CONTRIBUTING EDITORS

PREFACE

The second edition of "Van Nostrand's Scientific Encyclopedia," follows generally the premise of the original volume. The book encompasses the basic sciences and the applied fields of science and engineering. There have been added to this edition new sections on Electronics and Radio, Metallurgy, Meteorology, Photography and Statistics. The sections on Aeronautics, Engineering—Chemical, Civil, Mechanical, and Electrical—and Navigation have been expanded considerably. The entire book now takes into account the broad advances of science that have occurred in Aeronautics, Astronomy, Botany, Chemical Engineering, Chemistry, Civil Engineering, Electrical Engineering, Electronics and Radio, Geology, Mathematics, Mechanical Engineering, Medicine, Metallurgy, Meteorology, Mineralogy, Navigation, Photography, Physics, Statistics, and Zoology—twenty sections.

The responsibility for each science has been left largely in the hands of a single author in order to attain a unity impossible when many men contribute. However, although the responsibility rested largely in the one scientist of note, a number of men in each field have worked with the author, and a still larger group have consulted in an advisory capacity with the authors and publishers.

In this Encyclopedia over eleven thousand terms of scientific interest are arranged alphabetically, and an extensive system of cross-indexing has been developed to enable the reader to find all the facts that bear directly on each included topic. By this system, terms explained in this book are printed in bold face (black face) type wherever they are used significantly in the course of the articles on other terms. This practice makes it possible to turn readily to every article that has a bearing on the particular topic in which the reader is interested, as well as to obtain all supplementary information relative to any particular subject. Wherever bold face type appears within an article, the word or term appearing in this type is described in its alphabetical position. The user can gain a very comprehensive knowledge of each term if these references are consulted.

Naturally there are limits in the compiling of any one-volume book. These limits necessarily restrict the length of the article and the size of the illustration. However, the comprehensiveness of the book is noteworthy both in the scope of the terms covered and the breadth of the treatment in the individual article. The meticulous care of the authors, their advisers and their helpers, together with their systematic cross-referencing, have contributed in great measure to the inclusion of so much material within the covers of one book.

A feature of this Encyclopedia is the progressive development of the discussion of each topic, beginning with a simple definition expressed in the plainest terms and progressing to a final reflection of the more detailed scientific aspects of the topic treated. Articles dealing with simple concepts are, of course, treated in simple terms throughout the Encyclopedia, but those of a highly technical nature may be of value both to the inquiring layman and to the trained technician by a selection of their reading from the earlier or later portions of such an article.

The authors and the publishers will appreciate the indulgence of the reader for omissions. The exercise of judgment in the selection of material was unavoidable, and it was necessary to maintain a limit of difficulty beyond which it was impractical to go in attempting to cover so broad a field within the physical confines of one useful volume.

October, 1946

ACKNOWLEDGMENTS

Special articles appear under the authorship of individuals outside the consulting and contributing editor group. Acknowledgment is sincerely offered here to the following men and organizations who have prepared material or who have served in an advisory capacity: The Aluminum Company of America; The American Society for Metals; R. E. Barnard, R. S. Burns, M. E. Carruthers, G. H. Cole, and C. R. Taylor of the American Rolling Mills Company; J. S. Smart, Jr., of the American Smelting and Refining Company; Bakelite Corporation; E. B. Ferrell, Bell Telephone Laboratories, Inc.; J. A. Fizzell, Illinois Testing Laboratories, Inc.; Dr. Alexander Klemin; Polaroid Corporation; Dr. H. F. Olson, Radio Corporation of America; Dr. J. K. Robertson, Queen's University, Ontario; E. W. Saunders, University of Virginia; G. A. Roberts, Vanadium Alloy Steel Company.

Van Nostrand's
Scientific Encyclopedia

A

A SUPPLY. The source of the heating current for the cathode of an electronic tube. In the early days of radio the various voltages needed to operate a receiver were obtained from batteries, called A, B and C batteries, supplying the filament, plate and grid voltages respectively. These letter designations have carried over to the present-day sources, although the voltages are usually obtained now from an a-c source, either directly as in the case of the A supply or indirectly for B and C voltages. (See **Amplifiers** and **Radio Receivers.**) (L.R.Q.)

AA. An Hawaiian term introduced into geological nomenclature by C. E. Dutton, in 1883, and signifying the jagged, scoriaceous, blocky and exceedingly rough surface of some **basic** lava flows. (R.M.F.)

AARD-VARK. Mammalia, Tubulidentata. *Orycteropus.* African animals of peculiar form, including an Ethiopian and a South-African species. All are ant-eaters, feeding exclusively on ants and termites. The southern species has been called the ant-bear. (A.W.L.)

AARITE. Niccolite.

AARD WOLF. Mammalia, Carnivora. An African species, *Proteles cristatus,* superficially like the striped hyena. (A.W.L.)

AASVOGEL. South-African **vultures.** The name was applied by the Dutch colonists and means carrion-bird. (A.W.L.)

ABACA. The **sclerenchyma** bundles from the sheating leaf bases of *Musa textilis* (**Manila hemp**), a plant closely resembling the edible banana plant. These bundles are stripped by hand, after which they are cleaned by drawing over a rough knife. The fiber bundles are now whitish and lustrous, and from six to twelve feet long. Being coarse, extremely strong and capable of resisting tension, they are much used in the manufacture of ropes and cables. Since the fibers swell only slightly when wet, they are particularly suited for rope which will be used in water. Waste manila fibers from rope manufacture and other sources are used in the making of a very tough grade of **paper,** known as manila paper. The fibers may be obtained from both wild and cultivated plants, the latter yielding a product of better grade. The cultivated plants, propagated by seeds, by cuttings of the thick *rhizomes* or by suckers, are ready for harvest at the end of three years, after which a crop may be expected approximately every three years. (R.M.W.)

ABALONE. Mollusca, Gasteropoda. *Haliotis.* Marine species, mostly of the Pacific and Indian Oceans. The single broad shallow shell has a richly colored iridescent inner surface and is an important source of mother-of-pearl and blister pearls for costume jewelry. The flesh is palatable. It is eaten on our west coast, but much larger quantities are dried in California for shipment to the Orient. (A.W.L.)

ABAMPERE. The abampere, formerly called the "electromagnetic unit current," is the fundamental unit of the **c.g.s.** (centimeter-gram-second) electromagnetic system of **electrical units.** If a current of this magnitude flows in a circular loop of 1 cm. radius in a vacuum, the resulting magnetic field has an intensity, at the center of the circle, of 2π **oersteds;** which is the same as 1 oersted per unit length of wire. An equivalent statement is that if a current of 1 abampere flows in a straight wire across a magnetic field of 1 oersted intensity, at right angles to the magnetic intensity, the resulting lateral thrust, or "electric motor effect," is equal to 1 dyne for each centimeter of length of the wire. The (absolute) **ampere** is defined as $\frac{1}{10}$ of the abampere. (L.D.W.)

ABCOULOMB. Electric and Magnetic Units.

ABDOMEN. The abdomen is the posterior division of the body in many **arthropods.** It is the *posterior* portion of the trunk in **vertebrates.** In the vertebrates this region of the body contains most of the alimentary tract, the excretory system, and the reproductive organs. It contains part of the **coelom** and in mammals is separated from the thorax by the **diaphragm.**

The abdominal cavity of the human body is subdivided into the abdomen proper and the pelvic cavity.

The walls of the abdominal cavity are lined with a smooth membrane called the **peritoneum,** which also provides partial or complete covering for the organs within the cavity.

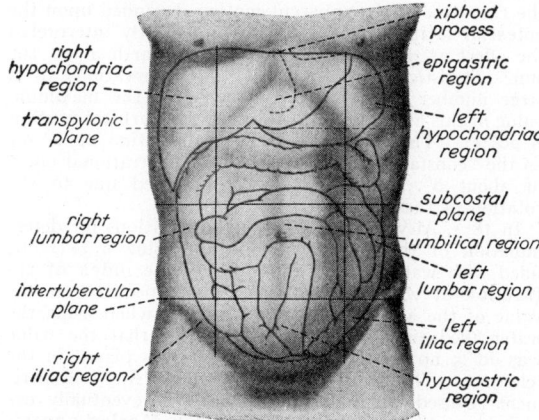

Planes of subdivision of the abdominal cavity and outline tracing of the liver, stomach, and intestine in relation to the anterior abdominal wall.
The oblique position of the stomach and the high position of the transverse colon are largely due to the fact that the subject was fixed in the horizontal position.
(*Cunningham, Textbook of Anatomy, Oxford Press.*)

The abdomen proper is bounded above by the diaphragm; below it is continuous with the pelvic cavity; posteriorly it is bounded by the spinal column, and the back muscles; and on each side by muscles and the lower portion of the ribs. In front, the abdominal wall is made up of layers of **fascia** and muscles. The surface of the abdomen is divided into sections. The mid-section above the navel between the angle of the ribs is known as the epigastric region; that portion around the navel, as the umbilical; below the navel and above the pubic bone, as the hypogastric region. It is further divided into right and left upper quadrants on each side above the navel, and right and left lower quadrants on each side below the navel. The lumbar region extends on either side of the navel posteriorly and laterally.

The principal organs of the abdominal cavity are the **stomach, duodenum, jejunum, ileum,** and **colon** or large intestine, the **liver, gall bladder** and biliary system, the **spleen, pancreas** and their blood and lymphatic vessels, **lymph glands,** and **nerves.**

The pelvic portion of the abdomen contains the urinary **bladder, uterus, Fallopian tubes,** and **ovaries** in the female, the sigmoid colon and **rectum,** and a portion of the small intestine. (A.W.L., R.S.M.)

ABERRATION OF LENSES. Chromatic Aberration; Spherical Aberration.

ABERRATION OF LIGHT. The apparent change of position of an object, due to the speed of motion of the observer, is known as the aberration of light. Care must be taken not to confuse this effect with that of **parallax.**

If a **telescope,** assumed to be stationary, is pointed at a source of light, the light which enters the object glass centrally and in the direction of the optic axis will pass through the telescope along that axis and emerge through the center of the eyepiece. If the telescope is in motion relative to the source, in any direction other than parallel to the optic axis, the light which enters centrally will emerge off the center of the eyepiece. If this light is to emerge centrally the telescope must be tilted forward in the plane containing the direction of motion of the instrument and the source. The amount of tilt will depend on the direction of the source and the ratio of the speed of the telescope to the speed of light.

This aberrational effect was first announced by Bradley in 1726. He noticed that stars had apparent periodic motions with a period of one **sidereal year,** and that the character of the apparent motion depended upon the **celestial latitude** of the star. He correctly interpreted the effect as due to the motion of the earth about the sun. Statistical discussions of the observations of a large number of stars have shown that the maximum value of this aberration due to the earth's **orbital motion** is 20″.47. This is known as the "aberration angle" or as the "constant of aberration." An aberrational effect of about 0″.3, at maximum, is observed due to the rotation of the earth on its axis.

In 1871, Airy made a series of observations for determination of the aberration constant using a telescope filled with water. Since the value of the **index of refraction** of water is about ¾, Airy expected that the value of the aberration would be 27″.3 when using the water-filled tube. He found, however, that the value was 20″.5 no matter what substance was placed in the telescope. The result of this so-called "Airy's Experiment" caused much discussion, but was eventually explained on the basis of the **Michelson-Morley experiment** and the **theory of relativity.**

All observations, in which the positions of the stars are involved, must be corrected for aberration of light if the results are to be accurate to within 20″. Both the motion of the earth about the sun and the rotation of the earth must be considered. The magnitude of the correction depends upon the **celestial coordinates** of the star, the position of the observer on the earth, and the date and time of observation. (W.K.G.)

ABFARAD. Electric and Magnetic Units.

ABHENRY. Electric and Magnetic Units.

ABIOGENESIS. The origin of living matter or living organisms from non-living material.

The ancients believed that living things, such as insects and mice, sprang from decaying organic matter or even

from mud in situations where they were sometimes seen in large numbers. Careful experiments finally showed that such highly organized creatures were produced only by others like themselves but the discovery of microorganisms again raised the question. In the experiments conducted by Pasteur and other scientists, it was at last proved that thoroughly sterilized materials gave rise to no living things unless they were later contaminated. Modern biology admits the possibility that an exceedingly simple type of living substance may arise from non-living materials but recognizes that living things as we know them are too complex to develop abruptly in this way. Even the origin of simple living substance has not actually been demonstrated. (A.W.L.)

ABLATION. From the Latin *ab* and *latio,* carried from, refers to the wasting away of the surfaces of rocks or **glaciers,** but principally used in the latter connection. Ablation deposits are the masses of **detritus** left after surface melting of glacial ice. (E.S.C.S., R.M.F.)

ABOHM. Electric and Magnetic Units; Ohm.

ABORAL OR APICAL SYSTEM. Part of the nervous system of the **echinoderms.** Unlike most nervous tissue it is developed from the middle germ layer. (A.W.L.)

ABORT. This term has two common meanings in medicine: 1. To check a disease or a condition during its early stages. 2. To expel the **fetus** during the first 4 or 5 months of pregnancy. (D.M.H.)

ABORTION. The expulsion of the **fetus** during the first half of **pregnancy,** which is always incompatible with the life of the fetus. Various descriptive terms are used to indicate the type of abortion as (1) accidental, (2) artificial or induced, i.e., intentional, (3) criminal, that is, induced illegally, (4) habitual, or where it occurs repeatedly with successive pregnancies in the same person, (5) incomplete, or where only a portion of the products of conception are expelled, (6) therapeutic, an abortion induced by a doctor as a protection to the life or health of the mother. Therapeutic abortions are commonly done when the mother has advanced **tuberculosis, kidney** or **heart disease.** (7) Threatened abortion, that is, the appearance of hemorrhage or labor pains early in pregnancy, may or may not develop into an actual abortion. (R.S.M.)

ABRASION. All metallic and non-metallic surfaces, no matter how smooth, consist of minute serrations and ridges which induce a cutting or tearing action when two surfaces in contact move with respect to each other. This wearing of the surfaces is termed abrasion. Undesirable abrasion may occur in bearings and other machine elements, but abrasion is also adapted to surface finishing and machining, where the material is too hard to be cut by other means, or where precision is a primary requisite. (H.C.H.)

ABRASIVES. Grinding, cutting, drilling, and polishing of various materials are often done with a wheel, sheet, or blast of abrasive, so-called, somewhat harder than the material treated. Accordingly, in the selection of an abrasive, the size of the individual grains and their specific hardnesses become important factors which determine the appearance of the finished surface or edge. The selected grains are bonded to a surface, such as sandpaper or emery paper, or formed into special shapes, especially wheels of various widths and diameters, or spotted on the face of drills as in sinking oilwells.

Many naturally occurring materials are used for this purpose, and for the harder abrasives specially manufactured materials are commonly utilized. The natural abrasives most used are sand, quartz, emery, diatomite, tripoli, pumice, and diamonds.

Sand, sandstone, quartzite and quartz, flint, and garnet are most used in sand-blasting of metals, in sawing-stones, grindstones, pulpstones, whetstones, in the grinding of ores in mills, and as sheets on paper or cloth. Emery and corundum are similarly used, practically all being imported as crude material and processed in this country.

Diatomite is used principally in polishes, but also as filter aid due to its high porosity, and as filler in special cases such as in plastics. It is highly resistant to heat, has low absorptive power for moisture, is chemically inert, has excellent electrical properties, and produces a good surface finish. It is stated that four leading brands of silver polishes contain 15 to 19% of diatomite as the sole abrasive. The principal states producing diatomite are California and Oregon. Tripoli is used for abrasive purposes, as an oilwell drilling mud, and also for foundry facing, and for filler in concrete. The production is centered around Newton County, Missouri, and adjacent Ottawa County, Oklahoma, and in Alexander County, southern Illinois. Pumice is used in cleansing and scouring mixtures and in hand soaps, and also for acoustic plaster and as an admixture for concrete.

Diamonds for abrasive purposes come from Brazil—these are called carbonados or black diamonds—and from the Union of South Africa—these are called bort. A beryllium-copper alloy is used successfully for cast-setting diamond core bits and reaming shells. These drilling bits are tough, strong, and hard, and the bond between the alloy and the diamonds is very close. Many small stones can be spaced over a comparatively small drill face.

Manufactured or "artificial" abrasives have made possible rapid working of very hard materials. Aluminum oxide glass ("Alundum") and silicon carbide crystals ("Carborundum") have proved outstandingly important. The total production manufactured for all uses, in 1941, in the United States, was aluminum oxide 150,-000 short tons, and silicon carbide 45,000 short tons, with additional 85,000 short tons of metallic abrasives. Tungsten carbide has attracted much attention as a very hard abrasive. It is typical of several metallic carbides, nitrides, and borides that have been suggested as abrasives. Some information on this point is contained in the following table:

streptococci are the common bacteria producing abscesses, although any organism may do so.

Abscesses may be single or multiple, primary or metastatic (see **Metastasis**), and may involve any organ or tissue in the body. The wall of an abscess is made up of inflammatory tissue which acts as a barrier to the spread of infection; the abscess cavity contains **pus** which consists of white blood cells, debris of tissue destruction, and **bacteria**, living and dead.

The physical signs associated with a superficially situated abscess are those of **inflammation**, swelling, increased heat, redness and tenderness over the involved area. Treatment in uncomplicated cases is usually surgical incision and drainage at the proper time. Chemotherapy with **sulfonamides** and **penicillin** may be necessary. (D.M.H.)

ABSCISSA OF A POINT. Rectangular Coordinates in a Plane.

ABSCISSION. This term is applied to the process whereby leaves, leaflets, fruits or other plant parts become detached from the plant. Leaf abscission is a characteristic phenomenon of many species of woody dicots and is especially conspicuous during the autumn period of leaf fall. Three main stages can be distinguished in the usual process of leaf abscission. The first is the formation of an abscission layer which is typically a transverse zone of parenchymatous cells located at the base of the petiole. The cells of this layer may become differentiated weeks or even months before abscission actually occurs. The second step is the abscission process proper which occurs as a result of a dissolution of the middle lamellae of the cells of the abscission layer. This results in the leaf remaining attached to the stem only by the vascular elements which are soon broken by the pressure of wind or the pull of gravity and the leaf falls from the plant. In the final stage of the process the exposed cells of the leaf scar are rendered impervious to water by lignification and suberization of the walls (see **Lignin; Suberin**). Subsequently other layers of corky cells develop beneath the outer layer. These layers eventually become a part of the periderm of the stem. The broken xylem elements of the leaf scar become plugged with **gums** or **tyloses** and the phloem elements become compressed and sealed off.

HARDNESS AND MELTING POINT OF VARIOUS CARBIDES, NITRIDES, AND BORIDES

> Greater than. Mohs' Values: Diamond 10, Corundum 9, Topaz 8, Quartz 7.

METAL	CARBIDES		NITRIDES		BORIDES	
	Hardness (Mohs' value)	Melting Point (°C.)	Hardness (Mohs' value)	Melting Point (°C.)	Hardness (Mohs' value)	Melting Point (°C.)
Chromium.........	>7	1890			8	
Columbium.......		3500	>8		>9	
Molybdenum......	7–9	2700			>9	
Tantalum........	>9	3875	>8	3100	>9	
Titanium.........	>8	3150	>8	2950	>9	
Tungsten.........	>9	2850			>9	
Vanadium........	>9	2830		2050	>9	
Zirconium........	>8	3530	>8	2980	>9	3000

(R.K.S.)

ABSAROKITE. A geologic term proposed by Iddings in 1895 for a **porphyritic basalt** containing **phenocrysts** of **olivine** and **augite** in a ground mass of smaller **labradorite** crystals. Type locality, Absaroka Rane, Yellowstone Park. (R.M.F.)

ABSCESS. A localized collection of pus usually formed in response to an infectious agent. **Staphylococci** and

In some kinds of plants an abscission layer is only imperfectly formed and in many others, especially herbaceous species, no abscission layer develops at the base of the petiole. In a few herbaceous species, of which coleus, begonia, and fuchsia are examples, an abscission layer develops. In the majority of herbaceous species, however, and in some woody species, there is no true abscission process. In such herbaceous plants most or all of the leaves are retained until the death of the plant. In the woody plants falling in this category (example: shingle oak, *Quercus imbricaria*) the leaves are shed only by mechanical disruption from the plant. Abscission of the fruits of apple and doubtless of many other species occurs in much the same manner as abscission of leaves. The abscission of apple fruits can be artificially retarded by spraying with certain **auxins**. (B.S.M.)

ABSINTHE. Artemisia.

ABSOLUTE HUMIDITY. The mass of water vapor in a specified volume. It can be expressed in any convenient units: ounces per cu. yd., grams per cu. meter. Example: 22 grams per cu. meter. (See **Humidity**.) (P.E.K.)

ABSOLUTE MAGNITUDE. The apparent brightness of a star, or any other luminous object, depends both upon the intrinsic brightness of the object and also upon its distance from the observer. In the case of the stars the apparent brightness, expressed as **stellar magnitude**, may be determined by any one of the standard methods of stellar **photometry**. In case the distance of the star is known the intrinsic brightness may be immediately calculated. Conversely, if we have any method available for determining the intrinsic brightness of a star independently of a knowledge of the distance, this distance may be computed from the ratio between the apparent and intrinsic brightness.

The absolute magnitude of a star is the apparent brightness, expressed on the magnitude scale, that a star would have if it were situated at a distance of ten **parsecs** from the sun or, in other words, if the **stellar parallax** of the star were $\frac{1}{10}$ of a second. Analytically, the absolute magnitude, M, of a star is connected with the apparent magnitude, mg, and the stellar parallax, π'', by:
$$M = mg + 5 + 5 \log \pi''.$$
On this scale we find the sun, with apparent magnitude -26.72 and parallax 206265″, to have an absolute magnitude of 4.85. **Antares** with parallax 0″.009 and apparent magnitude 1.22 is found to have an absolute magnitude of -4.0. On the basis of these absolute magnitudes and the defining relation of the magnitude scale, we find the brightness ratio of Antares to the sun to be 3470 or the star Antares is actually 3470 times as bright as the sun. (W.K.G.)

ABSOLUTE TEMPERATURE. Absolute Zero; Temperature Scales.

ABSOLUTE VALUE OF A REAL NUMBER. Number.

ABSOLUTE VARIABILITY. A measure of variability that is expressed in the unit of the **variable**. A measure of **relative variability** relates the absolute variability to an **average**, generally to the **mean**. (L.A.A.)

ABSOLUTE ZERO. A temperature at which bodies would possess no heat whatever. Prior to the discovery of the dynamic character of heat, no significance could be attached to a zero of temperature save that of a point arbitrarily chosen, such as the melting point of ice, from which temperatures might be reckoned both ways. But when it became known that heat is the kinetic energy of random molecular motion, it was at once possible to visualize, if not to realize, a condition of "absolute cold," merely by supposing the **molecules** of a substance to have come to rest relative to each other.

It was formerly customary to define the measure of **temperature** in such a way that a linear relation exists between the temperature θ and the pressure p of a gas (hydrogen) kept at constant volume; thus:
$$p = p_0 + a\theta \qquad (1)$$

in which p_0 is the pressure at the arbitrary zero of temperature. This relation is represented by the straight line in the accompanying figure. The slope of this line,

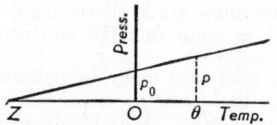

Relation between temperature and pressure of a gas at constant volume.

expressed by the constant a, corresponds to the change in pressure for each unit of temperature change (degree). Experiment shows that when the centigrade scale is used, this change is 0.0036626 of the pressure of the gas at the zero of that scale, viz., the melting point of ice; so that $a = 0.0036626$. (This is on condition that the pressure p_0 is 1000 mm.) Equation (1) accordingly may be written
$$p = p_0 + 0.0036626 p_0\theta = p_0(1 + 0.0036626\theta). \qquad (2)$$

Equation (2) now implies the possibility of causing the pressure to vanish altogether by reducing the temperature until the factor $1 + 0.0036626\theta = 0$; that is, until $\theta = -273.03°$ C. $= -459.45°$ F. (corresponding to Z in the figure). Since gas pressure depends upon the motion of molecules, it follows that, for the pressure to vanish, all such motion must cease. Hence we may suppose that the centigrade temperature $-273.03°$ C. is the absolute zero, so far as translational thermal energy is concerned. More recent studies employing the **Kelvin scale** of temperature give $-273.16°$ C. as absolute zero. There is reason to believe that a pressure would still exist in hydrogen even under these conditions. The lowest temperature so far experimentally attained is calculated to be within 0.005° of absolute zero. (See **Thermometry**.) (L.D.W.)

ABSORBENT COTTON. Absorbent cotton is prepared from the fibers of the **cotton** plant. These fibers are treated in such a way as to remove the natural waxy substances and the small amount of mineral matter present. Subsequent washing yields the product known commercially as absorbent cotton, which can absorb as much as eighteen times its own weight of water, and which has many commercial and medical uses. (R.M.W.)

ABSORPTION. This term has the widest significance in science and technology. In the organic world it denotes the process by which materials enter the living substance of which the organism is composed. Substances including food and oxygen are taken into special organs by ingestion and respiration but they must pass through the outer surface of the **cells** to become an integral part of the organism by absorption. The nature of the process is considered under **osmosis**.

The absorption of gases plays an important part in engineering. It is the frequent cause of **corrosion** by condensate (due to the high content of dissolved oxygen). Gaseous absorption is the basis of the absorption system of **refrigeration**. In this system a gas (or vapor) is absorbed in a suitable medium and is then separated by distillation, followed in some cases by liquefaction under pressure. The principles involved in the absorption of gases in liquids are treated in the article on **Solutions** and **Solubility**, and in the article on **Dissolving**. The adsorption of gases (as well as liquids and solids) is a related phenomenon that is treated in the article on **Adsorption**.

The quantity of heat absorbed by a substance is calculated from **specific heat** and **temperature**. The capacity of surface to absorb radiant heat is measured by its absorptivity (**Thermal Radiation**).

The absorption of mechanical energy by **dynamometers**, which convert the mechanical energy to heat or electrical forms, has lead to the use of the term "absorption dynamometers" to distinguish these machines. For the absorption of radiation see **Absorption Coefficient**. (F.T.M., A.W.L.)

ABSORPTION COEFFICIENT. A quantity used to express the rate at which a substance absorbs radiation passing through it. When light, x-rays, or other **electromagnetic radiation** enters a body of matter, it experiences in general two types of attenuation. Part of it is subjected to **scattering**, being reflected in all directions without essential change of character, while another portion is absorbed by being converted into other forms of energy. The scattered radiation may still be effective in the same ways as the original, but the absorbed portion ceases to exist as radiation or is re-emitted as secondary radiation. Strictly therefore we have to distinguish the true absorption coefficient from the scattering coefficient; but for practical purposes it is sometimes convenient to add them together as the total attenuation or extinction coefficient.

Accurate measurements upon radiation which has traversed various thicknesses of matter has established that any infinitely thin layer perpendicular to the direction of propagation cuts down the flux density by a fraction of its value proportional to the thickness of the layer, and that the flux density after having penetrated the medium to a distance x is

$$I = I_0 e^{-ax};$$

in which I_0 is the flux density just after entrance into the medium (i.e., for $x = 0$). For true absorption, the constant a is the absorption coefficient (commonly designated by μ). For scattering, which obeys the same law, a is the scattering coefficient. And for the total attenuation, including both, it is the extinction coefficient, which is the sum of the absorption and the scattering coefficients.

Another way of expressing the absorbing effect of a substance is to specify the "half-value layer," which is that thickness of the substance which will reduce the flux density to $\frac{1}{2}$ its original value, so that $I = \frac{1}{2}I_0$. This thickness is equal to $0.6931/a$. Thus if the absorption coefficient of copper for certain x-rays is 13.5 cm.$^{-1}$, the half-value layer for these rays is 0.0513 cm. thick. For many purposes it is convenient to use the mass absorption coefficient, which is the absorption coefficient of the substance divided by its density.

In general the absorption coefficient of a medium varies characteristically with the wavelength of the radiation, as illustrated by the absorption of x-rays in aluminum, tabulated below.

WAVE-LENGTH (X-units) *	ABSORPTION COEFFICIENT (Cm.$^{-1}$)
100	0.45
200	0.72
300	1.45
400	2.95
500	5.30
600	8.70
700	13.50
800	20.40
900	28.10
1000	38.00
	(L.D.W.)

* An X-unit is 10^{-11} cm. (See **X-Rays**; **Gas Absorption**.)

ABSORPTION SPECTRUM. The **spectrum** of radiation which has been filtered through a material medium. When white light traverses a transparent medium, a certain portion of it is absorbed, the amount varying, in general, progressively with the frequency, of which the **absorption coefficient** is a function. Analysis of the transmitted light may, however, reveal that certain frequency ranges are absorbed to a degree out of all proportion to the adjacent regions; that is, with a distinct

selectivity. These abnormally absorbed frequencies constitute, collectively, the "absorption spectrum" of the medium, and appear as dark lines or bands in the otherwise continuous spectrum of the transmitted light. The phenomenon is not confined to the visible range, but may be found to extend throughout the spectrum from the far infra-red to the extreme ultra-violet and into the x-ray region.

A study of such spectra shows that the lines or bands therein accurately coincide in frequency with certain lines or bands of the emission spectra of the same substances. This was formerly attributed to **resonance** of electronic vibrations, but is now more satisfactorily explained by **quantum theory** on the assumption that those quanta of the incident radiation which are absorbed are able to excite atoms or molecules of the medium to some (but not all) of the energy levels involved in the production of the complete emission spectrum.

A very familiar example is the spectrum of sunlight, which is crossed by innumerable dark lines—the **Fraunhofer lines**—from which so much has been learned about the constitution of the **sun**.

A noteworthy characteristic of selective absorption is found in the existence of certain anomalies in the refractive index in the neighborhood of absorption frequencies; discussed under **Dispersion**. (L.D.W.)

ABSORPTION TOWER. Gas Absorption.

ABSORPTIVITY. Thermal Radiation.

ABUTMENT. A bridge abutment is a **masonry** or **concrete** structure which functions both as a **pier** and as a **retaining wall**. It must support the end of the bridge and hold the abutting earth in position. The simple abutment consists of a **footing**, a breast or cross wall, a bridge seat, a back wall, and usually companion wing walls. The footing transfers the loads to the supporting soil, consequently the area in contact with the soil must be large enough to insure a safe bearing pressure. The breast wall must be large enough to withstand safely the combined effects of the bridge loads, its own weight and the pressure of the soil back of the abutment. The bridge seat is the surface which supports the end **bearings** of the bridge. The back wall supports the earth above the bridge seat. The wing walls are usually attached to both ends of the breast wall and are used to retain the side slopes of the fill at the end of the abutment. (C.W.C.)

ABVOLT. Electric and Magnetic Units.

ABYSSAL FAUNA. The animals found in the depths of the ocean below 600 fathoms. The abysses are characterized by darkness, low temperature, great pressure, and the absence of plant life due to the lack of light. Grotesque form and the extensive development of light-producing organs are frequent among abyssal animals. (A.W.L.)

ABYSSAL ROCKS. Proposed by Brögger as a general term for deep-seated **igneous** rocks, or those which have crystallized from **magmas** far below the surface of the earth, very slowly and under great pressure. **Granite** is a typical abyssal rock. The term **Plutonic** is synonymous. (R.M.F.)

ACACIA. Leguminosae: tribe Mimosae. A very large genus of trees and shrubs, particularly abundant in Africa and Australia. The small flowers are aggregated into ball-like or elongate clusters, which are quite conspicuous. The leaves are rather diverse in shape; quite commonly they are dissected into compound pinnate forms; in other instances, especially in Australian species, they are reduced even to a point where only the flattened petiole (see **Leaf**), called a phyllode, remains. This petiole grows with the edges vertical, a fact which some have been led to construe as a protective adaptation against too intense sunlight on the surface. Several species, particularly those growing in Africa and tropical Asia, yield products of commercial value. For example, from *Acacia Senegal* gum arabic is obtained; and from *A. Catechu*, a brown or black dye called cutch. Many species are valuable timber trees. Certain tropical American species are of particular interest because of the curious pairs of thorns, which are united at their base. These thorns are often hollowed out and used as nests by species of stinging ants. (R.M.W.)

ACANTHITE. Argentite.

ACANTHOCEPHALA. Worms with recurved spines at the anterior end, parasitic in the intestines of vertebrates. They are usually regarded as a class of roundworms (**Nemathelminthes**). (A.W.L.)

ACANTHUS. Acanthaceae. Acanthus is a small genus of Mediterranean plants largely grown for ornamental purposes. The flowers are white or various shades of red. The leaves of these plants are the source of the more or less conventionalized architectural design called the acanthus. (R.M.W.)

ACARINA. The order of **Arachnida** which includes the **mites** and **ticks**. (A.W.L.)

ACCELERATED FLIGHT. When the velocity of an airplane along its flight path contains elements of acceleration, the structure receives increments of inertia loading which may prove to be far more severe upon the structure than the loading imposed by the static weight of the airplane and its contents. Consequently, accelerated flight has been the subject of extensive analytical and experimental investigation. Acceleration of rectilinear velocity, as by increasing engine power in straight level flight, is of small import, since radial accelerations resulting from curvilinear flight at constant speed are so large as to be the critical influence. Cases of curved flight paths capable of accelerations of several g (acceleration of gravity, i.e., 32.2 ft. per sec.²) are quick pull-ups (or "zooms") from high-speed rectilinear flight, spins, steeply banked turns, loops. The magnitude of the effect of accelerated flight is well illustrated by the case of a flight path curved in a vertical plane. With a constant tangential speed of 120 m.p.h., our airplane will experience a radial acceleration of 4g, even though the radius of curvature be as great at 250'. (See **Load Factor**.) (F.T.M.)

ACCELERATION. The rate of change of the velocity with respect to the time is called acceleration. It is expressed mathematically by $\dfrac{dv}{dt}$, the vector derivative of the velocity, v with respect to the time, t. If the motion is in a straight line whose position is clearly understood, it is convenient to treat the velocity v and the acceleration $\dfrac{dv}{dt}$ as scalars with appropriate algebraic signs; otherwise they must be treated by vector methods.

Acceleration may be rectilinear or curvilinear, depending upon whether the path of motion is a straight line or a curved line. A body which moves along a curved path has acceleration components at every point. One component is in the direction of the tangent to the curve and is equal to the rate of change of the speed at the point. For uniform circular motion this component is zero. The second component is normal to the tangent and is equal to the square of the tangential **speed** divided by the **radius of curvature** at the point. This normal component which is directed toward the center of curvature also equals the square of the **angular velocity** multiplied by the radius of curvature. The acceleration due to gravity is equal to an increase in the

velocity of about 32.2 ft. per sec. per sec. at the earth's surface and is of prime importance since it is the ratio of the weight to the **mass** of a body. For examples of acceleration in both curved and linear motion, see **Kinematics**. (L.D.W.)

ACCELERATORS, RUBBER. Rubber and Accelerators.

ACCELEROMETER. The accelerometer is an instrument for determining the acceleration of the system with which it moves. Work in accelerometry is becoming increasingly important as means of transport continue to provide higher motive speeds for the use of mankind. The principal instrument in this field of work is the accelerometer. It has been used in airplane work to study the stresses that the airplane structure undergoes, and to determine how long these stresses last. The records can also be used to study pilots' ability, especially in landings and acrobatic maneuvers. Other uses for this instrument are the study of the oscillations of automobile springs, the pickup and braking power of automobiles, the side load on tires or rails when rounding curves, and study of vibrations of various sorts. The accelerometer should have a natural period of vibration which is considerably higher than that of any shocks it may experience. In addition, it should give a graphic, easily interpreted record, and it should be rugged, strong, and accurate. Not all these characteristics can be met by one design, and accelerometers suitable for measuring the accelerations produced in certain flight maneuvers are unsatisfactory for measuring landing shock accelerations. One style of the accelerometer is the seismograph type. Unfortunately, this instrument records displacements against an axis of time sequence, and accelerations are not read directly. The slope of the displaced curve is the rate of change of displacement with time; in other words, the velocity. If the velocity is determined and plotted, a similar measurement of slope gives the rate of change of velocity with time, and this is the acceleration. Thus the record of a seismograph type of accelerometer must be differentiated twice in order to obtain accelerations, for:

$$v = \frac{ds}{dt}$$

$$a = \frac{dv}{dt}$$

Hence: $a = \dfrac{d^2s}{dt^2}$

While this type of instrument has its certain uses, it has been superseded for most accelerometry work by more practical designs. (F.T.M.)

ACCESSORY NIDAMENTAL GLAND. A gland of the female reproductive system in the **squids** and allied species. (A.W.L.)

ACCOMMODATION. The power of altering the focus of the **eye** so that divergent light rays may be brought to a point on the retina. (See **Vision**.) (R.S.M.)

ACCOUCHEMENT. Confinement, or delivery of a baby, by an **obstetrician** or midwife. (R.S.M.)

ACCOUCHEUR. An obstetrician or midwife. (R.S.M.)

ACCRETION. Used in mineralogy and geology to define the process by which inorganic bodies grow larger by the addition of material onto the external surface. (R.M.F.)

ACCUMULATOR, HYDRAULIC. The hydraulic accumulator is a hydraulic machine consisting of vertical cylinders with weighted pistons or plungers under which a certain amount of water can be stored, and is, consequently, available for doing work at the pressure yielded by the weighted piston. The accumulator feature is obtained by virtue of the fact that while the water may be discharged rapidly, giving large hydraulic power for short periods of time, it may be refilled by a comparatively small and low-powered pump working a much longer time.

Another type of hydraulic accumulator is the pumped storage plant, now being looked on with considerable favor by electric power systems for the economic carrying of variable load. As employed in conjunction with steam generating stations, steam turbine-driven centrifugal pumps raise water from a lower to an upper pool with off-peak power. During the peak-load periods this water is released to the lower pool through a hydraulic turbo-generator as rapidly as is needed to give the required power. The hydraulic storage of power of this nature is essentially a high head development, low head equipment and hydraulic losses being too expensive. In favorable locations the over-all efficiency of conversion and storage may not need to be greater than 50% in order to justify the project. (F.T.M.)

ACCUMULATOR, STEAM. The steam accumulator is an effective means for smoothing out irregular steam demand into a uniform boiler output. Its operation is based upon the fact that the heat contained in water in a liquid form varies with the pressure of the water. Thus in a tank of water under pressure with the water at the saturation temperature, a decrease of pressure on the tank will be accompanied by a release of some of the heat energy held by the water, and a consequent flashing of a portion of the water into steam. This process can be continued with the production of steam at ever decreasing pressures until the lower pressure limit is reached. Since the heat required to evaporate a pound of water is much more than the heat of the liquid at the commonly used pressures, only 20–40% of the weight of water in a charged accumulator tank can be converted into steam.

An accumulator installation is shown by diagram in the accompanying figure. The irregular steam demand

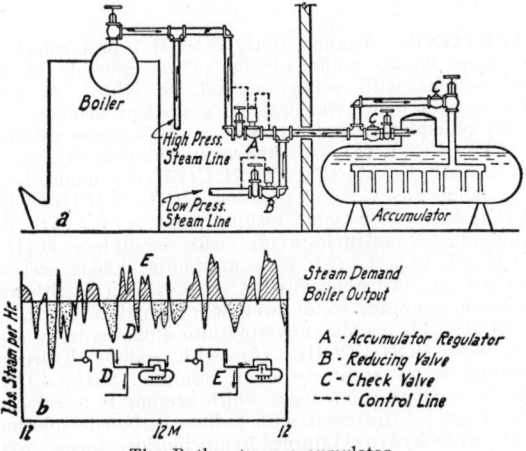

The Ruths steam accumulator.

is assumed to be that of an industrial process using considerable quantities of low-pressure steam, while the desired boiler output is shown as the straight line. The accumulator must absorb steam during the valleys which occur below the steady boiler output line and supply steam during the peaks. The direction of steam flow when charging the accumulator is shown at *D*, while at *E* is the flow diagram for accumulator discharging. (F.T.M.)

AC-DC RECEIVERS. These are radio receivers designed without transformers in the power supply so they may be connected to either alternating-current or direct-current circuits. The heaters or filaments of the

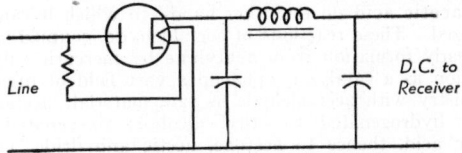

Plate supply of ac-dc receiver.

various tubes are connected in series with the proper series or parallel resistors to adjust the current to the correct value. The d-c voltage for the plates is obtained from a rectifier-filter circuit connected directly to the line. A simple half-wave type is shown. Such supplies prevent the use of a direct ground since the 110-volt line has one side grounded and there is always the possibility of connecting the receiver plug so it would short the line if the receiver were grounded. (L.R.Q.)

ACERDISE. See boehmite under **Lepidocrocite.**

ACETALDEHYDE. Acetaldehyde ($CH_3 \cdot CHO$) is a colorless, odorous liquid, boiling point 20° C., miscible with water, alcohol, or ether in all proportions. Acetaldehyde reacts with many chemicals in a marked manner, (1) with ammonio-silver nitrate ("Tollen's solution"), to form metallic silver, either as a black precipitate or as an adherent mirror film on glass, (2) with alkaline cupric solution ("Fehling's solution") to form cuprous oxide, red to yellow precipitate, (3) with rosaniline (fuchsine, magenta), which has been decolorized by sulfurous acid ("Schiff's solution"), the pink color of rosaniline is restored, (4) with sodium hydroxide, upon warming, a yellow to brown resin of unpleasant odor separates (this reaction is given by aldehydes immediately following acetaldehyde in the series, but not by formaldehyde, furfuraldehyde or benzaldehyde), (5) with anhydrous ammonia, to form aldehyde-ammonia ($CH_3 \cdot CHOH \cdot NH_2$), white solid, melting point 97° C., boiling point 111° C., with decomposition, (6) with concentrated sulfuric acid, heat is evolved, and with rise of temperature, paraldehyde ((C_2H_4O)$_3$ or $CH_3 \cdot CH {<}^{OCH(CH_3)}_{OCH(CH_3)}{>}O$), colorless liquid, boiling point 124° C., slightly soluble in water, is formed, (7) with acids, below 0° C., forms metaldehyde (C_2H_4O)$_x$ white solid, sublimes at about 115° C. without melting but with partial conversion to acetaldehyde, (8) with dilute hydrochloric acid or dilute sodium hydroxide, aldol ($CH_3 \cdot CHOH \cdot CH_2 \cdot CHO$) slowly forms, (9) with phosphorus pentachloride, forms ethylidene chloride ($CH_3 \cdot CHCl_2$), colorless liquid, boiling point 58° C., (10) with ethyl alcohol and dry hydrogen chloride, forms acetal, 1,1-diethyoxyethane ($CH_3 \cdot CH(OC_2H_5)_2$), colorless liquid, boiling point 104° C., (11) with hydrocyanic acid, forms acetaldehyde cyanhydrin ($CH_3 \cdot CHOH \cdot CN$), readily converted into alpha-hydroxypropionic acid ($CH_3 \cdot CHOH \cdot COOH$), (12) with sodium hydrogen sulfite, forms acetaldehyde sodium bisulfite ($CH_3 \cdot CHOH \cdot SO_3Na$), white solid, from which acetaldehyde is readily recoverable by treatment with sodium carbonate solution, (13) with hydroxylamine hydrochloride forms acetaldoxime ($CH_3CH:NOH$), white solid, melting point 47° C., (14) with phenylhydrazine, forms acetaldehyde phenylhydrazone ($CH_3 \cdot CH:N \cdot NH \cdot C_6H_5$), white solid, melting point 98° C., (15) with magnesium methyl iodide in anhydrous ether ("Grignard's solution"), yields, after reaction with water, isopropyl alcohol ((CH_3)$_2CHOH$), a secondary alcohol, (16) with semicarbazide, forms acetaldehyde semicarbazone ($CH_3 \cdot CH:N \cdot NH \cdot CO \cdot NH_2$), white solid, melting point 162° C., (17) with chlorine, forms trichloroacetaldehyde ("chloral") ($CCl_3 \cdot CHO$), (18) with hydrogen sulfide, forms thio-acetaldehyde ($CH_3 \cdot CHS$ or ($CH_3 \cdot CHS$)$_3$). Acetaldehyde stands chemically between ethyl alcohol on one hand—to which it can be reduced—and acetic acid on the other hand—to which it can be oxidized. These reactions of acetaldehyde, coupled with its ready formation from acetylene by mercuric sulfate solution as a catalyzer, open up a vast field of organic chemistry with acetaldehyde as raw material: acetaldehyde hydrogenated to ethyl alcohol; oxygenated to acetic acid, thence to acetone, acetic anhydride, vinyl acetate, vinyl alcohol. Acetaldehyde is also formed by the regulated oxidation of ethyl alcohol by such a reagent as sodium dichromate in sulfuric acid (chromic sulfate also produced). Reactions (1), (3), (14) and (16) above are most commonly used in the detection of acetaldehyde. (R.K.S.)

ACETALS. Organic compounds of the general formula $RCH(OR')(OR'')$. They are formed by the reaction of aldehydes with alcohols in the presence of small amounts of acids or certain inorganic salts. They are stable toward alkali, are volatile and insoluble in water but are decomposed into aldehyde by the action of acids. The last reaction is often used as a source of aldehydes. (R.K.S.)

ACETANILIDE. Aniline.

ACETIC ACID AND ACETATES. Acetic acid ($H \cdot C_2H_3O_2$ or $CH_3 \cdot COOH$) is a colorless liquid, melting point 16.6° C., boiling point 118° C., miscible with water, alcohol, or ether in all proportions. Acetic acid solution reacts with alkalis to form acetates, e.g., **sodium** acetate, **calcium** acetate; similarly, with some oxides, e.g., **lead** acetate; with carbonates, e.g., **sodium** acetate, **calcium** acetate, **magnesium** acetate; with some sulfides, e.g., **zinc** acetate, **manganese** acetate. **Ferric acetate** solution, upon boiling, yields red precipitate of basic ferric acetate. Acetic acid solution attacks many metals, liberating hydrogen and forming acetate, e.g., magnesium, zinc, iron. Acetic acid is an important organic substance, with alcohols forming **esters** (acetates); with **phosphorus** trichloride forming acetyl chloride ($CH_3 \cdot CO \cdot Cl$), which is an important reagent for transfer of the acetyl (CH_3CO-) group; forming **acetic anhydride**, also an acetyl reagent; forming **acetone** and calcium carbonate when passed over a suitable **catalyzer** (**barium** carbonate) or when calcium acetate is heated; forming **methane** (and sodium carbonate) when sodium acetate is heated with sodium hydroxide; forming mono-, di-, tri-chloroacetic (or bromoacetic) **acids** by reaction with chlorine (or bromine) from which **hydroxy-** and **amino-, aldehydic-, dibasic-** acids, respectively, may be made; forming **acetamide** when **ammonium** acetate is distilled. Acetic acid dissolves sulfur and phosphorus, is an important solvent for organic substances, and causes painful wounds when it comes in contact with the skin. Normal acetates are soluble, basic acetates insoluble. The latter are important in their compounds with lead, copper ("verdigris"). Acetic acid is made (1) by destructive distillation of wood. Dilute acid is obtained in the aqueous distillate, recovered by neutralization with **calcium** hydroxide, and then evaporation and recovery of calcium acetate; (2) from calcium or sodium acetate, acetic acid of high strength is made by distillation with concentrated **sulfuric acid;** (3) by the action of bacteria on dilute ethyl alcohol, containing the proper food materials for the bacteria, dilute acetic acid (vinegar) is produced. The vinegar contains, besides acetic acid and water, the materials characteristic of the alcohol and the process used; (4) by the reaction of acetaldehyde and air over a suitable catalyzer. Acetic acid is used as has been suggested by its reactions, (1) in the preparation of many organic substances, notably,

cellulose acetate, as a non-inflammable photographic film and also as a textile fiber; (2) in the preparation of many acetates and basic acetates and carbonates (white lead in conjunction with carbon dioxide); (3) as a weak, moderately cheap acid; (4) as a solvent when concentrated, for organic chemicals; (5) in pharmaceutical preparations, dyeing, rubber, artificial leather.

Esters (acetates of various alcohols) of note are:
Methyl acetate (CH_3COOCH_3), boiling point 57° C.
Ethyl acetate ($CH_3COOC_2H_5$), boiling point 77° C.
Propyl acetate ($CH_3COOC_3H_7$), boiling point 102° C.
Butyl acetate ($CH_3COOC_4H_9$), boiling point 125° C.
Amyl acetate ($CH_3COOC_5H_{11}$), boiling point 149° C.
Glycol monoacetate ($CH_3COOCH_2 \cdot CH_2OH$), boiling point 182° C.
Glycol diacetate ($CH_3COOCH_2 \cdot CH_2COOCH_3$), boiling point 190° C.
Glyceryl monoacetate (monoacetin) ($CH_2OH \cdot CHOH \cdot CH_2OOCCH_3$) decomposes upon heating.
Glyceryl diacetate (diacetin) ($CH_2OH \cdot CHOOCCH_3 \cdot CH_2OOCCH_3$), melting point 40° C., boiling point 176° C. at 40 mm. pressure.
Glyceryl triacetate (triacetin) ($CH_2OOCCH_3 \cdot CHOOCCH_3 \cdot CH_2OOCCH_3$), melting point −78° C., boiling point 259° C.
Glucose pentacetate ($C_6H_6(OH)(COOCH_3)_5$), melting point 113° C., sublimes.
Cellulose triacetate ($C_6H_5(OH)_2(COOCH_3)_3$).
Cellulose tetracetate ($C_6H_5(OH)(COOCH_3)_4$), softens at about 150° C.
Cellulose pentacetate ($C_6H_5(COOCH_3)_5$).
Cetyl acetate ($CH_3COOC_{16}H_{33}$), melting point 22° C., boiling point 200° C. at 15 mm. pressure.
Phenyl acetate ($CH_3COOC_6H_5$), boiling point 195° C.
Acetates may be detected by formation of foul-smelling cacodyl (poisonous) on heating with dry arsenic trioxide. (R.K.S.)

ACETOACETIC ACID ESTER. This is an important organic liquid of the formula $CH_3(CO)CH_2 \cdot COOC_2H_5$, which is used as a starting point for the synthesis of **ketones** of the type $CH_3(CO)CHR'R''$ and acids of the type $R'R''CHCOOH$ where R' and R'' are hydrocarbon **radicals**. The ethyl ester of acetoacetic acid is made by treating ethyl acetate (see **Esters**) with sodium. (R.K.S.)

ACETONE. Acetone ($CH_3 \cdot CO \cdot CH_3$) is a colorless, odorous liquid, boiling point 56° C., miscible in all proportions with water, alcohol, or ether. Acetone reacts with many chemicals in a marked manner, (1) with **phosphorus** pentachloride, yields acetone chloride (($CH_3)_2CCl_2$), (2) with **hydrogen chloride** dry, yields both mesityl oxide ($CH_3COCH:C(CH_3)_2$), liquid, boiling point 132° C., and phorone (($CH_3)_2C:CHCOCH:C(CH_3)_2$), yellow solid, melting point 28° C., (3) with concentrated **sulfuric** acid, yields mesitylene ($C_6H_3(CH_3)_3$) (1,3,5), (4) with **ammonia**, yields acetone amines, e.g., diacetoneamine ($C_6H_{12}ONH$), (5) with **hydrogen cyanide**, yields acetone cyanhydrin (($CH_3)_2CHOH \cdot CN$), readily converted into alpha-hydroxy acid (($CH_3)_2CHOH \cdot COOH$), (6) with sodium hydrogen sulfite, forms acetone sodium bisulfite (($CH_3)_2COH \cdot SO_3Na$), white solid, from which acetone is readily recoverable by treatment with sodium carbonate solution, (7) with **hydroxylamine** hydrochloride, forms acetoxime (($CH_3)_2C:NOH$), solid, melting point 60° C., (8) with **phenylhydrazine**, yields acetonephenylhydrazone (($CH_3)_2C:NNHC_6H_5 \cdot H_2O$), solid, melting point 16° C., anhydrous compound, melting point 42° C., (9) with **semicarbazide**, forms acetonesemicarbazone (($CH_3)C:NNHCONH_2$), solid, melting point 189° C., (10) with magnesium methyl iodide in anhydrous ether ("**Grignard's solution**"), yields, after reaction with water, trimethylcarbinol (($CH_3)_3COH$), a tertiary alcohol, (11) with ethyl thioalcohol and hydrogen chloride dry, yields mercaptol $((CH_3)_2C(SC_2H_5)_2)$, (12) with **hypochlorite, hypobromite,** or **hypoiodite** solution, yields chloroform $(CHCl_3)$, bromoform $(CHBr_3)$ or iodoform (CHI_3), respectively, (13) with most reducing agents, forms isopropyl alcohol $((CH_3)_2CHOH)$, a secondary alcohol, but with **sodium** amalgam forms pinacone $((CH_3)_2COH\cdot COH(CH_3)_2)$, (14) with **sodium** dichromate and sulfuric acid, forms **acetic** acid (CH_3-COOH) plus **carbon dioxide** (CO_2). When acetone vapor is passed through a tube at a dull red heat, **ketene** $(CH_2:CO)$ and **methane** (CH_4) are formed. Acetone is made (1) by heating **calcium** acetate at 400° C., calcium carbonate being simultaneously formed, (2) by passing **acetic acid** vapor over a heated catalyzer, e.g., **barium** carbonate, **manganese** carbonate, (3) by fermentation of **starch** by specific bacteria, normal-**butyl alcohol** being simultaneously produced, (4) in the water condensate (approximately 0.5% acetone) in the destructive distillation of wood, and, (5) in the urine of persons having **diabetes**. Acetone may be detected by the addition of acetic acid and sodium nitroprusside (trace). The appearance of a violet color in the interface between this solution and a layer of ammonium hydroxide indicates acetone. Acetone is used (1) as a solvent, e.g., for **acetylene**, (2) as a solvent for **cellulose** and glyceryl esters in the manufacture of celluloid, smokeless powders, airplane dopes, varnishes, (3) in the preparation of **chloroform, iodoform, sulfonal.** (R.K.S.)

ACETOPHENONE. Aldehydes, Ketones and Related Compounds.

ACETYL CHLORIDE. Chlorine.

ACETYLENE.
Acetylene, ethyne (C_2H_2 or $CH:CH$) is a colorless gas, of characteristic odor, moderately poisonous, boiling point —84° C., density, 1.17 grams per liter at 0° C. and 760 mm. (specific gravity 0.91, air equal to 1.00), slightly soluble in water or alcohol, very soluble in acetone (300 volumes of acetylene in 1 volume acetone at 12 atmospheres pressure), burns when ignited in air with a luminous sooty flame, requiring a specially devised burner for illumination purposes, forms an explosive mixture with air over a wide range (about 3% to 80% acetylene), explosive when compressed to 2 or more atmospheres, but safe when dissolved in acetone, of high fuel value (1455 B.T.U. per cubic foot). Acetylene reacts (1) with **chlorine**, to form acetylene tetrachloride ($C_2H_2Cl_4$ or $CHCl_2\cdot CHCl_2$) or acetylene dichloride ($C_2H_2Cl_2$ or $CHCl:CHCl$), (2) with **bromine**, to form acetylene tetrabromide ($C_2H_2Br_4$ or $CHBr_2\cdot CHBr_2$) or acetylene dibromide ($C_2H_2Br_2$ or $CHBr:CHBr$), (3) with **hydrogen chloride** (bromide, iodide), to form ethylene monochloride ($CH_2:CHCl$) (monobromide, monoiodide), and 1,1-dichloroethane, ethylidene chloride ($CH_3\cdot CHCl_2$) (dibromide, diiodide), (4) with water in the presence of a **catalyzer**, e.g., mercuric sulfate, to form **acetaldehyde** ($CH_3\cdot CHO$), (5) with **hydrogen**, in the presence of a catalyzer, e.g., finely divided nickel heated, to form **ethylene** (C_2H_4) or **ethane** (C_2H_6), (6) with metals, such as copper or nickel, when moist, also lead or zinc, when moist and unpurified. Tin is not attacked. Sodium yields, upon heating, the compounds C_2HNa and C_2Na_2. (7) With ammoniocuprous (or silver) salt solution, to form cuprous (or silver) acetylide (C_2Cu_2), dark red precipitate, explosive when dry, and yielding acetylene upon treatment with acid, (8) with **mercuric chloride** solution, to form trichloromercuric acetaldehyde ($C(HgCl)_3\cdot CHO$), precipitate, which yields with hydrochloric acid acetaldehyde plus mercuric chloride. Acetylene is made by reaction of **calcium** carbide and water, calcium hydroxide being simultaneously formed, and is formed when the gas in a Bunsen burner burns at the base of the burner; and when hydrogen is passed through a carbon **arc** (about 7% acetylene in the exit gas). Acetylene may be detected by the formation of explosive copper acetylide. Acetylene is used (1) as a fuel with oxygen for high temperature flames, (2) as an illuminant, (3) in the manufacture of **acetaldehyde**, from which a variety of chemicals is prepared, (4) in the manufacture of chloroderivatives, thus:

Acetylene → vinyl chloride by HCl, Cu_2Cl_2, NH_4Cl
        vinyl acetylene by same, which yields chloroprene
       → tetrachloroethane by Cl_2, which yields trichloroethylene ($CHCl:CCl_2$), pentachloroethane ($CHCl_2\cdot CCl_3$), perchloroethylene ($CCl_2:CCl_2$), hexachloroethane ($CCl_3\cdot CCl_3$). (R.K.S.)

ACETYLSALICYLIC ACID.
A **drug** commonly known as "aspirin." (See **Salicylic Acid.**) It is used for relief of milder forms of pain, especially joint and muscle pain. It also tends to reduce fever. It does not harm the heart, contrary to popular opinion. This drug is used in massive doses in acute rheumatic fever. (R.S.M.)

ACHENE.
An achene is a single-seeded **fruit**, which does not split when mature, and which has the seed free from the ovary wall except at the point of attachment. (R.M.W.)

ACHEULEAN. Paleontology of Man.

ACHILLES, TENDON OF.
In man the prominent tendon at the back of the ankle, extending from the muscle of the calf to the heel. Technically it is the tendon which attaches the gastrocnemius and soleus muscles to the calcaneum or heel bone. The name derived from human anatomy is used in relation to other vertebrates. (A.W.L.)

ACHLORHYDRIA.
Absence of **hydrochloric acid** in the **stomach.** This may occur normally in older people, and in certain diseases, such as **cancer** of the stomach, and pernicious **anemia.** (D.M.H.)

ACHOLIA.
Absence or lack of secretion of **bile.** (R.S.M.)

ACHONDRITES.
A form of stony meteorites without **chondri**, and having textures similar to those of some terrestrial rocks. (R.M.F.)

ACHRAS SAPOTA.
Sapodilla. Sapotaceae. A large tree native to the

Tendon of Achilles.

forests of Central and tropical South America, the fruit of which is an edible berry. Its greatest value is in its **latex** product, which yields chicle. The chicle-gathering industry is centered in Yucatan and Central America. The tapping is done in the rainy season. The tapper climbs to a height of 30–50', and with a machete cuts a series of connecting zig-zag diagonal gashes in the bark as he descends. At the bottom of this series of cuts he attaches a cup, into which the latex flows. The crude substance is collected, boiled down to eliminate much of its water and the coagulated product pressed into 20–25 lb. blocks. This substance, chicle, varies in quality from the best grade, which is milk-white in color, to pinkish or darker grades, which have received less care in preparation. Each tree yields about 2½ lbs. of chicle during one season and may be tapped every 6 years. The blocks of chicle are shipped largely to the United States, where they are melted and cleaned, flavored and sweetened, and then marketed as the familiar chewing gum. This use of the latex of the

Sapodilla is not new, since the Aztecs and their predecessors knew of it and used it. When first introduced into the United States it was tried as a rubber substitute, but proved unsuitable. (R.M.W.)

ACHROIT. Tourmaline.

ACHROMATISM. Chromatic Aberration.

ACICULITE. Aikinite.

ACICULUM. A strong internal **seta** found in the **parapodia** of annelid worms. (A.W.L.)

ACID ANHYDRIDES. Acids, Carboxylic.

ACID ROCK. According to A. Holmes an igneous rock which contains 66% silica is said to be acidic. The term is gradually going out of use. Since the geologist uses acid in a different sense from the chemist, Clarke has proposed persilicic for igneous rocks which are relatively rich in silica. (R.M.F.)

ACIDOSIS. A condition occurring in the body in which **acids** are absorbed or form in excess of their elimination or **neutralization**. The alkali reserve of the body is disturbed, the first step being a decrease in amount of bicarbonate (see **Carbon**) in the **blood**. The opposite condition results from excess formation or ingestion of **alkalis**, or from prolonged loss of acid from the stomach. The resulting condition is known as alkalosis. In this condition the alkali reserve is increased over the normal limit.

Ordinarily, excess acid or alkali formed or taken into the body does not cause either of these conditions. This is due to the ability of the body to protect itself automatically by preserving the acid-base **equilibrium**. This is accomplished by several mechanisms. In general, the balance is maintained by elimination, oxidation, excretion and neutralization. The buffer substances in the blood—the **salts, haemoglobin** and **protein**—act to lessen the change toward increased acid or alkali concentration. Further, there is a reserve of alkali to take care of any excess acid. This reserve of alkali consists of sodium bicarbonate, di-potassium phosphate and protein salts. Acidosis only occurs when the buffer substances and alkali reserve are depleted.

Excess acid or alkali can be eliminated by the kidneys and by the respiratory mechanism through its power to throw off greater or lesser concentrations of **carbon dioxide**. Neutralization of acid occurs with **ammonia** formed by the body **metabolism**. These are all normal bodily processes and the mechanism of equilibrium can still be maintained in abnormal conditions without producing an acidosis or alkalosis unless certain adverse factors enter into the picture.

Acidosis may occur in many conditions, usually those of serious nature. They are (1) starvation or inadequate intake of water or food (especially **carbohydrates**), or during acute **infections**, (2) after prolonged **anaesthesia**, (3) in diseases accompanied by severe **diarrhea** and vomiting, (4) in severe untreated **diabetes**, (5) in advanced kidney and heart disease.

Alkalosis may result from (1) prolonged vomiting with excess loss of **hydrochloric acid** from the stomach, (2) prolonged increase in the respiratory rate as is seen in certain disorders and in higher altitudes, (3) excess ingestion of alkalis by mouth as might occur in the treatment of peptic ulcer and in other conditions. This may also occur through prolonged use of alkali products by the laity due to the pernicious advertising of these products on the radio and in advertisements for treatment of imaginary and non-existent "acid conditions" and indigestion. (D.M.H.)

ACIDS, BASES AND SALTS. These are chemical compounds classified as **electrolytes**. Acids are electrolytes which furnish **hydrogen ions**, e.g., $HCl \rightarrow H^+ + Cl^-$. Bases are electrolytes which furnish **hydroxyl ions**, e.g., $NaOH \rightarrow Na^+ + OH^-$. Salts are electrolytes which furnish neither hydrogen nor hydroxyl ions, e.g., $NaCl \rightarrow Na^+ + Cl^-$. Salts are formed by the combination of equivalent weights of an acid and a base, a process called neutralization. The result is the formation of a salt and the combination of the hydrogen and hydroxyl ions to form water. $HCl + NaOH \rightarrow NaCl + H_2O$.

Water as an electrolyte occupies a unique position in that it furnishes both hydrogen ions and hydroxyl ions in equal amounts. In pure water the concentration of each of these ions is 10^{-7} moles per liter. The product of the hydrogen times the hydroxyl concentration is always constant and equal to 10^{-14}. When the hydrogen ion concentration is greater than 10^{-7} due to the presence of an acid, the hydroxyl ion concentration becomes less than 10^{-7} and the solution is said to be acidic. When the hydroxyl ion concentration is greater than 10^{-7} due to the presence of a base the hydrogen ion concentration adjusts itself to a value less than 10^{-7}. Thus the hydrogen ion concentration is a measure of the acidity or basicity of a solution. It is usually defined by stating the pH which is the negative logarithm of the hydrogen ion concentration (in moles per liter). The process of neutralization, whereby an acid and a base in **solution** react to form a salt—actually hydrogen ion of the acid and hydroxyl ion of the base react to form water leaving the **cation** of the base and the **anion** of the salt by recombination—is discussed elsewhere. (See **Reactions Involving Recombination of Ions.**)

Upon evaporation of the solvent, the salt is obtained as such, frequently as crystals, sometimes with, sometimes without water of crystallization. (See **Reactions Involving Water.**)

A salt can be defined as "a system built from oppositely charged ions which do not neutralize each other" (Kilpatrick, 1935). In this sense **hydrochloric acid** (H^+Cl^-) and **sodium** hydroxide (Na^+OH^-) are salts. Acids are those salts whose cation is hydrogen ion, H^+ (probably acid (H_3O^+) dissociating to proton (H^+) plus base (H_2O)), and bases those whose anion is hydroxyl, OH^- (probably acid (water) dissociating to proton (H^+) plus base (OH^-)). A salt (1) when dissolved in an ionizing solvent, e.g., sodium chloride in water, is a good conductor of electricity, and (2) when in the solid state forms a **crystal** lattice, e.g., sodium chloride crystals possess a definite lattice structure for both sodium cations (Na^+) and chloride anions (Cl^-), determinable by examination with x-rays.

A broader definition than that confined to solutions is demanded in some fields of chemistry, for example, in high temperature reactions of acids, bases, salts. In the formation of metallurgical **slags**, at furnace temperatures, **calcium** oxide is used as base and **silicon** oxide and **aluminum** oxide, as acids, and calcium aluminosilicate is produced as a fused salt. Sodium carbonate and silicon oxide when fused react to form the salt sodium silicate with the evolution of carbon dioxide. In this sense:

$$\left[\begin{array}{l}\text{Oxide of any}\\\text{element func-}\\\text{tioning as a}\\\text{metal, that is,}\\\text{as a base.}\end{array}\right] \text{plus} \left[\begin{array}{l}\text{Oxide of any}\\\text{element func-}\\\text{tioning as a}\\\text{non-metal, that}\\\text{is, as an acid.}\end{array}\right] \text{yields [Salt]}$$

Iron and **sulfur** when heated react to form the salt ferrous sulfide. In this sense:

[Metal] plus [Non-metal] yields [Salt]

Salts are, therefore, prepared (1) from solutions of acids and bases by **neutralization**, and separation by evaporation and crystallization, (2) from solutions of two salts by precipitation where the solubility of the salt formed is slight, e.g., **silver** nitrate solution plus sodium **chloride** solution yields silver chloride precipitate (almost all as solid) and sodium nitrate as sodium cations and nitrate anions in solution (recoverable as sodium nitrate solid by separation of silver chloride and subsequent evaporation of the solution), (3) from fusion of a basic oxide (or its suitable compound—sodium carbonate above) and an acidic oxide or its suitable compound—ammonium **phosphate** since ammonium and hydroxyl are volatilized as ammonia and water, thus:

ammonium sodium hydrogen phosphate $\begin{array}{c} NH_4 \diagdown \\ Na \!\!-\!\! \\ H \diagup \end{array} P^{5+}O_4$

yields sodium **metaphosphate** $Na—P^{5+}O_3$ upon heating), (4) from reaction of a metal and a non-metal.

Reactions of acids as such in solution without decomposition of anion, are dependent upon the presence of hydrogen cation (H^+) and the anion of the acid.

Reactions of bases as such in solution without decomposition of cation, are dependent upon the presence of the cation of the base and hydroxyl anion (OH^-).

Reactions of salts as such in solution, without decomposition of cation or anion, are dependent upon the presence of the cation and the anion of the salt.

Acids in general (exceptions are common) attack metals, **oxides, carbonates, sulfides, sulfites,** with the formation of a salt and, in the respective instances, hydrogen, water, **carbon dioxide, hydrogen sulfide, sulfur dioxide.** For Ionization Constants of Acids, see **Carboxylic Acids, Amines** and **Amides** for Nitrogen acids); of Bases, see **Amines** and **Amides** (for Nitrogen bases). (R.K.S.)

ACIDS, CARBOXYLIC, AND RELATED COMPOUNDS (Acid Anhydrides, Lactones, Lactides).

Carboxylic **acids** (containing carboxyl group—COOH) are of wide variety as to constitution, physical properties, methods of preparation, and uses, well illustrated by reference to some of the particular acids, such as formic, acetic, stearic, oleic, benzoic. Several hydroxy acids (containing hydroxyl group, —OH, and carboxyl group, —COOH) are found in important natural materials. Such acids are tartaric, citric, malic, lactic. The ionization constants of some organic acids and of **phenol,** which constants indicate the relative strength of these acids, are as follows, arranged in decreasing acidic strength:

Acid	Ionation Constant of Acid
Trichloroacetic	2×10^{-1}
Dichloroacetic	5×10^{-2}
Oxalic	4×10^{-2}
Malonic	2×10^{-3}
Chloroacetic	2×10^{-3}
Phthalic	1×10^{-3}
Tartaric	1×10^{-3}
Salicylic	1×10^{-3}
Citric	8×10^{-4}
Malic	4×10^{-4}
Formic	2×10^{-4}
Lactic	1×10^{-4}
Benzoic	7×10^{-5}
Succinic	7×10^{-5}
Acetic	2×10^{-5}
Carbonic	3×10^{-7}
Hydrocyanic	7×10^{-10}
Phenol	1×10^{-10}

Acids that are insoluble or slightly soluble in water may usually be titrated after dissolving in alcohol, and the amount of sodium hydroxide standard solution required to neutralize a given weight of the acid is characteristic, and an indication of the particular acid involved.

Substituted chloro-, bromo-, iodo-, amino- and cyano-, thio-, phospho-, acids will be found under the elements **chlorine, bromine, iodine, nitrogen, sulfur, phosphorus,** respectively.

Primary **alcohols** or **aldehydes,** upon **oxidation,** yield the corresponding carboxylic acids, and methyl **ketones** yield acetic acid, among other products. Regulated reduction of carboxylic acids yields the corresponding aldehydes or primary alcohols.

When the sodium or calcium salt of carboxylic acids is heated with **sodium** hydroxide or **calcium** oxide the **hydrocarbon** containing one less carbon atom than the acid is formed, e.g., sodium acetate yields **methane,** sodium benzoate yields **benzene** (and Na_2CO_3 or $CaCO_3$).

Acid **anhydrides** and acid or acyl **chlorides** are important organic reagents, e.g., acetic anhydride ($(CH_3CO)_2O$), benzoyl chloride (C_6H_5COCl).

When hydroxy- or amino-acids lose water or ammonia, respectively, characteristic reactions occur, as follows:

Hydroxy- or Amino Acid	Acid	Product
Alpha-hydroxy	$CH_3CHOHCOOH$	Lactide: $\begin{array}{c} H_3C \cdot CH—CO—O \\ \mid \qquad\qquad \mid \\ O—CO—HC \cdot CH_3 \end{array}$
Beta-hydroxy	$CH_3CHOHCH_2COOH$	Unsaturated acid: $CH_3CH:CH \cdot COOH$
Gamma-hydroxy	$CH_3CHOHCH_2CH_2COOH$	Gamma-lactone: $\begin{array}{c} CH_3CHCH_2CH_2CO \\ \mid\!\!—O—\!\!\mid \end{array}$
Alpha-amino	$CH_2NH_2 \cdot COOH$	Lactim: $\begin{array}{c} CH_2—NH—CO \\ \mid \qquad\qquad \mid \\ CO—NH—CH_2 \end{array}$
Beta-amino	$CH_2NH_2 \cdot CH_2 \cdot COOH$	Unsaturated acid: $CH_2:CH \cdot COOH$
Gamma-amino	$CH_2NH_2 \cdot CH_2 \cdot CH_2 \cdot COOH$	Gamma-lactam: $\begin{array}{c} CHCH_2CH_2CO \\ \mid\!\!—NH—\!\!\mid \end{array}$

SELECTED REPRESENTATIVE CARBOXYLIC ACIDS

Acid	Formula	Melting Point (°C.)	Boiling Point (°C.)
1. Carbonic	$(HO)_2CO$		
2. Formic	$H \cdot COOH$	8.5	100.5
3. Acetic	$CH_3 \cdot COOH$	16.6	118
4. Propionic	$C_2H_5 \cdot COOH$	−22	141
5. Normal-butyric (butanoic)	$C_3H_7 \cdot COOH$	−8	163
6. Iso-butyric	$(CH_3)_2CH \cdot COOH$	−47	154
7. Valeric (pentanoic)	$C_4H_9 \cdot COOH$	−59 appr.	187
8. Caproic (hexanoic)	$C_5H_{11} \cdot COOH$	9	202
9. Heptanic (oenanthylic)	$C_6H_{13} \cdot COOH$	17	260 appr.
10. Caprylic (octanoic)	$C_7H_{15} \cdot COOH$	16	237
11. Nonanoic (pelargonic)	$C_8H_{17}COOH$	12	254
12. Capric (decanoic)	$C_9H_{19} \cdot COOH$	31	269
13. Undecylic (undecanoic)	$C_{10}H_{21}COOH$	30	228 (160 mm.)
14. Lauric (dodecanoic)	$C_{11}H_{23} \cdot COOH$	48	225 (100 mm.)
15. Tridecylic (tridecanoic)	$C_{12}H_{25}COOH$	51	236 (100 mm.)
16. Myristic	$C_{13}H_{27} \cdot COOH$	58	250 (100 mm.)
17. Pentadecylic	$C_{14}H_{29} \cdot COOH$	52	257 (100 mm.)
18. Palmitic	$C_{15}H_{31} \cdot COOH$	64	340 appr. dec.
19. Margaric (heptadecanoic)	$C_{16}H_{33}COOH$	60	227 (100 mm.)
20. Stearic	$C_{17}H_{35} \cdot COOH$	69	383
21. Nondecylic	$C_{18}H_{37} \cdot COOH$	66	
22. Arachidic	$C_{19}H_{39}COOH$	75	
23. Behenic	$C_{21}H_{43}COOH$	83	
24. Lignoceric	$C_{23}H_{47}COOH$	80	
25. Cerotic	$C_{25}H_{51} \cdot COOH$	78	
26. Melissic	$C_{29}H_{59} \cdot COOH$	90	
27. Acrylic	$CH_2:CH \cdot COOH$	12	142
28. Crotonic (alpha)	$CH_3 \cdot CH:CH \cdot COOH$	72	185
29. Iso-crotonic (beta)	$CH_2:CHCH_2COOH$	15	172 dec.
30. 2-Methylacrylic	$CH_2:C(CH_3) \cdot COOH$	15	162
31. Vinylacetic	$CH_2:CH \cdot CH_2 \cdot COOH$	−39	163
32. Angelic (2-methycrotonic)	$CH_3 \cdot CH:C(CH_3) \cdot COOH$		
33. Oleic	$CH_3(CH_2)_7CH:CH(CH_2)_7 \cdot COOH$	14	286 (100 mm.)
34. Linoleic (linolic)	$CH_3(CH_2)_4CH:CHCH_2CH:CH(CH_2)_7COOH$	−18	230 (16 mm.)
35. Linolenic (3 double bonds)	$C_{17}H_{29}COOH$		
36. Propargylic acid (propiolic)	$CH:CCOOH$	9	144 dec.
37. Furoic (pyromucic)	$C_4H_3O \cdot COOH (2)$	131	231
38. Benzoic	C_6H_5COOH	122	249
39. Phenylacetic	$C_6H_5CH_2COOH$	77	265
40. Diphenylacetic	$(C_6H_5)_2CHCOOH$	148	
41. Triphenylacetic	$(C_6H_5)_3CCOOH$	265	
42. Cinnamic (beta-phenylacrylic)	$C_6H_5CH:CHCOOH$	133	300
43. Ortho-toluic	$CH_3C_6H_4COOH (2)$	102	259
44. Meta-toluic	$CH_3C_6H_4COOH (3)$	110	263
45. Para-toluic	$CH_3C_6H_4COOH (4)$	177	275
46. Oxalic	$COOH \cdot COOH$ {anhydrous / crys. $2H_2O$}	189 / 101	150 subl.
47. Malonic	$COOHCH_2COOH$	136	dec.
48. Succinic	$COOH(CH_2)_2COOH$	185	235
49. Glutaric	$COOH(CH_2)_3COOH$	97	304 dec.
50. Adipic	$COOH(CH_2)_4COOH$	151	265 (100 mm.)
51. Pimelic	$COOH(CH_2)_5COOH$	103	272 (100 mm.)
52. Suberic	$COOH(CH_2)_6COOH$	140	
53. Sebacic	$COOH(CH_2)_8COOH$	135	295 (100 mm.)
54. Camphoric	$HOOC\!-\!C(CH_3)_2\!-\!C\!-\!COOH$ dextro inactive; H_3C, CH_2, CH_2, H	187 / 202	
55. Fumaric (trans-ethylene dicarboxylic)	$COOHCH:CHCOOH$	287	290
56. Maleic (cis-ethylene dicarboxylic)	$COOHCH:CHCOOH$	130	135 dec.
57. Ortho-phthalic (ortho-benzene dicarboxylic)	$C_6H_4(COOH)_2 (1,2)$	191	dec.
58. Meta-benzene dicarboxylic (isophthalic)	$C_6H_4(COOH)_2 (1,3)$	330	subl.
59. Para-benzene dicarboxylic (terephthalic)	$C_6H_4(COOH)_2 (1,4)$	subl.	
60. 1,2,3-Benzenetricarboxylic (hemimellitic)	$C_6H_3(COOH)_3 (1,2,3)$	190	
61. 1,2,4-Benzenetricarboxylic (trimellitic)	$C_6H_3(COOH)_3 (1,2,4)$	216 dec.	

SELECTED REPRESENTATIVE CARBOXYLIC ACIDS—*Continued*

ACID	FORMULA	MELTING POINT (°C.)	BOILING POINT (°C.)
62. 1,3,5-Benzenetricarboxylic..... (trimesic)	$C_6H_3(COOH)_3$ (1,3,5)...................	350 subl.	
63. 1,2,3,4-Benzenetetracarboxylic.. (prehnitic)	$C_6H_2(COOH)_4$ (1,2,3,4)................	237 dec.	
64. 1,2,3,5-Benzenetetracarboxylic.. (mellophanic)	$C_6H_2(COOH)_4$ (1,2,3,5)................	238	
65. 1,2,4,5-Benzenetetracarboxylic.. (pyromellitic).............	$C_6H_2(COOH)_4$ (1,2,4,5)................	264	
66. Benzenehexacarboxylic........ (mellitic)	$C_6(COOH)_6$....................	dec.	
67. Glycollic (hydroxyacetic)......	$CH_2OH \cdot COOH$...............	65 appr.	dec.
68. Lactic (alpha-hydroxypropionic)	$CH_3CHOHCOOH$...............		122 (14 mm.)
sarcolactic	dextrolaevo	18	
paralactic	dextro	25	
69. Beta-hydroxypropionic........ (hydracrylic)	CH_2OHCH_2COOH......		dec.
70. Alpha-hydroxybutyric.........	$CH_3CH_2CHOHCOOH$..........	42	260
71. Beta-hydroxybutyric..........	$CH_3CHOHCH_2COOH$..........		130 (14 mm.)
72. Gamma-hydroxybutyric.......	$CH_2OHCH_2CH_2COOH$........	−17	
73. Alpha-hydroxystearic........	$C_{16}H_{33}CHOHCOOH$........	92	
74. Glyceric.................... (2,3-dihydroxypropionic)	$CH_2OHCHOHCOOH$........		
75. Tartaronic................... (2-hydroxypropandioic)	$COOHCHOHCOOH$........	158 dec.	
76. Malic (2-hydroxysuccinic).....	$COOHCH_2CHOHCOOH$.......dextrolaevo	133	150 dec.
	laevo	100	140 dec.
77. Tartaric (2,3-dihydroxysuccinic) racemic.................	$COOHCHOHCHOHCOOH$.........		
	dextrolaevo	205	
	dextro and laevo	170	
mesotartaric..........	inactive	140	
78. Arabonic....................	$CH_2OH(CHOH)_3COOH$...............	89	
79. Gluconic...................	$CH_2OH(CHOH)_4COOH$.......		
80. Glycuronic.................	$CHO(CHOH)_4COOH$.......		
81. Mucic.....................	$COOH(CHOH)_4COOH$.......	206 dec.	
82. Saccharic.................	$COOH(CHOH)_4COOH$.......	(lactone)	
83. Citric (hydroxytricarboxylic)...	$COOHCH_2C(OH)(COOH)CH_2COOH$......	153	
84. Ricinoleic..................	$C_{17}H_{32}(OH)(COOH)$.......	4	
85. Mandelic..................	$C_6H_5CHOHCOOH$.......	118	dec.
86. Salicylic..................	HOC_6H_4COOH (2).......	159	
87. Gallic....................	$(3,4,5)(HO)_3C_6H_2COOH$.......	235 dec.	
88. Glyoxalic (glyoxylic).........	$CHO \cdot COOH$.......	dec.	
89. Pyruvic...................	$CH_3COCOOH$.......	9	165 dec.
90. Acetoacetic...............	CH_3COCH_2COOH.......		100 dec.
91. Levulinic.................	$CH_3COCH_2CH_2COOH$.......	33	246
92. Mesoxalic.................	$COOH \cdot CO \cdot COOH \cdot H_2O$.......	120	
	or $COOH \cdot C(OH)_2 \cdot COOH$...............		

SELECTED REPRESENTATIVE ACID ANHYDRIDES

ACID ANHYDRIDE	FORMULA	MELTING POINT (°C.)	BOILING POINT (°C.)
1. Acetic.....................	$(CH_3CO)_2O$...............	−73	140
2. Propionic..................	$(C_2H_5CO)_2O$...............	−45	165 appr.
3. Normal-Butyric.............	$(C_3H_7CO)_2O$...............	−75	198
4. Iso-Butyric................	$((CH_3)_2CHCO)_2O$...............	−53	182
5. Benzoic...................	$(C_6H_5CO)_2O$...............	43	360
6. Cinnamic..................	$(C_6H_5CH:CHCO)_2O$...............	135	
7. Succinic..................	$(CH_2CO)_2O$...............	120	261
8. Maleic...................	$(CHCO)_2O$...............	57	202
9. Glycollic.................	$(CH_2OHCO)_2O$...............	130	
10. Lactic....................	$(CH_3CHOHCO)_2O$...............	260 dec.	
11. Salicylic.................	$(HOC_6H_4CO)_2O$...............	200 appr.	dec.

(*Continued on next page*)

SELECTED REPRESENTATIVE ACID ANHYDRIDES—*Continued*

Lactone and Lactide (Hydroxyacid Anhydrides)

Acid Anhydride	Formula	Melting Point (°C.)	Boiling Point (°C.)
1. Butyrolactone (gamma).......	$CH_2CH_2CH_2CO$ .... with O bridge		206
2. Lactide (dilactide)...........	$H_3C \cdot CH - CO - O$ .... $O - CO - HC \cdot CH_3$	125	255

(R.K.S.)

ACIDS, INORGANIC. See individual acids.

ACINIFORM GLANDS. The glands of **spiders** which produce the silk used to enclose the prey. (A.W.L.)

ACME. Screw Thread.

ACMITE-AEGIRITE. Acmite is a comparatively rare rockmaking mineral, usually found in nephelite **syenites** or other **nephelite** or **leucite** bearing rocks, as phonolites. Chemically it is a **soda-iron** silicate, and its name refers to its sharply pointed **monoclinic** crystals. Bluntly terminated crystals form the variety **aegirite**, named for Aegir, the Icelandic sea god.

Acmite has a hardness of 6 to 6.5, specific gravity 3.5, vitreous luster, color brown to green, transparent to opaque.

The original acmite locality is in Norway. Greenland furnishes fine specimens. United States localities are Magnet Cove, Arkansas, and Libby, Montana, where a variety carrying **vanadium** occurs. (E.S.C.S.)

ACNE. *Acne vulgaris* is a chronic infection of the **sebaceous glands** in the skin. The exact cause of this disease is not known. It is usually first seen in certain susceptible individuals with the onset of **puberty**. Because of its occurrence at this time, many observers believe that the primary cause is bound up with physiological and biochemical changes associated with sudden increase in function of the sex glands. It usually disappears on reaching adult life, although in certain individuals the disease becomes chronic. It is most often seen in people with oily skins. The external opening on the skin of the sebaceous gland becomes narrowed, the sebaceous material cannot escape and the retained material serves as a source of irritation, secondarily infecting the surrounding tissue. Acne may cause permanent scarring. It is most successfully treated by **x-rays** and **ultra-violet rays** or sunlight, although vaccines, local treatment of the skin, glandular and hygienic methods are also useful in certain cases. It is not contagious.

Acne rosacea is a much less-common skin disorder characterized by a red bulbous nose, and inflammation and infection of the flush area of the face. Its etiology is unknown. (D.M.H.)

ACOELA. An order of free-living flatworms in which the alimentary tract is without a cavity. (A.W.L.)

ACOELOMATA. Animals without a **coelom**. The term is applied especially to the flatworms, nemertine worms, and roundworms; these animals have attained the mesoderm in which the body cavity develops but it remains a more or less continuous mass with small spaces if any. (A.W.L.)

ACONINE. Alkaloids.

ACONITINE. Alkaloids.

ACONTIA. Filaments bearing stinging cells in **sea anemones.** They can be shot out through the mouth or through special pores when the animal is stimulated. (A.W.L.)

ACORN TUBE. A small vacuum **tube** in a glass envelope shaped somewhat like an acorn and designed to have a minimum of **interelectrode capacitance** and lead inductance. Such characteristics are essential for successful operation at very high frequencies. (See **Ultra High Frequencies.**) (L.R.Q.)

ACOUSTICS. In the broader sense, acoustics is the physics of **sound**, treated in all its aspects. Commonly, however, the term is restricted to a study of the transmission of sound through various media or in various enclosures or coduits, including the effects of reflection, refraction, interference, diffraction, and absorption.

Of especial importance is the acoustics of buildings and auditoriums. Sound from a source within an enclosure tends to build up to a maximum of intensity, limited only by leakage (as through open doors or windows) and by the dissipation of the energy through air viscosity and absorption. It is thus much easier to make one's self heard in a small, closed room than out of doors. In a large enclosure, an important factor is "reverberation," that is, the re-echoing of sounds among the various exposed solid surfaces, which, if too protracted, may impair the distinctness of audition. Sabine defines the "reverberation time" as the period in which the intensity falls to one millionth of its steady value after the source is suddenly silenced. He obtained an empirical formula for the time in seconds, as follows: $T = 0.164V/a$. Here V is the volume of the room in cu. meters, while a is the sum of the equivalent absorptions of the various exposed surfaces, each equal to the product of the area of the surface in sq. meters by its "acoustic absorptivity" (ratio of absorbed to incident sound energy). Sabine found the absorptivity of plaster to be 0.033, glass 0.027, wooden floor 0.061, linoleum 0.120, etc., and the absorption of an audience to be 0.44 sq. meter per person (equivalent to 0.44 sq. meter of a perfect absorber), of each empty wooden seat 0.008 sq. meter, of each upholstered seat 0.30 sq. meter, etc. It is thus possible to predict by Sabine's law the reverberation time of an auditorium before it is built, and to judge whether it will be good for piano music (for which the best value is about 1.1 sec.) or for speaking (for which a smaller value is desirable). Another factor in auditorium acoustics is concerned with "focusing" the sounds from the stage upon the audience by means of properly shaped walls and ceiling, a problem which Sabine studied by means of photographs of sound waves in small model enclosures.

The investigation of such problems as the passage of sound through simple tubes, or tubes having branches, or through cavities of various shapes, such as musical instruments or resonators, reveal certain remarkable analogies to the theory of a-c circuits. Thus we encounter the property known as acoustic impedance; with its components, acoustic resistance and acoustic reactance, the latter being dependent upon the acoustic inertance (analogous to inductance) and acoustic compliance (sometimes called acoustic capacitance). These correspond to analogous properties of a-c circuits, in which the electric impulses may be compared to acoustic waves with electricity as the medium. The acoustic resistance, inertance, and compliance depend, respectively, upon the viscosity, the density, and the elasticity of the medium. It is an interesting fact that by causing sound to pass through a suitably designed conduit, such as the two-branched or shunted "Quincke tube," certain frequencies may be suppressed, much as electric oscillations are suppressed by electric wave filters. This effect is due to the interference of wave trains following different routes. (L.D.W.)

ACRASPEDOTE. A medusa without a **velum,** such as the common **jellyfishes.** (A.W.L.)

ACRIDINE. Pyridine and Related Compounds.

ACRIFLAVINE. An acridine (see **Pyridine and Related Compounds**) **dye,** reddish-yellow in color, which is used in dilute solution in the treatment of infected or contaminated wounds. It posseses germicidal and antiseptic properties and, when used in the proper dilution, will not injure body tissues. In weak solutions it is also used for irrigation of diseased body cavities, as, for instance, the urinary **bladder** when **cystitis** is present. (R.S.M.)

ACROCYST. A chamber formed of the **blastostyle** of some species of **hydroids,** in which the eggs develop. (A.W.L.)

ACRODONT. Dentition.

ACROLEIN. Aldehydes, Ketones, and Related Compounds.

ACROMEGALIA or ACROMEGALY. A disease of the **pituitary gland** characterized by overgrowth of the skull, thickening and coarsening of the skin and facial features, enlargement and protrusion of the lower jaw, and enlargement of the flat bones, particularly the hands and feet. There is also a general enlargement of the heart and abdominal organs.

The etiology is an increase and overfunctioning of the cells staining with acid dyes, the acidophilic group of cells in the anterior pituitary gland, usually with the formation of an acidophilic adenoma. If the overfunction occurs before the growth of the long bones is complete, gigantism occurs; if it occurs in adult life, acromegaly is the result. (D.M.H.)

ACRYLIC ACID. Alcohols and Ethers.

ACTINIUM. Symbol: Ac. A radioactive element, and also a series of radioactive elements. (See **Radioactive Changes.**)

ACTINOLITE. The term for a **calcium-iron-magnesium amphibole,** the formula being $Ca_2(Fe,Mg)_5$-$(OH)_2(Si_4O_{11})_2$, but the amount of iron varies considerably. It occurs as bladed crystals or in fibrous or granular masses. Its hardness is 5–6, specific gravity 2.9–3.2, color green to greyish green, transparent to opaque, luster vitreous to silky or waxy. Iron in the **ferrous** state is believed to be the cause of its green color. Actinolite derives its name from the frequent radiated groups of crystals. Actinolite is found in **schists,** often with **serpentine** and in **igneous** rocks probably as the result of the alteration of **pyroxene.** The schists of the Swiss Alps carry actinolite. It is also found in Austria, Saxony, Norway, Japan, and Canada in the provinces of Quebec and Ontario. In the United States actinolite occurs in Massachusetts, Pennsylvania, Maryland, and as a **zinc-manganese** bearing variety in New Jersey. (E.S.C.S.)

ACTINOMYCOSIS. A disease of man and certain domestic animals ("lumpy jaw" of cattle) caused by a pathogenic **fungus** commonly known as the ray fungus.

The organism is not infrequently carried in the normal mouth without harm. Infection usually occurs by invasion from the mouth. Rarely, the source of infection may be outside the body. Actinomycosis is not contagious.

Clinically, several forms occur, depending on the site of localization. The cervico-facial involving primarily the face and neck, the abdominal, and pulmonary are the common types. Any viscus may be involved, however. The lesions are always multiple with chronic **abscesses** which burrow, spread, leave sinus tracts, and heal with the production of large amounts of fibrous tissue. The organism is found in the **pus** in actinomycotic abscesses as pin-point-sized yellow granules.

The mortality is high in the abdominal and pulmonary types; the commoner cervico-facial type has a better prognosis.

Treatment is surgical plus long continued use of **sulfonamide drugs.** In the past, iodides have been used, and recently **penicillin,** the latter with some success. (D.M.H.)

ACTINOMYXIDA. Sporozoa.

ACTINON. Symbol: An. A radioactive element of the actinium series. (See **Radioactive Changes.**)

ACTINOZOA. Anthozoa.

ACTINULA. A larval form of **hydroid,** resembling a **polyp** with a short stalk. (A.W.L.)

ACTION. In certain discussions of dynamics there is need of an expression for the product of twice the mean total **kinetic energy** of a system of particles, during a specified interval of time, by the duration of the interval. This product is called the "action." Mathematically, it is expressed by

$$S = 2\int_{t_0}^{t} E_K dt,$$

in which E_K is the kinetic energy and t_0 and t are the times of beginning and ending of the interval. The c.g.s. unit of action is the erg-second. The well known "Planck's constant h" is the common designation for the elementary quantum of action.

Maupertuis enunciated a law, known as the "principle of least action," which states that when a dynamic system is left to itself, unaffected by outside forces, so that its total energy cannot alter, any spontaneous change within the system takes place in such fashion that the action has the least possible value during the interval covered by the change. (L.D.W.)

ACTIVATION. Cathode.

ADAMANT. The term adamant was used by Theophrastus about 300 B.C. to mean **lodestone.** Chaucer as well as other medieval writers used it similarly. There seems to have been some confusion as to the exact nature of adamant; some regarded it as referring either to lodestone or **diamond** or to a mythical substance combining the properties of both. Curiously enough there is a rare variety of bort (black diamond) which is magnetic, due to the mechanical admixture of particles of **magnetite.** Whether this sort of black diamond was known to the ancients is wholly a matter of conjecture. The long continued association of the idea of hardness with the word adamant has led to our modern usage of the term as well as the derived, adamantine, referring to the luster of substances of high refractive indices. (E.S.C.S.)

ADAMANTINE SPAR. Corundum.

ADANSONIA DIGITATA. Baobab. Bombaceae. A tree of tropical Africa, the baobab has an extremely large trunk (the diameter sometimes exceeding 30'), but does not attain great height. The natives often excavate the light, easily worked wood and use the hollow so formed for shelter. The large fruit has a mucilaginous pulp which is often eaten; and the bark yields a fiber used for making cloth and rope. (R.M.W.)

ADAPTABILITY. A property of living matter through which it makes adjustments to its environment. It is accomplished chiefly through four fundamental properties: irritability, conductivity, contractility, and secretion. The first enables it to receive impressions from surrounding factors. Through the second all parts of the living body are in communication with the parts capable of receiving stimuli from without. The last two are the more common means of response. Contractility is the source of movement by which spatial adjustments of the body in relation to its surroundings are accomplished, and secretion is a means of producing special substances, such as digestive fluids, involved in the animal's reactions.

Adaptability is expressed in all of the actions of an organism, and in a more permanent way in the **adaptation** of the individual and species to its mode of life. (A.W.L.)

ADAPTATION. The process of modification of the living organism to adjust it to the conditions of its environment. Also an inherited character that enables the organism to meet certain environmental conditions.

All living things are adapted for a mode of life characteristic of their kind, under equally characteristic environmental conditions. They receive from previous generations a heritage (see **Heredity**) that fits them for this mode of life, and all characters in the hereditary complex that are of definite use are adaptive. Wings, for example, are an essential flight adaptation, and fins or some similar appendage are commonly found as adaptations for swimming. Other characters such as the colors and patterns of butterfly wings are usually of no apparent value and may be called non-adaptive or incidental.

Regardless of its adaptive heritage, however, each individual encounters some fluctuations in its environment to which it must adjust itself. The resulting changes in its body are adaptive, no less than its inherited structures. They are the acquired characters of biological literature, and have also been called individual adaptations. Human beings commonly experience two fine examples of this kind of adaptation in the calluses formed by the skin in response to friction, and the deposition of pigment, or tanning, as a protection against excessive ultra-violet light. A less evident result of exposure to ultra-violet light is a protective thickening of the epidermis, probably as important as the accompanying increase in pigmentation.

The relation of adaptations of both kinds to individual life is evident in any living thing. Beyond this field they are of great importance in theories of **evolution,** the one as the extensive result of evolutionary processes of the past and the other as a possible factor in the accomplishment of changes of evolutionary significance. (A.W.L.)

ADAPTIVE (RADIAL) EVOLUTION. Fossil Invertebrates and Fossil Reptiles.

ADAPTIVE RADIATION. A principle of evolutionary development formulated by Henry Fairfield Osborn. It assumes that a limited stock of animals in a restricted area tends to become broken up in the course of time into species adapted to various special habitats. Thus, as originally applied to the **mammals,** the ancestral stock is supposed to have been a small walking species with generalized teeth and omnivorous habits. From it the running and jumping species, burrowers and fliers, and species adapted to eat flesh or to eat vegetation are supposed to have evolved.

Although originally applied to the mammals the principle has been found to apply equally well to other groups, such as the insects. In some cases divergent or branching evolution has been noted, as well as radiating development from a common central stock. (A.W.L.)

ADDAX. Mammalia, Artiodactyla. An **antelope,** *Addax nasomaculatus,* of northern Africa and Arabia. (A.W.L.)

ADDENDUM. In a gear tooth, the distance from the pitch circle to the outer circle or top of the tooth. (H.C.H.)

ADDER. Reptilia, Serpentes. A term loosely applied to various **snakes,** both poisonous and non-poisonous. Among the poisonous species are the resplendent adders of Asia, which are near the American coral snakes, and the deadly **crait** and raj-samp of India, also known as the blue adder and banded adder, respectively. Australia also has two poisonous species, the death adders (*Acanthophis*), and the deadly puff-adder, *Bitis arietans,* is found throughout Africa. In North America the last name is applied to the harmless hog-nosed snakes, although spreading adder is perhaps more commonly used. (A.W.L.)

ADDICT. One who makes a habit of taking a narcotic **drug,** and comes to require it to maintain a sense of well-being. The narcotics chiefly used by addicts are **morphine, heroin,** and **cocaine.** (D.M.H.)

ADDISON'S DISEASE. A chronic disease due to destruction of the **cortex** of the **adrenal glands,** and diminution of the cortical **hormone.** It is characterized by marked weakness, wasting, brown pigmentation of the skin and **mucous membranes,** disturbances of the digestive tract and low **blood pressure.** The blood **electrolyte** balance is upset, with low chlorides, low sodium and high potassium. The disease is gradual, with progressive exhaustion and debility, eventually resulting in death if treatment is not given.

At autopsy, **tuberculosis** of the adrenal glands is the most common finding, although primary **atrophy,** or other destructive lesions may be found.

Treatment is with cortical hormone and salt—which produce a prompt remission in the acute disease; both of these must be continued throughout life much as insulin must be supplied to a diabetic. With adequate therapy, the patient with Addison's disease may lead a relatively normal life for years. (D.M.H.)

ADDITION. Addition is one of the fundamental operations with **numbers,** by which two or more numbers are combined to give a number; the result is called the sum.

Addition of numbers is subject to several fundamental rules or so-called laws: the commutative and associative laws, and in combination with **multiplication,** the distributive law.

The commutative law for addition of numbers is expressed by the formula

$$a + b = b + a$$

for any two numbers a and b; in words, the sum of any two numbers is the same in whatever order they are added. A similar statement applies to the sum of more than two numbers.

The associative law for addition of numbers is expressed by

$$(a + b) + c = a + (b + c)$$

for any three numbers a, b, c; in words, the sum of any three numbers is the same in whatever manner they are grouped. This is easily extended to more than three numbers.

The absolute value of a positive number is the number itself, and the absolute value of a negative number is its numerical value regardless of algebraic sign.

To add two numbers having like signs, add their absolute values and prefix the common sign.

To add two numbers having unlike signs, take the difference of their absolute values and prefix to it the sign of the number having the larger absolute value. (L.L.S.)

ADDITIVE COLOR PROCESS. A system of color photography in which the color synthesis is obtained by the addition of colors one to another in the form of light rather than as colorants. This color addition may take place (1) by the simultaneous projection of two or more (usually three) color images onto a screen, (2) by the projection of the color images in rapid succession onto a screen or (3) by viewing minutely divided juxtaposed color images. The principles of color analysis for a subject to be reproduced by an additive color process are discussed more fully under **color photography.** In the case of a 3-color process, three color records are made from the subject recording, in terms of silver densities, the relative amounts of red, green and blue present in various areas of the subject.

When the additive synthesis is made by simultaneous projection, positives are made from the color separation negatives and projected with a triple lantern onto a screen through red, green and blue filters. The registered color images give all colors of the subject due to simple color addition, red plus green making yellow, red plus blue appearing magenta, etc.

When the additive synthesis is made by successive viewing, the same three color images must be flashed onto the screen in such rapid succession that the individ-

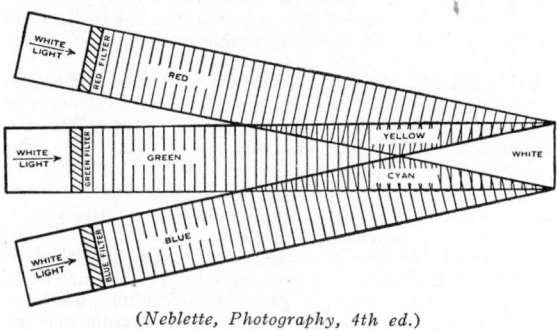

(*Neblette, Photography, 4th ed.*)

ual red, green and blue images are not apparent. Simple color addition is again obtained but this time use is made of the persistence of vision to "mix" the colors.

The third type of additive synthesis makes use of the fact that small dots of different colors, when viewed from such a distance that they are no longer individually visible, form a single color by simple color addition. The three color images in this type of process are generally side-by-side in the space normally occupied by a single image. The red record image will be composed of a number of red dots or markings of differing density which in their sum total will compose the red record image. Alongside the red markings will be green and blue markings, without any overlapping. When viewed at such a distance that the colored markings are at, or below, the limit of visual resolution, the color sensation from any given area will be the integrated color of the markings comprising the area—an additive color mixture.

All three types of additive color synthesis have been used commercially but, even with their high color fidelity, the expensive and inconvenient viewing or projecting devices have seriously limited the use of the first two types of addition. The third method, a color

mosaic formed photographically, has been the most successful of the additive color processes and has found widespread use in the mosaic screen processes and in the lenticular process. (H.C.C.)

ADENINE. Akaloids.

ADENITIS. Inflammation or infection of glandular tissue. The term generally is applied to inflammation of the lymph glands. Cervical adenitis is inflammation of the glands of the neck; axillary adenitis is inflammation of the glands of the axilla; inguinal adenitis is inflammation of the glands of the groin; mesenteric adenitis is inflammation of the glands in the mesentery of the small intestine. The infection can be local, originating in the glands or by drainage into the glands from a nearby inflammation, or it can accompany a systemic disease. (R.S.M.)

ADENOIDS. An increase in the normal **lymphoid** tissue in the nasopharynx (see **Pharynx**), which is sometimes referred to as the third or pharyngeal tonsil. The condition usually accompanies enlargement of the tonsils and, as with their enlargement, chronic infection plays an important part in its development. The disease, usually occurring during childhood, is easily recognized when fully developed by the dull facial expression, inability to blow the nose, mouth-breathing, snoring, and nasal twang to the voice. It is usually accompanied by frequent head colds and ear infections. The treatment is surgical removal; this is usually carried out at the same time as tonsillectomy. (D.M.H.)

ADENOMA. A benign **tumor** consisting of an encapsulated overgrowth of epithelial cells of a glandular structure. Adenomata may occur in the **endocrine glands,** the gastro-intestinal tract, the **respiratory system,** the breast, and wherever glandular **epithelium** occurs. (See **Gland.**) (D.M.H.)

ADHESION. This term is used both in medicine and in physics.

In medicine, adhesion refers specifically to an abnormal adherence of tissues of the body, either directly, surface to surface, or by bands of connective **tissue.** This most commonly occurs as a result of inflammation, as in **joints,** where adhesions between the lining membranes (synovial membranes) can produce stiffness and pain. In the peritoneal cavity following various degrees of peritoneal infection, bands or adhesions can occur between the peritoneal coats of solid organs or intestines. These occasionally cause intestinal obstruction requiring operative treatment. In the chest, as a result of various types of **pleurisy,** adhesions often form between the **pleura** covering the lung and the lining of the interior of the chest cavity.

In physics, the terms adhesion and cohesion designate intermolecular forces holding matter together. The tendency of matter to hold itself together or to cling to other matter is one of its most characteristic properties. Adhesion and cohesion are merely different aspects of the same phenomenon, which is apparently of the nature of an intermolecular attraction. We speak of cohesion as an interaction between adjacent parts of the same body and as acting throughout the interior of its substance, while adhesion refers to a similar interaction between the closely contiguous surfaces of adjacent bodies.

There is reason to believe that as two neutral molecules or atoms approach each other, their mutual potential energy reaches a minimum value at a certain equilibrium distance; so that work would be necessary either to push them closer or to pull them farther apart, because of forces which are probably electrical. (See **Least Energy Principle.**) The distribution of molecules, ions, or atoms in a solid is determined by this type of equilibrium, and the regular spacing of crystal structure and the architecture of the molecule itself are dependent upon it. Any force tending to diminish the equilibrium distance meets with the rapidly increasing reaction of compressive elasticity, while any force tending to increase it is opposed by cohesion, which increases at first and then rapidly diminishes toward zero as the point of fracture is reached.

The behavior of bodies which are aggregates of crystals or of fibers is complicated by the friction and the adhesion of the adjacent particles, so that the ultimate strength of a material is not a safe measure of its true cohesion. A filament of spun quartz may be much stronger when freshly drawn than later when crystallization replaces its initial cohesion by the adhesion between separate crystals; and yarn is not nearly so strong as the cotton or wool fiber composing it.

Adhesion increases with closeness of contact. This explains why one must bear down with a pencil to make a mark on paper, why fine dust adheres more firmly than coarse sand, and why a liquid or a gum usually sticks to a solid better than another solid does.

Cohesion in liquids is usually less, and in gases it is always much less, than in solids. Aside from the pressure in liquids due to external causes, there is presumably a very great internal or intrinsic pressure, due to intermolecular attraction, but not capable of direct measurement by means at our disposal. The clearest evidences of its existence are the work required for thermal **expansion** and the phenomenon of **surface tension**. (R.S.M., L.D.W., D.M.H.)

ADHESIVES.

Materials such as glue which cause two surfaces to become united are known as adhesives. There is a wide variety of these materials, and a large number make use of water as the medium in which they are used. These include glue, starch, dextrin, gum arabic, casein, sodium silicate, and rubber latex. Usually glue is used hot and is odorous, whereas casein can be used cold and is odorless. A typical preparation is made of 100 parts by weight of casein, 10 of alum, 5 of sodium carbonate in 500 of water. Many organic proteins, of which glue is one, are useful as adhesives, including the residues from the extraction of oils from seeds, such as soy bean.

Media other than water are used in the case of such adhesives as rubber. These are made with varying degrees of fluidity and of rubber content, depending upon their use, as for adhesive tape. The principal solvents are chlorinated hydrocarbons, carbon disulfide, coal tar hydrocarbons, petroleum hydrocarbons, and ethyl ether. In this order there is progressively decreasing solubility for rubber, and increasing viscosity with the same rubber content.

Adhesives are sometimes classified, according to the purpose for which they are used, into those (1) for bonding rigid surfaces, such as wood, metal, glass, and porcelain, where the bond does not have to be flexible. Water as a medium is generally satisfactory in such cases; and (2) for bonding flexible surfaces, such as paper, cloth, and leather, where it is important that the bond possess as high a degree of flexibility as the less flexible of the bonded surfaces. In the book-binding and shoe-manufacturing trades there is need to make temporary bondings in the processing, and rosin may be used to increase the tackiness even though rosin has a deleterious effect on rubber over a period of time.

Several of the solvents mentioned for rubber are also effective media for adhesives made of nitrocellulose, cellulose acetate, or numerous synthetic plastics, in which field great advances have been made. In cementing synthetic plastic transparent sheets the use of an organic solvent softens the surfaces sufficiently so that, when pressure is applied and the solvent is evaporated, the two surfaces adhere.

Resistance to temperature changes and to humidity conditions are frequently determining factors in the selection of the proper adhesive for given materials. (R.K.S.)

ADIABATIC PROCESSES.

Changes in matter which take place without transfer of heat. When heat is imparted to or withdrawn from a body of matter, the body generally experiences changes of temperature, pressure, and volume, and sometimes a change of state. These changes severally involve the absorption or the release of energy, which may be regarded respectively as positive and negative energy increments, and the algebraic sum of which is equivalent to the quantity of heat supplied or withdrawn. If the body in question could be provided with perfect thermal insulation, so that no heat could enter or leave it, then any change requiring energy, which might take place within the body, would necessarily be effected at the expense of energy yielded by other internal changes. But a rise of temperature might be caused by heat generated in compression. Processes of this sort, unaccompanied by any transfer of heat across the insulating boundaries of the body, are said to be adiabatic.

For an ideal gas, the pressure and the volume maintain, during an adiabatic expansion or compression, a relationship in which the pressure changes proportionately more with volume than for an isothermal change. This adiabatic relation, for a very slow change, is represented by the formula $pv^\gamma = $ constant, in which γ is the ratio of the **specific heat** of the gas at constant pressure to that at constant volume (for air, about 1.41). Corresponding formulas are readily obtained for pressure and temperature and for volume and temperature by utilizing the **ideal gas law**. It may be shown that such **reversible adiabatic processes** are also isentropic, that is, they take place without change of **entropy**. Adiabatic processes, though hardly realizable in practice, are often considered in thermodynamic reasoning. (L.D.W.)

ADIABATIC PROCESSES IN THE ATMOSPHERE.

When a parcel of air is moved from one position to another with respect to surrounding air, in such a manner that energy does not flow across the boundaries of the parcel, thermal changes taking place within the parcel are said to be adiabatic changes. Any process in the atmosphere occurring adiabatically is known as an adiabatic process.

Adiabatic processes during which the air involved remains unsaturated throughout the process are relatively simple. Adiabatic processes involving condensation or evaporation are considerably complicated by heat of condensation. The first law of thermodynamics applied to a parcel of unsaturated air of unit mass stipulates:

$$dq = c_v dT + A p dv$$

which, when combined with the gas equation, becomes

$$dq = (c_v + AR)dT - \frac{ART}{p}\,dp.$$

For the adiabatic process, this becomes

$$\frac{dT}{T} = \frac{AR}{c_p}\frac{dp}{p}$$

which, upon integration, becomes

$$\frac{T}{T_0} = \left(\frac{p}{p_0}\right)^{\frac{AR}{c_p}} = \left(\frac{p}{p_0}\right)^{.288}.$$

Temperature and dew-point changes within a moving parcel are of primary meteorological importance. Dry adiabatic horizontal transfer of a parcel from higher to lower or lower to higher pressure is of only minor consequence because of the comparatively small magnitude of pressure change. Dry adiabatic vertical transfer of a parcel, however, is one of the important meteorological processes. Temperature decrease in a rising and increase in a sinking parcel amounts to very nearly 9.8° C. per kilometer, or 5.4° F. per 1000'. Dew-point changes in a vertically moving unsaturated parcel are considerably less. Dew point decreases in rising air at a rate between 0.7° F. per 1000' and 1.0° F. per 1000', depending on the air temperature, and increases in sinking air at the same rate.

Adiabatic changes in a saturated parcel (as long as the parcel remains saturated) are known as moist adiabatic or pseudo-adiabatic changes in contrast to dry adiabatic changes in an unsaturated parcel of air. Pseudo-adiabatic changes are by no means constant or simple. In a parcel rising pseudo-adiabatically the temperature decrease is always less than the dry adiabatic temperature change by an amount depending on the weight of the water being condensed and the temperature at which condensation occurs. Condensation releases the latent heat of vaporization within the parcel which partially counteracts dry adiabatic cooling. The rate of cooling in rising saturated air varies from about 1.8° F. per 1000' in warm air at relatively high altitudes to 5.3° F. per 1000' in very cold air at sea level. This range of cooling rates in saturated air is the direct result of the variance in the amount of water resident in a given mass of air at full saturation. Very cold air can retain only a slight amount of water whereas very warm air can hold relatively large quantities. Values of resident water vapor at saturation range from 0.01% by weight in arctic air to 3% by weight in tropical air.

Sinking saturated air remains saturated only for a comparatively short distance during which it is heated pseudo-adiabatically at a rate determined by the amount of evaporation occurring within the parcel. As soon as it becomes unsaturated, the sinking parcel descends dry-adiabatically. **Foehn winds** are examples of both pseudo-adiabatic and dry-adiabatic changes. Air flowing uphill is cooled dry-adiabatically until saturated, after which it is cooled pseudo-adiabatically until it reaches the hilltop. On the lee side, the air descends dry-adiabatically. Observable results of the true foehn wind are abundant clouds and rain or snow on the windward side and clear, warm air on the lee side of a mountain range.

Dew points in saturated air rising pseudo-adiabatically decrease at the same rate as the temperature. Dew points in saturated sinking air increase at the same rate as the temperature until the air parcel is no longer saturated; then they rise slowly as previously described in connection with sinking saturated air.

A large percentage of all clouds and nearly all precipitation result from adiabatic ascent of air.

Assuming increasing positive values with altitude the following relations hold:

1. Dry adiabatic temperature change with altitude:

$$\frac{\partial t}{\partial h} = -\frac{gK}{R} = -9.8° \text{ C./km.}$$

where t = Temperature of parcel.
g = Gravitational constant.

$$K = \frac{c_p - c_v}{c_p} = .288.$$

R = Gas constant.

2. Dry adiabatic dew-point change with altitude

$$\frac{\partial t_d}{\partial h} = -1.71 \left[1 + \frac{2t_d}{237.3} - \frac{t}{273} \right] \text{ °C./km.}$$

where t_d = Dew-point temperature in °C.
t = Air temperature in °C.

3. Pseudo-adiabatic temperature change with altitude

$$\frac{\partial t}{\partial h} = -g \left(\frac{A + .621 \frac{e}{p} \frac{L}{Rt}}{c_p + .621 \frac{L}{P} \frac{de}{dt}} \right) \text{ °C./km.}$$

where t = Temperature of the parcel in centigrade.
g = The gravitational constant.
A = Heat equivalent of work.
e = Water vapor pressure.
p = Air pressure.
L = Heat of condensation.
c_p = Specific heat at constant pressure for air.

(P.E.K.)

ADIPOCERE. Waxy matter formed by the chemical transformation of bodies protected from air by burial in moist places or by submergence in water. Grave wax. (A.W.L.)

ADIPOSE FIN. A fleshy **fin** without supporting spines, occurring behind the dorsal fin in some fishes. (A.W.L.)

ADIPOSIS. An excessive accumulation of fatty tissue in the body. This condition is usually caused by overindulgence in food; in some cases heredity and disinclination to exercise play a part. In a few instances, disturbances in the endocrine glands, particularly the **pituitary** and **thyroid** are responsible for **obesity**. (D.M.H.)

ADJUSTED MOMENTS. Sheppard's Correction.

ADJUTANT. Aves, Ciconiiformes. *Leptoptilus.* A name applied to stork-like birds (**Aves**) of several species. They occur in Africa and the Oriental region where they are valuable as scavengers. From at least one species the soft downy feathers known as marabou are secured. (A.W.L.)

ADLERSTEIN. Geothite.

ADMIRALTY METAL. Brass and Bronze.

ADMITTANCE. Admittance is the reciprocal of **impedance**, hence is that quantity in an **alternating-current circuit**, by which the voltage may be multiplied to give the current. It may be computed from the circuit constants by the following equation:

$$Y = \sqrt{G^2 + B^2}$$

where G is the conductance (a positive number) and B the susceptance (positive number for capacitance and negative for inductance) of the circuit. These are computed from:

$$G = \frac{R}{R^2 + X^2} = \frac{R}{Z^2}$$

$$B = \frac{X}{R^2 + X^2} = \frac{X}{Z^2}$$

R being the **resistance** and X the **reactance** of the circuit.

In many types of calculations, particularly in communication circuits, it is frequently more convenient to use admittance rather than **impedance**. In parallel circuits the admittances of the various branches are added as shown to give the total admittance:

$$Y_0 = \sqrt{(G_1 + G_2 + \cdots)^2 + (B_1 + B_2 + \cdots)^2}$$

$$I_1 = VY_1, I_2 = VY_2, \text{ etc.}, I_0 = VY_0$$

where V is the impressed voltage and I the current in the circuit. (L.R.Q.)

ADOBE. An extremely fine-grained wind-blown **clay** particularly characteristic of the arid and semi-arid south-western United States, Mexico and South America. Used by the south-western Indians and Mexicans for huts and buildings from pre-historic times. (R.M.F.)

ADOLESCENCE. The period between **puberty** and maturity; usually between the ages of 12 and 20 years. (D.M.H.)

ADRENAL GLAND (SUPRARENAL GLAND).

One of the **glands** of internal secretion. The two adrenal glands are small, triangular structures situated on the upper portion of each kidney just below the diaphragm. They are composed of two portions: the internal portion, called the medulla; and the outer portion, called the cortex. Each portion has a separate secretion and function. Epinephrine is secreted into the blood stream by the medullary portion. This medullary portion of the gland is derived from the same embryonic cells that form the sympathetic **nervous system.** Epinephrine stimulates this nervous system and the reverse is also true; that is, stimulation of this nervous system also causes additional epinephrine to be poured forth into the system. Fear, anger, excitement, sudden physical exertion, etc., are stimulating factors, both to the sympathetic nervous system, and, in turn, to the secretion of epinephrine.

The cortex of the gland secretes a hormone concerned with the regulation of carbohydrate metabolism, and the metabolism of sodium and potassium. The adrenal cortex is necessary to life. **Addison's disease** occurs when its hormone production is insufficient. (D.M.H.)

ADRENALINE (EPINEPHRINE) ($C_6H_3(OH)_2$-$CHOHCH_2NHCH_3$).

One of the secretions of the **adrenal glands.** Adrenalin is the trade name for epinephrine. It is usually used either by local application or by hypodermic injection in 1:1000 solution. It is inactive when taken orally.

Normally, a sufficient amount of this substance is secreted by the gland to maintain normal tone of the blood vessels. During exercise and especially with fear, anger, danger or sudden muscular activity, the glands are stimulated and larger amounts are discharged into the blood stream. This stimulates the sympathetic nervous system with the following results: blood pressure is increased quickly and markedly, respirations are quickened and deeper, and smooth muscle is contracted. These actions are favorable for muscular exertion but unfavorable for digestion, as may easily be understood. The same reaction is seen when adrenalin is given by hypodermic injection.

Adrenalin is used medically to increase the blood pressure and restore tone to the blood vessels. It is the most rapidly acting circulatory stimulant of marked power, but owing to its transient action, it is used only in emergencies. It is used in cases of failure of the normal heart as may occur following electric shocks, etc., where it is given in dilute solution into the heart blood. Adrenalin is sometimes used to shrink mucous membranes of the nose when congestion is present. It is sometimes used to control bleeding from an oozing surface, through its action of constricting blood vessels.

A common use, which is specific, is to prevent or treat foreign **protein** reactions, as are seen when serum is given to a sensitive person, and in **hives,** hay fever and **asthma.** In the latter diseases its action is often dramatic, for the drug is a powerful dilator of constricted bronchi and the attack is usually aborted temporarily at least. (See **Hormones.**) (D.M.H.)

ADSORPTION.

Adsorption is a type of adhesion which takes place at the surface of a solid or a liquid in contact with another medium, resulting in an accumulation or increased concentration of molecules from that medium in the immediate vicinity of the surface. For example, if freshly heated charcoal is placed in an enclosure with ordinary air, a condensation of certain gases occurs upon it, resulting in a reduction of pressure; or if it is placed in a solution of unrefined sugar, some of the impurities are likewise adsorbed, and thus removed from the solution. Charcoal, when activated (i.e., freed from adsorbed matter by heating) is especially effective in adsorption, probably because of the great surface area presented by its porous structure. Its use in gas masks is dependent upon this fact. **Penicillin** is recovered in one stage of the process by adsorption on activated carbon.

When **colloidal** hydroxides, notably **aluminum** hydroxide, are precipitated in a solution of acidic **dyes,** that is, those containing the groups —OH or —COOH, the dye adheres to the precipitate, yielding what is termed a lake. The "adsorption" of dirt on one's hands results from the unequal distribution of the dirt between the skin of the hands and the air or solid with which the skin comes in contact. Water is frequently ineffective in removing the dirt. The efficacy of soap in accomplishing its removal is due to the unequal distribution of dirt between skin and soap "solution," this time favoring the soap and leaving the hands clean.

At a given fixed temperature, there is a definite relation between the number of molecules adsorbed upon a surface and the pressure (if a gas) or the concentration (if a solution), which may be represented by an equation, or graphically by a curve called the adsorption isotherm. The degree of adsorption depends upon (1) the composition of the adsorbing material, (2) the condition of the surface of the adsorbing material, (3) the material to be adsorbed, (4) the temperature, and (5) the pressure (if a gas). A notable case in point is carbon. Of the finely divided varieties of carbon there are important sugar charcoal, bone black or animal black, blood charcoal, wood charcoal, coconut-shell charcoal, activated carbon. The temperature of preparation of adsorbent charcoal is an important factor, high temperatures being deleterious, and the removal (or non-removal) of gases by passing steam over the heated carbon, which operation increases the adsorptive power. Bone black is used for removing the coloring matter from raw sugar solutions. Fusel oil is removed from whiskey and poison gases from air by adsorption with the proper form of carbon. By cooling carbon in a vacuum to the temperature of liquid air, the concentration of residual gas is greatly decreased. Dewar (1906) found that 5 grams of charcoal (presumably coconut-shell charcoal) at the temperature of liquid air reduced the pressure of air in a 1-liter container from 1.7 mm. to 0.00005 mm. This is now a standard method for the production of a high vacuum, as in the **neon** lighting tube. Travers (1906), using **carbon dioxide** gas and animal charcoal, found that the concentration of gas divided by the third power of the concentration of adsorbed gas is a constant. In other cases the power to which the concentration of adsorbed gas must be raised is approximately 3 to 2.

Besides carbon, other important adsorbents in use are infusorial or diatomaceous earth, fuller's earth, clay, activated silica, and activated alumina. All surfaces that behave indifferently towards non-electrolytes have the ability to adsorb **electrolytes.**

Adsorption plays an important role in the process of **dyeing,** and in contact catalytic processes such as the conversion of **sulfur dioxide** to trioxide, and of **nitrogen** plus **hydrogen** to **ammonia.** In the case of insoluble organic acids (containing —COOH group) and substances containing hydroxyl (—OH) groups on the surface of water, the film is oriented so that the —COOH or —OH groups are attracted into the surface of the water, while their hydrocarbon ends project away from the surface of the water showing no tendency to dissolve (Langmuir).

The heat of adsorption, or wetting in this case, of starch by water is 29 **calories** per gram of dry starch. The heat of adsorption of various vapors and adsorbents has been measured. Since increase of temperature reduces adsorption, the adsorption process is accompanied by the evolution of heat. It appears that the heat liberated for a given volume of liquid filling the capillary spaces of a given adsorbent is practically constant. The heat of adsorption of hydrogen is, on **nickel, palladium, platinum, copper,** 11,700, 18,000, 13,800, 9500 calories respectively per gram mol (2 grams) of hydrogen; and of **carbon monoxide** on platinum 35,000 calories per gram mol (28 grams) of carbon monoxide; and of **ethylene** on copper 9500 calories per gram mol (28 grams) of ethylene.

Occlusion is a type of adsorption, or perhaps more properly absorption, exhibited by metals or other solids toward gases, in which the gas is apparently incorporated in the crystal structure of the solid. Palladium thus occludes extraordinary quantities of hydrogen, with the simultaneous liberation of much heat. (L.D.W., R.K.S.)

ADULARIA. Feldspar.

ADULT.

A full-grown animal. In species with a pronounced **metamorphosis** the last stage is known as the adult or in insects as the imago. Sexual maturity is characteristic of most adults but some are without sexual functions and in some species reproduction may take place in an earlier developmental stage. (A.W.L.)

ADVECTION.

The transfer of air and air characteristics by horizontal motion. Fog drifts from one place to another by advection. Cold air moves from polar regions southward. Large-scale north-south advection is more prominent in the northern hemisphere than the southern, but west to east advection is prominent on both sides of the equator. (P.E.K.)

ADVECTION FOG.

Fog formed because of horizontal flow of air of high dew point over much colder surfaces. (See **Fogs**.) (P.E.K.)

ADVENTITIOUS BUDS.

Buds which appear elsewhere than in the leaf axils or above them. They may appear anywhere in the internode, or on roots or even on leaves, and develop either naturally or as a result of injury. The dense bunches of buds, which frequently appear on burls or in witches'-broom, may be adventitious. The buds which appear at the tops of thistle roots, particularly when the natural top of the plant is

Pollarded trees (*Catalpa*). Numerous branches developed from adventitious buds when the stem was cut.

cut off, are adventitious. So also are the buds which develop on the leaves of the Begonia and **Bryophyllum**. The practice of pollarding, or cutting off the branches of a tree in such a way as to leave only the main trunk or perhaps the stumps of a few large branches, results in the development of dense groups of adventitious buds. These grow into adventitious branches which in certain willows may be long and supple and so useful in manufacturing wicker furniture. The habit of forming adventitious buds on roots is of material value, since in consequence it is possible to propagate many plants by means of root cuttings.

Adventitious **roots** also exist. They may appear from the stem where they arise in the **pericycle**, or from other tissues of the plant. The roots which appear on slips or stem cuttings are adventitious. (R.M.W.)

ADVENTITIOUS ROOTS. Adventitious Buds.

AECIOSPORES, AECIA. Rust, Fungi.

AEOLIAN DEPOSITS.

Sediments and **sedimentary** rocks which are largely, if not entirely, composed of wind-blown material. Desert sands are typical aeolian sediments, characterized by relatively uniform, well-rounded particles whose surfaces are usually covered with microscopic pits due to their mutual bombardment during transportation. This pitting gives each sand grain a frosted appearance. Wind-blown sediments frequently show characteristic **cross-bedding**, ripple marks (miniature dunes) and wind-faceted pebbles (**glyptoliths**). Further evidence of their origin is the absence of fossils. Aeolian deposits are usually largely composed of **quartz** sand. An important fine-grained wind-blown deposit is **Loess**. Extensive desert deposits are also composed of **gypsum**, salt, etc. (R.M.F.)

AERATION.

Aeration is an artificial method in which water and air are brought into direct contact with each other. The purpose of aeration is to release certain dissolved gases, which often cause water to have obnoxious odors or disagreeable tastes. It is also used to furnish oxygen to waters which are deficient in this element. Aeration may be accomplished in many ways. One method consists of spraying the water into the air. The same result may be obtained by allowing the water to flow in thin sheets down a slope arranged as a series of horizontal steps.

The principle of aeration is used in the treatment of sewage by a method known as the activated **sludge** process. The **sewage** is allowed to flow into an aeration tank where it is mixed with a predetermined volume of sludge. Compressed air is introduced which agitates the mixture and furnishes oxygen which is necessary for certain biological changes which take place. Sewage may also be aerated by mechanically actuated paddles which rotate the liquid and constantly bring a fresh surface in contact with the atmosphere. (C.W.C., F.T.M.)

AERENCHYMA.

Spongy tissue occurring chiefly in the stems of many aquatic or marsh plants. This tissue is formed by the **phellogen** layer, and in many cases results from the separation of the cell walls, leaving extensive intercellular spaces, as in *Decodon*. Such porous tissue gives great buoyancy to the stem and so helps keep the leaves up in the air and permits gas diffusion within the plant. The term aerenchyma is frequently applied to any loose porous tissue found in plants, as for example that occurring in **lenticels**. (R.M.W.)

AERIAL MAP. Photogrammetry.

AERIAL PHOTOGRAMMETRY. Photogrammetry.

AERIAL SURVEYING. Photogrammetry.

AEROBE OR AEROBIC BACTERIA.

Most species of **bacteria** can grow only when there is available to them free oxygen. These are called aerobic bacteria. (Compare **Anaerobe**.) (R.M.W.)

AERODYNAMIC CENTER.

This term is used with reference to airfoils and wings. It is a point, lying on the aerodynamic chord, about which the pitching moment coefficient has the most constant value as attitude varies. Although different airfoils exhibit some variation in aerodynamic center, most of them have a center at or near 25 per cent of the chord aft of the leading edge. (F.T.M.)

AERODYNAMIC EFFICIENCY.

There is no single, rigid, and definite meaning ascribed to this term.

However, in general, it might be said that aerodynamic efficiency measures the ratio of the useful effect on a body exposed to air in motion as compared to the total effort required or to the undesirable reactions unavoidably incurred. Thus, one automotive vehicular shape would be more efficient than another when it transported, for example, five persons with equal comfort at the same speed but with less expenditure of power to overcome wind resistance than the other. Or, again, the aerodynamic efficiency of an airfoil, such as that employed in an airplane wing, can be measured by the ratio of the lift created to the drag entailed in creating this lift (see **Aerodynamics**). (F.T.M.)

AERODYNAMIC FORCES. Aerodynamics; Airfoil; Lift; Drag; Induced Drag; Trim.

AERODYNAMICS. Fluid mechanics is the study of reactions between fluids and solids immersed in them. Aerodynamics is a phase of the mechanics of fluids, its study being limited to the reactions caused by relative motion between the fluid and solid, with the fluid being air. Sometimes this strict definition is broadened so that *aerodynamics* may also include the reactions of gases other than air. The scope of the subject of aerodynamics is, nevertheless, broad. It encompasses the flow of gases in conduits, the effects of winds on static structures such as building, chimneys, and bridges, and the effect on moving bodies of the atmosphere through which they move. Thus the air resistance of automobiles, the flight of airplanes, and the sway of suspension bridges are examples of aerodynamic action. The field of study and application in which the subject attains its maximum importance and encounters its greatest problems is **aeronautics**. Consequently, the emphasis of the present writing is in the direction of aeronautics, particularly aviation. Normally, the object, such as wing, fuselage, wheel, etc., is in motion through still or relatively quiet air. The aerodynamic forces are the same as those prevailing had the object been at rest and the fluid moving past it at the same relative velocity. As this *inversion* is usually beneficial to comprehension of aerodynamics, it is common to employ it. This employment extends not only to visual picturization of aerodynamic forces created by air streams upon stationary airfoils and other objects, but also is customarily employed in testing these objects in moving streams. This is true of the **wind tunnel**.

As air is the common fluid of aerodynamics, a study of its properties is an excellent starting point. Most of the properties of **air** are described elsewhere in this volume. There is one physical concept of fundamental importance to the study of aerodynamics. It is *viscosity*. Possibly this will surprise the reader, who may have previously comprehended air as a frictionless fluid having zero viscosity—as indeed it is in many respects. Air is nearly a perfect gas. Its viscosity is only .00019 poises under standard conditions compared with .01 poise for water, and 1 poise for oil (light). There are many fields of aerodynamic study in which the air can be considered as frictionless; in fact, practically all aerodynamics except that associated with the conditions existing in a very thin layer immediately adjacent to the solid object may rest on this concept. In aerodynamic expressions the absolute viscosity, μ, is usually associated with mass density, ρ as a ratio. This is called the kinematic viscosity, ν, which has a magnitude of .000157 ft.² per sec. in standard sea level atmosphere. At subsonic air speeds it is customary to consider the air as incompressible, and to rely on **Bernoulli's Theorem** of the interchangeability of static head and velocity head. This permits the employment of the **principle of continuity** for flow, represented by the equation

$$AV = \text{constant},$$

which means, symbolically, that the product of the cross-sectional area of flow and the velocity is a constant at all points in the path taken by a finite quantity of the fluid. If the fluid were compressible, this continuity would be

$$\rho AV = \text{constant}.$$

A streamtube in air may be visualized as an imaginary conduit through which the condition of continuity of flow of the incompressible fluid takes place. Although the flow is steady the tube is not necessarily of uniform size, but where its sectional area decreases we would anticipate an increase of velocity; furthermore, a consideration of Bernoulli's Theorem would indicate that these diminished sections of the streamtube were regions of lower static pressure than elsewhere. Although a streamtube is ordinarily imagined to have a circular cross-section, it will be convenient here to think of a streamtube as having a rectangular cross-section. If we have enough of these streamtubes, all of the same width, but whose heights may vary, as velocity varies, we might imagine them stacked, one on another, as shown in the figure, and be thus encompassing the whole flow of air through a given region.
It will be convenient to think of the edges of these streamtubes as equivalent to the **streamline** pattern which is so useful in the study of the effect of air flows past solid objects. It follows that where the stream lines approach each other there is a region of lower pressure and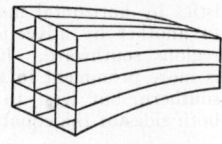

Fig. 1. Stacked streamtubes.

higher velocity; conversely when they diverge. Thus converging streamtubes (and stream lines) are like nozzles, while diverging are like diffusers. The reader may then judge fluid pressure upon viewing a streamline pattern by the density of stream lines. This is readily done, qualitatively, if not quantitatively.

It is in order now to state the nature of elementary streamline patterns and explain how a complex streamline pattern may be considered to be the result of the addition of two potential forces, each capable of producing a different character of stream line in the same region. A simple potential pressure can cause a plain rectilinear streamline pattern. A cylinder rotating in still air will drag with it a circulatory pattern of flow, the stream lines of which could be represented by a series of concentric circles. The circulation stream pattern is of great importance because an airfoil is a shape which automatically induces a circulating component of air flow, which when superimposed upon a rectilinear stream provides the typical stream line pattern around an airfoil. The lifting streamline pattern of the typical airfoil could be induced in the absence of the airfoil by imposing some specific circulatory pattern of stream flow upon a rectilinear field, and this fact has been put to useful service in theoretical consideration of the aerodynamics of airfoils.

In Fig. 2a, a rectilinear flow is passing a static cylinder. At some distance from the cylinder the flow is unaffected by the presence of the cylinder, and the streamtubes remain of constant size throughout, but it is obvious that between this undisturbed region and the cylinder, the intermediate tubes must be smaller in order that all can

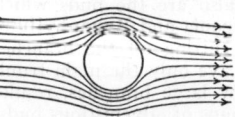

Fig. 2a. Cylinder at rest.

be accommodated in the space immediately above and below the cylinder. The action of the individual streamtube approaching a point abreast of the center line of the cylinder is, therefore, that of a nozzle which has increased the air speed and decreased its pressure. Hence, above and below the cylinder, static pressure is less than in the free, unaffected stream. If the cylinder were in rotation, tending to set up its concentric field of stream lines, the resulting streamline pattern would be like that shown in Fig. 2b, where, on one side of the cylinder, there is a reduction of static pressure due to crowding together of the streamlines, whereas on the other side there may be no change of pressure, or even an increase. Briefly, then, the presence of a circulatory component of

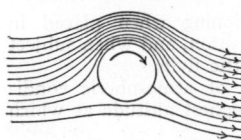

Fig. 2b. Rotating cylinder.

flow in a rectilinear field is to produce a net transverse force on the object since it will be subjected to unbalanced static pressure of the fluid against its surfaces. This is the force which causes a spinning tennis ball to assume a sharply curved trajectory, and this is the force which causes the lift to exist upon an airplane wing.

The influence of viscosity of air in aerodynamics is confined chiefly to the action of air in the **boundary layer.** Within this boundary layer (which, although extremely thin, constitutes the atmosphere immediately adjacent to the solid) the frictional qualities of air are of importance in determining the air flow. The thin boundary layer might be imagined subdivided into a great many lamina of air parallel to the surface and extending out to the limits of the boundary layer. A non-turbulent boundary layer is considered, although, as will be developed later, the boundary layer may become turbulent under certain conditions except for the lamina immediately adjacent to the stationary surface. As Fig. 3 displays, the laminar velocity, u, in-

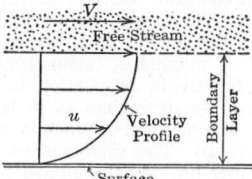

Fig. 3. Velocity profile.

creases above the surface until, at the edge of the boundary layer, it equals the free stream velocity. The smaller the viscosity of the fluid, the thinner the boundary layer will be. Consequently it is extremely thin for air, but what goes on in this boundary layer is of utmost importance in aerodynamics. The first layer of air sticks or adheres to the surface, and lamina above it successively slide on each other, exerting drags that are proportional to the viscosity. The rate of change of the velocity between adjacent lamina is a measure of the unit shearing force between them, and is the unit skin friction when the lamina considered is the one in motion nearest the stationary surface. This occurs at practically zero boundary layer thickness. A curve joining the tips of velocity vectors plotted for the different lamina in the boundary layer is called a velocity profile. If u is the variable velocity, increasing from O to V across the boundary layer, then du/dy is the rate of change of velocity (also the tangent to the velocity profile) and, through the application of one of the Newtonian principles, the unit shear stress,

$$\tau = \mu \left(\frac{du}{dy} \right)_{y=0}.$$

The proximity of streamlines to a stationary surface is now seen to produce a reaction resulting from the static pressure p existing in the flow nearest the surface, and a skin friction τ, which is, of course, tangential to the surface. The combination of these two is a force oblique to the surface. The foregoing statement premises an adherence of the streamlines to the surface, for, if they have separated from the surface, the region between will be filled with vortices of random motion, and the surface pressure cannot be definitely predicted from

the character of the streamline flow. It is apparent that aerodynamic reaction is a function of μ, ρ, V, and the extent of the surface area on which the reaction takes place. Usually some one dimension l is taken to describe the size of the surface. In Engineering units the aerodynamic force F is pounds, μ is poises, ρ is slugs per square feet, and l is feet. To discover the relationship existing between the reaction and its controlling factors, it is customary to set up the dimensional relationship:

$$|F| = |\rho^a \mu^b V^c l^d|$$

and employ the methods of dimensional analysis to discover the value of a, b, c, and d. It is found that

$$|F| = \left| \left(\frac{\nu}{Vl} \right)^b \rho V^2 l^2 \right| \quad \text{dimensionally}$$

and

$$F = K \left(\frac{\nu}{Vl} \right)^b \rho V^2 l^2 \quad \text{numerically,}$$

in which ν = Kinematic viscosity.

K = A dimensionless constant.

$\dfrac{Vl}{\nu}$ = **Reynolds number** (also dimensionless).

The product, $2K \left(\dfrac{\nu}{Vl} \right)^b$, is frequently given as a dimensionless coefficient C. The exponent b is of such magnitude that the Reynolds number has a minor (but not negligible) effect on C. The coefficient is affected primarily by the attitude of the body in the airstream. If it is completely symmetrical, as is a sphere, this variation is non-existent, and C is a true constant except for the above-mentioned minor effect of Reynolds number variation. However, the non-symmetrical shape of an airfoil causes large and typical variations of C, depending on the attitude with which the airfoil is presented to the airstream. The force F would then be $C \dfrac{\rho V^2}{2} S$, or CqS, in which q is dynamic pressure, and S is some significant surface of the object. (See **Angle of Attack.**)

Flow within a streamtube may be laminar or slightly turbulent. It was stated that, in laminar flow, drag per unit area is $\mu \left(\dfrac{du}{dy} \right)_{y=0}$, but turbulent flow produces higher skin friction because the velocity profile is fuller close to the surface on account of turbulent energy interchange between lamina. Though at the beginning of contact between airstream and surface true laminar flow existed, it has been found that at a certain critical value of Reynolds number, determined chiefly by the linear dimension of contact passed over, the boundary layer suddenly becomes turbulent and thickens. In spite of the higher skin friction, the drag on **bluff bodies** may actually become less with turbulence, because of an action, within the turbulent boundary layer, resisting separation of the stream lines from the surface. Viscous drag gradually slows down the fluid in a laminar boundary layer, and if contact with the surface is maintained over sufficient length, the innermost parts of the boundary layer will be brought to rest. This becomes the *separation point*. If a negative pressure gradient exists beyond the separation point, as it often does, there will be a reverse flow towards the separation point, all of which will produce a breaking away of the streamlines from the surface, leaving the surface between the two separation points, on either side of the body, in contact with air of random vortices and low pressure. Referring to Fig. 4, because of the vortices the pressure at n has failed to build up to the pressure at m (as it would have had the fluid been frictionless and without premature separation point), resulting in the creation of a downstream dragging force. This is characteristic of any body having a turbulent wake. Early turbulence in the

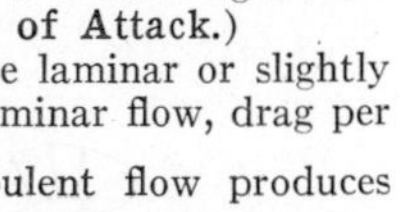

Fig. 4. Low-pressure wake.

boundary layer is desirable, so that the greater average momentum of boundary layer air can carry it against skin friction farther around the bluff surface, and so delay the separation point and narrow the turbulent wake. Thus the coefficient of drag for a sphere $\left(C \text{ of } Cq \frac{\pi d^2}{4} \right)$ is about .5 for laminar boundary layer and .1 for turbulent, with the transition occurring at a Reynolds number of approximately 300,000.

The applications of aerodynamics are varied and numerous, and stretch from scientific research to com-

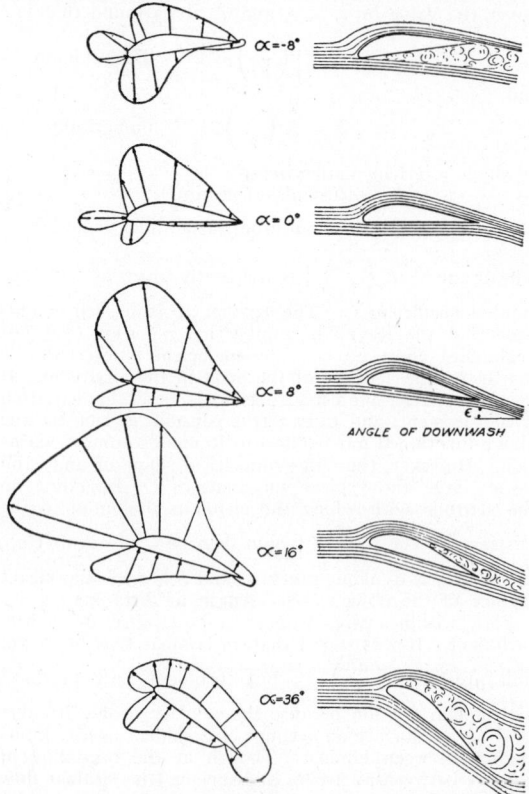

ANGLE OF DOWNWASH

Fig. 5. Distribution of pressure on a cambered wing at different angles of attack. Sketches to the right show, approximately, the manner of air flow at the different angles of attack.

mercial design. Many of them are described specifically in this volume, and the reader is invited to consult the following: **Circulation, Airfoil, Boundary Layer, Stagnation Point, Stream Line, Drag, Lift, Vortex, High Lift Devices, Kármán Vortices, Downwash.** (F.T.M.)

AEROLITE. A general term for **meteorites** which are richer in the basic silicates than in nickel and iron. (R.M.F.)

AERONAUTICAL ENGINES. An aeronautical engine is any engine which, by virtue of design or adaptation, successfully meets the special requirements of the propulsion of aircraft. These special requirements are:

1. Weight. Low weight per hp. developed.
2. Reliability. This applies especially in aviation, but is not to be disregarded in aerostation (see **Aeronautics**).
3. Stamina. Aircraft engines must be able to deliver continuously at least 75% of their full power without overheating or excessive wear.
4. Frontal Area. A minimum frontal area should be presented to the windstream.

5. Rotative Speed. The rotative speed cannot be increased above about 2000 rpm without adding reduction gearing because of the poor efficiencies of **propellers** operated at high speeds.
6. Ability to continue functioning when placed in other than normal horizontal position, and when subject to accelerations as during acrobatic flying.
7. Compensation for changes of atmospheric conditions at altitudes equal to the highest altitude at which the airplane can be flown.
8. Balance. Freedom from shaking and vibrating forces.
9. Thermal Efficiency. This affects the amount of fuel that needs to be carried.
10. Ease of maintenance and overhaul.

It has not been possible to achieve the best in all these requirements in any single engine, but many available aeronautical engines exhibit an excellent combination of many of them. Owing to the comparative rarity of **airships,** the airship engine is here accorded only passing notice. The blimps are powered by the smaller stock airplane type engines, but special engines are usually designed and built for dirigibles.

One of the most important requirements of the airplane engine is the necessity for achieving all the other requirements with due regard to keeping the weight at a minimum. Generally speaking, reliability, stamina, balance, etc., are opposed to low weight per horsepower, so that the design of an aeronautical engine becomes a compromise between the other desirable features and weight. The low weights of today's engines are achieved by the use of special alloy metals, careful designing, and extensive testing. The larger engines will be higher per unit output than the smaller ones, because the weight of auxiliaries, such as magnetos, carburetors, and the like, does not vary directly in proportion to output, but is nearly as much in small engines as in large ones. The weight of large aeronautical engines is usually between 1 and 2 lbs. per hp. The smaller sizes weigh between 2¼ and 3¼ lbs. per hp. As there must be a horsepower for at least every 25 lbs. of airplane weight, it can easily be seen that the engine is no small part of the weight of an airplane. As a matter of fact, 25 lbs. per hp. would represent an airplane greatly underpowered, judged from modern standards. Fifteen pounds per hp. is more nearly an average figure for the small sportsman or trainer type of airplane, 10–12 lbs. per hp. for the high-speed transports, and as low as 6–7 lbs. per hp. for high-performance military airplanes. This requirement of lightness has led to the perfection of the internal combustion engine of the **Otto cycle** type as the leading aeronautical engine. While Diesel and steam types of prime movers both have successfully flown aircraft, and may become very important to the aircraft of the future, the gasoline engine is preeminent in this field at the present time. Its commanding position is principally due to its superiority in weight per horsepower developed. In many other points it is inferior, as it uses a type of fuel that has high fire and explosion hazard, it requires auxiliaries which are delicate and easily thrown out of adjustment, and suffers a loss of output at increasing altitude. These disadvantages, however, only serve to emphasize the importance attached to weight.

Reliability in the **Otto engine** may readily be secured, but usually only at the expense of increasing the weight, and to secure the maximum possible reliability, the weight would be prohibitive. Hence the aircraft engine cannot be entirely reliable, although manufacturers have been able to achieve remarkable results in this direction. Airplanes acting as public carriers should, for night schedules, be multi-motored, and able to maintain altitude with one motor out. This feature is of less importance in daytime flying, as the pilot, confronted with engine trouble, will usually be able to select an emergency field and land safely, an achievement practically impossible at night. The spark plug element of the ignition system is one of the less perfect components of an engine, but the whole electrical ignition system is such that it is deemed best to duplicate it, so common practice on standard engines is to have two spark plugs in each cylinder, and two independent sources of voltage to these plugs. It has not been expedient from the standpoint of cost or weight to duplicate all the auxiliary systems of the gasoline engine.

Unlike the automobile engine, which may operate over long periods of time at only 25–35% of full power output, the aeronautical engine must deliver its "cruising" power steadily, and be able to deliver full power for a period of several minutes' duration. The airplane engine is called on for a full power output during the period of take-off and climb to some altitude at which the pilot feels it safe to throttle back. The throttled power output, however, is rarely less than 65% full power for cross-country work, and it is not usually possible to maintain altitude in an airplane with the power output less than 50% of rated. For these reasons, a high degree of stamina, which involves special attention to the cooling and lubricating systems, as well as to the wearing parts, must be incorporated in aeronautical engines.

The parasitic drag of an aeronautical power plant is dependent on its cooling system. Between 25 and 35% of the heating units in the fuel must be passed through the cylinder walls in order to keep them cool enough, and this heat must ultimately be dissipated to the atmosphere. If this is done by extending the outside surface of the cylinder to form fins and moving an airstream across them, the more effective the cooling, the greater the parasite **drag** involved. If a liquid cooling medium is interspersed between cylinder and atmosphere, as is the case in liquid-cooled engines, the engine may be completely enclosed in a well streamlined fuselage or nacelle, but a radiator must be placed in the windstream at some point, and becomes a source of drag. The radiator resistance of very large engines is less than the drag of air-cooled engines of equivalent power, and hence the liquid-cooled engine is favored for high-speed, high-powered airplanes. Special low drag types of radiators have been employed on high-speed racing planes, where neither engine drag nor radiator drag can be tolerated. These surface radiators, however, are altogether too fragile and expensive to be considered for any other service than the special one of high-speed racing, wherein the working life of the aircraft may not extend beyond a test period and a record-breaking flight over a measured mile or similar short course.

A most serious defect of the internal combustion engine is its decrease in **volumetric efficiency** and developed power when it is supplied with air of decreasing density. Such is the case when an airplane climbs steadily. It eventually arrives at an altitude where the decrease in engine horsepower leaves no excess power over that required to maintain horizontal flight, and no further climb is possible. The **airplane** is then said to have reached its "ceiling." This ceiling varies upward from 10,000 feet above sea level, depending upon the horsepower, weight, and wing area of an airplane. It may be increased several thousand feet with a **supercharger.** Since flight at high altitudes (i.e., stratosphere flight) holds forth much of promise for the aeronautical industry of the future, the supercharged internal combustion engine, or an external combustion cycle, such as the steam cycle, will be favored types.

The average thermal efficiency of the aeronautical engine is 25% at full rated power, and individual engines depart only slightly from this value. The gasoline tanks must hold sufficient gasoline to provide about .09 gal. for each horsepower used per hour.

Practically all cylinder arrangements that have been used for internal combustion engines are to be found in aircraft practice, but many of them are seldom seen.

The more common cylinder arrangements are shown in Fig. 1. The radial arrangement of cylinders is probably used more than any other at the present time.

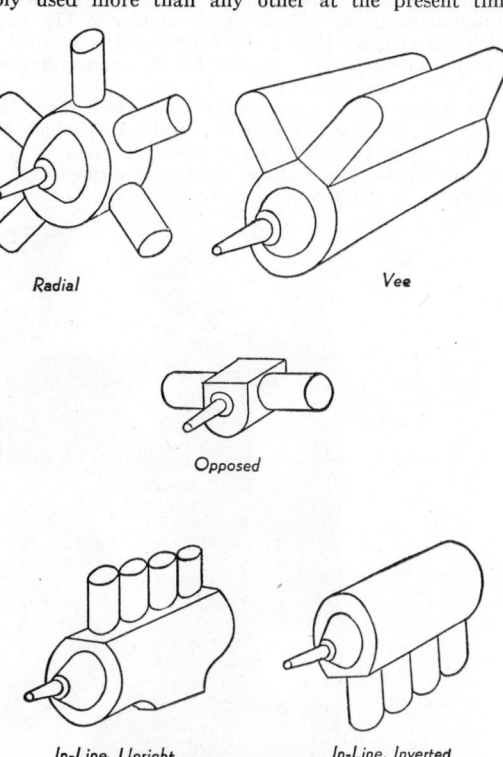

Fig. 1.   Aeronautical engine cylinder arrangement.

The radial aeronautical engine is always air-cooled. In fact, one of the principal advantages of the radial arrangement is the presentation of each cylinder, individually, to the cooling airstream, whereas in-line arrangements, while presenting a lower frontal area, have problems of cooling introduced by the fact that only the front cylinder meets an undisturbed airstream. The radial engine has a short cylindrical crankcase to which are attached, with axes radial to the crankcase, cylinders to the number of 3, 5, 7, or 9. The cylinders are nearly evenly spaced around the circumference of the crankcase. The crankshaft, which passes through the center of the crankcase, has a single throw and is supported on two main bearings and one thrust bearing. The front end of the crankshaft protrudes from the crankcase and on it is mounted the propeller hub. Through gearing, the crankshaft drives the valve lifting cams, the magnetos, the oil pump, and many other needed accessories. A radial engine of the four-cycle type must be built with an odd number of cylinders because it fires on alternate cylinders, and the four-stroke cycle requires two full revolutions before all cylinders have fired. Thus, in a seven cylinder engine, in which the cylinders are numbered consecutively 1, 2, 3, . . . 7, the firing order is 1-3-5-7-2-4-6. Radial engines using 6, 14 and 18 cylinders have been built, but they are really twin threes, twin sevens, or twin nines, and their crankshafts have two throws. Since some cylinders of a radial engine must project from the lower part of the crankcase, no integral oil sump is possible, but rather the oil is withdrawn from the lower part of the crankcase by a sump pump and delivered to an external oil tank where it is stored and cooled, and from which it is taken by the pressure oil pump and delivered back to the bearings and other parts of the engine. The piston and connecting rod arrangement is interesting in that there is one master rod which bears upon the crankshaft, while the remaining cylinders have articulated connecting rods, that is, they bear upon pins in the big end of the master connecting rod. All this is clearly shown in Fig. 2, in which the cylinder having the master rod is vertical, a customary arrangement. Except for the connecting rods the cylinders are identical.

The in-line cylinder arrangement, in which the cylinders are mounted in a line over a crankshaft which has as many throws as there are cylinders, is used in both air- and liquid-cooled engines. For engines of equivalent output, the in-line engine tends to be heavier because of

seats, and spark plugs. Aluminum bronze rings are inserted to form the valve seats as aluminum does not possess sufficient strength to resist the pounding action of the valves. Pistons are aluminum alloy, and connecting rods are duralumin or drop forged alloy steel. Crankcases are cast or forged aluminum, usually machined all over, while crankshafts are of special alloy steel, such, for instance, as nickel chromium steel. Valves offered difficult problems as they tended to warp, leak, pit, break, and stick under the severe service conditions of aeronautical engines. Engine manufacturers have, with the cooperation of parts manufacturers and metallurgists,

Fig. 2.

the longer crankcase and crankshaft, but presents a much lower frontal area and is more susceptible to streamlining than the radial type. Six cylinders in line, however, are about the maximum that can be successfully cooled, and these must be equipped with a special cowling containing air scoops and deflectors to direct cool air over the cylinders and to discharge heated air from the forward cylinders so it will not tend to heat those behind. In-line, air-cooled engines are frequently made inverted, that is, having the cylinders projecting downward from the crankcase. Advantages of improved forward visibility and high thrust center are secured by inverted designs. The upright engines may have self-contained oil sumps, but inverted engines must have external oil tanks.

Small 4- and 6-cylinder air-cooled engines are frequently made horizontally opposed. This type is excellent as a light plane engine. It rarely exceeds 150 hp., whereas the air-cooled in-line engine is found up to 500 hp., and the radial as high as 2200 hp., in stock models. The vee arrangement of cylinders is frequently used in multi-cylinder water-cooled engines, and occasionally in air-cooled engines. The present trend, however, is to employ liquid-cooled engines only in the larger powers.

The constant effort of aeronautical engine designers to reduce the weight per horsepower has not left much place for the commoner materials of construction, such as cast iron. While a few small air-cooled engines will have cast iron cylinders, standard practice in this field indicates the use of forged steel cylinder barrels, to which are attached cast aluminum alloy cylinder heads by bolting, screwing, shrinking, or combinations of these. Questions of cooling, volumetric efficiency, and manufacturing cost dictate the use of overhead valves in these engines, consequently the cylinder heads contain the ports, valve

succeeded so well in solving these problems that the latest engines produced satisfy, to a high degree, all the requirements mentioned in the introduction to this article. (See also **Gas Turbine, Rocket, Reaction Propulsion, Valve Gear, Carburetor Types, Ignition System, Otto Engine, Spark Plug, Airplane Engine.**) (F.T.M.)

AERONAUTICS. Aeronautics encompasses all that pertains to the study, design, manufacture, maintenance, testing, and use of aircraft. Since the term aircraft includes any machine or device for navigating the air, aeronautics is seen to be a term embracing a broad field of human activity. One general method of aeronautical classification is based on the method of sustention of the aircraft, namely:

1. Aviation, where lifting is obtained from the dynamic action of air in motion relative to wings.

2. Aerostation, in which lift is obtained by the buoyancy of a volume of light gas.

Aviation is therefore that branch of aeronautics which pertains to heavier-than-air craft, such as fixed wing **airplanes, gliders, autogiros, ornithopters, helicopters.**

Aerostation is that branch of aeronautics which pertains to **airships, free balloons,** and captive balloons.

Not included in either of the above categories is the rocket type of aircraft.

The following outline will serve to give the reader an idea of the activities in the whole field of aeronautics.

1. Study of aircraft, theoretical and practical.
    a. Education in aviation schools and colleges. Flying instruction, ground school work for mechanics and craftsmen, business and commercial studies, technological or engineering education.

b. Research, in developing new aircraft types, improving present types, investigating meteorological conditions, etc.
    (1) Governmental agencies (Army, Navy, National Advisory Committee for Aeronautics).
    (2) Universities.
    (3) Manufacturers.
2. Design. Design of aircraft, landing fields, hangars, aircraft equipment, communication systems, etc. This activity is found mainly in factories manufacturing aircraft, but is also, to a lesser extent, carried on by individual experimenters and various government agencies.
3. Manufacture.
    a. Processing raw materials. Manufacturing of special alloy metals, selection and seasoning of wood, manufacture of finishing materials, extraction of helium, etc.
    b. Fabricating. Shaping raw materials to the final form in accordance with design drawings.
    c. Assembling. Building the finished aircraft, hangar, radio transmitter, etc.
4. Testing.
    a. Ground testing of materials, methods, engines, etc.
    b. Flight testing of the aircraft structure, of instruments, of power plants, of transport schedules, and of communication systems.
5. Maintenance, inspection, and repair of flying equipment, and of ground equipment.
6. Use.
    a. Military. Observation, bombing, pursuit, patrol, direct ground offense.
    b. Civilian. Non-commercial. For personal transportation, pleasure flying, and air racing.
    c. Scheduled air transport. An important phase of the aeronautic industry, which acts as a public carrier of persons, mail, and cargo, on definite time schedules between specified points, and over specified routes, in so far as exigencies of weather conditions permit.
    d. Fixed base operation. Special for-hire charter trips, of both business and sporting nature.
    e. Miscellaneous.
        (1) Forest patrol.
        (2) Executive transport.
        (3) Advertising.
        (4) Freight transportation to otherwise inaccessible regions.
        (5) Map photography.
        (6) Crop dusting, etc.
7. Regulation and control. United States Civil Aeronautics Administration and State Aeronautical Commissions or Bureaus. Supervision of the aeronautical industry for the purpose of promoting general public safety against the hazards of aeronautical transportation, including:
    a. Setting up standards of design and construction of aircraft and other aeronautical equipment.
    b. Inspecting design and construction, and certifying equipment, i.e., licensing.
    c. Supervising the airworthiness of aircraft.
    d. Examining operating personnel, pilots and mechanics, for competence and physical condition.
    e. Developing and maintaining airways.
    f. Developing new and improved aeronautical equipment. (F.T.M.)

AERO-THREAD. Screw Thread.

AESCULAPIUS. The god of healing in ancient Greek mythology. (R.S.M.)

AESTIVATION. Summer dormancy, the antithesis of the more familiar hibernation. (A.W.L.)

AFRICAN CRESTED RAT. Mammalia, Rodentia. *Lophiomys.* A rare species of rat of northeastern Africa, related to the wood-rats of North America. (A.W.L.)

AFRICAN JUMPING HARE. Mammalia, Rodentia. *Pedetes.* A South-African species with large hind legs. (See **Rodentia.**) It belongs with the **jerboas** near the true mice and not with the hares as the name suggests. (A.W.L.)

AFRICAN LEMUR. Galago.

AFRICAN SLOW LEMUR. Potto.

AFTERBIRTH. The membranes and **placenta** expelled from the **uterus** a short time after the birth of the child. (R.S.M.)

AFTERGLOW. Electrodeless Discharge.

AGAMA. A large South African **lizard** of the genus *Agama.* (A.W.L.)

AGAMONT. A single-celled animal of a generation which reproduces asexually. (See also **Gamont.**) (A.W.L.)

AGAR-AGAR. A gelatine-like substance which is prepared from various species of red **algae** growing in Asiatic waters. The prepared product appears in the form of cakes, coarse granules, long shreds, or in thin sheets. It is used extensively alone or in combination with various nutritive substances, as a medium for culturing **bacteria** and various **fungi.** It is sometimes recommended as a mild laxative. (R.M.W.)

AGARICS. Agaricaceae. Fungi. This family of **fungi** is probably better known than any other, since it contains most of the plants commonly described by the names toadstool and mushroom, which popularly and mistakenly denote poisonous and edible fungi, respectively.

The Agarics are mostly fleshy fungi of that very definite structure, the familiar parasol-like toadstool. This is composed of convex pileus or cap, usually supported on an evident stalk. The underside of the cap shows a series of radiating plates, or gills, which are formed in agaric fungi only, and so serve to separate them from all others. The two sides of the gills are covered with the microscopic spore-bearing bodies called basidia, which are club-shaped or cylindrical cells bearing **spores,** generally four each. Agarics vary in size from delicate species with a cap a millimeter or so in diameter, supported by a slender thread-like stalk, to massive forms twelve inches in diameter: the larger species form millions of spores.

The spores float in the air for considerable distances, and finally come to rest on some solid substance. Should this be favorable for germination, the spore puts out a slender tube, which elongates rapidly and penetrates the substratum, from which it absorbs substances necessary for its continued growth. Gradually this thread-like body, known as the mycelium, spreads through extensive masses of substratum, branching frequently as it does so.

Finally there is accumulated in the mycelium a supply of reserve food sufficient for fruiting: then, if atmospheric conditions, such as moisture and temperature, be suitable, the familiar toadstool appears, it being only the reproductive stage of the fungus. Its rate of growth is often phenomenal, as is also the force it may exert in its growth. Seemingly delicate bodies not only break open the hard-packed surface of the ground, but also may push aside pebbles of considerable weight. Not infrequently whole rings of toadstools appear in a field, springing to maturity in a single night—these are the familiar fairy rings, resulting from the growing outward from a common source of the unseen mycelium and not from the dancing of fairy forms.

When the young fruit-body first comes up it is completely enclosed in a membranous skin known as the velum. As enlargement continues this skin is broken. Often traces of the velum remain in the form of flakes on the upper surface of the cap, and as a ring or annulus around the stalk. Attempts have been made to find in these characteristics a means for separating the edible from the poisonous species. However, no reliable distinction is found here. Actually, unless one is absolutely sure of the identity of a given species, the only safe rule is complete abstinence. Classification is based on the color of the spore-masses, and also the determination of

the way in which the gills are attached to the stalk, the color changes shown as a cut or broken surface dries, and various other means.

A few species of Agarics, notably *Agaricus (Psalliota) campestris,* are extensively cultivated, and justly esteemed as food. Other species are violently poisonous, particularly species of the genus *Amanita.* Yet certain Siberian people use *Amanita muscaria,* a poisonous species, to produce a form of intoxication. Another common mushroom, the Inky-cap, a species of

Sporophore of a mushroom, *Agaricus campestris.*

Coprinus, has been used as a writing fluid, the substance of the toadstool breaking down into a fluid mass containing vast numbers of black spores. (See **Basidomycetes.**) (R.M.W.)

AGATE. Agate is a variety of **chalcedony** whose variegated colors are distributed in regular bands or zones, in clouds or in **dendritic** forms, as in moss agate.

The banding is often very delicate with parallel lines of different colors, sometimes straight, sometimes undulating or concentric. The parallel bands represent the edges of successive layers of deposition from solution in cavities in rocks.

As agate is an impure variety of quartz it has the same physical properties as that mineral. It is named from the river Achates in Sicily where it has been known from the time of Theophrastus.

Agate is found in many localities; India, Brazil, Uruguay and Germany are notable for fine specimens.

Onyx is a variety of agate in which the parallel bands are perfectly straight and can be used for the cutting of cameos. Sardonyx has layers of sard (red carnelian) alternating with lighter-colored layers of onyx. (E.S.C.S.)

AGAVE. Amaryllidaceae. A large genus of plants, particularly abundant in Mexico, in which the thick rigid leaves form a basal rosette from the center of which rises the tall flower stalk. Because of the time required to store sufficient food reserves for flowering, certain species, notably *Agave americana,* are called **century plants** from the belief that they flower but once a century; actually flowering may occur in from 5 to 50 or more years. Once started, the flower stalk develops very rapidly, requiring immense quantities of sap. In Mexico, the flower stalk is cut off early in its formation and the stump scooped out to form a cup into which quantities of sweet sap exude. This sap is collected and fermented to form pulque, a strong drink with an unpleasant odor. Distilled pulque gives a more potent drink, mescal. From the leaves of several species, particularly *Agave sisalana* and *A. fourcroydes,* are obtained fibers. These fibers occur as **sclerenchyma** sheaths surrounding the vascular bundles in the leaves. To obtain the fibers, the leaves are cut off and the spiny tip and margin removed. Machines then heat and scrape the leaves and wash them until the clean fibers are obtained. These are then dried either in the sun or by artificial heat, and are ready for export under the name of sisal or henequen, according to the species from which they were obtained. Many species of Agave are cultivated for their ornamental value. (R.M.W.)

AGE OF THE EARTH. Chronology.

AGE-HARDENING. Precipitation Hardening.

AGGLOMERATE. A term proposed by Sir Charles Lyell in 1831 for coarsely graded volcanic ejectamenta similar in appearance to ordinary **conglomerates** or **breccias.** An extremely thick and widespread accumulation of so-called agglomerates occurs on the borders of the Yellowstone Park. These deposits, however, include numerous beds of water-laid pebbles, gravels and sands, the latter containing fossil plants of early **Tertiary** age. (R.M.F.)

AGGRADATION. In geology, the deposition of sediment by a river in its valley. (See also **Alluvial Fans.**) (R.M.F.)

AGGREGATE. The solid particles which form the major portion of the volume of **concrete** are called aggregate. The aggregate may be classed as fine or coarse depending upon the size of the individual particles. The specifications for the concrete on any project will give the limiting sizes which will distinguish between the two classifications. Fine aggregate generally consists of sand or stone screenings while crushed stone, gravel, **slag** or cinders are used for the coarse aggregate. The aggregate should be strong, clean and free from clay or organic matter since the strength of concrete depends upon the strength of the individual particles as well as the efficiency of the cement as a binder. (C.W.C.)

AGGREGATE GLANDS. Silk glands of **spiders** which produce the viscid spiral lines of the web. (A.W.L.)

AGONIC LINE. The line of no magnetic variation (declination) or deviation of the compass from true north. (See also **Geomagnetism** and **Isogonic Chart.**) (R.M.F.)

AGOUTI, AGUTI. Mammalia, Rodentia. Large rodents of the genera *Dasyprocta* and *Myoprocta* found in Central and South America and the West Indies, where they live chiefly in the forests. They are hunted for their flesh. (A.W.L.)

AGRANULOCYTOSIS (Agranulocytic Angina. Malignant Neutropenia, Granulocytopenia). A syndrome characterized by a marked decrease in white blood cells, particularly those of the granulocytic series (see **Blood**). The clinical picture is one of high fever and ulcerative lesions of the skin and **mucous membranes** which may become gangrenous. The mortality is high. There is an idiopathic type whose etiology is unknown, but the majority of cases are traceable to a toxic reaction to any of a number of drugs. Chief among these are aminopyrine, a common pain-relieving drug, arsenicals, gold salts, dinitrophenol, benzol and certain of the **sulfonamides**, particularly sulfanilamide. Treatment is supportive. (D.M.H.)

AGRICULTURAL CHEMISTRY. Agricultural chemistry is that branch of the science of chemistry which deals with the chemical processes connected with the growth of plants and animals, and the preparation of these and their products for market. Since plants are dependent upon air and soil, and upon water and climate for their growth, these are matters of prime importance in connection with crop cultivation. The maintenance of proper soil involves the study of the various chemical types of plant food, of fertilizers, and the effect of small percentages of various elements. The cultivation of plants also necessitates battling against injurious pests and diseases. The preparation of plant and animal product calls for examination of preparation methods as such, for example, in the case of milk products, butter, cheese, lactose, casein, and of the storage and transportation conditions, for example, meats, eggs, fruits, vegetables. (See **Soils: Fertilizers; Reactions Involving Recombination of Ions, Hydrogen Ion Concentration; Poisons; Foods; Photosynthesis; Biochemistry.**) (R.K.S.)

AGUE. Recurrent chills and **fever,** usually of **malarial** origin. (R.S.M.)

AI. Sloth.

AIGRETTE. The tuft of slender feathers found on the back of the **egret** during the breeding season. (A.W.L.)

AILANTHUS. Simarubaceae. Tree of Heaven, or Tree of the Gods. A genus of trees having large pinnate leaves, small flowers and winged fruit. One species, *Ailanthus glandulosa,* a native of China and Japan, has been extensively introduced into cities in the eastern United States, because of its ability to withstand the gas and smoke of the city, and is often found growing to large size in most unlikely habitats. An objectionable feature is the unpleasant odor of the male flowers. (R.M.W.)

AILERON. The aileron is one of the three **aerodynamic** surfaces of an **airplane** which are variable in attitude at the will of the pilot, and the purpose of which is to provide the required degree of maneuverability of the aircraft. The aileron is that surface which produces rotation about the longitudinal axis of the fuselage. This motion is known as roll, and is necessary to correct other rolls produced unintentionally, as by gusts, and is also employed to accomplish such maneuvers as banks or sideslips, both of which require a certain degree of roll executed under the control of the pilot.

The conventional type of aileron is a flap attached to a portion of the trailing edge of the **wing,** usually towards the extremities. This flap is rotatable around its forward axis, upwards and downwards, and in effect changes the **camber** of the airfoil. The result is a change of pressure on the wing much greater than that which could be obtained by pressure on the aileron alone. The principal defect of this type of aileron is that it becomes relatively ineffective when, for safety's sake, it is needed to be most effective; that is, when the airplane approaches the stall, or is stalled. With adequate control of roll during a **stall,** many serious accidents involving the tailspin could have been avoided. Except for this point, the trailing edge aileron is highly satisfactory. Its aerodynamic efficiency, simplicity, reliability, freedom from flutter, and balance are better than in other types, such as wing tip ailerons, spoiler ailerons, etc., which in consequence have not been widely accepted. (F.T.M.)

AIMLESS DRAINAGE. Type of drainage or stream pattern which occurs in low swampy lands. Particularly characteristic of glaciated regions of low relief. (R.M.F.)

AIR. The term air is frequently used as synonymous with "atmosphere of the earth" and it is in that sense that we shall use it here. The earth's atmosphere consists of a vast body of gases, vapors, and suspended matter of total mass about 5.1×10^{15} tons, or somewhat less than one-millionth part of the total mass of the earth. The height to which the air extends can be estimated only by the effects which it produces. On purely theoretical grounds, since according to the **kinetic theory** of gases some molecules attain velocities great enough to escape not only from the earth but also from the **solar**

system, we may say that the atmosphere of the earth extends out into space until it mingles with the interstellar gases of the **milky way.** The **twilight** arch has been observed to a height of about 45 miles, **meteors** become visible about 75 miles, on the average, above the surface of the earth, and **aurorae** have been observed up to more than 600 miles.

Most of our knowledge of the composition of air is based upon samples that have been taken at the surface of the earth. There air is composed chiefly of oxygen and nitrogen (and this is probably true to a height of 70 kilometers). Air also contains a variable proportion of water vapor, and small quantities of other gases. In many engineering calculations, air is considered to be composed only of nitrogen and oxygen, and the proportions assumed for such calculations are, by weight, 76.8% nitrogen and 23.2% oxygen; and, by volume, 79.1% nitrogen, and 20.9% oxygen. A more detailed analysis of air (without taking into consideration the variable amount of water vapor) is given below:

AVERAGE COMPOSITION OF AIR *

(At, or near to, the surface of the earth)

SUBSTANCE	PERCENTAGE (By volume of dry air)	PERCENTAGE (By weight of dry air)
Nitrogen	78.09	75.54
Oxygen	20.93	23.14
Argon	0.93	1.27
Carbon Dioxide	0.03	0.05
Neon	0.0018	0.0012
Helium	0.0005	0.00007
Krypton	0.0001	0.0003
Hydrogen	0.00005	0.000004
Xenon	0.000008	3.6×10^{-5}
Ozone	0.00005	1.7×10^{-6}

* Paneth, *Quart. J. R. Met. Soc.* **63,** 436 (1937).

Carbon dioxide is produced by the combustion of **carbon**-containing **fuels** and from the decay of organic matter. Its concentration in the air would be much greater if it were not consumed by vegetation in the process of **photosynthesis,** by which oxygen is liberated. In this way, the steady addition to the atmosphere of carbon dioxide by the combustion of fuels, and by the breathing of men and animals, is counteracted, and the oxygen-carbon dioxide ratio of the air is maintained.

The content of water vapor in the atmosphere varies greatly in amount, depending upon the locality, the season of the year, and the hour of the day, due to local

WATER VAPOR CONTENT OF NATURAL AIR

(Monthly Averages at Pittsburgh, Pa.)

MONTH	AVERAGE TEMPERATURE (°F.)	CONCENTRATION OF WATER IN AIR (Grains per cu. ft.)	VOLUME OF WATER (Gallons per hour at air rate of 20,000 cu. ft. per min.)
Jan.	37.0	2.18	87
Feb.	31.7	1.83	73
Mar.	47.0	3.40	136
Apr.	51.0	3.00	120
May	61.6	4.80	192
June	71.6	5.94	238
July	76.2	5.60	224
Aug.	73.6	5.16	206
Sept.	70.4	5.68	227
Oct.	56.4	4.00	160
Nov.	40.4	2.35	94
Dec.	36.6	2.25	90

and general states of the weather. On account of the great importance, meteorologically and industrially, of the amount and variation of water vapor in the atmosphere, extensive studies have been made by weather bureaus and various industries.

The amount of water vapor contained in the air may be expressed as the relative humidity, which is simply the fraction actually present of the amount of water required completely to saturate the air. The amount of water necessary to saturate 1 lb. of dry air is shown by the following table:

WEIGHT OF WATER NECESSARY TO SATURATE ONE POUND OF DRY AIR

TEMPERA-TURE (°F.)	WEIGHT OF WATER (Lb.)
40	.00520
45	.00632
50	.00765
55	.00920
60	.01105
65	.01322
70	.01578
75	.01877
80	.02226
85	.02634
90	.03108
95	.03662
100	.04305
105	.05052

The temperature of the air decreases, in general, with altitude up to about 10 or 12 kilometers within the layer, known as the "troposphere," in which convection currents and storms occur (see **Winds**). Above this is the "stratosphere," a region of unknown height within which a uniform temperature of about $-55°$ C. ($-67°$ F.) obtains, and in which there are no clouds, little wind, and no storms. Nearly all of the water vapor and dust are in the lower half of the troposphere; the stratosphere being very clear and excessively dry (see **Humidity**).

The various data cited above on the composition of the air are based upon samples taken at or near the surface of the earth. However, it is thought that at higher altitudes, notably above 70 kilometers, there is a marked increase in the concentration of the lighter gases, and it is believed that whatever air exists above 100 kilometers consists largely of hydrogen and helium. It is also thought that there is more ozone in the stratosphere than near the earth, and that it is due to the action of **ultra-violet** radiation on oxygen. Through some agency, perhaps **cosmic rays** or else **electrons** from the **sun**, the upper atmosphere is much more highly ionized than the lower (see **Ionosphere**). The earth's negative charge and the electricity in the upper atmosphere give rise to a vertical potential gradient, amounting to about 1.1 volts per cm. at the surface, but rapidly diminishing with altitude (see **Aurora Borealis** and **Lightning**).

By means of instruments carried by sounding balloons and from other sources of information such as, for example, the pioneer stratospheric flights, it is known that the atmosphere varies greatly with altitude in many respects such as (1) density, (2) pressure, (3) temperature, (4) motion, (5) composition, (6) electrical condition, and a number of other characteristics. The density, of course, is determined by the composition, pressure, and temperature. The pressure decreases with increasing altitude as is shown in the following table.

The standard density of air at 32° F. and 14.7 lbs. pressure is .081 lb. per cu. ft. Its gas constant is 53.4 cu. ft. per °F., and its composite molecular weight is 28.84. (See **Atmosphere**.) (F.T.M., R.K.S., L.D.W., W.K.G., P.E.K.)

VARIATION OF AIR PRESSURE WITH ALTITUDE

ALTITUDE (Ft.)	PRESSURE (In. of Mercury)	PRESSURE (Lbs. per sq. in.)
Sea Level	29.92	14.7
1,000	28.86	14.2
5,000	24.89	12.2
10,000	20.58	10.1
15,000	16.88	8.3
20,000	13.75	6.8
25,000	11.10	5.4
30,000	8.88	4.4
....	....	...
50,000	3.44	1.7

AIR-BLADDER. A pouch found in some fishes (**Pisces**), derived from the gut and filled with a mixture of gases, chiefly oxygen and nitrogen. It regulates the buoyancy of the body. Embryological development and structural relationship indicate a common evolutionary origin for the air-bladder and lungs. (A.W.L.)

AIR BRAKES. This could refer to air-operated mechanical brakes, or to surfaces whose aerodynamic drag provides a retarding force, such as the fan-type dynamometer.

Airplanes which may be expected to make very steep and prolonged dives are fitted with air brakes to limit the maximum diving speed. Dive bombers furnish the principal example of this. They need to approach their target at high altitude for safety from hostile gunfire, then dive on it almost vertically. The speed during the dive could rise to unsafe magnitude for an aerodynamically "clean" airplane. To control the terminal velocity in a dive, these aircraft have movable surfaces resembling large flaps (see **High Lift Devices**) which can be lowered, that is, set at a sharp angle to the wing chord, during a dive. But whereas flaps are designed to increase the maximum lift coefficient, and increased drag is incidental, air brakes are designed to produce as much drag as possible, hence are often perforated to assist the production of a turbulent wake.

Passenger automobiles and light trucks have **brakes** actuated by foot pedal pressure through the medium of mechanical or hydraulic linkage. The magnitude of brake pressure required is low enough so that the moderate mechanical advantage of these brake systems is adequate. Automotive buses, large trucks and truck-trailer vehicles are not readily and safely controlled with the aforementioned brakes, so frequently have an air brake system. This is generally a pressure system. A small engine-driven air compressor keeps a storage tank charged at moderately high pressure. The brake pedal operates a valve which admits the high-pressure air to one side of a diaphragm. Resultant distortion of the diaphragm furnishes a powerful force to operate either a mechanical or hydraulic brake system.

The safety of railway transportation is guarded by air brakes. In 1872 George Westinghouse patented his famous railway air brake which, with improvements made from time to time, has become the standard railway braking system. This air brake system applies the brakes when the air pressure is lowered. Air pressure is thus a force to release brakes and any accidental breakage of the air lines, including car-to-car couplings, will result in application of the brakes.

The elements of this air brake system will be described with reference to the accompanying figure. A steam-driven compressor on the locomotive is governed to maintain a constant air pressure in the locomotive storage tank, say about 90 lbs. per sq. in. The air brake line connects this tank with a "triple valve." This is a servo-piston-operated valve with openings to the train air line, to a car storage tank, to the brake operating cylinder, and to the atmosphere. When the brakes are

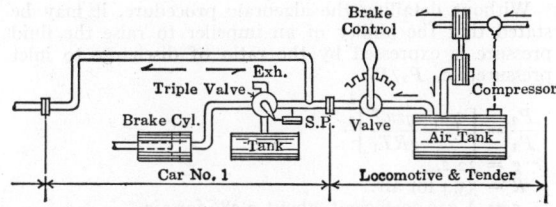

Railway air brake system.

off, line pressure is charging the car tank and the servo-piston has the triple valve turned so as to connect the brake cylinder with the atmosphere. To apply the brakes line pressure is relieved by operation of the brake control lever, whereupon the triple valve is turned to admit car tank air pressure into the brake cylinder.

This simple system has been considerably modified in practice as longer and heavier trains were formed. The need for uniform brake application along the length of the train, and of a graduated energy absorption by the wheel brakes, have required the addition of compensating elements. Some train systems are electropneumatic in nature. (F.T.M.)

AIR CHAMBER. Pumps.

AIR COMPRESSION.
The compression of air by mechanical means, and the raising of it to some desired pressure above that of the atmosphere, is effected, usually, by an approximate adiabatic change of state. The cycle of operation of an air compressor of the piston and cylinder form is briefly as follows: Beginning with the piston ready to start on the compression stroke, the piston compresses the air until the pressure rises to slightly above the discharge pressure, when the spring-loaded discharge valve opens and the remainder of the compression stroke is a delivery of the compressed air through the discharge valve at approximately constant pressure. At the end of the compression stroke, the discharge valve returns to its seat, but there is a small amount of high-pressure compressed air retained in the clearance space between the piston and the end of the cylinder. On the suction stroke of the piston this clearance air must first expand to the suction pressure before the spring-loaded inlet valve opens. The remainder of the suction stroke is then the induction of the air to be compressed in the cylinder. At the end of the suction stroke the inlet valve returns to its seat, and the piston is ready for another compression stroke.

If the ideal compression were possible, it would be represented by the following equation showing the relation between pressure and volume:

$$PV^{1.4} = \text{a constant.}$$

A compression of this nature may heat the air to temperatures which would interfere with reliable action of an air compressor and introduce lubrication difficulties, were there no provision for cooling the cylinder walls. Therefore, in compressors we find the cylinders to be externally finned or water-jacketed so that sufficient cooling is secured to keep the temperatures from becoming excessive. The extraction of heat from the cycle in this way modifies the conditions of compression from the ideal to some change more nearly represented by

$$PV^n = C,$$

in which n usually lies between 1.35 and 1.4. The ratio of the temperature before and after compression is expressed by the following equation, the temperatures being degrees Fahrenheit absolute.

$$\frac{T_2}{T_1} = \left[\frac{V_1}{V_2}\right]^{n-1}.$$

In compression to high pressures, the temperature rise may be too great to permit the compression to be car-

ried to completion in one cylinder, even though it is cooled as mentioned above. In high-pressure compressors, the compression is carried out in stages, with a partial increase of the pressure in each stage, and cooling of the air between the stages. Two- and three-stage compression is very common where pressures of 300–1000 lbs. per sq. in. are needed.

The volume of clearance air should be made as small as possible in order to improve the **volumetric efficiency** of the compressor, since the clearance air must expand to the suction pressure before the cylinder can begin to be charged.

The mechanical construction of air compressors varies with the amount of compression required. Piston and cylinder compressors are usually employed for the highest pressures. In small sizes these are frequently single-acting, but are made double-acting in larger sizes. High compression pressures also may be produced by centrifugal action. (See **Air Compressor**.) Light pressures, such as are required in draft and ventilating systems, are obtained most economically by the use of fans of the centrifugal or propeller type.

There is a type of compressor intermediate between the fan and the piston types. It is the rotary compressor or **blower**, which, operating by displacement, produces a positive air pressure, and, at the same time, is a compressor in which there are no reciprocating parts.

The compression of air and other gases may also be secured by the employment of steam jets if an admixture of vapor in the compressed gas is not undesirable. High-pressure steam is blown through nozzles which create a high-velocity jet. The gas to be compressed is led into the regions about the nozzle discharge where it is entrained in the steam jet. The mixture then travels into a diffuser for compression and attendant velocity reduction. Although the compression thus achieved is of limited magnitude, staging the compression in a series of nozzles, with intermediate coolers for partial condensation of vapor, allows moderate compression ratios to be achieved. (F.T.M.)

AIR COMPRESSORS. *Reciprocating.*
A common type of air compressor is the piston and cylinder compressor in which a reciprocating piston positively displaces the air from a cylinder during its discharge stroke. Compressors for charging tanks of air used to inflate pneumatic tires at the numerous automotive service stations are of the reciprocating type. Being of small capacity they are generally single acting and air cooled (by exterior fins) since those features are common in small compressors. Larger compressors, unless for extremely high pressure, are usually double acting and frequently cooled by water jackets. One or two cylinder arrangements are conventional, with a tendency to secure large capacity by increased bore and stroke rather than by multiple cylinders. Pistons are reciprocated by a crankshaft and connecting rod mechanism commonly deriving motion from the driving source by belt. Valves are spring-loaded to open upon slight differential pressures.

The compressor cycle is shown in Fig. 1. The discharge stroke which begins at A builds up the pressure to B where it exceeds the receiver pressure sufficiently to

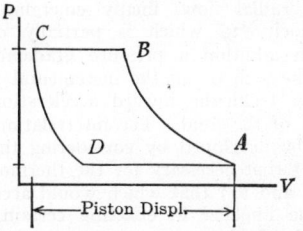

Fig. 1. Reciprocating compressor cycle.

open a discharge valve. Discharge then takes place at a constant control pressure from B to C. The volume C is the clearance volume of the compressor. The air in the clearance volume must expand to D during the suction stroke before the inlet valve will open. Thus the volume of air drawn in per stroke is only that from D to A. Obviously the compressor should have as small a clearance as possible in order to obtain good volumetric efficiency, especially at high discharge pressures. Small compressors may be operated with high compression ratios (8–12) if desired because cooling is more effective in small cylinders and mechanical strength is readily provided. Large volumes compressed to ratios exceeding 4 will need a multi-stage compressor to permit cooling between stages and to lessen the structural loads on the large first stage cylinders.

Rotating. Rotating compressors may be subdivided into four classes, known as fans, blowers, centrifugal compressors and turbo-compressors.

Fans are limited to low pressures where compression is negligible but large volumes are delivered.

Blowers are defined as machines for compressing air at pressures up to 35 lbs. per sq. in. gauge. Centrifugal and turbine type compressors may be built to deliver high pressures by providing sufficient stages; however, reciprocating compressors are commonly employed for pressures exceeding 50 lbs. per sq. in.

Centrifugal Compressor. A rotating impeller mounted in a casing and revolved at high speed will cause a fluid which is continuously admitted near the center of rotation to experience an outward flow and a pressure rise due to centrifugal action.

Assume that an impeller with radial blades of depth $r_2 - r_1$ is revolving at a speed of ω radians per minute. This is illustrated in Fig. 2. Consider that a compressible

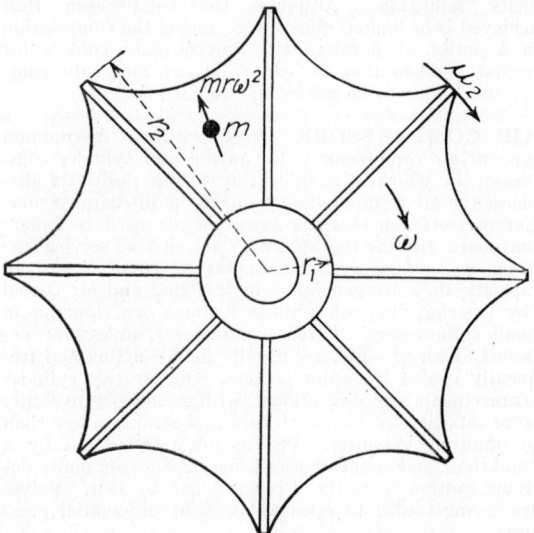

Fig. 2. Impeller for centrifugal compressor.

fluid (a gas) is admitted at the center and flows into the impeller radially. Relative to the impeller blades it has an outward radial flow, finally emerging with some absolute velocity v_2 which is partially diffused into pressure. In addition a pressure gradient must exist to balance the sum of all the incremental $mr\omega^2$ inertia forces arising from the inward acceleration $r\omega^2$ given each particle of the fluid. The interrelation of r, ω, P, can readily be developed by considering the power required as (1) that necessary for the thermodynamics of compression, and (2) that which would account for the action of the impeller in effecting certain momentum changes on the fluid.

Without detailing the algebraic procedure, it may be stated that the ability of an impeller to raise the fluid pressure is expressed by the ratio of discharge to inlet pressure, i.e., P_2/P_1

$$\frac{P_2}{P_1} = \left[1 + \frac{\eta z u_2^2}{gRT_1}\right]^{\frac{1}{z}}.$$

$g = 32.2$.
$R = 53.4$ for air.
$z = $ A gas coefficient, about 0.286 for air.
$T_1 = $ Inlet temperature, degrees Rankine.
$\eta = $ Energy coefficient. This would be 1.0 if the flow through the impeller were non-turbulent and frictionless. Typically, $\eta = .75$ to $.85$.
$u_2 = $ Impeller rim velocity, ft. per sec.

For pressure ratios higher than can be obtained from the action defined above, several impellers may be mounted on the same shaft and enclosed in a compound casing with passages arranged to lead the output from one impeller to the "eye" of the next. This multi-staging principle is used to produce pressures above the capability of a single impeller compressor. Multi-stage compression may have cooling between the stages so the overall compression may be more isothermal than adiabatic. If compression were isothermal

$$\frac{P_2'}{P_1} = e^{\frac{\eta u^2}{gRT}}.$$

(See also **Supercharger, Airplane Engine.**)

Turbo-compressors. This is a multi-stage **axial-flow** compressor, so called because of the resemblance, in action, to a reversed turbine. Fig. 3 shows a turbo-com-

Fig. 3. Turbo-compressor rotor. (*General Electric Co.*)

pressor, while Fig. 4 illustrates the compressing action. Air flows over a set of stationary airfoils (arranged circumferentially as fixed blading, with an angle of attack α. As the airfoil **aspect ratio** is small, the **downwash** created is considerable. Downwash turns the air stream through angle ϵ, which results in a flow channel

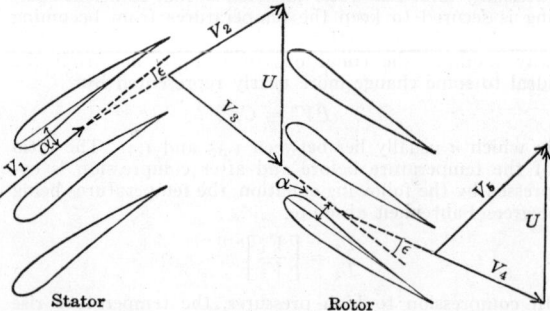

Fig. 4. Velocity relations in axial-flow compressor blading.

of increasing cross section. The diffusion thus effected slows down the air velocity to V_2 and increases pressure. The moving airfoils (rotor blades) receive this air stream at V_3 as a result of the vectorial combination of air stream velocity V_2 and blade speed U. Relative to the moving airfoils the downwash again causes a diffusion of velocity and another increase of pressure. The relative velocity V_4 leaving the moving blades produces an absolute final velocity of V_5 because of blade motion. Suitable combinations of U and a enable V_5 to duplicate V_1 and permit a similar following stage. The blade height in the following stage is decreased because of the smaller specific volume of the compressed air. Increasing a will increase the pressure rise without much decrease in volume. The turbo-compressor should be designed for conservative a's for two reasons. First, an operation near the stalling angle would be hazardous as small variations of a might occur which could **burble** the airfoils and cause an unstable, rough, or even hazardous condition to exist. Secondly, turbo-compressors may be employed under conditions where utmost efficiency is imperative (as in gas turbine power units) and a should create the optimum favorable balance between good downwash and minimum turbulent airfoil wake. Turbo-compressors have been built with energy efficiencies as high as 85%. They may be operated effectively at high speeds, i.e., 5000–10,000 rpm. (F.T.M.)

AIR CONDITIONING. Air conditioning is the artificial treatment of air in buildings to render the living conditions of persons within the building more comfortable and healthful, or to ensure better conditions for the production and storage of materials. Complete air conditioning involves adjustment and control of the following operations performed on the air supply of a building:

1. Heating or cooling.
2. Humidification or dehumidification.
3. Ventilation.
4. Cleaning.

While portions of this complete program of conditioning were often used in past years, it has only been recently that the importance of complete air conditioning has been fully understood. That a change of air in a room has beneficial effects is quite generally understood and the ventilation of buildings has been a subject of considerable study. Although it is quite obvious that dust and obnoxious fumes have no place in either industrial or domestic rooms, little has been done towards the filtering of air supplies. One new and important fact brought out within recent years has given great stimulus to more scientific handling of the air conditioning problem. It is that, in addition to temperature and air movement, the relative **humidity** is a very important factor in determining the comfort of the occupant of a room. Experimental research of the American Society of Heating and Ventilating Engineers indicated that air conditions which yielded equal degrees of warmth to human subjects plotted as straight lines on the common psychometric chart. These lines are called "comfort lines." The meaning of this fact is that a person may feel equally comfortable at two different temperatures, provided the humidity of the air at these two temperatures differs by the proper amount. The reason for the influence of humidity on comfort is that the nearer to a completely saturated state the air in a room becomes, the less tendency there is for evaporation to take place from the body. This evaporation plays no small part in determining hot weather comfort, and is a factor in surface cooling of the skin. On the other hand, in cold weather the atmosphere in an artificially heated room is very likely to prove deficient in moisture unless some special means for increasing the moisture content is provided. Some physicians believe that failure to humidify properly the air within heated buildings is an important

contributing cause of the common cold during the winter season.

Complete air conditioning, then, will involve the following equipment: a ventilating system for giving motion and circulation to the air; a furnace or heater to raise it to proper temperature in cold weather; a refrigerator, or cooler, to temper it for hot weather comfort; a humidifier, or dehumidifier; an air washer, or filter. This complete service is frequently applied to industrial buildings, theaters, auditoriums, etc., but has not been employed to any appreciable extent in homes. One reason for this, of course, is the cost of such equipment for the average home. Another is lack of information on the benefits realized from an air conditioning installation; another, the widespread use of vapor or hot water heat, which is not readily converted to an air conditioning system.

It is interesting to note the reversion to the once obsolete warm air heating system, on the basis of its adaptability to partial or complete year-round air conditioning. The warm air heating system, once discredited as obsolete and inferior to other types, is returning to a position of some considerable importance through the employment of positive fan-created air circulation—eliminating many real defects in the gravity system—and the use of thermostat-operated automatic controls preventing fluctuations of temperature. Moisture can be added to the air at the furnace in the winter time, and to the required degree. The same system may be employed in the summer time, either with refrigerating coils in the return circuit, or with water sprays, or simply with the induction of air from without the building. Under any of these operating conditions, of course, the air passes through filters for the removal of dust.

Just as the air conditioning of industrial and public buildings has proven to be practical and profitable, so will, in all probability, air conditioning be an important feature in the new homes of the future. (F.T.M.)

AIR GAP. This term is commonly used in connection with various **magnetic circuits** and denotes a gap left in the magnetic material. In the construction of various **chokes** and **transformers** used in communications circuits a short gap is usually left in the core material to prevent the material being saturated by the d.c. which often flows in such circuits. (See **Amplifiers** and **Power Supplies**.)

In rotating electrical machinery the rotating part of the magnetic circuit must, of course, be separated by a gap. In these machines this gap is kept as small as consistent with adequate mechanical clearance. In most instances the gap introduces no desirable electrical or magnetic characteristics and necessitates the application of additional electrical energy to overcome its **reluctance**. (L.R.Q.)

AIR HUNGER. Breathlessness or craving for air when insufficient **oxygen** is supplied to the **tissues** of the body. (R.S.M.)

AIR LIFT. An air lift is a water pumping method whereby water may be raised from a well through the medium of compressed air. The drop pipe in the well is supplied at the bottom with compressed air from a small air pipe, and the effect of mixture of air and water at the bottom of the drop pipe is to bring water to the surface. This is accomplished either by the water acting as pistons, trapping intermediate layers of air, the expansion of which drives the water pistons to the surface, or it may be accomplished by the mingling of air and water, forming a mixture which is sufficiently lighter than the undisturbed water in the well so that the mixture rises above the surrounding water. In order for this rise to reach the surface the discharge pipe must be submerged in the water of the well an amount varying from 100–300% of the actual lift. The **pumping**

efficiency of the system is very low, but it is very suitable for handling gritty or corrosive waters. (F.T.M.)

AIR LOCK. An air lock is an air-tight compartment in which the air pressure may be regulated to any desired intensity. When men are required to work in regions where the air pressure is above (or below) that of the atmosphere, an air lock must be provided to permit passage of the workmen from the open atmosphere to the pressure region. Thus, in the case of **caissons,** where workmen must labor under a high enough pressure to equalize the **hydrostatic** pressure existing at the bottom of the caisson, or in tunnels where flooding is avoided by forcing compressed air into it at sufficiently high pressures to hold the water back, the air lock is a feature essential to the maintenance of pressure during the admission of workmen. It is also used when entering boiler rooms of steamships where, to supply forced draft to the boilers, the entire boiler room is sometimes held at a pressure slightly above that of the atmosphere.

In construction work, the air lock is a chamber of

it travels. Modification continues until the air mass loses its identity in the general atmospheric circulation.

Classification of air masses begins from latitudinal consideration. There are four major zones which contribute primary classifications: 1. Arctic. 2. Polar. 3. Tropical. 4. Equatorial.

These are subdivided into Maritime (*m*) and Continental (*c*), depending upon the exact source region. Finally, each air mass must be classified as either cold (*k*) or warm (*w*). A cold air mass is one which is colder than the surface over which it is traveling and is therefore being heated from below. A warm air mass is one which is warmer than the surface over which it is traveling and is therefore being cooled from below. Capital letters are used to designate the zone location of an air-mass source: *A*, Arctic; *P*, Polar; *T*, Tropical; *E*, Equatorial. Small letters are used for appropriate subdivisions: *m*, maritime; *c*, continental; *w*, warm; *k*, cold. Small letters placed before and after the capital letter serve as descriptive symbols for the air masses.

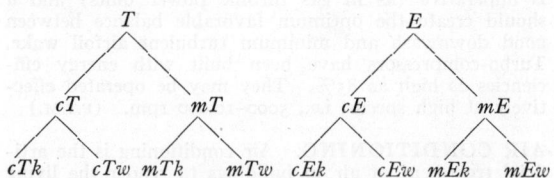

sufficient size to hold the number of men that must be accommodated in it at one time. It is provided with well braced doors having sealing-type edges and tightening locks. The chamber is equipped with valves for admitting and releasing air and with safety devices to prevent excessive pressures endangering the lives of the occupants. The air lock must have two air-tight doors, one leading to the atmosphere, the other leading to the pressure region. These doors open inward so that the pressure in the air lock tends to tighten them against the frame. To enter a pressure region, a caisson for example, the workmen enter the air lock, after which the door leading to the atmosphere is tightly fastened. Compressed air is then slowly admitted until the pressure in the air lock equals that in the caisson, after which the connecting door may be opened without trouble or loss of air from the working chamber. After the workmen enter the caisson, the air lock door is tightly closed, after which the air lock may be opened from the outside without affecting conditions within the caisson.

In contrast to air locks, the decompression chamber such as is used in deep sea diving has one door. The diver, on emergence from the water, is briefly subjected to atmospheric pressure, and must be rushed to the decompression chamber and quickly subjected to pressures approximating those encountered in diving. The pressure is then slowly released, the rate of decompression being such as to prevent the malady known as "the bends." Pressures in the decompression chamber are ordinarily much higher than those for which air locks need to be built. (F.T.M.)

AIR MASSES. Very large parcels of air, ranging from about 500–5000 miles in lateral dimensions and from several thousand feet to several miles deep, which have properties (temperature, humidity, thermal structure) that vary only slightly, or vary linearly, from point to point within the parcel, are known as air masses. Air masses develop over large relatively homogeneous geographical areas where air is stagnant for a sufficient period to acquire the characteristics of that region. These regions are either continental or maritime and are known as air-mass source regions. After an air mass begins to move from its source region it acquires modifying features characteristic of the surface over which

Over North America, except for extreme northern and southern parts, only Polar and Tropical air masses occur. Areas east of the Rockies are dominated by continental Polar and maritime Tropical air masses whose contrasting features are the cause of most of the stormy weather of this section. Pacific Coast regions are almost continually under the influence of maritime Polar and maritime Tropical air masses which are cool in summer and warm in winter.

Cold air masses tend to produce cumuloform clouds, showers and thunderstorms. The visibility is relatively unobstructed by pollution or haze; it often exceeds 50 miles at ground levels and several hundred miles from aircraft in flight. Fog is infrequent. Winds are frequently gusty from mid-morning to sunset. Flying air is rough up to considerable height. Precipitation, when it occurs, is showery, often with sunshine between bursts of rain or snow.

Warm air masses favor the formation of fogs, low clouds and drizzle. Visibility is more or less universally restricted by pollution and haze which are held in a shallow surface layer by the air mass stability. Cloud forms are generally stratified types ranging from fog to altostratus. (P.E.K.)

AIR MEDICINE. The study of physiological and pathological changes (anoxia, "bends," etc.) associated with flying. (D.M.H.)

AIR PLOT. A plot showing the movements of an airplane relative to the air is known, in **navigation,** as an air plot. It is similar in every respect to the old-fashioned method for obtaining the **dead reckoning** (DR) position of a sea-borne ship, disregarding the effects of ocean currents.

The air plot is made up of a series of distance **vectors,** added in the usual manner, the direction of each vector being a **heading** of the plane and the length is proportional to the distance moved through the air along the heading (distance = **air speed** × time on heading). The summation of the vectors gives a no wind (NW) position of the plane at any time. To obtain the DR position of the plane it is only necessary to add the total wind vector, whose direction is that of the motion of the air and whose length is the total movement (wind speed × total time since departure) of the air, to the

NW position. Such a method avoids the necessity of constructing the velocity-vector diagram to obtain **course** and speed along course for each heading, as is necessary in the standard method for obtaining a DR position in air navigation.

The air-plot method has been in use with British aviators for many years and is rapidly gaining favor in all air services. In cases where the plane is frequently changing headings, the air-plot method saves from 30–60% of the time involved in obtaining the DR position by standard methods. Furthermore, in the frequent cases where the predicted wind for high altitude is very inaccurate, the air plot provides a method for determining the wind.

The solution of the following problem indicates the advantages and value of the air plot. On a day when the predicted winds are uncertain, a plane is to search an area centered about a point in latitude $L = 42° 22'$ N and longitude $Lo = 35° 25'$ W. The plane, flying with true air speed TAS = 130 knots at altitude 10,000′, is over the point at 0500 and heads 150°. The altitude and air speed are maintained but the plane alters headings at the given times as follows: 0510 heading 240°, 0520 heading 330°, 0540 heading 060°, 0600 heading 150°, 0640 heading 240°. The air plot, properly labeled is shown in the figure from which the NW position at 0700 is found to be $L = 41° 15'$ N and $Lo = 35° 07'$ W.

The predicted wind at the altitude of flight is from 000° and speed 35 knots. From the 0700 NW position a vector in direction 180° and of length 70 miles is drawn. This is shown as a dotted line in the figure and repre-

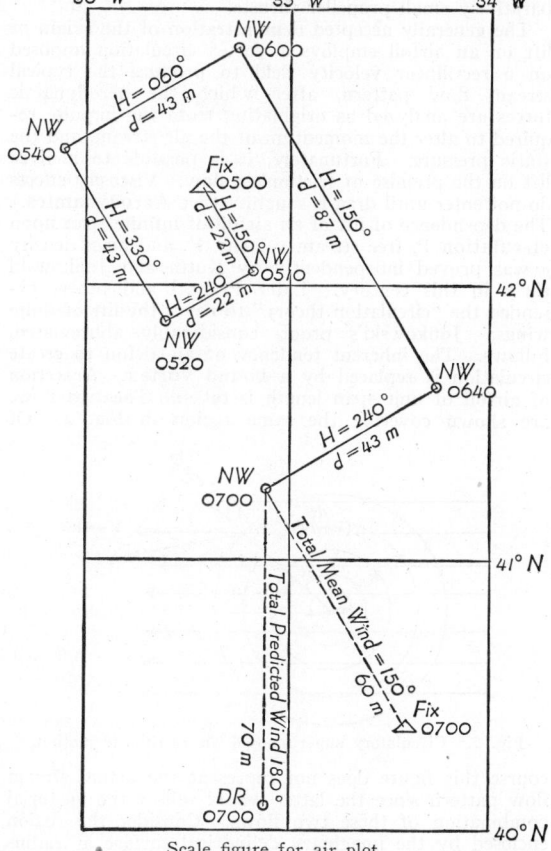

Scale figure for air plot.

sents the total movement of the plane due to the predicted wind in the two hours from 0500. This gives a dead-reckoning position in latitude 40° 05′ N and longitude 35° 07′ W.

A fix is obtained at 0700 in latitude 40° 23′ N and longitude 34° 27′ W. If a vector is drawn from the NW position to the fix, it will represent the actual movement of the air in the interval from 0500 to 0700. From the diagram this is found to be in the direction 150° and 60 miles long. Since this represents 2 hours total motion of the air, we have a mean wind from 330°, speed 30 knots.

To obtain the DR position, using the standard methods of dead reckoning, would have required the plotting of at least four velocity-vector diagrams to obtain the course and ground speed on each heading. Each of these would have taken at least twice as much time as is required to plot an air vector and, after the courses and ground distances had been obtained, the geographic-motion vector diagram would have had to be plotted. Then to obtain the mean wind, the assumed air motion would have had to be subtracted from the DR position to obtain the NW position, which has to be used with the fix to obtain the actual total wind motion. (w.k.g.)

AIR POCKETS. Bump, Air.

AIR PREHEATER. There are many devices of which the purpose is to heat air for some specific usage. However, in speaking of air preheaters, what is ordinarily meant is the heater employed for raising the temperature of air used for **combustion** of a fuel. This may occur in some industrial process such as preliminary heating of the air supplied to **blast furnaces**, but the most frequent use of air preheaters today is in connection with steam **boilers**. This type of air preheater is a heating surface installed between the boiler flue gas outlet and the stack. In arrangement the heating surface is composed either of tubes with flue gas inside and the air to be heated outside, or of rectangular plates spaced about one-half inch apart, leaving alternate gas and air passages. Its use is chiefly justified on economic grounds.

Two principles are employed for heat transfer in air preheaters. The recuperative principle implies transfer of heat through a separating partition, such as the walls of a tube, by continuously recuperating the cool side with conduction of heat from the hot side. Regenerative heaters are those which alternately heat and cool the same mass, regenerating it thermally by passing hot spent gas over its surface. Regenerative heaters are frequently used with blast furnaces, and are composed of two heating chambers in which are piled checkerworks of brick having sufficient heat storage capacity for the purpose. The burned gas leaving the furnace passes through one chamber, heating up the checkerwork, while in the other chamber the heated bricks are being cooled by air passing to the combustion region. When the air-heating chamber is thermally exhausted, valves shift the flow of hot gas through that chamber, and air is drawn through the hot one. (f.t.m.)

AIR PUMPS. The earliest air pump, constructed by von Guericke (1650), differed in no essential way from the ordinary suction water pump, and was capable of producing a vacuum of the order of 1 mm. of mercury. Pumps of this type held the field for two centuries, until Geissler (1855) evolved a practical means of utilizing the Torricellian principle illustrated by the vacuum above the mercury in a barometer. This was also the plan followed in designing the Töpler pump of 1862. The Sprengel mercury pump (1865) is quite different, utilizing the air-trap action of drops of mercury falling into a narrow funnel, after the manner of the ordinary water-jet pump used for filtering, etc. These pumps were used in many pioneer researches with vacuum tubes, etc., but are far too slow and fail to give sufficiently low pressure for much modern work.

Greatly improved forms of mechanical air pump have been devised in recent years. Among these are pumps of the scraping vane type, in which a solid cylinder rotates eccentrically within a hollow cylinder, carrying the air around by means of a blade or vane protruding radially from one and rubbing against the other; also a rotary mercury pump by Gaede, in which air is imprisoned by spiral compartments dipping into mercury, and forced out by their further rotation. Several so-called molecular pumps belong also to the mechanical class; in these the air is dragged along and expelled by the friction of a cylinder rotating rapidly inside a close-fitting casing. These rapid mechanical pumps are used for many purposes directly, and also as fore pumps or backing pumps for more effective types, in which case they serve to provide a fair vacuum into which the latter may discharge the last removable traces of gas from more highly evacuated enclosures.

The most efficient air pumps at present in use are those of the diffusion type. In most of these the pumping agent is mercury vapor issuing from a suitable boiler, the fast-moving molecules of which carry off the gas molecules diffusing into the enclosure from a side opening, somewhat as falling raindrops clear the atmosphere of dust. In Langmuir's condensation pump, the mercury vapor is prevented from entering the high-vacuum enclosure, or obstructing the diffusion, by a cold-water jacket which condenses the vapor. Such diffusion pumps, with the aid of a fore vacuum, are capable of very rapid exhaustion and are used extensively in the manufacture of x-ray tubes, radio tubes, and lamp bulbs. (L.D.W.)

AIR SPEED. Airplane Speeds.

AIR STANDARD EFFICIENCY. The actual thermal efficiency of **internal combustion** engines depends on many indeterminate factors which render the rational computation difficult, if not impossible. An efficiency may be computed, based on certain assumptions, as follows: first, that the internal combustion engine has no mechanical friction; second, that the compression and expansion in the cylinder are those of pure air, whereas actually the expanding gases are composed of nitrogen, oxygen, steam, carbon dioxide, and carbon monoxide, and the gas compressed is never pure air; third, that the compression and expansion are **adiabatic.** This would imply a heat insulation jacket around the cylinder, but the difficulties of successful lubrication have required all practical internal combustion engines to be positively cooled. The steady flow of heat from the cylinder to the cooling system destroys the possibility of an adiabatic compression or expansion. Based on these three assumptions, thermodynamic theory can be used to yield an equation of efficiency of the cycle, and such is termed the "air standard efficiency." The actual thermal efficiency will, of course, be considerably less than the air standard efficiency. Nevertheless, the air standard efficiency is useful as a measuring stick for the various designs, and it also shows the effect, on efficiency, of varying the compression ratio. The air standard efficiency of the **Otto,** or gasoline engine cycle, is given by the equation:

$$E = 1 - \frac{1}{r^{k-1}}.$$

For the **Diesel cycle** it is:

$$E = 1 - \frac{R^k - 1}{kr^{k-1}(R - 1)}.$$

In these formulae, the symbols have the following meanings: r is the **ratio of compression;** R is the cut-off ratio, i.e., roughly, the percentage of the stroke through which fuel injection occurs; k varies, often being assumed 1.3 for hot air, 1.4 for cold air. (F.T.M.)

AIRFOIL. An airfoil is any body whose shape causes it to receive a useful reaction from an air stream moving relative to it. This definition is broad and could include many shapes not ordinarily considered to be airfoils. The term is usually associated with a body of

the shape shown in Fig. 1. The cross-section of the airfoil is shown. The dimension perpendicular to this section is called the *span.* Many different airfoil

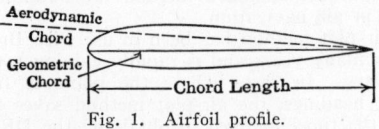

Fig. 1. Airfoil profile.

shapes have been used, proposed for use, or tested. Some of them have flat lower surfaces, while others have convex or concave surfaces. Systems have been devised for classifying and cataloging airfoils of different profiles. (See **Airfoil Classification.**)

The characteristic lifting airfoil profile has a maximum thickness of 10–15% of the chord at 20–40% of the chord aft of the leading edge. Early experiments were made with flat-plate airfoils, and later with thin curved-plate airfoils, but neither possessed as large a ratio of lift to drag as the double-surface cambered airfoil shown. Also an airfoil of finite thickness provides space for the foundation structure of a light weight **wing,** so removing the structural elements from the drag of the air stream.

The value of the airfoil shape to aviation resides in the magnitude and direction of the resultant air reaction on it when employed at attitudes below the stall (about 15° **angle of attack.**) The component of the air reaction normal to the free air stream is several times the magnitude of the parallel component. Thus lifting forces may be generated on wings by the employment of comparatively small propeller thrusts.

The generally accepted demonstration of the origin of lift on an airfoil employs a vortex circulation imposed on a rectilinear velocity field to produce the typical stream flow pattern, after which the aerodynamic forces are analyzed as originating from the impulse required to alter the momentum of the air stream, and the static pressure. Fortunately, it is possible to analyze lift on the premise of frictionless flow. Viscosity effects do not enter until drag is sought. (See **Aerodynamics.**) The dependence of lift of an airfoil of infinite span upon **circulation** Γ, free stream velocity V, and mass density ρ was proved independently by Kutta and Joukowski early in this century. Later Prandtl and others extended the "circulation theory" to cover the lift of finite wings. Joukowski's proof, considerably abbreviated, follows. The inherent tendency of an airfoil to create circulation is replaced by a **bound vortex.** A section of airfoil of unit span length is taken. The two flows are shown covering the same region in Fig. 2. Of

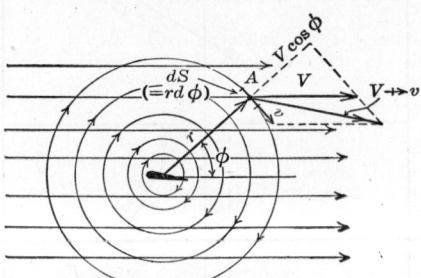

Fig. 2. Circulatory superimposed on rectilinear motion.

course this figure does not represent the actual stream flow pattern since the latter would follow the vectorial combination of these two flows. Consider the region enclosed by the imaginary cylindrical surface at radius r from the vortex center. Let r be large enough to have circulatory velocity v small compared to rectilinear velocity V and integrate for vectorial change in momentum in this cylinder and vertical component of static pressure acting against it.

The velocity of circulation, $v, = \dfrac{\Gamma}{2\pi r}$ and its combination with V gives the direction and magnitude of the stream line at point A. The mass of air leaving the cylindrical region at A is $\rho ds(V \cos \phi)$ and its vertical component of velocity is $v \cos \phi$. The net vertical momentum in the cylindrical mass of air, due to inflow and outflow is:

$$\oint (\rho r d\phi)(V \cos \phi) \left(\frac{\Gamma}{2\pi r} \cos \phi \right)$$

which simplifies to

$$\frac{\rho \Gamma V}{2\pi} \oint \cos^2 \phi d\phi.$$

Whence, by integration, the vertical impulse force is $\dfrac{\rho \Gamma V}{2}$. Since the change of momentum is downward the impulse force is upwards, i.e., it is the lift. An integration of the horizontal component of momentum yields zero net momentum.

Proceeding next to evaluate the static pressure, note that Bernoulli's Theorem applied at point A is covered by the following statement. Let p be the static pressure at A, and p_0 the free stream static pressure, then

$$p + \frac{\rho}{2} (V + v)^2 = p_0 + \frac{\rho}{2} V^2$$

After expanding $(V + v)^2$ and dropping all v^2 terms (since A was chosen to make v small relative to V)

$$p = p_0 - \rho V v \sin \phi.$$

Since the second term on the right-hand side of this equation represents the variation from free stream pressure at point A, the lift (if any) will be the line integral of its vertical component; that is,

Vertical pressure component

$$= \oint (\rho V v \sin \phi)(\sin \phi)(r d\phi)$$

$$= \rho \frac{V \Gamma}{2\pi} \oint \sin^2 \phi d\phi$$

$$= \rho \frac{V \Gamma}{2}.$$

The horizontal integral yields zero net pressure. The total lift per unit of span is the sum of that originating from momentum and that originating from pressure. From the preceding demonstration it is seen that each contributes equally to the total lift, which becomes $\rho V \Gamma$.

Theory and experiment show that Γ is always sufficient to cause the divided flow over the upper and lower surfaces to reunite at the trailing edge without reverse flow.

If the aerodynamic chord of the wing is parallel to the free stream velocity V, the lift is zero and obviously Γ is zero. As the angle of attack is increased Γ increases nearly proportionately. The strength of the circulation for an airfoil of chord c at aerodynamic angle of attack a_a is $\frac{1}{2}aVca_a$, where a is the proportionating factor. Since Lift, $L, = \rho V \Gamma b$ (b = span), it also equals

$$aa_a \frac{\rho V^2}{2} (bc).$$

Call aa_a the coefficient of lift, bc the area S, and $\dfrac{\rho V^2}{2}$ the dynamic pressure q, then the lift equation becomes:

$$L = C_L q S.$$

The variation of C_L with geometric angle of attack is seen in Fig. 3. The slope of the straight portion is a

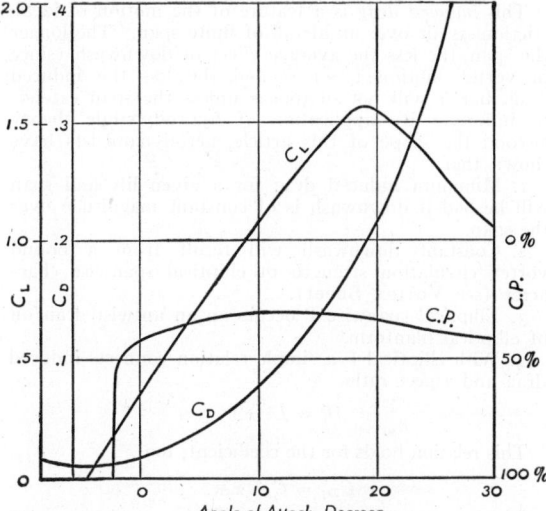

Fig. 3. Common aerodynamic characteristics of a Clark Y airfoil.

per radian. Theory indicates a value of 2π for a and tests agree well with the theory. The lift coefficient increases uniformly with angle of attack until the **stall** or **burble** point is reached, where it breaks and decreases rapidly with further increase of angle of attack because the high attack angle creates a bluff body and the **stagnation point** advances forward on the airfoil.

The full aerodynamic reaction upon an airfoil is a force of which the lift is the component normal to the free stream velocity. The other component is measured in the direction of the air velocity and is called drag. Drag coefficient has a characteristic parabolic variation, as Fig. 3 shows. The drag equation is analogous to that for the lift:

$$D = C_D q S.$$

If the resultant of all air reactions on an airfoil is consolidated into a single imaginary force, it must act at the **center of pressure** of the airfoil. The figure shows that the location of this center of pressure of an airfoil varies with its attitude.

If the airfoil has a finite span (i.e., definite tips), the **bound vortex**, assumed for the purpose of accounting for lift, is found to extend from the tips downstream as a free vortex. The effect of these tip vortices is to modify the air reaction since a **downwash**, w, is produced at the lifting line (see figure). The effect of

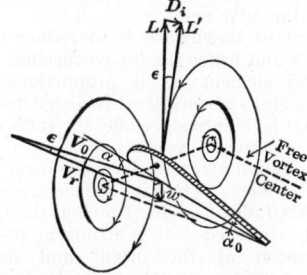

Fig. 4. Perspective of a section of airfoil subject to downwash from the free vortices at the tips.

downwash is to change the lift direction from L to L' since the relative wind, V_r, is now inclined. The angle ϵ is always small so L' may be considered of the same magnitude as L. However, L still remains the significant lift since the free stream velocity is V_0. Consequently, a drag D_i is introduced, although the original assumption of zero air viscosity remains in order.

This *induced drag* is a feature of the motion of ideal, frictionless air over an airfoil of finite span. The longer the span, the less the average effect of downwash (since in vortex motion $v \sim r^{-1}$) and the less the induced drag, but it will not disappear unless the span extends to infinity. By application of hydrodynamic theory beyond the scope of this article, aerodynamicists have shown that:

1. Minimum induced drag for a given lift and span will be had if downwash is of constant magnitude over the span.

2. Constant downwash will result from a bound vortex circulation strength of elliptical spanwise character (see **Vortex Sheet**).

3. Elliptical spanwise Γ occurs on an untwisted airfoil of elliptical planform.

4. With elliptical Γ a simple relation connects induced drag and aspect ratio.

$$D_i = L^2/\pi \mathcal{R}.$$

This relation holds for the coefficient, too.

$$C_{D_i} = C_L^2/\pi \mathcal{R}.$$

As

$$\frac{D_i}{L} = \frac{C_{D_i}}{C_L} = \frac{W}{V_0} \text{ (nearly),}$$

$$\alpha = \alpha_0 + C_L/\pi \mathcal{R}.$$

It appears that a finite wing possesses an induced drag of C_{D_i} not possessed by an infinite wing in ideal flow, also an induced angle of attack, requiring the angle of attack α of the finite span to be more than α_0 of an infinite span for the same lift per unit of span. The differences of induced drags of two airfoils of the same profile at the same lift coefficient but with different planforms follow naturally from the above:

$$C_{D_{i1}} - C_{D_{i2}} = \frac{C_L^2}{\pi}\left(\frac{1}{\mathcal{R}_1} - \frac{1}{\mathcal{R}_2}\right).$$

It is commonly assumed that the difference between the total drag coefficients of two airfoils of the same profile but different aspect ratios is the difference in induced drag. Likewise the difference in necessary angle of attack to the free stream velocity for the same C_L is:

$$\alpha_1 - \alpha_2 = \frac{C_L}{\pi}\left(\frac{1}{\mathcal{R}_1} - \frac{1}{\mathcal{R}_2}\right).$$

While these induced effects are premised on elliptical planform for the airfoil, they hold very well for rectilinear airfoils with elliptical tips and those with moderate taper ratio.

The lateral or spanwise distribution of lift on an airfoil depends on the planform and the distribution of downwash. Two cases are illustrated—an elliptical airfoil and a rectangular one.

1. The effect of downwash is to reduce the effective angle of attack and hence the lift coefficient. The lift of any lateral element dS is proportional to the product of C_L and c; hence the lift distribution of an elliptical wing is elliptical, while its induced drag, mirroring the downwash, is constant.

2. The downwash strength increases near the tips of a rectangular wing. The lift coefficient accordingly declines and lift falls off at the tip even though chord is maintained. Induced drag is strong at the tips.

This discussion of the origin and nature of aerodynamic reactions on the airfoil is concluded with a brief résumé of qualities of an airfoil which are sought for when it is employed in aviation.

1. High maximum C_L, in order to give low landing speed for a given size wing.

2. Low minimum C_D, so that the high speed, which occurs at small angles of attack, may be the greatest possible.

3. High ratio of C_L to C_D, so that an efficient, economical airplane will result.

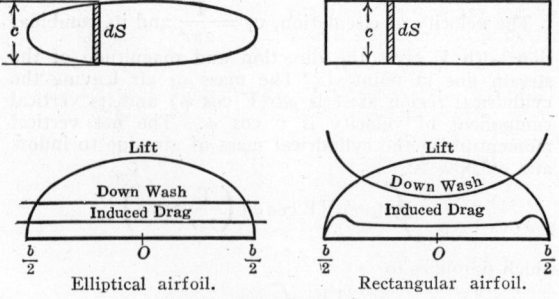

Elliptical airfoil.   Rectangular airfoil.

Fig. 5. Spanwise variations.

4. Minimum variation of the center of pressure, so that it will not be difficult to construct a stable airplane.

5. A shape well suited to the construction of a strong, but light-weight wing, at minimum cost. (F.T.M.)

AIRFOIL CLASSIFICATION. The shape of an airfoil depends on several controlling dimensions, and the number of variations possible is extremely great. The profiles which have been proposed, experimentally tested, or built into airplane wings, are sufficiently numerous to have caused some attempts at systematic classification. Elementary classifications such as (1) thin, (2) thick, (3) low drag, (4) high lift, (5) single camber, (6) double camber, (7) reflexed, etc., lack sufficient detail to be of much use.

During the evolution of aviation various unconnected classifications grew out of the identification of series of airfoil profiles by those who developed them or by the laboratories that tested and reported their characteristics. Thus there were the Munk series M_1, M_2, M_3, the Göettingen Laboratory series, an RAF series, etc.

From studies made by the **NACA** it was found that when a number of the better and more used airfoil profiles were "straightened," that is, their mean camber lines were made straight lines, and were proportionally reduced so that all had the same thickness, the profiles were almost identical. From this discovery was evolved the idea that an airfoil could be described by a standard thickness function and a camber line function. Two numerical classification systems have been devised, viz., a four-digit and five-digit system.

The **NACA** four-digit system of airfoil classification describes (1) the maximum camber of the mean line, (2) its position on the chord, (3) and (4) the maximum thickness. Thus the 4312 airfoil has a maximum mean camber of 4% of the chord at a position .3 of the chord from the leading edge and a maximum thickness, t, of 12% of the chord. The mean line is composed of two parabolas joined at the maximum camber point, as shown in the figure, with variations in p and m to produce different members of the four-digit family of air-

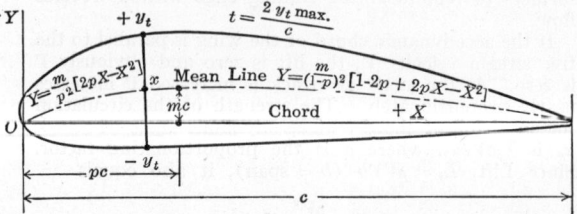

foils. The thickness function adopted for this classification is:

$$\pm y_t = [t/0.2][.2969x^{1/2} - .1260x - .3516x^2 + .2843x^3 - .1015x^4]$$

The radius of the leading edge is $1.1t^2$.

The five-digit system similarly employs a combination of numbers to describe camber line and thickness. The camber line in this series is composed of a cubic which becomes tangent to a straight line comprising the rear portion of the mean camber line, or, alternately, to a reflexed cubic. (F.T.M.)

AIRFOIL EFFICIENCY FACTOR. This is a correction factor used to reconcile the theoretical aerodynamics of induction (see **Airfoil**) with experimental data. Induction theory yields 2π for the value of dC_L/da, the rate of change of lift coefficient with angle of attack. Actually, $dC_L/da = 2\pi e$, where e is "airfoil efficiency factor," usually 90% or more. (F.T.M.)

AIRPLANE, FIXED WING. An airplane is an aircraft of the heavier-than-air type, deriving its sustention from aerodynamic reaction on the sustaining surfaces, and propelled by a force, usually the thrust of an air propeller, which is driven from some adequate source of power. Alternately, reaction propulsion of jets may be employed. The practical airplane of the present time consists of the following components:

1. Wing (or wing cellule, in the case of multi-planes), to provide the required lift.

2. Power unit, consisting of one or more prime movers, usually internal combustion types, attached to an air propeller by means of which a thrust sufficient to overcome the drag of the airplane is produced.

3. Fuselage, or body, which is required to house the crew, passengers, and cargo, and serves as the foundation structure of the airplane. The other components are usually, though not always, fastened directly to the fuselage.

4. Undercarriage for supporting the airplane when on the ground or on the water.

5. Controls for giving maneuverability, and allowing the pilot full control of the motion and performance of the airplane.

6. Stabilizing surfaces for imparting automatic aerodynamic stability to the airplane.

The foregoing components are assembled in a great many different ways in the aircraft of the present time, but a common arrangement has a single **wing**, or monoplane, attached directly to the **fuselage** and braced with **struts**, or more commonly attached without external supports, in which case the wing is designated a cantilever monoplane as will be indicated more fully later on. The power plant (see **Aeronautical Engines**) is mounted in the nose of the fuselage with an air **propeller** bolted directly on the engine **crankshaft**. The fuselage has its maximum section at the point which houses passengers and pilot, and tapers from there to a tail located some distance behind the wing. The tail consists of a horizontal surface and a vertical surface, each divided into fixed and movable portions. The movable tail surface, together with the **ailerons** in the **wings**, are actuated by controls operated from the cabin. When on the ground, this plane rests on two main wheels carried by a short undercarriage of struts and bracing, and upon a tail wheel at the rear. The location of the center of gravity is such that the main wheels carry most of the weight of the plane. In modern aircraft a popular variant of the landing gear is one in which a swiveling wheel is placed at the nose of the fuselage, and the two main wheels are placed just aft of the center of gravity. Such a landing gear, termed a tricycle landing gear, facilitates cross-wind landings.

If we were to imagine an aircraft of this type taking to the air, the operation would be somewhat as follows: The plane is at the end of a level straight runway about 2000' long, with the engine idling, but warmed up and ready to deliver full power. The plane is heading into the ground wind, if there is one. To **take off**, the pilot opens the throttle wide and operates the movable portion of the horizontal tail surfaces to lift the tail from the ground under the influence of the air blast from the propeller, and hold the longitudinal axis of the airplane horizontal. Meanwhile, under the influence of the propeller thrust, the plane is traveling on its landing gear, with increasing speed, down the runway, being guided in a straight line with the use of the rudder, which is the movable portion of the vertical tail surfaces. As the speed increases, aerodynamic reaction of lift on the wing surface builds up steadily, and after a run of from 600–1500', depending on the circumstances, the lift is nearly equal to the weight of the airplane, and could readily be made to exceed it by increasing the angle of attack on the wings. At this point the pilot may elect to continue the same take-off attitude until the lift equals the weight of the plane, and the plane leaves the ground, or he may elect to operate the movable horizontal tail surfaces, called the elevator, and thus increase the **angle of attack**, and the lift, so that the airplane will leave the ground at some definite point.

Once in the air, the plane has three degrees of freedom, and the maneuvering of the plane involves simultaneous coordination of controls for three basic airplane motions, the pitch, the roll, and the yaw. Referring to Fig. 1, the three mutually perpendicular axes about

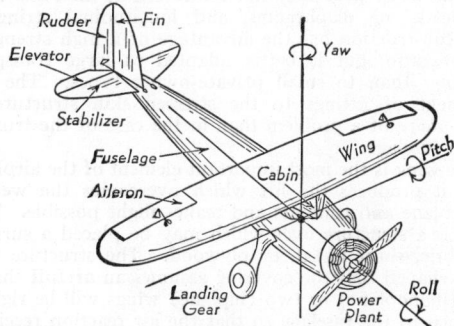

Fig. 1. Axes of an airplane.

which motion of an airplane takes place are the longitudinal axis of the airplane, the horizontal axis, which is parallel to the wing, and the vertical axis. These intersect at a common point, which is the center of gravity of the airplane. The rotation to the right or left around the vertical axis is called a yaw, and is obtained by the use of the rudder. Motion about the lateral axis is called pitch, and is controlled by the elevator. Rotation around the longitudinal axis is called roll, and is controlled by the ailerons. Parts of the vertical and horizontal tail surfaces are fixed for the purpose of giving longitudinal and directional stability to the airplane, whereas lateral stability is obtained by setting the wings at a slight dihedral angle.

Stability in an airplane is that ability which the airplane possesses of itself, and without aid from the pilot, of returning to a normal horizontal flight position when once displaced from that position by gusts. An airfoil, alone, is unstable. For instance, a slight unintentional increase of the angle of attack creates aerodynamic forces which tend to increase that angle rather than to decrease it. Nevertheless, stable airplanes can be built incorporating unstable airfoils through the medium of a fixed surface—a tail—placed at some distance behind the wing, or through a large amount of sweepback of the wings, a method employed in the tailless airplane. If, in the conventional arrangement of the components of an airplane, the center of gravity of the entire airplane is placed about one-third of the wing chord aft of its leading edge, and the tail surfaces between two and three chord lengths to the rear of the wings, the airplane will likely be stable, provided, of course, that the proper relation obtains between wing area and tail area.

The operation of the control surfaces, which involves their rotation around a hinge point through an angle of 10 to 20°, is accomplished from the pilot's seat by connecting horns on the control surfaces to controls at the seat by cables, torque tubes, or push-pull tubes. The rudder is operated by foot pedals, the ailerons and elevator by a control column having two degrees of freedom, or a column with one degree of freedom upon which is mounted a rotating wheel. In the wheel type of control, turning the wheel operates the ailerons, while moving the control column operates the elevator. In large airplanes, the force required of the pilot becomes very great unless the control surfaces are balanced by placing their hinges aft of the leading edge or by employing power boost to supplement manual effort.

Most airplanes can be classed structurally as trussed types or stressed skin types. The trussed structure used in the United States utilizes a triangularly framed fuselage of alloy steel or duralumin tubing, either welded or riveted at the joints. This has been a very popular modern type everywhere.

Stressed skin construction has no triangular framework, but rather relies upon the tube-like strength of the outer skin, which is oval or circular in cross-section, and which is made of thin sheet metal, or plywood. Thin sheets of the size required to cover a fuselage would be very weak were they not reinforced at intervals with bulkheads, or diaphragms, and longitudinal stringers. This construction has the advantage of a high strength-weight ratio, but is better adapted to large transport airplanes than to small private-owner types. The attachment of fittings to the stressed skin structure is much more of a problem than in the case of the trussed frame fuselage.

The wing is the most important element of the airplane since it produces the lift which overcomes the weight of airplane and contents and makes flight possible. The wing is a structure over which may be placed a surface of fabric, sheet metal, or plywood. The structure will be so shaped that the covering assumes an **airfoil** shape. One (monoplane) or two (biplane) wings will be rigidly fastened to the fuselage so that the air reaction received by the wing surface, and passed on to the wing structure, can act on the fuselage. The wing structure is sometimes purely cantilever, sometimes externally braced.

All parts of the airplane exposed to the air stream are subject to aerodynamic drag, the sum total of which must be overcome by the propeller thrust in order to maintain speed. The drag of all parts except the wings is parasitic drag, that is, it is the drag of objects which do not create any useful lift. The less this parasitic drag, the less power will be required to propel the airplane at a given speed. By streamlining the struts, concealing fittings, cowling the engine, filleting wing roots and other angular intersections, retracting landing gear, etc., airplane designers have endeavored to reduce parasitic resistance and increase performance. The modern airplane of the low-wing cantilever type, with retractable landing gear, is an example of low parasitic resistance. Speeds of over 400 m.p.h. are thus obtained.

Location of the power plant in an airplane is subject to great variation. The single-engined airplane of the past was almost invariably a tractor type, with the engine mounted in the nose of the fuselage. The present and future may see an increasing use of pusher airplanes, in which the engine is to the rear of the cabin, as this yields a superior arrangement from the standpoint of visibility and access to the cabin. It is also better from the standpoint of noise and odors. For reasons of clearance, the engine of the flying boat, or amphibian airplane, is mounted atop the wing, where the propeller will be in less danger from damage by spray or wave. The multi-motored installations, with two engines, usually have the engines mounted in the leading edge of the wing, one on each side of the fuselage. The third engine in trimotored design is placed in the nose of the fuselage. In four-motor designs, the engines have been spaced in the leading edge of the wing. The National Advisory Committee for Aeronautics found by research that the placing of the engine just ahead of the leading edge of a wing, coupled with careful streamlining of the engine frame into the wing, resulted in the best aero-

dynamic results—that is, the propulsive efficiency was high, and the aerodynamic characteristics of the wing were but little affected.

The structure for support of the airplane while at rest on the surface, or while taking off or landing, is naturally different in the case of land and water designs. Undercarriage for land planes consists of three or more wheels supporting the weight of the plane, and holding it sufficiently high off the ground, so that there is no danger of contact of the rotating propeller with the ground. As it is possible for the airplane to contact the ground upon landing in a very severe manner, the landing gear must be very rigid and strong, and should be provided with some means, such as pneumatic or hydraulic shock absorbers, to absorb some of the energy of the impact. Whereas two wheels and tail skid, or wheel, have been common practice in the past, a great deal of attention is being given now to three-wheel landing gear, in which the third wheel is ahead of the center of gravity, and the other two behind. The advantage of this arrangement is that brakes may be applied heavily to the two main wheels without any danger of turning the airplane over on its back, as was the case with the older design. It has already been mentioned that advantages of performance are to be secured by retracting landing gear into the fuselage after the airplane has left the ground. This is done on nearly all transport type airplanes, and on a great many sportsman planes as well where high performance at moderate power is desired. Many interesting mechanically, hydraulically, or electrically operated retracting landing gears have been devised and successfully applied. Every effort is made to have these retractable gear absolutely reliable, although it has been proven, in several instances, that an airplane may land on the belly of the fuselage with wheels retracted without serious injury to its occupants, and sometimes with only minor damage to the airplane.

Airplanes designed to land on the water are classed as seaplanes, or flying boats. The seaplane resembles a land plane with the undercarriage removed, and replaced by another of similar design, but with floats in place of wheels. The flying boat is structurally quite different. The fuselage, for instance, becomes a hull, and is shaped considerably like the hull of a speed boat. The tail is not carried by the hull, but by special tail-carrying structural members extending backward from the wing. The wing is usually above the hull, for water clearance, and the engine either in the leading edge of the wing, or mounted on a stand above it. Amphibians are essentially flying boats, with retractable landing gear to enable their landing on smooth fields when necessary. (F.T.M.)

AIRPLANE, JET PROPELLED.

An airplane "flies" by virtue of a lift produced by wings. These same wings, together with the remainder of the airplane, incur a **drag**. The latter includes the price paid for lift. It is the function of the propulsion system to produce a forward thrust at least equal to the drag and, sometimes, for accelerating or climbing maneuvers, to exceed it. The thrust can be derived (1) from an airscrew, (2) from the reaction to a fluid jet, (3) from a component of the aircraft's own weight, as in soaring (**Sailplane**), (4) from a tow as in the case of a "**glider.**" Number one is the conventional and usual method, but number two is the present subject. A **jet engine** will produce a reaction force on the engine body, which, being rigidly fastened to wing or fuselage, transmits that force as a propelling thrust. The principles of thrust production are covered under **Jet Propulsion,** so the present section will be confined to the airplane itself.

Most jet planes are powered by combustion **gas turbine** or turbo-jets and are not like rockets. Rocket-powered designs have been built for special military purposes, but cannot be considered to have shown much promise outside a few military applications. Also rockets have helped heavily loaded planes into the air and shortened the take-off run of others. (See **Rockets** and **Rocket Assist**.)

But the turbo-jet powers most of the jet airplanes. The airplane itself does not need to be radically altered to use jet propulsion. Stability and control remain about the same, and superficially the jet plane may appear about like any others. The differences are:

1. The jet plane fuel may be kerosene or some other fuel not possessing the fire hazard of gasoline.

2. Since propeller clearance is not a specification, ground clearance—which sometimes dictates the length of undercarriage—is reduced. This is accompanied by a whole group of advantages, structural and aerodynamic. Also, in bi-motored designs, engines can be set close to the fuselage, thus reducing the effect of unbalanced thrust when one engine is out of commission.

3. The front projected area of a large jet engine is smaller than that of the radial engine of equivalent horse-power.

4. Reciprocating engine type of vibration is eliminated.

5. High-powered jet engines are lighter than reciprocating types.

6. The jet plane performance holds up better at extreme altitudes.

7. At high speeds, especially, the jet plane is the more efficient.

No new flight problems are presented to a pilot by the jet engine; in fact controls are simplified. Since the jet plane has a propulsion system that can function near or above the speed of sound, airplane design is not limited to the subsonic range as it probably is with propeller type engines. The jet airplane of the future may have sharp or pointed forward shapes. Wings may no longer have the usual blunt leading edges, fuselages may be tapered to a sharp point, cabins may be wholly submerged in the smooth streamlined fuselage.

At low speeds and altitudes the jet plane compares poorly with the conventional airplane. In take-off, rate of climb, and efficiency in the use of fuel, the reciprocating engine and screw propeller prove to be superior. But let the flight be in the stratosphere at the high speeds of 500 or 600 m.p.h., and the jet is supreme. (F.T.M.)

AIRPLANE, ROTATING WING. Helicopter; Autogyro.

AIRPLANE EFFICIENCY FACTOR. This term, originally introduced by W. B. Oswald, is intended to reconcile the actual drag variation of a complete airplane with the ideal equation for drag coefficient, viz.,

$$C_D = C_{D_p} + C_{D_i}$$

in which C_D = Airplane drag coefficient.

C_{D_p} = Wing profile drag and airplane parasite drag coefficient.

C_{D_i} = Induced drag coefficient.

It has been found that the actual C_D vs. C_L (coefficient of wing lift) curve of an airplane is closely fitted by the ideal equation if an arbitrary "airplane efficiency factor" e is introduced, thus:

$$C_D = C_{D_p} + C_{D_i}/e \qquad \text{(F.T.M.)}$$

AIRPLANE SPEEDS. An airplane has three degrees of freedom, and in flight is sustained by a fluid medium which itself may be in motion relative to the ground. This combination provides possibilities for a variety of both critical and non-critical velocities. The speeds of the more important of these are described below.

Design Gliding Speed. Arbitrarily set value of maximum indicated airspeed used in determining pertinent structural loading conditions in commercial airplane design. Pilots should not exceed this speed in glides or dives. In the United States this speed is restricted to 150% of the maximum level speed, or less (depending on loading conditions).

Ground Speed. The horizontal component of the velocity of an aircraft, relative to the ground. Determines distance travelled.

Indicated Airspeed. True airspeed multiplied by

$$\sqrt{\frac{\text{atmospheric density}}{\text{std. sea level density}}}$$

Landing Speed. The minimum speed of an airplane at the instant of contact with the landing area in a normal landing. This would be stalling speed for some aircraft, while for others (3-wheeled undercarriage) it might be somewhat higher.

Maximum Level Speed. The indicated airspeed obtained in level flight using rated engine horsepower. Taken at the altitude where it reaches maximum value.

Maximum Vertical Speed. Terminal velocity in vertical flight path. A fictitious airspeed resulting from diving an unlimited distance in air of uniform density (assumed sea level conditions) with normal gross weight and zero propeller thrust.

Stalling Speed. The speed of an airplane in steady flight at its maximum coefficient of lift.

True Airspeed. The speed of an aircraft relative to the air. Determines aerodynamic reactions. (F.T.M.)

AIRSCREW. A term having the same meaning as air propeller. The rotating airfoil used to produce a forward propulsive thrust for aircraft is called *propeller* in the United States, but elsewhere airscrew will often be the accepted term. (F.T.M.)

AIRSHIP. The airship is that form of aircraft which derives its lifting power from aerostatic forces rather than the **aerodynamic** forces such as support the **airplane**. In other words, the lift of an airship is one of **buoyancy**, and this is derived from the difference between the density of the atmosphere and that of the lifting gas contained by the airship. The gases most commonly used are **hydrogen** and **helium**—the two lightest known gases. Hydrogen is cheaper than helium, but helium has the very distinct advantage of being noninflammable. The explosion hazard is reduced to a minimum in a helium-filled airship. The size of an airship is determined by the volume of lifting gas required to lift the weight of the ship and the load. The lifting power of helium is approximately .07 lb. per cu. ft. at 32° F., standard atmospheric pressure, and that of hydrogen is 6% more. The airship is equipped with a propulsive device and controls, so that its attitude, speed, altitude, and course are under the control of the pilot, whereas in the free balloon, altitude only is controllable, and velocity depends entirely upon the wind.

The requirements of navigability and **aerodynamic efficiency** have caused an effective aerodynamic shape to be given the airship. This shape is usually circular in cross-section, with a rounded nose and a tapering tail. The ratio of length to diameter varies between 5 and 8. Propulsion is obtained from air **propellers** driven by internal combustion engines, which are either mounted in **nacelles** attached to the hull, or are within the hull, and drive the propellors by means of shaft and gearing extending through the skin of the hull. The latter location is considered safe only in helium-filled airships.

There are three principal classifications of airships, i.e., the non-rigid, semi-rigid, and rigid. The shape of the non-rigid airship is maintained by the pressure of the lifting gas. However, the gas expands and contracts with temperature, and in the contracted state the airship would be limp (the popular name blimp is derived from a war time designation of this as the B-limp type), were it not for an air-filled ballonet which is built inside the main covering. As the lifting gas expands it forces air out of the ballonet, and when the lifting

Fig. 1. Dirigible airship.

Fig. 2. Semi-rigid airship.

gas contracts, air scoops fill the ballonet by virtue of the velocity of the airship. The variable volume ballonet can also be used by the pilot to regulate the altitude of the airship by using it to compress or expand the lifting gas, so changing its density and lifting power. The ballonet is emptied for ascent, and refilled for descent. The cabin and engine installations are carried in the car or in nacelles, which are suspended below the envelope.

The rigid, or dirigible airship, has a complete metal framework, and its shape is, accordingly, independent of the degree of inflation of the lifting gas cells which are contained within. Fixed equipment, like power plants and living quarters, is rigidly attached to the structural frame. The semi-rigid airship resembles the non-rigid in that its shape is maintained by gas pressure, but there is a structural keel extending longitudinally from the nose to the tail, with additional structural reinforcement at the nose and at the attachment of the control surfaces. In size, the blimps are the smallest, and the dirigibles the largest, airships.

The framework of the modern dirigible is made of girders running longitudinally connected by parallel circumferential rings. The circumferential rings must be absolutely rigid, and if the construction is not such that their shape is self-sustaining, they must be braced diametrically. One to three of the longitudinal members at the bottom of the hull are made especially heavy and rigid to form a keel. The keel serves to strengthen and integrate the ship fore and aft, provide main walkways for access to the interior of the ship, and support heavy equipment, such as cabins, engines, and control surfaces. (F.T.M.)

AIR-SPEED METER. An instrument used for measuring the speed of an airplane relative to the air is known as an air-speed meter.

One early type of air-speed meter employed a small propeller which was located out on the wing of the plane, well outside of the slip stream of the plane's propeller. The rpm of the small propeller is proportional to the speed of the plane through the air.

The propeller type of air-speed meter has been superseded by an instrument which employs the principle of the **Pitot tube.** The instrument is placed well out on the wing, beyond the influences of the slip stream of the propeller, and the leads from a **piezometer** tube and the Pitot tube are brought into different chambers separated by a pressure-sensitive element similar to that used in an aneroid **barometer.** The changes in thickness of this element are transmitted and amplified by a system of levers and gears to a dial, which will read directly in **knots** on naval planes, or statute m.p.h. on army or commercial planes. When the instrument is properly adjusted for standard atmospheric conditions of pressure and temperature, the reading of this dial is what is known as the indicated air speed.

To obtain the actual speed of the plane relative to the air, the indicated air speed must be converted to calibrated air speed by means of a calibration curve determined for the particular instrument being used. The calibrated air speed is then converted to true air speed by means of a formula involving the calibrated air speed, the temperature of the air, and the altitude at which the plane is flying. This final conversion from calibrated air speed to true air speed may be made rapidly on any one of a number of standard **navigational computers.** (W.K.G.)

AIRWORTHINESS. The quality of an aircraft denoting its fitness and safety for operation in the air under normal flying conditions. (F.T.M.)

AIRY'S EXPERIMENT. Aberration of Light.

AITKEN'S METHOD OF INTERPOLATION.
Given the arguments a, b, c, d and the corresponding values of the function u_a, u_b, u_c, and u_d, we desire the value of u_x, $a < x < d$. We use the simple method provided by A. C. Aitken. This method works for unequal intervals for the argument as well as equal intervals, for inverse interpolation as well as for direct interpolation, and it provides a way of determining the number of significant figures in u_x.

A table is constructed as follows to find u_x:

				Parts
u_a				$a - x$
u_b	$u_x(a, b)$			$b - x$
u_c	$u_x(a, c)$	$u_x(a, b, c)$		$c - x$
u_d	$u_x(a, d)$	$u_x(a, b, d)$	$u_x(a, b, c, d)$	$d - x$

We calculate the entries in the table

$$u_x(a, b) = \frac{\begin{vmatrix} u_a & a - x \\ u_b & b - x \end{vmatrix}}{b - a},$$

$$u_x(a, c) = \frac{\begin{vmatrix} u_a & a - x \\ u_b & c - x \end{vmatrix}}{c - a},$$

$$u_x(a, d) = \frac{\begin{vmatrix} u_a & a - x \\ u_d & d - x \end{vmatrix}}{d - a},$$

$$u_x(a, b, c) = \frac{\begin{vmatrix} u_x(a, b) & b - x \\ u_x(a, c) & c - x \end{vmatrix}}{c - b},$$

$$u_x(a, b, d) = \frac{\begin{vmatrix} u_x(a, b) & b - x \\ u_x(a, d) & d - x \end{vmatrix}}{d - b},$$

$$u_x(a, b, c, d) = \frac{\begin{vmatrix} u_x(a, b, c) & c - x \\ u_x(a, b, d) & d - x \end{vmatrix}}{d - c},$$

$$= u_x.$$

We may stop the calculations whenever the particular column $u_x(a, b, \cdots)$ shows an agreement to the number of significant figures desired. For example if $u_x(a, b, c)$ and $u_x(a, b, d)$ agree to 4 significant figures. then u_x would be that result to 4 significant figures. Aitken's method corresponds to polynomial interpolation, and in the illustration cited with 4 points a polynomial of the third degree is being used. The points (a, u_a), (b, u_b), (c, u_c) and (d, u_d) may be arranged in any order without affecting the value of u_x. (L.A.A.)

ALABAMINE. Chemical Composition, I. Elements.

ALABANDITE. Manganese sulfide, MnS. Associated with pyrite, sphalerite and galena as vein deposits. (R.M.F.)

ALABASTER. A fine-grained variety of the mineral gypsum, formerly much used for vases and statuary. It is usually white in color or may be of other light, pleasing tints.
The word alabaster is derived from the Greek name for this substance. (E.S.C.S.)

ALAGDAGA. Jerboa.

ALALITE. Diopside.

ALANINE. Aminoacids.

ALARY MUSCLES. Muscles which attach the heart of insects to the body wall and diaphragm. (A.W.L.)

ALBATROSS. Aves, Procellariiformes. A large marine bird (**Aves**) with unusual powers of flight, as is known from its habit of following ships for many hours without alighting. There are several species, belonging to *Diomedea* and allied genera. (A.W.L.)

ALBEDO. The term albedo is used astronomically to indicate the reflecting power of an object. Technically defined, albedo is the ratio of the radiation reflected from an object to the total amount incident upon it. For example, the albedo of the moon is 0.073 which means that the moon reflects that fraction of the sunlight which is incident upon it.
The value of the albedo of a planet is a measure of the quantity of **atmosphere** which surrounds the object. The higher the albedo the thicker the atmospheric layer. In the case of objects without atmosphere, as in the case of the moon, the albedo, combined with the color of the reflected light, may be used to make estimates of the character of the material making up the surface of the object. (W.K.G.)

ALBERT EFFECT. A reaction of light-sensitive photographic materials discovered by E. Albert in 1899. Albert found that if a wet collodion plate is considerably overexposed and the latent image destroyed with nitric acid, a positive image is produced upon exposure to white light and subsequent development. Albert's observations with wet collodion have been shown to hold also for gelatin emulsions by Luppo-Cramer, J. Precht and others. The effect is produced by chromic acid, ammonium persulfate and other substances which destroy the latent image as well as by nitric acid. (C.B.N.)

ALBERTITE. An oxygenated **hydrocarbon** which differs from **asphaltum** slightly in that it is not completely soluble in **turpentine**, nor can it be perfectly fused. Its hardness varies from 1 to 2; specific gravity 1.097; luster pitchy; color black.
It occurs as fissure filling in the **Carboniferous** rocks of Nova Scotia. (E.S.C.S.)

ALBINISM. Absence of pigmentation. The condition has been noted in occasional individuals of many species which are normally pigmented, including man. In some cases the term is applied to a partial lack of pigment, as in the white form of certain normally yellow butterflies; in this form the black markings characteristic of the species are fully developed. In contrast, the total lack of pigment in albino birds and mammals is shown by the pink eyes. In these organs pigment is functionally important but the albino fails to develop it, hence the color of the blood is seen through the tissues.
Albinism in man is known to be inherited as a recessive (see **Heredity**) to normal pigmentation. The term is also applied to normally green plants which fail to develop chlorophyll. As in man, albinism is inherited as a simple recessive. Albino plants die as soon as food is exhausted from the seed. Partial albinism in plants is known as variegation. (A.W.L., P.A.W.)

ALBINO. An individual without the normal pigmentation of its kind. (See **Albinism**.) (A.W.L.)

ALBITE. Feldspar.

ALBUMIN. An albumin is a member of a class of proteins (see **Aminoacids, Polypeptides, and Proteins**) which is widely distributed in animal and vegetable tissues. Albumins are soluble in water and in dilute salt solutions, and are coagulable by heat.
The repeated appearance of albumins in the urine (albuminuria) may indicate a diseased condition of the kidneys. (See **Nephritis**.) (D.M.H.)

ALBUMINOIDS. Aminoacids, Polypeptides, Proteins.

ALCOHOL. This is the name of a type of chemical compounds which are discussed under the heading, **Alcohols.** Ethyl alcohol, the member of the series in most common use and of most common occurrence, is often referred to by the group named, that is, simply as alcohol. It is discussed in this book under the heading **Ethyl Alcohol.** (R.K.S.)

ALCOHOLIC INSANITY. Ethyl Alcohol.

ALCOHOLISM. Ethyl Alcohol.

ALCOHOLS AND ETHERS. Alcohols (containing hydroxyl groups, —OH, attached to non-benzenoid carbon) are characterized by a wide variety of chemi-

cal reactions and uses, well illustrated by reference to particular alcohols, namely, methyl alcohol, ethyl alcohol, glycol, glycerol. There are three types of alcohols, illustrated as follows: (1) primary, CH_3—CH_2OH, ethyl alcohol, methyl carbinol, (2) secondary, $(CH_3)_2$=$CHOH$, isopropyl alcohol, dimethyl carbinol, (3) tertiary, $(CH_3)_3$≡COH, trimethyl carbinol, and characterized by the behavior upon oxidation, thus, (1) primary alcohols yield first **aldehyde**, and upon further oxidation yield **carboxylic acid**, each with the original carbon atom content, e.g., CH_3CHO, **acetaldehyde**, and CH_3COOH, **acetic acid**, (2) secondary alcohols yield initially **ketone** with the original carbon atom content, e.g., $(CH_3)_2CO$, **acetone**, dimethyl ketone, and

SELECTED REPRESENTATIVE ALCOHOLS

ALCOHOL	FORMULA	MELTING POINT (°C.)	BOILING POINT (°C.)
*1. Methyl alcohol (methanol)	$H \cdot CH_2OH$	−98	64.5
*2. Ethyl alcohol (ethanol)	$CH_3 \cdot CH_2OH$	−117	78.5
3. Normal-propyl alcohol (propanol-1)	$CH_3CH_2 \cdot CH_2OH$	−127	98
4. Iso-propyl alcohol (dimethyl carbinol, propanol-2)	$(CH_3)_2CHOH$	−86	82
5. Normal-butyl alcohol (butanol-1)	$CH_3(CH_2)_2CH_2OH$	−90	118
6. Methylethyl carbinol (butanol-2)	$\frac{C_2H_5}{CH_3}CHOH$		99.5
7. Trimethyl carbinol (2-methylpropanol-2)	$(CH_3)_3COH$	25.5	83
PENTANOLS			
8. Normal-primary-amyl alcohol (pentanol)	$CH_3(CH_2)_3CH_2OH$	−78	138
9. Iso-primary-amyl alcohol (iso-butyl carbinol)	$(CH_3)_2CHCH_2CH_2OH$	−117	130
10. Active-primary-amyl alcohol (secondary butyl carbinol)	$\frac{C_2H_5}{CH_3}CHCH_2OH$		128
11. Tertiary butyl carbinol	$(CH_3)_3CCH_2OH$	53	114
12. Methyl-normal-propyl carbinol	$\frac{CH_3}{CH_3(CH_2)_2}CHOH$		119
13. Methyl-iso-propyl-carbinol	$\frac{CH_3}{(CH_3)_2CH}CHOH$		114
14. Diethyl carbinol	$(C_2H_5)_2CHOH$		116
15. Dimethylethyl carbinol	$\frac{(CH_3)_2}{C_2H_5}COH$	−12	102
HEXANOL			
16. Normal-hexyl alcohol	$C_5H_{11}CH_2OH$	−52	156
HEPTANOL			
17. Normal-heptyl alcohol	$C_6H_{13}CH_2OH$	−35	176
OCTANOL			
18. Normal-octyl alcohol (caprylyl alcohol)	$C_7H_{15}CH_2OH$	−16	194
NONANOL			
19. Normal-nonyl alcohol	$C_8H_{17}CH_2OH$	−5	215
DECANOL			
20. Normal-decyl alcohol	$C_9H_{19}CH_2OH$	47	231
21. Lauryl alcohol (normal-dodecyl alcohol)	$C_{11}H_{23}CH_2OH$	24	259
22. Myristyl alcohol (tetradecyl alcohol)	$C_{13}H_{27}CH_2OH$	38	170 (15 mm.)
23. Cetyl alcohol (hexadecyl alcohol)	$C_{15}H_{31}CH_2OH$	49	344
24. Octadecyl alcohol	$C_{17}H_{35} \cdot CH_2OH$	59	
25. Eicosyl alcohol	$C_{19}H_{39} \cdot CH_2OH$	68	
26. Ceryl alcohol	$C_{25}H_{51}CH_2OH$	80	
27. Myricyl alcohol (melissyl alcohol)	$C_{29}H_{59}CH_2OH$	88	
28. Cyclohexanol	$CH_2(CH_2)_4CHOH$	24	162
29. Cycloheptanol (suberyl alcohol)	$CH_2(CH_2)_5CHOH$		185
30. Allyl alcohol	$CH_2:CHCH_2OH$	−129	97
31. Crotonyl alcohol	$CH_3CH:CHCH_2OH$		118
32. Phytol	$C_{20}H_{37}CH_2OH$		145 (0.03 mm.)
33. Propargyl alcohol	$CH:CCH_2OH$	−17	115
34. Furfuryl alcohol	$C_4H_3O \cdot CH_2OH(2)$		170
35. Phenyl carbinol (benzyl alcohol)	$C_6H_5CH_2OH$	−15	206
36. Diphenyl carbinol (benzhydrol)	$(C_6H_5)_2CHOH$	68	299
37. Triphenyl carbinol	$(C_6H_5)_3COH$	162	>360
38. Methylphenyl carbinol	$\frac{CH_3}{C_6H_5}CHOH$		205
39. Ethylphenyl carbinol	$\frac{C_2H_5}{C_6H_5}CHOH$		219

* Discussed separately under individual name.

SELECTED REPRESENTATIVE ALCOHOLS—*Continued*

ALCOHOL	FORMULA	MELTING POINT (°C.)	BOILING POINT (°C.)
40. Benzyl alcohol	$C_6H_5CH_2OH$	−15	206
41. Fluorene alcohol (diphenylene carbinol)	$(C_6H_4)_2CHOH$	156	
42. Cinnamyl alcohol	$C_6H_5CH:CHCH_2OH$	33	254
43. Salicyl alcohol (Saligenin)	$C_6H_4(OH)(1)(CH_2OH)(2)$	86	subl.
44. Terpineol	$C_{10}H_{17}OH$	35 appr.	220
45. Borneol (dextro-laevo) (camphol)	$C_{10}H_{17}OH$	210	subl.
46. Borneol (dextro) (Borneo camphor)	$C_{10}H_{17}OH$	209	213
47. Geraniol	$C_{10}H_{17}OH$	>−15	229
48. Menthol	$C_{10}H_{19}OH$	35 appr.	215
*49. Glycol (ethylene glycol, ethandiol)	$CH_2OH \cdot CH_2OH$	−17	197
50. Propylene glycol (propandiol-1,2)	$CH_3CHOHCH_2OH$		189
*51. Glycerol (propantriol)	$CH_2OHCHOHCH_2OH$	18	290
52. Erythritol	$CH_2OH(CHOH)_2CH_2OH$	126	331
53. Arabitol	$CH_2OH(CHOH)_3CH_2OH$	103	
54. Mannitol	$CH_2OH(CHOH)_4CH_2OH$	166	295 (4 mm.)
55. Dulcitol	$CH_2OH(CHOH)_4CH_2OH$	188	295 (4 mm.)
56. Sorbitol	$CH_2OH(CHOH)_4CH_2OH$	110 (anhydrous)	
STEROLS			
57. Cholesterol	$C_{27}H_{45}OH$	148	>300
58. Iso-cholesterol	$C_{27}H_{45}OH$	138	
59. Ergosterol	$C_{27}H_{41}OH$	160	

* Discussed separately under individual name.   (R.K.S.)

upon further oxidation yield carboxylic acids or aldehydes containing fewer carbon atoms than the original alcohol, e.g., CH_3COOH, acetic acid plus carbon dioxide, (3) tertiary alcohols yield no product with the original carbon atom content but yield carboxylic acids, aldehydes or ketones containing fewer carbon atoms than the original alcohol, e.g., CH_3COOH, acetic acid plus carbon dioxide. Upon reduction of these alcohols (as the haloid compound with magnesium in ether—Grignard's reagent—treated with water) the corresponding **hydrocarbons** are obtained, (1) $CH_3 \cdot$

CH_3, ethane, methyl methane, (2) $(CH_3)_2CH_2$, propane, dimethyl methane, (3) $(CH_3)_3CH$, trimethyl methane.

By loss of water, directly or by indirect reaction, alcohols form **ethers**, e.g., $(CH_3)_2O$ dimethyl ether, $((C_2H_5)_2O)$ diethyl ether, CH_2—CH_2 ethylene oxide.

$$CH_2OH \cdot CH—CH_2 \text{ glycide alcohol.}$$

By the removal of 1 mol of water from ethyl alcohol, **ethylene** is produced.

By reaction with acids, **esters** are formed in great variety.

SELECTED REPRESENTATIVE ETHERS

ETHER	FORMULA	MELTING POINT (°C.)	BOILING POINT (°C.)
1. Dimethyl ether (methoxymethane)	CH_3OCH_3	−138	−25
*2. Diethyl ether (ethoxyethane)	$C_2H_5OC_2H_5$	−116	35
3. Dipropyl ether (propoxypropane)	$C_3H_7OC_3H_7$	−122	91
4. Methylethyl ether (methoxyethane)	$CH_3OC_2H_5$		11
5. Methyl-normal-propyl ether (methoxypropane)	$CH_3OC_3H_7$		40
6. Methyl-iso-propyl ether	$CH_3OC_3H_7$		32 appr.
7. Ethyl-normal-propyl ether	$C_2H_5OC_3H_7$	−79	61
8. Ethyl-iso-propyl ether	$C_2H_5OC_3H_7$		54
9. Diallyl ether	$(CH_2:CHCH_2)_2O$		94
10. Methyl furfuryl ether	$CH_3OCH_2C_4H_3O$		135
11. Ethyl furfuryl ether	$C_2H_5OCH_2C_4H_3O$		150 appr.
12. Methylphenyl ether (anisole)	$CH_3OC_6H_5$	−38	156
13. Ethyl phenyl ether (phenetol)	$C_2H_5OC_6H_5$	−30	172
14. Diphenyl ether	$C_6H_5OC_6H_5$	27	259
15. Methylbenzyl ether	$CH_3OCH_2C_6H_5$		174
16. 1-Methoxy-4-propenyl benzene (anethole)	$C_6H_4(OCH_3)(1)(CH:CHCH_3)(4)$	22	235
17. Ethylene oxide (glycol oxide)	$(CH_2)_2O$	−111	11
18. Propylene oxide	CH_3CHCH_2 ⌞O⌟		35
19. Diphenylene oxide	$(C_6H_4)_2O$	86	288
20. Diphenylenemethane oxide	$C_6H_4\langle{}^{CH_2}_{O}\rangle C_6H_4$	105	315
21. Methyl ortho-hydroxyphenylene ether (guaiacol)	$CH_3OC_6H_4(OH)(2)$	28	205
22. Diethoxymethane	$CH_2(OC_2H_5)_2$		89
23. Diethylene oxide	$O(CH_2CH_2)_2O$	12	101
24. Diethyleneglycol	$(CH_2OH \cdot CH_2)_2O$	−6	245

* Discussed separately under Ether.   (R.K.S.)

ALCYONARIA. An order of the class Actinozoa in the phylum **Coelenterata.** The animals of this order are marine and because of the hard deposits formed in and around their bodies they are often known as **corals.** Among them are the **sea-pen,** the **sea-fan,** organ-pipe coral, and precious coral.

Alcyonaria are usually found in colonies. The individuals are **polyps** connected together by living structures and by the hard skeletal structures. They differ from the true corals in having only eight tentacles, pinnately branched. (A.W.L.)

ALDEBARAN. Aldebaran (α **Tauri**) is derived from an Arabic phrase indicating that the star is the "leader of the followers," i.e., the leader of the **asterism** known as the **Hyades,** which follow the **Pleiades** in their nightly journey across the sky. Astrologically, Aldebaran was a fortunate star, portending riches and honor. This star was one of the four royal stars of the Persians about 3000 B.C.

Aldebaran is one of the smaller stars whose diameter has been measured with the stellar **interferometer.**

The diameter is found to be about 33,000,000 miles, or 38 times the diameter of our sun. (W.K.G.)

ALDEHYDES, KETONES, AND RELATED COMPOUNDS (Acetals, Ketenes). Aldehydes (containing —CHO group) are characterized by a wide variety of chemical reactions, well illustrated by reference to two particular aldehydes, namely, **acetaldehyde** and **benzaldehyde.** Regulated oxidation of primary **alcohols** produces the aldehyde corresponding to the alcohol used, and vigorous oxidation of aldehydes produces the corresponding **carboxylic acid,** thus, ethyl alcohol ($CH_3 \cdot CH_2OH$), to acetaldehyde ($CH_3 \cdot CHO$), to **acetic acid** ($CH_3 \cdot COOH$). Regulated reduction of carboxylic acids produces the corresponding aldehyde, and vigorous reduction, the corresponding primary alcohol—the reverse of the first named reactions. Not only is the range of reaction of aldehydes wide, but also the range of applications, as illustrated under **formaldehyde, acetaldehyde, benzaldehyde, furfuraldehyde.** The reaction with Tollen's solution, as described under formaldehyde, acetaldehyde, benzaldehyde, is commonly used to classify a substance as an aldehyde.

SELECTED REPRESENTATIVE ALDEHYDES

ALDEHYDE	FORMULA	MELTING POINT (°C.)	BOILING POINT (°C.)
*1. Formaldehyde (methanal)	HCHO	−92	−21
*2. Acetaldehyde (ethanal)	CH_3CHO	−123	20
3. Propionic aldehyde (propanal)	C_2H_5CHO	−81	49
4. Normal-butyric aldehyde (butanal)	$CH_3(CH_2)_2CHO$	−99	76
5. Normal-amyl aldehyde (normal valeric aldehyde)	$CH_3(CH_2)_3CHO$		103
6. Iso-amyl aldehyde (iso-valeric aldehyde)	$(CH_3)_2CHCH_2CHO$	−51	92
7. Caproic aldehyde	$CH_3(CH_2)_4CHO$		130
8. Caprylic aldehyde	$CH_3(CH_2)_6CHO$		168 appr.
9. Capric aldehyde	$CH_3(CH_2)_8CHO$		208
10. Lauric aldehyde	$CH_3(CH_2)_{10}CHO$	45	185 (100 mm.)
11. Acrolein (acrylic aldehyde)	$CH_2:CHCHO$	−88	52
12. Crotonic aldehyde	$CH_3CH:CHCHO$	−69	102
13. Propargylic aldehyde	$CH:CCHO$		61
*14. Furfuraldehyde (furfural)	$C_4H_3OCHO(2)$	−39	162
15. Citral (gerianal)	$C_{10}H_{16}O$		110 (12 mm.)
16. Citronellal	$C_{10}H_{18}O$		206 appr.
*17. Benzaldehyde	C_6H_5CHO	−56	180
18. Cinnamic aldehyde	$C_6H_5CH:CHCHO$	−7	251
19. Glycol aldehyde (glycollic ald.)	CH_2OHCHO	97	
20. Glyceric aldehyde	$CH_2OHCHOHCHO$	138	
21. Glyoxal (oxalic aldehyde)	CHOCHO	15	50
22. Succinic aldehyde	$CHOCH_2CH_2CHO$		202
23. Vanillin (2-hydroxy-3-methoxybenzaldehyde)	$C_6H_3(OH)(2)(OCH_3)(3)(CHO)(5)$	81	
24. Benzil (yellow solid)	$C_6H_5COCOC_6H_5$	95	

* Discussed separately under individual name.

SELECTED REPRESENTATIVE KETONES

KETONE	FORMULA	MELTING POINT (°C.)	BOILING POINT (°C.)
*1. Acetone (dimethyl ketone) (2-propanone)	CH_3COCH_3	−94	56
2. Diethyl ketone (3-pentanone)	$C_2H_5COC_2H_5$	−42	102
3. Di-normal-propyl ketone (4-heptanone)	$C_3H_7COC_3H_7$	33	143
4. Di-normal-amyl ketone	$C_5H_{11}COC_5H_{11}$	15	226
5. Methylethyl ketone (2-butanone)	$CH_3COC_2H_5$	86	80
6. Methyl-normal-propyl ketone (2-pentanone)	$CH_3COCH_2CH_2CH_3$	−78	102
7. Methyl-iso-propyl ketone	$CH_3COCH(CH_3)_2$	−92	93
8. Ethyl-normal-propyl ketone (3-hexanone)	$C_2H_5COCH_2CH_2CH_3$		124

* Discussed separately under individual name.

SELECTED REPRESENTATIVE KETONES—*Continued*

KETONE	FORMULA	MELTING POINT (°C.)	BOILING POINT (°C.)
9. Ethyl-iso-propyl ketone................	$C_2H_5COCH(CH_3)_2$............	....	114
10. Lauryl ketone (laurone)...............	$(C_{11}H_{23})_2CO$...........	69	
11. Benzophenone (diphenyl ketone)........	$C_6H_5COC_6H_5$............	48	306
12. Dibenzyl ketone (diphenyl acetone).....	$(C_6H_5CH_2)_2CO$...........	34	330
13. Acetophenone (methylphenyl ketone).....	$CH_3COC_6H_5$.............	20	202
14. Methylbenzyl ketone.................	$CH_3COCH_2C_6H_5$.........	−15	217
15. Ethylphenyl ketone.................	$C_2H_5COC_6H_5$...........	21	218
16. Diphenylene ketone (fluorenone)........	$C_6H_4COC_6H_4$...........	84	342
17. Anthrone (9-oxyanthracene)...........	$C_6H_4{<}{CO \atop CH_2}{>}C_6H_4$.............	154	
18. Cyclopentanone....................	$CH_2(CH_2)_3CO$............		130
19. Cyclohexanone....................	$CH_2(CH_2)_4CO$............		155
20. Cycloheptanone (suberone)............	$CH_2(CH_2)_5CO$............		180
*21. Camphor.......................	$C_{10}H_{16}O$.............	179	209
22. Carone.........................	$C_{10}H_{16}O$.............		210
23. Carvone........................	$C_{10}H_{14}O$.............		228
24. Menthone.......................	$C_{10}H_{18}O$.............		207
25. Diacetyl........................	$CH_3COCOCH_3$............		88
(dimethyl diketone, 2,3-butandione)			
26. Acetylacetone (2,4-pentandione).........	$CH_3COCH_2COCH_3$............	−23	137
27. Benzoylacetone...................	$C_6H_5COCH_2COCH_3$.........	81	
28. Acetophenone acetone (phenacyl acetone)..	$C_6H_5COCH_2CH_2COCH_3$.......	....	162 (12 mm.)
29. Acetyl carbinol (alpha-hydroxy acetone)...	CH_3COCH_2OH.............	−17	146

* Discussed separately under individual name.

Ketones (containing =CO group) are in several reactions similar to aldehydes, but less marked than the latter in the variety, as illustrated by reference to the commonest ketone, namely, **acetone.** Regulated oxidation of secondary **alcohols** produces the ketone corresponding to the alcohol used, but vigorous oxidation ruptures the substance with the formation of two acids, one of which is found to be **acetic acid** (CH_3COOH) if a methyl ketone was used. Thus, dimethyl carbinol, isopropyl alcohol (($CH_3)_2CHOH$) to **acetone** (($CH_3)_2CO$), to acetic acid (CH_3COOH) plus carbon dioxide (CO_2). Regulated reduction of ketones produces the corresponding secondary alcohol, and vigorous reduction the corresponding **hydrocarbon,** thus being the reverse of the first named reactions. The range of applications of ketones is wide, as illustrated under acetone, **camphor.** The reaction with **hydroxylamine** to form oximes of characteristic melting point, and the absence of acidic characteristics in ketones are commonly used to classify a substance as a ketone.

SELECTED REPRESENTATIVE ACETALS

ACETAL	FORMULA	MELTING POINT (°C.)	BOILING POINT (°C.)
1. Methylene dimethyl ether................	$CH_2(OCH_3)_2$............	....	42
(methylal, dimethoxy methane)			
2. Ethylidene dimethyl ether................	$CH_3CH(OCH_3)_2$............	....	64
(1,1-dimethoxy ethane)			
3. Ethylidene diethyl ether................	$CH_3CH(OC_2H_5)_2$...........	....	104
(acetal, 1,1-diethoxyethane)			

SELECTED REPRESENTATIVE KETENES

KETENE	FORMULA	MELTING POINT (°C.)	BOILING POINT (°C.)
1. Ketene..........................	$CH_2{:}CO$.............	−151	−56
2. Methyl ketene (in ether solution)...........	$CH_3CH{:}CO$.............		
3. Dimethyl ketene....................	$(CH_3)_2C{:}CO$.............	−98	34
4. Diphenyl ketene....................	$(C_6H_5)_2C{:}CO$.............		265

Acetals are formed by reaction of aldehydes with **alcohols** (or with **carboxylic acids**). The aldehyde is so easily obtained from acetals that the latter may, in certain cases, be conveniently used as a source of aldehyde. With hydrochloric acid heated, the aldehyde and alcohol (or carboxylic acid) are readily formed. Acetals are relatively stable towards alkalis.

Ketenes are formed along with zinc bromide from alpha-bromo-substituted acetyl bromides by reaction with zinc, e.g., dimethylbromoacetyl bromide ($(CH_3)_2C$ $Br \cdot COBr$) yields dimethylketene ($(CH_3)_2C:CO$). Disubstituted ketenes or ketoketenes are reactive (1) with water, alcohols, ammonia, amines, phenylhydrazine, quinones, forming addition products, (2) with pyridine or quinoline forming ketene bases, (3) with olefin substances forming addition products. Ketenes do not form **phenylhydrazones** or **semicarbazones** as do ketones. The simplest ketene is $CH_2:CO$, which may be considered an anhydride of acetic acid, although it is best made by passing acetone vapor through a red hot tube. Carbon suboxide ($O:C:C:C:O$) is regarded as a diketene. (R.K.S.)

ALDER-FLY. Insecta, Neuroptera. A name given to the adults of a single subfamily of flies from their common occurrence on the alders bordering small streams. The larvae are aquatic. (A.W.L.)

ALDOL. Acetaldehyde.

ALEURITES (Euphorbiaceae). Tropical trees bearing small many-seeded fruits extremely rich in oil. *Aleurites triloba,* the candlenut of the orient, produces a fruit extensively used for food and for light. *Aleurites cordata,* a native of China, is the "varnish-tree." *Aleurites Fordii* yields tung, or nut oil (see **Fixed Oils**), a valuable drying agent used instead of linseed oil, especially in the preparation of waterproof varnishes. It is now widely cultivated in the southernmost parts of the United States. (R.M.W.)

ALEURONE GRAINS. Protein (see **Amino acids and Proteins**) reserves found in the seeds of several different kinds of plants. In many plants there is a special aleurone layer of definite thickness found in the **endosperm**. In corn the layer is a single cell in thickness and may contain a colored pigment; in **oats** it is two cells thick. In the **castor oil** plant the aleurone grains are not restricted to a single layer, but are distributed rather generally in the endosperm, and have a complex structure. (R.M.W.)

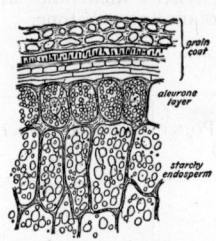

The aleurone layer in wheat. Part of a section of a wheat grain showing aleurone layer with cells filled with granules of protein.

ALEWIFE. Pisces, Teleostei. A common fish (**Pisces**) of the Atlantic coast, *Promolobus pseudoharengus,* related to the herring and shad. It enters the streams to spawn, and is found in lakes of New York. (A.W.L.)

ALEXANDRITE. A variety of **chrysoberyl**, originally found in the **schists** of the Ural Mountains. It absorbs yellow and blue light rays to such an extent that it appears emerald green by daylight but columbine-red by artificial light. It is used as a gem, and was named in honor of Czar Alexander II of Russia. (E.S.C.S.)

ALFALFA. *Medicago sativa.* Lucerne. A leguminous (see **Fruit**) plant probably native in southwestern Asia. It is an important perennial forage plant, which has been extensively cultivated since the Roman civiliza-tion. The plant has an extremely deeply-penetrating root system, reaching down 25' or more, and so is admirably adapted for growing in dry lands, where resistance to drought is important. Its fragrant purplish flowers are an important source of honey, especially in California. (R.M.W.)

ALGAE. (Sea weeds, Pond-scums, etc.) These are **Thallophytes** characterized by possessing chlorophyll, and so capable of elaborating their food by **photosynthesis**. Often the green pigment is completely concealed by other pigments, so that the plant is brown, red, or even black.

Algae are found in almost every habitat. In the oceans vast numbers of minute species float suspended in the upper levels of the water, while the shores are covered with many and varied forms from high tide level to depths of 30' or more. In fresh water they are equally abundant, but due to their smaller size are seldom so conspicuous as the marine forms; they occur in running water, in ponds, and in stagnant, often putrid water. Many species are found only in hot springs. They are found on the surface of the ground, on the bark of trees, on rocks, and even underground to a depth of several feet. A few species have found a favorable habitat within the bodies of higher plants and animals. In fact, wherever they find support and can obtain the necessary materials for growth, there algae may be found.

Algal plants offer a wonderful diversity of forms. In size they range from unicellular microscopic plants to structures having dimensions comparable to the larger land plants. The plankton forms, those free-floating plants often so abundant in both fresh and salt water, are nearly all unicellular; other free-floating forms, usually found near shore or in small fresh water ponds and streams, are multicellular organisms of various shapes, filamentous forms being especially common. Finally, attached marine forms often attain massive dimensions. The common kelp, or devil's apron, of the colder coastal waters of North America may grow to a length of 30' or more, and to a width of 2 or 3', while related species found in the Pacific Ocean far exceed them in size, reaching lengths of 100' and more.

Not only do algae vary greatly in size, but they also show almost every conceivable shape. Unicellular types are often adorned with a complex but beautifully symmetrical series of arms, or bristles, which may be of service in keeping them floating in the water. The filamentous forms may be simple or very much branched; often they are delicate plants of rare beauty. Other algae grow in flat sheets or membranes, either spreading over the substratum or rising gracefully in the water. The larger forms are of coarser habit, varying from irregular tumorous plants to long slender cords and broad flat fronds. Plants of the genus *Sargassum,* one of the brown algae found in warm regions, have an appearance very similar to that of flowering plants. Each plant has a slender branching stem, often 2' or more in length. From this stem flat lateral branches arise, which look very much like leaves, except for their brown color. Other short lateral branches end in small sub-spherical balls easily mistaken for fruits. Actually these structures are hollow bladders which help to keep the plant floating. Other branches, the real reproductive parts of the plant, are short cylindrical objects which might be mistaken for buds. At first the plant grows on rocks and other solid objects. It is easily broken loose, however, and floats about in the ocean currents. Thus, these plants are frequently washed up on northern beaches.

There are several systems of classification of algae, varying in details but all using the various pigments found in their cells as a basis for separation. Obviously such a classification is very artificial, but in the algae it seems to agree quite closely with natural systems based on such other criteria as the structure of the thallus or plant body, as the substances formed by the cells and stored in them, and especially as the reproductive processes which are found in the different groups.

Sargassum linifolium. (*From Altmanns Morphologie und Biologie der Algen, Gustav Fisher, Jena.*)

Separated according to pigments the algae fall into four large classes and several smaller ones. The four large classes are the Myxophyceae or blue-green algae, the Chlorophyceae or green algae, the Phaeophyceae or brown algae, and the Rhodophyceae or red algae. The minor classes are the Xanthophyceae (also called the Heterokontae, because of the two unequal cilia (see **Cilium**) which characterize them), the Chrysophyceae, the Bacillariphyceae or **diatoms**, the Cryptophyceae, Dinophyceae, Chloromonadineae and Eugleninae.

The blue-green algae are characterized by having within the cell, in addition to chlorophyll, a bluish pigment, phycocyanin. These pigments are not localized in a definite pigment-bearing body or plastid, but are frequently diffused in the outer zone of **protoplasm.** Surrounding the protoplasm is a definite cell wall of cellulose (see **Carbohydrates**). In most blue-green algae the outer part of the wall is modified and becomes a soft slimy substance which often forms a layer of considerable thickness. This slime substance may be of great value as an insulation against heat and desiccation, thus enabling blue-green algae to live in what seem to be most unfavorable environments. In the central portion of the **protoplast** are found many **chromatin** bodies, which, however, are not organized into a definite nucleus. The structure of the cell of these algae, with its absence of plastids and any definite nucleus, seems to indicate a relatively primitive organism, and is suggestive of the structure found in **bacteria.** Because they reproduce by simple fission, algae and bacteria are sometimes combined into a single group, the Schizophytes.

Many blue-green algae are single-celled organisms. The individual cells are often separate or they may be held together by the gelatinous outer wall in aggregates sometimes of considerable size. Representatives of this group are frequently observed in temporary puddles formed by a summer shower or in quiet shallow ponds in which the water often becomes very warm. Some of them are of considerable economic importance, when they appear in water supply reservoirs. In these they sometimes occur in numbers so great as to color the water and to be only too obvious to the most casual observer. Such occurrences are frequently described as "Water-blooms." Due to the products formed by the metabolism of the cells and liberated into the water, such "blooms" are real problems, for not only do these substances give to the water a distinctly unpleasant oily fishy taste and color, but several cases are recorded in which drinking of such water has been quickly fatal to live stock.

Other blue-green algae occur as simple filaments of cells. Single filaments may occur among other algae or they may exist in extensive masses covering considerable areas with a soft felt-like layer. In some genera many filaments are held together in a common gelatinous sheath. Many blue-green genera show what is called false branching. The rapid division of cells in the middle of a filament causes them to grow out laterally. Sometimes the filaments grow out as a single branch, as in *Tolypothrix,* or in pairs, as in *Scytonema.* True branching is found in a few genera.

In all the non-filamentous species the only method of reproduction is that of **cell division.** In the filamentous forms continued division may produce a filament of in-

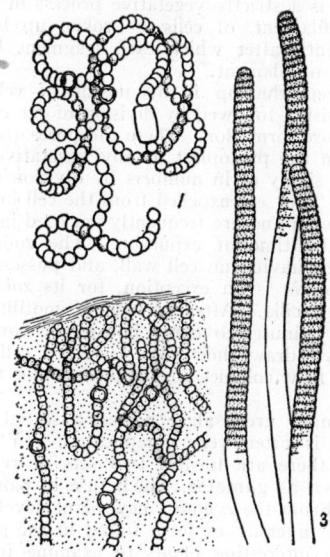

Three species of Myxophyceae. 1, Filaments of a species of *Anabaena.* 2, Filaments of *Nostoc* encased in the characteristic jelly. 3, *Oscillatoria.* (*1 after G. M. Smith, 2 after Fremy.*)

definite length. However, it eventually breaks up. This fragmentation may be due to animals feeding on cells of the filament, or to the death of certain cells. In some forms it is due to the development of cells which do not adhere tenaciously to the cells adjoining them. Specialized cells known as heterocysts are largely responsible for the last condition. They are large transparent cells which develop from ordinary cells.

Blue-green algae, particularly abundant in regions having a warm climate, develop elsewhere in great abundance, during the warmer seasons of the year. Besides being a source of trouble in water reservoirs, blue-green algae sometimes form unsightly stains by growing on the walls of stone buildings, especially if the latter be constantly wet. Blue-green algae are of little importance otherwise.

The green algae, or Chlorophyceae, are found in both salt and fresh water, where they often form conspicuous masses. They are characterized by a bright green color. The reserve food stored by these algae is starch, which is usually found around certain bodies known as pyrenoids, located in the **chloroplasts.**

Except that they are an important source of food for many animals, little importance can be ascribed to the green algae. They are, however, of considerable interest because of the possibility that from them the higher plants may have arisen, and because of the diversity of forms which are found within the group.

The protoplasts of the green algae are with very few exceptions enclosed within a rigid wall composed of two layers, an inner made up wholly or largely of cellulose and an outer layer of pectose (see **Carbohydrates**). Within the protoplast of the cell is located one or more conspicuous chloroplasts containing pigments approximately like those occurring in the plastids of higher plants. The chloroplasts of any single genus are usually very constant in appearance but in the different genera remarkable diversity of size and shape obtains. The primitive form seems to be the massive cup-shaped type such as occurs in many lower Chlorophyceae. The nucleus of all green algae is a definitely organized body possessing a nuclear membrane, one or more **nucleoli**, nuclear sap (karolymph), and a chromatin network. Many unicellular forms have one or more cilia which persist throughout their existence. The reproductive cells of most green algae have cilia. (See **Cilium**.)

Reproduction in green algae takes place in various ways. One is a strictly vegetative process in which the colony, or filament, of cells is broken up by various external agents, after which each fragment becomes a new colony, or filament.

Asexual reproduction is by means of cell division (nuclear division followed by division of the cytoplasm) or by zoöspore formation. These zoöspores are generally formed from the protoplast of any vegetative cell, and may appear singly or in numbers by division of a single protoplast. They are expelled from the cell in a manner as yet unknown and are frequently enclosed in a delicate vesicle at the time of expulsion. The zoöspores are naked bodies, having no cell wall, and possessing apical cilia. *Vaucheria* is an exception, for its zoöspores are covered with cilia. After periods of motility varying from a few minutes to many hours, zoöspores become quiescent, withdraw their cilia, secrete a cell wall and develop to new colonies or organisms like the parent form.

Finally, many green algae possess a sexual reproduction which is often very complicated. In sexual reproduction there are formed two sets of reproductive bodies known as **gametes**, which fuse in pairs to form a **zygote**. From the zygote a new plant develops.

The great diversity of form found in the green algae makes them interesting plants to examine for lines of **evolution** which have produced the many forms existent today. Several different lines have been found. One of these includes many of the forms which remain motile throughout their existence. A relatively simple unicellular organism is the starting point for such a line. *Chlamydomonas* is of this type. The plant is a small spherical or oval cell enclosed in a definite cell wall and having two apical cilia. Within the protoplast there is a single massive chloroplast which has a shape like that of a soft rubber ball pushed deeply in on one side. Within the hollow of the plastid the single distinct nucleus is located. At the apex of the cell near the origin of the two cilia there is a minute red body called an eye-spot. This is a light-sensitive organ which when stimulated causes the organism to move toward or away from the light source. In the apical end there is a pair of contractile vacuoles. *Chlamydomonas* reproduces asexually by means of zoöspores. These are formed by divisions of the protoplast to form 2, 4 or 8 daughter protoplasts, contained within the wall of the original cell. This wall softens and liberates these

naked cells. At once they become small replicas of the parent cell. They soon grow to the size of the original cell and again form a new group of zoöspores. In a suitable environment this is a very rapid method of reproduction. *Chlamydomonas* also reproduces sexually.

In sexual reproduction the protoplast of the cell divides to form motile bodies called gametes, which are quite like zoöspores, but smaller. On liberation from the parent cell wall, gametes from different cells unite in pairs, and form zygotes which develop into zygospores. A zygospore is a resistant spore with a thick wall which enables the organism to survive periods of adverse conditions. When favorable conditions return, the contents of the zygospore divide to form zoöspores, which behave as do similar spores from motile cells. The similarity of gametes and zoöspores indicates that one is derived from the other, that sex results from the transformation of asexual zoöspores into gametes.

The primitive character of *Chlamydomonas* is found in its contractile vacuoles, its eye-spot, its single massive plastid, and its cilia. From **Flagellates** it differs only in having a definite cell wall. Comparing other motile green algae with *Chlamydomonas* makes it possible to discover an interesting series of species of increasing complexity. First in this series is *Gonium sociale,* with colonies of four cells, and *Gonium pectorale,* with sixteen cells, held together loosely in a gelatinous matrix. All the cells of a colony are alike and any cell may form either zoöspores or gametes. All gametes are alike, but fusion occurs between gametes from different colonies. Next in the line of increasing complexity is *Pandorina,* in which a spherical colony is formed. In this genus the gametes are slightly different in size and behavior, some being small and active, while others, slightly larger, are more sluggish. This is an indication of a differentiation of sex.

Still further advance is shown by *Eudorina,* in which a colony is composed usually of 32 cells located in the peripheral portion of the gelatinous matrix. Each cell of the colony is like a *Chalmydomonas* cell and each is capable of reproducing asexually. But in sexual reproduction a very obvious difference in sexes is apparent. Some colonies are definitely female, the cells enlarging slightly and functioning as eggs. In other colonies, each cell divides to form 64 minute biciliate **sperm.** Fusion between an egg and a sperm produces an oöspore which gives rise to a new colony. In *Pandorina* then a very obvious distinction between sexes has appeared, but the vegetative cells remain alike. In *Pleodorina* each colony is composed of small, purely vegetative cells and larger reproductive cells. The smaller vegetative cells are formed in the anterior end of the colony. In *Volvox* we find the highest degree of differentiation exhibited in this line of motile algae. In this plant the number of cells in a colony is very great, in some species there being as many as 25,000. Of these cells only a few are reproductive, while thousands remain vegetative. Many *Volvox* colonies are so large as to be readily visible to the unaided human eye. In these the many cells form a single layer embedded in the gelatinous matrix. The center of the colony is either water or a thin gelatinous substance. The cells of the colony have the same structure as *Chlamydomonas* cells. They are joined together by fine protoplasmic strands. The beating of the cilia causes the large colony to roll rapidly about in the water, making it a fascinating object to watch. In asexual reproduction certain cells of the colony enlarge and move to the central region. There they lose their cilia, after which they divide rapidly to form new colonies which remain for some time within the parent. Often a single colony will contain a dozen or more of these small colonies. The latter are liberated by the disintegration of the parent colony. In sexual reproduction, cells in the posterior region of the colony differentiate. Some lose their cilia and become very large; these are eggs. Other cells, either in the same or different colonies, divide many times to form large numbers of minute biciliate sperm. These swim to the eggs. A single sperm enters an egg, its nucleus fusing with that of the egg. As a result a zygote is formed. This secretes around itself a thick wall and becomes an oöspore, capable of enduring protracted periods of unfavorable conditions. With the return of favorable conditions the thick wall of the oöspore breaks, the protoplast emerges and divides rapidly, forming a new colony. *Volvox* represents the climax reached in this line of evolution. There is not only a very great increase in the number of cells forming a colony, but also a distinct separation of vegetative and reproductive cells. The reproductive cells are of two kinds, eggs and sperm. But every cell of the plant retains the primitive character of the individual cell.

It is possible to build up other evolutionary series of green algae in which the vegetative cells are non-motile. *Chlamydomonas* often assumes a non-motile condition; the cells become embedded in a copious gelatinous matrix and lose their cilia. This condition is known as the palmella stage. *Tetraspora* is an alga which has the appearance of the palmelloid stage of *Chlamydomonas*. Another genus, *Palmella*, normally

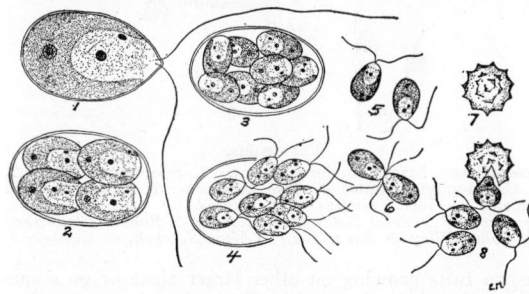

Chlamydomonas. 1, Mature cell; 2, Four young cells formed in asexual reproduction; 3, Eight gametes formed by the division of one cell; 4, Gametes escaping from old cell wall; 5, Free-swimming gametes; 6, Two gametes conjugating; 7, Resting zygote; 8, Four cells formed from the zygote. Somewhat diagrammatic.

exists as a shapeless colony of cells held together in a gelatinous matrix. Cilia are lacking, but may be developed by any cell in the colony. A ciliated cell escapes and swims about freely for a time, then settles down and divides to form a new colony. Asexual zoöspores are formed in *Palmella* and also isogametes, that is, gametes of equal size. *Palmella* shows the beginning of a non-motile habit, with a restriction of the motile stages to the reproductive cells. In *Geminella* the amorphous habit of the colony is lost; divisions take place in such a way that the resulting cells tend to exist in a single series, the individual cells being held together only by the gelatinous matrix around them. In *Ulothrix* further advance is made. In this plant the protoplast of the cell divides within the wall of the cell. But it does not escape therefrom; instead cross-walls are formed between daughter protoplasts which remain permanently joined. Repeated divisions in a single direction result in the formation of a long unbranched filament of cells. Asexual reproduction in *Ulothrix* is by zoöspores. These are very similar to the cells of *Chlamydomonas*, but each has 4 cilia. From each zoöspore new filaments are formed directly. Sexual reproduction is by biciliate gametes which are formed in the usual manner. Two gametes fuse to form a zygote, which develops a thick wall, becoming a zygospore, and on germinating produces zoöspores. In some species of *Ulothrix* the gametes are alike, while in others slight differences in size of the gametes produced from different cells indicate the beginning of sex differentiation.

Other genera of algae, related to *Ulothrix*, show greater differentiation in the vegetative cells. Branch-

ing occurs in many; certain cells produce the zoöspores or gametes. Other genera, notably the marine *Ulva*, the Sea Lettuce, have cell divisions in two planes, so

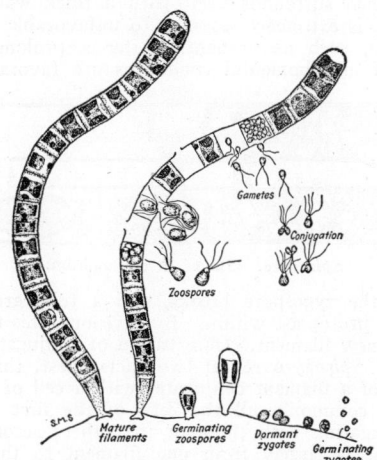

Ulothrix. Stages in development.

that extensive membranes are formed, often two meters or more in length.

The highest stage of development in this series is found in *Coleochaete*. This alga appears as small disks epiphytic on other aquatic plants. It consists of branching filaments which in some species grow out to form a flat shield-like thallus. Asexual reproduction is by biciliate zoöspores, which are formed singly from the protoplast of any cell. In its sexual reproduction, *Coleochaete* is especially interesting, exhibiting an advanced degree of differentiation of sex cells called oögamy. The eggs are formed in special cells at the tips of certain filaments. By continued growth, the vegetative cells form a layer enclosing the oögonium. From the oögonium a projection called the trichogyne grows out; in some species it remains short, but in one at least it becomes long and slender. The biciliate sperm are formed singly in special cells known as antheridia. A single sperm enters the egg through the trichogyne or papilla, through a soft place which forms in the wall. Following fertilization, or the union of egg and sperm, the fertilized egg enlarges greatly and secretes around itself a thick wall, in which condition it remains very resistant to external changes. On germination its contents divide to produce 16 or 32 cells, each of which produces a zoöspore. These give rise to new plants. In *Coleochaete* we have a highly developed end product to an advancing degree of differentiation of cells and development of sex.

Many other equally interesting lines of evolution can be found in green algae. One leads to *Vaucheria*, a branching filamentous plant in which cross walls are not formed, nuclear divisions occurring until an extensive multinucleate filament is formed. Many marine relatives of *Vaucheria*, especially abundant in tropical seas, have elaborate bodies, often thickly encrusted with lime. In the genus *Caulerpa* the thallus of some species appears to be differentiated into leaves, stems and roots; however, no differentiation of tissue occurs, the whole structure being essentially filamentous.

Another group of algae, found only in fresh water, is the Conjugales. This includes the frequently observed *Spirogyra*, a genus having spiral chloroplasts. Sexual reproduction in *Spirogyra*, as in all the Conjugales, is by a process called conjugation. When the cells of two filaments are to conjugate they come into contact one with another. From the cells of one filament projections grow outward to meet similar projections from the cells of the other filament. These projections meet, and the walls at their tips are absorbed, leaving a tube connecting two cells of opposite filaments. Through this tube the protoplast of one cell moves into the opposite cell, where fusion occurs. Following this the zygote thus formed surrounds itself with a thick wall. This zygospore is extremely resistant to unfavorable external conditions, such as drought. After a prolonged rest period, if environmental conditions are favorable, the

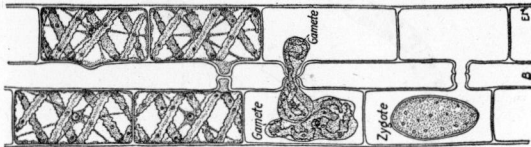

Spirogyra. Stages in conjugation.

wall of the zygospore breaks, and a tube grows out from the protoplast within. By divisions, this tube becomes a new filament. Observation of conjugating filaments of *Spirogyra* reveal two facts: first, that when one cell of a filament conjugates with a cell of another filament, commonly all the cells of the first filament are conjugating with those of the other; second, that movement is largely from one filament to the other, so that at the end of the process one filament is composed of empty cell walls and the other filled with zygospores. This may be conceived as a sexual condition, the empty cells having been male and the cells in which the zygospores formed, female. Many related algae in the Conjugales show no indication of sexuality, the protoplasts uniting in the tube joining the two cells. No ciliated cells of any sort occur in this order.

Related to *Spirogyra* is a family containing many species whose cells are very symmetrical and beautiful objects. This is the Desmidiaceae, a family whose members commonly occur as single cells composed of two symmetrical halves. In many species the halves are distinctly indicated by a deep constriction which leaves only a slender isthmus connecting them. Conjugation in the Desmids is essentially like that in the other Conjugales.

The brown algae, or Phaeophyceae, are distinguished from other algae by the presence of the brown pigment fucoxanthin, which masks the chlorophyll present and which imparts to them a brown color. They are nearly all marine plants. In this group there are no very simple primitive forms, comparable to *Chlamydomonas* of the greens. The simplest forms, such as *Ectocarpus,* are branched filamentous plants.

Other brown algae are very large, with tough bodies of many often complex shapes. They are most numerous on rocky coasts, where they grow attached to the rocks and are often exposed to the severest pounding of the tides. Brown algae are found in all regions, but are especially abundant in the cold temperate and arctic waters. One genus, *Sargassum,* occurs in great abundance floating in the Atlantic Ocean, forming the Sargasso Sea, through which Columbus passed so slowly on his voyage of discovery. Probably these plants are carried by ocean currents from the shores where they grow to the Sargasso Sea, where they float endlessly. The brown algae include the largest plants of this division of the plant kingdom, many of the kelps growing 30–40' long, and some of the plant forms of the Pacific Ocean attaining lengths of a hundred feet and more.

All brown algae are multicellular plants. Each cell contains a single distinct nucleus and several chloroplasts which contain chlorophyll, carotin, xanthophyll and fucoxanthin. The presence of the latter hides the other pigments. Photosynthesis in brown algae, as in other plants, results in the formation of sugar. This sugar however is changed into a compound, laminarin, instead of starch, which is never found in this group of algae.

The Phaeophyceae are divided into several orders. These show interesting differences in their life histories.

In most of them there is a very definite alternation of generations. Plants of the sexual generation form gametes; the asexual generation forms zoöspores. The motile cells of the brown algae are quite distinct from those of the green algae, having 2 unequal cilia which are attached laterally and extend in opposite directions.

One of the orders of this class is the Ectocarpales, of which *Ectocarpus* is a common form. It occurs as soft

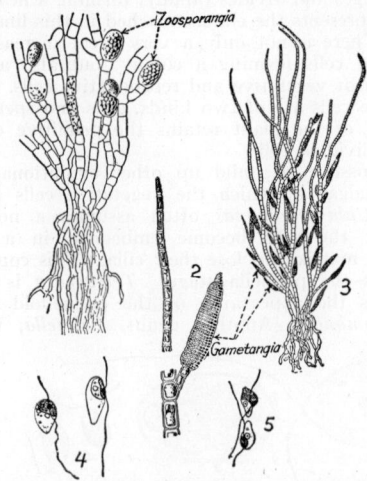

Ectocarpus.

Ectocarpus. Stages in development. 1, Sporophyte plant; 2, 3, Gametophyte plants; 4, Two zoospores; 5, Conjugation of gametes to form zygotes. (*Reprinted by permission from Textbook of General Botany, by Holman & Robbins, published by John Wiley & Sons, Inc.*) (*After Setchell, & Gardner.*)

brown tufts growing on other larger algae or on stones or woodwork in the water. Each plant is composed of slender branching filaments. In this genus there is a definite alternation of two generations which are indistinguishable in the vegetative condition. Plants of the sexual generation are haploid, that is, have the reduced number of chromosomes and bear gametangia. These are elongated structures composed of many small, cubical cells. Each of these cells forms a single gamete. The plants of this generation and the gametangia they bear all look exactly alike. The gametes which they produce also look very much alike. But in behavior they are different. Some are sluggish, moving but little, while others swim actively and are attracted to the sluggish ones from other plants. The active gametes are males, the others females; sometimes the female gametes are slightly larger than the males. One male gamete fuses with a female, forming a zygote. The zygote always forms an asexual plant which is **diploid.** There are two kinds of asexual plants, which look exactly alike, and also like a sexual plant. Each forms zoösporangia. One form of zoösporangium, called a plurilocular zoösporangium, is composed of many small cubical cells, each of which forms a single zoöspore. These zoöspores are liberated, swim about for a time, sink to the bottom and settle against any solid substratum and give rise to new asexual plants of the same type as that producing them. The other type of zoösporangium, often found on the same plant as the first, is composed of a single, usually much-enlarged, cell. The protoplast of this cell divides many times and gives rise to several zoöspores which are haploid. Therefore reduction division takes place in the sporangia, which produce these zoöspores. These zoöspores produce sexual plants. In some species of *Ectocarpus* the alternation of generations is not as regular as that described.

Another order of brown algae is the Cutleriales, of which the genus *Cutleria* is a well-known example. In *Cutleria* the two generations have a very different appearance, the sexual plants being much branched, and several inches tall, while the asexual plants are small lobed thalli growing prostrate on the substratum. So different are these two generations that they are often mistaken for different plants. The sexual plants are of two kinds, male and female, which differ very little. The sex organs are borne on the surface of the thallus. The male gametangia are elongate structures borne on branching filaments; the female gametangia are stouter and composed of few cells. The male gametes are minute and biciliate, the female, also biciliate, are many times larger. Many male gametes swim to a single female, one fuses with it, producing a zygote. The zygote develops into an asexual plant. The zoöspores, formed from small sac-like zoösporangia, which develop in large numbers on the upper surface of the thallus, are biciliate and very similar to gametes. They form sexual plants. In this order there is a very distinct difference in the two gametes, male and female, and a striking difference between sexual and asexual plants.

A third order is the Laminariales, which includes the largest algae known, commonly known as kelps. Some of them have a very striking appearance. *Postelsia*, for example, has a plant body composed of an erect stiff stalk often several inches in diameter. From its base many thick root-like outgrowths spread out and fix the plant firmly to the substratum. At the top of the stalk, long spreading branches are found. When the plant is seen growing in water it has much the appearance of a palm tree, whence the common name, sea palm. Another plant in this order is *Chorda filum*. Its thallus is a tough cord-like object 3–8' long and about ¼" in diameter. It looks very much like a coarse round leather shoestring. A very common genus is *Laminaria*, which has many species of various forms, mostly large. Some of them consist of long cord-like stalks which bear at their upper end a broad flat expansion often 6–12' long and 8–15" broad. Colloquially these are known as devil's aprons. These large plants are the asexual generation and so are diploid. The zoösporangia are formed in immense numbers on the surface of the thallus. Each zoösporangium is a cylindrical object which produces many small biciliate zoöspores. These zoöspores swim down to the sea bed, where they develop into haploid or sexual plants. The latter are minute, usually consisting of a few cells which form a branching filament. Some of the plants are male, others female. In the female plant, any cell may become a sexual cell; often the plant is only a single cell. This sexual cell is an oögonium and forms a large non-motile egg which remains in the parent plant. Any cell of the male plant may become sexual, producing minute biciliate sperms which swim to the egg and fuse with it, forming a zygote. The latter at once develops into an asexual plant. In this order, there is also a distinct alternation of generations, but the sexual generation contains the small plants, the asexual usually very large plants. The sexual cells are distinctly different: the large non-motile egg and the small biciliate sperm.

A fourth order of brown algae is the Fucales. Members of this order are tough, much-branched plants which are particularly abundant between the tide levels on rocky shores in regions where the water is cold. *Fucus*, the common rockweed or bladder wrack, is a common and well-known plant. In this, as in all members of this order, there is no asexual reproduction. Therefore no distinct alternation of generations can occur. A *Fucus* plant consists of a tough dichotomously branched frond, which is attached to the rock on which it grows by a disk-shaped holdfast. In many species hollow bladders develop along the frond and serve to bring the plant into an erect position at high tide. At the tips of the branches of the thallus the reproductive bodies, receptacles, are formed. In some species these tips are swollen to form hollow bladders, in others they are flat and little differentiated from the rest of the thallus. The reproductive cells are formed in spher-

ical cavities which are connected with the surface by small pores. Each cavity is called a conceptacle. Numerous branching filaments rise from the lower part

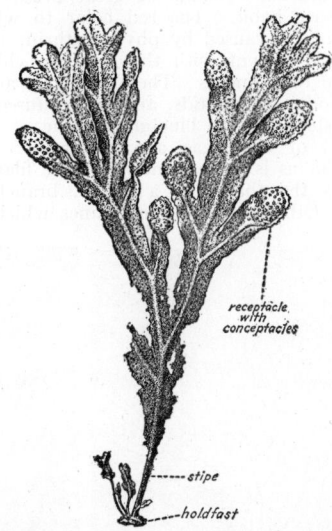

Fucus. Mature plant, ×⅓. No bladders are shown. (*From Morphologie und Biologie der Algen, Gustav Fischer, Jena.*)

of the conceptacle wall. Branches of these filaments bear the sexual organs. In some species the two sexes are borne in the same receptacle, in others they occur on different plants. The male sex organs or antheridia are oval sacs. The protoplast of each sac divides to form 64 cells, each of which becomes a laterally biciliate sperm. When mature these antheridia are extruded through the ostiole or opening of the conceptacle into the water. There the wall of the antheridium bursts, liberating the sperms. Each oögonium consists of a single cell. Its protoplast divides to form eight eggs. These also are extruded from the conceptacle, while still within the wall of the oögonium, and freed by the bursting of the same. Both types of conceptacle secrete a gelatinous matrix which surrounds the oögonia or antheridia and aids in bringing them to the surface of the receptacle. This matrix is squeezed out of the conceptacles by the partial drying of the frond at low tide. Each egg is a very large non-motile cell. Thousands of sperms are attracted to each egg and swim about it, causing it to revolve rapidly. Finally one sperm gains entrance to the egg and fertilizes it. The other sperms immediately swim away. The fertilized egg settles to the bottom, attaches itself to the substratum and at once starts to develop into a new plant. In the Fucales there is no alternation of generations. The gametes of the Fucales are very distinct, one being a large non-motile egg, the other a very small swimming sperm. One is tempted to arrange the various orders of brown algae in a series showing the way in which each may have evolved. However, such evolutionary relationships are purely speculative and not supported by any real evidence. It is impossible to trace the ancestry of the brown algae back to any simple ancestor, since no simple forms of brown algae are known.

The economic importance of the brown algae, while slight, is much greater than that of the green algae. Large quantities of these plants are gathered and used for fertilizer, wherever agriculture is carried on near the coast. From the ash produced by burning the larger forms, the kelps and Fucales, iodine and also potassium are obtained. In the Orient and in some of the north Atlantic islands, some of the brown algae are used as food, both for human beings and for live stock.

The red algae or Rhodophyceae form a very large group of plants, nearly all marine, of small to medium size. They are particularly abundant in warm coastal waters and are often plants of great beauty and extremely delicate habit. The red color to which they owe their name is caused by phycoerythrin, a red pigment which is present with the common chlorophyll-carotin group of pigments. These pigments are present in definite bodies or plastids, and not diffused through the protoplasts, as in the blue-green algae.

The forms of red algae are numerous. In many species the thallus is an extremely delicate filament. In other species the thallus is a tough, branched body 6–15″ long. Others are flat membranes which may be

Four red algae. 1, *Agardhiella tenera;* 2, *Dasya elegans;* 3, *Grinellia americana;* 4, *Chondrus crispus.* × ⅓. (*Photographs of herbarium specimens by Naylor.*)

a single cell in thickness or may be many cells thick. Some are thickly covered with a calcareous deposit, so that they are hard and stony, resembling corals. No motile reproductive cells are produced by members of this group. In sexual reproduction there is always a large female cell which is fertilized by a small male cell. This sexual reproduction is a rather complicated process. The antheridia are single-celled bodies; in some species the whole cell is liberated, in others the protoplast of the antheridium is freed. In either case the male cell floats in the water, carried only by the currents. The female reproductive organ is known as the procarp. In simpler forms this consists of a swollen basal portion called a carpogonium and a long slender portion called a trichogyne. Chance brings the male cell to the surface of the trichogyne, against which it sticks. The wall of the trichogyne is dissolved, allowing the nucelus of the male cell to enter the trichogyne. This nucelus passes down the trichogyne and enters the carpogonium, where it fuses with the female nucleus. From the fertilized carpogonium asexual spores called carpospores are formed, usually at

the ends of branches which grow out from the carpogonium or from cells which are formed from those surrounding the carpogonium. Into these carpospores, nuclei from the carpogonium pass. The carpospores of simpler red algae at once produce sexual plants. In most red algae, however, they produce asexual plants which may be identical with the sexual plants in appearance. These asexual plants bear reproductive cells called tetraspores, because 4 of them are borne in a single sporangium. Each tetraspore gives rise to a sexual plant. So in the majority of red algae there is a distinct alternation of generations.

Of the many species of red algae very few are of any importance. In northern waters of both coasts of the Atlantic, dulse, *Rhodymenia palmata,* is found. It is gathered, cleaned more or less, and dried. It is then sold as a food or a relish. Species of *Porphyra,* often called laver, are also eaten, especially by Oriental people. Irish moss, or corrageen, which is *Chondrus crispus,* is another red alga which is gathered for food. It is a small much-branched plant, commonly dark red in color and with a beautiful iridescent surface. The plants are gathered, thoroughly cleaned and dried. Drying bleaches them to a creamy white color. When thoroughly dry they are bagged and sold. The powdered plant is commonly boiled in milk, flavored and sweetened, and allowed to cool. It forms a firm smooth gel known as blanc-mange. From species of red algae growing in the Pacific Ocean, **agar agar** is obtained.

The origin of the red algae and their relationships with other algae is a matter of considerable speculation. A few trace them from the blue-green algae, finding in some of the more primitive red algae "connecting links" which support this view. There are no ciliated reproductive cells in either of the two groups. But their distinct well-developed nucleus, their plastids, and their complex reproductive process set the red algae off very clearly. In view of these facts, it is perhaps more logical to derive the red algae from the green, using a form like *Coleochaete.* If there is any relationship between the red and the blue-green algae, it must be rather remote. (R.M.W.)

ALGAL REEFS. Paleobotany.

ALGEBRA. As generally understood in an elementary sense, algebra is a branch of mathematics which deals with the operations with **numbers** by means of general symbols, such as letters, and with some of the simpler applications of these operations.

For the historical origin of algebra, we must go back to the time of the early Egyptians. The germs of algebra are found in the ancient papyrus of Ahmes (about 1700 B.C.). The ancient Greeks developed elementary **geometry** to a remarkable degree, but occupied themselves very little with algebra. It was not until the time of Diophantus (about 300 A.D.) that any Greek work in algebra was done to amount to anything. The early Hindus, however, cultivated elementary algebra to a surprising extent, but were little concerned with geometry.

The origin of the name "Algebra" is somewhat obscure. The Arabic mathematician *Mohammed ibn Mûsâ al-Khowârizmî* (about 825 A.D.) wrote a work entitled *"Al-jebr w'al-muqâbalah,"* which is sometimes translated as "restoration and equation," but the meaning is not exactly clear. This work treated of algebraic topics and its name is generally considered to have been the source of the name algebra. The modern Europeans became acquainted with algebra from the Arabs.

Some indication of the subject-matter usually considered as belonging to algebra may be found by consulting the following topics: **Number, Algebraic Operations, Algebraic Expressions, Factoring, Fractions, Involution, Evolution, Powers and Exponents, Radicals, Identities, Equations, Solution of Equations, Linear Algebraic Equations, Quadratic Equations, Polynomial Equations, Graphs, Inequalities, Binomial Formula, Logarithms, Progressions, Variation, Permutations, Combinations, Prbability, Determinants.** (L.L.S.)

ALGEBRAIC EQUATIONS. An algebraic equation is an **equation** in which both members are **algebraic functions.**

Algebraic equations in one unknown are classified into **polynomial equations** (rational integral equations), **fractional equations, radical** or **irrational equations;** polynomial equations are classified further. (L.L.S.)

ALGEBRAIC EXPRESSIONS. An algebraic expression is a symbol or combination of symbols that represents a number. If the expression consists of two or more parts connected by plus and minus signs, each of these parts with the sign preceding it is called a term. A monomial is an algebraic expression consisting of only one term, a polynomial is one consisting of more than one term; a binomial and a trinomial are polynomials of two or three terms respectively.

If two or more numbers or number symbols are multiplied together, each of them is called a factor of the product. Any factor of a product may be called the coefficient of the remaining part. (L.L.S.)

ALGEBRAIC FUNCTIONS. An algebraic function is a **function** which involves the **variable** in only the operations of **addition, subtraction, multiplication, division,** raising to **powers** with constant rational exponents, and extraction of roots (**involution**), a limited number of times.

An algebraic function may also be defined as a function $y = \phi(x)$ which satisfies an equation of the form $f(x,y) = 0$, where $f(x,y)$ is a **polynomial** in x and y.

Algebraic functions are classified into **rational functions** and **irrational functions,** and **power functions;** rational functions are further classified into sub-classes. (L.L.S.)

ALGEBRAIC NUMBERS. An algebraic number is a **number** which satisfies a **polynomial equation in one** variable with integral coefficients. (L.L.S.)

ALGEBRAIC OPERATIONS. Algebra is concerned with **numbers** and with various operations with numbers. The fundamental operations of algebra are: **addition, subtraction, multiplication** and **division.** Two other derived operations are **involution** (raising to a power), and **evolution** (extraction of roots). (L.L.S.)

ALGIN. A hydrophilic colloidal polysaccharide obtained from *Macrocystis pyrifera* and other species of **brown algae.** The term is used both in reference to the pure substance, alginic acid, extracted from the algae and also to the salts of this acid such as sodium or ammonium alginate, in which forms it is used commercially. The alginates currently find a large number of applications in the paint, rubber, pharmaceutical, food, and other industries. (B.S.M.)

ALGOL. Algol (β **Persei**) is one of the first **variable stars** to be recognized as such. The first scientific notice of this variability was made by Montanari in 1670, but it is quite evident that the changes in the

light of this star were noticed long before this time. In fact, the very name Algol, which signifies "Demon star," was probably assigned to the star because of its peculiar behavior. Astrologically, Algol was considered the most unfortunate star in the heavens.

Algol is an **eclipsing binary** and is the first star of this type to be explained. Because of its great brightness it has been extensively observed with all types of stellar **photometers** and the characteristics of its **light curve** are known with great precision. Algol is also a **spectroscopic binary,** and, from the determination of the **orbital elements** from the light variability as well as from the spectroscopic data, the physical characteristics of the component parts may be determined. (W.K.G.)

ALGONKIAN. A term applied to rock formations of late pre-Cambrian (**Proterozoic**) date in the Great Lakes region, and to the unit of time represented by these formations. Some American geologists use the term as a synonym for late pre-Cambrian. (See **Historical Geology.**) (R.M.F.)

ALIDADE. Plane Table.

ALIENIST. A legal term applied to a specialist trained in neurology and psychiatry who testifies in court on questions concerned with mental disease. (D.M.H.)

ALIGNMENT CHART. Nomograms or calculating charts used to represent formulae containing three or more variables. Graduated lines represent the variables;

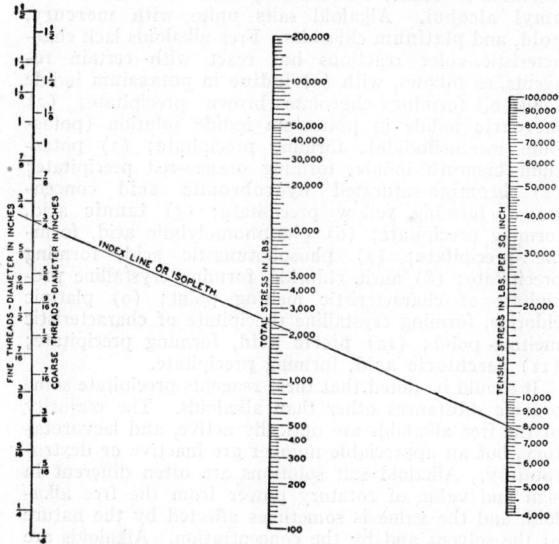

an index line or isopleth passing through points of two scales of known values will intersect the third scale to give the solution of the problem. The nomogram shown will give the solution of the equation $0.785D^2S = T$, where D is the root diameter of an American Standard screw, S is the unit tensile strength, and T is the total stress. (H.C.H.)

ALIMENTARY TRACT. The structures through which nourishment pass during the process of digestion and elimination. These include the mouth, pharynx, esophagus, stomach, the small intestine, which includes **duodenum, jejunum,** and **ileum,** and the large intestine, which includes the **cecum, colon, rectum** and **anus.** (See also **Anatomy; Digestive System.**) (R.S.M.)

ALIZARINE. Dyes.

ALKALI RESERVE. Acidosis.

ALKALI ROCKS. Igneous rocks which contain a relatively high amount of alkalis in the form of soda **amphiboles,** *soda* **pyroxenes,** or **felspathoids,** are said to be alkaline, or alkalic. Igneous rocks in which the proportions of both lime and alkalis are high, as combined in the minerals, **feldspar, hornblende,** and **augite,** are said to be calc-alkali. (R.M.F.)

ALKALIS. Acids, Bases, and Salts.

ALKALOIDS. Alkaloids are types of organic bases. They are generally colorless, odorless solids, of definite melting point, which decompose upon attempted distillation, of bitter taste (poisonous), insoluble in water, soluble in alcohol, ether, chloroform, carbon tetrachloride, amyl alcohol, benzene. Alkaloids are generally related to nitrogen ring compounds, e.g., **pyridine, quinoline, isoquinoline, pyrrole, pyrrolidine, purine,** and are found in certain plants (*Papaveraceae, Leguminosae, Ranunculaceae, Solenaceae, Rubiaceae* (cinchona). They are powerful poisons, many possess high medicinal value, some are habit-forming. Alkaloids of plants are usually found as alkaloid salts of **carboxylic acids,** e.g., of malic, citric, oxalic, succinic, quinic (**cinchona** alkaloids), meconic acid (**opium** alkaloids), in all parts of the given plant, but are generally accumulated in the fruit, seeds or bark. Alkaloids react as bases to form salts (especially used for crystallization purposes are **hydrochlorides, sulfates, oxalates**), which are generally soluble in water or alcohol, insoluble in ether, **chloroform, carbon tetrachloride, amyl alcohol.** Alkaloid salts unite with **mercury, gold,** and **platinum** chlorides. Free alkaloids lack characteristic color reactions but react with certain reagents, as follows, with (1) **iodine** in **potassium** iodide solution, forming chocolate brown precipitate; (2) **mercuric** iodide in potassium iodide solution (potassium mercuriiodide), forming precipitate; (3) potassium **bismuth** iodide, forming orange-red precipitate; (4) **bromine**-saturated **hydrobromic acid** concentrated, forming yellow precipitate; (5) **tannic** acid, forming precipitate; (6) phosphomolybdic acid, forming precipitate; (7) phosphotungstic acid, forming precipitate; (8) auric chloride, forming crystalline precipitate of characteristic melting point; (9) platinic chloride, forming crystalline precipitate of characteristic melting point; (10) **picric acid,** forming precipitate; (11) **perchloric acid,** forming precipitate.

It should be noted that these reagents precipitate some organic substances other than alkaloids. The majority of the free alkaloids are optically active, and laevorotatory, but an appreciable number are inactive or dextrorotatory. Alkaloid salt solutions are often different in sign and value of rotatory power from the free alkaloid, and the value is sometimes affected by the nature of the solvent and by the concentration. Alkaloids are usually extracted from the plant material by hydrochloric or sulfuric acid. This extract is then made slightly alkaline, thus precipitating the free alkaloid base, which may be separated by filtration and purified by further crystallization. It was suggested by Pictet as early as 1905 that the methylation (formation of —CH₃ group) of hydroxyl (—OH) or amino (—NH₂) groups, occurs in the plant by means of formaldehyde (HCHO), followed by rearrangement in which the methyl group enters the ring, thus increasing the ring by one additional carbon:

Pyrrole, by formaldehyde, into 1-methylpyrrole, rearranging into pyridine.

Indole, by formaldehyde into 1-methylindole, rearranging into quinoline.

More recently Robinson conducted the reaction of formaldehyde with 2,5-diaminopentanoic acid ($CH_2NH_2 \cdot$ $CH_2 \cdot CH_2 \cdot CHNH_2 \cdot COOH$) obtaining 1-methyl-2-hydroxypyrrolidine plus ammonia plus carbon dioxide. (See table, p. 45). (R.K.S.)

ALKALOSIS. Acidosis.

ALKYL. Radical of aliphatic **hydrocarbon.** (R.K.S.)

ALLANITE. Allanite is a rather rare **monoclinic** mineral of somewhat variable but quite complex chemical composition, perhaps represented satisfactorily by the formula $Ca_2(Al,Ce,Fe)_2(Al \cdot OH)(SiO_4)_3$. The color of the fresh mineral is black but it is usually brown or yellowish with a coating of some alteration product; often the altered crystals have the appearance of small rusty nails. It occurs characteristically in plutonic rocks like **granite, syenite** or **diorite** and is found in large masses in **pegmatites.** Localities in the United States are Essex and Orange Counties, New York, Franklin, New Jersey, Amherst County, Virginia, and Llano County, Texas. The slender prismatic crystals are sometimes called orthite. Allanite was named for its discoverer, T. Allan. Orthite was so named from the Greek word meaning straight, in reference to the straight prisms, a common habit of this mineral. (E.S.C.S.)

ALLANTOIN. Alkaloids.

ALLANTOIS. A sac-like outgrowth of the hind gut of the **embryo** found only in **reptiles, birds** and **mammals.** In reptiles and birds it serves as a respiratory organ and receives waste matter, and in mammals it forms part of the **placenta** through which all interchange with the blood of the mother during embryonic development is carried out. (A.W.L.)

ALL-DAY EFFICIENCY. In some applications of equipment, the load varies widely during the day and the machines have a no-load loss plus an additional loss which varies with the load. Hence the *efficiency* will vary from time to time during the day. The all-day efficiency is an over-all picture of the varying efficiency and is defined as the total energy output multiplied by 100 and divided by the total energy input in a day.

$$n = \frac{p_1 t_1 + p_2 t_2 + \cdots}{P_1 T_1 + P_2 T_2 + \cdots} \times 100$$

where p_1, p_2, etc., are the various output powers, and t_1, t_2, etc., are the corresponding times of use, while P and T represent the input powers and times. (L.R.Q.)

ALLELE, ALLELOMORPH. In Mendelian inheritance (see **Heredity** and **Evolution**), contrasting pairs of characters, as tall and dwarf, or the contrasting genes which produce them, are known as alleles or allelomorphs. The new term, allele, is preferable. (R.M.W., B.S.M.)

ALLERGY. The state of exaggerated susceptibility to a substance which is harmless in similar or greater amounts to most individuals. A person exhibiting this is said to be allergic to the substance causing the reaction. These substances are usually of **protein** nature, although physical agents such as heat, cold, and light can provoke an allergic response. The principal diseases of allergy are serum disease, serum accidents, **hay-fever, asthma, angioneurotic edema,** and **hives.** (R.S.M.)

ALLIGATOR. Reptilia, Crocodilia. A large freshwater **reptile.** Two species are known, one in China, *Alligator sinensis,* and the other, *A. mississippiensis,* in the southern United States. The American alligator, larger of the two, reaches a length of 16′ and a weight of 500 lbs. It has been hunted to some extent for its skin, which makes durable leather. (A.W.L.)

SELECTED REPRESENTATIVE ALKALOIDS

ALKALOID	FORMULA	MELTING POINT °C.	BOILING POINT °C.
1. Aconine....	$C_{25}H_{41}NO_9$....	132	
2. Aconitine....	$C_{34}H_{47}NO_{11}$....	204	
*3. Adenine....	$C_5H_5N_5$....	220 subl.	
4. Adrenaline (active principle of the hormone of the adrenal gland)....	$C_9H_{13}NO_3$....		
5. Allantoin....	$C_4H_6N_4O_3$....	235	
6. Apomorphine....	$C_{17}H_{17}NO_2$....	170 decom.	
7. Atropine (hydrolyzes to tropine plus tropic acid)....	$C_{17}H_{23}NO_3$....	118 subl.	
8. Belladonnine....	$C_{17}H_{21}NO_2$....		
9. Brucine....	$C_{23}H_{26}N_2O_4 \cdot 4H_2O$....	$\begin{cases} 105 \\ 178 \text{ anhyd.} \end{cases}$	
*10. Caffeine (theine)....	$C_8H_{10}N_4O_2 \cdot H_2O$....	$\begin{cases} 235 \text{ anhyd.} \\ 180 \text{ subl.} \end{cases}$	
11. Cinchonidine....	$C_{19}H_{22}N_2O$....	207	
12. Cinchonine....	$C_{19}H_{22}N_2O$....	264	
13. Cocaine....	$C_{17}H_{21}NO_4$....	98	
14. Codeine....	$C_{18}H_{21}NO_3 \cdot H_2O$....	155 anhyd.	
15. Coniine (2-normal propylpiperidine)....	$C_3H_7 \cdot C_5H_{10}N$....	−2	167
16. Ephedrine (1-phenyl-2-methylaminopropanol-1)	$C_{10}H_{15}NO$....	40	255
17. Ergotimine....	$C_{35}H_{39}N_5O_5$....	229	
18. Ergotoxine....	$C_{35}H_{41}N_5O_6$....	163	
*19. Guanine....	$C_5H_5N_5O$....	>360 decom.	
20. Homoatropine....	$C_{16}H_{21}NO_3$....	95–99	
21. Hydrastine....	$C_{21}H_{21}NO_6$....	235	
22. Hydrastinine....	$C_{11}H_{13}NO_3$....	116	
23. Hydroquinine....	$C_{20}H_{26}N_2O_2 \cdot 2H_2O$....	172 anhyd.	
24. Para-hydroxyphenylethylamine. (active principle of ergot)	HO〈 〉$CH_2CH_2NH_2$....	160	
25. Hyoscine (scopolamine)....	$C_{17}H_{21}NO_4$....	59	
26. Hyoscyamine....	$C_{17}H_{23}NO_3$....	107	
*27. Hypoxanthine....	$C_5H_4N_4O$....	150 decom.	
28. Laudanine....	$C_{20}H_{25}NO_4$....	166	
29. Lupanine....	$C_{15}H_{24}N_2O$....	99	
30. Lupinine....	$C_{10}H_{19}NO$....	69	
31. Morphine....	$C_{17}H_{19}NO_3 \cdot H_2O$....	230 decom.	
32. Narcotine....	$C_{22}H_{23}NO_7$....	176	
33. Nicotine. (1-methyl-2-beta-pyridylpyrrolidine)	$C_{10}H_{14}N_2$....		246 (730 mm.)
34. Novocaine (diethylaminoethyl ester of para-aminobenzoic acid hydrochloride)	$C_{13}H_{20}N_2O_2 \cdot HCl$.	156	
35. Papaverine....	$C_{20}H_{21}NO_4$....	147	
36. Paraconiine....	$C_8H_{15}N$....		169
37. Pilocarpine....	$C_{11}H_{16}N_2O_2$....	34	
38. Piperine (hydrolyzes to piperidine plus piperic acid)....	$C_{17}H_{19}NO_3$....	129	
39. Protopine....	$C_{20}H_{19}NO_5$....	208	
40. Pseudoaconitine....	$C_{36}H_{51}NO_{12}$....	211	
41. Pseudoephedrine....	$C_{10}H_{15}NO$....	116	
42. Pseudomorphine....	$C_{34}H_{36}N_2O_6 \cdot 3H_2O$....	327 decom.	
43. Pseudotropine....	$C_8H_{15}NO$....	108	
44. Quinine....	$C_{20}H_{24}N_2O_2$....	175 anhyd.	
45. Solanine....	$C_{44}H_{71}NO_{15}$....	244–250	
46. Strychnine....	$C_{21}H_{22}N_2O_2$....	286	270 (5 mm.)
47. Thebaine (paramorphine)....	$C_{19}H_{21}NO_3$....	193	
*48. Thebaine, iso....	$C_{19}H_{21}NO_3$....	203	
*49. Theobromine....	$C_7H_8N_4O_2$....	290 subl.	
50. Theophylline....	$C_7H_8N_4O_2 \cdot H_2O$....	269–272	
51. Thyroxine.... (present in the hormone of the thyroid gland)	HO〈 〉$-O-$〈 〉$CH_2 \cdot CHNH_2 \cdot COOH$ (with I substituents)		
52. Tropacocaine....	$C_{15}H_{19}NO_2$....	49	
53. Tropinone (tropanone)....	$C_8H_{13}NO$	41	224
54. Tropine ("tropanol")....	$C_8H_{15}NO$.	63	
*55. Veratridine....	$C_{36}H_{51}NO_{11}$....	180	
56. Xanthine....	$C_5H_4N_4O_2 \cdot H_2O$....	150 decom.	

(R.K.S.)

* See **Purine and Uric Acid Compounds.**

ALLIGATOR PEAR. Avocado.

ALLIUM. Liliaceae. A large genus whose species are found widely. Some 75 species are found in North America, especially in the western states. All are bulbous plants with flat or tubular leaves, and with spherical heads or umbels of variously colored flowers. Particularly important cultivated species are the onion, *Allium Cepa;* leek, *Allium Porrum;* garlic, *Allium ursinum;* and chives, *Allium Schoenoprasum.* One European species now extensively introduced in the United States is the field garlic, *Allium vineale,* which (if eaten by cows) noticeably flavors milk and butter. (R.M.W.)

ALLOCHTHONOUS. A term proposed by Gümbel in 1888 for **sedimentary** rocks whose constituents have been transported and deposited at some distance from their place of origin. The bulk of the sedimentary rocks are of this type. The term is now used most commonly in reference to masses of rock transported considerable distances by tectonic movements. (See **Overthrust.**) (R.M.F.)

ALLOPALLADIUM. Iridosmine.

ALL-OR-NONE LAW. A principle of reaction in living matter under which a structure responds to a stimulus to the maximum degree possible in its existing physiological state, regardless of the strength of the inciting stimulus. Thus any stimulus capable of exciting a nerve cell at a given moment will arouse the same degree of activity in that cell. First demonstrated by Bowditch for heart muscle, this principle has since been found applicable to single neurons and single muscle fibers when stimuli similar to those occurring in nature are applied. (A.W.L.)

ALLOTRIOMORPHIC. A term proposed by Rosenbusch in 1887 for minerals in **igneous** rocks which are not bounded by their typical **crystal** faces. Such minerals are said to be anhedral. (R.M.F.)

ALLOTROPES. Chemical Composition.

ALLOTYPE. An animal or plant fossil selected as a species or sub-species as illustrating morphological details not shown in the **holotype.** (R.M.F.)

ALLOWANCE. Interchangeable Manufacturing, Fit.

ALLOYS. Metals and Alloys.

ALLSPICE. *Pimenta officinalis.* Myrtaceae. Allspice is the dried fruit of a small tree native in the West Indies and Central America. The tree grows to a height of 30 or 40', has leathery leaves, fragrant with the oil they contain, and small white flowers borne in axillary cymes (see **Flower**). The fruits contain one or two seeds. Before they are ripe the fruits are gathered, rapidly dried and marketed under the name allspice. This name was given because early users thought the flavors of **cinnamon, cloves** and **nutmeg** were all found in this one spice. It is used for flavoring cakes, puddings and pies. (R.M.W.)

ALLUVIAL FAN. Also termed subaerial delta. Cone-shaped to delta-shaped deposits of coarsely graded **clastic** sediments deposited by intermittent streams that

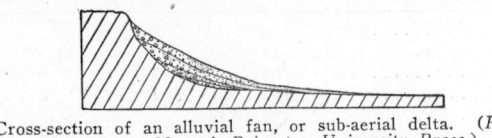

Cross-section of an alluvial fan, or sub-aerial delta. (*Field, Laboratory Manual, Princeton University Press.*)

debouch from steep valleys onto a relatively gentle slope or plain. Alluvial fans may extend for many miles, and confluent fans may eventually cover and fill relatively large intermontane basins. (R.M.F.)

ALLUVIUM. A general term used to designate the sand, silt and mud deposited by a stream, along its banks or upon its **floodplain,** during periods of high water. The word is derived from the Latin *ad,* to; and *luo,* wash. When alluvium is relatively fine-textured and contains sufficient organic matter it forms **soil.** Some of the oldest and richest agricultural regions are the great **delta** areas, such as the Nile, Euphrates, etc. (E.S.C.S.)

ALMAGEST. This is the name assigned by the Arabs to the great treatise on science compiled by **Ptolemy** during the 2nd century. The very name Almagest, which is a hybrid combination of the Greek superlative (μεγιστη) with the Arabic article (al), indicates the importance of this work to the early astronomers.

The Almagest is a collection of treatises on a variety of scientific subjects. In it is to be found the complete exposition of the Ptolemaic system for the structure of the universe. Perhaps the best known section of the Almagest is that dealing with the stars and the **constellations.** This section was taken from the works of Hipparchus and incorporated in the Almagest by Ptolemy with some improvements and additions. In this catalogue we first find the brightnesses of the stars divided into six **magnitudes,** a system which has persisted down to modern times. The positions of the stars given in the Almagest have proved of some little value in determining the constants of **precession** and also the **proper motions** of the stars. (W.K.G.)

ALMANAC. For the work of every person engaged in astronomy, whether as an astronomer in an observatory, a navigator on a ship at sea, or in the air, or a suveyor in the field, tables of certain astronomical data are indispensable. Many, in fact most, of these tables change from year to year. Among such material may be listed: the positions of the **sun, moon,** and **planets** for every day in the year, accurate positions of **stars** to be used for determination of local **time,** tables for computing **precession,** nutation, **aberration,** etc. Such material is published in almanacs which are computed and published several years in advance so that ships going off on long voyages can have the data at hand when they leave port.

In addition to the ephemerides and data listed above, almanacs also contain descriptions of such phenomena as **eclipses** of the sun and moon, **occultations** of stars by the moon, eclipses and configurations of the **satellites** of **Jupiter,** etc.

At the present time the computation of the material for the ephemerides of the different governments is a cooperative plan. The nautical almanac offices of the United States, Great Britain, France, Germany, and Spain do a share of the work. An examination of the preface for the American Ephemeris and Nautical Almanac for any year will show how the work for that particular year was distributed. (W.K.G.)

ALMANDITE. Garnet.

ALMOND. *Prunus Amygdalus.* Rosaceae. A medium-sized tree with pale pink or white flowers, probably native in western Asia and northern Africa. The fruit, a drupe (see **Fruit**), has the seed or kernel enclosed within a reticulated endocarp (see **Fruit**).

There are two kinds of almonds, bitter and sweet. Bitter almonds, used for flavoring, contain a high percentage of **hydrocyanic acid.** The sweet almond yields almond oil, and is used as a dessert and for confections. Almonds are grown incidentally in Mediterranean Europe, and in this country extensively in California. (R.M.W.)

ALOË. Liliaceae. A large genus of plants characteristic of drier parts of Africa, especially the southern part. Because of their ornamental appearance, with stiff habit and spiny-margined leaves, many of them are grown in cultivation. The rather small yellow or red flowers are born in large masses. Many species yield from the crushed leaves a purgative juice, which is called aloes, and which has been used extensively by eastern people. (R.M.W.)

ALOPECIA. Baldness—abnormal or natural loss of hair. It may be partial or complete, transient or permanent. It may be the natural accompaniment of old age or an indication of a toxic process in certain diseases. It can affect all body hair. (R.S.M.)

ALPACA. Mammalia, Artiodactyla. A South American domestic animal of the **camel** family, probably derived from the wild species known as the **guanaco,** *Lama huanaco.* It is the source of a long wool of fine quality and its flesh is excellent. Alpacas are kept at high altitudes in Bolivia and Peru. (A.W.L.)

ALPHA PARTICLE. Among the so-called nuclear particles, the **proton** and the alpha particle enjoy the distinction of not only appearing upon the disintegration of atomic nuclei (whether they had individual existence within the undivided nucleus or not), but of serving, by themselves, as the nuclei of certain atoms. Just as the atom of ordinary **hydrogen** (H^1) has the proton for its nucleus, so the nucleus of the **helium** atom is identical with the alpha particle, first recognized as composing the alpha rays emitted by many radioactive substances (see **Radioactive Changes**).

The mass of the alpha particle is very close to 6.644×10^{-24} g., and it carries a positive charge whose magnitude is exactly twice the elementary (electronic) charge (see **Electron**). This latter point, coupled with the fact that the mass is approximately four times that of the proton, leads one to question whether the alpha particle is a fundamental entity or is, like other atomic nuclei, made up of simpler structural units.

When alpha particles are given very high speed, either in natural radioactive emission or by artificial means, they prove to be extraordinarily effective in producing ionization in gases. (L.D.W.)

ALPHERATZ. Alpheratz (α **Andromedae**) is a star which was formerly allotted to the **constellation** of **Pegasus** by the Arabs. It is situated at the northeast corner of the great square of Pegasus. In **astrology,** Alpheratz portends honor and riches to all born under its influence. The star is a **spectroscopic binary** with a period of approximately 100 days. (W.K.G.)

ALSTONITE. Bromlite.

ALTAIR. Altair (α **Aquilae**) forms with β and γ of the same **constellation** the well-known line of stars which is a conspicuous feature of the early autumn sky and is sometimes referred to as the shaft of Aquila. This star was ill-omened in **astrology,** portending danger from reptiles. (W.K.G.)

ALTAZIMUTH. The altazimuth is the earliest type of mounting for astronomical telescopes. It is an instrument so mounted that it may be rotated about a horizontal and a vertical axis (i.e., rotated in **altitude** and **azimuth**). Perhaps the most familiar altazimuth instrument is the ordinary surveyor's transit or theodolite.

The great advantage of this type of instrument is the ease with which it may be set up. If the instrument has been properly constructed, the horizontal and vertical

axes will be strictly perpendicular to each other, and all that is necessary to adjust the instrument for use is to level the horizontal axis of the instrument for all azimuths.

The altazimuth instrument is used in the field for laying down azimuth lines and for determination of **latitude** and **longitude** by measuring altitudes of celestial objects. A few large, fixed altazimuth instruments are in use in observatories for accurate determination of **declinations** of stars but for this purpose the **meridian circle** is most commonly used.

For ordinary astronomical observing the altazimuth instrument is less convenient than the **equatorial** because of the fact that the diurnal motion of the celestial sphere is parallel to the equator rather than the horizon, with the result that the instrument has to be moved about both axes to follow the celestial objects. (W.K.G.)

ALTERNATING CURRENTS. Currents in which the electricity moves periodically back and forth. The usual types of generator are so constructed that the electromotive forces induced in the armature conductors are periodically reversed, and unless the machine is provided with a commutator or other type of **rectifier,** this alternating voltage will be impressed upon the external circuit, giving rise to an alternating current. (See **Electric Currents** and **Electric Circuits.**)

The electromotive force may be represented by an equation of the type

$$E = E_0 \cos 2\pi nt, \qquad (1)$$

in which E_0 is the maximum value of the e.m.f. ($t = 0$) and n is the frequency. (Some writers use the sine instead of the cosine in Eq. (1).) The resulting current follows a similar law, though it is in general out of phase with the e.m.f.

When a harmonic voltage such as that represented by Eq. (1) is impressed upon a circuit, the current is in general not derivable from the e.m.f. by a simple application of Ohm's law, but depends upon several factors. In the general case, account must be taken not only of the resistance R (ohms) of the circuit, but also of its **inductance** L (henrys), and its **capacitance** C (farads). (The circuit may include a series condenser, or may have enough **distributed capacitance** to have similar effect.) The current at any instant t in a series circuit is then given by the equation

$$I = \frac{E_0}{\sqrt{R^2 + \left(2\pi nL - \frac{1}{2\pi nC}\right)^2}} \cos(2\pi nt - \phi). \qquad (2)$$

Here ϕ is the "phase angle," the angular amount by which the phase of the e.m.f. exceeds that of the current. Its value is given by

$$\tan \phi = \frac{4\pi^2 n^2 LC - 1}{2\pi nRC}. \qquad (3)$$

ϕ reduces to zero if $L = 0$ and C is infinite (a noninductive, no-condenser circuit), or if the circuit and the frequency are so adjusted that $C = \frac{1}{4}\pi^2 n^2 L$; in which latter case the circuit is in **resonance** with the e.m.f. (as is a tuned radio circuit). Under either of these conditions also, the current is related to the e.m.f. by Ohm's law, that is, $I = E/R$. Large inductances tend to make ϕ positive (current lags behind voltage); small capacitances tend to make it negative.

The radical in Eq. (2), which takes the place of R in Ohm's law and which equals R in the case of a resonant or a non-inductive, no-condenser circuit, is called the impedance of the circuit, while the parenthesis containing L and C represents the reactance.

The currents in circuits having other than simple series arrangement of resistances, inductances, and capacitances are given by appropriate formulas depending upon the arrangement and derived from the laws of networks. (See **Kirchhoff's Laws.**)

Since the power at any instant in a circuit of resistance R is E^2/R, the average power over a complete period is proportional to the average value of E^2. This may be denoted by E_v^2, in which E_v is the effective or virtual e.m.f., equal to $E_0/\sqrt{2} = 0.707\ E_0$ and to 1.11 times the average e.m.f. Similar relations hold for effective, average, and maximum current. The average power is not in general the product of the effective voltage and the effective current, $\frac{1}{2}E_0 I_0$ (the apparent power), as it would be in a non-reactive circuit, but is equal to

$$P = \tfrac{1}{2}E_0 I_0\ \cos\phi, \qquad (4)$$

in which $\cos\phi$ is called the "power factor." The actual power is given by Eq. (4) in **watts**, as usual, while the apparent power is rated in "volt amperes" (effective volts $\times$ effective amperes). Many principles of ordinary alternating currents apply also to radio circuits of moderate frequency. (See **Radio Commuication**.) See also **Polyphase Currents** and **Transients**. (L.D.W.)

ALTERNATING-CURRENT CIRCUITS.

The alternating-current circuit is that type of circuit which carries electrical current which rapidly reverses its direction of flow. A circuit is any series of electrical connections which give a continuous path through the various component parts and back to the starting point. Most electrical circuits are now a.c., although at first d.c. was a formidable contender. A.c. has received great impetus because of the simple way in which it can be changed in voltage by a **transformer**. This is a great advantage in transmitting electrical energy over long distances, since the cost of the line decreases and the efficiency increases for a given power transmission when the voltage is increased. The advantages of the **induction motor** are also a point in favor of a.c. It is not, however, without defect since its use brings into prominence **inductance** and **capacitance**, factors which appear in d-c calculations only during switching and are usually of no importance in **direct-current circuits**.

The essence of a-c voltage can be simply pictured by a curve showing the relation of the voltage value to time. Such a curve is shown in Fig. 1a. The points on the curve above the time axis represent the voltage in one direction across the circuit and those below the line the voltage in the other direction. An ideal a-c generator would produce an output which would be represented by Fig. 1a; but **harmonics** are frequently present and modify the wave form of the output. Fig. 1b shows, for a

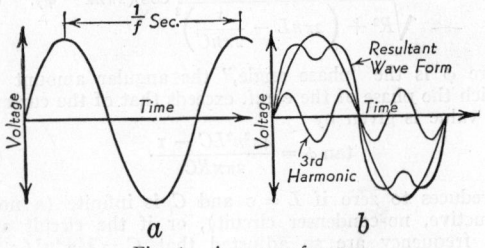

Fig. 1. A-c wave forms.

much exaggerated value, the effect on the wave of the third harmonic. The current would be distorted in a similar manner, the exact amount, however, depending on the amount of **reactance** in the circuit.

Since the value of the a.c. is continually changing it is necessary to define some sort of average value to use in calculations, measurements, etc. The unit is the ampere and the effective value of an a.c. is defined so that the same number of amperes of d.c. would produce the same heating effect in a resistance. This effective value is called the *r.m.s.* (**root mean square**) value and for a pure sine wave of current is the maximum value divided by $\sqrt{2}$. Voltage is similarly measured in r.m.s. volts (maximum value divided by $\sqrt{2}$). Unless otherwise specified a-c and voltage values are always

given in these r.m.s. quantities. The frequency (f) of an a.c. is the number of complete cycles per second (see Fig. 1a).

The sine wave of Fig. 1a could be formed by plotting against time the projection upon a vertical axis of the end of a **vector** rotating at $2\pi f$ radians per sec. Since any sine wave can be obtained in this manner it is customary to represent the a-c and voltage waves by the vectors which would be used to get the plot. This greatly simplifies the calculations involving the alternating quantities. In a purely resistive circuit the current and voltage are in phase, that is, the peaks of the current occur simultaneously with those of the voltage. The vectors representing these would coincide in direction as shown in Fig. 2. The current lags the voltage in time sequence if the circuit has a net inductive reactance and leads if the net reactance is capacitive. This means that peaks of the current wave would occur after or before those of the voltage. These would be represented by drawing the vectors for current and voltage displaced as shown in Fig. 2. The angle θ is the angle

Fig. 2. Voltage-current relations in a-c circuits.

lag or lead and is called the power factor angle. The component of the current which is actually in phase with the voltage multiplied by the voltage gives the **power**, while the component 90° to the voltage gives the "wattless" or reactive power. It will be noted that the smaller the angle θ the less the reactive and the more the active power. It will also be noted that inductance and capacitance displace the current in opposite directions so by proper proportioning of them in certain circuits their effects will cancel. Electrical power is always the voltage times the current component in phase with it. As seen in Fig. 2 this current component is $I\cos\theta$, and hence the power is $EI\cos\theta$ where E and I are expressed in r.m.s. volts and amperes. The $\cos\theta$ is an important quantity in a-c circuits and is called the power factor.

Series and parallel arrangements of a-c circuits are more difficult to solve than those of d-c circuits as they usually contain inductance and capacitance, necessitating the use of reactance as well as resistance in obtaining the electrical characteristics of the circuit. The basic equations and quantities for the solution of a-c series and parallel circuits are given below:

$$E = IZ$$

For a series circuit:

$$Z = \sqrt{R^2 + X^2}$$
$$X = X_1 + X_2 + X_3 \cdots$$
$$R = R_1 + R_2 + R_3 \cdots$$

where X_1, X_2, etc., are $2\pi fL$ for the inductive reactance and $-\dfrac{1}{2\pi fC}$ for capacitive reactance, L being the inductance in henries and C the capacitance in farads; Z is the **impedance** and R is the **resistance**.

For a parallel circuit:

$$Z = \frac{1}{Y_0}$$

where Y_0 is the total **admittance**.

While a.c. supplied to small consumers is usually single-phase, *alternators* are commonly wound for 3 phases. Sometimes 2 or 4 phases are used for distribution, and frequently 6, 12 or more phases are obtained by special transformer connections for supplying **rectifiers**. The single-phase circuit is obtained from a polyphase one by using only one phase of the system. Polyphase power has certain advantages over a single phase:

1. Transmission is more economical.
2. Equipment is smaller, less complicated and generally more efficient.
3. Power fluctuations are less. This is analogous to the difference between a single-cylinder and multi-cylinder automobile engine.

Three-phase is the most common system in use. The power is $\sqrt{3}$ times the power of a single-phase circuit having the same line voltage and current. (F.T.M., L.R.Q.)

ALTERNATING-CURRENT MOTOR. Motors.

ALTERNATION OF GENERATIONS. In the life history of many **Thallophytes** and of all plants in the divisions above the Thallophytes, there are two distinct phases in the life cycle which regularly alternate. In the **algae,** the alternating individuals are frequently indistinguishable until fruiting occurs, when it becomes apparent that one plant produces asexual **zoospores** which grow directly to form new plants, while the other plant (of the alternate generation) produces **gametes,** or sexual **cells.** These gametes fuse in pairs before growing to form new plants. Since two cells, and also their nuclei, fuse, it is obvious that there is a doubling of the nuclear substance. The plant having this double nuclear nature is called diploid, while the other generation is called haploid. In a mature diploid plant, there occurs at the time of spore formation a special type of division, **meiosis,** or reduction division, in which the double nuclear condition is reduced. While many algae have the two generations of similar appearance, others, notably the brown algae known as kelps, show a striking dissimilarity. The asexual plants are the familiar large brown seaweeds so frequently cast up on our coasts, while the sexual plants are minute and rarely seen. In kelps, and many other plants, these sexual plants are of two sorts, one, the male, producing minute biciliate sperms, the other, the female, forming oögonia with which the sperms unite. The alternate generations in the higher plants are separately described under each division (**Bryophytes, Pteridophytes,** and **Spermatophytes).** Among animals, alternation of generations is well marked in the **hydrozoan coelenterates,** where the **polyp** of many species is an asexual form capable only of securing nourishment, defending the colony, and producing new individuals by budding or some similar asexual process. The **medusa** is a sexual form produced in this manner. It gives rise to **germ cells** and through its sexual reproduction new individuals are formed which constitute another asexual generation. (R.M.W., A.W.L.)

ALTERNATOR. An electromotive force is generated in a conductor when it is moved so as to cut the lines of force between the poles of a magnet. The elementary principle of a simple two-pole, single-phase alternator is shown in Fig. 1. When the magnet revolves it will

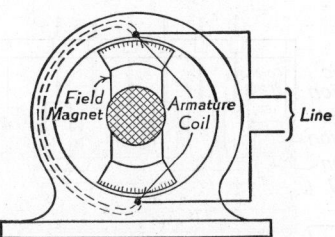

Fig. 1. Elementary alternator.

carry with it lines of force which will cut the conductor, which is a wire loop embedded in the stationary portion called the armature, and will generate a.c.

This elementary principle must be expanded in several

directions if a practical generator of a.c. is to be had. First, the rotating part, or rotor, must have magnetic strength in excess of that which could be obtained from a simple permanent magnet. In other words, the poles must be formed by electromagnets whose excitation, in the form of d.c., must be carried to the rotor through slip-ring connections. The rotor is called the field, and the current it uses is called the field current. In high-speed steam turbine-driven alternators as few as two poles are often used, while in slow-speed water-turbine units the number is frequently nearly 100. The stationary part, called the stator, or armature, usually has three sets of overlapping coils, connected in three separate circuits. These three circuits, or phases, are usually connected in one or the other ways shown in Fig. 2.

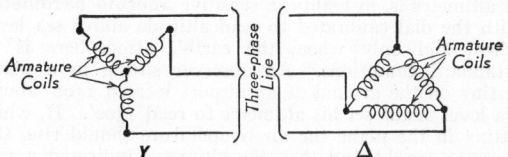

Fig. 2. Comparison of Y and Δ connections of three-phase alternator windings.

The Y connection is preferred because of the usefulness of the neutral point, and the fact that the line voltage is $\sqrt{3}$ times the phase voltage, whereas it is only equal to the phase voltage in Δ connection. The neutral point is connected to the fourth wire of a four-wire, three-phase system, and left unconnected, or grounded, in the three-wire system. Several advantages are realized by making the rotating part the **field,** and the stationary part the **armature.** The a.c. may be generated at very high voltages because it is not necessary to connect it through movable contacts as would be the case if the armature revolved. It is not necessary to conduct high-load currents through slip rings and **brushes** if the armature is fixed. The armature conductors can be very rigidly braced in position, and may be much better disposed than if they were required to be in the rotor.

Engine and hydraulic turbine-driven alternators are in the slow-speed class, and are characterized by large diameter, short length, and many poles. The steam turbine-driven alternator is a high-speed machine having a length larger than its diameter. Standard speeds of turbine-driven alternators range from 1200 to 3600 rpm, with 1800 rpm very common practice.

Basically, the alternator is a device for converting mechanical into electrical energy. While it is able to do this with a high degree of efficiency, it does suffer the following losses:

1. Friction and windage from bearings, brushes, and fan action of the rotor.
2. Core loss, which is the result of **eddy currents** and **hysteresis** in the iron **core.**
3. Resistance heating loss in the armature and field conductors.
4. Resistance loss of the field **rheostat.**
5. Exciter loss.
6. Ventilation loss.

Practically all generator losses appear as heat in and about the windings, and to maintain these at a safe working temperature, a cooling medium must be employed. Air has been the medium generally used. The rotor may or may not be able to produce its own fan action, depending on the size, speed, and construction of the rotor. Ventilating air frequently has to be brought through a duct and discharged through the alternator. Hydrogen is rapidly becoming a common cooling medium as it is much more effective and produces less windage loss. For its use the generator must be totally enclosed and gas-tight. To supply the d.c. for the field, a source of d.c. at 110–250 volts is necessary. This is delivered from a small d-c generator called the exciter. The exciter may be driven from an extension of the alternator shaft, or it may be driven by some independent means, such as motor or engine.

When alternators are operated in **parallel,** the division of load between them is not accomplished by changing the generated voltage, as in d-c generators, but by changing power input from the prime mover through the adjustment of the engine or turbine **governor.** Before being paralleled, two alternators must have the same phase sequence and frequency. Their voltages must be equal, and in phase; that is, with the peaks of wave forms coincident in time and direction. (F.T.M.)

ALTIMETER. An instrument used by aviators to determine the height of the plane above sea level is known as an altimeter. The most commonly used type of altimeter is, in reality, a sensitive **aneroid barometer** with the dial calibrated to read altitude above sea level when, and only when, the earth's atmosphere is in "standard conditions." An observer sitting in a plane resting on the ground at an airport located 1500′ above sea level might set his altimeter to read 1500′. If, while sitting in the plane the air temperature should rise, the observer would find that the altimeter indicated a rise of the plane; if the barometric pressure should fall, the altimeter would again indicate a rise of the plane.

On most altimeters there are at least two adjustments which must be made before take-off, and then checked constantly while in flight. The first is a so-called "altimeter setting" which corrects for the difference between the actual barometric pressure (reduced to sea level at standard temperature for the region in which the plane is operating) and the "standard barometric pressure" for the altitude of that region. Before take-off the aviator obtains the local barometric pressure, reduced to sea level and standard temperature, sets this figure on a scale visible on the face of the altimeter. He now makes the second adjustment by rotating the dial of his instrument until the indicator points to the altitude of the airport. As he climbs, the altimeter will register the so-called "pressure altitude" above sea level. This pressure altitude will be his actual altitude if, and only if, the decrease of temperature with altitude is that of the normal **lapse rate.** To obtain the actual altitude in case the lapse-rate temperature is not normal, involves a computation by means of an equation containing the observed pressure altitude and the actual temperature gradient from the ground. This computation may be made with an accuracy sufficient for most purposes by means of a scale and slide provided on most **navigational computers.** If the aviator now proceeds cross-country, the reading of the altimeter will give the pressure altitude if, and only if, the barometric pressure, reduced to sea level and standard temperature at the ground immediately below him, is the same as that at the locality where the first altimeter setting was made. The Weather Bureau and other agencies provide facilities for reporting the barometric pressure, reduced to sea level at standard temperature, at numerous stations. A careful aviator will obtain this altimeter setting at frequent intervals during a flight and alter the reading of the altimeter-setting scale of his own instrument whenever necessary. It should be remembered that the pressure-type altimeter gives the altitude above sea level after proper correction, not the altitude of the plane above the ground immediately below it. (See **Altitude; Instruments, Aviation.**)   (W.K.G.)

ALTITUDE. The term altitude is used as synonymous for height, or distance above the surface of the earth. The **altimeter** is used in aviation to measure the altitude of a plane.

In astronomy and **navigation** the altitude of a celestial object is that coordinate that is measured in the **horizontal system** of spherical coordinates that is measured in the plane of the vertical circle through the object from the **horizon** to the object. Altitude is probably the most frequently

measured of all celestial coordinates since it is universally used in **celestial navigation** to find **lines of position** for the determination of the location of a ship at sea, or a plane in the air. Altitude is also measured by geodetic surveyors for accurate determination of **latitude** and is used by astronomers for determination of the **declinations** of the stars.

At fixed stations, such as those established by the Coast and Geodetic Survey, an **altazimuth** instrument is used for measuring altitude. Navigators use the **sextant** for measuring altitude.

All observations for altitude must be corrected for atmospheric **refraction** and for **parallax.** Altitudes measured with the portable instruments must have additional corrections applied as discussed in the article on the **sextant.**   (W.K.G.)

ALTOCUMULUS.   Clouds.

ALTOSTRATUS.   Clouds.

ALUMINUM.   Symbol: Al. Atomic number: 13. Atomic weight: 26.97. Density: 2.70. Hardness: 3. Melting point: 660° C. Boiling point: 1800° C. No isotope, but of single atomic form: 27.

Aluminum is a silver-white metal, with a bluish tinge, capable of taking a high polish, ductile and malleable; a thin, protective, transparent film of oxide is formed upon exposure to air; upon heating to 580° C. in **oxygen,** burns with intense heat and light; soluble in **hydrochloric acid,** in **sulfuric acid** (of strength above 10%), and in concentrated or very dilute nitric acid, but made passive by other concentrations; insoluble in organic acids at room temperature; soluble in **sodium** hydroxide solution, forming sodium aluminate solution and **hydrogen** gas; reacts with dry **chlorine** upon heating to form aluminum chloride anhydrous. Isolated by Wöhler in 1827.

Aluminum occurs abundantly in all ordinary rocks, except **limestone** and **sandstone;** is third in abundance of the elements in the earth's crust (8.1% of the solid crust), exceeded only by **oxygen** and **silicon,** with which two elements aluminum is generally found combined in nature; present in igneous rocks and clays as aluminosilicates; in the mineral **cryolite** in Greenland as sodium aluminum fluoride (Na_3AlF_6); in the minerals **corundum** and **emery,** the gems ruby and sapphire, as aluminum oxide (Al_2O_3); in the mineral bauxite in Southern France, Hungary, Yugoslavia, Greece, the Guianas of South America, Arkansas, Georgia and Alabama of the United States, Italy, U.S.S.R., and Netherlands Indies as hydrated oxide ($Al_2O(OH)_4$); in the mineral **alunite** or alum stone in Utah as aluminum potassium sulfate ($Al_2(SO_4)_3 \cdot K_2SO_4 \cdot 4Al(OH)_3$).

Bauxite, the commercial source of aluminum metal, is treated to obtain pure aluminum oxide, and the oxide is electrolyzed from solution in fused cryolite. This method of producing aluminum metal was discovered independently by Hall in the United States and by Héroult in

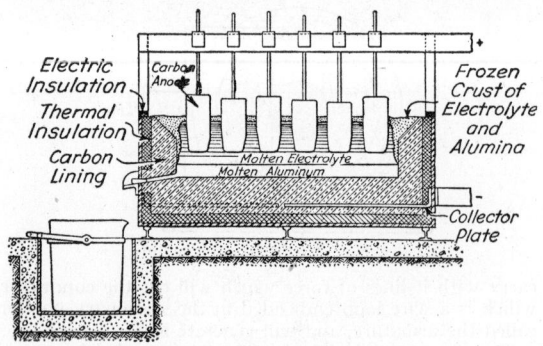

Hall process for aluminum manufacture.

France in 1886, and has been the sole method since that time. In present-day production and consumption of metals, aluminum stands fifth, and has shown a quantity increase in recent years greater than any other metal.

WORLD PRODUCTION

Year	Short Tons
1900	7,000
1910	48,000
1920	141,000
1930	293,000
1940	885,000

RECENT PRODUCTION OF LEADING METALS IN U.S.A.

Metal	1940	1941
Steel	134×10^9 lb.	148×10^9 lb.
Copper	3.69×10^9 lb.*	4.24×10^9 lb.†
Zinc	1.45×10^9 lb.‡	1.76×10^9 lb.‡
Lead	1.07×10^9 lb.	1.14×10^9 lb.
Aluminum	0.41×10^9 lb.	0.62×10^9 lb.
Ratio of steel to aluminum	330 to 1	240 to 1

* Primary copper 72% of this.
† Primary copper 66% of this.
‡ Primary zinc 93% of this.

Aluminum is largely used in articles of commerce and technology (1) where there is demanded lightness and strength, e.g., airplane, automobile, rail car, either pure or in alloy with silicon and another metal such as magnesium, (2) as electrical **conductor**, in competition with copper, (3) due to ease of workability and resistance to wear and corrosion, as cooking utensils, large and small containers, chemical apparatus, (4) for the production of local, very high temperature, or reduction of difficulty reducible oxides, e.g., aluminum plus **iron** oxide to form aluminum oxide glass, plus iron, and for iron oxide there may be substituted many other oxides, to produce the metal corresponding to the oxide used, (5) for paint, when finely powdered, (6) as a protective coating on steel, (7) in alloys, principally with copper, silicon, magnesium, manganese, nickel, zinc (**Aluminum Alloys**).

Aluminum compounds are generally made starting with bauxite, which is reactive with acids and with bases. With acids, e.g., sulfuric acid, any iron contained in the bauxite is dissolved along with the aluminum and silicon is left in the residue, whereas with bases, e.g., sodium hydroxide, any silicon is dissolved and iron left in the residue.

Acetate: Aluminum acetate ($Al(C_2H_3O_2)_3$), white crystals, soluble, by reaction of aluminum hydroxide and **acetic acid** and then crystallizing. Used (1) as a mordant in dyeing and printing textiles, (2) in the manufacture of lakes, (3) for fireproofing fabrics, (4) for waterproofing cloth.

Alums: Aluminum potassium sulfate, "alum" ($K_2SO_4 \cdot Al_2(SO_4)_3 \cdot 24H_2O$), white crystals, soluble, by crystallizing a solution of aluminum sulfate and **potassium** sulfate; other alums may be prepared, (1) by substituting for potassium sulfate, **sodium**, or **ammonium** sulfate, (2) by substituting for aluminum sulfate, **chromium** or **ferric** sulfate. The alums, therefore, are mixed salts of the type shown above, and do not necessarily contain aluminum; anhydrous aluminum potassium sulfate,

"burnt alum" ($K_2SO_4 \cdot Al_2(SO_4)_3$), is made by heating alum until water is removed.

Aluminates: Sodium aluminate ($NaAlO_2$), white solid, soluble, (1) by reaction of aluminum hydroxide and sodium hydroxide solution, (2) by fusion of aluminum oxide and sodium carbonate; the solution of sodium aluminate is reactive with carbon dioxide to form aluminum hydroxide. Used as a mordant in the textile industry, in the manufacture of artificial zeolites, and in the hardening of building stones. See silicates below and **calcium** aluminates.

Alundum: See oxide (below).

Carbide: Aluminum carbide (Al_4C_3), yellowish-green solid, by reaction of aluminum oxide and **carbon** in the **electric furnace**, reacts with water to yield **methane** gas and aluminum hydroxide.

Chlorides: Aluminum chloride ($AlCl_3 \cdot 6H_2O$), white crystals, soluble, by reaction of aluminum hydroxide and **hydrochloric acid**, and then crystallizing; anhydrous aluminum chloride ($AlCl_3$), white powder, fumes in air, formed by reaction of dry aluminum oxide plus carbon heated with chlorine in a furnace, used as a reagent in **petroleum** refining and other organic reactions.

Fluoride: Aluminum fluoride (AlF_3), white solid, soluble, by reaction of aluminum hydroxide plus **hydrofluoric acid** and then crystallizing ($2AlF_3 \cdot 7H_2O$), used in glass and porcelain ware.

Hydroxide: Aluminum hydroxide ($Al(OH)_3$), white gelatinous precipitate, by reaction of soluble aluminum salt solution and an alkali hydroxide, carbonate or sulfide (sodium aluminate is formed with excess sodium hydroxide but no reaction with excess ammonium hydroxide), upon heating aluminum hydroxide the residue formed is aluminum oxide. Used as intermediate substance in transforming bauxite into pure aluminum oxide.

Nitrate: Aluminum nitrate ($Al(NO_3)_3$), white crystals, soluble, by reaction of aluminum hydroxide and **nitric acid**, and then crystallizing.

Oleate: Aluminum oleate ($Al(C_{18}H_{33}O_2)_3$), yellowish-white powder, by reaction of aluminum hydroxide, suspended in hot water, shaken with **oleic acid**, and then drying, the product is used (1) as a thickener for lubricating oils, (2) as a drier for paints and varnishes, (3) in waterproofing textiles, paper, leather.

Oxide: Aluminum oxide, alumina (Al_2O_3), white solid, insoluble, melting point 2020° C., formed by heating aluminum hydroxide to decomposition; when bauxite is fused in the electric furnace and then cooled there results a very hard glass ("alundum"), used as an abrasive (hardness 9 Mohs scale) and heat refractory material. Aluminum oxide is the only oxide which reacts both in water medium and at fusion temperature, to form salts with both acids and alkalis.

Palmitate: Aluminum palmitate ($Al(C_{16}H_{31}O_2)_3$), yellowish-white powder, by reaction of aluminum hydroxide, suspended in hot water, shaken with palmitic acid, and then drying, the product is used (1) as a thickener for lubricating oils, (2) as a drier for paints and varnishes, (3) in waterproofing textiles, paper, leather, (4) as a gloss for paper.

Silicates: Many complex aluminosilicates or silicoaluminates are found in nature. Of these, clay in more or less pure form, pure clay, kaolinite, kaolin, china clay ($H_4Si_2Al_2O_9$ or $Al_2O_3 \cdot 2SiO_2 \cdot 2H_2O$) is of great importance. Clay is formed by the weathering of igneous rocks, and is used in the manufacture of bricks, pottery, porcelain and Portland cement. (See **calcium aluminosilicates; Ceramics; Cement, Portland.**)

Stearate: Aluminum stearate ($Al(C_{18}H_{35}O_2)_3$), yellowish-white powder, by reaction of aluminum hydroxide suspended in hot water, shaken with **stearic acid**, and then drying the product, used (1) as a thickener for lubricating oils, (2) as a drier for paints and varnishes, (3) in waterproofing textiles, paper, leather, (4) as a gloss for paper.

Sulfate: Aluminum sulfate ($Al_2(SO_4)_3$), white solid,

soluble, by reaction of aluminum hydroxide and **sulfuric acid,** and then crystallizing, used (1) as a clarifying agent in water purification, (2) in baking powders, (3) as a mordant in dyeing, (4) in sizing paper, (5) as a precipitating agent in sewage disposal; aluminum potassium sulfate, see **alums** and **alunite.**

Sulfide: Aluminum sulfide (Al_2S_3), white to grayish-black solid, reactive with water to form aluminum hydroxide and **hydrogen sulfide,** formed by heating aluminum powder and sulfur to a high temperature.

Aluminum in solution of its salts is detected by the reaction (1) with ammonium salt of aurin tricarboxylic acid ("aluminon"), which yields a red precipitate persisting in ammonium hydroxide solution, (2) with alizarin red S, which yields a bright red precipitate persisting in acetic acid solution. (R.K.S.)

ALUMINUM ALLOYS. The aluminum alloys, like those of other commercial metals, can be divided into two broad classes: (1) wrought alloys and (2) casting alloys. The wrought-alloy products include sheet, plate, foil, tubing, rods, bars, wire, structural and special shapes, rivets, forgings, screw machine products, paste and powder for paint, and impact extrusions, and the casting-alloy products comprise sand castings, permanent mold castings and die castings. Some wrought alloys are available in practically all of the commercial

where its resistance to corrosion and high thermal conductivity are desirable characteristics, and in the electrical industry, where its electrical conductivity of about 60% of that of copper and its light weight make it desirable for wire, cable and bus bars. Large quantities also are used for foil, to be used in packaging food products, for paste and powder to be used in paint and for collapsible tubes.

The addition of alloying elements to the commercially pure metal results in an increase in the strength and usually also has some effect on the other characteristics. Additions of silicon, for example, considerably improve the casting characteristics. The metals commonly used for alloying with aluminum are copper, magnesium, silicon, manganese, zinc, nickel, chromium, titanium and iron. Although they may be used singly, they more often are used in combination and, by properly selecting the combinations and amounts, may result in an alloy which can be heat treated to produce a tensile strength as high as 88,000–90,000 lbs. per sq. in. As with most other metals, the strength of any aluminum alloy can be increased by cold working but, if the alloy is heat-treatable, the strength usually can be increased much more by heat treatment than by cold working.

The first three wrought alloys in the accompanying table are not heat-treatable, but are given increased strength by cold mechanical deformation. These alloys

NOMINAL CHEMICAL COMPOSITION [1] AND TYPICAL PROPERTIES OF SOME COMMON ALUMINUM WROUGHT ALLOYS

	2S	3S	52S	Alclad 14S	17S	24S	Alclad 24S	61S	Alclad 75S
Nominal chemical composition	99% min. Alum.	1.2% Mn	2.5% Mg 0.25% Cr	4.4% Cu[2] 0.8% Si 0.8% Mn 0.4% Mg	4.0% Cu 0.5% Mn 0.5% Mg	4.5% Cu 1.5% Mg 0.6% Mn	24S with coating of high purity aluminum	1.0% Mg 0.6% Si 0.25% Cu 0.25% Cr	5.5% Zn[2] 2.5% Mg 1.6% Cu 0.2% Mn 0.25% Cr
Tensile strength psi...	A 13,000[8] H 24,000[8]	A 16,000 H 29,000	A 29,000 H 41,000	65,000	62,000	68,000	64,000	45,000	76,000
Yield strength psi[3]...	A 5,000 H 21,000	A 6,000 H 25,000	A 14,000 H 36,000	58,000	40,000	46,000	43,000	39,000	66,000
Elongation per cent in 2 in.	A 35 H 5	A 30 H 4	A 25 H 7	9	22	19	18	12	11
Modulus of elasticity [4]	10	10	10.2	.....	10.4	10.6	.....	10	.....
Brinell hardness [9]	23–44	28–55	45–85	.....	105	120	.....	95	.....
Melting range, °C	643–657	643–654	593–649	510–638	513–640	502–638	502–638	582–652	477–638
Melting range, °F	1190–1215	1190–1210	1100–1200	950–1180	955–1185	935–1180	935–1180	1080–1205	890–1180
Specific gravity	2.71	2.73	2.68	2.80	2.79	2.77	2.76	2.70	2.80
Electrical resistivity [5]	2.97	3.83	4.93	4.32	5.75	5.75	5.23	4.32	5.75
Thermal conductivity [6]	A 0.53	A 0.46	A 0.33	0.37	0.29	0.29	0.31	0.37	0.29
Coefficient of expansion [7]	23.6	23.2	23.8	23.0	23.6	23.2	23.2	23.6	23.6

[1] Aluminum plus normal impurities is the remainder.
[2] Core composition; coating is another alloy.
[3] 0.2% permanent set.
[4] Multiply by 10^6.
[5] Microhms per cu. cm. (room temperature).

[6] C.g.s. units (at 100° C.).
[7] Per °C. (20–100° C.); multiply by 10^{-6}.
[8] A = annealed; H = hard.
[9] 500 kg. load, 10 mm. ball.

products while others are special-purpose alloys and may be available in but one or two products. Similarly, some casting alloys are used only for one or two of the three usual types of castings.

Aluminum of commercial purity (99% minimum aluminum) is widely used for making wrought products but finds rather limited use in castings. In wrought forms it has a tensile strength of about 13,000 lbs. per sq. in. when in the annealed temper and of about 24,000 lbs. per sq. in. when in the cold-rolled hard temper. It is particularly useful in the food and chemical industries,

are usually classified as annealed, quarter hard, half hard, three-quarters hard and hard, depending on the amount of cold work done. Tensile strength increases with the hardness and ductility decreases. The range of mechanical properties is from 16,000 lbs. per sq. in. tensile strength with 30% elongation up to 41,000 lbs. per sq. in. with 7% elongation, depending on the alloy composition and the extent of deformation. A very important group of alloys is that in which the mechanical properties are greatly improved by thermal treatments. A typical example of an alloy in this group is Duralumin, developed by Wilm just before World War I. The alloy contained 4% copper, 0.5% magnesium and 0.5% manganese. After casting and rolling to sheet, the material was heated at 500° C. (932° F.), quenched in cold water and allowed to stand (age) for about 60 hours. This treatment increased the tensile strength to about 62,000 lbs. per sq. in. with an elongation of about 20%. A variety of alloys susceptible to heat treatment has since been developed. Differences in composition require different temperatures for the high-temperature heating known as solution treatment and differences both in time and temperature of aging. In some cases normal or room-temperature aging is sufficient, but in others a somewhat higher temperature (artificial aging) is necessary as the final operation. Tensile strengths vary in this group from about 40,000 lbs. per sq. in. to about 88,000 lbs. per sq. in. or even higher with elongations from 10 to

alloys, depending on the heat treatment used, develop tensile strengths of 25,000–46,000 lbs. per sq. in. and elongations of 1.5–14%. Several of the commercial heat-treatable alloys are particularly intended for use at elevated temperatures such as are encountered by the pistons and cylinder heads of an internal combustion engine. The oldest of these is an alloy containing 10% copper, about 1% iron and 0.2% magnesium. Another, often referred to as "Y Alloy," contains 4% copper, 2% nickel and 1.5% magnesium and a third contains 12% silicon, 2.5% nickel, 1% magnesium and 0.8% each copper and iron. A variety of heat treatments are used for these alloys in order to obtain specific characteristics, with the result that the tensile strength may vary from about 27,000 to 47,000 lbs. per sq. in. The typical properties of a number of casting alloys are shown in the accompanying table.

NOMINAL CHEMICAL COMPOSITION [1] AND TYPICAL PROPERTIES OF SOME ALUMINUM CASTING ALLOYS

Properties for alloys 195, B195, 220, 355 and 356 are for the commonly used heat treatment.

	13 [2]	43 [3]	108 [3]	A108 [4]	195 [3]	B195 [4]	214 [3]	218 [2]	220 [3]	355 [3]	356 [3]	380 [2]
Nominal chemical composition	12% Si	5% Si	4% Cu 3% Si	5.5% Si 4.5% Cu	4.5% Cu 0.8% Si	4.5% Cu 2.5% Si	3.8% Mg	8%Mg	10% Mg	5% Si 1.3% Cu 0.5% Mg	7% Si 0.3% Mg	8.5% Si 3.5% Cu
Tensile strength psi [5]	37,000	19,000	21,000	28,000	36,000	45,000	25,000	42,000	46,000	35,000	33,000	45,000
Yield strength psi [5]	18,000	9,000	14,000	16,000	24,000	33,000	12,000	23,000	25,000	25,000	24,000	25,000
Elongation per cent [5]	1.8	6	2.5	2	5	5	9	7	14	2.5	4	2
Brinell hardness [6]	.....	40	55	70	75	90	50	.....	75	80	70	.....
Melting range, °C	574–585	577–630	521–632	.....	549–646	527–627	580–640	540–621	449–621	580–627	580–610	521–588
Melting range, °F	1065–1085	1070–1165	970–1170	.....	1020–1195	980–1160	1075–1185	1005–1150	840–1150	1075–1160	1075–1130	970–1090
Specific gravity	2.66	2.69	2.79	2.79	2.81	2.78	2.65	2.53	2.58	2.70	2.68	2.76
Electrical resistivity	4.40	4.66	5.56	4.66	4.66	3.45	4.93	7.10	8.22	4.79	4.42	6.50
Thermal conductivity [7]	0.37	0.35	0.29	0.34	0.35	0.45	0.33	0.24	0.21	0.34	0.36	0.26
Coefficient of expansion [8]	20.0	22.8	22.8	22.7	23.9	22.8	24.8	24.0	25.4	22.8	22.8	20.0

[1] Remainder is aluminum plus minor impurities.
[2] Die cast.
[3] Sand cast.
[4] Permanent mold cast.
[5] For separately cast test bars.
[6] 500 kg. load, 10 mm. ball.
[7] C.g.s. units.
[8] Multiply by 10^{-6}. Per °C., for temperature range 20 to 200° C.

27%. Wrought alloys taken at the correct time in the treating cycle can be rolled, forged, drawn, stamped, extruded or spun.

Prior to about 1920, heat-treatable aluminum casting alloys were unknown and most castings were made from an alloy containing about 8% copper and known commercially as No. 12. This alloy, modified by the addition of silicon and iron, with or without zinc, is still used for many low-cost, general-purpose castings. As cast test bars have a tensile strength of about 24,000 lbs. per sq. in. when cast in sand and of about 29,000 lbs. per sq. in. when cast in permanent molds and an elongation of about 1.5%. Other alloys commonly used in the as-cast condition include the 5% and 12% silicon alloys, the latter chiefly for die castings, the 4% and 8% magnesium alloys and several alloys containing about 3–5% copper and about 4–8% silicon. Test bars of these alloys have a tensile strength of about 19,000–25,000 lbs. per sq. in. when cast in sand and of about 24,000–28,000 lbs. per sq. in. when cast in permanent molds. The elongation varies from about 2–9%. Die-cast test bars of alloys of this type have tensile strengths of 30,000–45,000 lbs. per sq. in. and elongations of about 2–7%. Several alloys containing zinc or zinc and magnesium, together with small amounts of several other elements, have attained some commercial importance in recent years and, in the as-cast condition, develop a tensile strength of 30,000–35,000 lbs. per sq. in. with an elongation of about 4–5%.

The more important heat-treatable casting alloys used for structural applications are the 4%-copper and the 10%-magnesium alloys and several alloys containing about 5–8% silicon with or without about 1–2% copper and small additions of magnesium. Test bars of these

In general, the alloys containing substantial amounts of silicon have the best foundry characteristics and a high resistance to corrosion, but they are not as easily machined as the alloys in which copper is the principal alloying element. The aluminum-magnesium alloys also have a high resistance to corrosion, can be machined with relative ease and will develop high strengths, but they require special foundry practices if the magnesium content exceeds about 4%. The aluminum-copper alloys, particularly if they also contain about 1% or more silicon, have good foundry characteristics, can be machined with relative ease and have a good resistance to corrosion, although they should be given paint protection if exposed to marine or severely-corrosive industrial atmospheres.

Aluminum alloys may be joined by any of the conventional welding methods, gas torch, electric spot welding and, for some applications, arc welding. Parts subject to heavy stresses are usually jointed by riveting since the heat of welding will destroy, to some extent, the beneficial effects of heat treatment. The physical characteristics of aluminum alloys, essentially their high electrical and thermal conductivity, their low melting points, and ease of oxidation make the welding techniques differ considerably from those for steel and require experience for successful operation. Brazing with aluminum-rich alloys produces excellent joints with some alloys. Soldering is rarely used and only when the joint can be provided with adequate protection against corrosion.

Most aluminum alloys are highly resistant to atmospheric attack. Under some conditions, as for example when the alloy is in continued contact with salt water, or under other corrosive conditions, some protection must be given. Some of the alloys are protected by cladding with a surface layer of pure aluminum or an alloy which is electronegative to the strong alloy core; these are known as alclad alloys and exhibit unusual resistance to attack. A protective coating can be applied by electrolytic oxidation (anodizing) which applies a thick protective oxide coating to the aluminum alloy. Under many conditions of service, the alloy may be protected by the use of paint, lacquer or enamel over the anodized surface or on the bare surface.

The principal applications of aluminum and its alloys are in the transportation industry—automobiles, buses and trucks, aircraft, railway cars and ships. Other large tonnages are used for metal products, electric conductors, cooking utensils, architectural uses, machinery and equipment for the chemical industries. (R. S. WILLIAMS, M.I.T.)

ALUMINUM BRONZES. Brass and Bronze.

ALUMS. These are double salts having the general formula $M_2SO_4X_2(SO_4)_324H_2O$. M is any univalent **cation** and X any trivalent **cation**, e.g., sodium alum, ammonium alum, chrome alum. (See **Aluminum.**) (R.K.S.)

ALUNITE or ALUMSTONE. The mineral alunite is a basic hydrous **sulfate** of **aluminum** and **potassium**; a variety called natroalunite is rich in **soda**. Alunite crystallizes in the **hexagonal** system and forms rhombohedrons with small angles, hence resembling cubes. It may be in fibrous or tabular forms, or massive. Hardness, 3.5–4; specific gravity, 2.58–2.75; luster, vitreous to pearly; streak white; transparent to opaque; brittle; color, white to grayish or reddish.

Alunite is commonly associated with acid **lavas** due to the sulfuric vapors often present; it may occur around **fumaroles** or associated with **sulfide** ore bodies. It has been used as a source of potash. It is found in Czechoslovakia, Italy, France, and Mexico. In the United States alunite is found in Colorado, Nevada, and Utah. (E.S.C.S.)

ALVEOLUS. 1. A minute sac-like chamber in a hollow organ, such as the air sacs of the lungs and the components of various glands. Its sac-like form distinguishes it from other chambers in which the walls are relatively thicker. 2. The cavity in the jaw in which the root of a tooth is fixed. (A.W.L.)

AMALGAM. An amalgam is an **alloy** of **mercury** with another metal. Usually amalgams are prepared by man, but there is a rare mineral, called amalgam, which is probably a mutual solid solution of silver and mercury, as the percentages of each vary to some extent. It crystallizes in the **isometric** system; hardness, 3–3.5; specific gravity, 13.75–14.1; luster, metallic; color, silver-white; streak the same, opaque.

Occurs in Bavaria, Czechoslovakia, France, Spain, Norway, Chile, and British Columbia. (E.S.C.S.)

AMANITA. Agarics.

AMARANTHS. A group of plants, including many coarse and obnoxious pigweeds. *Amarantus caudatus,* the familiar Love-lies-bleeding, and *A. hypochondriacus,* the Princess feather, are more attractive, widely planted garden annuals. (R.M.W.)

AMAZONITE. Feldspar.

AMAZONSTONE. Feldspar.

AMBER. Amber is a fossil resin which has been known since early times because of its property of ac-

quiring an **electric charge** when rubbed. In modern times it has been used largely in the making of beads, cigarette holders, and trinkets. Its amorphous nonbrittle nature permits it to be carved easily and to acquire a very smooth and attractive surface. Amber is soluble in various organic solvents, such as ethyl alcohol and ethyl ether.

It occurs in irregular masses showing a **conchoidal** fracture. Hardness, 2.25; specific gravity, 1.09; luster, resinous; color, yellow to reddish or brownish; it may be cloudy. Some varieties will exhibit **fluorescence.** Amber is transparent to translucent, melts between 250° and 300° C.

Amber has been obtained for over 2000 years from the **lignite**-bearing Tertiary **sandstones** on the coast of the Baltic Sea from Danzig to Memel, also Denmark, Sweden and the other Baltic countries. Sicily furnishes a brownish-red amber that is fluorescent.

The association of amber with **lignite** or other fossil woods as well as the beautifully preserved insects that are occasionally in it is ample proof of its organic origin. (E.S.C.S., R.M.W.)

AMBERGRIS. A fragrant waxy substance formed in the intestine of the sperm whale and sometimes found floating in the sea. It is used in the manufacture of **perfumes** to increase the persistence of the scent. (A.W.L.)

AMBLYGONITE. A rather rare compound of **fluorine, lithium, aluminum** and **phosphorus.** It crystallizes in the **triclinic** system; hardness 6; specific gravity 3.01–3.09; luster vitreous to greasy or pearly; color white to greenish, bluish, yellowish or greyish; streak white; translucent to sub-transparent.

Amblygonite occurs in **pegmatite** dikes and veins associated with other lithium minerals, is used as a source of lithium salts. Its name is derived from two Greek words meaning blunt and angle in reference to its cleavage angle of 75° 30′.

It is found in Saxony, France and Australia; and in the United States at Hebron, Paris, Greenwood, Rumford and Auburn, Maine; Branchville, Connecticut; Black Hills, South Dakota; and Pala, California. (E.S.C.S.)

AMBLYOPIA. Impairment of vision without organic disease of the eye proper. It may be due to **alcohols**—especially **methyl alcohol,** or various chemicals such as **arsenic** or **quinine,** etc. It may also occur in various diseases such as **nephritis** or **uremia** from any cause. It may also indicate certain diseases of the optic nerve or of the brain. (D.M.H.)

AMBLYPODS. Paleocene.

AMBULACRAL FEET. Tube Feet.

AMBULACRAL GROOVE. The groove along the lower or oral surface of the arm of a starfish, in which the tube feet are located. (A.W.L.)

AMBUSH BUG. Insecta, Hemiptera. Predacious **bugs** named from their habit of lying in wait for their prey in flowers where their colors conceal them. The most common species is *Phymata erosa.* (A.W.L.)

AMEIVA. Reptilia, Sauria. A lizard of Central and South America. The name is that of the genus, applied as a common name to the score of included species. (A.W.L.)

AMENORRHEA. Menstruation.

AMENT, or CATKIN. An inflorescence composed of many flowers, aggregated into long, often tassel-like masses. The perianth (see **Flower**) is completely lacking, or may be present in a scale-like form. The flowers of willows (pussy-willows), poplars, alders, beeches, oaks

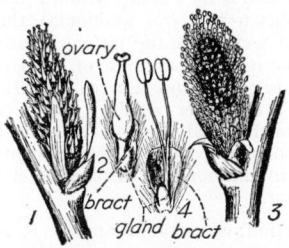

Flowers of willow, *Salix.* 1, pistillate catkin; 2, a single pistillate flower; 3, staminate catkin; 4, a single staminate flower.

and birches are familiar examples. Most of them are wind-pollinated flowers. (R.M.W.)

AMERICAN MONKEY. Mammalia, Primates. Any **monkey** of the family Cebidae, restricted to the New World. They differ from the apes and monkeys of the Old World in the broader nose and have sometimes been included with the marmosets in a group Platyrhini, based on this character. (A.W.L.)

AMERICAN STANDARD. Screw Thread.

AMERICIUM. Chemical Composition, I. Elements.

AMETHYST. Amethyst is purple- or violet-colored **quartz**, believed to be due to the admixture of a small amount of a **manganese** compound. Its physical characters are the same as quartz. An old superstition is that if worn as a gem it would cure intemperance. Oriental amethysts are purple **corundum.**

Amethysts are found at many localities, the Ural Mountains, India, Ceylon, Madagascar, Uruguay, Brazil, the Thunder Bay district of Lake Superior in Ontario, and Nova Scotia. In the United States amethysts are found in Michigan, Virginia, North Carolina, Montana and Maine.

The name amethyst is generally supposed to have been derived from the Greek word meaning not drunken. Pliny suggested that the term was applied because the amethyst approaches but is not quite the equivalent of a wine color. (E.S.C.S.)

AMICI PRISM. A direct-vision **prism**, that is, a prism combination by which a beam of light is dispersed into a spectrum without mean deviation. Such prisms are sometimes used in direct-vision **spectroscopes.**

The principle will be clear from the following example. Assume an inverted prism of crown glass with an angle

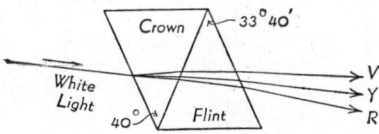

The deviation due to the crown glass prism is neutralized by that of the flint glass prism, for the middle of this spectrum (yellow) only.

of 40°, used with an erect prism of flint glass. Yellow sodium light (5893 A.) is deviated by the crown-glass prism through $+22°\,32'$ (upward). For the flint-glass prism to produce an equal negative (downward) deviation it must have an angle of 33° 40'. (Each prism is

supposedly set for minimum deviation.) Together they produce no deviation for this wavelength. But if light of 7682 A. (red) is used, the deviation of the crown glass is $+22°\,16'$, while that of the flint is $-22°\,10'$, giving a net deviation of $+6'$. And for the wavelength 4047 A. (violet), the deviations are, respectively, $+23°\,12'$ and $-23°\,44'$, giving $-32'$. There is thus, between the ends of the visible spectrum, a separation of 38'. Additional pairs of prisms may be used to increase the dispersion. (L.D.W.)

AMIDES. Amines and Amides.

AMIDOL. A developer used in photography. It is 2–4 diamino **phenol.** (R.K.S.)

AMIDOPYRINE. The chemical name for "pyramidon," often used as an **antipyretic** and to relieve pain— particularly of **joint** inflammations, headaches and in colds. In certain individuals who are sensitive to the **drug**, it diminishes the number of white **blood** cells, and leads to **agranulocytosis.** (R.S.M., D.M.H.)

AMINATION. Amination is the term used here to cover the processes of introducing the amino group ($-NH_2$) into organic compounds of the **amine** type. The principal methods in vogue are those of reduction (aniline from nitrobenzene) and of ammonolysis (aniline from chlorobenzene).

Reduction of groups other than the nitro group in the example named above is possible. Such groups as nitroso ($-NO$), hydroxylamine ($-NHOH$), nitrile ($-CN$), hydrazine ($-NH \cdot NH_2$), hydrazo ($-NH \cdot NH-$), and azo ($-N:N-$) yield amines by reduction, but the nitro group is principally employed. The production of aniline from nitrobenzene by reduction with iron and acid (usually hydrochloric) is an outstanding case. Only about 2% of the calculated amount of acid (to produce hydrogen by iron) is required, due to the fact that water plus iron in the presence of ferrous chloride solution (ferrous and chloride ions) functions as the real reducing agent. Insoluble ferroferric oxide, the chief iron-containing product that results from the reaction, is separated by filtration, and the aniline filtrate is made alkaline with sodium hydroxide and distilled. The oxide is also treated to recover its aniline content (it is stated that ½ the time consumed in this recovery is required to secure the last 10% on account of the mechanical quality of the sludge). Considerable heat is evolved in the reducing reaction, and provision must be made for withdrawing the heat in excess of that required to maintain the desired temperature in the apparatus.

Ammonolysis—the use of ammonia—is conducted for the replacement by $-NH_2$ of (1) halogen (chlorobenzene to aniline), (2) sulfonic acid group (sodium anthraquinone-2-sulfonate to 2-aminoanthraquinone), and (3) oxygen-function compounds (examples later). The production of aniline from chlorobenzene is conducted under high pressure, using excess ammonia—water in the presence of cuprous oxide catalyst at about 200° C. The resulting pressure is of the order of 900 lbs. per sq. in. These conditions of high temperature and high pressure demand specially constructed apparatus such as stainless steel now affords. Under the above conditions ammonia and water (as shown, not ammonium hydroxide) exist separately in the system, so that ammonia and water may each serve as reactants, the former producing aniline and the latter some phenol. Vorozhtzov, who studied this reaction in 1934, reported that 5 moles of ammonia (strength 32% NH_3, 68% H_2O) to 1 mole of chlorobenzene in the presence of 0.1 mole of cuprous oxide at approximately 200° C. yielded:

	IN THE ANILINE LAYER WEIGHT (%)	IN THE WATER LAYER WEIGHT (%)
Aniline............	81.6	4.9
Phenol.............	4.9	0.3
Diphenylamine......	0.9	0.0
Ammonia..........	Trace	13.8
Chloride ion........	Trace	8.8
Cuprous oxide......	Trace	2.9

which represents a yield of aniline 86.5% of theory, of phenol 5.2% of theory, and also 0.9% of diphenylamine. Addition of excess sodium hydroxide fixes the phenol as non-volatile sodium phenate and liberates free ammonia and free aniline. Upon distillation ammonia comes off first, and then the aniline—water fraction, from which aniline separates upon cooling. At 15.5° C. (60° F.) the water in the aniline layer is 4.5 weight per cent (density 1.026) and the aniline in the water layer 3.4 weight per cent (density 1.000). Aminoacetic acid is made by the ammonolysis of chloroacetic acid.

In the ammonolysis of benzenoid sulfonic acid derivatives an oxidizing agent is added to prevent the formation of soluble reduction products—$NaNH_4SO_4$ commonly resulting. When the sulfonate is used without an oxidizing agent the yield is low, of the order of 60% of theory. Lauer in 1932 studied the behavior of oxidizing agents on (a) sodium anthraquinone-1-sulfonate. The maximum yield, namely 92%, was obtained by using 1-nitrobenzene-3-sulfonic acid (in amount 120% of theory) as oxidizing agent, and (b) sodium anthraquinone-2-sulfonate. In this case he obtained the maximum yield, namely 83%, using potassium dichromate plus ammonium chloride (in amount 100% of theory).

Oxygen-function compounds that are subjected to ammonolysis are illustrated as follows:

Alcohols: Methanol plus aluminum phosphate catalyst yields mono-, di-, trimethylamines.
Phenols: *beta*-Naphthol plus sodium ammonium *sulfite* catalyst (Bucherer reaction) yields *beta*-naphthylamine.
Oxides: Ethylene oxide yields mono-, di-, triethanolamines.
Aldehydes: Glucose plus nickel catalyst yields glucamine.

Ketones: Cyclohexanone plus nickel catalyst yields cyclohexylamine.
Carbon dioxide: Urea formed at 190° C., 200 atmospheres pressure, in 2 hours from 1 mole of carbon dioxide plus 6 moles of ammonia (three times the theoretical weight of the latter) in silver- or lead-lined apparatus. A mixture of 51 parts by weight of urea, 12 of ammonium carbamate, 68 of ammonia, 15 of water results. By a succession of treatments the final result is 30 parts by weight of urea crystals and mother liquor containing 21 of urea, 12 of ammonium carbamate, 8 of ammonia, 15 of water.

The production of paraffin amines is conducted by heating under pressure the corresponding nitroparaffin with hydrogen gas, especially methyl, ethyl, propyl, butyl amines. Ammonolysis of the corresponding chloro-compounds is also satisfactory, especially with the higher members and with ethylene diamine from ethylene dichloride. (See **Amines.**) (R.K.S.)

AMINES AND AMIDES. Amines are derivatives of **ammonia** in which there is replacement of one or more **hydrogens** of ammonia (NH_3) by an alkyl group, e.g., methyl (—CH_3), ethyl (—C_2H_5) or an aryl group, e.g., phenyl (—C_6H_5), naphthyl (—$C_{10}H_7$). Aniline ($C_6H_5NH_2$) is a familiar and important amine, used in the dyestuff industry. (See **Aniline**, and **Dyes and Dyeing.**) Mixed amines contain at least one alkyl and one aryl group, e.g., methylaniline, methylphenylamine $\left(C_6H_5N\diagup^{H}_{\diagdown CH_3}\right)$. When one, two, and three hydrogens are thus replaced, the resulting amines are known as primary, secondary, tertiary, respectively. Methylaniline is, therefore, a secondary and dimethyl aniline a tertiary amine. Quaternary ammonium compounds, e.g., tetramethylammonium iodide, result from the reaction of a tertiary amine, e.g., trimethylamine and an alkyl haloid, e.g., methyl **iodide**. The corresponding hydroxide, tetramethylammonium hydroxide is a strong base, of the order of strength of sodium hydroxide. Amines have a characteristic odor, and many of the alkyl amines are soluble in water.

Amides are derivatives of ammonia in which there is replacement of one or more hydrogens of ammonia by an acyl group, e.g., acetyl (—$COCH_3$), yielding acetamide (CH_3CONH_2), benzoyl (—COC_6H_5) yielding benzamide ($C_6H_5CONH_2$). When one, two and three hydrogens are thus replaced, the resulting amides are known as primary, secondary, tertiary, respectively. Tribenzamide is, therefore, a tertiary amide.

STRUCTURAL GROUPING OF AMINES AND AMIDES

Ammonia	Primary Amine	Secondary Amine	Tertiary Amine	Quaternary ammonium compound
$N\diagup^{H}_{\diagdown II}^{\,H}$	$N\diagup^{CH_3}_{\diagdown H}^{\,H}$ Methylamine	$N\diagup^{CH_3}_{\diagdown H}^{\,CH_3}$ Dimethylamine	$N\diagup^{CH_3}_{\diagdown CH_3}^{\,CH_3}$ Trimethylamine	$H_3C\diagdown_{I}^{\diagup}N\diagup^{CH_3}_{\diagdown CH_3}$ Tetramethyl ammonium iodide
	$N\diagup^{C_6H_5}_{\diagdown H}^{\,H}$ Aniline (phenylamine)	$N\diagup^{C_6H_5}_{\diagdown H}^{\,C_6H_5}$ Diphenylamine	$N\diagup^{C_6H_5}_{\diagdown C_6H_5}^{\,C_6H_5}$ Triphenylamine	
		$N\diagup^{C_6H_5}_{\diagdown H}^{\,CH_3}$ Methylphenylamine (Methylaniline)	$N\diagup^{C_6H_5}_{\diagdown CH_3}^{\,CH_3}$ Dimethylphenylamine (Dimethylaniline)	

STRUCTURAL GROUPING OF AMINES AND AMIDES—*Continued*

	Primary Amide	Secondary Amide	Tertiary Amide	Quaternary ammonium compound
	$N \Big\langle \begin{array}{l} CO \cdot CH_3 \\ H \\ H \end{array}$ Acetamide	$N \Big\langle \begin{array}{l} CO \cdot CH_3 \\ CO \cdot CH_3 \\ H \end{array}$ Diacetamide	$N \Big\langle \begin{array}{l} CO \cdot CH_3 \\ CO \cdot CH_3 \\ CO \cdot CH_3 \end{array}$ Triacetamide	
	$N \Big\langle \begin{array}{l} COC_6H_5 \\ H \\ H \end{array}$ Benzamide	$N \Big\langle \begin{array}{l} COC_6H_5 \\ COC_6H_5 \\ H \end{array}$ Dibenzamide	$N \Big\langle \begin{array}{l} COC_6H_5 \\ COC_6H_5 \\ COC_6H_5 \end{array}$ Tribenzamide	
	$N \Big\langle \begin{array}{l} C_6H_5 \\ CO \cdot CH_3 \\ H \end{array}$ Acetanilide (phenylacetamide)	$N \Big\langle \begin{array}{l} C_6H_5 \\ CO \cdot CH_3 \\ CO \cdot CH_3 \end{array}$ Phenyldiacetamide		
	$N \Big\langle \begin{array}{l} C_6H_5 \\ C_6H_5 \\ CO \cdot CH_3 \end{array}$ Diphenylacetamide	$\begin{array}{l} CH_2CO \\ \| \\ CH_2CO \end{array} \Big\rangle NH$ Succinimide (butamide)		
		$C_6H_4 \Big\langle \begin{array}{l} CO \\ CO \end{array} \Big\rangle NH$ Phthalimide		

Amides of carbonic acid:

$OC \Big\langle \begin{array}{l} OH \\ NH_2 \end{array}$ Carbamic acid (Ethyl esters called Urethanes)	$OC \Big\langle \begin{array}{l} NH_2 \\ NH_2 \end{array}$ Urea $OC \Big\langle \begin{array}{l} NH_2 \\ NH \\ \end{array}$ $OC \Big\langle \begin{array}{l} \\ NH_2 \end{array}$ Biuret	$OC \Big\langle \begin{array}{l} \overset{2}{N}H\overset{1}{N}H_2 \\ \underset{4}{N}H_2 \end{array}$ ³ Semicarbazide	$HN : C \Big\langle \begin{array}{l} NH_2 \\ NH_2 \end{array}$ Guanidine (iminourea)	$HN : C \Big\langle \begin{array}{l} NHNH_2 \\ NH_2 \end{array}$ Aminoguanidine

The **ionization** constants of some **nitrogen** bases (also **silver** hydroxide) which constants indicate the relative strength of these bases, are as follows, arranged in decreasing basic strength:

BASE	IONIZATION CONSTANT OF BASE
Piperidine	2×10^{-3}
Diethylamine	1×10^{-3}
Dipropylamine (normal)	1×10^{-3}
Dimethylamine	7×10^{-4}
Brucine	7×10^{-4}
Triethylamine	6×10^{-4}
Ethylamine	6×10^{-4}
Tripropylamine (normal)	6×10^{-4}
Tetramethylenediamine	5×10^{-4}
Methylamine	5×10^{-4}
Propylamine (normal)	5×10^{-4}
Methyldiethylamine	4×10^{-4}
Silver hydroxide	1×10^{-4}
Ethylenediamine	9×10^{-5}
Trimethylamine	7×10^{-5}
Diethylbenzylamine	4×10^{-5}
Benzylamine	2×10^{-5}
Ammonium hydroxide	2×10^{-5}
Dimethylbenzylamine	1×10^{-5}
Hydrazine	3×10^{-6}
Quinine	2×10^{-7}

BASE	IONIZATION CONSTANT OF BASE
Pyridine	2×10^{-9}
Para-toluidine	2×10^{-9}
Phenylhydrazine	2×10^{-9}
Quinioline	1×10^{-9}
Meta-toluidine	6×10^{-10}
Aniline	5×10^{-10}
Ortho-toluidine	3×10^{-10}
Ortho-phenylenediamine	3×10^{-10}
Beta-naphthylamine	2×10^{-10}
Alpha-naphthylamine	1×10^{-10}
Semicarbazide	3×10^{-11}
Methyl red	3×10^{-12}
Thiazole	3×10^{-12}
Anthranilic acid	1×10^{-12}
Theobromine	5×10^{-14}
Caffeine	4×10^{-14}
Acetanilide	4×10^{-14}
Urea	2×10^{-14}
Acetamide	3×10^{-15}
Propylcyanide (normal)	2×10^{-15}
Thiourea	1×10^{-15}

The ionization constants of some **nitrogen** acids (also **carbonic** and **acetic** acids) are given for comparion, which constants indicate the relative strength of these acids, arranged in decreasing basic (increasing acidic) strength:

ACID	IONIZATION CONSTANT OF ACID	ACID	IONIZATION CONSTANT OF ACID
Hydrocyanic acid	7×10^{-10}	Acetic acid	2×10^{-5}
Cyanuric acid	2×10^{-7}	Hydrazoic acid	2×10^{-5}
Carbonic acid	3×10^{-7}	Barbituric acid	1×10^{-4}
Uric acid	2×10^{-6}	Hippuric acid	2×10^{-4}
Nicotinic acid	1×10^{-5}	Nitrous acid	4×10^{-4}
		Picric acid	2×10^{-1}

SELECTED REPRESENTATIVE AMINES
PRIMARY AMINES

AMINE	FORMULA	MELTING POINT °C.	BOILING POINT °C.
1. Methylamine	CH_3NH_2	−93	−7
2. Ethylamine	$C_2H_5NH_2$	−81	17
3. Propylamine (normal)	$C_3H_7NH_2$	−83	50
4. Propylamine (iso)	$C_3H_7NH_2$	−101	33
5. Butylamine (normal)	$C_4H_9NH_2$	−50	78
6. Butylamine (iso)	$C_4H_9NH_2$	−85	68
7. Amylamine (normal)	$C_5H_{11}NH_2$	−55	103
8. Amylamine (iso)	$C_5H_{11}NH_2$		95
9. Vinylamine	$CH_2 : CHNH_2$		56
10. Allylamine	$CH_2 : CHCH_2NH_2$		53
11. Aniline (phenylamine)	$C_6H_5NH_2$	−6	184
12. Benzylamine	$C_6H_5CH_2NH_2$		184
13. Para-phenylaniline (4-biphenylamine)	$C_6H_5 \cdot C_6H_4 \cdot NH_2$	51	302
14. Ortho-toluidine (2-methylaniline)	$(2)CH_3C_6H_4NH_2(1)$	−16	200
15. Meta-toluidine (3-methylaniline)	$(3)CH_3C_6H_4NH_2(1)$	−31	203
16. Para-toluidine (4-methylaniline)	$(4)CH_3C_6H_4NH_2(1)$	44	200
17. Naphthylamine, alpha	$C_{10}H_7NH_2(1)$	50	301
18. Naphthylamine, beta	$C_{10}H_7NH_2(2)$	111	306
19. 2,4,6-trimethylaniline (mesidine)	$(2,4,6)(CH_3)_3C_6H_2NH_2(1)$		229
20. Alpha-phenylethylamine	$C_6H_5 \cdot CHNH_2 \cdot CH_3$		187 (740 mm.)
21. Beta-phenylethylamine	$C_6H_5 \cdot CH_2 \cdot CH_2NH_2$		198
22. Dimethylenediamine (ethylenediamine)	$\left. \begin{matrix} CH_2NH_2 \\ \| \\ CH_2NH_2 \end{matrix} \right\}$	8	116
23. Trimethylenediamine	$(CH_2)_3 \begin{matrix} NH_2 \\ NH_2 \end{matrix} \Big\}$		135 (740 mm.)
24. Tetramethylenediamine	$(CH_2)_4 \begin{matrix} NH_2 \\ NH_2 \end{matrix}$		
25. Pentamenthylenediamine (cadaverine)	$(CH_2)_5 \begin{matrix} NH_2 \\ NH_2 \end{matrix} \Big\}$	9	179
26. Hexamethylenediamine	$(CH_2)_6 \begin{matrix} NH_2 \\ NH_2 \end{matrix} \Big\}$	42	204
27. 1,2-diaminopropane	$CH_3CHNH_2CH_2NH_2$		119
28. 4,4′-diaminobiphenyl (benzidine, para- para-prime-diaminobiphenyl) (crystallized from boiling water)	$(4')H_2NC_6H_4C_6H_4NH_2(4)$	218	200 (740 mm.)
29. Ortho-tolidine	$\begin{matrix} (4')H_2N \\ (3')H_3C \end{matrix} \Big\rangle C_6H_3C_6H_3 \Big\langle \begin{matrix} NH_2(4) \\ CH_3(3) \end{matrix} \Big\}$	129	
30. Ortho-phenylenediamine	$C_6H_4(NH_2)_2(1,2)$	103	257
31. Meta-phenylenediamine	$C_6H_4(NH_2)_2(1,3)$	63	285
32. Para-phenylenediamine	$C_6H_4(NH_2)_2(1,4)$	140	267
33. 1,2,3-triaminobenzene	$C_6H_3(NH_2)_3(1,2,3)$	103	336
34. 1,2,4-triaminobenzene	$C_6H_3(NH_2)_3(1,2,4)$	<100	340 appr.
35. Hexamethylenetetramine (urotropine)	$(CH_2)_6N_4$		
36. Ethanolamine ((beta-aminoethyl alcohol)	$NH_2CH_2CH_2OH$	171	

SECONDARY AMINES

AMINE	FORMULA	MELTING POINT °C.	BOILING POINT °C.
37. Dimethylamine	$(CH_3)_2NH$	−96	7
38. Diethylamine	$(C_2H_5)_2NH$	−39	55
39. Methylethylamine	$\begin{matrix} CH_3 \\ C_2H_5 \end{matrix} \rangle NH$		34
40. Dipropylamine (norm.)	$(C_3H_7)_2NH$	−40	110
41. Dipropylamine (iso)	$(C_3H_7)_2NH$		83 (743 mm.)
42. Dibutylamine (norm.)	$(C_4H_9)_2NH$		159
43. Dibutylamine (iso)	$(C_4H_9)_2NH$		139
44. Diphenylamine	$(C_6H_5)_2NH$	53	302
45. Ortho-aminodiphenylamine	$(2)NH_2C_6H_4NHC_6H_5$	79	
46. Para-aminodiphenylamine	$(4)NH_2C_6H_4NHC_6H_5$	66	
47. Methylphenylamine (N-methylaniline)	$C_6H_5NHCH_3$	−57	195
48. Ethylphenylamine ((N-ethylaniline)	$C_6H_5NHC_2H_5$	−64	204
49. Propylphenylamine (norm.) (N-normal-propylaniline)	$C_6H_5NHC_3H_7$		222

SELECTED REPRESENTATIVE AMINES—*Continued*

SECONDARY AMINES

AMINE	FORMULA	MELTING POINT °C.	BOILING POINT °C.
50. Propylphenylamine (iso)................ (N-iso-propylaniline)	$C_6H_5NHC_3H_7$................		
51. Butylaniline (norm.).................... (N-normal-butylaniline)	$C_6H_5NHC_4H_9$................		235 (720 mm.)
52. Butylaniline (iso) (N-iso-butylaniline).....	$C_6H_5NHC_4H_9$................		231
53. Allylaniline (N-allylaniline).............	$C_6H_5NHC_3H_5$................		218 (735 mm.)
54. Phenylbenzylamine (N-benzylaniline).....	$C_6H_5CH_2{>}NH$.	37	306
55. Benzalaniline (benzylideneaniline)........	$C_6H_5CH:NC_6H_5$.....	56	300 appr.
56. Dibenzylamine.....................	$(C_6H_5CH_2)_2NH$.....	−26	270 (250 mm.)
57. Diethanolamine (iminoethyl alcohol).....	$NH(CH_2CH_2OH)_2$................	28	270 (748 mm.)

TERTIARY AMINES

AMINE	FORMULA	MELTING POINT °C.	BOILING POINT °C.
58. Trimethylamine....................	$(CH_3)_3N$.	−124	3
59. Triethylamine....................	$(C_2H_5)_3N$.	−115	89
60. Methyldiethylamine....................	$CH_3N(C_2H_5)_2$.		66
61. Dimethylethylamine....................	$(CH_3)_2NC_2H_5$.		37
62. Tripropylamine (norm.)....................	$(C_3H_7)_3N$.	−93	156
63. Tripropylamine (iso)....................	$(C_3H_7)_3N$.		
64. Tributylamine (norm.)....................	$(C_4H_9)_3N$.		216
65. Tributylamine (iso)....................	$(C_4H_9)_3N$.		190
66. Dimethylaniline....................	$C_6H_5N(CH_3)_2$.	3	193
67. Diethylaniline....................	$C_6H_5N(C_2H_5)_2$.	−34	216
68. Methylethylaniline....................	$C_6H_5N{<}{CH_3 \atop C_2H_5}$		201
69. Dipropylaniline (norm.)....................	$C_6H_5N(C_3H_7)_2$.		241
70. Dipropylaniline (iso)....................	$C_6H_5N(C_3H_7)_2$.		
71. Dibutylaniline (norm.)....................	$C_6H_5N(C_4H_9)_2$.		263
72. Triphenylamine....................	$(C_6H_5)_3N$.	126	365
73. Dibenzylaniline....................	$C_6H_5N(CH_2C_6H_5)_2$.	70	>300
74. Diphenylbenzylamine....................	$(C_6H_5)_2NCH_2C_6H_5$.	86	
75. Methyldiphenylamine....................	$(C_6H_5)_2NCH_3$.		293
76. Ethyldiphenylamine....................	$(C_6H_5)_2NC_2H_5$.		297
77. Triethanolamine....................	$N(CH_2CH_2OH)_3$.	20	278 (150 mm.)

SELECTED REPRESENTATIVE AMIDES

PRIMARY AMIDES

AMIDE	FORMULA	MELTING POINT °C.	BOILING POINT °C.
1. Formamide....................	$H \cdot CONH_2$....................	2	193
2. Acetamide....................	CH_3CONH_2....................	82	222
3. Propionamide....................	$C_2H_5CONH_2$....................	79	213
4. Acrylamide....................	$CH_2:CH \cdot CONH_2$....................	84	
5. Methylacetamide (N)....................	$CH_3CONHCH_3$....................	28	206
6. Ethylacetamide (N)....................	$CH_3CONHC_2H_5$....................		205
7. Benzamide....................	$C_6H_5CONH_2$....................	130	290
8. Benzylacetamide (N) (acetylbenzylamine)..	$CH_3CONHCH_2C_6H_5$....................	60	>300
9. Dimethylbenzamide (N,N)....................	$C_6H_5CON(CH_3)_2$....................	41	272
10. Ethylbenzamide (N)....................	$C_6H_5CONC_2H_5$....................	70	299
11. Oxamide....................	$CONH_2 \atop {| \atop CONH_2}$.	418 dec.	
12. Malonamide....................	$H_2C{<}{CONH_2 \atop CONH_2}$.	170	
13. Succinamide....................	$H_2CCONH_2 \atop {| \atop H_2CCONH_2}$.	242	
14. Acetanilide (phenylacetamide) (N).......	$CH_3CONHC_6H_5$....................	113	305
15. Methylacetanilide (N).................... (acetylmethylphenylamine)	$CH_3CON{<}{C_6H_5 \atop CH_3}$.	103	253 (710 mm.)
16. Benzanilide (N-phenylbenzamide)........	$C_6H_5CONHC_6H_5$....................	163	118
17. Diphenylacetamide (N).................... (acetyldiphenylamine)	$CH_3CON(C_6H_5)_2$....................	103	subl.
18. Acetotoluide, ortho (N-tolylacetamide)....	$CH_3CONHC_6H_4CH_3(2)$.........	110	296
19. Acetotoluide, meta....................	$CH_3CONHC_6H_4CH_3(3)$.........	65	303
20. Acetotoluide, para....................	$CH_3CONHC_6H_4CH_3(4)$.........	153	306

(*Continued on next page*)

SELECTED REPRESENTATIVE AMIDES—*Continued*

SECONDARY AMIDES

AMIDE	FORMULA	MELTING POINT °C.	BOILING POINT °C.
21. Diacetamide..........................	$(CH_3CO)_2NH$..............	78	223
22. Dibenzamide..........................	$(C_6H_5CO)_2NH$..............	148	

TERTIARY AMIDES

23. Tribenzamide........................	$(C_6H_5CO)_3N$..................	207	subl.

IMIDES

IMIDES	FORMULA	MELTING POINT °C.	BOILING POINT °C.
24. Succinimide (butamide)..................	$\begin{array}{c}CH_2-CO\\ \mid \qquad\qquad >NH\\ CH_2-CO\end{array}$..............	125	287
25. Ortho-Phthalimide......................	$C_6H_4{<}^{CO}_{CO}{>}NH$..............	238	subl.

REPRESENTATIVE COMPOUNDS RELATED TO AMINES AND AMIDES

CARBAMATES

CARBAMATE	FORMULA	MELTING POINT °C.	BOILING POINT °C.
1. Carbamic acid (not isolated)			
2. Methyl carbamate.....................	$OC{<}^{OCH_3}_{NH_2}\}$..............	54	177
3. Ethyl carbamate (urethane).............	$OC{<}^{OC_2H_5}_{NH_2}\}$..............	49	184
4. Propylcarbamate (norm.)..............	$OC{<}^{OC_3H_7}_{NH_2}\}$..............	60	200
5. Phenylcarbamate.....................	$OC{<}^{OC_6H_5}_{NH_2}\}$..............	142	
6. Benzylcarbamate.....................	$OC{<}^{OCH_2C_6H_5}_{NH_2}\}$..............	86	dec.
7. Ethyl-N-methyl carbamate.............. (N-methylurethane)	$OC{<}^{OC_2H_5}_{NHCH_3}\}$..............		170
8. Ethyl-N-ethyl carbamate.............. (N-ethylurethane)	$OC{<}^{OC_2H_5}_{NHC_2H_5}\}$..............		175
9. Ethyl-N-normal-propyl carbamate....... (N-normal-propylurethane)	$OC{<}^{OC_2H_5}_{NHC_3H_7}\}$..............		192
10. Ethyl-N-phenyl carbamate.............. (N-phenylurethane)	$OC{<}^{OC_2H_5}_{NHC_6H_5}\}$..............	52	237
11. Ethyl-N,N-diphenyl carbamate........... (N,N-diphenylurethane)	$OC{<}^{OC_2H_5}_{N(C_6H_5)_2}\}$..............	72	>360
12. Thiourethane........................	$OC{<}^{SC_2H_5}_{NH_2}\}$	108	subl.

UREAS

UREA	FORMULA	MELTING POINT °C.	BOILING POINT °C.
1. Urea.................................	$OC{<}^{NH_2}_{NH_2}\}$..............	133	dec.
2. Methylurea..........................	$OC{<}^{NHCH_3}_{NH_2}\}$..............	101	dec.
3. Dimethylurea (sym.)....................	$OC{<}^{NHCH_3}_{NHCH_3}\}$..............	106	269

REPRESENTATIVE COMPOUNDS RELATED TO AMINES AND AMIDES—*Continued*
UREAS

UREA	FORMULA	MELTING POINT °C.	BOILING POINT °C.
4. Dimethylurea (unsym.)	$OC\begin{cases} N(CH_3)_2 \\ NH_2 \end{cases}$	183	
5. Ethylurea	$OC\begin{cases} NHC_2H_5 \\ NH_2 \end{cases}$	92	
6. Diethylurea (sym.)	$OC\begin{cases} NHC_2H_5 \\ NHC_2H_5 \end{cases}$	112	263
7. Diethylurea (unsym.)	$OC\begin{cases} N(C_2H_5)_2 \\ NH_2 \end{cases}$	74	
8. Normal-propylurea	$OC\begin{cases} NHC_3H_7 \\ NH_2 \end{cases}$	107	
9. Dipropylurea (sym., norm.)	$OC\begin{cases} NHC_3H_7 \\ NHC_3H_7 \end{cases}$	105	255
10. Dipropylurea (unsym., norm.)	$OC\begin{cases} N(C_3H_7)_2 \\ NH_2 \end{cases}$	76	
11. Tetramethylurea	$OC\begin{cases} N(CH_3)_2 \\ N(CH_3)_2 \end{cases}$		177
12. Tetraethylurea	$OC\begin{cases} N(C_2H_5)_2 \\ N(C_2H_5)_2 \end{cases}$		210 app.
13. Allyl urea	$OC\begin{cases} NHC_3H_5 \\ NH_2 \end{cases}$	85	
14. N-Acetylurea	$OC\begin{cases} NHCOCH_3 \\ NH_2 \end{cases}$	218	
15. Phenylurea	$OC\begin{cases} NHC_6H_5 \\ NH_2 \end{cases}$	147	160 dec.
16. Benzylurea	$OC\begin{cases} NHCH_2C_6H_5 \\ NH_2 \end{cases}$	147	
17. N-Benzoylurea	$OC\begin{cases} NHCOC_6H_5 \\ NH_2 \end{cases}$	214	
18. Diphenylurea (sym.) (carbanilide)	$OC\begin{cases} NHC_6H_5 \\ NHC_6H_5 \end{cases}$	238	261
19. Diphenylurea (unsym.)	$OC\begin{cases} N(C_6H_5)_2 \\ NH_2 \end{cases}$	189	
20. Ethylphenylurea (N,N')	$OC\begin{cases} NHC_6H_5 \\ NHC_6H_5 \end{cases}$	99	
21. Tetraphenylurea	$OC\begin{cases} N(C_6H_5)_2 \\ N(C_6H_5)_2 \end{cases}$	183	
22. Ethyleneurea	$\begin{matrix} CH_2-NH \\ | \quad\quad\;\; \rangle CO \\ CH_2-NH \end{matrix}$	131	
23. Ethylideneurea	$CH_3CH\begin{cases} NH \\ NH \end{cases}CO$	154	160 dec.
24. Glycollylurea (hydantoin)	$\begin{matrix} CH_2-NH \\ | \quad\quad\;\; \rangle CO \\ CO-NH \end{matrix}$	220	
25. Oxalylurea (parabanic acid)	$\begin{matrix} CO-NH \\ | \quad\quad\;\; \rangle CO \\ CO-NH \end{matrix}$	243 dec.	
26. Malonylurea (barbituric acid)	$H_2C\begin{cases} CO-NH \\ CO-NH \end{cases}CO$	245	260 dec.
27. Acetyonylurea (dimethylhydantoin)	$\begin{matrix} (CH_3)_2C-NH \\ \quad\quad\quad\quad \rangle CO \\ CO-NH \end{matrix}$	175	subl.
28. Glyoxyldiureide (allantoin)	$\begin{matrix} NH_2CONH-CH-NH \\ \quad\quad\quad\quad\quad\; \rangle CO \\ CO-NH \end{matrix}$	235	
29. Mesoxalylurea (alloxan)	$CO\begin{cases} CO-NH \\ CO-NH \end{cases}CO$	256 dec.	
30. Biuret	$NH_2-CO-NH-CO-NH_2$	192 dec.	
31. Acetylbiuret	$CH_3CONHCONHCONH_2$	107	
32. Thiourea	$SC\begin{cases} NH_2 \\ NH_2 \end{cases}$	181	
33. Methylthiourea	$SC\begin{cases} NHCH_3 \\ NH_2 \end{cases}$	118	
34. Ethylthiourea	$SC\begin{cases} NHC_2H_5 \\ NH_2 \end{cases}$	113	
35. Diethylthiourea (sym.)	$SC\begin{cases} NHC_2H_5 \\ NHC_2H_5 \end{cases}$	77	
36. Phenylthiourea	$SC\begin{cases} NHC_6H_5 \\ NH_2 \end{cases}$	154	
37. Diphenylthiourea (sym.)	$SC\begin{cases} NHC_6H_5 \\ NHC_6H_5 \end{cases}$	154	dec.
38. Benzylthiourea	$SC\begin{cases} NHCH_2C_6H_5 \\ NH_2 \end{cases}$	163	
39. N-Benzoylthiourea	$SC\begin{cases} NHCOC_6H_5 \\ NH_2 \end{cases}$	169	

(*Continued on next page*)

REPRESENTATIVE COMPOUNDS RELATED TO AMINES AND AMIDES—*Continued*

SEMICARBAZIDES

SEMICARBAZIDE	FORMULA	MELTING POINT °C.	BOILING POINT °C.
1. Semicarbazide	$OC\langle^{NHNH_2}_{NH_2}\rangle$	96	
2. Phenylsemicarbazide (1)	$OC\langle^{NHNHC_6H_5}_{NH_2}\rangle$	172	
3. Phenylsemicarbazide (4)	$OC\langle^{NHNH_2}_{NHC_6H_5}\rangle$	122	

SEMICARBAZONES

By reaction of hydrogens of number 1 nitrogen with oxygen of carbonyl group ($>CO$)

SEMICARBAZONE OF	MELTING POINT °C.
1. Formaldehyde	169
2. Acetaldehyde	162
3. Glyoxal	270
4. Acrolein	171
5. Furfural	202
6. Benzaldehyde	222
7. Salicylaldehyde	231
8. Cinnamaldehyde	215
9. Acetone	187
10. Methylethyl ketone	135
11. Methyl-n-propyl ketone	110
12. Methyl-iso-propyl ketone	113
13. Methyl-n-butyl ketone	122
14. Methyl-iso-butyl ketone	135
15. 2-Methylcyclohexanone	195
16. Diethyl ketone	139
17. Di-n-propyl ketone	133
18. Di-iso-propyl ketone	160
19. Cyclohexanone	166
20. Acetophenone	198
21. Mesityl oxide	164

tuted benzene sulfonamides, thus, ethylamine to form N-ethylbenzenesulfonamide ($C_6H_5SO_2NHC_2H_5$), soluble in sodium hydroxide, (4) with **chloroform** ($CHCl_3$) and a base, yielding **isocyanides** (Very poisonous!), (5) with nitric acid concentrated, yielding nitramines, thus, ethylamine to form ethylnitramine ($C_2H_5NHNO_2$). Primary amines may be formed (1) by reduction of **nitro-compounds** (aniline from nitrobenzene), **nitroso-compounds, hydroxylamines, cyanides, oximes or hydrazones**, (2) by the alkaline ($NaOH$) hydrolysis of **isocyanates** (sodium carbonate also formed) or isocyanides (sodium formate also formed), (3) by reaction with **bromine** followed by treatment with sodium hydroxide (Hofmann Reaction) (sodium carbonate and bromide also formed), (4) by reaction of sodium phthalimide plus alkyl halide, followed by heating with fuming hydrochloric acid (Gabriel reaction) (phthalic acid also formed), (5) with vapor of **alcohols** plus **ammonia** in the presence of a **catalyzer**, e.g., **thorium** oxide at 360° C., (6) from aminoacids, by living organisms, e.g., decomposition of fish, in the case of methylamine.

Secondary amines react (1) with nitrous acid, yielding nitrosoamines, yellow oily liquids, volatile in steam, soluble in ether. The secondary amine may be recovered by heating the nitrosoamine with **hydrochloric acid**, concentrated, or **hydrazines** may be formed by reduction of the nitrosamines, e.g., methylaniline forms

GUANIDINES

GUANIDINE	FORMULA	MELTING POINT °C.	BOILING POINT °C.
1. Guanidine	$HN:C\langle^{NH_2}_{NH_2}$		
2. 1,3-diphenylguanidine	$HN:C\langle^{NHC_6H_5}_{NHC_6H_5}\rangle$	147	
3. 1,1,3,3-tetraphenylguanidine	$HN:C\langle^{N(C_6H_5)_2}_{N(C_6H_5)_2}\rangle$	130	
4. 1,2,3-triphenylguanidine	$C_6H_5N:C\langle^{NHC_6H_5}_{NHC_6H_5}\rangle$	144	
5. 1,1,3-triphenylguanidine	$HN:C\langle^{NHC_6H_5}_{N(C_6H_5)_2}\rangle$	131	
6. Guanylurea	$HN:C\langle^{NH_2}_{NHCONH_2}\rangle$	105	160 dec.
7. Aminoguanidine	$HN:C\langle^{NHNH_2}_{NH_2}\rangle$	dec.	

Primary amines react (1) with **nitrous acid**, yielding (a) with alkyl amine, nitrogen gas plus alcohol, (b) with aryl amine warm, nitrogen gas plus phenol. The amino-group of primary amines is displaced by the hydroxyl group to form **alcohol or phenol**, (c) with aryl amine cold, **diazonium-compounds**, (2) with **acetyl chloride** or **benzoyl chloride**, yielding substituted amide, thus, ethylamine plus acetyl chloride to form N-ethylacetamide ($C_2H_5NHOCCH_3$), (3) with benzene-sulfonyl chloride ($C_6H_5SO_2Cl$), yielding substi-

methylphenylnitrosamine $\left(^{CH_3}_{C_6H_5}\rangle N \cdot NO\right)$, reduction yielding unsymmetrical methylphenylhydrazine $\left(^{CH_3}_{C_6H_5}\rangle NHNH_2\right)$

(2) with **acetyl** or **benzoyl chloride**, yielding substituted amide, thus, diethylamine plus acetyl chloride to

form N-N-diethylacetamide ($(C_2H_5)_2NOCCH_3$), (3) with benzene sulfonyl chloride, yielding substituted benzene sulfonamides, thus, diethylamine to form N-N-diethylbenzenesulfonamide ($C_6H_5SO_2N(C_2H_5)_2$), insoluble in sodium hydroxide, (4) with **phenol** warmed with **sulfuric acid** concentrated, then diluted with water and made alkaline with sodium hydroxide, yielding a blue to violet coloration.

Secondary amines may be formed (1) by alkylation of primary amines, using, for example, methyl chloride, dimethyl sulfate, (see **Esters**), **methyl alcohol**, heated, **formaldehyde** in acid medium heated, and may be recovered from mixtures with primary or tertiary amines by means of the nitrosamine reaction above, (2) from **aminoacids** by living organisms, e.g., decomposition of fish, in the case of dimethylamine.

Tertiary amines do not react with nitrous acid, acetyl chloride, benzoyl chloride, benzenesulfonyl chloride, but react with alkyl haloids to form quaternary ammonium haloids, which are converted by silver hydroxide to quaternary ammonium hydroxides. Quaternary ammonium hydroxides upon heating yield (1) tertiary amine plus alcohol (or, for higher members, olefin hydrocarbon plus water). Tertiary amines may also be formed (2) by alkylation of secondary amines, for example, by dimethyl sulfate, methyl alcohol heated, formaldehyde in acid medium heated, (3) from aminoacids by living organisms, e.g., decomposition of fish in the case of trimethylamine. (See **Amination**.)

Primary amides react (1) with **hypobromite** in sodium hydroxide, to form amines, of one carbon less than in the amides (sodium carbonate and bromide also formed). (Ureas yield nitrogen gas plus carbon dioxide with sodium hypobromite, (2) with **nitrous acid**, to form nitrogen gas plus the corresponding **carboxylic acid** (ureas yield carbon dioxide), (3) with **phosphorous** pentoxide, to form cyanides by loss of water, (4) with **sodium** hydroxide, to form ammonia gas plus the corresponding carboxylic acid. Primary amides are formed (1) by reaction of **acid chlorides**, **acid anhydrides** or **esters** with ammonium hydroxide (for urea, carbonyl chloride is used), (2) by heating the ammonium salt of the desired acid, e.g., ammonium acetate to obtain acetamide, (3) by reaction of **cyanides** with acids, e.g., hydrochloric acid concentrated cold. When amines instead of ammonium hydroxide are used with acid chlorides, acid anhydrides or esters, alkylated amides are obtained, e.g., N-methyl-acetamide ($CH_3CO \cdot NH \cdot CH_3$). Succinimide, it is to be noted, contains the ring of pyrrole, and may be converted into pyrrole by treatment with zinc and acetic acid, or with hydrogen in the presence of finely divided platinum heated.

Urea may be substituted, as to the hydrogens, (1) by alkyl or aryl, mono, di, tetra, most commonly di, symmetrical (1, 3), (2) by carbonyl groups (=CO) either (a) as open chain ureides, e.g., acetylurea ($NH_2CO \cdot NH \cdot OC \cdot CH_3$) or (b) as cyclic ureides, e.g.,

oxalylurea $\left(\begin{array}{c} CO—NH \\ | \quad\quad\quad >CO \\ CO—NH \end{array} \right)$. Purine and uric acid compounds contain two of these cyclic ureides, namely oxalylurea and malonylurea, interlocked. Biuret is formed by heating urea at 160° C., and its formation is commonly used as a test for urea. When sodium hydroxide plus **copper** sulfate is added to biuret a violet-red color is produced. Urea was the first artificially prepared product of vital processes, by Wöhler, in 1828, by heating ammonium cyanate. Urea is produced in the animal body from **proteins** of food. Mammals execrete urea in the urine, while many other animals excrete nitrogen in the form of uric acid.

Werner, in 1923, suggested the formula $HN:C\left\langle \begin{array}{c} NH_3 \\ | \\ O \end{array} \right.$ for urea.

Semicarbazide forms salts with **acids**, e.g., semicar-

bazide hydrochloride ($NH_2CONHNH_2 \cdot HCl$), melting point 173° C. dec., reacts with carbonyl group (=CO) of aldehydes and ketones to form semicarbazones, usually of sharp melting point, and useful in identification of **aldehydes** and **ketones**. Semicarbazide is formed by reaction of hydrazine hydrate and sodium cyanate.

Guanidine forms salts with acids, e.g., guanidine nitrate $HNC(NH_2)_2 \cdot HNO_3$. By heating at 120° C. for several hours a mixture of **ammonium** thiocyanate and dicyanidiamide, guanidine thiocyanate solution is obtained by extracting with water. Treating guandine with a mixture of nitric and sulfuric acids forms nitroguanidine $\left(HN:C\left\langle \begin{array}{c} NH \cdot NO_2 \\ NH_2 \end{array} \right. \right)$ which is reduced by zinc and acetic acid to aminoguanidine $\left(HN:C\left\langle \begin{array}{c} NH \cdot NH_2 \\ NH_2 \end{array} \right. \right)$. By treating aminoguanidine (1) with dilute acid or alkali, there is obtained, first, semicarbazide, finally hydrazine; (2) with nitrous acid, diazoguanidine $\left(HN:C\left\langle \begin{array}{c} NHN:NOH \\ NH_2 \end{array} \right. \right)$, which is decomposed by alkali into hydrazoic acid (HN_3) plus cyanamide ($H_2N \cdot CN$) plus water. (R.K.S.)

AMINOACIDS, POLYPEPTIDES, AND PROTEINS.

Aminoacids are formed by the reaction of water and proteins. Aminoacids correspond to hydroxyacids, the amino-group (—NH₂) occupying the position of the hydroxyl-group (—OH) of the latter. The simplest aminoacid is aminoacetic acid, glycocoll, glycine ($CH_2NH_2 \cdot COOH$) corresponding to hydroxyacetic acid, glycollic acid ($CH_2OH \cdot COOH$). Aminoacetic acid reacts (1) with **bases**, to form sodium aminoacetate ($CH_2NH_2 \cdot COONa$), (2) with **acids**, to form aminoacetic acid hydrochloride ($HCl \cdot CH_2NH_2COOH$), (3) with **formaldehyde**, to form $CH_2:NCH_2 \cdot COOH$, which can be **titrated** as an acid by a standard base, (4) with **alcohols** anhydrous plus **hydrogen** chloride, followed by treatment with **sodium** hydroxide solution cold, extraction with **ether**, and evaporation of ether, to form, for example the **ester** O-methylaminoacetate ($CH_2NH_2 \cdot COOCH_3$), boiling point 130° C., decomp. Such esters are useful in the separation of aminoacids, (5) with **benzoyl chloride**, to form N-benzoylaminoacetic acid, hippuric acid ($C_6H_5CO \cdot NHCH_2 \cdot COOH$), (6) with **acetyl chloride** in large excess at 0° to 20° C. plus phosphorus pentachloride, to form $HCl \cdot H_2N \cdot H_2C \cdot COCl$. Such compounds are useful in the synthesis of polypeptides, (7) with **nitrous acid**, to form glycollic acid plus nitrogen, (8) with phenyl isocyanate (C_6H_5NCO), to form phenyl ureidoacetic acid ($C_6H_5NHCONH CH_2 \cdot COOH$), slightly soluble. Such compounds from phenyl isocyanate and from alpha-naphthylisocyanate are useful in the separation of aminoacids, (9) with **copper** carbonate by boiling in aminoacetic acid solution, to form dark blue copper aminoacetate (($CH_2NH_2 \cdot COO)_2Cu \cdot H_2O$). N-methylaminoacetic acid, sarcosine $\left(\begin{array}{c} CH_3 \\ >N \cdot CH_2COOH \\ H \end{array} \right)$ is formed when creatine (meth-

ylguanidylacetic acid, $HN:C\left\langle \begin{array}{c} NH_2 \\ N\left\langle \begin{array}{c} CH_3 \\ CH_2 \cdot COOH \end{array} \right. \end{array} \right.$ of meat juice is warmed with barium hydroxide solution.

Aminoacids may be prepared (1) by reaction of halogen-substituted acids (or their esters) with ammonia, e.g., chloroacetic acid ($ClCH_2 \cdot COOH$) to form aminoacetic acid ($H_2NCH_2 \cdot COOH$), (2) by reaction of **aldehydes or ketones** with **hydrogen cyanide** to form aldehyde or ketone cyanhydrin, e.g., acetaldehyde cyanhydrin (CH₃·CHOH·CN), which by reaction with ammonia forms aminocyanide (CH₃CHNH₂·CN), and then by hydrolysis this forms the corresponding aminoacid (CH₃CHNH₂·COOH), (3) by reaction of proteins with hydrochloric acid, the aminoacid residues constituting the protein are obtained. The aminoacids thus obtained from proteins are alpha-substituted aminoacids. Some of the most important alpha-aminoacids in this connection are:

ide warm to form sodium sulfide, which with **lead** nitrate solution forms a black precipitate of lead sulfide.

When amino- or hydroxyacids lose ammonia, or water, respectively, characteristic reactions occur, as shown at top of page 77.

Aminoacids may be combined with one another, up to 19 aminoacid units, to form polypeptides, thus: alpha-aminopropionic acid, alanine (CH₃CHNH₂·COOH), by reaction with aminoacetic acid, glycine (CH₂NH₂·

SELECTED REPRESENTATIVE ALPHA-AMINO ACIDS

Alpha-Amino Acid	Formula
1. Aminoacetic acid (glycocoll, glycine)	CH_2NH_2COOH
2. N-methylaminoacetic acid (sarcosine)	CH_3NHCH_2COOH
3. Alpha-aminopropionic acid (alanine)	$CH_3CH_2NH_2COOH$
4. Alpha-amino-beta-hydroxy-propionic acid (serine)	$CH_2OHCHNH_2COOH$
*5. Alpha-aminoisocaproic acid (leucine)	$(CH_3)_2CHCH_2CHNH_2COOH$
*6. Alpha-amino-beta-methylvaleric acid (isolucine)	$\frac{C_2H_5}{CH_3}{>}CHCHNH_2COOH$
7. Alpha-aminosuccinic acid (aspartic acid)	$COOHCH_2CHNH_2COOH$
8. Alpha-aminoglutaric acid (glutamic acid)	$COOHCH_2CH_2CHNH_2COOH$
9. Alpha-amino-delta-guanidine-valeric acid (arginine)	$HN:C{<}^{NH_2}_{NHCH_2CH_2CH_2CHNH_2COOH}$
*10. Alpha-epsilon-diaminocaproic acid (lysine)	$CH_2NH_2CH_2CH_2CH_2CHNH_2COOH$
11. Bi-alpha-amino-beta-thiopropionic acid (cystine)	$CH_2S{-}SCH_2$ $CHNH_2 \quad CHNH_2$ $COOH \quad COOH$
*12. Alpha-amino-beta-phenylpropionic acid (phenylalanine)	CH_2CHNH_2COOH
13. Alpha-amino-beta-parahydroxyphenylpropionic acid (tyrosine)	CH_2CHNH_2COOH ... OH
14. Alpha-pyrrolidine carboxylic acid (proline)	$H_2C{-}CH_2$ $H_2C \quad CHCOOH$ NH
*15. Alpha-amino-beta-imideazolepropionic acid (histidine)	$N{-}CH$ $HC \quad CCH_2CHNH_2COOH$ N
*16. Alpha-amino-beta-indolepropionic acid (tryptophane)	CCH_2CHNH_2COOH CH N
*17. Alpha-aminoisovaleric acid (valine)	$(CH_3)_2CHCHNH_2COOH$
*18. Alpha-amino-beta-hydroxybutyric acid (threonine)	$CH_3CHOHCHNH_2COOH$
*19. Alpha-amino-gamma-methylthiolbutyric acid (methionine)	$CH_3·S·CH_2CH_2CHNH_2COOH$

* These acids have been shown by Rose to be essential in the diet of the growing rat or dog, based on experiments covering a period of about 6 weeks.

Phenylalanine, tyrosine, tryptophane are benzenoid compounds and consequently react readily with **nitric acid** to form yellow nitro-compounds. They are responsible for the yellow xanthroproteic reaction of proteins with concentrated nitric acid.

Tyrosine is a phenolic compound and consequently reacts with nitric acid to form upon boiling with nitrous acid plus **nitric acid** plus **mercurous** salt plus mercuric salt (Millon's solution) a brick-red color.

Tryptophane, when treated with glyoxylic (glyoxalic) acid (CHO·COOH), and stratified by concentrated **sulfuric acid** forms a violet color at the junction of the liquids. Glyoxylic acid is made by reduction of **oxalic acid** solution by **sodium** amalgam.

Cystine contains sulfur, reacts with **sodium** hydrox-

COOH) forms alanylglycine (CH₃CHNH₂—CO—NH—CH₂·COOH) in which the peptide group (—CO—NH—) serves to unite the two nuclei. In the reverse direction, alanylglycine can be caused to react with water to form alanine plus glycine, the disruption occurring at the peptide grouping.

Aminoacids are of great importance on account of being constituents, through the peptide group, of naturally occurring proteins of plants and animals. The percentage of **nitrogen** element found in proteins varies between the limits 15.0 and 17.6, of **sulfur** between 0.5 and 2.2, and of **phosphorus** between 0.4 and 0.9. Proteins are important **food** materials and are characterized by the above nitrogen content, being in this way distinguished from the other grand classes of food mate-

Amino- or Hydroxyacid	Formula	Product
Alpha-amino	CH_2NH_2COOH	Lactim: CH_2—NH—CO | | CO—NH—CH_2
Beta-amino	$CH_2NH_2CH_2COOH$	Unsaturated acid: $CH_2 : CHCOOH$
Gamma-amino	$CH_2NH_2 \cdot CH_2CH_2COOH$	Gamma-lactam: $CH_2CH_2CH_2CO$ $\lfloor$—NH—$\rfloor$
Alpha-hydroxy	$CH_3CHOHCOOH$	Lactide: H_3CCH—CO—O | | O—CO—$HCCH_3$
Beta-hydroxy	$CH_3 \cdot CHOH \cdot CH_2COOH$	Unsaturated acid: $CH_3CH : CH \cdot COOH$
Gamma-hydroxy	$CH_3 \cdot CHOH \cdot CH_2 \cdot CH_2 \cdot COOH$	Gamma-lactone: $CH_3CHCH_2CH_2CO$ $\lfloor$—O—$\rfloor$

rials, namely, fats, carbohydrates. Proteins are among the most important constituents of the living **cell**, and intimately connected with life processes. Each type of organism and each kind of cell within the organism possess their own characteristic proteins. Proteins are essential constituents of animal diet, and cereal grains, legumes, eggs, milk, cheese and meat are outstanding food sources of proteins. Wool, hair, silk, and skin are also proteins. Some proteins, such as keratin of hair and horn, are not broken down by the **enzymes** of the alimentary canal, and are thus not utilizable as foods. Certain proteins have important industrial uses. These are wool, silk, skins and hides for leather, hair, gelatin, casein for plastics, glue and casein for adhesives.

COMPOSITION OF CHIEF PROTEIN FOODS

	Protein, %	Fat, %	Carbohydrate, %
Wheat	8–17	1.5–4	65–79
Corn	7.5–13	3–7.5	65–76
Oats	12–14	5–8	67
Rice	8	2	77
Peas, dried	25	1	62
Beans, dried	23	2	60
Potatoes	1.4–2.8	0.2–0.4	15–29
Bananas	1.5	0.5	22
Apples	0.4	0.5	14
Peanuts	26	40	22
Milk	3.5	4	5
Butter	1	85	
Cheese, cream	26	37	3
Cheese, cottage	21	1	4
Beef, raw	19	19	
Eggs	15	11	

Proteins are complexes, which contain alpha-aminoacids united by peptide linkage (—CO·NH—). Unlike dissolved sugars, proteins do not diffuse through animal membranes, and, except for **aleurone** grains in the seeds of plants, are not found crystallized, and are only obtained crystalline with great difficulty. Proteins possess the acid-base characteristics of aminoacids, and the properties of protein solutions depend to a large degree upon the **hydrogen-ion** concentration. At the **isoelectric point**, which is characteristic of each protein, it is found that optimum conditions exist for coagulation of proteins by heating. Proteins are coagulated not only by heating, as in the familiar case of egg albumin, which coagulation is not reversible, but also by addition of **sodium, potassium** or **ammonium** salts, which, at a definite concentration of salt in solu-

tion, precipitate proteins. Thus precipitated, the protein retains its original propertes, and, by filtration and dissolving out of the salt, may be obtained in its original condition. The precipitation is thus reversible. The addition of certain substances, such as **alcohol, acetone, tannic acid,** or salts of heavy metals (for instance, mercuric chloride), causes irreversible precipitation.

The coagulation of proteins by heat is also affected by the presence of salts, small quantities of salts usually raise the coagulation temperature, and large quantities lower it.

A system of classification of proteins has been adopted by the American Society of Biochemists.

Proteins, in general, give positive reactions in many of the tests described under aminoacids, for example, (1) xanthoproteic acid reaction with **nitric acid** concentrated, the yellow color being changed orange by alkalis, (2) Millon's reaction, (3) glyoxylic acid reaction, (4) biuret reaction, by the reaction of protein in sodium hydroxide solution plus copper sulfate a violet red color is produced, (5) sulfur reaction of cystine. Proteins are precipitated in acid solution by phosphotungstic acid, **tannic acid, picric acid,** and hydroferrocyanic acid, and, in neutral or slightly alkaline solution, by salts of certain metals, such as **cupric, mercuric, lead, gold,** and **ferric.** Proteins are usually present in plant and animal materials in the colloidal state.

The data on page 78 on the composition of some proteins is selected from the work of Mitchell and Hamilton (1929). (R.K.S.)

AMINOAZO-COMPOUNDS. Azo-, Diazo-, and Related Compounds.

AMINOGUANIDINES. Amines and Amides.

AMITOSIS. Cell Division.

AMMETER. An instrument for measuring electric currents in amperes. D-c ammeters are usually of the moving-coil type, being similar in principle to the d'Arsonval **galvanometer.** A coil carrying the current to be measured turns between the poles of a permanent magnet against the torque of a hair-spring and causes a pointer to move over its dial. A-c ammeters commonly have two coreless coils in series, one turning in the field set up by the other which is fixed (electrodynamometer type). The reversal of current thus has no effect upon the direction of the torque. Some simple ammeters are of the hot-wire type, in which the longitudinal expansion of the wire carrying the current controls the movement of the pointer. For currents heavier than the coil or the hot wire can safely stand, a shunt may be provided in d-c ammeters, which allows only a prede-

PROTEIN	TRYPTO-PHANE %	GLUTAMIC ACID %	ARGININE %	CYSTINE %	PHENYL-ALANINE %	TYROSINE %	
Gliadin........	1	44	3	2	2	3	Proline 13%
Zein..........	....	31	2	1	8	6	
Milk albumin...	3	13	3	4	1	2	
Casein.........	....	22	....	....	4	7	
Egg albumin....	1	13	6	1	5	4	{ Leucine plus isoleucine 25%
Gelatin.........	....	6	9	....	1	....	{ Glycine 25%, hydroxyproline 14%

(R.K.S.)

termined fraction of the current to pass through the instrument; while a small transformer serves a similar purpose with a-c ammeters. Very sensitive ammeters, graduated in milliamperes, are called milliammeters. Very feeble currents are commonly measured by a microammeter of the **vacuum thermocouple** type. (L.D.W.)

AMMINES. Dry **ammonia** gas reacts with dehydrated salts of some of the metals to form solid ammines. Ammines, upon warming, evolve ammonia, sometimes with final decomposition of the salt itself, in a manner analogous to the decomposition of certain hydrates. The ammines of **chromic** (Cr^{+3}), **cobaltic** (Co^{+3}), **platinic** (Pt^{+4}) and other metals have been studied in detail. Two series of ammines are shown below, the first one in which the neutral ammonia group is replaced step by step by the negative nitro group (NO_2^{-1}), and the second one in which the neutral ammonia group is replaced step by step by the neutral H_2O group.

Number of Neutral Groups, e.g., (NH_3), on Metal, e.g., Co, varied from 6 to 0.

$$[Co(NH_3)_6]Cl_3$$
$$410$$

$$[Co(NH_3)_5(NO_2)]Cl_2$$
$$240$$

$$[Co(NH_3)_4(NO_2)_2]Cl$$
$$95$$

Number of Neutral Groups Constant, but Groups Varied	$[Co(NH_3)_3(NO_2)_3]$ 1.5
X = unit anion	
$[Cr(NH_3)_6]X_3$	$K[Co(NH_3)_2(NO_2)_4]$
$[Cr(NH_3)_5(H_2O)]X_3$	95
$[Cr(NH_3)_4(H_2O)_2]X_3$	$K_2[Co(NH_3)(NO_2)_5]$
$[Cr(NH_3)_3(H_2O)_3]X_3$	240
$[Cr(NH_3)_2(H_2O)_4]X_3$	$K_3[Co(NO_2)_6]$
	420

The neutral group of the complex may be replaced step by step by the following negative groups: Cl^-, Br^-, I^-, F^-, OH^-, NO_2^-, NO_3^-, CN^-, CNS^-, SO_4^{--}, CO_3^{--}, $C_2O_4^{--}$; or by the following neutral groups: H_2O, NO, NO_2, SO_2, S, N_2H_4, H_2NOH, CO, C_2H_5OH, C_6H_6. All neutral groups are of substances capable of independent existence.

In the ammines, trivalent metals, such as **cobaltic** and **chromic** above, possess a coordination number of 6, this number being the sum of the unit replacements on the metal in the complex ion. Since a regular octahedron has six corners equidistant from the center, it is assumed that the metal occupies the center and each of the six replacing groups occupies a corner of a regular octahedron. Support for this assumption is offered by the x-ray examination of these ammines. When there is only one of the six groups replaced by a second group, as in $[Co(NH_3)_5(NO_2)]Cl_2$, and in $[Cr(NH_3)_5$

$(H_2O)]X_3$ the octahedral placement of groups supplies only one form, but when two of the six groups are replaced by a second group, as in $[Co(NH_3)_4(NO_2)_2]Cl$, and in $[Cr(NH_3)_4(H_2O)_2]X_3$, two different octahedral corner arrangements are possible depending upon whether the two replacing groups are adjacent (cisform) or opposite (trans-form). Two substances differing in physical properties and corresponding to these two forms are known. Further, when three divalent groups, e.g., $3C_2O_4^{--}$ are present in the complex, two arrangements—not identical but mirror-images of each other—are possible. Two optically active substances are known in such cases corresponding to these two sterisomeric forms.

Six is the ordinary coordination number for metallic ammines and similar complexes. Additional examples are $K_2[Pt(NH_3)_2(CN)_4]$, $[Ni(NH_3)_6]Cl_2$, $K_4[Fe(CN)_6]$, $K_3[Fe(CN)_6]$, $K_2[Fe(Cn)_5(NO)]$, $K_2[SiF_6]$, $[Ca(NH_3)_6]Cl_2$. But, for the elements boron, carbon, and nitrogen four is the coordination number, e.g., $[BH_4]Cl$, $[CH_4]$, $[NH_4]Cl$, and in these substances the groups are assumed to occupy the corners of a regular tetrahedron; in $K_4[Mo(CN)_8]$ and $[Ba(NH_3)_8]Cl_2$ the

Square bracket contains the ion.
Equivalent electrical conductivity inserted below each compound.

coordination number is eight, and the groups are assumed to occupy the corners of a cube. (R.K.S.)

AMMONIA. Ammonia (NH_3) is a colorless gas, of characteristic choking odor, density 0.7710 gram per liter, 0° C., 760 mm., or 0.60 when air equals 1.00, melting point $-78°$ C., boiling point $-33°$ C., critical temperature 132° C., critical pressure 112 atmospheres, most soluble of the common gases (about 30% NH_3, 70% water, at room temperature and pressure).

Ammonia is used in solution (1) as an important **alkali** and (2) in producing ammonium compounds; as liquid (3) in refrigeration, and (4) as a **solvent** for certain substances; as gas (5) in local production of **hydrogen** by thermal decomposition for use in high temperature combustion, and (6) in the manufacture of **nitric acid** by incomplete combustion with air over heated platinum.

Ammonia is formed (1) in the destructive distillation of **coal** in the production of coke and coal gas, and is recovered from the gas by dissolving in water, and then

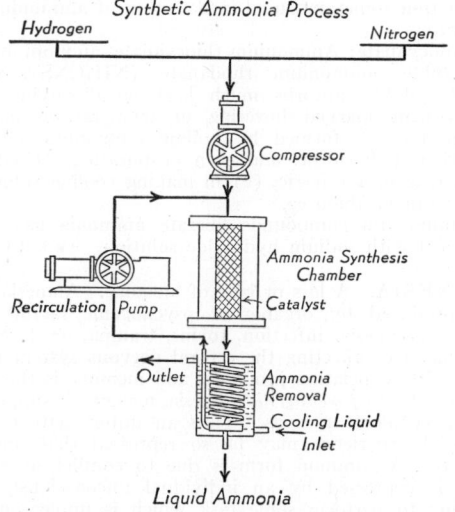

Synthetic Ammonia Process

Hydrogen — Nitrogen

Compressor

Ammonia Synthesis Chamber

Recirculation Pump — Catalyst

Outlet — Ammonia Removal

Cooling Liquid Inlet

Liquid Ammonia

Diagrammatic representation of the manufacture of ammonia from hydrogen and nitrogen.

distilling, (2) by reaction of **nitrogen** and hydrogen gases under pressure at elevated temperature in the presence of a **catalyzer** (see graph). Since the gases must be pure for this reaction to take place the resulting ammonia is of high purity, and may be directly liquefied, dissolved in water, or used as gas. When passed over heated **magnesium**, ammonia gas yields hydrogen gas and magnesium nitride (yellow solid, evolving ammonia

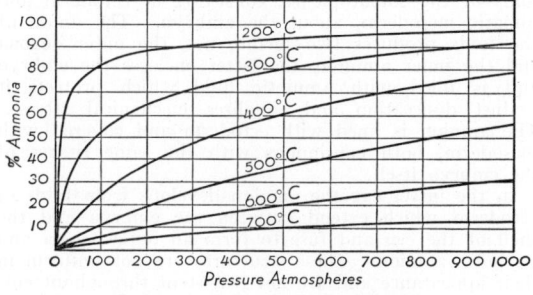

Graph showing ammonia equilibria.

with water), and when passed over heated cupric oxide, yields nitrogen gas, water, and finely divided **copper** metal. The reactions commonly assigned to ammonia as a base are given under **Ammonium.** (R.K.S.)

AMMONOLYSIS. Amination.

AMMONITE. Invertebrate Paleontology.

AMMONIUM. Radical: NH_4 The chemical radical ammonium (NH_4) is composed of **nitrogen** and **hydrogen** and it commonly behaves as a unit. Thus, it forms a series of ammonium salts which resembles the corresponding **potassium** salts. The radical should be distinguished carefully from ammonia (NH_3), which is a gas of separate existence (see **Ammonia**), whereas ammonium is encountered in compounds, such as ammonium chloride (NH_4Cl). When a concentrated water solution of ammonium chloride is electrolyzed, using a **mercury** cathode at $0°$ C., mercury ammonium amalgam, resembling sodium amalgam, is formed. When

warmed above $0°$ C., there are formed mercury, ammonia, and hydrogen.

Acetate: Ammonium acetate ($NH_4C_2H_3O_2$), white solid, soluble, formed by reaction of ammonia or ammonium hydroxide and **acetic acid,** reacts upon heating to yield acetamide.

Alum: Ammonium alums are those alums, such as **aluminum** ammonium sulfate ($Al_2(SO_4)_3 \cdot (NH_4)_2SO_4 \cdot 24H_2O$), **ferric** ammonium sulfate ($Fe_2(SO_4)_3 \cdot (NH_4)_2 SO_4 \cdot 24H_2O$), **chromium** ammonium sulfate ($Cr_2(SO_4)_3 \cdot (NH_4)_2SO_4 \cdot 24H_2O$) where ammonium sulfate is crystalized with the heavier metal sulfate.

Benzoate: Ammonium benzoate ($NH_4C_7H_5O_2$), white solid, soluble, formed by reaction of ammonium hydroxide and **benzoic acid.** Used (1) as a food preservative, (2) in medicine.

Borate: Ammonium borate, ammonium tetraborate ($(NH_4)_2B_4O_7 \cdot 4H_2O$), white solid, soluble, formed by reaction of ammonium hydroxide and **boric acid.** Used (1) in fireproofing fabrics, (2) in medicine.

Bromide: Ammonium bromide (NH_4Br), white solid, soluble, sublimes at $542°$ C., formed by reaction of ammonium hydroxide and **hydrobromic acid.** Used in photography.

Carbonates: Ammonium carbonate, sal volatile ($(NH_4)_2CO_3$), white solid, soluble, formed by reaction of ammonium hydroxide and **carbon dioxide** by crystallization from dilute alcohol, loses ammonia, carbon dioxide, and water at ordinary temperatures, rapidly at $58°$ C.; ammonium hydrogen carbonate, ammonium bicarbonate, ammonium acid carbonate (NH_4HCO_3), white solid, soluble, formed by reaction of ammonium hydroxide and excess carbon dioxide. This salt is the important reactant in the ammonia soda process for converting **sodium** chloride in solution into sodium hydrogen carbonate solid.

Chloride: Ammonium chloride, sal ammoniac, muriate of ammonia (NH_4Cl), white solid, soluble, sublimes at $520°$ C., formed (1) as a white smoke by reaction of ammonia gas and **hydrogen chloride** gas, (2) by reaction of ammonium hydroxide and hydrochloric acid, and then evaporating. Used (1) as an important nitrogenous fertilizer, (2) in dry cell electric batteries, (3) in soldering flux, (4) in the textile and tanning industries.

Chloroplatinate: Ammonium chloroplatinate ($(NH_4)_2PtCl_6$), yellow solid, insoluble, formed by reaction of soluble ammonium salt solutions and **chloroplatinic acid.** Used in the quantitative determination of ammonium.

Cobaltinitrite: Diammonium sodium cobaltinitrite ($(NH_4)_2NaCo(NO_2)_6 \cdot H_2O$) golden yellow precipitate, formed by reaction of sodium cobaltinitrite solution in acetic acid with soluble ammonium salt solution. Used in the detection of ammonium.

Cyanate: Ammonium cyanate (NH_4CNO), white solid, soluble, formed by fractional crystallization of **potassium** cyanate and ammonium sulfate (ammonium cyanate is soluble in alcohol), when heated changes into urea (the classical experiment by Wöhler in 1828).

Dichromate: Ammonium dichromate ($(NH_4)_2Cr_2O_7$), red solid, soluble, upon heating evolves nitrogen gas and leaves a green insoluble residue of **chromic** oxide.

Fluoride: Ammonium fluoride (NH_4F), white solid, soluble, formed by reaction of ammonium hydroxide and **hydrofluoric acid,** and then evaporating. Used (1) as an antiseptic in brewing, (2) in etching glass; ammonium hydrogen fluoride, ammonium bifluoride, ammonium acid fluoride (NH_4F_2), white solid, soluble.

Hydroxide: Ammonium hydroxide (NH_4OH), colorless solution, by solution of (1) ammonia gas in water, commercially of strength $26°$ Baumé (Specific gravity at $66°$ F., water at $60°$ F., 0.8974), 29.40% NH_3, (2) anhydrous liquid ammonia (in cylinders). Used (1) as a mild, moderately cheap alkali, e.g., laundering, (2) as a source of ammonium for many salts, (3) as an important chemical reagent. (See **Reactions Involving Recombination of Ions.**)

Iodide: Ammonium iodide (NH_4I), white solid, soluble, formed by reaction of ammonium hydroxide and **hydriodic acid,** and then evaporating. Used (1) in photography, (2) in medicine.

Linoleate: Ammonium linoleate ($NH_4C_{18}H_{31}O_2$). Used (1) as an emulsifying agent, (2) as a detergent.

Nitrate: Ammonium nitrate, nitrate of ammonia (NH_4NO_3), white solid, soluble, melting point 170° C., formed (1) by fractional crystallization of ammonium sulfate or chloride and **sodium** nitrate, (2) by reaction of ammonium hydroxide and **nitric acid,** and then evaporating. Used (1) in the preparation of nitrous oxide gas (by heating), (2) in explosives, pyrotechnics, (3) in freezing mixtures of salts.

Nitrite: Ammonium nitrite (NH_4NO_2) when ammonium sulfate or chloride and sodium or potassium nitrite are heated, the mixture behaves like ammonium nitrite in yielding nitrogen gas.

Oxalate: Ammonium oxalate (($NH_4)_2C_2O_4$), white solid, soluble, formed by reaction of ammonium hydroxide and **oxalic acid,** and then evaporating. Used as a source of oxalate; ammonium binoxalate ($NH_4HC_2O_4 \cdot H_2O$), white solid, soluble.

Perchlorate: Ammonium perchlorate (NH_4ClO_4), white solid, soluble, formed by reaction of ammonium hydroxide and **perchlorate acid,** and then evaporating. Used in explosives and pyrotechnics.

Periodate: Ammonium periodate (NH_4IO_4), white solid, moderately soluble.

Persulfate: Ammonium persulfate (($NH_4)_2S_2O_8$), white solid, soluble, formed by **electrolysis** of ammonium sulfate under proper conditions. Used (1) as a bleaching and oxidizing agent, (2) in electroplating, (3) in photography.

Phosphate: Diammonium hydrogen phosphate (($NH_4)_2HPO_4$), white solid, soluble, formed by reaction of excess ammonium hydroxide and **phosphoric acid,** and then evaporating at room temperature. Used (1) in fireproofing, (2) as a fertilizer supplying nitrogen and phosphorus, (3) in medicine; ammonium dihydrogen phosphate ($NH_4H_2PO_4$), white solid, soluble, formed by heating diammonium phosphate to 155° C.

Phosphomolybdate: Ammonium phosphomolybdate, (($NH_4)_3PO_4 \cdot 12MoO_3$ or similar composition), yellow precipitate, soluble in alkalis, formed by excess ammonium molybdate and nitric acid with soluble phosphate solution. Used as an important test for phosphate (similar product and reaction when arsenate replaces phosphate).

Salicylate: Ammonium salicylate ($NH_4C_7H_5O_3$), white solid, soluble, formed by reaction of ammonium hydroxide and **salicylic acid,** and then evaporating. Used in medicine.

Sulfate: Ammonium sulfate, sulfate of ammonia (($NH_4)_2SO_4$), white solid, soluble, formed by reaction of ammonium hydroxide and **sulfuric acid,** and then evaporating. Used (1) as an important nitrogenous **fertilizer,** largely obtained by recovery of ammonia from by-product coke oven operations, (2) as a soldering liquid, (3) in fireproofing fabrics, (4) in electric dry cell batteries, (5) as a source of ammonia; ammonium hydrogen sulfate, ammonium bisulfate, (NH_4HSO_4), white solid, soluble, melting point 147° C.

Sulfide: Ammonium sulfide (($NH_4)_2S$), colorless to yellowish solution, formed by saturation with **hydrogen sulfide** of one-half of a solution of ammonium hydroxide, and then mixing with the other half of the ammonium hydroxide. Dissolves sulfur to form ammonium polysulfide, yellow solution. Used as a reagent in analytical chemistry; ammonium hydrogen sulfide, ammonium bisulfide, ammonium acid sulfide (NH_4HS), colorless to yellowish solution, formed by saturation with hydrogen sulfide of a solution of ammonium hydroxide.

Tartrate: Ammonium tartrate (($NH_4)_2C_4H_4O_6$), white solid, moderately soluble, formed by reaction of

ammonium hydroxide and **tartaric acid,** and then evaporation. Used in the textile industry; ammonium hydrogen tartrate, ammonium bitartrate, ammonium acid tartrate ($NH_4HC_4H_4O_6$), white solid, slightly soluble, formation sometimes used in detection of ammonium or tartrate.

Thiocyanate: Ammonium thiocyanate, ammonium sulfocyanide, ammonium rhodanate (NH_4CNS), white solid, soluble, absorbs much heat on dissolving with consequent marked lowering of temperature, melting point 150° C., formed by boiling ammonium cyanate solution with sulfur, and then evaporating. Used (1) as a reagent for ferric, (2) in making cooling solutions, (3) to make thiourea.

Ammonium compounds liberate ammonia gas when warmed with sodium hydroxide solution. (R.K.S.)

AMNESIA. A loss or lack of memory. Amnesia may be produced by organic or psychogenic factors. In *organic* amnesia, infection, toxins, trauma, or degenerative changes affecting the central nervous system interfere with association processes, and memory is therefore impaired. In *psychogenic* amnesia, memory is suppressed for psychologic reasons. Thus an unfortunate or distasteful experience may be so repressed that amnesia results. A common form is due to conflict of wishes and is expressed by an individual unconsciously forgetting to perform some task which is unpleasant for him. A sudden and complete recovery of memory not infrequently occurs with psychogenic amnesia, but in the organic variety recovery is slow and rarely complete. (D.M.H.)

AMNION. An accessory embryonic membrane common to reptiles, birds and mammals and a superficially similar structure found in some insects.

To the three vertebrate classes, all fundamentally terrestrial, the amnion gives the name **Amniota.** In this group the amnion is usually formed by the growth of folds of the somatopleure, consisting of ectoderm and somatic mesoderm, about the **embryo.** The union of the folds produces two membranes, the outer **serosa** and the inner amnion; the latter encloses the embryo and is filled with amniotic fluid which protects it against desiccation and equalizes mechanical stresses. The amnion is lined with ectoderm and covered with mesoderm, both continuous with the same tissues of the embryo itself.

In the insect egg the amnion develops from folds of ectoderm which extend between the embryo and the shell of the egg and fuse to form an outer serosa and an inner amnion. These structures are not uniform in their appearance and are not persistent throughout embryonic development in all species in which they form, hence their functions are in doubt. (A.W.L.)

AMNIOTA. The classes **Reptilia, Aves,** and **Mammalia** of the vertebrates, in which an **amnion** appears during embryonic development. (A.W.L.)

AMOEBA. A genus of one-celled animals in which the body consists of a naked mass of **protoplasm** and the organs of locomotion are temporary blunt protuberances of cytoplasm (see **Cell**) known as pseudopodia. (See **Pseudopodium.**) (A.W.L.)

AMOEBIC DYSENTERY. Dysentery.

AMOEBULAE. Spores of amoeboid form which are produced by some 1-celled animals. (A.W.L.)

AMORPHOUS. As applied to minerals and rocks this term means: having no definite crystalline structure (texture). (R.M.F.)

AMORTISSEUR WINDING. Damper Winding.

AMOSITE. Amosite is a long fiber gray or greenish asbestiform mineral related to **anthophyllite**, found in South Africa. (E.S.C.S.)

AMPERE. The ampere is the practical unit of electric current. It is primarily an electromagnetic unit, the absolute ampere being defined as $\frac{1}{10}$ of the **abampere.** For reasons of expediency in standardization measurements, however, another basis has been adopted in the definition of the international ampere, which is one international **coulomb** per sec.; i.e., that current which would deposit silver at the rate of 0.001118 gram per sec. from a silver nitrate solution in a **coulombmeter** under prescribed conditions. This electrolytic unit of current is slightly less than the absolute ampere (by about 14 parts in 100,000), but the difference is commonly overlooked in practical measurements. Since the ampere is rather large for many laboratory purposes (a 50-watt, 110-volt lamp requires less than half an ampere), the milliampere (1/1000 ampere) and even the microampere (1/1,000,000 ampere) are often found convenient. (L.D.W.)

AMPERE-HOUR. An ampere-hour is a quantity of electricity, equal to 3600 **coulombs**, viz., the electricity flowing in one hour past any point of a circuit carrying one ampere. It is not a measure of quantity of electrical energy, since voltage does not enter into the product of amperes times hours. The use of the ampere-hour is confined largely to **storage battery** practice. (See **Storage Battery.**) Storage batteries are rated in ampere-hours to show the quantity of electricity that can be used without discharging the battery beyond safe limits. The quantity may be measured by an ampere-hour meter. (F.T.M.)

AMPÈRE'S LAW. This is a classic law of electromagnetism, useful in discussions of electrodynamics. It has been stated in two apparently distinct forms, which are, however, interconvertible.

One form, sometimes known as **Laplace's law**, states that the electric current i (**abamperes**), flowing along

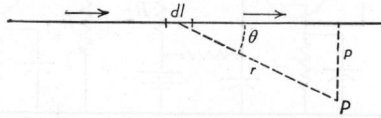

Current in element dl produces magnetic field at point P.

any line through an element of length dl, gives rise, at a point P distant r (cm.) from the element, to a magnetic field of intensity $dH = ipdl/r^3$ (**oersteds**), in which p is the perpendicular distance from P to the line of the element dl; or $dH = i \sin \theta dl/r^2$, in which θ is the angle between the line of the element and the line joining dl to P. The ultimate basis of the law is, of course, experimental. It may, for example, be deduced indirectly from the **Biot-Savart** law for the field about an infinitely long, straight wire. Ampère's law furnishes a basis for the solution of all problems relating to the magnetic fields produced by electric currents.

What is sometimes called the circuital form of Ampère's law may be tangibly expressed by saying that if a unit magnetic pole is carried completely around a conductor or system of conductors in which electricity is flowing, in such a way as to oppose the field set up by the currents, the work done, in ergs, is 4π times the algebraic sum of the currents, in abamperes. This is easily illustrated by a special case. The Biot-Savart law above referred to gives, as the magnetic intensity at a point distant r (cm.) from an infinitely long, straight wire carrying a current i (abamperes), the value $H = 2i/r$ (oersteds), directed, of course, along the circumference of the circle having r as its radius. The force acting upon a unit pole placed at this point

is therefore $2i/r$ (**dynes**). If now the pole is moved around the circle, against the field, the work done is

$$\frac{2i}{r} \times 2\pi r = 4\pi i \text{ (ergs).}$$

Maxwell pointed out that Ampère's law holds only for constant currents, and that when currents vary, the resulting changes of electric displacement in the surrounding space, giving rise to the radiation of energy in the form of electromagnetic waves, involves modifications embodied in the so-called Maxwell-Ampere law as expressed by the first of **Maxwell's equations.** (L.D.W.)

AMPHIBIA. The frogs, toads, newts, salamanders and related forms. A class of the phylum **Chordata.** Since these animals live only in moist places their distribution is restricted and they are among the less familiar vertebrates.

The amphibians are distinguished by: 1. Moist skin. 2. The absence of scales and claws. 3. Most species undergo a metamorphosis during development from an aquatic, gill-breathing larva to a semi-terrestrial, air-breathing adult stage.

While some of the salamanders are permanently aquatic and some of the tree frogs permanently terrestrial, most members of the class live near the water or in moist places and undergo the metamorphosis mentioned above.

The following orders of amphibians are recognized:

Order **Gymnophiona** (Apoda). Legless, worm-like animals, confined to the tropics of the Old and New Worlds.

Order **Urodela** (Caudata). Elongate animals with long tails and weak, short legs. The **salamanders, newts, efts, hellbender,** and **mud puppy.**

Order **Anura** (Salientia). Tailless species whose hind legs are the larger pair, more or less strongly developed for jumping. Most species have a larval stage known as the tadpole with a compact body and a long compressed tail but no legs until the onset of metamorphosis. The **frogs** and **toads.** (A.W.L.)

AMPHIBIAN AIRPLANE. Seaplane.

AMPHIBOLE. This is the name given to a closely related group of minerals all showing in common a prismatic cleavage of 54–56° as well as similar optical characteristics and chemical composition.

The amphiboles may be said to represent chemically a series of **metasilicates** corresponding to the general formula $RSiO_3$ where R may be **calcium, magnesium, iron, aluminum, titanium, sodium** or **potassium.** The crystals of the amphiboles may be in one of the three systems, **orthorhombic, monoclinic** or **triclinic.**

There is a clear parallelism between the amphiboles and the **proxenes.** The chief difference between the minerals of these two groups lies in the cleavage angles of 56° and 124° for amphibole and 87° and 93° for pyroxene. Amphibole crystals are usually long and slender and tend to be simple while pyroxene crystals tend to be complex, short and stout prisms.

Amphibole is common in both lavas and deep-seated rocks, though less so in the basic lavas than pyroxene. Many of the amphiboles may be developed as metamorphic minerals. The following members of the amphibole group are described under their own headings: **actinolite, anthophyllite, cummingtonite, glaucophane, grünerite, hornblende, riebeckite** and **tremolite.** Amphibole was so named by Haüy from the Greek word, meaning doubtful, because of the many varieties of this mineral. (E.S.C.S.)

AMPHIBOLITE. The amphibolites form a large group of rather important rocks of **metamorphic** character. As the name implies they are made up very largely of minerals of the amphibole group. There may be also a variety of other minerals present, such as **quartz, feldspar, biotite, muscovite, garnet,** or **chlorite** in greater or less amounts.

Depending upon the particular amphibole present these rocks may be light to dark green or black, the amphibole usually being in long slender prisms or laths, often quite coarse, sometimes in acicular or fibrous forms.

Because the mineral constituents are arranged parallel to the schistosity, amphibolites may have a strongly developed cleavage.

The occurrence of amphibolites accompanying gneisses, schists, and other metamorphic rocks of probable sedimentary origin strongly suggests a similar derivation. Yet some amphibolites cut other metamorphic rocks in the manner of dikes or sills. It is very likely that they have been derived from both original igneous and sedimentary rocks. Large masses of amphibolite suggest **gabbroic** stocks. Well known areas in which amphibolites are found are New England, New York State, Canada, Scotland, and the Alps. (E.S.C.S.)

AMPHILINIDEA. An order of **tapeworms** parasitic in fishes. (A.W.L.)

AMPHINEURA. The chitons and allied forms, a class of the phylum **Mollusca.** The more familiar members are flattened marine animals of oval outline. They have a shell composed of a series of separate plates, sometimes concealed within the body. The foot makes up most of the ventral surface and the limited **mantle** extends down about it to form a shallow groove. Nerve cells are in many cases distributed through the nerve cords so that the nervous system contains no ganglia (see **Ganglion**).

The class contains two orders:

Order **Aplacophora** (Solenogastres). Worm-like animals, somewhat cylindrical and elongate. Shell lacking.
Order **Polyplacophora**. The chitons. Flattened and oval, with shell plates. They are widely distributed, chiefly in the shallow waters. Sometimes used as food. (A.W.L.)

AMPHIOXUS. Commonly used to designate any of the primitive **chordates** called lancelets but more accurately a genus of these animals. **Cephalochordata.** (A.W.L.)

AMPHIPODA. An order of **crustaceans** including marine and fresh-water species. The beach-fleas are among the few which have a common name. (A.W.L.)

AMPHISBAENA. Reptilia, Lacertilia. The typical genus of a family of **lizards** which are snake-like because of the absence of legs. (A.W.L.)

AMPHOTERIC HYDROXIDES. These are hydroxides which can act as either **bases** or **acids,** e.g., **aluminum** hydroxide, **zinc** hydroxide, etc. (R.K.S.)

AMPLIFICATION FACTOR. This is a commonly used parameter in vacuum-tube work and is usually denoted by μ. In the conventional **tube** the current is controlled both by the voltage applied to the **grid** and that applied to the **plate**. Because the grid is closer to the **cathode** from which the **electrons** are drawn, a voltage applied to it is more effective in drawing the electrons across the tube than would be the same voltage applied to the plate (it is the passage of these electrons across the cathode-anode space which constitutes the tube current). Amplification factor is a measure of the relative effectiveness of voltages on the two electrodes and is defined as the negative of the ratio of the in-

finitesimal plate voltage change necessary to counteract a given infinitesimal change in grid voltage in order to keep the plate current constant.

$$\mu = -dE_p/dE_g, I_p \text{ constant}$$

where dE_p and dE_g represent very small increments of plate and grid voltages. (L.R.Q.)

AMPLIFIER. An amplifier is a device for increasing the magnitude of some quantity and as normally used in electrical engineering applies to a vacuum **tube** and its associated circuit arranged to reproduce in its plate circuit in greater magnitude a voltage or current which is applied in its grid circuit. The term amplifier is used to denote both a single stage or several stages in cascade. Amplifiers are classified in various ways according to their use, circuit, etc., but all types depend upon a few basic circuits. Thus they may be termed current amplifiers or voltage amplifiers, power amplifiers, audio-frequency amplifiers, radio-frequency amplifiers and various coupling types.

Current amplifiers are designed to give an amplified current in the plate circuit. However, since the vacuum tube is a voltage-controlled device, the input current is applied to the grid as a voltage drop that is produced in an impedance. This voltage drop controls the current in the plate circuit which is designed to give relatively large currents. Voltage amplifiers on the other hand have their plate circuits designed to produce large voltage drops.

In any vacuum tube amplifier circuit the fundamental operation is the controlling of the plate current by a voltage applied to the grid. This plate-current change is utilized in various ways, giving rise to several standard classifications of voltage amplifiers. Reference to the figures will aid in understanding the operation. Fig. 1*a* shows a simple resistance-coupled circuit. The a-c voltage to be amplified is applied across the grid re-

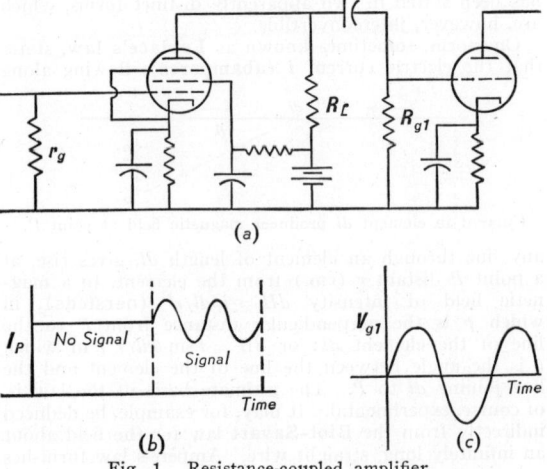

Fig. 1.   Resistance-coupled amplifier.

sistor r_g. This voltage alters the plate current which varies as the grid voltage varies. For no-signal voltage on the grid plate current is a constant value of d.c., but the a-c voltage on the grid causes this to vary as shown in Fig. 1*b*. This current produces a potential drop across the plate resistor R_L which has the same form as the current. This voltage is applied across the grid resistor of the following tube through the coupling **condenser**. However, since a condenser circuit responds only to varying voltages only the a-c component will appear across the resistor R_{g1} as shown in Fig. 1*c*. With proper choice of circuit components the voltage on the second grid is several times that impressed on the first grid. The ratio of these voltages is the gain or amplification of the stage.

While this gain varies with frequency, for the middle portion of the operating range it is given by:

$$A = \mu R'_L / (R'_L + R_p)$$

where R_p is the dynamic plate resistance of the tube, μ is the **amplification factor** of the tube, and R'_L is R_L and R_{gl} in parallel. **Pentode** tubes are normally used for resistance-coupled amplifiers because of their much greater gain.

Fig. 2 shows another method, transformer-coupled, of utilizing the current change in the plate circuit. Here

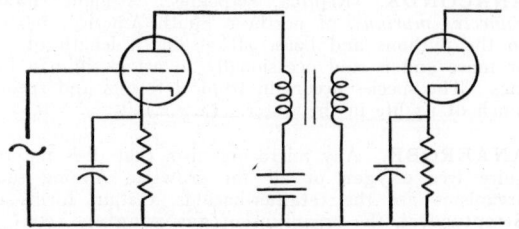

Fig. 2. Transformer-coupled amplifier.

the varying plate current produced by the alternating voltage on the grid produces a voltage across the secondary of the **transformer.** Since the transformer operates only on a.c. the voltage across the secondary is a reproduction of that applied to the grid of the first tube. The gain of such a stage is very closely given for the middle of its operating range by:

$$A = \mu n$$

where n is the **turns ratio** of the transformer (secondary/primary) and μ is the amplification factor of the tube. Because they give a wider frequency range with transformers than other types of tubes, **triodes** are normally used in such circuits. Fig. 3 shows a third

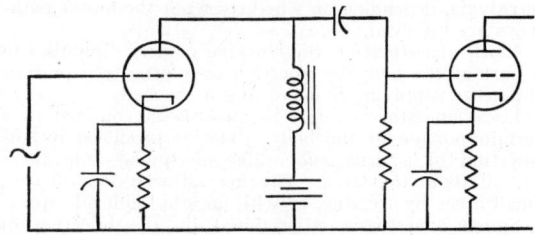

Fig. 3. Impedance-coupled amplifier.

method of coupling a voltage amplifier. This is very similar to the resistance-coupled case but the output voltage drop is across a **choke** or impedance coil. This is an impedance-coupled amplifier. Since the choke causes only a small voltage drop for d.c. very little of the supply voltage is lost and most of it appears at the plate of the tube, thereby increasing the gain by decreasing the dynamic plate resistance of the tube. The frequency response characteristic of this amplifier is usually not as good as for the resistance-coupled type. The gain in the middle range is given by:

$$A = \frac{\mu 2\pi f L}{\sqrt{R_p^2 + (2\pi f L)^2}}$$

where f is the frequency and L is the inductance of the choke in henries. Other symbols are as before.

The three circuits shown are commonly used in audio-frequency amplifiers which must operate over a wide frequency range. For radio-frequency use the range is much more restricted in terms of its proportion of the mid-frequency value. As a consequence such amplifiers can and ordinarily do use tuned circuits for coupling elements. Such an amplifier is shown in Fig. 4. While there are many variations of the tuned amplifier this is typical of those used in the radio-frequency portion of

the usual radio receiver. The response of a tuned amplifier varies markedly with frequency and the frequency at which the response is a maximum may be

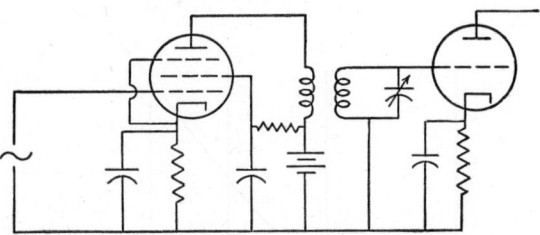

Fig. 4. Tuned amplifier.

adjusted by varying the condenser. This type amplifier, then, serves as a selective device to differentiate between stations and at the same time serves as an amplifier. The condenser shown is the tuning condenser controlled by the dial on the panel of the ordinary radio set.

For amplifying d-c voltages a direct-coupled amplifier must be used since the circuits shown before respond only to alternating voltages. Such an amplifier is shown in Fig. 5. It will be noted that the high voltage

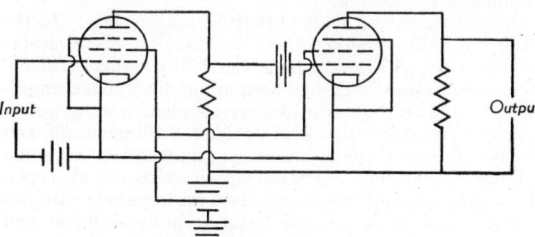

Fig. 5. Direct-coupled amplifier for d.c.

of the plate of the first tube would be applied to the grid of the second if it were not for the bucking action of the grid or C battery which must, therefore, be made large enough to give the correct net voltage on the grid. Unless the A, B, and C voltages are carefully regulated, direct-coupled amplifiers are rather unstable and are to be avoided except in cases where nothing else will serve satisfactorily.

The amplifiers discussed above are intended primarily for voltage amplification. However, in many applications it is necessary to get power from an amplifier unit so at least the final stage is usually adjusted to give power rather than voltage output. An amplifier so adjusted is called a power amplifier. In radio transmitting circuits or large power audio-amplifiers several stages may be power amplifiers. For this type of service the circuit is adjusted to give as large a current as possible through the load, which has much lower resistance than in voltage amplifiers. The tubes used are somewhat larger as a rule than those of voltage circuits and are specially designed for large current outputs. Power amplifiers are frequently classified as Class A, B, AB or C, depending upon just how they are operated. Class A amplifiers operate about a mean grid bias value such that the output wave is essentially the same as the input wave and the d-c plate current has a constant value (the voltage amplifiers are Class A). A Class B amplifier has the grid bias adjusted so that plate current flows only on the positive half-cycle of the input signal. Class AB is intermediate between A and B. In these three classes a subscript 1 is often used to denote that the grid does not take current, and a subscript 2 to denote that grid current flows. Since AB and B amplifiers using single tubes do not give output waves similar to the input waves they must be operated **push-pull** for audio-frequency work. Class C amplifiers are adjusted so plate current flows for less than a half-cycle of the input signal. They are used exclusively for radio-frequency work where a tuned circuit may be used for the load. (See **Tank Circuit.**) Fig. 6 indicates the rela-

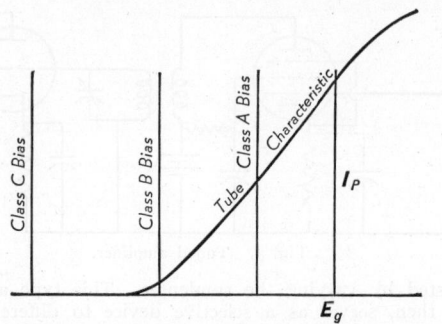

Fig. 6. Bias adjustments for various amplifier classes.

tive bias values for the different types of service. The efficiency becomes better but the distortion worse in the order A, AB, B and C. For audio-frequency work the power amplifiers are normally connected to the load (loud-**speaker**) through a transformer, while in radio-frequency work air-cored transformers and capacitance coupling are common.

Amplifiers have a wide variety of applications. In the ordinary radio **receiver** there are usually several stages of tuned amplifiers handling the high-frequency signal before **detection,** then one or more audio-voltage amplifiers, resistance- or transformer-coupled, and a power amplifier to drive the loud speaker. Phonograph and public address systems are other applications of audio amplifiers. Various industrial applications use all types. The radio broadcast station uses a wide variety—audio-voltage amplifiers in the studio, audio-voltage and power (ordinarily Class B) amplifiers at the transmitter, and usually several stages of Class C or C and B tuned amplifiers in the radio-frequency part of the transmitter. (L.R.Q.)

AMPUL or **AMPOULE.** A small sealed glass container for **drugs** that are to be given hypodermically. As they are completely sealed, the contents are kept in their original sterile condition. (R.S.M.)

AMPULLA. Any flask-like dilatation, such as the small saccular outgrowth of the water vascular system of starfishes, at the inner end of the tube foot, or the dilated portion of the semicircular canals of the vertebrate **ear.** (A.W.L.)

AMPULLIFORM GLANDS. Silk glands of **spiders** which produce the radial lines of the web. (A.W.L.)

AMPUTATION. This term is applied to the severance of a limb or other portion of the body. It may be traumatic or therapeutic. Congenital amputation is the removal of a part of the fetus in utero by a constricting band. (D.M.H.)

AMYGDALE. Amygdaloid.

AMYGDALIN. Glucosides.

AMYGDALOID. A vesicular rock, commonly a **lava,** whose cavities have become filled with a secondary deposit of mineral material such as **quartz, calcite, zeolites,** etc. The term is derived from the Greek word meaning almond in reference to the frequent almond-like appearance of the filled vesicles which are called amygdales or amygdules. (E.S.C.S.)

AMYGDULE. Amygdaloid.

ANABATIC WIND. A wind blowing uphill. In general, anabatic winds refer to winds originating in connection with surface heating, such as a breeze blowing up a valley when the sun warms the ground. (P.E.K.)

ANABOLISM. Metabolism.

ANACLINAL. Used in structural geology to define a direction opposite to the **dip** of the strata or formations. (R.M.F.)

ANACONDA. Reptilia, Serpentes. A giant **snake,** *Eunectes murinus,* of northern South America, related to the pythons and boas. It attains a length of 30' or more and is said occasionally to attack human beings. The species occurs in tropical forests and spends much of its life in the water. (A.W.L.)

ANAEROBE. Any micro-organism that does not require free **oxygen** or air for growth. Among such organisms are the **tetanus**-bacillus, certain forms of **Streptococci,** the organisms of **gas gangrene,** etc. A facultative anaerobe is a micro-organism which usually lives in air or oxygen, but which can, if necessary, live without it. (R.S.M.)

ANAEROBIC BACTERIA. Anaerobe.

ANAESTHESIA (ANESTHESIA). Loss of sensation or feeling in a part or a whole of the body. This occurs with interruption of sensory impulses of any portion of a nerve or the nerve pathways in the **spinal cord, brain,** or the centers in the brain. It can occur as the result of injury or severance, infection, circulatory disturbances, in or about nervous tissue. Any **drug** that depresses or inhibits nerve impulses, or **tumor** formation near or in nervous tissue may produce anaesthesia. Anaesthesia may or may not be accompanied by motor **paralysis,** depending on whether or not the motor pathways are involved.

Block anaesthesia is the blocking of nerve impulses to a part of the body by injection of a drug into or near the nerve supplying that portion of the body.

Local anaesthesia refers to anaesthesia confined to a certain portion of the body. This is produced by the injection of a drug (novocaine or similar substance) in and about that area. This may also be done in very small areas by freezing, as with an ethyl chloride spray.

Spinal anaesthesia is produced by the injection of novocaine or related drugs into the spinal fluid in the lower back region. By this method anaesthesia of the trunk and lower limbs may be obtained while the patient retains full consciousness. The level of anaesthesia (lower and upper abdomen and lower chest) may be regulated by the amount of drug, amount of injected fluid and position of the patient. There are certain cases in which spinal anaesthesia is indicated, although it is not used as frequently as inhalation anaesthesia.

Intravenous anaesthesia is a process in which certain drugs may be injected into the blood stream to produce anaesthesia for short periods. Such drugs as barbituric acid derivatives, for example, sodium amytal, "evipal," pentothal sodium, etc., may be given intravenously.

Basal anaesthesia is a very light anaesthesia that is produced by drugs given by mouth, by rectum, or by injection, which requires added inhalation anaesthesia but in much lighter concentration than if the preliminary medication had not been given. Avertin and other narcotic and sedative drugs are used for this purpose. The process known as twilight sleep falls in this category. An advantage of basal anaesthesia is the fact that the patient need not see the operating room, as the basal anaesthetic is given in his room.

General surgical anaesthesia refers to anaesthesia of the entire body accompanied by loss of consciousness and involving paralysis of all vital functions of the body except those governed by the respiratory and circulatory centers.

The anaesthetic agents most commonly used are as follows:

1. *Chloroform.* At present, chloroform is not commonly used in this country for anaesthesia. It is likely to cause ventricular fibrillation, a disturbance in heart rhythm which is fatal, and after prolonged or repeated use changes in the liver and kidneys may occur. It is contraindicated for children, for long operations, in patients who have diabetes, kidney, or liver disease and in general, when safer anaesthetics are available. Its anaesthetic qualities were first pointed out in 1847 by Flournes and by James Simpson.

2. *Diethyl Ether.* In non-expert hands diethyl ether is the safest of all anaesthetics. There are certain disagreeable after-effects following its use in any great quantity, such as nausea, vomiting and headache. It may be given by the open-drop method or through a closed machine in which it is mixed with oxygen or with nitrous oxide, ethylene, or cyclopropane. It also can be given by rectum, the diethyl ether being mixed with a bland oil. Diethyl ether (which is commonly referred to by the group name, that is, simply as "ether") was discovered in 1540 but it was over 300 years before it was used as an anaesthetic in surgery. Henry Hickman of England first made successful experiments with anaesthesia by inhalation in 1820, but no one was impressed with his discoveries. Crawford W. Long first used ether intelligently in 1842, but did not publicize its use until after Morton and Jackson claimed priority. William Morton, at the suggestion of Charles Jackson, a chemist, experimented with ether as an anaesthetic agent and in 1846 demonstrated it before the staff and students of the Harvard Medical School in the Massachusetts General Hospital. All modern methods date from this public demonstration. Of all anaesthetic agents, ether is the most commonly used. It should not be given in the presence of lung diseases.

3. *Nitrous Oxide (Laughing Gas).* This gas is the oldest of the inhalation anaesthetics. It has a rapid but weak anaesthetic effect and produces surgical anaesthesia only by asphyxia. It is therefore not used alone, but in combination with ether and oxygen, or with a basal anaesthetic such as avertin. If nitrous oxide were more potent it would be the perfect general anaesthetic since it is non-irritating and has no untoward side-effects.

4. *Cyclopropane.* This gas has been used as an anaesthetic since 1931. Its chief advantage lies in its effectiveness at a low concentration which permits its use with much higher concentrations of oxygen than can be given with ether, etc. It is therefore of special value in obstetrics and chest surgery. Although explosive, in the hands of an expert it is a safe and comparatively pleasant anaesthetic.

6. *Ethyl Chloride.* This is an inflammable liquid which vaporizes when sprayed on a mask. It is used for brief, light anaesthesia for minor procedures and as a preliminary agent to put a patient to sleep before giving him ether. Its use alone is not as safe as other anaesthetic agents. It is also used for local anaesthesia, being sprayed directly on the skin.

7. *Avertin (Tribromethylalcohol).* Avertin is a fluid preparation which, when given by rectum, produces a light anaesthesia. For surgical procedures of any degree, it must be supplemented by inhalation agents. Avertin, given first, reduces the amount of inhalation anaesthetics necessary to produce surgical anaesthesia. It also removes the element of fear before an operation since it is given while the patient is in his room in bed, and he usually arrives in the operating room sound asleep. (D.M.H.)

ANAL FEELERS. Posterior sensory appendages such as the anal cirri of **annelid** worms and the **cerci** of insects. (A.W.L.)

ANALCIME. Analcite.

ANALCITE. Analcite, sometimes called analcime, is a common **zeolite** mineral, a hydrous **soda-aluminum silicate.** It crystallizes in the **isometric** system, hardness, 5–5.5; specific gravity, 2.2; vitreous luster; colorless to white; but may be grayish, greenish, yellowish or reddish. It resembles **garnet** but is softer, and is distinguished from **leucite** only by chemical tests.

There are many excellent European localities. In the United States at Bergen Hill and West Paterson, N. J.; Keweenaw County, Mich.; and Jefferson County, Colorado. Nova Scotia furnishes beautiful specimens.

Analcite is a relatively common mineral and occurs with other zeolites in cavities and fissures in basic **igneous** rocks, occasionally in **granites** or **gneisses.** It seems to occur as a replacement and perhaps in some cases as a primary mineral crystallizing from a **magma** rich in soda and water vapor under pressure. The name analcite is derived from the Greek word meaning weak, in reference to the weak **electric charge** developed when heated or subjected to friction. (E.S.C.S.)

ANALOGY. Homology.

ANALYSIS OF COVARIANCE. Analysis of Variance.

ANALYSIS OF VARIANCE. The analysis of variance is a technique by which the sources of variation affecting a variable may be segregated and analyzed. It was introduced by R. A. Fisher in 1923. The method is of great importance in agriculture, biology, and industry. In all problems where the samples arise from normal populations having the same **variance,** the analysis of variance provides an effective and powerful technique. The analysis of variance is needed in the study of **randomized blocks** and in **Latin squares.** The analysis of **covariance** extends the analysis of variance to the problem of analyzing the sum of products into component parts. (L.A.A.)

ANALYTIC FUNCTIONS OF A COMPLEX VARIABLE. A complex variable w is said to be a **function** of a complex variable z when to each value of z in a certain region there corresponds a value of w. The function may be single-valued or multiple-valued.

At a given point of a region let Δz be an increment of the variable z and let Δw be the corresponding increment of a function $w = f(z)$, and form the ratio $\Delta w/\Delta z$. If this increment ratio $\Delta w/\Delta z$ approaches a unique limit when $\Delta z \to 0$, which is independent of the way in which $\Delta z \to 0$, then the function $f(z)$ is said to be differentiable at the point, and the limit is denoted by $f'(z)$ or dw/dz and is called the **derivative** of w.

An analytic function of a complex variable, $w = f(z)$, is a function which is differentiable everywhere in a certain region with the possible exception of a finite number of points, called singular points.

In order that a variable $w = u + iv$ may be an analytic function of the complex variable $z = x + iy$, it is necessary and sufficient that the following conditions,

$$\frac{\partial u}{\partial x} = \frac{\partial v}{\partial y}, \quad \frac{\partial u}{\partial y} = -\frac{\partial v}{\partial x},$$

called the Cauchy-Riemann differential equations, are satisfied.

If the function $w = u + iv$ is an analytic function of the variable $z = x + iy$, then both the real and imaginary parts u and v will satisfy **Laplace's equation:**

$$\frac{\partial^2 u}{\partial x^2} + \frac{\partial^2 u}{\partial y^2} = 0, \quad \frac{\partial^2 v}{\partial x^2} + \frac{\partial^2 v}{\partial y^2} = 0.$$

Two such functions u and v are called conjugate functions. (L.L.S.)

ANALYTIC GEOMETRY. Analytic geometry is essentially the study of geometric problems by use of algebraic (analytic) methods.

Elementary geometry had its beginnings with the ancient Greeks and most of its propositions were discovered by them. The first systematic treatment of this subject which we have was written by Euclid about 300 B.C.; our present geometry textbooks are merely modifications of this great work of Euclid. This development of geometry was attained by use of purely geometric methods and with almost no use of algebraic methods. Indeed, **algebra** was practically unknown to the ancient Greeks.

Elementary algebra originated among the ancient Hindus and was later developed further by the Arabs of the earlier Middle Ages. The stage of development indicated by our present elementary textbooks of algebra was not reached historically until after the close of the Middle Ages. Thus, elementary geometry and algebra developed historically as two separate branches of mathematics.

The development of elementary geometry made little progress beyond the stage of the ancient Greeks until the 17th century, when algebra was first applied systematically to the treatment of geometric questions. In 1637 the French mathematician and philosopher René Descartes published his great work *"La Géométrie,"* in which he showed how algebraic methods could be applied to the study of geometry. He thus became the recognized founder of analytical geometry. Almost simultaneously, another Frenchman, Pierre Fermat, also discovered the idea of applying algebra to geometry systematically, but his work was not published and recognized until much later.

While analytical geometry deals with a much more extensive subject matter than does elementary geometry, its especial value lies in its new method, by which geometric properties of figures are treated systematically by means of algebraic methods.

The methods of solution of problems and proofs of theorems in elementary geometry involve a great many special and ingenious devices, and no general and uniform procedure is apparent. Analytical geometry, however, furnishes simple, general procedures for the solution of problems, which greatly simplifies the study of geometry. These new methods also enable us to solve in a simple manner many problems of geometry not considered by the ancient Greeks or problems very difficult of solution by the methods of elementary geometry. Thus, analytical geometry gives a new powerful tool for the study of old and new problems of geometry.

The essence of the new method of analytical geometry lies in its representation of points by sets of numbers and its representation of geometric figures by algebraic equations.

Some idea of the subject matter usually treated in Analytic Geometry may be gained by consulting the following topics: Rectangular Coordinates, Polar Coordinates, Locus of an Equation, Equation of a Locus, Curves in a Plane, Graph of a Function, Slope of a Line, Straight Line in a Plane, Circle, Conic Sections, Ellipse, Parabola, Hyperbola, Higher Plane Curves, Parametric Equations, Trigonometric Curves, Exponential Curve, Empirical Equations, Transformation of Coordinates, Plane, Straight Line in Space, Surfaces, Sphere, Quadric Surfaces, Curves in Space. (L.L.S.)

ANALYTICAL CHEMISTRY. Analytical chemistry is that branch of the science of chemistry which deals with the detection or identification of a substance or a part of the same, either element or radical, by qualitative tests and the estimation of the same by quantitative tests. The difficulties encountered are in part caused by the chemical inertness of a given unknown substance itself, and in part to the other substances that may accompany a given unknown material. It is comparatively easy to analyze **air** for **oxygen,** qualitatively by heating **copper** or **iron** metal in air and observing the formation of copper or iron oxide on the surface of the metal, and quantitatively by subjecting an isolated measured volume of air to an oxygen-absorbent, such as a **cuprous** salt solution in **hydrochloric acid** or **ammonium** hydroxide, or **sodium** pyrogallate solution, and measuring the non-oxygen residue.

The reaction with or resistance to action of various selected reagents serves to indicate whether an unknown substance is or is not a known substance. Analytical chemistry, therefore, demands the knowledge of a vast range of reactions of individual substances, the ability to apply this knowledge in systematic sequence, and the experience to make correct observations and deductions.

Quantitative analysis may be said to be applied qualitative analysis, since the methods used in the quantitative estimation of a substance depend largely upon the accompanying substances. Furthermore, a good qualitative analysis should furnish evidence of the quantitative order of magnitude of each substance present.

Methods of treating an unknown substance, in order to establish its identity, are known as dry or wet. Dry methods involve the heat treatment at various temperatures either alone or with various reagents. Assay methods for some metals, e.g., **silver, gold,** and many organic methods are of this type. Wet methods involve first the problem of getting the substance into solution by some reasonable treatment, or its proved resistance to this treatment. Such simple and apparently inconsequential matters as the size of the particles of a substance tested or the time allowed for treatment may determine the correctness or incorrectness of a given test. The technique demands ability to attend to details.

In the detection, identification and estimation of substances methods must be invented to cope with gases, liquids, solids, with active and inactive, and with similar and dissimilar substances. The greater the differences in chemical behavior, the easier it is in general to accomplish the analytical purpose. The principles of procedure involve:

Solubility of substances in water:

Reactivity of substances	
(a) With acids.	(1) In water as a medium,
(b) With bases.	(2) By the application of
(c) With certain salts.	high temperature or
(d) With certain other chemicals.	electricity.

General, group, and specific (individual) tests:

Much information can be gained by such simple preliminary examinations as those concerned with (a) color, ordinary and if solid of the fine powder, (b) odor, (c) general and detailed appearance of the specimen, (d) hardness, if solid, (e) density, (f) approximate solubility, in water, (g) behavior on heating *without* free access of air (as in a test tube): melting, boiling, sublimation, transition, decomposition point, if any.

The accepted system of qualitative analysis in which 23 metals are separated and identified in *solution* of their **nitrates** well illustrates the principles under discussion. Arranging these in the order of group separation, the reaction serving to separate the group is stated.

In the case of Group II, a sub-group treatment with yellow ammonium sulfide (ammonium sulfide containing dissolved sulfur) is used to dissolve **arsenic, antimony** and **tin** sulfides. Followed by filtration.

Analytical Group Number	Metals of the Group	Precipitate	Reaction Used to Separate the Group
I.	Lead.......... Mercurous...... Silver.........	$PbCl_2$, white............. $HgCl$, white............. $AgCl$, white.............	Addition of hydrochloric acid to the cold solution resulting in the precipitation of these chlorides followed by filtration. The separation of lead is not complete.
II, A. II, B	Mercuric....... Lead.......... Bismuth........ Cupric......... Cadmium....... Arsenic......... Antimony...... Tin............	HgS, black................. PbS, black................. Bi_2S_3, black................. CuS, black................. CdS, yellow............... As_2S_3, As_2S_5, yellow......... Sb_2S_3, Sb_2S_5, orange......... SnS, brown, SnS_2, yellow.....	Addition of hydrosulfuric acid to the properly acidified (0.25 normal hydrochloric acid) filtrate from Group I resulting in the precipitation of these sulfides. Followed by filtration.
III.	Iron.......... Chromium...... Aluminum......	$Fe(OH)_3$, red-brown........ $Cr(OH)_3$, green............. $Al(OH)_3$, white............	Addition of ammonium hydroxide to the properly prepared filtrate from Group II, resulting in the precipitation of these hydroxides. Preparation involves removal of hydrogen sulfide by boiling, and the oxidation of ferrous to ferric. Followed by filtration.
IV.	Cobalt......... Nickel......... Manganese..... Zinc..........	CoS, black .... NiS, black................. MnS, pink................. ZnS, white.................	Addition of ammonium sulfide to the filtrate from Group III, resulting in the precipitation of these sulfides. Followed by filtration.
V.	Barium........ Strontium...... Calcium........	$BaCO_3$, white.............. $SrCO_3$, white.............. $CaCO_3$, white.............	Addition of ammonium carbonate to the properly prepared filtrate, from Group IV, resulting in the precipitation of these carbonates. Preparation involves removal of ammonium salts by evaporation and dry heating. Followed by filtration.
VI.	Magnesium..... Sodium........ Potassium...... Ammonium.....	....	Individual tests for these in properly prepared portions of the filtrate from Group V, except ammonium must be tested for in the original material.

The groups are separately treated for the final identification of the presence or absence of each member by the use of reactions whose effectiveness in separation and identification have been found satisfactory in each case.

The **anions** or the acid radicals of inorganic salts have been classified by Bunsen in 1878 in the following way:

Group I: Silver nitrate produces a precipitate insoluble in nitric acid while barium chloride produces no precipitate: chloride, bromide, iodide, cyanide, hypochlorite, ferrocyanide, ferricyanide, thiocyanide.

Group II: Silver nitrate produces a precipitate which is soluble in nitric acid while barium chloride produces no precipitate: sulfides, tellurides, selenides, nitrites, acetates, cyanates.

Group III: Silver nitrate produces a white precipitate soluble in nitric acid and barium chloride also produces a precipitate soluble in nitric acid: sulfites, selenites, tellurites, phosphites, carbonates, oxalates, iodates, borates, molybdates, tartarates, citrates, metaphosphates and pyrophosphates.

Group IV: Silver nitrate produces a colored precipitate soluble in nitric acid and barium chloride also produces a precipitate soluble in nitric acid: orthophosphates, arsenates, arsenites, vanadates, thiosulfates, chromates and periodates.

Group V: Both silver nitrate and barium chloride produce no precipitate: nitrate, chlorate, perchlorate, persulfate and manganates.

Group VI: Silver nitrate produces no precipitate while barium chloride produces a precipitate insoluble in nitric acid: sulfates, fluorides, fluosilicates.

Group VII: Non-volatile acids which form precipitates with both silver nitrate and barium chloride. Both precipitates are insoluble in nitric acid; silicate, tungstinate and anions of some rare elements.

After identifying the group into which the anion falls recourse must be made to confirmatory tests which can be found under the corresponding acid. (R.K.S.)

ANAMNIA. Vertebrates which do not develop an **amnion** during embryonic life. The group includes the cyclostomes (see **Cyclostomata**), **fishes,** and **amphibians.** (A.W.L.)

ANAMORPHISM. A term proposed by Van Hise in 1904 to designate the deep-seated constructive processes of **metamorphism** by which new complex (metamorphic) minerals are formed from the pre-existing simpler minerals, as contrasted with the surface alteration of rocks due to **weathering** and **cementation,** termed katamorphism. (R.M.F.)

ANASARCA. Generalized **Edema.**

ANATASE. Titanium Dioxide. (See Octahedrite.)

ANATEXIS. A term proposed by Sederholm in 1907 for the supposed end-processes of deep-seated **metamorphism** resulting in the partial or complete remelting of a specific type of rock in situ. (R.M.F.)

ANATOMY. In zoology, the structure of the body; also the science embracing our knowledge of structure. The term is usually applied to gross structure while the minute structure is treated under **histology,** but the two fields are sometimes distinguished as gross anatomy and microscopic anatomy. Another aspect of anatomy is

morphology. While this term applies properly to all structure it is usually employed in connection with external anatomical features of importance in **taxonomy.**

The simplest animals (**Protozoa**) are made up of a single **cell,** hence their structure is essentially that of the cell. Parts which are developed for the performance of special functions are known as organelles. With the grouping of cells to form more complex bodies a division of labor becomes possible and cells are specialized to perform different functions for the benefit of the individual. In the **sponges** this association is loose and the cells preserve a greater range of independent action than in other complex animals. The body is organized about a central cavity and is perforated by canals through which currents of water pass, carrying food and oxygen to the cells of the interior.

Above the sponges the cells of the body are more closely coordinated and are found in aggregates of varying complexity. The simplest aggregate is an association of similar cells for the performance of a special function; such a structure is called a **tissue.** One or more tissues may also enter into the formation of structures called **organs** which also perform special tasks, based on the associated functions of the component cells and tissues. Thus the stomach digests food through the motion produced by its muscular tissue, the secretion by its glandular lining, and the coordination of all parts by its nervous structures.

The initial differentiation of the body in animals above the sponges is the formation of two or three **germ layers** from which the various tissues, organs, and systems are derived. At the maximum eleven systems appear: the **skeletal system** provides support, the **muscular system** produces motion, the **integumentary system** is a protective covering which also takes part in interchange with the environment, the **digestive system** receives food and prepares it for absorption, the **respiratory system** provides oxygen, the **excretory system** removes wastes, the sensory (**sense**) **organs** receive stimuli from the environment and from the body itself and with the **nervous system** provide for coordination of the entire body, the **endocrine system** (see **Hormone**) provides internal chemical coordination, the **circulatory system** distributes materials throughout the organism, and the reproductive system (see **Reproduction**) perpetuates the species. There is some overlapping of these functions in many animals.

A consideration of the anatomy of any animal involves all of the details of development of these basic structures and systems. The anatomy of **vertebrates,** and human anatomy in particular, has become an intricate subject because of its relation to medicine. Some of the details of the general subject will be found under the various systems, under the principal groups of animals, under appendages, and under special anatomical terms too numerous to be listed here. (A.W.L.)

ANCHOR ICE. Ground Ice.

ANCHOR RING. Torus.

ANCHOVY. Pisces, Teleostei. Any of numerous small sardine-like fishes (**Pisces**), related to the herrings, whose richly flavored flesh is esteemed as a hors d'oeuvre. Especially the common anchovy, *Engraulis encrasicholus,* of the Mediterranean and the east Atlantic. (A.W.L.)

ANDALUSITE. An **aluminum silicate** corresponding to the formula Al_2SiO_5, this mineral is usually found in **metamorphic rocks** with **sillimanite, kyanite, garnet,** and **tourmaline.** It crystallizes in the **orthorhombic system,** developing coarse prisms of approximately square cross-section, but may be massive or granular. It shows a distinct cleavage parallel to the prism; hardness, 7.5; specific gravity, 3.16–3.20; vitreous luster; colorless to white, gray, brown, greenish or reddish; streak, white; transparent to opaque.

This mineral is named for its original locality, Andalusia, Spain. A variety of andalusite, chiastolite, has carbonaceous impurities so oriented that they produce a cross or a tesselated figure at right angles to the prism. Chiastolite comes from the Greek word meaning a cross. Localities are the Urals, the Alps, the Tyrol, The Pyrenees, Australia and Brazil. In the United States at Standish, Maine; Sterling and Lancaster, Massachusetts; Delaware County, Pennsylvania; and Madera County, California.

When clear it is used as a gem, and it has also been used to manufacture porcelain for spark plugs. (E.S.C.S.)

ANDESINE. Feldspar.

ANDESITE. A term originally applied to a porphyritic **lava** from the Andes Mountains by Leopold Van Buch, in modern terminology andesite is an extrusive **igneous rock,** the surface equivalent of **diorite.** In other words, it is composed chiefly of **plagioclase,** corresponding in chemical composition to **oligoclase** or **andesine** together with **biotite, hornblende,** or **pyroxene** in varying quantities.

Andesites are of rather widespread occurrence, being found in the Rocky Mountains, California, Alaska, South America, and at many other foreign localities. (E.S.C.S.)

ANDRADITE. Garnet.

ANDROGENESIS. The development of an egg after the entry of the male germ cell without the participation of the egg nucleus. (See also **Merogany.**) (A.W.L.)

ANDROMEDA. (Map, page 380.) The brighter stars of this **constellation** make an almost straight line between the constellations of **Perseus** and **Pegasus.** The most famous feature of the constellation is the great **nebula.** This is the only **spiral** which is actually visible to the naked eye and may be distinguished as a faint blur against a moonless sky close to the faintest star in the constellation which appears on the map (page 380). This is the largest of the spirals and is distant from the earth about 800,000 **light years.**

The bright star in Andromeda closest on the map to Perseus was called Almach by the Arabs and is a **double star,** one of the most beautiful in the sky in a small telescope. One component is a brilliant orange and the other a striking emerald color. Careful examination in a large telescope shows the green component to be also a double star. (W.K.G.)

ANDROMEDES. Andromedes is an alternative name for the **Bielid meteor shower** which is observed the latter part of November. This alternative name comes from the fact that the **radiant point** is in the **constellation** of **Andromeda.** (W.K.G.)

ANEMIA. A state of deficiency of either the number of circulating red **blood** cells or the amount of **haemoglobin** in the red cells.

The common symptoms and signs of anemia are weakness, easy fatigability and pallor of the skin and **mucous membranes.** When the condition is due to acute blood loss from a large hemorrhage, **shock** may be the outstanding sign. Pernicious anemia is frequently accompanied by a yellow tint of the skin, a smooth, red, sore tongue, digestive disturbances, and neurological disorders —numbness, tingling of the hands and feet, and weakness and paralysis of the legs.

The chief causes of anemia are (1) iron deficiency and consequent insufficient heamoglobin production, (2) blood loss, either acute or chronic, (3) increased destruction of red cells, (4) decreased production of red cells due to depression of the bone marrow.

Iron deficiency is the commonest cause of anemia. It is usually on a nutritional basis, associated with insufficient ingestion of iron-containing foods. Faulty absorption of iron from the gastrointestinal tract may also result in insufficient supply of iron. Nutritional deficiency is apt to occur when the need for iron is large, as in infancy and childhood, and during pregnancy. When the iron stores are inadequate, blood-cell production is interfered with, so that there is a reduction in the number of cells as well as a reduction of the cell haemoglobin content. Chronic blood loss results in gradual depletion of iron from the body and hence an iron-deficiency anemia. Abnormal bleeding from the uterus, from **hemorrhoids**, and from gastric or duodenal ulcers, (**peptic ulcer**), are the common causes of this.

Acute blood loss may be due to hemorrhage from any organ or tissue. Trauma, perforated peptic ulcer or other intra-abdominal hemorrhage, **abortion**, childbirth, are frequent causes.

Hemolytic anemia is characterized by increased blood-cell destruction. It occurs in an hereditary, familial form in which there is an abnormal fragility of the red cells. Other types are associated with reactions to chemicals and drugs, and with certain infections. (See **Hemolytic Jaundice**.)

Decreased blood formation due to depression of the bone marrow occurs as a response to certain chemicals (benzene, arsphenamine), to x-ray treatment, and to diseases which mechanically interfere with the bone marrow, such as tumor masses which crowd out the blood cells.

The cause of pernicious anemia is unknown. The disease characteristically attacks blue-eyed, prematurely gray-haired individuals, usually in the 4th decade of life. Dietary factors and the lack of certain substances in the stomach and liver play a part in the disease. Treatment with liver extract produces a prompt remission. The patient must receive this therapy for the rest of his life or he will again become anemic.

Treatment of the various anemias is dependent on their type. The iron-deficiency anemias respond well to iron given orally. Correction of conditions causing chronic blood loss eliminate the attendant anemia. Hemorrhage often requires **transfusion**. Removal of the spleen is effective in the treatment of familial hemolytic anemia. (D.M.H.)

ANEMOGRAPH. Meteorological Instruments.

ANEMOMETER. One of a variety of instruments for measuring the velocity of the wind. The most common type resembles a windmill in principle, having either a fan-like vane wheel or, in the widely used "cup anemometer," a whirligig arrangement of four cup-like

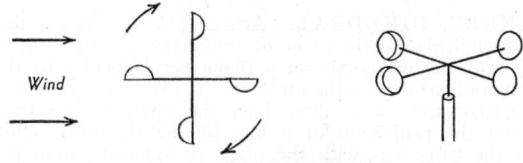

Design of cup anemometer.

vanes, the speed of whose revolution about a vertical shaft is proportional to the wind velocity if the velocity is not too great. Another type utilizes the pressure due to the wind blowing against a vertical surface, as measured by the compression of a spring or by a tube connected with a pressure gauge. A modification of this type consists of a metal plate suspended by its upper edge and blown into an oblique position by the pressure of the wind. All such instruments must be experimentally calibrated. (See **Meteorological Instruments**.) (L.D.W.)

ANEMOSCOPE. Meteorological Instruments.

ANEMOTROPISM. Orientation of the body in relation to the wind. (A.W.L.)

ANEROID BAROMETER. Meteorological Instruments.

ANEURYSM. A sac formed by dilatation of the walls of an artery which communicates with the interior of the artery and is therefore filled with blood. This occurs most frequently in the wall of the arch of the aorta, as a late manifestation of syphilis. Aneurysm of the abdominal aorta is due to arteriosclerosis and consequent weakness of the vessel wall.

Arteriovenous aneurysm is the term used when there is an abnormal connection between an artery and a vein. (D.M.H.)

ANGEL-FISH. Pisces. A species, *Squatina squatina*, intermediate between the **sharks** and **rays**. Also applied to several tropical genera of bony fishes. (A.W.L.)

ANGINA PECTORIS. A symptom complex characterized by recurrent pain in the chest following physical exertion. The pain is usually in the region of the sternum and may radiate to the left arm. It may be mild or so severe that the patient feels he is about to die. It is usually of short duration. The precipitating exertion may be in the nature of climbing stairs, or eating a heavy meal, or it may be an emotional outburst.

The attacks are believed to be caused by **anoxemia** of the heart muscle, causing a cramp-like pain. Any disease process which causes narrowing of the lumen of the coronary arteries which supply the heart with blood may cause angina pectoris. The most frequent cause is **arteriosclerosis**. Marked **anemia** may also cause angina.

The syndrome occurs usually after 40, most frequently in the 50's. It is four times as common in men as in women. The disease is thought by some to be increasing, but the increase may be only apparent since the span of life is now longer and diseases of the older age-groups are now seen more frequently.

Angina pectoris does not cause death. Frequently an individual may have a number of anginal attacks and finally die in one which seems more severe than the others. In such a case he has died from **coronary occlusion** in which the lumen of the coronary artery is completely blocked, a portion of the heart is deprived of its blood supply and consequently dies. In angina pectoris the artery is not completely shut off and the cardiac anoxia is reversible.

Treatment of an attack consists of rest (usually the pain itself causes the individual to seek rest), nitrites, nitroglycerin, or nicotinic acid (**Vitamin B**). These drugs dilate the coronary vessels and may thereby give relief. Alcohol, as whiskey or brandy, may be of help by the same mechanism. Nitrites, animophylline and nicotinic acid may also be of use prophylactically. (D.M.H.)

ANGIOMA. A tumor which is composed mainly of **blood** vessels (hemangioma) or of **lymph** vessels (lymphangioma). (R.S.M.)

ANGIONEUROTIC EDEMA. This allergic disease is characterized by localized transient swelling of the skin or a mucous membrane of the body. The swelling may persist for a few hours or days. If the swelling occurs in the throat, death may occur by obstruction of the air passages. One form is inherited. Other forms are due to sensitivity to some **protein**; the disease is really a form of giant hives.

The swellings may be accompanied by prickling, itching, and burning sensations. The lips, skin about the eyes, hands, and feet are commonly involved. If the mucous membrane of the gastrointestinal tract is involved in these localized swellings, severe abdominal pain with vomiting occurs and may simulate appendicitis, obstruction of the intestines, or some other abdominal accident.

Treatment is unsatisfactory as regards future attacks. **Adrenaline** by hypodermic aids during the acute attack. (D.M.H.)

ANGIOSPERMS. The angiosperms are **spermatophytes** in which the ovule matures to form a seed which is completely enclosed in an ovary, in contrast with the **gymnosperms,** spermatophytes which have the seed borne exposed on the surface of a scale. Angiosperms are more familiarly known as flowering plants, and the characteristic feature is the **flower.**

As a rule the angiosperms are land plants growing in a fixed position. A few of them have returned to the water as a habitat, but these are obviously reversions to an aqueous life and not primitive forms pointing the way along which angiosperms evolved. Tremendous diversity in size is found in this group; some of the so-called duckweeds are spherical masses of cells less than a millimeter in diameter; at the other end of the scale are the giant redwood trees (*Sequoia sempervirens*), many of which are over 300′ high. The variety of form shown in the angiosperms is nearly endless; each of the 130,000 and more species has a distinct appearance by which it can be distinguished. Some are tiny, herbaceous plants which live but a few weeks; others are giant trees living hundreds of years.

Included in the angiosperms are two types of plants. One, held to be the more primitive type, has a woody **stem** which has a much more complex structure than that found in the stems of Gymnosperms. The other has an herbaceous stem, a form of stem which dies to the ground at the end of the growing season. Herbaceous plants may live through to the next growing season by means of perennial roots and underground stems, or they may die completely, only the seeds surviving. This habit fits herbaceous plants especially to live in regions having growing seasons alternating with cold or dry periods. Herbaceous plants, or herbs, are most abundant in temperate and arctic regions.

The internal structure of angiosperm stems is much more specialized than that of gymnosperms. The xylem contains not only tracheids but also vessels and fibers. The vessels are open tubes of considerable length through which water is rapidly carried. The fibers give strength to the stem. In the **phloem** there are sieve tubes and companion cells, and also numerous fibers. The latter are often of great value to man. Linen, for example, is made from the phloem fibers of the flax plant. In the woody angiosperms and in many of the herbaceous forms there is a well developed **cambium.**

The leaves of angiosperms are of many shapes and sizes, but are typically thin and contain numerous veins. In the axils of each leaf there is a bud which may develop into a branch or a flower.

But all bear flowers at some time during their lives. The flower is the basis for classifying angiosperms. A flower is a special shoot or branch which is adapted to advance **pollination** and **fertilization.** Flowers may arise from the **axils** of ordinary leaves or be found in the axils of special modified leaves called bracts. A flower consists essentially of two organs: stamens in which the pollen grains are formed, and pistils in which the ovules are found. In addition to these two essential organs, there are usually accessory structures, which collectively make up the perianth. These accessory structures include the calyx, composed of separate parts called sepals, and the corolla, composed of petals. The latter are usually bright-colored and are assumed to attract insects or other animals which effect **pollination.**

The mechanical transfer of pollen grains from the stamens to the pistil is known as pollination. In many flowers the wind is the agent effecting pollination; in others, insects, and in a few, water or other agents.

Having reached the pistil, the pollen grain germinates, forming a slender tube called the pollen tube, which grows down through the tissues of the pistil until it reaches the ovule. The latter is enclosed in one or two layers called the integuments, in which there is a minute hole called the micropyle. The pollen tube grows through this micropyle and into the embryo sac. The embryo sac is typically a 7-celled structure with rather definite characteristics. At the end farthest from the micropyle there are three small cells called antipodal cells which are of little importance. At the opposite end, nearest the micropyle, there are 2 cells, the synergids, and a third cell, the egg cell, which in many species is reduced to an egg nucleus. The remainder of the embryo sac is a large cell in the center of which there are 2 polar nucei close together. These 2 nuclei soon unite, forming the fusion nucleus.

In angiosperms 2 male nuclei are discharged from the pollen tube into the embryo sac. One of these haploid nuclei unites with the egg nucleus, forming the diploid **zygote,** from which the **embryo** develops. The other nucleus unites with the fusion nucleus in the center of the embryo sac, forming the triploid endosperm nucleus. There are therefore two separate nuclear fusions in the embryo sac. This is characteristic in all angiosperms and is called double fertilization.

From the fertilized egg or zygote the embryo is formed. Usually this embryo is an elaborate body consisting of one or two seed leaves or cotyledons, a primitive root or hypocotyl and a primitive bud or epicotyl. Often the embryo is surrounded by a mass of nutritive material known as endosperm which develops from the endosperm nucleus. Surrounding this there are one or two seed coats, derived from the integuments. This whole structure is the **seed,** which is contained in the ripened ovary or **fruit.**

The angiosperms are separated into two large groups, the **dicotyledons** and the **monocotyledons.** The origin of the angiosperms is as yet unknown. They are known to have existed in the **Jurassic** period, but were not at all abundant until the **Cretaceous** period (see **Paleobotany**). The earliest **fossil** members of this group are well differentiated plants which give little indication as to their possible ancestry. Within the group, **evolution** seems to be from the woody type to the herbaceous, and from plants with flowers having an indefinite number of parts arranged in spiral manner and not fused. As evolution progressed the number of flower parts became reduced and definite and finally fused. In many cases great irregularity replaced the more primitive regularity. The angiosperms are the dominant land flora of the present day. (R.M.W.)

ANGLE, DELAY. Ignition.

ANGLE, DIHEDRAL. Acute angle between a line perpendicular to the plane of symmetry and the projection of the wing axis on a plane perpendicular to the longitudinal axis of the airplane. If the wing axis is not approximately a straight line, the angle is measured from the projection of a line joining the intersection of the wing axis with the plane of symmetry and the aerodynamic center of the halfwing on either side of the plane of symmetry.

Most airplanes have some dihedral in the wings, it being necessary to use more dihedral where the lateral stability of the airplane depends primarily upon the stabilizing action of dihedral. Dihedral is most pronounced in the low-wing monoplane; least in the parasol monoplane. (F.T.M.)

ANGLE, DOWNWASH. Airfoil; Downwash.

ANGLE, PROPELLER BLADE. Propeller Blade Angle.

ANGLE, STABILIZER. This is, structurally, the acute angle between the plane of the stabilizer and the longitudinal axis of an airplane. The attack angle of the stabilizer is aerodynamic in character and is defined as the acute angle between the plane of the stabilizer and the relative wind *at the tail*. It is seen in the accompanying figure that this involves the **downwash** from

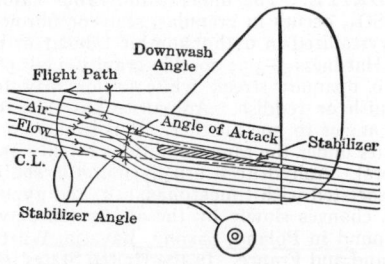

the wings, and consequently a slight positive stabilizer angle might be a negative angle of attack.

The stabilizer angle is sometimes made adjustable within a 5–10° range to allow the pilot to "trim" the airplane in flight. (F.T.M.)

ANGLE, STALLING. Burble.

ANGLE OF ADVANCE.
The degrees of crank angle traveled by the crankpin of an Otto (spark ignition) engine between ignition position and outer dead center. It represents a certain lead given to combustion, so that it may be nearly completed at the beginning of the power stroke. This helps to achieve the maximum possible **mean effective pressure**. Angle of advance needs to be greater the higher the rotative speed of the engine. It is seldom less than 10° (except during cranking) and may advance on some engines to 30° or 40° before dead center at maximum speed. (Compare with **Delay Angle** of the compression ignition engine. See **Ignition System**.) (F.T.M.)

ANGLE OF ATTACK.
The angle of attack is the acute angle included between the direction assumed by the wind relative to an **airfoil** and some basic reference **chord** of the airfoil. There are two of these reference chords. The one employed for general and structural use is the geometric chord, the other, the zero-lift or aerodynamic chord. (F.T.M.)

ANGLE OF INCIDENCE.
Acute angle between plane of the wing chord and the longitudinal axis of an airplane. The angle is positive when leading edge is higher than trailing edge. (F.T.M.)

ANGLE OF REPOSE.
The angle of repose is the maximum angle with the horizontal at which loose material such as grain, sand, coal, or stone will retain its position without tending to slide. The moisture content and the distribution of the fine and coarse particles have a marked effect on the value of this angle. The angle of repose is an important factor in the design of **retaining walls**, earth **dams**, and embankments and is particularly valuable in the design of storage bins and **bunkers** since the allowable surcharge as well as the active horizontal pressure depends upon its value. Tables giving the approximate value of the angles of repose for various materials will be found in most Civil Engineering handbooks. (C.W.C.)

ANGLE-WING.
Insecta, Lepidoptera. **Butterflies of** the genus *Polygonia*. Their wings are sharply angular but no more so than those of some other species. (A.W.L.)

ANGLER-FISH.
Pisces, Teleostei. Bottom-feeding fishes (**Pisces**) of the family Lophiidae, named from the tufted tentacle on the head which is said to attract prey. A common species is *Lophius piscatorius*. (A.W.L.)

ANGLES.
An angle is a fundamental mathematical concept, being one of the simple geometrical elements.

A plane angle is generated by the rotation in a plane of a half-line about a fixed point, called the vertex of the angle, from a position called the initial side of the angle to a final position called the terminal side of the angle. An angle generated by rotation in a counterclockwise sense is called positive, by clockwise rotation negative. Such an angle, in which a sense of rotation is distinguished, is called a directed angle, as shown by the two examples given (Figs. 1 and 2).

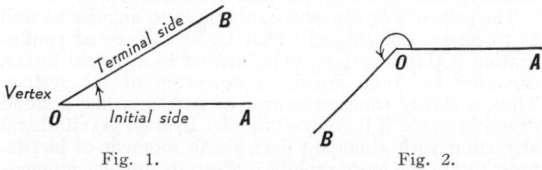

Fig. 1.   Fig. 2.

Co-terminal angles are angles which have the same initial and terminal sides. They differ in measure by a multiple of 4 right angles. The numerically smallest of a set of co-terminal angles is called the principal value of the set.

Complementary angles are angles whose sum is a right angle; supplementary angles are angles whose sum is 2 right angles.

Quadrantal angles are angles which are multiples of a right angle; if such an angle is placed with its vertex at the origin of a set of **rectangular coordinate** axes and its initial side along the positive X-axis (in so-called standard position), its terminal side will fall along one of the axes.

Besides plane angles, there are **dihedral angles** and **spherical angles**. (L.L.S.)

ANGLESITE.
The mineral anglesite is naturally occurring **lead** sulfate, crystallizes in the **orthorhombic** system and may be found mixed with **galena**, from which it is usually formed by **oxidation**. Hardness, 3; specific gravity, 6.12–6.39; luster, adamantine to vitreous or resinous; transparent to opaque; streak, white; colorless to white or green but may be rarely yellow or blue.

It is used as a source of lead. There are many foreign localities, and in the United States it has been found in large crystals in the Wheatley Mine, Phoenixville, Pa.; also in Missouri, Utah, Arizona and Idaho.

It is named from Anglesey, England. (E.S.C.S.)

ÅNGSTRÖM.
The angstrom, or angstrom unit, named for the pioneer spectroscopist Ångström, is the unit of length customarily used in expressing **wavelengths of light**. It is equal to 0.00000001 cm. or 10^{-10} meter, and is therefore sometimes called the "tenth-meter." The wavelength of sodium (yellow) light is about 5890 angstroms (A). The micron, or 0.001 mm., used for approximate designations of wavelength in the visible and infra-red, is 10,000 A. Both the angstrom and the micron are convenient also in expressing other very small lengths, such as the thickness of liquid films, etc. The unit above defined is the absolute angstrom. To obviate inconvenience due to variation of spectroscopic wavelengths with air pressure, another unit, the international angstrom, has been adopted, such that the wavelength of the red line of cadmium is exactly 6438.4696 of these units. The two units are equal only in a vacuum. (L.D.W.)

ANGUCLAST. Phenoclast.

ANGULAR ACCELERATION. Angular Velocity and Angular Acceleration.

ANGULAR MOMENTUM.
The product of moment of inertia and angular velocity. The analogy between

concepts relating to translational motion and to rotation is emphasized by reference to both linear and angular velocity, acceleration, and momentum. For rotational motion, **angular velocity** takes the place of linear velocity and **moment of inertia** takes the place of mass. Hence the angular momentum of a body with respect to a given axis of rotation is defined as the product of its moment of inertia with respect to that axis by its angular velocity about that axis. It must be regarded as a **vector** quantity, whose magnitude is that of the product just stated and whose direction is that of the angular velocity, determined by the axis of rotation.

The principle of conservation applies to angular as well as to linear momentum. That is, no change of configuration within a system, uninfluenced by external forces, can alter the total angular momentum of the system. Thus, a slowly rotating swarm of particles, like a cloud of gas in space, if it contracts under its own gravitational attraction with attendant decrease in moment of inertia, must rotate the more rapidly to keep its angular momentum constant. Again, if a person, whirling about on tiptoe, with arms extended, suddenly brings the arms down to the sides, he will as suddenly begin to whirl faster, the effect being more pronounced if he holds heavy weights in his hands. Angular momentum being a vector quantity, the principle applies as well to its direction as to its magnitude. The result is that any rotating body tends to maintain the same axis of rotation; a fact well illustrated by the spinning top and by the stabilizers used on some ocean vessels. (L.D.W.)

ANGULAR PERSPECTIVE. Perspective.

ANGULAR UNCONFORMITY. Unconformity.

ANGULAR VELOCITY AND ANGULAR ACCELERATION.
Quantities relating to rotational motion. While the use of the term "angular velocity" may be extended to any motion of a point with respect to any axis, it is commonly applied to cases of rotation. It is then the **vector**, whose magnitude is the time rate of change of the angle θ rotated through, i.e., $d\theta/dt$, and whose direction is arbitrarily defined as that direction of the rotation axis for which the rotation is clockwise. The usual symbol is ω or Ω.

Angular velocities, like linear velocities, are vectorially added; for example, if a top is spinning about an axis which is simultaneously being tipped over toward the table, the resultant angular velocity is the vector sum of the angular velocities of spin and of tipping. (This enters into the theory of **precession**.)

Angular acceleration is the time rate of change of the angular velocity, expressed by the vector derivative $d\omega/dt$. Only in case the direction of the axis remains unchanged can the angular velocity and angular acceleration be treated as scalars. The effect of torque applied to a body free to rotate about an axis is to give it angular acceleration, and the opposition offered by the body to this process gives rise to the concept of **moment of inertia**. (L.D.W.)

ANHARMONIC RATIO.
If we have given four points A, B, C, D on a straight line, their anharmonic ratio (or cross-ratio) is defined as the ratio $\dfrac{AC}{AD} \Big/ \dfrac{BC}{BD}$ in which the segments are to be regarded as positive or negative according to the order of the letters.

The anharmonic ratio of any four points z_1, z_2, z_3, z_4 in a complex plane is defined as the ratio

$$\frac{z_1 - z_3}{z_1 - z_4} \cdot \frac{z_2 - z_3}{z_2 - z_4},$$

in which the z's are complex numbers representing the points. As a special case of this we have the definition of the anharmonic ratio of four given real numbers. There are in general six distinct anharmonic ratios obtained by rearranging the given numbers or points in order. (L.L.S.)

ANHEDRAL. Allotriomorphic.

ANHYDRITE.
The mineral anhydrous **calcium** sulfate, $CaSO_4$, occurs in granular, scaly or fibrous masses, rarely crystallized in **orthorhombic** tabular or prismatic forms. Hardness 3–3.5; specific gravity 2.9–2.98; translucent to opaque; streak white; color may be white, gray, bluish or reddish. Anhydrite has three cleavages at right angles to one another. It is similar to **gypsum** and occurs under the same conditions often with the latter mineral. Anhydrite is usually found in **sedimentary** rocks associated with **limestones**, salt, and **gypsum**, into which it changes slowly by the absorption of water.

It is found in Poland, Saxony, Bavaria, Württemberg, Switzerland and France. In the United States in Niagara County, N. Y., West Paterson, N. J., and Nashville, Tenn. It occurs also in Nova Scotia and New Brunswick. (E.S.C.S.)

ANILIDES. Amines and Amides.

ANILINE.
Aniline, phenylamine, aminobenzene ($C_6H_5NH_2$) is a colorless, odorous liquid, melting point $-6°$ C., boiling point $184°$ C., slightly soluble in water, miscible in all proportions with alcohol or ether, poisonous, which turns yellow to brown in the air, is a weak base forming salts with acids, e.g., anilinehydrochloride ("aniline salt" $C_6H_5NH_2 \cdot HCl$) from which aniline is reformed by addition of **sodium** hydroxide solution. Aniline reacts (1) with **hypochlorite** solution, to form a transient violet coloration, (2) wth **nitrous acid** (a) warm, to form **nitrogen** gas plus phenol, (b) cold, to form **diazonium** salt (benzene diazonium chloride C_6H_5N—Cl), (3) with **acetyl chloride, acetic anhydride**, or **acetic acid** glacial, to form N-phenylacetamide $\left(\text{acetanilide, "antifebrin" } C_5H_6N\!\!\begin{smallmatrix}H\\OCCH_3\end{smallmatrix}\right)$, (4) with **benzoyl chloride**, to form N-phenylbenzamide $\left(\text{benzanilide, } C_6H_5N\!\!\begin{smallmatrix}H\\OCC_6H_5\end{smallmatrix}\right)$, (5) with benzenesulfonyl chloride, to form N-phenylbenzene sulfonamide ($C_6H_5SO_2NHC_6H_5$), soluble in sodium hydroxide, (6) with **chloroform** ($CHCl_3$) plus alcohol plus sodium hydroxide, to form phenyl isocyanide (C_6H_5NC) very poisonous, (7) with **sulfuric acid** at $180°$ to $200°$ C., to form para-aminobenzene sulfonic acid (sulfanilic acid, $H_2N \cdot C_6H_4 \cdot SO_2H(1,4)$), (8) with **nitric acid**, when the amine group is protected, e.g., using acetanilide, to form mainly para-nitroacetanilide ($CH_3CONH \cdot C_6H_4 \cdot NO_2$ (1,4)), from which para-nitroaniline ($H_2N \cdot C_6H_4 \cdot NO_2$ (1,4)) is obtained by boiling with concentrated hydrochloric acid, (9) with **chlorine** in an anhydrous solvent, such as chloroform or acetic acid glacial, to form 2,4,6-trichloroaniline ((1)$H_2N \cdot C_6H_2Cl_3$($2,4,6$)), (10) with **bromine** water, to form white solid 2,4,6-tribomoaniline ((1)$H_2N \cdot C_6H_2Br_3$($2,4,6$)), (11) with **potassium** dichromate in sulfuric acid, to form aniline black dye, and, by further oxidation, benzoquinone ($O:C_6H_4:O(1,4)$), (12) with **potassium** permanganate in sodium hydroxide, to form azobenzene ($C_6H_5N:NC_6H_5$) along with some azoxybenzene ($C_6H_5NO:NC_6H_5$), (13) with reducing agents, to form aminohexahydrobenzene (cyclohexylamine, $H_2N \cdot C_6H_{11}$), (14) with alkyl halides or alcohols heated, to form alkyl anilines, e.g., methylaniline ($C_6H_5NHCH_3$), dimethylaniline ($C_6H_5N(CH_3)_2$).

Aniline may be made (1) by the reduction, with iron or tin in **hydrochloric acid**, of nitrobenzene, and (2) by the amination of chlorobenzene by heating with ammonia to a high temperature corresponding to a pressure of over 200 atmospheres in the presence of a catalyst (a mixture of cuprous chloride and oxide). Aniline is the end-point of reduction of most mononitrogen substituted benzene nuclei, as nitrosobenzene, beta-phenylhydroxylamine, azoxybenzene, azobenzene, hydrazobenzene. Aniline is detected by the violet coloration produced by a small amount of sodium hypochlorite.

Aniline is used (1) as a solvent, (2) in the preparation of compounds as illustrated above, (3) in the manufacture of dyes and their intermediates, (4) in the manufacture of medicinal chemicals. (See also **Amines and Amides.**) (R.K.S.)

ANIMAL ASSOCIATIONS. While most animals are solitary, associating with others of their kind only incidentally or during the breeding season, others normally live in some relationship with members of the same or of other species.

The simplest association of members of the same species is gregariousness. Gregarious animals are not bound by the association but profit by it. Examples are the great herds of herbivorous animals such as the bison and the packs of predacious animals, such as wolves.

Colonial association may be accompanied by structural union between individuals, as in many marine polyps, or may be based on behavior, as in the social insects. The term merges with social organization. This type of association is accompanied by structural specialization of individuals for special tasks, except in human society where it depends on specialized training.

The association of individuals of different species may be the relatively loose type called commensalism in which both forms benefit but not in an essential way, or the indispensable symbiosis in which neither organism can persist without the other. An excellent example of symbiosis is the relation of **termites** with the **protozoa** found in their intestine; neither can live without the other.

An association in which one individual lives at the expense of the other is called **parasitism.**

Slavery is an association practiced by some of the social insects and by man; among the insects the slaves are of a different species.

Such relations as symbiosis and parasitism also occur among plants where they are exemplified by the combining of algae and fungi to form lichens and by the mistletoe, parasitic on trees. Symbiotic relations between animals and plants also occur. (A.W.L.)

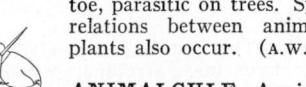
Bear animalcule.

ANIMALCULE. A minute animal. Applied to the **protozoa** and to such microscopic forms as the **rotifers.** (A.W.L.)

ANIONS. Anions are negatively charged atoms or radicals. (See **Ions.**) (R.K.S.)

ANISE. *Pimpinella Anisum.* Umbelliferae. An herb of the Umbel family (see **Carrot Family**) having a strong odor and bearing seeds from which is distilled the aromatic substance, oil of anise. It is sometimes cultivated as a medicinal plant (see also **Volatile Oils**). (R.M.W.)

ANISOGAMY. Heterogamy.

ANISOLE. Methylphenyl ether. (See **Alcohols and Ethers.**)

ANKERITE. Dolomite.

ANKLE. The slender part of the lower leg at its articulation with the foot. (A.W.L.)

ANKYLOSIS. Loss of motion in a **joint.** This may be caused by a disease process in or around the joint, producing either stiffening of the surrounding muscles, or the laying down of bone in the joint itself. In other cases, fibrous tissue and adhesions may cause the loss of

motion. Surgical ankylosis or arthrodesis is carried out as a treatment for certain diseases of joints, especially tuberculosis. (D.M.H.)

ANNABERGITE. The mineral Annabergite is a rather rare **nickel arsenate** with the formula $Ni_3As_2O_8 \cdot 8H_2O$ crystallizing in the **monoclinic** system. It is of secondary origin, resulting from the alteration of preexisting nickel minerals. It has been found in Saxony, France, and Cobalt, Province of Ontario, Canada. It was named from Annaberg in Saxony. (E.S.C.S.)

ANNEALING. Heating and cooling of metals to soften or effect other changes in properties. Ductile metals which have been work-hardened by cold rolling, cold drawing, or by other forms of plastic deformation may be softened by process annealing. The temperature required is relatively low compared with full annealing and the cooling rate is not critical. Full annealing for maximum softness and ductility requires heating to higher temperatures and in some cases the cooling rate is important. Most steels, for example, have a critical transformation temperature range, and they must be cooled slowly through this range for full annealing. On the other hand, austenitic stainless steels have no such range and are annealed by heating to a very high temperature followed by a rapid cool, often by quenching in water. Silicon steel sheets for transformer cores, motor and generator laminations, etc., require slow cooling in annealing to avoid internal stresses which impair their electrical properties.

Normalizing is a special treatment applied to steels to refine the grain structure and remove effects of hot rolling or other previous treatments. Steels are generally normalized by heating to approximately 100° F. above their critical transformation range followed by cooling in still air. (See also **Heat Treating.**) (R.H.H.)

ANNELIDA. The segmented worms, including **earthworms** and **leeches.** This phylum is biologically interesting because it shows in a primitive form the structural plan of the more complex animals.

The annelids are characterized by: 1. Metameric segmentation. 2. A closed tubular circulatory system in

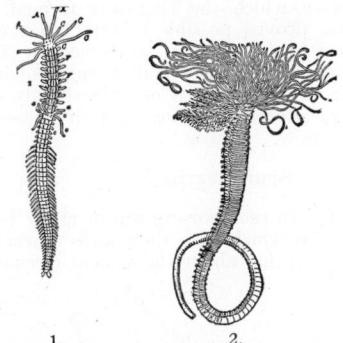

1.   2.

1. A sexual individual of *Autolytus* with male about to detach. (*From Verrill, Invertebrate Animals of Vineyard Sound.*) 2. Tufted worm (*Amphitrite ornata*). (*Drawn by Verrill.*)

most forms. 3. A coelom. 4. An excretory system with tubules opening from the **coelom** to the exterior in various segments. 5. The alimentary tract is tubular, with regions specialized for various functions. 6. The nervous system consists of a dorsal brain above the oesophagus connected by cords passing around the **oesophagus** with a ventral chain of ganglia (see **Ganglion**) below the alimentary tract. 7. **Setae** are present in many species.

The annelids are classified as follows:

Class **Archiannelida.** Small marine annelids without setae; with few to many segments.

Class **Chaetopoda.** Worms with setae. No suckers. External segmentation distinct and metameric. **Earthworms** and many aquatic species.

Class **Gephyrea.** Large marine worms, not segmented when adult.

Class **Hirudinea.** Flattened worms without setae but with a sucker at each end of the body. External segmentation consisting of 2–14 annuli to each metamere. Mostly aquatic, a few marine and a few terrestrial. Mostly blood-sucking parasites. The **leeches.** (A.W.L.)

ANNUAL. A plant which normally completes its life cycle, from seed to seed, in a single growing season. Typical annuals are corn, wheat, cucumber, and nasturtium. Annual plants are especially suited for life in regions where the growing season is short and alternates with an unfavorable cold period or dry season. (R.M.W.)

ANNUAL RING. A layer of wood added to the **stem** in one growing season.

In temperate climates stem growth occurs during the warm spring and summer months. The cells formed in spring when active growth is taking place are characteristically large, while during the summer only smaller cells are added. This alternation of cells results in the formation of definite concentric rings readily seen in cross-sections of woody stems. Actually the growth increment is in the form of a sheath continuous over the entire stem except at the growing tips. External conditions may have a profound effect on the appearance of the annual ring; favorable growing seasons with ample moisture result in broad rings, while seasons of drought produce narrow rings. Removal of surrounding overshading trees may result in a pronounced increase in the thickness of the annual ring. At times events such as severe defoliation by insects or cases of drought may produce two rings in one season; such rings are ordinarily not sharply distinct as are normal ones, and are called false annual rings. Counting of annual rings gives an accurate index of the age of the tree, while attention to details such as variable thickness of successive rings serves to indicate environmental changes. By careful comparison of different logs, even though they be largely reduced to charcoal, one may determine the actual year in which the ring was formed. By this means it has proved possible to establish the probable age of many ruins in the south-western states. In tropical countries having a continuous growing season, annual rings are not formed or only slightly developed. If, however, alternating rainy and dry seasons occur, then they appear. (R.M.W.)

ANNULAR. Spur Gearing.

ANNULUS. In the **sporangium** of many **ferns,** there is a ring of cells which have their walls characteristically thickened, and bring about the violent discharge of the

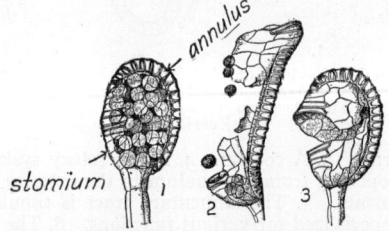

Fern sporangia. 1, unopened; 2, discharging spores; 3, empty.

spores within. This ring of cells is called the annulus. In **agarics,** the ring of tissue which is found around the stalk in many genera, is also known as an annulus. (R.M.W.)

ANOA. Oxen.

ANODE. The anode (also frequently called the plate) is the principal electrode for collecting electrons in an electron **tube,** and is, therefore, connected to the positive side of the circuit. Anode also designates the positively charged electrode in an electrolytic cell. (See **Electrochemistry.**)

Since most of the electrons which flow across the space between the electrodes of a tube are eventually collected by the anode and usually they are traveling at high speed when they strike it, there is an appreciable amount of energy to be dissipated by the anode. This energy is the kinetic energy which the electrons have acquired in traversing the tube under the influence of the various voltages applied to the tube. The electrons are stopped upon striking the anode and the energy they possess goes into heating it. It is constructed to dissipate this heat more or less readily, typical constructions involving radiating fins, water jackets, or simply blackening the metal to make it radiate more efficiently. Even with these precautions in construction the anodes of some tubes will be hot enough to glow. (L.R.Q., R.K.S.)

ANODE RAYS. Among the positively charged particles recognizable in a (so-called) **vacuum tube,** and mixed with the ionized molecules and atoms of the rarefied gas, are sometimes found ions which are traceable to the metallic anode or to impurities in it or upon its surface. The anode may be oxidized or have films or patches of metallic salts upon it which, in the operation of the tube, in some manner not fully understood, yield ions of the metal. If the anode is treated with alkali or alkaline-earth salts or oxides and strongly heated, very copious positive emission may result, serving as a convenient source of positive rays. In some instances occluded hydrogen seems to supply the ions. The experimental study of these rays is hampered by the uncertainty of the supply, which depends upon impurities often of unknown nature and amount; it is also complicated by the presence of the positive ions of the residual gas. The subject is intimately related to positive thermionic emission. **(Thermionic Phenomena.)** (L.D.W.)

ANODYNE. Any medicine which relieves pain or discomfort. The best-known anodynes are **morphine, codeine, hyoscine, atropine, ether, aspirin** and "pyramidon" **(amidopyrine).** (R.S.M.)

ANOLIS. Reptilia, Sauria. The name of a genus of **lizards** adapted also as a common name. Small, mostly brightly colored lizards of the warmer latitudes of the Americas. The little lizard sometimes sold under the name chameleon is the Carolina anolis, *Anolis carolinensis,* a common species of the southern United States and southward. (A.W.L.)

ANOMALISTIC YEAR. Year.

ANOMALODESMACEA. An order of **bivalve** mollusks, mostly burrowing marine species. (A.W.L.)

ANOMALOUS DISPERSION. Ordinarily the **refractive index** n of a medium decreases with increasing wavelength λ (see **Dispersion**). It often happens, however, that in the immediate vicinity of a certain wavelength λ_1 there is a break or discontinuity in the dispersion curve and the usual rule may be locally reversed (see figure). In some cases there are several such points, $\lambda_1, \lambda_2, \lambda_3, \cdots$. These discontinuities correspond to lines or bands in the **absorption spectrum** of the medium. In the Sellmeier dispersion formula,

$$n = 1 + \frac{A\lambda^2}{\lambda^2 - \lambda_1^2} + \frac{B\lambda^2}{\lambda^2 - \lambda_2^2} + \cdots,$$

the several fractional terms make provision for the respective discontinuities. The absorption wavelengths

λ_1, λ_2, $\cdots$ and the constants A, B, $\cdots$ must be determined experimentally. If there is pronounced absorption

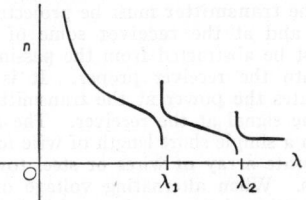

Variation of refractive index with wavelength, illustrating anomalous dispersion.

and anomalous dispersion in the visible range, the medium appears colored, as illustrated by transparent **dyes**. (L.D.W.)

ANONACEOUS FRUITS. *Anona* sp. Anonaceae.

The genus Anona contains shrubs and small trees, many of which bear fruits much esteemed by man. These fruits are composed of many individual ovaries which are more or less sunk in the fleshy receptacle and united to it and to each other. In some species these collective fruits are 5 or 6″ in diameter and so heavy as to drag down the branches. *Anona muricata* is the Soursop, whose somewhat acid white pulp is used in making sherbets and drinks. A native of southern Asia, *Anona squamosa*, the Sugar apple or Sweetsop has sweet fruits which are pleasantly fragrant and which are considered by many the most desirable of this group. Another species native in South America is *Anona Cherimola*, in which the fleshy carpels are completely fused. This species grows best at elevations of 4000–6000′, and is found in the Andean region. *Anona reticulata*, the Custard-apple, also a native of tropical America, is used locally in the West Indies and elsewhere. None of these fruits has as yet appeared in northern markets. (R.M.W.)

ANOPHELES. Mosquito; Malaria.

ANOPLURA.

The order of insects which includes the true or **sucking lice**. They are wingless parasitic insects with mouths formed for piercing and sucking. (See **Louse**.) (A.W.L.)

ANOREXIA.

Loss of appetite, or distaste for food. This condition often accompanies the onset of acute illness, fever—especially prolonged fevers, and particularly in chronic wasting disease such as **cancer**. (R.S.M.)

ANORTHITE. Feldspar.

ANORTHOCLASE. Feldspar.

ANORTHOSITE.

The name anorthosite was given by T. Sterry Hunt to rocks of **gabbroid** nature which were essentially free from pyroxene, hence almost wholly **plagioclase** *usually* labradorite. The term is derived from the French word for plagioclase, anorthose. Small quantities of pyroxene may be present as well as magnetite or ilmenite. The rock is commonly white to gray, bluish, greenish, or perhaps nearly black. A variety from the Province of Quebec is purplish-brown due to the inclusion of ilmenite dust within the feldspars. Although not a common rock in the ordinary sense of the word, occurrences of great areal extent are known in Canada, Norway, and Russia and in the United States in northern New York State and Minnesota. Opinions as to the origin of this rock differ. The development of anorthosite may have been due to the settling out of **labradorite** crystals from a gabbro magma as many believe, or there may have been an original anorthosite magma.

A study of anorthosite occurrences brings out two very curious circumstances, first, that there is no extrusive

(lava) equivalent of anorthosite, and second, that most anorthosite masses seem to be of pre-**Cambrian** age. (E.S.C.S.)

ANOXEMIA.

Deficiency in the **oxygen** content of the blood. This may be due to insufficient **aeration** of the blood as it passes through the lungs due either to disease of the lung or to circulatory failure, or to insufficient **haemoglobin** content of the blood. (D.M.H.)

ANSERIFORMES.

An order of birds including the **geese**, **swans**, **ducks**, mergansers, and related species. (A.W.L.)

ANT.

Insecta, Hymenoptera. Social **insects** of varied structure and habits. They may be distinguished from the related bees and wasps by the form of the slender

Ant.

petiole which connects **thorax** and **abdomen**; in the ants it is expanded above and looks more or less wedge-like in profile. (A.W.L.)

ANTARES.

Antares (α **Scorpii**) derives its name from two Greek words signifying that the star is "similar to" or "a rival of **Mars**," doubtless because of its distinctly reddish hue. In fact, this reddish color has always made the star an object of interest and importance in the ancient religions, and many of the Egyptian temples are so oriented as to indicate that Antares played an important part in their ceremonials. Antares was one of the four royal stars of the Persians about 3000 B.C., and some writers claim that it is the "lance star" referred to in the 38th chapter of the book of Job.

The diameter of Antares has been determined with the stellar **interferometer** and found to be about 390,-000,000 miles or slightly greater than the distance of Mars from the sun. It is a typical M **spectral type giant** star of very low density. (W.K.G.)

ANT-BEAR.

A name applied to the great **ant-eater**, *Myrmecophaga jubata*, of South America and to the **aard-vark** of South Africa. (A.W.L.)

ANT-BIRD.

Aves, Passeres. Several species of birds (*Aves*) found in the forests of Brazil, named from their fondness for ants. (A.W.L.)

ANT-EATER.

Mammalia. Any member of the class which is highly specialized for a diet of ants or termites. The specializations are a slender elongate snout, a long sticky tongue which aids in gathering a sufficient number of the small prey, and strong claws for tearing open ant nests.

The ant-eaters include the spiny ant-eaters or **echidnas** of the Australian region, which are monotremes, the

Echidna. (*N. Y. Zool. Soc.*)

aard-varks of Africa (Order Tubulidentata), the scaly ant-eaters or **pangolins** of the Oriental region and Africa and the **ant-bear**, tamandua (*Tamandua tetra-*

Giant ant-eater. (*N. Y. Zool. Soc.*)

dactyla), and 2-toed ant-eaters (*Cyclopes didactylus*) of Central and South America (Order Edentata). The banded ant-eater (*Myrmecobius*) is an Australian **marsupial** of more squirrel-like appearance than the more highly adapted species. (A.W.L.)

ANTECEDENT STREAM. A stream that has maintained its consequent course in spite of localized uplifts which, if they had proceeded rapidly in relation to the cutting power of the stream, would have caused diversion of the stream. A good example of an antecedent stream valley is one which cuts across a ridge or several ridges. Excellent examples occur in the valley and ridge province of the Appalachian Mountains. On the other hand, it has been suggested that the Appalachian antecedent stream valleys may be really **superimposed**. The accompanying diagram illustrates the origin of the pres-

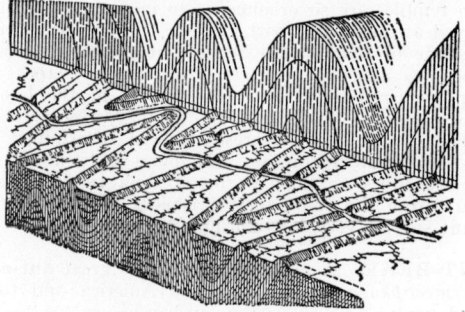

Block diagram illustrating the structural and erosional history of the Appalachian Range. (*After W. M. Davis.*)

ent topography and stream pattern of the Appalachians. It is postulated that the folds were reduced to a **peneplain** on which were flowing a few master streams. Uplift of the peneplain caused the rejuvenation of the master streams which were able to maintain their courses across the upturned edges of the more resistant strata, while the new tributary stream pattern was largely determined by the less resistant formations. (R.M.F.)

ANTELOPE. Mammalia, Artiodactyla. The antelopes are an extensive group between the oxen and the sheep and goats. Many species occur in Africa and some in India and Tibet. The prong-horn antelope of Western North America belongs to a separate group with hollow horns like the oxen; true antelopes have the horns almost solid.

The antelopes include many species with special names, such as the **eland**, the **kudu**, the **addax**, the **gemsbok**, the **oryxes**, the **gazelles**, the **wildebeests**, the **hartebeests** and others. (A.W.L.)

ANTENNA. In zoology an antenna is a jointed sensory appendage of the head found in several classes of **arthropods**. **Crustaceans** have two pairs, while in-

sects, **centipedes** and **millipedes** (see **Diplopoda**) have one pair.

In the process of radio communication the power generated in the **transmitter** must be projected or radiated into space and at the **receiver** some of this radiated energy must be abstracted from the passing radio wave and fed into the receiver proper. It is the antenna which radiates the power at the transmitter and which picks up the signal at the receiver. The antenna form ranges from a simple short length of wire for the receiver to an elaborate array of wires or steel towers for large transmitters. When alternating voltage of a high frequency is connected to a conductor which is open at the end a corresponding high-frequency a.c. will flow in the conductor and return to the voltage source through the capacitance between the conductor and the rest of the circuit. This rapid a.c. causes energy to be radiated into space from the conductor. This energy travels out from the conductor and does not return. The conductor in this case is the antenna (of course the various connecting wires of the transmitter also have high-frequency a.c. and hence will radiate to some extent but very inefficiently). The efficiency with which an antenna radiates is determined by its length and configuration, and its location with respect to the ground, surrounding objects, etc. In general, better radiation is obtained when the antenna length is an appreciable part of a wavelength of the radio signal. Thus they are usually such values as quarter-wave, half-wave, etc. In special cases they are made several wavelengths long but, where space is a major consideration, they may be made very short. It is found that a maximum signal is produced at the receiver by the **ground wave** for a vertical antenna slightly over 0.6 wavelength, but due to **fading** phenomena the best service is usually obtained by one between 0.5 and 0.6 wavelength. Other dimensions give best results for long-distance communication with the **sky wave**. It is to be noted that the characteristics of an antenna are governed by its dimensions in terms of a wavelength, so a very short antenna will have the same characteristics at a short wavelength (high frequency) as a much longer one would have at a correspondingly longer wavelength (lower frequency). Thus it is often more feasible to make a high-frequency antenna of the optimum dimensions than it is a low-frequency one. At the very low frequencies lumped inductance is frequently added to adjust the electrical length. The antenna will have the same characteristics used for receiving but, since in receiving, efficiency is not of such great importance as long as sufficient voltage is induced in the antenna by the passing radio wave to actuate the receiver, no great attention is paid to the design of ordinary receiving antennas. For commercial service, however, receiving antennae are specially designed. The usual receiving antenna with modern sets may be just a short piece of wire, but the picked-up signal will be greater for a moderately long one placed in the clear of absorbing (electrical) objects. However, the noise picked up will also increase and often a longer antenna results in less satisfactory reception because of the additional noise. (See **Directional Arrays.**) (L.D.W., L.R.Q.)

ANTENNA ARRAY. Directional Array.

ANTENNA, DUMMY. A dummy antenna is a substitute for an actual antenna used for test purposes. In making comparative tests, calibrations, etc., on **receivers** it is highly desirable to have conditions as near as possible to actual use conditions, yet have them standardized so they may be reproduced or the results on different units accurately compared. To do this a standard dummy antenna is used. The make-up of the antenna varies with different types of sets, being a series circuit with an inductance of 20 microhenries, capacitance of 200 micromicrofarads and resistance of 25 ohms for regular broadcast receivers. For auto radios, short-wave sets, etc., other circuit combinations are standard. The dummy antenna is connected between the set and the standard signal generator which supplies the radio-frequency test voltages. A dummy antenna consisting of just resistance is frequently used as a load on radio transmitters for making preliminary adjustments without radiating a signal. The output power of the transmitter is dissipated as heat in the resistance. (L.R.Q.)

ANTENNAL GLAND. Glands associated with the antennae of certain **crustaceans;** probably excretory. (A.W.L.)

ANTENNAL SCALE. The modified outer branch exopodite) of the second **antenna** in some **crustaceans.** (A.W.L.)

ANTHELMINTIC. A **vermifuge** or a remedy used to rid a patient of worms (see **Drugs**). (R.S.M.)

ANTHER. The terminal part of a stamen, containing the pollen sacs (see **Flower**). (R.M.W.)

ANTHERIDIUM. The structure which gives rise to the **sperm.** In the **algae** it is a single cell, the contents of which may become a single sperm or divide to produce many sperms. In the higher divisions of plants, the antheridium is a multicellular body which contains the sperms. (R.M.W.)

ANTHOCYANINS. Pigments in Plants.

ANTHOPHYLLITE. The mineral anthophyllite is an orthorhombic **amphibole** essentially (Mg, Fe)SiO_3 with **aluminum** sometimes present. This mineral corresponds to **enstatite** and **hypersthene** in the **pyroxene** group. It has a prismatic cleavage; hardness, 5.5–6; specific gravity, 2.8–3.2; luster, vitreous; color, gray, yellow, brown, green or brownish-green; transparent to translucent. Probably always a **metamorphic** mineral; very common in **schists.** Found in Norway, Austria, Greenland, Pennsylvania, Georgia and elsewhere. The name is derived from the Latin *anthophyllum,* clove, because of its usual brownish shades. (E.S.C.S.)

ANTHOZOA. The **sea anemones, corals, alcyonarians** and related forms. A class of the phylum **Coelenterata** in which the **polyp** form gains its highest development and the **medusa** is unknown.

Like the **hydrozoan** polyps, these animals have relatively thin walls, due to the thin middle layer (mesogloea), and are approximately cylindrical in form. The base is a disk by which the animal is attached to some support and the opposite end forms an oral disk bearing numerous hollow tentacles surrounding the mouth. The mouth leads into a long tube lined with ectoderm, known as the stomodaeum. In it **ciliated** grooves serve for the passage of currents of water into and out of the enteric cavity. In this cavity radiating partitions, the mesenteries, pass from the wall to the stomodaeum, which they hold in place. Others extend into the cavity from the wall without reaching the stomodaeum. The edges of the mesenteries bear mesenteric filaments with stinging cells. They are important in digestion and respiration. Reproductive bodies also develop in the mesenteries and slender acontia with many stinging cells arise from their edges. Muscle bands in the mesenteries and in the body wall contract the entire animal and close the margins of the oral disk in over the tentacles.

The class is divided into two orders:

Order **Alcyonaria.** Polyp with eight tentacles, pinnately branched. Colonial forms, usually supported by a hard skeleton. The **sea fans,** precious **coral,** and **sea feathers.**
Order **Zoantharia.** Colonial or solitary. Polyp with few to many tentacles, not pinnately branched.

Hard deposits formed under the basal disk in some species. The stony **corals** and **sea anemones.** (A.W.L.)

ANTHRACENE. Anthracene

$$\left(C_{14}H_{10} \text{ or } \right)$$

is a colorless solid, melting point 218° C., having blue fluorescence when pure, insoluble in water, slightly soluble in alcohol or ether, soluble in hot benzene, slightly soluble in cold benzene, transformed by sunlight into para-anthracene (($C_{14}H_{10}$)$_2$). Anthracene reacts (1) with oxidizing agents, e.g., **sodium** dichromate plus **sulfuric acid,** to form anthraquinone ($C_6H_4(CO)_2C_6H_4$), (2) with **chlorine** in water or in dilute **acetic acid** below 250° C. to form anthraquinol and anthraquinone, at higher temperatures 9,10-dichloroanthracene. The reaction varies with the temperature and with the solvent used. The reaction has been studied using, as **solvent,** benzene, chloroform, alcohol, carbon disulfide, ether, glacial acetic acid, and also without solvent by heating. **Bromine** reacts similarly to chlorine, (3) with concentrated **sulfuric acid** to form various anthracene sulfonic acids, (4) with **nitric acid,** to form nitroanthracenes and anthraquinone, (5) with **picric acid** (($_1$)$HO\cdot C_6H_2$-(NO_2)$_3$(2,4,6)) to form red crystalline anthracene picrate, melting point 138° C. Anthracene is obtained from coal tar in the fraction distilling between 300° and 400° C. This fraction contains 5–10% anthracene from which by fractional crystallization followed by crystallization from solvents, such as oleic acid, and washing with such solvents as **pyridine,** relatively pure anthracene is obtained. Anthracene may be detected by the formation of a blue-violet coloration on fusion with mellitic acid. Anthracene derivatives, especially anthraquinone, are important in **dye** chemistry. (R.K.S.)

ANTHRACITE. Anthracite is the "hard **coal**" of commerce. It contains usually less than 10% of volatile matter and more than 90% of **carbon,** hence burning with a smokeless flame. Anthracite has a high luster and, unlike "soft" or bituminous coal will not soil the fingers when handled. (E.S.C.S.)

ANTHRAQUINONE. Anthraquinone (9,10)

$$\left(C_6H_4 \begin{smallmatrix} CO \\ CO \end{smallmatrix} C_6H_4 \right)$$

is a yellow solid, melting point 286° C., can be sublimed, forms monoxime, melting point 224° C., by heating under pressure at 180° C. with hydroxylamine chloride, forms no phenylhydrazone with **phenylhydrazine,** with strong oxidizing agents reacts with difficulty to yield phthalic acid ($C_6H_4(COOH)_2$(1,2)), with reducing agents, such as **sodium** hyposulfite, **zinc** in sodium hydroxide solution, tin or **stannous** chloride in hydrochloric acid (but not sulfurous acid), anthraquinone is reduced to

anthraquinol $\left(C_6H_4 \begin{smallmatrix} COH \\ COH \end{smallmatrix} C_6H_4 \right)$,

anthrone $\left(C_6H_4 \begin{smallmatrix} CH_2 \\ CO \end{smallmatrix} C_6H_4 \right)$

dianthrol $\left(\begin{smallmatrix} C_6H_4 \begin{smallmatrix} COH \\ C \end{smallmatrix} C_6H_4 \\ C \\ C_6H_4 \begin{smallmatrix} C \\ COH \end{smallmatrix} C_6H_4 \end{smallmatrix} \right)$

di=anthrone $\left(\begin{array}{c} C_6H_4 \underset{}{\overset{CO}{<}} C_6H_4 \\ \text{C} \\ | \\ \text{C} \\ C_6H_4 \underset{CO}{\overset{}{<}} C_6H_4 \end{array} \right)$

depending upon the conditions. Anthraquinone is obtained by oxidation of anthracene using **sodium** dichromate plus **sulfuric acid,** and is purified by dissolving in concentrated sulfuric acid at $130°$ C. and pouring into boiling water, whereupon anthraquinone separates as pure solid, and is recovered by filtration. Further purification may be accomplished by sublimation or crystallization from nitrobenzene, aniline or tetrachlorroethane. Anthraquinone is used as the material from which many dyes are made, notably alizarin ($C_6H_4(CO)_2C_6H_2(OH)_2$) and related substances. These are vat dyes, that is, insoluble colored substances which are readily reduced to a substance having marked affinity for the fibre to be dyed and which upon exposure to the air are readily reoxidized to the original dye. Anthraquinone may be detected by the appearance of a red color on treatment with alkali, zinc powder and water. For quinones, see **Phenols and Quinones.** (R.K.S.)

ANTHRAXOLITE. A coal-like, metamorphosed **bitumen,** often closely associated with **igneous** rocks. (R.M.F.)

ANTIBIOTICS. Antibiotics are chemical substances secreted by microorganisms, principally fungi and bacteria, which are capable of inhibiting the metabolic activities, growth, or reproducton of bacteria. Some of them can be isolated in quantity and employed in the effective treatment of many stubborn microbial infections and diseases which have resisted earlier attempts to control them. **Penicillin,** for example, inhibits completely the growth not only of pneumococci, streptococci, and staphylococci, but also that of certain strains of gonococci which are resistant to the **sulfonamides.**

Thirty or more antibiotics are now known. Nearly all of them are bacteriostatic, inhibiting the growth of bacteria. Some are bacteriolytic, causing the death and disintegration of bacteria, others merely bactericidal, causing death of the organism without disintegration. Many, unfortunately, are useless in the treatment of disease because they are toxic to man.

Antibiosis is the opposite of **symbiosis.** When colonies of the Ascomycete, *Penicillium notatum,* are grown on the surface of nutrient agar near colonies of *Staphylococcus aureus,* the *Staphylococcus* fails to grow in the immediate neighborhood of the *Penicillium.* Realization of the significance of this particular example of antibiosis led to the isolation of penicillin. The spectacular success of this discovery led to the intensive investigation of many other fungi and bacteria. Several other substances have now been isolated. Many of them are specific in their action; they attack certain bacteria but have no effect on other organisms closely related to them. Some antibiotics interfere with the growth of organisms which cause plant diseases, others with the growth of useful organisms found normally in soil, in water, and in the digestive tract of animals. It is becoming popular to think of them in connection with water and sewage purification. They probably play an important role in the preservation of balance among soil organisms. But their great value lies in the fact that some of them control diseases without bringing harm to diseased tissue of the body. Along with the sulfonamides they provide adequate treatment for wound infection, blood poisoning, pneumonia, meningitis, sore throat, the common venereal diseases, and others; but not, unfortunately, for tuberculosis or the virus diseases. (P.A.W.)

ANTIBODY. When an animal is exposed to infection, there appear in his blood and body fluids soluble substances called antibodies, i.e., bodies acting against introduced substances. Antibodies are also produced on exposure to **antigens** other than bacteria. Any **protein** may act as an antigen and stimulate the formation of specific antibodies. All allergic phenomena come in the category of an antigen-antibody reaction.

In infection, antibodies have a protective and curative value. They act by neutralizing the effect of the antigen. The aim of immunization against disease is the establishment of antibodies in the body. Antiserums are **serums** high in antibacterial or antitoxic antibody content; they are used to combat the effects of the bacteria or their toxins in **scarlet fever, tetanus, and diphtheria.** (D.M.H.)

ANTICLINE. A **folded** structure involving **bedded rocks** in which the strata are arched upward so that the beds bend downward on either side. These downward-bending beds constitute the limbs of the fold.

The angle which the beds on the limbs of the fold make with the horizontal is spoken of as the dip. The term dip is also used to indicate the inclination of bedding in other structures.

Anticlinal arches may be broad and gentle or sharp with a steep dip, symmetrical or unsymmetrical, or may be complicated by minor folds on the limbs. Anticlinal folds may be of sufficient magnitude to be measured in miles, involving great thicknesses of sediments, or they may be so small as to be measured in inches.

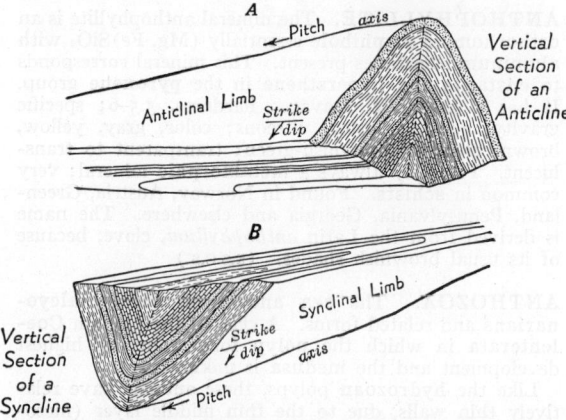

Diagrams illustrating parts of folds. (*After Willis, U. S. Geological Survey.*)

The direction of prolongation of the fold is termed the axis of the fold, and if not exactly horizontal the angle of inclination of the top bed of the anticline is called the pitch. Plunge is used as a synonym for pitch by some writers. (R.M.F.)

ANTICLINORIUM. A composite anticlinal structure of folded beds is called an anticlinorium; a composite

Diagrammatic section of an anticlinorium. (*After Van Hise.*)

synclinal structure is called a synclinorium. The latter term, however, should be applied only to the compressed sedimentary filling of a **geosyncline.** (R.M.F.)

ANTICYCLONE. A relatively large atmospheric eddy, whose dimensions vary from a few hundred miles to several thousand miles. It rotates in a clockwise manner in the northern hemisphere and a counterclockwise manner in the southern hemisphere when viewed from above. The barometric pressure within an anticyclone is high relative to its surroundings, and a pressure gradient exists from its center toward its periphery. A well-developed anticyclone is essentially an air mass. It is in general a region of slowly settling air, the rate of descent from aloft being from 300–1500′ per day. Anticyclones are migratory in the region north of 30–40° Lat. Their path is usually to the east and south. Seasonal semipermanent anticyclones develop over both North America and Eurasia during winter. A belt of permanent anticyclones lies between 10 and 40° Lat. with their centers usually over the oceans. Generally speaking, cloudiness is at a minimum in anticyclonic areas and storms are conspicuously absent. In street language, an anticyclone is known as a "high." (P.E.K.)

ANTIDOTE. A drug or drugs which counteract a poisonous dose of another drug. This may be accomplished by neutralization, as of an acid or alkali, or by changing the drug into an insoluble or non-toxic form. Some antidotes are drugs which give the opposite or antagonistic effect in the body to that produced by the poisoning drug. (R.S.M.)

ANTI-FRICTION BEARINGS. Bearings.

ANTIGEN. A substance, usually a protein, which stimulates the production of antibodies in the body. Bacteria, their toxins, pollens, dust, and many other substances may act as antigens. (D.M.H.)

ANTI-LOGARITHM. Logarithms.

ANTIMONY. Symbol: Sb (stibium). Atomic number: 51. Atomic weight: 121.76. Density: 6.7. Hardness: 3.0–3.3. Melting point: 630.5° C. Boiling point: 1380° C. (Isotopes: page 290.)

Antimony is a silver-white metal, brittle and easily pulverized; scarcely tarnished in dry air but oxidized slowly in moist air; burns at a red heat in air or oxygen with incandescence forming antimonous oxide; insoluble in hydrochloric acid; converted by nitric acid into antimonous oxide or antimonic oxide, depending upon the concentration of acid; by chlorine into trichloride or pentachloride, by sodium hydroxide solution into antimonite. Discovered by Valentine in 1450.

Antimony is used in alloys, with lead for storage battery plates, with lead and tin in type metals, with tin and copper in bearing or anti-friction metals. Antimony occurs chiefly as the sulfide (stibnite, Sb_2S_3) which is produced mainly in China, only small amounts in Mexico and Bolivia. Stibnite is (1) melted and reduced to antimony by iron metal and separated from fused ferrous sulfide; (2) is roasted in air and sublimed antimonous oxide collected and reduced by heating to fusion with carbon and sodium carbonate.

Acids: Antimonous acid (ortho, H_3SbO_3, pyro, $HSbO_2$), white solids forming antimonite salts; antimonic acid (ortho, H_3SbO_4, pyro, $H_4Sb_2O_7$, meta, $HSbO_3$), white solids, forming antimonate salts.

Antimonates: Sodium antimonate (pyro, $Na_2H_2Sb_2O_7 \cdot H_2O$, meta, $2NaSbO_3 \cdot 7H_2O$), white solids.

Antimonide: Silver antimonide (Ag_3Sb), black precipitate, by reaction of stibine and very dilute silver nitrate solution.

Antimonite: Sodium antimonite (meta $NaSbO_2 \cdot 3H_2O$) white solid.

Chlorides: Antimonous chloride, antimony trichloride, "butter of antimony" ($SbCl_3$) white solid, melting point 73° C., boiling point 220° C., by reaction of antimony upon heating with a deficiency of chlorine, reactive with water to form antimony oxychloride, antimonyl chloride ($SbOCl$), white solid, soluble in hydrochloric acid; oxychloride, white insoluble solid; antimonic chloride ($SbCl_5$), pale yellow liquid, boiling point 140° C., by reaction of antimony upon heating with an excess of

chlorine, reactive with water to form oxychloride, white insoluble solid.

Hydride. See Stibine.

Oxides: Antimonous oxide, antimony trioxide (Sb_2O_3), white solid when cold, but yellow when hot, formed (1) by burning antimony in air or oxygen, (2) by reaction of dilute nitric acid and antimony, melting point 656° C., sublimes at 1550° C. (in the absence of air); antimony tetroxide (Sb_2O_4), white solid, by heating antimony metal, trioxide, pentoxide, or trisulfide in air at 800°–900° C. for some time, decomposes above 900° C. to trioxide plus oxygen; antimonic oxide, antimony pentoxide (Sb_2O_5), pale yellow solid, by reaction of antimony metal or trioxide with concentrated nitric acid, decomposes on heating, forming tetroxide below 800° C.

Stibine, antimony hydride (SbH_3), colorless, odorless, very poisonous gas, by reaction of a solution of antimony-containing material with a metal (e.g., zinc), and hydrochloric or dilute sulfuric acid (but not with sodium hydroxide, thus differing from arsine). Stibine, (1) when heated in a glass tube yields a metallic mirror of antimony, (2) when passed into a very dilute solution of silver nitrate yields back precipitate of silver antimonide. Stibine burns in air with a faintly bluish-green flame forming antimonous oxide and water.

Sulfides: Antimonous sulfide, antimony trisulfide (Sb_2S_3), orange-red precipitate, by reaction of antimonous salt solution and hydrogen sulfide, soluble in concentrated hydrochloric acid, soluble in sodium or ammonium sulfide to form thioantimonite; antimonic sulfide, antimony pentasulfide, "antimony red" (Sb_2S_5), orange precipitate, by reaction of antimonic salt solution and hydrogen sulfide, soluble in concentrated hydrochloric acid, soluble in sodium or ammonium sulfide to form thioantimonate. Used (1) in the manufacture of matches and fireworks, (2) as a pigment, (3) in vulcanizing and coloring rubber.

Numerous organic compounds containing antimony have been prepared, among these are trimethylstibine ($Sb(CH_3)_3$), tetramethylstibonium hydroxide ($Sb(CH_3)_4 OH$), which is a strong base, triphenylstibine ($Sb(C_6H_5)_3$).

Solutions of antimony containing substances, when boiled with hydrochloric acid and iron metal, yield a black precipitate of antimony metal.

Tartrate: potassium antimonyl tartrate, "tartar emetic" ($K(SbO)(C_4H_4O_6) \cdot \frac{1}{2}H_2O$), white crystals, soluble, by reaction of antimony trioxide and potassium hydrogen tartrate, and then crystallizing. Used as a mordant in dyeing textiles and leather, and in medicine.

In alloys, antimony is easily detected by its formation of a white solid upon treatment with concentrated nitric acid and subsequent separation from tin, which is the only other metal thus forming a white solid. (R.K.S.)

REPRESENTATIVE ALLOYS CONTAINING ANTIMONY

Regulus of Venice........	49% Sb	51% Cu	
Type metal..............	20% Sb	75% Pb	5% Sn
Locomotive bearing metal..	14% Sb.	82% Sn	4% Cu
Stereotype metal.........	14% Sb	84% Pb	2% Sn
Antifriction metal (Babbitt metal)........	8% Sb	84% Sn	8% Cu
Britannia metal (can be spun for utensils)...........	5% Sb	93% Sn	2% Cu
Pewter....0–7% Sb	83–75% Sn	0–20% Pb	0–4% Cu

(R.K.S.)

ANTIPODAL CELLS. The three usually small cells which occur in the embryo sac of angiosperms at the end most distant from the micropyle. No known function has been ascribed to them. (R.M.W.)

ANTIPYRETIC. Any drug or physical agent which lowers the temperature of the body. Formerly antipyretic drugs were commonly used in fevers. Today they are little used, as fever is considered an index of the resistance of the body to infection, and is probably exerting a beneficial effect. The most common antipyretics are "aspirin" (acetylsalicylic acid), quinine, antipyrene, acetanilid and phenacetin. (R.S.M.)

ANTISEPTIC. Any substance inhibiting the growth of or killing microorganisms, without undue injury to bodily tissues. The chief antiseptics used are: iodine, ethyl alcohol, "merthiolate," "metaphen," "mercurochrome," phenol, picric acid, Dakin's solution, and azochloramid. (R.S.M.)

ANTI-SIDE-TONE. In the older types of telephone subscribers' sub-set the sound going into the transmitter (mouthpiece) can be heard in the receiver. This is known as side-tone and has some objectionable features. It causes the speaker unconsciously to lower the voice and thus reduces the energy transmitted, and it also increases the effect of any local noise by causing it to appear in the receiver. It is thus desirable to eliminate or at least reduce the side-tone and special circuits and equipment have been devised. These sub-sets are known as anti-side-tone sets. Fig. 1a shows a conventional

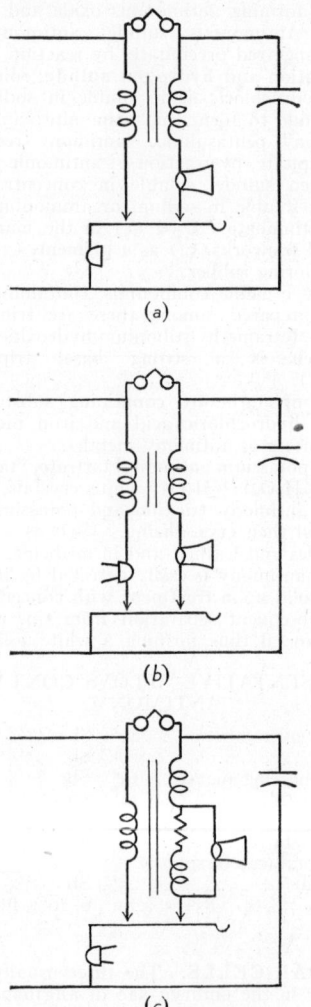

(a)

(b)

(c)

Fig. 1. Sub-set connections.

connection, 1b a side-tone reduction circuit, and 1c an anti-side-tone circuit. In the latter the values of the transformer windings and the resistances are so proportioned that the side-tone voltages cancel in the receiver

while the incoming signal from the line is impressed on the receiver. (L.R.Q.)

ANTITRADE WINDS. Above the northeast and southeast trade winds there frequently is present a westerly wind known as the Antitrades. Reversal of direction occurs as low as a few thousand feet or as high as 3–4 miles. (P.E.K.)

ANTLER. The large and complex horns of deer, consisting of bony outgrowth with no covering of keratin. When growing they are covered with skin, the velvet, which is soon lost. (A.W.L.)

ANT-LION. Insecta, Neuroptera. Immature insects of the family Myrmeleonidae which lie buried at the apex of conical pits in dry sand or dust. Ants or other small insects which enter the pit slide down the loose slope and are seized by the upturned jaws below. (A.W.L.)

ANT-LOVING CRICKET. Insecta, Orthoptera. *Myrmecophilia.* Small peculiarly formed crickets which live in ant nests. (A.W.L.)

ANURA. The frogs, toads, and allied species; a division of the Amphibia characterized by the absence of the tail. Also known as the Salientia from their jumping powers. (A.W.L.)

ANUS. The external opening of the rectum. (R.S.M.)

ANVIL. Forging

ANVIL HEAD, ANVIL CLOUD. Clouds.

AORTA. The main and largest blood vessel of the arterial blood system. It arises from the left ventricle of the heart, and arching over the root of the left lung, descends along the vertebral column, passing through the chest, and pierces the diaphragm into the abdominal cavity, finally dividing into the right and left iliac arteries in the pelvis. Many large and small blood vessels branch from it. (D.M.H.)

AORTITIS. Inflammation of the walls and lining of the aorta at its origin and along its ascending arch. The inflammatory process often spreads to involve the aortic valve, resulting in its dilatation and incompetency. Tertiary vascular syphilis is the commonest cause of aortitis. Arteriosclerosis of the aorta and the endocarditis of rheumatic fever are sometimes the etiologic factors. (D.M.H.)

AOUL. Mammalia, Artiodactyla. A gazelle of northeastern Africa. (A.W.L.)

APAR. Mammalia, Edentata. The 3-banded armadillo, *Tolypeutes,* of South America. (A.W.L.)

APATITE. The mineral apatite is a phosphate of calcium with either fluorine or chlorine or sometimes both, hence the distinction between fluor-apatite and chlor-apatite. Sometimes both fluorine and chlorine are present. Most apatite is, however, fluor-apatite.
Apatite crystallizes in the hexagonal system in prismatic and tabular forms. Hardness, 4.5–5; specific gravity, 3.17–3.23; luster, vitreous to resinous; transparent to opaque; streak, white; cleavage, imperfect basal and prismatic; color, white, green, yellow, red, brown and purple; sub-conchoidal fracture. The variety called asparagus stone is yellow-green and manganapatite which is a dark bluish-green may contain as much as 10% manganese dioxide replacing the calcium. Werner devised the name apatite from the Greek word meaning to deceive, as it was frequently mistaken for beryl and other species. Apatite has been found widely distributed

both geographically and petrologically as it occurs in many sorts of rocks, metamorphic **limestones, gneisses, schists, granites** and **syenites, pegmatite** veins and even with **iron** ores. It has been prepared artificially. It has been mined for the manufacture of fertilizers and to a slight extent for jewelry.

Apatite occurs extensively in Europe and America, especially in New England, New Jersey, New York, North Carolina, California, and in the provinces of Ontario and Quebec in Canada. (E.S.C.S.)

APE. Mammalia, Primates. Particularly any of the man-like apes, including the **gorilla, orang-utan, chimpanzee** and **gibbons** (Family Simiidae) but also applied to certain monkeys of another family, such as the Barbary ape. All are tailless. (A.W.L.)

"APE-MAN" OF JAVA. Paleontology of Man.

APERTURE, TELEVISION. In transmitting the **television** image it is necessary to break it down into elements of very small area. In early mechanical scanning methods this was done by "observing" the image through an opening or aperture which scanned the scene. In modern equipment the scanning is no longer done mechanically but the term aperture is used to denote the size of one of the elements into which the picture is broken for transmission. (L.R.Q.)

APHANITE. An aphanite is any fine-grained **igneous** rock whose constituents cannot be distinguished with the naked eye. The term is derived from the Greek, meaning invisible. The adjective aphanitic is applied to these rocks as well as their fine-grained groundmasses. (E.S.C.S.)

APHASIA. Diminution or loss of expression by the spoken or written word, or of the understanding of spoken or written language. It is due to injury or disease of the brain centers involving memory, hearing, speech or associated centers. (R.S.M.)

APHELION. Aphelion is the point in the **orbit** of a member of the **solar system,** except a **satellite,** where the object is most remote from the sun. It is the point on the line of apsides diametrically opposite to **perihelion.** (W.K.G.)

APHID, APHIS. Insecta, Homoptera. A plant louse. Small delicate **insects** with sucking mouths. They live on the sap of plants and many species are of economic importance. They are characterized by an intricate life cycle which results in a high rate of reproduction.

In the temperate zone aphids hatch in the spring from winter eggs; these individuals are females known as stem mothers. They bear living young without mating (**viviparity, parthenogenesis**) and these in turn are females capable of the same type of reproduction. Late in the season a generation known as the sexuparae bears both male and female offspring which mate to produce the eggs that pass the winter.

Winged aphids appear under conditions which demand migration from plant to plant. Experiments have shown that the appearance of wings is a response to definite environmental conditions, probably complex in nature.

Examples of the economic species are the melon aphid and the apple-grain aphid. Spraying with contact poisons such as nicotine sulfate is effective against all species. (A.W.L.)

APHIS-LION. Insecta, Neuroptera. The **larva** of the golden-eyes or lacewing flies (Family Chrysopidae), so called because they feed on aphids and other small insects. (A.W.L.)

APICAL GROWTH. Growth at the tip of an organ, as occurs in the roots and stems of all higher plants.

Examination of the stems of most plants will show that growth in length occurs only in the apical portion, and only for a relatively short period of time, usually a matter of a few weeks. This may be determined by observing the distances between successive leaves: near the growing tip the leaves are very small and close together; as one goes back along the stem the size of the leaves increases and also the distance between them; but after the leaf is mature little elongation of the stem occurs, as shown by the uniform distances between the mature leaves. Older stems increase constantly in diameter, but only exceptionally in length. One notable such exception, known as intercalary growth, is found in grasses and some other plants. Here a group of **cells** in the region of the **node** are capable for a time of active division and of causing increase in the length of older portions of the stem. Both here and in the tip of the stem increase in length is due to elongation of cells produced by an actively dividing tissue known as a meristem. (R.M.W.)

APICAL ORGAN. A ciliated structure found in the larvae of **annelid** worms and **Bryozoa.** (A.W.L.)

APLACOPHORA. An order of **Amphineura** made up of animals of worm-like form. (A.W.L.)

APLITE. This term is applied to fine-grained, sometimes sugary-textured **igneous** rocks, composed almost wholly of **quartz** and **feldspar.** Except for size of grain, aplites resemble **pegmatites** both in mineral composition and in mode of occurrence in dikes and veins, save that the rare minerals often present in pegmatites are wanting here. The word aplite is derived from the Greek word meaning simple, referring to its ordinarily simple mineral composition. (E.S.C.S.)

APODA. Gymnophiona.

APODEME. An internal projection of the hard outer covering (**exoskeleton**) of **arthropods.** Apodemes provide muscle attachments and in some species are extensively developed. They are collectively termed the **endoskeleton.** (A.W.L.)

APONEUROSIS. A sheet of tough white glistening **fascia** or membrane which surrounds muscle and muscle fibers or connects a muscle to the part which it moves. The tensile strength and resistance of muscle tissue is dependent upon the fascial tissue around the muscular fibers. (R.S.M.)

APOPHYLLITE. This mineral is a hydrous **silicate** of **potassium, calcium,** and **fluorine,** corresponding to the formula $(KFCa_4(Si_2O_5)_4)8H_2O$. It crystallizes in the **tetragonal system** in square prisms resembling cubes terminated by pyramids; may be tabular, sometimes massive. Cleavage is perfect, parallel to the base; hardness, 4.5–5; specific gravity, 2.3–2.4; luster, vitreous to pearly; transparent to translucent or nearly opaque; color may be white, grayish, greenish, yellowish or reddish. This mineral was named by Haüy from the Greek words meaning from a leaf, referring to its exfoliation when heated with the blow pipe.

Apophyllite is a secondary mineral found with the **zeolites** and has been classed with them by some writers, but it contains no **aluminum,** which element is understood to be an essential in a zeolite. It occurs in cavities in basalts and less often filling openings in granites or other crystalline rocks; it also is a gangue mineral in certain ore veins.

There are many localities for apophyllite: Bohemia, Trentino, Italy, the Hartz Mountains, and Iceland. Fine specimens have been obtained from the Ghats Mountains in India. The Triassic trap rocks of New Jersey, Connecticut and Nova Scotia have also furnished many specimens. (E.S.C.S.)

APOPHYSIS. In zoology, an apophysis is a protuberance or outgrowth of an organic structure, such as a process on a bone.

In geology, an apophysis is a tongue or other direct offshoot of a larger vein or **dike**. (A.W.L., E.S.C.S.)

APOPLEXY. Sometimes called "stroke." Sudden paralysis and coma following a vascular accident in or about the brain or brain stem. The vascular accidents occur as the result of hemorrhage from a cerebral vessel, or **thrombosis** of the blood within the vessel, shutting off the flow of blood to a portion of the brain. **Embolus** to the cerebral vessels occurring in the course of certain types of heart disease, or in chronic suppurative lung disease produces the same picture. Apoplectic seizures occur most commonly in the age group of 40–50 years. **Arteriosclerosis** of the vessels at the base of the brain and **hypertension** are the predisposing factors. Formerly it was thought that hemorrhage was from small arteries, but recent work indicates that it is the small veins which give way.

The apoplectic seizure follows immediately in the wake of the hemorrhage. Sometimes there is a premonitory period of dizziness, mental confusion, headache, disturbance of speech. Often the patient is overtaken suddenly, falls to the ground, and rapidly loses consciousness. Depending on the site and size of the hemorrhage, the severity of the seizure varies. At times, there is simply transitory confusion accompanied by the loss of motor power of a limb. If a wide area of brain tissue has been involved, deep **coma** develops. It may last for hours or days, and may be followed by transitory or permanent mental changes, apparent diminution in intelligence and loss of memory.

The paralysis which results from an apoplectic seizure is a hemiplegia, a loss of motor power of the muscles on one side of the body. The face, tongue, arm and leg are involved. At first the paralysis is complete, but gradually over a course of weeks, some power returns; with encouragement and helpful massage and exercise, the patient usually can walk again. The muscles on the affected side become stiff and spastic; contracture of the arm in the flexed position and fixation of the leg in extension tend to occur. (D.M.H.)

APOPYLE. The opening by which a canal in the wall of a **sponge** communicates with the central cavity. (A.W.L.)

APPALACHIAN REVOLUTION. Permian.

APPENDAGE. A supplementary structure attached to an organ or body. Externally many animals have appendages which serve for defense, for locomotion, for securing food, and as sensory organs.

The simplest external appendages are mere outgrowths of the body wall, either solid or hollow; the tentacles of **coelenterates** are an excellent example. In the **annelid** worms both tentacles and parapodia appear as external appendages, in the **echinoderms** the rays or arms may be radiating divisions of the body or appendages, and in the **mollusks** tentacles of complex structure occur. In all cases these structures allow greater facility of movement than is possible for the body as a whole and so compensate the lack of freedom which attends increased size and complexity or sessile habits.

The most elaborate appendages are the jointed appendages of the **arthropods** and **chordates**. Although they are not fundamentally related, the appendages of both phyla are similar in principle. They consist of a series of segments connected by movable joints with each other and with the body, and are provided with muscles which operate them as a series of levers. In arthropods the skeleton is external to the muscles, in the vertebrates internal.

The jointed appendages of arthropods are specialized in various species for swimming, walking, running, jumping and grasping, and in the form of antennae and palpi as sensory organs. In the jumping appendage powerful muscles result in the extension of one segment of the leg, as in the familiar grasshoppers. Grasping is accomplished by the folding back of one segment against another in the insects or by the development of a process on the next to the last segment which works against the terminal segment in a forceps-like relation; the latter is the chelate type of appendage. Arthropod appendages were originally metameric, one pair appearing on each segment of the body. In existing species they are variably limited as described under the term **Arthropoda**.

The **vertebrate** appendage is also specialized for various purposes, including jumping, swimming, flight, and burrowing. The primitive form appears to be the paired fins of the fishes (**Pisces**), and the arms and legs of man are good examples of fairly specialized appendages. Two pairs, the pectoral or anterior pair and the pelvic or posterior pair are typical; only in highly specialized forms such as the snakes is this number reduced.

The **vertebrate** appendage differs from the arthropod appendage in the terminal series of digits. These structures have been developed in some animals so that they can be opposed to each other for grasping, as in the opposition of the human thumb to the four fingers. In other forms they have been reduced in importance. (**Biramous Appendage, Hand, Fin, Foot, Antenna, Pentadactyl Appendage, Tail, Wing, Telson,** and **Mouth**.) (A.W.L.)

APPENDECTOMY. The surgical removal of the appendix. (R.S.M.)

APPENDICITIS. Appendicitis is an acute inflammatory process involving the **appendix**. It is characterized by abdominal pain, nausea and vomiting, tenderness and spasm of the muscles of the right side of the abdomen, fever, and an elevated white blood-cell count.

At the start of the 19th century the importance of the appendix was not recognized. Inflammation in the right side of the abdomen was believed due to infection about the **caecum** and was called "typhlitis" or "perityphlitis." Isolated reports in the early 19th century showed that the appendix could be involved in an acute process but operations and autopsies were quite rare. The real significance of the disease was recognized by Reginald Fitz, of Boston, in 1886, who first used the term "appendicitis."

Appendicitis is more frequent in highly civilized races than in primitive peoples and is more frequent in urban than in rural districts. The disease is most common between the ages of 10 and 35, although it may occur at other periods.

An attack of appendicitis may subside but, with a virulent infection or with obstruction of the blood supply to the organ, the appendix may become gangrenous or rupture, allowing grossly contaminated material to infect the abdominal cavity. This produces **peritonitis**. Nearly 25,000 individuals die from appendicitis in the United States yearly. In simple acute appendicitis, uncomplicated by abscess or peritonitis, the mortality is less than 1%. But once these complications develop, the mortality soars. Hence, the importance of early operation as soon as the diagnosis of acute appendicitis has been made. Delay and the taking of cathartics during an attack are responsible for the high mortality.

The most important and constant symptom of appendicitis is pain in the abdomen, most frequently in the right lower quadrant. Often the pain at first is around the navel and later shifts to the right lower quadrant of the abdomen. The pain may be constant or intermittent but is usually cramp-like. It varies considerably in severity, depending on many factors such as anatomical peculiarities of the appendix, severity of infection, and degree of obstruction of the appendix. Frequently, with rupture of the appendix the pain becomes lessened for a time—a dangerous sign. Other symptoms are nausea and vomiting—the latter following the onset of pain rather than preceding it as in simple gastroenteritis.

The most constant physical finding is tenderness localized over the region of the appendix. Fever may be absent, slight, or moderately high. The white blood count varies but is often either slightly or moderately elevated. It is a valuable confirmatory sign that at times may be misleading (a patient may have a normal blood count with a gangrenous appendix).

In children, the diagnosis of appendicitis is more difficult than in adults and it is for this reason that operation is usually delayed too long.

Treatment for acute appendicitis is always surgical. The earlier surgery is undertaken, the lower the mortality.

Chronic appendicitis is a much-abused term as many so-called cases are due to trouble in organs not having any relation to the appendix. Such cases are not relieved of symptoms by appendectomy. The term chronic appendicitis should be used only to describe an appendix which has been damaged by previous acute attacks of appendicitis. (D.M.H.)

APPENDICULARIAN. Chordata, Tunicata. A free-swimming tailed tunicate. These forms make up the class **Larvacea.** (A.W.L.)

APPENDIX (Appendix Vermiformis). A blind worm-like tubular portion of the intestine which arises from the base of the **cecum.** Its size and position vary greatly, the length averaging $3\frac{1}{2}''$, although it has been found to vary from $\frac{3}{4}$–$9''$. Its position often varies from the normal so that it has been reported in every possible situation in the **abdomen,** depending on the position of the cecum and the length of the organ and its attachments.

An appendix is found only in man, the higher **apes,** and the **wombat,** and possibly in some **rodents.** In herbivorous animals the cecum attains a very large size and it is thought by some that the appendix represents the degenerated remains of the herbivorous cecum. Others believe that it is a **lymph** organ functioning as other lymph glands in the body. By many it is considered to be in the process of gradual obliteration in the human species. It is subject to inflammatory processes because of its limited blood supply, and since it is a blind tube, it is subject to obstruction from fecal impaction. (R.S.M., D.M.H.)

APPLE. Rose Family.

APPLICATIONS OF CHEMISTRY. Chemistry, Applied Chemistry.

APPLIED CHEMISTRY. Chemistry.

APPROXIMATE INTEGRATION. When the function in the integrand of a **definite integral** is given analytically, the definite integral is usually evaluated by finding the corresponding **indefinite integral** by the methods of **integration technique** and substituting in this the given limits. But it sometimes happens that the indefinite integral cannot be found, and in this case a method of approximate integration must be used. When the integrand function is given empirically by a set of values, the integration technique is not applicable, and in this case also a method of approximate integration is ordinarily used. Formulas for the approximate evaluation of **definite integrals** are given by the following rules:

Suppose we wish to evaluate approximately the integral $\int_a^b f(x)dx$. Divide the interval (a, b) into n equal parts, each of length h, so that $h = \dfrac{b-a}{n}$. Let the successive

values of x in this subdivision be denoted by $x_0(=a)$, x_1, $x_2, \cdots, x_{n-1}, x_n(=b)$, and let the corresponding values of the function $y = f(x)$ be denoted by $y_0, y_1, y_2, \cdots, y_{n-1}, y_n$, so that

$$y_0 = f(a),\ y_1 = f(x_1),\ y_2 = f(x_2), \cdots,$$
$$y_{n-1} = f(x_{n-1}),\ y_n = f(b).$$

The trapezoidal rule is expressed by the formula:

$$\int_a^b f(x)dx \approx \tfrac{1}{2}h(y_0 + 2y_1 + 2y_2 + 2y_3 + \cdots + 2y_{n-1} + y_n),$$

where the coefficients of the y's are all 2 except the first and last. In this case, n may be any positive integer; in general, the larger n is taken, the closer the approximation.

For Simpson's one-third rule (or parabolic rule), the positive integer n must be taken as an even number; the rule is then expressed by the formula:

$$\int_a^b f(x)dx \approx \tfrac{1}{3}h(y_0 + 4y_1 + 2y_2 + 4y_3 + 2y_4 + \cdots$$
$$+ 2y_{n-2} + 4y_{n-1} + y_n)$$

Here also the approximation is in general closer the larger n is taken.

There are additional similar rules known for the approximate evaluation of definite integrals. (L.L.S.)

APPROXIMATIONS, NOTATION. The symbol $\approx$ is put between two expressions to indicate that the right-hand expression is an approximation to the left-hand expression.

Thus, we may write: $\frac{1}{3} \approx 0.33$, $\sqrt{2} \approx 1.414$, $\pi \approx 3.1416$. (L.L.S.)

APRICOT. Rose Family.

APSIDES. Orbit.

APTERYGIFORMES. An order of birds containing only the flightless kiwis of New Zealand. (A.W.L.)

APTERYGOTA. The primitive wingless **insects.** A subclass made up of the orders **Protura, Thysanura,** and **Collembola** in which the existing species are wingless and there is no evidence to show that wings have occurred in any ancestral form. (A.W.L.)

AQUA. Latin, water, previously used widely by chemists as a term for various solutions in water, e.g., aqua ammonia, solution of **ammonia** in water, ammonium hydroxide; aqua calcis, lime water, **calcium** hydroxide solution; aqua fortis, **nitric acid;** aqua regia, mixture of concentrated nitric and **hydrochloric acids,** named "royal water" because it was the only acid that would attack the noble metal gold. (R.K.S.)

AQUAMARINE. Beryl.

AQUARIUM. A water-tight container, usually with glass sides, for the maintenance of aquatic organisms in captivity. Also an establishment for the public exhibition of such displays.

Small aquaria for the home are often of the globular type known as goldfish bowls, blown in one piece, or one-piece rectangular glass vessels. Larger aquaria are made of glass plates set in a metal frame with a specially prepared aquarium cement. The latter require greater care and usually leak when first filled or when shifted later; if well constructed the flexible cement permits adjustment to the changed stresses and the leaks stop automatically. For laboratory purposes larger aquaria are made of a combination of stone slabs and glass plates.

Formerly aquaria for the home were limited to the display of goldfishes but in recent years many species of tropical fishes have been made available. In all cases the purity of the water used and the maintenance of an adequate supply of oxygen are important factors in the success of the aquarium. The maintenance of tropical fishes is further complicated by the need for higher temperatures than can be found in most homes. These aquaria cannot safely be allowed to drop below 70° F. and should be maintained at 75–80° F. Heating is best accomplished by electric heaters with **thermostatic** control, for which special equipment can be secured from dealers in aquarium supplies.

As a rule a good city water supply is safe for fishes. If not, water from any clear stream or lake where fishes thrive is likely to be safe for aquarium species, or a chemical purifier can be secured from dealers in fishes.

The oxygen supply can be maintained by changing the water frequently but it is better to establish a balanced aquarium by including some water plants such as **Elodea, Vallisneria,** or **Sagittaria.** The plants free oxygen and utilize the carbon dioxide produced by the fishes. Some plants float but others must be planted in a bed of sand or gravel in the bottom of the aquarium. With a little practice pleasing arrangements may be secured by the use of a few rocks and the careful grouping of the plants, and the quantities of plants and fishes may be adjusted so that changes need not be made for months at a time. A good rule to follow is "an inch of fish to a gallon of water." Plants need not be so carefully regulated, for they tend to crowd the aquarium and must usually be thinned from time to time, while fish increase only under the most favorable conditions.

Some common aquarium troubles are the accumulation of **algae,** diseases, and incompatibility of different species of fishes. Once algae are introduced it is difficult to keep them down without complete sterilization and renovation of the aquarium and any sand and rocks that it may contain. Plants should be replaced with clean stock but the fishes may be returned to the aquarium. For the recognition and treatment of diseases and for the proper stocking of the aquarium the fish-fancier should consult a reputable dealer or one of the excellent books on aquarium fishes now on the market.

A few snails and bottom-feeding fishes included in an aquarium play a useful part as scavengers. The flocculent wastes that inevitably accumulate at the bottom in the course of time can be siphoned off by means of a flexible tube. Some water is necessarily carried off, hence, after cleaning the aquarium, enough should be added to bring the contents to their normal level.

While convenience must often determine the location of an aquarium in the home, plants will not thrive without some sunlight so a balanced aquarium must be kept near a window. North light is by far the best. If the sun's rays must strike the aquarium directly, it should be only for a short time during the day.

Among the aquaria where large public exhibits of fishes are maintained may be mentioned the Shedd Aquarium in Chicago, Steinhart Aquarium at San Francisco, the Naples Aquarium, London Zoological Gardens, the Bermuda Aquarium, and the Honolulu Aquarium. The two last are famous for the brilliant tropical fishes which they exhibit from neighboring waters. (A.W.L.)

AQUARIUS. (The water bearer.) (Map, page 380.) This constellation is the eleventh sign of the **zodiac** and is of importance solely because of that fact. There is a theory that the constellation received its name because the sun is in this part of the sky during the rainy season in the Euphrates valley.

Though the constellation is relatively large it contains no bright or particularly striking features. (W.K.G.)

AQUEDUCT. An artificial conduit built to carry water is called an aqueduct. Generally speaking, aqueducts are built to convey a fresh water supply to congested districts from suitable sources more or less distant, and are therefore peculiar to cities. The first settlers in a place may depend upon local springs,

streams, and wells, but with the growth of population there comes a time when these will prove inadequate, and suitable distant water supplies may have to be tapped through the medium of the aqueduct.

An aqueduct may be either a pressure or grade-line type. Pressure conduits are commonly employed for small capacities or adverse topography, open or grade-line conduits for large capacity or favorable topography. Circumstances may require both types on the same project as the most economical combination. Pressure tunnels can convey water at pressures considerably above atmospheric, and are constructed with circular cross-sections. They are most frequently found in tunnel sections cut through hills and mountains, and in siphons. The principal distinguishing hydraulic characteristic of the pressure aqueduct is that it may depart from the normal open flow line both above and below the normal **hydraulic gradient.** However, siphon action, for sections above the hydraulic gradient, should be avoided wherever possible. Grade-line sections of aqueducts are usually built with open cut and fill construction following a hydraulic grade-line which will yield the requisite flow in the aqueduct at approximately atmospheric pressure, i.e., the fall per mile being just sufficient to overcome the friction loss in the same distance.

Some of the most important of the ancient aqueducts were those supplying the city of Rome, among which might be mentioned the Marcian, with a length of 58 miles, the Julian, a length of 17 miles, and the Claudian, with a length of 43 miles. These were high level aqueducts of the grade-line type, principally cut and fill where possible, with grade-line tunnels for piercing hills, and resting on multiple arches when spanning valleys. These older aqueducts rarely had cross-sections greater than 30 sq. ft. in area. The Catskill aqueduct which conveys the water of the Ashokan Reservoir to the City of New York, approximately 100 miles away, has a capacity of 500,000,000 gals. a day and is a splendid example of modern engineering on a large scale. It has in places cross-sections greater than 150 sq. ft. in area, inverted siphons, one going more than 1000′ below sea level, as well as a score of tunnels. (C.W.C., F.T.M.)

AQUEDUCT OF SYLVIUS. The portion of the central canal of the **nervous system** of **vertebrates** which lies in the mid-brain, connecting the third and fourth ventricles. (A.W.L.)

AQUEOUS VAPOR. Water vapor is known as aqueous vapor. It varies from 0.01–3.0% of the total atmosphere, depending on the temperature, source, and history of the air in question. Aqueous vapor enters the air by evaporation from oceans, lakes, rivers, and other small bodies of water, from precipitation, from water films on surface objects, from transpiration of plants. It exerts a partial pressure which is added to the total air pressure. (P.E.K.)

AQUIFER. Ground Water.

AQUILA. (The eagle.) (Map, page 380.) A constellation lying in the **milky way** and hence containing rich star fields when viewed with a low-powered **telescope.** The distinguishing feature of this constellation is the group of three stars almost in a straight line, with the bright star **Altair** between two fainter ones. Several **Novae** have appeared in this constellation, the most famous one being Nova Aquilae III of 1918. (W.K.G.)

AQUINO'S METHOD. Celestial Navigation.

ÆR. Symbol for **Aspect Ratio** of an object, especially an airfoil or wing. (F.T.M.)

ARABINOSE. Carbohydrates.

ARACHIS OIL. Esters.

ARACHNIDA. A class of the phylum **Arthropoda** including the **spiders, mites, ticks, scorpions,** pseudoscorpions, **whip scorpions,** sun spiders and **harvestmen.** Next to the insects this class is probably the best known among the invertebrates.

Arachnids differ from the other members of the phylum in one or more of the following characters: 1. The body is usually divided into two regions, a **cephalothorax** and abdomen. 2. Only simple eyes are present. 3. There are no **antennae.** 4. The **thorax** bears four pairs of legs in the adult. 5. The **abdomen** is often unsegmented and bears no appendages. 6. The first pair of appendages are **chelate** grasping organs. 7. Respiration is carried on by **tracheae** or **lung-books.**

Arachnids are almost exclusively terrestrial and are predominantly predacious or parasitic, although some of the mites are plant feeders and the harvestmen include vegetable materials among their food.

The development of poison glands is fairly general in the group. Spiders have such glands, opening in the jaws or **chelicerae,** and scorpions have a special sting at the tip of the abdomen. With the exception of the black-widow spider of the United States and a small scorpion found near Durango the poison is not known to be harmful to man. There is some probability that these two species may sometimes inflict fatal wounds.

The secretion of silk by spiders is another salient feature of the group. Silk glands are located in the abdomen, discharging through a group of spinnerets near the posterior end of the body. The silk is used to build webs of various forms for snaring prey, for the construction of cocoons to receive the eggs, as a lining for burrows, and in some cases as a vehicle to carry the animal on currents of air.

The economic importance of arachnids is rather limited. Spider silk has been woven but it is too delicate for extensive use and is valuable only as a source of cross-hairs for optical instruments. Aside from the poisons mentioned above, the principal harm from these animals is derived from the mites and ticks. The ticks do some damage to man and domestic animals by sucking blood but their greatest damage is due to the transmission of diseases. Texas fever of cattle, Rocky Mountain spotted fever of man, and other diseases are so conveyed. Mites living in the hair follicles, the **sebaceous glands,** and the tissues of the skin cause such diseases as scab in sheep and itch in man. Plant-feeding species of economic importance include the bulb mite, the pear blister mite and various gall mites.

The classification of the arachnids is briefly as follows:

Order Scorpionida. The **scorpions.** Abdomen divided into a preabdomen and a slender postabdomen bearing a claw-like sting at the tip.
Order Pedipalpi. The **whip-scorpions.** Anterior pair of legs slender and antenna-like.
Order **Solpugida.** The sun spiders. Head and thorax separate.
Order Chelonethida. Pseudoscorpions. Very small scorpion-like animals; no postabdomen nor sting.
Order Phalangida. **Harvestmen** or daddy longlegs; commonly regarded as spiders but have a segmented abdomen broadly joined to the thorax.
Order Araneina. **Spiders.** Abdomen unsegmented and joined to the thorax by a slender waist.
Order Acarina. **Mites and ticks.** Small to moderate species with a sac-like body showing no well-marked divisions. (A.W.L.)

ARAGONITE. The mineral aragonite is **calcium** carbonate, $CaCO_3$, chemically identical with **calcite** but crystallizing in the **orthorhombic system,** with acicular crystals. By repeated twinning, pseudo-hexagonal forms result. Aragonite may be columnar or fibrous, occa-sionally in branching **stalactitic** forms called flos-ferri (flowers of iron) from their association with the ores at the Carinthian iron mines. Its hardness is 3.5–4; specific gravity, 2.93–2.95; luster, vitreous to resinous; colors, white, gray, green-yellow or purple; transparent to translucent. Aragonite forms at temperatures of 80–100° C. and is relatively unstable at ordinary temperatures and pressures. It alters to calcite, although very slowly. There are many localities for aragonite in Europe, Bolivia, Pennsylvania, Iowa, Missouri, South Dakota, New Mexico, Arizona and Colorado. Its name is derived from Aragon in Spain. (E.S.C.S.)

ARANEINA. Arthropoda; Arachnida. The spider.

ARAPAIMA. Pisces, Teleostei. A large fish (**Pisces**), *Arapaima gigas,* of the rivers of northern South America. It attains a length of 15′ and a weight of more than 400 lbs. Also applied to related fishes of several genera. (A.W.L.)

ARAUCARIA. Coniferous trees found in southern South America, Polynesia and Australia. *Araucaria imbricata,* the "Monkey puzzle" tree of South America, reaches a height of 150′. *Araucaria excelsa,* the Norfolk Island Pine, is frequently seen in florists' windows and in collections of living plants, where it finds place because of its elegant symmetry. In its natural habitat it becomes a tree reaching a height of 200′ or more. (R.M.W.)

ARBOR. A shaft or stud, usually cylindrical or conical, on which a cutting tool, a tool holder, or a part to be machined is mounted or held. (H.C.H.)

ARBUTUS. Heath Family.

ARC BACK. This is the occurrence of an arc from **anode** to **cathode** in a gaseous rectifier **tube.** Normally such a tube has **electrons** flowing from the cathode to the anode but under certain conditions excessive heating of the anode, excessive voltage across the tube, or other effects may cause the anode to emit electrons and allow an arc discharge to take place in a direction opposite to the normal direction. Under many circuit conditions this may destroy the tube or it may merely open the protective devices. (L.R.Q.)

ARC DISCHARGE. The electric arc, so called because of the shape of the "flame," was discovered by Davy about 1808. It is a type of discharge between electrodes in a gas or vapor which is characterized by a relatively low voltage drop and a high current density. The two types which are of considerable practical importance are the arc in open air and the arc in gases at low pressure. The familiar carbon-arc (see **Arc Lamp**) and the electric-arc furnace are examples of the former. In this type the arc is started by impressing a voltage across the electrodes in contact and then separating them. At the instant of breaking contact the high field and current density initiate the arc. Thereafter, if the current is kept constant, the potential necessary is a linear function of the interelectrode distance. In its steady state the arc has an intensely hot cathode which emits a plentiful supply of electrons. The energy for heating the cathode is obtained from the high current density and from the bombarding positive ions. The arc is thus apparently a **thermionic phenomenon.** If the carbons are impregnated with a volatile metallic salt, the result is a "flaming arc," useful in producing the arc **spectrum** of the metal. If a d-c carbon arc is placed in parallel with a suitable condenser and an inductance, the circuit so formed may be made to oscillate, as discovered by Duddell, and to serve as a source of undamped electric waves. (See **Ionized Gases.**)

The arc is one of the most serious problems in switching electrical circuits, since the separation of the switch or **circuit breaker** contacts establishes an arc which must be extinguished in order to break the circuit. Many schemes have been developed to accomplish this. (See **Circuit Breaker**.)

The **mercury-arc tube** is the most important example of an arc in a gas at low pressure. Here the gas is the mercury vapor, the **cathode** is the mercury pool, and the **anode** is usually carbon. The arc is initiated by breaking contact between the mercury pool and a starting electrode. In gases at low pressure an arc may be established without breaking contact between the electrodes if the impressed voltage is high enough. If the voltage impressed across two electrodes separated by a low-pressure gas is gradually increased the current increases slowly at first due to the residual ions and electrons present in the gas. The electrons ionize the gas molecules upon colliding with them, thus giving rise to additional ions and electrons to carry the current and also to produce more ionization by collision. This process continues until suddenly the breakdown voltage is reached, when the current increases very rapidly, even if the voltage is lowered. This is the initiation of the glow discharge. If the circuit resistance does not limit the current the discharge progresses almost instantaneously into an arc discharge having the distinguishing features of the regular arc. The discharge in **thyratrons** and other gas-filled hot cathode tubes is often called an artificial arc discharge since it is characterized by low voltage and high current density. It is not a true arc however since the cathode heat energy is supplied by an external source and not by the discharge itself. (L.D.W., L.R.Q.)

ARC LAMP. The electric-arc lamp has, as its source of illumination, an electric arc struck between two electrodes. In contrast to the incandescent lamp, in which the illumination results from a heated filament, and vapor lamps, in which the illumination is derived from a vapor made luminous by electric current, the light from an arc lamp comes from the highly incandescent crater of one of the electrodes, and from the heated, luminous, ionized gases surrounding the arc. The positive electrode, having the crater, is mounted above the negative one. The light is largely directed downward from the crater. To start the arc the electrodes are brought into contact and then separated a short distance. The arc which follows vaporizes the electrode material slightly, forming a conducting ionized vapor which bridges the space separating the electrodes. The separation distance is maintained by a series-connected **solenoid**.

Arc-lamp electrodes.

The principal electrode material employed is carbon containing mineral salts which tend to intensify the flame between the electrodes. In any arc lamp using other than plain carbon electrodes, there is the problem of disposing of the fumes of the arc so that the enclosing glass globe will not be discolored, and in all such lamps, carbon or otherwise, there is a steady consumption of electrodes when the lamp is in use, necessitating their replacement every 50 to 150 hours of use. The light from arc lamps is very much more intense than that from the incandescent type lamp. From the standpoint of current consumption, the illumination is produced efficiently. Some arc lamps may not be operated on a.c., but all types are adaptable to d.c. A constant current series type circuit is used to operate street-light arc lamps, as that is found to be the most economical way to distribute the electrical energy over a large territory. (F.T.M., L.R.Q.)

ARC WELDING. Welding.

ARCH. An arch is a curved beam, made of **wood**, **brick**, **stone**, **concrete** or **steel**, whose supports are able to exert **lateral** as well as vertical forces to resist the action of any applied **loads**. These lateral forces are in the nature of thrusts which act inwardly toward the center of the arch span. The curvature of the beam must be in an upward direction in order to develop lateral **reaction** forces which will act in the required direction.

Brick arches are generally used to support walls above windows or doorways although they may be used for small **bridges** or **culverts**. In the past, brick arches of considerable magnitude have been constructed although at present brick is not commonly employed for major arch construction. **Masonry** arches are composed of stone blocks called voissoirs. They derive their load-carrying capacity from the fact that the shape of the arch ring is such that **compression** is the only type of **stress** caused by the **resultant** reaction on the end of any voissoir. Reinforced **concrete** arches, as the name implies, are composed of concrete and steel rods, the latter added primarily for the purpose of carrying any tensile stresses which may occur. Steel arches are subdivided into two classes: namely, the solid rib and open rib types. The solid rib arch is composed of structural shapes similar to those comprising a **plate girder** while the open rib arch is constructed of structural shapes forming triangles like those in a simple **truss**.

These structures may be further classified as fixed or hinged arches. A fixed arch is a structure which is rigidly connected to its supports in such a manner that they exert vertical and lateral reactions and prevent rotation. A two-hinged arch is one which is free to rotate about its supports, consequently they are able to exert only vertical and lateral reactions. Three-hinged arches have an additional hinge midway between the hinges at the supports. The tied arch is a structure in which the lateral forces are applied by means of a horizontal **tension** member connecting the ends of the arch. (C.W.C.)

ARCH BRIDGE. Bridge.

ARCH DAM. Dams.

ARCHAEOPTERYX. Fossil Birds.

ARCHAEOTHERIUM. Fossil Mammals.

ARCHEGONIATES. Those plants in which the female sex organ is an **archegonium** are called Archegoniates. (R.M.W.)

ARCHEGONIUM. The multicellular female sex organ characteristic of **Bryophytes**, **Pteridophytes** and **Gymnosperms**. It consists of a swollen basal portion called the venter, and an elongated neck. The venter may be a single layer of **cells**, but is often many cells thick. Within the basal portion is contained the single large egg, and a second, somewhat smaller, ventral canal **cell**, while the elongated neck contains a single row of cells which eventually dissociate, leaving an open canal through which the **sperm** may pass to reach the egg. (R.M.W.)

ARCHENTERON. Enteric Cavity.

ARCHEOCYTE. Cells of **sponges** which ingest and digest food, carry the products to other parts of the body, and form reproductive cells. They are amoeboid (see **Amoeba**) and are found in the mesenchyme (see **Mesenchymal Tissues**). (A.W.L.)

ARCHEOZOIC. (Archean.) The oldest of the five Eras of the earth's history. The rocks of this **System** are **metamorphosed** equivalents of all types of **sedimentary** and **igneous** rocks, but principally the latter. No undisputed fossils have been found in the Archean.

The lower Archean (Keewatin) of North America is composed of a preponderance of metamorphosed, **basaltic** lava flows and tuffs with some metamorphosed

A highly generalized section, about 25 miles long, showing the relations of the Archeozoic group of rocks in the Lake Superior-Lake Huron region of Canada. The Keewatin system was moderately folded and intruded by the Laurentian granite, after which there was deep erosion. Then the Timiskaming rocks were laid down, and later strongly folded and intruded by the Algoman granite, after which there was another period of profound erosion, marked by the upper surface.

sediments, such as **quartzite** and **slate**. The general character of the basal Archean proves that the oldest known rocks do not represent the original crust of the earth. The upper Archean (Laurentian) contains a preponderance of **granite, gneisses**, and **schists** in the form of **batholiths** intruding the Keewatin. The principal areas of Archeozoic rocks are in Canada, Finland, Scandinavia, Australia, Africa and northeastern South America. Many of the formations contain rich ore deposits, especially of gold and silver. Large amounts of graphite suggest the former existence of life. Length of time since the beginning of the Archeozoic, possibly 2000 million years. Owing to the lack of fossils, structural complexity, high degree of vulcanism and **metamorphism**, geologists have found great difficulty in deciphering the history of this earliest recognizable portion of the "crust" of the earth. The structural history of the Archean is therefore not so well known as that of the succeeding periods and intercontinental correlation is particularly difficult. On the other hand, the search for ore deposits has been an important stimulus to the study of the Archean formations, especially in Canada. (R.M.F.)

ARCHIANNELIDA. A class of **annelid** worms including small marine species of simple structure. (A.W.L.)

ARCHIMEDES' PRINCIPLE. Buoyancy.

ARCTIC AIR MASS. Air Masses.

ARCTIC SEA SMOKE. Fogs.

ARCTURUS, (α Bootes). Arcturus was probably one of the first stars to be named. It probably received its name because of its proximity to the constellation of **Ursa Major**, the name indicating that it is the "watcher of the bear." The name Arcturus is one of the few star names to be referred to in the Bible, being found in Job IX, but from the remainder of the verse there is evidence that the name Arcturus in this quotation actually refers to the constellation of Ursa Major rather than to the actual star itself. References to Arcturus are to be found in the writings of many of the ancient poets, including Virgil. **Astrologically,** the star portended honor and riches.

Arcturus is one of the few stars whose diameter has actually been measured with the stellar **interferometer.** The angular diameter is found to 0″.020 which, when combined with its **stellar parallax** of 0″.080, indicates a linear diameter of about 27 times that of our **sun.** (W.K.G.)

AREA, EQUIVALENT FLAT-PLATE. Flat-Plate Area, Equivalent.

AREA, SERVICE. This term is used to designate the coverage area of a broadcast station and is usually divided into a primary service area and a secondary

service area. The first is that area around the station in which reliable reception strong enough to override local noises, interference, etc., is obtained. The signal strength necessary to do this varies from 0.1 millivolt per meter for quiet rural locations to around 10 millivolts per meter for a metropolitan location. Coverage in this area is by the **ground wave** and hence is not subject to seasonal or diurnal variations. The secondary service area is the area where reception is fairly good but is not as perfect as the primary area. The signals here may be due to the ground wave or **sky wave.** Daytime reception in the secondary area is by ground wave but the night time area is often greatly extended by a useful sky wave signal. Excessive interference between the sky wave and ground wave may cause reduction of the night secondary area. (See **Fading.**) (L.R.Q.)

AREA ON A CURVED SURFACE. With a very few exceptions, the area of a portion or all of a given curved surface can only be found by use of methods of the **calculus.**

If an arc of a **curve** whose equation in **rectangular coordinates** is $y = f(x)$, between ordinates $x = a$ and $x = b$, is revolved about the X-axis, the area of the **surface of revolution** generated is given by

$$S = 2\pi \int_a^b y\,ds = 2\pi \int_a^b y \sqrt{1 + \left(\frac{dy}{dx}\right)^2} \cdot dx.$$

If an arc of a **curve** whose equation is $x = \phi(y)$, between **abscissas** $y = c$ and $y = d$, is revolved about the Y-axis, the area of surface generated is given by

$$S' = 2\pi \int_c^d x\,ds = 2\pi \int_c^d x \sqrt{1 + \left(\frac{dx}{dy}\right)^2} \cdot dy.$$

If the equation of a surface in rectangular coordinates is $z = f(x, y)$, the area of a portion S of the surface is given by the **double integral**

$$S = \iint_S \left[1 + \left(\frac{\partial z}{\partial x}\right)^2 + \left(\frac{\partial z}{\partial y}\right)^2 \right]^{1/2} dx\,dy.$$

If the equation of the surface is $F(x, y, z) = 0$, the area of S is given by

$$S = \iint_S \frac{\sqrt{\left(\frac{\partial F}{\partial x}\right)^2 + \left(\frac{\partial F}{\partial y}\right)^2 + \left(\frac{\partial F}{\partial z}\right)^2}}{\left[\frac{\partial F}{\partial z}\right]} dS.$$

(L.L.S.)

AREA UNDER A PLANE CURVE. One of the simplest interpretations of the **definite integral** is by the area under a plane curve.

If $y = f(x)$ is the equation in **rectangular coordinates** of a **plane curve**, the area between this curve, the X-axis and two **ordinates** $x = a$ and $x = b$, is given by the **definite integral**

$$A = \int_a^b y\,dx = \int_a^b f(x)\,dx.$$

If the equation in rectangular coordinates of the curve is $x = \phi(y)$, the area between this curve, the Y-axis and two **abscissas** $y = c$ and $y = d$ is given by

$$A = \int_c^d x\,dy = \int_c^d \phi(y)\,dy.$$

If the equation of the curve in **polar coordinates** is $r = f(\theta)$, the area bounded by the curve, two fixed radii vectors $\theta = \alpha$ and $\theta = \beta$ is given by

$$A = \tfrac{1}{2} \int_\alpha^\beta r^2\,d\theta.$$ (L.L.S.)

ARECA. Betel Nut.

ARENACEOUS. A textural term applied to sediments or **sedimentary** rocks which are composed of grains of sand. Psammitic has the same meaning. (R.M.F.)

ARGALI. Sheep.

ARGAND'S DIAGRAM. Complex Numbers.

ARGENTITE. The mineral argentite, sometimes called silver glance, is naturally occurring **silver** sulfide, corresponding to the formula Ag_2S. It crystallizes in the **isometric system** in cubes, octahedrons and dodecahedrons, or may be massive. Hardness, 2–2.5; specific gravity, 7.2–7.36; luster, metallic; streak, gray; color, black, blackish-gray or gray; opaque and sectile to such an extent that it cuts like wax with a knife. Heated upon charcoal it yields a malleable mass of silver. The name is derived from the Latin word for silver, *argentum.*

Foreign localities for fine crystals are Sonora, Mexico, and Freiberg, Saxony; in the United States at Butte, Montana; Tonopah, Nevada; and Aspen, Colorado.

Argentite is probably the most abundant ore of silver. Acanthite a rare **orthorhombic** silver sulfide has the same chemical composition and recent studies seem to show that it is a low-temperature form while argentite is the high-temperature form of this compound. (E.S.C.S.)

ARGENTUM. Silver.

ARGILLACEOUS. This term is used to designate **sedimentary** rocks composed of fine particles of the nature of clay or mud. Pelitic has the same meaning. (R.M.F.)

ARGILLITE. A dense, fine-grained, hard, **sedimentary** rock of various colors (usually white, gray or red). Composed of minute grains of both **clay** and **quartz.** Certain types of argillites are easily confused with certain types of fine-grained acid **lava** flows, such as felsites, unless studied microscopically. (R.M.F.)

ARGININE. Aminoacids.

ARGON. Symbol: A. Atomic number: 18. Atomic weight: 39.944. Density: 1.784 grams per liter, 0° C., 760 mm., or 1.380 when air is taken as 1.000. Melting point: −189.2° C. Boiling point: −185.7° C. Isotopes: 36 (0.31%), 38 (0.06%), 40 (99.63%).

Argon is a colorless, odorless gas, which does not react chemically under ordinary conditions. Discovered by Rayleigh and Ramsay in 1894, in ordinary air to the extent of 0.93%. The discovery was due to the investigation into the discrepancy of the density of **nitrogen** from the atmosphere and from chemical compounds. It appears that Cavendish a century earlier had made the observation that a small portion of air remained when nitrogen was sparked with oxygen. Argon shows a pale red glow in a vacuum electric discharge tube. Commercially, argon is used instead of nitrogen as the atmosphere of **tungsten** electric light bulbs. (R.K.S.)

ARGONAUTA. Mollusca, Cephalopoda. The genus to which the paper nautilus belongs. This species is not a true nautilus but is more closely related to the octopus. (A.W.L.)

ARGYRIA. Poisoning from the use of **silver** preparations over too long a period, causing a ghastly bluish discoloration of the skin over the entire body. (R.S.M.)

ARID REGIONS. Areas in which the rainfall is insufficient to support vegetation. Large parts of the western United States are semi-arid where sufficient rain falls only for a short period in any given year to support vegetation. (P.E.K.)

ARIES. (The ram.) (Map, page 380.) This **constellation** is far more famous from its classical significance than because of its appearance in the sky. It contains no bright stars and has no conspicuous features. Two thousand years ago the **vernal equinox** was located in the constellation of Aries and the symbol for the vernal equinox is the symbol for the constellation (i.e., the ram's head). **Precession** has caused the position of the vernal equinox to move backwards into the constellation of **Pisces** so that now the "sign of the first of Aries" is to be found in that constellation. (W.K.G.)

ARIL. In many plants there is formed in the developing fruit an outgrowth from the funiculus, or seed stalk, one which completely or partially surrounds the seed. In the litchi nut it is the thick translucent pulp surrounding the seed; in the nutmeg it is a mesh-like envelope which when removed and dried is known as mace. (R.M.W.)

ARISTOGENESIS. A principle of **evolution** formulated by Henry Fairfield Osborn, based upon observation of successive changes in fossil series. It postulates the origin of new characters in living things in adaptation to the environment and was expressed by its author as "the creative origin of the adaptive." (A.W.L.)

ARISTOTLE'S LANTERN. The masticating apparatus of the sea-urchin (**Echinoidea**). It consists of five jaws, each bearing a tooth, and five radial pieces (rotulae) which unite the bases of the jaws. (A.W.L.)

ARITHMETIC. Arithmetic, as generally understood, deals with the operations with numbers and their practical applications; it may be described as the art of computation and the applications of this art. (L.L.S.)

ARITHMETIC MEAN. In a **serial distribution** the arithmetic mean $\overline{X}$ (commonly called the **mean**) is the sum of the **variates,** x_i, divided by the number of variates, N.

$$\overline{X} = \frac{\Sigma x_i}{N} = \frac{x_1 + x_2 + x_3 + \cdots + x_N}{N}.$$

In a **frequency distribution** the formula for the mean becomes

$$\overline{X} = \frac{\Sigma f_i x_i}{\Sigma f_i} = \frac{f_1 x_1 + f_2 x_2 + f_3 x_3 + \cdots + f_N x_N}{f_1 + f_2 + f_3 + \cdots + f_N}$$

where f_i = frequency in the ith class and x_i is the class mark in the ith class. For computational purposes an arbitrary origin a is chosen and we have the formula

$$\overline{X} = a + c \frac{\Sigma f_i d_i^1}{\Sigma f_i}$$

where $d_i^1 = \frac{x_i - a}{c}$ and c is the **class interval.** In this method a is best chosen as the **class mark** of one of the classes in such a way as to make d_i^1 as small as possible.

The arithmetic mean is a **measure of central tendency** which summarizes the data. It lies between the least and the greatest variate in a distribution. The sum of the deviations from the mean is zero. The sum of the squared deviations about the mean is a minimum. The arithmetic mean should be used as the **average** unless special circumstances indicate the use of some other measure of central tendency. In a positively skewed distribution usually the mean is greater than the **median** which is greater than the **mode.** In a negatively skewed distribution these measures of central tendency are in reverse order.

The mean, $\overline{X}$, of the combination of r sets with means $\overline{X}_1$, $\overline{X}_2$, $\cdots$ $\overline{X}_r$ and numbers of variates N_1, N_2, $\cdots$ N_r, respectively, is

$$\overline{X} = \frac{N_1\overline{X}_1 + N_2\overline{X}_2 + N_3\overline{X}_3 + \cdots + N_r\overline{X}_r}{N_1 + N_2 + \cdots + N_r}.$$

The **standard deviation** of the mean, commonly called the standard error of the mean, $\sigma_{\overline{X}}$, measures the **variability** of sample means drawn from a **population** with **variance** σ^2, $\sigma_{\overline{X}} = \dfrac{\sigma}{\sqrt{N}}$, where N is the number of variates in the sample mean. An unbiased estimate of σ^2 is given by $\dfrac{N\sigma_S^2}{N-1}$, where σ_S^2 is the variance in a sample.   (L.A.A.)

ARITHMETIC PROGRESSION.

An arithmetic progression is a succession of **numbers** such that each member of the set differs from the preceding one by a constant called the common difference. The term "arithmetic progression" is often abbreviated A.P.

There are two fundamental formulae for arithmetic progressions—one for the general term and one for the sum of any number of terms.

The general term or nth term of an arithmetic progression whose first term is a and whose common difference is d is given by the formula

$$l = a + (n-1)d.$$

The sum of the first n terms of this arithmetic progression is given by

$$S = \frac{n}{2}(a+l).$$

The terms of an arithmetic progression between the first and last terms are called arithmetic means.

The arithmetic mean of two numbers is the middle term of an arithmetic progression whose first and last terms are the given numbers; it is given by half the sum of the given numbers. The arithmetic mean of two numbers is the same as the so-called arithmetic average of the two numbers.   (L.L.S.)

ARIZONITE.

A term proposed by Spurr and Washington in 1917 for a **dike** rock largely composed of **quartz** but with an appreciable amount of **orthoclase**. (R.M.F.)

ARKOSE.

Arkose is a relatively coarse-grained feldspathic **sandstone**, derived from the rapid disintegration of granite or other **feldspathic** rock. It is characterized by its content of fresh, unaltered, euhedral **feldspar**. The term was proposed by Brongiart in 1823, and has been in constant use ever since. Arkose is an important type of sediment especially in relation to the study of **unconformities, Paleoclimatology, "fossil" soils,** etc. (R.M.F.)

ARM.

An extended lobe or appendage of a body. The radiating lobes of the starfish are called arms or rays and the term is also applied to the branches of the **lophophore in brachiopods** and to other special structures. Its most familiar use is in application to the pectoral appendages of **vertebrates** when freed from the usual functions of support and locomotion, as in man and the other primates. In anatomy the term is restricted to the region from the shoulder to the elbow, to distinguish it from the forearm. In most of these species the arms are used for locomotion through the trees and some are still quadrupedal on the ground, but even in these species the arms can be used to some extent for handling objects. (A.W.L.)

ARMADILLO.

Mammalia, Edentata. Burrowing animals with many bony plates in the skin which form a more or less complete armor when the animal rolls up. Several species occur from Argentina northward through South America and one, the 9-banded armadillo, is found in Texas. They range in size from the 5-in. pichiciago to the 3-ft. giant armadillo. (A.W.L.)

ARMATURE.

The armature is one of the two essential parts of the dynamo electric machine. In a **generator**, the armature is the winding in which **electro-**

Armature of a d-c generator.

motive force is produced by magnetic induction. In the **motor** armature, conductors carry the input current which, in the presence of a **magnetic field**, produces a **torque** and effects the transmission of electrical into mechanical energy. In d-c machines it is the rotor, but the a-c armature may be rotor or stator. Larger size synchronous machines always have stationary armatures. The reluctance of the magnetic circuit to the **flux** of which the conductors of the armature must cut in order to generate electric energy, is decreased by providing a core of soft iron or steel, on the surface of which the conductors are embedded in slots suitably provided in the core. The armature windings of a d-c generator are terminated at the segments of a **commutator,** by means of which the alternating e.m.f.'s induced in the armature are rectified and transferred by **brushes** from the moving rotor to stationary terminals. The conductors must be separately insulated, as must be also the commutator segments, and must be well braced and anchored in their slots to resist the electromagnetic and mechanical forces which tend to displace them. (F.T.M., L.R.Q.)

ARMATURE REACTION.

This term refers to the reaction of the magnetic field produced by the current flowing in the armature conductors upon the main magnetic field of a dynamo machine. The result is a distortion of the magnetic field, the extent depending upon the reluctance of the magnetic circuit, the arrangement of the armature windings, the type field structure, and the phase angle between the armature voltage and current. In d-c machines the effect is to increase the flux at some pole tips and decrease it at others, while in a-c machines the effect depends upon the field structure and the phase angle of the armature current and voltage. The flux may be distorted as in the d-c machine, it may be changed in magnitude but undistorted in wave form or it may be changed in magnitude and shifted in position with respect to the field windings. Armature reaction is an important factor in the speed and voltage regulation of the machines. (L.R.Q.)

ARMY-WORM. Insecta, Lepidoptera. An economically important **caterpillar,** *Cirphis unipuncta,* named from its habit of migrating from field to field in large numbers. When severe outbreaks occur these insects completely strip fields of grain of all kinds. When migrating they are trapped in barrier ditches dug around the fields to be protected. They are also killed by poison baits. (A.W.L.)

ARNICA. Composite Family.

AROIDS. A large group of **monocotyledonous** plants, mostly tropical, having a characteristic flower habit. The numerous small inconspicuous flowers are borne on a fleshy stalk or spadix, which is surrounded, more or less completely, by a large, expanded, often brightly colored **bract** called a spathe. The spadix and spathe together are often but incorrectly considered to be the flower of the plant. The aroids are perennial plants, generally having tubers or **rhizomes** from which rise large leaves. Many tropical members are climbing plants. Well-known species are the Skunk Cabbage, whose foul-scented flowers appear so early in the spring, the Jack-in-the-Pulpit, and the wild arum, *Calla palustris,* of cold swamps, as well as the Sweet Flag, *Acorus Calamus,* of the marshes. The cultivated Calla Lilies are all aroids and not lilies at all; some of them are delightfully fragrant. On the other hand, in species of *Amorphophallus,* which are sometimes seen in collections of cultivated plants, the vile odor of the flower structure prevents them from becoming popular; the spathe and spadix of some of them are of gigantic size. In the tropics several species of *Colocasia* are cultivated for the edible rhizomes which appear under the name of dasheen or taro. (R.M.W.)

AROMATIC HYDROCARBONS. Benzenoid Hydrocarbons.

AROMATIC OILS. Volatile Oils.

ARRAY. A row or a column in a **correlation** table, or in an **analysis of variance** table. (See **Directional Antenna.**) (L.A.A.)

ARRHYTHMIA. Any variation in the normal **rhythm** of the **heart.** (See **Pulse.**) (R.S.M.)

ARROW WORM. Small marine animals sometimes classified with the **annelid** worms but more often included in the separate phylum **Chaetognatha.** (A.W.L.)

ARROYO. This term is applied to dry stream channels with nearly vertical walls and flat bottoms which are characteristic of semi-arid regions. They may suddenly become filled with torrential waters after heavy rains. The word arroyo is of Spanish origin. (E.S.C.S.)

ARSENIC. Symbol: As. Atomic number: 33. Atomic weight: 74.91. Density: gray 5.73; black 4.7; yellow 2.0. Hardness: 3.5. No isotope, but of single atomic form: 75.

Arsenic is a gray metal, brittle; sublimes on heating; is unchanged in dry air but a film of oxide is formed in moist air; heated in air at 180° C. forms arsenic trioxide of the odor of garlic, poisonous; insoluble in **hydrochloric acid** but soluble in concentrated **nitric** or concentrated **sulfuric acid** to form arsenic acid; soluble in hot **sodium** hydroxide solution; heated with chlorine forms arsenic trichloride; heated with metals forms metallic arsenides. When arsenic is heated in a tube and the vapor cooled (1) slowly (that is, in the hot part of the tube) black arsenic is formed, and this form is converted into the gray at 360° C., (2) rapidly (that is, in the cold part of the tube) yellow arsenic is formed, and this form is quickly converted into the gray by the action of light. Yellow arsenic is soluble in **carbon disulfide.** Arsenic element was discovered by Schröder in 1649.

Arsenic occurs as arsenide in many sulfide ores, e.g., of zinc, iron, cobalt, nickel, and as the mineral sulfides, **realgar** (arsenic monosulfide, AsS), red colored, **orpiment** (arsenic trisulfide, As_2S_3), yellow-colored—these two minerals when powdered are used as paint pigments—**arsenopyrite,** mispickel (iron arsenosulfide, $FeAsS$). The primary arsenic-containing material is arsenious oxide obtained by separation from roaster or smelter flue gases, and is most largely produced in Montana and Utah. Arsenic is obtained as a sublimate by heating the oxide with carbon. The use of free arsenic is practically limited to the manufacture of chilled shot—addition of arsenic (0.1%) lowers the melting point and increases the surface tension of lead so that chilling of the liquid drops yields spherical shot.

Acids: Arsenious acid is the name sometimes applied to solutions of arsenious oxide, but the acid has not been isolated; arsenic acid ($H_3AsO_4 \cdot \frac{1}{2}H_2O$), white solid, soluble, by heating arsenious oxide with **nitric acid** and crystallization, (1) ortho-arsenic acid (H_3AsO_4 or $As_2O_5 \cdot 3H_2O$), heating these crystals at 100° C., (2) pyro-arsenic acid ($H_4As_2O_7$ or $As_2O_5 \cdot 2H_2O$) by heating ortho-acid to 140°–180° C., (3) meta-arsenic acid ($HAsO_3$ or $As_2O_5 \cdot H_2O$) by heating pyro-arsenic to 200° C. While salts of these acids are known the only hydrates identified are $As_2O_5 \cdot 4H_2O$ and $3As_2O_5 \cdot 5H_2O$.

Arsenates: **Sodium** arsenate, sodium ortho-arsenate, trisodium arsenate ($Na_3AsO_4 \cdot 12H_2O$); sodium dibasic arsenate, disodium hydrogen arsenate ($Na_2HPO_4 \cdot 7H_2O$); sodium monobasic arsenate, sodium dihydrogen arsenate ($NaH_2AsO_4 \cdot H_2O$) and the corresponding potassium salts are white soluble solids; the former used as a mordant in textile dyeing and printing; **silver** arsenate (Ag_3AsO_4), reddish-brown precipitate, by reaction of silver nitrate solution and sodium arsenate solution, soluble in nitric acid or ammonium hydroxide; **magnesium** ammonium arsenate ($MgNH_4AsO_4$), white precipitate, by reaction of sodium arsenate solution and "magnesia mixture"—a reagent made by adding ammonium chloride to ammonium hydroxide and then to magnesium salt solution without the formation of a precipitate of magnesium hydroxide. Many pyro-arsenates and meta-arsenates have been described.

Arsenides: Metallic arsenides are formed (1) by heating the finely powdered metal with arsenic, e.g., iron arsenide ($FeAs_2$), (2) by reaction of the metallic salt solution and arsine, e.g., copper arsenide (Cu_3As_2).

Arsenites: Sodium arsenite, sodium meta-arsenite ($NaAsO_2$) and potassium arsenite ($KAsO_2$) are white soluble solids, the former is used as an insecticide, a weed killer, and poison; silver arsenite, silver ortho-arsenite (Ag_3AsO_3), yellow precipitate, by reaction of sodium arsenite solution and silver nitrate solution, soluble in nitric acid or ammonium hydroxide; copper arsenite ($Cu(AsO_2)_2$), green solid, used as the arsenite-acetate as an insecticide and poison ("Paris green").

Arsine: See Hydrides.

Chloride: Arsenious chloride, "butter of arsenic" ($AsCl_3$), colorless liquid, boiling point 130° C., by reaction of arsenic-containing solutions with concentrated hydrochloric acid, and applied in the analysis of arsenic containing substances by distilling in a current of hydrogen chloride gas below 108° C. (**Germanium** tetrachloride is volatilized similarly.)

Hydrides: Arsenic hydride, arsine (AsH_3), colorless gas, garlic odor, very poisonous by reaction of a solution of any arsenic containing material with (a) a metal, e.g., zinc, magnesium for iron, and hydrochloric or dilute sul-

SCHEME SHOWING INTERRELATIONSHIPS OF ARSENIC-FUNCTION ORGANIC COMPOUNDS

ARSINES

Methyl arsine
$CH_3 \cdot AsH_2$
B.P. 2° C.

Phenyl arsine
$C_6H_5 \cdot AsH_2$

Phenyl arsine dichloride
$C_6H_5 \cdot AsCl_2$

Phenyl arsenoxide
$C_6H_5 \cdot AsO$

ARSONIC ACIDS

Methylarsonic acid
$CH_3 \cdot AsO(OH)_2$
M.P. 158° C.

Atoxyl. Sodium salt of para-amino-phenylarsonic acid

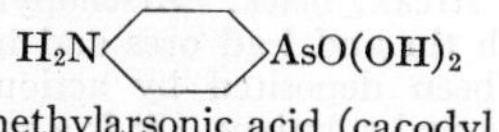

$H_2N\langle\quad\rangle AsO(OH)_2$

Dimethylarsine
$(CH_3)_2AsH$
B.P. 36° C.
(747 mm.)

Diphenyl arsine
$(C_6H_5)_2AsH$

Cacodyl oxide
$(CH_3)_2As\rangle O$
$(CH_3)_2As$

Dimethylarsonic acid (cacodyl acid)
$(CH_3)_2AsO(OH)$
M.P. 200° C.

Cacodyl chloride
$(CH_3)_2AsCl$

Cacodyl
$(CH_3)_2As$
$|$
$(CH_3)_2As$

Trimethyl arsine
$(CH_3)_3As$
B.P. 53° C.
Reacts with:
1. Oxygen
2. Chlorine
3. Sulfur
4. Methyl iodide

Triphenylarsine
$(C_6H_5)_3As$
M.P. 57° C.

Trimethyl arsine oxide
$(CH_3)_3AsO$

Quaternary arsonium
compounds
$[(CH_3)_4As]I$
$[(CH_3)_4As]OH$

Arsenobenzene
$C_6H_5As:AsC_6H_5$

Salvarsan
$HCl \cdot H_2N$ $NH_2 \cdot HCl$
$HO\langle\quad\rangle As:As\langle\quad\rangle OH$

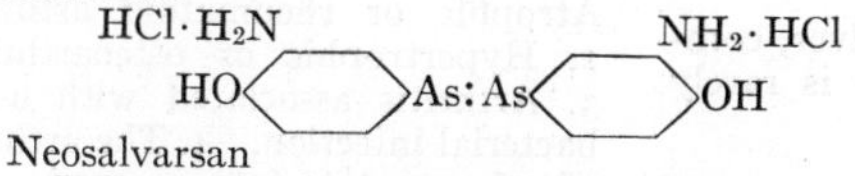

Neosalvarsan
Sodium salt of salvarsan condensed with
formaldehyde sulfoxylate (one or two
hydrosulfite (—OSONa) groups)

furic acid, or (b) a metal, e.g., zinc or aluminum, and sodium hydroxide solution. Arsine (1) when heated in a glass tube yields a metallic mirror of arsenic, (2) when passed into a solution of silver nitrate yields black precipitate of metallic silver and arsenious acid solution, (3) when passed over solid silver nitrate yields yellow solid $(Ag_3As \cdot 3AgNO_3)$, (4) when passed over solid mercurous chloride or solid mercuric chloride, or into mercuric chloride solution yields brown solid mercurous arsenide, (5) when passed into copper sulfate solution yields black precipitate of copper arsenide (Cu_3As_2). The formation and detection of arsine are important in estimating small amounts of arsenic (Marsh's test, Gutzeit's test, Fleitmann's test). Arsine burns in air with a bluish flame forming arsenious oxide and water; is unchanged in solution of sodium hydroxide; arsenic dihydride (As_2H_2), brown solid, formed by partial burning of arsine, and by electrolysis of water using an arsenic cathode.

Oxides: Arsenious oxide, arsenic trioxide, "white arsenic," "arsenic" (As_2O_3), white solid, moderately soluble, sublimes at 218° C., soluble in alkalis to form arsenites, known in three forms, (1) octahedral, by rapid cooling of the vapor (that is, in the cold part of the condenser), density 3.63, (2) amorphous, by slow cooling of the vapor (that is, in the hot part of the condenser), density 3.74, melting point 200° C., and three times as soluble as octahedral, (3) monoclinic (needle-like crystals), by heating either of the above forms to 200° C. for a long time, density 4.15. The oxide is formed by burning arsenic or arsenides in air. By-product in the fumes of metallurgical smelters. Used as insecticide, rat poison, preservatives of skins and hides, and as the principal source of arsenic compounds. Freshly precipitated ferric

hydroxide absorbs arsenious oxide, and is used as an antidote in cases of arsenic poisoning.

Sulfides: Arsenic monosulfide, arsenic disulfide (AsS or As_2S_2), red solid, by heating arsenious oxide and sulfur, melting point 307° C., used with sulfur and potassium nitrate in pyrotechnic powder for producing blue flame; arsenious sulfide, arsenic trisulfide (As_2S_3), (1) yellow precipitate by reaction of solution of arsenite and hydrogen sulfide, in the absence of acid, alkali or salt the precipitate does not coagulate but large quantities may be formed in yellow colloidal solution, insoluble in concentrated hydrochloric acid and formed in its presence, soluble in sodium or ammonium sulfide to form thioarsenite, (2) yellow solid by distilling arsenious acid and sulfur from a retort. Used as a paint pigment but the color is not permanent in the light; arsenic sulfide, arsenic pentasulfide (As_2S_5), yellow precipitate, by reaction of solution of arsenate and hydrogen sulfide (in the presence of concentrated hydrochloric acid).

Numerous organic compounds containing arsenic ("arsenicals") have been prepared; among these are salvarsan, neosalvarsan, atoxyl.

Nitric acid readily transforms all arsenic-containing substances (other than arsenic acid) to arsenic acid.

Solutions of arsenic-containing substances, when boiled with hydrochloric acid, deposit a gray film on a bright strip of copper. This film of arsenic or copper arsenide, when dried and heated in air in a glass tube, sublimes as arsenious oxide, tiny white crystals.

Arsenic in any form may be detected by the Marsh or Gutzeit tests. Arsenites give a yellow precipitate with silver nitrate solution; arsenates a chocolate-brown precipitate. (R.K.S.)

ARSENOPYRITE—MISPICKEL. The mineral arsenopyrite is a sulfarsenide of iron corresponding to the formula FeAsS. A variety in which some of the iron is replaced by cobalt is known as danaite. It crystallizes in the **orthorhombic system.** Its hardness is 5.5–6; specific gravity, 5.9–6.2; luster, metallic color, silvery-white to steel-gray, but usually with a yellow to gray tarnish; streak, black. Arsenopyrite is a common mineral with tin and lead ores and in pegmatites, probably having been deposited by action of both vapors and hydrothermal solutions. It is a widespread mineral, well known deposits occurring in Austria, Saxony, Switzerland, Sweden, Norway; Cornwall and Devonshire, England; Bolivia. In the United States at Roxbury, Connecticut; Franklin, New Jersey; Paris, Maine; Emery, Montana; and Leadville, Colorado. Danaite was first found in Franconia, New Hampshire, by J. D. Dana, for whom it was later named. Mispickel is an old German term whose exact derivation is unknown. (E.S.C.S.)

ARSPHENAMINE. "Salvarsan." An arsenic preparation given intravenously in the treatment of **syphilis, yaws, Vincent's Angina** (trench mouth), etc. It was first discovered by Ehrlich and is also known as "606." It is a yellowish powder, unstable when exposed to the air. Its chemical name is Diaminodihydroxyarsenobenzene dihydrochloride $(OH \cdot C_6H_3(NH_2 \cdot HCl)As)_2$. **Neoarsphenamine** is now more commonly used because it is less toxic and more soluble.

Marpharsen (arsenoxide) is believed to be the form in which arsphenamine is active in killing spirochetes in the body. Since only $\frac{1}{10}$ of injected arsphenamine forms arsenoxide, the dose of mapharsen is $\frac{1}{10}$ that of arsphenamine. This decrease in dose diminishes toxic reactions. In addition, mapharsen is stable in air and is more easily prepared than arsphenamine.

Silver arsphenamine is arsphenamine with silver, combining the therapeutic properties of both. It is rarely used. (D.M.H.)

ART MOBLIER. Paleontology of Man.

ARTEMISIA. Compositae. Many of the 280-odd species of Artemisia have been cultivated or used by man. Southernwood, *Artemisia Abrotanum,* is cultivated in gardens for its delicate foliage and aromatic odor. Another and a homely species, *Artemisia vulgaris* or Mugwort, is also frequently cultivated, as is *Artemisia Absinthium,* a native European perennial plant. All contain volatile oils. That from *Artemisia Absinthium,* oil of wormwood, is a powerful drug, capable of causing violent convulsions when taken even in small doses. It is used to flavor the alcoholic beverage absinthe, a liquor capable of producing much the same effects as the drug. The sage-brushes of the western United States are all species of Artemisia; like other species they contain an abundance of aromatic oil. From *Artemisia dracunculus,* a European species, tarragon is obtained. This is used as a condiment, and for flavoring vinegar and mustard. (R.M.W.)

ARTERIOSCLEROSIS. (Hardening of the **arteries.**) This term is used clinically to describe several varieties of pathological degenerative changes in the arteries. The large and medium-sized arteries show degeneration, deposition of fat and finally calcification. The small arteries (arterioles) show hyaline degeneration, replacement of normal tissue with a transparent acellular substance. The latter condition is almost always found in those who have had high blood pressure or **hypertension,** while the former condition may or may not be seen in hypertensives. Both types of arteriosclerosis occur most commonly after the age of 50. The symptoms due to this condition vary with the organs or structures involved.

The cause of arteriosclerosis is not known. Factors such as heredity, increasing age, and hypertension seem to contribute to its development. Diabetics are thought to be especially prone to develop arteriosclerosis.

The condition may be widespread throughout the body without producing any symptoms, but by local impairment of the blood supply, it may damage any organ. The heart, brain, kidneys and extremities are most frequently involved.

In the heart, arteriosclerosis commonly causes **angina pectoris, coronary occlusion, arrhythmias,** and **heart failure.** Kidney involvement results in decreased function which may terminate in **uremia.** Marked cerebral arteriosclerosis may be manifested by senile **psychoses,** or by occlusion of a cerebral vessel, **apoplexy,** ("a stroke") and hemiplegia. In the extremities, arteriosclerosis may cause cramps, numbness and tingling, and in advanced cases, **gangrene.**

The treatment of the disease is symptomatic, as well as being directed toward improvement of the circulation to the various organs involved.

The prognosis depends upon the organs affected. Arteriosclerosis causes the death directly or indirectly of the majority of individuals who live beyond the age of 60 years. (D.M.H.)

ARTERY. A vessel leading away from the heart. (See Circulatory System.) (A.W.L.)

ARTESIAN WELL. Ground Water. Hydrology.

ARTHRITIS. An acute or chronic **inflammation** of a joint. There are many varieties of the disease most of which fall into one of the four major types. 1. Atrophic or rheumatoid arthritis. 2. Hypertrophic or osteoarthritis. 3. Arthritis associated with acute bacterial infection. 4. The arthritis of **rheumatic fever, gout,** and certain systemic diseases, form a miscellaneous group.

Atrophic or rheumatoid arthritis (chronic infectious arthritis, arthritis deformans) is characterized by chronic inflammatory changes in the joints, and atrophy of the bone and surrounding musculature. Its etiology is unknown but certain features are thought to suggest an infectious element. The disease affects the whole organism, and such manifestations as weight loss, fever, **leukocytosis** are common. Women are attacked more often than men. The average age of onset is 30–35 years. Usually one joint alone is affected in the beginning, the metacarpo-phalangeal and phalangeal joints of the fingers being especially prone to develop fusiform deformities. In some cases the onset is acute and involves many joints at once. The course may be a mild one with frequent remissions, or there may be steady progression to multiple joint involvement and eventual crippling deformity. The treatment is not specific; general supportive measures—diet, rest, physiotherapy, prevention of contractures—are all important. Of the many drugs employed, gold salts intramuscularly have given the most promising results.

Hypertrophic osteoarthritis is a degenerative joint disease, common to almost all individuals in later life, its onset being as a rule in the 5th or 6th decades. It is characterized by degenerative changes in the joint cartilage with dense bony overgrowth in the form of spurs at

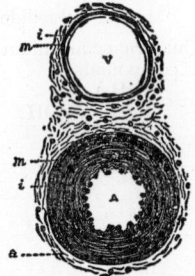

Transverse section through a small artery and vein, showing the relative difference in the thickness of their walls. *A,* artery showing *i,* the endothelial lining which with the thick layer of elastic tissue next to it constitutes the *intima.* The endothelial cells appear thick because the artery is contracted. *m,* the circular muscular coat or *media* constitutes the chief part of the wall of the vessel. Outside this is *a,* part of the outer coat or *adventita. V,* vein showing *i,* a thin endothelial membrane, *m,* a few circular muscle cells. *(From Kimber and Gray, Textbook of Anatomy and Physiology, Macmillan & Co.)*

the joint surface. The joints most frequently involved are fingers, knees, hips and vertebrae. The etiology is unknown, but trauma and excessive weight are predisposing factors. Treatment is symptomatic. The disease progresses gradually and may be attended by considerable joint pain, but deformity and serious loss of function are very rare.

The arthritis of acute infections is sometimes of the suppurative variety, joint infections with hemolytic streptococci or staphlyococci, which are secondary to penetrating wounds, or to a **septicemia** with these organisms. Arthritis also appears as a suppurative complication of **tuberculosis, pneumonia, scarlet fever, gonorrhea,** meningococcic **meningitis** and **typhoid fever.** In scarlet, meningococcus meningitis, brucellosis and **dysentery,** pain, swelling and sterile joint effusions may also occur.

The arthritis of **rheumatic fever** is characteristically fleeting and migratory, and leaves no permanent damage. **Gout** is a systemic disease of unknown etiology with paroxysmal attacks of arthritis as its chief manifestation. (D.M.H.)

ARTHROBRANCHIAE. **Gills** attached to the joint membranes of certain **crustaceans.** (A.W.L.)

ARTHRODIRE. Fossil Fishes.

ARTHROPODA. The largest and most diversified division of the animal kingdom, including **crustaceans,** horseshoe crabs (see **Xiphosura**), **insects, scorpions, spiders, centipedes,** millipedes (see **Diplopoda**) and other forms.

This phylum is characterized by the following structures: 1. The body is **triploblastic** and metameric (see **Metamere**), and is further subdivided into regions of which there may be a maximum of three: **head, thorax,** and **abdomen.** 2. Supporting structures are developed from the integument and constitute an exoskeleton made up of plates connected by flexible regions for freedom of movement. 3. Jointed appendages, fundamentally a pair to a segment, give the phylum its name. 4. The **circulatory system** is a combination of tubes and open spaces; the coalescence of the latter to form a haemocoel is accompanied by extreme reduction of the coelom. 5. The **eyes** are of a form peculiar to the group. 6. The **respiratory system** consists of air tubes or **tracheae,** of gills, or of **lung books** formed of leaf-like expansions of the body wall located in a cavity narrowly open to the exterior. Some species have no special respiratory organs.

The phylum is divided into several classes as follows:

Class **Onychophora.** Soft-bodied worm-like animals. Commonly called **Peripatus,** the name of one genus.
Class **Tardigrada.** The bear animalcules.
Class **Pentastomida.** Parasites known as pentastomids or linguatulids.
Class **Pycnogonida.** The sea-spiders.
Class **Crustacea.** **Crabs, lobsters, crayfishes, shrimps, barnacles, woodlice** or pillbugs, and other forms.
Class **Xiphosura.** The king crab or horseshoe crab. Also given the name Palaeostraca.
Class **Arachnida.** **Spiders, mites, ticks, scorpions,** etc.
Class **Diplopoda.** The millipedes.
Class **Pauropoda.** Rare forms allied to the preceding.
Class **Chilopoda.** The **centipedes.**
Class **Symphyla.** Rare forms allied to the preceding.
Class **Insecta.** The insects. (See also **Invertebrate Paleontology.**) (A.W.L.)

ARTICHOKE, GLOBE. *Cynara Scolymus.* Composite Family.

ARTICHOKE, JERUSALEM. *Helianthus tuberosum.* **Composite Family.**

ARTICULATION. This term has three common meanings. 1. A **joint,** such as the knee or elbow, etc. 2. Enunciation. 3. In dentistry, the contact of the teeth. (R.S.M.)

ARTIFACT. Paleontology of Man.

ARTIFICIAL HORIZON. Sextant.

ARTIFICIAL LINE. An artificial line is an electrical network consisting of **resistance, inductance** and **capacitance** so connected that it has the same characteristics (electrical) as the actual transmission **line.** Sometimes where the artificial line is not required to duplicate exactly the actual line the inductance or capacitance may be omitted. Such a line is very valuable for making laboratory tests as it makes possible connection at points corresponding to many miles of actual line. Artificial lines are also used widely in telephone and telegraph practice to balance actual lines to give desired operating characteristics in bridge type circuits. (L.R.Q.)

ARTIFICIAL PNEUMOTHORAX. **Tuberculosis** treatment. (R.S.M.)

ARTIODACTYLA. Hoofed animals which retain an even number of toes, the axis of the foot passing between the third and fourth digits. The species included are the **pigs, cattle** and related species, **antelopes, camels, hippopotami, giraffes, deer,** and some less familiar forms. The group constitutes an order of the class **Mammalia,** often called the even-toed **ungulates.** (A.W.L.)

ARYL. Radical of benzenoid **hydrocarbon.**

ASAFETIDA. *Ferula fetida.* Umbelliferae. Asafetida is a perennial herb found in Persia and Afghanistan. It grows 6–10' tall, and bears large compound leaves of bluish-green color, and large compound umbels of pale yellow **flowers.** The roots contain a milky juice which oozes out when they are cut, and hardens to a gummy substance bitter in taste. This substance, although extremely foul-smelling, is much used in the East and in France as a condiment. In small quantities its unpleasant odor is not apparent as, for instance, in "Worcestershire" sauce. As a drug, asafetida was earlier supposed to stimulate the nervous system. (R.M.W.)

ASBESTOS. This is the popular name for several fibrous minerals used for fireproofing and heat insulating material. Most of the commercial asbestos is a variety of **serpentine** called chrysotile. A fibrous kind of **amphibole** is called asbestus. Both terms asbestos and asbestus are derived from the Greek word meaning unquenched, which was applied to minerals that resisted fire. (E.S.C.S.)

ASCARIS. Nemathelminthes, Nematoda. *Ascaris.* Paratic roundworms of relatively large size found in the intestines of man and other animals. (A.W.L.)

ASCENT OF SAP. All of the organs of any terrestrial plant are dependent for their existence upon water absorbed from the soil. This water, which always contains traces of solutes and hence is often referred to as sap, moves in a generally upward direction through the plant. In some of the tallest known specimens of redwood trees (*Sequoia sempervirens*) the sap must ascend to heights exceeding 350' if the topmost branch is to be kept supplied with water. The upward movement of sap in plants occurs in the xylem, which in trees and shrubs corresponds to the wood. In the trunks or larger branches of trees sap movement is confined to a few of the outermost annual rings of wood. Sap movement occurs only through the vessels and tracheids of the woody tissue.

The earlier theories of the upward movement of sap in plants mostly invoked some vaguely conceived vital activity of the cells as furnishing the motive power for sap movement. Although the vessels and tracheids through which the water moves are dead, they are always in intimate contact with living wood parenchyma and wood ray cells and it is not inconceivable that these cells might in some way motivate the upward movement of sap. However, such theories receive very little support at the present time.

It is a common observation that sap may flow from the severed stems of many kinds of plants and that this flow ("bleeding") may continue for some time. This exudation of sap results from a pressure originating in the root called *root pressure,* and the exuded sap comes from the xylem tissues. Root pressures are also present in intact plants. For several reasons, however, root pressure can be considered only a secondary mechanism of water transport in plants. In the first place there are many species in which the phenomenon does not occur. In the second place the magnitude of measured root pressures seldom exceeds two atmospheres which could not cause a rise of sap of more than about sixty feet. In the third place known rates of sap flow under the influence of root pressure are inadequate to compensate for many known rates of transpiration. And finally, in woody plants at least, root pressures are usually present only in the early spring; during the summer period when transpiration rates and hence rates of sap movement are greatest, root pressures are negligible or non-existent.

The principal mechanism motivating the ascent of sap in plants is thought by most present day botanists to be dependent upon the property of *cohesion* (see Adhesion) in water. The cohesive forces between water molecules are very great. The evaporation of water from the mesophyll cells of the leaf during transpiration results in the movement of water molecules into these cells from the xylem (water-conducting tissue) of the veins. The xylem of the leaves is continuous with that of the stems which in turn is continuous with that of the root system out almost to the very tip of every rootlet. The water is apparently present in the cells and vessels of the xylem as continuous thread-like columns. As water molecules pass out of these water columns into the mesophyll cells the threads of water become taut throughout the plant. Eventually a tension of considerable magnitude may be set up in them which is transmitted from the top to the bottom of the plant. The water columns can sustain this tension only because of the high cohesive force of water. When the water in the xylem of the younger roots passes into a state of tension movement of water from the root cells into the xylem cells is induced. Loss of water from the root cells in turn causes absorption of water from the soil. Movement of water through the entire plant is thus brought about. Whenever transpiration rates are appreciable water does not, as a rule, enter the lower ends of the xylem ducts from adjacent root cells as fast as it passes from the upper ends into the mesophyll cells, hence the water is continuously under tension during periods of rapid transpiration and upward movement of sap. Calculations indicate that a cohesive force of between 30 and 50 atmospheres would be adequate to permit translocation of water to the very top of the tallest known trees by this mechanism. Experimentally determined values of the cohesive force of water are in excess of 300 atmospheres. (B.S.M.)

ASCHHEIM-ZONDEK TEST. A laboratory test which indicates the presence of pregnancy. During pregnancy the urine contains an ovary-stimulating or gonadotropic **hormone** produced by the chorionic cells of the placenta. This hormone is similar to the gonadotropic hormone of the anterior **pituitary** and is called A.P.L., or anterior-pituitary-like hormone. When urine from a pregnant woman is injected into immature mice or virgin rabbits, characteristic changes are produced in the animal's ovaries by the action of A.P.L. hormone. A positive test is obtained with urine excreted 8 days after the first missed menstrual period, when it is impossible to make a diagnosis by physical examination.

The Aschheim-Zondek test is also positive in pathological conditions in which there is a retention of placental tissue after delivery. (D.M.H.)

ASCIDIACEA. Chordata, Tunicata. The **tunicates,** sea squirts, or ascidians (**Ascidiacea**), constituting a class of the subphylum **Tunicata.** They begin life as **larvae** which resemble tadpoles in form and later become **sessile** animals invested in a covering called the test or **tunic.** Some species are solitary and others form colonies.

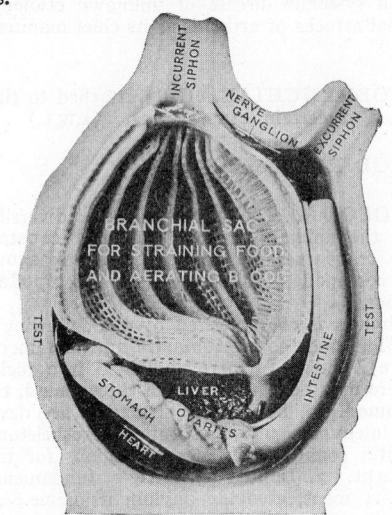

Internal anatomy of the adult sea squirt. (*American Museum of Natural History.*)

The class contains four orders:

Order Krikobranchia. Sessile colonial forms. Body elongate, usually with transverse constrictions.

Order Dictyobranchia. Mostly solitary species. Body without constrictions. Tunic translucent.

Order Ptychobranchia. Solitary or colonial. Tunic opaque.

Order Pyrosomida. Colonial, the entire colony free-swimming. (A.W.L.)

ASCITES. The presence of free fluid in the abdominal cavity, popularly called dropsy of the **abdomen.** Two causes of the condition are known; one is the obstruction of the venous return to the **heart** in diseases of the heart, **liver** and **kidney,** and the other is local inflammation of the **peritoneum,** as in tuberculous peritonitis. It is also seen when **cancer** has metastasized to abdominal structures and to the peritoneum. (R.S.M.)

ASCOMYCETES. Sac Fungi. Fungi. Many of the 40,000 species of **fungi** comprising the Ascomycetes are very common plants, but few are conspicuous. The great part of the species are small, often minute, while a few attain heights of 3 or 4", with a diameter of 1–2". Occasional individuals are even larger. All are characterized by the **ascus,** or spore-sac, commonly an elongate cylindrical body containing eight **spores.** In some species the ascus is spherical, or short cylindrical, while the number of spores may vary from two to many. Usually the asci are grouped together in a dense layer, called the hymenium. This may be composed entirely of asci, or may contain in addition numerous slender sterile filaments, called paraphyses. In some cases at least it seems the function of the paraphyses to protect the asci, since the outer tip of each paraphysis is a flattened cap which partially covers the ascus. Ascomycetes are found wherever suitable food-yielding materials exist. Many species are parasites, living on living plants; among these are species of great economic importance. Other species are saprophytes, wood-destroying species being particularly numerous.

The life-history of an Ascomycete comprises the very important but inconspicuous mycelium composed of slender branching septate hyphae which penetrate throughout the substratum, and extract from it nutrient materials which are elaborated and stored up within, and the fruiting stage in which the asci are formed. Two types of reproduction occur. One of these is the asexual type, in which asexual cells called conidia are cut off in various ways from the tips of hyphae, known as conidiophores. These conidia are single-celled spores which are disseminated by air currents. The other method of reproduction is sexual, and leads to the formation of asci. In *Pyronema confluens* this process has been carefully studied, and may be considered as typical in the main details for the process as it occurs in all the fungi of this class. The first step in this process is the formation of a multinucleate much-branched structure, which presently becomes septate. Some of the tips of this structure enlarge and become oögonia, called in this case ascogonia, while other tips become antheridia. From the oögonium a slender curved body called the trichogyne grows out. This is separated from the oögonium by a cross-wall. Since the oögonia and antheridia develop close together, the trichogyne comes in contact with the antheridium. All three bodies, oögonium, antheridium and trichogyne, are multinucleate. When the trichogyne comes in contact with the antheridium the walls between them at once break down, as does the wall between the trichogyne and the oögonium. The nuclei of the antheridium pass into the trichogyne, through it and into the oögonium. After this a new wall forms separating the trichogyne from a oögonium. In the oögonium the nuclei from the antheridium pair up with the nuclei of the oögonium, the nuclei of the trichogyne disintegrating early in the period of nuclear migration. Following the pairing of the nuclei in the ascogonium, coarse hyphae grow out from the latter. Into these the paired nuclei migrate. These coarse hyphae are the ascogenous hyphae, from which the asci eventually develop. In many Ascomycetes this process is considerably shortened, the acogenous hyphae arising directly from the mycelium, no sex cells being formed; while other species have sex cells but no fusion, the oögonium alone developing.

The life cycles of the various ascomycetes are remarkably uniform, suggesting that they are all derived from a common ancestor. Two different views are held by botanists as to what the ancestral form may have been. According to one group, they are derived from red algae; favoring this view is the very great similarity in the development of the ascogonium and that of the carpospore formation in the algae; another favorable point is the presence of the trichogyne and the behavior of the antheridial nuclei. On the other hand, the other group holds that the ancestors of the Ascomycetes are to be found in the Phycomycetes, basing this contention on the similarity of the Phycomycete sporagium and the ascus, the latter being merely a sporangium in which the number of spores has been greatly reduced, becoming stable at eight in most species.

Many members of this class of fungi are of great importance to man because of their destructive parasitic habit. A few species are of value as food, or in the production of foodstuffs, and other products used by man. Among the injurious species may be mentioned the Chestnut Blight fungus, *Endothia parasitica,* a disease probably introduced from China at the beginning of the twentieth century. In China the native chestnut trees had developed immunity; this the American trees did not have, so the fungus, which attacks the cambial tissue, was particularly destructive, nearly wiping out the native chestnut trees in a few years. Another disease caused by an Ascomycete is the Brown Rot of stone fruits, caused by *Sclerotinia cinerea.* This fungus is particularly destructive in wet seasons. Often infected fruits become shriveled up and dry, in which condition they are known as "mummies." A large group of Ascomycetes are known as Powdery Mildews, because of the abundant conidiophores which are formed by the mycelium on the surface of the leaves of infected plants. Often these are so abundant as seriously to impair the functional efficiency of the leaf.

Another group of Ascomycetes contains species which are destructive and also those which are commercially of great value; these are the ubiquitous blue and green molds, species of *Aspergillus* and *Penicillium.* The destructive species attack foodstuffs everywhere, causing rotting and spoilage. Citrus fruits become covered with the bluish-green conidial masses; as does moist bread, pie crusts and many other foodstuffs. Species of the genus *Penicillium* give to Camembert and Roquefort cheese their characteristic properties. Other species of this group are the casual organisms for skin diseases of animals, including man. Another species of this genus, Penicillium notatum, is the source of the important drug, penicillin. Since discovery of the bactericidal effects of penicillin, commercial culture of this mold has been undertaken on a large scale. The organism is grown in sterile culture on a liquid medium consisting of water with suitable organic and inorganic substances dissolved in it. In some procedures, bottles a quart or more in capacity are used as containers for the cultures; in others, large deep tanks are used. When the tank culture method is used, forced aeration of the cultures is necessary; in the bottle culture method the mold grows only on the surface of the liquid and thus obtains an adequate supply of oxygen. When the cultures have reached a suitable stage of development the mold colonies are filtered out of the liquid medium and discarded. During the growth of the organism the penicillin diffuses out of, or otherwise escapes from, the cells and hence is present in the medium. The penicillin is removed from the medium by extraction with organic solvents and suitably concentrated for medical use.

Among the largest of the Ascomycetes are species of truffles and morels, which are considered by mushroom fanciers to be particularly finely flavored. Truffles are fruit-bodies of the order Tuberales, and grow entirely underground. This makes it a matter of some difficulty to find them. Since they do not lend themselves to artificial cultivation, truffles must be sought in their wild habitat. To aid in locating them, man has trained dogs and pigs to find them by their superior sense of smell.

Another important Ascomycete is the genus *Claviceps,* which is parasitic on many grasses, including several cereal grains. This fungus forms a hard black sclerotium which is known as ergot, and which completely replaces the grain in the infected flower. The sclerotia are poisonous to livestock, causing the animals which have eaten them to become emaciated and covered with sores; another result is abortion in females.

Another very important group of Ascomycetes is the Yeasts. (R.M.W.)

ASCON. Porifera.

ASCUS.
The spore sacs characteristic of the Ascomycetes. typically each sac contains 8 spores, but many species are found where the ascus contains less than eight, while in others there are many more. The ascus develops from an ascogenous hypha. The latter is a septate hypha which grows in the form of a hook or

crosier. Each segment contains 2 nuclei; those of the apical hook-like segment divide simultaneously so that 4 nuclei are present. Septations cut off two of these nuclei, leaving a segment containing 2 nuclei which now fuse. The segment now elongates conspicuously and becomes the ascus while the fusion nucleus divides successively to form 8 nuclei which eventually become set off in 8 separate spores. Subsequent development of the segments cut off during ascus formation may lead to the development of additional asci from the same ascogenous hypha. (R.M.W.)

ASEPSIS. Absence of infection; sterility. (R.S.M.)

ASEXUAL REPRODUCTION. In asexual reproduction, a part of the parent organism, when removed, becomes an organism identical with the parent.

Different types of asexual reproduction are met in the different groups of plants, necessitating separate treatment. In green **algae** the characteristic method is by the formation of **zoöspores,** which are unicellular motile bodies formed from the **protoplast** of a single **cell.** Each zoöspore, after swimming around for a time, becomes quiet and grows to form a new individual of the parent type. In mosses and ferns, where a definite **alternation of generations** occurs, the unicellular **spores** which mark the end of one generation are an asexual means of reproduction. In flowering plants there are many methods of asexual reproduction, some of great importance to man. A common type is found in the strawberry, where long slender branches grow out from the short stem of the plant and take root at their tip. There a new plant forms. This reproductive structure is called a stolon. Many other plants produce similar branches which run along the surface of the ground (runners) or just beneath it (rhizomes), and send up one or more new plants from the nodes. Other plants, like the tiger lily, bear small buds in the axils of their leaves: the buds or bulblets are easily detached and readily grow to form new plants. Similar bulblets are formed at the base of many bulbs. Another very common method of asexual reproduction is through the formation of suckers, branches formed at the base of the parent

plants, and gradually growing to replace them. Man propagates extensively date palms, pineapples, and bananas, for example, by means of suckers. The familiar potato tuber, the swollen tip of a rhizome, is another type of asexual reproduction which is of tremendous value to man.

All the methods so far enumerated are from the stem of the plant. However, any part of the plant may be a means of asexual renewal. Many plants are known which reproduce by means of the leaves. Several of these are now cultivated extensively as objects of beauty or of curiosity. For example, many of the ornamental begonias are readily propagated by leaves; in several ferns, little plants form on the leaf surface, from which they fall to the ground and grow. Less widely known but equally interesting are species of *Bryophyllum* and *Kalanchoë.* In the notches of the leaves of these plants, while still attached to the parent plant or after being severed therefrom, tiny plants readily form and grow.

Many plants readily root when cut up into segments. Most people are familiar with the habit of the willow twig of striking root and growing when stuck in the ground. Equally well known is the geranium cutting, which is merely a branch removed from the parent plant and placed in a favorable environment.

Asexual reproduction is of tremendous importance to man. New and improved forms of plants are constantly being made: by asexual means they are reproduced in the great quantities necessary for commercial use. Without such reproduction their formation in quantity would be practically impossible. (See **Grafting and Budding; Propagation.**) (R.M.W.)

ASH DISPOSAL. All coal has ash, and since coal is the principal fuel for domestic and industrial use, the disposal of ashes is an important and widespread activity. The combustion of coal in a **furnace** results in the accumulation in an ash hopper, or ash pit, of the refuse, which, with proper technical handling of the **combustion,** will be principally ash, but may in some cases contain up to 20%, by weight, of carbon in the unburned state. All the ash should be in the ash pit or hopper, but as a matter of fact, from 5% to 40% may leave the

"Hydrojet" ash-handling system.

furnace in the flue gases—carried in suspension. While little can be done about this in domestic and industrial practice, the large amount of coal consumed by public utility plants is potentially capable of discharging from the stack so much ash to the surrounding territory that endeavor is made to reduce the percentage of ash leaving with the gas, and to separate the residual ash from the gas and return it to the hopper.

Removal of ashes, and their disposal, is no simple problem because, first, the ash is dusty, hence irritating and annoying to handle; second, it may contain clinkers which must be broken before being given to any reasonably sized conveying equipment; and third, it is abrasive and will wear all conveyor parts in contact with it if there is any relative motion. Ash disposal systems are designed for continuous or intermittent operation, as the case may be, and consist of some means of removing ash from the furnace, loading it on a conveyor system, unloading from the conveyor to storage, and a means of disposing of stored ash. Ashes can be raked from ash pits to boiler room floors, and then shoveled into wheelbarrows or cars, or raked to gratings where they will fall into a conveying system. The large furnace will often be designed so that the ash may be handled by gravity directly from the hoppers to cars or conveying system.

Present-day conveying systems variously employ bucket conveyors, scrapers, pneumatic conveyors, steam jets, and water jets. Of these, one of the most popular as well as interesting methods is the hydraulic or water jet system. It is essentially a large-plant system, but has the advantage of being clean and dustless. When the hopper is to be emptied, the ashes are hydraulically undercut with an oscillating feed jet. The mingled ash and water flows into a trench along which the ash is carried under the influence of jets of water directed along the run of the trench. At the end of the trench the flow is dropped into a sump where the ash settles to the bottom and the sluicing water overflows to a clear well, ready for recirculation. A wet ash pump or clam shell bucket will deliver the wet ash to an elevated ash bin, where it can be further dewatered. The ultimate disposition of industrial ash is to railroad cars, trucks or barges. An important factor in ash disposal is its use as a building material, particularly to form cinder blocks. In all larger cities this demand places a potential value on ash, which must be considered in the choice of fuel-burning equipment, since the ash recovery from pulverized coal furnaces would have much less value than from stokers. (F.T.M.)

ASP. Reptilia, Sauria. A poisonous **snake** found in

Asp. (*N. Y. Zool. Soc.*)

northern Africa and southwestern Europe, also called the southern viper and the Egyptian **cobra.** (A.W.L.)

ASPARAGUS. Liliaceae. A genus of about 125 species of liliaceous plants native to temperate and tropical regions of the Old World. All are characterized by having the leaves reduced to minute scales or bristles, while small, often very leaf-like branches called cladophylls function as leaves. The flowers are small, yellowish or white in color, and the fruit is a berry, often brightly colored. *Asparagus officinalis,* a native of the marshes of Europe, is the cultivated garden form with thick and fleshy young stems. Other species are widely grown for their delicate beauty, as the familiar Asparagus ferns, *Asparagus plumosus* and *Asparagus Spengeri* (which are not properly ferns at all), and the florist's smilax, *Asparagus asparagoides.* (R.M.W.)

ASPARAGUS BEETLE. Insecta, Coleoptera. An introduced European beetle, *Crioceris asparagi,* which is sometimes an important pest on asparagus. A related species, also introduced from Europe, is known as the 12-spotted asparagus beetle. They are held in check by hand-picking and by dusting plants with lead arsenate and lime. (A.W.L.)

ASPARAGUS STONE. Apatite.

ASPARTIC ACID. Aldehydes, Ketones, and Related Compounds.

ASPECT RATIO. Aspect ratio, one of the most important dimensions of a **wing,** measures its shape. It was long ago discovered that two wings of the same area, and built with the same airfoil section, did not have the same aerodynamic characteristics unless their area was of the same shape.

The term aspect ratio was then introduced to define this shape. The aspect ratio for a simple rectangular wing is the long dimension, i.e., the span perpendicular to the windstream, divided by the short dimension, the chord. However, many wings are not rectangular in plan form, and their aspect ratio is the span divided by the average chord. Since this average chord is the area of the wing divided by the span, it follows that for non-rectangular wings the aspect ratio is

$$\text{\AE} = \frac{\text{span}^2}{\text{wing area}}$$

In general, the larger aspect ratios are better from the aerodynamic standpoint, but poorer from the structural standpoint, since increasing the aspect ratio has the following results:

1. It decreases the drag.
2. It increases the rate of change of lift with the angle of attack.
3. It displaces the lateral center of pressure farther from the fuselage, and hence makes more difficult the problem of bracing rigidly the wing structure. (F.T.M.)

ASPECTS OF THE PLANETS. Planetary Motion.

ASPERGILLUS. Ascomycetes.

ASPHALT(UM). Asphaltum is a semi-solid mixture of several **hydrocarbons,** probably formed because of the evaporation of the lighter and more volatile constituents. It is amorphous, of low specific gravity, 1–2, with a black or brownish-black color and pitchy luster. Notable localities for asphaltum are the Island of Trinidad and the Dead Sea region, where Lake Asphaltites were long known to the ancients. (See also **Petroleum.**) (E.S.C.S.)

ASPHYXIA OR SUFFOCATION. This condition develops as a result of insufficient **oxygen** in the blood. The patient becomes **cyanotic** and unconscious, and if the causative factor continues, death results.

Local asphyxia is limited to a part of the body, and is due to an interruption of the blood supply to that part.

If total for any length of time gangrene of the part develops. (R.S.M.)

ASPIDOBRANCHIATA. GASTEROPODA.

ASPIDOCHIROTA. Holothuroidea.

ASPIRIN. Salycilates.

ASPLENIUM. Paleobotany.

ASS. Mammalia, Perissodactyla. Animals of several species related to the horses and zebras and belonging to the same genus, *Equus*. In addition to the domestic ass, *E. asinus*, wild asses are known in arid parts of Asia, from Persia to Mongolia, and in the deserts of northeastern Africa. (A.W.L.)

ASSASSIN-BUG. Insecta, Hemiptera. Any bug of the large predacious species constituting the family Reduviidae. (A.W.L.)

ASSOCIATION AND POLYMERIZATION; DISSOCIATION. The simplest formula for water is H_2O, molecular weight 18. By measurements of the **specific volume** and **surface tension** of liquid water at two different temperatures it is possible to estimate the molecular weight of this substance as liquid in terms of its molecular weight as vapor, and, consequently, the degree of association or dissociation, as the case may be. Water is estimated to be associated 2.7 to 2.0 times at 100° C.; 3.0 to 2.2 times at 60° C.; 3.6 to 2.4 times at 20° C. Dihydrol $(H_2O)_2$, and trihydrol $(H_2O)_3$, are believed to be the individual molecules concerned and to be present in proportions that vary with temperature. Substances believed to be associated as liquids to the degree of 2 to 3 times are: methyl **alcohol** (3.3), ethyl **alcohol** (2.5), **acetic acid** (3.2), **acetamide** (2.3), **benzamide** (2.2); and slightly: **acetone** (1.3), **phenol** (1.3). Normal liquids may be regarded as those whose molecules have a constant $\dfrac{\text{Molecular Weight} \times \text{Heat of Vaporization}}{\text{Absolute Temperature}}$ when observed under the condition of the same concentration of molecules in the vapor phase (Hildebrand, 1915), e.g., a concentration of 0.005 gram mol of vapor per liter. Such results are given in the following table:

NORMAL LIQUIDS	$\dfrac{\text{M.W.} \times \text{Ht.Vap.}}{\text{T.}}$
Nitrogen	27.6 (at 55° Abs.)
Oxygen	27.6 (at 75° Abs.)
Chlorine	27.8 (at 194° Abs.)
Pentane	27.0 (at 256° Abs.)
Hexane	27.2 (at 286° Abs.)
Benzene	27.4 (at 298° Abs.)
Mercury	26.2 (at 560° Abs.)
Average	27 approx.

ASSOCIATED LIQUIDS	$\dfrac{\text{M.W.} \times \text{Ht.Vap.}}{\text{T.}}$
Water	32.0 (at 325° Abs.)
Ethyl alcohol	33.4 (at 307° Abs.)
Ammonia	32.4 (at 200° Abs.)

Of two substances having the same elementary percentage composition but multiple or sub-multiple molecular weights, the substance of higher molecular weight is known as the polymer of the substance of lower molecular weight. Polymerization is a marked property of certain groups of chemicals; e.g., olefin **hydrocarbons**. Amylene (C_5H_{10}) in the presence of sulfuric acid or zinc chloride polymerizes to $C_{10}H_{20}$, $C_{15}H_{30}$, $C_{20}H_{40}$. Mono-

vinyl acetylene $(CH_2:CH—C:CH)$ is the polymer of acetylene $(CH:CH)$ and is made from the latter. **Formaldehyde** is polymerized spontaneously at ordinary temperature, and **acetaldehyde** upon the addition of a small percentage of hydrochloride acid, sulfuric acid, or zinc chloride.

Dissociation may be electrolytic or thermal. Electrolytic dissociation is discussed in articles on **Electrochemistry; Reactions Involving Recombination of Ions.** Thermal dissociation, an **equilibrium** phenomenon and not decomposition, is observed in the cases of such substances as the following:

Nitrogen tetroxide (N_2O_4) into nitrogen dioxide (NO_2)
    at 27° C., 80% N_2O_4, 20% NO_2 by weight.
    at 100° C., 20% N_2O_4, 80% NO_2 by weight.
Calcium carbonate $(CaCO_3)$ into calcium oxide (CaO)
    plus carbon dioxide (CO_2)
    at 600° C., 20 mm. pressure CO_2.
    at 920° C., 760 mm. pressure CO_2.
Mercuric oxide (HgO) into mercury (Hg) plus oxygen (O_2)
    at 400° C., 230 mm. pressure O_2.
    at 500° C., 800 mm. pressure O_2.
Trilead tetroxide (Pb_3O_4) into lead monoxide (PbO)
    plus oxygen (O_2)
    at 450° C., 5 mm. pressure O_2.
    at 635° C., 760 mm. pressure O_2.
Ammonium chloride (NH_4Cl) into ammonia (NH_3)
    plus hydrogen chloride (HCl)
    at 350° C., 40% NH_3, 40% HCl, 20% NH_4Cl by volume.
Sulfuric acid (H_2SO_4) into sulfur trioxide (SO_3) plus water (H_2O)
    at 450° C., 0.03% SO_3 by weight.
Sulfur trioxide (SO_3) into sulfur dioxide (SO_2) plus oxygen (O_2)
    at 600° C., 20% SO_2 to 80% SO_3 by volume.
    at 700° C., 50% SO_2 to 50% SO_3 by volume.
Carbon dioxide (CO_2) into carbon monoxide (CO)
    plus oxygen (O_2)
    at 1230° C., 0.04% CO_2 dissociated.
    at 1730° C., 1.77% CO_2 dissociated.
Water (H_2O) into hydrogen (H_2) plus oxygen (O_2)
    at 1230° C., 0.02% H_2O dissociated.
    at 1730° C., 0.59% H_2O dissociated. (R.K.S.)

ASSOCIATIONS, PLANT. Plants are not distributed in nature in a haphazard fashion, but in habitats in which certain species are present certain others usually occur also. Each such community of plants, composed of more or less the same group of species, is called a plant association. Some plant associations, such for example as the marginal rush or cattail association around a pond, may occupy only localized areas. Other associations, such as some of the grassland or desert shrub associations of western North America may occupy continuous areas totalling hundreds or thousands of square miles. In general, however, plant associations are the smaller units of vegetation occurring within a plant formation (see below) and their distribution is largely controlled by local soil and climatic conditions. Local differences in climate, in turn, are largely a function of topography. Some plant associations, such as a lichen association on a rock cliff, are relatively simple in organization. Others are relatively complex. The oak-hickory association of the eastern United States, for example, is named for the two prominent genera of trees present. Associated with the oaks and hickories, however, are occasional other large trees. In addition there are usually present smaller kinds of trees, species which constitute a shrub story, and herbaceous species which constitute a more or less continuous ground cover.

A larger unit of vegetation than the association is the formation. A *plant formation* usually occupies thousands of square miles and its limits are controlled primarily by climatic conditions. Some of the major plant formations of North America are the tundra, the boreal forest, the hemlock-hardwood forest, the deciduous forest, the grasslands, the western coastal forest, the western mountain forest, the semi-deserts, and the tropical forests. Within each formation there are usually many different plant associations. Most plant associations are not permanent, but in the phenomenon of **plant succession** one association gradually replaces another. Many successions are in progress in any plant formation. The end results of the successional replacement of one plant association in turn by another is, if the process goes to completion, the establishment of a *climax association*. Such an association is a stable plant community and is not succeeded by any other association; it is the apex of the successional process and, barring changes of climate, will continue to reproduce itself indefinitely. (B.S.M.)

ASSOCIATIVE LAW OF ALGEBRA. Addition and Multiplication.

ASTER. See Mitosis under **Cell Division.**

ASTEROID. The name asteroids (or planetoids) is given to a group of small objects which are members of the **solar system** and whose **orbits** lie, in general, between the orbits of **Mars** and **Jupiter.** The very name asteroid (star-like) describes adequately the appearance of the objects. Only one of them is ever bright enough to be seen with the naked eye and even this one, Vesta, is never a conspicuous object.

The first asteroid discovered, Ceres, was found accidentally by Piazzi on January 1, 1801. His attention was directed to it by noticing the motion of the object through the stars. As the object approached the position of the sun there was danger of its being lost, for the methods of orbit computation were not well developed at that time. The mathematician Gauss went to work on the problem and invented his well-known method for orbit computation, by means of which he was able to predict positions permitting the rediscovery of Ceres after it had passed the sun. Since the orbit was found to lie in the gap between the orbits of Mars and Jupiter, and the object was found to have a mean distance from the sun of 2.8 **astronomical units,** strong support was given by it to **Bode's Law.**

Up to the middle of the 19th century only five more asteroids were discovered, but, with the application of photography to astronomy, the discoveries became more and more frequent until at the present time more than 1300 are under observation. The objects are first detected by noticing the movement of a star-like object through the stars. Photographically, if the camera is arranged to follow the motions of the stars, the star images will appear as dots on the plate while the asteroid image will be trailed out into a short line. The most extensive program of search for asteroids was carried on by Wolf at Heidelburg during the period following 1891. From this time on through the first two decades of the present century Wolf and his assistants are credited with no less than 500 discoveries.

When an asteroid is first discovered it is designated with the year of discovery followed by two letters which indicate the half of the month in which the object was found and the chronological order within that half month. After the orbit of the object has been determined and it proves to be a new asteroid, it is assigned a permanent number, in chronological order of discovery, and the discoverer is privileged to name the object as he may choose. In general, asteroids are given Latinized names with the feminine endings.

Little is known regarding the physical characteristics of the asteroids themselves. The diameters of the four brightest, and presumably the largest, have been measured with large telescopes and found to run from 480 miles for Ceres down to 120 miles for Juno. Estimates of the sizes of the others may be made from the amount of sunlight which they reflect, after making certain assumptions regarding the reflecting power and shape of the objects. The results of the survey indicate that perhaps 150 asteroids have diameters greater than 50 miles, but the majority are between 50 and 20 miles in diameter, with some even smaller than that. Masses and densities can be estimated only from statistical studies but the indication is that the total mass of all of the objects combined cannot be more than $\frac{1}{500}$ part of the mass of the earth.

The problem of the determination of the shapes of the asteroids is still one of considerable interest, but for which there is no definite solution. It is well known that the reflected sunlight from many of these objects varies in a periodic manner which can only be adequately explained on the basis of a rotating object. In the case of Eunomia it has been quite definitely proven that the object must be spherical, and that the variation in light is due to different reflecting powers on different parts of the surface. On the other hand, **Eros** has been quite definitely proven to have a "dumb-bell" shape, with the light variations due to rotation of this irregular object.

The orbits of the asteroids have been studied with great zeal ever since the discovery of Ceres. In fact, this group of objects may be considered as a laboratory in which the workers in the field of **celestial mechanics** may test out various theories. Since the orbits lie between the orbits of Mars and Jupiter, and the masses of the asteroids are very small, the planets exert large **perturbations** on the asteroids, while they themselves are virtually unaffected by the asteroid attractions. Many of the methods of computing perturbations were developed as the result of the researches on the orbits of the minor planets. One particularly interesting result is found in the case of the so-called **Trojan Group** of asteroids, which is discussed elsewhere.

There are two theories for the origin of the asteroids. One is that these objects represent a planet that was "spoiled in the making," i.e., never developed into its solid form. The other theory postulates that the asteroids represent the remains of a planet which was formed but disintegrated later on. The latter theory is the older of the two but has gained some considerable strength on the basis of the so-called families of asteroids. Theoretically, it may be proved that, if a planet should be broken up by a series of explosions, the centers of the orbits of the asteroids which are products of any particular explosion, should lie along a line between the sun and Jupiter. In addition to the locations of the centers of the orbits along this line, the mean distances of the products of each explosion from Jupiter would be approximately the same, and also the inclinations of the orbit planes should be similar. As a result of a statistical study of the orbits of the asteroids, five such families of asteroids have been identified and the families each contain from 15 to 44 members. Within the past few years a theory has been gaining weight that the asteroids may be connected in some manner with comets, but as yet the evidence is far from conclusive. (W.K.G.)

ASTEROIDEA. A class of the phylum **Echinodermata.** The starfishes.

The starfishes are distinguished from other echinoderms by the presence of radiating arms or rays, usually five or in multiples of five, which contain part of the internal organs and are usually not sharply separated from the central disk. There are many species but the economic importance of the group is limited. They are sometimes serious pests in oyster beds since they feed largely on shellfish.

The class is divided into three orders: Phanerozonia, Spinulosa, and Forcipulata. (A.W.L.)

Common Starfish. (*N. Y. Zool. Soc.*)

ASTHENIA. Weakness, lack, or loss of strength. (R.S.M.)

ASTHENOSPHERE. A term proposed by Barrell, in 1914, for the zone beneath the relatively rigid **lithosphere.** The asthenosphere is considered to be the level of no strain in which there is maximum plasticity, and in which the **igneous** rock magmas are thought to originate. (R.M.F.)

ASTHMA. A term used to describe a characteristic "wheezing" type of respiration. Wheezing is due to obstruction caused by spasm of the bronchial muscles constricting the bronchi, and by swelling of the **mucous-membrane** lining of the respiratory passages. Asthmatic breathing may be paroxysmal or constant, and it occurs in many conditions. The term bronchial asthma is applied to the paroxysmal form, which appears as a definite clinical syndrome.

Asthma was first described by John Floyer in 1698, but the relation of **hay fever** and asthma to plant pollens and proteins in nature was observed for the first time in 1905 by Dunbar. Following this, the entire picture of the allergic diseases was correlated, and the role of protein sensitivity was demonstrated by means of pollen and protein skin sensitivity tests.

Bronchial asthma is a common disease. The allergic pattern and the asthmatic tendency is strongly familial; some member of the family of an asthmatic has allergic symptoms in at least 40% of cases. The clinical types are classified roughly as those in which the symptom complex is precipitated by extrinsic and intrinsic causes. The extrinsic form is traceable to the same group of substances, pollens and proteins, that usually cause hay fever. The common inhalants are the tree and plant pollens—notably ragweed, animal danders, vegetable powders, and house dust. Foods which are frequent offenders are milk, eggs, wheat and other cereals. Multiple sensitivities to many **antigens** at the same time are common. In adults, inhalants are the usual cause of asthma, and food is rarely so. The reverse is true in children.

Intrinsic asthma, or asthma originating from some cause within the body, is not seasonal and changes of residence or occupation are ineffective in its control. The exciting cause is difficult to determine in most instances. In some cases, however, chronic infection—particularly of the paranasal **sinuses** and the respiratory tract—play a role. Recently, asthma has come to be considered a **psychosomatic disease**, and psychogenic factors are known to precipitate attacks in individuals with an asthmatic tendency.

The symptoms of an acute attack of asthma are wheezy respiration with difficult inspiration and even more difficult long, noisy expiration. In a prolonged and severe attack all of the accessory muscles of respira-

tion are called into action. The attack may last minutes or hours. **Cyanosis** is common. Cough and sticky sputum are frequent. Dry squeaky râles or noises are heard over the lung fields and the breath sounds are distant. As with other allergic diseases, an increase in eosinophils is present in the **blood.** In the interval between attacks, the patient is asymptomatic, and the physical examination normal. Diagnosis is then made on history, and with extrinsic asthma, the offending substances can be identified by means of skin tests.

Treatment of the acute attack consists of the administration of **adrenalin** hypodermically. Stramonium in the form of powders or in cigarettes may alleviate the symptoms. In severe or intractable asthma, additional measures are necessary: aminophylline intravenously, rectal anesthesia (ether in oil or **avertin** in small doses), or the inhalation of 100% **oxygen**, or oxygen and **helium** mixtures may be required.

In the period between attacks, ephedrine combined with a barbiturate (**barbital**) is given as a prophylactic. Adrenalin in oil given intramuscularly has a slow, long-continued action and may help prevent attacks. In cases due to extrinsic allergens, a dust-free, pollen-free atmosphere is effective in decreasing the number of attacks. Similarly, the avoidance of foods to which the patient is known to be sensitive helps to keep the disease under control. In those cases associated with infection, eradication of the infection usually causes an improvement in the asthma.

Complications of asthma occur in long-standing cases. The patients are prone to chronic **bronchitis** and sometimes **bronchiectasis** develops. After years of increased resistance in the pulmonary tree, the right heart hypertrophies under the strain (cor pulmonale); if the strain becomes too great, congestive **heart failure** results. Emphysema and barrel-shaped chest are common. (D.M.H.)

ASTIGMATEA. Chordata, Thaliacea. An order of salpian **tunicates.** (A.W.L.)

ASTIGMATISM. When an optical system fails to bring the rays to a well-defined focus, it is said to be astigmatic. The term is, however, commonly applied to cases in which the fault is due to a lack of symmetry of the optical system about its axis, so that it has different focal lengths in different meridians. This would be true, for example, of a lens, one of whose surfaces is not a true surface of revolution but is slightly ellipsoidal or spoon-shaped. The defect is, unfortunately, quite common in the human **eye**, usually due to radial asymmetry of the **cornea.** When the patient looks at the optometrist's "clock face" or "wheel," the radiating lines along all but one diameter may appear blurred. The remedy is a spectacle lens having an astigmatism of its own at right angles to that of the eye, and to this end use is made of **cylindrical** or **toric** surfaces. (L.D.W.)

ASTROID. Hypocycloid.

ASTROLABE. The astrolabe is an ancient form of portable astronomical instrument invented during the 2nd or 3rd century B.C. probably either by Hipparchus or Appollonius. In its most common form the astrolabe consists of a circular disk which may be suspended by a ring so that it will hang in the plane of a **vertical circle.** A pointer, or **alidade**, is pivoted at the center of the disk and angular graduations are marked about the edge. In addition to the alidade and angular graduations many other astronomical materials such as **constellation** configurations, lists of **planets**, etc., are engraved on the disk. This ornamental engraving is very intricate and beautiful on many of the instruments and makes them interesting museum pieces.

The instrument was undoubtedly intended primarily for the purpose of measuring **altitude** of celestial bodies. For this purpose the ring was suspended by the thumb of one hand and the other fingers employed to steady the disk as the alidade was moved by the other hand until it pointed directly at the object under observation. The altitude could then be read directly on the disk.

The astrolabe was used by navigators for the determination of **latitude** from the 15th century down to the invention of the **sextant** in the 18th century. It has recently been revived for teaching purposes in elementary classes. (W.K.G.)

ASTROLOGY. Astrology is the ancient art of divining the future of human affairs from observations of the celestial objects. We find records of astrological methods back through the earliest recorded history and during the 14th and 15th centuries astrologers held important positions at many of the courts of Europe. Since observations of the celestial objects form an important part of astrology, the art may be considered as the parent of astronomy although the separation of the art of prophecy from the science of observation certainly took place long before the Christian era.

In the most commonly practised type of astrology a horoscope is drawn up for a person. Such a horoscope consists of a map of the heavens drawn at the instant of birth of the person interested. This map shows the aspects (see **Planetary Motion**) of the planets, the location of the **sun, moon,** and **planets** on the **zodiac** and other correlated phenomena. The signs of the zodiac, or "houses," in which the **various objects are located** at the time of birth are supposed to indicate all sorts of influence on the future of the person. Such words as lunatic, saturnine, ill-starred, etc., indicate the influence of astrology on general culture of past eras. (W.K.G.)

ASTRONOMICAL REFRACTION. In any type of astronomical observation the light from the distant object must pass through the atmosphere of the earth and suffer a change of direction known as **refraction.** The amount of change of direction depends upon two fundamental factors: the relative **refractive index** of the atmosphere and the angle which the ray from the distant object makes with the normal to the surface of the atmosphere. Since the normal to the atmosphere is the direction of the astronomical **zenith** the amount of refraction will depend upon the **altitude** of the object, being greatest when the altitude is least, or when the object is on the **horizon.** The effect of refraction is to make the altitude of an object appear greater than it would be if no atmosphere were present.

To calculate the amount of astronomical refraction the index of refraction of the atmosphere is needed and, unfortunately, this quantity varies with meteorological conditions. Various theoretical methods for computing the amount of astronomical refraction have been proposed but none of them are very satisfactory for altitudes less than 20°. A fair approximation to the true value may be obtained from the expression

$$R = \frac{983B}{460 + T} \cotan h$$

in which B is the reading of the **barometer** in inches, T is the **temperature** of the air in degrees **Fahrenheit,** h is the apparent altitude of the object, and R is the amount of refraction in seconds of arc. More accurate values may be obtained by using refraction tables such as those published in Bowditch American Practical Navigator. These tables give the amount of refraction in terms of observed altitude, and various meteorological conditions such as temperature and barometric pressure. This refraction must be subtracted from any observed altitude. In case changes due to refraction in other **spherical coordinates** than altitude are desired the **astronomical triangle** must be solved.

Sudden and irregular changes in astronomical refraction are produced by varying meteorological conditions and produce effects of twinkling in the stars. (W.K.G.)

ASTRONOMICAL TRIANGLE. The spherical triangle formed on the **celestial sphere** between the observer's **meridian,** the **hour circle,** and the **vertical circle** through an object, is known as the astronomical triangle. Since practically every problem of **nautical astronomy** deals with the solution of this triangle it will be well to have the definitions of the various parts

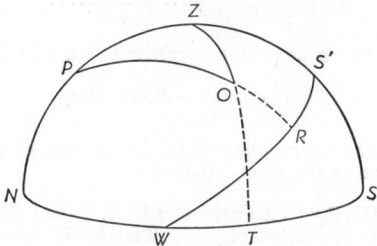

Astronomical triangle.

well in mind. In the figure (drawn for an observer in the northern hemisphere) we have:

N, W, and S, the north, west, and south points of the observer's **horizon.**
S', R, W is the celestial **equator.**
Z is the observer's astronomical **zenith.**
P is the north pole of rotation.
O is the object under consideration.
PZO is the astronomical triangle, the various parts of which are defined as follows:

The vertex angles are:

ZPO = the **hour angle** of the object.
PZO = 180°—the astronomical **azimuth** of the object.
ZOP = the parallactic angle.

The sides (measured in angular units) are:

PZ = 90°—observer's astronomic **latitude.**
ZO = 90°—**altitude** of the object = zenith distance of the object.
PO = 90°—**declination** of the object. (W.K.G.)

ASTRONOMICAL UNIT. The astronomical unit is a unit of distance principally employed in expressing distances within the **solar system,** but is also used to some extent for measuring interstellar distances. Technically defined, one astronomical unit is the mean distance of the **earth** from the **sun.** To express this in miles it becomes necessary to determine the distance of the earth from the sun in miles or, in other words, to determine the **solar parallax.** The value accepted at present for the length of the astronomical unit is 92,897,000 miles (149,504,000 kilometers). (W.K.G.)

ASYMMETRIC ATOM. Isomerism.

ASYMPTOTE TO A PLANE CURVE. An asymptote of a curve is a straight line which the curve approaches arbitrarily near as its tracing point recedes beyond all bounds.

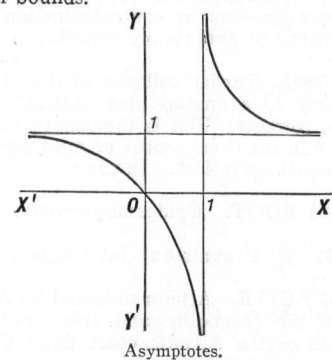

Asymptotes.

Thus, in the curve whose equation is $y = \dfrac{x-1}{x}$ as shown in the accompanying figure, the lines $x = 1$ and $y = 1$ are asymptotes. (L.L.S.)

ASYMPTOTIC SERIES. A divergent series of the form

$$A_0 + \frac{A_1}{x} + \frac{A_2}{x^2} + \cdots + \frac{A_n}{x^n} + \cdots$$

is said to be an asymptotic representation of a function $f(x)$ if

$$\lim_{x \to \infty} x^n[f(x) - S_n(x)] = 0,$$

for any value of n, where $S_n(x)$ is the sum of the first $n + 1$ terms of the series. (L.L.S.)

"ATABRINE." (Registered in U. S. Patent Office and used in U. S. Dispensary; "ATEBRIN" favored by *Chemical Abstracts* and by *British Chemical Abstracts;* "ATEBRIN" used in "Merck's Index"; and "ATE-BRINE" in Thorpe's "Dictionary of Applied Chemistry"). Atabrine is quinacrine hydrochloride, and the full chemical name 3-chloro-7-methoxy-9-(1-methyl-4-diethylaminobutylamino)-acridine dihydrochloride. It is used in the treatment of malaria. The production has been increased since its announcement in 1930, so that in 1942 the quantity was equivalent to the pre-World War II supply of quinine, and this has been increased several times since 1942. Atabrine is yellow in color, and its long continued use results in a harmless yellow pigmentation of all the tissues including the skin. (R.K.S., D.M.H.)

ATACAMITE. This mineral consists partly or entirely of **copper** oxychloride, $CuCl_2Cu(OH)_2$, probably, the exact formula is uncertain. Crystallizes in thin, orthorhombic prisms, may occur massive. Hardness, 3–3.5; specific gravity, 3.76–3.78; luster, adamantine to vitreous; color, green; streak, green; transparent to translucent.

It is a secondary mineral found associated with **malachite** and **cuprite;** originally found at Atacama, Chili, whence its name. Other localities are Bohemia, South Australia, and in the United States in Arizona, Utah and Wyoming. (E.S.C.S.)

ATAVISM. The appearance through **heredity** of characters which have not been developed in the parents of the organism in question. The strict meaning of the word is the reappearance of grandparental characters but it has been used also to designate the reappearance of characters from more remote generations. (A.W.L.)

ATAXIA. Lack of muscular coordination due to disease of the brain, particularly the cerebellum, or spinal cord. **Locomotor ataxia** is the result of degeneration of portions of the spinal cord occurring in the later stages of syphilis. (D.M.H.)

ATAXIC. A term applied by Keyes, in 1901, to all unstratified ore deposits in contradistinction to sedimentary, stratified or **eutaxic** ore deposits. (R.M.F.)

ATELECTASIS. Partial collapse of the **lung,** due most commonly to a mucous plug obstructing one of the bronchial passages. This is thought to be the preliminary stage in the development of most **pneumonias** following surgical operations. (R.S.M.)

ATHLETE'S FOOT. Epidermophytosis.

ATHODYD. Jet Propulsion; Jet Engine.

ATLANTIC SUITE. A term proposed by A. Harker, in 1896, for the chemically and structurally related **igneous** rocks of the Atlantic coast line. Chemically the rocks of this suite are described as **alkaline** and are represented by such types as **granite** and its magmatic relatives, as compared with the calc-alkali igneous rocks of the **Pacific Suite.** (R.M.F.)

ATMOSPHERE. The gaseous envelope surrounding any celestial object is known as the atmosphere of that object.

The general characteristics of the atmosphere of any object may be determined theoretically on the basis of the **kinetic theory of gases** and the mass, diameter, and surface temperature of the object. On the basis of this theory and the physical data regarding the various celestial objects, it is determined that the **stars** and all **planets,** except **Mercury,** should have atmospheres, while the **asteroids, moon,** and the satellites of the other planets should be without appreciable gaseous envelopes. The conclusions, drawn from purely theoretical considerations, can be checked by a variety of observations, such as spectroscopic, **albedo, twilight** arch, etc., and the observational results are in close agreement with the theory.

The compositions of the atmospheres of objects external to the earth can be determined from spectroscopic observations. For the stars the problem is relatively simple since the interior radiates a **continuous spectrum,** in most cases, and the cooler gases of the atmosphere produce their characteristic absorption lines. The analysis of planetary spectra is a far more complex problem. In the first place the light that we receive from these objects is sunlight which has penetrated the planetary atmosphere to an undetermined depth and has then been reflected to the earth. Before reaching the observer this reflected light must also pass through the earth's atmosphere. Hence the observed spectrum will be the normal solar spectrum crossed by the multitude of **Fraunhofer lines** due to solar and terrestrial atmospheres, plus lines due to any strange gases that may be present in that portion of the planetary atmosphere through which the sunlight has passed. The presence of any gases in the planetary atmosphere which are also in the solar or terrestrial envelopes will be manifested only by a slight intensification of the absorption lines over those present in the solar spectrum. At certain times the **orbital motion** of the planet has an appreciable component in the direction of the line of sight from the earth. Under these conditions there is a **Doppler shift** of the planetary lines from any corresponding ones in the solar spectrum. The application of high dispersion spectrographs to the study of planetary spectra has employed this Doppler shift and yielded the results given in the articles on the individual planets.

The atmosphere of the earth (see also **Air**) may be considered to extend out to interstellar regions since, in accordance with the kinetic theory, certain molecules may attain velocities greater than the velocity of escape from the earth and also from the solar system. **Auroral** effects have been observed to heights between 500 and 700 miles and we are entitled to say that the earth's atmosphere extends from the earth in sufficient amounts to permit ionization out to these distances. Of this, man has personally explored less than 10 miles; and directly, with instruments, not over 30 miles. What knowledge there is of the remainder is deductively derived from the behavior of radio emanation, auroras, sound waves, etc.

The diagrams presented herewith are self-explanatory. They show much of the interesting electronic and physical nature of our atmosphere, and establish the scale of man's penetration into it. Truly our aeronautical journeys to now have not ventured far from the "shores" of this aerial ocean. The variation of the physical nature of the atmosphere with altitude shows why this has been so.

Much as it must be regretted by men of science that the urge to destroy in warfare advanced in a few short

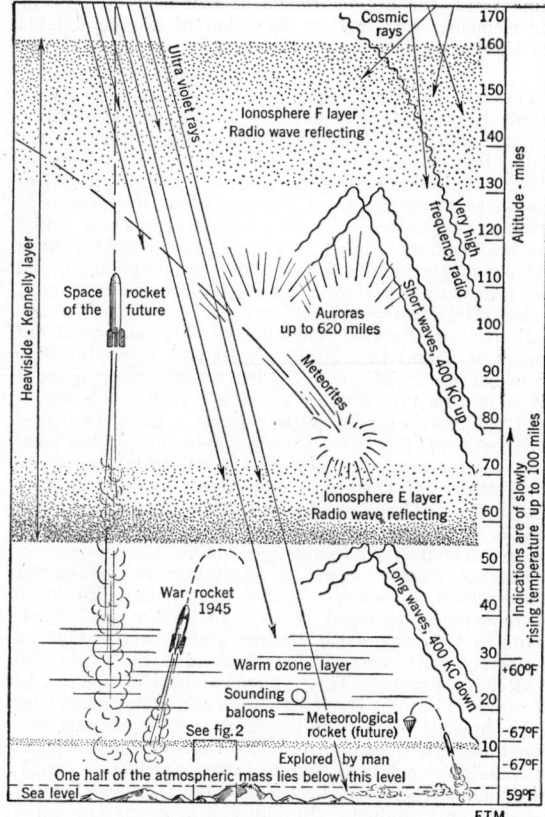

Fig. 1.   The lower atmosphere.   (Covers the small section indicated in Fig. 2a.)

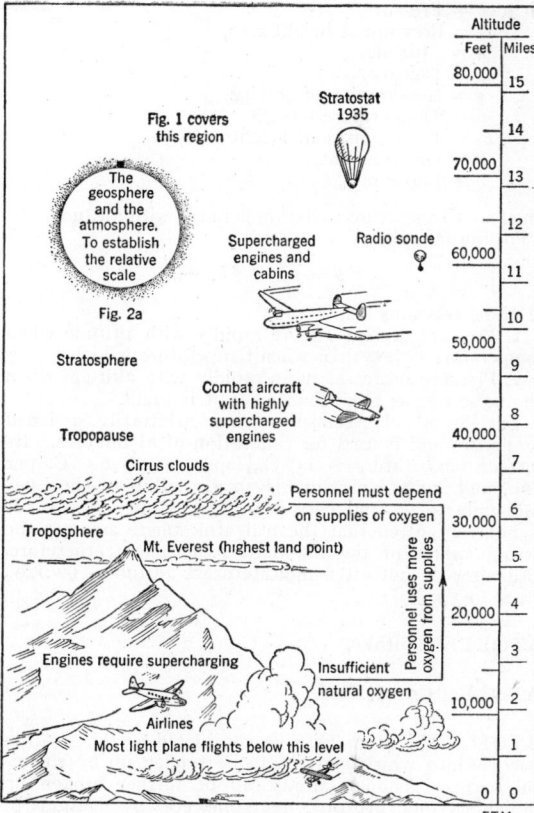

Fig. 2.   Detail in Fig. 1.

years the technology of rockets more than decades of peaceful investigation could accomplish, nevertheless they may look to the future with hope that adaptations of the epochal accomplishments of wartime rocketry will provide a means of scientific exploration of our upper atmosphere and of other worlds than ours.

The surveys thus far made indicate that the atmosphere of the earth is composed primarily of nitrogen and oxygen but also contains small amounts of other gases, from 0.01%–3.0% water vapor, dust, smoke, salt crystals, and small amounts of other impurities. Solid particles in the atmosphere account for the blueness of the sky for they scatter part of the sun's spectrum. One-half of the total mass of the air lies below approximately 18,000' with more than ¾ below 35,000'.

Three general divisions of the atmosphere are recognized. The lower layer is known as the **troposphere,** the middle layer, the **stratosphere,** and the outer layer is known as the **ionosphere.** All storms and practically all cloudiness are restricted to the troposphere. Temperature, in general, decreases from ground or water levels up to the top of the troposphere. There are relatively shallow layers in the troposphere in which the temperature increases, sometimes as much as 50° F., which are known as **inversions.** In general the stratosphere is **isothermal,** but in lower latitudes usually grows continually warmer with increase in altitude to the limit of observational data Very little is known about the thermal structure of the ionosphere from direct measurement, but indirect evidence indicates it is relatively warm. Air density at that level, however, is so small that the temperature would have no meaning in the usual sense. Temperature change with altitude, known as the **lapse rate,** is the principal criterion of differentiation between the

troposphere and the stratosphere whose boundary of separation is known as the **tropopause.**

Water vapor in the atmosphere has a varied distribution in the atmosphere being concentrated in greatest quantity in the lower parts of the troposphere. Water vapor enters the atmosphere from several sources, the principal ones being oceans and large lakes plus vegetation-covered land areas in summer. Tropics and subtropics are in general the large-scale source regions of water vapor. From source regions winds carry water vapor elsewhere over the entire globe. Transportation of water vapor is sufficiently localized in character so that tongues and islands of moist and dry air are present everywhere in the troposphere. In the northern hemisphere moist tongues usually flow from south or west, and dry tongues from some northerly direction, except that, particularly during summer, very dry air originating in the subtropical anticyclones often flows from the south and west. In the temperate zone moist tongues normally appear on the west side of anticyclones and on the east side of cyclones, and dry tongues appear on the east side of anticyclones and west side of cyclones.

Pressure-altitude relations in the atmosphere are mathematically precise and can be determined from the hydrostatic equation

$$\frac{\partial p}{\partial h} = -\rho g,$$

and the equation of state for air

$$p = \rho RT,$$

which leads to the relation

$$p = p_0 \left[\frac{T_0 - \lambda h^{g/R\lambda}}{T_0} \right],$$

where p = Pressure,
  p_0 = Pressure at height zero,
  h = Altitude,
  ρ = Density,
  g = Acceleration of gravity,
  T = Temperature (abs.),
  T_0 = Temperature at height zero,
  R = Gas constant,
  λ = Lapse rate,

when air temperature variation is linear with altitude.
  For isothermal air,

$$p = p_0 e^{-gh/RT}.$$

These relations state:
  1. Pressure decreases more rapidly with altitude when temperature is low than when temperature is high.
  2. Pressure decreases more rapidly with altitude when the lapse rate is large than when it is small.
  The Standard Atmosphere has arbitrarily assigned properties and is used for calibration of **altimeters**. Its surface temperature is 15° C., lapse rate is 6.5° C. per km., and surface pressure is 1013.2 millibars (in English units 59° F. and very nearly 3.5° F. per 1000'). It is not very often that the real atmosphere assumes the actual values of the Standard Atmosphere; therefore, altimeters do not often indicate exact altitude. (W.K.G., F.T.M., P.E.K.)

ATOKE. Epitoke.

ATOLL. Coral Reef.

ATOM. The atom may be considered as the smallest particle into which matter can be broken up by chemical means. Though atoms can be further broken up into **electrons, protons, neutrons,** etc., by methods of modern physics, they retain their individuality in chemical reactions and are used as fundamental units in the organization of theory and facts of chemistry. The atomic theory of chemistry is based on the following experimental laws:
  1. Law of Conservation of Mass—matter is not created nor destroyed in a chemical reaction.
  2. Law of Definite Proportions—compounds contain a definite fixed proportion by weight of the component elements—this proportion being characteristic of the compound.
  3. Law of Multiple Proportions—when an element combines with another to form more than one compound, the weights of one element which combine with a fixed weight of the other stand in the ratio of small whole numbers.
  On the basis of the above laws Dalton (1808) postulated the atomic theory which states that
  1. All matter consists of small indivisible particles called atoms.
  2. Atoms of the same element have the same weight, those of different elements different weights so that the atomic weight can be used to characterize an element.
  3. Chemical combination is the union of different elements. The combining weight of an element is therefore an integral multiple of the atomic weight.
  On the basis of recent discoveries modifications have to be introduced into the original theory of Dalton.
  1. Atoms are not indivisible. Some atoms can be made to decompose under the influence of processes involving enormous energies and others decompose spontaneously (radioactivity). Such phenomena do not occur in the ordinary chemical reactions and therefore the usefulness of the Dalton theory in chemistry is not impaired.
  2. Atoms of the same element often do not have the same atomic weight but may have several values which differ by several integral units from each other (isotopes). Since different isotopes usually occur in the

same ratio, the atomic weight is still characteristic of the element. For further discussion of atomic weights see **Chemical Composition.** (R.K.S.)

ATOMIC BOMB. On August 5, 1945, an atomic bomb, the first of its kind, was dropped on Hiroshima, Japan, with disastrous effects that fully justified the claim that it had more power than 20,000 tons of T.N.T. Atomic bombs involve the disintegration of the nuclei of atoms, not a rearrangement of atoms as in ordinary explosives. Their action is explained under the headings: (1) source of energy; (2) nuclear fission; (3) chain reactions; (4) materials; (5) detonation; and (6) radioactive products. (See **Radioactive Changes; Artificial Disintegration.**)
  Source of Energy. Fundamentally the source of energy is found in Einstein's relativity law, according to which energy of amount E has an equivalent mass m given by the law $E = mc^2$, where c is the velocity of light. Experiments in nuclear physics have confirmed the truth of this law, showing that mass associated with matter may disapper on a small scale giving rise to a release of energy of equivalent mass. Einstein's law shows that the loss of 1 unit of matter on the atomic weight scale releases 933 million electron volts (MEV) or approximately 0.001 unit = 1 MEV.
  Nuclear Fission. Such a disappearance of matter with release of energy occurs in uranium fission, a phenomenon discovered in 1938 by Hahn and Strassmann in Germany, with Meitner and Frisch playing an important part, and all but discovered by Curie and Savitch in France. This discovery showed that nuclei of uranium atoms (chiefly the U235 isotope as later work showed) by neutron bombardment disintegrate into two atoms of elements of intermediate atomic weights, such stable elements as barium (atomic weight = 137.4), lanthanum (138.9), iodine (126.9), and krypton (83.7) being typical products. In addition, neutrons (probably three) are liberated by the disintegration. As far as the atomic bomb is concerned, fission is of importance for two reasons: (1) a large amount of energy per fission is released; (2) since neutrons are liberated as a result of fission, and neutrons cause fission, there is the possibility of a chain reaction. The large release of energy is due to the fact that the average mass of protons and neutrons in the nuclei of elements of intermediate mass like barium and lanthanum is less than their average mass in the nucleus of the uranium atom. Hence the total mass associated with matter after disintegration by fission is less than the total mass before, with consequent release of 200 MEV per fission. In term of kilowatt-hours this is an extremely small amount of energy but, if all the atoms in 1 ounce of uranium could be disintegrated in this way, the release would be some 250 million kilowatt-hours.
  Chain Reaction. To establish a chain reaction there must be enough neutrons produced by fission to allow for the loss of those neutrons which do not cause fission and at the same time leave enough to continue the process. Neutron losses are due to (1) *escape* from the mass of uranium in which they are produced, (2) *capture* by uranium 238 atoms to form U239, and capture by atoms of impurities. The effect of escape is lessened by increasing the size of the unit containing the uranium, because escape is a surface effect, whereas production is a volume effect. Hence the larger the unit, the less the *relative* importance of escape. Loss of neutrons by escape is so important that neither a controlled nor an explosive chain reaction can be produced in a mass of material whose size is below a critical value. Escape is reduced by purifying the uranium and by increasing the concentration of the U235 isotope which undergoes fission many times more readily than U238.
  On December 2, 1942, a controlled chain reaction was established in a pile containing purified uranium with isotopes not separated. In this pile use was made

of two facts: (1) very slow or thermal neutrons cause fission in U238 atoms to an extremely slight degree compared with U235 atoms; (2) fast neutrons, the kind liberated by fission, can be slowed down by a *moderator,* that is, by certain substances of low atomic weight like hydrogen (in water, heavy water, or paraffin) and carbon. The pile consisted of a lattice-work of uranium and carbon. In the pile, stray neutrons, which are always about, started the process of fission; the fast liberated neutrons were slowed to thermal speeds by the passage through carbon before reaching an adjacent lump of uranium in which further fission was caused; and so the cycle was continued. Control of the process was possible through the use of strips of cadmium which could be moved in and out of the pile, the element cadmium having the property of absorbing slow neutrons to an exceptionally high degree.

For the successful operation of an atomic bomb two conditions must be fulfilled. (1) A chain reaction must take place so quickly that a large amount of material undergoes fission before the bomb flies apart. (2) There must be control of such a nature that the bomb can be detonated when wanted, but not before. The first condition is fulfilled by surrounding the bomb with an outer layer of heavy material called a *tamper,* which, because of its inertia, prolongs the time after detonation before the bomb actually bursts, and by the use of concentrated explosive material. The tamper has the further advantage that it reduces losses of neutrons by reflecting some of them.

Materials. Uranium 235 and plutonium are suitable materials because in each of these elements fission is brought about by the neutrous liberated by fission. U235 is obtained by separation of the isotopes of uranium, a problem which was one of the major fields of work preparatory to the successful construction of a bomb.

Plutonium is obtained from the U239 formed when atoms of U238 capture slow neutrons. U239 is an unstable isotope disintegrating with a half-period of 23 minutes as follows:

$$^{239}_{92}U = {}^{239}_{93}\text{neptunium} + \text{beta rays.}$$

Neptunium is also unstable, disintegrating with a half-period of 2.3 days, in the following manner:

$$^{239}_{93}\text{neptunium} = {}^{239}_{94}\text{plutonium} + \text{beta rays.}$$

Although plutonium emits a weak alpha radiation, its half-period is so long that it can be placed in the stable class. Its importance for an atomic bomb lies in the fact, already noted, that it undergoes fission by neutrons. The manufacture of plutonium for bomb material was the primary purpose of large production plants established at Oak Ridge, Tennessee, and at Hanford, Washington. In these plants, involving millions of kilowatts, U235 kept a chain reaction going, and capture of neutrons by the much more abundant U238 isotope provided the plutonium.

Detonation. In a bomb, explosion can be prevented by keeping units of active material below the critical size necessary for a chain reaction. The bomb can be detonated by bringing the fissionable units together quickly. Details of the mechanism used in detonation have not been released, but the report by H. D. Smyth issued by the U. S. War Department suggests that "the obvious method of very rapidly assembling an atomic bomb was to shoot one part as projectile in a gun against a second part as target."

Radioactive Products. The explosion of a bomb, as well as a controlled chain reaction, is accompanied by intense radioactivity. Although the end-products of fission are stable elements like barium, krypton, lanthanum, etc., these final products are the result of successive transformations of intermediate radioactive substances. For example, when lanthanum is an end-product, transformations like the following take place.

$$^{139}_{54}\text{xenon} \rightarrow {}^{139}_{55}\text{caesium} \rightarrow {}^{139}_{56}\text{barium} \rightarrow {}^{139}_{57}\text{lanthanum}$$

Since all isotopes of the elements preceding lanthanum are radioactive, it is evident that fission of any kind is accompanied by radioactivity. When the fisson is as concentrated as it is in a bomb, such activity is intense. (J. K. ROBERTSON, QUEEN'S UNIVERSITY.)

ATOMIC DISINTEGRATION. Radioactivity.

ATOMIC HEAT. Dulong and Petit's Law of Specific Heats.

ATOMIC NUMBER. Chemical Composition.

ATOMIC SPECTRA.

An atomic spectrum is the spectrum of radiation emitted by an excited **atom,** due to changes within the atom; in contrast to radiation arising from changes in the condition of a **molecule.** Such spectra are characterized by more or less sharply defined "lines," corresponding to pronounced maxima at certain frequencies or wavelengths, and representing radiation quanta of definite energy.

The lines are not spaced at random. In the spectrum of hydrogen, for example, there is a prominent red line (H_α) and, far from it, another (H_β) in the greenish-blue, then after a shorter wavelength interval a blue-violet line (H_γ), and after a still shorter interval another violet line (H_Δ), etc. One has only to plot the frequencies of these lines as a function of their ordinal number in the sequence, to get a smooth curve which shows that they are spaced in accordance with some law. In 1885, Balmer studied these lines, now called the Balmer series, and arrived at an empirical formula like the following:

$$w = 109678 \left[\frac{1}{2^2} - \frac{1}{(n+2)^2} \right].$$

This represents the wave number (number of waves per cm.) for any line, numbered n, in the series. As n increases, the wave number w approaches the "series limit" 27419, toward which the series "converges."

Other series have since been discovered in the hydrogen spectrum, including the Lyman series in the ultra-violet, represented by

$$w = 109678 \left[\frac{1}{1^2} - \frac{1}{(n+1)^2} \right],$$

and the Paschen series in the infra-red:

$$w = 109678 \left[\frac{1}{3^2} - \frac{1}{(n+3)^2} \right].$$

The coefficient 109678 (cm.$^{-1}$) is known as the **Rydberg constant** for hydrogen. It appears with only slightly differing values as a coefficient in the series formulae for all atomic spectra, a study of which has led to the **combination principle** of Ritz and its interpretation in terms of the **quantum theory;** though in general these formulae are not so simple as in the case of hydrogen. A study of the spectrum of a single element often reveals certain interesting relations between the most prominent series of lines composing it. For example, two series may have the same limit; or the limit of one may equal the limit of the other minus the wave number of the latter's first line, as is seen to be the case with the Paschen and the Balmer series above. Kossel and Sommerfeld noted that if an element is singly ionized, its spectrum resembles that of the element preceding it in atomic number; a fact explained by their having the same number of extra-nuclear electrons. (See **Hyperfine Structure, Pressure Shift and Broadening,** and **Doppler Effects.**) (L.D.W.)

ATOMIC STRUCTURE.

According to Bohr the **atom** is built up of two units—a positively charged nucleus and a number of negatively charged **electrons**. The nuclear positive charge is equal to the atomic number while the mass is equal to the atomic weight. The electrons have a negative unit charge and a mass $1/1840$ of the lightest nucleus. The number of electrons is equal to the charge on the nucleus measured in electron units of electrical charge, thus making the atom as a whole electrically neutral. The atom is essentially hollow with its mass concentrated at the nucleus and a cloud of orbital electrons revolving around it at various distances.

The inert gases occupy a unique position in the table in that the outer shell of electrons contains two electrons in the case of helium and eight in the case of the other inert gases. An outer shell of eight electrons is therefore correlated with chemical inertness. This correlation can be further extended to the other groups of the periodic table where we find in the same group of the periodic table the same arrangement of the electrons in the outer shell of all members of the group. In this way the chemical characteristics of an atom are associated with the number of electrons in the outer shell. (See **Valence; Spectra.**) (R.K.S.)

ATOMIC WEIGHT. Chemical Composition.

ATRACHEATA. In some systems of classification of plants, the **Bryophytes** are called Atracheata in distinction to all other plants above the Thallophytes, because they lack a definite **vascular system.** (R.M.W.)

ATRIUM. Literally an entrance chamber, and so applied to various organs. 1. The main part of the cavity of the middle **ear.** 2. The vestibule of the female genital passages. 3. A chamber into which the genital organs open in the flatworms. 4. A cavity formed of folds of the body wall in **Amphioxus** and the **tunicates,** which partially surrounds the **pharynx** and opens to the exterior by an atriopore. 5. The chamber at the end of an air tube in the lungs, with which the ultimate air sacs or alveoli communicate. 6. The chamber of the **heart** in **vertebrates** which empties into the ventricle. In this sense the term atrium is frequently replaced by auricle, although in strict terminology the auricle refers only to a small appendage of the atrium. (A.W.L.)

ATROPHY. Wasting of or decrease in size of a tissue. This occurs with any interference with the function or use of an organ or part. It may be temporary, as occurs following a fracture when for a period of time the affected limb can be but little used. It also occurs when nerves are injured or affected by disease, or as the result of circulatory accidents such as hemorrhage into the brain interfering with conduction along nerve pathways. If the damage is of permanent character, as results from infantile **paralysis, apoplexy,** or traumatic injury of nerves that cannot be repaired, the atrophy is progressive. With age tissues tend to atrophy, particularly skin, reproductive organs, and occasionally brain. (D.M.H.)

ATROPINE. An **alkaloid** derived principally from the leaves and root of **belladonna.** It is used to allay abnormal contraction and spasm of smooth muscle, especially the smooth muscle of the intestinal tract, and to check secretion of the skin, digestive and respiratory tracts. When given in adequate doses it produces dryness of the mouth and dilation of the pupil of the eye. Except in therapeutic doses, the drug is a poison. (R.S.M.)

ATTAR OF ROSES, OR ROSE OIL. An oil produced from a few kinds of roses, principally from *Rosa damascena.* Bulgaria leads other European countries in production.

In the preparation of the oil, the rose flowers are gathered early in the morning and immediately put into large copper stills. Water is added and then boiled until about 5 quarts have distilled over, carrying with it the rose oil. Fresh roses are added and distilled in the same water. The rose water thus obtained is redistilled separately, yielding the desired rose oil. Recently steam distillation has begun to replace this more primitive method. At best, about 2 tons of roses are necessary to produce one pound of the oil. Quite naturally, rose oil is very expensive, so adulteration by cheaper oils is a common practice. The oil is used principally in **perfumes.** (R.M.W.)

ATTENUATION FACTOR. This is the ratio of the input to the output current of a line or network. The attenuation factor of a series of connected circuits is the product of the individual attenuation factors. (L.R.Q.)

ATTENUATOR. The attenuator, often called a pad, is a network designed to introduce a definite loss in a circuit. It is designed so the **impedance** of the attenuator will match the impedance of the circuit to which it is connected, often being connected between two circuits of different impedance and serving as a matching network as well as an attenuator. It is distinguished from a simple resistance in that the impedance of an attenuator does not change for various values of its attenuation. It is a valuable unit in making many laboratory tests on communications equipment where it is used to adjust the outputs of two pieces of apparatus or for two different conditions so the relative merits may be determined from the attenuator setting. In much communication work it is desirable to transmit power at a higher level than will be used in order to overcome circuit noises, and then to reduce it to the proper value at the receiving end by a pad. It is usually calibrated in **decibels** and thus indicates the attenuation introduced by it. (L.R.Q.)

ATTRIBUTE. Qualitative Variable.

ATTRITION. From the Latin *attritio* meaning a grinding or rubbing down, is used in the terminology of geological science to refer to the grinding of particles through the transporting power of wind, running water, or by the movement of glaciers. (E.S.C.S.)

AUDIO AMPLIFIER. Amplifier.

AUDIO FREQUENCY. This term refers to the range of frequencies to which the human ear will respond. While the response of the ear varies from person to person, the range is approximately 20–20,000 cycles per second. Below 20 cycles the sound is normally heard not as a note, but as individual pulses. Most persons are unable to hear above 15,000 cycles, but a few people can hear even above 20,000. (L.R.Q.)

AUDITORY ORGANS. Organs sensitive to stimulation by sound waves. True auditory organs occur in **arthropods** and **vertebrates.** In the former they vary considerably but in the latter they are the ears and can be traced through their variations to a common structural foundation.

The simplest **arthropod** auditory organ is known as a chordotonal organ. It consists of a nerve ending with accessory cells connected with the body wall, which is apparently the immediate source of the vibrations to which the organ responds. More elaborate auditory organs are found in grasshoppers, **katydids, mosquitoes,** and related species. In the grasshoppers they are located on the sides of the first abdominal segment, in the katydids in the front tibiae, and in the mosquitoes at the base of the **antennae.** In all forms the **scolophore** is the essential sensory ending; accessory structures vary to a greater degree but usually include a modification of the **cuticula** which serves as a resonating membrane, or tympanum.

The essential auditory portion of the vertebrate ear is the cochlea, a spiral organ of elaborate structure containing terminations of the auditory nerve. This organ is part of the inner ear. In the mammals the outer ear includes the pinna, usually called the ear, and the external auditory canal leading inward to the tympanum or ear drum which vibrates in response to sound waves. Between these two regions lies the cavity of the middle ear, derived from the **pharynx** and connected with it by the **Eustachian tube.** The middle ear is bridged by a series of small bones, the hammer, anvil, and stirrup, which convey the vibrations of the tympanum mechanically to the liquid in the inner ear. These parts are variably developed in vertebrates below the mammals, all of which have simpler ears than described. The ears of **bats** play a unique part in the avoidance of obstacles during flight.

Vibrations ranging in frequency from 30 to 30,000 per sec. are perceived by man as sound. Other animals perceive higher or lower frequencies merging with variable pressures which must be regarded as tactile stimuli. (A.W.L.)

AUGEN-GNEISS. A gneissoid rock that contains **lenticular** crystals or mineral aggregates resembling "eyes." Derived from the German *augen,* eyes. (E.S.C.S)

AUGER BIT. Woodworking.

AUGITE. This mineral is a common **monoclinic** variety of **pyroxene** whose name is derived from the Greek word meaning luster, in reference to its shining cleavage faces. Chemically it is a complex metasilicate of **calcium, magnesium, iron** and **aluminum.** Color, dark green to black, may be brown or even white; hardness, 5–6; specific gravity, 2.93–3.49. Augite is important as a primary mineral in the **igneous** rocks and also as secondary mineral. The white augite is called leucaugite from the Greek word meaning white. Chemical analysis reveals this variety as containing little or no iron. Augite is of widespread occurrence. (E.S.C.S.)

AUGMENTATION. This term is used by astronomers to indicate the increase in apparent diameter of the **moon,** or any other object close enough to the earth to be observed as a disk, as the **altitude** of the object increases.

In Fig. (a) we have a representation of conditions for the object, *M,* on the horizon for an observer, *O;* while in Fig. (b) the object, *M,* is at the zenith on the

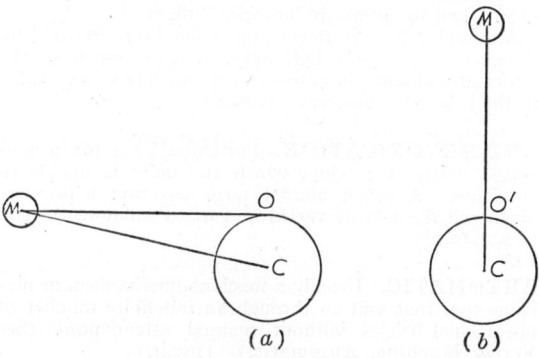

(a)   *(b)*

meridian for the observer, *O'.* In both figures, *C,* represents the center of the earth. The distance, *CM,* of the object from the center of the earth is assumed to be a constant. Examination of the figures will show at once that (a), conditions for rising object, gives the maximum distance of the object from the observer, while (b) gives the minimum value of this distance. Since apparent angular diameter of an object increases with decrease of distance, and since we define apparent size as the apparent angular diameter, the object is apparently larger under conditions (b) than in (a). In other words, the object is apparently larger on the **meridian** than when rising. Various attempts have been made, by psychologists and others, to explain the optical illusion that the contrary is the case and that the moon looks larger ("bigger than a house") when rising than when on the meridian.

For the **sun** and **planets** augmentation is too small to be considered except in the most refined observations of the altitude of a limb. However, in the case of the moon augmentation may amount to as much as 37". Failure to properly correct for this effect, when a limb of the moon is observed for determination of a **line of position** in **navigation,** might introduce an error as great as 0.3 miles in the position of the ship. (W.K.G.)

AUK. Aves, Charadriiformes. Marine birds (**Aves**) of several species related to the guillemots and puffins. They have a large compressed **beak** with oblique grooves toward the tip. One species, the **razorbill,** *Alca torda,* occurs on both sides of the Atlantic, nesting on rocky ledges. Another, the great auk or **garefowl,** was a flightless species which became extinct through wanton destruction about the middle of the 19th century. (A.W.L.)

AURA. A peculiar sensation which immediately precedes the onset of an epileptic **convulsion.** The aura most frequently arises in the epigastrium and is described as "a gone feeling" by patients. Numbness and tingling of an extremity or the face, palpitation, and dizziness are other forms. Psychic aura consists of feelings of apprehension, fright, horror. (D.M.H.)

AUREOLE. The contact **metamorphic** zone of varying width that often surrounds an **igneous** intrusion. Such areas of contact metamorphism often contain valuable ore deposits, especially when surrounding **batholiths** which have intruded sedimentary formations. (R.M.F.)

AURIC. GOLD.

AURICLE. 1. The outer ear (see **Auditory Organs**). 2. An appendage of the **atrium** of the mammalian **heart.** The term is frequently used as if it were synonymous with atrium. (A.W.L.)

AURICULAR CANAL. 1. The external auditory canal; **ear.** 2. The passage between the **atrium** and ventricle of the **heart.** (A.W.L.)

AURICULARIA. The form of dipleurula **larva** found in the **sea cucumbers.** It is more elongate and compact than the other types and the ciliated band follows an intricately curved path, outlining a conspicuous lobe before the mouth. (A.W.L.)

AURIGA. (The charioteer.) (Map, page 380.) This **constellation** is best known because it contains the bright star **capella** (the she-goat) and her kids. The kids are three fainter stars forming to the naked eye a small triangle and which always serve to distinguish Capella from other bright stars on a clear night. Capella is a bright star, yellowish in appearance, and of the same **spectral type** as our sun. The star, however, is so much larger than our sun that, in spite of its great distance (49 **light years**), it appears as 1st **magnitude,** whereas the sun at the same distance would be 6th magnitude, or barely visible to the naked eye on a clear moonless night. Capella is a spectroscopic **binary** with a period of 104 days. (W.K.G.)

AURIGNACIAN. Paleontology of Man.

AUROCHS. The wild ox of Europe, *Bos primigenius,* ancestor of domestic cattle. It is now extinct but in some parks of England are half-wild cattle supposed to be descended from this species. (A.W.L.)

AURORA BOREALIS. This well-known phenomenon of the upper atmosphere in middle and higher latitudes is now recognized as an electrical discharge in the ionized air, exhibiting, as it does, characteristic spectrum lines of the rarer atmospheric gases. The aurora appears in a variety of aspects, sometimes as a faintly luminous streak or arch, sometimes as bright streamers like a searchlight beam, sometimes resembling folds of a luminous curtain waved by the wind. Its intensity is greatest in an indefinite region apparently encircling the magnetic pole, toward which the streamers seem to converge. The occurrence of the phenomenon is intermittent, but with distinct evidence of several periodicities. Sunspot maxima, with their 11-year period, are always accompanied by maxima in the frequency and brightness of the aurorae. There are also smaller auroral maxima in March and October each year. Aurorae, like sunspots, are practically always attended by disturbances of terrestrial magnetism. A corresponding display in the southern hemisphere is called *aurora australis*.

The direct cause of the aurora is not known, but a widely accepted theory attributes it to electrons expelled with great velocity from the sun, especially during violent solar disturbances, and entering the upper terrestrial atmosphere, exciting the gases to luminosity like cathode rays in a Crookes or Geissler tube. Such high-speed particles would move in spiral paths along the lines of force of the earth's magnetic field, and hence converge toward the magnetic poles. (L.D.W.)

AURUM. Gold.

AUSCULTATION. This term is applied to the examination of the sounds within the chest, abdomen, heart, or larger blood vessels. It is carried out by listening with a stethoscope, or by applying the ear directly to the surface of the body. (R.S.M.)

AUSTEMPERING. Steel Austenite. Steel.

AUSTRALIAN RAT. Mammalia, Rodentia. Any of several species of the Australian region constituting the genus Hydromys. One large aquatic species is known as the beaver-rat. (A.W.L.)

AUSTRALOPITHECUS. Paleontology of Man.

AUTECOLOGY. The division of ecology that treats single factors or species in the association of organisms and environment, in contrast with synecology, in which the association of various factors and of species with species is considered. The subject of autecology is closely related with purely physical and chemical studies of environmental factors and with the physiological study of individual animals or plants. (A.W.L.)

AUTHIGENOUS, or AUTHIGENIC. A geologic term proposed by Kalkovsky in 1880, meaning generated on the spot, and referring particularly to the primary and secondary minerals of igneous rocks and the cements of sedimentary rocks. (R.M.F.)

AUTO STARTER. Starter.

AUTOCHTHONOUS. A geologic term proposed by Gümbel in 1888 for sedimentary rocks which have been formed in place. Now generally used to designate bedrock masses that have remained in place in a mountain belt, allochthonous masses that have been moved long distances. (R.M.F.)

AUTOCLASTIC. A term proposed by Van Hise in 1894 for crush breccias or fault breccias which have been fragmented in place. (R.M.F.)

AUTOCONVECTION LAPSE RATE. Undisturbed air will remain stratified even though the lapse rate exceeds the adiabatic rate of 5.5° F. per 1000′. If, however, the lapse rate becomes sufficiently large, density of the air will increase with altitude and will overturn. This critical lapse rate is 34.17° C. per km. or very nearly 19° F. per 1000′. Dust devils and whirlwinds result from this steep lapse rate which occurs at ground levels, particularly over concrete roads and sand and rock of deserts during the heat of day. (P.E.K.)

AUTOGAMY. A process of nuclear reorganization in protozoa in which the nucleus divides, each half undergoes a maturation, and the two persisting functional nuclei reunite. In the modified process known as paedogamy the individual forms a cyst within which it divides into two cells which reunite after the nuclear transformation is completed. (A.W.L.)

AUTOGENOUS. Self-generated—originating within the body. The term is usually applied to vaccines which are made from a patient's own bacteria as opposed to stock vaccines which are made from cultures grown from standard strains. (R.S.M.)

AUTOGYRO. The autogyro is an aircraft on which the lifting airfoil surface is not rigidly fixed to the body. In consequence of this fact, it has been possible to operate the airfoil with wind velocities greatly exceeding those of the airplane as a whole. As a result, the airfoil surfaces, or vanes, are a great deal smaller in area than those required for a conventional fixed-wing airplane. The autogyro was developed in Spain by Juan de Cievra, and first successfully flown in 1923.

The autogyro consists of a wingless fuselage mounting a pylon which contains a rotating head to which are affixed three or four balanced vanes of airfoil section. The angle of the rotor to the fuselage is controlled by the pilot and this takes the place of the normal control surfaces of the conventional airplane. The vanes rotate at speeds which give an average air velocity over them considerably in excess of the autogyro's air speed, and so the autogyro may be flown at speeds lower than the stalling speed of the airfoil section. This is a great advantage, as it permits the autogyro to land in small fields, with short landing runs, to descend almost vertically, and to approach "hovering" flight.

Although autogyro development has been retarded by recent successes with helicopter designs, much of the information gained in autogyro experiments was valuable to the helicopter pioneers. (F.T.M.)

AUTOINTOXICATION. Poisoning by a toxin generated within the body, which the body is unable to eliminate. A much abused term covering many undiagnosed diseases of varied and often unknown origin. (R.S.M.)

AUTOMATIC. Usually a machine, mechanism, or machine tool that will go through an indefinite number of operational cycles without manual attendance. (See Screw Machine, Automatic.) (H.C.H.)

AUTOMATIC FREQUENCY CONTROL. The automatic frequency control circuits are used in superheterodyne radio receivers to keep them tuned to a selected station in spite of random heating effects and other factors affecting the oscillator stability. Because of the method of reception used in these receivers, if the frequency of the local oscillator varies it produces a marked effect on the frequency response of the set,

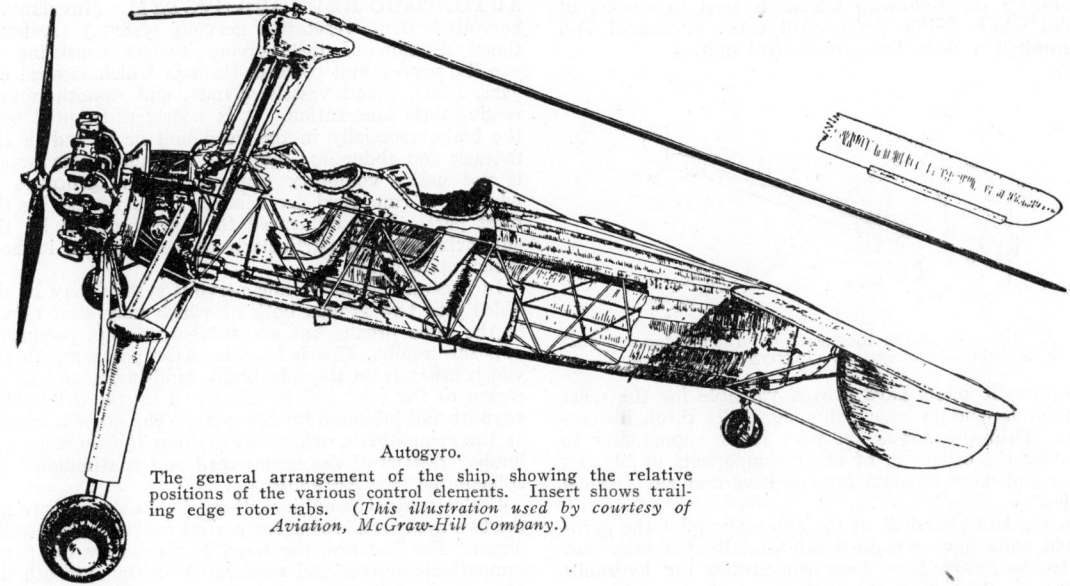

Autogyro.

The general arrangement of the ship, showing the relative positions of the various control elements. Insert shows trailing edge rotor tabs. (*This illustration used by courtesy of Aviation, McGraw-Hill Company.*)

Various random effects may cause this frequency to vary and hence the set apparently will get detuned. Automatic frequency control is designed to correct for these random effects and keep the local oscillator frequency at such a value that the intermediate frequency is kept constant. In this circuit the output of the intermediate frequency amplifier is fed into a **discriminator** as well as into the regular **detector**. The output of the discriminator is used to bias a **reactance tube** which is serving as part of the tuning circuit of the local oscillator. By proper adjustment of the circuit values the variation of the reactance tube can be made to keep the intermediate frequency constant within narrow limits and thus keep the receiver tuned. A similar automatic frequency control circuit is used to control the **carrier frequency** of some systems of frequency **modulation**. Here some of the output signal is fed into the discriminator which is adjusted so only the average or carrier frequency variation will produce a bias for the reactance tube. (L.R.Q.)

AUTOMATIC PILOT. This is a mechanism, the purpose of which is to control the movement of an airplane about its longitudinal, lateral, and vertical axes. The mechanism produces forces on the control system (see **Controls, Airplane Flight**) which act in parallel with those of the human pilot. The latter may at will overpower the automatic pilot, render it inoperative, or relinquish piloting of the airplane to it. The automatic pilot is primarily a mechanism designed to maintain the airplane in a constant heading (which may be set by the pilot) and in either level flight or constant climb or glide attitude. It does not perform flight maneuvers of an acrobatic character.

The automatic pilot has been brought to a high state of development, particularly in the United States. As might be expected from the complex nature of the action it performs, the mechanism is extensive and complicated. This, together with the demand for utmost reliability, raises design and manufacturing problems, the surmounting of which is to the credit of American scientific and mechanical talent. The mechanism is based on the principle of constancy of position of a universal mounted gyroscope and its action involves both pneumatics for gyroscope drive and primary signal, and hydraulics for the necessary large actuating forces. It is therefore seen to be a mechanical instrument.

A version of the automatic pilot made in large quantities in the United States consists essentially of two gyroscopic control units (one for pitch and bank control, the other for directional control), three servo units consisting of long stroke pistons in oil cylinders, one each for aileron, rudder, and elevator control, and numerous accessories for the supply of oil and air. The gyroscopes and their drives are mounted on the pilot's instrument panel where they provide a visual indication of the attitude and heading of the plane. The other elements are distributed where most convenient or practical, with connections between them being principally through pneumatic or hydraulic tubing.

A gyroscope is a spinning wheel which is universally mounted. A fundamental property is that the spinning axis tends to remain fixed in space, as is illustrated in Fig. 1. This property is called "rigidity," and is the

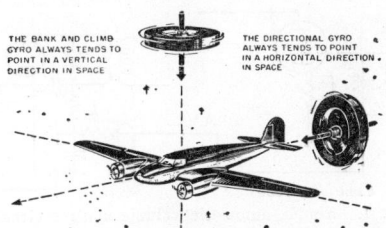

THE BANK AND CLIMB GYRO ALWAYS TENDS TO POINT IN A VERTICAL DIRECTION IN SPACE

THE DIRECTIONAL GYRO ALWAYS TENDS TO POINT IN A HORIZONTAL DIRECTION IN SPACE

Fig. 1.   The gyro in space.   (*Sperry Gyroscope Co.*)

basic principle in the operation of all automatic pilots. Any movement of the airplane about one of its three axes will move the case of the gyroscope but will not affect the position of its spinning axis. This relative movement between gyro and case is utilized to produce unbalanced air pressures on a diaphragm. The resultant diaphragm movement operates an oil relay valve which admits oil under pressure against the piston of a servo-motor, causing it to execute a stroke and move the proper control surface to a correcting position. Typical connections of the servo units into the flight control system are shown in Fig. 2, which, however, omits hydraulic lines for the purpose of clarity. In order to remove the applied control as the airplane is returning to its normal attitude, so that the control surface will be in its neutral position when the disturbance has been corrected, there must be a follow-up system which returns the mechanism to neutral position when the airplane has returned to the desired attitude.

In Fig. 2 the follow-up system is seen to consist of cables which follow the control cable movement and transmit it back to the gyro control unit.

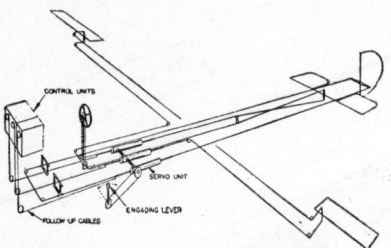

Fig. 2. Diagram of the incorporation of an automatic pilot in a flight control system. (*Sperry Gyroscope Co.*)

Automatic pilots have proven valuable for the relief of pilot fatigue on long flights, especially through overcast. Their use allows the pilot fuller opportunity to consider the operation of other components of his airplane and more freedom for executive and navigational duties.

In the latest versions of the automatic pilot the gyroscopic units have remained substantially the same but electronic means have been substituted for hydraulic means in picking up the indications of the gyroscope and converting its signals into actuation of the controls. Also, drive is electric rather than pneumatic. (F.T.M.)

AUTOMATIC VOLUME CONTROL.

An automatic volume control circuit is incorporated in a **radio receiver** to maintain the amplitude of the signal applied to the **detector** as nearly constant as possible and thus correct for variations in signal strength caused by **fading**. When a modulated radio-frequency wave is detected the output contains a d-c component which has a magnitude proportional to the amplitude of the **carrier** impressed on the detector. Other components (af and rf) are filtered out and this d-c voltage is applied as bias to the **variable mu** radio frequency amplifiers. Fig. 1 illustrates the essential components of such a system, the resistance and capacitance units

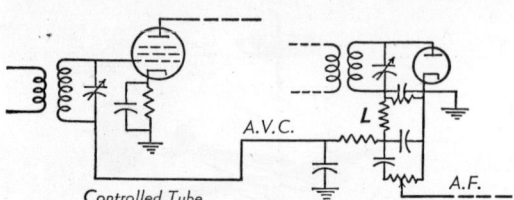

Fig. 1. Simple automatic volume control circuit.

indicated as *L* serving to separate the various components of the detector output and route them to the proper parts of the complete receiver. The d.c. for the avc is fed, as indicated, to the grid of the rf tubes, usually all being so biased although only one is shown. Thus a high value of applied signal gives a high value of d-c output which is applied to the variable mu tube, making its grid more negative and thus decreasing its gain and reducing the signal applied to the detector. A low value of carrier gives a lower value of d.c. which means less negative bias on the rf tube, hence more gain and increased signal to the detector. (L.R.Q.)

AUTOMOBILE. Motor Vehicle.

AUTOMOBILE ENGINE. Otto Cycle Engine.

AUTONARCOSIS.
Literally self-numbing. **Anaesthesia** resulting from accumulated products of the animal affected. (A.W.L.)

AUTONOMIC NERVOUS SYSTEM.
(Involuntary nervous system, vegetative nervous system.) A functional division of the nervous system consisting of ganglia, nerves, and plexuses through which visceral organs, heart, blood vessels, glands, and smooth muscle receive their innervation. It is widely distributed over the body, especially in the head and neck, and in the thoracic and abdominal cavities. The autonomic system is not under voluntary control and the processes in which it is concerned are beneath consciousness for the most part. It is influenced to a great degree by the **endocrine glands,** particularly the adrenal and its hormone, epinephrine.

In general the autonomic nervous system may be divided into two groups both of which may send nerves to the same organs but act antagonistically, producing opposite results. One is known as the parasympathetic, which arises from the mid-brain, hind-brain, and sacral region of the cord and is stimulated by the drug **pilocarpine** and inhibited by **atropine**. The other is known as the sympathetic, which arises from the thoracic and lumbar regions of the spinal cord and is stimulated by epinephrine (see **Adrenalin**).

Under normal conditions there is a balance between the two systems allowing for perfect function of a bodily organ. For instance, the heart is slowed by the parasympathetic system and accelerated by the sympathetic. Movement of the stomach is increased by the parasympathetic and is inhibited by the sympathetic. The pupil of the eye is contracted by the parasympathetic and dilated by sympathetic stimuli. (D.M.H.)

AUTO-OXIDATION. Oxidation.

AUTOPSY. Necropsy.

AUTO-ROTATION.
Auto-rotation, more commonly known as the tailspin, is a property of the motion of an **airplane** under a special set of circumstances. The attitude of an airplane, when spinning, is with the longitudinal axis rather steeply inclined to the horizontal; indeed, some airplanes spin almost "nose down." Furthermore, the airplane has a rolling motion which makes a vertical **helix** of the flight path during spinning. The peculiar significance of auto-rotation is that it is uncontrolled flight, and that recovery from it is not to be accomplished without the loss of 200–600′ of altitude; consequently, when spinning begins within this height above the ground level, a crash is very likely to ensue.

The action of auto-rotation is briefly described as follows. It is essentially a stalled maneuver, and can not be intentionally or unintentionally gotten into when the speed of the airplane exceeds that of its stalling speed. But when the airplane is flying at a speed equal to that at which it will **stall**, and a slight gust or change of angle of attack causes either a slight **roll** or a **yaw**, unless the pilot immediately and forcibly corrects the same to bring the ship again on the level keel with a slight increase of speed, the yaw will induce a roll (or the roll a yaw). It is the property of the conventional type of airplane that a roll produced by a yaw increases that yaw, and so results in an increasing roll. At the stalling speed **ailerons** become inoperative, and the pilot has no control which will permit him to correct this roll, so the airplane goes into a tight spiral dive which is almost immediately followed by the auto-rotative condition, in which the wings are operating at an **angle of attack** above that of the stall. Although the "twisted" wing (one whose angle of attack is not the same at all stations of the semispan) possesses some undesirable aerodynamic features, it has this virtue: if the center section has the greater incidence, it will burble before the tips, with sufficient loss of lift to cause the airplane to nose down to a greater speed and smaller angle of attack. The tips, remaining unstalled, tend to oppose the rolling tendency and any roll that develops may be counteracted by the ailerons which remain effective while the tip is unstalled. An opposite character of twist accentuates auto-rotation and may produce undesirable and dangerous spinning qualities in the airplane.

Recovery from auto-rotation is ordinarily accomplished by operation of the elevators and rudders, so as to decrease the angle of attack of the wings, allowing

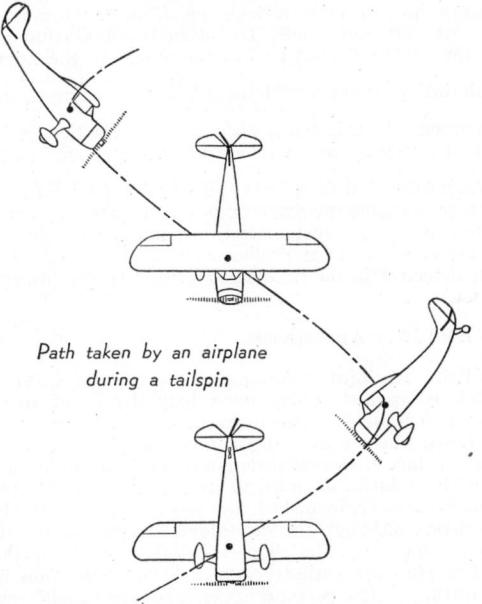

Path taken by an airplane during a tailspin

them to recover their lifting power, and giving to the tail surfaces the required maneuverable power to cause the pilot to be able to stop the rotation, and bring the airplane to an even keel. It has been possible to build airplanes which either will not spin, or which can be prevented from spinning at any time by the use of controls, especially of ailerons which are operative at the stall. (F.T.M.)

AUTOTOMY. Self-mutilation. Through the presence of a special modification near the base of the limb, some crustaceans and insects are able to drop off appendages by which they are seized. The autotomy of the arms of starfish and of the tail of lizards are other common examples. Autotomy is followed by regeneration. (A.W.L.)

AUTOZOOID. Members of **polyp** colonies whose function is to feed the colony. (A.W.L.)

AUTUNITE or CALCO-URANITE. This mineral is a hydrous **phosphate** of **calcium** and **uranium**, crystallizing in the **orthorhombic system**, usually in thin tabular crystals. Good basal cleavage; hardness, 2–2.5; specific gravity, 3.1; luster, subadamantine to pearly on the base; color, lemon yellow; streak, yellow; transparent to translucent; strongly fluorescent.

Originally from near Autun in France, whence the name, it is a secondary mineral associated commonly with **uraninite**. In the United States occurs sparsely in the pegmatites of Connecticut, New Hampshire and North Carolina. (E.S.C.S.)

AUXILIARY PROJECTION. Orthographic Projection.

AUXINS. The name given to a group of growth-regulating substances occurring in plants which have profound effects on the elongation of plant cells and other important growth phenomena. Three such compounds have actually been isolated from plant tissues: auxin a

$(C_{18}H_{32}O_5)$, auxin b $(C_{18}H_{30}O_4)$, and heteroauxin $(C_{10}H_9O_2N)$. In addition to these compounds known to be present in plants a number of synthetic compounds are known to exert similar or identical effects on plants. These are also often referred to as auxins and for convenience will be discussed under this heading in this treatment. The auxins exert their effects on the growth of plants when present in exceedingly minute quantities. Their effects are typical of **hormones** and the auxins are one of the important groups of plant hormones.

The first effect of the auxins to be discovered, and probably their most fundamental influence on plant growth, is that on the elongation of cells. During the growth process all plant cells go through the successive stages of cell division, cell elongation (or enlargement), and cell maturation. Auxins are absolutely essential for the occurrence of the elongation stage of growth. In their absence elongation or enlargement of cells ceases. In their presence elongation occurs if all other factors are also favorable for growth and, within limits, the elongation is proportional to the auxin concentration. Relatively high concentrations of auxin, however, inhibit cell elongation, the optimum concentration for cell elongation being different in different tissues. Concentrations which favor elongation of cells in stems, for example, may completely inhibit the elongation of cells in roots.

Another effect of the auxins is on root formation. More roots form, and more quickly, on the cuttings of many kinds of plants if they are first immersed in a dilute solution of one of the auxins before placing them in the rooting medium than if they are not so treated. Indolebutyric acid and alphanaphthalene acetic acid, both synthetic auxins, are especially effective in inducing root formation. The hormone may be applied to the cuttings in a solution, in a paste, as a vapor, or in a powder. Immersion for about 24 hours in a solution containing 5 to 20 mg. of either of these two substances per 100 cc. of water is effective in favoring root formation on many kinds of cuttings.

Fruit development on plants normally occurs only after **pollination** and subsequent **fertilization** have taken place. If pollination is prevented and any one of several of the auxins is introduced into the pistil by a suitable technique, development of a **parthenocarpic** fruit will result. Such fruits are seedless. During the winter season of low pollen viability treatment of greenhouse tomato flowers with indolebutyric acid results in a larger set of fruit.

The **abscission** of leaves, fruits and other plant parts from the stems can also be influenced by auxins. Artificially introducing auxin into the petioles of leaves at a time when they would normally abscise within a short period often appreciably delays the time required for the leaves to fall. Fruit fall can also be delayed by spraying with a solution of alphanaphthalene acetic acid shortly before harvest time. This practice is now widely followed in apple orchards. Application of the hormone spray delays abscission of the fruits and lengthens the period during which the apples can be picked from the tree.

Auxins also play an important part in the phenomena of phototropism and geotropism (see **Movement in Plants**). (B.S.M.)

AUXOSPORE. An auxospore is a special type of **spore** which occurs in **diatoms** and which seems to be a means of rejuvenating the cells. Rejuvenescence is necessary, since in the normal process of **cell division** one of the two daughter cells is always smaller than the parent cell. Consequently very small cells are ultimately formed. In some species of diatoms, auxospore formation is preceded by the escape of the protoplast from the walls of the cell. The free **protoplast** then enlarges and secretes about itself a wall. In time new valves more or less like those of the original diatom are formed. In other species of diatoms, auxospore formation is preceded by the union of the protoplasts of two similar diatom cells, the process being therefore sexual. (R.M.W.)

AVAHI. Mammalia, Primates. The woolly **lemur,** *Avahis laniger,* of Madagascar. (A.W.L.)

AVAILABLE ENERGY. Energy which can be converted into mechanical work by means at human disposal. While incalculable quantities of **energy** exist all about us, only an insignificant fraction of it is in such form that human invention has been able to utilize it for the performance of work. For example, water stored behind a dam has a supply of potential energy, some portion of which can be made to drive our machinery as it descends to the sea. But when it reaches the ocean level, though it still possesses energy, what is left is not available for use, because it cannot flow to a lower level as it might if the ocean basin were empty. Again, there is an abundance of the kinetic energy called **heat,** since the air, the ground, and bodies all about us are at temperatures far above **absolute zero.** But the second law of **thermodynamics** requires that to utilize any of this supply we must have a region colder than these bodies, into which heat would naturally flow from them. And even when such a region is at hand, as in the case of the relatively cool atmosphere surrounding an engine boiler, our best engines manage to capture only a small percentage of the thermal energy on its way from the hotter to the colder body.

A most disconcerting aspect of the subject is the fact that even if energy is available, it is not content to remain so until we are ready to use it, but takes every opportunity to escape and become unavailable. This is equivalent to the **least energy principle,** which may be expressed by saying that a system cannot be in stable **equilibrium** until it has got rid of all the available energy that it can. Any process in which available energy thus becomes unavailable is said to involve "degradation" or "dissipation" of energy. (See **Entropy.**)

Cosmic physicists long have recognized that the continued operation of this dissipative principle can result in only one ultimate condition, namely, that the entire universe will become in the end an absolutely cold, motionless lump of matter (if, indeed, the "proper energy" composing matter itself does not succumb, in which case there would be no matter). This fate, significantly expressed by the German term *Wärmetod* ("heat death") must surely overtake the cosmos unless, as some think, compensating influences are operating somewhere to prevent it. (L.D.W.)

AVENA. Oats.

AVERAGE. An average is a single value found by some mathematical process from a distribution of **variates.** It is in some way typical of the whole distribution.

Instances of simple averages are the **measures of central tendency.** More complex instances of averages are the **moments** of a distribution. (L.A.A.)

AVERAGE DEVIATION. Average deviation or mean deviation, M.D., is a measure of **absolute variability** in which the **deviations** are generally referred to the mean. It is defined by the formula

$$M.D. = \frac{\Sigma \mid x_i - \overline{X} \mid}{N}$$

$$= \frac{\mid x_1 - \overline{X} \mid + \mid x_2 - \overline{X} \mid + \cdots + \mid x_N - \overline{X} \mid}{N}$$

in a serial distribution where x_i are the variates, $\overline{X}$ the mean, and $\mid x_i - \overline{X} \mid$ is the absolute **deviation** of x_i from $\overline{X}$, and by

$$M.D. = \frac{\Sigma f_x \mid x - \overline{X} \mid}{\Sigma f_x}$$

$$= \frac{f_1 \mid x_1 - \overline{X} \mid + f_2 \mid x_2 - \overline{X} \mid + \cdots + f_N \mid x_N - \overline{X} \mid}{f_1 + f_2 + \cdots + f_N}$$

in a **frequency distribution** where f_i is the frequency in the class whose class mark is x_i. It shows on the average how far the variates are from the mean. Its uses are very restricted. The **standard deviation** is by far the preferred measure of dispersion. In the **normal probability function** $M.D. = \sqrt{\frac{2}{\pi}} \; \sigma_x = .7979\sigma_x$, $\sigma_x =$ the standard deviation. The average deviation is least when the deviations are referred to the median. (L.A.A.)

AVERAGE OUTGOING QUALITY LEVEL. The average outgoing quality level is the average per cent defective in the product after inspection, which shall not be exceeded no matter what may be the level of per cent defective in the product submitted to the inspector. (L.A.A.)

AVERTIN. Anaesthesia.

AVES. The birds. A class of the phylum **Chordata** which is marked chiefly by a high degree of specialization for flight. The great beauty of many species of birds, their songs, their interesting nesting habits, and the fact that few regions are so inhospitable as to be without birds, have led to wide interest in the group.

Birds are distinguished by several structural characteristics, although the first alone is sufficient for their recognition. 1. The skin is clothed with feathers. 2. The jaws are ensheathed in a horny **beak** and bear no teeth. 3. The pectoral appendages are usually modified for flight, forming **wings,** although they are rudimentary in some species and aid in swimming in some. 4. The **skeleton** is made rigid by the fusion of bones. 5. The **heart** is 4-chambered. 6. Birds are warmblooded (Homoiothermal).

As is true of all extensive groups, the birds are very diverse in habits. They are both herbivorous and carnivorous and are further specialized as seed-eating, fruit-eating, insect-eating, fish-eating, and other types. They are also specialized as swimming, wading, walking, running and diving forms, in addition to their usual ability in the air, and in a few cases they burrow effectively. Their nesting habits also vary remarkably and the construction of the nest is in many cases a source of wonder.

The seasonal **migrations** of birds are almost unique. No other group of animals is so generally characterized by this tendency. The subject has been widely studied and has aroused much speculation without being clearly understood. It is obviously correlated with seasonal variation in the food supply and with climatic conditions, and is made possible by high specialization for flight, but exact knowledge of cause and effect in migration is lacking.

The economic importance of birds is great, and is chiefly to their credit. Insect-eating species destroy countless pests and seed-eating species aid in checking the spread of many weeds, although they may also rob the farmer of a small part of his crops. Scavengers like the turkey **vultures** are useful, although the degree of their usefulness is difficult to estimate. On the other hand a few **hawks** and **owls**—and only a few—do some harm by destroying useful birds and the **crow** is given a very bad reputation by conservation experts as a robber of the nests of other birds. It is scarcely necessary to mention the value of birds as food and game. The domestic species, **chickens, ducks, turkeys, geese,** are too well known as food, and their eggs are too common a culinary material to be readily overlooked.

Probably because specialization for flight overshadows other adaptations, the classification of birds has been subject to some difficulty. The birds are divided into two subclasses by some writers, the Ratitae including flightless birds whose sternum is without the deep keel to which the powerful flight muscles are attached, and the Carinatae with a keeled sternum. These divisions are not, however, clean cut, hence the classification in twenty-five orders now in common use is given below.

Order Struthinioformes. Ostriches.
Order Rheiformes. Rheas.
Order Casuariiformes. The emu and cassowary.
Order Crypturiiformes. The tinamous.
Order Apterygiformes. Kiwis.
Order Sphenisciformes. Penguins.
Order Gaviiformes. The loons.
Order Colymbiformes. Grebes.
Order Procellariiformes. Albatrosses and petrels.
Order Pelecaniformes. Pelicans, gannets, darters, cormorants, etc.
Order Ciconiiformes. Herons, spoonbills, flamingos, storks, bitterns, etc.
Order Anseriformes. Swans, geese, ducks, etc.
Order Falconiformes. Eagles, falcons, hawks, vultures, etc.
Order Galliformes. Turkeys, pheasants, etc.
Order Gruiformes. Cranes, rails, gallinules, coots.
Order Charadriiformes. Gulls, plovers, curlews, auks, etc.
Order Columbiformes. Pigeons.
Order Psittaciformes. Parrots and related species.
Order Cuculiformes. Cuckoos, etc.
Order Strigiformes. The owls.
Order Caprimulgiformes. The goatsuckers, including the whip-poor-will and nighthawk.
Order Micropodiformes. Swifts and hummingbirds.
Order Coraciiformes. Kingfishers, hornbills, etc.
Order Piciformes. Woodpeckers, toucans, etc.
Order Passeriformes. An immense order including over half the known species of birds, among them the more familiar land species. The thrushes, sparrows, warblers, swallows, flycatchers, larks, wrens, titmice and many others. (A.W.L.)

AVIARY. A building or other enclosure for the maintenance of birds in captivity. (A.W.L.)

AVICULARIUM. A modified individual resembling the head of a bird, occurring in bryozoan colonies. (A.W.L.)

AVIGATION. The term avigation was introduced some years ago to describe the general subject of air navigation. The general methods of sea and air navigation are so closely interwoven at the present time that the term avigation has been virtually abandoned. In this encyclopedia the various topics of air and sea navigation are discussed together under the specific titles used by navigators in general. (W.K.G.)

AVITAMINOSIS. A state of malnutrition resulting from deficiency of vitamins in the diet, or inability to assimilate them through the gastrointestinal tract. Such deficiency of the different vitamins causes various diseases, such as beri-beri, scurvy, etc. Extreme avitaminosis, untreated, may cause death. (See Vitamins.) (D.M.H.)

AVOCADO or ALLIGATOR PEAR. *Persea americana.* Lauraceae. This tree, a native of the lowlands of tropical America, has been extensively cultivated in tropical and sub-tropical regions. It has been introduced into Florida and California, where, due to its nonhardy character, it is often injured by frosts. It is an attractive tree with large oval to ellipical leaves and small yellowish flowers. The large green to brownish fruit varies in shape from nearly spherical to pearshaped. It is extremely nutritious and rich in oil, and is becoming increasingly popular in American markets.

The thick yellowish flesh of the fruit has only a faint flavor, and therefore is usually served with salt, vinegar, or oil. (R.M.W.)

AVOCET. Aves, Charadriiformes. *Recurvirostra.* Wading birds (Aves) of several species found in the Old and New World. The beak is curved upward at the tip and the feet are fully webbed. (A.W.L.)

AVOGADRO'S LAW. The well-recognized principle known by this name was originally a hypothesis suggested by the Italian physicist Avogadro in 1811, to explain the puzzling rule of proportional volumes observed in chemical reactions of gases and vapors. It states simply that equal volumes of all gases and vapors at the same temperature and pressure contain the same number of molecules. Though this assumption accords with the facts and aids the kinetic theory of gases, just why it should be true is by no means self-evident; unless one starts with the much more recent Maxwell-Boltzmann law of equipartition of energy, which also requires proof. That Avogadro's law is true cannot be said to have been positively established until the experiments of J. J. Thomson, Millikan, Rutherford, and others determined the value of the electron as an electric charge and thereby made it possible to count the number of atoms of different elements in a gram. It is now known that 1 cc. of any gas at normal temperature and pressure consists of close to 2.687×10^{19} molecules (Loschmidt's number), and that the number of molecules in a mole of gas is 6.023×10^{23} (Avogadro's number). (See Chemical Composition.) (L.D.W.)

AWN. An awn is a slender projection found on the lemmas of many grasses. In some forms the awn arises from the tip of the lemma; in others from a point near the base. Commonly they are barbed, and not infrequently spirally twisted. They are particularly well developed in barley. (R.M.W.)

AXES OF AN AIRCRAFT. Three fixed lines of reference, usually centroidal and mutually perpendicular. The horizontal axis in the plane of symmetry, usually parallel to the axis of the propeller, is called the longitudinal axis; the axis perpendicular to this in the plane of symmetry is called the normal axis; and the third axis perpendicular to the other two is called the lateral axis. (See Airplane.) (F.T.M.)

AXIAL GRADIENT. An axis of organization characterized by progressive metabolic (see Metabolism) dominance. As formulated by C. M. Child the bilaterally symmetrical body is organized on a primary axis from head to tail, a secondary axis from dorsal to ventral and paired tertiary axes from the median plane to the lateral extremities. In any axis, beginning at the point mentioned first, and in the entire body proceeding from axis to axis in the order named, metabolic dominance is evident in functions and in development over lower levels or subordinate axes. (A.W.L.)

AXIAL LOAD. Load.

AXIAL ORGAN. An organ of peculiar structure and unknown function found near the axis of the body in all echinoderms except the sea cucumbers. (A.W.L.)

AXIAL SINUS. A portion of the body cavity in echinoderms into which the pores of the madreporite open. (A.W.L.)

AXIAL STRESS. Stress.

AXIL. The angle between the upper side of a leaf and the stem to which the leaf is attached is called the axil of the leaf. (R.M.W.)

AXILLA. The armpit. (D.M.H.)

AXINITE. This mineral is an **aluminum-boron-calcium silicate** with **iron** and **manganese.**

Crystallizes in the **triclinic system** yielding broad sharp edged forms, which has led to its name, derived from the Greek word meaning axe.

Axinite breaks with a **conchoidal** fracture; hardness, 6.5–7; specific gravity, 3.27–2.29; luster, vitreous; colors, brown, blue, yellow and gray. Transparent to translucent. Occurs in granites or more basic rocks along contacts and in cavities in Saxony, Switzerland, France, England, Tasmania and Japan. In the United States, in New Jersey, Pennsylvania, and California. (E.S.C.S.)

AXIS OF INSTANTANEOUS ROTATION. Dynamics of Rotation.

AXLE. An axle is a support for the rotation of wheels. In the dead axle, the wheel turns on the axle which is inserted in the hub portion, and forms the center of rotation. The contact surface between the wheel and the axle forms the bearing surface. In the live axle type the wheel is rigidly fixed to the axle which turns in bearings. The distinction between shaft and axle is this: a shaft is, in general, the support of

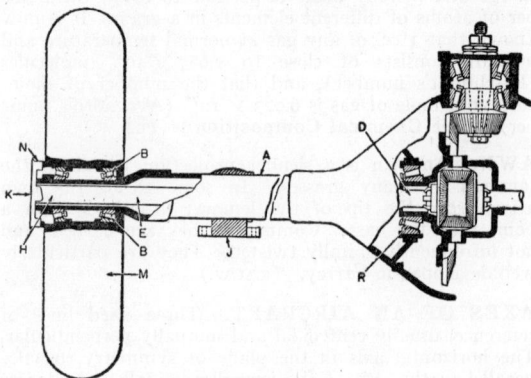

Full-floating axle.

rotating objects, whereas axle is more definitely applied to a shaft used with wheels.

The powered axle of an automobile offers an example of the live type. Three types of axles are shown in the illustration. They are, respectively, the full-floating,

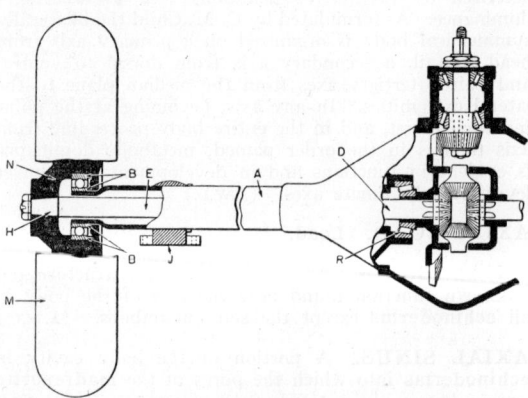

Three-quarter floating axle.

the three-quarter floating, and semi-floating axles. In the full-floating type the wheel is supported entirely on the axle housing, *A,* and the axle shaft, *E,* transmits **torque** only. The axle shaft, *E,* needs to be positively connected to the wheel, but not rigidly. The three-quarter floating axle has the wheel supported on one set of bearings on the axle housing, and the shaft *E*

must be rigidly fixed in the hub of the wheel to maintain alignment. Except when rounding turns, or when on roads which are not level, the shaft transmits pure torque only. The semi-floating axle is rigidly fixed in the wheel, and the axle shaft rotates on bearings in

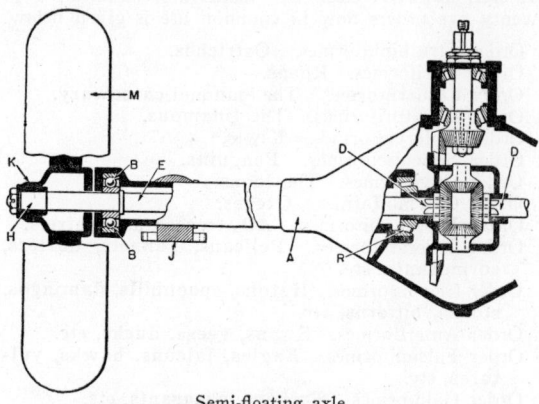

Semi-floating axle.

the axle housing. The axle shaft in this case is in bending as well as **torque,** since it must support all of the weight which is transferred to its wheel. (F.T.M.)

AXOLOTL. Amphibia, Urodela. A **salamander,** *Ambystoma tigrinum,* found near Mexico City which, although related to some of the terrestrial salamanders, retains its **larval** form throughout life, becoming sexually mature in this stage. Under experimental conditions the animal has been caused to undergo the usual **metamorphosis.** (A.W.L.)

AXON. Neuron.

AYE-AYE. Mammalia, Primates. A lemur, *Chiromys madagascariensis,* of Madagascar, resembling a squirrel in form, with large ears and a bushy tail. (A.W.L.)

AZALEA. Heath Family.

AZIDES. Hydrazoic Acid and Azides.

AZILIAN. Paleontology of Man.

AZIMUTH. In astronomy, the azimuth of a celestial object is that coordinate of the **horizontal coordinate system** which is measured in the plane of the **horizon** to the point where the **vertical circle** of the object cuts the horizon. Astronomical azimuth is measured from the south to the right (west) through 360°.

Astronomical azimuth may be computed by solving the **astronomical triangle,** provided three other parts are known. In most cases the **latitude** of the observer, the **hour angle,** and **declination** of the object are the known parts. In case the **longitude** of the observer is not accurately known, the **altitude** of the object may be obtained and combined with latitude and declination for computing azimuth.

In **navigation,** azimuth is of great importance for the determination of **compass corrections** and for plotting **lines of position** from celestial objects. Navigators do not use astronomical azimuth, but instead the **bearing** of the object, which is the **direction** measured from the **north.** Tables are published by the various governments, and by private individuals, which give the bearing of the sun in terms of local time, date, and latitude; and the bearings of other celestial objects in terms of declination, hour angle, and latitude.

In surveying, terrestrial azimuth of a mark is usually determined with an **altazimuth** instrument or surveyor's **transit.** The difference in azimuth between the vertical circle through some celestial object and the mark is measured and combined with the known azimuth of the object to obtain the azimuth of the mark. The object most commonly used for this purpose is **Polaris** (the north star), since this star is within 1° of the pole of rotation and its azimuth changes very slowly with time. Tables are published in the Nautical **Almanac** and a variety of other places which give the azimuth of Polaris in terms of local date and time. When using the north star for ordinary surveying purposes, the local time is only needed to within about 5 minutes, but for precise geodetic work the time must be known to within a few seconds.

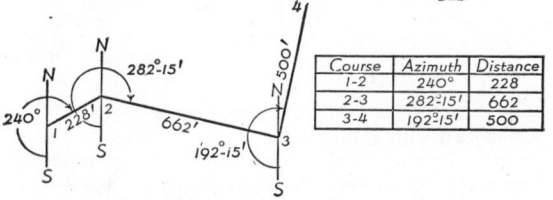

Course	Azimuth	Distance
1-2	240°	228
2-3	282°-15'	662
3-4	192°-15'	500

Plotting of a traverse.

Surveyors frequently run a **traverse** using azimuths and distances. The plotting of a traverse by this method is shown in the accompanying figure. (w.k.g.)

AZIMUTH CIRCLE. Alidade.

AZINES. Pyridine and Related Compounds.

AZO-, DIAZO-, AND RELATED COMPOUNDS.
Compounds related to aniline, either directly or by oxidation, and to nitrobenzene, by reduction are numerous and important. (See **Aniline; Hydrazines; Hydroxylamines; Nitro- and Nitroso-compounds.**) Azo- and Diazo-compounds are considered here. See table at end of this article.

When **nitrobenzene** is reduced in the presence of **hydrochloric acid** by tin or iron, the product is aniline (colorless liquid); in the presence of water by zinc, the product is phenylhydroxylamine (white solid); in the presence of methyl **alcohol** by sodium alcoholate or by **magnesium** plus **ammonium** chloride solution, the product is azoxybenzene (pale yellow solid); by **sodium** stannite, or by water plus sodium amalgam, the product is azobenzene (red solid); in the presence of sodium hydroxide solution by zinc, the product is hydrazobenzene (pale yellow solid). The behavior of other nitro-compounds is similar to that of nitrobenzene.

Hydrazobenzene is converted by oxygen of the air or by ferric chloride solution into azobenzene, and by strong acids into benzidine hydrochloride ($(4')H_2N \cdot C_6H_4 \cdot C_6H_4 \cdot NH_2(4) \cdot HCl$). Benzidine is prepared by reducing nitrobenzene to hydrazobenzene as above, and then treating the product with acid. Benzidine and its toluene relative, orthotolidine, are important intermediates for dyes. The counterpart of aniline is toluidine in its three forms, ortho, meta, para.

Azoxybenzene is converted by distillation with iron into azobenzene (ferrous oxide also formed), and by concentrated sulfuric acid warm into para-hydroxyazobenzene ($(4)HO \cdot C_6H_4N:NC_6H_5$), which is a dye.

Diazonium salts are usually colorless crystalline solids, soluble in water, moderately soluble in alcohol, and when dry are violently explosive by percussion or upon heating. These salts are generally used in cold (near 0° C.) acid solution, without separation of the salt, and are prepared by reaction of the desired benzenoid primary **amine** with **nitrous acid** (from **sodium** nitrite plus **hydrochloric acid**). Alkyl amines with nitrous acid yield the corresponding alcohol.

(A) By treatment of benzene diazonium chloride $\left(\begin{matrix} C_6H_5N-Cl \\ \cdots \\ N \end{matrix} \right)$ solution with **silver** oxide, or of the diazonium sulfate solution with **barium** hydroxide, the hydroxide (benzene diazo hydroxide, $C_6H_5N:N-OH$)

is obtained, which is intermediate in basicity between ammonium hydroxide and sodium hydroxide. Most diazo hydroxides are unstable, and are spontaneously transformed into nitrosoamines (group $-NH \cdot NO$), yellow neutral compounds. With sodium hydroxide, diazonium salt solutions yield sodium benzene-diazoate, more active chemically when first formed than upon standing, due to change from syn-diazoate $\left(\begin{matrix} C_6H_5N \\ \cdots \\ NaON \end{matrix} \right)$, which evolves nitrogen readily, to anti-diazoate $\left(\begin{matrix} C_6H_5N \\ \cdots \\ NONa \end{matrix} \right)$ which is more stable. Sodium benzene-syn-diazoate reacts with phenols in alkaline solution to give azo-dyes, e.g., para-hydroxyazobenzene, wherein hydrogen para (or ortho but not meta) to the hydroxyl group is reactive. Other reactions of diazonium salts are:

(B) Replacement of the diazo-group with loss of **nitrogen**, (1) by hydroxyl-group, forming **phenols** by warming with water, (2) by alkoxy-group, forming ethers, by warming with an **alcohol**, (3) by acyl-group, forming **esters** with an acid, (4) by **hydrogen**, forming **hydrocarbons** or substituted hydrocarbons, e.g., tribromobenzenediazonium chloride forms tribromobenzene, with alcohol or sodium stannite solution, (5) by **chlorine**, forming, for example, chlorobenzene, by warming with **cuprous** chloride (Sandmeyer's reaction), (6) by **bromine**, forming, for example, bromobenzene, by warming with cuprous bromide (Sandmeyer's reaction), (7) by **iodine**, forming, for example, iodobenzene, by warming a solution of the diazonium iodide, (8) by **cyanide**, forming, for example, cyanobenzene, by warming with cuprous cyanide (Sandmeyer's reaction), (9) by fluorine, forming, for example, fluorobenzene, by warming a solution of the diazonium borofluoride. Gatterman's modification of Sandmeyer's reactions above is to use copper powder plus the corresponding sodium or potassium salt instead of the cuprous salt. Nitro-compounds, since they are readily reduced to the corresponding amine, form an important group of compounds for the preparation of various derivatives by means of the diazo-reaction.

(C) Reduction, to form the corresponding **hydrazine**, for example, benzene diazonium chloride to form beta-phenylhydroxylamine hydrochloride.

(D) Bromination followed by treatment with **ammonia** to form azides, for example, benzene diazonium bromide plus bromine forms benzene diazonium perbromide ($C_6H_5Br_3$), which upon treatment with ammonia forms phenylazide

$$\left(C_6H_5N \diagdown \begin{matrix} N \\ \cdots \\ N \end{matrix} \right).$$

(E) Amines, (1) primary or (2) secondary amine to form diazoamino-compounds, e.g., benzenediazonium chloride (a) plus aniline forms diazoaminobenzene, benzene diazoaniline ($C_6H_5N:N-NHC_6H_5$), (b) plus methylaniline forms benzenediazomethylaniline

$$\left(C_6H_5N:N-N \diagdown \begin{matrix} C_6H_5 \\ CH_3 \end{matrix} \right).$$

Diazoamino-compounds readily change into aminoazo-compounds upon standing in alcohol solution or in the presence of amine hydrochloride, thus, benzenediazoaniline changes into para-amino-azobenzene, benzeneazoaniline-4 ($C_6H_5N:NC_6H_4NH_2(4)$), benzenediazomethylaniline changes into methyl-para-aminobenzene, benzeneazomethylaniline-4

$$\left(C_6H_5N:NC_6H_4N \diagdown \begin{matrix} H \\ CH_3 \end{matrix} (4) \right),$$

(3) tertiary amine to form aminoazo-compounds directly, e.g., benzenediazonium chloride plus dimethyl-

AZO-, DIAZO-, AND RELATED COMPOUNDS

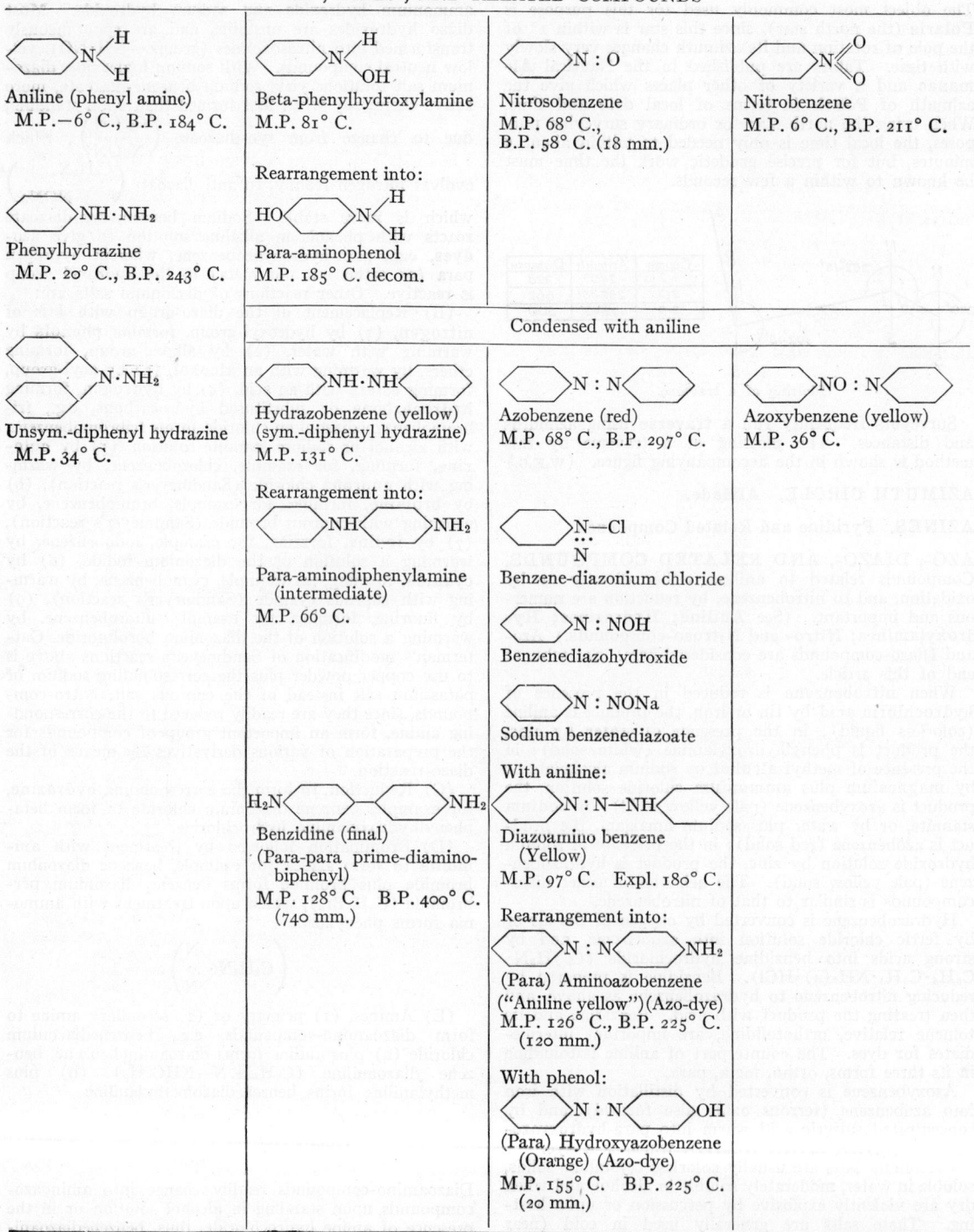

Aniline (phenyl amine)
M.P. −6° C., B.P. 184° C.

Phenylhydrazine
M.P. 20° C., B.P. 243° C.

Unsym.-diphenyl hydrazine
M.P. 34° C.

Beta-phenylhydroxylamine
M.P. 81° C.

Rearrangement into:

Para-aminophenol
M.P. 185° C. decom.

Nitrosobenzene
M.P. 68° C.,
B.P. 58° C. (18 mm.)

Nitrobenzene
M.P. 6° C., B.P. 211° C.

Condensed with aniline

Hydrazobenzene (yellow)
(sym.-diphenyl hydrazine)
M.P. 131° C.

Rearrangement into:

Para-aminodiphenylamine
(intermediate)
M.P. 66° C.

Benzidine (final)
(Para-para prime-diamino-
biphenyl)
M.P. 128° C. B.P. 400° C.
(740 mm.)

Azobenzene (red)
M.P. 68° C., B.P. 297° C.

Benzene-diazonium chloride

Benzenediazohydroxide

Sodium benzenediazoate

With aniline:

Diazoamino benzene
(Yellow)
M.P. 97° C. Expl. 180° C.

Rearrangement into:

(Para) Aminoazobenzene
("Aniline yellow")(Azo-dye)
M.P. 126° C., B.P. 225° C.
(120 mm.)

With phenol:

(Para) Hydroxyazobenzene
(Orange) (Azo-dye)
M.P. 155° C. B.P. 225° C.
(20 mm.)

Azoxybenzene (yellow)
M.P. 36° C.

---

aniline forms dimethyl-para-aminoazobenzene, benzene-azodimethylaniline-4 $\left(C_6H_5N:NC_6H_4N\genfrac{}{}{0pt}{}{CH_3}{CH_3}(4) \right)$. In this manner are prepared the azo-dyes, which contain the chromophore azo-group (—N:N—) plus an auxochrome amino-group $\left(-NH_2, -N\genfrac{}{}{0pt}{}{H}{CH_3}, -N\genfrac{}{}{0pt}{}{CH_3}{CH_3} \right)$.

The simplest azo-dyes are yellow, but by increasing the number of auxochrome groups or by increasing the percentage of carbon, the color darkens to red, violet, blue, and in some cases brown. Naphthalene residues darken to red, violet, blue and finally black. These amino-azo-dyes, together with the hydroxyazo-dyes (containing auxochrome hydroxyl-group —OH), are generally only slightly soluble in water. In order that the dye may be soluble it is desirable that it contain one or more sulfonic acid groups (—SO₂OH). This group may be introduced either by treating the dye with concentrated sulfuric acid, or by using sulfonic acid derivatives in preparing the dye, e.g., methyl orange, sodium dimethyl-para-aminoazobenzene-para-sulfonate $((4)$ $(CH_3)_2N$ $C_6H_4N:NC_6H_4SO_2ONa$ $(4))$ from dimethylaniline and diazotized sulfanilic acid (para-amino-benzene sulfonic acid, (1) $H_2N·C_6H_4·SO_2OH$ $(4))$, and then the sodium salt made from the product. Other azo-dyes are

chrysoidine $\left(C_6H_5N:NC_6H_3\diagdown^{NH_2(2)}_{NH_2(4)}\right)$

Bismarck brown $\left((3)H_2N·C_6H_4N:NC_6H_5\diagdown^{NH_2(2)}_{NH_2(4)}·HCl\right)$

Congo red $\left(^{(4)}_{(1)}{}^{HOO_2S}_{H_2N}\diagdown C_{10}H_5N:NC_6H_4.\right.$

$\left. C_6H_4N·NC_{10}H_5\diagdown^{SO_2OH(4)}_{NH_2(1)}\right)$

(R.K.S.)

AZOLES. Pyrrole and Related Compounds.

AZOXY- COMPOUNDS. Aniline; Azo-, Diazo- and Related Compounds.

AZURITE OR CHESSYLITE. This mineral is a basic **carbonate** of **copper**, $2CuCO_3·Cu(OH)_2$ so called from its beautiful azure-blue color. It is a brittle mineral with a **conchoidal** fracture; hardness, 3.5–4; specific gravity, 3.77–3.89; luster, vitreous; color and streak, blue; transparent to translucent.

Azurite like **malachite** is a secondary mineral, but far less common than that mineral. It has been formed by the action of carbonated waters on compounds of copper or solutions of copper compounds, probably most abundantly by rich solutions reacting with limestones. Azurite almost always occurs associated with malachite. Found in Siberia, Greece, Rumania, at Chessy, France, whence the name Chessylite, in South West Africa, Australia and elsewhere. Azurite occurs in the United States at Bisbee, Arizona and Kelly, New Mexico. It is used as an ore of copper. (E.S.C.S.)

B

BABBITT. Tin Alloys.

BABBLER. Aves, Passeriformes. Birds (**Aves**) of the Ethiopian and Indian regions, particularly those of the family Crateropodidae but sometimes more loosely applied. (A.W.L.)

BABINGTONITE. This mineral is a relatively rare calcium-iron-manganese silicate, occurring in small black triclinic crystals, found in Italy, Norway and in the United States at Somerville and Athol, Massachusetts and in Passaic County, New Jersey. It was named for Dr. William Babington. (E.S.C.S.)

BABIRUSA. Mammalia, Artiodactyla. A wild **pig**, *Babirusa alfurus,* of Celebes with unusually long curved tusks, resembling the horns of some deer, which are the source of the name. Babirusa means pig-deer. (A.W.L.)

BABOON. Mammalia, Primates. The true baboons are characterized by the elongate, dog-like muzzle with nostrils at the tip. The head is large and they walk on all fours. These animals constitute the genus *Papio* but the term baboon is also applied to the gelada baboon, *Theropithecus gelada,* of southern Ethiopia, also a dog-like species although the nostrils are some distance from the tip of the snout. Baboons are also called dog-faced monkeys.

Among the baboons several species are known by special names, including the mandrill, *P. sphinx,* the drill, *P. leucophaeus,* and the chacma, *P. porcarius.* All are found in the Ethiopian region.

Baboons include some of the most hideous of animals and are correspondingly ferocious in disposition. Individually they are able fighters and their defenses are augmented by their habit of living in bands. A group of males is said to be a good match for some of the larger predators. (A.W.L.)

BACILLARY DYSENTERY. Dysentery.

BACILLUS. Bacteria.

BACK BEARING. Bearing.

BACK-GOUDSMIT EFFECT. Zeeman Effect.

BACKSIGHT Differential.

BACK-SWIMMER. Insecta, Hemiptera. An aquatic **bug** of boat-like form which lives in an inverted position. The hind legs are broadened by fringes and are used like oars for propulsion. Family Notonectidae. (A.W.L.)

Back-swimmer.

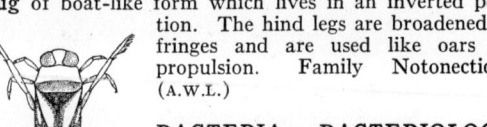

BACTERIA, BACTERIOLOGY. Bacteria, often called germs or microbes, are microscopic unicellular organisms only obscurely related to other organisms. Their unicellular form, lack of definite nucleus, and method of reproduction by fission, remotely suggest relationship with blue-green **algae,** from which most are, however, distinguished by the absence of pigments. This lack of pigments leads many to connect them with **fungi.** The absence of chlorophyll (see **Pigments in Plants,** also **Aminoacids** and **Proteins**) makes it necessary for bacteria to obtain their nutritional requirements from organic sources.

Many are parasites causing diseases of living plants and animals; others are **saprophytes** living on dead plants and animals, the chemical constituents of which are broken down by the action of the bacteria. Because of the results of their presence, bacteria, whether disease-producing or otherwise, are of tremendous importance to mankind.

The small size of bacteria makes them invisible to the unaided human eye. Although several early philosophers advanced the opinion that such invisible organisms must exist, it was not until the 17th century, with the invention of the **microscope,** that the Dutch lens-maker, Anton van Leeuwenhoek, actually saw these minute forms of life and reported them to the Royal Society of London, in 1683. After this the study of bacteria languished for a century. Throughout this early period little was known of the true nature of the organisms and the part they played in various processes, such as decay and **fermentation.** But during the 19th century the work of Pasteur definitely demonstrated the great biologic importance of bacteria.

The study of bacteria is closely connected with their culture, which means growing each form under such conditions as to exclude all other organisms, including other bacteria. To do this it is first necessary to clean thoroughly and sterilize all instruments and dishes to be used. Ordinary washing is insufficient. After thorough washing and rinsing, the dishes are covered or plugged with some substance which prevents the entrance of any solid particles, however small. A convenient material for plugging is ordinary cotton. All implements and vessels are next sterilized. This means exposure in an oven to a temperature as high as 170°–200° C. for 15 minutes or more, or treating with steam under pressure for 15 minutes or more at 120° C. If possible, such steam-pressure treatments are done in an autoclave. After sterilizing all objects to be used, it is necessary to prepare a suitable culture medium, which must be a substance on which the bacterium to be studied will grow satisfactorily. Many different media are used, such as meat broth and peptone, potato broth, etc. If gelatin is added, the medium becomes solid, and the organisms, growing on the surface, are more easily studied. However, the fact that gelatin liquefies at 37° C., and many bacteria grow best at higher temperatures, makes it desirable to find other substances which will remain solid at temperatures higher than 37° C. **Agar agar,** a vegetable product, is found to meet this requirement, and so is frequently used to prepare a solid medium on which to grow bacteria. The media are placed in suitably prepared dishes, sterilized, and then inoculated with the organism to be studied, precautions being taken to prevent the entrance of undesired organisms. On such solid media each species of bacterium forms characteristic masses or colonies.

Several methods are used for the study of bacterial cells. Living bacteria may be readily examined under the microscope by the hanging-drop method. To do this a drop of the culture, growing in a liquid medium, is transferred to a thin slip of glass (a cover slip) and inverted over a special glass slide in which a circular hollow has been ground. Bacteria grown on solid media may be transferred to a suitable liquid medium and mounted as a hanging drop. Such hanging drops allow one to study the living organism and to determine its motility, spore formation, and form.

Often it is desirable to study killed and stained bacteria. A thin film of bacteria is smeared on a clean glass slide and allowed to dry, after which it is fixed by passing through a flame, or by dipping in absolute alcohol, which is subsequently washed off. A suitable stain is then put on the smear of bacteria: gentian-violet is often used. After a few seconds the stain is washed off with water, and the preparation examined. Many methods of staining, some very elaborate, have been devised. One, known as Gram's method, merits description, since it is frequently used in describing bacteria. The stained organism is treated with an iodine solution and then washed in 95% alcohol; with this procedure many organisms give up the stain, and are said to be gram-negative; others retain the stain and are called gram-positive.

Bacteria are among the smallest of living organisms, varying from large species 1/250″ in length to minute forms only 1/250,000″ in length. The average dimensions of the species are 2 microns long and ½ micron in diameter (about 1/12,500″ by 1/50,000″).

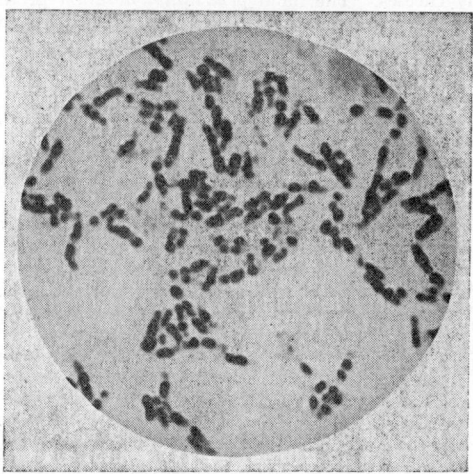

Bacillus diphtheriae. Magnification 2250×.

The minute size of the bacterial cell makes it difficult to determine its structure accurately. In recent years, however, the electron microscope, which is capable of magnifications of over 200,000 diameters, has been a valuable instrument in the study of bacterial morphology. With its aid, a distinct cell wall of bacteria can be made out, inclosing the inner cell protoplasm. Within the protoplasm spheroidal or discoidal granules are seen in many species, as well as areas of decreased density. Spores form within the protoplasm of sporulating species. Nucleoprotein has been demonstrated under certain conditions in what has been interpreted as a simple nucleus. Surrounding the cell membrane of some species (notably pneumococci, and bacilli of the Friedländer group) a thin gelatinous sheath is present, forming a capsule. Flagella are organs of motility which are seen on the outer surface of motile bacteria; they are minute lash-like structures arranged singly, or in tufts at the ends of the cell, or in some species, covering the entire surface of the bacterium.

Reproduction in bacteria seems to occur solely by the process of fission, or cell division. This process occurs with great rapidity, often requiring less than a half-hour for completion, and occurring in many cases as frequently as once an hour. Simple calculation will show that under such conditions a single bacterium may give rise by successive divisions in a single day to many millions of cells, and in a few days to unbelievable numbers.

Such rapid divisions do not continue indefinitely, however. The nutrient supply may become exhausted, or the products of bacterial growth accumulate and become toxic, and divisions cease. Often such conditions bring about spore formation, each cell generally becoming a spore. A bacterial spore is highly resistant, surviving prolonged desiccation, intense cold or high temperatures, as well as the presence of harmful chemicals. Spores should therefore be considered not as reproductive bodies, but as a means by which the organism survives unfavorable conditions which would be fatal to the ordinary cell. Not all bacteria are capable of forming spores. Fortunately many disease-producing bacteria are of this less resistant type.

Bacteria may be classified into three families, according to the shape of the single cell. The first contains all those whose cells are spherical; these are known as coccus forms—the family is Coccaceae and includes the genera *Streptococcus*, *Staphylococcus*, and *Neisseriae*. A second family, the Bacteriaceae, comprises all forms having rod-shaped cells, which may be either straight or slightly bent. Such cells are known as bacillus forms. This family has three genera, *Bacterium*, *Bacillus*, and *Pseudomonas*. The third family, the Spirillaceae, is a small one, and contains those forms in which the cell is in the form of a spiral, known as a spirillum-form. While coccus forms do not possess cilia, most spirillum forms are ciliated. Bacillus forms may or may not have cilia, and so belong to either the genus Bacillus, or Bacterium, respectively.

Bacteria may also be divided into two classes: **saprophytes** and **parasites**. Saprophytes are forms which grow and multiply best on dead tissue or organic matter, reducing it into simple chemical compounds which furnish food for plants, and thus, the saprophytes play an important part in the chemical cycle of living matter. They maintain the balance between the plant and the animal kingdoms. Ordinarily these forms do not cause disease in man unless potent toxins are produced by the

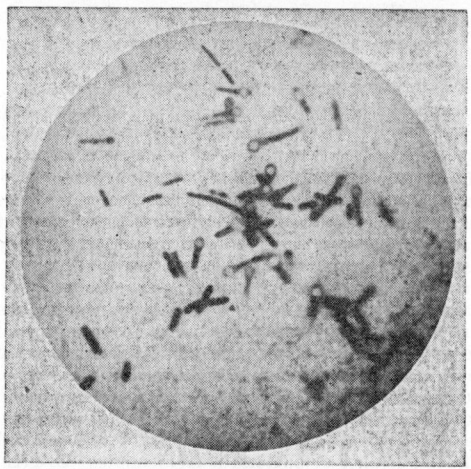

B. tetanus showing terminal spores. Magnification 1600×.

bacteria. The parasitic forms live off the living tissues of higher forms of life and, in general, they form the group of pathogenic or disease-producing bacteria.

Another classification of bacteria is based on their source of nutrition. Organic compounds of all sorts may be a suitable nutrient source. Complex nitrogenous bodies are attacked and broken down by many, the process being the familiar one of decay and decomposition. Simpler organic substances are attacked by other species. Autotrophic bacteria are those which obtain the carbon necessary for carbohydrate formation from carbon dioxide and their energy from the oxidation of inorganic compounds, such as ammonia, sulfur and sulfur compounds, nitrites (see Nitrous Acid), etc. *Beggiatoa* is a genus of this group. Heterotrophic bacteria are those requiring organic sources of both carbon and energy for their existence. Nitrogen-fixing bacteria are examples of heterotrophic bacteria.

Like all living organisms, bacteria are affected by environmental conditions, particularly by temperature, light, moisture, oxygen supply and food requirements. Many bacteria are capable of growing under remarkable temperature conditions. Some, like the hay bacterium (*Bacillus subtilis*) can divide at temperatures from 6° C. to 50° C., while others, such as the pathogenic species, may have a very narrow range of temperature in which they can grow. Bacteria are less sensitive to low temperatures than to high ones.

Light is another very important factor in the life of bacteria: the germicidal action of direct sunlight is well-known. The action of light on bacteria seems due to oxidation processes which are fatal to the well-being of the cell. This fatal effect of light is confined to the ultra-violet region of the spectrum, and is found both in sunlight and also in electric light.

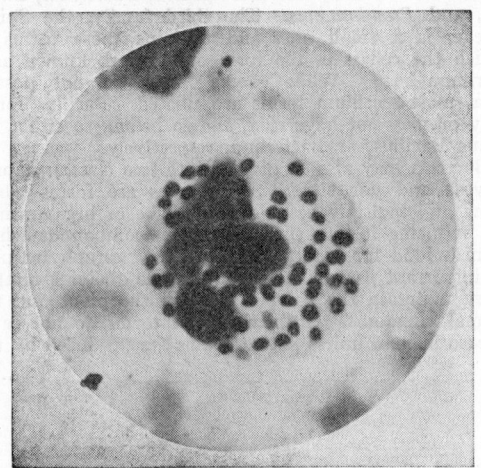

Gonorrheal pus showing the cocci within a leukocyte. Magnification 2250×.

Moisture is necessary for the continued existence of many bacteria: a short period of drying destroys a great many species in the vegetative phase, including fortunately most of the pathogenic forms. The spores of spore-producing forms are extremely resistant to desiccation.

Bacteria vary greatly in their oxygen requirements. There are many species which can exist only when free oxygen is available, such forms being known as obligate aerobes. Other species cannot exist except in the complete absence of free oxygen and are known as obligate anaerobes. Still other species are indifferent to the presence of oxygen and are called facultative anaerobes.

The importance of bacteria is tremendous. Of greatest significance to ordinary man are the many pathogenic species which, by their presence in the human body, cause disease. **Tuberculosis, leprosy, cholera, typhoid fever, diphtheria, pneumonia** and many other diseases are due to such organisms. Bacteria of this type are of highly specific nature, that is, can only attack a single host, or a small group of closely related hosts.

As bacteria may cause diseases in animals, so also are they the cause of many diseases in plants, although such diseases are not as numerous as those of animals. A common disease of plants which is caused by bacteria is Cucurbit Wilt, which appears in cucumbers, melons, pumpkins and squashes. In this case the bacteria gains entrance to the plant through wounds inflicted by insects such as the striped cucumber beetle. Once inside the plant body, the bacterium passes to the conducting strands of the stem, which are filled by the accumulating masses of rapidly dividing bacteria. These cut off the water supply of the leaves of the host plant so that wilting occurs and subsequently the death of the in-

fected parts. A somewhat similar disease is found in Brown Rot of Tomatoes. Black Rot of Cabbage, Bean Blight and Crown Gall of many plants are all bacterial diseases of some importance.

The disease-producing bacteria are known so much better than are other forms of bacteria as to lead to the idea that bacteria are eminently destructive. This, however, is far from the truth. Probably the beneficial types of bacteria many times outweigh the injurious forms. Some of these beneficial forms are used by man in the industries. The souring of milk is caused by the presence in the milk of certain bacteria which result in the formation of **lactic acid**. Bacteria present in cream during its ripening cause the butter presently made to have certain flavors. Many cheeses owe their particular characteristics to the presence of certain bacteria. Molds also play a very great part in the development of the much-sought flavors (and also odors) in the many cheeses. The mold *penicillium notatum* produces a substance known as **penicillin** which has proved to be one of the most effective of the **antibiotics**. Its use in the treatment of certain bacterial infections has saved many lives.

In the tanning of hides and in the curing of tobacco, bacteria are thought to play an important part. The oxidation of **alcohol** to **acetic acid** in such weak alcoholic liquids as cider and wine is due to the activities of different species of bacteria. In the preparation of **flax** fibers and also **hemp** bacterial action is very important, for by their action the cement substance binding the fibers together is broken down and the fibers freed. The action of certain bacteria is important for the proper preservation of ensilage.

Of particular importance to man is a certain group of several different bacteria which occur normally in the

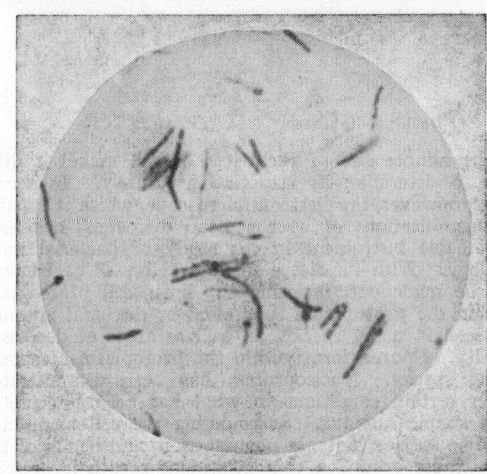

B. tuberculosis in sputum.   Magnification 2250×.

soil and are instrumental in making available to higher plants the very essential element **nitrogen**. All plants require a considerable amount of this element, which is available to them only in the form of certain soluble compounds, such as nitrates. The large percentage of free nitrogen in the atmosphere is unavailable, and few soils naturally contain a sufficient supply of usable nitrogen. Commercial **fertilizers** may be used to provide the necessary amount, but they are expensive. **Ammonia**, a product resulting from the breaking down of organic matter by bacterial action, is another possible source of nitrogen which is largely unavailable to higher plants. There are several species of bacteria which are able to convert ammonia into nitrates and other species which fix the free nitrogen of the air. These organisms are of very great importance to man. Their activities are collectively included in what is called the nitrogen cycle, and fall in three groups, those comprising the nitrogen-fixing bacteria, the nitrifying bacteria and the denitrifying forms. The nitrogen-fixing bacteria, in turn, fall into two groups, those living free in the soil and those which enter the roots of various plants. By their activities the fertility of a soil is built up greatly without any attention from man. The free-living forms in the soil obtain the energy necessary for their existence from various organic compounds and assimilate free nitrogen of the atmosphere so that it is subsequently

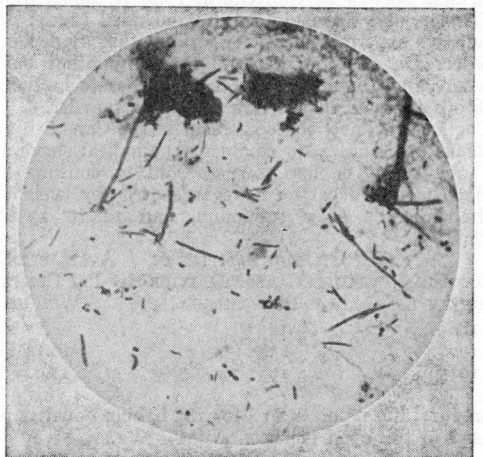

Bacterial flora from the mouth showing rods and cocci. Magnification 850×.

available to higher plants. Several species of bacteria are included in this group, including *Clostridium pasteurianum* and *Azotobacter chroococcum,* and they are most abundant in light, well-aërated soils. The nodule-forming bacterium, *Bacillus radicicola,* normally occurs in the roots of various plants, particularly legumes, such as clover, alfalfa, vetch, etc., causing small lumps or nodules to form at various places on the root. The

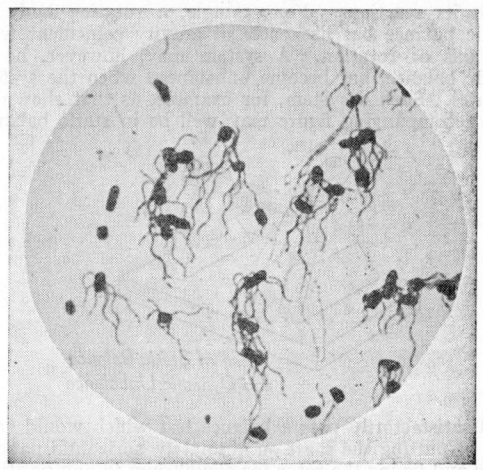

Bacillus typhosus showing flagella.   Magnification 150×.

bacteria are highly specific in the hosts they enter, each strain being restricted to a single species or to a small group of species. It is necessary, therefore, in spite of the widespread natural occurrence of the bacteria, to inoculate with the proper strain any new plant to be introduced. This is particularly noticed in the case of Alfalfa, where the seeds are covered with a suitable strain of bacteria before being planted. The bacterium enters the root through the root-hairs, a spot on the surface of the latter softening to permit their entrance.

Once inside the cell, the bacteria pass inward into the cortex, where they accumulate in great numbers in the vacuoles of the cells. Their presence causes definite enlargement of the infected cells, thus producing the conspicuous nodules. Within the nodules the bacteria become modified into irregular, frequently somewhat branched objects known as bacteroids. In the presence of abundant energy-supplying carbohydrates from the host plant the bacterioids assimilate free nitrogen, which is built up into complex nitrogenous compounds. These are either utilized by the host plant directly or liberated into the soil on the death of the host plant, so adding to the fertility of the soil.

In addition to the various nitrogen-fixing forms described above there are other bacteria which make available to higher plants the necessary nitrogen. These are collectively known as the nitrifying bacteria, and are of two kinds. The first, *Nitrosomonas,* converts ammonia into nitrites, and the second, *Nitrobacter,* converts the nitrites into nitrates.

The ammonia necessary for the development of the nitrite-forming *Nitrosomonas* results from the activities of another complex and very important group of organisms, largely bacteria, known as denitrifying organisms. These bacteria reduce nitrates to nitrites and ammonia, to oxides of nitrogen or to gaseous nitrogen. More important in the formation of ammonia are the various organisms causing decay of the dead bodies of plants and animals. These attack the complex organic substances which occur in dead bodies and break them down into simpler forms, eventually into ammonia, oxides of nitrogen or even free nitrogen. The process is one of great complexity and occurs in many stages, each organism attacking a single substance and reducing it to simpler forms, which will in turn be attacked by other organisms until the lowest stage is reached. The great result of the activities of these organisms is the disintegration of what otherwise would accumulate as vast quantities of dead matter, in which large amounts of chemicals necessary for living organisms would be bound. Putrefaction is thus a very necessary process for the continuation of life. (R.S.M., R.M.W., D.M.H.)

"BAD LANDS." The literal translation of the phrase *Mauvais Terre* of the French explorers who so described the highly dissected, relatively unconsolidated sandstones and shales such as occur in the western Great Plains near the Black Hills. Small areas also occur in the plateaus of the Rocky Mountain region. This type of topography develops in arid and semi-arid regions where the underlying formations are relatively soft, and, due to the climate, are not protected by a plant cover. (R.M.F.)

BADGER. Mammalia, Carnivora. Stoutly built and short-legged burrowing animals of several species, found throughout the northern hemisphere. The term is extended to the ferret-badgers, which are related to the skunks, and to the ratels of India and Africa, which are called honey-badgers.

Badgers are courageous and able fighters when attacked and are sometimes hunted for sport, but they are peaceful animals unless molested. The fur is moderately valuable and the long hairs, especially of the European species, *Meles taxus,* are used in making brushes. Badger-hair shaving brushes are a familiar example. (A.W.L.)

BAFFLE.   Mixing.

BAFFLES. In the engineering sense, a baffle is an object, usually a partition, placed for some specific purpose in the flow path of a fluid causing it to take some prearranged and circuitous path. Thus baffles are found in steam boilers to direct the hot gas properly back and forth over the tubes so that the gas will give up its heat to the required degree, and will not short-circuit directly from the furnace to the stack. For this service the baffle is composed of refractory material similar to firebrick and will be found in longitudinal or transverse arrangement. Transverse baffling is made by building the baffle perpendicular to the tubes. Longitudinal baffles are usually precast and laid upon the tubes of the boiler, forming a baffle whose surface is parallel to the tubes.

Baffles are built in coagulation basins to impede the flow of liquid, and are also found in exhaust **mufflers** where their purpose is to mix the flow of gases in adjacent exhaust puffs that they may emerge from the muffler in a silent steady stream. (F.T.M.)

BAGASSE. In the manufacture of sugar (see **Carbohydrates**) from sugar cane the crushed fibers from which the sap has been expressed are called bagasse. Its principal use is as a fuel to run the mills which crush the cane. For this purpose bagasse is mixed with petroleum oil. It is also used as a fertilizer and to some extent in manufacturing heavy insulation board and coarse paper. (R.M.W.)

BAG-WORM. Insecta, Lepidoptera. The **larva** of a **moth** which is encased in a covering of silk mixed with bits of leaves, twigs, etc. Only the head and legs protrude from the bag, hence the insect appears to be suspended from the twig on which it walks. The adult females are wingless and the eggs are deposited in the silken bag.

One species of bag-worm sometimes does great damage to evergreen trees, especially the cedars, and so has been named the evergreen bag-worm, *Thyridopteryx ephemeraeformis*. The larvae can be killed by spraying infested trees with lead arsenate and the destruction of the bags in the winter, when they contain eggs, is an important measure of control.

These insects constitute the family Psychidae. (A.W.L.)

BAILY'S BEADS. During an **eclipse** of the sun, at the instant when the moon's edge is just tangent to the edge of the sun, i.e., at either second or third contact, the thin crescent of the disappearing sun suddenly breaks up into a number of brilliant spots known as Baily's Beads. These are produced because the surface of the **moon** is very rough, and mountains on the moon will completely cover the sun's disk while the sunlight is still coming to the earth through the valleys. (W.K.G.)

BAINITE. Steel.

BAKING POWDERS. Baking powders are used instead of leavening or **fermentation** agents to make light those baking products made mainly from **flour.** Baking powder, when it is mixed with flour and other ingredients and the resulting mixture is thoroughly moistened and heated, evolves a gas, **carbon dioxide** (sometimes **ammonia**), thus producing a light porous product. To supply carbonate for the production of carbon dioxide, **sodium** hydrogen carbonate ("baking soda," $NaHCO_3$) is universally used. To supply acid, three classes of materials are in common use: (1) **potassium** hydrogen tartrate ("cream of tartar," $KHC_4H_4O_6$), and infrequently, **tartaric acid** ($H_2C_4H_4O_6 \cdot H_2O$), (2) **calcium** dihydrogen phosphate (calcium monophosphate, $Ca(H_2PO_4)_2 \cdot H_2O$), (3) sodium **aluminum** sulfate ($Na_2SO_4 \cdot Al_2(SO_4)_3 \cdot 24H_2O$). The proportions (approximate only because it is customary to use some excess of carbonate) are, with one part by weight of sodium hydrogen carbonate, (1) 2.2 parts by weight of potassium hydrogen tartrate, (2) 1.0 of tartaric acid, (3) 1.5 parts of calcium hydrogen phosphate, crystallized, or (4) 1.8 of sodium aluminum sulfate or ammonium aluminum sulfate. With 7 parts by weight of this finely powdered mixture, there is usually mixed about 3 parts by weight of starch to diminish the effect of moisture in storage. In some cases, dry powdered egg albumin is added to decrease the loss of carbon dioxide upon wetting the flour and baking powder mixture when used. For some purposes **ammonium** carbonate ($(NH_4)_2CO_3$) is used alone, since, upon heating, this material furnishes both ammonia and carbon dioxide gases to make the product light. These gases escape from the product in the baking process.

The referee board of consulting scientific experts of the United States government, consisting of Ira Remsen, chairman, Russell H. Chittenden, John H. Long, Alonzo E. Taylor, and Theobald Smith, concluded (1914) after exhaustive experimental investigations, that alum baking powders are no more harmful than any other baking powders, but that it is wise to be moderate in the use of foods that are leavened with baking powder.

The Inland Revenue department of the Canadian government has since 1889 conducted periodical surveys of the quality of baking powders sold to consumers in Canada. The department regards 10% by weight of carbon dioxide as a minimum, and 12 to 13% as normal.

DATE OF SURVEY	NUMBER OF SAMPLES IN WHICH CARBON DIOXIDE WAS DETERMINED	AVAILABLE GAS, PER CENT BY WEIGHT
1889	149	8.17
1908	158	10.24
1915	195	11.91

In 1917, the report stated for 185 baking powders purchased in the open market:

35 samples below standard (10% carbon dioxide)
68 alum-phosphate samples 10.78%
46 cream of tartar samples 11.01
31 acid phosphate samples 11.25
5 alum samples 12.40
(R.K.S.)

BALANCE. Mechanical balance consists of the equilibrium of masses, and can be divided into static and dynamic balance. Static balance occurs in a system when the **center of gravity** of the system coincides with its reactions. For example, a rotating body in static balance has its center of gravity coincident with its axis of rotation. A system may, however, be in static balance, but become unbalanced when the system rotates. Such a system, for example, as that shown in the accompanying figure may well be in static balance,

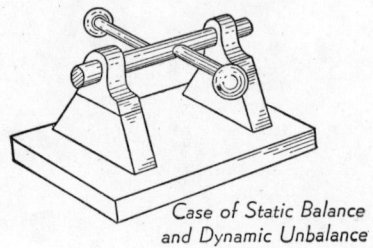

Case of Static Balance and Dynamic Unbalance

and satisfactorily pass a balance test which would consist of putting the shaft on absolutely horizontal parallel rails and trying the rotor for equilibrium in any position. But when this system rotates, the centrifugal forces of the two weights, not being in the same plane perpendicular to the axis of rotation, create a couple acting on the shaft. That couple rotates with the shaft and produces shaking forces at the journals, and vibrations in the foundation. The dynamic balancing of this system would involve the addition of a system of counter balances which, by themselves, would be in static equilibrium, but which in rotation would produce a couple equal in magnitude but opposite in direction to the one already considered.

Dynamic balance is especially important in high-speed or heavy rotating machinery, as the vibrating forces are proportional to the mass and the square of the speed of rotation. Manufacturers frequently use balancing machines for testing their product when it is especially important that it be perfectly balanced.

Reciprocating balance consists of opposing the shaking forces of a reciprocating mass by equal and opposite forces obtained from another reciprocating mass. One particularly difficult job of balancing occurs in a system consisting of both rotation and reciprocation, as exemplified by the **piston, connecting rod,** and **crank** mechanism. The difficulty lies in the fact that if perfect balance is secured in the direction of reciprocation by the employment of rotating counter balances, severe unbalance will result in a plane perpendicular to that direction. The solution of this difficulty is a compromise in which only part of the reciprocating mass is counterbalanced, thus reducing the maximum degree of unbalance, but producing a smaller unbalance in two planes.

An electric network is in a condition of balance when so adjusted that an e.m.f. in one branch produces no current in another branch. This is the case, for example, in a properly adjusted **Wheatstone bridge.**

The balance is a well-known instrument used in weighing. While any type of "scales" used for weighing may properly be called a balance, the term usually refers to the equal-arm balance familiar in every laboratory.

The dynamics of this instrument is relatively simple unless the pans are allowed to oscillate independently of the beam, a condition which should be carefully avoided while taking readings. The balance may then be treated as a gravity pendulum suspended from the central knife edge or pivot, the pans and their loads being regarded as concentrated at the end knife edges. Any slight excess weight Δw on one pan causes a change in the equilibrium position. The "sensitiveness" of the balance is appropriately expressed as the change in the equilibrium pointer reading per unit excess weight. More convenient in practical work is the reciprocal of the sensitiveness, which may be called the "stability." Thus if the change in pointer reading Δr is produced by a change of weight Δw on one pan, the stability is $s = \Delta w / \Delta r$. It is easily shown that unless the three knife edges are exactly in a straight line (which is seldom true), the stability not only depends upon the construction of the balance but is also a function of the load on the pans. If the load on each pan is W, the stability is expressed by a linear equation,

$$s = a + bW,$$

in which a and b are constants to be experimentally determined. The instrumental factors affecting the stability and upon which the constants a and b depend are: length of beam; relative lengths of pointer and of scale division; distances of central knife edge above center of mass and above or below line joining end knife edges; and total weight of moving parts, exclusive of load.

The pointer and scale are used in refined weighing to interpolate between the smallest weights in the set (commonly milligrams), for which purpose the value of s for the load on the pans must be known; it may be calculated from the above formula. Balances usually have also a rider scale, along which a small, hairpin-shaped weight may be moved to secure equilibrium, or to secure any desired pointer reading. The rider may be used in determining the stability constants a and b.

The arms of a balance, supposedly equal in length, are never exactly so. The result is that if the pointer stands at the zero of the scale with the pans empty, it will not do so when equal loads are placed on the two pans. This defect is compensated by the method of "double weighing." (See **Weighing Methods.**) (F.T.M., L.D.W.)

BALANCE COIL. A balance coil is a coil for supplying a three-wire circuit from a two-wire circuit. A 220-volt single-phase line, for example, can be used to supply two 110-volt circuits consisting of three wires, one of which is a common intermediate wire.

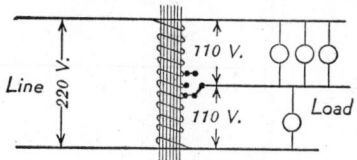

Balance coil for a.c.

The voltage between the intermediate wire and either of the outside ones is 110 volts. If the loads on the two 110-volt circuits are unequal, the voltage can be balanced by an adjustment of the central tap point at the balance coil. A balance coil is frequently an **autotransformer** having only one coil, a certain portion of which is used for both a high and low tension winding. The auto-transformer has connections at the ends of the coil, and an intermediate connection.

In d-c circuits the balance coil is used in conjunction with the **generator** to obtain a three-wire system. The generator windings are tapped at two diametrically opposite points and these taps are connected to the ends of a balance coil. The third wire for the system is obtained by connection to the center of the balance coil. Some manufacturers build the coil in the spider of the **armature** and bring out the center connection through a **slip ring** and **brush** while other manufacturers bring out the two tap connections on the armature through a pair of rings and brushes, mounting the balance coil outside the generator. (F.T.M., L.R.Q.)

BALANCED MODULATOR. A balanced modulator circuit is used to generate the **sidebands** of an amplitude modulated wave (see **Modulation**) and suppress the **carrier.** Since the power involved in a modulated wave is distributed at 100% modulation such that the carrier frequency component has ⅔ and each sideband ⅙ of the total, suppression of the carrier eliminates the necessity of supplying a major portion of the power usually needed. The output of such a modulator could be transmitted as only the sidebands, but in order to demodulate it a carrier wave would have to be introduced at the receiver. Such a system is known as a double sideband and suppressed carrier system. It is not used in practice because the carrier introduced at the receiver must have almost exactly the frequency of the one removed at the transmitter. Technical difficulties make this impractical. The balanced modulator output is normally fed through a filter circuit which cuts out one sideband and the other is transmitted. This gives single sideband, suppressed carrier transmission and is widely used, especially in **telephony,** since the frequency requirements for the carrier introduced at the receiver are not nearly so strict as the previous case. The reintroduced carrier needs to have an amplitude comparable to the received signal which is, of course, a small part of that needed at the transmitter of a radio system. The frequency band occupied by a single sideband, suppressed carrier system is somewhat less than half that needed for a full system. This is an important asset in telephone **carrier** circuits where the transmission line characteristics limit the total frequency band which can be transmitted. The single sideband system allows twice as many **channels** then on a given line. (L.R.Q.)

BALANCER SET. This consists of two shunt dynamos connected in series across a two-wire d-c distribution system to supply the third wire for a three-wire system. Either machine may act as **motor** or **generator**, the action being determined by the direction of unbalance of the load. For unbalanced loads one machine

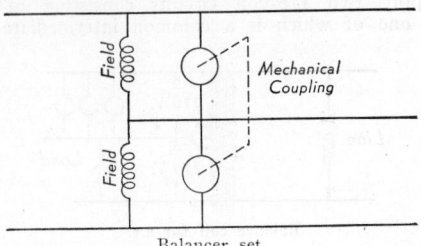

Balancer set.

motors and drives the other as a generator to tend to restore the voltage balance. (See Fig. 1.) (L.R.Q.)

BALANOGLOSSIDA. Enteropneusta.

BALANOGLOSSUS. A genus of worm-like marine animals belonging to the lower **chordates**. The name is also commonly applied to any of the similar animals of several genera. **Enteropneusta.** (A.W.L.)

BALAS RUBY. Spinel.

BALATA. Gutta Percha; Belting.

BALDNESS. Alopecia.

BALDPATE. Aves. 1. A North American **duck** *Mareca americana*. 2. A **dove** found in the West Indies. Both species are ineptly named from the white crown feathers. (A.W.L.)

BALL BEARING. Bearings.

BALLAST. Any material used for the purpose of providing stability is called ballast. Ballast is used in ships to bring the **center of gravity** of the vessel below a point called the **metacenter**, when there is a lack of cargo which would produce the same effect. **Balloons** and **airships** carry ballast which acts as a stabilizer and provides a means of controlling the rate of gain of rise as well as the altitude. Light vehicles which move at high speeds are often provided with ballast to lower the center of gravity and prevent overturning. Sand or water are very useful for ballast. Crushed stone which is placed under and between railroad ties to absorb impact and provide smooth riding conditions is also called ballast. (C.W.C., F.T.M.)

BALLAST TUBE. This is a resistance element sealed in a tube, usually in a hydrogen atmosphere, and is used as a series component of a circuit to maintain the current constant. This is accomplished by designing the ballast tube so the resistance increases and decreases rapidly with corresponding changes of current through it. (L.R.Q.)

BALLISTICS. This is the science which treats of the motion of masses projected into space, especially as associated with the motion of projectiles from guns and cannon. The complete path of a projectile is comprised of three separate and distinct phases. The first occurs in the bore of the gun, and the study of projectile motion here is that of interior ballistics. Secondly, there is the study of the path taken by the projectile as it flies through space from the gun to the target. This is exterior ballistics. Then, thirdly, there is the study of the penetration and the penetrating power of a projectile, which, for want of a better term, might be called penetration ballistics.

Interior ballistics is largely interwoven with the study of **thermodynamics**—the pressure, volume, and temperature of an expanding gas during travel of the projectile in the bore. It is concerned with the amount and combustion characteristics of gunpowder. The maximum pressures, and location of the same, stresses in the barrel, and the design of the barrel to resist these stresses, may also be said to be interior ballistics. The science of exterior ballistics might be said to have been rationally developed by Newton as a by-product of his study in **gravitation**. If the effect of air on the motion of a projectile is omitted, then the trajectory is **parabolic**, since as soon as the bullet or shell leaves the muzzle of the gun, force of gravity begins to pull it towards the earth. It is therefore impossible for a bullet to travel in a straight line, and if it is to return to a target at the same elevation as the muzzle, it must have an initial upward component given by aiming the barrel somewhat above the target. If the muzzle velocity is V, and the inclination of the barrel to the horizon is i, its upward component is $V \sin i$. If the action of gravity is wholly unresisted, the time it will take to reach the top of trajectory (at which point the upward component has been reduced to zero) is the same as that required by a freely falling body to attain a velocity equal to that of the vertical component at the muzzle. Assuming a simple case where the target is at the same level above the earth's surface as the gun, it would take another equal interval of time for the projectile to move from the top of its trajectory to the target. The distance to the target would be that covered by the horizontal component $V \cos i$ in the period of time taken by the projectile in reaching the top of its trajectory, and then returning to its original level.

But the effect of air forces cannot be neglected, practically speaking, and the simple equations derived from the mechanics of a freely falling body are not applicable without considerable modification. The retarding effect of air, and the effect of winds, as well as other matters, are factors which must be taken into account in modern ballastics.

There is but little known in respect to penetration ballistics, and laws which relate to the depth and character of penetration of bullets and shells into armor are not well understood. Empirical formulae are relied upon greatly at the present time in the field of ballistics. (F.T.M.)

BALLOON. The balloon is a lighter-than-air craft receiving its sustention from the **buoyancy** of the gas it contains. It is non-rigid. The shape, usually spherical, is formed by the internal pressure of the lifting gas. The lifting medium first used was air which was expanded and made lighter than the atmosphere by heat. The hot air balloon was first devised by the Montgolfiers in 1783. At the present time **hydrogen** is the principal lifting gas. Balloons may be thought of as free or captive. The captive balloon has some military use for observation, or fire control, but until recently the free balloon had little or no value other than sporting. The national and international balloon races gave impetus to this use of the free balloon, and many amazing records were hung up by sportsmen balloonists. The balloon drifts with the wind, and may be sent aloft by releasing **ballast**, or brought down by valving off the gas from the balloon. A knowledge of the science of meteorology is important in racing balloons, and there is a great deal of highly specialized knowledge required in connection with the operation of the balloon. Happy combinations of skill and weather conditions have enabled flights of more than 1000 miles to be made.

In recent times a new and scientific use of the free balloon has appeared. Several flights to the **stratosphere** have been made by specially-built balloons. The flights of these balloons to high levels may serve to advance scientific knowledge of cosmic rays and to study atmospheric conditions in the stratosphere. This information may become useful commercially. The problems of stratosphere ballooning are highly specialized, and the equipment quite expensive; consequently, stratosphere ballooning is undertaken largely by societies or governmental departments. Professor Picard of Belgium was one of the pioneers in this field, and ascended to over 60,000'. Whereas a racing balloon has a volume of some 80,000 cu. ft., the stratosphere balloons contain, when fully inflated, over ten times this volume. This extremely large volume is necessary in order to obtain sufficient buoyancy to lift the useful load to the stratosphere, where the atmosphere itself is very thin. Stratosphere ballooning is exciting, dramatic, and dangerous. It has

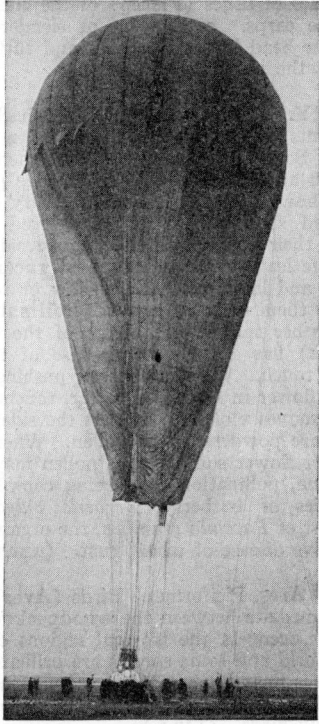

Balloon.

already taken its toll of human life, but has created one of the most thrilling modern chapters of the history of science.

In the winter of 1933 three Russians ascended to an unofficial height of 72,000'. Unfortunately, these intrepid men were killed during the descent, and no official mark was set. In the fall of 1934 Captains Stevens and Anderson and Major Kepner, of the United States, had a narrow escape from death. Having ascended to nearly 60,000', they were in the process of a normal descent when their balloon split at 5000'. Fortunately, all men parachuted safely to the ground, although the last man to leave was perilously low before clearing the gondola, which plummeted to the ground, destroying most of the scientific instruments carried. However, in November, 1935, Captains Stevens and Anderson again ascended, and this time to a new world's record. Leaving the ground at 7 A.M., they climbed to 73,000' in four hours, and obtained a great deal of scientific information on stratosphere photography, radio transmission, spore distribution, and air composition. Cosmic rays and the spectral distribution of sunlight were also studied during the hour or more that the stratosphere balloon (now being called stratostat) remained at the record height. During much of the remainder of the day the crew was busy governing the descent until the balloon was once again safely on the ground. In this case all instruments were intact and records were safely preserved. To give

an idea of the construction of a stratostat, the Explorer II used by Captains Stevens and Anderson had a volume of 3,700,000 cu. ft. fully extended, and this was used to lift a tiny globe-shaped gondola made of monel metal, just large enough to contain the two observers and their equipment. The globular shape is used because it must be sealed and made air-tight so that life will be possible in the stratosphere where the normal atmosphere would be almost instantly fatal. The globe is the best shape for resisting the bursting pressure of the air contained within. At the start of the ascent some 250,000 cu. ft. of **helium** were put into the Explorer II, and as it ascended the bag filled out as the air pressure outside grew less. (F.T.M.)

BALSA WOOD. *Ochroma Lagopus.* Bombacaceae. Balsa wood is obtained from a South American tree, *Ochroma Lagopus.* This wood is composed largely of wood **parenchyma,** a tissue characterized by thin-walled cells. Balsa wood is almost colorless and very light, having a specific gravity of 0.2, which is less than that of cork. The light porous nature of this wood leads to its use in life-preservers and floats, as an **insulating** material, and in the construction of airplane models and special furniture. (R.M.W.)

BALSAM. Resins.

BAMBOO. Gramineae. The tribe *Bambusae* comprises grasses which are particularly important in Oriental countries. The plants included in this group are of extremely variable nature, ranging from small inconspicuous species to the largest species of grass known, some having slender erect stems approaching a hundred feet in height. Many are clambering vines which form dense impenetrable masses. All these grasses are characterized by jointed hollow stems having solid nodes, familiar to Western people in the common bamboo fish pole or the fairly rare broomstick. Among the Eastern peoples, bamboo is much more commonly used. The hollow stems may serve as pipes for conducting water, or as containers for storing water and other substances. Split stems may be flattened and used in constructing shelters, or boats, or furniture. Some species yield a fibrous material which is used to manufacture a kind of paper. The young stalks of certain species become an important foodstuff. Indeed, in some regions the people have come to depend almost entirely on bamboo, so numerous are its uses. (R.M.W.)

BAMBOO RAT. Mammalia, Rodentia. Any member of several species of burrowing **rodents** of the Oriental and Malayan region, related to the mole rats but with the small eyes exposed and with a short tail. (A.W.L.)

BANANA. *Musa paradisiaca.* Musaceae. The banana plant is a striking tropical plant found wild in the Old World tropics. About thirty species are known. The plant has a thick underground **rhizome** from which rises what seems to be a thick erect stem. Actually this false stem is formed by the leaf bases which grow wrapped together to a height of 10'. From the upper portion, the large leaf blades spread out conspicuously, sometimes to a length of 10'. Up through the center of the tube formed by the leaf bases the flower stem pushes its way and bears bunches of inconspicuous small tubular flowers in the axes of large showy **bracts.** After fertilization of the flowers, the bracts fall off, and the whole cluster gradually hangs over due to the weight of the developing fruit. When green the banana fruit contains an abundance of starch, some of which changes to sugar as the fruit ripens. The banana was introduced at an early period into the New World tropics, where it has now become one of the most valuable crops. The banana is related to the **plantain.** (See **Parthenocarpy.**) (R.M.W.)

BANANA-QUIT. Honey Creeper.

BAND SPECTRUM. Molecular Spectrum.

BAND SPREAD. In the finer grade of radio **receivers,** such as used for communications purposes, it is desirable to be able to tune rapidly over a wide frequency range, yet be able to tune carefully to rather fine limits over a small range at any selected point in the wider range. This is accomplished by the band spreading tuning provided on such receivers. The usual arrangement is to have the wider range covered by conventional tuning **condensers,** but have these paralleled by small variable condensers which are tuned by the band spread dial. Thus the complete range can be covered by one rotation of the main condensers, while for any point at which they may be set the smaller ones can be used to give fine variation, thus giving, in effect, a spread-out tuning scale at this point. Sometimes other special circuit arrangements, such as tapping a condenser across part of the tuning coil, are used to give the same result. (L.R.Q.)

BANDED ANTEATER. Dasyure.

BANDICOOT. Mammalia, Marsupialia. Medium or small burrowing animals of the Australian region, related to the kangaroos but stoutly built and quadrupedal. There are several species of the genus *Perameles.* (A.W.L.)

BANDSAW. Woodworking; Saw.

BANK. 1. The position of an airplane when its lateral axis is inclined to the horizontal. A right bank is the position with the lateral axis inclined downward to the right. 2. To incline an airplane laterally, i.e., to rotate it about its longitudinal axis. (F.T.M.)

BANKET. A Dutch term originally applied to the gold-bearing **conglomerates** of the Witwatersrand, South Africa. (R.M.F.)

BANNER CLOUDS.. Clouds.

BANTAM. Aves, Galliformes. A small variety of the domestic fowl. Sebright bantams, the best-known breed, are characterized by the similarity of the sexes. (A.W.L.)

BANTING. Mammalia, Artiodactyla. The Javan **ox,** *Bibos sondaicus,* a species which also occurs in other East Indian islands and in India and the Malay Peninsula. Domesticated herds are kept in part of its range. (A.W.L.)

BANYAN. *Ficus benghalensis.* Moraceae. The banyan trees of India, sacred to the natives of that country, are of interest because of their habit of sending down from the spreading branches aerial roots, which enter the ground and increase in size until they resemble trunks. A single tree may have scores or, exceptionally, even hundreds, of these trunks, which enable it to spread out over a tremendous area. It has been stated that Alexander and his army of men took shelter under a single extensive banyan tree. (R.M.W.)

BAR CHART. A method of representing relative magnitudes by lines or bars the lengths of which are proportional to the quantities involved. Simple barographs show individual quantities; compound barographs show complete quantities on a single bar, and are often referred to as percentage bar charts. (H.C.H.)

BARABOO. A baraboo is a hill, mountain or other eminence which was once buried by the deposition of **sedimentary** material about it and has since been ex-

posed by the erosion of the younger beds. It is named from Baraboo, Wisconsin. (E.S.C.S.)

BARB. One of the divisions of a **feather** which branches from the shaft to form the vane. (A.W.L.)

BARBASTELLE. Mammalia, Chiroptera. A small European **bat** related to the California cave bat and to a few African and Asiatic species. Only one of the related species, the Himalayan barbastelle, receives the same name. (A.W.L.)

BARBEL. 1. Pisces, Teleostei. *Barbus.* Any of numerous species of fishes (**Pisces**) of the Old World, related to the carps. 2. A form of slender appendage found on the head in certain fishes and turtles, well illustrated by the **catfishes.** (A.W.L.)

BARBERRY. Berberidacae. The barberries are shrubby plants growing wild in the northern hemisphere, and also in South America. Many of them are spiny plants which are often cultivated as hedge plants. Examination shows that the leaves may vary from entire spiny-toothed structures to those reduced entirely to spines, only their position on the stem revealing the fact that they are leaves. The plants bear **racemes** of yellow flowers and later small red, yellow or black berries which make them very attractive. **Pollination** in barberries is rather interesting. Each of the six stamens (see **Flower**) has a spot at the base of the filament sensitive to touch. When an insect, pushing about the base of the flower in search of nectar, touches this spot, the stamen moves violently, so that the sides of the insect's head are powdered with pollen. When the insect visits another flower some of this pollen may be caught by the stigma, pollination then being completed.

One species of barberry, *Berberis vulgaris,* is the alternate host of *Puccinia graminis,* the organism causing the destructive disease of wheat **rust.** (R.M.W.)

BARBET. Aves, Piciformes. Birds (**Aves**) of several species intermediate between the **woodpeckers** and **toucans.** They occur in the tropical regions of both Old and New World and some species are brilliantly colored. The name was formerly applied to other members of the order found in tropical South America which are now known as puff-birds. (A.W.L.)

BARBITAL (Veronal). A commonly used drug which is effective as a **sedative, hypnotic,** and anticonvulsant. It is used in insomnia, hyperthyroidism, **chorea** and **epilepsy.** Other barbiturates are now more frequently prescribed. These are phenobarbital and the pentobarbitals. These drugs are more potent than simple barbital. The pentobarbitals have the advantage of acting rapidly; hence they are particularly useful in insomnia.

Poisoning with overdoses of barbiturates is common and suicides are frequently committed with these drugs. The acute intoxication is characterized by coma and depressed respirations. Death may be caused by paralysis of the respiratory center or vascular collapse. Chronic barbiturism causes lethargy, drowsiness, dizziness, and sometimes difficulty in walking and speaking. Withdrawal of the drug alleviates the symptoms completely. (D.M.H.)

BARCHAN. Dune.

BARITE. The mineral barite is **barium** sulfate, $BaSO_4$, crystallizing in the **orthorhombic system.** May occur as tabular crystals, in groups, or lamellar, fibrous and massive. Barite has two perfect cleavages, basal and prismatic; hardness, 2.5–3.5; specific gravity, 4.3–4.6; which has led to the term heavy spar, occasionally used for this mineral. Its luster is vitreous; streak, white; color, white to gray, yellowish, blue, red and brown;

transparent to opaque. Sometimes yields a fetid odor when broken or when pieces are rubbed together, due probably to the inclusion of carbonaceous matter. Barite is a frequently occurring gangue mineral and is found also in large masses in **sedimentary** rocks. Its name, barite, is derived from the Greek word meaning heavy. It is used as a source of barium compounds.

Barite localities are widespread, including many European occurrences, Czechoslovakia, Germany, France, Spain and England. In the United States barite is found in New York, Connecticut, Pennsylvania, Virginia, Michigan, Missouri, New Mexico, Oklahoma, Utah, Colorado, South Dakota, Georgia and Tennessee. In Canada it occurs in Ontario and in Nova Scotia. (E.S.C.S.)

BARIUM. Symbol: Ba. Atomic number: 56. Atomic weight: 137.36. Density: 3.5. Melting point: 850° C. Boiling point: 1140° C. (Isotopes: page 290.)

Barium is a silver-white metal, harder than lead, oxidizes rapidly in moist air, reacts with water yielding barium hydroxide and hydrogen gas, burns when heated in air, emitting a brilliant light and forming barium peroxide. Discovered by Scheele in 1774 and isolated by Davy in 1808.

Barium occurs chiefly as sulfate (**barite**, barytes, heavy spar, $BaSO_4$), and, of less importance, carbonate (**witherite**, $BaCO_3$). Georgia and Tennessee are the principal producing states. The sulfate is transformed into chloride, and the electrolysis of the fused chloride yields barium metal. Barium metal has been recently used as a "getter" in radio tubes, and as a nickel-barium alloy in automobile ignition systems.

Acetate: Barium acetate ($Ba(C_2H_3O_2)_2$), white crystals, soluble, by reaction of barium carbonate or hydroxide and **acetic acid**, and then crystallizing.

Carbide: Barium carbide (BaC_2), black solid, by reaction of barium oxide and **carbon** at **electric furnace** temperatures, decomposes water yielding **acetylene** gas and barium hydroxide.

Carbonates: Barium carbonate ($BaCO_3$), white solid, insoluble, (1) by reaction of barium salt solution and **sodium** carbonate or bicarbonate solution; (2) by reaction of barium hydroxide solution and **carbon dioxide,** excess carbon dioxide forms barium bicarbonate ($Ba(HCO_3)_2$), solution and this by boiling yields barium carbonate. Decomposes at 1450° C.

Chloride: Barium chloride ($BaCl_2 \cdot 2H_2O$), white crystals, soluble, by reaction of barium carbonate or hydroxide and **hydrochloric acid**, and then crystallizing.

Chromate: Barium chromate ($BaCrO_4$), yellow precipitate, by reaction of barium salt solution and **potassium** chromate solution.

Cyanamide: Barium cyanamide ($BaCN_2$) mixed with **cyanide** ($Ba(CN)_2$), by heating barium carbide at 800° C. with nitrogen gas. Fusion of this cyanamide-cyanide mixture with sodium carbonate converts entirely to cyanide.

Hydride: Barium hydride (BaH_2), white solid, by heating barium metal or amalgam in **hydrogen** gas at 1170° C., reactive with water yielding barium hydroxide and hydrogen gas.

Hydroxide: Barium hydroxide ($Ba(OH)_2$), white solid, (1) by reaction of barium oxide and water, (2) by precipitation of barium salt solution with **sodium** hydroxide solution, yields ($Ba(OH)_2 \cdot 8H_2O$) on crystallizing, decomposes upon heating at about 850° C. to form oxide (BaO) and water.

Nitrate: Barium nitrate ($Ba(NO_3)_2$), white crystals, soluble, by reaction of barium carbonate or hydroxide and **nitric acid,** and then crystallizing, used in pyrotechnics for the production of green light.

Oxides: Barium oxide (BaO), white solid, melting point about 1900° C., reactive with water to form barium hydroxide; barium peroxide ($BaO_2 \cdot 8H_2O$), white precipitate, by reaction of barium salt solution and hydrogen

or sodium peroxide, yields anhydrous barium peroxide (BaO_2) upon heating at 100° C. in a current of dry air. Anhydrous barium peroxide is also formed by heating barium oxide in air or **oxygen** under pressure (at somewhat over one atmosphere pressure) and temperature of 400° C. there is transformed into peroxide with air 70%, with oxygen 95%, of the total barium oxide taken. This reaction was applied in the Brin process for separation of oxygen from air, first forming the peroxide, and later heating the same to a higher temperature when oxygen was evolved and barium oxide remained for use again.

Oxalate: Barium oxalate (BaC_2O_4), white precipitate, by reaction of barium salt solution and ammonium **oxalate** solution.

Sulfate: Barium sulfate ($BaSO_4$), white precipitate, by reaction of barium salt solution and **sulfuric acid** or **sodium** sulfate solution, insoluble in acids, by heating with carbon yields barium carbonate. Present in a mixture with zinc sulfide, known as lithopone, a paint pigment, formed by the reaction of barium sulfide and zinc sulfate in water, followed by filtration and drying. The mineral barite when powdered is used as a paint pigment, as a filler for rubber goods, paper, and linoleum, only source of all barium containing substances.

Sulfides: Barium sulfide (BaS), grayish-white solid, by heating barium sulfate and **carbon,** reactive with water to form barium hydrosulfide solution; barium hydrosulfide ($Ba(SH)_2$), solution, (1) by reaction of barium sulfide and water, (2) by saturation of barium hydroxide solution with **hydrogen sulfide;** barium polysulfides are formed by boiling barium hydrosulfide with **sulfur.**

Volatile compounds of barium, such as the chloride, color the bunsen flame green. (R.K.S.)

BARK. All tissues of woody stems or roots which occur outside the **cambium** are collectively known as bark. In the earliest stages of its development, the **stem** is covered by a layer of thin epidermal cells, which may persist for some time. With increased age and growth, however, this epidermal layer is lost, and a new tissue is formed, either from the epidermal cells or from those cortical **cells** just beneath the epidermis. The cells which form this new protective layer are the cork cambium cells. The tissue which is formed by them is often called the cork or outer bark. If the divisions of the cork cambium cells occur fast enough the bark will remain smooth for some time. In most cases, however, an internal cork cambium forms in the deeper cortical tissues, and by producing secondary cortical tissue in isolated patches leads to the development of scales or patches of bark, with ever-deepening fissures as growth continues. In a few cases, as in the beech tree, the smooth condition persists throughout the life of the tree. In all young stems the continuity of the surface is broken by patches of loose cells, the lenticels, which permit an exchange of gases through the bark. The inner bark consists of phloem or bast, the food-conducting tissue of the plant. (B.S.M.)

BARK BEETLE. Insecta, Coleoptera. A small **beetle** of cylindrical form which burrows in the sapwood and inner bark of trees and logs, forming characteristic patterns. This habit also gives the name engraver beetle to these insects. Most of the numerous species infest forest trees but a few attack fruit trees. Together with the timber beetles and a few species which attack herbaceous plants they make up the family Scolytidae. (A.W.L.)

BARK LOUSE. Insecta, Homoptera. A scale **insect.** These insects are minute creatures with sucking mouths. They spend most of their lives attached to the leaves, stems, or roots of plants. The name scale insect is due to the common secretion of a scale which conceals the body of the insect. There are many species, of which some, like the oyster-shell bark louse, *Lepidosaphes ulmi,* and the San Jose scale, *Comstockaspis perniciosa,* are important enemies of fruit trees. Spraying with lime-sulfur and kerosene emulsion is a common method of control, and fumigation with hydrocyanic acid gas is extensively practiced in the citrus groves of California. Sprays can be used more effectively when trees are dormant in winter.

Some scale insects produce substances of commercial value, notably cochineal and shellac. (A.W.L.)

BARKHAUSEN EFFECT. A series of minute "jumps" in the magnetization of iron or other ferromagnetic substance as the magnetizing force is continuously increased or decreased; discovered by H. Barkhausen in 1919. The effect may be observed by winding on the specimen, along with the magnetizing coil, a secondary coil connected to some sensitive detector of current fluctuations, such as an **oscillograph** or an audio amplifier. As the magnetizing current is steadily increased, the current in the secondary circuit, instead of being constant, exhibits a succession of small, sharp peaks or maxima, which the amplifier reveals by a faint clicking or snapping sound. These discontinuities in magnetization are interpreted as indicating the existence in the ferromagnetic material of elementary magnets, called "domains," which are much larger than atoms, but not large enough to be identified with individual crystals. Another possible interpretation is that the discontinuties are caused by the accidental coincidence of orientations in elementary magnets of much smaller size, somewhat as the **Brownian movement** of a particle is caused by the accidental impact of many molecules upon it simultaneously. (L.D.W.)

BARKHAUSEN OSCILLATOR. This is a type of vacuum-tube **oscillator** which depends upon the **transit time** of the **electron** across the tube for its operation. The conventional oscillator operates well at frequencies low enough to make the transit time negligible compared to a period of the alternation. As the frequency is increased this condition can no longer be met and such oscillators become less and less satisfactory until they fail altogether. Fig. 1 shows in principle, a Barkhausen oscillator. The plate is maintained at a slightly lower

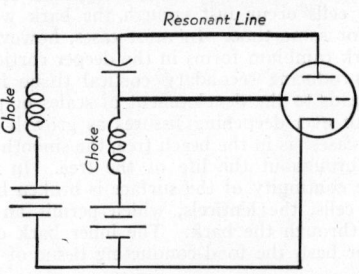

Resonant Line

Choke

Choke

Simple Barkhausen oscillator.

d-c potential than the cathode, while the parallel wires serve as a tuned circuit to which the electrons deliver energy. An electron starting from the cathode is accelerated by the positive grid voltage until it passes the grid, then is decelerated by the negative plate, reverses and goes back by the grid, then is again reversed in the grid cathode space and repeats until it finally hits an electrode and is removed. The electron thus oscillates about the grid. If the phase of these oscillations of the electron is correct with respect to the oscillations in the parallel line system, the electron delivers energy to the system. Fortunately the operation of the circuit is such that those with wrong phase are rapidly removed. However, the efficiency of the circuit is low. The frequency is determined largely by the tube dimensions and the

voltages applied, although other factors have a slight effect. (See **Klystron.**) (L.R.Q.)

BARLEY. *Hordeum vulgare* and *H. distichon.* Gramineae. Barley is a cereal grass with stalks up to three feet in height, and with the inflorescence a close spike. Many varieties are grown in cultivation, all probably derived from plants growing native in Asia. In the older varieties the flower possessed stout barbed **awns,** which were very disagreeable to those who handled the grain, and often the cause of serious trouble when the grain was fed to livestock. In the newer varieties, originating in Russia and widely cultivated and improved in the United States, these awns have been lost, leaving a smooth-fruited variety. In the mature barley grain the **aleurone** layer is usually three cells in thickness; the **embryo** is very small; and there is no **gluten.**

Primitive man in Europe used barley for food, as do many modern races. However, lacking gluten, it is impossible to make light bread from barley, a fact which prevents it from being popular with cultured races. It is frequently used in soups, and in the manufacture of breakfast cereals. Large quantities of the smooth grain are used as stock food. Its principal use, however, is in the manufacture of malt. For this purpose the grain is first soaked in water for 2 days or more at a temperature of about 55° F., with frequent changes of water to prevent excessive development of **bacteria.** The soaked grain is then spread out on the floor for twelve days, during which time **fermentation** occurs and germination starts. **Enzyme** action causes the starch (see **Carbohydrates**) of the grain to change to sugar. Before germination has led to a well-developed **coleoptile,** the grain is heated to stop growth and to drive off excessive moisture, which is reduced to about 2%. The sprouts are then removed and the malt, a pasty mass rich in starch and the enzyme diastase, is ready for use, principally in the manufacture of beer. (R.M.W.)

BARNACLE. Crustacea, Cirripedia. **Sessile** marine animals wholly unlike the more common **crustaceans** in appearance. Some are parasitic, some burrow in the shells of marine animals, and some attach themselves to submerged objects. The last habit is economically important since it results in the fouling of ships' bottoms and entails the expense of periodical cleaning. (A.W.L.)

BARNETT EFFECT. In 1915, S. J. Barnett discovered that a relatively long iron cylinder, when rotated at high speed about its longitudinal axis, developed a slight magnetization, the value of which was proportional to the angular speed. He found the magnetization to be about 1.5×10^{-6} c.g.s. electromagnetic unit per revolution per sec. for a cylinder about 7 cm. in diameter and 50 cm. long. The effect was attributed to the influence of the impressed rotation upon the revolving electronic systems within the atoms. The suggestion has been made that the magnetic moment of the earth and of the sun may be due in part to the rotation of these bodies. An inverse effect was discovered about the same time by Einstein and de Haas; viz., an iron cylinder, suspended vertically, was observed to rotate slightly when suddenly magnetized. (L.D.W.)

BAROCLINIC. When isobaric surfaces and equal-density surfaces do not coincide the atmosphere is said to be baroclinic. If the surfaces do coincide, the atmosphere is said to be barotropic. Local winds often arise when the atmosphere is highly baroclinic. (P.E.K.)

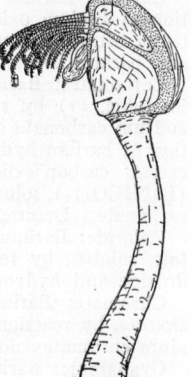

Barnacle.

BAROGRAPH. A recording **barometer**, commonly of the aneroid type. The pressure-sensitive unit consists of a pile of several elastic-walled evacuated boxes, connected through a suitable linkage to the stylus-arm, which carries the pen up and down on a slowly revolving drum. The variations of atmospheric pressure are thus traced on the paper, usually ruled in days and hours for the period of one week. Such instruments are extensively used by the U. S. Weather Bureau at its many observation stations. (See **Meteorological Instruments; Bar Chart.**) (L.D.W.)

BAROMETER. The barometer owes its origin to Torricelli, who, in 1643, first utilized a column of water to measure the atmospheric pressure and its variations. The obvious disadvantages of a barometer tube 34 feet long, with water vapor pressure above the liquid, soon brought about the substitution of mercury. This reduces the necessary length of the instrument to about 3 feet over all; and the vapor pressure of mercury at ordinary temperatures is negligible. To set up a mercurial barometer, one needs only to fill a clean glass tube, closed at one end, with pure mercury, heat it to expel air, and invert it in a small cup of freshly boiled mercury. At normal pressures the column will sink to about 76 centimeters, leaving a "Torricellian vacuum" above it. The scale must be made adjustable so that its zero may always coincide with the surface of the mercury in the cup. In some barometers the mercury is contained in a leather bag tied to the lower end of a short glass cylinder, making it possible to raise or lower the mercury level by a screw compressing the bag. The readings must of course be corrected for variations of temperature.

The aneroid barometer is merely a delicate pressure gauge, consisting essentially of a round, flat, air-tight,

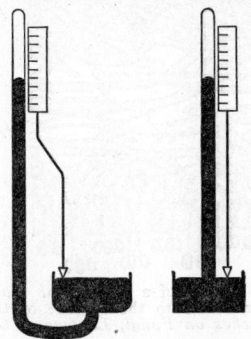

Two types of mercurial barometer (diagrammatic).

evacuated metal box, with elastic bases which respond to variations in pressure by springing in and out, and the movements of which are communicated to a pointer on a graduated dial. Such barometers are portable and quite sensitive, and are much used on shipboard, in airplanes, by engineers, etc. (L.D.W.)

BARRACUDA. Pisces, Teleostei. Vicious marine fishes (**Pisces**) of several species, found in coastal waters of temperate and tropical seas. They are said to be a more probable source of danger to bathers than the sharks; since many of them attain a length of 6' or more they are undoubtedly capable of inflicting severe injury.

The flesh is excellent and the fishes are caught both for sport and for the market.

The name barracuda is also applied to a New Zealand fish of the same order but of a different family. (A.W.L.)

BARRIER BEACH. A recently emerged coast (like the present Atlantic Coastal Plain) with shallow water extending for some distance from the shore will lend itself to the formation of barrier beaches. Waves tend to

stir up the bottom materials of sand and gravel to be redistributed which gradually build up a low ridge parallel to the coast. Such barrier beaches are also known as offshore bars and the quiet body of shallow water between the bar and the mainland is called a lagoon. (E.S.C.S.)

BARYCENTRIC PARALLAX. Barycentric parallax is a term given to a slight oscillatory motion of the **earth.** The center of gravity of the earth-moon system revolves about the sun in an **orbit** that is usually referred to as the orbit of the earth. Both the earth and **moon** are revolving about this center of gravity of the system as it in turn revolves about the sun. At the time of **conjunction** of the moon with the sun the moon is inside of the orbit of the system about the sun and the center of the earth is outside. At opposition this condition is reversed and the center of the earth is on the inside. Hence, the center of the earth oscillates slightly back and forth across the orbit with a period of one **synodic** month. This slight oscillatory motion of the earth is known as barycentric parallax.

Accurate measurements of barycentric parallax provide a method for the determination of the relative masses of the earth and the moon. In accordance with the principles of mechanical equilibrium the distances of the center of gravity of the earth-moon system from the centers of mass of the two individual objects are inversely proportional to the masses of the objects. Hence, if we can determine the distance of this center of mass of the system from the centers of the earth and moon we can at once determine the relative masses of the two objects. Accurate measurements of barycentric parallax determine the position of the center of mass of the earth-moon system as 2880 miles from the center of the earth. Since this is about $1/82.5$ of the distance of the moon from the earth, we find that the mass of the moon should be $1/81.5$ the mass of the earth. (W.K.G.)

BARYE. The barye, or bar as now commonly written, is the c.g.s. absolute unit of pressure, equal to one dyne per sq. cm. It is of course a very small unit, one atmosphere (760 mm. of mercury) being equal to about 1,013,246 baryes. Hence the megabarye or megabar, i.e., 1,000,000 baryes, is often more convenient. Some confusion exists in the use of these terms, some authors using "barye" to designate 1,000,000 dynes per sq. cm. and "microbarye" as synonymous with the barye above defined. Still more unfortunate is the term "barie," defined as 750 mm. of mercury, and hence very nearly equal to 1,000,000 dynes per sq. cm., though fixed on an altogether different basis. Acoustic engineers commonly express sound pressures in bars. (L.D.W.)

BARYSPHERE. Lithosphere.

BARYTES. Barite.

BARYTOCALCITE. This mineral is a **carbonate of barium** and **calcium** which crystallizes in the **monoclinic system** but occurs massive as well. It has a perfect cleavage parallel to the prism and one, less perfect, parallel to the base; fracture, sub-conchoidal; brittle; hardness, 4; specific gravity, 3.64–3.66; luster, vitreous; color, white or gray or may be greenish or yellowish; transparent to translucent. Barytocalcite is found in Cumberland, England, associated with **barite** and **fluorite.** (E.S.C.S.)

BASAL CONGLOMERATE. A conglomerate that lies on, or occurs just above, a plane of erosion or **unconformity.** Such a conglomerate constitutes the first **sedimentary** stage in a normal cycle of sedimentation. (R.M.F.)

BASAL METABOLISM. Metabolism.

BASALT. A fine-grained to dense, sometimes **porphyritic,** intrusive or extrusive **igneous rock,** black or greenish-black in color, characterized by a preponderance of calcic **plagioclase feldspars** and **pyroxene** together with minor amounts of accessory minerals such as **olivine.** Glass may be present. Amygdaloidal structure is common in such cavities and beautifully crystallized species of **zeolites, quartz** or **calcite** are frequently found.

The lava flows of the Plateau of the Deccan in India, the Columbia Plateau of Washington and Oregon States, as well as the Triassic lavas of eastern North America are basalts. Perhaps the most famous basalt flow in the world is the Giants Causeway on the northern coast of Ireland, in which the vertical joints give the impression of having been artificially constructed. Pliny used the word basalt and it is said to have had an Ethiopian origin, meaning a black stone. (R.M.F.)

BASANITE. Tephrite.

BASE. See individual bases under each metal; also **Acids, Bases and Salts.**

BASE CIRCLE. Involute.

BASE LEVEL. The ultimate physiographic feature of the processes of **denudation** is the reduction of the land surface to sea level, because it is at this level that the processes of river **erosion** are completely checked. As sea level is the datum plane below which stream erosion is impossible this may be taken as the theoretical lower limit of stream erosion and the level to which all the land surfaces must be brought if no other forces intervene. Actually, however, base level seems to be a limit which, however closely approached, may not in reality ever be reached. Therefore a broader use of the term base level is frequently made, base level being the lowest level to which a given stream can cut. A stream flowing into a body of water at an elevation above the ultimate datum, sea level, is said to be at a temporary base level. In any case when a stream has cut its channel to such a degree that it has just enough velocity to carry its load without further erosion being possible, it is then said to be at grade. Local or temporary base levels of relatively small area may be developed without reference to sea level. In this sense the term base level is not entirely comparable to the term **peneplain.** (E.S.C.S.)

BASE LINE. Base line is a **surveying** term employed to describe an accurately measured horizontal distance which is obtained by the use of a graduated steel or invar tape (see **Chain**). The purpose of this measurement is to establish accurately the horizontal distance between two points called stations, from which to proceed with the instrument surveying. The base line should, if possible, be located where the terrain intervening between its ends is fairly level so that the measurements may be simply yet accurately made. The purpose for which a base line is laid out determines the degree of precision which is to be used in the measurements of this line. The base line, laid out along the bank of a river in order to determine the width of the river by instrument sights taken from each end to a fixed point on the other bank, would not need to be as accurate as that for first order **triangulation** of the character needed in Coast and Geodetic surveying. The latter requires the use of a special tape standardized under certain conditions and made of invar, which has a very low coefficient of temperature expansion. When used in base line measurements for first order work the tape must be supported at intervals so as to eliminate, as far as possible, the effect of sag. Spring balances are required to reproduce the tension under which the tape was standardized. Thermometers are employed to record the temperature so that correction may be made for the thermal changes.

Another meaning of base line is connected with the methods of public **land subdivision** used principally in the western part of the United States. The primary unit of subdivision is the township, which is subdivided into sections. These units are referred to a pair of principal axes which have their origin at the intersection of a true **meridian** called the principal meridian, and the true parallel of **latitude** called the base line. (F.T.M.)

BASEMENT MEMBRANE. Epithelium.

BASIC ROCK. A term applied to **igneous rocks** whose content of silica is less than about 52%. It is not an exact term and is gradually going out of use. It may be convenient for field use. (E.S.C.S.)

BASIC SIZE. Interchangeable Manufacturing.

BASIDIOMYCETES. Nearly all the **fungi** commonly observed belong to this group, which includes toadstools, mushrooms, puff-balls, and many other forms. The characteristic feature which distinguishes them from other fungi is the basidium, typically a club-shaped structure bearing 4 **spores** at its apex. Members of this group are found wherever plant life can exist. The majority of them are **saprophytes,** living on dead wood and soil rich in humus; a few are parasites. The life-history of the common mushroom, one of the best known and most important, is fairly typical of the group.

The vegetative phase consists of a mycelium. This is a mass of slender much-branched threads, called hyphae, which grow throughout the substratum. The mycelium

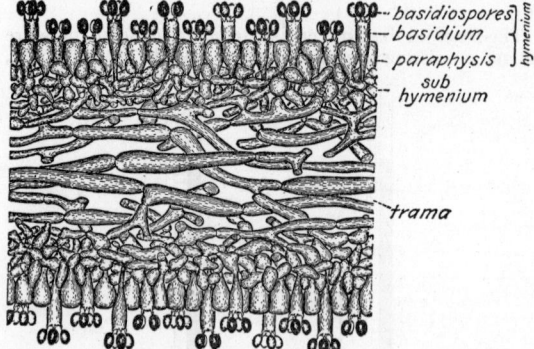

A section through the gill of a mushroom. *Coprinus Comatus.* Section cut perpendicular to the surface of the gill. (*From Buller's Researches on Fungi, Longmans, Green & Co.*)

is perennial. Each hypha is a long filament composed of many segments, each containing two nuclei. The hyphae absorb from the substrate the organic materials which the fungus needs in order to live, and convert it into other forms. Much of this food substance accumulates within the mycelium.

When sufficient material has been stored and conditions are suitable the fungus fruits. The fruit body first appears as a small round object rising from the substratum. At first a thin membrane completely envelops the fruit body. When the latter elongates, the membrane is broken, revealing the elongating stalk bearing at its tip an umbrella-shaped cap or pileus. Remnants of the broken membrane sometimes remain on the surface of the pileus or around its edge. Part of the membrane often remains at the base of the stalk, forming a cup-like volva. In some species of mushrooms the membrane is double. The inner portion, when broken, forms a ring, called the annulus, around the stalk. Both volva and annulus are found in many mushrooms; others have only one of them; in many no trace of either structure appears.

On the lower surface of the pileus there are numerous thin radiating plates called gills. The lateral surfaces of the gills are formed by hyphal tips which grow perpendicular to the surface and form a compact layer called the hymenium. The hyphal tips composing the hymenium are the basidia, the reproductive structures which

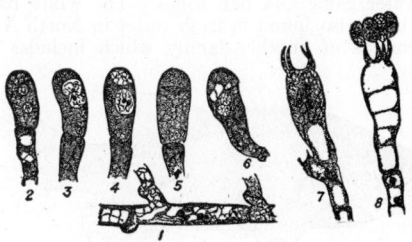

Development of basidium and formation of spores in mushroom. 1, Septate mycelius with binucleate cells; 2, Young basidium containing two unrelated nuclei; 3, 4, Fusion of nuclei in basidium; 5, 7, Development of fournucleate condition in basidium; 8, Formation of basidiospores that receive the four nuclei. (*Redrawn from Harper.*)

distinguish this group of fungi from all others. Each basidium is a cylindrical, binucleate (see **Nucleus**) body cut off from the tip of a hypha. The two nuclei fuse and immediately divide, usually twice, so that the basidium contains four haploid nuclei. From the outer end of the basidium four slender pegs called sterigmata develop. A small **spore** forms at the tip of each sterigma. Into each spore one of the nuclei of the basidium migrates. The spore is discharged from the sterigma and falls down between the gills and into the air. It is carried about by air currents and eventually falls to the ground. There the spore germinates, giving rise to a slender branching mycelium composed of uninucleate segments. This is the primary mycelium. Branches of two primary mycelia unite to form a secondary mycelium. The two nuclei present in each segment of a secondary mycelium have come from different spores. The manner in which they continue their identity during division is interesting. The two nuclei divide simultaneously. When division is about to occur a small bulge forms on the side of the hypha. One of the two nuclei enters this bulge, the other remains in the hypha. After division a cross wall forms, separating the two nuclei in the hypha. The protuberance containing the other nucleus continues to grow and forms an elbow-shaped structure, which joins the two cells of the hypha. It is called a clamp connection. One of the nuclei formed in this clamp returns to the original cell, the other passes through the clamp and into the other cell. By this means the two cells each receive a nucleus derived from one of the original nuclei. Nuclear fusion occurs only in the basidium. So there is in the mushroom an **alternation of generations** differing from that in most plants. The haploid or gametophyte phase consists of the primary mycelia. Following this a prolonged binucleate or dicaryon phase exists. Only in the basidium does nuclear fusion occur and reduction immediately follows. So the diploid phase or sporophyte is represented only by the basidium. There are no sex organs in this group of plants. The differences between the edible mushroom and other basidiomycetes is mainly in the structure of the fruit body, the location of the hymenium and the nature of the basidia.

The nature of the hymenium is the basis for classifying basidiomycetes. The Hymenomycetes are those in which the hymenium is exposed; in the Gasteromycetes it is formed within the fruit body. The principal order of Hymenomycetes is the Agaricales or **agarics**, or the gill fungi. These are commonly described as mushrooms and toadstools. While popular conception gives to these two terms very exact meanings, actually the terms are confusing and of little value. A common distinction is that a mushroom is edible and a toadstool poisonous. The question is then how to distinguish a harmless fungus from one that is poisonous. To eat them is perhaps a certain but not a safe way. There are many

definitions which aim to establish ways of distinguishing the noxious forms, based on the presence of volva and annulus, and other features. These do not always hold true. The only safe way is to learn to identify with certainty any fungus to be eaten and to reject all that are not known.

Another well-known order of basidiomycetes is the Polyporales; of these the polypores or Polyporaceae are best known. The distinguishing feature of these are the pits or tubes on the lower surface of the fruit body. The hymenium lines these pits. The fruit bodies of the polypores usually do not have a stalk and pileus, but form a layer spreading over the surface of rotting wood or grow out from the wood like a shelf. This shelf-like habit has given to these plants the name bracket-fungi. The fruit bodies of bracket-fungi live for many years and often show distinct growth layers. In some species these are a foot or more across and several inches thick.

The other large group of basidiomycetes is the Gasteromycetes, distinguished by having the hymenium lining irregular cavities in the fruit body. Until these are fully mature, the spore-bearing parts are completely enclosed by sterile tissue. There are many different kinds of Gasteromycetes. One of the best known is the puff-balls, or Lycoperdiales. The outer wall or peridium of the puff-ball surrounds the gleba or spore-bearing tissues. When the spores are mature the peridium breaks. In some genera the wall breaks up into irregular fragments and leaves the spore-mass exposed; in others a pore is formed at the apex of the fruit body. The slightest pressure against the peridium will cause clouds of spores to puff from the pore. The number of spores formed in a single puff-ball is tremendous; 7,000,000 has been given as the number from a good-sized sporophore. Some of the puff-balls are the largest fungi known, reaching a diameter of a foot or more. If gathered before the spores are formed, most of the puff-balls are edible. None are poisonous.

Geasters or earth-stars develop very much as puff-balls do. But when they are mature the outer peridium splits into sectors which bend outward, revealing the spore-bearing part within. These fungi are frequently found growing on dry sandy soil.

Another group of Gasteromycetes includes the Bird's-nest fungi. The spore-bearing structures here are the "eggs," small oval bodies resting in the bottom of an open cup of sterile tissue. A last and curious group of Gasteromycetes is the stinkhorns, vile-smelling fungi whose spores are included in a mass of sticky stinking tissue, formed by the disintegrated glebal substance. This attracts carrion flies and other insects, which carry the spores about.

Other important families of basidiomycetes are the **Rusts** and the **Smuts**.

There are several theories as to the origin of the basidiomycetes. Some maintain that they have evolved directly from certain primitive flagellates. Others derive them from the red **algae**. Many consider them to have descended from the **ascomycetes**. Adherents to this theory observe the dicaryon phase of the ascogenous hypha of the ascomycetes and note that if this phase were prolonged for some time it would be very similar to the secondary mycelium of the basidiomycete. They also note that the asexual reproduction by conidia occurs in the basidiomycetes very much as it does in the ascomycetes.

Dictyophora duplicata, Stinkhorn, a fleshy Basidiomycete.

The basidiomycetes include many important plants. Many are much sought as food. A few, and especially the field mushroom, *Agaricus campestris,* are extensively cultivated. This cultivation is commonly carried on in caves or cellars, where the requisite moisture conditions may be maintained and where sudden fluctuations in

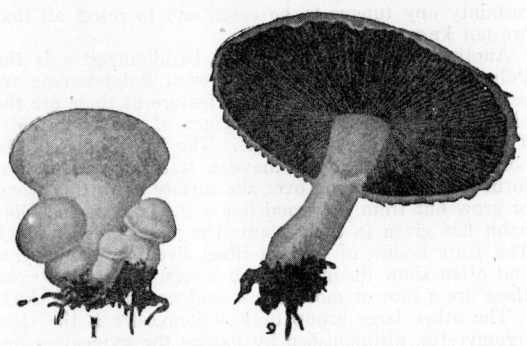

Agaricus campestris. The first figure shows the cultivated variety; the second, the wild form. (*From Farlow's Icones Farlow Herbarium, Harvard University.*)

temperature are not so apt to occur. The mushrooms are grown in beds. A mushroom bed is composed of a mixture of well-rotted manure, straw and other vegetable matter and earth. This mixture is allowed to stand for some time until heating has ceased and the temperature is constant. Then bits of mushroom spawn are buried in the mass. Mushroom spawn is a dense mass of healthy mycelium growing in a suitable substrate. The spawn grows rapidly and penetrates all parts of the bed. After about three weeks, fruiting begins, the young sporophores pushing up from the surface of the bed. They are gathered in the early stages of their development and kept in cool airy rooms until marketed.

Many basidiomycetes are serious disease-producing plants, attacking trees particularly. Often their presence is unnoticed until too late; the mycelium has permeated the tissues of the host. Other basidiomycetes attack dead plants, reducing them to simpler forms. These are the basidiomycetes which cause decay. They are saprophytes. (R.M.W., B.S.M.)

BASILAR MEMBRANE. A membranous partition in the auditory chamber (cochlea) of the inner ear forming the floor of the cochlear duct. Upon it lies the organ of Corti, which contains the sensory cells for hearing. The basilar membrane increases in width as it passes from the base of the cochlea towards the apex, thus influencing the character of the vibrations with which the basilar membrane responds to sounds of different frequency. (A.W.L.)

BASILAR PLATE. A plate derived from the ventral wall of the first body segment in millipedes (Diplopoda) which forms the most posterior component of the mouth parts. (A.W.L.)

BASILISK. Reptilia, Sauria. *Basiliscus.* A large tropical American lizard with crests on the back and tail which resemble the fins of fishes. Four species are known. (A.W.L.)

BASIN. In structural geology, a special type of folded structure in which the strata dip in toward a central point from all directions. As exposed at the surface the ground plan of a basin may be roughly circular or elliptical. Type locality, the Paris Basin, France. (R.M.F.)

BASIPODITE. Biramous Appendage.

BASOMMATOPHORA. An order of fresh-water snails with eyes located at the bases of the tentacles. (A.W.L.)

BASS. Pisces, Teleostei. A term so loosely applied that no specific definition can be given. The small-mouthed and large-mouthed black bass, Kentucky bass, and warmouth bass are highly valued North American fresh water game and pan fishes. The white bass and yellow bass, also found in fresh water in North America, are members of another family which includes mostly

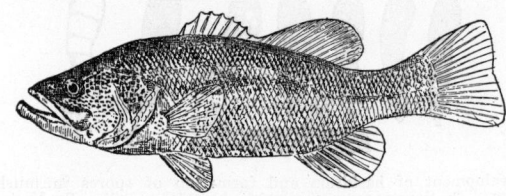

Large mouth black bass.   (*Slingerland-Comstock Publishing Co.*)

marine species known as sea bass. These are also valuable for food and some species are excellent game fishes. Among the marine species the striped bass is important as a game fish on the Atlantic coast and the black sea-bass or jewfish of the Pacific coast is among the largest of marine game fishes, attaining a weight of 500 lbs. or more. (A.W.L.)

BAST. As commonly used, the term bast is synonymous with phloem. (R.M.W.)

BAST FIBERS. In the phloem, pericycle, or cortex of many plants there occur long slender thick-walled cells with tapering ends. These cells, often occurring in groups of considerable size, are often called bast fibers. They give to the stem of the plant considerable strength. Many bast fibers are useful to man, especially in the manufacture of cloth and cordage. Linen is made from the pericyclic fibers of flax, while hemp and remie are coarser pericyclic fibers used in making coarse sacking and cordage. The phloem fibers of jute are also used for making rope and burlap, as well as the strong webbing used in upholstering furniture. Plant anatomists call them phloem fibers, pericyclic fibers, or cortical fibers. (R.M.W.)

BAT. Mammalia, Chiroptera. A flying mammal whose wings are formed of folds of skin, the patagia, stretching between greatly elongated digits of the fore limbs, the sides of the body and hind limbs, and thence to the tail. Since the wings are living tissue they form extensive radiating surfaces, a fact correlated with the much greater development of the group in warmer regions. Most species found in the temperate zones hibernate, although migration has been reported, and few species invade more northern limits of these zones.

It has recently been shown that the amazing ability of bats to avoid obstacles while flying rapidly in the dark is due to the reflection of sound waves from these objects as the animal approaches them. The bat emits a constant series of sounds so high-pitched that the human ear does not detect them, hence the use of suitable apparatus to bring them within range of human hearing is necessary to demonstrate their occurrence. The bat's ears are highly specialized to perceive these sounds and to determine by the elapsed time the direction of objects from which they are reflected and the distance of these objects.

The habits of bats vary greatly. The flying foxes or fruit bats of the Old World are fruit-eating species exclusively, the horseshoe and leaf-nosed bats are insectivorous, and some species of vampire and false vampire bats are blood-suckers. Among the bats with distinctive names are the barbastelle, pipistrelle, noctule and serotine, all found in Europe and other parts of the Old World. (A.W.L.)

BAT TICK. Insecta, Diptera. Highly specialized **flies** which live as **ectoparasites** on bats. Like other parasitic flies, their habits are accompanied by some structural resemblance to the true ticks, hence the inaccuracy of the common name. The bat ticks belong to two families, Streblidae and Nycteribiidae. All species of the latter are wingless. (A.W.L.)

BATAGUR. Reptilia, Chelonia. Large fresh-water **tortoises** of several genera found in India and the Malayan region. (A.W.L.)

BATHOLITH. Batholith, by some writers spelled bathylith, is derived from the Greek meaning deep, and stone. It is a very large intrusive **igneous** mass with steeply inclined contacts, enlarging downward to undetermined depths. Typical batholithic rocks have a relatively coarse and even texture, such as **granites and diorites.** Batholiths occur as the roots or cores of folded and faulted mountain ranges and are therefore closely associated with mountain-building, although probably not the cause of the deformation but rather consequent to it. The mode of emplacement of such large cross-cutting and relatively uniform igneous rock bodies has not yet been definitely determined. The term was proposed by Zuess in 1888. (R.M.F.)

BATHYSPHERE. A closed and air-tight chamber arranged to contain observers, and to be lowered at the end of a cable to great depths of water, is a bathysphere. It is principally used for scientific study of animal life and other conditions in the depths of the ocean. The bathysphere is spherical in shape. It has a small observation window and a telephone, the lead of which is unwound along with the supporting cable. The observer keeps in communication with the mother vessel and possibly dictates notes by means of his telephone. Suffocation is forestalled by the slow release of compressed oxygen from tanks. Depths of 3028′ have been reached by the bathysphere, and this, of course, is far beyond the maximum depth achieved by divers or submarines. (F.T.M.)

BATONETTE. Cell.

BATRACHIA. The frogs and toads.

BATTERY. The term battery refers to any group of duplicate units which are contributing individually to a common effect. A common usage of battery is in reference to a group of cannon forming, with the personnel, a military offensive unit. The word is also frequently applied to engineering usage, such as battery of boilers, which consists of a number of individual boilers on a common header supplying steam to a common service, or a battery of mixers, etc.

But by far the most common usage of the word is in reference to a collection of chemical cells for the production or storage of electrical energy. As such, the battery may be of the primary type, of which the individual unit is the primary **cell,** or it may be the ordinary **storage battery,** which is, strictly speaking, an electrical accumulator. The lead and sulfuric acid storage battery is the one most frequently found. A single cell of this type of battery has an e.m.f. of between 2 and 2½ volts when fully charged, and an ampere-hour capacity which is dependent upon the exposed area of the plates to which the electrolyte has access, as well as upon the quality of construction.

The electric system of the automobile is based upon the 3-cell, 6-volt storage battery. Farm lighting plants are frequently 32-volt, and require 16 cells. A 60-cell battery is standard for 110-volt service. To obtain these voltages, the cells must be connected in series. When a storage battery is used, the accumulated

energy is withdrawn through the conversion of chemical energy into electrical energy. Without some means of recharging, the battery will become completely dis-

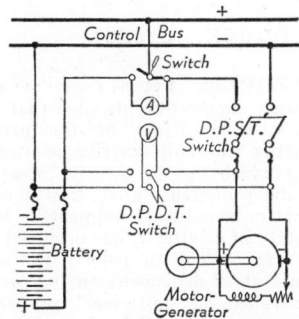

Battery and motor-generator parallel on control bus.

charged, and cease to deliver current. Automotive batteries are charged from a **generator** which is driven by the engine. Farm lighting batteries are recharged periodically by a special stationary engine-driven generator. Larger batteries, such, for instance, as those on 110-volt control buses, are usually kept charged by a motor-generator set or **trickle charger** which converts some available source of a.c. into d.c. The accompanying figure shows a connection frequently employed where the direct current bus can be energized either by a motor generator or a battery, and the same motor generator used to recharge the battery. (See **Electric Cell.**) (F.T.M., L.R.Q.)

BAUMÉ SCALE. A special **hydrometer** scale in industrial use, really two scales, one for liquids *lighter* than water and another for liquids *heavier* than water. The former is used in the petroleum industry and the latter in the acid and heavy chemical industry.

FOR LIQUIDS LIGHTER THAN WATER	DEGREES BAUMÉ	FOR LIQUIDS HEAVIER THAN WATER
. . . .	0	1.000
1.000	10	1.074
0.933	20	1.160
0.875	30	1.261
0.823	40	1.381
0.778	50	1.526
0.737	60	1.706
0.700	70	1.933
0.669	80	2.231
		(R.K.S.)

BAUXITE. The term applied, by Dufrenoy in 1847, to the amorphous mineral aluminum oxide dihydrate ($Al_2O_3 \cdot 2H_2O$), an important ore of **aluminum.** The term is also applied to aluminous **laterites** which contain commercial amounts of aluminum hydroxide. (See **Gibbsite.**) (R.M.F.)

BAY OIL. Volatile Oils.

BAYBERRY. *Myrica cerifera.* Myricaceae. This is a shrub which is found in marshy places in the United States. The simple leaves are very fragrant, especially when crushed, and are sometimes used in cooking. The fruit, a blackish drupe (see **Fruit**), is thickly coated with a mealy crust of wax. To obtain this wax, the fruits are boiled in water, causing the wax to melt and float to the top. The wax is pale green and very fragrant. It is used in making candles and also in scented soaps. The wax from this plant, and from other species of *Myrica,* has been used medicinally.

Sweet bay is the true laurel, *Laurus nobilis,* a small

tree found in the Mediterranean region. The leaves are very fragrant and are used as a condiment. From the berries a fragrant oil is obtained which is used medicinally. (R.M.W.)

BAYONET CHUCK. Chuck.

BAYES' THEOREM. Let $c_1, c_2, c_3 \cdots c_s$ be some s mutually exclusive random events such that one of them must happen and let $P(c_i)$ be the **probability** of c_i. These events we shall describe as causes. Let E be some event which can be directly observed and let $P(E \mid c_i)$ be the probability that E will occur on the assumption that c_i has already happened. Let $P(c_i \mid E)$ be the probability of c_i after E has occurred. The probabilities $P(c_i)$ are called "a priori" probabilities, i.e., probabilities not at all dependent on the event E, where $P(c_i \mid E)$ are called "a posteriori" probabilities since these depend on the event E having occurred. Bayes' theorem states

$$P(c_i \mid E) = \frac{P(c_i) \cdot P(E \mid c_i)}{\sum\limits_{j=1}^{s} P(c_j) P(E \mid c_j)}.$$

The theorem states exactly how the probability of a certain "cause" changes as different events actually occur. The theorem was published posthumously in 1764. When all conditions of the theorem are fulfilled there is no objection to Bayes' theorem. The difficulties in applying the theorem depend upon the fact that $P(c_i)$, the "a priori" probabilities, are not known and are assumed often to be equal in the absence of other knowledge. Bayes' theorem has been made the foundation of the theory of testing statistical hypotheses by some statisticians. However, it has been found to be unscientific, to give rise to various inconsistencies, and to be unnecessary. The modern theory of testing hypotheses makes no use of it. (L.A.A.)

BAYOU (Oxbow). A stagnant or abandoned course of a meandering river. Also a general term for a stagnant inlet or outlet of a lake or bay. (R.M.F.)

B BOARD. Telephony.

BDELLOIDEA. An order of rotifers, named for their fancied resemblance to leeches. (A.W.L.)

BDELLONEMERTEA. An order of **nemertine** worms of broad, flat form, named from their superficial resemblance to leeches. They live in the branchial chamber of **mollusks.** (A.W.L.)

BEAK. 1. A pointed protuberance. 2. The snout or rostrum of an animal, such as the beaked whale. 3. The horn-covered jaws of turtles and birds. The beak of birds is also called the bill. 4. The jaws of **cephalopod** mollusks.

Beaks of birds show a wide range of specialization for various uses, especially related to different types of food. Slender and elongate beaks are found in many wading species which must reach below the water for food and in the **hummingbirds** which visit deep-throated flowers. The broad beak of the **duck** has sieve-like structures at the sides and serves effectively to collect and strain out small particles from the water. The hooked beaks of birds of prey, small and slender beaks of insectivorous birds, and thick, strong beaks of seed-eating species are other examples.

The cephalopod mollusks, including the **squids**, **octopus** and related species, have two sharp jaws associated with the mouth which work together and resemble the beak of a bird. (A.W.L.)

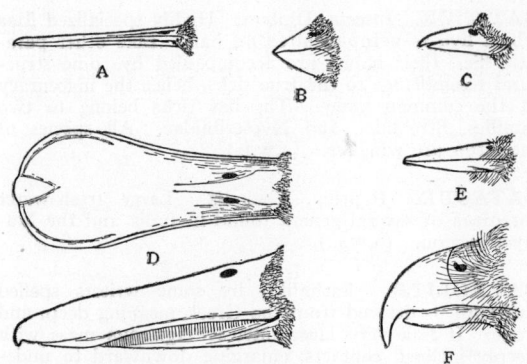

The beaks of birds. A, yellow-legs, a wader; B, cardinal, a seed-eater; C, flycatcher, an insect eater; D, the shoveler duck, dorsal surface above and side view showing the lateral sieve below; E, a woodpecker's chisel-tipped beak; F, hawk, a bird of prey.

BEAKED FISH. Pisces, Teleostei. A member of a group of African fishes (**Pisces**), some of which have the snout sharply elongated. Beaked members of other groups also occur, including the beaked salmon, which ranges from Japan to Australia. This fish is not a true salmon but constitutes a separate family. (A.W.L.)

BEAM. A beam is a straight or initially curved member which supports **bending loads** without the aid of **arch** action. (See **Cantilever Beam, Continuous Beam, Curved Beam, Fixed Beam, Indeterminate Structure, Simple Beam.**)

The term is also used to designate certain rolled steel sections such as the **I-beam** and the **wide flange beam** which may be used as beams or **columns.** (C.W.C.)

BEAM BRIDGE. Bridge.

BEAM POWER TUBE. This is a type of tetrode used for power amplification (see **Amplifiers**) in which the **electrons** passing from the **cathode** to the **anode** are concentrated into a beam which causes them to form a space charge near the anode and turn back any secondary electrons which are emitted from the anode. In this respect the tube behaves like a pentode (see **Tubes**), but, since it does not actually have the suppressor grid in the way of the electrons, it has a somewhat more desirable operating characteristic. (L.R.Q.)

BEAN. The fruits of many different plants and in many cases the plants themselves are called beans. Nearly all of them are members of the Leguminosae. In Europe and America, beans are principally plants of the genus *Phaseolus,* with *Phaseolus vulgaris,* the common Garden Bean, a most important species, with many varieties in cultivation. This plant is a tender annual, probably originally native in South America, and grows either as a low bush or as a twiner. Many of the varieties grown have been selected to yield a thick, rather fleshy pod with small seeds; these are string or snap beans. In some varieties chlorophyll (see **Pigments in Plants**) is either entirely or largely lacking in the pods, giving the Wax or butter bean. Many other varieties are grown primarily for the dried seeds, which have a very high food value and keep exceedingly well if dried and protected from insects. Common varieties used as dry beans are Pea, Yellow Eye and Red Kidney beans. Shell beans are those in which the nearly mature but green seed is eaten.

Related to the Garden Bean is *Phaseolus multiflorus,* the Scarlet Runner Bean, which is often grown as an ornamental plant because of its showy scarlet flowers. Another species, widely grown in warmer climates, is

Phaseolus lunatus, the Lima Bean, another plant from South America. The large flat pods of this plant contain a few large flat seeds. Varieties have been developed which can successfully mature in regions having a short growing season.

In Europe a common bean is the Broad Bean, also called Windsor or Horse Bean, *Vicia faba.* Its seeds are rich in nitrogenous compounds and rather hard to digest. These seeds are extensively used for horse food, as well as for human consumption.

In the Orient, the bean crop is almost entirely *Glycine Max,* the Soy Bean. The plant is an erect bushy annual with trifoliate leaves and bears fruit (pods) in great abundance. Tremendous quantities are grown in Manchuria. The plant has been introduced into the United States, where it is becoming increasingly popular, for several reasons. One is the fact that it will grow well on poor soils and, being a legume, will build up the fertility of the soil, because of its associated nitrogen-fixing bacteria. It is also an important forage crop. The seeds are rich in oil and have received much attention from chemists, with the result that large quantities are now used in the preparation of enamels, linoleum, inks, paints, soaps, etc. The cake remaining after the oil is pressed out is used as a stock food. Unquestionably, soy bean culture will increase in the United States in the future.

In Africa, species of *Dolichos* are used as food.

All the plants so far treated are legumes having "bean-" like fruit. But in the Vanilla bean one deals with an entirely different plant. Vanilla beans are the fruits of a climbing orchid growing wild in Central America and Mexico. Since prehistoric times these fruits have been used to flavor chocolate. At the present time (and with some difficulty) the plant is cultivated in several tropical countries, both in America and in the Orient. Like many other orchids, the flowers are pollinated only by certain specially adapted bees and perhaps the Hummingbirds. When introduced into regions where the necessary insects are lacking, pollination must be done by hand. The fruits, or pods, are slender elongate structures containing innumerable minute seeds. These seeds are germinated with difficulty, so new plants are usually obtained by cuttings of the old stem. Three years are required before the rooted cuttings will begin to bear fruit. In preparing the pods for market, they are first gathered and partially dried. After this they are stored in tight boxes during the night. This process is repeated daily throughout the period of "sweating." As a result of this treatment, the pods become deep chocolate brown, somewhat wrinkled objects, having a rich aromatic odor. The cured pods are packed in tight containers and shipped to the manufacturers of the extract. Vanilla is used mainly as a flavoring.

Many other plants contain vanillin, or similar substances, and so are often used in the manufacture of artificial vanilla. Among these are the seeds of a tropical South American plant, Coumarouna, which yield a product coumarin, used in perfumery. The seeds from which this coumarin is obtained are called Tonka Beans. (R.M.W.)

BEAN WEEVIL. Insecta, Coleoptera. The common bean weevil is a small beetle, *Mylabris obtectus,* which attacks growing beans and cowpeas in the pod. Methods of control are the same as for the pea weevil. The four-spotted bean weevil is a related species which attacks beans and peas both in the field and in storage. Control as above. (A.W.L.)

BEAR. Mammalia, Carnivora. Large, heavily built animals with rudimentary tail and plantigrade feet. In North America the polar bear, *Thallasarctus maritimus,* black bear, *Ursus americanus,* grizzly, *U. horribilis,* and Kadiak bear are recognized species and several less familiar names apply to species of brown bears of Alaska. The term brown bear itself does not designate a single species. Numerous varieties of the black and grizzly bears also occur.

Bears are found in every continent except Australia, although Africa and South America have only one species each. The African species, known as Crowther's bear, is found in the Atlas mountains and is related to the European brown bear. The South American spectacled bear, named from the light-colored rings around the eyes, lives in the Peruvian Andes. Among the more striking bears of Eurasia are the Himalayan bear, a black bear of moderate size with a white chevron on its breast, and the peculiar sloth-bear, *Melursus labiatus,* of India which has unusually large curved claws.

The term bear is also applied to the koala, which is known in Australia as the native bear for wholly superficial reasons. (A.W.L.)

BEARING. In navigation, both air and sea, the term bearing is used to indicate direction. It is the angle, measured at the observer, or some specified point, between two lines in the plane of the horizon. Careful distinction must be made between the meaning of the term bearing when used alone and the same term when qualified, e.g., relative bearing, compass bearing, etc.

The bearing of a given point from the observer is simply the true direction of that point, expressed in the conventional manner for expressing direction, i.e., to the right, through 360°, and in three digits.

If the compass is used for finding the bearing of a point from a ship, the value obtained will be the compass bearing and the compass corrections must be applied to obtain true bearing. Methods for obtaining bearings by using the pelorus are explained in the article on that instrument.

Frequently it is easier to use a ship's keel as a reference line, rather than north, for determining bearing.

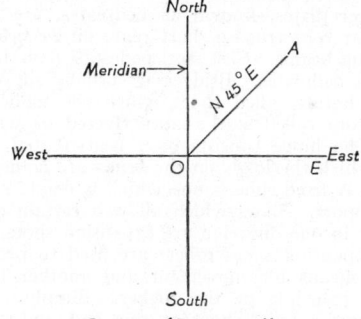

Bearing (Directional)

In this case relative bearing is obtained. This is defined as the angle between the keel of the ship and a line in the plane of the horizon to the object. This angle is measured from the forward end of the keel, to the right through 360°, and expressed in three digits. To find the bearing of the object when the relative bearing is known, the heading must be added to the relative bearing. For example, a ship is heading 326° and the relative bearing of a buoy is 124°. The bearing of the buoy is 326° + 124° = 450°, or 090°, and the buoy is due east of the ship. Radio bearings, taken with a radio direction-finder, are always given as relative bearings from the ship.

Lookouts on ships, gunners on planes, and other observers of this general type, may not have a pelorus available to determine bearings of a sighted object. Various approximate methods are in use for expressing relative bearings, depending upon the particular service. Lookouts on ships at sea frequently report relative bearings in terms of compass points. For example: an object sighted one point (11°.25) to the right of the bow would be reported as "one point off the starboard bow." An object with relative bearing 315° would be reported as "broad on the port bow," etc. In the air-services, clock numbers are sometimes used for reporting relative bearings, e.g., an object with relative bearing about 300° would be reported as "at ten o'clock." Complete descriptions of the various approximate methods used in the various services may be found in manuals of seamanship, regulations of the U. S. Army Air Corps, etc.

In surveying, bearing is usually expressed in terms of the acute angle measured either from north or south to the east or west. For example: a bearing of 135° would be expressed as S 45° E (south 45° east) ; a bearing of 290° as N 70° W, etc.

The latitude of a line is its orthographic projection on the meridian; the departure is its orthographic projection on a line perpendicular to the meridian.

$$\text{Latitude} = \text{Length} \times \text{cosine of bearing.}$$
$$\text{Departure} = \text{Length} \times \text{sine of bearing.}$$

Latitudes are positive for north bearings and negative for south bearings. Departures are positive for east bearings and negative for west bearings. In a theoretically closed **traverse** the algebraic sum of both the latitudes and the departures must be zero.

The bearing of any line such as *OA* referred to a meridian through *O* is a forward bearing; if referred to a meridian through *A* it is a back bearing or reverse bearing. Forward and back bearings differ by 180°.

Bearing is a mechanical term used to denote that part of a machine which bears the friction occasioned when parts are in contact and have relative motion. Bearings are to be found in a variety of forms, for example, lubricated and unlubricated, rotating and sliding, weight-carrying and thrust-carrying, etc. See also **Anti-friction Bearings, Bearing Friction,** and **Bearings.**

Bearing is a structural term denoting that part of a structure which transmits the loads to the supports. Rolled steel plates, known as bedplates, are used for bearings for roof trusses, short plate girder bridges and wall bearing beams. Cast steel pedestals have been used in place of bedplates. Bridges are usually supported on hinged pedestals called shoes which are made of cast steel or from rolled steel shapes riveted or welded together. The hinge consists of a heat-treated steel pin through which the loads on the **truss** are transferred to the shoe. A fixed shoe is one which is firmly connected to the support. Shoes which allow a certain degree of movement in one direction are expansion shoes. In one type of expansion shoes rollers are used to provide the necessary means for movement and another type employs the principle of the rocker. Simply supported bridges require fixed shoes at one end and expansion shoes at the other end to take care of the movement due to temperature changes.

The allowable pressure which may be exerted on a material is often called its bearing power. The force exerted by a plate on a rivet or pin is known as a bearing pressure. A compressive **stress** is frequently referred to as a bearing stress. (C.W.C., W.K.G.)

BEARING METALS. A great variety of alloys can be used as bearing metals. Probably the best known is **babbitt,** a tin-base alloy containing antimony and copper. Other so-called white metal bearing alloys have either lead, cadmium, or zinc as the principal element. Aluminum alloys are infrequently used for bearing purposes, although aluminum pistons cannot be discounted as bearing materials. The bearing plates for bridges and structures described under **bearing** are often a high tin bronze. Bronzes and leaded bronzes have many other applications as bearing surfaces in machines, particularly where the unit pressures are greater than can be withstood by babbitt and similar materials. (See **Brass and Bronze.**)

Internal combustion engine bearings range from babbitt, which is now used as a very thin coating on a

steel backing, to pure silver. Other important materials are leaded copper, the cadmium base alloys, and certain composite bearings consisting of a tin or lead base alloy bonded to steel by means of a bronze intermediate layer.

For the most exacting applications such as main bearings in aircraft engines, silver bearings are used. A thin film of indium is often applied to silver and other bearing surfaces to provide self-lubrication during the run-in period.

Cast iron is a very important bearing material. In addition to engine blocks, pistons, and rings, all of which require good bearing properties, cast-iron plain bearings perform satisfactorily with steel shafts where service conditions are not too exacting.

In the field of **anti-friction bearings,** hardened steel balls, rollers, and races form the bearing surfaces. Where even greater hardness, wear resistance, or precision is required, as in instruments and watches, jeweled bearings are used.

The requirements of a good bearing material obviously vary with the application but may include any of the following: wear resistance, compressive strength to resist extrusion or displacement, plasticity to accommodate slight misalignment or irregularities, shock resistance, fatigue resistance, low frictional characteristics to prevent galling or seizure, retention of strength and wear resistance at the temperatures developed in operation, high thermal conductivity to reduce operating temperatures, good bonding characteristics with the liner or container, and corrosion resistance under operating conditions. (R.H.H.)

BEARING MODULUS. The relationship Zn/p between the absolute viscosity Z of an oil, the bearing speed n in rpm, and the unit pressure p on the bearing, in lbs. per sq. in. of projected area. (H.C.H.)

BEARING STRESS. Stress.

BEARINGS. Bearings are employed to support, guide, and restrain moving elements. They may be classified as bearings for rotating and oscillating elements, and as bearings for reciprocating elements. Bearings for rotating or oscillating elements may be further classified as journal bearings and as anti-friction bearings.

A *journal bearing* is composed of two essential parts, the journal, which is the inner cylindrical or conical part and which usually rotates, and the bearing or surrounding shell, which may be stationary, as in the case of lineshaft bearings, or moving, as in a connecting rod bearing.

The simplest form of journal bearing embodies a shaft rotating in a hole in a frame or bracket. If any wear occurs in the bearing, it is necessary to replace the bracket or frame. For this reason, bearing holes are generally supplied with sleeves or bushings so that a comparatively inexpensive replacement is possible. The contact surfaces are lubricated through oil holes in the bushing.

Fig. 1 shows a split pillow block for transmission shafting, which has a babbitt-lined bearing surface.

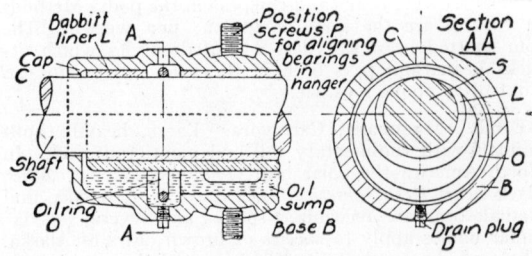

Fig. 1. Split pillow block.

The babbitt metal is cast into the cap and base of the bearing, and is locked in place by recesses or anchors in these members. This bearing is lubricated by a drop-feed oil cup or a grease cup which is screwed into the threaded hole in the cap. Split bearings are more expensive than solid or plain bearings but make it easier to remove and replace the shaft. Pillow blocks are usually stocked by manufacturers in sizes to fit standard transmission shafting.

Fig. 2 shows a split *ring-oiled bearing* for overhead transmission shafting. The entire bearing is sup-

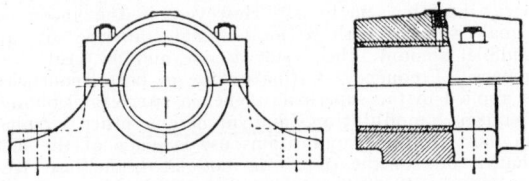

Fig. 2. Split ring-oiled bearing for overhead transmission shafting.

ported by two positioning screws *P* in the hanger of Fig. 3. The hangers are made of cast iron or pressed steel and are attached to wooden ceiling joists by

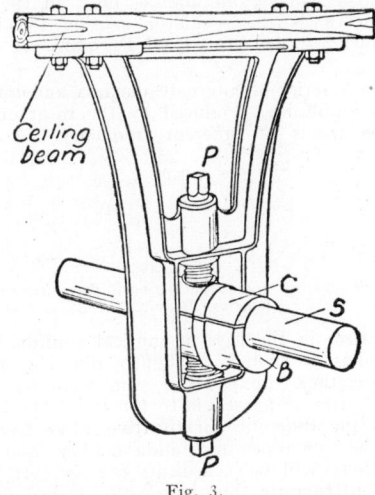

Fig. 3.

through bolts as illustrated, or by lag screws or hanger bolts. Hangers may also be attached to steel ceiling girders by steel girder clamps and bolts.

Thrust bearings are used for axial restraint, and are very important where heavy axial loads may prevail, as in vertical hydraulic turbo-generator sets, worm gearing, etc. In such instances, a collar on the rotating shaft rests on a series of freely pivoting segments attached to the stationary bearing. The rotation of the collar induces a wedge-shaped oil film between its lower surface and the top of the segments to provide thick-film lubrication. For light thrust loads, one or more thrust washers, restrained by separate collars or integral shoulders on the shaft, are satisfactory. *Pivot bearings* are used in instruments and light mechanisms; the shaft or spindle has either a ball or spherical end or a 60° conical end, which rotates in a mating conical hole. The resulting bearing carries both radial and thrust loads. (See **Kingsbury Thrust Bearing.**)

Fig. 4 illustrates several types of bearings for reciprocating slides. Both dovetail and rectangular slides are generally made with some form of gib or adjustable strip, so that the slide may be properly fitted and also to enable the slide to be clamped in the guide if desirable. The taper gib, which is adjusted by a double collar screw, is by far the most effective gib

but the guide must be planed with one tapered side. The flat gib is the least expensive but does not give as good contact with the slide as the other forms. The

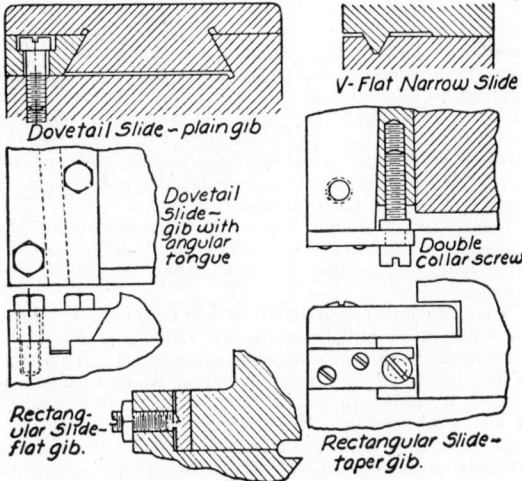

Fig. 4. Types of bearings.

V-flat narrow slide is employed on lathes and in installations where the length of the slide is less than the total width of the member.

Ball and roller bearings are known as *anti-friction bearings,* and have certain advantages over journal bearings. The actual bearing friction is less than in sliding bearings, and, as it is principally rolling friction, there is little danger of abrasion in machines that are frequently started and stopped under load. Rolling bearings will maintain relatively accurate alignment of parts over long periods of time, can carry heavy momentary overloads without failure, and are very easily lubricated.

Fig. 5 illustrates a single-row radial ball bearing and housing. The bearing has four elements: the outer

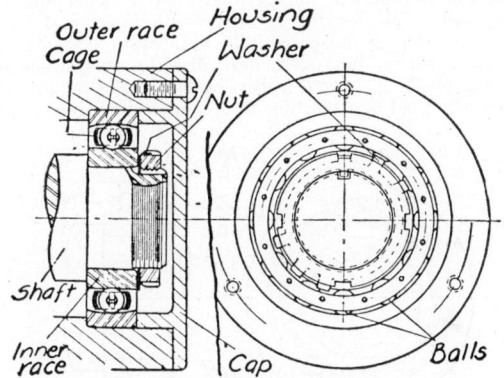

Fig. 5. Single-row radial ball bearing and housing.

race which fits in the housing; the inner race which fits on the shaft; and the cage or retainer which separates the balls and keeps them properly spaced about the periphery of the unit. Theoretically there is no reason why the balls could not roll on the shaft and in the housing, but the races are employed to maintain the proper fit and to provide satisfactory surfaces of the proper degree of hardness for the balls to roll on. In the figure, the bearing is resting against a shoulder on the shaft, and is held in position by a lock nut which may be locked at any twenty-fourth of a turn by the washer.

Ball bearings are generally supplied with bores, widths, and outer diameters in millimeters since the bearings were originally used in quantity in Germany, but bearings in standard inch sizes are also available at the present time. Radial ball bearings are made in three series—light, medium and heavy—and are numbered as follows: 205, 305 and 405, respectively. The bore, in millimeters, between sizes 204 and 213 in the light series, for example, is five times the last figure in the bearing number. Bearings 205, 305, and 405, for example, all have the same bore (25 mm.) but the medium and heavy series bearings, which are used for greater loads than the 205 bearings, have larger outer diameters and greater widths.

Double-row bearings have approximately twice the load-carrying capacity of single-row bearings of the same bore, and occupy less space than two single-row bearings.

Angular-contact ball bearings are designed to take a combination of radial and thrust loads, and should be used in pairs unless the load is pure thrust. This type of bearing is adapted to preloading, which consists of placing it under an initial load which is independent of the working load. Preloading tends to reduce the axial deflection under working loads, thus maintaining accurate alignment of the shaft or spindle elements. Self-aligning bearings are double-row bearings with a spherical surface on the inside of the outer race. This construction allows some deflection in the shaft without causing the bearings to bind.

Roller bearings have a greater load-carrying capacity but develop more friction than ball bearings of similar size. Cylindrical roller bearings are made in three series, similar to ball bearings, and have rollers whose diameters are approximately equal to their lengths. Needle bearings have cylindrical rollers of small diameter and considerable length, and operate without a cage or retainer. They occupy very little diametral space in relation to their load-carrying capacity, and are therefore coming into extensive use in gear mountings, and as piston pin bearings in large internal combustion engines.

Hyatt roller bearings have hollow cylindrical rollers that are made by winding strip steel into helical form. The hollow construction permits greater deflection under load. The bearing is made with inner and outer races but has been successfully applied to transmission shafting where the rollers bear directly on the surface of the shaft. Tapered roller bearings, Fig. 6, are extensively used for machine tool and automotive appli-

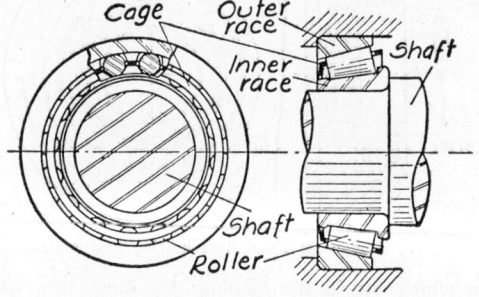

Fig. 6. Tapered roller bearings.

cations, and are capable of taking heavy unidirectional thrust loads in addition to large radial loads. The bearings are used in pairs for two-directional thrust. In some recently developed machine tools, rectilinear slides supported by two "chains" of bearing balls are used to provide practically frictionless table movement. Ball ways have been applied to cutter grinding machinery and radial drill presses, and will probably find further use in future developments. (H.C.H.)

BEAT FREQUENCY. When two signals of different frequencies are applied to a non-linear circuit, they will combine, or beat together, and give, among other components, one which has a frequency equal to the difference of the two applied frequencies. This difference frequency is known as the beat frequency. There are numerous applications of this effect, but two of the major ones in the communications fields are in the reception of continuous wave signals and in frequency shifting as in the superheterodyne receiver. Since continuous wave signals have no audio superimposed on them, they cannot be made audible by ordinary detection methods. However, if the incoming signal is beat with a local signal, differing by an audible amount, the result is an audible beat frequency. Frequency shifting by use of beat frequencies is applied in the superheterodyne, in carrier telephony, frequency modulation, and numerous other circuits. In most of these applications use is made of the fact that if one of the signals is modulated, the beat frequency signal will have the same modulation. (L.R.Q.)

BEAT FREQUENCY OSCILLATOR. This is any conventional oscillator whose function is to produce the signal to mix with a signal whose frequency is to be shifted. Thus in a continuous wave receiver, it is the oscillator causing an audible beat, in the superheterodyne it is the oscillator causing the intermediate frequency beat. (See **Beat Frequency.**) (L.R.Q.)

BEATS. A series of alternate maxima and minima in vibration amplitude, produced by the interference of two wave trains of different frequency. A familiar

Five coincidences in unit time between wave trains of frequencies 20 and 25.

example arises in the case of musical sounds. If two musical pipes or strings of slightly different pitch are sounded together, the result is a more or less distinct throbbing, often disagreeable to the ear. The beat frequency is the difference of the two wave frequencies. Thus, if the two tones are middle-c (256) and c-sharp (271.2), there will be 15.2 beats per sec. If the two tones are ultrasonic, but have a frequency difference within the audible range, the beats themselves may produce an audible "beat tone." A similar effect results from the simultaneous reception of two radio wave trains which are nearly, but not quite, synchronized. Thus if two stations are sending on carrier waves of 1000 and 998 kilocycles, the receiver will emit a shrill whistle of frequency 2000 cycles. This is the "heterodyne" effect, responsible for the annoying squeals and tremolos often heard in radio reception. The effect is utilized in heterodyne code receivers and in the frequency-conversion section of the modern radio. One type of radio "fading" may be regarded as a beat phenomenon of long period. (L.D.W.)

BEAUFORT WIND SCALE. In the days of sailing vessels, Admiral Beaufort introduced a wind scale for judging wind force on the sails of a vessel. Beaufort numbers have since then been correlated to a range of wind velocities, and the scale has continued in universal use for describing wind velocity.

CODE NUMBER	WIND VELOCITY (m.p.h.)	DESCRIPTION
0	0–1	calm
1	1–3	light air
2	4–7	light breeze
3	8–12	gentle breeze

CODE NUMBER	WIND VELOCITY (*m.p.h.*)	DESCRIPTION
4......	13–18	moderate breeze
5......	19–24	fresh breeze
6......	25–31	strong breeze
7......	32–38	moderate gale
8......	39–46	fresh gale
9......	47–54	strong gale
10......	55–63	whole gale
11......	64–75	storm
12......	over 75	hurricane

(P.E.K.)

BEAVER. Mammalia, Rodentia. An aquatic species with webbed hind feet and a broad flat tail. The largest of the North American **rodents**, reaching a length of 3½′ and a weight of 60 lbs. **Capybara.**

The beavers are noted for their extensive building operations. They construct dams of logs, branches and mud in order to form ponds in which their lodges are built. The lodge, like the dam, is a disorderly heap of sticks and mud rising above the water. It contains chambers with entrances below the surface of the water and may be occupied by one animal or by an entire family.

Beavers fell trees for construction and for food by gnawing round and round, although some observers have claimed that they gnaw deeper on the side toward which the tree is to be felled. They store up a supply of logs in their ponds from which the bark is eaten during the winter, although in the warm season they eat other green food.

The fur of the beaver has been among the most valuable on the market and has been the chief reason for a slaughter which threatened to destroy the species. With rigid protection they have become numerous in some parts of North America during the present century so the threatened extinction has apparently been averted.

The European beaver, *Castor fiber,* is similar to the North American species, *Castor canadensis;* whether the two are separate species has been a matter of dispute. (A.W.L.)

BECARD. Chatterer.

BECCAFICO. Aves, Passeriformes. A bird (**Aves**). This is an Italian name translated fig-eater or fig-pecker, said to apply to the European garden warbler, *Sylvia hortensis.* (A.W.L.)

BÊCHE-DE-MER. Sea Cucumber.

BECKE METHOD. A microscopic method of determining which of two minerals in contact has the higher or lower average index of refraction. (R.M.F.)

BECKMANN THERMOMETER. Liquid Expansion Thermometers.

BECQUEREL EFFECT. A photographic effect discovered by E. Becquerel (1895). Experimenting with the **daguerreotype** process, Becquerel found that a plate will produce a direct (positive) image if exposed first to diffuse daylight. (C.B.N.)

BED. A supporting member or frame, usually of arc-welded steel or cast iron, such as lathe and planer beds. (H.C.H.)

BEDBUG. Insecta, Hemiptera. A wingless, blood-sucking **bug** which hides during daylight in the crevices of beds and other furniture and about the woodwork of houses and seeks its victims at night. It has been found also about chicken roosts.

Corrosive sublimate (mercuric chloride) dissolved in alcohol and applied to the hiding places of the insects is one method of control. Severe infestations are usually handled by fumigation of the entire building with **hydrocyanic acid** gas, a procedure which demands the services of an expert since the gas is deadly.

This species, *Cimex lectularius,* gives the name bedbug to the family Cimicidae which also contains a few species that attack bats and birds.

Still another member of this order, the large bedbug, is sometimes found in beds. It is almost an inch long and is capable of inflicting a painful wound. This species is one of the **assassin bugs** (Family Reduviidae). (A.W.L.)

BEDDING. A term used by geologists to designate the natural layering or stratification usually characteristic of **sediments** and sedimentary rocks. Bedding is the result of the unequal rates of settling of particles of different sizes and specific gravities. In the case of very fine-grained sediments, or **shales**, the bedding may be shown by color bands. The thickness of a bed or stratum may vary from several feet to a fraction of an inch. Extremely thin beds are called laminae. (R.M.F.)

BEDEN. Ibex.

BEDROCK. The solid rock of the **lithosphere** which may be directly exposed or covered by loose unconsolidated materials such as sand, clay, soil, etc. (R.M.F.)

BEE. Insecta, Hymenoptera. **Insects** which gather nectar and pollen to provision their nests; they are provided with structural adaptations associated with this habit. The term means, to most persons, the **honey-bee,** of the genus *Apis,* unless it is further qualified, but there are many other kinds of bees and a large number of species. Among the solitary bees are the carpenter bees, the leaf-cutting bees, and a number of mining or burrowing species, while the social bees include the honey-bees, the bumblebees, and the stingless honey-bees of the tropical zone. Some species are parasitic in the nests of other bees.

The social life of bees shows structural specialization into castes and reproduction is restricted to a few individuals of the colony, as in other social insects. In the honey-bee colony three castes are recognized, the workers, drones, and queens. The first are abortive females, the second males, and the last functional females. Under normal conditions only a single queen is found in a colony.

Bees illustrate a complete transition from solitary habits to permanent social groups. Some are strictly solitary; some of the burrowing species build their nests independently but in groups; some form a common tunnel with which their individual nests connect; the bumblebees are social during the summer but hibernate as individuals, and the honey-bee colony is permanent.

Their association with flowers makes the bees important in the **cross-fertilization** of many plants. **Red** clover, for example, is cross-fertilized by the bumblebees and for the production of a good crop of seed these insects are necessary. (A.W.L.)

BEE FLY. Insecta, Diptera. **Flies** whose habit of visiting flowers for pollen and nectar is like that of the bees. They make up the family Bombyliidae. (A.W.L.)

BEE LOUSE. Insecta, Diptera. A minute, wingless, parasitic fly found attached to the queen and drones in honey-bee colonies. The few known species constitute the family Braulidae, containing the genus *Braula.* (A.W.L.)

BEE-EATER. Aves, Piciformes. Brightly colored birds (**Aves**) of the Old World. They have a long curved **beak** and are adept at catching insects in flight. The several species make up the family Meropidae. (A.W.L.)

BEE-MARTIN. Aves, Passeriformes. *Tyrannus.* The common kingbird of North America. The name is undeserved, since scientific investigation has disclosed that he eats very few bees. (A.W.L.)

BEE-MOTH. Insecta, Lepidoptera. A moth whose larva eats the wax and debris in old honeycombs, spinning a silken tunnel as it goes. These insects are found chiefly in weak colonies of bees and in stored combs. They may attack beeswax products, such as comb foundation in the supplies of the apiarist, but they cannot develop on a diet of pure wax; the organic waste in old brood combs provides the necessary nitrogenous material and furnishes a favorable breeding place. The best known form is *Galleria mellonella.*

In well-kept apiaries the bee-moth is rarely a serious pest. The maintenance of strong colonies of bees prevents its entrance and the protection of stored supplies against the entry of the adult moths safeguards them against damage. Fumigation of supplies is sometimes necessary. It may be effectively carried out with **carbon disulfide** but since this compound is highly explosive it must be used with due precautions. (A.W.L.)

BEESWAX. A tough wax formed of a mixture of several compounds, secreted by honey-bees in the form of thin scales from glands on the ventral surface of the abdomen and used in building the combs in which the young are raised and honey and pollen are stored. Commercially it is a compact mass varying from yellowish-white to brownish in color according to its purity. It has a high melting point, near 140° F.

Beeswax is used commercially to make fine candles, in polishing materials, as a component of modeling waxes, and in a variety of other products. See **Esters.** (A.W.L.)

BEET. *Beta vulgaris.* Chenopodiaceae. The many varieties of beets now in cultivation are perhaps all derived from the native *Beta maritima* of southern Europe.

The most important variety is the sugar beet, which in recent years has become an important rival of the sugar cane. As a source of sugar, beets were first utilized in Germany and in France about 1800. In the United States they became important commercially only after the World War. Their culture is still largely restricted to a few states, notably Colorado, Wisconsin, and California.

The sugar beet is a biennial plant which during its first year of growth forms a large tapering tap **root** and a rosette of leaves. At the end of this first year the plant is gathered for sugar production. If allowed to grow the second year, the plant forms a branching stem and an abundance of inconspicuous flowers, utilizing the sugar stored in the root to produce them.

The plant has an elongated tap root which tapers into a long slender root. This may penetrate 4–6' into the ground. In size and shape the root is extremely variable. Cut transversely, the root is seen to be composed of from 6 to 10 or even more concentric zones. Each zone comprises a ring of conducting cells outside which is a ring of small cells in turn surrounded by a ring of large cells. The small cells are rich in sugar, while the large cells are primarily water-storage cells. The formation of these zones is a consequence of the formation of a succession of **cambium** rings, each of which persists for a few weeks.

In preparing the beets for sugar manufacture, the roots are first lifted from the ground, the leaves cut off, and the roots hauled to the factory. There they are washed thoroughly and cut into thin slices. These slices are put into hot water, which extracts the sugar. The sugar solution is next treated with lime, and then precipitated with carbon dioxide: this removes many impurities which are filtered off. The purified liquor is bleached with sulfur dioxide, and then concentrated by boiling and crystallized under a partial vacuum. From this crude product the molasses is removed by centrifuging, leaving the sugar which is dried and granulated, after which it is ready for the market.

Many of the waste products of beet sugar production are utilized. The tops are used as a stock food either in the raw condition or after preserving as ensilage. The beet pulp left after extraction of the sugar is also used as a stock food, as is the molasses from the sugar. Often the pulp and molasses are mixed before feeding. Any refuse from the factory may be used as a fertilizer.

In addition to sugar beets, several other varieties of *Beta vulgaris* are known. One of them is the common table beet, which is eaten either boiled or pickled. When correctly grown it has a minimum of fibrous elements as well as a high sugar content. Most varieties of table beet are deep red in color, in contrast to the white-fleshed sugar beet. Another variety of beet is the Mangel-wurzel or Mangel, of which there are several varieties. They are of large size, have a sugar content varying from 4–8% and are developed principally as a stock food. (R.M.W.)

BEETLE. Insecta, Coleoptera. Any member of this order of insects. The term is often compounded with other words, as rove-beetle, click-beetle, and leaf-beetle. (See **Coleoptera.**) (A.W.L.)

BEGGIATOA. Filamentous forms of sulfur **bacteria** which are capable of converting **hydrogen sulfide** to **sulfuric acid.** **Sulfur** granules are stored in the cells, and may be oxidized to supply the cell with the necessary energy for life. (R.M.W.)

BEHEMOTH. A biblical name for a large animal, probably the **hippopotamus.** (A.W.L.)

BEISA. Mammalia, Artiodactyla. An **antelope,** *Oryx beisa,* of northeastern Africa, characterized by long sharp horns, ringed at the base and almost straight. With several other species it belongs to the genus *Oryx,* a name often used as a common name for any of the included animals. (A.W.L.)

BEL. This is a unit of power level defined by the following equation:

$$\text{bel} = \log_{10} \frac{P_1}{P_2}$$

where P_1 is the power being measured and P_2 is a reference power value (usually arbitrary). The bel is seldom used, as the **decibel** is the common unit for communication power work. (L.R.Q.)

BELEMMITE. Invertebrate Paleontology.

BELL AND SPIGOT. Pipe.

BELL CRANK. A means frequently used to transfer reciprocating motion at right angles is that of a rigid-angled arm pivoted to a fixed point at its vertex, and

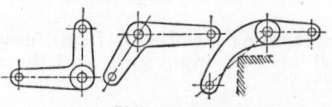

Bell cranks.

having hinge connections at its extremities. The bell crank really is a type of **lever,** and the motions of its end are not those of reciprocation, but rather of rotation, but if comparatively long rods are hinged to it, the other ends of those rods will have similar motion. The amplitude of the reciprocation transmitted by the bell crank is directly proportional to the radii from the pivot point to the joints. (F.T.M.)

BELLADONNA. *Atropa Belladonna.* Solanaceae. The plant grows as a native in Europe and in parts of Asia. It is commonly known as Deadly Nightshade. It is about 3′ tall, and has dull green leaves and purple flowers, which are followed by cherry-like red fruits. Every part of the plant contains the poisonous substance for which it is known. This drug, **atropine,** is a very poisonous **alkaloid** obtained principally from the roots and leaves. Belladonna is used medicinally most often as the tincture. Its action is that of atropine. (R.M.W., R.S.M.)

BELLADONNINE. Alkaloids.

BELL-BIRD. Aves. A term applied to several species of birds (**Aves**) which produce bell-like tones, including a **chatterer** in the Guianas and a **honey-sucker** in New Zealand. (A.W.L.)

BELTING. When the center distance between the axes of parallel shafts is comparatively large, flexible connectors or belts are generally used. Flat belts are generally made of leather, although folded canvas in "plies" impregnated with rubber or with balata gum (a substance obtained from South America) are sometimes used. Flat leather belts are made of oak-tanned or chrome-tanned leather strips cemented together to obtain the required length and thickness. Single leather belts are about $\frac{5}{16}''$ thick; double and triple belts are composed of two or three single plies, and are $\frac{5}{8}''$ and $\frac{7}{8}''$ thick. Belts may be endless, with cemented joints, or they may be joined or laced with wire, rawhide lace, or metal hooks.

Open and crossed belt drives are used for power transmission between parallel shafts; quarter-turn belts may be used when the shaft axes are not parallel. In order for a belt to stay on a pulley, the approaching side of the belt must approach the pulley perpendicular to the axis of rotation of the pulley. By this rule, it may be seen that a quarter-turn belt without an idler pulley will operate satisfactorily for one direction of rotation but will not stay on the pulleys if the rotation is reversed. (By proper use of an idler or "mule" pulley, reversing quarter-turn belt drives can be satisfactorily operated.) By the same rule, open and crossed belt drive shafts should be parallel, with the pulleys in alignment. Narrow belts are often kept on pulleys by means of flanges on the pulley face. Pulleys for wider belts are crowned or slightly greater in diameter at the center than at the edges. As the belt seeks the highest point on the pulley, the effect of crown is to keep the belt in a central position.

The speed ratio of a belt drive varies inversely as the pulley diameters. For purposes of estimate, the following, known as the Millwright's Rule, will give the horsepower a belt will transmit:

$$H = WV/K$$

where H is the power transmitted, V is the belt velocity in ft. per min., W is the belt width, in inches, and K is a constant equal to 800 for single and 500 for double belts.

Pulleys are made of cast iron with a rim of rectangular section, and with from 4 to 12 spokes. Two-piece pulleys which are bolted together at the rim and the hub are more expensive than one-piece or solid pulleys, but are easier to install. Steel pulleys built up of pressed steel rims, hubs, and arms, and welded or riveted, are lighter than cast-iron pulleys of the same diameter and face width, and can be run at higher speeds with safety. Pulleys of wood, compressed paper, and compressed fiber are lighter than metal pulleys and have a higher coefficient of friction, and therefore a greater ca-

pacity to transmit power for a given belt pull, than metal pulleys.

Vee-belts are made of cords impregnated with rubber, and are of trapezoidal form. They are made in five standard sections varying from section ·A which is $\frac{1}{2}''$ wide and $\frac{11}{32}''$ high, to section E which is $1\frac{1}{2}''$ wide and $1''$ high. The belts are endless and operate either in two grooved sheaves or in the grooves of a small sheave and on the face of a large pulley (V-flat drives). Vee-belt and V-flat drives offer a satisfactory solution for industrial transmission problems where the center distance is small and the velocity ratio is high, although these are not necessary conditions.

Round leather belt drives are used for light service as in sewing machine drives and similar applications. Like vee-belts, they may be used for misaligned shafting or for quarter-turn drives. Round belts are available commercially in diameters from $\frac{1}{8}''$ to $\frac{1}{2}''$. (H.C.H.)

BELUGA. Mammalia, Odontoceti. The white **whale,** *Delphinapterus leucas.* A small arctic species which attains a length of 16′ or more. The species has been captured for oil and for its flesh and the hide has been used for leather. (A.W.L.)

BENCH MARK. A definitely established point whose elevation is known in relation to some actual or arbitrary plane is called a bench mark. It is used in **surveying** as a vertical reference point when finding the **elevation** (with respect to an actual or arbitrary plane) of other points of a less permanent nature. The point may be the head of a spike or bolt driven into a tree in such a manner that the top of the head is as nearly horizontal as possible, the highest point on the top of a hydrant, or the top of a flat, non-corrosive plate set securely in stone or **concrete.** Bench marks may be temporary or permanent depending upon their use and should be located so that they are easily accessible for instrument work. (C.W.C.)

BENCH PHOTOMETERS. (See **Photometry.**) The principle underlying the usual forms of bench photometer depends upon finding a point so located between the two light-sources under comparison that the flux densities produced by them at that point are equal. The luminous intensities of the two sources are then proportional to the squares of their distances from that point. To this end, the two sources are mounted near the extremities of the scale of an optical bench, and on a movable carriage between them is some device, called the photometer "head," for receiving and comparing the illuminations from opposite directions.

Among the many types of bench photometer head in use, only four can be mentioned here. A crude form is the Joly "block" screen, composed of two blocks of opal glass or paraffin with an opaque partition between them. When the sides of the blocks are equally illuminated, the two front faces are lighted up equally by diffusion from within. One of the best known types utilizes the Bunsen screen, which is in effect a sheet of white paper with a grease spot at the middle. When the two surfaces are equally illuminated, the grease spot becomes indistinguishable. The Lummer-Brodhun "cube" screen somewhat resembles the Bunsen in principle, but with the paper and grease spot replaced by the interface between two right-angled prisms which are in optical contact only over a central area of some conventional shape. This central area and the area surrounding it merge into a uniform field when the illuminations are equal.

The distinctive feature of the "flicker photometer" head is that, by means of a rotating sector-disk, the two illuminations to be compared are presented to the observer in rapid alternation (but not too rapid), any difference between them being detected as a noticeable flicker. This type of photometer is especially useful when the sources are not of exactly the same color. (L.D.W.)

BENDING DEFORMATION. Deformation.

BENDING LOAD. Load; Flexure.

BENDING MOMENT. The external bending moment at any section in a beam is equal to the algebraic sum of the moments, about the gravity axis of the section. This definition assumes that all of the external forces are coplanar, that is, act in one plane. An internal resisting moment at any section is equal to the sum of the moments of the internal stresses about the gravity axis of the section. The external bending moment acting on any section is numerically equal to the internal resisting moment but acts in the opposite direction. External moments are positive or negative depending upon the direction in which they tend to rotate the section of the beam under consideration. This sign convention is entirely arbitrary although it is customary in beam analysis to assume that positive moments are those tending to shorten the top surface of the beam while negative moments are those which lengthen the top surface. The point on the longitudinal axis of a beam at which the bending moment changes sign is called the point of contraflexure or the point of inflection. Pure bending is a term used to denote the condition where the shear is zero over a finite length. It follows that the bending moment is constant over this length. Bending moments have a very important part in beam action since they cause the flexural stresses (see **Flexure**) and are the principal cause of deflections.

A graphical representation of the variation of bending moment on a beam is called a bending moment diagram. An illustration of a bending moment diagram is given below for an overhanging beam with a uniformly distributed load covering the entire length of the beam.

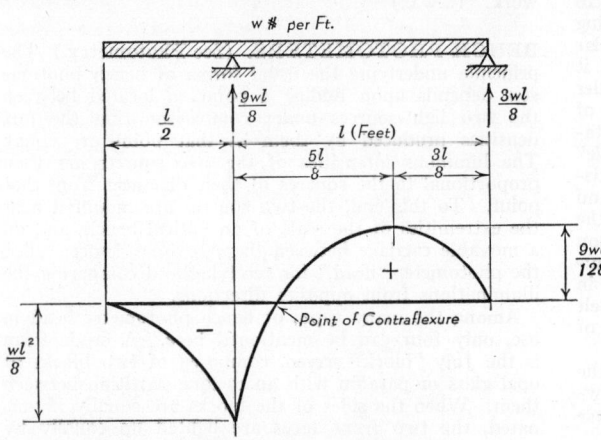

Bending moment diagram.

The maximum bending moments which are indicated on the diagram occur where the shear changes sign. (See **Shear** for the shear diagram for this beam.) (C.W.C.)

BENDING STRESS. Stress.

BENDS. Caisson Disease (Compressed Air Illness, "The Bends," Diver's Palsy).

BENEDICT'S SOLUTION. This is an alkaline solution of copper hydroxide and sodium citrate in sodium carbonate used either as a mild oxidizing agent or as a test for easily oxidizable groups such as aldehyde groups. The formation of cuprous oxide is a positive test, its color red but often yellow at first. (See **Carbohydrates**.) (R.K.S.)

BENIGN. Harmless; not malignant. This term is usually applied to tumors of non-cancerous character. (R.S.M.)

BENT. A transverse frame which forms an integral part of a structural unit or supports another structural unit is called a bent. Bents are designed to carry lateral as well as vertical loads, and are made of wood or structural steel. They are used principally in connection with viaducts and mill buildings. Viaduct bents consist of columns held firmly together by bracing in horizontal and vertical planes. The mill building bent is composed of the roof truss and the supporting columns which are connected by inclined members called knee braces. These knee braces stiffen the bent against the action of wind forces. Transverse mill building bents are connected by members called purlins and girts in addition to the necessary longitudinal bracing. The purlins rest on the top chord of the truss and support the roof while the girts which are connected to the columns are used to carry the siding. (C.W.C., F.T.M.)

BENTHOS. That part of the plant or animal world made up of individuals which rest on a solid support. The term is also applicable to free or attached forms living on the ocean floor. (See **Distribution**.) (A.W.L., R.M.F.)

BENTONITE. The term applied to altered fine-grained volcanic ashes which have been blown considerable distance from their origin and deposited in marine waters. The resulting material is usually a white, but sometimes a colored, clay-like sediment which may contain bits of volcanic glass but is composed mainly of colloidal silica which will absorb large quantities of water. Since bentonites are wind-blown deposits they are useful as definite datum planes in stratigraphy, especially in helping to determine the contemporaneity of the different facies of marine sediments. (R.M.F.)

BENZALDEHYDE. Benzaldehyde ("oil of bitter almonds," C_6H_5CHO) is a colorless liquid, boiling point 180° C., of characteristic odor, slightly soluble in water, miscible in all proportions with alcohol or ether, readily oxidized, even on standing in air, to benzoic acid. Benzaldehyde reacts with many chemicals in a marked manner, (1) with ammonio-silver nitrate ("Tollen's solution") to form metallic silver, either as a black precipitate or as an adherent mirror film on glass, but does not reduce alkaline cupric solution ("Fehling's solution"), (2) with rosaniline (fuchsine, magenta), which has been decolorized by sulfurous acid ("Schiff's solution"), the pink color of rosaniline is restored, (3) with sodium hydroxide solution, yields benzyl alcohol and sodium benzoate, (4) with ammonium hydroxide, yields tribenzaldamine (hydrobenzamide, $(C_6H_5CH)_3N_2$), white solid, melting point 101° C., (5) with aniline, yields benzylideneaniline ("Schiff's base" $(C_6H_5CH:NC_6H_5)$), (6) with sodium cyanide in alcohol, yields benzoin ($C_6H_5 \cdot CHOHCOC_6H_5$), white solid, melting point 133° C., (7) with hydroxylamine hydrochloride, yields benzaldoximes ($C_6H_5CH:NOH$), white solids, anti-oxime, melting point 35° C., syn-oxime, melting point 130° C., (8) with phenylhydrazine, yields benzaldehyde phenylhydrazone ($C_6H_5CH:NNHC_6H_5$), pink solid, melting point 156° C., (9) with concentrated nitric acid, yields metanitrobenzaldehyde ($C_6H_4CHO(NO_2)$ (2)), white solid, melting point 58° C., (10) with concentrated sulfuric acid, yields metabenzaldehyde sulfonic acid ($C_6H_4CHO(SO_3H)(2)$), (11) with anhydrous

sodium acetate and **acetic anhydride** at 180° C., yields sodium cinnamate (C_6H_5COONa), (12) with **sodium** hydrogen sulfite, forms benzaldehyde sodium bisulfite ($C_6H_5CHOHSO_3Na$), white solid, from which benzaldehyde is readily recoverable by treatment with sodium carbonate solution, (13) with **acetaldehyde** made slightly alkaline with sodium hydroxide, yields cinnamic aldehyde ($C_6H_5CH:CHCHO$), (14) with **phosphorus** pentachloride, yields benzylidine chloride ($C_6H_5CHCl_2$). Benzaldehyde is obtained (1) by boiling the glucoside amygdalin of bitter almonds with dilute acid, glucose plus hydrogen cyanide being formed simultaneously with benzaldehyde, (2) by heating benzal chloride with calcium hydroxide, (3) by heating a mixture of calcium benzoate and formate. Benzaldehyde may be detected by the appearance of a blue color on treating with acenaphthene and sulfuric acid, followed by heating. Benzaldehyde is used (1) as a flavoring essence, (2) in the production of cinnamic acid, (3) in the manufacture of malachite green **dye**. (R.K.S.)

BENZENE. Benzene ("benzol," C_6H_6 or ⬡) is a colorless, odorous liquid, melting point 5.5° C., boiling point 80.1° C., insoluble in water, miscible in all proportions with alcohol, ether, and many organic liquids, dissolves iodine, sulfur, oils, fats, rubber, resins, burns, when ignited, with a smoky flame, the vapor forms with air an explosive mixture, used as a fuel in **internal combustion engines**. Benzene reacts (1) with **chlorine**, to form (a) substitution products (one-half of the chlorine forms **hydrogen chloride**) such as chlorobenzene (C_6H_5Cl), dichlorobenzene ($C_6H_4Cl_2(1,4)$ and (1,2)), trichlorobenzene ($C_6H_3Cl_3(1,2,4)$), tetrachlorobenzene (1,2,3,5), and (b) addition products, such as benzene dichloride ($C_6H_6Cl_2$), benzene tetrachloride ($C_6H_6Cl_4$), benzene hexachloride ($C_6H_6Cl_6$). The formation of substitution products of the benzene nucleus, whether in benzene or its homologues is favored by the presence of a **catalyzer**, e.g., **iodine, phosphorus, iron,** (2) with concentrated **nitric acid,** to form nitrobenzene ($C_6H_5NO_2$), 1,3-dinitrobenzene ($C_6H_4(NO_2)_2(1,3)$), 1,3,5-trinitrobenzene ($C_6H_3(NO_2)_3(1,3,5)$), (3) with concentrated **sulfuric acid,** to form benzene sulfonic acid ($C_6H_5SO_3H$), benzene disulfonic acid ($C_6H_4(SO_3H)_2(1,3)$), benzene trisulfonic acid ($C_6H_3(SO_3H)_3$-(1,3,5)), (4) with methyl **chloride** plus anhydrous **aluminum** chloride (Friedel-Crafts reaction) to form toluene, monomethyl benzene ($C_6H_5CH_3$), dimethyl benzene ($C_6H_4(CH_3)_2$), trimethyl benzene ($C_6H_3(CH_3)_3$), (5) with acetyl chloride plus anhydrous aluminum chloride (Friedel-Crafts reaction) to form acetophenone, methyl phenyl ketone ($C_6H_5COCH_3$), (6) with **hydrogen** in the presence of a catalyzer, e.g., finely divided nickel, heated, to form dihydrobenzene, cyclohexadiene (1,3)(C_6H_8), a cyclic diolefin hydrocarbon, tetrahydrobenzene, cyclohexene (C_6H_{10}), a cyclic mono-olefin, hexhydrobenzene, cyclohexane (C_6H_{12}), a cyclo paraffin, (7) with ozone, to form benzene triozonide ($C_6H_6(O_3)_3$).
Benzene is obtained from coal tar and coal gas. Coal tar is distilled, the part of the distillate which is insoluble in water is collected up to about 210° C. By fractional distillation of this portion, benzene, **toluene**, **xylene** are obtained. From coal gas, benzene and toluene, in larger amounts than from coal tar, are obtained by "scrubbing," that is, passing the gas through a special non-volatile absorbing oil, and then distilling off the absorbed oil followed by fractional distillation of this portion. "Ninety per cent benzol" is a commercial benzene product, 90% of which distils over before 100° C. This product contains about 81% benzene, 15% toluene, 2% xylenes. Pure benzene and toluene are obtained by further fractionation. Benzene is used (1) as a **solvent** for many substances, such as oils, fats, resins for varnishes and lacquers, rubber, old paint, and in dry cleaning of fabrics, (2) as a motor fuel, (3) in the manufacture of nitrobenzene for **aniline,** of chlorobenzene for **phenol,** of benzene disulfonic acid for **resorcinol,** of azobenzene for **benzidine,** and of other organic chemicals, especially **dyes.**
Homologues of benzene, e.g., toluene, xylene, ethyl benzene, mesitylene, upon oxidation yield **carboxylic acids,** containing one carboxyl group (—COOH) for each side chain, and substituted side chains behave similarly. (R.K.S.)

BENZIDINE (44′ diaminodiphenyl). Benzidine is an important dye intermediate. (See **Dyes** and **Rearrangements.**) (R.K.S.)

BENZIL. Benzoin, Benzil, and Related Compounds.

BENZOIC ACID AND BENZOATES. Benzoic acid ($H \cdot C_7H_5O_2$ or $C_6H_5 \cdot COOH$) is a white solid, melting point 122° C., boiling point 249° C., insoluble in cold water, soluble in hot water in alcohol, and in ether. Forms benzoates; e.g., **sodium** benzoate, **calcium** benzoate, which, when heated with calcium oxide, yields benzene and calcium carbonate; forms with **phosphorus** trichloride benzoyl chloride (C_6H_5COCl) an important reagent for transfer of the benzoyl (C_6H_5CO—) group; forming meta-chlorobenzoic acid by reaction with **chlorine,** meta-nitrobenzoic acid by reaction with **nitric acid.** The following are esters of benzoic acid:

			°C.
Methyl benzoate	$C_6H_5COOCH_3$	boil. pt.	200
Ethyl benzoate	$C_6H_5COOC_2H_5$	boil. pt.	211
Glycol dibenzoate ...	$C_2H_4(COOC_6H_5)_2$	melt. pt.	73
Glyceryl tribenzoate..	$C_3H_5(COOC_6H_5)_3$	melt. pt.	76

Benzoic acid may be obtained (1) from some natural products, e.g., gum benzoin, **dragon's blood** resin, Peru and Tolu **balsams,** cranberries, urine of horses; (2) from benzotrichloride by reaction with water when heated; (3) as a by-product in the manufacture of benzaldehyde from benzal chloride or benzyl chloride. Dilute solutions of benzoic acid give a violet coloration with hydrogen peroxide (drop) and ferric chloride, on heating. (R.K.S.)

BENZOIN. *Styrax Benzoin.* Styracaceae. (For the chemical compound benzoin, see the article on **Benzoin, Benzil, and Related Compounds.**) Benzoin is a fragrant resin obtained from the bark of a tall quick-growing tree which is native in Sumatra and Java. The tree has alternate entire leaves, the lower surface of which is soft and hairy, the upper smooth. The flowers are borne in compound axillary **racemes;** the fruit is a drupe (see **Fruit**). The **resin** is obtained by making incisions in the bark. From these a thick white juice exudes and hardens. This is scraped off. It is a soft fragrant substance either white or of yellowish color. It is used in medicine as a soothing inhalant and for application to the skin. It contains about 25% **benzoic acid.**
Frequently confused with this is the North American shrub *Benzoin aestivale,* often called Spice bush, which blossoms very early in the spring. All parts of the shrub contain an aromatic substance which is very noticeable when the plant is bruised. (R.M.W.)

BENZOIN, BENZIL AND RELATED COMPOUNDS. (For the resin, benzoin, see preceding article—**Benzoin.**) When **benzaldehyde** (C_6H_5CHO) is warmed with **sodium cyanide** dissolved in alcohol, benzoin ($C_6H_5COCHOHC_6H_5$) white solid, melting point 137° C., is formed, and has the characteristics of a **ketone** and a secondary **alcohol.** Benzoin, (1) upon reduction with sodium amalgam (sodium dissolved in mercury), forms hydrobenzoin ($C_6H_5CHOHCHOHC_6H_5$) white solid, melting point 134° C., mixed with isohydrobenzoin, melting point 119° C. Hydrobenzoin yields an oxide of the formula ($C_6H_5 \cdot CH\text{—}CH \cdot C_6H_5$), which

$$\underset{O}{\diagdown}$$

is known in two stereoisometric forms, symmetrical (inactive) and unsymmetrical (two optically active forms plus a racemic form), (2) upon reduction with zinc plus **acetic acid** or **hydrochloric acid**, forms desoxybenzoin ($C_6H_5COCH_2C_6H_5$), (3) upon oxidation with **nitric acid**, forms benzil, dibenzoyl, diphenyl glyoxal ($C_6H_5COCOC_6H_5$), yellow solid, melting point 95° C., most common alpha-diketone. Two monoximes are known, namely, alpha-benzil monoxime

$$(C_6H_5\text{—}CO \cdot C_6H_5),$$
$$\underset{HO\text{—}N}{\overset{\|}{}}$$

melting point 140° C.; beta-benzilmonoxime,

$$\left(\begin{array}{c} C_6H_5C\text{—}COC_6H_5 \\ \| \\ N\text{—}OH \end{array} \right),$$

melting point 113° C.; and three dioximes, namely anti-benzildioxime $\left(\begin{array}{c} C_6H_5 \cdot C\text{—}C \cdot C_6H_5 \\ \| \quad \| \\ HO\text{—}N \quad N\text{—}OH \end{array} \right)$, melting point 243° C.;

syn-benzildioxime $\left(\begin{array}{c} C_6H_5C\text{———}CC_6H_5 \\ \| \quad \quad \| \\ N\text{—}OH \; HO\text{—}N \end{array} \right)$, melting point 206° C.; amphi-benzildioxime

$$\left(\begin{array}{c} C_6H_5C\text{———}CC_6H_5 \\ \| \quad \quad \| \\ N\text{—}OH \quad N\text{—}OH \end{array} \right),$$

melting point 164° C. When benzil is heated with sodium hydroxide dissolved in alcohol, benzilic acid, diphenyl-glycollic acid $\left(\begin{array}{c} C_6H_5 \cdot C \cdot C_6H_5 \\ \diagup \quad \diagdown \\ HO \quad COOH \end{array} \right)$, melting point 150° C., is formed.

The related hydrocarbons are as follows:

Symmetrical-diphenylethane...... $C_6H_5CH_2 \cdot CH_2C_6H_5$
(dibenzyl)
Symmetrical-diphenylethylene.... $C_6H_5CH:CHC_6H_5$
(stilbene)
Diphenylacetylene.............. $C_6H_5C:CC_6H_5$
(tolane) (R.K.S.)

BENZOPHENONE. Aldehydes, Ketones, and Related Compounds.

BENZOYL. Radicals.

BENZOYL CHLORIDE. Chlorine.

BENZYL. Radicals.

BERGAMOT OIL. An essential oil (see **Volatile Oils**) produced from the rind of the fruit of *Citrus bergamia,* a relative of the orange and lemon. The small trees are cultivated in southern Europe and bear small yellow fruits. From the skin of these the oil is expressed. This oil is also used to some extent as a clearing agent in the preparation of material for microscopic examination. (R.M.W.)

BERGIUS PROCESS. This is a process of catalytic **hydrogenation** of **coal** and **petroleum** under high pressure. (See **Catalysis.**) (R.K.S.)

BERGSCHRUND. Cirque.

BERI-BERI. A deficiency disease due to dietary lack of Vitamin B_1, thiamine (see **Vitamins**). Beri-beri has been known for centuries in the Orient, wherever polished rice is the staple diet. It is still a major cause of death in China, Japan and Brazil. In the United States it occurs chiefly in alcoholics who obtain most of their caloric requirements from alcohol and consequently eat a poor diet.

The symptoms of beri-beri are referable to the nervous and cardiovascular systems. The common manifestations are multiple **neuritis,** localized areas of altered skin sensitivity to touch in the extremities, and pain on pressure over large nerves. There is gradual loss of muscle strength, which may lead to paralysis of a limb.

The cardiovascular symptoms and signs are enlargement of the heart, dyspnoea, increased pulse rate, palpitation, and **edema.** If the edema is very extensive, the condition is called wet beri-beri.

Pathologically, degenerative changes are found in the nervous tissue, heart muscle, and gastrointestinal tract. These lesions, however, do not explain the pathologic **physiology** of the disease, which remains obscure.

Treatment is specific, with large doses of thiamine chloride. Since individuals with one vitamin deficiency are very likely to have other deficiencies, patients with beri-beri usually need treatment with other members of the B group. Untreated beri-beri is a fatal disease. (D.M.H.)

BERM. A relatively flat erosion surface brought to grade during a previous cycle of erosion. (R.M.F.)

BERNOULLI DISTRIBUTION. Bernoulli Probability Function.

BERNOULLI'S LAW. An important law relating to the flow of **liquids.** Let the attention be fixed upon a small portion of the liquid, whose motion is traced along the line of flow. Then if, at any instant, the elevation of this particle above an arbitrary datum is denoted by e (cm.), the pressure upon it by p (dynes per sq. cm.), the speed at which it moves by v (cm. per sec.), and the density of the liquid by ρ (grams per cm.^{3}, supposed constant), the total "head" of the liquid at this point is given by Bernoulli's equation as

$$e + \frac{p}{\rho g} + \frac{v^2}{2g} = H.$$

g is the acceleration of gravity (cm. per sec. per sec.). In hydraulics the three terms of the first member are commonly called respectively the "elevation head," the "pressure head," and the "velocity head." Bernoulli's law states that if the flow takes place without external interference, that is, without any work being done by or upon the liquid as it flows (an ideal condition, of course), then the total head H remains unchanged throughout the flow; a fact which follows from the principle of conservation of energy.

Special cases arise when any one of the three terms is kept constant. Thus if the flow is horizontal, so that e is constant, an increase in speed necessitates a decrease in pressure, a principle utilized in the Venturi meter. Many of the theorems of **hydrokinetics,** such as Torricelli's, may be derived from Bernoulli's law. (L.D.W.)

BERNOULLI PROBABILITY FUNCTION. If the **probability** of success, p, is constant from trial to trial and if these trials are independent, then we obtain a Bernoulli series and the distribution of successes is called a Bernoulli probability function. Let P_x be the probability of obtaining x successes in s trials, then $P_x = {}_sC_x p^x q^{s-x}$. The mean is sp, $\sigma_x = \sqrt{spq}$, $\alpha_{3:x} = \dfrac{q-p}{\sqrt{spq}}$, and $\alpha_{4:x}$

$= 3 + \dfrac{1 - 6pq}{spq}$. The mean of the proportion of successes, $\dfrac{x}{s}$, is p, $\sigma_{\frac{x}{s}} = \sqrt{\dfrac{pq}{s}}$, $\alpha_{3:x} = \alpha_{3:\frac{x}{s}}$, $\alpha_{4:x} = \alpha_{4:\frac{x}{s}}$. The Bernoulli probability function is of frequent occurrence. (See **Bernoulli's Theorem;** the **De Moivre-Laplace Theorem.**) (L.A.A.)

BERNOULLI'S THEOREM. Ths theorem was published posthumously in 1713. Let the **probability** of an event p be constant from trial to trial, and the probability of failure be denoted by q. Let the **relative frequency** be denoted by $\frac{x}{s}$ where x is the number of successes in s trials. Let $P\left(\left|\frac{x}{s} - p\right|\right)$ denote the probability of obtaining the absolute value of the deviation $\frac{x}{s} - p$. Usually this deviation will be small. Bernoulli's theorem states that as the number of trials s increases, the probability of a difference in absolute value more than a stated positive amount ϵ approaches zero. In symbols

$$\lim_{s \to \infty} P\left(\left|\frac{x}{s} - p\right| > \epsilon\right) = 0.$$

It should be clear that the theorem does not state

$$\lim_{s \to \infty} \frac{x}{s} = p.$$

The theorem states roughly that if s is sufficiently large usually $\frac{x}{s}$ will not differ much from p. However, cases may occur though quite infrequently when $\frac{x}{s}$ will differ considerably from p. Bernoulli's theorem is one of the "laws of large numbers." It is exceedingly important in all problems of modern sampling theory and is the basis of many practical problems, particularly when p is unknown and must be estimated in some way. (L.A.A.)

BERRY. A berry is a fruit in which the entire ovary becomes soft and fleshy, as in tomatoes and grapes. Frequently, as in currants and gooseberries, the calyx tube grows around this ovary wall and forms the skin of the berry. Strawberries, raspberries, mulberries, and so forth, are not berries in the botanical sense.

Many other fruits are closely related to berries, but are distinguished by special names. For example, melons, squashes and gourds are berries with a hard outer shell, and are called pepos, while a citrus fruit, such as an orange, lemon or grapefruit, etc., differs in having a tough leathery rind, and so is a modification of berry known as a hesperidium. (R.M.W.)

BERTRAND LENS. A small lens used in the petrographic microscope for magnifying the **interference figure.** (R.M.F.)

BERYL. The mineral beryl is a silicate of **beryllium** (glucinium) and **aluminum** corresponding to the formula $Be_3Al_2(SiO_3)_6$. Crystallizing in the **hexagonal system** the 6-sided prisms of beryl may be very small or range up to several feet in length and a yard or so in diameter. Terminated crystals are relatively rare. Its fracture is **conchoidal;** hardness, 7.5–8; specific gravity, 2.63–2.80; colors, emerald green, green, blue-green, blue, yellow, red, white and colorless; luster, vitreous; transparent to translucent. The mineral beryl has long been used as a gem, the emeralds being a rich green variety, colored probably by minute amounts of some chromium compound. A beautiful bluish sort is called aquamarine; morganite is pink, and the golden beryl is a clear bright yellow. Other shades like honey yellow and yellowish-green are common. Beryl is found in granite rocks and especially in pegmatites, but it occurs also in mica schists in the Urals. In addition to the many European localities as Austria, Germany, Ireland, etc., beryls of gem quality are found in Africa, Madagascar (especially for morganite), and Brazil. The most famous place in the world for emeralds is at Muso, Colombia, South America, where they form a unique occurrence in lime-

stones. Emeralds are also obtained in the Transvaal and near Mursinsk, in Siberia. In the United States, New England has furnished much beryl from its pegmatites and for a long time the huge crystals from Acworth and Grafton, New Hampshire, were the largest known. Recently, however, giant crystals even larger than those from New Hampshire were discovered in Albany, Maine, the largest of which was 18' long, 4' through, and weighed about 18 tons. Other localities are Paris and elsewhere in Oxford County, Maine; Royalston, Massachusetts; North Carolina, Colorado, South Dakota, and California. Metallic beryllium (glucinium) is obtained from beryl. Its lightness and strength make it very valuable for industrial purposes. Alloyed with copper in small amount, it confers extraordinary properties on the copper. The word beryl comes from the Greek meaning the gem beryl. (E.S.C.S.)

BERYLLIUM OR GLUCINIUM. Symbol: Be. Atomic number: 4. Atomic weight: 9.02. Density: 1.84. Melting point: 1350°. No isotope, but single atomic form: 9 (99.95%).

Beryllium is a silver-white, hard, malleable metal of markedly low density. Beryllium is only slightly affected by water or cold **nitric acid,** but is readily dissolved by hot nitric acid, by **hydrochloric** or dilute sulfuric acid, and by **sodium** hydroxide solution; and when ignited in air forms beryllium oxide. Chemically related to **magnesium** and **aluminum.** Discovered by Vauquelin in 1797. On account of the lightness, strength, hardness and resistance to corrosion of alloys of beryllium extensive investigations have been undertaken with a view to its more common use. Two per cent beryllium in copper is used as material for springs.

Beryllium occurs in beryl (11%–13% BeO) in the New England states, also in South Dakota and Colorado, in South Africa, Madagascar, Austria, and France. The Copaux method of extraction consists in heating the beryllium aluminosilicate (beryl) with **sodium** silicofluoride, and later extracting with hot water, whereupon the beryllium salts are dissolved and thus separated from the other constituents. **Electrolysis** of fused beryllium chloride at 700° C., a temperature below the melting point of beryllium metal, or of fused beryllium oxyfluoride mixed with barium fluoride at 1400° C., a temperature above the melting point of beryllium metal, are methods used to secure the metal.

Hydroxide: Beryllium hydroxide ($Be(OH)_2$), white, gelatinous precipitate by the reaction of **ammonium** hydroxide or sulfide with beryllium salt solutions. This precipitate, while formed similarly to **aluminum** hydroxide, differs from the latter in being formed also by boiling a solution of beryllate, thus furnishing a separation from aluminum by boiling in sodium hydroxide (of 5 normal concentration).

Oxide: Beryllium oxide (BeO), white solid, by ignition of the nitrate, hydroxide, or of the metal in air; an excellent refractory material, melting point 2450° C.; used in small amount in the mixture of oxides of **thorium** and **cerium** in the incandescent gas mantle.

Beryllium dimethyl ($Be(CH_3)_2$), snow-white solid, sublimes at 200° C., and beryllium diethyl ($Be(C_2H_5)_2$), colorless liquid, of melting point 12° C. and boiling point at 15 mm. pressure, 110° C., both compounds spontaneously inflammable in air, have been prepared by Gilman and Schulze (1927). (R.K.S.)

BERYLLONITE. A mineral composed of the **rare sodium beryllium phosphate,** which crystallizes in the **orthorhombic** system with short prismatic or tabular crystals. Cleavage, perfect basal; hardness, 5–5.6; specific gravity, 2.8; luster is vitreous to somewhat pearly on the base; colorless to white or yellowish. Attempts to make use of this mineral as a gem stone were unsuccessful owing to its relative softness. Beryllonite is found at Stoneham and Newry, Oxford County, Maine. (E.S.C.S.)

BESSEL FUNCTIONS. The differential equation

$$x^2 \frac{d^2y}{dx^2} + x\frac{dy}{dx} + (x^2 - n^2)y = 0$$

is called **Bessel's** equation. It cannot be solved in terms of elementary functions. It defines a new class of functions, the Bessel functions of the first kind, denoted by $J_n(x)$, and those of the second kind, denoted by $K_n(x)$. These functions are also sometimes called cylindrical functions.

The functions $J_n(x)$ may be defined by the convergent series

$$J_n(x) = \sum_{k=0}^{\infty} \frac{(-1)^k x^{n+2k}}{2^{n+2k} k! \Gamma(n+k+1)},$$

and the functions $K_n(x)$ by the convergent series

$$K_n(x) = J_n(x) \cdot \log x - \frac{1}{2} \sum_{k=0}^{n-1} \frac{(n-k-1)! x^{-n+2k}}{2^{-n+2k} k!}$$

$$- \frac{1}{2} \sum_{k=0}^{\infty} \frac{(-1)^k x^{n+2k}}{2^{n+2k} k! (n+k)!} \cdot [H_k + H_{k+n}],$$

where $H_k \equiv 1 + \frac{1}{2} + \frac{1}{3} + \cdots + \frac{1}{k}$. When n is an integer, the series for J_n may be written

$$J_n(x) = \frac{x^n}{2^n \cdot n!}\left[1 - \frac{x^2}{2(2n+2)} + \frac{x^4}{2 \cdot 4(2n+2)(2n+4)}\right.$$

$$\left. - \frac{x^6}{2 \cdot 4 \cdot 6(2n+2)(2n+4)(2n+6)} + \cdots\right].$$

When n is not an integer, the general solution of Bessel's differential equation is $y = c_1 J_n(x) + c_2 J_{-n}(x)$, where c_1 and c_2 are arbitrary constants. When n is an integer, the general solution is $y = c_1 J_n(x) + c_2 K_n(x)$, where c_1 and c_2 are arbitrary constants. (L.L.S.)

BESSEMER PROCESS. The Bessemer process is

one of four methods for making **steel**, the others being the crucible, the open hearth, and the electric processes. The pig iron (see **Blast Furnace**) used in the manufacture of steel must be refined to eliminate the impurities. The refining operation consists largely of oxidation of the impurities to effect their removal. In 1856 Sir Henry Bessemer gave the world his process for converting pig iron into steel. The Bessemer process is based on blowing compressed air through a mass of molten iron. On its way through the **iron**, the **oxygen** in the air combines with the **silicon, manganese,** and **carbon** in the iron, oxidizing them with the production of considerable heat. The "blow," as this oxidizing action is called, consumes only a short time, perhaps from 10 to 30 minutes. During this period the progress of the oxidation of the impurities is judged by the appearance of the flame issuing from the mouth of the converter. The use of photo-electric methods of recording the characteristics of the flame has greatly aided the "blower" in controlling the final quality of the product. After the blow, the liquid metal is recarburized to the desired point, and sometimes other alloying materials added, dependent on whether plain carbon steel or **alloy** steel is to be the final product. Spiegeleisen, a low grade ferromanganese, high in carbon, is largely used in the manufacture of Bessemer steel.

The Bessemer converter is often built in the form of a pear-shaped vessel having the top opening inclined towards the side. Tuyère openings for air are placed in

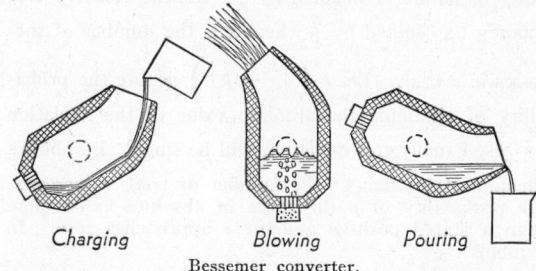
Bessemer converter.

or near the bottom of the converter, and the whole is mounted on trunnions so that it may be rotated into one or the other of three different positions, namely, inclined so that the opening is vertical for charging, upright during the blow, and horizontal for discharge of the finished product. The converter shell is made of steel plate and castings, and is lined with a **refractory** lining which may be either acid or basic, depending on the composition of the pig iron. Raw iron with low phosphorus and sulfur is readily refined, by the blowing process just described, in a converter lined with a siliceous material such as ganister. If the pig iron has considerable **phosphorus** and **sulfur** it is necessary to use a lining which has the proper fluxing characteristics, since phosphorus and sulfur are not readily removed by oxidation. Some of the materials for linings of the converter for the basic Bessemer process are mixtures containing limestone (**calcite**) and fire clay, **magnesite,** or calcined **dolomite.**

In the operation of the Bessemer converter it is necessary to have the whole charge and the converter intensely hot before beginning the blow. To this end the pig iron is either conveyed directly to the converter from the blast furnace, or is remelted in a cupola. The converter is prepared by firing with coal or oil until the lining is white hot. After the blow the iron is recarburized, and alloying elements are added and thoroughly mixed by stirring with a long iron bar and rocking the converter. (F.T.M., R.H.H.)

BETA FUNCTION. The beta function is a mathe-

matical expression; it is one of the transcendental functions which occurs first in the advanced part of **calculus.** It is also sometimes called an Eulerian integral of the first kind.

The **improper integral**

$$B(m, n) = \int_0^1 x^{m-1}(1-x)^{n-1}dx \quad (m > 0, n > 0)$$

defines a **function** of m and n which is called the beta function. It can be expressed in terms of the **gamma function** by

$$B(m, n) = \frac{\Gamma(m)\Gamma(n)}{\Gamma(m+n)}.$$

(L.L.S.)

BETA PARTICLES. Beta Rays. Radioactive Changes.

BETATRON. Cyclotron.

BETEL NUT. The fruit of the Areca palm, *Areca*

Catechu. The Areca palm, a plant native to Malaya, is extensively cultivated in the southern Asiatic countries and in the East Indian Islands. It is a slender unbranched tree which reaches a height of 40' with a trunk diameter of 3 or 4", and bears at its top a crown of 6–10 large leaves. The fruit is slightly more than an inch in diameter and has a fibrous rind surrounding the very hard seed. This seed is extensively chewed in Asiatic countries. To prepare them for chewing, the fruits are gathered just before maturity, boiled, sliced and dried. A piece of this dried fruit, together with a bit of lime, is wrapped in a betel leaf and the whole placed in the mouth. Chewing this preparation causes an abundant flow of saliva which is colored brick-red and stains the mouth, as well as blackening the teeth. The betel leaves are obtained from an entirely different plant, the Black **Pepper**, *Piper betel.* (R.M.W.)

BETELGUEZE. Betelgueze (*α Orionis*) is a contraction of an Arabic phrase indicating that this star is the "armpit of the central one," i.e., the armpit of Orion. Because of its rich reddish color the star has frequently been referred to as the "martial one," and in **Astrology** portends military or civic honors. Because it is the first star to rise of the brilliant and well known **constellation** of Orion, the title of "roarer" or "announcer" has been assigned to it by ancient writers.

Betelgueze is of great interest astronomically. It is an irregular **variable** star. It is also one of the first stars to have its diameter measured with the stellar **interferometer**. The diameter is found to be variable between 260,000,000 and 180,000,000 miles. At maximum diameter the star would extend out beyond the **orbit** of Mars if put in the sun's place. (W.K.G.)

BEVEL GEARING. Straight-tooth and spiral-tooth bevel gearing are used to transmit motion between shafts whose axes intersect. The operation of such units is analogous to that of friction cones, which may be considered to represent the pitch cones for the bevel gearing, and which correspond to pitch cylinders for spur gearing. Straight-tooth bevel gears have teeth of involute form, but the straight-line elements converge (if extended) at the intersection of the shaft axes, in contrast to the parallel-tooth elements of spur gears. There are several forms of bevel gearing; in the most important, the gear and pinion operate at a shaft axes angle of 90°. The unit is termed miter gearing if both gear and pinion have the same number of teeth, and angular gearing if the angle between the shaft axes is less than 90°. Bevel gearing in which the shaft axes angle is greater than 90° is also used to some extent. The pitch cone angles of the pinion and gear must be complementary for a shaft axes angle of 90°.

The pitch diameter of bevel gearing is measured at the large end of the pitch cones; the addendum and dedendum are not measured in the plane of the pitch

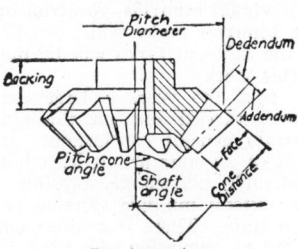

Bevel gearing.

circle, as in spur gearing, but are constructed perpendicular to the elements of the pitch cone on the surface of the back cone. The tooth shape is therefore dependent upon the magnitude of the back-cone radius, rather than the pitch radius, as in spur gearing.

Mortise gears have cast-iron rims with cored slots into which hard maple cogs or teeth are fitted and held in place by wedges at the back of the rim, and are designed to operate with cast iron cast tooth pinions. Cast-tooth gearing, however, is used only where the pitch-line velocity is low and where smooth action is not particularly important.

Spiral bevel gears have teeth cut in the arc of a spiral across the gear face, and bear the same relation to

straight bevel gears that helical gears do to spur gears. This construction results in a larger number of teeth in contact than in straight-tooth bevel gearing, and like helical gearing, permits higher pitch-line velocities and greater load-carrying capacities for the same occupied space. They are often used to replace straight-tooth bevel gear sets.

Hypoid gearing resembles spiral bevel gearing in general appearance, but is used for transmitting power between shafts whose axes are perpendicular, but nonintersecting. They find wide-spread application in automobile rear-axle drives, and in numerous other instances where bevel or worm gearing is not applicable. (H.C.H.)

BEVEL PROTRACTOR. **Measurement.**

BHARAL. Mammalia, Artiodactyla. The blue **sheep**, *Pseudois nayaur*, of Tibet. The species is closely related to the goats. (A.W.L.)

BIAS. A statistic θ' is said to be unbiased if the expected value of θ' equals the population value θ. A statistic which is not unbiased is biased. Bias in this sense should not be confused with the idea of a **consistent statistic**.

Bias is also sometimes used in the sense of an **error** in the observations, or in an error in the mean caused by persistent errors in the same direction. This notion is not the one usually considered when we say a statistic is biased.

In electrical work this term is used to denote a voltage whose principal function is to locate the operating point on the characteristic of a piece of apparatus. The term is most commonly applied to the grid voltage of a vacuum **tube**, in which case it means the d-c voltage (other than signal voltage) applied between the **cathode** and control **grid** of the tube. In this connection the term C-bias is also commonly used. The bias may be obtained from a source of d-c voltage, from the potential drop across a resistor (cathode resistor) in the cathode circuit or, when the grid carries current, from the drop across a resistor in the grid circuit. The first method is called fixed bias and the latter two self-bias. Bias is also used in **telegraphy** to indicate undesirable voltage additions to or subtractions from the code signals. (L.A.A., L.R.Q.)

BIAXIAL CRYSTALS. **Double Refraction.**

BICEPS. Any of several **muscles** with two heads. The principal examples are (1) the biceps femoris which,

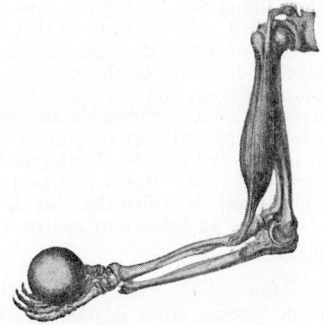

Biceps muscle contracting forearm.

in man, lies in the back part of the thigh and flexes the lower leg, and (2) the biceps of the upper arm which flexes the lower arm. (A.W.L.)

BICHIR. Pisces, Chondrostei. An air-breathing fish, *Polypterus,* found in the Nile, one of a few living representatives of a group which was once abundant and is supposed to have been ancestral to the bony fishes. It was formerly thought that *Polypterus* belonged to the Crossopterygii, a group of fish (**Pisces**) ancestral to the terrestrial vertebrates. (A.W.L.)

Bichir.

BICUSPID. A tooth with two cusps, especially the premolars of man. (A.W.L.)

BIELIDS. Bielids or Andromedes are the names applied to a **meteor shower** which is observed about November 24th of each year. The **radiant point** of the shower is in the **constellation** of **Andromeda.**

The history of the Bielids and their connection with **Biela's Comet** is one of the interesting chapters in the development of meteoric astronomy. Biela's Comet was first discovered in 1772 but was not found to be periodic. In 1826 Biela discovered the comet again and it now bears his name. Its **orbit** is elliptical with a period of about 6 years. In 1832 the comet passed very close to the earth. In 1845 the comet was observed to break in two, and in 1852, at the time of the predicted return, it was found that the two parts of the comet were both very faint and separated by over a million miles. They were unfavorably located relative to the sun for observation in 1859, and at the time of the return in 1866 they were not to be found.

The first mention of a swarm of meteors located on the orbit of Biela's comet is found in the display of December 5, 1741, when a brilliant shower was observed in Russia. They were observed during December in 1798, 1830, and 1838. By this time calculations of the radiant point had been made and it was first located, apparently, in **Cassiopeia.** By 1867 the date of the shower had shifted to November and it has always been observed in that month since that year. Up to 1885, many brilliant meteors were associated with the radiant point with from two to four observers on November 27 of that year observing no less than 39,546 meteors in 4 hours and 8 minutes. Since 1899, but very few members of the shower have been observed and it is evident that **perturbations** have shifted the orbit of the main swarm well outside the orbit of the earth. (W.K.G.)

BIENNIAL. Many plants, including some of the commonest cultivated ones, require 2 years of growth to complete their life-cycle. Such plants are called biennial. During the first year of growth they commonly form a close rosette of leaves growing from a very short stem and spreading out close to the ground, or form a head (as in the **cabbage**), and develop a thick tap root in which is accumulated a considerable amount for food reserves. During the second year of growth this reserve food is drawn upon to permit the development of a tall stem and flowers and fruit. **Beets** and **carrots** are examples of biennial plants in which the root rich in stored food reserves becomes an important source of food for man. (R.M.W., B.S.M.)

BIGHORN. Mammalia, Artiodactyla. *Ovis.* The mountain **sheep,** ranging from Alaska to Mexico and eastward into the bad lands of South Dakota. Several species and varieties are recognized in different parts of this range. (A.W.L.)

BILATERAL TOLERANCE. Interchangeable Manufacture.

BILE. Bile is a secretion of the **liver,** alkaline in reaction, yellow, brown or green in color. It consists of water, bile salts (sodium glycocholate and sodium tauro-cholate), inorganic salts, bile pigments (see **Pyrrole and Related Compounds**), **cholesterin,** and lecithin (see **Aminoacids, Polypeptides, and Proteins**). About a quart of bile is secreted every 24 hours, the quantity depending on the amount and kind of food eaten.

The bile salts are important; they stimulate the liver and have other functions of which little is known. After being secreted in the bile, the bile salts to a large extent are reabsorbed in the intestine and resecreted by the liver. The amount of bile pigments present determines the color of the bile.

The action of bile in the **digestion** of food is concerned with **fats.** Bile, plus the action of the **pancreatic** secretion, serves to break down the fats to **glycerin** and fatty acids. Fat soluble viatmins A, D, and K cannot be well absorbed without bile. Bile is also supposed to have an antiseptic action in the intestine, preventing putrefaction. Other functions of bile are not definitely known.

The bile is secreted by the liver into the biliary channels in the liver which enter the common bile duct. The bile duct enters the **duodenum** where a muscular **sphincter** controls the amount of flow into the intestine. Fatty foods and certain drugs cause the sphincter to open so that more bile escapes into the intestine.

If the flow of bile is prevented through obstruction, by stone formation, or by swelling of the bile duct walls from infection, the bile backs up and is absorbed by the blood stream and carried throughout the system. Such a condition results in **jaundice.** The same results follow if sufficient number of the smaller bile channels in the liver are obstructed. (R.S.M., D.M.H.)

BILL. The **beak** of a bird. (A.W.L.)

BILLFISH. Pisces, Teleostei. The garfish (**Pisces**), a marine species found from Massachusetts to Texas and ascending the rivers. Not to be confused with the garpikes or gars. (A.W.L.)

BINARY GRANITE. A granite containing only the essential minerals quartz and feldspar. Also used to describe granites which contain **muscovite** and **biotite** micas. (R.M.F.)

BINARY STARS. The term binary star was apparently first introduced by Sir William Herschel in 1802 to designate "a real double star—the union of two stars that are formed together in one system by the laws of attraction." At present, binary stars are classified under three headings: **visual binaries, spectroscopic binaries,** and **eclipsing binaries** under which headings the characteristics of the different types will be found discussed elsewhere in this work.

During the past 150 years a large amount of research has been carried out on the binary systems leading to certain general conclusions. It is believed that at least ¼ of all stars are at least binary systems, with a considerable percentage, possibly as great as 10%, of these systems multiple systems, i.e., containing more than two stars. There is a direct correlation between the period of revolution of a binary star and the eccentricity of its **orbit,** with the systems of short period having the smaller eccentricities. There is a regular gradation from pairs with short period in which the stars are practically in contact up to pairs so widely separated that the physical connection is only indicated by their common **proper motion** through space. Finally, in pairs in which the components are equal in brightness both stars have the same **spectral type,** while in systems where the brightnesses are different the fainter star is bluer if the brighter star is a giant; and redder if the brighter star belongs to the main sequence.

Since mass can be determined only from **gravitational** attraction and the only stars (with the exception of the **sun**) for which gravitational attraction can be determined are the binary stars, these objects form the one group from which the masses of stars may be determined. In the case of a visual binary star, after the orbit has been determined and the **stellar parallax** of the system obtained, the combined mass of the two stars may be obtained by a direct application of the **Keplerian Harmonic Law.** Unfortunately, it is impossible to obtain the complete orbit of a spectroscopic binary unless it is also a visual or an eclipsing binary; so from these objects a determination of mass is impossible except on the basis of a statistical discussion. In the case of those eclipsing binaries which are also spectroscopic binaries it is possible to make a complete solution for the specifications (i.e., masses, densities, sizes, luminosities, and approximate shapes) of both members of the system. From such objects, and only from such objects, may the complete characteristics of individual stars be determined. (W.K.G.)

BINNACLE. The stand for supporting and protecting the **compass** on board a ship is known as the binnacle. This stand is usually constructed of brass and is provided with a shaded light which illuminates the compass during the night, but does not shine in the eyes of the helmsman.

In addition to protecting the compass from effects of weather, the binnacle also contains a number of fixed and adjustable magnets and masses of soft iron for the purpose of partially compensating for the effects of the **magnetic field** of the ship. The process of adjusting the various compensating devices is known as compass adjusting and is usually carried on by experts while a ship is in port. (W.K.G.)

BINOCULAR. An instrument composed of two similar **telescopes,** one for each eye, usually with focusing tubes controlled by a common screw adjustment. The ordinary opera glass is a binocular utilizing Galilean telescopes. The field glass employs erecting telescopes of the spy-glass type. A well-known modern form is the "prism binocular." The special feature of this instrument is a pair of right-angled, **total reflection** prisms

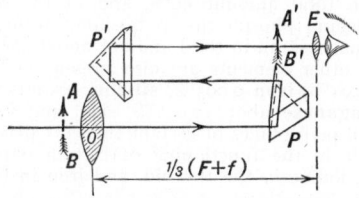

Optical system of prism binocular. (*Weld and Palmer, Textbook of Modern Physics, Blakiston.*)

in each telescope, which contribute three advantages. 1. The prisms, by means of two double total reflections in planes at right angles, accomplish the erection of the image without additional lenses. 2. The tube is rendered much shorter than in the ordinary field glass of equal power by the "doubling up" of the rays due to the reflections. 3. The objectives are by the same means set farther apart than the eyepieces, thus increasing the "stereo power" of the instrument as a binocular, so that objects can be seen to have depth or solidity at a greater distance than with the ordinary type. (See **Binocular Vision.**) (L.D.W.)

BINOCULAR VISION. The possession of two **eyes** set at a distance apart, but with approximately parallel axes, enables us to obtain two views from slightly different angles, and thus to become sensible of the solidity of single objects and to get an idea of the actual distribution of different objects in space. To become vividly conscious of this faculty, one has only to look about the room for a time with a hand cupped over one eye, and then suddenly to remove the hand. If the vision is reasonably normal, it will be noticed that with

one eye only the scene appears flat, like a photograph, but as soon as both eyes are used, objects spring into clear relief. In some manner the brain is able, through long experience, to blend the two different sensory pictures from the two different **retinal** images and to interpret the resulting sensation in terms of geometrical solidity. There is, however, a limit to the distance at which this impression is perceptible, and for very distant objects other factors must be relied upon, such as the apparent size (as of buildings or trees), or the opacity of the atmosphere (as in viewing distant mountains). In the absence of such factors, no estimate of distance can be formed; thus the stars all appear to be at the same distance. This limiting "stereoscopic radius" is for normal, unaided eyes, only a few hundred feet, but with a **binocular** telescope, and especially with a prism binocular, it is increased in a ratio called the "stereo power" of the instrument.

An interesting aspect of the subject is the use of binocular pictures and the stereoscope. Two photographs or drawings are prepared of the same group of objects from viewpoints approximately the same distance apart as the human eyes (say about $2\frac{3}{4}''$) and mounted side by side on a card so that each is viewed separately by the eye to which it corresponds; the observer gets the sensation of viewing a 3-dimensional scene. The observation is facilitated by a pair of lenses so designed as to allow of focusing the eyes for distance, and with a diaphragm set up between them to avoid seeing both pictures with either eye. This arrangement is the stereoscope. (L.D.W.)

BINOMIAL COEFFICIENTS. Binomial Formula.

BINOMIAL FORMULA. The binomial theorem gives a formula expressing any **power** of a binomial $a + x$ in terms of powers of a and x.

Binomial Theorem: If n is any positive integer, the expansion of $(a + x)^n$ is given by:

$$(a + x)^n = a^n + na^{n-1}x + \frac{n(n-1)}{2!}a^{n-2}x^2$$

$$+ \frac{n(n-1)(n-2)}{3!}a^{n-3}x^3 + \cdots$$

$$+ \frac{n(n-1)(n-2)\cdots(n-r+1)}{r!}a^{n-r}x^r + \cdots$$

$$+ nax^{n-1} + x^n,$$

where there are $n + 1$ terms on the right. If n is not a positive integer, the binomial expansion becomes an **infinite series:**

$$(a + x)^n = a^n + na^{n-1}x + \frac{n(n-1)}{2!}a^{n-2}x^2 + \cdots$$

$$+ \frac{n(n-1)(n-2)\cdots(n-r+1)}{r!}a^{n-r}x^r + \cdots$$

which is convergent for $|x| < |a|$.

The symbol $m!$, called factorial m, is the repeated product $1 \cdot 2 \cdot 3 \cdots m$ of all the positive integers from 1 to m inclusive; it is also sometimes denoted by $\underline{|m}$.

The rth term (general term) of the binomial expansion $(a + x)^n$ is:

$$\frac{n(n-1)(n-2)\cdots(n-r+2)}{(r-1)!}a^{n-r+1}x^{r-1};$$

sometimes the general term is taken as the $(r + 1)st$ term

$$\frac{n(n-1)(n-2)\cdots(n-r+1)}{r!}a^{n-r}x^r.$$

The coefficients in the binomial expansion are called binomial coefficients. They are often denoted by th symbols $1, \binom{n}{1}, \binom{n}{2}, \cdots, \binom{n}{r}, \cdots$, so that

$$\binom{n}{r} = \frac{n(n-1)(n-2)\cdots(n-r+1)}{r!}.$$

By the formula for **combinations**, it follows that $\binom{n}{r} = {}_nC_r$.

The binomial theorem is often written:

$$(1+x)^n = 1 + \binom{n}{1}x + \binom{n}{2}x^2 + \binom{n}{3}x^3 + \cdots + \binom{n}{r}x^r + \cdots$$

in terms of the binomial coefficient notation.

The binomial coefficients may be arranged in an interesting scheme called Pascal's triangle:

```
                    1
                  1   1
                1   2   1
              1   3   3   1
            1   4   6   4   1
          1   5  10  10   5   1
        1   6  15  20  15   6   1
      1   7  21  35  35  21   7   1
    1   8  28  56  70  56  28   8   1
```
. . . .

where each row of numbers represents the binomial coefficients of a certain power of a binomial.

The binomial formula gives the following important approximation:

$$(1 + x)^n \approx 1 + nx$$

if x is sufficiently small; special cases are:

$$(1 + x)^2 \approx 1 + 2x, \qquad (1 - x)^2 \approx 1 - 2x,$$
$$(1 + x)^3 \approx 1 + 3x, \qquad (1 - x)^3 \approx 1 - 3x,$$

$$\frac{1}{1+x} \approx 1 - x, \qquad \frac{1}{1-x} \approx 1 + x,$$

$$\sqrt{1+x} \approx 1 + \tfrac{1}{2}x, \qquad \sqrt{1-x} \approx 1 - \tfrac{1}{2}x,$$

$$\frac{1}{\sqrt{1+x}} \approx 1 - \tfrac{1}{2}x, \qquad \frac{1}{\sqrt{1-x}} \approx 1 + \tfrac{1}{2}x,$$

if x is sufficiently small. (L.L.S.)

BINOMIAL THEOREM. Binomial Formula.

BINTURONG. Mammalia, Carnivora. An Oriental species, *Arctitis binturong*, related to the **civets**. It is cat-like in appearance with tufted ears and a long, slightly bushy tail. (A.W.L.)

BIOCENOLOGY. The division of biological science which considers the factors binding populations of animals together as units in relation to environmental conditions. A subdivision of **ecology**. (A.W.L.)

BIOCHEMISTRY. Biochemistry is that branch of the science of chemistry which deals with the chemical processes and products of living organisms. The fundamental biochemical process upon which living organisms are dependent is that of **photosynthesis** which takes place in the green leaf of the plant. Under the influence of sunlight, the green leaves of plants, through the presence of **chlorophyll**—the green coloring matter of plants—and water, fix **carbon dioxide** of the atmosphere (about 0.03% carbon dioxide of the atmosphere) into **glucose**, and ultimately into **sucrose, starches, celluloses, carboxylic acids, tannins** and **fats**. By means of **nitrogen** supplied from the soil, certain plants synthesize **proteins** (in legumes, cereal grains, seeds) and many other nitrogen-containing organic substances, such as the **alkaloids**. The sugars, starches, celluloses, fats and proteins are the diet of herbivorous animals. These animals serve as food producers for their own and other kinds of animals, supplying these with protein and fat in the form of milk, eggs, and meat. See

Foods; Carbohydrates; Fats; Aminoacids and Proteins; Photosynthesis; Vitamin. Compounds of three carbon atoms form the common intermediary by which carbohydrates, fats and proteins are mutually interconvertible. Such three carbon compounds are lactic acid ($CH_3CHOHCOOH$), pyruvic acid ($CH_3COCOOH$), glyceric aldehyde ($CH_2OHCHOHCHO$), and pyruvic aldehyde (CH_3COCHO).

Chemical reactions of a specific character take place throughout the lifetime of each animal in each organ of its body. In the **stomach** there occurs acid digestion of foods, in the **intestines** alkaline digestion of foods, in the **lungs** absorption of oxygen into and liberation of carbon dioxide from the blood stream, in the **liver** production of glycogen and transformation of protein nitrogen into urea. The **thyroid** gland contains ten times as much iodine as any other organ of the body, and the thyroid secretion stimulates carbohydrate and calcium **metabolism** and exercises an effect on body fats. Over-secretion of the thyroid gland may result in **goiter**, whereas undersecretion may result in stunted growth or in great increase of weight from fat, accompanied by slow metabolism and retarded bone-growth.

The composition of the vegetation of the earth is estimated to consist of the four elements, **carbon, oxygen, hydrogen, nitrogen** in total amount 95%, and of the nine elements, **potassium, sodium, calcium, magnesium, silicon, sulfur, phosphorus, chlorine, iron** in total amount nearly 5%. These thirteen elements are stated above in the order of their decreasing abundance. The first three, carbon, oxygen and hydrogen, are derived from carbon dioxide of air and from water of air and soil. These, along with soluble nitrate nitrogen and the small amounts of the second group of nine elements supplied from the soil, are manufactured by the plant into its component materials, largely carbohydrates, fats and proteins. About 80 to 90% of the weight of the plant cell is water, and only in resting tissues, such as those of dried seeds, is the percentage of water small.

The composition of the human body is estimated to consist of the four elements, oxygen, carbon, hydrogen, nitrogen in total amount 96%, and of the seven elements, calcium, phosphorus, potassium, sulfur, sodium, chlorine, magnesium, in total amount about 4%. Traces of several other elements are also present in the body, **fluorine** 0.01%, **iron** 0.005%, **silicon, bromine, aluminum, manganese** about 0.001% each, and iodine still less. Small percentages of certain elements play a determining role in the functioning of certain parts of the body, e.g., iodine in the thyroid, and iron in the blood. (R.K.S.)

BIOECOLOGY. The division of biological science which treats the general relations of living things. (A.W.L.)

BIOGENESIS. The established principle that all living things spring from previously existing living things. **Abiogenesis.** (A.W.L.)

BIOGENETIC LAW. Evolution.

BIOHERM. A geologic term for beds or mounds of colonial and gregarious marine **fossils** with calcareous shells or skeletons. Present day bioherms are usually referred to as **coral reefs**. (R.M.F.)

BIOLOGICAL SURVEY. A governmental agency which gathers data on the wild life of the country, makes this information available to the public through official publications, takes an active part in conservation, and aids in the control and eradication of harmful organisms.

The United States Bureau of Biological Survey was a subsidiary of the Department of Agriculture primarily occupied with work on birds and mammals. The insects are dealt with by the Bureau of Entomology and Plant Quarantine and the fishes and many marine invertebrates by the Bureau of Fisheries. The work of the Bureau of Biological Survey and that of the Bureau of Fisheries was combined under the Fish and Wildlife Service of the Department of the Interior in 1940. This Service has an extensive organization reaching to all parts of the national domain and carries on the former work of the Bureau in the establishment of game refuges, the protection of game by all available measures, the extension of assistance in the propagation of game species, and the control of predatory animals.

The work of the United States Service is supplemented by that of state biological surveys. The latter, however, are limited chiefly to the study of wild life and the publication of records. Conservation activities in the states are usually in the hands of special bureaus with greater power and resources than the biological surveys. (A.W.L.)

BIONOMICS. The division of biological science which treats the relations of living things to each other and to the environment. (A.W.L.)

BIONOMY. Life habits of animals and plants. (R.M.F.)

BIOPHYSICS. The physics of biological processes or phenomena. So much has been learned in recent years as to the mechanisms of life functions that biophysics has taken its place along with **biochemistry** as an important area of material science.

Of course such matters as the dynamics of animal skeletons—with bones as levers or toggle joints—have always been cited in physics. Also the eye has long been recognized as an optical, the ear as an acoustic, instrument. More recent is the identification of nerve responses with electric currents. An extensive phase of biophysics is the role of **osmosis** in vital processes such as **secretion** and **respiration.**

Among the puzzling problems of biophysics are the phenomena of "bioluminescence," that is, the emission of light by living organisms such as the firefly and luminous fungi; and animal electricity, exhibited by such creatures as the **electric eel.**

The uses of physical techniques in medical practice are closely related to this field; such for example as the applications of **x-rays** and radioactivity to diagnosis and treatment, and the various phases of electrotherapy and **diathermy.**

Vaguely discernible on the horizon of biophysics are certain mysterious and controversial phenomena denoted as "telepathy" and "clairvoyance." While these have as yet not been divested of their associations with charlatanism, no one can say how soon they may emerge as demonstrable physical realities and take a legitimate place in biophysical science. (L.D.W.)

BIOPSY. A diagnostic procedure consisting of removal of a piece of tissue from a living patient and its examination under a microscope after sectioning and staining. This is an aid in diagnosis, especially in the differentiation of tumors, tuberculosis, and various types of chronic infections. (D.M.H.)

BIOSTROME. A geologic term for layers, beds, or strata composed of calcareous fossil shells which form **coquina,** or shell-limestone. The term biostrome is primarily intended to distinguish shell-limestone from **bioherms,** or typical coral reefs. (R.M.F.)

BIOTIC AREA. A geographical division characterized by certain environmental factors which determine the nature of its population. (A.W.L.)

BIOTIC FACTOR. Plants are profoundly affected by various environmental factors. Among these are the

effects of other plants and various animals. The living forces are the biotic factors, which severally affect every part of a plant. In the soil are countless numbers of **bacteria,** including the nitrogen-fixing group, which are of tremendous importance to the plant. Here also are earthworms and various soil-inhabiting animals. Above ground are the many animals which feed on the plant and also the numerous parasites which attack it. Large numbers of these parasites gain entrance to the plant tissues, where their presence may cause very great changes. Again, there is the effect which any plant may have on its neighbors in the constant struggle for existence. Many plants are much more aggressive than others, and so able to invade new regions, where they crowd out native plants already present. Plants of this type include many of the common **weeds.** Finally a very important biotic factor is man. He is constantly disturbing the established balance by his activities, often with serious consequences to himself. Wittingly or otherwise he introduces many new plants into regions where their presence is not desirable. He cuts off forests, and so leads to the destruction of shade-loving or shade-tolerant plants which grew under the trees. Many indeed are the **fungous** pests which he has carried to new regions, where the fungus has found new hosts which had not developed resistance and so readily succumbed to the parasite. (R.M.W.)

BIOTIC POTENTIAL. A quantitative expression of the dynamic significance of various inherent vital properties of living things as a factor in the establishment of external relationships. It summarizes the reproductive and survival potentialities of the organism. (A.W.L.)

BIOTIC RESISTANCE. The living factors in the environment of an organism which tend to hinder its normal increase. Parasitic and predacious enemies and organic food supply are outstanding examples. (A.W.L.)

BIOTITE. A common **silicate** mineral containing **potassium, magnesium, iron** and **aluminum,** sometimes called "iron mica," is found in granitic rocks, **gneisses, schists,** etc. Although actually **monoclinic,** it often assumes a pseudo-**hexagonal** form. Like the others of the **mica** group biotite shows a highly perfect basal cleavage. It has a hardness of 2.5–3; specific gravity, 2.7–3.1; luster, pearly to vitreous or sometimes submetallic when very black in color; cleavage sheets are elastic; color, greenish to brown or black; transparent to opaque.

Biotite is occasionally found in large sheets, especially in **pegmatite** veins and also occurs as a contact metamorphic mineral or the product of the alteration of **hornblende, augite, wernerite** and similar minerals.

Biotite occurs in the lavas of Vesuvius, at Monzoni, and many other European localities. In the United States it is found especially in the pegmatites of New England and of Virginia and North Carolina and the granite of Pike's Peak, Colorado. Biotite was named in honor of the French physicist, J. B. Biot. (E.S.C.S.)

BIOT-SAVART LAW. A law expressing the intensity of the **magnetic field** in the neighborhood of a long, straight wire carrying a steady **current.** If a permanent **magnet** is rigidly attached in any position to a rod or frame which is capable of rotation about such a wire as an axis, it is found that there is no resultant **torque** about the wire. From this it is readily shown that the field intensity varies inversely as the distance from the wire. If the current is i (abamperes) and the distance r (cm.), the intensity is given by the Biot-Savart law as $H = 2i/r$ (oersteds). **Ampère's law** is sometimes called by this name, since either of the two laws may be deduced from the other. (L.D.W.)

BIOTYPE. A group of individuals of similar hereditary organization. (A.W.L.)

BIPACKS AND TRIPACKS. Combinations of two or more **photographic emulsions** combined in such a way that they are all exposed together. The emulsions may be on separate supports in which case they are called separable bipacks and tripacks, or the emulsion may be coated on a single support when they are called integral bipacks and tripacks.

The simplest bipack consists of two emulsions on separate supports placed in the film or plate holder with the emulsions held together in intimate contact. Such a separable bipack has found wide application in making 2- and 3-color **motion pictures** and as a means of obtaining two of the necessary three color separation negatives in a 3-color camera. Bipacks may also be obtained by coating two photographic emulsions on opposite sides of a single support (two-sided bipack) or two emulsions coated on the same side of a single support (integral bipack). The type of bipack construction is generally dictated by the method of color image formation to be employed in the process concerned.

Tripacks may contain three separate emulsions on different supports (separable tripack) or three emulsions on two supports using an integral bipack on one and the other emulsion on a separate support. The most common tripacks today are integral and use but one support to carry all three emulsions. There may be an integral bipack on one side and the third emulsion on the other side (two-sided integral bipack) or all three emulsions may be on one side of the support (monopack).

The **photographic emulsions** employed in integral bipacks and tripacks are generally quite thin (.005–.02 mm.) and transparent to prevent too great light absorption in the upper emulsions and consequent loss of speed and diffusion. In the usual construction for a 3-color system the red sensitive emulsion is nearest the support with the blue recording layer being on the top surface. The top blue sensitive emulsion is generally separated from the green and red recording emulsions by a yellow separating layer to prevent any blue light from being recorded by these emulsions.

Integral bipacks and tripacks are now widely used as camera materials for taking pictures which are converted into color images by means of dye coupling development or **chemical dye destruction.** Integral tripacks are also used as a means of obtaining three-color prints by the same photographic processes. Present trends indicate a gradual replacement of most systems of **color photography** in which separate images are converted into the subtractive colorants by systems making use of the advantages of integral tripacks. (H.C.C.)

BIPARTITE CUBIC. This name is given to the curve whose equation in **rectangular coordinates** is

$$y^2 = x(x - a)(x - b) \quad (o < a < b).$$

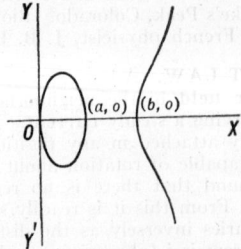

Bipartite cubic.

Its shape is shown by the accompanying figure. (L.L.S.)

BIPINNARIA. The form of **dipleurula** larva found in the starfishes (**Asteroidea**). It is bilaterally symmetrical and has a ciliated band. (A.W.L.)

BIPLANE, EFFICIENCY OF. Equivalent Monoplane.

BIPRISM. Young's Interference Experiment.

BIQUADRATIC EQUATIONS. Quartic Equations.

BIRAMOUS APPENDAGE. The primitive jointed appendage of the **arthropods,** still found in various form in the **crustaceans.**

The appendage consists of a single basal portion called the protopodite which is usually divided into a proximal coxopodite and a distal basipodite. It may bear on its outer margin one or several lobes called epipodites. From the protopodite two branches arise, an inner endopodite and an outer exopodite; this characteristic of the appendage is responsible for the name biramous. The endopodite is divided into five or less segments, named in order from the base the ischiopodite, meropodite, carpopodite, propodite, and dactylopodite. The exopodite is much less uniform and is often lacking.

These appendages have become modified and specialized for many functions in the existing crustaceans, as is nicely demonstrated by the appendages of the **crayfish** and **lobster.** In these animals they form sensory organs (antennae), mouth parts (jaws and accessory appendages), walking legs, swimmerets, accessory reproductive organs, and broad, flat swimming appendages. They are also regarded as the form from which the simpler jointed appendages of insects and other arthropods have been evolved. (A.W.L.)

BIRD. A warm-blooded feathered vertebrate. (See **Aves.**) (A.W.L.)

BIRD OF PARADISE. Aves, Passeriformes. Any bird (**Aves**) of numerous beautiful species which make up the family Paradiseidae. They are characterized by the gorgeous colors and bizarre forms of the plumage and are unsurpassed in splendor by any other group of birds, although they are fairly near to the crows in classification. Most of the species are found in New Guinea. (A.W.L.)

BIRD-LOUSE. Insecta, Mallophaga. A wingless ectoparasitic insect with biting mouth parts. Most species of bird lice live among the feathers of birds and eat

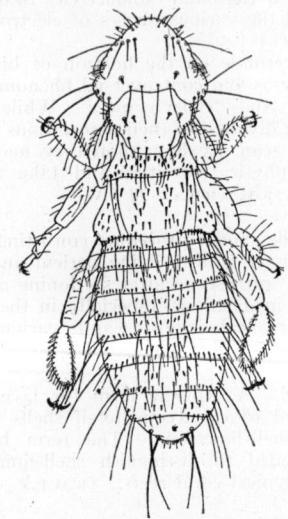

Bird-louse.

bits of feather and other debris. Although a few species are found on mammals the prevailing type of host has given its name to the entire order; the name biting lice is also distinctive.

The bird-lice which affect poultry are economically important. Even though they do not suck blood like other parasites the irritation resulting from their presence in large numbers is serious to the birds. Various measures of control have been devised, among them whitewashing roosts, oiling perches with kerosene, and the use of various insect powders in nests and on the birds themselves. The maintenance of clean surroundings for the flock is of the utmost importance in preventing severe infestation. (A.W.L.)

BIRD'S-NEST FUNGI. Basidiomycetes.

BIRMINGHAM WIRE GAUGE. Gauge Number.

BISHOP-BIRD. Aves, Passeriformes. Any bird (**Aves**) of several brightly colored species of African weaver-birds which make up the genus *Pyromelana*. The name has also been applied in the past to some of the brightly colored birds of North America, especially by the early settlers in Louisiana. (A.W.L.)

BISMITE. Stibiconite.

BISMUTH. Symbol: Bi. Atomic number: 83. Atomic weight: 209.00. Density: 9.8. Hardness: 2.5. Melting point: 271° C. Boiling point: 1450° C. No isotope, but of single atomic form: 209.

Bismuth is a white metal having a reddish tinge, lustrous, brittle, not very hard, somewhat malleable, not very ductile, when bent at 100° C. emits a creaking sound due to friction of the crystals; permanent in air or ordinary temperatures, burns to the trioxide upon heating to high temperature; insoluble in **hydrochloric** or dilute **sulfuric acid**, soluble in **nitric acid** to form nitrate; heated with **chlorine** yields chloride. Discovered by Valentine in 1450. Bismuth metal is used in alloys, for bearing or anti-friction metals, and for fusible metals. The salts of bismuth are frequently used in medicine, principally in (1) digestive disorders as a soothing protective to irritated mucous membranes, (2) in **x-ray** diagnostic examination where the opacity of bismuth makes simple the production of contrast pictures of the entire gastro-intestinal tract, (3) in the treatment of **syphilis** where it is given intramuscularly by hypodermic injection.

Bismuth occurs as native bismuth in Bolivia and Saxony and frequently is associated with lead, copper, and tin ores—the sulfide (**bismuthinite**, bismuth glance, Bi_2S_3) is also found in nature. Separation of bismuth from lead takes place during the **electrolytic** refining of the latter with bismuth remaining in the anode mud.

Chlorides: Bismuth chloride ($BiCl_3$), white crystals by reaction of (1) bismuth metal heated with **chlorine**, (2) bismuth oxide and **hydrochloric acid**, and then crystallizing, reactive with water to form oxychloride; bismuth oxychloride, bismuth subchloride (BiOCl), white solid, soluble in hydrochloric acid.

Hydroxide: Bismuth hydroxide ($Bi(OH)_3$), white precipitate, by reaction of bismuth salt solutions and alkalis, soluble in hydrochloric, sulfuric, or nitric acid.

Nitrates: Bismuth nitrate ($Bi(NO_3)_3 \cdot 5H_2O$), white crystals, by reaction of (1) bismuth metal or oxide and **nitric acid**, and then crystallizing, reactive with water to form oxynitrate; bismuth oxynitrate, bismuth subnitrate ($BiONO_3$), white solid, soluble in nitric acid, used in cosmetics, in pharmacy, in producing luster on metals, in enamels on ceramic ware.

Oxides: Bismuth oxide, bismuth trioxide (Bi_2O_3), pale yellow to brownish solid (1) by heating bismuth hydroxide or nitrate, (2) by burning bismuth at a high temperature in air, soluble in hydrochloric, sulfuric, or nitric acid; bismuth textroxide (Bi_2O_4), yellow to brown solid by reaction of bismuth trioxide and sodium hypochlorite solution; bismuth pentoxide (probably $Bi_2O_5 \cdot$

H_2O or $HBiO_3$), scarlet red, by electrolytic oxidation of bismuth trioxide in alkali, decomposes when heated at 150° C. yielding trioxide and oxygen.

Sodium bismuthate ($NaBiO_3$) is used in acid to oxidize manganous to permanganate in the identification of manganese.

Sulfates: Bismuth sulfate ($Bi_2(SO_4)_3$), white crystals, by reaction of bismuth trioxide and sulfuric acid, and then crystallizing, reactive with water to form bismuth oxysulfate; bismuth oxysulfate, bismuth subsulfate (($BiO)_2SO_4$), white solid, soluble in sulfuric acid.

Sulfide: Bismuth sulfide (Bi_2S_3), dark brown precipitate, by reaction of bismuth salt solution and **hydrogen sulfide**, soluble in hot dilute nitric acid, insoluble in **sodium** or **ammonium** sulfide. (R.K.S.)

REPRESENTATIVE ALLOYS CONTAINING BISMUTH

Fusible alloy, melting at 96° C.	53% Bi	32% Pb	15% Sm
Fusible alloy, melting at 91.5° C.	52% Bi	40% Pb	8% Cd
Fusible alloy, melting at 100° C.	50% Bi	30% Sn	20% Pb
Fusible alloy, melting at 70° C. (Wood's metal).	50% Bi	25% Pb	12.5% Sn 12.5% Cd
Fusible alloy, melting at 70° C. (Lipowitz' alloy)	50% Bi	27% Pb	13% Sn 10% Cd
Rose metal	50% Bi	27% Pb	23% Sn
Bismuth solder, melting at 111° C.	40% Bi	40% Pb	20% Sn

(R.K.S.

BISMUTHINITE (Bismuth Glance). A mineral containing a **sulfide** of bismuth and sometimes **copper** and **iron**; a variety from Mexico contains about 8% **antimony**. Bismuthinite is **orthorhombic** although its thin needlelike crystals are rare as it usually occurs in foliated or fibrous masses. It has one good cleavage parallel to the prism; hardness, 2; specific gravity, 6.4–6.5; metallic luster; streak, lead gray; color, similar but often with iridescent tarnish; opaque.

Bismuthinite is a rather rare mineral although somewhat widely distributed. European localities are in Norway, Sweden, Saxony, Rumania, and England. It is found also in Bolivia, Australia, and in the United States in Utah. It is used as an ore of bismuth. (E.S.C.S.)

BISON. Mammalia, Artiodactyla. A large hoofed animal with a prominent hump, short curved horns, and in the male sex a heavy mane. Two species occur, the European, *Bison bonasus,* and American, *B. americanus.* The former is a browsing forest species of northern Europe and the latter a grazing species, formerly very abundant on the plains of North America but now restricted to a few protected herds.

The romantic—and tragic—story of the American bison is known to everyone. A mainstay of the Plains Indians for food, hides for the construction of shelters, and fuel in the form of buffalo chips, it served the same purposes for early settlers, but the greater destructiveness of firearms and the availability of an Eastern market for hides soon threatened its extinction. A herd of about 6000 is now maintained in the Yellowstone National Park and a much larger number in Canada.

The European bison also goes by the names wisent and zubr but the name aurochs, sometimes applied to it, is not correctly used here. (A.W.L.)

BIT. Woodworking.

BITING LOUSE. Bird-Louse.

BITTERLING. Pisces, Teleostei. Fishes (**Pisces**) of several species allied to the carp. One, *Rhodeus amarus,* lives in Europe and the others in Eastern Asia. (A.W.L.)

BITTERN. Aves, Ciconiiformes. Wading birds (**Aves**) allied to the herons and egrets. They have moderately long legs and a straight beak which is strong and sharp. Two species, the American, *Botaurus lentiginosus,* and least, *Ixobrychus exilis,* bitterns, occur in North America, and several others are found on other continents. (A.W.L.)

BITUMEN. A general term for petroliferous substances ranging from true **petroleum** through the so-called mineral tars to **asphalt.** (R.M.F.)

BITUMINOUS COAL. Coal.

BIURET. Amines and Amides.

BIVALVE. A shell composed of two distinct parts or valves. Such shells are secreted by **brachiopods,** in which the valves are dorsal and ventral, and by certain **crustaceans** (Ostracoda) and **mollusks** (Pelecypoda) in which the valves are lateral. The most common examples of bivalves are among the edible mollusks, including **clams, oysters,** and **scallops.** (A.W.L.)

BLACK APE. Mammalia, Primates. A monkey, *Cynopithecus niger,* of Celebes whose rudimentary tail gives it some superficial likeness to the apes. It has an elongate muzzle and is related to the **macaques** and **baboons.** (A.W.L.)

BLACK BODY. This term denotes an ideal body which would, if it existed, absorb all and reflect none of the **radiation** falling upon it; its reflectivity would be zero and its absorptivity would be 100%. Such a body would, when illuminated, appear perfectly **black,** and would be invisible except as its outline might be revealed by the obscuring of objects beyond. The chief interest attached to such a body lies in the character of the radiation emitted by it when heated and the laws which govern the relations of the **flux** density and the **spectral energy distribution** of that radiation to the temperature.

The total emission of radiant energy from a black body takes place at a rate expressed by the **Stefan-Boltzmann** (fourth-power) **law;** while its spectral energy distribution is described by **Wien's laws,** or more accurately by **Planck's equation,** as well as by a number of other empirical laws and formulae. (See **Thermal Radiation.**)

The nearest approach to the ideal black body, experimentally, is not a sooty surface, as might be supposed, but an almost completely closed cavity in an opaque body, such as a jug. The laboratory type is usually a somewhat elongated, hollow metal cylinder, blackened inside, and completely closed except for a narrow slit in one end. When such an enclosure (called a *Hohlraum* in German) is heated, the radiation escaping through the opening closely resembles the ideal black-body radiation; while light or other radiation entering by the opening is almost completely trapped by multiple reflection from the walls, so that the opening usually appears intensely black. For this reason, black-body or "Planckian" radiation is also often called "cavity radiation." (L.D.W.)

BLACK BUCK. Mammalia, Artiodactyla. The Indian **antelope,** *Antilope cervicapra.* (A.W.L.)

BLACK COLOB. Mammalia, Primates. A species of thumbless **monkey,** *Colobus satanus,* found in western Africa. (A.W.L.)

BLACK DEATH. The great pandemic of **plague** which swept Europe in 1348. (D.M.H.)

BLACK PRINTER. A separate image used in the making of a color reproduction by subtractive synthesis. The black printer is often necessary because of the difficulty of reproducing grays and black from mixtures of the colorants used in making the color reproduction and also to impart sufficient density to the dark portions of the reproduction.

A negative for making the black printer may be made by using a yellow filter with a panchromatic emulsion or by using an infra-red emulsion and filter transmitting only infra-red light. Most photomechanical methods of color reproduction use a black printer and some photographic procedures, such as imbibition, occasionally make use of such an achromatic image. (H.C.C.)

BLACKBERRY. Rose Family.

BLACKBIRD. Aves, Passeriformes. 1. The ouzel of Europe. 2. Several species of North American birds of the genus *Agelaius* related to the orioles and grackles,

A North American red-winged blackbird. Glossy black, shoulders scarlet, edged with yellowish. Female streaked with no red.

and sometimes applied to the grackles as well. 3. In the West Indies applied to the ani, a member of the order Cuculiformes. (A.W.L.)

BLACKCOCK. Aves, Galliformes. The male of the Eurasian black **grouse,** *Lyrurus,* of which there are two species, one limited to the Caucasus. (A.W.L.)

BLACKFIN. Pisces, Teleostei. A small fish (**Pisces**), *Lythrurus atripes,* of southern Illinois and Iowa with black-marked dorsal and anal fins. Related to the shiners. (A.W.L.)

BLACKFISH. 1. Pisces, Teleostei. The black sea bass, *Centropristis striatus.* 2. Mammalia, Odontoceti. A black **whale** of moderate size, up to 20' long. It is found in both the Atlantic and Pacific Oceans and is widely distributed from north to south. Also called the pilot whale, *Globicephalus melas.* (A.W.L.)

BLACK-FLY. Insecta, Diptera. A minute fly whose small head and large **thorax** give it a hump-backed appearance. They are also called buffalo-gnats and the Indian name no-see-'em is sometimes used for the very small species. They constitute the family Simuliidae.

While some of these insects are harmless, others are among the most troublesome of our blood-sucking insects. Their bite is extremely irritating, considering its size, and the swarms are sometimes so numerous that their attack is serious to man and may cause the death of smaller animals, such as chicks. They are especially abundant in the woods, where campers and outdoor workers sometimes find it necessary to use oil of citronella, or one of the preparations containing tar, on exposed portions of the skin to prevent attack. (A.W.L.)

BLACK-GAME. Aves., Galliformes. The Eurasian black grouse. (A.W.L.)

BLACKHORSE. Pisces, Teleostei. A fish (**Pisces**) *Cycleptus elongatus,* of the Mississippi River system, also called the Missouri sucker. It attains a length of thirty inches and its flesh is excellent. (A.W.L.)

BLACKWATER FEVER. Malaria.

BLADDER. A thin-walled muscular sac which serves as a reservoir for urine. It is situated in the anterior portion of the pelvic cavity. The urine drains from the kidneys down the ureters, and the latter enter the bladder one on either side. The mucous-membrane linings of the kidneys, ureters and bladder are continuous.

   The term bladder is also used to describe similar sacs such as the **gall bladder** and the swim bladder of fishes. (D.M.H., A.W.L.)

BLADDER WORM. Platyhelminthes, Cestoda. An immature resting stage of **tapeworms** consisting of a bladder-like cyst in which one or more heads are inverted. Also known as the cysticercus stage. (A.W.L.)

BLADDERWORTS. Insectivorous Plants.

BLADE. Leaf.

BLANK. Sheet Metal Processes; Press-working.

BLAST FURNACE. The blast furnace is the chief means of reducing iron ore to pig iron. The reduction process is carried out at high temperature, and in the presence of a fluxing substance. A cross-section of a blast furnace is shown in the accompanying figure. The

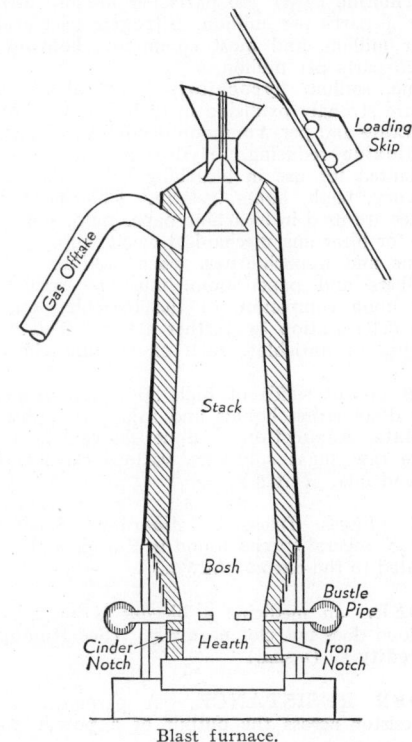

Blast furnace.

furnace may be 90–100′ high, and of varying diameters. It may be thought of as being composed of three sections: the lower section, called the hearth, in which the molten iron is accumulated; a short section above the hearth, called the bosh; and the stack, which is above the bosh. The hearth portion is constructed of fire-brick of the best quality with thick walls. The tapping hole is situated at some convenient point in the hearth, as is also the slag hole. Near the top of the hearth section tuyères are spaced around the circumference for the admission of air for combustion. Each tuyère receives a supply of air from the tuyère stock which is connected to a header called a bustle pipe. The latter is a large pipe which encircles the furnace and distributes heated air to the tuyères. The bosh, being just above the air admission point, encloses the hottest portion of the furnace. In order to keep the bricks from burning out, water-cooled plates are built into the brickwork of the bosh, and much heat is carried away by the circulating water. Extending upward from the bosh to the top of the furnace is the stack, made of plate steel and lined with fire-brick. At the top of the furnace is a double bell, which forms a sort of air lock for the admission of materials during continuous operation of the furnace.

   When in use, the blast furnace is first charged in the proper manner with alternate layers of **coke**, ore, and **limestone**. The coke is ignited at the bottom and is rapidly burned under the influence of a forced draft of air blown in through the tuyères. As the coke is burned away, the material moves downward in the furnace towards the hearth, but the stack is kept full by fresh charges admitted through the bells. The process of reduction is rather complicated chemically, but the net effect is the conversion of the ore, which may consist chiefly of **hematite** (Fe_2O_3), into liquid iron in the hearth, with slag, consisting of the limestone flux and the ash from the coke, together with compounds formed by reaction of the flux with substances present in the ore, floating in a molten state on the iron. This molten slag, or cinder, may be tapped off through the cinder notch, which is a slag opening. The iron is tapped at intervals through the iron notch. As the hot gases from the combustion region pass upward from the furnace they heat the fresh charges and pass out of the furnace through ducts which carry them to the purifying equipment. Since a reducing atmosphere is maintained in the blast furnace, the gases leaving it have some fuel value, due to their **carbon monoxide** content. These gases are passed through dust catchers and scrubbers until they are comparatively free from dust, after which part of the gas is used in gas engines which operate the blowers, and part of it is used in stoves or regenerative heaters, which preheat the air going to the bustle pipe. (F.T.M.)

BLASTING. Blasting is the rending or loosening of a hard or closely packed material by **explosives**. Blasting action can be one of shattering percussion or heaving, depending upon the type of explosive used, and the method by which it is placed. The principal field of blasting is in connection with earth or rock, although many other uses are found for it. For example, it may be used to loosen coal from the seam, to fell a large obsolete chimney, or to break up foundations which are to be replaced, etc. Blasting is very frequently used in preparing the right of way for a new road. Rocks must be shattered, boulders loosened and broken up, tree stumps blown out, and earth loosened and rendered suitable for rapid excavation. Quarry work offers a considerable opportunity for skilful removal of stone from the native mass by means of judiciously placed explosives. Ditches may be dug with explosives by transmitted blasting, and material of all sorts may be loosened so that loading by power shovel is rapid and economical.

   The explosives used for blasting constitute one very definite branch of all explosives. Dynamite and powder are chiefly used. While a great many high-powered explosives of different sorts are used in shells for offensive weapons, these are not suitable for general blasting service. Nitroglycerin dynamite and ammonia dynamite are the principal explosives for commercial blasting work. Blasting powders are also used where a permissible explosive is not required. A permissible explosive is a short-flame explosive which is considered safe for use in mines where the atmosphere may contain some combustible gas. This feature is not important, naturally, for blasting in stone, slate, and granite quarries, where blasting powder is in use. Dynamite is available in cartridges, in granular form, and powder form. Some dynamite, such as gelatin dynamite, is distinguished by plasticity, high density, and imperviousness to water. Other dynamites cannot be used in wet locations. In cold weather special low freezing formulae are required. The dynamite cartridges are made in strength varying from 15–60% dynamite so that the explosive power of a blast may be regulated. Dynamite and blasting powder are set off by means of blasting caps containing picric acid, tetryl and mercury fulminate, or lead azide. The blasting caps are ignited to produce an initial detonation, either by the heat of a fuse or by an electric current. A blasting machine is a hand-operated apparatus for generating current used in firing electric blasting caps.

There are several ways of using an explosive in blasting. A hole may be drilled in the earth or rock, and the dynamite cartridges placed at the bottom of the hole. The blasting cap, with fuse or wires attached, is placed on the top of the charge, and the hole is then "stemmed." Stemming consists of carefully tamping the hole above the explosive full of a material such as moist clay or earth, so that the explosive force of the dynamite will be confined and will produce the maximum shattering or loosening effect. Another way of using dynamite is to place it on top of the boulder or rock to be shattered, and cap it thoroughly with a layer of mud or clay well packed. The effectiveness of this kind of blasting is the result of the intensely fast action of dynamite, so rapid that the mud and atmospheric reactions are sufficient to give a shattering blow to rock or boulders having a brittle structure. This method of blasting tends to be wasteful of explosive, and is not used unless other methods, such as boring of holes, are impracticable. A blast of this character is known as a "dobie" shot. Considerable skill in the use of multiple charges in excavation or break-up of rock or coal is attained by those dealing regularly with dynamite. There are two methods of detonating multiple charges. With one, the charges are exploded simultaneously; with the other, one charge explodes the next one, and so on. For simultaneous explosion, electrical firing is a necessity, but where electrical firing is not used, multiple charges can be set off without difficulty if the ground is so thoroughly wet that the shock of an initial charge is transmitted through the ground to the adjacent one with sufficient intensity to detonate it. Blasting is comparatively safe in the hands of those trained and skilled in the use of explosives as long as they remain careful and cognizant of the hazards involved, but is not to be recommended for the inexperienced or casual user of explosives. (F.T.M.)

BLASTOCOELE. The first cavity formed during the **embryonic** development of animals. In many species the cleavage of the fertilized **ovum** gives rise to a hollow **blastula** of spheroidal form; the cavity of this structure is the blastocoele. (A.W.L.)

BLASTOMERE. Any of the **cells** resulting from the subdivision of the fertilized **ovum** during early embryonic development. (A.W.L.)

BLASTOMYCOSIS. A chronic infection caused by **fungi** of the blastomycetes group. It is characterized by multiple abscesses in the skin and at times of the internal organs. The disease is not common but more cases

are seen in the Middle West than elsewhere in the United States. Blastomycosis is chronic and prolonged with remissions. The systemic form is usually fatal. (R.S.M., D.M.H.)

BLASTOPORE. The external opening of the **gastrula.** (A.W.L.)

BLASTOSTYLE. A **polyp** from which medusae arise by budding in colonies of hydrozoan **coelenterates.** (A.W.L.)

BLASTULA. The stage in embryonic development which results from cleavage of the fertilized **ovum** and precedes the establishment of the germ layers. It is a hollow sphere in its primitive form but is modified in many animals, particularly in connection with the extensive storage of yolk in the egg, and in some of these modified forms the exact equivalent of the primitive blastula is difficult to determine. (A.W.L.)

BLAUBOK. Mammalia, Artiodactyla. *Hippotragus.* A South African **antelope** allied to the oryxes. It is said to be extinct. (A.W.L.)

BLEACHING SUBSTANCES, BLEACHING AND DECOLORIZING. Bleaching and decolorizing treatments are applied to **textile** fibers, **paper** pulp, **wheat** flour, **petroleum** products, **oils and fats,** raw **sugar** solutions. In some cases, the results are attained by chemical reaction of the reagent and material treated, and in other cases by absorption of undesirable material by the reagent followed by separation, e.g., by filtration, from the desired material.

The following gases are used in bleaching flour: **chlorine** 300 parts per million, chlorine (98%) plus nitrosyl chloride (2%) 300 parts per million, **nitrogen** tetroxide 4 parts per million, **nitrogen** trichloride 50 parts per million, and most commonly, **benzoyl peroxide** 120 parts per million.

Chlorine, sodium **hypochlorite** and calcium hypochlorite are strongly oxidizing in their behavior, **hydrogen peroxide** and per- compounds mildly oxidizing, and **sulfur dioxide** reducing. Hydrogen peroxide is specially adapted for use in bleaching hair, silk, feathers, straw, ivory, teeth, bones, gelatin. Chlorine and hypochlorites are used in bleaching paper pulp, and sodium peroxide for bleaching mechanical wood pulp.

Chlorine and hypochlorites, when used in bleaching textile fibers and paper pulp, must be scrupulously removed upon completion of the bleaching operation to avoid deterioration by further action. This is done by the use of antichlor, such as **sodium** thiosulfate solution.

Various porous solids of high adsorptive power are used for decolorizing liquids and solutions. (See **Collodial state, Adsorption.**) Bone charcoal is used to decolorize raw sugar solutions, various clays and gels for oils and fats. (R.K.S.)

BLEAK. Pisces, Teleostei. *Alburnus.* Small fishes **(Pisces)** of several species found in Europe and western Asia, related to the **carps.** (A.W.L.)

BLEEDER. In medicine a bleeder is an individual whose blood does not clot normally. (See **Hemophilia** and **Heredity.**) (R.S.M.)

BLEEDER RESISTANCE. A permanently connected resistor across the output of a **power supply.** (L.R.Q.)

BLEEDER TURBINE. Steam Turbine.

BLEEDING IN PLANTS. Often, particularly in spring, sap may be observed flowing from the broken ends of branches, from the fresh stumps left when stems are cut, or from cracks. The phenomenon is called "bleeding." The force causing this flow of sap from stems is called root pressure, since it seems due to osmotic forces in the living cells of the root. The flow of maple sap when the tree is tapped in the spring may be considered as a form of "bleeding." (See **Ascent of Sap.**) (R.M.W.)

BLENDE. Sphalerite.

BLENNY. Pisces, Teleostei. Marine and fresh-water fishes (**Pisces**) of wide distribution. Some move about on the bottom by means of the paired fins and others are able to leave the water and move about on the shore. About 50 species of blennies are known. (A.W.L.)

BLEPHAROPLAST. The basal body from which a cilium or flagellum grows forth. In many cases synonymous with **centrosome.** (A.W.L.)

BLESSBOK. Mammalia, Artiodactyla. One of the smaller species of South African **antelopes,** *Damaliscus albifrons*. (A.W.L.)

BLIND FLIGHT. Ordinarily, piloting of aircraft is accomplished by visual orientation on the part of the pilot with respect to fixed references such as horizon, clouds, and sky. Although some experimentation has been made with flying instruction starting with the hooded cockpit, this is still in its infancy, and piloting talent of the present time has been built up on the basis of visual orientation. Now when the aforementioned references are blotted out from view by fog, clouds, snow, or rain, the pilot cannot, by sense of gravity alone, maintain even keel and constant heading of the aircraft. Under these circumstances he is faced with the necessity of blind flight, that is maintenance of direction and equilibrium by the use of instruments located in the cockpit, which are read by the pilot, and used by him to create a mental image of the attitude of his plane.

When blind flight becomes necessary the only alternatives are to quit the airplane by parachute, or risk a crash. The ordinary flight instruments are not sufficient for blind instrument flight because of errors they introduce when the airplane undergoes circling flight. The ordinary bubble or bank indicator, pitch indicator, and compass may be sufficient for short stretches of blind flight if the air is comparatively calm, but are not to be thought of as blind flying instruments. In this classification come the artificial horizon, the turn and bank indicator, the sensitive altimeter, and various radio apparatus. The gyroscope possesses features which fit it admirably for incorporation into various types of flight instruments, particularly those which indicate the flight condition of the airplane. The artificial horizon, for instance, has a vertical gyroscopic axis in a frame mounted on gimbals so that it can rotate freely in the casing. The case is fixed to the airplane, and should the latter move out of the desired flight conditions, a small visible bar maintains a true horizontal position. A small painted airplane on the dial moves with the frame and the airplane, and its position with respect to the true horizontal bar gives the pilot an easily assimilated image of the position of the real plane with respect to the actual horizon. Thus he may instantly apply corrective measures to offset **rolls** or **pitches** caused by air conditions.

Since, in blind flight, the lack of sight of the ground prevents the visual observation of drift or other methods of determining the effects of wind, the radio must be depended on for the maintenance of the airplane on a course which is being flown blind. While meteorological information may be given by radio, a radio system designed for the specific purpose of indicating a given course has been devised and put into operation in the United States by the Civil Aeronautics Administra-

tion. This is known as the **radio range,** and forms the important element of the Federal airways.

The radio beacon sends out signals directively along predetermined courses coinciding with the established airways. Each radio range beacon provides four courses. The standard offcourse signals of dash-dot and dot-dash are transmitted in groups separated by the station identifying signal. "On-course" is indicated by the interlocking of "off-course" signals, thus forming a series of long dashes or continuous monotone signals which are also separated by the station signal. The sending antenna consists of two loops at an angle to each other. The signals sent out by these antennae form a pattern in a horizontal plane, whereby the course may be determined. It will be noted that though there are apparently four courses, the inclination of the sending loops is set to produce a narrow beam on the desired course leaving that at right angles too wide to be of much use. Reliable reception on aircraft in flight can normally be had up to distances approximately 100 miles from the station, and the width of the beam in which the dot-dash and dash-dot signals are completely interlocked is about 10 miles at a point 100 miles from the station. As long as the pilot is on course, he hears in the earphones a series of long dashes interspersed with a range beacon identifying signal. Should the plane get off to one side of the marked course, the dashes would gradually break up and be replaced by the off-course signal, either a dot-dash, or a dash-dot, depending on which side of the true course the deviation had been made. (F.T.M.)

BLIND LANDING. Radio has proven the means of overcoming many of the difficulties of flying under adverse conditions, and present developments indicate that it will soon bring the solution to the most difficult of all flying problems, that of landing an airplane when visibility is reduced to zero. For aircraft to become a safe and reliable means of transportation under all weather conditions it must be possible to land the plane safely when fog prevents the pilot seeing the landing strip from even the lowest altitude. Blind-landing or, more accurately, instrument-landing equipment has been under development for several years and several systems have been devised. However, the more successful ones differ more in the details than in the essentials of the system.

A system for making an instrument landing must provide the pilot with an accurate indication of his location with respect to the runway in three dimensions—laterally, vertically and longitudinally. This information is transmitted by means of a so-called localizer beam, two position fixes provided by "marker beacons" and a glidepath beam. While not the only system, the following one, developed under the sponsorship of the Civil Aeronautics Authority (later the Civil Aeronautics Administration) will indicate the general problem and how it may be solved.

The ground equipment calls for four **transmitters** for each landing direction while the plane equipment consists of three **receivers,** associated circuits and indicating instruments. The localizer beam, which lines the incoming plane up with the runway, is very similar to the **radio range** used to guide aircraft cross-country. A very high frequency transmitter produces two overlapping field patterns by means of directive antennae. (See **Directional Antenna.**) One of these patterns is modulated at 150 cycles and the other at 90 cycles. The course is indicated by the line where the overlapping patterns produce equal intensities. The glide path, which determines the path of the incoming plane in the vertical direction, is provided by another very high frequency (but frequency different from the localizer) transmitter producing a directional field pattern. This signal is modulated with 60 cycles. The glide path is a path in this field which gives constant signal strength at the airplane receiver. In order to get the exact sort of path desired the transmitter is located to one side of the runway and the beam is directed over it at an angle. This path is practically a straight line as the plane starts down and then becomes parabolic and meets the runway tangentially. Two marker beacon transmitters (fan markers), modulated with different frequencies are located at known distances from the end of the runway. These transmitters radiate relatively thin fan-shaped fields upward so they indicate well defined locations to an airplane flying through them. Fig. 1a shows the general location of the various transmitters and 1b shows the profile of the glide-path and marker patterns. The plane carries three receivers to pick up

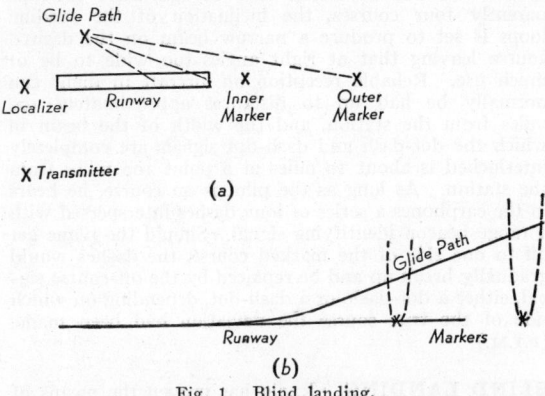

Fig. 1. Blind landing.

the signals radiated at three different frequencies by the ground stations, i.e., the localizer signals, the glide-path signals and the marker signals. Several types of instruments have been developed to combine these various signals so a single instrument indicates all the pilot needs to land the plane. In the so-called cross-pointer instrument the marker beacons cause colored lights to operate when the plane passes through their field, the localizer signals actuate a needle which is centered vertically for on-course and deflects to the appropriate side when the airplane gets off the course, and the glide-path receiver output actuates a horizontal needle which deflects up or down if the plane is not on the glide path. Thus when the plane is coming in along the correct path the two needles cross in the center of the instrument and the lights indicate two known points along this path. Other instruments give the same indications by means of spots or lines on the screen of a cathode ray tube.

Reference to Fig. 2 will show the path of a plane making an instrument landing by this system. As the

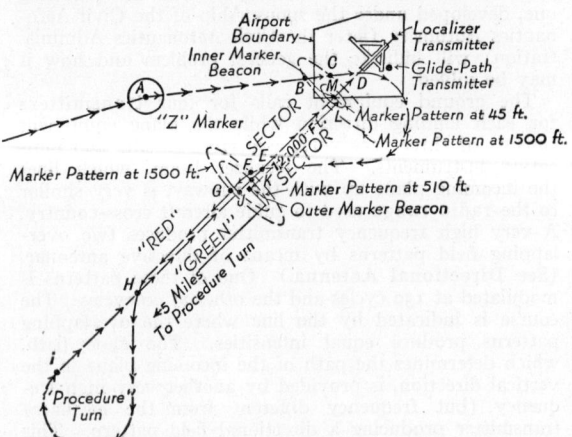

Fig. 2. Landing path. (*From Electrical Engineering, December 1940, p. 496.*)

pilot approaches the airport he first gets a fix by the marker at A, then flies at an altitude of 1500' directly towards the airport, cutting the field of the inner marker beacon. The appropriate light on his instrument lights while he is in the field, giving an indication of his angle with respect to the fan field. This instrument also indicates which side of the localizer beam he is on and he makes a turn to align himself with the localizer beam. He crosses the outer marker and continues to fly the beam for 5 miles and then makes a turn to begin the landing run. By keeping the plane on the localizer and glide-path beams as indicated by his landing instrument he passes through the marker fields again, the outer one at 500' and the inner one at 45'. Within a short interval after passing the inner marker contact is made with the runway. (L.R.Q.)

BLIND SPOT. That portion of the retina of the eye where the optic nerve enters; it is insensitive to light. (D.M.H.)

BLIND WORM. Amphibia, Gymnophiona. Slender worm-like amphibians with no trace of legs and with the tail and eyes rudimentary. They are also called caecilians. (A.W.L.)

BLIND-FISH. Pisces. Teleostei. Small fishes (Pisces) of which certain cave-inhabiting species have rudimentary eyes. Together with a few species whose eyes are normal they make up the family Amblyopsidae. (A.W.L.)

BLISTER BEETLE. Insecta, Coleoptera. Soft-bodied beetles of medium to large size. They are named from their blistering properties; when crushed on the skin even the common species are capable of raising a blister.

Blister beetles are of commercial importance as a source of a pharmaceutical preparation known as Spanish-fly, from the species of that name. This material is composed of the dried pulverized bodies of the insects and is used for producing blisters. Some of the North American species are also occasionally important enemies of plants, among them the old-fashioned potato beetle. They can be checked by the application of sprays containing arsenical poisons.

Several hundred species of blister beetles have been described. They constitute the family Meloidae. (A.W.L.)

BLIZZARD. Wind storms accompanied by low temperatures and blowing or falling snow. (P.E.K.)

BLOCK CHAIN. Chain.

BLOCK SIGNAL. A block signal is the mechanism employed to show the engineman, operating a train over a railroad protected by the automatic block signal system, whether there are any trains in the block ahead. In the operation of a railroad system the trackage is divided into blocks of from 1–5 miles' length, and the railway signalling system devised to prevent two trains occupying the block simultaneously. The block signal itself is a semaphore mounted on a mast and placed alongside the track in clear view of the enginemen. It is electrically energized from an a-c or d-c transmission line running along the right of way, and is actuated by electromagnets properly connected across the "blocks" of track. A standard system of signalling is the semaphore arm held vertical for "proceed" and horizontal for "stop." At night, red and green lights supplement the semaphore as a means of indicating the condition of the block. (F.T.M.)

BLOCKING CONDENSER. This is a condenser used at various points in an electrical circuit where it is desirable to pass a.c. and block d.c. It is commonly used in vacuum-tube circuits to prevent the a-c load from affecting the d-c plate voltage. (L.R.Q.)

BLOCKING LAYER. Photovoltaic Effects.

BLOOD. A liquid tissue whose chief function is the ready transportation of materials through the animal body. In some invertebrates, particularly insects, it is called haemolymph.

All bloods consist of two principal parts: the liquid intercellular material called the plasma, and the cells which are suspended in it. The plasma is a complex mixture containing absorbed products of digestion, waste products of the various tissues, special secretions, enzymes, and antibodies, and gases: **oxygen, carbon dioxide,** and **nitrogen.** The composition of plasma is roughly as follows: water, 90%, proteins, 9%, salts, 0.9%, sugar, urea, uric acid, creatin, etc., traces. All bloods contain white cells or leucocytes and the blood of vertebrates also contains red cells whose function is the transportation of oxygen. (See Plate facing p. 180.)

The transportation of dissolved materials by blood depends upon selective **osmotic** interchange with the tissues with which it comes into contact. The transportation of oxygen is, in some animals, of the same nature but in others, complex proteins containing copper or iron enter into combination with the gas. Haemocyanin (see **Protein**) containing copper and **hemoglobin** (see **Protein**) containing iron, are the two most common compounds known to act in this way. The former is found in some **arthropods** and **mollusks** and the latter in some **nemertine** and **annelid** worms, a few arthropods, the phoronids, and the **vertebrates.** In the vertebrates oxygen is contained in the red cells.

More is known of human blood than of that of any other species because of its importance in medicine. The average amount of blood in the human body is about ½₀ of the body weight or about 4–7 quarts. Arterial blood is bright red due to its relatively high oxygen content. Venous blood is darker in color due to depletion of its oxygen by the tissues. Blood is salty in taste, slightly heavier than water, has a peculiar odor and its normal temperature is about 100° F. In addition to extensive knowledge of the chemistry of the plasma, the normal characteristics of the cell content have been determined and the significance of deviations from the normal has become an important factor in clinical diagnosis. The human red cells or erythrocytes are biconcave disks averaging .0075 mm. in diameter. The enucleate condition is characteristic of mammals but in other vertebrates the nucleus is present. The red cells are filled with hemoglobin, which has the power of combining with oxygen easily and giving up the oxygen readily when the body cells require it. The color of the blood depends on its content of oxygen— the higher the oxygen content, the brighter the color. The principal function of red corpuscles is the transportation of oxygen to the cells and tissues. This function depends upon their hemoglobin content and their total number.

The number of red cells varies. In health the average is 4,500,000–5,000,000 per cu. mm. The hemoglobin content normally varies between 75–100%, or 12–15 grams. Any percentage lower than 75 with a total count of less than 4 million cells is classed as **anemia.**

The red blood cells are manufactured by the marrow cells of the long bones of the body. The average life of red cells is estimated from 50–70 days. They are destroyed, when useless, by the **liver, spleen** and **lymph** nodes.

The white cells are of three types: the lymphocytic, monocytic, and myeloid series. 1. Lymphocytes are 7–18 μ in diameter. Their nuclei are round to oval, deeply staining, and the cytoplasm is clear, with few or no granules. 2. Monocytes are 12–20 μ in diameter and have indented, kidney-shaped nuclei. Their cytoplasm

contains abundant fine reddish-blue granules. 3. The myeloid series comprise the granulocytes, which are classified as neutrophils, eosinophils, and basophils, according to whether their granules stain pink, red, or deep blue, respectively, with Wright's stain. All the granulocytes are approximately the same size, varying between 10 and 15 μ in diameter. Their nuclei are lobulated and, for this reason, they are called polymorphonuclear leucocytes.

The total number of circulating white blood cells in a normal adult is between 5000 and 10,000 per cu. mm. Of these 55–65% are polymorphonuclear neutrophils, 1–3% eosinophils, and 0–0.75% basophils; 25–33% are lymphocytes, and 3–7% monocytes. In addition to these cells, there are two other cellular elements, the macrophages—clasmatocytes or histiocytes—and the platelets. The macrophages are derived largely from the monocytic series of leucocytes; they are 15–80 μ in diameter. The blood platelets or thrombocytes are minute (2–4 μ) particles thought to arise from the giant megakaryocytes of the bone marrow. Their function is concerned with the initiation of blood clotting, by their liberation of thromboplastin.

The functions of the leucocytes are multiple, and some of them are as yet little understood. The cells are motile and one of their most important activities is **phagocytosis,** by means of which bacteria and cellular debris are destroyed. The neutrophils engulf bacteria and small particles of material, while monocytes and macrophages phagocytize larger particulate matter, **protozoa,** and red cells. **Enzymes** of various types are produced by the leucocytes, and these are probably responsible for the digestion of engulfed particles. Eosinophils are thought to play some role in detoxification, the disintegration and removal of protein. Their number is increased in allergic diseases and with infestations with the intestinal parasites. Nothing is known of the function of basophils. The function of the lymphocytes is obscure also, but they seem to be involved in some way in healing processes and in the body's reaction to foreign protein.

The total number of leucocytes and the ratio between neutrophils and lymphocytes deviates from the normal in the presence of certain diseases, particularly infections. With infections of bacterial rather than virus etiology, a marked increase in the total count (**leucocytosis**) and a high percentage of granulocytes occurs. Occasionally, in response to an overwhelming infection, or as a toxic reaction to a drug, the opposite occurs: the total count falls below 5000 per cu. mm. (**leucopenia**) and the percentage of granulocytes drops. This is an unfavorable prognostic sign. **Leukemia** is a fatal disease associated with enormous increases in the white cells of myeloid, lymphocytic, or monocytic series.

Diseases of the blood are numerous. In addition to the leukemias, the most important ones are the various anemias, polycythemia, hemophilia, hemorrhagic purpura, and agranulocytosis. These are discussed under their respective headings. (A.W.L., R.S.M., D.M.H.)

BLOOD GROUPS OR TYPES. There are four recognized groupings of human blood. The blood grouping of a person is hereditary and follows the principles of the Mendelian Law (see **Heredity**); it remains unchanged through life. The presence of blood groups in the human race is due to congenital antigenic differences in the red blood cells of individuals.

Landsteiner, in 1901, first accurately described three of the blood groups. In 1902, the fourth group was added by two of his pupils. Jansky in 1907, in further work, called the groups by numbers, thus, group I, II, III, and IV. In 1910, Moss classified the groups IV, II, III, and I so that Jansky I became Moss IV and Jansky IV became Moss I. This caused considerable confusion and led to the adoption of an international grouping arrangement naming the groups according to the antigen contained in the cells. Thus the groups are called Group A, B, and AB, and O, depending on whether antigens A, or B, or both are present in the red cells.

Group O is called the universal donor group, because the red cells contain no antigenic factor. However, a group O donor is never used in transfusing patients of other groups unless as an emergency procedure, when there is no time to obtain a donor of the same group.

Blood grouping is vital in the transfusion of blood. Not only must a donor of the same group be used, but his blood must be compatible or must cross-match with the patient's blood.

Lately, grouping has been used legally in establishing paternity. In certain cases it can be proved that the suspected individual could not have been the father. It never can prove that the individual is the father.

In 1940 a new antigen was found in human blood, the Rh factor. This factor is present in the red cells of 85% of the population, absent in 15%, and individuals are classified accordingly as Rh positive, or Rh negative.

Although antibodies to the Rh factor do not occur spontaneously in the serum of Rh-negative individuals, the Rh type is of importance in such people because by repeated transfusions with Rh-positive blood, such antibodies may be built up, and a fatal transfusion reaction will result if Rh-positive blood is then given. Rh-negative patients, therefore, should be transfused with Rh-negative blood only. This is of particular importance in obstetrical patients. If an Rh-negative woman is married to an Rh-positive man, 3 out of 4 of their children will be Rh positive. The blood of these children while in utero will stimulate the formation of anti-Rh immune bodies in the mother's blood; if she be given a transfusion with Rh-positive blood, the antibody will react with the Rh antigen, and a fatal transfusion reaction may result.

The Rh-positive fetus of an Rh-negative mother is also in danger: the anti-Rh immune bodies which are built up in the mother's blood diffuse through the placenta into the fetal circulation, and an antigen-antibody reaction takes place within the fetus, resulting in the agglutination and destruction of its red cells. Such an infant suffers from erythroblastosis fetalis, a disease whose etiology was unknown until the discovery of the Rh factor. (D.M.H.)

BLOOD PRESSURE. The pressure in the arterial system of the body. (See **Circulatory System.**) This pressure depends upon interrelations between a number of factors, the important ones being (1) the force of contraction of the left **ventricle** of the heart as it pumps forth blood; (2) the volume of blood forced into the **aorta** by the contraction of the left ventricle; (3) the peripheral resistance of the **arteries**, their elasticity and tone; and (4) the viscosity of the blood.

Both systolic and diastolic pressures are measured. The systolic indicates the pressure during **systole**, when the heart is contracting and forcing out blood, while the diastolic indicates the pressure during **diastole**, when the heart muscle is relaxed and its chambers are filling with blood. The normal systolic values are given as 90 to 140 mm. of mercury, and the diastolic 60 to 90 mm. of mercury. These figures vary with age, exercise, degree of obesity, and emotional stress. The normal is thus highly elastic. (D.M.H.)

BLOOD WEAVER-FINCH. Aves, Passeriformes. A group of **weaver-birds** of Arabia and Africa. Named from the prevalence of scarlet in their plumage. (A.W.L.)

BLOOD WORM. 1. **Annelida.** Certain marine worms whose bright red blood gives color to the entire body. 2. Insecta, **Diptera.** The aquatic larvae of certain **midges** which have hemoglobin dissolved in the plasma of the blood and so are red in color. (A.W.L.)

BLOODSTONE or **HELIOTROPE.** A massive variety of **quartz** of greenish color with small spots of red **jasper** somewhat resembling blood drops. It is used as a semi-precious stone. When placed in water in full sunlight bloodstone will frequently give a general reddish reflection, hence the term heliotrope, derived from the Greek words meaning sun and to turn. (E.S.C.S.)

BLOWERS. Compressors for compressing air or gas to pressures as high as 35 lbs. per sq. in. gauge are known as blowers. The principal types are the *reciprocating* (blowing tub), *centrifugal,* and *positive displacement rotary* (as exemplified by the Roots blower). A familiar example of the centrifugal blower is the airplane engine supercharger. This has a high-speed, radial-bladed impeller enclosed in a casing so that air, drawn from the atmosphere, may be caught by the impeller and compressed through a range of 20 to 30″ of mercury. The principle of pressure increase is the same as that explained for **air compressors,** *centrifugal.* Centrifugal blowers propelled by steam or gas turbines are called "turbo-blowers."

A Roots-type blower has two irregular, lobe-shaped impellers rotating (in synchronism) within a circular casing. Rotation of the impellers traps air between them and the casing, following this by an expulsion through the discharge port. Clearances between impellers and casing are small, but positive. (F.T.M.)

BLOW-FLY. Insecta, Diptera. **Flies** which deposit their eggs on meat. The name is applied to an entire

EXPLANATION OF PLATE A.

Fig. 1. The cells of normal blood reproduced from actual cells. Wright's stain ×1000 (1 mm. = 1 μ).

1, Red corpuscles and blood platelets. 2, Two lymphocytes. 3, Lymphocyte with azurophilic granules. This cell lay in a thin portion of a film and was exceptionally large. 4, Three endothelial leucocytes, one with fine cytoplasmic granules. The granules are rarely so distinct as here shown. 5, Polymorphonuclear neutrophils. 6, Eosinophils, one ruptured. The cells selected for drawing contained fewer granules than are usual. 7, Basophils.

Fig. 2. Leucocytes found in the blood in disease. All reproduced from actual cells stained with Wright's stain, excepting No. 15, which is copied from Pappenheim ×1000 (1 mm. = 1 μ).

8, Two myeloblasts, showing nucleoli. 9, Two promyelocytes. Note the blue edge of one. 10, Two mature neutrophilic myelocytes. 11, Eosinophilic myelocyte. 12, Basophilic myelocyte. Some of the granules have dissolved, leaving vacuoles and staining the cytoplasm. 13, Two lymphoblasts, one with lobulated nucleus (Rieder cell). 14, Turck's irritation leucocyte with vacuoles. 15, Plasma cell. 16, Degenerated nuclei, one a so-called "basket-cell." 17, Neutrophilic leucocyte with vacuoles (toxic change).

DESCRIPTION OF PLATE B.

Abnormal red-corpuscles. All drawn from actual specimens and all stained with Wright's stain except where noted. ×1000 (1 mm. = 1 μ)

Fig. 1. Variations in size, shape, and hemoglobin content; from cases of pernicious anemia and chlorosis.

Fig. 2. Polychromatophilia and basophilic granular degeneration; from cases of lead-poisoning and pernicious anemia.

Fig. 3. Normoblasts, reticulated red cells, and one microblast. The top row represents stages in the development of the normoblast. The two reticulated red cells are stained with brilliant cresyl blue.

Fig. 4. Megaloblasts from cases of pernicious anemia. Two show polychromatophilia and fairly typical nuclei, two have condensed nuclei, and one of these has basophilic cytoplasmic granules.

Fig. 5. Nuclear particles or "Howell-Jolly bodies." One cell also shows basophilic granular degeneration.

Fig. 6. Mitotic figures, two from myelogenous leukemia, one with polychromatophilic cytoplasm, from von Jaksch's anemia. The last was stained with Leishman's stain.

Fig. 7. Cabot's ring bodies, from a case of von Jaksch's anemia. Two cells also contain nuclear particles and one shows basophilic granular degeneration. Leishman's stain.

PLATE A

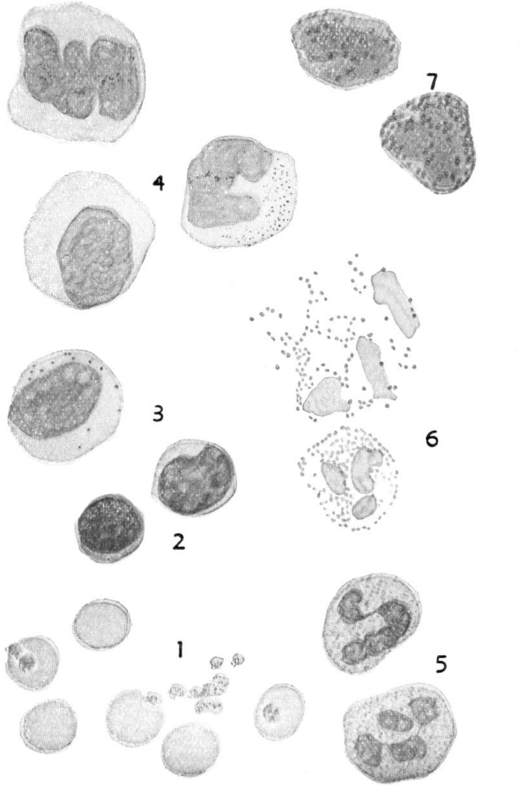

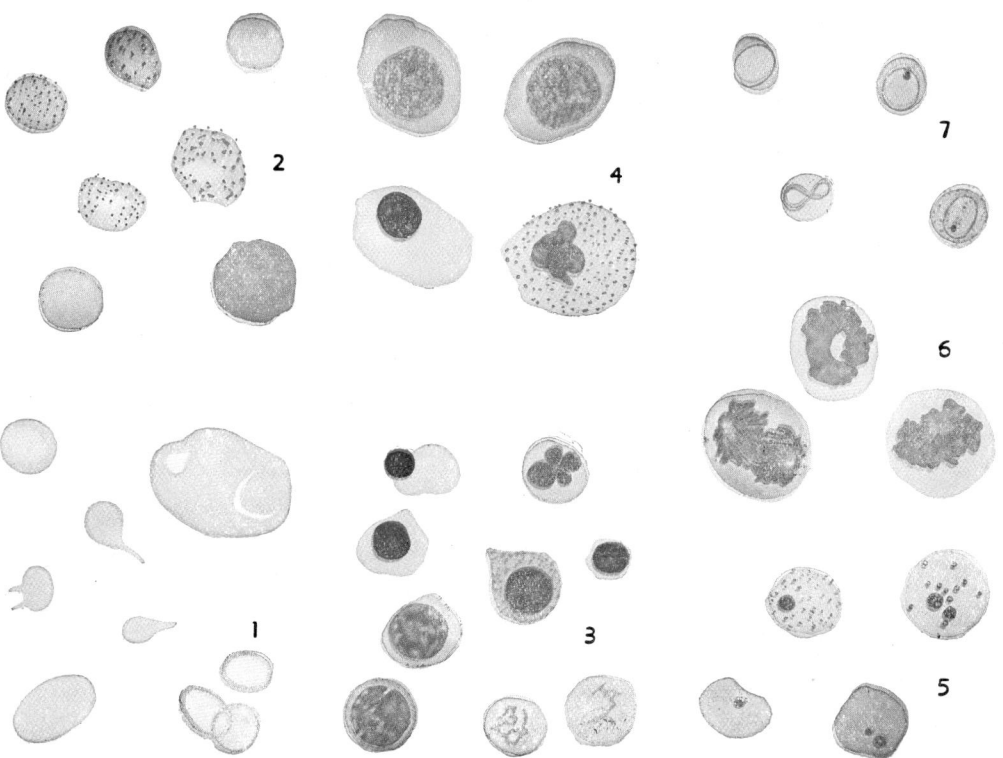

The cells of normal blood (scale: 1 mm. = 1 μ).

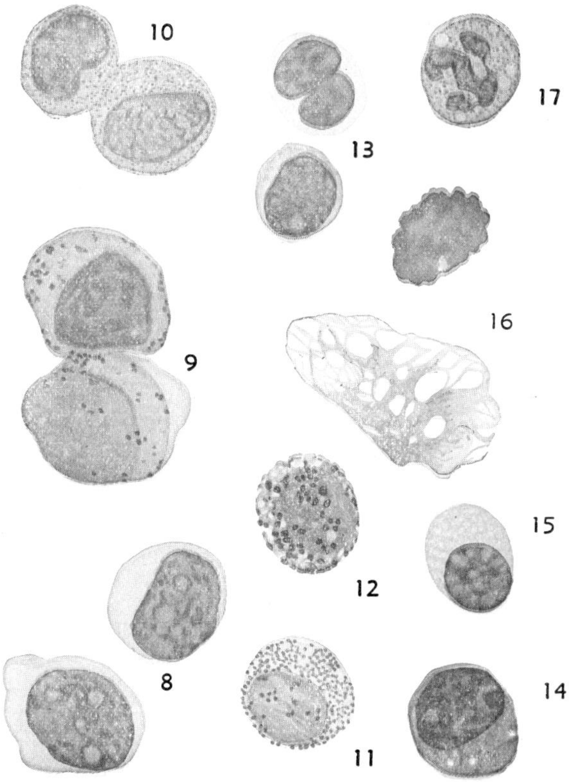

Leucocytes which appear in the blood in disease (scale: 1 mm. = 1 μ).   (J. W. Rennel, pinx.)

PLATE B

See page 180 for detailed discussion of these Plates. Reproduced from Todd and Sanford, *Clinical Diagnosis by Laboratory Methods*, courtesy of W. B. Saunders Co.

PLATE C

MALARIAL PARASITES.

Wright's stain. $\times$ 1000 (1 mm. $=1\mu$).

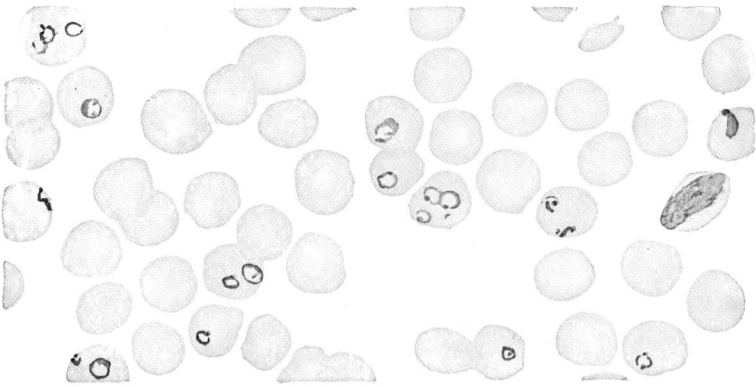

FIG. 1.—Estivo-autumnal malaria, exact reproduction of a portion of a field, showing an exceptionally large number of parasites.

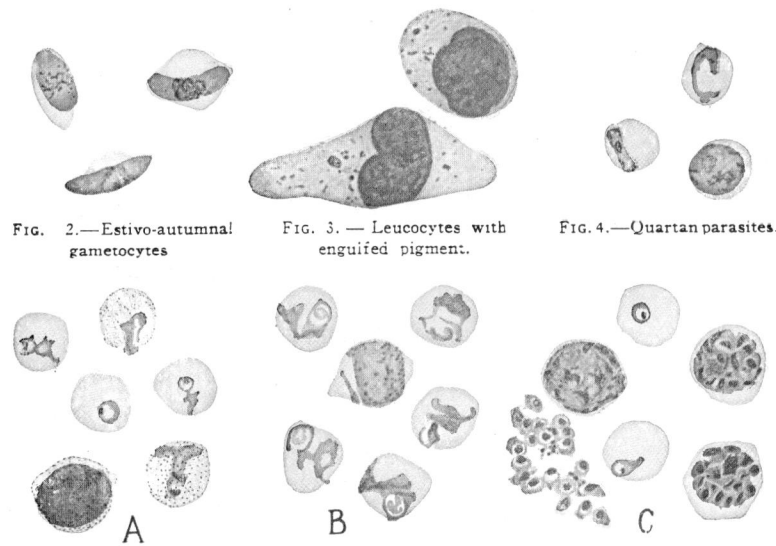

FIG. 2.—Estivo-autumnal gametocytes

FIG. 3. — Leucocytes with engulfed pigment.

FIG. 4.—Quartan parasites.

FIG. 5.—Tertian parasites. A, Eight hours after chill, showing malarial stippling, five young parasites, and one gametocyte, from two slides; B, twenty-four hours after chill, five half-grown parasites: one gametocyte; C, during chill, one presegmenter, two segmenters, a cluster of freshly liberated merozoites, and two very young parasites, from one slide.

(J. W. Rennel, pinx.)

family, however, containing other species which breed in dung, in wounds on living animals, and as blood-sucking parasites of nestling birds. The commoner species are also known as **bluebottle** flies. Family Calliphoridae. (A.W.L.)

BLOW-HOLE. A blow-hole is a defect in a **casting** caused by passage of gases through the metal of the casting during the process of solidification. Large numbers of blow-holes will materially weaken a casting. Even a single blow-hole may be sufficient cause for rejection of castings which are used to contain liquids or gases. The majority of blow-holes are due to the condition of the mold. The use of molding sand which does not vent the gases freely, improper ramming of the sand, or errors in making or placing the cores, are prolific sources of blow-holes.

Blow-holes in steel *ingots* are welded during rolling or forging unless oxidation of the internal surfaces occurs, as by exposure of subsurface blow-holes to the air. (F.T.M., R.H.H.)

BLUE. Insecta, Lepidoptera. **Butterflies** whose prevailing color is bright blue. The females are usually less blue than the males and some few species are not blue. With the coppers and hair-streaks they constitute the family Lycaenidae. (A.W.L.)

BLUE BRITTLENESS. Steels mechanically worked in the temperature range of 300–700° F. are less plastic than at lower or higher temperatures. Because a blue oxide surface film forms upon heating to about 550° F. the phenomenon has become known as blue brittleness. (R.H.H.)

BLUE EARTH. Kimberlite.

BLUE GROUND. Kimberlite.

BLUE MUD. A typical deep sea, fine-grained sediment containing an appreciable amount of **calcium** carbonate. The bluish-gray color is caused by organic matter and finely divided **iron** sulfide. (R.M.F.)

BLUE PRINT. Paper coated with ammonium **ferric** citrate or **oxalate** and **potassium** ferricyanide. On exposure to light the ferric ion is reduced to ferrous, and on subsequent treatment with water the ferrous ion forms ferrous ferricyanide which is blue while the ferric ion forms the brown ferric ferricyanide which is washed off the paper. (See **Drawing Reproduction**.) (R.K.S.)

BLUE RACER. Reptilia, Sauria. A variety of the **blacksnake** found in the prairie and plains states. Like many snakes, it has a bad reputation but is quite harmless. *Zamenis constrictor.* (A.W.L.)

BLUEBACK. Pisces, Teleostei. A commercially important **salmon**, said to be second only to the king salmon in this respect. It ranges from California to Alaska. Also called the redfish. Also a species of **trout** found in Maine and northward. (A.W.L.)

BLUEBERRY. Heath Family.

BLUEBIRD. Aves, Passeriformes. 1. Several species of North American **thrushes** whose prevailing color is blue. Of these the male mountain bluebird, *Sialia currucoides,* of the West is entirely blue while other species are marked with shades of reddish brown. 2. One of the **babblers** of the Oriental region. 3. Said to be applied to a South African **albatross** (Order Procellariiformes). (A.W.L.)

BLUEBOTTLE. Insecta, Diptera. Large **flies** of shining blue, green or purple color. They lay their eggs on meat and other foods and so are often seen in dwellings. (A.W.L.)

BLUEFIN. Pisces, Teleostei. A fish (**Pisces**) found in the Great Lakes and smaller lakes of the same region; one of the lake herrings, also called the blackfin. *Leucichthys nigripinnis..* (A.W.L.)

BLUEGILL. Pisces, Teleostei. A fish (**Pisces**), *Helioperca incisor,* related to the sunfishes and bass. Widely distributed east of the Rockies in lakes and the quieter parts of streams and esteemed as a pan fish. It attains a length of 10″ or more but in well-fished waters rarely attains this size. Although small, it rises readily to a fly and so ranks among desirable game fishes. (A.W.L.)

BLUETHROAT. Aves, Passeriformes. A European bird (**Aves**) related to the warblers. *Luscinia suecica.* (A.W.L.)

BLUFF BODY. An object immersed in fluid stream flow is said to be bluff (or blunt) if its shape promotes a rapidly increasing downstream pressure gradient in the streamline flow around it. A high adverse gradient assists the creation of a **stagnation point.** The streamline flow breaks loose from the surface of the body on either side, leaving a turbulent low-pressure wake. This wake causes the characteristically high drag of bluff bodies. (See **Aerodynamics**.) (F.T.M.)

BLUNT-NOSE. Pisces, Teleostei. A small fish (**Pisces**), *Hyborhynchus notatus,* found in streams west of the Alleghanies, related to the dace and chub. (A.W.L.)

BOA. Reptilia, Sauria. Any **snake** of the family Boidae. The United States has only two small species of the western states; the bulk of the family is made up of large arboreal snakes such as the **boa-constrictors,** *Boa constrictor,* of South America and the **pythons,** *Python,* of the Old World. One of these species, the **anaconda,** *Eunectes murinus,* attains a length of over 30′ and is the largest snake known. Boas are not poisonous. (A.W.L.)

BOAR. The male of any species of swine. (See **Pig**.) The term wild boar is also applied generally to two species found respectively in India, *Sus cristatus,* and in Europe and the adjoining areas of Asia and Africa, *S. scrofa.* (A.W.L.)

BOARD DROP. Forging.

BOBAC. Marmot.

BOBCAT. Wildcat.

BOBOLINK. Aves, Passeriformes. A widely distributed North American bird (**Aves**), *Dolichonyx oryzivorus,* of which the male, marked with black, white, and yellowish, is conspicuous in the prairies and mead-

Bobolink.

ows where the species breeds. The female is duller and plainer. The bobolink is noted for its cheerful song. (A.W.L.)

BOB-WHITE. Quail.

BODE'S LAW. In the latter part of the 18th century an empirical relationship was noticed between the mean distances of the various **planets** from the sun. This relationship was first published by Bode in 1772 and has since become known as Bode's Law, in spite of the fact that there is certain evidence that it was known and used by Titus a number of years previous to the time of its announcement.

Bode's Law may be stated as follows: write down a series of 4's; to the first one add 0, to the second one add 3, to the third one add $6 = 3 \times 2$, to the fourth one $12 = 6 \times 2$, to the fifth $24 = 12 \times 2$, etc.; the resulting numbers divided by 10 will give the approximate mean distances of the planets from the sun in **astronomical units.** The sequence is as follows:

PLANET	BODE DISTANCE	MEAN DISTANCE
Mercury	$4 + 0 = 4$	0.39
Venus	$4 + 3 = 7$	0.72
Earth	$4 + 6 = 10$	1.00
Mars	$4 + 12 = 16$	1.52
	$4 + 24 = 28$	
Jupiter	$4 + 48 = 52$	5.20
Saturn	$4 + 96 = 100$	9.53
Uranus	$4 + 192 = 196$	19.19
Neptune	$4 + 384 = 388$	30.07
Pluto	$4 + 768 = 772$	39.5

The value in the last column is the actual mean distance of the planet from the sun in astronomical units.

At the time that the law was first proposed the gap between Mars and Jupiter was not filled and no planets were known outside of Saturn. The law predicted distances and when Uranus was discovered with mean distance so close to the predicted value, Bode's Law was believed to be established. The discovery of the **asteroid** Ceres with a mean distance of 2.77 gave further support to the validity of the law. It is interesting to note that, in making the computations which led to the discovery of **Neptune**, Adams used the predicted Bode distance for the then unknown object.

During the 19th century many unsuccessful attempts were made to place Bode's Law upon a theoretical foundation. The failure of the law in the cases of Neptune and Pluto has convinced most astronomers that the law is a purely empirical relationship, more in the realm of coincidence than an actual physical law. (W.K.G.)

BODY CAPACITANCE. This refers to the capacitance effect produced when the operator's hand is brought near a vacuum-tube circuit. Since many of these circuits, especially those employing tuned circuits, are very sensitive to small changes in the capacitance coupling to surrounding objects, the bringing of the hand near any of the circuit conductors will materially alter the response of the circuit. Such effects can be minimized by shielding the circuit and by appropriate grounding. (L.R.Q.)

BOEHMERIA NIVEA. Urticaceae. A perennial Asiatic plant growing up to 7' in height and producing several crops of canes annually. Cultivation of the plant occurs principally in China, although smaller quantities are grown in other Oriental countries. The plant is grown for its pericyclic fibers, which are both long (up to 200 mm.) and of unusual strength, though they do not stand twisting very well. In the Orient these fibers are removed from the plant entirely by manual labor, the cortex is scraped off by drawing the stems over a coarse knife against which they are pressed. The fibers obtained by this process are dried, and in this condition are known as "China Grass." Repeated washing and drying is next used to remove the gummy substance which surrounds the fibers and to separate them, a slow tiresome process. Ramie fibers are considerably coarser than those of flax, and have great tensile strength, but are little used because of their lack of resistance to twisting. Cultivation has been attempted in the United States, but without much success, because of the difficulty of preparing the fibers for manufacturing processes. (R.M.W.)

BOEHMITE. Lepidocrocite.

BOG IRON. Limonite.

BOG MANGANESE. Wad.

BOIL. Furuncle. A localized purulent infection of the skin and underlying tissues, usually caused by the **Staphylococcus** group of micro-organisms. It is characterized by destruction of tissue in its center (core), with softening and discharge of pus. Treatment is surgical incision and drainage. (R.S.M., D.M.H.)

BOILER EFFICIENCY. By boiler efficiency is meant the measure of ability of a boiler to transfer the heat given it by the **furnace** to the water and steam. This will sometimes include **superheater** performance. Boiler and furnace are so much a unit nowadays that "boiler and furnace" efficiency is more important than boiler efficiency. In fact, boiler efficiency is often taken with the meaning of boiler and furnace efficiency; that is, the percentage of the higher heating value of the fuel which will be in the steam. Efficiency is also designated by "evaporation," which is simply the pounds of steam produced per pound of fuel fired. This evaporation may be either the actual evaporation or the equivalent evaporation that would be obtained with the same heat producing steam at 212° F. from feedwater at 212° F.

In the light of the second law of **thermodynamics,** it would hardly seem right to charge the boiler with being unable to absorb any of the heat that was at a lower thermal level than the boiler itself. The "true boiler efficiency" is an expression used to denote that portion of the heat above the saturation temperature which is absorbed by the boiler. It is obtained by dividing the heat absorbed by the boiler by the heat liberated in combustion, less the heat regained if the products of combustion were cooled from the boiler saturation temperature to atmospheric temperature. (F.T.M.)

BOILER HORSEPOWER. The term boiler horsepower was defined in 1876. At that time the average engine would operate on 30 lbs. of steam per horsepower per hour, and a boiler horsepower was defined as the capacity to generate steam at that rate. The average rate of evaporation, then, was 3 lbs. of steam per sq. ft. of heating surface per hour in a water tube boiler, so 10 sq. ft. of heating surface was accepted as manufacturer's rating for a boiler horsepower. The 10:30 ratio of heating surface to steam rate became obsolete when steam rates were reduced to less than 10 lbs. per horsepower hour through improvements in steam turbines and engines, and evaporative rates were increased from 10 to 15 lbs. per sq. ft. per hour by improvements in boiler design and manufacture. Yet the 10 sq. ft. per boiler horsepower persisted, so that now capacity can be expressed in terms of the manufacturer's horsepower rating only by misuse of the term "per cent rating." For instance, an 8000 sq. ft. boiler when producing 72,000 lbs. of steam per hour would be said to be operating on 300% rating. This might be construed to mean an overload of 200% on some equipment, but it would be the normal load for the boiler. When the significance of the term "per cent rating" is understood, there is no disadvantage to the use of boiler horsepower actually developed, as 1 boiler horsepower = 33,500 B.T.U. per hour transferred to the water. (F.T.M.)

BOILER TEST.

The performance of a steam generating unit cannot be gauged by visual inspection of the equipment in operation. Yet an understanding of where the heat units are going, what portion of them are usefully retained and what portion lost, and of those lost, how they are lost, and whether the losses are tending to increase, or whether they may be reduced, is highly desirable whenever there is an attempt to operate a boiler plant with the highest standards of technique and economy. Furthermore, a breakdown of costs often requires definite numerical data as to the performance of the steam generating unit. Then, too, there is always the natural curiosity of boiler operators as to the results they are getting from their equipment, and whether these results are the best that can possibly be obtained. This information can be secured by a test upon a boiler, although a complete test is a job of no little magnitude.

Most boiler tests are planned to obtain sufficient data to set forth the results in the form of the balance called for in the **boiler test code.** The test information that must be obtained is as follows: First, there is an analysis to determine the percentage of constituent gases present in the products of combustion. These gases are some or all of the following: **nitrogen, carbon dioxide, carbon monoxide,** and **oxygen.** Their relative proportions are obtained by a gas analyzer known as the Orsat apparatus. The temperature of the products of combustion (called flue gas) at the exit from the **boiler** must also be taken. The ultimate chemical analysis of the **fuel** in the state as fired is determined for a representative sample of the fuel used during the test. Then if there is refuse, such as the ash of a coal-fired furnace, that refuse must be analyzed for the presence of unburned combustible. The atmospheric conditions of temperature and humidity are also needed, although frequently the effect of moisture in the air is neglected, and humidity readings are omitted. The boiler test is made over an extended period of time to eliminate the effect of small irregularities of load, and no test of less than one hour's duration should be considered worth while. Tests are more frequently run for a period of three to twelve hours. The amount of steam produced during this period, as well as the weight of fuel fired, temperature of the feedwater, steam pressure and temperature, complete the information required to determine the distribution of the total heat available in the fuel, between useful heat and the various losses. The computations involved are quite beyond the scope of this article, but will be found adequately explained in any good book on heat engineering. (F.T.M.)

BOILER TEST CODE.

Many years ago the need for a set of rules and regulations standardizing the method of reporting the results of a steam boiler test was felt to be necessary for true comparison and evaluation of performance. The accepted code is a project of the American Society of Mechanical Engineers. The test code is, in reality, an arithmetical balance of the heat entering a steam generating unit against the heat leaving. The heat in the coal or other fuel fired into the furnace must be accounted for either as useful heat or as one or the other of several different losses. A **boiler test** is performed to obtain the information necessary to account for the heat produced by combustion of the fuel.

The boiler test code for stationary steam generating units has the following items:

1. Heat absorbed by the boiler.
2. Loss due to moisture in coal.
3. Loss due to water formed by burning of hydrogen.
4. Loss due to hygroscopic moisture in air.
5. Loss due to heat carried away in dry chimney gas.
6. Loss due to incomplete combustion of carbon.
7. Loss due to unconsumed combustible in refuse.
8. Loss due to unconsumed hydrogen and hydrocarbons, radiation, and other losses.

The sum of these eight items should be equal to the higher heat value of the fuel. (F.T.M.)

BOILERS.

A boiler is a pressure vessel designed to transmit heat developed by combustion of a fuel to water and steam contained in the boiler. In some instances the liquid in a boiler has been something other than water, but the purpose of this article is to discuss the **steam** boiler. The boiler should hold its contents safely and deliver the steam in the desired condition. It is desirable, also, from the standpoint of economy that the heat be transmitted with minimum loss to the atmosphere.

The steam boiler is an external combustion device, that is, the combustion takes place externally from the region of boiling water. This fact, then, implies the existence of a surface separating the interior of the boiler from the combustion zone. All of the heat which reaches the water must be transferred through this surface. It is called the **heating surface.** All but a very small portion of the surface encircling the water and steam region of a boiler is heating surface. Some boilers are built so that the heating surface encloses the furnace, while others have furnaces built as an auxiliary to the boiler. In the latter the boiler and furnace are enclosed in a refractory, heat-insulating casing called the setting.

Heat generated by **combustion** is transferred to the heating surface in two ways. The surfaces which might be said to "see" the incandescent region of combustion receive heat at a rapid rate by **radiation.** Other portions may receive heat by **convection** from the products of combustion which have been heated in the furnace and act as vehicles for transportation of heat to the boiler surface. The heat is transferred by **conduction** from the gas side of the heating surface through the metal to the water.

Beginning with the simple cylindrical shell type boiler, heated with flames applied to the outer surface, many and varied types of boilers have been evolved. Some have characteristics which fit them for marine service, others for land service, others for heating, etc. For a comprehensive picture of the whole field, it is well to classify the group in logical order. The following table illustrates various classifications.

Classification of Boilers

1. On the basis of the contents of the tubular heating surfaces.
   a. Fire tube, having flame and products of combustion in the tubes.
   b. Water tube, having the water and steam in the tubes, and the products of combustion outside.
2. On the basis of position of the furnace.
   a. Internally fired.
      The internally fired boiler has the furnace region completely surrounded by heating surface, as in the small portable vertical boilers, the Scotch boiler, or the locomotive boiler.
   b. Externally fired.
      All water tube boilers are externally fired, and one of the fire tube class, the horizontal return tubular, is externally fired.
3. On the basis of the shape and position of the tubular heating surface.
   a. The inclination of the tubes, as horizontal, inclined, or vertical.
   b. The form of the tubes—straight or bent.
4. On the basis of the drums.
   a. The number of drums, as single drum, two drum, or multiple drum.
   b. The position of the drum with respect to the tubes, that is to say, across the tubes or

parallel to the tubes, giving rise to the names cross-drum and long-drum.

5. On the basis of the type of headers employed to connect the tubes to the drums.

    a. Sinuous, forged steel headers having a number of individual sections. Each section serves the

bursting pressure. Holes are bored in the tube sheets, which are the ends of the cylinder, in such a manner that tubes may be passed through the shell and fastened tightly in the holes. These tubes are submerged in the water within the boiler, and hot gas passes through them. In some boilers the surface of the tubes comprises

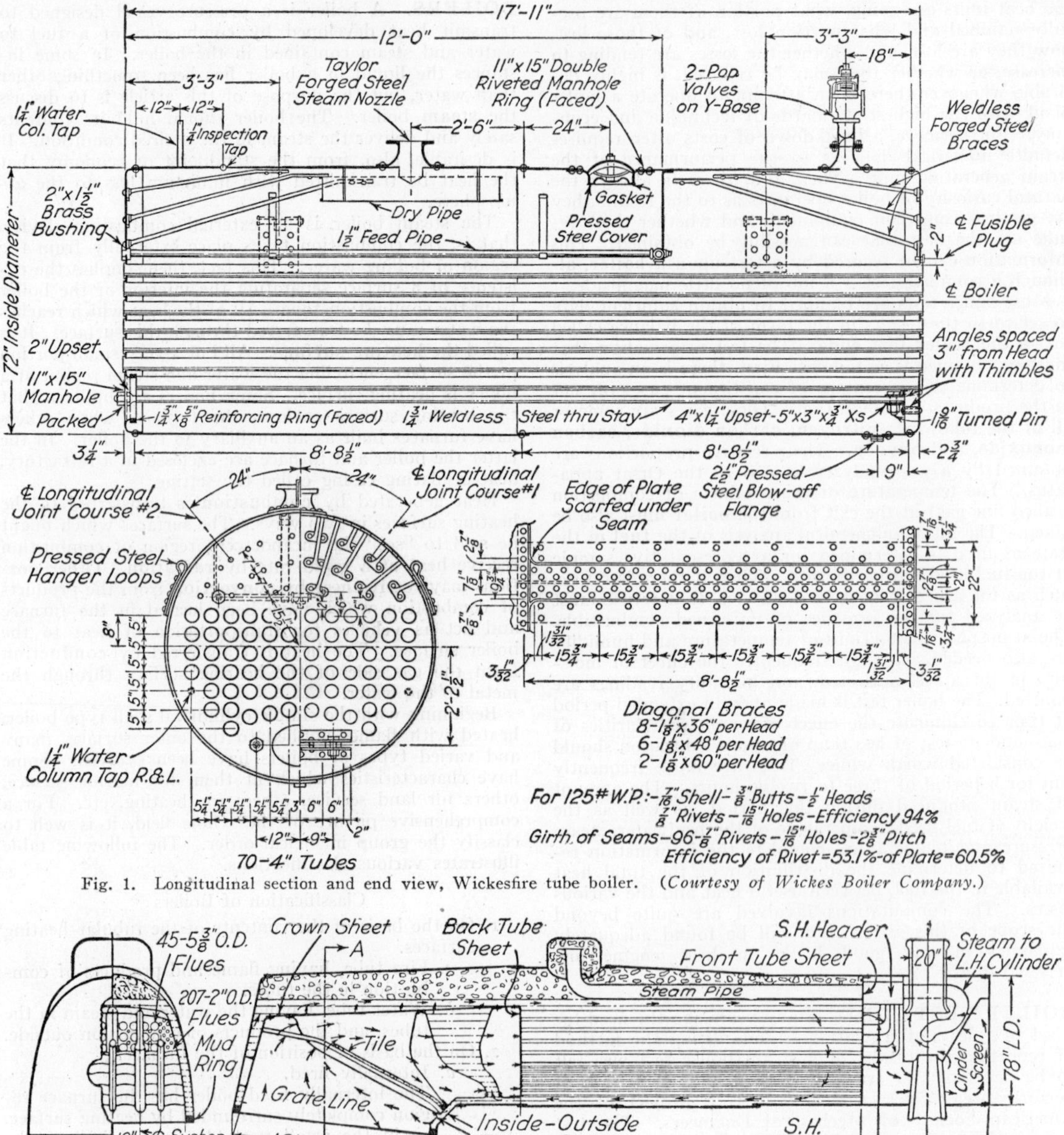

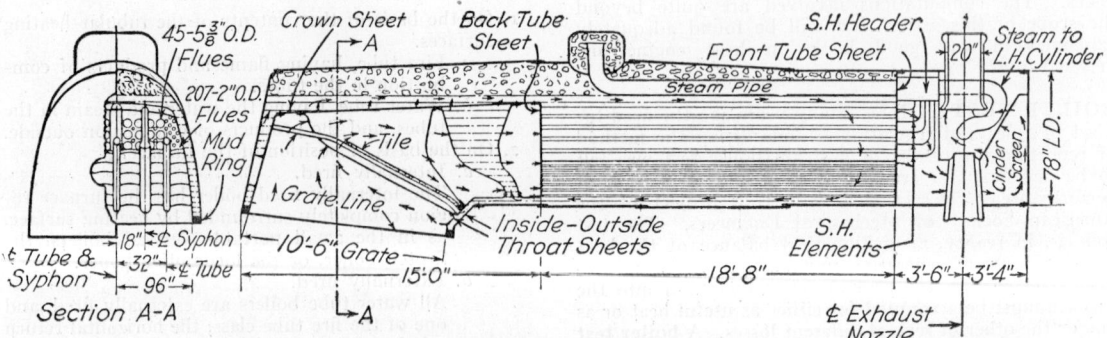

Fig. 1. Longitudinal section and end view, Wickesfire tube boiler. (*Courtesy of Wickes Boiler Company.*)

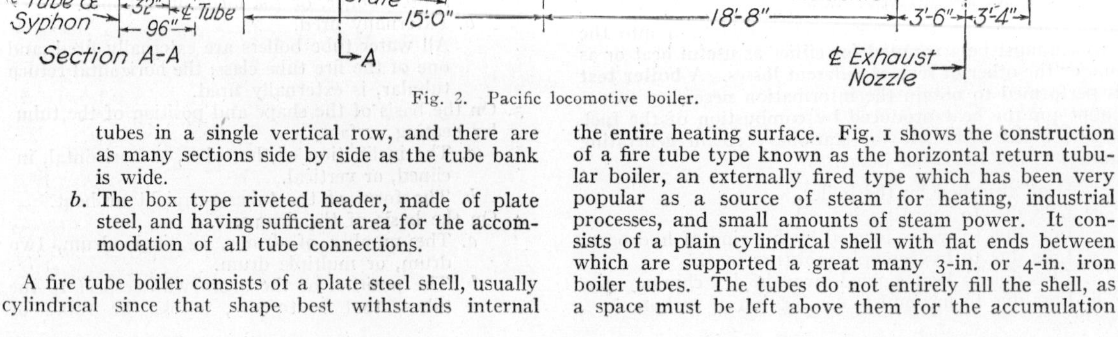

Fig. 2. Pacific locomotive boiler.

tubes in a single vertical row, and there are as many sections side by side as the tube bank is wide.

    b. The box type riveted header, made of plate steel, and having sufficient area for the accommodation of all tube connections.

A fire tube boiler consists of a plate steel shell, usually cylindrical since that shape best withstands internal

the entire heating surface. Fig. 1 shows the construction of a fire tube type known as the horizontal return tubular boiler, an externally fired type which has been very popular as a source of steam for heating, industrial processes, and small amounts of steam power. It consists of a plain cylindrical shell with flat ends between which are supported a great many 3-in. or 4-in. iron boiler tubes. The tubes do not entirely fill the shell, as a space must be left above them for the accumulation and storage of steam. The tube sheets above the tubes are braced against bulging by steel stay braces. The boiler is completed by the addition of proper taps for instruments, steam leads, feed water pipe, and safety devices, and with the provision for some means of supporting it, such as brackets or loops. The furnace is built below and external to the boiler shell, and both furnace and boiler are enclosed in a brick setting. The flames and hot gases play against the bottom of the shell, pass from the front to the back, and return to the front of the boiler through the tubes. The lower portion of the shell and the tubes form the heating surface.

Slightly different is the arrangement of the locomotive fire tube boiler, shown in Fig. 2, where the boiler is internally fired, and the flame and hot gases leaving the furnace region pass forward through the tubes. They emerge from the tubes into the smoke box, from whence they are blown to the atmosphere through a short stack, under the influence of a steam jet which receives steam from the exhaust of the locomotive cylinders. The heating surface consists of the tubes and the shell surrounding the furnace.

A water tube boiler is composed of drums and tubes, the tubes always being external to the drums, and a means of joining the tubes to the drums. The drums are used for storage of water and steam, and for connections of steam and water pipes. As they are not required to contain any tubular heating surface, they are much smaller in diameter than the shells of fire tube boilers, and can be built for higher boiler pressures than are possible in fire tube design. The heating surface is entirely in the tubes, the function of which is therefore simply to absorb the heat and generate the steam. The method of joining a tube to a boiler or header is to insert that tube into a hole having the same diameter as the outside of the tube (tubes are sized on their external diameter, whereas pipes are sized by the nominal internal diameter). The tube wall is forcibly expanded or rolled against the metal surrounding the hole, so as to make a tight joint. This process requires that the axis of the tube be perpendicular to the plane of the hole, or if the hole is bored in a curved surface, the axis of the tube must be radial to the curvature of the surface. If tubes are expanded directly into the drums, dispensing with an intervening header, the only row of tubes that may be absolutely straight is that row that lies in the surface which joins the center lines of two parallel drums. All other rows of tubes must be bent at their ends so as to enter the drum surface radially. Some headerless boilers are built entirely with straight tubes by connecting the tubes to the flat ends of cylindrical drums, but the more common construction is that shown in Fig. 3, where it can be clearly seen that the tubes are bent.

The common construction of a straight tube boiler involves the interposition between the tubes and the drum of a header which will provide a flat surface for the connection of the tubes, and which is connected to the drum by means of circulation tubes or sheet steel saddles. Fig. 4 shows the arrangement of surface in a straight tube box header boiler. As it is necessary to incline the tubes slightly to the horizontal in order to promote water circulation, the box type header must necessarily be inclined somewhat to the vertical, since the surface of the header must be perpendicular to the tubes. Sectional forged steel headers may be exactly vertical, since an inclined surface may be forged at the opening for each tube so that the hole will be perpendicular to the axis of the tube.

The materials from which boilers are constructed are steel and iron. The low-pressure steam boilers so frequently employed in steam heating systems are generally made up of cast iron sections. This material is entirely too weak and heavy for pressure boilers and rolled sheet iron and steel are employed in its place. The tubes are generally seamless iron or steel, and the drums are made

of rolled steel sheets which are fabricated by welding or riveting, or both.

In operation, a boiler is supplied with heat from the combustion of a fuel, and with water from a pump

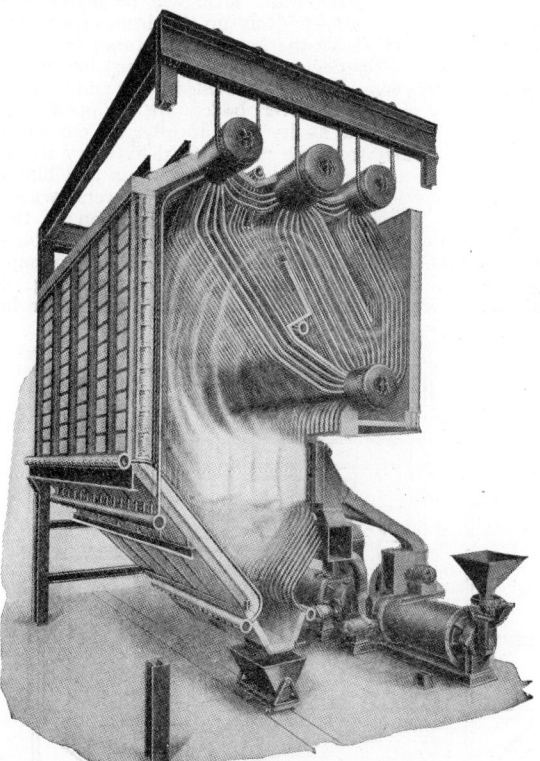

Fig. 3. Boiler equipped with water walls and fired by pulverized coal.

known as the boiler feed pump which is capable of overcoming the pressure existing in the boiler. Upon entrance to the boiler, the water absorbs heat until it reaches its boiling temperature, and is then boiled off to

Fig. 4. Arrangement of Longitudinal Drum Boiler. (*Edge Moor Iron Co.*)

form steam, which collects in the highest part of the boiler drum because of the difference in density of steam and water. Fresh feed water must be introduced to take the place of all steam generated and withdrawn from the boiler, so that the water level will be maintained at the proper position—midway up the drum of a water tube boiler, and above the level of the tubes in a fire tube boiler. Safety devices to protect against low water and high pressure must be provided. In addition to these safety devices, however, boilers are equipped with other auxiliaries, such as dry pipes to filter the entrained droplets of water from the steam and deliver dry steam at the steam nozzle, water level gauge, drain and blow-down connections, **superheaters**, soot blowers, **water walls**, and steam pressure gauges. Operation of the whole steam generating unit may also involve combustion equipment, draft fans, chimney, boiler feed pump, breechings, **economizer**, and **air preheater.** Fig. 5 shows a sectional drawing of a large power boiler in

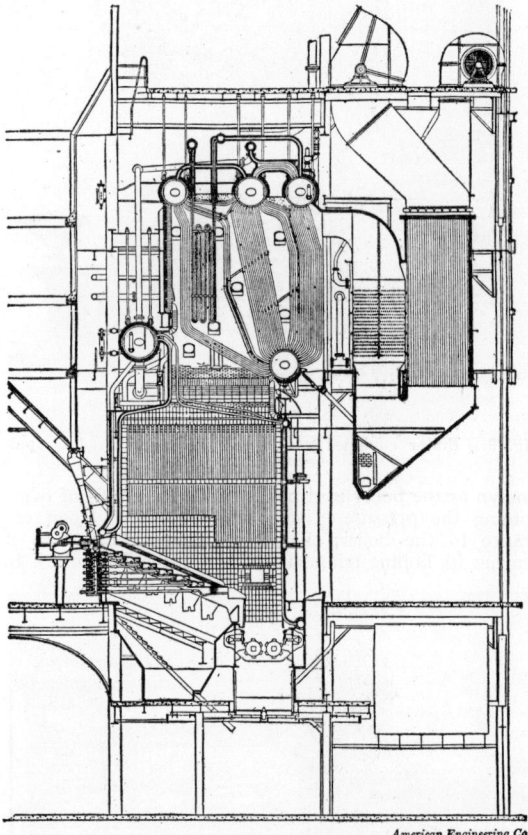

American Engineering Co.

Fig. 5. Cross-section of steam generating unit, Delray No. 3 Station, Detroit Edison.

which many of these auxiliaries are pictured. An air preheater and economizer, fans, water wall, stoker, and superheater may be noted.

The capacity of a boiler is frequently given in terms of horsepower. A rated **boiler horsepower** is 10 sq. ft. of heating surface for a water tube boiler, and 12 sq. ft. for a fire tube boiler. The rated heat transfer capacity of 10 sq. ft. of heating surface is 33,500 B.T.U. passed to the water in an hour, but since modern boilers can exceed this figure steadily without being overloaded, actual capacity is more than rated capacity, a situation which makes necessary the use of the term "per cent rating."

The selection of a boiler is likely to be greatly influenced by personal prejudice or previous experience of the owner. Given service conditions can usually be met

equally well by several boilers of different design. However, in spite of the wide variations in design, there are certain requirements which are fundamental to all water tube boilers. All boilers deserving consideration should meet these requirements, though, of course, not all the favorable points of different designs are covered. But it is nearly certain that if the boiler does not conform to them, operating difficulties will occur early in its life. First, there are the conditions governing the behavior of the water within the boiler. Most important of these is good water circulation. The process of evolution in boiler development has eliminated types with faulty circulation. The disengagement surface where the steam breaks through the surface of the water in the drum should be unrestricted. Priming of the steam with droplets of moisture may result from restricted disengagement surface. In order to control impurities which will be precipitated from the feedwater, it should discharge into the drum at a point where the circulation will deposit the precipitate in a settling chamber called the mud drum. For example, on a boiler in which the mud drum is at the bottom of the rear header, the feedwater should be introduced where the precipitate will be swept out of the drum and into a downcomer leading to the header. Adequate storage space for steam is a requirement indirectly connected with water conditions. The volume of steam storage should conform to the demands of the load served. Insufficient storage space has a bad effect on steadiness of steam pressure under variable load, and has been known to cause pulsations in the boiler and steam piping.

The path of the gases through the boiler should be so baffled and directed over the tubes that they give up heat to the required degree. This required degree is less when auxiliary heat transfer surface in economizer or air preheater are provided than when they are absent. Certain features of a boiler may result in undetermined thermal stresses being set up, such as the discharge of cold feedwater against the boiler shell setting up contraction stresses. Joints and seams should be well protected from the flames, and burners should never be set so the flames play directly against tube surfaces. To provide for intelligent and safe operation of the boiler a full complement of leads, gauges, and safety devices should be provided. These would include blow off, steam lead, feedwater lead, water gauge, pressure gauge, safety valve, and fusable plugs. By no means the least important requirement is the necessity of having an accessible boiler. This need is true of all boilers which are expected to be insured, in order that the insuring companies' inspectors may from time to time determine the state of the risk. Acessibility is also needed for maintenance, inspection, and repair by the regular boiler operating personnel. (F.T.M.)

BOILING POINT. The normal boiling point of a **liquid** is the temperature at which its maximum or "saturated" **vapor pressure** is equal to the normal atmospheric pressure, 760 mm. of mercury. If the pressure on the liquid varies, the actual boiling point varies in accordance with the relation between the vapor pressure and the temperature for the liquid in question. (See **Vapors.**) Water, for example, with a normal boiling point of 100° C. or 212° F., boils at ordinary room temperature when the pressure is reduced to about 17 mm.; and inhabitants of elevated regions often find difficulty in cooking food by boiling, because of the low boiling point. On the other hand, the boiling water and steam in a "pressure cooker" are so hot that such foods as meat and rice are cooked tender in a very short time. If a solid is dissolved in the liquid, or if another, less volatile liquid is mixed with it, the boiling point is raised to a degree expressed by the boiling point laws of van't Hoff, Raoult, and others. (See **Solutions.**)

A liquid does not necessarily begin boiling when the temperature reaches the boiling point. If kept perfectly quiet, and especially if covered with a film of oil, water may be raised several degrees above its normal boiling point, before it suddenly boils with explosive violence; it then returns to its true boiling point. (See **Ebullition**.)

Following is a brief table of normal boiling points:

Substance	B.P. (°C.)	Substance	B.P. (°C.)
Alcohol	78.3	Glycerin	291
Benzene (C_6H_6)	80.0	Helium	−272
Bromine	58.8	Hydrogen	−259
Carbon disulfide	46.3	Mercury	357
Chloroform	61.2	Turpentine	159
Ether	34.6	Water	100

(L.D.W.)

BOLE. A fine-grained, sticky, bright red **laterite**; the decomposition product of **basic igneous** rocks, such as **basalt**. (R.M.F.)

BOLIDE. Bolide is the term applied to **meteors** which are observed to explode in the air and break up into two or more fragments. Such objects are frequently described as having the appearance of an exploding rocket. Not infrequently following the explosion of a bolide a sharp detonation is heard. (W.K.G.)

BOLL WEEVIL. Insecta, Coleoptera. A **snout-beetle** or weevil, *Anthonomus grandis,* averaging about ¼″ in length, which damages cotton. The adult punctures the cotton squares to lay its eggs and thus prevents the formation of the boll, and later in the season deposits eggs in the bolls, where its larvae damage the seeds and lint.

The species entered the United States from Mexico in the early nineties and is now an established pest in practically all cotton-growing areas, causing annual loss estimated at $200,000,000. The problem is met by various methods of keeping the insect in check. The destruction of cotton plants after the crop is harvested kills many insects, and dusting the growing plants with **calcium arsenate** has been found effective. The powder is applied by hand dusters in small fields or by larger power dusters where necessary. Airplanes have been used in recent years for dusting on a large scale. (A.W.L.)

BOLLWORM. Insecta, Lepidoptera. The pink cotton bollworm is a small **moth** whose **larva**, the bollworm proper, lives in the flowers of cotton and usually prevents their maturing, and later enters the bolls and damages the seed and lint. The species lives on several other species of plants and is therefore difficult to check, although it is not yet as serious a pest as the boll weevil. The larva of another moth of larger size, more widely known as the corn earworm, also attacks cotton squares and bolls, as well as corn and tobacco, and so is known as the cotton bollworm. It is estimated to cause several millions of dollars damage each year. Fall plowing and disking are practiced to destroy this insect in the pupal stage, which is passed in the ground, and dusting as for the weevil is effective. (A.W.L.)

BOLOMETER. A very sensitive **thermometric** instrument of the metallic **resistance** type, devised by Langley and used for measuring feeble **radiation.** It consists of a slender strip of platinum mounted at the lower end of a long cylindrical tube having circular stops across it at intervals to screen off all radiation except that to be observed. A slight amount of radiant energy falling upon the strip causes a measurable deflection in a sensitive galvanometer in the resistance **bridge** coupled with it. The sensitivity of the instrument is of the same order as that of the **radiomicrometer** of Boys. When a bolometer tube is mounted as the receiving element of a spectroscope, instead of the usual observing telescope or camera, the instrument may be used to detect and measure the lines or bands of **infra-red** spectra. This arrangement is called a spectrobolometer. (L.D.W.)

BOLSON. An undrained basin in an arid region, which generally is partly filled with rock-waste washed by temporary streams from the bordering mountains. (R.M.F.)

BOLT. Screw Fastening.

BOLTZMANN CONSTANT. Ideal Gas Law.

BOLTZMANN'S PRINCIPLE. A somewhat general law relating to the statistical distribution of large numbers of minute particles subject to thermal agitation and acted upon by a **magnetic**, an **electric**, or a **gravitational field**, or by **inertia.** The number of particles per unit volume in any region of the field, when the system is in statistical **equilibrium**, is given by the equation

$$N = N_0 e^{-\frac{E}{kT}}$$

Here E is the **potential energy** of a particle in the given region, N_0 is the number per unit volume in a region of the field where E is zero, k is Boltzmann's constant (ideal gas constant per molecule), and T is the absolute temperature of the system of particles. Such an equilibrium may exist, for example, in a mass of electrified colloidal particles kept in suspension by their **Brownian movement** while acted upon by an electric field. A well-known special case is Laplace's "law of atmospheres," treated in the **kinetic theory** of gases. (L.D.W.)

BOMB, VOLCANIC. Lapilli.

BOMBARDIER BEETLE. Insecta, Coleoptera. A **ground-beetle** which discharges a strong-smelling volatile secretion in small jets when disturbed. Each discharge is accomplished by an audible report and a visible puff of vapor as the secretion evaporates. The numerous species make up the genus *Brachinus*. (A.W.L.)

BOND. Chemical bonds are discussed under **valence.** Electrical bonds are conductors connected between metal shapes, when those shapes do not constitute an effective electrical circuit. Railroad rails, for example, are used for the **block signal** circuit. In order to utilize them for that service, the rails must be electrically continuous. The bonding of a rail consists of attaching a conductor firmly between adjacent rail ends so that a positive path for flow of electric current will replace the haphazard and unreliable circuit that would be comprised by the ordinary rail construction in which rails are connected mechanically by fishplates. The contact resistance at these plates is too great, and so a short bonding connection is used. Bonding may be dispensed with when the rails are welded together. Bonding is also necessary where a radio is installed in a metal framework, some parts of which are bolted together. These joints, where the electrical contact is a surface contact, are frequently by-passed electrically by a bond. Aircraft and automobiles are often bonded this way.

Mechanical bonds play an important part in many structures.

The grip exerted by one material on another when in contact is called bond. The bond resistance is measured in lbs. per sq. in. of contact surface. The theory of reinforced **concrete** design takes into account the bond existing between the concrete and the reinforcing steel. This bond is the result of the molecular attraction called adhesion and the frictional force resulting from the shrinkage of the concrete around the rods. The intensity of the bond depends upon the proportions of the concrete, the age of the mix, the condition of the surface of the rods and the kind of rods used. Reinforcing rods which are made with irregular surfaces are called deformed rods and are able to exert a greater bond resistance than plain rods. (See **Grinding, Silicate Bond, Vitrified Bond.**) (F.T.M.)

BONDERIZING. Parkerizing.

BOND STRESS. Stress.

BONE. A rigid supporting tissue of which the skeleton of **vertebrates** is composed. Its rigidity is due to deposits of inorganic salts, chiefly **calcium phosphate** combined with some calcium carbonate, between the living components of the tissue. **Cells** are scattered throughout this hard matrix and at intervals it is penetrated by blood vessels and nerves which are necessary for the maintenance of its living parts. The matrix is also penetrated by a foundation of organic fibrils produced by the cells.

Bone occurs in two forms: compact bone such as the walls of the long bones of the skeleton, and cancellous bone composed of a reticular arrangement of slender parts which results in a spongy appearance. The latter is found in the ends of long bones, among other situations.

Compact bone is made up of thin plates called lamellae arranged in a definite plan. At the surface they follow the periphery of the bone but inside of the compact mass they are arranged concentrically about slender canals which run lengthwise of the bone. These cylindrical components are called Haversian systems; the central Haversian canal contains blood vessels and a small amount of connective tissue, together with the nerve fibers which supply these parts. Nourishment reaches the cells which lie among the surrounding lamellae by way of protoplasmic processes lying in minute canaliculi which radiate from the cavities containing the cells and communicate with the Haversian canals. Between the Haversian system are irregularly disposed layers known as ground or interstitial lamellae.

Bone is of **mesodermal** origin and develops chiefly from the **mesenchyme.** The vertebrae, certain bones of the skull, and the long bones are first formed in cartilage which is later replaced by (not transformed into) bone, while others, such as the thin bones of the skull and the bones of the face, are formed directly from mesenchyme; the former are called replacement bones and the latter membrane or dermal bones.

The term bone is also used to designate a single structural unit of the skeletal system. (A.W.L.)

BONGO. Mammalia, Artiodactyla. One of the harnessed **antelopes** or bush-bucks of western Africa. It lives in the forests. *Boöcercus euryceros.* (A.W.L.)

BONITO. Pisces, Teleostei. *Sarda.* Marine fishes (**Pisces**) of several species, related to the mackerel and tuna. (A.W.L.)

BONNER DURCHMUSTERUNG. The name Bonner Durchmusterung is applied to the monumental catalogue of 324,198 stars observed by that tireless observer F. W. A. Argelander. Accompanying the catalogue is an atlas of the heavens upon which each of the catalogued stars is shown by a dot, the size of the dot being proportional to the apparent brightness of the star. The catalogue contains practically every star brighter than the tenth **magnitude** north of **declination** —2°. The catalogue is commonly referred to as the B.D. and in many astronomical writings a particular star is referred to by its B.D. number (i.e., by the number assigned to it in the Bonner Durchmusterung).

The catalogue was continued by Schonfeld down to declination —23°, and Thome at Cordoba has extended it still further to —61°. It is hoped that the plan will be continued to the south pole.

In each of the catalogues stars are numbered in order of increasing **right ascension** within a particular zone of declination. Hence, a star known as CDM —48 1116 is the 1116th star in the Cordoba extension of the BD catalogue between declination —48 and —49. (W.K.G.)

BONTEBOK. Mammalia, Artiodactyla. A small South African **antelope,** *Damaliscus pygargus.* (A.W.L.)

BONY PIKE. Pisces, Holostei. A fish (**Pisces**). The garpike and **gars.** (A.W.L.)

BONY-TAIL. Pisces, Teleostei. A fish (**Pisces**) of the Colorado and Gila rivers, related to the minnows and chubs. *Gila elegans.* (A.W.L.)

BOOK SCORPION. Arachnida, Chelonethida. A European species of pseudoscorpion which is sometimes found in books and papers. (A.W.L.)

BOOK-LOUSE. Insecta, Corrodentia. A small **insect** found in old papers, books and rubbish and in collections of biological specimens. The order to which they belong is a small one containing winged species found on bark and lichens, and wingless species to which this name is applied.

Book-lice must be very numerous to do appreciable damage, and since they frequent damp situations, heating and drying rooms where they are found is usually a simple method of destroying them. Severe infestations can be checked by fumigation. (A.W.L.)

BOOM. A boom is a movable inclined arm of wood or steel used on some types of **cranes** or **derricks** to support the hoisting lines which carry the **loads.** The loads cause direct **compression** in the boom due to the manner in which the hoisting lines are connected to the member.

The word boom also describes a floating chain of logs, which is anchored in such a position in a body of water as to deflect or intercept saw logs, or to prevent floating debris from approaching water intakes to pipe lines, penstocks, etc. Nautically, a boom is a spar holding the foot of a fore and aft sail. (C.W.C., F.T.M.)

BOOMER. Mammalia. 1. The great gray **kangaroo** of Australia, a **marsupial.** 2. The mountain beaver, *Aplodontia,* a species of **rodent** which lives in the forests in the mountain ranges of the Pacific Coast. It is more closely related to the squirrels than to the true beavers. (A.W.L.)

BOOST. Supercharger; Airplane Engine.

BOOSTER. An electrical booster is inserted in series in an **electric circuit,** and increases the **voltage** of that circuit. There are several uses to which the booster can be put. It may be employed to compensate for a line voltage drop, or it may be employed to vary voltage in such a way that constant current is maintained. The boosting of d-c circuits is accomplished by rotating equipment called booster generators. If this booster is driven by an electric motor the set is called a motor-booster. The booster generator can be used to raise the line voltage at a feeder point on an electric traction system.

The booster transformer is sometimes used in **alternating-current circuits.** On a simple single-phase circuit it boosts the line voltage by connecting the primary of the **transformer** across the line, and the secondary in series. There are some disadvantages to this connection, however, since blowing of a fuse, or otherwise open-circuiting the primary, leaves the transformer connected as an open-circuited series transformer, and the open-circuit voltage on the primary winding may be excessive. The **induction regulator** is a form of booster transformer whose effect is varied by rotating one winding with respect to the other.

A mechanical booster is an auxiliary cylinder with which steam locomotives are sometimes equipped. It drives on the trailing truck or on a truck of the tender. The booster installation is used principally on freight locomotives, especially in the mountain sections where the extra tractive power is needed in starting the train. When starting a train from rest the locomotive is obliged to exert a much greater tractive effort than when the train is in motion, because the coefficient of static friction exceeds the coefficient of rolling friction. The booster is used only when starting the train, and is disengaged, usually automatically, when the train speed reaches something like 15 m.p.h. The booster drives through a geared crankshaft to the trailer truck. This extra power is available when starting by virtue of the fact that with a locomotive moving slowly, the main cylinders are not able to use all the steam that can be produced by the boiler. (F.T.M., L.R.Q.)

BOOTES. (The herdsman.) (Map, page 380.) While not in the zodiac, Bootes is one of the earliest recorded **constellations.** It is readily recognized in the early summer skies from the kite-shaped configuration of stars with the bright star Arcturus at the position of the tail of the kite.

The star **Arcturus** is the fourth brightest star visible in the northern latitudes. It is also a very interesting star from the astronomical point of view, being what is known as a **giant** star. In appearance the star is a reddish-yellow and the **spectral type** is such as to indicate that its temperature is slightly lower than that of the sun. Its angular diameter has been very carefully determined, and, since its distance is known, we find its linear diameter to be about 27 times that of the sun.

Many of the other brighter stars in Bootes are **double stars,** several of them forming interesting objects of study with relatively small instruments. (W.K.G.)

BORACITE. Boracite is a **borate** of **magnesium** containing some **chlorine.** It appears to be **isometric** but probably becomes so only at 265° C. below which temperature it is believed to be **orthorhombic.** Its hardness is 7; specific gravity, 2.9; luster, vitreous; color, white to gray, sometimes yellow or green; translucent to subtransparent. It occurs in beds with **gypsum** and salt in Germany, particularly at Stassfurt in Saxony. (E.S.C.S.)

BORAX. Boron.

BORE. The size of a hole in a pulley, gear, or bearing. As a verb, to enlarge and/or accurately finish a hole. (H.C.H.)

BORER. Insecta. Any **insect** which burrows into the tissues of plants. Among the more common pests of this kind are the peach-tree borer, *Synanthedon,* the squash borer, *Melittia satyriniformis,* and the wheat sawfly borer, *Cephus pygmaeus.* In the garden, borers frequently kill young plants by hollowing out the stem and they may even kill larger plants or weaken them so that they break off in the wind. The borers include insects of several orders and various habits, hence it is impossible to suggest general methods of treatment. A knowledge of the habits of the insect, particularly of the time of year when the eggs are laid or when the larvae enter the plants, is important since at other times the insect is beyond reach. In the home garden the hole made by the insect on entering the stem is often readily visible and by slitting the stem the invader can be found and killed, but unfortunately the weakening of the plant is the first indication of trouble. A little pyrethrum or rotenone powder introduced into the burrow will also prevent further damage. (A.W.L.)

BORING. The process of enlarging and/or accurately finishing a hole, usually with a single-point tool. Contrast **drilling, reaming, trepanning.** Boring may be effected on a **lathe** by a suitable tool held in the **tool** post; the work may be held in a chuck, clamped to the lathe faceplate, or mounted on the base of the cross-slide after the compound rest and cross-slide proper have been removed. In such cases, the boring tool is carried in a boring bar which is swung between the lathe centers and driven by a lathe dog.

For work of any variety either boring mills or boring machines are used. A boring mill is essentially a lathe set on end, and has a rotating horizontal table which is driven by a vertical spindle. Single-point cutting tools are carried in tool heads on rams that may be adjusted and fed vertically, and may be positioned or fed along a supporting cross-rail.

There are two important types of boring machines: those with a horizontal spindle, and those with a vertical spindle. Horizontal-spindle boring machines for large work are built with a base or floor plate level with the floor of the shop. Table-type horizontal-spindle boring machines have a table which is mounted on a saddle. Both the table and saddle are equipped with power feed and rapid traverse movements; the table moves on the saddle, in a direction perpendicular to the spindle, and the saddle moves parallel to the spindle, on the bed. The spindle rotates in the spindle head which may be power adjusted or fed along the vertical face of the column.

Boring, drilling and milling operations may be performed on horizontal-spindle boring machines. Milling cutters may be held in the spindle socket, or bolted directly to the spindle flange. Boring is handled with piloted bars whenever possible, but boring tools may be held in an adjustable boring head.

The vertical-spindle boring machine **or jig borer** is widely used for boring holes in jigs, fixtures, and dies, and has almost entirely replaced toolmakers' buttons or disks for accurate hole location.

Wash borings are frequently used to examine the subsurface formations in connection with deep **foundation** work. The equipment consists of a hollow pipe called a jet pipe and a larger hollow pipe called a casing. Water under pressure is forced down the jet pipe. This water washes the disintegrated material up through the space between the jet pipe and the casing to the surface where it may be retained for future examination. As the material at the bottom of the casing is washed away the casing is slowly forced downward. Where the precise character and formation of subsurface rock formations must be known, as in the case of foundations for important dams, core boring (sometimes called core drilling) is usually a necessity. A core drill consists of a hollow cylindrical bit with its cutting edge set with hard cutting particles (such as commercial diamond particles) connected to a hollow cylindrical drill shank. The whole is rotated by mechanical power and thus cuts out a vertical, or inclined, cylinder of the rock. These cores are periodically removed and when re-assembled constitute a clear and visible section of the rock structures pierced. (H.C.H., C.W.C., E.W.S.)

BORNEOL. Alcohols and Ethers.

BORNITE (Peacock Ore) (Horse-Flesh Ore). Named for the German mineralogist of the 18th century, Ignatius von Born, this mineral is a **sulfide** of **copper** and **iron** corresponding to the formula Cu_5FeS_4. It is **isometric** with a cubic habit, although crystals are rare, usually occurring as granular or compact masses. Its fracture is conchoidal to uneven; brittle; hardness, 3; specific gravity, 4.9–5.4; color, copper-red to reddish-brown (hence the name horse-flesh ore) when freshly fractured; it soon assumes an iridescent tarnish (hence the name peacock ore); luster, metallic; streak, grayish-black; opaque.

Bornite as a primary mineral has been observed in

pegmatite veins and in **igneous rocks** and is also a common secondary mineral.

Bornite crystals have been obtained in Austria and England. As an ore it is important in Tasmania, Chile, Peru and in Montana. Bornite has been found in Connecticut and in the Province of Quebec. (E.S.C.S.)

BORON. Symbol: B. Atomic number: 5. Atomic weight: 10.82. Density: 2.54 (for the crystalline form), 2.45 (for the amorphous form). Hardness: 9.5 (for the crystalline form). Melting point: 2300° C. Boiling point: 2550° C. Isotopes: 10 (20%), and 11 (80%).

Boron is (1) a yellowish-brown crystalline solid, (2) an amorphous greenish-brown powder. Both forms are unaffected by air at ordinary temperatures but when heated to high temperatures in air form oxide and nitride. Crystalline boron is unattacked by **hydrochloric** or **nitric acid**, or by **sodium** hydroxide solution, but with fused sodium hydroxide forms sodium borate and **hydrogen;** reacts with **magnesium** but not with **sodium.** Discovered by Davy and by Gay-Lussac and Thenard in 1808.

Boron occurs as **rasorite** or kernite (sodium tetraborate tetrahydrate, $Na_2B_4O_7 \cdot 4H_2O$) and **colemanite** (calcium borate, $Ca_2B_6O_{11} \cdot 5H_2O$) in California, as **sassolite** (boric acid, H_3BO_3) in Tuscany, Italy, and also locally in Chile, Turkey, and Tibet. The borates are transformed to boric acid, which is heated to form the oxide, and this last reduced to boron by ignition with **aluminum** powder. The aluminum is then dissolved by hydrochloric acid or sodium hydroxide solution, leaving boron residue.

Acids: Boric acid (ortho H_3BO_3 or $B_2O_3 \cdot 3H_2O$; pyro or tetra $H_2B_4O_7$ or $2B_2O_3 \cdot H_2O$; meta HBO_2 or $B_2O_3 \cdot H_2O$), white solid, soft and smooth to the touch, moderately soluble in the cold but very soluble in the hot, made by reaction of solution of a borate and hydrochloric, nitric, or sulfuric acid of proper concentrations, forms boron oxide glass upon ignition, forms borates by precipitation reactions or by fusion methods, used as an antiseptic, and in cosmetics and skin powders; hydrofluoboric acid (HBF_4), colorless solution, by reaction of boron fluoride with **hydrofluoric acid** solution forms fluoborates by neutralization method.

Borates: Sodium tetraborate, borax ($Na_2B_4O_7 \cdot 10H_2O$), white solid, soluble, used (1) in soaps and laundry starch glazes, (2) in ceramic glazes and enamels, and special glasses, (3) in metallurgical **fluxes,** (4) as source of boron-containing substances; **calcium** borate, white solid, insoluble; **silver** borate ($AgBO_2$), white precipitate, by reaction of borate solution and silver nitrate solution (silver oxide, brown solid, may accompany in this reaction); potassium fluoborate (KBF_4), white solid, moderately soluble, by reaction of hydrofluoboric acid and **potassium** carbonate, and then crystallizing; sodium perborate ($NaBO_3$), white precipitate, by reaction of sodium borate solution and **hydrogen peroxide,** used (1) as germicide and antiseptic, (2) as oxidizing and bleaching agent, (3) in cosmetics and soaps; methyl borate, and ethyl borate, see below.

Borides: Carbon boride (CB_6) and silicon borides (SiB_3 and SiB_6) are hard, crystalline solids, produced in the electric furnace; magnesium boride (Mg_3B_2), brown solid, by reaction of boron oxide and magnesium powder ignited, forms boron hydrides with hydrochloric acid; calcium boride (Ca_3B_2), forms boron hydrides and hydrogen gas with hydrochloric acid.

Chloride: Boron chloride (BCl_3), colorless fuming liquid, by reaction of boron, or boron oxide plus carbon, heated with **chlorine,** boiling point 12.5° C.

Fluoride: Boron fluoride, boron trifluoride (BF_3), colorless gas, by reaction of boron oxide, **calcium** fluoride and hot concentrated **sulfuric acid;** reacts with water to form boric acid plus hydrofluoric acid, and the latter, with excess boron fluoride, forms hydrofluoboric acid; combines with ammonia (e.g., $BF_3 \cdot NH_3$).

Fluoborate: Potassium fluoborate (KBF_4), white solid, slightly soluble, decomposed upon heating to 500° C.

Hydrides: Tetraboron hydride, tetraborane, borobutane (B_4H_{10}), colorless liquid 16° C., by reaction of **magnesium** boride and **hydrochloric acid,** followed by fractional purification of the gas; tetraboron hydride is the best known hydride; others lower and higher in the series are known.

Nitride: Boron nitride (BN), white solid, insoluble, reacts with steam to form **ammonia** and boric acid, formed by heating anhydrous sodium borate with **ammonium** chloride, or by burning boron in air.

Oxide: Boron oxide, boron trioxide, boric acid anhydride (B_2O_3), white, glassy solid, reactive with water to form boric acid and used as a dehydrating agent, melting point about 575°C., formed by heating boric acid to high temperature with loss of water, used as a nonvolatile acidic oxide.

Sulfide: Boron sulfide (B_2S_3), white solid, unpleasant odor, irritating to the eyes, reactive with water to form boric acid and hydrogen sulfide, formed by reaction of boron oxide plus carbon heated in a current of **carbon disulfide** at red heat.

Organic compounds: Methyl borate, trimethoxy boron ($B(OCH_3)_3$), colorless liquid, boiling point 65° C., by reaction of boric acid, even in very dilute solution, and methyl alcohol, and recognized by the green color of the flame when ignited; ethyl borate, triethoxy boron ($B(OC_2H_5)_3$), colorless liquid, boiling point 120° C., by reaction of boric acid, even in very dilute solution, and ethyl alcohol, and recognized by the green color of the flame when ignited.

Boron trimethyl ($B(CH_3)_3$) is a colorless gas; boron triethyl ($B(C_2H_5)_3$), a colorless liquid, boiling point 95° C.; boron diethyl dihydroxyl ($(C_2H_5)_2B(OH)_2$), sublimes at 40° C.

Borate solution when slightly acidified, or dilute boric acid solution imparts a characteristic reddish color to turmeric paper, intensified upon drying, and turned a characteristic bluish color upon moistening with ammonium hydroxide.

When borates are treated in a porcelain dish with methyl alcohol and concentrated sulfuric acid, the mixture stirred and ignited, a green-bordered flame will appear. (R.K.S.)

BORROW PIT. The section of ground from which earth is excavated for the purpose of being used as fill at some other point is called a borrow pit. Borrow pits are used extensively in connection with highway construction since it is often more economical to obtain fill from some nearby source than to carry it from some distant point along the line where there is an excess of cut. When it is necessary to ascertain the amount of material taken from a borrow pit the ground surface is divided into rectangles or squares before any material has been removed. This grid system is referred to an arbitrary **base line** and elevations are taken on each of the corners. After the required material has been removed the system of rectangles or squares is reproduced by means of the base line and elevations are again taken on the corners. The volume of material which has been excavated then consists of the sum of the volumes of the individual prisms. (C.W.C.)

BORT or **BOART.** Diamond.

BOSS. The term boss or stock is used to indicate a cross-cutting mass of **igneous rock** which has ascended into the crust of the earth and may or may not represent the roots of volcanic conduits. Bosses are roughly circular or elliptical in ground plan and usually of greater cross-sectional area than a volcanic neck and lack **pyroclastic** materials. Most probably bosses are the irregular upward extensions of **batholiths** the main parts of which are as yet unexposed.

Boss also designates a circular projection on a casting, usually serving as the seat for a bolt head or nut. (R.M.F., H.C.H.)

BOSTONITE. A rather rare rock type, dense, with an occasional feldspar **phenocryst** and grayish in color. It is composed almost wholly of alkaline feldspar, being analogous to **aplites.** The type locality is Salem Neck, Massachusetts, not many miles from Boston, for which it was named. (E.S.C.S.)

BOT FLY. Insecta, Diptera. **Flies** of two families whose larvae live as internal parasites in mammals. The adults as a rule have vestigial mouth parts and attack the host only to deposit their eggs.

Horses are attacked by two species of bot flies of the genus **Gastrophilus.** One deposits its eggs on the lips, whence the **larvae** reach the throat or stomach, and the other attaches them to the hairs of the forelegs, where they die unless the horse takes the larvae into its mouth by licking or biting the legs. The larvae develop in the alimentary tract and pass out when mature with the feces. These flies belong to the family Gastrophilidae.

The sheep bot adult, *Oestrus ovis,* deposits living larvae in the nostrils of the host, whence they work their way into the sinuses, into the horns, and even into the brain. Two other species known as **warble-flies** lay their eggs on the hairs and the young larvae migrate through the skin and ultimately to the **oesophagus.** Later they migrate again through the connective tissue to the back, where they complete their development in subcutaneous abscesses which produce the external lumps known as warbles. The skin over the abscess is perforated and the larva leaves its host and drops to the ground to pupate. These species make up the family Oestridae, together with numerous others which attack wild animals. (A.W.L.)

BOTANY. Botany is the science which deals with plants. It is divided into many sections, each dealing with a specific part of the subject. One section, which describes plants and arranges them in classes, is called taxonomy; another section, morphology, considers the form of the various parts of a plant, while its subsections include anatomy and histology, the study of the internal structure of plants, and cytology, the study of the cell and its parts. A third, physiology, deals with the functions of the parts. In addition, one may study plant geography, or the distribution of plants on the earth; ecology, the relations of plants to each other and to their environment; phytopathology, or the diseases of plants; paleobotany, the science of fossil plants; and economic botany, which considers the uses which man has found for plants and plant products.

The science of botany is very old. Since the welfare of man is closely connected with plants, it is natural that they should receive attention early. Undoubtedly plants were known and observed by men long before the period of Greek supremacy. Various recorded observations suggest that such is true. But only with the intellectual curiosity of the Greek mind do plants receive close attention. Aristotle (384–322 B.C.) studied them attentively and cultivated many species from widely separated regions. His disciple Theophrastus (371–287 B.C.) carried on the work and wrote about them in his "Enquiry into Plants," in which he describes some five hundred species and gives extensive and keen observations concerning them. In Rome another naturalist, Pliny the Elder (23–79 A.D.), writes extensively on Natural History, setting forth information on some thousand species of plants. His facts are largely drawn from sources other than the plants themselves and are often grossly exaggerated. His Natural History was of immense importance, however, and largely controlled the thought of botanists for many centuries. Another ancient naturalist, Dioscorides, also studied plants.

He was mainly interested in them because of the important place they held in the medical practice of that time. Indeed, the study of plants was for a long period of time considered the province of physicians and doctors, whose main interest was in plants as remedies or supposed remedies for various ills. After this, centuries followed in which little attention was given to plants; all knowledge thereof was drawn directly from the works of the ancient writers.

Beginning with the 16th century, however, interest in plants was revived. Men began observing the native plants around them and recording these observations, often accompanied by illustrations, in herb books or herbals. Such observations led to attempts to arrange and classify the various plants. Among the first herbals were those of Brunfels (1530) and Fuchs (1542), both of them containing excellent illustrations, but relying for their descriptions largely on the ancient writers of Greece and Rome. Hieronymus Bock (1498–1554) was another herbalist, who gave in his book extensive first-hand descriptions of the plants which he treats. William Turner and John Gerard published herbals treating of English plants. Valerius Cordus (1515–1544) gave even more complete and accurate descriptions of the plants in his books than Bock.

As a result of the work of these men and many others, came a need for a better understanding of plants and the necessity for arranging them in some sort of system other than that of size or of the alphabet. John Ray (1628–1705) advanced the problem considerably by introducing an exact concept of species, which he held to come from a single parent and to continue to produce like organisms, although he does allow some variation to occur. Ray separated flowerless plants from flowering, and divided the latter into **Dicotyledons,** with two seed leaves, and **Monocotyledons,** with only one.

The number of plants described was constantly increasing, rendering even more necessary a system of arranging them in order. Many systems were proposed, some having great merit. As early as 1583 Casalpino had eliminated any classification based on such variable organs as roots, stems or leaves, and had concluded that the flowers and fruit offered the only real basis. It remained for Carolus Linnaeus (1707–1778) to bring order to the situation. He invented the binomial system of nomenclature, by which each plant (and animal also) should be known by a name designating the genus and a qualifying adjective limiting the species named. His system of classification was purely artificial, being based on the number of stamens and pistils (see **Flower**), but did make it easy to refer to a description and so verify an identification. He also grouped plants and animals in larger divisions, the classes and orders. The present-day names of plants date from the time of Linnaeus. It has long been recognized that there seemed to be a natural grouping of plants; John Ray apparently understood some of the larger groups of plants. With the work of the French taxonomist A. L. de Jussieu came a definite knowledge of the natural relations of plants, which he grouped into fifteen classes with about a hundred orders.

While classification and description occupied a large place in the development of botany, other branches of the science were not neglected, although of necessity many of them waited on advancement in taxonomy. The anatomy of plants was studied by Nehemiah Grew (1641–1712) in England, and Marcello Malpighi (1628–1684) in Italy, while casual observations on the internal structures of some plant substances were made by Robert Hook. The finely illustrated writings of these men established the foundations for an understanding of the internal structure of plants. Subsequent workers in this field showed the similarities existent in the internal structures of plants, and the changes which have occurred during the evolution of plants. Out of this have come the later studies of cytology and histology.

Any knowledge of the way in which the plant lives and the functions of its various parts was slow to develop. The lack of definite organs connected with such functions as digestion, circulation, respiration, etc., made the problem even more difficult. Occasional observations had been made from time to time, often leading to erroneous conclusions. With Stephen Hales (1677–1761) plant physiology became established. He first used instruments to measure various physiological activities which he studied. His observations, recorded in his "Vegetable Staticks," published in 1727, show how attentively he studied the problem of nutrition in plants and the movements of liquids within the plant. Ingen-Housz (1730–1799) gained more exact knowledge of the problem of nutrition in plants, definitely showing that the **carbon** in plants came from the **carbon dioxide** of the atmosphere. He had an accurate knowledge of the role of gases in the life of the plant. Another worker, Andrew Knight (1758–1838), studied an entirely different field, being largely interested in the problem of direction of growth of root and stem. To him is due the use of a rapidly revolving wheel to which seedlings were attached. From this experiment he determined that roots grew away from the center of the revolving wheel and stems towards the center. Out of his studies came the study of tropisms (see **Movements in Plants**) in general. At the present time the study of physiology of plants is one of the most important and most fascinating, engaging the attention of many workers.

However, other branches of the science of botany have not been overlooked. The study of the distribution of plants has been pursued with great vigor, bringing to light many interesting problems, at times difficult to explain. Why should certain similar groups of plants appear in widely separated regions? At present this and many other questions are subjects for speculation and cause for further study.

Another branch of botany which occupies an important position today is that of plant pathology, which treats of the diseases of plants. When a single disease such as **wheat rust**, attacking a single crop, causes the loss of millions of dollars in reduced harvests, and with so many crops subject to numerous diseases, this must be recognized as a study of vital importance to man. Comprehensive and exact knowledge of the disease-producing organism is necessary. Often it is obtained only after prolonged, painstaking study. Then follows the problem of treatment leading to elimination of the disease, a study in itself. Sometimes this is impracticable; it is quicker to attack the problem in another way—to attempt to develop strains of plants which are resistant or immune to the disease. In this field new problems are constantly arising, or assuming greater importance, as for example was the case in the outbreak of the Dutch Elm Disease in recent years, or of the Chestnut Blight disease which so nearly wiped out the chestnut trees of America a few years ago.

These and many other problems show how close is the welfare of mankind tied up with the study of botany and the knowledge of the many sides of that science. (R.M.W.)

BOTHRIOLEPIS. Fossil Fishes.

BOTHRIUM. A projecting or grooved structure which takes the place of a sucker on the **scolex** of some **leeches.** (A.W.L.)

BOTRYOIDAL TISSUE. A peculiar kind of pigmented **tissue** with **cells** arranged end to end and containing intracellular capillaries filled with red liquid, found in **leeches.** (A.W.L.)

BOTULISM. A type of food poisoning resulting from the ingestion of toxins of *Clostridium botulinum*, a species of **anaerobic bacteria.** These organisms are gas-forming bacilli which grow in sealed or canned food. Commercially canned products formerly caused outbreaks but, since proper sterilization by heat has been employed by packers, commercial products have been safe. In recent years, it has been the home canned foods which are improperly sterilized and therefore contain the resistant **spores** of the organism which cause botulism.

The symptoms of poisoning appear 18 to 36 hours after eating spoiled food. The toxin has a special affinity for the nervous system. Double vision, difficulty in swallowing and speaking, are the most prominent features. Nausea and vomiting occur in only one third of the cases. The mortality rate is high—about 66%; death is usually due to respiratory paralysis. Treatment is with specific antiserum containing an antitoxin to neutralize the effects of the bacterial toxin. (D.M.H.)

BOUGAINVILLAEA. Nyctaginaceae. A small genus of plants, natives of South America, which are frequently cultivated in the tropics and to some extent as greenhouse plants outside the tropics. The flowers of the plants are small and inconspicuous, but are surrounded by showy bracts of various colors. The plants are generally grown for these brilliant-colored **bracts.** One species, *Bougainvillaea spectabilis*, is a clambering vine which is often grown in cultivation. (R.M.W.)

BOULANGERITE. A mineral compound of lead-antimony sulfide, $Pb_5Sb_4S_{11}$. Crystallizes in the monoclinic system; hardness, 2.5–3; specific gravity, 6.23; color, lead gray. (R.M.F.)

BOULDER. A large fragment of rock, usually rounded, which has been moved from its place of origin by a natural agency or has been formed in situ by weathering processes. Rather arbitrarily, 8″ has been set as the minimum diameter for a boulder. (E.S.C.S.)

BOULDER CLAY. Boulder clay is a glacial deposit of clay with subangular rock fragments of different sizes. (E.S.C.S.)

BOULDER TRAIN. Erratic.

BOUNDARY LAYER, FLUID FLOW. Motion of a fluid of low viscosity, such as air or water, around a stationary body or through a stationary conduit possesses the free velocity of an ideal fluid everywhere except in an extremely thin layer immediately next to the stationary body. Many of the phenomena of fluid flow may be studied and analyzed without consideration of this boundary layer but, thin as it may be (usually a few thousandths of an inch), its internal mechanics must be understood and evaluated in certain of the phenomena of fluid motion. Some of the more important of these are:

1. The magnitude of the maximum lift coefficient of the airfoils.

2. Profile drag of airfoils.

3. The drag of bluff bodies.

4. The large variations of drag coefficient at critical Reynolds number for laminar-turbulent transition.

5. The transfer of heat through surface films.

Many of the phenomena of the boundary layer are explainable on the basis of the theory advanced by Prandtl at the great University of Göttingen laboratory nearly half a century ago. In the same flow-research group were others, like Blasius, who broadened and experimentally confirmed the original hypotheses.

An elementary understanding of the effect of fluid viscosity will be had by considering a two-dimensional flow along the upper surface of a very thin flat-plate, as shown in Fig. 1. The thickness of the boundary layer, greatly exaggerated, is y_v; the free stream velocity is V, the variable velocity in the boundary layer is u,

A basic assumption of the theory is that a fluid layer of infinitesimal thickness resting against the plate "sticks" to it so shearing force of the next fluid layer on the sta-

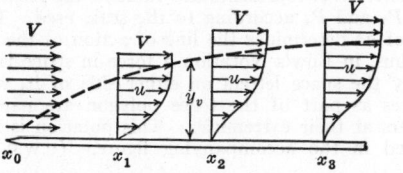

Fig. 1.   Velocity profiles in the boundary layer (dotted line is hypothetical upper edge of the boundary layer).

tionary layer determines the skin friction. Assuming that the boundary layer consists of lamina of fluid sliding on each other, the velocities of these lamina increase with y until, at the edge of the boundary layer, $u = V$. A series of boundary layer velocity profiles for stations x_1, x_2, x_3 are drawn to enable the reader to visualize the effect of friction on the momentum in the boundary layer and the thickening of it due to lower average u.

Note also the variation of the profile near the surface of the plate. Skin friction has steadily decelerated the individual fluid particles. The profile at x_3 indicates that the lower portion has come to rest. This is known as the stagnation point. Air-flow phenomena in this region are important in many ways, especially when there is an intended rising downstream pressure gradient, as in diffuser tubes or over the surface of airfoils.

Fluid flow in a divergent tube is illustrated in Fig. 2. On the lower profile, greatly enlarged velocity profiles are

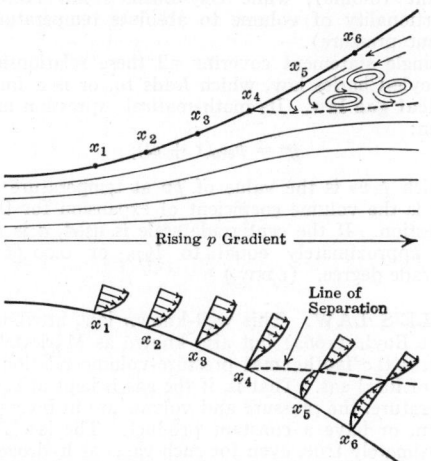

Fig. 2.   Effect of boundary layer viscosity on flow on a diffusor.

shown for the boundary layer at stations x_1, x_2, $\cdots x_6$. The x_4 profile indicates a stagnation point. Since the pressure gradient of the diffuser has a pressure at x_5 above that at x_4, a reverse flow is produced towards the stagnation point. Streamlines drawn in the upper half of the tube show what happens to the fluid flow. The region of the surface of separation between the reverse flow along the wall and the forward flow is unstable and breaks up into random vortices. Kinetic energy is irreversibly transferred to heat and the diffusion fails to produce the expected pressure gradient. Had the divergence of the conduit been sufficiently small, the pressure gradient would have been lowered per unit length and turbulence in free stream or boundary layer would have delayed the stagnation point.

For the airfoil with burbled or partly burbled air flow (Fig. 3), a similar explanation exists. Streamline flow over the upper surface of the airfoil increases in

velocity as angle of attack (and circulation Γ) increases. However, the surface friction in the boundary layer is decelerating the air particles next to the surface

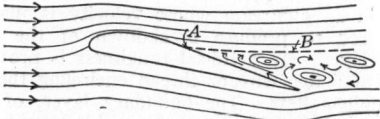

Fig. 3.   Partial burble on airfoil at high angle of attack. A—Stagnation point.   B—Line of separation.

and with high adverse pressure gradient (existing at high angle of attack) opposing the kinetic energy of the boundary layer the velocity profiles ultimately show a stagnation point toward the rear of the airfoil. The flow separates from the surface and vortices form a turbulent wake in place of the streamline wake previously existing. Once the separation surface moves onto an airfoil, minor increases of angle of attack bring it rapidly forward. High circulation strength is no longer needed to fulfill the requirement of unity of upper and lower flows at the trailing edge (Kutta's hypothesis) so Γ decreases sharply, and with it the lift. Boundary layer theory accounts in this manner for the maximum lift coefficient. At the same time the extended turbulent wake sharply increases the profile drag coefficient.

Now return to a view of the nature of flow in the boundary layer. It has been called laminar, and so it is for values of **Reynolds number** below a *critical* value. But for years, beginning about the time of Osborne Reynolds' experiments and revelations in the field of fluid flow, it has been known that the laminar property disappears and the flow suddenly becomes turbulent when the critical $\dfrac{Vl}{\nu}$ is reached. Usually flow starts over a surface as laminar but after passing over a suitable length the boundary layer becomes turbulent, with

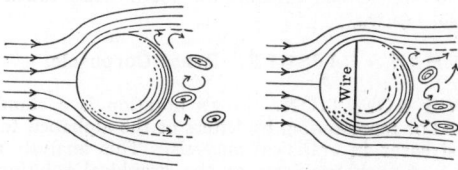

Laminar Flow   Turbulent Flow

Fig. 4.   Boundary layer turbulence reduces width of low-pressure wake.

a thin laminar sublayer thought to exist because of damping of normal turbulent components at the surface.

This transition has profound effects in all fluid dynamics, and certainly so in aerodynamics. The velocity profile in the boundary layer becomes fuller near the surface on account of the higher average kinetic energy of the layer created by turbulent energy exchange from layer to layer. The effective viscosity is therefore larger in turbulent than laminar flow, the turbulent boundary layer thickens more rapidly downstream, and skin friction increases.

The importance to aerodynamics is the beneficial effect of turbulence on the wake existing in the rear of **bluff bodies**. Even thin airfoils become bluff bodies at high angles of attack. A turbulent boundary layer has more kinetic energy than a laminar one. This carries the air farther toward the rear of the surface before a **stagnation point** is reached and so reduces the width of the wake and that part of the profile drag. This reduction of drag can be, and usually is, much larger than the increase of skin friction so turbulence has a good effect on both profile drag and maximum lift coefficient of an airfoil. The critical value of Reynolds number lies between 200,000 and 2,000,000, being affected by the initial turbulence existing in the air prior to meeting the surface. Decreasing initial turbulence increases the critical Reynolds number. The effect of turbulence on the boundary layer is strikingly displayed by observing the drag wake behind a smooth sphere mounted in an airstream whose velocity is just under that required to produce breakdown of the laminar boundary layer. An artificial roughness is provided in the form of a fine wire or thread encircling the sphere in the laminar flow. The boundary layer, of course, becomes turbulent downstream from this irregularity resulting in a delayed stagnation point and narrower wake. (F.T.M.)

BOUQUETIN. Ibex.

BOURNONITE (WHEEL ORE). An antimony-copper-lead sulfide corresponding to the formula $2PbS \cdot Cu_2S \cdot Sb_2S_3$. It is orthorhombic, and repeated twinning often produces crosses or wheel-shaped crystals. It is brittle; fracture, sub-conchoidal; hardness, 2.5–3; specific gravity, 5.7–5.9; luster, metallic; color and streak, dark gray to black; opaque.

Bournonite is found with **galena, chalcopyrite, sphalerite**, etc. There are many European localities; it was first found in Cornwall, England, by Count Bournon, for whom it was later named. Bournonite occurs in Bolivia and Peru and in the United States in Arizona, Montana, Nevada and Utah. (E.S.C.S.)

BOUTO. Dolphin.

BOWER-BIRD. Aves, Passeriformes. Birds (**Aves**) of several species found in the Australian region. They build runs roofed with grass or sticks and decorated with bright articles of all kinds. (A.W.L.)

BOWFIN. Pisces, Holostei. A species of fish (**Pisces**), *Amia calva*, found in lakes and sluggish streams in the eastern and central United States. Related to the gars and, like them, not valuable for food. Also called the dogfish. (A.W.L.)

BOWMAN'S CAPSULE. Renal Corpuscle.

BOW'S NOTATION. Bow's notation is a standard method of representing, by letters of the alphabet, forces and **stresses** in graphical analysis. This analysis may consist of such problems as the graphical solution of stresses in simple framed structures or the determination of the **resultant** of an independent system of unbalanced forces lying in the same plane and having a common point of application. The accompanying figure illustrates the method of applying Bow's Notation to the

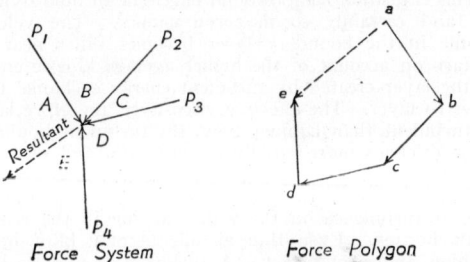

Force System   *Force Polygon*

latter system. Let P_1, P_2, P_3 and P_4 be a system of unbalanced forces lying in the same plane and having a common point of application. Denote the space between the line of action of each force by the letters *A, B, C* and *D*. Next construct a figure called a force polygon. This is accomplished by drawing a line parallel to P_1 and laying off its magnitude to a definite scale denoting the ends of the line by the letters *a* and *b*. **From point *b*** lay off *bc* equal in magnitude and parallel to P_2. Repeat the operation for the other forces. Upon comple-

tion of this graphical figure it will be found, in general, that the line representing P_4 will not pass through point *a*. The distance from point *a* to end of this line, which will be lettered *e*, represents the value of the resultant of P_1, P_2, P_3 and P_4 according to the scale used. The direction of *ae* determines the line of action of the resultant. Thus, in Bow's Notation a force in space is designated by the space letters on either side of it, whereas the forces as part of the force polygon are named by the letters at their extremities. This notation is further illustrated in the accompanying figure. (C.W.C.)

BOX TOOL. A tool with tangentially-cutting blades and a supporting rest, used on automatic screw machines. (H.C.H.)

BOX WRENCH. Wrenches.

BOXWOOD. *Buxus sempervirens. B. balearica.* Slow-growing trees or shrubs often planted for ornament because of their dark thick smooth leaves and compact habit. They are tender plants which must be treated as greenhouse subjects in cold climates. *B. balearica* is a tree growing to a height of 75′ or more. From it is obtained a dense yellow wood having a very close fine grain. This wood is used in wood engraving, in the manufacture of rulers and of musical instruments, and in inlay work. (R.M.W.)

BOYLE-CHARLES LAW. Boyle's law expresses the variation of pressure and volume of a body of ideal gas at constant temperature; Charles' law expresses the proportionality of pressure to absolute temperature (at constant volume), while Gay-Lussac's law states the proportionality of volume to absolute temperature (at constant pressure).

A single statement covering all these relationships is the Boyle-Charles law, which leads to, or is a form of, the **ideal gas law.** Its mathematical expression may be written:

$$pv = p_0v_0(1 + at),$$

in which p_0v_0 is the value of pv at temperature $t = 0$, and a is the volume coefficient of expansion for the gas in question. If the centigrade scale is used, a is for all gases approximately equal to $\frac{1}{273}$ or 0.003663 per centigrade degree. (L.D.W.)

BOYLE'S LAW. This well-known law, attributed to Robert Boyle (1662) but also known as Mariotte's law, expresses the **isothermal** pressure-volume relation for a body of ideal gas. That is, if the gas is kept at constant temperature, the pressure and volume are in inverse proportion, or have a constant product. The law is only approximately true, even for such gases as hydrogen and helium; nevertheless it is very useful. Graphically, it is

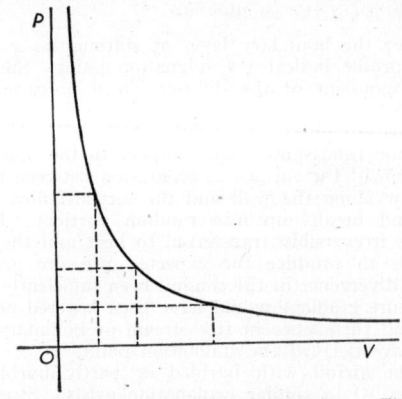

Equilateral hyperbola representing Boyle's law. The rectangular areas (pv) are all equal.

represented by an equilateral hyperbola. If the temperature is not constant, the behavior of the ideal gas must be expressed by the **Boyle-Charles law** or by the **ideal gas law.** (See **Characteristic Equation.**) (L.D.W.)

BRACHIAL. Pertaining to the arm, from the Latin term *brachium*. (A.W.L.)

BRACHIATION. Locomotion.

BRACHIOPODA. A phylum of marine animals which resemble the **bivalve** mollusks superficially. In the remote past they were much more abundant, as is shown by extensive fossil remains of many more forms than exist today.

The brachiopods are characterized by the following structures: 1. The body is enclosed by a shell consisting of dorsal and ventral valves. 2. The animal is **triploblastic** and coelomate but not segmented. 3. A **ciliated** organ called the **lophophore** projects about the mouth. It maintains currents of water which carry food and oxygen to the animal and wash the wastes away.

The phylum is divided into two orders:

Order Ecardines. Valves of shell not joined by a hinge. Anus present.
Order Testicardines. Valves of shell joined by a hinge. Alimentary tract without an anus.

(See also **Invertebrate Paleontology.**) (A.W.L.)

BRACHIOSAURUS. Fossil Reptiles.

BRACHYBLAST (OR SHORT SHOOT). In many plants, especially in the **Gymnosperms,** the display of leaves to light is considerably advanced by the formation of short lateral branches called brachyblasts. In the **larch** this short shoot is well developed. In this plant it persists, year after year, bearing at its tip a small group of leaves. It does not, however, increase in diameter even after several years of growth. The brachyblast develops from a bud formed in the axil of a leaf. The **maidenhair tree** or *Ginkgo* is another tree having well-developed short shoots. In both these plants and in many others the short shoot bears at its tip a terminal bud from which the leaves of the following year develop. In the **pines,** on the contrary, the short shoot is very much reduced and bears no terminal bud. In these plants the short shoot is reduced to a single bundle of leaves which persist for a year or two and then drop off completely. That this condition in pines is a reduced condition is clear from the condition found in fossil pines, which have a well-developed brachyblast, bearing many leaves and having a terminal bud. (R.M.W.)

BRACHYCEPHALIC. Short-headed. As applied to measurement of the human skull, with a width which is more than $\frac{4}{5}$ of the length. (A.W.L.)

BRACKELSBERG FURNACE. A horizontal rotating coal- or oil-fired melting furnace used principally for gray and white cast iron. (R.H.H.)

BRACKET FUNGI. Polyporaceae (also called Shelf fungi). A large group of **fungi** the fruit-body of which forms a characteristic shelf-like outgrowth from the trunks of trees. This fruit-body arises from a mycelium of fine hyphae, which penetrate throughout the woody tissue of the host plant, from which they derive nourishment and which they slowly destroy. The fruit-bodies are often perennial, showing on sectioning the successive growth-layers, which are added each year. (See **Basidiomycetes.**) (R.M.W.)

BRACT. In many flowering plants there is found at the base of the flower stalk a small leaf, often consider-ably modified: this is called a bract. In many plants its minute size causes it to be overlooked; in others it is a conspicuous object. In the Poinsettia, for example, the large showy red "flower" is really composed of bracts, as also is the conspicuous white petal-like structure surrounding the very small flowers of the Flowering Dogwood. (R.M.W.)

BRAGG'S LAW. The law expressing the condition under which a **crystal** will reflect a beam of **x-rays** with maximum distinctness, at the same time giving the angle at which the reflection takes place. For x-ray reflection it is customary to use the complement of the angle of incidence and reflection, that is, the angle which the incident or the reflected beam makes with the crystal planes, rather than with the normal. Let this "Bragg angle" be θ. If the planes or layers of atoms are spaced at a distance d apart, and if λ is the wavelength of the x-rays, Bragg's law is expressed by the equation

$$\sin\theta = \frac{n\lambda}{2d}.$$

The condition for an intensity maximum is that n must be a whole number. For example if the planes of rock salt parallel to the natural cubical faces are spaced at $d = 2.814 \times 10^{-8}$ cm. or 2814 x-units, and if the incident rays have a component of wavelength $\lambda = 714$ x-units, the above equation gives $\sin\theta = 0.1269n$. Then if the crystal is rotated slowly, there will be a distinct reflection when θ reaches $7° 17'$ $(n = 1)$, again at $14° 42'$ $(n = 2)$, also at $22° 23'$ $(n = 3)$, etc. (L.D.W.)

BRAIN. The principal central organ of the **nervous system,** consisting of a mass of nervous tissue lying in the head or in an equivalent position in animals which have no head.

A true brain appears only in bilaterally symmetrical animals. In its simplest form it is relatively small and is known also as the cerebral **ganglion.** Such a brain occurs in the **worms.** In cephalopod mollusks and in some of the **arthropods** the brain is larger and more complex but in no invertebrates does it attain the degree of development which is found in the vertebrates.

The **vertebrate** brain develops as a series of three expansions of the embryonic neural tube. Beginning at the anterior end these are known as the fore brain, mid brain, and hind brain, or as the prosencephalon, mesencephalon, and rhombencephalon. Later, the first and last subdivide, producing a total of five vesicles called the telencephalon, diencephalon, mesencephalon, metencephalon, and myelencephalon. These give rise to various adult structures as follows: The telencephalon produces two expansions, the cerebral hemispheres, which reach their highest development in man; they are the centers of reasoning and dominate all other regions. The telencephalon also gives rise to the **olfactory** centers and certain parts for which no simple description can be given. The diencephalon is important as the source of the **thalamus,** the retina of the **eyes,** the optic nerves, and the pars nervosa of the **pituitary gland.** The mesencephalon contains the centers of visual and auditory reflexes. The metencephalon gives rise dorsally to the cerebellum, which is a center of complex muscular coordination. The myelencephalon becomes the medulla oblongata, composed chiefly of great fiber tracts running between the higher centers and the spinal cord. In short, the human brain is made up of five portions: the cerebrum, cerebellum, midbrain, pons varolii and medulla oblongata.

The average weight of the human brain varies between 44 and 49 ounces. Growth of the brain ceases about the 19th to 20th year. It loses weight as age advances.

The surface of the brain is irregular, being divided into fissures and ridges (convolutions). The brain as well as the spinal cord is covered by three layers of

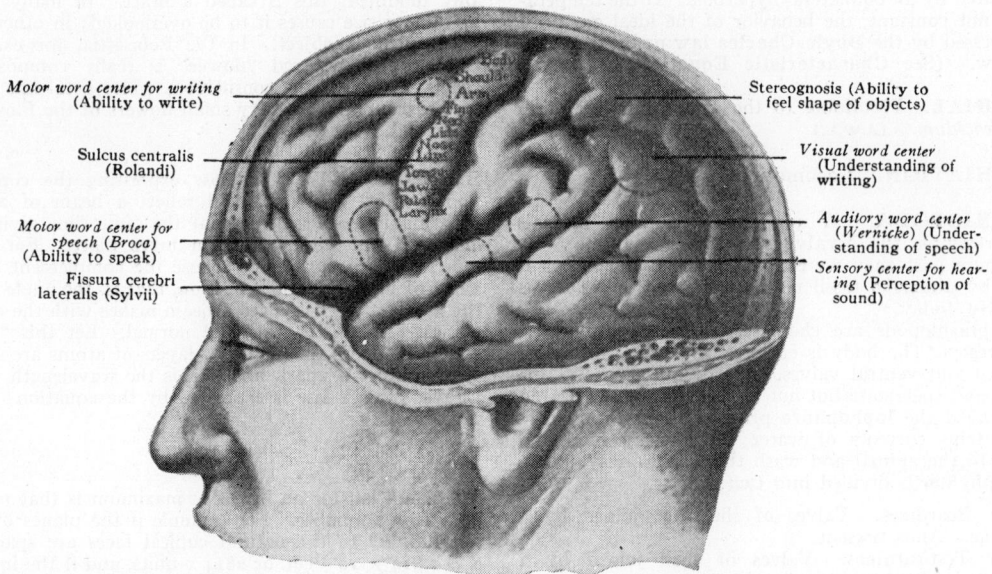

Main functional localizations in the cerebral cortex. (*From Callander's Surgical Anatomy, W. B. Saunders Co.*)

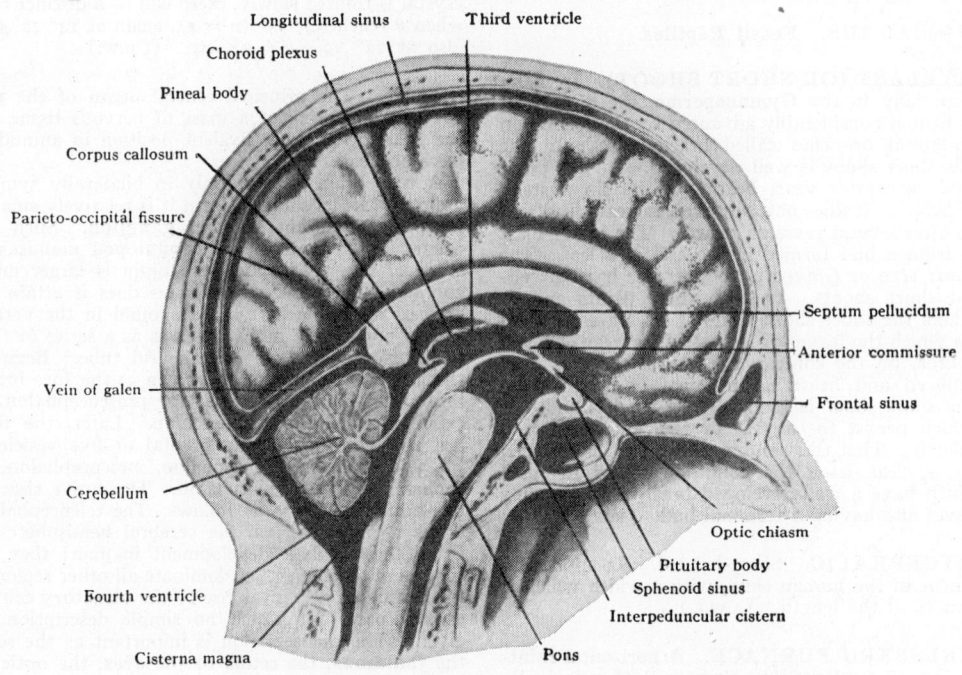

Ventricular and subarachnoid spaces viewed in median sagittal section of the head.
(*From Callander's Surgical Anatomy, W. B. Saunders Co.*)

membranes called from without inward, the dura mater, arachnoid, and pia mater. The cerebrospinal fluid surrounds the brain and spinal cord lying beneath the arachnoid layer. This fluid is produced in the cavities of the brain (**ventricles**) and serves as a protective and nutritive medium for the brain and spinal cord. (A.W.L., R.S.M.)

BRAKE. Bracken. *Pteris aquilina.* A common **fern** widely distributed in temperate regions. It has an extensive **rhizome** creeping deep in the ground and bearing solitary leaves at intervals along its length. These erect leaves, called fronds, have a petiole (see **Leaf**) from 1–4′ long and a much dissected blade. The

margin of this blade grows over and protects the spores born on the under side of the frond. Where it is at all abundant, as on abandoned rocky hillsides, the leaves may be gathered and used as bedding for livestock or even used as a coarse fodder in case of need. (R.M.W.)

BRAKE HORSEPOWER. The mechanical output of an engine, turbine, or motor is called the brake horsepower because one of the most common methods of testing for mechanical output is with the **Prony brake.** However, the output available at the shaft is called brake horsepower whether measured by brake or not. For example, when an **engine** drives an electric **gener-**

ator by direct connection, the horsepower available at the coupling between the machines is termed brake horsepower, but would not, under these circumstances, be measured by a brake. In the case of direct connected electric generating sets, the brake horsepower of the prime mover is found by dividing the electrical output from the generator, converted to horsepower, by the efficiency of the generator. The result is the generator input, which is the same as the mechanical output of the prime mover.

To measure brake horsepower by a Prony brake, readings of the weight registered on the scales and the speed of rotation of the brake drum are taken. In addition, the measured distance from the center of rotation to the force on the scales must be known. These quantities are substituted in the following formula to obtain brake horsepower:

$$\text{Brake horsepower} = \frac{2\pi W R N}{33,000}$$

W = net load on the scales, in pounds.
R = radial distance in feet to the line of action of the force registered on the scales.
N = rotative speed, in revolutions per minute.

The W of the above formula is less than the load actually recorded on the scales because the tare weight must be deducted from the scale reading to give net effective weight. The tare weight is an allowance for the fact that the arm of the Prony brake itself gives some reading on the scale due to its unbalanced position, relative to the center of rotation. The tare weight is that weight which, acting at the scales, will give the same moment about the center of rotation that the unbalanced dead weight of the brake would produce. (F.T.M.)

BRAKED LANDING. Landing, Types of.

BRAKES. A brake absorbs mechanical energy by transferring it into heat through frictional resistance. The most frequently seen brake at present is a mechanical type used to reduce the speed of a machine or to bring it to a state of rest. Complete control of automotive vehicles necessitates brakes adequate to bring the vehicle from maximum speed to rest in a reasonable distance. Stationary machines such as hoists or elevators have mechanical brakes, and numerous other applications will be found outside the automotive field.

The parts of a brake are the rotating part, which is generally, though not always, the drum, and a stationary part, the brake shoe. The shoe is pressed against the drum, generating a friction force at the contact surface. The brake shoe is frequently surfaced with a special material called brake lining, whose function is to pro-

vide for a high coefficient of friction, together with smooth braking action. The wear is largely taken by the lining. When not in use, the contact surfaces are separated by a light spring. Braking action may be applied by a mechanical linkage, such as lever or toggle joints, forcing the shoe against the drum, or by cables and pulleys, or by hydraulic means, incorporating a cylinder and plunger.

To bring a freely moving vehicle having velocity of V ft. per sec., and a weight of W lbs., to rest, requires the absorption of $\frac{1}{2}\frac{W}{g}V^2$ ft.-lbs. of work at the brake drum. This energy must be equaled by the product of the friction force acting at the braking surface multiplied by the speed of rubbing, multiplied by the time during which the brake is in action, i.e., the time taken to stop. According to the theory of friction, as long as the pressure between brake shoe and drum remains constant, the friction force generated is independent of the rubbing speed.

There are many possible arrangements of brake shoe and drum. Some of these are shown in the accompanying figure, which shows internally expanding and exter-

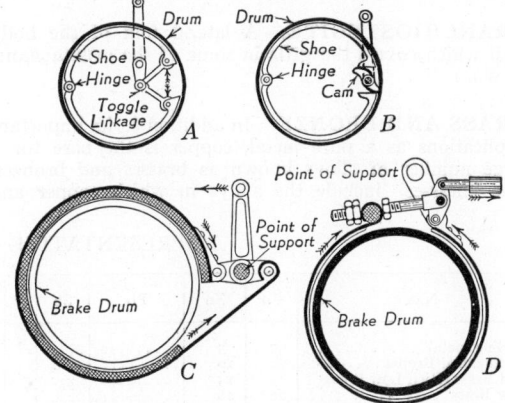

Types of brakes found mostly on machines.

nally contracting brakes, as well as methods of applying pressure between the friction surface. To lower a load at constant speed, the brake shoes can be mounted on the rotating shaft of the hoist drum, free to move outward under the centrifugal force set up by rotation. The brake drum is stationary. Any tendency of the load to increase the speed of rotation is counteracted by an increased centrifugal force pressing the brake shoes more firmly against the drum. (See also **Sheet Metal Processes, Press-working, Air Brakes.**) (F.T.M.)

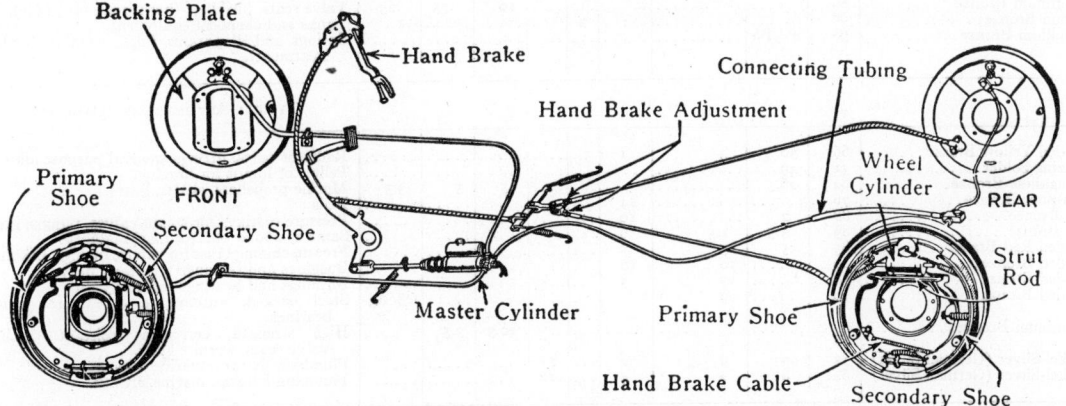

Passenger-car braking system comprising four-wheel hydraulic service brakes and rear-wheel mechanical parking brakes.

BRAMBLING. Aves, Passeriformes. A finch of the Old World which nests in northern Europe and Asia. *Fringilla montefringilla.* (A.W.L.)

BRANCHIAL. Pertaining to gills. (A.W.L.)

BRANCHIAL ARTERIES. Blood vessels which lead both to and from the gills in the fishes. Those which lead into the gills are the afferent branchials and those which lead out are the efferent branchials. (A.W.L.)

BRANCHIAL HEARTS. Muscular expansions of the branchial veins in the squids and related species which pump the blood through the gills. (A.W.L.)

BRANCHIAL VEINS. Branches of one of the large veins, the vena cava, of cephalopod mollusks which carry the blood to the gills on its way back to the heart from the anterior part of the body. (A.W.L.)

BRANCHIOPODA. A subclass of crustaceans. Not to be confused with Brachiopoda, the name of a phylum. (A.W.L.)

BRANCHIOSTEGITE. A lateral fold of the body wall which covers the gills in some of the crustaceans. (A.W.L.)

BRASS AND BRONZE. In addition to its important applications as a pure metal, copper is the base for a large number of alloys known as brasses and bronzes. The "brasses" include the alloys in which copper and zinc are the principal elements, although there are certain exceptions in which such alloys are called bronzes. The term "bronze" cannot be as definitely defined; it applies to the copper-tin alloys, and to many other copper-base alloys as indicated in the accompanying table.

The copper-base alloys have lower melting points than steels and are more readily produced from their constituent metals. They are comparable to steel in density, most copper alloys being only about 10% heavier, while those with considerable aluminum or silicon content may be slightly less dense than steel. Their strength properties, like those of steel, can be greatly varied by alloying and special processing, but in general they are at a lower strength level. The cost of copper-base alloys is generally higher than that of steels but lower than the nickel-base alloys.

Copper and its alloys have an infinite variety of uses reflecting their versatile physical, mechanical, and chemical properties. Some well-known examples are the high electrical conductivity of pure copper, the excellent deep-drawing qualities of cartridge case brass, the anti-friction properties of certain bearing bronzes, the resonant qualities of bell bronze, and the resistance to corrosion by sea water of several condenser-tube alloys. Representative compositions and typical applications of alloys generally produced in wrought forms such as sheets and plates, bars, shapes, wire, and tubing are listed.

The casting alloys, which are even more numerous than the wrought alloys, are used for bearings and other wearing surfaces, steam and water valves and fittings, electrical fittings, hardware, ornamental castings, and many applications requiring special corrosion-resisting qualities, pressure tightness, and good machinability.

REPRESENTATIVE BRASSES AND BRONZES

Name	Cu	Zn	Pb	Sn	Si	Ni	Al	Fe	Mn	Typical Applications—Wrought Forms
Gilding metal	95	5								Small arms ammunition.
Commercial Bronze	90	10								Jewelry, screw products, fly screens.
Red Brass (Rich Low Brass) *	85	15								Auto radiators, plumbing.
Low Brass	80	20								Wire cloth for paper making, flexible hose.
Malleable Brass (Spring Brass)	75	25								Springs for moderate loading.
Cartridge Brass	70	30	0.07 max							Cartridge cases, other drawn or spun products.
Common High Brass	66	34	0.30 max							Cheaper drawing grade than cartridge brass.
Muntz Metal *	60	40								Hardware, tubing.
Free Cutting Brass	61.5	35.5	3							Screw machine products, hardware.
Brass Forging Rod	60	38	2							Hardware, ammunition parts, plumbing.
Admiralty Metal	70	29		1						Marine condenser and heat exchanger tubes.
Naval Brass	60	39.25	0.20 max	0.75						Marine shafting, tube plates.
Manganese Bronze	58.5	39.0	0.20 max	1				1.4	0.1	Marine shafting, bolts, tie rods.
Aluminum Brass	76	22					2			High velocity condenser tubes.
Silicon Brass *	77	22			1					Refrigerator evaporators.
Copper-Nickel Alloy	70					30				Marine condenser tubes.
Nickel-Silver (German Silver)	72	10				18				Silver-plated flatware, jewelry.
Nickel-Silver (German Silver)	65	17				18				Silver-plated flatware, jewelry.
Nickel-Silver (German Silver)	66	24				10				Silver-plated flatware, jewelry.
Phosphor Bronze	95			5						Springs and clips, friction plates in clutches.
Phosphor Bronze *	90			10						Springs and clips, friction plates in clutches.
Aluminum Bronze	92						8			Bushings and bearings, machine parts.
Aluminum Bronze *	89						10	.75	.25	Valve seats, bushings (heat-treatable grade).
Silicon Bronze	96				3				1	Tanks and chemical equipment.
Beryllium Bronze	98	2% Be								Springs and diaphrams (high strength by heat treatment).

Name	Cu	Zn	Pb	Sn	Si	Ni	Al	Fe	Mn	Typical Applications—Castings
Leaded Yellow Brass	66	30	3	1						Radiator parts, fittings, general purpose alloy.
Brazing Solder	51	49								Pellets or lumps for brazing.
Manganese Bronze	62	26.5					5	3	3.5	Marine propellers, shafts, gears.
Phosphor Bronze	76			24						Bells
Tin Bronze	88	2		10						Pressure castings for steam valves, pumps, gears.
Tin Bronze	89	3.5	1.5	6						Same as above—better machinability.
Leaded Red Brass	85	5	5	5						Free machining brass for valves, fittings, bushings.
Leaded Bronze	80		10	10						Bushings and bearings for heavy duty.
Leaded Bronze	70		25	5						Bushings and bearings for heavy duty.
Leaded Bronze	70		29						1% Ag	Steel backed, automotive & airplane engine bearings.
Aluminum Bronze	86						10.5	3.5		High strength, corrosion resistant castings. Valve seats, worm wheels.
Nickel-Silver (German Silver)	57	20	9	2		12				Plumbing fixtures, marine fittings.
Nickel-Silver (German Silver)	64	8	4	4		20				Plumbing fixtures, marine fittings.

* Also used as casting alloys.

(R.H.H.)

BRASSICA. Cruciferae. A genus of plants, largely wild in Europe and adjacent Asia, which give us many of our garden vegetables. All have characteristic and often noticeable odors due to various sulfur compounds which they contain. Many of them become obnoxious weeds when introduced into new countries. Among the species cultivated are the following:

Brassica oleracea is a native of the coast of Europe from England to the Mediterranean. The wild plant is a stout branching biennial bearing large white or pale yellow flowers, and having no tendency to "head." From this have arisen, perhaps as mutants, cabbage and Brussels sprouts, in which the leaves form tight heads, kohlrabi with its swollen stem, the cauliflower and broccoli with their compact, swollen inflorescences. Another form is red cabbage, used mainly for pickling. Sauerkraut is finely cut cabbage fermented in a salt juice made from its own sap; the sour taste is caused by **lactic acid** resulting from fermentation.

Brassica rapa is the turnip, a yellow-flowered biennial, growing wild in Europe and western Asia. It is a plant adapted to cool climates and much used as human food. Ruta-bagas are a related species.

Brassica napus, or rape, is a biennial 2–3′ high. It is used as a green manure plant, being plowed under to improve the soil, and as a forage plant. From its seeds rape oil is obtained and rape cake.

Brassica nigra, or black mustard, an annual of Europe and Asia, is used as a garnish and as a salad plant. The ground seeds, mixed with water, give us table mustard. *Brassica alba* is very similar to black mustard.

Brassica chinensis and related species are Chinese Cabbages, annual plants used in salads. (R.M.W.)

BRAYTON CYCLE. Gas Turbine Plant.

BRAZIL-NUT. *Bertholetia excelsa.* Lecythidaceae. In the forests of northern Brazil grows the giant tree, the **seeds** of which are the familiar Brazil-nuts. These seeds are borne in a hard capsule 6–8″ in diameter and containing about two dozen seeds. The fruits develop high up in the air and fall to the ground without opening. This favors the gathering of the fruits, which are then split open and the seeds within removed. Immense quantities of the seeds are eaten yearly in Brazil and the United States. From the seeds an oil is obtained which is used in salad dressings, in paints and as a fine lubricant. (R.M.W.)

BRAZING. Brazing, according to the American Welding Society, is a group of welding processes wherein the filler metal is a non-ferrous metal or alloy whose melting point is higher than 1000° F., but lower than that of the metals or alloys to be joined.

At one extreme, brazing is very similar to **soldering** and is sometimes called hard soldering. A typical example of this is silver soldering or silver brazing using alloys containing 10–80% silver, balance principally copper and zinc. The melting points of these alloys are in the range 1175–1500° F., which is considerably higher than for soft solders. Silver solders are used to join both ferrous and the higher melting point non-ferrous alloys. A composition containing 15% silver, 5% phosphorus, 80% copper is often used to join copper and brass.

Joints to be silver-soldered are usually designed so as to require only a thin film of filler metal which is drawn into the joint by capillary action when the solder becomes molten. This principle is also used in copper brazing, a high-temperature process in which pure copper is used to join close-fitting steel parts. Whereas most other forms of brazing depend on local heating with a torch or by other means, copper brazing is generally done in a furnace in which the part is heated to about 2050° F. A reducing atmosphere is used to prevent oxidation, thus no flux is needed as in nearly all other types of brazing. The copper is applied in any convenient form such as wire or sheet and is so placed that it can flow into the joint when molten. A mechanically tight joint such as a forced fit between two cylindrical parts is the most effective in drawing the copper into the joint. The resulting thin film of bonding metal gives a high-strength connection.

Where somewhat higher temperatures can be tolerated than those used in silver soldering, brazing spelter can be used. This is an alloy containing about equal proportions of copper and zinc and melting at about 1600° F. Like silver solders it flows readily between contacting surfaces when properly fluxed to prevent oxidation. The parts to be joined may be heated to the melting point of the spelter as in soft soldering, or may be dipped in a bath of molten brazing alloy.

Another type of brazing is similar to welding in many respects and is sometimes called bronze welding. The joints are generally of the V or fillet types in which a bead of filler metal is deposited with a torch. The technique differs from welding in that the base metal is not melted but only raised to the "tinning" temperature at which bonding takes place between the base metal and the brass or bronze filler metal by slight interdiffusion or alloying. The filler metals most widely used are brasses of the 60% copper-40% zinc type with additions of tin, iron, manganese, or silicon. For some purposes special bronzes are used such as aluminum bronze, silicon bronze, and nickel silver. The melting points of all of these alloys are higher than those of the silver solders but lower than that of pure copper. They are applied with a torch as in gas welding. Bronze welding is particularly well adapted to the repair of worn or broken parts. (R.H.H.)

BREADFRUIT. *Artocarpus incisa.* Moraceae. The breadfruit tree is a native of the East Indian and Pacific Islands, but has been widely planted in tropical regions everywhere. It is an attractive shade tree with large leaves deeply cut into pinnate lobes. The large ovoid fruits have a rough surface. Each fruit is composed of many **achenes** each surrounded by a fleshy **perianth** and growing on a fleshy receptable. These fruits are very rich in starch. Before being eaten they are roasted or boiled. Some of the improved varieties bear seedless fruits. (R.M.W.)

BREAKDOWN VOLTAGE. This is the voltage necessary to cause the passage of appreciable electric current without a connecting conductor. It is commonly used to express the voltage at which an insulator or insulating material fails to withstand the voltage and ceases to behave as an insulator. (L.R.Q.)

BREAM. Pisces, Teleostei. Several species of freshwater fishes (**Pisces**) of the Northern Hemisphere, allied to the carps. Two of the European species are known as the zope and the zarthe and one of the North American species is called the golden shiner. (A.W.L.)

BREAST. 1. The front of the **chest.** 2. The **mammary gland.** (A.W.L.)

BRECCIA. Breccia, derived from the Latin meaning broken, is a rock formed of angular fragments in a **matrix** which may be of similar or of different material.

Fault breccias result from the grinding action of the two **fault** blocks as they slide past each other. Subsequent **cementation** of these broken fragments may occur by means of mineral matter introduced by the ground water. **Talus** slopes may become buried and the talus cemented in a similar manner.

Volcanic breccias result from the cementation of fragments that have been broken by volcanic action. Sometimes the surface of a lava flow will harden while the interior will be yet liquid; the fracturing of this surface material and its subsequent cementation by the uncooled lava produces a flow breccia.

The intrusion of **plutonic rocks** will often shatter the invaded country rock, forming a shatter breccia. In the case of plutonic rocks partly cooled and subsequently broken by further invasions of the magma, we have intrusive breccias. (E.S.C.S.)

BREED. A type of animal produced within the species by artificial selection, distinguished by definite hereditary characteristics but usually capable of interbreeding freely with other members of the species and so maintained only through artificial control of its propagation. Thus Guernsey and Angus cattle are breeds which would soon cease to exist in a state of nature, while the Michigan beaver and the Pacific beaver are self-maintaining in nature and are called subspecies. (A.W.L.)

BREEZE. Any light or moderate wind. Special names like sea breeze, lake breeze, land breeze, valley breeze, mountain breeze are used to describe particular breezes, usually of convective origin. (P.E.K.)

BREWSTER'S LAW. In 1815 Sir David Brewster discovered that for any dielectric reflector there is a simple relationship between the polarizing angle (see **Polarized Light**) for the reflected light of a particular wavelength and the **refractive index** of the substance for the same wavelength. The relationship is that the tangent of the polarizing angle is equal to the refractive index. For example, if the refractive index of flint glass for sodium light is 1.66 the polarizing angle for the reflection of sodium light by this glass is $50° 56'$.

An interesting consequence of this relationship may be deduced. If the angle of incidence is the polarizing angle p, and the corresponding angle of refraction at a boundary with refractive index, n, is ρ, then, by Brewster's Law we have

$$\tan p = \frac{\sin p}{\cos p} = n = \frac{\sin p}{\sin \rho}.$$

Hence $\cos p = \sin \rho$ or $p + \rho = 90°$. It follows that when the light is incident at the polarizing angle, the reflected and refracted rays are perpendicular to each other. The law might be used as a means for measuring the refractive index of highly absorbing **dielectrics** such as polished obsidian, pitch, etc. (L.D.W.)

BRIAR. Heath Family.

BRICK. Brick ordinarily refers to a rectangular prism of clay or shale which has been burned in a kiln. Clay is no longer the only material for brick manufacture, being supplemented by slag, cement, lime, etc. However, when other than the ordinary structural clay brick is meant, a descriptive term such as fire-brick, sand-lime brick, etc., is employed. The principal classifications of brick are for structural purposes in buildings, for paving, and for lining furnaces, the latter known as refractory brick, or fire-brick. Ordinary bricks are made from a selected clay soil first by preparation of the clay by grinding and thoroughly mixing with enough water to make the mud. The bricks are then formed in the required shapes by one of several methods. In the soft mud method the prepared clay is quite plastic and the bricks are molded to shape by hand or machine. This is the principal method for making bricks by hand. The commercial manufacture of bricks is more frequently by the stiff mud process, in which the mud is less plastic, and is extruded through a die by pressure and wire cut to the proper size. Brick made from clay that is hardly more than dampened must be formed in molds by application of a great deal of pressure. As hydraulic presses are frequently used, these dry-pressed bricks are sometimes referred to as hydraulic-pressed brick. This proc-

ess gives to the bricks a dense surface which makes them suitable for facing work.

After the bricks have been molded they are air dried and piled in the kiln for burning. It is quite difficult in any other than the continuously fired kilns, in which the bricks move slowly through the kilns on conveyors, to obtain uniform characteristics in all the brick, and so the product of the ordinary brick plant consists of various grades, ranging from hard brick to softer bricks, such as salmon brick. Hard-burned brick should be used for face work exposed to the weather, and soft brick for filling, for foundations, and the like. The standard brick measures approximately $2\frac{1}{4}'' \times 4''$ $\times 8''$, and has a crushing strength of between 1000 and 3000 lbs. per sq. in., depending on the quality. A highly impervious and ornamental surface may be laid on brick either by salt glazing, in which salt is added during the burning process, or by the use of a "slip," which is a glaze material into which the bricks are dipped. Subsequent reheating in the kiln fuses the slip into a glazed surface integral with the brick base.

A **refractory** brick is built primarily to withstand temperature. Good resistance to heat flow is not to be secured simultaneously with refractoriness. Indeed, the most refractory bricks usually have the highest thermal conductivities. It is important for the refractory brick to have high resistance to erosion by ash-laden gases and to the fluxing action of molten slag. It should not spall badly under rapid temperature changes, and its structural strength should hold up well under rapid temperature changes. Fire clay bricks are made from certain clays, including a plastic clay which binds the others into brick form. The firing in the kiln is carried out at a temperature such that the brick is partly vitrified. For special purposes they may be glazed by one of the methods previously described. The fire clay brick contains 30–40% alumina (see **Aluminum**), and about 50% silica (see **Silicon**). Progress in the art of combustion of fuels in furnaces has advanced the service requirements of refractory brick, sometimes to the point where they are so severe that a refractory superior to fire clay is needed. High alumina bricks containing 50–80% alumina, and correspondingly less silica, and silicon carbide, a product of the electric furnace, are typical of these super-refractories. Of course fire clay bricks are preferred wherever they give satisfactory service because they are lowest in cost of all the refractory bricks. The standard size of fire brick is $9'' \times 4\frac{1}{2}'' \times 2\frac{1}{2}''$. (See **Ceramics**.) (F.T.M.)

BRICK ARCH. Arch.

BRICKWORK. When laid, **bricks** are bedded into a mortar which, hardening, bonds the separate bricks into a brickwork unit. A solid brick wall of more than one layer thickness has the different layers of brick bonded into each other by the use of headers, that is, brick laid perpendicular to the face of the wall. There are different systems of bonding, each of which gives a somewhat different appearance to the wall. In the common **bond** every fourth or fifth course is composed entirely of headers. In the English bond, every other course is a header course, while in the Flemish bond headers and stretchers alternate in each course. (See **Masonry**.)

The strength and durability of brickwork depend on the quality of mortar and excellence of workmanship with which the brickwork is laid. The proportions of the mortar are from one to three parts of dry sand to one part of Portland **cement**, depending on the strength needed. The cement mortar is much stronger than lime mortar, but the addition of a small amount of lime (see **Calcium**) to cement mortar renders it more readily worked without materially impairing its strength. In estimating brickwork, one rule is to allow 1000 standard brick and $\frac{1}{2}$ cu. yd. of mortar for each 2 cu. yds. of brickwork in place. Some masons estimate number of bricks by assigning 7 to each superficial sq. ft. of area of wall 1 brick thick. Brickwork varies in weight from 1.5 to 1.9 tons per cu. yd., depending on the density of the bricks used. The maximum crushing strength to which brickwork should be subjected is 170 lbs. per sq. in. when set in cement mortar, although this may be increased to 250 lbs. if the effects of eccentric loading and lateral forces are fully analyzed. (F.T.M.)

BRIDGE, STRUCTURAL.

In civil engineering, a bridge is a structural unit or a series of structural units called spans designed primarily for the purpose of supporting moving loads, in addition to its own weight. The term bridge is generally associated with a structure which provides a means for foot, highway, or railroad traffic to pass over water, ground depressions or congested districts, although certain kinds of traveling cranes used for loading or unloading bulky materials such as ore or coal are sometimes referred to as bridges. All bridges are either stationary or movable. The so-called stationary bridges may be subdivided into spans of the **arch** type and those in which one end is allowed to move longitudinally so as to eliminate the horizontal force at the supports and also the stresses that would otherwise be induced by temperature changes. Movable spans are used in connection with low level bridges over navigable waters where these bridges interfere with shipping.

There are three general types of movable spans: namely, the bascule, the vertical lift, and swing bridge. The bascule bridge pivots about a horizontal axis or rolls back on circular segments. If the entire span rotates about a horizontal axis near one end, it is called a single leaf bascule. A double leaf bascule is one which is made up of two cantilevers, each of which rotates about a horizontal axis, forming a single span when closed. When the entire movable section may be lifted vertically, parallel to its original position, the bridge is called a vertical lift span. Swing bridges are those which turn in a horizontal plane about a vertical axis located at the center of the bridge. Movable bridges, when closed, are similar and perform the same service as the stationary types. Bridges may also be classified as framed **truss**, beam, or suspension bridges, depending upon the way in which they support the loads. All bridges are either straight or skew. When the end supports are not on lines at right angles to the longitudinal center line of the span, the resulting structure is called a skew bridge.

Bridges are usually constructed of **steel** or reinforced **concrete** although wood is sometimes used for temporary spans. Stone or **brick** are occasionally used for very short spans of the arch type. Reinforced concrete is particularly well adapted for use in connection with the beam or arched bridge since it can be molded into any desired form. The span length is limited, as in any bridge, by the strength of the component parts.

The ordinary framed bridge as illustrated in **Fig. 1** is composed of two vertical trusses, a floor system, upon which the roadway or railway is directly supported, a

certain amount of bracing and the end **bearings**. The floor system consists of longitudinal beams called stringers which transfer the effects of the moving loads to transverse beams known as floor beams. The floor beams are connected to the trusses at the lower intersection points of the truss members. Each intersection point is called a joint or panel point. The truss is composed of an upper and lower chord and web members which are joined together in the form of triangles. Fig. 2 shows some typical trusses. Since the loads are

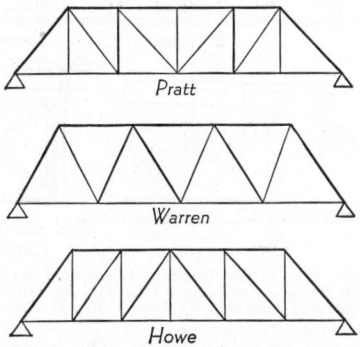

Fig. 2. Typical Framed Bridge Trusses.

applied to the truss at the panel points, the primary **stresses** will be axial. In this particular illustration the top chord will be in **compression** and the lower chord in **tension**. Some of the web members will be in tension, others in compression, but there are certain web members near the center of the span which may have either tension or compression depending upon the position of the moving loads. The bracing is usually made up of an upper and lower **lateral** system, sway frames, and portal bracing. These bracing systems resist the horizontal loads caused by wind, and, together with the floor system, tie the truss together forming a relatively rigid unit. Short-span framed bridges which do not require trusses sufficiently deep to allow for top chord, sway or portal bracing, because of interference with vehicular traffic are called pony truss spans. The trusses are the principal load-carrying components of the bridge, since they must support their own weight and the weight of the floor system in addition to the moving load and wind loads. The total load on each truss is transferred through a horizontal **pin** to the end bearings or shoes, generally castings, which distribute the load to the supporting masonry. At one end of the bridge the bearings will be firmly fastened to the masonry, but at the other end they will be of the expansion type, which allows a limited amount of longitudinal movement to take care of the temperature changes in the structure. The separate structural members of a truss bridge are composed of rolled steel shapes or built-up sections formed by riveting two or more rolled shapes together. The truss members are connected at their intersections by **gusset plates** or pins.

Beam bridges are composed of two or more beams laid parallel to the direction of traffic. The roadway or track may be supported directly on the beams or by a stringer and transverse floor beam system connected to beams called **girders**. The construction of a simple deck **plate girder** bridge, frequently used for short railway spans, is shown in Fig. 3. Each girder is composed of a steel plate called the web to which are riveted four angles. Additional plates called covers are riveted to the angles. The two top or bottom angles and the attached cover plate form the flanges. The girders are stiffened laterally by the lateral system and the cross frames. A through bridge is a span of the beam or framed truss type in which the floor system is placed between the girders or trusses usually near the plane of the bottom flange or chords. In deck spans the floor

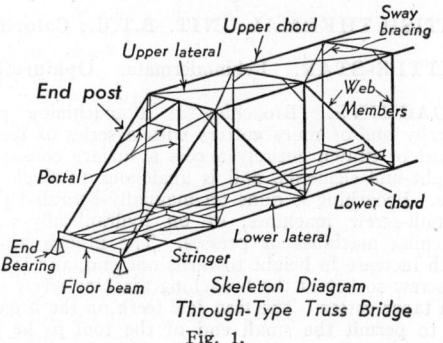

Skeleton Diagram
Through-Type Truss Bridge
Fig. 1.

system is placed between the girders or trusses near the plane of the top flanges or chords, or rests upon them.

The simplest type of suspension bridge, applicable for short spans, consists of a floor system connected by

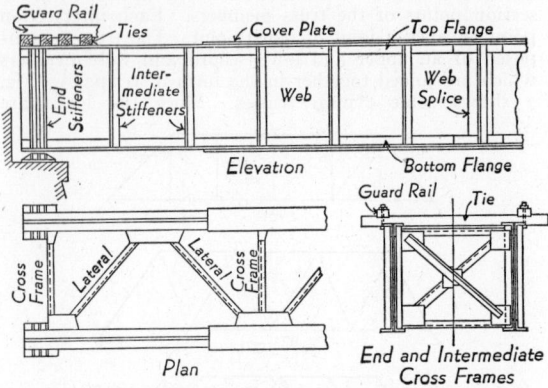

Fig. 3. Deck type girder bridge. (*Reprinted by permission from Modern Framed Structures, Part III, by Johnson, Bryan and Turneaure, published by John Wiley & Sons, Inc.*)

hangers to two cables or chains. The latter pass over towers and are firmly anchored at the end of the span. For longer spans it becomes necessary to connect the floor system to stiffening trusses which distribute the moving loads more uniformly to the hangers. This method of distribution reduces the distortion of the cable or chain. The hangers are formed of twisted wire ropes while the cable may consist of twisted wire ropes or a number of parallel wires securely bound together into a compact unit of circular cross-section. The chain is made up of a number of separate tension links called **eye-bars**. The floor system, stiffening truss and towers are constructed of rolled steel shapes.

In its simplest form the cantilever bridge consists of a suspended span and two anchor spans. Each anchor span which rests on two **piers** is made up of an anchor arm and a cantilever arm. The latter projects beyond the river pier to form a support for one end of the suspended span. Cantilever bridges may be either **trusses** or **plate girders**. The former are particularly well adapted to long-span construction.

A continuous bridge is one which rests on three or more supports and is capable of transmitting both **shear** and **moment** throughout its length. These bridges which are statically indeterminate may be plate girders or trusses. The continuous bridge is more rigid than the cantilever bridge but settlement of the supports has a marked effect on the stress distribution.

A bridge of two or more spans which is supported at each intermediate pier by a hinged quadrilateral is called a Wichert truss. This type of bridge is determinate (see **Determinate Structure**) and therefore the stresses are not affected by settlement of the supports.

A pontoon bridge is a floating roadway which is used to bridge narrow bodies of water. It consists of barges called pontoons which carry a roadway made up of beams which, in turn, support a plank floor. The pontoons must be firmly anchored so that they will not float out of position. The pontoon bridge is generally used for military purposes although there are instances in which this type of bridge has been constructed for ordinary vehicular and pedestrian traffic.

The type of bridge to be used at a particular location depends upon many considerations. Typical arched bridges are shown in Fig. 4. Satisfactory **foundations**, possible **pier** locations, and access to the bridge are some of the local conditions which will influence the selection of a particular type. From an architectural standpoint the type which is used should harmonize with the natural surroundings. When the cost of a bridge

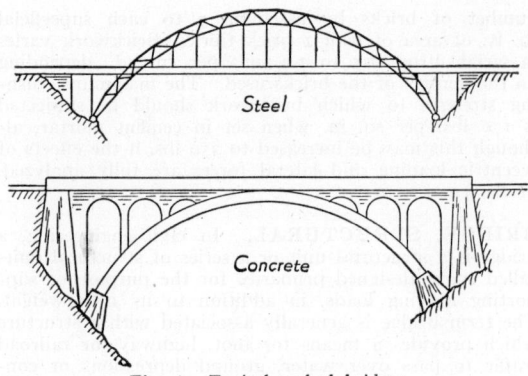

Fig. 4. Typical arched bridges.

project is limited to a predetermined amount certain types of bridges are automatically eliminated. (L.D.W., F.T.M., C.W.C.)

BRIDGING TRANSFORMER. Telephony.

BRIGHTNESS. The brightness of a surface giving out light, as viewed from a given direction, is the quantity of light flux emitted in that direction per unit area of cross-section perpendicular to that direction. This term may be applied either to a self-luminous body, as a white-hot metal plate, or to an illuminated diffuse reflector. If the emission is 1 lumen per sq. cm. of normal cross-section, the brightness is said to be one lambert; while 1 lumen per sq. ft., likewise, is called a foot lambert. (See **Photometry.**)

For a good diffuser (like a hot, sooty plate), the brightness is the same in all directions. Thus a white-hot cylindrical rod photographs as a uniformly white streak, as if it were a flat strip. The well-known **cosine emission law** may be deduced from this fact. (L.D.W.)

BRIGHT'S DISEASE. Nephritis.

BRILL. Pisces, Teleostei. A flat-fish, *Psetta laevis*, found in British waters and valued as food. The more familiar members of this group are the **flounders**, **turbot**, and **soles.** (A.W.L.)

BRINE-FLY. Insecta, Diptera. **Flies** whose larvae live in strong briny or alkaline waters. They belong to the family Ephydridae which also contains species whose larvae live in fresh water and one remarkable insect which lives in pools of crude petroleum in the California oil fields.

Large quantities of the **larvae** of certain species are washed ashore along some of the western alkaline lakes and are gathered by the Indians as food under the native name koo-tsabe. (A.W.L.)

BRINELL HARDNESS. Hardness.

BRITISH THERMAL UNIT. B.T.U.; Calorimetry.

BRITTLE-STAR. Echinodermata. Ophiuroidea.

BROACHING. Broaching is a machining process whereby one or more cutters with a series of teeth are pushed or drawn entirely across a surface consisting of straight-line elements, and is analogous to single-stroke filing. Broaching is done on manually-operated presses, on pull-screw machines, or on hydraulically-actuated broaching machines or presses. The broach has teeth which increase in height towards one end, and is held in the screw socket of the broaching machine screw or ram by a taper cotter. The first few teeth on the broach are low to permit the small end of the tool to be passed through the cylindrical hole in the work; the intermediate teeth remove most of the metal; and the last few teeth finish the surface to size. Each tooth removes an equal amount of metal. The broach is drawn or pulled through the work by a non-rotating screw which receives its axial motion from a rotating nut driven by a pulley.

Push broaching is usually performed on vertical machines with broaches that are comparatively short to insure stiffness. They are ordinarily employed where only a small amount of metal is to be removed. A set of several short push broaches may be required to remove the same amount of metal that one long pull broach will handle. Short broaches are used whenever possible, however, as they are easy to make, harden, and handle.

Broaches are extensively employed for machining square and hexagonal holes and splined shaft fittings. The figure illustrates a square hole broach which transforms the round hole shown at A to the square hole

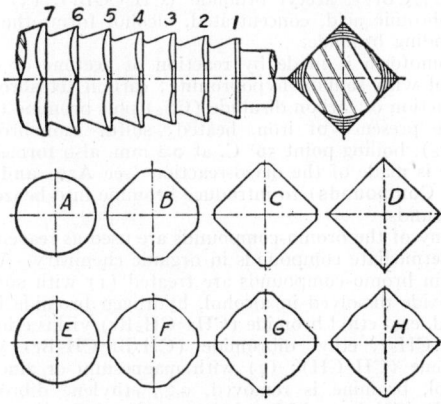

Square hole broach.

shown at D. Tooth number 1 fits the round hole A; B shows the cutting effect of tooth 3, and C the cutting effect of tooth 5, while D shows the finishing effect of tooth 7. (In actual practice, of course, more than seven teeth are required for an operation of this character; the intermediate teeth between 1 and 2, 2 and 3, etc., have been omitted for sake of clarity.) An alternate design and process are shown at E, F, G, and H. The hole at E is drilled slightly larger than the side of the square and the final teeth of the broach do not remove all the material in the corners. The design at H is practically as effective for power transmission as the design shown at D, and may be more easily broached. (H.C.H.)

BROADBILL. Aves, Passeriformes. Oriental birds (**Aves**) with a shallow but very broad beak. (A.W.L.)

BROADTAIL. Aves, Psittaciformes. A group of **parrots** or parraquets found only in Australia, Norfolk Island, and Tasmania. (A.W.L.)

BROCCOLI. Brassica.

BROCHANTITE. A mineral composed of basic **copper** sulfate corresponding to the formula $CuSO_4 \cdot 3Cu(OH)_2$, crystallizing in the **orthorhombic** system in needle-like prisms, or forming **druses** or masses. Hardness, 3.5–4; specific gravity, 3.9; vitreous luster; color, green; streak, green; transparent to translucent.

Brochantite is a secondary mineral occurring in the oxidized zones with other copper minerals, and is found in the Urals, in Rumania, in Sardinia; Cornwall, England; Chile. In the United States this mineral has been found at Bisbee, Arizona, Utah, in the Tintic District and in Inyo County, California. Brochantite was named for Brochant de Villiers. (E.S.C.S.)

BROCKET. Mammalia, Artiodactyla. *Cariacus.* Small South American **deer** of several species. Unlike most deer they are solitary in habits. (A.W.L.)

BROMIC ACID AND BROMATES. Bromic acid ($HBrO_3$) is a colorless liquid, miscible with water. A solution of bromic acid is decomposed upon boiling, but can be concentrated in vacuum up to 51% $HBrO_3$, with considerable decomposition. A strong oxidizing agent, e.g., **sulfides**, sulfur, sulfites to sulfuric acid, thiosulfates to tetrathionates (but depending upon the conditions), oxalates to carbon dioxide.

Prepared by reaction (1) of barium bromate solution and sulfuric acid, and filtering off barium sulfate.

Potassium bromate is formed (1) by **electrolysis** of **potassium** bromide solution, with stirring, heated, and over some hours, (2) by reaction of bromine and potassium chlorate solution, (3) by heating potassium **hypobromite** solution.

Metallic bromates are solids, soluble in water, except that **silver, lead, barium, thallous** bromates are slightly soluble. Basic bromates require the presence of a little free acid for solution. Bromates when heated, evolve oxygen and leave the bromide (or oxide) as residue.

Upon addition of hydrobromic acid (or bromide plus dilute sulfuric acid) bromate liberates bromic acid, which reacts with hydrobromic acid to form bromine. (R.K.S.)

BROMIDE. Bromine.

BROMINE. Symbol: Br. Atomic number: 35. Atomic weight: 79.916. Density: 3.12. Melting point: $-7.2°$ C. Boiling point: 58.8° C. (Isotopes: page 290.)

Bromine is a brown liquid, volatilizing at room temperature to a red vapor, which has a very irritating effect on eyes and throat. Liquid bromine causes painful sores upon contact with the flesh. Somewhat soluble in water, soluble in carbon tetrachloride to reddish-brown solution. Discovered by Balard in 1825. Used chiefly in photography, and in the preparation of organic chemicals, among which is ethylene dibromide, which is used in conjunction with lead tetraethyl in gasoline. Many bromides are used in medicine, especially those of potassium, sodium, ammonium, lithium, strontium and calcium. They are used chiefly as sedatives to lessen nervous excitability and to inhibit vomiting, to prevent or lessen the extent, discomfort or frequency of convulsive or spasmodic states, especially **epilepsy.** The drug is given orally or rectally. When given over too long a period, especially in sensitive subjects, a stubborn skin eruption may develop. Other toxic reactions include nervous system depression, coma, and sometimes psychotic manifestations. The severity of the bromide poisoning may be measured by determining the blood bromide level.

Bromine occurs as bromide in sea water (0.188% Br), in the mother liquor from salt wells of Michigan, Ohio, West Virginia, and in the potassium deposits of Germany and France. Separated from solutions by treatment with **chlorine**, and purified by distillation.

Bromine is soluble in water, 3.6 grams per 100 grams of water at 20° C., in ether, ethyl alcohol, carbon disulfide, chloroform, carbon tetrachloride, and potassium bromide solution. Bromine reacts (1) with **sodium** thiosulfate solution to form sulfate, (2) with **sulfurous acid** to form sulfuric acid, (3) with **hydrosulfuric acid** to form sulfur, (4) with unsaturated **hydrocarbons** to form organic bromo-compounds, e.g., ethylene bromide, acetylene tetrabromide, (5) with benzenoid **hydrocarbons** to form organic bromo-compounds, e.g., bromobenzene, dibromobenzene.

Acids: Hydrobromic acid (HBr); hypobromous acid (HBrO); bromic acid ($HBrO_3$). (See each acid.)

Bromate: (See **Bromic Acid.**)

Bromides: **Sodium** bromide (NaBr), **potassium** bromide (KBr), **ammonium** bromide (NH_4Br) are soluble bromides; **silver** bromide (AgBr), **mercurous** bromide (HgBr), **mercuric** bromide (HgBr₂), **lead** bromide (PbBr₂) are insoluble bromides. (See **Hydrobromic Acid.**) Organic bromo-compounds, see Bromine, Organic Compounds, below.

Chloride: Bromine chloride (BrCl).

Fluoride: Brominetrifluoride (BrF₃).

Iodide: Bromine iodide (BrI).

Hydride: Hydrogen bromide (HBr), colorless gas when pure (frequently contains free bromine as reddish gas), melting point —86° C., boiling point —67° C., very soluble in water yielding hydrobromic acid. Formed by reaction of (1) red **phosphorus**, bromine and water under the proper conditions, (2) bromine and **hydrogen** gases.

Hypobromite: (See **Hypobromous Acid.**)

Sulfide: (See **Sulfur**, Bromide.)

Organic bromo-compounds:

Paraffin and benzenoid **hydrocarbons**, e.g., ethane and benzene, respectively, react with bromine by *substitution* of bromine for hydrogen (hydrogen bromide also formed), e.g., ethane to yield ethyl bromide (C_2H_5Br) plus further substitution products; benzene, in the presence of a **catalyzer**, e.g., iodine, phosphorus, iron, to yield bromobenzene (C_6H_5Br) plus further substitution products; toluene, under like conditions to benzene, to yield ortho-bromotoluene and para-bromotoluene ($CH_3C_6H_4Br$) plus further substitution products, but at the boiling temperature, in sunlight, dry, and in the absence of a catalyzer, to yield paraffin-side-chain substitution products, benzyl bromide ($C_6H_5CH_2Br$), benzal bromide ($C_6H_5CHBr_2$), and benzotribromide ($C_6H_5CBr_3$).

Olefin, acetylene, and benzenoid **hydrocarbons**, e.g., ethylene, acetylene and benzene, respectively, react (1) with bromine by *addition*, e.g., ethylene dibromide ($C_2H_4Br_2(1,2)$), acetylene tetrabromide ($C_2H_2Br_4,1,1,2,2,$), benzene hexabromide ($C_6H_6Br_6$), also carbon monoxide yields carbonyl bromide (COBr₂), (2) with hypobromous acid by *addition*, e.g., olefins form, for example, ethylene bromohydrin ($CH_2Br·CH_2OH$), (3)

with hydrogen bromide by *addition*, to form, for example, ethyl bromide ($CH_3·CH_2Cl$) from ethylene. When the two olefin carbons have an unequal number of hydrogens attached, bromine of hydrogen bromide and hydroxyl of hypobromous acid attach to the carbon having the smaller number of hydrogens.

Oxygen-function compounds, e.g., **ethyl alcohol, acetaldehyde, acetone, acetic acid,** react (1) with bromine, to form *bromo-substituted* corresponding or related *compounds*, e.g., ethyl alcohol or acetaldehyde to yield bromal ($CBr_3·CHO$), acetone to yield bromoacetone ($CH_2Br·CO·CH_3$) acetic acid to yield, at the boiling temperature, dry, and in the absence of a catalyzer, monobromoacetic acid ($CH_2Br·COOH$), dibromoacetic acid ($CHBr_2·COOH$), tribromoacetic acid ($CBr_3·COOH$), the substitution taking place on the alpha-carbon (the carbon next to the carboxyl-group (—COOH)), (2) with **phosphorus** bromides, to form corresponding *bromides,* e.g., ethyl bromide (C_2H_5Br), ethylidene dibromide (CH_3CHBr_2), acetone bromide (($CH_3)_2CBr_2$), acetyl bromide (CH_3COBr), (3) with hydrobromic acid, concentrated, alcohol forms the corresponding bromide.

Bromoform is made by reaction of acetone or ethyl alcohol with sodium hypobromite; carbon tetrabromide by reaction of carbon disulfide (CS_2) plus bromine (Br_2) in the presence of iron, heated; sulfur monobromide (S_2Br_2), boiling point 56° C. at 0.2 mm. also formed.

Use is made of the diazo-reaction (see **Azo- and Related Compounds**) to introduce bromine into benzenoid compounds.

Many of the bromo-compounds are used as reagents or as intermediate compounds in organic chemistry. When paraffin bromo-compounds are treated (1) with **sodium** hydroxide dissolved in alcohol, hydrogen bromide is removed, e.g., ethyl bromide ($CH_3·CH_2Br$) yields ethylene ($CH_2:CH_2$), ethyl dibromide ($CH_2Br·CH_2Br$) yields acetylene ($CH:CH$); (2) with magnesium or **zinc** and alcohol, bromine is removed, e.g., ethylene dibromide ($CH_2Br·CH_2Br$) yields ethylene ($CH_2:CH_2$), acetylene tetrabromide ($CHBr_2·CHBr_2$) yields acetylene ($CH:CH$).

SELECTED REPRESENTATIVE ORGANIC COMPOUNDS OF BROMINE

Name	Formula	Melting Point °C.	Boiling Point °C.
1. Methyl bromide	CH_3Br		5
2. Ethyl bromide	C_2H_5Br		38
3. Normal-propyl bromide (1-bromopropane)	$C_2H_5·CH_2Br$		71
4. Iso-propyl bromide (2-bromopropane)	$CH_3·CHBr·CH_3$		60
5. Vinyl bromide (bromoethylene)	$CH_2:CHBr$		16
6. Allyl bromide	$CH_2:CH·CH_2Br$		16
7. Bromoacetylene	$CH:CBr$		—2
8. Alpha-bromostyrene	$C_6H_5CBr:CH_2$		160 (75 mm.)
9. Beta-bromostyrene	$C_6H_5CH:CHBr$		219
10. Benzyl bromide	$C_6H_5CH_2Br$		198
11. Glycol bromohydrin (ethylene bromohydrin)	$CH_2Br·CH_2OH$		150 (750 mm.)
12. Alpha-glycerol bromohydrin	$CH_2Br·CHOH·CH_2OH$		
13. (Mono)bromoacetic acid	$CH_2Br·COOH$	50	208
14. Alpha-bromopropionic acid	$CH_3·CHBr·COOH$	26	205
15. Beta-bromopropionic acid	$CH_2Br·CH_2·COOH$	63	
16. Bromosuccinic acid	$COOH·CH_2·CHBr·COOH$	160	
17. Acetyl bromide	$CH_3·COBr$		76 (750 mm.)
18. Benzoyl bromide	$C_6H_5·COBr$		219
19. Bromoacetone	$CH_2Br·CO·CH_3$		127
20. Alpha-bromocamphor	$C_{10}H_{15}O·Br(3)$	77	274
21. Cyanogen bromide	$CN·Br$	52	61 (750 mm.)
22. Bromofurane	$C_4H_3O·Br(2)$		101
23. Bromofuroic acid	$(3)Br·C_4H_2O·COOH$	128	
24. Bromobenzene (phenyl bromide)	C_6H_5Br		156
25. Ortho-bromotoluene (1,2-)	$C_6H_4(Br)(2)(CH_3)(1)$		182
26. Meta-bromotoluene (1,3-)	$C_6H_4(Br)(3)(CH_3)(1)$		184

SELECTED REPRESENTATIVE ORGANIC COMPOUNDS OF BROMINE—*Continued*

NAME	FORMULA	MELTING POINT °C.	BOILING POINT °C.
27. Para-bromotoluene (1,4-)	$C_6H_4(Br)(4)(CH_3)(1)$	29	184
28. Ortho-bromobiphenyl	$Br(2)C_6H_4 \cdot C_6H_5$		297
29. Para-bromobiphenyl	$Br(4)C_6H_4 \cdot C_6H_5$	90	310
30. Alpha-bromonaphthalene	$C_{10}H_7Br(1)$	5	281
31. Beta-bromonaphthalene	$C_{10}H_7Br(2)$	59	281
32. Ortho-bromophenol (1,2-)	$C_6H_4(Br)(2)(OH)(1)$	6	194
33. Meta-bromophenol (1,3-)	$C_6H_4(Br)(3)(OH)(1)$	32	236
34. Para-bromophenol (1,4-)	$C_6H_4(Br)(4)(OH)(1)$	63	238
35. Ortho-bromoaniline (1,2-)	$C_6H_4(Br)(2)(NH_2)(1)$	31	229
36. Meta-bromoaniline (1,3-)	$C_6H_4(Br)(3)(NH_2)(1)$	18	251
37. Para-bromoaniline (1,4-)	$C_6H_4(Br)(4)(NH_2)(1)$	63	
38. Ortho-bromonitrobenzene (1,2-)	$C_6H_4(Br)(2)(NO_2)(1)$	43	261
39. Meta-bromonitrobenzene (1,3-)	$C_6H_4(Br)(3)(NO_2)(1)$	56	256
40. Para-bromonitrobenzene (1,4-)	$C_6H_4(Br)(4)(NO_2)(1)$	126	255
41. Ortho-bromobenzoic acid (1,2-)	$C_6H_4(Br)(2)(COOH)(1)$	149	Subl.
42. Meta-bromobenzoic acid (1,3-)	$C_6H_4(Br)(3)(COOH)(1)$	154	....
43. Para-bromobenzoic acid (1,4-)	$C_6H_4(Br)(4)(COOH)(1)$	252	....
44. Alpha-bromoanthraquinone	$(1)Br \cdot C_6H_3(CO)_2C_6H_4$	188	Subl.
45. Beta-bromoanthraquinone	$(2)Br \cdot C_6H_3(CO)_2C_6H_4$	204	....
46. Methylene bromide	CH_2Br_2		97
47. Ethylene dibromide (1,2-dibromoethane)	$CH_2Br \cdot CH_2Br$		131
48. Acetylene dibromide, cis	$CHBr:CHBr$		110 (754 mm.)
49. Acetylene dibromide, trans	$CHBr:CHBr$		108
50. Ethylidene bromide (1,1-dibromoethane)	CH_3CH_2Br		110
51. Benzal bromide (benzylidene bromide)	$C_6H_5 \cdot CHBr_2$		140 (20 mm.)
52. Dibromoacetic acid	$CHBr_2 \cdot COOH$	49	233 (decom.)
53. Dibromosuccinic acid	$COOH \cdot CH_2 \cdot CBr_2 \cdot COOH$	166	180 (decom.)
54. Carbonyl bromide	$COBr_2$	....	64
55. Ortho-dibromobenzene (1,2-)	$C_6H_4Br_2(1,2)$	2	221
56. Meta-dibromobenzene (1,3-)	$C_6H_4Br_2(1,3)$	−7	219 (755 mm.)
57. Para-dibromobenzene (1,4-)	$C_6H_4Br_2(1,4)$	87	218
58. Dibromoanthracene (9,10-)	$C_6H_4(CBr)_2C_6H_4$	221	Subl.
59. Bromoform	$CHBr_3$		151
60. Methyl bromoform (1,1,1-tribromoethane)	$CH_3 \cdot CBr_3$		....
61. Benzotribromide (phenyl bromoform)	$C_6H_5 \cdot CBr_3$		....
62. Tribromoethylene	$CHBr:CBr_2$		163
63. 1,2,3-Tribromopropane	$CH_2Br \cdot CHBr \cdot CH_2Br$	16	220
64. Tribromoacetic acid	$CBr_3 \cdot COOH$	135	245 (decom.)
65. Tribromophenol, sym	$C_6H_2(OH)(Br)_3(2,4,6)$	96	....
66. Bromal	$CBr_3 \cdot CHO$		174
67. Bromal hydrate	$CBr_3 \cdot CH(OH)_2$	53	....
68. Carbon tetrabromide	CBr_4	92	189
69. Bromotrichloromethane (trichlorobromomethane)	$BrCCl_3$		104
70. Acetylene tetrabromide (1,1,2,2-tetrabromoethane)	$CHBr_2 \cdot CHBr_2$		151 (54 mm.)
71. Tetrabromoethylene	$CBr_2:CBr_2$	56	226
72. Naphthalene tetrabromide	$C_{10}H_8Br_4$		
73. Tetrabromofluorescein (Eosin)	$C_{20}H_8O_5Br_4$		
74. Hexabromoethane	$CBr_3 \cdot CBr_3$	200 (decom.)	....
75. Benzene hexabromide, alpha, trans	$C_6H_6Br_6$	212	
76. Hexabromobenzene	C_6Br_6	316	....

(R.S.M., D.M.H., R.K.S.)

BROMLITE. Alstonite. The mineral Bromlite or Alstonite is a rare **barium-calcium carbonate** that may contain small amounts of **strontium** carbonate as well. It occurs at Bromly Hill, near Alston, Cumberland, England (whence the two names) in association with **calcite** and **witherite**. (E.S.C.S.)

BROMOIL PROCESS. A process of photographic printing based upon the selective adherence of an oil pigment to an image in tanned gelatin. It was once widely used by pictorialists because of the control over tone values which it affords but is now nearly obsolete. A print on bromide paper is bleached in a solution, the chief ingredients of which are usually potassium ferricyanide—or copper sulfate—and potassium bichromate. In this solution the silver image is bleached and the gelatin in contact with the silver image tanned, the degree of tanning being more or less proportional to the amount of silver present. After fixing and washing to remove the silver image, the water adhering to the surface of the print is blotted off and a thick ink of the type used by lithographers is applied locally with a brush. The ink is accepted by the tanned gelatin but repelled by the water contained in the unhardened gelatin. Thus an image in pigment is built up on the differentially hardened gelatin. By repeated application of the ink to certain portions and by changing its consistency the bromoil worker is able to change the tone values of the original photographic image to meet his own artistic ends.

A development of the bromoil process known as bromoil transfer consists in transferring the pigment from the finished bromoil to a plain drawing paper by passing the two together in contact through a transfer press. Bromoil transfer results in an image of pigment on plain paper thus eliminating all traces of the photographic base. (C.B.N.)

BRONCHIECTASIS. A chronic progressive disease of the bronchi or bronchioles due to weakening or destruction of the muscle layer of their walls. The condition may be caused by repeated bronchial infections, or by the plugging of a bronchus by mucus or a foreign body, with stasis and chronic infection. **Whooping cough** may be followed by bronchiectasis. Destruction of the muscle results in the formation of cylindrical saccular or fusiform dilatations of the bronchi, with stagnation of secretions in them and consequent infection. The disease is characterized by chronic cough productive of large amounts of foul-smelling sputum. Except for **tuberculosis,** it is the most common cause of **hemoptysis.** The course is a chronic one with acute flare-ups of fever and constitutional symptoms. Surgical removal of the affected area of the lung offers the only cure. **Sulfonamide** therapy and **penicillin,** with postural drainage, are effective in controlling the element of infection. (D.M.H.)

BRONCHITIS. An inflammation of the bronchial tubes usually associated with a similar process in the trachea (tracheobronchitis) (see **Respiratory System**). The disease may be primary but most often is associated with an upper respiratory infection, a "cold." It may also be part of the disease picture of **tuberculosis, whooping cough, measles, influenza, typhoid fever,** and **asthma.** Pneumonia is the chief and most dangerous complication; bronchopneumonia is often ushered in as an acute bronchitis.

The etiology of bronchitis in the course of an upper respiratory infection is obscure. Although many pathogenic organisms may be found in the bronchial secretions there is no evidence that they play a significant role in the genesis of the condition. The possibility that the disease is caused by a filterable **virus**—similar to the agent which causes the common cold—has been considered, but not proven.

In acute bronchitis the most prominent symptom is cough, dry at first, but later productive of sputum. Low-grade fever may be present for several days. Soreness of the chest and a sensation of tightness beneath the sternum are common. The general lassitude and discomfort subside with the fever, but the cough may persist for several weeks. Treatment of the acute disease consists of steam inhalations and drugs to control involuntary coughing and to liquefy the sputum.

Chronic bronchitis usually effects old people, men more than women. It follows repeated attacks of acute bronchitis, and often is associated with **bronchiectasis,** pulmonary **emphysema,** and **asthma.** The only symptom is persistent cough, usually worse in the winter months. The general health is not impaired. The treatment is directed toward the underlying cause if the bronchitis is secondary; drugs which keep the sputum liquid and help control the cough may be useful. (D.M.H.)

BRONCO or **BRONCHO.** Mammalia, Perissodactyla. **Horses** of western North America, descended from stock originally brought in by the Spanish conquerors. They are noted for endurance and spirit. There is no certain distinction between the terms broncho and mustang, and both are embraced by the name Indian pony. (A.W.L.)

BRONZE. Brass and Bronze.

BRONZE AGE. Paleontology of Man.

BRONZITE. Enstatite.

BROOD SAC. A pouch in which the eggs are retained during development in some of the **crustaceans.** (A.W.L.)

BROOKITE. Brookite, composed of **titanium** dioxide, TiO$_2$, is an **orthorhombic** mineral of the same chemical composition as **rutile** and **octahedrite.** It was named for the English mineralogist H. J. Brooke. (E.S.C.S.)

BROOM CORN. *Andropogon Sorghum* var. Gramineae. A variety of a tropical grass, **sorghum,** characterized by having an inflorescence in which the branches are very long and slender, growing in loose **panicles.** The plant is extremely drought resistant and so adapted for regions having an arid climate. The close-bunched stiff inflorescent-branches account for the principal use of broom corn, i.e., the manufacture of brooms and whisk-brooms. Large quantities are grown in Oklahoma. (R.M.W.)

BROWN AND SHARPE. Gauge Number; Taper.

BROWN BODY. In the **bryozoa** the entire active portion of the individual, known as the polypide, degenerates in certain conditions to form a compact mass known as the brown body. The enclosing sheath or zooecium forms a new polypide whose stomach sometimes encloses the brown body representing the previous individual. In such cases the mass is later discharged by way of the anus. (A.W.L.)

BROWN PRINT. Drawing Reproduction.

BROWN ROT. Ascomycetes and Bacteria.

BROWNIAN MOVEMENT. The random movement observed among microscopic particles suspended in a fluid medium. The phenomenon was observed in 1827 with suspensions in liquids, **colloids,** by Robert Brown, English botanist, who is said to have attributed it to living organisms. Not until the kinetic theory was developed was it generally understood to be due to the thermal agitation of the suspending medium. A smoke particle floating in the air, for example, is battered on all sides by the high-speed air molecules. The resultant displacement is for the most part nearly zero, but there are statistical inequalities which now and then reach such magnitude as to produce motions visible in a high-powered microscope, and which result in an irregular migration of the particle. In fact, such particles may be regarded essentially as huge molecules, with mean square speeds of thermal motion proportionately smaller as their masses are larger than that of the true molecules of the surrounding medium. A mathematical analysis of the problem by Einstein in 1905 led to an equation connecting the observed motions with the **Boltzmann constant,** the development being based upon the law of **equipartition of energy;** and the agreement between the predictions of this theory and experimental results is very satisfactory. (L.D.W.)

BROWN-TAIL MOTH. Insecta, Lepidoptera. A European species, *Euproctis chrysorrhea,* related to the tussock moths of this continent. It was introduced into Massachusetts during the last century and together with the **gypsy moth,** another introduced species, has become an important pest in the New England states and the adjacent Canadian provinces, attacking shade and fruit trees. Spraying with **lead** arsenate in the late summer has been found an effective method of control, as well as the collection and destruction of the winter nests in which the caterpillars hibernate.

The hairs of the **larva** are very irritating to the human skin. (A.W.L.)

BRUCELLOSIS. Undulant Fever.

BRUCINE. Alkaloids.

BRUCITE. The mineral brucite is **magnesium** hydroxide corresponding to the formula $Mg(OH)_2$; **iron** and **manganese** may occasionally be present. The crystals are usually tabular **rhombohedrons** of the **hexagonal** system; it may also occur fibrous or **foliated.** Brucite has one perfect cleavage parallel to the prism base; hardness, 2.5; specific gravity, 2.38–2.4; luster, pearly to vitreous; commonly white but may be gray, bluish or greenish; transparent to translucent. Brucite is a secondary mineral found with **serpentine** and **metamorphic** dolomites. It has been found in Italy, Sweden, and the Shetland Islands; and in the United States in New York, Pennsylvania, Nevada and California. Brucite was named in honor of Archibald Bruce, an American physician. (E.S.C.S.)

BRUNTON COMPASS. Specially designed for geologists and fitted with a clinometer and other devices for reading both horizontal and vertical angles. (R.M.F.)

BRUSH. A brush is a device for conducting current to or from a rotating part. The brush is stationary, and is held and guided by a fixed brush holder in which it slides freely. There may be several brushes side by side to form a single-brush set. The rotating member may be the **commutator** of a d-c **generator** or **motor,** or it may be the slip **rings** of an a-c motor or generator. Examples of brushes might also include those used in **magnetos** and static electricity machines.

Brush material is usually carbon, though spring-leaf copper and copper gauze are used in special cases. Circuit connections to carbon brushes are made directly to the brush by means of short flexible cables from the external circuit because the contact surface between brush holder and brush is an unreliable conductor.

Brushes wear and must be replaced periodically. Also, their ends should fit commutators so as to make good contact over the entire brush surface. They may sometimes need periodic redressing with sandpaper in order to maintain the proper arc for contact. The most serious fault of a brush is the formation of electric **arcs** between the rotating member and the brush. This may be due to condition of the brush, vibration of the brush in the holder, or improper setting of the brush. The position of the brushes on the commutator is adjusted so the coils or turns being shorted by the brushes will have a minimum voltage. This means that the coil sides will be in the position of minimum flux, which position will depend upon the load unless correction is applied. In modern machines interpole windings are used to compensate for the distortion of the field caused by the armature current so the brush position is opposite the center of the poles. In machines without some form of compensation the position will be to one side of this and is adjusted to give minimum sparking under normal load. (F.T.M., L.R.Q.)

BRUSH DISCHARGE. Ionized Gases.

BRUSH TURKEY. Aves, Galliformes. Large dull-colored birds (**Aves**) of several species found in New Guinea and Australia. They deposit their eggs in large heaps of decaying vegetation. (A.W.L.)

BRUSSELS SPROUTS. Brassica.

BRYOPHYLLUM CALYCINUM AND OTHER SPECIES. Crassulaceae. A genus of tropical plants which are frequently seen in cultivation because of their habit of reproducing by means of their leaves. If a mature leaf is removed from one of the plants and placed on damp sand, within a short time there appear in the notches of the leaf margin tiny roots and later small green plants which soon become independent of the parent leaf. Infrequently the little plants appear in the notches while the leaf is still attached to the parent plant.

The explanation of this uncommon habit is that in the development of the leaf certain cells in the leaf notches remain permanently **embryonic.** For some reason their

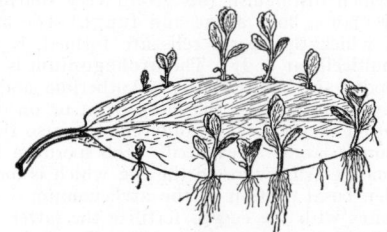

Vegetative reproduction of *Bryophyllum calycinum.* New plants develop in the notches of the leaves.

development is inhibited so long as the leaf remains attached. Severing the leaf removes the inhibition and the embryonic cells resume active development. (R.M.W.)

BRYOPHYTES. The second subdivision in the Plant Kingdom, comprising Mosses and Liverworts (or Hepatics), is a small group of plants, having some 20,000 species. Most of these are terrestrial plants. The bryophytes inhabit a wide range of habitats, from dry barren rocks to submerged objects, but are most frequent where an abundance of moisture is assured. They are found on trunks and branches of trees, on the soil, and even on the leaves of some tropical plants.

The body of these plants is small and without much structural complexity. Rhizoids, slender outgrowths which serve mainly to attach the plant to its substratum and which serve only slightly as absorbing organs, are common. They may be single celled or multicellular, but are colorless. The habit of the plant body is diverse. In many hepatics it is a thin flat

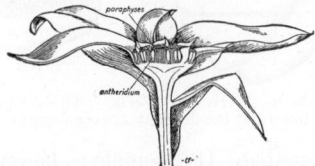

The tip of a plant of *Mnium* which bears antherida, cut lengthwise to show the antheridia. Diagrammatic.

thallus, one to several cells thick. Often the edge of the thallus is so lobed that it appears to be differentiated into a central stem and lateral leaves. In the mosses this differentiation is much greater. There is an erect central stem which bears many thin radiating leaves. However complex the structure may be, the cells of the plant-body of a bryophyte show very little differentiation. The central cells may be longer; other cells may have thicker walls; cells nearer the surface contain more **chloroplasts,** but there is never any real modification of cells to form **vascular elements.** The latter are entirely wanting in this group, which are therefore sometimes called Atracheata, that is, plants without any vascular cells, to distinguish them from the ferns and

The tip of an archegonial plant of *Mnium,* cut lengthwise to show the archegonia. Diagrammatic.

seed plants, the Tracheata, characterized by a highly developed tissue.

The sex organs of the bryophytes are highly developed
objects, which distinguish this group very sharply from
the lower plants, both algae and fungi. The **antheridium**, in which the sperm cells are formed, is a club-
shaped multicellular body. The **archegonium** is a flask-
shaped body, also multicellular. Antheridia and archegonia may be formed on the same plant or on different
plants; often they appear at different times so that self-
fertilization is largely prevented. The sperm, a biciliate
actively motile cell, swims to the egg which is located in
the swollen basal portion of the archegonium. There a
sperm unites with the egg to fertilize the latter and in-
cite growth of a new generation. But the plant resulting
from this fertilized egg is one entirely unlike the parent
plant. Commonly it is a well-developed plant, but less
conspicuous than the one on which it is usually entirely
dependent for its food supply. When mature it forms
a large mass of small spherical cells called spores, which
are freed from the body in which they are formed and
carried away by currents of air. On reaching a suitable
environment each spore germinates and eventually form
a plant like that which bore the sexual organs. There is
then in the bryophytes a very definite **alternation of
generations**, a haploid sexual generation called the
gametophyte, bearing male and female sex organs and a
diploid asexual generation called the sporophyte, in
which are formed haploid asexual spores. The gametophyte is green and carries on **photosynthesis.** In most

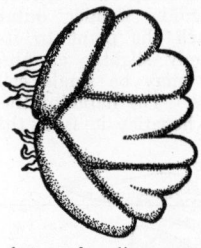 

Gametophytes of a liverwort, *Riccia*. The bodies embedded
in the right-hand plant are sporophytes.

cases it is terrestrial. The sporophyte, however, depends
on the gametophyte for its food supply and water.

Bryophytes are separated into two classes, the Hepaticae or liverworts and the Musci or mosses. Each
class is subdivided into three orders, as follows:

Class I. Hepaticae.
   Order 1. Marchantiales.
       " 2. Jungermanniales.
       " 3. Anthocerotales.
Class II. Musci.
   Order 1. Sphagnales.
       " 2. Andreaeales.
       " 3. Bryales.

The liverworts are generally considered lower in the
scale of evolution than the mosses. The thallus, or
plant body, of the liverworts, is prostrate and flat.
When it forks, the two branches are equal, a method
of branching known as dichotomous. In the second
order of liverworts the thallus is so divided as to appear
leafy, the order often being called the leafy hepatics.
All liverworts have unicellular **rhizoids** borne on the
lower side of the thallus by which they are anchored
firmly to the substratum. The sex organs are borne
embedded in the body of the thallus or in special outgrowths called gametophores, which rise from the
thallus. The small sporophytes are dependent on the
gametophyte for their nutrients. In these plants an
asexual reproduction occurs by means of small masses
of cells which develop from the gametophyte to which
they are attached by a very slender stalk. These small
bodies, called gemmae, are easily separated from the

parent plant. They develop into a new gametophyte
when they are carried by wind or water to a suitable
environment.

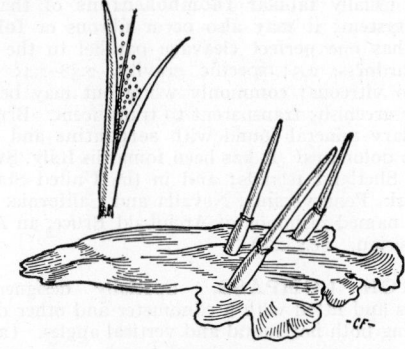

Anthoceros, one of the horned liverworts. The "horns" are
sporophytes. A portion of a sporophyte is shown separately
to illustrate the method of liberation of spores.

The three orders of the liverworts form an interesting
series, of increasing complexity. The Marchantiales include forms which have a prostrate thallus, often showing a structure of considerable complexity. The sporophyte is very simple. Growth of the thallus is by
repeated divisions of a single apical cell which itself
sometimes forms two such cells, whose continued divisions form a dichotomous branching of the thallus.
As the thallus increases in length at the apical end,
death of the cells occurs at the other end, so that the
plant slowly grows ahead until in time a fork is reached
and the two halves separate by progressive disintegration of the older portions.

One of the simplest members of this group is *Riccia*,
a small plant found either floating on still waters or
growing on wet mud. Some species are thick and
fleshy, others are slender much-branded bodies, having
a very evident median groove.

From the lower surface single-celled rhizoids grow
downward. From this surface also, thin scales or plates
are developed, forming an overlapping row along the
middle of the thallus. Both antheridia and archegonia
are formed on the upper surface along the midrib.
The antheridia have a wall a single cell in thickness,
and contain many sperm mother cells, each of which
divides to form two biciliate sperms. The archegonial
wall is also a single cell in thickness, and encloses a

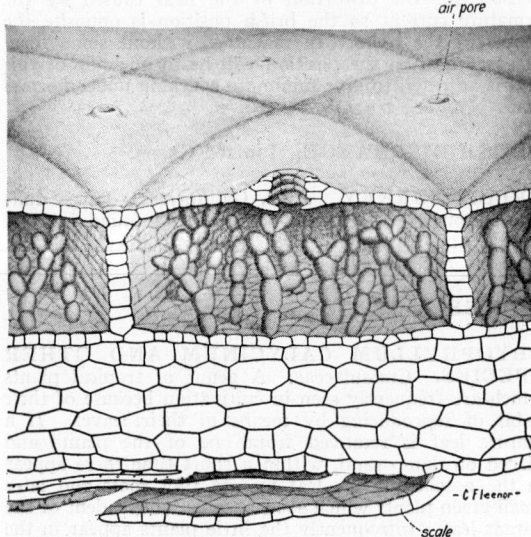

Marchantia. Section through a thallus showing air chambers.

row of six cells, four of which are the canal cells, the other two a ventral cell and an egg cell. When the latter is mature, the other five disintegrate, while at the same time the apical cells of the archegonium split apart, forming a canal through which the sperm swims to fuse with the egg. The sporophyte which develops from the fertilized egg remains embedded in the gametophyte thallus; when mature it is nearly all sporogenous tissue enclosed in a thin-walled capsule. A more complex member of this order is *Marchantia.* In this the gametophyte thallus is several inches long and from a half an inch to an inch broad. Its lower surface bears rhizoids and scales, or lamellae, quite like those of *Riccia.* The upper part of the thallus, just beneath the upper surface, contains a number of large chambers, each connected with the outside air by a large pore. Gemmae are produced in cup-like organs growing out of the upper surface of the thallus. The sexual organs in this plant are not formed in the thallus, but are borne on special erect branches called gametophores. Antheridia and archegonia are borne on different plants, the plants thus being dioecious. They are quite like the antheridia and archegonia of *Riccia.* The sporophyte is considerably larger than that of *Riccia,* with a well-developed foot attaching it to the gametophyte, a short thick stalk which pushes the spore-containing capsule out from the tissues of the gametophore. Not only does this capsule contain large numbers of spores, but also, scattered among the spores, slender elongate cells called elaters, whose walls have spiral thickenings. These are affected by differences in humidity which cause the elater to twist about, apparently to stir up and loosen the spores. In *Marchantia* the gametophyte is very elaborately developed; the sporophyte, very simple.

The second order of liverworts is divided into two suborders, the Anacrogynae, in which the archegonia are borne on the upper surface of the thallus, and the Acrogynae, in which they are borne at the apex. In the species of the Acrogynae the thallus is so incised

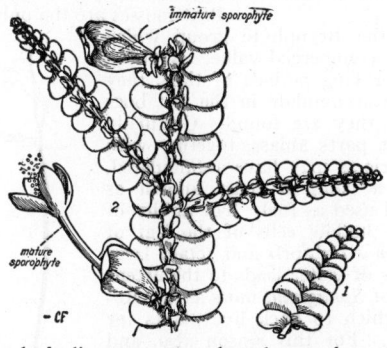

Porella, a leafy liverwort. 1, a branch seen from the upper side. 2, a portion of a plant seen from the lower side; sporophytes also are visible attached to archegonial branches, each partially enclosed in a perianth.

on its margins that it appears to bear two rows of small leaves. The Jungermanniales are small plants, many of them very delicate, growing in wet places, either on the ground or on rocks and tree trunks. Cellular differentiation in the thallus is very slight. In them the sporophyte is much more highly developed than in the first order. It has a long slender erect stalk which bears the capsule. When the latter is mature, it splits into four valves, which spread apart and free the spores within. Members of this suborder also form a series more or less as do the Marchantiales. The third order, the Anthocerotales, is a small group containing three genera. In all of them the gametophyte thallus is of a very simple type, with no great cellular differentiation, and with the sex organs always embedded. In this order the sporophyte is a most interesting object, far advanced in comparison with those of the other

liverworts. It is an erect, slender more or less cylindrical object composed of a basal foot and a long capsule. The central portion of the capsule is a rod of sterile tissue called the columella. Around it is the sporogenous tissue. The spores are formed in zones alternating with narrow bands of sterile tissue. Outside the sporogenous tissue is a wall of sterile tissue composed of chlorophyll (see **Pigments in Plants**) containing cells. The epidermal portion of this wall contains many stomata (see **Stoma**). The basal portion of the sporophyte, just above the foot, is composed of meristematic cells, which by their divisions cause the capsule to elongate. When mature this capsule splits into two valves which pull apart, resembling horns. These plants, with their very simple gametophyte and very elaborate sporophyte, contrast strikingly with the Marchantiales, with their elaborate gametophyte and simple sporophyte.

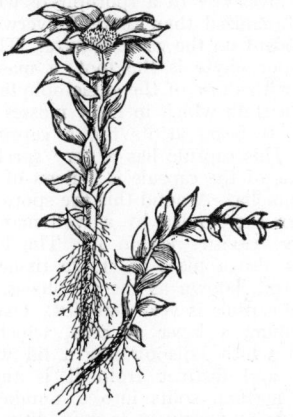

Mnium, a common genus of mosses.

The liverworts are of no economic importance, but are of interest because they suggest what may have been the habit of those plants which first left the water and grew on land. They have never become independent of water, since it must be present if fertilization is to occur.

The number of mosses known is much larger than the number of liverworts. Mosses are found in many different regions, being much more abundant than liverworts, and able to grow under a wide range of environments. They are found in greatest abundance in moist shaded regions where they often cover an extensive area. Other species grow on the trunks of trees, often well above the ground, where they are exposed for long periods of time to desiccation. In tropical forests mosses often clothe not only the trunks but also the branches of trees with a thick green covering; they may even succeed in growing on the thick evergreen leaves which characterize many tropical trees. Some species of mosses grow well on the exposed surfaces of barren rocks. A few are aquatic, living entirely submerged in running water throughout their existence.

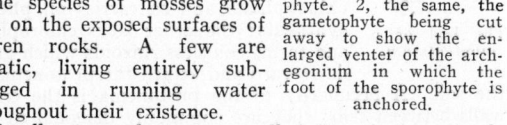

The sporophyte of a moss *Mnium.* 1, the sporophyte attached to the gametophyte. 2, the same, the gametophyte being cut away to show the enlarged venter of the archegonium in which the foot of the sporophyte is anchored.

Usually moss plants are small (often tiny), and seldom exceed a few inches in length. A few genera, such as *Fontinalis,* which grows in water, and several tropical members, grow to lengths of 10-15", which is very unusual in this group. Moss plants show a much higher development than hepatics. Usually the gametophyte, the part ordinarily seen and called a moss, has a very distinct, often erect stem, which bears many small radiating leaves. In each leaf there is generally a fairly evident midrib. There are many **rhizoids** growing from the lower part of the stem and attaching the plant to the ground. The rhizoids of mosses are longer than those of liverworts and are multicellular. There is no true vascular system in any moss, though the cells of the central portion of the stem are often much longer and more slender than those surrounding them. The sexual organs of mosses are very similar to those of hepatics and are borne at the tips of the stem or branches. Biciliate sperms are formed which must have water in which they can swim to the egg. The fertilized egg gives rise to a sporophyte which is much more highly organized than that of liverworts but still entirely dependent on the gametophyte. The basal portion of the sporophyte is the foot, a mass of cells in close contact with those of the gametophyte. Above the foot there is a stalk which in most mosses is very long and slender. It bears at its top a capsule or spore-bearing sac. This capsule has a very specialized structure. The axis of the capsule is a mass of sterile tissue called the columella. Around this the sporogenous tissue occurs, in turn surrounded by a wall many cells thick and with large cavities within it. The basal part of the capsule is also a mass of sterile tissue, often considerably swollen, known as the apophysis. The apical portion of the capsule is very complex. Over its surface is the operculum, a layer of cells, which completely covers it and which falls off like a lid when mature. Beneath this and distinct from it is the peristome, which, when mature, splits into a number of slender teeth which react to changes in humidity, rolling back when dry and closing together when wet. Surrounding the developing capsule and remaining around it for some time is a loose jacket of cells in no way connected with it. This is the calyptra, formed from cells of the gametophyte which were originally cells of the archegonial

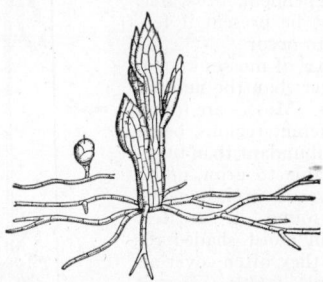

Moss protonema, showing the production of the buds which become the upright, leafy shoots.

wall. The ripe spores of mosses are shaken out through the apical opening and scattered by currents of air. On germinating, these spores do not give rise to a new moss plant directly. Instead they form a slender branching filamentous structure called a protonema which very much resembles certain kinds of **algae.** There are two types of cells composing the protonema: one contains many **chloroplasts** and so carries on **photosynthesis:** the other lacks chloroplasts and forms colorless rhizoids which grow downward and attach the protonema to the soil. A peculiarity of the protonema is the cross-walls between cells: they are commonly **diagonal** to the long axis of the filament rather than at right angles. From the cells of the protonema short erect branches

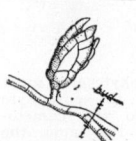

The bud of a moss *Mnium.*

ending in small buds are formed. These buds develop into erect moss plants. The life-history of a moss plant shows a very distinct **alternation of generations.** In addition there is the juvenile phase of the gametophyte, the protonema.

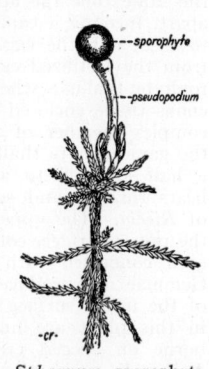

Sphagnum sporophyte attached to a leafy plant.

The first order of mosses, the peat- or bog-mosses, contains the single genus *Sphagnum* with many species of world-wide distribution, always growing in low, wet bogs. These mosses are considered very primitive. The gametophyte has an erect stem from which arise numerous branches, all of two kinds, either spreading, or pendent against the stem. The many leaves are small and but a single cell in thickness. Some of the cells are small and elongated, forming a fine anastamosing network in the leaf. These are living cells containing chloroplasts. The openings of the network are filled by very large in-

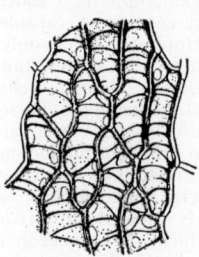

A portion of a leaf of *Sphagnum,* the peat moss. Surface view.

flated cells with thin walls and no protoplasm. Large pores in the walls of these dead cells permit free passage of water into the cell cavity. The small sporophyte has a spherical capsule, which is black or dark brown, and a very short stalk. The sporophyte is borne at the end of a specialized structure called a pseudopodium which lifts the sporophyte above the tuft of branches at the top of the gametophyte. The protonema of *Sphagnum* is a small flat thallus resembling that of the Anthocerotales.

Peat-mosses are the only members of the Bryophyte group which have any commercial value. Growing slowly for long periods of time, they gradually accumulate in the wet bogs in which they are found. Gradually the lower parts amass, together with such debris as may have accumulated, forming a compact mass known as peat, and used as fuel. The ability of the large hollow cells of the leaf of *Sphagnum* to absorb and retain large quantities of water leads to the extensive use of *Sphagnum* moss as a material in which to pack live plants for shipment. For this reason also, and because *Sphagnum* is naturally a sterile substance, harboring few bacteria, certain species have been used as surgical dressings, especially in times of great need.

The Andreaeales is a small group of small mosses growing on the surfaces of siliceous rocks. They are unimportant.

Most mosses belong to the third order, the Bryales, which are the true mosses.

As a group the Bryophytes are of little importance. They are recognized as primitive plants which developed from some simple ancestral forms from which they have gradually diverged independently along several different lines. The existing forms do not form a single series, representing stages in the development of the most advanced forms, nor are they plants from which the higher plants have taken their origin. In this group the gametophyte appears in its most advanced form. (See also **Paleobotany.**)   (R.M.W.)

Individual plant of *Bryum.* Leafy axis bearing young sporophyte, with calyptra on top.

BRYOZOA. One of the smaller divisions of the animal kingdom, made up of **sessile** animals which are mostly colonial, forming branching or encrusting colonies or large jelly-like masses. The appearance of certain branching colonies is the source of the name Bryozoa, which means moss-animals. Both marine and fresh-water forms are known.

These animals are distinguished by the following characters: 1. The individual consists of two parts, a sheath-like zooecium or body wall and an enclosed polypide consisting of the alimentary tract, tentacle sheath and tentacles. 2. The body is coelomate but not segmented. 3. A circlet of **ciliated** tentacles occurs at the free end of the body; in most species they can be retracted when the animal is disturbed or at rest. 4. The alimentary tract is sharply bent, mouth and anus opening near each other at the free end.

The phylum is divided into two classes:

Class Endoprocta. Primitive bryozoans whose tentacles cannot be retracted but are folded inward and covered by a fold of the body wall. The anus opens inside the circlet of tentacles.

Class Ectoprocta. Tentacles retractile. Anus opening outside the circlet of tentacles. (A.W.L.)

B SUPPLY. Power Supply.

B.T.U. Also Btu. Abbreviation for **British Thermal Unit.** (See **Calorimetry.**)

BUBBLE CAP. Distillation.

BUBBLE OCTANT. Sextant.

BUBO. Swelling of a lymphatic gland, particularly in the groin. Buboes develop in response to infection and are usually tender and painful; they may break down and discharge **pus.** They are most commonly associated with **lymphogranuloma inguinale, chancroid** and **plague** (bubonic plague), although they occur in the course of other diseases such as **gonorrhea, syphilis, tularemia.** (D.M.H.)

BUCCAL CAVITY. The cavity of the mouth. (A.W.L.)

BUCCAL FUNNEL. The funnel-like depression leading to the mouth in the **lampreys.** (A.W.L.)

BUCK. The male **deer** and sometimes the male **sheep.** Also applied to entire species in some cases, as the water-buck and prongbuck. (A.W.L.)

BUCKLING. Columns.

BUCKWHEAT. *Fagopyrum esculentum.* Polygonaceae. Outside of the grasses, buckwheat is the only plant used to any extent as a cereal in the United States. It is an erect branching plant from one to four feet tall, with a small root system and a smooth rather weak stem at each node of which is borne a single heart-shaped leaf. The **inflorescence** is a many-flowered **raceme,** the individual flowers being white or pink-tinged. The **calyx** lobes, five in number, are colored; there is no **corolla.** There are eight **stamens** and a single one-celled **ovary** which bears three curved **styles.** Cross-pollination is brought about by the numerous insect visitors attracted by the pleasant fragrance of the flowers. The mature fruit is a triangular brown or black **achene.** The single seed within contains an abundance of white **endosperm** high in starch content.

Buckwheat is an Asiatic plant which is cultivated in widely scattered regions. It grows well in cool climates, on poor soils, and where the growing season is short. The plant is often used as a green manure, being turned under to enrich the soil. Pancake flour is made from buckwheat seeds. The grain is also used as food for poultry and other domestic animals, either whole or divested of the hulls. Buckwheat flowers are an important honey source, producing a very finely flavored product. (R.M.W.)

BUD. In zoology, a bud may be defined briefly as an outgrowth from the body which develops into a new individual. In botany a bud is an undeveloped shoot and normally occurs in the axil of a leaf or at the tip of the stem. Once formed a bud may remain for some time in a dormant condition, or may develop into a shoot immediately. Various factors may cause dormant buds to grow, such as removal of the apical bud or of the part of the stem above the bud.

The buds of many woody plants, especially in temperate or cold climates, are protected by a covering of modified leaves called scales which tightly enclose the more delicate parts of the bud. Many bud scales are covered with a gummy substance, which serves as added protection. When the bud develops, the scales may enlarge somewhat but usually drop off, leaving on the surface of the growing stem a series of horizontally elongated scars. By means of these scars one can determine the age of any young branch, since each year's growth ends in the formation of a bud, the development of which causes the appearance of an additional group of bud scale scars. Continued growth of the branch causes these scars to be obliterated after a few years so that the total age of older branches cannot be determined by this means.

In many plants scales are not formed over the bud, which is then called a naked bud. The minute undeveloped leaves in such buds are often excessively hairy. Such naked buds are found in shrubs like the **Sumach** and **Viburnums** and in herbaceous plants. In many of the latter, buds are even more reduced, often consisting of undifferentiated masses of cells in the axils of leaves. A head of cabbage (see **Brassica**) is an exceptionally large terminal bud, while Brussels sprouts are large lateral buds.

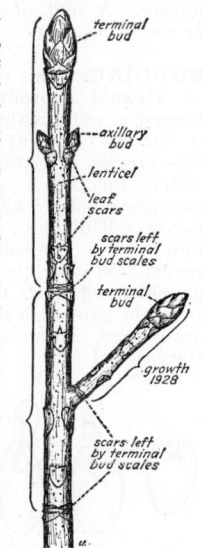

Twig of buckeye, *Aesculus glabra,* showing two years' growth.

Since buds are formed in the axils of leaves, their distribution on the stem is the same as that of leaves. So we find alternate, opposite and whorled buds, as well as the terminal bud at the tip of the stem. Whorled buds are found in many coniferous trees, as the fir or spruce. In many plants buds appear in unexpected places: these are known as **adventitious buds.**

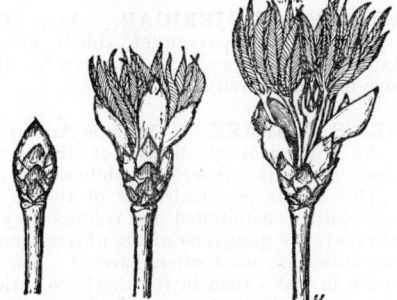

A series of stages in the growth in the Spring of a bud of buckeye, *Aesculus glabra.*

Often it is possible to find in a bud a remarkable series of gradations of bud scales. In the Buckeye, for example, one may observe a complete gradation from the small brown outer scale through larger scales which on unfolding become somewhat green to the inner scales of the bud, which are remarkably leaf-like. Such a series suggests that the scales of the bud are in truth leaves, modified to protect the more delicate parts of the plant during unfavorable periods. (A.W.L., R.M.W.)

BUDAN'S THEOREM.

Budan's theorem is of help in locating the real roots of a polynomial equation; it is stated as follows:

Let $P(x) = 0$ be a polynomial equation of degree n with real coefficients. Let a and b be real numbers, neither of which is a root of $P(x) = 0$, and suppose $a < b$. Let V_a denote the number of variations of sign of

$$P(x), \ P'(x), \ P''(x), \ \cdots, \ P^{(n)}(x)$$

(successive derivatives) for $x = a$, after vanishing terms have been deleted, and similarly V_b the number of variations for $x = b$. Then $V_a - V_b$ is either the number of real roots of $P(x) = 0$ between a and b or exceeds the number of these roots by a positive even integer. A root of multiplicity m is here counted as m roots. (L.L.S.)

BUDDING.

This term is used to designate a process of asexual reproduction in which the young are formed as outgrowths of the parent body. It is limited to animals or plants of relatively simple structure. In this process a portion of the wall of the parent cell softens and pushes out. The protuberance thus formed enlarges rapidly while at this time the nucleus of the parent cell divides. One of the resulting nuclei passes into the bud. Presently the bud becomes cut off from its parent cell and the process is repeated. Often the daughter cell starts to bud before it becomes separated from the parent, so that whole colonies of adhering cells are formed. Eventually cross walls cut off the bud from the original cell.

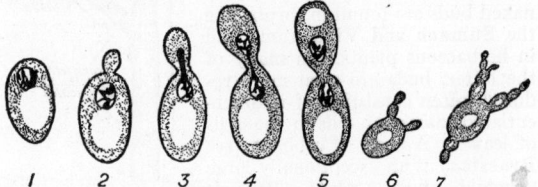

1, 2, 3, 4, and 5. Reproduction by budding. Yeast cells stained to show nucleus. (*Redrawn from Guillermond, The Yeasts, John Wiley & Sons, Inc.*) 6 and 7. Colony formation by rapidly growing yeast. Nuclei not stained.

The term budding is also applied to a process of embryonic differentiation in which new structures are formed by outgrowth from preexisting parts.

A third use of the term budding is discussed in the article on grafting and budding. (A.W.L., R.M.W.)

BUDGERIGAR, BUDJERIGAR.

Aves, Psittaciformes. An Australian parraquet, chiefly green with a blue tail and yellow face. Also known as the Australian love-bird in captivity. (A.W.L.)

BUERGER'S DISEASE.

(Thrombo-Angiitis Obliterans.) An inflammatory disease of the peripheral blood vessels which is followed by obliteration of their lumina. This results in impairment of the circulation to the extremities, manifested by coldness, cyanosis, pain, and eventually gangrene of the affected part. The lower extremities are most often affected. The disease is commoner in males than in females by a ratio of 75 to 1, and over half the cases occur in Jews. The onset is usually between the ages of 20 and 45. Treatment is directed toward improving the circulation of the extremities by special exercises, elevation of the affected part, drugs, and occasionally surgical measures which remove vessel-constricting impulses by interfering with their nervous pathways. (D.M.H.)

BUFFALO.

Mammalia, Artiodactyla. Hoofed animals of several species, some of which have been domesticated. They are strongly built, with a short heavy neck and a broad muzzle. The horns are hollow and are not twisted or branched.

The Cape buffalo, *Syncerus caffer*, is an African species with short curved horns whose bases are broadly expanded over the upper part of the head. It is found in swampy ground. In western Africa is another species, the short-horned buffalo or bush cow. The Indian buffalo, *Anoa bubalis*, is another water-loving species whose sweeping curved horns extend back toward the withers. This species is domesticated and in the Philippines is called the carabao. Another smaller species of the Philippines is known as the tamarao, *Anoa mindorensis*.

The term buffalo does not apply correctly to the American bison. (A.W.L.)

BUFFALO CARPET-MOTH.

Insecta, Coleoptera. More properly known as the carpet beetle but misnamed through the similarity of habits of its larva and those of the clothes moths.

The true carpet beetle is an introduced European species, *Anthrenus scrophulariae*, whose larva eats woolen materials of all kinds as well as furs and feathers. The adult is a compact oval insect about one-eighth inch long and marked with brick-red, black and white, and the larva is a brown hairy grub. A number of other species, native to North America, have the same habits and may be equally troublesome. The adults frequent flowers and eat pollen, hence they may migrate readily to houses. They are important pests in museums.

In the home, good housekeeping methods are usually an adequate safeguard against these pests. In special cases fumigation is necessary to destroy them but as a rule the use of sprays now supplied commercially for application to clothing and other fabrics is the only unusual measure required. Fumigation with carbon disulfide is the method commonly used in museums to destroy them; the explosive nature of this fluid and its vile odor do not recommend it for home use. (A.W.L.)

BUFFALO FISH.

Pisces, Teleostei. *Ictiobus*. North American food fishes (Pisces) of three species, related to the suckers. (A.W.L.)

BUFFALO GNAT.

Black Fly.

BUFFER AMPLIFIER.

This term is commonly applied to an amplifier stage whose main function is to isolate the oscillator of a transmitter from the main power amplifiers. The frequency of an oscillator depends, among other factors, upon the load which is applied to it. Since, for satisfactory communication, it is highly desirable that the oscillator frequency remain constant, buffer amplifiers are always used in broadcast transmitters and are usually used in others. Such an amplifier usually operates Class C (see Amplifiers) with the grid drive so low the grid does not draw current from the exciting circuit and hence doesn't load it. The output circuit of the buffer can then supply the normal grid load of a conventional Class C amplifier without this load affecting the oscillator. Sometimes more than one buffer stage is used to make certain that no load is reflected back to the oscillator. (L.R.Q.)

BUFFING.

A finishing process for producing a lustrous surface, usually on sheet metal. Buffing is usually effected by using a buffing wheel composed of layers of cloth sewed together to which a cake abrasive is periodically applied. (H.C.H.)

BUFFLEHEAD. Aves, Anseriformes. *Charitonetta albeola.* A small North American duck of wide distribution. (A.W.L.)

BUG. Insecta, Hemiptera. Insects with sucking mouth parts usually arising near the front of the head, with antennae usually long but few-jointed, and with wings, when present, thicker at the base and membranous at the tip, overlapping when folded to form a more or less conspicuous X on the back. The bugs are so diverse that no concise definition can be generally adequate. The term bug is not synonymous with insect. (A.W.L.)

BUHR-STONE. Relatively porous, calcareous, and siliceous sandstones with sharp or angular grains, used in making millstones. (R.M.F.)

BULB. A thick short stem which grows many thick leaves in which food reserves are stored. In many bulbs the leaves are closely wrapped together, forming a compact body called a tunicated bulb, as is the case in the onion. In other bulbs the fleshy leaves are loosely arranged to form a scaly bulb, such as the Easter Lily. Bulbs are particularly common in monocotyledonous plants, and aid the plants greatly in surviving long dry seasons. So regions subject to regularly recurrent dry seasons are particularly rich in bulb-forming plants. Commonly the term bulb is erroneously applied to any fleshy underground plant part, regardless of its nature, so that rhizomes, corms and tubers are all popularly classed as bulbs. The dahlia "bulb," for instance, is really a fleshy root. (R.M.W.)

BULBIL (OR BULBLETS). In a few plants there occur small reproductive bodies called bulbils. An example is found in the small black objects growing in the axils of the leaves of Tiger Lilies. These are really buds in which the scales are very much swollen; when mature the whole body falls to the ground and under favorable conditions puts out roots and in time grows into a new plant. Similar bodies are found in the familiar onion sets, and in several sedges. Serving the same purpose are the small globose bodies which develop on the leaves of several species of ferns, as for example, *Cystopteris bulbifera.* All such bodies form one method of vegetative propagation. (R.M.W.)

BULBUL. Aves, Passeriformes. Birds (Aves) of several species found in Africa and the Oriental region, related to the babblers. They are said to be melodious singers. (A.W.L.)

BULBUS ARTERIOSUS. A muscular expansion of the ventral aorta at its origin from the heart in bony fish. Distinguished from the bulbus cordis or conus arteriosus by the lack of cardiac muscle. (A.W.L.)

BULBUS CORDIS. A term closely linked with conus arteriosus and rather loosely used. Where the ventral aorta of vertebrates joins the ventricle of the heart, especially in the 2-chambered stage in the fishes and in embryos of higher classes, a prominent, somewhat conical expansion of the tube is found. This expansion may contain valves and its wall may be composed in part of cardiac muscle. In the former case it appears to be an an expansion of the artery and is properly termed the conus arteriosus. In the latter it seems to be a specialized portion of the ventricle and is called the bulbus cordis. The two characteristics may be combined to constitute a bulbo-conus. This term is sometimes applied to the corresponding portion of the embryonic circulatory system, which is better designated as the truncus arteriosus. (A.W.L.)

BULK MODULUS. Elasticity.

BULKHEAD. A bulkhead is a partition or a transverse strengthening frame. Ships' bulkheads are the important transverse partitions which subdivide the hold into separate water-tight compartments, being built from the keel to the bulkhead deck. They must be not only water-tight, but have sufficient structural strength to resist the bursting pressure to which they will be subjected when one bulkhead space is filled with water, while the adjacent one is empty.

In construction, any wall used to restrain fluid or semi-fluid pressure, such as that resulting from water or saturated earth in foundation excavations, is called a bulkhead or bulkhead wall. (F.T.M., E.W.S.)

BULL. The male of certain animals, as domestic cattle, the bull elephant, and the bull alligator. (A.W.L.)

BULLFINCH. Aves, Passeriformes. *Pyrrhula.* Birds (Aves) of northern Europe and Asia, related to the grosbeaks. (A.W.L.)

BULLFROG. Amphibia, Salientia. *Rana.* Large frogs, closely related to the more familiar grass frogs and leopard frogs. The bullfrogs reach a length of 8″. The flesh of the bullfrog is delicious. (A.W.L.)

BULLHEAD. Pisces, Teleostei. Small catfishes native to the streams and lakes of the eastern and central United States. Excellent food fishes. The common bullhead is also called the horned pout, *Ameiurus nebulosus.* (A.W.L.)

BUMBLEBEE. Insecta, Hymenoptera. Stoutly built hairy bees of moderate to large size. Some species are colonial, building nests on the surface of the ground, while others live as parasites in the nests of other bumblebees.

Unlike the honey-bee, bumblebees are not permanently colonial in temperate regions. Only the queen lives through the winter. When she emerges from hibernation in the spring she builds a nest or occupies an abandoned nest of a bird or mouse, and in it makes waxen cells in which she lays eggs and a waxen honey pot in which to store surplus food. She feeds her young until they mature, and only when they emerge as worker bees does the colony take on an organization like that of the honey-bee. In the fall, males and queens appear and the colonies break up. The queens mate before hibernating. Most authorities include all of these bees in the genus *Bombus.*

The parasitic bumblebees, making up the genus *Psithyrus,* enter the nests of other bumblebees and lay their eggs to be cared for by the hosts. Their chief structural difference is the lack of pollen-gathering organs in the females.

Bumblebees are important in the cross-fertilization of red clover and other deep-throated flowers. (A.W.L.)

BUMP, AIR. An airplane flying horizontally into an upward gust of air experiences a sudden change of direction of relative wind. This induces a temporary excess of lifting power of the wing which thereby accelerates the airplane upwards. The effect on the occupants is not unlike that of a surface vehicle passing over a bump in the roadway, hence the term "air bump." If the gust is downward, a downward acceleration is produced by the sudden decrease of angle of attack. This is sometimes referred to as an "air pocket" because the occupants feel as if the airplane had entered a region in which there was no air to sustain the lift. Actually, of course, there is practically no difference in the density of air in or out of the regions of air "bumps" and "pockets." (F.T.M.)

BUMPINESS OF AIR. Turbulence.

BUNCHER. Klystron.

BUNDLE. Also often called vascular bundle or fibro-vascular bundle. In most **vascular** plants the vascular tissues are arranged in the form of a cylinder. In many cases, notably in woody plants, this cylinder is a solid mass of cells. But in many plants, particularly in herbaceous **dicotyledons** and **monocotyledons,** the vascular tissues occur in strands which are more or less distinctly separated from one another, and are called vascular bundles. Such bundles appear as discrete objects when seen in a cross-section of the stem. However, they really form a continuous conducting system which extends from a single bundle in the root through the stem and into the leaves and other parts, and becomes an elaborate system of inter-connecting parts.

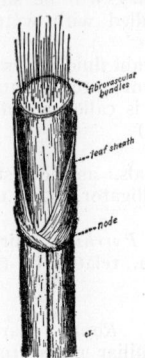

Portion of corn stem (*Zea mays*) with the vascular bundles protruding; illustrating the structure of a typical monocotyledonous stem.

In the axis of the plant each bundle consists of masses of **xylem** and **phloem** cells which may appear in various arrangements; frequently the xylem and phloem cells appear in radially adjoining masses, forming a collateral bundle; less frequently one kind of cells is surrounded by cells of the other kind, forming a concentric bundle; in **roots** a third arrangement is found; the xylem occupies the center of the single bundle, nearly surrounded by strands of phloem. In cross-section the xylem of the young root (see illustration, page 1256) is in the form of a cross with four or five arms (Dicotyledons), or many arms (Monocotyledons), and the phloem strands occupy the positions between the radiating arms of the xylem. Such bundles are called radial bundles.

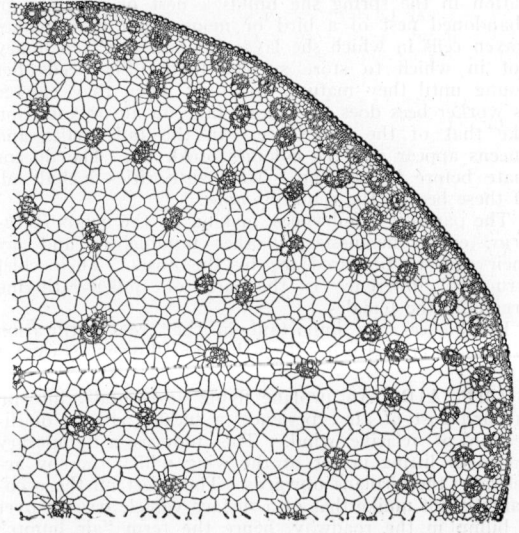

Cross-section of corn stem showing the distribution of fibro-vascular bundles in a typical monocotyledon.

While a bundle consists normally of associated masses of xylem and phloem cells, in many plants there appears with them strands of fibers whose presence may give protection to the bundle, and additional rigidity to the stem. Because of the frequent occurrence of such fibrous masses as part of a bundle, the term fibro-vascular bundle has been used to describe them. In many plants the fibrous cells form a mass on the outer side of the bundle, between it and the surface of the stem; while in other plants, particularly in monocotyledons, the fibers form a sheath completely encircling the bundle.

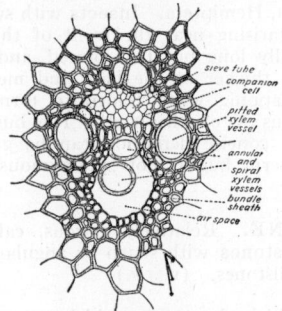

Cross-section of a single fibro-vascular bundle of corn, *Zea mays.*

Bundles are often described as open or closed. Open bundles are those in which **cambium** cells are found, so that by the repeated division of the cambium cells, the size of the bundle constantly increases. Closed

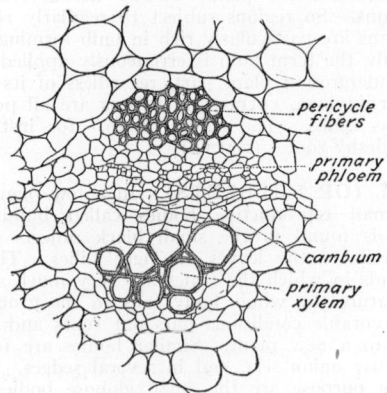

A cross-section of a single fibro-vascular bundle of a young sunflower stem.

bundles are those in which no cambium occurs, they being composed entirely of primary tissues and so once formed remain constant in size. (R.M.W., B.S.M.)

BUNION. A swelling of a **bursa** or fluid-filled sac on the outer part of the large joint of the great toe. This condition may be caused by too tight or too short shoes. It frequently accompanies the condition of hallux valgus, a lateral deflection of the great toe. (R.S.M., D.M.H.)

BUNKER. A bunker is any large bin, but its technical meaning is confined to receptacles for the storage of fuel or loose materials in bulk. The bunker is frequently constructed of plate steel which, for some purposes, must be lined with a substance such as **concrete** or wood to protect the metal. One prevalent type, the Berquist suspension bunker, has a profile of such a shape that **tension** is the only **stress** produced in the steel. Since the bunker is usually placed so that the material is deposited from the top, and withdrawn by gravity from the bottom, a suspension type is particularly advantageous. One of the most common uses for bunkers is to hold the coal required for the combustion equipment of **power plants.** It may be able to hold from three to five days' supply for a small plant, but space limitations will govern the capacity for large power plants. However, the bunker should hold enough to supply the plant during minor repairs to the coal conveying equipment. (F.T.M.)

BUNSEN SCREEN. Bench Photometers.

BUNTER SANDSTONE. Triassic.

BUNTING. Aves, Passeriformes. Birds (**Aves**) related to the finches and sparrows, including the British yellowhammer, not to be confused with the North American woodpecker which bears this name, the ortolan, and the snow bunting. In North America the indigo bunting, *Passerina cyanea*, is the most widely distributed species, ranging over the eastern half of the country and sometimes to western Texas. The lazuli, varied, painted, and lark buntings are characteristically western birds. The painted bunting, *Passerina ciris*, is also called the nonpareil. (A.W.L.)

BUOY. A buoy is a floating object, usually attached to a specific location on the bottom of the sea or to some submerged object, which is used by navigators. Buoys are used to locate channels, dangerous rocks or shoals, mooring positions, submerged wrecks, and a variety of other purposes. The various countries have each devised a system of shapes and colors to be used on buoys for specific purposes. (W.K.G.)

BUOYANCY. The familiar lifting effect of a fluid upon a body wholly or partly submerged in it, known as buoyancy or buoyant force, was first closely studied by the Greek philosopher Archimedes in the 3rd century B.C. What is now known as the "principle of Archimedes" states that the buoyant force is equal to the weight of that body of the fluid which the submerged body displaces, and may be treated as a single force acting vertically upward through the center of gravity of the displaced fluid (center of displacement). This statement applies whether the submersion is partial or complete. The principle readily follows from the consideration that if the submerged body were withdrawn and the resulting cavity allowed to fill with the fluid, the latter would be in equilibrium under the joint action of its own weight and the external forces formerly exerted by the surrounding fluid upon the submerged body.

If the buoyant force equals the weight of the submerged body and acts through its center of gravity, the body will be in equilibrium. This might be true if the body had exactly the same mean density as the fluid. It would then remain at rest when completely submerged at the proper level. A balloon, for example, may rise to a certain height and remain suspended or drift along horizontally. But a solid body completely submerged in a liquid does not come so readily to stable equilibrium, because of the very slight compressibility of liquids, the presence of highly compressible gases in pores of the body or in bubbles clinging to it, and the unequal coefficients of expansion of the body and the liquid.

If, however, the body has a lower mean density than the liquid, it will "float," partly submerged to a level, and in a position with reference to the vertical, determined by the Archimedes principle. An important phase of flotation, especially in ship design, is the degree of stability. This may be expressed in terms of the position of the "metacenter." When a boat is tipped very slightly in a given plane, the center of displacement shifts to one side. There is one point of the boat, called the metacenter, which remains vertically above the center of displacement. This point must be higher than the center of gravity, as the resulting couple then tends to restore the boat to its normal position; if it is lower, the boat is unstable and will capsize. The

height of the metacenter above the center of gravity (metacentric height) is a measure of the stability in the given vertical plane. It is in general different for different planes; for example a boat has a transverse and a longitudinal metacenter and metacentric height, corresponding, respectively, to its rolling and its pitching. Thus one may usually change seats in a rowboat without danger, while an equal shift across the boat might overturn it. (See **Weighing Methods**.) (L.D.W.)

BURBLE. Burble describes the action of air over the upper surface of an **airfoil** when the **angle of attack** has been increased to the point where the air stream no longer follows the profile of the airfoil but breaks away from it. The space between the airfoil and the detached air stream is filled with eddying, burbling air, and the lift is largely lost. The airfoil is then said to have reached the burble point. This is synonymous with "stalled" in wing terminology. (See **Stagnation Point**.) (F.T.M.)

BURBOT. Pisces, Teleostei. The fresh-water representative, *Lota lota*, of the codfish family. It is found in streams and lakes of Europe and North America, more commonly north of latitude 40°. The species also goes by the names eel-pout, ling and lawyer. (A.W.L.)

BURDOCK. *Arctium lappa.* Composite Family.

BURIN. Paleontology of Man.

BURLAP. Jute.

BURNER. An arrangement for mixing a fluid **fuel** with air and burning it is called a burner. Burners are commonly provided for liquid and gaseous fluids, but **pulverized coal**, when mixed with a certain amount of air which lubricates it, flows much as a fluid, and is burned by burners. Burner designs for gaseous fuels are the simplest because gas and air are easily mixed and a gas burner may consist of little other than a means to subdivide the total flow of the fuel into a large number of small jets which will present a large surface exposure to the combining oxygen. A liquid fuel will approach the efficiency of gas firing after it is sufficiently atomized or vaporized so as to present a large surface for combustion. This can be done by wicks, by contact with heated metal as in wickless kerosene burners, by the use of high-pressure sprays with very fine holes, or by the use of steam jets. Fuel oil burners of large capacity accomplish an atomization of the oil mechanically by giving the oil a whirl in the burner tip, and discharging it into the furnace through a small orifice. When steam is used, the oil is broken up by a continuous discharge of steam under pressure within the burner and at the tip outlet. Preparation of the oil for use in the burner may include filtering, heating, and pumping. Fig. 1 shows typical oil burners and tips.

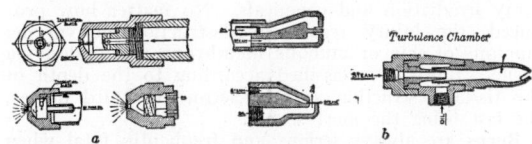

Fig. 1. a. Typical mechanical pressure atomizer tips. b. Typical outside-mixing and inside-mixing steam atomizing burners. (*Handbook of Oil Burning, American Oil Burner Association.*)

Pulverized coal offers much more of a problem in burner design than do oil and gas, even through the coal is pulverized to such an extremely small size that 90% of it will pass through a screen having 100 openings to the inch. Small as the coal particles are, they average 25,000 times larger than the oxygen **molecule**. The pulverized coal is floated or air-borne to the burner on a stream of air amounting to some 10% or 20% of the total combustion requirements. This is called primary air. The remainder of the air needed, called secondary air, is admitted directly to the burner, the function of which then becomes one of properly proportioning the fuel and air, and thoroughly mixing them. There are two general types of pulverized coal burners, called long and short flame burners. The long flame is produced by moderate tip velocities from a simple form of burner, coupled with an admission of the secondary air through openings in the furnace setting located along the traverse of the flame. Long flame burners are gradually being abandoned in favor of the short flame. The short flame burner produces complete mixture of fuel and air by violently whipping the secondary air through the primary air and fuel. This type is essentially a high-capacity, forced-draft type of burner, whereas the long flame burner is better adapted to induced draft.

Fig. 2 shows the primary and secondary air inlets into the short flame burner. The rate of combustion is regu-

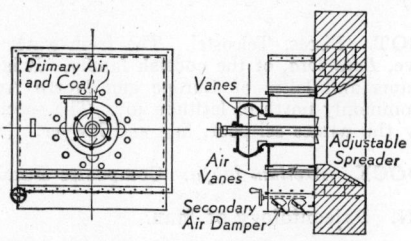

Fig. 2. Circular burner. (*Riley Stoker Corp.*)

lated externally by the rate of feed of coal to the burner. Proper proportioning of air is accomplished by a damper in the secondary air passage. Air vanes cause rotation of the secondary air moving into the furnace. A fan type spreader sets the primary air and coal whirling in the opposite direction. With this turbulence and in the presence of a furnace temperature higher than that required for ignition, the flame will propagate itself back into the incoming air fuel stream at from 15 to 40 ft. per sec. The velocity of the mixture leaving the burner must exceed this in order to prevent flarebacks, and since friction slows down parts of the incoming stream, burner exit velocities of 80 to 130 ft. per sec. are used. (F.T.M.)

BURNISHING. A surface-hardening or surface-finishing process for metals, effected by the application of a roller or blunt rod under pressure to the surface. It is used for gear-tooth finishing to some extent. The work is rotated between three hardened and ground burnishing gears. Burnishing will not correct errors but serves to compress the surface of the teeth and provides a slight surface hardness. As a finishing operation prior to hardening, it can also be used to remove burrs and bruises. (H.C.H.)

BURNS. Burns may be produced by heat, electricity, x-ray irradiation and chemicals. No matter how produced, the injury results in destruction of varying amounts of skin or mucous membrane and underlying tissue. Burns are classified according to the depth of the tissue destruction as first, second or third degree, the last being the most severe.

Burns are always serious and frequently fatal when more than a small part of the body surface is involved. The cause of the marked systemic reactions associated with burns is not known, but several factors are believed to be involved. The patient with severe burns exhibits immediately a typical picture of profound shock, accompanied by loss of body heat and extreme pain. There is a loss of blood plasma into the burned area, and the surrounding tissues become edematous. A greatly diminished circulating blood volume with a relative increase in haemoglobin results. The edema fluid contains a high per cent of chlorides, and there is

a corresponding decrease in blood chlorides. The great loss of body fluids plus the low blood chlorides are thought to be responsible for the so-called toxemia which severely burned patients show. Infection is a serious complication of burns, and is the cause of many late deaths. Because bacteria thrive well in devitalized tissue, the burned area is open to bacterial invasion. Fever, local redness, swelling and eventually pus formation follow. Kidney damage and marked anemia also occur in severe burns.

The present treatment of burns consists of relieving pain with morphine, combating shock by blood or preferably plasma transfusions, and preventing infection. Tannic acid has been abandoned because it has been shown to be ineffectual and is thought by some to produce liver and kidney damage.

First and second degree burns do not destroy the regenerative layer of the skin, and consequently heal without scarring. Third degree burns destroy the regenerative layer, and produce marked scarring. This can be minimized by the use of skin grafts, which involves the transplantation of normal skin from some part of the patient's body to the burned area. (D.M.H.)

BURR. 1. A ragged edge of metal particles resulting from cutting or grinding operations. 2. Fine-tooth rotary milling cutters used in die-sinking. (H.C.H.)

BURRO. Mammalia, Perissodactyla. A small variety of the domestic ass or donkey, used as a beast of burden in the West. The burro is incredibly tough and subsists on a diet of almost anything edible, hence it has been valuable in the exploration of the more arid areas. (A.W.L.)

BURROWING. The habit of living underground and also the preparation of runways and living quarters beneath the surface. Many animals, such as wolves, that are not specially developed for burrowing, prepare burrows or dens for the birth of their young and for hiding places. Others are normally at home in the ground and come to the surface only under certain conditions; here examples range from the worms, of which the earthworm is particularly well known, to highly specialized mollusks, insects, and vertebrates.

The less specialized burrowing animals cut away the earth by means of structures whose origin is not associated with this use, such as the claws of reptiles, birds, and mammals, while in many of the more specialized forms these parts are highly developed and modified and other special adaptations are evident. The earthworm merely eats its way through the earth, and some insects are capable of burrowing into the tissues of plants in the same way. The shipworm, a mollusk, has the valves of the shell adapted for cutting burrows into wood.

The moles are highly developed for burrowing by the powerful build of the fore limbs and by their large claws, and are further adjusted to life underground by their fine moisture-resisting fur. The poorly developed eyes are also correlated with conditions underground, since these animals rarely enter the light.

Mole crickets are in some ways like the moles. Their front legs are broadened and provided with clawlike processes and the body is covered with a downy vestiture which repels moisture.

Burrowing offers the animal greater safety than can be enjoyed on the surface of the earth, since no predator can compete with the highly developed burrowing animal on equal terms below the surface. (A.W.L.)

BURSA COPULATRIX. A pouch found in the females of some species of insects which receives the seminal fluid of the male during copulation, before it passes to the spermatheca. (A.W.L.)

BURSAE. Small sacs lined with synovial membrane (see **synovia**) which lie near joints. Their function is protective: they act as lubricating buffers between moving parts, between the tendons and bone, or bone and joint capsule, or between skin and bony structures. There are in the neighborhood of 1000 bursae in the human body. (D.M.H.)

BURSITIS. Inflammation of the synovial lining of the **bursae.** This is caused by overuse of a joint, trauma, or infection. Bursitis may be acute or chronic. The acute form is extremely disabling, the involved area being swollen, tender, and very painful on motion. The bursal sac is often distended with fluid in chronic bursitis. Immobilization, heat, and **diathermy** are used in the treatment of acute bursitis; surgery may be necessary in the chronic form.

The common sites for bursitis are the acromial and subdeltoid bursae around the shoulder joint, the olecranon (elbow) bursa, and the pre-patellar bursa ("house-maid's knee"). (D.M.H.)

BUS, AUTOMOTIVE. This automotive term was derived from *omnibus,* a public vehicle for use on public highways, designed to carry a relatively large number of passengers. The present automotive bus represents the outgrowth of a great deal of specialized engineering effort. The service rendered by an automotive bus varies from the comparatively short-haul service for cities and metropolitan areas to the long-distance transcontinental type bus service, and naturally the problems of speed, acceleration from a standstill, passenger comfort, fuel capacity, top speed, etc., will vary.

Since the bus is usually part of a transportation system involving many similar units, specialized overhaul and maintenance service can be worked out applicable to the fleet of buses. Repair centers, emergency stations, etc., may be established by a large system which is then in a much better position to take advantage of new mechanical developments than is the case for the ordinary run of business. Thus a bus transportation system may employ the **Diesel engine** which has been developed to the point of serviceability as a source of automotive power, but whose progress in this field has suffered from the large investment in equipment and talent to build and service the **Otto** type engine. (F.T.M.)

BUSH-CRICKET. Insecta, Orthoptera. **Crickets** of several species, most of which are found chiefly in shrubby vegetation. (A.W.L.)

BUSHING. In mechanical terminology, to bush is to reduce the size of a hole. A bushing is a hollow cylinder used as a renewable liner for a bearing or a drill jig.

A bushing is also a pipe fitting employed to reduce the size of pipe to a smaller size. When pipe is employed to contain electrical wires, the open end from which the wires emerge is often capped by a bushing which substitutes a smoothly rounded surface for the sharp edges of an unbushed conduit. The sharp edges would tend to abraid the insulation on the wires.

In electrical work, where a **conductor** at high voltage emerges from one insulated condition to another, an intermediate support must be provided. An electrical bushing is needed to provide the support and insulation between the conductor and the supporting surface. For example, where the conductor leaves the insulated interior of a **transformer** case, a bushing is provided to support the terminal where it passes through the case, and to insulate the voltage difference between the terminal and the grounded case. A bushing is also required at terminals of oil **circuit breakers**, and at potheads where the conductors of a multi-conductor **cable** are separated and brought out from the cable sheath for external connections. To obtain sufficient dielectric strength for very high voltage bushings without having the physical dimension of the bushing become excessive, the oil-filled bushing or the condenser-type bushing was developed. The condenser bushing is made of thin layers of tin foil wound between concentric layers of **insulation**. It is possible in this way to give uniform potential drop through the thickness of the bushing. (F.T.M., H.C.H.)

BUSHMASTER. Reptilia, Sauria. A large and deadly **snake**, *Lachesis mutus,* found in the forests of Central and northern South America. It is a pit-viper, like the rattlesnakes, water moccasin, and copperhead of North America, but it reaches a length of 12′ and may have fangs over an inch long, hence it is more to be feared than any but the largest of the related species. (A.W.L.)

BUSTAMITE. Rhodonite.

BUSTARD. Aves, Gruiformes. Large birds (**Aves**) of numerous species found chiefly in Africa, although some occur in Europe and Asia. They are chiefly terrestrial in habits but are powerful fliers. One of the African species is called the hubara and those of India are known as floricans. The bustards are related to the rails and cranes. (A.W.L.)

BUSTARD-QUAIL. Aves, Galliformes. *Turnix.* Small birds (**Aves**) related to the pigeons and rails, as well as to the gallinaceous birds. They are widely distributed in the Old World. Also called hemipodes. (A.W.L.)

BUTANOL. Alcohols.

BUTTE. Mesa.

BUTTER. Esters.

BUTTER YELLOW. Dyes.

BUTTERCUP FAMILY. Ranunculaceae. This family, also called the Crowfoot family, is a large family consisting chiefly of perennial herbaceous species distributed throughout the world. It belongs to the order Ranales which is of importance as the most primitive order of Dicotyledons and probably gives rise to the Monocotyledons on the one hand and the two large branches of Dicotyledons on the other. The family is important commercially because it includes many familiar and widely-grown ornamental genera, notably the buttercup (*Ranunculus*), peony (*Paeonia*), larkspur (*Delphinium*), virgin's bower (*Clematis*), and columbine (*Aquilegia*). It also includes the genera *Aconitum* and *Hydrastis*, **aconite** and goldenseal, important medicinal plants. The flowers vary from those having the primitive, symmetrical corolla of the buttercup and peony to those with irregular tubular corollas with long spurs as in columbine and larkspur. *Delphinium* is poisonous to cattle and has been responsible for losses on the western ranges. (P.A.W.)

BUTTERFISH. Pisces, Teleostei. A small species of fish (**Pisces**), also called the gunnel. (A.W.L.)

BUTTERFLY. Insecta, Lepidoptera. An insect with four large wings, usually completely covered with scaly vestiture. Distinguished from most other members of the order (moths and skippers) by the terminal club of the **antennae**. (A.W.L.)

BUTTERWORT. *Pinguicula species.* **Insectivorous Plants.**

BUTT JOINT. Riveted Fastenings.

BUTTRESS. Screw Thread.

BUTTRESS ROOTS. Roots.

BUYS BALLOT'S LAW. Professor Buys Ballot at Utrecht in 1857 stated this law: "Standing with back to wind, low pressure is to left and high pressure to right in northern hemisphere, with the reverse being true in the southern hemisphere." This law was one of the earliest statements of the principles of winds now accepted as common meteorological knowledge. (P.E.K.)

BUTYRIC ACID. Acids, Carboxylic.

BUZZARD. Aves, Falconiformes. Birds (**Aves**) of prey of several species belonging to the genus *Buteo*. The North American representatives are commonly called hawks, as Swainson's hawk. The same may be said of the nearly related rough-legged buzzards; American representatives of the genus are the rough-legged hawks. The name is incorrectly, although commonly, applied to the turkey buzzard, which is a vulture. (A.W.L.)

BY-PASS CONDENSER. This is a **condenser** placed in an electrical circuit to allow a.c. to flow around some circuit component and cause d.c. to flow through the component. The most common usage is in by-passing various voltage-dropping resistors used in vacuum-tube circuits to adjust the voltages applied to the several parts of the circuits. These resistors are by-passed so there will be no, or very little, alternating signal voltage drop to produce undesirable feedback. The **reactance** of the condenser should be low compared to the **resistance** of the resistor being by-passed. (L.R.Q.)

BYSMALITH. A plug-like **igneous** intrusion related to a laccolith but bounded laterally by faults due to upward "punching" rather than "pushing" of the **magma** as it forces its way into a series of stratified rocks. (R.M.F.)

BYSSUS. An organ of attachment formed by a special byssus gland in the foot of many **bivalve** mollusks. It consists of thread-like processes which are at first adhesive. (A.W.L.)

BYTOWNITE. Feldspar.

C

CAA. Civil Aeronautics Administration. A department of the United States Government, whose duties consist in approving the designs of aircraft to be used in interstate commerce, certifying the airplanes as airworthy, examining and licensing pilots for competence, examining planes in use for continuance of airworthy condition, maintaining federal airway and communications facilities, and engaging in needed aviation research and development. Its position in the federal system has been varied. In 1945 it was found to be a division of the Department of Commerce. (F.T.M.)

CABANE. A cabane is a part of the **airplane** structure. A monoplane which has a wing not attached directly to the fuselage, but mounted above it, requires a cabane structure to carry the load from the wing to the fuselage. A cabane is also needed for the biplane. A cabane is composed of **struts** triangularly framed to remain rigid against fore and aft loads and braced with wire for rigidity laterally. At its upper end it is connected to the wing root fittings and at its lower end to the fuselage. If the main structural members of the wing are not continuous from tip to tip the outboard sections are hinged at the cabane joint, and there is a short center section. (F.T.M.)

CABBAGE. Brassica.

CABBAGE BUTTERFLY. Insecta, Lepidoptera. A white **butterfly,** *Pieris rapae,* with black-tipped forewings and two or three black spots on the wings of each side. Introduced from Europe about the middle of the nineteenth century, this species spread rapidly until at present it is found throughout the United States and much of Canada. The caterpillar feeds on all cruciferous plants, but is especially important as a pest on cabbage and cauliflower, whence it receives its common name.

The insect can be held in check by spraying with Paris green. Cabbage heads formed on sprayed plants do not contain enough poison to affect human beings. (A.W.L.)

CABINET REPRESENTATION. Pictorial Representation.

CABLE. A cable is a strong rope composed of several strands of fiber or wire. In nautical terms, large ropes usually 10″ or more in circumference are known as cables, while smaller ropes are called hawsers. However, by common usage, a cable is any strong rope. Structural cable, made of straight or twisted steel wires, is used for suspension **bridges.** Twisted wire cables are also used as guy lines and lifting lines for some types of **derricks.** (See **Cable, Electrical.**) (F.T.M.)

CABLE, ELECTRICAL. An electrical cable is one or more conductors surrounded by an insulating medium and a protective sheath. Such cables are used for the transmission of electric power and for transmission of communication signals. The power cables have relatively few conductors of heavy gauge and are insulated for high voltages. Such cables are frequently filled with oil to increase the insulation strength. The outer sheath is commonly of lead, although for submarine work this in turn is often further strengthened by a second sheath of steel strands. Communications cables usually contain many pairs of small-gauge copper conductors, paper-insulated, surrounded by a lead sheath. Sometimes the entire cable is nitrogen-filled under pressure. The

various pairs of conductors are arranged by twisting and placing to minimize pick-up between them (see **Cross-Talk).** Common practice is to include two extra pairs for spares for each hundred active pairs. Cables used for submarine circuits have fewer pairs and are heavily insulated and armored to withstand the severe strains to which they may be subjected in laying and by ocean currents. The coaxial cable is a special type in which the pair of conductors is formed by a center wire and the outer sheath. In this case the sheath is copper and the insulation is often a gas with solid dielectric spacers at intervals to hold the inner conductor centered. This coaxial cable may in turn be enclosed with others in a lead sheath for protection. Coaxial cables have a wide usable frequency range and hence are used for transmission of television programs. They are also often used for radio-frequency transmission lines as they have no external field. (L.R.Q.)

CABLEWAY. A suspended steel **cable** acting as a track for aerial hoisting and conveying devices is a cableway. While occasionally used for transporting persons across deep gorges, where the amount of traffic does not warrant the building of a **bridge,** the cableway, in its more common application, handles construction material for building of **dams,** etc., or has a permanent use in connection with the handling of material such as rock or gravel which is taken from open pits. Clear spans up to a half-mile in length are possible in a cableway. The carriage which operates on the cableway may or may not have provision for carrying pasengers, depending on the purpose of the cableway. (F.T.M.)

CACAO. Esters; Theobroma Cacao.

CACHALOT. Mammalia, Odontoceti. A name for the sperm **whale** from the French. *Physeter catodon.* (A.W.L.)

CACHEXIA. A state of extreme malnutrition with resultant weakness and wasting of body tissues. It is frequently seen in the terminal stages of chronic disease —particularly **cancer.** (D.M.H.)

CACOMISTLE. Mammalia, Carnivora. Small animals related to the raccoons but of more slender build. Several species, ranging from Mexico to Oregon. Also called **civet cat** and ring-tailed cat. (A.W.L.)

CACTUS FAMILY. Cactaceae. The cactus is known to all as a prickly inhabitant of dry American deserts. In popular parlance, the name "cactus" applies to any fleshy spine-covered plant. But not all spiny plants are cacti, nor are all cacti characterized by spines.

With the exception of a single genus *Rhipsalis,* some of whose members are said to occur in Ceylon and Madagascar, all cacti are natives of America, where they are found widely scattered from latitude 59° in North America through the tropics to the southern Andean region and Argentina. They are particularly conspicuous features of the flora of dry desert regions where they are found in a wide variety of forms and sizes.

A few genera, and especially *Pereskia,* are very like ordinary **mesophytic** plants, having well-developed ovate leaves borne alternately on a long slender stem. But in nearly all of the Cactus Family the leaf surface is very much reduced, the leaves appearing as small fleshy bodies which last but a brief time before dropping off. In many species leaves of any recognizable kind are never formed, the green fleshy stem taking over the function of leaves completely. In these fleshy stems large amounts of water are stored, a feature which enables these plants to survive in the arid regions in which they so frequently grow. Due to the mucilaginous nature of the cell contents and to the greatly reduced surface of the plants the contained water is held most tenaciously and lost very slowly.

In their natural environment cactus plants have an extensive system of long fibrous roots which not only extend outward from the plant to considerable distances, but also penetrate the soil deeply. In cultivated plants the root system is usually greatly reduced. The stems of cacti show a variety of forms. In addition to the normal-stemmed *Pereskia,* there are the Prickly Pears, species of *Opuntia.* In most of these the stem is a series of flattened, fleshy joints often abundantly protected with bristling bunches of barbed spines. In species of *Mammillaria* and *Cereus* the stem is a cylindrical or globular body, often conspicuously ridged, and armed with numerous spines. In *Phyllocactus* and *Epiphyllum* the stem is flattened and largely unarmed, the small weak spines being borne in notches along the edges of the stem. The familiar night-blooming "cereus" is of this type.

One of the best known and largest of all the cacti is the sahuaro or giant cactus (*Cereus giganteus*). This species is a native of southern Arizona, northern Sonora, and extreme southeastern California. This massive cactus may grow to a height of 50′ with many side branches. Some individuals probably attain an age of 200 years. In some parts of its range extensive "forests" of this species have developed.

The flowers of most species of cactus are large and brightly colored. They are regular, although in some species a definite tendency toward zygomorphic flowers is seen. The flowers are borne singly. The perianth is composed of a large number of separate members which show a gradual transition from the outer small sepals (see **Flower**) through to large brightly colored petals. The stamens are likewise numerous and have long filaments. The single compound pistil contains many ovules and in fruit becomes a many-seeded berry. In many species the fruit is edible. Species of *Opuntia* are frequently planted in rows to form an impenetrable barrier against intruders. These plants were early introduced into the Old World and later into Australia, where in many places they have become a troublesome and almost worthless weed. Cactus plants are frequently seen in cultivation, being especially sought by those who like the bizarre effect they give. Somewhat similar in appearance are many species of *Euphorbia* from tropical Africa, and of *Stapelia,* a genus of the Milkweed Family, and likewise native to Africa. The flowers of these plants are quite unlike cactus flowers, however, so that the plants are readily distinguished when they bloom.

One genus of cactus, the Living Rock Cactus, *Lophophora williamsii,* is used by the native American Indian to produce a state of extreme exhilaration. It is used in the form of dried disks or buttons, sections of the stem, known as peyotl or mescal, a curious-tasting object which gives to the user the most agreeable sensations of color and pleasure. While it is not habit-forming, its use is frowned upon by Americans. (R.M.W., B.S.M.)

CADDIS OR CADDICE FLY.

Insecta, Trichoptera. The adult of any species of this order. The caddis flies are slender **insects** with four wings, sometimes clothed with hair-like scales which give them a moth-like appearance. The mouth parts are formed for biting but are vestigial. Since the **larvae** are aquatic the caddis flies are much more abundant in the vicinity of water, but they are attracted to light, often at some distance. (A.W.L.)

CADDIS WORM.

Insecta, Trichoptera. The aquatic **larva** of a **caddis fly.** They are noteworthy for the silken webs and cases which they build, some for protection and some to catch prey. Some species spin silken nets attached to rocks in the bottom of a stream in such a position that the current washes into the wide mouth and passes out through a web at the smaller end. In this way the **insect,** which lives in a tube near by, snares its food. Many of the caddis worms live in cases from which only the head and legs protrude. These cases are formed of many different materials, held together by silk. Some are made of small flat pebbles, some of bits of leaves, and some of small snail shells. They are economically of some value as food for fishes. (A.W.L.)

CADELLE.

Insecta, Coleoptera. A small brown **beetle,** *Tenebroides mauritanicus,* which infests granaries, feeding on other insects as well as on grain.

Fumigation with **carbon disulfide** and **hydrocyanic acid** gas has been used to destroy this and other granary pests. Both methods are dangerous and should be employed only by experts. It has been found effective also to heat the entire building to a temperature of 120–140° F. for several hours. In mills which have heating plants of ordinary efficiency for winter use these temperatures can be secured in the summer. (A.W.L.)

CADMIUM.

Symbol: Cd. Atomic number: 48. Atomic weight: 112.41. Density: 8.6. Hardness: 2. Melting point: 320.9° C. Boiling point: 767° C. (Isotopes: page 290.)

Cadmium is a silver-white metal, malleable and ductile, but at 80° C. becomes brittle, remaining lustrous in dry air and only slightly tarnished by air or water at ordinary temperatures, may be sublimed in a vacuum at a temperature of about 300° C., when heated in air burns to form oxide, dissolves slowly in hot dilute **hydrochloric** or **sulfuric acid,** and more readily in **nitric acid.** Discovered by Stromeyer in 1817.

Cadmium metal is used (1) as a protective coating for iron and steel, (2) in alloys such as solder, "bearing metal," and low melting point "fusible" alloy.

Cadmium occurs as the sulfide (greenockite, CdS) in Greenland, Scotland and Pennsylvania, but chiefly in zinc ores (frequently 1 part cadmium to 400 parts zinc), from which it is separated during the process of manufacture by fractional distillation—cadmium is more volatile than zinc and is collected in the initial distillate.

Bromide. Cadmium bromide ($CdBr_2$), white soluble solid. Used in photography.

Chloride: Cadmium chloride ($CdCl_2 \cdot 2\frac{1}{2}H_2O$), white soluble crystals by reaction of cadmium metal and **hydrochloric acid,** and crystallizing.

Hydroxide: Cadmium hydroxide ($Cd(OH)_2$), white precipitate by the reaction of cadmium salt solution and **sodium** hydroxide solution; soluble in acids and in ammonium hydroxide, but insoluble in sodium hydroxide solution.

Nitrate: Cadmium nitrate ($Cd(NO_3)_2 \cdot 4H_2O$), white, soluble crystals, by reaction of cadmium metal and **nitric acid,** and crystallizing.

Oxide: Cadmium oxide (CdO), brownish-yellow solid, by burning cadmium metal in air, or by ignition of cadmium hydroxide, carbonate or nitrate.

Iodide: Cadmium iodide (CdI_2), white, soluble solid, by reaction of cadmium oxide, hydroxide, or carbonate with **hydriodic acid,** soluble in alcohol (most iodides are insoluble in alcohol), forms cadmium potassium iodide, soluble, by excess of potassium iodide and crystallizing.

Sulfate: Cadmium sulfate ($3CdSO_4 \cdot 8H_2O$), white soluble solid, by reaction of cadmium metal and dilute **sulfuric acid,** and crystallizing. Used in making the Weston standard electromotive force cell.

Sulfide: Cadmium sulfide, "cadmium yellow" (CdS), yellow precipitate, by reaction of cadmium salt solution and **hydrogen sulfide.** Used as a yellow paint pigment and when mixed with ultramarine for green pigment, in coloring soap, paper and rubber, and in ceramic glazes.

Tungstate: Cadmium tungstate ($CdWoO_4$), yellow insoluble solid. Used in fluorescent paint. (R.K.S.)

REPRESENTATIVE ALLOYS CONTAINING CADMIUM

Fusible alloy, melting at 145° C............	18% Cd	50% Sn	32% Pb
Fusible alloy, melting at 70° C. (Wood's metal)..	12.5% Cd	50% Bi	25% Pb 12.5% Sn
Fusible alloy, melting at 70° C. (Lipowitz' alloy).	10% Cd	50% Bi	27% Pb 13% Sn
Fusible alloy, melting at 91.5° C............	8% Cd	52% Bi	40% Pb

Alloys melting at:

420° C............	90% Cd	10% Ag
610° C............	70% Cd	30% Ag
760° C............	50% Cd	50% Ag
280° C............	90% Cd	10% Zn
295° C............	70% Cd	30% Zn
327° C............	50% Cd	50% Zn

(R.K.S.)

CADUCEUS. The wand of Hermes, or Mercury, used as a symbol of the medical profession and of the Medical Corps of the U. S. Army. (R.S.M.)

CAECILIA. Gymnophiona.

CAECUM. Any blind pouch, especially a sac-like appendage of the alimentary tract (see **Anatomy**). The blind pouch with which the large intestine begins is a caecum; it bears the vermiform appendix in man and the anthropoid apes. Some of the fishes have pyloric caeca at the point of union of the stomach and intestine and in some insects the rectum bears such pouches. A pair of blind branching diverticula associated with the vestigial intestine of the starfishes are called rectal caeca. (A.W.L.)

CAESAREAN SECTION, OPERATION. Removal of the child from the **uterus** by means of an incision through the abdominal and uterine walls.

It is generally thought that Julius Caesar was born in this way and that he obtained his name from the manner of his delivery (a *caeso matris utero*). However, this could hardly be correct since his mother lived for many years after his birth and also Julius was not the first to have the name Caesar. The correct derivation of the name for the operation comes from the Roman law by which it was required that such an operation be performed upon women who died just before the termination of pregnancy. At first this law was known as the *lex regia* but under the emperors its name was changed to the *lex caesarea* and the operation henceforth was known as the caesarean operation.

As to when the first caesarean operation was performed on a living woman there is no definite information, although certain passages in the Talmud may indicate that it was carried out in the early Christian era.

The first record of an authentic caesarean operation on a living woman was in 1610 by Trautmann, of Wittenberg. There were probably other operations done before this but we have no record of them. Up to 1882 no sutures were used in the incision in the pregnant uterus and most of the women perished from hemorrhage. Even after sutures were used the mortality was so great from peritonitis, shock, and hemorrhage, that in 1887 out of 11 operations performed in New York City only one mother survived. Since then, however, with improvements in the whole field of surgery, the operation has become relatively safe. The mortality at present in the hands of capable surgeons is very low.

The operation is performed under the following circumstances: 1. When there is deformity and narrowing of the bony pelvis which does not permit delivery through the vaginal route. 2. If a pelvic **tumor** blocks the birth canal. 3. In certain patients with advanced heart disease, **tuberculosis, nephritis** and other conditions which may make labor dangerous to the health of the mother. 4. In patients who have previously been delivered by caesarean section. (R.S.M., D.M.H.)

CAFFEINE, OR TRIMETHYLXANTHINE. An **alkaloid** prepared usually from **tea-leaves.** It is present also in **coffee,** Kola and guarana. It is a frequently used and valuable drug in medicine, where it serves as a stimulant and **diuretic.** (R.S.M.)

CAFFRE CAT. Mammalia, Carnivora. A **cat** which ranges throughout Africa and into Asia. It resembles the domestic cat and is supposed to represent the ancestral stock from which some strains of the latter are derived. (A.W.L.)

CAGE. Bearings.

CAIMAN. Reptilia, Crocodilia. *Caimein.* Large **reptiles** of five species inhabiting the waters of Central and tropical South America. They resemble the crocodiles but are less dangerous. Called by the natives jacare. (A.W.L.)

CAIRNGORM STONE. The name given to the smoky brown variety of **quartz,** particularly when transparent, from Cairngorm, Scotland, a well-known locality. (E.S.C.S.)

CAISSON. A caisson is a large, usually bottomless box or cylindrical shell, made of wood, **concrete or steel,** which is used in the construction of **foundations** located in water or below the ground-water level. It is also useful when quicksand or other unstable soils are encountered during the excavation for a deep foundation. Caissons for the foundations of buildings or other land structures are forced down through the soil by means of weights while the underlying material is being excavated. A water jet is often used to aid the sinking action of the weights by reducing friction and by breaking up the material under the edge of the caisson. As the sinking progresses additional sections are added until the required depth is obtained. When the foundation is to be constructed under water, the caisson is built on land, floated to the site and sunk in place by means of weights. If it is to be sunk to bed rock through the mud or sand at the bottom of the body of water, the operation is carried on in a manner similar to that described for building caissons. In order to withstand **hydrostatic** or earth pressure the caisson should be properly braced or reinforced.

The box caisson is a water-tight structure used for **bridge** foundations which are under water. It is open at the top so that concrete or stone may be deposited to provide the necessary weight to sink it to the correct position, the bottom having been previously prepared so that a firm footing is assured.

An open caisson is a rectangular or cylindrical shell, usually open top and bottom, although it may have a bottom with wells for removing the excavated material. It is provided with a cutting edge to penetrate the underlying soil and thus facilitate the sinking operation. Double walls or **ballast** pockets are used so that at least a part of the permanent filling may be placed subsequent to sinking and in this way reduce the amount of temporary loading. In land operations the bottom of the caisson is assembled over the site of the foundation and the earth within the walls excavated. As the material is removed the weight of the caisson and the additional loads, if required, cause it to sink. The water jet, whose action has been described, may be used to assist gravity action. When the caisson reaches its final position and the remaining earth has been excavated, a concrete seal is poured through the water that is in the bottom of the caisson. This makes the bottom watertight. The water is then pumped out and the interior filled with concrete. Open caissons may be used for bridge piers if they are constructed in such a manner that the upper end projects above the surface of the water. The depth of the seal must be large enough to provide a mass of concrete that will counteract the effects of the uplift of water which may seep through the sand or mud at the bottom of the caisson. If the depth of the concrete seal is made about one-half the depth of the water, measured from the surface to the bottom of the caisson, the caisson will be safe against flotation, even for 100% uplift. Open caissons are generally used for deep building foundations which are carried below the ground-water level. The depth to which the shell may be sunk depends upon the friction developed on the sides and upon the material which is encountered. This type of caisson has several disadvantages. The character of the bottom cannot be readily ascertained due to the fact that it is under water; the bearing area cannot be properly prepared to receive the caisson and the concrete seal must be placed under water. It also has some advantages. The use of compressed air, which is a necessary factor of the pneumatic caisson, is not required. It may be used for greater depths than the pneumatic caisson when the ground-water level is high and for some types of foundations it is more economical. The genesis of the open caisson method was originated and developed by the Chinese probably as early as 1500 B.C. in connection with the sinking of shallow wells in soft material where other methods of construction were unsuitable or impractical. In modern times this ultimately led to the development of the pneumatic caisson.

The pneumatic caisson consists of an inverted box or cylinder which utilizes compressed air to counteract the hydrostatic pressure. The essential parts of the caisson are a working chamber, open shafts and **air locks**. The working chamber is the place where excavation is carried on under compressed air. The shafts provide a means for workmen to enter or leave the caisson and are also used as outlets for the excavated material. The air lock is a room where the air pressure is gradually adjusted as the workmen enter or leave the working chamber. When the pneumatic caisson is to be used for land foundations, where the water level is high, the bottom containing the working chamber is constructed first. This bottom has a cutting edge similar to that for the open caisson. The material in the working chamber is dredged out and the caisson gradually sinks into the soil. When the ground water is encountered compressed air must be used to keep the water out of the working chamber, and as the sinking progresses, the pressure must be increased to balance the hydrostatic pressure. As the caisson is lowered cofferdams are used to keep out earth or water. Concrete, which is deposited inside the cofferdam, furnishes additional weight for sinking. After the caisson has reached its final resting place all loose material is excavated. A concrete seal is then used to exclude water. The air pressure is reduced to atmospheric pressure, the air locks removed and the remainder of the working chamber and shafts filled with concrete. Pneumatic caissons, used for foundations located in water, are usually built on shore and floated into position. If the top is made air-tight the air in the working chamber will add to the buoyancy. Additional buoyancy may be obtained by means of a temporary water-tight and air-tight bottom. The sinking of the caisson through the water is accomplished by pouring concrete inside the walls of the cofferdam. When the caisson reaches the bottom the procedure is similar to that described for the pneumatic caisson sunk through earth. The sinking of the pneumatic caisson may be controlled more readily than that of the open caisson. Obstructions such as boulders or sunken logs can be easily removed and the surface under the caisson properly prepared and inspected. The concrete seal does not have to be poured through water as in the case of the open caisson. One serious objection is that the depths is controlled by the fact that there is a limit to the intensiy of the air pressure in the working chamber which workmen can stand safely. (C.W.C.)

CAISSON DISEASE (COMPRESSED AIR ILLNESS, "THE BENDS," DIVER'S PALSY). This disease occurs in individuals who, having been exposed to increased air pressure, are subjected to too abrupt reduction in the pressure.

At present, with engineering operations in digging tunnels and underwater supports, and in deep-sea diving, there are numerous opportunities for this disease. Great precautions, however, are taken to prevent its occurrence. On entering or leaving the caisson where air pressures of 30 to 40 lbs. are present, air locks are provided where gradual compression and decompression are produced. In spite of these precautions a number of cases of caisson sickness develop. The same disease occurs from rapid ascension to a great height, as in high-altitude flying. Descent to a higher air pressure relieves all symptoms.

The cause of this disorder is the rapid liberation of nitrogen from the fluids and tissues of the body in going too rapidly from high to low pressures. The presence of this nitrogen is due to the increased solubility of the gas under the higher pressures as in the caisson. The liberated gas forms bubbles in nerve tissue, muscles, brain, and other body tissues.

The symptoms produced are localized pain in the abdomen or extremities, vertigo, sensory or motor disturbances, dyspnea, collapse and unconsciousness.

Cases treated early usually make a complete recovery. Those that reach the stage of collapse usually do not recover.

The only treatment is recompression and slow decompression. The victim is placed as quickly as possible in an air lock and the air pressure increased until all the air bubbles go into solution. The pressure is then very gradually lowered until normal air pressure is realized. (R.S.M., D.M.H.)

CALAMARY. Mollusca, Cephalopoda. The true squids, *Loligo,* including the common squid of the Atlantic Coast and related species. (A.W.L.)

CALAMINE. The mineral calamine is zinc silicate (H_2ZnSiO_5) occurring in tabular and prismatic **orthorhombic** crystals although often in massive and fibrous forms. There is a perfect cleavage parallel to the prism; it is brittle with a sub-conchoidal fracture; hardness, 4.5–5; specific gravity, 3.40–3.50; luster, vitreous; color, white, tending toward light bluish or greenish shades; streak, white; transparent to translucent. Calamine differs from **willemite**, also a zinc silicate, in that the former contains considerable water which may be driven off when heated to a high temperature.

There are many localities for calamine in Europe, fine specimens having come from Saxony, Sardinia; Cumberland, Alston Moor and Derbyshire, England. It is found in Siberia, Algeria and Mexico. In the United States calamine has been found at Sterling Hill, New Jersey; in Lehigh County, Pennsylvania, and in Virginia, Missouri, Montana, Colorado, Utah, New Mexico and Nevada.

The name calamine is said to have been derived from the Latin *calamus,* a reed, in reference to its occurrence in slender stalactitic forms. It is an important ore of zinc.

Calamine has also been called hemimorphite because of the tendency to form doubly terminated crystals showing a different grouping of faces at either end. The name is derived from the Greek meaning half and form. (E.S.C.S.)

CALAMITES. Paleobotony.

CALAMISTRUM. A comb-like structure on the fourth leg of some **spiders.** It is formed of stiff hairs and is used to manipulate silk for certain purposes. (A.W.L.)

CALANDRA. Aves, Passeriformes. A European **lark,** noted for its song. The name is sometimes applied to other related species of the Old World. (A.W.L.)

CALANDRIA. The part of a vacuum-evaporating apparatus wherein the liquid being concentrated is circulated through tubes that are surrounded by steam to furnish heat. (R.K.S.)

CALAVERITE. Sylvanite.

CALCAREA. A class of the phylum **Porifera** containing **sponges** whose spicules are calcareous. They are marine animals exclusively.

The sponges of this class include the simplest of the entire phylum. Some are of the ascon type, with canals passing completely through the body wall, and others are sycon sponges with two sets of canals, the incurrent leading into the body wall from the exterior and the radial leading from the body wall to the interior of the sponge. Common sponges are available to illustrate both forms, *Leucoselenia* representing the former and *Grantia* the latter. These structural differences and the examples mentioned characterize the two orders into which the class is divided:

Order Homocoela. Ascon sponges; *Leucoselenia.*
Order Heterocoela. Sycon sponges; *Grantia.* (A.W.L.)

CALCAREOUS. Limestone.

CALCAREOUS ALGAE. Paleobotany.

CALCAREOUS TUFA. Tufa.

CALCIFEROUS GLAND. Oesophageal Gland.

CALCINATION. The process of calcination involves the subjection of a substance to a high temperature below its **fusion** point, usually to make the substance friable. Material so treated may (1) lose moisture, e.g., the heating of **silicic acid** or **ferric** hydroxide resulting in the formation of silicon oxide or ferric oxide, respectively, (2) lose a volatile constituent, e.g., the heating of **limestone** (calcium carbonate) resulting in the formation of **carbon dioxide gas** and **calcium** oxide residue—destructive distillation of many organic substances is of this type—(3) be oxidized or reduced, e.g., the heating of pyrite (iron disulfide) in air resulting in the formation of **sulfur dioxide** gas and **ferric** oxide residue. When the calcination involves oxidation, as in the preceding case, the operation is termed roasting. When heating involves reduction of metals from their ores, with separation from the gangue of the liquid metal and slags the process is termed smelting. (See **Smelting.**) (R.K.S.)

CALCIPHILE. Plants which require an abundance of **calcium** in the soil for satisfactory growth are known as calciphiles. Many important agricultural plants, as for example **alfalfa** and many of the clovers, are of this sort. This has led to the practice of liming the soil when these crops are planted. (R.M.W.)

CALCIPHOBES. Plants which cannot tolerate **calcium** are called calciphobes. Such plants, moreover, prefer an extremely acid soil. Members of the heath family are of this group. Rhododendrons, for instance, so admired for their beautiful showy flowers and thick evergreen leaves, often fail to grow satisfactorily due to an excess of lime or calcium in the soil. Working into the soil around the plants an abundance of leaves often helps to correct this condition. The effect may be due to the toxic action of the calcium, or may be because the calcium prevents the absorption of other elements necessary to the plant. (R.M.W.)

CALCITE. The mineral calcite, **carbonate** of **calcium** corresponding to the formula $CaCO_3$, is one of the most widely distributed minerals. Its crystals are **hexagonal-rhombohedral** although actual calcite rhombohedrons are rare as natural crystals. However, they show a remarkable variety of habit including acute to obtuse rhombohedrons, tabular forms, prisms, or various **scalenohedrons.** It may be fibrous, granular, lamellar or compact. The cleavage in three directions parallel to rhombohedron is highly perfect; fracture, conchoidal but difficult to obtain; hardness, 3; specific gravity, 2.7; luster, vitreous in crystallized varieties; color, white or colorless through shades of gray, red, yellow, green, blue, violet, brown, or even black when charged with impurities; streak, white; transparent to opaque; it may occasionally show phosphorescence or fluorescence. Calcite is perhaps best known because of its power to produce strong **double refraction** of light such that objects viewed through a clear piece of calcite appear doubled in all of their parts. A beautifully transparent variety used for optical purposes comes from Iceland, for that reason being called Iceland spar.

Acute scalenohedral crystals are sometimes referred to as dogtooth spar.

Calcite represents the stable form of calcium carbonate; aragonite will go over to calcite at 470° C.

Calcite is a common constituent of **sedimentary** rocks, as a vein mineral, and as deposits from hot springs and in caverns as stalactites and stalagmites.

Among many localities noted for fine crystals of calcite are: Saxony; Cumberland, England; Guanajuato, Mexico; St. Lawrence County, N. Y.; West Paterson and Bergen Hill, N. J.; and Joplin, Mo. (E.S.C.S.)

CALCIUM. Symbol: Ca. Atomic number: 20. Atomic weight: 40.08. Density: 1.54. Hardness: 1.5. Melting point: 810°. Boiling point: 1170° C. Isotopes: 40 (96.76%), 42 (0.77%), 43 (0.17%), 44 (2.30%).

Calcium is a silver-white metal, somewhat malleable and ductile; stable in dry air but in moist air or with water reacts to form calcium hydroxide and hydrogen gas; when heated burns in air to form calcium oxide emitting a brilliant light. Discovered by Davy in 1808.

Calcium occurs generally in rocks, especially limestone (average 42.5% CaO) and igneous rocks (average 5.0% CaO); as the important minerals **limestone** (calcium carbonate, $CaCO_3$), **gypsum** (calcium sulfate dihydrate, $CaSO_4 \cdot 2H_2O$), **phosphorite,** phosphate rock (calcium phosphate, $Ca_3(PO_4)_2$), apatite (calcium phosphate-fluoride, $Ca_3(PO_4)_2$ plus CaF_2), **fluorite,** fluorspar (calcium fluoride, CaF_2); in bones and bone ash as calcium phosphate, and in egg shells and oyster shells as calcium carbonate. Calcium is the fifth element in abundance, constituting 3.6% of the earth's solid shell; is the most abundant metallic element of the human body (2.0% Ca), is the third most abundant metallic element of vegetation (0.6% Ca, exceeded by potassium and by sodium).

Calcium metal is produced in small amount by **electrolysis** of anhydrous fused calcium chloride, finding a limited use as a reducing metal, and alloyed with lead (0.04% Ca) for sheathing electric cables. Calcium compounds are among the most common metallic compounds encountered in industry.

Acetate: Calcium acetate, "acetate of lime," "lime pyrolignite" ($Ca(C_2H_3O_2) \cdot H_2O$), white solid (technical is "gray" or "brown"), solubility: at 0° C., 27.2 grams; at 40° C., 24.9 grams, at 80° C., 25.1 grams of anhydrous salt per 100 grams saturated solution, formed by reaction of calcium carbonate or hydroxide plus **acetic acid**. (A common source of acetic acid is "pyroligneous acid," a crude product from wood distillation.) Used (1) as one of the sources of acetic acid, (2) as one of the sources of acetone, (3) in dyeing and printing cotton goods.

Aluminates: Calcium aluminates, four in number, have been prepared by high temperature methods and identified ($3CaO \cdot Al_2O_3$, at 1535° C. decomposes with partial fusion; $5CaO \cdot 3Al_2O_3$, melting point 1455° C.; $CaO \cdot Al_2O_3$, melting point 1590° C.; $3CaO \cdot 5Al_2O_3$, melting point 1720° C.). These aluminates are important in connection with Portland cement.

Aluminosilicates: Calcium aluminosilicates, two in number, have been prepared by high temperature methods and identified ($2CaO \cdot Al_2O_3 \cdot SiO_2$, gehlinite; $CaO \cdot Al_2O_3 \cdot 2SiO_2$, anorthite). Calcium aluminosilicate mixtures are important in connection with Portland **cement**, and metallurgical **slags**.

Arsenate: Calcium arsenate, arsenate of lime ($Ca_3(AsO_4)_2$), white precipitate, formed by reaction of soluble calcium salt solution and **sodium** arsenate solution. Used as an insecticide.

Arsenite: Calcium arsenite, arsenite of lime ($Ca_3(AsO_3)_2$), white precipitate by reaction of soluble calcium salt solution and **sodium** arsenite solution. Used as an insecticide and germicide.

Borates: Calcium borates are found in nature as the minerals **colemanite** ($Ca_2B_6O_{11} \cdot 5H_2O$), **borocalcite** ($CaB_4O_7 \cdot 4H_2O$), and pandermite ($Ca_2B_6O_{11} \cdot 3H_2O$).

Bromide: Calcium bromide ($CaBr_2 \cdot 6H_2O$), white solid, soluble, formed by reaction of calcium carbonate or hydroxide plus **hydrobromic acid**, and then evaporation. Used in medicine and in photography.

Carbide: Calcium carbide, "carbide" (CaC_2), grayish-black solid, reacts with water yielding **acetylene** gas and calcium hydroxide, formed at **electric furnace** temperature from calcium oxide plus **carbon** (coke). Used as the starting point for many organic chemicals (1) from acetylene, (2) from calcium cyanamide.

Carbonate: Calcium carbonate ($CaCO_3$), found in nature as **calcite**, Iceland spar, marble, limestone, coral, chalk, shells of **mollusks, aragonite**. This is the most widely distributed compound of calcium in nature, and the source of calcium for many manufacturing processes, since it is (1) readily dissolved by acids forming the corresponding calcium salt, (2) converted to calcium oxide upon heating. Aragonite is an unstable form at room temperature, although no change is observable until heated, when at 470° C. it is quickly converted into calcite; calcium hydrogen carbonate, calcium bicarbonate, $Ca(HCO_3)_2$, colorless solution, formed by reaction of calcium carbonate and carbonic acid, either in a test tube or in limestone regions where this action accounts for the caves and holes encountered. When the solution loses carbon dioxide by exposure to air or by warming, calcium carbonate is precipitated, e.g., in steam boilers or stalactites and stalagmites of limestone caves.

Chloride: Calcium chloride ($CaCl_2 \cdot 6H_2O$), white solid, soluble, absorbs water from moist air, formed by reaction (1) of calcium carbonate or hydroxide plus **hydrochloric acid**, and then evaporation, (2) of calcium hydroxide plus ammonium chloride in the reaction for the production of ammonia gas. The latter is the commercial source, where calcium chloride is a final product of the "ammonia-soda" process for **sodium** carbonate. Obtainable commercially as fused anhydrous, and as flakes. Used as dust preventative, as brine for refrigeration, in freezing mixtures and solutions, as a drying agent for gases (not ammonia), as a dehydrating agent, **as a wood** preservative.

Chromate: Calcium chromate ($CaCrO_4$), yellow solid, soluble, formed by the reaction of chrome iron ore and calcium oxide heated to a high temperature in a current of air and then extraction with water. Used as a pigment.

Citrate: Calcium citrate ($Ca_3(C_6H_5O_7)_2 \cdot 4H_2O$), white solid, solubility: at 18° C. 0.085 gram crystals per 100 grams of water, at 30° C. 2.2 grams, former by reaction of calcium carbonate or hydroxide plus citric acid solution. Used in effervescent beverages.

Cyanamide: Calcium cyanamide ($CaCN_2$), white solid, formed (1) by fusing cyanamide or urea with calcium oxide, sublimes at 1050° C., (2) industrially, by heating calcium carbide at 1100–1200° C. in a current of **nitrogen**. The mass is black due to the separation of carbon. Used (1) as a nitrogenous **fertilizer**, (2) as the starting point for the manufacture of various organic chemicals, and sodium cyanide.

Fluoride: Calcium fluoride, **fluorite**, fluorspar (CaF_2), white precipitate, formed by reaction of soluble calcium salt solution plus **sodium** fluoride solution. Used (1) as a flux in the production of metallurgical slags, (2) as a source of hydrofluoric acid by reaction with concentrated sulfuric acid.

Formate: Calcium formate ($Ca(CHO_2)_2$), white solid, solubility: at 0° C. 13.90 grams, at 40° C. 14.56 grams, at 80° C. 15.22 grams of anhydrous salt per 100 grams saturated solution, formed by reaction of calcium carbonate or hydroxide plus **formic acid**, and then evaporation. Calcium formate, when heated with a calcium salt of a carboxylic acid higher in the series, yields an aldehyde.

Furoate: Calcium furoate ($Ca(C_4H_3O \cdot COO)_2$), formed by reaction of calcium carbonate or hydroxide plus furoic acid. Used for preparing fungicides and bactericides.

Hydride: Calcium hydride (CaH_2), white solid, reacts with water yielding **hydrogen** gas and calcium hydroxide; when electrolyzed in fused potassium lithium chloride, hydrogen behaves like chloride and is liberated at the anode. Used as an easily regulated reducing agent in organic chemistry.

Hydroxide: Calcium hydroxide, hydrated lime, slaked lime, milk of lime, lime water ($Ca(OH)_2$), white solid, soluble slightly (about 0.2 gram per 100 milliliters of water), formed (1) by reaction of calcium oxide and water, with the accompanying evolution of much heat, (2) by precipitation of soluble calcium salt solution and **sodium** hydroxide solution. Used (1) as an important and cheap alkali, and so used in many chemical reactions to form calcium salts or to liberate **ammonia** gas from **ammonium** salts, (2) in lime, mortar, and cements, frequently mixed with sand.

Hypochlorite: Calcium hypochlorite, "H.T.H." (High Test Hypochlorite) ($CaOCl_2$) white solid, emits a marked odor in air, contains 60%–65% "available chlorine" and sufficient calcium hydroxide to stabilize, formed by reaction of calcium hydroxide and **chlorine** and evaporating at low temperature. Used as **bleaching** agent for fabrics, as a disinfectant, for the preparation of dilute solution of sodium hypochlorite in laundering, and in the treatment of wounds; "chloride of lime," "bleaching powder" ($CaOCl_2$), white solid, contains 36%–38% "available chlorine." Used as above.

Hypophosphite: Calcium hypophosphite ($Ca(H_2PO_2)_2$), white solid, formed (1) by boiling calcium hydroxide suspension in water and yellow **phosphorus** (phosphine gas, poisonous, evolved), and then evaporation, (2) by reaction of calcium carbonate or hydroxide plus **hypophosphorous** acid, and then evaporation. Used in medicine.

Iodide: Calcium iodide (CaI_2), yellowish-white solid, soluble, formed by reaction of calcium carbonate or hydroxide plus **hydriodic acid**, and then evaporation. Used in photography.

Lactate: Calcium lactate ($Ca(C_3H_5O_3)_2 \cdot 5H_2O$), white solid, solubility: at 0° C. 3.1 grams, at 30° C. 7.9 grams of anhydrous salt per 100 grams water, formed by reaction of calcium carbonate or hydroxide plus **lactic acid,** and then evaporation. Used in medicine.

Malate: Calcium malate ($CaC_4H_4O_5 \cdot 2H_2O$), white solid, solubility: at 0° C. 0.670 gram, at 37.5° C. 1.011 grams of anhydrous salt per 100 grams saturated solution. Formed (1) by reaction of calcium carbonate or hydroxide plus **malic acid,** (2) by precipitation of soluble calcium salt solution and sodium malate solution.

Nitrate: Calcium nitrate, "Norway salpeter" ($Ca(NO_3)_2 \cdot 4H_2O$), white solid, soluble, formed by reaction of calcium carbonate or hydroxide plus **nitric acid,** and then evaporation. Used widely as a nitrogenous fertilizer (initially manufactured in Norway by neutralizing the dilute nitric acid produced in the fixation of atmospheric **nitrogen** by the electric arc process). Also used in pyrotechnics and as a source of nitrate for other nitrates.

Oxalate: Calcium oxalate (CaC_2O_4), white precipitate, insoluble in weak, soluble in strong, acids, formed by reaction of soluble calcium salt solution and **ammonium** oxalate solution. Solubility: at 18° C. 0.0056 gram anhydrous salt per liter of saturated solution.

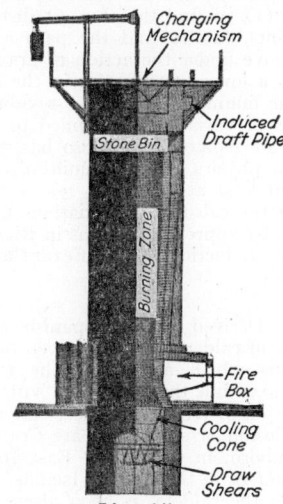

Lime kiln.

Oxides: Calcium oxide, lime, burnt lime, quick lime, live lime, caustic lime (CaO), white solid, melting point 2570° C.; reacts with water to form calcium hydroxide with the evolution of much heat; reacts with water vapor and carbon dioxide of the atmosphere to form calcium hydroxide and carbonate mixture (air slaked lime); formed by heating limestone to a high temperature (800° C.) and removal of carbon dioxide also formed. An important and cheap alkali, which is used (1) as calcium hydroxide with water, (2) as calcium carbonate for high temperature reactions (carbon dioxide evolved as gas); calcium peroxide (CaO_2), white solid, only slightly soluble in, but reacts slowly with, water, formed by reaction of soluble calcium salt solution and sodium peroxide. Used as a germicide.

Phosphates: Tricalcium phosphate, tertiary phosphate of lime, phosphorite, bone ash ($Ca_3(PO_4)_2$), white solid, the source of practically all phosphate and phosphorus-containing substances; insoluble in water; reactive with **silicon** oxide and carbon at electric furnace temperature yielding **phosphorus** vapor (later condensed under water) plus carbon monoxide plus calcium silicate molten slag; reactive with sulfuric acid to form, according to the proportions used, the two following calcium phosphates or **phosphoric acid,** dicalcium hydrogen phosphate, secondary phosphate of lime, reverted phosphate (CaHPO₄), white solid, insoluble, used as a source of acid in **baking powders** for leavening; calcium dihydrogen phosphate, primary phosphate of lime, superphosphate ($Ca(H_2PO_4)_2 \cdot H_2O$), white solid, soluble, formed

(1) by reaction of tricalcium phosphate and sulfuric acid, yielding calcium dihydrogen phosphate plus calcium sulfate, the mixture constituting the widely produced and utilized "superphosphate" **fertilizer,** (2) by reaction of tricalcium phosphate and sulfuric acid yielding phosphoric acid plus calcium sulfate, and, after separation, the resulting phosphoric acid is treated with tricalcium phosphate yielding calcium dihydrogen phosphate, known as "treble superphosphate"; used also as a source of acid in baking powders for leavening.

Silicates: Calcium silicates, four in number, have been prepared by high temperature methods, and identified, $3CaO \cdot SiO_2$, prepared by heating the constituents to a temperature below the melting point (melting point is 1700° C. but substance unstable); $2CaO \cdot SiO_2$, melting point 2080° C., but upon slow cooling changes to forms of different volume; $3CaO \cdot 2SiO_2$, melting point 1475° C.; $CaO \cdot SiO_2$, wollastinite, melting point approximately 1400° C. These silicates are important in connection with Portland cement.

Sulfate: Calcium sulfate, **gypsum** ($CaSO_4 \cdot 2H_2O$), plaster of Paris ($CaSO_4 \cdot \frac{1}{2}H_2O$), **anhydrite** ($CaSO_4$), white solid, soluble slightly (about 0.2 gram per 100 milliliters of water), formed by reaction of soluble calcium salt solution and sodium sulfate solution. Used (1) in plaster, stucco and cement, frequently mixed with fillers.

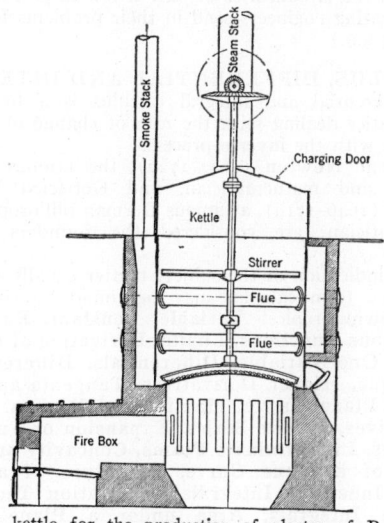

Calcining kettle for the production of plaster of Paris from gypsum.

Dry plaster is gypsum which has been heated until about one-fourth of the original water remains. This dry plaster, when mixed with water, can be worked or shaped but soon sets to a coherent and adherent solid, which is composed of interlacing crystals of gypsum, (2) as a filler for **paper** and **paints.**

Sulfides: Calcium sulfide (CaS), grayish-white **solid,** reactive with water, formed by reaction of calcium sulfate and **carbon** at high temperatures. Used as a depilatory, and in luminous paints; calcium hydrogen sulfide, calcium bisulfide ($Ca(HS)_2$), formed in solution by saturating calcium hydroxide suspension with hydrogen sulfide. Used as a depilatory.

Sulfites: Calcium sulfite ($CaSO_3$), white precipitate, formed by reaction of soluble calcium salt solution and **sodium** sulfite solution, or by boiling calcium hydrogen sulfite solution; calcium hydrogen sulfite, calcium bisulfite, sulfite liquor ($Ca(HSO_3)_2$), formed in solution by saturating calcium hydroxide or carbonate suspension with **sulfurous acid.** Used as cooking liquor (with excess sulfurous acid) in converting wood chips into paper pulp.

Tartrate: Calcium tartrate ($CaC_4H_4O_6 \cdot H_2O$), white solid, solubility: at 0° C. 0.0875, at 80° C. 0.180 gram anhydrous salt in 100 milliliters saturated solution, formed (1) by reaction of calcium carbonate or hydroxide and **tartaric acid,** (2) by precipitation of soluble calcium salt solution and **sodium** tartrate solution. (R.K.S.)

CALCULATOR, NETWORK.
The network calculator is an elaborate layout of **inductance, capacitance,** and **resistance** units, and **generators** used for duplicating the electrical characteristics of a power system so various system conditions may be studied. These components are connected to panels where, by means of switches or plug connections, a network of them may be wired up relatively easily. This network is given the same electrical characteristics as the system to be analyzed, that is, the lumped values of the calculator are set for the values equivalent to the actual system values. Then by connecting electrical **instruments** in the network the electrical quantities in any part of the network may be measured. These measurements give the results which would be obtained in the real transmission system if a similar operation had been performed on it. Thus by duplicating system conditions in the laboratory with the calculator and measuring the electrical results of various operations the solutions to many involved problems in transmission work may be found without laborious mathematical calculations. These calculators are set up in the laboratories of only a few large companies and operating engineers send in their problems for solution. (L.R.Q.)

CALCULUS, DIFFERENTIAL AND INTEGRAL.
The differential and integral calculus is a branch of mathematics dealing with the rate of change of a function and with the inverse process.

Sir Isaac **Newton** (1642–1727), the famous English scientist and mathematician, and Gottfried Wilhelm Leibnitz (1646–1716), a famous German philosopher and mathematician, are considered the founders of the calculus.

Some indication of the subject matter usually included under the term calculus may be found by consulting the following topics: **Variable, Constant, Functions, Continuous Functions, Limits, Derivative of a Function of One Variable, Differentials, Differentiation Technique, Higher Derivatives, Tangents and Normals to Plane Curves, Maxima and Minima, Partial Derivatives, Infinite Series, Expansion of Functions in Series, Indeterminate Forms, Concavity and Convexity of a Plane Curve, Curvature of a Plane Curve, Indefinite Integrals, Integration Technique, Definite Integrals, Area under a Plane Curve, Length of Plane Curve Arc, Double Integral, Triple Integral, Volumes by Double Integrals, and by Triple Integrals, Line Integrals, Differential Equations.** (L.L.S.)

CALCULUS OF FINITE DIFFERENCES.
The calculus of finite differences is a study of methods of treating problems involving differences of functions.

If we have given a sequence $u_0, u_1, u_2, \cdots, u_n, \cdots$, and if we form the differences of consecutive terms of this sequence, as $u_1 - u_0, u_2 - u_1, \cdots$, we call these differences the first differences of the original sequence and denote them by $\Delta u_0, \Delta u_1, \Delta u_2, \cdots, \Delta u_n, \cdots$, so that $\Delta u_n = u_{n+1} - u_n$. We may then form the first differences of these first differences and obtain second differences, denoted by $\Delta^2 u$, so that $\Delta^2 u_n = \Delta(\Delta u_n) = u_{n+2} - 2u_{n+1} + u_n$; and similarly we may form third and higher differences.

The study of the properties of successive differences is applied to problems in **interpolation** and quadratures (**approximate integration**), to problems in the summation of series, and also to the solution of **difference equations.** (L.L.S.)

CALCULUS OF VARIATIONS.
The calculus of variations is concerned with the problem of maxima and minima of functions of functions. In the ordinary differential calculus, the problem of extreme values of functions is dealt with, in which a function of one or more independent variables is examined to determine the values of the variable or variables for which the function takes maximum or minimum values. The calculus of variations is a natural extension of this problem. In the calculus of variations, a curve or surface (or corresponding function) plays the role of the independent variable of the former problem.

The function to be minimized or maximized in the calculus of variations is essentially of the form

$$\int_a^b f(x,y,y')dx,$$

where y is a function of x and $y' = dy/dx$, or some generalization of this. The central problem of the calculus of variations is then to find the function $y = F(x)$ or the corresponding curve which will make the above integral take a minimum value or a maximum value.

A few of the simplest problems of the calculus of variations are (1) to find the shortest distance between two given points, (2) to find the path along which a particle will move under the action of gravity from one given point to a lower given point in the shortest time, (3) to find the minimum surface of revolution.

The calculus of variations is applied to many important problems of geometry, and also has many valuable applications to physics, as in Hamilton's principle and the principle of least action.

Problems in the calculus of variations may be solved either directly, by approximation, as in Ritz's method, or indirectly, by reduction to **differential equations.** (L.L.S.)

CALDERA.
Derived from a Spanish word meaning caldron, the term caldera has been given to great craterlike depressions which are either the result of subsidence of lava within the body of a **volcano** or of an explosive eruption of terrific violence. Examples of these craters of explosion or subsidence are Crater Lake, Oregon, Mt. Tamboro, in the Dutch East Indies, and the original *La Caldera* in the Canary Islands. Crater Lake, which occupies the caldera, is 2000' deep and about 25 sq. mi. in area, surrounded by cliffs whose maximum height is 2000' above the lake. (E.S.C.S.)

CALEDONIAN DISTURBANCE. Silurian.

CALENDAR.
The problem of time-keeping has always been a vexing one to mankind. There are three "natural" units, the solar **day,** the lunar **month,** and the tropical **year.** The normal or true solar day had to be abandoned with the improvement of mechanical time-keeping devices and the mean solar day has been adopted as the standard short unit for keeping records. The task of the calendar builder is to combine this unit with the two longer units and, since the three are mutually incommensurable, a rigorous solution of the problem is impossible and compromises must be made.

The fact that the economic world is largely dependent upon agriculture introduces one important restriction on the freedom of the calendar builder. The seasons should remain at approximately the same place in the completed calendar from year to year. The date upon which the sun apparently passes through the **vernal equinox** is of fundamental importance to the agriculturalist and for many centuries was considered as the time of starting a new year. One of the earliest calendars on record started the year on this date and then proceeded through ten lunar months. Such a calendar would cover only 295.3 mean solar days, while the period from one passage of the sun through the vernal equinox to the next is 365.2422 days. The period between the end of one year to the beginning of the next was determined by the priesthood and politicians with hopeless confusion resulting.

The first step toward the modern calendar was made by Julius Caesar with the advice of the astronomer Sosigenes. The so-called Julian Calendar discards the lunar month and adopts 365.25 days as the length of the year. This year is divided into twelve periods (months) of 30 or 31 days. The normal year was 365 days in length but, to make up the extra ¼ day, an extra day was intercalated (i.e., put into the normal calendar) every four years.

Running parallel with the Julian Calendar we find the far more ancient calendar of the Jewish and Mohammedan peoples, which holds rigorously to the lunar month. Division of the number of days in the tropical year by the days in the lunar month will indicate that there are 12.36 lunar months in a tropical year. To retain the synchronism between the calendar and the seasons, this calendar is variable in the number of months which it contains and the process of intercalating months becomes very complicated. However, the Eastern calendar exerts a powerful effect upon the calendar of the Western world, because of the fact that the date of Easter is fixed by a date on the Eastern calendar.

In A.D. 325 the Christian Church took its first step in calendar building and at the Council of Nice made two decrees: a decree that the sun should pass through the vernal equinox on the 21st of March on the Julian Calendar, and a second decree relative to the date for the celebration of Easter. The latter of the two decrees was within the province of the Church and can be followed; the former, however, applies to factors beyond the control of man.

It should be noted that the length of the tropical year is 0.0078 day less than the 365.25 days of the Julian Calendar. This means that after the lapse of 1000 years the sun will pass through the vernal equinox 7.8 days earlier than the 21st of March, assuming it was at the vernal equinox on this date in the first place. By 1582 the date of the vernal equinox was the 11th of March instead of the 21st and Pope Gregory decided to return the sun to its proper date and to modify the calendar in such a way that the error would not reappear. The Gregorian Calendar is identical with the Julian except in the fact that only such century years are leap years as are divisible by 400. This is equivalent to dropping 3 days every 400 years leaving an average length for the year of 365.2425 days which differs from the tropical year by only 0.0003 day. This calendar was immediately adopted by all Catholic countries but the Greek Church and most Protestant countries refused to recognize it. The confusion following this change persisted well down into the present century (Rumania used the Julian Calendar until 1919) and is still felt by historians in reading records of the early years of this country when both calendars were in use.

Within the past 20 years a strong movement has been on foot to modify the calendar in the attempt to have dates and days of the week agree in successive years. Any such scheme involves the necessity of introducing one day each year without date or day of the week, and two such days on leap years, if the year and the seasons are to retain the present synchronism. This intercalation of a day will break the 6-day sequence between Sabbaths, an idea which is abhorrent to many religious sects. The scheme which has the most general support is one in which the year is divided into four equal quarters of 3 months each. In each quarter the first month has 31 days and the second and third, 30 each. This gives exactly 13 weeks in each quarter, and 52 weeks in each year. The days are to be intercalated without date or day of the week between December 30 and January 1 each year and between June 30 and July 1 every leap year (e.g., the normal calendar would read Saturday, Dec. 30; New Year's Day; Sunday, Jan. 1). (w.k.g.)

CALF. 1. The young of cattle and of related wild species. 2. The fleshy mass formed by the muscles at the back of the leg below the knee. (A.W.L.)

CALICHE. An important natural fertilizer occurring in Chile as a deposit containing sodium nitrate and other soluble salts. These deposits were of great economic importance, prior to the development of present-day methods for the fixation of atmospheric nitrogen. Geologists more commonly apply the term to a calcareous deposit which in semi-arid regions accumulates at and near the surface of the ground through evaporation of capillary water. (R.M.F.)

CALIPASH. A structure beneath the carapace of turtles containing a greenish fat, sometimes regarded as a delicacy. (A.W.L.)

CALIPEE. A structure lying neor the plastron of the turtles, similar to the calipash. (A.W.L.)

CALIPERS. Measurement.

CALKING. Forging.

CALL INDICATOR. Continuity and speed of service are among the primary objectives of the telephone companies. This would be fairly simple if all the phones on a system were manual or all automatic (see Telephony), but where a given exchange area involves both types of service the problem is more involved. In order to avoid the necessity of a dial subscriber having to call the operator and give her the desired number, the call indicator has been developed to give the manual operator the number automatically as it is dialed. Thus when a subscriber on an automatic office calls one on a manual office, he may simply dial the desired number. The call indicator circuit then causes the dialed number to appear on a board before the operator who reads the called number and completes the connection. In this way, it is not necessary for the subscriber to make any distinction between calls to other automatic subscribers or manual subscribers. (L.R.Q.)

CALLA LILIES. Aroids.

CALLIER QUOTIENT. The ratio of the density measured by diffused light to the density measured by specular (direct) light. (C.B.N.)

CALLUS. There are five common meanings with which this term is used throughout biological science.

1. Callus is a thickened horny mass found in the outer layer of the skin, which is the result of continued pressure or friction.

2. Callus designates a thickened spot near the umbilicus in the shells of certain snails.

3. Callus also denotes a tissue formed about the fragments of a broken bone. This tissue eventually develops into bone, and by this means the fracture heals.

4. Callus is a term applied to colorless glistening pads of substance which close the sieve plates of sieve tubes which have ceased to function. The formation of these callus pads usually ends the activity of the cell, although they may be dissolved later and activity resumed.

5. Callus designates a certain protective tissue which occurs widely in plants. When the root or stem of a gymnospermous or a dicotyledonous plant is wounded, exposing the tissues within, the cambium cells around the wound begin to divide rapidly, forming a protective mass of soft parenchymatous tissue. These living cells are called callus, or wound tissue, and in time will entirely close the wound if the latter is not too extensive. Callus tissue may even grow over such a wound as results when a tree is girdled, that is, when a ring of bark is removed completely round the stem. Ordinarily,

however, such a wound is fatal. After the tissue is formed, cell differentiation goes on and a new **phellogen** layer may be formed, as well as the other tissues composing the cortex of the **stem**. The **cambium** becomes once more a continuous layer. When wounds are made in pruning, that is, when a branch is cut off, callus tissues gradually form a ring which spreads over and finally completely closes the wound. (A.W.L., R.S.M., R.M.W.)

CALM. The absence of horizontal air motion is recognized as a calm. (P.E.K.)

CALOMEL. Mercury.

CALORESCENCE. This term refers to the production of visible light by means of energy derived from invisible **radiation** of frequencies below the visible range. Tyndall found it possible to raise a piece of blackened platinum foil to a red heat by focusing upon it infra-red radiation from an arc or from the sun, the visible wavelengths having been filtered out. It is to be noted that the transformation is indirect, the light being produced by heat and not by any direct stepping up of the infra-red frequency. A somewhat analogous phenomenon is the production of visible sparks or the glowing of a fine platinum wire in a resonant circuit energized by long-wave Hertzian radiation. (L.D.W.)

CALORIE. The gram calorie is the quantity of heat required to raise the temperature of one gram of pure water from 15 to 16° C. Since the specific heat of water is not quite constant, it is necessary to specify the degree interval used in defining the calorie, this one being chosen because water has its average specific heat at this point. The calorie is equal to about $\frac{1}{252}$ B.T.U., and is dynamically equivalent to about 41,855,000 ergs of energy. The heat unit thus defined is the one most commonly employed in physics. For some purposes it is convenient to use a larger unit, the kilogram calorie, which is equal to 1000 gram calories. The calorific values of foods are, for example, usually expressed in terms of this unit. (L.D.W.)

CALORIFIC VALUE. Heating Value.

CALORIMETRY. The measurement of quantity of heat; any apparatus used for the purpose being called a "calorimeter." The most obvious result of the application of heat is the rise in temperature of the body to which it is applied; hence it is natural that this readily measurable effect should be utilized in heat measurement. Since the application of heat melts solids and vaporizes liquids, these effects also furnish the basis of valuable calorimetric methods.

The common units of heat are based upon the temperature-raising effect. Thus we have the gram **calorie**, the kilogram **calorie**, and the British thermal unit. The latter, equal to about 252 gram calories, is the quantity of heat required to raise the temperature of one pound of water one Fahrenheit degree. Gas companies sometimes use a large unit called the "therm," viz., 100,000 B.T.U., but this name has also been applied to various multiples of the calorie and is hence ambiguous.

The so-called water calorimeter is essentially a thermally insulated metal cup containing water and furnished with a thermometer. The quantity of heat to be measured is applied to the water, and the rise of temperature resulting, multiplied by the mass of the water, gives the number of heat units received. Corrections must of course be made for the heat absorbed by the cup, thermometer, and other accessories, the "water equivalent" of which is in the computation simply added to the mass of actual water. Allowance must also be made for heat lost or unintentionally introduced through radiation and conduction; for which purpose **Newton's law of cooling** is commonly used.

Since the **heat of fusion** of ice is known to be very nearly 79.71 calories per gram, the heat to be measured may be applied to the melting of ice without change of temperature, and the mass of ice melted, multiplied by the heat of fusion, gives the quantity of heat. Bunsen, Lavoisier and Laplace, Black, and others devised calorimeters based upon this principle.

A steam calorimeter was perfected by J. Joly (1886) and used for the accurate determination of specific heats of solids, liquids, and gases. In principle this apparatus consists of a balance, with the specimen hung from one pan and surrounded by an enclosure which can be flooded with steam. The mass of moisture condensing on the specimen,

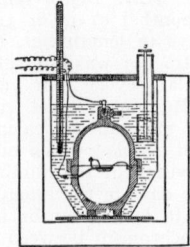

Bomb calorimeter for determining the heating value of solid and liquid fuels.

multiplied by the heat of vaporization of water, gives the quantity of heat imparted to the specimen.

Other calorimetric methods have been proposed and tested, and many special calorimeters have been designed for measuring heats of combustion of fuels, food values, heats of chemical reaction, heat from electric currents, etc.; most of these methods depend upon one or another of the principles mentioned above. (L.D.W.)

CALORIZING. Production of a protective coating of iron-aluminum alloy on iron or steel. The articles are ordinarily coated by heating to a high temperature in a closed container packed with powdered aluminum. Other processes include impregnation at high temperature with an aluminum chloride vapor and spraying with molten aluminum from a spray gun and then heating to a high temperature. When the aluminum coating is held at high temperatures, an iron-aluminum alloy forms which is resistant to oxidation and corrosion by hot combustion gases, especially those containing sulfur compounds which are particularly corrosive to bare iron or steel.

Steel sheets are aluminized by a hot-dip process similar to galvanizing. The principal applications for such a product are furnaces and ovens, automobile mufflers, and other equipment requiring heat and corrosion resistance. When a sheet which has been coated with aluminum by a hot-dip process is exposed to a temperature over 1000° F., the aluminum forms an iron-aluminum alloy which is heat- and corrosion-resistant. (R.H.H.)

CALYX. For the use of this term in botany, see **Flower**. In zoology, a calyx is a cup-shaped or funnel-like structure, such as the body of a **sea-lily** and the chambers branching from the principal cavity of the vertebrate **kidney**. (A.W.L.)

CAM. A cam is a rotating or sliding member which imparts a desired motion or series of motions to another member. Cams are used whenever a desired motion is of such character that it cannot be obtained by using cranks or linkages. There are two important forms of cams: radial cams where the follower moves in a plane perpendicular to the axis of the shaft, and cylindrical cams where the follower moves in a plane parallel to the axis of the shaft. Each of these types may be classified further as positive-motion cams in which the reciprocating motion of the follower is definitely controlled by the cam, and nonpositive-motion cams in which the follower is returned to its starting point by spring or gravity action.

Fig. 1 shows a radial cam with a flat follower or cam tappet. The cam is integral with the cam shaft. The cam profile is composed of two circular arcs connected by tangent lines. Cylindrical, helicoidal, and plane surfaces are used for cam faces whenever possible, since they are more easily and accurately manufactured than irregular curves.

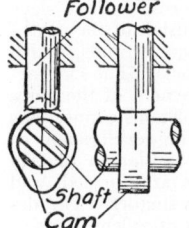

Fig. 1. Radial cam.

The radial disk cam, at the right of Fig. 2, is similar to the cam of Fig. 1. Roller followers are preferred to flat followers because the line contact between the roller and the cam is of a rolling nature, since the sliding is transferred to the pin that carries the roller. The face cam, at the left of Fig. 2, is a positive-motion cam, but is much more difficult to manufacture than a disk cam because the cam groove must be of accurate uniform width. This face cam has a cast iron disk on which the inner and outer hardened steel plates are screwed and dowelled.

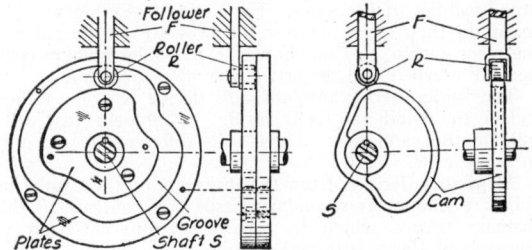

Fig. 2. *Left:* Positive-motion cam. *Right:* Radial disk cam.

Fig. 3 shows a solid cylindrical cam with a bell-crank or lever follower for the thread-controlling function on moderate-speed sewing machines. A development or

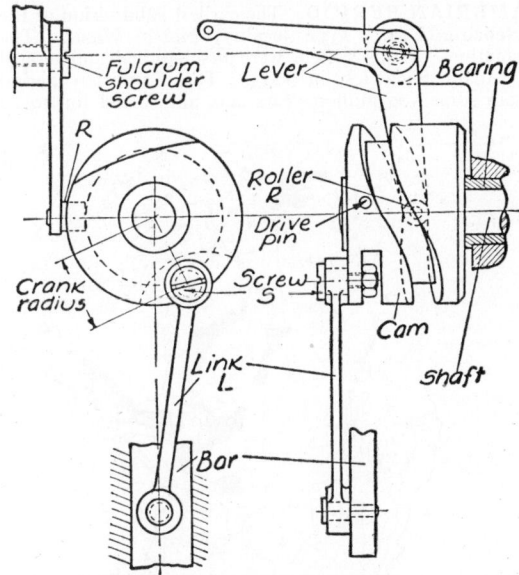

Fig. 3. Solid cylindrical cam.

layout of a portion of a cylindrical cam is shown in Fig. 4. This development shows uniform or straight-line motion of the roller, modified by an arc equal to the roller radius at the beginning and end of each phase of motion, to permit gradual acceleration and to provide roller clearance. The drum cam may have positive motion and will therefore require a cam strap on either side of the roller, or it may be constructed with a single strap, in instances where the inertia of the slide is great enough to enable the roller to remain at rest unless acted on by the cam strap. (H.C.H.)

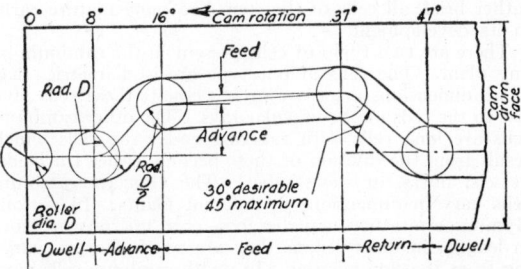

Fig. 4. Development of portion of cylindrical cam.

CAMBER. Camber is the upward curvature which is given to **bridge** trusses with theoretically horizontal lower **chords**, bridge girders with theoretically horizontal bottom **flanges** and **beam bridges** to compensate for the actual **deflection**. Although these deflections are small in a properly designed structure, they may be objectionable from the standpoint of appearance. Due to an optical illusion these structures appear to have a pronounced downward deflection. This term is also used to denote the initial curvature which occurs in steel beams as the result of rolling.

In short-span trusses camber is obtained by lengthening the top chords $\frac{1}{8}$ to $\frac{3}{16}$" for each 10' of length. No change is made in the lower chords and **verticals** but the length of the **diagonals** must correspond to the new outline. Long-**span** trusses are cambered by increasing the geometrical length of the **compression members** and decreasing the geometrical length of **tension members**. The change in length is based on the calculated longitudinal **deformation** of the members under **dead load** and partial or full **live load**.

It is not customary to camber short-span girders. Long-span girders are cambered by fabricating them with an upward curvative corresponding to a predetermined amount of deflection. This is accomplished by using two or more plates for the **web**, spliced in such a way as to produce this curvature approximately. The straight flange angles and cover plates are then bent to the desired curvature during the fitting-up operation.

Camber may be obtained in a beam bridge by placing the beams so that the initial curvature due to rolling is upward. The initial curvature may be increased by heating the flange on the concave side with a torch. (C.W.C.)

CAMBIUM. In **Gymnosperms** and **dicotyledonous Angiosperms**, a large part of the tissues of the stem is derived from a special layer of cells known as the cambium. The cambium originates from certain cells of the procambial strand. In the procambial strand of the stem (that part of the growing tip in which **cell** differentiation first takes place), cell differentiation commences at the tangential edges of the strand and progresses towards the center, forming primary xylem cells towards the center of the stem and primary **phloem** cells towards the surface. Some of the cells in the middle portion of the procambial strand do not differentiate into **xylem** or phloem, but become meristematic cells, dividing actively. These are the cambium cells. Often they begin to divide before the other cells of the procambial strand have ceased elongating.

At first the cambium is a vaguely defined layer of cells occupying the middle portion of the procambial strand. In **roots** the cambium appears on the inside of the primary phloem strands, which alternate with the primary xylem strands.

Gradually additional cells are formed laterally, either from those cambium cells already formed or by differentiation of parenchyma cells of the medullary ray, until a complete cylinder of cambium exists. Once formed, the cambium of woody plants persists throughout the life of the plant; in herbaceous plants its existence is rather brief, all cells of the stem becoming mature early in its development.

There are two types of cells present in the cambium of any plant. The cells of one type are isodiametric, that is, all dimensions are more or less equal; these cells give rise to the cells of the vascular rays. The other cambium cells are long cells with tapering ends; the cells which result from the division of these become either tracheids, vessels, fibers, or sieve tubes. The elongate cambium cells vary in dimensions in different plants. In various Gymnosperms they may be 3000–4000 microns or more in length; in dicotyledons they are much shorter, varying from 100–800 microns. In width cambium cells vary in different plants from 20–40 microns, and in thickness, or radial dimension, 5–15 microns. Cambium cells have a dense **cytoplasm** in which **vacuoles** are either lacking or very minute. Each cell of the cambium has a single **nucleus** which is usually elongated. The walls, especially the tangential ones, are very thin. Division of the cambium cells occurs in a longitudinal tangential plane, that is, the cell divides lengthwise to form two slender elongate cells, one of which lies outside the other, towards the outside of the stem or root. It is certain that the division is always mitotic (**mitosis**). One of the cells resulting from this division soon begins to change its form. If this differentiating cell is on the inside of the cambium cylinder it may elongate even more, its ends sliding by and between those of other cells about it. Presently thickening of the wall occurs through deposits of cellulose which are laid down on the primary wall. The cytoplasm of the cell gradually disappears. When mature, this cell, now a **tracheid**, is a long slender tapering cell with thick wall and no protoplasm. In the wall are numerous **simple** or **bordered pits**, which are continuous with pits of adjoining cells. In Gymnosperms, all elongate cells derived from the cambium become tracheids, except in those forms which have wood parenchyma cells. In these, transverse divisions occur to form a linear row of short cells. In angiosperms, other types of cells are formed. One of these, the fiber, differs little from the tracheid except that it has a thicker wall, in which there are few small pits. The other type is quite distinct. The cambium derivative which is going to form one of these does not elongate noticeably, but does increase greatly in diameter. As it increases, a large central vacuole forms, and the nucleus moves to a position near the middle of the end wall. At that stage, the vessel appears as a series of very large vacuolate cells separated from one another by distinct end walls. When full size is reached, secondary wall thickening occurs. Then the end wall breaks down, leaving a series of cells forming a long open tube; in many plants perforations are formed in the end wall, so that direct continuity from cell to cell exists. The tremendous increase in diameter of the vessel cells causes the cells around it to be flattened and crowded into angular shapes and irregular arrangements. Once the walls have formed and the cell matured, no further change takes place. Its structure is fixed permanently.

The cells which are formed externally to the cambium become phloem cells. The manner of differentiation is not so well known in these cells as in the xylem cells. Apparently divisions of these phloem mother cells, cut off from the cambium cells, are much more frequent than are divisions of the xylem mother cells. Phloem parenchyma results from the transverse division of one of these cells to form a longitudinal series. In angiosperms each phloem mother cell divides unequally, cutting off a very small cell from the corner of the mother cell. The larger cell forms part of a sieve tube, the smaller becomes a companion cell. Often the companion cell divides again to form two or more companion cells associated with a single sieve tube. The cytoplasm of the companion cells remains dense, develops few vacuoles, and always has a well-developed nucleus. In the sieve tube, on the contrary, the cytoplasm becomes peripheral, and

there is a large central vacuole. The nucleus has disappeared in the mature sieve tube. The end walls of the sieve tube cells are characterized by the presence of porous places called sieve plates. The pores of these sieve plates result from the enlargement or fusion of the protoplasmic strands, known as plasmodesma strands, which connect the protoplasts of adjoining cells. The enlargement of these strands causes an enlargement of the pores through which they pass, so that conspicuous connections are formed between adjacent cells. The development of sieve tubes in Gymnosperms is very similar to that in angiosperms, but no companion cells are formed, and the pores in the sieve plates are much smaller. The development of phloem fibers is like that of xylem fibers.

It is obvious that with continued formation of xylem cells inside the cambium and consequent increase in stem diameter, the cambium is constantly being pushed outward and stretched. Gliding growth of cambium cells and those cut off from them causes increase in circumference of the cambium cylinder and so prevents any breaking of the same. For a time the phloem cells maintain their shape against the pressure of the enlarging stem within. In time, however, the older phloem cells become crushed and distorted beyond recognition.

The isodiametric cambium cells divide to form either xylem or wood ray cells inside, or phloem ray cells outside the cambium. These cells differentiate directly into ray cells.

All tissues derived from the divisions of the cambium cells are known as secondary tissues, in contrast to the primary tissues, which are formed by differentiation of the cells of the procambial strands.

Another cambium, the cork cambium (formerly called phellogen), arises in the pericycle of roots and in the outer cortex of stems. It produces cork and secondary cortex. (See **Bark**.) (R.M.W., B.S.M.)

CAMBRIAN PERIOD. The earliest subdivision of the **Paleozoic Era.** Type locality, North Wales. The formations of this system were first studied and named by Adam Sedgwick in 1835. The Cambrian period began some 500 million years ago, and lasted for 70–80

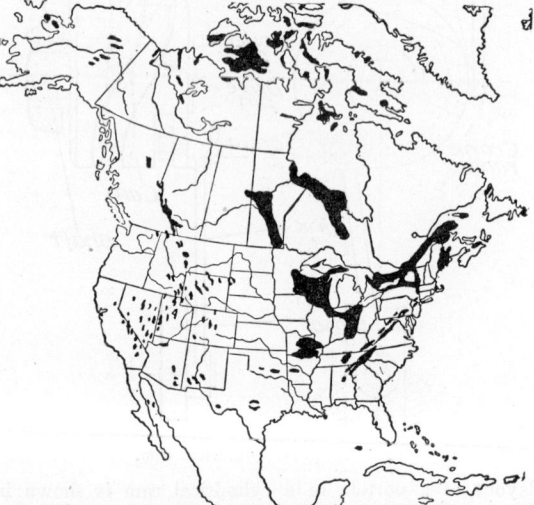

Map showing known areas of outcrops (surface distribution) of Cambrian, Ordovician, and Silurian strata in North America.

million years. Cambrian formations are well exposed in North America in the Appalachians and Rocky Mountains. Important lower Cambrian beds containing the oldest known faunas occur in British Columbia. Other countries in which the Cambrian is well exposed are Sweden, Britain, Spain, Scandinavia, France, Germany, eastern China, northeastern Siberia, India (Himalayas and Salt Range), Morocco, Australia, Argentina and Antarctica. Cambrian sediments represent the earliest evidence of deposition in well defined geosynclines, the principal types being **sandstones, shales** and **limestones.** Tillites indicate continental glaciation. The maximum thickness of 40,000′ of Cambrian strata occurs in North America. The oldest known invertebrate fossils occur in this period, the principal types being **trilobites, chitinous brachiopods,** and primitive **graptolites;** all of which had a marine habitat. It is interesting to note that the paleontological record begins with such highly developed organisms as trilobites, whose ancestors are entirely unknown in the pre-Cambrian formations. The "father" of American Cambrian stratigraphy and paleontology is Charles D. Walcott. The base of the Cambrian has not been clearly defined and in some places it probably passes without hiatus into the lower **Proterozoic.** In eastern North America the top of the Cambrian is also under debate, the upper beds forming the base of a new system called the Ozarkian by E. O. Ulrich. (R.M.F.)

CAMEL. Mammalia, Artiodactyla. Animals with long legs and necks and a conspicuously humped back. They are adapted for life in arid regions, including sand deserts, by the broad feet and slit-like nostrils; internally the development of cells for the retention of water in one part of the stomach is especially important for life in such regions.

The two existing species of camels, the Arabian, *Camelus dromedarius,* with one hump and the Bactrian, *C. bactrianus,* with two, are found in the Old World. The Arabian camel is found both in Africa and Asia and the Bactrian in the desert regions of Central Asia. Both species have been domesticated. They are used for riding and as beasts of burden, and also supply foods in the form of milk and flesh. Their hair is used in making fabrics.

The Arabian camel was introduced into the southwestern states by the United States government in 1856 but after many years of apparent success the experiment was abandoned. The animals which were freed died out after persisting for some years.

The group is represented in the New World by the **llama** and related species. (A.W.L.)

CAMEL-CRICKET. Insecta, Orthoptera. Wingless insects related to the **katydids.** They live in dark moist places and are dull colored. These facts together with the strongly humped back give them their name. They are also called cave-crickets. (A.W.L.)

CAMERA, COLOR. Color Camera.

CAMERA, MOTION-PICTURE. Motion-picture cameras may be divided into two classes: (1) those in which the film moves intermittently, being at rest during the actual exposure (intermittent movement) and (2) those in which the film moves continuously (continuous movement). Cameras of the second class are used chiefly for high-speed cinematography and are described under the heading of **high-frequency photography.**

In a motion-picture camera with intermittent movement (see Fig. 1) the film is drawn from the roll at the top and delivered to the take-up reel at the bottom by a sprocket which is driven by a spring motor, or, in professional cameras, by an electric motor. The film passes around the sprocket then through the film gate and finally around the opposite side of the sprocket and thence to the take-up spool. The supply spool (unexposed film) and the take-up spool (exposed film) are driven in the directions shown by the same mechanism as the sprocket. Since the sprocket operates continuously while the claw pull-down produces an intermittent

movement of the film in the film gate, it is necessary to leave a loop on either side to prevent damage to the performations of the film.

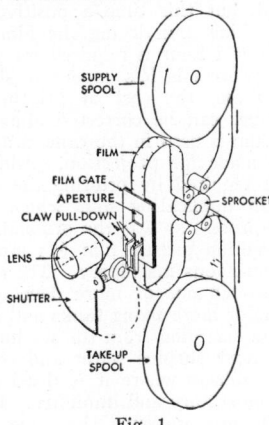

SUPPLY SPOOL

FILM

FILM GATE
APERTURE
CLAW PULL-DOWN

SPROCKET

LENS

SHUTTER

TAKE-UP SPOOL

Fig. 1.

A rotating shutter, the shaded part of which is opaque, placed between the lens and the exposing aperture exposes the film. The shutter is synchronized with the pull-down mechanism so that the film is stationary during the exposure—that is, while the lens is uncovered by the shutter—and when the opaque portion of the shutter covers the lens, the perforations of the film are engaged by the claw of the pull-down and the film is moved down into position for the next picture. At an exposure speed of 16 pictures per sec., the actual exposure is from $\frac{1}{30}$–$\frac{1}{40}$ sec. on cameras using 16 mm. film.

In simple cameras designed principally for home movies, the exposure of the film is controlled by the lens diaphragm; cameras intended for professional use, however, are usually provided with shutters in which the open portion may be varied to regulate the exposure (variable shutters). (See **Shutters, Photographic.**)

Film and Film Processing. In America, motion-picture film and motion-picture equipment are available in three sizes: the so-called standard (35 mm.), 16 mm., and 8 mm. In the first-mentioned the pictures are $\frac{3}{4}''$ high or 16 to a foot as compared with 40 pictures to the foot of 16 mm. film and 80 to the foot of 8 mm. film. A 100-foot roll of standard 35 mm. film as used in motion-picture theaters requires but 67 secs. to pass through the projector, while a roll of 16 mm. film of the same length requires 4 full minutes, and 8 mm. film 8 minutes. The two smaller sizes—8 and 16 mm.—were originally intended for the making of home movies by the amateur. The use of the smaller size is confined almost entirely to the amateur but the larger (16 mm.) is now widely used in the making of advertising, educational and other films which are not designed to be shown to large audiences.

Most 8 and 16 mm. film is of the reversal type and after exposure in the camera is returned to the manufacturer for processing. The film is processed on continuous machinery consisting of tanks containing the various processing solutions and racks with rollers over which the strand of film is led through the different processing solutions and to the drying cabinets from which it emerges wound on a reel. In reversal processing, the film is developed as a negative first, after which the developed silver of the negative image is removed in a bleaching bath. The film is then cleared of the bleaching solution and the undeveloped silver halide remaining is resensitized and exposed to light to render it developable. To correct for errors in exposure this second exposure is varied in accordance with the amount of undeveloped silver halide remaining, the exposure being controlled automatically by a thermopile which gives each scene on the film the exposure required to produce a positive image of the proper density and contrast for projection. After this second exposure, the film is again developed, but this time a positive is obtained. After fixing, washing and drying the film is ready for projection. About 1 hour is required for processing.

The reversal process is less expensive since only one film is required and the cost of printing is avoided. Errors in exposure can be corrected almost as well as when negative film is used in the camera and a print on positive film is made for projection. Additional copies can be made if required by printing the positive on a special fine-grain duplicating film which is developed by the reversal process as was the original.

Professional motion picture film (35 mm.) is usually processed by continuous machinery. The strand of film is conducted through one or more developing tanks, a rinsing tank, one or more fixing tanks and several washing tanks. Upon emerging from the washing tanks, it is squeegeed to remove surplus water and conducted into air-conditioned cabinets where it is dried under proper conditions of temperature and humidity. Both negative and positive film are processed the same way but the time of processing is less for positive than for negative film and the speed of the film through the processing machine is usually higher.

Elon-hydroquinone borax developers are generally employed for the development of the negative and development is to a relative low **gamma** while the positive is developed to gammas ranging from 1.9 to 2.4 using a concentrated elon-hydroquinone developer. The general practice is to develop both the negative and the print to predetermined gammas thus placing on the cameraman the responsibility for lighting and exposing so as to obtain the results desired. The processing of both negative and positive film is controlled by developing sensitometric strips at frequent intervals and measuring the densities to determine the exposure-density relationship and the gamma. (See **Sensitometry.**)

Printing. Positive prints for projection are made on motion-picture positive film which corresponds to negative film in size but is coated with a slow, fine-grain, contrast emulsion. Printing is ordinarily by contact but projection printing (optical printing) is used for special effects and for reducing 35 mm. film to 16, or enlarging 16 mm. to the larger size.

In contact printing the developed negative and the positive film are brought together in contact and exposed by moving the two together at a uniform speed, across an illuminated slit.

The exposure of the positive film is controlled by varying the voltage on an incandescent tungsten lamp, by an adjustable diaphragm or back shutter, which controls the amount of light reaching the exposing aperture or by means of what is termed a timing strip or travelling matt. These are made by exposing and developing positive film so as to leave a clear strip down the center or to produce a given density. The width of the clear strip, or the density, is varied according to the amount of light the strip is to transmit. In printing, this timing strip is placed between the light source and the printing aperture so that the light reaching the negative must first pass through this strip. The timing strip moves with the negative—although usually at a lower rate—so that by splicing together timing strips with the proper light transmission and of the proper lengths, the exposure will be changed automatically at the beginning of each scene to compensate for any difference in negative density. (See also **Projectors, Motion Picture.**) (c.b.n.)

CAMERA, PHOTOGRAPHIC. No one knows precisely when it was first discovered that light, admitted to a darkened room through a small aperture, will form an image on the opposite wall of objects beyond the aperture. The first unmistakable reference appears in the *Problemata* of Aristotle. Later, between the 10th and the 15th centuries the discovery was again made and the term *camera obscura* (Latin, dark enclosure or dark room) came into use. Still later, portable models were devised for sketching and through the use of these by Niepce, Talbot and Daguerre, who developed the first practical processes of photography, the word camera became a part of photographic terminology.

All photographic cameras consist of (1) a light-tight enclosure for the sensitive material, (2) a lens, or other means of forming an image, (3) a holder for the sensitive material, and (4) a means of controlling the time during which light is permitted to reach the sensitive material (i.e., a shutter). It may have, in addition to these essentials, (1) a finder to show what will be included in the picture, (2) means for changing the sensitive material, i.e., roll holder, film or plate holder, film cartridge, or film pack, (3) means of focusing the image, e.g., a collapsible bellows, or a tube with a sliding or screw motion, combined with a focusing scale, or focusing screen, (4) a diaphragm to control the amount of light admitted by the lens, (5) a range finder to assist in focusing the image, (6) a depth-of-focus scale, and (7) movements, or swings, which enable the sensitive material to be placed at different angles to the optical axis of the lens.

Cameras, other than those designed for specialized fields of photography, such as photoengraving, photolithography, photomicrography, etc., may be divided into seven classes: (1) box, (2) folding roll film, (3) hand cameras with ground-glass focusing, (4) miniature cameras, (5) reflex cameras, (6) twin-lens cameras, (7) view cameras.

The typical box camera uses roll film; has a single lens with a maximum opening of about f/14; a shutter with one "instantaneous" speed (usually from $\frac{1}{25}$ to $\frac{1}{40}$ second); two reflecting or one direct-vision finder; and may or may not have an adjustable diaphragm. It is designed to meet the needs of those wishing a simple and inexpensive camera.

Considerable ingenuity has been shown by manufacturers in the design of small roll-film cameras and there is considerable variety in this group. The most popular sizes are 2¼ x 2¼, 2¼ x 3¼ and 2½ x 4¼. Most cameras of this type now have self-erecting fronts, the front springing into position automatically when the camera is opened. Focusing is by scale and is accomplished most often by partially unscrewing the front element of the lens, but in some cameras the lens and shutter may be moved forward or backward as a unit. Inexpensive models are fitted with single lenses and simple shutters; the more expensive with large aperture, high-quality anastigmat lenses and precision shutters. In some cases, the lens is coupled to a range finder to permit direct focusing. Reflecting finders have largely given way in recent years to direct-vision finders. The convenience and compactness of the folding roll-film camera make it one of the most popular types for amateur use.

The term miniature as applied to cameras is rather loosely employed. Here the term miniature will be restricted to cameras utilizing 35 mm. film in cartridges. The typical miniature camera has a metal body and a lens mounted in a metal tube which is extended for focusing by a milled screw. In the less expensive models a between-the-lens shutter is generally used; focal-plane shutters, however, are generally used on the more expensive models to enable lenses to be changed readily. In practically all cases direct-vision finders are employed. Some of these have framing devices so that the field included in the finder will correspond to that of other lenses of different focal length. Many cameras of this type are fitted with coupled range-finders so that the distance of the subject is determined and the lens focused in one operation. In some cases a range-finder is provided—or is available as an attachment—but is not coupled to the lens. In practically all cases the film spools are carried in light-tight magazines (cartridges) and the film movement is limited by an automatic stop which is responsible for the spacing of the exposures on the roll. On most cameras an interconnecting mechanism prevents the operation of the shutter, after an exposure has been made, until the film has been changed, thus preventing double exposure.

Considerable enterprise has been shown by the promoters of such cameras in providing attachments for specialized work. So much has been done in this direction that some miniature cameras are more nearly universal in their application than any other type of camera.

Hand cameras with ground-glass focusing, using sheet film, glass plates or film-pack are popular with both press and advanced amateur photographers. The most popular of such cameras is fitted with a coupled range-finder and a direct-vision finder and is used much like the miniature camera. Generally ground-glass focusing is resorted to only for copying and other work beyond the capabilities of the range finder. For press photography the camera is usually equipped with a synchronizer for use with flashlamps (see **Flashlamps**).

In the *reflex* camera, a mirror placed at an angle between the lens and the sensitive material reflects the image to a focusing screen at the top of the camera where it is seen right side up and in full size up to the moment of exposure. When the exposure lever is depressed to make a picture, the mirror swings out of the way and covers the focusing screen. It then releases the focal-plane shutter (see **Shutters, Photographic**), exposing the film. The reflex camera is favored by those who wish to study the image on the ground glass but object to the inconvenience of the other types of ground-glass focusing cameras. It is the only camera with ground-glass focusing suitable for moving objects. Ground-glass focusing appeals to many not only because the image is seen full size but the relative sharpness of different planes of the subject (see **Depth of Field**) can be determined. This can only be guessed at when using a range-finder or when focusing by scale.

The *twin-lens* camera is another type of ground-glass focusing camera and has two lenses of the same focal length mounted side by side one of which forms an image on the ground-glass focusing screen while the other, mounted in a shutter, is used in making the picture. The twin-lens camera has many of the advantages of the reflex but is simpler mechanically.

The *view* camera is used chiefly by professional and advanced amateur photographers. It is designed for use on a tripod and does not have either a view-finder or a focusing scale. Ordinarily it is made only in sizes 4 x 5″ or larger—5 x 7″ and 8 x 10″ being two of the usual sizes—and has a long bellows extension with vertical and horizontal adjustments to the back, and in some cases to the front, for dealing with difficult subjects such as tall buildings, cramped interiors and industrial products requiring unusual perspective.

Cameras designed for professional portrait photography are similar to the view camera except that they are much larger and heavier. Ordinarily they are not made in sizes smaller than 8 x 10″ and are fitted with a sliding carriage so that the loaded film holder is placed on one side and the slide withdrawn before the image is focused. When everything is ready, the shutter is closed and the carriage shifted to place the film in position for the exposure. This reduces the time elapsing between focusing and the exposure and assists greatly in making lifelike portraits. The camera is mounted on a substantial stand which enables it to be raised or lowered as required. (C.B.N.)

CAMPANULARIAE. Coelenterata, Hydrozoa. An order made up of colonial species with two forms of **polyps,** one nutritive and the other reproductive, known respectively as hydranths and blastostyles, and both enveloped partially in a cupped extension of the sheath of the colony. The common and widely distributed genus *Obelia* is an example. (A.W.L.)

CAMPHOR. *Cinnamomum camphora.* Lauraceae. Camphor is a crystalline compound occurring in various parts, as the wood and leaves, of the camphor tree, a large evergreen tree with light green leaves, growing in many warm regions of southeastern Asia, but particularly on the Island of Formosa. It has been introduced into California, Florida and other warm parts of the United States as an ornamental tree. Its abundance in Formosa formerly gave Japan virtual control of the natural supply of camphor.

The old method of extracting camphor involved chopping down the tree and cutting it up into small chips which were distilled by primitive means. The crude product thus obtained was redistilled to remove the oil of camphor present and the crystalline camphor obtained.

Camphor

$$\left(C_{10}H_{16}O, \text{ or } C_9H_{16}{:}CO, \text{ or } CH_3 \cdot C \underset{CH_2{-}CH_2}{\overset{CO{-}CH_2}{<}C(CH_3)_2>}CH \right)$$

is a white solid, melting point 179° C., boiling point 209° C., of characteristic pleasant odor, insoluble in water, soluble in alcohol or ether. Camphor (1) when heated with **phosphorus** pentoxide, yields cymene ($CH_3 \cdot C_6H_4 \cdot C(CH_3)_2(1,4)$), (2) when heated with **iodine**, yields carvacrol ($CH_3 \cdot C_6H_3(OH) \cdot CH(CH_3)_2(1,2,4)$), (3) when treated with **nitric acid**, yields camphoric acid

$$\left(CH_3C \underset{CH_2{-}CH_2}{\overset{COOH \quad HOOC}{<}C(CH_3)_2>}CH \right),$$

and then, by further oxidation, camphoronic acid

$$\left(CH_3 \underset{CH_2{-}COOH}{\overset{COOH}{<}}C(CH_3)_2{-}COOH \right),$$

(4), when treated with **hydroxylamine**, yields camphoroxime ($C_9H_{16}{:}C{:}NOH$), solid, melting point 120° C., (5) when treated with **nitrous acid** (amyl nitrite plus sodium alcoholate), yields isonitrosocamphor

$$\left(CH_3C \underset{CH_2{-}CH_2}{\overset{CO{-}HON{:}C}{<}C(CH_3)_2>}CH \right),$$

solid, melting point 153° C. Camphor may be made synthetically by converting pinene into bornyl chloride with hydrogen chloride, thence to isobornyl acetate, thence to isoborneol, and finally oxidizing borneol to camphor.

Camphor is detected by formation of the **oxime** and determination of its melting point. Camphor is used (1) in medicine as a heart stimulant and for other purposes, (2) in celluloid and lacquers, (3) in insecticides and moth preventives. Camphorated oil is a solution of camphor in cotton-seed oil. (See **Aldehydes, Ketones, and Related Compounds**.) (R.K.S., R.M.W.)

CAMPTONITE. A dark **basaltic dike** rock of the essential mineralogical composition of a **diorite,** requiring, however, microscopical examination for proper identification. It was named from the type locality, Campton, New Hampshire. (E.S.C.S.)

CANADA BALSAM. A slightly yellow, transparent, fluid **resin** procured from a North American species of silver fir tree. Used for mounting thin sections of rocks, and of tissues of plants and animals for microscopic examination, between glass slides, and for cementing glass in optical instruments. The **refractive index** of Canada balsam after it has been heated varies between 1.534 and 1.540, according to A. Johannsen. (R.M.F.)

CANADIAN. For the geological significance of this term, see **Ordovician.**

CANAL. In zoology, a canal is a tubular structure or passage, with specific applications in many groups of animals among which are the following: 1. The passages in the wall of a **sponge.** 2. Slender diverticula of the enteric cavity in **coelenterates** and **ctenophores.** 3. The stone canal, ring canal, and other parts of the water vascular system in **echinoderms.** 4. The inguinal canal through which the testis descends from the abdomen into the scrotum in **mammals.**

In engineering, a canal is an artificial channel conveying or holding water. There are canals built primarily for navigation, and canals whose purpose is to convey water from one point to another. Examples of the former type are to be found in some of the famous ship canals of the world, and examples of the latter type in hydraulic power canals and irrigation canals.

The construction of irrigating canals represents one of man's earliest major engineering achievements. Indeed such construction goes back to earliest recorded history, or even beyond, and the importance of such works is evidenced by present-day remnants and ruins of such structures. Mesopotamia and Egypt afford remains furnishing evidence of some of the earliest irrigation canal construction. Some of the larger of these were doubtless used for navigation as well. At a somewhat later date certain portions of India were covered with an extensive system of irrigation canals. Some of these old irrigation systems have been rehabilitated under modern governments and returned to their original purpose. Construction of extensive irrigation canals in the New World (in Peru, Central America, and the states of Arizona and New Mexico) considerably antedate its discovery in 1492.

Navigable canals constitute the first solution of the problem of transporting freight in bulk overland. Prior to the early development of the railroad, about 1835, there was considerable canal development in Europe and some in the United States. However, the competition of the railroads prevented further development and even rendered obsolete much prior canal construction. At present such navigable canals can compete with railroads only under certain favorable economic circumstances and character of freight.

Navigable canals may be classified according to the relative levels of the two bodies of water which they connect. The Suez Canal is a simple channel connecting the Mediterranean and the Red Sea at practically the same level. The New Orleans Industrial Canal connects a higher and a lower body of water by a simple gradient, but the Panama Canal crosses a continental divide, making the middle some 85′ above the ends which have about the same elevation. A canal such as the Sault Ste. Marie follows a river valley and by-passes the portions of the river which cannot be navigated. The Barge canal is a hybrid type composed partly of artificial canal and partly of canalized river.

Where the profile of a navigable canal is on a gradient, locks must be employed to divide the waterway up into a number of adjacent steps. The lock is a short section of the waterway just large enough to accommodate the longest boat. By means of gates the lock section may be shut off from both adjoining sections of the canal. Valves control the inlet of water to the lock, the operation of which is as follows: A boat proceeding upstream in the canal approaches a lock. The water level in the lock is the same as that in the section containing the boat, so the gates between the boat and the lock are opened, permitting the boat to enter the lock at the lower water level. The gates are then closed, and water is admitted from the higher level. As the water rises in the lock it carries the boat with it. When the water level in the lock equals that in the adjacent canal section, the gates are opened on that end, and the boat is then free to proceed along the canal at the higher level.

Canals built for the purpose of transporting water from one point to another must, for economy, convey the water as rapidly as possible, but not with a velocity great enough to cause erosion. While navigable canals are usually simply excavated in earth and rock, and have little or no water movement within them, water-conveying canals may be eroded if the water exceeds a certain critical velocity. On the other hand, if the water flows too slowly, in some cases there may be a silting up of the canal by earth material released from suspension. There is a critical velocity, however, at which neither silting nor erosion will occur. Many canals having a flow of water are lined with concrete or rock. The rate of flow of water in a canal depends upon the cross-section of the canal, the character of the wetted surface, and the gradient of the bottom. The friction loss in a canal or natural channel is equal to the fall of the surface level of the stream, provided the velocity remains constant. The velocity of water can be found by a number of empirical formulae. Probably the most widely accepted of these is the Kutter-Chezy formula:

$$v = C\sqrt{RS}.$$

v is the mean velocity, ft. per sec., S is the slope, fractional fall per ft. of length, C is a coefficient involving the channel roughness and the hydraulic radius, as well as the slope, R is the hydraulic radius of the cross-section of the canal, and is equal to the cross-sectional area of flow divided by the wetted perimeter. A canal cross-section in the shape of a semicircle is ideal from the standpoint of minimum friction, and concrete lined canals are often so built. However, canals excavated in earth or rock are ordinarily trapezoidal in shape, and velocities of from 1–4 ft. per sec. are employed, depending upon the nature of the soil. (A.W.L., C.W.C., F.T.M.)

CANAL RAYS. This term is a bad translation of the German *Kanalstrahlen.* A more accurate designation would be "tunnel rays" or "perforation rays." They consist of positive particles in a vacuum tube which escape through tunnels or holes bored in the cathode. Positive ions originating in the gas near the cathode move toward it with great speed, ordinarily striking it and causing the surface disintegration and **sputtering** of the metal soon observable; also doubtless releasing **cathode-ray** electrons. But if the cathode is perforated with holes so placed that the positive rays can enter them, some of the particles pass through into the space behind the cathode. In especially constructed canal-ray tubes, this space is elongated so that the positive rays, now isolated from the cathode rays, can be studied separately. (L.D.W.)

CANANGA OIL. Volatile Oils.

CANARY. Aves, Passeriformes. A finch, *Serinus canarius,* native to the Canary Islands, which has been extensively used as a cage bird. The wild species is brownish with yellow markings but in captivity pure yellow strains have been developed.

The goldfinch of North America and to a lesser extent the yellow warbler are called wild canaries from their similar yellow color. (A.W.L.)

CANCER. 1. *Astronomy.* (The crab.) (Map, page 380.) Cancer is the name of a small and poorly marked **constellation** of faint stars that is of importance principally because it is the fourth sign of the **zodiac.** In this constellation is found the fine **cluster** known as Praesepe (or the Beehive). The stars are not so numerous as in some other star clusters, but are of sufficient brightness to make this an interesting object in a small telescope. Galileo counted 36 stars with his telescope but observers using modern equipment have counted over 300. On a clear moonless night the object appears as a faint glow of light, and is frequently used by astronomers as a test of the transparency of the **atmosphere.**

2. *Medicine*. (Carcinoma.) Any malignant **tumor** arising from **epithelial** tissue. In its broader sense it includes all disease associated with the presence of such a tumor.

Hippocrates, the 5th century B.C., probably first used the Greek word crab to describe a spreading cancerous growth. From the Romans came the word cancer, meaning crab.

Cancer is one of the most ancient of diseases and its occurrence appears in the earliest records of man. It is one of the most universal and common of all diseases and not only attacks man but all forms of life, including plant life. No age group is exempt. The majority of cases, however, occur in individuals between 40 and 50.

The total deaths from cancer in the United States have been estimated to be around 150,000 annually (1 out of every 8 deaths)—a rate of 125 per hundred thousand. No accurate figures can be given as only about 1% of those dying in this country are autopsied, and many die from unrecognized cancer, the death being attributed to terminal disease such as pneumonia and various circulatory diseases.

In 1900, cancer rated sixth as a cause of death. This increase to second and almost first place is probably accounted for by more accurate diagnosis of cancer as a cause of death, and increase in the average life-expectancy. In other words, more people live longer than previously so that they reach an age when cancer is more likely to develop.

Cancer begins as a simple body cell or group of cells which suddenly starts to multiply without restraint, growing independently of the rest of the body, not serving any useful purpose, invading the surrounding tissues or organs by direct spread, or spreading throughout the body by means of **metastasis** (small groups of cells breaking off from the primary growth and being carried by the blood or lymph channels to far distant tissues

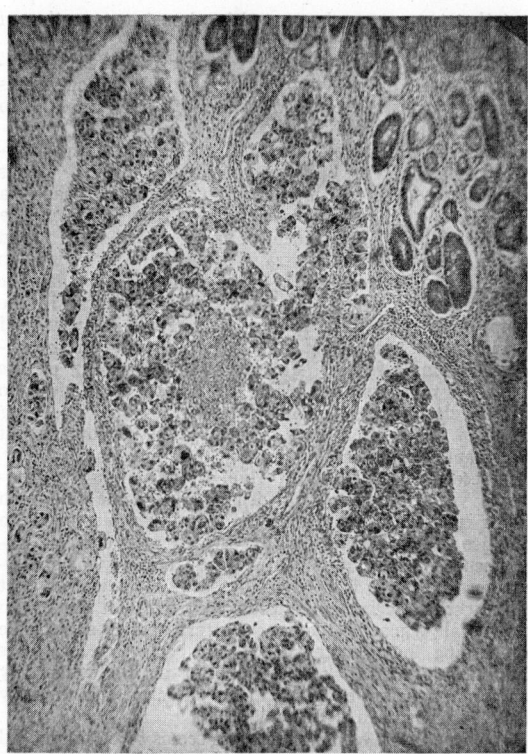

Fig. 2. Section through a cancerous tumor of the stomach, magnified one hundred times. Remnants of normal gland tissue similar to Fig. 1 may be seen in upper right hand corner. The remainder of the illustration shows nests of cancer cells invading the muscular layer of the stomach wall. (*Fortune Magazine.*)

where they in turn produce growths similar to the original or primary tumor). It is this characteristic of spread or metastasis that distinguishes a malignant from a benign tumor. Benign tumors differ from malignant tumors in that (1) they do not metastasize or invade tissue, (2) they usually (but not always) grow slowly, (3) they interfere with function only through pressure or obstruction of surrounding structures, (4) they usually threaten life only by compression of some vital structure because of their increasing size. This is especially true when they grow in a confined or closed space. In most instances it is easy to distinguish between benign and malignant tumors. In other cases, the growth may be borderline between the two classifications, and it may be difficult for an expert to distinguish between the two. In this borderline group an accurate diagnosis cannot be made by physical examination of the patient or any other method except surgical **biopsy** and microscopic examination of the tissue removed.

In spite of observations and theorizing for the past 2000 years, the causes of cancer are still unknown. The Greeks and Romans attributed cancer to a disturbance of the humoral balance of the body, and today various theories are similar, in that they presuppose alterations in the secretions of the endocrine glands as a partial cause for the abnormal cell growth. Modern research has demonstrated that chemical irritation and the irritation of chronic inflammation often are starting points for at least some malignant tumors.

One theory of long standing is that of Cohnheim, assuming that tumors start from misplaced embryonic **cells** which may remain dormant for years. With some stimulus that is unknown, these **embryonic** cells start their unrestrained growth and become a malignant tumor. Such embryonal displacement of cells is not uncommon and may well account for the start of some tumors. A serious objection to this hypothesis is the

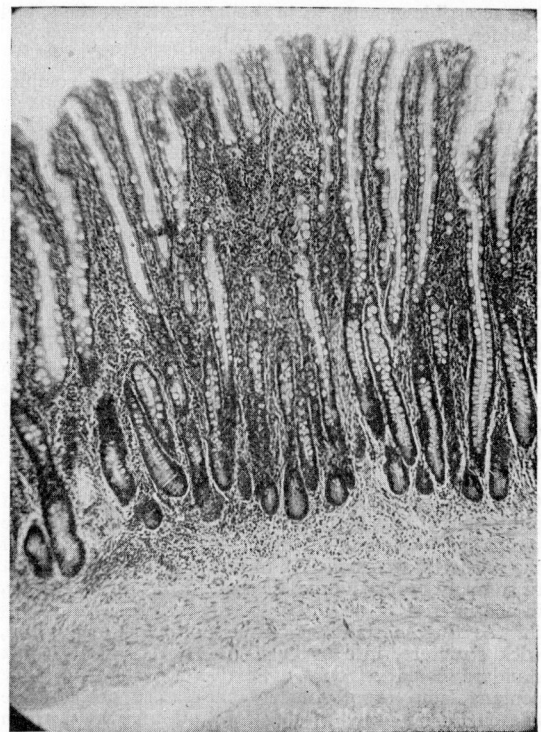

Fig. 1. Photomicrograph of normal stomach wall magnified eighty times. The portion at the top represents stomach cavity. The strand-like processes projecting upward represent the glands and their ducts which secrete the gastric juice. The lighter area below the glands is the muscular wall of the stomach. (*Fortune Magazine.*)

fact that tumors can be caused in almost any portion of the body by chronic irritation.

Another theory assumes that cancer is due to the action of some unknown parasite, a **virus** which causes irritation of tissues or invades the cell and causes it to proliferate abnormally.

It is true that in experimental animals, certain malignant tumors have been produced by viruses. Rous demonstrated the transmission of chicken sarcoma by a filtrate of tumor tissue which had passed through a Berkfeld filter. More recently, an agent has been found in the milk of certain strains of mice which develop spontaneous breast cancer. This "milk factor" has the properties of a virus. It induces breast cancer in the young suckled by their mothers, and also in the young of certain normal strains suckled by mice whose milk contains the milk factor. However, if the young of the cancerous mothers are suckled by normal mice, they do not develop cancer. Thus ingestion of the infectious agent in the milk is necessary for the development of malignant tumors in these susceptible animals.

A recent theory, that of Gye, is that an ultramicroscopic virus, or group of viruses, plus some unknown factor inherent in the animal's tissues are necessary to produce the malignant tumor. Much of the recent experimental work supports the virus etiology of malignant growths. It is believed by some investigators that the infectious agent is acquired very early in life by an individual whose hereditary pattern plays some role in making him susceptible. The infection itself is thought to remain latent until later in life, when any one of a number of physical or chemical factors may cause its activation.

The occurrence of cancer in man seems to be influenced by heredity in some instances, but the fact that every family tree shows malignant disease in some of its members, if adequate search is made, makes the apparent greater susceptibility in some families hard to evaluate. Considerable work has been done on inheritance of cancer in mice by Maud Slye, who has developed a strain all the members of which have spontaneous cancer.

The relationship between an injury and the occurrence of cancer is sometimes based on faulty or superficial observation, or accidental coincidence. A single injury cannot be said to initiate a cancerous growth, but chronic irritation has long been observed to be related to new growths. Examples of this are cancer of the lip in pipe smokers, skin cancers in x-ray workers and chimney sweeps, who develop the disease in sites exposed to irritation.

In a very small proportion of malignant tumors spontaneous disappearance of the growth has occurred. This has been estimated to happen not oftener than once in 100,000 cases.

The clinical picture presented by a patient with a malignant tumor varies widely depending on the site of the tumor. The first sign of cancer on the body surface or in the breast, tongue, or lip is a lump or slight thickening of the tissue. Cancer in a vital internal organ is particularly difficult to diagnose early. Pain is an extremely late symptom. In intestinal, stomach, or rectal cancer, the first symptom may be persistent indigestion or abdominal distress which is treated by the patient with frequent doses of soda and other alkalies until it is too late for proper treatment. Lack of appetite and disturbances of taste may be other signs.

Cancer in other parts of the body may be first marked by unusual discharges or slight bleeding. Sores or ulcers that refuse to heal may have undergone malignant change. Bleeding from the nipple, or the rectum, or unusual discharge of blood from the female genital organs are suggestive of cancer as well as other diseased conditions. Diagnosis can be made by a competent physician or surgeon. Cancer of the rectum is often treated as hemorrhoids or constipation. Cancer of the bone may cause vague aching pain early, and is often treated as "rheumatism." X-ray is of utmost value in diagnosis of bone cancer.

Diagnosis of cancer requires a complete physical examination and, depending on the location, x-ray or other special diagnostic procedures. When a positive diagnosis cannot be made, as is sometimes the case, it is necessary to do a **biopsy**. By this means an exceedingly small piece of tissue is removed, cut and stained and then examined under the microscope by an expert pathologist.

The earlier cancer is found, the more difficult it is to recognize. When diagnosis is easy the disease often is too far advanced for any but palliative treatment.

Often the doctor is at fault, through not noticing salient points in the history of a case, not doing a thorough examination or making use of special examinations. But more often it is the patient who is at fault, either through ignorance, fear, or use of home remedies or remedies advertised for constipation, piles, indigestion, etc.

Treatment of cancer involves surgery, radium or x-ray or a combination of surgery and radiation. Different forms and location of the tumor may require one or the other or both.

Cancer of the skin or of the **cervix** of the uterus is often best treated by radium. Palliative treatment of certain foci of inoperable cancer is done by use of x-ray or radium.

In general, cancer of the internal organs is best treated by surgery. Radiation by x-ray or radium in some cases is used before operation and after operation. Cancer is curable but only when it exists as a primary localized growth. When it has begun to spread over the body death usually results sooner or later, although the patient may live for several years or more with but few symptoms and in relative comfort.

In certain forms of cancer electrosurgery is used in the form of the endotherm knife (radio knife) for coagulation and removal of tumors. It is particularly used in less accessible growths as in the larynx, brain, chest, and bladder. (W.K.G., R.S.M., D.M.H.)

CANCRINITE. The mineral cancrinite is a complex orthosilicate (see **Silicon**) corresponding approximately to the formula $3H_2O \cdot 4Na_2O \cdot CaO \cdot 4Al_2O_3 \cdot 9SiO_2 \cdot 2CO_2$. It is **hexagonal,** with prismatic cleavage; hardness, 5–6; specific gravity, 2.42–2.50; color, white to gray or may be greenish, bluish, yellow, or flesh red; colorless streak; luster, subvitreous to greasy; transparent to translucent. Cancrinite is found only in the **nephelite-syenites** and related rock types and commonly associated with **soda-lite.** Cancrinite is believed to be in part primary, having crystallized direct from the **magma,** and in part secondary as a result of alteration of nephelite by solutions of calcium carbonate. Cancrinite is found in the Ilmen Mts. of Russia, in Rumania, in Norway and in Canada in Hastings County, Ontario, and in the United States in Kennebec County. This mineral was named for Count Georg Cancrin, a Russian statesman who died in 1845. (E.S.C.S.)

CANDLE POWER. Chief among the problems of **photometry** has been the development of a suitable standard of luminous intensity. In the earlier stages of the science it was sufficient to rate a lamp as equivalent to so many ordinary candles. Later it became necessary, in the interest of accuracy, to agree upon specifications for the English standard candle, as to its composition, wick structure, rate of burning, etc. At best such a standard could be only approximately uniform. The pentane lamp (approximately one candle power), the French Carcel lamp, burning colza oil, and the German Hefner lamp, burning amyl acetate, as well as standardized gas flames, were subject in less degree to the same limitations.

In 1921 the International Commission on Illumination adopted the international candle, which had for some years been in practical use in the national standards laboratories of the United States, Great Britain, and France. It was defined in terms of carbon-filament lamps constructed and operated under specified conditions. The United States candle was thus fixed in reference to a group of 45 lamps kept at the National Bureau of Standards, and has been our standard until 1940.

J. Voille long ago proposed as a unit of intensity 1 sq. cm. of surface of molten platinum at its melting point. Waidner and Burgess improved on this idea by replacing the platinum surface by a **black-body** radiator, and their suggestion has now been embodied in the "new candle," adopted in 1937 by the International Committee on Weights and Measures to be put into effect shortly. This standard is defined as $\frac{1}{60}$ of the luminous intensity of 1 sq. cm. of the black-body radiator at the melting point of platinum. (See **Photometry** and **Illumination**.)   (L.D.W.)

CANDLEFISH. Pisces, Teleostei. An oily fish, *Thaleichthys pacificus,* belonging to the **smelt** family. It is found from Oregon to Alaska and ascends the coastal streams. It is an important food fish and takes its name from the fact that it can be burned as a candle when dried. (A.W.L.)

CANDLE-FLY. A southern name for **moth,** equivalent to the northern miller or moth-miller. (A.W.L.)

CANE CUTTER. Cottontail.

CANE-RAT. Mammalia, Rodentia. An African burrowing species more closely related to the **porcupines** than to the true rats. (A.W.L.)

CANGA. A Brazilian term for an iron-rich conglomerate or breccia in which the pebbles or **anguclasts** are **hematite** and **itaberite** cemented by hematite or **limonite.** (R.M.F.)

CANINE. 1. Pertaining to dogs. 2. An elongate conical tooth between the incisors and premolars of each half-jaw. The eye-teeth of man in the vernacular. (A.W.L.)

CANIS MAJOR. (The great dog.) (Map, page 380.) Both this **constellation** and its companion Canis Minor, or the little dog, have been named from remote antiquity as the dogs of **Orion. Sirius** in Canis Major and **Procyon** in Canis Minor are both well-known stars, Sirius being the brightest observable. References to these stars are to be found in nearly all ancient classical literature. Sirius, in particular, was of great importance to the Egyptians because it rose with the sun at the period when the waters of the Nile were due to rise and was considered as a herald of the returning fertility of the valley. Sirius is not only the brightest, but also the closest star visible to the naked eye which can be observed in the **latitudes** of Europe or North America. Intrinsically Sirius has a brightness more than 20 times that of our sun. Both Sirius and Procyon have faint companions, that of Sirius being particularly famous as the first of the **white dwarfs** discovered. (W.K.G.)

CANKER-WORM. Insecta, Lepidoptera. **Insects** of two chief species, the spring canker-worm, *Paleacrita vernata,* and fall canker-worm, *Alsophila pometaria,* respectively, which attack apple trees and are sometimes important pests. The spring canker-worm is found throughout the United States and the fall canker-worm in the northeastern quarter of the country.

The caterpillars eat leaves and so can be destroyed by spraying with arsenate of lead. Since the females are wingless, bands of tanglefoot around the tree trunks prevent their ascending the trees to deposit their eggs.

Canker-worms of three other species occur in North America. (A.W.L.)

CANNABIS INDICA. A variety of common **hemp** from which is procured the so-called hashish and marijuana, etc.—habit-forming narcotic **drugs.** They are not commonly used in medical practice. (D.M.H.)

CANNEL COAL. A form of bituminous or soft **coal** which ignites easily and burns with a hot "candle-like" flame, due to its high percentage of volatile matter. Especially desirable for open fires and metallurgical processes. Most cannel coals owe their desirable peculiarities to their large content of paraffine-like fossil spore cases which are highly volatile. The peculiar chemical and organic composition of this type of coal strongly suggests that it originated as a lake muck (sapropel) under open water rather than typical peat bog conditions. (R.M.F.)

CANTALOUPE. Gourd Family.

CANTHARIDES. Spanish Fly.

CANTILEVER BEAM. A beam which is rigidly connected at one end to a fixed support and free to move at the other end is called a cantilever beam. This theoretically fixed condition rarely occurs because of deformation of the supporting material. The maximum **bending moment** and maximum **shear** occur simultaneously at the face of the support. The usefulness of this type of beam is demonstrated in structures such as canopies, unbraced airplane wings and cantilever **retaining walls.** (C.W.C.)

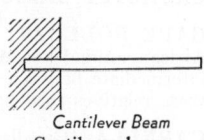

Cantilever Beam
Cantilever beam.

CANTILEVER BRIDGE. Bridge.

CANVAS-BACK. Aves, Anseriformes. A large North American **duck,** *Nyroca valisneria,* similar to the redhead but with a dusky crown and face. Its flesh at the proper season is superior to that of other ducks. (A.W.L.)

CAP SCREW. Screw Fastenings.

CAPACITANCE. Also often called "electric capacity," the capacitance of an electrically conducting body is the ratio of the quantity of electricity imparted to the body to the resulting change in its electric potential. It is usually expressed in coulombs of charge per volt of potential change, that is, in terms of the **farad,** or its submultiples; the fundamental c.g.s. electromagnetic unit being, however, the "abfarad." (See **Electrical and Magnetic Units.**)

If a conductor is completely isolated, that is, far removed from other conductors, including the earth, and is surrounded by a homogeneous, perfectly insulating **dielectric,** its capacitance depends only upon the size and shape of its external surface and upon the **dielectric constant** of the surrounding medium. For some bodies of definite geometrical form, the capacitance may be calculated. For example, the capacitance of an isolated sphere, in farads, is equal to $kr/(9 \times 10^{11})$, in which r is the radius in cm. and k is the dielectric constant of the surrounding medium. The capacitance of a conductor may be greatly increased by bringing it near to other conductors (in which case it forms part of a **condenser**).

Very long electric circuits, especially when the wire is surrounded by a conducting sheath, as an ocean cable, have considerable capacitance because of the condenser-like action of wire and sheath with the insulation between them acting as dielectric. The same is true of insulated wire wound in a close coil, adjacent turns of which, being at slightly different potential, act as the conductors of a condenser; an effect which may be partially avoided by a criss-cross or "honey-comb" winding or by a "banked" winding (in flat spirals). The capacitance of a circuit, whether thus "distributed" or intentionally introduced by means of condensers, acts as a condenser in parallel with the conductor, and may have marked effect upon alternating or variable currents traversing it. (See **Electric Circuits** and **Alternating Currents.**) (L.D.W.)

CAPACITIVE LOAD. An **alternating-current circuit** in which the current drawn leads the voltage in phase is said to provide a capacitive load. Capacitive loading may be the result of actual **condensers**, or of virtual condensers in the form of long transmission lines, or over-excited synchronous rotating equipment. Most electrical apparatus, such as **motors**, **coils**, etc., draws from the line a current which lags the voltage, and the use of some capacitive load is desirable in order to bring the total current and voltage more nearly in phase, and thus raise the power factor. (F.T.M.)

CAPACITY, DISTRIBUTED. Coil.

CAPACITY, ELECTRIC. Capacitance.

CAPE POLECAT. Mammalia, Carnivora. A South African animal resembling the **skunks** and apparently intermediate between them and the true **polecats.** Like these relatives they have a foul odor. (A.W.L.)

CAPELLA. Capella (a **Aurigae**) is the third brightest star visible in the northern latitudes, and the fifth brightest star on the celestial sphere. It is closer to the pole than any of the other bright stars. Capella has always played an important part in mythological writings and we find it referred to on an old tablet dating back to 2000 B.C. **Astrologically,** Capella portended civic and military honors and wealth.

Astronomically, Capella is particularly interesting for it is a star which is a **spectroscopic binary** with a period of 104 days, and the angular distance between the components has been measured with the interferometer. From the complete solution of the **orbit,** the physical characteristics of the object may be found. It is a **giant** star of the same **spectral class** as our sun. (W.K.G.)

CAPERCAILLIE. Aves, Galliformes. A large woodland **grouse** of northern and central Europe and Asia. The species is also known as the capercally, capercailizie, wood-grouse and cock-of-the-wood. *Tetrao urogallus.* (A.W.L.)

CAPILLARITY. The name given to a class of phenomena, of which the elevating or depression of liquids in fine tubes is representative. When the interface between a liquid and a gas, or between two liquids, is intercepted by a solid surface, an equilibrium is established at the junction among the forces acting along the three surfaces of contact. For example, let a plate of solid S be dipped into a liquid L having gas G above it (see Fig. 1). A molecule at the junction O is

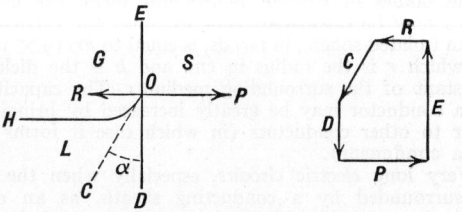

Fig. 1. Capillary force, large adhesion.

acted upon by the adhesive attraction P, by the forces which give rise to the three surface tensions along the interfaces OH, OE, and OD, and by the reaction R of the plate S against which it is drawn by the adhesion.

(Its weight may be considered negligible.) The flexible interface OH adjusts itself so that these forces come into equilibrium; unless, indeed, one of them, E, exceeds the sum of D and C, in which case the liquid "creeps" indefinitely along the surface as oil does over a glass or tin container. The equilibrium polygon at the right is labeled in each case to correspond with the figure representing the surfaces. The "angle of contact" a, between the liquid surface at O and the solid surface OD, is determined by the aforesaid forces acting at O. For most liquids against glass it is acute; for mercury against glass it is obtuse. (Fig. 2.) In special cases it may be 90°, and in others it reduces to zero.

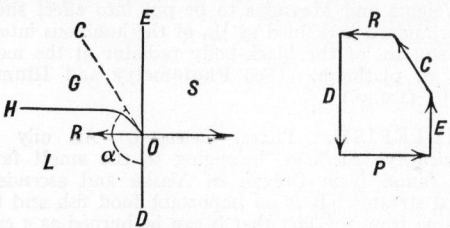

Fig. 2. Capillary forces, small adhesion.

If the interface between two media A and B (Fig. 3) is curved, A being on the concave side, the pressure in A is greater than in B on account of the surface tension; much as the pressure inside a rubber balloon is greater than outside. We can now understand why water rises in a capillary tube. For, to secure equilibrium, the liquid must rise until the pressure inside the surface at B, plus the pressure due to gravity at depth h, makes the pressure at L equal to that at the surface level outside; that is, to the atmospheric pressure. Similar reasoning applies to the depression of mercury in a glass tube. For a circular tube of internal radius r, the distance h to which capillarity will elevate (or depress) a liquid of density ρ and surface tension T (against air) is readily shown to be

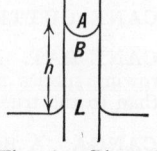

Fig. 3. Rise of liquid in capillary tube.

$$h = \frac{2T\cos\alpha}{r\rho g},$$

where g is gravity. (See also **Electrocapillarity.**) (L.D.W.)

CAPILLARY. 1. Hair-like, especially in application to fine tubes. 2. A minute blood vessel intervening between the arteries and veins. (See **Circulatory System.**) (A.W.L.)

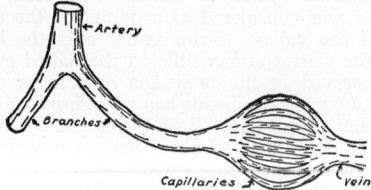

Diagram to illustrate variations in velocity of bloodflow. If a vessel divides into two branches, these will be individually of less cross-section than the main trunk, but united they will exceed it. Linear velocity will be lower in the branches than in the parent stock. The sum of the cross-sectional areas of the capillaries is greater than that of the artery or vein. (*Kimber and Gray, Textbook of Anatomy and Physiology, Macmillan & Co.*)

CAPILLARY ELECTROMETER. Electrocapillarity.

CAPILLARY FRINGE. Above the zone of saturation, capillary pores may exist which, if filled with water, form a zone or fringe of moisture higher than the true water table. This is the capillary fringe or zone of capillarity. (E.S.C.S.)

CAPITULUM. 1. A protuberance on **mites and ticks** which appears to be a small head but is formed only of the mouth parts. 2. Any small head, or small, bony, articular eminence, such as the capitulum of a rib. (A.W.L.)

CAPRICORNUS. (The sea-goat.) (Map, page 380.) This is a **constellation** of small stars, not at all striking in appearance, but important because it is the tenth of the **zodiac.** The star Alpha (named Giedi) is one of the more remarkable stars, it being actually made up of six components. The two larger can be seen by some keen-eyed persons as separate, but it is easily resolved by means of an opera glass. Each of these two components may be still further resolved into three component parts with appropriate telescopes. (W.K.G.)

CAPRIMULGIFORMES. An order of nocturnal birds (**Aves**) with very wide mouths. The **nightjars,** including the **whip-poor-will and nighthawk,** the goat-suckers, and the **oil birds.** (A.W.L.)

CAPSAICIN. Potato Family.

CAPSICUM, CAPSAICIN. Potato Family.

CAPSTAFF EFFECT. A photographic effect discovered by J. G. Capstaff in 1921. Sulfurous acid followed by a weak alkali is capable of extending the sensitivity of blue sensitive films and plates to longer wavelengths. The effect appears to be due to the formation of a form of colloid silver and is of no practical value, although materials sensitive to the entire visible region (panchromatic) can be produced by bathing successively in sodium bisulfite (2%) and a weak solution of sodium carbonate, then washing in running water. (C.B.N.)

CAPSTAN. A capstan is a machine, usually rotary, used chiefly in connection with shipping for hoisting heavy weights. Although formerly turned by hand with a capstan bar, the capstan is now frequently operated by a steam engine or electric motor. The capstan may have a reduction gear so that it can be operated at two different speeds, and thus be suited to the loads to be hoisted. (F.T.M.)

CAPSULE. Fruit.

CAPTACULA. Filaments with suckers at the tips, found on the dorsal surface of the head in the tooth-shells, marine **mollusks** of the class Scaphopoda. (A.W.L.)

CAPUCHIN MONKEY. Mammalia, Primates. Any of several species of South American **monkeys** of the genus *Cebus,* which are characterized by the moderately long prehensile tail. They differ from the other species with prehensile tails in having this organ fully covered with hair and not bare on the lower surface near the tip. Also called sapajous. (A.W.L.)

CAPYBARA. Mammalia, Rodentia. A large South American species, *Hydrochoerus capybara,* of the **cavy** family. These animals attain a length of 4′ and a weight of almost 100 lbs. They are semiaquatic in habits and while their food usually consists of water plants and other vegetation they sometimes make inroads on cultivated crops. They are also called capivara and carpincho. (A.W.L.)

CARABAO. Buffalo.

CARACAL. Mammalia, Carnivora. A member of the cat family found in India and thence into Africa where it is widely distributed. It is related to the **lynxes** and like them is marked by tufted ears. Sometimes known as the Persian lynx. *Lynx caracal.* (A.W.L.)

CARACARA. Aves. Falconiformes. South American birds (**Aves**) of several species related to the hawks. They eat carrion but also catch living prey and sometimes rob other birds of their prey. One species, Audobon's carcara, *Polyborus cheriway,* occurs in the extreme southern parts of the United States. (A.W.L.)

CARACUL. Karakul.

CARAPACE. A shield-like covering of the upper part of the body. In the crustaceans it is the body wall of the thorax and in the **turtles and tortoises** it is a complex structure made up of bony plates, including flattened ribs and vertebrae, covered with thin horny plates. The armor of the **armadillo,** composed of many bony plates developed from the skin and covered with horny plates, is also called a carapace. (A.W.L.)

CARAPATO. Arachnida, Acarina. Ticks of two species, found in tropical Africa and Central America, respectively. The African species is also called the tampan.

The wounds produced by these creatures are severe in themselves but their transmission of the germs of **relapsing fever** is a much greater danger. (A.W.L.)

CARAWAY. Carrot Family.

CARAWILA. Halys, Pit Viper.

CARBAMIC ACID. Amines and Amides.

CARBAZOLE. Pyrrole and Related Compounds.

CARBIDES. Carbon.

CARBINOLS. Alcohols.

CARBOHYDRATES. Carbohydrates are polyhydroxy **aldehydes or ketones,** or those products which are chemically related to such simple units by combination of two or more of these units through the loss of water. These higher units are thus **anhydrides.** They yield the simple units, referred to above, by treatment with acids, such as **hydrochloric or sulfuric,** or particular **enzymes.**

Carbohydrates respond to the alpha-naphthol test as given under glucose, reaction number (12) below, when heated (1) in air, carbohydrates burn to form **carbon dioxide** plus water, evolving heat, (2) in the absence of air, combustible gases, watery and tarry distillates, and charcoal residue.

Carbohydrates may be classified as follows:

Monosaccharides (sugars).

Crystalline solids, soluble in water, sweet taste. Those that occur in nature are fermentable by certain enzymes.

Tetrose, $C_4H_8O_4$

  1. Erythrose

Pentoses, $C_5H_{10}O_5$

  2. Arabinose

    By boiling gum arabic, cherry gum, corn pith, elder pith with dilute sulfuric acid.

  3. Xylose

    By boiling substances mentioned under arabinose above.

  4. Ribose

  5. Lyxose

Hexoses, $C_6H_{12}O_6$

  Aldohexoses

    6. Glucose, dextrose ("grape sugar"), melting point 146° C. (anhydrous). With the enzyme zymase (of yeast) yields ethyl alcohol plus **carbon dioxide.** Specific rotatory power—see glucose below.

    7. Galactose

      Specific rotatory power + 83.9°.*

    8. Mannose

      Specific rotatory power + 14.1°.

9. Gulose
10. Idose
11. Talose
12. Altrose
13. Allose
Ketohexoses
14. Fructose, laevulose ("fruit sugar"), melting point 95° C. Specific rotatory power — 88.5°.
15. Sorbose
16. Tagatose
Disaccharides (sugars), $C_{12}H_{22}O_{11}$
Crystalline solids, soluble in water, sweet taste.
17. Sucrose ("cane sugar," "beet sugar"), melting point 170–186° C. (decomposes). With the enzyme invertase, yields glucose plus fructose. Specific rotatory power + 66.4°.
18. Lactose ("milk sugar"), melting point 202° C. (anhydrous). With the enzyme lactase yields glucose plus galactose. Specific rotatory power + 52.4°.
19. Maltose ("malt sugar"), melting point of $C_{12}H_{22}O_{11} \cdot H_2O$: 100° C. With the enzyme maltase yields glucose plus glucose. Specific rotatory power + 138.5°.
20. Melibiose
   With enzymes or dilute acid yields glucose plus galactose.
21. Cellobiose
   With the enzymes maltase, or cellase, yields glucose plus glucose.
22. Trehalose.
Trisaccharide, $C_{18}H_{32}O_{16}$
Crystalline solid, soluble in water, tasteless.
23. Raffinose, melitose, melting point 118° C. (anhydrous). With the enzyme invertase, yields fructose plus melibiose. With the enzyme emulsin, yields sucrose plus galactose.
Polysaccharides (non-sugars), $(C_6H_{10}O_5)_n$
Non-crystalline solids, insoluble in water, tasteless.
24. Starches
   With the enzyme diastase yield maltose.
25. Celluloses
   With hydrochloric acid, heated, yield glucose. With acetic anhydride plus concentrated sulfuric acid, yield cellobiose.
26. Dextrin
   With the enzyme diatase yields maltose. With the enzyme maltase or with acids yields glucose.
27. Inulin, melting point 178° C. (decom.) $(C_6H_{10}O_5)_n$
   With the enzyme inulase (but not with diastase) yields fructose.
28. Glycogen, melting point 240° C.
   With the enzyme diastase (or ptyalin), yields glucose plus maltose.
29. Pentosans

Glucose may be called the key carbohydrate, since it is the leading member of the aldohexose group, and is formed as one of the products or the only product when the following carbohydrates are hydrolyzed, sucrose, lactose, maltose, cellulose, glycogen. Glucose is a colorless solid ($C_6H_{12}O_6$), less sweet than sucrose, soluble in water from which it may be crystallized ($C_6H_{12}O_6 \cdot H_2O$), melting point 86° C., slightly soluble in alcohol from which it may be crystallized anhydrous of melting point 146° C. Either of these crystalline forms, upon examination of the solution in the polariscope, shows a specific rotary power of +110° when the solution is freshly prepared, but the solution changes, gradually upon standing or rapidly upon heating, to a constant value of +52.5°, a phenomenon known as mutarotation. Glucose reacts (1) with alkaline **cupric** salt solution (**Fehling's solution or Benedict's solution**) to form **cuprous** oxide, (2) with ammonio-silver salt solution

(Tollens' solution) to form finely divided or mirror film of silver, (3) with **phenylhydrazine** in acetic acid, to form glucose phenylhydrazone ($CH_2OH(CHOH)_4CH:NNHC_6H_5$), white solid, melting point alpha 159–160° C., beta 140–141° C., with excess phenylhydrazine to form glucosazone ($CH_2OH(CHOH)_3C:NNHC_6H_5 \cdot CH:NNHC_6H_5$) yellow solid, melting point 205° C. decom., (4) with acetic **anhydride,** to form glucose pentacetate ($C_5H_6(OOCCH_3)_5CHO$), melting point alpha 112 to 113° C., beta 131 to 134° C., (5) with **sodium** amalgam, to form sorbitol ($CH_2OH(CHOH)_4CH_2OH$), (6) with **hydriodic acid,** to form 2-iodo-normal-hexane ($CH_3(CH_2)_3CHICH_3$), (7) with **sodium** hydroxide solution, to form yellowish-brown solutions upon warming, (8) with **calcium** hydroxide solution, to form calcium glucosate ($CH_2OH(CHOH)_4COCa(OH)$), slightly soluble solid from which glucose is recoverable by action of **carbon dioxide** (calcium carbonate formed simultaneously). **Strontium** hydroxide and **barium** hydroxide react similarly. Any of these three reactions may be utilized to recover glucose, with the limitation that barium soluble compounds are poisonous, (9) with **hydroxylamine** hydrochloride, to form glucoseoxime ($CH_2OH(CHOH)_4CH:NOH$), melting point 138° C., (10) with hydrocyanic acid, to form glucosecyanhydrin ($CH_2OH(CHOH)_4CHOHCN$), (11) by oxidation, to yield with **bromine** gluconic acid ($CH_2OH(CHOH)_4COOH$), and with **nitric acid** saccharic acid ($COOH(CHOH)_4COOH$), (12) with alpha-naphthol dissolved in chloroform and then forming a layer of concentrated sulfuric acid beneath the mixture, to form a red coloration at the junction of the two liquid layers (Molisch's test for carbohydrates). Upon standing, the color changes to purple. (13) With methyl **alcohol** in the presence of **hydrogen chloride,** to form methyl glucoside (methyl ether of glucose)

$$(CH_2OH \cdot CH(CHOH)_3CH(OCH_3)),$$

specific rotatory power, alpha +158°, beta —34°. Alpha and beta methyl glucoside (an ether) correspond to alpha and beta forms of glucose

$$(CH_2OH \cdot CH(CHOH)_3CH(OH)),$$

specific rotatory power, alpha +110°, beta +17.5°. Alpha-glucose may be prepared by crystallization of glucose from acetic acid plus water solution at ordinary temperatures, and beta-glucose from glacial acetic acid at higher temperatures. Ordinary glucose is chiefly the alpha form. Haworth suggests the formulas:

Alpha-glucose

Beta-glucose

The following methylglucosides have been prepared and used in the study of the structure of glucose, namely, 2,3,4,6-tetramethylglucose or galactose

$$(CH_2OCH_3 \cdot CH(CHOCH_3)_3CHOH),$$

2,3,6-trimethylglucose

$$(CH_2OCH_3 \cdot CH \cdot CHOH(CHOCH_3)_2CHOH),$$

2,3,4-trimethylglucose

$$(CH_2OH \cdot CH(CHOCH_3)_3CHOH),$$

1,3,4,6-tetramethylfructose

$$(CH_2OCH_3 \cdot CH(CHOCH_3)_2C(OH) \cdot CH_2OCH_3).$$

Glucose possesses four asymmetric carbon atoms (2, 3, 4, 5), making possible 16 different stereoisomerides. These are represented by the two forms (dextro and laevo) of each of the 8 aldohexoses listed. In structure these differ solely in the arrangement of the groups —H and —OH about the four asymmetric carbon atoms. Similarly, the aldopentoses consist of 4 pairs, or 8 in all, stereoisomerides, namely, dextro- and laevo-arabinose and the pairs of each of the three sugars, which follow in the list.

Glucose and fructose are present in sweet fruits, such as grapes and figs, and in honey. These two are the only hexoses found in nature in the free state. Glucose is normally present in human urine to the extent of about 0.1%, but in the case of those suffering from diabetes glucose is secreted in large amount. Glucose is formed, as previously mentioned, by the reaction of polysaccharides and water, the reaction with starch in the presence of very dilute hydrochloric acid serving as the industrial source (the hydrochloric acid acts as a catalyzer, and the small percentage present is later neutralized to form sodium chloride). The solution is evaporated to a syrup or to crystallization, and is used in the manufacture of sweets, and (usually) alcohol, and in foods. The reaction of glucosides with water, by enzymes or acids, produces glucose as one of the products. With sodium hydroxide, under carefully defined conditions, glucose forms lactic acid. Glucose is used as food and for the production of alcohol of wines from fruit juices. Glucose may be detected by formation of glucosazone, and determination of its melting point.

Fructose is present with glucose in sweet fruits and honey, may be obtained free by reaction of inulin of dahlia tubers or artichokes with water, and with glucose by reaction of sucrose with water, the product being known as invert sugar. Fructose differs from glucose in structure in being a pentahydroxy-2-ketone, CH_2OH $(CHOH)_3COCH_2OH$ instead of aldehyde. The specific rotatory power of fructose is —88.5°. Fructose forms the same identical osazone as glucose, and sorbitol plus mannitol by reduction. Fructose may be used as sugar by diabetic patients to advantage instead of glucose or sucrose. Fructose is detected by the violet color its alkaline solution gives with meta-dinitrobenzene.

Sucrose is a colorless solid, when heated melts at 170° to 186° C., and on cooling forms barley sugar which gradually crystallizes, upon heating above the melting point forms caramel, a brown liquid, with decomposition. Caramel is used in confectionery, and in coloring beverages and foods. At higher temperatures decomposition into gaseous and tarry substances occurs, finally leaving a residue of carbon ("sugar charcoal"). Other sugars behave similarly. Sugars are also carbonized by concentrated sulfuric acid. Sucrose is very soluble in water, and is obtained from solution by crystallization, usually by vacuum evaporation. The solution has a specific rotatory power of +66.4°, does not exhibit mutarotation, but is converted by acids or invertase into invert sugar (glucose plus fructose), specific rotatory power —19.7°. Sucrose forms with calcium hydroxide calcium sucrosate, a 1% solution of sugar dissolves about 18 times as much calcium hydroxide as does pure water. This behavior is utilized to recover sugar from solutions, as in the case of glucose, and also to determine free calcium oxide in burnt lime, due to the reactivity of calcium hydroxide and non-reactivity of calcium carbonate. Sucrose is non-reactive with dilute sodium hydroxide, with phenylhydrazine, with ammonio-silver salt solution, but when inverted to glucose plus fructose, these reactions may be obtained. Sucrose forms with acetic anhydride sucrose octaäcetate. The suggested structural formula is

$$CH_2OHCHOHCH(CHOH)_2CH-O-\overset{\displaystyle \overset{CH_2OH}{|}}{C}\ (CHOH)_2\overset{\displaystyle \overset{CH_2OH}{|}}{CH}$$

Glucose residue   Fructose residue

Sucrose is an important food preservative, food flavor, and a raw material for confectionery and for industrial alcohol.

Sucrose is extensively distributed in the seeds and leaves of plants, and is the most abundant of the sugars. The commercial sources of sucrose are the stems of sugar-cane (11 to 16% sucrose, average 13%), the root of the sugar-beet (average 16% sucrose, selection having raised the sucrose content from 5% to a maximum of 20%), the sap of the sugar maple, and the stems of sorghum-cane. Sucrose is pressed from the stems of sugar-cane or sorghum-cane, and extracted with the water from the sliced roots of sugar-beets. The solutions are purified, evaporated and crystallized to such a degree that commercial sucrose is practically chemically pure (about 99.8% sucrose). The purity of sugar and the concentration or strength of sugar solutions is determined by the rotatory power of the solution, the special polariscope usually used being called a saccharimeter. Sucrose is reduced with Fehling's solution only after inversion.

The sugar content of some common fruits have been reported by Kulisch:

	SUCROSE	HEXOSES
Apple	1.0–5.4	7.0–13.0
Apricot	6.0	2.7
Banana, ripe	5.0	10.0
Pineapple	11.3	2.0
Strawberry	6.3	5.0

Lactose is obtained from the residual water solution ("whey") of milk after removal of fat and casein for butter and cheese. Milk contains about 4.5% of lactose. Lactose forms hard gritty crystals ("sand sugar") $(C_{12}H_{22}O_{11} \cdot H_2O)$, loses water at 140° C., melting point 202° C. (anhydrous) with decomposition; is less sweet than sucrose, reduces ammonio-cupric salt solution, ammonio-silver salt solution, forms osazone, melting point 200° C., turns yellow when warmed with sodium hydroxide solution. Lactose is the source of galactose, and undergoes, with the proper enzymes, fermentation into lactic acid and butyric acid.

Maltose is found in soy bean, and is produced by the action of the enzyme diastase of germinated barley (malt) on starch at 50° C., and is thus an intermediate product in the transformation of starch into alcohol. Maltose $(C_{12}H_{22}O_{11} \cdot H_2O)$, melting point 100° C., when rapidly heated, may be crystallized from the concentrated malt syrup after removal of proteins and insoluble material. Maltose reduces ammonio-cupric salt solution, and forms osazone.

Starch is a white powder, odorless and tasteless, insoluble in cold water, forming an emulsion ("starch paste") or gel with hot water, the consistency of which depends upon the ratio of starch to water used. When boiled starch emulsion is cooled and treated with a solution of iodine in alcohol or **potassium** iodide, a blue coloration is produced, which is a sensitive and characteristic test. The blue color is associated with the adsorption of iodine on the surface of the starch, and disappears in the presence of alkalis. When boiled with dilute acid, starch is first changed into a soluble gummy mixture known as dextrin, and finally into glucose. When starch, either alone or in the presence of a slight amount of nitric acid, is heated to 120° to 200° C., dextrin is formed; at higher temperatures starch behaves similarly to sucrose. With concentrated nitric acid, starch forms esters, similar to cellulose nitrates. By the action of the **enzyme** diastase, starch is converted into maltose, which with the enzyme maltase yields glucose. Starch is non-reactive with ammonio-**cupric** salt solution, and with **phenylhydrazine**. Starch is extensively distributed as granules in many plants. Those that serve as important reserves of starch are the seeds of wheat, corn, barley, rye, rice, the tubers of potato, cassava (tapioca starch), the pith of the sago palm. The form and size of the starch granules are characteristic of the plant in which it is found. When viewed in the microscope, the grains are seen to be cells made up of an inner nucleus, surrounded by concentric layers. True starch or amylose is present in the interior of the starch cell, while pseudo-starch or amylopectin is contained in the walls of the cell. The formation of starch paste is associated with the presence of the latter substance, which appears to contain a small proportion of combined phosphoric acid.

Starch is obtained by coarsely crushing the raw material, e.g., corn kernels or potatoes, the starch grains are washed free from cellulose and other materials by water. The suspension of starch is allowed to settle and the water drawn off. In the case of corn kernels, the oily germs float on the surface of the suspension and are separated for subsequent extraction of oil. Starch consists of a number of glucose units attached to each other by oxygen linkages to the 3,6 carbon atoms. Starch is an important food material commonly consumed with the **proteins** of cereal grains, rice and potatoes. Starch is a raw material for the brewing industry, for industrial alcohol, for dextrin, and is used in laundering textile goods as a stiffening agent and finish, in sizing paper, and as a thickening agent and adhesive.

Cellulose is a white solid, odorless and tasteless, insoluble in cold or hot water, chemically non-reactive except when drastically treated. Cellulose (1) when heated with water at 260° C. under pressure, dissolves completely with decomposition, (2) when treated with concentrated **sulfuric acid** dissolves, the solution upon dilution and boiling yielding glucose, (3) when treated with **sodium** hydroxide (15% to 25% NaOH) the fibers swell up and upon washing and drying possess a lustrous appearance (mercerized cotton), (4) with **iodine** in **potassium** iodide solution plus **zinc** chloride (Schulze's solution) produces a dark blue color, (5) with 80% **sulfuric acid** and rapidly washed and dried yields parchment surface. Simple cellulose is represented in nature by the cottonseed hair. Compound celluloses are widely distributed in plants, the two principal types being:

(a) Lignocelluloses, of woods, cereal straws, jute. These cellulose materials yield lignin by treatment (1) with 43% **hydrochloric acid**, cold for 12 hours, (2) with 8% to 12% sodium hydroxide at 140° to 160° C. for 6 to 10 hours, (3) with 72% sulfuric acid at ordinary temperature for 18 hours (the common method). Wood yields about 25% of lignin by the last treatment. The composition of lignin is not known.

(b) Pectocelluloses, of flax, hemp, ramie. These cellulose materials yield pectic substances by treatment with oxalic acid or ammonium oxalate at 85° C. for 24 hours, followed by carefully defined treatment with **alcohol**, **acetic acid** and **calcium** chloride. Pectic substances are most abundant in leaves, e.g., ivy, sycamore, and in apples or oranges, especially the white peel of the latter.

Cellulose of cotton fiber is supplemented industrially by that made from wood of **coniferous** trees, e.g., spruce, hemlock, into pulp, mainly for paper, by treatment (1) with **calcium** hydrogen sulfite plus sulfurous acid solution—the sulfite process—heated under pressure for several hours, (2) with **sodium** hydroxide solution—the soda process—under similar conditions, (3) with **sodium** sulfide plus **sodium** hydroxide solution—the sulfate process—under similar conditions. Other industrial sources of cellulose are flax, hemp, ramie, jute, Manila hemp, cereal straw, used mainly for textiles, and cordage.

Cotton cellulose is an important textile fiber. Cotton and wood cellulose may be transformed into esters, three of which are of outstanding importance. (1) Cellulose nitrates (see Fig. 1), for explosives, plastics, photo-

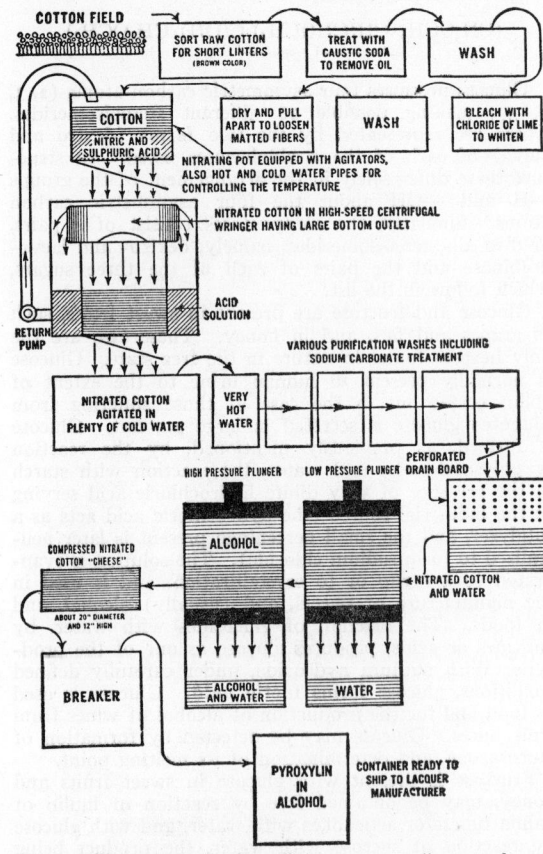

Fig. 1. Flow sheet of nitrocellulose or pyroxlin manufacture from cotton and nitric-sulfuric acid mixture.

graphic films, lacquers, collodion, rayon (made noninflammable by denitrifying with ammonium sulfide solution), (2) cellulose acetate, for plastics, photographic films, rayon, lacquers, (3) cellulose xanthate, used in making synthetic textile fiber or translucent wrapping paper called viscose. Viscose is made by impregnation of cellulose with sodium hydroxide (17.5% NaOH) solution, sodium cellulose ($C_6H_{10}O_5 \cdot 2NaOH$) approximately being formed (see Fig. 2). Addition of **carbon disulfide** (one mol CS_2 to one mol $C_6H_{10}O_5$, approximately) results in the formation of yellow sodium xanthate (possibly $C_6H_8(ONa) \cdot O \cdot CS \cdot SNa$), which after ripening and purifying is exuded as fibers or sheets into a bath consisting of dilute sulfuric acid plus sodium hydrogen sulfate plus sugar (many recipes for bath liquor are known) at 30° C. The precipitated material is desulfurized by treatment with very dilute **sodium** sulfide solution at 40° to 50° C., or by exposure to air, leaving the fiber or sheet as translucent cellulose. Mixed with metallic dust and coloring matter, viscose is made into artificial leather.

Cellulose, when treated with ammonio-**cupric** hydroxide solution (Schweitzer's solution), dissolves, and, when

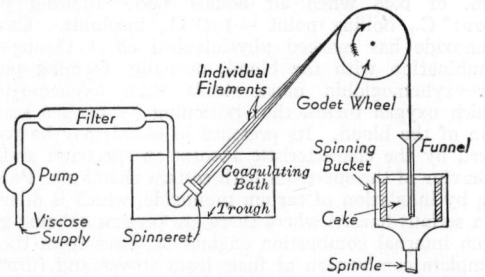

Fig. 2. Diagram showing the formation of rayon filaments and thread. (*duPont Rayon Co.*)

the resulting solution is acidified, translucent flocculent cellulose is precipitated. By exuding the ammonio-cupric cellulose solution through fine orifices into a bath of acid artificial fibers of cellulose are produced—cupra-ammonium process.

Ground wood pulp or paper turns yellow on exposure to sunlight, and with phloroglucinol dissolved in alcohol produces a red color. Lignin complexes of wood are destroyed by treatment with **chlorine or bromine** at ordinary temperatures, leaving cellulose residue. Ground wood pulp is bleached by sodium peroxide. Wood powder ("sawdust") when heated at 220° C. with sodium hydroxide plus potassium hydroxide yields sodium potassium oxalate, from which oxalic acid may be obtained. Destructive distillation of wood yields gases, watery distillate—containing **acetic acid**, methyl alcohol, acetone, tar distillate and charcoal residue.

Dextrin is a white to yellow solid, forming an adhesive with water, non-reactive with ammonio-cupric salt solution, reactive with iodine in alcohol or potassium iodide, usually forming red, brown, or blue color. Formed when starch is (1) heated to 120° to 200° C. either alone or in the presence of a slight amount of nitric acid. Dextrin is formed when bread is toasted and is present in well-baked bread crust, and on the surface of starched goods that have been ironed hot. Dextrin is used in adhesives.

Inulin is a white solid, soluble in warm water, specific rotatory power —40°, with iodine in alcohol or potassium iodide gives yellow color. Inulin is present in tubers of dahlia to the extent of about 10%. Inulin reacts with water in the presence of the enzyme inulase or of acids to form fructose. The enzyme diastase does not produce this change.

Glycogen, or animal starch, is a white solid, soluble in water, specific rotatory power +197°, with iodine in alcohol or potassium iodide solution, forming brown color. Glycogen is found as reserve carbohydrates in the animal body, more particularly in the liver. Horseflesh, oysters and beef are sources of glycogen.

Pentosans are polysaccharides which may be considered as anhydrides of pentose sugars, after the manner of the hexosans, sucrose, starch, from glucose, fructose. When pentosans or pentoses are heated with hydrochloric or sulfuric acid, furfural ($C_4H_3O \cdot CHO$) is formed, and addition of aniline produces a red color. Pentosans are present in gummy carbohydrates, in bran of wheat seed, and in woods.

By means of the cyanhydrin reaction higher sugars of the heptose, octose and nonose types have been prepared. A monosaccharide such as an aldohexose may be converted into the next lower monosaccharide, such as an aldopentose, by oxidation to the acid, which corresponds to the aldohexose, then treating the calcium salt solution of this acid with a solution of ferrous acetate plus hydrogen peroxide. Carbon dioxide is evolved and aldopentose formed. (R.K.S.)

CARBON. Symbol: C. Atomic number: 6. Atomic weight: 12.010. Density: (1) Diamond, 3.52, (2) Graphite, 2.25, (3) Amorphous, 1.88. Hardness: (1) Diamond, 10, (2) Graphite, 0.5 to 1.0. Melting point: >3500° C. Boiling point: 4200° C.

The chemical element carbon is known in three forms, namely, (1) diamond, (2) graphite, (3) amorphous, such as charcoal, coke. When heated in excess air or **oxygen**, carbon burns to form carbon dioxide with the evolution of a definite amount of heat from each form. See **Chemical Composition, Allotropic Elements.** Discovery prehistoric. Isotopes: 12 (99.3%), 13 (0.7%). The uses of carbon are dependent upon the form and variety, diamonds for jewels and as abrasive, graphite in lubricants and as an electrical conductor, cocoanut charcoal for adsorbing gases at low temperature in an enclosed space to produce a high vacuum, activated carbon to absorb color from solutions and to remove color from water, coke and wood charcoal as fuels.

Carbon occurs (1) free, as **diamond** and graphite of local distribution, e.g., diamonds in South Africa, **graphite** in Ceylon, and (2) combined, as carbonate rocks, such as **limestone** and **dolomite**, as **hydrocarbons** in petroleum, natural gas, coal, and as plant and animal constituents generally distributed.

Microscopic diamonds have been made by crystallization of carbon dissolved in iron under high pressure of the solidifying iron. Graphite is made artificially from anthracite coal by heating to the temperature of the electric furnace. Various varieties of carbon are prepared by heating certain organic materials, e.g., cocoanut charcoal for gas adsorption, bone charcoal and activated car-

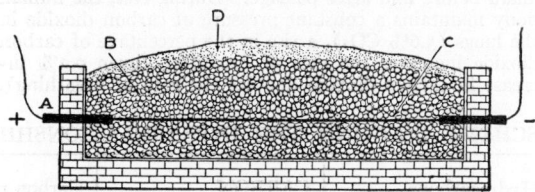

Graphite production in electric resistance furnace by Acheson process. A, Electrodes of graphite; B, Charge of coarse grains of coke; C, Core of carbon rod; D, Covering of sand and coke mixture.

bon for decolorizing solutions of organic substances, lamp black or soot for paint pigment by incomplete combustion of natural gas or petroleum, gas carbon by coking the residue in the distillation of petroleum, coke and wood charcoal by the destructive distillation of coal and wood, respectively. (See **Destructive Distillation.**)

The behavior of different forms of carbon towards strong oxidizing agents, e.g., **nitric-sulfuric acid** mixture or sulfuric acid plus **potassium** chlorate, serves to distinguish them. Diamond is unacted upon by these reagents, graphite upon prolonged treatment and heating yields graphitic acid, yellow, of the original external form of the graphite, and amorphous carbon forms are dissolved yielding humic acid and finally mellitic acid ($C_6(COOH)_6$), melting point 286° C.

Acids, carboxylic.

Carbides. **Calcium** carbide (CaC_2) and **sodium** carbide (Na_2C_2) yield **acetylene** with water; **aluminum** carbide (Al_4C) yields **methane** with water; **silicon** carbide (SiC) and **titanium** carbide (TiC) do not react with water; silicon carbide and **tungsten** carbide are used in machining operations on account of their hardness (see **Abrasives**); carbides of iron (Fe_3C), **manganese** (Mn_3C), and **chromium** (Cr_4C, Cr_3C_2), and many other metals, also of **boron** (B_6C) have been reported.

Chlorides. (See **Chlorine.**)

Hydrides. (See **Hydrocarbons.**)

Oxides. Carbon dioxide, carbonic acid anhydride, "carbonic acid gas" (CO_2) is a colorless gas, density 1.9769 grams per liter, 0° C., 760 mm. or 1.53 when air equals 1.00, melting point —56.6° C. at 5.2 atmospheres, solid sublimes at —79° C. at 1 atmosphere, critical temperature 31° C., critical pressure 73 atmospheres, soluble about one volume in one volume of water at 15° C. and 1 atmosphere. Carbon dioxide is used in solution under pressure for carbonated beverages, as a liquid for extinguishing fires and making solid carbon dioxide, and as a solid ("carbon dioxide snow," "dry ice") in the maintenance of low temperatures, and as a gas quickly generated, for extinguishing fires and making carbonates, and is the source of carbon (in the atmosphere) for plant life. Carbon dioxide is chemically a remarkably permanent gas, requiring such treatment as burning magnesium to separate carbon (plus magnesium oxide); it is, however, reduced to carbon monoxide at a red heat by carbon or iron; and dissociates into carbon monoxide plus oxygen above 1300° C. (69% unchanged at 2400° C.). Carbon dioxide is formed (1) in the combustion of fuels when burned with excess air, (2) in the respiration of animals by oxidation of foods, (3) by the action of acids on carbonates, (4) by heating carbonates, such as sodium hydrogen carbonate (not sodium carbonate), **calcium** carbonate, or **magnesium** carbonate; and is recovered from (5) **fermentation** processes, and (6) many gas wells. The atmosphere contains about three parts by volume of carbon dioxide in 10,000 volumes of air.

Carbon dioxide is estimated by absorption in **sodium** or **potassium** hydroxide solution and measuring the volume of the gas before and after absorption; by absorption with solid sodium hydroxide or soda-lime and weighing before and after absorption; or by passage through **barium** hydroxide solution and titration of the alkali before and after passage. During rest, the human body maintains a constant pressure of carbon dioxide in the lungs (5.6% CO_2), a rise in the percentage of carbon dioxide increases the rate of breathing (even 0.2% increase in carbon dioxide doubles the rate of breathing),

and a fall in the percentage decreases the rate of breathing with danger of cessation.

Carbon monoxide, carbonic oxide (CO) is a colorless, odorless gas, density 1.2504 grams per liter, 0° C., 760 mm. or 0.98 when air equals 1.00. Melting point —207° C., boiling point —192° C., insoluble. Carbon monoxide has marked physiological effect through its combination with the blood, probably forming purple carboxyhemoglobin more stable than oxyhemoglobin which oxygen forms, thus poisoning by oxygen starvation of the blood. Its presence in blood may be recognized by the characteristic absorption spectrum and the behavior of the spectrum with certain chemicals. Poisoning by inhalation of carbon monoxide, which is odorless, is a serious hazard where there are present exhaust gases from internal combustion engines or gases from the incomplete combustion of fuels from stoves and furnaces. (See **Carbon Monoxide Poisoning.**) Chemically, carbon monoxide is (1) combustible to form carbon dioxide, accompanied by a transparent blue flame and the evolution of heat, but the fuel value is low (320 B.T.U. per cu. ft.) (2) reactive with **chlorine**, forming carbonyl chloride ($COCl_2$) in the presence of light and a **catalyzer**, (3) reactive with **sulfur** vapor at a red heat, forming carbonyl sulfide (COS), (4) reactive with **hydrogen**, forming methyl alcohol (CH_3OH) or methane (CH_4) in the presence of a catalyzer, (5) reactive with **nickel** (also **iron, cobalt, molybdenum, ruthenium**) to form nickel carbonyl ($Ni(CO)_4$) (and carbonyls of the other metals named), (6) reactive with fused **sodium** hydroxide, forming sodium formate (HCOONa), (7) reactive with **cuprous** salt dissolved in either **ammonium** hydroxide or concentrated **hydrochloric acid**, which solutions are utilized in the estimation of carbon monoxide in mixtures of gases, e.g., flue gases of combustion, coal gas, exhaust gases of internal combustion engines, (8) reactive with **iodine** pentoxide at

SCHEME SHOWING THE INTERRELATIONSHIPS OF SOME CARBON-CONTAINING SUBSTANCES

Hydrocarbons	CARBON	Carbon monoxide	Carbon suboxide	Carbon dioxide
Natural gas, petroleum, coal, lignite, shale in nature.	Diamond, graphite in nature. Graphite, charcoals, coke artificially.	Formic acid Acetic acid	Malonic acid Oxalic acid	Carbonic acid
		Metallic acetates	Metallic oxalates	Metallic carbonates Calcite, dolomite, magnesite, strontianite, witherite, malachite, azurite in nature.
	Formaldehyde Acetaldehyde Acetone		Insoluble Except sodium, potassium ammonium.	
	Methyl alcohol Ethyl alcohol Glycol Glycerol	Organic acetates	Organic oxalates	Insoluble Except sodium, potassium, ammonium. Organic carbonates
	Fatty oils, fats, waxes. In plant and animal substances.			
	Carbohydrates In plant and animal substances.			
	Proteins In plant and animal substances.			

150° C. For the reaction of carbon monoxide with oxygen to form carbon dioxide—finely divided iron, or **palladium** wire is a catalyzer; for the reaction of carbon monoxide with water vapor to form carbon dioxide plus hydrogen ("water gas reaction") important studies have been made of the conditions; and for the reaction of carbon dioxide plus carbon heated similar important studies have been made (at 675° C., 50% CO_2 plus 50% CO; at 900° C., 5% CO_2 plus 95% CO). Carbon monoxide is formed by the incomplete combustion of **fuels**. The reaction of carbon plus oxygen at such a temperature as produces carbon monoxide (say 900° C., 95% CO plus 5% CO_2) evolves heat, while the reaction of carbon plus carbon dioxide, producing carbon monoxide at the same temperature absorbs heat. Accordingly, it is possible to arrange the oxygen (free or as air) and carbon dioxide supply ratio in such a way that the desired temperature may be continuously maintained. The reduction of carbon dioxide by iron forms carbon monoxide plus ferrous oxide.

Carbon suboxide (C_3O_2) is a colorless gas, of unpleasant odor, poisonous, boiling point 7° C., burns with a blue smoky flame producing carbon dioxide. When condensed to liquid, carbon suboxide slowly changes at ordinary temperature to a dark red solid, soluble in water to a red solution. Reacts with water to form malonic acid, with **hydrogen chloride** to form malonyl chloride, with **ammonia** to form manolamide. Made by heating malonic acid or its ester at 300° C. under diminished pressure, and separation from simultaneously formed carbon dioxide and ethylene by condensation and fractional distillation.

Compounds of carbon are discussed as follows:

Antimony-containing. (See **Antimony, organic compounds.**)

Arsenic-containing. (See **Arsenic, organic compounds.**)

Boron-containing. (See **Boron, organic compounds.**)

Bromine-containing. (See **Bromine, organic compounds.**)

Chlorine-containing. (See **Chlorine, organic compounds.**)

Fluorine-containing. (See **Fluorine, organic compounds.**)

Halogen-containing. (See **Individual Halogen.**)

Hydrocarbons.

Iodine-containing. (See **Iodine, organic compounds.**)

Metal-containing. (See **Individual Metal.**)

Nitrogen-containing. (See **Nitrogen, organic compounds,** but first consult key below.)

Oxygen-containing. (See **Oxygen, organic compounds,** but first consult key below.)

Phosphorus-containing. (See **Phosphorus, organic compounds.**)

Silicon-containing. (See **Silicon, organic compounds.**)

Sulfur-containing. (See **Sulfur, organic compounds.**)

Organic compounds are discussed as follows:

Sulfur-containing organic compounds are listed under **Sulfur.**

Phosphorus-containing organic compounds are listed under **Phosphorus.**

Chlorine-containing organic compounds are listed under **Chlorine.**

Bromine-containing organic compounds are listed under **Bromine.**

Iodine-containing organic compounds are listed under **Iodine.**

Fluorine-containing organic compounds are listed under **Fluorine.**

Nitrogen-containing organic compounds are listed under **Nitrogen.**

Organic compounds containing carbon and hydrogen only are listed under **Hydrocarbons.**

Organic compounds containing carbon and oxygen only are listed under **Carbon, oxides.**

Organic acids containing carbon, hydrogen and oxygen only are listed under **Acids, carboxylic.**

Organic compounds containing oxygen, *except those cited above,* are listed under **Oxygen.**

Organic compounds are grouped as follows:

Division Number	Chemical Elements	Generic Groups	
I	C, H	Hydrocarbons.......	See **Hydrocarbons**
II	C, O	Oxides, carbon......	See **Carbon oxides**
III	C, H, O	Acids, mono, di, tri, basic..............	See **Acids, carboxylic**
		Aldehydes, mono, di..	See **Aldehydes**
		Ketones, mono, di....	See **Aldehydes**
		Quinones...........	See **Phenols**
		Alcohols, prim., sec., tert...............	See **Alcohols**
		Phenols, mono, di, tri	See **Phenols**
		Acid, anhydrides.....	See **Acids, carboxylic**
		Esters.............	See **Esters**
		Ethers.............	See **Alcohols**
		Acetals............	See **Aldehydes**
		Carbon-hydrogen oxides...........	See **Alcohols**
		Lactones...........	See **Acids, carboxylic**
		Lactides...........	See **Acids, carboxylic**
		Hydroxy aldehydes...	See **Carbohydrates**
		Hydroxy ketones....	See **Carbohydrates**
		Hydroxy acids......	See **Acids, carboxylic**
		Aldo-acids.........	See **Acids, carboxylic**
		Keto-acids.........	See **Acids, carboxylic**

For further reference, see **Oxygen,** for scheme showing interrelationships of oxygen-function organic compounds, and alphabetical list of compounds.

Division Number	Chemical Elements	Generic Groups	
IV	C, N With or without H, O Without S	Nitrate............	See **Nitric acid**
		Nitro..............	See **Nitro**
		Nitrite............	See **Nitrous acid**
		Nitroso............	See **Nitro**
		Nitrosamines.......	See **Nitro**
		Hydroxylamines.....	See **Hydroxylamines**
		Oximes............	See **Hydroxylamines**
		Hydrazines.........	See **Hydrazines**
		Hydrazones........	See **Hydrazines**
		Osazones..........	See **Hydrazines**
		Pyrrole............	See **Pyrrole**
		Pyridine...........	See **Pyridine**
		Amines............	See **Amines**
		Quaternary ammonium compounds...	See **Amines**
		Cyanides..........	See **Hydrocyanic acid**
		Isocyanides........	See **Hydrocyanic acid**
		Amides............	See **Amines**
		Anilides...........	See **Amines**
		Aminoacids........	See **Aminoacids**
		Polypeptides.......	See **Aminoacids**
		Proteins...........	See **Aminoacids**
		Cyanates..........	See **Cyanic acid**
		Isocyanates........	See **Cyanic acid**
		Fulminates.........	See **Cyanic acid**
		Cyanamides........	See **Cyanamides**
		Carbamates........	See **Amines**
		Ureas.............	See **Amines**
		Ureides...........	See **Amines, and Purine**
		Purines...........	See **Purine**
		Semicarbazides.....	See **Amines**
		Semicarbazones.....	See **Amines**
		Guanidines........	See **Amines**
		Aminoguanidines....	See **Amines**
		Hydrazo...........	See **Hydrazines, and Azo**
		Azo...............	See **Azo**
		Azoxy.............	See **Azo**
		Diazo (nium).......	See **Azo**
		Aminoazo..........	See **Azo**
		Hydroxyazo........	See **Azo**

For further references see **Nitrogen,** for alphabetical list of compounds and scheme showing interrelationships of nitrogen-function organic compounds.

Division Number	Chemical Elements	Generic Groups	
V	C, S With or without H, O, N	Sulfonic acids.......	See **Thioalcohols**
		Sulfonyl...........	See **Thioalcohols**
		Sulfones..........	See **Thioalcohols**
		Sulfinic acids......	See **Thioalcohols**
		Sulfinyl..........	See **Thioalcohols**
		Sulfoxides........	See **Thioalcohols**
		Thiophene........	See **Thiophene**
		Penthiophene......	See **Thiophene**
		Thioalcohols......	See **Thioalcohols**
		Thioethers........	See **Thioalcohols**
		Tertiary sulfonium compounds.......	See **Thioalcohols**
		Thiophenols.......	See **Thioalcohols**
		Thioaldehydes.....	See **Thioaldehydes**
		Thioketones.......	See **Thioketones**
		Thioic acids......	See **Sulfur, acids**
		Thiocyanates......	See **Thiocyanic acid**
		Isothiocyanates....	See **Thiocyanic acid**
		Thiocarbonic acid..	See **Thiocarbonicacid**

For further references see **Sulfur,** for alphabetical list of compounds, and scheme showing interrelationships of sulfur-function organic compounds.

CARBON SCHEME SHOWING TYPES OF CARBON ARRANGEMENTS.

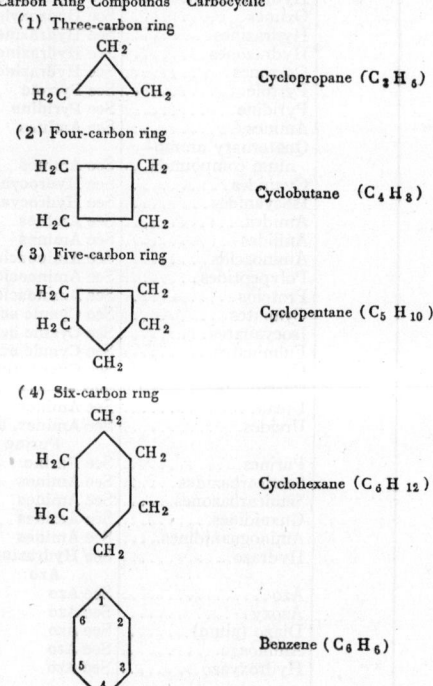

A. Carbon Chain Compounds

(1) Straight chain

$$H_3\underset{3}{C} - \underset{2}{CH_2} - \underset{1}{CH_3} \quad \text{Propane } (C_3H_8)$$

(2) Forked or branched chain

Trimethylmethane (C_4H_{10})
(2-Methylpropane)

Tetramethylmethane (C_4H_{10})
(2,2-Dimethylpropane)

B. Carbon Ring Compounds   Carbocyclic

(1) Three-carbon ring

Cyclopropane (C_3H_6)

(2) Four-carbon ring

Cyclobutane (C_4H_8)

(3) Five-carbon ring

Cyclopentane (C_5H_{10})

(4) Six-carbon ring

Cyclohexane (C_6H_{12})

Benzene (C_6H_6)

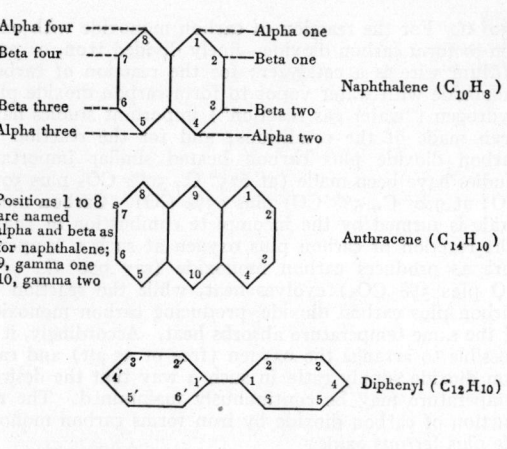

Naphthalene ($C_{10}H_8$)

Positions 1 to 8 are named alpha and beta as for naphthalene; 9, gamma one 10, gamma two

Anthracene ($C_{14}H_{10}$)

Diphenyl ($C_{12}H_{10}$)

C. Combinations of Carbon Chain and Carbon Ring

(1) Six-carbon plug

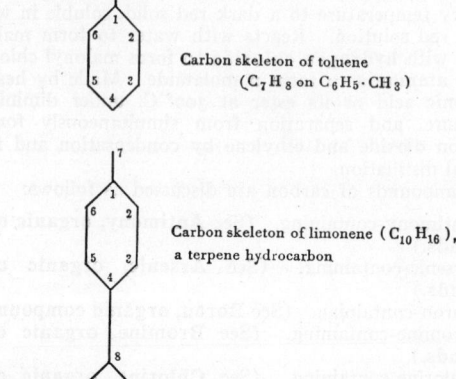

Carbon skeleton of toluene
(C_7H_8 on $C_6H_5 \cdot CH_3$)

Carbon skeleton of limonene ($C_{10}H_{16}$), a terpene hydrocarbon

(2) Bridged ring (3 bridges from No. 2 to No. 4, (1) via No. 3, (2) via No. 7, (3) via No. 1,6,5. Therefore, "bicyclo [3.1.1] heptane" Numbers inside the square brackets denoting the number of carbons in each bridge - not counting the piers (No. 2,4)

Carbon skeleton of pinene ($C_{10}H_{16}$), a terpene hydrocarbon

D. Heterocyclic Compounds

(1) Oxygen-carbon ring

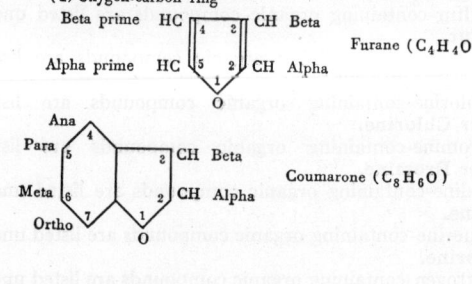

Beta prime HC   CH Beta
Alpha prime HC   CH Alpha

Furane (C_4H_4O)

Ana
Para
Meta
Ortho

CH Beta
CH Alpha

Coumarone (C_8H_6O)

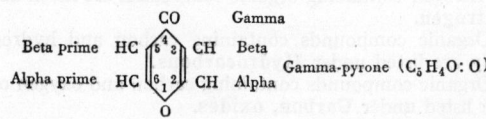

CO   Gamma
Beta prime HC   CH Beta
Alpha prime HC   CH Alpha

Gamma-pyrone ($C_5H_4O{:}O$)

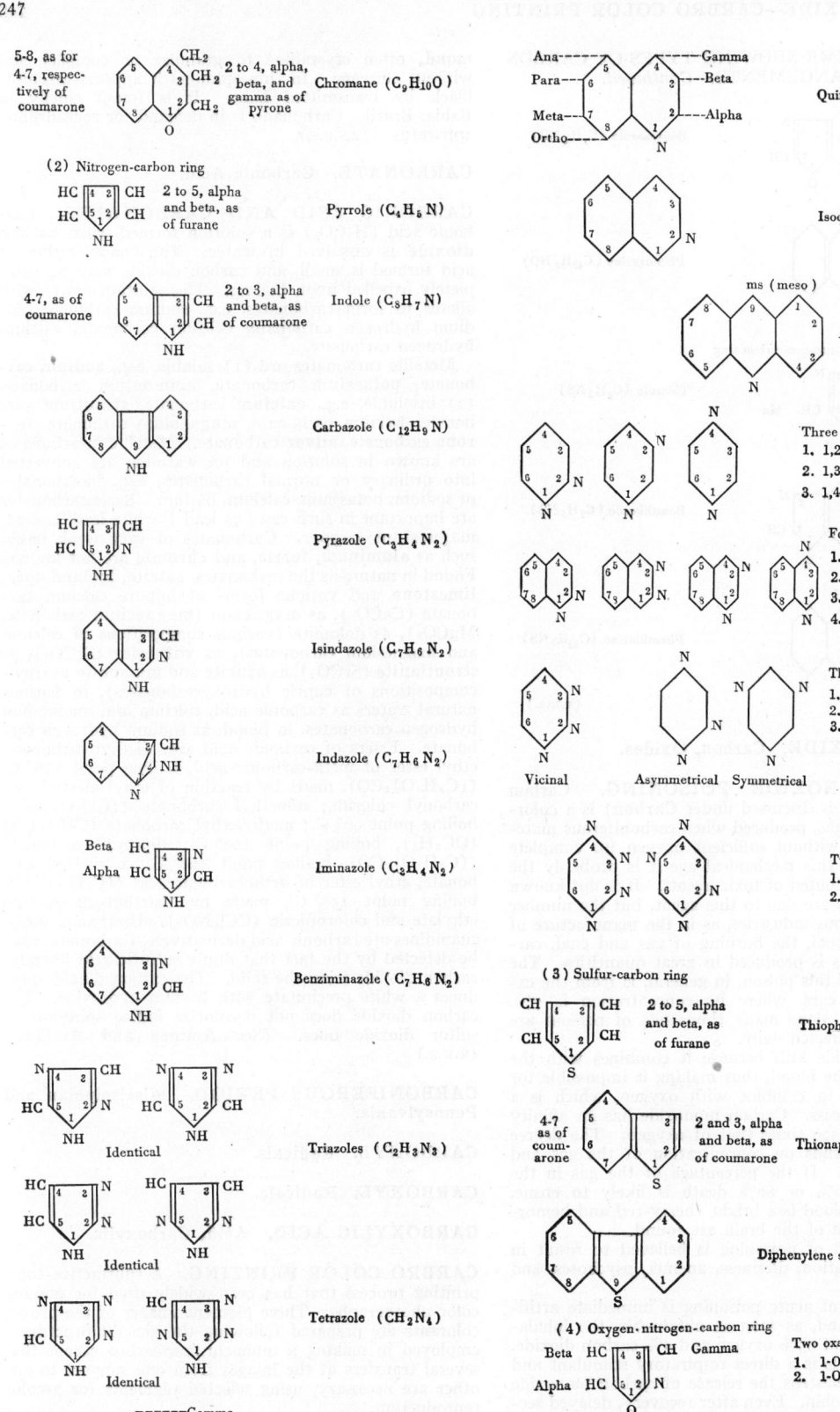

5-8, as for 4-7, respectively of coumarone — 2 to 4, alpha, beta, and gamma as of pyrone — Chromane ($C_9H_{10}O$)

(2) Nitrogen-carbon ring

2 to 5, alpha and beta, as of furane — Pyrrole (C_4H_5N)

4-7, as of coumarone — 2 to 3, alpha and beta, as of coumarone — Indole (C_8H_7N)

Carbazole ($C_{12}H_9N$)

Pyrazole ($C_3H_4N_2$)

Isindazole ($C_7H_6N_2$)

Indazole ($C_7H_6N_2$)

Beta HC / Alpha HC — Iminazole ($C_3H_4N_2$)

Benziminazole ($C_7H_6N_2$)

Identical — Triazoles ($C_2H_3N_3$)

Identical — Tetrazole (CH_2N_4)

Gamma / Beta / Alpha — Beta prime / Alpha prime — Pyridine (C_5H_5N)

Ana — Gamma
Para — Beta
Meta — Alpha
Ortho —
Quinoline (C_9H_7N)

Isoquinoline (C_9H_7N)

ms (meso) — Acridine ($C_{13}H_9N$)

Three diazines ($C_4H_4M_2$)
1. 1,2 (pyridazine)
2. 1,3 (pyrimidine)
3. 1,4 (pyrazine)

Four benzodiazines
1. 1,2 (cinnoline)
2. 2,3 (phthalazine)
3. 1,3 (quinazoline)
4. 1,4 (quinoxaline)

Three triazines
1. 1,2,3
2. 1,2,4
3. 1,3,5 (cyanidine)

Vicinal — Asymmetrical — Symmetrical

Two tetrazines
1. 1,2,3,4
2. 1,2,4,5

(3) Sulfur-carbon ring

2 to 5, alpha and beta, as of furane — Thiophene (C_4H_4S)

4-7 as of coumarone — 2 and 3, alpha and beta, as of coumarone — Thionaphthene (C_8H_6S)

Diphenylene sulfide ($C_{12}H_8S$)

(4) Oxygen-nitrogen-carbon ring

Beta HC — Gamma / Alpha HC — Two oxazoles (C_3H_3NO)
1. 1-O-2-N
2. 1-O-3-N

Beta HC / Alpha HC — CH — Mu

CARBON SCHEME SHOWING TYPES OF CARBON ARRANGEMENTS.—*Continued*.

Benzoxazole (C₇H₅NO)

Phenoxazine (C₁₂H₉NO)

(5) Sulfur-nitrogen-carbon ring

Thiazole (C₃H₃NS)

Benzthiazole (C₇H₅NS)

Phenthiazine (C₁₂H₉NS)

(R.K.S.)

CARBON DIOXIDE. Carbon, Oxides.

CARBON MONOXIDE POISONING. Carbon monoxide (which is discussed under **Carbon**) is a colorless and odorless gas, produced when carboniferous materials are burned without sufficient oxygen for complete **combustion**. In this mechanical age it is probably the most widely distributed of toxic agents. It is not known how many deaths are due to this agent, but the number is great. In various industries, as in the manufacture of steel, mining of coal, the burning of gas and coal, carbon monoxide gas is produced in great quantities. The greatest source of this poison, in general, is from the exhaust of motor cars, where its concentration is 7%. From this source alone many thousands of persons are to some degree affected daily.

Carbon monoxide kills because it combines with the haemoglobin of the blood, thus making it impossible for the haemoglobin to combine with oxygen, which is a necessary life process. Carbon monoxide has an affinity for haemoglobin 300 times that of oxygen. The degree of poisoning depends on concentration of the gas and time of exposure. If the percentage of the gas in the blood rises to 70% or 80% death is likely to ensue. After death the blood is a bright cherry-red and hemorrhages and **edema** of the brain are found.

A chronic form of poisoning is believed to result in headaches, palpitation, dizziness, anemia, psychoses, and neuritis.

The treatment of acute poisoning is immediate artificial respiration, and, as soon as obtainable, the inhalation of a mixture of 95% oxygen and 5% carbon dioxide. The carbon dioxide is a direct respiratory stimulant and the pure oxygen hastens the release of carbon monoxide from the haemoglobin. Even after recovery, delayed secondary complications such as paralysis and cerebral hemorrhage may occur. (R.S.M., D.M.H.)

CARBONADO (BLACK DIAMOND). The mineral carbonado is an opaque massive black variety of dia-

mond, often crystalline to granular or compact and without cleavage. In thin splinters it appears greenish-black by transmitted light. It is found chiefly in Bahia, Brazil. Carbonado is in demand for rock-drilling apparatus. (E.S.C.S.)

CARBONATE. Carbonic Acid.

CARBONIC ACID AND CARBONATES. Carbonic acid (H_2CO_3) is a solution formed when **carbon dioxide** is dissolved in water. The concentration of acid formed is small, and carbon dioxide may be completely expelled upon boiling. The solution reacts with alkalis to form carbonates, e.g., sodium carbonate, sodium hydrogen carbonate, calcium carbonate, calcium hydrogen carbonate.

Metallic carbonates are (1) soluble, e.g., **sodium** carbonate, **potassium** carbonate, **ammonium** carbonate, (2) insoluble, e.g., **calcium** carbonate, **strontium** carbonate, **barium** carbonate, **magnesium** carbonate, **ferrous** carbonate, **silver** carbonate. Metallic bicarbonates are known in solution and on warming are converted into ordinary or normal carbonates, e.g., bicarbonates of sodium, potassium, calcium, barium. Basic carbonates are important in such cases as lead ("white lead"), **zinc**, **magnesium**, **copper**. Carbonates of very weak bases, such as **aluminum**, **ferric**, and **chromic** are not known. Found in nature as the carbonates, **calcite**, **Iceland spar**, **limestone** and various forms of impure calcium carbonate ($CaCO_3$), as **magnesite** (**magnesium** carbonate, $MgCO_3$), as **dolomite** (various compositions of calcium and magnesium carbonates), as **witherite** ($SrCO_3$), as **strontianite** ($SrCO_3$), as **azurite** and **malachite** (various compositions of cupric hydroxycarbonates), in various natural waters as carbonic acid, calcium and magnesium hydrogen carbonates, in blood, as sodium hydrogen carbonate. Esters of carbonic acid are: diethyl carbonate, ethyl ester of meta-carbonic acid, boiling point 126° C. (($C_2H_5O)_2CO$), made by reaction of ethyl alcohol and carbonyl chloride; dimethyl carbonate (($CH_3O)_2CO$), boiling point 90° C.; methylethyl carbonate ($CH_3O)CO$ (OC_2H_5), boiling point 109° C.; dipropyl carbonate (($C_3H_7O)_2CO$), boiling point 168° C.; tetraethyl carbonate, ethyl ester of orthocarbonic acid (($C_2H_5O)_4C$), boiling point 158° C., made by reaction of sodium ethylate and chloropicrin (CCl_3NO_2). Urethanes, ureas, guanidines are carbonic acid derivatives. Carbonates may be detected by the fact that dilute sulfuric acid liberates carbon dioxide from the solid. The carbon dioxide produces a white precipitate with barium hydroxide. The carbon dioxide does not decolorize iodine solutions as sulfur dioxide does. (See **Amines and Amides**.) (R.K.S.)

CARBONIFEROUS PERIOD. Mississippian and Pennsylvania.

CARBONYL. Radicals.

CARBOXYL. Radicals.

CARBOXYLIC ACID. Acid, Carboxylic.

CARBRO COLOR PRINTING. A subtractive-color printing process that has been widely used for making color photographs. Three pigment images of the proper colorants are prepared following the general procedures employed in making a monochrome carbro except that several transfers of the images from one support to another are necessary, using selected pigments for 3-color reproduction.

Suitable bromide prints are prepared from each of the separation negatives and then each is contacted, gelatin surfaces together, with the suitably colored pigment paper which has been previously soaked in the sensitizing solution. Sensitizing solutions vary widely but generally contain an acid chromate, ferricyanide and bromide. The combining of the pigment papers with the appropriate bromide prints is generally carried out in a roller wringer. After the silver in the bromide prints is bleached, the pigments are squeegeed onto waxed celluloid or other suitable supports so that the paper backing and soluble gelatin-bearing pigment may be removed in hot water leaving on each celluloid a positive colorant image of pigmented gelatin. These images are dried, then resoaked and contacted, one at a time, with paper which has a clear gelatin coating soluble at a relatively low temperature. When the 3-colorant pigment images are superimposed, they are resoaked in water and placed in contact with the final support which may be almost any hardened gelatin surface. When they have partially dried together the temporary support paper is stripped off in hot water leaving the color print on its final support. (H.C.C.)

CARBUNCLE.
In geology this term is applied to that variety of **garnet, almandine,** which was much used formerly for jewelry, when cut en cabochon. It is derived from the Latin, *carbunculus,* a small spark, in reference to the glowing effect of that style of cutting. In the early part of the Christian Era, the term seems to have been used for red stones of all sorts.

In medicine, a carbuncle is a severe localized infection, usually occurring in the **subcutaneous** tissues. It starts in the same way as a boil, but then spreads down to the subcutaneous tissues where it may form a diffuse lesion discharging to the surface by several—instead of one— openings. It is a more serious infection than a boil and a longer time is required for it to subside. Fever and constitutional symptoms are frequently marked and in some instances death may result from **septicemia.** The causative organism is usually a **staphylococcus** *aureus.* Treatment is surgical incision and chemotherapy with one of the **sulfonamides** or **penicillin,** if necessary. (E.S.C.S., R.S.M., D.M.H.)

CARBURETION.
The fuel for an **internal combustion engine** must be thoroughly mixed with the air needed for combustion. This is very true of the **Otto cycle** engine, since thorough distribution of particles of fuel in the air is essential to the rapid and complete explosive combustion of the fuel in that cycle. One of the most effective means of mixing the particles of a liquid fuel with air is by vaporization. The vaporizing and mixing of a liquid fuel with air in the correct proportions is called carburetion, and the device to accomplish this, a carburetor.

The automobile engine offers an instance of applied carburetion, also one of the more difficult carburetive problems, since the automobile engine is expected to operate smoothly and economically with wide variation of power, using fuels of varying quality. The **Venturi tube** principle forms the basis of most commercial car-

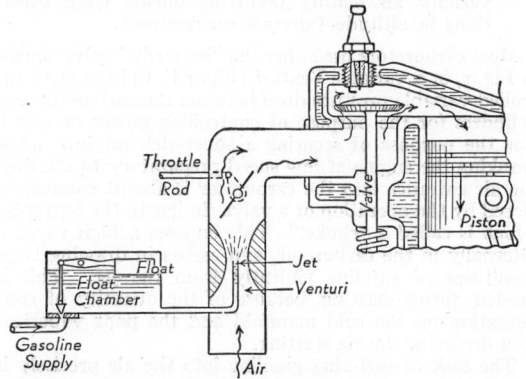

Induction system of the gasoline engine showing elementary form of carburetor.

buretors. In its simplest form it consists of a Venturi tube in the passage leading from the atmosphere to the cylinder. As the air passes this tube, it is speeded up at the throat due to the constriction in flow area, and the static pressure of the air is reduced. A gasoline chamber with a gasoline level which is maintained slightly below that of the level of the throat of the Venturi, supplies gasoline through an interconnecting tube to a jet whose opening is located in the area of reduced pressure at the Venturi throat. Since the surface of the gasoline in the float chamber is subject to atmospheric pressure the pressure difference causing gasoline to be sprayed out of the jet is the same as that which exists between the atmosphere and the throat of the Venturi tube. Thus both air and gasoline are subjected to the same driving pressure, and when one increases the other increases, and vice versa. If the discharge of gasoline from the jet, and of air through the Venturi increased in the same ratio, this simple carburetor would be successful for automotive service, but unfortunately the rate of increase of gasoline flow with increasing pressure difference exceeds that for air.

The mixture delivered by this carburetor is given by the following equation, in pounds of air per pound of gasoline:

$$\text{Mixture} = \frac{C_a A_a}{C_g A_g} \sqrt{\frac{d_a}{d_g}}.$$

C's are coefficients of discharge.
A's are areas of flow.
d's are densities of the fluids.

With ordinary gasoline this mixture should be **maintained** at a theoretical 15.2 lbs. of air per lb. of gasoline. The simple Venturi tube carburetor will not do that, but rather will give an increasingly richer mixture, as the velocity of air through the Venturi increases. The principal reason for this characteristic will be found in the variability of the coefficients of discharge. The coefficient of discharge for air through a Venturi tube is approximately constant, but that of gasoline from a jet increases with increasing pressure differential across the jet.

The commercial Venturi-type carburetor contains modifications which offset this basic undesirable characteristic so that a suitable mixture will be maintained at all loads. The simplest method employed to oppose enrichment of mixture was to provide an auxiliary valve admitting air to the mixture before it reached the cylinder. This auxiliary air valve opening was closed by a spring-loaded valve which was held partially open by the suction in the manifold, and admitted fresh air to dilute the overrich mixture. The carburetion obtained by the auxiliary air valve principle was inferior to other methods, and has been abandoned.

If the gasoline supply tube is constricted between the float chamber and the jet, the pressure drop across the constriction will increase at increasing rates of flow, and the result will be a tendency for starvation of the jet, and weakening of the mixture at high air velocities. By combining properly a jet of this type with one of the unrestricted type, a compound jet arrangement may be designed which will give fairly uniform mixture at varying loads. The use of a restricted jet, together with extra high speed jets that come into play when the air speed through the Venturi exceeds a certain value, is another modification of the simple Venturi carburetor. Some carburetors have a variable orifice on the gasoline jet created by the use of a throttle-operated metering pin; others have a Venturi tube with variable opening.

In one type of carburetor the density of the gasoline issuing from the jet is, in effect, varied by bleeding air bubbles into the vertical stem of the jet in such a way that the amount of air mixed with the gasoline increases with increasing load, thus tending to offset the enriching tendency of a plain Venturi jet carburetor. This air bleeding of the jet is called the plain-tube principle of carburetion.

Although a mixture of 15.2 lbs. of air per lb. gasoline is theoretically correct, mixtures as rich as 9:1 or as lean as 20:1 are explosive. However, the rich mixtures are uneconomical, and the lean mixtures must be employed cautiously at maximum power to prevent **detonation** and overheating. Thus rich mixtures are indicated near full power operation. Also, on account of the proportionately high dilution of incoming fuel stream by unscavenged products of combustion at the previous cycle at light loads (under quantity control), a rich mixture is needed at low power output. Between these needs for rich mixture, the carburetor should deliver a fairly uniform lean mixture for the sake of fuel economy. The accompanying graph illustrates a desirable performance for the carburetor of a gasoline engine. By employing

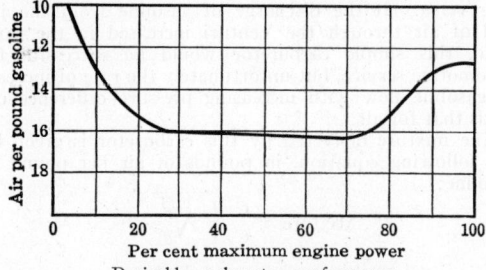

Desirable carburetor performance.

multiple jets, adjustable orifices and other intricacies, commercial carburetors attain mixture control approximating this desired performance, at considerable sacrifice of the simplicity of the principle of Venturi tube metering. (F.T.M.)

CARBURETION, AIRCRAFT ENGINE. Carburetor Types.

CARBURETOR TYPES. A carburetor is any device for securing the **carburetion** function described in the article of that name. It is a necessary device for any engine, furnace, kiln, etc., where a volatile liquid fuel is prepared for combustion by vaporizing into the flow of combustion air. The possibility of accomplishing carburetion satisfactorily in many different ways is attested by the great variety of commercial carburetors in use. The types vary from simple to extremely complicated, depending upon the degree to which the following requirements are present, and the perfection with which they must be met:

1. Variable power over a range from idling to maximum horsepower at full rated speed.
2. Speed variations ranging from nearly constant speed under governor control to speed controllable from idling speed to a maximum speed of several thousand revolutions per minute. Rapid acceleration at varying rates may also be an imposed requirement.
3. Desired thermal efficiency, which may vary from a factor of little importance for some types of engines to one of major importance in others.
4. Required operation when idling may vary from rough and irregular to perfectly smooth.
5. Requirements imposed by the mobility of the engine. If stationary, there is no trouble with surge of fuel in supply chambers. Some engines, as on automobiles, have the carburetors subject to moderate accelerations, but remaining in approximately level position. Aircraft engine carburetors are subject to great accelerations, and those which are employed on acrobatic or combat aircraft may be required to operate in all possible physical positions.
6. The fuels to be carbureted may be of varying volatility.

Most carburetors expect to secure vaporization by the use of a low-pressure spray of fuel (gasoline) into a rapidly moving stream of air from a stationary jet. The purpose of the spray is to expose a large surface of the volatile gasoline for auto-evaporation. As the heat evaporated must come mainly from the fluids themselves, the carbureted mixture undergoes a cooling of several degrees. This is helpful in improving **volumetric efficiency**, although promoting, under some conditions, formation of ice in the body of the carburetor. Other carburetors employ a moderate pressure spray and are adaptable to less volatile fuels than gasoline. Still other carburetors have been devised which, instead of a spray, expose a liquid surface in a pool over which is swept the air which it is desired to carburet. Such have been called puddle carburetors, but are deemed to be suitable to meet the requirements mentioned in the preceding paragraph only in the most limited fashion.

A comprehensive and understanding survey of carburetor types may well revert, for background, to the equation for the air-fuel mixture supplied by a simple, low-pressure jet, Venturi carburetor. The equation, as set forth in **carburetion,** shows a "coefficient of discharge" ratio $C_a C_g$ multiplying orifice area of the gasoline and air flow, and densities of the air and gasoline. When a simple Venturi carburetor is operated at increasing rates of air flow through it, the numerical magnitude of the coefficient ratio tends, by natural phenomena, to decrease, thus increasing the richness of the mixture. To offset this, and obtain approximately uniform mixture under variable air flow, A_a might be increased, or A_g might be decreased. There can be little control over the density of the air, but the effective density leaving the gasoline jet is controlled in some types of carburetors by mixing it with variable quantities of air bubbles. Now these three possibilities have all been employed in carburetors. The "variable Venturi" carburetor in effect varies the area of the air orifice. The "metering pin" carburetor is arranged to produce a variable gasoline discharge orifice. The "restricted air bled jet" produces an effect analogous to a change of density of the gasoline. Other compensating principles employed in carburetors involve ingenious uses of extra jets to supplement the main jet, while others have been designed with auxiliary air valves arranged to admit air to the over-rich carbureted mixture, and so dilute it approximately to the desired richness.

The principal parts of a carburetor might be functionally classified as:

1. A device to meter the fuel into variable air flow to obtain the desired fuel-air ratio.
2. A regulated pressure supply of fuel to the metering equipment.
3. A means for varying the flow of the air-fuel mixture to meet variable demand for power.
4. Auxiliary device to provide the additional perfection of smooth idling, temporary richness for acceleration, and mixture adjustment to compensate for variable air density occurring during large variations in altitude (aircraft carburetors).

Most carburetors use either the "butterfly" valve shown in Fig. 1, or a variable Venturi (Fig. 1A) to impose a controlled variable pressure drop between the carburetor and cylinders for the purpose of controlling power or speed. For the purpose of securing a super-rich mixture while cranking the engine at low speed preparatory to starting, the air entrance into the carburetor is almost completely closed by the operation of a valve similar to the butterfly, which is called a "choke." This imposes a high vacuum internally in the carburetor, and assists in drawing large quantities of gasoline violently from the jet. This is needed during starting, because of the quantity of condensation on the cold manifold and the poor vaporization occurring during starting.

The task of metering gasoline into the air precisely is a difficult one, at best, and for its success, the metering equipment must be supplied with the gasoline under constant head or pressure. Between the original source, such as a gasoline supply tank, and the jets, there must be imposed a means for regulating the gasoline supply to the metering devices. Many carburetors use a float-con-

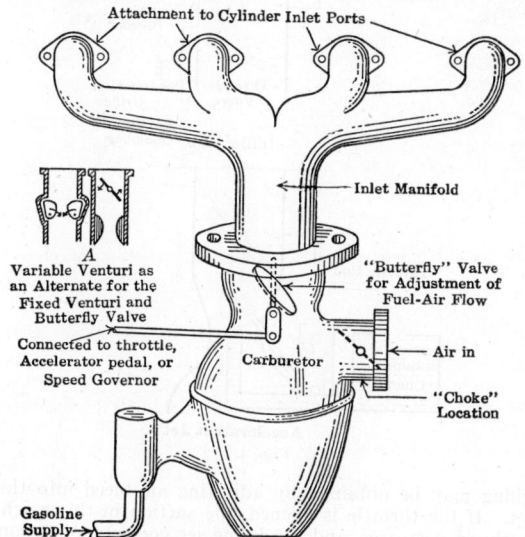

Fig. 1. Elements of the carburetor induction system.

trolled valve arranged to maintain a steady liquid level in a float chamber. This arrangement is subject, however, to the fault of surge and splash, and disruption of the desired level if the carburetor is not maintained approximately horizontal, or if it is subject to severe acceleration. Float control is quite successful for automobiles, trucks, stationary engines, and many other applications, includ-

ing even aircraft engines where the required operating conditions are not excessively severe. Where the float control is unsatisfactory, constant pressure supply is possible by using a spring-loaded diaphragm whose motion will control the valve admitting fuel from the source of supply. Both of these methods of regulating the supply chamber are illustrated in Fig. 2. The success of any car-

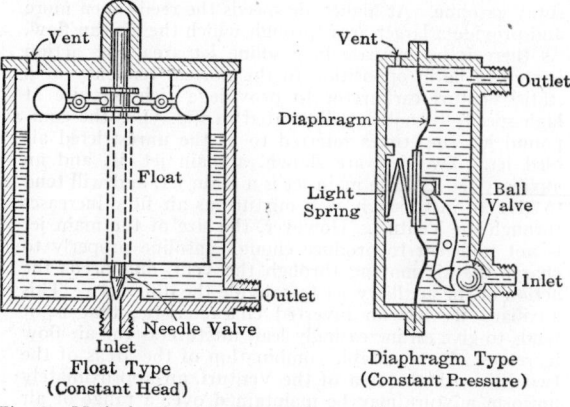

Fig. 2. Methods of regulating gasoline supply to metering devices.

buretor depends on how good its metering system is. There has been previous mention of methods of metering. Some of these will be described in more detail by reference to Fig. 3. The auxiliary air valve type displaces a light spring-loaded valve opening into the carburetor body on the discharge side of the Venturi. The simple jet as shown will tend to produce too rich a mixture at the higher air velocities through the carburetor. These high velocities are accompanied by lower pressures in the region around the auxiliary air valve,

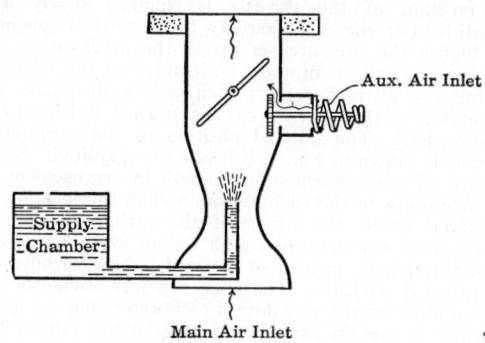

A   Auxiliary Air Valve

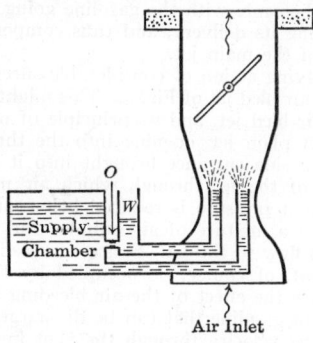

B   Compound Jet

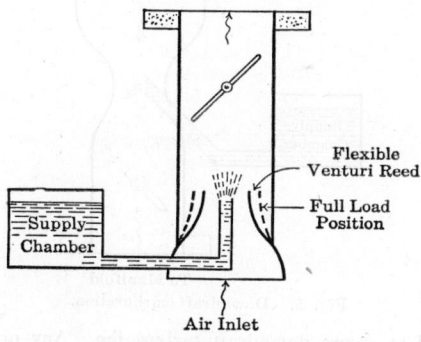

C   Variable Air Venturi

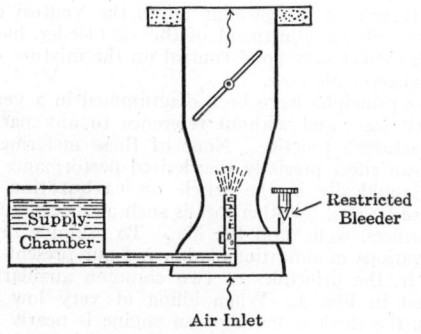

D   Air Bled Jet

Fig. 3. Methods of metering.

which is thereby induced to open to an extent roughly proportional to the decrease of pressure, admitting uncarbureted air and effecting a dilution of the mixture. The variable Venturi carburetor employs a Venturi section (of somewhat imperfect design) whose sides are built of springy strips, or reeds. At slower air speeds the reeds are nearly closed, producing a high vacuum, and improving the tendency of the jet to spray gasoline. At higher air speeds the reeds open more and provide a larger area through which the air can flow. As there is no increase in gasoline jet area, this action is seen to be in opposition to the normal tendency of a static Venturi carburetor to provide a rich mixture at high speeds. Another (illustrated in Fig. 3) is the compound jet, sometimes referred to as the unrestricted air bled jet. Two jets are shown, a main jet M, and an auxiliary jet A. The main jet is a plain jet, and will tend to give an increasingly rich mixture as air flow increases through the Venturi. However, the size of the main jet is not sufficient to produce enough gasoline properly to charge the air moving through the vent, and needs the action of the auxiliary jet to provide the difference. This auxiliary jet has an inverted characteristic, that is, it tends to give an increasingly lean mixture as the air flow increases. By a suitable combination of the areas of the two jets with the area of the Venturi, an approximately uniform mixture may be maintained over a range of air speed. The inverted characteristic is secured by feeding jet A from a well W which receives gasoline from the supply chamber through an orifice O. The higher the rate of gasoline flow through A, the more of the available driving pressure there will be consumed by the orifice O. Consequently, the discharge of gasoline from jet A will not increase proportionately faster than the air speed, as is the case with M, but will even fail to keep pace with the air flow. Thus, at low air speeds, the gasoline level in W will be nearly up to the constant level of the supply chamber, but will lower as air speed increases, and finally, at maximum speeds, will uncover the orifice, and air will then mix with the gasoline going to the jet, greatly reducing its delivery, and thus compensating for the increase of the main jet.

A jet modifying action of considerable success is illustrated by the air bled jet of Fig. 3. This might be termed a restricted air bled jet, and its principle of operation is as follows. A plain jet, opening into the throat of the Venturi, has a side entrance brought into it somewhere near the tip of the jet through which air may inflow, when the throat pressure is reduced below atmospheric, thus providing a mixture of air bubbles and gasoline in the ascending flow to the jet. Space occupied by the air bubbles cannot, of course, be occupied by the gasoline also, and hence the effect of the air bleeding is to choke the quantity of gasoline that can be discharged from the jet. As the air velocity through the vent increases, and the pressure driving the gasoline from the supply chamber up the jet thereby is increased, the induction effect of the air through the bleeder is likewise increased. Suitable proportioning of the gasoline jet to the Venturi orifice, coupled with an adjustment of the air bleeder, has been found to effect very good control on the mixture over a wide range of air flow.

These principles have been diagrammed in a very elementary way, and without reference to any particular manufacturer's practice. None of these metering principles can effect precisely the desired performance as set forth graphically in the article on **carburetion** unless they are assisted by other means such as special auxiliary jets, orifices, wells, and the like. To examine in detail these various modifications is beyond our present scope; however, the principles of two common auxiliaries are pictured in Fig. 4. When idling at very low engine speeds, the throttle valve of an engine is nearly closed, and the flow of air through the Venturi is so slow that no adequate or reliable jetting pressure may be expected. An idling jet is led to an outlet point on the cylinder

side of the closed throttle valve, where the suction vacuums are high. The suction will draw up gasoline from the supply to this jet, and a mixture for smooth

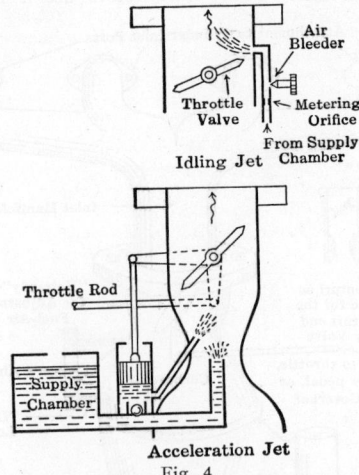

Acceleration Jet
Fig. 4.

idling may be obtained by adjusting air bleed into this jet. If the throttle is opened, the suction in this neighborhood decreases, and the idling jet goes out of action. In order to meet the requirements of temporary enrichment of mixture for rapid acceleration, most carburetors designed for applications where this requirement must be met, are provided with some sort of acceleration jet. The operation of the throttle rod may be used to cause a piston to descend in the accelerating well, forcing an extra supply of gasoline from the accelerating jet. The piston fits loosely in the well or cylinder, and has little tendency to pump gasoline out of the accelerating jet if the position of the throttle is changed slowly and smoothly, but the more rapidly the throttle is opened, the higher the pressure set up in the accelerating jet, and the more the temporary enrichment of the mixture.

Thus far, all carburetor principles described have incorporated vertical flow of the air upward and open top gasoline jets. The implied position of the carburetor, then, as is shown in Fig. 1, is below the manifold. Some definite advantages are secured, both in arrangement of the manifolds, in the carburetion, and in the location of auxiliaries about the engine if the carburetor can be placed above the manifold, with the air sweeping downward through it, instead of upward. Such carburetors are called downdraft. The open top jets obviously are not suitable for the downdraft carburetor, but no great difficulty is met in carbureting horizontally out of the jets. In Fig. 5 an idea is conveyed of the system em-

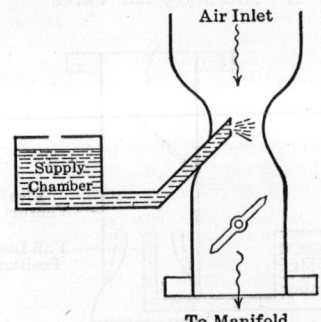

To Manifold
Fig. 5. Downdraft carburetion.

ployed to secure downdraft carburetion. Any or all of the metering modifications heretofore discussed may be applied in downdraft carburetion.

The reader is asked to refer once again to the equation for mixture, and note that if all other terms in the equation were constant, the mixture will again become too rich, because of the decrease of air density, if an engine were employed at increasingly greater altitudes above sea level. Surface vehicles powered by gasoline engines may have their carburetors adjusted from time to time for operation in regions of different altitude, usually by needle valve adjustments which impose orifice-like restrictions in the flow to the main jet. Aircraft may travel into regions of widely varying air density frequently and without the possibility of a static adjustment to compensate for variable air density. Thus, carburetors which are installed on aircraft capable of attaining altitudes where the change of air density would affect

to the Venturi), the jetting pressure could be cut back and a compensation effected for the decreasing density of the air at higher altitudes. This is called back suction control.

The problem in securing good carburetion reaches its maximum difficulty in the supply to airplane engines which are required to engage in acrobatic maneuvers and operate at high altitudes as, for example, fighter airplanes. As has been mentioned, float control of supply would have to be replaced by some other system, also vapor lock and ice formation might occur in a conventional carburetor at high altitudes. These difficulties will be overcome by injection carburetion, and, in addition, it might be possible to use fuels of lower volatility. A direct solution of this problem would be to inject metered amounts of

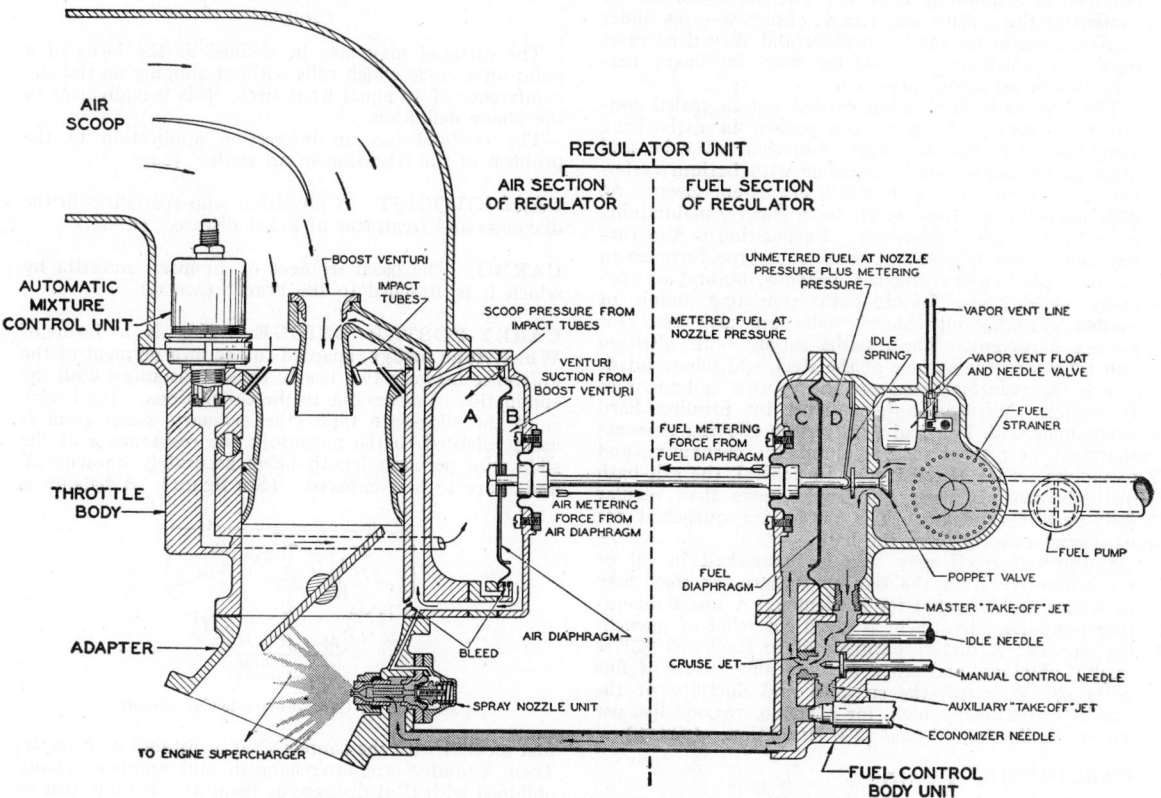

Fig. 6. Diagram of injection carburetor. (*Bendix Aviation Corp.*)

the carburetion in a detrimental way, are equipped with mixture control which may be operated manually or may be automatically adjusted by the atmospheric pressure itself. This mixture control is usually either of the *needle valve* or *back suction* type. In needle valve adjustment what amounts to a variable orifice is imposed in the flow line leading to the gasoline jet, and the setting of this orifice may be adjusted by the engine operator through the operation of a mechanical linkage leading to the carburetor. Alternately, the needle valve adjustment may be obtained automatically by spring-loaded bellows whose expansion is a reflection of the existing air pressure, and hence indirectly of the air density. It will be seen in the various figures of this article that the gasoline supply chamber is vented to the atmosphere. If it could be arranged partially to close this vent, and connect the region above the liquid level with the throat of the Venturi, the normal pressure differential delivering gasoline through the jet could be interfered with, and, in fact, completely dispersed at will. By regulating the degree of interference through adjustments of the vent opening (and possibly also the valving of the connection

fuel directly into the cylinder by injection valves. This relieves the manifold of fuel distribution problems but distributes the fuel system about the engine, whereas there are advantages of keeping the fuel supply system as compact as possible. An injection carburetor which may be a single unit, and thus obtain the simplicity and compactness of the float carburetor, has been built and is in use on military airplanes. This injection carburetor injects a metered spray of fuel into the air inlet at about 5 lbs. pressure. This is sufficient to atomize the gasoline finely and mix it with the air, and the carburetion will be completed while the mixture flows through the supercharger and inlet manifolds into the cylinders. The carburetor is floatless, has automatic mixture control, and allows the power of the engine to be controlled by a throttle rod exactly as in the case of the float carburetor. (F.T.M.)

CARBURIZING. Machine parts requiring high strength, hardness, and toughness can often be made by either of two methods, one based on the use of a medium-carbon steel (.30–.50% carbon) heat treated to the required properties, and the other based on the use of a low-carbon steel (.08–.25% carbon) carburized to give a high-carbon surface layer and then heat treated. The carburized part will have a harder more wear-resistant surface and a tougher core than the heat-treated medium-carbon steel. Transmission gears, camshafts, and piston pins are typical parts which can be made advantageously of carburizing grade steels.

The process consists of heating the fully machined part in an atmosphere rich in carbon monoxide or hydrocarbon gases at a temperature in the range 1650–1800° F. Reactions at the surface of the metal liberate atomic carbon which is readily dissolved by the steel and diffuses inward from the surface. In a typical carburized case a depth of penetration of .05″ was obtained in 4 hours at 1700° F. The maximum carbon content at the surface was 1.10%. Shallow cases under 0.02″ are useful for many purposes and very deep cases over 0.10″ thick are required for gears for heavy machinery and for armor plate.

The process is most often carried out in sealed containers in which the parts are packed in carburizing compound consisting of a mixture of charcoal, coke, and other carbonaceous solids, together with barium carbonate and other compounds which act as energizers. At high temperatures these solids burn slowly, maintaining a supply of carbon monoxide. Carburizing is also carried out in batch-type and continuous-type furnaces in an atmosphere of natural gas, propane, butane, or specially mixed gases. Liquid baths consisting mainly of molten cyanide and chloride salts are also used for surface hardening. These baths supply both nitrogen and carbon to the surface of the steel, and where nitrogen is the principal hardener the process is known as cyaniding. Nitrogen hardens steel by forming hard compounds with iron and with certain alloying elements that may be present such as aluminum, chromium, and vanadium. (See **Nitriding**.) In general, the salt-bath methods give shallower but harder cases than regular solid-pack carburizing. The pieces are quenched for hardening directly from the bath.

Carburized steels may also be **quenched** in oil or water directly from the box or furnace, or they may be cooled and reheated for hardening. A low temperature **tempering** treatment is given for relief of quenching stresses. A surface hardness of 60 Rockwell "C" is readily obtained, and when medium alloy steels of fine grain size are used, the strength and ductility of the core is exceptionally high, for example, 165,000 lbs. per sq. in. tensile strength and 18% elongation. (R.H.H.)

CARCINOMA. Cancer.

CARDAMOMS. *Elettaria Cardamomum.* Zingiberaceae. Cardamoms are the seeds of a leafy-stemmed perennial **monocotyledon** growing from 5–9′ in height. The flowers, white with purple-striped perianth parts, are borne on leafless stems which rise from the thick fleshy **rhizomes** apart from the leafy stems. The angular seeds are borne in 3-celled fruits. The dried seeds are used in India and elsewhere in tropical Asia as a highly flavored spice. (R.M.W.)

CARDIAC. Pertaining to the heart. (D.M.H.)

CARDINAL TEETH. In the shells of some **bivalve** mollusks, the interlocking prominences of the two valves just below the umbo. (A.W.L.)

CARDIOID. The cardioid is a type of mathematical curve which received its name from its heart-shaped form.

The cardioid may be defined geometrically as follows: From any point O on a circle of diameter a draw any secant OS cutting the circle at B, and extend OB to P so that $BP = a$; as this secant line rotates about O, the point P describes the cardioid.

If we take O as the pole, and OX through the center of the circle as polar axis and the circle to the left of O, the polar equation of the cardioid is $r = a(1 - \cos \theta)$.

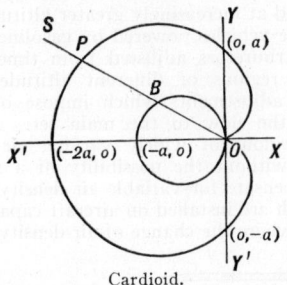

Cardioid.

The cardioid may also be defined as the locus of a point on a circle which rolls without slipping on the circumference of an equal fixed circle; this is equivalent to the above definition.

The cardioid has an interesting application to the problem of the trisection of an angle. (L.L.S.)

CARDIOLOGIST. A physician who specializes in the diagnosis and treatment of **heart** disease. (R.S.M.)

CARDO. The basal segment of an insect **maxilla** by which it is attached to the head. (A.W.L.)

CAREY-FOSTER BRIDGE. This is a form of **Wheatstone bridge,** adapted to the measurement of the difference between two nearly equal resistances, with the elimination of errors due to the connections. The bridge is of the slide-wire type (the ordinary 4-gap form is easily adapted to the purpose); the resistance ρ of the slide wire per unit length being accurately known. X and S are to be compared. (See figure.) A balance is

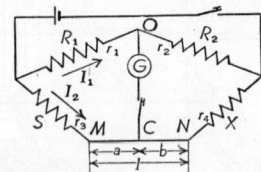

Diagram of **Carey-Foster** bridge circuit.

first secured with the contact C at a distance a_1 from M. Then X and S are interchanged, and another balance obtained with C at distance a_2 from M. It may then be shown that

$$X - S = (a_1 - a_2)\rho.$$

The Callendar and Griffiths bridge is a special type of Carey-Foster bridge used with **resistance thermometers.** (L.D.W.)

CARIBOU. Mammalia, Artiodactyla. *Rangifer.* **Deer** with partly flattened branching antlers in both sexes; North American relatives of the European reindeer. The several species inhabit the more northern parts of the continent, although two come into the United States. Caribou are found in northern Maine and in several of the northwestern states. (A.W.L.)

CARIES. Decay or death of a portion of a bone, due to a chronic inflammatory disease. Dental caries is decay of the enamel, dentin or pulp of a tooth. Spinal caries is Pott's disease (tuberculosis of the spinal column). (R.S.M.)

CARINA. 1. A sharp ridge, comparable with the keel of a boat, such as is found on the shells of certain **snails.** 2. The median dorsal plate of the shell of a

barnacle. 3. The deep ridge on the breastbone of most birds (**Aves**), more often called by the English equivalent, keel. (A.W.L.)

CARINATAE. A division of the birds (**Aves**) including all of the flying species which have the breastbone provided with a deep keel for the attachment of the powerful flight muscles. The division includes a few species which do not fly and which have only a rudimentary keel. The classification of birds as Carinatae and Ratitae is now rather generally abandoned in favor of subdivision into orders without such grouping. (A.W.L.)

CARNELIAN. The mineral carnelian is a red or reddish-brown **chalcedony**; the word is derived from the Latin word meaning flesh, in reference to the flesh color sometimes exhibited. (E.S.C.S.)

CARNIVORA. An order of **mammals** made up largely of flesh-eating species, mostly predacious in habits, although some are omnivorous and some eat carrion. They have four or five toes on each foot, armed with claws, the canine teeth are prominent, and the premolar and molar teeth are formed for cutting. Common examples of the several families into which the order is divided are the **bears**, the **wolves**, the **raccoons**, the **weasels**, **minks**, and **skunks**, and the **cats**. (See also **Pinnipedia**.) (A.W.L.)

CARNOT CYCLE. An ideal cycle of four reversible changes in the physical condition of a substance; useful in thermodynamic theory. Starting with specified values of the variable temperature, specific volume, and pressure, the substance undergoes in succession (1) an isothermal (constant temperature) expansion, (2) an adiabatic expansion (see **Adiabatic Processes**), and (3) an isothermal compression to such a point that (4) a further adiabatic compression will return the substance to its original condition. These changes are represented on the

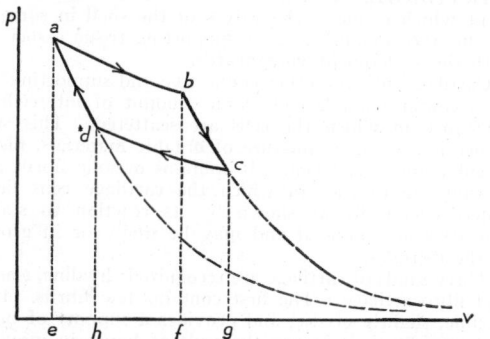

Carnot cycle on v-p diagram. ab and cd, isothermals; bc and da, adiabatics, which for some theoretical purposes, are produced to infinity.

volume-pressure diagram respectively by *ab, bc, cd,* and *da* in the accompanying figure. Or the cycle may be reversed: *a d c b a.*

In the former (clockwise) case, heat is taken in from a hot source and work is done by the hot substance during the high-temperature expansion *ab;* also additional work is done at the expense of the thermal energy of the substance during the further expansion *bc.* Then a less amount of work is done on the cooled substance, and a less amount of heat discharged to the cool surroundings, during the low-temperature compression *cd;* and finally, by the further application of work during the compression *da,* the substance is raised to its original high temperature. The net result of all this is that a quantity of heat has been taken from a hot source and a portion of it imparted to something colder (a "sink"), while the balance is transformed into mechanical work represented by the area *abcd.* If the cycle takes place

in the counter-clockwise direction, heat is transferred from the colder to the warmer surroundings at the expense of the net amount of energy which must be supplied during the process (also represented by area *abcd*). The operation of the cycle thus illustrates the second law of **thermodynamics.** (L.D.W.)

CARNOTITE. The mineral carnotite is a **vanadate** of **potassium** and **uranium** with small amounts of **radium.** Its formula may be written $K_2O \cdot 2UO_3 \cdot V_2O_5 \cdot 2H_2O$. The amount of water, however, seems to be variable. It occurs usually as a lemon-yellow earthy powder disseminated through **sandstones,** rarely as **orthorhombic** shales. It was mined in Colorado and Utah, as a source of radium. Other localities are in Arizona, Pennsylvania, and the Belgian Congo. (E.S.C.S.)

CAROTENE. Pigments in Plants.

CARP. Pisces, Teleostei. Large **fishes** belonging to the same family as the minnows, dace, and chubs. The common carp is an introduced species, indigenous to eastern Asia but now thoroughly at home in the rivers and lakes of North America and Europe. It is coarse and bony but it is widely used as food. The **goldfish** or golden carp is a related species native to China and Japan. Many strange varieties have been developed in captivity and the species is thriving in some lakes and streams in the eastern United States. Several other species of carp occur in Europe and Asia. (A.W.L.)

CARP SUCKER. Pisces, Teleostei. *Carpiodes.* Fishes (**Pisces**) of several species related to the buffalo and suckers. They are found in streams and lakes of the United States and Mexico. Most species are known by this name but one member of the genus is called the **quillback.** (A.W.L.)

CARPAL BONES. The small bones of the wrist in man and the similar group in the forelegs of animals. (A.W.L.)

CARPEL. Flower.

CARPENTER MOTH. Insecta, Lepidoptera. **Moths** whose larvae bore in the trunks of trees, entering the solid wood. A few species, including the locust borer, are of large size, and because of their narrow wings and long bodies may be mistaken for sphinx moths. These insects make up the small family Cossidae. (A.W.L.)

CARPENTER-BEE. Insecta, Hymenoptera. A bee which excavates its nest in wood. The small carpenter-bee, *Ceratina dupla,* of North America merely digs out the pith or soft wood of a plant, such as sumac, while the large carpenter-bees, *Xylocopa,* of which there are several species, bore into solid wood, even attacking unpainted wood in construction. The larger bees are not unlike **bumblebees** in appearance. (A.W.L.)

CARPINCHO. Capybara.

CARPOPODITE. The third segment from the tip of the **crustacean** appendage. (A.W.L.)

CARRIAGE. Lathe.

CARRIER. A person who carries the specific organisms of any disease, and who can, though showing no signs of the disease himself, distribute disease through contact with others. Virulent hemolytic streptococci, diphtheria bacilli, and meningococci, may be carried in the throats of healthy individuals; typhoid and the dysentery bacilli may be present in the intestinal tract of healthy carriers. (D.M.H.)

CARRIER CURRENT. Carrier current is used in connection with both power and communication circuits but, basically, the principle is the same for both systems. The term refers to the use of a relatively high-frequency a.c. superimposed on the ordinary circuit frequencies in order to increase the usefulness of a given transmission line. Thus in the case of power systems, carrier currents of several kilocycles frequency are coupled to the 60-cycle transmission lines. These carrier currents may be modulated to provide telephone communication between points on the power system or they may be used to actuate relays on the system. This latter use is known as carrier relaying. Carrier currents have greatly extended the usefulness of existing line facilities of the telephone and telegraph companies. Several carrier frequencies may be coupled to the lines already having regular voice or telegraph signals on them. Each of these carrier frequencies may be modulated with a separate voice or telegraph channel and thus a given line may carry the regular signals plus several new carrier channels, each of which is equivalent to another circuit at regular frequencies. At the receiving end the various channels are separated by filters and the signals demodulated and then fed to conventional phone or telegraph circuits. The number of carrier channels which may be applied to a given line depends upon the characteristics of the line, varying from one or two for some lines to several hundred for the coaxial cable. (L.R.Q.)

CARRIER FREQUENCY. Carrier Current; Radio Communication.

CARRIER RELAYING. Carrier Currents.

CARRIER WAVE. Radio Communication.

CARRION BEETLE. Insecta, Coleoptera. Moderate to large beetles which are found about decaying flesh and to some extent about other decaying matter. Applied to members of the family Silphidae, although many other beetles breed in decaying matter and are found in it, both as adults and as larvae. (A.W.L.)

CARRION BIRD. Aves, Falconiformes. The English equivalent of the Dutch name aasvogel, applied to some of the larger South African vultures. (A.W.L.)

CARROT FAMILY. Umbelliferae. The plants of this family, mostly found in the north temperate zone, are nearly all annual or biennial herbs, with a few trees or shrubs. They are characterized by having stems hollow except at the nodes, alternate leaves which are either pinnately or ternately compound, and large numbers of small flowers produced in simple or compound umbels. The distinctive umbel is an inflorescence in which the pedicels of the individual flowers arise from a very short axis and are of nearly equal length, producing a flat-topped or rounded cluster.

The single flowers are small and regular, the calyx being either absent or adnate to the ovary, and five-toothed, the corolla composed of five separate petals, the stamens five and the single pistil with inferior ovary, with two one-seeded carpels. The flowers are insect-pollinated. The fruit is a form known as a schizocarp, a dry fruit having two carpels which separate at maturity into two mericarps, each of which usually has five longitudinal ridges. Between the ridges are found longitudinal oil canals containing the volatile oils which give the odors characteristic of many members of this family.

Many members of this family are important food plants to man and domestic animals. Among these are plants grown for their roots, the carrot, *Daucus Carota*, and parsnip, *Pastinaca sativa*, both natives of Europe; also celery, *Apium graveolens*, grown for its leaf stalks, and parsley, *Apium Petroselinum*, grown for the much dissected leaves which are used as a garnish and for flavoring. A large number of plants in this family are

frequently cultivated, especially in Europe, for flavorings and medicinal use, though possibly their therapeutic value is rather overrated. Among these are anise (*Pimpinella anisum*), caraway (*Carum Carvi*), coriander (*Coriandrum sativum*), fennel (*Foeniculum officinale*), and dill (*Anethum graveolens*). Many of these are also used in flavoring confections. *Ferula fetida* produces a drug, asafoetida.

In contrast to these, which are useful to man, are many members of the family which contain violent poisons. *Conium maculatum*, the Poison Hemlock, is reputed to have been the source of the poison which Socrates drank. *Cicuta maculata* and related species are sometimes eaten by livestock with fatal results. (R.M.W.)

CARTESIAN COORDINATES. Cartesian coordinates of a point are either its rectangular coordinates or its oblique coordinates. They are called Cartesian after the French mathematician and philosopher Descartes, who first introduced the idea of coordinates as a basis for the analytic study of geometry. (L.L.S.)

CARTESIAN OVAL. This curve is defined as the locus of a point P that moves so that its distances from the origin $O(o,o)$ and the point $A(a, o)$, in rectangular coordinates, satisfy the relation $AP = \pm b \cdot OP \pm c$, where b and c are given positive constants. (L.L.S.)

CARTHAMUS TINCTORIUS. Safflower. Compositae. This plant, a native of the East Indies, is now widely cultivated in tropical Asia and Egypt, and to a limited extent in southern Europe. It is a low annual plant with yellowish-red flowers which have tubular corollas.

Of commercial importance are the flowers, which are washed in water to remove the yellow pigment present, dried, and ground to a powder. This powder is mixed with starch and talc to make rouge. (R.M.W.)

CARTILAGE. 1. The internal structure of the ligament which connects the valves of the shell in some of the bivalve mollusks. 2. A supporting tissue associated with the skeleton of vertebrates.

Cartilage, like the other connective and supporting tissues, contains a relatively large amount of intercellular substance in which the cells are scattered. This substance is a complex mixture of organic materials, bluish in color and translucent. It contains organic fibrils and around the cavities in which the cartilage cells lie it differs chemically as shown by its reaction to stains. The cells are rounded and may lie singly or in groups in the capsules.

Three kinds of cartilage are recognized: hyaline, elastic, and fibrocartilage. The first contains few fibrils. It is flexible, slightly elastic, and provides a support of moderate rigidity. It covers the ends of bones in movable joints as the articular cartilages, forms the rings of the trachea, and occurs in other parts of the body where such qualities are required. Elastic cartilage is similar to hyaline but has many elastic fibers in the intercellular matrix. It occurs in the pinna of the ear, where its qualities provide support and elasticity, the latter very necessary in a delicately formed projecting structure of this kind which might otherwise be easily broken. Fibrocartilage contains many inelastic white fibers which give it extreme toughness. It is associated with some joints and forms the intervertebral disks of the backbone. These disks provide very firm connections between the separate vertebrae and at the same time cushion the series.

The term cartilage is also applied to separate skeletal units formed of this material. Each cartilage is surrounded by a tough connective tissue sheath called the perichondrium.

Cartilage is a primitive skeletal material of the vertebrates. It precedes bone in embryonic development and persists in the adult skeleton in the sharks and related fishes. It is not transformed into bone but is replaced by bone in the formation of some of the parts of the skeleton. It may become rigid through the deposition of calcareous material in its matrix, particularly in old age. This calcified cartilage, while rigid like bone, does not have the minute structure of that tissue. (A.W.L.)

CARYOPSIS. Fruit.

CASCARA SAGRADA. The bark of *Rhamnus Purchiana*, a shrub growing in western North America. It is a **drug** used as a laxative and cathartic. (R.S.M.)

CASE HARDENING. Carburizing, cyaniding, and **nitriding** are the principal methods of case hardening steels. (R.H.H.)

CASEIN. Casein is a protein present in milk that is used to make **plastics.** (See **Aminoacids, Polypeptides and Proteins, Adhesives, Pigments, Paints, Varnishes.**) (R.K.S.)

CASHEW. *Anacardium occidentale.* The cashew **nut** is the fruit of a Brazilian tree of moderate size. The kidney-shaped nut grows at the end of a curiously enlarged fleshy **peduncle** which is juicy and bright yellow or red. This fleshy portion is much eaten in tropical America. The nut itself contains a biting caustic oil which is driven off by roasting. The single kernel of this fruit is the familiar cashew nut so widely known as a confection. The oil is used to a limited extent, **to** protect book bindings and wood against the attack of **termites.** (R.M.W.)

CASING. The casing is that part of a machine which encloses or encases the working portions. Not all machines are said to have a casing; the term is usually applied in instances where a nearly complete enclosure is made, such, for example, as the fan casing and the turbine casing.

The pipe which is used to line a well is called casing pipe, and is a prominent feature of oil and gas wells. It is usually characterized by light weight, and joints with fine pitched threads. In sinking a deep well the casing constituting the first section may be 6″ in diameter. When the 6-in. casing has been sunk as deep as practicable, the size will be stepped down to, say, 5″, which is lowered into the well inside the 6-in. casing. Thus progressively the size of the casing is decreased as the well deepens. (F.T.M.)

CASINGHEAD GAS. A large amount of technically pure **gasoline** termed casinghead or "aviation gas" is now extracted by condensation from vapors present in certain **natural gases** as they flow from the well. This product is also used to blend with gasoline which has been refined from **petroleum.** (R.M.F.)

CASSAVA. Spurge Family.

CASSEGRAINIAN. Telescope.

CASSIA. *Cinnamomum cassia.* Lauraceae. Cassia, or Chinese Cinnamon, is one of the earliest used of all spices, and is mentioned in the Old Testament as a spice. It is a bushy plant native to Cochin China, from which region it had spread in cultivation to much of southeastern Asia in early times. Its presence in the island of Ceylon led the Portuguese to seize that island in 1636. That and other spices were the main cause of the bitter and bloody struggles among various European nations for possession of tropical Asian lands.

In cultivation the much-branched plant is ready for harvest about six years after planting. In harvesting, the bark on the young stems is split longitudinally and peeled back in cylinders about 16″ long. The adherent **periderm** is planed off. With continued drying the bark

rolls up rather tightly and turns dark brown in color. This is the product used mainly as a spice in pastry making. Immature fruit may be dried and also used for spice.

By distillation, oil of cassia, mainly cinnamic **aldehyde,** is obtained from the bark and leaves. This oil is used mainly as a flavoring for candies. Some of it is used in making scented soaps, and in medicines, where its principal value is in concealing unpleasant tastes.

The name Cassia is also applied to two unofficial drugs, Cassia pulp, which is the fruit pulp of *Cassia Fistula;* and American Senna, the dried leaves of *C. marilandica,* both from the family Leguminosae. (R.M.W., B.S.M.)

CASSIOPEIA. (Map, page 380.) This is one of the most widely known and striking **constellations** of the northern latitudes. It is easily recognized by the 5 bright stars forming an irregular W, some observers seeing not only a W but also a chair. Since this object is circumpolar (i.e., remains above the **horizon** at all hours every night) for most northern countries, and is easily recognized, it is frequently used as a rough indicator of sidereal **time.** The leading bright star of the W (the star Beta Cassiopeiae) lies almost in zero hours **right ascension.** Hence a line drawn through **Polaris** and Beta Cassiopeiae must pass close to the vernal equinox. The **hour angle** of this line must be equal to sidereal time. Hence when Beta Cassiopeiae is on the meridian directly above the pole the sidereal time is zero, when on the meridian directly below the pole the sidereal time is 12 hours, etc.

One of the brightest **novae** on record appeared in this constellation in 1572 and was observed and recorded by Tycho Brahe. (W.K.G.)

CASSIQUE. Aves, Passeriformes. South American birds (**Aves**) of several species, related to the Old World starlings. (A.W.L.)

CASSITERITE—TIN STONE. The mineral cassiterite, chemically **tin** dioxide, SnO_2, is almost the sole ore of tin. It is a noticeably heavy mineral crystallizing in the **tetragonal** system, as low pyramids, prisms, often very slender, and as twinned forms. It is a brittle mineral, hardness, 6.0–7.0; specific gravity, 6.8–7.1; luster, adamantine; color, generally brown to black, but may be red, gray to white, or yellow; streak whitish, grayish, or brownish; may be almost transparent to opaque. A fibrous variety somewhat resembling wood is called wood tin. Cassiterite occurs in widely scattered areas, but deposits of a size to be commercially important are few. It is associated with **granites** and **rhyolites** and is believed to be, in part at least, the result of **pneumatolytic** action. The weathering of the cassiterite bearing rocks often permits the distribution of it along stream beds where it is known as stream tin. The Malay Peninsula deposits are stream tin. Other productive regions are Australia, Africa, and South America. In the United States cassiterite has been mined in the Black Hills of South Dakota and elsewhere but the deposits are not of great consequence. The tin veins, now about exhausted, in Cornwall, England, were known to the Phoenicians before the Roman invasions. Saxony and Bohemia were formerly important producing regions also.

The word cassiterite is of Greek origin. (E.S.C.S.)

CASSOWARY. Aves, Casuariiformes. *Casuarius.* A large flightless bird (**Aves**) of Australia, New Guinea, and adjacent islands. The several species are all forest birds, unlike the related **emus.** The naked skin of the head and neck is brightly colored with blue, green and red, and the head is surmounted by a bony crest, in some species of large size. The wings are rudimentary. (A.W.L.)

CAST IRON. Cast iron is the product of remelting and casting **pig iron**. The melting furnace is usually a **cupola** but other types of melting equipment including **electric-arc furnaces** may be used for special grades. In the melting process the composition can be changed from that of the pig iron by additions to the charge of steel and sometimes of alloying elements. However, the iron is not refined as in steel-making; that is, no considerable amount of phosphorus, sulfur, silicon, or manganese is removed from the charge. The total of these elements plus the carbon content may be 7% in a typical cast iron compared to 1% in a typical steel. The high carbon content of pig iron is largely retained in cast iron, about 3.25% being normal for general purpose cast irons. By the addition of considerable steel scrap to the charge, and by other means, the carbon can be reduced to between 2 and 3%, with considerable improvement of the strength of the castings. A large part of the carbon is present in the form of flakes of **graphite** (see photomicrograph), which accounts for the relatively low tensile strength and the characteristic gray appearance of the fracture of high-carbon gray cast irons.

The silicon content, or the silicon-carbon balance, plays an important part in determining the structure and properties of cast iron. Silicon promotes graphitization of the carbon, hence reduced silicon content favors higher strength. When both carbon and silicon are low, it is possible to cast a graphite-free iron known as white cast iron. This product is not, however, high in strength because much of the carbon is present as **cementite** which is very brittle; thus the hardness and wear-resistance is high but the tensile strength is low. Most white cast iron is converted to malleable cast iron by annealing at high temperatures; however, certain hard alloy grades (see Ni-Hard in table) are used in the as-cast state for abrasion-resisting applications.

In the annealing of white iron castings for malleablizing, a considerable part of the carbon is removed by oxidation at the surface and the remainder of the cementite is converted to graphite and ferrite. The nodular graphite, or temper carbon, of malleable cast iron (see photomicrograph) can readily be distinguished from the flake graphite of gray cast iron. Whereas gray iron is relatively brittle, good malleable iron has a ductility of about 20% in 2" and a tensile strength of 50,000 lbs.

TYPICAL CAST IRONS

No.	Composition, %									
	C	Si	Mn	P	S	Cr	Ni	Mo	Cu
1	3.25	2.25	0.65	0.15	0.10	....	....	....	....
2	3.20	2.15	0.70	0.15	0.10	0.50	0.50	0.50	....
3	3.35	2.00	0.70	0.15	0.10	0.35	1.75	....	....
4	3.50	2.90	0.65	0.50	0.06	....	....	....	....
5	3.25	1.75	0.50	0.50	0.10	....	....	....	....
6	3.60	1.75	0.50	0.80	0.08	....	....	....	....
7	3.35	0.65	0.60	0.35	0.12	....	....	....	....
8	3.25	2.00	0.60	....	....	....	....	0.50	1.00
9	3.00	1.00	0.75	....	....	0.35	1.25	....	....
10	2.50	2.50	1.00	....	....	....	1.00	1.00	....
11	1.50	0.95	0.70	....	....	0.45	....	....	1.75
12	2.50	0.85	0.30	0.15	0.05	....	....	....	....
13	3.25	1.00	0.40	0.15	0.10	1.5	4.5	....	....
14	2.90	1.60	1.25	0.20	0.08	2.5	13.5	....	6.0
15	0.75	13.0	0.5	....	....	....	....	....	....

No.	Application
1	Automobile cylinder blocks, plain iron
2	Automobile cylinder blocks, alloy iron
3	Automobile cylinder blocks, alloy iron
4	Piston rings, individually cast
5	Heavy section (1.5" thick) castings
6	Light section sand cast water pipe
7	Car wheels
8	Brake drums
9	Machine tool beds
10	Crankshafts
11	Crankshafts (Ford analysis)
12	White iron for malleablizing

Special Alloy Compositions

No.	
13	Ni-Hard—a very hard white iron for crushers, grinders, ball mills, etc.
14	Ni-Resist—a corrosion and heat resistant cast iron for processing equipment, furnace castings, manifolds, etc.
15	Duriron—acid resisting castings

(R.H.H.)

per sq. in. Specially treated malleable iron known as pearlitic malleable has even better mechanical properties. Malleable irons are used for small castings requiring greater toughness and ductility than ordinary gray cast irons; for example, automobile brake levers, pump parts,

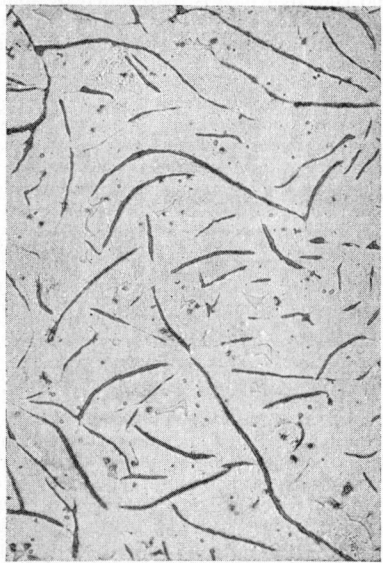

Fig. 1. Graphite flakes in cast iron at 100 times magnification. Not etched. (See **Metallography.**)

and numerous parts for agricultural implements. Pipe fittings are another important application.

A white cast iron structure can be produced locally in a gray iron casting by the use of chills in the mold if the composition is properly balanced; thus chilled cast iron freight-car wheels and cast iron brake shoes have hard white iron wearing surfaces and gray iron

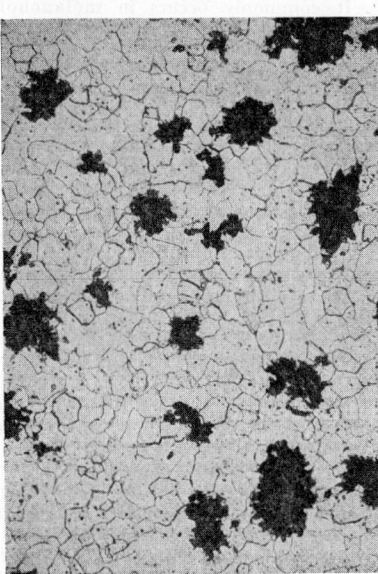

Fig. 2. Temper carbon in matrix of ferrite grains in malleable cast iron. 100 times magnification. Etched in a solution of nitric acid in alcohol.

cores. If an entire wheel were white iron it could not be bored or machined by ordinary methods and would fail in service because of excessive brittleness. On the other hand, if the tread were gray iron it would be too soft for good wear-resistance in this type of service.

Cast iron is ordinarily thought of as a cheap low-strength material. In recent years the foundry industry has found methods for increasing its strength without sacrificing its inherent good machinability, ease of casting, and excellent bearing qualities. These methods include (1) lowering the carbon and silicon content by the use of steel scrap in the charge, (2) addition of alloying elements to a properly balanced base mixture, (3) superheating the molten metal and inoculating with a graphite-forming agent in the ladle before casting. High-strength Meehanite cast irons are made by a patented process in which a graphitizing ladle addition is made to a melt which would otherwise be white or mottled in structure. Whereas the tensile strength of an ordinary gray iron casting may be only 20,000 lbs. per sq. in., various higher strength grades are made with up to 60,000 lbs. per sq. in. tensile strength.

High-alloy irons such as Ni-Resist and Duriron have special corrosion-resisting qualities.

CAST STEEL. Steel.

CASTING. Casting is the process of producing metal shapes by pouring molten metal into molds of the required form where it is allowed to solidify. The metal part formed as a result of this operation is called a casting. The art of casting is one of the oldest methods for making metal parts and is still extensively used in spite of more modern developments such as **forging.**

The production of a casting involves the use of a pattern, usually of wood or metal, which is similar in shape to the desired finished piece and slightly larger in all dimensions to allow for shrinkage of the metal upon solidification. The pattern is bedded down in a special damp sand by an operation called molding. When the pattern is removed it leaves an impression of the shape of the desired casting. This impression is completely surrounded by sand and provided with openings called gates through which the molten metal enters. After pouring and cooling the mold is broken open and the casting removed. All adhering sand particles together with any extraneous projections such as those left by the gate system are removed after which the casting is machined to the required finish.

The term is also applied to the casting of **pig iron** in **blast furnace** practice and the casting of **ingots** in steel-mill practice.

Centrifugal casting is applicable to the production of pipe and tubing, wheels, gear blanks, and other castings having rotational symmetry. While the mold is rotated on a horizontal axis for pipe and tubing, and on a vertical axis for wheels and gear blanks, a measured amount of molten metal is added. The mold may be sand or water-cooled metal for more rapid solidification. Centrifugal castings have good structure and density.

Metal molds are also used for making die castings and permanent mold castings. In the latter process a permanent metal mold is filled by gravity in the usual manner, while in die casting considerable pressure is exerted on the molten metal, insuring rapid and complete filling of the mold. Die-casting machines are highly mechanized for rapid and nearly automatic operation. The product is characterized by high dimensional accuracy and clear reproduction of mold details including screw threads, holes, and intricate sections, all of which greatly reduces the machining required. The process is limited in its application by the high cost of making alloy steel dies or molds. The lower melting **zinc alloys** and **aluminum alloys** are most successfully die cast; however, certain **brasses and bronzes** can also be die cast. Tin- and lead-base alloys are easily die cast but have limited application.

The zinc-base die-casting alloys are the most widely used. A typical composition is 1.0% copper, 3.9% aluminum, 0.06% magnesium, balance zinc. This alloy has a strength of about 45,000 lbs. per sq. in. with 3%

elongation in 2". Typical applications are carburetors, fuel pumps, tools, typewriter frames, instrument cases, and hardware which is often finished by chromium plating.

The investment or "lost wax" process has lately been revived as a method of making precision castings of metals such as steel, etc., having too high a melting point for die casting. A wax pattern is made in a die-casting machine, sprayed with a highly refractory slurry, dried and imbedded in sand. The mold passes through a furnace where the wax is melted out and recovered, and the mold baked. The casting is then poured into the cavity left by the melting out of the wax, resulting in castings which rival die castings for dimensional accuracy. (C.W.C., R.H.H.)

CASTOR. Castor (α Geminorum) is the fainter star of the twins. Since these two stars are always considered together in the ancient literatures, the history and astrological significance will be found discussed under **Pollux,** the brighter of the two.

Astronomically, Castor is a very remarkable star. It was discovered in 1719 to be a visual **binary,** with the **magnitudes** of the components 2.8 and 2.0. The separation is about 6" and the star is certainly a true binary, but the period has not yet been accurately determined. The period is probably of the order of magnitude of 350 years. Each of the two components of the binary system is also a **spectroscopic binary** so Castor is a quadruple system. Castor has a faint companion separated from it by about 72" but having the same **parallax** and **proper motion.** This companion is also a spectroscopic binary with a period of slightly less than 1 day. (W.K.G.)

CASTOR OIL. *Ricinus communis.* Euphorbiaceae. Castor oil is obtained from a short-lived perennial tree which occurs wild in tropical Africa and perhaps in India. Cultivation of the tree is widespread not only in the tropics but also in temperate regions, where it is often grown as an ornamental plant. In the tropics it becomes a tree 36' tall, with large coarse leaves often of reddish color, and green flowers. An annual herbaceous variety is grown widely and produces a superior oil. The seeds, borne three in each of the smooth or prickly capsules, have a hard mottled shell. These seeds are ejected violently from the mature fruit.

The principal use of the plant is for the oil which is contained in the seeds. This oil is pressed out without heating the seeds. It is used as a lubricant for airplane and marine engines, either in the pure state or mixed with mineral oils. Much castor oil is used in preparing various dyes used in coloring textiles. Certain soaps, also, are prepared from castor oil. (See **Esters.**) Not a little has been used as a lubricant for the human system where, however, its action has proved mildly irritating. The seeds contain a violent poison which must be removed before the oil can be used medicinally. India produces most of the castor bean crop of today. (R.M.W., B.S.M.)

CASUARIIFORMES. An order of flightless birds (**Aves**) containing the **emus** and **cassowaries** of Australia and New Guinea. (A.W.L.)

CAT. Mammalia, Carnivora. Although commonly used in the unqualified form only to designate the domestic cat, *Felis domestica,* this term properly indicates any member of the large family Felidae, which includes the **lion,** the **tiger, leopards, lynxes,** and many species of smaller size.

Among the distinctive characters of the family are the simple dentition and the sharp, curved, retractible claws. In most species the claws can be withdrawn completely into sheaths.

The larger cats of the Old World include the lion, tiger, leopards, and the moderately large serval, jungle-cat, and **caracal,** all African and Indian species. In North America the only large form is the **puma** or mountain lion, also known as the panther, painter, catamount, and cougar, which is represented by four species ranging from Florida to the west and northward in the Rocky Mountains and in the coastal area. South America has the **ocelot** or tiger cat, the **jaguar,** and the **jaguarondi** cat or eyra, all of which enter Texas by way of Mexico. The lynxes include a European species and in North America the **wildcat** or bobcat, the Canadian lynx, and a few less familiar species. (A.W.L.)

CATABOLISM. Metabolism.

CATACLASTIC. As proposed by Teall in 1887 this term has the same meaning as **crush breccias.** This term is also applied to the deformation and granulation of minerals such as may take place during dynamic metamorphism. (R.M.F.)

CATACLYSM. Before the doctrine of uniformitarianism (the uniformity of all geologic processes throughout all time) had been promulgated by Sir Charles Lyell, it was believed that mountains, mountain ranges, deep valleys and rugged highlands or other features of the earth's surface had been produced by great and sudden convulsions called cataclysms. The word cataclysm is derived from the Greek words denoting downward and to wash away, i.e., floods; the implication being that great floods or violent earthquakes had been instrumental in producing the relief features of the surface of the globe, now known to be the result of the slow but ever-continuous action of the processes of weathering and **erosion.** (E.S.C.S.)

CATALEPSY. A nervous seizure, marked by absence of voluntary motion and of sensibility. The trance lasts from a few minutes to several days. During the attack, the body is cold and pale, and the pulse and breathing are slow. It commonly occurs in **melancholia** and schizophrenia. (R.S.M., D.M.H.)

CATALYZER. Catalysis.

CATALYSIS. This is a phenomenon observed in chemical reactions whereby the reaction between two or more substances is influenced by the presence of a third substance (the catalyst) which remains unchanged in the process. For instance **hydrogen** and **oxygen** gas do not react with each other appreciably at room temperature. The introduction of **platinum** powder produces an instantaneous union of these gases to give water. The platinum is chemically unchanged in this process and if proper precautions are taken can convert an infinite amount of hydrogen and oxygen into water. This is an example of positive catalysis. There are cases known of negative catalysis in which the rate of a reaction is repressed by the presence of a foreign body. An example is the inhibition of the **hydrogen peroxide** decomposition by traces of **acetanilid.** In the case of gaseous reactions in the presence of solid catalysts, the reaction takes place on the surface of the solid and the adsorption of the gaseous reactants and products plays an important role. It has been found that certain substances (called promoters), having themselves but little catalytic ability, when added to a catalyst enhance the latter's catalytic ability. An example is the use of **alumina** to promote the catalytic ability of **iron** for the reaction of **hydrogen** and **nitrogen** to form **ammonia.** Since the catalytic reactions take place on the surface of the catalyst the latter must be prepared in as highly subdivided a state as possible. This is accomplished by using the catalyst in a colloidal state or dispersing it on inert carriers. Charcoal, alumina, silica gel, kieselguhr, etc., are com-

mon catalytic carriers. Many catalysts lose their catalytic activity completely when exposed to the action of
minute amounts of certain substances called poisons.
Thus arsenic compounds poison platinum catalysts. The
following are a few of the more important catalytic
reactions.

1. **Hydrogen** and **oxygen** react to form water in the
   presence of **platinum** and other finely divided
   metals.
2. Hydrogen and **nitrogen** react to form **ammonia** in
   the presence of iron, and alumina (promoter).
3. Ammonia and oxygen (air) react to form **nitric
   oxide** in the presence of smooth platinum or
   platinum-rhodium in the form of fine wire gauze
   to furnish large surface to weight ratio.
4. **Sulfur dioxide** and oxygen (air) react to form
   sulfur trioxide in the presence of **platinum** or of
   vanadium oxide.
5. **Carbon** monoxide and hydrogen react to form
   methane in the presence of nickel.
6. Carbon monoxide and hydrogen react to form
   methanol in the presence of **zinc** and **chromium**
   oxides.
7. **Carbon** monoxide and hydrogen react to form a
   synthetic gasoline (Fischer-Tropsch Process) in the
   presence of cobalt-thoria or nickel-thoria supported
   on kieselguhr.
8. Liquid fats and hydrogen react to form solid fats in
   the presence of nickel.
9. Coal and hydrogen react to form **petroleum**
   (Bergius Process), in the presence of various
   catalysts.
10. Carbon monoxide and oxygen react to form carbon
    dioxide in the presence of copper and manganese
    oxides. (See also **Chemical Changes** and **Hydro-
    genation.**)   (R.K.S.)

CATALYST. Catalysis.

CATAMOUNT. Mammalia, Carnivora. The **puma**
or mountain lion. Sometimes applied to the **wildcat.**
(A.W.L.)

CATARACT. Any opacity which develops in the
crystalline lens of the **eye** or in the capsule of the lens.
Cataracts may be primary or secondary to a systemic
disease. They may be partial or complete, stationary or
progressive, hard or soft. The most common predisposing cause is age, hence senile cataracts are frequently
seen. Other causes, either direct or predisposing, are
heredity, faulty intra-uterine development, general diseases of the eye.
The treatment of a cataract uncomplicated by other eye
disorders is very successful. Useful and even perfect
vision follows in the great majority of cases. The
procedure is surgical removal of the lens when vision
is impaired. After removal of a cataract, the patient
must wear strong convex glasses to compensate for loss
of the lens of the eye. (D.M.H.)

CATARRH. An inflammation, usually chronic, of any
mucous membrane. The term generally applies to the
mouth, throat, or nose; it is, however, vague and rarely
used at the present time. (D.M.H.)

CATAWBERITE. The term applied by Lieber to a
metamorphic rock chiefly composed of **magnetite** and
talc. (R.M.F.)

CATBIRD. Aves, Passeriformes. 1. A common
North American bird, *Damatella carolinensis,* related to
the mockingbird and the thrashers. Although quietly
colored in slate gray it is a welcome resident because of
its fine singing. 2. An Australian **bower-bird.** (A.W.L.)

CATCHER. **Klystron.**

CATENARY. The catenary is the locus of the equation

$$y = \frac{a}{2}\left(e^{x/a} + e^{-x/a}\right) = a \cosh\frac{x}{a}.$$

It is the shape of the curve assumed by a uniform, heavy
flexible cord freely suspended
from its extremities. (L.L.S.)

CATERPILLAR. The larval
form of the **butterflies** and
moths. (A.W.L.)

CATFISH. Pisces, Teleostei.
Fishes (**Pisces**) of many species without scales, although the
skin has bony plates in some species. Barbels occur on
the head. They are represented in the fresh waters of
all continents except Australia and gain great diversity
in the Americas.

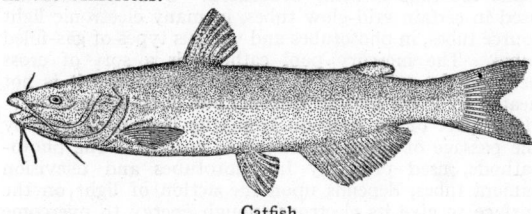

Catfish.

The catfishes are important food fishes in some parts
of the world. In North America the channel cats are
especially desirable. They reach a large size in some of
the larger rivers and lakes.
Some of the species included in the group whose
names do not indicate the association are the **bullheads,**
horned pout, goujon or mud cat, mad toms, and the
wels of Europe. (A.W.L.)

CATGUT. Sheep's intestine which has been treated,
made **aseptic,** and prepared into strands of various
strengths for use in surgery as an absorbable **suture**
or **ligature** material. Chromic catgut is treated with
chromium trioxide and takes a longer time to be absorbed by the body than plain untreated catgut. (D.M.H.)

CATHETOMETER. Comparator.

CATHODE. In the nomenclature of the **electrolytic
cell,** the negative electrode is designated as the cathode.
In a gas-filled or vacuum tube of two elements, the filament is called the cathode.
The cathode might well be called the heart of the
electron **tube** since it is the source of the primary
electrons. It is the electrons which are controlled
in various ways to give the multitude of types and
applications of such tubes. Cathodes may logically be
divided into two groups according to their condition
of operation: thermionic or hot cathodes, and cold
cathodes. The thermionic type is used in all vacuum
tubes and in many gas-filled tubes. It is heated by
passing an electrical current directly through it, in
which case it is called a directly heated or filament type
cathode, or by passing the current through a special
heating element enclosed in the cathode, in which case
it is called an indirectly heated cathode. The filament
type is used in most large radio transmitting and industrial **vacuum tubes** and **thyratrons,** while the indirectly
heated type is used in most radio receiving tubes and
many smaller industrial types. In either type, however,
the mechanism is the same, the so-called free electrons
in the cathode are given enough energy by thermal
means to overcome the restraints of the cathode surface and pass beyond the boundary, i.e., they are emitted.

Once free of the cathode they may be controlled in various ways to give us the many types of tubes which are now available. The amount of energy necessary to overcome the surface restraint is called the work function of the cathode. Various materials have been developed to give more electrons for a given amount of heating energy, but only three are used to any great extent at present. These are pure tungsten which is suitable for operation at very high voltages but requires rather high temperatures for appreciable emission, thoriated tungsten which is a combination of tungsten, thorium oxide and thorium, and finally a combination of barium and strontium oxides. Both of these composite cathodes are more efficient than pure tungsten but cannot be operated satisfactorily at as high voltages (voltage between the cathode and other tube electrodes), so are used in medium- and low-voltage tubes. Both must be put through special processes, called activation, to get the desired characteristics. This is done after the cathode is assembled in the tube and while the tube is being evacuated. Cold cathodes are used in certain grid-glow tubes, in many electronic light source tubes, in phototubes and various types of gas-filled tubes. The mercury pool cathode is a sort of cross between the cold and hot cathode types since it is not heated by extra heating current, but does get an appreciable part of its emission from the heat generated by the passage of the main discharge current. The photocathode, used primarily in phototubes and television camera tubes, depends upon the action of light on the surface to give its electrons enough energy to overcome the surface forces. These cathodes are usually composite surfaces prepared to give electron emission with the relatively low amount of energy available from the light. In general, emission from cold cathodes is due to photo-electric action, high electric field emission and bombardment by charged particles. (L.R.Q.)

CATHODE RAYS. An emission from the cathode in a **vacuum tube,** which becomes more conspicuous as the tube is cleared of gas molecules with diminishing pressure. At pressures of 0.01 mm. of mercury or lower, the rays leave the cathode normally to its surface and move in straight lines across the tube, as shown by the early experiments with the **Crookes tube.** By using a concave cathode, they may be brought to a focus, and any obstacle placed at the focus becomes intensely hot. For some time doubt existed as to the nature of these rays, but Sir J. J. Thomson finally proved that they are negatively electrified particles with a **charge-mass ratio** now known to be of about 1.76×10^8 coulombs per gram; and that they move with speeds varying with the voltage but commonly of the order of $\frac{1}{3}$ the speed of light. We now know that they are **electrons,** released from the metal of the cathode and driven away by its negative voltage.

Lenard (1898) showed that cathode rays will penetrate through thin aluminum or gold leaf and can thus be allowed to pass outside the tube. Electrons so escaping are called Lenard rays. Recent technique due to Coolidge and others has produced Lenard rays of such high speed as to be comparable to the **beta rays** from radium. (L.D.W.)

CATHODE-RAY TUBE. This is a special form of vacuum tube used in various electronic applications, e.g., as the picture tube of the television receiver and as an oscilloscope tube. The tube consists fundametally of three sections enclosed in an evacuated bulb. The first is the electron gun which produces and projects down the tube a beam of electrons. This gun has a **cathode,** usually indirectly heated, which emits a plentiful supply of electrons, then a **grid** which controls the number of electrons drawn towards the anode and finally the **anode.** Both the grid and anode are different in structure from the corresponding elements of the conventional vacuum

tube. They consist of metal cylinders, coaxial with the cathode and closed at the end except for a small circular hole. Many of the electrons which are drawn towards the anode under the influence of the grid and anode voltages pass through these holes and are projected as a beam into the main part of the tube. Here they are focused upon the screen by electric or magnetic fields in a manner analogous to the focusing of light rays by a lens system. In passing down the tube the beam passes between two sets of plates, called deflecting plates, which are perpendicular to one another. A voltage applied to either of these sets will deflect the beam towards the positive plate. Sometimes the plates are omitted and coils outside the tube substituted. A current through the coils will set up a magnetic field and hence cause deflection of the beam. Finally the beam hits the screen which is a coating on the end of the tube of some material which will fluoresce under the impact of the electrons. Reference to Fig. 1 will show the relative positions of the various parts of the tube. A brief description of the

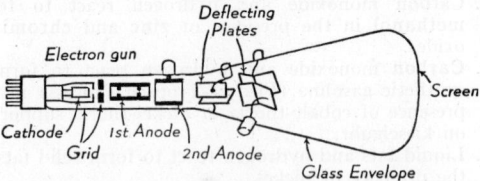

Fig. 1. Typical cathode-ray oscilloscope tube.

use of the tube as an **oscilloscope** will indicate its possibilities. The oscilloscope is used primarily for studying the current or voltage wave forms present in a circuit which is being investigated. If the voltage being studied is connected to one set of the deflecting plates it will cause the electron beam in the tube to move back and forth as the voltage varies. This will cause the fluorescent spot on the screen to move, and due to the persistence of the screen it will appear as a line. If now, another voltage is placed on the other set of plates it will move the beam and hence the screen spot, the resultant being a line pattern which is a combination of the effects of both voltages. Usually the second voltage is a sawtooth wave which gives linear (with respect to time) deflection of the spot and hence the pattern will be the same as if the first wave were plotted on conventional rectangular coordinates. For current study either the voltage drop, caused by the current through a resistor, is used or the current is passed through the coils of the magnetic deflection type. For use as a television picture tube the picture signals are applied to the various electrodes so the fluorescence of the screen is a reproduction of the original scene being televised. (L.R.Q.)

CATHODE SPOT. **Mercury-Arc Tube.**

CATIONS. Cations are positively charged **atoms** or **radicals.** (R.K.S.)

CATKIN. **Ament.**

CATLINITE. A red, siliceous **clay** occurring in Minnesota. (R.M.F.)

CATNIP. **Mint Family.**

CAT'S-EYE. This name is applied to varieties of several mineral species that enclose fine fibers or cellular structures in parallel arrangement, causing, particularly when cut and polished *en cabochon,* a band of reflected light to play on the surface of it. Because of fancied resemblance to the eyes of cats, such stones are called cat's-eyes, and the effect is referred to as chatoyancy. The stone is said to be chatoyant. True cat's-eye is a variety of **chrysoberyl,** but **tourmaline** and **quartz** are also found which show this same effect. Ordinary quartz cat's-eyes are a pale yellowish or greenish, but a beautiful golden-yellow sort is known from South Africa called tiger's-eye which probably represents a replacement of **crocidolite** by quartz. (E.S.C.S.)

CAT-TAILS. *Typha latifolia* and related species. Typhaceae. These are well-known plants growing in marshy places and along the margins of ponds and slow-flowing streams. Usually they form extensive stands, crowding out nearly all other plants. The cat-tail plant has a thick horizontal **rhizome** which grows along the surface of the ground or just beneath it, generally in several inches of water. From this rhizome the long linear leaves grow in erect bunches. These leaves have widely overlapping bases and are from 3–6′ long. The flower stem rises stiffly erect in the center of the bunch of leaves and is from 3–8′ tall. Near its tip the cat-tail bears two dense cylindrical spikes of flowers, one above the other, which are unisexual. The **pistillate** flowers are found below the **staminate** flowers. Each staminate flower consists of from 2 to 5 or more stamens surrounded by a number of hairs. Soon after the pollen grains are shed the staminate flowers drop off, leaving the naked tip of the stem projecting above the pistillate spike. Each pistillate flower consists of a single pistil surrounded by a group of long hairs. Cat-tails are entirely wind-pollinated. After pollination the **ovaries** develop to 1-seeded **achenes** surrounded by the fine hairs. These fruits form the familiar black or dark-brown cat-tail of late summer and fall. The seeds are blown about by the wind, the long hairs greatly aiding in distribution.

The dried leaves of cat-tails were formerly used in making the seats of rush-bottomed chairs. The hair-covered seeds have been used to a slight extent for stuffing for pillows and small things. The entire fruiting stem is often used as an ornament. (R.M.W., B.S.M.)

CATTLE. Mammalia, Artiodactyla. Broadly applied to all bovine animals, including oxen, buffaloes, sheep, goats and others but more commonly restricted to the **buffaloes, oxen,** and related species and especially to the domesticated races. (A.W.L.)

CAUCHY-RIEMANN DIFFERENTIAL EQUATIONS. Analytic Functions of a Complex Variable.

CAUCHY'S THEOREM ON ANALYTIC FUNCTIONS. If $f(z)$ is an **analytic function of a complex variable** z which has no singularities within or on a given closed curve C, then $\int_C f(z)\,dz = 0$, where the integral is extended over the entire contour C. (L.L.S.)

CAUDAL FILAMENT. A median jointed appendage resembling an antenna and sensory in function, found in some of the primitive **insects.** (A.W.L.)

CAUDATA. Urodela.

CAULDRON-SUBSIDENCE. A term proposed by E. B. Bailey and other Scottish geologists for the sinking of the portion of the roof or cover of a deep-seated **igneous intrusion,** aided by circumferential **faults.** (R.M.F.)

CAULIFLOWER. Brassica.

CAUSAL VARIABLE. Let a variable Y depend directly on a variable X, then X is said to be a causal variable. Naturally there must be a logical connection between the two variables and not merely an association. (L.A.A.)

CAUSTIC. Term applied to bases or alkalis. Caustic potash, **potassium** hydroxide; caustic soda, **sodium** hydroxide; caustic line, **calcium** oxide; lunar caustic, **silver** nitrate. (R.K.S.)

CAUSTIC EMBRITTLEMENT. Water, Utilization.

CAUSTICS. Spherical Aberration.

CAVE. A cave is a natural opening in the earth's surface, and from time immemorial caves have attracted much attention because of the inherent mystery suggested by them or from the fact that they have been used as the refuge of robbers, as depositories for their plunder, and as the first habitations of primitive man. Caves are chiefly developed in **limestone** regions where because of the easy solubility of **calcium** carbonate the ground water succeeds in dissolving and carrying away large quantities of this otherwise physically resistant rock. The water enters through the **joint** cracks or

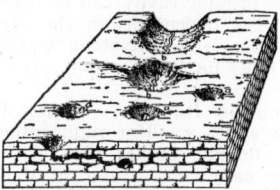

Diagram illustrating the origin of caves (in black), sink holes (a), and natural bridges (b) in limestone. (*After H. F. Cleland.*)

bedding planes, passing downward by gravity until it becomes saturated with calcium carbonate or reaches the ground water level where more or less complete saturation exists. With the continued solution of the limestone, large channels and even great underground chambers are formed. The steady removal of the limestone in solution thus tends to weaken the whole formation with the result that the roofs of underground channels or chambers frequently collapse, forming depressions varying in size from a few feet in depth and of small area to those of many acres and 100′ or more in depth. Such fallen-in areas are called sink holes or simply sinks. If they contain water they are then referred to as sink-hole lakes. Sometimes after the continued collapse of the roofs of caverns a small portion will remain, thus forming a natural bridge, the classical example of which is the Natural Bridge, Virginia.

Wherever limestones occur, if there is a sufficient supply of ground water, underground drainage will develop.

Among the more famous caverns of the United States are the Luray and Shenandoah Caverns in Virginia, Mammoth Cave, Kentucky, and Carlsbad Caverns in New Mexico.

Certain parts of Florida abound in sink holes and sink-hole lakes. An important feature in caverns is the so-called rock icicles or stalactites and their associated stalagmites. (E.S.C.S.)

CAVE-CRICKET. Camel Cricket.

CAVENDISH EXPERIMENT. Gravitation Constant.

CAVIARE, CAVIAR. The roe of **sturgeons,** preserved in brine and used especially as an hors d'oeuvre. (A.W.L.)

CAVITATION. It is possible to operate the blades of a water screw propeller or a hydraulic turbine at a condition such that cavities are formed in the wake or about the blade, resulting in vibrations, loss of efficiency, or corrosion of the blades. The chief cause of trouble with **hydraulic turbine** runners is corrosion and attendant pitting of the blades or buckets. This occurs chiefly in regions of high vacuum, as, for instance, near the **draft tube,** and is probably due to the liberation of oxygen from the water at low vapor pressure. The use of too-high **specific speed** for the head available is thought to aggravate this trouble.

When the engines or turbines of a ship drive the propeller blade through the water at excessive speeds the blade may move through the water faster than the water can close in around it. The cavities that are thus formed cause a great loss of propulsive efficiency, and should be avoided. To avoid cavitation the blades of a marine propeller are made short and wide. (F.T.M.)

CAVITY RESONATOR. Just as the ordinary transmission **line** may be shorted and adjusted to multiples of quarter-wavelengths and used as resonant circuit, the **wave guide** may be closed at the end and, if the dimensions are correct, made to serve as a resonant circuit at **ultra high frequencies.** When used in this manner the guide is known as a cavity resonator. Generally speaking, any hollow metallic cavity can be made to resonate at some frequency dependent on the shape, size and mode of oscillation. The **klystron** makes use of this property of cavities for its operation. (L.R.Q.)

CAVY. Mammalia, Rodentia. Stoutly built **rodents** with short legs and a small or rudimentary tail, represented by the common **guinea-pig.** Among the other species are the **capybara,** largest of the rodents and aquatic in habits, although most of the species are small and terrestrial. All are native to South America. They constitute the family Caviidae. (A.W.L.)

CAYMAN. CAIMAN.

C-BIAS. Bias.

CECOSTOMY. An artificial opening made through the abdominal wall into the cecum.. This is done temporarily or permanently to provide an artificial **anus** when the large intestine is obstructed or is the site of certain chronic diseases such as **colitis.** It is also a frequent operation for cancerous ulceration of the large intestine, as a preliminary stage before removal of the tumor. (D.M.H.)

CECUM, CAECUM. A sac-like, blind pouch of the large intestine, situated below the level of the junction of the small intestine into the side of the large intestine. At the lower portion of the cecum, but variable in position, is the **appendix.** (R.S.M.)

CEDAR. Under this name many trees, of several genera, are included. All are characterized by having in the woody tissue an aromatic **volatile oil** which persists for a long time after the tree is cut down and dried. The wood of many cedars is very resistant to rotting, therefore cedar was a favored material for rail fences. The oil present in the wood is repulsive to insects. Among the many cedar trees are *Thuja plicata,* the Western Red Cedar, much used for shingles; *Juniperus virginiana,* Eastern Red Cedar, an important wood used in the making of cedar chests and also lead pencils; *Cedrela odorata,* Spanish Cedar, a tree related to the true **Mahogany** tree, and used in the manufacture of furniture and for cabinet work, also for cigar boxes. In these and in its other uses, Spanish Cedar is frequently cut into thin sheets and applied as a veneer over cheaper woods. Cedar of Lebanon, *Cedrus libanotica,* and White Cedar, *Chamaecyparis occidentalis,* are frequently planted for ornamental purposes, wherever they are hardy. (R.M.W.)

CEILING. Total distance from ground or water vertically to the base of the lowest cloud layer covering more than ½ of the sky. It is one of the most often measured and critical observations in aircraft operation because it is the clearance an aircraft has above terrain before going on instruments.

The term ceiling is used to designate vertical operating limits of an **airplane.** An airplane flies by virtue of expenditure of a certain amount of horsepower by an

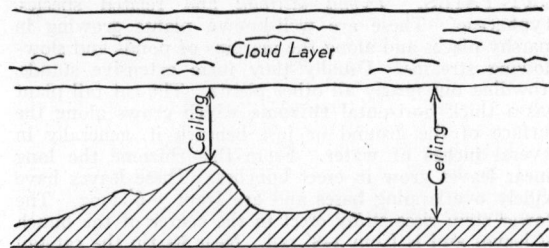

engine. When the horsepower thus available from the engine exceeds that required to overcome the air resistance due to motion of the plane, the excess may be employed by the pilot for increasing the velocity or for climbing at constant velocity. However, as the altitude is increased the performance of the internal combustion engine is affected by the rarefied air, with the result that the horsepower available from the engine will have diminished until it is just sufficient to maintain horizontal flight. This altitude is the absolute ceiling of the plane, above which it is not possible to climb. Since theoretically it takes a plane an infinite time to reach absolute ceiling, the service ceiling is a more practical measure of performance. This is the altitude at which the rate of climb has diminished to 100 ft. per min. (P.E.K., F.T.M.)

CELERY. *Apium graveolens.* Umbelliferae. A biennial plant growing wild in the marshes of western Europe. The wild plant has a rank taste and strong characteristic odor, but cultivation has resulted in the elimination of these objectionable features, and has produced an enlarged leaf-stalk extensively used as a salad plant. In another form, celeriac, the root stock and stem are much enlarged. (R.M.W., B.S.M.)

CELESTIAL MECHANICS. The term celestial mechanics is applied to that field of astronomical study and research which deals with the motions of two or more bodies in space under the influence of their mutual gravitational attractions. The fundamental elements of the subject are found in the **Newtonian** law of universal gravitation, the laws of motion, and the **Keplerian** laws of planetary motion. In the classical theory we find space of three dimensions treated, with time considered as an independent variable. Within recent years some slight modifications of the classical theory, particularly when the time interval is very long or velocities and accelerations are very high, have become necessary on account of the **Theory of Relativity.** Under the general heading of celestial mechanics we find such problems discussed as the development of the various methods for **orbit** computation, methods for computing **perturbations,** and solutions of the **problem of three bodies.** (W.K.G.)

CELESTIAL NAVIGATION. Navigation, on sea or in the air, by means of observations of celestial objects, is known as celestial navigation.

In the year 1837, Captain Thomas Sumner discovered what has since been known as the **Sumner Line,** and modern celestial navigation may be said to date from that discovery. The methods for determining terrestrial **latitude** (L) and **longitude** (Lo), from **sextant** observations of **altitude** of celestial objects, are briefly described in the article on Sumner Line. In the years that have elapsed since 1837 a great deal of research has been carried out on the theory of the celestial **line of position,** and many methods for calculating the data necessary to plot that line have been developed.

Before discussing these methods, it will be well to understand just what is meant by a celestial line of position, in the modern usage of the term. At any instant any celestial object is directly at the zenith for some particular spot on the surface of the earth. This point was formerly known as the subsolar, sublunar, or substral point, depending upon whether the object observed was the sun, the moon, or a planet or star. At present the U. S. Navy uses the term Ground Position (GP) to designate that point, no matter what celestial object is used. The terrestrial coordinates of GP may be expressed in terms of celestial coordinates of the object: the latitude being equal to the declination of the object, and the longitude equal to its Greenwich hour angle. At present both of these quantities are tabulated for the sun, moon, planets, and navigator's stars, in the Nautical Almanac, or the Air Almanac, both of which are issued by the Nautical Almanac Office of the U. S. Naval Observatory. These tabulations are given in terms of Greenwich Civil Time, and this time is used by all navigators for the recording of the times of sextant observations for altitude.

At the GP the altitude of an object is 90°. If an observer moves out from the GP, in any direction, the altitude of the object decreases in minutes by an amount equivalent to the distance the observer has moved in nautical miles. Now assume that an observer obtains the geocentric altitude (h_t) of an object at a given Greenwich Civil Time. From the data given in the Almanac the terrestrial coordinates of the GP may be determined and that point plotted on a sphere representing the earth. Then with this point as center, and with a radius proportional to the number of minutes in $90° - h_t$, a circle is drawn on the sphere. Since the observer is on this circle it is a line of position, by definition of the term.

This method for laying down a line of position on a sphere is very simple in theory, and has been understood for many years. However, to obtain an accuracy in plotting, consistent with the accuracy attainable in sextant observations, requires the use of a sphere of unwieldy dimensions. It is very difficult to draw a line with width less than 0.05''. On the sphere recently proposed for use in the so-called "Spherographic System of Navigation" such a line would have a width equivalent to about 10 miles, and sextant observations are attainable to a much higher accuracy than this.

In modern practice the line of position is drawn on a **Mercator** Chart, plotting sheet, or **small-area plotting sheet** by employing the geometric proposition that a radius of a circle is always perpendicular to an arc. At some particular GCT the altitude (h_s) of a celestial object is obtained with the **sextant** or bubble octant. Then, using the **dead-reckoning** position, or some position close to it which leads to simplified computations, the values of the altitude (h_c) and the **bearing** (Z_n) that the object would have at the assumed position and GCT of observation are computed, or taken from suitable tables. The dead-reckoning, or assumed, position is now set down on the plotting sheet and a line drawn through it in the direction Z_n. This line is a section of the radius of a circle drawn about the GP, which, in reality, is usually off the plotting sheet. Next the difference between the computed and observed altitudes ($h_c - h_t$) is taken and called the "intercept." If the intercept is zero the ship must be on a line of position perpendicular to the bearing line and passing through the plotted position. If the intercept is plus ($+$) the line of position must pass through a point ($h_c - h_t$) minutes of arc, or nautical miles, away from the GP along the bearing line, and if the intercept is minus ($-$) the line must be between the plotted position and the GP. In either case the line of position must be perpendicular to the bearing line.

In spite of the fact that the line of position is actually a circle with radius $90° - h_t$ miles, nevertheless, in practically all cases, the line may be drawn as straight. If we assume that the altitude of the object is 80° the value of ($90° - h_t$) is 10° or 600 miles. In this case a straight line 60 miles long, perpendicular to the radius, will differ from the actual circle by less than a mile at its extremities. Accordingly, the assumption of a straight line will not lead to appreciable errors, if the altitude is less than 80° and the drawn line is less than 60 miles long.

Most of the developments in celestial navigation since 1837 have been in improving the methods for obtaining h_c and Z_n. For computation, the so-called "cosine-haversine" formula is used when the tables of H.O. 9 (Bowditch American Practical Navigator) are available. This formula is:

$$\text{hav } (90° - h_c) = \text{hav } \theta + \text{hav } (L - d)$$

$$\text{hav } \theta = \cos d \cos L \text{ hav } t$$

in which L is latitude of the DR, or assumed position, d is the declination of the observed object, t is the local hour angle ($t = \text{GHA} - \text{Lo}$, with Lo the longitude of the plotted point) and θ is an auxiliary angle. The value of Z_n may be obtained directly from any one of a number of published azimuth tables, or may be computed by means of a variety of formulae (e.g., $\sin Z_s = \sin t$ $\cos d \sec h$ and $Z_n = Z_s \pm 180°$). Various individuals, among them Admiral Marq Saint Hilaire, Admiral Ogura, and Captain De Aquino, have transformed the fundamental equations for solving the **astronomical triangle** to obtain tabular methods for solving for h_c and Z_n. The tables developed by Ageton (H.O. 211), or those by Dreisonstok (H.O. 208), were standard for many years in the U. S. Navy. Still more recently, tables by Weems and another by Ageton, have been privately printed. H.O. No. 214 and No. 218 (the former computed by the U.S.W.P.A. and the latter by the British Admiralty Office) tabulate the values of h_c and Z_n for every degree of latitude, every degree of t, and every 30' of d. For most of the tabular methods and for H.O. 214 and 218, much time may be saved, and very little accuracy sacrificed, by using as a plotted point a position as close as possible to the DR position; but with Latitude an even degree and longitude so selected that (GHA − Lo) will give a value of t some even degree. Such a position is referred to by navigators as an assumed position.

Navigators spend much valuable time in arguing the relative advantages of the different methods for laying down a line of position by celestial navigation. Time, convenience, and personal preference form the topics for debate. Using a DR position and the cosine-haversine method, a line of position can be plotted with from 4 to 5 minutes of computational work. The tables of Bowditch (H.O. 9) are all the equipment needed, and this method will yield good results in practically all conditions. The Ageton or Dreisonstok tables require from 2 to 3 minutes of computational work, and the tables are small and convenient. However, to attain the stated speed, an assumed position must be used. H.O. 214 is published in eight large volumes, each covering 10 degrees of latitude, and, using an assumed position, the line may be laid down in about 1 minute. H.O. 218 is published in 18 volumes, each covering 5 degrees of latitude, and using an assumed position, the line may be obtained in slightly less than 1 minute. H.O. 218 has additional advantages over H.O. 214 when working with the navigator's stars.

Simple statistical analysis shows that the point on the line of position closest to the dead-reckoning point is the most probable position that can be obtained for the ship from a single observation of altitude. In air navigation this position is referred to as the estimated position (EP). Care must be taken not to confuse this EP with the EP obtained by dead-reckoning navigation in marine navigation. This most probable position, or EP, can be obtained without plotting the line by any of the above methods if the DR position instead of the assumed position is used, since the intercept gives the shortest distance from the DR position to the line.

An example of the use of celestial navigation at sea is given in the following practical case.

During the night of August 18, 1943, the navigating officer of a ship on passage from England to the United States wishes to check the dead-reckoning position of the ship. The two stars Alpheratz and Altair are well placed for observation. When the navigating officer's watch reads 23h 40m 10.0s the sextant altitude of Alpheratz is 50° 34'.3. For the purpose of checking the deviation of the steering compass the bearing of the star is taken by this compass and found to be 121°. At watch reading 23h 46m 15.4s the sextant altitude of Altair is 50° 20'.7. The watch times must be corrected to obtain Greenwich Civil Time (GCT) and the sextant altitudes corrected for instrumental errors, dip, and refraction to obtain the geocentric altitude (h_t). These corrected results are:

Star	GCT	h_t
Alpheratz	Aug. 19 02h 39m 34.0s	52° 27'.4
Altair	Aug. 19 02h 45m 39.4s	50° 13'.8

Using the average of the watch times, the dead-reckoning position of the ship at Aug. 18, 2345 is found as Latitude 43° 24'.6 N and Longitude 48° 27'.4 W. Since the ship is proceeding at only 16 knots and the DR position is probably somewhat in error, this position is used for computing the altitudes and bearings that the stars should have at the GCT's of observation. These values are found to be:

Star	Bearing	h_c
Alpheratz	098°.4	52° 32'.5
Altair	216°.2	50° 07'.4

The intercepts ($h_c - h_t$) are now found: for Alpheratz +5'.1 and Altair −6'.4. These yield two "most probable" positions of the ship, one 5.1 miles from the DR position in the direction 278°.4 (i.e., away from the GP of Alpheratz), and the other 6.4 miles in the direction 216°.2 (i.e., toward the GP of Altair). To determine the fix the DR position is plotted, the bearing lines to the two GP's drawn through it, the intercepts measured off in the proper direction, and the two lines of positions drawn through the intercepts perpendicular to the lines to the GP points. The point of intersection of the lines of position is the fix at 2345. The actual plotting, on small area plotting sheet, is shown in Fig. 1. From the

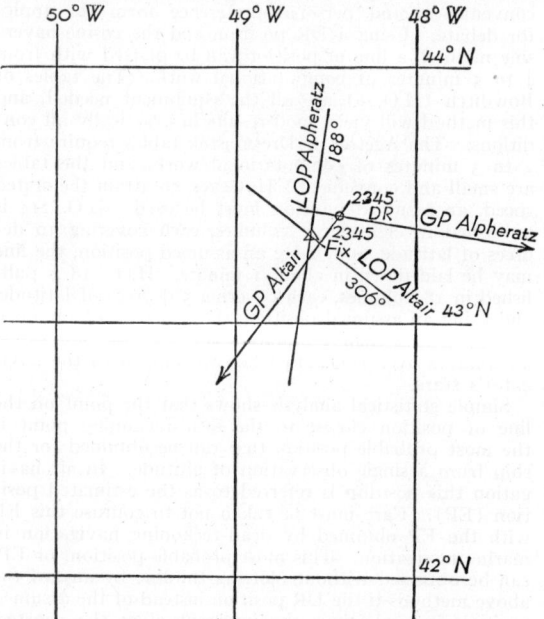

Fig. 1. Scale drawing for celestial navigation example.

figure the fix is found to be in Latitude 43° 20'.7 N and Longitude 48° 35'.2 W. To determine the compass deviation, the difference between the observed compass bearing of Alpheratz and the computed bearing is found to be 121° − 098° = 23°. Since the variation in this region is 26° W, the deviation must be 3° E.

Because of the rapid motion of aircraft the most rapid methods for obtaining the data for plotting the line of position must be employed. The bubble octant is used for measuring the altitude of the celestial object. After this is corrected to obtain geocentric altitude, a position is assumed as close as possible to the DR position that is in some even latitude and in such longitude as to make the local hour angle an even degree. Then the data for plotting the line is obtained from the Air Almanac and H.O. 214 and 218. Under the most favorable conditions a rapid computor can have the line of position plotted about 5 minutes after the observation is begun. If two objects are available for observation, about 5 minutes will elapse between the observations, if they are made with maximum accuracy attainable with that instrument. In modern planes a motion of greater than 15 miles between observations is common and the method of running fix (see **Radio Navigation**) must be employed.

Proper selection of stars to be observed will yield data of extreme importance to the pilot and navigator. For example, if the object is nearly ahead or astern of the plane, the line of position will cross the course nearly at right angles and the length of the intercept will provide a check on the ground speed being made good. On the other hand, if the object is in a direction approximately perpendicular to the course the value of the intercept will indicate the accuracy of the wind correction angle.

In many cases altitudes and bearings of celestial objects may be computed in advance of the actual observing. These are known as predetermined altitudes and have many uses in air navigation. If a plane is to depart at a definite time and follow a specified course, the altitudes and bearings to be expected at indicated times may be computed before the plane leaves the ground. The course to be followed is plotted on a chart, the predetermined DR positions for indicated times are marked, and the bearings of the GP of the object are indicated by lines drawn through the DR positions. The precomputed altitudes are geocentric but they may be transformed into those expected to be read on the octant at the specified times, by applying the various corrections with reversed signs. The navigator measures the altitude at an indicated time, obtains the difference between his value and that predicted, and lays off this distance along the drawn bearing line either toward or away from the GP. In this way an EP is determined in a few seconds after the observation is completed, and no computing is required during the flight. The pilot is notified to alter heading and air speed to bring the plane back to schedule. If, due to unforeseen conditions, the plane gets so far off scheduled position that the intercepts are more than 150 miles, the predetermined altitudes must be abandoned and regular celestial navigation adopted.

Under some conditions it may be necessary for a plane to make an accurate landfall (e.g., locate a small island, life raft, etc.) under conditions where celestial navigation must be relied upon. In such cases the use of precomputed altitudes gives great assistance. In such cases an estimated time of arrival (ETA) is obtained. Then, using the latitude and longitude of the landfall, a series of altitudes and bearings of a celestial object is computed. The interval of time between computed values depends somewhat on the rapidity with which the values are changing, but is usually about 10 minutes. The series begins at least half an hour before ETA and extends beyond that value. Two curves are now drawn on graph paper showing altitude and bearing as a function of time. The plotted altitudes are those expected with the octant, i.e., with corrections applied to computed geocentric values. If possible an object that is approximately ahead or astern of the plane should be selected. About half an hour before the predicted ETA the pilot alters heading 10° or 15° to the right or left of that predicted for the true course, so that there will be no question as to which side of the landfall he is approaching, and the navigator begins taking altitudes of the object. The navigator plots his observed values, as a function of time of observation, on the same graph as that showing the predetermined values, and obtains a curve of observed values. At the instant that the observed curve intersects that obtained from precomputation the plane must be on a line of position running through the landfall. The bearing of the celestial object at this instant is read off the plotted bearing curve, the line drawn at right angles to this bearing through the destination, and the pilot instructed to alter heading to run down the line.

The "Star Altitude Curves," computed and published in 1940 by the Weems System of Navigation, are, in reality, nothing more or less than predetermined altitudes of selected objects. An appropriate solution of the astronomical triangle may be made to give the latitude and local sidereal time at which a celestial object will be at a specified altitude. On a **Mercator graticule**, using local sidereal time instead of longitude for the invariable coordinate, the positions at which a specified object will have a given altitude may be plotted. A curve drawn through such points will be a curve of equal altitude and, hence, a celestial line of position. In a series of volumes, each covering 10° of latitude and 24 hours of local sidereal time, such equal altitude curves are printed. The printed values of altitude have had the refraction correction applied with reverse sign for standard atmospheric conditions. On each page of the "Star Altitude Curves" three stars are shown, being selected so that they will be favorably situated for observation and give lines of position intersecting at angles of about 60°. The curves are printed for each 10′ of altitude for each star, with the even degrees of altitude indicated by a heavier line with the value of the altitude printed along the line. To use the curves most efficiently the navigator should have a timepiece keeping Greenwich Sidereal Time (GST). Using the proper volume which contains his DR latitude, the navigator then turns to the page covering his local sidereal time (LST), finding this quantity from his DR longitude by the relationship GST − Lo = LST. On the page he will find curves for three stars and he takes sextant altitudes of one or more of these. If his instrument has no appreciable instrumental corrections he may use the observed value directly and the printed curve with the observed altitude is his line of position. Two such lines will provide a fix in terms of latitude and local sidereal time. The longitude of the fix is determined from its local sidereal time by the relationship above Lo = GST − LST. The determination of celestial line of position by means of the Star Altitude Curves is undoubtedly the most rapid of all methods. However, it requires the transportation of several volumes of curves, and the sky must be clear enough to permit the observation of certain particular objects. (W.K.G.)

CELESTIAL SPHERE. The concept of a sphere, on which all of the so-called "fixed stars" are projected, persisted in all descriptions of the structure of the universe from the earliest historical records down through the 17th century. Even in modern times such a concept is very convenient for discussing the common motions of the stars, and the individual motions of the different members of the solar system.

This celestial sphere may be defined as a sphere of infinite radius with the center located within the solar system. The reference frames for all systems of astronomical **spherical coordinates** are established on the celestial sphere.

When projecting the different members of the solar system onto the celestial sphere it becomes necessary to restrict the location of the center to some particular point within the solar system. If the center of the celestial sphere is considered as a point on the surface of the earth we have systems of **apparent coordinates**; with the center at the center of the **earth** we have **geocentric coordinates**; at the center of the sun, **heliocentric coordinates**; at the center of **Jupiter**, Jovicentric coordinates; etc.

Due to the fact that the earth is actually rotating about an axis, the celestial sphere is apparently rotating, as seen from the earth, about an axis parallel to the axis of the earth and with the same angular velocity as the earth, but, of course, in the opposite direction. For purposes of convenience it is customary to refer to this apparent rotation simply as the rotation of the celestial sphere. All systems of spherical coordinates which are established on the sphere, such as the **equatorial**, **galactic**, **ecliptic**, etc., rotate with the sphere, while the **horizontal** system apparently remains fixed in space. (W.K.G.)

CELESTITE or CELESTINE. The mineral celestite is composed of **strontium** sulfate, $SrSO_4$, occasionally with **calcium** and **barium**. It crystallizes in the **orthorhombic** system in tabular or prismatic crystals. More rarely it may be pyramidal or simply fibrous or granular. Two essentially perfect cleavages may be observed, one parallel to the base, the other parallel to the prism. Its fracture is uneven; hardness, 3–3.5; specific gravity, 3.95–3.97; luster, vitreous; color, white, but may be slightly reddish or bluish; transparent to translucent. Celestite may occur with **gypsum** and salt associated with beds of limestone, or by itself in large commercially important veins. It sometimes occurs with sulfur in volcanic localities and is often a gangue mineral in veins of **galena**, **sphalerite** and similar metallic minerals. In Europe there are many localities for fine crystals, especially in England. In the United States celestite is found in New York, Pennsylvania, West Virginia, Tennessee, Kansas, Colorado, and California. The first celestite described was the delicate blue material from Blair County, Pennsylvania. Its "celestial" tints suggested the name. Celestite resembles barite very strongly. (E.S.C.S.)

CELIAC DISEASE. Sprue.

CELL. For the cell as a source of electricity, see **Electrical Cell.** In biology a cell is: 1. Essentially a small chamber, and so applied in the designation of cells in honey-comb, the water cells in the stomach of the camel, and other structures. 2. Also the unit of living matter. Early observation of cells was based on plant **tissue** in which the cell walls were the most prominent structure, hence the units were regarded as small compartments. Even after the discovery that these units are small masses of living substance and that the enclosing wall is merely one component which is not always present, the name cell was retained with the new meaning. The first sound expression of the cell theory is credited to Schleiden and Schwann, in 1838 and 1839.

Cells are divided into two principal parts, a nucleus which is usually central or nearly so and an enveloping cytosome, more often known by the name of the material composing it, cytoplasm. Together the nucleus and cytosome make up a living unit which has been named the protoplast or energid; it is enveloped by various membranes which complete the cell. In some cells the nuclear constituents are more numerous, including several separate parts which may be similar and may differ visibly, as in the micronuclei and macronuclei of certain single-celled animals. All of these major parts of the cell are themselves complex.

The cytosome is the part of the cell in which are differentiated the structures by which its special functions are carried out, hence cells are much more varied in cytoplasmic structure and in general form than in their nuclei. Cells often show a clear peripheral zone of **cytoplasm** called the ectoplasm and an inner mass containing various formed constituents and called endoplasm. The more common of the formed constituents are mentioned in the following paragraphs.

The nucleus differs greatly in form. Although most often spheroidal to slightly elongate it is sometimes of complex shape. It is invested by a delicate nuclear membrane and contains a ground substance called the karyolymph or enchylema in which are two formed components, the linin or supporting network and the chromatin. In the resting cell chromatin is usually dispersed in granules or in a reticular form. It is named for its strong affinity for stains, usually for the basic dyes, which leads it to color readily in preparation for microscopic observation. Since it sometimes takes acid stains the tendency to distinguish the chromatin according to its staining reaction led to the terms oxychromatin and basichromatin, and these led in turn to the suggestion of another collective term, karyotin.

The nucleus also contains knots of chromatin known as karyosomes and other chemically different bodies called nucleoli or plasmosomes. In some cells the two are closely united to form amphinucleoli.

All evidence points to the chromatin as an essential substance which controls the differentiation and constructive action of the cell and as the bearer of hereditary characteristics. The chief evidence of these functions is found in **cell division** and **heredity,** although the removal of nuclei from living cells by microdissection has disclosed the importance of this body in metabolism. Enucleate 1-celled animals continue their activities as long as they have the energy but are unable to assimilate more energy-bearing materials.

The cytoplasm contains an important body known as the central body or centrosome which is active during cell division and is often self-perpetuating. In some cases a structure of similar functions lies within the nucleus. The centrosome contains a central granule or minute rod, sometimes paired, which is called the centriole. This structure is the most common constituent; it is always present, whether or not the surrounding structures appear, and is the active part during cell division. When the investing portion of the centrosome is large it has been called attraction sphere, centrosphere, periplast, and idiosome in various cases but the term centrosome is commonly adopted. The centriole may also be associated with motile appendages of the cell, forming the basal body from which the central filament of a flagellum or cilium arises; it is then called a blepharoplast.

The cytosome of many cells contains differentiated fibrils which are very fine thread-like structures of various functions. Good examples are the myofibrils of muscle cells and the neurofibrils of nerve cells; the former are contractile and the latter may be conductors.

Plastids are relatively large self-perpetuating bodies which are frequent in plant cells but rare in those of animals. The chloroplasts found in the green parts of plants are the most abundant example.

Chondriosomes are specialized bodies of various forms, varying from granular mitochondria to rod-like chondrioconts. Mitochondria sometimes associate in linear groups called chondriomites or in rounded chondriospheres. The functions of all such bodies are uncertain. According to one view they have the power to act upon special chemical constituents of the cytoplasm.

Vacuoles are globular cavities filled with liquid. In the 1-celled animals they contain food and water during digestion, and in the case of the contractile vacuoles the waste products in solution.

The Golgi apparatus is one of the most baffling of all cell structures. Since it is revealed by definite methods of preparation it seems to have definite chemical properties, yet it is extremely diverse and is of uncertain function. It has also been named Golgi bodies, trophospongium, internal reticular apparatus, dictyosomes, and canalicular system.

The Golgi apparatus is either a somewhat reticular structure, usually near the central body and nucleus, or a scattered lot of reticular or granular parts. The separate bodies of which it is composed are of many shapes and have been designated as batonettes, among other terms.

Many cells contain cytoplasmic granules of various kinds whose functions are well understood. They include pigment granules, granules of material for storage, secretory granules which are transformed into the special secretion of the cell during periods of activity, and other special forms.

The membranes surrounding the cytosome in animal cells are thin. They include a delicate plasma-membrane which is regarded as a peripheral layer of the cytoplasm itself, in addition to which there may be an outer cell wall. The latter is secreted by the cell but is not a part of the living cytosome; it is formed of cellulose in many plant cells and in animal cells is represented most strikingly by such highly developed structures as cuticular layers.

Other facts concerning cells are discussed under **Tissues, Cell Division, Gamete,** and **Gametogenesis.** (A.W.L.)

CELL DIVISION. The subdivision of cells is a fundamental expression of the power of reproduction inherent in all living matter. All living cells are capable of dividing to form others or are the products of such division.

Cell division is of two kinds, amitosis and mitosis. The former is an apparently simple elongation, constriction and division of the cell, including its nucleus, and is of relatively rare occurrence. It takes place chiefly in cells of temporary value, as in certain accessory reproductive structures.

Mitosis is a much more complex process of cell division which prevails in the organic world and by its orderly sequence of stages indicates the precise maintenance of essential constituents of the cell. It is conveniently divided into four stages, the prophase, metaphase, anaphase, and telophase.

During the prophase in animal cells the centrioles separate, or if only one is present it divides and the two halves separate, and a series of radiating lines, the astral fibers, forms about each. As these two asters continue to draw apart similar fibers appear between them in the form of a spindle. The whole structure constitutes the achromatic or mitotic figure.

In plant cells (which have no centrioles) the spindle is formed from the transparent protoplasm (karyolymph) of the nucleus and small amounts of cytoplasm on opposite sides of the nucleus. The spindle, in either case, is a polarized gel which, in some unknown way, guides the daughter chromosomes (chromatids) to opposite ends of the cell.

At the same time the chromosomes, which have existed in the resting nucleus as long, thin threads, contract. Each chromosome now appears in the form of two parallel threads (chromatids). These reach their minimum length and maximum width at the end of prophase. The membrane and nucleoli now disappear and a special portion of each chromosome, the centromere (spindle attachment constriction), migrates to the equator of the spindle. This is metaphase.

In the anaphase stage the chromatids, now called chromosomes, migrate to the poles of the spindle. The centromere of each chromosome starts first toward the pole so that each chromosome takes the shape of the letter V or J depending upon the location of the centromere.

In the telophase this migration is completed, the cytoplasm constricts and divides, or is separated by the formation of a partition between the two groups of chromosomes, and the two daughter cells undergo a return to the resting form.

Since the differentiation exhibited by chromosomes is

longitudinal, this process of cell division apparently results in the formation of daughter chromosomes exactly equivalent to those of the parent cell, and guarantees to each daughter nucleus a similar assortment of these bodies. With the exception of the centriole the cytoplasmic constituents appear not to be evenly divided.

The type of cell division described here is called somatic mitosis if it involves the diploid number of chromosomes, equational mitosis if the nucleus has the haploid chromosome number. **Meiosis** is a series of two rapidly succeeding mitoses which reduces the diploid to the haploid number and occurs during **gametogenesis** in animals and sporogenesis in the higher plants. (A.W.L., P.A.W.)

CELLULITIS. Diffuse inflammation of the cellular tissues, marked by swelling, redness, heat and tenderness. During this stage of inflammation, **pus** is absent, although an **abscess** is usually surrounded by an area of cellulitis. (R.S.M.)

CELLULOID. Plastics.

CELLULOSE. This is a polysaccharide having the formula $(C_6H_{10}O_5)_n$. (See **Carbohydrate**.) It is the substance which composes the greater part of the **cell** walls of plants. **Cotton** fibers are over 90% pure cellulose. Cellulose is insoluble in water, but when dry it soaks up water readily until it is saturated. Water passes readily through the cellulose walls of plant cells. In many cells the cellulose wall is strengthened by the addition of lignin, forming lignocellulose. Being insoluble in water and not affected by digestive juices, cellulose is not digestible and so is available as food only for herbivorous animals (horse, cow, goat, rabbit, etc.) in which the digestive tract is capable of retaining it long enough for digestion by bacteria, or for such insects as the cockroach and termite. Digestion is accomplished in such insects by ameboid protozoa harbored in the digestive tract.

In many cells, especially in those of seeds and in fungi, the walls are formed of hemicellulose, a substance resembling cellulose in many of its properties, but more soluble. Hemicelluloses, of which there are many kinds, are apparently a food reserve of the plant and may be used when need arises. (For further details on cellulose and its chemistry, see **Carbohydrates**.) (R.M.W.)

CELSIUS SCALE. Temperature Scales.

CEMENT. Cement is a finely powdered substance which possesses strong adhesive powers, when combined with water. Gypsum plaster (see **Calcium**), common lime, hydraulic limes, Puzzolan, natural and Portland cements are a few of the materials which are used for cementing purposes.

Portland cement, which is the most important of these materials since it is a basic ingredient of concrete, was first manufactured in England in the early part of the nineteenth century. It derived its name from the fact that this newly discovered cement resembled a building stone that was quarried near Portland, England.

There are three fundamental stages in the process of manufacture of Portland cement, namely, (1) preparation of the raw mixture, (2) production of the clinker, (3) preparation of the cement. Whether the process used is wet or dry, the raw materials are selected, analyzed, and mixed so that, after treatment, the product, or clinker, has a desired, narrowly specified composition. A factory analysis of slurry, where the wet process is in use, is as follows: **calcium** oxide 44%, **aluminum** oxide 3.5%, **silicon** oxide 14.5%, **ferric** oxide 3%, **magnesium** oxide 1.6%, loss on ignition about 33% (largely carbon dioxide), showing that the composition of the resulting burned clinker is essentially a calcium aluminosilicate. The system calcium oxide-

aluminum oxide-silicon oxide has been determined by Rankin and co-workers. In some places the composition of the rock is practically of the desired composition, and in other places clay and limestone are mixed in the desired proportions.

The raw mixture is heated in a continuously operated, long, almost horizontal, slowly rotated furnace or kiln at a high temperature. The temperature is regulated so that the product consists of sintered but not fused lumps. This is clinker. Too low a temperature causes insufficient sintering, and too high a temperature results in a molten mass or glass, the product in either of these cases being valueless for cement purposes. Clinker is unaffected by water, and may be stored indefinitely without detriment. In 1824, Joseph Aspin, an English bricklayer, took out a patent for the manufacture of an improved cement, which he called Portland cement because, after hardening, it looked like Portland stone, a famous English building stone. The cement thus produced was not what is known as Portland cement as the temperature of burning was not sufficiently high. The value of burning at a temperature sufficiently high to cause incipient fusion was soon afterwards discovered.

In order to obtain the desired setting qualities in the finished cement, there is added to the clinker about 2% of gypsum (calcium sulfate, $CaSO_4 \cdot 2H_2O$), and the mixture is pulverized very finely. For every ton of Portland cement shipped, over two and one-half tons of raw materials *and* cement clinker must be ground to the fineness of flour, and in addition one-half ton of coal or equivalent fuel is burned. Since most of the coal so used is powdered, the total weight of materials to be ground per ton of cement is about 3 tons. Furthermore, some 6 million fire-bricks are used annually in the United States for relining cement kilns.

When Portland cement is mixed with water the product sets in a few hours and hardens over a period of weeks. The initial setting is caused by the interaction of water and tricalcium aluminate ($3CaO \cdot Al_2O_3$), present in the cement, accompanied by the separation of gelatinous hydrated product. The later hardening and the development of cohesive strength are due to the interaction of water and tricalcium silicate ($3CaO \cdot SiO_2$), also present in the cement, accompanied by the separation of gelatinous hydrated product. In each case the gelatinous material surrounds and cements together the individual grains. The hydration of dicalcium silicate ($2CaO \cdot SiO_2$), also present in the cement, proceeds still more slowly than that of the above compounds. The ultimate cementing agent is probably gelatinous silica (SiO_2), and it is thought by some that the value of the aluminate lies in its action as a flux in the burning of the clinker.

The analysis of the finished cement made from the above-noted slurry is as follows: calcium oxide 64%, aluminum oxide 5.5%, silicon oxide 21%, ferric oxide 4.5%, magnesium oxide 2.4%, sulfate 1.6%, loss of ignition about 1% (largely water).

Deductions regarding the mechanism of setting and hardening, and the identity of the substances concerned are the results of extensive studies, involving the use of the microscope in the examination of thin sections, on the individual compounds, the clinker, and the resulting concrete. Elaborate researches have also been conducted to determine the best way of incorporating the ingredients of concrete, the nature of the aggregate (sand, gravel, crushed rock) to be used, and the proportions of cement, water, and aggregate, in order that the resulting concrete, really an artificial rock, shall possess the greatest possible strength.

Special applications of cements and lutes are frequently demanded, as for example in floor covering, in tank lining, and in the closure of joints. The difference between a cement and a lute is that the former sets to a rigid solid mass whereas the latter retains some plasticity so that some movement of the lute is possible

ANALYSIS OF MATERIALS USED FOR MANUFACTURE OF LIME AND CEMENT

MATERIAL	FROM	SiO$_2$	Fe$_2$O$_3$ Al$_2$O$_3$	CaO	MgO	CO$_2$	SO$_3$	USED FOR
Limestone.......	Annville, Pa..........	0.36	0.45	54.45	0.54	43.24		Portland cement
Limestone.......	Glens Falls, N.Y......	3.30	1.30	52.15	1.58	40.98		" "
Limestone.......	Mitchell, Ind.........	0.74	0.13	52.94	1.87	43.68		" "
Marl...........	Bronson, Mich.......	1.78	1.21	49.55	1.30	40.35		" "
Cement rock.....	Nazareth, Pa........	13.44	6.60	41.84	1.94	32.94		" "
Cement rock.....	Martin's Creek, Pa....	11.11	6.31	42.51	2.89	36.57		" "
Clay...........	Alpena, Mich.........	61.09	26.97	2.51	0.65	....	1.42	" "
Clay...........	Suisun, Cal..........	58.44	26.50	1.70	1.88	....		" "
Cement rock.....	Rondout, N.Y........	15.37	11.38	25.50	12.35	34.20		Natural cement
Limestone.......	Union Bridge, Md.....	0.89	0.47	54.68	0.32	43.44		Lime
Limestone.......	Woodville, Ohio.......	0.78	0.48	31.15	20.78	45.76		"
Oyster shells.....	Long Island Sound....	3.30	0.25	52.14	0.25	41.61		"

ANALYSIS OF PORTLAND CEMENTS *

WHERE MADE	MADE FROM	SiO$_2$	Fe$_2$O$_3$	Al$_2$O$_3$	CaO	MgO	SO$_3$	Loss
New Jersey.........	Cement rock and	21.82	2.51	8.03	62.19	2.71	1.02	1.05
Pennsylvania........	limestone	21.94	2.37	6.87	60.25	2.78	1.38	3.55
Michigan...........	Marl and clay	22.71	3.54	6.71	62.18	1.12	1.21	1.58
Ohio..............		21.86	2.45	5.91	63.09	1.16	1.59	2.98
Virginia...........	Limestone and clay	21.31	2.81	6.54	63.01	2.71	1.42	2.01
Missouri...........		23.12	2.49	6.18	63.47	0.88	1.34	1.81
Pennsylvania †......		23.56	0.30	5.68	64.12	1.54	1.50	2.92
Illinois.............		22.41	2.51	8.12	62.01	1.68	1.40	1.02
Germany...........	Blast furnace slag	20.48	3.88	7.28	64.03	1.76	2.46	
Belgium............	limestone	23.87	2.27	6.91	64.49	1.04	0.88	
France.............		22.30	3.50	8.50	62.80	0.45	0.70	
England............		19.75	5.01	7.48	61.39	1.28	0.96	
Germany ‡.........	Iron ore and limestone	20.5	11.0	1.5	63.5	1.5	1.0	

* From Meade's "Portland Cement." † White Portland cement. ‡ Sea-water cement.

without cracking. A lute must have support in order that it be retained in position.

A somewhat crude though convenient classification can be made on the basis of the principal ingredients, thus, (1) Portland cement, (2) high alumina cement, (3) sodium silicate, (4) magnesium oxychloride plus copper powder, (5) litharge or red lead plus glycerol, (6) rubber latex, and (7) synthetic resins. Supplementary materials to be considered are asbestos, white lead, plaster of Paris, sulfur, graphite, sand, pitch, tar, rosin, and boiled linseed oil.

The choice to be made depends upon the kind of material to which the cement or lute must adhere; what it must withstand in the way of acid, base, sulfate, or organic liquid; also what temperature is involved; and finally the matter of resistance to vibration and shock.

Portland and high alumina cements do not withstand acids but are resistant to bases. High alumina cement attains its maximum strength more quickly than Portland, and has the extra advantage that it withstands solutions of sulfates.

Sodium silicate cement does not withstand bases, but is resistant to acids except hydrofluoric. This cement sets to a very rigid solid, so that when subjected to mechanical shock or to temperature change it is liable to crack.

A cement containing 90% magnesium oxychloride and 10% copper powder is strong, resistant to abrasion, and can be bonded to Portland cement.

Litharge or red lead plus glycerol is very commonly used, especially for pipe joints.

Rubber latex cement withstands dilute acids and dilute bases, and adheres well to ceramic materials such as stoneware. This cement remains somewhat pliable, thus resisting mechanical shock and temperature change. Organic liquids in general attack this cement.

Synthetic resin cements withstand hydrochloric acid, dilute nitric acid, dilute sulfuric acid, and dilute bases, and are frequently more resistant to organic liquids than is rubber latex cement. The adherence to ceramic materials is good, and the liability to cracking less than for sodium silicate cement.

As for the miscellaneous ingredients mentioned, some are used as fillers or extenders as in the case of sand, and some are used in their own right as when pitch, tar, rosin, molten sulfur, or packed asbestos can be used. (R.K.S.)

CEMENTATION. In geology, cementation is the process of deposition from solution of mineral matter in the interstices of rocks, and is an important factor in the consolidation of coarse-grained elastic rocks such as sandstones and conglomerates or breccias. This action is continually going on in the ground water zone, so much so that the term zone of cementation has come into common use.

Cementation may occur in fissures or other openings of the rocks and in time all such spaces will be closed to further deposition or entrance of ground water. In metallurgy, cementation is the process by which one substance is caused to penetrate and change the character of another, by the action of heat, at temperatures below the melting points. (See **Carburizing, Nitriding, Calorizing, Sherardizing, Chromizing,** and **Siliconizing.**) (E.S.C.S., R.H.H.)

CEMENTED CARBIDE TOOLS AND DIES. Extremely hard and abrasive carbides of tungsten, tantalum, titanium, and columbium cemented together by a tough and ductile binding agent, generally cobalt or nickel.

The product is made by the methods of **powder metallurgy,** using a combination of sintering and pressing and a finish grinding operation. In cutting-tool applications cemented carbide tips are inserted in steel holders by brazing in order to conserve the cutting material and provide a tool of suitable shock resistance.

The original material (Carboloy) consists of tungsten carbides in a matrix of cobalt. The cobalt content may be as low as 3% for the hardest varieties and as high as 20% for applications requiring less abrasive hardness but considerably higher strength and shock-resistance.

The die applications include drawing dies for wire, rods, tubing, and special shapes, and dies for extrusion, forming, and deep drawing operations. Such dies are surpassed in hardness only by diamond dies, which have restricted use because of size limitations and cost. Practically all ferrous and non-ferrous alloys can be worked in cemented carbide dies.

As cutting tools, cemented carbides are particularly suited to machining cast iron, aluminum alloys, babbitt, brasses and bronzes, and many non-metallic materials such as hard rubber, Bakelite, asbestos, porcelain, and Masonite. Many of these materials are not readily machined with hardened steel tools because of their abrasiveness. For cutting steel, tantalum carbide and mixed carbide types are used because they resist chip welding and cratering much better than the straight tungsten carbide types. Heat-treated steels of high hardness which are unmachinable even by **high-speed tool steel** can often be machined by cemented carbide tools.

The carbide tools are relatively brittle and must be very rigidly mounted. Best results are obtained with specially built machine tools with capacity to take deep cuts at high cutting speeds.

Other applications of cemented carbides for wear- and corrosion-resistance include lathe centers, gauges, valve stems and seats, sandblast nozzles, and **Brinell hardness** balls. (R.H.H.)

CEMENTITE. Iron carbide, Fe_3C, a compound which is present in nearly all alloys of iron and carbon such as steel and cast iron. It is very hard and brittle and weakly magnetic. At certain elevated temperatures it is dissolved by the iron, forming a **solid solution,** and in the quenching of steels it is essentially retained in solution, but upon **tempering** (reheating to low temperatures) it is precipitated in a very finely divided form, giving structures having high strength and toughness and known as tempered martensite and sorbite. The cementite present in cast irons and steels having appreciable silicon contents can be decomposed at elevated temperatures to iron and carbon in the form of **graphite.** (See **Malleable Cast Iron.**) (R.H.H.)

CENOGENESIS. The development during **evolution** of **adaptations** which persist only during early stages in the life of the individual, giving way finally to the more primitive adult characters. Thus among the insects the **beetle, fly,** or **butterfly** appears to be the original form of the species and the **larva** is a special adaptation which enables the individual to meet special conditions during its growth without giving up the advantages of its adult form. (A.W.L.)

CENOZOIC. The latest major subdivision of the geologic time-scale. The Era of "recent" life, or age of mammals and modern flora. Subdivided, from the base up, into the following Periods: **Paleocene, Eocene, Oligocene, Miocene, Pliocene, Pleistocene, Recent.** The term Tertiary is applied to the first five periods. The Quaternary begins with the Pleistocene. The Era was characterized by: vanishing of **Archaic Mammals** (Eocene); rise of higher mammals (Oligocene); culmination of mammals (Miocene); transformation of the ape-like ancestor into man (Pliocene); periodic glaciation and rise of man (Pleistocene). The Cenozoic Era began 60–70 million years ago. (R.M.F.)

CENTAURUS. (The centaur.) (Map, page 380.) This is a large and brilliant **constellation** of the southern sky, being invisible to observers in North America or Europe. Centaurus has two bright stars which are frequently spoken of as the "southern pointers" since the line through them passes through the southern cross (the constellation Crux). The brighter one of these two (Alpha Centauri) is not only the third brightest star in the entire sky, but is also the nearest bright star. The distance of this star is 4.3 **light years,** the only star thought to be closer than this to the earth being a faint star known as Proxima Centauri. Alpha Centauri is a **double star** and its brighter component is interesting in that it is almost a duplicate of our **sun** in size, temperature, and other physical characteristics. (W.K.G.)

CENTER DRILL. Counterbore.

CENTER GAUGE. A small gauge used for checking lathe center point angles and threading tool points; also used for setting single-point threading tools on a lathe. (H.C.H.)

CENTER HEAD. Measurement.

CENTER OF BUOYANCY. Hydrostatics.

CENTER OF CURVATURE OF A CURVE. Curvature of a Plane Curve.

CENTER OF DISPLACEMENT. Buoyancy.

CENTER OF GRAVITY. Center of Mass.

CENTER OF MASS. This is a more accurate term for what is commonly referred to as the center of gravity. If we imagine a body divided into infinitesimal particles or elements of **mass,** and if each of these elements is acted upon in the same direction, chosen at random, by a force proportional to its mass, it is easily shown that, whatever the direction of this set of parallel forces, their resultant always passes through a certain point, which is the center of mass of the body. Since the weights of the particles of a small body constitute approximately such a system of forces, it has become customary to call this point the center of gravity, though it is in general not strictly correct to do so.

If the body is given a linear acceleration in any direction without rotation, since the **inertia** of each particle is proportional to its mass and acts in direct opposition to the acceleration, the resultant inertia of the whole body acts in a line through the center of mass; which is therefore properly called also the center of inertia.

If any plane is passed through the center of mass of a body, it divides the body into two parts which have the property that their mass **moments** with respect to the plane are equal but of opposite sign. This means that if the mass of each particle is multiplied by its distance from the plane, and the products added, the sum is numerically the same for both parts of the body. This principle may be put into mathematical form and the position of the center of mass calculated therefrom. It may be shown, for example, that the center of mass of a homogeneous right circular cone is on its axis at a distance from the apex equal to ⅔ of the altitude. For a continuous body of homogeneous material, the center of mass coincides with the **centroid.** (L.D.W.)

CENTER OF PRESSURE, HYDROSTATIC. The point of application of the resultant of all the pressure forces acting upon an exposed area is called the center of pressure. Water and air create pressures which are of importance in phases of engineering. The center of pressure on a horizontal plane immersed in water to a depth h is the center of the area, that is, the point corresponding to the center of mass of a flat, uniform plate coinciding with the area.

More generally, if a plane surface is completely immersed in a liquid at an angle θ with the horizontal, the center of pressure upon that surface lies at a depth below the top of the liquid, given by the equation $d = \sin \theta \cdot I/M$; in which I is the areal **moment of inertia** of the plane figure with respect to the line of intersection of its plane with the plane of the top of the liquid, and M is the areal **moment** of the figure with respect to the same intersection. For a vertical surface ($\theta = 90°$), $d = I/M$. Thus, in the case of a vertical rectangle whose upper and lower edges are horizontal and at depths h_1 and h_2, and whose width is b, we have $I = b/3 (h_2{}^3 - h_1{}^3)$, while $M = b/2(h_2{}^2 - h_1{}^2)$; hence the depth of the center of pressure is

$$= \frac{2}{3} \frac{h_1{}^2 + h_1 h_2 + h_2{}^2}{h_1 + h_2}.$$

In particular, if the upper edge is at the top of the liquid, and if the altitude of the rectangle is a, so that $h_1 = 0$ and $h_2 = a$, $d = \frac{2}{3}a$. A rectangular dam 12′ high, and completely filled with water, would thus have a center of pressure at a depth of 8′.

Water which is assumed by engineers to weigh 62.5 lbs. per cu. ft. exerts a horizontal pressure varying uniformly from zero at the water surface to $62.5h$ at a depth h. The total horizontal water pressure acting on the vertical rectangle mentioned above at $\frac{2}{3}a$ below the water surface is $31.25h^2b$ lbs. in which b is the other principal dimension or the vertical width of the rectangle. (See **Hydrostatics**.)   (F.T.M., L.D.W.)

CENTER PUNCH. A hand tool with a sharp conical point, used for layout work and locating centers for drilling.   (H.C.H.)

CENTERLESS GRINDER. Grinding.

CENTIGRADE SCALE. Temperature Scales.

CENTIMETER. C.G.S. System; Electric and Magnetic Units.

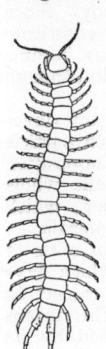

CENTIPEDE. Arthropoda, Chilopoda. Worm-like animals with segmented bodies, a distinct head, and a pair of jointed legs on most segments of the body. They have a pair of **antennae** and the first pair of legs are modified as poison claws for the capture of prey.

The centipedes differ from the nearly related millipedes (**Diplopoda**) in the presence of poison claws, in having only one pair of legs to a segment, and in the more flattened body. Each segment has a dorsal and a ventral plate connected by softer tissue and has the legs joined to the sides of the body. (A.W.L.)

Centipede.   **CENTRAL CONICS.** Conic Sections.

CENTRAL LIMIT THEOREM. The distribution of sample **means** usually approximates a **normal probability function.** The approximation becomes more accurate as the size of the **sample** increases. This is particularly remarkable in that the shape of the distribution of the **population** from which the sample was drawn has little influence over the shape of the distribution of sample means. The necessary and sufficient condition that sample means be distributed normally constitutes the central limit theorem of probability. We shall state a sufficient condition proved by Liapounoff which is extremely general. Let $x_1, x_2, \cdots, x_n$ be independent variables with means zero, possessing absolute third **moments.** $\mu_3{}^{(1)}, \mu_3{}^{(2)}, \cdots$

$\mu_3{}^{(n)}$, respectively. (An absolute third moment is defined as the expected value $|x|^3$.) Let $\sigma^2 = \sigma_{x_1}{}^2 + \sigma_{x_2}{}^2 + \cdots + \sigma_{x_n}{}^2$, $\sigma_{x_i}{}^2$ being the **variance** of x_1 and $\alpha_3 = \sum\limits_{i=1}^{n} \mu_3{}^{(i)}/\sigma^2$. If $\lim\limits_{n \to \infty} \alpha_3$ approaches zero, then the distribution of sample means, $\overline{X}$, approaches a normal distribution. A certain bell-shaped symmetrical probability function called Cauchy's distribution furnishes an example in which the conditions of the central limit theorem are not satisfied, since the Cauchy distribution possesses no mean. The distribution of sample means from a Cauchy distribution reproduces exactly the same Cauchy distribution which does not approach normality no matter how large the size of the sample becomes.

The central limit theorem has many applications. Recently a remainder term has been obtained providing rather accurate estimates of the closeness of the approximation. Similar theorems may be proved for the sample variance when each item is drawn from the same population.   (L.A.A.)

CENTRAL LOAD. Load.

CENTRAL OFFICE. Telephony.

CENTRAL STATION. A central station is a place where electrical energy is produced from an original source such as **fuel** or water power, and from which the energy is sent out over a network of transmission lines radiating from the station to the load served. The central station is illustrative of the fact that large-scale production is the most economical way of providing electrical energy to the individual consumer, for energy thus produced may usually be sold to the individual for less than he could generate it himself, in spite of the fact that there will be transmission losses on the line between the central station and the customer.

A typical steam central station will consist primarily of a **boiler, turbine,** and **condenser,** the boiler to produce steam, the turbine to utilize the steam for the production of mechanical power, and the condenser to receive the exhaust steam and reduce it to liquid for returning to the boiler, and at the same time provide a vacuum for the more economical operation of the turbine. A **generator** directly connected to the turbine effects conversion of mechanical into electrical energy with a minimum of loss. The boiler-turbine-condenser group must be serviced by auxiliary equipment which might, for clarification, be grouped under two heads,—that auxiliary equipment which might be said to be connected with the flow of air or flue gas in the combustion portion of the plant, and that having to do with the proper conditioning of the condensate produced by the condenser, so that it possesses the requisite heat, pressure, and purity for boiler feed. Included in the first of these groups will be such equipment as **draft fans, stokers, air preheaters,** etc., while in the second group there would be found **feedwater heaters** and **pumps,** feedwater treatment, etc. Between the generator or generators and the outgoing transmission lines, there is contained in the central station no inconsiderable amount of electrical equipment needed for the control and electrical protection of the generators, the supply of auxiliary power, and switching equipment. The accompanying figure is typical of the type of station just described. It shows a boiler room containing the fan and combustion equipment as well as the boiler, a turbine room, which contains little other than the turbines with their generators, but below which will be a basement well filled with condenser and feedwater auxiliaries. Electrical galleries and control rooms on the opposite side of the turbine room from the boiler complete the picture as far as the central station itself is concerned. However, the switching and control as done in the plant are frequently the remote control of high voltage switching

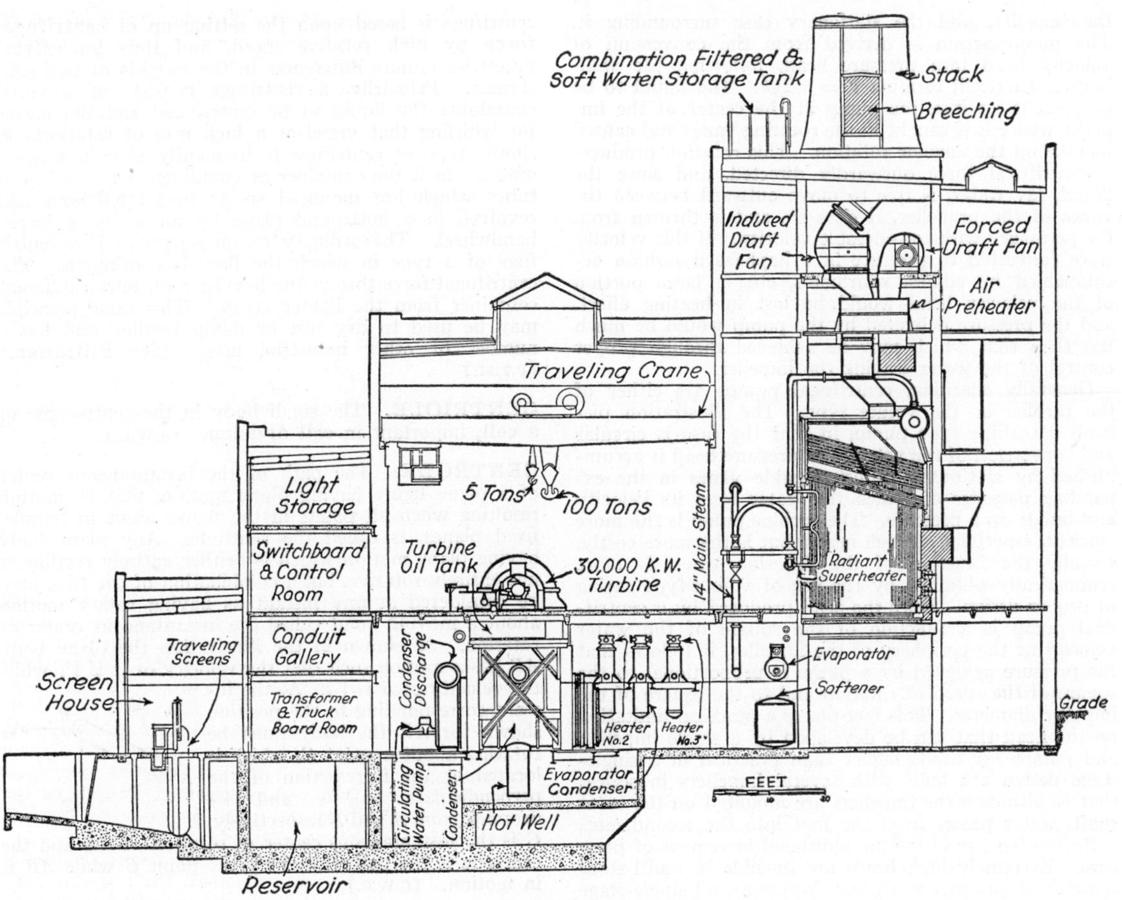

Combination Filtered &
Soft Water Storage Tank

Stack

Breeching

Induced
Draft
Fan

Forced
Draft Fan

Air
Preheater

Traveling Crane

5 Tons

100 Tons

Light
Storage

Switchboard
& Control
Room

Turbine
Oil Tank

30,000 K.W.
Turbine

14" Main Steam

Radiant
Superheater

Traveling
Screens

Conduit
Gallery

Condenser
Discharge

Evaporator

Softener

Screen
House

Transformer
& Truck
Board Room

Heater
No.2

Heater
No.3

Grade

Circulating
Water Pump

Condenser

Hot Well

Evaporator
Condenser

FEET

Reservoir

Central station.

equipment which is located in an outdoor type **substation** usually placed adjacent to the plant. (F.T.M.)

CENTRECHINOIDEA. Echinodermata, Echinoidea. An order of sea urchins with a central mouth about which are gills. The commoner sea urchins of North America belong in this order. (A.W.L.)

CENTRIFUGAL CASTING. Foundry Practice.

CENTRIFUGAL FORCE. A manifestation of the **inertia** of a body moving in a curved path, the effect being that of a **force** directed radially toward the convex side of the curve. The stresses developed in a rotating rigid body which may cause it to fly asunder, the sensations experienced upon rounding a corner in an automobile, and the action of cream separators, centrifuges, and centrifugal driers, are well known examples.

When a particle moves in a circle of radius r with uniform speed V, it is subject to an **acceleration** toward the center equal to V^2/r. This can be produced only by the application of a **force** in that direction, whose value in absolute units is mV^2/r, m being the mass of the particle. The particle may therefore be regarded as pulling or pushing outward against its constraint with a force of this magnitude. In the case of a rigid body of finite size, it may be shown that the centrifugal force is the same as if the body were concentrated at its **center of mass**.

The centrifugal force involved in the motion of a vehicle around a curve makes desirable the "superelevation" or "banking" of the roadway at an angle dependent upon the curvature and the speed, in order that the wheels of the vehicle may push perpendicu-

larly against the pavement or track. If the speed is V and the radius of the curve is r, then the roadbed should be inclined at an angle s given by the formula: $\tan s = V^2/gr$, where g is the acceleration of gravity. For example, if $V = 40$ ft. per sec. and $r = 1000$ ft., the value of g being 32.15 ft. per sec. per sec., the formula gives $s = 2° 51'$; which on a standard-gauge railroad track would require the outer rail to be 2.8″ above the inner. (L.D.W.)

CENTRIFUGAL PUMP. Although this type of **pump** is now widely used, it is a comparatively new development. The centrifugal pump operates at higher speeds than were easily obtainable before the day of the steam turbine or the electric motor, but with the advent of these high-speed devices, high efficiencies were made possible in the centrifugal pump field, and its advantages, insofar as size, simplicity, and quietness of operation are concerned, have emphasized the possibilities of this type of pump. The figure shows that the essential parts of a centrifugal pump are a rotating member **called**

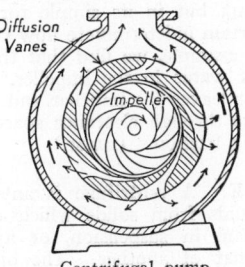

Diffusion
Vanes

Impeller

Centrifugal pump,

the impeller, and the stationary case surrounding it. The pump action is derived from the conversion of velocity head into pressure head, following the well-known Torricelli formula $V = \sqrt{2gh}$. The liquid to be pumped is let into the pump at the center of the impeller where it is caught in the rotating vanes and caused to take up the angular rotation. This rotation produces a centrifugal force outwardly directed, and since the liquid, say water, is free to move outward between the vanes of the propeller, it does so, and is thrown from the periphery with considerable velocity. If this velocity were converted to pressure in a haphazard fashion accompanied by eddies, whirlpools, etc., a large portion of the velocity head would be lost in heating effect, and the pressure delivered by the pump would be much less than that which may be achieved through proper control of the water leaving the impeller.

Generally speaking, centrifugal pumps are either of the turbine or the volute type. The illustration pictures a turbine type pump, in that the case is circular and conversion of the velocity to pressure head is accomplished by stationary diffusing guide vanes in the expanding passages of which the water loses its velocity and builds up a pressure. The turbine pump is the more efficient, especially on high heads, but is the more costly. Usually the function of the diffusion vanes is more economically obtained by the use of volute-type casing of proper design. Since the head produced by a centrifugal pump is a function of the square of the water velocity at the periphery of the impeller, it follows that the pressure achieved by a pump is proportional to the square of the speed of rotation and to the square of the impeller diameter. This fact places a practical limitation on the head that can be developed by a single impeller, and pumps for heads higher than practical in a single-stage design are built with several impellers in series; that is, although the impellers are mounted on the same shaft, water passes from the first into the second, etc., each impeller producing an additional increment of pressure. Extremely high heads are possible in multi-stage centrifugal pumps. With few exceptions, all single-stage pumps are the volute type, but builders of high-pressure pumps frequently employ diffusers or guide vanes.

Standard lines of pumps are constructed with impellers of different diameters, and a standard casing. Different capacities are obtained either by changing the impeller diameter or the speed of rotation. At constant speed the head is proportional to the square of the impeller diameter, the capacity is proportional to the impeller diameter, and the power required is proportional to the cube of the impeller diameter. A pump using a given impeller, but having speed variable, will show that the head is proportional to the square of the speed, that the capacity is proportional to the speed, and that the power is proportional to the cube of the speed. The efficiencies of centrifugal pumps vary with the size of the pump and the number of stages; that is, the head pumped against. Very small sizes may have efficiencies as low as 25%. However, for most normal installations the efficiency for small pumps should not be less than 50–60% and for large pumps against a low head may be as much as 80–90%. Modern centrifugal pumps are considerably more efficient than those of twenty or more years ago. A workable theory has been derived for the centrifugal pump, but in its simple form it is incomplete, due to certain indeterminate flow conditions in the propeller. For example, we find that there are friction and eddying of water in the impeller, and shock loss where the water leaves the impeller and enters the case. There is leakage through clearance spaces, and there is disk and bearing friction. (F.T.M.)

CENTRIFUGE. A centrifuge is any machine used to separate liquids from solids which might normally be held by them in suspension, or a liquid of one density from that of another. The operation of the centrifuge is based upon the setting up of **centrifugal force** by high rotative speed, and thus intensifying otherwise minute differences in the weights of two substances. Primarily, a centrifuge consists of a vessel containing the liquid to be centrifuged and the means for whirling that vessel at a high rate of rotation. A simple type of centrifuge is frequently seen in medical offices. In it the container or containers are simply test tubes which are mounted so as to be balanced and revolved in a horizontal plane by means of a geared handwheel. The ordinary cream separator is a centrifuge of a type in which the flow is continuous. The centrifugal force throws the heavier milk into a different container from the lighter cream. This same principle may be used to dry wet or damp textiles, and has a number of other industrial uses. (See **Filtration**.) (F.T.M.)

CENTRIOLE. The small body in the centrosome of a **cell**, important in **cell division**. (A.W.L.)

CENTRODE. The path of the instantaneous center of a plane figure having plane motion, that is, motion resulting when all points in the figure move in parallel fixed planes, is called the centrode. Any plane body having plane motion which is neither entirely rectilinear nor entirely rotative, but a combination of the two, may be considered at any instant as having rotary motion about a moving point called the instantaneous center of rotation. As shown in the illustration the plane body AB has a motion such that the velocity of A is V_1 while the velocity of B is V_2. At the instant corresponding to the position shown for AB the body must be rotating about a point C, which is located at the intersection of the perpendiculars to V_1 and V_2 dropped from A and B respectively. C is the instantaneous center of rotation of AB and the centrode is the path traced by the point C while AB is in motion. (C.W.C.)

CENTRODORSAL. The central or basal plate (ossicle) of a free-swimming **crinoid**. (A.W.L.)

CENTROID. The centroid of a given geometrical figure (curve arc, portion of a plane or curved surface, or solid) is the point whose **coordinates** are the **mean values** of the coordinates of the points of the given figure; it is independent of the choice of axes. The centroid of a geometrical figure corresponds to the center of gravity (or center of mass) of a material body of similar form.

For a plane curve arc, the centroid is given by formulae:

$$\bar{x} = \frac{\int_a^b x\,ds}{L}, \quad \bar{y} = \frac{\int_a^b y\,ds}{L},$$

$$ds = \sqrt{1 + \left(\frac{dy}{dx}\right)^2} \cdot dx = \sqrt{1 + \left(\frac{dx}{dy}\right)^2} \cdot dy,$$

where L is the length of the arc.

For a plane area, the centroid is given by:

$$\bar{x} = \frac{\iint_A x\,dS}{A}, \quad \bar{y} = \frac{\iint_A y\,dS}{A},$$

where A is the area of the given region.

For a solid, the centroid is given by:

$$\bar{x} = \frac{\iiint_V x\,dV}{V}, \quad \bar{y} = \frac{\iiint_V y\,dV}{V}, \quad \bar{z} = \frac{\iiint_V z\,dV}{V},$$

where V is the volume of the given region. (L.L.S.)

CENTROSOME. A small body in the cell. (A.W.L.)

CENTROSPHERE. A specialized structure in the cell. (A.W.L.)

CENTRUM. Earthquakes.

CENTURY PLANT. Agave americana, the Maguey or American aloe, is often called the century plant. It has a very short thick stem which bears a rosette of thick fleshy leaves, in which are stored food reserves. Each year the plant adds three or four leaves to the rosette. After a period varying from 5 to many years (perhaps a century) vegetative growth ceases and flowering occurs. A gigantic terminal flower bud is formed and grows rapidly. It bears many flowers. After the fruit has formed the whole plant dies. It is this habit of bearing flowers only after a prolonged period of vegetative growth that has given this plant the common name, century plant. It is widely grown in cultivation, and seldom flowers under such conditions. The plant propagates vegetatively by means of suckers which grow out from the base of the stem. (R.M.W.)

CEPHALASPIS. Fossil Fishes.

CEPHALIC GLAND. A structure lying in the anterior end of nemertine worms. (A.W.L.)

CEPHALOCHORDATA. Chordata. A subphylum containing a number of species of small marine animals called lancelets. From one of the included genera the name Amphioxus has become common in laboratories to designate any lancelet, although the animals so designated more commonly belong to the genus *Branchiostoma*.

The lancelets differ from the other groups of lower **chordates** in their fish-like form. They are small, usually from 1½ to rather more than 2″ long, and taper toward both ends; the body is laterally compressed. Like other members of the phylum they have a dorsal nerve cord, below it a stiffening longitudinal rod, the **notochord,** and a **pharynx** perforated by many slits. A depression called the vestibule leads to the mouth. It is surrounded by a circlet of slender **cirri. Cilia** about the mouth carry food particles into the alimentary tract where they are caught by the mucus in a ventral groove, the endostyle, carried forward, upward, and then back to the intestine, all by ciliary action.

The lancelets are important to the evolutionist in several ways. Since they swim actively in the tides and bury themselves in the sand at other times, taking their food like sessile animals, they illustrate possible ancestral conditions for the tunicates on the one hand and the fishes on the other. The peculiar food-concentrating mechanism described above is also important; it occurs in larval lampreys and provides one of the few well-marked evidences of relationship between the vertebrates and lower chordates.

Although of great scientific interest the lancelets have been of little economic importance. They are caught in large numbers for food by the Chinese. (A.W.L.)

CEPHALONT. An attached form of certain parasitic 1-celled animals (**Protozoa, Sporozoa**). (A.W.L.)

CEPHALOPODA. A class of the phylum **Mollusca** containing the **squids, cuttlefish, octopus** or devilfish, and **nautilus.** The class is relatively small but it includes the most highly developed of unsegmented invertebrates and is of some economic importance. All of the species are marine.

These mollusks are distinguished by the following characteristics: 1. The head bears a pair of large eyes which are of the camera type like those of man although they differ fundamentally in structure. 2. The mouth is surrounded by a group of tentacles provided with many cup-like suckers and sometimes with hooks. 3. The **mantle** cavity opens to the exterior by a broad entrance and by a slender tube, the funnel or siphon; by forcing jets of water from the cavity through the siphon the animal propels itself through the water in the opposite direction. 4. The foot is modified to form the siphon and possibly the tentacles. 5. The mouth is provided with a sharp beak superficially like that of a bird of prey. 6. Some species have an ink-sac from which a dark fluid is discharged for concealment. 7. The shell is internal or absent in most species.

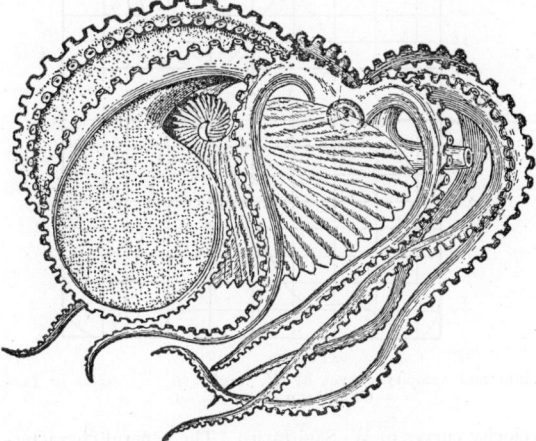

Female *Argonauta argo.* (*Lull, Organic Evolution, after Claus-Sedgwick. Macmillan & Co., Ltd.*)

Squids and cuttlefish are eaten by some peoples. The latter species furnishes the cuttlebone so familiar in bird cages and the true sepia pigment. Squids are of importance in North America chiefly as bait in the fisheries of the Atlantic Coast. The octopus has been credited with amazing feats of destruction at sea which are probably entirely imaginary. Giant squids are known to occur, with a total length of 50′, and are more likely subjects of these tales, but there is no satisfactory evidence that even these creatures are greatly to be feared by man.

The class is divided into two orders. The first was abundant in the past but is now almost extinct.

Order Tetrabranchiata. Shell external, spiral, and divided into a series of chambers of increasing size, in the last of which the animal lives. Once abundant, now limited to four known species in the genus *Nautilus.*

Order Dibranchiata. Shell internal or lacking and usually straight. The **squids** and **cuttlefishes** have ten tentacles, the **octopus** or devilfish and the **argonaut** or paper nautilus only eight. (See also **Invertebrate Paleontology.**) (A.W.L.)

CEPHALOTHORAX. The anterior body region of **spiders, crustaceans** and some other **arthropods,** consisting of the compactly associated head and thorax. (A.W.L.)

CEPHEID. Examination of the curve in the article on **variable stars,** showing the number of variable stars of different periods, indicates that there are two groups of periodic variables with periods less than 50 days. The term Cepheid, derived from the typical short period variable δ **Cephei,** is applied to all variables with periods under 50 days. The group with periods less than one day are frequently referred to as cluster-type variables, because of the fact that they are most numerous in globular star **clusters,** while for purposes of differentiation Cepheids with periods greater than 5 days are referred to as "typical Cepheids."

The mean **light curve** of a typical Cepheid, together with the mean **velocity curve** plotted on the same time scale, is shown in the accompanying figure. (Light and

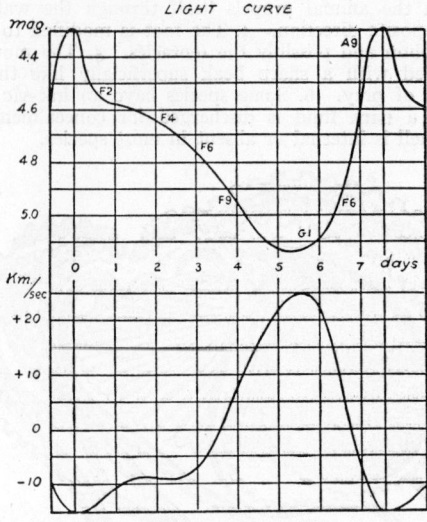

LIGHT CURVE

VELOCITY CURVE

Light and velocity curves of W. Sagittarii. (*Curtiss in Lick Observatory Bulletins.*)

velocity curves of W. Sagittarii.) The general characteristics of the curve are typical for all Cepheids with the rapid rise to maximum followed by a more gradual decline to minimum. The descending portion of the curve is characterized in typical Cepheids by certain irregularities.

With the change in brightness of a Cepheid there is also a change in **spectral type,** the stars being somewhat bluer at maximum as shown by the letters along the light curve. Also correlated with change in brightness of the typical Cepheid is a change in **radial velocity,** with the star approaching the observer at maximum and receding at minimum.

The cause of the variability of the Cepheids has been a mystery ever since the study of this type of variables was begun. For many years it was believed, because of the variations in radial velocity, that the objects were **spectroscopic binaries.** Shapley, Eddington, and others have conclusively shown that this cannot be the true explanation, and have proposed an alternative hypothesis that the stars are pulsating. On the basis of this theory the star when approaching its minimum is relatively cool and is contracting under the influence of gravitational attraction. As the mass of gas contracts the increased pressure in the interior produces a rise of temperature and eventually outward pressure due to high temperature overbalances the gravitational force and the star suddenly expands. The sudden expansion produces cooling until gravitation overbalances the outward force and contraction takes place again. Since at the time of expansion, the outer surface will be approaching the observer and also the temperature is higher than during the contracting stage, we find the correlations between brightness, spectral type, and radial velocity all accounted for on this theory. Many details remain to be accounted for before we are entitled to say that the problem of Cepheid variation is completely solved, but the pulsation hypothesis certainly seems to be a step toward the final solution.

In the course of the study of typical Cepheids in the Small Magellanic cloud, Miss Leavitt, of the Harvard College Observatory, found a direct correlation between the average **magnitude** of the Cepheids and the period of variation. Since all variables in the cloud are at approximately the same distance from the earth, it

appeared that the period of variation was directly correlated with intrinsic brightness or, in other words, with average **absolute magnitude.** Shapley, after an exhaustive study of numerous Cepheids with known distances and hence known average absolute magnitudes, was able to show that the correlation between period and luminosity was apparently characteristic of all Cepheids and, by plotting period against average absolute magnitude, obtained what is commonly known as the period-luminosity curve. After this curve is established, it is possible to determine the absolute magnitude of any Cepheid, no matter how distant, by simply determining the period. With the absolute magnitude and the apparent magnitude known, the calculation of the distance of the Cepheid is a relatively easy task. The period-luminosity relationship has been extended to the cluster type Cepheids and, while there is still some question as to the "zero point" of the various curves, it has served a very valuable purpose in determining distances of globular clusters. Cepheid variables are also found in distant extragalactic **nebulae** and hence they may be used to determine distances of objects outside of our own **galactic system.** For this reason the Cepheids have been referred to as "the measuring rods of the universe." The explanation of the period-luminosity relation has not yet been found, but it may very probably be connected with the **mass-luminosity** relationship found by Eddington and others for all stars. (w.k.g.)

CEPHEUS. (Map, page 380.) This **constellation** is particularly interesting because it contains the remarkable **variable star** Delta Cephei. This star is the typical **Cepheid** variable, a class of stars which is most valuable in providing a measuring rod for probing the remote regions of space. (w.k.g.)

CERAMICS. The requisite raw material for making ceramic materials is clay. Clay is essentially an **aluminosilicic acid** ($H_2Al_2Si_2O_9$), containing more or less foreign matter such as (1) **ferric** oxide (Fe_2O_3), which contributes the reddish color frequently associated with clay, (2) **silica** (SiO_2) as sand, (3) **calcium** carbonate ($CaCO_3$) as **limestone.** Since clay is formed by the decomposition of igneous rocks, followed by transportation of the fine particles by running water and later deposition of these particles by sedimentation when the flow of water diminishes in speed, the quality of clays shows a wide range. When clay is wet it is plastic and can be shaped according to the desire and skill of the operator. The shape is retained on drying, and subsequent heating produces a coherent, hard mass, which suffers in the process more or less shrinkage and deformation depending upon the composition of the raw materials, and the method and temperature of treatment. Common bricks are made of crude materials without careful regulation of the conditions of treatment.

Bricks and plain clay products possess an earthy surface and fracture, are porous, and the strength depends upon the materials and treatment. Porcelain, on the other hand, possesses a glass-like or vitreous surface and fracture, and is not porous. Porcelain is made by mixing with the clay some powdered feldspar mineral potassium aluminosilicate ($KAlSi_3O_8$ approximately). At the temperature of firing, feldspar undergoes a gradual change from the crystalline to the glassy state, the rate depending upon the time of heating and the temperature to which it is subjected. The fusion point of **feldspar** is of the order of 1300° C., whereas that of **kaolin** (pure clay) is of the order of 1700° C. Subjection of the porcelain raw material to the latter temperature would result in the formation of a glass. But when the temperature used is below the melting point of the clay portion and about the melting point of the feldspar, the latter produces a glass cement which binds together the particles of the former. When ground **quartz** (SiO_2) is added to the original clay mixture the shrinkage of the material in the processes of drying and firing is reduced, the resistance to deformation during firing is increased, and the temperature coefficient of expansion of the product is affected.

The range of clay, feldspar, and quartz, as to the ratios in the mixture and as to individual composition of each (see tables 1 and 2), as well as the available range of temperature of firing makes possible the production of products of a wide variety of physical structure. There has been proposed an arbitrary line of demarcation, namely that the unglazed product, such as has been described, which absorbs not more than 1% of its weight upon and after immersion in water, shall be termed porcelain, otherwise it shall be called earthenware. Such a non-porous material as porcelain which includes chinaware is also distinctly translucent in thicknesses of a few millimeters, whilst earthenware is non-translucent and somewhat porous.

Materials that are to be glazed are dipped in a slip (the mixture of raw materials and water), dried, and refired. The glaze mixture is made up so that its fusion temperature is lower than that of the body of the ware, and the firing temperature is such that a surface of glass is formed over the body of the ware.

Designs and colors may be placed, as is commonly done, on the glaze and refired, or, as less commonly and more recently with fine effect, directly on the body under the glaze, in which case the glaze when produced covers and protects both the body of the ware and the decoration.

The range of ceramic products is great, as is illustrated by the following materials, namely, common bricks, hollow bricks, pipes, tiles, terra cotta, fire-bricks resistant to high temperatures, stoneware, crockery, earthenware, porcelain, chinaware. Fire-bricks are in-

TABLE 1

RAW MATERIALS	CHINAWARE	EARTHEN-WARE
Total clay, usually blended..	46.5%	50.5%
Feldspar....................	15	13.5
Quartz.....................	36	36
Dolomite..................	2.5	....
Porosity average..........	0.5	8

TABLE 2
TABLE SHOWING ANALYSES OF CERTAIN CERAMIC RAW MATERIALS

	SILICA	ALUMINA	LIME	MAGNESIA	IRON OXIDES	SODA	POTASH
	%	%	%	%	%	%	%
Kaolin...........	58	29	0.2	0.3	1	1	1
Fire clay........	61	26	0.3	0.4	1	1	1
Common brick clay	58	14	7	1.5	4	3	3
Feldspar........	71	16	0.3	0.0	0.5	4	7
Quartz...........	100	....	....	....	....	....	....

dispensable in high-temperature industries that use electric furnaces, blast furnaces, reverberatory furnaces, converters, that produce ceramics, Portland cement, glass, and in the high-temperature furnaces for producing steam and for subjecting coal to destructive distillation. (R.K.S.)

CERARGYRITE (Horn Silver). The mineral cerargyrite, silver chloride, AgCl, crystallizes in the iso-

metric system but is usually massive appearing like wax or horn, hence the name horn silver. It has no cleavage, is highly sectile, yielding bright surfaces; hardness, 1–1.5; specific gravity, 5.55; luster, resinous to adamantine; color, gray, white to colorless, may be blue, violet-brown after exposure to light; transparent to translucent. Cerargyrite is largely a secondary mineral, usually associated with other silver minerals as well as with compounds of **lead, zinc,** and **copper.** Saxony and the Harz Mountains are European localities. The Broken Hill district of New South Wales is a well-known occurrence, but probably the most important deposits are found in Atacama, Chile. It is found also in Bolivia and Mexico. In the United States cerargyrite comes from Colorado, Idaho, Utah, Nevada, Arizona and New Mexico. The name cerargyrite is derived from the Greek meaning horn and silver. (E.S.C.S.)

CERATA. Projecting respiratory organs found on the upper surface of the body in some of the marine **snails.** (A.W.L.)

CERATITE. Invertebrate Paleontology.

CERCARIA. A larval form of the **flukes** (parasitic flatworms) with a compact body and slender tail (A.W.L.)

CERCOPOD. Slender projections at the posterior end of the body in certain crustaceans (Phyllopoda). (A.W.L.)

CERCUS, CERCI ANALES. Paired jointed appendages at the posterior end of the body in some **insects.** (A.W.L.)

CEREALS. Grass Family.

CEREBELLUM. A division of the vertebrate **brain.** (A.W.L.)

CEREBRAL GANGLION. 1. The simple brain of many invertebrates. 2. Either member of the anterior pair of **ganglia** in the molluscan nervous system. 3. The upper posterior component of the brain of **cephalopod mollusks.** (A.W.L.)

CEREBRAL TUBE. A ciliated tube associated with the brain in certain marine worms (Sipunculids). (A.W.L.)

CEREBROSPINAL FLUID. The fluid which circulates in the ventricles of the brain and in the subarachnoid spaces of the brain and spinal cord. (See **Brain.**) Its function is the mechanical protection of the brain and cord; possibly it takes part in metabolic activities, although there is no proof of this. In appearance, the fluid is normally water clear; it contains sodium chloride in slightly higher amounts than in blood; its glucose content is approximately half that of blood, and protein is present only in minute quantities. There are 0 to 8 cells per cu. mm., all lymphocytes. Lumbar puncture, the insertion of a needle between the lumbar vertebrae and the withdrawal of a small amount of fluid for examination, is a safe and valuable diagnostic procedure in diseases of the central nervous system. In **meningitis,** the fluid is cultured, and examined directly for **bacteria.** Alterations in cell count, pressure, and chemistry are helpful in the differential diagnosis of tumors as well as infections. (D.M.H.)

CEREBRUM. A region of the vertebrate **brain.** (A.W.L.)

CERES. Asteroid.

CERIUM. Symbol: Ce. Atomic number: 58. Atomic weight: 140.13. Density: 6.8. Melting point: 640° C.

Types of compounds: Ce_2O_3, white cerous oxide; CeO_2, pale yellow ceric oxide. Color of salts: cerous, colorless; ceric, yellow. Discovered by Berzelius and Hisinger in 1803. (Isotopes: page 290.)

Cerium occurs in **monazite** sand, a cerium phosphate (50%–75% ceria), **cerite**, a basic silicate of **calcium** and **iron** (50%–70%), and **allanite**, a hydrated silicate (up to 50%).

The cerium sub-group of the rare earth metals (see **Yttrium**) consists of the elements scandium, lanthanum, cerium, praseodymium, neodymium, illinium, samarium, europium, and gadolinium, all of whose potassium sulfate compounds are relatively insoluble in water. Cerium is the most abundant of the rare earth metals, and each member of the cerium sub-group, except samarium, is said to be more abundant than yttrium. Cerium metal is prepared by electrolysis of the fused chloride, and is used as a pyrophoric alloy (70% Ce with iron) in gas and tobacco lighters.

Nitrates: Cerous nitrate ($Ce(NO_3)_3 \cdot 6H_2O$), pink to colorless crystals, soluble; ceric nitrate ($Ce(NO_3)_4$), reddish-yellow crystals, soluble.

Oxide: Cerous oxide (Ce_2O_3), grayish-green powder; ceric oxide (CeO_2), pale yellow powder, soluble in sulfuric acid, used in gas mantles (1% CeO_2 with thorium oxide), and in coloring glass and ceramic ware.

Sulfates: Cerous sulfate ($Ce_2(SO_4)_3$), green to pink, depending upon the amount of water of crystallization; ceric sulfate ($Ce(SO_4)_2$), yellow to orange-red solid, soluble, used in **sulfuric acid** solution as a strong oxidizing agent. (R.K.S.)

CERUMEN. The waxy secretion that collects in the external **ear**. (R.S.M.)

CERUSSITE. The mineral cerussite, **lead** carbonate, $PbCO_3$, is **orthorhombic** with tabular, prismatic and pyramidal crystals, with twinned forms very common. If not in crystal aggregates it may occur in granular or compact masses. Cerussite is very brittle with a conchoidal fracture; hardness, 3–3.5; specific gravity, 6.46–6.57 (a heavy mineral); luster, adamantine but may be vitreous to resinous, pearly or even submetallic. Its color is variable, white to gray, grayish-black or blue or green, transparent to translucent.

Cerussite is of secondary origin being found associated with other lead minerals, and is widely distributed. There are many European and American localities. Fine crystals have been obtained from Phoenixville, Pennsylvania; Joplin, Missouri; Leadville, Colorado; Pima County, Arizona, and Dona Ana County, New Mexico. It is an ore of lead, and frequently carries values of **silver**. Derived from the Latin *cerussa,* white lead. (E.S.C.S.)

CERVALCES. Pleistocene.

CERVICAL GROOVE. A groove marking the boundary between head and thorax in the **crustaceans**. (A.W.L.)

CERVIX. Any narrow or neck-line portion of an organ. The term is usually used in reference to the narrow end of the **uterus** that projects into the **vagina**. (R.S.M., D.M.H.)

CESIUM. Symbol: Cs. Atomic number: 55. Atomic weight: 132.91. Density: 1.87. Melting point: 28.5° C. Boiling point: 670° C. No isotope, but of single atomic form: 133.

Cesium is a silver-white, very soft metal, possibly the softest of all the metals; tarnishes instantly on exposure to air, soon igniting spontaneously with flame to form oxide; preserved under kerosene; reacts vigorously with water forming cesium hydroxide solution and hydrogen gas. Discovered by Bunsen and Kirchhoff in 1860 by means of the spectroscope.

Cesium occurs in **pollucite** (cesium aluminosilicate, 35% Cs_2O), **lepidolite** (lithium aluminosilicate), and traces in certain mineral waters. The localities having the highest known concentrations of cesium are in Maine, Black Hills, South Dakota, and the island of Elba. Cesium salts may be recovered from the mother liquor upon crystallization of lithium salts, but are separable from rubidium salts with great difficulty. Cesium metal is obtained (1) by **electrolysis** of fused cesium **barium** cyanide mixture out of contact with air, (2) by distillation of cesium chloride with **calcium** metal.

Chloride: Cesium chloride (CsCl), white deliquescent solid, melting point 646° C., soluble.

Hydroxide: Cesium hydroxide (CsOH), white, deliquescent solid, melting point 272° C., soluble.

Oxide: Cesium oxide (Cs_2O), red solid, by heating cesium metal in oxygen or dry air, reactive with water to form soluble cesium hydroxide; several higher and lower oxides have been reported.

Other soluble salts: Cesium sulfate (Cs_2SO_4); cesium nitrate ($CsNO_3$); cesium carbonate (Cs_2CO_3); cesium fluosilicate (Cs_2SiF_6).

Slightly soluble salts: Cesium perchlorate ($CsClO_4$), insoluble in alcohol; cesium chloroplatinate (Cs_3PtCl_6); cesium permanganate ($CsMnO_4$).

Volatile cesium salts, such as the chloride, color the bunsen flame violet. (R.K.S.)

CESTIDA. An order of ctenophores (**Ctenophora**). (A.W.L.)

CESTODA, CESTOIDEA. The tapeworms. A class of the phylum **Platyhelminthes**. The tapeworms, like other members of the phylum, are flat-bodied. The body consists of two regions, a head or scolex usually bearing hooks, suckers, or both, and a strobila which, in all but the simplest species, is formed of a series of segments called proglottids. Tapeworms are parasitic in vertebrates.

In addition to the characters mentioned, tapeworms are distinguished by a complex life cycle. The adults live in the intestine of the host, absorbing food through the wall of the body since they have no alimentary tract, and produce a long succession of **proglottids** which break off as they mature and pass out with the faeces of the host, break off and remain in the intestine, or mature while still attached to the worm. They are reproductive bodies containing the organs of both sexes, rarely those of only one. The fertilized egg becomes a simple **embryo** with six hooks called the onchosphere. In this stage it is taken into the alimentary tract of a new host, migrates into the blood vessels, and after drifting along the blood stream lodges in some part of the body and develops into another form, usually vesicular, called the bladder worm. In this stage it remains inactive unless the tissue containing it is eaten by another animal, in which case it attaches itself to the intestinal wall of the new host and develops into an adult tapeworm.

Tapeworms are among the important parasites of man. Some of these species live in hogs and cattle and become established in man as the result of eating imperfectly cooked meat and one of the most dangerous species is found in the dog during its adult stage and in domestic animals and man in the bladder worm stage. The elimination of tapeworms from the human body requires the careful attention of a physician.

The tapeworms are classified as follows:

Subclass Cestodaria. Parasitic in fishes as adults and in **annelid** worms and **mollusks** in the early stages No distinct **scolex** and no segments.
Order Amphilinidea. Species of leaf-like form.
Order Gyrocotylidea. Leaf-like, with a projecting organ of attachment at the posterior end.
Subclass Cestodes. Usually segmented. Proglottids with organs of both sexes.

Order Tetraphyllidea. Scolex without retractile projections (proboscides), with four suckers or bothria. Parasitic in cold-blooded vertebrates.

Order Tetrarhynchidea. Scolex with proboscides. Parasitic in fishes.

Order Pseudophyllidea. No proboscides; only two suckers or bothria. In vertebrates of all classes.

Order Cyclophyllidea. Four suckers and usually hooks. Body elongate, with distinct segments, often numerous. Principally in warm-blooded vertebrates. (A.W.L.)

CETACEA. The whales, dolphins, porpoises and related mammals. An order, now replaced by two, **Odontoceti** and **Mystacoceti**, for the toothed whales and whalebone whales, respectively. (A.W.L.)

CETANE NUMBER. Ignition Quality.

C.G.S. SYSTEM. The **metric system** of physical units is based primarily on the standard **meter** and the standard **kilogram**, preserved at Sèvres. Experience has shown, however, that the usage of physics is often better served by founding its measures upon the centimeter and the **gram** (along with the mean solar second), rather than to use the meter and the kilogram directly. The "c.g.s." (centimeter-gram-second) system is the basis of nearly all present-day physical measurement, except in certain fields still commonly employing English units. (See **M.K.S. System.**)

Following are listed the c.g.s. units of certain familiar magnitudes, together with their dimensional makeup in terms of the fundamental units of the system (see **Physical Magnitudes and Physical Equations**):

MAGNITUDE	C.G.S. UNIT	DIMENSIONS
Area	Square centimeter	cm.^{2}
Volume	Cubic centimeter	cm.^{3}
Speed	Centimeter per second	cm. sec.$^{-1}$
Acceleration	Centimeter per second per second	cm. sec.$^{-2}$
Momentum	Gram-centimeter per second	g. cm. sec.$^{-1}$
Force	Dyne	g. cm. sec.$^{-2}$
Torque	Dyne-centimeter	g. cm.^{2} sec.$^{-2}$
Pressure	Barye or Bar	g. cm.$^{-1}$ sec.$^{-2}$
Energy and Work	Erg	g. cm.^{2} sec.$^{-2}$
Power	Erg per second	g. cm.^{2} sec.$^{-3}$

(L.D.W.)

CHABAZITE. The mineral chabazite is a member of that group of hydrous **silicates, the zeolites,** and corresponds to the formula $CaAl_2Si_6O_{16} \cdot 8H_2O$ with sodium sometimes replacing a part of the **calcium. Potassium, barium** and **strontium** may be present in very small amounts. Chabazite is **hexagonal,** usually in rhombohedrons that tend to resemble cubes. It has a rhombohedral cleavage; is brittle; hardness 4–5; specific gravity 2.08–2.16; luster vitreous; color white to flesh-red; streak white; translucent to transparent. Chabazite is found in the amydaloidal cavities of **basalts** often associated with other zeolites. It is occasionally found in such crystalline rocks as **syenites, gneisses** and **schists.** Chabazite is a rather common zeolite, being found in many localities in Europe. In the United States it occurs in the Triassic traps of New Jersey and Maryland. The **Triassic** lavas of Nova Scotia have yielded fine specimens. The name chabazite is derived from the Greek word meaning a precious stone. (E.S.C.S.)

CHACHALACA. Guan.

CHACMA. Mammalia, Primates. The largest of the baboons, *Papio porcarius,* resident in the extreme southern part of Africa. Also called the pig-tailed baboon. (A.W.L.)

CHAETA. A slender pointed structure secreted by cells of the **integument** in many invertebrates, such as the **annelids.** A seta. (A.W.L.)

CHAETOGNATHA. The arrow-worms, a group of small marine animals sometimes included in the phylum Annelida but now more often regarded as a separate phylum.

Arrow-worms are usually elongate transparent animals. They have two or three pairs of horizontal fins, one forming a caudal fin at the end of the body. The head bears a pair of eyes and a group of spine-like jaws which give the name to the phylum. The alimentary tract runs through the body as a straight tube to an anus near the caudal end. The body cavity is divided into three chambers by two transverse septa. Although the phylum includes only about 30 species, arrow-worms are found from the surface to great depths and in all of the oceans. (A.W.L.)

CHAETONOTOIDEA. An order of rotifers. (A.W.L.)

CHAETOPODA. A division of the **annelid** worms including the forms which have setae set in pockets in the integument. The **coelom** is well developed and is at least partially divided into **metameric** chambers, and the external segmentation of the body is also metameric. Most of the marine annelids such as the lobworm and **clam worm** and the **earthworms** and freshwater annelids are included here. The **leeches** and a few more primitive worms make up the rest of the phylum.

By some authorities this division is called a class and is divided into two orders:

Order **Polychaeta.** Free-swimming and sedentary worms, mostly marine, with lobed appendages (**parapodia**) and many **setae.** Usually a head with sensory organs. This group is sometimes regarded as a class and is then divided into two orders, the Errantia with similar body segments, most species free-swimming or burrowing, and the Sedentaria with specialized body regions, living in tubes in the bottom of the ocean, or between the tides.

Order **Oligochaeta.** Fresh-water and terrestrial worms with few setae and with neither parapodia nor sensory appendages. The earthworms are common examples. Sometimes ranked as a class. (A.W.L.)

CHAFER. Insecta, Coleoptera. A name applied to certain plant-eating **beetles,** including the rose chafer of the United States. The adults of this species are sometimes a troublesome pest on small fruits, especially grapes. They damage the fruit itself. Spraying with **lead** arsenate is recommended for their control. (A.W.L.)

CHAFFINCH. Aves, Passeriformes. Birds (**Aves**) of several species found in Europe and western Asia. The brambling is a member of the same genus, *Fringilla.* (A.W.L.)

CHAIN. A flexible connector composed of metal links, used for hoisting or for power transmission. *Coil chain* is used for hoisting and haulage, and consists of oblong links of circular sections, usually of welded wrought iron or steel. Coil chain with a stud or bridge across the center of the coil is preferred to plain coil chain in some instances, since the studs tend to prevent stretching and kinking.

Chain used for power transmission is shown in the figure. *Detachable link chain* is used for low-speed and light-load power transmission, and for conveyors and elevators of moderate capacity and length. The links can be easily detached and replaced, as illustrated. *Pintle chain* is from two to four times as strong as detachable link chain and can be used with the same sprockets.

Both types of chain are usually made up of malleable iron unmachined links. They can be supplied with integral pin, plate, or scraper attachments.

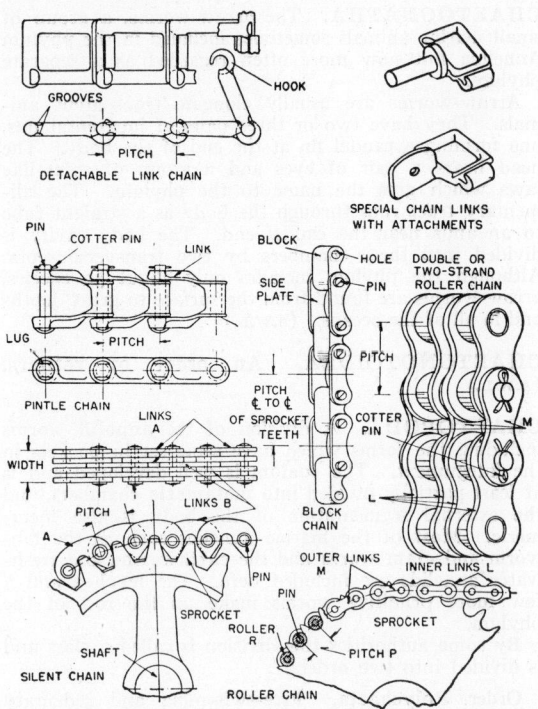

Steel block, roller, and silent chains are used where an exact speed ratio is desired and the center distance of the shafts is too large for gearing. *Block chain* is used for comparatively slow speeds and consists of blocks linked together by connecting links and pins. *Roller chain* consists of alternate links L and M held by connecting pins which are fastened by cotters. The pins also serve to carry the rollers which bear on the sprocket teeth. Roller chain can transmit more power than block chain and can operate at chain velocities up to 1200 ft. per min. For power requirements too great for single chains, double-, triple-, or quadruple-strand roller chains may be employed.

Silent chain is composed of alternate flat steel links A and B connected by pins. The links have straight faces in contact with the sprocket, and rotate slightly on the pins as the chain bends around the sprocket. Silent chain is used for heavy loads at speeds up to 1600 ft. or more per min. The silent chain is not actually quiet in operation but is much less noisy than other types of chain in use at the time of its adoption.

The speed ratio of a power chain depends upon the numbers of teeth in the driving and driven sprocket wheels; velocity ratios up to 7:1 are satisfactorily employed. Short-center drives with high ratios are usually more economical if fine-pitch chain is employed, while narrow large-pitch chain is cheaper for low-ratio long-center drives. (H.C.H.)

CHAIN BLOCK. Chains and sheaves may be employed in combination to produce an unusually powerful lifting mechanism. The best of these is the differential chain block, the action of which is explained in connection with the accompanying diagram. The mechanism consists of two sheaves, A and B, A being double sheave having diameters R and r. It will be shown that the multiplying power of this mechanism depends upon the ratio of these diameters. If they are equal, the pull P will not move

the weight, and the efficiency of the mechanism will be 0%, but the theoretical mechanical advantage is infinity. A slight difference in radii will produce a very large lifting effort, although the efficiency may still be very low. The sheaves are made with link pockets so that the chain fits nicely into the circumference, and is restrained from slipping. Furthermore, the chain is endless, and the mechanism is self-locking by virtue of the friction intentionally allowed on the journals.

In explanation of the chain block, if the pull P revolves sheave A one revolution, the vertical chain at a is lowered through a distance to $2\pi r$, while the side b is raised the distance $2\pi R$. The net vertical displacement of the sheave B is $\pi(R-r)$ upward. With no friction considered, the work of lifting W through this distance must be equal to the work done by the pull P moving through $2\pi R$. Solving this equation for advantage W/P,

$$\frac{W}{P} = \frac{2R}{R-r}.$$

Applying the mechanical efficiency e to this equation, the actual mechanical advantage is

$$\frac{W}{P} = \frac{2Re}{R-r}.$$

These chain blocks are built in different sizes for hoisting loads from ¼ ton to 3 or 4 tons, by hand. On account

Chain block hoist.

of the self-locking feature depending on friction, the average mechanical efficiency of this device is only about 30%. (F.T.M.)

CHALCEDONY. Chalcedony is one of the cryptocrystalline varieties of the mineral **quartz**, having a waxy luster. It may be semi-transparent or translucent and is usually white to gray or grayish-blue or some shade of brown, sometimes nearly black. Other shades have been given different names. A clear red chalcedony is known as **carnelian** or **sard**; a green variety colored by nickel oxide is called **chrysoprase**. **Prase** is a dull green. **Plasma** is a bright to emerald-green chalcedony which sometimes is found with small spots of **jasper** resembling blood drops; it is then referred to as **blood stone** or **heliotrope**. The term chalcedony is derived from the Greek word meaning Chalkedon, a town in Asia Minor. (E.S.C.S.)

CHALCID FLY. Insecta, Hymenoptera. Minute **insects** of many species, usually shining or metallic and with very simple wings showing a single vein. A few feed on plants, and the fig insects, *Blastophaga psenes*, are essential for the fertilization of Smyrna figs, but most species are parasitic on other insects during their larval life. The host species are attacked by these parasites in all stages of metamorphosis, including the egg.

A few species are important pests in wheat but the chief economic importance of the group lies in the destruction of other pests by the parasitic species. It is difficult to estimate the importance of such natural checks. (A.W.L.)

CHALCOCITE (COPPER GLANCE). The mineral chalcocite is **cuprous sulfide**, Cu_2S, crystallizing in the **orthorhombic** system, often in pseudo-**hexagonal** forms. Above a temperature of $91°$ C. chalcocite changes into an **isometric** form. It has conchoidal fracture; hardness 2.5–3; specific gravity, 5.5–5.8; metallic luster; color dark gray to blackish-gray, frequently with bluish-green tarnish. Chalcocite is of widespread occurrence and a valuable copper ore. It seems in some cases to be definitely secondary in origin, in other cases primary. It may have been formed from bornite by the action of alkaline solutions. It sometimes carries valuable amounts of silver.

Among the many European localities might be mentioned Cornwall, England, the Ural Mountains, and Rumania. It occurs also in the French Congo, South West Africa, Peru, Mexico, and Alaska. In the United States it is found at Bristol, Connecticut, in fine crystals, Montana, Tennessee, Arizona, Nevada, and California.

The word chalcocite is derived from the Greek word meaning copper. (E.S.C.S.)

CHALCOPYRITE (COPPER PYRITES). The mineral chalcopyrite is a **sulfide** of **copper** and **iron** corresponding to the formula $CuFeS_2$. Its **tetragonal** crystals are often complex with repeated twinning; massive chalcopyrite is common. It has an uneven fracture; is brittle; hardness, 3.5–4; specific gravity, 4.1–4.3; luster, metallic; color, brass-yellow, may be iridescent from tarnish; streak, greenish-black; opaque. Chalcopyrite is the most common copper-bearing mineral known and it is the most important ore of copper. It is a primary mineral in many igneous rocks and from it a host of secondary copper minerals have been derived. Among the many localities where fine specimens of this mineral have been obtained might be mentioned: Freiburg, Saxony; Alsace; Rio Tinto, Spain; Cornwall, England; Australia; Chile, Peru, and Bolivia, South America; and in the United States, Ellenville, New York; Chester County, Pennsylvania; Joplin, Missouri; Gilpin County, Colorado; Arizona, Montana, Utah, Nevada, California, New Mexico and Tennessee. In Canada there are notable deposits of chalcopyrite in the Provinces of British Columbia, Ontario, and Quebec. The name chalcopyrite is derived from the Greek word meaning copper, and the word pyrites. (E.S.C.S.)

CHALK. Chalk is a soft, porous **limestone** of white, grayish-white or buff color made up of the minute shells of **foraminifera** and fragments of **cocospheres**. It occurs extensively in England and France and less so in the United States. (E.S.C.S.)

CHALYBITE. Siderite.

CHAMELEON. Reptilia, Sauria. Any member of several genera of lizard-like **reptiles** of very peculiar form, occurring in Africa, the Oriental region, and about the Mediterranean. They are arboreal species with grasping feet, a crested head, and a long extensile tongue with a clubbed sticky tip which is used to catch insects. They are able to change color readily. The most common genus is *Chamaeleon*.

The little lizard commonly sold at street fairs in the United States is not a true chameleon but is more closely related to the iguanas. (A.W.L.)

CHAMOIS. Mammalia, Artiodactyla. A goat-like European **antelope**, *Rupicapra tragus,* found in the forests of mountainous regions and to a limited extent under alpine conditions. Noted for its agility. (A.W.L.)

CHANCRE. The first manifestation of **syphilis**. A chancre is a single, hard, elevated sore, which appears on the penis, vagina or external genitalia 3 to 4 weeks after exposure. Occasionally an extragenital lesion—usually on the lips—occurs. The lesion lasts 6–8 weeks.

Although its clinical characteristics are striking, a positive diagnosis of syphilis cannot be made unless the spirochetes can be demonstrated in the secretions of the sore by examination with a dark-field microscope. Early diagnosis is extremely important because the earlier treatment is begun, the better is the prognosis. (See **Syphilis**.) (D.M.H.)

CHANCROID. A veneral disease characterized by multiple soft **ulcers** on the genitalia. It is sometimes called soft **chancre**, in contrast to hard chancre of **syphilis**. The ulcers of chancroid contain the causative organism, Ducrey's bacillus. The regional lymph nodes in the groin may form **buboes,** ulcerate, and discharge pus. Treatment is local, plus sulfonamides orally. (D.M.H.)

CHANNEL. Compression Member; Tension Member; Gauge Line.

CHANNEL, FREQUENCY. This term denotes the band of frequencies which is associated with a single unit of intelligence in a communications system. Thus it applies to the band of frequencies radiated by a broadcast station, or to the band of frequencies which must be handled by a carrier system to handle a single conversation. In the various systems the application of intelligence to a given frequency will generate certain other frequencies which are then associated with the original in some manner to convey the intelligence to the receiver. This band of frequencies then determines the response characteristics which the receiver (or other units of the system) must have for satisfactory results. Thus in conventional broadcasting the various stations use channels about 10 kc. wide; in frequency modulation the present channel is about 200 kc.; in television it is 5 or 6 megacycles; in carrier telephony it is only about 3 kc. (L.R.Q.)

CHAPARRAL. The name applied to a **plant association** occurring over wide areas in western North America and composed of a mixed population of low-growing shrubs. Some stands of chaparral are dense; others are open. Chaparral is usually found between a lower zone of **sagebrush** or grassland and an upper zone of woodland or forest. This association is found principally in the foothills of the Coast Ranges and Sierra Nevada in California, of the southern Rocky Mountains, and other ranges in Utah and Arizona. Some types of chaparral, such as that occurring on the Coast Ranges of southern California, are composed largely of evergreen shrubs; others, such as that occurring in the southern Rocky Mountains, largely of deciduous shrubs. (B.S.M.)

CHARACTERISTIC CURVE. Gamma, Exposure-Density Relationship.

CHARACTERISTIC EQUATIONS. A class of equations connecting those variables, such as temperature, pressure, and volume, which define the physical condition of a given substance and are called variables of state.

The **ideal gas law** and the **Boyle-Charles** law represent approximately the behavior of all gases, but if one wishes to be accurate, some modification of these must be sought which will take account of the differences between individual gases. The best known characteristic equation for gases is that of van der Waals. Using the same notation as for the ideal gas law, this may be written

$$\left(p + \frac{a}{v^2}\right)(v - b) = RT.$$

(Here, however, the volume is not in liters but is in terms of the volume of the same body of gas at standard temperature and pressure; and the pressure is in atmospheres.) a and b are constants characteristic of the gas in question

They are very small; if they were zero we should have the ideal gas law. Following are their approximate values for certain gases:

Gas	a	b
Ammonia	0.00831	0.00165
Helium	0.00007	0.00106
Hydrogen	0.00049	0.00119
Nitrogen	0.00277	0.00175
Oxygen	0.00271	0.00142

Clausius modified van der Waals' equation as follows:

$$\left(p + \frac{a}{T(v+c)^2}\right)(v-b) = RT,$$

employing three empirical constants a, b, c; while the equation of Dieterici has an exponential factor:

$$p \cdot e^{\frac{a}{RTv}} \cdot (v-b) = RT.$$

None of these equations represents the behavior of all gases equally well.

Beattie and Bridgman have proposed a characteristic equation for fluids in general, as follows:

$$pv^2 = RT\left(1 - \frac{c}{vT^3}\right)\left[v + B\left(1 - \frac{b}{v}\right)\right] - A\left(1 - \frac{a}{v}\right),$$

in which a, b, c, R, A, and B are empirical constants to be determined for each fluid. (L.D.W.)

CHARACTERISTIC IMPEDANCE. This is the impedance which a transmission line would present at its input terminals if the line were infinitely long. If, instead of the line being actually infinite in length, it is finite and is terminated by an impedance equal its characteristic impedance it will behave, as far as the input is concerned, as if it were infinite. This means that there will be no reflected electrical wave, with the attendant losses, etc., at the terminal point. In electrical circuits this is an extremely important consideration as reflection means some energy which would otherwise go to the load is reflected back down the line to cause losses on the line, objectionably high voltages, echo effects and other undesired conditions. When a line is terminated in the characteristic impedance it is said to be matched, or the load matches the line. While not always attainable it is a condition highly desirable. (L.R.Q.)

CHARACTERISTIC OF A COMMON LOGARITHM. Logarithms.

CHARACTERISTIC SPEED. Specific Speed.

CHARACTERISTIC TEMPERATURE. Dulong and Petit's Law of Specific Heats.

CHARADRIIFORMES. A large order of shore and wading birds (**Aves**) including the **gulls, terns, auks, puffins, plovers, sandpipers, snipes,** and many others. (A.W.L.)

CHARGE-MASS RATIO. This term refers to the relationship between the electric charge of a particle and its mass, so important in the physics of **electrons, ions,** and other electrified bodies of molecular order.

The earliest information on the subject followed from the researches of Faraday on electrochemical equivalents. From his results it appears that in the electrolysis of chlorine, for example, one **coulomb** of negative electricity is carried by 0.00037 gram of this element, and hence that the carriers or ions have a charge-mass ratio of about 2700 coulombs or 8.1×10^{12} electrostatic units of electricity to the gram. Similarly, one coulomb of positive electricity is carried by 0.0000104 gram of hydrogen, which gives about 95,700 coulombs

or 2.87×10^{14} e.s.u. to the gram for hydrogen ions. This is 35 times the ratio for chlorine ions. But the atomic masses of hydrogen and chlorine are in the ratio 1:35, which means that if the carriers are atoms, the charge per carrier is the same for both elements. Bivalent elements, on the other hand, carry twice this charge per ion.

When J. J. Thomson applied a magnetic field to a stream of hydrogen **canal rays,** and then neutralized the resulting deflection by means of an electric field, he was able to calculate the charge-mass ratio of these particles from the curvature of the magnetically deflected stream and the values of the two field intensities. This he found to be either 95,700 coulombs per gram as in the electrolysis of hydrogen, or ½ that value, which indicated that some of the ions were atoms and some were molecules carrying the same charge as the atoms. But when a similar test was applied to the **cathode rays** in a **Crookes tube,** the ratio was found to be about 5.303×10^{17} e.s.u. per gram, or 1850 times that for hydrogen atoms, whatever the nature of the cathode. We know now that this enormous difference is one of mass, not of charge; and that these experiments were the first direct revelation of the identity of the **electron,** which has a mass only 1/1850 of that of the hydrogen atom. (L.D.W.)

CHARLES' LAW. Although the coefficients of expansion of different solids or of different liquids are notably different, the coefficients of expansion of all gases are nearly the same, namely, about ½73 of the volume at 0° C. per centigrade degree. The law, stated by Charles in 1787 and independently by Gay-Lussac in 1802 (hence sometimes called Gay-Lussac's law) is not strictly true. Regnault obtained the following values of the volume coefficient for various gases:

Air	0.0036706
Hydrogen	0.0036613
Carbon dioxide	0.0037099
Sulfur dioxide	0.0039028
Carbon monoxide	0.0036688
Nitrous oxide	0.0037195
Cyanogen	0.0038767

None of these is far from ½73 = 0.003663, which is therefore commonly taken as the expansion coefficient for gases; especially as the value for hydrogen, commonly used in the standard gas thermometer, is very near it. If the pressure as well as the volume is allowed to vary, the behavior of the ideal gas must be expressed by the **Boyle-Charles law** or the **ideal gas law.** (L.D.W.)

CHARLIER COEFFICIENT OF DISTURBANCY. The Charlier coefficient of disturbancy, C, is defined as

$$C = 100 \frac{\sqrt{\sigma^2 - spq}}{sp}$$

where s is the number of trials, p is the **probability of** success in a single trial, and σ^2 is the **variance** of the distribution of the number of successes. (L.A.A.)

CHARNOCKITE. Charnockite is a granular variety of **hypersthene granite** which was first described from the gravestone of Job Charnock, who founded the city of Calcutta, India, whence the derivation of the name charnockite. (E.S.C.S.)

CHARR. Pisces, Teleostei. *Salvelinus.* A lake fish (**Pisces**) of the British Isles, related to the trout. (A.W.L.)

CHART. A chart is a **map** drawn to show those details which are of primary importance to navigators. Since the **rhumb line** and **great circle** are curves on the surface of the earth most used by navigators, most charts are drawn on either the **Mercator** or **Great Circle** projections in order that these curves may appear as straight lines. A few charts of harbors or small islands will be found drawn on the **polyconic projection.** Within recent years the **Lambert projection** has superseded all others for use by air navigators.

Charts differ from maps by emphasizing those features and details which are of importance in **navigation** and in omitting many features of interest to the general public. The parallels of **latitude** and meridians of **longitude** are always clearly marked, and **compass** "roses," frequently showing both true and magnetic **directions,** are printed on the charts in convenient places. The coast pilot is interested in such things as prominent landmarks to be watched for when approaching from seaward, such as high mountains not too far inland, church spires, Coast Guard stations, light houses, light ships, beacons, **buoys, range marks, soundings,** etc. The off-shore navigator is interested in the depth of water, character of the sea bottom, compass variation, prevailing currents, positions of **radio aids,** etc. Aviation charts emphasize those surface features of most importance to the air navigator such as **contour lines,** heights of mountains, rivers, railroads, power transmission lines, water towers, radio stations, radio beams, airports, etc.

Before using any chart the navigator must study the legend carefully to determine such things as the date of issue, interval of contours, units of soundings, and similar data. No navigator, either over land or sea, should ever use a map published for commercial purposes. To do so is to court disaster since, for example, railroad companies frequently publish maps showing their lines as straight for long distances, and moving towns from their correct locations so that they appear to be on the straight route. (W.K.G.)

CHAT. Aves, Passeriformes. Birds (**Aves**) of several species, usually designated by a compounded word, as the stonechat and whinchat of Europe, the yellow-breasted chat of North America (*Icteria virens*), and several North African species. The European wheatear and hedge warbler are also chats. (A.W.L.)

CHATOYANCY. Cat's-Eye.

CHATTERER. Aves, Passeriformes. South and Central American birds (**Aves**) making up the family Cotingidae. They are quite varied, some with strangely formed plumage and some beautifully colored. The family includes the umbrella bird, **bell-birds, cotingas,** manakins, and **cocks-of-the-rock.** A single species, the xantus becard, *Platypsaris aglalae,* enters the United States near the Mexican border.

These birds are near the flycatchers. (A.W.L.)

CHATTER MARKS. Moon-shaped scratches or gouges on the bed rock which are supposed to be caused by the "chattering" action of angular boulders which are carried in the bottom of a glacier. (R.M.F.)

CHAULMOOGRA OIL. *Teraktogenos (Hydnocarpus) Kurzii.* Flacourtiaceae. Chaulmoogra oil is expressed from the seeds of a tall tree native to the jungles of northern Burma. The tree has a smooth light-brown bark, large leathery evergreen leaves and inconspicuous flowers. The fruits have a soft hairy surface and contain several large seeds.

The oil has long been known and used in Burma and other Malasian countries in the treatment of various skin afflictions and **leprosy,** often with very favorable results. Modern science has advanced this problem by expressing the oil and purifying it, so that the painful effects associated with the use of the crude oil have been eliminated. Thus, ethyl esters of the fatty acids in the oil are prepared. This product is given by intra-muscular injection into the buttocks. Treatment is continued over long periods of time—3–5 years or more, with intervening periods of rest without treatment. (R.M.W., R.S.M.)

CHEEK. The lateral wall of the oral cavity. (A.W.L.)

CHEETAH, CHETAH, CHITA. Mammalia, Carnivora. A large **cat** found in Africa and India. It is marked by its slender build and is tawny with black spots, thus somewhat resembling the leopards. The species is known as the hunting leopard and has been tamed for use in the chase. (A.W.L.)

CHELA. A form of grasping appendage found in **lobsters, crabs** and other **arthropods.** The large pincers of these species are the most familiar examples.

All chelate appendages have the next to the last segment prolonged into a process against which the terminal segment works to form a forceps-like organ. (A.W.L.)

CHELICERAE. The first pair of appendages in **spiders** and related animals. They are associated with the mouth and are formed for chewing and in some cases for grasping, as in the **scorpions.** (A.W.L.)

CHELIPED. An appendage of the **thorax** formed for grasping, in the **crustaceans.** The chela or pincher of the lobster and crayfish. (A.W.L.)

CHELLEAN. Paleontology of Man.

CHELONIA. Testudinata; Fossil Reptiles.

CHEMICAL CHANGES. The chemical composition of a substance is subject to various changes under various conditions, depending upon (1) the nature of the individual substance, (2) the nature of other substances present, and (3) the conditions. The vast majority of reactions take place between two substances—occasionally one substance only, and sometimes three or more different substances. There are many cases where simple contact of the substances is sufficient to bring about the chemical change, e.g., the rusting of iron in oxygen. In many other cases, the change is not spontaneous, but must be induced, frequently by raising the temperature, as in the burning of fuels. The conditions that are considered important and fundamental are (1) temperature, (2) pressure, (3) medium, if any, (4) catalyzer, if any, (5) electric direct current, (6) light. In a given reaction, the change in composition of the substance or substances involved is inherently connected with a change in energy. Thermal, electrical or light energy of a certain potential or intensity and in definite amounts, is requisite to initiate and carry on the reaction, and thermal, electrical or light energy of definite amount is liberated or consumed in the reaction. Every reaction, properly speaking, has both a matter and an energy aspect. While the energy aspect is frequently neglected directly, the conditions must always be in accord with the energy demand, even if apparently not considered. (See **Thermochemistry; Electrochemistry; Photochemistry.**) Chemical changes require consideration of three topics, namely, I. Natural rate of chemical reactions, II. Acceleration of the natural rate in the presence of a catalyzer, and III. The end-point of chemical reactions.

I. *Natural Rate of Chemical Reactions.* Various factors operate to affect the rate of chemical reactions. By natural rate is understood the rate of a reaction in the absence of a **catalyzer.** Excluding electrochemical and photochemical reactions, and giving attention to thermochemical reactions only, there are four factors or conditions to be considered, namely, (1) concentration of constituents, (2) temperature, and (3) pressure—important where a gas is involved, (4) nature of the medium, if any.

(1) Relation between concentration of reactants and rate of reaction. The rate of a given reaction, at constant temperature and pressure under stated conditions of concentration of the reacting substances, is quantitatively expressible by a velocity constant, which is the fraction of the substances transformed in a unit of time. Many reactions occur instantaneously—true for most reactions in solution in inorganic chemistry—and many others are complicated by subsidiary reactions, so that the velocity constant is measurable in comparatively few cases. The principle, however, holds as stated, whether or not the desired value can be ascertained experimentally.

A simple reaction that was studied by Wilhelmy (1850), and since then by various investigators, is the transformation (hydrolysis) of sucrose ($C_{12}H_{22}O_{11}$) in water solution into glucose ($C_6H_{12}O_6$, a polyhydroxy aldehyde) plus fructose ($C_6H_{12}O_6$, a polyhydroxy ketone), which proceeds at a measurable, steady rate in the presence of acid (hydrogen ion). The rate of reaction at any instant is found to be proportional to the amount of sucrose present at that instant.

When a dilute water solution of an ester, such as methyl acetate, is similarly hydrolyzed in the presence of hydrogen ion, the reaction is of the same type. And this statement also applies to the decay of radioactive elements. One of the important radioactive constants is the period, that is, the time required for the decay of one-half of the element present at a given instant.

The preceding cases are instances of monomolecular reactions—in the case of sucrose and of methyl acetate the concentration of water, when it is in large excess, remains constant throughout the reaction. Only in monomolecular reactions is the velocity constant independent of the initial concentration.

The rate of bimolecular and higher types of reactions depends upon the concentration of each reactant, and is specific in each case.

When a reaction involves two different phases, that is, when the system is not homogeneous but heterogeneous, as in reactions between a solid phase, such as zinc or calcium carbonate, and a liquid phase, such as hydrochloric acid solution, the rate of reaction involves consideration of (1) the area of the surface of contact of the solid with the solution, and (2) the rate of diffusion from the surface of the solid, as well as (3) the concentration of hydrogen ion of the acid solution.

When the rate of a chemical process is dependent upon (1) two or more *consecutive* reactions, the observed rate is limited by the rate of the slowest reaction in the series, (2) two or more *concurrent* reactions, the products are in the same ratio at any instant only when the reactions themselves are of the same rate.

(2) Relation between temperature of reactants and rate of reaction. The rate of chemical reaction is increased two or three times for a rise in temperature of 10° C. The rate of decay of radioactive elements has been found to be unaffected by temperatures from the lowest to the highest attainable.

(3) Relation between pressure of reactants, if gaseous, and rate of reaction. Since pressure changes amount to concentration changes in such systems, the behavior is as described above under concentration.

(4) Relation between nature of the medium and rate of reaction. Very slight changes in the nature of the medium greatly affect the rate of a chemical reaction, but attempts to relate any physical property of a solvent with the effect observed on the rate of a given reaction appear to have proved unsuccessful.

Summarizing, the rates of chemical reactions are subject to highly specific influences in each case, as has been abundantly demonstrated by experimental investigations, and recognized in numerous legal battles in chemical patent suits.

II. *Acceleration of the Natural Rate of Chemical Reactions* in the presence of a positive or negative catalyzer (see **Catalysis**). When, in the presence of a given substance, the natural rate of a chemical reaction is changed, either increased or decreased, the given substance is called a catalyzer. Examples are numerous. 1. When a gas-lighter of the type known as platinum black, the active part of which consists of very finely divided platinum, is held in a stream of hydrogen or city gas, the gas is ignited in air. Platinum is a catalyzer for this reaction, and causes ignition to take place at a temperature much lower than by subjecting to fire. 2. The changing of sulfur dioxide into sulfur trioxide is accomplished by passing a mixture of sulfur dioxide and air (one-fifth oxygen) over asbestos coated with finely divided platinum. The temperature required is much lower by the use of platinum catalyzer than without its use. 3. Solutions of sulfites are subject to oxidation to sulfates by oxygen upon allowing to stand in air. The addition of sugar or glycerol retards the speed of this reaction. These substances act in this case as negative catalyzers. 4. The combination of nitrogen and hydrogen gases under high pressure to form ammonia gas is accomplished at a lower temperature in the presence of a catalyzer than in its absence, thus increasing the yield of ammonia (see **Equilibrium**). One of the catalyzers is composed of iron, intimately mixed with 1% aluminum oxide and 1% potassium oxide. Iron is a catalyzer for this reaction, but is more active as such in the presence of aluminum oxide and potassium oxide, which are spoken of as promoters, a sort of catalyzer of a catalyzer. 5. The hydrogenation of liquid fatty oils and of oleic acid is conducted in the presence of finely divided nickel as a catalyzer. 6. Enzymes are very specific catalyzers, "the most selective and delicate of all known catalysts (Hilditch)," at ordinary temperatures, say 25° to 30° C. Dextro-glucose is converted into ethyl alcohol in the presence of the enzyme (zymase) of yeast, and ethyl alcohol into acetic acid (vinegar) in the presence of the enzyme of *Mycoderma aceti*. 7. Nitric acid reacts slowly with copper metal, but the rate of reaction is accelerated more and more as nitrogen tetroxide (catalyzer) is formed in the solution. This is an example of autocatalysis, wherein the reaction brings about the formation of its own catalyzer. 8. Arsenic-containing substances are extreme negative catalyzers, called inhibitors or poisons, of platinum catalyzer.

When the catalyzer is a solid substance, the greatest difficulty in use is to maintain a clean surface. The presence of a positive catalyzer enables a reaction to proceed more rapidly at a lower temperature than corresponds to the natural rate of the reaction. This increases the amount of substances converted in a given time, decreases the demands as to temperature resistance of materials of construction of the apparatus, and frequently makes possible a state of equilibrium (below) more favorable to the yield of desired material.

III. *The End-point of Chemical Reactions.* If a chemically reactive system is isolated from the rest of the universe at a constant temperature and pressure, a definite end-point is often attained short of the complete transmutation of reactants into resultants. In order to be certain that this end-point (short of complete transmutation) is what is known as the equilibrium point, the equilibrium must be approached from both directions, e.g., $A + B \rightarrow C + D$ and $C + D \rightarrow A + B$. If the value of the equilibrium constant (below) is the same when approached from both directions, then the reaction is one of true chemical equilibrium. Such equilibrium reactions are also referred to as balanced or reversible reactions. In such reactions the extent of the chemical change is proportional to the concentrations of all the reactants—reactants and resultants being interchangeable, depending upon the direction of the reaction. (Generalization of Guldberg and Waage, 1864, called Law of Mass Action, or more correctly Law of Concentration Effect. Reaction studied by Guldberg and Waage (1867): Barium sulfate plus potassium carbonate plus barium carbonate plus potassium sulfate.)

A classical case, frequently cited, is that investigated by Berthelot in 1863. When 1 mol (60 grams) of acetic acid (CH_3COOH) and 1 mol (46 grams) of ethyl alcohol (C_2H_5OH), both of which substances are soluble in water, are mixed, a reaction takes place which results in the formation of water and ethyl acetate ester, which is likewise in the ratio of 1 mol (18 grams) of water, and 1 mol (88 grams) of ethyl acetate ($CH_3COOC_2H_5$). On the other hand, when 1 mol of water and 1 mol of ethyl acetate ester are mixed, a reaction takes place which results in the formation of acetic acid and ethyl alcohol in the ratio of 1 mol of acetic acid and 1 mol of ethyl alcohol. Three important observations have resulted from the detailed study of this reaction, namely, (1) the reaction between acetic acid and ethyl alcohol as reactants proceeds at such a rate that the fraction 0.00575 of the amount present at any instant reacts, at 6° to 9° C., in 1 day to form equivalent amounts of water and ethyl acetate ester, (2) the reaction between water and ethyl acetate ester as reactants proceeds at such a rate that the fraction 0.00144 of the amount present at any instant reacts, at 6° to 9° C., in 1 day to form equivalent amounts of acetic acid and ethyl alcohol, and (3) the end-point of each reaction is the same, that is, the reaction is one of true chemical equilibrium, and the resulting equilibrium mixture contains, in each case, 0.33 mol acetic acid plus 0.33 mol ethyl alcohol plus 0.67 mol water plus 0.67 mol ethyl acetate ester. This system attains practical equilibrium, at 6° to 9° C. in about 1 year, at 100° C. in about 8 days, and at 200° C. in about 24 hours.

The equilibrium constant is calculated numerically as follows:

Equation:	CH_3COOH	$+ C_2H_5OH$	$\rightleftarrows HOH$	$+ CH_3COOC_2H_5$
Reaction weights:	60	46	18	88
Molar ratio at equilibrium:	0.33	0.33	0.67	0.67
Weights at equilibrium:	0.33×60	0.33×46	0.67×18	0.67×88

$$\left.\begin{array}{c}\text{Equilibrium}\\\text{constant at } 9° \text{ C.}\end{array}\right\} = \frac{\text{conc. } HOH \times \text{conc. } CH_3COOC_2H_5}{\text{conc. } CH_3COOH \times \text{conc. } C_2H_5OH}$$

$$= \frac{0.67 \times 0.67}{0.33 \times 0.33} = 4.$$

Knowing the equilibrium constant at any stated temperature enables one to calculate the equilibrium endpoint at that temperature for any ratio of reactants. Thus, when 1 mol (60) grams of acetic acid and 10 mols (460) grams of ethyl alcohol at 9° C. are taken:

$$\text{Equilibrium constant at } 9° \text{ C.} = 4 = \frac{X \times X}{(1 - X) \times (10 - X)}$$

where X is the number of mols of water and also the number of mols of ethyl acetate ester formed (1 mol of each is formed by reaction of 1 mol acetic acid plus 1 mol ethyl alcohol). Solution of this equation shows $X = 0.97$. Therefore, by taking the above ratio of acetic acid (1 mol) to ethyl alcohol (10 mols) 0.97 (or 97%) of the acetic acid, the excess reactant being ethyl alcohol (9 mols), is converted at equilibrium into water plus ethyl acetate ester. In practice, the reaction is conducted by the use of a catalyzer, e.g., sulfuric acid concentrated, zinc chloride.

In cases where one of two resultants can be separated from the reactants and the other resultant, by precipitation as a solid, by condensation as a liquid, or by volatilization as a gas or vapor, the yield of the desired substance from a given amount of reactants can

sometimes be materially increased. In the case of heterogeneous systems (those that are not homogeneous) whenever a solid participant is present, the *concentration* of said solid is considered constant. The precipitation and solution of solids are in this category, as well as the reactions between a gas and a solid, e.g., the system ferroferric oxide plus hydrogen gas plus iron plus water vapor.

The effect of change of temperature on a system in chemical equilibrium is that the equilibrium point is shifted (1) towards the side that *away* from that which evolves heat when the temperature is *raised,* and (2) towards the side which evolves heat when the temperature is lowered. It is *as if* the amount of heat were a *material* reactant and its concentration (temperature or intensity of heat) increased, in respect to the *direction* of the shift of the equilbrium point. The amount of the shift at constant pressure can be calculated in cases where one possesses the proper data. (See **Thermochemistry.**)

The effect of change of pressure on a system in chemical equilibrium is that the equilibrium point is shifted (1) towards the side possessing the smaller aggregate volume when the pressure is increased, and (2) towards the side possessing the larger aggregate volume when the pressure is decreased. The amount of the shift at constant temperature can be calculated by means of the equilibrium constant (above) recalling that increase of pressure is equivalent to increase of concentration of gases (temperature constant). When the volume of resultants equals the volume of reactants, no effect is produced on the equilibrium point by change of pressure. (See **Equilibrium.**)

Chemical Equations. In calculating the equilibrium constant (above) the equation of the reaction was presented and the reaction weights inserted. Every chemical reaction between pure substances (individual chemicals) with the formation of pure substances is capable of being represented in the form of a chemical equation, wherein the formula of each pure substance has the significance stated in chemical composition, section II, Compounds and related topics. The only difficulty which should present itself to the uninitiated is that of always recalling that the formula of each chemical substance stands for a definite chemical unit weight (or volume, if a gas). In the vast majority of reactions encountered in practice the writing of the chemical equation of a reaction is comparatively simple when one knows (1) what substances are reactants, (2) what substances are resultants, (3) the formula of each. The sum of the weights (masses) of reactants equals the sum of the weights (masses) of the resultants. Without this information, equations cannot be written.

Grand Groups of Chemical Reactions. Certain grand groups of reactions are recognized. The classification may be carried further with extensive additions as desired.

1. **Reactions involving recombination of ions**
2. **Reactions involving water**
   (a) Consumption of water
   (b) Production of water
   (c) Water as catalyzer
3. **Reactions involving oxidation—reduction**
   (a) In solutions of electrolytes
   (b) Not in solutions of electrolytes. (R.K.S.)

CHEMICAL COMPOSITION. The following outline will make clear the subjects discussed under chemical composition of substances. For the chemical changes undergone by such substances see **Chemical Changes.**

I. Elements, symbols, atoms, allotropes (allotropy), isotopes (isotopy), isobares, atomic constants.
II. Compounds, formulas, molecules, allotropes (allotropy), molecular constants, isomers (isomerism, stereoisomerism).

III. Radicals.

IV. Equivalents, valency, graphical formulas.

I. Elements and Related Topics

At the present time there are recognized by the Committee on Atomic Weights of the International Union of Chemistry 86 chemical elements. For list see **Chemistry**. The most recent additions to the list are **hafnium** in 1927, **rhenium** in 1929, and **protactinium** in 1937. Some elements have been known since prehistoric times, e.g., the common metals; some have been proved to be elements that were previously considered not elements, e.g., **chlorine** (Davy, 1810); some have been discovered by new methods of investigation, e.g., by the spectroscope (Bunsen and Kirchhoff); **cesium** (1860), **rubidium** (1861), **helium** in the spectrum of the sun by Janssen and Lockyer in 1868 and in certain minerals of the earth by Ramsay and by Cleve independently in 1895, by the

ago with some confidence. The Moseley atomic number is probably the best peg on which to hang the concept of atom as used practically. This constant, along with the Mendeléeff placement of the elements, makes it clear that the elements form a system. Other data confirm this. Symbol weights—commonly known as atomic weights—of the elements are relative numbers for the mass of each element based on the standard of reference, oxygen equal to 16.0000. The alphabetical table of elements on page 288 gives the name, symbol, atomic number and atomic weight.

Other constants of certain atoms are given (pp. 288, ff.): namely, atomic volume, atomic heat, atomic radius, atomic compressibility. Constants in frequent use are hardness, crystal form, melting point, boiling point, vapor pressure; also important, spectrum, both visible and x-ray.

Atomic volume is the volume in milliliters occupied by 1 atomic weight in grams of the solid element, there-

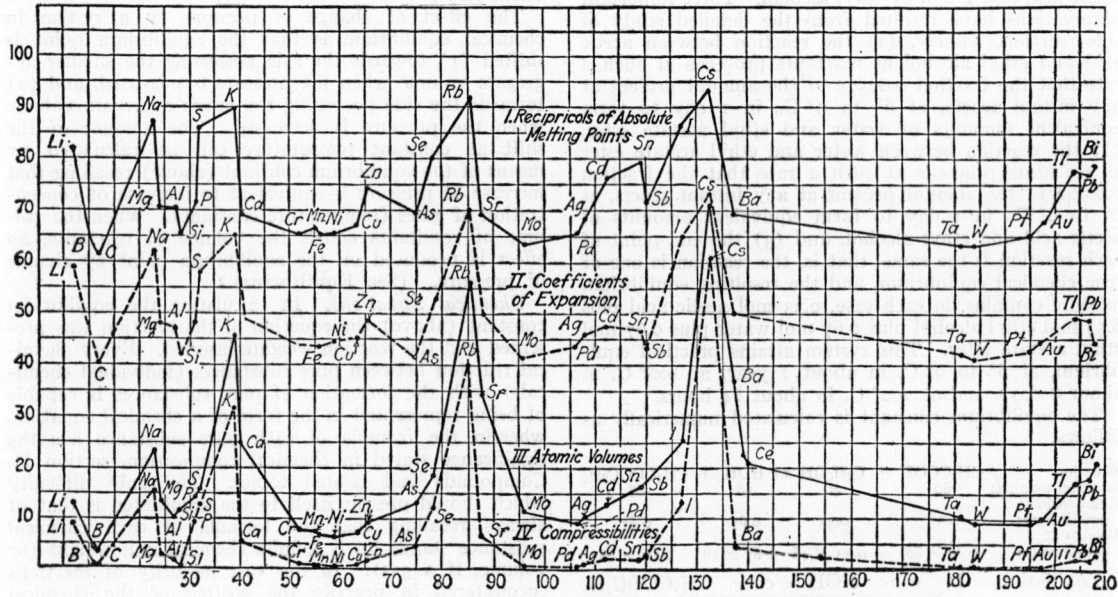

Graph showing periodicity of the physical properties of the elements.

x-ray spectra (Moseley, 1914), e.g., hafnium by Coster and Hevesy in 1923, and rhenium by Walter Noddack and Ida Tacke in 1925.

The introduction of a systematic relationship among the chemical elements is due to Mendeléeff in 1869. His pronouncement was that the properties of elements, as well as the forms and properties of their compounds, are in periodic dependence on the atomic weight of the elements. A full account is given in the Faraday Lecture of the Chemical Society (London) on "The Periodic Law of the Chemical Elements" delivered by Mendeléeff on June 4, 1889. This periodic arrangement is the one in common use at the present time. Extension and experimental verification of the periodic arrangement are due to the work on x-ray spectra of the chemical elements by Moseley in 1914, who was able to assign a serial number, called the atomic number, to each of the elements from hydrogen 1 to uranium 92. As was stated above, the identity of 86 of these has been officially recognized, and some of the remaining are known to exist in the radioactive series of elements, namely, uranium X₂ of atomic number 91, mesothorium 2 of atomic number 89, and several of atomic numbers 81, 82, 83, 84, 86, 88, 90, 92. (See **Radioactive Changes**.)

It is not attempted here to define a chemical element in a simple manner, as could have been done some years

fore, atomic weight (grams) divided by density (specific gravity) of the solid element equals atomic volume (milliliters).

Atomic heat is the amount of heat required to raise the temperature 1° C. of 1 atomic weight in grams of the solid element at a given temperature, therefore, atomic weight (grams) multiplied by specific heat of the solid element equals atomic heat (calories). The atomic heat values given are not exactly comparable— the temperature is as stated in each case. Omitting **beryllium, boron, carbon, sodium, silicon, potassium, titanium**, the remaining 23 elements show a value within 1% of 6.1 (Dulong and Petit's generalization of constancy of atomic heat (1819)). (For Molecular heat, see following section II, Compounds and related topics.) Atomic radius as determined by Neuburger (1936) using x-ray studies. Atomic compressibility is the fractional change in volume of the solid element corresponding to a given change in pressure at a constant temperature.

Due to the investigations initiated by J. J. Thomson (1913) using **neon** gas in a low vacuum discharge tube and examining the effects of **anode** rays, the discovery was made that neon consists of two (later three) masses of atoms about 90% of the 20 and about 10% of the 22 mass (later, 20 (90%), 21 (0.27%), 22 (9.73%)). The work in this field has progressed since 1919 in the

The following arrangement, using the accepted symbols (see **Atomic Weights**, page 288), summarizes the above:

PERIODIC CLASSIFICATION OF THE CHEMICAL ELEMENTS
ACCORDING TO MENDELÉEFF AND MOSELEY

Number	Group O	I	II	III	Sub-Group iv	v	vi	vii	viii			i	ii	iii	Group IV	V	VI	VII
I		H																
2–9	He	Li	Be	B											C	N	O	F
10–17	Ne	Na	Mg	Al											Si	P	S	Cl
18–35	A	K	Ca	Sc	Ti	V	Cr	Mn	Fe	Co	Ni	Cu	Zn	Ga	Ge	As	Se	Br
36–53	Kr	Rb	Sr	Y	Zr	Cb	Mo	(43)	Ru	Rh	Pd	Ag	Cd	In	Sn	Sb	Te	I
54–85	Xe	Cs	Ba	La–Lu Hf	Ta	W	Re	Os	Ir	Pt	Au	Hg	Tl	Pb	Bi	(84)	(85)	
				57–71 (Below)														
86–96	Rn	(87)	Ra	(89)	Th	Pa	U	**(93)**	**(94)**	(95)	(96)							

57–71 of Group III: La   Ce   Pr   Nd   (61)   Sa   Eu   Gd   Tb   Dy   Ho   Er   Tm   Yb   **Lu**

NOTES:

Element 43, masurium, announced by Walter Noddack and Ida Tacke 1925.

Element 61, illinium, announced by Harris, Yntema and Hopkins, 1926.

Element 84, radioactive element known.

Element 85, alabamine, announced by Allison and Murphy, 1929.

Element 87, virginium, announced by Allison and Murphy, 1929, and by Papish and Wainer, 1931.

Element 89, radioactive element, mesothorium 2.

Element 93, neptunium, a product of atomic bomb research, formed by radioactive change spontaneously from uranium 239 by emission from the latter of a beta particle (high energy electron).

Element 94, plutonium, a product of atomic bomb research, formed by radioactive change spontaneously from neptunium 239 by emission from the latter of a beta particle. (See **Atomic Bomb**.)

Element 95, americium, announced by Glenn T. Seaborg and co-workers in 1946. Possesses six 5F electrons corresponding to europium which has six 4F electrons.

Element 96, curium, announced by Glenn T. Seaborg and co-workers in 1946. Possesses seven 5F electrons corresponding to gadolinium which has seven 4F electrons.

Groups by name:

Group O, rare or noble gases
Group I, Li–Cs, alkali metals
Group II, alkaline earth metals
Group III, Sc–Lu, rare earth metals
Group viii, Ru, Rh, Pd, Os, Ir, Pt, rare or noble metals
Group VII, halogens

---

hands of Aston and co-workers to such a degree of accuracy that the atomic weight of **hydrogen** by this method of the mass-spectrograph is, since 1929, the accepted value. Substances which are recognized from their Mendeléeff-Moseley properties as chemical elements and which are *not* made up of atoms which are all of

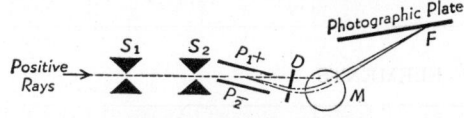

Aston's mass-spectrograph.

the same weight, as neon above, are said to consist of isotopes. Isotopes of a given element possess the same chemical properties but different masses. That is, neon 20, 21, 22 are chemically indistinguishable isotopes. Physical confirmation of the existence of isotopes has been obtained by separation of neon into two fractions of different densities by effusion (Aston, 1920), (Hertz, 1932). The partial separation of isotopes of a given element has been successfully accomplished also in the case of **mercury** by Brönsted and Hevesy (1920) using fractional **distillation**; of **chlorine** by Brönsted and Hevesy (1920) and by Harkins and Hayes (1921) by fractional diffusion through pipeclay; and of **hydrogen** by Urey (1932). Washburn and Urey (1932) found that the residual water of electrolytic cells, which had

been operated for years without drawing off the caustic solution, contained a marked increase in abundance of mass 2 hydrogen relative to mass 1 (see **Water**).

Isotopes of radioactive elements are discussed under **Radioactive Changes.**

A table of isotopes of the elements, arranged in two columns, the one containing the odd-numbered, the other the even-numbered elements, is presented. (Page 290.) Odd-numbered elements have at the most two isotopes (Aston). Of 38 odd-numbered elements, 19 are single; of 41 even-numbered elements, 4 only are single. Isobares are elements of the same mass but different chemical properties, e.g., titanium 50 and chromium 50.

The method of positive or anode-ray analysis by the mass-spectrograph involves the use of a gaseous or volatilizable element or compound of the element. The blank spaces in the preceding table are accounted for by the fact that no satisfactory volatile compound has yet been found for the elements in question.

Allotropic elements are those which present different physical forms of the same elementary chemical composition. The energy content and the volume of the different forms are not the same, and frequently a given temperature determines the transformation of one form to the other. In the case of **oxygen** and ozone the composition O_2 and O_3, respectively, makes it easy to understand the differences in behavior, but in other cases, different internal arrangements of the substances must be assumed.

INTERNATIONAL TABLE OF ATOMIC WEIGHTS

1944

	Symbol	Atomic Number	Atomic Weight		Symbol	Atomic Number	Atomic Weight
Aluminum	Al	13	26.97	Molybdenum	Mo	42	95.5
Antimony	Sb	51	121.76	Neodymium	Nd	60	144.27
Argon	A	18	39.944	Neon	Ne	10	20.183
Arsenic	As	33	74.91	Nickel	Ni	28	58.69
Barium	Ba	56	137.36	Nitrogen	N	7	14.008
Beryllium	Be	4	9.02	Osmium	Os	76	190.2
Bismuth	Bi	83	209.00	Oxygen	O	8	16.0000
Boron	B	5	10.82	Palladium	Pd	46	106.7
Bromine	Br	35	79.916	Phosphorus	P	15	31.02
Cadmium	Cd	48	112.41	Platinum	Pt	78	195.23
Calcium	Ca	20	40.08	Potassium	K	19	39.096
Carbon	C	6	12.010	Praseodymium	Pr	59	140.92
Cerium	Ce	58	140.13	Protactinium	Pa	91	231
Cesium	Cs	55	132.91	Radium	Ra	88	226.05
Chlorine	Cl	17	35.457	Radon	Rn	86	222
Chromium	Cr	24	52.01	Rhenium	Re	75	186.31
Cobalt	Co	27	58.94	Rhodium	Rh	45	102.91
Columbium	Cb	41	92.91	Rubidium	Rb	37	85.44
Copper	Cu	29	63.57	Ruthenium	Ru	44	101.7
Dysprosium	Dy	66	162.46	Samarium	Sm	62	150.43
Erbium	Er	68	167.2	Scandium	Sc	21	45.10
Europium	Eu	63	152.0	Selenium	Se	34	78.96
Fluorine	F	9	19.00	Silicon	Si	14	28.06
Gadolinium	Gd	64	157.3	Silver	Ag	47	107.880
Gallium	Ga	31	69.72	Sodium	Na	11	22.997
Germanium	Ge	32	72.60	Strontium	Sr	38	87.63
Gold	Au	79	197.2	Sulfur	S	16	32.06
Hafnium	Hf	72	178.6	Tantalum	Ta	73	180.88
Helium	He	2	4.003	Tellurium	Te	52	127.61
Holmium	Ho	67	164.94	Terbium	Tb	65	159.2
Hydrogen	H	1	1.0080	Thallium	Tl	81	204.39
Indium	In	49	114.76	Thorium	Th	90	232.12
Iodine	I	53	126.92	Thulium	Tm	69	169.4
Iridium	Ir	77	193.1	Tin	Sn	50	118.70
Iron	Fe	26	55.85	Titanium	Ti	22	47.90
Krypton	Kr	36	83.7	Tungsten	W	74	183.92
Lanthanum	La	57	138.92	Uranium	U	92	238.14
Lead	Pb	82	207.22	Vanadium	V	23	50.95
Lithium	Li	3	6.940	Xenon	Xe	54	131.3
Lutecium	Lu	71	174.99	Ytterbium	Yb	70	173.04
Magnesium	Mg	12	24.32	Yttrium	Y	39	88.92
Manganese	Mn	25	54.93	Zinc	Zn	30	65.38
Mercury	Hg	80	200.61	Zirconium	Zr	40	91.22

ATOMIC CONSTANTS OF ELEMENTS

		Atomic Volume of Solid	Atomic Heat of Solid at ° C.	Atomic Radius 10^{-8} cm. Neuburger	Atomic Compressibility at 20° C. $\times 10^7$
1.	Hydrogen	13.21	....	0.37	
2.	Helium				
3.	Lithium	11.8	6.6 (50°)	1.50	9.0
4.	Beryllium	5.3	4.5 (0–300°)	1.11	
5.	Boron	4.5	5.5 (900°)	0.7	0.3
6.	Carbon	5.6 (graphite)	5.5 (900°) (graphite)	0.77	3.0 (graphite)
7.	Nitrogen	13.6		0.53	
8.	Oxygen	11.2		0.60	
9.	Fluorine	....	....	0.68	
10.	Neon	....		1.60	
11.	Sodium	22.9	6.8 (20°)	1.86	15.6
12.	Magnesium	14.0	6.0 (20°)	1.60	2.9

ATOMIC CONSTANTS OF ELEMENTS—*Continued*

	ATOMIC VOLUME OF SOLID	ATOMIC HEAT OF SOLID AT °C.	ATOMIC RADIUS 10^{-8} CM. NEUBURGER	ATOMIC COMPRESSIBILITY AT 20° C. $\times 10^7$
13. Aluminum...................	10.2	5.8 (20°)	1.48	1.47
14. Silicon...................	11.4	4.7 (14°)	1.17	0.32
15. Phosphorus...................	16.9 (yellow)	5.9 (90°) (yellow)	1.08	9.2 (yellow)
16. Sulfur...................	15.3 (rhombic)	5.7 (15–96°) (rhombic)	1.06	12.9 (rhombic)
17. Chlorine...................	....	....	0.97	
18. Argon...................	....	....	1.91	
19. Potassium...................	45.3	7.0 (14°)	2.27	31.7
20. Calcium...................	25.9	5.8 (0–20°)	1.97	5.7
21. Scandium...................	....	....	1.51	
22. Titanium...................	9.3	5.4 (0–100°)	1.45	
23. Vanadium...................	8.8	....	1.31	
24. Chromium...................	7.7	5.8 (18–100°)	1.25	0.9
25. Manganese...................	7.4	6.6 (20–100°)	1.24	0.84
26. Iron...................	7.1	6.0 (20°)	1.24	0.63
27. Cobalt...................	6.8	5.9 (20°)	1.25	
28. Nickel...................	6.6	6.2 (20°)	1.24	0.40
29. Copper...................	7.1	5.9 (15–100°)	1.28	0.75
30. Zinc...................	9.2	6.2 (0–100°)	1.33	1.7
31. Gallium...................	11.8	....	1.22	2.09
32. Germanium...................	13.6	....	1.22	
33. Arsenic...................	14.8	6.2 (0–100°)	1.25	4.5
34. Selenium...................	16.5	....	1.16	12.0
35. Bromine...................	....	....	1.13	
36. Krypton...................	....		2.0	
37. Rubidium...................	56.2	....	2.43	40.0
38. Strontium...................	34.5	....	2.14	
39. Yttrium...................	19.5	....		
40. Zirconium...................	14.0		1.58	
41. Columbium...................	12.7		1.43	
42. Molybdenum...................	10.7	....	1.36	0.46
43.				
44. Ruthenium...................	8.3	....	1.32	
45. Rhodium...................	8.5	....	1.34	
46. Palladium...................	9.0	....	1.37	0.54
47. Silver...................	10.3	6.0 (20°)	1.44	1.01
48. Cadmium...................	13.0	6.2 (28°)	1.49	2.1
49. Indium...................	15.1	....	1.57	
50. Tin...................	16.3	6.4 (18°)	1.51	1.9
51. Antimony...................	18.2	6.1 (20–100°)	1.44	2.4
52. Tellurium...................	20.4	....	1.44	
53. Iodine...................	5.9	6.6 (20°)	1.35	13.0
54. Xenon...................	...	....	2.2	
55. Cesium...................	70.4	....	2.62	61.0
56. Barium...................	39.0		2.17	
57–71.				
72. Hafnium...................	13.4	....	1.59	
73. Tantalum...................	10.9	....	1.46	0.53
74. Tungsten...................	9.8	....	1.41	0.27
75. Rhenium...................	8.8		1.38	
76. Osmium...................	8.5	....	1.34	
77. Iridium...................	8.6	....	1.35	
78. Platinum...................	8.7	6.3 (20°)	1.38	0.38
79. Gold...................	10.2	6.2 (0–100°)	1.44	0.64

(*Continued on next page*)

ATOMIC CONSTANTS OF ELEMENTS—*Continued*

		ATOMIC VOLUME OF SOLID	ATOMIC HEAT OF SOLID AT °C.	ATOMIC RADIUS 10^{-8} CM. NEUBURGER	ATOMIC COMPRESSIBILITY AT 20° C. $\times 10^7$
80.	Mercury				
81.	Thallium..........................	17.2	....	1.71	2.3
82.	Lead..............................	18.2	6.3 (20°)	1.75	2.33
83.	Bismuth...........................	21.3	6.1 (20°)	1.82	3.0
84.					
85.					
86.	Radon				
87.					
88.	Radium				
89.					
90.	Thorium...........................	19.2	....	1.82	
91.	Protoactinium				
92.	Uranium...........................	12.8			

STABLE ISOTOPES OF THE ELEMENTS AND THEIR PERCENTAGE

At. No.	ELEMENT	Mass Number of Isotope and Percentage — Odd No. Elements		Even No. Elements		At. No.	ELEMENT	Mass Number of Isotope and Percentage — Odd No. Elements		Even No. Elements	
1	Hydrogen.........	1	99.98			22	Titanium..........	...	.....	46	8.5
		2	0.02							47	7.8
2	Helium............	...	.....	4	100					48	71.3
3	Lithium...........	6	7.9							49	5.5
		7	92.1							50	6.9
4	Beryllium.........	...		9	99.95	23	Vanadium.........	51	100		
5	Boron.............	10	20			24	Chromium.........	...	.....	50	4.9
		11	80							52	81.6
6	Carbon............	...	.....	12	99.3					53	10.4
				13	0.7					54	3.1
7	Nitrogen..........	14	99.62			25	Manganese........	55	100		
		15	0.38			26	Iron..............	...	.....	54	6.5
8	Oxygen............	...	.....	16	99.76					56	90.2
				17	0.04					57	2.8
				18	0.20					58	0.5
9	Fluorine..........	19	100			27	Cobalt............	57	0.2		
10	Neon..............	...	.....	20	90.00			59	99.8		
				21	0.27	28	Nickel............	...	.....	58	66.4
				22	9.73					60	26.7
11	Sodium...........	23	100							61	1.6
12	Magnesium........	...	.....	24	77.4					62	3.7
				25	11.5					64	1.6
				26	11.1	29	Copper...........	63	68		
13	Aluminum........	27	100					65	32		
14	Silicon............	...	.....	28	89.6	30	Zinc..............	...	.....	64	50.4
				29	6.2					66	27.2
				30	4.2					67	4.2
15	Phosphorus........	31	100							68	17.8
16	Sulfur............	...	.....	32	96					70	0.4
				33	1	31	Gallium...........	69	61.2		
				34	3			71	38.8		
17	Chlorine..........	35	76			32	Germanium........	...	.....	70	21.2
		37	24							72	27.3
18	Argon.............	...	.....	36	0.31					73	7.9
				38	0.06					74	37.1
				40	99.63					76	6.5
19	Potassium.........	39	93.4			33	Arsenic...........	75	100		
		40	0.01			34	Selenium..........			74	0.9
		41	6.6							76	9.5
20	Calcium...........	...	.....	40	96.76					77	8.3
				42	0.77					78	24.0
				43	0.17					80	48.0
				44	2.30					82	9.3
21	Scandium.........	45	100			35	Bromine..........	79	50.6		
								81	49.4		

STABLE ISOTOPES OF THE ELEMENTS AND THEIR PERCENTAGE—*Continued*

At. No.	Element	MASS NUMBER OF ISOTOPE AND PERCENTAGE				At. No.	Element	MASS NUMBER OF ISOTOPE AND PERCENTAGE			
		Odd No. Elements		Even No. Elements				Odd No. Elements		Even No. Elements	
36	Krypton	...	.....	78	0.35	54	Xenon	...	.....	124	0.09
				80	2.01					126	0.09
				82	11.53					128	1.90
				83	11.53					129	26.23
				84	57.11					130	4.07
				86	17.47					131	21.17
37	Rubidium	85	72.8							132	26.96
		87	27.2							134	10.54
38	Strontium	...	.....	84	0.5					136	8.95
				86	9.6	55	Cesium	133	100		
				87	7.5	56	Barium	...	.....	130	0.16
				88	82.4					132	0.015
39	Yttrium	89	100							134	1.72
40	Zirconium	...	.....	90	48.0					135	5.7
				91	11.5					136	8.5
				92	22.0					137	10.8
				94	17.0					138	73.1
				96	1.5	57	Lanthanum	139	100		
41	Columbium	93	100			58	Cerium	...	.....	140	89
42	Molybdenum	...	.....	92	14.2					142	11
				94	10.0	59	Praseodymium	141	100		
				95	15.5	60	Neodymium	...	.....	142	25.95
				96	17.8					143	13.0
				97	9.6					144	22.6
				98	23.0					145	9.2
				100	9.8					146	16.5
43						61				148	6.8
44	Ruthenium	...	.....	96	5					150	5.95
				98	?						
				99	12	62	Samarium	...	.....	144	3
				100	14					147	17
				101	22					148	14
				102	30					149	15
45	Rhodium	101	0.1	104	17					150	5
		103	99.9							152	26
46	Palladium	...	.....	102	0.8					154	20
				104	9.3						
				105	22.6	63	Europium	151	50.6		
				106	27.2			153	49.4		
				108	26.8	64	Gadolinium	...	.....	155	21
				110	13.5					156	23
47	Silver	107	52.5							157	17
		109	47.5							158	23
48	Cadmium	...	.....	106	1.5					160	16
				108	1.0	65	Terbium	159	100		
				110	15.6	66	Dysprosium	...	.....	161	22
				111	15.2					162	25
				112	22.0					163	25
				113	14.7					164	28
				114	24.0	67	Holmium	165	100		
				116	6.0	68	Erbium	...	.....	166	36
49	Indium	113	4.5							167	24
		115	95.5							168	30
50	Tin	...	.....	112	1.1					170	10
				114	0.8	69	Thulium	169	100		
				115	0.4	70	Ytterbium	...	.....	171	9
				116	15.5					172	24
				117	9.1					173	17
				118	22.5					174	38
				119	9.8					176	12
				120	28.5	71	Lutecium	175	100		
				122	5.5	72	Hafnium	...	.....	176	5
				124	6.8					177	19
51	Antimony	121	56							178	28
		123	44							179	18
52	Tellurium	...	.....	122	2.9					180	30
				123	1.6	73	Tantalum	181	100		
				124	4.5	74	Tungsten	...	.....	180	0.2
				125	6.0					182	22.6
				126	19.0					183	17.3
				128	32.8					184	30.1
				130	33.1					186	29.8
53	Iodine	127	100								

(*Continued on next page*)

STABLE ISOTOPES OF THE ELEMENTS AND THEIR PERCENTAGE—*Continued*

At. No.	Element	Mass Number of Isotope and Percentage — Odd No. Elements		Even No. Elements		At. No.	Element	Mass Number of Isotope and Percentage — Odd No. Elements		Even No. Elements	
75	Rhenium..........	185	38.2							201	13.17
		187	61.8							202	29.56
76	Osmium..........	...	.....	184	0.02					204	6.72
				186	1.58	81	Thallium..........	203	29.4		
				187	1.64			205	70.6		
				188	13.3	82	Lead..............	...	.....	204	1.5
				189	16.2					206	23.5
				190	26.4					207	22.7
				192	40.9					208	52.3
77	Iridium...........	191	38.5			83	Bismuth..........	209	100		
		193	61.5			84					
78	Platinum..........	...	.....	192	0.8	85					
				194	30.2	86	Radon				
				195	35.3	87					
				196	26.6	88	Radium				
				198	7.2	89					
79	Gold.............	197	100			90	Thorium..........	...	.....	232	100
80	Mercury..........	...	.....	196	0.15	91	Protoactinium......	...	.....		
				198	10.11	92	Uranium..........	...	.....	234	0.006
				199	17.03					235	0.7
				200	23.26					238	99.3

SOME ALLOTROPIC ELEMENTS AND THEIR DISTINGUISHING PROPERTIES

Element	Crystal Form	Density	Melting Point	Transformation Temperature, °C.	Heat of Combustion, Calories Per Gram
Carbon:					
Diamond.............	Cubic	3.51 (20° C.)	....		7870
Graphite, natural.......	Hexagonal	2.25 (20° C.)	....	Stable form 400–800	7854
Amorphous.......... (wood charcoal)	....	....	....		8080
Oxygen:					
Oxygen...............	(Gas)	1.105 (air 1)	−218	....	Zero } Heat of
Ozone................	(Gas)	1.658 (air 1)	−251	....	−720 } formation
Phosphorus:					
Yellow (white)........	Hexagonal....	1.82 (20° C.)	44		
Red.................	Cubic	2.20 (20° C.)	590 (43 atm.)		
Black...............	Rhombohedral	2.69			
Sulfur:					
Rhombic..............	Rhombic	2.07	113	Stable below } 95.6	{ 2220
Monoclinic............	Monoclinic	1.96	119	Stable above }	2240
Iron:					
Alpha-ferrite (magnetic).	Cubic, body centered....	....	....	Stable below } 769	
Beta-ferrite...........	Cubic, body centered....	....	....	Stable above }	
Beta-ferrite...........	....	....	....	Stable below } 906	
Gamma-ferrite........	Cubic, face centered....	....	....	Stable above }	
Gamma-ferrite........	....	....	....	Stable below } 1404	
Delta-ferrite..........	Cubic, body centered....	....	1535	Stable above }	
Cobalt................ (Magnetic below 1150° C.)	....	....	....	444 1150	
Nickel (Magnetic below 350° C.)	....	....	....	350	

SOME ALLOTROPIC ELEMENTS AND THEIR DISTINGUISHING PROPERTIES—*Continued*

ELEMENT	CRYSTAL FORM	DENSITY	MELTING POINT	TRANSFORMATION TEMPERATURE, °C.	HEAT OF COMBUSTION, CALORIES PER GRAM
Zinc:					
Alpha.................	....	....	....	Stable below } 174	
Beta..................	....	....	....	Stable above }	
Beta..................	....	....	....	Stable below } 322	
Gamma...............	....	....	....	Stable above }	
Tin:					
White, ordinary.......	Tetragonal	6.55	....	Stable above } 18–20	
Gray.................		5.80	....	Stable below }	
Rhombic, white.......	Rhombic	7.20	....	Stable above Ord. stable below } 195	

II. Compounds and Related Topics

When two or more chemical elements are present in a substance possessing properties, e.g., melting point, boiling point, density, percentage composition by elements, that are constant and individual for the substance, the given substance is regarded as a compound.

Water and carbon dioxide are two of the most important substances of everyday life that are definite compounds. See **Water,** and **Carbon, oxides,** for full description of these compounds, also **Hydrogen Peroxide** for another compound which contains the same elements, namely, hydrogen and oxygen, as water, and carbon, oxides, for other compounds which contain the same elements, namely, carbon and oxygen, as carbon dioxide—these are carbon monoxide and carbon suboxide.

The formulae of gaseous compounds are obtained from a study of the composition by elements and the density, by a method introduced by the Italian chemist, Cannizzaro in 1858. Later, in 1872, in the course of his Faraday Lecture before the Chemical Society (London) on the subject "Some Points in the Teaching of Chemistry" Cannizzaro stated that "Symbols and formulas, in my opinion, constitute the introduction, preparation, and base of the study of the transformations of matter, which is the true object of our science." The simplest way to understand the method is to arrange in tabular form, (1) the individual gases, (2) the weight in grams

DISPLAY OF DATA TO ILLUSTRATE THE CANNIZZARO METHOD OF ARRIVING AT THE SYMBOL AND SYMBOL WEIGHT OF CHEMICAL ELEMENTS, AND THE FORMULA AND FORMULA WEIGHT OF CHEMICAL COMPOUNDS

GAS	GRAMS PER 1 STANDARD LITER	PERCENTAGE COMPOSITION BY CHEMICAL ELEMENTS	GRAMS PER 1 STANDARD LITER BY CHEMICAL ELEMENTS					
			Hydrogen	Oxygen	Carbon	Nitrogen	Sulfur	Chlorine
1. Hydrogen chloride.	1.639	Hydrogen 2.76% Chlorine 97.24	0.045	....	....	....	....	1.594
2. Ammonia.........	0.771	Hydrogen 17.75 Nitrogen 82.25	0.137	....	....	0.634		
3. Carbon dioxide....	1.977	Oxygen 72.73 Carbon 27.27	....	1.438	0.539			
4. Carbon monoxide..	1.250	Oxygen 57.14 Carbon 42.86	....	0.714	9.536			
5. Methane..........	0.717	Hydrogen 25.14 Carbon 74.86	0.180	....	0.537			
6. Ethylene..........	1.260	Hydrogen 14.38 Carbon 85.62	0.181	....	1.079			
7. Acetylene.........	1.173	Hydrogen 7.75 Carbon 92.25	0.091	....	1.082			
8. Oxygen...........	1.429	Oxygen 100.00	....	1.429				
9. Hydrogen.........	0.090	Hydrogen 100.00	0.090					
10. Nitrogen..........	1.251	Nitrogen 100.00				1.251		
11. Chlorine..........	3.214	Chlorine 100.00	....	....	....	....		3.214
12. Sulfur dioxide.....	2.927	Oxygen 49.95 Sulfur 50.05	....	1.462	....	....	1.465	
13. Hydrogen sulfide...	1.539	Hydrogen 5.91 Sulfur 94.09	0.091	....	....	....	1.448	
14. Nitrous oxide......	1.978	Oxygen 36.35 Nitrogen 63.65	....	0.719	....	1.259		
15. Nitric oxide.......	1.340	Oxygen 53.32 Nitrogen 46.68	....	0.715	....	0.625		
Minimum weight (approximate).........	....	....	0.045	0.715	0.538	0.626	1.45	1.60

of 1 liter (at 0° C., 760 mm. of mercury pressure) of each gas, (3) the weight in grams of *each element* present in the above volume (1 standard liter) found by exact analysis (percentage composition by chemical elements using the methods of analytical chemistry), as on the preceding page.

Careful examination of the figures in the last six columns reveals the experimental fact that (1) in each separate vertical column the figures represent a minimum weight or a small multiple (approximately) of this weight, (2) the smallest of the six minimum weights is that for hydrogen, namely, 0.045 gram in 1 standard liter of hydrogen chloride gas.

The next step involves changing 0.045 gram of hydrogen to exactly 1.000 gram and finding arithmetically the volume of hydrogen chloride containing this weight (1.000 gram hydrogen). The volume is found to be 22.2 standard liters.

Therefore, 1.000 gram minimum weight of hydrogen is contained in 22.2 standard liters of hydrogen chloride.

of atomic weights. See preceding section I, Elements and related topics.

One formula volume of a gas is 22.2 liters. It is necessary to state that actual gases under ordinary conditions show some variation from this value, so that for accurate work the records should be consulted in each case. The volume, 22.4 liters (1% larger than used in the above deductions) is commonly used.

Summarizing, the formula "HCl" states that "36.5 grams of hydrogen chloride gas occupies a standard volume of 22.2 liters and is composed of 1 gram of hydrogen element chemically united with 35.5 grams of chlorine element." The reason for the formulae of the simple gases, oxygen, O_2, hydrogen, H_2, nitrogen, N_2, chlorine, Cl_2, is apparent from the general method of deduction. The formula O_2 represents 22.2 liters or 32 grams of oxygen *gas*, whereas O represents 16 grams of oxygen *element* in any substance.

Other constants of **molecules** are similar in kind to those referred to under atoms (see preceding section I,

---

Using this standard volume of 22.2 liters, the next step is to ascertain the minimum weight of the other elements in this volume.

Chemical Element	Hydrogen	Oxygen	Carbon	Nitrogen	Sulfur	Chlorine
Approximate minimum weight in grams of each of the six chemical elements in the standard volume, 22.2 liters.................................	1 gram	16 grams	12 grams	14 grams	32 grams	35.5 grams

---

Then, the abbreviation is introduced of representing:

SYMBOL WEIGHTS OF EACH ELEMENT BY THE SYMBOLS

1 gram	of hydrogen	by the symbol	H
16 grams	of oxygen	" " "	O
12 grams	of carbon	" " "	C
14 grams	of nitrogen	" " "	N
32 grams	of sulfur	" " "	S
35.5 grams	of chlorine	" " "	Cl

By setting up again the second half of the table for the 15 gases, this time for 22.2 standard liters instead of 1 standard liter, the results obtained may be observed in the table on the following page.

Thus, it is seen, the chemical formulae and formula weights (last column) of 15 gaseous chemical compounds have been arrived at, using the Cannizzaro method, by purely experimental and rational means, involving no theoretical considerations. Extension of the method serves to ascertain the chemical formula of all gases and vaporizable substances. For compounds which are neither gases nor vaporizable, other methods are available. Of these the most used are those of Raoult depending upon the depression of the freezing point or the elevation of the boiling point of a compound dissolved in a given **solvent**. (See **Solutions**.)

It remains to be noted that, when there is no method available for ascertaining the formula weight of a compound, the *simplest* formula, based on chemical analysis and the use of symbol weights of the contained elements, is used, e.g., ferric oxide, Fe_2O_3, ferroferric oxide, Fe_3O_4, ferrous oxide, FeO, cupric oxide (black copper oxide), CuO, cuprous oxide (red copper oxide), Cu_2O. The customary formula of water is H_2O, which is correct at temperatures above 100° C.—actually, liquid water is mainly dihydrol $(H_2O)_2$.

It should be understood from the above discussion that a chemical formula is no chance throwing together of chemical symbols, but represents the results of careful analysis, and the scrutiny and deduction of the most skillful workers in the field. On this score alone, chemical formulae demand the greatest respect in understanding and use.

Symbol weights and atomic weights are used synonymously. Formula weights and molecular weights are used synonymously. Unless otherwise stated, symbol weights and formula weights are expressed in grams, and the numbers used are those taken from the accepted list

Elements and related topics), namely, hardness, crystal form, melting point, boiling point, vapor pressure, molecular volume, molecular heat (see below), molecular dimensions of gases, allotropy and transformation temperatures. Molecular properties of great significance in organic chemistry are those of **isomerism** and stereoisomerism, and molecular dissociation, and **association**. A molecule of a chemical compound may be defined as the smallest unit which retains the characteristic properties of the compound.

It has become customary in chemical literature to use the formula of a substance as an accepted abbreviation for the name of the substance, especially in cases of frequent repetition.

The molecular heat of a solid compound is equal to the sum of the atomic heats (see preceding section I, Elements and related topics) of the elements of the compound. This generalization is the result of investigations by Kopp (1864). Since several elements that are commonly found in solid compounds are not themselves solids or are not, at ordinary temperature, in agreement with the atomic heat constant, 6.1, calculated values from experimental observations of the molecular heat of solid compounds have been assigned to them. Nernst recommended that the following values be used:

Oxygen............	4.0	Phosphorus........	5.4
Hydrogen..........	2.3	Sulfur..............	5.4
Carbon............	1.8	Boron..............	3.7
Silicon............	3.8	Fluorine...........	5.0

The term molar fraction is used to express the ratio of the number of gram mols of one compound present in a system to the total number of gram mols of all the compounds present in that system. Thus, formaldehyde solution is approximately 40% formaldehyde (HCHO, formula weight 30) and 60% water (H_2O, formula weight 18)

$$\frac{\text{number of mols HCHO}}{\text{number of mols HCHO} + \text{number of mols } H_2O} = \frac{\frac{40}{30}}{\frac{40}{30} + \frac{60}{18}}$$

$$= \frac{1.33}{4.66} = 0.285, \quad \text{molar fraction of formaldehyde.}$$

This value multiplied by 100 gives the molecular percentage, namely, 28.5 molecular per cent formaldehyde

DERIVATION OF FORMULAE AND FORMULA WEIGHTS OF GASES, HAVING GIVEN THE PERCENTAGE COMPOSITION BY CHEMICAL ELEMENTS OF EACH GAS AND THE SYMBOLS AND SYMBOL WEIGHTS OF THE ELEMENTS CONTAINED

Gas.... { Symbol Weight / Symbol......	IN 22.2 LITERS						FORMULA OF GAS	GRAMS OF SAME IN 22.2 LITERS
	1 g. H	16 g. O	12 g. C	14 g. N	32 g. S	35.5 g. Cl		
1. Hydrogen chloride...	1					1	HCl	36.5
2. Ammonia..........	3			1			NH₃	17
3. Carbon dioxide.....		2	1				CO₂	44
4. Carbon monoxide....		1	1				CO	28
5. Methane....:......	4		1				CH₄	16
6. Ethylene..........	4		2				C₂H₄	28
7. Acetylene..........	2		2				C₂H₂	26
8. Oxygen...........		2					O₂	32
9. Hydrogen.........	2						H₂	2
10. Nitrogen..........				2			N₂	28
11. Chlorine..........						2	Cl₂	71
12. Sulfur dioxide.......		2			1		SO₂	64
13. Hydrogen sulfide.....	2				1		H₂S	34
14. Nitrous oxide.......		1		2			N₂O	44
15. Nitric oxide........		1		1			NO	30

solution (40 weight per cent of formaldehyde solution. For "volume per cent" see **Ethyl Alcohol**).

III. Radicals

In many chemical compounds there are groups of two or more elements that frequently have the properties of or enter into chemical reaction as a unit. Of those which are of outstanding importance the following are cited:

(1) **Ammonium** (NH₄—) behaves as a unit in ammonium compounds and in some of these compounds is very similar to **potassium** (K—) in potassium compounds.

(2) Hydroxyl (—OH) which behaves as a unit in bases (e.g., **sodium** hydroxide, NaOH), **alcohols** (e.g., methyl alcohol, CH₃OH), and **phenols** (e.g., phenol, C₆H₅OH).

(3) Anion-groups of acids, their salts and their esters: **Sulfate** (>SO₄), sulfite (>SO₃), nitrate (—NO₃), nitrite (—NO₂), phosphate (≥PO₄), perchlorate (—ClO₄), chlorate (—ClO₃), chlorite (—ClO₂), hypochlorite (—OCl), carbonate (>CO₃), formate (—CHO₂), acetate (—C₂H₃O₂), palmitate (—C₁₆H₃₁O₂), stearate (—C₁₈H₃₅O₂), oleate (—C₁₈H₃₃O₂), oxalate (>C₂O₄), lactate (—C₃H₅O₃), malate (>C₄H₄O₅), tartrate (>C₄H₄O₆), citrate (≥C₆H₅O₇), benzoate (—C₇H₅O₂), cinnamate (—C₉H₇O₂), phthalate (>C₈H₄O₄), salicylate (—C₇H₅O₃). See **Acids, Carboxylic**, for others.

(4) Alkyl- and aryl-groups of alcohols, phenols, their esters and their alcoholates and phenolates: (a) Alkyl (non-benzenoid)-methyl (CH₃—), ethyl (C₂H₅—), propyl (C₃H₇—), butyl (C₄H₈—) and similar radicals of alcohols (see **Alcohols** for others); (b) Aryl (benzenoid)-phenyl (C₆H₅—), tolyl (C₇H₇—), xylyl (C₈H₉—), naphthyl (C₁₀H₇—) and similar radicals of phenols. (See **Phenols** for others.)

(5) Acyl-groups of organic acids: acetyl (CH₃CO—), benzoyl (C₆H₅CO—). (See **Acids, Carboxylic**, for others.)

(6) Miscellaneous radicals, for example, cacodyl ((CH₃)₂As—), celebrated on account of the investigations by Bunsen (1838).

All of the above radicals are associated with a corresponding radical or element in a compound. While a radical frequently and rather generally enters into chemical reaction as a unit, it is not implied that this is always so, the stability in each case is characteristic of each radical and each reaction in which it is involved. Thus, **ammonium** hydroxide (NH₄OH) yields **ammonia** gas (NH₃) and water (H₂O) at room temperature; ammonium nitrate (NH₄NO₃) is decomposed, upon heating, with the accompanying disruption of both the ammonium and nitrate radicals to yield **nitrous oxide** (N₂O) gas and water (H₂O).

Radicals enter widely into reactions involving electrolytic dissociation of salts, acids, bases in water solution. (See **Reactions Involving Recombinations of Ions**.)

Except in rare cases, free radicals are not encountered. Gomberg (1900), by treating triphenylmethyl chloride in an atmosphere of **carbon dioxide**, with zinc, silver, or mercury, obtained the free radical, triphenylmethyl. On dissolving the colorless solid in organic solvents a yellow solution is obtained, and the reactivity (due to unsaturation) of the yellow solution is marked towards **oxygen**, dissolved **iodine, ether**. Triphenylmethyl is present in solution in two forms, (1) monomolecular ((C₆H₅)₃C) yellow, in equilibrium with (2) dimolecular ((C₆H₅)₃C)₂ colorless. But tribiphenylmethyl ((C₆H₅—C₆H₄)₃C) occurs only in the monomolecular form, purple. The action of alkali metals on **ketones** in some cases produces metallic ketyl (Schlenk, 1913) thus: $\frac{R'}{R''}$>C—ONa, which is a free radical, or contains trivalent carbon as does monomolecular triphenylmethyl.

IV. Equivalents, Valency, Graphical Formulas

In the preceding section on radicals, each radical is written with one or more accompanying lines, characteristic for each, (a) on the right-hand side for ammonium, alkyl, aryl, acyl, and (b) on the left-hand side for hydroxyl, anion groups. The metals in their salt compounds would be similarly written (a) on the right-hand side, thus, sodium (Na—), calcium (Ca<), aluminum (Al<), (also hydrogen (H—) of water, acids, etc.), and the non-metals in their salt compounds would be similarly written (b) on the left-hand side, thus, chloride (—Cl), bromide (—Br), iodide (—I), sulfide (>S), nitride (≥N), (also oxygen (>O) of water, oxides, etc.). Compounds that are known to consist of the union of these (a) and (b) radicals and elements are found to be united "bond for bond," thus water,

(H_2O) ($H-OH$ or $H_2{>}O$); hydrochloric acid, (HCl) $(H-Cl)$; sulfuric acid, (H_2SO_4) ($H_2{>}SO_2$ or $O_2{=}S(-OH)_2$); nitric acid, HNO_3 ($H-NO_3$ or $O_2{=}N-OH$); phosphoric acid, H_3PO_4 ($H_3{>}PO_4$ or $O{=}P(-OH)_3$); acetic acid, $HC_2H_3O_2$ ($H-C_2H_3O_2$ or CH_3CO-OH or $H-OOC\cdot CH_3$ or $H_3C-C(O)(-OH)$); oxalic acid, $H_2C_2O_4$ ($H_2{>}C_2O_4$ or $O{=}C-OH \,/\, O{=}C-OH$); acetyl chloride, (CH_3CO-Cl or $H_3C-C(O)(-Cl)$); silver chloride, $Ag-Cl$, cupric chloride, $CuCl_2$, green soluble solid ($Cu{<}Cl_2$); cuprous chloride, $CuCl$, white insoluble solid, or Cu_2Cl_2 (vapor) ($Cu-Cl$ or $Cu-Cl \,|\, Cu-Cl$); cupric oxide, CuO, black insoluble solid ($Cu{=}O$); cuprous oxide, Cu_2O, red insoluble solid ($Cu_2{>}O$); magnesium oxide, MgO, white insoluble solid ($Mg{=}O$); magnesium nitride, Mg_3N_2, yellow solid yielding ammonia with water ($Mg_3{:}N_2$); ammonium nitrate, NH_4NO_3 (NH_4-NO_3 or $H_2(H){>}N-N{=}O_2$); ethyl alcohol, C_2H_5OH (C_2H_5-OH or H_3C-CH_2-OH). Also, oxygen gas, O_2 ($O{=}O$); chlorine gas, Cl_2 ($Cl-Cl$); nitrogen gas, N_2 ($N{\equiv}N$); hydrogen gas, H_2 ($H-H$); cyanogen gas, $(CN)_2$ ($CN \,|\, CN$ or $C{\equiv}N \,|\, C{\equiv}N$).

VALENCY OF CERTAIN CHEMICAL ELEMENTS

Positive 7	6	5	4	3	2	1	o	Negative 1	2	3	4
						H^{1+} In water, acids, methane			O^{2-} In water, oxides, bases, alcohols, aldehydes		
		N^{5+} In nitrogen pentoxide, nitrates, nitro-compounds		N^{3+} In nitrogen trioxide, nitrites, nitroso-compounds	N^{2+} In nitric oxide					N^{3-} In ammonia, ammonium compounds ($N^{1+,\,4-}$)	
		P^{5+} In phosphorus pentoxide, phosphates		P^{3+} In phosphorus trioxide, phosphites		P^{1+} In hypophosphites				P^{3-} In phosphine, phosphonium compounds ($P^{1+,\,4-}$)	
	S^{6+} In sulfur trioxide, sulfates		S^{4+} In sulfur dioxide, sulfites						S^{2-} In sulfides		
Cl^{7+} In perchlorates		Cl^{5+} In chlorates		Cl^{3+} In chlorites		Cl^{1+} In hypochlorites		Cl^{1-} In chlorides			
			C^{4+} In carbon dioxide, carbonates	$C^{3+,\,1-}$ In carbon monoxide, formic acid	$C^{2+,\,2-}$ In formaldehyde				$C^{1+,\,3-}$ In methyl alcohol		C^{4-} In methane
			Si^{4+} In silicates	Al^{3+} In aluminum compounds	Ca^{2+} In calcium compounds	Na^{1+} In sodium compounds	Free elements $Na°, Ca°, Al°, Si°, Cu°, Fe°, Mn°, Cr°, C°, S°, P°$				
					Cu^{2+} In cupric compounds	Cu^{1+} In cuprous compounds					
				Fe^{3+} In ferric compounds	Fe^{2+} In ferrous compounds						
Mn^{7+} In permanganates	Mn^{6+} In manganates		Mn^{4+} In manganese dioxide	Mn^{3+} In manganic compounds	Mn^{2+} In manganous compounds						
	Cr^{6+} In chromates			Cr^{3+} In chromic compounds	Cr^{2+} In chromous compounds						

(See **Reactions Involving Oxidation, Reduction** (B) (2) for more complete list.)

The last formula in each of the above examples is known as the graphical formula of the compound. Graphical formulae are of great significance in the field of organic chemistry. They are frequently partially abbreviated, thus, ethyl alcohol, above, is often written CH_3—CH_2OH.

The number of lines shown accompanying each radical or element is a measure of its valency. Univalent radicals are equivalent. Higher valent radicals are equivalent "bond for bond." The univalent radicals, ammonium (NH_4—) and nitrate (—NO_3) are equivalent in *combining power* (ammonium nitrate NH_4NO_3). The divalent radical sulfate (>SO_4) and the univalent radical nitrate (—NO_3) are equivalent 1 to 2 in *replacing power* (copper sulfate $CuSO_4$, copper nitrate $Cu(NO_3)_2$). These values have been experimentally determined in each case, based on certain elementary assumptions, principally two, namely, (1) oxygen in most chemical compounds has a valency of minus two (>O^{2-}), (2) hydrogen in most chemical compounds has a valency of plus one (H—$^{1+}$). The cases where oxygen is zero (—$^{1+}O^{1-}$—) and hydrogen minus one (—H^{-1}) are rare (hydrogen peroxide

$$\begin{pmatrix} H^{+1}—{}^{-1}O^{+1} \\ \phantom{H^{+1}—}| \\ H^{+1}—{}^{-1}O^{-1} \end{pmatrix} \text{ and calcium hydride } \begin{pmatrix} Ca^{+2} \diagdown \begin{matrix} H^{-1} \\ H^{-1} \end{matrix} \end{pmatrix}$$

respectively.)

Grand Groups of Chemical Substances—Elements and Compounds. Certain grand groups of substances are recognized. The classification may be carried further with subdivisions extended as desired.

1. Acids, bases, salts (see **Acids, Bases, Salts; Individual Acids; Acids, Carboxylic**).

2. Metals, non-metals, alloys (see Individual elements: **Alloys**).

3. Oxides, chlorides, sulfides—most important grand group of 2-element compounds (see **Oxygen**, oxides; **Chlorine**, chlorides; **Sulfur**, sulfides; **Thermochemistry**).

4. Organic compounds (see **Chemistry, B. Organic Chemistry**). (R.K.S.)

CHEMICAL COORDINATION. The coordination of complex bodies by the reaction of various parts to substances whose occurrence within the body is conditioned by its own processes. A simple example is the regulation of the rate of respiration by the concentration of **carbon dioxide** in the blood. The more striking cases of chemical coordination concern special secretions produced by glands within the body and called **hormones**. (A.W.L.)

CHEMICAL DYE DESTRUCTION. A process of image formation used in processes of color photography. Such a process is based upon the chemical destruction of a dye present in the photographic emulsion, the quantity of dye destroyed being dependent quantitatively on the presence or absence of a silver image.

If a photographic emulsion is dyed with a suitable dye and, after exposure and processing to a silver image, is treated with a solution of hydrogen peroxide, the dye will be selectively destroyed. The finely divided particles of the silver image act as a catalyst and enable the peroxide to attack the dye. The dye image remaining will be the inverse of the silver image—dye-positive image obtained by treatment of a negative silver image. With suitable dyes and processing reagents, chemical dye destruction may be controlled by either silver or silver halides. Recent advances make it unnecessary for the colored dye to be put into the emulsion during manufacture, since colorless dye formers may be incorporated in the emulsions that may be converted into colorants shortly before the chemical step of dye destruction takes place.

Chemical dye destruction has been chiefly employed in photography as a method of color formation in integral tripacks (see **Bipacks and Tripacks**). The dyes used are generally quite stable and of excellent hue and saturation. (H.C.C.)

CHEMICAL ELEMENTS. Chemical Composition.

CHEMICAL ENGINEERING is that branch of engineering concerned with the development and application of manufacturing processes in which chemical or certain physical changes of materials are involved. These processes may usually be resolved into a coordinated series of unit physical operations and unit chemical processes. The work of the chemical engineer is concerned primarily with the design, construction, and operation of equipment and plants in which these unit operations and processes are applied. Chemistry, physics, and mathematics are the underlying sciences of chemical engineering, and economics its guide in practice. (D.E.M.)

CHEMICAL EVENTS. See **History and Evolution of Chemistry**.

CHEMICAL SOCIETIES AND PUBLICATIONS

1663 Royal Society (London) founded.
1665 Philosophical Transactions of the Royal Society first published. (Volume 200, 1903.)
1699 Memoires de l'Académie des Sciences de l'Institut de France, Paris, first published. Académie founded 1666, recognized by letters patent 1713.
1780 American Academy of Arts and Sciences founded. Memoirs 1785– Proceedings 1846–
1785 Chemical Society of the University of Edinburgh in existence. Date of founding not known. Students of Joseph Black.
1789 Annales de Chimie first published.
1792 Chemical Society of Philadelphia founded by James Woodhouse.
1818 American Journal of Science first published.
1831 British Association for the Advancement of Science founded.
1835 Comptes rendues de l'Académie des Sciences (Paris) first published. (Volume 201, 1935.)
1840 Chemical News (London) first published.
1841 Chemical Society (London) founded. Chartered 1848.
1848 American Association for the Advancement of Science founded. Journal of the Chemical Society (London) first published.
1857 Société Chimique de France founded. Bulletin 1858–
1863 National Academy of Sciences (U.S.A.) founded.
1867 Deutiche Chemische Gesellschaft founded.
1868 Berichte der Deutschen Chemischen Gesellschaft first published. (Volume 68, 1935.)
1869 Nature (London) first published.
1871 Oil, Paint and Drug Reporter founded.
1876 American Chemical Society founded. Analyst (London) first published.
1878 Institute of Chemistry (British) founded.
1879 Journal of the American Chemical Society first published. American Chemical Journal first published. Philosophical Magazine (London) first published.
1881 Society of Chemical Industry (British) founded.
1882 Journal of the Society of Chemical Industry first published.
1883 Science first published. (New Series 1895.)
1897 Journal of Physical Chemistry first published.
1902 Electrochemical Society (U.S.A.) founded. Transactions published. Chemical and Metallurgical Engineering founded as Electrochemical Industries.
1904 Annual Reports of the Progress of Chemistry first published.
1905 Journal of Biological Chemistry first published. Transactions Faraday Society first published.

1906 American Society of Biological Chemists organized. American Men of Science first published.

1907 Chemical Abstracts first published.

1908 American Institute of Chemical Engineers founded. Transactions published.

1909 Industrial and Engineering Chemistry first published (as Journal of Industrial and Engineering Chemistry).

1912 Eighth International Congress of Applied Chemistry, Washington and New York.

1914 Chemical Industries founded as Chemical Markets.

1915 Société de Chimie Industrielle founded. First Chemical Exposition in New York.

1916 National Research Council (U.S.A.) founded. Annual Reports of the Progress of Applied Chemistry first published. Chemical Engineering Catalog founded.

1918 Chimie et Industrie first published.

1919 Proceedings of Chemical Engineering Group of Society of Chemical Industry first published.

1921 American Chemical Society Monographs established.

1923 Chemistry and Industry (London) first published. Institution of Chemical Engineers (London) founded. Transactions published. American Institute of Chemists founded. Chemical and Engineering News founded as News Edition of Industrial and Engineering Chemistry.

1924 Chemical Reviews first published. Journal of Chemical Education first published.

1928 Food Industries founded.

1929 World Power Congress, Tokio. Metals and Alloys founded. Analytical Edition of Industrial and Engineering Chemistry founded.

1933 Journal of Chemical Physics first published.

1934 Ninth International Congress of Pure and Applied Chemistry, Madrid.

1935 National Farm Chemurgic Council organized.

1936 Chemical Engineering Conference of World Power Congress, London. Journal of Organic Chemistry first published.   (R.K.S.)

CHEMICAL FORMULAE. The formulae of chemistry constitute a shorthand notation used to represent the composition by weight, the **molecular** properties, the characteristic chemical reactions or at times even the ordering of the **atoms** in space of the elements which go to make up the chemical compound. Chemical formulae are classified into empirical, molecular, structural or configurational, the order given being that of increasing content of information. The following steps indicate the type of chemical experiments necessary to establish the different kinds of formulae. The first step consists in the isolation of a pure chemical compound. Chemical purification can be obtained by **crystallization, distillation, adsorption, sublimation,** etc. Some of the criteria of purity which a substance must satisfy are constancy and sharpness of melting point and boiling point on repeated purification. As an example let us assume that we have succeeded in purifying a solid compound which we shall call **tartaric acid** and whose formula we wish to determine.

The second step consists in a qualitative and quantitative analysis of the compound. In the case of tartaric acid qualitative analysis tells us that the compound contains carbon, oxygen and hydrogen, while quantitative analysis shows that the proportions are 48 parts by weight of carbon, 96 of oxygen, and 6 of hydrogen. To obtain the empirical formula one divides each proportion by the atomic weight of the particular element, obtaining in this way a set of numbers which can be represented by a ratio of small integers. The simplest ratio of integers is commonly used to indicate as subscripts on the right of the chemical symbol of the element to represent the empirical formula. In the case of tartaric acid the atomic weights are 12 for **carbon,** 16 for **oxygen** and 1 for **hydrogen.** Dividing the percentages as determined by analysis by the atomic weights we get

$$\text{Carbon} \quad 48/12 = 4.00$$
$$\text{Oxygen} \quad 96/16 = 6.00$$
$$\text{Hydrogen} \quad 6/1 = 6.00$$

The set of numbers is 4,6,6 and can be represented in this case by the ratio of integers 2:3:3. The empirical formula is therefore $C_2O_3H_3$. Empirical formula is thus only a convenient method for representing the percentage composition by weight of the different elements in the compound. The third step is the determination of the molecular weight of the compound in question. This allows us to assign to the compound a molecular formula. The molecular weight can be determined in a variety of methods such as by the determination of the weight of 22.4 liters of the vapor of the substance at 1 atmosphere pressure and 0° C., temperature. Other methods are based on the differences in the boiling point or freezing point of solutions of known concentration and those of the pure solvent. To determine the molecular formula from the knowledge of the empirical formula and the molecular weight the following procedure must be followed. Multiply the atomic weight of each element by its subscript as indicated in the empirical formula and add the result. On comparison of such a sum with the molecular weight it will be found that the molecular weight is equal to the sum times an integer. To obtain the molecular formula multiply each subscript in the empirical formula by this integer and obtain a new set of subscripts. We found the empirical formula of tartaric acid was $C_2O_3H_3$. The sum mentioned above is

$$12 \times 2 + 16 \times 3 + 1 \times 3 = 75$$

The molecular weight determined experimentally is 150. The integer multiple is 2 and the molecular formula becomes $C_4O_6H_6$.

The molecular weight of the compound can be obtained from the molecular formula by summing the products obtained by multiplication of the atomic weights of the elements times their subscripts in the molecular formula. The latter contains all the information that the empirical formula contains but in addition specifies the number of atoms in the molecule and also the molecular weight of the substance. The chemical formulae met in practice are molecular formulae.

Structural formulae have a twofold purpose: they attempt to show which atom is attached to which in the molecule and also to summarize the more important chemical reactions of the molecules. The cornerstone of the structural formula theory are the assumptions of definite valency for each element, the ability of certain atoms, especially carbon, to unite with each other to form chains and rings, and the formation of multiple valence bonds between atoms in a molecule. Most of the evidence for the manner in which atoms are attached to each other in the molecule is circumstantial. Yet all the circumstantial deductions of the organic chemist have been substantiated by direct evidence of spectroscopy. The structural formula is also a shorthand notation for the important chemical reactions of the compound. It can be considered as being built up of a group of organic **radicals,** i.e., groups of atoms which retain their individuality in the course of certain reactions. Each radical has reactions which are characteristic of its presence in the molecule. For instance, the **carboxyl** radical —C—OH with double bond O will react with alkali such as **sodium** hydroxide to form salts —C—ONa, with **phosphorus** pentachloride to form acid **chlorides** —C—Cl; with **alcohols** to form **esters;** with reducing agents under certain conditions to form successively the **aldehyde** radical —C—H and the **alco-**

hol radical. Any compound which undergoes such reactions is said to contain a carboxyl group. The number of such carboxyl groups in a molecule can be determined by studying the above reactions quantitatively. On the other hand if the compound will react with sodium to give off hydrogen; with phosphorus trichloride to give a halogen substitution product which can be reduced to **hydrocarbon**; with an oxidating agent to give an **aldehyde** or ketone, with organic **acids** to form **esters;** with **alcohols** to form **ethers;** then the molecule is said to contain a hydroxyl group —OH. Analogously there are similar characteristic reactions for a variety of radicals. It often happens that the presence of one type of a radical near another type mutually influences their reactivity, but one can consider to the first approximation that the radicals act independently of each other. The structural formula is considered completely established if one can synthesize the compound by simple clear-cut reactions involving no **rearrangments** on the basis of the proposed formula. In the particular example of tartaric acid the third step in the determination of the structural formula would be to determine what radicals are present in the molecule and their number. The results show that there are two carboxyl and two hydroxyl groups. The only structural formula involving these groups, satisfying valence requirements, possessing the proper molecular formula and consistent with the synthesis of tartaric acid is the following one

$$O{=}C{-}OH$$
$$HC{-}OH$$
$$HC{-}OH$$
$$O{=}C{-}OH.$$

To conclude, the structural formula purports to give an idea as to how the individual atoms are attached to each other in the molecule, to give a résumé of the chemical reactions, and to suggest methods of synthesis. Configurational formulae are discussed in **Isomerism and Stereoisomerism.** (R.K.S.)

CHEMICAL NOMENCLATURE AND PRONUNCIATION.

CHEMICAL NOMENCLATURE AND PRONUNCIATION. Based on the "Report of the Commission on the Reform of the Nomenclature of Organic Chemistry," from the translation published in the Journal of the American Chemical Society, Vol. 55, No. 10, p. 3905, October, 1933.

The subject is treated under the following general headings: I. General. II. Hydrocarbons. 1. Saturated Hydrocarbons. 2. Unsaturated Hydrocarbons. 3. Cyclic Hydrocarbons. III. Fundamental Heterocyclic Compounds. IV. Simple Functions. V. Complex Functions. VI. Radicals. VII. Numbering.

I. General

1. As few changes as possible will be made in terminology universally adopted.

2. For the present, only the nomenclature of compounds of known constitution will be dealt with; the question of substances of imperfectly known constitution is postponed.

3. The precise form of words, endings, etc., prescribed in the rules should be adapted to the genius of each language by the subcommittees.

II. Hydrocarbons

4. The ending *ane* is adopted for saturated hydrocarbons. Open-chain hydrocarbons will have the generic name *alkanes*.

The name "alkane" is better and shorter than "paraffin," especially since the latter term is now so commonly applied to a solid mixture.

5. The present names of the first four normal saturated hydrocarbons (methane, ethane, propane, butane) are retained. Names derived from the Greek or Latin numerals will be used for those having more than four atoms of carbon.

6. Branched-chain hydrocarbons are regarded as derivatives of the normal hydrocarbons; their names will be referred to the longest normal chain present in the formula by adding to it the designations of the side chains. In case of ambiguity, or if a simpler name would result, that chain which admits of the maximum of substitutions will be selected as the fundamental chain.

If there are two or more choices for the longest chain, then that one should be chosen in which there is the greatest number of substitutions (the reason being that the substituting radicals, while more numerous, will be of simpler structure). Example:

$$CH_3CH_2CH_2CH_2CHCH_2CH_2CH_2CH_3$$
$$CH(CH_3)CH(CH_3)CH_3$$

By the principle of the "longest chain" the name would be 5-(1,2-dimethylpropyl)- nonane; but according to the rule the name 4-butyl-2,3-dimethyloctane (which avoids a branched side chain) is the one to be chosen if it seems simpler.

7. In case there are several side chains, the order in which such chains are named will correspond to the order of their complexity. The chain having the greatest number of secondary and tertiary atoms will be considered the most complex. The alphabetic order may also be followed in such cases.

8. In the names of open-chain unsaturated hydrocarbons having one double bond the ending *ane* of the corresponding saturated hydrocarbon will be replaced by the ending *ene;* if there are two double bonds, the ending will be *diene,* etc. These hydrocarbons will bear the generic name *alkenes, alkadienes, alkatrienes,* etc. Examples: propene, hexene, etc.

9. The names of triple-bond hydrocarbons will end in *yne, diyne,* etc. They will bear the generic name *alkynes.* Examples: propyne, heptyne, etc.

10. If there are both double and triple bonds in the fundamental chain the endings *enyne, dienyne,* etc., will be used. The generic names of these hydrocarbons will be *alkenynes, alkadienynes,* etc.

11. Saturated monocyclic hydrocarbons will take the names of the corresponding open-chain saturated hydrocarbons, preceded by the prefix *cyclo.* They will bear the generic name *cycloalkanes.*

12. When they are unsaturated, rules 8–10 will be applied. However, in the case of partially saturated polycyclic aromatic compounds the prefix *hydro,* preceded by *di-, tetra-,* etc., will be used. Example: dihydroanthracene.

13. Aromatic hydrocarbons will be denoted by the ending *ene* and will otherwise retain their customary names. However, the name *phene* may be used instead of *benzene.*

III. Fundamental Hetrocyclic Compounds

14. The endings of customary names, endings which do not correspond to the function of the substance, will undergo the following modifications, so far as they are in accord with the genius of each language: (a) The ending *ol* will be changed to *ole.* Example: pyrrole. (b) The ending *ane* will be changed to *an.* Example: pyran.

The change from -*ol* to -*ole* is obviously for the purpose of reserving -ol as an ending for the names of alcohols and phenols; similarly, the change from -*ane* to -*an* is made in order to reserve -ane for saturated parent compounds.

15. When nitrogenous heterocycles not having the ending *ine* give basic compounds on progressive hydro-

genation, such derivation will be indicated by the successive endings *ine, idine*. Examples: pyrrole, pyrroline, pyrrolidine; oxazole, oxazoline.

16. The ending *a* is adopted for hetero atoms occurring in a ring. Oxygen will accordingly be indicated by *oxa*, sulfur by *thia*, nitrogen by *aza*, etc. The letter *a* may be elided before a vowel. Examples: thiadiazole, oxadiazole, thiazine, oxazine.

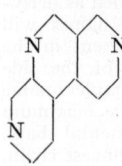

While the universally accepted names of heterocyclic compounds are retained, the names of other heterocyclic compounds are derived from that of the corresponding homocyclic compound by adding to it the names of the hetero atoms ending in *a*. Example: 2,7,9-triazaphenanthrene.

IV. Simple Functions

17. Substances of simple function are defined as those containing a function of one kind only, which may be repeated several times in the same molecule.

That is to say, a compound which is an acid, an alcohol or an aldehyde and only that, is defined as a substance of simple function, while one which is at the same time an alcohol and an acid, or an acid and an aldehyde, is said to be a substance of complex function.

18. When there is only one functional group, the fundamental chain will be selected so as to contain this group. When there are several functional groups the fundamental chain will be selected so as to contain the maximum number of these groups.

19. Halogen derivatives will be designated by the name of the hydrocarbon from which they are derived, preceded by a prefix indicating the nature and number of the halogen atoms.

20. Alcohols and phenols will be given the name of the hydrocarbon from which they are derived, followed by the suffix *ol*. In accordance with rule 1 names universally adopted will be retained, as: phenol, cresol, naphthol, etc.

This nomenclature may also be applied to heterocyclics. Example: quinolinol.

21. In naming polyhydric alcohols or phenols, one of the forms *di, tri, tetra*, etc., will be inserted between the name of the parent hydrocarbon and the suffix *ol*. Example: CH_2OHCH_2OH, 1,2-ethanediol.

22. The name *mercaptan* as a suffix is abandoned; this function will be denoted by the suffix *thiol*. Examples CH₃SH, methanethiol; C_6H_5SH, benzenethiol; CH_2SHCH_2SH, 1,2-ethanedithiol.

23. Ethers are considered as hydrocarbons in which one or several hydrogen atoms are replaced by alkoxy groups. However, for symmetrical ethers the present nomenclature may be retained. Examples: $CH_3OC_2H_5$, methoxyethane; CH_3OCH_3, methoxymethane or methyl ether.

24. Oxygen linked, in a chain of carbon atoms, to two of these atoms will be denoted by the prefix *epoxy* in all cases where it would be unprofitable to name the substance as a cyclic compound. Examples: ethylene oxide = epoxyethane; epichlorohydrin = 3-chloro-1,2-epoxypropane; tetramethylene oxide = 1,4-epoxybutane.

25. Sulfides, disulfides, sulfoxides and sulfones will be named like the ethers, *oxy* being replaced by *thio, dithio, sulfinyl* and *sulfonyl*, respectively. Examples: CH_3SO_2 C_2H_5 methysulfonylethane; $CH_3SC_3H_7$, methylthiopropane; $CH_3CH_2CH_2SOCH_2CH_2CH_3$, 1-(propylsulfinyl) butane.

26. Aldehydes are characterized by the suffix *al* added to the name of the hydrocarbon from which they are derived; thioaldehydes, by the suffix *thial*. Acetals will be named as 1,1-dialkoxyalkanes.

27. Ketones will receive the ending *one*. Diketones, triketones, thioketones will be designated by the suffixes *dione, trione, thione*.

28. The name *ketene* is retained.

"Ketene" (or, as spelled by some, "keten") is accordingly recognized as a name for the parent compound $CH_2{=}CO$.

29. For acids the rule of the Geneva nomenclature is retained. However, in cases where the use of that nomenclature would not be convenient the carboxyl group will be considered as a substituting group and the name of the acid will be formed by adding to the name of the hydrocarbon the suffix *carbonique* or *carboxylic*, according to the language.

30. Acids in which an atom of sulfur replaces an atom of oxygen will be named according to the Geneva nomenclature. Example: ethanethioic, -thiolic, -thionic, -thionothiolic. If the carboxyl is considered as a substituent the compounds will be named *carbothioic* acids. The suffix *carbothiolic* will be used if it is certain that the oxygen of the OH group is replaced by sulfur; the suffix *carbothionic* if it is the oxygen of the CO group; the suffix *carbodithioic* will be used if both oxygen atoms are replaced. Examples of the two systems of names: CH_3COSH or CH_3CSOH (either one), ethanethiolic acid, methanecarbothiolic acid; CH_3CSOH, ethanethionic acid, methanecarbothionic acid, CH_3CSSH, ethanethionothiolic acid, methanecarbodithioic acid.

31. The existing conventions will be retained for salts and esters. Examples: Sodium butanoate or sodium salt of butanoic acid; diethyl, 1,2-ethanedicarboxylate or diethyl ester of 1,2-ethanedicarboxylic acid; sodium acetate; methyl succinate.

32. Acid anhydrides will retain their present mode of designation according to the names of the corresponding acids. For names formed in accordance with the Geneva nomenclature, the amides, amidoximes, amidines, imides and nitriles will be named like the acids by adding to the name of the corresponding hydrocarbon the endings *amide, amidine, amidoxime, imide* and *nitrile*, respectively, while the halides will be named by combining *chloride*, etc., with the name of the radical. Examples: C_3H_7COCl, butanoyl chloride; $C_3H_7CONH_2$, butanamide; etc.

If the carboxyl is considered as a substituent the endings *carbonamide, carbonamidine, carbonamidoxime, carbonimide, carbonitrile* will be used. Examples: C_3H_7COCl, propanecarbonyl chloride; $C_3H_7CONH_2$, propanecarbonamide; etc.

33. The ending *ime* is reserved exclusively for nitrogenous bases. The present nomenclature of monoamines is retained. For polyamines, the name of the hydrocarbon will be followed by the suffixes *diamine, triamine*, etc.

For aliphatic compounds containing quinquivalent nitrogen the ending *ine* will be changed to *onium*. For cyclic substances containing quinquivalent nitrogen in the ring the ending *ine* will be changed to *inium;* for those with the ending *ole*, this will be changed to *olium*. Examples: pyridine, pyridinium; imidazole, imidazolium.

In accordance with the first sentence of this rule the spelling of names of non-bases ending in -ine should be changed; thus glycerine becomes glycerol, dextrine becomes dextrin, propine becomes propyne (see rule 9). Examples of names of amines: CH_3NH_2, methylamine; $(CH_3)_2NH$, dimethylamine; $(CH_3)_3N$, trimethylamine; $H_2NCH_2CH_2NH_2$, 1,2-ethanediamine; $C_6H_4(NH_2)_2$, benzenediamine.

34. The nomenclature of the derivatives of phosphorus, arsenic, antimony and bismuth, being very complicated, requires special consideration.

35. Oximes will be named by adding the suffix *oxime* to the name of the corresponding aldehyde, ketone or quinone. Examples: $C_2H_5ONH_2$, ethoxyamine; C_2H_5 NHOH, ethylhydroxylamine.

36. The generic term *urea* is retained; it will be used as a suffix for the alkyl and acyl derivatives of urea. Examples: butylurea, $C_4H_9NHCONH_2$; butyrylurea,

$C_3H_7CONHCONH_2$. The bivalent radical —NHCONH— will be named *ureylene*.

37. The generic name *guanidine* is retained.

38. The name *carbylamine* is retained.

39. Isocyanic and isothiocyanic esters (RNCO, RNCS) will be named *isocyanates* and *isothiocyanates*.

40. The name *cyanate* is reserved for true esters which on saponification yield cyanic acid or its hydration products. The name *sulfocyanate* will be replaced by *thiocyanate*.

41. Nitro derivatives: no change in the present nomenclature. That is, the group NO_2 is always indicated by the prefix *nitro*, never by a suffix. Nitroso compounds are treated similarly (see rule 52). Examples: nitrosobenzene, 2,4,6-trinitrophenol.

42. Azo derivatives: the forms *azo, azoxy* are retained.

43. (a) Diazonium compounds, RN_2X, are named by addition of the suffix *diazonium* to the name of the parent substance (benzenediazonium chloride).

(b) Compounds having the same empirical formula but containing trivalent nitrogen will be named by replacing diazonium with *diazo* (benzenediazohydroxide).

(c) Substances of the type RN_2OM will be named *diazoates*.

(d) Compounds in which the two nitrogen atoms are united to a single carbon atom will be designated by the prefix *diazo* (diazomethane, diazoacetic acid).

44. Hydrazines are designated by the name of the alkyl radicals from which they are derived, followed by the suffix *hydrazine*. In cases where the amino group, of carbonamides is replaced by the hydrazino group, the suffix *hydrazide* will be used. Hydrazo derivatives are regarded as derivatives of hydrazine. Examples: CH_3NHNH_2, methylhydrazine; $C_2H_5NHNHC_3H_7$, 1-ethyl-2-propylhydrazine; $C_3H_7CONHNH_2$, butyrohydrazide or propanecarbohydrazide.

45. Hydrazones and semicarbazones are named like the oximes. The term *osazone* is retained.

46. The name *quinone* is retained.

47. Sulfonic and sulfinic acids will be designated by adding the suffixes *sulfonic* and *sulfinic* to the name of the hydrocarbon.

The analogous acids of selenium and tellurium will bear the names *alkaneselenonic* and *-seleninic* acids; *alkanetelluronic* and *-tellurinic* acids. Examples: $C_2H_5SO_3H$, ethanesulfonic acid; $C_{10}H_6(SO_2H)_2$, naphthalenedisulfinic acid.

48. Organometallic compounds will be designated by the names of the organic radicals united to the metal which they contain, followed by the name of the metal. Examples: dimethylzinc, tetraethyllead, methylmagnesium chloride.

However, if the metal is united in a complex manner it may be considered as a substituent. Example: $ClHgC_6H_4CO_2H$, chloromercuribenzoic acid.

49. The nomenclature of cyclic derivatives having side chains requires special consideration.

50. If it is necessary to avoid ambiguity, the names of complex radicals will be placed in parentheses. Examples: (dimethylphenyl)amine = $(CH_3)_2C_6H_3NH_2$; dimethylphenylamine = $C_6H_5N(CH_3)_2$.

V. Complex Functions

51. For compounds of complex function, that is to say, for compounds possessing different functions, only one kind of function (the principal function) will be expressed by the ending of the name. The other functions will be designated by appropriate prefixes.

52. The following prefixes and suffixes will be used for designating the functions.

FUNCTION	PREFIX	SUFFIX
Acid and derivatives.	carboxy	carbonylic, carbonyl, carbonamide, etc., or oic, oyl, etc.

FUNCTION	PREFIX	SUFFIX
Alcohol	hydroxy	ol
Aldehyde	oxo, aldo (for aldehyde O) or formyl (for CHO)	al
Amine	amino	amine
Azo derivative	azo	....
Azoxy derivative	azoxy	....
Carbonitrile (nitrile)	cyano	carbonitrile or nitrile
Double bond	....	ene
Ether	alkoxy	....
Ethylene oxide, etc	epoxy	....
Halide	halogeno [halo]	....
Hydrazine	hydrazino	hydrazine
Ketone	oxo or keto	one
Mercaptan	mercapto	thiol
Nitro derivative	nitro	....
Nitroso derivative	nitroso	....
Quinquivalent nitrogen	....	onium, inium [olium]
Sulfide	alkylthio	....
Sulfinic derivative	sulfino	sulfinic
Sulfone	sulfonyl	....
Sulfonic derivative	sulfo	sulfonic
Sulfoxide	sulfinyl	....
Triple bond	....	yne
Urea	ureido	urea

For the order used in the *Chemical Abstracts* indexes, see Patterson and Curran, *Journal of the American Chemical Society*, **39**, 1624 (1917).

53. The names of derivatives of fundamental heterocyclic substances will be formed according to the preceding rules. Example: Hydroxyquinolinecarbonamide, not quinolinolcarbonamide.

VI. Radicals

54. Univalent radicals derived from saturated aliphatic hydrocarbons by removal of one atom of hydrogen will be named by replacing the ending *ane* of the hydrocarbon by the ending *yl*.

Examples: methyl, ethyl, pentyl (or amyl), etc. Since isopropylidene is recognized (rule 56) it was no doubt the intention of the Committee to recognize isopropyl similarly.

55. The names of univalent radicals derived from unsaturated aliphatic hydrocarbons will have the endings *enyl, ynyl, dienyl*, etc., the positions of the double or triple bonds being indicated by numerals or letters where necessary.

Examples: $CH_2=CH—$, ethenyl (or vinyl); $CH\equiv C—$, ethynyl; $CH_2—CH=CH—CH_2—$, 2-butenyl; $CH_2=CH—CH=CH—$, 1,3-butadienyl.

56. Bivalent or trivalent radicals derived from saturated hydrocarbons by removal of 2 or 3 hydrogen atoms from the same carbon atom will be named by replacing the ending *ane* of the hydrocarbon by the endings *ylidene* or *ylidyne*. For radicals derived from unsaturated hydrocarbons, these endings will be added to the name of the hydrocarbon. The names isopropylidene and methylene are retained.

57. The names of bivalent radicals derived from aliphatic hydrocarbons by removal of a hydrogen atom from each of the two terminal carbon atoms of the chain will be ethylene, trimethylene, tetramethylene, etc.

Only saturated radicals are provided for: —CH_2CH_2—, ethylene; —$CH_2CH_2CH_2$—, trimethylene, etc.

58. Radicals derived from acids by removal of OH will be named by changing the ending carboxylic to *carbonyl* or, if the Geneva nomenclature is used, oic to oyl. Examples: CH_3CO, ethanoyl or methanecarbonyl (or acetyl).

59. Univalent radicals derived from aromatic hydrocarbons by removal of a hydrogen atom from the ring

will in principle be named by changing the ending *ene* to *yl*. However, the radicals C_6H_5 and $C_6H_5CH_2$ will continue provisionally to be named phenyl and benzyl, respectively. Moreover, certain abbreviations sanctioned by usage are authorized, as *naphthyl* instead of *naphthalyl*. Examples: $CH_3C_6H_4$—, tolyl (instead of toluyl), anthryl (instead of anthracyl), phenanthryl, fluoryl.

60. Univalent radicals derived from heterocyclic compounds by removal of hydrogen from the ring will be named by changing their endings to *yl*. In cases where this would give rise to ambiguity, merely the final *e* will be changed to *yl*. Examples: pyridine, pyridyl; indole, indoyl; pyrroline, pyrrolinyl; triazole, triazolyl; triazine, triazinyl.

61. Radicals formed by removal of a hydrogen atom from a side chain of a cyclic compound will be regarded as substituted aliphatic radicals. Examples: $C_6H_5CH_2CH_2$—, (2-phenylethyl); $C_6H_5CH{=}CHCH_2$—, (3-phenyl-2-propenyl).

62. In general, special names will not be given to multivalent radicals, derived from cyclic compounds by removal of several hydrogen atoms from the ring. In this case prefixes or suffixes will be used. Examples: triaminobenzene or benzenetriamine; dihydroxypyrrole or pyrrolediol.

63. The order in which prefixes or radicals are stated (alphabetical order or conventional order) remains optional.

VII. Numbering

64. In aliphatic compounds the carbon atoms of the fundamental chain will be numbered from one end to the other with the use of arabic numerals. In case of ambiguity the lowest numbers will be given (1) to the principal function, (2) to double bonds, (3) to triple bonds, (4) to atoms or radicals designated by prefixes. The expression "lowest numbers" signifies those that include the lowest individual number or numbers. Thus, 1,3,5 is lower than 2,4,6; 1,5,5 lower than 2,6,6; 1,2,5 lower than 1,4,5; 1,1,3,4 lower than 1,2,2,4. The Committee has left full latitude on the position of numbers.

65. Positions in a side chain will be designated by numerals or letters, starting from the point of attachment. The numerals or letters will be in parentheses with the name of the chain. Examples: $(CH_3)_2CH$—, (1-methylethyl) or isopropyl; $CH_3CHClCH_2$—, (2-chloropropyl). The rule equally permits Greek letters, ordinary letters, primed numbers ($1'$, $2'$), numbers with indices (4^1, 4^2) or other designations.

66. In case of ambiguity in the numbering of atoms or radicals designated by prefixes, the order will be that chosen for the prefixes before the name of the fundamental compound or side chain of which they are substituents.

67. The prefixes, *di, tri, tetra,* etc., will be used before simple expressions (for example, diethylbutanetriol) and the prefixes *bis, tris, tetrakis,* etc., before complex expressions. Examples: bis(methylamino)propane: $CH_3NH(CH_2)_3NHCH_3$; bis(dimethylamino)ethane, $(CH_3)_2NCH_2CH_2N(CH_3)_2$. The prefix *bi* will be used only to denote the doubling of a radical or compound; for example, biphenyl. (R.K.S.)

CHEMISTRY. On account of the vast scope of the field of chemistry, the subject has, for convenience, become divided into many branches. It should be emphasized that these branches are not, cannot be, and are not desired to be mutually exclusive. The field is a unit, covering the composition and changes in composition of matter, and the accompanying energy phenomena. The natural tendency is toward unification and removal of artificial barriers; the artificially created demand for systematic treatment has necessitated the erection of boundaries, but the more penetrable the

boundaries, the better for the healthy unification and growth of science.

I. Classification of the subject, wherein the primary emphasis is on *Matter Changes.*

II. On *Energy Changes.*

A. *Inorganic Chemistry.* Study of chemical elements and their compounds, their properties, chemical behavior, preparation, and applications. List of elements accepted by the committee on Atomic Weights of the International Union of Chemistry:

Aluminum	Fluorine	Mercury	Silicon
(Ammonium)		Molybdenum	Silver
Antimony	Gadolinium		Sodium
Argon	Gallium	Neodymium	Strontium
Arsenic	Germanium	Neon	Sulfur
	Gold	Nickel	
Barium		Nitrogen	Tantalum
Beryllium	Hafnium		Tellurium
Bismuth	Helium	Osmium	Terbium
Boron	Holmium	Oxygen	Thallium
Bromine	Hydrogen		Thorium
		Palladium	Thulium
Cadmium		Phosphorus	Tin
Calcium	Indium	Platinum	Titanium
Carbon	Iodine	Potassium	Tungsten
Cerium	Iridium	Praseodymium	
Cesium	Iron	Protactinium	Uranium
Chlorine		Radium	
Chromium	Krypton	Radon	Vanadium
Cobalt		Rhenium	
Columbium	Lanthanum	Rhodium	Xenon
Copper	Lead	Rubidium	
	Lithium	Ruthenium	Ytterbium
Dysprosium	Lutecium		Yttrium
		Samarium	
Erbium	Magnesium	Scandium	Zinc
Europium	Manganese	Selenium	Zirconium

With the following exceptions the compounds of the above elements are treated in this book alphabetically by elements. The exceptions are in the cases of bromine, carbon, chlorine, fluorine, hydrogen, iodine, nitrogen, oxygen, phosphorus, sulfur, where the number and variety of substances to be discussed necessitates the extension of the discussion to additional separate articles. The titles of the articles related to each element are, however, stated under the given element.

B. *Organic Chemistry.* Among the chemical elements, carbon stands alone in the large number and complexity of its compounds. Carbon-containing substances are grouped and treated separately under the heading of organic chemistry. Traditionally, certain of these substances are commonly treated in inorganic chemistry, namely, the various forms of carbon element, carbon oxides, cyanogen, carbonic acid, formic acid, acetic acid, oxalic acid, hydrocyanic and related acids. The sections devoted to organic phases of chemistry are selected by topics, by important groups, and by individual substances. (See **Carbon, Key to Organic Compounds.**)

(a) Hydrocarbons.

(b) Oxygen function compounds. The following are found under **Oxygen:** Furane, pyrone; alcohols, phenols; aldehydes, ketones, quinones; carboxylic acids. The following are found in this book under their individual names:

Acetals	Carbohydrates	Glucosides
Acid anhydrides	Cellulose	
Aldehydeacids	Coumarone	Hydroxyacids
Anthocyanins		Hydroxyaldehydes
	Esters	Hydroxyketones
Benzil	Ethers	
Benzoin	Fats	Ketenes
		Ketoneacids

Lactones   Phthaleins   Sugars
Lactides

               Saccharides   Tannins
Oils, fatty   Starch
Oxides   Sterols   Waxes

(c) Nitrogen function compounds. The following are found under **Nitrogen**: Nitro-, nitroso-, hydroxylamines, hydrazines, amines; pyrrole, pyridine; amines; amides, cyanides and isocyanides; cyanates and isocyanates, ureas. The following are found in this book under their individual names:

Acridine	Indazole
Alkaloids	Indigo
Aminoacids	Indole
Aminoazo-compounds	Isoquinoline
Aminoguanidines	
Anilides	Lactams
Azides	Lactims
Azines	
Azo-compounds	Nitrates
Azoles	Nitriles
Azoxy-compounds	Nitrites
	Nitrobenzene
Biuret	Nitrolic acids
	Nucleic acids
Carbamic acid	
Carbazole	Osazones
Chlorophyll	Oxazoles
Cyanamides	Oximes
Cyanuric acid	
	Phenylhydrazones
Diazoamino-compounds	Piperidine
Diazo-compounds	Polypeptides
	Porphyrins
Esters	Proteins
	Purines
Fulminates	Pyrazole
	Pyrrolidine
Guanidines	
	Quinoline
Hydrazo-compounds	
Hydrazones	Semicarbazides
Hydrazoates	Semicarbazones
Hydroferricyanides	
Hydroferrocyanides	Tetrazoles
Hyponitrites	Triazoles
Hydroxyazo-compounds	
	Ureides
Imides	Urethanes
Imines	Uric acid

(d) Sulfur function compounds may be found by reference to the article on **Sulfur**. They include thiophene, penthiophene, thioalcohols, thiophenols, thioic acids, thiocyanates, thioureas, thiocarbonic acid, and related compounds; sulfinic acids, sulfinyl compounds, sulfonic acids, sulfonyl compounds.

C. *Analytical Chemistry*. Qualitative and quantitative analysis of elements, radicals and compounds, or by groups of these. (See **Analytical Chemistry**.)

D. *Synthetical Chemistry*. Qualitative and quantitative synthesis of compounds and radicals, or by groups of these.

E. *Principles of Chemistry*—Pure Chemistry. See below.

F. *Applications of Chemistry*—Applied Chemistry. See below.

G. *Radioactive changes* and ultimate structure of matter. (See **Radioactive Changes**.)

H. *Biochemistry*. Plant and animal substances and processes, both normal and abnormal. (See **Biochemistry**.)

I. *Geochemistry*. Composition of and changes in the atmosphere (see **Air**), hydrosphere (see **Water**),

and lithosphere. By extension, astral chemistry. (See Geochemistry.)

J. *Domestic Chemistry*. Chemical aspects of **air, water, foods**, clothing, shelter, cleansing, transportation, communication, and sanitation.

K. *Agricultural Chemistry*. **Soils**, crops, fertilizers, agricultural **poisons** and live stock.

L. *Metallurgical Chemistry*. Recovery and purification of metals (see each metal), properties of metals and alloys.

M. *History and Evolution of Chemistry*. (See **History and Evolution of Chemistry**.)

N. *Studying and Teaching of Chemistry*.

O. *Chemical Nomenclature*. (See **Chemical Nomenclature**.)

The PRINCIPLES OF CHEMISTRY have been formed around various nuclei as the following arrangement illustrates:

(a) Matter in various physical states: (See *Each* of the following states of matter)
   (1) **Gases**
   (2) **Liquids**
   (3) **Solutions**
   (4) **Solids**
   (5) **Colloidal state**

(b) *Chemical composition* of matter: (See **Chemical Composition**)
   (1) Elements, symbols, atoms, allotropes, isotopes
   (2) Compounds, formulae, molecules, allotropes, isomers
   (3) Radicals
     (a) in combination
     (b) as ions (See **Reactions Involving Recombination of Ions**)
     (c) free
   (4) Equivalents, valency, graphical formulae

(c) Changes in composition of matter: (See **Chemical Changes**)
   (1) Reactions
     (a) Natural rates of reactions
     Concentration, temperature, pressure, medium
     (b) Artificially controlled rates of reactions
     Catalyzers
     (c) End-point of reactions—Equilibrium reactions
     Concentrations, equilibrium constant, temperature, pressure

(d) Changes in energy in chemical reactions.
   (1) **Thermochemistry**
   (2) **Electrochemistry**
   (3) **Photochemistry**

(e) Grand groups of chemical substances:
   (1) Acids, bases, salts (see **Reactions Involving Recombinations of Ions**)
   (2) Metals, non-metals, alloys
   (3) Oxides, chlorides, sulfides (see each; **Thermochemistry, Heat of Formation**)
   (4) Organic (classification given on preceding pages)

(f) Grand groups of chemical reactions
   (1) **Reactions involving recombinations of Ions**—Hydrogen-ion concentration; pH; indicators
   (2) **Reactions involving water**
     (a) Consumption of water—including acid and alkaline mediums—Hydrolysis; saponification of esters
     (b) Production of water—Neutralization; esterification; nitration; sulfonation
     (c) Water as catalyzer
   (3) **Reactions involving oxidation—reduction**

(a) In solutions of electrolytes
(b) Not in solutions of electrolytes, including oxygenation and deoxygenation; hydrogenation and dehydrogenation; chlorination and dechlorination; bromination and debromination; nitro-reduction; and sulfonyl-reduction

(4) **Würtz-Fittig**—sodium-organic halide reaction

(5) **Friedel-Crafts** — anhydrous aluminum chloride-alkyl-aryl halide-benzenoid hydrocarbon reaction

(g) Grand groups of phenomena and processes:

(1) Isotopy (see **Chemical Composition**); allotropy (see **Chemical Composition**); **Isomerism and Stereoisomerism**

(2) **Association and Polymerization; Dissociation; Rearrangement; Passivity**

(3) Role of water; **deliquescence; efflorescence;** desiccation; drying; water as catalyzer (see **Reactions Involving Water**)

(4) **Equilibrium** including Phase rule

(5) **Adsorption**

(6) Vapor pressure

(7) Heating of materials
(a) With access of air: roasting (see **Calcination**); smelting
(b) Without access of air: **calcination; destructive distillation;** sublimation; **distillation, evaporation and drying** (see **Distillation**)

(8) Solid-liquid-gas treatments
Including precipitation; crystallization; sedimentation; filtration; dissolving solids; extraction

APPLICATIONS OF CHEMISTRY. Applied chemistry both industrial and non-industrial, concerned with (1) chemical materials, their composition, location, transportation, storage, (2) chemical processes naturally occurring or artificially operated, their matter and energy changes, the conditions, both favorable and unfavorable, of temperature, pressure, concentration, medium and catalyzer for these changes to take place, including apparatus, (3) chemical products, their quality (composition or specifications), quantity, applications, uses, and value.

Industrial chemistry, including chemical engineering, economic aspects of chemistry plus industry, embraces a part of the field of applied chemistry. Those product industries which have attained large proportions and general recognition are among the following:

Abrasives
Acetaldehyde
Acetic acid
Acetone
Acetylene
Acids. (See individual acids, e.g., **sulfuric, nitric, hydrochloric, phosphoric, oxalic, phthalic.**)
Adhesives
Alcohols
Alkalis. (See individual alkalis, e.g., **sodium** hydroxide and carbonates, **potassium** hydroxide and carbonates, **ammonium** hydroxide, **calcium** oxide and carbonate.)
Alloys
Aluminum and its compounds
Amines
Ammonia
Ammonium compounds
Aniline
Anthracene
Arsenic compounds
Artificial fibers and films. (See **Carbohydrates.**)

Baking powders
Barium compounds
Benzene
Bleaching substances, bleaching and decolorizing
Boron compounds
Bricks. (See **Ceramics.**)
Bromine and its compounds

Calcium compounds
Carbon dioxide
Casein. (See **Aminoacids; Plastics.**)
Cellulose. (See **Carbohydrates.**)
Cement, Portland
Ceramics
Chlorine and its compounds
Chromium and its compounds
Coal. (See **Fuels; Destructive distillation** products; **Hydrogenation.**)
Coal Tar products and intermediates
Coke. (See **Fuels; Destructive distillation** products; **Calcium** carbide.)
Copper and its compounds

Decolorizing. (See **Bleaching** substances.)
Destructive distillation products
Dextrin. (See **Carbohydrates; Adhesives.**)
Disinfectants. (See **Poisons.**)
Drugs
Dyes and dyeing, textile fibers

Earthenware. (See **Ceramics.**)
Essential Oils. (See **Hydrocarbons,** terpenes.)
Esters
Ethers. (See **Alcohols.**)
Ethyl alcohol
Explosives

Fats. (See **Esters.**)
Fertilizers
Flavors. (See **Esters.**)
Foods
Formaldehyde
Fuels
Fungicides. (See **Poisons.**)

Gases. (See individual gases, e.g., **oxygen, acetylene, ammonia, chlorine, carbon dioxide, hydrogen, nitrous oxide, sulfur dioxide; Fuels.**)
Gasoline. (See **Hydrocarbons,** paraffin.)
Germicides. (See **Poisons.**)
Glass
Glazes. (See **Glass.**)
Glycerol
Gold and its compounds

Hydrochloric acid
Hydrogen
Hydrogenation products. (See **Catalysis; Hydrogenation.**)

Inks. (See **Tannins.**)
Intermediates. (See **Coal Tar** products.)
Insecticides. (See **Poisons.**)
Iron and its compounds.

Ketones. (See **Aldehydes.**)

Lead and its compounds
Leather and tanning. (See **Tannins.**)
Lime. (See **Calcium** oxide.)
Lubricants

Magnesium and its compounds
Mercury and its compounds
Metals. (See individual metals, e.g., **iron, copper, lead, zinc, aluminum, magnesium, nickel, tin, mercury, silver, gold.**)
Methyl alcohol

Naphthalene
Nickel and its compounds
Nitric acid
Nitrobenzene
Nitrocellulose. (See **Explosives**.)
Nitrogen and its compounds. (See also **Fertilizers; Explosives**.)
Nitroglycerine. (See **Explosives**.)
Nitrous oxide. (See **Nitrogen**, oxides.)

Oils. (See **Esters** for fatty oils; **Hydrocarbons**, for petroleum products, and for essential oils.)
Oxalic acid
Oxygen

Paints. (See **Pigments**.)
Perfumes
Petroleum. (See **Hydrocarbons**, paraffin.)
Phenol
Phosphate rock. (See **Phosphoric acid; Fertilizers**.)
Phosphoric acid
Phosphorus and its compounds
Photography
Phthalic acid
Pigments
Plaster. (See **Calcium** sulfate.)
Plastics
Poisons
Potassium compounds. (See also **Fertilizers**.)
Pottery. (See **Ceramics**.)

Rayon. (See **Carbohydrates**.)
Refractories
Rubber and accelerators

Salts. (See individual salts.)
Silica. (See **Silicon**, oxide)
Silver and its compounds
Slags. (See **Glass**.)
Soaps. (See **Esters**.)
Sodium compounds
Solvents
Starch. (See **Carbohydrates**.)
Steel. (See **Iron**.)
Stoneware. (See **Ceramics**.)
Sugars. (See **Carbohydrates**.)
Sulfur and its compounds
Sulfur dioxide. (See **Sulfur**, oxides.)
Sulfuric acid

Textile fibers. (See **Dyes**.)
Tiles. (See **Ceramics**.)
Tin and its compounds
Toluene
Turpentine. (See **Hydrocarbons**, Terpenes.)

Varnishes. (See **Pigments**.)

Waxes. (See **Esters**.)

Zinc and its compounds.   (R.K.S.)

CHEMORECEPTOR. Sense Organ.

CHERRY. Rose Family.

CHERT. An impure, flinty hard rock composed chiefly of cryptocrystalline silica. Chert varies in color from gray through brown to black according to the kind and amount of coloring matter. It occurs principally as concretions, nodules or bands in **limestones** and **dolomites**, and unlike **flint** its fracture tends to be splintery instead of conchoidal. A great deal has been written on the occurrence and origin of chert and there is no doubt but that it may be formed in several different ways. Many of the nodular and concretionary cherts have grown around siliceous sponge spicules or **radiolaria**.

Chert may be either **sygenetic** or **epigenetic**. The former type is supposed by some authors to be chemically precipitated from river waters on the bottom of the sea as a colloid contemporaneously with the limestones or dolomites. On the other hand certain cherts are obviously secondary although they may have been formed previous to the final lithification of the formations in which they occur. Cherts which contain relatively large amounts of iron are called **Jasper**. (R.M.F.)

CHESTNUT BLIGHT. Ascomycetes.

CHEVROTAIN. Mammalia, Artiodactyla. The mouse-deer of Asia and Africa, a group of several species whose appearance is like that of very small deer but whose structure differs in several important details. *Tragulus* is found in Asia, *Dorcatherium* in Africa. (A.W.L.)

CHEWING GUM. Achras Sapota.

CHEZY FORMULA. The Chezy formula is an important formula for friction loss in large water conduits. The formula is stated as follows:

$$V = C\sqrt{RS}.$$

R is the **hydraulic radius** of the cross-section of flow. It is the cross-sectional area of flow divided by the wetted perimeter. S is the friction loss, in fractional foot head per ft. length of conduit for a conduit running full of water, or it is the slope of the water surface, fractional foot per ft. of length for open flow. C is a coefficient. A widely accepted formula for C is Kutter's formula, as follows:

$$C = \frac{41.65 + \dfrac{1.811}{n} + \dfrac{.00281}{S}}{1 + \dfrac{n}{\sqrt{R}}\left(41.65 + \dfrac{.00281}{S}\right)}.$$

In this formula, R and S have meanings already defined, n is a channel roughness coefficient. Typical values of n are as follows:

Wood stave penstocks	.010
Steel penstocks	.017
Lumber flumes	.014
Earth canal	.030
Natural river channel	.05 to .10

The original expression for C of the Kutter formula did not contain the term in S. It is said that this term was inserted at a later date to make the Chezy formula give results consistent with actual data for the very flat slopes of the Mississippi River where slopes as flat as $\frac{1}{5}''$ per mi. were encountered. For normal use the term in S has little effect on results computed by the formula. (E.W.S.)

CHIASTOLITE. Andalusite.

CHICKADEE. Aves, Passeriformes. Birds (**Aves**) of several species found in various parts of North America, all quietly colored in grays with some black markings

Carolina Chickadee.

and in some species a little white and brown. The common widely distributed species is also called the black-capped **titmouse**, *Penthestes atricapillus*. (A.W.L.)

CHICKENPOX (Varicella). A common highly contagious disease of childhood caused by a filterable **virus**. The disease has been distinguished from smallpox since 1553. It is spread by direct contact, the incubation period being 14 to 16 days.

In children the disease is usually quite mild. The first symptom is often the eruption which begins on the face or trunk as small, red, widely scattered itchy spots, and spreads to the rest of the body, including the scalp and **mucous membranes** of the mouth and **pharynx**. The eruption is often most abundant on the neck and shoulders. The red spots increase in size and soon are capped by small blisters of clear fluid, which progress to scab formation. The rash characteristically appears in crops over a period of 3–4 days so that early and late lesions may be seen side by side. A fever up to 101° F. may be present during the first day or two. Usually, if not infected by scratching, the lesions heal without leaving scars.

In adults, chickenpox is often a more serious disease, with severe systemic reactions and high fever.

Treatment is supportive. The patient should be isolated, and efforts made to prevent secondary infection. (D.M.H.)

CHICLE. Achras Sapota.

CHICORY. Cichorium Intybus. Composite Family.

CHIGGER, CHIGOE, JIGGER. 1. Arachnida, Acarina. A harvest **mite** whose **larva** is very irritating to the human skin, causing intense itching. North American, sometimes locally abundant. 2. Insecta, Siphonaptera. A small **flea** found in the tropical parts of Africa and the New World. After mating the female burrows into the skin of a warm-blooded animal and becomes distended as her eggs develop until she is as large as a pea. When they attack man they usually enter the skin of the foot and cause a troublesome sore. (A.W.L.)

CHILARIA. A pair of rudimentary appendages on the first abdominal or pregenital segment in the king crab (**Xiphosura**). (A.W.L.)

CHILBLAIN. Inflammation, accompanied by swelling, painfulness, itching, and redness of the hands or feet due to exposure to cold. (D.M.H.)

CHILDBED FEVER. Puerperal Sepsis.

CHILDRENITE. The mineral childrenite is a complex hydrous compound of **aluminum, iron** and **phosphorus**. **Manganese** may replace the **iron**, this variety being called **eosphorite**. Childrenite is known only in orthorhombic crystals; massive sorts have never been found. Its hardness is 4.5–5; specific gravity, 3.18–3.24; luster, vitreous to resinous; color, pale yellowish through yellowish-brown to nearly black; streak is white or yellowish; translucent. Childrenite is found in Saxony; Devonshire, England; and Oxford County, Maine. Eosphorite occurs in Branchville, Connecticut; Oxford County, Maine, and Bavaria. Childrenite was named for J. G. Children, an English mineralogist. Eosphorite is derived from the Greek word meaning dawn-bearing, because of the pinkish color of this mineral. (E.S.C.S.)

CHILL. A paroxysm of shaking or shivering, accompanied by a sense of cold and pallor of the skin. During a severe chill the temperature becomes elevated. A chill may indicate the onset of a disease, often a severe infection, notably lobar **pneumonia** and **malaria**.

Nervous chill is shaking or shivering due to excitement, fear, or anger, and is not accompanied by any rise in temperature. (D.M.H.)

CHILOPODA. The centipedes, usually considered as a class of arthropods related to but distinct from the millipedes and a few rare forms, but sometimes ranked as an order in the class Myriapoda, containing all of these forms.

These animals are elongate and slender with numerous segments, most of them bearing a single pair of appendages. They have one pair of **antennae** and the first pair of legs is modified to form a pair of poison claws with which they catch their prey. They are terrestrial, breathing by air tubes (**tracheae**) which open separately on the various segments. The body is flattened and the segments are composed of dorsal and ventral plates connected by softer lateral walls which bear the legs. Centipedes have poison glands opening through the poison claws but there is no evidence to show that they are ever dangerous to man. (A.W.L.)

CHIMACHIMA. Aves, Falconiformes. A bird (**Aves**) of prey found from Panama to southern Brazil. One of the caracaras. (A.W.L.)

CHIMANGO. Aves, Falconiformes. A South American bird (**Aves**) of prey found in Tierra del Fuego and the southern part of the continent. A caracara. (A.W.L.)

CHIMNEY. A chimney is a vertical tubular structure of masonry, steel, or reinforced concrete, built for the purpose of enclosing a column of hot gas, to produce thereby a draft. Combustion requires **oxygen**. Air is needed to supply it. To move this air through the fuel bed, and to produce a flow of the gaseous products of combustion through the furnace and boiler, or through a stove, requires a difference of pressure, called draft. The chimney is built primarily to produce a certain available draft, although sometimes a chimney may have to be high for reasons entirely foreign to draft. In addition to the useful draft it produces, a chimney must also overcome the friction loss in the chimney itself. These losses are proportional to the cross-sectional area of the stack. Hence the problem of chimney diameter is more than the assumption of a velocity comparable to that used in actual practice; it should be such that the diameter and height it indicates result in a chimney of least cost. Ordinarily, the economic chimney gas velocities range between 20 and 40 ft. per sec.

A chimney produces a draft by virtue of an extremely simple principle of thermodynamics. When gas is heated it expands in volume and decreases in density, in which condition it may be displaced by a more dense gas. The light, hot gas is confined by the chimney. The tendency of hot gas to move up the chimney is proportional to the height of the stack, since the difference of weight of equivalent columns of air and flue gas (and this is the draft) is greater the higher the columns.

The draft of a stack is, in an elementary way, expressed by:

$$D = \text{Height multiplied by difference in density of flue gas and air.}$$

Many empirical formulae are advanced for computing the height of a chimney required for a given boiler. A rational scientific approach would necessarily be based on the above equation, as it truly represents the physical action actually creating the draft. When the elementary equation is written for a chimney of 100 ft. incorporating certain factors needed to convert draft to inches of water, and allowing for cooling and friction in the stack, it has the form:

$$D = K(d_a - d_f) - 0.0148 d_f \sqrt{\frac{V^5}{F}}$$

where D = effective draft per 100 ft. of stack, in inches of water.

K = 17.3 for masonry stacks, and 15.4 for steel stacks.

d_a = density of air, in pounds per cubic foot.

d_t = density of flue gas, in pounds per cubic foot.

V = gas velocity in the stack, in feet per second.

F = gas flow, in cubic feet per second.

The second term in this equation allows for friction in the stack itself, so that D represents the effective draft per 100 ft. of stack. The actual height of a chimney is obtained by dividing required draft by D and multiplying by 100. The required stack draft is the sum of all friction losses external to the stack, plus the impact loss (of gas discharged from stack), and less the effective draft furnished by fans or jets. The impact loss is $0.003V^2 d_f$ in. of water, the symbols having meaning as given above.

The use of perforated radial bricks has found much favor in chimney construction. Their dead air space acts somewhat as heat insulation, and they are lighter than ordinary bricks. Comparatively short stacks are frequently made of plate steel. These are lined for a portion of their height with refractory lining, and, unless very short, are braced by suitable guy wires. A tall masonry chimney must have sufficient area in any transverse section to distribute the superimposed weight sufficiently to prevent the unit pressure from exceeding the safe bearing power of the masonry material. The effect of a transverse applied load due to wind pressure cannot be neglected. The effect of wind pressure is to increase the compression in the masonry on the leeward side and decrease it on the windward side. Since tension is not permissible in masonry, not only the thickness of the wall, but the external diameter of the chimney, must be selected with due respect to strength and stability under the condition of maximum wind load. It is very important that the foundation of a tall chimney be absolutely firm and unyielding, as a very small settling on one side would throw the top of a tall chimney several inches out of line and induce an unexpected eccentric loading in the structure.

To be satisfactory, chimneys for residences must, like all other chimneys, have sufficient height for the required draft, and sufficient area to carry off the volume of gases produced. Deficiencies in either or both of these needed characteristics will be sure to cause lazy fires and smoky furnaces. Insofar as possible, a chimney should be straight and perpendicular. Necessary bends should be reduced to the minimum, and corners should be well rounded. The use of a flue lining made of jointed sections of tile is an aid to draft as well as a safeguard against the risk of fire due to defective brickwork in the chimney. (F.T.M.)

CHIMPANZEE. Mammalia, Primates. One of the large man-like **apes** of tropical Africa. This species has been found the most easily kept and tractable of the apes in captivity and for this reason has been a favorite subject for the study of these near relatives of the human species. A number of chimpanzees have become famous for their ability to learn human habits and have displayed the rudiments of true intelligence. (A.W.L.)

CHINA-CLAY. A commercial term, more or less identical with **kaolin**, as applied to the relatively pure clay concentrated by washing from a thoroughly kaolinized **granite**. England is the chief exporter of China-clay. France has unique clays from which are made the famous Sèvres and Limoges potteries. At the present time the United States imports only about $\frac{1}{3}$ of the China-clay which it consumes, most of its domestic supply coming from the **Cretaceous** clays of New Jersey. (R.M.F.)

CHINAWOOD OIL. Esters.

CHINCH BUG. Insecta, Hemiptera. A small black bug, *Blissus leucopterus*, with whitish wing membranes. Although its total length is only $\frac{1}{4}''$ it is one of the most serious pests of wheat and damages other cereals as well. The methods of control are complex, including means of trapping the bugs during migration from field to field, the destruction of rubbish in which they spend the winter, and the use of decoy plots planted at such a time as to be most attractive when the eggs are being deposited. These plots are later plowed under to destroy the insects. (A.W.L.)

CHINCHILLA. Mammalia, Rodentia. Small squirrel-like **rodents** related to the porcupines. They live at high altitudes in the mountains of South America. The fur of the common chinchilla is valuable. There are two genera, *Chinchilla* and *Lagidium*. (A.W.L.)

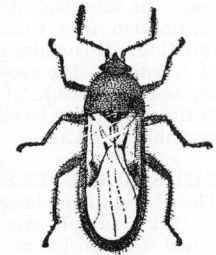

Chinch bug.

CHINE. Seaplane.

CHINOOK. Pisces, Teleostei. The king or quinnat **salmon,** most important of all North American food fishes. It is found from California to Alaska.

In meteorology, a warm wind blowing down the east slopes of the Rockies. It is similar to the **foehn wind.** (A.W.L., P.E.K.)

CHIP. Cutters.

CHIPMUNK. Mammalia, Rodentia. Small burrowing **rodents** of squirrel-like appearance but with the tail shorter and not bushy. They are brown or grayish with longitudinal stripes on the back or sides. They are omnivorous and while not usually troublesome they sometimes destroy flowering bulbs during the winter.

Chipmunks belong to several genera and numerous species and subspecies. Some of the western species are called golden chipmunks or rock squirrels and others antelope chipmunks or ground squirrels. They are related to the ground squirrels and gophers. While predominantly North American, chipmunks are also found in Siberia. (A.W.L.)

CHIROPTERA. The **bats.** An order of mammals highly specialized for flying. (A.W.L.)

CHIRU. Mammalia, Artiodactyla. The Tibetan **antelope,** *Panthalops hodssoni*, a species of moderate size with long horns, ringed in the basal half. (A.W.L.)

CHISEL. Lathe, Woodworking.

CHISEL-JAW. Pisces, Teleostei. A small fish (Pisces) of West African rivers. It has very strong teeth. (A.W.L.)

CHISEL-MOUTH. Pisces, Teleostei. A North American fish (Pisces), *Acrocheilus alutaceus,* related to the carps and dace, found in the lower Columbia River and its tributaries. (A.W.L.)

CHI-SQUARE. Chi-square, χ^2, is defined

$$\chi^2 = \sum \frac{(f_o - f_t)^2}{f_t}$$

where f_o is the observed frequency and f_t is the theoretical frequency calculated in accordance with the hypothesis which is being tested. The summation extends over all classes. (L.A.A.)

CHI-SQUARE PROBABILITY FUNCTION. The chi-square **probability function** is fundamental in modern **statistics.** Its equation is

$$P_{\chi^2} d\chi^2 = \frac{1}{\Gamma\left(\frac{n}{2}\right)} e^{-\frac{\chi^2}{2}} \left(\frac{\chi^2}{2}\right)^{\frac{n}{2}-1} d\left(\frac{\chi^2}{2}\right), \quad 0 \leq \chi^2 < \infty.$$

The χ^2 test of goodness of fit and the χ^2 test of independence in **contingency tables** depend upon it. **Confidence limits** for the **population variances**, when the **variates** are **normally** distributed, are based upon it also. In addition, it has many other uses. The χ^2 distribution is a special case of **Pearson's Type III function.** It contains only one parameter n, the **degrees of freedom.** The expected value of χ^2 is n, its variance $2n$, and its measure of **skewness** $\alpha^2{}_{3:\chi^2} = 8/n$. Tables of the probability levels of the χ^2 distribution are readily available. (L.A.A.)

CHI-SQUARE TEST OF GOODNESS OF FIT.

The χ^2 test of goodness of fit is used to test the hypothesis that a **sample** arose from a given **population.** The actual frequencies are compared with the theoretical frequency by means of **chi-square**

$$\chi^2 = \sum \frac{(f_o - f_t)^2}{f_t}.$$

If χ^2 is "large" it is assumed that the hypothesis is not tenable. Just how large χ^2 should be to reject the hypothesis is determined by the χ^2 distribution and the level of significance adopted, such as the 5% or 1% level. In most applications the theoretical frequency f_t should be 5 or more. Furthermore, the degrees of freedom n must be determined in using the χ^2 table, where the degrees of freedom equal the number of classes whose theoretical frequency is 5 or more less the number of **statistics** used in finding the theoretical frequencies less 1. Essentially the same methods are used in testing **contingency tables** for independence. (L.A.A.)

CHITAL. Mammalia, Artiodactyla. The Indian spotted **deer** or axis deer, *Axis axis.* (A.W.L.)

CHITIN. An essential constituent of the **cuticula** of **arthropods.** Also found to a lesser extent in most of the invertebrate groups. A material of variable and intricate chemical composition, this substance has been the subject of habitual misstatement especially among entomologists. It is a principal component of the insect exoskeleton and appears in both the rigid and flexible parts, which are usually said to be chitinized or not chitinized; the degree of rigidity has recently been said to depend upon other materials deposited with the chitin. Chitin is inelastic, hence the arthropods shed the exoskeleton at intervals during growth to permit expansion during the formation of a new covering. Chitin is chemically very inactive. (A.W.L.)

CHITRA. Chital.

CHIVES. Allium.

CHLAMYDOSPORE. A single spore or reproductive cell enclosed in a spore case, formed by some of the **slime molds.** (A.W.L.)

CHLOANTHITE. Smaltite-Chloanthit.

CHLORAL HYDRATE ($CCl_3 \cdot CHO \cdot H_2O$). A valuable and powerful sedative prepared by passing chlorine gas through absolute alcohol (see **Ethyl Alcohol**) and precipitating with water. It occurs in crystalline form, is soluble in water, has a bitter caustic taste and penetrating odor.

With therapeutic doses, quietness and drowsiness are produced. Pain is not diminished. With poisonous doses there is profound stupor, coma and collapse which may result fatally.

Chloral should never be given with alcohol as the alcoholic solution which is formed constitutes the well-known powerful "knockout drops," a quick-acting and dangerous depressant.

Chloral is used medically to quiet nervousness, irritability, and to produce sleep. It is contraindicated in threatened failure of the circulation or respiration or in the presence of marked liver, kidney, or heart disease. (See **Acetaldehyde** and **Drugs.**) (D.M.H.)

CHLORENCHYMA CELLS. Parenchyma.

CHLORIC ACID AND CHLORATES. Chloric acid ($HClO_3$) is a colorless solution; fairly stable but slowly decomposes into (a) **perchloric** and **hydrochloric acids,** (b) chlorine dioxide and **oxygen,** (c) hydrochloric acid and oxygen, (d) **chlorine** and oxygen (when of strength as high as 40% $HClO_3$). Can be concentrated in vacuum up to 52% $HClO_3$ with considerable decomposition. A powerful oxidizing agent, e.g., **sulfur** and all forms of sulfur compounds (other than **sulfates**) to **sulfuric acid**—but **persulfates** oxidize chlorate to perchlorate.

Prepared by reaction (1) of **barium** chlorate solution and sulfuric acid, and filtering off barium sulfate, (2) **potassium** or **sodium** chlorate and sulfuric acid.

Potassium chlorate is formed (1) by electrolysis of potassium chloride solution, stirring and heating over some hours, (2) by heating potassium hypochlorite solution.

Metallic chlorates are solids, soluble in water, except that those of bismuth, tin and mercury require a little free acid. Chlorates, when heated, evolve oxygen and leave the chloride (or oxide) as residue. The process of evolution of oxygen from chlorates liberates heat and may take place with explosive violence. The addition of a **catalyzer,** e.g., **manganese** dioxide, causes a rapid evolution of oxygen at the melting point of the chlorate. Potassium chlorate explodes violently, upon friction or percussion, with sulfur, phosphorus, charcoal, sugar and many organic substances. When potassium chlorate (melting point about 350° C.) or sodium chlorate (melting point 260° C.) is heated for some time somewhat above the melting point, the respective perchlorates and chlorides are formed.

Upon addition of hydrochloric acid, chlorate liberates chloric acid, which reacts with hydrochloric acid to form chlorine. (R.K.S.)

CHLORIDE. Chlorine.

CHLORINE. Symbol: Cl. Atomic number: 17. Atomic weight: 35.457. Density: 3.214 grams per liter, 0° C., 760 mm. or 2.486 when air equals 1.000. Formula of chlorine gas: Cl_2. Melting point: —100.6° C. Boiling point: —34.6° C. Critical temperature: 144.0° C. Critical pressure: 76.1 atmospheres. Isotopes: 35 (76%), 37 (24%), have been separated by diffusion, and have been identified and quantitatively estimated by the mass spectroscope.

Chlorine is a pale greenish-yellow gas, of marked odor, irritating to the eyes and throat, poisonous. Somewhat soluble in water, soluble in carbon tetrachloride to a colorless solution. Discovered by Scheele in 1774, but identified as a chemical element by Davy in 1810.

Chlorine occurs as sodium chloride in ocean water (2% Cl), in salt beds, salt brines, salt lakes, e.g., Stassfurt, Germany, the Dead Sea, the states of New York, Michigan, Louisiana, Utah, and many other places. Practically all the chlorine manufactured in the United States is made by electrolysis of sodium chloride solution under special conditions. Chlorine is marketed as liquid in steel cylinders up to large sizes, or as "bleaching powder" (calcium hypochlorite) or as sodium hypochlorite, and is used (1) for the purification of drinking water, (2) as a disinfectant, (3) in the preparation of many chemicals, (4) for bleaching paper pulp and textiles.

Chlorine does not react with oxygen gas, but is reactive with explosive violence with **hydrogen** gas when exposed to sunlight; reacts vigorously with many metals such as **copper, iron, aluminum, arsenic, antimony;**

with some non-metals such as **phosphorus, sulfur;** with many organic compounds such as **ethylene, acetylene** to form ethylene dichloride, acetylene tetrachloride, respectively, with warm **turpentine** to form **carbon** plus **hydrogen chloride** violently, with benzenoid compounds to form chloro-benzenoid substitution products plus hydrogen chloride. Chlorine reacts with **sulfurous acid** to form **sulfuric acid;** with **hydrosulfuric acid** in deficiency of chlorine to form **sulfur,** in excess of chlorine to form sulfuric acid.

Azide: Chlorazide (ClN_3).

Acids: Hydrochloric acid (HCl); hypochlorous acid (HOCl); chlorous acid ($HClO_2$); chloric acid ($HClO_3$); perchloric acid ($HClO_4$). See each acid. Also chloro-organic acids, see below.

Bromide: Chlorine bromide (ClBr).

Chlorate: (See **Chloric Acid.**)

Chlorides: Chlorides are known of many elements. Metallic chlorides are (1) Soluble, e.g., **sodium, potassium, ammonium** chlorides (NaCl, KCl, NH_4Cl); (2) Reactive with water, e.g., **aluminum** chloride anhydrous, **magnesium** chloride crystals when heated, titanium tetrachloride, **stannic** chloride anhydrous; (3) Insoluble, e.g., **silver** chloride (AgCl), **lead** chloride ($PbCl_2$) (soluble in hot water), **mercurous** chloride (HgCl); and (4) organic chlorides are noteworthy, e.g., methyl chloride (chloromethane, CH_3Cl), chlorobenzene (phenyl chloride, C_6H_5Cl), carbon tetrachloride (CCl_4), carbonyl chloride ("phosgene," $COCl_2$). Hydrochloric acid and soluble chlorides are important chemical reagents for the precipitation of insoluble chlorides. (See **Hydrochloric Acid.**) Organic chloro-compounds, see Chlorine, organic compounds, below.

Chlorites: (See **Chlorous Acid.**)

Hydride: Hydrogen chloride (HCl), colorless gas, of characteristic odor, melting point $-111°$ C., boiling point $-85°$ C., density 1.27 (air equal to 1.00), very soluble in water yielding **hydrochloric acid.** Formed by reaction of (1) sodium chloride and concentrated sulfuric acid upon heating, (2) hydrogen and chlorine gases under regulated conditions (explosive in sunlight or magnesium light). Reacts with ammonia gas to yield ammonium chloride, white smoke.

Hypochlorites: (See **Hypochlorous Acid.**)

Iodides: Chlorine iodide (ClI); trichloroiodine (Cl_3I).

Oxides: Chlorine monoxide (Cl_2O), pale orange-yellow gas, boiling point 5° C., odor resembling but distinguishable from chlorine, soluble in water to yield hypochlorous acid, reactive with explosive violence with sulfur, phosphorus, and many carbon compounds. Formed (Hazard!) by reaction (1) of chlorine gas and dry finely divided mercuric oxide at 400° C., (2) chloramine and water in the presence of **calcium** chloride; chlorine dioxide, chlorine peroxide (ClO_2), reddish-yellow gas, melting point $-79°$ C., boiling point 11° C., unpleasant odor, soluble in water and upon cooling chlorine dioxide octahydrate crystallizes, dangerously explosive alone as solid, liquid, gas or with organic matter or phosphorus; reactive with sodium hydroxide solution to form sodium chlorite plus sodium chlorate, and with sodium peroxide

to form sodium chlorite plus oxygen gas. Formed by reaction of **potassium** chlorate and concentrated sulfuric acid (Hazard!) and slight warming (chlorine dioxide gas and perchloric acid are formed); chlorine heptoxide (Cl_2O_7), colorless volatile oil, formed by reaction of perchloric acid and phosphorus pentoxide at $-10°$ C., for one day, and then distilling at the boiling point 82° C. of the heptoxide, explosive by percussion or by flame, but unreactive with paper or wood, reactive with water to yield perchloric acid.

Perchlorates: (See **Perchloric Acid.**)

Sulfides: (See **Sulfur,** chlorides.)

Organic Chloro-compounds:

Paraffin and benzenoid **hydrocarbons,** e.g., **ethane** and **benzene,** respectively, react with chlorine by *substitution* of chlorine for hydrogen (hydrogen chloride also formed), e.g., ethane to yield ethyl chloride (C_2H_5Cl) plus further substitution products; benzene, in the presence of a catalyzer, e.g., iodine, phosphorus, iron, to yield chlorobenzene (C_6H_5Cl) plus further substitution products; toluene, under like conditions to benzene, to yield ortho-chlorotoluene and para-chlorotoluene ($CH_3C_6H_4Cl$) plus further substitution products, but, at the boiling temperature, in sunlight, dry, and in the absence of a catalyzer, to yield paraffin-side-chain substitution products, benzyl chloride ($C_6H_5CH_2Cl$), benzal chloride ($C_6H_5CHCl_2$) and benzotrichloride ($C_6H_5CCl_3$).

Olefin, acetylene and benzenoid **hydrocarbons,** e.g., **ethylene, acetylene,** and **benzene,** respectively, react (A) with chlorine by *addition*, e.g., ethylene dichloride ($C_2H_4Cl_2(1,2)$), acetylene tetrachloride ($C_2H_2Cl_4(1,1,2,2)$), benzene hexachloride ($C_6H_6Cl_6$), also carbon monoxide yields carbonyl chloride ($COCl_2$), (B) with hypochlorous acid by *addition* to form, for example, ethylene chlorohydrin (CH_2ClCH_2OH).

Oxygen-function compounds, e.g., **ethyl alcohol, acetaldehyde, acetone, acetic acid,** react (A) with chlorine, to form *chloro-substituted* corresponding or related *compounds*, e.g., ethyl alcohol or acetaldehyde to yield chloral (CCl_3CHO), acetone to yield chloroacetone ($CH_2ClCOCH_3$), acetic acid to yield, at the boiling temperature, dry, and in the absence of a catalyzer, monochloroacetic acid ($CH_2Cl·COOH$), dichloroacetic acid ($CHCl_2COOH$), trichloroacetic acid (CCl_3COOH), the substitution taking place on the alpha-carbon (the carbon next to the carboxyl group (—COOH)), (B) with phosphorus chlorides, to form corresponding *oxygen-function chlorides*, e.g., ethyl chloride (C_2H_5Cl), ethylidene dichloride (CH_3CHCl_2), acetone chloride (($CH_3)_2CCl_2$), acetyl chloride (CH_3COCl).

Chloroform is made by reaction of **acetone** or **ethyl alcohol** with calcium or sodium hypochlorite; carbon tetrachloride by reaction of carbon disulfide (CS_2) plus chlorine (Cl_2) in the presence of iron heated (sulfur monochloride, S_2Cl_2, boiling point 139° C., also formed, and separated by fractional distillation); trichloroethylene by reaction of acetylene tetrachloride and dilute alkali. The diazo-reaction (see **Azo- and Related Compounds**) may be used to introduce chlorine into benzenoid compounds.

SCHEME SHOWING THE INTERRELATIONSHIPS OF CHLORINE-CONTAINING SUBSTANCES

Hydrogen chloride	Chlorine	Chlorine monoxide		Chlorine dioxide	Chlorine heptoxide
Hydrochloric acid		Hypochlorous acid	Chlorous acid	Chloric acid	Perchloric acid
Metallic chlorides. Of sodium, potassium, magnesium in nature		Metallic hypochlorites	Metallic chlorites	Metallic chlorates	Metallic perchlorates
Organic chlorides. See next page.		Organic hypochlorites.			

SCHEME SHOWING THE INTERRELATIONSHIPS OF CHLORINE-FUNCTION ORGANIC COMPOUNDS

RELATED TO	METHANE	ETHANE OR PROPANE	TOLUENE	BENZENE
—CH₂OH >CHOH ⪴COH of alcohols and phenols	Methyl chloride	Ethyl chloride	Benzyl chloride	Phenyl chloride (chlorobenzene)
		Ethylene dichloride		Ortho-dichlorobenzene
		Chloroalcohols: Ethylene chlorohydrin		Para-dichlorobenzene 1,2,4-trichlorobenzene 1,2,3,5-tetrachlorobenzene Hexachlorobenzene Benzene hexachloride
			Ortho-chlorotoluene Para-chlorotoluene 2,4-Dichlorotoluene 2,4,6-Trichlorotoluene....	Ortho-chlorophenol Para-chlorophenol 2,4-Dichlorophenol 2,4,6-Trichlorophenol
—CHO >CO of aldehydes and ketones	Methylene chloride	Acetylene tetrachloride	Benzal chloride	
		Aldo- and keto-chlorides: Acetaldehyde chloride (ethylidene dichloride) Acetone chloride Chloroaldehydes and -ketones: Chloral Chloroacetone		
H—COOH Carboxylic acids	Chloroform	Hexachloroethane	Benzotrichloride	
		Acid chlorides: Acetyl chloride Chloroacids: Monochloroacetic acid Dichloroacetic acid Trichloroacetic acid	Acid chlorides: Benzoyl chloride Chloroacids: Meta-chlorobenzoic acid	
HO–COOH Carbonic acid	Carbon tetrachloride Carbonyl chloride			

ORGANIC COMPOUNDS OF CHLORINE

ORGANIC COMPOUNDS OF CHLORINE	FORMULA	MELTING POINT °C.	BOILING POINT °C.
1. Methyl chloride...................	CH_3Cl....................	....	−24
2. Ethyl chloride....................	C_2H_5Cl...................	....	12
3. Normal-propyl chloride (1-chloropropane).	$C_2H_5 \cdot CH_2Cl$..............	....	46
4. Iso-propyl chloride (2-chloropropane)......	$CH_3 \cdot CHCl \cdot CH_3$.............	....	37
5. Vinyl chloride (chloroethylene)...........	$CH_2 : CHCl$.................	....	−12
6. Allyl chloride....................	$CH_2 : CH \cdot CH_2Cl$............	....	45
7. Chloroacetylene..................	$CH : CCl$...................	....	Expl.
8. Alpha-chlorostyrene................	$C_6H_5CCl : CH_2$...............	....	199
9. Omega-chlorostyrene...............	$C_6H_5CH : CHCl$.............	....	199
10. Bornyl chloride...................	$C_{10}H_{17}Cl$.................	161	....
11. Benzyl chloride (phenyl chloromethane)....	$C_6H_5 \cdot CH_2Cl$..............	....	179
12. Cinnamyl chloride..................	$C_6H_5CH : CH \cdot CH_2Cl$.............	....	
13. Glycol chlorohydrin (ethylene chlorohydrin)	$CH_2Cl \cdot CH_2OH$..............	....	129*
14. Alpha-glycerol chlorohydrin.............	$CH_2Cl \cdot CHOH \cdot CH_2OH$.........	....	139 (18 mm.)
15. Beta-glycerol chlorohydrin............	$CH_2OH \cdot CHCl \cdot CH_2OH$.........	....	146 (18 mm.)
16. Chloroacetic acid..................	$CH_2Cl \cdot COOH$..............	....	190
17. Alpha-chloropropionic acid.............	$CH_3 \cdot CHCl \cdot COOH$.............	....	186
18. Beta-chloropropionic acid.............	$CH_2Cl \cdot CH_2 \cdot COOH$.............	61	203
19. Chloromalonic acid.................	$CHCl(COOH)_2$.................	133	
20. Ethyl chloroformate................	$ClCOOC_2H_5$.................	....	94

*Constant boiling mixture, 42.5% water, 96° C.

ORGANIC COMPOUNDS OF CHLORINE (continued)

Organic Compounds of Chlorine	Formula	Melting Point °C	Boiling Point °C.
21. Alpha-chloroacrylic acid.................	$CH_2 : CCl \cdot COOH$.............	65	
22. Beta-chloroacrylic acid.................	$CHCl : CH \cdot COOH$.............	85	
23. Acetyl chloride.....................	$CH_3 \cdot COCl$.................		51
24. Benzoyl chloride....................	$C_6H_5 \cdot COCl$.................		197
25. Chloroacetone.....................	$CH_2Cl \cdot CO \cdot CH_3$.............		121
26. Alpha-epichlorohydrin............	$CH_2Cl \cdot CH-CH_2$.............		117
(chloropropylene oxide)	$\quad\quad\quad \underset{\displaystyle O}{\underline{\quad\quad}}$		
27. Cyanogen chloride.................	$CN \cdot Cl$.................		13
28. Chlorofurane.....................	$C_4H_3Cl(2)O$.................		77 (744 mm.)
29. Furoyl chloride...................	$C_4H_3O \cdot COCl(2)$.............	0	170
30. Furfuryl chloride.................	$C_4H_3O \cdot CH_2Cl(2)$.............		49 (26 mm.)
31. Chlorofuroic acid (3).............	$C_4H_2Cl(3)O \cdot COOH(2)$............	149	
32. Chlorofuroic acid (5).............	$C_4H_2Cl(5)O \cdot COOH(2)$............	179	
33. Chlorobenzene (phenyl chloride)......	C_6H_5Cl.................		132
34. Ortho-chlorotoluene (1,2).........	$C_6H_4(Cl)(2)(CH_3)(1)$........		160
35. Meta-chlorotoluene (1,3).........	$C_6H_4(Cl)(3)(CH_3)(1)$........		162
36. Para-chlorotoluene (1,4).........	$C_6H_4(Cl)(4)(CH_3)(1)$........	8	162
37. Ortho-chlorobiphenyl.............	$(2)Cl \cdot C_6H_4 \cdot C_6H_5$........	34	267
38. Meta-chlorobiphenyl.............	$(3)Cl \cdot C_6H_4 \cdot C_6H_5$........	89	
39. Para-chlorobiphenyl.............	$(4)Cl \cdot C_6H_4 \cdot C_6H_5$........	75	282
40. Alpha-chloronaphthalene.........	$C_{10}H_7 \cdot Cl(1)$........		259
41. Beta-chloronaphthalene.........	$C_{10}H_7 \cdot Cl(2)$........	56	265
42. Ortho-chlorophenol (1,2).........	$C_6H_4(Cl)(2)(OH)(1)$........		175
43. Meta-chlorophenol (1,3).........	$C_6H_4(Cl)(3)(OH)(1)$........	28	214
44. Para-chlorophenol (1,4).........	$C_6H_4(Cl)(4)(OH)(1)$........	41	217
45. Ortho-chloroaniline (1,2).........	$C_6H_4(Cl)(2)(NH_2)(1)$........	0	210
46. Meta-chloroaniline (1,3).........	$C_6H_4(Cl)(3)(NH_2)(1)$........	−10	230 (767 mm.)
47. Para-chloroaniline (1,4).........	$C_6H_4(Cl)(4)(NH_2)(1)$........	71	231
48. Ortho-chloronitrobenzene (1,2).....	$C_6H_4(Cl)(2)(NO_2)(1)$........	32	245
49. Meta-chloronitrobenzene (1,3).....	$C_6H_4(Cl)(3)(NO_2)(1)$........	44	236
50. Para-chloronitrobenzene (1,4).....	$C_6H_4(Cl)(4)(NO_2)(1)$........	83	242
51. Ortho-chlorobenzoic acid (1,2).....	$C_6H_4(Cl)(2)(COOH)(1)$........	141	
52. Meta-chlorobenzoic acid (1,3).....	$C_6H_4(Cl)(3)(COOH)(1)$........	158	
53. Para-chlorobenzoic acid (1,4).....	$C_6H_4(Cl)(4)(COOH)(1)$........	242	subl.
54. Ortho-chlorobenzaldehyde (1,2).....	$C_6H_4(Cl)(2)(CHO)(1)$........	11	208 (748 mm.)
55. Meta-chlorobenzaldehyde (1,3).....	$C_6H_4(Cl)(3)(CHO)(1)$........	18	213
56. Para-chlorobenzaldehyde (1,4).....	$C_6H_4(Cl)(4)(CHO)(1)$........	48	213 (748 mm.)
57. Alpha-chloroanthraquinone.........	$(1)Cl \cdot C_6H_3(CO)_2C_6H_4$....		
58. Beta-chloroanthraquinone.........	$(2)Cl \cdot C_6H_3(CO)_2C_6H_4$....	208	
59. Methylene chloride.................	CH_2Cl_2.................		42
60. Ethylene dichloride (1,2-dichloroethane)...	$CH_2Cl \cdot CH_2Cl$.................		84
61. Acetylene dichloride (1,2-dichloroethene)...	$CHCl : CHCl$.................		48
62. Ethylidene chloride (1,1-dichloroethane)...	$CH_3 \cdot CHCl_2$.................		58
63. Benzal chloride (benzylidene chloride).....	$C_6H_5CHCl_2$.................		214
64. Acetone chloride (2,2-dichloropropane)....	$(CH_3)_2CCl_2$.................		70
65. Dichloroacetic acid.................	$CHCl_2 \cdot COOH$.................		194
66. Chloroacetyl chloride.................	$CH_2Cl \cdot COCl$.................		105
67. Carbonyl chloride (phosgene).......	$COCl_2$.................		8
68. Ortho-dichlorobenzene (1,2).........	$C_6H_4Cl_2(1,2)$.................		179
69. Meta-dichlorobenzene (1,3).........	$C_6H_4Cl_2(1,3)$.................		172
70. Para-dichlorobenzene (1,4).........	$C_6H_4Cl_2(1,4)$.................	53	174
71. Dichlorophenol (2,4).........	$HO \cdot C_6H_3 \cdot Cl_2(2,4)$.........	45	210
72. Dichloroanthracene (9,10).........	$C_6H_4(CCl)_2C_6H_4$.........	209	
73. Chloroform.....................	$CHCl_3$.................		61
74. Methylchloroform (1,1,1-trichloroethane)...	$CH_3 \cdot CCl_3$.................		74
75. Benzotrichloride (phenyl chloroform).....	$C_6H_5 \cdot CCl_3$.................		221
76. Trichloroethylene.................	$CHCl : CCl_2$.................		88
77. 1,2,3-Trichloropropane.................	$CH_2Cl \cdot CHCl \cdot CH_2Cl$.................		158
78. Trichloroacetic acid.................	$CCl_3 \cdot COOH$.................	58	195
79. Trichlorophenol.................	$C_6H_2(OH)(Cl)_3(2,4,6)$.........	68	
80. Chloral (trichloroacetaldehyde).........	$CCl_3 \cdot CHO$.................		98 (768 mm.)
81. Chloral hydrate.................	$CCl_3 \cdot CH(OH)_2$.................	52	96
82. Carbon tetrachloride.................	CCl_4.................		76
83. Acetylene tetrachloride.................	$CHCl_2 \cdot CHCl_2$.................		146
(1,1,2,2-tetrachloroethane)			
84. Tetrachloroethylene.................	$CCl_2 : CCl_2$.................		121
85. Naphthalene tetrachloride.............	$C_{10}H_8Cl_4(1,2,3,4)$.................		
86. Hexachloroethane.................	$CCl_3 \cdot CCl_3$.................	187 (sealed tube)	185 (777 mm.)
87. Benzene hexachloride, alpha, trans........	$C_6H_6Cl_6$.................	158	
" " beta, cis...........	$C_6H_6Cl_6$.................	297 (decom.)	
88. Hexachlorobenzene.................	C_6Cl_6.................	230	309 (742 mm.)

(R.K.S.)

Many of the chloro-compounds are used as reagents or as intermediate compounds in organic chemistry, and as non-inflammable solvent liquids for the extraction of fats and oils, and in cleaning textile materials.

CHLORITE SCHIST. A schist whose color and foliation are chiefly due to the mineral chlorite. Other minerals common in this type of schist are **quartz** and **epidote.** Garnet and **magnetite** sometimes occur as **idiomorphic** crystals giving the schist a **porphyroblastic** texture. (R.M.F.)

CHLORITOID. A mineral which occurs as tabular crystals, probably **triclinic**, foliated masses or scattered scales and plates of a greenish-gray to greenish-black mineral that is characteristic of the less intensely altered **metamorphic** rocks such as **phyllites** and **quartzites.** Chemically it is a hydrous **iron-aluminum silicate.** **Ottrelite** contains some manganese as well. Chloritoid was originally noted as from the Ural Mountains and named for its greenish color from the Greek word meaning green. Ottrelite was named from Ottrez in Luxemburg. (E.S.C.S.)

CHLOROFORM. Chlorine; Anaesthesia.

CHLOROMONADINA. An order of single-celled animals of green color. (See **Mastigophora.**) (A.W.L.)

CHLOROPHYLL. Pyrrole and Related Compounds; Pigments in Plants.

CHLOROPLAST. Pigments in Plants.

CHLOROPLATINIC ACID. Platinum.

CHLOROUS ACID AND CHLORITES. Chlorous acid ($HClO_2$) is a yellow solution, unstable, and of characteristic odor of chlorine dioxide. A strong oxidizing agent.

Formed by the spontaneous decomposition of **chloric acid** into chlorous and **perchloric acids.**

Sodium chlorite is formed (1) by the reaction of chlorine dioxide (ClO_2) and **sodium** hydroxide, with the simultaneous formation of sodium chlorate, and (2) by the reaction of chlorine dioxide and sodium peroxide with accompanying evolution of oxygen gas.

Chlorites are generally yellow. The chlorites of **sodium** and **potassium** are deliquescent, those of **lead, silver, mercurous, mercuric** (red) are insoluble. (R.K.S.)

CHOANOCYTE. The collar cell of **sponges,** bearing a high ridge surrounding a **flagellum** at the free end. They are located in cavities in the sponge and produce currents of water through the passages in the body wall. (A.W.L.)

CHOCOLATE. Theobroma Cacao.

CHOKE COIL. This term is applied to various types of inductances used in electrical circuits primarily to present high reactance at certain frequencies. Such coils usually have high reactance compared to their resistance and offer impedance to the flow of alternating currents by the induced counter electromotive force. Since this **impedance** will vary directly with frequency the choke may be designed to let certain lower frequencies through and stop or impede higher ones. An air-core choke coil is often used in electrical power circuits to block high-frequency transients produced by lightning surges. In communications circuits air-core chokes are used extensively to block radio frequencies from audio-frequency circuits or from d-c parts of the circuit. In these applications they are often called radio-frequency chokes. Iron-cored chokes are frequently used in audio circuits

in a similar manner. Iron-cored chokes are also important components of power supply filters as well as many wave filters. (L.R.Q.)

CHOLANGITIS. An acute or chronic infection of the **bile** ducts, usually associated with infection of the **gall bladder,** or obstruction of the common duct. While it usually occurs with **cholecystitis** or **cholelithiasis,** rarely it may complicate **pnuemonia, typhoid fever,** or other acute infections.

The symptoms of acute cholangitis are those of an acute cholecystitis with sepsis. **Jaundice** is present due to the obstruction of the common bile duct. **Leukocytosis,** chills, high intermittent fever, colic, vomiting, and severe pain in the abdomen may be present. **Septicemia** may complicate the picture.

Chronic cholangitis is characterized by jaundice, and sometimes by recurring attacks of fever. The degree of jaundice depends on the degree of obstruction.

The treatment in certain cases is surgical drainage of the common bile duct and the gall bladder. (D.M.H.)

CHOLECYSTITIS. Infection of the **gall bladder** occurring in an acute and chronic form. Cholecystitis rarely develops as a primary infection. It is usually secondary to **cholelithiasis,** or is part of a general **cholangitis.**

The **bacterial** organisms usually found in cholecystitis are of the colon-typhoid group. Less frequently the *Streptococcus, Staphylococcus* and *Pneumococcus* are found. The organisms gain access to the organ either by the blood stream or ascend from the **duodenum** through the bile.

The acute form is accompanied by fever, right-sided abdominal pain, vomiting and severe prostration. Jaundice is not seen unless the common bile duct is involved in the acute infection (**cholangitis**). In severe infections of the gall bladder, perforation may occur. This usually results fatally unless surgical intervention is undertaken early.

Chronic cholecystitis is the more common form. It may follow the acute form, or develop insiduously over a period of time. Usually, stones are present in the gall bladder and predispose to infection sooner or later.

The symptoms of chronic cholecystitis are practically the same as those of cholelithiasis. (See **Cholelithiasis.**)

Diagnosis is verified by means of **cholecystography.** The treatment of cholecystitis is surgical. This is usually done in the chronic stage, and consists of removal of the infected gall bladder. In mild cases, or when operation is contraindicated, a medical regime is followed. (See **Cholelithiasis.**) At times, operation becomes imperative in the acute stage when danger of perforation, or **empyema** of the gall bladder occurs. (R.S.M., D.M.H.)

CHOLECYSTOGRAPHY. X-ray visualization of the **gall bladder.** This procedure is in routine use in diagnosis of gall-bladder disease. It is accomplished by giving the halogen salts of **phthaleins** by mouth, or less often by vein. The salts are excreted by the liver into the **bile** and concentrated in the gall bladder. This renders the gall bladder opaque to x-ray. By this method the position, size, shape and often the presence of stones can be demonstrated. Later a fat meal is given. This causes the gall bladder normally to contract, causing a disappearance or diminution of the shadow on the x-ray plate. When inflammation of the gall bladder or obstruction of the cystic duct is present, the gall-bladder shadow may be absent. (R.S.M., D.M.H.)

CHOLELITHIASIS (GALLSTONES, BILIARY CALCULUS). Gallstones usually form as the result of infection and stasis of **bile.** Changes in the **metabolism** of the body, causing an increase in the **cholesterol** content of the blood or other changes in the composition of the bile have been observed in patients with cholelithiasis, but the significance of these is unknown since the same changes occur without the formation of stones. In some cases neither infection nor stasis is present and the precipitating factors remain obscure. It is believed by some that in certain cases infection follows stone formation.

Certain predisposing factors are definitely known. The majority of patients are obese, and are usually over 35 or 40 years of age. Females are affected more often than males. Women are prone to cholelithiasis, perhaps because of the increased cholesterol content of the blood during pregnancy. **Typhoid** infection seems to dispose to stone formation.

The chemical composition, number, size, and structure of gallstones varies greatly. The largest number reported in one **gall bladder** is said to be 7802. Some stones are very large. In shape, they may be round, irregular or sharply faceted. Solitary stones are often found. Chemically, the stones are commonly of four types: 1. Pure cholesterol. Such stones present, in section, a radiating crystalline structure. 2. Alternate layers of cholesterol and calcium salts. 3. Mixtures of cholesterol and bilirubin-calcium. Such stones are usually soft, noncrystalline, and small in size. 4. Pure bilirubin-calcium.

The symptoms caused by gallstones depend on their position, and the amount of infection (**cholecystitis**) present. The common symptoms seen in most cases are usually vague digestive disturbances, such as dull pains, sour eructation, fullness of the upper abdomen and flatulence. These symptoms are aggravated by eating rich foods, especially those of a fatty nature.

The severe pain or **colic** that occurs in recurring attacks is due to a stone lodging in the cystic or the common bile duct, to spasm of the gall-bladder walls due to irritation by stones or the infection, or to distention and stretching of the gall bladder by the obstructed bile. If the common bile duct becomes obstructed so that no bile can pass, **jaundice** develops.

The pain accompanying a stone impacted in the system of biliary ducts is often excruciating, and is accompanied by nausea, vomiting and fever—the latter depending on the amount of infection present. The paroxysms of biliary colic tend to recur at more frequent intervals and to become more severe. A large gallstone escaping into the intestine may cause **intestinal obstruction.**

Complications of cholelithiasis are chronic cholecystitis, pancreatitis (inflammation of the **pancreas**), perfora-

Cholesterin-bilirubin-calcium stones, outer layer pure cholesterin. (*Nelson Loose-Leaf Surgery, Thomas Nelson & Sons.*)

tion of the gall bladder with resulting **peritonitis,** intestinal obstruction from large gallstones, infection of the **liver** (hepatitis), and adhesion formation around the gall bladder and neighboring organs.

Surgical removal of the gall bladder gives the best results in cholelithiasis. If the patient does not wish or cannot withstand the operation, various medical measures may offer some relief of symptoms. The elimination of fatty and spicy foods and alcohol from the diet, plus the administration of bile salts may help to avoid attacks. Nitroglycerin or amyl nitrite may relieve the muscle spasm during the acute attacks. **Morphine** may be necessary for the relief of colic. (R.S.M., D.M.H.)

CHOLERA. An acute infection involving primarily the terminal portion of the small intestine, the **ileum.** It is a disease of antiquity, and great epidemics have occurred in India, the Orient, Egypt and Europe. Cases have occurred in the eastern part of the United States as well. During the 1879–1883 pandemic, the causative agent was demonstrated to be a small rod, *vibrio cholerae,* which is spread in the same manner as *typhoid* bacilli, i.e., by the intestinal-oral route.

Cholera usually has a sudden onset with profuse **diarrhea,** vomiting, muscular cramps, suppression of urine, **uremia, acidosis,** and collapse. Many of the symptoms are due to extreme dehydration. Death occurs in as high as 50 to 75% of cases. Early treatment with adequate salt and water intravenously, and alkali to prevent acidosis and uremia, reduce the mortality to 25 or 30%. **Penicillin, sulfonamides,** and **antiserum** may have some beneficial effect but this has not been proved. (D.M.H.)

CHOLESTEROL. Alcohols and Ethers.

CHONDRI. Rounded and ellipsoidal grains of silicates which are characteristic of meteorites. When viewed in their section the aggregate looks like grains of wheat. (R.M.F.)

CHONDRIOCONT. A structure found within some cells. (A.W.L.)

CHONDRIOSOME. A structure within the cell. (A.W.L.)

CHONDRITE. A term proposed by Rose in 1864 for meteoric stones which contain spheroidal aggregates of basic minerals (**pyroxene** and **olivine**), **oligoclase feldspar,** and varying proportion of **nickel-iron.** The percentage of free nickel increases as the percentage of nickel-iron decreases. (R.M.F.)

CHONDROSTEI. The **paddle-fishes** and **sturgeons.** An order with the skeleton made up largely of **cartilage** but with some bony components. (A.W.L.)

CHONOLITH. A term proposed by R. A. Daly in 1905 for irregular **igneous** intrusions which according to their shapes and field relationships cannot be classified as **dikes, laccoliths, batholiths, bysmaliths,** etc. (R.M.F.)

CHORD. The word chord has three well-defined engineering meanings. In railroad and highway practice a chord is a straight line joining the ends of an arc of any given curvature. The sharpness of **circular curves** used on railroad lines is defined by the number of degrees in a central angle subtended by a chord of 100'.

The principal boundary members of the framed **truss** used in bridges or buildings are designated as chord members or chords. Refer to **bridges** for an illustration of chord members.

In aviation terminology, the dimension of an **airfoil** surface parallel to the air stream is called the chord. (F.T.M.)

CHORD, AIRFOIL. The chord is a basic reference axis for the geometric or aerodynamic properties of an **airfoil.** It is normal to the span and lies in the plane of the airfoil. There are two of these reference chords. The one used for general and structural reference is the *geometric* chord. The other is an *aerodynamic chord,* being an imaginary line through the airfoil parallel to the free air stream at zero lift and passing through the trailing edge. The length of this chord is of no importance. It is useful mainly in aerodynamic studies because the lift varies directly with the **angle of attack** of the aerodynamic chord.

If the airfoil has a flat lower surface an element of this surface is taken as the geometric chord. The chord length is the over-all projection of the profile on this chord. In double-cambered airfoils the geometric chord is taken as the longest straight line possible between leading and trailing edges, or as a straight line joining the ends of the profile median line. The angle of attack to the geometric chord at zero lift is the angle between these chords. This may be discovered by wind tunnel tests, although empirical constructions have been devised which locate the aerodynamic chord surprisingly well. If the wing is tapered there is a tip chord and a root chord. The location of the intermediate chord on which the aerodynamic forces could be assumed to act is of importance in studies of airplane balance and stability. When the coefficient of lift may be assumed to be constant over the semi-span, the mean aerodynamic chord coincides with the mean geometric chord (i.e., the centroid of the semi-wing planform). This simplification is in error if the wing has twist, or if it is rectangular, in which case the uneven downwash causes decreased lift coefficient near the tips. It is an excellent approximation for untwisted tapered wings and exact for untwisted elliptical wings. (F.T.M.)

CHORD, MEAN AERODYNAMIC. Chord, Airfoil.

CHORDATA. The **vertebrates** and a few marine animals of simpler form, including the **tunicates, salpians** and **lancelets.** Although the true vertebrates make up most of this phylum the inclusion of the other forms is scientifically accurate. With these limits the distinctive characters of the phylum are few. The animals are **triploblastic, coelomate,** and **metameric** like the higher invertebrates but differ from them in three points: 1. The skeleton is internal. In its primitive state it consists of a slender longitudinal rod lying above the alimentary tract and called the **notochord.** This structure is present at some stage in development in all of the included species. 2. The **nervous system** is entirely dorsal in position, lying in the body wall above the notochord. 3. The alimentary tract includes a chamber, the **pharynx,** just behind the oral cavity, whose walls are perforated by openings associated with respiration and called the gill slits or pharyngeal clefts. These openings appear or are indicated only in the **embryos** of terrestrial species.

It is difficult to estimate the relative importance of the phylum since man himself is one of the included species. The chordates include the most highly developed animals from the scientific point of view, and from the practical point of view they are equally important as the source of most of our animal foods, furs, feathers, wool, leather, and as beasts of burden. Man has depended on the vertebrates, indeed, for much of his progress, and has taken his domestic animals from this group.

The classification of the phylum is briefly as follows:

Subphylum **Hemichordata.** Worm-like marine animals. **Balanoglossus.** Also named Enteropneusta.
Subphylum **Urochordata.** Sessile or free-swimming forms, marine, with larvae resembling tadpoles in which the characters of the phylum are evident. Also named Tunicata.
   Class **Larvacea.** Small floating animals with the larval form of the subphylum.

   Class **Ascidiacea.** The tunicates. Sessile or free, named from the investing test or tunic which encloses them.
   Class **Thaliacea.** The **salpians.** Free-swimming.
Subphylum **Cephalochordata.** The lancelets. Small fish-like animals which swim freely and also burrow in the sand.
Subphylum **Vertebrata.** The skeleton includes cartilaginous or bony components in addition to the notochord. Also named Craniata.
   Class **Cyclostomata.** The round-mouthed **eels:** lampreys and hags.
   Class **Pisces.** The fishes. Sometimes placed in two classes: Elasmobranchii, containing the **sharks, rays,** etc., and Pisces, containing the remaining fishes.
   Class **Amphibia.** The **salamanders, frogs, toads,** etc.
   Class **Reptilia.** The **lizards, snakes, turtles, crocodiles,** etc.
   Class **Aves.** The birds.
   Class **Mammalia.** Popularly called animals without further qualification. They secrete milk for the nourishment of their young and the skin usually bears some hair, often a complete coat which may be in the form of fur or wool. Mice (**mouse**), **horses** and **cattle, monkeys,** man, and many other forms. (A.W.L.)

CHORDOTONAL ORGAN. An organ for the perception of vibrations, found in the insects where it may exist singly or in association with complex auditory organs. It consists of a nerve ending with accessory cells connected directly to the body wall or to some modified derivative of the body wall in an organ of hearing. (A.W.L.)

CHOREA (St. Vitus' Dance. Sydenham's Chorea). Acute chorea is an affection of the central nervous system characterized by involuntary writhing movements and coarse muscular twitchings. It occurs primarily in children, girls more often than boys, and is frequently associated with acute **rheumatic fever.** The cause is obscure; in some respects chorea behaves as an infection, but no bacterial or virus agent has ever been isolated in the disease. It runs a course of weeks or months and may recur. Treatment is rest and sedation. (D.M.H.)

CHORION. 1. An accessory structure formed during embryonic development in **mammals.** It provides the connection with the tissues of the mother through which all interchange of materials between her blood and that of the **embryo** is carried on prior to birth.

The chorion is a composite structure formed of the **serosa** and the **allantois,** although the name is sometimes erroneously applied to the serosa alone. In some mammals a specialized **placenta** develops from part of the chorion as the persistent connection with the mother. Like all other extraembryonic membranes, this structure is discarded at birth.

2. The shell of an insect egg. (A.W.L.)

CHOROLOGY. Zoogeography.

CHOUGH. Aves, Passeriformes. *Pyrrhocorax.* Eurasian birds (**Aves**) related to the crows and resembling them in form and color. A few other Asian birds are known as chough-thrushes. (A.W.L.)

CHROMATIC ABERRATION. The indistinct color effects observed along the edges of images formed by a simple **lens** constitute what is known as the chromatic aberration of the lens. This aberration is due to the fact that the glass, or any other substance, out of which the lens is constructed produces **dispersion** (i.e., refracts light of different colors by different amounts). In a convergent glass lens the focal length is greater for red light than for blue, while in a divergent (concave surface) glass lens the blue focus is longer than the red. The effect for the convergent (convex surface) lens is shown, highly exaggerated, in the figure, in which the location of the image of a source, S, is shown for red

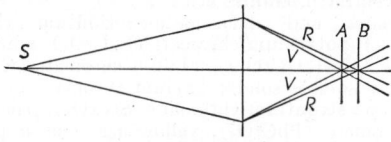

Chromatic aberration.

light, R, to be in the plane, B, and for blue light, V, to be in the plane, A. An observer using an **eyepiece** focused for the plane A will see a sharp image of the source in blue light surrounded by the margin of a confused set of images of greater size in other colors.

Chromatic aberration is very bothersome to users of lenses either for telescopic, microscopic, or photographic purposes and ever since optical instruments came into use attempts have been made to design "achromatic" lenses. For relatively short focus instruments, such as **cameras** or field glasses, practically complete achromatism can be obtained by using, instead of a simple convex lens, a combination of a convex and a concave lens, the two lenses being constructed of glasses of different dispersive powers. Prospective purchasers of field glasses, opera glasses or binoculars should always examine them carefully to determine whether or not the lenses are properly achromatized. A simple test is to examine the edge of a white building, which is in full sunlight, through the instrument under consideration. Move the image of the edge well over to one side of the field of view, and, if the lenses are properly figured, no color effects will appear. If, however, chromatic aberration is present, the image of the edge of the white building will be found to be bordered with a bright-colored fringe which increases in width as the image is moved closer to the edge of the field. In passing it might be said that the same test will indicate whether or not the field glasses are properly corrected for **spherical aberration** and other aberrations, for in a poor lens the image of the edge will become blurred and curved when moved to the side of the field of view.

In long focus instruments, such as astronomical **telescopes,** complete achromatism is virtually impossible. Partial achromatism may be obtained in such instruments by employing the combination of the divergent and convergent lenses of glass of different dispersive powers. (See **Dispersion.**) Two types of partial achromatism are employed, depending upon the purposes for which the telescope is designed. In a telescope to be used for photographic purposes the colors which are most active photographically, i.e., the greens and blues, are all brought to the same focal point, whereas the reds and oranges are thrown well out of this focal plane.

In a so-called visual telescope the yellowish-green light is all brought to one sharp focus, while the blues and violets are bent well inside this visual focus. On looking at the image of a very bright object, such as the **moon,** a **planet,** or a bright **star,** a halo of bluish light can be observed due to the out of focus photographic light, but this halo is so diffuse that it is not objectionable when working with objects of the brilliance for which the instrument is designed. Instruments designed for visual observing (visual refractors) cannot be used satisfactorily for celestial photography without employing yellow sensitive plates and color filters to eliminate the out of focus blue and green light.

The image formed by a concave metal or silver on glass mirror is free from chromatic aberration since all colors are reflected in the same direction. This is the most important optical advantage of the reflecting telescope over the refracting type. (W.K.G.)

CHROMATIN. A substance found in the **nucleus** of the **cell,** named for its pronounced affinity for biological stains. It is the bearer of **hereditary** qualities of the organism. It usually occurs in the form of small granules distributed throughout the nucleus. When the nucleus is about to divide the chromatin particles mass together to form the **chromosomes.**

Chromatin of special qualities has received special names in several cases. Basichromatin takes basic stains, oxychromatin, acid stains. Trophochromatin governs the vegetative activities of the cell and idiochromatin its reproductive functions. (A.W.L., R.M.W.)

CHROMATOPHORE. The chromatophore, also called a chromoplastid, is a definite body occurring in the **protoplast.** It is characteristic of a chromatophore that it should have a definite color, due to the pigment or **pigments** present in it.

In plants the most common chromatophore is the **chloroplastid.** Other chromatophores are of various colors, including yellow, brown, orange, and red. Not all plant pigments are found in chromatophores, however. Many occur dissolved in the cell sap.

Chromatophores found in the skin of animals of several groups, including **arthropods, mollusks,** and **vertebrates,** are large cells, often extensively branched. They are well developed in the **octopus** and **squid,** in many **amphibians** and fishes, and in **reptiles.**

Rapid changes in color such as those of squids and some lizards have been ascribed to a contraction of the chromatophores but it now seems evident that the cell itself does not contract although the pigment within it may undergo a considerable change in distribution, revealing itself when widely distributed and otherwise concealed from view. (R.M.W., A.W.L.)

CHROMIDIA. Granules in the cytoplasm of **cells** which stain deeply with basic stains and are derived from the nucleus. In some cells they constitute a nucleus of scattered parts, as in **bacteria.** (A.W.L.)

CHROMIOLE. A granule within a **chromomere.** (A.W.L.)

CHROMITE. An important mineral in the chromite series of multiple oxides. The dominant compound is $FeCr_2O_4$, but most chromite also contains magnesium (Mg) and aluminum (Al). Crystallizes in the isometric system. Hardness, 5.5; specific gravity, 4.5–4.8; color, black. Associated with **periclotite** and **serpentine** (metamorphised peridotite). Chromium (Cr) is a highly important metal in the making of steels which are particularly resistant to rust and heat. It is also used in decorative and wear-resisting plating; the manufacture of refractories, such as furnace brick and cement; and in the manufacture of chemicals used in dyeing, tanning, and pigment industries. Commercial amounts occur as placer deposits in serpentine areas. In the United States chromite has been mined in Maryland, Pennsylvania, California, Montana, Oregon, Wyoming and North Carolina. Important deposits occur in Quebec, Canada. Valuable deposits occur in Asia Minor, Southern Rhodesia, New Caledonia, Cuba, India, and the Philippines. Also in New South Wales, and in the Urals associated with platinum. (R.M.F.)

CHROMIUM. Symbol: Cr. Atomic number: 24. Atomic weight: 52.01. Density: 7.1. Hardness: 9. Melting point: 1615° C. Boiling point: 2200° C.

Chromium is a slightly grayish metal, hard, and capable of taking a brilliant polish; not appreciably ductile or malleable; soluble in **hydrochloric** or **sulfuric acid** (dilute); made passive by **nitric acid** (dilute or concentrated) or sulfuric acid (concentrated); not affected by air or water at ordinary temperatures; not affected by fused alkalis; when heated to 200° C. in air forms chromic oxide. Discovered by Vauquelin in 1798. (Isotopes: page 290.)

Chromium metal is used (1) in special steels—less than 1% Cr in chrome or nickel-chrome steels greatly increases the hardness, (2) in corrosion resistant or "stainless" steels—8%–14% Cr, (3) in electroplating, as a protective and ornamental coating for less resistant metals—electrodeposited chromium is the hardest form of the metal, being almost as hard as diamond, and very resistant to corrosion, but the very thin layers deposited electrolytically may be easily damaged due to the relative softness of some supporting metals, (4) in special alloys, e.g., wire of high electrical resistance for heating units, nichrome or chromel wire (11%–25% Cr, 50% or more nickel, remainder iron).

Chromium occurs chiefly as chromite (ferrous chromite, $Fe(CrO_2)_2$) in southern Rhodesia, Union of South Africa, U.S.S.R., New Caledonia, India, Philippine Islands, Japan, Turkey, Greece, Cuba, and California. (1) Heating chromite in the electric furnace with carbon yields ferrochrome for alloys, and (2) when chromite is heated with sodium carbonate and nitrate, sodium chromate is formed, which is then extracted with water. This is the substance from which chromium compounds are obtained.

Acetates: Chromous acetate ($Cr(C_2H_3O_2)_2$), brownish-violet solid; chromic acetate ($Cr(C_2H_3O_2)_3 \cdot H_2O$), grayish-green solid. Both are soluble and used as mordants in dyeing and printing textiles, and in tanning.

Alum: See Sulfate below.

Chlorides: Chromous chloride ($CrCl_2$), white solid, by heating chromic chloride in dry hydrogen gas, and obtained as a blue solution by reduction of chromic chloride solution with zinc metal, or by reaction of chromium metal and hydrochloric acid in the absence of air; chromic chloride ($CrCl_3$), exists in two forms, (1) green, deliquescent crystals, soluble in water, formed by reaction of chromium hydroxide and hydrochloric acid and subsequent crystallization, (2) reddish-violet crystals, anhydrous, insoluble in water or dilute or concentrated acids, although the presence of a very small amount of chromous or stannous chloride renders the chromic chloride soluble, formed by heating chromic oxide in sulfur chloride above 400° C. Solutions of chromic chloride, concentrated or acidified, are green, when diluted violet; chromyl chloride (CrO_2Cl_2), red liquid, boiling point 118° C., by heating anhydrous sodium or potassium dichromate with sodium chloride and concentrated sulfuric acid.

Chromates and dichromates: Sodium chromate (Na_2CrO_4), potassium chromate (K_2CrO_4), ammonium chromate ($(NH_4)_2CrO_4$), calcium chromate ($CaCrO_4$), are yellow soluble solids; barium chromate ($BaCrO_4$), pale yellow, strontium chromate ($SrCrO_4$), pale yellow, lead chromate ($PbCrO_4$), yellow (used as a pigment, "chrome yellow"), zinc chromate ($ZnCrO_4$), yellow, used as a pigment, mercurous chromate (Hg_2CrO_4), yellow to red to brown, silver chromate (Ag_2CrO_4), reddish-brown, are insoluble solids. Sodium dichromate ($Na_2Cr_2O_7$), potassium dichromate ($K_2Cr_2O_7$), readily crystallized, ammonium dichromate ($(NH_4)_2Cr_2O_7$) (forming nitrogen gas and green chromic oxide solid upon heating), are red soluble solids, of important application as oxidizing agents, e.g., sulfurous acid causes reduction to chromic; silver dichromate ($Ag_2Cr_2O_7$), red insoluble solid, changing to silver chromate (Ag_2CrO_4) upon boiling with water. Solutions of chromate in the presence of acid are changed to the corresponding dichromate.

Dichromates: See Chromates. Solutions of dichromate in the presence of alkali are changed to the corresponding chromate.

Hydroxides: Chromous hydroxide ($Cr(OH)_2$), brown precipitate, formed by reaction of chromous salt solution with sodium hydroxide solution; chromic hydroxide ($Cr(OH)_3$), grayish-green precipitate, formed by reaction of chromic salt solution with sodium or ammonium hydroxide or sulfide solution, soluble in excess of sodium hydroxide but reprecipitated upon boiling, soluble in acids.

Oxides: Chromic oxide, chromium sesquioxide (Cr_2O_3), green solid relatively insoluble in acids after having been heated to 500° C., a good refractory material, formed by ignition of ammonium dichromate or chromium hydroxide, reduced to chromium metal (1) by heating

REPRESENTATIVE ALLOYS CONTAINING CHROMIUM

25–30% Cr		75–70% Fe	Resistant to oxidation at high temperature and to nitric acid. Not very ductile.
25% Cr	12% Ni	63% Fe	Tougher than "18–8" below. Heat resistant up to 1150° C. High resistance to oxidation.
20% Cr	67% Ni with C, Si, Mn approximately 1% each		Heat resistant up to 1200° C. Electrical resistance wire for heating.
18% Cr	8% Ni	74% Fe	"18–8" for castings, hot and cold-rolled bars, forgings, cold-drawn bars, plates, sheets, tubes, strips, and wire. For chemical apparatus. Especially resistant to dilute and concentrated nitric acid, and to sodium hydroxide (up to 20% NaOH); refrigeration equipment; furnace parts.
16–20% Cr		84–80% Fe	For castings, forgings, hot-rolled sheets, bars, plates, strips, and tubes. Heat-resistant up to 870° C. Resistant to ammonia and to nitric acid.
12–16% Cr		88–84% Fe	(a) with C maximum 0.10%, as "Stainless iron" for moderately corrosive conditions. (b) with C 0.1 to 0.3%, as "Cutlery stainless iron," more resistant to corrosion than (a). (c) with C 1 to 2%, as "Die stainless iron," resistant to abrasion, but less resistant to corrosion than (a) or (b).

(R.K.S.)

with carbon in the electric furnace, or (2) by ignition with **aluminum** powder; chromium trioxide, chromic anhydride, "chromic acid" (CrO_8), brownish-red crystals, soluble in water forming chromic acid (H_2CrO_4 or $H_2Cr_2O_7$), formed by addition of concentrated **sulfuric acid** to dichromate solution, forms chromic oxide and oxygen upon being heated to 190° C., a powerful oxidizing agent in which reaction chromic compound is formed. Used as a mordant in dyeing wool and silk.

Sulfate: Chromium sulfate, chromic sulfate (Cr_2 (SO_4)$_3 \cdot 18H_2O$), violet crystals, formed by reduction of dichromate with sulfurous acid in sulfuric acid solution, and crystallization. Used as a mordant in the textile industry, in the photographic fixing bath, in tanning, and in ceramics; chromium potassium sulfate, "chrome alum" ($Cr_2(SO_4)_3 \cdot K_2SO_4 \cdot 24H_2O$). Used as chromium sulfate.

Sulfide: Chromium sulfide (Cr_2S_3), brownish-black powder, formed by heating chromium metal in **carbon disulfide** vapor, not formed by wet methods because with water forms chromium hydroxide plus hydrogen sulfide.

Chromic acid and chromates are detected by the formation with lead acetate solution of a yellow precipitate (lead chromate) soluble in nitric acid, but insoluble in acetic acid.

For the detection of chromous and chromic compounds see **Analytical Chemistry**. (R.K.S.)

CHROMIZING.
Production of a high chromium content surface layer on iron and steel by heating at high temperatures in a solid packing material containing chromium powder, or in an atmosphere containing chromium chloride. The surface layer is formed by diffusion of chromium into the iron in the same manner as carbon diffuses into iron in **carburizing**; however, the process is much slower and requires higher temperatures than carburizing. A similar result can be obtained by high-temperature diffusion of electrodeposited chromium. Chromized coatings have corrosion resistance and elevated temperature oxidation resistance similar to the high chromium types of **stainless steels**. The principal use of chromizing has been in Europe. (R.H.H.)

CHROMOGENIC COUPLERS.
Couplers are used in secondary color development. (See **Color Development**.) The term "coupler" is applied to a large number of organic compounds which combine with a limited number of chromogenic developers to produce dye images with a wide range of color and intensity. Couplers are dye intermediates but differ from chromogenic developing agents in that they do not as a rule have the ability to develop a silver image.

Couplers may be hydroxy or amine derivatives of aromatic groups, as benzene, naphthalene or anthracene. Phenol, naphthol, analine, cresol, para-aminophenol and dimethylparaphenylenediamine are examples. With these compounds coupling takes place with the hydrogen atom which is in the ortho or para position to the hydroxy or amino group on the coupler.

Compounds having active methine (=CH₂) groups, with strong polar groups for other valences as the cyano (—CN), the carborryl (—CO), the aceto (CH₃CO—), the acid ester (—COOC₂H₅), and the phenyl, (—C₆H₅) groups will couple. Acetoaceticester and paranitrophenylacetonitrile are illustrations.

The methine group may be part of a ring structure as in coumarine and indoxyl, or may be part of a heterocyclic ring attached to a phenyl group as in 1-phenyl-3 methyl-5 pyrazolone.

Compounds having a N in the ring will couple if there is a methyl (—CH₃) group, attached to the ring in the alpha position to the nitrogen. Two illustrations for this type are picoline and 2-methyl-thiazole.

Because secondary color development has been particularly successful in the development of color materials, and because of the ease by which it is possible to

control color by this method, the field of couplers has expanded rapidly. (See **Chromogenic Developing Agents** and **Color Development**.) (S.M.T.)

CHROMOGENIC DEVELOPING AGENTS.
Chromogenic developing agents are selective organic reducers which not only have the ability to reduce exposed silver halide but also become dye intermediates by oxidation during the developing process. Color results when the oxidized developing agent, or product from reduction, unites, condenses, or couples with a molecule of the unoxidized developing agent or with another organic compound, known as a coupler, to form a dye.

Developing agents capable of forming dyes by addition, or condensation with the oxidized product, are known as *class 1* developers. Examples of these developers are para and ortho derivatives of dihydroxy, diamino, and hydroxy and amino benzene, or naphthalene. It is believed that the reaction is possible because these developers are readily oxidized to quinone and quinonimid structures, as

Quinone

Quinonimid

Among the developers showing these characteristics are: para-aminophenol, orthoaminophenol, catechol, pyrogallol, alphanaphthol, 4-methoxyalphanaphthol, and hydroxycoerulignon. Certain alpha hydroxyketones and alpha aminoketones also show the same property. Among the better known are indoxyl and thioindoxyl.

Developing agents that require the addition of dissimilar organic compounds to form dyes belong to *class 2*. Typical of this class are paraphenylenediamine and the paraphenylenediamine derivatives in which the hydrogens of one of the amine groups are mono- or disubstituted. Dimethylparaphenylenediamine, diethylparaphenylenediamine and 2-amino-5-diethylaminotoluene are excellent developers in this class. In a number of instances the N of the amine group may be a part of the heterocyclic ring structure as: 6-amino-N-methyl-penomorphine, 4-hydroxy-isocarbstyril, and 1-phenyl-3 methyl-4 amino-5-pyrazolone.

Leuco derivatives of certain vat dyes can be prepared in stable form. These act as developers in alkaline solution forming insoluble colored compounds on oxidation. Indigo white, a leuco base, is an example. (See **Chromogenic Couplers; Color Development**.) (S.M.T.)

CHROMOMERE.
One of the masses into which the chromosome is differentiated longitudinally. (See **Cell**.) (A.W.L.)

CHROMONEMA. A thread of **chromatin** in the resting **cell.** (A.W.L.)

CHROMOPHORE. Dyes.

CHROMOSOME. A unit into which the **chromatin** of the cell forms during **cell division.** Important in **heredity.** (A.W.L.)

CHROMOSOME NUMBER, PLOIDY. Chromosome number is of importance in **heredity** and **taxonomy.** Each species has a characteristic, though not necessarily different, chromosome number. The term ploidy is applied to the number of chromosomes, and other terms derived from it are used to designate modifications in the characteristic number as they normally occur during reproduction or as they are artificially produced. The body cells in each species have two complete sets of chromosomes in each nucleus. One set constitutes the haploid complement, the two sets the diploid complement. Chromosomes occur in pairs, the haploid or monoploid set consisting of one member of each pair found in the diploid nucleus. The diploid number is reduced to the haploid number by **meiosis,** restored by **fertilization,** and maintained as a constant number in every cell of the body by mitosis during **cell division.** In man, for instance, the diploid number is 48; eggs and sperm produced by meiosis have the haploid number 24, and the fertilized egg two sets of 24 each, one set contributed by each parent. Two of the chromosomes in the female, in man and certain other animals, are called allosomes or x-chromosomes and are concerned with sex determination, the male having only one allosome paired with a y-chromosome which has nothing to do with sex and is missing in many species. Chromosomes other than allosomes are called autosomes.

Further variation in chromosome number is common in plants, although less common in animals, and may be induced experimentally. Even haploid plants have been found, proving that the haploid set constitutes the basic, irreducible complement of each species. Variation may take the form of simple multiplication of the haploid number, a condition known as euploidy (true ploidy). If N is taken as the haploid number, $2n$ would be diploid, $3n$ triploid, $4n$ tetraploid, $5n$ pentaploid, $6n$ hexaploid, $7n$ heptaploid, $8n$ octoploid, $9n$ enneaploid, and $10n$ decaploid. Multiples above $10n$ are referred to as 11-ploid, 12-ploid and so on. The tetraploid number is common among cultivated plants. This and other multiples have been induced by the use of **colchicine** which prevents the consummation of normal mitosis or meiosis and doubles the diploid number. Even-numbered euploids breed true, the gametes of a tetraploid, for example, being diploid. Odd-numbered euploids often give rise to aneuploid plants having numbers such as $2n + 1$, $2n + 2$, etc. The trisomic condition, $2n + 1$, has been of use in the identification of the hereditary effect of whole chromosomes, one at a time, since the extra chromosome is an extra member of a pair. In the Jamestown Weed, **Datura,** which has 24 chromosomes, all possible primary trisomics, 12, have been induced and studied in detail. The various types mentioned in this paragraph are called collectively autoploids.

Another type of variation is induced when species are crossed with one another. These contain two different haploid complements; no chromosome finds a homologous chromosomes with which to pair at meiosis with the consequence that spore formation and consequently gamete formation are interfered with and species hybrids are usually sterile. Even before colchicine and other treatments had been used to double the chromosome number of the species hybrid and so make normal reproduction possible, amphidiploids (double diploids) had been found. The classic example is the cross between the radish (18 chromosomes) and the cabbage (18 chromosomes). The hybrid had 9 radish chromosomes and 9 cabbage chromosomes and was almost completely sterile. A few seeds were found one of which gave rise to a true-breeding hybrid which had 18 radish and 18 cabbage chromosomes. This is the amphidiploid condition and the plant produces gametes with the original haploid complement of each of the parent species. It is now possible to double the chromosome number of sterile hybrids thus producing amphidiploids almost at will. A very powerful tool for the production of entirely new varieties of plants is thereby made available. The phenomenon may also explain the origin of certain species. Ploidy of this type is known as alloploidy. (P.A.W.)

CHROMOSPHERE. The chromosphere is the layer of **atmosphere** of the sun which is composed principally of hydrogen, helium and calcium. It lies at a distance of several hundred **miles** above the **photosphere** and merges into the **reversing layer** below and the **corona** above. Since the elements which compose the chromosphere contain strong spectral lines in the red, it is usually visible at a total **eclipse** of the sun as a brilliant red envelope about the sun, and gets its name from this fact. The chromosphere may be observed with specially designed instruments, similar to those used in the study of **prominences,** even without a total eclipse. (W.K.G.)

CHRONOGRAPH. As the word itself implies, a chronograph is an instrument for writing time. In the usual type of instrument a drum carrying a sheet of paper is rotated by clockwork at the rate of 1 rpm. A pen, attached to the armature of an electromagnet, is carried parallel to the axis of the drum by a screw and rests on the paper, tracing a spiral line. The electromagnet is connected in an electric circuit with a standard clock in such a manner that every second the pen makes a short lateral movement which graduates the spiral line into seconds. A key is connected in the circuit in such a manner that each time the key is closed an extra graduation is made on the spiral. When the sheet is removed from the drum a scale may be used to measure the distance of the marks made by pressing the key from the marks made by the clock. In this way the time of pressing the key may be readily measured to 0.01 of a second. Where observation times are required with greater accuracy, the drum may be rotated more rapidly and a tuning fork used to make the time graduations. Other types of chronograph use a steadily moving strip of paper on which the pen traces its record. On the printing type of instrument a set of type wheels is rotated by a standard clock and each time a key is pressed a piece of paper is pressed against the type and the instant printed to the nearest hundredth of a second. Chronographs have many uses, for example, in apparatus designed to record vibrations from **earthquakes.** (W.K.G.)

CHRONOLOGY. The study of the measurement of time. This term is used by the geologist to include the methods employed in determining the age of the earth, and also the sequence of events. The principal methods that are employed are: 1. **Salinity** of the oceans. Assuming that the oceans were originally fresh water and that their present saltiness is due to river-borne solutions, it is computed that the oceans are from 90–600 million years old. 2. **Erosion.** It has been estimated that, to produce a major **unconformity,** having an approximate area of 3 million sq. miles, would take 9 million years. Therefore the unconformity (plane of erosion) which forms the boundary between the Archeozoic and the Proterozoic is a fundamental factor in the computation of **pre-Cambrian** time alone. 3. **Deposition.** The rate of deposition or time which it has taken to produce the stratigraphic sequence of sedimentary formations may be taken as a measure of the age of the earth. It is estimated that the post-**Proterozoic** stratigraphic column is approximately 350,000'. If this amount of sediment was the result of continuous depositions, at the present rate of deposition for the Nile delta, it would intimate that the earth was only 12 million years old. (Compare with "radioactive methods.") 4. **Radioactive** elements. The study of the radioactive minerals offers a method of determining the time at which the minerals crystallized from the **magma**. Since igneous rocks intrude sedimentary **formations** which may be dated by **stratigraphic** methods, it is possible to date both the intrusive and the intruded sedimentary rocks in terms of the appropriate period of the **geologic time-scale**. According to the radioactive method the oldest known rocks were formed approximately 1800 million years ago. (R.M.F.)

CHRONOMETER. For many types of field observing where accurate time is required, the use of the pendulum clock is not practicable. This is particularly true on shipboard or in countries like Japan, where earthquakes frequently disturb the pendulum. The chronometer is an accurate type of escapement timekeeper which is spring driven. The movement is similar to that in the ordinary pocket watch but much more massive. Various devices to compensate for changes in temperature and for changes in the tension of the spring are incorporated in the instrument. When properly handled, the rate of a chronometer may be relied upon to within a few hundredths of a second per day. With modern methods of obtaining clock comparisons by radio at frequent intervals, the chronometer may be used for all except the most refined astronomical observations. Many chronometers are equipped with devices for making or breaking an electric circuit so that they may be used in conjunction with the **chronograph**. (W.K.G.)

CHRYSALIS, CHRYSALID. The third stage in the development of **butterflies**, also properly called the pupa. The caterpillar of a butterfly spins no cocoon but hangs itself by a silken button or by a belt and button. The skin of the pupa into which it changes is often brightly colored or protectively colored and marked, unlike the mahogany-colored pupae of most moths. Although some moths form similar naked pupae the term chrysalis is applied only to those of the butterflies. (A.W.L.)

CHRYSOBERYL. (CYMOPHANE — GOLDEN BERYL). The mineral chrysoberyl, an **aluminate** of **beryllium** corresponds to the formula $BeAl_2O_4$, crystallizes in the **orthorhombic** system with both contact and penetration twins common, often repeated resulting in rosetted structures. Hardness 8.5; specific gravity, 3.5–3.84; luster vitreous; color various shades of green, sometimes yellow. A variety which is red by transmitted light is known as **alexandrite**. Streak colorless; transparent to translucent, occasionally opalescent.

Chrysoberyl occurs in granitic rocks, **pegmatites** and mica **schists**; often is found in alluvial deposits. The Ural Mountains yield **alexandrite**; other localities for chrysoberyl are Czechoslovakia; Ceylon; Southern Rhodesia; Brazil and Madagascar where it occurs of gem quality in the pegmatites of that island. In the United States it is found in Maine, Connecticut and New York. The word chrysoberyl is derived from the Greek words meaning golden, and beryl. **Cymophane** has its derivation also from the Greek words meaning wave, and appearance, in reference to the opalescence exhibited at times. (E.S.C.S.)

CHRYSOCOLLA. This mineral, a hydrous **silicate** of **copper** probably corresponding to the formula $CuSiO_3 \cdot 2H_2O$, is perhaps a mineral gel for it usually appears as an amorphous mass, in veins, or as incrustations. It is very rarely found in small acicular crystals which are either **hexagonal** or **tetragonal**.

Its color is generally some shade of blue or green but if impure may be brown or black. It has a characteristic conchoidal fracture; hardness, 2–4; specific gravity, 2–2.2; vitreous to dull luster; translucent to opaque.

Chrysocolla is a secondary mineral and associated commonly with other copper minerals of similar origin. Among the localities for excellent specimens may be mentioned Cornwall and Cumberland, England; Belgian Congo; Chile; Lebanon and Berks Counties, Pennsylvania; the Clifton-Morenci Globe and Bisbee districts in Arizona; Dona Ana County, New Mexico, and the Tintic district, Utah.

The word chrysocolla is derived from the Greek words meaning gold and glue, formerly the name for gold solder. Chrysocolla is one of the less important ores of copper and has a minor use as a gem stone. (E.S.C.S.)

CHRYSOLITE. Olivine.

CHRYSOMONADIDA. One-celled animals of yellow or brown color, or colorless, constituting an order of **Mastigophora**. (A.W.L.)

CHRYSOPRASE. Chalcedony.

CHRYSOTILE. A delicately fibrous variety of **serpentine** which separates easily into silky flexible fibers of greenish or yellowish color. Its name is derived from the Greek words meaning gold and fibrous. Most of the common asbestos of commerce is chrysotile. It is mined in Thetford, Province of Quebec, and in South Africa. (E.S.C.S.)

CHUB. Pisces, Teleostei. A European fish (**Pisces**), *Leuciscus cephalus,* related to the dace, and several North American species. One, an excellent food and game fish, is the fallfish, *Leucosomus corporalis,* or chub of northeastern streams and lakes. Another, the horned dace, *Semotilus atromaculatus,* creek chub, or common chub, is widely distributed in the United States and Canada with the exception of the far west. In small streams it becomes abundant and sometimes rises to the fly as freely as the game fishes. Its flesh is usually lacking in flavor. Other species of the group are sometimes called chubs. (A.W.L.)

CHUB SUCKER. Pisces, Teleostei. A small **sucker,** *Erimyzon sucetta,* of rather stout build, common throughout the eastern half of the United States. (A.W.L.)

CHUCK. A rotating vise which may be attached to the spindle of a machine. There are two important varieties of lathe chucks, independent and universal. In general, the *independent chuck* has four jaws each of which is separately actuated and adjusted. It may be employed for almost any type of work, cylindrical, square, or irregular. In turning cylindrical work, it is necessary to adjust the jaws very carefully, and test the concentricity of the work and spindle axes with some form of indicator. When a 4-jaw independent chuck is employed for repetitive work, only two adjacent jaws are actuated as each new part is placed in the chuck after the initial adjustment and alignment of the jaws have been obtained.

Three-jaw *universal* or self-centering chucks are employed for cylindrical and hexagonal bar stock. The jaws are simultaneously advanced or retracted by turning the scroll plate in which the jaw teeth fit. On account of the curvature of the scroll, it is necessary to have separate sets of jaws for inside and outside clamping, in contrast to the independent chuck where the jaws may be reversed. Combination chucks combine an independent chuck for holding odd-shaped work; a universal chuck for self-centering and gripping round or square work; 2-jaw universal chucks have jaws to which special adapters may be fitted and are employed principally in turret lathe work.

Independent and universal chucks may be attached to a threaded adapter which screws on the spindle nose. These chucks may also be obtained with adapters to fit heavy-duty taper spindle noses; the chuck is drawn on the spindle nose by the engagement of the "pull-on" nut with the externally-threaded hub of its adapter.

Air-operated chucks have a body with two jaws and work-holding adapters with an actuating wedge which closes and opens the jaws by moving parallel to the chuck axis. The actuating wedge is threaded so that a draw-rod may be attached. The other end of the draw-rod is attached to a piston operating in an air cylinder which is attached to the lathe headstock. The piston is double-acting so that the chuck jaws may be both opened and closed by the action of compressed air. The chuck is operated by a valve convenient to the machine operator. Air- and oil-operated chucks are generally employed for production work, as in turret and chucking lathes.

Draw-in chucks and collets are used for bar work, and are designed to fit on the heavy-duty spindle nose if the live center is removed. The collets, which are of various sizes, fit in the chuck and are clamped with a removable key or wrench. Drawing in the spring collet forces its outer surface against the taper on the inside of the chuck; releasing the collet causes its jaws to open by their spring action. Bars of any length may be held in the chuck, extending entirely through the hole in the spindle if necessary. Collets for all standard sizes of circular rod are available as well as collets for hexagonal bar stock and cylindrical metric sizes. Magnetic chucks of both the electrically-actuated and the permanent-magnet type may also be employed on the lathe. (H.C.H.)

Three-jaw universal chucks are used on drill presses. Rotating the outer sleeve by hand or by a key fitted to the sleeve gear teeth, opens and closes the three self-centering jaws. The arbor hole in the drill chuck fits the tapered end of the drill press spindle. *Bayonet chucks* are single-purpose chucks and are used on automatic drilling machinery. They are equipped with a bayonet slot for rapid attachment and release.

CHUCK-WALLA. Reptilia, Sauria. A common lizard, *Sauromalus obesus*, of the southwestern deserts, ranging into Utah and Nevada. It attains a length of 11″ and is sometimes eaten. (A.W.L.)

CHUCK-WILL'S-WIDOW. Goatsucker.

CICADA. Insecta. Homoptera. Large **insects of** many species. They are stoutly built and have two pairs of membranous wings which are folded roof-like over the body when at rest. They are best known for the loud songs of the males, which are produced by a pair of elaborate organs located on the under surface at the base of the abdomen. These organs have a vibrating structure controlled by special muscles and thin resonating parts which result in a peculiarly penetrating sound.

The female has a powerful **ovipositor** with which she punctures the twigs of trees and shrubs to deposit her eggs. The young cicada does not remain in the twig but drops to the ground and burrows, feeding on the roots of plants. In the case of one species, *Tibicina septendecim*, the duration of the larval period is unusually long and has resulted in the name seventeen-year locust. Locust is inaccurately but very commonly applied to these insects and some are called harvest-flies.

Great damage is sometimes done to young orchard trees by the breaking of twigs where the eggs have been deposited, especially after one of the great broods of the seventeen-year locust, or periodical cicada, has passed. The only effective protection is to cover young trees with inexpensive cloth when such a brood is imminent; the years of emergence are known by economic entomologists and can readily be learned for any part of the country. (A.W.L.)

CICHLID. Pisces, Teleostei. Fishes (**Pisces**) of a large family, including several hundred species, found in tropical and subtropical waters of India, Africa, and the New World. They occur in both fresh and brackish water.

A number of species of cichlids are offered by dealers in tropical fishes for aquaria. They include beautifully colored fishes and are interesting for their breeding habits; the eggs are carried in the mouth or **pharynx** until they hatch. (A.W.L.)

CICONIIFORMES. An order of long-legged wading birds (**Aves**) including the **herons, flamingoes, bitterns, storks** and others. (A.W.L.)

CIDAROIDA. An order of sea-urchins without gills around the mouth. (A.W.L.)

CIENEGA. A type of spring which occurs in intermontane basin deposits or **bolsons**, especially of semiarid to arid regions. When the underground waterbearing stratum, or **aquifer**, is blocked by cemented gravels the water may be forced by hydrostatic pressure to the surface, forming the type of spring called a cienega. (R.M.F.)

CILIARY JUNCTION. An association of separate gill filaments in certain **bivalve** mollusks, which is characterized by interlocking **cilia**. (A.W.L.)

CILIATA. One-celled animals which have **cilia** during adult life but are without suctorial tentacles. They constitute a class of this name in the subphylum **Ciliophora.** (A.W.L.)

CILIOPHORA. A subphylum of the phylum **Protozoa** containing species which have cilia or cirri during some stage of life. They vary greatly in form and habits. Some are sessile, some free swimming, and some parasitic.

The following is a brief summary of a classification of ciliates now widely used:

Class Ciliata. With cilia or cirri throughout life.
  Subclass Protociliata. Leaf-like species with two to many nuclei. Parasitic in the intestines of fishes and amphibians. *Opalina* and related forms.
  Subclass Ciliata (Euciliata). With two kinds of nuclei, large and small (macronucleus and micronucleus).
    Order Holotrichida. **Cilia** uniformly distributed. *Paramecium* and many other genera.
    Order Heterotrichida. With a zone of cilia of larger size, or membranelles, associated with the mouth. *Stentor*.
    Order Oligotrichida. With cilia about the mouth but few on the body.
    Order Hypotrichida. Flattened, with cilia or cirri on the under surface. *Stylonychia*, etc.
    Order Peritrichida. Oral end of body enlarged, ciliated, many species stalked. *Vorticella*, etc.
Class Suctoria. With cilia only during early life. Adults sessile, with tentacles for ingesting food and for piercing. (A.W.L.)

CILIUM. A slender hair-like process of minute size on the surface of a cell. It is part of the living **cytoplasm.** In association with their minute size cilia occur in relatively large numbers and act in unison to produce aggregate effects. They are capable of waving movement. Commonly they bend consecutively in the same direction so that a wave of movement passes along the ciliated surface, followed by the return of the cilia to the resting position and this again by their bending. This type of movement is said to be metachronal. The successive waves follow each other closely so that several may be apparent at the same time.

Cilia are found in many species of Protozoa and give the name Ciliophora to one subdivision of the phylum. Among the multicellular animals they occur on the surface of the body in a few groups (e.g., **coelenterates, turbellarian** worms and **molluskans**) during adult life and in many larvae such as those of the **echinoderms,** the **annelid** worms, coelenterates, and others. They are also found on epithelia of limited distribution in many complex animals, as in the mantle cavity of mollusks and the trachea of man.

Cilia on the surface of small animals, such as the **Protozoa** and **larvae** of greater complexity, are able by their action against a surrounding liquid to propel the animal and so serve as organs of locomotion. In animals of larger size, in those which are not surrounded by a liquid medium, on sessile forms, and where the direction of their movement is contrary to the movements of propulsion, they act to set up currents either in liquids secreted by the body or in the surrounding medium. Thus some cilia in Protozoa, coelenterates, and mollusks direct a current bearing food and oxygen into the gullet and in man cilia carry toward the throat the mucus secreted by glands in the lining of the trachea. This secretion may bear foreign particles which have been inhaled. (A.W.L.)

CINCHONA. *Cinchona Calisaya* and other species. Rubiaceae. This is a small tree (30–40' high) with light green leaves and pink flowers. It is native in Bolivia. There, because of the destructive habit of chopping the trees down in order to obtain the bark, the wild tree has almost disappeared. In 1859 the plant was introduced into the East Indies, where it was so widely cultivated as to produce a tremendous decrease in the price of the bark.

The bark is the source of the drug quinine, a bitter alkaloid used as a preventive and cure for **malaria.** When the trees are seven or more years old, they are felled, and the bark removed. Other methods of collection include uprooting the tree to secure the root bark in addition to the stem bark and scraping the outer bark from living trees. The yield of quinine is increased by protecting the stem of the tree from sunlight, as by covering trunk and branches with moss. Coppicing is also practiced (see **Cinnamon**). Production for world trade at present is centered largely in the Dutch East Indies, particularly in Java. (R.M.W., B.S.M.)

CINCHONIDINE. Alkaloids.

CINGULUM. 1. A girdle, such as the outer ciliated ring at the anterior end of a **rotifer.** 2. A bundle of association fibers in the mammalian **brain,** partially encircling the corpus callosum not far from the median plane. 3. The basal ridge of a tooth. (A.W.L.)

CINNABAR. The mineral cinnabar, mercuric sulfide (HgS) occurs in small and often highly modified hexagonal crystals, usually of rhombohedral or tabular habit. It is found chiefly in crystalline crusts, granular or simply massive. The fracture of cinnabar is subconchoidal; hardness, 2–2.5; specific gravity, 8–8.2; luster, adamantine tending toward metallic, sometimes dull. This mineral has a characteristic cochineal-red color

which, however, may be brownish at times, occasionally dull lead gray. The streak is scarlet; it is transparent to opaque.

Cinnabar occurs in veins or may be in masses in **shales, slates, limestones** and similar rocks due to the impregnation by mineral-bearing solutions or as replacements. Russia, Czechoslovakia, Bohemia, Bavaria, Italy and Spain have furnished excellent specimens. The most important of the world's mercury deposits is at Almadin in Spain. Italy, Peru, Dutch Guiana, China and Mexico have commercially valuable occurrences of cinnabar. In the United States this mineral is found in California (most important deposit), Nevada, Utah, Texas and Oregon. Cinnabar is the chief ore of mercury. Its name is supposed to be of Hindu origin. (E.S.C.S.)

CINNAMON. Cinnamomum zeylanicum. Lauraceae. Cinnamon is obtained from a small tree native to Ceylon and India, where the plant is now extensively grown in cultivation. The trees grow from 25–40' high, have shining dark green leathery leaves, small whitish flowers having a rather disagreeable odor, and dark purple fruits. The bark of young twigs is smooth and somewhat mottled; in older branches and the main stem, the bark becomes thick, rough and of little value. To insure the desideratum of many young branches, the limbs are severed so that many slender branches will form, a practice known as coppicing. From these slender stems the bark is removed by lengthwise splitting and partial loosening from the stem; as it dries it rolls back. It is then removed from the stem, the dry useless **periderm** scraped off, and the inner bark remaining allowed to dry completely. During drying its color changes from pale yellow to deep brown. The tight rolls of dried bark are packed together in bundles, called pipes, and are ready for marketing.

This bark contains considerable amounts of a powerful drug which in large doses is a dangerous poison. The principal use of cinnamon is as a spice for pastries. By distillation of cinnamon stems and leaves there is obtained oil of cinnamon (see **Volatile Oils**), used in flavoring candy and in scenting soap. (R.M.W., B.S.M.)

CIPOLIN. A metamorphic rock transitional between a **marble** and mica-**schist** in which the principal **mica** is **phlogopite.** (R.M.F.)

CIRCLE. A circle is a plane curve such that all of its points are at a fixed distance, called the radius, from a fixed point called its center.

The length of the circumference of a circle of radius r is given by $C = 2\pi r$, its area by $A = \pi r^2$.

The area of a circular sector is given by $A = \frac{1}{2}r^2\theta$, where r is the radius of the circle and θ is the **radian measure** of the angle of the sector.

The area of a segment of a circle (between a circle and a chord) is given by $A = \frac{1}{2}r^2(\theta - \sin\theta)$, where r is the radius of the circle and θ is the radian measure of the angle of the segment.

In **rectangular coordinates,** the equation of the circle of radius r with center at the origin is

$$x^2 + y^2 = r^2;$$

and the equation of the circle of radius r with center at the point (h, k) is

$$(x - h)^2 + (y - k)^2 = r^2.$$

Any equation, in rectangular coordinates, of the form $Ax^2 + Ay^2 + Bx + Cy + D = 0$ represents a circle. For the circle whose equation is $x^2 + y^2 + Dx + Ey + F = 0$, the center is $(-\frac{1}{2}D, -\frac{1}{2}E)$ and the radius is $\frac{1}{2}\sqrt{D^2 + E^2 - 4F}$ if this latter expression is real. A circle is determined by three conditions.

In **polar coordinates,** if the center is on the polar axis and the circle passes through the pole, its equation is $r = 2a \cos\theta$ (if the center is to the right of the pole) or

$r = -2a \cos \theta$ (if the center is to the left of the pole), where a is the radius of the circle. If the center is on a line perpendicular to the polar axis through the pole and the circle passes through the pole, the equation of the circle is $r = \pm 2a \sin \theta$, where the $+$ or $-$ sign is to be used according as the center is above or below the pole. If the center is at the pole, the equation of the circle is $r = a$. If the circle passes through the pole and has X- and Y-intercepts a and b, its polar equation is $r = a \cos \theta + b \sin \theta$.

The simplest **parametric equations** of a circle are: $x = a \cos \theta$, $y = a \sin \theta$, where a is the radius and θ is the angle XOP, and P is the point (x, y).

If $C_1 = 0$ and $C_2 = 0$ are the equations of two given circles, then $C_1 + kC_2 = 0$ is the equation of the system of all circles passing through the intersections of the given circles, if k takes all positive and negative values.

If $x^2 + y^2 + D_1x + E_1y + F_1 = 0$ and $x^2 + y^2 + D_2x + E_2y + F_2 = 0$ are the equations, in rectangular coordinates, of two given circles, then the straight line whose equation is $(D_1 - D_2)x + (E_1 - E_2)y + (F_1 - F_2) = 0$ is called the radical axis of the given circles. It is perpendicular to the line of centers of the given circles. If the given circles intersect (or touch) each other, the radical axis is their common chord (or common tangent). The radical axis of two circles is the locus of all points from which the lengths of the tangents to the circles are equal.

If three circles are given, and the three radical axes of these circles taken in pairs are found, it will be found that they intersect in a common point, which is called the radical center of the three given circles.

The length along the tangent from a point (x_1, y_1) to a circle of radius r and center (h, k) is given by

$$t^2 = (x_1 - h)^2 + (y_1 - k)^2 - r^2.$$

The length along the tangent from a point (x_1, y_1) to the circle $x^2 + y^2 + Dx + Ey + F = 0$ is given by

$$t^2 = x_1^2 + y_1^2 + Dx_1 + Ey_1 + F.$$

The equation of the **tangent** to the circle $x^2 + y^2 = a^2$ at the point (x_1, y_1) is $x_1x + y_1y = a^2$.

The equation of the tangent of **slope** m to the circle $x^2 + y^2 = a^2$ is $y = mx \pm r\sqrt{1 + m^2}$.

The circles $x^2 + y^2 + 2D_1x + 2E_1y + F_1 = 0$ and $x^2 + y^2 + 2D_2x + 2E_2y + F_2 = 0$ are orthogonal (intersect at right angles) if $2D_1D_2 + 2E_1E_2 = F_1 + F_2$. (L.L.S.)

CIRCLE OF POSITION. Line of Position.

CIRCUIT BREAKER.

A circuit breaker is a special switching device for opening electrical power circuits under load, the operation being initiated either by an attendant, or automatically, in the event of abnormal circuit conditions. It is a safety device installed in those portions of an electric circuit where it is desired to open the circuit under load upon presence of abnormal current, abnormal voltage, high temperature, grounds, etc. There is no more important element of an electrical power system than the circuit breakers and their control equipment. They are used in generator leads, feeders, bus ties, large motor leads, etc. Ordinary switches cannot be used for such applications as they are incapable of extinguishing the resultant arc.

Circuit breakers may be classified as air break or oil break types. An air break type, commonly a carbon circuit breaker, consists of stationary contacts mounted on a vertical panel, a movable contact which is closed against a heavy spring by a handle, a trip to release the latch, and carbon auxiliary contacts which take the arc when the breaker opens. These breakers are mostly closed by hand, and opened by series overload trip coils. They are infrequently used above 600 volts. In recent years, a type of air circuit breaker called the Deion breaker has been developed. By means of magnetic fields the arc which tends to follow upon opening of a circuit under load, is deflected into a deionizing chamber where it is broken up into a number of short arcs

which are deionized and quenched during a few cycles. The Deion breaker is applicable to a-c circuits.

The carbon circuit breaker is a type used principally for d-c circuits. The oil-immersed circuit breaker is standard with a.c. because the arc can be oil quenched as the voltage passes through the zero point. The accompanying figure shows the elements of an oil circuit

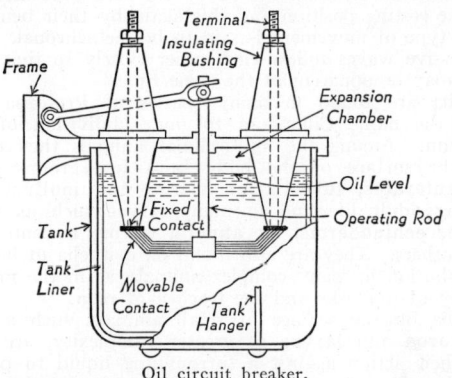

Oil circuit breaker.

breaker. The contacts are opened under oil by means of a rod extending through the cover of the case. The operating arm is actuated by hand, motor, or solenoid, and when in closed position is tensed by a heavy spring whose purpose is to produce a rapid opening of the contacts when the operating arm is released. The Deion principle is also applied to oil breakers. The latch which locks the operating arm in closed position against spring pressure may be tripped free by energizing a small solenoid called the trip coil, and the breaker has a protective device which is intended to be automatically opened by a trip coil actuated either directly by the current passing through the breaker, or indirectly by special trip circuits utilizing **relays**.

Recently special types of breakers have been developed. These employ various means such as oil impulses, pressure chambers, etc., to extinguish the arc. (F.T.M., L.R.Q.)

CIRCULAR CURVES.

From a mathematical standpoint a circular curve is an arc having a constant radius but it is used in Civil Engineering as a general heading to cover simple, compound and reversed curves.

A circular arc joining two tangents (straight lines) is called a simple curve. Large radius simple curves are used in **highways** to provide a means of gradually changing the direction of the center line of a roadway.

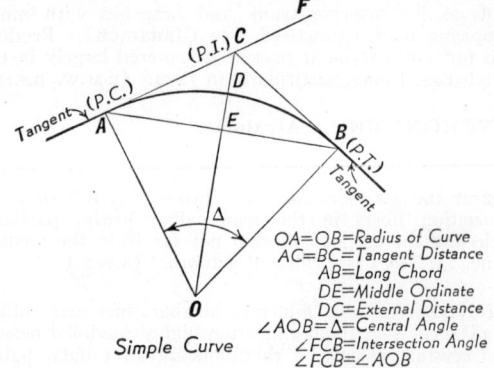

Simple Curve

OA=OB=Radius of Curve
AC=BC=Tangent Distance
AB=Long Chord
DE=Middle Ordinate
DC=External Distance
∠AOB=Δ=Central Angle
∠FCB=Intersection Angle
∠FCB=∠AOB

Simple curves connected by tangents were formerly used on railroads but they have been superseded by the combination of **spiral** and simple curves. The accompanying figure shows the elements of a simple curve

Point *A* is called the point of curvature (P.C.) and point *B* the point of tangency (P.T.). Point *C* is known as the point of intersection (P.I.) of the tangents. In highway practice the length of the curve is generally represented by the length of the circular arc but in railroad practice it is given in terms of **chord** lengths.

A curve made up of two or more simple curves, each having a common tangent point at their junction and lying on the same side of the tangent, is called a compound curve. Compound curves have an advantage over simple curves since they may be easily adapted to the natural **topography** of a particular location.

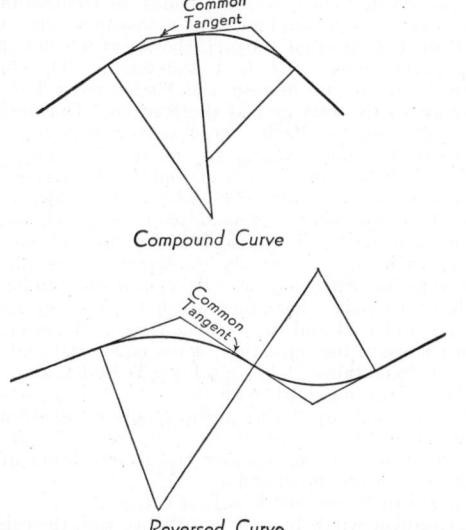

Common
Tangent

Compound Curve

Common
Tangent

Reversed Curve

A curve made up of two simple curves, having a common point of tangency at their junction and lying on opposite sides of the common tangent is called a reversed curve. This type of curve is advantageous for use in connection with railroad cross-overs and spur tracks but should never be employed for main lines. Reversed curves are used in highway location when the alignment requires an abrupt reversal in direction. (c.w.c.)

CIRCULAR FUNCTIONS. Trigonometric Functions.

CIRCULAR MEASURE OF ANGLES. Radian Measure of Angles.

CIRCULAR MIL. This is a unit of area, employed to designate the cross-sectional area of electrical conductors. It replaces measurement in square inches, and is more convenient since it is not necessary to multiply by the factor π. A mil is $1/1000$ of an inch. The area of a circle whose diameter is M mils is simply M^2 circular mils.

Wires are sized by their area in circular mils or by the American wire gauge. Despite the arguments in favor of one system using diameters in mils as size numbers, the American wire gauge is commonly used for wires sized from 40 to 0000 American wire gauge. Wires larger than 0000 (212,000 circular mils) are always sized in circular mils. (F.T.M.)

CIRCULAR MIL FOOT. Resistance.

CIRCULAR PITCH. Spur Gearing, Worm Gearing.

CIRCULAR POLARIZATION. Polarized Light.

CIRCULAR SECTOR. Circle.

CIRCULAR SEGMENT. Circle.

CIRCULATION. The distribution through the body of a liquid medium known as **blood** or haemolymph which conveys necessary materials to the tissues and carries wastes away from them.

In the simplest animals materials are merely transmitted from cell to cell but in the more complex forms the bulk of tissues which is removed from direct contact with the centers of interchange with the environment is too great to be served adequately by this method. In such animals a special **circulatory system** is developed for the more rapid distribution of materials. Circulation in such a system may be accomplished by the mixing of liquid contents in open cavities such as the **coelom** or it may be the function of special organs in closed tubular systems.

In a closed tubular system such as that of the **vertebrates**, including man, pressure is applied to the blood stream by a muscular organ, the heart, and by the muscular walls of the larger vessels. The blood leaves the **heart** under its greatest pressure and enters the arteries whose elastic walls expand and then, returning to their contracted state, partially equalize the fluctuating pressure supplied by the heart. The blood is at a lower pressure in the minute capillaries where some of it escapes into the spaces between the cells of other tissues; vessels of this kind occur in abundance at all points where interchange between the blood and the tissues takes place. From the capillaries the blood flows back to the heart in veins. In total, the pressure diminishes steadily from the time the blood leaves the heart until it returns again, a necessary condition according to the laws of hydraulics.

The relations existing at any point in the body between the liquids in the cells and intercellular spaces and the blood itself make possible osmotic (see **osmosis**) interchange in both directions which serves the needs of the tissues and maintains a normal concentration of materials in solution in the blood.

The paths of circulation in the bodies of various animals involve the discussion of the **circulatory system**. (A.W.L.)

CIRCULATION, AERODYNAMIC. A mass of air in rotary motion is said to be in circulatory flow if its velocities at various radii from the center of rotation are of the proper magnitude to induce radial equilibrium of the circulating mass. Consideration of the requirements for equilibrium consists in balancing centrifugal forces against static pressures derived from **Bernoulli's Theorem** and results in the specification that velocity of a particle must be inversely proportional to its radius from the center of rotation. Note that this is not like the motion of portions of a wheel, whose velocities are proportional to their radii. A simple case of circulation could be visualized as the flow pattern (concentric circles) induced by a rough cylinder rotating rapidly in still air.

If air is in circulation and no object such as a cylinder occupies the central core the velocity at the center of rotation reaches a theoretical value of infinity. This is impossible and the center of circulatory flow must be occupied by a small core of air in simple rotary motion. Air in this condition is described as a free **vortex**. Exceptionally high velocities may exist at the edge of the rotary core, and in vortices of high circulatory strength such as tornadoes the atmospheric energy may reach destructive proportions.

The strength of aerodynamic circulation (usually called the *circulation* and designated by symbol Γ) is the line integral of the tangential component of the velocity along any closed line encircling the vortex center, or some object.

$$\Gamma = \oint v \, dl.$$

It has been proven that Γ has the same value for any closed path through the flow around the object (such as an airfoil, cylinder, etc.) and this fact is of value in various phases of aerodynamic analysis. A simple case of a flow pattern of concentric circles illustrates the constancy of Γ. The velocity at radius r is $\dfrac{K}{r}$ and $\Gamma = \oint \dfrac{K}{r}\,dl$

$= \oint \dfrac{K}{r}\,r\,d\theta = 2\pi K$. Therefore Γ is not a function of r; its magnitude is constant throughout the flow pattern. (F.T.M.)

CIRCULATION OF THE ATMOSPHERE.

When averaged over long periods of time, local and small-scale irregularities in the atmosphere's motions disappear and a generalized pattern of winds is manifest. There are five latitudinal belts in each hemisphere into which generalized winds can be classified.

1. The Doldrum belt which extends roughly from the equator to 10 or 15° north and south is a belt of light variable winds.

2. The Trade Wind belt extends from 10 or 15° north and south to approximately 30° north and south. Trade winds blow from the northeast to east-northeast in the northern hemisphere and from the southeast to east-southeast in the southern hemisphere and are known respectively as the Northeast Trades and the Southeast Trades.

3. A narrow and drifting belt of light variable winds extends about the earth at approximately 30° north and south. This belt is known as the Horse Latitudes.

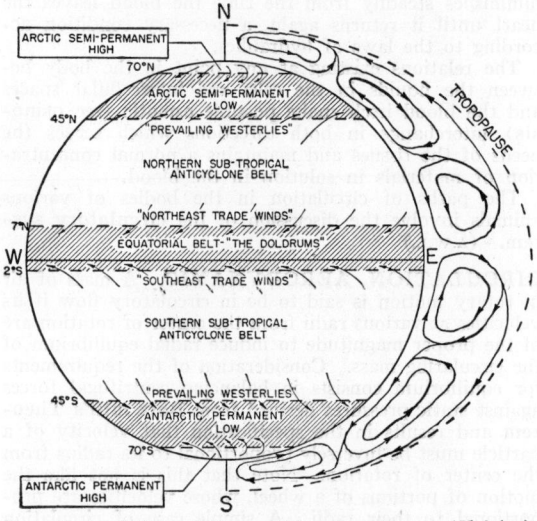

General circulation of the earth's atmosphere. (*Halpine's A Pilot's Meteorology.*)

4. Westerly winds with a slight component from the south blow in a relatively wide band from approximately 30° north and south to 60 or 70° north and south. They are known as the Prevailing Westerlies but in the southern hemisphere are more commonly known as the Roaring Forties because of their stronger and more steady character.

5. The polar area winds tend to blow anticyclonically with an easterly component over each region.

Surface atmospheric pressure belts fit these surface wind zones. There are four distinct zones or belts of high and low pressure.

1. Pressure in the doldrums is low.

2. Pressure in the horse latitudes is high.

3. Pressure between the temperate-zone westerlies and the polar winds is low.

4. Pressure over the polar caps is high.

Two well-developed semi-permanent cyclones are present in the northern hemisphere, one located over the North Atlantic near Iceland, known as the Icelandic Low, and the other in the Aleutian area of Alaska, known as the Aleutian Low. In the southern hemisphere there is a more or less continuous belt of low pressure around the world at the corresponding latitude.

Because pressure decreases more rapidly with altitude when air is cold, surface-wind characteristics disappear very slowly with altitude and at considerable height the circulation is generally westerly.

There is considerable change in the general circulation from winter to summer in so far as continents and oceans are concerned. These seasonal winds, or changes in the general circulation, are known as monsoon winds. Over the United States and Canada, the average wind directions during winter are west and north. This brings relatively warm air in over the Pacific coast but cold Polar air to the area east of the Rockies. During summer, winds on the Pacific coast remain from the west and north bringing relatively cool air to this region. East of the Rockies, winds are west and south bringing hot and often humid air over the entire section. Because of these monsoon winds, seasonal temperature changes in the eastern parts of North America are considerably greater than changes in solar radiation would indicate. Monsoons are extremely well developed over India and southeastern Asia where they are hot and rainy during summer and cool and dry during winter. A major seasonal change is the migration of the equatorial and subtropical circulations (doldrums, horse latitudes, and trade winds) north and south with the sun. This also causes weakening of the westerlies during summer and strengthening during winter; weakening of the Icelandic and Aleutian Lows during summer and a weakening of the Polar circulations in summer.

Winds can be divided into four categories:

1. Gradient winds blow in accordance with the existing pressure gradient centrifugal force and Coriolis Force.

a. Cyclonic winds blow counterclockwise about regions of relatively low pressure in the northern hemisphere and clockwise in the southern hemisphere.

b. Anticyclonic winds blow clockwise about regions of relatively high pressure in the northern hemisphere and counterclockwise in the southern hemisphere.

2. Geostrophic winds blow in accordance with the pressure gradient, but only where the pressure gradient is balanced by the Coriolis Force. They are, therefore, winds which blow in straight or nearly straight lines over the earth. Geostrophic winds are not possible at the equator because there is no Coriolis Force present.

3. Cyclostrophic winds blow cyclonically in both hemispheres in wind systems where the pressure gradient is balanced by centrifugal force in the absence of the Coriolis Force. Cyclostrophic winds occur near the equator as hurricanes and other local less intense vortices.

4. Antitriptic winds are small-scale, short duration winds which blow, in general, along the pressure gradient. Land and sea breezes are of this type.

In general, winds are mainly gradient winds.

Many strictly local winds blow over relatively small regions. Most of these occur where there is sharp contrast in surface temperature over a relatively small distance or where terrain is highly irregular. Sea breezes blow from cool water to heated land during the heat of day. Land breezes blow from cooled land to warmer water during the cool of the night. Valley breezes blow upslope in valley-hill terrain during sunny days, and mountain breezes blow downhill in a reverse manner during darkness. Mountain breezes often become very strong and extremely variable as a result of large-scale eddies and Venturi effects in mountain passes. (P.E.K.)

CIRCULATION THEOREM.

If the atmosphere is baroclinic, that is, if the surfaces of equal pressure and equal density intersect at any angle whatsoever, there is a tendency for a circulation to develop in such a manner that the atmosphere will become barotropic, that is, the surfaces of equal pressure and equal density will coincide. The atmosphere is normally baroclinic. Sea and land breezes, mountain and valley breezes are results of well-defined baroclinic states. Direction of circulation is always such that cold air flows toward warm air at the

contracticle vessels are principally the dorsal vessel and a series of five pairs encircling the **oesophagus** and known as hearts. The blood flows forward in the dorsal vessel, down through the hearts, and back in the ventral vessels, reaching the dorsal vessel again after following various routes through the tissues which it serves.

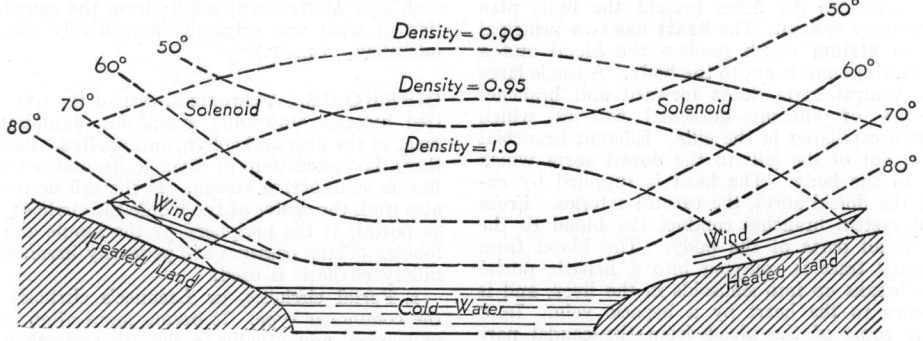

Circulation theorem and solenoids.

base of the circulation pattern, and warm flows toward cold at the top of the pattern. Air sinks in the cold air region and rises in the warm air region. Thus the sea breeze blows along the surface of the earth from cold water to heated land, rises, then returns seaward and sinks.

It is possible to compute the magnitude of the circulation from a given baroclinic state if temperature and pressure are known. Suppose a vertical plane were erected perpendicular to a shore, extending from the cold water to heated land. Lines of equal pressure and temperature drawn in this plane will produce a field of approximate parallelograms which are known as solenoids. A tendency for circulation exists about the perimeter of each solenoid, but in the field as a whole this tendency is nullified in adjacent solenoids. There is no nullification along the border solenoids of the whole field and it is here that the circulation springs up. The number of solenoids in the field is a measure of the expected strength of the resulting circulation. (P.E.K.)

CIRCULATORY SYSTEM. The system of passages and chambers through which materials are distributed in the body in a liquid mixture called **blood** or haemolymph.

Animals of very simple structure and small size secure adequate distribution by the diffusion of materials from cell to cell and so have no need for a special circulatory system. In more complex bodies, however, the principal centers of interchange, such as the alimentary tract, respiratory organs, and excretory system, are far removed from some of the parts that they serve; here more rapid transportation is necessary.

This need is met in some animals by the extension of centers of interchange. In the flatworms, for example, the alimentary tract branches through the body and no other structure is far from some part of it.

Other forms, as the roundworms, have extensive spaces within the body in which liquid contents are moved to some extent by the movements of the body. This type of circulation extends to animals with a true body cavity or **coelom** but here it is associated with the development of a closed tubular circulatory system, a condition which exists in the earthworm.

In this simple state the tubular system consists of a longitudinal vessel above the alimentary tract and others at different levels in the ventral part of the body. Other tubes running around the alimentary tract associate the longitudinal vessels, and muscular walls of certain regions, together with valves in the cavities of the tubes, propel the blood which they contain. In the earthworm the

In such systems as this several types of vessels are developed. Those which lead from the central pumping organ receive the blood under its highest pressure and have the strongest walls, containing elastic and muscle fibers. These vessels are called arteries. They lead into

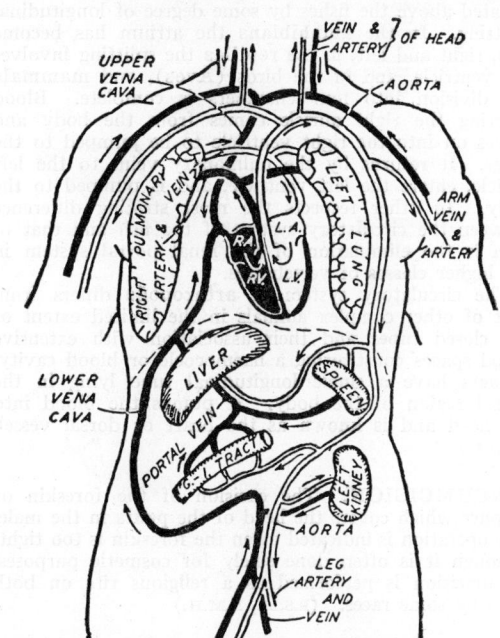

Circulatory system. (*Carlson & Johnson's The Machinery of the Body, University of Chicago Press.*)

smaller branches known as arterioles and these in turn into minute vessels with very delicate walls made up chiefly of a single layer of thin cells. In the human body these delicate **capillaries** are from $\frac{1}{200}$ to $\frac{1}{90}$ mm. in diameter. Some of the blood-fluid passes between the cells of their walls into spaces in the surrounding tissues where interchange with the various cells is possible. In some parts of the body, as in the **liver** of vertebrates, the ultimate tubular passages are even simpler than the capillaries and are called sinusoids. Their walls are, at least in part, merely the surrounding tissues; whether a special lining exists in some parts is disputed. From such small vessels the blood is collected into veinlets and these converge to form veins which carry it to the heart. The veins have strong walls of complex structure but in vessels of the same caliber the walls are thinner in veins than in arteries.

Some of the fluid from tissue spaces is gathered into a type of vessel resembling veins but more delicate; it flows into trunk vessels which rejoin the veins. This system of vessels is known as the lymphatic system.

In the vertebrates the fishes present the basic plan of the circulatory system. The **heart** has two principal chambers, an **atrium** which receives the blood and a **ventricle** which pumps it out to the body. A single large artery, the ventral aorta, leads forward and branches into a series of afferent branchial arteries which break up into capillaries in the gills. Efferent branchial arteries lead out of the gills to the dorsal aorta which runs back to the body. The head is supplied by extensions of the dorsal aorta, the carotid arteries. From the arterial system branches conduct the blood to the capillaries of all parts of the body. The blood from the alimentary tract is collected into a hepatic portal vein which breaks up into sinusoids in the liver, and is carried thence to the heart by a hepatic vein. In a like manner some of the blood from the caudal part of the body is conveyed to the **kidneys** by a renal portal vein before entering the veins which carry it to the heart. From all other regions of the body the blood is collected by veins which flow directly to the heart.

The **vertebrate** heart is a specialized region of the tubular system whose subdivision into chambers is complicated above the fishes by some degree of longitudinal splitting. In the **amphibians** the atrium has become two, right and left, in the **reptiles** the splitting involves the ventricle, and in the birds (**Aves**), and **mammals** the division into four chambers is complete. Blood entering the right auricle comes from the body and passes on into the right ventricle to be pumped to the lungs. It returns by the pulmonary veins to the left auricle, enters the left ventricle, and is pumped to the body. In other respects the most striking difference between the circulatory system of the fish and that of man is the elimination of the renal portal system in the higher classes of vertebrates.

The circulatory system of **arthropods** differs from that of other complex animals in the limited extent of the closed tubes and their association with extensive blood spaces constituting a haemocoele or blood cavity. **Insects** have a single longitudinal tube lying in the dorsal region of the body. It pumps the blood into the head and is known as the heart or dorsal vessel. (A.W.L.)

CIRCUMCISION. The excision of the foreskin or prepuce which covers the head of the **penis** in the male. The operation is indicated when the foreskin is too tight, although it is often done solely for cosmetic purposes. Circumcision is performed as a religious rite on both sexes by some races. (R.S.M., D.M.H.)

CIRQUE. Topographic feature produced by a mountain glacier. A mountain glacier usually starts in some sheltered ravine, slightly below the top of the mountain. Névé (firn) is the name of the granular ice which gradually develops from the original snow. The névé or accumulation of ice is restricted to that altitude at which the average summer temperature is 32° F. This line (summer **isotherm**) may vary from sea level, in the polar regions, to 20,000′ in the tropics. After the accumulation of granular ice on a slope reaches a certain thickness, the mass moves slowly downward under its own weight, by a process of plastic flow. In the névé region, where the snow and ice bank rests against the sloping cliff, the ice tends to work away from the rock, forming a crevice called the bergschrund. Frost action in the region of the bergschrund causes the recession of

the cliff, the broken material from which is frozen into the base of the ice and serves as tools to scour out circular basins called cirques. Small lakes and ponds which occur in cirques are called tarns. Where several cirques are developed near the summit of a mountain, the side walls of the cirques are called combs; the ridges between the cirques, cols; and the elevated terminations of the comb ridges, monuments. A triangular mountain peak, such as a Matterhorn, results from the complete cirquation of what was originally a relatively smooth-topped mountain. (R.M.F.)

CIRRHOSIS. A chronic disease of the **liver** characterized by gastrointestinal symptoms, **jaundice,** enlargement of the liver and spleen, and **ascites.** Pathologically, there is destruction of liver cells and overgrowth of fibrous **connective tissue.** If the cell destruction radiates from the center of the lobule the cirrhosis is classified as portal; if the periphery of the lobule and the interlobular biliary passages are primarily involved, the term biliary cirrhosis is used.

1. Portal cirrhosis, or Laennec's cirrhosis, is by far the commoner type. It occurs more often in men than in women, and usually in the 4th and 5th decades. It was formerly believed to be due to excess ingestion of alcohol ("gin-drinker's liver") since approximately 75% of cases occur in alcoholics. Now, however, **Vitamin B** deficiency is thought to be the major factor, and alcohol is considered to be secondary in importance. In a certain number of cases, where vitamin B deficiency plays no apparent role, a variety of etiologic agents may be involved. These include infections such as schistosomiasis (fluke infestation) and **syphilis,** and chemical poisons, including **carbon** tetrachloride, cinchophen, **phosphorus** and **arsenic.**

The symptoms and signs of portal cirrhosis are variable. Rarely, they may be absent in well-marked cases. The contraction of the fibrous tissue of the liver causes obstruction of the venous blood from the abdominal organs, and consequently congestion of the gastric and intestinal mucosa. As a result, **anorexia,** nausea, and vomiting are common. As the obstruction of the vessels progresses, the veins dilate to form varices, and may rupture into the stomach and produce a massive, frequently fatal, **hemorrhage.** Bleeding from **hemorrhoids,** which are also the result of back pressure from obstruction, is common. Patients with cirrhosis are sallow, anemic and, late in the disease, jaundiced. **Ascites** occurs in 80% of cases and as many as 12–15 liters of fluid may distend the abdomen. Peripheral **edema** and enlargement of the liver and spleen are common. Many patients have a low-grade fever.

The treatment of portal cirrhosis consists of a high-caloric, high-protein diet, and supplementary vitamin B complex, as brewer's yeast or thiamine chloride and nicotinic acid. Liver extract is also recommended. If ascites is present, fluid may have to be removed by repeated abdominal taps and drainage. If the disease is treated before ascites appears, the prognosis is better, although marked improvement may occur even after ascites has developed. The disease usually runs 3 to 6 years, but 2 years is the average duration of life after signs of liver failure such as ascites, jaundice, and hematemesis appear.

2. Biliary cirrhosis. Two forms of biliary cirrhosis occur, a primary and secondary form. The first is uncommon and is characterized by enlargement of the liver and the **spleen** with the presence of **jaundice.** Infection is the most probable cause. The disease cannot be arrested and the average duration of life is 5 or 6 years.

The secondary biliary cirrhosis is an obstructive type of cirrhosis caused by occlusion of the bile duct by **gallstones** or **tumor** formation. Jaundice is present with recurring attacks of chills and high fever. Treatment is surgical. (R.S.M., D.M.H.)

CIRRIPEDIA. Crustaceans of which the only commonly known representatives are the **barnacles.** The name applies to a subclass characterized by adaptations for sessile life; the included species are unlike other crustaceans in appearance.

Economically the barnacles are important because they attach themselves to the bottoms of ships and necessitate occasional removal.

Classification:

Order Thoracica. The common **barnacles.**
Order Acrothoracica. Small forms of which the females live in cavities in the shells of mollusks.
Order Apoda. One simplified form, parasitic in a common barnacle.
Order Rhizocephala. Parasitic, usually on crabs and related crustaceans. The adult forms root-like growths in the body of the host and is otherwise degenerate in form.
Order Ascothoracica. Parasitic in corals; less degenerate than the preceding. (A.W.L.)

CIRROCUMULUS. Clouds.

CIRROSTRATUS. Clouds.

CIRRUS. 1. In Protozoa a spine-like organ of locomotion formed of aggregated **cilia.** 2. In **annelid** worms a slender fleshy process on the **parapodia.** (See **Clouds.**) (A.W.L.)

CISCOE. Pisces, Teleostei. *Leucichthys.* Important food fishes (**Pisces**) of the Great Lakes and a few smaller lakes. Also known as the lake herrings. Included are two species known as **long-jaws,** the **bluefin** or blackfin, one species which is called the cisco, *Leueichthys hoyi,* and the **tullibee** which is found in smaller lakes in the Great Lakes region. (A.W.L.)

CISSOID OF DIOCLES. The cissoid is a type of mathematical curve which was first studied by the ancient Greek mathematician Diocles.

The cissoid is a **plane curve** which may be defined geometrically as follows: Draw a circle of radius *a* with its center on the *X*-axis and passing through the origin of a system of **rectangular coordinates.** Draw the **tangent** to the circle at *A*, where the circle cuts the *X*-axis. Let any secant line *OS* cut the circle at *R* and the tangent at *A* in the point *Q*; then lay off $OP = RQ$. The locus of the point *P* as the secant rotates about *O* is the cissoid.

Cissoid.

In **polar coordinates,** the equation of the cissoid is $r = 2a \sin^2 \theta / \cos \theta$. In **rectangular coordinates,** the equation is

$$y^2 = \frac{x^3}{2a - x} \quad \text{or} \quad x^3 + xy^2 - 2ay^2 = 0.$$

The cissoid can be used to "duplicate a cube," i.e., to find the edge of a cube whose volume is twice that of a given cube. (L.L.S.)

CITRAL. Aldehydes, Ketones, and Related Compounds.

CITRIC ACID AND CITRATES. Citric acid ($H_3 \cdot C_6H_5O_7$ or $COOH \cdot CH_2 \cdot C(OH)(COOH) \cdot CH_2 \cdot COOH \cdot H_2O$) is a white solid, water of crystallization lost at 130°, melting point 153° C., decomposes at higher temperatures, soluble in water or alcohol, slightly soluble in ether. Citrates (like tartrates) in solution change silver of ammonio-**silver** nitrate into metallic silver. **Calcium** citrate on account of its solubility characteristics, is of importance in the separation and recovery of citric acid. Calcium citrate plus dilute sulfuric acid yields citric acid plus calcium sulfate, and the latter may be separated by filtration. Citric acid may be obtained by evaporation of the filtrate. Citric acid may be obtained (1) from some natural products, e.g., the free acid in the juice of citrus and acidic fruits, often in conjunction with **malic** or **tartaric** acid, and the juice of unripe lemons (approximately 6% citric acid) is a commercial source; (2) by the **fermentation of glucose;** (3) by synthesis. Citric acid is a tribasic acid, that is, three series (mono-, di-, tri-) of salts and esters are known. Citric acid is used (1) in effervescent beverages, and effervescent medicinal salts, e.g., "citrate of magnesia," consisting of citric acid, sodium hydrogen carbonate and magnesium sulfate, (2) as ferric ammonium citrate in **blue printing.** (R.K.S.)

CITRINE. The mineral citrine is a yellow variety of quartz sometimes used as a gem. It is often marketed under the name topaz and may mislead the unwary. Brazil and Madagascar have furnished material of excellent quality. (E.S.C.S.)

CITRON. Citrus Fruits.

CITRONELLA. Volatile Oils.

CITRONELLAL. Aldehydes, Ketones, and Related Compounds.

CITRUS FRUITS. Rutaceae. The various Citrus fruits are obtained from thorny shrubs or small trees, having smooth leathery evergreen leaves with numerous internal resinous glands which contain an aromatic oil. In most of the species the petiole of the leaf has winged margins and is articulated both to the branch and to the blade. The flowers are extremely fragrant and are usually pure white, or in some species purplish pink. They are borne singly or in small **cymes** in the axils of the leaves. The **calyx** has 3–6 small teeth, the **corolla** is composed of 4–8 thick petals, while there are numerous **stamens** and a many-celled **ovary.** The fruit of these plants is a modified **berry,** known as a hesperidium.

The roots of Citrus plants lack root hairs entirely, absorption being directly through the fine terminal fibrous roots. The wood is hard and fine grained.

All Citrus plants are native to the Old World. The principal species in cultivation are as follows:

Citrus trifoliata, or the trifoliate orange, has trifoliate leaves, and hairy useless fruit. Its importance is due to its hardiness, making it valuable for hybridizing with more desirable species. It can be grown outdoors as far north as southern New York State.

Citrus sinensis, the Orange, originating in China, and with many varieties, is widely cultivated. In this country two principal growing regions exist, with characteristic varieties in each. Florida produces sweet, thin-skinned juicy oranges, which ripen in early winter. They are picked before fully ripe and matured in storage. California produces thick-skinned navel oranges (see **Parthenocarpy**), which have a pleasing acid pulp, and ripen during the winter and spring months. The source of these navel oranges was a sport arising in Brazil, and later carried to California, where it was extensively propagated. California also produces the many-seeded Valencia oranges, which ripen from June through October, producing a crop when other varieties are not in bearing.

Citrus Limonia, the Lemon, is probably a native of India. The trees are small and have many stout thorns. The flowers are large and purplish. The plants are not at all hardy, so are grown only in regions entirely free from frost. The fruits are produced continuously throughout the year and are picked green. To allow them to ripen on the tree causes them to become bitter and unmarketable. As soon as they attain the size demanded by the market they are picked and stored, often for several months. Coloring gradually develops during storage, or may be hastened by placing them in rooms heated above 90° F., where the yellow color develops in four to five days.

Citrus paradisi, Grapefruit, is probably derived from a bitter-fruited Malaysian tree. This is the largest fruited of all the Citrus plants. The trees are somewhat hardier than orange trees, and are principally grown in Florida.

Citrus nobilis, King oranges, Mandarin oranges, Tangerines. This species is characterized by the ease with which the skin or rind is removed from the sweet pulp.

Citrus aurantifolia, Lime, are very thorny shrubs or small trees, which are not at all hardy, and so are grown mainly in tropical countries and, in the United States, in California and the southern tip of Florida. The white flowers are small, as is the very acid thin-skinned fruit.

Citrus medica, Citron, is rarely grown in the United States. The rind of the fruit is treated with brine to remove the bitter oil it contains, then boiled in a sugar solution to which glucose (see **Carbohydrates**) has been added. The glucose is used to keep the product from becoming brittle. After boiling for a short time the rind remains in the syrup for a few weeks. It is then removed, boiled once more in pure sugar solution and dried. It is now the commercially-marketed citron.

Citrus Aurantium, Bitter or Sour Orange, is an ornamental tree, which is cultivated as a stock on which to bud other more desirable forms.

Citrus Bergamia, bergamot.

Related to the Citrus group is the genus *Fortunella,* with several species. They are the kumquats. They are small evergreen shrubs often planted for ornament and because of the small yellow fruits, which are either eaten raw or used in preserve making. In addition to the species enumerated, many hybrids have been developed.

The citrus industry of the United States has received many setbacks during its history. An ever-present source of worry is the danger of frosts, which are fatal to the present crop, and often do serious damage to the growing trees. At times frost injury is so serious as to menace the success of the industry. The use of smudge pots, producing clouds of smoke, increasing the humidity of the air by irrigation and other procedures are a partial insurance against damage.

Recently Florida has been beset with the Mediterranean **fruit fly,** which attacks a variety of fruits, in addition to the citrus varieties. This pest is now declared to be eradicated. Citrus canker, a bacterial scourge, is another source of destruction of the Florida crop. In California, very undesirable **scale insects** are present.

The uses of citrus fruits are many and varied. A very large part of the crop is consumed as raw fruit or as a drink. The rich **vitamin** content they possess has done much to further their consumption. Citrus oils and **citric acid** are by-products from culled fruit, which are becoming of commercial importance in the United States. The oils are largely used in flavoring extracts, and to some extent in perfumes and soaps. (R.M.W.)

CITY SURVEYING. Surveying.

CIVET. Mammalia, Carnivora. Animals of slender build with long tails and short legs, somewhat like the weasels in appearance. They occur in Africa, southern Europe, and southeastern Asia and constitute the family Viverridae including, in addition to the true civets, species known as **genets, linsangs,** palm-civets, the **binturong,** the mongooses, hemigales, and a few related forms.

The name is also applied to the **cacomistle** and locally to the small spotted **skunk** of North America. (A.W.L.)

CLADOCERA. The water-fleas, an order of small crustaceans. (A.W.L.)

CLADOSELACHE. Fossil Fishes.

CLADOSICTIS. Miocene.

CLAIRAUT'S DIFFERENTIAL EQUATION. Ordinary Differential Equations of First Order and Higher Degree than the First.

CLAM. Mollusca, Pelecypoda. Any member of numerous species of **bivalve** mollusks; a mussel. Some species are commonly called mussels, others clams, and in some

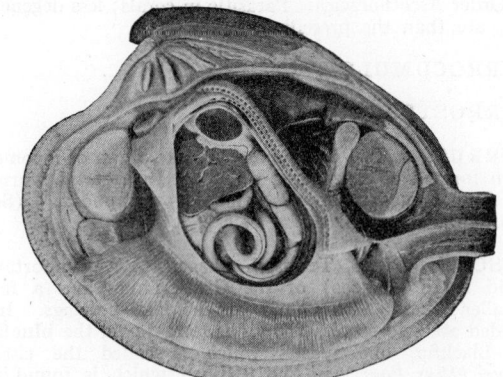

Anatomical model of hard shell clam. (*American Museum of Natural History.*)

cases the names are freely interchanged. Thus the edible mussel and other members of the same family receive this name only, the little-neck clam is always a clam, and the fresh-water species are called freely both clams and mussels.

The clams are moderately important as food. Occasionally valuable pearls are found in them and the shells of fresh-water species are the foundation of the important pearl button industry of the Mississippi Valley. (A.W.L.)

CLAM WORM. Annelida, Polychaeta. Marine worms of the genus *Nereis.* (A.W.L.)

CLAPEYRON'S EQUATION. This is a very useful differential equation involving the variables associated with the transition of a pure substance from one state to another; that is, from solid to liquid, liquid to vapor, or solid to vapor, and vice versa.

If the variables of state used to represent the condition of the substance are the absolute (Kelvin) temperature T, the pressure p, and the specific volume v, the equation may be written

$$\frac{\partial p}{\partial T} = \frac{L}{T(v_2 - v_1)}, \tag{1}$$

in which L is the heat of transition, that is, the heat evolved or absorbed, per unit mass, during the transition. v_1 and v_2 represent the specific volumes of the substance before and after the transition, respectively. If we use temperature, **entropy** (ϕ), and specific volume as the variables, the equation takes the form

$$\frac{\partial \phi}{\partial v} = \frac{L}{T(v_2 - v_1)}. \tag{2}$$

For a transition taking place at some specified fixed temperature T, the common second member of these equations becomes a known constant, whose value represents the slope of either the T-p or the v-ϕ transition curve, at the point of transition. In other words, it shows the rate at which one variable must change with respect to the other in order to maintain equilibrium between the two states at the given temperature.

For example, we may use Eq. 1 to calculate the change of the boiling point of a liquid with pressure in the vicinity of its normal boiling point.

Take the case of water near 100° C. The absolute temperature corresponding to 100° C. is $T = 373°$. The heat of vaporization of water at this temperature is $L = 540$ cal./g. $= 2.26 \times 10^{10}$ ergs/g. (see **Mechanical Equivalent of Heat**). The specific volumes of steam and of water at this temperature are respectively $v_2 = 1671$ cm.^{3}/g. and $v_1 = 1$ cm.^{3}/g. Substituting these values in (1) we obtain $\partial p / \partial T = 36{,}260$ ergs/deg. cm.$^3 = 36{,}260$ dynes/cm.^{2} deg. $= 27.2$ mm. Hg/deg. The reciprocal of this, $\partial T / \partial p = 0.368$ deg./mm. Hg, is the quantity required. That is, near normal pressure the boiling point of water rises at the rate of 0.368° per mm. of pressure. This result agrees closely with experiment.

Many other applications of Clapeyron's equation are encountered in the thermodynamic treatment of changes of state. (L.D.W.)

CLARAIN. A term proposed by Marie Stopes in 1919 for a finely banded variety of "bright" or shiny bituminous **coal**. In thin sections, under the microscope clarain is seen to be composed of disintegrated plant substances including bands of **spore** cases, which impart a yellowish to reddish color to the substance. (R.M.F.)

CLARIFICATION. Filtration.

CLARK CELL. Standard Cell.

CLASS. The second subdivision of the animal kingdom: major subdivisions or **phyla** are divided into **classes.** (A.W.L.)

CLASS INTERVAL. Class Limits.

CLASS LIMITS. Class limits are of two types, open and closed. The open class limits are determined by the degree of accuracy with which the **variable** is measured. If measurements of the variable have been made to the nearest ¼ inch, then the open class limits might be 10–14.75 inches, 15–19.75 inches, etc. The open part, λ, in this case ¼ inch, is the inaccuracy in the measurements. The closed class limits are found by subtracting $\lambda/2$ and adding $\lambda/2$ respectively to the lower open class limit and the upper open class limit. In our example the closed class limits would be 9.875–14.875″, 14.875–19.875″, etc. The class mark in either open class limits or closed class limits is the sum of class limits divided by 2. The class interval, c, is the difference between successive class marks, or the difference between successive lower closed class limits, or it may be found in various other ways. The class interval should be kept constant, if possible, in a **frequency distribution.** (L.A.A.)

CLASSIFICATION. Classification is the separation of a mixture of substances of various sizes and specific gravities into two or more divisions, the cuts being made both with reference to size and to specific gravity. There are three general classes:

1. Sorting classifiers; using a relatively dense aqueous suspension as the fluid medium. These species of equipment make more than two cuts and use hindered settling as the means. The cuts contain material settling at the same rate in the dense medium. Each cut contains material of more than one density and more than one size. The relative distribution is coarse-light and fine-heavy. This makes excellent tabling feed. Feed is from 0.05 mm. to 5 mm. (or larger for coal). They have little use outside the metallurgical field.

2. Sizing classifiers; make only a single separation, coarse from fine, but often this is a separation of densities also, as clay from sand. They are merely settling tanks. The fine particles overflow and the coarse settle. Provision is made for removing the coarse particles and cleaning them by several turn-overs. This process is often used in connection with closed circuit grinding. The feed can vary over wide limits. Drag classifiers dewater and raise the coarse material, thereby doing three jobs at once.

3. Air classifiers; produce two grades of material, coarse and fine. Utilizing centrifugal force and gravity, the coarse material is made to fall into receivers while the fine is blown out. This classifier is also used largely in connection with closed circuit grinding.

Efficiencies of classifiers based on screen analyses range from 50 to 80% but would be much higher if based on settling rates. Cost of treatment is about 1 cent per ton of material. Classification, except in closed circuit grinding, has been replaced to a large extent by flotation.

Tabling is a classification based on the stratification of particles in a liquid stream. Coarse-heavy and fine-light come off together, the opposite of hindered settling, so that these two processes in series can make good separations. 10-micron to 10-mesh particles can be treated. Larger sizes can be worked if the material is coal. Density, size, and shape of particles are the determining factors. The average cost of processing is 5 to 10 cents per ton of material.

Jigging is the consolidated result of three effects; hindered settling, differential acceleration at beginning of fall, and consolidation trickling at end of fall. The result is that an almost complete separation with respect to densities can be obtained. Jigs fall into two general classes, depending on how they are constructed, namely, (1) the movable screen type, and (2) the vibrating liquid type. For materials too fine for regular jigging, the same results can sometimes be obtained by putting a layer of fine shot on the screen which acts as the classifying layer. Jigs are used only on fairly coarse material, and

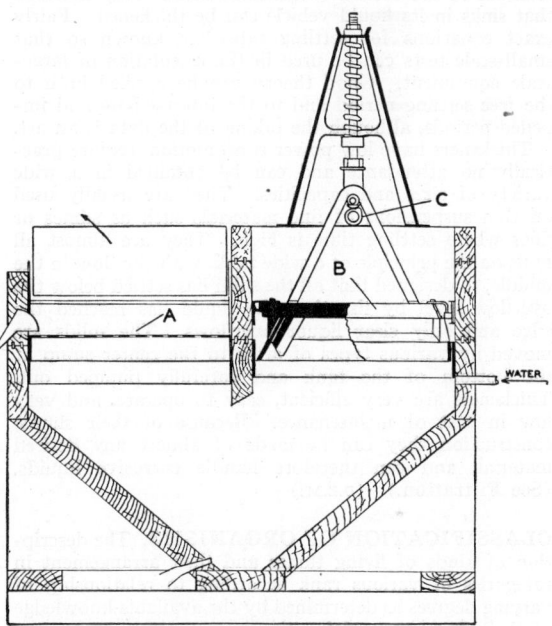

A. SCREEN   B. PLUNGER   C. ECCENTRIC CAM
Hydraulic jig.

so are well adapted to treating coal. Their use in other fields is largely being replaced by flotation. They are cheap and simple of operation but require much water and fairly expert handling.

Flotation is the best way (when it is possible) of separating different substances. The results are based on preferential wetting or non-wetting of the materials. The separation is accomplished by allowing one material to sink (wetted) and the other to float off (non-wetted) as foam. The wetted material must be coarse enough to sink and the non-wetted must be fine enough to be floated. As both materials are ground together, a proper size balance must be maintained. In general, with the exception of coal, the material is fairly fine. The foam is made by air bubbles, the air being either beaten in by stirrers or blown in below the liquid level. The maximum concentration of material in the feed varies with the material in question, the highest being about 40%. Several ingredients are added to assist in the separation, thus: 1. Frothers—to make foam (0.025–0.25 lbs. per ton). 2. Collectors—to give a preferential wetting effect (0.05–2.5 lbs. per ton). These collectors are usually large organic molecules partly soluble in water, and partly insoluble. 3. Hydrocarbon oils—to protect films (1.0–5.0 lbs. per ton). 4. pH regulators—(0.5–10 lbs. per ton). 5. Activators—to change the surface effects so that the collector will work better (0.5–2.0 lbs. per ton). 6. Depressants—to reduce the activity of one substance to the advantage of another (0.05–1.01 lbs. per ton). 7. Deactivators—when the activator works on both materials, a deactivator is used to destroy its effect on one of the materials (0.05–0.5 lbs. per ton).

These reagents are seldom all used at the same time but various combinations may be found desirable. Theoretically, flotation can be applied to any two materials, it is necessary only that the right flotation reagents be found. Except for a few generalizations, the reagents are determined largely through trial and error. The usual flotation process is a combination of several cells in series and in parallel, arranged to give the best separation.

Thickening is merely the settling out, by gravity, of the solid material in a slurry, followed by the mechanical removal of the settled (thickened) solid. Any material that sinks in its liquid vehicle can be thickened. Fairly exact equations for settling rates are known so that small-scale tests can be used in the calculation of large-scale equipment. Good theory can be applied both to the free settling period and to the intermediate and impeded periods, although the taking of the data is an art.

Thickeners have low power consumption, require practically no attendance and can be obtained in a wide variety of sizes and capacities. They are usually used on thin suspensions of fine materials such as slimes or flocs where settling time is high. They are almost all built on the principle of a wide tank, with the flow in the middle, so designed that all the solid has settled below the overflow level by the time the liquid has reached the edge and only clear liquor overflows. The solids are moved by various types of arms to the center sump at the bottom of the tank and carefully pumped out. Thickeners are very efficient, easy to operate, and very low in cost of maintenance. Because of their simple construction they can be made of almost any desired material, and can therefore handle corrosive liquids. (See **Filtration**.)   (D.E.M.)

CLASSIFICATION OF ORGANISMS. The description of kinds of living things and their arrangement in categories of various rank according to relationship of varying degrees as determined by the available knowledge of their structure and functions.

The classification of organisms is the function of the science of **taxonomy**. By common agreement under the principles of this science the organic world is now divided into plant and animal kingdoms, and in the animal kingdom the following series of subdivisions is recognized:

Phylum
  Class
    Order
      Family
        Genus
          Species

Various other divisions are introduced into some classifications to express other degrees of relationship. Terms used in this way are subclass, suborder, section, superfamily, subfamily and group.

Agreement is lacking on a few small divisions of the animal kingdom, but most of the phyla into which it is divided are generally accepted. The following is a classification commonly used.

Phylum **Protozoa.** The 1-celled animals.
Phylum **Porifera.** The **sponges.**
Phylum **Coelenterata.** The **hydroids, jellyfishes, corals, sea anemones,** etc.
Phylum **Ctenophora.** The comb jellies and sea walnuts.
Phylum **Platyhelminthes.** Flatworms, including flukes and tapeworms.
Phylum **Nemertea.** The nemertine worms.
Phylum **Nemathelminthes.** Roundworms, including horsehair worms.
Phylum **Rotifera.** Rotifers or wheel animalcules.
Phylum **Bryozoa.** The moss animals.
Phylum **Brachiopoda.** Lamp shells.
Phylum **Phoronidea.** Phoronids.
Phylum **Chaetognatha.** Arrow worms.
Phylum **Echinodermata.** Starfishes, brittle stars, sea urchins, sea cucumbers, sea-lilies, basket stars.
Phylum **Mollusca.** Snails, chitons, mussels, squids, cuttlefish, nautilus.
Phylum **Annelida.** The segmented worms.
Phylum **Arthropoda.** Crustaceans, spiders, scorpions, insects, etc.
Phylum **Chordata.** Fishes (**Pisces**), **amphibians, reptiles,** birds (**Aves**), **mammals,** and a few other forms.

Similar principles have been adopted for the classification of plants. (See **Paleobotany** and **Nomenclature**.) (A.W.L.)

CLASTIC ROCK. A sedimentary rock that is entirely or chiefly composed of fragmental material. **Sandstones** and **conglomerates** are typical clastic rocks. (R.M.F.)

CLAVICLE. The collar bone of man. The ventral anterior bone of the **pectoral girdle** of **vertebrates.** (A.W.L.)

CLAY. Clay is a very fine-grained unconsolidated rock material which is plastic when wet, but becomes hard and stony when heated to redness. Mineralogically, clay consists chiefly of hydrous **silicates** of **aluminum** together with numerous impurities, as **hematite, limonite,** etc., which often impart various colors to it. (E.S.C.S.)

CLAY GALLS. Thongallen.

CLAYDEN EFFECT. A photographic effect discovered by A. W. Clayden (1900). A photographic material which is first given a partial exposure to diffuse light, produces a reversed image if exposed a second time to a brilliant source of light. The effect is observed frequently in photographing lightning. If the lens is left open to await a flash in the proper position, other flashes occurring in the meantime (street lights, etc.) produce a general exposure. Then when a brilliant flash occurs the shutter is closed and the image developed. The image of the flash may develop as a positive or as a negative depending upon the relation of the two exposures. (C.B.N.)

CLEARANCE. The distance by which one object clears or misses another is called the clearance. Although this term might be applied to almost any number of situations, there are some cases of special importance.

Clearance in a piston and cylinder mechanism, such as found in the **engine**, is the space left in the end of the cylinder when the piston is in dead center position towards the end of the cylinder for which the clearance is determined. This space in volumetric units of measurement is the clearance volume. Clearance is commonly stated as a percentage, being the ratio of the clearance volume to the piston displacement during a stroke. The per cent clearance is a characteristic of no little importance to the performance of steam engines and internal combustion engines. (See **Ratio of expansion.**)

The unobstructed space which must be allotted for the occasional removal of parts of physical equipment such as the rotors of motors or turbines, the tubes of a boiler, etc., may be accurately referred to as clearance. (See **Fit; Gearing.**) (F.T.M.)

CLEAR-WINGED MOTH. Insecta, Lepidoptera. A **moth** whose wing membranes are largely free from scales and therefore transparent. A majority belong to the families Aegeriidae, containing the squash-borer and other borers of economic importance, and Sphingidae or hawk-moths, of which the genus *Hemaris* includes such species. (A.W.L.)

CLEAVAGE. In biology, cleavage is the subdivision of the fertilized **ovum** which precedes the formation of germ layers as the first stage of **embryonic** development. In the simplest type of cleavage the egg and each cell thereafter split completely through; this process is called holoblastic cleavage. With the accumulation of yolk in the egg its division is hampered until in the eggs of birds and reptiles the living matter lies on one side of the yolk and cleavage merely subdivides this small mass into a layer of cells; this is meroblastic cleavage. In still other eggs with much yolk, such as those of the insects, cleavage gives rise to a layer of cells completely enclosing the yolk. The last type is called superficial cleavage.

In geology, cleavage is the tendency of crystalline minerals to split more easily in certain definite directions, with the development of more or less smooth surfaces called cleavage planes. Cleavage planes can only be developed parallel to some possible crystal face. Cleavage may be described either in terms of the ease with which it is developed, or its direction. **Slaty cleavage,** as the term implies, is the tendency for slaty rocks to split in relatively thin, flat plates. Slaty cleavage is the result of the **metamorphism (foliation)** of a sedimentary rock (**shale** or mudstone) and is not to be confused with mineral cleavage. (A.W.L., R.M.F.)

CLICK BEETLE. Insecta, Coleoptera. **Beetles** with a peculiar junction between the first and second thoracic segments which permits them to be moved with a convulsive snap. When laid on its back the beetle uses this method of righting itself, throwing itself into the air by these snapping movements of the body. Most members of the family Elateridae and some of the Eucnemidae are click beetles. (A.W.L.)

CLICKS, KEY. Keying.

CLIMACTERIC. That time or portion of a person's life during which the body undergoes a radical change. The term refers particularly to puberty and the **menopause.** (R.S.M.)

CLIMATE. Average weather of any region. Averages need not be for a whole year but can be computed from data for months, weeks, or days. It is necessary that sufficient data be available to remove the element of irregularity from climatic values. (P.E.K.)

CLIMB CUTTING. There are two methods of machining surfaces with rotating multi-toothed cutters; in the first method, known as climb cutting or cutting down, the work moves or feeds in the same direction as the periphery of the cutter teeth at the line of contact; in the second, known as cutting up, the work feeds in a direction opposite to that of the periphery of the cutter teeth at the line of contact. In cutting up action, each tooth of the cutter takes a wedge-shaped chip whose initial thickness is zero and whose final thickness depends upon the rate of feed. The cutting action has a tendency to lift the work from the surface on which it is placed, and work must therefore be restrained vertically as well as longitudinally.

Climb cutting produces a chip of maximum thickness at the beginning and zero thickness at the end of the cut. As a result, the machined surface has a better finish than can be obtained by cutting up.

Climb cutting has one major disadvantage. The cutting action has a tendency to draw the work in towards the cutter. This tendency is manifested when a heavy feed is used or when the machine is not sufficiently rigid or powerful for the size of the cut. In such a case, the cutter tends to climb over the work and may thus ruin the work, the cutter, and possibly the machine. For this reason climb cutting is not employed for medium and heavy-duty service, particularly where comparatively old machine tools are used. (H.C.H.)

CLIMBING PERCH. Pisces, Teleostei. An Indian fish (**Pisces**), *Anabas scandens*, not related to the true perch, which is able to move along the ground by using its paired fins. It and allied species secure oxygen from the air by means of a special organ associated with the gills. It has been said to climb low trees but it climbs rarely, if at all. (A.W.L.)

CLINKSTONE. Phonalite.

CLINOGRAPHIC REPRESENTATION. Pictorial Representation.

CLINOMETER. A pendulum-like instrument for measuring angle of slope, usually in degrees. Used by geologists for measuring the **dip** of formations. (See **Brunton Compass.**) (R.M.F.)

CLITELLUM. A swollen glandular region occupying several segments near the anterior end of the **earthworm.** It is an accessory reproductive structure. (A.W.L.)

CLITORIS. The organ in the female corresponding to the penis in the male. It is a small cylindrical organ about an inch long, situated in the anterior angle of the **vulva,** capable of erection. Unlike the penis the clitoris is not traversed by the **urethra.** (R.S.M.)

CLOACA. A chamber at the end of the gut which receives the ducts of the reproductive and excretory systems. A cloaca occurs in the **rotifers** and in many **vertebrates** but in most **mammals** it is present only during embryonic life. (A.W.L.)

CLOCK. Time.

CLONE. The descendants of a single individual produced through continued asexual reproduction. As parts of the parental body, these individuals may be expected to have the same heritage. (A.W.L.)

CLOSED TRAVERSE. Traverse.

CLOT. A firm mass which results from coagulation, or change from a fluid or semi-fluid state to a semi-solid soft mass. This occurs in blood or lymph when it escapes from the blood or lymph vessels. For a discussion of the mechanism of blood clotting, see **coagulation.** (D.M.H.)

CLOTHES MOTH. Insecta, Lepidoptera. A **moth** whose **larva** eats dead and dry animal matter, especially fur, feathers, and wool. Three species are known. The

A, clothes moth. B, larva of clothes moth. (*After Riley. U. S. Dept. Agr.*)

case-bearing clothes-moth larva, *Tinea pellionella,* lives in a case made of bits of the material on which it lives, spun together with silk. The tube-building or tapestry moth, *Trichophaga tapetiella,* makes a gallery of silk and fragments as it works. The common or naked clothes-moth larva, *Tineola biselliella,* does not spin until it makes its cocoon. The adults of all three species are small, expanding usually about ½″. As in the case of the **buffalo carpet-moth** good housekeeping is the best remedy for these pests. When present they may be killed by the use of commercial sprays or in heavy infestations by fumigation.

Frequently when moths become over abundant in a house it is due to some forgotten breeding place, such as feathers or old garments in storage. A search in unexpected places is often the best cure for such attacks. (A.W.L.)

CLOUD CHAMBER. An enclosure containing air or other gas saturated with water vapor, the cooling of which by a sudden expansion results in the formation of fog droplets upon particles of dust or other nuclei. That **ions** in the gas are capable of serving as condensation nuclei, even when no dust is present, was demonstrated by the experiments of C. T. R. Wilson. Thus the clouds produced are much more dense if the gas is traversed by some ionizing emission like **x-rays** or **alpha rays.** Sir J. J. Thomson utilized this effect in his early measurements of the electronic charge. One of the most striking phenomena of the Wilson cloud chamber is exhibited when single ionizing particles, such as alpha or beta particles, are allowed to traverse it just before the expansion. The path of each particle is marked by a visible white streak or "track" of mist, sometimes several cm. in length, which soon diffuses and disappears. The study of photographs of such cloud tracks has in recent years afforded much information as to the nature and the movements of the particles producing them. (L.D.W.)

CLOUD TRACK. Cloud Chamber.

CLOUDBURST. Torrential rainfall, usually associated with thunderstorms. Very frequently an excessive number of large drops occur in a cloudburst. (P.E.K.)

CLOUDED LEOPARD. Mammalia, Carnivora. A **cat,** *Felis nebulosa,* of southeastern Asia, neither closely related to the true leopard nor like it in markings. (A.W.L.)

CLOUDS. Large numbers of water droplets **or ice** crystals virtually suspended in the atmosphere. Actually the water or ice in a cloud occupies only a small fraction of the total space appearing as a cloud. Light is well reflected from the droplets or crystals and the cloud body appears as an opaque drifting object. Clouds can be classified in several ways. The two most common are in regard to form, stratified or billowed; and in regard to height, high, medium, or low. The basic cloud forms

are internationally recognized but there are many variations of each form.

Cirrus is a high cloud composed of ice crystals and, therefore, lies entirely above the freezing level. Cirrus is never lower than about 4 miles in the tropics but may be near ground levels in the polar areas. In appearance they are usually thin, wispy, often in streaks, and always whitish without shadows. Cirrus often forerun storms but not all cirrus are associated with storms. They cannot be used as a foolproof indication that a storm is approaching until considerable experience in cloud observation is attained.

Cirrostratus is a cloud veil of more or less uniform texture composed of ice crystals and therefore, like cirrus, lies entirely above the freezing level. Cirrostratus varies from white to gray, is usually translucent, and partially obscures the sun and moon. There are no shadows but often mock-suns or mock-moons which are images of the real celestial bodies. Cirrostratus often heralds the approach of a cyclonic storm, particularly in the temperate zone.

Cirrocumulus is a small billowed cirrus-type cloud composed of ice crystals. This type cloud indicates some instability in the layer at and above the cloud level which permits rising currents to form the cloud parcels and descending currents to create clear spaces between them. Cirrocumulus frequently occurs in advance of a cyclonic storm.

Altostratus is a translucent to opaque cloud composed of water droplets through thin layers of which the sun or moon might appear as seen on a ground-glass screen. It is a middle-level cloud in contrast to the high cirrus forms but may lie either above or below the freezing level. Very frequently, the top part of a layer of altostratus is a cirrus-type cloud, although this is not observable from below. Altostratus cast very little if any shadow but usually appear as a dull, drab, grayish sheet. Altostratus following cirrus and cirrostratus is an almost certain indication that a cyclonic disturbance is approaching.

Altocumulus is a billowed cloud of small cumuli that generally form in layers. It is composed of water droplets, although it may lie either above or below the freezing level. It is a middle-level cloud. Altocumulus casts shadows and varies in color from pure white to nearly black. In general they are whitish with darker shadows. "Mackerel sky" is an appropriate description for many altocumulus bands. The cloud may or may not be associated with cyclonic storms.

Stratus is a uniform layer of dull, grayish, low-level cloud. It is not the low-level equivalent of altostratus, rather it is more a cloudlike fog. Fragments of stratus torn by wind or remnants of clearing stratus are usually known as fractostratus.

Nimbostratus is a type of stratus from which rain or snow appears to fall. Usually the stratus layer is merged with altostratus or altocumulus, and the total cloud layer is very deep. More frequently there is a clear break between the two layers. Nimbostratus is dull and grayish and usually ragged.

Stratocumulus is a layer consisting of large lumpy masses of cloud. It is billowed, often appears in rolls with occasional blue showing between rolls. The cloud casts considerable shadow with shades varying from very whitish in thin spots to very dark in thick spots. Types vary from almost true cumulus to almost true stratus.

Cumulus clouds are billowed heaps with flat bases and tufted tops. They have considerable shadow and often are very dark on the underside. Size and shape vary from flat small balls of cloud-cotton to great towers with valleys and ravines along the sides. The cloud is a low type but can be found with bases from 500–1000′ and tops as high as 20,000′. It is composed of water droplets and may produce rain if well developed. Flat, fair-weather types are known as cumulus humilis and the well-developed variety as cumulus congestus.

Cumulonimbus is the thunderstorm cloud. It is tall, billowed, full of contrast from brilliant white to inky black. Most of the cloud is composed of water droplets but the top which penetrates above the freezing level is composed of ice crystals. False cirrus often develops at the top or anvil head. Cumulonimbus occur with bases from 500–15,000′ and tops from 10,000–50,000′.

There are also several minor cloud forms. Scud is generally wind-torn nimbrostratus. A lenticular cloud is a feather-shaped cloud usually formed over the peak of a mountain or terrain where the wind blows uphill and the air reaches the condensation level near the top of its flow. Lenticular clouds also form in mid-air if there is considerable undulation in the horizontal winds and if the air rising on the crest of a wave reaches the condensation level. Clouds of this type do not travel with the wind but remain practically stationary. Crest clouds form, covering ridges and mountain tops in a similar fashion, but are more extensive and do not usually have the lenticular form. They develop in a current of air rising along the ridge or mountain when air reaches its condensation level along the slope. Mammatocumulus sometimes are associated with severe thunderstorms. They appear as small inverted cumulus with the bulges extending downward. They are indications of extreme turbulence. Fracto clouds are wind-torn clouds. They are usually low clouds, cumulus or stratus. (P.E.K.)

CLOVER. Pea Family.

CLOVES. Eugenia aromatica. Myrtaceae. Cloves

are the dried flower buds of a beautiful evergreen tree growing 25–40′ high. Its native home is the Molucca Islands. In cultivation it has spread throughout tropical Oriental islands. The trees grow best when near the coast. The clove tree has thick shining leaves, a smooth gray bark and flower buds and flowers of deep red color. The fragrant oil is located mainly in the leaves and flower buds, and causes the air surrounding the tree to be richly scented. Flower buds are first formed when the tree is 4 or 5 years old. In gathering, the branches are pulled down and the flower buds picked off by hand, or they may be pounded from the trees with bamboo sticks. After gathering, the pedicels, or short flower stems, are picked off, and the buds dried. These buds may now be used as a spice, either whole, or ground to a powder. They are also used to improve the breath; indeed, in ancient times it is recorded that their use for this purpose was required before royalty could be addressed, a commendable requirement. In addition to their use as spice, an oil is distilled off and condensed by cooling. This oil, which is rich in eugenol, is much used in making artificial vanilla. Clove oil (see Volatile Oils) is also used in scenting soaps and perfumes. It is also much used in the preparation of sections which are to be examined by the microscope, it being a means of clearing the sections. (R.M.W.)

CLUB FOOT (Talipes). A deformity of the foot,

usually congenital in origin, characterized by marked flexion, extension, inversion, or eversion.

Heredity is a factor in some cases. In others the condition results from an intra-uterine accident or maldevelopment, firm pressure or constriction due to deficient amniotic fluid, or tumor in or around the uterus, interlocking of the feet, constriction of the umbilical cord, pressure of twins, etc.

Treatment is effective in childhood. In older children the prognosis for a useful foot is not as good. Treatment may be manipulative or operative, depending on the type of deformity and the age of the patient. (R.S.M., D.M.H.)

CLUB MOSSES. Lycopodiales; Paleobotany.

CLUSTERS. An examination of the night sky will

indicate at once that the stars are not uniformly distributed. Certain clusters may be readily distinguished by the unaided eye, such as the Pleiades, the Hyades, the constellation of Coma Berenices, Praesepe, and the double cluster in Perseus. The application of the telescope to astronomy revealed a number of other clusters and the application of photography further augmented the number until at present several hundred such objects are listed. Star clusters are studied under two main types, galactic clusters and globular clusters.

A galactic cluster contains between a few hundred and a thousand stars far enough apart to be distinguished as individuals. The name galactic is derived from the fact that all except two or three of the brightest and closest of these objects lie within 15° of the galactic plane. The method of determination of the distances and sizes of those galactic clusters which are close enough to be classified as moving clusters are discussed elsewhere. For the more distant ones the spectral classes of individual stars are plotted as abscissae and apparent magnitudes as ordinates and the resulting diagram compared with that obtained for giant and dwarf stars. From this comparison an approximate determination of absolute magnitudes may be made and the distance determined. From the distance and observed angular diameter of a cluster its linear diameter may be found. By such methods a few galactic clusters are found to be within a few hundred light years of the sun, while the great majority are between 1500 and 15,000 light years away. The shapes of the galactic clusters are quite irregular.

The globular clusters contain thousands of stars and are characterized by their globular form and concentration of stars in their central regions. In a moderately large telescope these objects are strikingly beautiful to see. Many attempts have been made to determine the total number of stars contained within a globular cluster, but the concentration at the center is so great that the images all run together. Certainly, in the brighter objects of this class there are more than 50,000 stars. Shapley finds, from an extensive study of a number of clusters, that there are stars of all spectral types, but there is a very definite correlation between spectral class and brightness, with the fainter stars all being redder than the brighter ones.

The distances of globular clusters may be approximately determined from the cluster type Cepheid variables which are found in large numbers in these objects. Shapley finds that the globular clusters are all contained within a vast ellipsoid with a maximum diameter of approximately 250,000 light years, having its center about 75,000 light years from the sun in the direction of the constellation of Sagittarius. While the principal plane of this ellipsoid practically coincides with the galactic plane, nevertheless practically none of the globular clusters are observed close to the milky way. The only plausible explanation seems to be found in the hypothesis that there is some sort of absorbing material in this region of space.

A study of the shapes of the globular clusters indicates that they are somewhat ellipsoidal as though they were rotating, but no rotation has ever been definitely observed. From the estimated distances and angular diameters the diameter of the clusters is found to be of the order of magnitude of 100 light years. In the central dense portion of a globular cluster there must be about 1500 times as many giant stars as there are in an equal volume of space in the vicinity of the sun. However, even with this relatively close packing of stars, the average distance between the individuals must be of the order of magnitude of 1 light year. (W.K.G.)

CLUTCH. The function of a clutch is to effect the

coupling of two working parts in such a way as to permit connection or disconnection at will, and without the necessity of bringing both parts to rest. Almost always a clutch is a coupling between two shafts which are in line, one being the driving, the other the driven, shaft. Under these conditions the function of the clutch upon being engaged is to pull the driven member up to the speed of the driving member, and to transmit the required amount of power without **slip**.

Clutches may be classified according to the method of transmitting the torque:

1. Friction clutches are operated by the surface friction created when two surfaces are pressed together.

2. Magnetic clutches make use of the attraction of a magnet for its armature. This type has attractive features, but has made little headway because of the development of the friction clutch.

3. Jaw clutches are those in which positive drive is obtained by the use of projecting lugs. If this type of clutch is engaged when the drive is in motion, there will be a sudden shock to the driven mechanism unless some degree of flexibility is imposed between clutch and driven machinery by the use of springs, rubber, or the like.

Most clutches are of the friction type, since it has been possible to build this type to transmit almost any amount of power desired with moderate-sized clutches which are comparatively trouble-free. The friction clutch may be engaged either by an axial or by radial movement. Cone and disk clutches are the former type; internal-expanding shoe and external-contracting bands are examples of the latter.

The **internal combustion engine** is not easily started under load, and therefore when used to power self-propelled vehicles, it must be disengaged from the axle for starting purposes. For this reason every automobile, truck, or bus must have a clutch. The two clutch forms frequently used for this service are the cone clutch and the disk clutch.

In explanation of the cone clutch, the reader is asked to refer to Fig. 1. In this figure at A the clutch is

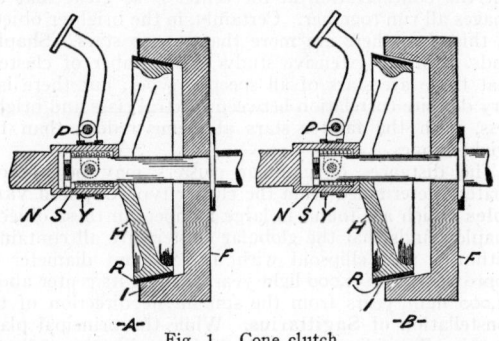

Fig. 1. Cone clutch.

"in"; at B it is "out." The flywheel is connected to and rotates with the engine crankshaft. A conical surface is machined on the inside of the rim, and into this the male member of the clutch, also conical, is pressed by the action of the spring S. The rim of the external cone is faced with a friction material such as leather. Other friction surfaces employed are asbestos fiber, steel with cork insets, or plain steel against steel, or steel against bronze. In B the foot pedal is shown depressed, and by means of the collar Y, the friction cone is withdrawn, compressing the spring S. The spring pressure required is that needed to produce the force at the friction surface which is necessary to transmit the required horsepower. If the mean diameter of the friction surface is D ft., and P hp. is to be transmitted at N rpm., then the sum of all the friction forces acting in the friction surface can be expressed by the following formula:

$$F = \frac{33,000P}{2\pi N \times \frac{D}{2}} = \frac{33,000P}{\pi ND}.$$

The fundamental law of friction states that frictional force equals coefficient of friction times pressure normal

to the surface, hence, calling the pressure normal to the surface Q,

$$Q = \frac{F}{f},$$

in which f is the coefficient of friction of the contacting surfaces.

The conical clutch surface is used because a small spring pressure acting along the axis of the shaft will produce a large force if the cone angle is small. If the central angle of the cone is α and the force exerted axially by the spring is S,

$$Q = \frac{S}{\sin \alpha}.$$

The cone clutch is either metal to metal, running in an oil bath, or leather to metal, running dry. Since the leather must be kept soft and pliable, but free from grease, necessitating a certain amount of inspection and maintenance, this type of clutch has been superseded, for automotive uses at least, by the disk clutch. The dry-plate disk clutch operates without lubrication, gaining thereby so high a coefficient of friction that one plate held between two disks usually will develop sufficient friction force. Fig. 2 shows a multiple disk clutch of

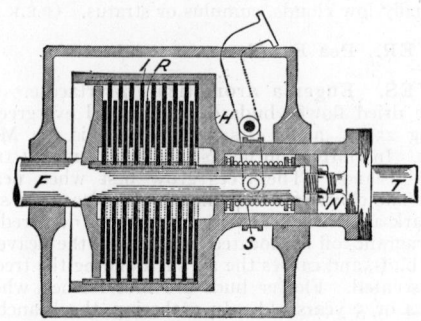

Fig. 2. Multiple disk clutch.

the type frequently used in the automobile. F is an extension of the crankshaft to the engine, and carries with it the series of disks similar to R. The driven shaft, T, is expanded inside the clutch to a large diameter drum having mounted internally a series of disks similar to I, alternating with the R disks. In each case the disks must move with their respective shafts, but are permitted some endwise movement through the method of attachment. The clutch is shown in disengaged position. Should the clutch pedal be released, the spring S will push on the member H and press together the fixed and moving disks. The friction set up thereby will pull the driven shaft up to speed, and then transmit to it the power for which the clutch is designed. Such clutches are sometimes metal to metal, immersed in oil, or metal to friction material if run dry. (F.T.M.)

CNIDOBLAST. A stinging cell of the **coelenterates**, found in all forms, **polyps, jellyfishes,** and **sea anemones.** It produces **nematocysts** which are discharged in defense and in securing food. (A.W.L.)

CNIDOCIL. A projection on the free surface of a **cnidoblast** which is sensitive to external stimuli. (A.W.L.)

COAGULATION. 1. The clotting of **blood** or **lymph.** 2. The changes produced in tissue by the application of increased temperatures or by certain chemicals.

The mechanism of the clotting of blood is not completely understood. It is now generally believed that two reactions occur as part of the process, both of them involving **enzyme** systems. The substances concerned in the formation of fibrin and therefore clotting are certain blood proteins (fibrinogen, prothrombin, thrombin) and thromboplastin, or tissue factor, which is widely distributed as an intracellular substance and is liberated when the blood platelets (see **Blood**) or other cells are injured. **Calcium** also takes part in blood clotting. All these substances are postulated to interact in the following manner:

1. Thromboplastin + calcium ions, react with prothrombin to produce thrombin.

2. Fibrinogen + thrombin produces fibrin, which forms the clot.

Failure of the clotting mechanism anywhere along the line, results in various blood diseases, such as **hemophilia**, **purpura hemorrhagica**, etc. Vitamin K is necessary for prothrombin formation, and its deficiency results in a bleeding tendency. (See **Vitamins**.)   (D.M.H.)

COAITA. Mammalia, Primates. The red-faced spider-monkey of the Lower Amazon. (A.W.L.)

COAL. A combustible substance of organic origin which occurs as beds or "seams" and which has a variable physical and chemical composition, including variable amounts of mineral or non-combustible matter. Although the combustible constituents of coal are of organic origin, the geological processes involved in its formation act so slowly and over such a long period of time that coal is, economically speaking, non-reproducible.

It is known that the origin of coal is to be sought in the vast and luxuriant vegetation which flourished over portions of the earth's surface in past geological ages, for the imprint of leaves of giant tree ferns of a form similar to those now found only in tropical jungles can often be seen on the face of blocks of coal at the mine. (See **Paleobotany**.) Undoubtedly the somewhat higher mean temperature of the earth, coupled with the prevalence of a more steamy atmosphere, produced conditions which led to very rapid growth of vegetation. Such conditions are today approximated in hothouses, so it is probable that the coal-forming epochs are definitely of the past, and that such coals as are used constitute a depletion of a fixed natural resource. Through the centuries of the **carboniferous** age there was produced a rapidly accumulating mass of decaying vegetation, a great deal of which was undoubtedly also preserved from complete decay by having air excluded from it either by its being partially submerged or being covered with a considerable thickness of superimposed matter. The process of decay soon converted the **cellulose, lignin**, and **proteins** of the original vegetable debris into peat bogs which, due to some subsequent earth movements, were either rapidly or gradually covered with inorganic material which, in time, produced sedimentary rocks such as shale or sandstone. Coal is found today in bedded deposits, or seams, in company with **shale, sandstone, limestone**, and other sedimentary products. The pressure of the overlying strata, together with some possible small amount of residual decay, created an increase of temperature in the peat which slowly accomplished its conversion into coal. The fact that under conditions of controlled pressure and temperature a material of vegetable origin may today in the laboratory be converted into a black powder resembling coal, lends credence to this theory of the origin of coal. It is entirely conceivable that the variations in overlying strata, in condition of the bed when overlaid, and in subsequent rise of temperature, are accountable for the differences in coal as it is brought from the mines today. During the process of this chemical change, percolation of water through the incipient coal seams probably carried into the coal many of the minerals which, in the aggregate, produce the ash which is common to all coals. (Of course, some of the ash can be attributed to the minerals present in the vegetation from which the coal was formed.) The classification of coal is primarily in terms of its origin both as to (1) composition, or the original types of vegetable material composing it, and (2) the degree of **metamorphism** which it has undergone

since burial. A simple but practical classification of coal according to its origin and chemical composition is as follows: (1) Peat. Partially carbonized vegetable matter, such as accumulates in a bog. Because of its high water content, peat is not an economical fuel when either lignite or coal is available. (2) Lignite. Brown coal and lignite are formed by the burial and consequent compression of peat by overlying sediments (formations), the peat substances being thus compacted and changed, first to brown coal or **lignite**, and then to black lignite or sub-bituminous coal. Although lignites are thus the parent forms of the higher grades of coals, even pure lignite is a relatively poor form of fuel because of its high water content and consequent low calorific value. Calculated on an ash-free basis, lignite yields moisture, 43.4%; volatile matter, 18.8%; fixed carbon, 37.8%; heat value in **British Thermal Units**, 7400 B.T.U. per lb. (3) Bituminous Coal. Soft Coal (bituminous) varies in heating value from 9720 to 15,360 B.T.U. per lb. The various ranks of bituminous contain roughly 70–80% of fixed carbon and 30–20% of volatile matter. When most bituminous coals are heated in a closed retort this volatile matter is driven off as gas and **coal tar**. The residue of fixed carbon may form a coherent porous mass called **coke**. Coking coals are particularly valuable as they yield the essential fuel for the smelting of iron ores. Other by-products of bituminous coals are a great series of organic compounds which are used extensively in the dye and other industries. (4) **Anthracite** (Hard Coal). When bituminous coals have been heated and compressed (by natural geological causes) so that they have changed from the finely jointed and crumbly "Soft Coals" to the massive and hard form, they are called anthracite, which is relatively hard, clean and moisture free; low in volatile matter and high in fixed carbon. Anthracite grades (through **metamorphism**) to **graphite** which is high in fixed carbon, combustible only at high temperatures, and classed as a refractory. High-grade anthracite, calculated on the ash-free basis, yields moisture, 3.2%; volatile matter, 1.2%; fixed carbon, 95.6%; and heat value, 14,400 B.T.U. per lb. The geologic distribution of coal is as follows: Pre-Paleozoic. No coal beds but abundance of graphite in the **Archeozoic** crystalline rocks. Paleozoic. Important bituminous and anthracite coal deposits in England and eastern United States of **Pennsylvanian Period**. (See **Historical Geology**.) Mesozoic. A few beds of **cannel** and bituminous coals occur in the **Triassic**. In the **Cretaceous** occur a great series of bituminous coal beds second only in importance to those of Pennsylvanian age in the eastern United States.

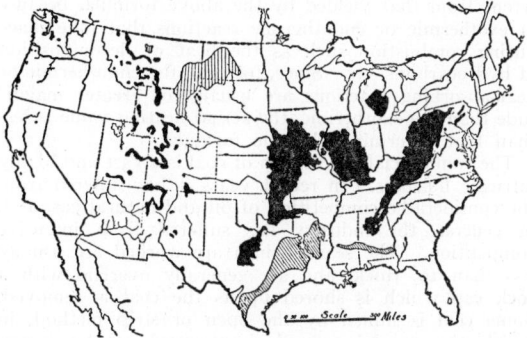

Map of the United States showing the principal coal fields. The lined areas represent lignitic coals. All solid black areas east of the Rocky Mountains, excepting the very small ones of Triassic age in Virginia and North Carolina, represent fields of Pennsylvania coal. All coal in the western United States is Cretaceous and Tertiary. (*Modified after U. S. Geological Survey*.)

Cenozoic. Poor grades of lignite and bituminous coals occur in the **Tertiary**. These coals are of some value for local use only. The chief coal reserves of the world are located in the United States, England and the territory contiguous to Germany and France.

Coal contains **carbon, hydrogen, oxygen, nitrogen, sulfur,** and various mineral substances such as **silica, alumina,** and iron (ferric) oxide. All of the mineral substances are included in the descriptive term **ash.** Hydrogen is present in coal in two forms, called *free* and *combined,* although actually it is in two methods of combination. Hydrogen is combined with oxygen, forming the moisture in coal, and all coal contains more or less moisture. Hydrogen is also combined with carbon, forming the hydrocarbon, or volatile portion of coal. A very important difference, however, exists in these two combinations of hydrogen. In one, water, it is completely oxidized, and must be considered incombustible; in the other, the hydrocarbons, the hydrogen is combustible. Occurrence of ash in coal is more or less accidental. Coal rarely contains more than 4% of sulfur, or 2% of nitrogen. Only a small portion of the carbon content of coal is accounted for in the volatile portion (the hydrocarbons); in fact, the largest portion of the heating value of coal is obtained from the free or "fixed" carbon. A sample analysis of anthracite coal is: 2% moisture, 5½% volatile material, 86½% fixed carbon, 6% ash. Typical semi-bituminous coal, the principal steaming coal, would be covered by the following analysis: 3% moisture, 18% volatile material, 75% fixed carbon, 4% ash.

The analysis of coal on the basis just described is known as the **proximate analysis,** but studies in combustion, which is essentially a chemical reaction, necessitate analyzing the coal into its elements. This analysis is known as the **ultimate analysis.** Since the heating value of hydrogen is roughly five times that of carbon, it follows that the coal high in volatile material possesses a higher calorific value.

The combustion of free carbon liberates 14,500 B.T.U. per lb., of free hydrogen 62,000 B.T.U. per lb., and of sulfur 4000 B.T.U. per lb. One pound of coal has theoretical heating value of

$$HV = \frac{14{,}540C + 62{,}000H + 4000S}{100} \text{ (B.T.U.)}$$

$C =$ the per cent of carbon in the coal. $S =$ per cent of sulfur in the coal. $H =$ per cent "free" hydrogen in the coal, understood to be that portion of the total hydrogen in an ultimate analysis which is not already combined with oxygen. The actual heating value of coal as determined by calorimeter may be slightly different from that yielded by the above formula, because of exothermic or endothermic reactions that take place during combustion such as the heat of decomposition of hydrocarbons, etc. Since, for coal, the endothermic or heat-absorbing reactions are usually of greater magnitude than the exothermic, the actual heating value is less than that determined by the formula.

The commercial production of coal is a vast and highly intricate business. In recent years it has suffered from the considerable competition of oil and natural gas, and, in general, the industry has suffered from intensive competition. The seams which are worked are usually less than 10′ thick, and are generally overlaid with a rock cap which is shored up as the coal is removed. Some coal is mined by the open or strip method, in which the overlying earth is stripped off by power shovel. Coal is undercut or blasted out of the face of a seam and loaded onto cars which convey it to the plant, where the slate is removed before the coal is broken and graded.

Bituminous coals are produced in the following sizes:

Run of Mine. This coal is mined, unscreened, and varying in size from large lumps to slack.
Lump. Coal which passes over a 1¼″ screen.

Nut. Cut coal of a size which passes through a 1¼″ screen, but is retained on a ¾″ screen.
Slack. All coal passing the ¾″ screen.
Anthracite coal, in addition, is produced in the finely graded sizes called Pea and Buckwheat.

The firing qualities are important characteristics of coals when selection of coal or coal-burning equipment is under consideration, for, if it were not for the effects of firing qualities and impurities, coal could be purchased on the basis of heating value alone. The firing characteristics of coal result from (1) its volatile content, (2) moisture content, (3) sulfur content, (4) ash composition. Some coals retain their original shape during combustion, being gradually reduced in size as combustion proceeds; others soften and fuse into a mat or crust which blankets off the air supply; others soften but do not form a crust. This quality influences the stoker selection. The sulfur and ash content will have much to do with the types of clinkers that are formed. The fusion temperature of the ash becomes an important characteristic when low excess air and high furnace temperatures are employed. The hardness and way in which a coal shatters when pulverized may influence the selection of pulverized fuel equipment. Some coals are fast-burning, others slow. That characteristic influences shape and thickness of fuel bed carried, or type of burner installed. The tendency of a coal to disintegrate and become slack will affect the type of grates and the draft used.

The bulk of coal is transported by boat or railroad, although these are supplemented, to a limited extent, by trucks and wagons. The individual user of coal usually puts considerable quantities of it in storage as insurance against complete shut-down of a plant occurring from failure of normal fuel supplies to arrive. When coal is piled in storage it "weathers." It has a tendency to become slack; the surface oxidizes and heat is liberated. When loosely piled in shallow piles or large lumps, natural circulation of air currents carries the heat away rapidly enough so that the temperature does not rise dangerously. Fines with their larger surface exposure must be watched more carefully than coarse lumps. Two methods of preventing spontaneous combustion are practiced. One of them is to promote sufficient circulation of air to carry off heat at low temperatures. This is done by shallow piling and by use of ventilating ducts. The other is to store in such a way that the oxidizing air is excluded from the coal.

Coal is the source of a great many important and useful by-products which are derived either from distillation or carbonization of bituminous coal. Amongst the more important of these might be mentioned the following: illuminating and fuel gas, coke, ammonia, analine and other dyestuffs, explosives, pitch, cresoline and paint compounds, antiseptics, tar. (R.M.F., F.T.M.)

COAL BALLS. Concretions composed of mineralized plant fragments preserved as **petrifactions.** Because the original structure of the plants has been so well preserved, the coal-ball flora has been of great aid to **paleobotanists** in determining the character of the **carboniferous** flora. (R.M.F.)

COAL MEASURES. Upper Carboniferous.

COAL TAR PRODUCTS AND INTERMEDIATES. When coal tar (see **Destructive Distillation, Coal**) is subjected to distillation, the various condensates contain substances of value for intermediates, which with the light oil recovered from coal gas, may be considered the immediate raw materials for making dyestuffs. While the exact treatment may vary with each coal tar obtained, general results are probably satisfactorily summarized as follows:

FRACTION COLLECTED	NAME OF FRACTION	PERCENTAGE OF COAL TAR
Up to 170° C........	Light oil	0.2
170–230° C.........	Carbolic oil	10
	⅓ phenol	
	⅔ cresylic acid	
230–270° C........	Creosote oil	10
270–350° C........	Anthracene oil	25
Residue at 350° C....	Pitch	55
Residue at 1200° C....	Retort carbon	17

The light oil fraction plus light oil recovered from coal gas (the latter the main source of benzene and **toluene**) contains:

FRACTION OF LIGHT OIL	FRACTION COLLECTED	DISTILLATION CHARACTERISTICS		
Crude benzene..	Up to 95° C.	90% recovered at 100° C.		
Crude toluene...	95–125° C.	5%	" "	100° C.
		90%	" "	120° C.
Crude solvent naphtha.....	125–170° C.	5%	" "	130° C.
		90%	" "	160° C.
Heavy naphtha .	170–200° C.	5%	" "	160° C.
		90%	" "	200° C.

Residue: Wash oil (for benzene absorption from coal gas), naphthalene and phenols. Drawn off and cooled, whereupon naphthalene crystallizes and is separated by filtration.

When these fractions are further fractionally distilled, pure benzene, pure toluene, xylenes and solvent naphtha are recovered. (See also **Hydrocarbons.**)

Benzene, toluene, xylene, naphthalene, anthracene, are the important hydrocarbons used for intermediates.

In addition to phenol there are used as intermediates the following **phenols:** cresols, resorcinol, alpha- and beta-naphthol. In addition to aniline, the following amines: ortho- and paratoluidine, dimethyl aniline, diethylaniline, diphenylaniline, benzidine, toluidine, alpha- and beta-naphthylamine, acetanilide; in addition to chlorobenzene: para-dichlorobenzene, benzyl chloride; in addition to **benzoic acid: phthalic acid,** phthalic anhydride, dibutylphthalate, **salicylic acid,** cresylic acid (see **Acids, Carboxylic**); also diphenylguanidine, ortho-ditolylguanidine (see **Guanidine**), phenylglycine, phenylglycocoll, $C_6H_5NHCH_2COOH$. (See **Aminoacids,** tricresylphosphate, benzaldehyde.)

There are about five largely used processes for producing intermediates. The products may be classified as:

1. *Sulfonated Products.* Wherein sulfuric acid concentrated is used to produce sulfonic acid products (group —SO_3H) by elimination of water (—OH from sulfuric acid plus H— from the benzenoid compound taken react to form H_2O). Sulfanilic acid (1-amino-4-benzene sulfonic acid) naphthionic acid (1-amino-4-naphthalene sulfonic acid and others). (See **Thioalcohols.**)

2. *Nitrated Products.* Wherein nitric acid concentrated (usually with sulfuric acid concentrated) is used to produce nitro-compounds (group —NO_2) by elimination of water (—OH from nitric acid plus H— from the benzenoid compound taken react to form H_2O). (See **Nitrocompounds** (nitrobenzene, 2- and 4-nitrotoluene, meta-dinitrobenzene, 2,4-dinitrochlorobenzene, 2,4-dinitrochlorotoluene, 2,4-dinitrophenol, nitroanilines); **Nitrobenzene; Toluene; Phenol.**)

3. *Chlorinated Products.* Wherein chlorine is used to produce chloro-compounds by elimination of hydrogen chloride (—Cl from chlorine plus H— from the compound taken react to form HCl). (See **Chlorine,** organic compounds.)

4. *Hydroxylated Products.* (See **Phenols.**)

5. *Aminated Products.* (See **Amines; Aniline.**) (R.K.S.)

COATI. Mammalia, Carnivora. *Nasua.* Animals related to the raccoons but with a long snout. They occur in Mexico, Central America, and South America. (A.W.L.)

COAXIAL LINE. This is a type of transmission **line** in which one conductor completely surrounds the other, the two being coaxial and separated by a continuous solid dielectric or by dielectric spacers with gas as the principal insulating material. Such a line is characterized by no external **field** and by having no susceptibility to external fields from other sources. It is extensively used for radio-frequency transmission lines and is being installed as a multi-channel telephone **carrier** and television program line. (L.R.Q.)

COBALT. Symbol: Co. Atomic number: 27. Atomic Weight: 58.94. Density: 8.9. Melting point: 1480° C. Boiling point: 2900° C. (**Isotopes:** page 290.)

Cobalt is a silver-white metal, harder and stronger than iron or nickel, not very malleable but a small amount of carbon markedly increases the malleability and ductility, magnetic below 1150° C., soluble in nitric acid. Compact cobalt is not oxidized on exposure to air at ordinary temperature, and does not react with alkalis. Discovered by Brandt in 1735.

Cobalt occurs as arsenide and sulfide (**smaltite,** $CoAs_2$; **cobaltite,** CoAsS), generally associated with iron, nickel, copper and silver minerals, the principal sources being Ontario and Belgian Congo. The ore is treated in a **blast furnace** and recovered as arsenide of the above-contained metals, which product is roasted with sodium chloride or treated with sulfuric acid; cobaltic hydroxide is obtained by a succession of treatments and then ignited to tricobalt tetroxide. From the oxide cobalt metal is obtained by heating with carbon.

Cobalt is used in alloys, (1) ferro-cobalt (35% Co) as a permanently magnetized steel, (2) "carboloy," tungsten carbide and cobalt, of high hardness and used as in cutting steel, (3) "stellite," cobalt-chromium alloy used in high-speed cutting steels, and in heat-resistant jet engine turbines.

Acetate: Cobalt acetate, cobaltous acetate ($Co(C_2H_3O_2)_2 \cdot 4H_2O$), red-violet solid, soluble, used as a dryer for paint and varnish oils.

Chloride: Cobalt chloride, cobaltous chloride ($CoCl_2$), red crystals, soluble.

Cobaltinitrites: Sodium cobaltinitrite ($Na_3Co(NO_2)_6$) soluble, and potassium cobaltinitrite, "cobalt yellow" ($K_3Co(NO_2)_6$), insoluble, are yellow solids, the latter used as a pigment.

Hydroxides: Cobalt hydroxide, cobaltous hydroxide ($Co(OH)_2$), rose-red precipitate by reaction of boiling cobalt salt solution and **sodium** hydroxide solution; cobaltic hydroxide ($Co(OH)_3$), brownish-black precipitate by reaction of cobalt salt solution and sodium hypochlorite solution.

Nitrate: Cobalt nitrate, cobaltous nitrate ($Co(NO_3)_2 \cdot 6H_2O$), red crystals, soluble.

Oxides: Cobalt monoxide, cobaltous oxide (CoO), gray to greenish solid, formed by heating cobaltous hydroxide or carbonate; used in the production of various colors with other oxides in **ceramics** such as blue, purple, green, red, yellow; cobalt sesquioxide, cobaltic oxide (Co_2O_3), black solid, formed by heating cobaltous nitrate at 180° C.; tricobalt tetroxide, cobalto-cobaltic oxide (Co_3O_4), black solid, formed by heating in air to a red heat any of the other oxides or compounds which yield these oxides by heating.

Sulfate: Cobalt sulfate, cobaltous sulfate ($CoSO_4 \cdot H_2O$), red crystals, soluble.

Sulfide: Cobalt sulfide, cobaltous sulfide (CoS), black precipitate, formed by reaction of cobalt salt solution with **ammonium** sulfide solution, relatively insoluble (after precipitation) in **hydrochloric acid.**

Cobalt soluble salts are pink, in solid or solution, but when heated gently to dehydration the hydrated salts, preferably chloride, turn blue, and the blue turns pink upon absorption of water vapor from the atmosphere. (R.K.S.)

COBALTITE. The mineral cobaltite is a sulfarsenide (see **Sulfur** and **Arsenic**) of **cobalt**, corresponds to the formula CoAsS, crystallizing in the isometric system as cubes or pyritohedrons, also may be massive. Cobaltite has a very good cleavage parallel to the cube faces; uneven fracture; brittle; hardness, 5.5; specific gravity, 6–6.4; metallic luster; color, silvery-white to reddish, sometimes steel gray or violet to grayish-black; streak, grayish-black. Cobaltite is found with cobalt and **nickel** minerals deposited commonly by metasomatic processes. It is found in Sweden, Norway, England and the Province of Ontario. It is an ore of cobalt. (E.S.C.S.)

COBEGO. Dermoptera.

COBIA. Pisces, Teleostei. A fish (**Pisces**) of the family Elacatidae, especially the species *Elacate canada*. This species is taken on the Atlantic Coast of North America. Also called the crab-eater and the sergeant-fish. (A.W.L.)

COBRA. Reptilia, Sauria. Poisonous **snakes** of several species found in Africa and southern Asia. Distinguished by the ability to inflate the neck when aroused. The common cobra of India, *Naja naja,* is the best-known species. It attains a length of 6′, half the size of the giant cobra or hamadryad, *Naja hannah,* which is less common. The asp or Egyptian cobra, *Naja haje,* occurs in Africa.

As is true of other poisonous snakes, the effect of the bite of a cobra depends on its own supply of venom, on the location of the bite, on the condition of the victim when bitten, and on his tolerance for poison. If the fangs penetrate well, however, and the snake has a good supply of venom, the bite of any of these species is said to be almost certainly fatal. The venom acts upon the nervous system. (A.W.L.)

COCA. *Erythroxylon Coca.* Erythroxylaceae. Coca is a drug which is found in leaves of a shrub, *Erythroxylon coca,* growing wild in Peru. This shrub grows about 6′ high, and has small ovoid leaves which are whitish on the under side, small yellow flowers growing in the axils of the leaves, and scarlet fruit. Cultivation has been carried on in Peru and adjacent regions since prehistoric days. The trees have been introduced into Java where they are now extensively cultivated. The leaves of the plant are highly prized by the native South American Indian for chewing. For this purpose they are mixed with lime, or ashes, and when chewed in this form enable the masticator to endure hunger and fatigue. The active principle is a poisonous alkaloid, **cocaine,** which is extracted from the leaves. (R.M.W.)

COCAINE. One of the alkaloids obtained from **coca** leaves. It is a valuable local anaesthetic when applied to skin or **mucous-membrane** surfaces, or infiltrated into tissues. This is the basis of its medical use. Other effects include stimulation of the central nervous system, increasing the heart rate, increasing the blood pressure, and increasing the body temperature.

The disadvantage of cocaine is that because of its central nervous system effects, it is habit-forming. Addiction usually occurs only in individuals who are psychopathic personalities. The drug is taken as snuff ("snow"), by application to the gums, by mouth, or by hypodermic injection. A feeling of **euphoria,** great mental and physical strength ensues. Often **hallucinations** and ideas of persecution make the addict dangerous; an innocent person may be killed because of his delusions.

Many synthetic local anaesthetics have been developed. These have largely replaced cocaine because of their greater potency and diminished toxicity. The commonly used ones are procaine ("novacaine"), butyn, nupercaine, and metycaine. (D.M.H.)

COCCIDIA. An order of parasitic Protozoa. (See **Sporozoa.**) (A.W.L.)

COCCIDIOIDOMYCOSIS (Coccidiodal granuloma. Valley Fever). A fungus infection due to *Coccidiodes immitis.* The disease is prevalent in southern California and the Southwest and appears to be spreading eastward. It is contracted usually by the inhalation of **spores,** although the agent may be introduced through an abrasion in the skin. The primary infection is often so mild that it passes unnoticed; if it is associated with signs and symptoms, the picture is one of an atypical bronchopneumonia (see **Pneumonia**) which heals completely in a few weeks. The granulomatous form of the disease, with widespread involvement of bones, joints, skin, subcutaneous tissues, and internal organs may rarely follow the acute primary infection, or it may develop years later. It is always fatal, whereas the primary pneumonic form has an excellent prognosis. There is no specific treatment. (D.M.H.)

COCCIDIOMORPHA. An order of parasitic Protozoa in some classifications, including **Coccidia** and **Haemosporidia.** (See **Sporozoa.**) (A.W.L.)

COCCOLITH. A microscopic calcareous unicellular marine plant. An important constituent of **chalk.** (R.M.F.)

COCCOSTEUS. Fossil Fish.

COCCUS. One of a family of the order of Eubacteriales which includes bacteria whose cells are spherical in form. (See **Streptococcus, Staphylococcus.**) (D.M.H.)

COCCYX. The lower end of the spinal column. It is composed of four rudimentary small vertebrae which are usually fused together. (R.S.M., D.M.H.)

COCHLEA. The auditory portion of the inner **ear of vertebrates.** (A.W.L.)

COCK OF THE ROCK. Aves, Passeriformes. *Rupicola.* Birds (**Aves**) of several species found in tropical South America. The males are crested and brilliantly colored. (See **Chatterers.**) (A.W.L.)

COCKATIEL. Aves, Psittaciformes. A small Australian **parrot** related to the cockatoos. (A.W.L.)

COCKATOO. Aves, Psittaciformes. Any member of a family of crested **parrots** whose beaks are transversely ridged on the under surface of the hook. The tail is short and broad. Cockatoos occur in the Australian and Oriental regions. (A.W.L.)

COCKCHAFER. Insecta, Coleoptera. A European beetle, *Melolontha vulgaris,* related to the May beetles of North America. (A.W.L.)

COCKLE. Mollusca, Pelecypoda. *Cardium.* Marine **bivalve** mollusks of several species. (A.W.L.)

COCKPIT. An open space in an airplane for the accommodation of pilots or passengers. When enclosed completely, such a space usually is called a cabin (F.T.M.)

COCKROACH. Insecta, Orthoptera. Flattened oval **insects,** usually brown in color. The head is almost concealed by the broad margins of the **thorax** and in winged species the wings overlap above the body.

Cockroaches are widely known from a few species which inhabit houses, especially where quantities of food are available. They eat almost anything used by man as food and often damage other things, such as articles made of cloth containing sizing or paste. They are especially troublesome in restaurants. The two most important pests are the Croton-bug, *Blatella germanica*, and the Oriental cockroach, *Blatta orientalis*. They can be destroyed by sprinkling borax (see **Boron**), **sulfur**, or **pyrethrum** powder liberally about their hiding places or by the use of commercial roach pastes.

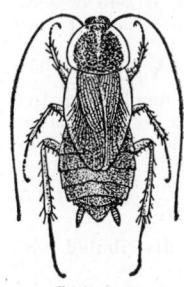

Cockroach.

The cockroaches gain their greatest development in the tropics, where species with a normal length of more than two inches are found. (A.W.L.)

COCOA. Theobroma Cacao.

COCOA BUTTER. Theobroma Cacao.

COCONUT. *Cocos nucifera.* Palmaceae. To the natives of tropical islands the coconut is a most valuable tree; it provides him with shelter, with dishes, with food and drink. To the inhabitant of north temperate regions, the plant is equally valuable—some half million tons of coconut oil are used annually.

The coconut tree is a striking plant having a columnar trunk 60–80' high, bearing at its top a crown of bright green pinnate leaves, each leaf varying from 15–20' long. The fruit is a large nut composed of three united **carpels**. The shell of the coconut has three distinct layers; the outer or epicarp is thin, smooth and brown; the middle, or mesocarp, is thick and fibrous; it is the source of the fiber, coir. Usually these two layers are removed before the coconuts are shipped, but not infrequently the entire fruit is exhibited as a curiosity by dealers in fruits. Within is the hard brown shell or **endocarp**. This contains the seed, which consists of an **embryo**, located under one of the germ pores at the end of the endocarp, and the **endosperm**, which is of two parts, one the familiar white meat of the coconut, the other the milk which partially fills the cavity of the seed. It is within the meat that the embryo is embedded. The thin brown skin immediately investing the meat and sticking to it when it is removed from the shell is the inner seed-coat. The hard shell of the coconut is much used as a dipper or as a vessel for storage of various substances.

The leaves of the tree yield a fibrous material used by the native islanders, and are also used at times as a thatch for shelters. The trunk of the tree is sometimes used by cabinet makers for ornamental work.

Commercially the most valuable product of the coconut is copra, the dried meat of the nut. This is obtained by splitting the nut and drying the meat, preferably by the sun. Artificial drying is done in regions of great humidity. From the meat is obtained coconut oil, which forms about 63% of the meat. The natives obtain this oil by various means. The hot sun of the tropics may cause it to dry out of the pounded meat. Crude presses are often used to squeeze the oil from the dried meats. Or again the crushed dried meats may be placed in hot water, the melted oil rising to the top and being skimmed off. Any one of these methods is sufficient to supply the moderate needs of the natives, but utterly inadequate to meet the requirements of the civilized nations.

For modern treatment, copra is shipped to large factories. Here it is cleaned and ground, then heated and pressed. The meal is then ground up once more,

cooked in water and pressed by powerful hydraulic presses. By this method nearly all the oil contained in the copra is extracted.

The principal use of the oil is in the making of **soap**, especially in soaps which will produce a copious lather. Considerable quantities are also used in making butter and lard substitutes. Some oil finds use in salad oils and in confectionery. (See **Esters**.)

The copra cake remaining after the oil is expressed may be used as a stock feed.

Many coconuts are used to make shredded coconut, while some 20,000,000 are sold fresh each year in the United States. Another product is **coir**. (R.M.W., B.S.M.)

COCOON. A case containing eggs or young, or a developing animal in an inert stage. In the **earthworms** a cocoon to contain the eggs is secreted by a special region of the body, the clitellum, as an external structure, but most cocoons are spun of silk, like those formed by the spiders to contain their eggs and those which some **larval** insects spin about themselves when they are ready for their final transformation. (A.W.L.)

CODEINE. An alkaloid, one of the most valuable and frequently used **drugs**. It is one of the components of **opium** (the dried opium contains 0.5 to 1% of codeine). It is a much weaker narcotic than **morphine** and it has less power to produce sleep, or to allay pain. However, its use does not readily cause addiction. It is a component of nearly all good cough remedies because of its ability to diminish cough and allay discomfort. It is also very often used to control mild pain, and promote sleep, especially in combination with other sedatives. In chronic conditions codeine is much preferred to morphine, due to the ease of addiction to the latter. (R.S.M.)

CODFISH. Pisces, Teleostei. Fishes **(Pisces)** of a large number of species, chiefly marine. They occur

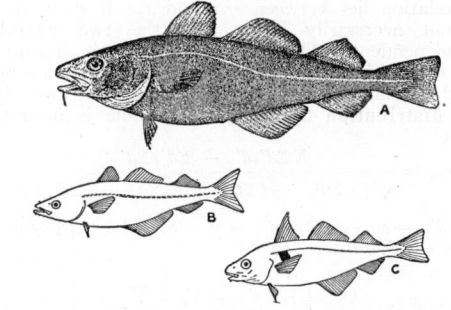

A, cod. (*U.S.B.F. Manual.*) B, pollock. C, haddock. (*Nichols and Breder, N. Y. Zool. Soc.*)

in the northern hemisphere and are the most important of all food fishes.

In the cod family (Gadidae) are included the **hakes, haddocks, whiting, pollacks,** and coal fish, and in the fresh waters of northern North America a single species, the ling, **burbot,** or lawyer, is found.

The common cod, *Gadus morrhua*, has been ranked as the most important food fish. The annual catch amounts to more than one billion pounds. It is abundant in the North Atlantic. (A.W.L.)

CODLING MOTH. Insecta, Lepidoptera. The moth, *Cydia pomonella*, whose **larva** lives in apples. An important economic species.

This insect is so serious an enemy of the apple that the successful production of the fruit depends on a program of spraying which has been carefully worked out for all parts of the country. **Lead** arsenate is an effective poison but its application must be regulated according to the entrance of the caterpillars into the fruit. Once they have penetrated the surface they are beyond reach of sprays. The principle of spraying is to apply a first spray when the petals of the flowers fall, and a second when the eggs of the next generation of insects are hatching later in the summer. Economic entomologists of the various states are prepared to furnish the proper information for various localities, according to the conditions of the year. (A.W.L.)

CODLIVER OIL. Esters.

COEFFICIENT OF ALIENATION.
The coefficient of alienation is $\sqrt{1 - r^2_{xy}}$, where r^2_{xy} is the **coefficient of correlation** between x and y. (L.A.A.)

COEFFICIENT OF ASSOCIATION.
The coefficient of association, Q, is defined by

$$Q = \frac{ad - bc}{ad + bc}$$

where a, b, c, d are the frequencies in a 2×2 contingency table as illustrated. Its use is not recommended.

a	b
c	d

Two by two contingency table.

(L.A.A.)

COEFFICIENT OF CONTRACTION. Orifice.

COEFFICIENT OF CORRELATION.
The coefficient of correlation, r_{xy}, is defined by

$$r_{xy} = \frac{\Sigma d_x d_y}{N \sigma_x \sigma_y}, \quad d_x = x - \overline{X}, \quad d_y = y - \overline{Y},$$

σ_x is the **standard deviation** of x, σ_y is the standard deviation of y and N is the number of paired **variates** (x_1, y_1), (x_2, y_2), $(x_3, y_3) \cdots (x_N, y_N)$, $\overline{X}$ = mean of the x's, $\overline{Y}$ = mean of the y's. It measures the dependence of x on y or y on x when the **regression** is linear. The coefficient of correlation lies between -1 and 1. If $r_{xy} = 0$, this does not necessarily imply that the two **variables** are independent. There may be strong dependence when $r = 0$ since the **correlation** may be non-linear instead of linear. For computational purposes in a **serial distribution** the following formula is more convenient

$$r_{xy} = \frac{N\Sigma d'_x d'_y - \Sigma d'_x \Sigma d'_y}{\sqrt{[N\Sigma d'^2_x - (\Sigma d'_x)^2][N\Sigma d'^2_y - (\Sigma d'_y)^2]}}$$

where $d'_x = x - a$, and $d'_y = y - b$. The regression line of y on x is determined by

$$y' = r\frac{\sigma_y}{\sigma_x}(x - \overline{X}) + \overline{Y}$$

and of x on y by

$$x' = r\frac{\sigma_x}{\sigma_y}(y - \overline{Y}) + \overline{X}.$$

The **standard error of estimate** $\sigma_e = \sigma_y\sqrt{1 - r^2_{xy}}$. As r_{xy} approaches 1 or -1, $\sigma_e \to 0$. Another method of defining r then is

$$r^2_{xy} = 1 - \frac{\Sigma(y - y')^2}{\Sigma(y - \overline{Y})^2}$$

where y' is given by the regression line.

In a correlation table let f_{xy} = frequency of an item in a cell, then the formula becomes fo r_{xy}

$$r_{xy} = \frac{\Sigma f_{xy} d'_x d'_y - \Sigma f_x d'_x \cdot \Sigma f_y d'_y}{\sqrt{[N\Sigma f_x d'^2_x - (\Sigma f_x d'_x)^2][N\Sigma f_y d'^2_y - (\Sigma f_y d'_y)^2]}}$$

where $N = \Sigma f_{xy}$, $d'_x = \frac{x - a}{c}$, $d'_y = \frac{y - b}{g}$, x and y are class marks, c is the class interval for the x's and g is the class interval for the y's. The coefficient of correlation is

unchanged if we substitute $\frac{x - a}{c}$ for x and $\frac{y - b}{g}$ for y. If the variates are distributed in a normal bivariate probability distribution, we may test the hypothesis that ρ, the population coefficient of correlation, is zero or some other value ρ_1, by methods due to R. A. Fisher. To test the hypothesis $\rho = 0$ we calculate $t = r\sqrt{\frac{n}{1 - r^2}}$, where n the degrees of freedom is $N - 2$, N the number of paired variates, r is the sample coefficient of correlation and refer t to the tables of **Student's t distribution**. To test the hypothesis $\rho = \rho_1$, we calculate

$$W = \frac{1}{2}\log_e\frac{1 + r}{1 - r}, \quad \overline{W} = \frac{1}{2}\log_e\frac{1 + \rho_1}{1 - \rho_1} \quad \text{and} \quad \sigma_W =$$

$$\frac{1}{\sqrt{N - 3}}$$ and assume W is **normally distributed** with mean $\overline{W}$. In reality W is only approximately normal. The standard error of r, $\sigma_r = \frac{1 - \rho^2}{\sqrt{N}}$, has sometimes been used to test the hypothesis $\rho = \rho_1$. Its usefulness is questionable. (L.A.A.)

COEFFICIENT OF COUPLING. Mutual Inductance.

COEFFICIENT OF DETERMINATION.
The coefficient of determination is defined as r^2_{xy}, the square of the **coefficient of correlation**. (L.A.A.)

COEFFICIENT OF DISCHARGE.
Water discharged from an **orifice, weir, pipe**, etc., theoretically has a velocity which is directly proportional to the square root of the head of water causing flow through the opening. Actually, however, contractions in the stream, surface roughness, and other causes, result in the actual velocity being smaller than the theoretical. The theoretical velocity is identical with that of the velocity attained by a freely falling body. It is $\sqrt{2gh}$, wherein g represents the acceleration of gravity, and h the height of fall corresponding to the pressure head creating the discharge. The ratio of the actual velocity to the theoretical is the coefficient of discharge. The coefficient of discharge from circular orifices is affected by the diameter of the orifice, the head, the sharpness of the edge, the velocity of approach to the orifice, and other minor factors. (F.T.M.)

COEFFICIENT OF DISPERSION.
Given n series containing s trials each, and let m_1, m_2, m_3, $\cdots$, m_n, represent the number of successes in each of these series. Let p be the **mean probability** in all $N = ns$ trials without specifying in any way the nature of the trials as either independent or dependent, and let $q = 1 - p$. Then we define the coefficient of dispersion D as the **expected value** of Q, $D = E(Q)$,

$$Q = \frac{\sum\limits_{i=1}^{n}(m_i - sp)^2}{Npq}.$$

Usually D is called the Lexis ratio. The empirical value of D' obtained by using observed frequencies m'_1, m'_2, $\cdots$, m'_n is given by

$$D' = \frac{\sum\limits_{i=1}^{n}(m'_i - sp)^2}{Npq},$$

assuming p is known. If p is unknown we obtain

$$D'' = \frac{N\sum\limits_{i=1}^{n}\left(m'_i - \frac{sM}{N}\right)^2}{M(N - M)}, \quad M = \Sigma m'_i$$

on the assumption that all trials are independent and the unknown value of p may be approximated by M/N. We have mainly followed the exposition of Uspensky.

If the events are independent, three important cases arise. The probabilities may be the same within a series but may vary from series to series. This is the Lexis case, $D > 1$,

$$D = 1 + \frac{s-1}{npq} \sum_{i=1}^{n} (p_i - p)^2,$$

p_i = the probability in the ith series, and

$$p = \frac{p_1 + p_2 + \cdots + p_n}{n}.$$

The Lexis case is called a Lexis series of trials.

If the probability γ_i for the ith trial in each series is the same, we have a Poisson series of trials, $D < 1$,

$$D = 1 - \sum_{i=1}^{s} \frac{(\gamma_i - p)^2}{spq}, \quad p = \frac{\gamma_1 + \gamma_2 + \cdots + \gamma_s}{s}.$$

In a Bernoulli series of trials the probability is constant from trial to trial throughout all the series and $D = 1$. If $D = 1$ the series is said to have **normal dispersion**, $D > 1$ **supernormal dispersion**, $D < 1$ **subnormal dispersion**. In practical cases D'' is calculated since D is not available. Usually deaths, births, marriages, suicides and divorces form series which are Lexian if these series are taken over large regions or over different periods in a smaller locale. A Poisson series of trials seldom occurs in practice. (L.A.A.)

COEFFICIENT OF EXPANSION. Expansion.

COEFFICIENT OF FRICTION. Friction.

COEFFICIENT OF HEAT TRANSFER. Heat Transfer.

COEFFICIENT OF MEAN SQUARE CONTINGENCY.
Let n_{rc} be the frequency in the cell formed by the intersection of the rth row and the cth column of an $r \times c$ contingency table, n_r = sum of the frequencies in the rth row and n_c = sum of the frequencies in the cth column and $N = \Sigma n_r = \Sigma n_c$, the total frequency in the table. We calculate

$$x^2 = N\left[\left(\sum \frac{n^2_{rc}}{n_r n_c}\right) - 1\right]$$

the summation being extended over all cells. The coefficient of mean square contingency, C.C., is

$$C.C. = \sqrt{\frac{x^2}{N + x^2}}$$

Its use is not recommended, unless the x^2 test shows lack of independence when the test of independence in a **contingency table** is applied. (L.A.A.)

COEFFICIENT OF MULTIPLE CORRELATION.
The coefficient of multiple correlation $r_{1.23}$ (in the case of three **variables** x_1, x_2, x_3) is defined by

$$r^2_{1.23} = 1 - \frac{\sigma^2_{e1}}{\sigma^2_{x1}}, \quad \sigma^2_{e1} = \frac{\Sigma(x_1 - x'_1)^2}{N},$$

$$\sigma^2_{x1} = \frac{\Sigma(x_1 - \overline{X})^2}{N},$$

$\overline{X}$ is the **mean**, N triplets of values $(x_{11}, x_{21}, x_{31}), \cdots (x_{1N}, x_{2N}, x_{3N})$, $x'_1 = b_1 + b_2x_2 + b_3x_3$. By permuting x_1, x_2, and x_3 we obtain similar formulas for $r^2_{2.13}$ and $r^2_{3.12}$. The coefficient of multiple correlation measures the

dependence of x_1 on x_2 and x_3, $0 \leq r^2_{1.23} \leq 1$. Always $r^2_{1.23} \geqq r^2_{12}$ or r^2_{13}. Another formula for $r^2_{1.23}$ is

$$r^2_{1.23} = \frac{r^2_{12} + r^2_{13} - 2r_{12}r_{13}r_{23}}{1 - r^2_{23}}.$$

To test the hypothesis $R^2_{1.23} = 0$, $R_{1.23}$ being the population value of the multiple coefficient of correlation we calculate

$$F = \frac{\Sigma(x'_1 - \overline{X}_1)^2/2}{\Sigma(x_1 - x'_1)^2/N - 3}$$

with **degrees of freedom** $n_1 = N - 2$, $N - 3$. Then F is referred to tables of **Snedecor's F distribution**. If the value of F so determined is greater than the value of F at the 5% level of significance from Snedecor's tables, we reject the hypothesis $R^2_{1.23} = 0$.

For m variables $x_1, x_2, \cdots, x_m, r^2_{1.23 \cdots m}$ is defined by

$$r^2_{1.23\cdots m} = 1 - \frac{\sigma^2_{e1}}{\sigma^2_{x1}}$$

where σ^2_{e1} and σ^2_{x1} are defined as before and $x'_1 = b_1 + b_2x_2 + \cdots + b_mx_m$. The test of significance becomes

$$F = \frac{\Sigma(x'_1 - \overline{X})^2/m - 1}{\Sigma(x_1 - x'_1)^2/N - m - 2}$$

$$= \frac{r^2_{1.23\cdots m}/m - 1}{(1 - r^2_{1.23\cdots m})/N - m - 2}$$

$n_1 = m - 1, n_2 = N - m - 2$. The easiest computational process for finding $r^2_{1.23\cdots m}$ is presented in the **Doolittle method of solving normal equations**. (L.A.A.)

COEFFICIENT OF NON-DETERMINATION.
The coefficient of non-determination is $1 - r^2_{xy}$, where r_{xy} is the **coefficient of correlation**. (L.A.A.)

COEFFICIENT OF PARTIAL CORRELATION.
The coefficient of partial correlation $r_{12.3}$ is defined as the **correlation** between x_1 and x_2 when x_3 is kept constant in a certain way or more precisely let $x'_1 = b_1 + b_3x_3$, $x'_2 = c_1 + c_3x_3$ then $r_{12.3}$ is the correlation between $x_1 - x'_1$ and $x_2 - x'_2$. This definition gives us the formula

$$r_{12.3} = \frac{r_{12} - r_{13}r_{23}}{\sqrt{(1 - r^2_{13})(1 - r^2_{23})}}, \quad -1 \leq r_{12.3} \leq 1.$$

Similar formulas are evident for $r_{13.2}$ and $r_{23.1}$. The partial coefficient of correlation always lies between -1 and $+1$. In the case of m **variables** $x_1, x_2, \cdots, x_m$, $r_{12.34\cdots m}$ is defined as the correlation between $x_1 - x'_1$ and $x_2 - x'_2$ where

$$x'_1 = b_1 + b_3x_3 + b_4x_4 + \cdots + b_mx_m$$

and

$$x'_2 = c_1 + c_3x_3 + \cdots + c_mx_m.$$

It is found most simply by the formula

$$r_{12.34\cdots m} = \sqrt{\beta_{12.34\cdots m} \cdot \beta_{21.34\cdots m}}$$

$\beta_{12.34\cdots m}$, and $\beta_{21.34\cdots m}$ are the **regression** constants

$$t_1 = \beta_{12.34\cdots m}t_2 + \beta_{13} \cdots t_3 + \cdots + \beta_{1m} \cdots t_m$$
$$t_2 = \beta_{21.34\cdots m}t_1 + \beta_{23} \cdots t_3 + \cdots + \beta_{2m} \cdots t_m$$
$$t_i = \frac{x_i - \overline{X}_i}{\sigma_{xi}}.$$

The most efficient computational technique in finding $r_{12.34\cdots m}$ is given under **Doolittle's method of solving normal equations**. To test the hypothesis $r_{12.34\cdots m} = 0$ we apply **Student's t test**

$$t = \frac{r}{\sqrt{1 - r^2}} \sqrt{N - m}$$

with **degrees of freedom**, $n = N - m$ where r is the sample value of the coefficient of partial correlation. (L.A.A.)

COEFFICIENT OF PERFORMANCE. Refrigeration.

COEFFICIENT OF SKEWNESS. Skewness.

COEFFICIENT OF VARIATION. The coefficient of variation of a distribution, V, is defined as the standard deviation divided by the mean, $V = \frac{\sigma_x}{\overline{X}}$. Ordinarily its value is less than 1. Some authors define it as $V = 100 \frac{\sigma_x}{\overline{X}}$ in order to avoid decimals. (L.A.A.)

COEFFICIENT OF VELOCITY. Orifice.

COELATA. A group of free-living flatworms which have an intestine. (See **Turbellaria**.) (A.W.L.)

COELENTERATA. The **hydroids, jellyfishes, sea anemones, corals,** and related animals. These forms make up a major division of the animal kingdom of simple structure. The body is developed from only two germ layers and is radially symmetrical. The alimentary track is sac-like, with a single opening, but sometimes has complex tubular branches. The nervous system is a scattered network of cells connected by their slender processes. Coelenterates have peculiar stinging cells (**cnidoblasts**) which discharge irritating **nematocysts.** Most species are marine and all are aquatic.

Two forms of individuals occur in this phylum, the **polyp** or hydroid and the **medusa.**

The economic importance of the phylum is limited principally to the corals. Precious coral is sold in considerable quantities and the rock corals have built up many of the oceanic islands.

The principal subdivisions of the phylum are the following:

Class **Hydrozoa** (Hydromedusae). Both polyps and medusae occur in these species, usually alternating in a reproductive cycle. Species are often colonial. In colonies individual polyps or hydranths are borne on branches of a common stalk which also gives rise to reproductive individuals or **blastostyles** from which medusae arise. All individuals are joined by the continuous digestive tract. Most species are small but the **Portuguese man-of-war** and a few others are large.

Class **Scyphozoa** (Scyphomedusae). The **jellyfishes.** Usually free-swimming species of moderate to large size. All individuals are medusae. Mostly marine.

Class **Actinozoa** (Anthozoa). **Sea anemones, sea feathers, corals** and allied species. Solitary or colonial marine species. The individuals are polyps. Colonies of some species build up massive hard deposits inside or outside of the body. (A.W.L.)

COELOM. The true body cavity, formed by the splitting of the middle **germ layer** of the body (mesoderm) and lined with a definite layer of cells.

The coelom is well developed in **annelid** worms, where it appears as a series of **metameric** chambers. It is of limited extent in other invertebrates but in the vertebrates gains great development. Here it is divided into a thoracic and an abdominal cavity and in the terrestrial species the thoracic cavity is further divided into pleural cavities containing the lungs and a pericardial cavity containing the heart. The abdominal or peritoneal cavity contains principally the greater part of the digestive tract.

Excretory organs open from the coelom in the annelids and lower **vertebrates,** and the liquid which it contains is apparently supplementary to the blood. In all cases the cavity furnishes a space into which developing organs may expand in complex animals. (A.W.L.)

COELOMATA. Animals which have a **coelom.** The term embraces the phyla **Bryozoa, Brachiopoda, Phoronidea, Chaetognatha, Echinodermata, Mollusca, Annelida, Arthropoda,** and **Chordata.** (A.W.L.)

COELOMODUCT. A tubule opening at one end into the coelom and at the other on the surface of the body. It occurs in a simple form in some **annelid** worms as an excretory organ and in other species is associated with the **nephridial** tubule to form a more complex excretory structure. Coelomoducts are regarded as the evolutionary forerunners of the excretory tubules of **vertebrates.** (See **Excretory System.**) (A.W.L.)

COELOSTAT. In many types of astronomical research it is desirable to have the main instrument stand still. To accomplish this purpose and also allow for the apparent motion of the **celestial sphere** it is necessary to reflect the light by a moving mirror from the object in question into the instrument. Such a device which "makes the sky stand still" is known as a coelostat. In the coelostat a mirror is mounted parallel to the polar axis of an **equatorial** mounting. The axis is rotated by clock work from east to west at such a rate that it would complete one rotation in 48 hours of **sidereal time.** Since the celestial sphere rotates from west to east once in 24 hours of sidereal time, and reflection doubles the angle, this rotation of the polar axis will compensate for the apparent rotation of the celestial sphere. With one single mirror mounted in the manner described the direction in which the light will be reflected will depend entirely upon the declination of the object observed. To obtain any desired direction of reflection a second mirror is employed to send the stationary beam from the first mirror in the desired direction.

In case it is desired to hold an image of the sun apparently stationary the polar axis must be rotated once in 48 hours of solar time. Such an instrument is known as a heliostat. Other types of mounting, used for particular purposes, are known as siderostats. (W.K.G.)

COENENCHYME. The middle and outer tissues of certain coelenterates (**Alcyonaria**). (A.W.L.)

COENOCYTIC CELLS. There are many plants which are composed of cells each containing many **nuclei.** In some cases the entire plant is a single cell, no cross-walls being formed until **reproduction** takes place. Multinucleate cells of this sort are called coenocytic cells. The **Phycomycetes** contain many plants of this kind, as do also the green **algae.** (B.S.M.)

COERCIVE FORCE. Magnetism.

CO-FACTOR OF A DETERMINANT. Determinants.

COFFEE. *Coffea arabica.* Rubiaceae. Probably the the coffee plant is a native of Ethiopia. In early historical times it was recognized as a valuable drink by the Arabs, who were responsible for its introduction into Europe. The cultivation of coffee slowly spread throughout the Old World and later, beginning in 1720, in the New World. The plant is a small tree with dark green leaves and fragrant white star-like flowers borne in axillary clusters. The fruits are about ½″ long, have a deep crimson skin and a yellow pulp surrounding the two seeds within. These seeds are the coffee beans.

The trees grow best in well-drained soil in regions entirely free from frost. In cultivation, the trees are grown from seed and begin bearing fruit in 4 or 5 years, continuing to bear up to 30 years. Seed production, once started, is more or less continuous throughout the year. The berries are gathered from the trees by threshing, or by hand, a costly operation in regions where labor is expensive. The berries are separated from any dirt present by washing, the heavy berries sinking to the bottom and the trash washing away. After this the pulp surrounding the seeds is largely removed by machines, and the seeds allowed to ferment slightly. They are then spread out to dry in the sun, or sometimes they are dried artificially. This permits the removal of any remaining pulp and of the thin inner skin which surrounds the seeds. The seeds are then ready to be bagged and shipped.

Before they are ready for the consumer the coffee beans must be carefully roasted. Roasted coffee contains from 1–2% of the alkaloid **caffeine**, which acts as a diuretic and as a cerebral stimulant. The aroma of coffee is due to an oily substance, caffeol, which is quickly oxidized when exposed to the air. By means of modern vacuum packing, this substance is preserved unchanged. Coffee also contains **glucose, dextrin,** and **protein.**

Brazil is one of the greatest coffee-producing countries in the world. The only important product of the coffee tree is the widely used beverage.

In many Eastern countries other species of coffee, as *Coffea liberica* and *robusta,* are planted to some extent, because of their greater resistance to diseases which have seriously reduced the better *Coffea arabica* plantations. (R.M.W.)

COFFER FISH. Pisces, Teleostei. Strangely shaped tropical marine fishes (**Pisces**) whose bodies are enclosed in bony plates. (A.W.L.)

COFFERDAM. A cofferdam is a structure of a temporary nature, used to exclude water from an otherwise submerged area, for the purpose of preparing for **foundations** or for other subaqueous construction. Cofferdams are frequently required both above and below the site of a permanent dam in order to by-pass the stream through a temporary channel during the construction period. The simplest cofferdam is an earth dyke which should be used only in shallow water where there is little or no current. Facing an earth cofferdam with sand bags or constructing it entirely of sand bags will make it serviceable when there is a current which would wash away loose materials.

If the depth of the water to be held back by the cofferdam exceeds that for which the earth dam is practicable (approximately 5′) other types must be constructed. Sheet piling (see **Piles**) can be used to form a cofferdam around a site by erecting it in double parallel walls, the space between being filled with sand or gravel and clay. This mixture is thoroughly tamped in order to form a solid filling. This materially adds to the stability of the structure. Where a small area is to be unwatered, a single wall of sheet piling, internally braced, may be erected around the site.

When cofferdams, constructed of sheet piling, are to rest on hard bottom, a timber framework is required to hold the sheeting in place. These frames are generally built on shore, floated into position and sunk into place. Sheet piling is then placed around the outside and banked with earth.

After the cofferdam has been completed it must be unwatered by pumps. As sheet piling is not entirely water-tight, leakage must be pumped out in order to keep the interior as dry as possible. (C.W.C., F.T.M.)

COG. Obsolete term for gear tooth. (See **Spur Gearing.**)

COHERENCE. Interference.

COHESION. Adhesion.

COIL. This term applies to one or more turns of conductor when wound as a definite unit of an electrical circuit. Thus we have the **choke coil,** or as it is sometimes called, impedance coil, as a number of turns of wire forming a coil used primarily for its reactance effect. The **transformer** is a unit of one or more coils used for transferring electrical energy by magnetic induction. Coils are particularly important in communications circuits where they serve in the above capacities but also form parts of the tuned circuits which make possible our complex systems. While the coil is ordinarily used for its inductive properties it inherently has both resistance and distributed capacity. The former is because of the resistance of the wire of which it is wound. The latter is due to the potential difference between turns which are separated by the turn insulation. At high frequencies this distributed capacity becomes extremely important and limits the usefulness of a given coil. Various special winding schemes have been used to minimize this effect. Electrical machines have coils as essential components; thus we have field coils, armature coils, etc. (L.R.Q.)

COIL CHAIN. Chain.

COINCIDENCE METHOD. Physical Measurements.

COINING. Press-Working.

COIR. The outer husk of the **coconut** is composed of coarse rough brown fibers, and is known as coir. These fibers are light, tough, and extremely resistant to heavy wear, as well as to wetting. They are therefore much in demand for making doormats, cordage, coarse matting, and stuffing for upholstery. Coarse brushes are made from shorter fibers. India and Ceylon supply the greater part of the crop. (R.M.W.)

COITUS. Sexual intercourse, **copulation.** (R.S.M.)

COKE. Coke is a solid product obtained by heating **coal** (in practice, bituminous coal) in a furnace without access of air. Coal is composed of moisture, ash, "fixed" carbon, and hydrocarbons which are volatile, in that they are distilled from the coal by the application of heat. When coal is subjected to intentional and controlled distillation, the volatile matter and moisture are driven off and the residue, consisting of the fixed carbon and the ash-forming substances, is commercial coke. The volatile products of the distillation are comprised of water, coal tar, and gas. The coal tar is a complex mixture, containing many substances of value to industry, especially to the chemical industry. The gas is widely used as an industrial fuel.

Since the smoke-producing constituents are driven off during the "coking" of the coal, the coke forms a desirable fuel for stoves and furnaces in which conditions are not suitable for the complete combustion of the bituminous coal itself. Coke may be burned with little or no smoke under combustion conditions which would result in a large amount of smoke were bituminous coal the fuel. Coke is the standard fuel for metallurgical purposes, i.e., cupola and blast furnace heating.

The solid residue remaining from the refining of petroleum by the "cracking" process is a form of coke. Petroleum coke has many commercial uses besides being a fuel, and is employed in the manufacture of dry cells, electrodes, etc. Gas works engaged in the manufacture of coal gas also have a coke end-product. This is called gas house coke. (F.T.M.)

COKITE. The term applied by Lacroix in 1917 to natural coke, the result of the **contact metamorphism** of coal beds. (R.M.F.)

COL. A relatively small area about midway between two cyclones and two anticyclones where the pressure gradient is very weak and winds are usually light and variable. It is the point of intersection between a trough line and a wedge line. (P.E.K.)

COLA. *Cola acuminata,* and other species. Sterculiaceae. Cola nuts are the cotyledons of trees growing native in the West African forests. The natives chew the fresh nuts, finding therein stimulation to prevent fatigue. From the dried nuts is made a beverage used in Africa.

The nuts contain up to 3% caffein and 2% tannin, as well as small amounts of theobromine. They are used in the manufacture of beverages. (R.M.W.)

COLCHICINE. An alkaloid obtained from the seeds or corm of the autumn-crocus, *Colchicum autumnale.* Liliaceae. The plant grows wild in Europe and North Africa and has been introduced into cultivation. It should not be confused with *Crocus,* a member of the Irish Family (Iridaceae), several species of which are grown widely as early-spring-flowering or as autumn-flowering perennials. These are tiny plants with narrow leaves and flowers on short stalks. The autumn-crocus is a much larger plant with leaves up to 12″ long and 2″ wide. The leaves appear in the spring on a stalk which also bears the ripening fruit. The fruit was formed in the previous fall and has remained underground on the top of the corm. Seeds are collected for the market in August as the leaves dry up and die. In September several flowers grow out of the corm. The flower has no stalk and consists of a corolla tube 4 or 5″ long partly underground and a limb of six segments often 4″ across. Corms are collected for the drug market in early August, cut into thin slices and quickly dried to kill the cells and prevent metabolic activity which destroys the alkaloid.

Colchicine is very poisonous, producing violent gastro-enteritis when taken internally, even in small amount. Its use has been mainly in the treatment of gout. The sudden emergence of the drug from relative obscurity is due to the recent discovery that it interferes with cell division by destroying the spindle mechanism. The two chromatids which represent one chromosome at the metaphase stage fail to separate and do not migrate to the poles (ends) of the cell. Each chromatid becomes a chromosome *in situ.* The entire group of new chromosomes now form a resting nucleus and the next cell division reveals twice as many chromosomes as before. The cell has changed from the diploid to the tetraploid condition. Applied to germinating seeds or growing stem tips in concentrations of about 1 gram in 10,000 cc. of water for 4 or 5 days, colchicine may thus double the chromosome number of many or all of the cells, producing a tetraploid plant or shoot. Offspring from such plants may be wholly tetraploid and breed true. Tetraploid plants are larger than diploid plants and often more valuable. The alkaloid has also been used to double the chromosome number of sterile hybrids produced by crossing widely separated species of plants. Such plants, after colchicine treatment, contain in each cell two complete diploid sets of chromosomes, one from each of the parent species, and become fertile, pure-breeding hybrid species. (P.A.W.)

COLD FRONT. Any boundary surface separating cold air from warmer air along which cold air is actively displacing the warm air. (See Fronts.) (P.E.K.)

COLD SECTOR. Temperate-zone cyclones usually involve two air masses. That part of the cyclone occupied by the cold air is known as the cold sector (in contrast with the area occupied by the warm air which is the warm sector). Cold sectors constitute more than ½ and often practically all the area covered by a cyclone. (P.E.K.)

COLD WORKING. An important property of ductile metals such as iron, copper, and aluminum is their ability to undergo severe deformation at normal temperatures without rupture. Cold rolling, forming, stamping, extrusion, and wire drawing are among the many processes involving cold working of metals. With the exception of metals that recrystallize at or near room temperature, for example, lead, cold working increases the hardness and strength and reduces the ductility. The strength thus developed in processing is often used to meet the strength requirements of the part. (See Temper.) Alloys processed in this manner are called work-hardening type alloys—in contrast with heat-treating alloys which are given a final heat treatment to develop required hardness or strength. (R.H.H.)

COLEMANITE. The mineral colemanite is a borate of calcium corresponding to a formula which is perhaps best represented as $Ca_2B_6O_{11} \cdot 5H_2O$. It occurs either as massive deposits or in monoclinic crystals. It has a subconchoidal fracture; hardness, 4–4.5; specific gravity, 2.42; vitreous to adamantine luster, may be colorless to milky white, grayish or yellowish; transparent to translucent. Colemanite was found originally in Death Valley, Inyo County, California, and has since been found rather widely distributed in San Bernardino, Los Angeles, Kern and Ventura Counties, California, as well as in Clark, Esmeralda and Mineral Counties in Nevada.

Colemanite was, until the discovery of kernite, the chief source of borax. Kernite, $Na_2B_4O_7 \cdot 4H_2O$, because of its easy solubility in water, has displaced very largely other boron-bearing minerals as a source of borax. Colemanite, kernite and inyoite (probably $2CaO \cdot 6B_2O_3 \cdot 13H_2O$) are lake deposits associated with other and rarer boron minerals, laid down during periods of volcanic activity or resulting from the leaching of the adjacent Tertiary sedimentary formations. Colemanite was named for Mr. William T. Coleman of San Francisco; Kernite and Inyoite were named from Kern and Inyo Counties, California. (E.S.C.S.)

COLEOPTERA. The beetles. An order of insects usually recognizable by the thickened wing covers which meet in a straight line down the middle of the back. These wing covers, or elytra, are modified fore wings. In most species of beetles they are thickened or horny but in some they are soft. In some species they are divergent and in some they are short, leaving much of the abdomen exposed. The typical condition of the elytra is found outside of this order only in the earwigs. Beetles have biting mouth parts and a complete metamorphosis in which the larval stage is often a grub.

This order of insects is the largest group of its rank in the animal kingdom, with almost 200,000 described species. It embraces almost the entire range of adaptation of the class, although very few beetles are parasitic. Many species are of economic importance. (A.W.L.)

COLEOPTILE. In the seeds of grasses the primitive bud or plumule is enclosed in a protective sheath called the coleoptile. During germination of the seed this coleoptile elongates, pushing its way out of the seed and up through the soil. It is very sensitive to light, growing directly towards a beam of light. (R.M.W.)

COLIC. This is a general term denoting abdominal pain which comes on quickly, is sharp and penetrating in character, intermittent, brief, and cramp-like. Biliary colic is a sharp severe pain which occurs with the passing of gallstones through the bile passages. Lead or Painter's colic occurs in lead poisoning and is associated with increased intestinal peristalsis. Renal colic occurs with the passing of a stone through the ureter. Treatment of colic is symptomatic during an attack. Atropine is used to relieve muscle spasm, and morphine may be necessary for severe pain. (D.M.H.)

COLITIS. Inflammation of the wall of the **colon** which may be due to a number of causes. Amoebic colitis or amoebic dysentery is colitis caused by the parasite *Entamoeba Histolytica* (see **Dysentery**). Mucous colitis is a chronic disorder of the colon commonly seen as a manifestation of a **psychoneurosis**. It is characterized by overactive **peristalsis** of the intestinal tract, the passage of mucus, constipation or diarrhea, and spasm of the walls of the colon. Ulcerative colitis is widespread infection of the colon with ulceration of its walls accompanied by diarrhea with blood and **pus**, fever, and weight loss. Its exact cause is unknown, but it belongs to the group of **psychosomatic diseases**. The disease is chronic with acute exacerbations, and responds poorly to treatment. **Sulfonamides** plus general supportive measures—high-caloric high-vitamin diet—are used. (D.M.H.)

COLLAPSE THERAPY. A method of treatment in **tuberculosis** where the diseased lung is put at rest to promote healing. (See **Pneumothorax, Thoracoplasty**.) (R.S.M.)

COLLAR. A fold or ridge of tissue more or less completely encircling the body behind its anterior end. In the snails, cuttle fishes, and related mollusks the ventral edge of the mantle is called the collar and in *Balanoglossus* (**Chordata**) the region of the body between the proboscis and the trunk is so named. (A.W.L.)

COLLAR CELL. A cell bearing a **flagellum** at one end, surrounded by a high membrane. Some of the 1-celled animals and the choanocytes of sponges have this form. (A.W.L.)

COLLECTOR. Klystron.

COLLEMBOLA. The spring-tails. An order of primitive wingless insects characterized by a forked appendage at the tip of the body which is used in leaping. This

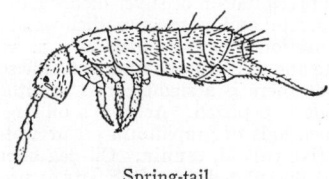

Spring-tail.

appendage is bent forward beneath the body and when released snaps sharply down and back, projecting the animal into the air.

Spring-tails are small and delicate. They are found mostly in moist places on the ground or on bark, though a few species live in dry hot situations. The snow flea, which sometimes appears in large numbers on the surface of snow, is a spring-tail. Some species are found on the surface of water. (A.W.L.)

COLLENCHYMA. Stem.

COLLES' FRACTURE. A fracture of the arm near the wrist, in which the radius is broken in its lower quarter. It is one of the most common fractures and occurs usually from a fall on the outspread hand, or a direct blow against the wrist. (D.M.H.)

COLLET. Drilling.

COLLETERIAL GLANDS. Glands associated with the female reproductive system of **insects**. They secrete materials which cement the eggs together or form a protective covering over them. (A.W.L.)

COLLIMATOR. An optical arrangement for producing parallel rays of light. A common form consists of a converging **lens**, at one of whose focal points is placed

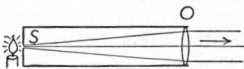

Divergent rays from slit S rendered parallel by objective O.

a small source of light, usually a pinhole or narrow slit upon which light is focused from behind. Rays diverging from this focal point emerge from the objective **lens** in a parallel beam. The slit or other source is viewed through the collimator without parallax, since it appears at an infinite distance. The arrangement is very generally used on **spectroscopes** and spectrometers. (L.D.W.)

COLLISION. As used in physics, this term refers to any encounter between free bodies in which they come near enough to exert a mutual influence, generally with exchange of energy. It does not necessarily imply actual contact. The process is subject to conservation of momentum, and in an "elastic collision," also to conservation of energy. In the latter case, if the initial velocities are given, the velocities of the bodies after collision can be calculated by applying these two conservation principles. The subject is of special significance in the case of **atoms, molecules,** etc. Collisions of this type between two particles A and B are denoted as of the "first class" or the "second class" according as (1) particle A loses kinetic energy and thereby affects B in some way, or (2) particle A communicates to B energy which it has by virtue of excitation or ionization. In either case B may receive additional kinetic energy or may be excited or ionized by the energy given to it by A. (L.D.W.)

COLLOBLAST. An adhesive cell used by the comb jellies (**Ctenophora**) in catching prey. Lasso cell. (A.W.L.)

COLLODION PROCESSES. The use of collodion in photographic emulsions (see **Photographic Emulsions**) dates from 1851 when Frederick Scott Archer published the details of his wet collodion process. While this process is no longer used in general photography, it is still widely used in making the half-tone negatives required in photoengraving. A brief description of the wet collodion process follows: A clean glass plate is first coated with collodion containing potassium iodide and potassium bromide. It is next sensitized by immersion in a solution of silver nitrate. It is then placed in a plate holder—specially designed for the handling of the wet plate—and the exposure made. After exposure it is developed in a solution of ferrous sulfate and fixed in potassium cyanide, or in hypo, washed and dried. The wet collodion process, as it is used by the photoengraver, results in a negative of high density and extreme contrast, high resolution and with an extremely fine grain. These characteristics render wet collodion well adapted to the requirements of photoengraving and it is only with recent years that gelatino-bromide emulsions of similar characteristics have been made available. These are now rapidly replacing the less convenient and less reliable wet collodion process.

Collodion printing-out paper was introduced by Obernetter of Munich in 1867 and was for many years the favorite printing process of the portrait and professional photographer. It was in general use until the early years of the present century when it was gradually replaced by developing-out paper. (C.B.N.)

COLLOIDAL STATE. All living matter is built up of colloidal materials, and almost all of our food, clothing and shelter materials are colloidal. The colloidal state is determined by the size of the particles of the substance, being intermediate in size between visibly suspended particles and invisible molecules. Turbid suspensions, visible to the naked eye or by means of the ordinary microscope, consist of particles of 250 millimicrons (250×10^{-7} cm.) or larger diameter, and molecules invisible by the **ultramicroscope** are of 5 millimicrons (5×10^{-7} cm.) or smaller diameter. True suspensions, such as fine clay in water, settle on standing, but colloidal suspensions exhibit continuous random motion—so called Brownian movement—of the suspended particles. **Diffusion** through animal membranes or parchment paper is negligible for substances in the colloidal state, whereas for true solutions diffusion is marked. Phenomena related to vapor pressure, freezing point, osmotic pressure of solvent or dispersion medium are marked in the case of true **solutions** but negligible for emulsions, like milk, emulsoids like gelatin, boiled starch, silicic acid, and suspensoids like **arsenious** sulfide sol, **sulfur** sol, **gold** sol, **ferric** hydroxide sol, clay (**ammonium** hydroxide) sol. In the case of emulsoid sols, the **surface tension** is lower than that of the dispersion medium, the **viscosity** is greater than that of the dispersion medium, and, in the case of organic emulsoid sols, the solid may be recovered from and redispersed in the medium at will, that is, they are reversible—for example, gelatin-water sol—and in the case of inorganic emulsions the system is non-reversible—for example, silicic acid-water sol. Particles of colloidal dimensions possess an **electric charge**, either positive or negative, and by treatment with certain **electrolytes** or colloids or by action of the electric current the electric charge is neutralized and the particles are coagulated.

The subdivided phase, corresponding to the solute in solutions, is called the dispersed phase, and the enveloping phase, corresponding to the solvent in solutions, the dispersing or continuous phase. Where both phases are continuous, the distinction between dispersed and dispersing phases merges into an interlacing system. Most of the stable colloidal systems are dependent for their stability on the presence of a protective colloid or stabilizer. Thus gelatin is added to milk and to ice-cream to stabilize the colloidal system, and mustard flour (1.0%) and egg yolk (8.0%) serve as stabilizers in mayonnaise dressing where as much as 75% of oil is dispersed in 16% of water phase, which phase contains acetic acid and salt about 1.0 and 1.5 parts per 100 of the total mayonnaise mixture (Corran, 1935).

A jelly is a completely transparent elastic mass; a gel is a flocculent and gelatinous precipitate. Jellies may be obtained without marked interference with the **equilibrium** in the solution; formation of gels disturbs radically the equilibrium in the solution. Fruit jellies and soap jellies are familiar examples; silicic acid gel is produced by addition of hydrochloric acid to sodium silicate solution. In using **pectin** for the production of jellies, the practical conditions are: maximum concentration of pectin 0.97% of the weight of the finished jelly; pectin-sugar ratio definite for a given pH (see **Reactions Involving Recombination of Ions**); sugar 50% to 70%; maximum pH 2.9 to 3.1; with 50% dry material at least 1.2% pectin, with 70% dry material 0.6% pectin.

Finely divided clay forms a colloidal suspension—is peptized—in water when treated with ammonium hydroxide or with tannin. Acheson made use of this principle when he used tannin to peptize finely divided graphite for lubricants. He relates that upon looking up the literature, he could find only one instance of the use of vegetable matter in clay working. That was in the Bible where the Egyptians used straw in the making of bricks. Since straw contains no tannin he wondered what the effect could be. Upon boiling straw with water he found that about half of the straw dissolved in water, and that the solution produced the same results with clay that tannin produced. He states that he found, in one case, that a sun-dried brick made of

treated clay was of greater tensile strength than a burned one made of the untreated clay.

Particles of a given substance may be brought into the colloidal state in one of two ways, namely, (1) by comminution or dispersion from macroscopic size—visible to the eye—to ultramicroscopic or colloidal size, (2) by precipitation or condensation from sub-ultramicroscopic, molecular or solution size to colloidal size, either without or with accompanying particles large enough to settle out of the medium.

Dispersion Processes. 1. The simplest method of accomplishing dispersion is by grinding the solid (or liquid) material with the liquid medium until particles of the required size are ultimately obtained. The colloid mill (Plauson, 1921) is used for such purpose, as in mixing paints and pastes, regenerating milk from milk powder, dispersing cellulose in **sodium** hydroxide and **carbon disulfide** for the production of xanthates for viscose, and in emulsifying fats and waxes. 2. Zinc sulfide, **cupric** ferrocyanide, **stannic** acid, **silver** chloride are examples of precipitates which, when washed on the filter paper until the accompanying soluble electrolyte has been removed, form colloidal solutions and pass through the pores of the paper. Since this is usually to be avoided in practice, the washing is then done with an electrolyte which does not conflict with the treatment to follow. Frequently **ammonium** nitrate solution is used. 3. A peptizing agent is frequently employed. **Tannin** is peptized by water, and by glacial **acetic** acid. Soaps are peptized by water. Gelatin swells in cold water but is not peptized, but is peptized in warm water. Starch, although insoluble in cold water, behaves similarly to gelatin with warm water (63° to 74° C., depending upon the kind of starch). Cellulose nitrate swells in ethyl alcohol and not in ether, but is peptized in ethyl alcohol-ether mixture. Clay is peptized by ammonium hydroxide, and it is held by some that the action of sodium hydroxide on **zinc, aluminum,** and **chromium** hydroxides is one of peptization. 4. Water-peptizable colloidal substances such as gelatin, dextrin, gum arabic, and soap peptize many precipitates, and are often called protective colloids. Gelatin in the solution prevents the precipitation of silver dichromate upon mixing silver nitrate and potassium dichromate solutions. (See Condensation Processes, below.) 5. When dilute **silver** nitrate and dilute **potassium** bromide solutions are mixed so that there is a slight excess of either solution, silver bromide is peptized. Acheson's oil-dag and aquadag are suspensoids of graphite in oil or water containing a protective colloid, **tannin.** Oil-dag contains about 15% of a "deflocculated graphite," and is used in dilute solution in lubricating oil (about 0.1% graphite). Bearings gradually become coated with a thin layer of graphite.

Condensation Processes. 1. When a solution of **ferric** chloride is poured into a relatively large volume of boiling water, colloidal ferric hydroxide is formed. The ferric hydroxide sol does not react with hydrogen sulfide nor with **potassium** ferrocyanide, and like all colloidal substances does not pass readily through animal membranes or parchment. 2. When **hydrogen sulfide** is passed into a solution of **arsenious** oxide, arsenious sulfide sol is formed which in the absence of an electrolyte may be made of the high concentration of 60 grams of arsenious sulfide per 100 grams of water. Upon addition of **hydrochloric acid,** arsenious sulfide coagulates and is precipitated. 3. When hydrochloric acid is added to **sodium** silicate solution either silicic acid sol or silicic acid gel is formed. 4. When hydrogen sulfide solution is treated with an oxidizing agent, for example, the proper concentration of nitric acid, sulfur sol is formed. 5. When **gold** chloride very dilute solution (0.01% to 0.001% of gold chloride) is made slightly alkaline (say by the addition of magnesium oxide) and then treated with a reducing agent, for example, formaldehyde or sodium hydrosulfite ($Na_2S_2O_4$), red gold sol is formed. 6. Use of a protective colloid in solution prevents the formation of the ordinary and expected precipitate in many cases and causes the formation of the expected substance as colloidal sol. Silver nitrate (0.6 gram per liter) and potassium dichromate (0.5 gram per liter) to one of which is added 0.1 volume of hot gelatin solution (2 grams per 100 milliliters of water) are mixed with stirring silver dichromate sol is formed. 7. When an electric arc is formed under water between two metallic rods, particles of the metal of colloidal size are formed along with more or less separation of free metal. A protective colloid increases the stability. If the metal vaporizes and then condenses to the colloidal state this is strictly speaking a condensation process, if otherwise, a dispersion process.

The disappearance of the colloidal state of a substance may be accomplished in either of two directions, namely, by the colloid passing into solution or into suspension. Practically, the latter is the more important method. Coagulation, agglomeration or precipitation is readily brought about by discharge of the electric charge on the particles. Ions carrying a charge of opposite sign to that carried by the colloidal particles are active precipitants, and the higher the valency of the ion the more effective (Linder-Picton-Hardy). When the colloidal particles are made neutral the conditions are least favorable to their stability. For colloidal arsenious sulfide, which is negatively charged in water, the coagulating power of potassium iodide (K^+I), calcium chloride ($Ca^{2+}Cl_2$), aluminum chloride ($Al^{3+}Cl_3$) is in the ratio of $1:80:1500$ (Svedberg); and for colloidal ferric hydroxide, which is positively charged in water, the coagulating power of potassium chloride (KCl^-), potassium sulfate ($K_2SO_4^{2-}$) is in the ratio of $1:45$. The active ion is carried down with the precipitated particles. Oppositely charged colloids, e.g., arsenious sulfide and ferric hydroxide, when mixed, precipitate each other. Other methods of coagulation are by migration of colloidal particles to and their discharge at electrodes, and by heating, as in the case of egg albumin. Coagulation is usually irreversible, especially when caused by electrolytes.

An interesting case, operating on a large scale in nature, of the precipitation of a colloidal system by an electrolyte is that of the action of sea water on the mud and silt of river water entering the ocean. When river water flows into the ocean the former, on account of its lower specific gravity, tends to flow over the latter and spread out in widening range. As the current diminishes some of the suspended mud and silt settles out, but the finer colloidal particles are coagulated by the electrolyte of the sea water and form deltas at the mouths of rivers.

Scope and Importance of the Colloidal State. As was stated in the beginning all living matter, whether animal or plant, is made up of colloidal materials and is sustained by colloidal processes. Of similar importance is colloidal chemistry in everyday living, in almost all of our foods, such as proteins and starches, in our clothing, whether of natural or synthetic origin, and in our shelter materials, such as wood, bricks, concrete. When there is added to these other common things and operations of everyday life, such as pottery and porcelain, paper, rubber and leather, and cooking and washing, where colloidal matter and processes operate, it is evident how broad is the scope and how great is the importance of the field. To these there must also be added other applications in the realm of industry, such as dyeing, printing, photography, water purification, smoke prevention, ore flotation, sewage disposal and soil preparation, paints, varnishes and lacquers, plastics, adhesives, and innumerable other operations and materials. (See Adsorption.) (R.K.S.)

COLLUM. 1. The dorsal plate of the first body segment in the millipedes (Diplopoda). 2. Any neck-like part or structure. (A.W.L.)

COLOB. Mammalia, Primates. *Colobus.* African thumbless monkeys of several species. (A.W.L.)

COLOCOLLO. Mammalia, Carnivora. A small cat of northern South America. It is light gray marked with black. (A.W.L.)

COLOCYNTH. Citrullus colocynthus. Gourd Family.

COLOGARITHMS. Logarithms.

COLON. The large intestine, which extends from the cecum to the rectum. It is divided into several parts, although the colon forms a continuous hollow muscular

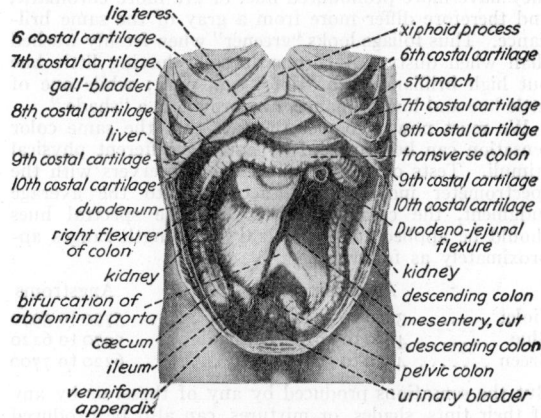

The abdominal viscera after the removal of the jejunum and ileum. (*Cunningham, Textbook of Anatomy, Oxford Press.*)

tube. The ascending colon extends from the lower right side of the abdomen at the termination of the small intestine, upward to the under surface of the liver, where it turns to the left and runs across the abdomen to the lower border of the spleen as the transverse colon. Beneath the spleen it bends downward, descending along the left side of the abdomen as the descending colon. As it enters the pelvis the colon makes a double curve, similar to the letter S. This portion is known as the sigmoid colon. The end of the sigmoid colon terminates in the rectum.

The functions of the colon are (1) final absorption of the products of digestion, (2) absorption of fluid from the feces so that this material becomes semi-solid, (3) removal of the fecal waste products into the rectum.

The principal diseases of the large intestine are cancer, tuberculosis, diverticulitis, polyp formation, and colitis. (R.S.M., D.M.H.)

COLONY. A group of individuals of the same species living together for mutual benefit. They may be structurally united or separate and may be alike in form or of different types suited for various functions. (A.W.L.)

COLOR. This vast subject is complicated by the distinction between the physical basis of colors and the sensations produced by them; and still further by a somewhat confused and unsettled vocabulary.

If one looks at two surfaces of the same size and shape, equally illuminated, and at the same distance, his only means of distinguishing them is by what is called their "color." To a physicist, the color means the spectral energy distribution of the light emitted or reflected by the surface. One may, however, specify it more simply, though less accurately, in terms of a limited number of variables employed in color measurement or colorimetry.

The practical standard "white" light is direct noon sunlight. A "perfectly white" surface would reflect white light completely without any alteration. No such surface exists. Even snow does not reflect white completely, though it does reflect all visible wavelengths in the same proportion. Its color is one of the "grays" or achromatic colors, of very high "brilliance"; while that of a lead-pencil mark is a much feebler achromatic color. A "black" surface would reflect no light at all (see **Black Body**). Most colors, however, are chromatic, that is, they exhibit "hue," because their spectral energy distribution differs so much from that of white or gray that they look "reddish," "bluish," etc. Some colors of the same hue are more "brilliant" than others. Just as snow is of a more brilliant gray than graphite, so bright red is more brilliant than dark red. Further, some colors have greater purity or "saturation" than others; that is, they have more pronounced hue, or are more chromatic, and therefore differ more from a gray of the same brilliance. Thus foliage looks "greener" when freshly washed than when dusty. A chromatic color having little hue but high brilliance is a "tint," e.g., pink; while one of little hue and low brilliance, like brown, is a "shade."

We must now recognize the fact that the same color sensation can be produced by entirely different physical stimuli. Tests of a large number of observers with the spectrometer indicate that, according to the average judgment, the common names of pure spectral hues should be applied to the several wavelength ranges approximately as follows:

	Angstroms		Angstroms
Violet	3900 to 4550	Yellow	5770 to 5970
Blue	4550 to 4920	Orange	5970 to 6220
Green	4920 to 5770	Red	6220 to 7700

But the sensations produced by any of these, or by any of their tints, shades, or mixtures, can also be produced in a variety of other ways. For example, red and green light may be mixed to produce a good imitation of spectral yellow light, though no yellow wavelengths are present in the mixture.

According to the Young-Helmholtz theory, the human vision has three separate color sensations, each capable of stimulation in various degrees. It is thought that, if stimulated separately, they would prove to be the sensations produced by red, blue, and green light, respectively. But they always act together, and every color sensation is the effect of their joint stimulation in some definite proportion. A result of this is that any color can be successfully imitated by adding together red, blue, and green light with suitable relative intensities. These are therefore called "additive primaries." If added in equal intensities, they produce a sensation of white. White may, however, be produced also by adding in suitable proportions various pairs of pure spectral hues, which are "complementary" to each other; thus:

Angstroms	Angstroms
6562 and 4921	5671 and 4645
6077 and 4897	5644 and 4618
5853 and 4854	5636 and 4330
5739 and 4821	

(The third pair, for example, is a certain yellow and a certain blue.)

Color sensations are also commonly produced by removing certain components from white light, as by the use of filters. Pigments such as paint and colored inks act in this way, being selectively reflective. The complementary hues of the three additive primaries are the "subtractive primaries" blue-green (for red), yellow (for blue), and purple (for green).

Maxwell devised an ingenious graphical scheme, called the "color triangle," for representing mixtures of colored lights (not pigments). The three additive primaries red, blue, green, are at the vertices of an equilateral triangle, with their complementary subtractives blue-green, yellow, purple on the sides opposite. Each altitude, as PG, is taken as 100%. Any hue is represented by a point

H, whose distances r, b, g, from the three sides represent the percentages of the three primaries R, B, G which must be added to imitate it ($r + b + g = 100$). Thus

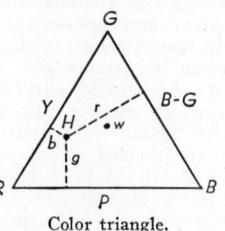

Color triangle.

for the point H in the figure, $r = 60$, $b = 10$, $g = 30$, giving a reddish-yellow sensation. The point W, at the center of the triangle, corresponds to the sensation of white. (L.D.W.)

COLOR BLINDNESS. Vision.

COLOR CAMERAS. Cameras designed to take color separation negatives simultaneously from the subject. Any camera used in normal black and white photography may be used to take a set of color separation negatives, but the time necessary to change filters and emulsions rules out all but inanimate subjects. Certain subjects can be satisfactorily photographed by means of a repeating back adapted to an ordinary camera. Such a mechanical device shifts filters and emulsions rapidly enough so that the total exposure time lapse is only a few sec-

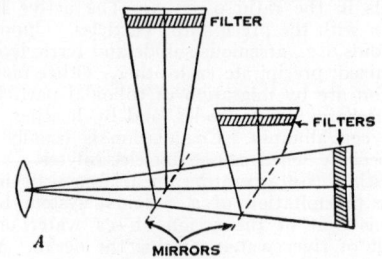

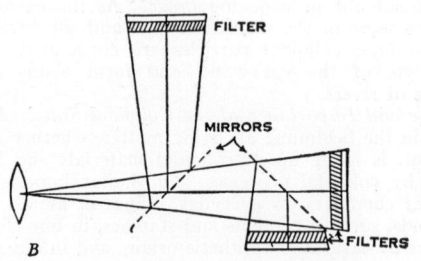

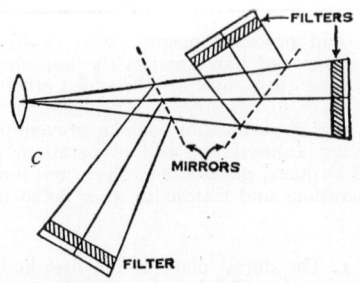

Double-mirror color cameras. (The mirrors shown are partially reflecting-transmitting.) (*Neblette's Photography*.)

onds. Whenever any movement is likely to be encountered, however, a color camera is necessary.

A color camera divides the light coming from the subject at any time into two or more beams which may be simultaneously recorded on different photographic emulsions. The optical and beam-splitting devices that are currently in use consist of a single lens followed by one or two partially reflecting mirrors, or a partially reflecting prism of glass. (See Figs.) With a double-mirror

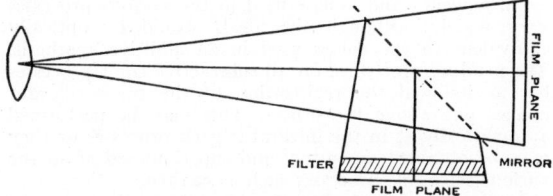

Single-mirror color camera using bipack emulsion. (*Neblette's Photography.*)

camera three appropriate emulsions and color filters yield the three-color separation negatives. With a single-mirror camera a **bipack** must be used to record two of the color records while the third is recorded on a single emulsion. A slight amount of diffusion is obtained with the latter system because of the light passing through the front emulsion of the bipack before reaching the second film exposed at that aperture.

Cameras of the above types have enjoyed considerable popularity among illustrative color photographers because of the color accuracy obtainable in color prints made from direct color separation negatives. Such cameras are often termed "one-shot cameras." (H.C.C.)

COLOR CORRECTION. The steps taken in improving the reproduction in a color photograph. Color correction, or masking, is necessary if accurate color reproduction is desired because the colorants used in subtractive color processes do not fulfill the theoretical requirements. All known cyan dyes, inks, or pigments, for example, are rather poor transmitters of blue and green light, the colors to which they should be completely

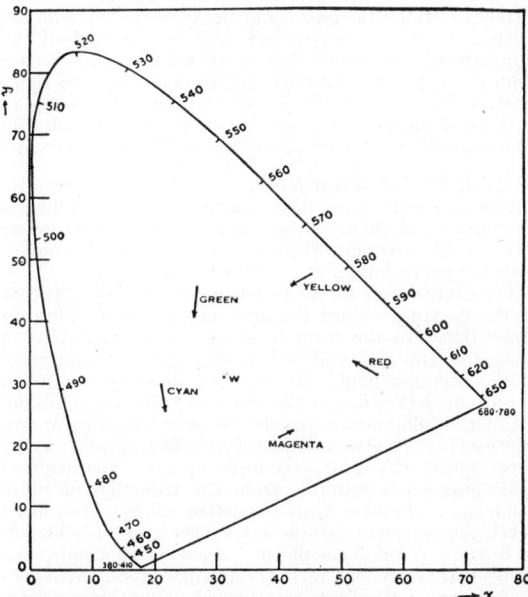

Five colors (dots) and their reproductions (arrow heads) plotted on a chromaticity diagram. (*Neblette's Photography.*)

transparent. In a similar manner most magenta colorants only transmit about 50% of the incident blue light. The use of such colorants leads to color reproduction shifts that may be generalized as reds, magentas and yellows reproduced too light; cyans, greens and blues reproduced too dark. Reds and magentas shift toward yellow; cyans and greens reproduce as somewhat colder hues. With these colorants it is also difficult to obtain a satisfactory scale of neutrals and the use of a supplementary black-printer image is often necessary.

Masking, or color correction, may be carried out photographically by combining with a given negative, or positive color record, a positive or negative which is a record of another color. It is common practice, for example, to apply to the green-record separation negative a positive mask which has been obtained from the red-record separation negative. This positive mask, of relatively low contrast, prevents magenta from printing in those areas of the reproduction that contain cyan, thus resulting in brighter cyans, blues and greens. There have been a great many systems of masking used in commercial photography but they usually involve a positive made from the red record used as a mask on the magenta printer and a positive mask made from the unmasked magenta printer used in correcting the yellow printer. The relative contrasts of negatives and masks determines the extent of the correction for each printer. Masking of this type directly on the color separation negatives greatly improves the color accuracy obtainable in a reproduction from a given set of subtractive colorants.

When it is desirable to make a reproduction directly from a positive or negative **color transparency** by photographing it and printing, a somewhat different form of color correction becomes necessary. Because of the incorrect absorptions of most subtractive colorants, the recording of a color transparency by a green filter and panchromatic emulsion, for example, yields an image not only of the transparency's magenta dye layer but also an image of the cyan dye layer. Since the magenta layer in the transparency resulted from the original green-record negative obtained when the transparency was made, this is the only image desired when photograph-

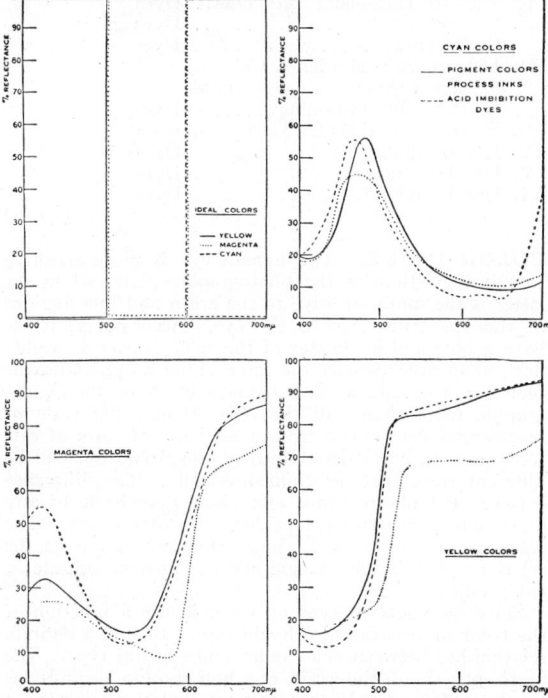

Ideal and real reflectance values for subtractive printing colors. (*Neblette's Photography.*)

ing the transparency. The recording of the unwanted cyan image may be avoided by combining with the transparency before it is rephotographed a negative or positive cyan or silver image that will partially or completely neutralize the positive or negative cyan dye image. Then a green filter separation may be obtained of the transparency's magenta dye layer without interference from the cyan image. Such masks may be made separately and combined with the transparency, or they may be formed during the processing of the transparency in which case they are part of the integral tripack. (H.C.C.)

COLOR DEVELOPMENT. Color development, also known as chromogenic development, or dye coupling, is a process which involves the conversion of a latent image into metallic silver and the deposition of a dye in contact with the silver image.

The silver image serves to localize, or fix, the position of the dyestuff. Since the amount of dye produced is proportional to the metallic silver, the amount of dye formed in any area will be proportional to the intensity of the exposing light. Removal of silver by reduction or solvent action leaves the dye image in the emulsion.

Color development is possible because a number of dye intermediates also function as developing agents. Color development may take place in two ways. The oxidized developing agent resulting from the reduction of silver halide may combine with a portion of the unoxidized developing agent to form a dye. This method is known as primary color development. Developing agents, like pyrogallol, catechol, methoxyalphanaphthol, hydroxycoerulignon, indoxyl and thioindoxyl, show this capacity. In practice this method has not been very successful because the images formed are relatively unstable, the highlights are usually stained, and the range of color is limited.

In the second method, another dye intermediate, known as the coupler, is added to the developer. The dye forms when the added dye intermediate couples, or condenses, with the oxidized developing agent. The term coupler is applied to the second dye intermediate used. Phenols, aromatic amines, substituted methines, and azomethines having strong polar groups are satisfactory for dye coupling. This process, using two dye intermediates for dye formation, is termed secondary color development. Among the dyes used for secondary color development are: the indophenols formed by the reactions of para-aminophenol and a phenol; the indoanalines, by reaction of a paraphenylenediamine and a phenol; the indiamines with a phenylenediamine and an amine; the indothiophenols by paraphenylenediamine and a thiophenol; and the azomethines from the reaction of a para-aminophenol or a phenylenediamine with a methine containing active or strong polar groups. Secondary color development has been used successfully for the production of prints in color and for color transparencies with integral tripack processes because a wide variety of colors is possible by using different couplers with a limited number of developing agents or with a single developing agent. Couplers may either be incorporated in emulsions or added during processing.

The formation of a dye image by color development may take place simultaneously with the reduction of the silver image or in two steps. A molecule of the developing agent becoming oxidized as a result of the reduction of the silver halide molecule, and the coupling of the oxidized developing agent with the coupler which requires the reduction of these silver halide molecules. In either method four atoms of silver are reduced for every molecule of dye formed.

Latent images, bleached silver images, or chromate images, may be used for the production of dye images by color development. (See **Chromogenic Developing Agents, Chromogenic Couplers.**) (S.M.T.)

COLOR IMAGE FORMATION. Processes of color photography using an additive synthesis usually obtain their color images by means of neutral silver images and the placement of color filters in the light beams. Subtractive synthesis, however, requires that the image itself be colored, and the many ways of forming these images, both chemically and physically, have led to the numerous photographic **subtractive color processes** now in existence.

Three-color subtractive processes use the colorants cyan, magenta and yellow in printing from the red, green and blue analysis separation records. The exact hues of cyan, magenta and yellow used in the various processes vary widely, although the ideals would be optically equivalent to the colors used in an additive synthesis. The printing procedure for all subtractive color processes has as its goal the registration of the three colorant images one above the other. This may be performed automatically as in the integral tripack processes or they may be obtained separately and superimposed as in the various separable processes such as **carbro.**

A classification of photographic subtractive processes shows six distinct types of colorant image formation. The lines of demarcation between the various methods cannot be drawn too clearly since many commercial processes make use of more than one method of colorant image formation. In the accompanying table the processes are listed together with the usual colorant material employed with each type of image formation. (See **Additive Color Process, Bipacks and Tripacks.**)

SUBTRACTIVE PRINTING PROCESSES

Process	Medium
I. Toning Processes	
A. Metallic Toning	Colored Metallic Salts
B. Dye Toning or Mordanting	Dyes, usually basic
II. Gelatin-Relief Processes	
A. Tanning Development	Dyes or Pigments
B. Dichromated Colloids, Hardening Brought About by:	
1. Light	Pigments
2. Silver	
a. Carbro	Pigments
b. Hardening Bleaching	Dyes
c. Dyebro	Dyes
C. Etching	Dyes
III. Differentially Hardened Colloids —relief formation not essential	
A. Differential Staining	Dyes
B. Differential Water Absorption	Inks
IV. Dye Coupling	Dyes
V. Dye Destruction	Dyes
VI. Dye Bleaching	Dyes

(H.C.C.)

COLOR INDEX. The human eye is more sensitive to red light than is the photographic plate, while the latter is the more sensitive to the green and blue regions of the **spectrum** than is the eye. Accordingly, if we have a blue and a red star of the same apparent brightness, or in other words, the same visual **magnitude,** the blue star will make a much stronger image on the photographic plate than will the red. Hence, the scale of magnitudes determined from a sequence of stars of different colors, when determined photographically, will be different from that determined visually. The difference between the photographic and visual magnitude of any given star is known as the color index of the star. The algebraic sign of the color index is determined from the relation: photographic magnitude − visual magnitude = color index.

Since the **spectral type** of a star is also a function of the color of the star, we should expect to find a definite relationship between color index and spectral type. The "zero point" of the scale of photographic magnitudes has been defined in such a manner that the color index is zero for an Ao type star between 5.5 and 6.5 magnitudes. For the other spectral classes the values of the color index are approximately:

B	A	F	G	K	M
−0.24	0.00	+0.28	+0.56	+1.00	+1.35

In accordance with the theory of **black-body radiation**, the apparent color of a radiating gas should be a function of the temperature of the gas and color index may be used as an approximate method for the determination of the temperatures of the stars. (W.K.G.)

COLOR NEGATIVE. A photographic image in which not only the brightness of the various parts of the scene is reversed, but the color of each area is complementary to that in the original scene. (H.C.C.)

COLOR PHOTOGRAPHY. The creation, by photographic means, of sensations in the brain of the observer approximating as closely as possible the sensations he would have experienced in oberving the colored subject directly.

The steps necessary to form the sensation may best be classed as analysis and synthesis. The proces of analysis involves the recording on two or more photographic emulsions of the proportions of the visible spectrum reflected or emitted by the subject. For example, a color-sensitive emulsion exposed behind an orange-red filter will record the areas of the subject photographed which reflect or emit red light, these areas being shown on the developed negative as silver densities. The actual densities will bear a definite relation to the amount of orange-red light from each subject area. Photographic density thus becomes a quantitative measure of the orange-red light present in the subject. If the same procedure is employed with another color-sensitive emulsion but this time using a blue-green filter, a two-color analysis of the original subject will have been obtained in the form of two photographic negatives, one a "red-orange record," the other a "blue-green record." If three negatives had been made using red, green and blue filters, a three-color analysis would have resulted.

The synthesis, or recombination, of the photographic analysis may take many different physical or chemical forms, depending on whether the synthesis is to be additive or subtractive and on the method of image formation. In its simplest additive form, the synthesis consists of the simultaneous projection, in register, from two projectors, of the black and white positive images made from the analysis negatives. Since densities in the red-orange record indicate the presence of red-orange in the subject, a positive or lantern-slide image is reversed in the sense that the silver densities now indicate absence of red-orange while the clear areas of the positive indicate the presence of red-orange. A similar situation exists for a positive made from the negative recording the blue-green in the subject.

If a red-orange filter is inserted in the light beam of the projector containing the first positive, only red-orange light will reach the screen. The position and quantity of this light is controlled by the silver positive. Image areas that correspond to those areas of the subject completely lacking in orange-red will be dark, while the more orange-red there was in the subject the lighter the positive will be, thus transmitting more orange-red onto the screen. If the second positive is projected onto the screen in register with the first by means of another projector employing a blue-green filter, the additive synthesis will be complete. A three-color additive synthesis may be carried out in a similar manner with three positives and projectors equipped with red, green and blue filters.

The observer viewing the image of such a synthesis receives a complex stimulus through the medium of his eye The similarity of his sensation to that obtained when viewing the subject directly is indicative of the accuracy of the photographic color reproduction.

The majority of color photographs today are made by some form of subtractive synthesis. This differs from an additive synthesis where light is added to light on a screen to form the final color image. In the subtractive synthesis colorants are used to subtract from white light the desired color; i.e., a yellow dye (colorant) is used to subtract blue light from white. An ideal subtractive synthesis gives the same sensation as that obtained from the additive synthesis. (See **Additive Color Process, Subtractive Color Process.**) (H.C.C.)

COLOR TRANSPARENCY. A color photograph made for viewing by transmitted light. As the light only passes through the colorants of a transparency once, their contrast is about double that necessary in a print on an opaque support which is viewed by reflection. The colorant images may either be the components of an integral tripack or may be simply three colorants superimposed after their preparation by subtractive synthesis. (H.C.C.)

COLORADO POTATO BEETLE. Insecta, Coleoptera. A leaf-eating beetle, *Leptinotarsa decemlineata,* native to the western United States. Originally it fed on a native plant but about the middle of the nineteenth century it became troublesome in potato fields and rapidly spread across the country. Both larva and adult eat the leaves of potato plants. They can be checked by the use of Paris green or lead arsenate used as a spray. (A.W.L.)

COLORATION. The coloration of animals embraces both the colors that appear in their bodies and the patterns in which they are arranged. In many species color appears to be incidental but in others coloration has an important bearing on the life of the individual.

Colors are due in some cases to the presence of compounds which are important for other reasons. The blood of insects, for example, may be green, that of certain mollusks blue, and that of the vertebrates and annelid worms red because of their chemical composition (see **Protein**), but these colors have no important bearing on individual life. The black pigment found in eyes is a protection against random light rays but any impervious layer would be as effective; black color is unimportant in itself.

In many animals colors of this kind are supplemented by special pigments located in the superficial layers of tissue and in the vestiture, such as hair, feathers, and scales. The surface of the integument in insects is often formed so that it breaks up light rays and reflects only a portion of them. The colors produced in this way are called physical colors and are usually metallic or glassy. All of these colors, whether physical or pigmental, are often arranged in intricate patterns characteristic of the species. In this arrangement details may be observed in some animals which seem to have definite value in relating the individual to its environment.

The white tail patch found in some **deer** and **rabbits** has been interpreted as a signal mark. It may serve for ready recognition in poor light and may catch the attention of other members of a group when danger is near.

Warning colors are exemplified by the brilliant red and yellow of the **coral snakes.** Poisonous or distasteful animals are usually avoided as food and conspicuous appearance makes it all the easier for other animals to see and avoid them.

Concealing colors render the animal less conspicuous in its normal environment, and are essentially the same in the hunter and in the hunted. The white winter coats of the prairie hare and of the weasel are equally inconspicuous against the snow, but the one animal is helped to escape its enemies and the other to catch its prey by this means. Concealing colors may also be in patterns. The black and tawny stripes of the tiger, for example, are said to conceal it admirably in tall grasses lighted by the sun, where vertical shadows are conspicuous. This aspect of coloration, however, merges with protective mimicry, and in mimicry physical form is involved as well as coloration. (A.W.L.)

COLORS. For list of approved food, drug and cosmetic colors see reference in **Dyes and Dyeing Textile Fibers.**

COLOR-SENSITIZING DYES. The color-sensitizing dyes are those which added to the silver halide emulsions cause them to become more sensitive not only to the ordinary range of silver halides in gelatin but increase the spectral range, even in the infra-red or long-wave part of the spectrum. The dyes used are, generally, selective with respect to a particular range, hence, it is common practice to use more than one dye for general photography. These dyes not only increase the spectral range but permit the photographer to obtain better results where only short-time exposures are possible. The following examples may serve as an illustration.

Corallin is the dye which was originally used to prevent halation in silver halide-collodion emulsions, and accidentally the general principle of sensitization of photographic emulsions was discovered.

The tetraiodo derivatives of the dye, selena and telluraxanthones, are said to be good sensitizers for the spectral range of 520–610 mμ. It is doubtful, however, if they can be used in place of the simple cyanines or simple merocyanines.

The cyanines are probably the most important sensitizing dyes, although the merocyanines have come into considerable prominence in the last 10 years. The cyanines comprise the largest number of satisfactory sensitizers used in present-day emulsions.

Very good sensitizers for the yellow are found in the benzooxazoles and also benzthiazoles. Kryptocyanine is an example of the quinoline derivatives and is important, since its introduction opened up the field of extreme red and near infra-red sensitivity. It does not sensitize beyond 800 mμ.

The dicarbocyanines or pentamethines have a maximum absorption at about 670–700 mμ. The tricarbocyanines or heptamethines are sensitizers for the range 650–1050 mμ. The tetra and pentacarbocyanines have been prepared and contain a linking chain of 9 and 11 carbon atoms respectively. Only a few of these long-chain dyes are of practical value for the extreme infrared. (RALPH BRADEN, ROCHESTER INSTITUTE OF TECHNOLOGY.)

COLOR-SEPARATION NEGATIVES. Photographic negatives which record the relative intensities of the primary colors used in the analysis necessary to reproduce a subject by means of color photography. In three-color photography, for example, the separation negatives are records, in terms of silver densities, of the amounts of red, green and blue light received at the camera from the subject.

A set of color-separation negatives may be prepared by photographing the subject three times on separate color-sensitive emulsions so that each is a record of one of the primary colors. A panchromatic emulsion is generally employed with a set of tricolor filters, the colors of the primaries. It is only necessary, however, to obtain the color records on separate negatives so it is also possible to use for each record any combination of color filter and emulsion sensitivity that will record one of the primary colors. A set of color-separation negatives may be made by exposing (1) each one in turn in a camera, (2) by the use of a color camera which will expose them simultaneously, or (3) in a tripack.

It is common practice to balance a set of color-separation negatives, by altering the exposure and development times, so that a gray scale will be recorded equally on each negative. The particular densities desired are dependent on the method of color synthesis to be employed. (H.C.C.)

COLOSTRUM. The fluid secreted by the mother's breasts a few days before and after the birth of the child, before the secretion of true milk begins. It is a thin watery fluid containing considerable albumin. (D.M.H.)

COLUMBIFORMES. The order of birds (Aves) containing the pigeons and doves. (A.W.L.)

COLUMBINE. Buttercup Family.

COLUMBITE. A mineral oxide of iron, manganese, columbium and tantalum. Crystallizes in the orthorhombic system. Hardness, 6; specific gravity, 5.20±; color, red to brown. (R.M.F.)

COLUMBIUM or NIOBIUM. Symbol: Cb. Atomic number: 41. Atomic weight: 92.91. Density: 8.4. Melting point: 1950° C. No isotope, but of single atomic form: 93.

Columbium is a slightly bluish metal; ductile, malleable, and when polished resembles platinum; burns upon being heated in air; insoluble in hydrochloric or nitric acid, but soluble in hydrofluoric acid or a mixture of hydrofluoric and nitric acids. Discovered by Hatchett in 1801.

Columbium occurs, usually with tantalum, in columbite ($Fe(CbO_3)_2$, 80% Cb_2O_5), pyrochlore (50% Cb_2O_5), samarskite (50% Cb_2O_5), chiefly found in western Australia, and South Dakota. Recovered along with tantalum by fusion with potassium bisulfate, and obtained in the residue after subsequent extraction with water. Columbium and tantalum are separated by fractional crystallization of the potassium fluorides, columbium concentrating in the mother liquor and tantalum in the crystals. Chemically related to vanadium and tantalum. Columbium is used (1) in steel, especially *stainless steel,* to stabilize the carbon present (as carbide), and (2) for columbium carbide (see **Abrasives**) for dies and cutting tools. Ferro-columbium alloy is used as a source of columbium for addition to corrosion-resistant steel.

Fluoride: Columbium potassium fluoride ($K_2CbF_5 \cdot H_2O$).

Chlorides: Columbium trichloride ($CbCl_3$); columbium oxychloride ($CbOCl_3$), white solid; columbium pentachloride ($CbCl_5$), yellow crystals, melting point 194° C., boiling point 240° C.

Columbates: Columbium of valence plus 5.

Oxides: Columbium monoxide (CbO) (sometimes called dioxide, Cb_2O_2); columbium dioxide (CbO_2) (sometimes called tetroxide, Cb_2O_4); columbium pentoxide (Cb_2O_5). (R.K.S.)

COLUMELLA. 1. The axis of the spiral shell of a snail. 2. A bone in the middle ear of amphibians, reptiles and birds. (A.W.L.)

COLUMN. A slender structural compression member, standing vertically, is a column. The ratio of the length of the column to the least radius of gyration of its cross-section is called the slenderness ratio. This ratio affords a means of classifying columns. A short steel column is one whose slenderness ratio does not exceed 50; an intermediate length steel column has a slenderness ratio ranging from 50 to about 200, while a long steel column may be assumed as one having a slenderness ratio greater than 200. A short concrete column is one having a ratio of unsupported length to least dimension of the cross-section not greater than 10. If the ratio is greater than 10 it is a long column. Timber columns may be classed as short columns if the ratio of the length to least dimension of the cross-section is equal to or less than 10. The dividing line between the intermediate and long timber columns cannot be readily evaluated. One way of defining the lower limit of long timber columns would be to set it as the smallest value of the ratio of length to least cross-sectional dimension that would just exceed a certain constant K of the material. Since K depends upon the **modulus of elasticity** and the allowable compressive **stress** parallel to the grain it can be seen that this arbitrary limit would vary with the species of timber. The value K is given in most structural handbooks.

If the load on a column is applied through the center of gravity of its cross-section it is called an axial **load**. A load at any other point in the cross-section is known as an eccentric load. A short column under the action of an axial load will fail by direct **compression** but a long column loaded in the same manner will fail by buckling (bending), the buckling effect being so large that the effect of the direct load may be neglected. The intermediate length column will fail by a combination of direct stress and bending.

In the middle of the 18th century a mathematician named Euler derived a formula which gives the maximum axial load that a long, slender ideal column can carry without buckling. An ideal column is one which is perfectly straight, homogeneous and free from initial stress. This maximum load, sometimes called the critical load, causes the column to be in a state of unstable equilibrium, that is, any increase in the loads or the introduction of the slightest lateral force will cause the column to fail by buckling. The Euler formula for columns is given below.

$$P = \frac{K\pi^2 EI}{l^2}$$

in which P = Maximum or critical load.
       E = Modulus of elasticity.
       I = Moment of inertia of cross-sectional area.
       l = Unsupported length of column.
       K = A constant whose value depends upon the conditions of end support of the column. For both ends free to turn $K = 1$; for both ends fixed, $K = 4$; for one end free to turn and the other end fixed $K = 2$ approximately, and for one end fixed and the other end free to move laterally $K = \frac{1}{4}$

Examination of this formula reveals the following interesting facts with regard to the bearing power of columns. First, that elasticity and not compressive strength of the materials of the column determines the critical load. Secondly, the critical load is directly proportional to the moment of inertia of the cross-section. The strength of a column may therefore be increased by distributing the material so as to increase the moment of inertia. This can be done without increasing the weight of the column by distributing the material as far from the principal axes of the transverse section as is possible consistent with keeping the material thick enough to prevent local buckling. This bears out the well-known fact that a tubular section is much superior to a solid section for column service. Another bit of information that may be gleaned from this equation is the effect of length upon critical load. For a given size column doubling the unsupported length quarters the allowable load. The restraint offered by the end connections of a column also affects the critical load. If the connections are perfectly rigid, the critical load will be four times that for a similar column where there is no resistance to rotation (hinged at the ends).

Since the moment of inertia of a surface is its area multiplied by the square of a length called the **radius of gyration**, the above formula may be rearranged as follows. Using the **Euler** formula for hinged ends, and substituting Ar^2 for I the following formula results:

$$\frac{P}{A} = \frac{\pi^2 E}{\left(\frac{l}{r}\right)^2}$$

$\frac{P}{A}$ is the allowable unit stress of the column, and the quantity $\frac{l}{r}$ is known as the slenderness ratio. This ratio is an important function of a column, since columns having a slenderness ratio below certain definite values (determined from the material of the column) cannot be considered to be in the Euler range, as the compressive stress may reach the yield point before failure by bending can occur.

Since the structural column is generally an intermediate length column and it is impossible to obtain an ideal column, the Euler formula has little practical value for ordinary design. Consequently, various empirical column formulae have been developed to agree with test data, all of which embody the slenderness ratio.

In architecture, the column is a round vertical structural member which may, however, sometimes be more ornamental than utilitarian. As such columns are frequently made of inelastic material, the above analyses of critical loads would not necessarily apply to all architectural columns. (C.W.C., F.T.M.)

COLUMN RESTRAINT. Column.

COLUMNAR STRUCTURE. Prismatic columns which develop in basic lavas, due to the stress-strain relationships set up by rapid chilling. A relatively frequent phenomenon in basalt flows, such as those of the Giants' Causeway, Ireland. (R.M.F.)

COLY, COLIES. Aves, Cuculiformes. *Colius*. African birds (**Aves**) of several species, also known as mouse birds. They have four toes, directed forward. The beak is strong and slightly curved, the head crested, and the tail long. (A.W.L.)

COLYMBIFORMES. An order of swimming and diving birds (**Aves**) with lobed toes. The grebes. (A.W.L.)

COMA. A state of complete unconsciousness from which the individual cannot be aroused even by powerful stimuli. It occurs following severe head injury, in cerebro-vascular accidents (**apoplexy**), in acute alcoholism, in uncontrolled **diabetes** (diabetic **acidosis**), in terminal **uremia**, and just before death in many diseases. Coma also follows overdosage with certain drugs, particularly central nervous system depressants like **morphine, barbital** derivatives, and **chloral hydrate.** (D.M.H.)

COMANCHEAN. Cretaceous.

COMB RIDGE. Cirque.

COMBINATION DIE. Die Casting, Press Work.

COMBINATION PRINCIPLE. The principle, first recognized by Ritz, that the many frequencies exhibited by the spectrum of a substance can be regarded as differences between a comparatively few terms characteristic of the substance, taken two at a time in their various possible **combinations**. Ritz's statement was quite empirical, but we now understand that these terms correspond to the different possible energy states of the atom or molecule, and that the much more numerous spectral frequencies correspond to "jumps" or transitions from one state to another with consequent release or absorption of radiation quanta. For example, if an atom had twenty possible energy states or "levels," the number of possible transitions releasing energy would be theoretically $20 \times 19/2! = 190$.

It does not follow, however, that all of the corresponding frequencies are actually found in the spectrum; some may be for some unknown reason apparently "forbidden," or of such rare occurrence as not to produce observable spectrum lines. When the principle is applied to certain **molecular spectra**, slight discrepancies are found which may be explained by assuming that some of the energy levels are not single but are close doubles. Such a discrepancy is known as a "combination defect." (See **Quantum Theory**.)   (L.D.W.)

COMBINATION SET.   Measurement.

COMBINATION TONES.
No acoustic or electroacoustic communication system or receiving apparatus is free from **distortion**. By this is meant that the wave form of the acoustic output is not strictly similar to that of the acoustic input, so that the quality of the sound is somewhat altered. (See **Musical Sounds**.) The human ear is no exception to this rule. The fact is that the vibrations received in the cochlea, with the consequent sensations, do not accurately represent the periodic variations of air pressure in the auditory canal, and that we hear subjectively sounds that have no external physical existence. The most important aspect of this auditory distortion is the fact that when two pure musical tones are sounded, which are not of too nearly the same pitch, many persons can distinctly hear other tones in addition to the two actual ones. These subjective sensations are called combination tones, from the fact that they correspond in pitch to tones having frequencies equal to the difference, to the sum, or to other simple combinations of the two actual frequencies.

By far the most conspicuous of these subjective tones is the difference tone. Thus if one plays on a piano the two notes e (333) and c (528), many listeners hear, in addition to these, a note in the key of g (195), in the octave below, which is equal to $528 - 333$. It was formerly supposed that the difference tone was the effect of **beats**, but Helmholtz detected sum tones and showed, further, that the difference tone sensation is too loud to be attributed to beats.

The cause of distortion in the case of the ear is believed to lie in the fact that the ear drum, or tympanic membrane, does not vibrate symmetrically. Mathematical analysis shows that if the tympanum moves farther inward than outward from its normal position during vibration, the waves passed on from it to the cochlea will contain other frequencies than those making up the original sound, and that in the case of a sound composed of two pure tones of frequencies n_1 and n_2, there will arise also the frequencies $n_1 - n_2$, $n_1 + n_2$, $2(n_1 - n_2)$, etc.

The existence of difference tones is of value in electroacoustic apparatus such as the telephone and amplifier systems, and also in the pipe organ, the deficiencies of which in the lower frequency range are partially offset by the low difference tones that the listener thinks he hears. There is evidence that the "striking note" of certain chime bells is really a difference tone and does not exist among the partials of the complex sounds of these bells. The occurrence of combination tones can hardly fail to be of importance also in the effect on the ear of chords in musical harmony.   (L.D.W.)

COMBINATIONS.
Each of the groups or selections which can be made by taking some or all of a set of things, without regard to order of arrangement, is called a combination.

The number of combinations of n things taken r at a time is denoted by $_nC_r$ or $C_{n, r}$ or $C(n, r)$.

The number of combinations of n different things taken r at a time is

$$_nC_r = \frac{n(n-1)(n-2)\cdots(n-r+1)}{r!} = \frac{n!}{r!(n-r)!}.$$

Also,

$$_nC_r = {_nC_{n-r}}.$$

The total number of combinations is:

$$_nC_1 + {_nC_2} + {_nC_3} + \cdots + {_nC_n} = 2^n - 1;$$

that is, the total number of combinations of n things taken 1, 2, 3, $\cdots$, n at a time is $2^n - 1$.

Combinations are closely related to **permutations**. Each combination containing r elements can yield $r!$ permutations by rearranging its elements in different orders. Hence, $_nP_r = r! \cdot {_nC_r}$.   (L.L.S.)

COMBINED STRESS.   Stress.

COMBINING WEIGHTS, LAWS OF.
The laws of combining weights are:

1. Fixed Proportions. The elements in a chemical compound are combined together in a fixed and definite proportion by weight.

2. Multiple Proportions. When two chemical compounds contain the same elements, then the fixed proportions in each are related to each other by a simple ratio. For example, in carbon dioxide 12 parts by weight of carbon element combined with 32 parts by weight of oxygen element, in carbon monoxide 12 parts by weight of carbon element combined with 16 parts by weight of oxygen element, therefore, in the two oxides the ratio of oxygen is 32:16 or 2:1 on the same carbon weight basis.

3. Reciprocal Proportions. The proportions in which two elements or radicals combine with a third element or radical are in a simple ratio to the proportions in which they combine with each other (if they combine) or with a fourth element or radical (Richter, 1792). Therefore, each element or radical has a characteristic equivalent weight.   (R.K.S.)

COMBUSTIBLE.
Any substance which will enter rapidly into chemical union with *oxygen*, and produce light or heat in useful quantities, is combustible. Practically all of the combustible fractions of useful commercial fuels, including solid, liquid, and gaseous fuels, are composed of carbon and hydrogen, either in the elementary or combined form (hydrocarbons). Even carbon monoxide (in water gas) and the alcohols are but oxygen compounds of carbon, and of carbon and hydrogen, respectively. Nearly 100% of natural gas or **fuel** oil is combustible. In **coal** there may sometimes be found a considerable amount of non-combustible matter, which is chiefly moisture and an aggregation of various mineral compounds known commonly as ash. Ash-free and moisture-free coal is known as combustible.   (F.T.M.)

COMBUSTION.
Combustion is the rapid chemical union of a **fuel** with **oxygen**, accompanied by the liberation of useful heat energy. (See **Reactions Involving Oxidation-Reduction**; and **Thermochemistry**.) When a coal is subjected to combustion conditions it first absorbs the heat necessary to bring it to the volatilization temperature. This will include both sensible heat and latent heat necessary to vaporize its moisture content. Further heating evolves the volatile hydrocarbons, leaving behind the fixed carbon and ash. It is the hydrocarbons that must be most carefully handled to insure freedom from smoke and attendant incomplete combustion. A major function of any furnace and stoker, grate, or burner installation is the proper mixing of oxygen with this evolved volatile material, and the further subjecting of the mixture to the ignition temperature, with sufficient time being given for complete combustion. The fixed carbon will then burn freely with an intense heat, and under correct conditions will be completely consumed to carbon diox-

ide. Each small piece of incandescent carbon has its surface atoms burned to carbon dioxide upon contact with an oxygen atmosphere. The piece will then be blanketed with carbon dioxide, an inert gas, unless additional air is applied in such a manner as to scrub the particles of carbon free of carbon dioxide and expose fresh surface to combustion. Stokers and grates accomplish this by permitting the motion of air past a stationary fuel bed at high velocity induced by draft. Pulverized coal **burners** are designed for the turbulent mixing of air and fuel, both of which are moving into the furnace about the same velocity.

The requirements for combustion are:

1. Thorough mixing of fuel and air in proportions which will insure complete combustion.

2. Exposure of fuel particles to oxygen throughout a period of time sufficient for their combustion.

3. Maintenance of the combustion zone at a temperature above that of ignition of the fuel.

The air required for combustion may be considered as composed of two parts: that which is used to supply oxygen for union with the fuel during the process of combustion; and that which is supplied in excess so that there will be a certainty of having an oxygen atom adjacent to a fuel atom when required. Combustion equations of typical fuel elements or compounds are shown in the accompanying table. The following equations illustrate how the information may be obtained. Taking the combustion of carbon to carbon dioxide for example:

Chemical reaction: $C + O_2 = CO_2$
Combining mols: $1 \text{ mol} + 1 \text{ mol} = 1 \text{ mol}$
Combining weights: $1 \times 12 \text{ lbs.} + 1 \times 32 \text{ lbs.} = 1 \times 44$ lbs.

Thus, each 12 lbs. of carbon require 32 lbs. of oxygen for complete combustion. Taking into consideration the fact that air has only 23.2% by weight of oxygen, a pound of oxygen is contained in $\frac{1}{.232}$ lbs. of air; consequently, in the combustion of carbon to carbon dioxide, 12 lbs. of carbon require $\frac{32}{.232}$ lbs. of air, or 1 lb. of carbon requires $\frac{32}{12 \times .232} = 11.5$ lbs. of air, theoretically.

Correspondingly, hydrogen needs 34.5 lbs. of air per lb., and sulfur 4.3 lbs. per lb. The air theoretically required to burn a pound of coal is

$$Air = 11.5C + 34.5\left(H - \frac{O}{8}\right) + 4.35S \text{ lbs. of air per lb. of coal.}$$

The symbols represent fractional proportions of the elements in the ultimate analysis. The $\left(\frac{O}{8}\right)$ correction allows for the hydrogen which is already combined and for that which will be burned with free oxygen in the fuel.

An account of the dynamic conditions surrounding furnace combustion, more air than might be indicated by the above equation will be necessary for complete combustion. The excess air is the ratio of the amount of air actually used over and above the theoretical amount to the theoretical amount. It varies from about 10% in large furnaces fired by burners to 40 or 50% in well-operated stokers, and 100% or more in hand-fired furnaces. Otto cycle engines operate on an excess of practically 0%, Diesel cycle engines may use 100% or more of excess air.

Since atmospheric air is composed of only 20.9%

COMBUSTION PROPERTIES OF FUELS

Fuel	Molec- ular Weight	Molecular Equation	Approx- imate Ignition Temper- ature, °F.	Lbs. of Air Required per Lb. of Fuel	Gravimetric Basis			Combining Weights	Heating Value, B.T.U. per Lb.	
					Products of Combustion, Lbs.					
					CO₂	H₂O	N₂		Higher	Lower
C to CO₂	12.0	$C + O_2 = CO_2$	752	11.5	3.67	....	8.8	1 lb. C + 2.67 lb. O₂ = 3.67 lb. CO₂	14,540	......
C to CO	12.0	$2C + O_2 = 2CO$	752	5.7	2.33*	....	4.4	1 lb. C + 1.33 lb. O₂ = 2.33 lb. CO	4,430	......
CO to CO₂	28.0	$2CO + O_2 = 2CO_2$	1253	2.4	1.57	....	1.8	1 lb. CO + 0.57 lb. O₂ = 1.57 lb. CO₂	4,380	......
H₂	2.016	$2H_2 + O_2 = 2H_2O$	1090	34.5	....	9.0	26.5	1 lb. H₂ + 8 lb. O₂ = 9 lb. H₂O	62,000	52,100
S	32.0	$S + O_2 = SO_2$	470	4.3	....	2.0†	3.3	1 lb. S + 1 lb. O₂ = 2 lb. SO₂	4,000	......
CH₄	16.03	$CH_4 + 2O_2 = CO_2 + 2H_2O$	1240	17.2	2.75	2.25	13.2	1 lb. CH₄ + 4 lb. O₂ = 2.75 lb. CO₂ + 2.25 lb. H₂O	23,850	21,375
C₂H₄	28.03	$C_2H_4 + 3O_2 = 2CO_2 + 2H_2O$	1124	14.8	3.14	1.29	11.4	1 lb. C₂H₄ + 3.43 lb. O₂ = 3.14 lb. CO₂ + 1.29 lb. H₂O	21,450	20,035

Fuel	Molec- ular Weight	Molecular Equation	Approx- imate Ignition Temper- ature, °F.	Cu. Ft. of Air Required per Cu. Ft. or per Lb.	Volumetric Basis (60° F., 14.7 Lb. per Sq. In.)			Combining Volumes	Heating Value, B.T.U. per Cu. Ft.	
					Products of Combustion					
					CO₂, Cu. Ft.	H₂O, Lbs.	N₂, Cu. Ft.		Higher	Lower
C to CO₂	12.0	$C + O_2 = CO_2$	752	151.7	31.7	....	120.0	1 lb. C + 31.7 C. F. O₂ = 31.7 C. F. CO₂	......	......
C to CO	12.0	$2C + O_2 = 2CO$	752	75.5	31.7*	....	59.7	1 lb. C + 15.8 C. F. O₂ = 31.7 C. F. CO	......	......
CO to CO₂	28.0	$2CO + O_2 = 2CO_2$	1253	2.4	1.0	....	1.9	1 C. F. CO + 0.5 C. F. O₂ = 1 C. F. CO₂	323	......
H₂	2.016	$2H_2 + O_2 = 2H_2O$	1090	2.4	....	0.048	1.9	1 C. F. H₂ + .5 C. F. O₂ = .048 lb. H₂O	328	275
S	32.0	$S + O_2 = SO_2$	470	56.8	....	11.9‡	44.9	1 lb. S + 11.9 C. F. O₂ = 11.9 C. F. SO₂	......	......
CH₄	16.03	$CH_4 + 2O_2 = CO_2 + 2H_2O$	1240	9.6	1.0	0.095	7.6	1 C. F. CH₄ + 2 C. F. O₂ = 1 C. F. CO₂ + .095 lb. H₂O	1004	900
C₂H₄	28.03	$C_2H_4 + 3O_2 = 2CO_2 + 2H_2O$	1124	14.4	2.0	0.095	11.4	1 C. F. C₂H₄ + 3 C. F. O₂ = 2 C. F. CO₂ + .095 lb. H₂O	1581	1476

  * Carbon monoxide.   † Sulfur dioxide.   ‡ C. F. Sulfur dioxide.   C. F. = cubic feet.

oxygen by volume, the remaining 79.1% (nitrogen chiefly, but with minute amounts of argon, carbon dioxide and other gases) is inert and merely absorbs heat, causing a loss of efficiency. From these proportions, and the fact that a cu. ft. of oxygen burns to a cu. ft. of carbon dioxide, we see that the theoretical maximum of carbon dioxide in stack gas will be 21% with perfect combustion, no excess air and no hydrogen in fuel. Actual coal-burning conditions lower this to a point at which 14% is considered to represent very good combustion. (F.T.M.)

COMBUSTION CHAMBER. An isolated region, arranged especially to promote the combustion of a fuel in it is a combustion chamber. The **furnace** is one type of combustion chamber. The space into which the piston of an internal combustion engine compresses the air, or air-fuel mixture, is called the combustion chamber. A description of engine combustion chambers should begin with a classification into those for spark ignition and those for compression ignition engines.

Spark Ignition Engines. Combustion chambers for these engines are comparatively simple. Details vary as to location of spark plugs and position of valves. As compression ratios were increased, prevention of **detonation** became a major problem. Much research has been carried out to determine in what manner detonation may be a function of combustion-chamber shape. This has resulted in various arrangements of spark-plug location, as well as special shaping of the interior curves of the cylinder head surface in order to promote turbulence, reduce length of flame travel from the spark, and secure good heat transmission from gas to cylinder head.

Compression Ignition Engines. Diesel-engine combustion chambers exhibit great differences, especially in the field of the high-speed automotive engine. The best enactment of the Diesel cycle of operations is secured at low engine speed but at the expense of heavy

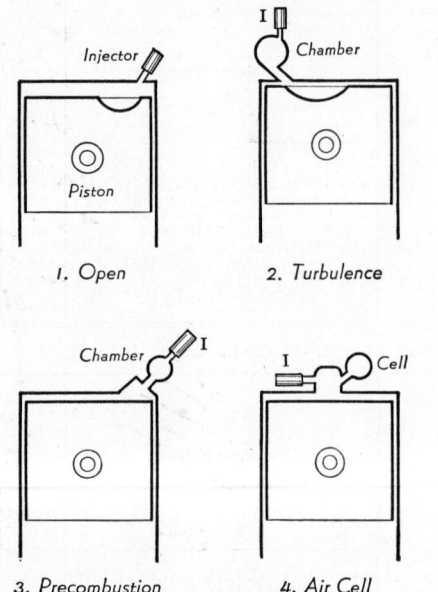

1. Open   2. Turbulence

3. Precombustion   4. Air Cell
Various types of Diesel combustion chambers.

construction. In the compression ignition engine only air is compressed by the piston, after which fuel must be injected into the air. The air pressure is high and a fuel jet of good atomization will have poor penetration, and consequently poor mixing with the air. The same would be true of a solid jet which, while possessing good penetration power would expose little surface and

fail to obtain mixing. In a slow-speed engine a fuel jet of moderate atomization and good penetration will be granted sufficient time for complete combustion without detonation in a plain or "open" combustion chamber. It is in high-speed engines that the variations in combustion chamber design are most noticeable. These might be classified as *turbulence precombustion,* and *air-cell* combustion chambers. Such cylinder heads have auxiliary chambers or pockets incorporated for the purpose of promoting a high degree of turbulence. The fuel spray is received in both the turbulence and precombustion chamber, but incomplete combustion predominates in the latter since approximately ⅓ of the air in the combustion chamber is driven into the precombustion chamber, whereas over ¾ of it is compressed into the turbulence chamber. However, in either case the fuel is sprayed into the auxiliary chamber, whereas the main chamber receives the fuel in an air-cell type. The air cells usually contain even less of the compressed air than a precombustion chamber will. As soon as piston travel lowers combustion chamber pressure, the air-cell contents are ejected into the main chamber, creating turbulence. (F.T.M.)

COMBUSTION CONTROL, AUTOMATIC. Good combustion of fuel in a furnace, efficiently, smokelessly, and at the desired rate, is difficult to achieve by manual adjustments if the desired rate of heat liberation is subject to variation. Systems of automatic combustion control have been developed which, when installed, relieve the furnace operator of the need of making some or all of those adjustments, depending on the degree of automatic control supplied.

As the steam boiler furnace furnishes an example of as complex a combustion control requirement as any, it is used as an example of the application of automatic combustion control. The control consists of simultaneous and proportionate variation of fuel and air to meet the load variability and is exercised through change of fan speed, damper position, and fuel feed. In order that a complex control system be most successful, application of the control regulators is given consideration at the time of choosing the drive for fans, fuel feeders, etc. Combustion control is of special benefit in plants where the highly variable demand makes it next to impossible to maintain a uniform steam pressure manually. Furthermore the carbon dioxide can be continuously maintained at the desired value. The ideal combustion control equipment should accomplish the following objectives:

1. Divide the load equally among boilers.
2. Regulate rate of firing in accordance with load.
3. Maintain steam pressure within predetermined limits.
4. Maintain proper proportioning of fuel to air.
5. Regulate the furnace vacuum.

A simple form of control wherein steam pressure automatically adjusts the draft is applicable to the smallest plants. The damper regulator substitutes mechanical control for the frequent opening and closing of dampers by competent firemen, or for the open position often maintained by poor firemen. The rate of fuel feed and adjustment of over-fire air remains under manual control; however, this phase of combustion control can be better judged by visible indication than may air supply.

A complete combustion control system is illustrated. It is necessarily restricted in use to large installations, where initial cost and operating conditions justify the expense of completely automatic control. The steam pressure is transmitted to a master regulator which modifies the signal supply medium (shown by dot-dash) to a variable signal (shown by dash). The individual controllers upon receiving this signal act to change motor speeds, damper openings, coal feeds, etc. During the process of initial adjustment of the system the controllers have been corrected so that the signal transmitted by the master regulator results in just the change in motor speed, damper position, etc., needed by the boiler unit at that particular rate of combustion. The signalling medium may be air under either plenum

tracted a great deal of attention. They were formerly regarded with superstitious awe and supposed to portend all sorts of calamities to the earth and its inhabitants. For these reasons the appearance of a comet

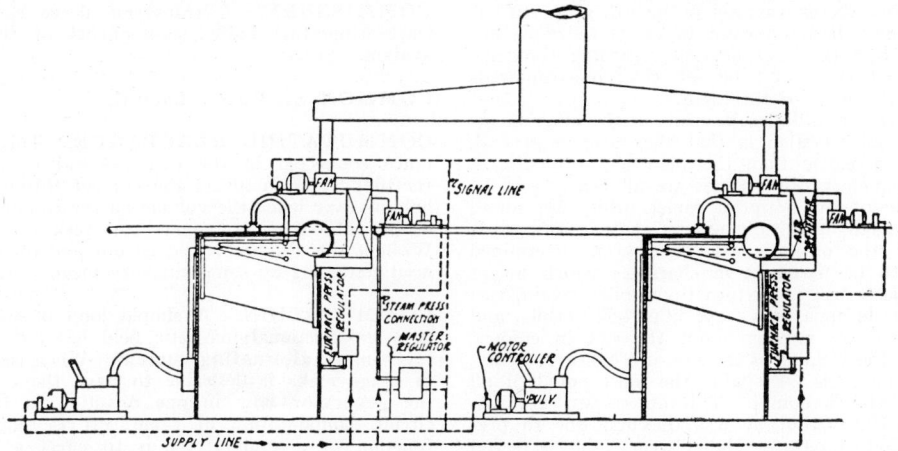

Elements of a fully automatic control system. (*Morse's Power Plant Engineering and Design.*)

or vacuum, liquid, or electricity. The furnace pressure controller is independent of steam pressure, for a predetermined vacuum should be maintained in the furnace at all ratings. (F.T.M.)

COMBUSTION JET VELOCITY. When fuels are burned in a combustion chamber, the exit of which is a well-shaped nozzle, the products of combustion attain a high jet velocity since combustion supports a chamber pressure of considerable magnitude. Like any other gas jet (see Jet, Gas), the velocity is a function of combustion chamber pressure. Rockets are burdened with the need of carrying an oxidizing agent as well as a fuel. The product of mass per second of the jet and its velocity is the principal factor in rocket action—the greater this product the more powerful the rocket. The total rocket weight for a shot of definite thrust and duration can be made smaller if the fuel used has a higher jet velocity.

The ideal maximum velocity for a fuel is based on the assumption that all the lower heating value is transferred into kinetic energy of the jet. A pound of hydrogen which yields up some 52,400 B.T.U. per lb. produces 9 lbs. of jet matter when completely burned with pure oxygen. Nine pounds moving at 17,100 ft. per sec. represents kinetic energy equivalent to this heat; consequently, that is the ideal maximum velocity for hydrogen. Other fuels have jet velocities as follows:

Fuel	Maximum Exhaust Velocity, Ft. per Sec.
Acetylene	15,900
Gasoline	14,500
Kerosene	14,500
Alcohol	13,300
Gunpowder	7,000
Smokeless powder	9,600

(F.T.M.)

COMET. Comets are objects which are moving in space under the influence of the sun's gravitational field and which occasionally come close enough to the earth to be observed. The name comes from the descriptive Latin phrase stellae comatae (hairy stars), and aptly describes the appearance of this class of celestial objects. Their appearance is so different from all other objects in the night sky that they have always at-

was always recorded by the ancient scribes and we have records of comet observations running back over more than a thousand years prior to the Christian era.

Only a very small proportion of the comets which are discovered ever become visible to the naked eye, and there must be many comets which are never discovered at all. From a study of ancient records there has been, on the average, about one naked eye comet per year. The year 1911, in which there were four such objects, apparently holds the record for maximum number, while there are many years in which no comet at all is visible to the unaided eye.

In a comet which becomes visible to the naked eye there are three general parts: the nucleus, the coma, and the tail. The relative sizes and appearances of these parts change radically as the comet approaches **perihelion.** It is believed that the nucleus (or head) of a comet is a mass of more or less condensed material. It appears much like an ordinary star when the comet is a long way from the sun. As the comet approaches the sun the nucleus increases in brightness and apparently shrinks in size, although this latter characteristic may be an optical illusion since actual measures of the diameter of the head are practically impossible. With the approach to the sun a hazy shell makes its appearance about the nucleus and this so-called coma increases in size. When quite close to the sun the coma seems to stretch out in the direction away from the sun and the tail develops in this direction. The real glory of a comet to the naked eye is this tail. It should be carefully noted that the tail points in a direction directly away from the sun and does not, in general, trail out behind the comet in its motion. The sizes and shapes of the tails of different comets vary in all characteristics. Many comets apparently never develop a tail at all. The tails which do develop have a variety of different shapes and curvature running from short, sharply curved tails, to long straight streamers pointing away from the sun.

Comets are the largest members of the **solar system.** The nucleus may have a diameter up to 10,000 miles, the coma diameters from 10,000–50,000 miles, and the tails have been observed with length as great as several hundred thousand miles. Hence, the volume of a large comet may be greater than the volume of all other members of the solar system combined. In contrast to the enormous bulk of these objects, their masses are so small that they have never been accurately measured.

The only available method for determining the mass of a comet is by means of the **perturbations** which they might produce in the **orbits** of the **planets** or **asteroids**. Many cases of very close approach of comets to objects of known mass have been observed but, while the comet orbit itself may be enormously perturbed, no perturbations have ever been observed in orbits of other objects. The best that can be said regarding cometary masses is that they must be less than one hundred-thousandth the mass of the earth.

Orbits of comets differ from orbits of the other members of the solar system in that they are, in general, much more eccentric than the planetary orbits, and, while the planetary orbit planes are all nearly parallel to the plane of the **ecliptic**, comet orbits are found inclined through practically every angle. The great majority of the orbits which have been determined are found to be parabolic in character which means that the objects will not return to visibility again, two or three objects apparently have hyperbolic orbits, and the remainder are moving about the sun in ellipses, returning to the vicinity of the sun and hence visibility, at periodic intervals. Probably the most noted of all comets is **Halley's comet**. This object was not discovered by Halley, but he was the first one to predict the period of return. Halley's comet has a period of approximately 75 years and has been observed and recorded at practically every return for over a thousand years.

Comparatively little is known regarding the composition and origin of comets. The **spectra** have been carefully studied over a period of years. From these observations we find that comets shine partly by reflected sunlight and partly by radiation from gases, largely hydrocarbon in character, which are excited to incandescence by radiant energy from the sun. There is no valid explanation as to why the sun's radiation should stimulate this activity in comets and not in any of the other members of the solar system. From a careful study of the orbits the preponderance of evidence points to the fact that comets are now and always have been members of the solar system and have evolved and developed with it.

A definite connection has been established between comets and **meteors**. In one well-authenticated case a comet was observed to be disintegrating at several successive returns and finally failed to reappear as a comet. In its place there was a meteoric shower. In several other cases we have meteoric **radiant points** following identical orbits with comets. These observations have led to the hypothesis that the head of a comet may be nothing more or less than a densely packed mass of meteoric material. Why this material should be stirred to activity by the sun's radiation so that it develops a coma or tail is not as yet explained.

There is practically no danger to the earth from comets. There used to be a belief that the tail of a comet contained poisonous gases which might cause wholesale death on the earth if the earth ever came close to the tail. Such belief is entirely without foundation for the gas is in such a highly diffused condition that, even if it were poisonous, a lethal concentration would be highly improbable. During the return of Halley's comet in 1910 the earth passed through the tail and not the slightest effects could be noted either in diminution of sunlight or change in the chemical constitution of the atmosphere. A direct collision with the head of a comet would undoubtedly have a very destructive effect over the portion of the earth where the collision took place. However, from the number of comets and the average orbital characteristics, such a collision is highly improbable. (w.k.g.)

COMMENSALISM. An association between individuals of different species which is beneficial to both but not indispensable. Ants and plant lice are sometimes associated in this way. The plant lice are guarded by the ants and sometimes carried to a good food supply and the ants receive the sweet honey dew secreted by their charges. (A.W.L.)

COMMISSURE. A transverse nerve cord or fiber tract connecting paired components of the **nervous system**. (A.W.L.)

COMMON SEWER. Lateral.

COMMUTATING REACTANCE. This is a reactance connected in the cathode lead of mercury-arc **rectifier** units to insure the current through the tube holding over when the voltage on the conducting **anode** drops until the next anode can pick up conduction. Without this the arc would go out and the tube would need restarting by some auxiliary means. (L.R.Q.)

COMMUTATION. A simple loop of wire rotating in a unidirectional magnetic field has induced in it a reversing or **alternating current**. When the conditions of usage make it desirable to have the current from the **generator** flow in one direction in the external circuit, commutation of some sort is required. The function of a commutator is to effect a reversal of the current induced in the winding of the generator at the proper point to produce a direct current in the connected circuit.

A single loop would generate a current which, when commutated, was pulsating in nature because of the varying rates at which the conductor, during rotation, cuts the lines of force of the **magnetic circuit**. The generator is composed of many such loops properly connected at their ends to the commutator. The commutator for a single loop would be simply a split ring with the two halves insulated not only from each other, but from the frame and shaft of the machine as well. In an actual practical generator having a multiplicity of windings, the commutator consists of a large number of segments of copper assembled around a hub which is attached to the shaft. The segments are thoroughly insulated from each other, usually with mica, and to them the ends of the armature coils are soldered. **Brushes** bearing against the commutator conduct the current away from the generator. (F.T.M., L.R.Q.)

COMMUTATIVE LAW OF ALGEBRA. Addition and **Multiplication**.

COMPANDOR. This is the term used to designate the combination of a volume **compressor** and a volume **expander**. (L.R.Q.)

COMPANION CELL. This is a very small elongate **cell** always found in close association with a **sieve tube**. It contains a dense **protoplasm** with a prominent **nucleus** and very small **vacuoles**. It is assumed that in some way they function with the sieve tubes.

For development, see **cambium**. (R.M.W.)

COMPARATIVE Anatomy, Physiology, etc. The comparative treatment in any division of biological science focuses attention upon a limited subject, such as **anatomy**, but introduces into its treatment data drawn from many species. This method of study has been applied widely to the vertebrates, hence comparative anatomy is likely to mean comparative anatomy of the vertebrates unless otherwise qualified.

The comparative method is valuable in determining the evolutionary development of organs and the relationship of species. (A.W.L.)

COMPARATOR. An instrument for the accurate measurement of moderately small lengths or distances. The feature common to various forms is a reading microscope or telescope arranged to travel along a scale, its axis remaining parallel to a fixed line. A typical form consists of a low-power microscope mounted on a carriage movable forward or backward by a **micrometer** screw. The two points or lines whose distance apart is to be measured are brought into the focal plane of the microscope, the cross-hairs of which are adjusted first upon one and then upon the other. For example, the distance to be measured may be that between two star images on an astrographic plate, or the images of two spectrum lines taken by a spectograph; in either case the plate is simply mounted on the stage of the comparator microscope. Another familiar type is the "cathetometer," consisting of a telescope sliding on a vertical scale and provided with a **vernier**. This instrument is used to measure heights of liquid columns, or other differences of level. (L.D.W.)

COMPASS. While this term is sometimes used to indicate an instrument for drawing circles (pair of compasses), we shall confine this article to descriptions of various instruments used for finding **direction on the surface of the earth**.

The oldest and most commonly used is the magnetic compass. The directive forces for this instrument are the horizontal component of **terrestrial magnetism** and, in a properly designed compass, the effect of the vertical component must be reduced as much as possible. The simplest form of the modern instrument is that used by surveyors and consists of a light, thin **magnet** (compass needle), pivoted so that it can turn freely about an axis perpendicular to its length, and with the north-seeking end clearly marked. A **compass card** is mounted in the plane parallel to that in which the compass needle can turn. In using this or any other type of magnetic compass, care must be taken to see that the instrument is level with the axis of rotation vertical. With the instrument properly adjusted, the north-seeking end of the needle will point toward compass north, and true **north** can be determined by the application of the **compass corrections** for deviation and variation. This simple instrument is perfectly satisfactory for surveyors or for any purpose where it can be mounted on a firm foundation. However, the needle is very sensitive and will swing back and forth for some time before settling down to compass north.

For use on ships and airplanes a more stable system than the compass needle must be employed. On seagoing ships the marine magnetic compass is used. This consists, essentially, of a large number of compass needles bound together in bundles, and these bundles attached by a system of fine, non-magnetic wires to the compass card. Various arrangements of the bundles of needles are employed by different designers in order that the maximum directive force may be obtained. This direction-seeking system is supported above its center of gravity on a jewelled pivot and is mounted in a heavy bowl of non-magnetic material known as the compass bowl. The compass bowl is covered with a heavy glass plate and is completely filled with liquid which reduces friction on the pivot by partially floating the moving system, and also tends to damp out short vibrations. The bowl is heavily weighted on the bottom and contains an expansion chamber to compensate for alterations in the volume of the liquid with changes of temperature. A conspicuous line, known as the lubber's line, is painted on the inside of the bowl. The bowl is mounted on gimbals in the ship's **binnacle** with the lubber's line parallel to the keel of the ship. An **alidade** is frequently used with the marine compass for taking **bearings** of distant objects for purposes of **piloting,** or of celestial objects for the determination of compass corrections. While the liquid tends to damp out short vibrations of the direction-seeking system, nevertheless, the inertia of the system

and liquid produces a slow oscillation of the compass card across the magnetic meridian, particularly when the ship is frequently changing headings or is in a heavy sea. This swinging is not too troublesome to the helmsman on an ocean-going vessel for he can watch the compass and obtain a good average reading.

In an airborne ship, where the pilot can only glance at the compass along with a number of other dials, this periodic swing is very troublesome. Furthermore, in airplanes the frequent and severe accelerations make the use of gimbals practically impossible and without them the compass card will frequently rub against the cover of the bowl. The magnetic compass first used by aviators has a floating direction-seeking system similar to that used in the marine type, but the bowl is deeper and the compass card is marked on the outside of a portion of a spherical surface with the graduation visible through a window in the side of the bowl. This instrument is set in the cowl of the plane in front of the observer with the 000° (north) graduation on the south-seeking side of the card so that when the plane is heading north (the pilot looking at the south side of the compass card in front of him) the compass will read 000°. Great care must be taken in using this instrument for taking bearings because of the 180° rotation of the graduations on the card. While this instrument is more convenient for aviators than the standard marine compass, nevertheless, it does swing across the equilibrium position with a slow periodic motion.

The aperiodic (without period) magnetic compass is an improvement on the standard marine instrument. Instead of the numerous bundles of needles, small and very strong magnets are used. These are suspended from the compass card by very fine wires which do not exert appreciable drag on the compass liquid. The whole direction-seeking system is very light and hence has small inertia. The fine wires produce eddy currents in the liquid and these damp out the troublesome swinging back and forth across the equilibrium position. The result is that the card swings quickly into the proper direction and comes to a stop. To avoid the necessity of continually reading the graduations on a moving card, the direction-seeking system carries a single heavy line parallel with the north-seeking direction, with the north end clearly marked. The degree graduations are on a grid ring which can be rotated until any desired graduation coincides with the lubber's line on the bowl. The grid ring carries two conspicuous wires relatively close together, parallel to each other, and in the direction of the 000°–180° graduations on the ring. To follow a desired compass **heading**, the pilot merely rotates the grid until the heading value is at the lubber's line. He then heads the plane so that the "needle" on the direction-seeking unit is parallel to the grid wires. In case a sudden maneuver or air pocket sets the plane off the desired heading, the pilot does not have to look at the graduations but merely brings the plane onto such heading that the "needle" is again parallel to the grid wires. In this, as well as other forms of magnetic compass which cannot be mounted in gimbals in a plane, banking on turns brings the vertical component of the earth's magnetic field into play. (See northerly turning error under **Compass Corrections.**)

The gyro-flux-gate compass is a recent development of a compass which can be used in modern fast and quick maneuvering planes. Two fundamentally important developments are incorporated in this instrument. In the first place, a gyroscopic stabilizer holds the direction-sensitive element horizontal during maneuvers of the plane. In the second place, an **electronic amplifier** increases several hundredfold the strength of the horizontal component of the earth's magnetic field. This increase in field strength is sufficient to permit the use of the gyro-flux-gate **compass** within 300 miles of the earth's magnetic poles, whereas the ordinary magnetic compasses are unreliable within 1200 miles of these points. Furthermore, the field strength is sufficient to allow the use of repeaters in various locations about the plane. A repeater is a compass card and lubber's line, whose directive force is obtained by a complicated electrical system from a master compass at a distance. An example of the importance of these repeaters is found in a large bombing plane where the pilot, navigator, and bombardier, all have instruments giving identical readings for the heading of the plane. If separate compasses were used in each location, the chance that they would all agree in reading would be very small, because of unavoidable differences in deviation correction.

The earth-inductor compass was designed a number of years ago for use on aircraft, but has been rendered virtually obsolete by the aperiodic and gyro-flux-gate instruments.

All of the above types of compass depend upon a magnetic field for directive force. This field in practice is the resultant of the earth's magnetic field and the magnetic field of the ship. Compass corrections which are frequently variable must always be used with these instruments.

The gyro-compass was developed for the purpose of avoiding the troublesome corrections necessary with the magnetic compass. The first practical installation for seagoing vessels was completed about 1911, and during World War I the gyro-compass was perfected. The detailed theory is too complicated to be undertaken here. The essential feature is a heavy gyroscope driven at high speed by electric power. The frame of the gyro rotor is mounted in gimbals and a "ballistic tube" is attached which causes the axis of rotation of the gyro to set itself parallel to the earth's axis of rotation provided the instrument is stationary or moving in the east-west direction. Since the axis of rotation of the earth is in the plane of the meridian, the axis of the rotor will point true north on a stationary ship. A compass card may be attached to the gyro, but usually the gyro is used to operate repeaters in various locations about the ship.

On modern ships a master gyro is installed in some well-protected part of the ship and repeaters are located where they will be of greatest value. The electrical circuit which operates the repeaters may be used for a variety of other purposes, such as keeping a constant record of the heading of the ship, and for operating the steering mechanism so that the ship will be held on a predetermined heading. The application of this automatic steering device, frequently referred to as "metal Mike," does not eliminate the necessity of having a helmsman constantly on duty. In cases of emergency such as the appearance of an unexpected obstruction to navigation, the "metal Mike" must be disconnected and the ship steered by hand. A good helmsman can anticipate the action of waves and by constant vigilance and long practice can "touch" the helm in time to prevent a large wave from forcing off the bow. "Metal Mike" has no such powers of observation and the bow would be forced off by a large wave. The gyro would immediately swing the ship back toward proper heading and the next wave would throw the ship off again. The result is a continual strain on the hull of the ship which a good helmsman can avoid. On most ships at present, the steering is done by hand, using a gyro repeater in the binnacle.

On airplanes the rapid accelerations of the ship prevent the use of a true gyro-compass. However, a gyro pilot device may be used to hold a plane on constant heading and altitude, when these factors have been set by standard instruments.

While the gyro-compass is not subject to the compass corrections necessary with the magnetic instrument, nevertheless, it is subject to other corrections when the ship is in motion. These corrections depend on the course and speed of the ship, and the latitude in which the ship is located. On modern gyro-compasses, devices are incorporated which, when properly set for course, speed, and latitude, will automatically correct the readings of all repeaters to true values. The gyro-compass is a complicated piece of equipment and the multitude of electric contacts and connections frequently cause trouble. Every ship, no matter how complete a gyro-compass installation is aboard, must carry a standard magnetic compass for which the corrections are known at all times, so that it can be used whenever the gyro-compass is out of order. (W.K.G.)

COMPASS CARD. The compass card is a light cardboard or paper disk or annulus which is securely attached to the directing system of a mariner's com-

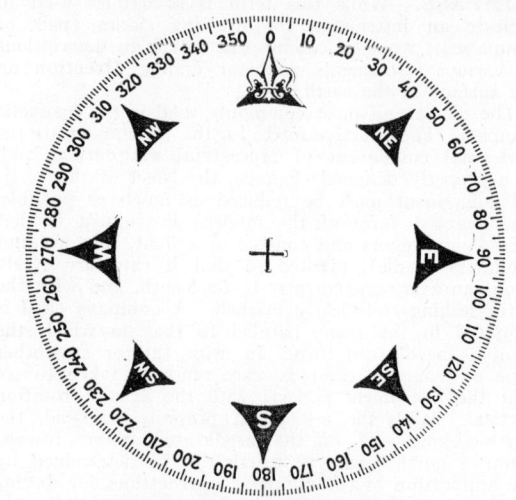

Compass Card. U. S. Navy Standard.

pass. On this card are painted the "points" and other systems of markings for reading the compass.

During the sailing-ship era, when it was difficult for a helmsman to hold the ship within two or three degrees of a given heading, the point and quarter-point system of marking was sufficiently accurate for graduating the compass card. With the increase in ease of steering, the degree system, such as had long been in use on surveyor's compasses, made its appearance on the mariner's compass. Two such degree systems are in common use. One system has zero both at the north and south points of the card and reads in degrees both right and left from the zero's to the east and west points. In such a system the north-west point would be referred to as north 45° west and the north-east point as north 45° east. The present standard U. S. Naval system of marking has north marked zero and reads to the right through the east through 360°. The marking of the card for a naval ship's compass is shown in the accompanying figure. On such a system the north-west point is referred to as 315° and the north-east as 045°. (W.K.G.)

COMPASS CORRECTIONS. The directive force of any magnetic compass is the horizontal component of the magnetic field in which the instrument is located. This magnetic field is the resultant of the field of terrestrial magnetism and any other fields in the immediate vicinity. The direction in which the north-seeking end of the compass points is known as compass north, and to convert directions indicated by the magnetic compass to those referred to true north, the compass corrections must be applied.

The angle between compass and magnetic **north** is known as deviation. It is named east or west depending upon whether the north-seeking end of the compass is east or west of magnetic north. Deviation is the resultant of a large number of local magnetic fields and these are subject to unpredictable changes. A steel ship has a strong magnetic field that is set up in the hull at the yard where the ship is constructed. The direction of this field relative to the keel at the time of launching will depend upon the orientation of the keel during construction. At sea, the direction of this field relative to the keel of the ship may be altered by heavy pounding of the ship by waves, by salvos of gun fire, by severe temperature changes, and a variety of other causes. In addition to the magnetic field of the hull itself, other fields are introduced by such things as cargo, electric currents, degaussing coils, etc. Since deviation is produced by **magnetic** fields connected with the ship, its value will change with different headings. The ship's **binnacle** contains a number of movable magnets which may be adjusted to reduce deviation to a minimum value. In port, a professional compass adjuster will swing the ship (put the ship on various headings completely around the compass) and determine the deviation on each heading. He then adjusts the magnets in the binnacle. After the compass adjusting is complete the adjuster will issue a calibration card or table showing the value of the deviation for each 15° of heading of the ship. At sea, the navigator must check this correction on different headings whenever possible by taking compass **bearings** of distant objects of known **azimuth**, or of celestial objects and computing their azimuths. In case the observed values do not agree with those given on the calibration card, the navigating officer must swing the ship and redetermine the entire curve just as soon as possible.

Correcting the compass heading for deviation will give the magnetic heading of the ship. This must be further corrected for variation to obtain the true heading. The angle between magnetic north and true north is known as variation. Variation is named east or west depending on whether magnetic north is east or west of true north. (N.B. Unfortunately a confusion of terminology is in common use for describing this compass correction. Most surveyors, some geologists, and a few other scientists use the term **declination** instead of variation for this angle. The standard practice in the American air services, and in the U. S. Navy and Merchant Marine, is to use the term variation.) Variation differs from place to place on the surface of the earth and from day to day in any particular locality. As a result of long and painstaking research on terrestrial magnetism, it is possible to plot on **charts** the value of the variation for any locality at some particular date and the rate of change of variation. Curves, known as isogonic lines, may be drawn through places of equal variation. Whenever old surveys are being used, it is important to determine whether the directions given are magnetic or true. If directions are given as magnetic, it is necessary to obtain the variation for the year of the survey to obtain the true directions.

The standard U. S. Navy practice in correcting directions from compass to true is to first apply deviation and then variation. East deviations and variations are marked plus, and west values are marked minus. With this convention in signs we have compass + deviation = magnetic. Magnetic + variation = true. The summations are to be made with due regard of the algebraic signs. A convenient phrase for remembering this method is *Can Dead Men Vote Twice.* There are many other conventional methods employed by different services such as Army Air Corps, Civil Aeronautics Administration, etc., for compass correction. No one of these can be called "the best method." The im-portant thing is to know what you are doing and do it properly.

The above corrections are based on the assumption that the only force effective on the direction-seeking unit of the compass is the resultant of the earth's field and the fields of the ship. In all airplane compasses, other than the new gyro-flux-gate type, banking the plane brings the vertical component into play. If the plane is turning there will also be a centrifugal effect due to the turn. These two factors are in opposite directions and are of greatest magnitude in the northern hemisphere when a plane is turning from a north heading. Under certain conditions the compass may indicate a right turn when the plane turns left, and sometimes may indicate first a left and then a right turn. There are no set rules of correction for this so-called northerly turning error, but lives have been lost because of the failure of pilots to realize and allow for it. Certain gyroscopic devices have been developed to indicate the actual turning of a plane, but the magnetic compass should never be used for this purpose. (W.K.G.)

COMPASS PLANTS. Several plants are called compass plants. It is characteristic of them that all their leaves extend in a north-south direction, regardless of the position in which the leaves are borne on the stem. This arrangement of leaves is brought about by the bending of the petioles of the leaves. *Silphium Scariola,* a relative of garden lettuce, exhibits this peculiar habit when grown in a dry open place. This uncommon response of the leaves protects them from excessive sunlight at midday, but assures maximum exposure in early morning and late afternoon, when radiation is not so intense. (R.M.W.)

COMPENSATION SAC. A reservoir opening to the exterior in certain **Bryozoa** which fills with water when the tentacles are extruded and empties when they are retracted. (A.W.L.)

COMPENSATOR. While this term may serve to designate any device used for compensation, there are two specific applications of its use.

In physics the term compensator is ordinarily restricted to an arrangement for measuring the phase difference between the two components of elliptically **polarized light.** This is accomplished by introducing a known, opposite phase difference of equal magnitude, which reduces the existing phase difference to zero. The

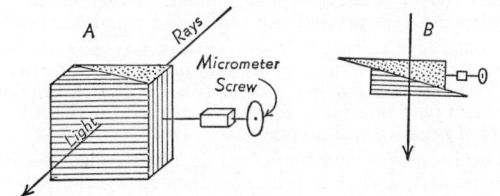

A. Diagrammatic sketch of Babinet Compensator: angle of wedges much exaggerated. Hatching and strippling indicate direction of crystal axes. B. Shows wedges displaced.

most familiar form, devised by Babinet, consists of two quartz wedges, with thin optic axes at right angles to each other. When passed through this apparatus and a **Nicol prism** set to extinguish light plane-polarized at 45° to either axis, any given elliptically polarized light produces a system of parallel dark bands. Plane-polarized light is first used (zero phase difference), then the elliptic light of unknown phase difference, and the relative displacement of the wedges necessary to restore the bands to their original position gives the phase difference required.

Another use of the term compensator is in electrical engineering. When a motor is to be started other than by direct connection to the supply line, some means must be incorporated in the circuit to limit the current taken during the starting period. A method frequently used to start a-c motors which are too large to be directly connected for starting, is to decrease the impressed voltage during the starting period. The term compensator usually refers to the automatic transformers, switches, and wiring employed to limit the starting current for the squirrel-cage type of a-c motor. This equipment is enclosed in a case through which projects an operating handle which has three positions, "off," "starting," and "running." This handle controls a double-throw switch. When in starting position, the switch is thrown so as to connect the motor to the line through an auto-transformer, which reduces the voltage to an amount suitable for starting purposes. When in the running position, the switch connects the motor directly to the line through fuses, and full line voltage is impressed upon the motor windings. (See **Starter, Motor.**) (L.D.W., F.T.M.)

COMPETENT. As defined in terms of **structural geology** see **Incompetent.** (R.M.F.)

COMPLEMENTAL MALE. 1. In certain **barnacles** which are normally **hermaphrodite** a few individuals lack female organs and are called complemental males. 2. In colonies of white ants, termite reproduction is carried on principally by a highly specialized king and queen. Other sexually mature individuals resembling the immature insects are known as the second reproductive caste. The males of this caste are complemental males. (A.W.L.)

COMPLEMENTARY COLORS. Color.

COMPLEMENTARY FUNCTION OF A LINEAR DIFFERENTIAL EQUATION. Linear Differential Equations.

COMPLEX FRACTIONS. A complex **fraction** is one whose numerator or denominator or both are fractions or mixed expressions. A complex fraction may be reduced to simpler form by multiplying every term of the numerator and denominator by the **least common multiple** of the denominators of the subsidiary fractions. (L.L.S.)

COMPLEX NUMBERS. An imaginary number is a square root of a negative real number. Every imaginary number can be expressed as a product of a real number by the so-called imaginary unit $\sqrt{-1}$; thus, $\sqrt{-4} = 2\sqrt{-1}$.

Sometimes the term imaginary number is used for what we call complex numbers; to emphasize the distinction, the term pure imaginary is sometimes used for the square roots of negative real numbers.

The imaginary number $\sqrt{-1}$ is often called the imaginary unit, and it is generally denoted by the letter i for brevity; its fundamental property is expressed by $i^2 = -1$. In electrical engineering work, the letter j is often used in place of i for the imaginary unit.

The successive powers of the imaginary unit are: i, $i^2 = -1$, $i^3 = -i$, $i^4 = +1$, $i^5 = +i$, $i^6 = -1$, $i^7 = -i$, $i^8 = +1$, and so forth in cycles of four.

Complex numbers are numbers of the form $a + bi$, where a and b are real numbers and i is the imaginary unit; they consist therefore of a real part a and a pure imaginary part bi.

The complex numbers $a + bi$ and $a - bi$ are said to be conjugate to each other.

Two complex numbers are defined as equal when their real parts are equal and their imaginary parts are equal. When any two expressions involving real and imaginary parts are equal, we may equate their real and imaginary parts separately.

The four fundamental operations with complex numbers are expressed by the formulae:

$$(a + bi) \pm (c + di) = (a \pm c) + (b \pm d)i,$$

$$(a + bi)(c + di) = (ac - bd) + (ad + bc)i,$$

$$\frac{a + bi}{c + di} = \frac{ac + bd}{c^2 + d^2} + \frac{bc - ad}{c^2 + d^2} i.$$

That is, to add (or subtract) complex numbers, add (or subtract) the real and imaginary parts separately; to multiply two complex numbers, multiply them as though they were ordinary binomials and substitute $i^2 = -1$; to divide two complex numbers, multiply numerator and denominator by the conjugate of the denominator.

Since a complex number $x + yi$ depends on two real numbers x and y, complex numbers require a 2-dimensional field, such as the plane, for their graphic representation.

Argand's diagram represents a complex number as $x + yi$ graphically by the point whose **rectangular coordinates** are x and y, or by the corresponding **vector** from the origin to this point, as shown in Fig. 1.

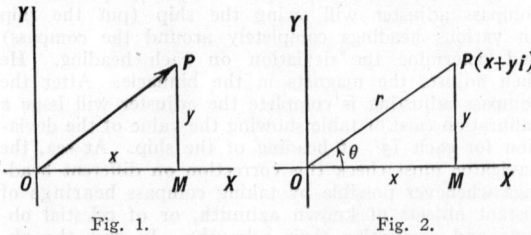

Fig. 1.   Fig. 2.

Let the complex number $x + yi$ be represented graphically by the point $P(x, y)$ in rectangular coordinates, by Argand's diagram, and let the **polar coordinates** of P be (r, θ). (See Fig. 2.) Then $r = OP = \sqrt{x^2 + y^2}$, and $\theta = \angle XOP = \arctan(y/x)$. Then the radius vector r is called the absolute value (or modulus) of the complex number and is denoted by the symbol $|x + yi|$, and the vectorial angle θ is called the amplitude (or argument) of the complex number and is occasionally denoted by $am(x + yi)$. The complex number $x + yi$ may then be represented in terms of r and θ by the expression

$$r(\cos \theta + i \sin \theta),$$

(sometimes abbreviated into $r \operatorname{cis} \theta$), which is called the polar (or trigonometric) form of the complex number.

The product and quotient of two complex numbers in polar form are expressed by the formulae:

$$r_1(\cos \theta_1 + i \sin \theta_1) \cdot r_2(\cos \theta_2 + i \sin \theta_2) =$$
$$r_1 r_2 [\cos (\theta_1 + \theta_2) + i \sin (\theta_1 + \theta_2)],$$

$$\frac{r_1(\cos \theta_1 + i \sin \theta_1)}{r_2(\cos \theta_2 + i \sin \theta_2)} = \frac{r_1}{r_2} [\cos (\theta_1 - \theta_2) + i \sin (\theta_1 - \theta_2)].$$

(L.L.S.)

COMPOSITE BEAM. A **beam** which is composed of two materials properly bonded together and having different modulii of elasticity (see **Modulus of Elasticity**) is called a composite beam. The reinforced concrete beam and wood beams reinforced with steel plates are typical examples.

The analysis of composite beams depends on the assumption that a plane section before bending remains plane after the load is applied. Therefore the two materials must be connected in such a way that they will act as a unit. This condition is realized in the reinforced concrete beam by means of the bond (see **Bond Stress**) between the reinforcing rods and the concrete. In the case of reinforced wood beams the parts are connected by bolts properly spaced to resist the shearing forces (see **Shear**) between the plates and the beam.

The flexure formula is applicable to composite beams if the beam is transformed into an equivalent homogeneous section by means of the transformed area method which is found in texts on strength of materials and reinforced concrete design. (C.W.C.)

COMPOSITE COURSE. The shortest **track** between two points on the surface of the earth is a great circle if we neglect the slight oblateness. However, the following of such a track has two fundamental disadvantages: 1. Such a course is a **rhumb line** only in the particular cases of two points both on the **equator,** or two points on the same **meridian** of longitude. 2. Such a course will frequently lead the vessel into impossible positions (e.g., if the two points are in the same **latitude** but differ by 180° in longitude the great-circle track between them would lead over the nearest pole).

A composite course is a combination of great-circle and rhumb-line courses designed to carry a ship from one point to another by the shortest practicable path. In case the great-circle track does not lead the ship into impossible positions the composite course is usually a series of rhumb-line courses to successive positions along the great-circle track, the rhumb-line distances so figured that the course of the ship will be altered at convenient times (e.g., the changing of the watch). In case the great-circle course leads the ship into danger the problem of computing the composite course is one of a number of compromises which are different for every problem. A good example of such a composite course may be obtained by examining the steamer lanes across the Atlantic Ocean which will be found on many terrestrial globes. (W.K.G.)

COMPOSITE FAMILY. Compositae. This is the largest family within the plant kingdom and contains over twelve thousand species in over eight hundred genera. Furthermore, it represents the highest evolutionary attainment among dicotyledonous plants. Composites are mostly herbaceous plants, distributed practically all over the earth. The few members which are shrubs or trees are largely limited to tropical regions, especially to island floras. The family is divided into two groups, distinguished by the nature of its flowers. The plants of one group have latex vessels usually containing a white **latex,** those of the other group lack latex, but in most cases contain oil-canals with acrid or bitter watery contents.

The leaves of most composites are alternate, often entirely radical, although in a few cases they are opposite, as in the sunflower, *Helianthus annuus,* or whorled,

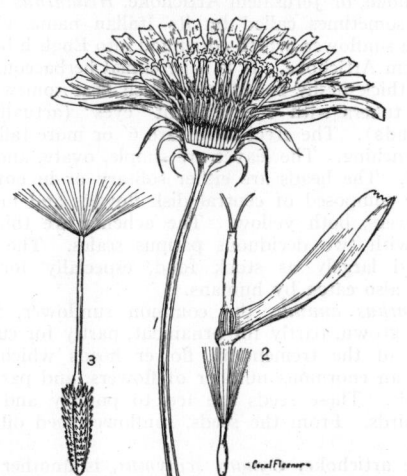

Flower-head and flowers of the dandelion, *Taraxacum.* 1, the inflorescence or head, composed of many flowers upon a flattened stem. 2, a single flower more enlarged. 3, a single fruit.

as in species of *Eupatorium.* Stipules are seldom found in this family. The root is most commonly a tap-root, frequently much thickened, as in the dandelion. The inflorescence is of the type known as a head, or capitulum. Often the heads are aggregated in larger inflorescences of various types, as panicles or cymes or spikes. Commonly the single head is inaccurately regarded as a flower, rather than as an inflorescence. Surrounding the head is a group of bracts, making up the involucre. These **bracts** are usually green and serve to protect the flowers before they are mature and also to protect the maturing fruit. The flowers of a head are arranged on the enlarged end of the stem or axis, called the receptacle. This receptacle may be flat and disk-shaped, as in the common sunflower, conical as in the yellow daisy, *Rudbeckia hirta,* or otherwise. It may be smooth or covered with hairs or scales.

The individual flowers show considerable difference in structure. In many species the flowers of a head are

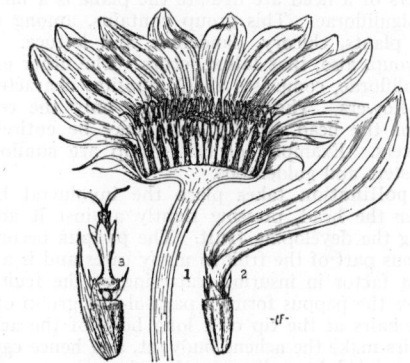

Flowers and flower-heads of a sunflower, *Helianthus.* 1, the flowerhead, cut so as to show the relation of the ray and disk flowers to the end of the stem. 2, a single ray-flower. 3, a single disk flower.

all alike and all perfect. In other species they are of two kinds, one called ligulate and the other tubular. Ligulate flowers, also called ray-flowers, are irregular, but bilaterally symmetrical; tubular flowers, also called disk flowers, are regular. Both types occur in the head of the sunflower, where the tubular or disk flowers occupy the larger part of the receptacle, the ray-flowers being the conspicuous yellow flowers forming a ring around the periphery of the receptacle, just inside the involucral bracts. Each disk flower is perfect and regular. The **calyx** appears in different genera as bristles, barbs, scales, or teeth, and sometimes is completely lacking. It is known as the pappus and occurs at the apex of the inferior **ovary.** Often the pappus becomes a very important structure in the dissemination of the fruit. The **corolla** is tubular and 5-lobed, and inserted on the apex of the ovary. The short filaments of the five **stamens** are inserted on the base of the corolla tube, while the **anthers** are attached to each other by their edges, forming a tube which surrounds the style. The pollen is discharged into the anther tube. The single **pistil** has an inferior ovary containing a single erect **ovule,** a simple style which splits into two parts, the inner surfaces of which are stigmatic. The fruit is usually of the type called an achene. Nectar is secreted in a ring-shaped **nectary** which surrounds the base of the style, located at the base of the tubular corolla. This nectar attracts insects, but only those with mouth parts sufficiently long to reach to the bottom of the corolla tube can obtain it. When the flower opens the **pollen is** shed into the anther tube. At the base of this tube the style, as yet unforked, occurs. This style elongates, pushing like a ramrod against the pollen above it, and causing an accumulation of pollen at the upper end of the anther tube. Insects seeking the nectar necessarily come in contact with the pollen masses, which are thus likely to be transferred to another flower as the insect goes about collecting nectar. However, should insect-pollination fail, self-pollination is assured in many species, by the behavior of the styles, which emerge from the anther tube, protrude considerably, and then split and coil backwards so that the inner stigmatic surface rolls down into contact with the pollen. In some cases this self-pollination is almost the only method. Many species are self-sterile, thus requiring cross-pollination by insects.

The ligulate or ray flowers differ from the disk flowers in having the corolla of five united petals forming a tubular base which gradually emerges into a flat strap-shaped lateral structure with five teeth at its tip. Growing from the tube of the corolla are the small **stigma** and the five coalesced stamens. In many species where the rays flowers are marginal in the head they are entirely sterile or are pistillate.

The Composites are divided into two groups. If all the flowers of a head are ligulate the plant is a member of the Liguliflorae. This group contains, among other common plants, chicory, dandelion, and lettuce. It is in this group, also, that **latex** occurs. The other group, the Tubuliflorae, comprises all Composites characterized by disk flowers. These may occupy only the central portion of the head, or the latter may be entirely of disk flowers. Examples of this group are sunflowers, daisies, asters and goldenrods.

After **pollination** takes place the involucral bracts close over the head, pressing tightly against it and so protecting the developing **fruit**. The **pappus** becomes a conspicuous part of the fruit in many cases and is a very important factor in insuring scattering of the fruit. In some cases the pappus forms a parasol-like group of fine radiating hairs at the tip of a long beak of the achene. These hairs make the achene buoyant, and hence capable of being carried long distances by air currents. Often the base of the hair is **hygroscopic**, responding to changes in moisture, so that the parasol-like structure opens and closes with decrease or increase of humidity, which may help to loosen the achene from the receptacle, and also to shove it along over the surface of the ground, or to push it into a crack in the soil. In *Bidens,* often called beggar-ticks or beggar-lice, the pappus is in the form of stiff usually downwardly barbed bristles which catch into the hair of passing animals or the clothing of man and so are carried about, finally breaking off and falling to the ground. In the burdock, *Arctium,* the involucral bracts form recurved hooks which serve in the same way, the entire head often breaking off and being transported. In thistles the pappus takes the form of a tuft of long silky hairs which enable the achene to float readily through the air. In many cases members of the composite family have no special pappus development, seed dispersal being entirely accidental.

The reasons which suggest that the Composite Family is highest in rank are found in the massing of the individual flowers into a compact head surrounded by protecting bracts, the structure of the individual flower with its inferior ovary, its united petals forming a corolla tube, the united anthers forming the anther tube, and the reduction of the number of carpels to two with but one ovule developing.

For so large a family comparatively few of its species are used by man. Several are extensively cultivated for ornament, the flowers often becoming very large and double and extremely showy, as in the case of the *Chrysanthemum* and *Dahlia.* Less showy ornamentals are the *Aster, Bellis, Tagetes* and *Calendula,* among others. Some composites yield oils or other substances useful to man, as *Arnica, Artemisia, Tanacetum, Calendula, Chrysanthemum* and *Helianthus.* A few are used as food, such as Lettuce, Artichokes, Endive, Chicory, Salsify and Dandelion.

Lettuce, *Lactuca sativa,* is an annual herb which is native to Europe, Asia and northern Africa. The plant is very leafy and contains a milk-white **latex.** In young plants the leaves are crowded on a short stem, forming a close rosette; in older plants the stem elongates greatly and bears a panicle of heads of yellowish flowers. The achenes are flat, ribbed, and contracted into a slender beak bearing numerous soft white or brownish pappus hairs which radiate outward like a parasol. Many varieties have been developed in cultivation. Lettuce is used almost entirely as a salad plant.

Endive, *Cichorium endivia,* is an annual or biennial herb having many basal leaves which in cultivated forms have become very much dissected and crisped. Mature plants are tall-stemmed and have purple, rarely white, flowers, all ligulate. The plant is used either as a salad plant or as a pot-herb. It is a native of India, and has long been cultivated in European countries. It is becoming more popular in the United States.

Chicory, *Cichorium intybus,* also called succory, is one of the many European plants which has become a persistent weed on introduction to North America. It is a perennial plant having a deep tap-root and a stiff tough stem two or three feet tall. Root leaves are numerous, forming a dense basal rosette; stem leaves are small, of various shapes, and clasping the stem. The flowers are usually blue, sometimes pink, or white. The plant is used as a salad plant or as a pot-herb, often being mistaken for dandelion. Its roots have been dried, ground and roasted and used as a substitute for coffee. There are several varieties in cultivation, one of which, Witloof chicory, finds considerable favor as a salad plant.

Salsify, *Tragopogon porrifolius,* or Oyster plant, is a plant indigenous to southern Europe. It is a hardy biennial having a thick tap-root 8–12″ long and an inch or two in diameter. This root is formed during the first year of growth when it bears a crowd of leaves. During the second year's growth the rather succulent branching stems grow up two or three feet tall. The stem leaves are alternate, entire and clasping, and have a smooth waxy surface. The flower heads are borne on long hollow stalks, or peduncles, and have purple flowers, all ligulate. The achenes are linear and have a long slender beak with radiating pappus hairs at its tip. The roots of the plant have a flavor suggestive of that of oysters, and are used as a cooked vegetable. A yellow-flowered species also occurs, and has been widely introduced into the United States.

Scolymus hispanicus, another southern European member of the Composite Family, is also known as an Oyster plant, or salsify, usually being designated as Spanish salsify. *Scorzonera hispanica,* Composite Family, is black salsify.

Artichoke, or Jerusalem Artichoke, *Helianthus tuberosus,* is sometimes called by its Italian name, girosole, meaning sunflower, which corrupted into English becomes Jerusalem Artichoke. It is a perennial herbaceous plant having thick fleshy rootstocks which bear somewhat irregular tubers with very evident "eyes" (actually dormant **buds**). The erect stems are 6′ or more tall, stout and branching. The leaves are simple, ovate, and long-petioled. The heads are either solitary or in **corymbs,** and are composed of central disk flowers and marginal ray flowers, both yellow. The achenes are thick and hairy, with two deciduous pappus scales. The tubers are used largely as stock food, especially for hogs, but are also eaten by humans.

Helianthus annuus, the common sunflower, is frequently grown, partly for ornament, partly for curiosity because of the tremendous flower heads which often contain an enormous number of flowers, and partly for the seeds. These seeds are fed to poultry and larger caged birds. From the seeds, sunflower seed oil is expressed.

Globe artichoke, *Cynara scolymus,* is another Composite which is grown for food. It is a herbaceous perennial native in northern Africa. The flowers form large globular heads surrounded by several rows of fleshy

bracts. The basal portions of each bract and the thick fleshy receptacle are cooked and eaten. The blanched leaves of a related species, *Cynara Cardunculus,* or cardoon, are often eaten like celery.

Dandelion, *Taraxacum officinale,* is a stemless perennial herb, having a thick tap-root and a rosette of basal leaves which grow close to the ground. Contraction of the root each year keeps the leaves of that year's growth at the ground level. The heads, composed of yellow ligulate flowers, are borne singly on hollow peduncles. When the **achenes** are mature the **peduncle** elongates greatly, lifting the achenes well above the ground. Each achene is beaked and has a crown of pappus hairs which aid it in floating through the air. The dandelion is frequently used as a pot-herb, and is grown extensively. Wild plants, often pestiferous weeds in lawns, are much gathered in the spring for greens. The root contains a glucoside taraxirin, and is used medicinally as a tonic. Another species of dandelion, *Taraxacum kok-saghyz,* is extensively cultivated, especially in Russia, as a source of rubber.

Several other members of the Composite Family have medicinal value. The young roots of the common burdock, *Arctium lappa,* are used in southern Europe and Asia as a drug plant. The stem tips and young leaves of *Eupatorium perfoliatum,* a native North American plant, are used as a stimulant and tonic, and in the making of compounds to relieve colds and fevers. Species of *Grindelia,* native in western United States, are used in preparations to relieve asthma. *Matricaria chamomilla,* a native of Europe, is a mild stimulant and tonic plant, as is also **Anthemis nobilis.**

Certain species of **Artemisia** are used in several ways. *Arnica montana,* of western United States, is another stimulant and tonic plant. *Calendula officinalis,* a strong-smelling garden annual frequently grown for its brilliant yellow or orange flowers, contains a volatile oil, and serves a variety of uses in the East and in Europe. Species of Chrysanthemum are grown extensively for their ornamental flower heads. *Chrysanthemum parthenium* is a febrifuge, popularly used to cure mild fevers. The entire flower heads of *Chrysanthemum roseum, Chrysanthemum cinerariaefolium* and *C. (Anacyclus) pyrethrum* are ground up and used as insect powders, under the name of pyrethrum.

Several members of this family contain poisonous substances which cause serious consequences when eaten by stock. *Helenium autumnale,* the sneeze-weed, is a perennial herb with smooth stem 2–6' tall, pointed lanceolate leaves, and bright yellow flowers in heads containing both ray and disk flowers. It occurs widely scattered in North America. Eaten in quantities it causes fatal results in cattle, horses and sheep. White Snakeroot, *Eupatorium urticaefolium,* another perennial with stem 1–5' tall, with opposite ovate leaves and dense masses of small heads of white flowers, occurs widespread in northern North America. It also produces fatal results if eaten by stock.

Ragweed, *Ambrosia artemisiifolia* and related species, is another widely known member of the family. It is a native plant, widely distributed and usually very abundant. It is a branching annual, ordinarily growing 2–4' tall. The leaves are finely divided and rather thin. The flower heads are of two sorts, the staminate heads are borne in elongated racemes, while the pistillate are borne in clusters. Flowers are produced in late summer and autumn and, unlike most composites, are wind-pollinated. The pollen of these plants is one of the principal causes of hay fever. (See **Allergy.**) (R.M.W.)

COMPOSITE SET. Telegraphy.

COMPOSITION, CHEMICAL. Chemical Composition.

COMPOSITION OF VECTORS. Vector Addition.

COMPOUND, CHEMICAL. Chemical Composition.

COMPOUND CURVE. Circular Curve.

COMPOUND ENGINE. A steam engine in which the total pressure drop of the steam between throttle and exhaust conditions is divided into two or more stages is said to be compounded when the expansion in each stage is performed in a separate **cylinder.** By common usage the engine in which steam is successively expanded in two cylinders is called a compound engine. If three cylinders are used, the engine is triple expansion; if four cylinders, quadruple expansion, etc. There are three very definite advantages attending compounding an engine. Any one cylinder does not have the big difference in temperature between incoming and outgoing steam that exists in a simple steam engine. The amount of initial condensation of the steam consumed in warming the cylinder and ports at the beginning of each stroke is thereby lessened. The confining of the highest pressure steam to the small high-pressure cylinder makes it possible to design the low-pressure cylinder, which is always the largest, against much lower loading forces than in the case of the simple engine, where one cylinder must not only meet the requirements of the highest pressure, but also contain the greatest volume of the steam. Furthermore, the flow to the low-pressure cylinder may be split between two identical low-pressure cylinders, thus reducing the cylinder size.

A compound engine may be tandem-compounded or cross-compounded. Tandem compounding refers to the arrangement wherein the different cylinders are in line and their pistons are fastened to the same piston rod. One crank and one connecting rod are required in a tandem compound engine. The cross-compounded engine has cylinders set with axes parallel, each cylinder having its own piston rod and connecting rod, the latter bearing on separate cranks of the crankshaft. There is more mechanical friction in the cross-compounded engine, but the cranks may be set at an angle to each other, and the rotative effort made more uniform. Steam exhausted from the high or intermediate pressure cylinder in a tandem engine may be passed directly to the next lower pressure cylinder through short interconnecting piping since the pistons move in synchromism. A steam receiver is necessary between the cylinders of a cross-compounded engine because of the angle between the cranks. The compound engine was at one time built in very large sizes, and a quadruple expansion engine was no uncommon thing, but while the engine was efficient, it was bulky, slow-speed, and extremely costly, and has been rendered obsolete by the steam turbine. There are, however, many multiple expansion engines in service at present in excellent working condition, and they will probably continue to give good service for many years, but new installations at present are made with either the simple steam engine or the steam turbine. (F.T.M.)

COMPOUND FRACTURE. Fracture.

COMPOUND GEAR. Gear Train.

COMPOUND GENERATOR. Generator.

COMPOUND PROBABILITY. Probability.

COMPOUND TURBINE. A compound turbine is one in which there are two casings—high- and low-pressure. The steam is partially expanded in the high-pressure casing, then exhausted to the low-pressure casing. The rotor arrangements may be either tandem- or cross-compound. Two generators must be supplied to cross-compounded turbines. The principal advantages of compounding a turbine are the reduction in physical size of any one casing, the confinement of the highest pressures to the smaller casing, which may be built of steel, and the possibility of divided flow in the low-pressure casing for the purpose of equalizing end thrusts. (F.T.M.)

COMPRESSIBILITY, AERODYNAMIC. The "incompressible fluid" theory of classical hydrodynamics has proved most useful for the estimation of aerodynamic parameters, and when applied to problems of low-speed flight has yielded sufficiently accurate results. It has been found, however, that the flow pattern about a body moving through the air at high speeds is affected to a large degree by changes in density resulting from compression or expansion of the fluid. Consequently, aerodynamic coefficients based on an incompressible-flow theory are in considerable error when applied to airplanes moving at high speeds. An understanding of compressible flows is, therefore, of the utmost importance to the designer of high-speed aircraft.

In the study of compressibility phenomena as applied to aerodynamic problems, airplane speeds are classified according to their relation to the speed of sound in air. The speed of sound is taken as a reference velocity because it is a function of fluid elasticity. As applied to compressible flows this means that the amount of pressure necessary to effect a given change in density in any fluid is proportional to the speed of sound in the fluid. Since the pressure is proportional to the square of the velocity, the velocity which a body may attain before appreciable density changes occur is also proportional to the velocity of sound in the fluid. It is apparent, therefore, that the flow pattern about a body will be altered by density changes to a degree dependent upon the ratio of the velocity of the body to the velocity of sound. This ratio is known as the *Mach number* and is taken as an index of the effects of compressibility on the flow pattern. A curve showing the variation of the speed of sound with altitude is presented in Fig. 1.

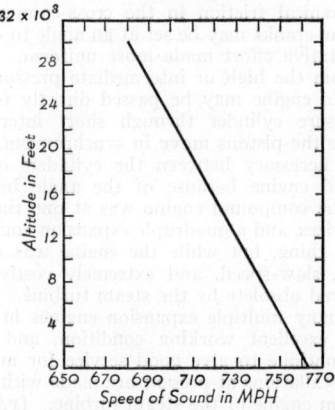

Fig. 1. Speed of sound as a function of altitude (sea-level temperature 60° F.).

In order to show quantitatively the changes in air density associated with increasing Mach number the following table has been prepared. This table is based on an adiabatic flow through a converging nozzle. M is the stream Mach number and ρ/ρ_s is the ratio of the corresponding air density to the density at static conditions.

M	ρ/ρ_s
.2	.981
.4	.925
.6	.841
.8	.742
1.0	.635

The preceding discussion fully accounts for the effects of compressibility only at Mach numbers less than one,

or at subsonic speeds. As air speeds approach and attain the velocity of sound, radical changes occur in the flow pattern which do not result entirely from changes in air density. The flow pattern in a perfect incompressible fluid is instantaneously influenced at all points by pressure changes occurring at any point in the flow field. A consideration of the theory of elasticity as applied to fluids, however, indicates that the effects of small pressure changes in a real fluid are transmitted throughout the fluid in the form of waves which travel at the speed of sound. It may be seen, then, that the effects of a pressure change which occurs behind the critical point at which the speed of sound has been reached cannot influence the flow field ahead of the point.

Since at the critical point the forward motion of the pressure waves is completely arrested by an air stream velocity equal to the velocity of wave propagation, a wave front is formed at the critical point. This wave front constitutes a sharp discontinuity in the flow with which is associated large increases in pressure, density and temperature and a decrease in velocity. Such a wave front with its attendant discontinuities is known as a shock wave. The flow field about a body traveling at or near sonic velocities will be radically different from that at low speeds.

Once the velocity has increased beyond the sonic range and attained a sufficiently high supersonic value, the shock wave will be forced downstream and the flow at the original point of shock will be comparatively smooth. However, the flow pattern at supersonic speeds will bear little resemblance to that which obtains at low speeds.

High-speed flight is greatly complicated by the compressibility phenomena which have been described. An attempt will be made to outline the more important considerations in the aerodynamics of high-speed flight.

The change in aerodynamic characteristics which results from fluid compressibility is greatest when shock waves form on some part of the airplane such as an airfoil or cowling. The forward velocity of the airplane which corresponds to the formation of shock waves is called the critical speed. The critical airplane speed is always less than the velocity of sound since the local velocity over the airfoil surfaces and other components is in excess of the forward speed of the airplane. The critical Mach number is the ratio of the critical airplane speed to the speed of sound.

The airfoil section usually predominates as the factor which controls the effects of compressibility on the characteristics of high-speed airplanes. The variation in

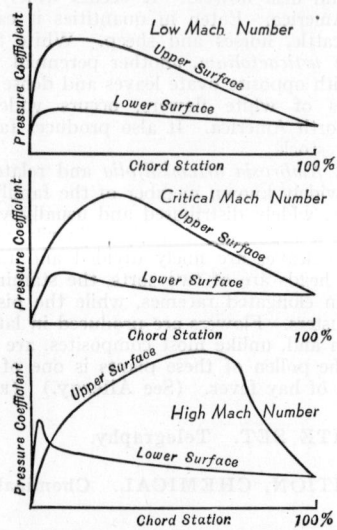

Fig. 2. Pressure distribution of an airfoil section at three Mach numbers.

pressure distribution around an airfoil as the Mach number is increased is shown in Fig. 2. Both flight and wind-tunnel tests have shown that this change in pressure distribution affects the following important airfoil parameters:

1. The drag coefficient, C_D.
2. The slope of the lift curve $\dfrac{dC_l}{d\alpha}$.
3. The pitching-moment coefficient, C_M.
4. The maximum lift coefficient, $C_{l\ max}$.

The drag coefficient suffers most, its value increasing tremendously as the critical Mach number is reached. The large increase in drag results not only from energy loss in the shock wave but also from the large positive pressure gradient existing across the shock. Such a pressure gradient causes boundary-layer separation which results in a wide, turbulent wake with its attendant form-drag. A curve is presented in Fig. 3 which shows

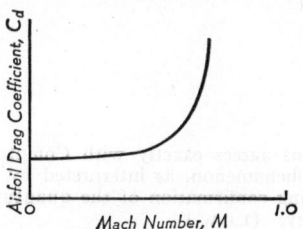

Fig. 3. Variation of airfoil drag coefficient with Mach number.

qualitatively the variation of profile-drag coefficient with Mach number. In terms of airplane performance, the increase in drag which occurs at the critical speed indicates that a tremendous amount of power would be required for an airplane to fly through the sonic range of speeds and reach speeds above the speed of sound.

A loss in maximum lift coefficient occurs as the critical speed is approached. However, airplanes do not operate

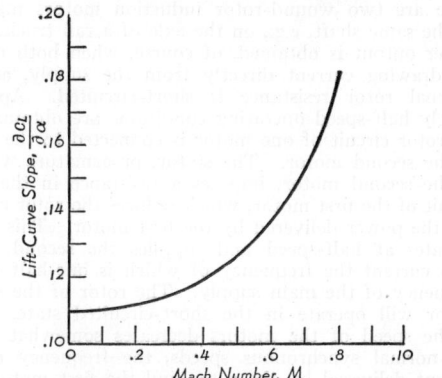

Fig. 4. Change of lift-curve slope with increasing Mach number.

at high lift coefficients when traveling at high speeds. The changes in lift-curve slope and pitching-moment coefficient which occur with increasing Mach numbers are shown in Figs. 4 and 5. The variation of lift-curve

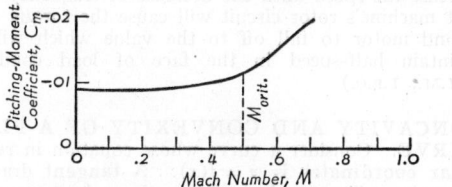

Fig. 5. Variation of pitching-moment coefficient with increasing Mach number.

slope with Mach number was calculated from Glauert's approximately correct theoretical relation which states that the change in lift-curve slope is proportional to $\dfrac{1}{\sqrt{1 - M^2}}$, where M is the free-stream Mach number.

The changes in pitching-moment coefficient and lift-curve slope indicate a considerable variation in the external forces acting on the lifting surfaces of airplanes traveling at high speeds. Unless the changes in C_M and $\dfrac{dC_l}{d\alpha}$ are taken into consideration in the design of high-speed airplanes, complete loss of control and stability may result at high speeds. The large turbulent wake discussed in connection with the drag of bodies at shock speeds also has an adverse effect on the control and stability. The angle of downwash and thus the trim of the airplane will change instantly when the wake widens and the plane may be thrown completely out of control. The large turbulent wake may also cause serious control-surface buffeting.

Knowledge of compressible flows is still very limited. The most fruitful methods of obtaining information about compressibility phenomena have been found in the field of wind-tunnel testing. Reliable data are very hard to obtain at speeds in the sonic range, however, and it is in this range that quantitative results are most needed. Tremendous mathematical difficulties are associated with problems involving compressible flows. It is to be hoped, however, that the mathematician working with those engaged in experimental research will eventually develop a reasonably complete body of knowledge regarding compressible flows. (F.T.M.)

COMPRESSION. Compression is descriptive of the decrease of volume of a compressible substance due to the application of pressure. The common example of compression is found in cylinders filled with gas being compressed by the motion of a piston moving in the cylinder. The pressure increase required for any given reduction of volume of a gas depends upon the particular condition surrounding the act of compression; for example, if the cylinder is thoroughly insulated against heat transmission, the compression will be of a type known as **adiabatic**, and the temperature of the gas will rise during compression because the mechanical work expended on the piston in obtaining the compression is converted into heat energy in the gas. On the other hand, if the cylinder is a good conductor of heat, and passes heat to the atmosphere as rapidly as it is received by the gas, the temperature at the end of compression may be the same as that at the beginning. In this isothermal compression the pressure at the end of compression is less than that of an equivalent adiabatic compression. In general, actual compressions are neither strictly adiabatic nor isothermal. (See **Air Compression, Compression Ratio, Polytropic.**)

In structural engineering compression is used to denote the type of stress which causes the fibers of a member to compress. (F.T.M.)

COMPRESSION MEMBER. Any member of a structure which is subjected to a primary compressive stress is called a **compression** member. The analysis or design is the same as that for a **column.**

When the member consists of two or more separate elements, known as ribs, the parts are connected to produce unity of action and also to prevent excessive distortion during fabrication, shipping or erection. Tie plates and lacing are used for this purpose. The tie plates, also called stay plates or batten plates, which are rectangular in shape are placed at the ends of the member and at intermediate points where the lacing is interrupted. The lacing bars, also known as lattice bars, are plates, angles or channels placed at an angle to the longitudinal axis of the member. Lacing may be single or double as shown in the accompanying illustration of a compression member made up of two channels which are rolled steel sections. Single lacing should make an angle of not less than 60° with the longitudinal axis of the

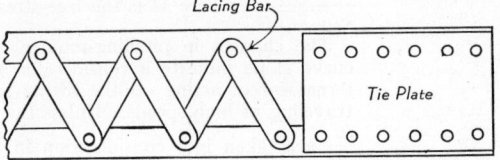

Lacing Bar

Tie Plate

Single Lacing

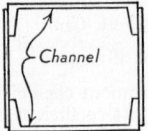

Channel

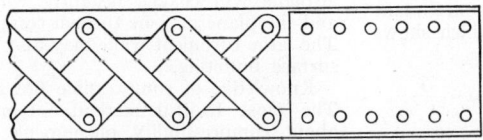

Double Lacing

member. The angle for double lacing should be not less than 45°. The purpose of these minimum angles is to reduce the buckling tendency of the individual segments. (C.W.C.)

COMPRESSION RATIO. Ratio of Expansion.

COMPRESSION SPRING. Springs.

COMPRESSOR, VOLUME. In amplitude **modulation** systems of **radio communication** the amount of intelligence volume which can be modulated upon the **carrier** is limited to an amount which will give 100% modulation. Since the percentage of modulation depends directly upon the volume of the sound, it follows that, in order not to exceed the allowable modulation on very loud sounds, the percentage on most sounds will be rather low. Since the maximum use is made of the power and a higher signal to noise ratio is obtained for high degrees of modulation it is very desirable to keep the level of modulation as high as possible. To do this the volume range of the original sound is compressed into a much smaller range. Thus, while a symphony orchestra may have a volume range of 100–110 db., the range is compressed to about 40 db. for broadcast purposes. In recordings the maximum volume which may be recorded is limited by the thickness of the groove walls so the volume range is reduced here also. An **expander** may be used in the reproducing system of the radio circuit or the phonograph to restore the original range. (L.R.Q.)

COMPTON EFFECT. The well-known increase in wavelength of **x-rays** scattered by the **electrons** of the lighter **atoms.** Professor A. H. Compton found in 1923 that when a beam of homogeneous x-rays enters carbon, for example, and is scattered at various angles θ with the incident direction, the wavelength of the scattered radiation, in angstroms, exceeds that of the incident by $0.0243(1 - \cos\theta)$. This he explained as due to collisions of the x-ray quanta with free electrons in the carbon, assuming conservation of momentum and conservation of energy as in ordinary elastic **collision.** The quantum is in general deflected obliquely (at angle θ), while the electron moves off in another direction, taking part of the energy $h\nu$ of the quantum. This lowering of the quantum energy, when the proper relativistic formulae for momentum and kinetic energy are used, corresponds accurately with the reduction in frequency and **increase** in wavelength actually observed.

The recoiling electrons were later detected by Professor C. T. R. Wilson by means of the cloud tracks which they produce in water vapor, and the speed of these electrons agrees exactly with Compton's analysis. The whole phenomenon, as interpreted by Compton, is a most striking confirmation of the **quantum theory** of radiant energy. (L.D.W.)

CONCATENATION. Variable speed control is more of a problem in a-c **motors** than it is in motors operating on d.c. The wound-rotor motors, the speed of which is controllable by insertion of resistance in the rotor circuit, nevertheless suffer from the disadvantage of losses occasioned by conversion of electrical energy into heat in the resistors. A method of speed reduction without heating loss frequently used in railway service is known as concatenation. This form of connection is also called cascade control. Concatenation may be employed when there are two wound-rotor induction motors mounted on the same shaft, e.g., on the axle of a rail truck. Full power output is obtained, of course, when both motors are drawing current directly from the supply, and all external rotor resistance is short-circuited. Approximately half-speed operating conditions are obtainable if the rotor circuit of one motor is connected to the stator of the second motor. The stator, or armature winding of the second motor, imposes a resistance in the rotor circuit of the first motor, which reduces the rotor current and the power delivered by the first motor. This motor operates at half-speed and supplies the second motor with current the frequency of which is one-half of the frequency of the main supply. The rotor of the second motor will operate in the short-circuited state. Now as the speed of the motors decreases somewhat below half-normal synchronous speeds, the frequency of the current delivered by the rotor and the first motor rises somewhat over the half-speed point, and increases the magnetic slipping of the second motor, which is then receiving current at a frequency higher than half that on the supply mains, and is operating at slightly less than one-half normal synchronous speeds. The result is an increase of torque in the second machine, which will increase the speed until the decrease of frequency in the first machine's rotor circuit will cause the torque of the second motor to fall off to the value which will just maintain half-speed in the face of load conditions. (F.T.M.; L.R.Q.)

CONCAVITY AND CONVEXITY OF A PLANE CURVE. Consider a curve whose equation in **rectangular coordinates** is $y = f(x)$. A **tangent** drawn to the curve will turn clockwise as the point of tangency moves from A to C, and counterclockwise as the point moves from C to E. The curve from A to C is called concave downward and from C to E concave upward; the point C is called a point of inflection. At a point of inflection the curve changes from concave upward to concave downward, or vice versa.

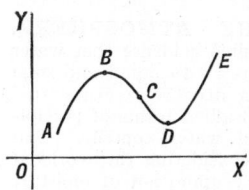

A curve whose equation is $y = f(x)$ is concave downward at points where $f''(x)$ (i.e., d^2y/dx^2) is negative, and is concave upward where $f''(x)$ is positive; if $f''(x) = 0$ at $x = x_0$ and if $f''(x)$ changes sign as x increases through x_0, then the curve has a point of inflection at $x = x_0$. (L.L.S.)

CONCENTRATION. Concentration is the amount of a given substance in a stated weight or volume of material. The most commonly used method of expressing concentration is by stating the percentage, that is, parts by weight of the given substance in 100 parts by weight of the stated material. There are settled exceptions to this method of expressing concentration, and the units used should be carefully recorded or observed according as the reader is operator or reader respectively. Ethyl alcohol, in water mixtures, is commonly reported as a stated per cent by volume, where 50.0% by volume is equivalent to 42.47% by weight. (See **Ethyl Alcohol.**) Gases in a mixture are commonly reported by volume per cent, thus nitrogen in air 78.0% by volume (equivalent to 75.5% by weight). (See **Air.**)

The concentrations of substances in solution are expressed variously, thus, per cent by weight of the actual material stated, per cent by weight of a material calculated chemically from the actual material, grams of the actual material (or a material calculated chemically from this) per 100 milliliters of solution, gram mols (the formula weight taken in grams) of the actual material per liter of solution (this is molar or formal concentration and the abbreviations, M or F, respectively, are used to express it), gram equivalents (the equivalent weight taken in grams—see **Chemical Composition, IV.** Equivalents) of the actual material per liter of solution (this is normal concentration and the abbreviation, N, is used to express it).

Since the concentration is proportional in many individual cases to an easily determined physical constant, such as specific gravity (e.g., of solutions), index of refraction, specific rotatory power (e.g., sugar solutions, terpenes), such constants are frequently used to ascertain and express concentration data. (R.K.S.)

CONCH. Mollusca, Gasteropoda. Any of numerous species of large marine mollusks (**Mollusca**). The spiral shells have a long aperture, in many species beautifully colored. The shells are used for ornaments and as horns, and the animals are sometimes eaten. (A.W.L.)

CONCHIOLIN. A horny material which forms the outer layer of the shell mollusks (**Mollusca**). (A.W.L.)

CONCHOID OF NICOMEDES. The conchoid is a mathematical curve, first discussed by the ancient Greek mathematician Nicomedes. The name given to it is due to its supposed shell-like shape.

The conchoid may be defined geometrically as follows: With a set of rectangular axes, draw a line AB parallel to the X-axis, above it and at a distance a from it. Draw a secant line OS cutting AB at Q, and extend OQ in each direction a fixed distance b to points P and P'. As the secant line rotates about O, the points P and P' describe the conchoid.

In **polar coordinates,** the equation of the conchoid is $r = a \csc \theta \pm b$; in **rectangular coordinates,** the equation is $(x^2 + y^2)(y - a)^2 = b^2y^2$.

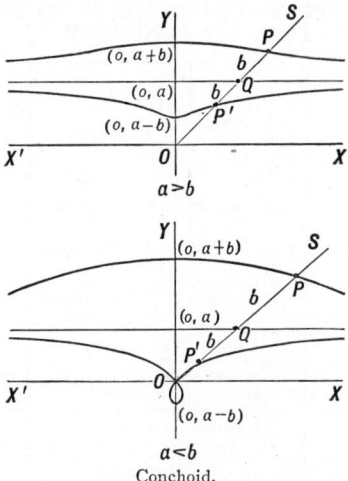

Conchoid.

The conchoid can be used for the "trisection of an angle," also to "duplicate a cube." (L.L.S.)

CONCHOLOGY. The study of molluscan shells. The word is sometimes applied to the study of mollusks generally but this field is more accurately called malacology. (A.W.L.)

CONCRETE. Concrete is a mixture of fine and coarse aggregates firmly bound into a monolithic mass by a cementing agent. The cement ordinarily employed for concrete is the standard Portland cement. The aggregates are sand and crushed stone or gravel. Crushed slag or cinders are used in special kinds of concrete. The formation of concrete can be thought of as a process in which the voids between the particles of coarse aggregate are filled by the fine aggregate, and the whole is cemented together by the binding action of the cement. The nature of Portland cement in this respect is described under **cement.**

Due to its strength, permanency, and low cost, concrete is one of the most important building materials employed in modern construction. It is widely used for **foundations** of all types, buildings, **bridges, dams, retaining walls, highways,** and other purposes too numerous to mention. However, the success of concrete in meeting any particular set of conditions, depends upon the proper correlation of many factors bearing on the selection and mixing of the materials, the placing of the concrete, and the original design. Concrete is strong in compression, but relatively weak in tension. Therefore structures in which the concrete is likely to be in **tension** must be reinforced with steel rods, which carry the tension and relieve the concrete of tensile stress. The design of plain and **reinforced concrete** structures has been placed upon a semi-rational technical basis. The uniformity of commercial cement enables the engineer to design an economical concrete structure—one having a minimum of unused excess strength. For strong permanent concrete, the aggregates should be clean, coarse, and well graded. River or coarse sand is better than pit sand, and should always be used where possible. The table on page 370 gives typical concrete mixes, with the characteristics of each.

This table is suitable for preliminary estimates or for small amounts of concrete, but it should be remembered that research and development in the science of concrete proportioning have advanced to the point where little short of a laboratory analysis can establish the best and most economical mix for a given condition. Type and gradation of aggregate, moisture content of the sand, and water-cement ratio are typical factors taken into account

DATA ON CONCRETE MIXES TO YIELD 1 CU. YD. CONCRETE

Mixture	Cement, Sacks	Sand, Cu. Yd.	Stone, Cu. Yd.	Application	Weight, Tons per Cu. Yd.	Safe Comp. Stress, Tons per Sq. Ft.
1:2:3	7	.51	.77	Roofs, sills, tanks, tunnels	2	35
1:2:4	6	.44	.88	R. C. floors, beams, and columns	2	30
1:2½:4	5.6	.52	.83	Building walls	2	25
1:3:5	4.7	.52	.86	Foundations and footings	2	20
1:2:4	6.6	.49	Cinders, Cu. Yd. .98	R. C. floors	1.5	14
1:2:4	6.6	.49	Slag, Cu. Yd. .98	R. C. floors	1.6	14

in a complete analysis for the specification of large amounts of concrete work.

Concrete should be transported rapidly from the mixer to the forms, so that no initial set will have occurred before the concrete is placed in its final position. It is necessary to place concrete in the forms with care to prevent segregation of the lighter and heavier parts. This precludes dropping the concrete into place from any height. After the "green" concrete has been poured, it should be cured, or hardened, slowly over a period of about a week, during which time it should be protected from vibration, freezing, and a too-rapid rate of drying out. Strengths of concrete are usually classified on the basis of 28-day strength. Most of its strength will be acquired in this time but there is a slow increase in strength for a much longer period thereafter. (F.T.M.)

CONCRETION. Used in petrology and geology to designate spheroidal or discoidal aggregates formed by the secretion of silica, calcium carbonate, gypsum, or other chemical compounds around an original nucleus. (See also Oölites.) (R.M.F.)

CONCURRENT FORCES. Statics.

CONCUSSION. Concussion of the brain is a loose term used to describe certain events following a blow on the head. These consist of momentary loss of consciousness and general cessation of physiologic processes; the pulse is imperceptible, respirations cease, and there is no response to sensory stimuli. Recovery takes place rapidly. The patient regains consciousness and has no recollection of the blow. There is no satisfactory explanation of the pathologic physiology of concussion. Death rarely follows simple concussion and, when it does, gross changes in the brain are not found. (D.M.H.)

CONDENSATE. A liquid in the vapor state may be reduced to the liquid by removal of such portion of the latent heat of evaporation as it may contain. The act is called condensation, and the liquid is condensate. It is the property of vapors that the condensate is dense as compared to the vapor, which, in condensing, formed it; that is, there is considerable shrinkage of volume upon a reversion to the liquid state. The thermal condition of the condensate immediately upon formation is that of saturated liquid at the temperature of vapor. In other words, it has the temperature of the heat of liquid

corresponding to saturation conditions at the vapor pressure. (F.T.M.)

CONDENSATION IN THE ATMOSPHERE. Clouds and precipitation are visible evidence that water vapor in the atmosphere condenses into liquid and solid water. Moisture in cloud form may be re-evaporated into the air, but rain, snow, and allied forms of precipitation actively lessen the total water content. This moisture loss by precipitation, considering the world at large, is replaced by equivalent evaporation of moisture into the atmosphere.

Nuclei upon which condensation can take place are absolutely necessary. Non-ionized and pollution-free air will become up to 400% supersaturated before condensation occurs. It also is known that clouds sometimes form before air becomes 100% saturated. Nuclei upon which condensation begins at the approach toward saturation are highly hygroscopic and encourage condensation. Minute salt crystals, primarily **sodium, magnesium** and **calcium** chlorides, carbonates, and sulfates, smoke particles, and ions serve as condensation nuclei. Lowering of the temperature of air below its dew point, or adding water vapor beyond the holding capacity of the air, is a second requirement for the formation of clouds. At high temperatures, air is able to hold up to 3% water vapor before becoming saturated but, at low temperatures, this amount may be as low as .01%. When air is cooled, therefore, it slowly loses its capacity to hold water vapor and that excess vapor condenses into cloud droplets (or ice crystals). Cooling is achieved in nature by one of four methods.

1. Cooling occurs by contact with surfaces cooler than the air. If clouds or precipitation are to result from this type of cooling, it is necessary that cooling proceed beyond the dew point of the air in question. Contact cooling is effective in forming fogs, dew, or hoarfrost, but as to precipitation, causes only drizzle.

2. Cooling occurs by lateral mixing of one air parcel with another at a lower temperature. The cooler parcel, however, is warmed somewhat. Lateral mixing is virtually noneffective in forming any extensive clouds, fogs, and does not cause precipitation.

3. When a layer of air having a lapse rate less than the adiabatic rate is mixed vertically, the top of the layer is cooled and the base warmed. This transfer of heat results from the establishment of an adiabatic lapse rate in the thoroughly mixed layer in contrast to its previously nonadiabatic rate. Cooling at the top of the layer, if there is sufficient water vapor present, will produce a cloud layer and may cause some small amounts of precipitation. Vertical mixing, however, is not the cause for more than slight amounts of precipitation even though it is the cause for considerable cloudiness.

4. Lifting of whole layers of air or parcels of air causes excessive cooling; cooling in turn causes cloudiness and precipitation. Lifted layers and parcels of air cause nearly all precipitation. Lifting occurs on windward sides of sloping terrain, along frontal surfaces, and in convection currents; these three taken together are the real causes of the world's precipitation.

Lifted particles or layers of air while unsaturated, cool dry-adiabatically until the temperature and dew point coincide. The level at which the temperature and dew point become the same is known as the lifting condensation level because it is at this level that condensation into cloud droplets or ice crystals occurs. Condensation levels depend on the temperature and dew point of the lifted air, being high for dry air and low for nearly saturated air. (P.E.K.)

CONDENSATION PUMP. Air Pumps.

CONDENSER, ELECTRICAL. An electrical condenser is an arrangement of conductors and **dielectrics** used to secure an appreciable capacitance, sometimes one of specified value. The essential feature of all condensers is a system of two or more conductors, separated by layers of dielectric. The potential difference between the conductors, when charged, is limited by the electric polarization in the dielectric. This makes it possible to accumulate large charges at comparatively small voltages. The oldest form of condenser is the **Leyden jar,** still often used where heavy electric discharges are desired. Many modern condensers consist of alternate metal and dielectric plates or sheets, sometimes of metal foil and paraffin paper strips rolled in a compact bundle. Condensers in which the dielectric is air, usually of adjustable capacitance, are much used in radio and oscillatory circuits. Standard condensers, of accurately known capacitance, are employed in electrical measurements. The capacitance of a condenser depends upon the total area a and the thickness d of the dielectric and upon its dielectric constant k. If the dimensions are in centimeters, the capacitance in a condenser of flat plates is approximately given in electrostatic units by the formula $C = ka/4\pi d$ and in microfarads by $C = 8.84 \times 10^{-8} \, ka/d$. Thus if there are 21 metal plates 10 cm. sq., separated by 20 sheets of mica 0.01 cm. thick and of dielectric constant 6, the capacitance is about 0.106 microfarads. The capacitance is often made adjustable by varying the distance d or by arranging the plates to move past one another so as to vary the area a of dielectric subject to the electric field between them. (F.T.M., L.D.W., L.R.Q.)

CONDENSER, STEAM. A steam or vapor condenser is a device which performs the operation of reducing a vapor to a liquid through the extraction from it of the heat of evaporation it may contain. Industry and research use the condenser extensively to recover a liquid which has been separated from another liquid of higher boiling point or from impurities by the evaporation method. The condenser is an important part of the modern steam power cycle. The purpose of the steam condenser is to lower the back pressure on the prime

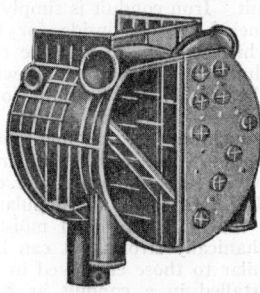

C. H. Wheeler Radial-Flow Condenser. End view with half of water box cover removed and cross-section at middle.

mover, allowing the steam a larger pressure and temperature drop, thereby increasing both efficiency and capacity. A secondary purpose may be the supply of quantities of warm water or the collection of condensate for boiler feed water. Condensers are applicable to both engines and turbines, though especially to the latter because its thermodynamic advantages occur chiefly in the low-pressure range. Engines are ordinarily operated condensing up to 26″ Hg vacuum and turbines up to 29.5″. Size of the equipment required to create the necessary piston displacement is the limiting factor in reduction of engine back-pressure. Furthermore, the larger the temperature difference between inlet and exhaust steam in the ordinary counter-flow engine, the larger the initial condensation loss. Economies effected by condensing operation may be from 10% to 30% higher than for non-condensing.

Condensers can be divided into *contact* and *surface* types. Steam and condensing water are intimately mixed in the contact type. Then the condensate and condens-

ing water are withdrawn thoroughly mixed together. A dividing surface is interposed between steam and water in the surface condenser. Heat is transferred through this partition surface at rates varying from 300 to 800 B.T.U. per sq. ft. per hour per degree Fahrenheit. The steam and condensing water never come into direct contact and are withdrawn from the condenser separately. The dividing surface is ordinarily a tube with condensing water circulating inside and steam outside.

The contact condensers (jet condensers) may be further subdivided into those from which the water is extracted directly by pumps, those from which it is forced by the ejector action of a high-velocity water jet, and those from which it flows by virtue of a head of water maintained in a long tail pipe. Elements of the jet condenser are: (1) nozzles or distributors for the condensing water, (2) steam inlet, (3) mixing chamber, (4) hotwell, (5) in some cases, a diffusing chamber or a tail pipe.

Surface condensers are classified as horizontal or vertical by the position of their tubes. They are single-pass or double-pass according to whether the water passes the length of the condenser once or twice. Three-pass condensers are infrequently used at present. They are also classified upon a basis of the shape of the shell, as, for instance, cylindrical, heart-shaped, U-shaped, oval; and upon the location of the air-cooling surface—internal or external. The elements of the surface condenser are (1) cooling surface for both air and steam, usually ¾, ⅞, or 1-in. copper alloy tubes from 10 to 25 ft. long, (2) tube sheets into which the tubes are expanded or packed, (3) intermediate tube supports (sheets bored similar to the tube sheets but in which the tubes

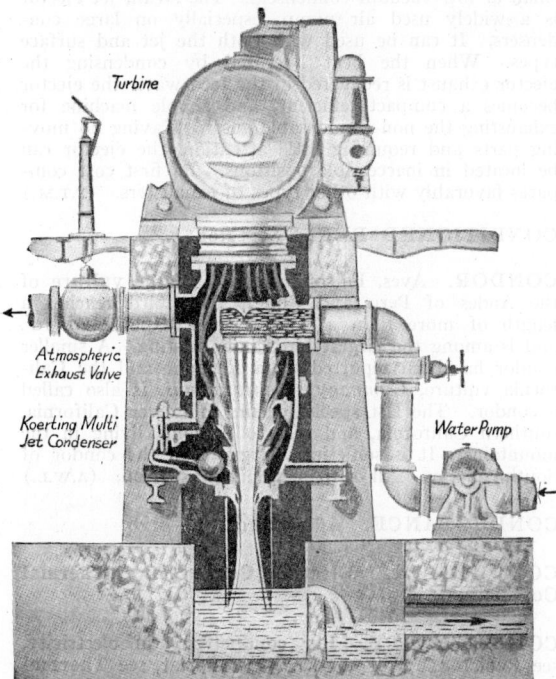

Typical arrangement of multi-jet condenser under turbine.

fit loosely), (4) water boxes provided with circulating water connections and enclosing the space furnished for water flow to and from the tubes, (5) steam inlet, air and condensate outlet, (6) hotwell in which the condensate collects (in some plants deaeration of the condensate is done in a deaerating hotwell, while in a few cases reheating hotwells are installed to offset undercooling of the condensate), (7) condenser shell enclosing and supporting the other elements.

Heat transfer action in a surface condenser is hindered by the presence of non-condensable gases which mix with the film of condensate on the tube surface. The sources of air and other non-condensable gas leakage are numerous. Some may come over with the boiler steam, or leak in through turbine packing gland or exhaust nozzle connection. Air leakage will seriously affect the heat transfer and every effort is made to minimize it.

The auxiliaries required for condensers can be arranged under two heads: first, those connected with the flow of water; second, those connected with the vacuum. The condensate pump serving a large condenser is always of the centrifugal type but reciprocating pumps may be used for small surface condensers and jet condensers. The head on it is the vacuum plus friction of the piping to the surge tank plus the velocity head plus the difference in elevation between the discharge to the surge tank and the condenser hotwell.

A high-vacuum surface condenser requires around a hundred pounds of water per pound of steam condensed. Supply of circulating water is often a deciding factor in station location and a limiting factor in extension of existing plants. These large amounts of circulating water make the circulating water system of considerable importance. In case of limited supply of circulating water it may be necessary to resort to cooling towers and ponds to cool the water for recirculation.

The auxiliary equipment having to do with vacuum includes the vacuum pump, atmospheric relief, vacuum breaker, and manometer. Vacuum pumps may be classified as reciprocating, rotary, or ejector types. The reciprocating type, similar in principle to the air compressor, becomes inconveniently large for other than small or low-vacuum condensers. The steam jet ejector is a widely used air pump, especially on large condensers. It can be used with both the jet and surface types. When the heat liberated by condensing the ejector exhaust is recovered in the feed water the ejector becomes a compact, efficient, and simple machine for exhausting the non-condensable gases. Having no moving parts and requiring little attention, the ejector can be located in inaccessible positions. Its first cost compares favorably with other types of exhausters. (F.T.M.)

CONDITIONED REFLEX. Reflex.

CONDOR. Aves, Falconiformes. A large **vulture** of the Andes of Peru and Chile. This bird reaches a length of more than 4′ and a wing expanse of 10′, and is among the largest birds now existing. A smaller condor has been reported from Ecuador and the California vulture, *Gymnogyps californianus,* is also called a condor. The last species occurs in Lower California, southern California, and east to Arizona, living in the mountains. It is sometimes larger than the condor of South America. All of the condors eat carrion. (A.W.L.)

CONDUCTANCE. Admittance.

CONDUCTION. Electric Conduction; Thermal Conduction.

CONDUCTIVITY. For conductivity of electricity, see **Resistance;** for conductivity of heat, see **Thermal Conductivity.**

In biology, conductivity is a fundamental property of living matter through which impulses generated by some form of stimulus are transmitted from one part to another of the individual body. It attains its highest expression in the nervous tissue of animals. Here impulses set up by the reception of stimuli by sense organs travel through the nerve fibers and arouse other impulses in associated nervous structures or activate muscle fibers or other organs capable of appropriate action. A stimulus may also arise in the nervous system itself, conditioned by memory, to bring about some action.

The nature of conduction in living substance is not completely understood. In nerve fibers the simultaneous occurrence of electrical and chemical changes during the passage of impulses has been demonstrated. (F.T.M., A.W.L.)

CONDUCTOR. A material which, when placed between terminals having a difference of electrical potential, will readily permit the passage of an electric current, is an electrical conductor. Different materials have different degrees of conductivity, and their effectiveness in this respect is computed as the **conductivity.** The best conductors are the metals, such as silver, copper, aluminum, platinum, mercury, etc., but non-metallic substances such as carbon, saline solutions, and moist earth also are sufficiently conductive so that this property becomes of significance under certain circumstances. By virtue of their cost-conductivity characteristic, copper and aluminum are the most widely used conductors. They will usually be found as wires or **buses.** Copper is used more commonly than aluminum, the use of which is still largely confined to high-voltage transmission lines, where its lighter weight is of definite advantage. Steel as a conductor is inferior to the other two materials mentioned, but its greater strength and resistance to wear have led to its adoption as a conductor of special purposes, such as that of third rail service on electrified railways, and as an inner core of copper or aluminum cables. The resistance of a conductor is its resistivity multiplied by its length and divided by its cross-sectional area. When the length is expressed in feet, and the area in **circular mils,** the resistivity of aluminum is 17.48; of copper, 10.35; and of iron, 58. This property varies with the temperature of the conductor, and the constants quoted apply only at 20° C. (F.T.M., L.R.Q.)

CONDUIT. In engineering, a conduit is ordinarily taken to be a means for holding and enclosing electrical wires. The most widely used electrical conduit is the rigid iron conduit. Iron conduit is simply wrought-steel pipe well enameled on the inside for the protection of the wires, which are drawn into the conduit. Fiber and brass conduits are used when all wires of an a-c circuit cannot be run in one conduit, for the installation of one conductor in an iron conduit would greatly increase the inductance of the line and would also make of the body of the conduit a circuit in which would circulate currents similar to the core loss of a transformer. The reason for the popularity of a rigid iron conduit is that it is fire- and moisture-proof, reliable, and mechanically strong. It can be laid rapidly by methods similar to those employed in plumbing.

Wires are installed in a conduit as follows: A fish tape or wire, a tempered steel wire of rectangular cross-section, is pushed through the conduit until it appears at the farther end. A draw line is then attached to it and by withdrawing the fish tape, the line is drawn through the conduit. The wires are in turn attached to the draw line and drawn into position. This method of installation requires (1) that the conduit interior be smooth and uninterrupted, (2) that the bends be of long radius and limited in number. If the conduit were not smooth internally there might be difficulty in pushing the fish tape through it, and the roughness would doubtless damage the insulation on wires being drawn in. When conduit is cut and threaded the ends should be reamed to remove burrs. The ends should be set up well into the coupling and should be leaded unless in a dry location. Ordinary pipe elbows are not used in conduit wiring. The pipe itself is bent to a long radius or long radius elbows are used. Due to the snubbing action of bends on wire being drawn through the conduit, not more than four equivalent 90° bends are permitted between pulling points, and many prefer to limit the number to three. In size, conduits smaller than ½-in. iron pipe size are not to be used, while conduits larger than 4-in. size are seldom required. Fiber ducts or tunnels are favored over the large iron conduit. Manufacturers have developed, in place of ordinary pipe elbows, tees, etc., lines of special conduit fittings designed to satisfy all requirements for outlets, junctions, etc. (F.T.M., L.R.Q.)

CONDYLARTHS. Paleocene.

CONDYLE. A rounded prominence on a bone, associated with a joint. (A.W.L.)

CONE. A cone is one of the simple and basic type forms of geometric solid.

The surface generated by a straight line turning around one of its points and intersecting a given curve is called a conical surface, and the solid bounded by such a surface and a plane cutting it is called a cone.

The volume of a cone is equal to ⅓ the product of the area of the base by the altitude. For a right circular cone of altitude h, radius of base r and slant height s, the volume is given by $V = \frac{1}{3}\pi r^2 h$, and the lateral area by $A = \pi r s$.

A frustum of a cone is the portion of a cone between two parallel planes. The volume of a frustum of a cone is given by the formula

$$V = \frac{1}{3}h(A_1 + A_2 + \sqrt{A_1 A_2}),$$

where h is the altitude of the frustum (perpendicular distance between cutting planes or bases), and A_1 and A_2 are the areas of the bases.

The **locus** of a second-degree equation in x, y, z which is **homogeneous** in these variables is a conical surface with vertex at the origin.

The standard equation of a cone is

$$\frac{x^2}{a^2} + \frac{y^2}{b^2} - \frac{z^2}{c^2} = 0. \qquad \text{(L.L.S.)}$$

CONE OF SILENCE. Radio Range.

CONE, WIND. Wind Indicator.

CONE-IN-CONE STRUCTURE. Probably a **con-cretionary** structure frequently observed in **limestones, dolomites** and other **sedimentary** rocks which originated as fine-grained sediments. As the term suggests the structure is formed by a series of concentric cones, the result of radial **crystallization** around a common axis, probably due to differential pressures. (R.M.F.)

CONFIDENCE LIMITS. Confidence limits are intervals, different for each **sample**, set up in successive samples, which cover the **population parameter** a certain proportion of times, called the confidence coefficient. Usually the confidence coefficient is chosen as .95, .99, or .995, depending on the particular level of confidence desired. When the confidence coefficient is increased, the confidence interval becomes larger. It should be realized that each sample gives another confidence interval and that after a confidence interval is determined that either the population parameter is covered by the interval or it lies outside the interval. To find confidence intervals the distribution of the **statistic** must be known. Confidence intervals have been obtained in cases where the statistic is distributed according to **Student's** t distribution, in the **Bernoulli probability function** for p, in the **Poisson probability function** for m, and for many other population parameters. Confidence intervals should not be confused with probability limits within which a chance **variable** lies a certain proportion of the time. A population parameter is a constant and does not vary from sample to sample. A confidence interval may be con-

sidered as the estimate of a population parameter by an interval instead of by a point. (L.A.A.)

CONFLUENT. Used by geologists and physiographers to designate a stream that unites with another; especially applied to streams nearly equal in size. (R.M.F.)

CONFOCAL CONICS. If two **central conics** have the same foci, they are said to be confocal.

The equation

$$\frac{x^2}{a^2 + k} + \frac{y^2}{b^2 + k} = 1 (a > b),$$

where k is an arbitrary constant, represents the system of confocal conics consisting of the **ellipses** and **hyperbolas** whose common foci are at the points $(\pm\sqrt{a^2 - b^2}, 0)$. When $k > -b^2$, the conic is an ellipse, when k is between $-a^2$ and $-b^2$, the conic is a hyperbola.

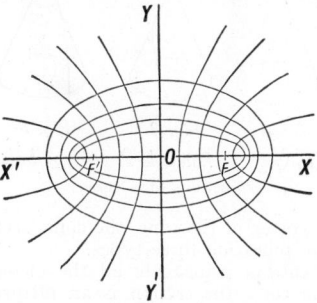

Confocal conics.

Through any given point there pass just one ellipse and just one hyperbola of the confocal system. The two conics of the system which pass through a given point intersect at right angles. (L.L.S.)

CONFOCAL QUADRICS. Confocal quadrics are geometric surfaces having a common focus.

The system of surfaces represented by the equation

$$\frac{x^2}{a^2 + k} + \frac{y^2}{b^2 + k} + \frac{z^2}{c^2 + k} = 1,$$

in which k is an arbitrary constant, is called a system of confocal quadrics. The principal sections of the system (sections by the coordinate planes) are **confocal conics**.

Through every point there pass three quadrics of the system, namely, an **ellipsoid**, an **hyperboloid of one sheet** and an **hyperboloid of two sheets**. These three surfaces are orthogonal, that is, their **tangent planes** are mutually perpendicular. (L.L.S.)

CONFORMABLE CONTACT. Used by **stratigraphers** and **structural** geologists to designate strata in parallel contact. (See also **Unconformity**.) (R.M.F.)

CONGENITAL. A condition or deformity that exists at or before birth. (D.M.H.)

CONGER. Pisces, Teleostei. Large marine **eels**, *Leptocephalus*, with sharp cutting teeth. They are widely distributed. Also applied to the **Congo snake**. (A.W.L.)

CONGLOMERATE. Conglomerate, called in older writings "pudding-stone," consists of aggregates of gravel or pebbles with a matrix of sand and cement. The proportion of pebbles and matrix may vary considerably both as to amount and actual or relative size of the component material. The common cementing materials are silica, **calcite**, and iron oxide.

Consolidated glacial debris called **tillite**, may consist of boulders of considerable size in an heterogeneous mixture of pebbles, clay and sand. (R.M.F.)

CONGO RED. Dyes.

CONGO SNAKE. Amphibia, Urodela. A slender aquatic **salamander**, *Amphiuma means*, of the southeastern states. The legs are very small, hence the animal is much like a snake in appearance. Also called the blind eel, Congo eel, and conger. (A.W.L.)

CONIC SECTIONS. The curve of intersection of a right circular **cone** by any plane is called a **conic section** or simply a **conic**. If the cutting plane does not pass

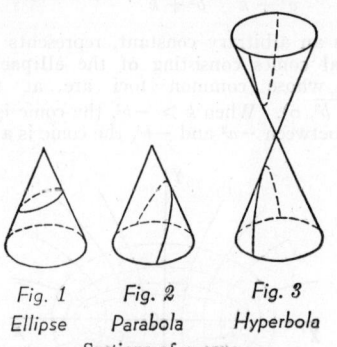

Fig. 1   Fig. 2   Fig. 3
Ellipse   Parabola   Hyperbola
Sections of a cone.

through the vortex of the cone, the conic section belongs to one of the following three types.

1. If the cutting plane cuts all the elements of one nappe of the cone, the section is an **ellipse**, including the **circle** as a special case.

2. If the cutting plane is parallel to an element of the cone, the section is called a **parabola**.

3. If the cutting plane cuts both nappes of the cone, the section is called a **hyperbola**.

The conic sections were discussed by the ancient Greek mathematicians.

When a variable point moves so that its distance from a given fixed point to its distance from a given fixed line has a constant **ratio**, the **locus** is a conic. The given fixed point is called the focus, the given fixed line the directrix, and the ratio of distances the eccentricity e of the conic.

If $e = 1$, the conic is a parabola; if $e < 1$, the conic is an ellipse; if $e > 1$, the conic is a hyperbola.

The ellipse and hyperbola are called central conics, since they each have a center of symmetry.

The general equation of the second degree, in **rectangular coordinates**,

$$ax^2 + 2hxy + by^2 + 2gx + 2fy + c = 0,$$

represents:

(1) a hyperbola if $h^2 - ab > 0$,
(2) a parabola if $h^2 - ab = 0$,
(3) an ellipse if $h^2 - ab < 0$;

it represents a proper conic (non-degenerate) if

$$\Delta = \begin{vmatrix} a & h & g \\ h & b & f \\ g & f & c \end{vmatrix} \neq 0.$$

If $\Delta = 0$, the equation is represented by a pair of straight lines. An equation of the form $ax^2 + by^2 + 2gx + 2fy + c = 0$ will represent an ellipse if a and b are of the same sign, and a hyperbola if a and b are of opposite signs.

The equation of a conic in **polar coordinates** is

$$r = \frac{ep}{1 - e \cos \theta},$$

where e is the eccentricity and p is the distance from the focus to the directrix, if the pole is at the focus and the polar axis is perpendicular to the directrix.

The equation of the **tangent** to the conic

$$ax^2 + 2hxy + by^2 + 2gx + 2fy + c = 0$$

at the point (x_1, y_1) is

$$ax_1x + h(x_1y + xy_1) + by_1y + g(x + x_1) + f(y + y_1) + c = 0$$

The tangents and **normals** to conics have many interesting properties. A few of them will be mentioned.

The tangent and normal to an ellipse bisect, respectively, the external and internal angles formed by the focal radii of the point of contact. (See Fig. 4.) There is a similar theorem for the hyperbola.

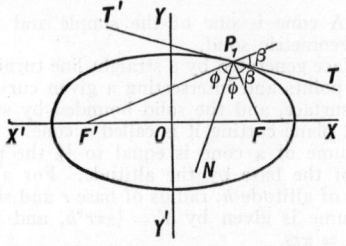

Fig. 4. Reflecting property of ellipse.

This theorem has an interesting application to drawing a tangent and normal to an ellipse at a given point on the curve.

The phenomenon observed in "whispering galleries" depends on this property. Thus, in Fig. 5, let the

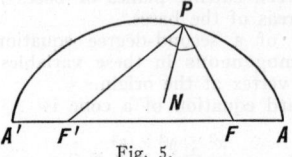

Fig. 5.

elliptic arc $A'PA$ be a vertical section of such a gallery. The waves of sound from a voice at focus F will, after meeting the ceiling of the gallery at P, be reflected in the direction PF'. The law of reflection of sound waves is that the angles of incidence and reflection are equal. Hence sound waves emanating from F in all directions will converge at F'. A whisper at F, which would not carry over the distance FF', might consequently, through reflection, be audible at F'.

The tangent and normal to a parabola bisect respectively the internal and external angles formed by the focal radius of the point and the line through that point parallel to the axis. (See Fig. 6.)

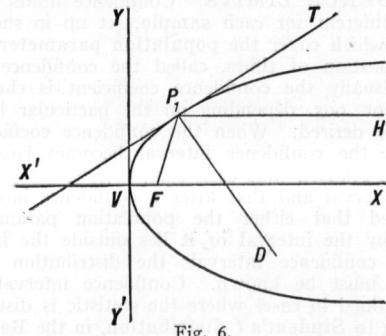

Fig. 6.

This can be applied to the construction with ruler and compass of the tangent and normal to a parabola.

The principle of parabolic reflectors depends upon this property. The reflecting surface of such a reflector is obtained by revolving a parabolic arc about its axis. If now a light be placed at the focus, the rays of light which meet the surface at P_1 in the figure will be re-flected in a direction making with the normal P_1D an angle equal to the angle FP_1D. But this direction is, by the above property, parallel to the X-axis (axis of parabola). Reflecting telescopes operate on this prin-ciple. (L.L.S.)

CONICAL PROJECTIONS.

The term conical pro-jection is applied to any one of that class of map pro-jections in which the surface of the earth is projected from a point within the earth to the surface of one or more cones, and then developed on a plane.

In the simple conical projection a cone is placed tangent to the surface of the earth along the central parallel of latitude for the region to be mapped. The surface features are then projected onto the cone, the cone is cut along that element which represents the meridian of longitude 180° from the central longitude of the mapped region, and the cone is rolled out on a plane. The resulting graticule will show parallels of latitude as circles concentric at the pole of rotation (usually outside the limits of the map), and meridians of longitude as straight lines converging on the pole.

Along parallels of latitude and along the central meridian the scale of distance is uniform. The map is not conformal except close to the central parallel and the distortion of shape increases rapidly with distance from that parallel. The simple conical projection is sometimes used for maps of small countries. For rela-tively large areas the polyconic and Lambert projec-tions are preferable conical projections. (W.K.G.)

CONICAL SPRING. Springs.

CONIDIA.

A conidium is an asexual spore, character-istic of many fungi. Commonly it is formed at the tip of a hyphal branch, which is called a conidiophore. Conidia may be one- to many-celled. (R.M.W.)

CONIFERS, OR CONIFERALES.

This group con-tains some 400 of the 500 species of gymnosperms. It includes the dominant evergreen trees of the north tem-perate zone, as well as many tropical species. Conifers are at their best in regions where severe winters occur. Nearly all conifers are trees, often of great size, and may attain great age, as for ex-ample the redwoods of California and the Douglas firs of Washington and Oregon.

While they are dominant plants in the northern forests of today, in Mesozoic times they were much more numerous and often of much greater size. The petrified forest of Arizona contains a fossil Gym-nosperm, *Araucarioxylon arizoni-cum*, from the Triassic period. As recently as the Miocene age, coni-ferales were much more widely scattered than at present, species of Redwood and Cypress growing in regions as far north as Green-land; today they are found only in warmer climates and often in a very restricted range there. Though naturally restricted in habitat, many of them are easily introduced into new, often distant regions, where they thrive.

The plant body of the conifers varies from low straggling shrubs to large trees. The several parts of the plant also show great diversity. Nearly all of them have tall straight stems extending to the very top of the tree, and numerous lateral branches which are progressively shorter from bottom to top of the trees, which therefore have an attractive conical shape. In many species there

Leaf clusters from dif-ferent species of pine. 1, *Pinus Murrayana;* 2, *Pinus ponderosa;* 3, *Pinus flexilis.*

is a central tap root extending deep into the ground, from which smaller lateral roots arise. In other species extensive lateral roots spread out near the surface of the ground. The leaves of conifers of the northern hemi-sphere are either slender and needle-shaped or short and scale-like. Those of the pines are formed in fascicles or bunches of 2–5 which grow from a very short lateral branch. In spruces and firs the leaves are borne singly around the stem. In white cedar and certain other conifers the leaves are reduced to short pointed scales which are formed in pairs of opposite sides of the stem, and are pressed tightly to it. In the southern hemi-sphere there are several conifers with broad leaves very similar in outward appearance to those of many angio-sperms. The leaves of conifers have an epidermis of thick-walled cells which give stiffness to the leaves. The stomata are sunk deep in grooves. The chloren-chyma of the leaves, the cells containing the chloro-plasts, is formed of cells the walls of which are curiously infolded. These cells surround the central vein, in which are found the vascular bundles. Some species have one, others two bundles in a leaf. In the chlorenchyma numerous resin canals are found. These are also present in the bark and in the wood of many gymnosperms. The leaves of most gymnosperms re-

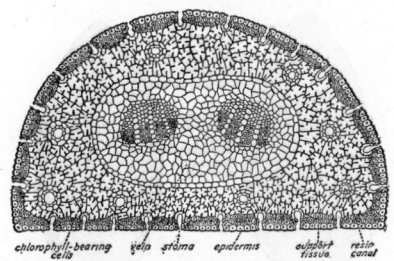

Cross-section of a pine leaf.

main on the tree from 2 to 5 or more years, and even, in some species, persist for as long as 20 years. Those of the larches are deciduous, falling in the autumn, and leaving the branches bare through the winter.

The reproductive structures of the conifers are the cones or strobili, which are of two kinds, staminate and ovulate cones. Usually these are found on the same tree, which is therefore monoecious. The staminate cones, too, often called the male cones, are small and short-

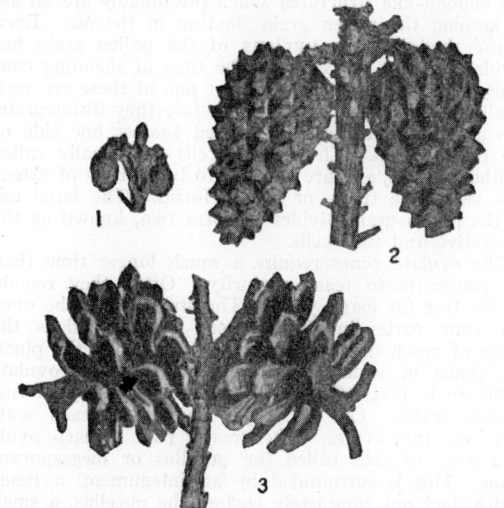

Lodgepole pine, *Pinus Murrayana,* showing development of ovulate strobili. 1, Pair of strobili shortly after pollination; 2, pair of strobili one year old; 3, pair of strobili two years old after the seeds have been shed.

lived. They are generally found in clusters near the tips of the branches. A staminate cone consists of a central axis and a series of spirally arranged microsporophylls. Each microsporophyll bears two pollensacs or microspororangia on its lower surface. Each microsporangium contains numerous cells called microspore mother cells which divide by meiosis to form

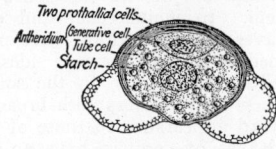

Austrian pine, *Pinus laricio.* Microspore developing into pollen grain. (*Chamberlain's Elements of Plant Science, McGraw-Hill Book Co., Inc.*)

haploid microspores, or pollen grains, four from each microspore mother cell. The number of pollen grains produced is tremendous; often they are liberated from the sporangia in such quantities as to produce what are known as "sulfur showers," which cover the ground with a layer of yellow pollen, or cause the water of ponds to become turbid with the pollen grains. The

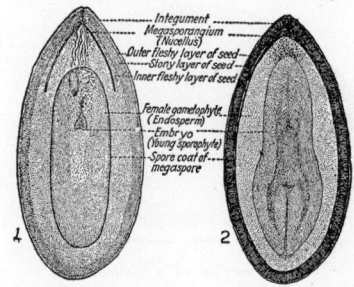

Development of pine seed. 1, two embryos have started and the suspensors have pressed one of them deep into the endosperm; 2, single mature embryo, its mate having failed to develop to maturity. (*Chamberlain's Elements of Plant Science, McGraw-Hill Book Co., Inc.*)

pollen is carried about by the wind, often to distances of many miles. The wall of a pollen grain is composed of two layers, an inner, called intine, and an outer, exine layer. In some conifers the outer layer is separate from the inner in two places, forming conspicuous balloon-like structures which presumably are an aid in keeping the pollen grain floating in the air. Even before shedding, the **nucleus** of the pollen grain has divided, so that the grain at the time of shedding contains three or more cells. All but one of these are very small and last but a short time before they disintegrate, leaving thin disk-like cells pressed against one side of the pollen grain. These small cells are usually called prothallial cells, and are thought to be vestiges of extensive vegetative tissue or earlier forms. The large cell of the pollen grain divides to form two, known as the generative and tube cells.

The ovulate cones require a much longer time than the staminate to reach maturity. Often they remain on the tree for many years. The structure of the ovulate cone varies in the different genera, and is the cause of much discussion in many cases. In the pines, the genus in which the development of the ovulate structure is best known, the cone is made up of numerous scales. On the upper surface of each scale there are two ovules. The greater part of each ovule is a mass of cells called the nucellus or megasporangium. This is surrounded by an integument, a tissue which does not completely enclose the nucellus, a small opening known as the micropyle being left. It is through this opening that the pollen grains reach the surface of the megasporangium. As the ovule develops,

one or sometimes more of the cells in the nucellus becomes distinct from the other cells because of its larger size and denser protoplasm. This is the megaspore mother cell. It divides by meiosis to form a row of

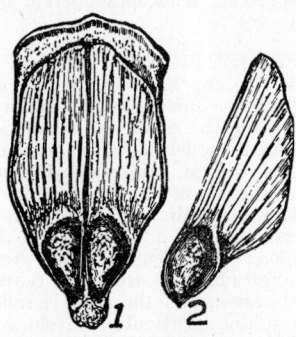

1, megasporophyll and two seeds; 2, a single seed with wing. (*Curtis, Nature and Development of Plants, Henry Holt & Co.*)

four cells, one of which becomes the megaspore, the others degenerating. The single megaspore divides into many cells which form the female gametophyte or megagametophyte. At the end of the megagametophyte nearest the micropyle several **archegonia** develop. Each archegonium contains a single very large egg cell.

The pollen grain, carried by the wind, comes in contact with a small drop of fluid which has been secreted by cells in the region of the micropyle. As this evaporates the pollen grain is drawn down into the micropyle to the surface of the nucellus. There the pollen grain puts out a pollen tube which grows through the

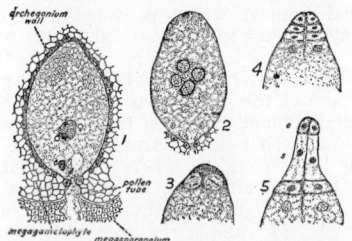

Stages in the early development of a pine sporophyte. 1, fertilization; 2, zygote nucleus has divided into four nuclei which pass to the end of the sac farthest from the micropyle and divide again; 3 and 4, walls have formed between the nuclei; 5, suspensor cells (s) elongate, pushing pro-embryo cells (e) into the megagametophyte. (*Curtis, Nature and Development of Plants, Henry Holt & Co.; after Ferguson and Coulter and Chamberlain.*)

nucellus tissue until it reaches the tip of the megagametophyte. During this development a final division of the generative nucleus has occurred, and two male gametes or sperm nuclei are formed. When the tip of the tube reaches an archegonium these nuclei are discharged into the egg. One of the nuclei passes to the egg nucleus and fuses with it; the other disintegrates. The time required for the pollen tube to reach the egg varies greatly in different genera; in some, like the spruce and the hemlock, it is a matter of a few weeks; in others, like the pine, it is nearly a year.

The nucleus of the fertilized egg passes to the basal end, dividing twice, and forms a rosette of four nuclei. Each of these divides again, forming two tiers of four nuclei. Walls then begin to form, separating the apical tier from the other. Subsequent nuclear divisions increase the number of tiers to four, each composed of four cells. The apical tier presently divides and forms the **embryo**; the second tier, called the suspensor, elongates greatly, shoving the embryo down into the gametophyte tissue. There the embryo absorbs food substances from the tissues around it and matures.

The mature seed of a gymnosperm consists of a hard seed coat, developed from the integument, the nucellar tissue, and the embryo. The latter consists of a straight slender **hypocotyl** and two or more **cotyledons**, and a very small **epicotyl** or plumule. During its development this embryo has been surrounded by a mass of tissue called the **endosperm**, which is megagametophyte tissue.

The seeds of different conifers vary greatly in size. In some species of pine, they are large enough to be used as food for human beings. In the southwestern United States there are several species of pine which bear edible seeds. These are called piñon nuts, and are used in the same way as peanuts are. In southern Europe other species of pine yield edible seeds. One of these, called pignolea nuts, is frequently used in confectionery.

The conifers yield many other valuable products. The **wood** of many of them is extremely valuable, forming the principal timber used in construction work. Its use in this work is largely due to its composition. Each annual ring is composed of two distinct layers, an inner soft layer and an outer hard layer often much darker colored than the other; these give to the wood great strength and also flexibility, and allow easy driving of nails into the wood without splitting it. The wood is also used for **paper pulp**. **Turpentine** and **resin** are obtained from many of the conifers. **Amber** is a fossil resin coming from an extinct conifer. Many conifers are grown as ornamental trees, often in regions far from their natural habitat.

Several genera of conifers are well known for various reasons. Perhaps the best known are the Redwoods, two species of the genus *Sequoia*, which is found in California. *Sequoia gigantea*, the giant redwood, grows in scattered groves on the western slopes of the southern Sierra Nevada mountains in California. The trees may reach a height of 300′ and have massive trunks 12–36′ in diameter. Many of them are 3000 years old or more. Fortunately the wood of this species is a poor timber, so that they are reasonably safe from cutting for lumber. The other species, *Sequoia sempervirens*, grows north of San Francisco near the coast. It also is a tall tree, but more slender than the "big trees," averaging 15′ in diameter. It grows more rapidly, reaching maturity in 1000 years or less. Its soft red wood is an excellent timber and much used.

Another conifer native in the western United States is the Douglas Fir, *Pseudotsuga Douglasii*. This tree, like the redwood, attains great size, growing 300′ or more in height, and 6–10′ in diameter. It is a valuable timber tree.

A third conifer, the cedar of Lebanon, *Cedrus libanotica*, is an attractive tree with numerous spreading branches which grows 70–100′ tall and 5–7′ in diameter.

The Cypress, *Taxodium distichum*, is a tree of the wet lowlands of southern United States and Mexico. It is a large tree, often more than 100′ high and 4–6′ in diameter. When growing in wet swamps, erect woody protuberances known as cypress "knees" grow out from the roots. These have a long conical shape and may be several feet high. The wood of the cypress is soft, straight-grained, and has a peculiar somewhat unpleasant odor. The heartwood varies from red to very dark brown; the sapwood is light. The wood is very resistant to exposure to water, and also not liable to damage by insects. It is used in greenhouse construction, in making vats, and other containers for liquids, in furniture-making and for many other purposes requiring a soft, durable, non-shrinking wood. One of these trees, growing in southern Mexico, is very large and, while undoubtedly very old, is still sound. This is the "Big Tree of Tule," which has a trunk diameter of 50′.

Many species of Firs (*Abies*) and Pine (*Pinus*) also reach great size. But others are remarkable for their ability to endure adverse conditions and thrive. Under these conditions they may remain very small. Specimens of *Pinus Thunbergia*, for example, are much grown in cultivation by the Japanese, who give to them a bizarre yet very attractive shape, while keeping them very small. Some of these dwarf trees may be many years old and yet only a foot or less in height. Grown normally, it is an irregularly branched tree 100′ or more high and 6–7′ in diameter. Other conifers are also used by the Japanese in pot cultures and dwarf gardens (see **Paleobotany**). (R.M.W., B.S.M.)

Isolated tree of the Western Yellow Pine, *Pinus ponderosa*.

CONIINE. Alkaloids.

CONJUGATE FOCI. In photography the interdependent distances between object and lens and lens and image are known as the conjugate distances. As the distance from the object to the lens increases, that from the lens to the image decreases and vice versa.

The formula expressing the geometrical relation between these distances is:

$$\frac{1}{f} = \frac{1}{v} + \frac{1}{u}$$

where f is the focal length of the lens, v the lens-to-image distance and u the object-to-lens distance. Thus:

$$u = \frac{f \times v}{v - f}$$

$$v = \frac{f \times u}{u - f}$$

$$f = \frac{u \times v}{u + v}$$

The ratio of the conjugate distances determines the scale of the image. Thus:

$$R = \frac{v}{u}$$

where R = the relative size of the image to the object and u and v are the object and image distances as before. When u is greater than v the object is larger than the image (reduction); when v is the greater then the image is larger than the object (enlargement).

Other formulae follow:

$$R = \frac{f}{u - f} \qquad R = \frac{v - f}{f}$$

$$f = \frac{u \times R}{R + 1} \qquad f = \frac{v}{R + 1}$$

$$u = \frac{v}{R} \qquad u = \frac{f}{R} + f \qquad u = v \times R \qquad u = (R + 1) \times f$$

$$v = u \times R \qquad v = (f \times R) + f \qquad v = \frac{u}{R} \qquad v = \frac{f}{R} + f$$

D = total image-to-object distance.

$$D = \frac{v \times (R + 1)}{R}$$

$$D = u \times (R + 1)$$

(C.B.N.)

CONJUGATE GEARING. Spur Gearing.

CONJUGATION. A process of interchange of nuclear material between two individual 1-celled animals. The cells come together in a close association and undergo nuclear divisions. One nuclear component of each passes over into the other conjugant and unites with a nuclear

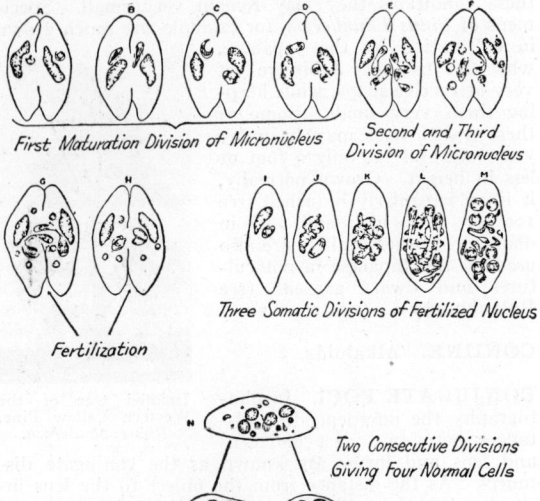

First Maturation Division of Micronucleus

Second and Third Division of Micronucleus

Three Somatic Divisions of Fertilized Nucleus

Fertilization

Two Consecutive Divisions Giving Four Normal Cells

Conjugation in *Paramecium caudatum*. (*After Calkins.*)

derivative of that individual to form a new nucleus. The completion of the process takes place after the separation of the two cells. It is equivalent to **fertilization**. (See **Algae** for conjugation in plants.) (A.W.L.)

CONJUNCTIVA. The thin membrane which covers the eyeball and is continued as a lining of the eyelids. (D.M.H.)

CONJUNCTIVITIS. Inflammation of the **conjunctivae**. This may be caused by wind, dust, or other foreign bodies, or infection. **Staphylococci, gonococci** and other bacteria may be the etiological agents. (D.M.H.)

CONNATE WATER. A term proposed by A. C. Lane in 1908 for underground water which is preserved as the water, fresh or saline, in which the **sediments** were originally deposited. (R.M.F.)

CONNECTING ROD. The common connecting rod is one of the four elements of the mechanism known as the slider crank chain. This mechanism consists of a base which carries two members, one of which rotates while the other reciprocates. The connecting rod connects the reciprocating and rotating elements by means of pinned or hinged joints at its ends, the same constituting the connecting rod bearings. The importance of this mechanism, and of its elements, including the connecting rod, is that it is the basis for a large number of machines of great importance to modern civilization. This is probably due to the fact that in so many cases a reciprocating motion is produced where rotary is desired, and vice versa. Among the more common illustrations of the machines of which the connecting rod is a vital and important part, are **engines, pumps, compressors, punches,** etc. A familiar example is the connecting rod of the gasoline engine. This engine derives its power from the push of the exploding gas against the reciprocating piston. One end of the connecting rod is joined to the piston by the wrist pin on which it has bearing. The other end of the connecting rod has a bearing on the rotating crank pin. Thus the connecting rod has a composite motion: one end of it reciprocates, while the other rotates. It is subject not

only to tension and compressive stress, but also, by virtue of its inertia, to transverse bending. Due to this latter factor, high-speed connecting rods are carefully designed so that their mass will not only be as small as possible, but so placed as to cause the least shaking forces. Commonly used sections are the solid rectangular, the I-beam, and the tubular. Almost all materials, including cast iron, brass, wood, steel, aluminum alloy, have at one time or another been used for connecting rods. (F.T.M.)

CONNECTIVE TISSUE. The connective **tissues,** as the name suggests, are primarily those which bind together other structures. They are derived from the loosely arranged **mesenchyme** of the embryo and are characterized in the adult by the presence of various kinds of cells and of much intercellular substance, also in various forms.

In addition to **bone** and **cartilage** the connective tissues of the adult **vertebrate** include the loose irregularly arranged tissue which underlies the skin and occupies spaces between other organs. In this tissue lie cells which produce its own structures, blood cells, fat cells, and cells which become active in the repair of wounds. Between the cells is a soft matrix containing two kinds of fibers: white and elastic. The white fibers give tensile strength to the tissue and the elastic fibers give elasticity.

The other connective tissues are made up of certain of these structures. Thus **tendons** and **ligaments** are composed principally of parallel white fibers. Fibrous membranes may be either elastic or tough according to the fibers composing them.

Some of the connective tissues have other functions which are not associative. Adipose tissue, for example, is composed largely of cells in which fat is stored.

Some minute parts of many organs are held together by a network of reticular tissue whose fibers run in all directions among the cells of the organ.

The development of these tissues is so extensive that if all other components of the body could be removed, its gross form would still be evident. (A.W.L.)

CONODONTS. Invertebrate Paleontology.

CONRADSON CARBON TEST. This test is an indication of the percentage of carbon residue in an oil, either fuel or lubricating. Where lubricating oil is exposed to high temperatures during use, a certain amount of carbon remains after the volatile parts of it have been volatilized. This is true of the lubrication of the cylinders of the internal combustion engine, one of the objectionable features of which is the deposit of carbon in the cylinder head after continued use. Likewise with fuel oils, especially the Diesel fuel oil, carbon residue is an important characteristic. Although not very well understood at present, even by many of those associated with Diesel power development, recent research indicates that the carbon residue as indicated by tests such as the Conradson, may be expected to be an important factor in the selection of a fuel oil.

The Conradson test consists of heating, under special conditions, a weighed sample of the oil at a sufficiently high temperature to volatilize all of the volatile matter of the oil. Once the distillation is completed, there remains the carbon residue, which is determined by weighing. Although it is objected that this test fails in many important respects to duplicate the conditions under which a carbon residue is formed in internal combustion engines, it is a reasonably successful indicator of the value of an oil with respect to its carbonizing qualities. (F.T.M.)

CONSANGUINITY. As used by geologists this term implied the "blood relationship" of **igneous** rocks. Proposed by Iddings in 1892 to designate a related group of igneous rocks which have been derived from a common **magma** and thus form a distinct **petrographic province.** (R.M.F.)

CONSEQUENT STREAMS. The type of **drainage** pattern which develops on the initial slopes of a land surface newly exposed to **erosion** by running water.

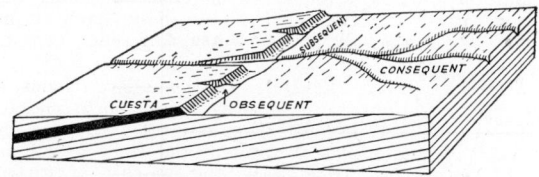

Block diagram to illustrate the meaning and relations of consequent and obsequent streams, in a youthful stage of the normal cycle of erosion, in a region of tilted strata. The more resistant rock layer (in black) stands out in the form of a cuesta, against the erosion.

Some such surfaces, as in coastal plains, are underlain by sedimentary strata gently tilted in the direction of surface slope. If one of the parallel series of tilted formations is more resistant to erosion than the others, a cliff or **cuesta** will be developed. Stream valleys which develop parallel to the cuesta are called subsequent, and their tributaries which cut back into the cuesta are called obsequent. (R.M.F.)

CONSERVATION OF ENERGY. Energy.

CONSISTENT STATISTIC. A statistic is said to be consistent if when calculated from the whole **population** it gives the correct value of the **population parameter.** More precisely as the size of the sample N increases then

$$\lim_{N \to \infty} P\{ |\theta_1 - \theta| > \epsilon \} \to 0,$$

that is, the probability that the statistic θ_1 differs from the population parameter θ by more than ϵ in absolute value approaches zero. The least one should demand of a statistic is that it be consistent. (L.A.A.)

CONSTANT BOILING MIXTURE. Distillation.

CONSTANT CURRENT TRANSFORMER. This is a specially constructed **transformer**, sometimes called a tub transformer, built so the **primary** and **secondary** can move relative to one another under the influence of the forces set up by the load current. The currents in the primary and secondary react on one another to produce a repelling force between the two windings, the greater the current the greater the force. The secondary is movable along the core and is partially balanced by a counter-weight which is adjusted so the current forces will produce just the right reaction to hold the current essentially constant. An increase of current increases the force and hence the separation, which in turn increases the leakage flux between the windings and lowers secondary voltage. The lowered voltage causes a reduction of the current to a value which, when equilibrium is reached, is practically constant. Reduction of the load current causes an opposite sequence of reactions. These transformers are used in series street-lighting circuits to maintain constant current when the lamps short out because of trouble. (L.R.Q.)

CONSTANT OF INTEGRATION. Indefinite Integral.

CONSTANTS. A constant is a symbol for a single number.

An absolute constant is one which has always the same value. The symbol for such a constant may be a particular number-symbol, as 5 or ⅓ or $\sqrt{2}$, or it may

be a special symbol always representing the same number, as $\pi \approx 3.1416$.

An arbitrary constant is one which has only one particular value in a given case, but may have another value in another case.

Arbitrary constants are often represented by letters, as a, b, c, etc., from the first part of the alphabet. (L.L.S.)

CONSTANTS, LINE. Line, Transmission.

CONSTELLATIONS. In astronomy this term is used to designate certain groupings of the **stars.** From earliest recorded history we find that the larger star groups (constellations), the smaller groups (asterisms such as the **Pleiades**), and the individual stars have received names symbolizing meteorological, religious, or mythological beliefs. The idea that the constellation names and myths are of Greek origin has been quite completely disproved. It seems highly probable that they are of Semetic or Pre-Semetic origin and that they found their way into Greece through contact with the Phoenicians (sailors who used the stars constantly in their profession).

The oldest record of actual constellation listing is found in the Creation Legend in about 650 B.C. This Legend was recorded on Cuneiform from even earlier records. From this time onwards frequent references to the constellation legends are to be found both in poetical and historical writings. The basis for the modern constellation division is to be found in the list of 48 constellations published by Ptolemy in about 150 A.D. This list of Ptolemy is based upon the writings of his predecessors, notably Hipparchus.

The boundaries of Ptolemy's constellations were very indefinite and many visible stars were left out entirely. Furthermore, his list only covered that portion of the

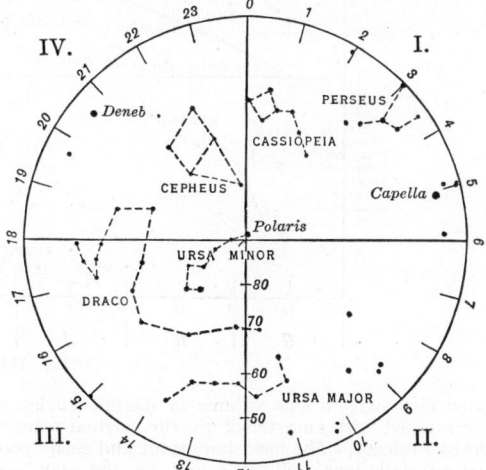

Fig. I. Circumpolar stars.

heavens visible from the southern Mediterranean regions. In the 1800 years since Ptolemy's time the list has been added to and the boundaries defined until at present all stars are included in some one of the constellations. The International Astronomical Union placed the matter of defining the constellation boundaries in the hands of a special committee and in 1930 the final list was published.

Several attempts have been made to supplant the ancient mythological names with more modern ones (e.g., the Coelum Stellatum Christianum of Julius Schiller in 1627 in which we find the ancient names replaced by names of various Church dignitaries) but none of the attempts have been successful.

The following list contains the names of all of the constellations now used; those in bold-face type are

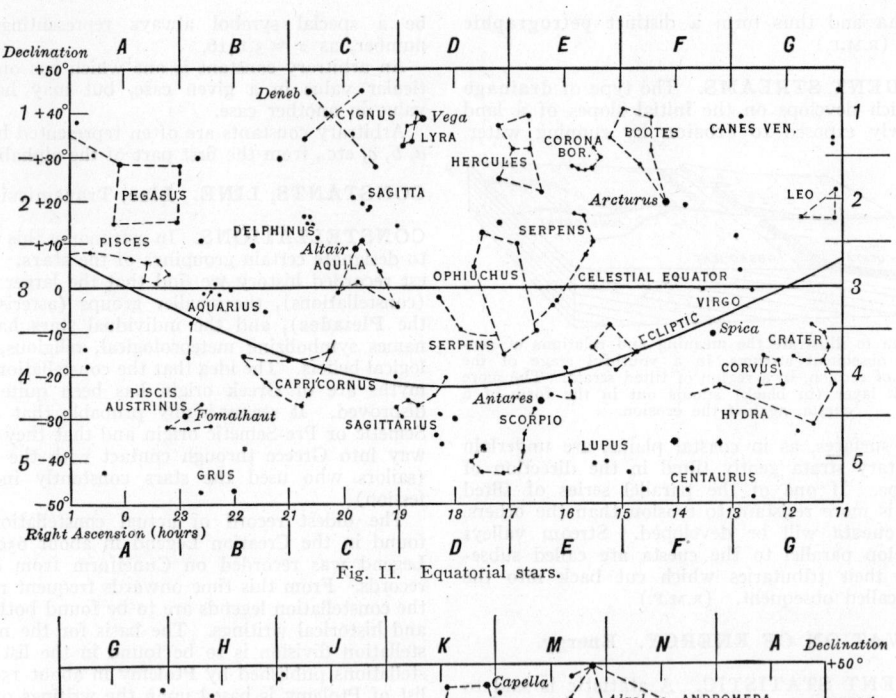

Fig. II. Equatorial stars.

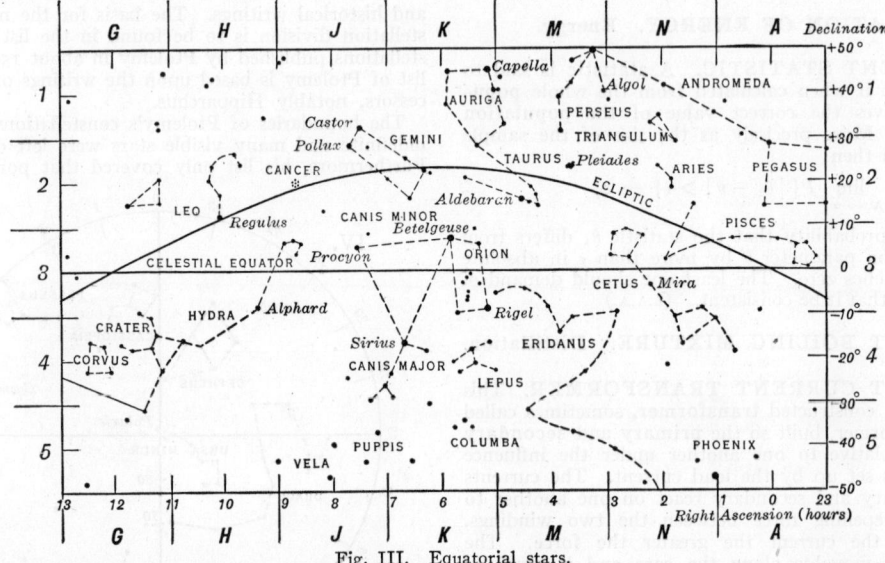

Fig. III. Equatorial stars.

treated elsewhere in this volume in special articles, and those marked with an asterisk are the original constellations of Ptolemy. The most important and easily recognized constellations will be found on the star maps, at the top of the page, and the numbers (e.g., Aries—Fig. III, 2, N) refer to the particular map and location on that map where the constellation will be found.

LIST OF CONSTELLATIONS

North of the Ecliptic

* **Andromeda**—Fig. III, 1, N
* **Aquila**—Fig. II, 3, C
* **Auriga**
* **Bootes**—Fig. II, 1, F
  Camelopardalis
  Canes Venatici
* **Cassiopeia**—Fig. I, I
* **Cepheus**—Fig. I, IV
  Coma Berenices
* **Corona Borealis**
* **Cygnus**—Fig. II, 1, C

* **Delphinus**
* **Draco**
  Equuleus
* **Hercules**—Fig. II, 2, E
  Lacerta
  Leo Minor
  Lynx
* **Lyra**—Fig. II, 1, D
* **Ophiuchus**
* **Pegasus**
* **Perseus**—Fig. III, 1, M

* **Sagitta**
* **Serpens**
* **Triangulum**

Zodiacal

* **Aquarius**—Fig. II, 3, B
* **Aries**—Fig. III, 2, N
* **Cancer**—Fig. III, 2, J
* **Capricornus**—Fig. II, 4, B
* **Gemini**—Fig. II, 2, K
* **Leo**—Fig. III, 2, H
* **Libra**—Fig. II, 4, E

* **Ursa Major**—Fig. I, II
* **Ursa Minor**—Fig. I, III
  Vulpecula

* **Pisces**—Fig. III, 3, A
* **Sagittarius**—Fig. II, 4, D
* **Scorpius**—Fig. II, 5, E
* **Taurus**—Fig. III, 2, M
* **Virgo**—Fig. II, 3, F

South of Ecliptic

  Antilia
  Apus
* **Ara**
  Caelum
* **Canis Major**—Fig. III, 4, K
* **Canis Minor**

  Carina
* **Centaurus**—Fig. II, 5, F
* **Cetus**
  Chamaeleon
  Circinus
  Columba
* **Corona Australis**

* Corvus—Fig. II, 4, G	Norma
* Crater	Octans
* Crux	* Orion—Fig. III, 3, K
Dorado	Pavo
* Eridanus	Phoenix
Fornax	Pictor
Grus	* Piscis Austrinus
Horolgium	Puppis
* Hydra	Pyxis
Hydrus	Reticulum
Indus	Sculptor
* Lepus	Scutum
* Lupus	Sextans
Mensa •	Telescopium
Microscopium	Traingulum Australe
Monoceros	Tucana
Musca	Vela
	Volans

(w.k.g.)

CONSTIPATION. Infrequent or delayed passage of the intestinal contents. One of the most common causes in a normal person is neglect of the impulse to move the bowels. The stimulus gradually lessens or is lost entirely. Two factors are involved which prevent formation of a habit cycle of a daily bowel movement at a certain regular time. They are laziness and failure to adjust the time of impulse to the time of evacuation.

Constipation is common in those who take no exercise, and consequently suffer from an impairment of muscular tone. In other cases diet is at fault. It may be too concentrated, so that it leaves too little residue or bulk to enable the intestinal muscles to function properly. On the other hand, the diet may contain too much roughage, producing an irritated spastic colon. Certain individuals suffer from a spastic colon due to reflex nervous causes and normal intestinal function is prevented. Instead, the large intestine is usually in a state of spasm or contraction preventing normal passage of intestinal contents through it. Laxatives only increase the spasm; many functional symptoms accompany this condition with complaints of indigestion, abdominal pain and headache predominating.

The question of habit is an important one in the subject of constipation. Frequently the cause is a continued lack of bowel habit training beginning in infancy. Again, there are many individuals who believe, in some cases due to the pernicious influence of cathartic advertising, that they must "help Nature along." Gradually the artificial stimulus has to be increased and the dose must be increased or a stronger cathartic taken. Eventually the various drugs have little or no effect. The same applies to the enema habit.

In individuals with generally poor muscle tone there is insufficient tone to the intestinal muscles so that there is always a stasis with delay in passage of the intestinal contents, and constipation.

There are many local causes of constipation. These include weakness of the abdominal muscles, hemorrhoids or other painful disorders causing spasm of the sphincter muscles at the lower rectum, tumors either within the intestine or causing pressure from without, adhesions, bands or any structural defect or abnormality causing a partial obstruction of the intestines.

The symptoms of constipation depend largely on the cause of the condition. Many mistaken ideas about constipation have gained wide popular acceptance. Intestinal "auto-intoxication" is largely a myth. Symptoms from constipation occur chiefly in those whose nervous systems, especially the autonomic system, are easily disturbed—this is true particularly of those who suffer from spastic colitis. It must be remembered that certain individuals may normally have one, two, or three bowel movements a day, others may normally have a movement every other day. In other words, the frequency of

bowel movements depends on habit. Certain individuals may have a daily bowel movement and still be constipated—producing a hard irritating movement instead of a normal semi-solid one.

Treatment of constipation involves correction of the cause and considerable habit training coupled with a satisfactory diet and avoidance of all laxative preparations and drugs. (r.s.m., d.m.h.)

CONSTRAINED MOTION. Kinetics.

CONSTRAINT. Constraint is that property which distinguishes a mechanism from other mechanical linkages. A mechanism has constrained motion in that a motion of one part is followed by a predetermined motion of the remainder of the mechanism. To determine whether a mechanical linkage is a mechanism or not, Klein advocates applying the criterion of constraint, which he writes as follows:

$$J = \frac{3N - 4 + \gamma - P}{2}$$

in which J is the number of joints in the mechanism, N is the number of links in the mechanism, γ is the number of independent prismatic chains, that is, those whose joints are of the sliding type, $P =$ the number of point or line type of contact joints in the mechanism. When this equation yields an identity, the mechanism is said to be constrained for all dimensions. (f.t.m.)

CONSUMER'S RISK. Suppose a consumer accepts or rejects lots on the basis of **samples** chosen from such lots. The consumer's risk then is the **probability** of accepting lots on the basis of the sample inspection when such lots should in reality be rejected. The consumer's risk arises from the **error** of accepting lots which should be rejected. (l.a.a.)

CONSUMPTION. Tuberculosis; Pulmonary Diseases.

CONTACT FLYING. Conducting an airplane from place to place by constant reference to surface features on the earth is known as contact flying. The Civil Air Regulations of the Civil Aeronautics Administration are very specific in defining the amount of cloud, the range of visibility, and the height of ceiling with which contact flying may be attempted. Adherence to these rules will result in the view of enough surface features to permit the safe navigation of the plane without reference to instruments of navigation such as the **compass, radio,** etc., although Civil Air Regulations specify that certain of these instruments must be carried on the plane and in good condition. (w.k.g.)

CONTACT METAMORPHISM. Metamorphism.

CONTACT POTENTIAL DIFFERENCE. In his experiments with electroscopes, Volta found that when pieces of two different metals, otherwise insulated, are brought into contact, they acquire opposite charges and maintain a difference of electrical potential even while still touching. This potential difference he found to be characteristic of the given pair of metals. Thus when the metals are iron and copper, the iron has a potential about 0.15 volt higher than the copper, while for tin and iron the difference is 0.31 volt, tin being the higher. Volta listed a series of several metals, viz., zinc, lead, tin, iron, copper, silver, gold, such that when any two are put in contact, the one first named is at the higher potential. "Volta's law," which he was not in position to demonstrate but which was established much later, states that the potential difference between any two metals in direct contact is the sum of the potential differences between intervening metals of the series. Thus for tin and copper (above) it is 0.31 volt + 0.15 volt = 0.46 volt; and it makes no difference whether the tin and copper are in direct contact or have other metals intervening between them.

A distinction must be made between the contact potentials in air and the so-called "intrinsic" contact potentials in a vacuum with all adsorbed gases removed. According to Millikan, the intrinsic potential difference between two metals A and B is expressed by $V_{AB} = h(\nu_A - \nu_B)/e$, in which h in **Planck's constant**, ν_A and ν_B are the critical frequencies of photoelectric emission for the two metals (see **Photoelectric Phenomena**), and e is the electronic charge. In any case, if the electronic **work functions** of the metals are p_A and p_B, the contact potential difference is $V_{AB} = (p_A - p_B)/e$. The work functions, and hence V_{AB}, are in general dependent upon the medium surrounding the metals. Accurate measurements of these potentials are, unfortunately, very difficult. (L.D.W.)

CONTACT PRINTING, PHOTOGRAPHIC.

In contact printing the photographic paper is exposed in contact with the negative. During exposure the exposing light passes through the back of the negative to the coated side of the paper; emulsion side of negative facing emulsion side of the print. Absolute contact at all points between negative and paper is necessary, if a sharp print is to be obtained. Contact printers should be checked frequently for uniform pressure over entire printing surface.

Prints with white margins are obtained by placing the negative on opaque, or non-actinic, masks, or the mask may be placed between negative and print. The latter position is usually adopted when printing glass plates.

Printing is controlled either by varying the length of the time of the exposure, keeping the intensity constant or varying the intensity of illumination and keeping the exposure time constant. The illumination is varied by increasing or decreasing the distance between light source and negative, or by using lamps of different power. The color temperature of the lamps has an effect on the time of exposure and on print contrast, but this is not critical with contact papers as these are sensitive primarily to the blue, violet and ultra-violet portions of the spectrum.

The simplest method of exposing prints is with a lamp and printing frame, as shown in the accompanying illustration. The printing frame is loaded by placing

(*Neblette's Photography.*)

glass-side down on table, loosening the springs on the back, and lifting the back out. The glass, which must be free from imperfections, is cleaned, and the frame

dusted and a mask with the desired border is placed on glass first, then the negative with the emulsion-side up. To this is added, in proper safelight illumination, a sheet of sensitive paper, emulsion-side down. The back is replaced and springs are clamped down. While this method of printing is slow, it does offer excellent opportunities for shading of light areas and printing-in of dark portions of the negative.

The exposure on different parts of the negative, or "dodging," is accomplished by placing tissues and cutouts on glass plates under the printing surface or, on some machines, by raising the printing head and exposing local areas with a small spotlight held in the hand. (S.M.T.)

CONTAGION.

The communication of disease by (1) immediate contact, (2) contact with excretions or secretions of a sick person, e.g., saliva by coughing, sneezing or talking, etc. (R.S.M.)

CONTINGENCY TABLE.

A contingency table consists of r rows and s columns. It is similar to a correlation table but one of the **variables** (and usually both) is **qualitative** instead of **quantitative**. (L.A.A.)

CONTINUITY, PRINCIPLE OF.

In brief, this is the principle that mass is indestructible and may be completely accounted for at different points of a steady flow even though physical changes have occurred. Also the mass of chemical elements which undergo chemical change during flow, becoming disassociated from or associated with other elements, may be identified at two states in a flow through application of the principle of continuity.

If an incompressible fluid flows steadily along a conduit past two stations where the cross-sectional areas of the fluid are found to be a_1 and a_2, respectively, then the mean fluid velocities v_1 and v_2 are related thus:

$$a_1 v_1 = a_2 v_2.$$

Had the fluid been compressible, then the continuity of mass would be exemplified by

$$a_1 v_1 d_1 = a_2 v_2 d_2$$

wherein d is the *density* of the fluid.

Again, suppose a fuel containing C lb. of carbon per lb. fuel is burned with air and the resulting products of combustion are analyzed for X lb. of CO_2 and Y lb. CO per lb. of gases. X and Y may be mathematically broken down and their carbon contents C_x and C_y established. Then $C_x + C_y =$ weight of carbon per lb. gases. "Carbon continuity" implies that $C_x + C_y$ was continuously derived from C and so useful knowledge of quantity of gaseous products per lb. fuel burned may be obtained by expressing this continuity mathematically.

$$\frac{\text{Lbs. of gaseous products}}{\text{Lb. fuel}} = \frac{C}{C_x + C_y}.$$

This principle is susceptible of useful application to numerous other problems in the field of mechanical and chemical engineering. (F.T.M.)

CONTINUOUS BEAM.

A continuous beam is one which has more than two supports. A beam which is continuous over several supports offers more difficulty in analysis than would a series of freely supported beams covering the same overall span. Because of the restraint at the intermediate supports, the continuous beam can carry a greater load than a simple beam of the same size and span. It is quite important to provide a firm **foundation** for the intermediate supports, as small **deflections** due to sinking of an intermediate support may introduce **stresses** of an entirely different nature and magnitude from those used in designing the beam. (F.T.M.)

CONTINUOUS FUNCTION.

A function of one variable $f(x)$ is said to be continuous at a value $x = c$ when $f(c)$ has a definite finite value which is equal to the limit of $f(x)$ when $x \to c$:

$$\lim_{x \to c} f(x) = f(c).$$

This may also be expressed as follows: $f(x)$ is continuous at $x = c$ if $f(c)$ exists, and if for any arbitrary number $\epsilon > o$ there exists another number δ such that $\left| f(c) - f(x) \right| < \epsilon$ for all values of x such that $\left| x - c \right| < \delta$.

A function $f(x)$ is said to be continuous in an interval (a, b) when it is continuous at every point of the interval, it being sufficient at the end points that

$$\lim_{x \to a^+} f(x) = f(a),$$

and

$$\lim_{x \to b^-} f(x) = f(b).$$

A function is said to have a discontinuity at a point where it is not continuous. The usual type of discontinuity is a point at which the function becomes infinite, or where it has a finite jump.

Important properties of a continuous function are the following:

If $f(x)$ is continuous in a closed interval (a,b), then among the different values of $f(x)$ in (a, b), there is a greatest value M and a least value m.

If $f(x)$ is continuous in an interval (a, b), then between every two values x_1 and x_2 of x in (a, b), $f(x)$ takes at least once every value between $f(x_1)$ and $f(x_2)$.

A function of two variables, $f(x,y)$, is said to be continuous at a point (a, b) if

$$\lim_{\substack{x \to a \\ y \to b}} f(x, y) = f(a, b). \qquad \text{(L.L.S.)}$$

CONTINUOUS SPECTRUM.

Light or other radiation may have such composition that, when analyzed with a **spectroscope** it presents apparently an unbroken continuity of wavelength over a wide range. Such, for example, is the light from an ordinary lamp filament. Any incandescent solid, liquid, or gas under high pressure will radiate with a continuous **spectrum**. In contrast to this, the spectrum of a glowing gas, as in a **neon** tube, is made up mainly of sharply defined maxima ("lines") or of more diffuse bands with dark spaces between them. Sunlight appears to have a continuous spectrum until analyzed carefully, when the continuous spectrum is found to be crossed by a multitude of dark **Fraunhofer lines**.

A continuous spectrum may extend without interruption from the extreme **infra-red** to the extreme **ultraviolet**. Study with a **spectrophotometer** will indicate that the relative intensity of the radiation differs in different wavelengths, having a well marked maximum at some particular wavelength. The position of this maximum intensity is a function of the temperature of the source, and the study of **spectral energy distribution** in stellar spectra provides a method for the determination of the temperatures of the stars. (L.D.W.)

CONTINUOUS VARIATE.

A continuous variate is one which may have any value either integral, fractional, or irrational within the range of variation. Examples of continuous variates are the weights of objects, the ages of people, etc. A **variate** which is not continuous is said to be **discrete**. (L.A.A.)

CONTINUOUS WAVE.

Continuous wave transmission in radio is the standard type now for radio telegraph communication. It means a wave which is constant in amplitude, as opposed to the varying amplitude of a modulated wave, and which changes only during **keying**. For transmission of the dots and dashes of the telegraph code the wave is transmitted at constant amplitude for a time corresponding to the length of the dot or dash. This type transmission when combined with modern receiving systems produces much better reception than the previously used **damped waves** or **interrupted continuous waves**. A much narrower **frequency channel** is required for this type than for the other types. Reception is by the heterodyne principle of beating another locally generated signal (see **Beat Frequency Oscillator**) to produce an audible note. (L.R.Q.)

CONTOUR.

A contour is an outline or a boundary line, or it is a line passing through a series of points, all of which have some one characteristic in common. The contour is a useful means of displaying on one plane described by x and y coordinates, the interrelation of three variables, x, y, and z. Each contour line drawn upon the xy plane represents a single value of the variable z.

In surveying terms, a contour is a means by which the three dimensions necessary to represent a point in space can be shown on a map. These three variables are, length measured north and south, length measured east and west, and height, above some arbitrary reference plane. On a **topographical map** height is shown by contours. All points on a given contour line will be at the same elevation. The accompanying figure shows

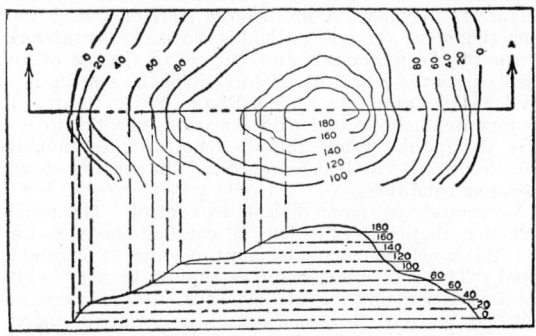

Contours.

the relation between height of a small irregular hill, and the corresponding contour lines as drawn upon a map. Occasionally special maps, instead of showing elevation as the third variable, give line of constant magnetic **declination**. These lines are called isogonic lines and the map is known as an isogonic chart.

While the contour line has its greatest use in land surveying and mapping, there are other technical uses. For example, the efficiency of a centrifugal pump varies both with the pumping head and with the discharge. Accordingly, centrifugal pump characteristics are often given with the efficiency displayed as contour lines upon a head-discharge plane. (F.T.M.)

CONTRACTILITY.

The fundamental property of living matter on which its power of movement depends. In the simplest forms of living things it is evident in a flowing movement of the material of the **cell**. In more complex forms the property is centralized in muscle tissue. Muscle cells are elongate and are so arranged that necessary movements result from their shortening when stimulated. Chemical explanations of contractility have been formulated but they are theoretical. (A.W.L.)

CONTROL.

An experiment or test done to confirm or to rule out error in clinical or experimental observations. (R.S.M.)

CONTROL COLUMN.

A lever having a rotatable wheel mounted at its upper end for operating the longitudinal and lateral control surfaces of an airplane. This type of control is called "wheel control."

A control stick is the vertical lever by means of which the longitudinal and lateral control surfaces of an airplane are operated. The elevator is operated by a fore-and-aft movement of the stick; the ailerons, by a side-to-side movement. (F.T.M.)

CONTROL, MOTOR.
Two major types of control of motors are needed in the usual industrial applications. For any but the smallest motors some type of control must be provided for starting the motor without throwing excessive load on the line or excessive current surges on the motor. For this type see starter, motor. The second type of control which is often needed is speed control. This may be accomplished in several ways, depending upon the type motor and the required degree of control. D-c motors are more easily controlled than a-c ones and consequently are frequently used where a high degree of control is required. Shunt- or compound-wound machines are normally used and in either case the speed may be increased by putting resistance in the shunt-field circuit. This gives an efficient means of speeding up the motor since the field current is small and hence the loss in the resistance is small. To slow down such a motor it is necessary to decrease the armature voltage. This may be done by inserting resistance in the armature circuit, but this is very inefficient since the armature current is high, causing the loss in the resistance to be high. A much more efficient and at the same time more versatile method is the use of thyratrons to control the motor armature current or the use of an auxiliary motor-generator set to control the voltage applied to the armature. (See Phase Shift Control.) Motors are also operated with armatures in series to reduce the speed. While this does not give continuous speed control by itself it does give an efficient low-speed operating condition.

A-c motors are more difficult to control. The usual induction motor is essentially a constant-speed motor and the synchronous motor must operate at constant speed. The a-c series motor is subject to very wide speed variation under load, just as its d-c counterpart, and may be controlled, inefficiently, by series resistance. The most common method of controlling the speed of an a-c motor is the rotor resistance control of a wound-rotor induction motor. The rotor windings are connected to the outside circuit through slip rings and brushes so extra resistance may be inserted. For fixed load the speed of the motor varies with the rotor resistance so this gives a method of adjusting the speed. It corresponds to the d-c control with armature resistance and is inefficient. Special types of motors have been developed and elaborate control systems devised but their use is not too common. Concatenation is another method of getting reduced speed operation where continuous control is not necessary. By a combination of concatenation and rotor resistance control better efficiency with complete control is obtained. (L.R.Q.)

CONTROLLABILITY.
The quality of an aircraft that determines the ease of operating its controls and/or the effectiveness of displacement of the controls in producing change in its attitude in flight. (F.T.M.)

CONTROL SURFACE, AIRPLANE.
The control surfaces of an airplane are movable airfoils arranged to be moved by the pilot to change the attitude of the airplane during flight.

The primary surfaces of a monoplane are (1) ailerons, one on each wing, (2) a rudder, and (3) an elevator. Operation of these impresses (1) rolling, (2) yawing, and (3) pitching motions on the airplane. Roll and yaw are not independent and a roll may induce a yaw without the rudder being operated; similarly, yaw with rudder may induce a roll.

The secondary surfaces are those designed to reduce the force necessary to operate the primary surfaces. These consist of trim and balance tabs.

Besides the above, flaps and air brakes might be considered as control surfaces.

The customary position of these control surfaces is:

1. Ailerons. Form a hinged rear section of the wing, occupying 10–20% of the chord and 20–50% of the span.

2. Rudder. Hinged to the rear edge of the vertical fin.

3. Elevator. Hinged to the rear edge of the stabilizer. (F.T.M.)

CONTROLS, AIRPLANE FLIGHT.
Flight controls of an airplane are installed to give the pilot control over the three types of motion possible around its center of gravity, viz., pitch, roll, and yaw. The primary flight controls are those which operate elevator, aileron, and rudder. Secondary flight controls operate the trim tabs or stabilizer adjustment, and flaps.

The pilot gives his principal attention to the primary controls which are actuated by a control stick, or column, and pedals. Principles of both wheel and stick control are illustrated. It is observed that the wheel or stick

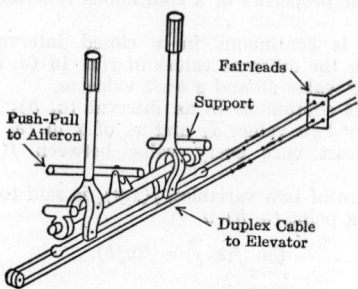

Fig. 1.   Stick control (tandem).

controls only ailerons and elevator, for in any case the rudder is controlled by foot pedals. Wheel and column control may be applied to all sizes of airplanes, but stick control is restricted to the smaller airplanes. There are four possible dual control arrangements as either wheel or stick control could be used with either

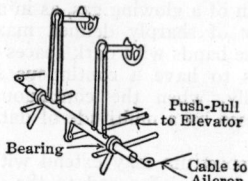

Fig. 2.   Wheel control column (side by side).

tandem or side-by-side seating. Only two of these are illustrated. The control motions required are logically consistent with the resulting airplane motion. A backward pull on wheel or stick and the nose rises, a push forward and the nose drops into a glide or dive. Simi-

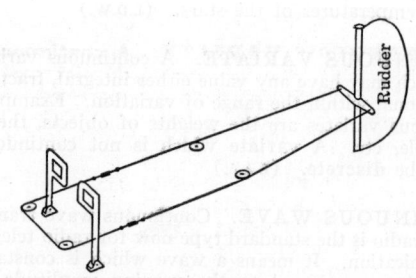

Fig. 3.   Simple pedal and cable rudder control.

larly, a left movement of the stick, or counter-clockwise turn of the wheel, rolls the airplane counter-clockwise (as viewed by the pilot), while a push on the left rud-

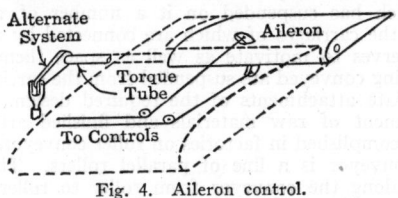

Fig. 4. Aileron control.

der pedal swings the airplane into a left yaw. The airplane may be pitched without engaging in roll or yaw, but if it is rolled it will also yaw and vice versa. Hence lateral stick or wheel motion must be correlated with pedal action for smooth banks and turns. Some attempts have been made to simplify the controls to two elements. Theoretically either aileron or rudder control might be omitted if the airplane were strongly stable laterally or directionally, since a banked turn may be accomplished by use of ailerons or rudder alone. Whether the sacrifice of maneuverability is worth the advantage of simplified control is debatable.

Three common methods of transmitting motion from controls to the surfaces they actuate are (1) flexible cables, (2) push-pull rods and tubes, (3) torque tubes. Where transmission distances are great as in transport planes, cables are favored because of low weight, freedom from vibration, and absence of lost motion. Pulleys, rollers, fairleads (guides of a non-abrasive material), bellcranks, and levers are required to support, guide, and change the direction of the tubes or cables. The details of their assembly vary with every make and model of airplane.

Secondary controls are principally those operating trimming devices, the most common of these being a **tab** located in the trailing edge of the control surface. These are only infrequently operated in flight. They are often not essential but provide for pilot comfort by allowing the airplane to be trimmed for "hands-off" (of the primary controls) flight in a predetermined heading. Tabs are generally cable-operated from cranks, wheels, or dials, located convenient to the pilot.

The maneuvering forces developed by the control surfaces are many times the force required on the controls by the pilot. It is not uncommon to have 10 lbs. pullback on the control stick produce 100–200 lbs. force on the tail surfaces. Only a small part of this multiplying action can be credited to the mechanical advantage of the control linkage. Furthermore, a relatively small portion of the incremental maneuvering force is developed on the control surface itself. One looks to the effect of the control surface displacement on the surface to which it is attached for an understanding of the magnitude of the forces involved. The rotation of a hinged surface, such as an elevator, about its hinge line changes the effective *camber* of the entire surface of the stabilizer. Similarly, an aileron movement changes the effective camber of the wing ahead of it. As camber of an airfoil is increased the lift on it increases. A small movement of the elevator may therefore produce large forces, which act principally on the fixed stabilizer surface.

The control forces exerted by the pilot are therefore only a small part of the total maneuvering force acting; however, they may be uncomfortably large if the airplane is a big one. Larger airplanes develop heavy forces on the control surfaces and only by trim adjustment could the pilot manually control the airplane without excessive fatigue. Such craft are usually equipped with an **automatic pilot** which will relieve the pilot of much of the effort required in piloting larger airplanes. (F.T.M.)

CONURE. Aves, Psittaciformes. Small **parrots** of numerous species found from Mexico into South America. Their prevailing colors are green and yellow. The Carolina paraquet of North America is a member of this group. (A.W.L.)

CONUS ARTERIOSUS. Bulbus cordis.

CONVECTION. Thermal Convection.

CONVECTION CURRENTS. Air currents that travel vertically. Convection can be either mechanical or thermal. Orographical and frontal lifting are mechanical convections of whole layers of air; whirlwinds, dust devils, air mass, cumulus clouds and thunderstorms are results of small-scale convection. (P.E.K.)

CONVERGENCE AND DIVERGENCE. If an imaginary box is erected in the atmosphere near the earth's surface in such a manner that its base, top, and

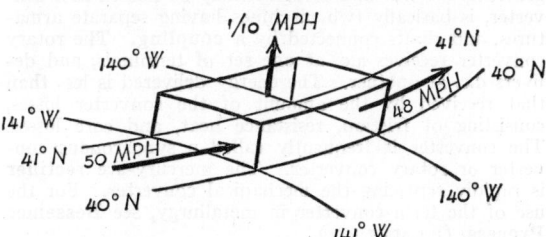

sides are parallel to the air flow, i.e., the winds, it is possible to illustrate the effects of convergence and divergence. When air flows uniformly through this box there will be no accumulation or diminution of air inside the box. If, for any reason, however, more air flows into one end of the box than flows out the other, there is an accumulation of air which must seek an outlet. Because the pressure is less at the top of the box than at the bottom, this accumulated air flows upward out of the box. If we followed a cube-shaped unit mass of air through this flow it would become distorted into a rectangular prism elongated vertically. This process is called convergence and results in a field of rising air. Converging, and therefore rising air can, if the process endures over a sufficient period of time, produce clouds and precipitation. It also tends to destabilize the air. If, in the same box, less air flows into one end than flows out the other, there is an air diminution and space is available at the top of the box for more air. One unit cube of air will be flattened into a rectangular prism elongated horizontally. This process is called divergence and results in a field of sinking or subsiding air. Divergence, therefore, and subsiding air tend to stabilize the air. Clouds and turbulence diminish in regions of divergence and subsidence. (P.E.K.)

CONVERSION EFFICIENCY. This term is applied to vacuum tube **amplifier** and **oscillator** circuits and denotes the a-c (usually radio frequency) output divided by the d-c input to the plate circuit. It is also called the plate efficiency. It varies quite widely with different classes of amplifiers, running as high as about 80% in some class C amplifiers, from 30 to 70% (depending upon conditions) in class B and being of the order of 20% in the usual class A amplifier. (L.R.Q.)

CONVERSION TRANSCONDUCTANCE. This is one of the vacuum-tube coefficients which is valuable in predicting the behavior of the **tube** in a circuit. It is applied to the mixer tube (first detector or converter) of a **superheterodyne** receiver and may be defined as the intermediate frequency current in the plate circuit divided by the applied radio-frequency voltage at the control **grid**. It is also used in connection with detectors where it gives a measure of the audio-frequency output to be expected for a given modulated radio-frequency input. (L.R.Q.)

CONVERGENCY OF SERIES. Infinite Series.

CONVERTER. In certain fields of engineering the nature of the desired electrical power may be different from that available. For example, an electric railway having d-c motors propelling its rolling stock may obtain its electrical energy in the form of alternating current. This situation calls for the conversion of a.c. to d.c. with little loss. The converter accomplishes this. If a conversion is made from d.c. to a.c., we have an inverted converter. The most common type of rotating converter at present is the rotary converter, which is essentially an alternator and d-c generator combined in one machine having a single armature and a single-field circuit. This is not to be confused with a motor-generator set which, although it may be classed as a converter, is basically two machines having separate armatures, and shafts connected by a coupling. The rotary converter receives a.c. at one set of terminals, and delivers d.c. at another. The energy delivered is less than that received by the amount of the converter losses, consisting of friction, resistance heat, and core losses. The converter is frequently called a synchronous converter or rotary converter. The mercury-arc rectifier is rapidly replacing the mechanical converter. For the use of the term converter in metallurgy, see Bessemer Process. (F.T.M., L.R.Q.)

CONVEYOR. Strictly speaking, a conveyor is any device which is capable of moving material from one point to another. The material may be moved intermittently, that is to say, in a succession of separate loads on the conveyor system, or continuously. The term furthermore embraces the conveying both in horizontal or vertical directions, or a combination of the two. This general definition of a conveyor would include a vast number of cases which it is not proposed to treat here. Ordinarily, a conveyor is thought of as a mechanical conveyor constituting a definite installation fixed in position except for its belts, buckets, sprockets, and other moving parts. Equipment for vertical conveying would come more properly under the designation of hoist. In many instances materials are conveyed at an angle to the horizontal, in which case the conveyor system partakes of the nature of a hoist. A conveyor may be designed either for handling loose bulk material such as coal, sand, ore, etc., or for handling individual parts during or after production.

One of the simplest and most-used conveyors consists of a belt, usually horizontal (although it is possible to use a belt conveyor at small angles), continuous over two end pulleys, one of which drives it. The belt is supported at intermediate points on rollers. For transporting material in bulk, the rollers are arranged so that the belt forms a trough, but for handling packages, as on a factory conveying system, the belt would be flat. Another simple form of conveyor is the drag conveyor, which, in its primitive form, consists of a flat, heavy chain which is caused to move along the bottom of a trough, and which drags with it the material being conveyed. The friction losses are comparatively heavy in this type. The flight conveyor is an improvement, since vertical wood or steel plates are attached to the chain and nearly fill the trough from side to side. These effect a much more thorough movement of material, and a reduction of friction and wear. The flight conveyor may be used at fairly steep angles.

Mass-production methods depend largely upon the successful application of conveyor systems to the enterprise. The mass production of automobiles is one of the best examples. Economies effected by maintaining the workman at a station and moving the material past him on conveyors have made possible the production of higher quality vehicles at moderate prices. The trolley conveyor is frequently used where parts are being fabricated on an assembly line. It is an overhead steel monorail which has suspended on it a number of 2-wheel trolleys the carriages of which are connected by a chain which serves to motivate as well as space them. The parts being conveyed are suspended from the carriages by appropriate attachments of the required design.

Movement of raw materials and finished articles is often accomplished in factories on roller conveyors. The roller conveyor is a line of parallel rollers. The load travels along the conveyor from roller to roller under the influence of a gravity force component derived from a slight slope of the conveyor. Articles having flat surfaces, such as boxes, may be loaded directly on such a conveyor.

An apron conveyor consists of two chains upon which are mounted a number of flat metal plates or boards which are small enough to allow the chains to pass over sprockets at the ends of the conveyor, but which between sprockets form a continuous flat moving apron upon which are placed the articles to be conveyed.

Bucket conveyors consist of a number of V-buckets which are fastened between two moving chains. In the plain bucket conveyor the buckets are fixed to the chains. Therefore, on horizontal runs, they must be carried through troughs and act much like flight conveyors. If the buckets are pivoted so that they always maintain a level position, they will carry the material on horizontal runs rather than scrape it. A long spiral worm revolving in a closely fitting trough forms a type known as the screw conveyor. Bucket elevators may accomplish vertical turns, trolley and roller conveyors may accomplish horizontal turns, but the others are essentially straight run conveyors. On them, turns must be made by dumping from one conveyor on to another running in the new direction. Some use has been made of pneumatic conveyors, in which loose material may be drawn along with the stream of air, or in which articles may be sent in special containers which, fitting the walls of the conveyor conduit rather closely, are moved along somewhat as air pressure would move a piston in a cylinder. Modern conveyors are usually driven by electric motors operating through reduction gearing in order to operate the conveyor at an appropriate speed. If inclined, the conveyor will be equipped with a device to prevent it from starting backwards when stopped. Various automatic or semi-automatic conveyor discharging devices are employed. Power requirements are minimized by the use of anti-friction bearings on rollers, sprockets, or pulleys. Wear is minimized by the use of hardened wearing parts, or allowed for by the use of renewable wearing parts. (F.T.M.)

CONVULSION. A violent spasm or involuntary contraction (generally a series of them) of a group or of all the body muscles. Convulsions may be associated with high fever due to any cause in children, and with infectious diseases of the central nervous system such as encephalitis and meningitis. Convulsions are commonly and regularly seen in epilepsy, eclampsia, tetanus, strychnine poisoning, uremia and brain injuries. Convulsive seizures may also be seen in hysteria. (D.M.H.)

CONY, CONEY. Mammalia. 1. Rodentia. The pika, *Ochotona*, a small stoutly built animal found at high altitudes in the northern hemisphere. Numerous species. 2. Hyracoidea. The hyraces, small rodent-like animals, form an order of their own. They occur in Syria, Arabia, and Africa. Most of the toes are armed with broad nails and the superficial resemblance to rodents is due to the long cutting teeth and the compact tailless body. (A.W.L.)

COOLANT. Lubrication.

COOLIDGE TUBE. X-Rays.

COOLING, INTERNAL COMBUSTION ENGINE.

The products of combustion in an engine are necessarily at a high temperature when formed. Upon attainment of thermal equilibrium in the combustion chamber the gas may have a temperature of from 1000° to 1400° F. If the cylinder walls are not maintained well below 500°, overheating and breakdown of the piston-cylinder lubrication film can ensue. A deliberate and purposeful cooling of the outside of the cylinder wall is imperative. Usually about a third of the heating value of the fuel consumed must be abstracted by a system of cooling which therefore becomes an essential and major element of any internal combustion engine.

The heat received by the cooling system is dissipated to the atmosphere in the case of land mobile units such as automobiles, railcars, etc.; to the water on which engine-propelled vessels float; and either to the atmosphere or to large natural bodies of water or flowing streams in the case of stationary engines. Cooling systems may be divided into (1) those using air directly, and (2) systems employing a liquid for the cooling of the cylinder and combustion chamber. Liquid cooling is accomplished by surrounding the cylinder with a cooling jacket containing the liquid. Water is ordinarily employed, but there are two exceptions. One is met where cooling systems of an idle engine are exposed to water-freezing temperatures. Liquids of lower freezing point, either pure or in solution with water, are then substituted for water. Alcohol and ethylene glycol are two such liquids. There is another reason for using the latter as a cooling system liquid. Its boiling temperature is about 380° F., therefore the jacket liquid may be heated to a possible 250° instead of the 180° common in circulating water systems. This materially reduces both the required flow of coolant and the size of the radiator.

Direct air cooling is obtained by circulating air over the exterior of the cylinder. Heat transfer from cylinder to air is so much slower than to water that it becomes necessary to extend the exposure of metal surface by casting or machining cooling fins on the exterior of the cylinder and combustion chamber. Air- and liquid-cooled cylinders are compared in Fig. 1.

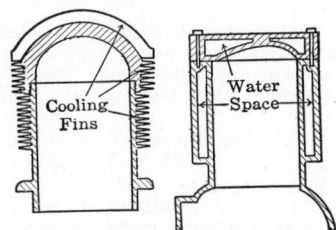

Fig. 1. Air and liquid cooled cylinders compared. (Valves and plugs omitted.)

Liquid-cooled systems may be further subdivided into evaporative and circulating types. Evaporative cooling is found on small stationary engines having hopper cooling. The water jacket is extended into an open hopper, or reservoir, so that a considerable quantity of water may be held. During operation of the engine the water is boiling. Heat is conducted away from the engine in the form of steam. Each pound of water evaporated will remove as much heat as 15 or 20 lbs. of water circulated but, unlike a circulating re-cooling system, water must be added periodically to the hopper to replace the evaporation. Most liquid-cooled engines employ the circulating system wherein the liquid is cooled externally and readmitted to the engine. A steady flow is maintained through the cooling jackets by means of pumps or thermo siphon action.

Methods of cooling the liquid are:

1. Pass the liquid through a "radiator" where, by conduction and convection (but very little radiation) heat

may be removed by air blown through or across the radiator.

2. Pass the liquid through a surface heat exchanger and absorb the heat in raw water which may then be re-cooled or wasted.

3. Pass the liquid through a cooling tower or spray pond to cool it evaporatively.

A typical water cooling system for a gasoline tractor engine is shown in Fig. 2. Water is circulated from the

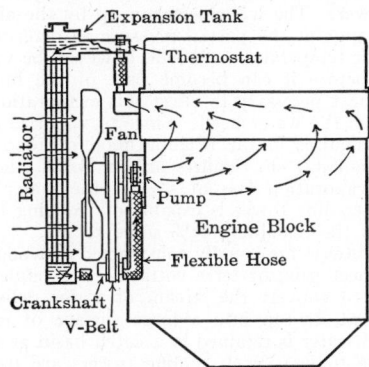

Fig. 2. Forced circulation water cooling system.

engine jackets to a radiator core where it is cooled, then withdrawn by an engine-driven centrifugal pump and sent through the cylinder jackets. From there it flows into the jackets surrounding the cylinder head and then back to the radiator. Since the radiator must be of sufficient size to obtain adequate cooling in hot weather, a thermostat is inserted in the flow line to maintain suitable engine temperature in cold weather.

Direct air-cooled engines are used in many power fields. Some have blowers to force the air to circulate over the cooling fins, while others depend on the motion of the engine itself to create the circulation. A motorcycle engine is illustrative of the self-cooled principle. The airplane engine is another. However, the latter is essentially a high-output engine installed under conditions demanding utmost efficiency of cooling. The air is carefully guided over its cooling fins by cowling so that the air used is the least possible and its usage is accomplished with a minimum of disturbance to the surrounding air stream. Effective air cooling of an airplane engine is so dependent on the cowling that it cannot be described except in conjunction with the cowling. Air-cooled engines habitually operate at higher temperatures than liquid-cooled, which gives them some thermodynamic advantage. It is common to consider the rear spark plug gasket temperature as the cylinder temperature and to embed a thermocouple in it for the activation of a remote reading thermometer. Typically the average temperature is 420° F., although 50° variation is possible among the cylinders of a multi-cylinder engine. The amount of heat to be removed by the cooling system is related to engine operation variables as follows:

1. Increase of engine power increases required cooling.
2. Increase of air fuel ratio decreases required cooling.
3. Increase of combustion air temperature increases required cooling.

There is an optimum ignition timing at which required cooling is minimum. Much research has been directed to the problem of fin size, spacing, heat transfer coefficients, etc., most of which was stimulated by the needs of the large air-cooled radial engine designer. The ratio of piston displacement to surface area of the cylinder increases with increase in cylinder bore, making it progressively more difficult to cool the engine adequately as larger sizes are attempted. In modern air-cooled aeronautical engines of high specific output, cooling is often the limiting factor in performance. (F.T.M.)

COOLING TOWER. The cooling tower is a means for cooling water through the medium of partial evaporation. The action that takes place is one whereby the stream to be cooled is broken up into a fine mist or rain, thus exposing a very large surface area. In this condition the water is mixed with air which is brought into the cooling tower by natural convection currents, or which is forced in by fans. If that air is not already saturated with water vapor, it will become so by coming in contact with the large area of moisture exposed in the cooling tower. The water is taken up by the air in the form of vapor at the partial pressure as determined by atmospheric temperature. It will, however, be vaporized to steam before it can become part of the humidified air. The heat necessary to effect this evaporation comes largely from the water itself. Since to vaporize a pound of water requires in the neighborhood of 1000 B.T.U., while to cool it 1° F. requires only 1 B.T.U., it follows that an evaporation loss of 5% of the water passing through a cooling tower is capable of reducing the temperature of the remaining 95% some 50° F.

Cooling towers are usually constructed of wood having wooden, sheet iron, or terra cotta interior baffles so arranged as to convert the stream of water delivered to the tower at its top into a large surface of exposure. The cooled water is retained in a catch basin at the bottom of the tower. Such cooling towers are frequently employed to cool the jacket water of internal combustion engines, and the circulating water of condensers of various types, i.e., in connection with refrigerating plants. Long, narrow cooling towers are more effective than squarely built ones. Higher cooling towers are more efficient than low ones. The efficiency of a cooling tower is stated as follows:

$$\text{Eff.} = \frac{\text{Hot water temperature} - \text{cold water temperature}}{\text{Hot water temperature} - \text{wet bulb temperature}}$$

A cooling tower in good condition should have an efficiency of at least 60%. (See **Absorption.**)   (F.T.M.)

COOPERAGE. Wood.

COORDINATE PAPER. Rectangular Coordinates in a Plane, and Polar Coordinates in a Plane.

COORDINATES OF A POINT. Coordinates of a point are **numbers** which determine the position of the point.

In a plane, the position of a point is usually determined by either **rectangular coordinates** or **polar coordinates**.

In space, the position of a point is usually determined by either **rectangular coordinates, polar coordinates, spherical coordinates** or **cylindrical coordinates.** (L.L.S.)

COORDINATION. The regulation of an organism so that its various parts act cooperatively for the benefit of the whole.

Coordination depends upon the reaction of living structures to surrounding conditions, both inside and outside the body. Reaction to substances present in the body is classed as chemical coordination and reaction to stimuli which cause the transmission of impulses through the living substance is nervous coordination. The separation between the two is not sharp. In general the maintenance of a normal state in the body is due to the transformation and distribution of substances within it, or to chemical coordination. These processes are, however, subject to nervous regulation. The rapid transmission of nerve impulses likewise is conditioned by substances within the body. The two kinds of coordination are conspicuously illustrated by the action of **hormones** and of the **nervous system.** (A.W.L.)

COOT. Aves, Gruiformes. Birds (**Aves**) of several species occurring in Europe and Asia, North America, and Africa. They are waders and swimmers, with lobed toes. The plumage is dull in the adult, in contrast with the beak and part of the head. In the American species, *Fulica americana,* the beak is ivory-white. Although sometimes eaten, the coots do not rank with the ducks as food and game birds.   (A.W.L.)

COOTER. Reptilia, Testudinata. A turtle of the genus *Pseudemys.* The several species are found chiefly in the eastern and southern United States. Also called sliders.   (A.W.L.)

COPAL. Copal is the hardened **resin** derived from several tropical trees. One of these, *Trachylobium Hornemannianum,* is a large white-flowered tree of tropical east Africa. Another *Hymenaea Courbaril,* a tree with large white or purplish flowers, is a native of tropical South America.

The resin may be obtained from living trees, in which case it is a soft substance naturally slow to harden. In Zanzibar, copal resin occurs in fossil form, masses of resin closely resembling amber being dug from the ground. This is the best grade of copal, known as Zanzibar copal. Similarly in South America the resin may be dug from the ground at the base of the tree, where it slowly accumulates. Copal is used in the making of high-grade varnishes. Of late it is being supplanted with synthetic cellulose lacquers. Several other trees also yield copal resin.   (R.M.W.)

COPALITE or **COPALINE.** The mineral copalite or "Highgate resin" is a fossil **resin** found in irregular fragments in the **blue clay** of London, England. It resembles copal, the resin of certain modern tropical trees. Copalite is pale yellow to greenish or brownish, and emits an aromatic odor when broken. It has a hardness of 1.5; a specific gravity of 1.046; burns with a very smoky yellow flame.   (E.S.C.S.)

COPEPODA. A subclass of small **crustaceans.** Some species are free-swimming and others live as parasites on fishes. The latter are called fish-lice.   (A.W.L.)

COPPER. Insecta, Lepidoptera. In zoology, coppers are small **butterflies** whose prevailing colors are coppery shades, often with metallic luster. With the **blues** and the **hairstreaks** they make up the family Lycaenidae.

Copper is the metallic element—Symbol: Cu (cuprum). Atomic number: 29. Atomic weight: 63.57. Density: 8.92. Hardness: 2.5–3.0. Melting point: 1083° C. Boiling point: 2310° C. (**Isotopes:** page 290.)

Copper is a yellowish-red metal, very malleable and ductile, soft, good **conductor** of electricity, but traces of certain impurities markedly decrease the **conductivity;** unattacked by dry air, but in moist air containing **carbon dioxide** a protective greenish film of basic carbonate is formed; dissolved best by **nitric acid,** not attacked by cold dilute **hydrochloric** or **sulfuric acid,** but in hot hydrochloric acid dissolves to yield cuprous chloride, in hot concentrated sulfuric acid to yield copper sulfate; attacked by **chlorine,** especially when heated, to form cuprous and cupric chlorides; only slight action by **hydrogen sulfide** or **sulfur dioxide** at ordinary temperatures in the absence of air. Very thin sheet copper is translucent and transmits greenish-blue light. Discovery prehistoric. Probably the first metal to be used by mankind. Copper is the second most largely used metal, being exceeded only by iron, and considerable scrap metal is recovered.

Copper is largely used (1) in construction and apparatus where workability is demanded of and definite resistances to corrosion supplied by the metal, (2) as an electrical conductor (99.95% Cu) commonly in the form of wire, (3) as a constituent of various **alloys,** especially brass and bronze, coins, **aluminum** alloys, **nickel** alloys, and **steel,** (4) as a catalyzer when finely divided, for certain chemical reactions.

Copper occurs as native copper particularly in the region south of Lake Superior (often 99.9% Cu), as sulfides (**chalcocite,** copper glance, cuprous sulfide, Cu_2S; **chalcopyrite,** CuFeS), as oxide (**cuprite,** cuprous oxide, Cu_2O, red); as basic carbonates (**malachite,** $CuCO_6 \cdot Cu$ $(OH)_2$, green; **azurite,** $2CuCO_3 \cdot Cu(OH)_2$, blue). The copper content of its ores varies from 0.3% to 8% Cu and the average is of the order of 2.5%. The value depends largely upon the content of silver and gold. The area of production is widely distributed, in the United States, Montana, Utah, New Mexico, Arizona, Michigan, Tennessee, in Canada, Mexico, Chile, Peru, Africa, Spain, Portugal, Japan. (1) Native copper ore is crushed, concentrated by washing with water, smelted, and cast into bars. (2) Oxide and carbonate ores are treated with **carbon** in a smelter. (3) Sulfide ore treatment is complex, but in brief, consists of smelting to a matte of cuprous sulfide, ferrous sulfide, and silica, which molten matte is treated in a converter by the addition of lime and air is forced under pressure through the mass. The products are blister copper, ferrous calcium silicate slag, and sulfur dioxide gas. Refining is conducted by electrolysis, and the anode mud is treated to obtain the gold and silver.

Acetates: Copper acetate, cupric acetate (Cu $(C_2H_3O_2)_2 \cdot H_2O$), greenish-blue solid, soluble, formed by reaction of cupric oxide (or copper plus oxygen of the air) and **acetic acid,** and then crystallizing. Used as an insecticide, fungicide; basic copper acetate, copper subacetate, "verdigris," green and blue solids, insoluble, color depending upon the ratio of copper oxide to copper acetate present, used (1) as paint pigment, (2) as insecticide and fungicide, (3) in dyeing and printing fabrics. (See **Arsenic**—Arsenites, and Arsenite below.)

Acetylide: Cuprous acetylide ($Cu_2C_2 \cdot H_2O$), red precipitate (explosive when dry), formed by passing **acetylene** into cuprous chloride solution in ammonium hydroxide.

Arsenite: Copper arsenite, cupric arsenite, "Scheele's green" ($CuHAsO_3$), light green solid, insoluble, poisonous, formed by reaction of soluble copper salt solution and **sodium** arsenite solution, used as paint pigment and insecticide; copper acetoarsenite, cupric acetoarsenite, "Paris green" ($Cu(AsO_2)_2 \cdot Cu(C_2H_3O_2)_2$), green powder, insoluble, used as (1) paint pigment, (2) insecticide, (3) in wood preservatives.

Bromide: Cuprous bromide (CuBr), white solid, insoluble, formed by reaction of cupric bromide solution and copper metal, or of excess copper metal with **bromine;** copper bromide, cupric bromide ($CuBr_2$), brownish-black solid, soluble, formed by reaction of cupric oxide and **hydrobromic acid,** and then crystallizing. Color of solution depends markedly upon the concentration, e.g., concentrated is dark brown, dilute is blue, intermediate is green.

Carbonates: Basic copper carbonates, green precipitate by reaction of soluble copper salt solution and sodium carbonate solution, or superficially by oxidation of copper metal in moist air containing **carbon dioxide.** **Malachite** (green) and **azurite** (blue) occur in nature.

Chlorides: Cuprous chloride (CuCl), white solid, insoluble, formed by reaction of cupric chloride solution and copper metal, or of excess copper metal with **chlorine;** copper chloride, cupric chloride ($CuCl_2 \cdot 2H_2O$) green crystals, soluble, formed by reaction of cupric oxide (or copper plus oxygen of the air) and **hydrochloric acid** and then crystallizing. Crystals become anhydrous and brownish-yellow at 110° C., the chloride melts at 500° C., and changes to cuprous chloride and chlorine gas at higher temperature.

Cyanides: Cuprous cyanide (CuCN), white solid, insoluble, formed by decomposition of cupric cyanide, upon heating with water, very poisonous **cyanogen** gas being evolved at the same time. Used in certain organic reactions, i.e., **Sandmeyer's** for benzenoid cyanides; cupric cyanide ($Cu(CN)_2$), brownish-yellow precipitate by reaction of soluble cupric salt solution and **sodium** cyanide solution. Upon heating with water, very poisonous cyanogen gas is evolved, and cuprous cyanide.

Ferrocyanide: Cupric ferrocyanide ($Cu_2Fe(CN)_6$), reddish-brown precipitate, formed by reaction of soluble cupric salt solution and **potassium** ferrocyanide solution. This is a very delicate test for copper, and enables the detection of as little as one part of copper in a million parts of solution.

Hydroxides: Cuprous hydroxide (CuOH, formula doubtful), yellow precipitate, by reaction of cuprous salt solution and **sodium** hydroxide solution, soluble to colorless solution by hydrochloric acid or ammonium hydroxide; cupric hydroxide ($Cu(OH)_2$), blue gelatinous precipitate, by reaction of cupric salt solution and sodium hydroxide solution, soluble to deep blue solution by ammonium hydroxide, cupric hydroxide is changed to black cupric oxide upon boiling with water, or upon standing.

Iodide: Cuprous iodide (CuI), white solid, insoluble, formed by reaction of soluble cupric salt solution and **potassium** iodide solution with separation of **iodine** at the same time. Used in certain organic reactions, i.e., Sandmeyer's for benzenoid chlorides.

Nitrate: Copper nitrate, cupric nitrate ($Cu(NO_3)_2 \cdot 6H_2O$), blue crystals, soluble, formed by reaction of cupric oxide (or copper) and **nitric acid,** and then crystallizing. Copper nitrate is decomposed, upon being heated, leaving cupric oxide residue and evolving nitrogen tetroxide and oxygen gases.

Oxides: Cuprous oxide, red copper oxide, (Cu_2O), red solid, insoluble, formed (1) by reaction of cupric salt solution in alkaline medium with a reducing solution such as glucose or arsenite. The cupric salt solution in sodium hydroxide is maintained by the addition of sodium tartrate or citrate. Fehling's test is the application of this reaction; (2) by superficial oxidation of copper upon heating to moderate temperature; cupric oxide (CuO), black solid, insoluble, formed (1) by the oxidation of copper upon heating in air, (2) by heating cupric hydroxide, carbonate or nitrate to red heat.

Sulfate: Copper sulfate, cupric sulfate, "blue vitriol," "blue-stone" ($CuSO_4 \cdot 5H_2O$), blue crystals, soluble, formed by reaction of cupric oxide (or copper plus oxygen of the air) and **sulfuric acid,** and then crystallizing. Crystals lose water upon being heated and the white residue of anhydrous copper sulfate is used to detect the presence or absence of water in certain organic liquids (water causes blue coloration).

Sulfide: Cuprous sulfide (Cu_2S), black solid, insoluble, formed by reaction of cuprous salts, e.g., cuprous chloride, with **hydrogen sulfide** or **sulfur** upon boiling; cupric sulfide, copper sulfide (CuS), black precipitate, formed by reaction of cupric salt solution and **hydrogen** or **sodium** or **ammonium** sulfide, soluble in dilute nitric acid.

Cupric salts (the common copper salts) dissolve in water to give a beautiful blue color; the crystalline salts are blue solids (e.g., sulfate, nitrate), green solid (e.g., chloride), brownish-black solid (e.g., bromide) and the oxide is black. The salt solutions yield, with excess ammonium hydroxide, a dark blue solution. Various forms of this solution are applied in special reagents, e.g., Fehling's, where reducing agents form cuprous oxide, and Schweitzer's, where cellulose is dissolved, and upon making the solution acid, cellulose is precipitated, as in the earliest artificial fiber process.

Cuprous salts are insoluble in water; the salts are white (e.g., chloride, bromide, iodide) and soluble in concentrated hydrochloric acid to brownish solution, and in ammonium hydroxide to colorless solution (the blue coloration usually encountered is due to the presence of cupric). These solutions readily absorb **oxygen** or **carbon monoxide,** and are so utilized in the analysis of gases. Cuprous oxide or hydroxide is formed of various colors ranging from yellow to red. (R.K.S.)

COPRA. Coconut.

COPROLITES. Paleontology.

COPULATION. The act of sexual union by which the seminal fluid, containing the reproductive cells of

REPRESENTATIVE ALLOYS CONTAINING COPPER

Alloy	% Cu				
Phosphor bronze........	91.6	8.25% Sn	0.15% P		
Statue bronze..........	91.4	5.5% Zn	1.7% Sn	1.4% Pb	
Gun metal.............	90	10% Sn			
Gilding metal..........	90	10% Zn			
Hardware bronze........	89	9% Zn	2% Pb		
Aluminum bronze.......	88–96	10.5–2.3% Al, Fe, Sn			
Manganin.............	85	12% Mn	3% Ni		
Red brass.............	82	15% Zn	3% Sn		
Bearing bronze.........	82	2% Zn	16% Sn		
Bell metal.............	80	20% Zn			
Nickel coinage, U.S.A....	75	25% Ni			
Cartridge brass.........	70	30% Zn			For ductility and strength.
Admiralty metal........	70	29% Zn	1% Sn		
Yellow brass...........	67	33% Zn			
65:35 brass............	65	35% Zn			For wire, tubing, and cold pressing.
Nickel silver...........	65–55	17–27% Zn	18% Ni		
Rivet metal............	63	37% Zn			For cold pressing.
Muntz metal; yellow metal	60	40% Zn			For sheet and hot working.
Manganese bronze; high tensile brass..........	58	38% Zn	4% Mn, Fe, Al, Ni, Sn		For casting, extrusion, hot stamping; high tensile strength.
Hot stamping brass......	58	40.5% Zn	1.5% Sn		For casting, extrusion, hot stamping, machining, free cutting.
German silver..........	50	25% Zn	25% Ni		
Palladium gold substitute.	40	31% Au	19% Ag	10% Pd	
White solder...........	40	60% Zn			
Monel metal..........	25–35	60–70% Ni	1–6% Fe		
Antifriction metal.......	12.5	75% Sn	12.5% Sb		
Silver coinage, U.S.A....	10	90% Ag			
Standard sterling silver...	7.5	92.5% Ag			

COPPER LOSS. This term is frequently used to denote the resistance loss in the conductors of electrical circuits or machines. In most machines there are two types of electrical losses, those caused by winding resistance, i.e., the copper loss, and those caused by the magnetic core, i.e., core loss. Copper loss is given by

$$P = I^2 R$$

where I is the current and R the resistance. (L.R.Q.)

COPPER OXIDE RECTIFIER. Rectifier.

COPPERHEAD. Reptilia, Sauria. A poisonous snake, *Agkistrodon mokasen,* of the eastern United States, extending locally into the central states. It is a pit viper, related to the rattlers and water moccasin. The species frequents rocky uplands more than wet ground and is nocturnal and retiring in habits.

The reddish head of this snake has led to confusion with certain harmless species, including the hog-nosed snake. The copperhead can be distinguished with certainty by the pit between the eye and the nostril. It is not a large snake, usually reaching a length of less than four feet, and is not as dangerous as supposed because of its shy nature. Its bite is treated like those of the other **pit vipers.** (A.W.L.)

COPPERSMITH. Aves, Piciformes. A small Indian bird, *Xantholaema haematocephala,* one of the **barbets.** It is green above and yellow with green markings below. (A.W.L.)

the male, is transferred to the genital passages of the female.

The germ cells are adapted for locomotion through liquids, hence many aquatic species need only discharge them into the surrounding water simultaneously to enable them to come together for **fertilization.** If the egg is to develop in the body of the mother, however, or if the animal is entirely terrestrial, the liquid medium in which fertilization occurs is secreted by the body and a direct transfer from male to female is usually necessary. Artificial insemination of animals has been accomplished.

In some animals copulation is accomplished merely by the apposition of the orifices of the genital ducts, but in most cases the terminal portion of the female organs becomes a vagina for the reception of a male intromittent organ. This organ varies greatly. In some of the **rotifers** the pointed end of the body serves for the introduction of the germinal material, although some have a special projecting organ called the penis. Some of the roundworms have a pair of copulatory **setae,** which project from the alimentary tract. In the **crustaceans** certain paired appendages are modified for introduction into the female and in the **spiders** the male discharges the seminal fluid onto a web and takes it up into his **palpi,** which are modified for the transmission of the material to the female ducts.

Among the vertebrates most intromittent organs are in the form of a penis developed either as a projecting fold in the wall of the **cloaca** or as a protrusible organ associated with the urogenital passages at their caudal extremity. In the **shark** the pelvic fins sometimes bear lobes which are thrust into the cloaca of the female during copulation. Copulation is associated with the involved process of **mating**. (A.W.L.)

COQUETTE. Aves, Micropodiformes. A **humming-bird** of a small group of species found from southern Mexico to southern Brazil. They are distinguished by the crested head and conspicuous frills at the sides of the neck. (A.W.L.)

COQUINA. This is a Spanish word meaning little shells. It is a coarse and highly porous **limestone** made up of shells and shell fragments loosely cemented. It is being formed at present along the coasts of Florida, where it is frequently referred to as "beach rock." Only a few of the limestone formations of former geological periods are true coquina. In Bermuda coquina, largely of **Aeolian** origin, is sawed into blocks and used as a building material. (R.M.F.)

CORACIIFORMES. An order of birds including the hornbills, kingfishers and **rollers**. (A.W.L.)

CORAL. Coelenterata, Actinozoa. The hard deposit built up by minute colonial animals called coral polyps which occur in the warmer oceans. The deposit consists principally of **calcium** carbonate.

The term coral is applied to the deposits of animals of two orders, **Alcyonaria** and **Zoantharia,** of different form and habits. Those of the alcyonarians are made up of minute spicules formed within the tissues, occasionally compacted in a hard central rod running through the entire colony and sometimes supplemented by an external covering. Red **or** precious coral is the

Coral. (*American Museum of Natural History.*)

hard axis of such a form and organ-pipe coral is made up of the connected tubes which once surrounded the living animals. Zoantharian corals build up hard deposits externally beneath the basal disk which attaches them to the ocean floor. As new individuals arise from the edge of the living tissue their deposits become continuous with those already laid down and so large colonies produce extensive masses of coral rock. The form of these deposits varies. Some are slender and branching and others rounded and massive. They have received common names such as staghorn coral and brain coral.

Precious coral is secured principally in the Mediterranean and is the foundation of a considerable industry in Italy. Several thousands of persons in that country work coral into beads and other ornaments and make it into jewelry.

The formation of coral islands in the warmer oceans has resulted in many habitable land masses, and in the same waters submerged reefs of this material are serious obstacles to navigation. (A.W.L.)

CORAL REEF. A complex, **ecological** association of benthonic (bottom-living) and attached, calcareous, shelly marine **invertebrates**, forming either fringing

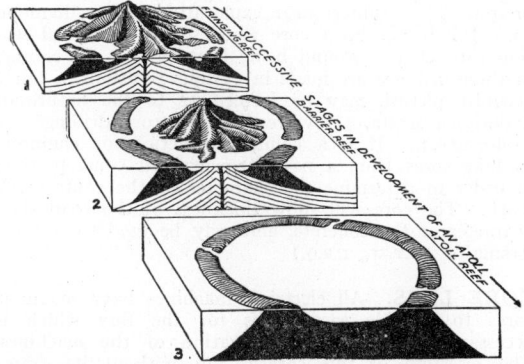

Block diagrams illustrating the successive stages in the development of an atoll during the subsidence of a volcanic cone. (*Field, Laboratory Manual, Princeton University Press.*)

reefs, barrier reefs, or atolls. The lagoons of barrier reefs and atolls are important loci for the deposition of fine-grained calcium carbonate mud called drewite. Fossil reefs include all types of organic reefs which show a distinct ecological and structural evolution from the earliest known fossiliferous limestones to the typical atolls of the South Pacific Oceanic Islands. (R.M.F.)

CORBINO EFFECT. Hall Effect.

CORDAITALES. Paleobotany.

CORDIERITE (IOLITE) (DICHLORITE). The mineral cordierite, composition $Mg_2Al_4Si_5O_{18}$, is an **orthorhombic** mineral frequently seen, however, in pseudo-**hexagonal** forms, as well as massive. It is brittle, with a subconchoidal fracture; hardness, 7–7.5; specific gravity, 2.60–2.66; luster, vitreous; color, blue of varying shades; translucent to transparent. Cordierite exhibits pleochroism (or dichroism) being dark blue, light blue and light yellow when examined by transmitted light in different directions. Hence it is frequently called dichroite. It is occasionally used as a gem. Cordierite is found as a primary mineral in the igneous rocks. It is, however, found ordinarily in **gneisses, schists** and in areas of contact metamorphism. Localities for good specimens are numerous in Europe, including Bavaria, Finland, Norway. It is found in Greenland, Madagascar and Ceylon from which latter place come the rolled pebbles of a rich blue color known as saphir d'eau and prized as a gem. In the United States it is found principally in Connecticut. Named for the French geologist, Pierre Louis Antoine Cordier, this mineral has also been called iolite from the Greek word meaning violet, and stone, as well as dichroite from the Greek meaning *two-colored*. (E.S.C.S.)

CORE. A magnetic core is an important element in many applications in electrical design. Electromagnetic equipment, as exemplified by the **transformer, the motor,** and the **generator,** have electrical circuits, usually of copper conductors, and **magnetic circuits.** The magnetic circuit follows a path largely contained in a core composed of iron or iron alloys. The purpose of the core metal is to offer the best path for the magnetic lines of flux, and its success in this respect is measured by its permeability. Cores are usually composed of a large number of thin metal laminations which are fabricated by punching from thin sheets of metal, and after being enameled are assembled to form a core. The enamel forms an insulation between laminations which reduces the eddy currents induced in the metal of the core by transformer action. Normal oxidation scale is frequently sufficient insulation for this.

A drilled core of suitable size is frequently removed from a material to be used as a sample for inspection or test. Thus, subterranean exploration with a diamond core drill brings up a core which gives a vertical section through the ground being investigated. If enough of these drillings are made in various positions, a fairly accurate picture may be formed as to the nature of underlying strata. However, not all core drillings are subterranean. It is a common practice for engineers to take cores from a newly constructed concrete road in order to determine the character of the contractor's work. The core will show thickness of the road slab, character of the concrete, and may be used to test for strength. (F.T.M., L.R.Q.)

CORE LOSS. All electrical machines have magnetic cores to provide easy paths for the flux which is necessary for the proper operation of the machines. The insertion of this core is not without its drawbacks, however, as it introduces additional losses known as core losses. In spite of the additional loss present in the magnetic material the over-all effect of the core is a tremendous increase in the efficiency of the machine because of the smaller currents needed to produce the desired magnetic flux. The core losses are composed of eddy current loss and hysteresis loss. The former is caused by the currents which are induced in the core material by the changing flux through it. This is, of course, much more pronounced in a-c than in d-c machines because of the much greater rate of change of the flux in the former. In order to reduce this loss the core of all a-c machinery and much d-c machinery is laminated or composed of thin sheets. The hysteresis loss is due to a sort of molecular friction when the molecules of the core try to align themselves with the magnetic field. While the two types of losses do not follow the same laws the total loss varies approximately as the square of the frequency if other factors are held constant, or about as the 1.5 power of the flux density if it is the only variable. (L.R.Q.)

CORIANDER. Carrot Family.

CORIOLIS EFFECTS (Acceleration or Force). Any object moving above the earth with constant space **velocity** is deflected relative to the surface of the rotating earth. This deflection was first discussed by the French scientist Coriolis about the middle of the last century, and is now usually described in terms of the Coriolis **acceleration** or the Coriolis **force**. The deflection is found to be to the right in the northern hemisphere and to the left in the southern.

As a first approximation to the problem we assume that an observer is at the center (C) (Fig. 1) of a disk

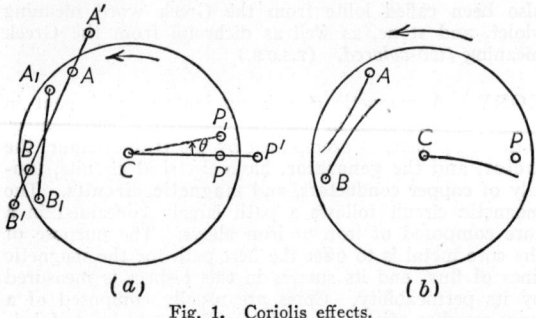

(a)   *(b)*

Fig. 1. Coriolis effects.

that is rotating with constant angular velocity ω. In Fig. 1(a) the observer can see objects off the disk and is conscious of the rotation. At a given instant, when the object P on the disk is directly in line with

the point P' off the disk, the observer fires a shot at P. During the time (t) that is required for the shot to move over the distance CP at speed v ($CP = vt$), the disk will turn through the angle θ ($\theta = \omega t$). The observer notices that the bullet misses the point P, but that it hits P'. In Fig. 1(b) the observer's "world" is limited to the rotating disk and he has no way of knowing that his world is in rotation. Under these conditions when he fires a shot at P he will notice, as before, that the bullet misses P, and he will also determine that the shot follows a curve, similar to that shown, in his world. The same effects will be observed no matter where the observer is located in his world or in what direction he aims his shot (e.g., from A in direction BB', or from B toward AA'). The deflection is always to the right with the direction of rotation indicated in the figure, and would be to the left if the direction of rotation were reversed.

In accordance with fundamental definitions a curved motion represents an acceleration (a) and an acceleration is the result of the action of a force (f) which is proportional to the **mass** (m) of the object and the speed (v) with which it is moving. Analysis of the motions shown in the figure show that the acceleration relative to the disk is given by $a = 2v\omega$, or, since $\omega = \frac{2\pi}{T}$, in which T is the period of rotation, $a = \frac{4\pi}{T} v$. The force producing this acceleration will be given by $f = k \frac{4\pi}{T} mv$, in which k is a factor of proportionality whose value is determined by the units employed.

The conditions on the rotating earth in the vicinity of either pole of rotation are comparable to the "world conditions" on the rotating disk. The sense of rotation will be opposite at the two poles and the deflection will be to the right at the north pole and to the left at the south. The angular velocity of rotation of an area at any place on the earth in **latitude** L is given by $\omega = \frac{2\pi}{T} \sin L$, in which T is the **sidereal** period of rotation of the earth or 86,163.4 **seconds**. Hence the Coriolis acceleration is $a_c = \frac{4\pi}{T} v \sin L$, and the Coriolis force

$$F_c = k \frac{4\pi}{T} mv \sin L.$$

The Coriolis effects must be considered in a great variety of phenomena in which motion over the surface of the earth is involved. Among these may be listed: 1. Rivers in the northern hemisphere should scour their right banks more severely than the left, and the effect should be more evident for rivers in high latitudes. Studies of the banks of the Mississippi and Yukon rivers indicate the predicted results. 2. The motions of air over the earth are governed, to an appreciable extent, by the Coriolis force. (See **Wind Equations**.) 3. A term, due to the Coriolis force, must be included in the equations for exterior **ballistics**. While the effect is negligible in small-arms fire, nevertheless, it is very important in the fire-control material for long-range guns. 4. Any **level** bubble, which is being carried on a ship or plane, will be deflected from its normal position. The deflection will be perpendicular to the direction of motion of the ship or plane. The correction for this effect may amount to several **miles** in the determination of a position of the ship or plane by methods of **celestial navigation** if the bubble octant is used in the necessary observations. (See **Sextant**.) (W.K.G.)

CORK. Cork cells are found in the outer **bark** of most woody-stemmed plants, but in amounts too small and with too brittle walls to be of any use to man. But in *Quercus Suber,* the Cork Oak, the cork cells become a very large part of the tissue of the bark, and have been used for centuries by man. The cork oak tree is a medium-sized tree, seldom much over 50′ in height, growing in nearly all countries bordering the Mediterranean Sea. The evergreen leaves are small, 1½–3″ long, and about an inch wide, with slightly toothed margins. The bark of the tree soon becomes rough and deeply furrowed, but is of little value except as ground cork or as a source of **tannin.** When the tree is about 20 years old this first formed bark is removed, care being taken not to injure the phloem and **cambium** layers. Within 10 years a new cork layer has formed. This layer is the first of many layers which are removed once every 10 years or so throughout the life of the tree. Removal is generally done in the early summer at a time when hot dry winds will not cause injury to the unprotected phloem and cambium.

After removal, the cork is air-dried for a time, then boiled to soften it and to remove some of the tannin. The outer part of the bark is scraped off, and the rest pressed out flat and dried. It is then ready to ship.

The physical properties of cork account for its many uses. It is very light and buoyant, more than 50% of its volume being air, and hence is used in the manufacture of floats, life-preservers, and so forth. Since the living **protoplasm** of the cork cells dries up early in their development, leaving hollow cells, each containing a small mass of air which expands after compression, cork is very resilient, and is frequently used as a core on which to wind yarn or string in the manufacture of baseballs. In the early stages of their formation, the walls of cork cells are **cellulose,** but this is soon impregnated by a waterproof and non-absorbent lipoid substance, **suberin.** Therefore cork is used in making handles for fishing rods, shoe-soles, and cork stoppers. Since the hollow cork cells are poor conductors both of heat and sound, cork is much used as insulating material. For this use cork is ground up and then pressed into sheets with various binding materials, giving much larger sheets than can be obtained from the tree. Ground cork is also an important constituent of linoleum.

Cork is traversed by lenticels, loose masses of porous tissue, which appear as dark spots or holes in stoppers. Usually in making stoppers the bark is cut so that these will be transverse in the stopper. In making stoppers, the forms are first punched out as cylinders, and then trimmed down by machine to the required tapering shape. (R.M.W., B.S.M.)

CORK CAMBIUM. Phellogen.

CORLISS. Engine.

CORM. In botany a corm is a very short, thick, subterranean stem distinguished from the **rhizome** by erect instead of horizontal growth. Its surface shows more or less distinct **nodes.** From the nodes of the upper portion **buds** develop. Generally roots are formed in the lower part of the corm. In many plants the corm is surrounded by scaly leaves or leaf bases. The crocus and the gladiolus produce corms.

In zoology the corm is the median branch or endopodite of the appendage of certain **crustacea** together with the common basal portion. When the exopodite is reduced these parts sometimes appear as the principal axis of the appendage. (R.M.W., A.W.L.)

CORMIDIUM. A group of individuals of various forms budded from the parent stalk of certain floating marine coelenterates. **(Siphonophora.)** (A.W.L.)

CORMORANT. Aves, Pelecaniformes. A large bird (**Aves**) of slender build with a moderately long neck and slender beak, slightly hooked at the tip. The feet are webbed and the birds are strong swimmers and divers. They live entirely on fish. The numerous species are widely distributed, the common cormorant,

Phalacrocorax, occurring in eastern North America, Europe, Asia, and northern Africa.

In Japan and China cormorants are kept in captivity to be used for fishing. A ring or strap around the bird's neck keeps it from swallowing the fish that it catches, although some are said to be so well trained that they bring fish to their owners without this check.

Cormorants are eaten in some parts of the world. (A.W.L.)

CORN. This term is used in physiology and in botany.

In physiology, a corn is a hard callus or horny thickening of the skin on or between the toes, caused by constant pressure or friction. Beneath the hardened callus there is often a small sac or bursa filled with fluid. A soft corn is one found usually between the fourth and fifth toes, and is formed of thickened moist skin and fibrous tissue.

In botany, corn is a well known plant, Indian Corn or Maize, *Zea mays.* Gramineae. There has been considerable speculation as to the origin of this plant. Everything indicates that it is native to America, probably originating in Mexico. The claim has been made, however, that the Chinese knew the plant before America was discovered, and that the grain was carried from Asia to America during the periods of migration of man from one continent to the other. The evidence supporting this claim is not very strong. The plant is not known to occur in the wild state, but a native Mexican grass, teosinte, *Euchlaena mexicana,* is a closely related grass with which corn **hybridizes** freely. Some botanists hold that teosinte is the ancestral grass from which corn originated.

There are many kinds of corn in cultivation, ranging from dwarf forms less than 3′ high to giant plants 15′ or more in height. All kinds have an extensive fibrous root system, the individual roots not only occupying the surface portion of the soil but also extending downward to depths of 6′ or more. In addition to these normal roots, which all arise from the basal portions of the very young stem, there develop from the lower **nodes** of the older stem prop roots. These prop roots are coarse outgrowths which radiate outward and downward until they reach the surface of the ground. During their growth in the air, their tips are protected from drying by an abundant slime coating; once they have entered the ground they branch abundantly and become like normal roots. They serve to support the plant. The stem of the corn plant is coarse and, unlike other grasses, solid throughout its length. The leaves, borne alternately on the stem, have large broad blades at the base of which is a conspicuous ligule, a membranous outgrowth which tightly invests the stem and so may serve to prevent water entering between the stem and the leaf-sheath. The corn plant is **monoecious,** that is, both staminate (**stamen**) and pistillate (**pistil**) flowers are borne on the same plant. However, they are usually not borne in the same inflorescence. The staminate inflorescence or tassel appears at the top of the plant, and matures some time before the pistillate flowers do. The male flowers produce immense quantities of pollen which when ripe is shed in the air, to be carried by wind currents or gravity to the pistillate flowers. Corn pollen is a cause of hay fever (see **Allergy**) to many people. The pistillate inflorescence, or ear, is a modified branch developing in the axis of a leaf. This branch has a fleshy axis or cob on which are borne rows of pistillate flowers. These occur in two flowered **spikelets,** the lower flower usually being abortive, but its **bracts,** the lemma and palea persisting, as do the two short **glumes** which subtend the entire spikelet. The ovary bears a long style, commonly known as the silk of the corn. When first developed the silk is green and has a sticky surface; after pollination has occurred, the silk turns brown and dries up. The entire pistillate inflorescence is enclosed in many overlapping modified leaves known as husks, from the tip of which the silk protrudes. The mature corn grain is variously shaped according to the kind of corn. In most species it is a flattened object with a shallow groove on one side, indicating the location of the **embryo**. This embryo is on one side of the grain, the rest of which is filled with a starchy substance, the **endosperm**. This endosperm is usually separable into two parts, one hard and horny, called the horny endosperm; the other less firm and of lighter color, known as the starchy endosperm. The horny portion contains more protein than does the starchy and has its starch grains more densely packed together. The corn grain will not germinate unless the temperature is above 40° F., and sprouts best when the temperature is about 90° F.

Corn is grown most successfully in regions having a deep, warm, well-drained loam soil, an abundance of rainfall and a growing season of ninety days or more, depending on the kind of corn. Improved varieties have been developed which grow satisfactorily in regions having a shorter growing season. Corn is principally grown in the Americas, especially in the United States and Argentina. It has been introduced in European countries and into Africa, but is not grown there to any great extent. In the United States the so-called corn belt grows more than half the entire world crop. This corn belt comprises the states from Ohio west to South Dakota and south to Kansas, a region having climatic conditions most favorable for this crop.

The uses of the corn crop are many and varied. The green plants are fed to stock directly or are stored in silos. For storage in silos the entire plant is cut down and chopped into small pieces. These are compactly stored in large tight structures of wood, concrete or other material, in which partial fermentation occurs, forming a product called ensilage. This is an important part of the ration of dairy cows.

Corn grains form a most important food product for man and his domesticated animals. The ripe ears are picked from the plants and allowed to dry. The grain is then removed from the cob and used directly as stock and poultry food or ground into coarse particles known as cracked corn, much used in feeding poultry. Ground somewhat finer, but still consisting of coarse particles, white corn becomes grits, much used in southern states. More finely ground, corn becomes corn meal, used in making corn bread and various kinds of puddings. Rarely corn is finely ground to flour. Another corn preparation is hulled corn. In making this the grain is soaked in lye which loosens the **pericarp** or outer portion of the grain. This is then removed and the remaining grain cooked soft. As a breakfast food corn appears principally in the form of corn flakes. In making these, clean corn grains are steamed and the hulls and embryo removed, leaving the endosperm. This is sweetened and flavored, and then cooked by steam under pressure. Following this cooking, the grains are partially dried and then passed between heavy rollers, which make them flakes. These flakes are carried to huge ovens where they are quickly toasted. Cooling follows, after which the product is packed in waterproof cases and is ready for the consumer.

In addition to use as food for man and beast, corn yields many important secondary products, such as corn starch, glucose (see **Carbohydrates**) and corn oil. Starch is prepared from corn by soaking the grains in slightly acidulated water for several days, after which they are broken up, the embryos removed and the grain ground. Following the grinding, the whole is passed through sieves which remove the pericarp. The resulting paste is allowed to flow slowly over tilted tables, which allows the starch to settle. This starch is washed and dried, then pulverized for use.

From corn starch is prepared glucose, or corn sirup, a thick substance which results from the partial hydrolysis of starch with acid. The acid is neutralized with sodium carbonate, and the liquor resulting from neutralization is filtered. This liquor is evaporated, and again filtered, emerging as a clear thick sirup, which is boiled down even further. It now becomes commercial glucose, a thick sirupy substance about half as sweet as cane sugar, and with little flavor. It is used in making jellies and preserves, and in blending with other sweets such as cane sirup and maple sirup.

From the embryos of corn, corn oil is prepared. Corn oil is used as a cooking oil and also in making soaps and paints. The cake remaining after the oil is pressed from the embryos becomes a stock food. (See **Esters**.)

Corn stalks have been used experimentally in manufacturing paper. However, the abundance of objectionable non-fibrous material will probably prevent any extensive use of the stalks in this industry. The cobs left after shelling the grain are used as a fuel to a slight extent and also in making cob pipes. The dried husks form a stuffing for a particularly disturbing kind of mattress, and are also used to some extent as a stuffing in upholstery.

The principal varieties of corn include pod corn, a rarely grown form in which the glumes, lemma and palea of each floret are particularly well-developed, surrounding the kernel; pop corn, in which the sudden explosion of the moisture within the grain turns the latter more or less inside out; sweet corn, characterized by the high sugar content of the grains, and largely grown in home gardens; flint corn, with very hard grains containing a large amount of horny endosperm; and dent corn, so-called because the floury endosperm extends to the end of the kernel and is surrounded laterally by horny endosperm. On maturing the floury endosperm shrinks, causing an obvious dent to appear at the apex of the kernel.

The description above applies to Indian corn or maize, commonly called corn in America and Australia. In Europe the term corn is applied to various cereal grains, in England to wheat, in Scotland and Eire to oats. In its original use corn meant a hard seed or grain. (R.S.M., R.M.W.)

CORN ROOTWORM. Insecta, Coleoptera. Larvae of two species of **beetles**. The adult of one is yellowish green with six black spots on each wing cover. The **larva** of this species, the southern corn rootworm, damages the roots and lower stems of various grains and grasses and sometimes kills the plants. The other species is the western corn rootworm. The adult is entirely yellowish-green and the larva burrows inside the roots of corn and stunts or kills the plant. Both beetles are about ⅜″ long.

The most effective protection against these pests is proper crop rotation. Since the southern rootworm lives on plants other than corn this treatment is supplemented by planting early or late to avoid the most serious attack of the larvae. (A.W.L.)

CORNCRAKE. Aves, Gruiformes. A small shore bird (**Aves**) with long legs and short beak, found in Europe and Asia and occasionally in North America. It is also called the land rail and is related to the Carolina rail of North America. (A.W.L.)

CORNEA. The transparent outer layer of the front of the eyeball. (See **Eye**.) It is a complicated structure composed of five layers. The cornea may be the site of inflammation or ulceration. (D.M.H.)

CORNEAGEN CELL. A kind of cell found in the eyes of some insects. It produces the transparent lenticular cornea at the outer surface of the **eye**. (A.W.L.)

CORNER. A corner is the point of intersection of adjacent property lines. Landed property is ordinarily bounded by broken lines meeting at the "corners" of the property. A land survey is generally a **traverse** with the transit stations at the corners and with the traverse lines coinciding with the property lines. The corners are indicated on a survey plat. They are frequently marked by monuments, either artificial or natural. When subdividing land in accordance with the scheme of the United States **land subdivision,** the corners are designated by standard monuments whose character and markings are governed by certain specific regulations.

If it is impossible to set a monument at a corner, permanent markers called "witness corners" are placed on all lines intersecting at the corner. When a property line intersects a body of water the intersection is marked by a permanent monument known as a "meander corner." (C.W.C., F.T.M.)

COROLLA. Flower.

CORONA. Diffraction of light from either the sun or moon through cloud water droplets produces a corona or series of colored rings which appear about the celestial body but actually are at the cloud height. Corona differs from halo in that the latter is due to refraction by ice crystals. Reddish colors always occupy the outer part of the ring. (P.E.K.)

CORONARY OCCLUSION (Coronary thrombosis). Obstruction of a coronary artery, or one of its branches, which supplies blood to the heart muscle. Such obstruction results in death of the heart muscle supplied by the involved vessel.

This vascular accident is a common cause of death between the ages of 45 and 70, although it may occur earlier, in the late 20's or in the 30's. It occurs 3 times as frequently in men as in women. The cause is bound up with the etiology of arteriosclerosis and **hypertension.** Most of the patients give a previous history of **angina pectoris,** and many of them have high blood pressure.

Pathologically, the coronary artery is seen to be occluded by a thrombus, or plug, formed by an arteriosclerotic process in the lining of the vessel. Gradual narrowing of the vessel along its course may also result in coronary occlusion.

Clinically, the chief symptoms and signs are those of pain and **shock.** The pain is sudden, severe, and constricting, located beneath the sternum or over the heart. Unlike the pain of angina, it is not related to exertion or activity, and instead of being brief, it lasts for hours or occasionally, days. Shock is manifested by an ashen-gray color, cold sweating skin, thready pulse, and a fall in blood pressure. Nausea and vomiting may occur. The diagnosis may be confirmed by an **electrocardiogram** which demonstrates characteristic patterns.

Treatment in the acute phase consists of the administration of **morphine** to relieve the pain and the anxiety (fear of impending death is common); aminophylline intravenously and **oxygen** inhalation are also useful.

The immediate mortality is about 40%. Many patients who die suddenly are said to have "acute indigestion" because the pain may be epigastric, and nausea and vomiting often occur. If the patient survives the first few weeks, sufficient recovery will usually take place to allow him to lead a life of restricted activity. (D.M.H.)

CORONATAE. An order of jellyfishes (**Scyphozoa**) with a lobed margin. Found in the open ocean. (A.W.L.)

CORPUS CALLOSUM. A broad band of nerve fibers within the **brain,** connecting the two cerebral hemispheres. (A.W.L.)

CORPUS LUTEUM. Sex Hormones; Menstruation.

CORPUSCLE. A term applied to many minute bodies. 1. The **cells of the blood.** 2. The excretory unit of **vertebrates,** consisting of a small knot of blood vessels enveloped by a capsule. **Renal corpuscle.** 3. Bone cells are called corpuscles of Purkinje. 4. Many nerve endings including Pacinian or lamellar corpuscles, Grandry's corpuscles, tactile corpuscles. These and all other sensory corpuscles consist of nerve endings enveloped by accessory cells of various forms. They are sensory organs. (A.W.L.)

CORRASION. This term is applied to the mechanical wearing away of rocks through the agency of running water and the rock fragments which it carries in suspension. (R.M.F.)

CORRELATION. Two **variables** are said to be correlated if there exists any dependence between them. If the **probability function** of the two variables $P_{x_1,\ x_2}$ is related to the probability function of each variable x_1 and x_2 by the relationship

$$P_{x_1,\ x_2} dx_1 dx_2 = P_{x_1} dx_1 \cdot P_{x_2} dx_2,$$

then the two variables x_1 and x_2 are independent. Otherwise the variables are correlated. The nature of the dependence may be so complete that knowing the value of one variable we may predict the value of the other variable, or the dependence may be of any amount from the case of complete dependence to that of independence. It should be noted that neither variable is more important than the other variable but that the relationship of dependence is mutual. Correlation may indicate either a causal relationship or may be merely one of association. The idea of correlation may be extended to any number of variables. Types of correlation are **linear, multiple, curvilinear.** The functional relationship $y = f(x_1, x_2, x_3)$ in the case of three variables is said to be the **regression** of y on x_1, x_2 and x_3. The degree of correlation may be measured by the **coefficient of correlation,** the **index of correlation,** the **coefficient of multiple correlation,** the **index of multiple correlation,** and the **correlation ratio.** (L.A.A.)

CORRELATION RATIO. The correlation ratio of y on x, η_{yx}, is defined by

$$\eta_{yx}{}^2 = 1 - \frac{\sigma_{ey}{}^2}{\sigma_y{}^2}, \quad \sigma_{ey}{}^2 = \frac{\Sigma(y - \bar{y}_x)^2}{N},$$

$\sigma_y{}^2$ is the **variance** of y, N = number of paired **variates,** and $\bar{y}_x$ is the **mean** of a column for a particular value of x. This may be written as

$$\eta_{yx}{}^2 = \frac{\Sigma N_c (\bar{y}_x - \bar{Y})^2}{\Sigma(y - \bar{Y})^2},$$

$\bar{Y}$ = mean of the y's, N_c the number of items in a column for a particular value of x. For computational purposes

$$\eta_{yx}{}^2 = \frac{N\left[\sum_{c=1}^{c} \frac{\left(\sum_1^{N_c} f_{xy} d'_y\right)^2}{N_c}\right] - (\Sigma f_y d'_y)^2}{N \Sigma f_y d'_y{}^2 - (\Sigma f_y d'_y)^2},$$

c is the number of columns, $\sum_1^{N_c}$ is a summation over a column, $N = \sum_{c=1}^{c=c} N_c$, d'_y is a **deviation** from an arbitrary origin in class intervals, f_{xy} is the frequency in a cell. The range of values for η_{yx}^2 is $0 \leq \eta_{yx}^2 \leq 1$ and always $\eta_{yx}^2 \geqq r_{xy}^2$. It should be noted that the **index of correlation** is defined in essentially the same way as η_{yx} but η_{yx} is always obtained from a correlation table. We test the hypothesis that $\eta_{yx}^2 = r_{xy}^2$, r_{xy} being the linear **coefficient of correlation**, by use of the **analysis of variance**

$$F = \frac{\eta_{yx}^2 - r_{xy}^2/c - 2}{1 - \eta_{yx}^2/N - c}$$

with $n_1 = c - 2$, $n_2 = N - c$ and refer to tables of **Snedecor's F distribution**. To find η_{xy}^2, x and y are interchanged. (L.A.A.)

CORRELATION SURFACE. The correlation surface $z = P(x_1, x_2)$, $z \geqq 0$, is a surface depending on the two **variables** x_1 and x_2. The volume under the surface $P(x_1, x_2)\Delta x_1 \Delta x_2$ represents approximately the **probability** that x_1 shall have a value between x_1 and $x_1 + \Delta x_1$ simultaneously with x_2 a value between x_2 and $x_2 + \Delta x_2$. This generalizes for more than two variables, say m, to $z = P(x_1, x_2, x_3, \cdots, x_m)$, $z \geqq 0$, and the volume $P(x_1, x_2, \cdots, x_m)\Delta x_1 \Delta x_2 \cdots \Delta x_m$ is interpreted as the probability that x_i will have a value between x_i and $x_i + \Delta x_i$, $i = 1, 2, \cdots, m$. The **moments** of $x_1, x_2, \cdots, x_m$ and the **higher moments** would be found in the usual way, depending on the **discrete** or **continuous** nature of the **variates**.

We briefly describe the normal probability function or surface in two variables. The corresponding generalization in m variables, the multivariate distribution will not be discussed. In the case of two variables in **standard units** t_1, t_2

$$z = P(t_1, t_2, r) = \frac{1}{2\pi\sqrt{1-r^2}} \exp. - \frac{t_1^2 - 2rt_1t_2 + t_2^2}{2(1-r^2)},$$

$$-\infty < t_1 < \infty - \infty < t_2 < \infty.$$

$$t_1 = \frac{x_1 - \overline{X}_1}{\sigma_{x_1}}, \quad t_2 = \frac{x_2 - \overline{X}_2}{\sigma_{x_2}}, \quad r \text{ the \textbf{coefficient of}}$$

correlation between t_1 and t_2. Then the **simultaneous** probability that $a_1 \leq t_1 \leq a_2$, and $b_1 \leq t_2 \leq b_2$ is given by the double integral

$$\int_{t_1 = a_1}^{t_1 = a_2} \int_{t_2 = b_1}^{t_2 = b_2} P(t_1, t_2, r) dt_1 dt_2.$$

The **regression** of t_1 on t_2 is given by $t_1 = rt_2$, i.e., the locus of the **mean** value of t_1 is a straight line. Similarly $t_2 = rt_1$ is the regression of t_2 on t_1. The plane $t_2 = a$ intersects $P(t_1, t_2, r)$ in a **normal curve**, and similarly for a plane $t_1 = b$. A plane $z = c$, $0 \leq c \leq \dfrac{1}{2\pi\sqrt{1-r^2}}$ intersects $P(t_1, t_2, r)$ in an ellipse. When $r = 0$

$$P(t_1, t_2, r) = P(t_1) \cdot P(t_2) = \frac{1}{\sqrt{2\pi}} e^{-\frac{t_1^2}{2}} \cdot \frac{1}{\sqrt{2\pi}} e^{-\frac{t_2^2}{2}}.$$

In the limit as r approaches 1, $P(t_1, t_2, r)$ approaches a normal curve in the plane $t_1 = t_2$. Volumes under the surface $P(t_1, t_2, r)$ have been calculated. It should be noted that $P(t_1, t_2, r) = P(t_2, t_1, r) = P(-t_1, -t_2, r) = P(t_1, -t_2, r) = P(-t_1, t_2, -r)$. (L.A.A.)

CORRESPONDING STATES. Critical State.

CORRODENTIA. An order of minute **insects** containing the **book lice** and psocids. They have biting mouths and some species bear four wings. Some are found among plants and on bark and others frequent books and papers, especially in damp buildings. (A.W.L.)

CORROSION. The most common kind of corrosion, with which everyone is familiar, is that of *rusting*. This is but a special case of a general classification known as atmospheric corrosion, wherein the oxygen of the atmosphere reacts with the material in question. Most metals, with the exception of the noble metals, such as gold, can be oxidized by atmospheric oxygen. In the usual case, however, water vapor must be present before any appreciable oxidation can take place. With iron, for example, 40% humidity is needed at ordinary temperatures before rusting will occur.

Corrosion which takes place when materials are immersed in solutions cannot be explained by any single rule. Several different explanations are required for a complete understanding of the phenomena. All substances are soluble to some extent in water. When a piece of iron is put into water it will immediately begin to dissolve in the water in the form of actual metallic iron (Fe) in solution. If there is oxygen present in the water, the oxygen will immediately attack the iron in solution oxidizing it to ferric hydroxide. The hydroxide will precipitate, thereby removing the iron from the solution, which allows more iron to dissolve. Obviously, in order to have the process continue to the ultimate wearing out of the iron, the supply of oxygen must be continually renewed.

The acidity of the water has a great effect on the speed and the degree to which the material can be dissolved. In this regard, carbon dioxide plays an important rôle. It is dissolved in the water from the atmosphere and forms carbonic acid, which makes the water slightly acidic. Because the acidity is quite low, its concentration, if put in usual chemical terms, would be insignificant. The pH system of expressing concentration, wherein the pH of a solution is the logarithm of the reciprocal of the hydrogen ion concentration, is commonly used. Using this system, the pH number is somewhere between 1 and 14. Neutral solutions have a pH of 7. A pH less than 7 means that the solution is acidic and more than 7 means that it is alkaline.

When a metal becomes corroded because of the acidity of the solution in which it is immersed, the process of solution is different from the one mentioned previously. In this case the substance goes into solution because there is an interchange of hydrogen ions in the solution with the atoms of the metal in the piece. The metal goes into solution and hydrogen tends to plate out on the piece. As soon as the piece has a film of hydrogen on its surface the dissolving of the metal will cease. Here again, oxygen plays an important part, because the oxygen dissolved in the water will react with the film of hydrogen to eliminate it by forming water and allow the corrosion reaction to proceed.

Another type of corrosion is called concentration cell corrosion. It is known that two solutions of different concentrations will set up an electrical potential between them similar to that produced by a battery. If oxygen is present in the liquid and it is continually being replenished by contact with the air, then the oxygen concentration in the liquid will remain substantially constant. Any liquid that is contained in small holes or cracks on the metal surface will not be able to obtain oxygen from the main bulk of the solution, so when the supply in the holes and cracks is exhausted, no more oxygen can get in to replace it. Therefore the oxygen concentration in the cracks is different from that of the main bulk of the solution and a concentration cell is set up. This minute electrical effect is sufficient to make corrosion proceed quite rapidly.

A similar cell type of corrosion is that called galvanic or two-metal corrosion. Two different metals in contact will set up an electrical potential between them. If the two metals are surrounded by an electrolyte so that a closed circuit can be obtained, corrosion takes place. The magnitude of the electrical potential and therefore the speed and extent of the corrosion will depend upon the types of metals in each pair. In general, pairs farther apart in the electromotive series will corrode faster than those closer together.

Certain alloys consist of more than one kind of crystal. When these alloys are exposed to a conducting solution, corrosion tends to take place wherever two dissimilar crystals are together and in contact with the solution. Here the size of the crystals is an important factor in whether or not corrosion will take place. Because particle size can be changed by proper heat treatment, it is sometimes possible to make an alloy more corrosion-resistant by giving it a heat treatment before use. Very large or very small crystals will not corrode as fast as intermediate sizes.

When a piece of metal is under strained conditions its properties are usually different from those when it is not strained. There are certain cases where corrosion has taken place in a metal which was homogeneous, but corrosion started at the point where the metal was known to be in the most strained condition. It is believed that here also an electrical potential may be set up between the strained and the unstrained metal.

The obvious way to prevent oxidation is to prevent the oxygen from getting into contact with the surface. In a steam boiler this is done by heating the water before it enters the boiler, thereby driving out the dissolved oxygen.

Because the acidity of a solution contributes largely to the speed of the corrosion, the acidity may be reduced by the addition of alkaline agents, or the solution may be "buffered" by the addition of certain salts. Buffering is the process of depressing the hydrogen ion concentration by using the proper concentrations of other substances.

If oxygen cannot be eliminated from the solution and if the acidity must remain at a certain fixed value, then occasionally inhibitors can be found to prevent corrosion. Inhibitors are substances which are put into a solution in a small amount to stop the corrosion. They fall in two general classes, namely, (1) substances which show a "preferential wetting" effect, that is to say, the inhibitor will wet the surface of the metal easier than will the solution. Therefore the inhibitor will spread over the surface of the metal and effectively keep the solution from attacking it; and (2) substances which either stabilize the natural protective coating on the metal or which react with the metal to form a coating. For example, if the metal has a tendency to form an oxide film, then a solution which has a small amount of oxidizing material in it such as a dichromate will allow oxide film to persist, thereby protecting it against corrosive action. When a natural protective film such as an oxide is on a metal it is spoken of as being in the "passive" state; when the solution is such that it destroys the protective film then the metal is said to be in the "active" state.

All metals with the exception of the noble metals form oxide coatings. Some of these, such as the coating on aluminum, lead, or copper give protection from corrosion. Under the proper conditions an oxide film can be put on iron to give a considerable measure of protection. This process is called "bluing." Information on the protection of metals by paint and other artificial coatings may be found under **Protective Coatings, Pigments, Paints, and Varnishes.**

The term is applied by mineralogists and petrologists to the resolution and modification of the crystal form of **phenocrysts** during their growth in the parent **magma.** The same term is also used by some geologists and **physiographers** to denote the mechanical erosion of the surface of the earth, particularly by rivers and glaciers. A better term for this purpose, however, is **corrasion.** (D.E.M., R.M.F.)

CORROSION, PROTECTION AGAINST. In combating corrosion, the surface of the metal may be protected by **paints,** plates, or other surface coatings. Or again, the metal may be **alloyed** in such a way as to render it corrosion-resistant. In most cases corrosion is merely a slow, obvious destruction of investment in a metal, but there are instances where the potential destructive action greatly exceeds that caused by the corrosion alone. For example, in pressure vessels such as the steam boiler, corrosion in the boiler itself may hold potential threat of destruction for an entire plant, and endanger lives. It is possible for corrosion to occur at many places in the piping leading to boilers or heaters, but usually it occurs in the boiler itself. The trouble is ordinarily found to be due to an acid condition of the boiler feed water, or to dissolved oxygen contained in it. The raw water used may be acid from surface pollution or from sub-surface drains. Usually this can be detected and readily remedied. A more serious factor is the oxygen dissolved in water. Under the high-temperature conditions existing in the boiler itself, this oxygen becomes extremely active in attacking metal surfaces. The operators of large high-pressure boilers well know the necessity of removing oxygen from feed water through the employment of deactivators or deaerators.

The ordinary user of a metal has at his disposal several means for minimizing corrosion. One of the more common is the application of a protecting surface of corrosion-resisting metal. **Zinc** is one such metal. Galvanized iron is iron with a thin layer of zinc applied to it by dipping in molten zinc or by electrical means. Cadmium, nickel, tin, and chromium are metals often used as protective coatings, generally applied by **electroplating.** (F.T.M.)

CORROSIVE SUBLIMATE. Mercury.

CORSITE. An orbicular **diorite** resulting from the segregation, in rounded concentric forms, of ferro-magnesian minerals (see **Iron** and **Magnesium**). It derives its name from its occurrence on the Island of Corsica, and is also sometimes called Napoleonite. (E.S.C.S.)

CORTEX. This is the outer portion of a **stem** or **root,** bounded externally by the epidermis, and internally by the cells of the pericycle. It is composed mostly of **cells** which are very little differentiated. Usually these are rather large, thin-walled **parenchyma** cells. The outer cortical cells often acquire irregularly thickened cell walls, contain **chloroplasts,** and carry on **photosynthesis.** These are the collenchyma cells.

In zoology, a superficial layer of an organ. Included are such organs as the kidney, the adrenal gland, the ovary, the thymus and portions of the brain. Among these examples the cerebral cortex of the brain is the most familiar. The term designates no common characteristic of origin or structure, but only the existence of a distinctive layer at the surface of the organ involved. (B.S.M., A.W.L.)

CORTLANDTITE. Peridotite.

CORUNDUM (Ruby) (Sapphire) (Emery). The mineral corundum, Al_2O_3, **aluminum** oxide, occurs as well-developed **hexagonal** crystals which may display prismatic, rhombohedral, pyramidal or tabular habits. The larger crystals are often rounded or barrel shaped. Corundum shows both basal and rhombohedral partings; the fracture is conchoidal; hardness, 9; specific gravity, 3.95–4.10; luster, vitreous to adamantine, may be pearly on base; transparent to translucent. Common corundum is gray, grayish-blue or brown, but may be red, yellow or whitish; it is sometimes called adamantine spar. Transparent corundum may be colorless or of various tints. The highly prized ruby is deep red, the sapphire, blue. Transparent yellow corundum is known as oriental topaz; if violet, oriental amethyst; if green, oriental emerald.

Emery is a mixture of granular corundum of dark color, magnetite and hematite, sometimes with spinel. **Quartz** may be present. For a long time emery was supposed to be an ore of iron. Until the introduction of artificial abrasives emery was much used for such purposes.

Corundum is found as an accessory mineral in the crystalline rocks such as crystalline limestones and dolomites, gneisses, schists as well as in the igneous rock types granite and syenite. Corundum syenites are found in Canada, especially in the Province of Ontario. Rubies have long been mined in Upper Burma; both rubies and sapphires are found near Bangkok, Siam. Numerous localities in India furnish gem stones of high quality.

In the United States common corundum is found in New York, New Jersey, Pennsylvania, Virginia, North Carolina, South Carolina and Georgia. Sapphires of gem quality near Helena, Montana, associated with alluvial gold in the Missouri River. From the crystalline limestones and schists of the islands of Naxos and Samos in the Grecian archipelago most of the emery of commerce comes. Other deposits are near Ephesus in Asia Minor, and in the town of Chester in Massachusetts. The word corundum comes from the Hindu, *kurand;* emery is derived from the Greek name for this substance. (E.S.C.S.)

CORVUS. (The raven or crow.) (Map, page 380.) Corvus is a small **constellation** containing no particularly bright or interesting stars. This group of stars has long been a friend to lovers of the sea because of its resemblance to the "fore and aft" sail of a cutter. For this reason the constellation is frequently referred to by sailors as "the cutter's mainsail." On a clear moonless night the resemblance to the sail is very remarkable, even the "step" of the mast and a small "pennant" flying from the gaff being discernible. (W.K.G.)

CORYDALIS. Insecta, Neuroptera. A large gray **insect** with four membranous wings and in the male sex with very long slender jaws. The adult of the **hellgrammite.** Also called dobson fly.

In botany, the term Corydalis is applied to a genus of plants occurring in the north temperate regions and in South Africa, which are sometimes found in cultivation. All are herbs with small, somewhat irregular flowers. (A.W.L., R.M.W.)

CORYMB. Flower.

CORYZA. Acute coryza or the common cold is a catarrhal inflammation of the upper respiratory tract of unknown origin but probably caused by a filterable virus (see **Filterable Virus**). Various descriptive terms are used, depending upon the particular portion of the tract involved, such as acute rhinitis, when the nose is involved primarily, acute pharyngitis or acute laryngitis.

Experiments indicate that chilling may precipitate an attack. Susceptibility to infection is increased when the body is exposed to such sudden changes in temperature as may produce localized cooling of nose and throat.

Diseases of the upper respiratory tract are often highly contagious and spread rapidly. There is a definite seasonal variation of incidence, the peak usually occurring during the winter months. The common cold in itself is not dangerous, but measles is frequently ushered in by an acute coryza, and pneumonia may be preceded by it. (D.M.H.)

COSINE EMISSION LAW. A law relating to the emission of **radiation** in different directions from a radiating surface. If a small, white-hot metal plate is viewed from a great distance, its apparent candle power, measured by a photometer, is greatest when it is perpendicular to the line of sight, and reduces to practically zero when it is turned edgewise. If the observer now moves nearer, he finds that this change is due to the smaller angle subtended by the surface, that is, the smaller cross-section of the beam proceeding from it

in his direction; and that the apparent **brightness** of the surface is the same however it is turned. To apply this, let the radiating surface, of area *a,* be emitting a luminous flux *L* (lumens) in the normal direction (see figure), and, in any other direction making an angle

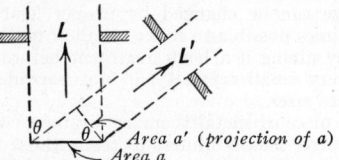

Illustrating cosine emission law.

θ with the normal, the smaller quantity *L'.* Then since the apparent brightness is unchanged, $L'/a' = L/a$. This gives $L'/L = a'/a$. But $a'/a = \cos \theta$, hence $L' = L \cos \theta$; which means that the energy emitted in any direction is proportional to the cosine of the angle which that direction makes with the normal. This is the "cosine emission law" of Lambert. It applies to thermal radiation as well as to light, and to diffusely reflected as well as directly emitted radiation. The law is true only for a perfectly diffusing surface, strictly, for a **black body.** (L.D.W.)

COSINE-HAVERSINE FORMULA. Celestial Navigation.

COSMETICS. The term cosmetic means (1) articles intended to be rubbed, poured, sprinkled, or sprayed on, or introduced into or otherwise applied to the human body or any part thereof for cleansing, beautifying, promoting attractiveness, or altering the appearance, and (2) articles intended for use as a component of any such articles; except that such term shall not include soap. Face powders, lipsticks, nail polishes, cleansing creams, bath powders are cosmetics. The distinction between a cosmetic and a drug is brought out by this part of the definition of a drug, namely, articles (other than food) and components of such articles that are intended to affect the structure or any function of the body of man or other animals. Some commonly accepted cosmetics qualify, therefore, as drugs as well, for example, antiperspirants since they affect the functions of the sweat glands, and hair preparations that claim to treat dandruff, since the claim made for an article has an important legal bearing. Furthermore, a cosmetic preparation is deemed to be adulterated, and consequently to be in violation of the Federal Food, Drug, and Cosmetic Act of 1938, if it contains any "poisonous or deleterious" substance, a statement that clearly excludes the use of mercury-, arsenic-, cadmium-, selenium-, or thallium-containing substances, and amines (not ethanolamines).

Face powder is a blend of (1) white pigments, (2) sufficient coloring materials to give the desired tint, and (3) perfumes. By virtue of its covering power, face powder hides the color of the part of the body to which it is applied, and gives this part the color of the face powder. Some covering agents used are zinc oxide (22), titanium oxide (10), magnesium carbonate (20), magnesium stearate (20), zinc stearate (20), kaolin (20) or clay, talc (80), calcium carbonate (20) or precipitated chalk, and rice starch (40), where the numbers in parentheses signify the relative amounts required to produce the same shade with a given amount of coloring material (thus talc is the poorest and titanium oxide the best covering material). In the matter of adhesiveness or staying-on power, some combination of stearates, palmitates, or myristates of magnesium, zinc, and aluminum are frequently used—a common recipe is about 4% of magnesium or zinc stearate. To produce smoothness in face powder, both in use and in handling, talc is unique, and is usually used to the extent of one-half or more of the products. White pigments that are used to absorb perfume are precipitated chalk, starch, magnesium carbonate, and kaolin, which pigments may be thus used up to 10% of the product. The coloring materials are either natural ground earth colors or else organic colors certified under the previously mentioned Act of 1938. (See **Dyes and Dyeing Textile Fibers.**) Color is considered a primary factor in face powders. Perfumes (see same) are added after the talc has been "prefixed" by a special treatment involving a chemical treatment followed by ageing.

Talcum powders are essentially like face powders, but baby talcum powder is borated by the addition of powdered boric acid, and also treated with an antiseptic material. So-called "liquid powder" requires the presence of some mucilagenous substance for adherence, and bentonite clay to aid in maintaining the solids in suspension and to prevent caking on standing.

In the preparation of cold cream, part of the fatty acid component of beeswax is allowed to react with borax in hot solution to form a soap, essentially sodium cerotate, plus boric acid. An oil-in-water emulsion of the excess beeswax or other oil or fat results. A very general type of recipe uses approximately 1 part by weight of beeswax, 2 parts of water, and 3 parts of other oil or fat, such as mineral oil, almond oil, spermaceti, ceresin, or lanolin. Borax is used to react with some 5% of the beeswax, say 1 part borax to 16 parts beeswax. Instead of beeswax there are used such materials as triethanolamine, the glycerides or glycols of fatty acids, for example, glyceryl or glycol monostearate, or fatty acid ester of mannitol or sorbitol. Perfumes, antioxidants, and preservatives are added ingredients. Vanishing cream contains stearic acid (approximately 15%), sodium or potassium stearate (5%), glycerol (8% or less), and the remainder water.

Lotions are (1) water plus mucilage emulsions containing up to say 3% of mucilage of quince seed, tragacanth, karaya, psyllium, Irish moss, or linseed with a small percentage (say 3 to 10) of alcohol; or (2) alcohol solutions containing 35% or more of alcohol for use as astringents or body rubs. About 2 parts by weight of zinc sulfocarbolate (phenolsulfonate) in 50 parts of alcohol and 50 parts of water, with perfume added, is a popular astringent lotion. Hand lotion is commonly 25% glycerol and 75% rose water, but the addition of such materials as triethanolamine (1%) or sorbitol stearate (1%) is practiced.

The essential ingredients of lipstick are (1) an oil, (2) a wax to make the oil semisolid, and (3) coloring materials, usually of two kinds (a) soluble dibromo- or tetrabromofluorescein, called bromo acid, and (b) insoluble color lakes certified in the Act of 1938. Castor has been the favorite oil used, but the fatty acid esters of glycerol and of glycol seem to have some advantage over castor oil in the matter of increased solubility for bromo acid, as well as less likelihood to produce allergic results. The bromo acid content is commonly of the order of 1 to 3%, and the colored lakes about 10%. The product is blended so that it is easily applied, does not separate or sweat, does not crack or crumble, and stays on satisfactorily for several hours. It must not be irritating to the flesh, and should have a softening point not lower than 48° C. (118° F.) nor higher than 63° C. (145° F.), the favored range being 55 to 60° C. (131 to 140° F.). Mineral oil, 15% in amount, imparts gloss, but has no other useful function. The flavor and odor are also important, and are decidedly limited in choice—coumarin, methyl coumarin, vanillin, and ethyl vanillin for flavor, and rose, jasmin, and orange blossom for odor—with the total amount of both about 2% only.

Nail lacquer and nail lacquer remover are two cosmetics closely related to larger chemical industries, since the lacquer is not unlike quick-drying **lacquers** based on nitrocellulose dissolved in the proper organic solvent, and the lacquer remover performs the same function as an ordinary lacquer remover such as acetone or ethyl acetate. Addition is required of appreciable amounts of softener or plasticizer, even in amount equal to the nitrocellulose content, to avoid brittleness of the film; of resin to improve the adhesion of the film to the nail, and to improve gloss; and of coloring materials, as discussed under other cosmetics. (R.K.S.)

COSMIC RAYS. A highly penetrating radiation apparently reaching the earth in all directions from outer space. The existence of this radiation was first suspected from the discharge of **electroscopes** in air free from all known ionizing influences, the rate of discharge increasing as the electroscope was carried to higher altitudes. (See **Ionized Gases.**) The atmosphere apparently absorbs a measurable portion of the rays, but traces of their effect persist even at depths of many feet below water or below the ground. Professor R. A. Millikan and his associates, who have studied the distribution and absorption of the rays very thoroughly, find that the rays are of nearly the same intensity from all parts of the **celestial sphere,** and that, as they proceed through an absorbing medium, they show evidence of unequal absorption rates. Professor A. H. Compton and many others have expressed the view that the rays are not of any one type, being probably in part true radiation and in part corpuscular, like radioactive emissions, and that they are influenced by the earth's magnetic field.

Observations on the cosmic rays have been extended to nearly all parts of the earth's surface and many miles up into the atmosphere. The most common detector is the electroscope, special forms of which, with automatic recording devices, have been perfected for the purpose. Also used are various types of **counting tube** or ion counter and the Wilson **cloud chamber.** Dr. T. H. Johnson and others have devised ingenious arrangements of counters by means of which it is possible to trace the actual path of a single cosmic particle or photon. At present it is not possible to be sure whether a supposed cosmic particle is really part of the original rays or is a result of cosmic-ray action upon atmospheric molecules. (See **Secondary Emission.**)

The origin of the cosmic rays is still entirely unknown. Attempts to trace them to the **sun,** the **stars,** or to other recognizable celestial objects, have been equally unsuccessful. Millikan advanced the theory that the rays are a result of the synthesis of certain types of **atom** in outer space and that they represent the "mass defect" energy of these atoms. This was when the rays were supposed to be purely **electromagnetic radiation.** More recently Millikan and his associates have ascribed the rays to the transformation of the "rest mass" of interstellar atoms (see **relativity**) into electron pairs of enormous energy. At present their complex character renders the question of their origin exceedingly difficult. (L.D.W.)

COSMOGONY. Milky Way; Historical Geology.

COSTAL. This term means pertaining to a rib or the region of the ribs. Thus, costal cartilage is the cartilaginous portion of the ribs which joins them to the **sternum.** The term intercostal refers to the space between the ribs. (R.S.M.)

CO-TERMINAL ANGLES. Angles.

COTINGA. Aves, Passeriformes. A group of Brazilian **chatterers** closely related to the bell birds. (A.W.L.)

COTTER. A wedge-shaped metal piece for fastening two parts subjected to reciprocatory motion. The connecting rod bearing of a steam engine is often constructed with a removable cap held in place by a cotter. A cotter pin (see Fig. 2, screw fastenings) is a split pin made of semi-cylindrical bar stock, bent so that the flat surfaces are in contact, to permit insertion in a drilled hole. The cotter pin is used as a locking medium for nuts and bolts; since the pin is made of a soft steel, the split ends can be spread after it has been inserted in the hole. (H.C.H.)

COTTON. *Gossypium* species. Malvaceae. Many species of cotton plants are known, some native to warm regions of America, others growing wild in tropical Asia. Cotton seeds are surrounded by an abundance of soft white fibers which when mature are very conspicuous. These fibers are unicellular and have been known and used by man for years. Cultivation of cotton in India has continued through twenty-six centuries. Even before the discovery of America the natives of tropical America were growing Sea Island cotton and using its fibers. At present the United States leads the world in cotton production, but other countries are rapidly increasing their production as demand for the product grows. Brazil, Egypt, and India are important producers.

The cotton plant of cultivation is a woody annual growing three or four feet tall. It bears large palmate leaves and showy white flowers which turn pink as they grow older. Outside the 5-parted **corolla** and **calyx** is an involucre composed of three large green **bracts** having very irregular margins. These bracts persist throughout the growth of the fruit. The **stamens** of the cotton flower have their filaments united to form a hollow tube through which the stigmas must grow; this stamen structure is characteristic of the mallow family, or Malvaceae. The fruit is a large dehiscent capsule of ovoid shape. On dehiscence, or splitting open, at maturity, the soft white fibers which surround the five or six large dark brown or black seeds become visible. The fibers are of two kinds, one relatively long, called lint, the other short and called fuzz. The capsule of the cotton plant is generally called the cotton boll. The length of the lint fibers varies in different species: Sea Island cotton, the native American species *Gossypium barbadense,* cultivated on coastal areas of the southern states, has the longest fibers of any cotton ($1\frac{1}{2}$–$2\frac{1}{2}''$ long). Egyptian cotton fibers are slightly shorter than these. Next comes upland cotton, which is mainly derived from *Gossypium hirsutum.* This is the principal cotton plant of the southern states. In these the length of the fibers varies considerably, but averages about $1''$. Asiatic cottons have still shorter fibers.

Cotton cultivation is limited to regions having a growing season of 6 months or more of continuous high temperature. Any soil having a proper moisture content is suitable for cotton, but a deep, well-drained loam soil is best. One requirement is that excessive rainfall shall not occur during the period when the bolls are opening, since the cotton fibers are injured by moisture at that time. During the growing season the cotton fields must be kept clear of weeds. From 5 to 6 months is required before the production of fibers begins, after which fiber production continues for another 3 months or more.

Picking of the lint and contained seeds is done almost entirely by hand, and requires much cheap hand labor during the picking season. Recently machines have been introduced which bid fair to replace human labor. The soft light fibers, after picking, are carted to the gins. Previous to the invention of the cotton gin by Eli Whitney in 1793, the lint was removed from the seeds by hand, a slow and laborious process. The invention of the gin greatly advanced the cotton-manufacturing industry. The cotton gin of today is essentially the same as that of Whitney, a steel grate with narrow slits through which reach thin notched saws. The rapid rotation of the latter causes lint to catch on their teeth. The lints are pulled through the grating, after which brushes remove them from the saws. This ginned cot-

ton is then pressed into bales weighing about 500 lbs. each.

The uses of cotton are many. First among them is the manufacture of cloth. In this, cotton ranks first among textiles, far exceeding any other fiber. Cloth may be woven entirely of cotton, as much is, or may be of cotton mixed with other fibers, as silks, linen and wool. Cotton fabric forms an important part of automobile tires, being the base on which the rubber is held.

Cotton fibers, especially the fuzz, are frequently used to stuff mattresses, pads and upholstered furniture. Treated with chemicals which remove the thin coating of waxy substances which cover the fibers, the latter becomes **absorbent cotton,** which is capable of absorbing many times its weight of water.

Cotton treated in this way is almost pure **cellulose,** and so is in great demand by those industries using cellulose. The pure cellulose of the fiber may be dissolved and then precipitated in sheets, giving the familiar thin transparent cellophane. Or the dissolved cellulose may be pressed through fine holes and solidified, giving rayon or artificial silk. If treated with concentrated caustic soda, cotton fibers take on a high degree of luster, resembling silk. The product of this process is called mercerized cotton, after John Mercer, its discoverer.

Treated with nitric acid under various conditions, cotton yields a long series of by-products. Some of them are plastic substances known as pyroxylins. If highly nitrated, cellulose becomes guncotton, used in the manufacture of explosives. So-called artificial skin or collodion is one of these nitrated products. Many varnishes and lacquers of recent development are made from cotton cellulose.

Not all the derivative products of cotton come from the fibers. Some, for example, are obtained from the seeds. In preparing these, the hulls are first removed from the kernel within. These hulls are used as fuel in the ginning mill, as food for cattle, and as fertilizer. The kernels are heated and pressed to remove the oil in cotton. During this pressing the kernels are wrapped in cloth to prevent anything but oil from being expressed. The oil is purified to a soft white substance very similar to lard in appearance. Cottonseed oil is used in making salad oils, butter substitutes, and soap. (See **Esters.**) After the oil is expressed, the seed cake may be used as food for stock or as a fertilizer. (R.M.W.)

COTTON STAINER. Insecta, Hemiptera. *Dysdercus.* A bug of Florida and adjacent states which punctures the bolls of cotton and causes staining of the fiber. It develops in groups which can be jarred from the plants into vessels containing a little kerosene. (A.W.L.)

COTTONMOUTH. Reptilia, Sauria. A poisonous snake, *Agkistrodon piscivorus,* of the southeastern United States. Named from the whitish color of the inside of the mouth. It frequents wet places, especially swamps, and is sometimes called the water moccasin.

This species is a pit viper, related to the copperhead and rattlers. It reaches a length of 6′ and is more aggressive than the copperhead but is said to be less deadly than popularly supposed. Treatment of its bite is the same as for other **pit vipers.** (A.W.L.)

COTTON-MOUTON EFFECT. Electric and Magnetic Double Refraction.

COTTONTAIL. Mammalia, Rodentia. *Sylvilagus.* Wild **rabbits** of several variable species, distributed from southern Florida to Mexico and north into Canada. They are gray to brown and most forms have the tail white beneath. An included species is the pontoon or marsh rabbit of Dismal Swamp. The cane cutter or swamp rabbit of the south central states is another species that inhabits wet ground. (A.W.L.)

COTYLEDON. Seed.

COTYLOSAUR. Fossil Reptiles.

COUCAL. Aves, Cuculiformes. Ground birds (**Aves**) of several species, black or red and black in color, found in the Oriental, Australian, and Ethiopian regions. Cuckoos. (A.W.L.)

COUDÉ. The Coudé is a modification of the **equatorial** form of mounting for an astronomical **telescope** designed for the purpose of providing a maximum amount of comfort for the observer. The telescope itself is mounted in bearings parallel to the axis of rotation of the earth and forms the polar axis of the instrument. The eyepiece is at the upper end and usually projects into a closed room. Below the object glass a mirror is mounted in such a manner that it may be rotated about an axis perpendicular to the optic axis of the telescope. Hence the mirror may be rotated about its own axis parallel to an **hour circle** (in the coordinate of **declination**) and is carried along with the rotating tube parallel to the coordinate of **hour angle.** As the observer is seated in a comfortably heated room he looks down into the eyepiece and sees the field of view slowly rotating about the **optic center** of the field.

The fundamental difficulty with the Coudé mounting may be traced to the fact that the mirror will introduce distortion. Furthermore, unless the silver coating on the mirror is very perfect there will be a large amount of light loss. For ordinary "star-gazing" purposes the comfort of the observer makes the instrument a very popular one. (W.K.G.)

COUGAR. Mammalia, Carnivora. Large North American **cats** of several species. The Florida cougar, *Felis coryi,* lives in wild lands of central Florida, another species in Louisiana, and the western species, *F. oregonensis,* is still found from the Rocky Mountains to the coast. The species once found throughout the eastern half of the continent, *F. concolor,* is now limited to the wild mountainous areas of the east or is entirely extinct. The western species is called the mountain lion or puma and the eastern species was also called panther, painter and catamount. (A.W.L.)

COULOMB. The coulomb is the practical unit of quantity of electricity. It is primarily an electromagnetic unit, the absolute coulomb being $\frac{1}{10}$ of the abcoulomb or **abampere**-second. The "international coulomb," however, is defined electrochemically, as the quantity of electricity transferred during the electrolytic deposition of 0.001118 gram of silver from a silver nitrate solution in a **coulombmeter.** It is thus the basis of the international **ampere.** The international coulomb is about 0.99995 of the absolute coulomb.

The coulomb, while convenient in reference to electric currents and electrochemical processes, is far too large for electrostatic purposes, being approximately equal to 3×10^9 e.s.u. of electricity. If one could perform the impossible feat of charging two small spheres with one coulomb each, these charges, if placed one meter apart, would attract or repel each other with a force of about 9×10^{14} dynes or well over a million tons, and the potentials in the near vicinity of the charges would reach hundreds of billions of volts. (L.D.W.)

COULOMBMETER or COULOMETER. An electrolytic cell used for the measurement of the quantity of electricity passing through a circuit; also called "voltameter." A standard form, based upon the definition of the international **coulomb**, and depositing silver, has been developed at the Bureau of Standards. (*Bulletin Bureau of Standards,* 13, 479, 1916.) The more practical form for laboratory use employs copper electrodes

in a bath of copper sulfate. The thin copper cathode, between two heavy copper anodes, is removable for weighing; and since one coulomb deposits 0.000329 gram of copper, the weight of the copper deposit enables one to determine the quantity of electricity in coulombs. A solution recommended for this cell consists of 15 grams of crystalline copper sulfate, 5 grams of pure sulfuric acid, and 5 grams of pure alcohol, dissolved in 100 grams of distilled water. (L.D.W.)

COULOMB'S LAWS. This term usually refers to the familiar laws of interaction between two concentrated electric charges and between two magnetic poles. The force of attraction or repulsion between two charges (or poles) is directly proportional to the product of the charges (or pole strengths) and inversely to the square of the distance between them. The similarity to the Newtonian law of **gravitation** will be readily noted.

The experimental proof of these laws was carried out by means of the **torsion balance,** employing small, electrified balls or the poles of slender bar magnets. In the electrical case, a far more precise proof of the inverse-square feature is based on the mathematical analysis of the experimental fact that at every point within a charged, hollow, spherical conductor, the electric field intensity due to the surface charge is exactly zero. (See **Friction.**) (L.D.W.)

COUMARIN. Bean.

COUMARONE. Furane and Related Compounds.

COUNTERBALANCING. Counterbalancing means simply the application of extra **mass** to a system in order to produce balance for the system as a whole, and to offset the unbalance arising from some particular part. Rotating machinery, especially high-speed machinery, needs to be counterbalanced if the center of gravity of the rotating mass does not lie on the axis of rotation. Hoists are frequently counterbalanced so that a descending weight will supply some of the energy required for hoisting the non-useful load. Numerous examples of counterbalancing will be found in everyday practice, but those cases associated with the counterbalancing of high-speed rotating machinery are the most imperative of solution. (See **Balance.**) (F.T.M.)

COUNTERBORE. Counterboring, spot-facing, countersinking, and center drilling are hole-enlarging operations generally performed on a drill press, although they are actually reaming operations. Counterbored holes are those with cylindrical enlargements, for fillister head screws; spot-faced holes are counterbored to a depth just sufficient to clean up the surface. Countersunk holes have conical enlargements; center drilling is essentially countersinking, using an integral drill and countersink for machining center holes for turning operations on lathes.

Both counterboring and countersinking are performed with piloted cutters, but usually (except for center drilling) require separate operations for drilling the body hole before countersinking or counterboring. A subland drill is a twist drill of special design, constructed so as to provide separate cutting edges and flutes for each hole diameter, so that counterbored holes may be produced in one operation. (H.C.H.)

COUNTERPOISE. This is a network of wires placed near the ground to serve as the ground side of the circuit of a radio radiating system, the other side being the **antenna.** It is used primarily where the actual **ground** has such poor conductivity that satisfactory operation without it cannot be obtained. This network should extend well beyond the antenna if it is to be fully effective. (L.R.Q.)

COUNTERSHAFT. Shaft.

COUNTERSINK. Counterbore.

COUNTING TUBE. A type of **ionization chamber** adapted to detect the arrival of individual ionizing particles, such as **beta rays** or **cosmic-ray** particles. The enclosure is commonly cylindrical with its thin metal wall serving as one electrode, preferably surrounded by a protecting envelope of glass. The other electrode may be a needle-point projecting into the cylinder, as in the Geiger counter, or a fine, straight, axial wire, as in the Rutherford-Geiger and the Geiger-Müller counters. These electrodes are maintained at a potential difference just too small to ionize the air between them without some additional assistance. When an ionizing particle enters through the tube wall, the resulting ionization causes a momentary current. This may be amplified to operate a suitable signal or even to record itself on a drum or tape. (L.D.W.)

COUNTRY ROCK. The general term used for the main mass of rock in which occur the **veins, dikes** or ore bodies which are of particular interest, or which are described in detail. (R.M.F.)

COUP DE POING. Paleontology of Man.

COUPLE. As a usual thing the action of two forces on a body can be duplicated by a single force, equal to their **resultant,** acting at their **center of pressure.** Two parallel forces of equal magnitude but opposite direction cannot be reduced to a single force. They form a couple. The effect of a couple upon a body is independent of the location of that couple with respect to the body. The net action of several couples all in the same plane on a body is equal to the algebraic sum of the moments of the couples, the sign being determined from the direction of rotation which the couple tends to give. The moment of a couple is the product of the perpendicular distance between the forces and one of the forces. The action of any force, acting at any particular point on a body, upon another point lying in the plane of the force can be resolved into another force of the same magnitude acting at the desired point plus a couple. For example, let F be any force acting at point a, and b any other point at distance d from the line of action of F.

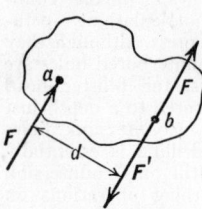

At point b place two equal and opposite forces F and F' which are parallel to the direction of the original force F. Then F' at b and F at a form a couple Fd, leaving F which acts through point b. The effect on point b of the latter force and the couple is the same as the effect of the original force F acting at a. Thus it is seen that it is possible to replace a force acting at a with an equal force acting at b, and the couple Fd, where d is perpendicular distance from b to the force F in its original position. (F.T.M.)

COUPLED CIRCUIT. While any group of circuits which are so connected or related that effects in one produce effects in the other constitute a coupled circuit, the term is usually used to designate circuits related so a-c effects are transferred but steady state d-c effects are not. The two most common classifications of coupled circuits are the inductive and the capacitive coupled circuits, so named because of the primary method of transferring the effects. Capacitance coupling is used quite extensively in various vacuum tube amplifier circuits, in thyratron circuits, and similar applications where it is desired to block d-c effects and transfer a.c. Since

the **condenser** does this it may be used as the common element between the two circuits. The so-called resistance coupled amplifier is really capacitance coupled. Fig. 1 shows examples of this type. Inductive coupling is the most widely used type since it is used extensively in the power field as well as in the communications and electronics fields. The ordinary power **transformer** is the

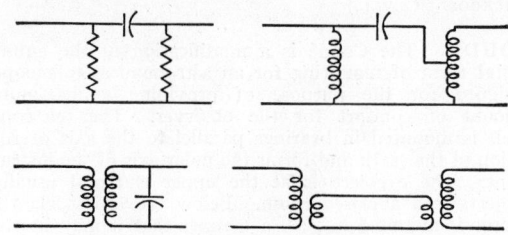

Fig. 1. Capacitance-coupled circuits (top).
Fig. 2. Inductance-coupled circuits.

means of inductively coupling two power circuits. The various transformers, tuning coils, etc., of radio circuits are other examples. Fig. 2 shows some typical circuits.

For two circuits to be inductively coupled an inductance element in one circuit must be so related to an inductance in the other that flux set up by one links the other. Thus a current flowing in circuit number 1 produces a flux which, in part at least, links the other. When this flux changes it produces a voltage in the second coil, and if the second circuit is closed a current flows. This flux which links both circuits is the mutual flux and the effect gives rise to **mutual inductance.** Mutual inductance may be defined as the flux linkages in one circuit per ampere in the other,

$$M = N_2 \phi_m / I_1$$

where M is the mutual inductance, $N_2 \phi_m$ the flux linkages by the mutual flux in circuit 2 and I_1 is the current in circuit 1 which caused the flux ϕ_m. M is also expressible in terms of the self-**inductance** of the coupled coils,

$$M = k \sqrt{L_1 L_2}$$

where k is the coefficient of coupling and the L's are the respective self-inductances. The manner in which the current in the secondary circuit varies is a rather complicated function of the various circuit parameters, frequency, and primary current. However, there is a certain value of coupling which will produce a maximum secondary current for fixed values of the other parameters. This value of coupling is called the critical coupling. In the usual inductive coupled circuit used at radio frequencies where the circuits are tuned, the secondary current plotted as a function of frequency presents a single peak, increasing in value, up to the critical coupling, then presents a double peak with no increase in value as the coupling is made still closer.

Not only are coupled circuits used to transfer a-c energy from one circuit to another, but by proper design may be made to match the impedances of the connected circuits also. (See **Impedance Matching.**) (L.R.Q.)

COUPLING. A means for joining two parts. There are a number of specific uses of the term. A pipe coupling is a hollow cylinder with internal pipe threads, used to join two sections of externally threaded pipe, either of the same or of different sizes; in the latter case, the unit is referred to as a reducing coupling. Shaft couplings are used to connect rotating shafts, and are of two types—rigid and flexible. The sleeve coupling is an example of the former type, and consists of a hollow cylinder, usually provided with a keyway and set screws to prevent relative rotary and axial motion of the shafts. Another type of rigid coupling, the flange coupling, consists of a disk and hub, on each shaft, connected by through bolts. Since perfect alignment of two theoretically collinear shafts is difficult to attain, some form of flexible coupling is usually employed for moderate

Flange coupling.

or heavy duty transmissions, such as motor-generator sets, motor-driven pumps, and the like, to prevent the transmission of shock and eliminate stress reversals. There are numerous commercial forms of flexible couplings; one type is similar to a flanged coupling, but employs laminated steel pins instead of through bolts for transmitting power from one flange to the other. In another form, two sprockets of equal size are mounted— one on each shaft—and connected by means of a roller or silent **chain**. In other forms, the connection between the shaft flanges is effected by springs or by bolts or pins mounted in rubber.

For connecting shafts whose axes are slightly out of alignment, but approximately parallel, the Oldhams or cross-keyed coupling is used. This device consists of two coupling halves fastened to the shafts; each half has a groove or slot cut in it, and the two halves are arranged so that the grooves are perpendicular. The halves are connected by a central member with perpendicular tongues that engage the coupling half slots. In some instances, cross-keyed couplings are used as flexible couplings; in such cases, the central member is made of fiber or has leather-faced contact surfaces.

A universal joint is a rigid coupling for connecting shafts whose axes will intersect if prolonged, and for applications where the angle between the shaft axes may vary during operation. The device is usually composed of two forked coupling halves, with a central block free to oscillate about two mutually perpendicular axes lying in a plane perpendicular to the shaft axes. Universal joints operate satisfactorily when the shaft coincidence error does not exceed ten to fifteen degrees; beyond this range, the joint is likely to be quite inefficient.   (H.C.H.)

COURLAN. Aves, Gruiformes. A large Brazilian bird (**Aves**) which resembles the rails in appearance and habits. Also called the limpkin. (A.W.L.)

COURSE. The term course is used in a number of contradictory meanings by different writers on the general subject of **navigation**. During the past few years the U. S. Navy has adopted two standard meanings for this term. Course is the **direction** that a navigator desires his ship to follow for a given period of time, and course is also the direction that a navigator hopes his ship has followed for a given period. To be more specific: In the first place, the navigator knows or assumes that his ship is in a given location at a given time and he wishes to proceed to another specified location. By graphical methods, on a **chart** or **small-area plotting sheet**, or by any one of a number of standard computational methods, such as **plane sailing, middle-latitude sailing**, Mercator sailing, great-circle sailing, composite sailng, etc., the navigator obtains the direction and length of the line joining the two points. These are known as the predicted course and distance between the two locations. In the second place, the navi-

gator knows, or assumes, that his ship is at a given location at a certain time. With the ship proceeding on a given **heading** with a known speed, the navigator may determine, by graphical or computational **dead-reckoning** methods, a position of the ship at the end of a definite period of time. The direction and length of the line joining the two locations are the assumed course and distance between the two points, or the dead-reckoning course and distance made good. (See **Track**.) (W.K.G.)

COURSER. Aves, Charadriiformes. Long-legged birds (**Aves**) related to the plovers, found in Africa and southern Europe. One species is a desert bird. (A.W.L.)

COURTSHIP. The special behavior of animals in seeking mates.

Courtship varies from the complex behavior of birds and mammals to the random association of the sexes in many simpler animals, where meeting under proper conditions inevitably results in mating without evident preliminaries.

In its more complex aspects courtship almost invariably consists of some display on the part of the male which influences the female to accept or reject his advances, and is often accompanied by rivalry between males which may be settled by combat. The singing of male birds during the breeding season and the display of brilliant plumage by some species are well-known examples, and unusual evolutions in flight are little less familiar. Such behavior also occurs among the invertebrates, particularly among the spiders and some of the insects. Some of the South American birds, the **tinamous**, reverse the usual type of courtship. In these species the females court the males. (A.W.L.)

COVALENCE. Valence.

COVARIANCE. The covariance is defined as $\Sigma(x_i - \overline{X})(y_i - \overline{Y})N$. It equals $r\sigma_x\sigma_y$, where $\overline{X}$ is the **mean** of the x's, $\overline{Y}$ = mean of the y's, r is the coefficient of correlation, σ_x and σ_y are the **standard deviations** of x and of y, respectively. (L.A.A.)

COVELLITE. The mineral covellite, **cupric sulfide**, CuS, is **hexagonal**, usually in thin platey crystals, but may be massive. It has a hardness of 1.5–2; specific gravity, 4.6; luster, submetallic to resinous; color, dark indigo blue, sometimes showing a purplish tarnish, or if moistened may appear purple in color. Its streak is dark gray to black; it is opaque. Covellite is found associated with **chalcopyrite, bornite, cholcocite**, etc., and is believed to be chiefly of secondary origin. Covellite occurs in Serbia, Saxony, Sardinia, Argentina, Chile, Bolivia and Peru, and in the United States at Butte, Montana, and in Colorado, Wyoming and Utah. This mineral was named for Covelli, who discovered it in the lavas of Mt. Vesuvius. (E.S.C.S.)

COWBIRD. Aves, Passeriformes. 1. The yellow **wagtail** of England. 2. A dark-colored bird, *Molothrus ater,* of southern Canada and the United States, related to the blackbirds, and several species of the same genus extending from Texas into South America. Most of these species deposit their eggs in the nests of other birds like the European cuckoo. They are named from their frequent association with cattle and before the settlement of North America the common species was called the buffalo bird. (A.W.L.)

COWLING. Cooling, Engine.

COWLING, ENGINE. Cowling is removable covering placed over or around an engine. Its purpose may be simply that of a weather guard—or it may serve the more complex function of streamlining or engine cooling. The *hood* of an automobile engine is a simple form of cowling.

It is in connection with air-cooled airplane engines that cowling is of the greatest technical interest since there it becomes part of high-performance design. Any inferiority of cowling design will severely penalize airplane performance.

At first, radial air-cooled engines were installed without cowling of any description; later the crankcase and accessory sections were enclosed in a streamlined aluminum cowling, leaving cylinder barrels and heads exposed. Then it was found that a narrow ring (Townend ring) around the cylinder heads increased speed by confining turbulence. However, the ring cowl tended to produce high cylinder-head temperatures in engines of high specific output. During the decade from 1930 to 1940 the whole subject of cowling and cooling of airplane engines was investigated in considerable detail by private and governmental organizations, with the result that much better cooling is obtainable with far less aerodynamic drag than was possible before. Most of this progress lay in the field of scientific cowling design. While most of this work was directed toward the radial air-cooled engine, reduction of the drag of liquid-cooled engines was also studied, especially that originating in the coolant radiator and the air ducts and scoops associated with it.

It was shown by F. W. Meredith that if air is admitted at high speed to a diffuser passage, then heated and expanded through an efficient nozzle, a jet thrust is produced which has capabilities of offsetting the drags of the system. Liquid-cooled engine advocates were quick to see the advantages of improved design of cooling air ducts and were able to produce airplanes with extremely low cooling-system drags. Later, in the radial engine field, improved designs of cowling were able to accomplish almost the same effect.

In the United States the NACA undertook comprehensive investigation of cowling and cooling of large radial engines. The NACA cowl is a close-fitting covering, usually cylindrical, placed around the engine and extended as a skirt some distance rearward of the cylinders. Its nose section is designed to divide the air flow into a small portion which enters the cowl, and a much larger portion which flows smoothly over the outside surface. Equally important is the inner cowling, or cylinder baffling, which prevents wastage of cooling air. The cowling skirt ends in a slot between the cowl and fuselage or nacelle. The heated air from the engine leaves through this slot and mingles with the main air flow outside the cowl.

Diagrams of the inner engine cowling (baffles) are shown for different types of air-cooled engines. These

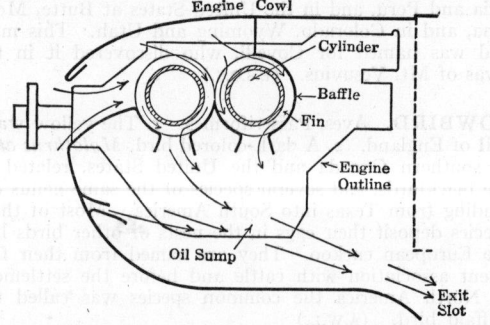

Fig. 1. Cowling and pressure baffles for horizontally opposed engine.

baffles confine the air flow to the region of the cooling fins, thereby using the minimum air for the necessary cooling. A pressure differential ΔP exists across these baffles because of the impact pressure existing upstream and the low pressure of the exit slot. Expressed as a

percentage of the **dynamic pressure** corresponding to airplane speed, this ΔP may be 60–80%. It can be controlled by making the exit slot variable, as with

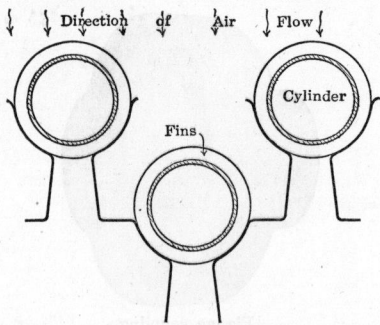

Fig. 2. Development of pressure baffle arrangement for two-row radial engine.

cowl flaps or sliding skirt. A fixed exit slot of sufficient size to cool a large engine at low speed promotes unnecessarily high cooling-system drags at higher speeds. This effect is not as noticeable for small engines.

Propellers are expected to provide an air inflow to the cooling system when the airplane is on the ground.

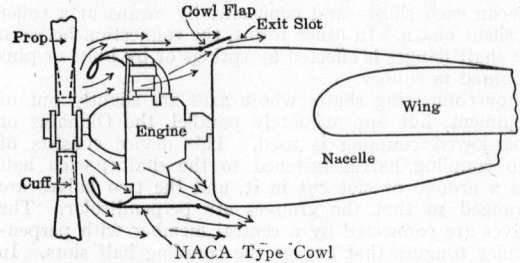

Fig. 3. Cooling air flow in NACA cowl.

The shanks of large metal propellers have evolved into nearly cylindrical sections possessing but little fan action. To provide cooling air for idling, cuffs of airfoil section are fitted to the propeller blade shanks. Engine-driven axial blower fans have been used for the same purpose.

Improvements continue to be made in the cooling and cowling of aircraft engines. Ample opportunity for improvement still exists since the engine delivers to the cooling system more energy than it does to the propeller. In time it may be possible to make much of this energy available as reaction propulsion. (F.T.M.)

COWPER'S GLAND. A gland of the male reproductive ducts of mammals whose secretion is part of the seminal fluid (see **Reproduction**). (A.W.L.)

COWRY. Mollusca, Gasteropoda. Compactly oval shells with a long narrow aperture, smooth surface, and often bright colors.

There are many species of cowries, especially in the Pacific and Indian Oceans. The shells of some are used as money and for decorations. (A.W.L.)

COXA. The segment by which the leg of an arthropod is joined to the body. (A.W.L.)

COXAL GLAND. Excretory glands opening on the fifth segment of the body in the king crab and scorpions. Coelomoducts. (A.W.L.)

COXOPODITE. Biramous Appendage.

COYOTE. Mammalia, Carnivora. Wolves of several smaller North American species, ranging from Iowa to

California and from Mexico into Canada. The common coyote or prairie wolf, *Canis latians,* occurs in the northern prairie area, the plains coyote, *C. nebracensis,* throughout the great plains, and the mountain coyote, *C. lestes,* in the western mountain areas.

Coyotes sometimes destroy chickens and other small domestic animals. They have been relentlessly killed for their depredations and remain common only in sparsely settled regions. (A.W.L.)

COYPU. Mammalia, Rodentia. A large aquatic rodent of South America whose habits are like those of the muskrat. Its fur is the nutria of commerce. (A.W.L.)

CRAB. Crustacea, Decapoda. **Crustaceans with a** short broad **cephalothorax** and a small abdomen bent below it. The large pinchers and four pairs of legs are

Above, male spider crab. Left, female spider crab. Right, ghost crab. Center, mud crab. (*From Mayer, Seashore Life, N. Y. Zool. Soc.*)

the only conspicuous appendages. Most species of crabs are found in or near the ocean but some are terrestrial and others live in fresh water. The land crabs deposit their eggs in the water. The many species of crabs constitute a division of the order named the Brachyura from the short abdomen.

A number of species of crabs are used for food. Of these the edible crab of the Atlantic, which ranges from Cape Cod to Louisiana, and the edible crab of the Pacific Coast, are the most important species.

Many crabs have received common names which apply to one species or to a group of similar species. Among these names are spider crab, hermit crab, fiddler crab and land crab. (A.W.L.)

CRACKING PROCESS. Gasoline, as one of the products of crude oil, is required today in such quantities for motor vehicles that were the crude oil to be separated into its commercial components (such as petroleum ether, gasoline, kerosene, etc.) by **distillation** or **fractionation** through heating and condensation, there would be an uneconomic excess of the products other than gasoline. For petroleum, the natural crude **oil,** is made up of a complex mixture of **hydrocarbons,** which differ in composition and structure. When petroleum is distilled, the more volatile fractions, such as gasoline, are found to be composed, in general, of hydrocarbons of simpler molecular structure—that is, those having molecules which contain a comparatively small number of atoms. Speaking broadly, the hydrocarbon molecules having larger numbers of atoms are more largely concentrated in the less volatile petroleum products.

In the cracking process these larger molecules are broken up to produce, in part, a greater yield of gasoline and other low-boiling fractions. One of the simplest methods of cracking is by heating under pressure. By efficient cracking, gasoline yields of 60% or more have been obtained from certain crudes. (For further discussion of cracking, see **Hydrocarbons; Gasoline.**) (F.T.M.)

CRAG IMPLEMENTS. Paleontology of Man.

CRAIT. Reptilia, Sauria. A poisonous **snake** of India, *Bungarus caeruleus,* related to the cobras and adders. It is said to conceal itself in houses, and to this fact must be attributed its record of killing more human beings than any other snake, with the exception of the cobra.

The name crait has also been applied to several related species of Oriental snakes, including the banded adder. (A.W.L.)

CRANBERRY. Heath Family.

CRANE. Aves, Gruiformes. Large birds (**Aves**) with long legs and neck. They are superficially like the larger herons and the name crane is sometimes inaccurately applied to the latter birds, especially to the great blue heron. Cranes are found in Europe, Asia, North America and Africa. (See **Lifting Crane.**) (A.W.L.)

CRANE FLY. Insecta, Diptera. **Insects** which resemble mosquitoes in form but are usually much larger. They have a V-shaped groove across the thorax and have no scales on the wings.

The **larvae** of some species live in the ground and are sometimes injurious to the roots of grasses and grains. These larvae are called meadow maggots or leather jackets. They come to the surface at times and can be destroyed by the use of poison baits. (A.W.L.)

CRANE, LIFTING. In engineering, a crane is a hoisting machine in a particular category. It is not a simple matter to point an exact distinction between **hoists** and cranes, as they have not been clearly distinguished in common usage. However, cranes might be classified as those types of hoisting machines where vertical lift is not the primary purpose (as it is, for example, in the case of an elevator) but where the lift is only that sufficient to allow proper clearances for those parts being transported by the crane. The primary purpose of a crane is to lift a piece from one place and set it down in another, and so a degree of horizontal freedom is essential. The horizontal movement may be accomplished in two ways, viz., the rectilinear motion of a traveling crane mounted on parallel rails, or the rotary motion of a pivoted swinging crane. The ordinary traveling crane is that most frequently found inside buildings. The rails are laid on the walls or

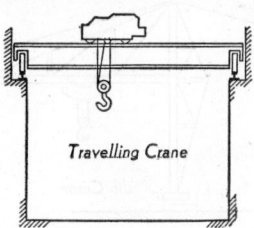

Travelling Crane

on a special steel framework inside the walls. Rolling on the rails on each side is a wheeled carriage which carries the end of a **girder** bridging the distance between the rails. This girder must be sufficiently strong to support by beam action the lifting capacity of the crane. A rolling carriage containing the hoist is mounted on top of the girder so that it gives motion in a direction perpendicular to the main rails. Thus every part of **a** rectangular floor area inside the walls may be reached by the crane. This type of crane is widely used in factories and other plants where heavy objects must be transported about in a rectangular area. If the crane is seldom used, it may be a hand-operated type, having chains reaching down to the floor level to be pulled by hand—one to roll the crane back and forth along the main rails, the other to pull the carriage from side to side and to operate the hoist. Electrically operated traveling cranes usually have two motors, one to roll the crane along the main supports, the other to move the carriage on the crane girder and operate the hoist. The control can be accomplished remotely from an electrical control panel, or by an operator stationed on a cab which travels with the crane. The latter is generally the case when the crane is in almost continuous use. In outdoor locations, where it is not convenient to erect a structure to carry the overhead rails, the crane can be equipped with legs which support it on rails at the ground level. It is then a Gantry crane.

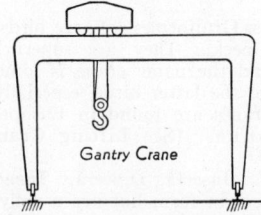

Gantry Crane

The post crane is a simple type of rotary crane frequently used for unloading material from one transportation system and placing it on another. The area

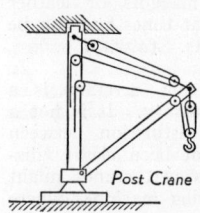

Post Crane

which can be served by a crane of this type is more limited than in the case of the traveling crane, but it is a simpler, cheaper type. As shown in the figure, the post is supported on pivots at the top and bottom. It would be necessary to support the top by guy wires out of doors. The principal members are the post and the **boom**. In the type shown, the radius of the circle in which the hook travels can be varied, but this is fixed in some post cranes, where the post and boom are connected by a simple **tie rod** member. If the lower pivot of a post crane were made large and heavy, the top support could be eliminated. The load would be carried by bending in the post, which would then be shortened and become the pillar of a pillar-type crane.

The jib crane is better adapted for an all-steel rotary crane where the radius to be covered is larger than that to which the post or derrick type crane is suited. The

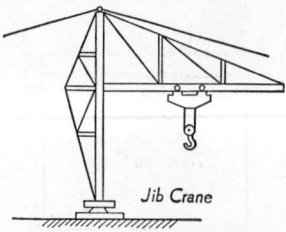

Jib Crane

radius is varied by changing the position of the trolley on the jib beam. The post and jib cranes may be swung about either by hand or by motors. Cranes having a further degree of mobility are the locomotive and crawler types, which are self-contained units mounted, in the case of a locomotive crane, on a railway carriage, or in the case of the crawler crane, on caterpillar treads. They are motivated by internal combustion or steam

engines, and in structure are somewhat similar to a pillar crane. (F.T.M.)

CRANIATA. Vertebrata.

CRANIUM. Skeletal System.

CRANK. Crankshaft, Bell Crank.

CRANKSHAFT. A crank is a bent arm which moves with rotary motion about its unbent end. In order to provide a support for this rotation the crank is mounted on a crankshaft. The crank and crankshaft form one of the important basic units of mechanism. There has been found no simpler, more efficient way of transforming rotary to reciprocating motion, or vice versa. The crank alone does not accomplish this, but in conjunction with the **connecting rod** and slider, it forms the basis of many such important machines as engines, pumps, compressors, and a host of other mechanisms. A simple crankshaft of the overhung type is readily made from an arm (or disk) and a crank pin which is set into it at some radial distance from a crankshaft, which is the center of rotation. This type is widely used where a crankshaft is to accommodate but one crank, and where the crankshaft bearing surface is entirely on one side of the crank. Multiple cranks on the same crankshaft are obtained by fitting the crank pin between two crank arms, so that the connecting rod may swing freely without interference with the crankshaft. Multiple throw crankshafts are made by building up the crank pin between two cranks, and by machining from a single forged piece. It is the function of a crank to transmit the force at the crank pin into a torque at the crankshaft, the torque being equal to the component of crank pin pressure perpendicular to the crank radius multiplied by that radius. Since crank pin pressure is not always exactly perpendicular to crank radius, it follows that a crank may carry some compression or tension as well. It should never be loaded with bending by forces parallel to the crankshaft. (F.T.M.)

CRASPEDON. Velum.

CRASPEDOTE. Provided with a craspedon or **velum.** Applied to the medusae of many species of **Hydrozoa** in contrast with the jellyfishes. (A.W.L.)

CRAYFISH. Crustacea, Decapoda. Fresh-water **crustaceans** resembling small lobsters, with which they are closely related. They are widely distributed. The term is applied to a limited extent to certain marine species which are more closely allied to the lobsters than to the true crayfishes. In the United States **Cambarus** is a common genus East of the Rockies, and *Astacus* to the west.

Crayfishes are eaten in Europe and to a limited extent in the United States. (A.W.L.)

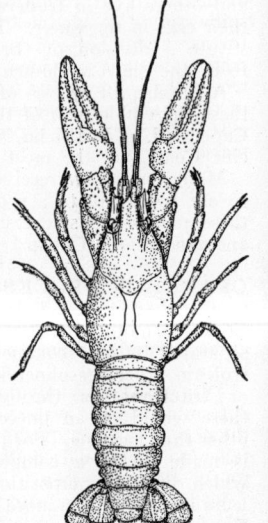

Crayfish.

CREAM OF TARTAR. Tartaric Acid.

CREEP. Creep of metals is slow plastic deformation under stress. When lead is subjected to any considerable stress, deformation occurs and continues indefinitely or until rupture takes place. On the other hand, steel and most structural metals do not creep or flow plastically until the stress level is above a critical point such as the **yield point** in mild steel. At elevated temperatures, e.g., 900° F., creep occurs at much lower stresses and must be carefully considered in the design of equipment such as boilers and tubes which are intended to operate at elevated temperatures over long periods of time. Alloy steels have been developed having higher creep strengths than carbon steels.

Creep strength is the unit stress which will produce deformation at a specified rate at a specified temperature; for example, the creep strength of a certain 0.15% carbon open-hearth steel at 1000° F. is 6100 lbs. per sq. in. for a rate of 0.01% elongation in 1000 hours. Other values for this material are 6900 lbs. per sq. in. for a rate of 0.1%, and 7800 lbs. per sq. in. for a rate of 1.0% elongation in 1000 hours. Creep strength values can only be determined by long-time laboratory tests under carefully controlled conditions of temperature and loading.

The term is used by geologists to designate the slow movement of soil and rock waste down a slope. This movement is due to the combined influence of gravity, frost, and **ground water.** Creep has been proved to be an important factor in soil erosion even on relatively flat slopes. (R.H.H., R.M.F.)

CREEPER. Aves, Passeriformes. Small insectivorous birds (**Aves**) which cling to the trunks of trees or cliffs in seeking food. They are found in the Northern Hemisphere and are represented in North America by the brown creeper, *Certhia familiaris.* (A.W.L.)

CREODONT. Fossil Mammals.

CREOSOTEBUSH. *Larrea tridentata.* Zygophyllaceae. An evergreen shrub, sometimes growing to a height of 12′ but usually smaller, which is a prominent species of large parts of the semi-desert regions of California, Arizona, New Mexico, western Texas, and northern Mexico. In parts of its range the plant may occupy large areas as a practically pure stand. Creosotebush bears small yellow flowers and has a characteristic odor of creosote. (B.S.M.)

CREPUSCULAR. Active during twilight.

CREST GATE. The principal items of equipment to be used in conjunction with a dam, for the maintenance of desired water conditions, are crest gates, sluices, and spillways, and sometimes, by legal requirement, by-passes for fish or for logs. In order to control the head closely, crest gates are provided on the spillway sections, but where the spillway is a natural rock channel at one end of the dam it is convenient to effect the control by sluice gates. The different types of crest gates in use are summarized as follows:

1. Sliding gates.
2. Tilting gates.
3. Rolling gates.
4. Taintor gates.
5. Stationary flash boards.

Sliding gates are limited in capacity by the effort that is required to move them against frictional drag set up by heavy water pressure. When equipped with roller guides, as in the Stoney roller gate, the capacity per gate is much increased.

Tilting gates usually are so constructed that a predetermined rise of water against them automatically tilts them to the open position, then, after the water has fallen to its normal elevation, they automatically return to the closed position.

Rolling gates are supported on a cylindrical drum which can be rolled up a steep incline by various means. Chains, wrapped around the cylinder ends, are connected to hoists which roll the gates up an incline of 20° to the vertical.

The Taintor gate has been widely used in this country for crest control. The face of the Taintor gate is the

circumference of an arc the center of which is at the pivot point. This causes the resultant of water pressure against the gate always to pass through the pivot so

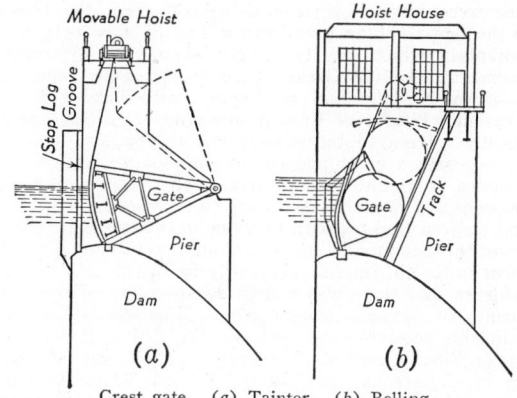

Crest gate. (*a*) Taintor. (*b*) Rolling.

that, with the exception of pivot friction, it is necessary to handle only the weight of the gate itself when opening or closing it.

When no other control is exercised, it is good practice to use temporary flash boards on the spillway crest. Generally these are constructed of wood and held in place by iron pins which will bend over and release the flash boards should the water rise to a point where the maximum spillway capacity is needed to pass the flood. (F.T.M.)

CRETACEOUS PERIOD. The last major division in the **Mesozoic Era** of the geologic time-scale. Type locality, chalk (creta) cliffs of the English Channel. The period was named by A. d'Halloy in 1822. In 1877, Hill proposed that the Lower Cretaceous be erected as a separate system which he called the Comanchean. The Comanchean period began about 135 million years ago and lasted about 25 million years. The Cretaceous period began about 110 million years ago, and lasted about 50 million years. The greatest thicknesses of Cretaceous strata in the United States occur in the Rocky Mountain region and in California. During the Cretaceous Period there was also an extensive overflow of the waters of the Gulf of Mexico toward

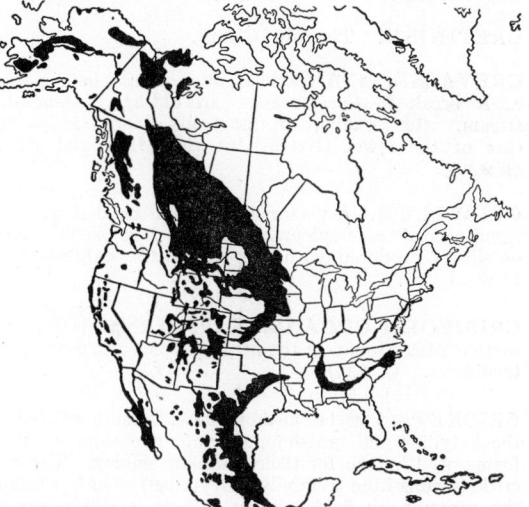

Map of North America showing the surface distribution (areas of outcrops) of Cretaceous strata. The large and small areas in the western interior of the continent very largely represent Upper Cretaceous deposits.

the southern interior of the United States, depositing there a great series of overlapping sediments. This is the type region of the Lower Cretaceous or Comanchean. While the Atlantic and Gulf continental margins were invaded by the Cretaceous Sea, the region of the **Appalachian geosyncline** had been reduced to a **peneplain** which may have been overlapped by marine sediments. In the Great Plains region were deposited fresh-water shales and sandstones with local swampy areas in which were formed numerous **lignites**, subsequently altered to bituminous coal. The Middle Cretaceous saw a great marine invasion of western North America from the Gulf of Mexico to the Arctic Ocean. Deposits of Cretaceous Age are well represented in central and western Europe, with fairly continuous marine deposition throughout the entire period. Cretaceous deposits occur in South America, especially in Brazil. As further evidence that there was widespread invasion of the continents by oceanic waters during this period, marine sediments occur in southwestern Asia, China, Himalayas, Japan, Siberia, and Africa. Important mineral resources are of Cretaceous Age. In the United States the Cretaceous formations contain numerous important aquifers which underlie great semi-arid areas. Bituminous coal beds of average value occur in Alaska, Australia, British Columbia and Germany. Important oil pools occur in the Gulf region. The Cretaceous clays of the eastern United States are extensively used in the manufacture of china and building materials. The famous sulfide copper ores of Butte, Montana, occur in igneous rocks of Cretaceous and Early Tertiary age. Lower Cretaceous plants and animals were only slightly different from those of Jurassic time. Ferns, **cycads**, **ginkgoales** and conifers still predominated, and the principal marine invertebrates were **ammonites** and **belemnites**. During the Cretaceous there was a great expansion of the **pelecypods**, **gastropods**, and the modern types of fishes. The Mesozoic reptiles had reached their climax in the early Cretaceous and, except for a few large and bizarre forms, were on their way to extinction. The most remarkable types were **Triceratops** (horned dinosaur), **Tyrannosaurus** (tyrant dinosaur), **Tylosaurus** (marine lizard), and **Pteranodon** (crested pterodactyl, or flying reptile). Since uppermost marine Cretaceous does not occur in North America, there appears to have been a pronounced period of uplift and erosion, accompanied by mountain building, with the combined growth of the Cordilleran ranges. This period of diastrophism, which closed the Mesozoic Era in the western hemisphere, is called the **Laramide Revolution.** (R.M.F.)

CRETINISM. Thyroid Gland.

CREVASSE. 1. A crack or open fissure in a glacier. 2. A break in the levee of an old-age, meandering stream, often leading to disastrous floods, as in the case of the lower stretches of the Mississippi River. (R.M.F.)

CRIBELLUM. A perforated plate associated with the spinnerets of some spiders. Through it a peculiar form of silk is produced with the aid of the **calamistrum.** (A.W.L.)

CRIBIFORM ORGANS. Series of thin vertical calcareous plates found in the arms of some starfishes (**Asteroidea**). (A.W.L.)

CRICKET. Insecta, Orthoptera. **Insects** related to the katydids and grasshoppers and, like some of these forms, well known for their resonant singing. The true crickets constitute a family (Gryllidae) which contains the common or field crickets and in addition several other forms more or less different in appearance. The field crickets are black or brown species, some of which enter houses. Tree crickets are usually green with broad transparent wings. They frequent trees and shrubs. Mole crickets are thick-bodied brown insects whose forelegs are strongly developed for burrowing. In addition to these and a few other forms of crickets several insects belonging with the katydids among the long-horned grasshoppers bear this name. They are the cave or camel crickets, the sand cricket, and the Mormon cricket. (A.W.L.)

Cricket.

CRINOIDEA. The sea lilies, feather stars, and basket stars, a class of the phylum **Echinodermata.** There are now only a few hundred species of these animals although several thousand fossil species are known.

The crinoids are distinguished from the starfishes and other echinoderms by the following characteristics: 1. The body consists of a disk, arms, and a stalk. 2. The mouth and anus are both directed upward. 3. The arms bear small lateral branches and in many species fork repeatedly.

Most of the living species lose the stalk when mature and become free-swimming. These forms, known as comatulids, are found in shallower waters of the ocean, where they swim or creep by means of the arms. The stalked species are found in deep water.

The classification of crinoids is of little interest save to specialists. Many families are recognized and they are sometimes grouped in three orders: Holopodida, Ptilocrinida, and Comatulida. The first includes a primitive attached form, the second the stalked crinoids, and the last the free species. (A.W.L.)

CRISIS. 1. The turning-point of a disease, either toward recovery or death. In certain diseases this may occur dramatically—particularly in **pneumonia**, where the high fever suddenly drops, and the acutely ill, delirious patient within a short time becomes quiet, of better color, breathes more easily, and seems to be definitely on the way to recovery. 2. A painful paroxysm or seizure seen in **syphilis** of the spinal cord (tabes dorsalis) where it usually occurs as a "gastric crisis" characterized by severe, intractable abdominal pain. (D.M.H.)

CRISTOBALITE. Tridymite.

CRITICAL ANGLE. Total Reflection.

CRITICAL COUPLING. Coupled Circuits.

CRITICAL FREQUENCY. A wave radiated from the **antenna** of a radio **transmitter** spreads in various directions, the exact nature of its spread being determined by the directional characteristics of the antenna. The part which travels towards the outer atmosphere goes into the **ionosphere** where it acts upon the ionized particles, principally **electrons**, which absorb energy and re-radiate it. The net result of this action is an effective change of the index of refraction of the medium through which the wave is traveling. This causes the wave to be bent back towards the earth, the extent of the bending varying with the index. It is this which causes radio waves to be returned to the earth to give reception at points very remote from the transmitter. However, this change of index of refraction varies with frequency in such a manner that the waves are bent less and less as the frequency is raised. A critical frequency is finally reached where the wave is not bent enough to return to the earth, even for the most glancing angle of incidence which it is possible to obtain. As the frequency of the signal is progressively raised the critical frequencies for the various layers of the ionosphere are reached in turn, the wave penetrating each at its critical frequency and being refracted back to the earth by the next layer until finally all layers are penetrated and there is no returning signal. This occurs in the vicinity of 40 megacycles. (L.R.Q.)

CRITICAL GRID VOLTAGE. Thyratron.

CRITICAL POTENTIAL. Electrons are commonly set into motion by an electric field, the energy thus acquired being proportional to the **potential difference** traveled through. If, when moving very rapidly, an electron encounters an atom or a molecule, the latter may be ionized or it may be merely excited; either process involving a change in the relation of one or more of the atomic or molecular electrons to the atom or molecule to which it belongs. Such a change requires a certain minimum amount of energy and unless the moving electron has at least this energy, the ionization or excitation does not take place. The critical potential corresponding to such a process is the potential difference necessary to give the moving electron the requisite energy. For example, the least energy which a cathode particle must have to excite the $K\alpha$ x-radiation of nickel (wavelength 1.662 angstroms) is 7426 **electron volts,** that is, the energy which would be given to it by applying a potential difference of 7426 volts to the x-ray tube. Hence 7426 volts is the critical potential for this excitation. (L.D.W.)

CRITICAL PRESSURE. When a fluid flows through an orifice or a nozzle, the volume of flow from a fixed initial pressure depends on the final pressure as long as that final pressure is in excess of a certain critical pressure. If the final pressure is lower than the critical pressure, the amount of fluid passed will be the same as though the final pressure had been equal to the critical pressure. The critical pressure is a constant proportion of the initial pressure. For example, for air flow it is 53%, for saturated steam, 58%, and for superheated steam, about 56%.

Critical pressure also denotes the pressure of a vapor at its critical point, the latter occurring at a temperature above which the vapor can exist only in the form of a gas. Water, for example, has a critical pressure of 3226 lbs. per sq. in. (See **Critical State.**) (F.T.M.)

CRITICAL SPEED. Because of lack of homogeneity of a body caused by manufacturing difficulties and variations in material densities, it is impossible to distribute the mass of a rotating body about its geometric center. If the shaft on which the body rotates deflects under load, the center of mass may move from the axis of rotation of the shaft. Shaft rotation will then begin about the geometric axis but at some speed the centrifugal force caused by the displacement of the center of mass will equal the deflecting forces on the shaft. The deflecting force may be opposite to and balance the centrifugal force, or the two may be in phase, causing a periodic variation in the radial force on the shaft, which induces a series of vibrations. Since the magnitude of the centrifugal force depends upon the angular velocity and the mass, the vibrations will attain a maximum value at some speed, which is called the critical speed of the shaft. Above the critical speed a state of equilibrium may again be attained in which the body virtually rotates about its mass center. Second, third, and fourth critical speeds are also possible but the amplitudes of the shaft vibration are progressively less. (H.C.H.)

CRITICAL STATE. For every pure substance there is a definite temperature (the critical temperature) and a definite pressure (the critical pressure) at which the liquid and the vapor are indistinguishable. If the liquid

substance is enclosed in a hermetically sealed tube with nothing above it but its own vapor, and the temperature is raised, the liquid expands and also evaporates, adding to the vapor; so that the density of the liquid decreases while that of the vapor increases. Ultimately the two densities become equal (the common value being the critical density and its reciprocal the critical specific volume). The surface separating liquid and vapor then becomes invisible; though apparently the surface tension does not altogether vanish until a slightly higher temperature is reached. The pressure attained in this process may be very great; for water it is about 218 atmospheres, the corresponding temperature being 374° C. Above this temperature the substance is said to be a gas, and no amount of compression will cause it to separate into two fluid phases. (See **States of Matter.**)

For certain purposes it is convenient to express the absolute temperature, the pressure, and the specific volume of a substance in terms of their respective critical values, i.e., as abstract ratios T/T_c, p/p_c, v/v_c, called "reduced variables of state." If two gases have two reduced variables of state in common, the third variable will also be approximately the same for the two gases; a principle known as the "law of corresponding states." (L.D.W.)

CRITICAL TEMPERATURE. Critical State.

CROCIDOLITE (BLUE ASBESTOS). The mineral crocidolite may be considered as a fibrous variety of the **monoclinic amphibole**, riebeckite. It is also known as a massive mineral. Its hardness is 4; specific gravity, 3.2–3.3; luster, silky to dull; color, blue or bluish-green. It is found in Austria, France, Bolivia, South Africa (the variety known as tiger's-eye), and in the United States in Massachusetts and Rhode Island. The name crocidolite is derived from the Greek, meaning woof, in reference to its fibrous appearance. (E.S.C.S.)

CROCODILE. Reptilia, Crocodilia. Large aquatic **reptiles** of elongate slender form, with short legs, a long compressed tail, and skin armored with bony plates. The crocodiles differ from the related **alligators** and **caiman** in the more tapering snout, although their jaws are much wider than those of the **garial**. Crocodiles are found in southern Asia, Africa, northern Australia, tropical South America, and in southern Florida. Some species occur only in fresh water and others in salt marshes and estuaries. The most common genus is *Crocodilus.*

Crocodiles are much more aggressive and vicious than alligators and are dangerous to man (see also **Fossil Reptiles**). (A.W.L.)

CROCODILIA. An order of large **reptiles** of long slender build. The skin is armed with bony plates and the long jaws bear many conical teeth set firmly in bony sockets. The animals are chiefly aquatic. They are found only in warmer regions and chiefly in fresh water, although some enter the ocean.

The order includes **crocodiles, alligators, caimans,** and the **garial.** (A.W.L.)

CROCOITE. The mineral crocoite, **lead** chromate, corresponds to the formula $PbCrO_4$, and forms prismatic **monoclinic** crystals, often acicular. It is also found in columnar or granular masses. It has a rather distinct cleavage parallel to the prism, and a less distinct cleavage parallel to the base. It has a conchoidal fracture, is sectile; hardness, 2.5–3; specific gravity, 5.9–6.1; luster, adamantine to vitreous; color, red; streak, orange-yellow, translucent. Crocoite is a secondary mineral believed to be formed by waters containing chromic acid acting upon lead minerals like galena, with which it is associated. It is found in Russia, Rumania, Tasmania, Brazil, the Philippines, and Arizona. It is not of com-

mercial value. The name crocoite is derived from the Greek word for saffron in reference to the color of the powdered mineral. (E.S.C.S.)

CROOKES TUBE. Sir William Crookes was a pioneer in the study of electric discharge in gases. In the **vacuum tubes** which he used, and certain forms of which still bear his name, the pressure was reduced to such a point that the bright glow observed at higher pressures practically disappeared. The **cathode rays**, obstructed by but little residual gas, shot straight across the tube and, impinging upon the opposite wall, caused it to glow with greenish fluorescence. By placing an obstacle, such as a metal plate shaped

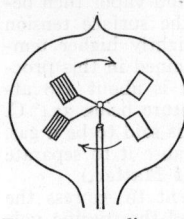

Four-vane radiometer in bulb.

like a Maltese cross, in the path of the rays he was able to demonstrate their rectilinear character by the shadow on the fluorescing surface. In one type of tube he interposed a light paddle-wheel of metallic vanes in the path of the rays, and found it driven at high speed as if by a stream of air. It is, however, probable that the greater part of this effect is due to the heating of the bombarded surfaces and that the paddle-wheel really operates as a Crookes **radiometer**. (L.D.W.)

CROP. A thin-walled expanded portion of the alimentary tract (see **Digestive System**) used for the storage of food prior to digestion. Crops are found in many animals, including **earthworms, insects,** and **birds.** (A.W.L.)

CROSS MODULATION. This is an effect produced in radio receivers which results in the **modulation** from one **carrier** being impressed on another carrier. If two modulated carriers are applied to the **grid** of a vacuum **tube** which has appreciable curvature in its operating characteristic, there will be numerous frequency components in the output of the tube. However, following selective or tuning circuits will cut out many of these but there will be certain ones which will not be cut out. Some of these represent the sidebands of the modulation of the undesired signal modulated on the desired carrier. Since they occupy the same frequency band as the desired carrier and its normal modulation there is no way in which they may be cut out and hence upon detection the audio output will contain both the desired and the undesired audio signals. (L.R.Q.)

CROSS PRODUCT OF TWO VECTORS. Vector Product of Two Vectors.

CROSS-BAR SWITCHING. This is a type of automatic telephone switching used for very large service areas. It is characterized by freedom from brushes and sliding contacts, the operation being through magnetically operated mechanisms. (L.R.Q.)

CROSS-BEDDING. Oblique lamination of certain beds in **aeolian** or water-laid **sediments**, caused by current action, is called cross-bedding. Cross-bedded sediments are found especially in river and stream deltas, alluvial fans and cones, river sand bars and marine sand deposits, and are also characteristic of windblown deposits of all kinds. The different types of cross-bedding are useful criteria for helping to determine the physical (including climatic) conditions under which certain types of clastic sediments were deposited. In regions where the formations have been highly deformed, and possibly overturned, cross-bedding may also be used by the **stratigrapher** and structural geologist to determine the original order in which sedimentary strata were laid down. (R.M.F.)

CROSSBILL. Aves, Passeriformes. *Loxia.* Small seed-eating birds (**Aves**) of the northern hemisphere whose mandibles cross at the tip when the mouth is closed. This adaptation enables them to open seeds and fruits, such as cones, very readily. (A.W.L.)

CROSS-FIRE. In nearly all communications circuits the problem of interference from adjacent communication or power circuit must be dealt with. This interference may be between lines of a like nature, between communications lines of different types, or between the communication line and a power circuit. In any event, the net effect is to produce interference which may be merely objectionable noise, or it may be severe enough to cause false operation. When the interference is between two telegraph circuits it is known as cross-fire. The effect is for the signals in one circuit to produce signals and hence false operation in the other. (L.R.Q.)

CROSSHEAD. This is a machine part having reciprocating motion and mounting a pin for the attachment of a connecting rod. In the crank and connecting rod mechanism which is employed to interchange reciprocating and rotating motion, the crosshead is the reciprocating part. It slides between crosshead guides which are attached to the frame. One end of the **connecting rod** is jointed to the crosshead by the wrist pin.

Piston and cylinder engines which are single-acting, that is to say, which have the gas or vapor on one side of the piston only, may have the wrist pin set directly in the piston. A crosshead is essential to a double-acting engine. The piston rod of a double-acting engine must go through a stuffing box or a packing gland in the end of a cylinder, a thing which is impossible for the swinging connecting rod. Some large single-acting engines also have a crosshead so that the wear due to the side thrust of the connecting rod will occur on the easily adjustable crosshead guides instead of the cylinder walls. (F.T.M.)

CROSSOVER. Heredity.

CROSS-RATIO. Anharmonic Ratio.

CROSS-SLIDE. Lathe.

CROSS-STAFF. Both the cross-staff and the **astrolabe** were used by navigators for the purpose of measuring **altitude** of celestial objects in the period prior to the invention of the **sextant** in the 18th century. The astrolabe was the more compact of the two, but the use of the cross-staff was simpler and the results slightly more accurate, particularly for small altitudes.

The principle and use of the instrument is illustrated in the accompanying figure. A "cross" with a peep

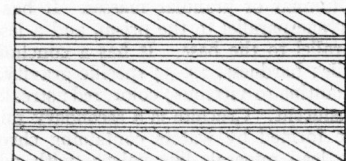

Cross-bedding (false bedding). (*Field, Laboratory Manual, Princeton University Press.*)

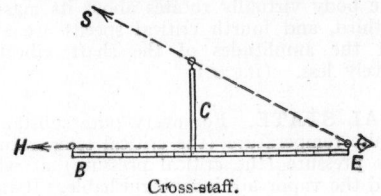

Cross-staff.

sight in its upper end slides along a rod *EB*. Holding the cross vertical, the observer sights from *E* along the rod toward the horizon at *H*, and slides the cross along until the object under observation appears through the peep sight in the cross. The graduations on the rod indicate directly the value of the angle *HES,* which is the desired apparent altitude. The instrument may also be used to measure the angular distance between any two objects.

Within recent years the instrument has been used to some extent for teaching purposes in elementary courses in astronomy. (W.K.G.)

CROSSTALK. This is the undesirable operating condition of a telephone communication system where the electrical effects of one circuit induce an undesired current in another. In the case of a **phantom** telephone circuit, crosstalk will be troublesome unless the lines with loading coils, terminal equipment, etc., are perfectly balanced electrically. Considerable care must be taken in the manufacture of the loading units in order to minimize crosstalk. (See also **Cross-Fire** and **Morse Thump.** (F.T.M., L.R.Q.)

CROTON. Spurge Family.

CROTON BUG. Insecta, Orthoptera. A European **cockroach,** one of the two chief pests of this kind in the United States. Its name is said to be due to its association with water pipes from the Croton reservoir in New York City. (A.W.L.)

CROTONIC ACID. Alcohols and Ethers.

CROTONIC ALDEHYDE. Aldehydes, Ketones, and Related Compounds.

CROUP. A catarrhal inflammation of the **larynx** occurring in children, characterized by cough, hoarseness, and difficult, noisy inspiration (stridor). The disease usually occurs in attacks at night, and by morning the child is entirely well.

Membranous croup is a synonym for laryngeal **diphtheria.** (D.M.H.)

CROWN. Belting.

CROWN SHEET. Locomotive.

CROWS. Aves, Passeriformes. Large birds (**Aves**) of the northern hemisphere and Africa. They are black or black and white in color. Together with the ravens,

rooks, magpies, jays, and other species they make up the **crow** family (Corvidae).

The importance of the common crow of North America, *Corvus brachyrhynchus,* has been debated. He does some good in destroying vermin and insects, but the farmer and conservationist are both against him for his destruction of fruit and grain and the eggs and young of other birds. (A.W.L.)

CRUCIBLE. A crucible is a vessel made of heat-resistant material and employed to hold a material that in itself is at high temperature, or is to be subjected to high temperature. Crucibles will range in size from the small laboratory types to large ones having a capacity of several tons of molten metal. They are roughly cup- or barrel-shaped, and are made of some material such as clay. In the laboratory, platinum, iron, and porcelain crucibles are used. In the iron and steel industry, clay crucibles have been used, but at the present time most manufacturers employ the graphite crucible, which is made half and half from graphite and fire-clay, well-mixed, molded, and burned to vitrification. The crucible may be used to hold a material being melted or burned, as in some processes for making steel, where the raw material is put in the crucible, which is then set in a hot furnace until the contents are melted. Or a crucible may be used to receive molten metal, which has been produced elsewhere, as from a brass furnace or cupola, in which case it is the means for conveying it from the point of melting to the point of casting, serving thus as the intermediate reservoir between the furnace and the mold. (See **Casting.**) (F.T.M.)

CRUDE OIL. Petroleum; and Hydrocarbons.

CRUSH BRECCIAS. Autoclastic.

CRUSHING AND GRINDING. Machines for the mechanical subdivision of solids are usually classified as coarse crushers, intermediates and fine grinders.

The best known types of coarse crushers are jaw crushers and gyratory crushers. A typical jaw crusher, the Blake, consists of a set of vertical jaws, one jaw being

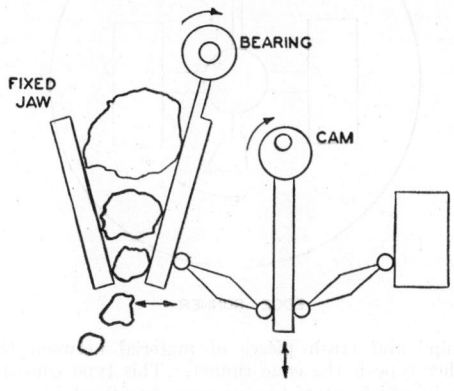

JAW CRUSHER

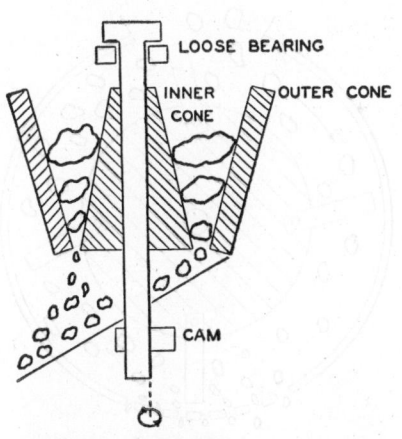

GYRATORY CRUSHER

fixed while the other moves back and forth by a cam and pitman arrangement. The jaws are farther apart at the top than at the bottom so that the material is progressively crushed smaller as it travels down the jaws, falling through the bottom when it becomes small enough. The movement of the jaw may therefore be small since the complete crushing is not performed with one stroke. A gyratory crusher consists of inner and outer vertical crushing cones. The inner cone, apex upward, is inverted with respect to the outer. The inner cone has a slight circular movement, but does not rotate. The movement is given to the cone shaft by a cam arrangement. Its crushing action is similar to that of the jaw crusher.

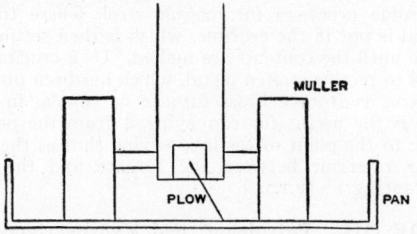

One type of intermediate crusher is a set of crushing rolls. A set consists of two horizontal cylinders placed close together and rotating in opposite directions so as

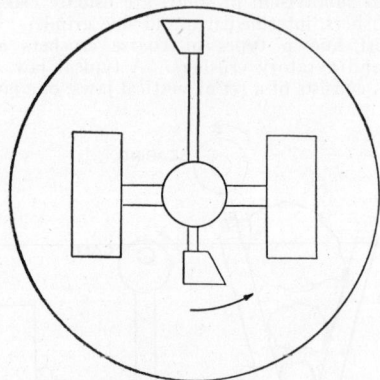

EDGE RUNNER

to "nip" and crush pieces of material between them. Another type is the edge runner. This type consists of a pan in which rotate two or more heavy wheels known as mullers. The material is shoved under the wheels by

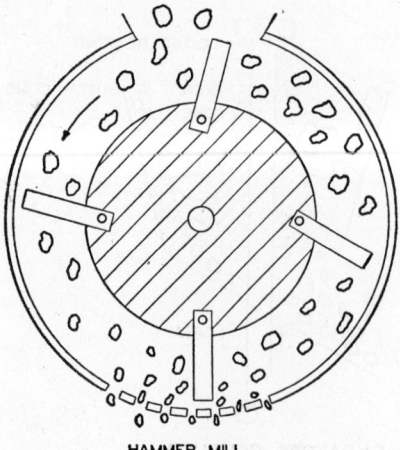

HAMMER MILL

plows. Grinding and mixing are often performed together, particularly on soft materials such as clay.

Hammer mills involve the principle of impact rather than pressure. Bars, hinged to disks attached to a horizontal rotating shaft, strike the material, enclosed in a cage, until it is fine enough to drop through the bottom openings. This type is usually used on soft material such as coal.

The ball mill is a typical fine grinder. A slightly inclined or horizontal rotating cylinder containing balls,

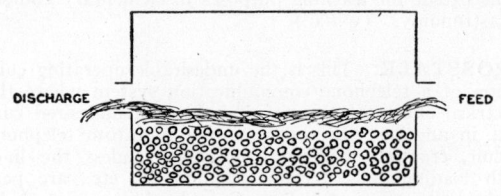

BALL MILL

usually stone or steel, grinds the material to the necessary fineness by the rubbing and impact of the tumbling balls. The feed is at one end and the discharge at the other. This is largely used in the Portland cement industry. Buhrstone mills, also for fine grinding, are similar to the old-fashioned flour mill. With modern mechanical refinements these are used in the paint industry. (D.E.M.)

CRUSTACEA. The lobsters, crabs, barnacles, shrimps, and many other species, constituting a class of the phylum **Arthropoda**. A large majority of the 25,000 known species are aquatic.

The Crustaceans are distinguished from other classes of the phylum by the following characters: 1. The body is divided into **cephalothorax** and abdomen. 2. The eyes are compound. 3. Two pairs of **antennae** are present. 4. Jointed appendages are found on the abdomen in many species. 5. Respiration is usually accomplished by gills.

The principal economic importance of crustaceans is due to the value of some species as food. Lobsters, shrimps, and some of the crabs are caught in large numbers for the market and are regarded as delicacies. Although formerly cheap, the lobster has been caught in such large numbers on the New England coast that it now has to be protected by law and brings high prices. The smaller crustaceans are important as food for fishes.

Barnacles have long been a nuisance for their part in the fouling of ship bottoms.

The following brief outline indicates the complexity of classification of the crustaceans.

Subclass **Branchiopoda.** At least four pairs of broad fringed appendages on the thorax. **Carapace** usually present.
Order Phyllopoda. Appendages similar, numerous. Fairy shrimp.
Order Cladocera. Usually with a bivalve carapace. **Water fleas.**
Subclass **Ostracoda.** With a bivalve carapace. Thorax with only two pairs of appendages or less. *Cypris.*
Subclass **Copepoda.** No carapace. Thorax with five pairs of appendages. *Cyclops* and the parasitic fish-lice.
Subclass **Cirripedia.** Sessile as adults. Barnacles and parasitic forms.
Subclass **Malacostraca.** Carapace normally present. Many appendages on thorax and abdomen.
Order Leptostraca. Like Phyllopoda but abdominal appendages slender.
Order Hoplocarida. Larger predacious species. **Mantis shrimps.**

Order Syncarida. A small group of rare forms.
Order Mysidacea. Also very limited.
Order Cumacea. A few small marine species.
Order Tanaidacea. Small marine order, sometimes included in the following.
Order Isopoda. Marine, fresh-water and terrestrial forms. Oval and flattened. No carapace. The sow-bugs, pill-bugs or wood-lice and other species.
Order Amphipoda. Related to the preceding but body high and narrow.
Order Euphausiacea. Cephalothorax covered with a shield. Marine species. Thoracic appendages biramous.
Order Decapoda. Cephalothorax with shield. Thoracic appendages with one branch. Lobsters, crabs, crayfishes, shrimps, etc. (A.W.L.)

CRYOLITE. Cryolite, sodium aluminum fluoride, Na_3AlF_6, crystallizes in the monoclinic system but in forms that closely approach cubes and isometric octahedrons. It is usually found massive. Cryolite has an uneven fracture, is brittle; hardness, 2.5; specific gravity, 2.95–3.0; luster, vitreous to greasy; color, snow-white but may be colorless, reddish or brownish; translucent to transparent. The only considerable occurrence of cryolite is at Ivigtut, Greenland, where veins of this mineral are associated with granites and gneisses. Small occurrences of cryolite have been noted in the Ilmen Mountains, Russia, and at Pike's Peak, Colorado. Cryolite has its chief use in the electrolytic production of aluminum, but small amounts are employed in the manufacture of opalescent glass.

The name cryolite is derived from the Greek words meaning frost (ice) and stone in reference to its translucency. (E.S.C.S.)

CRYOLOGY. That branch of hydrology which is concerned with snow and ice. (R.M.F.)

CRYPTOBRANCHUS. Hellbender.

CRYPTOCRYSTALLINE. When the texture of a rock is so finely crystalline (that is, made up of such minute crystals) that its crystalline nature is but vaguely revealed even in a thin section by transmitted polarized light, the rock is said to be cryptocrystalline. Among the sedimentary rocks, chert and flint are cryptocrystalline. Lava flows, especially of the acidic type such as felsites and rhyolites, may have a cryptocrystalline ground mass as distinguished from pure obsidian (acidic), or tachylite (basic), which are natural rock glasses. (R.M.F.)

CRYPTOMONADIDA. An order of 1-celled animals containing species of constant body form, with flagella. Mastigophora. (A.W.L.)

CRYPTOZOA. Paleobotany.

CRYPTURIFORMES. An order of ground birds (Aves) resembling the game birds. They range through South America and north into Mexico, and are called partridges by inhabitants of these countries. The tinamous (tinamu). (A.W.L.)

CRYSTAL. There are two types of crystals which are used in communications work, the rectifying and the piezo-electric. Among the first may be mentioned galena, silicon and silicon carbide. All have the property of passing current in one direction and not in the other, or in some cases unequally in the two directions. Since demodulation or detection is fundamentally a process of rectification these crystals may be used as detectors of modulated radio signals.

The piezo-electric crystal has the peculiar property of giving a voltage across certain faces when a mechanical pressure is applied to other faces. Among crystals exhibiting this effect are Rochelle salt and quartz, the first being by far the most active but the second being much stronger mechanically. These crystals are used in numerous ways in communications circuits. Rochelle salt is used for microphones where the sound waves striking the crystal produce corresponding voltages which are amplified for use, for phonograph pickups where the pressure is produced by the needle linkage (see Crystal Pickup), for loud-speakers and head-phones where use is made of the reversibility of the process, i.e., the voltage is applied and it causes the other faces physically to move. Quartz is widely used for frequency control of radio transmitters (see Oscillator), for filters in telephone carrier circuits and for very sharp tuning in receivers. (See also Piezo- and Pyro-Electric Phenomena; Crystalography.) (L.R.Q.)

CRYSTAL ANALYSIS. Crystal Structure.

CRYSTAL AXES. Crystallography.

CRYSTAL CONTROL. Oscillator.

CRYSTAL FORM. Crystallography.

CRYSTAL OSCILLATOR. Piezo-Electricity and Pyro-Electricity.

CRYSTAL PICKUP. Since piezo-electric crystals produce electrical voltages when subjected to mechanical stresses they offer possibilities for various electromechanical processes. One of these is the phonograph pickup, where the phonograph needle operating in the groove of the record must transmit mechanical motion to something which will convert it into electrical effects so vacuum-tube amplifiers may be used. The crystal is one of the most sensitive and at the same time one of the highest fidelity devices for doing this. While many crystals exhibit the piezo-electric effect, Rochelle salt is the most sensitive and is used for pickups, microphones, etc. Through a mechanical linkage the motion of the needle is transmitted into mechanical stresses on the crystal and hence produces electrical effects which may be amplified and then converted to sound by the loud-speaker. (L.R.Q.)

CRYSTAL PLANE. Crystallography.

CRYSTAL STRUCTURE. Long before the epochal experiments performed at the suggestion of von Laue by Friedrich and Knipping, it was inferred that the structure manifestly existing in crystals is due to an orderly arrangement of atoms or molecules. The ideal crystal is traversed by systems or "families" of parallel cleavage planes in various directions, three or more of which determine the normal external shape of the natural crystal. It is easy to imagine that these planes form a space lattice of equal polyhedral cells fitting together to fill the space, much as do the cells of a honeycomb or an egg-crate; and further, that each of the individual cells represents a characteristic unit grouping of particles, a structure built of atoms or of ions, perhaps the molecule itself.

Dr. Max von Laue (Munich) in 1912 hit upon a method of verifying and elaborating this theory. He knew from the density and atomic weights, that the number of atoms in a cc. of rock salt, for example, is about 4.488×10^{22}, and that therefore, if they are equally spaced in all three directions, their distance apart is 2.814×10^{-8} cm. or 2.814 angstroms. Certain quantum theory calculations had already indicated that x-rays have wavelengths of this order. It occurred to von Laue that if a beam of x-rays were directed upon a crystal and the crystal turned into a suitable position, one might observe interference maxima analogous to those produced with light by a **diffraction grating.** This proved to be the case, and the result verified beyond question the existence of regular spacings between reflecting planes of some sort, presumably plane arrays of atoms or ions.

Subsequent analysis of the problem by Bragg resulted in a formula analogous to that for interference of light reflected by thin plates. If the reflecting layers are spaced at equal distances d, and if the wavelength of the incident x-rays is λ, the angle θ between rays and layers necessary for an interference reflection maximum is given by **Bragg's law,** viz., $\sin \theta = n\lambda/2d$; in which n is an integer. By slowly turning the crystal, the various plane-families are brought into suitable orientations for the production of maxima. The result is a Laue pattern of black spots on the photographic plate placed beyond the crystal to catch the reflections. Another method, developed by Hull and by Debye and Scherrer, secures the necessary angular variation by crushing the crystal to powder and relying upon the fortuitous orientation of the fragments; the pattern in this case being a system of concentric rings.

While it is very easy, when one knows the structure of the crystal and the wavelength of the rays, to predict the diffraction pattern, it is quite another matter to deduce the crystal structure in all its details from the observed pattern and the known wavelength. This can be done, however, and we now have determined the arrangement and spacings of atoms in nearly all known crystals. Stereoscopic models of many typical crystals are to be found in an excellent album prepared by von Laue and Mises, entitled *Stereoscopic Drawings of Crystal Structure*. (L.D.W.)

CRYSTAL SYSTEMS. Crystallography.

CRYSTALLINE CONE. A glassy body between the cornea and the sensory portion of each component of the compound **eye** in some species of insects. (A.W.L.)

CRYSTALLINE STYLE. A digestive secretion of certain **bivalve** mollusks. It consists of protein bearing an amylolytic ferment and is produced in a continuous rod by a pouch of the intestine. One end projects into the stomach and is worn away by contact with a gastric shield and mixed with the food. (A.W.L.)

CRYSTALLIZATION. The phenomena, which are involved in the production of crystals, suggest consideration of the following topics:

1. Crystals proper. (See **Crystallography.**)
2. The process of crystallization. (See below.)
3. The converse process of crystal disappearance. (See **Dissolving.**)
4. The prevention of crystal formation. (See **Colloidal State, Condensation Processes** (6).)
5. Solutions proper. (See **Solutions and Solubility.**)

The Process of Crystallization. Crystals are formed (1) from solution, (2) from fusion, (3) by sublimation.

(1) The formation of crystals from solution, starting with an unsaturated solution, takes place when a solution is evaporated or cooled below the saturation point, except as retarded by supersaturation. Supersaturation is prevented by the addition of seed crystals of the substance. Since, in the case of the majority of soluble substances, solubility increases with increase of temperature cooling below the saturation point favors the formation of crystals. In very few cases, such as sodium sulfate above $32.4°$ C., **calcium** sulfate (slightly soluble), calcium hydroxide (slightly soluble), solubility decreases with increase of temperature, and the above statement would not apply. A case of wide scope and great importance is that of crystal formation by **precipitation** upon mixing two solutions. Actually, this is the same as for substances of greater solubility, since the sub-

stance precipitated is first formed in solution and the excess above the saturation point separates as precipitate. As a rule the crystals are larger and more perfect the slower their growth. Conversely, when small crystals are desired, rapid stirring and quick cooling are practical. The smaller the crystals of a given substance, the purer the material generally is. Small crystals may be increased in size by allowing them to stand in the mother liquor before separation.

(2) The formation of crystals from fusion takes place when the melted substance is cooled sufficiently slowly near and below the fusion point. If the cooling is rapid the fusion may result in the formation of an undercooled liquid of rigidity corresponding to a solid. Glasses, whether artificial, such as glass, vitreous enamels, and slags, or natural, such as vitreous rocks and minerals, e.g., obsidianite, are under-cooled liquids. Rocks and minerals which have cooled sufficiently slowly from fusion form crystals, for example, granite.

(3) The formation of crystals by **sublimation** takes place when the vapor of a substance is condensed as a solid without passing through the liquid phase in so doing. This occurs when the temperature of the condenser is below that of the melting point of the substance. (See **Vapor Pressure.**)

The heat of crystallization is in amount the same as the heat of solution of a given substance but of opposite sign. (R.K.S.)

CRYSTALLOBLASTIC. Designating the textures of metamorphic rocks resulting from recrystallization under differential and directed pressures and high viscosity. (See also **Riecke's Principle.**) (R.M.F.)

CRYSTALLOGRAPHY. This branch of physical science deals with the external shapes of crystals and with the geometrical relations between the atomic planes within them. If a solid crystal is broken, it is found to have separated along certain "cleavage planes" into polyhedral fragments. Even when crushed to powder, the minute grains show this characteristic. Measurements upon the variously shaped pieces reveal that if they were fitted together again, the planes would all be found to belong to one or another plane-family, the members of any one of which are all parallel.

In most crystal systems each of the more prominent crystal faces belongs to one of three plane-families intersecting along what are called the crystal axes. (In the hexagonal system there are four.) These may be conveniently used as coordinate axes, X, Y, Z, though they are not generally at right angles. Haüy discovered that if the ratio of the intercepts of two crystal planes on one of these axes is a simple fraction, such as $\frac{3}{5}$, the ratios of the intercepts on the other axes are likewise simple. This suggests that the two intercepts on any one axis are multiples of a common unit. The units are, however, generally different for the different axes, bearing to each other ratios called the axial ratios.

It is more convenient to use the reciprocals of the intercepts. For example, a plane might have intercepts equal to 10,000, 15,000, and 6000 of the respective units.

The reciprocals have the ratios $\dfrac{1}{10,000} : \dfrac{1}{15,000} : \dfrac{1}{6000}$,

which in lowest terms are $3:2:5$. These smallest integers are the Miller indices of the family to which this plane belongs, and the family is thus designated (325). The family (201) is parallel to the Y axis but intersects the X and Z axes. (The hexagonal system has four Bravais-Miller indices for each plane-family.)

Close study of the angles, indices, and axial ratios long since made it clear that every crystalline substance has a structure built upon a space "lattice" characteristic of the substance. We now know that this is due to the regular arrangement of the atoms, molecules, or ions composing the substance. (See **Crystal Structure.**) The same study shows that the lattice structures of all crystals may be classified into fourteen types, which are divided into six "systems" as follows:

1. **Isometric.** Three axes of equal length which intersect at right angles, with axial ratios all unity.

2. **Tetragonal.** Three axes which intersect at right angles. Two horizontal axes of equal length, and a vertical axis which is either longer or shorter than the horizontals. (Only one axial ratio is unity.)

3. **Hexagonal.** Four axes which intersect at right angles, three coplanar axes at 120°, and one at right angles to them. The two horizontal axes of equal length, and the vertical axes either longer or shorter than the horizontals.

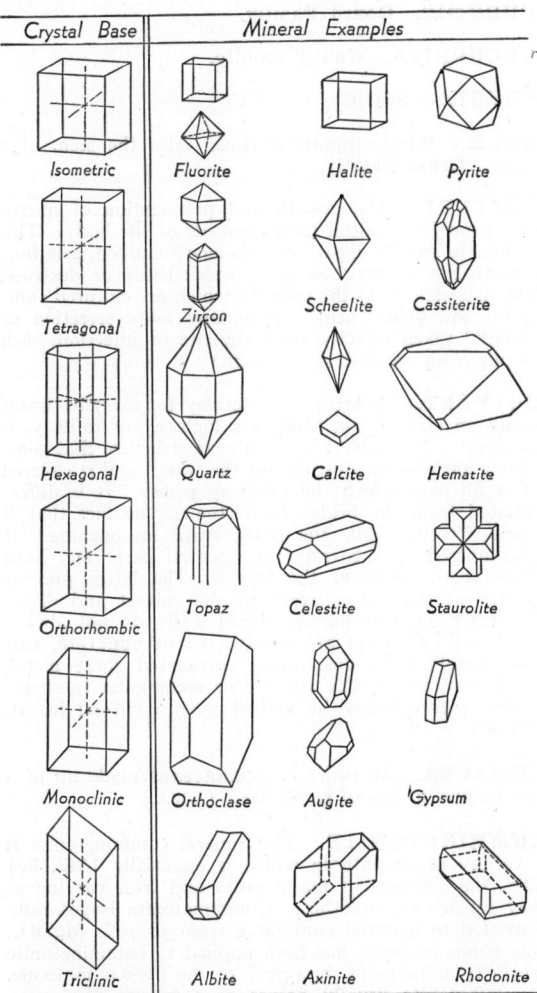

Crystal Base	Mineral Examples
Isometric	Fluorite Halite Pyrite
Tetragonal	Zircon Scheelite Cassiterite
Hexagonal	Quartz Calcite Hematite
Orthorhombic	Topaz Celestite Staurolite
Monoclinic	Orthoclase Augite Gypsum
Triclinic	Albite Axinite Rhodonite

Illustrating the six crystal systems and specific examples of each system. (*Field, Outline, Barnes & Noble.*)

4. **Orthorhombic.** Three axes of different lengths which intersect at right angles, and none of the axial ratios unity.

5. **Monoclinic.** Three axes of different lengths, two of which intersect at right angles, the third oblique to one of the others.

6. **Triclinic.** The axes of unequal length and all oblique to one another, and none of the axial ratios unity. The angles between axes are less than 90°, and are unequal.

Practically all minerals are crystalline, although perfect natural crystals are seldom, if ever, found. Because of the laws of crystallography, however, a crystallographer can usually determine the crystal form of a known spe-

cies from a fragment of the original crystal, provided that at least two of the crystal faces are visible. Crystalline aggregates are said to be **cryptocrystalline** when the individual particles are proved to have crystalline structure but their crystal faces are exceedingly small or indistinguishable. A mineral is **pseudocrystalline** if its external form does not correspond with its crystalline structure. (L.D.W., R.M.F.)

CTENIDIAL FILAMENTS. Slender processes of the **ctenidia** of mollusks (**Mollusca**). (A.W.L.)

CTENIDIUM. A projection of the body wall of mollusks (**Mollusca**), lying in the mantle cavity and usually serving as a gill. Ctenidia are slender filaments or leaf-like plates. (A.W.L.)

CTENOPHORA. The comb jellies or sea walnuts, constituting a small phylum of marine animals related to the coelenterates and sometimes included in that phylum. The phylum includes the peculiar **Venus' girdle**, a transparent ribbon-like animal whose longitudinal axis is across the middle of the slender body.

The ctenophores differ from the coelenterates in the following structures: 1. The alimentary tract opens to the exterior at the end of the body opposite to the mouth. 2. The body bears rows of ciliated (see **Cilium**) plates, the combs which give the animals one of their common names. These are organs of locomotion. 3. **Colloblasts** or adhesive cells take the place of **nematocysts**.

Classification:

Class Tentaculata. With a pair of long tentacles.
  Order **Cydippida.** Body spherical to cylindrical. Tentacles long.
  Order **Lobata.** Tentacles replaced in adult by fringe of short tentacles around mouth.
  Order **Cestida.** Body ribbon-like. Venus' girdle.
Class Nuda. Tentacles absent. Body conical to ovoid. (A.W.L.)

CUBEB. Piper Cubeba, Pepper.

CUBIC EQUATIONS. A cubic equation in one unknown is a **polynomial equation** of the third degree, and has the general form

$$a_0x^3 + a_1x^2 + a_2x + a_3 = 0.$$

The general cubic equation $y^3 + py^2 + qy + r = 0$ may be reduced by the substitution $y = x - \frac{1}{3}p$ to the normal form $x^3 + ax + b = 0$, where $a = \frac{1}{3}(3q - p^2)$, $b = \frac{1}{27}(2p^3 - 9pq + 27r)$.

The normal form $x^3 + ax + b = 0$ has roots x_1, x_2, x_3 given by

$$x_1 = A + B, \quad x_2, x_3 = -\frac{1}{2}(A + B) \pm \frac{1}{2}\sqrt{3}i(A - B),$$

where $i = \sqrt{-1}$, and

$$A = \sqrt[3]{-\frac{b}{2} + \sqrt{\frac{b^2}{4} + \frac{a^3}{27}}}, \quad B = \sqrt[3]{-\frac{b}{2} - \sqrt{\frac{b^2}{4} + \frac{a^3}{27}}}.$$

If a and b are real, and

(1) if $\frac{b^2}{4} + \frac{a^3}{27} > 0$, there are one real root and two conjugate complex roots,

(2) if $\frac{b^2}{4} + \frac{a^3}{27} = 0$, there are three real roots of which two are equal,

(3) if $\frac{b^2}{4} + \frac{a^3}{27} < 0$, there are three real and unequal roots.

In case (3), where $\frac{b^2}{4} + \frac{a^3}{27} < 0$, the above formulae for the roots are not in convenient form for numerical calculation; a trigonometric transformation reduces them to better form, thus: if $\cos\phi = \sqrt{\dfrac{b^2}{4} \div \left(-\dfrac{a^3}{27}\right)}$, then the roots are

$$x_k = \mp\,2\sqrt{-\frac{a}{3}}\cos\left(\frac{1}{3}\phi + k\cdot 120°\right)\ (k = 0, 1, 2),$$

where the upper sign is to be used if $b > 0$ and the lower sign if $b < 0$.

If $\dfrac{b^2}{4} + \dfrac{a^3}{27} > 0$, the real root is given in convenient form for numerical calculation by the trigonometric form

$$x = \pm 2\sqrt{\frac{a}{3}}\cdot\cot 2\phi,$$

where

$$\tan\phi = \sqrt[3]{\tan\psi}\quad\text{and}\quad \cot 2\psi = \sqrt{\frac{b^2}{4} + \frac{a^3}{27}},$$

and where the upper sign is to be used if $b > 0$ and the lower sign if $b < 0$.

If $\dfrac{b^2}{4} + \dfrac{a^3}{27} = 0$, the roots are

$$x = \mp 2\sqrt{-\frac{a}{3}},\ \pm\sqrt{-\frac{a}{3}},\ \pm\sqrt{-\frac{a}{3}},$$

where the upper sign is to be used if $b > 0$ and the lower if $b < 0$.   (L.L.S.)

CUBIC FUNCTIONS. A cubic function is a **polynomial function** of the third degree, which is therefore of the form $ax^3 + bx^2 + cx + d$, where a, b, c, d are **constants** (independent of x) and x is the **variable.** (L.L.S.)

CUBICAL PARABOLA. This is a curve obtained by plotting the **graph** of the equation $a^2y = x^3$. Its shape is shown in the accompanying figure. (L.L.S.)

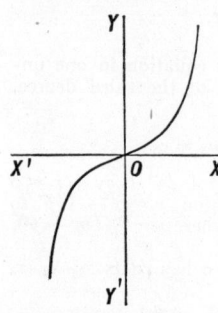
Cubical parabola.

CUBOMEDUSAE. An order of **jellyfishes** of roughly cuboidal form. (A.W.L.)

CUCKOO. Aves, Cuculiformes. Birds (**Aves**) of many species found in all continents. They take their common name from the call of the European cuckoo, *Cuculus canorus.* This species, like some others of the family, is well known for its habit of leaving its eggs in other birds' nests.

Some species are known by other names, including koel, coucal and malkoha. North America has two species, the yellow-billed, *Coccyzus americanus,* and black-billed,

Yellow-billed cuckoo. A long slender bird, grayish brown above, whitish below, with white marks on the ends of the tail feathers; lower part of beak yellow.

C. erythrophthalmus, and the related ani and roadrunner of the South and Southwest. (A.W.L.)

CUCKOO WASP. Insecta, Hymenoptera. Small **wasps** which lay their eggs in the nests of solitary wasps and bees. Usually metallic blue or green. Family Chrysididae. (A.W.L.)

CUCULIFORMES. The order of birds (**Aves**) which includes the **cuckoos,** anis, **road-runners,** quezal, and related species. (A.W.L.)

CUCUMBER. *Cucumis sativus.* **Gourd Family.**

CUCUMIS. Gourd Family.

CUCURBITA. Gourd Family.

CUESTA. Scarp.

CULM. Mississippian Period. Also the stem of a grass. **Grass Family.**

CULTURE. The growth and propagation of microorganisms by artificial means outside of the body. This is done in various media such as agar, bouillon, gelatin—either plain or enriched with blood, tissue or dextrose. The materials from the patients which are cultured commonly are blood, urine, tissue, any body secretion or material taken directly from the site of infection, such as pus from an abscess. (D.M.H.)

CULVERT. An artificial waterway for carrying water under an obstruction, such as a highway or railway, is a culvert. The culvert is a drainage structure that completely encloses the opening for the water, and is covered by a fill over which the roadway passes. It is differentiated from the bridge by virtue of the fact that it conveys water while the latter spans an opening. It is composed of a barrel (the essential part) and head walls at the ends of the barrel. The latter prevent erosion of the soil adjacent to the culvert and divert the water into the barrel. Head walls are not always employed. Culverts are constructed of **concrete,** cast iron, **brick,** terra cotta, and corrugated sheet metal. Small ones are usually circular or rectangular in shape. Larger culverts may be arched over a curved invert. (F.T.M.)

CUMACEA. An order of **crustaceans** made up of a few marine species of small size. (A.W.L.)

CUMMINGTONITE. The mineral cummingtonite is a variety of **amphibole** which is essentially (Mg, Fe) SiO_3, the amounts of **magnesium** and **iron** varying as they replace one another. Cummingtonite is generally restricted to material containing from 50–70% $MgSiO_3$. The name *grünerite* has been applied to cummingtonite which contains more than 50% of the $FeSiO_3$ molecule. Cummingtonite usually occurs as a brown fibrous to lamellar mineral. It derives its name from Cummington, Massachusetts. (E.S.C.S.)

CUMULATIVE FREQUENCY DISTRIBUTION. The cumulative frequency distribution is found by adding all the **variates** below a certain point, generally the closed **class** limits. It is useful in finding any measure of position and is needed to construct the **ogive.** We note that the **distribution function** is the mathematical counterpart of the cumulative frequency distribution. The cumulative frequency distribution may also be found (for variates above any point) by adding all the variates above a certain point. (L.A.A.)

CUMULATIVE METHOD. Physical Measurements.

CUMULONIMBUS. Clouds.

CUMULOSE. The term proposed by Merril in 1897 for **sediments** composed almost, if not entirely, of **carbonaceous** material, such as peat, lake mucks, etc. (R.M.F.)

CUMULUS. Clouds.

CUPOLA. In geology, cupola is the term proposed by R. A. Daly in 1911 for a subsidiary dome-like protrusion in the roof of a **batholith.** Cupolas are supposed to be the reservoirs for the concentrated rising gases of the batholithic **magma** and may serve as the loci for **volcanoes.** (R.M.F.)

CUPOLA FURNACE. The cupola is the furnace commonly employed in foundries for the melting of cast iron. In principle, the cupola is similar to, but simpler than, the **blast furnace.** Like the latter, it has a sheet-metal shell and is lined with refractory brick. It is similar, also, with respect to the admission of air through tuyeres, but unlike the blast furnace, it rarely uses pre-heated air, nor is there any attempt to save the heat in the waste gases. Accordingly, it is not necessary to provide the air lock at the upper end, as on the blast furnace. The top of a cupola is simply a stack discharging the products of combustion to the atmosphere. The cupola is fed with materials from an opening in the side above the line to which the charging is carried. It has an iron notch and a slag spout, and a hearth much like the blast furnace.

When it is ready to be charged, the hearth is lined with a refractory material such as sand, containing sufficient clay to bind it together to keep it from floating on the iron. A certain amount of kindling wood is then charged, on which is placed a bed layer of coke to an amount sufficient so that when the kindling has burned out and the coke has settled to the bottom of the cupola, it will form a bed extending above the tuyeres. As soon as the bed charge is well lighted additional charges of coke, scrap, pig iron, and limestone are added until the upper limit of charge depth is reached. This is allowed to burn some time under the influence of a small quantity of air so as to "soak" thoroughly the contents with heat. The cupola is then ready for the "blast." A blower is started, and air is rapidly forced through the tuyeres, resulting in a very high rate of combustion, and the production of a high temperature just above the tuyeres. As the coke burns out, and the iron sinks into this high-temperature region, it is melted, falls in drops and streams to the hearth, and runs out through the iron notch. The cupola operator allows a certain amount of this iron to run out at the beginning of a heat, as it is usually not sufficiently liquid to pour well. He then closes the iron notch with a bott, which is a cone of clay designed to cork the opening. The melting continues and molten iron is accumulated in a pool over the hearth. As successive charges sink lower inside the cupola, fresh charges are added to the extent of the iron required for that particular run. The operator judges this amount of iron accumulated by knowing the characteristics of his cupola, and noting the elapsed time of melting between successive tappings. When tapping a cupola, the clay bott is carefully cleaned out until only a thin wall remains between the iron notch and the interior pool. This is broken with a long tapping chisel. As the molten iron gushes forth it is caught in **crucibles,** in which it is carried to the molds. When a crucible has been filled the flow is stopped by inserting another bott in the tapping hole. This succession of tappings goes on until all the molds have been poured. Some cupolas are sufficiently large to melt at a rate equal to that at which the iron can be used, and the iron notch is not plugged. Iron is allowed to run continuously from the spout, fresh crucibles being placed under the spout as soon as the preceding one is filled. Although a blast furnace is often operated continuously for days at a time, a heat in an average cupola in a jobbing foundry usually consumes only 2 or 3 hours; and in a production factory rarely exceeds 8 to 10 hours. (F.T.M.)

CUPR(IC), (OUS). Copper.

CUPRITE. The mineral cuprite, **cuprous** oxide, Cu_2O, occurs as **isometric** crystals, usually **octahedrons,** but may be cubes, **dodecahedrons** or modified combinations. It also is found as a massive, earthy material. Its fracture is conchoidal to uneven; brittle; hardness, 3.5–4; specific gravity, 5.85–6.15; luster, submetallic to earthy; color, red; nearly transparent to nearly opaque. Its streak is shining brownish-red. Cuprite is a secondary mineral resulting doubtless from the oxidation of copper sulfides. It is often found associated with native copper, **malachite** and **azurite.**

Cuprite is a fairly common mineral, and of the many localities in which it occurs may be mentioned the Province of Perm, Russia; Chessy, France; Broken Hill, New South Wales; Corocoro, Bolivia; Andacollo, Chile; Bisbee, Arizona; and Del Norte County, California. The name cuprite is derived from the Latin cuprum, copper. (E.S.C.S.)

CUPRUM. Copper.

CURARE POISON. Strychnine.

CURASSOW. Aves, Galliformes. Birds (**Aves**) of a few species found in northern South America. They are about the size of turkeys and are arboreal in habit. Excellent as food and sometimes domesticated. (A.W.L.)

CURETTAGE. The scraping of an organ, a bony cavity or some other portion of the body with a curet— a spoon-shaped sharp-edged instrument. It is commonly spoken of in relation to the uterus, where the lining of the organ is removed with a curet either for diagnostic examination, or as a procedure to terminate pregnancy. (R.S.M., D.M.H.)

CURIE. The curie is a unit used in measuring the amount of emanation from **radium.** It is that amount which is in radioactive equilibrium with 1 gram of radium. (R.S.M.)

CURIE POINT. Curie-Weiss Law.

CURIE-WEISS LAW. It is well known that iron and other ferromagnetic substances do not exhibit magnetic properties at high temperatures. The change with rising temperature is gradual, but at a certain temperature, known as the Curie point, occurs a transition from ferromagnetic to paramagnetic properties, and likewise a change in the dependence of the magnetic susceptibility upon the temperature. (See **Magnetism.**) P. Curie stated in 1895 that above this point the susceptibility varies inversely as the absolute temperature. But this was found to be not generally true, and was modified in 1907 by P. Weiss to state that the susceptibility of a paramagnetic substance above the Curie point varies inversely as the excess of the temperature above that point. At or below the Curie point, the Curie-Weiss law does not hold. (L.D.W.)

CURIUM. Chemical Composition, I. Elements.

CURL OF A VECTOR FUNCTION. Let **v** be a **vector function** of position, with rectangular components v_1, v_2, v_3. The vector expression

$$\nabla \times \mathbf{v} = \hat{\mathbf{i}}\left(\frac{\partial v_3}{\partial y} - \frac{\partial v_2}{\partial z}\right) + \hat{\mathbf{j}}\left(\frac{\partial v_1}{\partial z} - \frac{\partial v_3}{\partial x}\right) + \hat{\mathbf{k}}\left(\frac{\partial v_2}{\partial x} - \frac{\partial v_1}{\partial y}\right)$$

$$= \begin{vmatrix} \hat{\mathbf{i}} & \hat{\mathbf{j}} & \hat{\mathbf{k}} \\ \dfrac{\partial}{\partial x} & \dfrac{\partial}{\partial y} & \dfrac{\partial}{\partial z} \\ v_1 & v_2 & v_3 \end{vmatrix}$$

is called the curl of **v**, and is denoted by curl **v**. It is also sometimes called the rotation of **v** and denoted by rotation **v**.

If $\mathbf{r} = x\hat{\mathbf{i}} + y\hat{\mathbf{j}} + z\hat{\mathbf{k}}$ is a variable vector and **a** is a constant vector, then

$$\text{curl } \mathbf{r} = \nabla \times \mathbf{r} = 0,$$
$$\text{curl } (\mathbf{r} \times \mathbf{a}) = \nabla \times (\mathbf{r} \times \mathbf{a}) = -2\mathbf{a}.$$
$$\text{curl } (r\mathbf{a}) = \nabla \times (r\mathbf{a}) = \frac{\mathbf{r} \times \mathbf{a}}{r}.$$

Let **u** and **v** be vector functions of position, and let u be a scalar function of position, then

curl $(\mathbf{u} + \mathbf{v}) = \nabla \times (\mathbf{u} + \mathbf{v}) = \nabla \times \mathbf{u} + \nabla \times \mathbf{v} = \text{curl } \mathbf{u} + \text{curl } \mathbf{v}$,

curl $(u\mathbf{v}) = \nabla \times (u\mathbf{v}) = (\nabla u) \times \mathbf{v} + u(\nabla \times \mathbf{v})$,

curl $(\mathbf{u} \times \mathbf{v}) = \nabla \times (\mathbf{u} \times \mathbf{v}) = \mathbf{u}(\nabla \cdot \mathbf{v}) - \mathbf{v}(\nabla \cdot \mathbf{u}) + (\mathbf{v} \cdot \nabla)\mathbf{u} - (\mathbf{u} \cdot \nabla)\mathbf{v}$,

curl (curl $\mathbf{v}$) = curl$^2\mathbf{v} = \nabla \times (\nabla \times \mathbf{v}) = \nabla(\nabla \cdot \mathbf{v}) - (\nabla \cdot \nabla)\mathbf{v}$ = grad (div $\mathbf{v}$) − $\nabla^2\mathbf{v}$,

curl (grad u) = $\nabla \times (\nabla u) = 0$,

div (curl $\mathbf{v}$) = $\nabla \cdot (\nabla \times \mathbf{v}) = 0$.

Let **v** be a vector function of position, let δ be a small region of space and also its volume, surrounding a point P, and let ω be the bounding closed surface of δ and let $d\sigma$ be an element of ω; let $\hat{\mathbf{n}}$ be the unit normal vector (outward drawn) at any point of ω. Then the curl of **v** is defined by

$$\text{curl } \mathbf{v} = \lim_{\delta \to 0} \frac{1}{\delta}\int_\omega \hat{\mathbf{n}} \times \mathbf{v}\, d\sigma.$$

(L.L.S.)

CURLEW. Aves, Charadriiformes. Shore birds (**Aves**) of several species with long legs, a moderately long neck, and a long curved beak. Related to the snipes. (A.W.L.)

CURRANT. Berry.

CURRENT CORRECTION ANGLE. The angle between the **heading** of a ship and the **course** of the ship relative to the earth is known as the current correction angle.

In sea **navigation** the actual motion of the water relative to the surface of the earth is known as current. The **direction** in which the water is moving is known as the set, and the speed of motion is called the drift. Three main types of currents are recognized by navigators: general ocean currents, tidal currents, and currents due to wind. General ocean currents are discussed in various publications of the Hydrographic Office and are plotted on the monthly Pilot Charts issued by that office. Tidal currents are predicted for any particular locality at any time by the U. S. Coast and Geodetic Survey in an annual publication known as Tide Tables. Tide tables for foreign shores are published by different governments. Currents due to winds are very difficult to determine with any exactness. Experience has shown that a wind, which has been blowing steadily for several hours, will produce a surface current which sets about 40° to the right of the direction toward which the wind is blowing in the northern hemisphere, and about 40° to the left in the southern hemisphere. The drift of wind currents is between 1 and 3% of the speed of the wind producing it. Many variables such as gustiness, steadiness of direction, etc., enter into the determination of wind currents and only long experience will give anything approaching accurate set and drift. In the article on **Dead Reckoning** a so-called "current" is discussed

which is the "catch-all" for various errors in dead-reckoning navigation.

If a ship is to make good a specified course in a region where a current is known to exist, the so-called current correction angle should be determined to find the proper heading for the ship. The determination of this angle may be made either by graphical or computational methods. Either method is sufficiently accurate and the graphical method is more commonly used. However, the computational method is fully as rapid as the graphical and does not require the space and paraphernalia needed for constructing the vector triangle. The graphical method is practically identical with that discussed under **Wind Correction Angle.**

The simplicity of the computational method is indicated in the following case: A ship is operating, with cruising speed of 15 knots, in a region where a current is known to set 160° with drift of 3 knots. The navigating officer wishes to determine the proper heading for the ship in order that its actual motion relative to the surface of the earth shall be 270° (due west), and he also wishes to know the speed the ship will make good along this course. A free-hand sketch is helpful, but by no means necessary. Such a sketch is shown in Fig. 1.

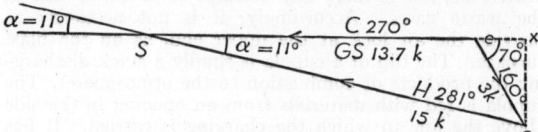

Fig. 1. Determining current correction angle.

The angle xec is known since the current sets 160° and the course is to be 270°. The side cs is known since the speed of the ship is 15 knots. Then by means of traverse tables, slide rule, or any form of three-place computing, we have:

$$\begin{array}{llll} & 70° & ec = 3.00\text{ k} & ex = 1.03\text{ k} & xc = 2.82\text{ k} \\ esc = & 11° & cs = 15.00\text{ k} & xs = 14.73\text{ k} & xc = 2.82\text{ k} \\ & & & es = 13.7\text{ k} \end{array}$$

Hence we have the heading to be used given by 270° + 11° = 281° and the predicted speed along the course to be 13.7 knots. This work was done in about 2 minutes using nothing other than a pencil, pad of paper, and the traverse tables.

Any one of the **dead-reckoning computers**, designed for air navigation, may be used for solving problems of this sort. When using these, care must be taken to remember that they are designed for use with wind, and that wind direction is given as opposite to that in which the air is actually moving; while the set of the current is properly given as the direction in which the water is moving. (W.K.G.)

CURRENT DENSITY. This is the current in a conductor divided by the total area of the conductor. In this usage an ionized gas is considered a conductor. (L.R.Q.)

CURRENT TRANSFORMER. Transformer.

CURRENTS. Current Correction Angle.

CURURO. Tucotuco.

CURVATURE OF A PLANE CURVE. The curvature of a plane curve is defined by $\dfrac{d\phi}{ds}$, where ϕ is the inclination angle of the tangent to the curve and ds is the differential of arc length.

The reciprocal of the curvature is called the radius of curvature at the point.

If the equation of the curve in rectangular coordinates is $y = f(x)$, the radius of curvature is given by

$$R = \left[1 + \left(\frac{dy}{dx}\right)^2 \right]^{3/2} \Big/ \frac{d^2y}{dx^2}.$$

If the equation of the curve is $r = f(\theta)$ in **polar coordinates**, the radius of curvature is given by

$$R = \frac{\left[r^2 + \left(\frac{dr}{d\theta}\right)^2 \right]^{3/2}}{r^2 - r\frac{d^2r}{d\theta^2} + 2\left(\frac{dr}{d\theta}\right)^2}.$$

The center of curvature of a curve at a given point is a point on the normal on the concave side of the curve at a distance from the point of contact equal to the radius of curvature. The circle of curvature is a circle with center at the center of curvature and radius equal to the radius of curvature.

The coordinates of the center of curvature are, in rectangular coordinates:

$$\alpha = x - \frac{dy}{dx}\left[1 + \left(\frac{dy}{dx}\right)^2 \right] \Big/ \frac{d^2y}{dx^2},$$

$$\beta = y + \left[1 + \left(\frac{dy}{dx}\right)^2 \right] \Big/ \frac{d^2y}{dx^2}$$

(L.L.S.)

CURVE FITTING. Empirical Equations.

CURVE TRACING. Locus of an Equation.

CURVED BEAM. A curved beam is one which has a finite **radius of curvature** before and after the **bending loads** are applied. Theoretically the **flexure** formula is not applicable to curved beams because the unit **deformation** does not have a straight line variation over the depth of the beam due to the difference in the length of the various fibers between any two radial planes. Formulae for curved flexural members are given in texts on advanced strength of materials. Reliable values of extreme fiber stresses may be found by means of correction factors, applied to the flexure formula, which are also in these texts. The correction factor depends on the radius of curvature and the shape of the cross-section. However, a member must have a considerable amount of curvature before there is an appreciable difference between the stresses found by the straight beam (flexure) theory and the curved beam theory. Consequently a correction factor of 1 is frequently used when the radius of curvature is large.

The curved beam theory or a modification of this theory should be applied to the analysis of any curved flexural member of small radius of curvature even though the member cannot be classed as a beam. Hooks and chain links are typical examples. (C.W.C.)

CURVES, IN A PLANE. A plane curve may be represented by a single **equation in two variables** interpreted as **rectangular coordinates** or as **polar coordinates.** Instead of this, it is sometimes desirable to use two equations which express the coordinates of a variable point on the curve in terms of a third variable or parameter; these equations are then called the **parametric equations** of the curve.

To find the points of intersection of two curves whose equations are given: solve the given equations as a simultaneous system, arrange the real solutions in corresponding pairs, and these will be the coordinates of all the points of intersection. (L.L.S.)

CURVES, IN SPACE. Curves in space which do not lie in one plane are called skew curves, or twisted curves, or space curves. Such curves are frequently thought of as the intersection of two **surfaces.**

Space curves are generally represented by **parametric equations**, by which the **rectangular coordinates**

x, y, z of any point of the curve are expressed as **functions** of a parameter.

The **cylinders** whose elements intersect a given curve and are parallel to one of the coordinates axes are called projecting cylinders of the curve.

If a given space curve is represented by parametric equations $x = f(t)$, $y = g(t)$, $z = h(t)$, the tangent line to the curve at a point $P_1(x_1, y_1, z_1)$ corresponding to $t = t_1$ has the equation

$$\frac{x - x_1}{f'(t_1)} = \frac{y - y_1}{g'(t_1)} = \frac{z - z_1}{h'(t_1)}.$$

The length of arc of this space curve is given by

$$s = \int_{t_0}^{t_1} [f'(t)^2 + g'(t)^2 + h'(t)^2]^{1/2} dt.$$

(L.L.S.)

CURVILINEAR CORRELATION. If two variables x and y are connected by a relationship which is not linear, then we say they are correlated curvilinearly. If $y' = b_0 + b_1 x + b_2 x^2$, then y' is determined by a quadratic function of x. However x' may be determined by y in a linear relationship, such as $x' = ay + b$, or by some curvilinear function. This theory may be generalized to any number of variables. It may also be used to determine the partial curvilinear correlation between x and y when any number of other variables are kept constant.

The coefficient of curvilinear correlation ρ is given by the relationship

$$\rho^2 = 1 - \frac{\Sigma(y - y')^2}{\Sigma(y - \bar{y})^2}$$

where the y's are actual values and y' is determined by an equation $y' = f(x_1, x_2, \cdots, x_r)$, where the function f is not of first degree in all of the variables $x_1, x_2, \cdots, x_r$. Here $\Sigma(y - \bar{y})^2 = N\sigma_y^2$. The **regression** $y' = f(x_1, x_2, \cdots, x_r)$ is usually found by the method of **least squares.** The significance of a curvilinear coefficient of correlation is determined by the **analysis of variance.** Naturally the regression $y' = f(x_1, x_2, \cdots, x_r)$ is not restricted to polynomial functions. (L.A.A.)

CUSCUS. Phalanger.

CUSP. A tapering projection, such as the projections on the crown of a tooth or one of the pointed segments of a cardiac valve. (A.W.L.)

CUSPATE FORELAND. A coastal **headland** of triangular shape, with its apex seaward and its sides concave. (R.M.F.)

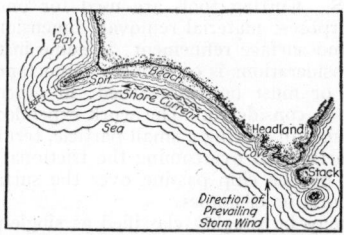

Sketch map illustrating erosion of a headland and development of shore current, beach, spit, bar, cove, and stacks, and cuspate foreland. (*After W. H. Hobbs.*)

CUTICLE. For the use of this term in botany, see **Leaf.** In zoology, cuticle is the outermost layer of the integument. (See **Integumentary System.**) As applied to the skin of the vertebrates this layer is composed of cells of ectodermal origin but in the invertebrates it indicates a noncellular layer secreted by the underlying cells. A cuticle of the latter type occurs in the parasitic flatworms, the rotifers, the roundworms and annelid worms, and the arthropods. The cuticle of insects is usually called the **cuticula.** (A.W.L.)

CUTICULA. 1. The thickened plate at the free end of some epithelial cells. 2. The layer of scales covering a hair. 3. The noncellular layer covering the bodies of some invertebrates. Some entomologists call the outer and inner layers of the chitinous cuticula of insects the epidermis and dermis. The terms **epidermis, dermis, cuticle** and **cuticula** are loosely used. (A.W.L.)

CUT-LIP. Pisces, Teleostei. A small fresh-water fish, *Exoglossum maxillingua,* common from the St. Lawrence to Virginia. The nigger chub. (A.W.L.)

CUT-OFF. The term cut-off generally refers to a particular point in a cycle at which a mechanism cuts off some flow to or from the cycle. The steam engine cut-off is that % of stroke accomplished by the piston when the inlet valve closes and prevents more steam entering the cycle from the boiler. Up to the point of cut-off, the pressure of the steam against the piston is fairly constant, but when cut-off has occurred, the steam acts expansively with decreasing pressure. The cut-off of a Diesel cycle is the fraction of the stroke accomplished when supply of fuel oil to the cylinder is stopped. (F.T.M.)

CUT-OFF BIAS. This is the voltage which must be applied to the **grid** in order to stop the flow of anode or plate current in a vacuum **tube.** It is a valuable reference point in the discussion of vacuum tube characteristics as much of the tube behavior is determined by where the bias voltage is set with respect to the cut-off value. (See **Amplifiers.**) (L.R.Q.)

CUT-OFF FREQUENCY. While this term may be used to designate the frequency at which any electrical device ceases to function in the normal manner it is more commonly used in connection with electric wave **filters** and **wave guides.** When referring to filters it means the frequency at which the attenuation begins to increase sharply. In the ideal filter the attenuation would go to infinity at the cut-off frequency, but in a practical filter the rise in attenuation is not so abrupt and never reaches infinity, but does usually go to a very high value. When applied to a wave guide, the cut-off frequency is the lowest value of frequency which will be propagated down the guide without attenuation. This is determined by the dimensions of the guide. (L.R.Q.)

CUTS AND FILLS. Earthwork.

CUTTERS. Cutting tools are used for one or more of three purposes: material removal, dimensional or size accuracy, and surface refinement; in some instances one of these considerations is of primary importance and the others are, or must be, disregarded; in other cases all three must be considered. The work expended in metal cutting consists of tearing a small particle, termed a chip, from the metal, and overcoming the frictional resistance engendered by the chip passing over the surface of the tool as it curls or crumples.

Cutters may be roughly classified as single-point cutters and as multi-toothed cutters. The most familiar examples of single-point cutters are the inserted bits used for lathe, planer, and shaper tools; multi-toothed cutters are used for milling and broaching, although both drills and reamers fall into this classification. In metal and wood cutting, the cutting speed is the velocity, usually in in. or ft. per min., with which the cutter passes over the work. The *feed* is the relatively slow motion of either the cutting tool or the work in metal cutting. In drilling, the feed is generally stated as so many thousandths of an inch per revolution of the drill; in milling, the feed is generally given as so many

inches of work movement per min. Single-point cutting tool nomenclature is illustrated in Fig. 1. This tool is similar to those used on planers and shapers, and moves

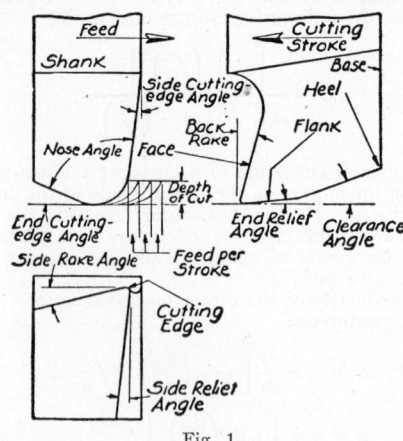

Fig. 1.

along a straight line and cuts on the forward stroke only. The cutting action takes place in two planes, and multiple rake and relief angles are provided.

Fig. 2 illustrates the action of and the difference between profile-type and form-type multi-toothed cutters. The profile-type cutter is sharpened by grinding along

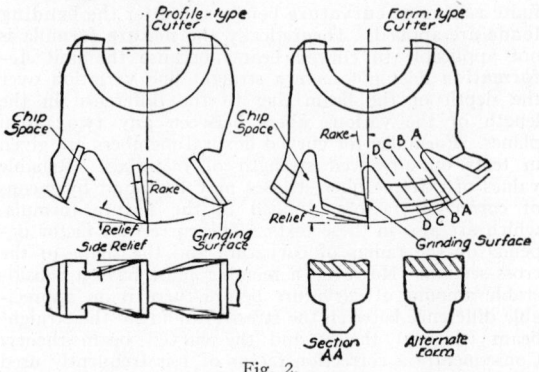

Fig. 2.

the indicated edge, and is therefore limited to straight-line cutting edges. The form-type cutter is sharpened by grinding the face of each tooth. Since the teeth are of unvarying shape—sections *BB, CC* and *DD* being exact replicas of section *AA*—these cutters may be ground without change of form if an equal amount is removed from each tooth of the cutter.

The profile-type cutter is the more efficient of the two types since it has considerably more chip space between the teeth. It is also possible to provide side relief for this type. The form-type cutter shown has no side relief since this would affect the tooth shape after grinding. Sometimes a change in the design of the work will permit an alteration in the cutter profile, as illustrated in the alternate form, thus permitting clearance or side relief.

If the edge of the tooth moves perpendicular to the plane of the paper simultaneously with the direction indicated, the minute serrations of the cutting edge, on account of their saw-like action, aid in the removal of the chip. This principle is employed in rotating cutters by making the tooth edges helical instead of straight lines parallel to the cutter axis. Some of the shock of tooth contact is eliminated and a cleaner, smoother finish generally results. (H.C.H.)

CUTTING OUT. Forging.

CUTTING SPEED. Cutters.

CUTTING UP. Climb Cutting.

CUTTINGS. Propagation, Asexual Reproduction.

CUTTLEFISH. Mollusca, Cephalopoda. Mollusks, related to the squids but forming a separate family (Sepiidae). Used as food in the Oriental region. The shell is the cuttlebone of commerce and the ink sac secretes the pigment sepia. (See also **Invertebrate Paleontology**.) (A.W.L.)

CUTWORM. Insecta, Lepidoptera. **Larvae** of many species of **owlet moths** (Noctuidae). They hibernate when partly grown and in the spring often cut off young plants at the ground. In severe attacks they can be killed by poisoned baits. (A.W.L.)

CUVIERIAN ORGAN. A defensive organ found in some sea cucumbers (**Holothuroidea**). It is a modified part of the respiratory tree attached to the **cloaca**, and made up of tubes covered with sticky material. When irritated the animal contracts violently and ejects this structure through the ruptured wall of the cloaca. In the sea water the sticky material forms long adhesive threads which entangle the enemy. (A.W.L.)

CUXIO. Mammalia, Primates. **A monkey of the** Amazon valley, also called the black saki. (A.W.L.)

CW. Continuous Wave.

CYANAMIDES. Cyanamide (NC·NH₂ or HN:C: NH) is a white solid, melting point 44° C., boiling point 140° C. at 20 mm. pressure, transformed at 150° C. into cyanuramide, tricyantriamide ((NC·NH₂)₃). Cyanamide reacts (1) as a base with strong acids forming salts, (2) as an acid forming metallic salts, such as calcium cyanamide (CaCN₂). Cyanamide is formed (1) by reaction of cyanogen chloride (CN·Cl) plus **ammonia** (ammonium chloride also formed), (2) by reaction of thiourea plus **lead** hydroxide (lead sulfide also formed).

When calcium cyanamide is boiled with water, dicyandiamide ((NC·NH₂)₂), melting point 207° C. is formed (along with calcium hydroxide). Fusion of dicyandiamide with **sodium** carbonate plus **carbon** produces **sodium** cyanide plus **ammonia** (also some tricyantriamide). Diethylcyanamide ((C₂H₅)₂N·CN) is a colorless liquid, boiling point 189° C. at 748 mm. pressure, and when hydrolyzed yields diethylamine ((C₂H₅)₂NH) plus ammonia plus carbon dioxide. Diphenylcyanamide (C₆H₅N:C:NC₆H₅) when hydrolzyed yields **aniline** plus carbon dioxide. Benzylcyanamide (C₆H₅CH₂NH·CN) is a white solid, melting point 43° C. (R.K.S.)

CYANIC ACID AND CYANATES, AND RELATED COMPOUNDS, cyanic acid, isocyanic acid, fulminic acid, cyanuric acid, cyamelide, fulminuric acid. Cyanic acid (HCNO or HOCN) is a colorless, odorous liquid; soluble in water and in ether; volatile with decomposition when heated; passing at ordinary temperature into a mixture of cyanuric acid ((HNCO)₃) and cyamelide ((CONH)ₓ), white solid, which on vaporizing yields cyanic acid; when cyanic acid vapor is rapidly cooled in a freezing mixture, cyanic acid, liquid, unstable, is obtained, when the vapor is condensed above 105° C., cyanuric acid

$$\left((HNCO)_3 \text{ or } CO \diagup_{\displaystyle NH-CO}^{\displaystyle NH-CO} \diagdown NH \right)$$

is obtained. Cyamelide dissolves in **sulfuric acid** unchanged and addition of water causes precipitation of

cyamelide; passes into cyanuric acid when warmed with concentrated sulfuric acid, finally into **carbon dioxide** plus **ammonia;** dissolves in **sodium** hydroxide solution forming sodium cyanate. Sodium cyanate is prepared by heating **sodium** cyanide and an oxide such as **lead** monoxide (PbO), trilead tetroxide (Pb₃O₄), or lead dioxide (PbO₂), addition of water and separation of the sodium cyanate solution from the lead oxide by filtration. Sodium cyanate solution upon boiling changes into sodium carbonate plus urea (CO(NH₂)₂).

Ammonium cyanate (CNONH₄), white solid, formed by reaction of sodium cyanate and **ammonium** sulfate solutions is transformed to **urea** upon being heated at 100° C. This reaction was carried out in 1828 by Wöhler, and is the first record of a so-called inorganic substance being transformed outside a living organism into a so-called organic substance. The following esters are known:

Methyl iso-cyanate (CH₃NCO), boiling point 44° C.
Ethyl cyanate (C₂H₅OCN), decomposes on heating.
Ethyl iso-cyanate (C₂H₅NCO), boiling point 60° C.
Phenyl iso-cyanate (C₆H₅NCO), boiling point 166° C.

Ethyl cyanurate $C_2H_5O \cdot C \diagup_{\displaystyle N=C(OC_2H_5)}^{\displaystyle N-C(OC_2H_5)} \diagdown N$

Ethyl iso-cyanurate $CO \diagup_{\displaystyle N(C_2H_5)-CO}^{\displaystyle N(C_2H_5)-CO} \diagdown N(C_2H_5)$

Fulminic acid (HONC) and the fulminates are violently explosive. Utilizing this property, mercuric fulminate (Hg(ONC)₂)·½H₂O) is used as a detonator for other explosives. Mercuric fulminate is made by the reaction of **ethyl alcohol** and **mercuric** nitrate in excess of **nitric acid**, from which insoluble mercuric fulminate separates. Silver fulminate (Ag(ONC)) is more explosive than mercuric fulminate, and is used in the manufacture of fire-crackers. Free fulminic acid may be obtained by reaction of potassium fulminate and excess of ether. It volatilizes with the ether upon distilling, and changes rapidly to metafulminic acid. Related to fulminic acid, is fulminuric acid ((HONC)₃ or NO₂·CH(CN)·CONH₂). (R.K.S.)

CYANIDING. Carburizing.

CYANITE. Kyanite.

CYANOGEN. Cyanogen ((CN)₂) is a colorless gas of marked characteristic odor, very poisonous, density 1.8 (air equal to 1.0), melting point —28° C., boiling point —20° C., soluble, when passed into water at 0° C., cyanogen forms **hydrocyanic acid** plus **cyanic acid,** but at ordinary temperatures the reaction is complex. With **sodium** hydroxide solution, there is formed with cyanogen sodium cyanide plus sodium cyanate, with dilute sulfuric acid oxamic acid (COOH·CONH₂), oxalic acid (COOH·COOH). By reaction with tin and hydrochloric acid, cyanogen is reduced to ethylene diamine (CH₂·NH₂·CH₂·NH₂). Cyanogen reacts with hydrogen to form hydrocyanic acid, and with metals, e.g., zinc, copper, lead, mercury, silver, to form cyanides. Cyanogen, (1) when burned in air produces a violet flame forming **carbon dioxide** and **nitrogen** in the outer part and carbon monoxide and nitrogen in the inner part, (2) when exploded with **oxygen** produces carbon dioxide or carbon monoxide and nitrogen depending upon the ratio of oxygen to cyanogen (2 volumes oxygen plus 1 volume cyanogen yields 2 volumes carbon dioxide plus 1 volume nitrogen; 1 volume oxygen plus 1 volume cyanogen yields 2 volumes carbon monoxide plus 1 volume nitrogen). The flame spectrum contains characteristic bands in the blue and violet. By means of the electric spark, the electric arc or a red hot tube, cyanogen is decomposed into carbon plus nitrogen. When heated at ordinary pressure at about 300° C., or under 300 atmospheres pressure at about 225° cyanogen is converted into paracyanogen, a brown powder, also formed when mercuric cyanide is heated. Cyanogen is prepared (1) by reaction of **sodium** cyanide and **copper** sulfate solutions, whereby one half the cyanogen is evolved as cyanogen gas and one half remains as cuprous cyanide. From the filtered cuprous cyanide, by treatment with ferric chloride solution, cyanogen is evolved with accompanying formation of ferrous chloride, (2) by heating mercuric cyanide solid, or a mixture of mercuric chloride and sodium cyanide solutions, mercury and mercurous, respectively, being formed, (3) by heating ammonium oxalate $(COONH_4 \cdot COONH_4)$ with phophorus pentoxide, water being abstracted. Small amounts of cyanogen are present in blast furnace gas and raw coal gas. (R.K.S.)

CYANOSIS. A blue color of the skin and mucous membranes, most marked in the lips, nose, cheeks, ears, hands, and feet. It is due to the presence of abnormally large amounts of reduced haemoglobin (i.e., haemoglobin which has given up its oxygen to the tissues) in the blood. This occurs when there is a failure of aeration of the blood as it passes through the lungs so that insufficient oxygen is picked up by the red cells, or in circulatory failure and stosis of venous blood in peripheral vascular beds. It also occurs when there is abnormal communication between the venous and arterial sides of the circulation as in certain congenital malformations of the heart. Cyanosis is commonly seen in severe heart disease, particularly the congenital type, pneumonia, severe infections, asthma, and emphysema. It is also seen as a result of poisoning with gases, or drugs which interfere with repiration or the rate of absorption of oxygen by the blood. (D.M.H.)

CYANURIC ACID. Cyanic Acid and Cyanates.

CYBOTAXIS. A condition in which certain liquids, under **x-ray** examination, give evidence of structure resembling that of crystals. By passing a beam of x-rays through various alcohols and other organic liquids, G. W. Stewart and his collaborators have obtained one, two, or even three diffraction maxima or halos, somewhat like the diffraction rings produced by powdered crystals. (See **Crystal Structure.**) These suggest that molecules are temporarily arranged in rows, layers, or stacks like bricks in a pile and that they have one, two, or even three different dimensions or spacings, corresponding, in accordance with **Bragg's law,** to the different angles of diffraction observed.

A closely related property is exhibited by certain substances known as "liquid crystals," which appear to be intermediate between merely cybotactic liquids and true crystals. In these there appear to be large groups of molecules which, though able to move and turn about, retain their structural arrangement. Such mesomorphic substances manifest even some of the optical properties of crystals, which the former type do not. (L.D.W.)

CYCADLES. Paleobotany.

CYCADOFILICALES. Paleobotany.

CYCADS. A group of **Gymnosperms** containing 9 genera and less than 100 species. They first appeared in late **Paleozoic** times, became a dominant group almost cosmopolitan in distribution in the **Mesozoic** period. Cycads are now limited to tropical or subtropical regions, often with a very restricted range. Some of the genera are found only in the New World, in Mexico and the West Indies; others occur only in the Old World, in Australia, Africa and various Islands of the

Pacific Ocean. Bescause of their decorative habit they are frequently grown in cultivation in places outside their natural range. Many are grown as greenhouse plants in temperate regions.

The appearance of the plants is uniform in all genera. The stem is either columnar and rarely branched or

A cycad, *Dioon edule*. *(After Chamberlain.)*

underground, and much enlarged. Columnar stems are usually from 6–10′ tall, but some species grow much higher. These stems are generally thickly clothed in the persistent leaf-bases, which sometimes give an indication of the age of the plants. As determined by the number of leaf bases many are several hundred years old. Internally the stem contains a very large pith and a thick **cortex** with a narrow cylinder of wood between them. The leaves form a large crown at the top of the stem. They are pinnate except in one Australian genus, *Bowenia,* which has bipinnate leaves which are thick and leathery. The primary root is large and extends deep into the ground.

All Cycads are **dioecious** plants. The ovulate cones are usually very large. The different genera have cones which show a very distinct series, ranging from those of *Cycas revoluta* with loosely arranged leaf-like **sporophylls** to the compact cones of Zamia. The large **ovules,** or megasporangia, are covered by a single thick integument. The male cones are much smaller and always formed of compactly massed sporophylls, each of which bears many sporangia, or pollen sacs. The pollen grains are very numerous and light. Pollination is effected by wind, though insects are frequently seen on the male cones, and may play some part in the pollen transfer.

The pollen grains are caught in a sticky fluid, which covers the micropyle of the ovule. As this sticky substance dries, it shrinks, drawing the pollen grains down through the micropyle. Each pollen grain then puts out a pollen tube which digests its way through the mass of nucellar tissue which surrounds the **female gametophyte,** and reaches a small chamber which is formed at the micropylar end of the gametophyte and the **nucellus.** Meanwhile two sperm cells have been forming in the pollen tube. Each is very large and has a spiral band of cilia wound about its anterior end. Freed from the pollen tube, these sperm pass to the gametophyte, where union of one sperm with the egg occurs.

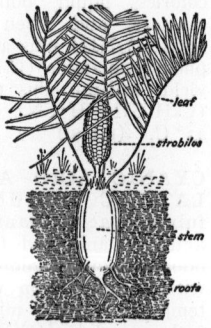

A cycad. Mature sporophyte of *Zamia* bearing a carpellate strobilus. *(Smith, Overton, Gilbert, Denniston, Bryan and Allen, Textbook of General Botany. Macmillan Co.)*

Sperm and egg nucleus presently unite, and the egg is fertilized. The fertilized egg divides to form a mass of cells which is known as the proembryo. At the base of this proembyro is a group of cells which becomes the true **embryo**. The embryo is pushed down into the tissue of the gametophyte by the elongation of a group of cells known as the suspensor. The mature seed of a Cycad has an outer fleshy coat which is variously colored. Inside this there is a hard stony layer which in turn surrounds another layer which is fleshy at first but soon becomes thin and dry. Within is the gametophyte which contains the embryo. Cycad seeds germinate as soon as they are mature.

Apart from their value as decorative plants or as curiosities, few products of importance are obtained from the Cycads. Their leathery leaves remain green for some time after removal from the plant. They are therefore often used on Palm Sunday and for funeral purposes. The seeds of many of them are edible, as is also the central portion of the stem of *Cycas* species. Cycads are sometimes confused with palms.

Cycads probably originated in very early times from some primitive ferns. Even today certain cycads closely resemble ferns. Cycads are "living fossils" which continue to exist in a very restricted range. (See **Paleobotany**.)   (R.M.W.)

CYCLE. A series of changes executed in orderly sequence, by means of which a mechanism, a working substance, or a system is caused periodically to return to the same initial condition, constitutes a cycle. Many complicated machines or assemblages of machines work in definite cycles. An important form of cycle is the heat engine cycle, in which a series of thermodynamic changes in a working medium periodically return the system to the same thermodynamic level. This working medium may be a gas, as in the **Otto** and **Diesel** cycles, or a vapor, as in the steam cycle. See also **Carnot cycle**, for an example of a general ideal cycle. A vapor cycle is so named from the fact that it is conceived as using the same vapor over and over, passing it around what might be thought of as a closed loop of equipment, and subjecting it to various thermodynamic changes by means of which useful mechanical energy is produced from heat. The distinction between vapor and engine cycle should be recognized. An engine cycle considers only the changes occurring within an engine, but a vapor cycle involves, in addition, all changes in the vapor state from the point of leaving the engine until it is again ready to enter it. (See **Rankine Cycle, Regenerative Cycle, Reheat Cycle**.)   (F.T.M.)

CYCLE OF EROSION. The work of rivers and streams is erosional, transportational and depositional. The erosional work of rivers and streams sculptures the surface of the earth into a variety of forms. So

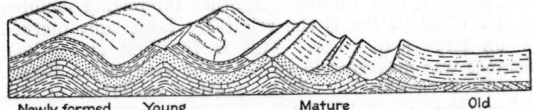

Diagram to illustrate successive stages in the normal cycle of erosion in a region of folded rocks. (*After G. H. Ashley*.)

long as the land surface remains above sea level it is subject to such agents of denudation as wind, rivers and glaciers. The principal agent of denudation is running water (rivers and streams). The river **erosion** pattern of any region will depend upon (1) climate, (2) relative hardness and solubility of the formations, (3) structure, (4) the degree to which the erosive process has completed its work, with or without interruptions caused by **diastrophism**. The stream pattern of a region is not only indicative of the structural control, but also of the stage in its erosional history. The ideal complete cycle of erosion begins with uplift of a region with low altitude and ends with reduction of the uplifted region to a **peneplain**.   (R.M.F.)

CYCLE OF STRESS. The stress variation on a particular plane through a specific point in a body which is subjected to a **repeated load** is called a cycle of stress. If the stress varies alternately between **tension** and **compression**, the variation is known as reversal of stress. The reversal is complete when the alternate stresses are equal in magnitude.

The algebraic difference between the maximum and minimum stresses of a cycle is the range of stress. The **endurance limit** depends on the range of stress.   (C.W.C.)

CYCLOHEXANE. Cyclohexane, hexahydrobenzene, hexamethylene (C_6H_{12}) is a colorless liquid, boiling point 81° C. Cyclohexane is formed by reaction of **benzene** and **hydrogen** in the presence of a **catalyzer**, finely divided nickel, heated. Intermediate products of the hydrogenation are cyclohexadiene, dihydrobenzene, (C_6H_8) and cyclohexene, tetrahydrobenzene (C_6H_{10}). By regulated oxidation of cyclohexane, cyclohexene, or cyclohexadiene, benzene is formed.   (R.K.S.)

CYCLOID. If a circle moving in a plane rolls along a straight line, a point on the circumference describes a curve called the cycloid.

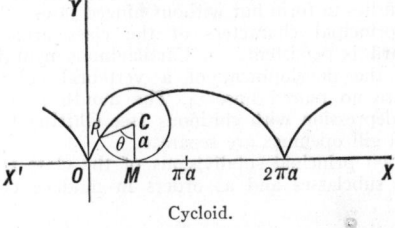

Cycloid.

If the given line is the X-axis, and the origin is a point on this line at which the fixed point on the circle touches the X-axis, the **parametric equations** of the cycloid are

$$x = a(\theta - \sin \theta), \quad y = a(1 - \cos \theta),$$

where θ is the angle turned through by the radius of the circle and a is the radius of the circle. The rectangular equation of the cycloid is

$$x = a \cos^{-1}\left(1 - \frac{y}{a}\right) \pm \sqrt{2ay - y^2},$$

which is not ordinarily convenient for use.

The area of one arch of the cycloid is $3\pi a^2$ and the length of one arch of the cycloid is $8a$, where a is the radius of the rolling circle.

The teeth of gears are often cut with faces which are arcs of cycloids, so that there is rolling contact when the gears are in mesh.

The inverted arch of a cycloid has the following two mechanical properties:

1. If two particles sliding without friction start from any two points of the curve at the same time, they will reach the lowest point at the same instant.

2. A particle sliding without friction will travel from O to B in less time than along any other curve connecting

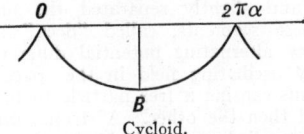

Cycloid.

O and B. Hence, the cycloid is sometimes called the curve of quickest descent.   (L.L.S.)

CYCLONE. A wind system around low atmospheric pressure. If a barometric depression is sufficiently low to be classed as a cyclone, it develops or has associated with it winds which flow around its center counterclockwise in the northern hemisphere and clockwise in the southern hemisphere. The intensity of a cyclone depends on the pressure gradient between the system's center and periphery, and on the characteristics of the air masses involved. Cyclones are divided into two groups.

1. Tropical cyclones are vortices with indefinite or short-lived frontal structures which occur over the tropical and subtropical regions, particularly over the western half of the Atlantic and Pacific and over the Indian Ocean. They are known as hurricanes, typhoons, and baguios.

2. Extra-tropical, or wave cyclones are composed during development, youth and maturity of definite parts including warm and cold sectors and warm and cold fronts, but beyond maturity approach vortex structure. Nearly all the storms of the temperate zone are wave cyclones. (See **Wave Cyclones; Fronts; Winds.**) (P.E.K.)

CYCLOPROPANE. Anaesthesia; Hydrocarbons.

CYCLOSTOMATA. The round-mouthed eels, **hag fishes** and **lampreys.** A class of the phylum **Chordata** made up of marine and fresh-water species resembling slender fishes in form but without hinged jaws.

The principal characters of the class are: 1. The **notochord** is persistent. 2. Cartilaginous neural arches indicate the development of a vertebral column. 3. There are no paired fins. 4. The mouth is a funnel-shaped depression with chitinous (see **Chitin**) teeth. 5. External gill openings are separate.

The two principal subdivisions of the class are listed both as subclasses and as orders in modern classifications.

> Subclass Myxinoidea (Hyperotreta). The **hag fishes.** Marine species with a poorly developed oral depression. Gill openings far behind head.
> Subclass Petromyzontia (Hyperoartia). **Lampreys.** Marine and fresh-water species with a well-developed oral funnel. Gill openings immediately behind head. (A.W.L.)

CYCLOSTROPHIC WIND. Winds which blow as a result of a pressure gradient and centrifugal force, but in the absence of Coriolis Force. They are, of necessity, cyclonic and restricted to equatorial zones which is the only place Coriolis Force is zero, or nearly zero. The cyclostrophic component of a wind is the difference between the gradient and the geostrophic winds. Hurricanes are largely cyclostrophic winds until they travel north or south sufficiently to be affected by Coriolis Force. (P.E.K.)

CYCLOTRON. This name has been given to the magnetic resonance accelerator, a device developed in 1931 by Lawrence and Livingston for imparting very great velocities to heavier **nuclear particles** without the necessity of excessive voltages. The electrified particles (such as protons or helium nuclei) are released in the region between two large, flat, hollow, semicircular segments S, S, of thin metal placed with their diametric edges closely parallel, as if one had cut a pill-box in two along a diameter and slightly separated the halves. (See figure.) These segments, called "dees," are given a high-frequency alternating potential difference, producing a rapidly oscillating field in the space I between, them, and thus causing a free particle to be pulled first one way and then the other. A strong, uniform magnetic field is applied perpendicular to the plane of the segments. The result is that, as a particle darts into one of the segments, it follows a semicircular path of radius proportional to the speed (as in a **mass spectrograph**) and re-enters the interspace I on the other side of the center. The time required for this semicircular journey depends only upon the intensity of the magnetic field,

and does not change with the velocity of the particle and the radius of its path. Now if the field is adjusted so that this time equals ½ the electric oscillation period, the particle will always emerge into an electric field so directed as to pull it in the direction it is already going, and in this way its speed increases at each crossing of the interspace. Thus, starting near the center, the particle spirals outward and speed increasing each half-turn, until it finally escapes into a receptacle K near the outer edge. The apparatus must, of course, be in a vacuum.

The first cyclotron imparted over 1,200,000 **electron volts** of energy to protons with an applied voltage of only 4000 volts; more recent installations have produced

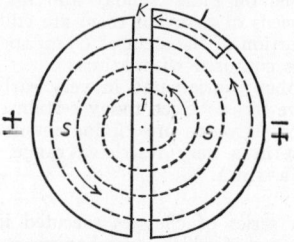

Diagrammatic plan of cyclotron. Ions released in interspace. I, traverse hollow sectors S in semi-circles, and are finally utilized or allowed to escape at K.

5,000,000 electron volts. Such fast-moving particles are of great value in nuclear research, being used to bombard the atoms of substances placed in the receptacle K. An apparatus which serves a similar purpose with electrons (to which the cyclotron is not well adapted) has been developed by Kerst and Serber (1941) and given the name "betatron." (L.D.W.)

CYDIPPIDA. An order of **Ctenophora.**

CYGNUS. (The swan.) (Map, page 380.) Cygnus is one of the most striking and interesting **constellations** of the northern sky. It represents a swan flying with outstretched wings and legs trailing out behind. It is also frequently referred to as the northern cross. Lying, as it does, in one of the most impressive portions of the northern **milky way,** the constellation contains many interesting objects. The brightest star (alpha Cygni) known as Deneb, is one of the most distant of all of the bright stars; its intrinsic brightness being about 1000 times that of the sun. Beta Cygni is one of the most striking of all of the **double stars,** both components being bright, one blue and the other orange, and easily separated by a small telescope. No physical connection has been determined between the components and it probably is not a true **binary.** Another interesting member of this constellation is the relatively faint star, marked 61 on large star maps. This is the first star whose distance was measured in 1838 by the astronomer Bessel. Bessel's determination of the **stellar parallax** of this object opened a field of astronomic research which has done much to solve many of the problems of general cosmogony. (W.K.G.)

CYLINDER. A cylindrical surface is a **surface** generated by a straight line which moves parallel to itself and intersects a given **curve.** A cylinder is a solid bounded by a cylindrical surface and two parallel planes which cut all the elements of the surface. The volume of a cylinder is equal to the area of its base times its altitude; for a right circular cylinder of altitude h and radius of base r, the volume is $V = \pi r^2 h$. The lateral area of a cylinder is equal to the perimeter of a right section times its lateral edge; for a right circular cylinder of altitude h and radius of base r, the lateral area is $A = 2\pi rh$. The **locus of an equation in rectangular coordinates** of higher degree than the first in which one variable is lacking is a cylindrical surface whose elements are parallel to the axis along which that variable is measured.

Because it is comparatively easy to form, by usual means of manufacture, and because its shape is very well adapted to the resisting of internal bursting pressure, the cylinder is a very common engineering shape. While, of course, anything of cylindrical shape might truly be called a cylinder, it is customary to apply the term to that part which, in conjunction with a closely fitting internal **piston** will provide an enclosed space the volume of which may be varied by motion of the piston. Expansion of volume of a working medium is the basis of all commercially employed power cycles, and the cylinder and piston have an important place in this field. Engine-type prime movers have cylinders and pistons, and the widespread employment of the **internal combustion engine** for personal transportation in the United States has made of the cylinder a part familiar to many persons. The cylinder of an internal combustion engine or an air compressor needs to be cooled while the machine is in operation in order to keep the interior wall temperature low enough so that a lubricating oil film may be maintained between the piston and the cylinder. Air cooling and water cooling are resorted to for this purpose, the former for finned, the latter for jacketed, cylinders. Water **pumps** and steam **engines,** by the nature and temperature of the working substance enclosed in their cylinders, do not need to be cooled. The cylinder should be made of some hard metal having good wearing characteristics and be machined and honed to a bright finish and exact size. Nickel, cast iron, and steel have been extensively used for internal combustion engine cylinders. Cylinders may be made of brass, bronze, or other special materials when the working medium is corrosive to iron or steel. (L.L.S., F.T.M.)

CYLINDRICAL COORDINATES. The cylindrical coordinates of a point in space are three numbers which determine the position of the point; they are useful in mathematical problems dealing with cylinders and cones.

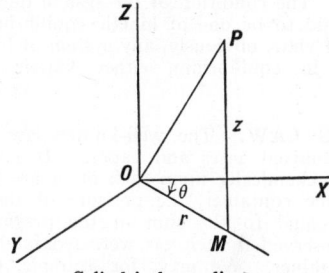

Cylindrical coordinates.

Let P be any point in space, and referred to a system of **rectangular coordinate** axes. Let M be the projection of P on the XY-plane; let $r = OM$, $\theta = \angle XOM$ be **polar coordinates** of point M in the XY-plane, and let $z = MP$ be the third rectangular coordinate of P.

The cylindrical coordinates of point P are the three numbers r, θ, z.

The relation between rectangular and cylindrical coordinates of a point in space are given by

$$x = r \cos \theta, \quad y = r \sin \theta, \quad z = z.$$

(L.L.S.)

CYLINDRICAL HARMONICS. Bessel Functions.

CYME. Flower.

CYMOPHANE. Chrysoberyl.

CYPRESS. Conifers.

CYST. 1. A capsule containing semi-solid or fluid material which may be under pressure. The contents of a cyst may be retained normal secretions, e.g., the sebaceous cyst which contains the products of a plugged sebaceous gland; or they may represent a fluid collection associated with a parasitic infection, e.g., echinococcus cysts of the brain, liver or other organs. Many tumors undergo cystic degeneration, especially carcinomas of the ovary, breast and uterus. Benign cysts occur in the ovary, spleen, lungs, kidney and liver where they are often congenital. Other congenital cysts result from fetal malformations and failures of development, such as the several cysts which occur in the neck. 2. The resting form of an organism in a protective covering. Significant examples in medicine are the cysts of *Endamoeba histolytica* which are found in the stools of patients with **amoebic dysentery**, and the encysted larva of **Trichinella spiralis** which can be demonstrated in the muscles of individuals who have had **trichiniasis**. 3. The vesicular portion of the bladder-worm stage of a **tapeworm**. (D.M.H., A.W.L.)

CYSTICERCOID. A small **bladder worm** in which the scolex fills the interior. (A.W.L.)

CYSTICERCUS. Bladder Worm.

CYSTINE. Aldehydes, Ketones, and Related Compounds.

CYSTITIS. Infection of the urinary bladder. This may be acute or chronic, and may be caused by a variety of bacteria, the most common ones being *B. coli* and allied organisms. **Streptococci** and **staphylococci** are next in frequency; specific infections with the tubercle bacillus and the gonococcus also occur.

Cystitis may be an ascending infection through the **urethra**, or it may be secondary to infection higher in the urinary tract in the **ureters** or **kidneys**. Obstructive and traumatic factors may play a role.

The symptoms of cystitis are frequency and urgency of urination with burning and pain. Treatment of *B. coli* and coccal infections with **sulfonamides** usually results in prompt cure. (D.M.H.)

CYSTOIDS (Cystids). Invertebrate Paleontology.

CYSTOSCOPY. Examination of the inner surface of the urinary bladder by means of a cystoscope, an instrument which is passed through the **urethra** into the bladder, where, by use of lights and lenses, a view of the inner bladder wall may be obtained. (D.M.H.)

CYTOGENIC GLAND. An organ which produces and discharges **cells**, such as the reproductive glands. (A.W.L.)

CYTOPLASM. Cell.

CYTOSOME. Cell.

D

DAB. Pices, Teleostei. *Limanda.* British and American flat fishes (**Pisces**) of several species. Edible. (A.W.L.)

DABCHICK. Aves, Colymbiformes. The little **grebe**, *Podicipes fluviatilis,* an aquatic bird of the Old World, and the pied-billed grebe, *Podilymbus podiceps,* of the New World. (A.W.L.)

DACE. Pisces, Teleostei. Various fishes (**Pisces**) of the family Cyprinidae, which contains also the **minnows** and **carps.** Among the North American species are a minnow called the red-bellied dace, *Chrosomus erythrogaster,* the common **shiner, redfin,** or dace, *Luxilus cornutus,* and two other small fishes, the long-nosed dace, *Rhinichthys cataractae,* and the black-nosed dace, *R. atronasus.* (A.W.L.)

DACITE. The name of a somewhat variable group of extrusive **igneous** rocks similar to the **rhyolites** but richer in **plagioclase feldspar.** Typical dacites are felsitic to **porphyritic** in texture. Dacites are the extrusive equivalents of the quartz-rich varieties of diorites and are sometimes classified as quartz-bearing **andesites.** The porphyritic types usually occur toward the center of the thicker dacite flows, **dikes** and **sills,** as well as the marginal zones of **laccoliths.** Dacites are common in the Cordilleran province of North, Central and South America. The term, dacite, was proposed by G. Stache of Austria for lavas in the old Roman province of Dacia. (R.M.F.)

DACTYLOPODITE. Biramous Appendage.

DACTYLOZOOID. A form of **polyp** found in colonial **Hydrozoa** which captures prey and brings it to the mouth. (A.W.L.)

DADO. A groove of rectangular cross-section, used in woodworking, cut across the grain. A plough cut is similar to a dado cut, but is cut with the grain. A rabbet is a dado cut at the edge of a piece of work. Mortise and tenon joints consist of two pieces—one piece has a slot or hole, called the mortise, cut in it, in which a corresponding piece, called the tenon, fits. A blind mortise and tenon joint is one in which the mortise hole is not cut through. Mortise and tenon joints may be of square, rectangular, or circular section. Other varieties of wood joints include crosslap joints, dovetails, etc. (H.C.H.)

DADDY LONGLEGS. Arachnida.

DAGUERREOTYPE. The first practical process of photography was invented by Louis J. M. Daguerre of Paris in 1837, although the details of the process were not published until 1839. The process was used chiefly for portraiture and became obsolete within a few years after the introduction of the wet collodion process in 1851. Although the first commercially successful process, modern photography is based on the negative-positive methods introduced the same year by William H. Fox-Talbot of England. In the daguerreotype process a light-sensitive layer of silver iodide is formed on a silver plate by contact with iodine. After exposure in the camera, a positive image is produced when the image is exposed to mercury and heated. The mercury, by attaching itself to the unexposed portions,

forms a positive image. The silver iodide remaining was removed at first with a solution of sodium chloride (salt) which was soon replaced, however, with sodium thiosulfate (hypo), the properties of which had been discovered by Herschel in 1819. The daguerreotype image so produced is very weak. In 1840 Fizeau described a process of toning with gold which greatly increased the strength of the image and was generally adopted.

At first, from 5 to 10 minutes' exposure was required on open landscapes and street scenes. The invention of a fast, large-aperture portrait lens by Petzval in 1841 and the discovery by Goddard in London (1840) of the superior sensitivity of silver bromide reduced the time of exposure to a few seconds. (C.B.N.)

DAKIN'S SOLUTION. An antiseptic aqueous solution of **sodium** hypochlorite, modified with **sodium** bicarbonate, used in the irrigation of infected wounds. The solution was introduced during World War I by Henry Dakin, a New York chemist. (R.S.M.)

D'ALEMBERT'S PRINCIPLE. The principle, first pointed out by d'Alembert in 1742, that **Newton's third** law (see **Newton's Laws of Dynamics**) holds for forces acting upon bodies entirely free to move as well as upon fixed bodies in stationary equilibrium. In the former case the "reactions" concerned are due solely to inertia. Thus, in the act of throwing a ball, one pushes upon the ball with a certain force, and the inertia of the ball causes it to push back on the hand with an equal force. The condition of the system during such a process is said to be one of kinetic equilibrium. From this point of view, obviously, any system of bodies must always be in equilibrium, either kinetic or static. (L.D.W.)

DALTON'S LAW. The well-known law of partial pressures in mixed gases and vapors. If several gases not reacting chemically upon each other are introduced into the same container, the pressure of the resulting mixture is equal to the sum of the pressures which would be observed if each gas were separately enclosed in that container. We may, for example, regard the **atmospheric** pressure as the sum of a nitrogen pressure, an oxygen pressure, an argon pressure, a carbon dioxide pressure, a water-vapor pressure, etc. The same principle holds for mixtures of the saturated vapors of two or more liquids evaporating in the same closed space, provided one liquid does not dissolve the vapor from the other (as water dissolves ammonia). Like other gas laws, this law is approximately valid only within limits. (L.D.W.)

DAM. The primary function of the dam is to fill the gap in the natural reservoir line left by the stream channel. The most desirable sites are usually those where this gap becomes a minimum for the required storage capacity. Topographical and geological conditions at the site may be expected, in most cases, to dictate the choice of the type of dam. The best economic arrangement is often a composite structure such as a masonry dam flanked by earth embankments. The dam further serves to divert the water to the intake works.

Dams may be classified as follows:

1. Timber dams.
2. Rock-fill dams.

3. Earth dams.
   Plain.
   Core wall.
   Hydraulic fill.
4. Masonry dams.
   Gravity, solid and hollow.
   Arch, single and multiple.

Timber Dams.—The timber dam is rarely used because of its short life and the limitation in height to which it may be carried. It is conceivable that in a location where timber is plentiful and cement costly and difficult to transport, and where only a submerged diversion dam is contemplated, the timber dam would be the most economical to construct, even taking due account of its lack of permanence.

Rock-Fill Dams.—The rock-fill dam is an embankment of loose rock with either a water-tight upstream face of concrete slabs or timber, or a water-tight core. Where suitable rock is at hand in plentiful amount, a minimum of transportation of materials can be realized with this type of dam. Like the earth embankment, it resists damage from earthquake shock very effectively.

Earth Dams.—The earth dam is constructed as (1) a simple homogeneous embankment of well-compacted earth, (2) the same, but with a water-tight core or an

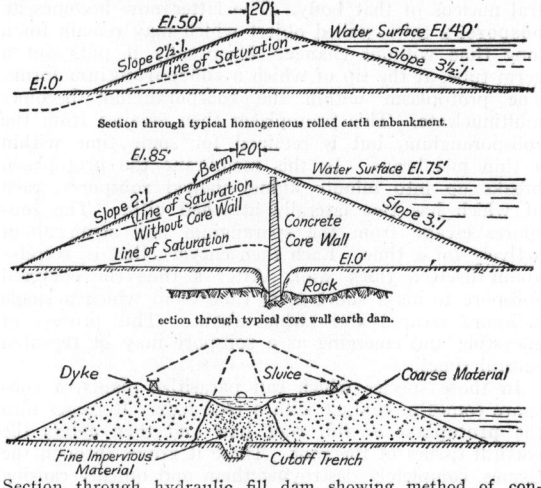

Section through typical homogeneous rolled earth embankment.

Section through typical core wall earth dam.

Section through hydraulic fill dam showing method of construction.

Earth dams.

upstream face pavement, (3) a **hydraulic fill** in which hydraulic segregation is relied upon to produce a water-tight core.

Masonry Dams.—Masonry dams are of either the gravity or arch type. Stability is secured in the gravity dam by making it of such a shape and size that it will resist overturning, sliding and crushing at the toe. In the arch dam stability is obtained by a combination of arch and gravity action. If the upstream face is vertical the entire weight of the dam must be carried to the foundation by gravity while the distribution of the normal **hydrostatic** pressure between vertical cantilever and arch action will depend upon the stiffness of the dam in a vertical and horizontal direction. When the upstream face is sloped the distribution is more complicated. The normal component of the weight of the arch ring may be taken by arch action while the normal hydrostatic pressure will be distributed as explained above. Hence, for the gravity type, good impervious foundations are essential, but, for the arch type, firm, reliable support at the abutments (either buttress or canyon side wall) is more important. The most desirable site for an arch dam is a narrow canyon with steep side walls of sound rock. When situated on a suitable site, the gravity dam inspires more confidence in

the layman than any other type. It has mass that lends an atmosphere of permanence, stability, and safety. When built upon a carefully explored foundation with stresses calculated from completely evaluated loads, the gravity dam probably represents the art of dam building at its highest point of development. This is an attribute of no mean significance because, due to flood disasters and their tremendous potentialities, fear of flood is a keenly developed human instinct. This factor has led to the adoption of the gravity section in some instances where an arch dam would have been the more economical construction.

(a) *The Solid Gravity Dam*

W is large enough, with respect to *P*, to incline *R* sufficiently to fall within the middle third.

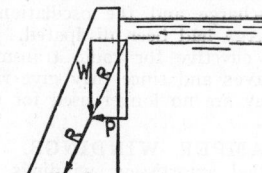

(b) *The Hollow Gravity Dam*

The slab is inclined enough to produce a vertical water pressure P_v, which inclines *P* sufficiently to overcome the effect of a small *W*, so that *R* falls within the middle third.

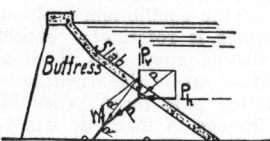

(c) *The Arch Dam*

Water pressure on the upstream face has the effect of shortening the dam, thereby creating resisting compressive stresses within the dam and tightening it against the abutments.

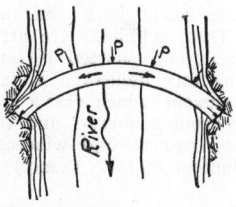

Comparison of stabilizing forces in dams.

Gravity dams are classified as *solid* or *hollow*. The solid type is the more widely used of the two, although the hollow dam is frequently the more economical to construct. Most forms of hollow dams have been patented by Ambursen and others. The gravity dams can also be classified as *overflow* and *non-overflow*. If the dam is to serve as a spillway section, its downstream face is ordinarily made an ogee curve with the curvature such that there will be no tendency of the water to leave the surface of the concrete, even with maximum water elevation at the crest.

Two types of single-arch dams are in use; namely, the constant angle and the constant radius dam. The constant radius type employs the same face radius at all elevations of the dam, which means that as the channel grows narrower, as at the bottom, the central angle subtended by the face of the dam becomes smaller. In a constant angle type of dam, this subtended angle is kept a constant and the variation in distance from abutment to abutment at various levels taken care of by varying the radii. The safety of an arch dam is dependent on the strength of the side wall abutments, hence the arch should not only be well seated on the side walls, but the character of the rock in bearing be carefully inspected to determine its ability to take the enormous thrust that will be set up as the water rises. The multiple-arch dam consists of a number of single-arch dams with concrete buttresses as the supporting abutments. The multiple-arch dam does not require as many buttresses as the hollow gravity type, so is very economically constructed. It requires good rock foundation because the buttress loads are heavy. (F.T.M.)

DAMMAR. Resins.

DAMPED WAVES. This is ordinarily used to designate electric waves which decrease in amplitude with time. In any oscillatory circuit which contains resistance (and all practical ones will) the oscillations will be dissipated in resistance losses and the amplitude of the oscillations will gradually decrease unless energy is continually added to the circuit. When energy is added to overcome this dissipation and maintain the amplitude constant, continuous waves result. A condenser discharging through an **inductance** will give rise to damped waves and this was the basis of the old spark radio transmitters where the spark gap initiated the discharge and the oscillations continued until all the energy had been dissipated. Since these waves are not as effective for radio transmission as the **continuous waves** and since they give rise to spurious frequencies they are no longer used for this purpose. (L.R.Q.)

DAMPER WINDINGS. These are windings, also called amortisseur windings, placed in the pole faces of synchronous **motors** and generators to counteract the tendency these machines have to hunt or oscillate about the equilibrium position. When the synchronous machine is rotating at synchronous speed and is not oscillating about this speed as a mean value, the a-c field set up by the armature is fixed with respect to the rotating pole faces. (See **Motors.**) There is, then, no change of magnetic flux across the windings in the pole face. However, if the machine hunts it oscillates about the synchronous speed so the windings in the pole faces are swept back and forth across the armature flux. These windings are actually bars placed in the pole faces and shorted by an end connection so the changing flux induces a voltage, resulting in a circulating current which reacts to damp out the hunting. In synchronous motors and condensers this winding is used as a squirrel-cage winding to start the machine as an induction motor. (L.R.Q.)

DAMPING. This term usually refers to the checking of a motion due to friction or similar cause. It is of especial significance in connection with the diminishing amplitude of an oscillation, as that of a pendulum swinging in the air or that of the electricity vibrating in an oscillating circuit. Unless energy is supplied during each cycle, the amplitude of such a vibrator falls off at each successive oscillation by an amount commonly expressed in terms of the decrement or damping factor, which is the ratio of any one amplitude to that next succeeding it in the same sense or direction. In so-called logarithmic damping, this decrement is constant; in an oscillating electric circuit, its value is an exponential e^δ, in which δ, the logarithmic decrement, is a constant depending upon the effective resistance, inductance, and capacitance of the circuit. (See **Electric Oscillations and Waves.**)

An important instance of damping is found in the reading of an oscillating index, like a balance pointer, on a scale. If one may assume that the amplitude falls off by equal amounts at each swing ("linear" damping), in order to find the equilibrium position one has only to average an even number of readings at one extreme and an odd number at the other, and then find the mean of the two averages.

When metals are stressed repeatedly within the elastic range they dissipate energy into heat because of "internal friction" or the "mechanical hysteresis effect." This characteristic is also known as "damping capacity" and may be expressed as the amount of work dissipated into heat by a unit volume of the material during a completely reversed cycle of stress. (See **Hysteresis.**)

The damping capacity of metals and alloys can be determined by the torsion pendulum method and by other methods in which the metal specimen is vibrated and the rate of damping-out of the vibrations observed.

Cast iron has been found to have a high damping capacity, a characteristic which is considered desirable in lathe beds and castings for other machine tools where rigidity is required. Materials with high damping capacity are also believed to be less sensitive to fracture at notches and other surface irregularities when subjected to cyclic or repeated stresses. For a given material, increase in hardness by **heat treatment** or **cold working** would normally be expected to reduce the damping capacity, however many exceptions have been noted. (L.D.W., R.H.H.)

DAMPING-OFF. This is a disease of plants, especially young seedlings, caused by several species of **fungi,** one of which is *Pythium.* This fungus has a slender, branching, non-septate **mycelium** containing many minute nuclei. At the apex of a branch of this mycelium an **oögonium** is formed. This is a spherical body containing many nuclei. One of these nuclei stays at the center of the oögonium; the others migrate to the periphery. From the tip of another branch of the mycelium a multinucleate **antheridium** is formed. The antheridium grows to the surface of the oögonium, and develops a slender conjugation tube which penetrates the oögonial wall. Through this tube a nucleus enters the oögonium and fuses with the central nucleus of that body. The latter now becomes an **oöspore,** a thick-walled object which may remain for a long time without change. Eventually it puts out a germ tube, at the tip of which a **zoösporangium** forms. The protoplasm within the zoösporangium becomes multinucleate. This protoplasm then escapes from the zoösporangium, but is retained for some time within a thin membrane. In this membrane the protoplasm breaks up into minute kidney-shaped zoöspores, each of which has two laterally attached **cilia.** The zoöspores escape from the sporangium and swim about actively for a time. Each then encysts, that is, secretes about itself a thick wall. After a time the encysted zoöspore forms a short germ tube from which a single zoöspore escapes and swims about. This process of encysting and emerging as a zoöspore may be repeated several times.

In those species which can parasitize plants, a zoöspore comes in contact with a root hair. It passes into the **protoplast** of this root hair and then enters the cortical tissues of the root. There it grows through the tissues, completely destroying them and quickly causing the death of the plant. Because of the extremely rapid growth of the parasite, once it enters the tissues of the host, the latter dies very suddenly, the top dropping over and collapsing. Often whole flats of seedlings succumb in a very few hours after the fungus appears. Adequate moisture is necessary for the fungus to grow, abundant moisture favors it greatly, and insufficient moisture retards it. Control of this pest is therefore largely a matter of reducing the moisture available as much as possible. The danger of losing valuable seedlings by damping-off may be prevented by steam sterilization of the soil and chemical sterilization of the surfaces of seeds. (R.M.W., B.S.M.)

DAMSEL FLY. Odonata.

DANBURITE. The mineral danburite, $CaB_2(SiO_4)_2$, **calcium-boron silicate,** crystallizes in the **orthorhombic** system in prismatic forms somewhat resembling the mineral topaz. Its fracture is subconchoidal; brittle; hardness, 7–7.2; specific gravity, 2.97–3.02; color, colorless, yellowish-white, yellow, dark wine yellow and brownish-yellow; luster, vitreous to greasy; translucent to transparent. It is found at Danbury, Connecticut, from whence its name was derived, Saint Lawrence County, New York, Switzerland, Japan and Madagascar. (E.S.C.S.)

DANDELION. *Taraxacum.* Composite Family.

DANDRUFF. Seborrhea.

DARLINGTONIA. Insectivorous Plants.

D'ARSONVAL. Instruments, Electrical.

DART SAC. A structure associated with the female genital duct of **snails.** It secretes a calcareous dart which is shot by muscular contraction into the body of another snail when the two approach each other prior to mating. (A.W.L.)

DARTER. 1. Pisces, Teleostei. Small fishes **(Pisces)** constituting the family Etheostomidae. They have a small swim bladder or none, hence they rest on the bottom when not in active motion. 2. Aves, Pelecaniformes. *Anhinga.* Long-necked diving birds **(Aves)** with long sharp beaks, found in all continents. Also known as snake birds, anhingas and snake necks, and certain species as the wryneck and water turkey. They resemble the cormorants. (A.W.L.)

DASH-POT. The dash-pot is a device for effecting a quick, jerking, mechanical motion. In its usual form it consists of a **cylinder** with a closely fitting **piston.** As there are no inlet ports on the cylinder, motion of the piston increasing the volume creates a **vacuum** in the cylinder. This vacuum, amounting to 10–13 lbs. pull on every sq. in. on the piston, will, in the case of a large piston, produce a pull on the piston rod sufficient to overcome considerable resistance, and literally dash the piston back to the point from which the vacuum was started. Although a spring could be considered potentially capable of performing the same service as a dash-pot, it is inferior for some jobs. There is always a certain amount of resilience or springiness not possessed by the dash-pot, a tendency for fatigue of the metal, and of stiffness when used for large forces. (F.T.M.)

DASYURE. Mammalia, Marsupialia. *Dasyurus.* Arboreal pouched animals **(Marsupialia)** of the Australian region. They are long-tailed forms with short legs resembling the civets and are carnivorous. The dasyure family includes also the thylacine or Tasmanian wolf, the Tasmanian devil, the phascologales, the pouched mouse, and the banded anteater. All are carnivorous or insect-eating. (A.W.L.)

DATE. *Phoenix dactylifera.* **Palm.**

DATE LINE. International Date Line.

DATOLITE. Datolite, **calcium boroxy silicate,** $Ca(BOH)SiO_4$, occurs in **monoclinic** crystals of varied habit, mostly short stout prisms, but often in highly modified forms. Datolite reveals no cleavage, its fracture is conchoidal to uneven; brittle; hardness, 5–5.5; specific gravity, 2.9–3.0; luster, vitreous to dull; color, white to gray or may be greenish, yellowish, or brownish. It has a white streak and is transparent to translucent usually, but has been observed opaque. Datolite is a secondary mineral being found in veins and cavities associated with **zeolites** and **calcite,** particularly in the **basic igneous** rocks. It has been found in the Harz Mountains, Germany; in the Trentino district, Italy; in Norway and Tasmania. In the United States it has been found in the Triassic **traps** of the Connecticut River Valley in Massachusetts and Connecticut, and from similar rocks in New Jersey. In Michigan Datolite has been found associated with the copper-bearing rocks of Keweenaw County. This mineral derives its name from the Greek word meaning to divide, in reference to the granular structures of some of the massive varieties. (E.S.C.S.)

DATUM. Earthwork.

DAVISSON-GERMER EXPERIMENT. In 1927, Davisson and Germer conducted a research, the results of which furnished a remarkable confirmation of the basic postulate of **wave mechanics.** De Broglie had suggested about 1925 that **electrons** have in some respects the characteristics of waves, and deduced, for the wavelength equivalent to a moving electron, the expression $\lambda = h/mv$, in which m and v are the mass and speed of the electron and h is Planck's constant. If the electron is moving, for example, with a speed corresponding to 65 **electron volts** of energy, the corresponding "de Broglie wavelength" is 1.52 angstroms, which is in the x-ray range. This led Davisson and Germer to see whether electrons might be reflected from crystals after the manner of x-rays. They used a single crystal of nickel cut parallel to the (111) planes, and upon varying the electron speed at a fixed angle of incidence, they found not only a distinct "regular" reflection but also a series of diffraction maxima strikingly similar to those obtained with the same crystal for x-rays of varying wavelength. The differences observed were satisfactorily explained as due to the refraction of the nickel for the electron waves. (L.D.W.)

DAW. Jackdaw.

DAY. Time.

DB. Decibel.

D.C. Direct Current.

DDT. An insecticide that has attracted widespread attention. It is currently made by reaction of 1 mol chloral $(CCl_3 \cdot CHO)$ and 2 mols chlorobenzene $(Cl \cdot C_6H_5)$ in the presence of concentrated sulfuric acid. The principal product of this reaction is DDT but accompanied by other isomers and by-products. DDT is 2,2-bis-(4-chlorophenyl)-1,1,1-trichloroethane, of formula

of melting point 103° C. minimum for "purified," of setting point 88° C. minimum for "technical." The technical grade is that specified by and produced for the U. S. Government, and the specifications for both purified and technical grades were adopted July 10, 1945, by the Manufacturing Chemists' Association of the U. S. "Freon 12" (dichlorodifluoromethane) liquid under pressure, forms a mist with DDT present, when released to the atmosphere, sufficient to kill the insects in the surrounding area—this is the principle of the "insect bomb." The U. S. Department of Agriculture terms it the most potent single insecticide ever discovered.

Dusted into clothing, it protects against the body louse, which carries epidemic typhus. (See **Rickettsial Diseases.**) In contrast to World War I, in which typhus accounted for many thousands of deaths among troops, the disease has been under complete control in World War II. Mosquito control with DDT is also highly effective; walls sprayed with the material will remain lethal to mosquitoes and flies for several months. The toxic effects of DDT on man, and its effect on wild plant and animal life, have not been completely worked out. As far as experimental work has proceeded, it appears that the agent is extremely potent and relatively safe. (R.K.S., D.M.H.)

DEAD CENTER. In machine tools, the term applies to the stationary center on which work rotates while being machined. In a **lathe**, the dead center is mounted in the tailstock. In heavy-duty lathes, the pressure on the pivot bearing effected by the center holes and the dead center is often so great that the dead center, or the center hole in the work, may wear excessively, and a rotating dead center, consisting of a center mounted in ball bearings carried in the shank which fits the tailstock sleeve, is often used. In grinding machines both centers are frequently stationary.

A steam engine is said to be on dead center when the piston is at one end of its stroke and the crank, connecting rod, and piston rod are in alignment. In this position, the steam pressure does not exert a rotative force on the crank, since it is transmitted directly to the shaft and bearings. (H.C.H.)

DEAD LOAD. Load.

DEAD MEN'S FINGERS. Porifera.

A branching sponge, *Chalina arbuscula,* found off the Atlantic coast of the United States. Its branches are rounded finger-like projections of white or light gray color. (A.W.L.)

DEAD RECKONING.

The meaning of this **navigation** term can be best understood from a consideration of its probable origin. Early navigators deduced their positions at any time by using the distances and directions of motion of their ship since leaving some previously known position. Such positions were determined at frequent intervals and early log books contained a column for entering them and this column was headed by the abbreviation "ded. pos." The calculating, or "reckoning" necessary to obtain the entries for this column, was known as "ded reckoning" and is now written as dead reckoning.

Dead reckoning is fundamental for the methods of modern navigation and should be clearly understood. **Pilotage, radio navigation, celestial navigation** and other methods for determining positions of a ship or plane, all involve the use of dead-reckoning methods. This is particularly true in the case where two **lines of position** are not observed simultaneously, and **the fix** must be found by moving one or both of the lines by dead reckoning.

There is a conflict of opinion among modern sea navigators as to just how much material should be included in the dead-reckoning (DR) position. It is certainly true that originally only the **heading** of the ship and the distance run on that heading were included. Some modern sea navigators cling to the original meaning of the term. The effects of **current** and **leeway** may be computed independently and the DR position corrected to an estimated position (EP). Some sea navigators include all known or suspected motions of the ship in their dead reckoning and call the resulting position either DR or EP.

It is an established procedure in air navigation to include the movement of the plane through the air (heading and air distance), and the movement due the motion of the air itself (wind direction and speed), in determining a DR position. In air navigation the term estimated position (EP) is used in connection with celestial navigation and is discussed in the article on that subject.

Before entering upon a discussion of actual methods of dead reckoning, it should be clearly understood that certain errors are bound to appear in the data employed in the calculations. These may be listed under five groups: **compass corrections,** steering, **patent logs** or **air-speed meters,** leeway of a seaborne ship, and predicted current or wind. Under the most favorable conditions, the dead-reckoning position is somewhere within a circle whose center is the DR position and whose radius is between 1 and 2% of the distance run. In conditions of rough sea, or turbulent air, the radius may increase to more than 5% of the distance run.

The graphical method of solution is used by practically all air navigators, and by a majority of those on the sea. This is essentially the graphical addition of **vectors.** In the determination of the old-fashioned DR position, the vectors used are total motions within given periods of time. These vectors are drawn to proper scale of distance on a **mercator chart,** mercator plotting sheet, or **small-area plotting sheet,** with the tail of the first vector resting on the last-known position, known as the point of departure, and the head of the final vector at the DR or EP position. All vectors must be properly labelled; the direction of each being the heading and the length proportional to the distance run.

The graphical method of solution is illustrated in the following case: At 1200 a ship is in latitude 43° 35' N and longitude 34° 38' W. At 1200 the ship is on heading 150°, speed 12 knots, at 1430 heading altered to 125°, speed 12 knots, at 2000 heading altered to 030°, speed 12 knots, at 2400 heading altered to 320°, speed 12 knots. The vector diagram is drawn on a small-area plotting sheet and properly labelled as shown in Fig. 1. The

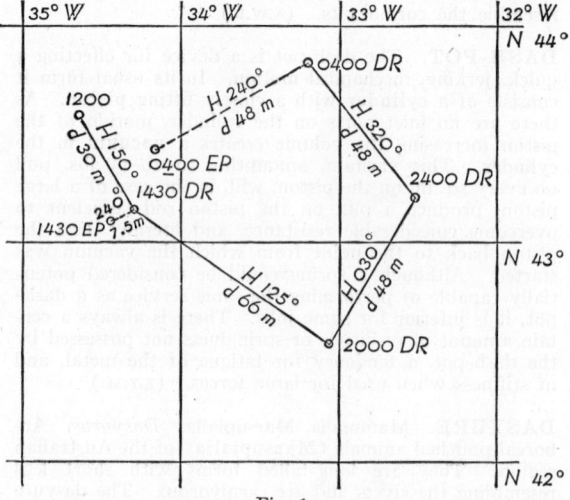

Fig. 1.

various distances are obtained for the vectors by simply multiplying the speed by the time. From this diagram the DR positions for any desired times can be read off directly. For example, the position at 1430 L = 43° 09'.0 N & Lo = 34° 17'.7 W and at 0400 L = 43° 49'.5 N & Lo = 33° 12'.0 W.

In case a current is predicted for this region, an additional vector may be added to any DR position to obtain the estimated position. For example, if in the above situation we assume a current to set 240° with drift 3 knots, vectors (shown dotted in the figure) in direction 240° and lengths 7.5 and 48 miles are added at the 1430 and 0400 DR positions respectively. This yields a 1430 EP L = 43° 05.3 N & Lo = 34° 26.6 W and 0400 EP L = 43° 25.5 N & Lo = 34° 09.4 W. If leeway is present it should be treated as an additional compass correction to each heading, as explained in the article on leeway.

The graphical method of solution for dead-reckoning problems requires a considerable amount of paraphernalia, e.g., plotting sheet, parallel rulers or protractor, scale, either pad and pencil or a **dead-reckoning calculator** for arithmetic computing, and most important of all, a flat surface for doing the drawing. On large ships or planes the navigator has the needed space, and the chart table is frequently equipped with a drafting machine which simplifies the plotting. On small boats or planes sufficient space for plotting is difficult to find. Computational methods are just as rapid as those involving plotting and they certainly yield more accurate results. However, it should be emphasized that the advantage of computational methods is only that of convenience in space, for the plotting methods are fully as accurate as the data warrants.

In the computational method the first step is to solve the right triangle in which the heading vector is the hypotenuse, with the difference of latitude (DL), and the departure (dep) the two legs. The DL is added algebraically to the latitude (L_1) of the point of departure, calling DL + when north, and − when south. The DR latitude (L_2) is then $L_2 = L_1 + DL$. The departure must be converted from nautical miles to difference in longitude (DLo) in minutes of arc by the method of **middle-latitude sailing,** using for the mid-latitude $L_m = \dfrac{L_1 + L_2}{2}$. The DLo is added algebraically to the longitude (Lo_1) of the point of departure, calling DLo + when west, and − when east. The DR longitude (Lo_2) is then $Lo_2 = Lo_1 + DLo$. Three-place computing, using slide rule or traverse tables is sufficient to attain all accuracy warranted by the data. To illustrate the computational method: At 0815 a ship is in latitude 41° 42′ N & Lo = 35° 25′ W. The ship proceeds along heading 243° at 14 knots, and the DR position at 1200 is desired.

L_1	41° 42.0 N	H	243°	Lo_1	35° 25′.0 W
DL	23.8 S	d	52.5 m	DLo	1 02′.5
L_2	41 18.2 N	dep	46.8 W	Lo_2	36 27′.5 W
		L_m	41°.5		
		DLo	62.5 W		

The actual numbers were read directly from traverse tables and the total time required for solution was 2 minutes. The equipment used was a pencil, scratch pad, and traverse tables.

In more complicated cases where a number of different headings and distances are used, and in a region where a current is predicted, we set up what is known as a traverse form in which the columns N, S, E, and W refer to the direction of the individual DL's and departures. At 0400 a ship is on heading 200°, speed 15 knots, and is in latitude 40° 21′.3 N and longitude 124° 34′.6 W. At 0723 the heading is altered to 300°, speed 15 knots, at 1136 heading altered to 030° speed 15 knots, at 1314 heading altered to 340° speed 15 knots. In this region there is a current setting 140° 2 knots. The traverse form used in determining the DR position at 1600 is given below:

TRAVERSE FORM

H	Time	d	N	S	E	W
200°	3 h 23 m	50.8 m		47.7		17.4
300	4 13	63.3	31.8			54.9
030	1 38	24.5	21.2		12.2	
340	2 46	41.5	39.0			14.2
		Sums	92.0	47.7	12.2	86.5
		Total	44.3			74.3

L_1	40° 21′.3 N	dep	74.3 m W	Lo_1	124° 34′.6 W
DL	44.3 N	L_m	40°.7	DLo	1 37.9 W
L_2	41 05.6 N	DLo	97.9 W	Lo_2	126 12.5 W

To this DR position the current motion must be added to attain the estimated position. The current motion in 12 hours is 24 miles along 140°, for which DL = 18.4 S, dep = 15.4 m E, and in mid-latitude 40°.7 DLo = 20.4 E. Applying these to the DR position we have EP L = 40° 47′.2 N & Lo = 125° 52.1 W.

Whenever a fix is obtained, a DR position is also obtained. The difference between the two positions is always attributed to "current" by sea navigators, never to errors in their data or obtaining the DR position. Here the difference in opinion as to whether or not leeway and current should be included in obtaining the DR position introduces confusion. If an old-fashioned DR position (i.e., without current or leeway included) is used, then the difference between fix and DR may be the actual current. If an EP is used, the difference between this position and the fix is a residual current, supplemental to that already used in the determination of the EP. In either case the difference is called current with set equal to the direction from the DR position to the fix, and drift the distance between the two positions divided by the number of hours elapsed from the point of departure for the dead reckoning. These so-called currents are entered in the ship's log book and later forwarded to the Coast and Geodetic Survey or the Hydrographic Office where statistical discussions yield improvements to previously predicted values.

In air navigation a slightly different procedure for finding DR position is employed. In the first place, the work is always done by constructing the vector diagram. Before plotting the total distances from the point of departure, a velocity-vector diagram is used to determine the **course** and ground speed made good along this course. The heading and air speed of the plane are drawn as one vector, and to this is added the vector representing the direction toward which the wind is blowing and the wind speed. The vector sum is a vector giving course and predicted ground speed. Unfortunately, the first man who erected a weather vane so balanced it that the arrow pointed into the wind instead of with it. Still more unfortunately, the weather bureaus have retained this archaic method of publishing wind directions, and air navigators must be particularly careful in constructing their diagram. After the course and ground speed along the course have been obtained, then a DR position can be obtained by plotting these in the same manner that the sea navigator plots heading and ship's speed. At 0916 a plane is known to be in latitude 41° 52′ N & longitude 56° 34′ W. The plane is heading 243° with true air speed 110 knots and the predicted wind is from 305°, speed 35 knots. The pilot wishes to determine the course and ground speed of the plane and the DR position at 1045. The vector-velocity diagram is shown in Fig. 2(a). The wind vector, *ew*, must be

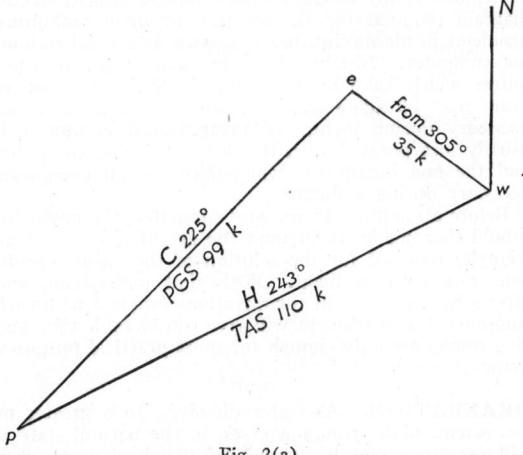

Fig. 2(a).

drawn in the direction that the wind will move the plane, i.e., wind direction − 180°; to this is added the heading air-speed vector, *wp*, and the sum, or vector, *ep*, is the course and speed made good. In this case the course is found to be 225° and the ground speed 99 knots. With these quantities determined, the DR position is found by plotting the distance-vector diagram shown in Fig. 2(b). The course 225° is laid off from the 0916 position on a Mercator plotting sheet or a small-area plotting sheet, and the distance 99 × 1 h 20 m = 146 m is measured off on the proper scale, giving the 1045 DR position. If the heading is altered, a new velocity-vector diagram must be constructed and a new course and ground speed obtained. Then a distance

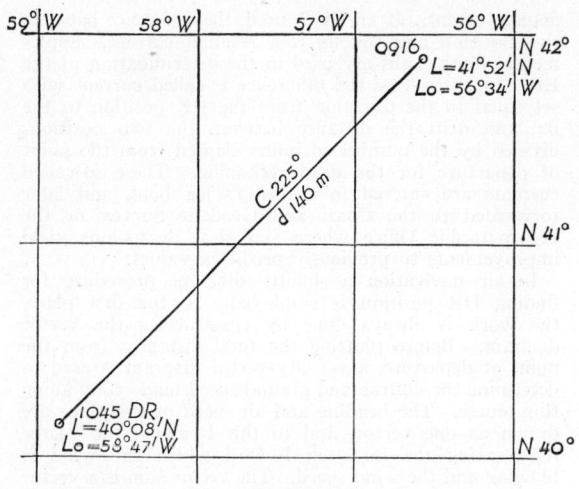

Fig. 2(b).

vector would be added to that already found, and the DR position found at any desired time.

The solution of the velocity-vector diagram is facilitated by the use of any one of a number of **dead-reckoning computers.**

In case a reliable fix is obtained at the same time a DR position is found, and the two do not coincide, the air navigator always assumes that the difference is due to an error in the predicted wind. The line between the point of departure and the fix is known as the track if but one heading is used, or the track made good if the plane has changed heading since leaving the point of departure. A discussion of this problem will be found in the article on **Air Plot.** (w.k.g.)

DEAD-RECKONING COMPUTER. A mechanical device for rapidly setting up the complete velocity-vector diagram required for the solution of **dead-reckoning** problems in air **navigation** is known as a dead-reckoning computer. Nearly all of the many types of computers which have been developed within the past 10 years may be adapted to the solution of all triangles necessary during flight. A **navigational computer** is usually incorporated with the dead-reckoning computer, and the one instrument will suffice for all computing necessary during a flight.

Before attempting to use any computer, the navigator should thoroughly understand the actual drawing of all triangles required for the solution of the various problems met with in flight. With that understood, and after a careful study of the directions supplied with each computer, the instrument will save considerable time and give results accurate enough for most practical purposes. (w.k.g.)

DEAERATION. As water dissolves, to a greater or less extent, many common gases, in the natural state it will contain a certain amount of dissolved gases, such as oxygen and carbon dioxide. Deaeration is the removal of this dissolved gas. Deaeration at the present is practiced where the gas that water contains would have undesirable effects. Often dissolved oxygen is objectionable because of its corrosive action. This is true in the case of the high-pressure steam boiler, where a small amount of oxygen dissolved in the feed water may become quite active in attacking the boiler metal under the high pressure and temperature conditions there experienced. Steam boiler operators often treat their boiler feed water in deaerators to remove this oxygen. These deaerators are either of the deactivating or the heating type. Deactivating types employ chemical means of deaeration. A representative deactivating deaerator consists essentially of a tank containing large surfaces of

scrap iron. When water containing dissolved oxygen is brought into contact with this iron, most of the **oxygen** combines with the **iron,** and is thereby removed from the water. Deaerating action in a heating-type deaerator is obtained by first reducing the solubility of the gas through heating the water (under pressure); second, reducing the pressure and producing explosive boiling; and third, controlling the agitation of the water subsequent to the second action in a partially evacuated region. The figure shows a heating-type deaerator. Water

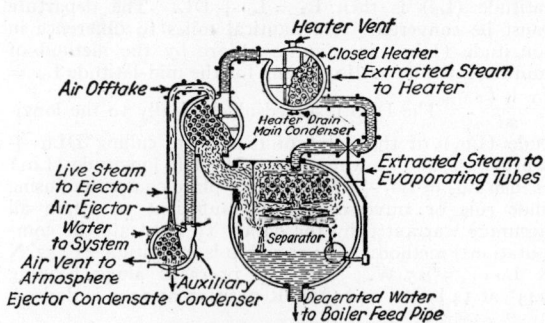

Elliott type "J" deaerator.

drawn from a supply line circulates through the tubes of the auxiliary and main condensers. It then flows under the heater where it is heated to the desired temperature. The heated water flows down into the separator, where it boils violently, giving off the dissolved gases. Vapor and gases are carried off through the condenser, in which the vapor is condensed for return to the deaerator. The gases are removed by an air pump. (f.t.m.)

DEATH. The termination of vital processes in the organism. The physiology of death has been studied extensively and various explanations have been offered of its causes without furnishing true understanding. It often occurs through accident, but under entirely normal conditions also it takes place after a lapse of a period characteristic of the species. Following maturity the metabolic processes of the individual become slower through senility until at last some vital part fails completely. Only animals that reproduce by binary fission, notably the 1-celled species, are immune from normal death. In these forms the identity of the parent is merged with that of its offspring and death is always accidental. (a.w.l.)

DEATH WATCH. Insecta, Coleoptera. A small beetle of the family Anobiidae which burrows in solid wood in buildings. By striking its head against the walls of the burrow it produces a sharp sound which can be heard in quiet places. It has been supposed, superstitiously, to foretell death. (a.w.l.)

DEATH'S HEAD MOTH. Insecta, Lepidoptera. A large European **sphinx moth** whose **thorax** bears a light mark shaped like a skull. (a.w.l.)

DEBRIDEMENT. The treatment of wounds, especially traumatic, dirty, crushing wounds, by means of excising all injured, contaminated, or devitalized tissue. (r.s.m.)

deBROGLIE WAVE. Davisson-Germer Experiment; Wave Mechanics.

DECADE BRIDGE. Bridge.

DECALAGE. The wings of a biplane are usually set parallel to each other. This is not absolutely necessary, and some designers have preferred characteristics obtained when the wings are set at a slight angle to each other. Decalage is the angle between the **chords** of the two wings of the biplane. It is rare for decalage to amount to more than 2 or 3°. (F.T.M.)

DECAPODA. 1. The shrimps and **prawns, lobsters, crayfishes,** and **crabs,** constituting a large and important order of **crustaceans.** The **thorax** is covered by a **carapace** and bears five pairs of appendages, the first pair chelate grasping structures and the remaining four formed for walking. 2. The **cuttlefish, squids** and related forms, constituting an order of **cephalopods.** They possess ten arms, with stalked suckers provided with horny rims, and have a well-developed internal shell. (A.W.L.)

DECARBURIZATION. Reduction in carbon content at the surface of **steel** or **cast iron** by heating in air or other oxidizing or reducing gases. In heating for hot rolling, **forging,** or **heat treatment,** decarburization is usually objectionable, and specially prepared neutral furnace atmospheres may be used to reduce or eliminate it. Molten salt or lead baths are also effective in protecting the surface during heat treatment.

In the case of heat-treated machine parts, surface decarburization is objectionable because it reduces **fatigue** strength and lowers the wear-resistance of **bearing** surfaces. Important surfaces of hardened steel parts are often finish-ground, in which case a limited amount of decarburized skin can be removed. **Tool steels** for cutting tools, punches, chisels, etc., are usually ground sufficiently to remove all decarburization; however, many tools and dies are machined to finish dimensions before hardening and extreme care must be taken to protect the surface.

Decarburization is intentional in the processing of low-carbon sheet steels for electrical applications. In the production of **malleable cast iron** by annealing white cast iron decarburization is beneficial. (R.H.H.)

DECAY COEFFICIENT. Certain processes studied in physics progress at a rate diminishing in accordance with an **exponential** function of the time; such, for example, as phosphorescence and **radioactive** emission. The falling off or "decay" of such a process may be represented by an equation giving the intensity at time t as $I = I_0 e^{-Ct}$, in which I_0 is the intensity at the beginning of the time t and C is the "decay coefficient."

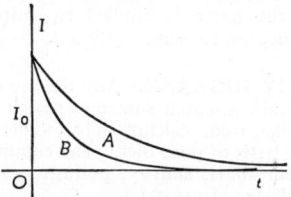

Typical exponential decay curves. The decay coefficient in B is greater than in A.

Closely related to C is the half-value period, which is the time required for I to fall to ½ its original value I_0; it is equal to $0.69315/C$. Thus, if the half-value period of radium B is 1608 sec., its decay coefficient C is

$$\frac{0.69315}{1608 \text{ sec.}} = 0.000431/\text{sec.}$$

This means that approximately 0.000431 of the substance existing at any instant disintegrates during the ensuing second. The reciprocal of C, called the "decay modulus," represents the time required for I to diminish to $1/e$ or 0.3697 of its original value I_0. It is equal to 1.4427 times the half-value period; for the decay of radium B its value is therefore 2320 sec. (L.D.W.)

DECCA NAVIGATION. Decca navigation is a system of **hyperbolic navigation** which employs low-frequency continuous-wave radiation. The system has been under experimentation for a number of years, but has not as yet proved to be entirely satisfactory in practice. It does have certain definite possibilities, and is worthy of consideration.

The master and any slave station radiate continuous waves whose frequencies are related by a simple fraction, say one fourth. The radiations from the two will be in phase when the distances from the stations differ by even multiples of a specified unit. This unit is a function of the wavelengths radiated.

In practice a master and two slave stations are used. The receiving set reduces all three to a common frequency and phase meters indicate the relative phase of each slave to the master. The accuracy of the setting of the phase meter is such that differences in distance may be determined with an accuracy of the order of magnitude of 100 feet, and this is independent of distance from the base line. However, there is complete ambiguity of position since there is no positive method of determining the number of complete phase changes between the observer and either station. Attempts are being made to overcome this difficulty by modulating envelopes on the carrier wave.

In spite of the ambiguity mentioned above, the system is of great accuracy and value when used in proceeding to some specified objective. A **line of position,** involving the master and one of the slave stations, is selected which passes through the desired objective. The pilot must get his ship onto this line and then set his phase meter. If he then proceeds so that the phase meter setting remains constant he must be following the hyperbola directly to his objective. The hyperbolic lines from the master and the other slave will intersect the hyperbolic track along which the ship is proceeding. The pilot computes the number of complete phase changes that are to be expected, between the point of departure and the objective, along the line that he is following. When this number, plus any remaining fractional phase change, has been completed the pilot must be directly over the objective. (W.K.G.)

DECIBEL. A commonly used unit of relative power, especially in expressing acoustic or electric power outputs in terms of some arbitrarily specified standard. The decibel is one tenth of a **bel.** Devices emitting sound energy, such as **loud-speakers,** or those which transform sound energy into electrical energy, such as **microphones,** are commonly rated in decibels above or below some reference device of similar type.

The decibel is a power ratio, equal to the tenth root of ten or about 1.259. Thus if a radio receiver, playing at a certain "volume" level, is turned up until the acoustic power output has increased by 25.9%, its new level is said to be one decibel above the original level. To raise the level 2 decibels would require a power step-up of $1.259 \times 1.259 = 1.585$; that is, of $\sqrt[10]{10} \times \sqrt[10]{10} = \sqrt[5]{10}$. An increase of 10 decibels would require a 10-fold output of power, since $(\sqrt[10]{10})^{10} = 10$. This last increase is called a **bel,** a term derived from the name of Alexander Graham Bell, inventor of the telephone. Twenty decibels or 2 bels corresponds to a 100-fold increase of power; and so on. To be quite accurate one should say, not that 3 decibels equals 3 times 1 decibel, but rather that it is the cube of 1 decibel.

If two power outputs to be compared are denoted by p_1, p_2 and their difference of level, in decibels, is denoted by d, then from the above definition,

$$\frac{p_2}{p_1} = (\sqrt[10]{10})^d = 10^{\frac{d}{10}}.$$

The relation is more conveniently applied in the logarithmic form,

$$\log \frac{p_2}{p_1} = \frac{d}{10},$$

or

$$d = 10 \log \frac{p_2}{p_1}.$$

For example, the electric power output of a certain microphone is 18 decibels higher for a musical tone of 1500 cycles per sec. than for a 500-cycle tone of the same intensity. This means that, for the same acoustic power input, the electric power output of the microphone at 1500 cycles exceeds that at 500 cycles in the ratio $\log^{-1} 18/10$, or about 63 to 1.

Generators of voltage are also often expressed in decibel measure, especially if the voltage has an acoustic origin. But since the power of a generator in a given circuit is proportional to the square of the voltage, $p_2/p_1 = (E_2/E_1)^2$, or $\log p_2/p_1 = 2 \log E_2/E_1$. Therefore, for such cases,

$$d = 20 \log \frac{E_2}{E_1}.$$

It is an interesting circumstance that the smallest change of sound intensity that the normal human ear can detect is approximately 1 decibel, whatever the original intensity. It may have been this fact that first suggested such a unit. (L.D.W.)

DECIDUOUS PLANTS. Plants which drop their leaves at the end of the growing season are called deciduous plants. (R.M.W.)

DECILE. A decile is a measure of position. The nine deciles $D_1, D_2, \cdots, D_9$ are the values of the **variates** which divide a distribution into ten equal parts. They are found by observation in a **serial distribution** after the items are arranged in order of magnitude, and in a **frequency distribution** by interpolation in the **cumulative frequency distribution**. (L.A.A.)

DECK BRIDGE. Bridge.

DECKEN STRUCTURE. Used by structural geologists to designate a series of great overthrust folds with nearly parallel and horizontal axial planes. (R.M.F.)

DECLINATION. The declination of a celestial object is the coordinate in the **equatorial** system of **spherical coordinates** measured in the plane of the **hour circle** through the object from the equator to the object. In case the object is between the equator and the north celestial pole the declination is said to be north or positive (+), otherwise the declination is south or negative (−). Declination is ordinarily measured either with a **meridian circle** or an **altazimuth** instrument.

The term declination is used by surveyors, and a few others, in place of **variation** to describe the angle between true and magnetic **north**. Navigators, both air and sea, always use the term variation for this angle. (See **Compass Corrections**.) (W.K.G.)

DECOMPOSITION OF VECTORS INTO COMPONENTS. Vector Addition.

DECOUPLING FILTER. In most multistage **amplifiers** there are certain circuits, such as voltage supplies, etc., common to more than one stage. Since these common circuits provide a path through which energy may be fed from the output back into the input of some stages, serious feed-back problems would result if something were not done to prevent them. The usual remedy is to insert a decoupling filter in those plate and grid leads which connect to points common to other plate or grid leads. These filters are frequently resistances in series with the lead and a **by-pass** condenser from the plate or grid side of the resistor to ground. The resistance used must be low enough not to cause a serious loss of voltage and the condenser should have a reactance which is low compared with the resistance at the lowest frequency for which the circuit is designed. Where the resistance would produce too much d-c voltage drop or where it does not give enough filtering action an inductance is sometimes used. For still more effective filtering a second resistance and condenser in cascade may be used. (L.R.Q.)

DEDENDUM. In a gear tooth, the distance from the pitch circle to the root circle or bottom of the tooth space. (H.C.H.)

DEDIFFERENTIATION. A process of change from a more specialized to a less specialized condition. (A.W.L.)

DEER. Mammalia, Artiodactyla. Hoofed animals which have solid bone antlers in the male or in both sexes. These antlers are shed each spring. The deer constitute the family Cervidae.

The typical deer are represented by the red deer of Europe, *Cervus elephas,* and the **elk** or wapiti, *C. canadensis,* and Virginia or white-tailed deer of North America, *Odocoileus virginianus.* The group also includes Asiatic species known as the shou and maral. Asia is the home of many species of deer which fall into several groups known as the sambar group, the fallow deer group (*Dama*), the muntjacs (*Cervulus*), and the tufted deer (*Elephodus*). Several species not associated with these groups are also found in Asia, among them the chital or Indian spotted deer, *Axis axis,* the **reindeer**, *Rangifer,* and the Chinese water deer, *Hydropotes.*

The North American fauna includes the **caribou**, *Rangifer,* related to the reindeer, and the **moose**, *Alces americana,* which is related to the elk of Europe, in addition to the mule deer, *Odocoileus hemionus,* of the west and the Virginia deer, *O. virginianus.* Mexico and Central and South America are the home of the brockets and guemals and of several other kinds of deer.

Many species of deer are among the finest of game animals. They have been hunted to extinction in some areas and have retreated readily before the advance of man, but through proper protection and management they are being maintained in many sections of the United States in satisfactory numbers. The flesh is excellent. (A.W.L.)

DEER FLY. Insecta, Diptera. Small **horse flies** with banded wings which are abundant in the eastern woods. In the west the name is applied to **snipe flies**. All species are annoying to man. (A.W.L.)

DEFICIENCY DISEASE. Any disease due primarily to lack of certain essential substances such as **vitamins**, or minerals like iron, calcium, etc. The deficiency is usually on the basis of poor diet. The common deficiency diseases are **beri-beri, scurvy, pellagra,** alcoholic **neuritis,** and rickets. (D.M.H.)

DEFINITE INTEGRAL. Let $f(x)$ be a **continuous function** in an interval (a, b). Suppose this interval to be divided and redivided into parts in any manner such that as the process is continued the lengths of the parts all approach zero as a limit. At any stage of the process let $\Delta x_1, \Delta x_2, \cdots, \Delta x_n$ denote the parts, and let z_1 denote any point in $\Delta x_1,$ z_2 any point in $\Delta x_2, \cdots, z_n$ any point in $\Delta x_n.$ Then form the sum of the products:

$$\sum_{i=1}^{n} f(z_i)\Delta x_i = f(z_1)\Delta x_1 + f(z_2)\Delta x_2 + \cdots + f(z_n)\Delta x_n.$$

As $n \to \infty$ and each $\Delta x_i \to 0$, this sum approaches a **limit**, which is independent of the mode of division of the interval (a, b) into parts Δx_i and the choice of the z_i in the subintervals Δx_i. This limit is denoted by the symbol

$$\int_a^b f(x)dx,$$

and is called the definite integral of $f(x)$ between the **limits** a and b; a is called the lower limit and b the upper limit of the integral.

The definite integral $\int_a^b f(x)dx$ has the properties:

(1) $\int_a^b f(x)dx = -\int_b^a f(x)dx,$

(2) $\int_a^b f(x)dx = \int_a^c f(x)dx + \int_c^b f(x)dx.$

If $\int_a^x f(x)dx = F(x)$, then $\frac{d}{dx}F(x) = f(x)$, so that $F(x)$ is an integral of $f(x)$.

If $F(x)$ is any known integral (**indefinite integral**) of $f(x)$, then

$$\int_a^b f(x)dx = F(b) - F(a).$$

This is the fundamental formula for the evaluation of a definite integral in terms of the indefinite integral found by inverse **differentiation**.

If the substitution $x = \phi(t)$ is made in the definite integral $\int_a^b f(x)dx$, we obtain

$$\int_a^b f(x)dx = \int_{t_1}^{t_2} f[\phi(t)]\phi'(t)dt,$$

where t_1 and t_2 are values of t such that $a = \phi(t_1)$, $b = \phi(t_2)$, and $\phi'(t)$ is the derivative of $\phi(t)$ with respect to t. (L.L.S.)

DEFINITE PROPORTIONS, LAW OF. The proportions by weight of the different elements which make up a chemical compound are the same in every sample of this compound. (See **Chemical Composition.**) (R.K.S.)

DEFLATION. In geology, deflation is the action of the wind in removing unconsolidated fine-grained **sediments** from a land surface. In 1895 dust fell in Missouri which must have come entirely from western Kansas and Nebraska, since the intervening country was covered with ice and snow. Dust from the Sahara has been blown over Germany and England (transported by air 2000 miles). (R.M.F.)

DEFLECTION. In engineering there are two common uses of the term deflection—deflection under load and deflection angle in surveying.

In general loads acting on an elastic structure cause a linear displacement of the several parts of the body, relative to their original position. This is known as deflection. Deflection is characteristic of all structures since all materials are elastic to a certain extent. Fig. 1 represents a beam which has been bent by the

Fig. 1. Deflection of a bent beam.

action of an external load. The deflections are the ordinates between the original and final positions of the **elastic curve.** The amount of deflection depends upon the load, stiffness of the material and the dimensions of the beam. The stresses cause the top fibers to compress and the bottom fibers to elongate. Since Hooke's Law states that **stress** is proportional to **strain** (deformation) as long as the stress is below the **proportional**

limit, the stress in the beam due to bending will be proportional to the deformation of the fibers. The beam will come to rest or be in a state of equilibrium after the application of a load, when at every section the **moment** of the internal stresses equals the moment due to the external load.

Deflection is an important element in the design of a load-carrying structure. If strength is the limiting condition for which a beam is designed, the design should be tested for deflection to make certain that the displacements are within allowable limits. Deflection rather than strength often governs the design. This is especially true in the design of beams for buildings, where small deflections, only, are permissible, because of the tendency of the deflection of the beams to crack terrazzo floors, plastered ceilings, etc. Solid rib and trussed bridges are also subject to deflection. Large steel bridges and building trusses are always cambered (see **Camber**) to counteract the effect of deflections.

The numerical value of the deflections of beams may be obtained by such methods as Double Integration, Conjugate Beam, Moment Areas and Work. The deflections in rib and framed structures may be computed by methods of Work, Elastic Weights, etc. The Williot-Mohr Diagram furnishes a graphical means for obtaining the deflection of trussed frames. These methods are given in any standard structural textbook.

Maxwell's Law of Reciprocal Deflections is very useful in the analysis of **indeterminate structures.** This theorem, which may be applied to any loaded structure, states that the deflection of a point A in an arbitrary direction AC due to any **load** P applied at B in a direction BD is equal to the deflection of B in the direction BD when the load P is applied at A in the direction AC. (See Fig. 2.)

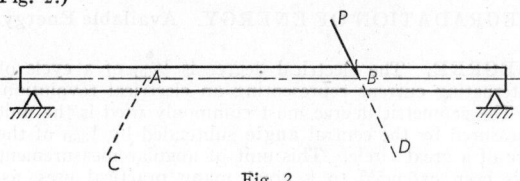

Fig. 2.

In surveying, the angle between a line and the extension of the preceding line is called the deflection angle. See Fig. 3. When the survey is a closed **traverse**

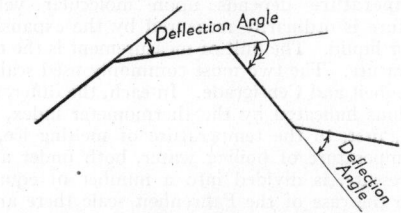

Fig. 3. Deflection angle.

the sum of the deflection angles must equal 360 **degrees. Circular curves** are frequently laid out by the method of deflection angles which is illustrated in Fig. 4.

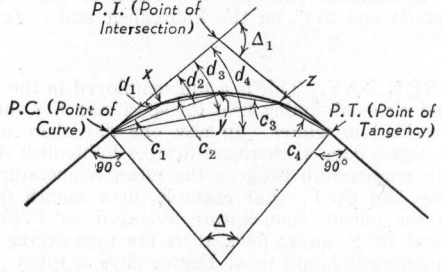

Fig. 4. Layout of curve by deflection angles.

Before it is possible to lay out points on the curve it is necessary to locate the P.C., P.I., and P.T. and obtain the central angle Δ which is numerically equal to the deflection angle Δ_1. Having the value of Δ_1, the deflection angles d_1, d_2, etc., may be calculated from the assumed chord lengths c_1, c_2, etc. The transit is first set up over the P.C. and sighted on the P.I. The deflection angle d_1, which gives the direction of the line from P.C. to x, is then turned off by means of the transit. The chord length c_1 is next measured on the ground by a steel tape (see Chain), which definitely fixes the positions of point x. Point y may be located by turning off the deflection angle d_2, and measuring the chord c_2 from the point x. Other points on the curve may be set in a similar manner. From the geometry of the figure it can be seen that the deflection angle (d_4) to the P.T. is equal to one-half of the central angle. (C.W.C., F.T.M.)

DEFLECTIVE FORCE. Wind Equations.

DEFORMATION. The change in the shape of a body which accompanies a stressed condition is called deformation or strain. The total amount of deformation in one direction is the total deformation. Unit deformation is the deformation per unit of length. Permanent deformation is known as set. If an axial load is applied to a body, the length and lateral (cross-sectional dimensions, are changed. **Poisson's** ratio is the ratio of lateral unit deformation to longitudinal unit deformation.

Deformation which is the result of a flexural stress (see **Flexure**) is called bending deformation. Shearing or shear deformation is caused by **shearing stress.** (C.W.C.)

DEGRADATION OF ENERGY. Available Energy.

DEGREE. The electrical degree is $\frac{1}{360}$ of a cycle of alternating current representing an electrical revolution.

The geometric degree most commonly used is the unit measured by the central **angle** subtended by $\frac{1}{360}$ of the arc of a great circle. This unit of angular measurement has been extended to a great many practical uses, as, for example, the reading of compass **bearings** in degrees. A right angle contains $90°$; a minute is $\frac{1}{60}$ of a degree, and a second is $\frac{1}{60}$ of a minute.

The thermal degree represents molecular activity, **in** that **temperature** depends upon molecular velocity. Temperature is ordinarily measured by the expansibility of a gas or liquid. The unit of measurement is the degree of temperature. The two most commonly used scales are the Fahrenheit and Centigrade. In each, the difference in the positions indicated by the thermometer index, when subjected, first, to the temperature of melting ice, then to the temperature of boiling water, both under atmospheric pressure, is divided into a number of equal degrees. In the case of the Fahrenheit scale there are 180 equal degrees, but for the Centigrade there are only 100. Thus a Centigrade degree is a larger unit than the Fahrenheit, the ratio being as 9 to 5. The melting temperature of ice is called zero on the Centigrade, and $32°$ on the Fahrenheit scale, and the boiling point of water at standard atmospheric pressure is $100°$ on the Centigrade and $212°$ on the Fahrenheit scale. (F.T.M., L.R.Q.)

DEGREE DAY. This is a unit employed in the heating and air conditioning field for specifying the nominal heating load in winter. In any one day there are as many degree days as there are degrees Fahrenheit difference in temperature between the mean temperature for the day and $65°$ F. For example, in a month during which the outside temperature averaged $20°$ F. for 10 days and $35°$ F. for 20 days, there are 1050 degree days. This number is found thus: Degree days $= 10(65 - 20) + 20(65 - 35) = 1050$.

Some years ago a statistical study determined that the fuel consumption for heating residences varied directly as the difference between some datum temperature and the outside temperature. A suitable datum was found to be $65°$ F. If the degree days are totaled for a heating period, the fuel consumption during that period, as compared with some other heating period, will be in the same proportion as the number of degree days in the two periods.

This is also a convenient unit for expressing the fuel consumption of a heating plant, since it eliminates weather as a variable. Operators sometimes compute fuel burned per degree day by the heating plant in order to make a comparison with previous operation of the plant. If the fuel quantity is divided by the degree days and by the amount of radiation, building volume, or some other unit of heating load, the result is comparable with similar figures from other heating plants without considering weather conditions. (F.T.M.)

DEGREE OF INDETERMINACY. Redundancy.

DEGREE OF REDUNDANCY. Redundancy.

DEGREES OF FREEDOM. This term has reference to the various ways in which a system may alter in respect to the configuration of its parts. For example, a system composed of three dimensionless particles has 9 degrees of freedom; for it takes nine independent coordinates to specify the positions of the particles in space, and their arrangement may therefore be changed in nine different ways. A single rigid body, on the other hand, has 6 degrees of freedom, since it may have motions of translation in three coordinate directions and it may also rotate about any one of the three coordinate axes through its center of mass. Any actual motion of the body is in general made up of all six, its linear motion being the resultant of three linear components and its rotation the resultant of three angular components. Each molecule of a diatomic gas has 7 degrees of freedom; viz., the six just mentioned for the molecule as a whole (regarded as a rigid body), and, in addition, one corresponding to the possible vibration of the two atoms toward and from each other. If the body is not rigid, the number of degrees of freedom may be virtually infinite. It should however be added that, because of restrictions imposed by the **quantum theory,** not all of the possible degrees of freedom can in general be expected to participate in changes of molecular energy. (See **Equipartition of Energy.**)

In statistics, the concept of the degrees of freedom, an extremely difficult one to explain mathematically, but very easy to understand intuitively, was introduced by R. A. Fisher in the **analysis of variance,** the χ^2 test of significance, **tests of independence in contingency tables,** and in **Student's** t test for the difference of two means. In the χ^2 test **for goodness of fit** the degrees of freedom n is found by the formula $n =$ number of classes in which the theoretical frequencies are 5 or more, less the number of **statistics** calculated from the **sample,** less one. It is evident that the last one must be subtracted since, when all the first class frequencies are inserted, the last one is uniquely determined in order to give the proper total frequency. We are subtracting the number of linear restraints when we subtract the number of statistics used from the sample in forming the theoretical frequencies. In an $r \times s$ contingency table the degrees of freedom will ordinarily be $(r - 1)$ $(s - 1)$. In Student's t test for the difference of 2 means the degrees of freedom will be $N_1 + N_2 - 2$, where $N_1 =$ number of **variates** in one sample, and N_2 the number of variates in the second sample, if the variates are not paired. In estimating a population variance σ^2_p we use $\dfrac{N\sigma^2_s}{N-1}$, where σ^2_s is the sample variance, N is the size of the sample, and $N - 1$ the degrees of freedom. In a **randomized block** experiment of r blocks and s varieties in each block, the degrees of freedom for the total sum of squares will be $rs - 1$, for the sum of squares among variety means $s - 1$, among block means $r - 1$, and for error $(r - 1)(s - 1)$. The degrees of freedom are important in all analysis of variance problems. Essentially the degrees of freedom represent the number of independent classes or conditions in certain statistical problems. (L.D.W., L.A.A.)

DEGU. Mammalia, Rodentia. A small animal of Chile and Peru. It resembles the rat but has moderately long ears and a tufted tail. (A.W.L.)

DEHUMIDIFICATION. Removal of saturated vapor from a gas is known as dehumidification. Most dehumidification is concerned with the removal of water vapor from air. This may be done chemically by the exposure of air to a dehydrating chemical such as calcium chloride. This method is rarely practical unless small quantities of air are to be treated.

As air can contain only a definite amount of moisture when completely saturated at some particular temperature, and since the amount of this vapor held at saturation decreases with the temperature, air that is nearly saturated with water vapor may have its water content greatly reduced if its temperature can be reduced. Reduction of temperature may be accomplished through surface cooling by passing the air over cold surfaces or through condensers, or cold water and air may be mixed together, as in the spray-type dehumidifier. (F.T.M.)

DEHYDROGENATION. Dehydrogenation is a chemical reaction involving removal of **hydrogen** from a compound. (See **Reactions Involving Oxidation-Reduction.**) (R.K.S.)

DE-ICERS, AIRCRAFT. Under certain atmospheric conditions ice may form on various parts of aircraft in flight. (See **Icing, Aircraft.**) When this happens to airplanes, flight may be difficult, if not impossible, to maintain, and tragic consequences are a possibility. Fortunately, there have been developed acceptable methods of ridding airplanes of ice accumulations which would interfere with flying characteristics. De-icing equipment might be classified as mechanical, thermal, or chemical. Wings, struts, empennage, and other parts tending to accumulate ice on a leading edge may be protected with mechanical de-icers consisting of rubber tubes that can be inflated and deflated for the purpose of cracking loose accumulations of ice. Propellers and windshields are better treated by heating or with liquid antifreeze solutions. A great amount of developmental work has been done in special refrigerated wind tunnels. Equipment was developed which, after extensive flight testing, has been applied as standard equipment on commercial and military airplanes.

Rubber de-icers contain a series of parallel tubes that can be inflated with compressed air. In operation, tubes of each de-icer are inflated and deflated rapidly, with a rest period between inflations. When not in operation, the tubes collapse against the surface of the airfoil section. While the tubes are flat on the surface, a coating of ice is permitted to form. After a coating is formed, the center tube is inflated, cracking the ice along the leading edge. As the center tube deflates, the outer tubes inflate, loosening and raising the cracked ice from the surface, so that the air stream can get under it and blow it away. Thus the aerodynamic lift of the wing is restored. The air is supplied from air pumps driven from the airplane's engines, with motor-driven distributor valves and pilot control valve inserted in the circuit. The de-icing equipment, consist-

ing of the rubber tubes and the air supply, may add between .1 and .3% to the weight of an airplane. When deflated, the de-icer tubes lie snugly on the leading edge, and do not greatly increase the drag. Only about 35% of the drag of an airplane is profile drag of the airfoils,

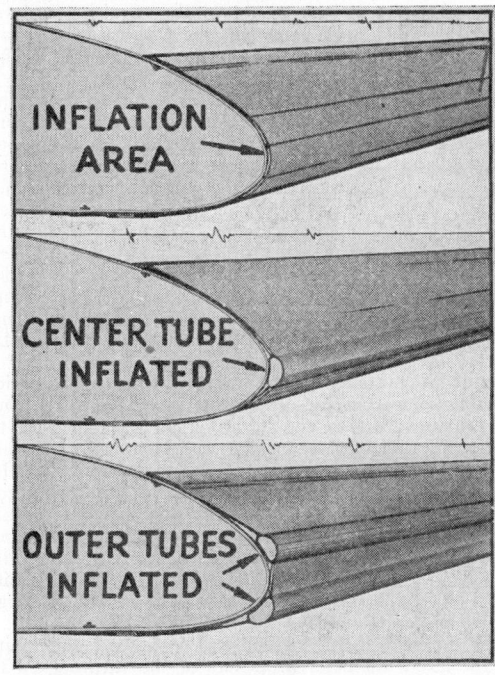

Fig. 1. Rubber de-icer tubes for the wing leading edge. (*B. F. Goodrich Co.*).

and the effect of de-icer equipment is to increase the profile drag about 10%. Hence, drag may be increased 3½%, and speed reduced by the square root of 3½%. When the de-icers are in operation, the drag varies; however, because of the nature of the inflation cycle and

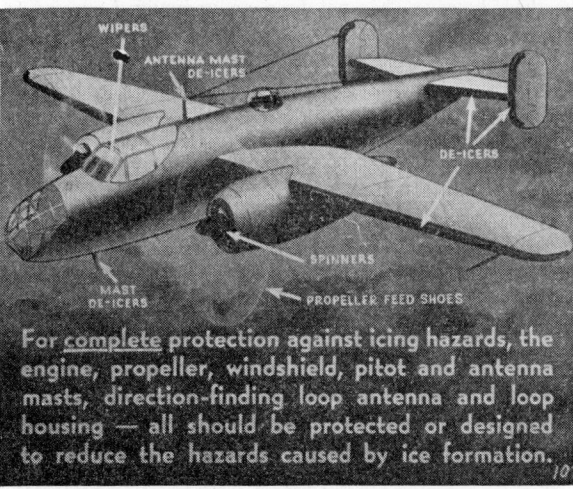

For **complete** protection against icing hazards, the engine, propeller, windshield, pitot and antenna masts, direction-finding loop antenna and loop housing — all should be protected or designed to reduce the hazards caused by ice formation.

Fig. 2. Points of application of localized de-icing. (*B. F. Goodrich Co.*)

the airplane's momentum, the speed is not greatly affected, and experience shows that during operation of the de-icers, the speed is reduced only about 1½%.

In spite of a considerable standardization on mechanical wing de-icers, experimentation with thermal de-icing has continued. This system appears to become more attractive as flight speeds increase and the slight protuberance of the inflation tubes more severely penalizes aerodynamic performance. Hot air, ducted to the interior of metal skin wings, can rid the wing of ice and not affect the airfoil profile in the least. Some extra drag must be accepted for this system uses an air scoop in the propeller stream to obtain the requisite ram for circulating hot air. When combined with cabin air heating (a necessity on airliners) the thermal de-icing equipment is estimated to increase airplane weight from ½% to 1½%. The air ducts add another item to the lengthy list of equipment the wing designer is struggling to incorporate inside the wing of the modern multi-engined airplane, and still find room for his primary structure.

Removal of ice from the wings and empennage is not enough. Other serious problems are those of propeller, carburetor, and windshield icing. Were one to list the icing up problems in order of the importance of their solution to safe scheduled airline operation, the order would probably be: (1) induction system (air scoop, duct, carburetor); (2) propeller; (3) wings and empennage; (4) windshield; (5) airspeed head (pitot-static tube).

Induction systems are commonly protected from ice plugging by carburetor air heaters which transfer from the exhaust gases to the incoming air enough heat to prevent the adiabatic expansion, and cooling by gasoline evaporation, from lowering the air temperature to the point where ice would begin to deposit on carburetor venturi, jets, throttle, or other protuberances in the air stream. For the protection of propellers, feed shoes distribute anti-icing fluid from a slinger-ring along the propeller leading edges. Ice is thrown off in small bits before it can cause unbalance or consolidate in large enough pieces to damage the fuselage.

It is doubtful if this method will prove to be adequate for the larger diameter propellers of the future (say 15 ft. diameter). Electric heating of propellers is possible and may be the solution to problems of preventing ice accumulations on the large-diameter, slow-speed propeller.

Powerful windshield wipers, when assisted by liberation of antifreeze, have been successful in keeping the forward windows in the pilot's compartment sufficiently free of ice to allow the pilot clear vision forward. This method, however, has not been an unqualified success. In the search for improvements, hot air heating applied to the interior of a double-glazed windshield has been used with satisfactory results. However, this is complicated, and the search continues for a better system of insuring clear forward pilot vision. Airspeed head openings if plugged with ice render the instrument inoperative. What is more hazardous is a partial blockage, or a build-up of ice on the air speed strut. These will cause a false airspeed reading, of which the pilot may be unaware. Airspeed heads are usually electrically heated where their accurate functioning is of prime importance, as in airliner piloting. This method is entirely successful. (F.T.M.)

DEIONIZATION TIME. Thyratron.

DELIQUESCENCE AND WATER ABSORPTION.
When a substance absorbs moisture upon exposure to the **atmosphere**, the substance is said to be deliquescent, and the phenomenon is known as deliquescence. At ordinary temperatures the vapor pressure of water varies as shown in the following tabulation.

If the solution in water of a substance has a lower water vapor pressure than corresponds to that of the atmosphere at the given temperature, water vapor condenses in the solution from the atmosphere until the water vapor pressure of the solution equals the water vapor pressure of the surrounding atmosphere.

Substances that are ordinarily deliquescent are **sulfuric acid** concentrated, **glycerol**, **calcium chloride**

TEMPERATURE, °C.	WATER VAPOR PRESSURE, IN MM. OF MERCURY	
	At Saturation	At 50% Humidity
0	4.6	2.3
10	9.2	4.6
20	17.5	8.8
30	31.8	15.9
40	55.3	27.7

crystals, **sodium** hydroxide solid, **ethyl alcohol** 100%. In an enclosed space these substances deplete the water vapor present to a definite degree. Other substances are used to accomplish this end by chemical reaction, e.g., **phosphorus** pentoxide (forming phosphoric acid), **boron** trioxide (forming boric acid).

Water is absorbed from non-miscible liquids by addition of such substances as anhydrous **sodium** sulfate, **potassium** carbonate, anhydrous **calcium** chloride, solid **sodium** hydroxide.

See **Efflorescence** for the converse phenomenon. (R.K.S.)

DELIRIUM. A mental disturbance characterized by disorientation, **hallucinations**, incoherent rambling speech, excitement and usually extreme restlessness. It occurs with high fever, especially in children, or with the **toxemia** of any severe illness. It is often present in various forms of mental disease and also may result from drugs or abuse of alcohol. Delirium tremens occurs in alcoholics; its chief features are hallucinations, agitation, and tremors of the extremities. It may occur in the chronic alcoholic following a temporary excess or sudden withdrawal of alcohol. It is particularly liable to develop in the alcoholic with **pneumonia**, in **erysipelas**, or following trauma. In uncomplicated delirium tremens the mortality is about 5–15%, pneumonia being the usual immediate cause of death. (R.S.M.)

DELTA. The terminal deposit of river-borne sediment in a lake or bay. So called because of its triangular

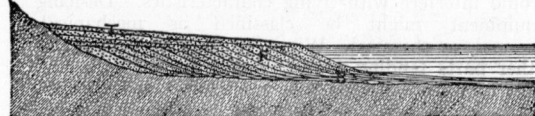

Ideal structure section of a delta. T, top-set beds; B, bottom-set beds; F, fore-set beds. (*Modified after G. K. Gilbert.*)

or delta-like ground plan. The cross-section or structure of a typical delta is shown in the accompanying sketch. Except in the case of small deltas only the top-set beds can be observed. In the case of **Paleozoic**, **Mesozoic**, and **Cenozoic** deltas it is extremely difficult to distinguish between the top-set, fore-set, and bottom-set beds, and the ultimate determination that a **sedimentary** formation is of delta origin depends largely on tracing the original source and areal distribution of the sediments, and the presence or absence of marine and terrestrial fossils. (R.M.F.)

DELTA CONNECTION. The Delta connection is one of the two most frequently used ways of connecting a three-phase **alternating-current circuit.** The other is the Y connection. A three-phase machine has three coils. These coils have six ends which must, in some way, be connected to the three wires of a three-phase circuit. The Delta connection, as illustrated in the accompanying figure, has the coils connected at three points corresponding to the three-phase circuit. When this is compared with the Y connection, it will become apparent that the line voltage in Delta connection equals the coil voltage, and that the line current in Y connec-

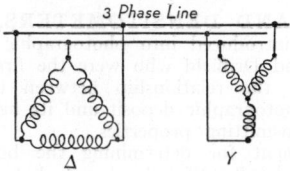

Δ-connection and Y-connection compared.

tions equals coil **current.** For a balanced Delta load the line current is $\sqrt{3}$ times the phase (or coil) current. (F.T.M., L.R.Q.)

DELTA RAYS. When various substances are bombarded with **alpha rays,** they are found to give off **electrons,** which sometimes move with great speed, and to which Sir J. J. Thomson has given the name delta rays. The delta electrons are thought to originate in a type of **ionization** of the bombarded substance, though by some it has been held that they are really thermions given off as the result of the very brief and very intense local heating of the body when the alpha particle strikes it. The rays appear to be emitted mostly in directions at right angles to that of the alpha-ray beam. Delta particles are themselves capable of causing ionization, as shown by the cloud tracks produced when alpha particles traverse a gas in a **cloud chamber.** (L.D.W.)

DELUSION. Hallucination.

DEMAND. In the nomenclature of economics, demand is the number of units of a commodity which will be purchased at a given price. It is the correlation of desire for the article, and ability to purchase it.

In a narrower technical sense, demand is that number of commodity units a source of supply is caused to produce, manufacture, or otherwise create. The maximum demand is an important index, especially in an industry unable to warehouse its product, and which, therefore, is required to have manufacturing capacity at least equal to the maximum demand. (F.T.M.)

DEMAND FACTOR. With reference to the electric service industry, demand factor is the ratio of a customer's maximum demand to his connected load, i.e., to the sum of the full-load ratings of all the electrical equipment he has connected to the supply line. (F.T.M.)

DEMANTOID. Garnet.

DEMENTIA. Mental deterioration. (See **Psychosis.**) (D.M.H.)

DEMENTIA PRAECOX. Schizophrenia.

DEMODULATION. Detection.

DE MOIVRE'S THEOREM. De Moivre's theorem is a rather remarkable mathematical result discovered by the French mathematician Abraham De Moivre (1667–1754).

De Moivre's theorem gives any power of a **complex number** in polar form:

$$[r(\cos\theta + i\sin\theta)]^n = r^n(\cos n\theta + i\sin n\theta).$$

This formula holds when n is a positive or a negative integer. It also holds when n is fractional, but may then be written in the more general form:

$$[r(\cos\theta + i\sin\theta)]^{1/n} =$$
$$r^{1/n}\left[\cos\frac{\theta + k\cdot 360°}{n} + i\sin\frac{\theta + k\cdot 360°}{n}\right],$$

where k takes the values of 0, 1, 2, $\cdots$, $n-1$ and where $r^{1/n}$ denotes the principal n^{th} root of r. This formula gives the n, n^{th} roots of any number. (L.L.S.)

DE MOIVRE-LAPLACE THEOREM. Let us consider a **Bernoulli probability function,** p = **probability** of success, $q = 1 - p$ = the probability of failure in a single trial, s = the number of trials, P_x = probability of x successes in s trials, $P_x = {}_sC_xp^xq^{s-x}$. When s is large it is difficult to evaluate P_x directly or to solve such problems as

$$\sum_{x=a}^{x=b} P_x, \quad 0 \leqq a \leqq x \leqq b \leqq s.$$

De Moivre overcame this difficulty by showing that as $s \rightarrow \infty$, the **probability function** P_x approaches a **normal curve** with **mean** sp and **variance** spq. Hence

$$P_x \approx \frac{1}{\sqrt{spq}\sqrt{2\pi}} e^{-\frac{(x-sp)^2}{2spq}}$$

or by areas

$$P_x \approx \int_{t_1}^{t_2} \phi(t)dt, \quad t_1 = \frac{x - \frac{1}{2} - sp}{\sqrt{spq}}, \quad t_2 = \frac{x + \frac{1}{2} - sp}{\sqrt{spq}},$$

$$\phi(t) = \frac{e^{-t^2/2}}{\sqrt{2\pi}},$$

the **normal probability function in standard units.** Similarly

$$\sum_{x=a}^{x=b} P_x \approx \int_{t_1}^{t_2} \phi(t)dt,$$

$$t_1 = \frac{a - \frac{1}{2} - sp}{\sqrt{spq}}, \quad t_2 = \frac{b + \frac{1}{2} - sp}{\sqrt{spq}}.$$

These approximations improve as s increases, when p and q are nearly equal and are better in the neighborhood of the mean than in the two tails. De Moivre gave additional terms and for greater accuracy the **Gram-Charlier Type A** gives generally as much accuracy as desired. Of course $\sum_{x=a}^{x=b} P_x$ may be found by means of the **incomplete Beta function** which is exact. However, generally the tables do not extend far enough if s is large. In some cases continued fractions possess definite advantages but at the expense of somewhat longer computations. (L.A.A.)

DEMOSPONGIAE. A class of **sponges** (Porifera) of complex structure, including the sponges of commerce and the fresh-water sponges.

The members of this class have siliceous spicules which are never 6-rayed, a spongin skeleton, or a combination of spongin and siliceous matter. Some species have no skeleton. The body plan is of the **rhagon** type.

Three orders are recognized:

Order Myxospongida. Simple sponges without skeletal structures.

Order Tetraxonida. Skeleton siliceous (see **Silicon**), sometimes with spongin. Fresh-water sponges are included in this order with many marine forms.

Order Keratosa. Skeleton of spongin fibers. **Spicules** absent. Commercial sponges belong here. (A.W.L.)

DENDRITE. A tree-like crystal formed during solidification of metals or alloys. Dendrites generally grow inward from the surface of the mold, extending branches from a central trunk in a manner resembling a fir tree. In alloys, the central portions of a dendritic crystal are richer in higher melting point constituents, while the outer portions consist of lower melting point material

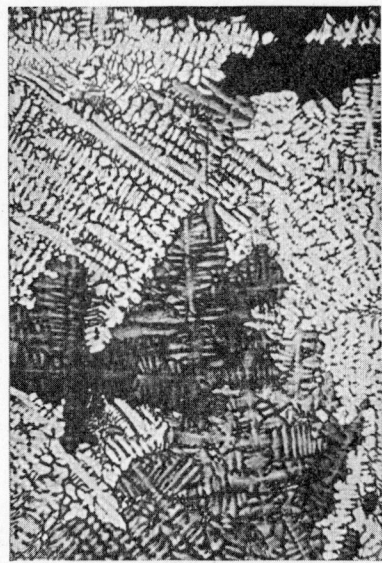

Dendrites in a brass casting (67% copper, 33% zinc) at 25 times magnification. Etched with a solution of $NH_4OH—H_2O_2$. (See *Metallography*.)

which is last to solidify. This form of segregation can usually be eliminated by diffusion during subsequent mechanical working and heat treatment. (See **Neuron**.) (R.H.H.)

DENDROCHIROTA. Holothuroidea.

DENEBOLA. Denebola (β **Leonis**) received its name because of its position in the **constellation** of Leo, the name Denebola being derived from an abbreviation of an Arabic phrase meaning "tail of the lion." In astrology, Denebola is one of the unfortunate stars, portending misfortune and disgrace. (W.K.G.)

DENGUE. "Breakbone Fever" or "Dandy Fever." An acute fever due to a filterable **virus** which is present throughout the circulation for one day preceding and 2 or 3 days following the onset of the disease. It is common in many tropical countries, and sudden outbreaks or epidemics occur in subtropical or even temperate climates. Several years ago there was a serious outbreak in the southern states, particularly in Florida. In 1928 there were widespread outbreaks in Greece and Egypt. In World War II, cases have occurred in Hawaii and New Guinea.

Dengue is transmitted by mosquitoes of the species *Aedes Aegypti,* which have become infected by biting a dengue patient during the first 48 hours of his fever. Twelve days are required for the mosquito to be able to transmit the disease to another human, who in turn becomes ill 4–10 days later. **Yellow fever** is transmitted by the same mosquito, and since both diseases are often prevalent in the same locality, some have speculated that similar organisms are responsible for the two diseases. However, in certain regions of the eastern hemisphere, dengue is common where yellow fever is unknown.

The disease lasts 7–8 days. It is characterized by the sudden onset of an acute fever with chill, severe headache, marked pain in the muscles and joints and profuse sweating. A skin eruption usually appears on the third day; the symptoms rapidly disappear and the patient seems well on the road to recovery. However, 48 hours later the symptoms reappear and last for 36 or more hours. Convalescence is protracted and mental depression is marked. Complications are rare, and the disease seldom causes death. One attack confers immunity for 6 months to several years. There is no specific treatment. The disease can, however, be prevented by anti-mosquito measures. (R.S.M., D.M.H.)

DENSITY AND DENSITOMETERS. The term *density* was introduced into photographic terminology by Hurter and Driffield who were the first to express quantitatively the relationship between the mass of silver in a photographic deposit and its light-absorbing and light-transmitting properties.

An instrument for determining the blackening, or photographic density, of a developed photographic plate or film is known as a densitometer. There is a great variety of types of densitometers, but the underlying principle of all of them is the same. The intensity of the radiant energy from a source which is maintained as nearly constant as is possible, is compared by any of the methods of **photometry** with the intensity of the same radiation after it has passed through the portion of the photographic plate whose density is required.

For the determination of the density of a very small area on a photographic plate, such as the image of a spectral line or a star, the instrument employed is usually referred to as a "microdensitometer." In such instruments the physical type of photometer, e.g., one employing a **photoelectric cell** or a **thermopile**, is frequently employed. (C.B.N., W.K.G.)

DENSITY AND SPECIFIC GRAVITY. The density of a substance is its mass per unit volume, usually expressed in grams per cubic centimeter. The specific gravity of the substance is the ratio of its density to that of water, usually at 4° C., or 20° C., or 60° F., in the same units, and is therefore an abstract number independent of units.

To determine the density of a given substance, it is necessary only to ascertain the volume of a specimen whose mass is known by weighing. This may be obtained from measurements on the dimensions of the specimen, or, in the case of a liquid, by the use of a **pycnometer** or specific gravity bottle. For solids a more precise method is to measure the buoyant force, upon the specimen, of a liquid of known density in which it is immersed, or by enclosing it in a specific gravity bottle and determining the volume by displacement. The Mohr-Westphal balance is especially designed to give densities of liquids by the buoyant force on a solid sinker of known volume. The **hydrometer** may also be used for quick determinations of liquid densities. The density of a gas is best obtained by weighing a specimen of it in a large, light bulb of known capacity, concurrently observing the temperature and pressure to which the gas is subjected, much as the pycnometer is used for liquids.

A brief table of densities, all in grams per cubic centimeter, is appended. For gases the densities are at standard temperature and pressure.

Substance	Density	Substance	Density
Air	0.001293	Gold	19.3
Alcohol	0.794	Hydrogen	0.0000899
Aluminum	2.70	Iron	7.86
Carbon dioxide	0.001977	Lead	11.3
Chlorine	0.003214	Mercury	13.55
Copper	8.90	Nitrogen	0.001251
Cork	0.24	Oxygen	0.001429
Gasoline	0.67	Platinum	21.45
Glass	2.4–2.8	Silver	10.5
Glycerine	1.27	Water (4° C.)	0.999973

The term density is also applied to the blackness of the image on a photographic plate or film. (See **Density and Densitometer**.) (L.D.W.)

DENSITY OF AIR. Total mass of air per cc. is the density of air. It is given by the relation

$$density = \frac{0.0012930}{1 + 0.00367t}\left[\frac{B - 0.378e}{760}\right]$$

where t = Temperature, °C.
B = Barometric pressure expressed in mm. of mercury.
e = The partial pressure of water vapor in the air.

Air density at standard conditions of 0° C. and 760 mm. of Hg is 0.0012930 grams per cc. of air free from water vapor. (P.E.K.)

DENSITY RANGE. A term commonly used in color photography to denote the range of densities found in a given negative or positive. Measures of density to obtain the range are generally, although not always, made on achromatic areas. The difference between the maximum and minimum densities in a particular negative or positive is its density range. Under special conditions a measure of the density range gives the same information as **gamma** but this correlation must not be assumed to hold unless complete sensitometric data in any individual case show it to be true. (H.C.C.)

DENTARY APPARATUS. Aristotle's Lantern; Tooth.

DENTINE. Tooth.

DENTITION. The form and arrangement of the teeth in **vertebrates.** Teeth are so intimately related to the food that they are involved in the fundamental adaptations of the animal. In connection with the study of the many adaptations of teeth, terms have been coined which apply in some cases either to the **tooth** itself or to the entire dentition, while others apply to the dentition in general.

The primitive form of tooth is apparently that of the **sharks,** which has a principal sharp flattened point and in some cases fairly prominent lateral points. These teeth are arranged in several rows and are renewed as needed. In other fishes and in the amphibians teeth are also developed in large numbers and in some forms occur elsewhere in the mouth than on the jaws. They are named for the part of the skull with which they are associated, as the vomerine teeth.

In the **reptiles** and **mammals** the simpler condition of a row of teeth along each jaw prevails. The teeth may be entirely conical as in the reptiles or of various types, as in the mammals, and may be indefinitely renewable or limited to one or two sets. Where teeth are of more than one kind the dentition is said to be heterodont. The forms of teeth include the sharply conical canines, the sharp-edged cutting incisors, and the broad grinders, which include premolars and molars. Renewal is unlimited in the reptiles but in the more highly specialized mammals only one or two sets appear normally. Monophyodont dentition consists of one set and diphyodont includes a temporary set of milk teeth which is replaced by a set of permanent teeth, as in man.

The numbers of teeth of different kinds are expressed in a dental formula as a distinctive characteristic of mammals. In this formula the teeth of one-half of each jaw are listed in this order: incisors, canines, premolars and molars, and those of the upper jaw are placed above those of the lower jaw. Thus the dental formula of man is 2123/2123 and that of the woodchuck is 1023/1013. The zeros in the latter formula indicate the absence of canines.

The position of the teeth in the jaw is also sometimes indicated by a special term. When placed along the edge of the jawbone the dentition is said to be acrodont, and when placed along the inner margin, pleurodont. (A.W.L.)

DENUDATION. Cycle of Erosion.

DEPARTURE. Any **course** and distance covered by a ship may be resolved into two components at right angles to each other. One of these components will be parallel to a **meridian** while the other will be along a parallel of **latitude** and is known by the term departure. If the lengths of these components are expressed in **nautical miles** the component along the meridian may be immediately converted into difference of latitude because a minute of arc of latitude is practically equal to one nautical mile. The component along the parallel of latitude, expressed in **nautical miles** and referred to as easting or westing, is known as the departure.

For the solution of various problems in **dead reckoning** and the **sailings** it becomes necessary to convert departure into difference of **longitude.** The accompanying figure illustrates the problem. In the figure

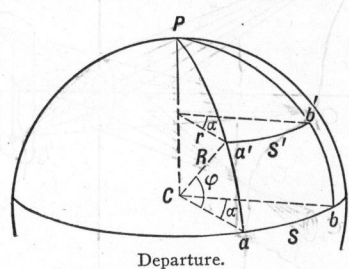

Departure.

we have S' a departure measured along the parallel of latitude ϕ between the meridians $Pa'a$ and $Pb'b$. The arc, S, of the **equator** represents the difference of longitude corresponding to the departure S'. Planes are passed through S' and S perpendicular to the axis of the earth, PC, and radii are drawn in these planes to include the angles subtended by S' and S. Call R the radius of the earth and r the radius of the arc S'. The angles subtended by S and S' must be equal and we have at once $S/R = S''/r$. By the definition of latitude (assuming the earth as spherical) we have the plane angle $aCa' = \phi$ and, from the definition of the trigonometric functions, $R/r = \sec \phi$. Therefore we have $S/S' = R/r = \sec \phi$; or $S = S' \sec \phi$. Since S' is expressed in nautical miles, which are practically equivalent to a minute of arc on the equator, we have at once $S = S' \sec \phi$ as the difference in longitude corresponding to the departure S'; the difference in longitude being thus expressed in minutes of arc. (W.K.G.)

DEPENDENT SYSTEMS OF LINEAR ALGEBRAIC EQUATIONS. Linear Algebraic Equations, Systems of.

DEPENDENT VARIABLE. Functions.

DEPHLEGMATOR. A partial condenser used in distillation. It also acts as a partial theoretical plate. (See **Distillation.**) (D.E.M.)

DEPRESSED EQUATION. Polynomial Equations.

DEPRESSION. A region over which atmospheric pressure is lower than surrounding regions. A depression, of necessity, has cyclonic winds. (See **Cyclone.**) (P.E.K.)

DEPTH OF FIELD. In photography the term depth of field refers to the distance over which satisfactory definition is obtained when the lens is in focus for a certain distance. If, for example, a lens is in focus for an object at a distance of 25′ and the definition is satisfactory on objects from 20–40′, the depth of field extends from 20–40′. Depth of field is frequently but incorrectly termed depth of focus, which is the range of image distances corresponding to the range of object distances covered by the depth of field.

The depth of field depends upon: 1. The standard adopted for "satisfactory" definition. 2. The distance of the plane on which the lens is in focus. 3. The focal length of the lens. 4. The relative aperture (f/number). So far as the last three are concerned we may note (1) that the depth of field increases with the distance of the plane on which the lens is in focus and (2) that it becomes less as the focal length or (3) as the aperture increases.

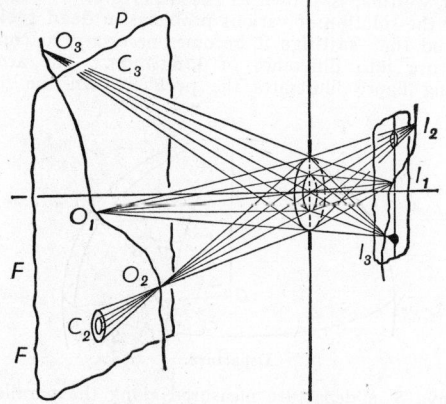

Let O_3, O_2, O_1 represent an object, parts of which are at different distances from the lens L. Suppose the lens to be focused on O_1, a point image I_1 will be formed on the focusing screen, or sensitive plate. With the focusing screen, or the sensitive plate, at this point it is clear that the image of O_2 is formed at I_2 *behind* the screen and that of O_3 at I_3 in *front* of the screen. In other words, the position of the image point varies with the distance of the object point and the lens cannot produce a point for point image upon a plane surface, such as the sensitive film, unless the subject itself is a plane. When this is not the case, the image of points nearer or further from the lens is a circular disk rather than a point.

Any disk, however, will appear to the eye as a point if the viewing distance is sufficiently great. At a distance of 10", for example, a circle with a diameter of $\frac{1}{100}$" appears as a point to the average eye. On an angular basis this corresponds to about 2 minutes of an arc.

Since any unsharpness in any part of the negative is increased when an enlargement is made, the disk, or circle of confusion, in an enlargement is equal to the diameter of the disk in the negative multiplied by the degree (times) of enlargement. Thus, if the largest circle allowable for sharp definition is assumed to be $\frac{1}{100}$" and the print is a $5\times$ enlargement, the maximum diameter of the circle of confusion in the negative is $\frac{1}{100} \times 5$ or $\frac{1}{500}$" and the depth of field should be calculated accordingly.

If it is assumed that the viewing distance is the distance at which the proper perspective is obtained, i.e., a distance equal to the focal length of the lens, then the circle of confusion can be expressed as a fraction of the focal length of the lens. The value generally used is 0.00058 or 1/1720 of the local length.

If u = distance focused on,
   θ = angular size of circle of confusion (1/1750 of the focal length),
   τ = effective diameter of lens or the focal length divided by the f/number,

then the nearest point sharply defined in front of the plane in focus is

$$\frac{u^2 \tan \theta}{\tau + u \tan \theta}$$

and the greatest distance beyond the plane in focus is

$$\frac{u^2 \tan \theta}{\tau - u \tan \theta}$$

Tables of depth of field are included in most camera manuals and in reference books. (C.B.N.)

DERIVATIVE OF A FUNCTION OF ONE VARIABLE. The rate of change of a function is expressed by the derivative of the function.

Let $y = f(x)$ be a given **function** of one **variable** and let x_1 be a chosen value of x, and let Δx be an increment of x to be added to x_1, then Δy will denote the corresponding increment of y:

$$\Delta y = f(x_1 + \Delta x) - f(x_1).$$

Form the increment ratio $\Delta y/\Delta x$:

$$\frac{\Delta y}{\Delta x} = \frac{f(x_1 + \Delta x) - f(x_1)}{\Delta x}.$$

Let $\Delta x \to 0$, then $\Delta y/\Delta x$ will usually approach a limit which is called the derivative of y with respect to x at the value $x = x_1$. Hence:

The derivative of a function $f(x)$ is the limit approached by the ratio of the increment of $f(x)$ to the increment of x when the increment of x approaches the limit 0.

The process of finding the derivative of a function $y = f(x)$ is called differentiation.

The derivative of $y = f(x)$ with respect to x is denoted by various symbols; sometimes $D_x y$ is used, sometimes $\frac{dy}{dx}$, or $f'(x)$, or y'.

If $\lim_{\Delta x \to 0} (\Delta y/\Delta x)$ is formed at the point $x = x_1$, it is called the derivative at $x = x_1$; but if the limit is taken for the general point x, the result is sometimes called the derived function of $f(x)$.

The rate of change of any function $f(x)$ with respect to the independent variable x is defined as the derivative $D_x y$ or dy/dx.

Special interpretations of the derivative are:

1. As the **slope** of a curve: If x and y represent rectangular coordinates of a variable point, and if a function $y = f(x)$ represents a curve, then the slope of the tangent to the curve at a point $x = x_1$ is represented by the value of the derivative of $f(x)$ at $x = x_1$.

2. As the speed of a moving particle: If s denotes the distance traversed by a particle moving in a straight line in time t, then a function $s = f(t)$ represents the law of motion, and the derivative of $f(t)$ at $t = t_1$ represents the speed of the particle at the instant t_1.

For a function $y = f(x)$, the ratio $\frac{dy}{dx}/y$ is called the relative rate of increase of y with respect to x; the product of this by 100 is sometimes called the percentage rate of increase.

If $f'(x)$ is positive throughout an interval (a, b), then $f(x)$ continually increases as x increases from a to b. If $f'(x)$ is negative throughout (a, b), then $f(x)$ continually decreases as x increases from a to b. (L.L.S.)

DERIVED CURVES. Higher Derivatives.

DERIVED FUNCTION. Derivative.

DERMAPTERA. The earwigs. An order of **insects** made up of species whose forewings, when present, are short leathery wing covers, and whose abdomen bears a pair of appendages like the jaws of a forceps at the tip. The common name is based on the supposition that they enter the ears of human beings. They are plant-eating, insect-eating, and probably in some species scavengers. (A.W.L.)

DERMATITIS. An irritation or inflammation of the skin due to infection, allergic reactions to a variety of agents, **drugs**, heat, cold, **x-rays**, or radium. Dermatitis may also be seen in metabolic and deficiency states. (D.M.H.)

DERMATOLOGY. The study, diagnosis, and treatment of the skin and its diseases. (D.M.H.)

DERMESTID. Insecta, Coleoptera. Any of the small **beetles** of the family Dermestidae, including the **buffalo carpet moth.** They damage clothing, woolen articles, and museum specimens. (A.W.L.)

DERMIS. 1. The corium or inner layer of the **skin** of vertebrates. 2. The inner layer of the **cuticula** of insects. (A.W.L.)

DERMOLITH. A term proposed by T. A. Jaggar in 1917 for ropy, wrinkled or **pahoehoe** type of basic lava, such as occurs in the volcanic islands of Hawaii. (R.M.F.)

DERMOPTERA. An order of mammals containing only the **flying lemurs** of the Malayan region. They differ from other mammals in dentition. Like the flying squirrels, they have folds of skin along the sides which support them in the air. Also called cobego, kaguan, and other names. (A.W.L.)

DERRICK. A derrick is a device for lifting heavy objects. It is used in all types of construction work when heavy material must be raised into position. Derricks are used extensively in structural steel erection.

The simplest form of derrick is the gin pole which consists of a vertical or nearly vertical mast held in

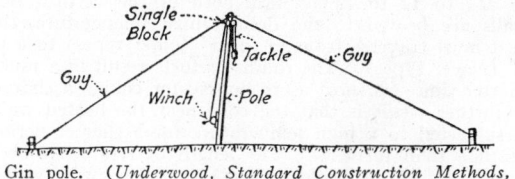

Gin pole. (*Underwood, Standard Construction Methods, McGraw-Hill Co.*)

position by at least four twisted wire cables or ropes known as guys. A pulley or some other form of hoisting tackle attached to the top of the mast completes the equipment. The gin pole is frequently used in the erection of steel frames for small buildings when all of the members can be placed in position without moving the derrick, as a whole, vertically, to some point on the frame.

The shears derrick is essentially a gin pole derrick which is constructed in such a manner that only two

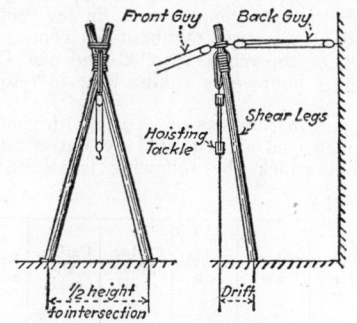

Shears. (*Underwood, Standard Construction Methods, McGraw-Hill Co.*)

guys are necessary. The "mast" is a V-shaped frame made of two poles which may, in some cases, be connected by transverse bracing. Since this form gives it lateral stability only two guys are required. This derrick is often used for the erection of heavy machinery

which is lifted by means of hoisting tackles connected to the "mast" at the point of intersection of the poles.

The guy derrick is made up of a **boom** and a mast that are connected at the top by a **cable** which allows

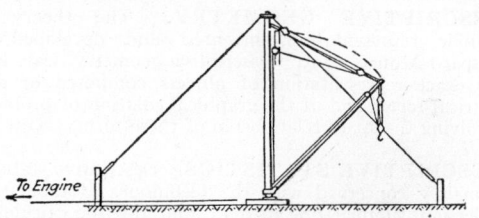

Engine-operated guy derrick. (*Underwood, Standard Construction Methods, McGraw-Hill Co.*)

the boom to be raised or lowered. The mast, which is longer than the boom, is held firmly in a vertical position by guys although the derrick, as a whole, may rotate through a horizontal angle of 360°. The load is applied to the boom by means of another cable which may be used to raise or lower the load without moving the boom in a vertical plane. This type is used in the erection of large steel building frames since it can be readily set up at any floor level and has a large range of movement.

The stiff-leg derrick is similar to the guy derrick except that it does not require guy wires or ropes to

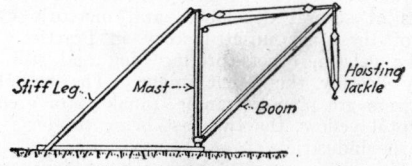

Engine-operated stiff-leg derrick. (*Underwood, Standard Construction Methods, McGraw-Hill Co.*)

support the mast and the boom has a horizontal swing of only about 240°. The mast is supported by struts set at right angles to each other. The lifting action is the same as that for the guy derrick.

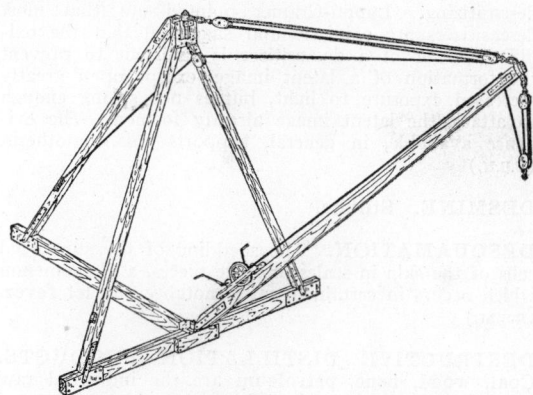

A Jinniwink—hand operated. (*Underwood, Standard Construction Methods, McGraw-Hill Co.*)

The A-derrick or Jinniwink is a modification of the stiff-leg derrick. It is used for lifting comparatively light loads. (C.W.C.)

DESCARTES' RULE OF SIGNS. If in passing from one coefficient of a **polynomial equation** to the next, there is a change of sign from plus to minus or from minus to plus, this is called a variation of sign; a succession of two like signs, either both plus or both minus, is called a permanence of sign.

Descartes' rule is: The number of positive **roots** of a polynomial equation $P(x) = 0$ with real coefficients is not greater than the number of variations of sign in the polynomial $P(x)$, and the number of negative roots is not greater than the number of variations of sign in the polynomial $P(-x)$. (L.L.S.)

DESCRIPTIVE GEOMETRY.

The theory of graphic representation, invented and developed by Gaspard Monge, 1795. Descriptive geometry deals with the exact representation of objects composed of geometrical forms and of the graphical solution of problems involving the space relationship of these forms. (H.C.H.)

DESCRIPTIVE STATISTICS.

Descriptive statistics is mainly concerned with the techniques of tabulation, collection, summarizing of data, and also the calculation of measures which adequately describe a **sample**. It is not concerned with the theory of statistical inference. (L.A.A.)

DESENSITIZING.

In 1920 Dr. Luppo-Cramer discovered that bathing an exposed film or plate in pheno safranine reduced its sensitiveness to light to such an extent that even high-speed orthochromatic materials may be developed without danger of fog in a bright yellow or orange light. With panchromatic materials phenosafranine was less effective but other desensitizers for use with panchromatic materials were soon found.

The number of desensitizing substances is fairly large. In addition to phenosafranine the list includes picric acid, aurantia, chrysoidine, methylene blue, the oxidation products of several developers and mercuric cyanide. Many of these are unsatisfactory in practice. Some stain the gelatin, others produce fog and still others partially destroy the latent image. The best known desensitizers are phenosafranine, pinakryptol green and pinakryptol yellow, the two last being products of the I.G. Farbenindustrie.

Desensitizers are used either as a preliminary bath prior to development or as an addition to the developer. Usually the desensitizer is more effective when added to the developer. Some, however, cannot be used in the developer; either the desensitizer is precipitated or it is destroyed by the sulfite contained in the developer.

Little is known of the chemical reactions involved in desensitizing. Luppo-Cramer pointed out that most desensitizers are oxidizers and suggested that the oxidizing power of a desensitizer is sufficient to prevent the formation of a latent image, except upon greatly increased exposure to light, but is not strong enough to attack the latent image already formed. The evidence available, in general, supports this hypothesis. (C.B.N.)

DESMINE. Stilbite.

DESQUAMATION.

The shedding of the superficial cells of the skin in scales or large pieces, a phenomenon which occurs in certain diseases, notably **scarlet fever**. (R.S.M.)

DESTRUCTIVE DISTILLATION PRODUCTS.

Coal, **wood**, bone, **petroleum** are the industrial raw materials most commonly subjected to the process of destructive distillation. In this process the materials are heated to various degrees of temperature without access of air (pyrolysis) but with provision for recovery and collection of the products desired. The reactions are complex due in large part to the complexity of composition of the material treated. For the treatment of petroleum see **Hydrocarbons**. The destructive distillation of coal, wood, bone and carboniferous shale will be discussed under the topics (1) Raw material, (2) Conditions of temperature and time of treatment, (3) Products.

Coal, Destructive Distillation. 1. Destructive distillation of coal is conducted primarily to produce coke or coal gas, and secondarily for coal tar, **benzene** and

toluene from the coal gas and coal tar, and **ammonia** from the coal gas and water condensate. Bituminous coals are found to be most suitable for this purpose. These coals have a fuel value of 12,000 to 15,000 B.T.U. per lb., that is 24 to 30 million B.T.U. per short ton (2000 lbs.). Reports of analyses made by the United States Bureau of Mines on samples of coking coal as received from 23 coal fields of the United States show:

Moisture	1–10%
Volatile matter	25–38
Fixed carbon	71–47
Ash	2.5–9
Nitrogen	1.4–2
Sulfur	0.5–1.5

From such coals there would be obtainable:

Furnace coke	60–80%	of the weight of coal
Coal gas	2700–3500	B.T.U. per lb. of coal
Surplus after heating retorts	1600–2100	B.T.U. per lb. of coal
Coal tar	6–12	gals. per ton of coal
Benzol	1.5–4	gals. per ton of coal
Nitrogen		
As ammonium sulfate	17–25	lbs. per ton of coal
As sodium cyanide	1.8 (appr.)	lbs. per ton of coal

2. Conditions. Since coal and coke are poor conductors of heat, it is necessary that the thickness of the layer heated be such that the coking temperature, at the places in the charge that are most distant from the source of heat, shall be attained in an economical period of time. In the history of the industry the horizontal width of retort has been gradually diminished from 24 to 22″ to 12 to 14″. Since both of the vertical side walls are heated to the desired high temperature, the heat must traverse from 12 to 11″ (older types) to 6 to 7″ (newer types). The thinner retorts result in a much shorter time—as short as 10 hours—for coking a charge. A further result is that the coke near the heated walls is subjected to a high temperature for a shorter period of time than formerly. The length of retort has been extended from 20′ (older types) to 40′ (newer types). The length is determined by mechanical considerations of removing the charge economically. The height has been increased from 6′ (older types) to 13′ (newer types). The retorts are charged from the top, and discharged by pushing the coke out of one end by a pusher entering the other end.

The heat is supplied by burning gas in flues between each retort, and batteries containing up to 90 retorts are in use. In 1919 the United States Bureau of Standards reported tests made in Koppers retorts. Temperature of 1225° C. maximum to 1200° C. minimum were observed in the heating flues and the operating temperature is close to these temperatures; in the center of the charge, and at distances of about 2′ from the bottom and the top of the retort 1010° C. and 980° C., respectively, in 10.5 hours—the coking time in retorts about 14″ thick.

The changes in composition of gas with time of heating of a given coal are reported by Bacon and Hamor (1922) from which the following is selected or calculated:

Hours Coal Carbonized	Methane	Hydrogen	Ethylene	Carbon Monoxide	Carbon Dioxide	Benzene	B.T.U. per cu. ft.
2	36.7	42.5	4.0	0.9	3.3	1.8	605
4	34.5	48.8	3.1	2.9	2.3	1.1	570
6	33.6	50.1	2.8	3.0	2.2	1.0	560
8	33.7	53.8	2.5	3.4	1.4	0.6	560
10	31.2	47.1	1.8	2.8	2.3	0.4	480
12	33.4	50.7	1.9	4.2	1.5	0.4	525
15	33.2	53.4	1.8	3.9	1.9	0.3	525
19	26.1	55.8	1.2	4.7	1.1	0.0	450

(24 hours time of carbonization. The amount of gas evolved at each stage of heating is not reported.)

The recent types of retorts use **silica** brick for the walls. These have been found most satisfactory on the basis of thermal, mechanical and economic considerations. Two-thirds of the heat consumed passes through the walls of the retort to the charge, and one-third is lost from the flues, the total actually required being about 1225 (1100–1400) B.T.U. per lb. of coal carbonized. Beehive ovens, in which coal was coked with loss of other products, and the small retorts used specially in making coal gas have been largely displaced by by-product recovery retorts as described.

3. Products. *a*. Coke. The quality of coke depends upon the kind of coal used, the amount of moisture present in the coal, the fineness of the coal (the coal is pulverized to a fineness approximately 85% passing through ⅛″ mesh), the width of the retort, and the temperature of the walls of the oven. The time of coking is in practice determined by the above factors. The ash present in the coal remains in the coke, and the percentage of ash is therefore higher in the coke than in the coal in proportion to the loss of volatilized matter. For metallurgical purposes the **sulfur** and **phosphorus** contents of the coke are considerations. As to the sulfur, this is present in three forms in the coal, namely, as organic sulfur compounds (from 0.5 to 2% of the coal), **iron** disulfide (pyrite) (from 0.1 to 8%), and **sulfates**. About two-thirds of the total sulfur is retained in the coke, and one-third removed in the volatilized matter. Data regarding **phosphorus** is regularly examined by some operators. **Nitrogen** exists in coal in two forms, namely, **amino** nitrogen (group —NH₂) and nitrogen-ring compounds

$$\left(\text{for example, } \textbf{pyridine} \hexagon_N\right). \text{ The total nitrogen in}$$

coal is 1.4 to 2%, of which 40 to 50% remains in the coke (possibly as a nitride of **carbon**), 18% is recovered from the volatilized matter as ammonia, 3% as pyridine and 1% as **cyanide**, while 25–30% escapes as free nitrogen in the gas. The supporting or crushing strength of coke is an important consideration for metallurgical purposes. *b*. Coal gas. The composition at various stages of heating, and the surplus amount after heating the retorts have been noted. **Ammonia** is recovered by passing the gas through water and then distilling the solution, and the **benzene** and **toluene** by passing the gas through a non-volatile absorbing oil and then distilling the solution. The benzene-toluene distillate makes up 90–95% of light oil, so called, the remaining 10 to 5% coming from the coal tar by distilling up to 170° C. **Sulfur** compounds are removed by passing the gas through layers of hydrated **ferric** oxide spread over a large surface such as wood shavings, or by passing the gas through a solution of **sodium** carbonate. *c*. Light oil. The light oil above mentioned, furnishes when distilled approximately the following materials and amounts:

	GALLONS PER TON OF COAL CARBONIZED
Pure benzene	1.6 to 2.0
Pure toluene	0.4 to 0.6
Xylenes and light solvent naphtha	0.4
Unsaturated hydrocarbons	0.3
Heavy hydrocarbons and naphthalene	0.25
Wash oil, for benzene absorption from coal gas	0.35

d. Coal tar. Upon distillation, coal tar furnishes, in addition to the above-mentioned light oil, up to 170° C. (0.2% of coal tar), approximately 10% of its weight distilling between 170 and 230° C. (one-third **phenol**, two-thirds **cresylic acid**), 10% between 230 and 270° C. (**anthracene** oil), and 55% pitch at 350° C. When the distillation is conducted to 1200° C. about 17% coke or retort carbon remains. (See **Coal Tar Products and Intermediates**.)

Wood, Destructive Distillation. 1. Destructive distillation of wood is conducted primarily to produce charcoal, **acetic acid** or **methyl alcohol** and secondarily for wood tar. Certain deciduous or hard woods have been found most suitable for this purpose. Freshly cut or green wood usually contains about 60% water calculated on the weight of dry wood but this may reach as high as 100%. Dry wood is mainly lignocellulose with little ash.

ANALYSES OF SAMPLES OF WOOD DRIED IN OVEN AT 105° C.

Kind of Wood	Cellulose	Lignin	Ash
Redwood	48.5%	34.2%	0.2 %
Western yellow pine	57.4	26.7	0.45
Longleaf pine	58.5	....	0.4
Western white pine	59.7	26.4	0.2
Sugar maple	60.8	....	0.4
Douglas fir	61.5	....	0.4
White spruce	61.9	....	0.3

The United States Forest Products Laboratory, Madison, Wisconsin, has conducted extensive researches in connection with wood, its treatment, and its uses. Carbonization of wood in piles partially covered with earth or in kilns where by-products are not recovered has been largely displaced by by-product recovery retorts. The production of acetic acid from **acetylene**, of acetone from acetic acid and from starch, and of methyl alcohol from **carbon monoxide** and **hydrogen** has been instrumental in furnishing severe competition for the wood distillation industry.

2. Conditions. Klason (1908–10) reported very little decomposition of wood by heating rapidly to temperatures below about 250° C. But chemical changes having notable effects on the strength of wood occur even below 100° C. when wood is thus exposed over a long period of time. Destructive distillation begins at 250° C., is exothermic (i.e., evolves heat) at 280° C., and is finished at about 350° C., when practically all of the acetic acid, wood alcohol and wood tar have been evolved. Above 350° C. the charcoal decomposes into gas and tar, leaving about 30% charcoal residue. It is interesting to note that **cellulose** forms no methyl alcohol upon destructive distillation. High vacuum rapid distillation of wood yields 44% wood tar, transparent, of light color and 20% charcoal; in contrast the yield when heated 14 days at atmospheric pressure is 2% wood tar, dark, and 40% charcoal.

The yield of materials upon the destructive distillation of certain hard woods is as follows:

Kind of Wood	Acetic Acid	Methyl Alcohol	Wood-tar	Charcoal
Beech	6.0%	2%	10%	41%
Maple	5.5	2	11	40
Ash	5.0	2	10	41
White oak	4.5	1.5	7	47

Coniferous or soft woods, upon destructive distillation, yield characteristic wood oils and wood tars; the recovery is complete from the contained volatile oils, about 75% complete from the contained rosin and slight from the fiber of the wood.

About 50% of wood ash is **potassium** carbonate.

3. Products. *a.* Charcoal is used as **fuel** and a reducing agent in metallurgy, where its low ash content and the absence of sulfur and phosphorus are important considerations. Its porosity is great, and crushing strength low. *b.* Wood gas. The composition of the gas evolved when wood is subjected to destructive distillation is of no commercial value. It consists of about 60% **carbon dioxide,** 30% **carbon monoxide,** 3% each **hydrogen** and **methane,** and about 8000 cu. ft. are obtained per cord (128 cu. ft.) of wood, that is, about 240 cu. ft. per 100 lbs. of wood. The gas (about 135 B.T.U. per cu. ft.) is burned to furnish heat to the retorts. *c.* Water condensate. In this is contained the **acetic acid,** wood alcohol (methyl alcohol to acetone in the ratio of about 100 to 8 parts by weight), other **ketones,** some **acetaldehyde** and ill-smelling oils. The amount of crude pyroligenous acid, so called, is from 200 to 250 gals. per cord of wood. Crude wood alcohol is obtained by fractional distillation, either with or without previous neutralization of pyroligneous acid, and the acetic acid then recovered by direct fractional distillation or by evaporation of the calcium acetate liquor. *d.* Wood tar. From hard woods 7 to 10% of the weight of the wood, and from pine about 14%. The tar is steam distilled to recover the acid contained. The residue from this treatment is either burned as fuel for the retorts or distilled for various fractions of wood oils, wood tar, creosote (boiling point 200 to 220° C.), and wood tar pitch 50 to 65%.

Bone, Destructive Distillation. 1. Bones, after removal of fat, consist of the two-thirds inorganic bone material, mainly **calcium** phosphate (56% calcium phosphate, 8% calcium carbonate, 1% each calcium fluoride and **magnesium** phosphate), and one-third organic bone material (osseine, **nitrogen**-containing). 2. Conditions. When subjected to destructive distillation there is formed, gas, water condensate, bone oil, and bone charcoal (bone black) residue. 3. Products. *a.* Gas contains **ammonia.** *b.* Water condensate contains **ammonium** carbonate, sulfide, thiocyanate and cyanide. An impure grade of ammonium sulfate may be obtained. *c.* Bone oil, 3 to 5% of the weight of bone, is dark brown to black, of offensive odor, and begins to distil when heated to 80° C. **Pyridine** (C_5H_5N, boiling point 115° C.) is present in the fraction collected below 120° C. and **pyrrole** (C_4H_5N, boiling point 131° C.) in the fraction below 150° C. Pyridine is recovered by separation as the picrate. **Quinoline** (C_9H_7N, boiling point 238° C.) is also obtainable. *d.* Bone charcoal (bone black) consists of the **calcium** phosphate ash of bone impregnated with the residual carbon (10% carbon and 1% nitrogen). This material is used in decolorizing solutions, especially of raw sugar, on account of its adsorptive power for such colored materials, and as a black paint pigment.

Carboniferous Shale, Destructive Distillation. 1. Carboniferous shale, representing one of the greatest reserve supplies of fuel material, occurs in certain regions, for example, Colorado, in vast quantities. 2. Conditions. These have been extensively investigated and yields of 64 gals. (slightly more than 1.5 barrels) per ton of shale are reported. Some 1500 tests were made by R. D. George (1921) of the Colorado Geological Survey. S. D. Kirkpatrick has contributed a series of articles on the subject (Chemical and Metallurgical Engineering, 1924). J. H. Ginet (1923) reported cost of 87 cents per ton of shale, ⅔ of which is cost of mining. The refined products obtainable from Colorado crude shale oil are stated to be as follows: 3. Products. *a.* Gasoline, 15–17% of the crude shale oil, *b.* Kerosene, 30–32%, *c.* Gas oil, 18–26%, *d.* Light lubricating oils, 15–18%, *e.* Heavy lubricating oils, 10–12%, including paraffin wax. Valuable contributions have been made by the Colorado School of Mines and the United States Bureau of Mines. (R.K.S.)

DETECTION. This is the process of separating the intelligence from the **carrier** upon which it was modulated for transmission by radio or wired carrier. (See **Modulation.**) For the detection of amplitude modulated signals some sort of rectification must be used since the rapid alternations back and forth of the radio signal are too fast for the loud **speaker** to follow, and even if it did follow them the ear could not respond. Among the earliest radio detectors were various **crystals** such as galena, silicon and silicon carbide. These materials, when properly mounted, will pass current in one direction and not in the other, so by impressing on them the modulated signal they may be made to serve as detectors. The resulting current is a pulsating d.c. which varies in magnitude with the amplitude of the modulation on the original carrier. Crystals are no longer used for ordinary radio reception but are widely used for ultra high frequency detection where the capacitance of the vacuum tube is too high. For radio reception the most common detectors are vacuum **tubes** and at present only two types of detectors are used to any extent. These are the bias detector which uses a grid-controlled vacuum tube and the diode detector which uses a simple diode. Of these the diode is preferable since it is more linear, can handle higher signals without distortion and is suitable for **avc.** On the other hand the bias detector is more sensitive. Figure 1 shows a typical diode detector circuit. The modu-

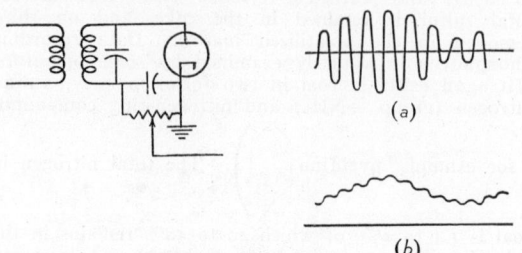

(a)

(b)

Fig. 1. Diode detection.

lated signal is impressed through the coupling **transformer** from a preceding circuit. Current flows through the diode only when its plate is more positive than its cathode, hence it rectifies the signal. This current charges up the **condenser,** which in turn discharges through the resistor when the diode is not passing current. The condenser does not completely discharge but has a voltage across it similar to that shown in Fig. 1*b*, while Fig. 1*a* shows the impressed modulated carrier. It is seen that the voltage across the condenser follows the modulation closely, and by adding additional condenser-resistance units even the slight ripples shown on it may be removed. This voltage may be coupled to the **grid** of an audio amplifier tube and further amplified before being impressed on the **speaker.** The bias detector uses a grid-controlled tube and is biased to or nearly to **cut-off.** Fig. 2 shows a circuit. A signal

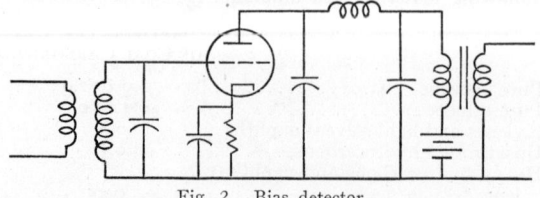

Fig. 2. Bias detector.

impressed upon the grid causes plate current to flow only when the signal is positive and hence the plate current flows in pulses. By means of the filter shown in the plate circuit this is smoothed out until it is a reproduction of the original audio wave.

Carrier transmission in telephone practice, while using the conventional vacuum tube detectors in several types, uses a copper oxide rectifier in the more modern systems. These rectifiers convert the modulated carrier into a varying d.c. which follows the modulation and which may be separated easily by a transformer into the d-c and audio-frequency components. These oxide units are not sufficiently free from distortion to be suitable for radio work, but are highly satisfactory for telephone service.

In the reception of continuous wave radio signals a heterodyne detector is often used. For such signals to be audible they must be beat with another signal to produce an audible frequency since they have no modulation and hence, even if detected by the usual means, cannot produce a varying output signal. When the continuous-wave radio signal is mixed in a vacuum-tube circuit with a locally generated signal which differs in frequency by some audio amount from it, the resultant output of the tube contains various component frequencies. Among these is one equal to the difference between the two frequencies mixed in the input. By proper adjustment of the frequency of the local signal this difference can be made any value desired and is usually from 500 to 1000 cycles. This gives a pleasing tone for the output. Such reception is called heterodyne detection. A somewhat similar process is used in the first detector of the **superheterodyne** receiver where the difference is made fairly high (this is the intermediate frequency and is always well above the audible range). When the heterodyning signal is generated by the detector serving as an **oscillator** at the same time it is often called autodyne detection.

For frequency modulation detection see **Discriminator**. (L.R.Q.)

DETECTOR. Detection.

DETERGENTS. Surface-Active Compounds.

DETERMINANTS. A determinant is a certain type of mathematical expression which plays an important part in the study of systems of linear equations and related topics. Determinants were first introduced by the German mathematician and philosopher G. Leibnitz (1646–1716).

A determinant of the second order is denoted by a symbol as

$$\begin{vmatrix} a_1 & b_1 \\ a_2 & b_2 \end{vmatrix},$$

and is defined as representing the expression $a_1b_2 - a_2b_1$.

A determinant of the third order is denoted by a symbol as

$$\begin{vmatrix} a_1 & b_1 & c_1 \\ a_2 & b_2 & c_2 \\ a_3 & b_3 & c_3 \end{vmatrix},$$

and is defined by the expression

$$a_1b_2c_3 + a_2b_3c_1 + a_3b_1c_2 - a_1b_3c_2 - a_2b_1c_3 - a_3b_2c_1.$$

A determinant of the nth order is defined as follows: Take a set of n^2 numbers, called elements, and arrange them in the form of a square array, with n columns and n rows, thus:

$$\begin{matrix} a_1 & a_2 & a_3\cdots a_n \\ b_1 & b_2 & b_3\cdots b_n \\ c_1 & c_2 & c_3\cdots c_n \\ \cdots\cdots\cdots\cdots \\ \cdots\cdots\cdots\cdots \\ l_1 & l_2 & l_3\cdots l_n, \end{matrix}$$

where the letter indicates the row and the subscript the column in which any particular element occurs.

(1) With the elements of such an array form all products that can be formed by taking as factors one element

and only one from each row and from each column of the array.

(2) In each product arrange the factors so that the letters are in the same order as in the alphabet and then count the inversions of the subscripts. If their number is even (or zero), give the product the plus sign; if odd, the minus sign.

(3) Take the algebraic sum of all these plus and minus products, and represent it by the array itself with vertical bars at either side of it, thus:

$$(a) \qquad \begin{vmatrix} a_1 & a_2\cdots a_n \\ b_1 & b_2\cdots b_n \\ \cdots\cdots\cdots\cdots \\ l_1 & l_2\cdots l_n \end{vmatrix}$$

The expression determined by (3) is called the determinant of the array, and it is denoted by the symbol (a). When there are n rows and n columns, the determinant is said to be of the nth order.

By an inversion in the preceding definition, we mean any case in which a larger subscript precedes a smaller one when the letters are arranged in their natural order.

This definition includes those given in the preceding for the determinants of the second and third orders.

The diagonal of elements $a_1, b_2, c_3, \cdots, l_n$ is called the principal diagonal of the determinant.

The number of terms in the expression which defines the determinant of the nth order is $n!$

The product of all the elements $a_1, b_2, \cdots, l_n$ lying in the principal diagonal is called the principal term. All the other terms can be formed in order from the principal term by permuting the subscripts in all possible ways.

Some of the principal properties of determinants are the following:

1. The value of a determinant is not changed if its corresponding rows and columns are interchanged.

2. If two rows (or columns) of a determinant are interchanged, the sign of the determinant is changed, but its absolute value remains unchanged.

3. If all the elements of a row (or column) of a determinant are zero, the value of the determinant is zero.

4. If two rows (or columns) of a determinant are identical, its value is zero.

5. If all the elements of a row (or column) of a determinant are multiplied by the same number, the determinant is multiplied by this number.

6. If each element of a row (or column) is expressed as the sum of two or more numbers, the determinant can be expressed as the sum of two or more determinants; thus, for a determinant of the third order:

$$\begin{vmatrix} a_1 + a'_1 & a_2 & a_3 \\ b_1 + b'_1 & b_2 & b_3 \\ c_1 + c'_1 & c_2 & c_3 \end{vmatrix} = \begin{vmatrix} a_1 & a_2 & a_3 \\ b_1 & b_2 & b_3 \\ c_1 & c_2 & c_3 \end{vmatrix} + \begin{vmatrix} a'_1 & a_2 & a_3 \\ b'_1 & b_2 & b_3 \\ c'_1 & c_2 & c_3 \end{vmatrix}.$$

7. The value of a determinant is not changed if to each element of any row (or column) there be added the corresponding elements of any other row (or column) each multiplied by the same number.

If in any determinant we strike out both the row and column in which any particular element lies, and then form the determinant of the remaining elements without changing their relative positions, the new determinant so formed is called the minor of that element.

The evaluation of determinants is generally based on an expansion of the determinant in terms of minors by use of the following rule: a determinant may be expressed as the sum of the products of each of the elements of any one of its rows (or columns) by their corresponding minors, each product having a positive or negative sign according as the sum of the row-number and column-number for the particular element considered is even or odd.

If H_k is the minor of an element h_k of a determinant, where the element lies in the hth column and k^{th} row, then $\overline{H}_k = (-1)^{h+k}H_k$ is called the co-factor of h_k.

The expansion of the determinant may then be written in any of the forms:

$$D = a_1\overline{A}_1 + a_2\overline{A}_2 + a_3\overline{A}_3 + \cdots + a_n\overline{A}_n$$
$$= b_1\overline{B}_1 + b_2\overline{B}_2 + b_3\overline{B}_3 + \cdots + b_n\overline{B}_n$$
$$\cdots\cdots\cdots\cdots\cdots\cdots\cdots\cdots\cdots$$
$$= l_1\overline{L}_1 + l_2\overline{L}_2 + l_3\overline{L}_3 + \cdots + l_n\overline{L}_n$$

Any sum, such as $b_1\overline{A}_1 + b_2\overline{A}_2 + \cdots + b_n\overline{A}_n$, obtained by adding the products of the elements of any row (or column) with the co-factors of the corresponding elements of any other row (or column), is zero.

The evaluation of determinants is based on the use of the preceding properties of determinants, particularly 7, and the expansion by minors, reducing the given determinant in order successively by one until its order is 3 or 2, when it can be directly evaluated by the definition of the second or third order determinant.

The product D' of two determinants D_1 and D_2 of the same order is obtained thus: multiply the elements of the ith row of D_1 by the corresponding elements of the kth column of D_2; the sum of the products thus obtained is the element in the ith row and k^{th} column of the product determinant D'.

One of the chief uses of determinants is in the solution and study of the properties of **systems of linear equations.** (L.L.S.)

DETERMINATE STRUCTURE. Any structure in which the reactions and stresses can be found by means of the equations of statics only is a determinant structure. If a sufficient number of such equations cannot be set up from known conditions, the structure is not statically determinate. (C.W.C.)

DETONATION. The term detonation in ordinary usage denotes any explosion. In the **explosives** industry, however, a detonator is an explosive or device used to initiate the explosion of another explosive, which can thus be less sensitive, and more safely handled and transported. Also, detonation is the term describing the knock or ping which, under certain circumstances, occurs in the cylinders of an internal combustion engine. Detonation is undesirable. It reduces power output, causes overheating, unduly stresses the cylinder head, and is generally objectionable from the noise and vibration standpoint. It decreases **thermal efficiency.**

In spark ignition engines, detonation appears to be due to spontaneous ignition of the explosive gasoline-air mixture in the cylinder head. Observers have noted that with no detonation, ignition of the charge starts at the spark plug and travels rapidly, but with a definite wave front, through the charge. However, when the engine is knocking, the inflammation proceeds normally only part way from the source of ignition, and then suddenly the remainder of the charge is ignited simultaneously at all points, accompanied by a sharp rise in pressure due to almost instantaneous expansion. The result is a heavy pressure wave striking the cylinder head hammer-like blows causing it to vibrate audibly. The explanation advanced for this sudden ignition of the entire charge is that the initial combustion increases the pressure of the rest of the charge, and its temperature, until it passes the point where ignition will take place spontaneously due to **adiabatic** compression.

Conditions affecting detonation might be discussed under three heads: viz., fuel characteristics, cycle characteristics, and engine characteristics. The rate at which a fuel burns when ignited is the first of the fuel characteristics. Since the ratio of air to fuel in the combustible mixture also affects the rate of burning, it has a bearing on detonation. The spontaneous ignition temperature of the fuel is another important characteristic. Turning next to the cycle upon which the **engine** is operating, we find the compression ratio and the time of ignition to be of great importance. The overcoming of detonation is of increasing importance, as compression ratios of engines are raised in the effort to increase thermal efficiency. The driver of an automobile with manually controlled ignition knows detonation may be offset by retarding the spark. But this is undesirable, since it is accompanied by loss of power and overheating. One of the more important engine characteristics affecting detonation is the material of the combustion chamber. The more rapidly heat is conveyed away from the cylinder head, the less will be the chance of the initial combustion raising the pressure above that for spontaneous ignition. Thus aluminum cylinder heads and improved cooling efficiency offset detonation in high compression engines. Manufacturers' provisions for thus conducting heat away rapidly will be nullified by a heavy layer of heat-insulating carbon deposit in the cylinder head. Other engine characteristics of importance are the location of the spark plug and the shape of the combustion chamber.

It is known that certain substances have the ability to suppress this kind of detonation. How they act is not fully known. Amounts of suppresser as small as one molecule to 100,000 molecules of explosive mixture are effective in eliminating detonation. The detonating quality of a gasoline is one of its comparable characteristics. A scale called octane rating has been devised to measure this quality. The octane rating is the percent by volume of isooctane in a heptane-isooctane mixture (see **Hydrocarbon**), which exactly matches in antidetonating property, an actual fuel under test in a standard test engine at standard conditions.

Lead tetraethyl in small quantities is effective as a knock suppresser. To prevent a lead deposit from forming in the cylinder head, ethylene dibromide is added to form a lead bromide which is powdery and is blown out through the exhaust ports. Gasoline so treated is called ethyl gasoline.

The compression ignition engine also experiences a knock from the combustion of improperly selected fuels. It is believed that combustion knock in a **Diesel engine** is caused by delay in **ignition** during the first part of fuel injection, causing accumulation of fuel which burns simultaneously with the fuel injected during the latter part of the injection. Detonation in these engines is aggravated by the slow-burning characteristic that is anti-knocking in spark ignition engines. Good ignition quality, as represented by high **Cetane Number,** minimizes the delay period and the tendency of a compression ignition engine to knock. (F.T.M.)

DETRITUS. General term for unconsolidated **sediments** derived from pre-existing rocks by natural agencies. Derived from the Latin word meaning worn. (R.M.F.)

DEUTERIC. Used by **petrologists** to describe those alterations in an **igneous rock** which occur during the later stages of its solidification. (R.M.F.)

DEUTERIUM. Symbol: D. Atomic number: 1. Atomic weight: 2, the second isotope of hydrogen, 0.02% of ordinary hydrogen. Discovery by Urey in 1932.
Oxide: Deuterium oxide (D$_2$O), heavy water.
(See **Hydrogen; Water.**) (R.K.S.)

DEUTEROCEREBRUM. The second region of the **brain** of an **arthropod,** derived from a pair of ganglia (see **Ganglion**) in the corresponding segment of the head. (A.W.L.)

DEUTERON. Nuclear Particles.

DEUTOPLASM. Inert material stored in **eggs.** Yolk. (A.W.L.)

DEVELOPERS. Developing Agents; Photography.

DEVELOPING AGENTS. Developing agents are selective inorganic or organic reducers which have the ability to distinguish between exposed and unexposed silver halide grains in an emulsion and have only sufficient energy to reduce exposed silver halide grains to metallic silver, thereby converting the latent image into a visible one.

Certain chemical reducers are too energetic; they reduce all the silver halide particles in an emulsion, causing complete fog. These are unsuited for photographic use. The energy of any developing agent is subject to the variations in emulsions, in exposure, and in the composition of the developing solution in which it is incorporated. The developing agent that is satisfactory in one case may not be suitable in another.

Developing agents can be broadly divided into two general classes: (1) those in which a metal undergoes a change in valence, and (2) those which do not contain a metal or in which the valence of the metal does not change. Inorganic compounds and metallic salts of organic acids belong to the first class, while the organic developing agents and certain inorganic compounds, as peroxide, hydroxylamine and sodium-hydrosulfite, are members of the second.

Inorganic Developing Agents. Among the small number of inorganic substances that have been found to act as photographic developing agents are ferrous fluoride, ferrous oxalate, ferrous citrate, copper chloride, cuprous bromide, cuprous iodide, sodium hydrosulfite, hydrogen peroxide, hydroxylamine, and hydrazine. Most of these compounds are of academic interest only. Many of the agents listed produce by-products or gases which are harmful to emulsions. Ferrous oxalate, however, was used for many years in negative developers. Sodium hydrosulfite, although unstable, is about the only inorganic developer that is used today. Inorganic developing agents have given way to organic compounds because the latter have more desirable properties, keep longer in solution, and are easier to control.

Organic Developing Agents. Several hundred organic compounds have been described in photographic literature as possessing developing power. Of these, only about a dozen are of practical use in black and white photography. The number used for processing color materials probably does not exceed twice the number of black and white agents. In some instances the same agent can be employed for both black and white and color photography. The remaining agents are, therefore, of scientific and research interest.

Organic developing agents can be considered to be organic derivatives of hydrogen peroxide HO—OH, hydroxylamine H_2N—OH and hydrazine H_2N—NH_2. For consideration, these may be divided into three groups:

1. Derivatives in which one or both of the hydrogens of an amine group, —NH_2, are substituted.

2. Derivatives in which organic radicals are inserted between the hydroxy and amino groups.

3. Some aliphatic compounds containing the characteristic hydroxy or amino groups also act as developing agents.

Common Developing Agents. Because of the flexibility of developing agents, it is possible to obtain a great variety of results with a small number of compounds.

By far the greater number of developing solutions today make use of only two agents, metol and hydroquinone. In some instances a third agent, such as glycin, chlorhydroquinone or pyrogallol, is added. Some solutions, notably process developers, make use of only one agent—hydroquinone. By varying the type of alkali and amount of restrainer almost any control may be obtained.

Amidol (Acrol, Diamol). Chemically the agent is known as 2-4 diaminophenol. Amidol is capable of development without an alkali. With sulfite it forms a soft working developer adapted to the development of bromide papers. It deteriorates rapidly in solution becoming inactive in a day or two. The keeping qualities of the solution may be improved by additions of boric, gly-

collic or lactic acids. Best results are obtained by mixing the developer immediately before use. Combined with sulfite, metabisulfite and bromide, it has been a popular developer for lantern slides and opal plates. The agent is usually sold in the form of the dihydrochloride, $2\text{-}4\ (NH_2)\cdot C_6H_3OH\cdot 2HCl$, for stability.

Catechol (Pyrocatechin). Catechol, also known as ortho hydroquinone, is a 1-2 dihydoxy benzene, 1-2 $(OH)_2\cdot C_6H_4$. It is often substituted for hydroquinone in formulae. It is specified in developers which harden emulsions during development and has been used in the pinatype process for making imbibition prints.

Chlor Hydroquinone (Adurol, Chlorquinol), 3-Cl, 1-4 $(OH)_2\cdot C_6H_3$. In behavior chlor hydroquinone is similar to hydroquinone but is more energetic especially at low temperatures. It is less sensitive to the action of restrainers and does not show as great a tendency to fog or stain as hydroquinone. It is combined with metol for negative development. It is also a component in some fine-grain developers. In combination with hydroquinone in carbonate developers it produces warm tones on chloride and chlor bromide papers.

Glycin (Athenon, Monazol, Iconyl). Glycin, or para-hydroxyphenylamino-acetic-acid, has the formula $HO\cdot C_6H_4\cdot NH\cdot CH_2COOH$. The agent is insoluble in water but dissolves readily in sodium sulfite solutions. It is a slow acting but powerful developing agent, producing fine-grained silver images with a marked freedom from fog even in the absence of alkaline bromides. Glycin, unlike most developing agents, is not oxidized by air, nor does it produce a stain image. It keeps well in solution and is recommended for deep-tank developing. Glycin is usually used in combination with metol, with paraphenylene diamine for fine grain developing, or with hydroquinone for warm tones ranging from warm blacks to browns, depending on dilution. In combination with metol and hydroquinone it produces the rich jet blacks particularly desired in salon prints.

Hydroquinone (Quinol). Hydroquinone, 1-4 $(OH_2$ C_6H_4 1-4 dihydroxybenzene, is the second oldest developing agent discovered by Abney in 1881. Hydroquinone is used to a greater extent than any other agent. It has a low energy, is very sensitive to restrainers, and is practically inactive at temperatures below $59°$ F. Because hydroquinone produces density rather than detail in images, it is used as an agent for producing contrast. Unless the developers containing hydroquinone have sufficient sulfite and bromide, the images are heavily stained and fogged.

Hydroquinone, in strong alkaline solutions as carbonate or hydroxide, is an excellent developer for line copy work. Combined with paraformaldehyde it produces the extremely high densities and contrasts desired in graphic arts.

For general photographic purposes it is combined with metol to produce a great variety of results, ranging from fine grain to process negatives and as standard developer for prints.

In developers having low concentrations of sulfite, the oxidation products of hydroquinone harden the gelatin in contact with the silver image. Reliefs made in this manner can be dyed and used to transfer dye images in color photography.

Metol. (Common trade names are Elon, Veritol, Genol, Pictol, Photol, Rhodol, Errol, Scalol, Satrapol, and Mirol.) N-Monomethyl-para-amino-phenol-sulfate, $HO\cdot C_6H_4\cdot NH\cdot CH_3\cdot \frac{1}{2}H_2SO_4$, or metol, since its introduction by Hauff in 1891, has been one of the most popular developing agents. Practically every photographic manufacturer and photo supply-house packages the developing agent under their own trade name.

Metol is a rapid, soft working developer that develops detail. It is not very sensitive to bromide or restrainers. It has good keeping qualities and a long useful life. Used alone, it will develop a larger quantity of sensitive material than any other agent without becoming exhausted. Although metol will develop images without an alkali it is usually incorporated in solutions containing carbonates, borax or borates. It is usually necessary to add some bromide to metol developers to prevent fog.

Since metol produces detail and hydroquinone builds contrast, the two agents can be combined in various proportions with different alkalies to produce almost any result desired, a fact which explains the wide popularity of the M-Q developers.

Other combinations as metol-pyro, metol-glycin and metol-hydroquinone-glycin are also popular.

Developers containing only metol and sulfite or metol, sulfite and bisulfite produce fine-grained images with excellent shadow detail and contrast.

Para-aminophenol. The free base $HO \cdot C_6H_4 \cdot NH_2$ is only slightly soluble in water; hence salts such as the hydrochloride, sulfate, oxalate or tartrate are employed. The compound is the parent substance from which all of the paraaminophenol developing agents are derived.

In solution with alkaline carbonates, paraaminophenol is a rapid, soft working developer. It produces images which are remarkably free from fog, even at temperatures above normal. For this reason it is recommended for both tropical and high-temperature development. Paraaminophenol developers can be prepared in highly concentrated solutions which have exceptional keeping properties.

Paraphenylenediamine (Diamine, Dianol, Paramine, P.P.D.). The free base $H_2N \cdot C_6H_4 \cdot NH_2$ is very slightly soluble in water. It is alkaline and may be used without an alkali. The hydrochloride $C_6H_4(NH_2)_2 \cdot 2HCl$ is more soluble in cold water than in hot. The salt is acid and must be made alkaline before it will develop. Both the base and the salt keep in the dry state if they are protected from air.

Both forms of the developer are toxic and produce painful skin eruptions on persons who are sensitive or allergic. Persons affected should wear rubber gloves when working with either the developer or its solutions.

The fine-grain properties of paraphenylenediamine developers were observed by Lumiere and Seyewetz in 1904. Fine-grain development is due to the low reduction potential of paraphenylenediamine and to its solvent action on silver halides; a condition that produces low effective emulsion speeds and necessitates increased exposures.

Silver images resulting from paraphenylenediamine developers are dichroic; they appear black by transmitted light and cream colored by reflected light. In reflected light the negatives often appear as positives. Because paraphenylenediamine produces stained images, negatives which appear quite thin produce prints with good quality and contrast.

To overcome the need for prolonged exposures, other developing agents having higher reduction potentials, as metol, glycin, hydroquinone, chlor hydroquinone and pyrogallol, are combined with paraphenylenediamine.

Paraphenylenediamine has also been used to produce warm-tone fine-grain images on lantern slides and motion-picture film.

Paraformaldehyde, a polymerized form of formaldehyde, is a solid. This agent, combined with hydroquinone in strongly alkaline developers, produces process or line developers with extreme contrast for graphic arts.

The energy of the solutions increases, after mixing, to a maximum in 4 or 5 days and then gradually decreases over a period of several weeks. It is necessary, therefore, to adjust the time of development according to the age of the developer.

Pyrogallol (Pyro, Piral). Pyrogallol, 1-2-3 $(OH)_3$ C_6H_3, is the oldest known organic developing agent. It was discovered independently by both V. Regnault and S. Archer in 1851.

Pyrogallol, in pure form, is a small, fine, white crystalline needle-like powder which is extremely light and floats through the air with the slightest movement. Pyrogallol is also obtained in heavy white flake crystals which are easier to handle and dissolve more rapidly during mixing. The crystal form is more stable and does not oxidize as rapidly as the powder.

Pyrogallol is a slow, soft working developing agent which forms both a stain and a silver image. The combination of the yellow stain and silver produces brown images that print more contrasty than they appear visually. While the stain formation may be inhibited by additions of sulfite and other preservatives, the developer is subject to oxidation by exposure to air and by oxidation at increased temperature and agitation.

The objections to using pyrogallol are the difficulties in controlling the stain formation and in obtaining negatives with a uniform stain deposit. The keeping qualities of the pyrogallol developers are poor. It is generally advisable to use a fresh bath with each batch of film. In order to obtain greater contrast and shorter times of development, pyrogallol is combined with metol, hydroquinone and an alkaline carbonate as an alkali. Metol-pyrogallol developers are popular for portraiture. Combinations of pyrogallol and paraphenylenediamine are used in fine-grain developers. (s.m.t.)

DEVELOPMENT. Developing Agents; Photography; Sheet Metal Processes; Press Working.

DEVIATION. The deviation of x from a is defined as $x - a$. If a is the **mean** we define $d = x - \overline{X}$. If x is any other point we define $d' = x - a$. If a is the **median** we have a deviation from the median. The absolute value of the deviation of x from a is defined as $|x - a|$. (See **Compass Correction**.) (L.A.A.)

DEVIATION RATIO. Modulation.

DEVIL FISH. 1. Mollusca, Cephalopoda. Any octopus. 2. Pisces, Plagiostomi. The giant **ray**.

Both of these forms are said to be dangerous to man, and when we consider that they are predacious and that some reach large size we can readily believe that they may attack divers. One octopus of the Pacific Ocean reaches a diameter of more than 20' and the ray attains an equal width. (A.W.L.)

DEVITRIFICATION. The process by which the natural rock glasses, such as **obsidian** and **tachylyte**, develop minute but definite minerals, usually **quartz** and **feldspar**. (R.M.F.)

DEVONIAN. The name of a geologic period. Type locality, Devonshire, England. The formations of this period were first studied and described by R. I. Murchison in 1839. The Devonian period began 330 million years ago and lasted for 50 million years. The Devonian formations are well exposed in eastern North America and parts of the North American Cordilleran. In the **Appalachian Geosyncline** the Devonian is largely represented by an immense thickness of red and brown shales and sandstones of deltaic and estuarine origin, and the transition between the sediments of this system and that of the underlying Silurian is so gradual that the boundary is extremely difficult to locate by physical means alone. In Britain the Devonian is represented by a marine limestone (facies) in the type locality, and a red non-marine sandstone (facies) to the north. This red sandstone (facies) is referred to as the "Old Red" by British geologists. The Scottish "Old Red" contains the famous fossil "fishes" described by Hugh Miller in 1851. These fish include two distinct groups, **Ostracoderms** and **Ganoids**, the latter being the supposed ancestors of the Amphibia or first terrestrial vertebrates.

The first undoubted evidence of terrestrial plants occurs in the Devonian, the late Devonian types being the progenitors of the **Carboniferous** forms. In England,

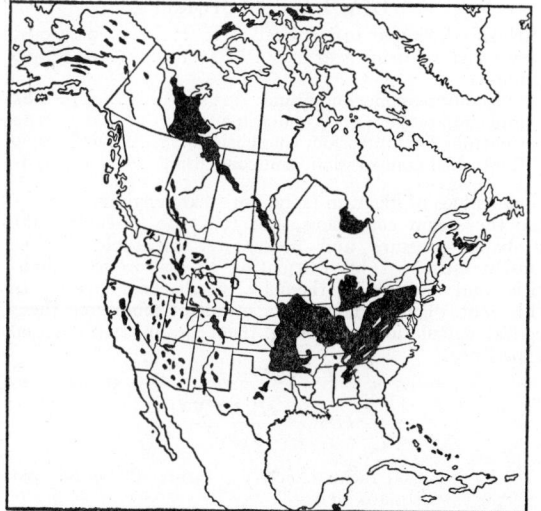

Map showing known areas of outcrops (surface distribution) of Devonian, Mississippian, and Pennsylvanian strata in North America.

Scotland, Spitzbergen, western Russia, and Norway, occur great thicknesses of terrestrial, intermontane clastic sediments similar to those found in the **Proterozoic**. Highly fossiliferous marine sediments, including sandstones, shales and limestones are particularly well exposed in New York State. Many of the limestone formations contain reefs composed principally of compound corals, **Bryozoa** and **Calcareous Algae**. Among the marine invertebrates **goniatites** and **Eurypterids** are particularly representative. Other common marine types are corals, **Bryozoans** (reefs) and **Echinoderms, Pelecypods** and **Trilobites**. The spire-bearing Brachiopods (**Spirifers**), which started in the Silurian, reach their maximum development in genera and species in the Devonian. The only fossil evidence of a terrestrial vertebrate rests upon a footprint (probably that of an amphibian) found in the Upper Devonian of western Pennsylvania. Beginning with the middle and ending with the period, mountain building occurred in the New England States. The principal economic products derived from the American formations are petroleum and natural gas, first exploited in 1859 in western Pennsylvania, New York, Ohio and West Virginia. (R.M.F.)

DEVONITE. The name given by Johannsen, in 1910, for a variety of **porphyritic basalt** containing **phenocrysts** of **potassium-rich plagioclase**. Type locality, Mt. Devon, Massachusetts. (R.M.F.)

DEW. If air in contact with any surface is cooled at the surface to a temperature below its dew point, some of the water vapor present in the air will condense onto the cool surface as liquid water or dew. When temperatures are below freezing, hoarfrost forms instead. (P.E.K.)

DEW CLAW. Small hoofs of the rudimentary toes, found just above the functional hoofs in some of the even-toed hoofed animals. (A.W.L.)

DEWLAP. The fold of skin which hangs below the neck in cattle. (A.W.L.)

DEW-POINT HYGROMETER. Hygrometers; Dew Point.

DEW-POINT TEMPERATURE. The temperature to which air must be cooled to become saturated is its dew-point temperature, but the cooling must be carried out at constant pressure. It is the temperature at which the

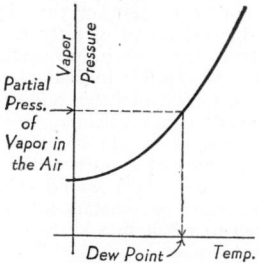

Graphical location of dew point from vapor pressure curve.

maximum vapor pressure of water would be equal to the actual partial pressure of the water vapor in the atmosphere (see **Dalton's law**). For example, let the temperature of the air be 20° C., and the relative humidity 60%. Then since the maximum vapor pressure of water at 20° C. is 17.4 mm., the actual water vapor pressure is 0.6 of this or 10.4 mm. The temperature at which 10.4 mm. is the maximum vapor pressure of water is 12°. Hence if the air is cooled to 12° C., it will reach saturation and under suitable conditions dew will form; 12° C. is the dew point. Likewise if the dew point is known to be 12° when the air is at 20°, it follows that the relative humidity is 60%. The dew point **hygrometer** depends upon this principle. (L.W.D., P.E.K.)

DEXTRIN. Carbohydrates.

DIABASE. Dolerite.

DIABETES MELLITUS. A disease of **metabolism** in which those processes concerned with the formation and utilization of **carbohydrates** are disturbed. An outstanding feature is the deficient supply of **insulin**, a **hormone** produced by the **pancreas**. Insulin is necessary for the utilization of carbohydrates, and when it is decreased in amount in the body, the metabolism of carbohydrates, and secondarily proteins and fats, is upset. As a result, large amounts of unused sugar appear in the blood and overflow into the urine. Intermediary products in the metabolism of fats and proteins, ketone or acetone bodies, also appear in the blood and urine, and their presence in large amounts is associated with diabetic **coma** and **acidosis**.

It is estimated that there are 660,000 diabetics in the United States and the incidence of the disease is increasing steadily. This is due in part to its hereditary nature and to improved treatment. Diabetic children now live to adult life, marry, and may produce diabetic children.

The cause of diabetes is obscure but many factors are known to play a part. The most important are (1) heredity, (2) endocrine gland disturbances—chiefly involving the **pituitary**, the **adrenal** cortex, and the **thyroid**, (3) obesity, (4) a varied group of factors such as age, sex, race, infection, etc.

1. The inheritance of diabetes as a Mendelian recessive (see **Heredity**) is well established. If both parents are diabetic, all of their children will have diabetes. If one parent has diabetes and the other neither has it nor carries a gene for it, none of their children will have the disease. Diabetes could, therefore, be bred out of the species over a number of generations by the consistent union of diabetics with individuals from nondiabetic families.

2. The interaction of the endocrine glands in the control of metabolism is complex and not fully understood. It has been shown that a derangement of the normal relationships of the glands may result in diabetes. The

pituitary under certain circumstances can manufacture a diabetogenic (diabetes-producing) factor. The adrenal cortex is important in carbohydrate metabolism but its exact action is not known.

3. Obesity is associated with diabetes in several ways. Although the popular conception that increased sugar consumption and obesity can directly cause diabetes is incorrect, it is true that diet influences the appearance of the disease in a susceptible individual. This is borne out by the fact that mild diabetes can be controlled by diet; and if a patient requires insulin, increasing the diet increases the amount of insulin necessary. If an individual has a low pancreatic reserve, the strain on the insulin-producing tissue which would exist over a long period under the dietary circumstances leading to obesity, might be expected to result in a situation where the insulin supply is inadequate.

4. The time of life at which diabetes is most likely to occur is in the 4th and 5th decades, and the disease is far more common in women than in men. The Teutonic races are more susceptible than the Latin. The incidence in the Jewish race is greater than in any other; this is due in part to inbreeding, heredity, and the tendency to obesity in middle life. Occupation can be correlated with diabetes in that the incidence is much greater in the professional groups, those of sedentary employment, and in the higher income groups.

The symptoms of diabetes are weakness, weight loss, excessive appetite and thirst, and the excretion of excessive amounts of urine. Treatment with insulin, which was isolated by Banting and Best in 1921, is life-saving in severe diabetics. In mild cases, dietary control alone may be sufficient. With the use of diet and insulin diabetics can now lead a normal, productive life. (D.M.H.)

DIAGENESIS. A term proposed by Gumbel in 1888 for the gradual and successive chemical physical changes which take place in **sediments** previous to or during their **consolidation.** Diagenesis may also include the numerous processes of **lithification** but is a useful term only when particularly applied to the more or less contemporaneous chemical alteration of sediments. (R.M.F.)

DIAGONAL. Any inclined **web member** of a **truss** other than the **end post** is known as a diagonal. Under various **live load** conditions certain web members of a **bridge** truss may be subjected to a **reversal of stress.** Stiff diagonals are web members designed to carry either **tension** or **compression.** Tension diagonals are assumed to be incapable of resisting any appreciable amount of compression.

When the diagonals are designed to take tension only (are not stiff) two are required, sloping in opposite directions, in a panel where either one, acting alone, would have its stress reversed. When one of these diagonals is acting the other is out of action for all practical purposes. The one which carries the **dead load** stress when no other loads are on the bridge is called the main diagonal. The other which comes into play when the main diagonal goes out of action is the counter. (C.W.C.)

DIAGRAM FACTOR. This is a factor relating particularly to **piston** and **cylinder** engines. Although it has a meaning applied to internal combustion engines, diagram factor is principally a dimension of the **steam engine.** Certain analyses of the steam engine, especially those concerned with predicting the performance of a given unit under stated steam conditions, are most easily made with the use of the diagram factor. This factor is defined as the ratio of the actual **mean effective pressure** to the theoretical effective pressure; also as the actual work to the theoretical work. The theoretical case is that of a steam engine having no compression, no **wire drawing,** and no clearance. (See **Rankine Cycle.**) The

ratio of the area of this cycle to that of the actual **engine** operating between the same pressure limits, is typical for any one class of engine. The accompanying table gives

DIAGRAM FACTORS

High-speed, simple automatic	0.70–0.85
Low-speed, releasing gear	0.80–0.90
Uniflows	
Full compression, condensing	0.75–0.85
Full compression, non-condensing	0.70–0.80
Controlled compression, condensing	0.85–0.90
Controlled compression, non-condensing	0.80–0.85

some values of diagram factor for steam engines. Knowing the steam conditions, and the type of engine, the probable pressure and horsepower realizable can be closely estimated by calculating the theoretical quantities and multiplying them by the diagram factor. In this light, diagram factor is a means of modifying theoretical calculations to bring them in line with actual experience.

$$P_a = P_t \times f$$

$$IHP = \frac{2 P_a L A N}{33,000}$$

P_a = actual mean effective pressure, in pounds per square inch.
P_t = theoretical mean effective pressure, in pounds per square inch.
f = diagram factor.
IHP = indicated horsepower (steam engine).
L = stroke in feet.
A = piston area, in square inches.
N = rotative speed, in revolutions per minute. (F.T.M.)

DIALLAGE. The mineral term for a **calcium-iron pyroxene,** similar in chemical composition to **diopside** but richer in iron oxide. In addition to the typical prismatic cleavage of the pyroxene group diallage has a marked "cleavage" parallel to the vertical **pinacoids,** known as diallage parting. Diallage is a common constituent of **gabbros.** The term Diallagite was proposed by Cloiseaux in 1845 for rocks particularly rich in diallage. The term diallage is derived from the Greek meaning difference, and referring to the peculiar cleavages of this variety of **monoclinic** pyroxene. (R.M.F.)

DIALYSIS. This is a process of removing crystalloids from **colloids** by means of semi-permeable membranes. (R.K.S.)

DIAMAGNETISM. Magnetism.

DIAMETERS OF CONICS. The **locus** of the midpoints of all chords parallel to a given chord of a **conic section** is called a diameter of the conic; it is a straight line in each case.

The diameter of the **parabola** $y^2 = 2px$ which bisects all chords of **slope** m is the line $y = p/m$ which is parallel to the axis of the parabola.

The diameter of the **ellipse** $\frac{x^2}{a^2} + \frac{y^2}{b^2} = 1$ which bisects all chords of slope m passes through the center and has the slope $m' = -b^2/a^2m$. The diameter of the **hyperbola** $\frac{x^2}{a^2} - \frac{y^2}{b^2} = 1$ which bisects all chords of slope m passes through the center and has the slope $m' = b^2/a^2m$.

Two diameters of an ellipse or hyperbola are called conjugate diameters if each bisects the chords of the given curve that are parallel to the other. For the ellipse $\frac{x^2}{a^2} + \frac{y^2}{b^2} = 1$, the slopes of two conjugate diameters are related by $mm' = -b^2/a^2$; for the hyperbola $\frac{x^2}{a^2} - \frac{y^2}{b^2} = 1$, the relation between the slopes of conjugate diameters is $mm' = b^2/a^2$. (L.L.S.)

DIAMETRAL PITCH. Spur Gearing; Worm Gearing.

DIAMOND. In any enumeration of gem stones the diamond unquestionably stands in first place because of its remarkable brilliancy and great hardness as well as because of its relative rarity. It is an allotropic form of **carbon.** Its luster is also "hard" in quality from which we have derived the term adamantine luster, through the Greek meaning, adamant, applied to diamonds and possibly other very hard substances, literally "unconquerable." In this connection it is interesting to note that the ancients confused hardness with strength or toughness, not understanding that minerals of considerable hardness may split or cleave easily in certain directions. This unfortunate fallacy is believed to have been responsible for the unwitting destruction of more than one diamond.

When colorless, water-clear, and flawless, diamonds are referred to as of the "first water" and are of the greatest value. "Off-color" stones, those with a slight tinge of yellow, are far less desirable although stones of a bright yellow, or of odd colors such as pink or steely blue, command large premiums, as they are highly prized.

The diamond fields of India, now seemingly exhausted, yielded quantities of stones, probably all of those known previous to 1725 when diamonds were discovered in Brazil. That country has produced many fine but mostly small stones.

Diamonds were discovered in 1867 along the Orange River in South Africa, and since then Africa has been preëminent in the production of diamonds; in the seventies and eighties occurred a series of amazing discoveries of diamond fields and stones of extraordinary size. Diamonds have also been found in Australia, Borneo, British Guiana, and Arkansas.

Much as the matter has been studied there is no general agreement as to the genesis of the diamond. It is found in **alluvial** deposits, both unconsolidated and consolidated, indicating that the erosion of rocks containing diamonds not only during the present era but also in past geologic time. In Africa diamonds are mined in a dark basic rock of the general nature of **peridotite** called **kimberlite** from the town of Kimberley. The kimberlite occurs in vertical "pipes," resembling what once may have been volcanic necks or other types of igneous rock conduits. It is supposed that the diamonds have been formed in and brought to the surface by the magma which was of the general nature of peridotite. Undoubtedly high pressures, and possibly high temperatures as well, are necessary for the development of crystallized carbon in the form of diamonds. (E.S.C.S.)

DIAMOND DIE. Die; Wire Drawing.

DIAPHRAGM. 1. The thin layers of muscle which suspend the insect heart within the body. 2. The muscular partition between the thoracic and abdominal cavities of mammals.

The mammalian diaphragm is important in **respiration.** (A.W.L.)

DIARRHEA. Increased frequency of bowel movements, the stools tending to be liquid. Diarrhea occurs commonly with infections of the gastro-intestinal tract (e.g., the **dysenteries**) or with any irritation of the bowel. (D.M.H.)

DIASPORE. The mineral diaspore is a hydrous **oxide** of **aluminum** corresponding to the formula $AlO(OH)$ occurring in prismatic **orthorhombic** crystals, usually somewhat flattened, or massive. It displays good cleavage; conchoidal fracture; is brittle; hardness, 6.5–7; specific gravity, 3.3–3.5; luster, vitreous to pearly; color, white, grayish, greenish, yellowish, brownish or colorless; transparent to translucent. Diaspore is found associated with **corundum, emery** and **bauxite,** being probably an alteration product of the oxide. It has been made artificially. Diaspore has been found associated with emery in the Ural Mountains, in Asia Minor, in the Island of Naxos, Greece, and in the United States at Chester, Massachusetts. Its name is derived from the Greek word meaning to scatter, because of its decrepitation upon heating. (E.S.C.S.)

DIASTASE. Enzymes.

DIASTEM. A term proposed for a slight hiatus, or loss of record, during the deposition of sediments. As diastems must be contemporaneous with sedimentation they are not to be confused with **disconformities.** (R.M.F.)

DIASTOLE. The stage of dilation of the heart or relaxation of the **heart** muscle. It is during this stage that the chamber of the heart fills with blood. (R.S.M.)

DIASTROPHISM. A general term for all types and modes of deformation of the crust of the earth. (See **Epeirogeny; Orogeny.**) (R.M.F.)

DIATHERMANCY. Thermal Radiation.

DIATHERMY. The generation of heat in the bodily tissues by means of high-frequency electric currents. Two electrodes are placed on opposite sides of a portion of the body. The heat is generated by the resistance provided by the body tissues between these electrodes.

Medical diathermy is used in **neuritis,** sprains, **arthritis,** etc. Surgical diathermy (endothermy) is used operatively. Sufficient heat is produced at the skin surface either to actively cut the tissues with the current or to coagulate and kill tissue cells. By use of this method in certain conditions, warts, malignant growths, etc., there is less bleeding and less danger of spreading malignant cells. This method can, however, only be used in certain conditions. (R.S.M.)

DIATHESIS. Congenital predisposition toward any disease. (R.S.M.)

DIATOM OOZE. Oceanic Deposits.

DIATOMS. Bacillarieae. Diatoms are **algae** which are very commonly found in both fresh and salt waters. Often they occur in immense numbers, especially in the ocean. The feature which distinguishes them from all other algae is the siliceous (see **Silicon**) wall which encloses them. This is composed of two halves or valves, one of which fits over the other much as a cover fits onto a box. These siliceous walls are often beautifully marked with the finest and most regularly arranged patterns which make these algae objects of great beauty when seen under a microscope. Because of the regularity of these marks, certain species are used for testing the resolving power of a **microscope.** Within the wall, the simple **protoplast** contains several **chloroplasts** which may contain a brown pigment that masks the **chlorophyll** present.

Many diatoms are distinctly unicellular organisms; others stick together to form long chains or are included in a gelatinous sheath which forms extensive branching aggregations. There are two large orders of diatoms, separated according to the shape of the cell. One, the Centrales, comprises those diatoms which are radially symmetrical; the other, the Pennales, those which are not radially symmetrical. Many of the Pennales are bilaterally symmetrical, others irregular. Many members of the Pennales move about in a gliding manner.

Two methods of reproduction are found in diatoms. The more common method is asexual. In this process the protoplast of the cell enlarges, pushing the two

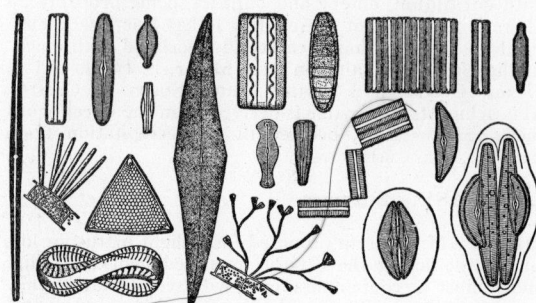

Collection of diatoms showing variation in appearance. (*Kerner's Natural History of Plants, Blackie & Son.*)

valves apart. Then the protoplast divides into two parts, each of which occupies one of the two valves. A new valve is formed over the exposed surface of the protoplast but inside the original valve of the cell. As a consequence the new valve is smaller than the original. As repeated divisions occur the size of the valve gradually decreases. This does not continue indefinitely however. To bring the cell back to its original size, auxospores are formed. When this happens the protoplast enlarges tremendously and escapes from its walls. It then divides and secretes about itself new walls which have the size of the original cell. Auxospore formation is often a result of sexual fusions. Sexual reproduction is accomplished by the fusion of two diatoms. There are several variations in the method in which this process occurs. Sometimes it is a simple fusion of two protoplasts to form one; in other forms, two amoeboid gametes are formed by each protoplast. **Gametes** from different protoplasts fuse, forming **zygospores**. Several other variations are known. Some of the Centrales form small bicilate gametes which fuse.

The diatoms are a very important group of algae. Occurring in immense numbers as they often do, they are the main food substance of many animals, which in turn become food for higher organisms, including man. When a diatom dies and its protoplast disintegrates, the siliceous shell sinks to the bottom of the water. Gradually immense accumulations of diatom valves are formed on the ocean bottom. These may eventually be buried beneath other deposits. They become diatomaceous earth, often forming beds hundreds of feet thick and covering large areas. Diatomaceous earth after removal of organic matter is used as an **abrasive** and scouring agent, as a **filter**, and in many other ways. (R.M.W.)

DIATREME. A general term for volcanic pipes and circular vents, the result of the explosive action of **magmatic** gases. (R.M.F.)

DIAZO PROCESS. The use of organic compounds as light-sensitive media for photographic processes is best exemplified by diazo-compounds. The bleaching of these dyes, or dye intermediates, finds its most important application in plan copying papers, as Ozalid, Oce, and Primuline.

The diazo process is based on the fact that exposure causes diazonium salts to undergo photochemical changes which affect their capacity to form dyes. Dyes are formed when diazonium compounds, or diazotized amines, couple with phenols or aromatic amines in a neutral or alkaline medium. During exposure the diazonium compound undergoes decomposition and is thus rendered incapable of coupling and dye formation. Several types of papers are possible. The undecomposed

diazo-compound can be developed by immersing the paper in an alkaline solution containing a coupler. The decomposed diazo-compound resulting from exposure couples with the undecomposed compound when the paper is made alkaline. Both the diazo-compound and the coupler are coated together on the paper but are prevented from reacting due to the presence of an organic acid. Papers of this type are dry developed by contact with ammonia fumes.

In the Ozalid process based on the patents of Kogel in 1924, use is made of diazo-anhydrides in a neutral medium because these do not couple until they are alkaline.

The coated paper, which can be handled in subdued daylight, is placed beneath a paper tracing and exposed to light. The exposed portions become bleached while the unexposed portions retain their yellow color. After exposure, the paper is passed over a tank of ammonia vapors which convert the unexposed portions to a dye image. Papers which produce various colored images as black, blue, brown, maroon, purple and green are available. Recently, papers capable of rendering a continuous scale of tones have been perfected. Large-size colored images printed from photographic positives are used for posters and advertising. Foils or film coated with diazo-solutions have high resolution and are finding numerous applications in the fields of microfilming, sound recording and in inter-office, plant, and radio communications. (S.M.T.)

DIAZOAMINO-COMPOUNDS. Azo-, Diazo-, and Related Compounds.

DIAZO-COMPOUNDS. Azo-, Diazo-, and Related Compounds.

DICHOTOMY. A system or method of branching in which the main axis divides into two branches, which may in turn branch in the same manner, as for example the **thallus** of an **alga** or an **hepatic**, or the root or stem of a **club moss**. (R.M.W.)

DICHROITE. Cordierite.

DICKCISSEL. Aves, Passeriformes. A small American bird, *Spiza americana*, related to the buntings. It is found in open country and is distinguished by its yellow breast and black throat patch. (A.W.L.)

DICOTYLEDONS. The larger of the two subclasses of **angiosperms** is the dicotyledons, containing over 100,000 species. The plants of this subclass have leaves, flowers and seeds distinctly different from those of **monocotyledons**. The leaves are generally broad and have netted veins; the parts of the flowers occur most frequently in fours or fives or multiples of these numbers. That is, there are 4 or 5 sepals, 4 or 5 petals, 4 or 5 (8 or 10, or more) stamens, and one to many pistils. The embryo of the seed has two cotyledons or seed leaves.

Many important food plants are members of this group; for example, **potato, cabbage, carrot, apples, oranges,** and **peanuts.** Other dicotyledons yielding economic products of great value are the various **rubber** plants, **cotton, flax, sugar** beets, and soy beans. The number cultivated as ornamental plants is too great to enumerate.

Dicotyledons vary in size from tiny annuals an inch or less in height to giant trees 350′ tall. They include herbaceous and woody members, annuals, biennials, and perennials. They are found all over the world, wherever plants can grow. (See also **Paleobotany**.) (R.M.W.)

DICTYOKINESIS. A process of subdivision of the Golgi apparatus of the **cell** during cell division, followed by the distribution of the resulting parts to the daughter cells. (A.W.L.)

DICTYOSOME. Cell.

DICTYOSTELE. Stele.

DIDELPHIA. Mammalia.

DIE. There are several commonly accepted usages of this term. In thread-cutting, a die is a tool resembling a slotted nut, and is used for cutting external threads on screws, bolts, and pipe. In wire drawing, a die is a device used to procure plastic flow of metal; the die is usually, although not necessarily, made with a bell-mouthed circular hole, into which the end of the wire is introduced. In the past, hardened steel dies for wire drawing were extensively used; at the present time, either tungsten-carbide or diamond dies are used, since the wear and abrasion on the die are considerably reduced. (See also **Die Casting, Drop Forging, Plastic Molding, Press Working.**) (H.C.H.)

DIE CASTING. Die castings are produced by forcing molten metal under pressure into a steel die. The pressure is maintained until solidification is complete. The process is essentially a further development of gravity-feed casting, but the pressure function entails finer detail and better finish. While gravity-feed casting tonnage is greater than that of pressure casting, the latter has a wider field of application and is more important in the quantity production of precision parts. Zinc alloys are generally used for die castings, although aluminum alloys, brass alloys and other non-ferrous metals are used to a considerable extent.

The process of die casting is entirely automatic and requires the following elements: a die-casting machine to hold the molten metal under pressure; a metallic mold or die capable of receiving the molten metal, and designed to permit easy and economical ejection of the solidified product; and a casting alloy that will produce a satisfactory product with suitable physical characteristics.

There are two types of die-casting machines: The first, or air-operated machine, forces the material into the die by high pressure on the surface of the molten metal in a special ladle or goose; and the second, or plunger type machine, forces the material into the die by means of a cylinder and piston which are submerged in the molten metal.

Die-casting dies are constructed in different styles for various production requirements. A single die contains an impression of only one part; a multiple die contains two or more impressions of any one part; a combination die contains one impression only of two or more parts; and a combination-multiple die contains a number of impressions of each of two or more parts. Single dies are comparatively cheap and are used for small-lot production, since they reduce the tool investment to a minimum for any one part. Combination dies, when properly planned, will reduce the total die cost for a given set of castings to a minimum. They are applicable to parts that will always be used in the same quantities and of the same alloy. These parts should be of the same general character and weight. Multiple dies are usually slower to operate than single dies but will give higher production rates for the same labor costs.

Die-casting dies are often vented by permitting air to escape through the clearance in the ejector and core pin bearings. The problem of venting is considerably more important than in sand casting because the mold has no porosity. Sometimes dies are vented by grinding shallow grooves on the parting surfaces of the dies; in other instances plugs with suitable vent grooves are added to the die. (H.C.H.)

DIE SINKING. The process of cutting a recess or cavity in a die for drop forging, press working, or plastic molding. (See also **Hob.**) (H.C.H.)

DIELECTRIC ABSORPTION. The persistence of electric polarization in some **dielectrics** after the removal of the polarizing electric field. When a **condenser** with glass plates is connected with a **battery,** the charging current may last, though gradually decreasing, for some minutes or hours; and when the charged condenser is short-circuited, the discharge current may not cease entirely with the first rush. After a **Leyden jar** has been discharged and allowed to stand disconnected for a time, another, smaller spark can usually be obtained from it. This is called a "residual" charge.

By melting mixtures of wax and allowing them to harden in a strong electric field, Eguchi (Japan, 1925) succeeded in obtaining dielectric absorption which persisted almost undiminished for several years. Such a permanently polarized body, singularly analogous to a permanent magnet, has been termed an "electret." (L.D.W.)

DIELECTRIC CONSTANT. Dielectrics.

DIELECTRICS. A dielectric is a body through which, or a medium in which, electric attraction or repulsion may be sustained. Thus glass is a dielectric, because unlike charges on opposite sides of a plate of glass attract each other; likewise two charged bodies immersed in oil or in nitrogen exhibit mutual electric force (though less than in a vacuum), hence these substances are dielectrics. Dielectrics are always insulators; a good conductor completely screens off an electric field (see **Electric Screening**). The **condenser** is the usual form of apparatus for studying and comparing dielectrics.

The explanation of most dielectric phenomena is found in what is known as "electric polarization." When a dielectric is placed in an **electric field,** it is believed that there is a slight shifting of the negatively charged particles (electrons), of which it is partly composed, in one direction and of the positive particles in the other, so that the body as a whole now has an electric moment where before it had none; the electric moment per unit volume being the measure of the polarization. This may result partly from a rectilinear shift of electrons and atomic nuclei, and in many cases partly also from the re-orientation of "polar" molecules, i.e., molecules which have a permanent electric moment of their own.

In dielectrics of this latter type, the polarization is greater in the liquid than in the solid state, presumably because the polar molecules turn about more easily. (The whole phenomenon bears a close analogy to the magnetization of iron; an analogy made still more striking by **dielectric absorption,** q.v.)

A notable result of the polarization of a dielectric is the reduction of the electric intensity due to any given distribution of electric charges. It is shown in dielectric theory that if the intensity in a vacuum, called the electric displacement, is denoted by D, that in a dielectric whose polarization is P is $E = D - 4\pi P$. The polarization P is apparently proportional to D. Hence E is proportional to D, and the ratio D/E, which is greater than unity, is called the dielectric constant of the substance, and usually denoted by k. It varies widely for different substances, as will be noted in the brief table below:

Dielectric	k	Dielectric	k
Air	1.0006	Mica	5.7
Alcohol	28.4	Paraffin	2.1
Carbon dioxide	1.001	Porcelain	5.7
Glass (flint)	9.9	Rubber	2.2
Hydrogen	1.0003	Sulfur	4.0
Linseed oil	3.3	Water	81.1

A convenient measure of the dielectric constant is the ratio of the **capacitance** of a condenser filled with the dielectric in question to that of a similar condenser with the dielectric removed, leaving a vacuum. The electric energy per unit volume in the polarized dielectric, for a given electric field intensity, is also proportional to the dielectric constant; therefore, if the dielectric constants of glass and air are in the ratio 10:1, a condenser with glass plates charged, say, to 1000 volts, has ten times as much stored electrical energy as the same condenser charged to the same voltage but with air substituted for the glass.

With non-polar dielectrics the dielectric constant k depends, for a given substance, upon the density ρ (and hence upon the temperature), increasing as ρ increases in accordance with a formula due to Clausius and Mosotti:

$$k = \frac{2a\rho + 1}{1 - a\rho};$$

in which a is a constant for the given dielectric. For different substances it depends upon a property of the molecule known as its "polarizability," which is the increase of electric moment produced in the molecule by unit electric field intensity. (See also **Electric Insulation**.) (L.D.W.)

DIESEL CYCLE. Although modified from the inventor's original conception, the modern Diesel cycle retains the most important feature, namely that of compression of air to the ignition temperature, followed by timed introduction of fuel. This cycle is shown in the accompanying diagram. The solid line indicates a theo-

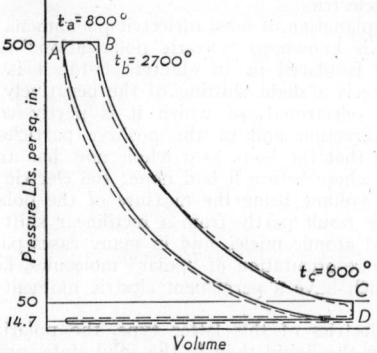

Showing the departure of the actual cycle from the theoretical, or air standard cycle.

retical cycle, the dotted line shows how a slow-speed actual cycle may depart from the theoretical. Typical temperatures are also indicated. Beginning with point D on the cycle, imagine that a cylinder filled with air is closed at the end by a tightly fitting position. The piston is moved to compress the air without addition or loss of heat through the cylinder walls. As the air is decreased in volume, the pressure rises adiabatically, and it arrives at the condition corresponding to point A. The piston is then reversed in direction, and starts to move so as to increase the volume of the air. The air is very hot due to its adiabatic compression. In fact, it is well above ordinary ignition temperatures of petroleum products. As the piston starts to move, carrying the cycle from point A, fuel is injected or sprayed into the cylinder just rapidly enough so that its combustion will keep the pressure up while the volume is being increased, at least up to point B. At B when the outward stroke is partially completed, the fuel is cut off, and the products of combustion expand adiabatically from B to C, giving work to the piston as they do. At C the exhaust valve opens, and the pressure drops to D. The line extending horizontally from D represents the theoretical exhaust and suction stroke. Adiabatic expansion and compression are not possible in a cylinder which

must be well cooled in order to maintain a lubricating oil film. Therefore an actual cycle will not be expected to follow the adiabatic. Another difference between actual and theoretical cases is the composition of the gas within the cylinder. Theoretical studies are made assuming pure air in the cylinder. Actually there is a little burned gas present during the compression, and a great deal of it during the expansion strokes. However, by assuming no friction loss, adiabatic compression and expansion, and air, only, in the cylinder, an expression may be derived for the efficiency of the cycle $ABCD$. (See **Air Standard Efficiency**.) (F.T.M.)

DIESEL ENGINE. In a patent dated 1892, Dr. Rudolf Diesel, a German engineer, described an engine to operate on the **Carnot cycle**. Coal dust was the fuel, and it was to be fed rapidly enough so that isothermal expansion would result. After fuel cut-off, an adiabatic expansion would continue, followed by a compression made isothermal by the injection of water into the cylinder. An adiabatic compression then brought the cycle back to its beginning. A further claim of the patent covered the use of liquid fuels and the spray valve. Early attempts to build this engine resulted in the adoption of a modified cycle which, after much experimentation, was built into a successful working engine. Since then the Diesel has slowly but surely established for itself a secure position as a prime mover.

The reasons for continued growth of Diesel engine power are to be found in the advantages of the Diesel over other prime movers for certain classes of service which abound in this country, and to the comparative low cost of high-grade petroleum fuel.

The Diesel applications at present may be divided into *mobile* and *stationary*. Marine and locomotive service absorbs a large proportion of the annual output of engine builders. Stationary Diesels are to be found in all kinds of factories, especially small factories. A great many of them are used for pumping oil in pipe lines, and from wells, and for pumping water both for drainage and irrigation. They are also in service in mines for pumping, for compressing air, and for electrical service.

The Diesel is an excellent prime mover for electrical generation in capacities of from 100 to 5000 horsepower. As such, it is widely used by private industry, hotels, utility companies, and municipalities, especially in the water and light plants of the latter.

The advantages of the Diesel engine are:
1. Low fuel cost.
2. No long warming-up period.
3. No standby losses.
4. Uniformly high efficiency of all sizes.
5. Simple plant layout.
6. No large water supply needed.

The Diesel can extract more work out of each heat unit than any other engine in the world. For that reason it becomes an attractive prime mover wherever first cost is written off slowly enough so that operating costs are influential.

Where fuel prices favor oil, where prices of all fuels are extraordinarily high, where water supply is either of poor quality or inadequate, where loads are relatively small and reliable transmission line service is not available at competitive rates, and where industrial power rates are too high, the Diesel will continue in popularity as an efficient, small capacity, reliable prime mover.

The stationary Diesel engine is usually a heavily built engine having a piston which reciprocates in a cylinder. A connecting rod is either pinned directly into a trunk type piston, or to a crosshead to which the piston is connected by a piston rod. The connecting rod bears on a crank, which has bearings in the main frame. Added to these parts are the auxiliaries such as valves and valve gear, fuel injection systems, water circulation systems, starting systems, etc. The Diesel is more heavily built than the gasoline engine, and is a relatively slow-moving machine. Rotative speeds are commonly 100 to 750 rpm, except that automotive Diesels are

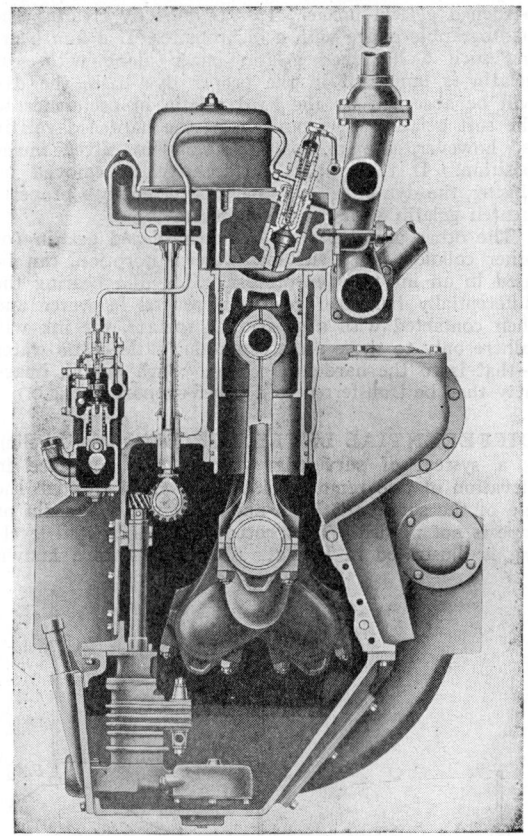

Cross-section of Diesel engine. (*Caterpillar Tractor Co.*)

designed for 2000 rpm and over. The fuel ordinarily employed is a product from crude petroleum. The low cost of good-grade petroleum fuel in the United States has, to the present time, precluded the general use of any other type of Diesel fuel. **Fuel oil** is the residue left when the distillation has removed the gasoline, kerosene, and light distillates from the crude. The heavier residues are also removed from the better grades of fuel oil. This fuel is pumped into the cylinder during the first part of the power stroke, correctly metered so that its combustion tends exactly to offset the drop of pressure which would otherwise be experienced. During combustion of the fuel in the Diesel engine the pressure remains approximately constant. After a small portion of the power stroke is completed, the fuel is cut off, and the products of combustion do work on the piston expansively.

The principal and important difference between the oil and the gasoline engine resides in the method of **ignition**. Compression of the air trapped in the cylinder of a Diesel engine is employed as its means of ignition. The compression is carried to much higher pressures in the Diesel than in the Otto cycle; consequently, the temperature at the end of compression is higher in the Diesel cycle. In fact, compression is carried high enough so that the temperature of the compressed gas exceeds the ignition temperature of the fuel. This is compression ignition. It may require the volume after compression to be only one-fourteenth of that before compression, whereas it is only a fifth or a sixth in the gasoline engine. It is not possible to use compression ignition in the ordinary Otto cycle engine, because an

inflammable charge is compressed, and the compression to the ignition temperature would cause spontaneous, uncontrolled, unregulated ignition. However, the charge compressed in the Diesel is fresh air—incombustible. This air is compressed until its pressure is over 500 lbs. per sq. in., and its temperature around 800° F. Then the oil is injected into the hot compressed air whose temperature is sufficiently high to cause immediate ignition of the spray.

The Diesel engine is built in both **two-cycle** and **four-cycle** types. The fuel is injected in one of two ways, either by being blown in and atomized by a high-pressure air jet (air injection), or sprayed in through a fine nozzle tip under the influence of an extremely high oil pressure created by pumps (solid injection). Thus there are four possible combinations. The slow-speed types in general use are either the two-cycle solid injection, or the four-cycle air injection. With few exceptions, high-speed Diesels are four-cycle, solid injection. Governing is accomplished by control of the fuel oil pump, more or less oil being delivered per stroke, depending on the load. For electrical generation even the multicylinder engines must be equipped with a heavy flywheel to prevent cyclic variation of speed.

A common method of starting large Diesels is by compressed air. In addition, the air-injection types require an injection air system and some airless-injection types have a scavenging air system. Both injection and scavenging air pumps are integral with the engine.

The temperatures existing in the stationary Diesel engine would soon break down the film of lubricating oil on the cylinder liners and otherwise put the engine out of service by warping of valves, pistons, etc., were the engine not cooled by circulating water through jackets surrounding the heated parts.

The heart of a Diesel engine is the **combustion chamber** end of its cylinder. The shape of the cylinder head and face of the piston, and the design of the nozzle and the injection system must be carefully considered. To obtain the complete combustion necessary to good efficiency, a fuel must be thoroughly mixed with the air charge, so that all particles of it will be burned. There must be good penetration of the oil spray into the highly compressed dense air, and there must be turbulence to insure mixing of the oil spray with the air. These two important characteristics are secured by special design, both of the cylinder head and the spray valve.

In the two-cycle engine there is no valve gear. The absence of this feature is, indeed, the virtue of the two-cycle principle. In the four-cycle engine the exhaust and inlet valves are mechanically operated from a camshaft. Since the Diesel engine is commonly rather large, the valves are correspondingly large in girth, and are operated from a massive camshaft. Although the Diesel engine is basically a slow-speed type, refinements of design, especially in the combustion chamber, have enabled builders to produce medium- and high-speed compression ignition engines. The illustration shows a cross-section of a medium-speed, four-cycle, solid injection engine. Many of its mechanical features are similar to the common automobile engine. The fuel system consists of an *injection pump* (one per cylinder), cam driven, which delivers a metered quantity of fuel oil at an extremely high pressure to an *injection nozzle*. The latter opens into a pre-combustion chamber into which some of the combustion air has been compressed. Partial combustion occurs in this chamber, then the swirling, gasified, fuel-air mixture is ejected into the main combustion chamber where nearly complete combustion will be achieved. (See **Injection, Fuel.**) (F.T.M.)

DIFFERENCE EQUATIONS. A difference equation may be described as an **equation** connecting values of an unknown **function** at two or more equally spaced values of the independent variable, as, for example, $u(x + h) - u(x) = 2x + 1$, or $u(x + 2h) + 2xu(x + h) - u(x) = x^2$.

A linear difference equation is one in which the unknown function $u(x)$ occurs linearly and the coefficients are functions of the independent variable only. A homogeneous difference equation contains no term independent of the unknown function u. (L.L.S.)

DIFFERENTIAL, AUTOMOBILE.

The differential drive is an important element in the automobile. As a 4-wheel vehicle rounds a corner, the outer wheels travel a greater distance than the inner. The wheels on a wagon are mounted on a dead axle, so that they turn independently of each other. On a live axle some device which will permit them to revolve at different speeds to compensate for the difference in travel when rounding a curve is necessary. The ordinary automobile differential is illustrated in Fig. 1. The driveshaft has

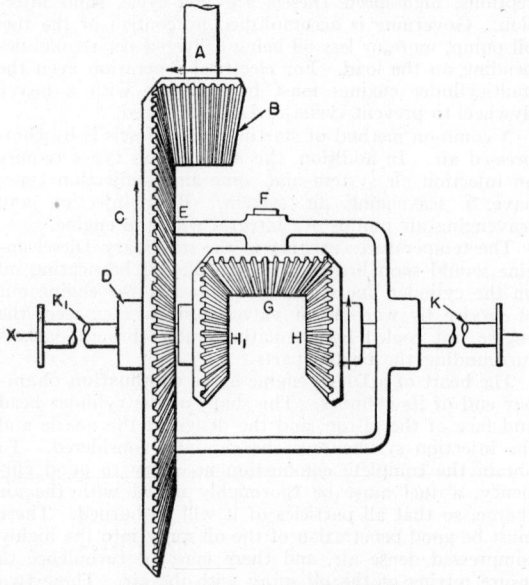

Fig. 1. Bevel gear differential.

mounted on it pinion B, which drives gear C. If it were not for the necessity of rounding curves, gear C could be rigidly fixed to the live axle KK. The differential action is obtained as follows: Gear C is not keyed to the axle. The spider E is rigidly fastened to the gear and has mounted on it, free to turn, the bevel gear G. Gear G meshes with gears H_1 and H, each of which is keyed to a half of the axle. When traveling straight ahead, gears G, H_1, and H revolve with the spider, but do not have any motion relative to each other. When rounding a curve, one wheel must travel faster than the other. The difference in rotation of the axle is compensated for by rotation of the differential gear G on its pin F. Any accelerated motion of one wheel is offset by a retarded motion of the other. (F.T.M.)

DIFFERENTIAL EQUATIONS.

An **equation** which involves **derivatives** (or **differentials**) is called a differential equation.

If **partial derivatives** occur in the equation, it is called a partial differential equation; if not, it is called an **ordinary differential equation**. (L.L.S.)

DIFFERENTIAL HARDENING.

A method of preparing printing plates or matrices used in **subtractive color** photography. The value of this process is dependent on the fact that (1) certain dyes exhibit a marked affinity for hardened gelatin as compared to unhardened gelatin, and (2) that hardened gelatin will adsorb a relatively small volume of water as compared to soft gelatin.

The first fact is made use of by employing a photographic silver image to control the position of the hardened gelatin image. This is done by treating the photographic image with a dichromate. If a film bearing such a hardened gelatin image along with soft gelatin is immersed in the proper dye bath, the dye will be absorbed by the hard gelatin in preference to the soft gelatin. The position of the hardened gelatin is, however, the original photographic silver image position. If the surplus surface dye is removed by rinsing, the dye image may be transferred to a properly treated gelatin surface by imbibition.

The other useful property of a hardened gelatin (or other colloid) image, its low water absorption, can be used in an ink-transfer process. If a film bearing the differentially hardened gelatin is soaked in water and then contacted with a greasy ink surface, the ink will adhere only to those areas containing little or no water —that is to the hardened gelatin. Such an ink image may then be transferred to a paper support. (H.C.C.)

DIFFERENTIAL LEVELING.

Differential leveling is a system of surveying whereby the difference in elevation of two remote points is obtained through the use of the surveyor's **level** and **level rod**. A chain or tape is not needed. The procedure in differential leveling is illustrated in Fig. 1. BM_1 represents a known

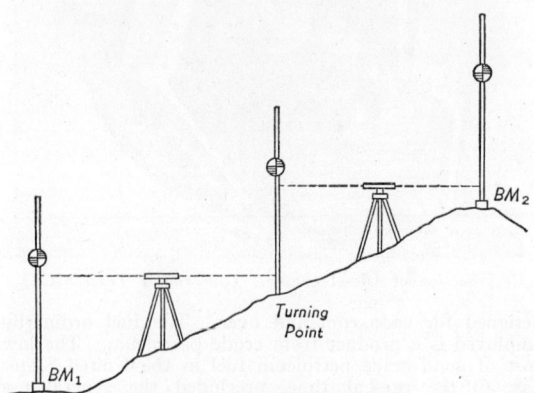

Fig. 1. Differential leveling.

bench mark. The elevation of BM_2 is to be found. The rod is held on BM_1 and the level set up so as to take a back sight on the rod. The rodman then advances to a turning point chosen by the instrument operator, and the telescope is swung around for a foresight reading on the rod. The levelman then advances the instrument to a new position, from which he takes a back sight on the rod, which is still at the turning point. This procedure is continued until the rodman reaches the site of BM_2. The back sight reading, added to the elevation of BM_1, gives the elevation of the level at the first station. The foresight reading, subtracted from the instrument elevation, gives the elevation of the turning point. In this way, by additions and subtractions of back sight and foresight readings, the total difference of elevation between BM_1 and BM_2 is determined.

The field work which is necessary for determining the elevation of points along a given line such as the center line of a railroad or highway is called profile leveling. Rod readings are taken at regular intervals and also at points of abrupt change of slope. The outline of a vertical section through the center line is called a profile. The profile is obtained by plotting elevations which are the result of profile leveling. (F.T.M.)

DIFFERENTIAL MANOMETER. A device for measuring small pressures. The device is best explained by referring to the figure. A U-tube, equipped with an

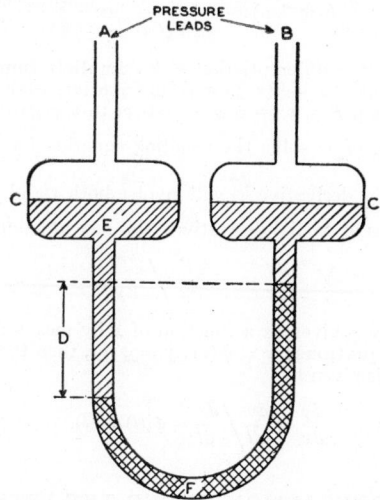

Fig. 1. Differential manometer.

enlarged section (c) at the top of each side, has in it two immiscible liquids, a lower (heavier) liquid F and an upper (lighter) liquid E. If a pressure difference be set up across A and B, the liquid F will change position giving a head D to compensate for the pressure. Because of the enlargements in the upper tubes the top level of liquid E changes a negligible amount. The head equivalent to the pressure varies inversely as the difference between the densities of liquids E and F. By selecting the proper liquids, the difference can be made very small, making for a large head. Thus a pressure which would give only a small reading on an ordinary manometer can be made to produce a large reading on a differential manometer, thus increasing the accuracy of measurement. (D.E.M.)

DIFFERENTIALS. A differential is a fundamental mathematical concept closely associated with the idea of the rate of change of a function.

For a function of one variable, the differential is defined as follows: Let $y = f(x)$ be a given function; assign an arbitrary increment Δx to x, then the function y takes an increment Δy given by $\Delta y = f'(x) \cdot \Delta x + \epsilon \cdot \Delta x$, where $f'(x)$ is the **derivative** of y and ϵ is a variable which approaches o as $\Delta x \to$ o. Then the first term $f'(x) \cdot \Delta x$ is called the differential of $y = f(x)$, and is denoted by dy or $df(x)$. The differential of the **independent variable** x is the same as Δx, i.e.: $dx = \Delta x$; but dy is not equal to Δy, where y is the function. We may regard dy or $f'(x) \Delta x$ as an approximate value of Δy.

Since $dy = f'(x) dx$, it follows that $f'(x) = dy/dx$, so that the derivative may be regarded as the quotient of dy by dx, which justifies the notation dy/dx commonly used for the derivative.

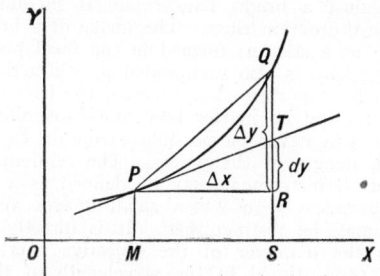

In the accompanying figure, PT is the tangent to the curve $y = f(x)$ at the point P. Then $PR = \Delta x$, $RQ = \Delta y$, and $RT = dy$.

Since dy is an approximation to Δy, if x has a small error Δx, then Δy will be the corresponding error in y, and dy may be taken as the approximate small error in y.

The differential of a function of several variables is defined as follows: Let $u = f(x, y)$ be a function of two variables, and let x and y take arbitrary increments Δx and Δy, then the function u takes an increment Δu given by

$$\Delta u = \frac{\partial u}{\partial x} \cdot \Delta x + \frac{\partial u}{\partial y} \cdot \Delta y + \epsilon_1 \cdot \Delta x + \epsilon_2 \cdot \Delta y,$$

where ϵ_1 and ϵ_2 approach o as $\Delta x \to$ o and $\Delta y \to$ o. Similarly for a function of three or more variables. The total differential of the function $u = f(x, y)$ is defined as the principal part of the total increment Δu:

$$du = \frac{\partial u}{\partial x} \cdot \Delta x + \frac{\partial u}{\partial y} \cdot \Delta y.$$

We define $dx = \Delta x$ and $dy = \Delta y$ for the independent variables. Hence

$$du = \frac{\partial u}{\partial x} \cdot dx + \frac{\partial u}{\partial y} \cdot dy.$$

Similarly for functions of three or more variables.

For the calculation of small errors in a function of several variables $u = f(x, y)$, we can use du as an approximation to Δu, so that if $\Delta x = dx$ and $\Delta y = dy$ are small errors in the independent variables, then $du = \frac{\partial u}{\partial x} \cdot dx + \frac{\partial u}{\partial y} \cdot dy$ is the approximate small error in u. (L.L.S.)

DIFFERENTIATION. In geology, differentiation refers to the general process of formation of different types of **igneous** rocks from a common parent magma.

In mathematics, the use of the term differentiation is discussed in the articles on **Derivative of the Function of a Variable**; and **Differentiation, Technique of (Mathematical)**.

In biology, differentiation is the development of the varied structures within living units through which special functions are performed. Differentiation is expressed within the **cell** in its various parts and in 1-celled organisms further differentiation gives rise to the **organelles** which carry on locomotion, digestion, and other functions. In the complex body during its development this process results in the appearance of the different kinds of cells, tissues, and organs which ultimately compose the individual. (A.W.L., R.M.F.)

DIFFERENTIATION, TECHNIQUE OF (MATHEMATICAL). General **differentiation** rules, and differentiation rules for **algebraic functions**:

1. The **derivative** of a **constant** is zero: $\frac{d}{dx}(c) = $ o.

2. The derivative of an algebraic sum of any number of functions is equal to the algebraic sum of their derivatives:

$$\frac{d}{dx}(u + v + w) = \frac{du}{dx} + \frac{dv}{dx} + \frac{dw}{dx}.$$

3. The derivative of a constant times a function is equal to the constant times the derivative of the function:

$$\frac{d}{dx}(cu) = c \cdot \frac{du}{dx}.$$

4. The derivative of the product of two functions is equal to the first factor times the derivative of the second plus the second factor times the derivative of the first:

$$\frac{d}{dx}(uv) = u \frac{dv}{dx} + v \frac{du}{dx}.$$

5. The derivative of the quotient of two functions is equal to the denominator times the derivative of the numerator minus the numerator times the derivative of the denominator, all divided by the square of the denominator:

$$\frac{d}{dx}\left(\frac{u}{v}\right) = \frac{v\frac{du}{dx} - u\frac{dv}{dx}}{v^2}.$$

6. The derivative of a constant **power** of a function is equal to the **exponent** times the function with its exponent diminished by one times the derivative of the function:

$$\frac{d}{dx}(u^n) = nu^{n-1}\frac{du}{dx}.$$

7. The derivative of a function of a function is expressed by: If y is a function of z and z is a function of x, then

$$\frac{dy}{dx} = \frac{dy}{dz} \cdot \frac{dz}{dx}.$$

8. If $x = \phi(y)$ is the **inverse function** to $y = f(x)$, then the derivatives of these two functions are **reciprocals** of each other: $\frac{dx}{dy} = 1/\frac{dy}{dx}.$

For transcendental functions, the differential rules are:
9. The derivative of the **exponential function** is given by:

$$\frac{d}{dx}(a^u) = a^u \log_e a \frac{du}{dx} \quad (a \text{ constant}),$$

$$\frac{d}{dx}(e^u) = e^u\frac{du}{dx}, \quad \frac{d}{dx}(e^x) = e^x.$$

10. The derivative of the **logarithmic function** is given by:

$$\frac{d}{dx}(\log_a u) = \frac{1}{u} \cdot \log_a e \cdot \frac{du}{dx},$$

$$\frac{d}{dx}(\log_e u) = \frac{1}{u} \cdot \frac{du}{dx},$$

$$\frac{d}{dx}(\log_{10} u) = \frac{M}{u} \cdot \frac{du}{dx}, \quad \text{where} \quad M = \log_{10} e,$$

$$\frac{d}{dx}(\log_e x) = \frac{1}{x}, \frac{d}{dx}(\log_{10} x) = \frac{M}{x}.$$

11. The derivatives of the **trigonometric functions** are given by:

$$\frac{d}{dx}(\sin u) = \cos u \frac{du}{dx}, \qquad \frac{d}{dx}(\cos u) = -\sin u \cdot \frac{du}{dx},$$

$$\frac{d}{dx}(\tan u) = \sec^2 u \frac{du}{dx}, \qquad \frac{d}{dx}(\cot u) = -\csc^2 u \frac{du}{dx},$$

$$\frac{d}{dx}(\sec u) = \sec u \tan u \frac{du}{dx}, \quad \frac{d}{dx}(\csc u) = -\csc u \cot u \frac{du}{dx}.$$

12. The derivatives of the **inverse trigonometric functions** are given by:

$$\frac{d}{dx}(\sin^{-1} u) = \frac{1}{\sqrt{1 - u^2}} \cdot \frac{du}{dx},$$

$$\frac{d}{dx}(\cos^{-1} u) = -\frac{1}{\sqrt{1 - u^2}} \cdot \frac{du}{dx},$$

$$\frac{d}{dx}(\tan^{-1} u) = \frac{1}{1 + u^2} \cdot \frac{du}{dx},$$

$$\frac{d}{dx}(\cot^{-1} u) = -\frac{1}{1 + u^2} \cdot \frac{du}{dx},$$

$$\frac{d}{dx}(\sec^{-1} u) = \frac{1}{u\sqrt{u^2 - 1}} \cdot \frac{du}{dx},$$

$$\frac{d}{dx}(\csc^{-1} u) = -\frac{1}{u\sqrt{u^2 - 1}} \cdot \frac{du}{dx}.$$

13. For the differentiation of an **implicit function in two variables**, say $F(x, y) = 0$: differentiate each term of the equation $F(x, y) = 0$ with respect to x, regarding y as a function of x; solve the resulting equation for $\frac{dy}{dx}$. In general, the derivative $\frac{dy}{dx}$ will involve both x and y.

In terms of partial derivatives, this rule amounts to:

$$\frac{dy}{dx} = -\frac{\partial F}{\partial x}\Big/\frac{\partial F}{\partial y}.$$

14. If y is given as a function of x by means of **parametric equations**: $x = \phi(t)$, $y = \psi(t)$, then the derivative of y is given by:

$$\frac{dy}{dx} = \frac{dy}{dt}\Big/\frac{dx}{dt} = \psi'(t)/\phi'(t).$$

(L.L.S.)

DIFFERENTIATION UNDER THE INTEGRAL SIGN. For the differentiation of a definite integral of a function $f(x, \alpha)$ containing a parameter α, when the limits of the integral are constants a and b, we have the formula

$$\frac{d}{d\alpha}\int_a^b f(x, \alpha)dx = \int_a^b \frac{\partial f}{\partial \alpha} \cdot dx,$$

and when the limits of the integral are functions of α: u and v, we have the formula

$$\frac{d}{d\alpha}\int_u^v f(x, \alpha)dx = \int_u^v \frac{\partial f}{\partial \alpha} dx + f(v, \alpha)\frac{dv}{d\alpha} - f(u, \alpha)\frac{du}{d\alpha}.$$

(L.L.S.)

DIFFRACTION. A class of phenomena arising from the interruption of a wave-train, as of light, by one or more opaque obstacles. For example, if light from a point source passes the edge of a postcard and falls upon a white screen, the shadow of the edge is not sharply defined, but deepens to darkness gradually on one side, and is bordered by very narrow alternate bright and dark **interference** fringes (diffraction bands) on the other.

The flux density at any point P in an uninterrupted wave field is either constant or varies progressively with the position of P. But if the waves have to pass obstacles before reaching the point P, there not only will be shadows, but also the flux density at P will be subject to interference effects due to phase differences in the waves reaching it from different parts of the advancing wave front (see **Huygen's Principle**). The **diffraction grating** and the **zone plate** are dependent upon well-recognized diffraction principles.

The image of a minute opaque speck under magnification against a bright background is surrounded by concentric diffraction rings. The image of a bright object, such as a star, as formed in the focal plane of a converging lens, is also surrounded by diffraction rings. If two such images are close together the fringe systems will overlap and no matter how much magnification is applied it will never be possible to obtain clear, well-separated, images of the points. The resolving power of an optical instrument may be defined as a measure of the sharpness with which small images very close together may be distinguished. It is directly proportional to the diameter of the objective aperture and inversely proportional to the wavelength of the light.

Diffraction thus limits the resolving power, and hence the practicable magnification, of optical instruments. (L.D.W.)

DIFFRACTION GRATING.

A series of very fine, closely spaced parallel slits, or of very narrow, parallel reflecting surfaces, which, when light is incident upon it at a definite angle, produces a succession of spectra. The complete optical theory is somewhat complicated, but the action of a plane transmission grating may be explained approximately as follows.

A plane, monochromatic light wave W, incident at angle i (see figure), reaches the slits at different times. A

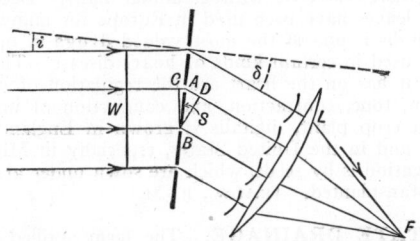

Diagram showing diffraction by a plane grating.

lens L receives the waves emerging from any two adjacent slits, A and B (among many others), after they have traveled paths differing by $CA + AD$; that is, by $S \sin i + S \sin \delta$, in which $S = AB$. If the lens is so placed that this path difference is a whole number of wavelengths, $n\lambda$, the successive wave-trains will reach it in the same phase, so that when they are brought to the focus F, they will be in synchronism and will produce a bright image of the distant source. Therefore any angle δ for which this result is possible is subject to the condition

$$S \sin i + S \sin \delta = n\lambda,$$

or

$$\sin \delta = \frac{n\lambda}{S} - \sin i.$$

Bright images will be produced for those angles δ which correspond to $n = 1, 2, 3, 4, \cdots$; the numbers denote the "orders" of the images. It is easily shown that for any order the total deviation $(i + \delta)$ is least when $\delta = i$ and therefore when $\sin \delta = \frac{n\lambda}{2S}$. If the incident light is composed of various wavelengths, the corresponding images of any order will appear at different points, since δ varies with λ; and the result is a **spectrum**. In short, the grating acts as a **dispersion** piece, and as such is of great value in **spectroscopes**.

For high dispersion the slits must be very fine and very close together (S small), and for high resolving power (sharpness of spectral lines) the total number of slits must be large. Gratings having several thousand slits to the inch of width are common. They may be made by ruling fine scratches with a diamond point on glass, or, with reflecting gratings, on polished metal. If the rulings are not spaced with absolute regularity, false lines, called "ghosts," appear in the spectrum.

Rowland was the first to rule reflection gratings on concave metal surfaces. Such gratings eliminate the necessity of the spectroscope collimator or focusing lenses, as they take light direct from the spectroscope slit and form the spectral-line images like a concave mirror. The **echelon** is another special type of grating. (L.D.W.)

DIFFUSER.

A diffuser is a passage so shaped that it will change the characteristics of a fluid flow from a certain pressure and velocity to a lower velocity and a higher pressure. The diffusion must be carried out in a well-streamlined passage having smooth interior surfaces, and sides not diverging at so great an angle as to cause the fluid to leave the sides of the diffusing chamber. By reducing the velocity through increasing the cross-sectional area of flow, the pressure may be built up as the velocity head is diminished. Diffusers are applied to centrifugal **fans, centrifugal pumps,** jet pumps, **wind tunnels,** and other equipment where it is required to conserve energy by efficiently converting velocity head into pressure. (F.T.M.)

DIFFUSION OF FLUIDS.

The **molecules** of a gas or of a liquid wander about ceaselessly, colliding frequently and exchanging kinetic energy, but maintaining a certain aimless progress. If an enclosure contains two gases, the lighter initially above and the heavier below, the gases at once begin to mingle because of their molecular motion. The same is true of a dense solution (as of sugar) and pure water; both the sugar and the water molecules wander across the boundary, so that in the course of time the whole body of liquid attains nearly uniform concentration. The process whereby this is effected is called diffusion. In the case of fluids of different color, its progress may be easily watched.

The rates at which different gases diffuse at a given temperature are inversely proportional to the square roots of their molecular weights. Thus, hydrogen diffuses four times as fast as oxygen. This follows, according to the **kinetic theory,** from the fact that the molecules of various kinds have the same mean kinetic energy and hence their mean square speeds are in the inverse ratio of their masses. In the case of a solution of non-uniform concentration, the diffusion of the solute from the more to the less concentrated regions takes place in accordance with Fick's law, expressed by the equation

$$\frac{dm}{dt} = -DS\frac{dc}{dx}.$$

This gives the mass of solute diffused per unit time through a cross-section S, in terms of the concentration gradient dc/dx in the direction x perpendicular to the cross-section. D is a constant for the given solute and solvent at a given temperature, and is called the diffusion coefficient. For any one pair of substances, D is found to be proportional to the absolute temperature. It should be stated that these laws apply only to non-electrolytic solutions.

Both gases and solutions exhibit selective diffusion through suitable porous partitions or membranes. Partitions may be used which will allow the smaller but not the larger molecules to pass through. There results an increase of pressure on the side where the larger molecules are. The phenomenon is called osmosis, and the osmotic pressures thus developed play an important part in many physiological processes. (L.D.W.)

DIFFUSION PUMP. Air Pumps.

DIGESTER.

In the process of paper making, the wood is first reduced to chips, which are then reduced to a pulp by cooking with a solvent in a digester. In the two principal methods of chemical pulp manufacture, either soda or sulfite liquor is mixed with the chips in the digester tank in definite proportions. The whole is heated for several hours by high pressure steam. During this process the wood chips are disintegrated, freeing the cellulose for further use in the paper-making process. (F.T.M.)

DIGESTION.

The process of mechanical treatment and chemical transformation by which food is prepared for absorption by the body.

The **foods** of animals consist of the complex organic materials, **proteins, carbohydrates,** and **fats.** Proteins are abundant in lean meat and in some plant products, including beans. Carbohydrates include starches and sugars and fats include both the animal fats and vegetable oils such as occur abundantly in nuts. With the exception of a few simple sugars (grape sugar) none of these materials can be absorbed into the animal body without first being transformed into simpler compounds.

In the **digestive system** the foods are mixed with secretions containing **enzymes** which bring about the chemical changes. The mixing is accomplished by mechanical processes dependent upon muscular action. Some animals chew their food to break it up into smaller particles and others grind it in a special region of the tubular tract. Once the food enters the tubular alimentary tract of the more complex animals such as man, it is propelled and mixed by peristalsis.

The enzymes in the digestive secretions change the food by hydrolysis, a process in which molecules of food and water together are split into molecules of simpler composition. Ultimately proteins are broken up into amino acids, the carbohydrates into simple sugars, and the fats into fatty acids and glycerol. These end-products are absorbed into the tissues of the body and there used to synthesize the necessary proteins, carbohydrates, and fats, differing chemically from those originally taken in as food. (A.W.L.)

DIGESTIVE SYSTEM. The organic system which receives food and prepares it for absorption.

One-celled animals and **sponges** take food into the cell to be transformed by a process of intracellular digestion. While this process persists to a limited extent in **coelenterates,** flatworms, and mollusks (**Mollusca**), these and all other animals also have some form of digestive system or alimentary tract in which food is retained for extracellular digestion preceding absorption. Secretions are discharged into this tract by the cells of its lining and are mixed with the food. The cavity is lined with endodermal tissue, in many cases supplemented by ectodermal ingrowths at both ends.

The simplest form of digestive system is the **enteric cavity** of coelenterate **polyps.** It is little more than a sac with one opening through which food enters and undigested wastes are discharged. In the flatworms a similar condition prevails, but in both the **jellyfishes** and in some flatworms the cavity is complex, extending throughout the body in a system of canals or branches which distribute the food as well as absorb it. From the roundworms through the remainder of the animal kingdom the system is tubular, opening at one end by the mouth and at the other by the anus.

In the tubular digestive tract specialization of digestion reaches a maximum. Here food passes successively through different regions instead of being mixed indiscriminately, hence each region may subject it to special treatment. The chief regions are those which aid in securing food, simple passages, storage reservoirs, grinding structures, and digestive regions which include chambers and tubular regions. In addition, glandular derivatives of the lining are so highly developed that they become separate organs associated with the tubular tract by slender ducts.

Some of the worms have a very simple tract with a muscular **pharynx** which aids in securing food and a long simple intestine, in which it is digested. Other animals, including the leeches, insects, and birds, have a crop in which food is stored prior to digestion. The **mastax** of rotifers and the **gizzard** of birds are examples of grinding structures.

The mammalian alimentary tract is a good example of regional specialization. The oral cavity with its teeth provides for chewing and some digestion, the **pharynx** and **oesophagus** furnish a passage to the **stomach,** where food is stored and slightly digested, the small **intestine** completes digestion and absorbs the end-products, the large intestine absorbs water, and the rectum stores the remaining wastes for periodical discharge by way of the anus. Glands associated with this tract are the **salivary glands,** the **liver,** and the **pancreas.** (A.W.L.)

DIGITALIS. *Digitalis purpurea.* Foxglove. Scrophulariaceae. The foxglove is a biennial often grown as an ornamental plant. The first year of growth produces only the long basal leaves, while in the second year the erect leafy stem 2–5′ tall is developed. The flowers are borne in a **raceme** which through the bending of the **peduncles** or individual flower stalks becomes one-sided. The purple flowers have a 5-parted **calyx;** a tubular bell-shaped **corolla** obscurely five-lobed, five **stamens** and a single **pistil.** They are pollinated mainly by bees. The fruit is a 2-celled capsule.

The drug digitalis is prepared mostly from leaves of the second year's growth. These are rather coarse ovate leaves covered with glandular hairs. Decoctions of the leaves have been used in Europe for many years.

Digitalis is one of the most valued **drugs** in medicine and is used in certain kinds of **heart** disease. The chief effects it has on the heart are the regulation of its rate, rhythm, tone, contraction, and conduction of impulses.

As a crop plant, digitalis is grown in England, Germany, and in the United States, especially in Michigan. Propagation is by seeds, which are sown under glass and later transplanted. (R.M.W., R.S.M.)

DIGITATE DRAINAGE. The term applied to the finger-like pattern of stream valleys. Such a stream pattern usually develops only when the underlying formations are relatively horizontal or if folded, faulted, or **metamorphosed,** are relatively of equal hardness and solubility. (R.M.F.)

DIHEDRAL. The dihedral of an airplane measures the amount of tilt of the wings upward from a normal horizontal axis. Though the right and left wing of an airplane may, to the casual observer, appear to be rigged in a straight line, actually there is a slight angle between them. This is for the purpose of procuring lateral stability. The low-wing monoplane is inherently less stable than the high-wing monoplane or the biplane, and one will note that its wings are set at a more pronounced angle. Dihedral is measured in degrees from the normal horizontal axis to the plane of the wing. It is ordinarily from 1–4° in magnitude. (F.T.M.)

DIHEDRAL ANGLE. A dihedral angle is formed by two intersecting planes; it is measured by the corresponding plane angle formed by drawing a plane perpendicular to the intersection of the planes. (L.L.S.)

DIKE (DYKE). A tabular, intrusive mass of **igneous** rock which cuts across other igneous rock bodies, such as batholiths; or cuts across the **bedding** (stratification) of lavas or **sedimentary** formations. Not to be confused, in the chemical and mineralogical sense, with **veins;** or, in the structural sense, with **sills.** (R.M.F.)

DILL. Carrot Family.

DILUTION, DISPOSAL OF SEWAGE BY. This refers to the method of sewage disposal in which the sewage wastes are discharged into bodies of water of such extent that the natural purification processes always operative in bodies of water may dispose of the sewage wastes without the creation of a nuisance. For successful ultimate disposal all sewage wastes, liquid or solid, must undergo decomposition accompanied by active bacterial action. If there is a deficiency of oxygen present during the decomposition process putrescent conditions will ensue accompanied by the emission of foul and noxious odors and the creation of unsightly floating masses, i.e., a nuisance will be created. If ample oxygen is present the decomposition will proceed without developing the objectionable characteristics of the putrescent stage of organic decomposition.

If sufficient diluting water is provided to supply the oxygen requirements of the decomposing sewage wastes and still leave a remainder, perhaps 40% of saturation, to support fish and other aquatic life, the sewage may be successfully disposed of by dilution; otherwise a nuisance is likely to be created.

The minimum ratio of diluting water to sewage depends primarily upon the dissolved oxygen content of the diluting water, the oxygen demand of the sewage, and the residual dissolved oxygen desired. For average conditions a fresh-water flow of 8 to 10 c.f.s per thousand population should permit successful disposal by dilution. (E.W.S.)

DILUVIUM. Derived from the Latin **diluo,** to wash apart, through **diluvialis,** flood. A relatively obsolete geologic term formerly applied to certain water-laid deposits within or bordering the glaciation regions of Europe and North America. Most of the sediments which were previously thought to have been the result of the "flood," and which are now known to be stratified drift, were called diluvium. In Germany, the term corresponds to our **Pleistocene.** (R.M.F.)

DIMENSION FORMULAE. Physical Magnitudes and Physical Equations.

DIMETRIC REPRESENTATION. Pictorial Representation.

DIMORPHISM. The occurrence of individuals of two forms in the same species.

In its broadest application dimorphism includes the alternation of forms such as the **polyp** and **medusa** of the **coelenterates.** It is commonly used to indicate less fundamental differences due to the conditions attending the development of essentially similar individuals. Thus some of our **butterflies** differ noticeably in the generation developed during the summer and that which emerges in the spring after passing the winter in an immature stage. This condition is seasonal dimorphism, as also is the occurrence of wet- and dry-season forms in some tropical species. Conspicuous difference between the sexes other than reproductive adaptations is sexual dimorphism. This term is also sometimes applied to the occurrence of two forms in a single sex, like the black and yellow females of the common yellow swallowtail butterfly.

The term is used by mineralogists to describe the phenomenon of certain natural compounds crystallizing in two different forms. (See **Calcite** and **Aragonite.**) (A.W.L., R.M.F.)

DINGO. Mammalia, Carnivora. A wild **dog,** *Canis dingo,* found in forested areas of Australia. It is probably descended from dogs introduced long ago.

The dingo is a serious enemy of sheep and has been killed in large numbers for its depredations. (A.W.L.)

DINOFLAGELLIDA. An order of 1-celled animals. Chiefly marine species, whose body is surrounded by an envelope of **cellulose,** often beautifully figured. **Mastigophora.** (A.W.L.)

DINOSAURS. Fossil Reptiles.

DIODE. Tube, Electronic.

DIOECIOUS. With separate sexes. Organs of only one sex developed in each individual.

There are many species of plants in which the flowers are unisexual, that is, have either stamens or pistils, but not both. If these two types of flowers occur on different plants, the plants are said to be dioecious. The willow is such a plant. (R.M.W.)

DIOPSIDE. The mineral diopside is a **monoclinic pyroxene** corresponding to the chemical formula $CaMg(SiO_3)_2$, **calcium magnesium silicate.** Its crystals, like other pyroxenes, tend to be short stout prisms of square or octagonal cross-section. Compact, granular, lamellar and fibrous varieties are often found. The prismatic cleavage is characteristic, cleavage planes intersecting at angles of 87° and 93°. A basal parting is often noted, but should not be confused with the cleavage. The hardness of diopside is 5–6; specific gravity, 3.2–3.3; uneven fracture tending toward conchoidal; luster, vitreous to dull; sometimes pearly on the base; color, light or dark greens, but may be colorless, gray, yellow or blue, although the latter color is rare. Diopside is a primary mineral in rocks like diorites, gabbros and the like, but is also found in schists, and, as the result of contact metamorphism, in such rocks as crystalline limestones and dolomites. Diopside is found in association with **vesuvianite, garnet, spinel, scapolite, tremolite, tourmaline** and similar minerals. It is a rather widespread mineral, important localities being found in the following European countries: Finland, Sweden, Switzerland, Italy; it is found in eastern Siberia near Lake Baikal. In Canada diopside localities are in Lanark and Hastings Counties, Province of Ontario, and in the United States in Lewis and St. Lawrence Counties, New York, and in Maine.

Two varieties of diopside, **malacolite** and alalite, both of a leaf-green color, have been somewhat used as gem stones. The word diopside is derived from the Greek meaning double and appearance, referring to its double refraction. Malacolite is also from the Greek, meaning soft, because of being softer than feldspar found with it. Alalite is from the Ala Valley, in the Italian Piedmont. (E.S.C.S.)

DIOPTASE. The mineral dioptase is a rather rare **copper silicate** corresponding to the formula H_2CuSiO_4 occurring in prismatic crystals of the **hexagonal** system, tri-rhombohedral in form. It may be found in crystalline aggregates or simply massive. Dioptase displays a conchoidal to uneven fracture; hardness, 5; specific gravity, 3.28–3.35; luster, vitreous; color, a beautiful emerald green. It has been found in Russia, French Congo, Belgian Congo, South West Africa, Chile, and in the United States in Arizona. The name is derived from the Greek words meaning through and to see, because cleavage was observed by looking through the crystals. (E.S.C.S.)

DIORITE. Diorite is a deep-seated **igneous rock** composed dominantly of **sodiaplagioclase feldspar** with **hornblende, biotite,** and (or) **augite. Orthoclase** may be present in small amounts, also **quartz.** Any considerable proportion of the latter mineral produces a quartz-diorite. With increasing amount of orthoclase, we have granodiorite, which is generally understood to be a rock intermediate in character between quartz-diorite and granite. If quartz is absent and there are essentially equal amounts of orthoclase and plagioclase the rock is then known as a monzonite from the type locality, Monzoni, in the Tyrol. There are quartz monzonites and, where the deficiency of silica is great enough, nephelite monzonites. Rocks of the latter sort have been reported from Madagascar. A variety of quartz, diorite, containing both hornblende and biotite, is called tonalite from the Tonale Alps, although the rock found there is more nearly a granodiorite. The word diorite is derived from the Greek meaning distinctive or defining, in contrast to the deceptive dolerites. The diorites are of widespread occurrence. (E.S.C.S.)

DIOSPYROS EBENUM. Ebony. Ebonaceae. The ebony tree is a native of India and Ceylon. It is a large tree with entire leathery leaves and axillary flowers. The wood of the tree is divided sharply into a soft white sapwood of little value and a hard very dark heartwood. The latter is much used for inlay work, for black piano keys, for musical instruments, and for handles of various instruments. Many other species of *Diospyros* have dark woods used as a substitute for true ebony. The wood of several other trees, especially that of the pear tree, are frequently stained to imitate ebony.

Other species of *Diospyros* are esteemed for their fruits. Especially so are *Diospyros virginiana,* the American persimmon, and *D. Kaki,* the Japanese persimmon. *Diospyros virginiana* is a large American tree, 60–100' high, with rather thick ovate-oblong leaves and pale yellow axillary flowers. The fruit is a large globular berry an inch or more in diameter, orange-yellow in color, and very astringent until fully ripe. The astringent quality is due to the presence of much soluble **tannin,** which is gradually formed into an insoluble compound as the fruit ripens, so that the mouth-puckering quality is nearly lost. Frost action has been considered by many to be the cause of the change in the fruit. The American persimmon is hardy as far north as Rhode Island. The Japanese persimmon is a smaller tree, seldom growing more than 40' tall, and is less hardy. Its fruits are larger than those of the American tree, and of reddish color. Both trees have a hard dark wood. (R.M.W.)

DIOTOCARDIA. An order of mollusks, mostly marine species. **Gasteropoda.** (A.W.L.)

DIP. Anticline; Horizon.

DIP FAULT. Fault.

DIP NEEDLE. More properly called an inclinometer. The instrument consists essentially of a magnetic needle poised to swing on a horizontal pivot and thus to

Dip needle or magnetic inclinometer.

indicate the "dip" or inclination of the earth's magnetic field. (See **Terrestrial Magnetism.**) The zero diameter of the vertical graduated circle should be carefully leveled and adjusted to the magnetic meridian. In order to correct for errors of level, balance, magnetization, and eccentricity, the circle should be reversed north to south, the needle axis should be reversed in its bearings, the magnetization should be reversed, and for each of these positions both ends of the needle should be read on the circle. A complete observation is thus the mean of 16 circle readings. (L.D.W.)

DIP OF HORIZON. Horizon.

DIPHENYL. Hydrocarbons.

DIPHTHERIA. An acute, contagious disease caused by the diphtheria bacillus. The organisms usually gain entry to the body through the respiratory passageways, and here they grow, multiply, and form a membrane— a thick, grayish-white exudate, which becomes necrotic.

The bacteria produce a potent toxin which is liberated and absorbed at the site of the membrane; the toxin is responsible for the constitutional symptoms of the disease.

Different types of diphtheria occur, depending on whether the membrane appears in the nose, **pharynx,** or **larynx.** Nasal diphtheria is often mild; it is characterized by a bloody, yellow nasal discharge. Pharyngeal diphtheria is usually the most severe because of the large area over which the membrane may spread and the consequent rapid absorption of large amounts of toxin. Laryngeal diphtheria is dangerous because of the possibility of asphyxiation as the membrane spreads and blocks the air pathways; when this occurs, **intubation** or tracheotomy are carried out. Tracheotomy consists of incision of the trachea low in the mid-line through the neck, and insertion of a metal tube which keeps the airway open; this operation has been largely replaced by intubation, the passing of a metal tube through the mouth down past the obstruction, where it remains until the edema and membrane have receded.

The signs and symptoms of diphtheria appear after an incubation period of 5 days following exposure. Sore throat and mild fever commonly usher in the disease. The membrane may appear early and spread rapidly and may be accompanied by tremendous swelling of the neck, resulting in the so-called "bull neck." In the laryngeal form, cough and noisy, difficult respirations occur. The severity of the constitutional symptoms depends on the amount of toxin absorbed. In a severe pharyngeal type, prostration may be extreme, and death may occur within several days.

Great advance in the treatment of diphtheria came with the development of a specific antitoxin which neutralizes the circulating toxin. (See **Serum.**) The mortality has been cut from 35% (90% in the laryngeal form) to 5%. If the antitoxin is given early, and in large amounts, the prognosis is good even in the most severe cases. The effect on the local reaction can be observed within 24 hours of administration: the membrane stops spreading, the swelling diminishes, and the patient's constitutional signs improve markedly also.

In spite of adequate treatment and recovery from the acute phase of diphtheria, late complications due to the effects of the toxin may occur; early treatment, however, does cut down their incidence. The commonest late complications are peripheral **neuritis** which may progress to paralyses, and acute **inflammation** of the heart muscle, which may be severe enough to cause **heart failure** and death. These occur usually in the second or third week after onset of the disease. The neuritis heals without treatment, although the period of recovery may extend over several months; the same is true of the heart lesion except in those cases which are fatal in the first few days after the development of cardiac symptoms.

The control of diphtheria has been one of the most successful programs in modern medicine. This has been accomplished by the immunization of the susceptible population, i.e., young children, with small doses of inactive toxin. The incidence of the disease in children has been cut down tremendously, and the cases now occurring are more often in the young adult population. This group represents individuals who have never been immunized, or who have lost their immunity over a period of years. The Schick test is used to detect susceptible individuals: a minute amount of toxin is injected into the skin, and the site is observed for the appearance of redness and swelling. If the person tested is immune, the antitoxin in his blood will neutralize the injected toxin, and no reaction will occur; but if he is susceptible, the reaction will appear and persist for several days. Shick positive individuals are then immunized.

Diphtheria **carriers** represent a public-health problem, since they may harbor virulent diphtheria organisms in their throats and nasal passageways, and spread the disease through a community. It is often difficult to correct the carrier state. (D.M.H.)

DIPLEURULA. A bilaterally symmetrical **larva** of the **echinoderms**. It is formed from the **gastrula** by the breaking through of a depression to form the mouth and by the assumption of a form peculiar to the several classes, usually with projecting lobes. The cilia of the outer surface become arranged in a band in most classes. The **auricularia**, **bipinnaria**, and **pluteus** are forms of dipleurula larvae. (A.W.L.)

DIPLOBLASTIC. Derived from two embryonic germ layers. The first step in the formation of tissues in the developing multicellular animal is the formation of two layers, one covering the outside of the body and the other lining a cavity within it. These are the ectoderm and endoderm, respectively. The bodies of **sponges**, **coelenterates**, and possibly **ctenophores** develop by the further differentiation of these two layers alone. (A.W.L.)

DIPLOPODA. The millipedes. A class of the phylum **Arthropoda**.

Millipedes are worm-like animals with a head and segmented body. Each segment bears two pairs of legs, ex-

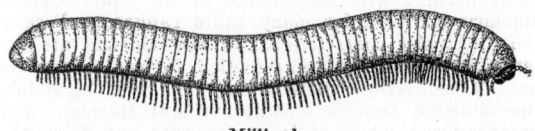

Millipede.

cepting the first and last few. The head bears a pair of antennae and in some species a pair of eyes. Most millipedes have a cylindrical body.

These animals live in moist places, usually among rubbish on the surface of the ground, and eat decaying organic matter or plant tissues. Some attack roots and are therefore of economic importance.

The class is divided into two orders, Pselaphognatha and Chilognatha. All common species belong to the latter. (A.W.L.)

DIPLOSOME. (See **Cell.**) A centrosome with two centrioles. (A.W.L.)

DIPNOI, DIPNEUSTI. The lungfishes. An order of the class **Pisces** made up of a few species found in Australia, Africa, and South America. They have an air sac opening from the **pharynx** which serves as a lung.

These fishes live in marshes and intermittent streams. When the water becomes stagnant they thrive by breathing air at the surface and when it dries up completely they form cells in the mud at the bottom with a vent leading to the surface and lie dormant until the pond is renewed. (See also **Fossil Fishes.**) (A.W.L.)

DIPPER. Aves, Passeriformes. The water **ouzel.**

DIPSOMANIA. Inordinate and uncontrollable drinking usually due to a chronically disturbed state of the mind or personality which causes the individual to have constant recourse to alcohol. It is almost always a conscious or unconscious attempt to obtain forgetfulness or escape from the realities or failures of life. (R.S.M.)

DIPTERA. Flies, mosquitoes, midges, gnats and other **insects.** An order characterized by sucking and sometimes piercing mouths and the presence of a single pair of wings. The hind wings are represented by the halteres, slender clubbed appendages, often inconspicuous. A few species lack wings. The metamorphosis

is complete and the larvae of many species are known as maggots.

This is one of the largest orders of insects, with about 50,000 described species, and in variety of adaptations it is exceeded by no other. Some species suck the juices of plants, some eat the tissues during larval life, some visit flowers for nectar, some suck blood, some are parasitic inside or outside the bodies of warm-blooded animals, and many are parasitic on other insects or are predacious. Many are scavengers, living on decaying organic matter or on the wastes of animals.

Species of economic importance include some of the plant-feeders, such as the **Hessian fly**, the blood-sucking **horse flies** and **mosquitoes**, and the parasitic **bot flies**. (A.W.L.)

DIPTERUS. Fossil Fishes.

DIRECT MOTION. Planetary Motion.

DIRECT STRESS. Stress.

DIRECT-ACTING PUMP. The direct-acting pump is a steam-driven **pump** of the piston and cylinder type, not having crankshafts, flywheels, or similar rotative apparatus. It is a simple, inexpensive, and reliable piece

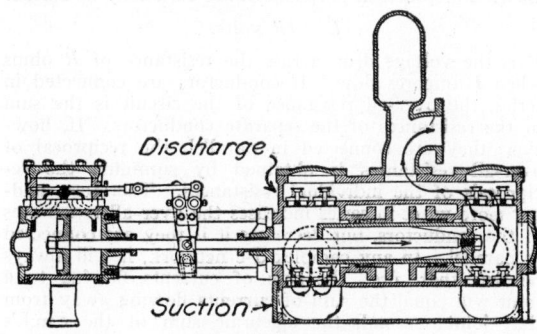

Direct-acting pump.

of equipment—but inefficient as a pumping unit. However, as the heat of the exhaust steam can often be recovered in feed water the low thermal efficiency is not of much importance. In construction it consists of steam and water cylinders, the pistons of which are rigidly connected by a pump rod. The steam piston must have a larger diameter than the water piston when the pump is used in boiler feed service. If p is the boiler gauge pressure, p_h the static plus friction head on the water end, p_f the friction drop in the steam pipe, $A_w p_i$, the pounds excess push required to overcome pump friction, water inertia, etc., A_w, D_w, A_s, D_s area and diameter of steam and water pistons:

$$\frac{D_s}{D_w} = \sqrt{\frac{p + p_h + p_i}{p - p_f}}$$ (all p's being expressed in lbs./sq. in.)

p_i involves the design and operating characteristics of the pump. In order to insure ample operating pressure D_s/D_w is made large enough to include considerable margin of reserve, being about 1.6 for ordinary boiler feed service and 2.5 for low-pressure feed service. Control of capacity is exercised through speed variation by throttling the steam line (i.e., adding a throttle pressure drop to p_f). In practice, units are rated at maximum piston speeds of 100 ft. per min., but should actually be operated at between 25 and 40 strokes per min. The pulsation of delivery is absorbed by an air compression chamber placed in the discharge line.

The direct-acting steam pump consumes from 100 to 300 lbs. of steam per horsepower hour. Thermal efficiency is so low as to have no comparative meaning and in its place is substituted pump duty; that is, the foot pounds work done in the pump cylinders per million B.T.U. chargeable to the steam end. The high steam consumption is caused by non-expansive use of the steam. Were the steam expanded the pump would stall before reaching the end of its stroke unless the ratio D_s/D_w were extremely large and a flywheel provided to steady the speed. (F.T.M.)

DIRECT-CURRENT CIRCUITS. Unidirectional current is produced from batteries, from dynamo machinery equipped with **commutators**, or by means of **rectifiers.** The great disadvantage of d.c. is the fact that until recently it has not been commercially expedient to transform it from low voltage to the high voltage necessary for long-distance transmission of electrical power. Difficulties of **commutation** prevent generation at high voltages. Recently the use of electron tubes, such as the **thyratron**, has opened up new transmission possibilities for d.c.

In lighting and heating apparatus there is not much difference between d.c. and a.c. D-c **motors** are more expensive than a-c of equivalent power rating, but they have better operating characteristics and simpler speed control. Inductance and capacitance are not important factors in d-c transmission.

The basis for most d-c circuit calculations is **Ohm's Law.** For practical purposes Ohm's Law may be stated:

$$E = IR \text{ volts.}$$

E is the voltage drop across the resistance of R ohms when I amperes flow. If conductors are connected in series, the over-all resistance of the circuit is the sum of the resistances of the separate conductors. If, however, they are connected in parallel, the reciprocal of over-all resistances is obtained by summing the reciprocals of the individual resistances. Therefore, adding more wires in series increases the over-all resistances of the conductors, but decreases it if they are connected in parallel. In any complex d-c network, it will always be true that, first, the sum of currents flowing to a joint will equal the sum of currents flowing away from that joint; second, the algebraic sum of the e.m.f.'s in any circuit will be equaled by the sum of the voltage drops in that same circuit. The electrical power flowing in a d-c circuit is found by multiplying the voltage by the current, the unit of power being watts. Heat which is generated by electrical current flowing through a resistance of R ohms for T seconds is:

$$\text{Heat} = I^2RT \text{ watt-seconds.}$$

The watt-second is a unit of electrical energy so small that 1055 watt-seconds are required to equal one B.T.U. (See **Electric Circuits.**) (F.T.M., L.R.Q.)

DIRECTION. The term direction is used in **navigation** to describe the angle between **north** and any given line. The modern practice is for navigators to express directions in degrees of angle measured from the north to the right (i.e., through the east) through 360°. The standard practice is to use three digits for expressing directions. For example, a direction 4° east of north is expressed as 004°, east is expressed as 090°, west as 270°, etc. Several approximate methods for expressing the directions of bearings are in use by lookouts on ships, by aviators, and in cases where rapid description of approximate direction is desired. (See **Bearings, Azimuth.**) (W.K.G.)

DIRECTION COSINES OF A LINE IN SPACE. Let a set of **rectangular coordinate** axes in space be chosen, and let l be any line in space. Through the origin O of the coordinate system draw a line l' parallel to the given line l. Let α, β, γ be the angles which line l' makes with the X-, Y-, Z-axes respectively. Then these angles α, β, γ are called the direction angles of the given line l, and $\cos \alpha$, $\cos \beta$, $\cos \gamma$ are called the direction cosines of the line l. (L.L.S.)

DIRECTIONAL ANTENNA. For ordinary radio-broadcast service it is desirable to transmit the radio signal in all directions equally, but for special broadcast services such as international short wave, it is often desirable to direct the radiation in some specific direction and avoid radiation in other directions. The need for directed radiation is even more pronounced in other types of radio service. The radio signals may be directed by the use of directional antennae, or, as often called when consisting of more than one element, directional arrays. Any **antenna** is directional to a certain extent, e.g., the common tower antenna for broadcast stations does not radiate directly upward, but in the sense used here a directional antenna is one having marked characteristics of this type. Basically the directional antennae all depend upon radiation from two or more components adding vectorially (see **Alternating-Current Circuits**). If the waves radiated from various elements add in a certain direction the signal will be strong in that direction, while if they tend to cancel, or subtract, in a given direction the signal will be zero or weaker in that direction. One of the simplest directional antennae is the loop such as used with many portable receivers. Here the two elements whose effects add vectorially are the two vertical sides of the loop. The result is a figure 8 radiation pattern, i.e., if lines are drawn to scale in various directions so their lengths represent the strength of the signal in each direction, the ends will all lie on a figure 8 curve with the antenna at the center. These antennae are used for many **radio ranges**, and, since the directional characteristics of any antenna are the same for transmission and reception, for **radio compass** use. The directional pattern may be altered by adding the radiation from a separate vertical antenna. For broadcast use where it is necessary to cut down the signal in certain directions, usually to avoid interference with another station, systems consisting of two or more vertical antennae are quite common. By proper spacing of the elements and proper choice of the phase of the currents (which can easily be adjusted by circuit values) a wide range of radiation patterns may be obtained. For international short-wave broadcasts and for point-to-point communication more elaborate extensions of this same principle are used. Since the more elements in an array, the sharper the pattern, the radiation may be beamed at will, the type service and the economics being the usual limiting factors. By stacking systems one above the other in a vertical plane the radiation may be directed vertically as well as horizontally. It should be mentioned that it is not desirable to have the beam too sharp even for point-to-point service since the variations in the **ionosphere** may cause the signal to miss the receiver if the beam is too sharp. Sometimes the elements of an array are not all fed directly from the transmitter, but some are fed and others pick up energy radiated by the first and reradiate it. By proper choice of the spacing and the antenna dimensions these various radiations may be made to give the desired pattern. The fed antennae are often referred to as driven antennae or elements and the others as parasitic antennae. There are many other types of directional antennae, such as rhombic, V, herring bone, etc., but all depend upon vector addition of the radiation to give the pattern. By the use of directional arrays the signal transmitted in a given direction may be increased manyfold over its value for the same transmitter with a non-directional antenna. A measure of this is the gain of the array which is the signal with the array divided by the signal in the same direction for one element of the array serving as antenna. Where the type service permits their use, directional arrays are the most economical means of obtaining increased signal strength at the receiver. In reception the directional antenna allows the reception of a signal from the desired direction and suppresses signals and noise from other directions. (L.R.Q.)

DIRECTIONAL DERIVATIVES.

A directional derivative is an important mathematical concept which expresses the rate of change of a function in any given direction.

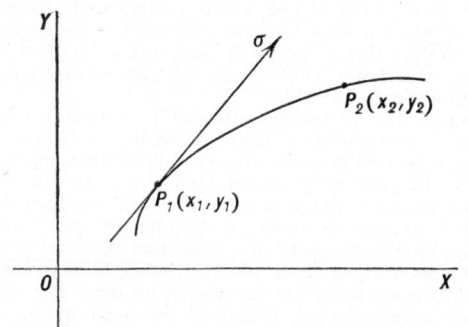

Let $u(x, y)$ be a **continuous function** in a region S of the plane, let $P_1(x_1, y_1)$ be a fixed point in S and let C be a curve passing through P_1. Take another point P_2 on C, and let $\Delta u = u(x_2, y_2) - u(x_1, y_1)$. Let ΔS be the arc P_1P_2 of the curve C, and form the ratio $\dfrac{\Delta u}{\Delta s}$. Let the point P_2 approach the point P_1 along curve C, so that $\Delta s \to 0$, then the **limit**

$$\lim_{\Delta s \to 0} \frac{\Delta u}{\Delta s}$$

is called the directional derivative of u at the point P_1 along the curve C. If σ is the direction of the **tangent** to C at P_1, then the above limit may also be called the directional derivative of u at P_1 in the direction σ. The usual notation for this directional derivative is either $\dfrac{\partial u}{\partial s}$ or $\dfrac{du}{ds}$.

This directional derivative may be evaluated in terms of **partial derivatives** by the following formula:

$$\frac{\partial u}{\partial s} = \frac{\partial u}{\partial x}\cos \alpha + \frac{\partial u}{\partial y}\sin \alpha,$$

where α is the inclination angle (with the X-axis) of the direction σ in which the derivative is taken.

The directional derivative of a function $u(x, y, z)$ of three variables in any direction in space may be defined in a similar way. For such a derivative we have the following formula:

$$\frac{\partial u}{\partial s} = \frac{\partial u}{\partial x}\cos \alpha + \frac{\partial u}{\partial y}\cos \beta + \frac{\partial u}{\partial z}\cos \gamma,$$

where α, β, γ are the **direction angles** of the direction σ in which the derivative is taken.

The directional derivative of a function $f(x, y, z)$ at a given point in the direction of the normal to the surface $f(x, y, z) = c$ (a constant) through this point is called the normal derivative of f, and is denoted by $\dfrac{\partial f}{\partial n}$ or $\dfrac{df}{dn}$. It is the greatest value of the directional derivative in any direction. (L.L.S.)

DIRECTRIX OF A CONIC. Conic Sections.

DIRIGIBLE. Airship.

DISACCHARIDES. Carbohydrates.

DISCHARGE, GASEOUS.

The gaseous discharge is the basis of operation of many of the electron tubes in common use. If two electrodes have a gas at low pressure between them and a gradually increasing voltage is applied across them, a series of events takes place as the voltage is raised. First a very small current, of the order of microamperes, will flow as the ions and electrons are attracted to the electrodes. These charged particles are present because various cosmic radiations, radioactive radiations, etc., which are always present except in specially shielded enclosures, ionize the gas molecules. As the voltage is raised the current finally begins to increase rapidly because the electrons being attracted towards the positive electrode have gained sufficient energy to ionize atoms of the gas and thus generate more carriers of the current. Suddenly the current increases extremely rapidly, and at the same time the voltage across the tube drops. The value of the voltage at which this occurs is the breakdown voltage and the gas has broken into a self-maintaining discharge called a glow. If the current is not limited by circuit resistance it continues to increase, almost instantaneously, while the voltage drops to a low value and the discharge becomes an arc. If the circuit has insufficient resistance the current will reach an enormous value with probable damage to the tube and other circuit elements. The glow discharge is characterized by the ability to pass moderate currents at moderate values of voltage, while the arc will pass very large currents at low values of applied voltage. (L.R.Q.)

DISCONFORMITY.

The geological meaning of this term is treated under **unconformity**.

DISCONTINUITY OF A FUNCTION. Continuous Function.

DISCONTINUITY SURFACES. Fronts.

DISCRETE VARIABLE.

A discrete **variable** is one which occurs only in terms of integral values of a certain unit. Examples are the number of votes cast in an election, or the number of Beta particles emitted by a radioactive substance. A variable which is not discrete is said to be continuous. (L.A.A.)

DISCRIMINANT OF A QUADRATIC EQUATION. Quadratic Equations in One Unknown.

DISCRIMINATOR.

This is the detector (see **Detection**) for a frequency modulation receiver. In frequency modulation the intelligence is impressed or modulated upon the **carrier** by causing variations in frequency. However, our **loud-speakers** operate upon a change of magnitude of the current through them so some method must be provided in the frequency modulation receiver to convert the frequency variations into amplitude variations. This is the function of the discriminator. It is a special type of balanced rectifier or detector which gives no output at the frequency for which it is tuned (the carrier frequency), but gives an output voltage whose magnitude and polarity are determined by the amount and direction of the frequency deviation of the input signal. As the received signal is varied back and forth in frequency by the **modulation**, the discriminator gives an output voltage whose magnitude follows these frequency changes. This output voltage may then be amplified and used to drive the loud-speaker. The discriminator is also used with appropriate circuits to tune automatically **receivers** having automatic frequency control. It is used in certain frequency modulation transmitter circuits as part of the frequency-stabilizing circuit. (L.R.Q.)

DISEASES OF PLANTS.

Plants are attacked by a great many diseases. A plant disease may be defined as any variation from a normal condition, either in structure or function. Disease may be due to an unfavorable environment, one lacking in water or in the mineral requirements of the plant, or to attacks by various organisms. Among the many agents which cause diseases in plants are **bacteria** and **fungi**, and many animals, as well as the **viruses**. The results of infection are many and varied. Often the afflicted plant is stunted, its leaves become wrinkled and irregular, and the normal green color spotted or mottled with white or yellow patches. Parts may die and drop off. Again, various abnormalities, such as **galls** and **witches' brooms,** may be formed.

If the disease is caused by a fungus or a bacterium, the disease-causing organism must gain entrance into the plant. There are many ways in which this may occur. The older stems of higher plants are covered by a thick layer of cork cells, the walls of which become suberized (see **Suberin**). The leaves and young stems are protected by a thinner epidermal layer of cells, the outer surface of which is often further protected by a layer of cutin. In the epidermis of the leaf, however, there are many minute openings or **stomata** through which gases pass. In the outer layer of stems there are lenticels, masses of loosely packed cells which permit a ready movement of gases from the outer air into the stem. Through these openings disease-causing organisms may enter the plant. A fungous **spore**, falling onto a leaf, germinates if moisture is available, and pushes out a slender tube or hypha which grows through a **stoma** into the body of the leaf. There conditions are more favorable, and the fungus grows rapidly, spreading through the leaf.

But entrance to the plant is more frequently through accidental openings. A branch is broken off, or bruised. Sucking insects or other organisms make a puncture through the outer layers. In any case, a favorable condition is then established for the rapid growth of the disease-producing organism. Careless pruning often provides the means of entrance for disease.

In many cases the growth of the parasite causing the disease is very slow. Years may elapse before the real damage appears. On the other hand, in some diseases the infection spreads rapidly and soon shows its presence. The spores or cells which cause disease are formed in immense numbers. They are of small size and easily carried about by wind or by water, and deposited everywhere. Only chance brings them to a suitable host. Insects also carry disease-producing organisms from one plant to another. In this stage these organisms are extremely resistant to adverse conditions around them and usually able to live unharmed for long periods of time. When conditions around them become favorable, infection quickly follows. One of the most important requirements is moisture. When moisture is abundant, therefore, epidemic outbreaks of the diseases may occur.

There are many ways of combating these various diseases. First among these is prevention. Since mechanical injury offers a means of entrance, it should be avoided as much as possible. When such injury is necessary, as in pruning, care should be taken to cover the cut surfaces with protective substances. Another means of combating disease is by spraying, spreading over the plant a film of substance which is toxic to the disease-causing organisms. Not only may most insects be controlled by this means, but also the growth of many fungi may be inhibited. Bordeaux mixture, sulfur dust, and many different chemical mixtures are much used in controlling fungous pests. Once a plant has become infected it becomes necessary to remove the infected part and destroy it, or to destroy the entire plant, to prevent the disease from spreading to other plants.

In recent years a new method of combating disease has been utilized. This is based on the observation that many plants, even of susceptible species, become immune. If the plant is a perennial and the disease attacks only one part, it is possible to graft resistant forms onto the plant. An example of this treatment is found in the European grape. The roots of the European grape are attacked by the **Phylloxera** insect, and succumb rapidly. This pest threatened to wipe out the grape industry of Europe. The American grape resists the insects, so that by grafting European grape stocks onto American roots, the disease was controlled and the industry saved. Another method is to breed immune strains by selecting resistant plants and crossing them with plants of desirable quality, but poor resistance. By selection among the offspring of such a cross a new form may be developed which combines the resistant quality of one plant with the superior or desired qualities of the other. Nature itself seems to develop such resistant varieties. Often one may use such naturally immune plants directly.

Among the commoner diseases of plants are fireblight, also called pear blight, caused by *Bacillus amylovorus,* a bacterium which attacks the **cambium** and inner bark. The disease is spread by insects. It causes the leaves, blossoms and young twigs to wilt and turn dark-colored as though damaged by fire. Citrus canker, caused by *Pseudomonas citri,* is another bacterial disease. It attacks various citrus plants, causing brown spots to appear on leaves, young shoots and on the fruit. Apple scab is a widespread disease of the apple trees. The organism causing the disease is an **ascomycete,** *Venturia inaequalis,* introduced into this country from Europe. The fungus grows on the leaves and fruit, forming small grayish spots which become scab-like. The fungus hyphae grow in the tissues just beneath the epidermis, spreading slowly. The parasite survives the winter on fallen leaves or on the young twigs. Another common and widespread disease is black knot of plum trees. It is caused by an **ascomycete,** *Plowrightia morbosa,* which attacks the **cambium, phloem** and cortical tissues of young twigs. The presence of the fungous **mycelium** stimulates the tissues of the plum to increased growth, forming conspicuous black warty growths. These may develop on one side of the stem only, in which case the branch continues to grow and fruit, or the warty growth may completely surround the stem, causing its death. Ergot, caused by an **ascomycete,** *Claviceps purpurea,* is a disease of rye and other grasses. The fungus grows in the young ovary, forming a dense mass of **mycelium** which replaces the ovary tissues and causes a pronounced hypertrophy of the ovary. Later the fungus forms hard, black, horn-like bodies which project from the **glumes** of the **floret.** These black bodies are called sclerotia, or ergot spurs. It is through them that the fungus survives the winter. In the spring the sclerotium develops a brownish outgrowth with an enlarged tip. In this tip flask-shaped cavities called perithecia are formed. In them the asci each containing eight ascospores are found. From these ascospores conidia are cut off. These conidia may infect developing rye ovaries. Ergot spurs are very poisonous, causing serious trouble when eaten by cattle. The animals suffer from grave digestive disturbances, and, in severe cases, serious impairment of the nervous system and a sloughing off of hoofs, and loss of teeth and hair. Bread made from ergot-infested rye causes similar effects in human beings. Ergot is used medicinally.

Other important plant diseases are caused by **rusts** and **smuts.** (R.M.W.)

DISHED HEAD. Pressure Cap.

DISINFECTANT. Any agent, physical or chemical, which destroys infective organisms. (R.S.M.)

DISK GRINDER. Grinding.

DISLOCATION. The displacement of any part from the normal position. This term is used especially in reference to the **joints** of the body. In a compound dislocation, a joint is penetrated by a wound. In a fracture dislocation, there is a fracture existing with the dislocation. A traumatic dislocation is one that is caused by violence. (R.S.M., D.M.H.)

DISORIENTATION. A confused state of mind in which the normal relationship of identity, time and place is lost. (R.S.M.)

DISPERSION. The selective deviation of an emission in accordance with some variable characteristics; such as the refraction of light at different angles for different frequencies. When an emission is so dispersed, the result is a **spectrum.**

The dispersion of light may be accomplished through **refraction** by a **prism, diffraction** by a grating, or other means. Refractive dispersion is due to the fact that the velocity of light in a given medium, and hence the **refractive index,** vary with the frequency. In any case, if the deviation Δ produced by the dispersing apparatus is expressible as a function of the wave length λ, then the measure of the dispersion may be taken as $D = \dfrac{d\Delta}{d\lambda}$. For example, for a plane **diffraction grating** at normal incidence, the deviation in the first order is given by $\sin \Delta = \lambda/s$, in which s is the grating space; from which it follows that the dispersion is $D = (s^2 - \lambda^2)^{-\frac{1}{2}}$.

Refractive dispersion is not so simply expressed. For a single refraction at angle of incidence i and with refractive index n, it may be shown that the dispersion $d\Delta/d\lambda$ is equal to

$$D = \frac{\sin i}{n\sqrt{n^2 - \sin^2 i}} \cdot \frac{dn}{d\lambda}.$$

Various attempts have been made to express n as a function of λ, for example, by the empirical dispersion formula of Cauchy:

$$n = A + \frac{B}{\lambda^2} + \frac{C}{\lambda^4} + \cdots;$$

or by that of Sellmeyer (see **Anomalous Dispersion**):

$$n = 1 + \frac{A\lambda^2}{\lambda^2 - \lambda_1^2} + \frac{B\lambda^2}{\lambda^2 - \lambda_2^2} + \cdots.$$

Different media are commonly compared through some arbitrarily defined "dispersive power"; such as $(n_F - n_C)/(n_D - 1)$, in which the n-terms are subscripted to indicate the refractive indices for the F, C, and D **Fraunhofer lines.**

In statistics, dispersion is that property of a distribution indicating the spread of the items of the distribution. If the **variates** are very much alike, we say the dispersion is small. If the items are very different or very dissimilar, we have a large amount of **variability.** Measures of dispersion are the **variance,** the **standard deviation,** the **average deviation,** and the **range.** Two distributions may have exactly the same mean, but may differ in their standard deviations. The variance or the standard deviation is the preferred measure of dispersion. (L.D.W., L.A.A.)

DISPLACEMENT. Displacement in a piston and cylinder mechanism is the volume swept out by the piston face. It is assumed that the face of the piston is coplanar. Given the bore and stroke as D and L, the number of cylinders n, the displacement is:

$$\frac{\pi D^2 L n}{4}.$$

The portion of a ship which is immersed displaces a certain weight of water. According to Archimedes' principle, a body immersed in water is buoyed up by a force equal to the weight of water displaced by the body.

Hence the displacement of a ship in tons of water is equal to the weight of the ship and of its contents. (For electric displacement, see **Dielectrics.**) (F.T.M.)

DISPLACEMENT CURRENT. Electric Currents.

DISPLACEMENT LAW. Wien's Laws.

DISSEMINATION OF SEEDS. Fruit, Seed.

DISSEPIMENT. A partition or septum. (A.W.L.)

DISSOCIATION. For electrolytic dissociation see **Electrochemistry** and **Reactions Involving Recombination of Ions;** for thermal dissociations see **Equilibrium; Association and Polymerization.** See also **Rockets.**

DISSOLVING. The limiting amount of a gas or solid dissolving in a liquid at a given temperature and pressure is termed the solubility. (See **Solutions and Solubility.**) The method of attaining the limit is to expose the liquid, which is generally water, to the gas or solid for a sufficient length of time, until no more of the gas or solid dissolves at the given temperature and pressure. In order to make the time as short as possible, vigorous mixing of the solution is demanded and a relatively large surface of contact between the gas or the solid and the solution.

These ends are accomplished in the case of a gas by the use of the principle of countercurrent flow, by which the gas is passed first through the almost saturated solution, consecutively thereafter through less and less saturated solution, and finally through the pure solvent. The gas is commonly introduced at the bottom of a tower into the top of which the solvent is fed. The tower may be charged with inert material or contain bubbling plates to furnish the desired large surfaces of contact and the mixing. **Ammonia, hydrogen chloride, sulfur dioxide, carbon dioxide** are common soluble gases which are subjected to such a process. The process of dissolving one gas from a gas mixture is often called scrubbing. The liquid or solution used in scrubbing may be one with which the gas reacts chemically.

The solution of a solid is accomplished by allowing the liquid or solution to pass over a sufficient surface of the solid, or the solid is suspended in the upper part of the liquid, in which case the solution as formed being itself denser than the liquid, descends and allows fresh contact.

An interesting application of dissolving on a large scale is that in which the sulfite liquor is prepared for cooking wood chips to pulp for **paper.** A tower, sometimes about 100' in height, is charged with lumps of **limestone** (calcium carbonate). Water is run into the top of the tower, flows down over the limestone, meets sulfur dioxide which enters at the bottom, and the resulting solution of **sulfurous acid** dissolves the limestone in the lower portion. Carbon dioxide gas passes out at the top of the tower, and the cooking solution, consisting of calcium hydrogen sulfite and sulfurous acid, is recovered at the bottom.

The heat change accompanying solution is characteristic of the substances. Specific examples are given in the following table. For solubility of the gases listed see **Solutions and Solubility.**

GAS		UPON DISSOLVING		HEAT EVOLVED IN GRAM-CALORIES
Ammonia	17	grams in 200 $\times$ 18 grams water		8,460
Hydrogen chloride	36.5	" " 200 $\times$ 18 " "		17,440
Sulfur dioxide	64	" " 300 $\times$ 18 " "		8,550
Carbon dioxide	44	" " 300 $\times$ 18 " "		4,760
SOLID				
Sodium hydroxide	40	" " 200 $\times$ 18 " "		9,940
Sodium chloride	58.5	" " 100 $\times$ 18 " "		−1,180
Sodium nitrate	85	" " 200 $\times$ 18 " "		−5,030
Sodium sulfate decahydrate	322	" " 400 $\times$ 18 " "		−18,760

Heat is evolved upon solution of all gases. Heat is absorbed upon the solution of the majority of solids. (The heat of solution in an almost saturated solution is negative in the case of solids whose solubility increases with increase of temperature.)   (R.K.S.)

DISTILLATION, EVAPORATION, AND DRYING.

Distillation, evaporation, drying, sublimation are processes that involve the vapor pressure of liquids. The vapor pressure of a given pure substance is a constant varying with temperature. In a solution of two or more liquids, the vapor pressure of each concentration ratio is definite. Many systems have been studied. For the separation of two liquids in solution, use is made of the fact that the ratio of the two substances in the vapor and liquid phases is usually different.

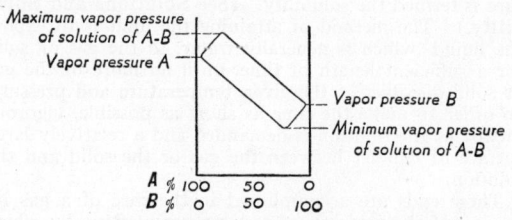

Fig. 1.   Diagram showing vapor pressure relations of a solution of two liquids.

SYSTEM: ETHYL ALCOHOL-WATER
Total Pressure, 760 mm.

Mol Percent in Liquid Phase		Temperature of Volatilization, °C.	Mol Percent of Ethyl Alcohol in Vapor (and, therefore, in condensate)
Ethyl Alcohol	Water		
0	100	100	0
1.90	98.10	95.5	17.00
7.21	92.79	89.0	38.91
12.38	87.62	85.3	47.04
26.08	73.92	82.3	55.80
39.65	60.35	80.7	61.22
51.98	48.02	79.7	65.99
67.63	32.37	78.7	73.85
89.43	10.57	78.15*	89.43

* Minimum boiling-point mixture.

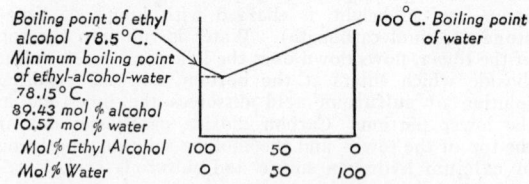

Fig. 2.   Diagram showing boiling-point relations of ethyl alcohol-water solutions.

EFFECT OF CHANGE OF PRESSURE ON BOILING POINT AND COMPOSITION OF DISTILLATE OF CONSTANT MINIMUM BOILING-POINT MIXTURE OF ETHYL ALCOHOL-WATER

Constant Minimum Boiling-Point Mixture, °C.	Pressure, mm.	Mol Percent of Ethyl Alcohol
78.15	760	89.43
62.8	400	91.4
87.8	1100	89.3

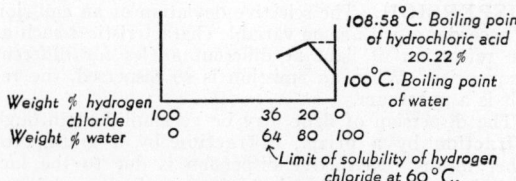

Fig. 3.   Diagram showing boiling-point relations of hydrochloric acid solutions.

Constant minimum boiling-point mixtures are also known of the systems: ethyl alcohol-carbon tetrachloride, ethyl alcohol-carbon disulfide, ethyl alcohol-chloroform, isopropyl alcohol-water, isopropyl alcohol-chloroform, acetone-methyl alcohol, etc.

SYSTEM: HYDROGEN CHLORIDE-WATER
Total Pressure, 760 mm.

Constant Maximum Boiling Point Mixture, °C.	Pressure, mm.	Density at 25° C.	Weight Percent of Hydrochloric Acid
108.584	760	1.0959	20.222
107.859	740	1.0962	20.268
106.424	700	1.0966	20.360
110.007	800	1.0955	20.155

Constant maximum boiling-point mixtures are also known of the systems: hydrobromic acid-water, hydriodic acid-water, hydrofluoric acid-water, nitric acid-water, sulfuric acid-water, formic acid-water, acetone-chloroform, etc.

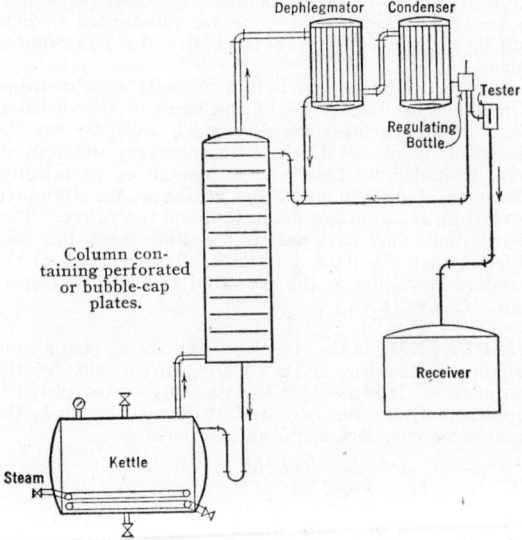

Fig. 4.   Intermittent still used in the distillation and separation of liquid mixtures.

Plain Distillation is conducted in an apparatus consisting of three essential parts connected in series, namely, (1) the still or retort, (2) the condenser, (3) the receiver. The material is heated in the still or retort, its vapor passes into the condenser in which the vapor is cooled, and the liquid condensate is collected in the receiver. Modifications of this set-up are introduced affecting (1) the pressure in the apparatus and (2) the degree of refluxing to which the vapor is subjected.

The system may be closed to the atmosphere and the pressure within increased or diminished as desired, resulting in pressure or vacuum distillation respectively.

The system may be subjected to refluxing by the introduction of a fractionating column between the still and condenser. The column may contain (1) any of various styles of packing material such as coke fragments, or Raschig rings, or (2) a series of horizontal plates that are either (a) perforated or (b) have bubble caps, through which the vapor ascends. The plates also have overflow tubes for the liquid to descend from plate to plate. The desideratum is always the intimate mixing of ascending vapor and descending liquid throughout the whole column. Assuming that the rate of flow is constant, the upward rate of flow of the vapor, V, in mols per sec., minus the downward rate of flow of liquid, L, in mols per sec., equals the condensed output, C, from the top of the column, in mols per sec., and $\frac{L}{V}$ is called the reflux ratio. Fractional distillation is practiced on a large scale in the production of **alcohols, benzene hydrocarbons,** and **petroleum** fractions.

When a substance such as **aniline** has a high boiling point but an appreciable vapor pressure at 100° C. and is practically insoluble in water, use is frequently made of steam to distil the substance. The total or atmos-

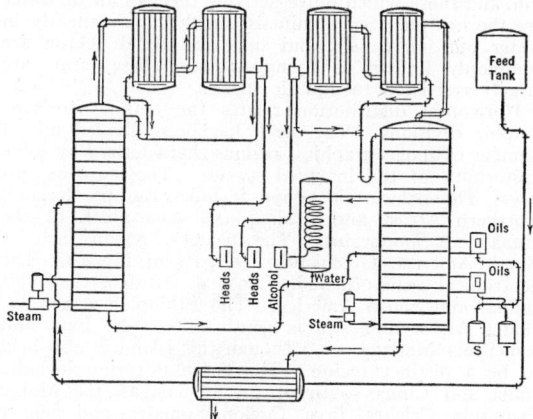

Fig. 5. Continuous still used in the distillation and separation of liquid mixtures.

pheric pressure equals the sum of the partial pressure of the substance, somewhat below 100° C., and of the partial pressure of water vapor at that temperature. The distillate contains the substance and water approximately in the weight ratio: the product of the vapor pressure of the substance and its molecular weight *to* the product of the vapor pressure of water (733 mm. at 99° C., 707 mm. at 98° C.) and 18 (the molecular weight of water).

Evaporation is in principle the same operation as plain distillation, with the modifications in practice that (1) the vapor may or may not be recovered, (2) the residue in the evaporator may or may not contain solids, and (3) vacuum evaporation is frequently used in a single compartment or in multiple stages with each successive stage operated at an increasing vacuum utilizing the heat of condensation of the vapor from the preceding stage. In multiple stage evaporators there is a saving in the cost of heat and an increased expenditure for apparatus. Vacuum evaporation is frequently utilized to lower the temperature to which a substance is subjected and thus avoid decomposition by passing a current of warm dry air over the substance. Combined high-vacuum and very low temperature evaporation or drying is practiced in the final removal of water vapor from frozen penicillin, due to the heat-sensitive nature of this material. Water vapor passes from the place of higher concentration, that is, the substance, to the place of lower concentration, that is, the air, and is thus

removed from the substance. If oxygen of the air reacts with the substance, an inert gas such as nitrogen may be substituted for air.

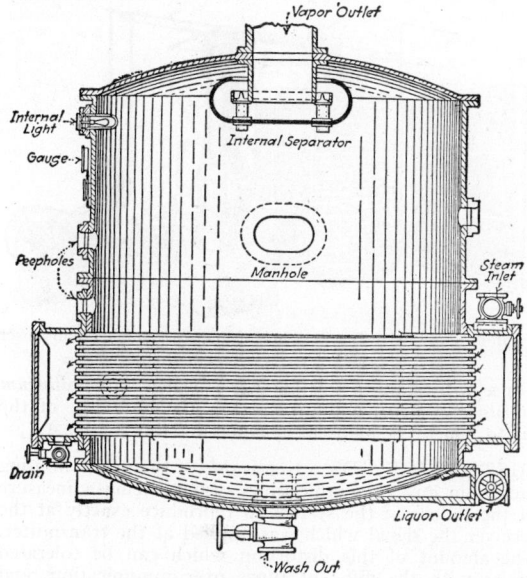

Zaremba horizontal tube evaporator.

Drying may be carried out in two different ways, either *batch* or *continuous*. In batch drying, the drier is filled with wet material and run until the material is dry, when the material is removed and the process is repeated. In continuous drying, the material is moved slowly through a tunnel. The hot gas enters at one end and the wet stock enters at the other. By the time the material has passed through the tunnel it is dry. Many devices are used to move the stock through the tunnel, such as carts equipped with trays, or conveyor belts. Finely divided material can be moved forward through a drier by having the drier itself made in the form of a slowly rotating, slightly inclined hollow cylinder. Such a rotary drier has the advantage of turning over the material and rolling it along at the same time, thus keeping fresh surfaces continually exposed.

For any substance, the drying rate can be determined by the following equation:

$$W = kA\Delta H,$$

where W = weight of water evaporated per unit time,
A = area available for drying,
H = humidity difference = humidity of air saturated with water minus humidity of air being used,
k = a constant characteristic of the material. k remains constant as long as the surface is completely wet but usually becomes smaller as the material approaches dryness.

The effect of air velocity, air **humidity,** temperature, and dimensions of the solid being dried have been studied in a quantitative manner for a number of materials, such as wood in the seasoning process, clay in ceramic goods, and leather. The higher the air velocity while the material is wet the more rapid the drying, but this effect diminishes as the material dries, due to the preponderance of the capillarity effect in transfer of water from the interior, through the interstices of the material to the air.

It is interesting to note that a great many materials ordinarily considered dry actually contain an appreciable amount of water. Dry glue, for example, becomes quite sticky on humid days but may become very brittle when subjected to long periods of low humidity. The swelling and shrinking of wood with changes in the weather are other examples. The amount of moisture which a sub-

Vacuum shelf drier installation (open) with surface condenser (center) and dry vacuum pump (right).

stance will retain in this manner is called its *equilibrium moisture content*, and depends on the humidity of the air to which the material is exposed. (R.K.S., D.E.M.)

DISTORTION. Distortion is one of the major limiting factors in any communications system, being a measure of the failure of the system to reproduce exactly at the receiver the signal which was applied at the transmitter. The amount of this distortion which can be tolerated varies with the different types of communication, and both for electrical and for economic reasons a system is usually not made much better than necessary.

Distortion may be divided into three types: frequency distortion, which is caused by the system (or some part of it) not responding to all frequencies equally; amplitude distortion, which is caused by the system not responding to amplitude of signal linearly; and finally, phase distortion which is caused by different frequencies being shifted in phase by varying amounts. The first two are important in sound communication while all three are important in **television**. Frequency distortion is determined by the circuit's ability to respond to a wide frequency band. The average human ear will respond to a range of about 20 to 15,000 cycles, but all of this range is not necessary for most services so the systems are not designed to respond to such a wide range. For telephone service the range from about 250 to 2800 cycles gives ample articulation and is the usual band employed. A wider band is not necessary and would greatly increase the cost of apparatus and the difficulties from interference. Radio broadcast service requires a much wider band since it is designed for entertainment and must give a fair reproduction of the original music. Amplitude modulation systems are usually limited to an upper value of about 5000 cycles to keep the side bands within the allowed **channel** and to avoid excessive noise pick-up which a wider band would introduce. Frequency modulation, being inherently more noise-free, uses a much wider band, going to the upper limit of audibility. When any system fails to reproduce the applied frequency range linearly it gives frequency distortion, which may or may not be objectionable. Amplitude distortion usually results from some component saturating when a large signal is applied, frequently the distortion occurring for one polarity of the signal current or voltage and not the other. This has the effect of introducing new frequencies which are harmonics of the original ones. Phase distortion is not important for sound since the ear is not sensitive to it, but in picture work it produces very noticeable distortion in the reproduced picture. It is caused by circuit elements offering different **imped-ances** at different frequencies and is one of the most difficult forms to eliminate when the frequency range of the system is very wide. Often compensating networks are introduced in the line-up of a system to counterbalance certain distortions and give an over-all response free from distortion. (L.R.Q.)

DISTRIBUTED CAPACITANCE. The capacitance which is inherent in any coil because of the adjacent turns, layers, windings, etc., which are separated by some dielectric material and which have voltage differences between them. The result of this is a capacitance action which lowers the effective **inductance** of the coil. This capacitance is often considered as lumped and in parallel with the true inductance of the coil. (L.R.Q.)

DISTRIBUTED LOAD. Load.

DISTRIBUTION. The portions of the earth's surface occupied by different kinds of animals are determined by the relation of environmental conditions and inherited adaptations and result in a spatial arrangement of living forms known as distribution.

In vertical or altitudinal distribution animals may be surrounded either by water or by air, and may either rest on the solid surface, float in the surrounding medium, or move actively through this medium. Those which rest on the solid support constitute the faunal group known as the benthos, those which float are the plankton, and those which move actively through air or water are the nekton. Since animals can float permanently in water but not in air, and since the aerial nekton are commonly known as flying animals, these terms are largely restricted to aquatic species.

Horizontal distribution relates the animals to geographic divisions. The earth has been divided into a number of zoogeographical regions characterized by some uniformity of the included species. These are as follows: The palaearctic region includes Europe, Iceland, Northern Africa and Arabia, and Asia north of the Himalayan mountains. The nearctic region includes North America, Greenland, and part of Mexico. The neotropical region includes most of Mexico, the West Indies, and South America. The Ethiopian region includes Africa and Arabia south of 20° N. Lat., and sometimes Madagascar, although this island is also held to be a distinct region. The Oriental region includes India and China south of the Himalayas, the Malay Peninsula, Celebes, Java, Ceylon, Sumatra, and smaller adjoining islands. The Australian region includes Australia, New Guinea, New Zealand, and the remaining Pacific islands with the exception of the Hawaiian group, which constitute a small separate region. (See also **Gaussian Distribution.**) (A.W.L.)

DISTRIBUTION FUNCTION. Given the **probability function of a discrete variate**, P_x, $a \leq x \leq b$, the distribution function $H(x)$ is defined as

$$H(x) = \sum_{x=a}^{x=x} P_x.$$

Always $H(a) = 0$, $H(b) = 1$. Given the probability function of a **continuous variate** $P_x dx$, $a \leq x \leq b$, the distribution function $H(x)$ is defined as

$$H(x) = \int_{x=a}^{x=x} P_x dx.$$

In this case also $H(a) = 0$, $H(b) = 1$. The distribution function generally has an S shape (Fig. 1). The distribution function is the mathematical counterpart of the **cumulative frequency distribution.**

The distribution function is important since it gives the probability of x being less than or equal to a particular value x_0. (L.A.A.)

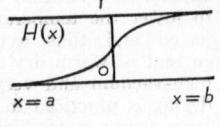

DISTRIBUTIVE LAW OF ALGEBRA. Multiplication.

DISTRIBUTOR. This is a rotary electrical switch used to distribute electric current from one source to a number of separate circuits. A common example is the distributor of the automobile engine **ignition system.** High-tension ignition current suitable for use by spark plugs is delivered to the distributor at its revolving switch arm. Around the periphery of the travel of this distributor arm are mounted contact points which are connected to the various circuits to which the current is to be distributed. On the multi-cylinder gasoline engine, each circuit leads to a separate spark plug, and there are as many contact points as there are spark plugs. The rotation of the distributor arm brings it successively opposite the contact points. There may be actual contact between the distributor arm and the contact points, or the current may be required to jump a small gap between them. Its voltage must be high enough to do this. The contact points are mounted in a case of insulating material such as bakelite. The distributor arm is mounted on a rotating shaft driven from the crankshaft. In the four-cycle engine this shaft revolves at one-half crankshaft speed. It revolves at crankshaft speed in the two-cycle engine. (F.T.M., L.R.Q.)

DISTURBANCE. Used by structural geologists to designate an **orogenic** deformation of less area and intensity than a **revolution.** (R.M.F.)

DITHIONIC ACID AND DITHIONATES. Dithionic acid ("hyposulfuric acid" $H_2S_2O_6$) is a colorless solution, formed by reaction of **barium** dithionate solution and dilute **sulfuric acid,** and filtering off barium sulfate. The resulting solution may be evaporated in vacuum to specific gravity 1.35 beyond which point decomposition occurs, with resulting formation of sulfuric acid and **sulfur dioxide.** Dithionic acid does not react in the cold with **chlorine, sulfur, nitric acid, permanganate** or **hypochlorite,** but **sodium** peroxide oxidizes it to **sulfate.**

Sodium dithionate ($Na_2S_2O_6$) is made by reaction (1) of barium dithionate and sodium sulfate, and filtering off barium sulfate, (2) of dithionic acid and sodium hydroxide, (3) of **sodium** sulfite solution and **iodine,** some sulfate being formed.

When **sulfurous acid** is allowed to come in contact with suspensions of **manganese** dioxide, **ferric** hydroxide, **cobaltic** hydroxide, but not barium peroxide, or sodium sulfite and lead dioxide, in the cold, the corresponding dithionate is formed, and this may be conveniently converted into barium dithionate by reaction with barium hydroxide, and filtering. The solution of barium dithionate is then evaporated to **crystallization.** (R.K.S.)

DIURETIC. Any **drug** or substance which increases the urine volume. Water is the most common diuretic. In medicine, mercury compounds are frequently used to produce diuresis in **edema** and **ascites** due to heart disease. (D.M.H.)

DIURNAL CHANGES. In the course of a normal 24-hour period, atmospheric features change in a cyclic manner, returning to the approximate value again at the end of each 24-hour period. Some types of clouds and weather reach a maximum or a minimum during certain parts of a day. Daily changes are known as diurnal changes.

1. Surface pressure undergoes two definite periods of increase and two of decrease. Mean maximum pressure occurs approximately at ten o'clock local time in the morning and evening, and mean minimum pressure occurs at four in the afternoon and morning. In the tropics, this surge and ebb of pressure are very pronounced and highly rhythmic.

2. Temperature tends to become a maximum about 2–3 hours after local noon and a minimum at sunrise. Over water there is a minimum diurnal change as small as a fraction of a degree and over sandy and rocky desert a maximum which sometimes amounts to 100° or more.

3. Relative humidity tends to become a maximum about sunrise and a minimum in the afternoon.

4. Over land, cumulus-type clouds tend to be a maximum during afternoons and a minimum at night.

5. Fogs tend to be a maximum at and shortly after sunrise and minimum in the afternoon.

6. Rough flying air tends to be a maximum in mid-afternoon and a minimum at night.

Diurnal changes are primarily associated with the apparent movements of the sun. (P.E.K.)

DIVER. Aves, Gaviiformes. Large diving birds (**Aves**) with strong pointed beaks and webbed feet, found throughout the northern part of the world. The **loons.** The most common species is the great northern diver. (A.W.L.)

DIVERGENCE. Convergence and Divergence.

DIVERGENCE OF A VECTOR FUNCTION. The divergence of a vector function is a certain type form of mathematical expression which occurs very frequently in discussions of mathematical physics.

Let **v** be a **vector function** of position, with rectangular components v_1, v_2, and v_3 (in magnitude). The **scalar** expression

$$\nabla \cdot \mathbf{v} = \frac{\partial v_1}{\partial x} + \frac{\partial v_2}{\partial y} + \frac{\partial v_3}{\partial z}$$

is called the divergence of **v**, and is denoted by div **v**.

A fundamental property of the divergence is the so-called **divergence theorem.**

If **r** is a variable vector of the form $\mathbf{r} = x\hat{i} + y\hat{j} + z\hat{k}$, and if **a** is a constant vector, then

$$\operatorname{div} \mathbf{r} = \nabla \cdot \mathbf{r} = 3,$$

$$\operatorname{div} (\mathbf{r} \times \mathbf{a}) = \nabla \cdot (\mathbf{r} \times \mathbf{a}) = 0,$$

$$\operatorname{div} (r\mathbf{a}) = \nabla \cdot (r\mathbf{a}) = \frac{\mathbf{r} \cdot \mathbf{a}}{r}.$$

If **u** and **v** are vector functions of position and u is a scalar function of position, then

$$\operatorname{div} (\mathbf{u} + \mathbf{v}) = \nabla \cdot (\mathbf{u} + \mathbf{v}) = \nabla \cdot \mathbf{u} + \nabla \cdot \mathbf{v} = \operatorname{div} \mathbf{u} + \operatorname{div} \mathbf{v},$$

$$\operatorname{div} (u\mathbf{v}) = \nabla \cdot (u\mathbf{v}) = (\nabla u) \cdot \mathbf{v} + u(\nabla \cdot \mathbf{v}),$$

$$\operatorname{div} (\mathbf{u} \times \mathbf{v}) = \nabla \cdot (\mathbf{u} \times \mathbf{v}) = \mathbf{v} \cdot (\nabla \times \mathbf{u}) - \mathbf{u} \cdot (\nabla \times \mathbf{v}),$$

$$\operatorname{div} (\operatorname{grad} u) = \nabla \cdot (\nabla u) = \nabla^2 u \quad \text{(Laplacian of } u\text{)}.$$

An alternative definition of the divergence is the following:

Let **v** be a vector function of position, let δ be a small region of space and also its volume, surrounding a point P, and let ω be the bounding closed surface of δ and let $d\sigma$ be a surface element on ω; let **n** be an outward drawn normal unit vector to any point on ω. The outward flux through the element $d\sigma$ is $\mathbf{n} \cdot \mathbf{v}\, d\sigma$. The total flux outward through ω is the integral of this expression, and represents the net quantity of outward flow from δ per unit time. The limit approached by the quotient of net flow by the volume δ, as $\delta \to 0$, gives the rate of diminution of density. This is

$$\operatorname{div} \mathbf{v} = V \cdot \mathbf{v} = \lim_{\delta \to 0} \frac{1}{\delta} \iint_\omega \mathbf{n} \cdot \mathbf{v}\, d\sigma.$$

(L.L.S.)

DIVERGENCE THEOREM. Green's Theorem in Space.

DIVERGENCY OF SERIES. Infinite Series.

DIVERSITY. That characteristic of public consumption of a good specifically produced and marketed for public use, known as diversity, arises from the diversification of use of the commodity by individual customers. In a public utility system marketing a commodity which may be stored, diversification is not of as much importance as in services such as transportation and electric power, where warehousing the salable commodity is either impossible or impractical. Thus, in a water or gas system, a certain amount of storage may be interposed between supply and demand which would effect the result that only a diversification of usage can achieve in the electric service system.

In the case of an electric utility system, the fact that customer A requires 5 kilowatts during some part of the day, customer B, 8, and C, 7 kilowatts, does not mean that at some time during the day 20 kilowatts will be drawn from the supply line. The diversity of usage between customers would so stagger their periods of maximum demand that the feeder capacity could be considerably less than the sum of the individual maximum demands. Taking into account a certain amount of diversity between feeders themselves, and between substations supplying these feeders, the maximum demanded load of a power plant is likely to be only a small fraction of the sum of the individual customer's peak loads. Diversity factor is the ratio between individual maximum demands of parts of a system and their combined simultaneous maximum demand. It is defined as the maximum simultaneous demand of a system or part of a system, divided by the sum of the individual maximum demands of the subdivision, taken as they may occur. A low diversity factor is a desirable loading condition. (F.T.M., L.R.Q.)

DIVERSITY FACTOR. Diversity; Group Drive.

DIVERSITY RECEPTION. Fading has been found to vary from place to place at a given time. Thus if a radio signal is received simultaneously at points separated by a few wavelengths' distance it is found that the outputs of the receivers do not all fade together. Diversity reception is a method of utilizing this effect to minimize the fading. Basically such a system consists of 2 or more (3 is quite common) antennae separated by several wavelengths (at least 10 times the wavelength of the received wave is desirable and 3 antennae placed at the vertices of an equilateral triangle give the best positioning) feeding separate radio-frequency receiver channels. The outputs of these channels are then combined to give a single output. By means of automatic gain control circuits the antenna receiving a non-faded signal supplies most of the output and as the signals at the different antennae fade out and back in, the control system acts to maintain a constant output level. While such a system, because of its complexity, is not suitable for home reception, it is widely used for reception of foreign broadcasts for rebroadcasting in this country. It is also used for transoceanic telephone reception. (L.R.Q.)

DIVIDERS. Measurement.

DIVI-DIVI. Tannins.

DIVISION. Division is the inverse operation to multiplication. The result of dividing one number by another is called their quotient. The quotient a/b of two numbers a and b is that number c such that $b \cdot c = a$ (provided $b \neq 0$). It follows that $(a/b) \cdot b = a$ and $(a \cdot b)/b = a$. That is, division undoes the effect of multiplication.

The reciprocal of a number a is the quotient of 1 by that number, namely $1/a$. The reciprocal of a fraction is that fraction inverted.

The quotient of two numbers of like sign is positive, that of two numbers of unlike sign is negative, the absolute value being the quotient of the absolute values of the numbers.

To find the quotient of two polynomials, arrange each in descending powers of some common letter involved. Divide the first term of the dividend by the first term of the divisor; the result is the first term of the quotient. Then multiply the whole divisor by the first term of the quotient and subtract the product from the dividend. Consider the remainder thus obtained as a new dividend and repeat the operation. Continue in this manner until a remainder is obtained which is either zero or an expression whose first term does not contain the first term of the divisor as a factor. (L.L.S.)

DOBSON FLY. Corydalis.

DOCTRINE OF SIGNATURES. This curious belief came into existence during the Middle Ages. According to its proponents, every plant was created for a purpose, and, more than that, was marked so that its purpose could be known. The most able interpreters of this doctrine were those gifted with an imagination capable of discovering the signature.

If the shape of the leaf suggested that of the human heart or liver, then obviously that plant, or its leaves, was meant to be used to cure diseases of the heart or the liver. A little plant, which has small white flowers with a conspicuous dark spot in the center, is known as Eyebright. It was quite clear that this plant was marked as a plant which should be used to treat eye trouble. Common walnuts were seen to resemble a skull—the meat within was very like the human brain in appearance. Surely here was a remedy for any trouble which originated in the brain. A certain lichen, *Usnea barbata,* commonly known as Old Man's Beard, grows on dead branches of trees, from which it hangs in slender branching threads. A decoction of this lichen was therefore used to promote growth of the hair.

Sometimes the marks or signs were extremely obscure. One lichen, for example, commonly grows on barren rock surfaces. This lichen, *Parmelia saxatilis,* will also grow on old bones, including skulls, if the latter happen to be in a favorable spot. Surely anything growing on a skull is valuable; so here is a plant which is a cure for epilepsy and also a healing salve for wounds. (R.M.W.)

DODO. Aves, Columbiformes. A large clumsy flightless bird, *Didus ineptus,* once common on the island of Mauritius but now extinct. (A.W.L.)

DOG. Mammalia, Carnivora. A member of certain species of the dog family (Canidae), characterized by the slender build, long legs, elongate muzzle, and blunt claws which cannot be retracted.

The family is represented in all continents but the wild species are known as dogs only in the case of a few which inhabit Asia, Africa, and South America, including the raccoon dog and the Siberian wild dog. The species that bear special names are the wolves, coyote, kaberu, jackals, dingo, foxes and fennecs.

The name is best known as applied to the domestic dog, *Canis familiaris,* one of the most highly diversified of animals as a result of long selection and controlled breeding. (A.W.L.)

DOG, LATHE. Lathe.

DOGFISH. Pisces. 1. Plagiostomi. Several species of small sharks, all marine. 2. Holostei. The bowfin, also called the fresh-water dogfish. (A.W.L.)

DOLDRUMS. Circulation of the Atmosphere.

DOLERITE. The term dolerite, derived from the Greek meaning deceitful, was originally applied to all dark, heavy, fine-grained **igneous rocks** of doubtful character. It is now used to indicate gabbroid or basaltic types occurring as dikes or sills whose mineralogical composition is **plagioclase, feldspar, hornblende** or **pyroxene** or both, **olivine** and perhaps **biotite, magnetite** or **ilmenite** and **pyrite.** Included in the dolerites are the diabases, which display plagioclase laths in a somewhat radial arrangement, and from this circumstance we have the textural term diabasic which is synonymous with ophitic.

Both terms dolerite and diabase have been used interchangeably but the suggestion of Kemp that diabasic refers to rocks in which the feldspar is in excess and the augite occupies the interstices between the feldspar laths is one which should receive more attention. (E.S.C.S.)

DOLICHOCEPHALIC VS. BRACHIOCEPHALIC.
Paleontology of Man.

DOLOMITE. The mineral dolomite, the **carbonate** of **calcium** and **magnesium,** corresponds to the formula $CaMg(CO_3)_2$ and closely resembles **calcite.** Its crystals, **rhombohedral** in habit, fall in the **hexagonal** system. Like calcite it may be massive or granular, some marbles being dolomite rather than calcite. It displays a perfect cleavage parallel to the rhombohedron; sub-conchoidal fracture, brittle; hardness, 3.5–4; specific gravity, 2.8–2.9; luster vitreous to pearly; color varies widely, white, reds, greens, black, browns, yellows or colorless; transparent to translucent. Unlike calcite, dolomite dissolves very slowly if at all in dilute cold hydrochloric acid; powdered dolomite will dissolve in warm acid. This is the common test for the two minerals. Much dolomite occurs as stratified rocks where it is believed to have been formed by a secondary process, probably by the action of waters charged with magnesium compounds. Dolomite also is found as a vein mineral, as is calcite. **Iron** or **manganese,** rarely **zinc** or **cobalt,** may replace some of the magnesium. **Ankerite** is the name given to a mineral whose composition is essentially a calcium-magnesium-iron carbonate. Among the many noted localities for dolomite may be mentioned the following: Saxony, Switzerland, Italy, France, Spain, Brazil, Mexico and in the United States at Roxbury, Vermont; Lockport, New York; Phoenixville, Pennsylvania; Alexander County, North Carolina; Hancock County, Illinois, and the Joplin District, Missouri. It was named for Deodat de Dolomieu who first described its characteristics. (E.S.C.S.)

DOLOMITIC LIMESTONE or **DOLOMITE.**
Many of the **carbonate** rocks consist largely of the mineral dolomite, the double carbonate of **calcium** and **magnesium,** thus differing from ordinary **limestone** which is essentially carbonate of calcium. Such rocks are called dolomites. It should be noted, however, that the term dolomite is sometimes loosely used to mean a magnesian limestone. The term dolomite is properly restricted to the rock made up of the double carbonate, $CaMg(CO_3)_2$. Dolomite occurs in bedded deposits probably as a chemical precipitate either as the result of inorganic or organic agencies, or in certain cases because of leaching-out of the calcium with the accompanying concentration of the magnesium, or from the carbonates of decomposed shells of marine animals. The replacement of calcium carbonate by dolomite is also of importance. Dolomite is used to some extent as a building stone, as a refractory and in the production of heat insulating materials. (E.S.C.S.)

DOLPHIN. 1. Pisces, Teleostei. Marine game fishes (Pisces), *Coryphaena,* with a deep head, short snout, and long tapering body. They look like the mammalian dolphins in general form. 2. Mammalia, Odontoceti.

Small toothed **whales,** attaining a length of about eight feet. There are many species of several characteristic forms. One of the more peculiar is the narwhal, *Monodon monoceros,* which has a single spirally twisted ivory tusk. Several species of dolphins live in large rivers of the Old and New World tropics, among them the Gangetic dolphin or susu of India and the Amazonian dolphin, *Inia geoffroyensis,* also called the inia or bouto, of South America. (A.W.L.)

DOME. As used by the geologists this term has several meanings. Principally applied to mounds of viscous lava which are squeezed out of volcanoes and solidify without forming lava flows. When portions of the older lavas or ashes are pushed up by the pressure of later lavas the resulting structure is called a volcanic dome. (R.M.F.)

DOMESTIC CHEMISTRY. Chemistry.

DOMESTIC HEATING. Heating.

DOMITE. The term proposed by Von Buch for the **trachyte** lavas of the famous volcanic Puy de Dôme district of France. More specifically, trachytes which contain appreciable amounts of **oligoclase** and **hematite.** (R.M.F.)

DOOLITTLE METHOD OF SOLVING NORMAL EQUATIONS. We shall show how the normal equations (1) may be solved readily by the Doolittle technique as simplified by P. Dwyer.

$$
\begin{aligned}
a_{11}w_1 + a_{21}w_2 + a_{31}w_3 &= a_{41} \\
(1) \qquad a_{12}w_1 + a_{22}w_2 + a_{32}w_3 &= a_{42} \\
a_{13}w_1 + a_{23}w_2 + a_{33}w_3 &= a_{43}
\end{aligned}
$$

where $a_{ij} = a_{ji}$. The solution is given in tabular form with a check column and an explanation at the right hand side.

w_1	w_2	w_3		Check	Explanation
a_{11}	a_{21}	a_{31}	a_{41}	a_{51}	$a_{5i} = a_{4i} + a_{3i} + a_{2i} + a_{1i}$,
	a_{22}	a_{32}	a_{42}	a_{52}	remembering $a_{ij} = a_{ji}$
		a_{33}	a_{43}	a_{53}	
a_{11}	a_{21}	a_{31}	a_{41}	a_{51}	$b_{i1} = \dfrac{a_{i1}}{a_{11}}$
1	b_{21}	b_{31}	b_{41}	b_{51}	$b_{51} = 1 + b_{21} + b_{31} + b_{41}$
	$a_{22.1}$	$a_{32.1}$	$a_{42.1}$	$a_{52.1}$	$a_{i2.1} = a_{i2} - a_{i1}b_{21}$
	1	$b_{32.1}$	$b_{42.1}$	$b_{52.1}$	$b_{i2.1} = \dfrac{a_{i2.1}}{a_{22.1}}$
					Check = sum of all items in row
		$a_{33.12}$	$a_{43.12}$	$a_{53.12}$	$a_{i3.12} = a_{i3} - a_{i1}b_{31} - a_{i2.1}b_{32.1}$
		1	$b_{43.12}$	$b_{53.12}$	$b_{i3.12} = \dfrac{a_{i3.12}}{a_{33.12}}$
					Check = sum of all items in row
w_1	w_2	$b_{43.12}$			

(2)

In the last row $w_3 = b_{43.12}$, $w_2 = b_{42.1} - b_{32.1}w_3$, $w_1 = b_{41} - b_{31}w_3 - b_{21}w_2$. Finally the values of w_1, w_2, w_3 are checked in (1). While we have illustrated the method for three equations, $a_{ij} = a_{ji}$, the method is perfectly general for n equations, $a_{ij} = a_{ji}$.

We demonstrate how $r^2_{4.123}$ may be found, assuming the predictive equation, the **regression** of t_4 on t_1, t_2, and t_3, is given in **standard units**

$$ t_i = \frac{x_i - \overline{X}_i}{\sigma_i} $$

$$ t_4 = \beta_{41.23}t_1 + \beta_{42.13}t_2 + \beta_{43.12}t_3. $$

We must evaluate $\beta_{41.23}$, $\beta_{42.13}$, $\beta_{43.12}$. The normal equations are

$$\beta_{41.23} + \beta_{42.13}r_{12} + \beta_{43.12}r_{13} = r_{14}$$
$$(3) \qquad \beta_{41.23}r_{12} + \beta_{42.13} + \beta_{43.12}r_{23} = r_{24}$$
$$\beta_{41.23}r_{13} + \beta_{42.13}r_{23} + \beta_{43.12} = r_{34}$$

Here r_{ij} is the **coefficient of correlation** between x_i and x_j. This set of equations (3) corresponds to (1) with $\beta_{43.12} = w_1$, $\beta_{42.13} = w_2$, $\beta_{43.12} = w_3$, $r_{ij} = a_{ij}$, and $r_{ii} = a_{ii} = 1$. After we have found $w_1, w_2, w_3, r_{4.123}$ is given by

$$r_{4.123} = \sqrt{a_{41}b_{41} + a_{42.1}b_{42.1} + a_{43.12}b_{43.12}}.$$

Thus both the regression equation for t_4 and $r_{4.123}$ are determined simultaneously. Let $\sigma_{e4.123}$ represent the **standard error of estimate** of x_4 based on the regression of t_4, then $\sigma_{e4.123} = \sigma_4\sqrt{1 - r_{4.123}{}^2}$, $\sigma_4 =$ the **standard deviation** of x_4. The regression of x_4 on x_1, x_2, and x_3 will be

$$x_4 = B_{41}x_1 + B_{42}x_2 + B_{43}x_3 + B_4$$

$$B_{41} = \beta_{41.23}\frac{\sigma_4}{\sigma_1}, \quad B_{42} = \beta_{42.13}\frac{\sigma_4}{\sigma_2}, \quad B_{43} = \beta_{43.12}\frac{\sigma_4}{\sigma_3}$$

$$B_4 = \sigma_4\left(\frac{\overline{X}_4}{\sigma_4} - \frac{\overline{X}_1}{\sigma_1} - \frac{\overline{X}_2}{\sigma_2} - \frac{\overline{X}_3}{\sigma_3}\right)$$

$\overline{X}_i =$ the **mean** of x_i.

To find $r_{43.12}$, the **coefficient of partial correlation** between x_4 and x_3 when x_1 and x_2 are kept constant, we have

$$r_{43.12} = \sqrt{\beta_{34.12}\beta_{43.12}}.$$

We have shown how $\beta_{43.12}$ is obtained and $\beta_{34.12}$ is found similarly by solving another set of three equations in three unknowns. In exactly the same way $r_{41.23}$, $r_{23.14}$, and $r_{42.13}$ are found.

The methods have been illustrated for four variables but are perfectly general. (L.A.A.)

DOOR-KNOB TUBES. These are vacuum **tubes** especially built for **ultra high frequency** use where the lead inductance and the capacitance between electrodes of the conventional tubes prevent their use. The door-knob tube gets its name from its resemblance to a door knob, the peculiar shape being the result of an effort to get the leads as short as possible and have a minimum amount of dielectric material in the construction. These tubes do not have a conventional base, the leads being brought out directly through the glass and made heavy enough to serve as connections. (L.R.Q.)

DOPPLER EFFECTS. The effects upon the apparent frequency of a wave train produced (1) by motion of the source toward or away from the stationary ob-

Doppler effect of motion of source. λ' *is the altered wave length.*

server, and (2) by motion of the observer toward or from the stationary source; the motion in each case being with reference to the (supposedly stationary) medium.

(1) It is easy to see that when the source moves, the waves are crowded together on the side toward which it moves, and are more widely separated on the opposite side, thus producing, respectively, an apparent increase and an apparent decrease in frequency. If the speed of the waves is V and that of the source is u, and if the true frequency is ν, the apparent frequency is

$$\nu' = \frac{V}{V + u}\nu; \qquad (1)$$

u being $+$ or $-$ according as the distance is increasing or decreasing. Thus if a whistle of actual frequency 250 per

sec. is moving away from the listener with a speed of 11 ft. per sec., and the speed of sound is 1100 ft. per sec., the apparent pitch is lowered to $\frac{1100}{1111} \times 250$ per sec. $= 247.52$ per sec. (This effect is often observed with the bell or the whistle of a passing locomotive.)

(2) If the observer is moving with speed u, which is $+$ or $-$ according as his distance from the stationary source is increasing or decreasing, the apparent frequency is given by

$$\nu'' = \frac{V - u}{V}\nu. \qquad (2)$$

Then if the listener moves away from the whistle at 11 ft. per sec., the pitch is lowered to $\frac{1089}{1100} \times 250$ per sec. $= 247.50$ per sec. The two effects are thus very nearly but not quite equal; frequently both are operative at once.

The Doppler effects are of great importance in the case of light (for which it is quite impossible to distinguish between them). The slight abnormality (Doppler shift) in the positions of the spectrum lines from a star, for example, affords a fairly accurate value of the relative speed with which the star and the earth are approaching or receding from each other (the **radial velocity**). Many **double stars** (**spectroscopic binaries**) are recognized as such only by the doubling of their spectrum lines due to the components moving in opposite directions. The spectrum lines of gases are often broadened because of the various speeds of the molecules. (L.D.W.)

DORAB. Pisces, Teleostei. A slender marine fish (**Pisces**) of the Oriental region, of large size and vicious habits. (A.W.L.)

DORMOUSE. Mammalia, Rodentia. Small arboreal **rodents** of the Palaearctic and Ethiopian regions. They are somewhat like squirrels in appearance, with long hairy or bushy tails. Also called sleepers. The common dormouse is *Muscardinus avellanarius.* (A.W.L.)

DORSAL LAMINA. In the **tunicates,** a ciliated ridge along the middle line of the dorsal wall of the **pharynx.** (A.W.L.)

DORY, JOHN DORY. Pisces, Teleostei. Marine fishes (**Pisces**) of ugly form and world-wide distribution. The John dory, *Zeus faber,* is a species of the northern hemisphere which is valued as food. Family Cyttidae. (A.W.L.)

DOT PRODUCT OF TWO VECTORS. Scalar Product of Two Vectors.

DOTTEREL. Aves, Charadriiformes. *Eudromias.* Several Old World species of birds (**Aves**) related to the plovers. (A.W.L.)

DOUBLE-CURRENT GENERATOR. If a synchronous converter is driven mechanically both a.c. and d.c. may be taken from it, the first from the **slip ring** connections and the second from the **commutator** connections. (L.R.Q.)

DOUBLE INTEGRAL. A double integral is a fundamental mathematical concept extensively used in **Calculus** for the treatment of geometric and physical problems, particularly those dealing with plane figures.

Let $f(x, y)$ be a **continuous function** in a region S of the XY-plane. Let the region S be divided in any manner into n sub-regions $\Delta S_1, \Delta S_2, \cdots, \Delta S_n$; denote any one of these sub-regions by ΔS_k and also denote its area by ΔS_k. Let (x_k, y_k) by any point within or on the boundary of ΔS_k. Then form the sum

$$\sum_{k=1}^{n} f(x_k, y_k) \cdot \Delta S_k = f(x_1, y_1)\Delta S_1 + \cdots + f(x_n, y_n)\Delta S_n.$$

Let $n \to \infty$ and at the same time let the greatest diameter of each sub-region $\Delta S_k \to 0$. Then the **limit** of this sum is defined as the double integral of $f(x, y)$ extended over the region S, and we write:

$$\lim_{\Delta S_k \to 0} \sum_{k=1}^{n} f(x_k, y_k)\Delta S_k = \iint_S f(x, y)dS.$$

A double integral may be interpreted geometrically as follows: If $z = f(x, y)$ is interpreted as the equation of a **surface**, the double integral $\iint_S f(x, y)dS$ represents the volume bounded by this surface, the XY-plane and a right **cylinder** standing on the boundary of the region S and perpendicular to the XY-plane.

Let $f(x, y)$ be continuous in a region S of the XY-plane bounded by the lines $x = a$, $y = b$, the X-axis and a curve $y = \phi(x)$, in **rectangular coordinates**. The integral

$$\int_a^b \left(\int_0^{\phi(x)} f(x, y)dy \right)dx,$$

which is called an iterated, or repeated, double **integral**, is defined as follows: hold x constant and integrate $f(x, y)$ with respect to y, substitute the limits $\phi(x)$ and 0 for y and subtract, then integrate this function of x with respect to x from a to b.

This iterated integral is sometimes written

$$\int_a^b \int_0^{\phi(x)} f(x, y)dy dx, \quad \text{or} \quad \int_a^b dx \int_0^{\phi(x)} f(x, y)dy,$$

and occasionally as

$$\int_a^b \int_0^{\phi(x)} f(x, y)dx dy.$$

The fundamental theorem on double integrals is the following: Let S be a region bounded by the lines $x = a$, $x = b$, the X-axis, and a curve $y = \phi(x)$. Then the following double integral and iterated integral are equal:

$$\iint_S f(x, y)dS = \int_a^b \left(\int_0^{\phi(x)} f(x, y)dy \right)dx;$$

i.e., the double integral may be evaluated by the iterated integral (using successive integrations).

In terms of **polar coordinates**, a double integral $\iint_S f(r, \theta)dS$ may be evaluated by an iterated integral of the form

$$\int_\alpha^\beta \left(\int_{r_1(\theta)}^{r_2(\theta)} f(r, \theta)rd\tau \right)d\theta.$$

(L.L.S.)

DOUBLE REFRACTION. If a crystal·of **calcite** (calcium carbonate, a common mineral) is held between the eye and a pinhole in a card, two bright dots are seen. If the crystal is rotated around the line of sight, one dot travels in a circle around the other, which remains fixed. Evidently* there are two refracted rays; with the light normally incident on the natural crystal face, they make (within the crystal) an angle of about $6° 9'$. The two refracted rays reveal a difference in **refractive index**. The one which remains fixed as the crystal revolves, called the ordinary ray, corresponds to a greater index than does the other, called the extraordinary ray. For sodium light the two indices of calcite are, respectively, 1.658 and 1.486. A simple test shows that the two rays consist of plane-**polarized light,** one vibrating at right angles to the other. By cutting the calcite into plates making various angles with the nat-

ural faces, a divergence as high as $6° 16'$ may be obtained for normal incidence; but it is generally less. For one direction (and, for calcite, only one) there is no divergence at all; the light is then said to be traveling along the optic axis of the crystal. Quartz also exhibits double refraction (though with much less divergence); but in this case the extraordinary ray corresponds to the greater refractive index (ordinary, 1.544, extraordinary, 1.553, for sodium light).

These phenomena were explained by Huygens (1678) as due to the fact that the ordinary wave has a spherical wave front, traveling with the same speed in all directions, just as if the medium were isotropic; while the extraordinary wave has either a maximum or a minimum speed along the optic axis, so that its wave front is either an oblate (door-knob-shaped) or a prolate (football-shaped) spheroid, externally or internally tangent to the spherical ordinary wave front at the two points on the axis. (See figure.) Along the "equators" of the two

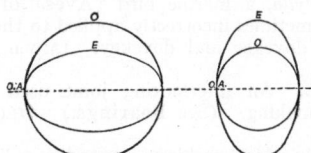

wave fronts the difference in speed is greatest. Here the extraordinary vibration is parallel to the axis, the ordinary vibration perpendicular to the axis; and the two vibration components, proceeding together, experience a progressively greater difference of phase. A "quarter-wave plate" is a layer cut just thick enough for this phase difference to be ¼ of a cycle.

Many crystals have two optic axes. In this case the double wave surface is a much more complicated system in which there are four points of intersection, corresponding to the two axes. Its section in the X-Z plane is a circle inside an ellipse and not touching it; in the Y-Z plane, an ellipse within a circle; and in the X-Y plane, which contains the axes, an ellipse intersecting a circle at four points. Crystals of **mica**, borax, and **topaz** are biaxial. (See also **Electric and Magnetic Double Refraction**, and **Photoelasticity**.) (L.D.W.)

DOUBLE SHEAR. Rivet; Shear.

DOUBLE STAR. (See **Binary** star.) There are numerous cases where two stars are so nearly in the same direction, as seen from the earth, that they appear as single stars to the unaided eye but may be separated into two components by the use of a telescope. Such a pair of stars is referred to as a double star. Double stars may be either one of two kinds. In cases where the two stars are only apparently close to each other (i.e., lie in approximately the same direction from the earth, but are separated by a great distance in the radial direction) the pair is known as an optical double. In the great majority of cases, however, the stars are actually close enough together to exert strong gravitational attractions on each other and are in orbital motion relative to each other. Such physically connected stars form what is known as a **binary star.** Optical doubles may be distinguished from binaries by observing the pair in a telescope over a period of years. In case the distance between them changes progressively over a long period of time while the position angle (see **Filar Micrometer**) remains constant it may be safely assumed that the motion is due to **proper motion** alone and that an optical double is under observation. Position angle changes progressively and distance oscillates between a maximum and minimum in the case of a binary.

The first recorded discovery of a double star was by Riccioli in 1650 when Zeta Ursae Majoris was announced to be a double star. Since that time the search for double stars has been carried on very thoroughly and every star, down to the 10th **magnitude,** has been carefully examined. Dr. R. G. Aitken, who has completed one of the most comprehensive surveys of variable stars thus far attempted, makes the following statement: "At least 1 in every 18, on the average, of the stars in the northern half of the sky which are as bright as 9.0 magnitude is a close **double star** visible with the 36-in. refractor." (W.K.G.)

DOUC. Langur.

DOUGLAS FIR. Conifers.

DOUROUCOLI. Monkey.

DOVE. Pigeon.

DOVEKIE. Aves, Charadriiformes. The black guillemot, *Uria grylla,* a marine bird (**Aves**) of the north Atlantic. Sometimes incorrectly applied to the little auk. Also spelled dovekee and dovekey. (A.W.L.)

DOVETAIL. An interlocking joint used in joinery and cabinet-making. (See **Bearings.**) (H.C.H.)

DOWEL. In woodworking, a wooden cylindrical pin for fastening two or more parts. In machine work, a cylindrical or tapered pin for locating and aligning parts. Dowels are sometimes used as transverse keys or driving pins where small torques are transmitted. (H.C.H.)

DOWNWASH. From an airfoil of finite span in flight there are a multitude of vortices trailing, forming a **vortex sheet.** As a result of circulatory flow about each vortex filament a net downwash flow of air is produced, having an important effect upon the air reactions. If the airfoil is untwisted, and elliptical in planform (or tapered in approximation of an ellipse), the vortices bound to the airfoil will produce a circulation of approximately elliptical strength distribution over the span. This produces a uniform downwash of air in the region of the wing. The combined effect of the forward motion of the airfoil and the downwash will be better understood upon reference to the accompanying diagram.

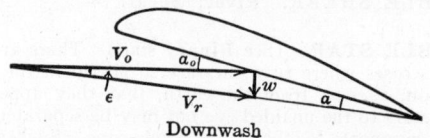

Downwash

The forward speed V_0 and the vortex-induced downwash w produce a resultant air velocity of V_r (substantially equal to V_0 in magnitude) which meets the airfoil at angle of attack α. The angle ϵ is called the downwash angle. Without downwash, the lift produced would be that corresponding to α_0 at velocity V_0, but downwash reduces the angle of attack to α thus decreasing the lift. To lift the same this airfoil would require α with downwash to be larger than α_0 without downwash. The effect of downwash is now seen to be deleterious since, for the same V and lift, it increases the needed α and produces extra drag. Furthermore this drag (**induced drag**) is not associated with aerodynamic cleanliness but with **aspect ratio.** Disposing the area of the wing as a slender wing of large wing span (i.e., high aspect ratio) is the way to reduce this form of drag. (F.T.M.)

DOWNWASH VELOCITY. Downwash.

DRAFT. Draft of a ship or boat is its depth of flotation. It is also the minimum depth of water in which navigation is possible.

In making a **pattern** which is to be molded in sand, a certain small taper is necessary where the pattern is of such shape as to be rather deeply imbedded in the sand. When the pattern is rapped or jarred loose from the sand, a slight taper, called, in foundry practice, draft, allows the pattern to pull free during its removal from the mold, without disturbing the side walls.

As applied to a gaseous system, *draft* is a pressure-differential that operates to move the gases. **Combustion** requires oxygen—and therefore air. To move this air through the fuel bed and to produce a flow of the gaseous products of combustion out of the furnace, then through the boiler, economizer, etc., requires a difference of pressure equal to that necessary to accelerate the gases to their final velocity, plus friction head losses. This difference of pressure is called draft whether measured above or below atmospheric pressure. The range of pressures required is most easily measured by manometers reading in inches of water.

A manometer measures pressures in terms of a displaced column of water. The pressure may be obtained from the manometer reading by employing the factor relating a head of water to the pressure produced at its base. It requires 2.31 ft. of water to result in a pressure of 1 lb. per sq. in. The ordinary draft gauge is a variation of the U-tube manometer.

Requisite draft can be obtained by use of **chimneys, fans,** steam or air **jets,** or combinations of these. The chimney is probably the most common, but the least understood of any of them. At one time the chimney was universally used as the sole means for producing a draft and even now it is relied on entirely in many small plants and partially in most of the large ones.

Mechanical draft may be classified as *forced* or *induced,* the former having the combustion air placed under a **plenum,** the latter referring to gas movement into a region of partial **vacuum.** With forced draft alone, furnace gases seep outward through cracks in walls, and blow through opened doors and ports. Induced draft alone allows considerable undesirable dilution of the products of combustion unless furnace, casings, ducts, etc., are maintained airtight. A logical compromise is to use both systems in a *balanced draft* adjusted to maintain atmospheric pressure (or a slight vacuum) in the furnace. There, expansion cracks, opened doors, etc., will not cause undesirable gas or air flows. (See **Foundry Practice; Watershed.**) (F.T.M.)

DRAFT GAUGE. As draft is always an extremely small pressure, special gauges have been devised for its measurement. An ordinary U-tube manometer is unsatisfactory for moderate draft since the liquid displacements are small. Inclined leg manometers (see **Manometer**) magnify the reading and place it on a single scale but need to be carefully levelled; also, the liquid is subject to evaporation, thus affecting the accuracy. Dry type draft gauges have been developed which operate from a spring-loaded diaphragm to one side of which is piped the pressure to be measured. Since a water-loaded U-tube manometer is a primary standard of reference, draft gauges are customarily calibrated to read "inches of water" no matter what their operating principle or fluid may be. (F.T.M.)

DRAFT TUBE. **Hydraulic turbines** frequently discharge the water with considerably more velocity than would be economical from the efficiency viewpoint, were it not possible to recover a great deal of that energy by the proper use of a diffusing chamber at the outlet. The diffusing chamber or tube is known as the draft tube, and there are a variety of types. However, the main objective is to convert the velocity head residing in the water leaving the turbine into pressure head. If this can be done efficiently, the turbine can be set somewhat below normal tailwater level.

The greater the **specific speed** of a turbine runner the higher will be the velocity of the water discharged into the draft tube, and the more important the recovery of this velocity by draft tube design. The draft tube is to take the water from the turbine at a point where the pressure is considerably less than atmospheric, and, by efficiently reducing the velocity, convert it into pressure head so that it can emerge smoothly into the tailrace at atmospheric pressure. By "efficiently" is meant without shock or whirl loss. Not all the velocity head can be recovered, for the water must be given to the tailrace at normal tailrace velocity to prevent its backing up into the turbine. Also, whatever friction loss occurs in the draft tube adds to this reduction of useful head. (F.T.M.)

DRAG, AERODYNAMIC. (See **Aerodynamics, Airfoil, Boundary Layer.**)

An object subjected to an air stream is acted on by a resultant air pressure. Drag is the component of air reaction which is parallel to the air stream. Its origin is profile impact of molecules of air against the face of the object, the skin friction of the molecules of air as they slide along the object, the vortices and eddying air currents set up in an otherwise undisturbed air stream by the presence of the object, and the induction effects of **downwash**, if any. Drag is that quantity which imposes limitations upon the top speed of vehicles, missiles, and so forth. As it is proportional to the square of the velocity, its magnitude mounts rapidly as velocities are increased.

There are two kinds of drag—drag on surfaces which obtain a useful reaction from the air stream as well as a drag, and drag upon surfaces where the only reaction is the drag. A wing has both drag and lift. Drag on the wing is the price paid for lift, and is so accepted. A strut, or wire, or wheel, creates no lift, and the drag is wholly undesirable. A drag of this type is called **parasite drag**. Careful streamlining and reduction of parts exposed to the air stream are ways of reducing parasite drag. The drag of a wing may be divided into profile drag and induced drag. **Induced drag** depends upon the lift and the **aspect ratio** of a wing. During a test, profile and induced drag are not separable. Parasite drag depends upon the surface roughness and shape of the object.

Profile drag of airfoils is the result of skin friction and turbulent wake. It is greatly influenced by **Reynolds number** (scale effect) and by initial turbulence in the air stream. Although some authors have proceeded on the basis that this is the sole influence on profile drag coefficient, others have found that the profile drag coefficient has a small variation with coefficient of lift. However, the generally accepted theories of origin of aerodynamic forces imply that scale effects modify profile and parasitic drags only and are absent in lift and induced drag. Experience confirms the validity of theory except near the **burble** attitude where viscosity has an effect on lift.

Experiments in the **wind tunnel** show that factors affecting reaction of air on airfoils are:
1. The relative velocity of the air and airfoil.
2. Extent of the surface area.
3. Density of the air.
4. The angle of inclination of the airfoil to the air stream.

All this may be stated somewhat as follows:

$$F \sim R^n \alpha \rho S V^2.$$

F is wind reaction, α is the **angle of attack** of the wing, ρ is the density of the air, S the surface area, V the air velocity, R the **Reynolds number**.

If a proper constant K be inserted, the similarity can be made into an equality. K may also be made to include the effect of Reynolds number, angle of attack and air density. The equation is then simplified to:

$$F = KSV^2.$$

This reaction is neither perpendicular nor parallel to the wind stream. It is convenient to divide it into its components of lift and drag. Letting K_L and K_D be the corresponding coefficients,

$$\text{Lift } L = K_L S V^2,$$

$$\text{Drag } D = K_D S V^2.$$

There will be a different value of K for each angle of attack and each R, although the effect of the latter is usually minor.

The defect of this equation is that K is not dimensionless, as a true coefficient should be; also, the formula is not flexible with respect to air density. If it were rewritten:

$$R = C \rho S V^2,$$

C would be dimensionless. The equation is usually written thus for drag,

$$D = C_d \frac{\rho}{2} S V^2,$$

because $\dfrac{\rho V^2}{2}$ is the pressure necessary to give air of density ρ a velocity of V. This is called the dynamic pressure. The drag is also $C_d q S$, where q stands for dynamic pressure. The drag coefficient C_d varies with angle of attack, but it is dimensionless. The wind tunnel test is used to establish the law of variation, and the results are plotted as a drag curve, with the coefficient of drag as the ordinate, and either coefficient of lift or angle of attack as the abscissa.

A few typical coefficients of drag are:
1. Flat-plate normal to air stream. Average value 1.28 often used, although subject to variation from R and R.N.
Flat-plate parallel to air stream. C_D = skin friction coefficient as there is no turbulent wake. Varies. 0.002–0.008 depending on initial turbulence and R.N.
2. Sphere (S = projected area), about 4×10^5 critical R.N. Laminar boundary, 0.45; turbulent 0.1.
3. Cylinder, axis transverse to air stream (S = projected area), about 4×10^5 critical R.N. Laminar boundary, 0.7; turbulent, 0.3.
4. Thin wing, infinite R, R.N. 3.5×10^6. C_{Dp} = .01.
5. Clark Y wing, R 6, R.N. 3.5×10^6. C_D = .009 + .06 C_L^2.
6. Airplane. $C_D = C_p + C_L^2 / \pi R$. C_p varies greatly as it depends on aerodynamic cleanliness (fairing, retraction of wheels, cowling of engine, etc.). (F.T.M.)

DRAG FOLDS.

The smaller folds included in major folds. Produced by the shearing stresses set up within a fold by the relative movements of strata parallel to their bedding planes. (R.M.F.)

DRAG LINE.

The drag-line excavator consists of a turntable on wheels or caterpillar treads, which supports the excavating machinery. A long **boom** is pivoted at its lower end to the turntable, and guyed at its outer end by a rope **sheave**. Extra long booms are characteristic of the drag-line excavator because the loading of the bucket is by scraping action in contrast to the shoveling action of a grab bucket. By the use of the boom, to which is attached the scraper, with the intermediary of a block and tackle, the operator can place the scraper bucket at a distance of 25–100′ from the position of the excavator. The scraper has attached to it a drag line which is wrapped around a powered drum in the cab. As this drag line is reeled in, it drags the scraper bucket and fills it, after which it is hoisted and dumped wherever wished. The drag-line excavator has been built in very large capacities, and is especially suitable for such work as levee building, **borrow pit** work, etc.

The drag-line scraper is a means for stocking-out bulk material to storage and reclaiming the same. It is comparatively low in first cost, and adaptable to a storage lot of irregular area. A head post and machinery house is located at the point from which stocking-out begins, and to which reclamation moves. A movable tail tower may be operated at different points along the edge of the storage lot farthest from the head post. Between the tail tower and head post is an endless wire cable, to which is attached a scraper bucket. This passes over the drive sheave in the head tower. The drive may

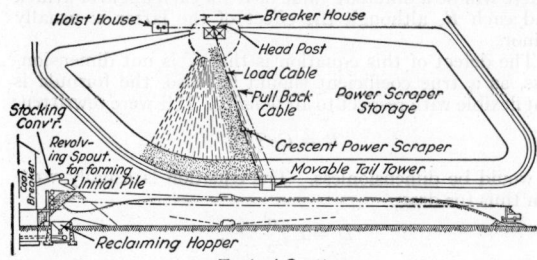

Typical Section

Storage and reclamation by drag scraper.

be reversed, and the bucket caused to move out on the stock pile and then be returned, scraping up a full load of loose material as it comes. Typical arrangement of the drag scraper is shown in the accompanying figure. (F.T.M.)

DRAGON FLY. Insecta, Odonata. Large insects of powerful flight which eat other insects caught in the air. The immature insect is aquatic and predacious. The order is composed of dragon flies and damsel flies only. All have four slender net-veined wings and long slender bodies, but the dragon flies have fore and hind wings of slightly different shape and hold them extended when at rest. Also called Devil's darning needles. (A.W.L.)

DRAGON'S BLOOD. Resins.

DRAGONET. Pisces, Teleostei. *Callionymus.* Brightly colored, marine shore fishes (**Pisces**), limited to the Palaearctic region with the exception of a few tropical species in the Pacific Ocean. (A.W.L.)

DRAIN. In surgery, this term refers to any appliance placed in a wound to provide for escape of discharge or drainage. A common drain for small infections is a small piece of rubber or gauze. In abdominal cases, where drainage is required, the cigarette drain is most commonly used. This is made up of several layers of gauze surrounded by a sleeve of thin rubber sheeting. A Penrose drain is a thin tube of rubber sheeting without any gauze filling. Soft rubber tubing of all sizes is also commonly used. (R.S.M.)

DRAINAGE. The removal of water from any particular locality constitutes the action of drainage. Natural drainage is embodied in the rivulets, streams, and rivers to which surface and subsurface water is drained. Artificial drainage is used for a wide variety of purposes. Drainage in pipes is very common, especially in removing the water-floated wastes from human habitations. The drainage from sinks, lavatories, baths, and toilets constitutes a special phase of plumbing. All such drainage is accommodated in pipes. The principal drainage arteries for such a system are cast iron pipes with bell and spigot joints which are made water-tight by calking. Outside the building such drainage systems connect to the regular terra cotta or other type sewer drains of the sewerage system. Secondary drains are usually 1¼″ or 1½″ screwed iron pipe.

Artificial drainage of surface waters for the purpose of drying, or partly drying, a certain area of land, or for the purpose of intercepting water on its natural way toward a particular tract, can be accomplished by the use of ditches which are pitched in the direction in which the water is to be removed. Swampy, marshy land is often drained by this method.

Subsurface drainage systems are used where a ditch would be undesirable, as on a playing field. The subsurface drain can be a French drain or a pipe drain. A French drain is a deep ditch filled to a short distance below the surface with loose rock, the larger rock being at the bottom. Due to the larger friction loss in a drain of this type, it must be laid to a steeper gradient than that used in the pipe drain. In a subsurface pipe drain terra cotta drain tile is laid end to end wth slightly open joints. A fill of loose materials, such as crushed stone or gravel, is laid around the open joints so that water which enters this porous area will seep through the open joints into the pipe drain. (F.T.M.)

DRAVITE. Tourmaline.

DRAWING. One use of the word drawing is in reference to an operation performed on metal. The metal is originally in the form either of sheets, solid blanks, or pierced billets. These are formed in suitable dies into many different shapes, such as hollow cylinders, cuplike parts, or solid parts of various shapes. Pipe, wire, various structural shapes, are often formed by drawing. Drawing tends to reduce the thickness of the metal. (See **Wire Drawing, Cold Working, Stamping, Ductility, Forming, Press Working.**) (F.T.M.)

DRAWING REPRODUCTION. Modern draftingroom practice requires penciled drawings either on tracing paper, or on tracing cloth, which is made of cotton, and treated and sized to provide a suitable working surface. Since original drawings or tracings are never sent into the shop, and are, as a matter of fact, infrequently used in the drafting department itself, some method of reproduction is necessary. Blueprinting is one of the oldest methods, and is effected by placing the tracing on a sheet of paper or of cloth sensitized with iron salts and exposing the two to sunlight or an electric-arc lamp. After exposure, the print is washed in water and dried. To insure permanency and intensification of color, blue prints are sometimes washed with a potassium bichromate solution. The resulting print has a dark blue background and white lines and figures. Brown prints, having white lines on a dark background, are made in a manner similar to blue prints, but Vandyke paper is used. The prints require fixing in a "hypo" solution.

Black-line and blue-line positive prints, having dark lines on a white background, may be made from blueprint or Vandyke originals, but may also be produced directly from tracings by using specially prepared paper, which is developed by covering one side of the paper, after printing, with a developing and fixing solution. Ozalid prints are positive prints made by printing directly from tracings and developed dry in the presence of ammonia vapor. The prints may be obtained with red, blue, or black lines on a white background. Reproduction on transparent foil sheets can also be obtained by the process.

The Photostat process is a direct copying process whereby photographs of maps, drawings, prints, and tracings may be made, developed and fixed. The copy may be enlarged or reduced as desired. When made from original pencil or ink drawings or tracings, Photostat prints have a dark background and white lines. Positives may be obtained by making a second print from the first (negative) Photostat. (H.C.H.)

DREDGE. An excavating machine for use in river, harbor, or drainage work, is a dredge. A characteristic application of the dredge is in submarine excavation. Dredges are usually floated on water-tight hulls. There are several types, which may be classified according to the way they excavate. The dipper dredge, which is the more common type, is very similar to a locomotive crane, except that it is mounted on a floating hull. Its digging equipment consists of a dipper or grab bucket, which is able to dip as deep as 50′ below the water surface. This type of dredge may be used in dredging channels or in cutting a wide drainage ditch provided the ditch is large enough to float the dredge. This dredge will dig its own waterway through land, and is particularly useful for drainage work. A dredge where the excavating is done by a number of buckets placed on an endless chain is known as an elevator dredge. A bucket chain is able to elevate the spoil considerably higher than other types. It is frequently used to raise sand and gravel from a stream bed.

The hydraulic dredge digs by suction of the spoil material from the bottom. A large pipe is lowered to the area to be dredged. A large power-driven rotary cutter is rigged in front of the entrance end of this pipe. Operation of the cutter breaks up the soft material of the bottom so that it may readily be transported by a current of water. A water pump attached to the upper end creates a powerful flow of water through the pipe which picks up and carries the loose material near the mouth of the pipe. Large, specially designed centrifugal pumps create the flow. Dredges are usually steam-driven. The larger sizes contain living quarters for the crew. (F.T.M.)

DREIKANTER. Literally a 3-cornered or 3-edged pebble. A term of German origin signifying a pebble that has been sculptured or faceted by natural sandblasting. Such pebbles are usually considered to be proof of the semi-arid or even desert condition under which they have been formed, but they may also be formed in pluvial climates provided that the regolith is composed of porous and shifting sands such as compose glacial sand plains and coastal beaches. They are also called gibbers or glyptoliths. (R.M.F.)

DRESSING. Grinding.

DREWITE. A term proposed by R. M. Field in 1918 for pure calcium carbonate muds of organic chemical origin. Drewite probably forms the bulk of the fine-grained unfossiliferous limestone from the pre-Cambrian to the present. Named after G. H. Drew, a pioneer in the study of marine bacteria. (R.M.F.)

DRIER. Pigments, Paints, Varnishes.

DRIFT. In mechanics a drift is a hand tool somewhat resembling a punch, but used to drive pins in or out of deep holes. The drift is made slightly smaller than the diameter of the hole.

In structural engineering the term drift pin refers to a tapered steel pin which is used during the fabrication of a member to hold the individual parts together before the fitting-up bolts (bolts which are necessary to draw the parts together subsequent to riveting) are inserted in the rivet holes. Drift pins are also used together with bolts in steel erection to fasten the members to the gusset plates or other connections before the field rivets are driven.

In geology, the use of the term drift came about in the following way. Before Louis Agassiz propounded his theory that extensive deposits of sand, gravel, boulder clays, etc., of Northern Europe and North America were the result of the action of great continental ice sheets, it had been suggested that during some period when the land had stood at a lower level, ice bergs were swept from the north over the continents and, melting, dropped their loads of detritus. Thus some of the material which we now know to be of glacial origin was called drift, because it was believed to have been "drifted" to its

place through the agency of ice. This was the theory of Sir Charles Lyell as a substitute for the still older theory of the diluvialists who believed that these glacial deposits were positive evidence of the "deluge." Geologists have retained the term drift to designate all unconsolidated sediments which are determined to be of glacial origin, and still further divide drift into stratified drift and unstratified drift or till. For "ether drift," see **Ether and Michelson-Morley Experiment.** (F.T.M., R.M.F.)

DRIFT METER. The instrument used in air navigation to measure the angle between the heading of a plane and the track being made good, is called a drift meter.

The simplest form of drift meter consists of a circular plate of heavy glass set in the floor of the cockpit in front of the pilot. The plate may be rotated within a ring on which degrees of angle are marked to the left and right of a zero mark. This zero point is in the direction of the forward end of the longitudinal axis of the plane. The plate has a series of parallel lines ruled on it. With the plane in level flight the pilot can look down through the plate and rotate it until objects on the ground are moving parallel to the lines. Under these conditions the lines on the plate will be in the direction of the track being made good, and the angle between the heading and this track may be immediately read on the scale.

Many modern and complicated types of drift sights have been devised but all of them operate on the fundamental principle described above. In some modern drift sights, a gyroscopic stabilizing system holds the grid lines level even though the plane is not flying level. Astigmatizers are frequently incorporated to assist in measuring drift angle, particularly when flying over water.

In some modern drift sights a system is incorporated so that ground speed may be determined. A pair of wires is marked on the grid, perpendicular to those set parallel to the apparent motion of the ground. The time required for an object on the ground to move from one of these wires to the other will be proportional to the ground speed. The distance of the plane from the ground must be accurately known, and the objects observed must be directly below the plane to obtain an accurate value of ground speed. (W.K.G.)

DRIFTING. Forging.

DRILLING. Originating a cylindrical hole with a one- or two-lipped straight-fluted or twist drill. (Contrast **Boring, Reaming, Trepanning.**) The twist drill is the most widely used tool for cylindrical holes. In drilling a hole, the point of the drill is forced into the work, crushing the material immediately beneath the web, and thereby allowing the two lips or cutting edges to cut. A two-lipped twist drill will originate a true cylindrical hole if both lips are of the same length and at the same angle with the axis of the drill, and if the axis of the drill and its axis of rotation are coincident. The point angle of a twist drill is generally 118° for drilling cast iron or steel; the lip relief angle is from 12 to 15°, and the chisel edge angle from 120 to 135°. The helical flutes provide rake for the cutting lips so as to curl the chips; they furnish a path for the escape of the chips and serve as channels for lubricants or coolants to reach the point of the drill. The drill is guided primarily by the two lips but also by the margin or full-diameter edge. That portion of the drill immediately behind the margin is reduced in diameter to minimize the friction between the drill and the walls of the hole. The web is the central section that connects the two outer helical portions of the drill.

There are three general types of drilling machines: the bench drill press, the upright drill press, and the radial drill press. In all three types, the spindle rotates in a sleeve or quill which does not rotate but is free to move axially to provide the necessary feed for the drill. In the bench drill, both the table knee and the head carrying the spindle are adjustable for various classes of work. In the upright drill press, the spindle sleeve supports are fixed, and all adjustment for different classes of work is made by moving the table, which is accomplished by turning the elevating crank. The table can be moved in a horizontal plane, clamped at any point or, if desired, swung out of the way so that large work may be placed on the base. The machine is equipped with a ratchet lever for hand-feeding the drill. A hand wheel is fastened to a worm shaft whose worm engages a worm gear on the pinion feed shaft, giving a motion much finer than that obtained by using the hand lever. Speed changes in bench drills are effected by cone pulleys; in upright drill presses either cone pulleys or a geared head is used.

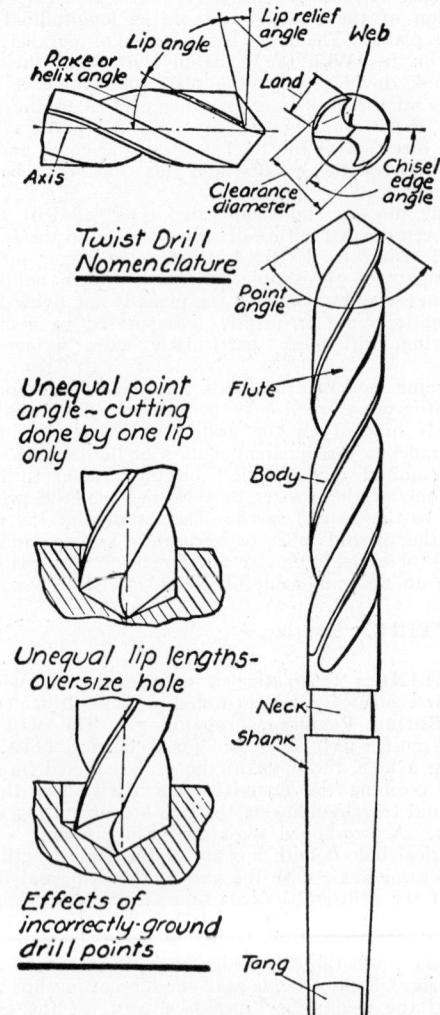

Twist Drill Nomenclature

Unequal point angle - cutting done by one lip only

Unequal lip lengths- oversize hole

Effects of incorrectly-ground drill points

The radial drill has a column which may be rotated about the base. On the column is a radial arm which moves in a horizontal plane with the column, but may also be moved in a vertical plane. A head carries a spindle, and may be moved radially along the arm. The radial drill is therefore adapted to heavy work where it is easier to move the drill than the work. Spindle speed and feed changes are effected by gearing. A sensitive

drill is a vertical or upright machine of comparatively light construction adapted to very high speeds and used for delicate work.

Drill-press spindles generally have a Morse taper hole in the spindle. Twist drills with taper shanks may be fitted directly to the tapered spindle hole, but in many cases the taper shank of the drill is smaller than the spindle socket. In such an instance, a collet or sleeve is employed. A drill collet is a conical sleeve which has an inside taper fitting the shank of a twist drill, and an outside taper fitting the socket in the drill-press spindle.

Straight-shank twist drills which have a cylindrical shank of the same diameter as the drill itself are generally held either in 3-jaw or 2-jaw **chucks**. Both types of chucks are generally held by inserting an arbor in a tapered hole in the body of the chuck, which in turn is held by inserting its tapered shank in the spindle socket.

For wood drilling, the work can ordinarily be held by hand. Except where small holes are drilled in comparatively large or heavy parts, however, metal drilling requires so much torque that the work should be held in a vise or be clamped to the drill press table.

Two-lipped twist drills with either taper or straight shanks are satisfactory cutting tools for holes in cast iron and steel. Straight-fluted drills are used for copper or brass; they manifest less tendency to "dig in" than the helically fluted tool. For heavy feeds and comparatively deep holes, oil-hole drills are employed. These drills have holes drilled from the solid metal through the body of the drill so that lubricants can be pressure-fed to the drill points. The lubricant not only assists in the cutting operation, but also facilitates the removal of the chips by forcing them out along the flutes of the drill. Oil-hole drills, with holes for lubricant in the body of the drill, may be employed in drill presses but they are generally used in automatic machinery where the drill is stationary and the work rotates.

Drilling speeds vary from 40 ft. per min. for cast and alloy steels to 300 ft. per min. for brass and bronze. Drilling feeds vary from .002″ per revolution for $\frac{1}{8}$″ diameter drills to .015″ per revolution for drills 1″ in diameter and over; a representative figure is .001″ per $\frac{1}{16}$″ diameter of the drill. (H.C.H.)

DRIVEN ANTENNA. Directional Antenna.

DRIVING FORCE. Gas Absorption.

DRIZZLE. Hydrometeors.

DROMEDARY. Mammalia, Artiodactyla. An Arabian camel of a breed used for riding. Incorrectly applied to the Bactrian camel. (A.W.L.)

DRONGO. Aves, Passeriformes. The king crows of southern Asia and Africa. Birds (**Aves**) of several species, mostly black, forming a family not closely related to the true crows. (A.W.L.)

DROP FORGING. Forging.

DROPSY. Edema.

DROSOPHILA. Insecta, Diptera. The name of a genus of fruit flies or pomace flies. Because of the extensive use of these insects, and especially *Drosophila melanogaster,* in the study of heredity the name of the genus has become more commonly used than the vernacular names. (See **Heredity.**)

These flies are small and usually light-colored insects which frequent decaying or fermenting fruit. They are attracted by vinegar, cider, and fruit wastes even when fresh and so are often common about the house during the canning season. The **larvae** live in the fermenting material. (A.W.L.)

DRUGS. A wide variety of sources has been tapped to furnish materials to combat disease and its effects and to promote health. Many of the products of vegetable, animal and mineral origin in use were recognized many centuries ago as possessing therapeutic value. Great improvements have been made in this field in recent decades. The desired products have been obtained in greater purity and of greater efficacy. In many cases individual or group substances have been isolated and synthesized. The role played by particular drugs in promoting recovery or improvement of health has been studied. Many drugs are used, not for their function in combating directly the causative agent of a disease, but for their indirect action (1) upon its toxic products, (2) by inducing a biochemical reaction, which causes the body organism to produce substances to combat the disease organisms or their toxic products, (3) by inducing physiological reactions which enable the patient to resist those reactions caused by the disease.

Vegetable drugs may occur in all parts of plants, but are particularly abundant in such storage regions as seeds, roots and bark. Many used by man are found in the leaves, and a few in the flowers. Their function in the plant is a debated one; perhaps they are by-products of the plant's metabolism, or they may be energy-yielding reserves stored until needed. The poisonous nature of many of them suggests the thought that they may be designed for protection against attacks by insects and other animals. Drug substances in the plant fall into several classes. One of these contains the alkaloids, nitrogenous organic substances which react with acids. Many of these alkaloids are powerful poisons; all of them are toxic to animals. In another group of drug substances are the glucosides, complex organic substances which, when acted upon by certain **enzymes,** break down into **sugars** plus other compounds. Other medicinal substances are found in **resins, tannins, gums, and oils.**

In former times the number of plants used medicinally was very great. Many were known by primitive man and the knowledge of their value handed down from generation to generation. In modern times the number of plants used medicinally has diminished greatly, partly because men learned how absurd was the reason for their use and how slight their value, and partly because other drugs, many of them prepared synthetically, replaced the natural products. But many continue to be of great benefit to man.

Every part of the world yields drug plants. In a very few instances, these plants are cultivated. Mostly, however, the plants are gathered wild. Usually all that is necessary to prepare the crude drug for market is to gather and dry the plants. They are then shipped to the wholesale buyers. If the plant itself is to be used as a drug, it is thoroughly cleaned and then ground. More frequently the preparation is an elaborate process calculated to free the pure drug from the tissue in which it occurs.

Common vegetable drugs are **belladonna, ipecac, quinine, coca, digitalis, strychnine, opium, castor oil, camphor,** and **menthol.**

The narcotics include very valuable drugs which, taken in moderate doses, properly prescribed, relieve pain and produce sleep, but in large doses cause stupor, or coma, and commonly convulsions. Among the habit-forming drugs are **opium** and its derivatives, **coca, and** hashish from **hemp.**

The pioneer workers in pharmacology acquired considerable empirical information about methods of extraction of drugs from natural materials by water, hot water, alcohol and other solvents, and combinations of solvents. In the more purely scientific field, some outstanding discoveries are the **hypnotic,** chloral hydrate (1867), the **antipyretic,** acetanilide (1885), the anti-**syphilitic,** salvarsan (1910), the **thyroid** principle, thyroxin, insulin (used in **diabetes),** sulfanilamide (first commercially produced in 1937) and other "sulfa" drugs, **Atabrin** and **plasmoquin** to combat malaria, and **penicillin.**

Refinements in the technique of biological assay have contributed largely to the progress in the study of **vitamins, enzymes,** and **hormones.** By means of the methods that have been developed in this branch of science the effects of complex mixtures of substances are measured by their physiological reaction on test animals. In this way the direction of physiological research is controlled, even though the chemical nature of the substance under examination is entirely unknown.

Some substances that are mentioned elsewhere may be classified as drugs, since they are of the greatest importance in treating disease and promoting health. These are (1) the serums and **vaccines** usually produced in animals, directly or indirectly, by biochemical reactions of **bacteria.** In some cases the bacteria produce an attenuated strain which is utilized. Vaccines are used to induce a mild attack of the disease or a related disease. By this means protective substances are formed in the system of the patient. The protective substances are added directly to the system in other cases. (2) The glandular products, of which **adrenalin,** the active principle of the **suprarenal gland,** and **thyroxin,** the active principle of the **thyroid gland,** have attracted widespread attention, have been studied physiologically on test animals, and on man, and both have been synthesized. (3) The **vitamins** have been shown to be necessary for the normal functioning of the organism as well as necessary for the avoidance and cure of certain deficiency diseases.

Representative drugs may be classified upon the basis of physiological function as follows:

1. *Antiseptics.* Sodium hypochlorite, iodine, alcohol, silver nitrate, tannic acid, picric acid, potassium permanganate, iodoform, chloramine T, dichloramine T, hydrogen peroxide, 4-hexylresorcinol (urinary disinfectant), phenol, mercuric chloride solution (very poisonous), boric acid, argyrol (colloidal silver), menthol, phenyl salicylate (salol), triphenylmethane dyes, both basic (very poisonous) and acidic, flavine dyes (proflavin, acriflavin, rivanol —active in the presence of proteins).

2. *Anesthetics.* *a.* General: ether, chloroform, nitrous oxide, ethylene, cyclopropane, avertin, alcohol. *b.* Local: cocaine, novocaine, stovaine, production of cold by evaporation of ethyl chloride.

3. *Soporifics, hypnotics.* Bromides of sodium, potassium, ammonium, calcium, lithium, chloral hydrate, sulfonal, trional, tetronal, ethyl carbamate, hedronal, barbital (Veronal), phenobarbital (Luminal), diallybarbituric acid, allylisopropylbarbituric acid, and other barbituric acid compounds.

4. *Antipyretics, analgesics.* Acetanilide, acetphenetidin (Phenacetin), acetyl salicylic acid (Aspirin), antipyrine, pyramidon, Atophan, Novatophan.

5. *Purgatives, laxatives, cathartics.* Citrates, tartrates, acetates, phosphates, sulfates of magnesium, sodium, potassium, phenolphthalein, castor oil, croton oil, cascara, rhubarb, aloin, podophyllin, and other plant principles.

6. *Circulatory depressants and stimulants.* *a.* Depressants: glyceryl trinitrate, erythritol tetranitrate, mannitol hexanitrate, isoamyl nitrite, sodium nitrite, benzyl nitrite, aconite, veratrine. *b.* Stimulants: digitalis, digitoxin, digitalin (not digitonin), caffeine, adrenalin, ammonia, camphor, cocaine.

7. *Parasympathetic nerve paralyzers and stimulants.* *a.* Paralyzers: atropine, homatropine. *b.* Stimulants: pilocarpin, physostigmine, histamine.

8. *Mydriatics and myotics.* *a.* Mydriatics (causing dilation of the pupil): belladonna, atropine, strychnine, scopolamine (hyoscine). *b.* Myoptics (causing contractions of the pupil): pilocarpine, eserine.

9. *Dieuretics.* Caffeine, theobromine, theophyline (theocin), hexamethylenetetramine.

10. *Emmenagogues.* Ergot, apiol, rue, tansy.

11. *Emetics.* Ipecac, potassium antimony tartrate (tartar emetic), apomorphine, mustard.

12. *Expectorants.* Ammonium chloride, ammonium anisate, woodtar creosote, eucalyptus, terebene.

13. *Astringents.* Tannic acid, alum.

14. *Counterirritants.* Mustard, capsicum, turpentine, chloroform, cantharides.

15. *Antiarthritics.* Lithium citrate, cinchophen, salicylates and iodides of sodium, potassium, ammonium, calcium, strontium, lithium.

16. *Alteratives.* Iodides of sodium or potassium, colchicum, taraxacum.

17. *Antacids.* Sodium bicarbonate, magnesium hydroxide and oxide, bismuth basic carbonate.

18. *Tonics, bitters.* Cinchona, nux vomica, ferric citrate, sodium hypophosphite, glycerophosphates of sodium, potassium.

19. *Narcotics.* Morphine, codeine, heroin, scopolamine (hyoscine).

20. *Demulcents.* Glycerol, olive oil, gelatin.

21. *Drugs for specific diseases.* For counteracting: *a.* malaria: quinine, Atebrin, plasmoquin. *b.* syphilis: arsphenamine (Salvarsan, Ehrlich 606), neoarsphenamine (Neosalvarsan). *c.* leprosy: esters and sodium salt of chaulmoogric acid. *d.* sleeping sickness: tryptane blue (Bayer 205). *e.* acute rheumatic fever: salicylic acid compounds.

22. *Enzymes.* Pepsin, pancreatin.

23. *Hormones.* Adrenalin (active principle of the suprarenal gland), thyroxin (active principle of the thyroid gland), insulin (active principle of the pancreas), pituitary extract (active principle of posterior lobe of pituitary gland).

24. *Vitamins.* (See **Vitamins**.) (R.K.S., R.M.W.)

DRUM. Pisces, Teleostei. Fishes (**Pisces**) of numerous species, mostly marine, constituting the family Sciaenidae. The **sheephead** is a North American fresh water species of some food value. (A.W.L.)

DRUMLIN. A drumlin is a hill composed of glacial material of unstratified and heterogeneous character usually about 100′ in height and ¼–½ mile in length, oval in shape, and with its long axis parallel with the general direction of ice movement. Sometimes a mass of bed rock seems to have been the anchor about which the glacial till was deposited. Drumlins are known, however which contain no bed rock core. Drumloid-shaped hills similar to the smaller glaciated rock features called **roches moutonnée** are sometimes called rock drumlins. (R.M.F.)

DRUPE. Fruit.

DRUSE. A cavity, usually in a **sedimentary** rock, the walls of which are encrusted with minerals which have been derived, through underground solutions, from the rocks in which the cavities were formed, by **solution**. (R.M.F.)

DRY PIPE. Steam Purifier.

DRYING. Distillation, Evaporation and Drying.

DRYOPITHECUS. Paleontology of Man.

DUANE AND HUNT'S LAW. The quantum-energy law for the generation of x-rays. When a cathode particle in an x-ray tube strikes the target, its energy may all be transformed into heat; or it may excite an atom of the metal to emit a quantum of x-radiation. (See **X-rays and Quantum Theory**.) If all the cathode particles have the same energy Ve (voltage times electronic charge), some may give rise to low-frequency quanta and have energy left to heat the target, others may excite quanta of higher frequency and have less

energy to spare. But no quantum can be emitted of frequency and energy higher than that corresponding to the original energy of the electrons. This is Duane and Hunt's law. It is expressed by the so-called Planck-Einstein equation. If the highest frequency emitted is ν_{max}, and h is Planck's constant, the equation may be written $h\nu_{max} = Ve$. It follows that for a given voltage V, the **x-ray spectrum** must terminate abruptly at the frequency Ve/h. (L.D.W.)

DUCK. Aves, Anseriformes. Swimming birds (**Aves**) of moderately large size with heavy bodies, short legs, webbed feet, and broad flattened beaks with sieve plates at the sides. In common usage the term includes the closely related **teals, sheldrakes,** and **mergansers.** With the exception of the mergansers, which have narrow beaks with serrate edges and live chiefly on fish, the ducks eat rice and other plant products, with some insects and other small animals. They are among the leading game birds and many wild species have very palatable flesh. Many species have been domesticated and interbred with the common domestic duck.

Among species which have received special names are the **gadwall, widgeon, mallard, shoveler, pintail, bufflehead, scoter,** and **canvas-back.** (A.W.L.)

DUCKBILL. Mammalia, Monotremata. *Ornithorhynchus.* A primitive egg-laying mammal of Australia, also called the platypus and the duck-mole. It is about 18″ long and has close fur somewhat like that of the moles. The muzzle is broad, flat, and naked, resembling the beak of a duck but not horny. The feet are broad and webbed. Duck-moles are found in the streams of southern and eastern Australia. They nest in burrows in the banks and deposit two small eggs at a time. The young are nourished with milk secreted by the mother. (A.W.L.)

DUCT. A tube-like structure providing for the passage of secretions or excretions from any organ. (D.M.H.)

DUCTILITY. Ductile metals are capable of undergoing considerable amounts of plastic deformation before developing shear cracks or failing by rupture in tension. Ductility is generally associated with tensile properties or the ability to be cold drawn, as in wire drawing or sheet stamping operations. Per cent elongation in the **tension test** is the usual measure of ductility. (See **Malleability; Metals, Physical Properties of.**) (R.H.H.)

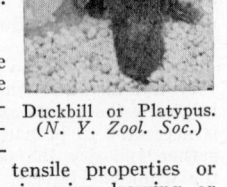

Duckbill or Platypus. (*N. Y. Zool. Soc.*)

DUCTLESS GLAND. Endocrine Gland.

DUCTUS COMMUNIS. A passage connecting the ducts of all female organs of reproduction with the genital **atrium** in the flatworms. (A.W.L.)

DUGONG. Mammalia, Sirenia. A marine **mammal,** *Halicore,* found along the shores in the Oriental region. Related to the American manatee. It has a blunt muzzle, a broad horizontal tail, and pectoral flippers. These animals eat seaweed and are caught for their flesh and oil. (A.W.L.)

DUIKERBOK. Mammalia, Artiodactyla. Small South African **antelopes.** Daintily built, with large ears and, in the male, short straight horns. (A.W.L.)

DULONG AND PETIT'S LAW OF SPECIFIC HEATS. It has long been known that the atomic heats of the great majority of elements have nearly the same value at room temperature; in fact, the thermal capacity of a gram atom of most elements is not far from 6 calories per degree. Dulong and Petit expressed this by stating that the specific heats of elements are in inverse proportion to their atomic weights.

That this should be the case for gases, easily follows from the **kinetic theory** and the principle of **equipartition of energy.** For example, if the same mean energy per molecule is necessary to raise the temperature of oxygen and of hydrogen $1°$, the same is true per atom, and weights of these gases having the same number of atoms will have equal thermal capacities.

For solids the matter is not quite so simple, and the more exacting theories of Einstein, Debye, and others show that the atomic heat should be expected to vary with the temperature. According to Debye, there is a certain characteristic temperature for each crystalline solid at which its atomic heat should equal 5.67 calories per degree. Einstein's theory expresses this temperature as $h\nu_m/k$, in which h is **Plack's constant,** k is **Boltzmann's constant,** and ν_m is a frequency characteristic of the atom in question vibrating in the crystal lattice (see **Crystal Structure.**)   (L.D.W.)

DUMMY ANTENNA. Antenna, Dummy.

DUNE. Dunes are elliptical or crescent-shaped mounds of sand which may reach a height of a hundred or more feet. The windward slopes of dunes are gentle, the lee sides steep, if crescentic in shape the convex side faces the direction from which the wind is blowing. The crescentic-shaped dunes are called barchans. Sand

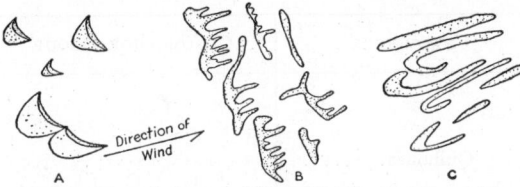

Types of sand dunes. (*After Walther and Cornish.*)

blown up the windward side drops down the lee slope causing the dunes to migrate slowly. Dunes are common and characteristic phenomena of lake and sea beaches, semi-arid regions, and particularly deserts, but also may develop on any dry, sandy soil in a relatively humid region where there is a lack of vegetation. Ancient geologic formations which may have originated as dunes may be identified by their peculiar type of **cross-bedding,** and by the peculiar texture of the sand. (R.M.F.)

DUNITE. Peridotite.

DUNLIN. Aves, Charadriiformes. *Tringa.* A common European shore bird; a **sandpiper.** Also called the ox bird. (A.W.L.)

DUNNOCK. Aves, Passeriformes. The European hedge sparrow, *Prunella modularis.* (A.W.L.)

DUODENUM. That portion of the intestinal tract immediately beyond the stomach. It begins at the **pylorus** and extends to the beginning of the **jejunum.** Horseshoe-shaped, it forms a curve around the head of the pancreas. The secretions of the duodenum are alkaline. Into it empty the bile duct from the liver and two ducts from the pancreas. The greater part of the digestion of food begins here.

The duodenum is a common site for ulceration. (See Peptic Ulcer.)   (D.M.H.)

DUODIODE. Tube, Electronic.

DUPLEX PUMP. The kind of **direct-acting pump** most often found has two steam cylinders arranged side by side. The water cylinders are also two in number, arranged side by side. The duplex pump is in reality a twin-cylinder **pump.** It has two steam pistons, two water pistons, and two piston rods. The special advantage of the twin arrangement comes from the convenience of valve operation in a twin-cylindered direct-acting pump. The steam valve of one of the cylinders is caused to reciprocate properly in its valve chest with a motion derived from the travel of the piston rod of the other.

The mechanism is quite simple. A collar block attached to the piston rod operates an oscillating type lever to one end of which is attached the steam valve. The motion of that piston rod moves the valve in the other cylinder, which then executes a pumping stroke as the steam enters and pushes against the piston. As that piston rod moves, it similarly operates the valve in the first steam cylinder, causing it to effect a pumping stroke. Thus in a duplex pump the valve action on one cylinder is derived from the stroke of the other, and pumping strokes are consummated alternately, first in one cylinder, then in the other. (F.T.M.)

DURAIN. The term proposed by Marie Stopes in 1919 for an ingredient of bituminous **coal,** occurring in bands. In thin, translucent sections durain is seen to be composed of reddish **spores** in a gray, granular **matrix.** (R.M.F.)

DURALUMIN. Aluminum Alloys.

DURCHMUSTERUNGEN. Bonner Durchmusterungen.

DURIRON. Alloys.

DUST DEVIL. Whirlwinds over sandy areas which pick up dust and sand, swirling the particles upward. They sometimes become fairly strong and reach hundreds of feet upward. (P.E.K.)

DUST STORM. Over dusty, dry, or desert areas strong winds which have picked up considerable dust and carried it to great heights. Sand storms are similar to dust storms. Three factors are necessary in the development of a dust or sand storm, (1) a dusty or sandy surface, (2) strong winds which can strike at the ground and pick up dust particles, and (3) a steep lapse rate, i.e., unstable air, which permits the dust to be carried to great heights.

Visibility in dust storms is usually only a few yards and sometimes reduced to night-like darkness. (P.E.K.)

DYADICS. Two **vectors** placed in juxtaposition neither with a dot · nor a cross × between them constitutes a dyad, as for example **ab.**

Any polynomial of dyads, as $a_1b_1 + a_2b_2 + \cdots + a_nb_n$, is called a dyadic.

A dyadic is sometimes regarded as a **tensor** of the second rank.

The first vector of a dyad is called its antecedent and the second its consequent. The antecedents and consequents of a dyadic are respectively the antecedents and consequents of its constituent dyads.

In general, a dyadic is usually denoted by bold-face capital Greek letters, as Φ.

If the order of the vectors in each dyad of a dyadic Φ is reversed, the resulting dyadic is called the conjugate of Φ.

Two dot products of a dyadic $\Phi = A_1B_1 + \cdots A_nB_n$ by a vector v are defined thus:

$$\Phi \cdot v = A_1(B_1 \cdot v) + A_2(B_2 \cdot v) + \cdots A_n(B_n \cdot v),$$
$$v \cdot \Phi = (v \cdot A_1)B_1 + (v \cdot A_2)B_2 + \cdots + (v \cdot A_n)B_n,$$

both being vectors. In $\Phi \cdot \mathbf{v}$, the dyadic Φ is said to act as pre-factor, in the other case, as post-factor.

A dyadic acting as pre-factor or post-factor upon any vector produces a **linear vector function** of this vector.

Any dyadic $\Phi = \mathbf{A}_1\mathbf{B}_1 + \cdots + \mathbf{A}_n\mathbf{B}_n$ can be reduced to the trinomial form $\mathbf{a}_1\mathbf{b}_1 + \mathbf{a}_2\mathbf{b}_2 + \mathbf{a}_3\mathbf{b}_3$, where $\mathbf{a}_1$, $\mathbf{a}_2$, $\mathbf{a}_3$ are chosen arbitrarily.

Any dyadic Φ can be expressed in the form

$$\Phi = a_{11}\hat{\mathbf{i}}\hat{\mathbf{i}} + a_{12}\hat{\mathbf{i}}\hat{\mathbf{j}} + a_{13}\hat{\mathbf{i}}\hat{\mathbf{k}}$$
$$+ a_{21}\hat{\mathbf{j}}\hat{\mathbf{i}} + a_{22}\hat{\mathbf{j}}\hat{\mathbf{j}} + a_{23}\hat{\mathbf{j}}\hat{\mathbf{k}}$$
$$+ a_{31}\hat{\mathbf{k}}\hat{\mathbf{i}} + a_{32}\hat{\mathbf{k}}\hat{\mathbf{j}} + a_{33}\hat{\mathbf{k}}\hat{\mathbf{k}}.$$

This is called the nonion form of the dyadic.

A dyadic which is identical with its conjugate is called a symmetric dyadic; for it, we have $a_{ij} = a_{ji}$, i.e., $a_{12} = a_{21}$, $a_{13} = a_{31}$, $a_{23} = a_{32}$.

A dyadic is called anti-symmetric or skew-symmetric when it is equal to the negative of its conjugate; then $a_{ij} = -a_{ji}$, i.e., $a_{11} = a_{22} = a_{33} = 0$, $a_{12} = -a_{21}$, $a_{23} = -a_{32}$, $a_{31} = -a_{13}$.

If Φ is symmetric, and $\mathbf{v}$ is any vector, then $\Phi \cdot \mathbf{v} = \mathbf{v} \cdot \Phi$.

A dyadic which, acting as pre-factor or post-factor upon a vector, produces the vector itself, is called an idemfactor or unit dyadic. Its standard form is

$$\mathbf{I} = \hat{\mathbf{i}}\hat{\mathbf{i}} + \hat{\mathbf{j}}\hat{\mathbf{j}} + \hat{\mathbf{k}}\hat{\mathbf{k}}.$$

Any dyadic can be resolved into a sum of a symmetric part and a skew-symmetric part.

Any symmetric complete dyadic Φ can be reduced to the normal form:

$$\Phi = a\hat{\mathbf{i}}\hat{\mathbf{i}} + b\hat{\mathbf{j}}\hat{\mathbf{j}} + c\hat{\mathbf{k}}\hat{\mathbf{k}}.$$

where the coefficients a, b, c are positive or negative (scalar) constants, by a transformation of axes. (L.L.S.)

DYADS. Dyadics.

DYAS. The western European term for the **Permian** system where it is readily divisible into two parts. (R.M.F.)

DYE BLEACHING. A method of forming a subtractive color synthesis to be used in making a color photograph. In a mixture of colorants such as cyan, magenta and yellow, each colorant tends to be bleached by light of its complementary color since such light is the only light absorbed by that colorant. Thus, the cyan colorant tends to be bleached by red light and not by blue or green light.

If an integral tripack composed of gelatin layers colored with such light-sensitive dyes is exposed to light from an original subject or a color transparency, each colorant layer will be bleached by its complementary color, yielding a dye image in which the colors correspond to those in the original. Such a simple process of color reproduction has long been sought but without notable practical success. The difficulty of making the process sensitive enough and at the same time stable after the images have been formed, has prevented any successful commercialization. (H.C.C.)

DYE MORDANTING. Dye Toning.

DYE TONING. A method of colorant formation used in subtractive processes of color photography. The positive silver images are converted into a substance, such as potassium iodide, which has a strong affinity for certain dyes. Such substances are mordants and will hold the dye in the same position and in the same relative density as formerly occupied by the silver image. Dye toning is, therefore, often termed dye mordanting.

The mordant image is treated with a dye which is then held by the mordant while the film or plate is washed free of unattached dye. The mordants generally used work well with basic dyes. Such dyes are very brilliant although usually not very stable or fast to light. (H.C.C.)

DYES AND DYEING TEXTILE FIBERS. With few exceptions the dyestuffs in use for dyeing textiles, coloring foods, as chemical indicators, biological stains, antiseptics, and photographic sensitizers, are made industrially from the benzenoid **hydrocarbons, benzene, toluene, naphthalene, anthracene.** These hydrocarbons are colorless substances when observed in natural light. They show, however, slight absorption of light of wavelength outside the range of easy visibility to the naked eye, at the extreme edge of the spectrum. Such simple derivatives of these hydrocarbons are pure **phenol, hydroquinone,** the naphthols, anthranol, chlorobenzene, dichlorobenzene, benzenesulfonic acid, **aniline** and hydrazobenzene are likewise colorless to the naked eye, and some show slight absorption of light beyond the visible spectrum.

But nitrobenzene is of a pale yellow color, and azoxybenzene pale yellow. Thus, these two substances show some absorption of light in the visible spectrum. Finally, benzoquinone and azobenzene are highly colored substances, thus showing marked absorption in the visible spectrum. Quinones are the only carbon-hydrogen-oxygen compounds that as a class possess marked color. Azo-nitrogen compounds possess marked color. Some nitro-compounds are decidedly colored, others slightly, some scarcely. The table on page 487 is intended to make these results graphic.

The groups which distinguish these and other highly colored organic substances—chromogenes—are called chromophors (color producers).

Chromogene	Chromophor Group
Quinones.........	
Azo-compounds....	—N : N—
Nitro-compounds...	—NO₂

Chromogenes, although highly colored, are not dyes. Colored substances, in order to be classed as dyes, must be capable of permanent attachment to some other substance, such as cotton and linen, wool and silk, rayon, in such a way as to withstand the action of water and soap. Dyes for ordinary use should also be fast to sunlight.

Chromogenes are converted into dyes by the introduction of a secondary group, called an auxochrome (color fixing) group, such as (1) amino group ($-NH_2$, also for example, $-N\begin{smallmatrix}H\\CH_3\end{smallmatrix}$, $-N\begin{smallmatrix}CH_3\\CH_3\end{smallmatrix}$, $>NH$), (2) hydroxyl group ($-OH$), (3) carboxyl group ($-COOH$). Picric acid (2,4,6-trinitrophenol) is a dye, while T.N.T. (2,4,6-trinitrotoluene) is not a dye, although highly colored. Para-aminoazobenzene ($C_6H_5-N:N-C_6H_4(NH_2)(4)$) and para-hydroxyazobenzene ($C_6H_5-N:N-C_6H_4(OH)(4)$) are dyes, while azobenzene ($C_6H_5-N:N-C_6H_5$) and para-para-prime-azotoluene ($(4')(CH_3)C_6H_4-N:N-C_6H_4(CH_3)(4)$) are not dyes although highly colored. Alizarin (1,2-dihydroxyanthraquinone) is a dye, while anthraquinone, like naphthoquinone and benzoquinone, is not a dye although highly colored.

TABLE SHOWING HIGHLY COLORED DERIVATIVES OF BENZENOID HYDROCARBONS

Note: Highly colored substances (not dyes) are underlined once, dyes twice

Hydrocarbon	Selected Derivatives of Hydrocarbons				
Benzene	Phenol Dihydroxybenzenes (1, 2; 1, 3; 1, 4) Aniline Benzidine Chlorobenzene Dichlorobenzene Benzene sulfonic acid	Benzoquinone Hydrazobenzene	Azobenzene Aminoazobenzene Hydroxyazobenzene	Azoxybenzene	Nitrobenzene 4-Nitrophenol Sodium-4-nitro-phenolate 2,4,6-Trinitro-phenol 2,4,6-Trinitro-aniline
Toluene	Cresols Orcinol Toluidine Tolidine Chlorotoluenes Toluene sulfonic acids	Toluquinone Hydrazotoluene	Azotoluene		Nitrotoluenes 2,4,6-Trinitro-toluene
Naphthalene	Naphthols Dihydroxynaphthalenes Naphthylamines Chloronaphthalenes Naphthalene sulfonic acids	Naphthoquinone			Nitronaphthalenes
Anthracene	Anthrols and Anthranol Dihydroxyanthracenes Anthramine	Anthraquinone 1,2-Dihydroxy-anthraquinone (alizarin)			

Auxochromes increase the solubility of the dye over that of the chromogens, and the presence of the sulfonic acid group is of great importance in this respect (see **Thioalcohols**). Dyes must usually possess a basic or acidic nature. Basic dyes contain the amino auxochrome (forming salts, e.g., hydrochlorides), feebly acid dyes the hydroxyl chromophore, and acid dyes the sulfonic acid (more commonly) or carboxyl (less commonly) auxochrome (forming salts, e.g., sodium salt). The solubility of acid dyes is frequently attained by the use of the sodium salt.

The effect of increasing the weight of the dye molecule by addition of methyl ($-CH_3$), ethyl ($-C_2H_5$), amino ($-NH_2$) groups is to darken the color. For example, aminoazobenzene is pale yellow, diaminoazobenzene orange, and triaminoazobenzene brown.

As to the basic or acidic nature of the fibers to be dyed, wool and silk are proteins (see **Aminoacids**) and behave as acids towards bases, and as bases towards acids (that is, they are amphoteric) and are readily dyed directly by dyes of a basic or acidic nature. Cotton and linen, on the other hand, are cellulose (see **Carbohydrates**) and noted for their chemical inertness. These are scarcely affected by basic or acidic dyes. Rayons of cellulose are similar to cotton and linen in respect to dyeing, and rayons of cellulose acetate require special dyes.

The cellulose fibers are dyed directly by some dyes such as benzidines, primulines (both are amino dyes),

curcumines (hydroxy dyes), by the use of **sodium chloride** or sodium sulfate solution. But generally these fibers are dyed by the use of mordants. In this process dyestuffs containing either hydroxyl ($-OH$) or carboxyl ($-COOH$) groups in their composition are used with a solution of **chromium, aluminum, iron, tin** or **copper** salt. Other dyes are used as mordant dyes, and other mordanting materials, for example, **tannic acid,** are used with basic dyes. With these salts the mordant dyes form a colored insoluble substance called a lake, which is permanently fixed in the fiber under the conditions of dyeing. The color is dependent upon the mordant used for a particular dye. **Cellulose** acetate fibers are dyed by ionamines, which are insoluble azo or anthraquinone dyes, temporarily converted into soluble derivatives by the introduction of the group $-CH_2SO_3Na$, and later split off in the dyeing process, leaving the insoluble dye on the fiber. Insoluble azo or anthraquinone dyes are also maintained in suspension in the dye bath by means of sulfonated castor oil. These dyes do not adhere to cotton, linen, or viscose, and therefore permit the operation of selective dyeing when other fibers are used with cellulose acetate.

In vat dye processes an insoluble dye is converted into a weakly acidic, alkali-soluble, colorless or leuco-compound. Upon immersion of the fiber in the dye bath and removal of the immersed fiber, the dye is fixed in the fiber by oxidation in the air. Indigo and anthraquinone dyes are typical vat dyes. **Sodium hydro-**

CHEMICAL INDICATORS.

(See **Reactions Involving Recombination of Ions.**)

	pH Range.		pH Range.
Methyl violet	0–2	Bromcresol green	3.8–5.4
	5–6	Methyl red	4.4–6.0
Meta-cresol purple	1.2–2.8	Ethyl red	4.5–6.5
	7.4–9.0	Para-nitrophenol	5–6
Thymol blue	1.2–2.8	Bromcresol purple	5.2–6.8
	8.0–9.6	Alizarin	5.5–6.8
Cresol red	2–3		10.1–12.1
	7.2–8.8	Phenol red	6.8–8.4
2,6-Dinitrophenol	2–4	Rosolic acid	6.9–8.0
Bromthymol blue	2.8–4.6	Cyanin	7–8
	6.0–7.6	Orange II	7.2–8.6
Bromphenol blue	3.0–3.6	Cresol red	7.2–8.8
Methyl orange	3–4	Phenolphthalein	8.3–10.0
Congo red	3–5	Thymolphthalein	9.4–10.6
Ethyl orange	3.5–4.5	Sodium indigosulfonate	12–14
		1,3,5-Trinitrobenzene	14–14.3

sulfite ($N_2S_2O_4$) is frequently used to reduce the dye to the leuco form.

Selected classes of dyes are discussed as follows:

(1) Azo- dyes, (2) Disazo- dyes, (3) Triphenylmethane dyes, (4) Xanthene dyes, (5) Acridine dyes, (6) Thiazine dyes, (7) Alizarin and Anthraquinone dyes, (8) Indigo dyes.

The azo- dyes contain one or more azo- groups (—N:N—) as in azobenzene,

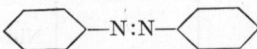

They are formed mainly by diazotization (see **Diazo-Compounds**) of primary amines and coupling with:

1. An **amine**, in acid solution. Reaction slow, accumulation of acid may be prevented by gradual addition of sodium carbonate.
2. A **phenol**, in alkaline solution. Reaction rapid, accumulation of acid is prevented by presence of alkali.

Coupling takes place at the carbon para to —NH₂ and —OH (—ONa) group; if the para carbon is occupied by another group, then coupling takes place at the carbon ortho to —NH₂ and —OH(—ONa), not at the carbon meta to —NH₂ and —OH but beta-naphthol and beta-naphthylamine at carbon number 1, never at 3 or 4. Aminonaphthols in (a) acid solution, at carbon ortho to —NH₂ group, (b) alkaline solution, at carbon ortho to —OH(ONa) group. Diamines undergo only meta coupling.

Colors certified for use under the Food, Drugs and Cosmetics Act of 1938 (see U. S. Code: Acts of the 75th Congress, page 990) (a) in foods, drugs and cosmetics: green numbers 1 to 3, yellow numbers 1 to 6, red numbers 1 to 4 and 32, orange numbers 1 and 2, blue numbers 1 and 2; (b) in drugs and cosmetics taken internally: green numbers 4 to 8, yellow numbers 7 to 11, red numbers 5 to 31 and 33 to 39, orange numbers 3 to 17, brown number 1, blue numbers 4 to 9, violet numbers 1 and 2, black number 1; and (c) in externally applied drugs and cosmetics: green number 1, yellow numbers 1 to 6, red numbers 1 to 13, orange numbers 1 and 2, blue numbers 1 to 5, violet numbers 1 and 2, black number 1. In the year 1940 certificates were issued for 1,610,000 lbs. of colors in the (a) group, 292,000 lbs. in the (b) group, and 9600 lbs. in the (c) group.

Special applications of dyes have attained a position of great importance. Some of these applications and the dyes used are given below. Data given for various dyes that are used for such special purposes as coloring food, chemical analysis, biological investigation, and photographic sensitizing. There are special dyes for such materials as gasoline and synthetic resins.

CERTIFIED BIOLOGICAL STAINS

The following dyes are certified by the Commission on Standardization of Biological Stains.

NAME OF STAIN	APPLICATION	SOLUBILITY grams per 100 ml.
Aniline blue, water soluble	Cytoplasm (acid)	
Bismarck brown Y		1.4
Brilliant cresyl blue		
Brilliant green		
Carmin	Nuclear (basic)	
Congo red	Cytoplasm (acid)	
Cresyl violet		0.4
Crystal violet		1.7 (chloride)
Eosin, bluish		39.1 (Sodium salt)
Eosin, yellowish	Cytoplasm (acid) Blood	
Ethyl eosin		0.03
		1.13% sol. in alcohol (95% strength)
Fast green FCF		16.0
Fuchsin, acid	Cytoplasm (acid)	
Fuchsin, basic	Nuclear (basic) Bacterial	
Hematoxylin	Nuclear (basic)	
Indigo carmine		1.7
Janus green B		5.2
Jeuner's stain		
Light green S F, yellowish	Cytoplasm (acid)	20.4
Malachite green		
Martius yellow		4.6 (Sodium salt)
Methyl green	Nuclear (basic)	
Methyl orange		0.5
		0.015 (acid)
Methyl violet		2.9
Methylene azure		
Methylene blue	Nuclear (basic) Vital Blood Bacterial	3.6 (chloride)
Methylene violet		
Neutral red	Cytoplasm (acid) Vital	5.6 (chloride)
Nile blue A		
Nigrosin		
Orange G	Cytoplasm (acid)	10.9
Orange II		11.4
Phloxine		50.9 (Sodium salt)
Pyronin		9
Rose bengal		36.3
Safranin O	Nuclear (basic)	5.5
Sudan III	Fat	Insol.
		0.15% sol. in alcohol (95% strength)

Name of Stain	Application	Solubility grams per 100 ml.
Sudan IV	Fat	Insol. 0.09% sol. in alcohol (95% strength)
Tetrachrome stain (MacNeal)		
Thionin	Nuclear (basic)	0.25
Toluidine blue		3.8
Wright's stain		

PHOTOGRAPHIC SENSITIZERS

Eosin, erythrosin	For yellow-green
Pinaflavole, orthochrome T	For green
Pinachrome, pinaverdol, acridine orange	For orange
Ethyl red	For orange-red
Pinacyanole (6800), naphthocyanole (7500)	For red of wavelength (Angstrom units) specified
Kryptocyanine (8000), dicyanine (9000), neocyanine (10,000)	For infrared of wavelength (Angstrom units) specified (R.K.S.)

DYNAMIC CHARACTERISTIC. This refers to the characteristics of a piece of equipment under actual operating conditions. As applied to vacuum **tubes** it designates the characteristics obtained when operating into a load with a varying or a-c signal present in the circuit. Static characteristics of such tubes are taken with fixed voltages, which are adjusted by steps, on the various electrodes. Examples of these are the mutual characteristic which gives the relation between grid voltage and plate current, and the plate characteristic which gives the relation between plate voltage and plate current. Often these characteristics do not begin to approach the behavior under regular operating conditions because of circuit impedances and varying voltages and currents present in the circuit. Fortunately, the dynamic characteristics may be obtained with sufficient accuracy from the static characteristics when the circuit constants have been given. The tube circuit may then be analyzed. (L.R.Q.)

DYNAMIC COOLING AND HEATING. Adiabatic Processes in the Atmosphere.

DYNAMIC METAMORPHISM. Metamorphism.

DYNAMIC PRESSURE, FLUID. The pressure necessary to accelerate a fluid from rest to a speed of V is the dynamic pressure equivalent to that speed. If ρ is taken as mass density of the fluid,

$$\text{Dynamic pressure, } q = \frac{\rho V^2}{2}.$$

This conception is useful in aerodynamics as the dynamic pressure represents the unit air pressure acting on a surface increment in atmospheric air moving with velocity V over the surface. By **Bernoulli's theorem**,

$$p = p_0 - \frac{\rho v^2}{2} = p_0 - q.$$

The vacuum caused by air in motion over a surface is greatest when this imaginary dynamic pressure is greatest since q represents the vacuum. (F.T.M.)

DYNAMICAL PARALLAX. Dynamical parallax is an indirect method for the determination of the distances of **binary stars**. In the determination of **orbits** of binary stars the semi-major axis of the relative ellipse is determined in angular units. This length cannot be expressed in any linear units unless the distance of the system is known. With this distance known the orbit may be solved, and the combined masses of the two stars determined in terms of the sun's mass as unity. For the systems thus far solved it is found that the majority of the stars are between ⅕ and 10 times the mass of the sun. Hence, as a first order of approximation we may assume that for any binary system the combined mass will not differ greatly from twice the mass of the sun.

The rigorous expression for the so-called harmonic **Keplerian Law of Planetary Motion** shows the period or revolution of one object about the other to be dependent upon the combined masses of the objects and also their distance apart. The combined mass of the earth and the sun is known, and the period of revolution of the earth about the sun is one year. Hence, assuming the combined mass of the binary system to be twice that of the sun, and knowing the period of revolution of the system in years, we may find the mean distance between the components of the system in astronomical units. Since from the orbit we know the distance between the components in angular units, we can immediately find the angular distance subtended by one astronomical unit at the distance of the star, or, in other words, we can find the stellar parallax of the system.

The parallax thus determined is based on the assumption that the combined mass is twice that of the sun. However, with the parallax thus computed and the apparent **magnitudes** of the stars known, we can then calculate the **absolute magnitude** of the system. With this absolute magnitude we go to the **mass-luminosity** relation and get a second approximation to the mass of the system. With this improved mass the process is repeated and an improved value of stellar parallax obtained. By a sufficient number of approximations, values of the so-called dynamical parallax may be obtained with an accuracy approaching that of the direct trigonometric determinations.

Within recent years several methods have been devised for obtaining dynamical parallaxes for systems which are moving so slowly that orbit determination is impossible. While such methods do not lead to results of great accuracy, nevertheless, they are valuable for statistical discussions. (W.K.G.)

DYNAMICS OF GASES. The laws pertaining to the forces of gas pressure and to the flow of gases are based ultimately upon the **kinetic theory**, but certain principles can be stated without analyzing their origin to that extent. To a first approximation, the **ideal gas law**, or the **Boyle-Charles law**, represents the dynamics of gases at rest. At a given temperature, the pressure of a body of gas varies inversely as its volume, and hence directly as its density (**Boyle's law**); and at a fixed volume, the pressure is a linear function of the temperature, varying at the same rate (1/273 per centigrade degree) for all gases (**Charles's law**). But dynamic processes in a gas are complicated by the fact that change in volume is in general accompanied by change in temperature, so that simple dynamics is overshadowed by **thermodynamics**. It was for this reason, for example, that the correct formula for the speed of **sound** in air proved, for a time, elusive. A gas is highly compressible, and this property affords ready opportunity for the energy of mechanical impulses, which would be merely transmitted by a non-compressible fluid, to be transformed into heat, or for the gas to use its thermal energy to create impulses of its own. The same circumstance complicates the effect of gravity. The **atmosphere** is not an ocean of uniform density and definite depth; its pressure and density are logarithmic functions of the altitude. The forces associated with moving gases form the subject-matter of **aerodynamics**. (See also **Ther-**

mal Convection, Winds, Pressure Gauges, Air Pumps, etc.) (L.D.W.)

DYNAMICS OF ROTATION. A body is said to rotate when all of its particles move in circles about a common axis with a common **angular velocity.** This motion may be either free or constrained, as illustrated, respectively, by the earth turning on its axis, and by a flywheel or a pendulum.

If one twirls an umbrella about its handle, it tends to open. This is because the **centrifugal forces** exert torques tending to throw the stays outward on their pivots. Through any point of a rigid body there are at least three lines, mutually perpendicular, about which the body would rotate without any such centrifugal torque. It may be shown that the **moment of inertia** of the body with respect to any one of these lines is either a maximum or a minimum as regards all lines through the given point. They are called principal axes. In general there is only one line about which a free body will rotate permanently; it is the principal axis of greatest moment of inertia through the **center of mass.** A body constrained to rotate about an arbitrary axis will, when released, tend to change its motion so as to rotate about this permanent axis, but the adjustment is complicated by precession, so that the body may "wobble" like a badly thrown discus.

If a free body, at rest, is given a sudden push along some line not through the center of mass, it begins to rotate about some other line beyond the center of mass and perpendicular to the applied force. This line is the axis of instantaneous rotation. It is only a temporary axis, the rotation being at once transferred to an axis through the center of mass. The line mutually perpendicular to the instantaneous axis and to the line of the force passes through the center of mass, and its intersections with the other two lines are conjugate points, having the same relation as the center of oscillation and the center of suspension of a rigid **pendulum.** If the push is given in line with the center of mass, the axis of instantaneous rotation is at infinity, and the motion is then one of pure translation.

A torque applied so as to tend to change the axis about which a body is rotating results in the peculiar behavior known as **precession.** The **angular momentum** of a rotating body is the product of its angular velocity by its moment of inertia about the axis of rotation. The **kinetic energy** associated with rotational motion is equal, in absolute units, to ½ the product of the moment of inertia by the square of the angular velocity—a formula analogous to that for kinetic energy of linear motion. (L.D.W.)

DYNAMITE. Explosives.

DYNAMO. Dynamo refers to a general class of machines capable of transformation of electrical into mechanical energy, or vice versa. The word is a shortened form of dynamo-electric. A feature of all dynamo machines is the employment of magnetic induction in effecting the transformation. The essential parts of an ordinary dynamo are the **armature** and the field. One of these is mounted on a rotating shaft, and the other is stationary. Theoretically, the dynamo is perfectly reversible, that is, it may be used either as a generator or a motor. Actually, this is not always possible. (See also **Generators, Motors.**) (F.T.M., L.R.Q.)

DYNAMOMETER. A dynamometer is an instrument for measuring force, such as a spring balance. Most writers, however, apply the term to certain devices for the measurement of mechanical power. The principal classification is derived from the fact that some types of dynamometers absorb all of the power, which is converted into heat, whereas others transmit the power they receive to some other absorber of power,

measuring it during the process. These are called, respectively, absorption and transmission dynamometers.

In the absorption dynamometer class there are types which convert the mechanical to heat energy through the medium of mechanical friction. They are all similar to the Prony brake. (See **Brake Horsepower.**) The friction surfaces are variously wooden blocks against metal drums or pulleys, bands with wooden cleats, ropes, or friction-surfaced brake bands. Also there are hydraulic dynamometers which absorb the power by fluid friction. One common arrangement is similar to a **centrifugal pump,** except that the casing, instead of being rigidly fixed to a bed plate, is freely supported on the propeller shaft. It is restrained from rotating by an attached arm. The restraining moment in the brake arm is measured by platform or spring scales. The energy absorbed appears as a heating of the water in the dynamometer. To prevent it boiling it must be steadily renewed. Thus the energy is carried off in a stream of water entering the dynamometer cool and leaving warm. Air friction has also been set to use in the fan brake absorption dynamometers.

One of the most convenient means for measuring power is to convert it to electrical power (watts). In an electrical dynamometer a **generator** is slightly modified. The stator is mounted, free to revolve, but restrained from revolving by a brake arm which is attached to it, and to which are fastened weighing scales. The tendency of the casing to rotate with the rotor which is connected to the source of the power is opposed by the brake arm. The force shown on the scales becomes a **torque** when multiplied by its lever distance from the center of rotation. Since power is torque multiplied by rotative speed, the only other reading necessary from the dynamometer is the speed of the rotor shaft. In all absorption dynamometers the casing is mounted free to revolve under the action of mechanical friction, fluid friction, or magnetic drag. Actual rotation is prevented by the attached brake arm. Power is measured as a torque operating at the rotative speed of the driven shaft.

A transmission type dynamometer is illustrated by the torsion type, in which a shaft delivering power is twisted through a small angle by the torque. Such a shaft may be calibrated at rest by measuring the torsional deflection obtained under known torque loadings. This dynamometer has its greatest field of usefulness where the other types are impractical. Measurement of power output from a large marine engine is typical. (F.T.M.)

DYNAMOTOR. This is a double **armature** rotating electrical machine. One of the armatures is wound for low voltage direct current, and serves as a motor armature. The other armature is wound for a high d-c voltage and serves as a generator winding. The machine is used for supplying plate voltage to portable radio equipment, being operated from storage batteries on the motor end and supplying the high voltage d.c. from the generator winding. (L.R.Q.)

DYNATRON. This term is applied to a vacuum **tube** operated in such a manner that its plate characteristic has a negative resistance section, i.e., the plate current increases while the plate voltage decreases. The most common application is for stable **oscillators.** The screen-grid or tetrode vacuum tube exhibits this characteristic when the screen voltage is higher than the plate voltage. (L.R.Q.)

DYNE. The c.g.s. absolute unit of **force,** defined as the force required to give a free mass of one gram an acceleration of 1 cm. per sec. per sec. Since the weight of one gram mass, or one gram of force, would give it an acceleration of about 980 cm. per sec. per sec., a dyne is about $\frac{1}{980}$ of this weight, or only a little more than a milligram. Nevertheless it is a fundamental dynamic unit of physics, and is the basis of the units of **energy (erg)**, of **power (watt)**, of **pressure (bar)**, etc. (L.D.W.)

DYSENTERY. A term applied to intestinal disorders characterized by frequent watery stools which may contain **blood, pus,** and mucus. Abdominal pain, tenesmus, and constitutional symptoms may accompany the diarrhea. Dysentery may be caused by **bacteria, parasites,** and chemical irritants. There are two important forms—amoebic and bacillary.

Bacillary dysentery is world wide in distribution, but occurs most commonly in tropical and subtropical climates. Lack of sanitation and crowding are predisposing factors. Flies are known to spread the disease. It is always a problem in armies and cases occurred on all fronts in World War I and II. The causative bacteria are varieties of *Bacillus dysenteriae—Shiga,* and the Flexner group, including Schmitz, Sonne, and the paradysenteriae are the commoner types. *B. dysenteriae Shiga* causes the most severe disease. Symptoms vary according to whether the disease is mild, fulminating, relapsing or chronic. The mild forms may be marked by diarrhea only. In more serious types, the onset may be sudden, with severe abdominal cramps and copious diarrhea. Fever, prostration, and dehydration may be extreme. **Arthritis, neuritis** and **endocarditis** may occur as late complications after the acute phase has passed. Chronic dysentery and healthy dysentery **carriers** occur.

The treatment of bacillary dysentery is with **sulfonamides,** sulfadiazine being the most effective at present. Sanitary disposal of feces, fly control, water, milk, and food inspection and personal cleanliness are all important in the control of the disease.

Amoebic dysentery is an infection of the large intestine caused by the protozoan, *Entameba histolytica.* Its geographical distribution is the same as that of bacillary dysentery. A common mode of spread is by carriers—healthy individuals who pass the **cysts in their** stools. Bloody diarrhea with relatively little pus is characteristic. The disease tends to be recurrent—many remissions and exacerbations occurring. It may be difficult to distinguish amoebic from bacillary dysentery in the acute stage, except by the detection of amoebic cysts in the stools of the patient with the amoebic form. Complications of amoebic dysentery are metastatic **abscesses,** usually in the liver, but other organs may be the site.

The treatment of the acute disease is with various drugs, chiefly carbarsone, an arsenical, and **emetine,** which is a specific in amoebic liver abscess. In intestinal amoebiasis, carbarsone and emetine are used frequently in courses, the one following the other. Chiniofon, an iodine compound, is also used. (D.M.H.)

DYSMENORRHEA. Menstruation.

DYSPAREUNIA. Painful or difficult **coitus.** (R.S.M.)

DYSPEPSIA. This term does not represent any disease entity, but is merely a group of symptoms due to various physiological or **pathological** activities of the **gastro-intestinal** tract which may indicate some form of impairment of the normal digestive function. (R.S.M.)

DYSPNOEA. Dyspnoea means difficult breathing. More specifically, the term indicates discomfort arising because of consciousness of the necessity to exert effort to breathe. This occurs normally after violent exercise, and is a frequent symptom of heart disease, **pneumonia,** and **emphysema;** any condition which is associated with **acidosis, oxygen** lack, or mechanical hindrance to the respiratory movements may cause dyspnoea. (D.M.H.)

DYSPROSIUM. Symbol: Dy. Atomic number: 66. Atomic weight: 162.46. (**Isotopes:** page 290.) Type of compound: Dy_2O_3, white. Color of salts: Yellow. Discovered by Boisbaudran in 1886. A member of the **yttrium** sub-group of the rare earth metals. (R.K.S.)

E LAYER. Ionosphere.

e, THE NUMBER. The number e may be defined by:

$$e = \lim_{n \to \infty} \left(1 + \frac{1}{n}\right)^n,$$

or by

$$e = \lim_{x \to 0} (1 + x)^{1/x},$$

and is represented by the infinite series

$$\dot{e} = 1 + \frac{1}{1!} + \frac{1}{2!} + \frac{1}{3!} + \frac{1}{4!} + \cdots + \frac{1}{n!} + \cdots$$

It has the approximate value: $e \approx 2.71828$.

The number e is an **irrational number** and is also a **transcendental number.** It is used as the base of the system of natural (or Napierian) **logarithms.** (L.L.S.)

EAGLE. Aves, Falconiformes. Large birds (**Aves**) of prey with strong hooked beaks, large curved claws, and powerful flight. The American golden eagle, *Aquila chrysaetus,* is an example of the typical members of the group, which also includes the harpy eagles, hawk eagles, harrier eagles, sea eagles, and others. The bald eagle, *Haliaeetus leucocephalus,* national bird of the United States, is one of the sea eagles. (A.W.L.)

EAR. Auditory Organ.

EARTH. (See tables of planetary data in the article on **Planet.**) The earth is third planet in point of distance from the sun and is unique in being the only planet which is known to carry an **atmosphere** capable of supporting human life.

The earth is approximately spherical but, due to the fact that it is rotating on an axis, the actual shape is an oblate **spheroid.** The actual shape is determined by making accurate determinations of astronomic **latitude** at a number of different stations and then determining the linear north-south distance between the two stations measured along the surface of the earth. If the earth were a perfect sphere, a degree of latitude would have the same linear length everywhere. Observations show that a degree of astronomic latitude is longer the higher the latitude on the earth. Accurate measurements indicate a difference of 26.70 miles between the equatorial and polar diameters of the earth, the polar being the shorter.

A theory that the earth is in rotation was advanced by Copernicus in the 15th century but it was not until the 19th century that any proofs independent of the motions of the heavenly bodies were available. The most familiar of these is the so-called **Foucault pendulum.**

In accordance with the Copernican Theory of the structure of the universe, the earth should be revolving about the sun, but a proof of this motion was not available until 1725, when Bradley demonstrated that the **aberration of light** which he had previously observed could only be explained by the revolution of the earth about the sun. There are several other proofs of the revolution of the earth, but none of them can be demonstrated without the use of refined telescopic observations. The best known of these proofs is the annual variation in the **radial velocities** of the stars and **stellar parallax.** (W.K.G.)

EARTH, ELECTRICAL. Ground, Electrical.

EARTH RADIATION. Heat Balance in the Atmosphere.

EARTHNUT OIL. Esters.

EARTHQUAKES. The lithosphere, or so-called crust of the earth, is continuously undergoing deformative movements which are expressed at the surface in folds, faults and volcanic activity. When the adjustments beneath the surface are sudden a vibration is expressed in the form of an earth tremor or earthquake, the result of a sudden fracture in the lithosphere.

An earth fracture is called by geologists a **fault,** and the surface along which realignment of the crustal blocks takes place is called a fault plane. The intersection of the fault plane with the surface of the earth is called the trace of the fault plane. The relative movement of the fault blocks may be vertical, horizontal or oblique. Pronounced vertical movements of the fault blocks may produce fault scarps, often the loci of intermittent earthquakes. Incipient fault scarps may be, however, obliterated or greatly reduced by erosion. Earth tremors which are too gentle to be recorded by the senses alone are called **microseisms.** Instruments for measuring earthquake waves are called **seismographs** or seismometers and the study of earthquakes is called **seismology.**

The first seismograph was used in Italy in 1841, but many different types of instruments have been invented since. In the simple type of instrument, as illustrated,

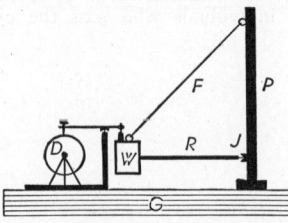

Diagram showing the principle of a seismograph. G, ground; P, post set in the ground; W, weight; R, rigid support contacting the post with a free-moving sharp point at J; F, flexible wire; D, recording drum revolved by clockwork; and the marker extends from W to D. When the ground shakes, the suspended weight, due to its inertia, scarcely moves, but the shaking motion is transmitted to the marker which leaves a record on the drum.

a delicately balanced horizontal pendulum is attached to a mast from the heavily weighted end. An arm attached to the pendulum magnifies and records the movements of the pendulum on a revolving drum or **chronograph** which automatically registers both the time and magnitude of the shock. The resulting records are called **seismograms.** A seismoscope is an instrument which detects an earthquake but does not record it. The various types of true seismographs may be classified as horizontal, vertical, inverted, electromagnetic, and torsional.

The principal machines now in use are the Milne-Shaw, McComb-Romberg, Wenner, Benioff, Wood-Anderson, Galitzin, Wiechert and Mainka. Of the American operating stations there are 46 in the United States, one in Puerto Rico, two in Alaska, and 6 in Hawaii. Many more are needed, and the same holds true of the rest of the world, especially on the oceanic islands.

The point of origin of an earthquake, at varying distances beneath the surface of the earth, is called the **centrum.** Directly above the centrum, on the surface of the earth, is the **epicentrum,** or center of the maximum shock.

Earthquake waves are classified as follows: 1. Preliminary. These waves pass through the earth following a curved path below the straight line from the earthquake (**epicenter**) to the recording instrument (**Seismograph**). There are two types of these waves, probably following the same path but at different velocities. The first preliminary wave, designated *P* for **primus,** is also known as (a) longitudinal, because of its back and forth vibration in direction of progress, and (b) compressional, because of its effect on the medium through which this type of wave passes. When these waves are reflected at the surface of the earth, or other reflecting surface, they are known as reflected waves. All of these waves are also known as body waves. 2. Surface waves or long waves. Much of the energy of an earthquake goes into surface or long waves. These are mainly of two types, (a) Rayleigh, with vibration in the form of a vertical ellipse back and forth in the direction of progress, and (b) Love, with only transverse-horizontal movement at right angles to the direction of progress.

The intensity or destructive activity of an earthquake, in populated areas, is determined by the acceleration, amplitude, period of vibration, length of time of vibration, and character of the surficial materials in the region of the epicenter.

Amplitudes vary from a fraction of an inch to several inches. The destructive phase of earthquakes varies from 1 minute to only a few seconds. It is estimated that for the whole earth, there are over 10,000 earthquakes per year, but the majority of them occur in regions of recent mountain building. A major earthquake occurs, approximately, once a week. The total number of all earthquakes, with definite epicenters, was 548, between 1925 and 1930; one about every 16 hours.

Dutch East Indies, Caribbean Sea, Mexico, South America, and Alaska. (R.M.F.)

EARTHWORK. This term includes work, the object of which is to alter the surface of the earth to serve some useful constructional purpose. In addition to excavation, building of embankments and trimming of slopes, earthwork also includes the clearing and grubbing of rough land, grading, etc. Excavation of rock and loose rock is usually considered earthwork. Among the most common forms of earthwork are preparation of subgrade for railways and highways, building of embankments for hydraulic work and construction of open drainage systems. In the preparation of a roadbed, the original surface of the ground is altered to the required degree by cuts and fills. As far as possible, cut should equal fill, so that the material excavated may be hauled a short distance and used to fill depressions in the proposed roadway. Where the amount of cut is insufficient for filling, the deficiency must be made up by hauling from **borrow pits.** An excess of cut is deposited on spoil banks.

The vertical dimension of the cut or fill at any **station** is found by leveling (see **Differential**). An engineer's **level** is set up near the station. The height of the instrument, which is the vertical distance between the line of sight of the level and a reference plane called the datum, is obtained by taking a rod reading (see **Level Rod**) on a **bench mark** or other point of known elevation above the datum. After the height of instrument is determined a rod reading is taken on the ground at the station. This is the ground rod. The difference between the height of instrument and the known **grade** elevation at the station is the grade rod. The cut or fill is the algebraic difference between the grade rod and the ground rod. The point where the side slope of a cut or fill will intersect the ground surface is marked by a stake called a slope stake. Slope stakes are set before construction is begun.

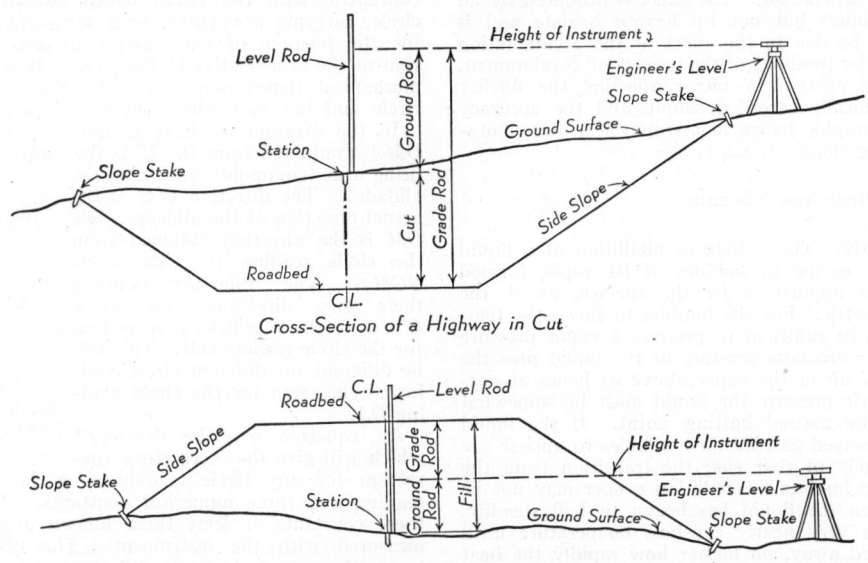

Cross-Section of a Highway in Cut

Cross-Section of a Highway in Fill

It has been estimated that over 3 million people have been killed by earthquakes of intensities 9 and 10 from the 6th century to 1927. Tremors of the ocean bottom cause seismic seawaves, called *Tsunami* (Japanese term). Seismic seawaves have been known to rise 100′ or more, and when such waves break upon a densely inhabited coast, they cause great destruction of life and property. Approximately 224 seismic seawaves are known to have occurred, chiefly on the coasts of Japan,

To measure the amount of cut or fill in earthwork, transverse cross-sections of the cut or fill are measured at regular intervals. These sections are then plotted on paper, and the area computed. Sections are taken close enough so that the volume of earthwork between them is considered that of a prism of length equal to the distance between stations, and area equal to the average of the two sections. This is known as the end area method. A more accurate result is obtained by the use of the prismoidal formula, which states that the volume of the earth equals $\frac{1}{6}(a_1 + 4a_m + a_2)D$, a_1 and a_2 are the end areas, a_m is the mid-section area, D is the length of the section being measured. Earthwork is usually done on a contract basis and paid for on the basis of volume excavated, or volume of fill. A fill of earth usually shrinks 10–20% upon compacting. Rock may be expected to occupy 15–30% greater volume after excavation than before.

If the volume of cut or fill between any two successive stations is given an algebraic sign (+ for cut and — for fill), the algebraic sum of the volumes starting at an initial station, may be represented by a continuous curve called a mass diagram. The abscissa for any point on the curve is its horizontal distance (in units of 100′) measured from the initial station; the ordinate is the algebraic sum (in cu. yds.) of the volumes up to the point. Plus-ordinates (cut) are plotted above the x-axis and minus-ordinates (fill) below. (F.T.M., C.W.C.)

EARTHWORM. Annelida, Oligochaeta. Terrestrial segmented worms of many species. They burrow in earth containing organic matter on which they live, coming to the surface only in damp cloudy weather and at night. Their activity in loosening and mixing the soil in fields is estimated to be valuable in crop production. (A.W.L.)

EARWIG. Dermaptera.

EBERHARD EFFECT. A photographic effect studied by Eberhard (1912). Eberhard gave a photographic plate a uniform exposure through a metal plate with openings of various sizes and found that the density of the exposed areas varied, the density decreasing with the size of the area. The amount of the variation in density increases with the thickness of the emulsion coating and is decreased by an exposure of the background or by general fog. The effect is produced by all organic developers but not by ferrous oxalate and is considered to be due to the effect of the accumulation of restraining by products of the process of development. The Eberhard effect is a factor affecting the fidelity of photographically recorded sound, and the accuracy of the photographic image in astrophysical and photometric investigations. (C.B.N.)

EBONY. Diospyros Ebenum.

EBULLITION. The boiling or ebullition of a liquid is due to the escape of bubbles of its vapor formed where heat is applied below the surface, as at the bottom of a kettle. For the bubbles to form, the temperature must be sufficient to produce a vapor pressure equal to the hydrostatic pressure of the liquid plus the pressure of the air or the vapor above it; hence at normal atmospheric pressure the liquid must be somewhat hotter than the normal boiling point. If the liquid contains a dissolved gas, the first bubbles to appear are composed largely of that gas; the transition from the escape of these bubbles to ebullition proper may not be marked. When the liquid has begun to boil steadily, it maintains a very nearly constant temperature until it has all boiled away, no matter how rapidly the heat is applied.

The formation of vapor bubbles appears to require some sort of nucleus; so that if the liquid is pure and perfectly free from suspended matter, and the vessel very smooth, the liquid may become superheated before ebullition suddenly begins. If the solid surface imparting heat to the liquid is very hot, a film of vapor forms between it and the liquid and the liquid escapes quietly from under the liquid without forming bubbles. This is the so-called "spheroidal state," observable when drops of water glide silently off a hot stove. (L.D.W.)

ECARDINES. Brachiopoda.

ECCENTRIC. The eccentric is a machine element employed to convert rotating to reciprocating motion. Its function is similar to that of the **crank**. The eccentric is used chiefly for short throws, where it would be undesirable to break the shaft, as is necessary in the case of a crank. It consists of a disk mounted on a shaft in such a way that the geometric center of the disk does not coincide with the center of rotation. The

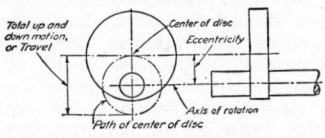
Simple eccentric.

distance between the center of rotation and the geometric center of the eccentric is the throw. This corresponds to the crank-arm distance of an equivalent crank. The eccentric must be used in conjunction with an eccentric strap which surrounds the eccentric, and which transmits the reciprocating motion to an eccentric rod rigidly attached. The eccentric is chiefly used to drive auxiliaries such as valve gear, and where reciprocation of small magnitude is needed. The **cam** and **crank** may be employed to provide similar motion. (F.T.M.)

ECCENTRICITY. This term is used in astronomy with two different significances. The term as descriptive of the shape of an **ellipse**, and hence as one of the elements of an **orbit**, is discussed elsewhere. We shall limit this article to a brief description of the correction for eccentricity that must be applied to many types of instruments used for **angular** measurement. Instruments for this purpose usually consist of a circle graduated in angular units, with an arm, assumed to be concentric with the circle, which sweeps around the circle, carrying a **vernier**, or a measuring microscope, for the purpose of determining accurately the direction of the arm relative to the circle. It is practically a mechanical impossibility to make the centers of the circle and the measuring arm exactly coincident.

In the diagram we have C the center of the circle OMA, graduated from O. C' is the center of the measuring arm (commonly known as the alidade). The direction $C'M$ is the actual direction of the alidade, while OM is the direction obtained from the circle reading (i.e., the angle OCM). The difference between these two directions, the angle $C'MC$, is the eccentricity correction for the circle reading OM. This will be different for different circle readings, being zero for the circle reading OA.

An equation may be developed which will give the eccentricity correction for any circle reading as a function of the reading and three numerical constants. To determine these constants at least three known angles must be measured with the instrument. The differences between the values obtained with the instrument and the known values of the angles are the eccentricity corrections for the circle readings. These eccentricity corrections are then used for the solution of three equations for the three constants. With the constants determined the equation, giving the eccentricity correction for any circle reading, may be written down. The results are usually tabulated, or plotted on a curve and supplied by the maker of the instrument. (W.K.G.)

Eccentricity correction for circle with alidade.

ECCENTRICITY OF A CONIC. Conic Sections.

ECDYSIS. The molting or shedding of the **cuticula** by **arthropods.**

Since the cuticula of these animals is also the rigid skeletal support of the body and is inelastic, it is shed at intervals during growth and a new covering of larger dimensions is formed. In preparation for molting the insect or other arthropod becomes inactive for a time, then by crawling movements crowds forward in the old integument, which splits down the back and allows the animal to emerge. During the resting period preparation is made by the secretion of fluid from the molting glands of the cellular layer and the loosening of the under part of the **cuticula.** Following the shedding of the old cuticula, a new layer is secreted during a further period of inactivity. All cuticular structures are shed at ecdysis, including the terminal linings of the alimentary tract and of the air tubes if they are present.

The **molting** of reptiles is sometimes called ecdysis. (A.W.L.)

ECHELETTE. Echelon.

ECHELON. A highly specialized form of **diffraction grating,** devised by Michelson. It consists of a row of glass plates of exactly equal thickness, packed together to form a miniature stairway of equal risers. The light enters normally to the largest plate at one end (see figure) and emerges at various deviations through

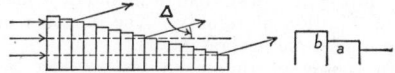

the low "risers." It is easily shown that if the thickness of the plates is a, the height of the "risers" b, and the refractive index of the glass n, the equivalent path difference between successive emergent streams for any angle of deviation Δ is $na - a\cos\Delta + b\sin\Delta$; or since Δ is in practice always small, $\cos\Delta = 1$ and $\sin\Delta = \Delta$ (in radians), giving $(n-1)a + b\Delta$. This must be equal to an integral multiple, N, of the wavelength λ for any spectrum line, the deviation of which is therefore

$$\Delta = N\frac{\lambda}{b} - (n-1)\frac{a}{b}.$$

The smallest value N can have (for $\Delta = 0$) is $(n-1)\frac{a}{\lambda}$, which, since a is usually several mm. and $(n-1)$ is 0.5 or more, is of the order of several thousand. The **dispersion,** viz.,

$$D = \frac{d\Delta}{d\lambda} = \frac{N}{b} - \frac{a}{b}\frac{dn}{d\lambda},$$

is correspondingly large. The echelon is thus especially adapted to the study of the **hyperfine structure** of spectrum lines.

A kind of reflection echelon of very small steps, called an "echelette," has been ruled on metal by R. W. Wood. (L.D.W.)

ECHIDNA. Mammalia, Monotremata. The spiny ant-eaters of the Australian region. Egg-laying mammals from 1′ to 20″ in length, belonging to several species. The body is covered with hair and spines and has a slender snout, short legs, and strong claws. They are burrowing animals. There are two genera, *Tachyglossus* (*Echidna*) and *Zaglossus* (*Proechidna*). (See also **Fossil Mammals.**) (A.W.L.)

ECHINODERA. Kinorhyncha.

ECHINODERMATA. A large division of the animal kingdom including the starfishes, sea cucumbers, brittle stars, sea lilies, sea urchins, and basket stars, all marine animals.

This phylum is characterized by the following structures: 1. The adult is almost perfectly radially symmetrical, although the young are bilateral. 2. The wall of the body contains a hard skeleton in most forms, made up of calcareous bodies called ossicles. 3. The **coelom** is well developed. 4. A water vascular system is present, consisting of a closed series of tubes opening to the exterior at one point on the body and bearing many delicate sacs, the tube feet or tentacles, which protrude at the surface of the body. 5. There is no special excretory system.

The echinoderms are divided into several classes which fall into two subphyla:

Subphylum Eleutherozoa. Without a stalk.
    Class **Asteroidea.** The starfishes.
    Class **Ophiuroidea.** The brittle stars.
    Class **Echinoidea.** The sea urchins, sand dollars, etc.
    Class **Holothuroidea.** The sea cucumbers.
Subphylum Pelmatozoa. With a stalk at least when young.
    Class **Crinoidea.** The feather stars, basket stars, and sea lilies.

(See also **Invertebrate Paleontology.**) (A.W.L.)

ECHINOIDEA. The sea urchins, keyhole urchins, and sand dollars. A class of the phylum **Echinodermata.**

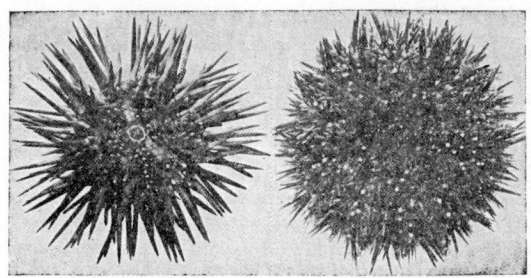

Echinoidea. (*N. Y. Zoological Society.*)

The members of this class are distinguished by the following characters: 1. The body is circular, varying from almost globular to thin disks. 2. The tube feet are suckers. 3. The surface bears long spines and **pedicellariae.** 4. There are no radiating arms. 5. The ossicles are closely associated to form a shell.

Sea urchins live on organic matter of all kinds, including small animals, plant tissues, and waste matter. They are of little economic importance but in some of the Mediterranean countries and to a limited extent in the Orient they are used as food.

The class is divided into the following orders:

Order Cidaroida. Sea urchins without gills around the mouth.
Order Centrechinoida. Gills present around the mouth.
Order Exocycloida. With indications of bilateral symmetry.

Many species flattened. Sand dollars, etc. (A.W.L.)

ECHINORHYNCHOIDEA. Acanthocephala.

ECHIURIDA, ECHIUROIDEA. Gephyrea.

ECHO SUPPRESSOR. When an electric wave on a **line** encounters a discontinuity or point at which the impedances do not match (see **Impedance Matching**) some of it is reflected. This reflected wave may return to the sending end of the line with sufficient amplitude to be objectionable. This is especially true in telephone service. While an effort is made to prevent reflections, there are cases where energy is fed back along the line and returns to the sender as an echo. In certain systems two lines (4 wires) are employed for transmission in the two directions and in such systems echo suppressors may be used to suppress the returning wave. This is accomplished by using a relay to short one line when there is a signal on the other. Thus if party A is talking and sending a voice signal to B, the voice currents on the line from A to B operate a shorting relay across the line from B to A so party A does not receive his own voice as an echo. (L.R.Q.)

ECLAMPSIA.
A disease marked by convulsions during the latter half of pregnancy. (See **Toxemias of Pregnancy.**) (R.S.M.)

ECLIPSES.
The term eclipse is applied to the darkening of a heavenly body due to the presence of another object. There are two cases to be considered, depending upon whether the object in question is self-luminous or is shining by reflected light. In the case of a self-luminous object, an eclipse or **occultation** takes place when an opaque object passes between the object and the observer. An object which is shining by reflected light is eclipsed when an opaque body passes between the object under consideration and its source of light. Eclipses of the first type are illustrated by an eclipse of the sun where the moon passes between the sun and the observer, in the case of an occultation of a star by the **moon**, and in the cases of **eclipsing binary** stars. A typical example of the second case is found in an eclipse of the moon where the earth passes between the sun and the moon.

Eclipses of the sun and moon have always been regarded with much superstitious awe and we find references to them in all of the ancient literatures. Because of their comparatively infrequent occurrence at any one point of the earth, records of eclipses of the sun may be used by historians to fix accurate dates for events. As an example of this reference may be made to an Assyrian tablet which states: "Insurrection in the city of Assur. In the month of Sivan the sun was eclipsed." This undoubtedly refers to the solar eclipse of June 15, 163 B.C. This is the same eclipse referred to in Amos VII, 9: "I will cause the sun to go down at noon, and I will darken the earth in the clear day."

The accompanying figure illustrates the shadow cast by the opaque body O; the body S being assumed luminous. An observer in the region A, the umbra

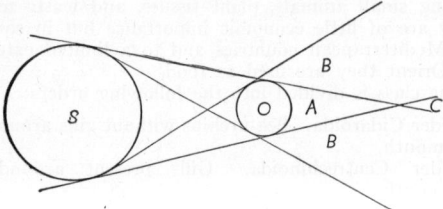

Fig. 1. Eclipse of sun or moon.

of the shadow, will be unable to see any portion of S and will be in darkness; in B and C, the penumbra, he will be able to see portions of S, in B observing the disk of S with a circular segment cut out, and in C observing S as an **annulus**, i.e., a ring of light. From a knowledge of the diameters of S and O and the distance between the two objects, the dimensions of the various parts of the shadow may be calculated.

For an eclipse of the sun we consider S (Fig. 1) as the sun and O the moon. Because of the fact that the **orbit** of the earth-moon system about the sun is eccentric the length of the umbra cone, A, varies be-

tween 236,000 miles at **aphelion** and 228,000 miles at **perihelion,** while, owing to the eccentricity of the moon's orbit, the distance of the moon from the surface of the earth varies between 217,800 miles at perigee and 247,500 miles at apogee. Examination of these numbers indicates that with the earth at aphelion and the moon at perigee, the surface of the earth will be 18,200 miles inside of the apex of the umbra. Under these conditions, the most favorable for an eclipse of the sun, the shadow of the moon on the earth will be a spot about 170 miles in diameter and within this area a total eclipse of the sun may be observed. Surrounding the spot of totality there will be a region of about 3000 miles radius within which the sun will be partially eclipsed. With the earth at perihelion and the moon at apogee the surface of the earth will be 19,500 miles beyond the apex of the umbra cone and, while an annular eclipse may be observed, no totality is possible.

As the moon revolves about the earth in its orbit the shadow sweeps along the plane of the moon's orbit with a velocity of about 2100 m.p.h. to the eastward. The earth is rotating at such a rate that a point on the **equator** is moving to the east with a velocity of about 1040 m.p.h. Accordingly under the most favorable conditions for a solar eclipse (i.e., the earth at aphelion, the moon at perigee, and the eclipse taking place at noon for an observer on the equator) the shadow will pass the observer with a speed of 2100 − 1040 = 1060 m.p.h. from west to east. The spot will pass the observer in slightly less than 8 minutes, which is the maximum duration of totality. The duration of a partial eclipse may be several hours.

For an eclipse of the moon we consider S, Fig. 1, the sun and O the earth. From the relative dimensions and distances we find that even under the most unfavorable conditions, i.e., with the earth at perihelion and the moon at apogee, the shadow of the umbra cone of the earth will extend well out beyond the distance of the moon. Hence an annular eclipse of the moon is impossible although partial eclipses, i.e., eclipses when the moon passes through the earth's shadow far enough off the central line to pass outside the umbra, are quite common.

Because of the fact that the plane of the moon's orbit is inclined to the plane of the **ecliptic,** an eclipse of either the sun or the moon may occur only when the moon is close to one of the **nodes,** i.e., close to the plane of the ecliptic, and must also be in **conjunction** (for an eclipse of the sun) or in opposition (for an eclipse of the moon). Hence eclipses of the sun occur with the moon in new **phase,** and eclipses of the moon with the moon in full phase. Since the earth-moon system revolves about the sun once each year the line of nodes would pass through the sun twice in each year if the direction of that line were fixed in space. However, due to a **perturbation** known as regression of the moon's node, the line actually passes close to the sun three times each year. The period when the line of nodes is close to the sun is known as an eclipse season. Two solar eclipses, either total or partial, must occur each year, and five may take place. No lunar eclipse need occur in any year, although three are possible. The minimum number of eclipses in any year is two, both solar, while the maximum number is seven, five of the sun and two of the moon, or four of the sun and three of the moon. Considering the earth as a whole, solar eclipses are more common than eclipses of the moon. However, since each eclipse of the moon is visible over a large portion of the earth's surface while eclipses of the sun are observed only over very restricted areas, for any particular locality eclipses of the moon are more common than eclipses of the sun. The sequence of eclipses may be determined by an ancient method known as the **Saros.**

The progress of an eclipse of the sun is designated by a series of "contacts": first contact coming when the edge of the penumbra *B* first touches the sun, second contact when the sun first passes into the umbra *A*, third contact when the umbra leaves the sun, and fourth contact when the last edge of the penumbra leaves the sun. Accurate recording of the times of the contacts gives valuable information regarding the complicated motions of the moon.

For many years the only information regarding solar **prominences**, the **flash spectrum, Baily's beads,** and the **solar corona** that was available was obtained from observations taken at the time of total solar eclipses. At the present time the prominences may be carefully and completely studied at any time by means of the **spectroheliograph** and related instruments. An instrument has already been developed which will permit photography of the inner corona without an eclipse, and the instrument is still in the process of improvement toward the photography of the entire corona. However there are still several problems, e.g., the flash spectrum, which can only be obtained at the time of a solar eclipse and expeditions will probably be dispatched to regions of totality for many years to come. An eclipse of the moon is of comparatively little scientific importance. At the time of such an eclipse the moon is not completely dark, but is illuminated by light which is refracted into the umbra by the atmosphere of the earth, and the moon is visible with a dull reddish light. (W.K.G.)

ECLIPSING BINARY. When the orbit plane of a **binary star** lies so nearly in the line of sight of the observer that the components undergo mutual **eclipses** the object is known as an eclipsing binary. In case the binary is also a **spectroscopic binary** and the **parallax** of the system is known we have one of the most valuable specimens for stellar analysis. Eclipsing binaries are **variable stars**, not because the light of the individual components vary, but because of the eclipses. The most notable of the eclipsing binaries is the star **Algol** (β **Persei**), so named the "demon star" by the Arabs in all probability because they noticed the variation in light.

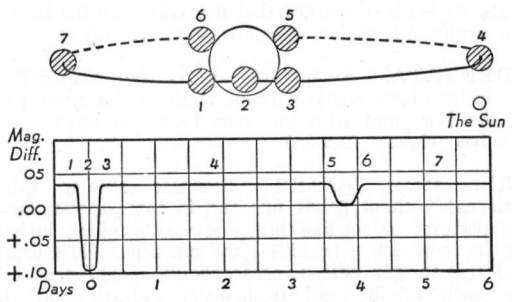

Apparent relative orbit and light curve of the eclipsing binary 1H. Cassiopeiae. (Curve and orbit determined by Joel Stebbins from his observations with the photoelectric photometer at the University of Illinois.)

The **light curve** of an eclipsing binary is characterized by periods of practically constant light with periodic drops in intensity. In the figure we have a characteristic light curve of such an object. In this case the eclipse of the larger and brighter primary star by the secondary is **annular**, while the eclipse of the secondary by the primary is total.

The **orbit** of an eclipsing binary may be determined from a study of the light curve. In addition to the seven **elements** of the orbit it is also possible to determine the relative sizes of the individual stars in terms of the radius of the orbit. In the determination of the orbit of a spectroscopic binary it is impossible to determine the semimajor axis, *a*, and the inclination of

the orbit plane, *i*, independently; but a quantity (*a* sin *i*), expressed directly in linear units (i.e., miles or kilometers) may be determined. If a star is both an eclipsing and spectroscopic binary we can determine all seven elements of the orbit, including *a* and *i*, in angular units from the light curve, and the quantity (*a* sin *i*) in linear units from the spectroscopic data. Hence, we can determine the radius of the orbit in linear units and then get the sizes of the individual stars in linear units. Since from the period we can get the relative masses of the two stars, and can get the relative sizes from the combination of the photometric and spectroscopic orbits, we are able to determine the densities of the individual stars. (W.K.G.)

ECLIPTIC. The great circle cut out on the **celestial sphere** by the plane containing the **orbit** of the earth is known as the ecliptic. The ecliptic is the fundamental plane for the system of **spherical coordinates** in which celestial **latitude** and **longitude** are measured. The ecliptic is also the reference plane to which the planes of the orbits of all the members of the **solar system** are referred.

The plane of the ecliptic is inclined to the plane of the **equator** by an angle of approximately 23°.5, known as the obliquity of the ecliptic. The two planes intersect in a line known as the line of **nodes**. The points where this line of nodes intersects the celestial sphere are known as the **equinoxes**. The apparent motion of the sun in the ecliptic about the earth, due to the actual motion of the earth in its orbit, causes the sun to pass through each one of the equinoxes once each year. The point where the sun crosses the equator from south to north is known as the **vernal equinox**, and the opposite extremity of the line of nodes is the autumnal equinox. Due to **precession** the direction of the line of nodes is continually changing relative to the stars. At present the vernal equinox is in the **constellation** of Pisces and is continually moving along the ecliptic in a direction contrary to the annual motion of the sun at such a rate that it will complete one revolution of the ecliptic in approximately 25,000 years. (W.K.G.)

ECLOGITE. This is a coarse, granular rock composed chiefly of **garnet** and **pyroxene** with subordinate amounts of various minerals such as **rutile, magnetite, apatite**, etc. **Hornblende** sometimes is present, replacing the pyroxene, often to the extent that a garnet **amphibolite** is produced. The origin of eclogites is obscure, they may result in part from the deep-seated **metamorphism** of gabbroic rocks, but some may have resulted from crystallization of a primary **basic magma** under conditions of great pressure. They may represent segregations in a highly basic magma analogous to segregations of basic minerals in **granites** and other common **igneous** rocks. Seemingly confirmatory evidence of this idea is found in the chunks of eclogite-like material found in the **kimberlite** of South Africa. (E.S.C.S.)

ECOLOGY. The biological science that deals with the relations of organisms and the environment, including relations with other organisms. The relations of the individual are the subject matter of **autecology** and those of groups belong to **synecology**. In all of its divisions the science is now complex.

The living organism is inevitably exposed to certain physical factors which vary over the surface of the earth. The force of gravity causes stresses in its body and influences the pressure under which it must live. It is immersed in a fluid medium, either water or air, whose pressure on a given unit of surface depends on the amount of material above and so upon the altitude at which the organism lives. The pressure of water in the ocean is roughly one ton per sq. in. per mile of depth. That of air is approximately 15 lbs. per sq. in. at sea level and lessens rapidly at higher altitudes. The oxygen content of these media influences respiration and their physical characteristics are important in locomotion. Many animals float and swim in water but relatively few fly and none can float permanently in the air. On the ground the friction between the body and the earth necessitates other types of locomotion and burrowing forms also must be specially adapted.

The temperatures to which the animal is subjected are extremely important, since they affect its physiological processes directly. The adaptation of animals to temperatures is often complex.

Light is important both for its effects on the living substance and as a factor in vision.

Since water is an essential constituent of living matter, the presence of an adequate supply is of fundamental importance to living things. On land variation is extreme. Between the dearth of water in arid regions and its abundance in marshes, and between the rainy and dry seasons of some areas, many factors interact in maintaining a supply which is sufficient only for animals that are adapted to conserve it. Even in such regions occasional years of drought impose severe conditions.

The food supply, like water, is a fundamental need. Since animals depend upon organic food this relationship is essentially one of living things with each other.

As a result of all of these factors, animals are restricted to parts of the earth where tolerable conditions prevail. Any species is found in its characteristic environment where it is associated with others that require similar conditions. These groups form the communities of the ecologist. By common agreement four main types are recognized: land communities, communities of waters and shores, communities of the seashore, and fresh-water communities. Land communities are those of deciduous forests, evergreen forests, grasslands, deserts, arctic and alpine regions, and minor subdivisions. In the sea are found the pelagic communities of the open waters, benthonic communities of the bottom, and littoral or shore communities, and in fresh water similar subdivisions are recognized in addition to those of still and running water.

Some of the major results of the adjustment of animals to the environment are expressed under **distribution**. (A.W.L.)

ECONOMIZER. Any device, the presence of which in a machine or cycle of machinery is nonessential, but which effects a saving, usually of the raw material, may be, with reason, named an economizer. Thus an attachment for the carburetor of a gasoline engine, designed to increase the energy delivered per unit of fuel used, may be called an economizer.

The economizer of a steam power plant is a heat exchange surface the purpose of which is to recover waste heat in the flue gas by absorbing it in the boiler feed water. Such economizers often form an integral part of the **boiler** surface, with heating surface in the form of tubes. Heat is recovered from flue gas by passing it through the tube surface into the feed water stream, which circulates inside the tubes. Ordinarily, no steam is produced in the economizer, though steaming economizers have occasionally been built. (F.T.M.)

ECTOCYST. Zooecium.

ECTODERM. Germ Layers.

ECTONEURAL SYSTEM. A portion of the nervous system of echinoderms which forms a plexus under the ectoderm and a radial nerve along each arm or equivalent radius. (A.W.L.)

ECTOPROCTA. A class including most members of the phylum **Bryozoa**. The included species live in colonies of many forms and are characterized by the retractile tentacles and by the anus lying outside of the circlet of tentacles.

The class is divided into two orders:

Order Gymnolaemata. **Lophophore** circular. Mouth usually closed by a flap called the operculum. Marine.
Order Phylactolaemata. Lophophore horseshoe-shaped or oval. Fresh-water species. (A.W.L.)

ECZEMA. A loose term indicating any inflammatory skin disease characterized by watery discharge with formation of scales and crusts. The term covers many skin diseases, but is used most commonly to describe the allergic dermatitis of infants and children. (R.S.M.)

EDEMA. Swelling due to excessive tissue fluids. The causes of edema are many. Physiologically the balance of the body fluids between the cells (intracellular fluid) the fluid bathing the cells (extracellular fluid), and the **blood** plasma is upset. Fluid is drawn from the blood into the tissues when there is a higher osmotic pressure in the tissues than in the blood (see **Osmosis**). This higher pressure may be due to an actual increase (e.g., in salt retention due to impaired kidney function) or it may be a relative increase, as in edema associated with low serum proteins in the blood due to nutritional deficiency. Obstruction to venous or lymph flow also results in edema, by the mechanical factor of increased pressure in the capillaries. This occurs with **varicose veins** and **elephantiasis**. Capillary damage due to infection, bacterial toxins, or to trauma will allow the passage of fluid from the blood into the tissue spaces and produce edema. Exudation of fluid into the extracellular spaces is part of the general process of **inflammation** and is found at the site of any localized inflammatory reaction or infection.

The common conditions characterized by edema are congestive **heart failure, nephritis, varicose veins, cirrhosis,** and allergic phenomena such as **angioneurotic edema.** In congestive failure, the fluid tends to collect in dependent portions, the feet, legs, and over the sacrum. In nephritis these areas plus the loose tissue around the eyes and other easily distensible tissues become edematous. If the edema is so marked that it involves all the tissues, the condition is spoken of as anasarca. (D.M.H.)

EDENTATA. Ant-eaters, sloths and armadillos. An order of mammals without teeth in the front part of the jaws and with no enamel on the teeth. The feet bear claws. (A.W.L.)

EEL. Pisces, Teleostei. Elongate slender fishes (**Pisces**) without pelvic fins and in some species lacking pectoral or median fins. Several families, including the true eels (*Anguilla*), the muraenas (*Muraena*), and the **conger eels** (*Leptocephalus*). The true eels are both marine and fresh-water animals and the muraenas and conger eels are marine. The muraenas are also called morays in some places. They are found in the warmer seas, especially about coral reefs.

Eels are eaten but they do not rank among the important food fishes. (A.W.L.)

EEL GRASS. *Zosterna marina.* Najadaceae. Eel grass is a common plant of sandy or muddy ocean shores. The plant has a creeping somewhat fleshy stem which roots freely at the nodes and has short erect branches. Usually it grows in salt water from a foot to over 4′ deep, and is frequently found in tidal pools. The long linear leaves have sheathing bases and float more or less erect in the water. The very peculiar flowers consist of a single **carpel**, containing a single **ovule**, two flat **stigmas**, and a **stamen**, which has two half **anthers** joined by a slender connective. These flowers are borne on a long flat **spadix**. The pollen

grains are thread-like and have a specific gravity equal
to that of the salt water in which they grow. Thus
when mature they float in the water, neither rising
nor falling, as they are carried by the currents to
the large flat stigma. The fruit is an **achene**.

Quantities of eel grass are raked up and dried. It is
used for packing, for stuffing for various objects, in
the manufacture of certain kinds of wall board and for
insulation in walls.

The plants suffer periodically from a certain disease
which seriously depletes their numbers. In some regions
the natural growth of eel grass has been practically
wiped out by this disease.

Fresh water eel grass, **Vallisneria spiralis,** is an entirely different plant, with an interesting method of
pollination. (R.M.W.)

EELWORM. Nemathelminthes, Nematoda. A roundworm, *Ascaris lumbricoides,* 6–15″ long, parasitic in the
adult stage in the small intestine of man and the domestic
animals and during development in other tissues of the
body. They sometimes occur in large numbers in
children, who take in the young **larvae** in water or on
raw foods, such as fruits and vegetables. (A.W.L.)

EFFECTIVE HEIGHT. Antenna.

EFFECTIVE RESISTANCE. Resistance.

EFFICIENCY. The general significance of this term
as applied to a device or machine may be expressed as
the ratio of output to input of energy or of power. If a
d-c motor, for example, is operating on 4 amperes at 100
volts (the power input is 400 watts), and if the motor
actually delivers only 280 watts of mechanical power, its
efficiency at that load is 280 watts ÷ 400 watts, or 70
per cent. In general, the efficiency of a machine varies
somewhat with the conditions under which it operates.
Usually there is a load for which the efficiency is a
maximum. This may be illustrated by a heavy block-
and-tackle. For a small load the efficiency would be
very low, because of power wasted in bending the ropes;
for an excessive load it would again be low, on account
of the large friction which would then develop; while for
intermediate loads, higher efficiencies would prevail.

The concept may be extended to other than purely
mechanical systems. Thus, the efficiency of an electric
lamp may be expressed in candles or lumens of luminous
flux (output) per watt of electric power (input); or that
of an automobile horn in watts of acoustic power (noise)
per watt of electric input. Various types of heat engine
exhibit different thermodynamic efficiencies, i.e., the ratio
of the work derived in the engine to the heat energy applied to it. (See **Thermal Efficiency.**) (L.D.W.)

EFFICIENT STATISTIC. Among consistent statistics, a **statistic** is said to be efficient if in large
samples it is **normally distributed** with a minimum
variance. Thus, an efficient statistic T_1 must first of
all be consistent, secondly it must be normally distributed
in large samples, and finally variance of $T_1 \leqq$ variance
of T_2 in large samples, where T_2 is any other consistent
statistic. There may be more than one efficient statistic
but as the size of the sample increases, efficient statistics
give essentially the same results since their variances
approach equality. We state some properties of efficient
statistics. If T_1 and T_2 are both efficient statistics of the
parameter θ, and if the distribution of $(T_1 + T_2)\sqrt{N}$
approaches normality in the limit as N increases, then
$\lim_{N \to \infty} r_{T_1 T_2} = 1$, where r is the **coefficient of correlation**

between T_1 and T_2. Let T_1 be an efficient statistic, and
T_2 denote a consistent statistic which is not efficient such
that

$$\lim_{N \to \infty} \frac{\sigma^2_{T_1}}{\sigma^2_{T_2}} = E < 1, \quad \text{then } r_{T_1 T_2} = \sqrt{E}.$$

A **sufficient** statistic T which is consistent and normally
distributed in the limit as N becomes large, is an efficient
statistic. Clearly, in large samples, efficient statistics
should be used wherever possible. (L.A.A.)

EFFLORESCENCE AND LOSS OF WATER.
When a substance evolves moisture upon exposure to
the **atmosphere,** the substance is said to be efflorescent, and the phenomenon is known as efflorescence.
At ordinary temperatures the vapor pressure of water is
as follows:

TEMPERATURE, °C.	WATER VAPOR PRESSURE, IN MM. MERCURY	
	At Saturation	At 50% Humidity
0	4.6	2.3
10	9.2	4.6
20	17.5	8.8
30	31.8	15.9
40	55.3	27.7

If the substance has a higher water vapor pressure
than corresponds to that of the atmosphere at the
given temperature, water vapor is evolved from the
substance until the water vapor pressure of the substance equals the water vapor pressure of the surrounding atmosphere.

Substances that are ordinarily efflorescent are **sodium** sulfate decahydrate, **sodium** carbonate decahydrate,
magnesium sulfate heptahydrate, **ferrous** sulfate heptahydrate.

When the saturated solution of a substance in water
has a water vapor pressure greater than that of the
surrounding atmosphere, evaporation of the solution
takes place, leaving the substance.

See **Deliquescence** for the converse phenomenon.
(R.K.S.)

EFFLUENT STREAM. A stream that flows out of
another stream or out of a lake. Also a stream whose
upper surface is below the surface of the local **ground
water table.** (R.M.F.)

EFFUSIVE. The term applied by geologists to molten material (lava) which has been poured out on the
surface of the earth from a vent or fissure, as distinguished from ejected volcanic material (ashes and
bombs) and injected **magmas** (plutonic rocks).
(R.M.F.)

EGG. The female reproductive cell or **gamete.**

EGGPLANT. Potato Family.

EGRET. Heron.

EIDER. Aves, Anseriformes. *Somateria.* **Ducks** of
several species which breed in the far north along
rocky coasts. They are noted for the fine down with
which they line their nests. This material has a high
commercial value and is collected for market. (A.W.L.)

EINSTEIN EQUIVALENCE PRINCIPLE. One
of the interesting features of the theory of general
relativity emphasizes the fact that the weights of
bodies and the **forces** which they oppose to acceleration are proportional, each being in direct ratio to the
masses of the bodies. The relativity theory points out
that they are really indistinguishable. Everyone is
familiar with the sensations of increased or reduced
weight caused by upward or downward acceleration in
a passenger elevator. If the elevator moved without
noise or jar and if one could not see out, an accelerated

motion in any direction would occasion forces which one would be unable to distinguish from that due to a gravitational field. A cream separator or a centrifuge, or a bucket of water whirled about the head without spilling it, illustrate how **centrifugal force** (also due to inertia) may imitate gravity. It is this relationship which is enlarged upon in relativity theory as the principle of equivalence. (L.D.W.)

EINSTEIN SHIFT. According to the relativity theory, when radiation quanta leave a massive source like the sun or a **star,** they are retarded by the gravitational attraction and hence lose energy. This means that they lose **frequency** and that the wavelength λ increases. For a star of radius R and mass M, the fractional increase in wavelength is

$$\frac{\Delta\lambda}{\lambda} = \frac{G}{c^2} \cdot \frac{M}{R},$$

in which G is the **gravitation constant** and c is the **electromagnetic constant** (speed of light). The coefficient $G/c^2 = 7.414 \times 10^{-29}$ cm. per gram. For the sun, $M = 2.3 \times 10^{33}$ grams and $R = 1.394 \times 10^{11}$ cm. Then $\Delta\lambda/\lambda = 1.23 \times 10^{-6}$, so that each solar spectrum line should be shifted toward the red by a little over a millionth of its own wavelength.

Measurements of this precision are hardly possible at present. But there are other stars (in particular, the **white dwarfs**) so massive and so dense that the shift has actually been observed and found to be of the correct order of magnitude. (L.D.W.)

EINSTEIN THEORY. Relativity.

EJA. Reptilia, Sauria. The desert saw viper of Egypt, a vicious poisonous **snake.** (A.W.L.)

EJACULATORY DUCT. The portion of the male genital duct between the duct of the **seminal vesicle** and the **urethra.** (A.W.L.)

EJECTAMENTA. Volcanism.

EJECTOR. Any mechanism or device which can in some fashion remove an object or a material from a certain position, is rightfully called an ejector. Thus many machines are equipped with ejector elements the function of which is the removal of a part from one position to another. The pump which moves a fluid from some place where it is not desired is an ejector. A device employed to remove non-condensable gases from a steam condenser is called an air ejector. The steam-jet ejector is a widely used type, especially on large condensers. It consists of a steam jet receiving steam at a pressure from 100–250 lbs. per sq. in. The steam issues from the jet with high velocity, picking up such particles of air or other gas as may be entangled in the steam jet. The jet is then decreased in velocity, and a pressure built up. In this way, such gas as is entangled with the steam jet is compressed. (F.T.M.)

ELAEOLITE. Syenite.

ELAIOPLAST. A minute body in the cytoplasm of some **cells,** supposed by some cytologists to be a form of plastid about which fat is deposited. (A.W.L.)

ELAND. Mammalia, Artiodactyla. The largest African **antelopes.** The body is like that of a cow but the head is smaller and the horns are moderately long, straight, and spirally twisted. There are several species of the genus *Taurotragus.* (A.W.L.)

ELASTIC AXIS. Flexure.

ELASTIC CURVE. The curve of the **neutral** surface of a structural member subjected to **loads** which cause bending is called the elastic curve. The ordinates between this curve and the original position of the neutral surface represent the **deflections** due to bending. (C.W.C.)

ELASTIC LIMIT. The maximum unit **stress** which can be obtained in a structural material without causing a permanent **deformation** is called the elastic limit. (See **Ultimate Strength.**) (C.W.C.)

ELASTICITY. The property whereby a body, when deformed, automatically recovers its normal configuration as the deforming forces are removed. Each of its several types is probably due to the action of intermolecular forces which are in equilibrium only for certain configurations.

Deformation or, more briefly, strain is of various kinds; in each case its measure is a certain abstract ratio. For example, the elongation of a rod under tension is expressed as the ratio of the increase in length to the unstretched length. Linear compression is the reverse of elongation. They are both accompanied by a fractional change in diameter, the ratio of which to the elongation is called the Poisson ratio. Shear is a strain involving change of shape, such that an imaginary cube traced in the unstrained material becomes a rhombic prism. The measure of shear is the tangent of the angle through which the oblique edges have been made to depart from their original perpendicular direction. Volume strain is the ratio of a decrease in volume to the normal volume. **Flexure or** bending, and torsion or twisting, are combinations of these more elementary strains. A straight rod bent into a plane curve undergoes elongation on the convex side and linear compression on the concave side, while there is an intermediate neutral layer which suffers neither.

For every strain there arises, in an elastic substance, a corresponding stress, which represents the tendency of the substance to recover its normal condition. Stress is expressed in units of force per unit area. Tensile stress, for example, is the ratio of the force of tension to the normal cross-section of the rod subjected to it. Shearing stress is the force tending to push one layer of the material past the adjacent layer, per unit area of the layers. Pressure, expressed in like units, is the stress corresponding to volume compresion, etc.

For each type of strain and stress there is a modulus, which is the ratio of the stress to the corresponding strain. In the case of elongation or linear compression, it is commonly called Young's modulus; we also have the bulk modulus and the shear modulus or rigidity. (See **Tension Test.**)

In engineering design Young's modulus is used for tension and compression and the rigidity modulus for shear, as in torsion springs. Nominal values in lbs. per sq. in. are given for the principal classifications of structural metals (and their alloys).

Material	Young's Modulus	Rigidity or Shear Modulus
Magnesium.....	6.5×10^6	2.4×10^6
Aluminum......	10.2×10^6	3.6×10^6
Copper.........	14.5×10^6	6.1×10^6
Steel..........	30×10^6	11.6×10^6

(L.D.W., R.H.H.)

ELATERITE. A variety of **bitumen** which is elastic when fresh but becomes hard and brittle when exposed to the air. (R.M.F.)

ELECTRET. Dialectric Absorption.

ELECTRIC AND MAGNETIC DOUBLE REFRACTION.

In 1875 Kerr discovered that glass and many other isotropic, transparent solids and liquids exhibit double refraction like crystals, when placed in a strong electric field; and in 1905 Cotton and Mouton, after some preliminary results by Kerr and others, demonstrated the corresponding phenomenon with a magnetic field. These are now known respectively as the Kerr electro-optical effect and the Cotton-Mouton effect. In both cases the magnitude of the effect, as measured by the phase difference produced per unit thickness of medium, is, for a given substance, wavelength, and temperature, proportional to the square of the field intensity. The optic axis of the doubly refracting substance corresponds to the direction of the imposed field.

Of the two phenomena the Kerr effect is much more pronounced and is as yet the only one of practical importance. The Kerr cell, in which nitrobenzene, a liquid, is commonly employed because of its large and quick response to the electric field, has in recent years been extensively used as an electro-optical control or shutter for light beams, for example, in the recording of sound pictures. (L.D.W.)

ELECTRIC AND MAGNETIC UNITS.

The measures of electrical quantities are based upon two quite distinct principles, giving rise to two distinct systems of electric units, which differ not only in size but in their physical makeup or dimensions; just as oil is measured and sold either by the gallon (volume) or by the pound (mass).

1. In the electrostatic system, the fundamental unit is that of quantity of **electricity**, defined as the charge which, concentrated upon a small sphere, repels a similar equal charge at unit distance (in a vacuum) with unit force. The c.g.s. electrostatic unit charge is the "statcoulomb," in which the unit distance is 1 cm. and the unit force is 1 **dyne**. It is equal to about 3.3×10^{-10} coulomb. Based on this is the seldom-used e.s.u. current or "statampere," which is 1 statcoulomb per sec.

2. The electromagnetic system, on the other hand, is founded upon the unit electric current, viz., the **abampere**, from which is derived the absolute **ampere**. The abcoulomb is 1 abampere per sec., and the absolute coulomb is $\frac{1}{10}$ of it. Aside from these are the international coulomb and ampere, defined electrochemically.

For some purposes it is convenient to use an e.s.u. of electric potential or of electromotive force, viz., the "statvolt," which is that potential difference through which the transference of 1 statcoulomb of electricity requires the expenditure of 1 erg of energy. But the practical e.m.f unit is the **volt**, similarly defined but in terms of the coulomb and the joule. The abvolt depends on the abcoulomb and the erg in the same way. There is also an international volt.

The practical unit of electrical **resistance** is the **ohm** (either absolute or international), equal to 10^9 abohms. The resistivity of a substance is commonly expressed in terms of the ohm-centimeter; i.e., as the resistance between opposite faces of a 1-cm. cube of the substance.

The practical unit of **capacitance** is the farad, which is 1 coulomb/volt, and is thus electromagnetic. The "abfarad" is the fundamental c.g.s. unit, 1 abcoulomb/abvolt or 10^9 farads. The microfarad (0.000001 farad) is commonly used in actual measurements.

The practical and fundamental c.g.s. units of **inductance** (or mutual inductance) are respectively the henry (defined under **Inductance**) and the abhenry, equal to 10^{-9} henry. The millihenry (0.001 henry) is usually more convenient.

The purely magnetic units begin with the concept of the unit magnetic pole (see **Magnetism**). The **oersted** and the **gauss** are respectively the units of magnetic intensity and magnetic induction. Often more convenient than the maxwell or "line" as a unit of **magnetic flux** is the "volt-second" or "weber," which is 10^8 maxwells. Magnetomotive force and magnetic potential may be expressed either in gilberts or in ampereturns (see **Magnetic Circuit**). (L.D.W.)

ELECTRIC CELL.

The electric cell is composed of two dissimilar metals or materials which are immersed in a solution capable of acting upon them chemically. A primary cell is one composed of two dissimilar electrodes which are immersed in a solution of an acid which acts more readily on one of the electrodes than on the other. The result of this chemical reaction is the production of an electromotive force. If the electrodes are connected by an external circuit, the electromotive force will maintain a flow of current through the circuit composed of the external connections, the electrodes, and the electrolyte. As current flows the chemical energy resulting from the action of the acid on the electrodes is converted into electrical energy.

A secondary cell is one in which electric current must first be passed between the electrodes of the cell under the influence of an externally generated electromotive force. The chemical action resulting makes the cell capable of giving off electric current by means of a secondary reversed chemical action. A group of secondary cells is a storage **battery**.

The physicist Volta discovered in 1800 that when two dissimilar metals were immersed in a solution such as sulfuric acid, they produced an electromotive force. By experimenting with different metals, he was able to arrange them in an electrochemical series which will show the relative electromotive force, and which of two elements used in a cell will be the positive, and which the negative electrode. A typical electrochemical series is as follows (for further details, see **Reactions Involving Oxidation-Reduction**):

> Zinc
> Iron
> Tin
> Lead
> Copper
> Silver
> Platinum
> Carbon (non-metal)

If any two of these metals are immersed in sulfuric acid the one nearest the top of the electrochemical series will be the positive electrode, and current will flow from it in the solution and to it in the external circuit.

A Voltaic cell is shown in Fig. 1. A simple cell of this type would soon become polarized. As current flows around the circuit, the following reaction takes place:

$$Zn + H_2SO_4 = ZnSO_4 + H_2.$$

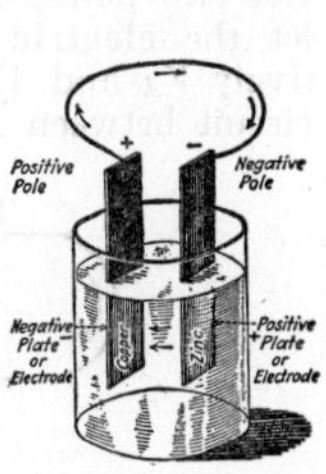

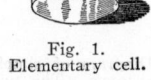

Fig. 1.
Elementary cell.

The hydrogen thus liberated by the chemical reaction gathers on the negative electrode and, as a result, the electromotive force diminishes and finally disappears. This action is known as **polarization**, and must be overcome in any practical cell.

There are many types of primary cells, amongst which might be mentioned the Daniell cell with zinc in sulfuric acid, and copper in copper sulfate: the LeClanche cell, using zinc and carbon in an aqueous solution of ammonium chloride; the Weston cell, with mercury and cadmium in a solution of cadmium sulfate. (See **Standard Cell.**) Depolarization in the Daniell cell is accomplished by absorption of the hydrogen by the copper sulfate, which is separated from the acid-filled portion of the cell by a thin porous partition, through which the hydrogen may pass. In the LeClanche cell, manganese dioxide, a substance which gives off oxygen to combine with the hydrogen, is used as a depolarizer. The Weston cell is a voltage standardizing apparatus, and is not designed to produce a flow of current, but rather to yield a standard voltage for comparison. Its voltage is 1.0183 volts at 20° C.

The most widely used primary cell is the dry cell, whose popularity is the result of its portability and simplicity. Having no liquid electrolyte, the dry cell offers no danger from spilled acids; moreover, it is inexpensive and reliable, and long-lived when used for supplying intermittent currents, as for ringing doorbells, or with telephones, etc. The common dry cell (Fig. 2) has an electromotive force of about $1\frac{1}{2}$ volts. The electrodes are zinc and carbon, with the former being in the shape of a cup some $2\frac{1}{2}''$ in diameter by $6''$ high for the standard type, although smaller sizes are used in flashlights. The carbon electrode is centralized in the zinc cup, and the space between carbon and zinc filled with the necessary electrolyte and depolarizer. The electrolyte is a damp sal ammoniac compound, and the depolarizer is manganese dioxide. The open end of the zinc cup is then sealed off with pitch or cement and binding posts are connected to the zinc and carbon. (F.T.M.)

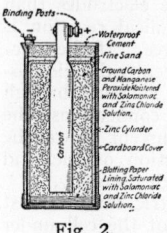

Fig. 2.
Dry cell.

ELECTRIC CIRCUITS. The simplest type of electric circuit may be considered as beginning at any point and continuing without branching until that point is reached again. Along this closed path electricity moves because in some portion or portions of the path an **electromotive force** E is applied to it, doing work on the electricity and causing it to progress against the **resistance** as an **electric current.** It is convenient to divide the total resistance R of the circuit into two parts, the internal resistance R_i, viz., the resistance of that part in which the electromotive force is applied, and the external resistance, R_e, of that part in which no electromotive force is active. The current is the same through the whole of such a simple circuit, and its value, if constant, is given by **Ohm's law:**

$$I = \frac{E}{R} = \frac{E}{R_i + R_e}. \qquad (1)$$

Let two points A and B be selected on the circuit and let the **electric potentials** at those points be respectively V_A and V_B. If the resistance of the part of the circuit between A and B is R_{AB} and if the part of the

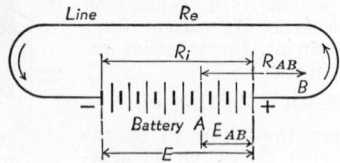

Diagram of simple circuit showing internal and external resistance. The two points A and B are selected at random; part of battery may or may not be included between them.

total electromotive force E applied to the circuit between these points is E_{AB}, then for the section AB, Ohm's law takes the special form

$$I = \frac{V_A - V_B + E_{AB}}{R_{AB}}, \qquad (2)$$

in which E_{AB} is $+$ or $-$ according as it acts with or against the current. (The latter is illustrated by the part of a circuit containing a storage battery being charged or a motor exerting its back electromotive force.) We may equate (1) and (2), since the current is the same in all

parts of the circuit; the potential difference between any two points A and B may then be calculated, thus:

$$V_A - V_B = \frac{R_{AB}}{R} E - E_{AB}. \qquad (3)$$

If there is no active electromotive force in the part AB, then (2) gives $I = (V_A - V_B)/R_{AB}$. For any two pairs of points, A, B and C, D, since I remains the same, we then have

$$\frac{V_A - V_B}{V_C - V_D} = \frac{R_{AB}}{R_{CD}}, \qquad (4)$$

the so-called "law of potential drop."

If the circuit has appreciable **inductance**, Ohm's law in its simple form (1) applies only so long as the current is constant. (See **Alternating Currents** and **Alternating-Current Circuits.**) Also, a circuit may be broken by a **condenser,** or have a condenser in parallel with it, or it may have appreciable **distributed capacitance**; in any such case its behavior depends upon the application of an electromotive force departs from Ohm's law. Divided circuits and networks, especially if they involve inductances or capacitances and carry variable currents, present somewhat complicated problems. (See **Kirchhoff's Laws of Networks.**) (L.D.W.)

ELECTRIC CONDUCTION. The conduction of electricity material substances is of two general kinds. 1. A migration of ionized (and hence electrified) atoms or molecules, as in **ionized gases** and electrolytes. 2. A process in which the atoms are in the main stationary, as in metals. This article deals with the second type only.

The electric conductivity of solids has an almost unbelievable range. For silver and sulfur (the best and the poorest among elements), the ratio is something like 1000 billions of billions to 1. Their resistivity is, of course, in the inverse ratio. (See **Resistance.**) Metals, as a class, aside from being almost immeasurably better conductors, differ in several respects in their conduction from non-metals. For example, the conductivity of pure metals consistently decreases with rising temperature, while the non-metallic solids, including carbon, generally have maximum conductivity at one or more temperatures. Selenium has most extraordinary properties (see **Photoconductivity**). Some metals exhibit **superconductivity** near the **absolute zero** of temperature. The alloying of metals, and even the admixture of small quantities of impurities, often profoundly affects the conductivity. Thus "constantan," an alloy of copper and nickel, shows almost no change of conductivity with temperature. The **Wiedemann-Franz law** expresses the remarkable relation which exists between the electric and the thermal conductivities of metals. The various **thermoelectric phenomena, thermionic phenomena,** the **Hall Effect,** etc., must also be taken into account by any theory of metallic conduction.

The idea proposed by Weber (long before electrons were known) and elaborated by Drude and Lorentz (with the aid of electron theory) has, until recently, completely dominated our thinking about metallic conduction. It supposes that good conductors contain electrons free to move about among the relatively fixed atoms; that when an electric field is applied to the conductor, the electrons drift along through it, colliding with the atoms and producing heat; and that poor conductors are such because few of their electrons are free. Tolman and Stewart (1916) tried suddenly stopping a metal rod in rapid longitudinal motion, and found that it became negatively charged at the forward end, as if the electrons had piled up there. A large number of experimental facts are in qualitative agreement with this theory. But two great difficulties are encountered: lack of even approximate quantitative agreement in many details, and the existence of certain stubborn anomalies; such, for example, as superconductivity and the conflict between **Dulong and Petit's law of specific heats** and the principle of **equipartition of energy.** At present it appears that the solution of the difficulty may be a revision of our concept of the **electron** and its behavior, in terms of **wave mechanics.** (L.D.W.)

ELECTRIC CONDUCTIVITY. Resistance.

ELECTRIC CURRENTS. Electricity is a tangible thing, with some of the properties of ordinary matter, and among them the ability to move. Whenever it moves, we have an electric current. Since electricity may exist as separate electrons, protons, or positrons, or in collected charges of these entities, the electric current presents a variety of aspects. Usually we think of the progressive motion of electrons in a conductor, called **electric conduction,** or the two-way traffic of charged ions called "electrolytic conduction"; but the current may consist of a flight of electrons or other charged particles through a vacuum, or it may be a bodily movement of an electric charge, like that on one of the carriers of a **static machine.** Also there is that singular process assumed by Maxwell to take place in a **dielectric** or even in a vacuum, to account for the behavior of electromagnetic fields and radiation, and called the "displacement current." A current may be continuous, that is a progressive movement of electricity in one direction (the so-called **direct current**), or it may consist of temporary surges called **transients,** or be merely a vibration of the electricity with very short amplitude, as in **alternating currents** and electric oscillations. In conduction currents, contrary to the usual impression, the actual net progress of the electricity is very slow, seldom more than a few centimeters a minute. This is because of resistance and the continuous dissipation of energy in accordance with **Joule's law.** The speed of the electrons in a vacuum tube may, on the other hand, reach thousands of miles per sec. Electric impulses, or "compression waves," in a conductor also travel with great speed (in a telephone wire, many thousands of miles per sec.), and it is this property that gives electric communication the lightning-like rapidity commonly associated with electric currents. To drive electricity against its inertia or against the opposition of obstacles such as massed atoms or molecules requires **electromotive force.** If the electricity moves through any region where there is such opposition but with no battery or other source of electromotive force within that region to supply energy (as in a resisting wire), its motion is maintained at the expense of its own energy. The **electric potential** changes as the electricity progresses, by an amount proportional to the potential energy which the electricity must thus give up. This leads to the familiar "law of potential drop." (See **Electric Circuits, Ohm's Law, Ampere, Electromagnetism,** etc.) (L.D.W.)

ELECTRIC DEGREE. Degree.

ELECTRIC DISPLACEMENT. Dielectrics.

ELECTRIC EEL. Pisces, Teleostei. A true eel, *Electrophorus electricus,* of the rivers of northern South America. It reaches a length of 6' and is capable of giving a powerful electric shock. (A.W.L.)

ELECTRIC FIELD. Electrostatics; Fields of Force.

ELECTRIC FURNACE. An electrically heated furnace for the melting and alloying of metals. It is especially successful in the production of alloy steels, but is also used for non-ferrous alloys, particularly in the brass industry. The effect of the electricity upon the metal is purely to supply heat, there being no electrochemical action. The heating effect of the modern fur-

nace is secured either by induction or by the use of the arc.

The high-frequency induction furnace may take several forms, the most usual being a refractory crucible surrounded by a water-cooled copper coil. Power in the form of high-frequency a.c. varying from 60 to 60,000 cycles per sec. is supplied to the coil from a suitable power supply such as a motor generator set which converts 60-cycle current to a higher frequency, a widely used value being 960 cycles per sec. Heat is produced in the metallic charge by means of eddy currents induced by the high-frequency field surrounding the coil. The

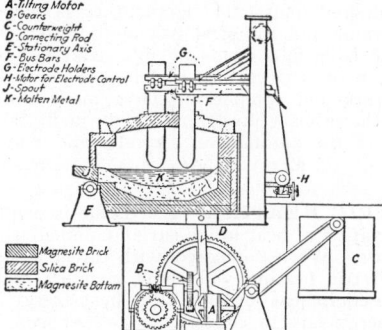

A - Tilting Motor
B - Gears
C - Counterweight
D - Connecting Rod
E - Stationary Axis
F - Bus Bars
G - Electrode Holders
H - Motor for Electrode Control
J - Spout
K - Molten Metal

▨ Magnesite Brick
▧ Silica Brick
▤ Magnesite Bottom

Fig. 1.　Electric-arc melting furnace.

capacity of this type of furnace varies from a few pounds to as much as ten tons. Refractories used may be made of either quartz sand as in the "acid" crucible or magnesite as in the "basic."

The furnace is primarily a remelting furnace, very little refining or purifying of the metallic charge being accomplished. Close control of composition is accomplished by selection of raw materials, the speed of melting being such that control by chemical methods is impractical. Violent stirring of the molten charge as indicated by the arrows in the figure is produced by the interaction of

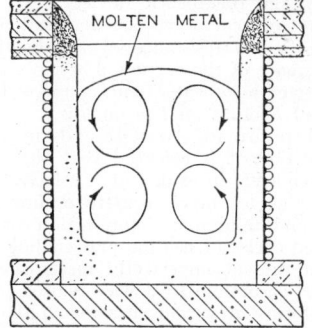

MOLTEN METAL

Fig. 2.　Section of induction furnace showing stirring of the melt.

the eddy currents with the high-frequency field. Since lower frequencies are most efficient for stirring, many modern furnaces have two frequencies applied to the coil, a high frequency for rapid melting, and a low frequency for rapid stirring.

The type of arc furnace in most general use is the three-phase arc furnace invented by Heroult and bearing his name. The heat is produced in this furnace by three arcs striking directly to the bath from the electrodes which are usually made of carbon or graphite. Power is supplied to the arcs from large transformers and is regulated by automatic equipment designed to give minimum carbon contamination of the melt from the electrodes. The capacity of these furnaces varies from about ½ ton to in excess of 100 tons. These furnaces are primarily refining furnaces and may be either "acid" where the refractories are predominately silica, or "basic" where the refractories are predominately magnesite. The basic furnace is the more widely used.

The operation of such a furnace is usually divided into two periods. First there is an "oxidizing" period in which the metallic charge is melted and held under a slag containing considerable iron oxide. This serves to eliminate carbon, silicon, phosphorus, and some sulfur. The "oxidizing" slag is then raked and poured off and a second or "reducing" slag usually very low in iron oxide and high in lime content is placed on the bath. This slag serves to eliminate sulfur and some other undesirable non-metallic material. The composition is adjusted to the desired limits, the temperature is raised to the proper value and the steel is poured into a ladle and from the ladle into ingots or into sand molds for the production of steel castings. Complete control of both temperature and composition have made this type of furnace the primary producer of high-quality alloy steels for use in the automotive, aircraft, and machine tool industries. (C. R. TAYLOR, AM. ROLLING MILLS.)

ELECTRIC HEATING. Electricity is employed for the generation of heat in industrial, commercial, and domestic use. When compared with direct heating from fuels, electric heating will show a much higher cost, since the equipment for converting heat energy into electrical energy fails to convert, at its best, some 70% of the heat units. Nevertheless, since electrical heating apparatus produces the heat at the very spot desired, and exactly in the desired amount, the difference in cost is not so pronounced, because of the economy of electric heating apparatus. For localized heating, such as cooking, soldering, and welding, electric heating is especially good; also it is used where cost is not an important item, viz., apartments, shipboard, hospitals, etc.

Heat may be obtained from electrical energy by means of resistance heating, induction heating, and the arc. Resistance heating is used in most domestic appliances, industrial ovens, and the like. Arcs are used for welding, lighting, and furnaces. (See **Electric Furnace, Arc Lamp.**) Induction heating has been employed in welders, furnaces, and therapeutic devices.

The amount of heat developed in resistance heating is proportional to the resistance of the heating element, and to the square of the current flowing. The electrical energy converted into heat when I amperes flow through a resistance of R ohms for t seconds is $I^2 \times R \times t$ watt-seconds. There are 1055 watt-seconds in a B.T.U. In resistance heating, it is essential that the resistor have little or no tendency to oxidation, a relatively high melting point, a high resistance so as to minimize the length of resistor, and minimum deterioration or embrittling upon repeated cycles of heating and cooling. Resistance wire used at present imperfectly meets these specifications. Nickel chromium steel wire is used in round and flat ribbon forms in various heating appliances. Insulation is usually by porcelain supports or mica. If, due to the convenience of application of heat, and the close control which can be exercised over temperature in electric heating, the product being subjected to heat can be made better, or with less labor, or with less wastage and spoilage, the higher cost of electric heating may often be more than overcome by savings in other lines. This has been true in cases such as ironing in laundries, cooking in candy manufacture, maintaining of glue pots, soldering irons, etc., at constant temperature, the even browning of loaves in bakers' ovens, the heating of presses and matrices, and a host of other heating applications.

In recent years advances made in domestic heating offer possibilities of electric heating for such purposes becoming economically feasible. Among these may be mentioned the system known as radiant heating. The first is based on the theory that most of the heat lost from the body is by radiation to colder surroundings.

If, then, the walls of a room are heated there will be no absorption of radiation from the body and hence there will be no sensation of cold. While pipes carrying steam or hot water are used for this purpose, electric heating elements embedded in the walls offer enticing possibilities. (F.T.M., L.R.Q.)

ELECTRIC IMAGE. Electrostatics.

ELECTRIC INDUCTION. Electrostatics.

ELECTRIC INSULATION. Any dielectric is an electric insulator, but experience has demonstrated the value of certain solids, such as glass, porcelain, rubber, mica, silk, paraffin, etc., for practical use. Oil and air also are often used where the cooling of the conductor is necessary or where very high voltages are employed. Insulating material is commonly applied to a conducting surface for the prevention of current leakage, and to protect materials and persons in the vicinity of the surface. Electrical insulation is sometimes applied continuously to a conductor in the form of a complete covering like the rubber insulation on wires; or else the insulation may be applied at regular intervals like the porcelain insulating supports on an electrical bus-bar. With solid insulators, mechanical strength is often a consideration; as are also incombustibility, flexibility, non-hygroscopic character, high surface resistance, the ability to withstand high temperatures, the possibility of being machined or molded, and moderate cost. Low dielectric constant would also be desirable in cases where distributed **capacitance** is to be minimized.

The primary requirement of an insulator, however, concerns its insulating strength, that is, the maximum voltage per unit thickness which the material will sustain without electric breakdown or sparkover. (This is quite apart from its resistivity.) In reckoning this with alternating voltages, the maximum or peak voltage must be used, which is $\sqrt{2}$ or 1.41 times the effective voltage at which the service is rated. (Thus the insulation on a 11,000-volt line must be able to sustain 15,600 volts.) Solid insulators are sometimes subject to surface leakage, or conductivity due to films of moisture or other impurities; as well as to true conductivity of the insulator itself. The latter is not serious except in the case of the mountings of electrometers, ionization chambers, etc., where quantitative accuracy is desired, and for which one of the best insulators is found to be sulfur.

Vulcanized rubber compound and varnished cambric are the principal materials for insulating wires. For standard low-voltage wiring, the rubber is usually about $\frac{3}{64}''$ thick. Rubber-insulated wires are covered with a protective braid. Slow-burning wires are insulated with layers of braid impregnated with fire-resistant compound. Fine wire used in low-voltage coils is commonly insulated by simply applying a suitable varnish or enamel directly to the bare wire. (L.D.W., F.T.M.)

ELECTRIC MOMENT. Electrostatics.

ELECTRIC MOTOR. Motor, Electric.

ELECTRIC NETWORKS. Kirchhoff's Laws of Networks.

ELECTRIC OSCILLATIONS AND ELECTRIC WAVES. The most important early researches in this field were carried out by Heinrich Hertz, who, about 1888, discovered that when an electric discharge takes place in a circuit having suitable inductance and capacitance, the resulting oscillations of the electricity therein give rise to radiation, usually some meters in wavelength. The existence of the radiation was proved by its inducing oscillations in a similar circuit set up at a distance. He found that this Hertzian radiation (as it is now called) can be reflected by metal surfaces and refracted by large blocks or prisms of dielectric material, just as light is reflected and refracted, and that it exhibits corresponding interference phenomena. The applications of Hertzian waves in radio are treated elsewhere. (See Radio Communication.)

The natural frequency of an oscillatory circuit or "Hertz oscillator" having resistance R (ohms), inductance L (henrys) and capacitance C (farads) is given by the equation

$$f = \frac{1}{4\pi} \sqrt{\frac{4}{LC} - \frac{R^2}{L^2}} \text{ (cycles per sec.).}$$

In using such formulae, the high-frequency values of R, L, and C must be employed, these being in general different from their low-frequency or steady-current values. The tuning condensers and inductance coils used in obtaining currents of desired frequencies are familiar to every radio amateur. High-frequency currents produced in this way range from 10 kilocycles per sec. for the longest radio waves (30 kilometers) to at least 75,000,000 kilocycles per sec. for the ultra short waves (4 mm.), detected some years since by Nichols and Tear and bordering on the infra-red. (See Electromagnetic Radiation.)

Electric waves may be propagated on long wires, somewhat as sound waves travel through a long tube. By terminating the wire in a small capacitance, the waves may be reflected and made to form interference nodes as do sound waves in an organ pipe. The wave length may be thus determined by what is known as the Lecher oscillator method, using two parallel wires.

It is often desirable to introduce into a communication line, such as a telephone circuit, a combination of inductances, or of inductances and capacitances, so designed as to form an effective barrier to currents of certain frequencies or frequency ranges, while others are allowed to pass through. Such arrangements, called wave filters, find many applications in modern electro-acoustic technology. (L.D.W.)

ELECTRIC POTENTIAL.

If a charge of electricity is moved from one region of space to another, it encounters, in general, electric forces which either help or hinder the transfer and which therefore add to or subtract from the potential energy of the charge. Let us suppose that a positive unit charge has been brought into the region A from a region so remotely beyond the borders of the material universe that no electric forces exist there. In general, a certain amount of work, V_A, has been done against the electric forces encountered on the way in; consequently V_A may be regarded as the potential energy which the unit charge has acquired in the process. This work per unit charge, V_A, is called the absolute electric potential of the region A. It is a scalar quantity, and may be either positive or negative; for example if A is in the vicinity of a large negative charge, the unit positive charge has been attracted and has done work, or lost potential energy, during its journey, and V_A is therefore negative. If a second region B is at a potential V_B, less than V_A, the unit charge would lose potential energy in moving from A to B, in the amount $V_A - V_B$. Thus, in an electrolytic cell, a positive ion migrates from the high-potential to the low-potential electrode, and does work in heating the solution. Negative charges tend to migrate from lower- to higher-potential regions; this is illustrated by the electrons in a wire.

The absolute zero of potential is of course that at an infinite distance from the universe; but for practical purposes an arbitrary zero is used, commonly that of the earth's surface (which is by no means constant); or that of some other large conductor, such as the metallic base or case of the apparatus being used. The ordinary unit of electric potential is the volt. (See Electromotive Force.) (L.D.W.)

ELECTRIC POWER.

Electric power is the product of electric current and electromotive force; that is, multiplication of current flowing by voltage forms the basis of the calculation of electric power. In a d-c circuit, the current measured in amperes, multiplied by the voltage between wires, is the power in watts. A thousand watts constitutes the kilowatt, a larger and more frequently employed unit of electric power.

The voltage and current may not be in phase with each other in an a-c circuit and, while the instantaneous power is the product of the instantaneous voltage and current, this out-of-phase relation causes the power to fluctuate between positive and negative values. Hence for the average power (which is usually what is desired) this factor needs to be taken into account in determining electric power in an a-c circuit, for it is only that component of the current which is in phase with the voltage that contributes to the average electrical power. The out-of-phase component produces the "wattless power." The power factor measures the fraction of the current that is in phase and available for true power. It is equal to the cosine of the phase difference between voltage and current. In a single-phase a-c circuit having current of I amperes, voltage of E volts, and power factor f, the true power is EIf watts. In a balanced three-phase circuit, it is $\sqrt{3} EIf$ watts. (See Alternating Current, Direct Current.) (F.T.M.)

ELECTRIC POWER TRANSMISSION.

If power could be generated for the same cost at any point in the country, there would be no need for electric power transmission. But since the larger the electrical generating unit, the lower the unit cost of production, there has developed the method, so extensively used at present, of generating electric current in large central stations. From this has arisen the need for electric power transmission to carry the energy so produced to the users. The system of distribution extending from the generating station to the user is of varying complexity, depending on the number of customers and their location relative to the plant. From the standpoint of economy in power transmission, it is desirable to have the plant near the center of the load served, but other factors, such as suitability of sites, proximity of fuels, real estate costs, water supply, etc., are prominent influences bearing on the location of the generating station.

The physical system of conductors, whereby the generators are connected with the customers' lines, may be separated into two parts. These are shown in the figure. The primary distribution system generally consists of a transmission line carrying three-phase a.c. from the switchboard of the plant to a substation located near the load served. A secondary distribution system extends from the substation to the customer's site. The primary system is generally known as the transmission line; the secondary system as the distribution network. The first is characterized by relatively high voltage; the second by medium and low voltage.

Energy may be transmitted electrically in overhead wires or underground cables as an electronic flow under pressure. The flow is measured in amperes. The voltage is the electric pressure. Electrical energy is proportional to the product of these two quantities. This energy can not be transmitted without some losses, the principal one being a resistance loss, which depends upon the current flow and the size of the wire. In transmitting a given amount of energy, this loss may be reduced by increasing the voltage, since any voltage increase will allow a decrease of current. Or, from another viewpoint, for a given amount of energy, and a given permissible loss, higher voltages will permit the use of smaller wires. The foregoing should serve to show that, where practicable, electric power should be transmitted at high voltages. For various reasons this is not feasible in the distribution networks, but is so on transmission lines. In consequence, the operating voltages of the latter range up-

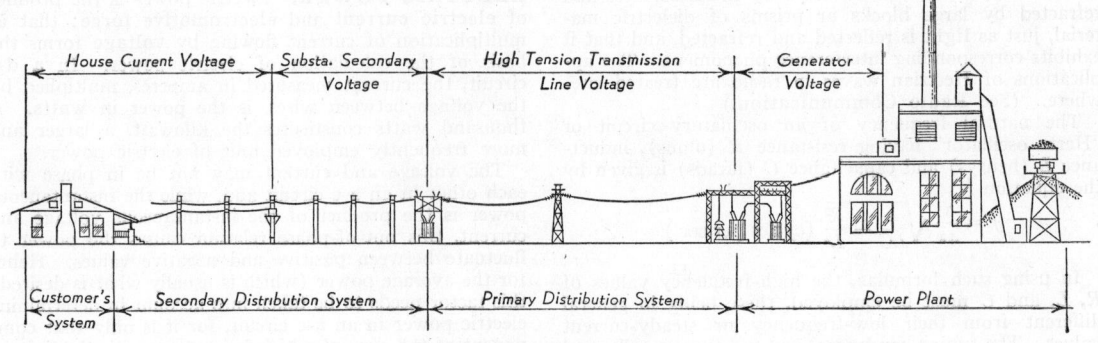

Elements of a power system.

wards from 6600 volts to nearly 300,000 volts. Commonly employed system voltages are 33,000, 66,000, 132,000, and 220,000. However, the higher voltages are strictly for overhead lines, cable transmission usually being limited to voltages lower than 40,000 volts.

These economic transmission voltages are higher than those under which power generating and utilizing apparatus can be operated, and a voltage transformation is essential if high transmission line voltages are to be used. By far the simplest and most effective means of accomplishing this is to employ a static **transformer.** The transformer may be used only with **alternating current,** and this goes far in explaining the prevalence of the a-c method of power transmission, because in many other respects high voltage d.c. is a superior means of transmission. As is shown in the diagram, the generator voltage is increased in the switchyard transformer to some economic transmission line value. The transmission line extending from station to load center may be of any length, as dictated by the geography of the locale; however, it does not commonly exceed 200 miles. Above some such distance the transmission line losses grow to where it becomes financially more attractive to build the generating station closer to the load and carry the energy to it by rail or water in the form of fuel. The longer lines are usually from hydro developments where the availability of water power offsets the line cost. At the receiving end of the transmission line, the voltage-decreasing transformer reduces the voltage to that which may be handled readily by the distribution network. This is often 2200 volts. At this voltage the energy flows through the network to the pole type transformers located close to the customer, or group of customers, it serves. At these transformers the voltage is reduced to 110 or 220 volts, in accordance with the customer's needs.

The energy loss occasioned by current flowing against the line resistance is not the only loss of energy in power transmission. The long stretches of parallel conductors have a capacitive effect, causing them to draw a current much as a **condenser,** even though the switches at the far end of the line are open. Furthermore, at high voltages, the air surrounding the conductors becomes partially ionized, and there exists a brush or **corona** discharge which represents a leakage of energy. The latter, as a matter of fact, is a limiting factor in the raising of electrical pressure on the transmission line and on the size conductor, and tends to offset the savings due to low current flowing when high voltages are used. Good transmission line design requires proper coordination of voltage, wire size, and line losses, so that the desired power may be transmitted at a minimum total annual cost.

When current flows on a long transmission line, the **inductance** of the line itself, plus its **capacitive** effect, combine with the **resistance** loss to give a voltage at the discharge end which varies with the load, even though the voltage is constant at the generator end. This feature is known as transmission line regulation, and is susceptible of prediction and analysis by theories developed by the electrical engineer. From the standpoint of the equipment served, it is desirable that this line regulation be offset so that lamps, heaters, etc., may operate on standard voltage. **Induction regulators** and tapchanging **transformers** are among the means employed to offset transmission line regulation. Furthermore, it is not difficult to demonstrate that the economy of power transmission varies as the square of the **power factor,** making it extremely desirable to operate the transmission line at unity power factor. On account of the induction characteristics of most electrical loads, and the capacitive effect of a transmission line itself, the current on a transmission line will tend to vary from lagging to leading, or vice versa with load, although customarily it tends to be lagging. For this reason, the synchronous motor is frequently employed for power factor correction, it being well known to have the electrical feature of drawing leading or lagging current, depending upon the amount of field excitation. Where this is not done, static capacitors are often added to improve the power factor.

The sheer physical extent of the ordinary transmission line makes it a likely victim of lightning, and no extensive transmission line could be successfully operated without adequate lightning protection. This may consist of **lightning arrestors** strategically placed, or an overhead grounded guard wire. Direct strokes are usually immediately dissipated by flash-over on the insulators, the lightning then finding its way down the pole to the ground. High-frequency induced waves (2000 to 5000 cycles) may not possess flash-over potential, and will travel along the line until relieved by an arrestor, or dissipated in resistance loss. Aside from lightning, a line should be protected against overload and short-circuits. This is commonly the function of transformer fuses, substation **circuit breakers,** and generator circuit breakers.

Turning from electrical to mechanical characteristics of electric power transmission, we find the material employed as conductor is universally either copper or aluminum frequently with a steel core for mechanical strength. The former is used in most cases, but the latter is in use for the very high voltage lines where, because of the larger diameter permitted by the lower density, it is effective in reducing corona loss. Except for very low-voltage distribution network lines, the wires are bare of insulation, and are carried by insulators of porcelain or glass of a design suitable for the voltage employed. Sometimes these insulators are mounted rigidly on the cross arm of the poles, but for the higher voltages they depend from the poles, and are known as suspension insulators. On high-voltage lines, each insulator is a chain of separate units, so that the voltage from line to pole may have a uniform gradient across the insulator. The distribution network and the low-voltage secondary transmission lines are generally supported by wooden poles having cross arms. The pole most commonly used at present is of Southern pine, well

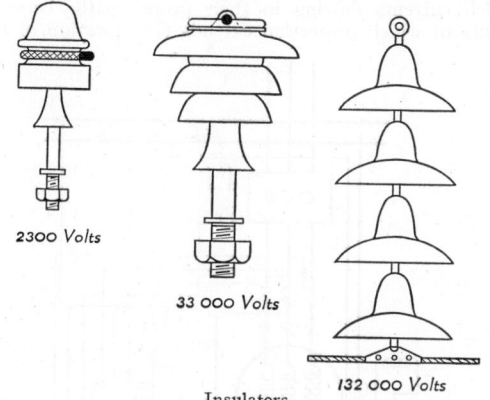

2300 Volts

33 000 Volts

132 000 Volts

Insulators.

creosoted. These poles should be sufficiently high so that, at the middle of the span between them, the bottom of the wire sag has a minimum clearance over the ground. As specified in the safety code governing the industry, this clearance varies with the voltage and with the surroundings; for example, in rural districts, for line voltages of 15,000 volts, 18′ clearance should be maintained.

The spacing of poles must be chosen with due regard for temperature and sag conditions. Larger sags in the wire between poles give lower stresses in the wire, and permit the use of longer spans, but at the expense of the use of taller poles, needed to maintain the minimum clearance previously mentioned. Consequently, the observer may see extremes of design representing individual designers' viewpoints, varying from relatively low, closely spaced poles, with little sag, to long spans where tall poles are spanned by wire having a much greater sag. At the same time, the effect of contraction caused by lowering of temperature in winter, and the loads suffered during storms, particularly sleet storms, must be guarded against. While high voltage lines are frequently constructed on exceptionally high wooden poles(i.e., 50–80′), the steel pole or steel tower is the more frequently used. A common arrangement for high-tension transmission lines is the four-legged steel tower, as diagrammed in the accompanying sketch. This tower is conveniently arranged to support two three-phase cir-

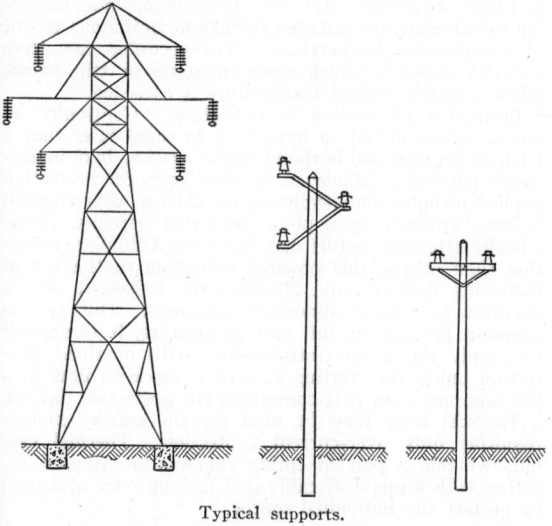

Typical supports.

cuits, and is frequently built as high as 100′, which permits extremely long spans because of the considerable amount of sag permissible. One hundred and ten thousand-volt transmission lines, so supported, are frequently run with no more than 6 or 8 towers per mile. (F.T.M.)

ELECTRIC RATES. An electric utility company derives its revenues directly from the customers it serves on the basis of monthly billings. The customers' meter readings are put into the rate structure and the amount due from that customer determined. The apparent simplicity of the process is misleading, for the establishment of the rate structure that will fulfill the requirements of a successful working rate is a matter of no inconsiderable difficulty. From the public's standpoint the rates should meet the following conditions:

1. Rate schedules should be simple. The problem of setting up a schedule that will fairly distribute the costs is aggravated by the necessity of its being comprehensible to the public as well as to the rate expert.

2. Rate schedules should be uniform over large territorial areas. There is much yet to be accomplished here. Persons in one community frequently are paying on one basis, and those in the neighboring community on another which is so different as to be unintelligible to the first.

3. Direct service from producer to consumer. This requires the elimination of the energy jobber, subcontractor, or middleman.

4. Distribution of costs in such a way that persons creating a desirable and relatively inexpensive type of load may enjoy the full use and benefit of electrical appliances.

Scientific electric rate-making might be said to have originated with Dr. John Hopkinson, an Englishman who lived in the last half of the nineteenth century. The Hopkinson rate theory was based on two charges, one a fixed annual charge per kilowatt of maximum demand, the other a small unit charge against each kilowatt-hour of energy used. Other early leaders in rate-making theory were W. J. Green and Arthur Wright in 1896, and H. L. Doherty in 1900.

The following elements enter into the cost of electrical energy to the consumer:

> Fixed element.
> Energy element.
> Variable load element.
> Customer element.
> Investors' profit.

The first of these is governed by the extent of plant investment and the current financial rates. It remains a fixed sum regardless of the amount of energy sold. The second is directly proportional to the plant output. The third, the variable load element, is governed by the characteristics of the load served, the load factor, the abruptness of the peaks, etc. The customer element will be proportional to the number of customers and nearly independent of both the plant investment and its kilowatt hour production. The profit is that which any sound normal business is expected to make. (F.T.M.)

ELECTRIC SCREENING. The experiments of Faraday revealed that any region completely enclosed by metal or other good conductor, however thin, is entirely free from electrostatic fields due to anything going on outside the enclosure. This is not because conductors are impervious to electric fields, but is due to the fact that the free electrons in the conducting shell surrounding the enclosure instantly adjust themselves so as to offset the effect of any electrostatic force that would otherwise penetrate the interior. Even a cage of fairly coarse screen wire is quite effective. Since electromagnetic radiation involves electric fields, we have here an explanation of why metals are opaque to it.

The screening effect of conductors is utilized in many kinds of electrical apparatus, as by enclosing electroscopes and the wires leading to them in metal cases or

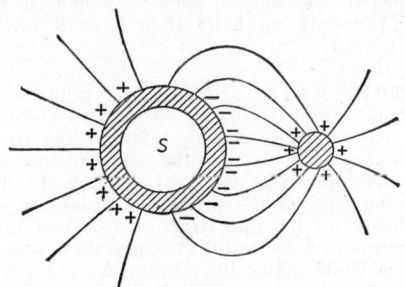

Showing induction and external field. Space S is completely shielded.

conduits, the placing of metal covers over radio tubes, etc. Whole buildings are sometimes covered with sheet iron to prevent induction sparks due to lightning from setting fire to inflammable contents, such as gasoline or explosives. (See **Electrostatics; Coaxial Cables and Wave Guides.**) (L.D.W.)

ELECTRIC SYSTEM PROTECTION. The protection of **alternating current** cable networks and overhead transmission lines is a large and important field of electrical power engineering. The need for protection should be obvious. The nature of electrical phenomena is such that trouble may develop and lead to considerable damage to equipment before attendants are aware of it or before they can perform the protective operations manually. Not only will failure to protect the electrical equipment endanger the equipment itself; what is more important, it may also endanger life. Failure of generating station equipment involves more than the cost of repairs to such. It may mean complete shutdown or restricted output, bringing loss of revenue, complaints, and ill will of customers. Though often complicated, the protective system is not excessively expensive compared to the value of the equipment it protects.

Protection is mainly against sustained overloads, high temperatures, and internal faults. In the case of **generators** or **transformers**, sustained overloads and high temperatures can usually be taken care of by the control room staff, guided by instruments indicating the magnitude of these abnormalities. They may be assisted by automatic relays giving audible or visual warning of these abnormalities when a predetermined value is exceeded. But upon internal fault developing, the equipment should be quickly and automatically disconnected. The internal faults may be short-circuited turns, open circuits, phase-to-phase short-circuits, or grounds.

The protective system should be satisfactory from the following points: 1. Selectivity. The protective system should be able to distinguish which portion of a sectionalized system is in trouble and to disconnect it only, leaving the remainder free to function normally. 2. Perception. It should distinguish between normal and abnormal conditions with a high degree of accuracy. 3. Sensitivity. It should not allow unduly large growth of the fault before the protective actions are completed.

The simplest protective apparatus is the fuse. The multitude of low-capacity, low-voltage power applications have made the fuse the most common protective device. All other protective systems are built around the circuit breaker.

The reactance of **alternator** windings or of added bus reactors is generally such that they can withstand a severe external fault such as a phase short-circuit without damage. Hence there is little value in protecting them against external faults. Alternator protection has resolved itself into protection against internal faults only, at least for large central station generators. Protection

against internal faults is ordinarily restricted to the **armature** circuit because of the low voltage of the field windings and the rarity of trouble therein.

A differential scheme of alternator protection is one in which currents flowing in their proper paths between points to which protection extends (i.e., location of the

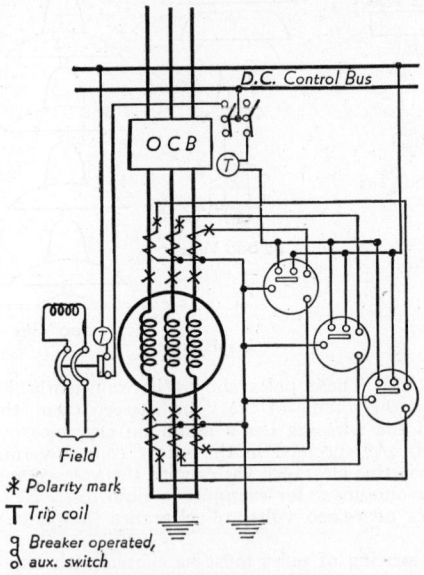

* Polarity mark
T Trip coil
ɋ Breaker operated,
aux. switch

Differential generator protection.

current transformers), are so balanced against each other as to leave the relays unaffected. A diversion of current from the proper path, as through a ground fault, will divert current through the relay and cause it to function so as to clear the apparatus in trouble. Note that the amount of current flow has no effect on the relays as long as it is confined in the proper path.

Transformers represent a large investment. They are essential equipment, they can hardly be installed in duplicate, and therefore an adequate protection system is a necessity. The transformer should be guarded against overload, excessive temperature, and internal faults. As in the case of an alternator, protection against internal faults should be positive and rapid in action. Overheating and overloading generally manifest themselves by the same symptom—rise of temperature. The transformer may be protected against such by alarms which will warn the attendants of high oil temperature, incorrect flow of cooling water, etc. In some instances recording instruments are installed to take a continuous record of transformer temperature. The recorder can have auxiliary contacts which close an alarm or trip circuit when a predetermined temperature is exceeded.

Protection of motors is important. Continuity of service is considered so important in some cases that a burnout on overload is risked rather than to take unnecessary trip-outs. However, in other cases the protection applied includes under-voltage, no voltage, over-current, heating, grounds, open phase, and short-circuits. Periodically attended motors may have much simpler protection than others, this possibly extending to the use of indicating devices only, allowing the judgment of the operator to handle abnormal situations. Due to the common practice of full voltage starting, it is difficult to supply the motor satisfactorily with overload protection unless the starting current is made to flow in a line separate from that containing the protective devices.

Thermal fuses may be used for the smaller motors requiring only over-current protection. Thermal cutouts wherein a pair of spring contacts are secured together with a metal of calibrated fusibility are also used to protect the individual motor.

Station bus short-circuits and grounds are not common occurrences, yet they may be so damaging both to station and connected network that every effort is put forth to confine the evil effects to the locality in trouble and to disconnect it from the rest of the system as rapidly as possible. For this purpose buses are sectionalized by reactors, by automatically operated oil circuit breakers, or by both, or, being non-sectionalized, they can be protected by over-current relays or by differential protection.

Circuit breakers are installed at strategic points in a transmission system to open the circuit and remove a faulty section of the system before damage can occur to the rest of the equipment. The operating relays of these breakers have a wide variety of possibilities and the type used in connection with any breaker depends upon the system, the particular part of the system, and just how it is to be protected. Thus a simple over-current relay with a breaker might be used to isolate completely an entire system from the generator and remove the voltage from the fault. However, in most power systems, as has been pointed out, it is highly undesirable to interrupt service, and so various sectionalizing breakers with special types of tripping relays are used. (F.T.M., L.R.Q.)

ELECTRICAL BRIDGE.

The electrical bridge is a term referring to any one of a variety of electric networks, one branch of which, the "bridge" proper, connects two points of equal potential and hence carries no current when the circuit is properly adjusted or "balanced." This is well illustrated by the **Wheatstone bridge** and the **Carey-Foster bridge** for measuring resistances. Among many other important special types may be mentioned the following: the decade bridge, of the Wheatstone type, in which the ratio coils are decimal multiples of an ohm; the percentage bridge, in which a change of one division on the slide-wire scale corresponds to a change of 1% in the ratio of the compared resistances; the Callendar and Griffiths bridge, a special adaptation of the Carey-Foster type; the Thomson (Kelvin) double bridge, having eight arms and used for comparing low-resistance standards; the Wien bridge for a-c capacitances, the Nernst high-frequency capacitance bridge, and the farad bridge which reads capacitances directly in farads; the inductance bridge and the Heaviside mutual inductance bridge; the resonance and the Maxwell bridges for comparison of inductance with capacitance; and the frequency bridge, resembling the Wheatstone bridge but used for the measurement of alternating current frequencies. For details, the reader should consult works on advanced electrical measurement. (L.D.W.)

ELECTRICAL CONDENSER.

An electrical condenser is an arrangement of conductors and **dielectrics** used to secure an appreciable capacitance, sometimes one of specified value. The essential feature of all condensers is a system of two or more conductors, separated by layers of dielectric. The potential difference between the conductors, when charged, is limited by the electric polarization in the dielectric. This makes it possible to accumulate large charges at comparatively small voltages. The oldest form of condenser is the **Leyden jar,** still often used where heavy electric discharges are desired. Many modern condensers consist of alternate metal and dielectric plates or sheets, sometimes of metal foil and paraffin paper strips rolled in a compact bundle. Condensers in which the dielectric is air, usually of adjustable capacitance, are much used in radio and other oscillatory circuits. An electrolytic condenser consists of an electrolytic cell in which a current has deposited a very thin layer of nonconducting material on one of the electrodes. This layer acts as the condenser dielectric. Such condensers have very high capacitance in proportion to their size and weight, and are much used in

electric filters and other electronic apparatus. Standard condensers, of accurately known capacitance, are employed in electrical measurements. The capacitance of a condenser depends upon the total area a and the thickness d of the dielectric and upon its dielectric constant k. If the dimensions are in centimeters, the capacitance for a condenser of flat-plates is approximately given in electrostatic units by the formula $C = ka/4\pi d$ and in microfarads by $C = 8.84 \times 10^{-8} ka/d$. Thus, if there are 21 metal plates 10 cm. sq., separated by 20 sheets of mica 0.01 cm. thick and of dielectric constant 6, the capacitance is about 0.106 microfarads. The capacitance is often made adjustable by varying the distance d or by arranging the plates to move past one another so as to vary the area a of dielectric subject to the electric field between them. (L.D.W.)

ELECTRICAL INSTRUMENTS.

Instrument, Electrical.

ELECTRICITY.

The electrification of amber by rubbing with wool or fur was observed many centuries ago. Not until the work of Volta, late in the 18th century, was electricity recognized through any but electrostatic phenomena, and investigations on the properties and applications of **electric currents** were among the most brilliant features of 19th-century physics. Even in the 1890's physicists were still asking, "What is electricity?" It had then long been known that an appropriate application of energy will separate electricity into two components, designated as positive and negative; that bodies charged with these components attract each other; and that the energy of separation is yielded upon the reunion of the two components. It remained for J. J. Thomson to recognize the **electron,** and for the recent analysis of **atomic structure** to identify the proton and the positron. As a physical magnitude, methods have been devised for measuring quantities of electricity, not only in terms of the elementary electronic charge, but in larger units determined by electrostatic, electromagnetic, or electrochemical effects. (See **Electric and Magnetic Units.**) Among the most important aspects of the subject are the magnetic properties of moving electricity (**electromagnetism**) and the incorporation of both components of electricity in the structure of all atoms and molecules of matter. (See **Electrostatics.**) (L.D.W.)

ELECTROCAPILLARITY.

The **surface tension** between two conducting liquids in contact, such as mercury and a dilute acid, is sensibly altered when an **electric current** passes across the interface. As a result, when the contact is in a capillary tube, the pressure difference on the opposite sides of the meniscus is affected by a current traversing the capillary column, to an extent dependent upon the direction of the current across the boundary. This has been utilized in different forms of capillary electrometer. In the Dewar type, two small

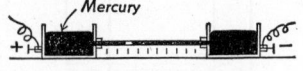

Capillary electrometer.

vessels of mercury are joined below the mercury level by a horizontal capillary tube, the mercury in which is interrupted by a short space filled with dilute acid. Upon applying a small potential difference to the two bodies of mercury, the equilibrium is disturbed and the drop of acid moves toward the low-potential end until the resultant capillary pressure is balanced by the hydrostatic pressure of the mercury. Since this effect is approximately proportional to the potential difference, the apparatus serves as a sensitive indicator for potentials of a few hundred millivolts. (See **Capillarity.**) (L.D.W.)

ELECTROCARDIOGRAM. A tracing on photographic film of the pathway traveled by the wave of electrical activity which sets off the contraction of the heart muscle with each beat of the heart. The electrocardiogram is a valuable diagnostic aid in heart disease.

Normally, the electrical current sweeps down from the heart's pacemaker in peristaltic waves over the auricles and by special conduction tracts and fibers into the ventricles. The electrocardiograph, which is essentially a string **galvanometer** with a photographic attachment, records this normal pathway as a series of waves or deflections of definite pattern. Deviations from the normal pattern occur in the form of changes in rate and rhythm, and in abnormalities of shape, direction, and amplitude of the individual waves. Such changes are often indicative of heart disease. **Arteriosclerosis** of the heart vessels and **coronary occlusion** result in heart muscle damage which may block the normal pathway and the electrical impulse must therefore pursue a different course in traversing the heart. Various disturbances in rhythm may follow interference with the normal conduction system. These are commonly associated with arteriosclerotic and **rheumatic heart disase.** (R.S.M.)

ELECTROCAUTERY. A cautery consisting of a platinum wire in a holder, which may be heated to any desired degree when electric current is supplied. (R.S.M.)

ELECTROCHEMISTRY. When substances are subjected to the **electric current** different results are observed, depending upon the nature of the substance. Solids such as metals and certain forms of non-metals, e.g., graphitic carbon, conduct the current without undergoing chemical change, whereas solutions of **salts, acids, bases** in water, and fused salts, acids, bases undergo chemical change at the **electrodes** upon passage of electric **direct current.**

The extent of chemical change is dependent in each case upon the amount of current which passes. One equivalent (see **Chemical Composition**) of **element, compound or radical**, is liberated per 96,500 coulombs of current (1 coulomb equals 1 ampere-second) (Faraday, 1833), at each electrode. The positive electrode *in the solution* is called the anode, and the reaction at the anode occurs by loss of negative charge of the anion. The negative electrode *in the solution* is called the cathode, and the reaction at the cathode occurs by loss of positive charge of the cation. The process itself is called electrolysis.

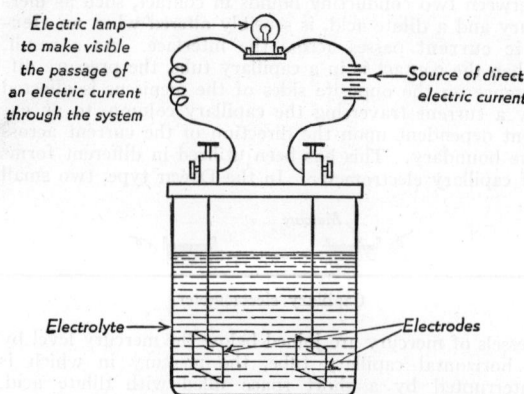

Diagram showing electrolysis of solutions.

In electrolysis of dilute solutions of the alkali and alkaline earth (Groups I an II, see **Chemical Composition**) salts or bases, oxygen (1 volume) is evolved at the anode and hydrogen (2 volumes) at the cathode. This is equivalent to the electrolysis of water. When the above salts are **chlorides** and the electrolyte not dilute, the reaction is more complicated at times. More or less **chlorine** is evolved with or without **oxygen,** according to the composition, concentration and mixing of the solution, the temperature, and the substance of which the anode is composed. Thus, with **sodium** chloride solution (1) chlorine is separated at the anode and **sodium** hydroxide at the cathode when a semi-porous diaphragm is used to separate mechanically but not electrically the anode and cathode compartments, (2) sodium **hypochlorite,** when the solution without a diaphragm is mixed by rapid stirring, (3) sodium **chlorate,** when the warmed solution without a diaphragm is mixed and the current is passed a long time, and, with sodium chloride fused, sodium metal at the cathode and chlorine at the anode. Graphite or amorphous carbon is used as a resistant anode, iron or nickel as cathode, and asbestos cloth as a diaphragm.

In electrolysis of many metallic salts, other than those of the alkali and alkaline earth metals, the metal is deposited at the cathode. **Copper or lead** is refined or deposited on another metal or on graphite by electrodeposition, **nickel, silver, gold, zinc, cadmium** or **chromium** is deposited in electroplating of objects, and **aluminum, magnesium, sodium** or **calcium** is produced by electrolysis of a fused electrolyte.

The electrical conductivity or conductance of an electrolyte depends upon the composition and concentration of the electrolyte, and on the temperature. Its value is expressed as reciprocal ohms, or mhos, and the equivalent conductivity is the conductivity of one equivalent weight of the substance at the given concentration and temperature when measured in a cell having two parallel platinized electrodes 1 cm. apart. In practice, the conductivity cell has two **platinum** electrodes each about 1 sq. cm. in area and is calibrated by the use of a standard solution.

ELECTRICAL CONDUCTIVITY OF POTASSIUM CHLORIDE SOLUTION AT 18° C.

GRAM MOLS PER LITER OF SOLUTION	CONDUCTIVITY (MHOS)	DIFFERENCE IN CONDUCTIVITY BETWEEN SUCCESSIVE DILUTIONS (MHOS)
1	98.2	
0.1	111.9	13.7
0.01	122.5	10.6
0.001	127.6	5.1
0.0001	129.5	1.9

The equivalent conductivity at zero concentration or infinite dilution is obtained by extrapolation of the curve plotted from the above values, and is in the case of **potassium** chloride at 18° C., 130.1 mhos. For **hydrochloric acid** the value is 379.5 and for **potassium** hydroxide 236.6.

The equivalent conductivity for each of the common anions and cations has been determined:

Hydrogen (H^+)	314	mhos
Sodium (Na^+)	43.5	
Potassium (K^+)	64.4	
Ammonium (NH_4^+)	64.5	
Hydroxyl (OH^-)	172	
Chloride (Cl^-)	65.5	
Nitrate (NO_3^-)	61.7	
Sulfate ($\frac{1}{2}SO_4^{--}$)	68	
Acetate ($C_2H_3O_2^-$)	34.6	

From these values the equivalent conductivity of acids, bases, salts may be ascertained, and in the case of electrolytes of low conductivity, for example, **acetic acid** and **ammonium** hydroxide, furnishes the accepted values.

The conductivity depends upon the velocity of each **ion** and upon the viscosity of the solution. Since the total current passing a given cross-section of the solution is carried partly by anions and partly by cations, the more rapidly moving ions (anions or cations, as the case may be) carry more than half of the current. The fraction of the current carried by anions and by cations is proportional to the relative speed of each group. The result is that the faster ions crowd around their electrode to a greater degree than the slower ions around their electrode—the discharge at the electrode is strictly proportional to the current passing and is equivalent at each electrode. Under a potential gradient of one volt per cm. at 18° C. the speed of ions is as follows:

H^+	10.8	cm. per hour
Na^+	1.26	" " "
K^+	2.05	" " "
OH^-	5.6	" " "
Cl^-	2.12	" " "
NO_3^-	1.91	" " "

These values are practically in the same ratio as the equivalent conductivities of the ions.

The modern theory of electrolytic dissociation was announced by Arrhenius (1887) and extended by Milner (1912), Debye and Huckel (1923) and Onsager (1926). In the case of strong **electrolytes** like salts, **hydrochloric, hydrobromic, nitric acids, sodium, potassium,** quarternary **ammonium** (not ammonium), **calcium, barium** hydroxides, the dissociation into ions is complete. Changes in conductivity with changing concentration are due to the electrical effects of the ions upon each other. Each ion of positive or negative charge is surrounded in the solution by an ionic atmosphere of negative or positive charge, respectively. Weak electrolytes include ammonium hydroxide and many organic bases (see **Amines**), and practically all of the organic and some inorganic acids. These are slightly dissociated into ions.

Conductivity ratios at 18° C. of some weak acids are as follows:

Acid	Concentration	Ratio of Conductivity at Given Concentration to Conductivity at Zero Concentration at 18° C.
Carbonic.......	0.1 molar	0.002
Acetic..........	1.0 normal	0.004
Hydrosulfuric...	0.1 molar	0.0007
Hydrofluoric....	1.0 normal	0.07
Boric..........	0.1 molar	0.001

Such acids are slightly ionized and the values given in the last column above represent practically the equilibrium fraction ionized at the corresponding concentration and temperature thus:

$$\text{Non-ionized Acid} \rightleftarrows H^+ + Anion^-$$

$$\text{Concentration: } \frac{1-x}{v} \qquad \frac{x}{v} \qquad \frac{x}{v}.$$

Where 1 is the original amount, $1 - x$ the amount nonionized under the conditions, and v the volume of solution.

Ostwald showed that, in each case of a weak electrolyte

$$\frac{\frac{x}{v} \times \frac{x}{v}}{\frac{1-x}{v}} = \frac{x^2}{1-x} \times \frac{1}{v} = \text{Constant}.$$

This constant is called the ionization constant of the acid or the base. (See **Acids, Bases, Salts.**) The less the ionization of the acid (or base), the nearer the constant approaches in value the product of the concentration of the two ions.

The intensity aspects of electrochemical reactions are discussed in the article on **Reactions Involving Oxidation-Reduction.** (R.K.S.)

ELECTRODE. In an electric circuit, part of which is composed of other than the usual conductor of copper, or other metal, the terminal connecting the conventional conductor and the conducting substance is an electrode. Examples of electrodes are to be found in the electric **battery**, where they dip in the electrolyte; the **electric furnace**, where the electrodes connect the external circuit with the heating arc, and in the electric chair, where the electrodes connect the external shocking circuit with the body. For electrode potentials, see **Reactions Involving Oxidation-Reduction.** (F.T.M.)

ELECTRODE POTENTIAL. Reactions Involving Oxidation-Reduction.

ELECTRODELESS DISCHARGE. There are two ways in which a current may be maintained in a rarefied gas without the introduction of **electrodes** into the gas. 1. A tube containing the gas may be placed between external metal plates having a rapidly alternating, high-potential difference. The tube then acts as the **dielectric** in a **condenser**, and the gas may become luminous with a discharge across the tube similar to that with internal electrodes. 2. The tube may be surrounded by a **helix** through which a high-frequency current is passing. In this case the luminosity takes the form of a ring, inside the tube, coaxial with the turns of the helix. This is due to the alternating electric intensity induced by the current in the helix. The discharge has the characteristics of the positive column in an ordinary discharge tube, except that it forms a closed ring. Striations sometimes appear, in radial planes. If the oscillations in the helix are intense, the ring discharge is confined to the space immediately inside the tube wall; if less so, it extends farther inward. The discharge is facilitated by **ultra-violet** radiation traversing the gas. Volatile impurities in the tube, such as sulfur or phosphorus, impair the discharge and may stop it. With some gases there is a distinct **phosphorescence**, called the "afterglow," persisting for some seconds after the helix oscillations cease. There is evidence that this effect is associated with impurities and is of chemical origin. (See **Ionized Gases.**) (L.D.W.)

ELECTROENCEPHALOGRAM. A record of the electrical potentials of the brain, which appear as waves of characteristic rate, amplitude, and rhythm. Abnormalities of these pictures aid in the diagnosis of such conditions as **epilepsy** and brain tumor. (D.M.H.)

ELECTROLYSIS. The process of decomposition of an **electrolyte** by the passage through it of an **electric current** is electrolysis. In electrolysis, the body of the electrolyte must surround two **electrodes**, one called the anode, the other, cathode. Current flows from the anode to the cathode, and the dissociated parts of the electrolyte are liberated at the surface of the electrodes, that moving toward the anode being the anion, and, to the cathode, the cation. For example, a solution of copper sulfate, commonly called blue-stone, is an electrolyte. When a current is passed through it between two electrodes, the copper sulfate is broken down into metallic copper, which is deposited on the cathode in the form of a copper plate. The molecule of water is caused to liberate its oxygen, which appears at the anode, while the hydrogen unites with the sulfate iron to form a solution of sulfuric acid. Electrolysis of a damaging, uncontrolled nature sometimes occurs in buried water pipes or the reinforcement of concrete. It originates from stray lighting or traction currents. A damp soil is essential to action of this nature. (See **Electrochemistry; Reactions Involving Oxidation-Reduction.**) (F.T.M.)

ELECTROLYTE. This term is commonly applied to substances which either in the molten state or in solution conduct electricity by transfer of **ions**. The term is not applied to the metals in elementary form. Another usage is to apply the term electrolyte to the conducting solution itself. The more important electrolytes are solutions of salts, acids, or bases, usually in water. (See **Acids, Bases, and Salts**; and **Electrochemistry**.) (R.K.S., F.T.M.)

ELECTROLYTIC DISSOCIATION THEORY.
Reactions Involving Recombination of Ions.

ELECTROMAGNET. A magnet whose field is produced by an electric current, and which is largely demagnetized upon cessation of the current, is an electromagnet. In order to obtain the strongest field possible, highly permeable soft iron or steel is employed for the **core** of electromagnets. In an electromagnet the current flows through a solenoid, which is a conductor wound in the form of a **helix**, and which produces a strong magnetic field coaxial with the helix. The core is placed inside the helix in order to give a magnetic path of the least reluctance. Electromagnets are found in a number of different forms, such as the plain solenoid with cylindrical core, or the horseshoe electromagnet, much used in electric bells, telegraph instruments, and telephones. Very powerful electromagnets are often used to move masses of iron, such as scrap iron, and have the advantage that the loading or unloading of the crane to which the magnet is attached is simply a matter of applying or disconnecting the electric current.

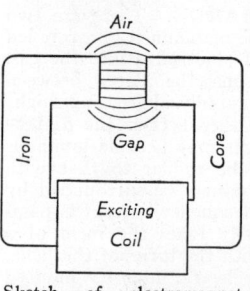

Sketch of electromagnet having two poles; one of a large variety of designs.

Typical electromagnet, carrying load. (*Electric Controller & Manufacturing Co.*)

For practical calculations we may write the Bosanquet law $\phi = \mathfrak{M}/\mathfrak{R}$ for the flux in a **magnetic circuit** in more useful approximate form:

$$\phi = \frac{nI}{5l/2\pi\mu a} = \frac{nI}{0.796l/\mu a}.$$

This refers to a simple, closed magnetic circuit of length l (cm.), uniform permeability μ, and uniform cross-section a (sq. cm.), excited by a magnetomotive force

nI (ampere-turns). The denominator is the reluctance of the circuit. If either the permeability or the cross-section is not uniform, this reluctance must be separated into parts, each having its own value for l, μ, and a. For example, let us calculate the ampere-turns nI required to produce an induction of 2200 gausses in the air between the poles of an electromagnet having an iron core of permeability 1500 and cross-section 40 sq. cm., the distance around through the iron being 78 cm. and the single air gap 6 cm. wide. The required flux is 40 sq. cm. $\times$ 2200 gausses = 88,000 lines or maxwells. The reluctance of the iron part of the circuit is 0.796 $\times$ 78 cm. $\div$ 1500 $\times$ 40 sq. cm. = 0.001 reciprocal cm., and that of he air gap, 0.796 $\times$ 6 cm. $\div$ 1 $\times$ 40 sq. cm. = 0.119 reciprocal cm. (taking the permeability of air as 1); so that the total reluctance is 0.12 reciprocal cm. That is, 88,000 maxwells = $nI \div$ 0.12 reciprocal cm., or nI = 10,560 ampere-turns. The result, though only roughly approximate, is sufficiently accurate for practical purposes. It is to be noted that since the air gap contributes by far the larger part of the reluctance, the flux is very sensitive to variations in its width. (F.T.M., L.D.W.)

ELECTROMAGNETIC CONSTANT. The speed of propagation of electromagnetic waves such as **light**, in a vacuum, commonly denoted by c, appears so frequently in physical formulae that it has become one of the most important of all physical constants. This is especially true since the advent of multitudes of formulae of **relativity** involving this factor. The precise determination of its value has justly, therefore, engaged the attention of the ablest experimenters.

Early attempts to measure the speed of light failed from lack of any adequate conception of the magnitude of the quantity to be measured. The first inkling of the truth was arrived at by Roemer in 1675, when he surmised that certain irregularities in the observed recurrence of the eclipses of Jupiter's satellites were due to the fact that time is required for the light to traverse the varying distance from Jupiter to the earth. The value thus calculated was not far from those obtained by the best experimental methods. The first direct determination was made by Fizeau (1849), who sent out a beam of light between the teeth of a rapidly revolving cogwheel, to be reflected by a distant mirror and returned in time to be stopped by the next tooth. Foucault (1854) improved upon this by using a revolving mirror, the beam finally reflected from which, after its return from a distant fixed mirror, was deviated because of the slight rotation of the revolving mirror during the journey of the light out and back. This method was later perfected and refined by Michelson and others. Michelson's final value (1930) was 299,772 kilometers per sec. The most probable value, as of 1941, is given by Birge as 299,776 kilometers per sec.

Aside from direct determinations of c as the speed of light, the constant has been deduced from measurements of standing electric waves on wires (Mercier, 1923), the result of which was 299,782 kilometers per sec.; and from the measured ratio of the abcoulomb to the c.g.s. electrostatic unit charge, the numerical value of which (according to Rosa and Dorsey, 1906) is 2.99781×10^{10}, giving $c = 299,781$ kilometers per sec. A careful comparison of the results of the direct and the indirect methods by Edmondson (1934) and others has led to the suspicion that the former are subject to some unexplained periodic influence, possibly associated with changes in the long base lines used in the measurements. The agreement and constancy of the indirect results are considerably better. (L.D.W.)

ELECTROMAGNETIC FIELD. It is commonly stated that a wire carrying an **electric current** is surrounded by a **magnetic field** whose lines of force are circles with the wire as their axis. This statement implies that the magnetic field is directly traceable to the moving electricity in the wire. There is, however, another aspect of the matter. Each electric particle projects into space a radiating field of electric force, whose lines may be thought of as bristling out from the wire like the fur on the tail of an angry cat; and as the particles move along the wire the lines of force move with them. According to the theory of Maxwell, it is the motion of these lines of electric force that sets up the magnetic field transverse to them. More generally, a variable electric field, with moving lines of electric force, is always accompanied by a transverse magnetic field; and conversely, a variable magnetic field, with moving lines of magnetic force, is accompanied by a transverse electric force. The joint interplay of electric and magnetic forces here described is what is called an electromagnetic field, and is considered as having its own objective existence in space apart from any electric charges or magnets with which it may be associated. An essential feature of the theory is that this process, whatever it is, represents a flow of energy at right angles to both electric and magnetic components. The flux density of this energy (corresponding to the intensity of radiation) is represented by what is known as the Poynting vector. Electromagnetic radiation is, on this theory, the propagation of these electric and magnetic stresses through space with the speed of light, somewhat as the much slower waves of elastic stress are propagated through steel. The conditions in an electromagnetic field are expressed mathematically by the well-known Maxwell's equations.

When an electric charge is set into motion, it builds about itself an electromagnetic field, and this implies a distribution of energy throughout space. The density of this energy at any point of the field is proportional to the product of the electric and magnetic vector components and the sine of the angle between them (vector product). The total field energy can be obtained by suitable integration, and is greater than that of the purely electric field of a stationary charge. Maxwell's theory treats this excess as kinetic energy, thus endowing the moving charge with an "electromagnetic mass" and an "electromagnetic momentum" inherent in its electrical character. That this may be the nature of all mass and momentum at once suggests itself when we consider the electrical constitution of matter, discovered since Maxwell's time. (L.D.W.)

ELECTROMAGNETIC INDUCTION.

Probably the most noteworthy of the many scientific contributions of the renowned Michael Faraday was his discovery in 1831 of electromagnetic (or more logically, magneto-electric) induction. As exhibited in the usual experimental arrangements, this phenomenon is the setting up, in a circuit, of an electromotive force by reason of the variation of the magnetic flux linked with the circuit; the magnitude of that electromotive force being, as Faraday found, proportional to the rate at which the flux through the circuit, or the "linkage," varies. If the flux linkage with the circuit, in maxwell-turns, is expressed by $N\phi$ (the actual flux, ϕ, times the number of turns, N), the electromotive force generated by its variation, in volts, is:

$$E = \frac{N}{10^8} \frac{d\phi}{dt}.$$

(See Volt.) The electromotive force is positive (counterclockwise) when $\frac{d\phi}{dt}$ is positive, that is, when the flux is increasing, negative when it is decreasing; as viewed by one looking in the direction of the magnetic induction.

Another aspect of the matter is that if a conductor moves through a magnetic field, or if a magnetic field sweeps over a conductor, in such a way that the conductor cuts across the lines of force, the electricity in the conductor experiences forces at right angles to the field and to the (relative) motion. More general still is the Maxwell concept that when magnetic lines of force move sidewise, their movement results in an electric field at right angles to the magnetic lines and to their motion. (See Electromagnetic Field.)

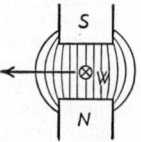

Wire (W) moving to the left across an upward magnetic field has induced in it an e.m.f. away from the observer.

Faraday's discovery was almost accidental. Happening to thrust a bar magnet into a coil connected with a galvanometer, he noted a momentary deflection of the needle. If the north pole is thrust downward into the coil, so as to increase the flux linked with the coil, the current will be counterclockwise as viewed from above, and reverses on drawing the magnet out again.

The far-reaching consequences of this simple observation can hardly be overestimated. It was the forerunner of the invention of electric generators and alternators, of the original Bell telephone, of the induction coil, of the transformer, of the induction motor, of magnetic damping devices, and of many other electric appliances. It is the basis of Lenz's law and of the Wilson experiment, and the explanation of eddy currents. The volt and the henry are definable in terms of it. The phenomenon is called mutual induction when the variation of current in one circuit causes a variation of magnetic flux through, and hence an electromotive force in, another coupled circuit. It is a curious fact that in a vacuum the linkage through one circuit A due to unit steady current in a neighboring circuit B is equal to the linkage through B due to unit current in A; hence the mutual inductance of two circuits is the same, whichever is the primary circuit. If the circuits are closely coupled and have high self-inductance, and if an alternating electromotive force E be applied to A, the resulting a.c. induces an electromotive force in B approximately equal to $\frac{n_B}{n_A} E$; in which N_A and N_B represent the numbers of turns in the respective circuits. The principle is utilized in induction coils and in potential transformers. (L.D.W.)

ELECTROMAGNETIC MASS. Electromagnetic Field.

ELECTROMAGNETIC RADIATION.

A century ago light was believed to be a transverse wave motion in an ether behaving like an elastic solid. Intensive study of the electromagnetic field, however, led James Clerk Maxwell (1873) to substitute electric and magnetic for elastic forces in the theory of light propagation. According to his view, light is the result of vibrating electric charges. These set up alternating electric and magnetic fields at right angles to each other and to the direction of propagation, which pass on the energy from one portion of the ether to the next as an electromagnetic wave. (Poynting's theorem states that the rate of this energy transfer is proportional to the product of the electric and magnetic intensities.) The theory was very successful in explaining many of the electrical and magnetic properties of light, such as the Faraday effect and the Kerr effect.

Maxwell suggested that it should be possible to produce waves of much longer wavelength than light by causing electricity to oscillate in a conductor. This was a forecast of the Hertzian radiation, upon which radio transmission depends, and which exhibits many of the characteristics of light, such as reflection, refraction, diffraction, interference, polarization, etc., but on a gross scale. Other researches showed that the infra-red and ultra-violet radiations have these same properties. It was therefore natural to classify them together as the same phenomenon in different frequency ranges. X-rays for a time could not be identified with this group, but their diffraction by crystals finally demonstrated their wave character and they now take their place next to the ultra-violet. Meanwhile the quantum theory put a new aspect on the whole matter, and incidentally added the gamma rays to the radiation family. The table below gives the approximate wavelength ranges of these various types of radiation. It is quite probable that the very penetrating cosmic rays contain radiation of still higher frequency than the gamma rays. (See also Thermal Radiation.)

TABLE OF ELECTROMAGNETIC RADIATION

Type of Radiation	Approximate Range of Wavelength (in Cm.)
Gamma Rays	10^{-10} to 10^{-9}
X-rays	10^{-9} to 10^{-7}
Ultra-violet	10^{-7} to 4×10^{-5}
Visible Light	4×10^{-5} to 7.7×10^{-5}
Infra-red	7.7×10^{-5} to 0.4
Hertzian Radiation	0.4 to 3×10^{6}

(L.D.W.)

ELECTROMAGNETIC UNITS. Electric and Magnetic Units.

ELECTROMAGNETISM. The pioneer discovery of the magnetic effect of the electric current was made by Oersted at Copenhagen in 1820. In experimenting with battery currents, he happened to bring a compass needle near a wire in which there was an electric current, and noted that the needle was deflected. Such a wire is surrounded by a magnetic field so that, to one looking along the wire in the direction from the positive to the negative battery-terminal (the so-called "direction of the current"), the direction of the field, as indicated by the north pole of the compass needle, is clockwise (Ampère's rule).

If the wire carrying the current is placed in a magnetic field perpendicular to its direction, this field reacts with that due to the current in such a way as to give the wire a lateral thrust, perpendicular to both the wire and the field in which it is placed. For a wire of length l (cm.) carrying current I (amperes) and placed across a field of intensity H (oersteds), this lateral force is given by the equation $f = HIl/10$ (dynes); which follows from the definition of the ampere. An electric motor is driven by forces thus produced.

If the wire is bent into a circular loop of radius r (cm.), still carrying current I (amperes), there is produced at its center, perpendicular to the plane of the loop, a magnetic field of intensity $H = \pi I/5r$ (oersteds). This, and the statement in the preceding paragraph, may be shown to be interdependent. If more loops are added, forming a coil of n equal turns close together, the resulting field is n times as great. By winding the n turns along a cylinder, forming a "helix" of radius r and axial length a, one obtains something greatly resembling a bar magnet, the ends of the helix corresponding to the poles. The field intensity at the center of the axis of this helix (without any core) is

$$H_o = \frac{2\pi nI}{5\sqrt{4r^2 + a^2}} \quad \text{(oersteds)}.$$

If we now insert an iron core, we have an electromagnet, and the helix supplies the magnetomotive force nI ampere-turns for a magnetic circuit composed partly of iron and partly of air.

More general calculations of electromagnetic effects are based upon Ampère's law, the Biot-Savart law, and Maxwell's equations. (L.D.W.)

ELECTROMETERS. Electroscopes and Electrometers.

ELECTROMOTIVE FORCE. Various means have been discovered whereby electricity may be propelled against any opposition such as the resistance of a conductor or the electromagnetic reaction of a motor. Thus we have batteries, generators, thermels, photovoltaic cells, etc., each of which has this electrical driving influence called electromotive force. The process of driving electricity against opposition requires the expenditure of energy. The electricity must have work done on it, and it, in turn, does work on something else. In an automobile headlight circuit, the electricity is able to do work on the lamp because, as it passes through the battery it receives a supply of potential energy. The electric potential where it emerges from the battery is therefore different from that at the entrance, and it is for this reason that there is a current through the lamp. The measure of the electromotive force of the battery is the work done on each unit of electricity as it passes through the battery. If this work is one joule per coulomb, the electromotive force is one volt. Frequently electromotive force is identified with difference of potential. This is permissible when the electricity gives up none of the energy while the work is being done, so that the resulting potential difference does actually correspond to the accession of energy. But the terminal potential difference of a generator and its electromotive force are in general not equal. (See Electric Circuits.) And it would be perfectly possible to arrange a circuit in which the electricity does work as fast as it receives energy and therefore moves along at constant potential; this is the case, for example, in a symmetrically placed, circular transformer secondary of one closed turn. (L.D.W.)

ELECTROMOTIVE SERIES. Reactions Involving Oxidation-Reduction.

ELECTRON. The first clear indication that electricity is composed of equal, elementary charges was afforded by electrolysis. (See Charge-Mass Ratio.) This gave, as the charge carried by a chlorine ion, 1.6×10^{-19} coulombs (negative); for the hydrogen ion, 1.6×10^{-19} coulombs (positive); for oxygen 3.2×10^{-19} coulombs (negative); for aluminum, 4.8×10^{-19} coulombs (positive); etc. We here have a common factor 1.6×10^{-19} coulombs or 4.8×10^{-10} electrostatic units. Sir J. J. Thomson (1897) succeeded in making a rough measurement of the charges carried by the ions of gases. Millikan used a much more precise method. It consists in measuring the speed with which a minute liquid drop is dragged along in an electric field when it picks up one or more stray air ions; the force acting on it being then deduced from the speed, and the charge, in turn, calculated from the force and the field intensity. He found this charge to be always a multiple of 4.77×10^{-10} electrostatic units—the same as that revealed by electrolysis. The most probable value at present (1945) is 4.8025×10^{-10} e.s.u.

Cathode rays and the beta rays from radium, when subjected by Thomson to measurement by a type of mass spectograph method, were found to be composed of particles having a negative charge of this value and a mass of about 9×10^{-28} gram. This mass is only about 1/1847 that of the lightest atom known,—hydrogen. For many years, no other particle was discovered having a mass comparable with it. The name "electron," suggested by Johnstone Stoney, is now given to any negative particle of this charge and mass, regardless of its origin. There is every evidence that not only are cathode and beta rays composed of electrons, but that swarms of these particles pervade all matter and play an important part in many physical processes such as electric conduction, thermal conduction, magnetism, photoelectric phenomena, thermionic phenomena, the emission of light, etc., as well as in chemical reactions.

In recent years it has been necessary to modify the concept of the electron as a mere electric particle and to recognize that it has wave characteristics, with a frequency and a wavelength. According to **wave mechanics**, an electron traveling with speed v is associated with a "de Broglie wave" train of wavelength $\lambda = h/mv$ (in which m is the electronic mass and h is Planck's constant), traveling with a speed greater than the speed v of the electron or even that of light, c, and equal to c^2/v. This concept gives the same frequencies for radiation emitted by atoms as the older Bohr theory. (See **Davisson-Germer Experiment**.) (L.D.W.)

ELECTRON GUN. Cathode-Ray Tube; Electronics.

ELECTRON LENS. Cathode-Ray Tube.

ELECTRON MICROSCOPE. The most outstanding achievement of electron optics (see **Electronics**) is the electron microscope. This instrument may be defined simply as a **microscope** in which streams of **electrons** function in essentially the same way as rays of light do in an ordinary optical microscope. When it is recalled that moving electrons exhibit properties characteristic of trains of waves (see **Wave Mechanics**), the analogy becomes still more striking; while the very short wavelength associated with electron waves serves to give the electron microscope extraordinary resolving power, thus making available **magnifying powers** far beyond the reach of any optical instrument.

The electron-optical system of this new microscope, as now usually constructed, consists of an electron gun (corresponding to a source of light) and a series of magnetic electron lenses which serve as condenser (to converge the rays on the specimen), objective, and projecting lens. Of course there is no eyepiece, since electron rays are not visible. So they must be made to form an image by projection, either on a fluorescent screen, which may be viewed through small windows in the case, or on a photographic plate, where the image is permanently recorded.

Magnetic lenses are of different types, and usually consist of a coil with iron pole pieces, so shaped as to give a "stray" field having lines of force with radial components symmetrically disposed about the axis. Unlike light rays, electron rays directed by magnetic lenses move in helical paths, so that the image is sometimes rotated many degrees with respect to the specimen. Focusing is accomplished, not by longitudinal movement as in optical instruments, but by control of the currents in the magnetizing coils, thus varying the focal lengths of the lenses. This accomplishes the result with great nicety.

Direct magnifications of as much as 30,000 diameters with resolving power to match, are entirely practicable. The enlarged photographic images thus obtained may then be still further magnified by ordinary optical means, so that over-all magnifying powers of 100,000 or more are now available. This brings within observation objects, such as very minute disease organisms, that have hitherto eluded detection. (L.D.W.)

ELECTRON MULTIPLIER. High-speed electrons impinging upon a metallic surface may cause other electrons to come from the surface. These secondary electrons as they are called may far outnumber the original or primary electrons. There are present in any conductor many so-called free electrons, that is, they are free to move at random in the material but not to leave it. If, however, they acquire more than a certain amount of energy, known as the work function, they may break through the surface restraints and leave the parent material. The high-speed primary electrons upon hitting the surface really penetrate it slightly and

by colliding with the free electrons inside give them enough energy to overcome the surface attraction and leave. Since the primary electron may have several times the work function energy it may divide it among several electrons and thus cause several to be emitted. In certain electron tubes this principle is utilized to give current amplification. The electrons carrying the original current strike a surface which has been treated to enhance the effect and cause many secondary electrons to be emitted. These in turn are accelerated and directed against a surface, and so on, until after several stages many thousands of times more electrons have been obtained than were in the original current. Such tubes are called electron multipliers. (L.R.Q.)

ELECTRON OPTICS. Electronics.

ELECTRON VOLT. This is a convenient unit of energy for calculations in **electronics** and in connection with **ionization** or excitation of atoms or molecules. When an electric charge e is transferred from a region where the **electric potential** is V_1 to one where the potential is V_2, its potential energy changes by an amount equal to $e(V_1 - V_2)$. If the charge e is the electronic charge 1.602×10^{-19} coulomb (as it is when the transferred particle is an **electron, a proton, a positron**, etc.), and if the potential difference $V_1 - V_2$ is one volt, the corresponding change in energy is equal to 1.602×10^{-19} joule or 1.602×10^{-12} erg, and is called an electron volt. Thus if a doubly ionized positive oxygen molecule moves in an electric field through a potential-drop of 500 volts, it receives $2 \times 500 = 1000$ electron volts or 1.602×10^{-9} erg of energy; and since the mass of the oxygen molecule is about 5.31×10^{-23} grams, this energy would give the molecule, if unimpeded, a speed of about 7.77×10^{6} cm. per sec. or 48.1 miles per sec. Careless writers often abbreviate electron volt to volt, a practice which should be discouraged. (L.D.W.)

ELECTRONIC ENGINEERING. This is the specialized branch of electrical engineering which deals with the design and application of electron **tubes**. While it is one of the newest branches of engineering it is rapidly becoming of great importance in the field. (L.R.Q.)

ELECTRONIC SWITCH. This is an **electron tube** device for alternately switching between two input signals. It consists of two amplifier channels, each fed by one of the inputs, arranged so they are alternately biased to cut-off (i.e., become inoperative) and feeding a common output. Thus while one channel is cut off, the other is amplifying its signal and feeding it into the output circuit. Then the first channel is cut off and the second becomes operative. One of the principal uses of the instrument is with the cathode ray **oscilloscope** where the switch is used to switch between two signals so rapidly that both signal waves appear on the oscilloscope screen. (L.R.Q.)

ELECTRONICS. The activity and the control of **electrons** have in recent years developed into an important field of physical science, called electronics. Cathode rays, thermionic phenomena, and photoelectric phenomena are separately treated elsewhere, as are also **electric conduction, thermoelectric phenomena**, and the **Hall** effect and allied effects in metals.

It was early learned that cathode rays are deflected by either a magnetic or an electric field. This has lately been utilized in various devices dependent upon the control of electronic motion, such as the **oscillograph** and the "electron gun" (a source of electrons with electrodes so charged and disposed as to project a sharply defined electron stream along a prescribed path). The fundamental facts are:

1. An **electric field** accelerates a free electron in the direction antiparallel to the electric intensity and with a force equal to $10^7 Xe$ (dynes); in which X is the intensity in volts per cm. and e the electronic charge, in coulombs. The acceleration is therefore $10^7 Xe/m$, where m is the electronic mass. Since for an electron the **charge-mass ratio** e/m is 1.76×10^8 coulombs per gram, the acceleration of an electron in a field of intensity X (volts per cm.) is $1.76 \times 10^{15} X$ (cm. per sec. per sec.).

2. If there is a component X of the field at right angles to the electronic motion, the electron moving with speed v (cm. per sec.), follows a curve whose radius of curvature is $r_X = 5.66 \times 10^{-16} v^2 / X$ (cm.).

3. A magnetic field produces no acceleration in its own direction, but if there is a component of it, H (oersteds), at right angles to the electronic motion, the electron is deflected at right angles to both its motion and to the component H, following a curve of radius $r_H = 5.66 \times 10^{-8} v/H$ (cm.).

By designing electrodes and coils or magnetic pole-pieces so as to produce non-uniform electric or magnetic fields in various desired configurations, it has been found possible to act upon streams of electrons very much as lenses and optical instruments act upon rays of light, bending them, diverging them, or converging them to a focus. We thus have **electron microscopes** and electron image projectors, indeed, a whole new field of "electron optics." (L.D.W.)

ELECTROPHORUS. The simplest of all **static machines**; devised by Volta in 1816. It consists of a slab of some resinous substance, such as sealing wax or vulcanite, which is negatively charged by rubbing with fur. A metal plate provided with an insulating handle is placed upon the electrified slab. The contact is localized at a few points, so that instead of taking the negative charge off the slab, the metal plate becomes charged by induction, positively on the under side and negatively on the upper. The negative induced charge is now removed by grounding with the finger, and upon being lifted by means of the handle, the plate becomes positively charged all over, often strongly enough to yield bright sparks. Very little of the negative charge on the slab is removed in this process, and it may thus be used over and over to induce an indefinite number of positive charges. The energy is of course furnished by the operator in pulling the metal plate away from the slab. If a slab of glass is used, and rubbed with silk, it becomes positive and the induced charges on the plate are then negative. The instrument is useful in lecture-table demonstrations, in charging electrometers, etc. (L.D.W.)

ELECTROPLATING. The coating of an object with a thin layer of some metal through electrolytic deposition is known as electroplating. (See **Electrochemistry; also Reactions Involving Oxidation-Reduction.**) The process is widely used in numerous industries, either for the purpose of rendering a lustrous or non-corrosive finish on some article, or, as in electrotyping, being the principal part of the process. In electroplating, the general object is to employ the article to be plated as the **cathode** in an **electrolytic** bath composed of a solution of the salt of the metal being plated. The other terminal, the **anode**, may be made of the same metal, or it may be some chemically unaffected conductor. A low-voltage current is passed through the solution, which electrolyzes and plates the cathodic articles with the metal to the desired thickness. In this way table utensils are **silver** plated, various parts are made weatherproof by **cadmium** or **chromium** plating, and a high finish may be imparted through **nickel** plating. **Copper, zinc,** and **gold** are also plated. As the plating proceeds, the strength of the solution must be kept up by the addition of crystals of the plating salt, or a renewal of the anode if it is of the plating metal. A firm bond

between the anode and the deposited metal is to be secured when the two metals are of a type which tends to alloy. If this is not the case, some intermediate metal, which will alloy between the base and the plate, is first deposited. For example, in silver plating, the iron would otherwise form a poor bond with the silver, so a thin layer of copper is first deposited on it.

Because of the excellent conducting properties of a metallic salt solution, only a low voltage is required. As this must be d.c., the process of electroplating calls for a supply of current from special low-voltage d-c generators. The voltage will be of the order of 6 volts or less between anode and cathode.

Some of the solutions used to plate various metals are as follows: for silver or gold plating, double **cyanide** of the metal and **potassium;** copper plating, copper sulfate; nickel plating, nickel ammonium sulfate. (See also **Galvanizing.**) The articles to be plated must be thoroughly and effectively cleaned of all grease and dirt by washing in caustic or acid solutions. While the above is a brief outline of the process of electroplating, in commercial operations there are many troublesome angles which would not be suspected from the foregoing. Irregularity of the plate, poor surface graining, "trees," insufficient bond, and other troubles creep in. The overcoming of these requires the use of various expedients, such as careful control of temperature, or the addition of certain colloids and other compounds which have been found effective in preventing formation of defects on the plated articles.

A particular and specialized branch of electroplating is the preparation of plates from printers' type, artists' engravings, etc. This art is known as electrotyping, and is an important phase of electroplating. To make a book plate, for example, melted wax is run over the printers' type as it is locked up in its frame, so that an impression of the type is made in wax. The surface of this wax impression is then made conducting by coating it with **graphite.** This is then put in an electroplating bath of copper sulfate, and a thin shell of copper built up over the wax mold. The shell can then be separated from the mold and backed up with type metal to give it body. Finally it is mounted on a wooden block for rigidity. (F.T.M.)

ELECTROPOLISHING. Production of a smooth **surface** on metals by **electrochemical** means.

In all **electroplating** processes metals (and hydrogen) are deposited on the cathode and dissolved from the anode, except when insoluble anodes are used in which case oxygen is liberated at the anode. Electropolishing is the reverse of electroplating. The work is made the anode and tends to be dissolved. The operating conditions are controlled so that atomic oxygen forms continuously and reacts with the metal surface. Part of this oxygen may be liberated as gas. According to one theory, the high points of the metal surface are most readily oxidized and this oxidized material is thereupon dissolved in the electrolyte or otherwise removed. In any case selective solution of the high points of the surface tends to give a very smooth finish comparable or superior to a mechanically buffed surface.

A wide variety of electrolytes is used. A typical one for **stainless steels** contains phosphoric acid and butyl alcohol.

All mechanical methods of polishing, including those used for **metallographic** samples, produce a thin surface layer of work-hardened metal. Electropolishing produces a strain-free surface which is especially suitable for microscopic examination.

An important commercial application of the process is the polishing of stainless steel parts of irregular contour which would be difficult or impossible to buff. Copper and its alloys, Monel metal, aluminum, and many other alloys can be electropolished. (R.H.H.)

ELECTROSCOPES AND ELECTROMETERS.

These are instruments for detecting small charges of **electricity**, or for measuring small voltages, or sometimes, indirectly, very small **electric currents**. One of the earliest sensitive electroscopes consists of two narrow strips of gold-leaf hanging together in a glass jar. Upon being charged, they stand apart on account of their mutual repulsion. One leaf may be replaced by a stiff strip of brass, so that only the remaining leaf can move. The Wilson electroscope has a single gold-leaf which, on being charged, is attracted by a grounded metal plate tipped at such an angle as to give maximum sensitivity. If the movement of the gold-leaf in an electroscope is observed through a microscope whose ocular is provided with a calibrated scale, the instrument becomes an electrometer, capable of measuring potential differences in microvolts. (Some forms of electrostatic **voltmeter** operate on the same principle.) If the **capacitance** of the charged system is known, the rate of movement of the electrometer index may be used to measure the current from the discharging body; ionization currents are often thus measured.

The quadrant electrometer has a thin, oblong, metal plate suspended horizontally in the interior of a flat,

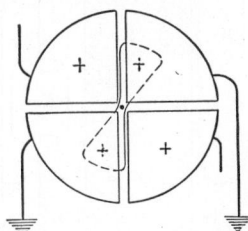

Quadrants and strip of quadrant electrometer.

circular metal box cut into four quadrants. One pair of opposite quadrants and the suspended strip are connected to the source of potential, the other pair of quadrants is grounded. This causes the strip to turn toward the grounded pair against the torsion of the suspending wire. Several electrometers have been designed, depending upon the lateral deflection of a lightly stretched, silvered or platinized quartz fiber; they are called string electrometers. The Wulf electrometer employs two such fibers side by side; on being charged, they bulge apart. The displacement of the fibers is observed in a micrometer microscope. Some of the special electrometers used for work with **cosmic rays** are of this type. Recently the thermion vacuum tube (see **Triode**) has been adapted to the amplification of exceedingly feeble discharge currents, thus serving as an electrometer. (See **Capillary Electrometer**.) (L.D.W.)

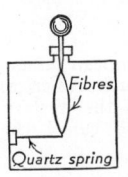

Quartz fiber electroscope or electrometer.

ELECTROSTATIC DEFLECTION. Cathode-Ray Tube.

ELECTROSTATIC UNITS. Electric and Magnetic Units.

ELECTROSTATICS.

While moving **electricity** has certain properties peculiar to its motion (see **Electric Currents, Electromagnetism**), in electrostatics we are concerned with phenomena exhibited by electricity whether in motion or at rest. The outstanding elements of the subject are electric charges, electric fields, electric induction in conductors, and electric polarization in dielectrics.

Owing to the tremendous mutual repulsion of all electricity for electricity of the same kind, it is impossible to gather together any considerable quantity of free electricity, positive or negative, in a limited space. To place a single coulomb of either sign on a metal sphere 10 cm. in diameter would require 4.5×10^{17} ergs of energy, equal to the output of a 100-hp. motor running continuously for a week; and its sudden release would rival the explosion of a carload of dynamite. But we can collect and experiment with very small charges. It is found that their attractions and repulsions obey **Coulomb's law** of inverse squares, and are also definitely dependent upon the dielectric constant of the surrounding medium. Faraday showed that any charge, imparted to a conductor, at once seeks the outside surface and so distributes itself there as to produce no influence anywhere inside. The **electric potential** of such a conductor is uniform both over its surface and throughout its interior.

The space outside in the neighborhood of the charge is, however, an electric field, as shown by the fact that a small charged body placed anywhere in it is urged by a definite force in some definite direction. Such a field may be mapped out by lines of force as in the case of a **magnetic field**. (See **Fields of Force**.) The electric intensity at any point of an electric field is measured by the force exerted upon a free unit charge placed at that point, and its direction is that of the force on a positive charge.

If a pair of equal, opposite charges at a fixed distance apart, called an "electric dipole," is placed in an electric field, it experiences in general a torque (like a bipolar magnet in a magnetic field). The maximum torque thus produced by a field of unit intensity is called the "electric moment" of the dipole; its magnitude is the product of either charge by the distance between them.

When a neutral conductor is placed in an electric field, it develops charges on opposite sides, the positive charge being on the side toward which a free positive charge is urged. The conductor has thus acquired an electric moment. This is thought to be due to a shifting of the **electrons** in the conductor until the region inside the conductor again attains the condition of zero intensity and uniform potential (see **Electric Screening**); and the process is called electric induction. The **electrophorus**, for example, utilizes this principle, as do other **static machines**. When a concentrated charge is brought near a large conducting surface, the charge thereby induced on the latter has in some respects the effect of a second concentrated charge, of opposite sign to the first, and lying behind the conducting surface; which gives rise to the idea of an "electric image," sometimes useful in electrostatic calculations.

If the object placed in the electric field is a dielectric or non-conductor, while there is no true induction, something like a stress, called an electric polarization, develops within the dielectric. This condition is the essential feature in the operation of a **condenser**, and its existence profoundly affects the **capacitance** of the conductors whose charges are responsible for the field. (L.D.W.)

ELECTRUM.

Electrum is a native alloy of **gold** and **silver** in which the latter metal may be present in quantities up to 40%. Electrum from the Urals is said to carry 20% **copper**. The color of electrum is a pale yellow or yellowish-white and the name is derived from the Greek word mentioned in the "Odyssey," meaning a metallic substance consisting of gold alloyed with silver. This same word was also used for the substance **amber**, doubtless because of the pale yellow color of certain varieties. (E.S.C.S.)

ELEMENT, CHEMICAL. Chemical Composition.

ELEMENTARY CHARGE. Electron

ELEMENTARY QUANTUM OF ACTION. Planck's Law.

ELEPHANT. Mammalia, Proboscidea. The largest existing land animals. Characterized by massive structure and by the elongation of the nose and upper lip to form a long prehensile proboscis or trunk. The two upper incisors develop into long tusks in the male and the broad grinding molar teeth grow into position gradually as they are worn off during the life of the animal.

Two species of elephants are recognized, the Indian, *Elephas maximus,* and African, *Loxodonta africana.* The former averages 8–9′ in height and the latter about 10′, although occasional specimens of both species considerably exceed these figures. The African elephant has much larger ears than the Indian and its tusks average somewhat heavier.

The Indian elephant is tamed for use as a beast of burden and for handling heavy materials, such as timbers. It does not breed freely in captivity, hence wild herds are the source of supply. The animals are caught or trapped by various methods.

Elephant tusks are the chief source of ivory, single tusks running from 90 to almost 200 lbs. in weight. (A.W.L.)

ELEPHANTIASIS. Filariasis.

ELEPHAS. Fossil Mammals.

ELEVATOR. A movable auxiliary airfoil, the function of which is to impress a **pitching moment** on the aircraft. It usually is hinged to the stabilizer. (F.T.M.)

ELEVATORS. Elevators are hoists for lifting passengers and freight within the confines of a building, usually by means of a car operating vertically in an elevator well—an open vertical shaft extending from top to bottom of the building. The direct acting hydraulic elevator with a long **piston** working in an upright **cylinder** set deep in the ground, carrying on top of it the car, has been superseded by electrically driven elevators because it had limitations as to height of the building that it could serve, and operating speed.

The electric elevator may be of the winding drum type, or of the traction-sheave type. Both of these are illustrated in the accompanying figure. The winding drum elevator has a drum with spiral groove on its face, on which is wound a turn or two of wire rope. The ends of this rope are connected respectively to the counterweight and to the car. In actual practice two ropes are attached to each, this having the advantage of smaller rope size, permitting greater flexibility. The counterweight moves in a direction opposite to the car, and has a mass sufficient to balance the weight of the car plus 30–40% of its maximum live load. The winding drum is driven by an electric motor through a reduction gear consisting of helical gears, or worm and spur gears. The reduction on freight elevators is larger than passenger elevators, since they are required to carry heavier loads, and high speed is not as essential. Since the rope winds around the winding drum, whose surface is grooved, it moves back and forth on the surface of the drum as the elevator is operated, and the face of the drum then is proportional to the height of the building. This limits the height to which winding drum elevators can be applied, and the high-lift, high-speed passenger elevators necessary for tall office buildings are of the traction-sheave type, also illustrated.

The car of a traction elevator is moved by friction existing between traction sheave and rope. A secondary, or idler sheave, near the traction sheave, provides for the return of the rope around the traction sheave a second time, as this has been found necessary in order to develop the high degree of friction required. The traction sheave may be used without reduction gearing between motor and sheave. The gearless traction elevators are driven by large slow-speed motors having the traction sheave mounted on the same shaft with arma-

ture and brake pulley. These electric elevators may be used in the tallest buildings, since the face of the traction sheave is independent of building height. The

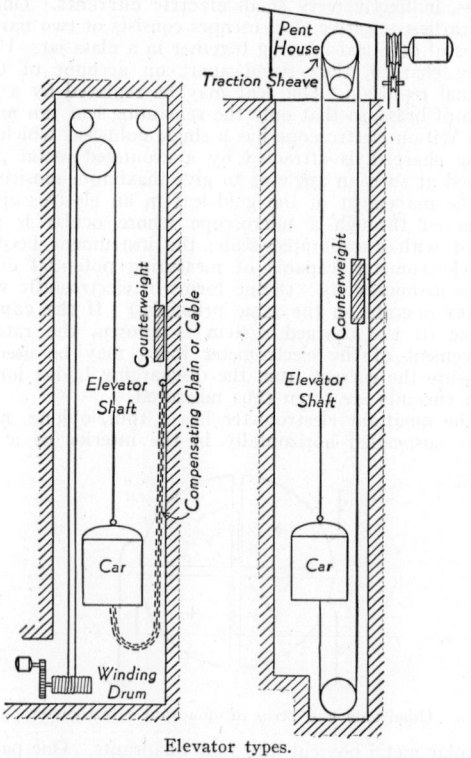

Elevator types.

motor and sheave must be mounted atop the elevator well. Elevator speeds, formerly limited to 100 to 200 ft. per min., have, with the gearless traction type, been advanced as high as 1000 ft. per min. (F.T.M.)

ELINVAR. Nickel Alloys.

ELK. Mammalia, Artiodactyla. In European usage, *Alces,* the animal called the moose in North America. In American usage, *Cervus,* a large **deer** of the western and northern parts of the continent, also called the wapiti. (A.W.L.)

ELLIPSE. An ellipse is one of the class of curves called **conic sections.** It may be obtained by cutting a right circular **cone** by a plane which cuts all the elements of the cone but which is not perpendicular to the axis of the cone.

An ellipse is the **locus** of a point which moves so that the sum of its distances from two fixed points remains constant. The fixed points are called the foci, the point midway between them is called the center. The line through the foci and the line through the center perpendicular to the preceding line are axes of symmetry, and the curve is symmetric about the center. (See Fig. 1.) The line through the foci cuts the curve in two points

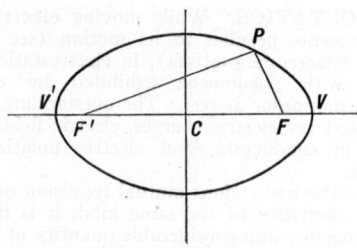

Fig. 1. Ellipse.

called the vertices; the segment (or length) joining the vertices is called the major axis; the distance between the points where the line through the center perpendicular to the line through the foci cuts the curve is called the minor axis. The major axis is equal to the constant sum of the focal radii in the definition of the curve.

The length of the semi-major axis is usually denoted by a, and the length of the semi-minor axis by b; the distance between the center and a focus is usually denoted by c.

The chord through either focus perpendicular to the major axis is called the latus rectum (or focal width) of the ellipse.

The eccentricity of the ellipse is the ratio $c/a = e$ of the distance between center and focus to the semi-major axis; it is a proper fraction ($e < 1$). It determines the shape of the curve; if e is near 0, it is nearly circular, and if e is near 1, it is elongated (narrow).

The standard equation of an ellipse in **rectangular coordinates** is

$$\frac{x^2}{a^2} + \frac{y^2}{b^2} = 1,$$

where a and b are the semi-major and semi-minor axes, respectively, and the center is at the origin and the major axis is along the X-axis.

The relation between the semi-axes a and b and the distance c between center and focus is $b^2 + c^2 = a^2$.

The two lines $x = \pm \dfrac{a}{e}$ are called the directrices of the ellipse $\dfrac{x^2}{a^2} + \dfrac{y^2}{b^2} = 1$. They have the property that the ratio of the distance of any point on the ellipse from a focus to the perpendicular distance of the point from the corresponding directrix is equal to the eccentricity.

The equation $\dfrac{(x-h)^2}{a^2} + \dfrac{(y-k)^2}{b^2} = 1$ represents an ellipse with center at (h, k) and axes of symmetry $x = h$ and $y = k$.

The ellipse may be constructed geometrically in several ways.

One method is the following, by continuous motion: Place two tacks in a drawing board at the foci F and F' and wind a string about them as indicated in Fig. 2.

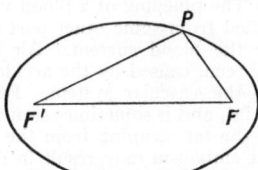

Fig. 2. Construction of ellipse.

If a pencil is placed in the loop FPF' at the point P and is moved so as to keep the string taut, then P describes an ellipse. If the major axis is to be $2a$, the length of the loop FPF' must be $2a + 2c$, where $2c$ is the distance between the foci.

A construction by ruler and compasses is as follows (Fig. 3): Draw circles on the major and minor axes AA'

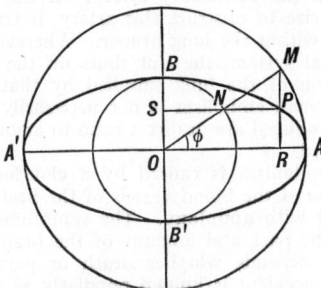

Fig. 3. Construction of ellipse.

and BB' as diameters. From the center O draw any radius intersecting these circles in M and N respectively. From M draw a line MR parallel to the minor axis, and from N a line NS parallel to the major axis. These lines will intersect in a point P on the ellipse.

The angle AOM is called the eccentric angle for the point P. The two circles are called the major and minor auxiliary circles.

The simplest **parametric equations** of an ellipse are:

$$x = a \cos \phi, \quad y = b \sin \phi,$$

where ϕ is the eccentric angle of the point $P(x, y)$, and a and b are the semi-axes of the ellipse.

The equation of the **tangent** to the ellipse $\dfrac{x^2}{a^2} + \dfrac{y^2}{b^2} = 1$ at the point (x_1, y_1) is $\dfrac{x_1 x}{a^2} + \dfrac{y_1 y}{b^2} = 1$.

The equation of the tangent with slope m to the ellipse $\dfrac{x^2}{a^2} + \dfrac{y^2}{b^2} = 1$ is $y = mx \pm \sqrt{a^2 m^2 + b^2}$.

The area of an ellipse is $\pi a b$, where a and b are the semi-axes.

Some of the applications of the ellipse are:

The orthogonal projection of a circle on a plane oblique to the plane of the circle is an ellipse.

Elliptical gears are used in machines, such as hay presses and power punches, where a slow, powerful motion is needed in a part, only, of each revolution.

The arches of stone and of concrete bridges are frequently constructed in the form of semi-ellipses.

The **orbits** in which the **planets**, including the earth, revolve around the sun, are ellipses.

A crescent, such as the crescent **moon**, is bounded by a semi-circle and a semi-ellipse. (L.L.S.)

ELLIPSOID. The **surface** represented in **rectangular coordinates** by the equation $\dfrac{x^2}{a^2} + \dfrac{y^2}{b^2} + \dfrac{z^2}{c^2} = 1$ is an ellipsoid; it is one of the **quadric surfaces**. It has a center at the origin, three principal axes (of symmetry) along the coordinate axes, all elliptic sections, and semi-axes a, b and c. It is a closed surface.

The volume of an ellipsoid with semi-axes a, b, c is $\frac{4}{3}\pi abc$.

If $a = b$, one set of sections consists of circles, and the surface is called an ellipsoid of revolution, which may be

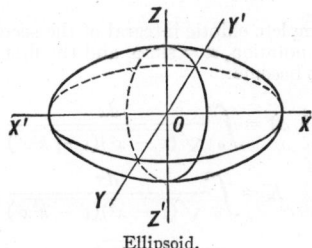

Ellipsoid.

generated by revolving an **ellipse** about its major or minor axis.

The ellipsoid of revolution obtained by revolving an ellipsoid about its minor axis is called an oblate spheroid; the one obtained by revolving about the major axis is called a prolate spheroid. (L.L.S.)

ELLIPSOIDAL HEAD. Pressure Vessels.

ELLIPTIC COORDINATES. Elliptic coordinates are three numbers which determine the position of a point in space; they are used in certain special types of problems of geometry and mathematical physics.

The **surfaces represented by the equation in rectangular coordinates:**

$$\frac{x^2}{\lambda - a} + \frac{y^2}{\lambda - b} + \frac{z^2}{\lambda - c} - 1 = 0,$$

where λ is a variable parameter, with $a > b > c > 0$, form a family of **confocal quadrics**. Through each point (x, y, z) of space there pass three surfaces of the family. The equation above has three roots $\lambda_1, \lambda_2, \lambda_3$ for λ for a given set of values of x, y, z. These three roots $\lambda_1, \lambda_2, \lambda_3$ determine the point (x, y, z), and are called the elliptic coordinates of the point. (L.L.S.)

ELLIPTIC GEOMETRY. Non-Euclidean Geometry.

ELLIPTIC HYPERBOLOIDS. Hyperboloids.

ELLIPTIC INTEGRALS AND FUNCTIONS. Any **integral** of the type $\int R(x, y)dx$, where R is a **rational function** of x, y, and where y is a square root of a third or fourth degree **polynomial function** of x, is called an elliptic integral.

These integrals may be expressed in terms of elementary functions and one or more of the following type forms of elliptic integrals:

$$F(k, \phi) = \int_0^\phi \frac{d\phi}{\sqrt{1 - k^2 \sin^2 \phi}} \quad (0 < k < 1)$$

called the elliptic integral of the first kind in Legendre's form;

$$E(k, \phi) = \int_0^\phi \sqrt{1 - k^2 \sin^2 \phi}\, d\phi \quad (0 < k < 1),$$

called the elliptic integral of the second kind in Legendre's form;

$$\Pi(k, n, \phi) = \int_0^\phi \frac{d\phi}{(1 + n \sin^2 \phi)\sqrt{1 - k^2 \sin^2 \phi}} (0 < k < 1),$$

called the elliptic integral of the third kind in Legendre's form.

The number

$$K = \int_0^{\pi/2} \frac{d\phi}{\sqrt{1 - k^2 \sin^2 \phi}}$$

is called a complete elliptic integral of the first kind, and the number

$$E = \int_0^{\pi/2} \sqrt{1 - k^2 \sin^2 \phi} \cdot d\phi$$

is called a complete elliptic integral of the second kind.

In Jacobi's notation, $x = \sin \phi$, and the first two elliptic integral types become:

$$F(k, \phi) = \int_0^x \frac{dx}{\sqrt{(1 - x^2)(1 - k^2 x^2)}},$$

with

$$K = \int_0^1 \frac{dx}{\sqrt{(1 - x^2)(1 - k^2 x^2)}},$$

$$E(k, \phi) = \int_0^x \frac{\sqrt{1 - k^2 x^2}}{\sqrt{1 - x^2}}\, dx,$$

with

$$E = \int_0^1 \frac{\sqrt{1 - k^2 x^2}}{\sqrt{1 - x^2}}\, dx.$$

The constant k is known as the modulus of the integrals, and the number k' defined by $k^2 + k'^2 = 1 (0 < k' < 1)$ is called the complementary modulus.

Tables of the elliptic integrals have been published.

If we put

$$u = \int_0^x \frac{dx}{\sqrt{(1 - x^2)(1 - k^2 x^2)}}, \quad (-1 < x < 1),$$

this equation defines u explicitly as a function of x. The **inverse function** x regarded as a function of u is called the

sine amplitude of u and is written $x = \sin am\ u$ or $x = sn\ u$. Two other functions are defined by $\sqrt{1 - x^2} = \cos am\ u$ or cn u, and $\sqrt{1 - k^2x^2} = \Delta\ am\ u = dn\ u$.

These three functions are type forms of elliptic functions.

An elliptic function in general is defined as a single-valued doubly periodic **analytic function of a complex variable,** whose only singularities in the finite part of the plane are poles. Any elliptic function can be expressed in terms of certain standard types of elliptic functions, of which the main types are those of Jacobi and those of Weierstrass. The Jacobi elliptic functions sn u, cn u, dn u have simple poles only, while the fundamental Weierstrass function $p(u)$ has double poles.

Elliptic functions are natural generalizations of the trigonometric functions and exponential functions, which are singly periodic functions. (L.L.S.)

ELLIPTIC PARABOLOID. Paraboloids.

ELLIPTIC POLARIZATION. Polarized Light.

ELONGATION. Planetary Motions; Tension Test.

ELUVIUM. General term for unconsolidated, residual sediments. (R.M.F.)

ELYTRON. The front wing of an **insect,** modified to form a leathery or rigid wing cover which folds above the body and conceals the hind wings at rest. They usually meet in a straight line down the middle of the back. Elytra are characteristic of the beetles and earwigs. They are sometimes short, sometimes cover the entire posterior part of the body, and are sometimes united to form an immovable shield. (A.W.L.)

EMBIIDINA. A small order of rare **insects** which live in nests and galleries of silk under objects lying on the ground. The few known species are found in warm regions, including the southwestern United States. (A.W.L.)

EMBOLISM. The plugging of a blood vessel by a clot, or embolus, carried from some other part of the circulatory system by the blood current. Air embolism is a condition, rarely seen, caused by the accidental introduction of air into the vascular system. Fat embolism is caused by oil or fat, and is sometimes seen following fractures of bones—the fat escaping from the bone marrow. Both air and fat embolism may result in death from the plugging of the capillaries of the brain or lung. Pulmonary embolism is the plugging of the pulmonary arteries by a blood clot. This is usually fatal, death resulting almost instantly. The condition may occur postoperatively, or with phlebitis or infection of some vein of the body, usually in the pelvis or legs. When such an infection occurs, a soft clot forms in the vein. If a piece of this clot should become dislodged, it goes, in the normal course of circulation, toward the lung, and thus may lodge in the pulmonary artery. If the clot is not of sufficient size to obstruct this artery, it travels along its branches within the lung proper. Wherever it lodges in the arterial system, the clot shuts off the circulation to the portion of the lung supplied by that particular branch. A small embolism is not necessarily fatal; the symptoms produced are in direct ratio to amount of lung tissue involved.

Cerebral embolism is caused by a clot lodging in a branch of one of the blood vessels of the brain. It may be associated with **apoplexy.** The symptoms produced depend on the part and amount of the brain involved, and on this depends whether death or paralysis, etc., result. This accident is known popularly as a "stroke." (R.S.M.)

EMBRITTLEMENT. Metals may become brittle under many different conditions. Ordinary **steel and wrought iron,** as well as **zinc alloys** and **magnesium alloys** suffer a reduction in impact **toughness** at subnormal temperatures. The reason for this form of brittleness is obscure. The effect is only temporary, full recovery of toughness occurring upon return to normal temperatures. Austenitic **stainless steels, brasses** and **bronzes, nickel alloys, aluminum alloys,** and **lead alloys** are not subject to severe embrittlement at low temperatures. Nickel additions to ordinary steels have a favorable effect.

Hydrogen embrittlement of iron and steel is caused by absorption of atomic hydrogen in **electroplating** processes or in **pickling** baths. After such exposure the normal toughness can usually be restored by prolonged aging or a short period of heating at a slightly elevated temperature, as in a steam bath.

Season cracking of high zinc brasses is a severe form of embrittlement resulting in cracking or disintegration. Somewhat similar forms of stress-corrosion cracking occur in many other metals and alloys. **Embrittlement of boiler plate** may be considered a special case.

Steels and ingot iron may be embrittled by any heat treatment that deposits films of either oxides or carbides in the grain boundaries. (See carbide embrittlement of **Stainless Steel.**)

Fatigue fractures are brittle in appearance but fatigue failures cannot properly be ascribed to a brittle condition of the metal. (R.H.H.)

EMBRITTLEMENT OF BOILER PLATE. This is the rarest of all **boiler** "diseases," yet it cannot be said to be so rare as to be unimportant. A serious feature of embrittlement is that when failure occurs it may come as a disastrous explosion, because embrittlement affects the drums and its presence is not detectable except on minute scrutiny. Embrittlement is attributed to the presence of a certain concentration of sodium hydroxide in the absence of inhibiting agents. The steel loses its toughness and cracks appear along the seams below the water line. They generally run from rivet to rivet, following the intercrystalline structure. In cases of embrittlement it has always been found that the feed water was high in sodium bicarbonate which broke down into sodium carbonate in the boiler and partially hydrolyzed. The reaction

$$Na_2CO_3 + HOH \rightarrow CO_2 \uparrow + 2NaOH$$

assumes considerable proportions at the elevated temperature of the boiler water. Prevention of embrittlement consists of reducing the causticity or adding an inhibiting agent such as sodium sulfate or phosphate. The A.S.M.E. recommendation consists of maintaining a ratio $\frac{sodium\ sulfate}{sodium\ carbonate}$ of 1 up to 150 lbs. gauge pressure, 2 up to 250 lbs., and 3 for 250 lbs. and over. No boilers meeting these ratios have experienced embrittlement. Recent developments have indicated that probably the more accurate expression for the inhibiting ratio is

$$\frac{sodium\ sulfate + sodium\ carbonate}{sodium\ hydroxide}.$$

Embrittlement or cracking of metal need not be feared in boilers provided this ratio is maintained in excess of 2. (F.T.M.)

EMBRYO. The developing individual between the union of the germ cells and the completion of the organs which characterize its body when it becomes a separate organism.

The term is difficult to limit because some development takes place after birth or hatching and in some species a considerable period of growth intervenes between the completion of the essential structures of the individual and its assumption of separate life. In the latter stage the organism is called a fetus if it is a

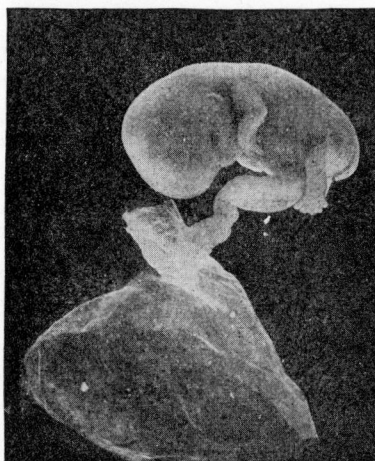

Human embryo at six weeks. (*Photo by Newton Millc*)

mammal, but this term is not applied to the similar period of birds and reptiles. For the embryo in botany, (see **Seed**). (A.W.L.)

EMBRYO SAC. Flower.

EMBRYOLOGY. The science which deals with the development of the individual from the union of the germ cells to the completion of its bodily structure. Although the term **embryo** cannot be precisely limited, the science of embryology is concerned with all development prior to birth or hatching.

Development of the fertilized ovum begins with the process of **cleavage.** Following cleavage a process of gastrulation gives rise to two or three **germ layers** and from this point the development of specialized tissues and organs goes on by gradual steps, all based on the subdivision and differentiation of many cells.

The processes of change by which germ layers give rise to other structures are varied. In some cases masses of cells grow out in solid protuberances from an existing source. This process is called budding and is exemplified by the appearance of legs and other appendages on the surface of the body. Other structures are developed by the pushing in or out of layers of cells. If the new part pushes in the layer it is said to invaginate, and if it pushes out, to evaginate. Hollow organs may also be formed by the splitting of solid masses and parts may separate by splitting from such masses; either process is delamination. A good example of evagination is the pushing out of a blind sac from the embryonic pharynx of vertebrates to form the respiratory system, and invagination is illustrated by the pushing in of ectoderm to form the stomodaeum which becomes the oral cavity in part. The formation of the vertebrate excretory tubules as solid knots of tissue whose cavities arise by internal splitting is a case of delamination.

The details of development of any species or group of animals are complex. Vertebrate embryology has been worked out in great detail and is fairly uniform but the number of invertebrate forms is so great that their embryonic development cannot be concisely summarized.

In the vertebrates, once the germ layers are formed their further development is the formation of organs and tissues and in some species **extraembryonic membranes,** with the exception of the mesoderm. This layer gives rise to diffuse mesenchyme and its compact portions differentiate into three regions, the dorsal, intermediate, and lateral or ventral mesoderm. The first subdivides into two longitudinal series of metameric masses, the mesodermal somites, flanking the middle line of the body where the notochord lies. This skeletal primordium is independently derived from the same source as the mesoderm. The lateral mesoderm splits to form an outer somatic layer associated with the body wall and an inner splanchnic layer which envelops the viscera. The split forms the coelom or body cavity. From this point the mesoderm, like the other germ layers, gives rise directly to organs of the body. The organs and systems developed from each embryonic tissue are listed under **germ layers.**

In the field of experimental embryology an effort has been made to learn of the controlling factors in development by subjecting embryos and ova to various unusual conditions. By exposure to chemical stimuli, unusual temperatures, radiation, and the effects of centrifuging, many abnormal results have been recorded. It is evident from these results that development, like the life of the organism, is conditioned by a delicate balance of environmental factors. The response of inherited potentialities to this balance in the development of normal organic structure links embryology very closely with the subject of **heredity.**

It has become evident that one of the important factors in the embryological differentiation is the interaction between parts of the embryo which have come close to each other in the course of development. This action is known as induction, the formation of a nervous system in the amphibian, for instance, being induced by the notochord. The term organizer is also used in this connection. (A.W.L.)

EMBRYONIC FISSION. The subdivision of a single **ovum** at some stage in its development into parts which give rise to complete **embryos.** Polyembryony.

As a result of this process a single egg of many insects (parasitic **Hymenoptera**) and of some **rotifers** develops into several or many individuals. (A.W.L.)

EMBRYOPHYTA. All plants which are not **thallophytes** are sometimes grouped together and called embryophytes. It is characteristic of these plants that for a time at least the developing plant or **embryo** remains dependent on the tissue of the **gametophyte.** The embryophyta are also generally characterized by the existence of an **archegonium,** a multicellular female sex organ in which the egg is contained. In the **angiosperms,** however, the reduction of the gametophyte has resulted in the loss of the archegonium.

The embryophyta include all the **Bryophytes, Pteridophytes,** and **Spermatophytes.** (R.M.W.)

EMERALD. This beautiful green variety of the mineral beryl has been known since ancient times and always prized as a gem, both because of its color and relative rarity. It is frequently cloudy or flawed, hence the expression "rare as an emerald without a flaw." The original source of emeralds seems to be the so-called Cleopatra's mines in Egypt, where in a range of low mountains about 15 miles from the Red Sea, they are found in **schists.** The quality of these emeralds is not high, but there is much evidence of considerable workings in a former period.

Although emeralds are found in the Urals and to some extent elsewhere the most important locality for emerald today is at Muso, Colombia, South America, about 75 miles northwest of Bogota. These mines are believed to be in part at least the source of the emeralds which Cortez and the Spanish conquistadores ruthlessly seized and which were believed for a long time to have come from Peru.

The word emerald is probably derived from the Persian. (E.S.C.S.)

EMETIC. A substance or **drug** that induces vomiting, either by direct action on the stomach or indirectly by action on the vomiting center in the brain. (R.S.M.)

EMETINE. A drug used in the treatment of amoebic **dysentery.** It is one of the alkaloids of **ipecac,** which is obtained from the dried roots of *Psychotria ipecacuanha* or *acuminata,* plants native to Brazil and Central America, but also cultivated in India. Emetine is used in the acute diarrhea stage of amoebiasis and is particularly useful in liver abscess due to entameba histolytica. It is given subcutaneously. (D.M.H.)

EMISSIVE POWER. Thermal Radiation.

EMMERMAN PROCESS. The Emmerman process, sometimes called the Direct **Person Process,** is a method of producing masks directly on photographic prints during processing for the purpose of obtaining good tone separation. The Emmerman process is an automatic masking method for use in projection printing. With this method, a silver mask is formed on paper by partially exposing a sheet of sensitive paper soaked with developer, on a glass plate on enlarger easel and allowing it to develop out fully before completing the exposure and developing the print in a tray.

The mask formed protects the shadow or dark areas of print so they do not become blocked up on the second exposure. The method is of particular value in making prints from negatives of subjects having an extreme brightness range, or an exposure scale too long for the print emulsion.

Three requirements are necessary for success with the Emmerman process. The developer must not fog or stain the wet paper while the mask is being made. The emulsion should not show reversal tendencies during the second exposure. It is advisable to use a paper grade one degree more contrasty than the negative normally requires. (S.M.T.)

EMPENNAGE. Tail; Airplane.

EMPHYSEMA. A structural change in the lungs, in which the smaller air sacs, the alveoli, are excessively distended at rest; they are permanently fixed in the inspiratory position thus reducing the vital capacity, or possible total exchange of air with inspiration and expiration. The **thorax** is altered in shape, and the patient with long-standing emphysema has a "barrel chest" which moves little, if at all, with respiration. **Cyanosis** and **dysponea** are common, and eventually **heart failure** occurs due to long-continued strain on the right side of the heart which pumps blood into the lungs.

Emphysema is the end picture in long-standing inflammatory and obstructive diseases of the lungs, in chronic **bronchitis** and bronchial **asthma,** and as a compensatory change when some parts of the lungs are not functioning. It is a common finding in individuals in certain occupations, being seen in glass-blowers and players of wind instruments. It is also a frequent finding in old age. (D.M.H.)

EMPIRICAL DATA. Empirical Equations.

EMPIRICAL EQUATIONS. A table of pairs of values of empirical or statistical data may be represented graphically by plotting the corresponding pairs of values as **rectangular coordinates** (or **polar coordinates**) of points in a plane and joining these points by a broken line or by a smooth curve.

The general problem in empirical equations is to find an equation which will represent the given data as accurately as possible. The general form of the required equation may be known in advance from theoretical considerations, but in other cases nothing may be known about the form of equation in advance.

If the given data are plotted as rectangular coordinates, and if the resulting points tend to lie along or very near a straight line, we may assume a linear law $y = mx + b$. To determine the coefficients m and b in this equation, several methods are available: the method of average points, the method of average equations, and the method of least squares.

The straight-line law may be tested by use of the following theorem: If the variable x has constant first differences Δx (that is, differences between consecutive pairs of values of x), and y is a **linear function** of x, then y will have constant first differences Δy and conversely. If the given table of data has the first variable at equidistant intervals, and if the first differences of the second variable are found by calculation from the table to be constant or very nearly constant, we may assume a linear law.

In fitting a straight line to a set of empirical data by the method of average points, we first divide the set of points representing the data into two groups and find an average point for each group, i.e., one whose coordinates are averages of the respective coordinates; then we find the equation of the line through these two average points.

In the method of average equations, we substitute each pair of values of the empirical data in the assumed straight-line equation $y = mx + b$, and thus obtain as many so-called observation equations as there are pairs of corresponding values. We then divide these equations into two groups as nearly equal in number as possible. Then we add the equations of each group, thus obtaining two equations in m and b. Solving these two equations for m and b gives the required linear law $y = mx + b$.

The values of the coefficients obtained by the method of averages depend on the way the given points or the observational equations are grouped. In accurate scientific work, the values of the coefficients are found by the method of least squares; this method may be somewhat longer than the method of averages.

The method of least squares as applied to the linear law may be explained as follows: Assume the required equation to be of the form $y = mx + b$, where m and b are to be determined. Let $(x_1, y_1), (x_2, y_2), \cdots, (x_n, y_n)$ be the given tabulated pairs of values of data. Let (x_k, y_k) be any pair of values of the data; then corresponding to the value of x_k of x the corresponding value of y as calculated from the equation is $mx_k + b$, but the tabular value of y is y_k. The difference between these two values of y, tabular and calculated, is $r_k = y_k - (mx_k + b)$, which is called the residual of the point (x_k, y_k) with respect to the line. Corresponding to the n points there are n residuals. Geometrically they represent the vertical distance between each point and the required line.

Let us form the sum of the squares of these residuals: $r_1{}^2 + r_2{}^2 + \cdots + r_n{}^2$, which is represented more briefly by the symbol Σr^2.

The basic principle of the method of least squares in this case may be stated thus: The values of m and b determined by the method of least squares are those that make the sum of the squares of the residuals Σr^2 as small as possible.

It is found that the values of m and b that make Σr^2 a minimum are determined by the equations $\Sigma y = m\Sigma x + bm$, $\Sigma xy = m\Sigma x^2 + b\Sigma x$. To form these equations, we may proceed as follows: We write down two sets of equations. The equations of the first set are formed by substituting the given pairs of values of x and y in the equation $y = mx + b$. The equations of the second set are formed by multiplying each equation of the first set by the coefficient of m in it. We obtain thus the following sets of equations:

$$y_1 = mx_1 + b \qquad x_1 y_1 = mx_1{}^2 + bx_1$$
$$y_2 = mx_2 + b \qquad x_2 y_2 = mx_2{}^2 + bx_2$$
$$y_3 = mx_3 + b \qquad x_3 y_3 = mx_3{}^2 + bx_3$$
$$\text{etc.} \qquad\qquad \text{etc.}$$

If we add the members of the equations in each of the two sets, we obtain two equations in m and b, which can be solved for these two unknowns, and these values put in the form $y = mx + b$ give the required equation in the sense of least squares.

If the graph of the given data does not indicate a linear law, or if the test of the data by first differences does not show a linear law, it may be that the data may be fitted by a quadratic law. This may be tested as follows:

If a variable x has constant first differences, and if the related variable y is a **quadratic function** of x, then y will have constant second differences $\Delta^2 y$ (that is, successive differences of the first differences Δy), and vice versa. Hence, if $\Delta^2 y$ is found constant or nearly so, we may assume a quadratic law $y = ax^2 + bx + c$.

If we decide that the given data may be fitted by a quadratic (or parabolic) law $y = ax^2 + bx + c$, we may determine the coefficients a, b, c by the method of averages or by the method of least squares.

In the method of average points, we divide the data into three groups nearly equal in number, and find three average points, one for each group. Substituting the coordinates of these three average points in the above equation, we obtain three equations with three unknowns, which may be solved for a, b, c.

By the method of average equations, we substitute the given data in the assumed equation, divide the resulting equations into three groups nearly equal in number, add the equations of each group, then solve the resulting three equations for a, b and c.

In the method of least squares for the quadratic law, we proceed as follows: Form three sets of equations, the first set being the equations obtained by substituting the given data in the general parabolic law equation $y = ax^2 + bx + c$; the second set is obtained by multiplying each equation of the first set by the coefficient of b in it, and the third set by multiplying each equation of the first set by the coefficient of a in it. Add the corresponding members of each set of equations, obtaining thus three new equations in three unknowns a, b, c. Solving these equations for a, b and c gives the coefficients of the required equation.

If a linear or quadratic law does not fit the given data, we may try plotting the data on **logarithmic paper**. If the resulting points tend to lie on a straight line, an equation of the power-law type: $y = ax^n$ is indicated.

For, if we assume the law $y = ax^n$, and take logarithms, we obtain $\log y = \log a + n \log x$. Put $\log y = Y$, $\log a = A$, $\log x = X$, and the equation becomes $Y = A + nX$, a linear law, giving a straight-line graph. If X and Y (i.e., $\log x$ and $\log y$) are plotted on ordinary squared paper, or if x and y are plotted on logarithmic paper, we should get a straight line. The equation of this straight line may be found by one of the previous methods for the linear law, obtaining A and n. Then a will be the anti-logarithm of A, and the coefficients a and n are known, which may then be substituted in $y = ax^n$.

If, when plotted on logarithmic paper, the data does not give a straight line, we may try plotting it on **semi-logarithmic paper**; if these points tend to lie on a straight line, an equation of exponential type: $y = ae^{bx}$ is indicated.

For, if we assume the equation $y = ae^{bx}$, and take (common) logarithms, we get $\log_{10} y = \log_{10} a + bx \log_{10} e$. Put $\log_{10} y = Y$, $\log_{10} a = A$, $b \log_{10} e = m$, and the preceding equation becomes $Y = A + mx$, a linear law. Hence, if x and Y ($= \log y$) are plotted on ordinary cross-section paper, or if x and y are plotted on semi-logarithmic paper, we should obtain a straight line. The equations of this straight line can be found by one of the previous methods for the linear law, giving A and m. Then a may be found as the anti-logarithm of A, and

b may be found from $m/\log_{10} e$. These values substituted in $y = ae^{bx}$ give the required equation.

It may be in some cases that a polynomial law $y = a + bx + cx^2 + dx^3 + \cdots$ will fit the data better than any of the preceding types; the equation of lowest degree is desired. It is treated similarly to the quadratic law.

Other forms of equations are sometimes useful. The following forms may be reduced to the straight-line law by suitable transformations.

The law $y = a + bx^2$ may be reduced to a straight-line law by the substitution $x^2 = X$, $y = Y$.

The law $y = a + \dfrac{b}{x}$ reduces to the straight-line law by the substitution $1/x = X$, $y = Y$.

The law $y = \dfrac{x}{ax + b}$ can be written $\dfrac{x}{y} = ax + b$, so that if we put $x/y = Y$ and $x = X$, the equation becomes $Y = aX + b$, a straight-line law. (L.L.S.)

EMPIRICAL PROBABILITY. Probability.

EMPYEMA. A collection of pus in any cavity or organ of the body. The term is generally used in describing a collection of pus in the **pleural** cavity. This is a common complication of **pneumonia**. In pneumonia a simple **pleurisy** with fluid develops which later may become infected, accompanied by formation of pus. The treatment is surgical. Empyema of the gall bladder may occur when there is obstruction to the cystic duct in the presence of infection of the organ. (R.S.M.)

EMU, EMEU. Aves, Casuariiformes. Large, flightless herbivorous birds (**Aves**) of Australia. They have very small wings are fleet of foot.

On some islands of the Australian region the emeus have been exterminated and in Australia itself they are said to be restricted to the wild interior and to be scarcer year by year. The best-known species is *Dromaeus novae-hollandiae*. (A.W.L.)

EMULSION. Colloids.

ENAMEL. Ceramics; Pigments, Paints, Varnishes; Tooth.

ENANTIOMORPHISM. Isomerism.

ENARGITE. A mineral sulfosalt composed of copper, arsenic and antimony, $Cu_3(As, Sb)S_4$. Crystallizes in the orthorhombic system. Hardness, 3; specific gravity, 4.451; color, gray to black with metallic luster. From the Greek, meaning **distinct** cleavage. (R.M.F.)

ENCEPHALITIS. Inflammation of the brain. The causative agents are injury, infection, or the action of toxic substances on the brain tissue. Epidemic encephalitis is considered elsewhere as a separate disease.

Encephalitis may complicate diseases such as **syphilis, malaria, measles, mumps, influenza, typhus,** and many others. It may be caused by metals, (e.g., lead encephalitis), carbon monoxide, caisson disease, fracture of the skull with injury to the brain, etc.

The symptoms may occur suddenly or gradually with lethargy, fever, headache and stiffness of the neck. Often the temperature is excessively high. **Paralysis,** deafness and other nervous manifestations are frequently present.

The mortality is high in recognized cases, although many cases are not diagnosed, especially when of lesser degree. Aftereffects such as deafness, **epilepsy, dementia,** changes in personality, etc., are found in many cases that survive. (R.S.M., D.M.H.)

END AREA METHOD. Earthwork.

END MILL. Milling.

END POST. Bridge.

END STIFFNER. Bridge.

ENDEMIC. A term used in reference to a disease that is prevalent in a particular locality, recurring frequently but affecting only a small part of the population at one time. (See **Epidemic**.) (D.M.H.)

ENDIVE. *Cichorium Endivia.* **Composite Family.**

ENDOCARDITIS. Inflammation of the tissue that lines the various cavities of the **heart,** and of the heart valves. The disease may occur as part of a symptom complex of **rheumatic fever,** or as a complication of many acute infections, such as septicemia due to the *Staphylococcus, Streptococcus, Pneumococcus, Gonococcus,* or *Influenza bacillus.* Endocarditis is a complication that usually results fatally.

Subacute bacterial endocarditis is a form of the disease usually caused by the streptococcus viridans. It occurs only on valves previously damaged by rheumatic disease or deformed by congenital anomalies. The disease was formerly fatal within two years, but recent work with **penicillin** suggests that many cases may be cured by large doses of this drug. (D.M.H.)

ENDOCARP. Fruit.

ENDOCRINE GLAND. A gland whose secretion is carried by the blood stream. Also called ductless glands. These organs produce substances known as **hormones** which condition the action and development of other parts of the body either by activation or by inhibition.

The principal ductless glands of the vertebrates are the thyroid, parathyroids, pituitary, adrenal or suprarenal, ovary and testis, and pancreas; the thymus and pineal were at one time classified as endocrine glands, but modern research has so far failed to substantiate this concept. The first two develop as outgrowths of the **pharynx.** These endodermal parts lose their connections with the pharynx and become associated with mesodermal tissue in the adult. The pituitary gland is made up of an anterior lobe derived from an ingrowth of ectoderm and a posterior lobe derived from the brain. In the adult it is connected with the under side of the brain. The adrenals lie near the kidneys or in contact with them and consist of an outer cortex and an inner medulla, both endocrine in function. The cortex arises from the lining of the body cavity and the medulla from cells of the sympathetic nervous system. The ovary and testis are developed from mesoderm. Although their primary function is the production of germ cells, they also produce hormones in some of their parts. The pancreas is a digestive gland derived from and connected with the intestine. Among its digestive cells are small islets of Langerhans which have no connection with its duct but which produce a hormone.

The action of these glands is considered under **Hormone.** (A.W.L.)

ENDODERM. Germ Layers.

ENDOGENETIC. As used by geologists to denote processes originating within the earth. (R.M.F.)

ENDOMORPHISM. That phase of **contact metamorphism** which takes place in the intrusive **magma** rather than in the walls of the rock mass which it invades. (R.M.F.)

ENDOPHRAGMAL SKELETON. An internal framework found in some **crustaceans.** It is made up of **apodemes** derived from the exoskeleton. (A.W.L.)

ENDOPODITE. Biramous Appendage.

ENDOPTERYGOTA. Insects whose wings are concealed beneath the integument of the **larva** during development. Such insects have complete metamorphosis, hence this term is synonymous with **Holometabola.** (A.W.L.)

ENDOSKELETON. Skeletal System.

ENDOSPERM. Flower.

ENDOSTERNITE. A skeletal plate lying beneath the anterior part of the alimentary tract in some **crustaceans.** Mesodermal (see **Germ Layers**) in origin. (A.W.L.)

ENDOSTYLE. A groove lying in the median line of the ventral wall of the **pharynx** in the **tunicates,** lancelets (**Amphioxus**), and larval lampreys (**Cyclostomata**). It is ciliated (see **Cilium**) and secretes mucus. Food particles swept into the pharynx are caught by the mucus and carried forward to peripharyngeal grooves which run around the pharynx to a dorsal median hyperbranchial groove. In this groove the food is carried to the intestine.

The thyroid gland of vertebrates evolved from the endostyle. (A.W.L.)

ENDOTHELIUM. The delicate lining of the organs of **circulation.** It is one cell in thickness and is continuous throughout the closed passages with the exception of the **sinusoids.** The walls of capillaries are made up of little more than the endothelium. (A.W.L.)

ENDOTHERMIC REACTION. Thermochemistry.

ENDURANCE, AIRPLANE FLIGHT. Range and Endurance; Airplane.

ENDURANCE LIMIT. Fatigue.

ENERGY Energy is probably to be regarded as the most fundamental of all physical entities, though within the past few years our concepts relating to it have undergone quite revolutionary changes.

From the elementary and older point of view, energy is thought of as an intangible something transferred to bodies of matter when **work** is done upon them, and delivered up by such bodies whenever they do work upon other bodies. Thus, let a free body be acted upon by a steady force of 6 lbs. until it has moved from rest a distance of 100 ft. in the direction of the force. The work thereby done upon the body is 600 **foot-pounds,** and since the moving mass is now capable of doing 600 foot-pounds of work upon other bodies by collision or otherwise, we think of this 600 foot-pounds as something which has been transferred and conserved throughout the process.

Considerations of this kind have led, in the past, to regarding energy primarily as that which is thus communicated and conserved whenever work is done by one body upon another. But in the nineteenth century, Joule and others discovered the equivalence of **heat** and energy. And when it was established that **light** is a form of energy traveling through space apparently independent of any matter, it became necessary either to construct a dynamics for light and to invent an **ether** in which to carry on its operations, or else to accept energy, at least radiant energy, as having an objective existence of its own, co-ordinate with that of matter itself.

In the measurement of energy, we still adhere to the elementary concept and utilize its familiar relationship with the dynamic magnitude, work. We even express energy in work units; though there is today no more logical reason for doing so than for adhering to the old water-temperature **calorie** in measuring heat. Thus the

practical energy units are, in pure physics, the **erg** or centimeter-dyne, in engineering, the foot-pound.

Whenever energy is obviously associated with the motion of masses of matter—the form in which we sense its existence most readily—it is called **kinetic energy.** But since work may be done in such a way as not to affect the motion of matter, but to alter its situation or condition in other ways (as when a battery is charged or a clock wound), we must also recognize the existence of latent or **potential energy.** Much of the activity going on in the material world involves the continual transformation of energy from one of these states to the other.

Early experimenters observed that when a **machine** is so constructed as to operate with negligible friction, the work done by the machine is equal to that done upon it. Thus, if a weight of 20 lbs. is lifted 7 ft. by means of a single movable pulley, the operator, though obliged perhaps to exert only 10 lbs. of force on a cord, must draw that cord upward 14 ft. In each case the work done is 140 foot-pounds. The extension of this principle to a multitude of complex cases finally led to the doctrine of conservation of energy, recognized in various aspects by different physicists toward the middle of the nineteenth century. This doctrine, in concise form, states that the total quantity of energy in existence remains unaltered throughout all the changes which take place in the material universe. It is thus analogous to the earlier doctrine of the conservation of **mass.**

Modern research in **relativity** and atomic physics has cast some doubt upon the literal truth of both of these doctrines. Einstein, in studying the relativistic connection between matter and energy, arrived at the conclusion that energy, like mass, possesses inertia and gravitational attraction, in such degree as to make the unit of energy equivalent to $1/c^2$ units of mass (in which c is the speed of light); so that 1 gram of matter has the same inertia and gravitational attraction as c^2 ergs, or 8.986×10^{13} joules, of energy. Subsequent observation has supported not only this view, but the further theory that in certain circumstances matter may be actually converted into energy, and perhaps energy into matter, in the same ratio. Thus if the sun emits its radiant energy without drawing upon any source of supply other than that arising from the transformation of its mass, the Einstein proper energy principle leads to the conclusion that its mass must be diminishing at the rate of some 4,600,000 tons per sec. If all the **stars** are likewise sacrificing their mass to keep up their radiation output, and if this energy is not somewhere re-created into matter, it is clear that the conservation doctrine is not universally valid, unless applied to the totality of both matter and energy together as a single entity. (See **Available Energy.**) (L.D.W.)

ENERGY LEVELS. Quantum Theory.

ENERGY LOAD. Energy.

ENGINE. In common usage, the term engine is used widely for devices which produce motion. In stricter technical sense, an engine is said to transform energy, especially heat energy, into mechanical work. Among the prime movers, those in which the power originates in a piston and cylinder are classed as engines, while those with purely rotative motion are known as turbines. (See **Diesel Engine, Otto Engine,** and **Steam Engine.**) (F.T.M.)

ENGINE LATHE. Lathe.

ENGINE PERFORMANCE, INTERNAL COMBUSTION. The performance of any prime mover is of interest to the owner, operator, and designer, as is also the variation in performance created by changes in mode of operation. While many features of prime-mover action might be designated as performance, the items usually selected are (1) **power** output, (2) **thermal efficiency,** (3) speed, usually rotative.

As internal combustion engine performance is influenced by more independent performance factors than any other commercial prime movers, internal combustion performance factors are chosen for description here.

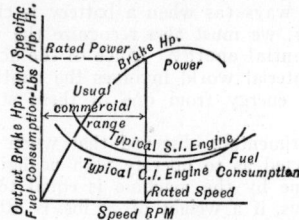

Fig. 1. Full throttle performance.

The output of such an engine is its "shaft horsepower," sometimes called brake horsepower because it used to be frequently measured by a brake-type of absorption **dynamometer.** Since the purpose of obtaining and using an engine is for its power, the brake hp. is of first concern in any event. The operating expense of an engine consists mainly of the cost of fuel; hence fuel consumption (i.e., quantity per unit time at a specified output), specific fuel consumption (lb. per brake hp.-hr.), or thermal efficiency is important, it being the measure of the direct cost of producing power. Rotative speed, say rpm, affects size and weight of the engine; also, its application to specific driven machinery, since direct drive, if possible, is usually simplest and best.

The internal combustion engine combines fuel and air in combustion for the production of a gas pressure against the piston. A basic factor of performance is, therefore, atmospheric air. But this may vary greatly both in temperature and density, especially in the aeronautical field. To combat loss of power due to diminishing atmospheric density at altitudes, airplane engines are often provided with **superchargers** to compress that air toward sea-level density. However, should this compression also be applied to sea-level air, the resulting charge would be so powerful as to damage the engine by overstraining and overheating cylinders, pistons, bearings, etc. It is necessary to throttle the intake so that the cylinder **mean effective pressure** will be limited to a safe value. This adds manifold pressure to other factors affecting power output.

Because friction and cooling heat losses tend to remain constant though output varies, thermal efficiency is not constant but decreases at part capacity. Correspondingly, the specific fuel consumption increases.

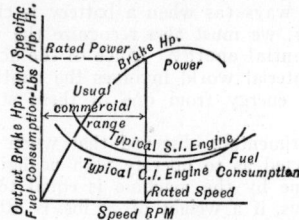

Fig. 2. Constant speed performance.

This variation (with constant rpm) is visualized in Fig. 2, part load being carried by throttling the inlet so as to reduce the fuel admitted per cycle. If the inlet is unrestricted, but the output is varied by adjustments of the connected load, so that speed increases, the output will first rise (as shown in Fig. 1), then decrease as a speed is reached where **volumetric efficiency** is decreasing more rapidly than number of power cycles increase. Likewise, fuel consumption is increased at slow speeds because more heat is transferred to the cooling system as a result of the relatively longer time the hot gases are exposed to the cylinder walls. Again, it is increased at high speeds because of increasing mechanical friction. Some intermediate speed will exhibit an optimum combination of cooling and friction, producing minimum fuel consumption and maximum thermal efficiency.

Ruggedness, reliability, and long life are features that cannot be expressed by factors shown in these figures. High speeds can result in damaging centrifugal and reciprocating stresses. Excessive supercharging of engine cylinders forces quantities of fuel and air into the combustion chamber, where excessive pressures and temperatures can be produced.

Cylinder mean effective pressure, or the hypothetical *Brake Mean Effective Pressure,* as has been mentioned, is controlled by throttling the induction system. The same power may be produced by various combinations of induction manifold pressure and rotative speed lying between the extremes of maximum allowable mean effective pressure with low speed and maximum speed with low mean effective pressure.

Fig. 3 shows a typical altitude-power curve involving rpm and manifold pressure as variables. At full

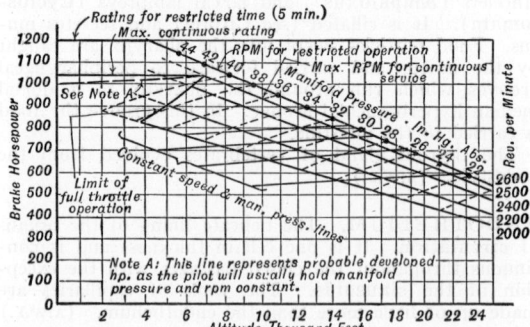

Fig. 3. Aircraft engine calibration curves.

throttle the power decreases uniformly with increase of altitude. The critical altitude is the maximum altitude attainable with a given rpm and manifold pressure combination. Full throttle operation is implied. For example, the critical altitude shown by Fig. 3 for normal rpm (2500) is 6800'. Thus the curve of the limit of full-throttle operation is also one of critical altitude vs. rpm. At superior altitudes power falls off. The lines showing power vs. altitude at constant manifold pressure and rpm rise slightly since constant manifold pressure on the inlet, and decreasing atmospheric pressure on the exhaust, tend to increase mean effective pressure; also, lower atmospheric temperature at higher altitudes has some tendency to increase volumetric efficiency. If the air intake is located so that it obtains some air **ram** there will be an increase of critical altitude. (F.T.M.)

ENGINEERING STATISTICS. Industrial Statistics.

ENGINES, AIRCRAFT. Aeronautical Engines.

ENOL. Tautomerism.

ENRICHMENT. Also "secondary enrichment." The term applied by students of ore deposits to the natural processes by which the lower levels of an ore deposit are enriched at the expense of the upper levels, or the original protore. Particularly applied to **lodes** in which the **sulfide** ores have been concentrated by the leaching of the upper levels of the **vein** and redeposition below the **groundwater** table. Important ore minerals belonging to this type are **chalcocite** and **argentite.** (R.M.F.)

ENSIGN FLY. Insecta, Hymenoptera. Small parasitic **insects** whose abdomen is elevated on a slender stalk above the thorax. It has been likened to a flag and gives the common name to the group. The en-

Ensign fly.

sign flies make up the family Evaniidae. In all species whose habits are known the larvae are parasitic in the eggs of cockroaches. (A.W.L.)

ENSILAGE. Corn.

ENSTATITE. The mineral enstatite is an **ortho-rhombic pyroxene,** rarely in distinct crystals, usually found as fibrous or lamellar masses or perhaps compact. It has one easy cleavage parallel to the prism; brittle with uneven fracture; hardness 5.5; specific gravity 3.1–3.3; luster pearly to vitreous, sometimes somewhat metallic in **bronzite,** a variety of enstatite carrying up to 15% ferrous oxide FeO. Color grayish to greenish or yellowish-white, green and brown. Chemically enstatite is a silicate of magnesium, $MgSiO_3$. It occurs in igneous rocks which are high in magnesium content, like **gabbros, diorites, pyroxenites,** etc., and less commonly in metamorphic rocks. Meteorites of both the stony and metallic types have been shown to contain enstatite. It has been found at many places in Europe, Czechoslovakia, Austria, Bavaria, Germany, Norway and South Africa. In the United States it occurs in Putnam and St. Lawrence Counties, New York; Lancaster County, Pennsylvania; Jackson County, North Carolina, and near Baltimore, Maryland. The name enstatite is derived from the Greek word meaning *opponent,* in reference to its refractory nature; it is almost infusible. (E.S.C.S.)

ENTERIC CAVITY, ENTERON. The digestive cavity. This cavity forms by the splitting or invagination of the inner **germ layer** early in embryonic (see **Embryology**) development and persists as a sac with one opening to the exterior in the **coelenterates** and flatworms. In this form it is also called the **archenteron.**

In animals with a tubular alimentary tract the enteric cavity becomes the primitive gut. Its endodermal lining becomes the glandular digestive tissue and gives rise to large glandular masses in some species, and in the terrestrial vertebrates also produces the respiratory system.

The tubular enteric cavity is supplemented by invaginated tubes of ectoderm at each end, the **stomodaeum** and **proctodaeum.** (A.W.L.)

ENTERITIS. Any inflammation of the small intestine, usually accompanied by fever, pain in the abdomen, **diarrhea** and other constitutional symptoms. (R.S.M.)

ENTEROCOELA. Animals whose bodies contain no other cavity than that used for digestion. **Coelenterates, ctenophores, flatworms, nemerteans, roundworms** and **rotifers.** (A.W.L.)

ENTEROPNEUSTA. Hemichordata.

ENTEROSTOMY. An artificial opening made surgically in the intestine so that it may communicate with the outside through the abdominal wall or communicate with another portion of the gastrointestinal tract. It may be temporary or permanent in character. (R.S.M.)

ENTEROZOA. Animals which have a digestive cavity, the **enteric cavity.** All animals but the **protozoans** and **sponges.** (A.W.L.)

ENTHALPY. The enthalpy of a substance is defined as the sum of its internal and potential energy above some datum. Thus $h = u + pv$. A theoretical datum would be absolute zero of temperature. However, enthalpy tables for the practical use of engineers have, for the zero datum of u, a temperature of 32° F. for steam, −40° F. for refrigerants, and 80° F. for gases.

The incremental change of enthalpy for all gas processes is $c_p\,dt$ where c_p is the isobaric specific heat. This is not true of vapor processes unless a high degree of superheat exists. (F.T.M.)

ENTLADUNGSSTRAHLEN. This German term (meaning "discharge rays"), used by Wiedemann and adopted into English, refers to certain types of radiation emitted by electric sparks, apparently in the extreme **ultra-violet** range. At atmospheric pressure, short condenser sparks give radiation of wavelength approximately 900 **angstroms,** while for long sparks from an induction coil the wavelength is shorter, probably between 400 angstroms and 900 angstroms. At reduced pressures the wavelength is still shorter. These radiations produce ionization in the air and are strongly absorbed by **fluorite,** which is thereby rendered thermoluminescent (an effect called radio-thermoluminescence; see **Luminescence**). Experiments on the reflection of *Entladungsstrahlen* indicate their **total reflection** by glass or by celluloid with a critical angle of about 82°, which corresponds to a refractive index of 0.99. Work with these rays is difficult, as elsewhere in the far ultra-violet, on account of the absence of simple means of detecting them and because of their rapid absorption in air. (L.D.W.)

ENTOMOLOGY. The science that deals with all facts pertaining to **insects.** Because of the large number of insect species and their frequent economic importance the principal divisions of the science have been systematic entomology and economic entomology. Classification is difficult and intricate, and demands the constant service of specialists. The economic field is of such importance, especially to agriculture, that the national and state governments maintain organizations for the scientific study of insects which also assist in their control. (See **Insect Pests.**)

Although entomologists may specialize in insect **morphology** or physiology, work of this type is less extensive than more practical studies and is of more general biological interest, hence it takes its place largely in the subsidiary sciences of general biology. (A.W.L.)

ENTOMOSTRACA. A division of the class **Crustacea** formerly used to include all but the subclass **Malacostraca.** (A.W.L.)

ENTOPROCTA. A class of **Bryozoa.**

ENTRENCHED (or INCISED) MEANDER. A river valley which has a distinctly meandering old-age pattern (longitudinal profile), and a V-shaped or canyon-shaped, youthful transverse profile. The meandering course of the river is inherited from the time when it flowed at, or close to, **base level,** that is on a relatively flat surface. Subsequent uplift of the region quickens the flow of the stream, hence its downcutting power, without necessarily altering its inherited meandering course. Entrenched meanders are therefore physiographic evidence of the rejuvenation of the erosive power of an old-age stream. Entrenched meanders

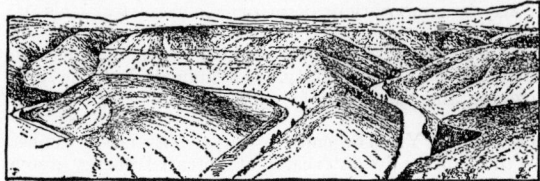

A rejuvenated region showing entrenched meanders. Yakima Canyon, Washington. (*Hobbs, after G. O. Smith.*)

may suggest the first stage in a new **cycle of erosion,** but usually imply an interruption in the normal cycle due to relatively sudden uplift of the region before the entire area has been reduced to a **peneplain.** (R.M.F.)

ENTROPY. The physical interpretation of this thermodynamic term, introduced by Clausius, has proved somewhat difficult.

In the mathematical treatment of thermodynamic processes there occurs very often a quantity, now relating energy to absolute temperature, now associated with the **probability** of a given distribution of **momentum** among molecules, and again expressing the degree in which the energy of a system has ceased to be **available energy.** Its mathematical form suggests that these are all aspects of a single physical magnitude. Application of the "second law" of **thermodynamics** leads to the conclusion that if any physical system is left to itself and allowed to distribute its energy in its own way, it always does so in a manner such that this quantity, called "entropy," increases; while at the same time the available energy of the system diminishes. This law applies to the universe as a whole, hence the proposition that the total entropy increases as time goes on. An interesting conclusion as to entropy in the vicinity of absolute zero is expressed by the Nernst heat theorem; viz., that all physical and chemical changes in this region take place at constant entropy. Any process during which there is no change of entropy is said to be "isentropic." This is true, for example, of an **adiabatic process** in which there is no dissipation of energy, i.e., one which is also a **reversible process.** In thermodynamic discussions entropy is commonly classed, along with temperature, pressure, and volume, as one of the variables defining the state of a body, and is often graphed as such on thermodynamic diagrams. (L.D.W.)

ENVELOPES. If the curves of a family are tangent to the same curve or group of curves, the name envelope of the family is applied to the curve or group of curves.

The envelope of the family of curves $f(x, y, a) = 0$, where a is the variable parameter of the family, is given by the pair of equations:

$$f(x, y, \alpha) = 0 \quad \text{and} \quad \frac{\partial f}{\partial \alpha} = f_\alpha (x, y, \alpha) = 0.$$

We may regard these as giving **parametric equations** of the envelope, or we may eliminate a between the two equations. (L.L.S.)

ENVIRONMENT. The assemblage of material factors and conditions surrounding the living organism and its component parts.

Environment includes both external and internal factors. In the external environment inanimate objects and the forces associated with them constitute the physical environment, and the living things and their derivatives with which the animal may be associated constitute its organic environment. Within its body it maintains an organization which constitutes an internal environment to which all of its parts respond directly, whether or not they also have external contacts. (See **Ecology, Distribution** and **Habitat.**) (A.W.L.)

ENZYME. An organic **catalyst,** i.e., a substance produced by a living organism that conditions some chemical action within the body and is not permanently altered in the process. The inorganic catalysts of chemistry and the enzymes of biology are similar in many ways. Both bring about specific chemical changes at a rate depending on the amount of active material present, and both influence the transformation of many times their own weight of material. Enzymes are more complex than inorganic catalysts, although their composition is not definitely known. They are regarded as disperse **colloids** in combination with **proteins.**

Enzymes have been classified in three principal groups: 1. Hydrolytic enzymes. 2. Oxidases, reductases, and zymases. 3. Catalase. A large number have been studied in some detail.

The hydrolytic enzymes facilitate the simplification of organic compounds by **hydrolysis** and are important in digestion, in the coagulation of **blood,** and in the formation of waste products and other substances in the body. Their action (hydrolysis) is the splitting of molecules of the compound affected in association with molecules of water to form compounds of simpler molecular structure, in some cases as one step in a series of similar enzymatic actions. For example, **starch** is hydrolyzed by the enzyme ptyalin in the saliva to produce maltose, and this **sugar** is hydrolyzed by maltase in the small intestine to form the simple sugar, **glucose,** which can be absorbed into the body.

Some of these enzymes act on **fats** and allied substances and are called esterases, some act on starches and sugars and are called carbohydrases, and some act on **proteins** and related compounds and are grouped as proteases, amidases, etc.

The oxidases and related enzymes bring about oxidation, reduction, and similar actions. Among them tyrosinase acts on the substance tyrosine, an **amino acid,** to produce the pigments known as melanins, while glycolase acts on sugars to produce lactic acid, and zymase, also acting on sugars, produces **alcohol** and **carbon dioxide.**

The enzyme catalase breaks down **hydrogen peroxide** into water and oxygen.

Enough is known of the specific enzymes and their action to show that they play an extensive part in the chemical processes of the body. It is entirely probable that many processes not clearly understood at present are enzymatic in nature, and that even the processes of reproduction and heredity are fundamentally of this kind. The literature on the subject records many established and many conjectural processes.

Enzymes are also used in various manufacturing processes, including the preparation of fibers and fabrics in the weaving industry, the removal of hair and subcutaneous tissue from hides preparatory to tanning, several stages of brewing, cheese making, and a number of processes in less familiar fields.

Because the enzymes are associated with colloidal substances they are very difficult to isolate and purify; nevertheless pepsin, urease and some others have been isolated in the crystalline state. Enzymes are very unstable substances and often lose their catalytic ability without the concomitant loss of their other characteristics (denaturation). They are as a rule inactive at 0° C., increase in activity with rise in temperature, and are destroyed at high temperatures. There is an optimum **hydrogen ion** concentration range for each enzyme. Many of them require activators, coenzymes or accelerators for their action and are inhibited in their catalytic activity by minute traces of certain substances such as **hydrogen cyanide.** The following is a list of the more important enzymes and the reactions they catalyze:

Enzyme	Reaction
Lipase	Hydrolysis of **fats**
Diastase	Hydrolysis of **starch**
Maltase	Hydrolysis of maltose and alpha **glucosides**
Emulsin	Hydrolysis of beta **glucosides**
Lactase	Hydrolysis of **lactose**
Pepsin	Hydrolysis of **proteins** in acid medium to peptones
Trypsin	Hydrolysis of **proteins** to amino acids
Urease	Hydrolysis of **urea**
Rennin	Coagulation of **casein** in the presence of calcium
Tyrosinase	Oxidation of **tyrosine**
Luciferase	Oxidation of luciferin to produce light (fire-fly)
Catalase	Decomposition of **hydrogen peroxide**
Zymase	Alcoholic **fermentation**
Invertase	Inversion of cane **sugar.**

(R.K.S., A.W.L.)

EOCENE. A subdivision of the **Tertiary** of the geologic time-scale. Type locality, near Paris, France. Term first proposed by Lyell in 1832. The Eocene began approximately 60 million years ago and lasted for about 20–30 million years. The greatest thickness of formations of this period occur in Wyoming. The principal areas of deposition in the United States are: 1. The unconsolidated marine gravels, **glauconite** sands and clays which overlap the Cretaceous marine sediments of the Atlantic Border. 2. The marine limestones, terrestrial sandstones and **lignites** of the Gulf Coast. 3. The marine sediments of the Pacific Coast. 4. The terrestrial intermontane deposits of the Western interior. The plants of this period suggest world-wide warm climate. The fossil plants include many of the modern genera, such as the beeches, dogwoods, walnuts, maples and elms. Fossil vertebrate skeletons show that the mammals are now dominant, although many of the existing orders of reptiles and birds also lived at this time. The mammalian fauna may be divided into two principal groups: 1. The archaic types which did not survive the Eocene. 2. The progenitors of the modern mammals, including the ancestors of the camels, pigs, horses, rats, and primitive monkeys. The principal surviving archaic forms are the creodonts (primitive flesh-eaters), uintatheria (hippopotamus-like forms), and zeuglodons (marine mammals). The mineral resources of this period are described under the Tertiary. (R.M.F.)

EOLITHS (Kentish). Paleontology of **Man.**

EOSIN. Dyes.

EOSPHORITE Childrenite.

EÖTVÖS BALANCE. Gravitation and Gravity.

EOZOÖN. The name of a "problematical" **fossil** in the Grenville limestone (marble) of Canada. Since the Grenville formation represents one of the oldest known **sedimentary** formations, Eozoön has aroused considerable discussion among geologists and **paleontologists.** Originally described as a giant **foraminifera,** it is now known to be of inorganic origin, probably the result of **contact metamorphism** between the interbedded basic lavas (now **serpentine**) and the limestones. The term Eozoön is derived from the Greek, meaning dawn animal. (R.M.F.)

EPEIROGENY. Signifying broad and relatively widespread or continental uplift as compared with mountain building or **Orogeny.** (R.M.F.)

EPHEDRA. Gymnosperms.

EPHEDRINE. Drugs and Alkaloids.

EPHEMERIDA. The may-flies, also known locally as shad-flies, salmon-flies, June bugs and Canadian soldiers. The adults are sluggish insects with slender filaments at the caudal end of the body and large triangular front wings. The hind legs are much smaller, in some species rudimentary. The immature insect is aquatic and in most species feeds on decaying vegetable matter. It may live for several years, while the adult stage lasts only a few days.

May-flies of one species emerge as adults in large numbers within a short period and are sometimes very abundant near favorable bodies of water. They fly at twilight and can sometimes be seen in large gray clouds at a distance of more than a mile over the islands of Lake Erie, where they are especially abundant. In the cities bordering the lake they are attracted to lights and their dead bodies are sometimes swept up in bushels after a heavy flight. Under such conditions they are a nuisance but not a serious pest. Their value as food for fishes more than offsets what little harm they do. (A.W.L.)

EPHEMERIS. Almanac.

EPHIPPIUM. The thickened covering of the **carapace** of certain **crustaceans** which is thrown off as a case for the winter eggs. (A.W.L.)

EPHYRA. A saucer-shaped jellyfish **larva** with deeply notched margin which is produced by the segmentation of the primary **scyphistoma** larva and develops directly into the adult **jellyfish.** (A.W.L.)

EPICARDIUM. The thin covering of the vertebrate **heart,** continuous with the lining of the pericardial cavity. (A.W.L.)

EPICENTRUM. Earthquakes.

EPICYCLIC TRAIN. Combinations of **gears** having a motion resulting from rotation about an axis which, in itself, is in rotation, are known as epicyclic trains. A simple epicyclic gear train, consisting of three gears and an arm, is shown in the figure. Mecha-

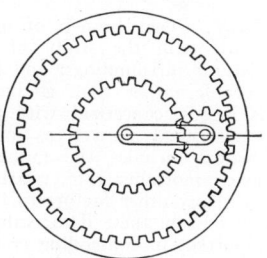

Epicyclic gear train.

nism of this nature is sometimes used for speed reducers. The ratios of speed of the driven and driving elements are found by the following simple rule: consider, first, the gears locked and the arm turned, then the arm locked and the gears turned. The algebraic sum of the separate motions thus determined will give the speed ratio. (F.T.M.)

EPICYCLOID. An epicycloid is a certain type form of mathematical curve, which may be defined as follows:

If a circle of radius b rolls upon the exterior of a circle of radius a, a point on the first circle traces an epicycloid. Its **parametric equations** are:

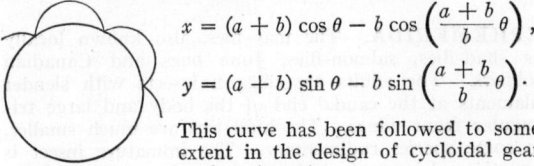

$$x = (a + b) \cos \theta - b \cos \left(\frac{a + b}{b} \theta \right),$$

$$y = (a + b) \sin \theta - b \sin \left(\frac{a + b}{b} \theta \right).$$

This curve has been followed to some extent in the design of cycloidal gear teeth. (L.L.S.)

Epicycloid.

EPIDEMIC ENCEPHALITIS. This term was first applied to the von Economo type of encephalitis, or "encephalitis lethargica" which occurred in a worldwide epidemic beginning in 1917 and was most prevalent in North America between 1919 and 1926. Curiously, it has never reappeared. The disease was a serious one with a mortality rate of 40–100%. The etiological agent was never isolated though a **virus** was suspected. The clinical picture was characterized by extreme lethargy, coma, and weakness and paralyses, particularly of the cranial nerves.

In recent years, the **arthropod**-borne encephalitides have formed a group of growing frequency and importance. At present, infections produced by the following viruses are classified in this group: Western and Eastern equine encephalitis, Japanese B, St. Louis, Russian spring-summer, and louping ill of sheep. The virus in these encephalitides is transmitted by the bite of an insect—a species of mosquito or tick, depending on which disease is involved. In Western and Central United States, western equine (which causes encephalitis in horses as well as man), and St. Louis viruses are present in birds and domestic fowl; although these animals remain well, mosquitoes may acquire the virus by feeding on them, and in turn transmit the disease to man. These two types of encephalitis are of greatest importance in the United States. Eastern equine is rare, and Japanese B and Russian spring-summer do not occur in North America so far as is known.

The clinical picture in the arthropod-borne encephalitides is similar but varying in severity. Fever, headache, stiff neck, paralyses and stupor may occur. Fatality rates for St. Louis and Western equine infections vary between 5 and 22%. Treatment is non-specific, and residual paralyses and mental deterioration are common. A **vaccine** made from infected chick embryos conveys temporary immunity to eastern and western equine varieties. (D.M.H.)

EPIDEMIOLOGY. That branch of medical science which is concerned with the study of disease as it appears in its natural surroundings, and as it affects a community of people rather than a single individual. Epidemiology is largely concerned with infectious diseases; it has a statistical as well as an experimental side. The statistical includes the gathering of facts about the incidence, mortality rate, relation of climate, age, sex, race and many other factors to the appearance of a given disease. From these data, valuable information which is important in controlling epidemics of disease is obtained. In experimental epidemiology, the investigator may produce epidemics in laboratory animals for the study of certain problems, or he may go out in the field and study epidemics in man by laboratory means. Such work is important in elucidating the cause, the modes of transmission, and the insect vectors possibly concerned, as well as other facts of fundamental importance in public health. (D.M.H.)

EPIDERMIS. In insects, the outer layer of the noncellular cuticula; here the cellular layer is called the hypodermis. In other invertebrates the cellular layer covering the body is called the epidermis. In vertebrates,

the epidermis is the outer cellular layer of skin. The human epidermis consists of four layers, from without inward as follows: (1) a layer of horny flattened cells, (2) a layer of transparent cells, (3) layers of granular cells, (4) a layer of rounded pigmented cells. (R.S.M., A.W.L.)

EPIDERMOPHYTOSIS (Athlete's Foot). A contagious **fungus** infection most commonly seen on the foot, especially between the toes. Mild cases show only cracked skin between the toes, but in the severe forms the skin may be macerated, fissured, blistered and infected secondarily with bacteria. The disease may spread to the dorsum and sole of the foot, the fingers, and other parts of the body. Itching is usually marked. Allergic eruptions to the fungus, especially on the hands, may occur; these are called epidermophytids.

The fungus flourishes in damp places such as showers, swimming pools and Turkish baths, and the infection is thought to be spread by contact with contaminated objects in these places.

Thorough drying between the toes is the best preventitive. Local applications such as sulfur-salicylic acid ointments and dilute iodine are effective in curing the chronic disease, but gentler treatment must be employed in the acute phase. (D.M.H.)

EPIDIABASE. Epidiorite.

EPIDIORITE. A term applied to gabbros, **dolerites**, and **diabases**, the **augite** of which has been partly altered to **hornblende**, thus approaching a **diorite** in mineral composition. The term is derived from the Greek meaning upon, plus diorite. (E.S.C.S.)

EPIDOTE. The mineral epidote is a basic orthosilicate of **calcium, aluminum** and **iron** whose formula may be written $HCa_2(Al,Fe)_3Si_3O_{13}$. The ratio of aluminum to iron ranges from 6:1 to 3:2. Epidote is found in prismatic **monoclinic** crystals, which may be acicular to fibrous. Fine granular and compact masses are common. The mineral displays one good cleavage, an uneven fracture; is brittle; hardness, 6–7; specific gravity, 3.25–3.5; luster, vitreous to resinous; typical color, pistachio green, but may be yellowish- to brownish-green, sometimes red, yellow, gray, white or colorless. Colorless to grayish streak; transparent to opaque. The characteristic color of ordinary epidote makes it usually an easily identified mineral. It occurs commonly in metamorphic rocks as gneisses and schists; however, it seems probable that under certain conditions it may appear as a primary mineral, for example in granitic rocks. The Urals, Austria, Switzerland, Italy, France and Norway are known for their occurrences of fine epidote crystals. In the United States epidote has been found in excellent specimens at Franconia and Warren, New Hampshire; Huntington, Massachusetts; Willimantic and Haddam, Connecticut; Chaffee County, Colorado, and Riverside County, California. The word epidote is derived from the Greek. The name pistacite, from the Greek word meaning the pistachio nut, has been occasionally applied to this mineral. It has been used as a gem stone but is in little demand for this purpose. (E.S.C.S.)

EPIGENETIC. A term used by **petrologists** to denote physical and chemical changes, particularly in **igneous** and **sedimentary rocks**, which are clearly secondary to (later in time) the conditions under which the rock originated. This term is commonly used by the students of ore deposits to designate minerals formed after the enclosing wall rocks, in contrast to those minerals formed contemporaneously with the wall rocks. The latter minerals are said to be syngenetic. In the case of the sedimentary rocks the term is used to describe textures, structures and mineral aggregates, of non-metamorphic origin, which have originated during the post-lithification history of the formation. Thus flint, chert, concretions, etc., may be described as being either epigenetic or syngenetic. (R.M.F.)

EPIGLOTTIS. A flap developed from the floor of the **pharynx** of terrestrial vertebrates which covers the opening to the **respiratory system** (glottis) while the animal swallows. (A.W.L.)

EPIGYNUM. A plate above the opening of the reproductive duct of the female **spider.** (A.W.L.)

EPIGYNY. Fruit.

EPILEPSY. A symptom complex characterized by recurrent convulsive seizures. In approximately 75% of cases, no cause can be assigned to the attacks (idiopathic or genetic epilepsy) while in 25%, described as acquired epilepsy, the causes are brain injury, congenital defect or birth injury to the brain, brain tumor, infections of the brain, or interference with cerebral blood supply due to vascular disease. The **pathology** is thus varied, except in the idiopathic type, where no organic changes can be demonstrated at autopsy.

Epilepsy occurs in roughly 0.5% of the population of the United States—or in about 600,000 individuals. Although epilepsy per se is not inherited, the predisposition to the disease is inherited. The pattern of the brain waves as shown on an **electroencephalogram,** a record of the electrical **potentials** of the brain, is an hereditary trait. Abnormal patterns, called cerebral dysrhythmia, are characteristic of an individual who is predisposed to epilepsy even though he may never develop seizures. If electroencephalograms are taken of the members of an epileptic's family, it will be found that a large proportion of them have cerebral dysrhythmia and are, therefore, potential epileptics.

Attacks of epilepsy usually begin in childhood or adolescence in the idiopathic type. Both sexes are equally affected. The seizures vary from transient loss of consciousness (petit mal) to violent **convulsions** (grand mal). Petit mal occurs without warning, and is characterized by sudden loss or impairment of consciousness which comes on abruptly, lasts a few minutes and ceases abruptly. Psychomotor seizures are similar, but longer attacks, marked by complete **amnesia** and loss of contact with the environment, without loss of activity. The individual may perform purposeful acts, but after the confusion clears, he has no recollection of what happened to him. Grand mal begins with a warning **aura** in about 50% of cases. The aura consists of various strange indescribable sensations, dizziness, and sometimes discomfort in the abdomen. It may be very brief and followed immediately by loss of consciousness, rigidity of the muscles, and cessation of breathing. If the patient is standing, he falls to the floor. His face becomes a livid blue and the veins engorge; perspiration and saliva flow. When breathing begins again, a period of violent intermittent muscle jerkings begins. At the same time frothy saliva is blown from the lips as deep respiratory movements begin. A severe convulsion is followed by heavy sleep, lasting several hours. On awakening, the individual often vomits, complains of headache and muscular soreness, and is depressed in spirits. Convulsions occur at varying intervals, sometimes only once a year, or as often as many times a day. If the seizures follow one another in rapid succession without the regaining of consciousness between attacks, the condition is described as status epilepticus. One-sided seizures, or those localized to one part of the body, are in general due to acquired epilepsy, or some recognizable brain lesion.

The treatment of acquired epilepsy is the removal, if possible, of the cause. In idiopathic epilepsy, it resolves itself into attempts to prevent the seizures by various anti-convulsant drugs. Of these phenytoin sodium (dilantin sodium), phenobarbital, and bromides are the most effective, and in the order named. During a convulsion care should be taken to protect the individual from injury. It is important to put some soft substance between the teeth to prevent biting of the tongue. It is useless to attempt to cut short a single attack, although status epilepticus may be shortened by phenobarbital hypodermically.

A certain proportion of epileptics, after years of seizures, deteriorate mentally. The number so affected is not as great as was formerly believed; probably not more than 10% of severe epileptics ever exhibit marked deterioration, and 25% mild changes. The unpleasant traits in the so-called epileptic personality—moodiness, egotism, quarrelsomeness, occur primarily in the group showing some deterioration; they are rare in the mentally normal.

The prevention of epilepsy is a problem in **eugenics.** Ideally, not only the epileptic, but all individuals with cerebral dysrhythmia should marry partners with normal brain waves in order to reduce the existing chance (about 1 in 40) of their children having epilepsy. (D.M.H.)

EPINASTY. Movement in Plants.

EPIPHARYNX. A fold on the inner surface of the upper lip (labrum) of the insect mouth. (A.W.L.)

EPIPHRAGM. A tough membrane of calcified mucus with which some **snails** close the aperture of the shell during periods of drought. (A.W.L.)

EPIPHYTES. A striking feature of tropical forests is the abundance of plants which grow attached to other plants. These attached plants are called epiphytes, which means plants growing on other plants. They are found both on the main trunk and on the branches, often far above the ground. In many cases the epiphyte grows on the under side of a branch to which it is firmly fastened by its roots. Epiphytes gain nothing but support, and a more favorable position of growth because of better light conditions and other environmental factors; they do not obtain any nutrients from the supporting plants, as parasites would. Particularly noteworthy among epiphytes are many **ferns, aroids** and especially **orchids.** Frequently as in the orchids, the **roots** of the plants are so modified as to absorb water directly from the atmosphere.

In the strictest meaning of the word, many lower plants including **algae, fungi, lichens** and **mosses** are epiphytes, since they are often found growing on other plants, not only in tropical regions but in temperate regions as well. (R.M.W.)

EPIPODITE. Biramous appendage. The epipodites of some **crustaceans** are filamentous or branching gills. (A.W.L.)

EPIPODIUM. A lateral ridge on the molluscan foot. (A.W.L.)

EPISTASIS. The masking of one hereditary character by another which is genetically independent. (See **Heredity.**) (A.W.L.)

EPISTAXIS. Hemorrhage from the nose. Nosebleed. (R.S.M.)

EPISTOME. A projection above the mouth of some **bryozoans.** (A.W.L.)

EPITHELIUM. Tissue which covers surfaces and lines hollow organs, and the derivatives of these tissues, whether solid or hollow. All epithelial tissues are made up of closely associated cells with very little intercellular material and most of them have one surface free and the other connected with an underlying tissue.

Epithelia, are classified according to the form of **cells**, the number of layers, and the embryonic origin and location. In flat, pavement, or squamous epithelium the cells are much thinner than their diameter. Cuboidal epithelium is made up of cells approximately as thick as their width. They are not strictly cuboidal but are polyhedral prisms. Columnar epithelium contains cells that are much higher than their diameter. Glandular epithelia are usually of the two latter forms. Any of these forms of cells may occur in more than one layer as a stratified epithelium, but flat epithelia are more often stratified and thicker cells usually form a single layer, or a simple epithelium. In some cases the epithelium is made up of cells of several forms in two or three layers, which change with movements of the part. Such a tissue is called transitional. Others appear to have several layers of cells but all are attached to the underlying tissue, rising to various heights. This is a pseudo-stratified epithelium.

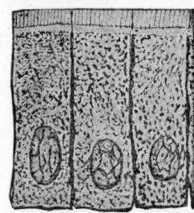

Simple columnar epithelium from intestinal lining. (*Kimber and Gray, Textbook of Anatomy and Physiology, Macmillan Co.*)

According to origin and position two kinds of epithelia derived from the mesoderm are recognized. Of these, endothelium lines the circulatory organs and mesothelium lines the body cavity. Other special types of epithelium are derived from each of the three germ layers. The free surfaces of some bear cilia.

The cells of many epithelia produce special secretions and in some cases these glandular layers are highly developed to form massive structures known as **glands**.

Goblet cells from epithelium lining large intestine. (*Kimber and Gray, Textbook of Anatomy and Physiology, Macmillan Co.*)

Most epithelia rest on a thin basement membrane or membrana propria derived from the connective tissues. (A.W.L.)

EPITOKE. The posterior sexual portion of the body in certain **annelids**. In one of the Pacific species this part of the worm breaks off when mature and rises to the surface. In this stage it is the palolo worm of the Pacific Islands. The anterior asexual portion (atoke) remains in the burrow and produces another epitoke. (A.W.L.)

EPOCH. A subdivision of the geologic time scale which is divided as follows: Era, Period, Epoch and Age. (R.M.F.)

EPSOM SALTS. Magnesium Sulfate.

EQUAL ARM BALANCE. Balance.

EQUALITIES. An equality is a statement that two mathematical expressions are equal. Equalities are of two kinds: identical equalities or **identities**, and conditional equalities or **equations**. (L.L.S.)

EQUALIZER. It is very difficult to construct apparatus and particularly **lines** for communication circuits which will pass all the necessary frequencies equally. Often it is necessary to insert networks whose function is merely to restore the original relation of the various frequencies. These are called equalizers or, sometimes,

inverse networks. Usually they are loss circuits which will cause more loss in those frequencies which had been attenuated less and the opposite for those attenuated the most. Thus the equalizer shows a frequency response which is the inverse of the system it is intended to equalize so the result of connecting it in the circuit is to restore the over-all response to a flat characteristic. The term equalizer is also used for the series of connections sometimes made in d-c machines to insure equal current distribution in the conductors. (L.R.Q.)

EQUATION OF A LOCUS. The equation of a given locus is the **equation** which is satisfied by the **coordinates** of all points of this locus, and of only such points. (L.L.S.)

EQUATION OF STATE FOR AIR. For any gas the equation of state per mol is:

$$PV = RT;$$

where
V = volume;
P = pressure;
R = universal gas constant;
T = abs. temperature.

For one gram this becomes

$$PV = \frac{RT}{m},$$

where m is the molecular weight of the gas. For G grams this becomes,

$$PV = \frac{GRT}{m}.$$

This equation is valid for each of the constituent gases of the atmosphere.
For nitrogen,

$$P_n V = \frac{G_n RT}{m_n}.$$

For oxygen,

$$P_o V = \frac{G_o RT}{m_o}.$$

For argon,

$$P_a V = \frac{G_a RT}{m_a}.$$

For water vapor,

$$P_w V = \frac{G_w RT}{m_w}.$$

When there is no water vapor present in the atmosphere, these equations can be combined as follows:

$$P_t V = (P_n + P_o + P_a)V$$

$$= RT \left[\frac{G_n}{m_n} + \frac{G_o}{m_o} + \frac{G_a}{m_a} \right] = RT \left(\frac{G_t}{m_t} \right).$$

In these equations P_t is the total pressure of the nitrogen, oxygen, and argon. Also G_t is the total mass of the gases, and m_t is the molecular weight of the mixture. m_t has a numerical value of 28.97.

Because water vapor is always present in varying quantities in the atmosphere, corrections in the equation of state must be made in accordance with the amount of water vapor present. Procedure is as follows:

$$PV = (P_t + P_w)V$$

$$= RT \left[\frac{G_t}{m_t} + \frac{G_w}{m_w} \right] = RT \left[\frac{G}{m_t} - \frac{G_w}{m_t} + \frac{G_w}{m_w} \right].$$

In these equations P is the total pressure of the air gases plus the water vapor, and G is the total mass of the air gases plus the water vapor.

This equation can be rearranged and simplified:

$$PV = RT \frac{G}{m_t} \left[1 - \frac{G_w}{G} \left(1 - \frac{m_t}{m_w} \right) \right].$$

m_t has a value of 28.97 and m_w has a value of 18.00. This equation is easily reduced to

$$PV = RT \frac{G}{m_t} \left[1 + 0.6 \frac{G_w}{G} \right].$$

Virtual temperature of the air is defined as,

$$T' = T \left[1 + 0.6 \frac{G_w}{G} \right].$$

Virtual temperature is in effect the temperature of a mass of dry air having the same density of another mass of air containing water vapor. Virtual temperature is always greater than real temperature except when G_w is nil.

The equation of state for real air becomes

$$PV = \frac{RT'G}{m_t}.$$

If R, the universal gas constant, is made into a specific gas constant for air by letting $\frac{R}{m_t} = R_a$, then for one gram of air the equation of state becomes $PV = R_a T'$.   (P.E.K.)

EQUATION OF TIME. Time.

EQUATIONS.

An equation is an **equality** in which both members are equal for certain particular values of the symbols involved but not for all values. To emphasize this, an equation is sometimes called a conditional equality or a conditional equation.

An equation is indicated by putting the equality sign = between the two sides or members of the equation.

An equation with one unknown is an equality which holds for certain particular values of the unknown but not for all values of the unknown.

Equations are classified into many types. There are equations with one or more unknown **numbers** and equations with one or more unknown **functions**. Equations with one or more number unknowns are classified into **algebraic equations** and **transcendental equations**; each of these two classes is further subdivided into other sub-classes. Equations with functional unknowns are classified into **differential equations**, **integral equations**, **difference equations**, functional equations, etc.; each of these classes is also divided into subclasses.   (L.L.S.)

EQUATIONS, CHEMICAL. Chemical Changes.

EQUATIONS OF MOTION. Kinematics.

EQUATOR.

A plane perpendicular to the axis of rotation of the **earth** and passing through the center of the earth will intersect both the surface of the earth and also the **celestial sphere** in great circles. These great circles are known as the terrestrial and celestial equators. Some navigators refer to the celestial equator as the equinoctial.   (W.K.G.)

EQUATORIAL AIR. Air Masses.

EQUATORIAL COORDINATES.

Equatorial coordinates are a system of **spherical coordinates** in which the origin may be the eye of the observer (in which we have the apparent system of coordinates), the center of the earth (geocentric system), or the center of the sun (heliocentric system). The fundamental line in this system is the line joining the poles of rotation of the earth which cuts the **celestial sphere** in its poles of rotation. The plane perpendicular to the fundamental line through the origin is the celestial **equator**. The fundamental direction in the plane may be either the point of intersection of the **local meridian** with the celestial equator, which is above the horizon, or the **Vernal equinox**.

To locate an object in this system of coordinates, a plane is passed through the object and the line joining the poles of rotation and this plane will cut out a great circle, known as an **hour circle**, on the celestial sphere perpendicular to the plane of the equator. The **declination** of an object is the angular distance of the object north (+) or south (−) of the celestial equator measured in the plane of the hour circle through the object. The **hour angle** of the object is the angular distance measured in the plane of the equator from the point of intersection of the meridian above the horizon to the point of intersection of the hour circle through the object in the direction of apparent rotation (west) of the celestial sphere. The right ascension of the object is the angular distance, measured in the plane of the equator from the vernal equinox to the point of intersection of the hour circle in a direction (east) contrary to the direction of apparent rotation of the celestial sphere. For purpose of convenience both right ascension and hour angle are frequently expressed in units of hour, minutes and seconds of time, rather than the more common angular notation of degrees, minutes, and seconds of arc.

Due to the fact that the local meridian remains fixed as the celestial sphere apparently rotates, the hour angle of an object is continually changing. Since both the vernal equinox and the hour circle rotate with the celestial sphere both the right ascension and declination of the object remain fixed as the sphere rotates. However, both right ascension and declination change slowly due to **precession** and **nutation**. In tabulating these coordinates in **star catalogues** the values are given for the position of the equinox for some particular date and the corrections necessary to reduce the positions to the present date must be applied.   (W.K.G.)

EQUATORIAL FRONT. Fronts.

EQUATORIAL TELESCOPE.

A telescope so mounted that it may be moved parallel to the **equatorial coordinates** of **hour angle** and **declination** is known as an equatorial telescope, or simply as an equatorial. In this form of mounting one axis, known as the polar axis, is parallel to the axis of rotation of the earth, and the other axis, about which the telescope may be rotated, is perpendicular to the polar axis and known as the declination axis. From this arrangement of the axes, as the telescope itself is rotated about the declination axis it must move in a plane perpendicular to the **equator**, hence parallel to hour circle or in the direction of declination. The rotation of the declination axis about the polar axis, with the telescope remaining fixed in declination, will cause the telescope to move parallel to the equator, hence in the coordinate of hour angle.

The majority of equatorials are carried on a single pier. This form of mounting has the difficulty that the telescope will frequently run into the pier when the hour angle is close to zero. To avoid this several other methods of supporting the polar axis have been devised. Perhaps the most common is the so-called English mounting, in which the two ends of the polar axis are supported on separate piers, with the telescope free to pass through zero hour angle.

The most difficult adjustment of the equatorial is to get the polar axis strictly parallel to the axis of rotation of the earth. When this has been accomplished the instrument is by far the most convenient of all forms of mounting. If the instrument is rotated about the polar axis from east to west at exactly the same rate that the earth is rotating about its axis from west to east, the telescope lens will remain fixed relative to objects on the celestial sphere. Hence once the instrument is set on a star, clock work may be devised to keep the telescope "following."

Many ingenious modifications of the equatorial mounting, such as the Coudé, have been devised. Most such modifications have the comfort of the observer in mind and sacrifice both light and good definition to attain this end. (w.k.g.)

EQUIANGULAR SPIRAL. Logarithmic Spiral.

EQUIDAE. Pleistocene.

EQUILATERAL HYPERBOLA. Hyperbola.

EQUILIBRIUM, BIOLOGICAL. The state of coordination which maintains an animal in normal posture.

Equilibrium of aquatic animals such as the fishes is maintained by the resistance of the surrounding water in relation to specialized body form, by muscular movements of body and fins, and by the gas-filled swim bladder. The bodies of most fishes are heavier above, as is shown by their floating back downward when dead, but the combination of these factors maintains their erect position.

Terrestrial animals maintain their posture by constant muscular adjustment in response to stimuli received by sensory organs in the soles of the feet and in the muscles and tendons. A portion of the inner ear of vertebrates is also a center of equilibrium. End organs in the semicircular canals of this organ are stimulated by movement in the liquid filling the canals when the animal moves. The three canals lie in the three planes of space so that at least one is activated by any movement. The results of their reaction are transmitted to one of the lower brain centers, whence the proper impulses are relayed to the muscles.

Equilibrium in flight demands very delicate coordination of essentially the same type. In insects and bats it is supposed to be accomplished partly through delicate sense organs located in the wings. (a.w.l.)

EQUILIBRIUM, CHEMICAL. The fundamental law of chemical equilibrium is that enunciated by Le Chatelier (1884), and may be stated as follows:—If any stress or force is brought to bear upon a system in equilibrium, the equilibrium is displaced in a direction which tends to diminish the intensity of the stress or force. This is equivalent to the "principle of least action." Its great value to the chemist is in that it enables him to predict the effect upon systems in equilibrium of changes in temperature, pressure, and concentration.

The chemical system, **hydrogen-nitrogen-ammonia**, furnishes a notable example of the application of the principle.

Nitrogen + Hydrogen $\rightleftarrows$ Ammonia + Heat
1 vol.   3 vol.   2 vol.   12,000 calories
                                        per mol ammonia
———————————————
        4 vol.

At the temperature 700° C. and pressure 1 atmosphere, the equilibrium percentage of ammonia is 0.03 in the above system, and at 100 atmospheres 2.5. Increase of pressure shifts the equilibrium towards the side of the smaller total volume, at a constant temperature. Decrease of pressure shifts the equilibrium towards the side of the larger total volume, at a constant temperature. Systems of the same initial and final volumes are unaffected, as to equilibrium amounts of materials, by change of pressure.

At the pressure 100 atmospheres, and temperature 700° C., the equilibrium percentage of ammonia is 2.5 in the above system, at 600° C. it is 5, at 500° C. it is 10. Increase of temperature shifts the equilibrium in the direction which absorbs heat, at a constant pressure. Decrease of temperature shifts the equilibrium in the direction which evolves heat (van't Hoff's principle, 1884).

At constant pressure and temperature, the equilibrium is shifted away from the side subjected to an increase in concentration of any constituent, or towards the side subjected to a decrease in concentration of any constituent. (See **Chemical Changes**.) For the qualitative effect of temperature change, one may visualize the heat of an equilibrium reaction as material, and an increase of temperature (heat intensity) as operating to increase the concentration of "heat material" thus shifting the equilibrium away from the side of its increased concentration, and conversely. It is possible, knowing the heat of reaction, Q, on the assumption that the heat of reaction is constant between two given (absolute) temperatures, T_1 and T_2, to calculate the equilibrium constant K_2 (at T_2) when the equilibrium constant K_1 (at T_1) and the gas constant, R (equals 2 calories per mol), are known, by the application of van't Hoff's equation:

$$\log_{10} K_2 - \log_{10} K_1 = \frac{Q}{2.3 \times R} \left(\frac{1}{T_2} - \frac{1}{T_1} \right).$$

In this way the quantitative effect of temperature change on the state of equilibrium may be calculated.

In reactions of the ammonia synthesis type, to which **sulfur trioxide** from **sulfur dioxide** plus **oxygen** also belongs, the rate of reaction decreases with lowering of the temperature as the conversion is increased. There is, in such types of reactions, a limit to the practicable lowering of the temperature. The finding of a positive **catalyzer** for a given reaction of this sort permits the operation to gain the advantage of equilibrium conversion at the lower temperature as well as the increased rate of reaction at that temperature due to the presence of the catalyzer. (See **Chemical Changes**.) The time yield of product is, therefore, very important, and, with a catalyzer, the space-time yield.

Systems in equilibrium are divided into two great divisions, according to whether they are (A) homogeneous, that is, chemically and physically uniform throughout, or (B) heterogeneous, that is, not uniform throughout but consisting of two or more phases. Each phase is a homogeneous, physically distinct, and mechanically separable portion of a system. For example, ice, water, water vapor are three different phases (solid, liquid, gas) of the substance water. There can be only one gas phase of a system, and only one liquid phase where a *single* homogeneous solution is present. But the number of liquid and of solid phases in general is limited by the number of components (not constituents) of a system. The number of components is the least number of constituents, *independently* variable, and requisite to compose each and every phase. For example, the system consisting of saturated solution in water (H_2O) of **sodium** sulfate (Na_2SO_4) plus solid sodium sulfate decahydrate ($Na_2SO_4 \cdot 10H_2O$) plus water vapor consists of three phases (a) gas, (b) solution, (c) solid sodium decahydrate. The *least* number of constituents, independently variable in amount *and* requisite to compose each and every phase is two, namely, Na_2SO_4 and H_2O. These, therefore, are the two components of this system. Since zero and negative as well as positive amounts of compounds are permitted in expressing the composition of each phase of any system, the three phases of this system are composed of the following components:

(a) Gas phase,   zero Na_2SO_4 plus H_2O
(b) Liquid phase, Na_2SO_4 plus H_2O
(c) Solid phase, Na_2SO_4 plus H_2O

The number of components in the ice-water-water vapor system is one, namely, H_2O.

To systems in which equilibrium depends solely upon the following variables, namely, (1) composition of each and every phase, (2) temperature, and (3) pressure, the phase rule (Willard Gibbs, 1874) applies: The number of variables, that is (1) the number of components, C, plus (2) temperature plus (3) pressure, above, equals the number of phases, P, plus the number of degrees of freedom, F. The number of degrees of freedom of a system is the least number of the above variables which must be arbitrarily fixed in order to define the condition of the system.

$$C + 2 = P + F.$$

The phase rule applies to true equilibrium systems, where the equilibrium can be reached from either side, and, furthermore, takes no account of the time involved to attain equilibrium. The phase rule is a qualitative statement, whereas the law of mass action (concentration effect) is quantitatively applicable to those equilibrium systems where the reaction which occurs may be considered to take place in a homogeneous system, e.g., gas phase, or solution phase. (See **Chemical Changes.**)

In a one-component system, $P + F = 3$, and physical changes only occur. When only one phase is present, for example, liquid water (no vapor, no solid) the system is bivariant, that is, two variables—temperature and pressure—may be independently changed over a range. When a second phase, either vapor or solid appears through a sufficient change of temperature or pressure or both, or when two phases are originally present, the system is univariant, that is, one variable—either temperature or pressure—may be independently changed over a range. When the third phase appears or when three phases are originally present, the system is invariant, that is, a change of either temperature or pressure destroys the equilibrium, and the disappearance of one of the phases occurs. A system of one component in three phases is invariant and the conditions are represented by a point known as the triple point. The triple point for water is 0.007° C., 4.6 mm. mercury pressure. When the total pressure is one atmosphere (760 mm.) the equilibrium temperature of water-ice is 0.000° C., and when the water vapor pressure is one atmosphere the equilibrium temperature of water-water. vapor is 100.000° C.

If, in dealing with any system, the gas phase or pressure may be neglected, on account of constancy or slightness of effect, the phase rule is simplified for practical purposes to $C + 1 = P + F$, and, if both may be neglected, to $C = P + F$.

Many two- and three-component systems have been studied and recorded in detail. The **iron-carbon** system is one that has attracted much attention and been of great value in iron metallurgy. (R.K.S.)

EQUILIBRIUM MOISTURE CONTENT. Drying.

EQUILIBRIUM OF FORCES. A state of balance between or among forces. The much used term equilibrium is here confined to its dynamical sense; such subjects as thermal equilibrium, radioactive equilibrium, etc., are treated in appropriate places elsewhere. Unless otherwise specified, the term refers to that set of conditions to which a system of forces must be adjusted in order that a free body acted upon by them will experience no acceleration. This is somewhat illogically termed "static equilibrium," to distinguish it from the "kinetic equilibrium" with which **D'Alembert's principle** is concerned.

Two conditions are necessary for the equilibrium of a set of forces. 1. The vector sum of the forces must be zero (see **Vector Addition**); then if they are resolved into rectangular components, the algebraic sums of the X, the Y, and the Z components must separately reduce to zero. 2. The algebraic sum of the torques of the forces about each of any three mutually perpendicular axes must be zero; the body then has no tendency to rotate about any axis. (See **Statics and Graphical Statics.**)

A body or a set of bodies under the action of forces may be in stable, unstable, or neutral equilibrium. For any one of these, a very slight displacement or change

of relative position may be regarded as taking place without change of energy, a fact known as the "principle of virtual work." If the energy is really constant, the equilibrium is neutral; if it is at a minimum, the equilibrium is stable (see **Least Energy Principle**); if it is at a maximum, the equilibrium is unstable. The degree of stability in the second case may be expressed as the amount of energy which must be supplied to render the system unstable. These various ideas may be illustrated by a body suspended by a string, a sphere resting on a horizontal plane, a pencil balanced on its point; and by blocks, pyramids, etc., on a table. (L.D.W.)

EQUINOCTIAL. Equator.

EQUINOX. The line of intersection of the plane of the earth's **equator** with the plane of the **ecliptic** (the **line of nodes** of the earth) intersects the **celestial sphere** in two diametrically opposite points known as the equinoxes. As seen from the earth the sun apparently passes through each of the equinoxes once each year, passing through the vernal equinox on approximately March 21st and the autumnal equinox on approximately September 21st.

The great circle passing about the **celestial sphere** through the equinoxes and the pole of the ecliptic is known as the equinoctial colure. (W.K.G.)

EQUIPARTITION OF ENERGY. If a great number of perfectly elastic, rapidly moving particles are turned into an enclosure together and are allowed time to mingle, darting about and striking or otherwise encountering each other, the kinetic energy which they possess becomes distributed in accordance with the famous principle of equipartition of energy, or Maxwell-Boltzmann law, as enunciated by Boltzmann. Each particle has a number of **degrees of freedom**, determined by its character. (For the molecules of a diatomic gas, such as nitrogen, the effective number is 5; plus some others not ordinarily concerned with thermal energy.) The equipartition principle states that the average energy taken up by motions in each of the several degrees of freedom is the same, and is independent of the sizes and masses of the particles. (For a gas it is equal to $\frac{1}{2}$ the product of the Boltzmann constant by the absolute temperature of the gas.) Thus when heat energy is imparted to a pure diatomic gas, $\frac{1}{5}$ of it goes into each degree of molecular freedom which heat can affect. Three of these degrees of freedom are concerned with motions of translation, so that $\frac{3}{5}$ of the energy takes this form. And indeed, when we calculate the change in translational energy due to rising the temperature of one gram of the gas one degree (see **Kinetic Theory**), it is found to be almost exactly $\frac{3}{5}$ of the specific heat as measured at constant volume, which represents the total imparted energy. One of the strongest supports of the principle comes from quantitative observations on the **Brownian movement.** Many other examples occur in physics. (L.D.W.)

EQUIVALENT CIRCUIT. This term is applied to an electrical circuit which is electrically equivalent to another circuit, or sometimes, to a mechanical device. Equivalent circuits of mechanical systems or electromechanical systems such as loud-**speakers** enable the designer to apply methods of circuit analysis and often obtain a solution easily which would be very difficult if not impossible otherwise. The equivalent circuit method is used extensively in the analysis of communications circuits, particularly those involving vacuum **tubes.** The tube itself may be replaced, for instance, by a generator having a generated voltage equal the amplification factor of the tube times the applied grid voltage and an internal resistance equal the dynamic **plate resistance** of the tube. While this does not give the d-c solution of the tube circuit it does allow the circuit to be simplified

for alternating currents and these are usually the ones of interest. Similarly, many other types of electrical circuits may be simplified in terms of equivalent circuits, sometimes giving all the necessary solutions, sometimes giving solutions for limited conditions, but, in most cases, greatly decreasing the labor involved in analyzing the circuit or equipment. (L.R.Q.)

EQUIVALENT EQUATIONS. Two equations in one unknown are said to be equivalent when they have the same **roots.**

An equation is changed to an equivalent equation by the following transformations:

1. When the same number or expression is added to or subtracted from each side of the equation.

2. When both sides of the equation are multiplied or divided by the same number or expression, provided that number is not 0 and the expression does not contain the unknown. (L.L.S.)

EQUIVALENT MONOPLANE. The monoplane wing which would duplicate some of the important aerodynamic characteristics of a multiplane (such as a sesquiplane, a biplane or a triplane) is termed an "equivalent monoplane." In studies of performance and stability it is much simpler to use the air stream reactions on an equivalent monoplane than to work with the interacting forces on two or more wings. The span of the monoplane of equivalent **induced drag** is kb, where b is the multiplane wing span and k is the "apparent span factor." In former years k was extensively investigated, but in recent aviation history the trend has been away from multiplanes on account of their inferior aerodynamic performance. The air forces acting on the equivalent monoplane were considered to be localized on an "equivalent mean aerodynamic chord" and theories were derived for estimating the location of this imaginary chord. (F.T.M.)

EQUIVALENT PARASITE AREA. Flat-Plate Area.

EQUIVALENTS, CHEMICAL. Chemical Composition.

EQUUS. Fossil Mammals.

ERA. One of the five major subdivisions of the **geologic time-scale.** The formations that belong in any one era are spoken of as a group. (R.M.F.)

ERBIUM. Symbol: Er. Atomic number: 68. Atomic weight: 167.2. Density: 4.77. (**Isotopes:** page 290.) Type of compound: Er_2O_3. Color of salts: red. Discovered by Mosander in 1842. A member of the **yttrium** sub-group of the rare earth metals. (R.K.S.)

EREPSIN. Enzymes.

ERG. The c.g.s. absolute unit of **work** and of **energy.** It is the centimeter-dyne; that is, the work done, or the energy transferred, when the continuous exertion of one dyne of force upon a body is accompanied by a displacement of one centimeter in the direction of the force. The erg is so small that often the more convenient **M.K.S. unit,** called the joule, is used; 1 joule equals 10^7 ergs. One foot-pound is approximately 1.355 joules. (L.D.W.)

ERGOMETER. Power.

ERGOSTEROL. Alcohols and Ethers; Vitamins.

ERGOT. A **fungus** which grows upon and replaces the grain or rye. (See **Diseases of Plants.**) Ergot in the body contracts smaller arteries and smooth muscle fibers —especially that of the **uterus.**

Preparations of ergot are used medically to cause contraction of the uterus after childbirth to check hemorrhage. It is questionable whether it has any abortifacient properties in early pregnancy.

During recent years a form of ergot has been used for the treatment of migraine.

Continued use of ergot produces chronic spasm of the small arteries in the hands and feet, producing gangrene. In the past gangrene commonly occurred in certain European countries from eating bread made from ergot-infected rye. (See **Ascomycetes.**) (R.S.M., D.M.H.)

ERGOTIMINE. Alkaloids.

ERGOTOXINE. Alkaloids.

ERIOMETER. Diffraction.

ERITHRITE. The mineral erythrite is a rather rare mineral of secondary origin. Chemically it is a hydrous **cobalt arsenate** corresponding to the formula $Co_3As_2O_8 \cdot 8H_2O$, the cobalt being replaced at times by **nickel, iron** or **calcium.** It is found in **monoclinic** crystals, masses, and in crusts with other cobalt minerals. It is a soft mineral; hardness, 1.5–2.5; specific gravity, 2.95; luster, adamantine to vitreous although massive varieties may be dull to earthy; color, red, occasionally gray; transparent to nearly opaque. It is known from Bohemia, Saxony, Baden; Chalantes, France; Tunaberg, Sweden; Cornwall and Cumberland, England; Cobalt, Province of Ontario, Canada; and in the United States from Nevada and California. Erythrite takes its name from the Greek meaning red. (E.S.C.S.)

ERMINE. Mammalia, Carnivora. The white winter phase of various species of **weasels** and the stoat of the Old World. These animals turn white only in colder northern latitudes and always retain the black-tipped tail with which royal ermine is spotted. The fur is fine but its value is less than is popularly supposed. (A.W.L.)

EROS. Of all of the more than 1300 **asteroids** which have been discovered up to the present time, Eros has probably attracted the most attention. The interest in this object comes from the fact that when this object comes to **opposition** at the time of **perihelion** passage it is distant from the earth by only 13,840,000 miles, or is closer to the earth than any other member of the **solar system** whose orbit is accurately known, except the moon. At the time of such close approach the **parallax** of Eros is nearly 60″ and is larger than that of any other member of the solar system. Hence, observations made of this object when at close approach provide accurate determinations of its parallax, and, with the **orbit** of the planet accurately determined, **solar parallax** may be more accurately determined than from any other known object. These favorable close oppositions of Eros come at rather widely separated intervals, those since discovery coming in 1901 and 1931. At both of these times many accurate observations, both visual and photographic, were made and from them a number of important astronomical constants have been determined.

Because of the close approach of Eros to the earth the orbit of Eros suffers large **perturbations** from the earth. A study of these perturbations provides data for the determination of the ratio of mass of the earth and the sun. With this mass ratio determined the solar parallax may be accurately determined.

Eros itself is a small object, between 15 and 20 miles in diameter. The light from it varies at times with a period of about 5 hours, and the regularity of the light variation can only be explained on the hypothesis that the object is in rotation with this period. At the time of the close opposition of 1931 observations made in South Africa indicated that Eros is not spherical, but is "dumb-bell shaped."

Several other asteroids have been discovered which come closer to the earth than does Eros. These objects are very minute, and at present the observational material is too meager to permit of accurate determination of their orbits. (W.K.G.)

EROSION. The general term referring to the reduction of the land surface toward sea level by the various agencies of weathering, stream action, glacial action, wind action, etc. A similar term is denudation. The individual and combined processes of erosion are the primary surficial agencies in the sculpturing of topographic features and the development of scenery. Erosion also provides the material from which are formed the bulk of the fragmental, or **clastic, sedimentary rocks.** (R.M.F.)

ERRANTIA. A division of marine segmented **worms** including chiefly free-swimming species. Sometimes ranked as an order and sometimes as a division of the order Polychaeta. (A.W.L.)

ERRATIC. In geology, an ice-carried **boulder** or block, sometimes weighing many tons, which because of its lack of similarity to the bedrock or formation on which it rests, and the peculiarity of its position, must have been transported to its present resting place by a glacier or an iceberg. When erratics occur in sufficient quantity to form relatively pronounced topographic features these are called moraines. When erratics of similar or identical **lithology** show a well-defined lineal distribution from the parent outcrop they are called boulder trains. (R.M.F.)

ERROR. The term "error" is employed in many ways in statistics. It may refer to an error in calculation, but more often to an **error of observation,** or an **error of sampling.** Other uses of the term occur in the standard error of estimate, the **standard error** of a statistic, the types of error in the **Neyman-Pearson** theory of testing hypotheses, and the **experimental error** in the **analysis of variance.** Thus the term "error" generally is used in a technical sense and not in its ordinary meaning of a "mistake" or an "incorrect value." (L.A.A.)

ERROR OF ESTIMATE. The standard error of estimate is

$$\sigma_e = \sqrt{\frac{\Sigma(y - y^1)^2}{N}},$$

where y is the actual value of the **variable** and y^1 is the value predicted by an equation, $N =$ number of items in the **sample.** The error of estimate in a particular instance is $y - y^1$. (L.A.A.)

ERROR OF OBSERVATION. Let y be an observed value of a **variate** and y^1 its correct value. Then we define an error of observation as $y - y^1$. Errors of observation are caused by experimental conditions. As long as all such errors are not in the same direction, no serious **bias** should arise. Usually we assume this to be the case. (L.A.A.)

ERROR OF SAMPLING. An error of sampling depends on the particular **sample** chosen. Each sample gives rise to various **statistics.** The difference between a statistic and the population value may be termed the sampling error. A sampling error is not a "mistake" in the common meaning of that term. It is an unavoidable consequence of the fact that samples vary. (L.A.A.)

ERRORS OF MEASUREMENT. Much of the routine work in any experimental research in the physical sciences is concerned with the eliminating, minimizing, or compensating for observational errors. By the error of any measurement is meant the result of the individual measurement minus the true value of the quantity measured. It may thus be either positive or negative. Errors may be broadly classified into two types: persistent or systematic errors, due to causes which endure throughout the whole series of observations and which, therefore, affect each individual observation alike; and accidental or chance errors, due to a combination of random influences which may be changed completely from observation to observation. These two classes may be illustrated, respectively, by measurements of a length with a ruler which is too long or too short, and by measurements of an angle with a surveyor's **transit** which is being shaken by the wind. Mistakes or blunders, due to the misreading of an instrument, etc., are not considered as errors of observation, but rather as errors of the observer.

Various standard devices for dealing with persistent or systematic errors will be found discussed under **physical measurements.** The treatment of accidental errors is discussed briefly in the article on **least squares.** (L.D.W.)

ERUPTION. A breaking-out of the skin, marked by various degrees of redness, swelling, and elevation. There are many types of eruption, some of which are characteristic of the disease which they accompany and are important in the diagnosis of these diseases. A drug eruption is one occurring after the use of a **drug** to which the individual is allergic. A plant eruption occurs after exposure of a sensitive subject to irritating substances found in certain plants, poison ivy for example. (D.M.H.)

ERUPTIVE. This term has the same general geological meaning as **effusive,** but is sometimes used in the much more general sense as synonymous with **igneous.** (R.M.F.)

ERYOPS. Fossil Amphibia.

ERYSIPELAS. An acute **inflammation** of the skin caused by infection with *streptococcus hemolyticus.* It is accompanied by high fever, **leucocytosis,** and sometimes **septicemia** and severe prostration. Infection usually takes place at the site of injury, sometimes a break in the skin too small to be visible. The face is the most common starting point of the infection. A small area of skin over the cheek becomes red and swollen. The margin of the lesion is raised and well demarcated; it spreads rapidly to cover the nose and involve the other cheek, giving the characteristic butterfly appearance. **Abscess** formation at the site of infection may occur as a complication. The most successful treatment is with sulfadiazine or other **sulfonamides** or **penicillin.** Relapses and recurrences in the same site are not uncommon. (D.M.H.)

ERYTHROCYTE. Blood.

ERYTHROXYLON. Coca.

ESCAPE VELOCITY. Rocket Flight.

ESCARPMENT. Used by physiographers and structural geologists to denote a line of steep slopes or cliffs. (See also **Cuesta** and **Scarp.**) (R.M.F.)

ESKER (OSE; plural, OSAR). Certain long, often winding, ridges formed of stratified sands and gravels, which occur within the glaciated regions of Europe and North America are called eskers. These pronounced topographic features are frequently several miles in length and because of their peculiar and uniform shape somewhat resemble railway embankments. Eskers represent the deposits of glacial streams which flowed within and under glaciers. After the retaining ice walls melted away the stream deposits remained as long winding ridges. (R.M.F.)

ESOPHAGEAL GLAND.

Glands associated with pouches on the sides of the **esophagus** in earthworms. They secrete **calcium** carbonate. Also called calciferous glands. (A.W.L.)

ESOPHAGUS.

The tube-like passage which connects the lower end of the **pharynx** with the stomach. It is about 10″ long, lies in front of the spine as it descends through the chest, and passes through the **diaphragm** just before it enters the stomach. Its walls are made up of circular and straight muscle fibers which allow for wave-like contractions to progress from above downwards. The inner surface contains many glands which secrete mucus for lubrication of its walls. The function of the esophagus is to mechanically pass food from the pharynx into the stomach. The main disorders affecting it are **tumors,** diverticula and stricture. Stricture usu-ally results from the swallowing of corrosive substances. (R.S.M.)

ESSENCES. Extracts.

ESSENTIAL OILS. Hydrocarbons; Volatile Oils.

ESSEXITE. Gabbro.

ESTABLISHMENT OF A FORT. Tides.

ESTERIFICATION.

Esterification is illustrated by the union of an alkoxy group such as C_2H_5O— and an acetyl group such as CH_3CO— in which case ethyl acetate ester $CH_3COOC_2H_5$ results. This is commonly accomplished by the use of an alcohol to supply the RO— group and an acid for the RCO— group, with the simultaneous formation of water HOH. The reaction takes time and is not instantaneous, and an equilibrium state is also established when sufficient time for this is allowed to elapse. The equilibrium investigated by Berthelot (1863) between acetic acid plus ethyl alcohol *and* ethyl acetate plus water is reported in the article on **Chemical Changes,** III. The End-point of Chemical Reactions. The rates of esterification and the equilibrium points for various acids and alcohols were studied by Menschutkin during the period 1879 to 1909, and the results are presented in the two following tables.

RATE OF ESTERIFICATION AND EQUILIBRIUM POINT FOR VARIOUS ALCOHOLS AND PHENOLS WITH ACETIC ACID IN EQUIVALENT AMOUNTS AT 155° C.

ALCOHOL		% OF ESTER OBTAINED	
Name	Formula	In 1 Hour	At Equilibrium
Primary:			
Methyl..........................	CH_3OH	56	70
Ethyl............................	$CH_3 \cdot CH_2OH$	47	67
n-Propyl.......................	$CH_3 \cdot CH_2 \cdot CH_2OH$	47	67
n-Butyl........................	$CH_3 \cdot CH_2 \cdot CH_2 \cdot CH_2OH$	47	67
Allyl............................	$CH_2:CH \cdot CH_2OH$	36	59
(propene-2-ol-1)			
Benzyl..........................	$C_6H_5 \cdot CH_2OH$	39	61
Secondary:			
sec-Butyl.......................	$(CH_3)_2CHOH$	27	61
(dimethylcarbinol)			
Methylethylcarbinol...............	$(CH_3)(C_2H_5)CHOH$	23	59
Diethylcarbinol...................	$(C_2H_5)_2CHOH$	17	59
Diallylcarbinol...................	$(CH_2:CH)_2CHOH$	10	50
Menthol.........................	(see structure)	15	61
(2-methyl-5-isopropylcyclohexanol-1)			
* *Tertiary:*			
Trimethylcarbinol.................	$(CH_3)_3COH$	1.4	6.6
Dimethylethylcarbinol..............	$(CH_3)_2(C_2H_5)COH$	0.8	2.5
Methyldiethylcarbinol..............	$(CH_3)(C_2H_5)_2COH$	1.0	3.8
Dimethyl-*n*-propylcarbinol..........	$(CH_3)_2(CH_3CH_2CH_2)COH$	2.2	0.8
Dimethyl-*iso*-propylcarbinol.........	$(CH_3)_2((CH_3)_2CH)COH$	0.9	0.9
Phenol..........................	C_6H_5OH	1.5	8.6
Thymol..........................	$C_6H_3(CH_3)(5)((CH_3)_2CH)(2)OH(1)$	0.6	9.5
(5-methyl-2-isopropylphenol)			

* Tertiary alcohols, upon heating, lose water, leaving unsaturated hydrocarbons, and give low yields of esters.

RATE OF ESTERIFICATION AND EQUILIBRIUM POINT FOR VARIOUS ACIDS WITH ISOBUTYL ALCOHOL IN EQUIVALENT AMOUNTS AT 155° C.

ACID		% OF ESTER OBTAINED	
Name	Formula	In 1 Hour	At Equilibrium
Formic.................................	$H \cdot COOH$	62	64
Acetic..................................	$CH_3 \cdot COOH$	44	67
Propionic...............................	$CH_3 \cdot CH_2 \cdot COOH$	41	69
(methylacetic)			
n-Butyric...............................	$CH_3 \cdot CH_2 \cdot CH_2 \cdot COOH$	33	70
(ethylacetic)			
Phenylacetic............................	$C_6H_5 \cdot CH_2 \cdot COOH$	49	74
Phenylpropionic.........................	$C_6H_5 \cdot CH_2 \cdot CH_2 \cdot COOH$	40	72
(benzylacetic)			
* iso-Butyric...........................	$(CH_3)_2 CH \cdot COOH$	29	70
(dimethylacetic)			
* Methylethylacetic.....................	$(CH_3)(C_2H_5)CH \cdot COOH$	22	74
* Trimethylacetic.......................	$(CH_3)_3 C \cdot COOH$	8	73
* Methyldiethylacetic...................	$(CH_3)(C_2H_5)_2 C \cdot COOH$	3	74
* Benzoic..............................	$C_6H_5 \cdot COOH$	9	73
* p-Toluic.............................	$C_6H_4(CH_3)(4)COOH(1)$	7	77
Cinnamic...............................	$C_6H_5 \cdot CH : CH \cdot COOH$	12	75

* The rate of esterification is relatively slow for branched chain aliphatic and for benzenoid acids. The equilibrium point is about 70 per cent (64 to 77 for the above acids) with isobutyl alcohol under the specified conditions.

Norris showed that the esterification reaction proceeds as stated above by use of the two following reactions:

$$CH_3 \cdot CO \cdot SH + H \cdot OC_2H_5 \rightleftharpoons CH_3 \cdot COOC_2H_5 + HSH$$

$$CH_3 \cdot CO \cdot OH + H \cdot SC_2H_5 \rightleftharpoons CH_3 \cdot COSC_2H_5 + HOH$$

and this has been confirmed using acid and alcohol containing the heavy isotope (18) of oxygen.

The esterification reaction can be speeded up by the use of a catalyst. Such a catalyst is hydrogen ion say from hydrochloric or sulfuric acid. Side reactions may occur, hydrochloric acid furnishing some organic chloride, and sulfuric acid causing dehydration of the alcohol. Phosphoric acid generally avoids both these results. Salts that hydrolyze to furnish hydrogen ions are also used, e.g., zinc chloride, aluminum sulfate, ferric chloride, sometimes by the addition of acid to these salts.

The equilibrium point can be displaced to produce more ester than is shown in the tables by increasing the relative amounts of either the acid or the alcohol as desired.

A complicating factor of considerable significance when recovery of the ester is to be made by distillation is the existence of 2-component (binary) azeotropes of constant boiling points with (1) ester and water and (2) ester and alcohol, and 3-component (ternary) azeotropes with (3) ester and water and alcohol, selected data on which are contained in the following table.

The industrial importance of esterification may be understood from the following remarks on esters, all of the esters made synthetically.

The annual production in the U. S. is approximately: normal, secondary, and isobutyl acetates 100,000,000 lbs., ethyl acetate 95,000,000 lbs., normal, secondary, and isoamyl acetates 3,000,000 lbs., methyl acetate 1,000,000 lbs. These acetates are used largely in the lacquer industry. Cellulose nitrate (nitrocellulose) is used in the plastics (30,000,000 lbs.), lacquer, and industrial and military explosives industries, cellulose acetate in the plastics (20,000,000 lbs.) and lacquer industries, glyceryl trinitrate (nitroglycerin) in the industrial (20,000,000 lbs.) and military explosives industry, cellulose xanthate (viscose) is an important synthetic fiber for textiles. In the specialized field of plasticizers for the plastics and lacquer

ESTER [Alcohol]	B.P. °C.		AZEOTROPE		
			B.P. °C.	Ester %	Other Component %
Ethyl acetate [Ethyl alcohol	77.1 78.3]	a	70.4	93.9	6.1 water
		b	71.8	69.2	30.8 alcohol
		c	70.3	83.2	{ 7.8 water { 9.0 alcohol
n-Propyl acetate [n-Propyl alcohol	101.6 97.2]	a	82.4	86.0	14.0 water
		b	94.7	49.0	51.0 alcohol
		c	82.2	59.5	{ 21.0 water { 19.5 alcohol
iso-Propyl acetate [iso-Propyl alcohol	91.0 82.5]	a	77.4	93.8	6.2 water
		b	81.3	40.0	60.0 alcohol
		c	....	....	....
n-Butyl acetate [n-Butyl alcohol	126.2 117.8]	a	90.2	71.3	28.7 water
		b	117.2	53.0	47.0 alcohol
		c	89.4	35.3	{ 37.3 water { 27.4 alcohol
n-Amyl acetate [n-Amyl alcohol	148.8 137.8]	a	95.2	59.0	41.0 water
		b	....	....	....
		c	94.8	10.5	{ 56.2 water { 33.3 alcohol

The ternary azeotrope has the lowest boiling point and in distillation, as long as the three components are present, it distills off first, then the binary azeotrope distills off as long as its components are present. Since esters are not very soluble in water, water can be added to the distillate, and then water plus alcohol are removed. Upon redistillation the ester soon distills off pure.

Many variations in details are used in applying the principles of esterification.

industries numerous synthetic esters are used, as examples, butyl stearate, diamyl phthalate, dibutyl oxalate, dibutyl phthalate (also for smokeless powder), dibutyl sebacate, dibutyl tartrate, diethylene glycol monostearate, diethylene glycol distearate, diethyl phthalate, dimethyl phthalate, diphenyl phthalate, glyceryl tripropionate, isobutyl phthalate, tributyl borate, tributyl citrate, tributyl phosphate, tricresyl phosphate, triethylene glycol dihexoate, triethylene glycol dioctoate, triethyl citrate, triethyl phosphate, triphenyl phosphate. Methyl methacrylate ester is an important plastic. Ethyl silicate is used to cover concrete, brick and stone with a coating of silicic acid to resist water penetration. Dioctyl phthalate is used as a plasticizer in cable and wire insulation, and dimethyl phthalate as an insect repellant. (See Esters.) (R.K.S.)

ESTERS, INCLUDING OILS, FATS, WAXES.

An ester is a compound, which by reaction with water, acid or alkali, forms an alcohol plus an acid (a salt of the acid is formed when alkali is used). The reverse reaction, namely, reaction of an alcohol plus an acid, accompanied by the elimination of water, is the most important general method of formation of an ester. Of the alcohols concerned with ester composition or formation, the most common are methyl alcohol, ethyl alcohol, glycol, glycerol, cellulose. The range of acids important in the study of esters is wider than that of the alcohols, and the importance is determined largely by the point of view. On the scientific side, esters of all acids are important in extending the knowledge of the acid, frequently important in the determination of unknown acids and some are individually important. On the industrial side, esters of nitric acid are important in explosives, e.g., glyceryl trinitrate (nitroglycerin, dynamite), cellulose hexanitrate (nitrocellulose, guncotton), glycol dinitrate, in plastics (pyroxylin and celluloid), photographic films, lacquers, rayon, and in collodion; esters of acetic, propionic, butyric acids in plastics, photographic films, rayon, lacquers, e.g., cellulose acetate, ethyl butyrate; esters of stearic, palmitic, oleic, linolic acids, in vegetable and animal oils and fats for food, in paints and varnishes, for soaps, fatty acids, glycerol; esters of certain acids, e.g., salicylic acid methyl ester (oil of wintergreen), in odoriferous substances and perfumes. The esters of each acid are listed under the acid, as follows: (1) carboxylic acids: acetic, benzoic, carbonic, citric, formic, lactic, malic, oleic, oxalic, palmitic, salicylic, stearic, tartaric, (2) nitrogen acids: nitric, nitrous, hydrocyanic, cyanic, (3) sulfur acids: sulfuric, sulfurous, thiocyanic, (4) phosphorus acids: phosphoric, hypophosphoric, phosphorous. In addition, in the cases of those acids not discussed individually, the following esters are noted:

olefin acid), ricinoleate (a hydroxy-olefin acid). A few alcohols other than glycerol are found in the esters of animal waxes, e.g., cetyl alcohol (as palmitate) in the head oil of the sperm whale, myricyl alcohol (as palmitate) in beeswax. There are various methods of classification of oils, fats and waxes. The classification presented on the opposite page gives a selected list of members, accompanied by selected physical and chemical constants.

Saponification Number represents the number of milligrams of potassium hydroxide (KOH) required, under specified conditions, to saponify one gram of the sample.

Iodine Number represents the number of grams of iodine absorbed, under specified conditions, by 100 grams of the sample. Hübl solution contains iodine and mercuric chloride in absolute alcohol. Wijs solution contains iodine monochloride in glacial acetic acid.

Titer Test. From the sample, the free fatty acids are made, and then cooled slowly, the temperature being observed by means of a thermometer whose bulb is immersed in the cooling liquid. When turbidity appears in the liquid the temperature remains stationary or rises slightly. The temperature of the top point of the rise is the titer of the sample.

Reichert Meissl Number represents the number of milliliters of decinormal alkali required to neutralize the volatile fatty acid distillate, obtained, under specified conditions, from 5 grams of the sample. For most oils and fats the number is small, of the order of 0.5, but in the case of butter fats is high, between 20 and 34. Coconut oil gives a value between 6.5 and 7.5, palm kernel oil between 5 and 7, and palm oil between 1 and 2.

Drying oils are those which, upon exposure to the air, gradually form a permanent dry film. Esters of linolic and linolenic acids are present in those oils, and the di- and triolefin acid radicals absorb oxygen from the air and produce this film. Linseed and similar oils are used mainly in paints and varnishes. Some fish oils, e.g., salmon oil, also are used in the paint industry.

Edible oils for food and cooking are mainly non-drying or semi-drying oils and fats. The consumption in this

	ESTER	FORMULA	MELTING POINT	BOILING POINT
1.	Methyl propionate	$C_2H_5COOCH_3$		80° C.
2.	Ethyl propionate	$C_2H_5COOC_2H_5$		99° C.
3.	Methyl normal-butyrate	$C_3H_7COOCH_3$		102° C.
4.	Ethyl normal-butyrate	$C_3H_7COOC_2H_5$		121° C.
5.	Methyl iso-butyrate	$(CH_3)_2CHCOOCH_3$		93° C.
6.	Ethyl iso-butyrate	$(CH_3)_2CHCOOC_2H_5$		112° C.
7.	Methyl normal-valerate	$C_4H_9COOCH_3$		127° C.
8.	Ethyl normal-valerate	$C_4H_9COOC_2H_5$		145° C.
9.	Methyl iso-valerate	$(CH_3)_2CHCH_2COOCH_3$		116° C.
10.	Ethyl iso-valerate	$(CH_3)_2CHCH_2COOC_2H_5$		135° C.
11.	Dimethyl malonate	$CH_2(COOCH_3)_2$		180° C.
12.	Diethyl malonate	$CH_2(COOOC_2H_5)_2$		199° C.
13.	Dimethyl succinate	$(CH_2COOCH_3)_2$	20° C.	195° C.
14.	Diethyl succinate	$(CH_2COOC_2H_5)_2$		216° C.
15.	Methyl cinnamate	$C_6H_5CH : CHCOOCH_3$	33° C.	263° C.
16.	Ethyl cinnamate	$C_6H_5CH : CHCOOC_2H_5$	12° C.	271° C.

Many esters are found in natural materials, and have uses. Such esters are those found in animal fats, e.g., beef fat for tallow, hog fat for lard, fish oils, and in vegetable oils, e.g., olive oil, corn oil, cottonseed oil, linseed oil. Fatty oils are stored in the plant mainly in the seed, e.g., cottonseed, flax seed (linseed oil), or in a portion of the seed, e.g., corn germ, coconut meat (copra when dried), and in the animal organism localized as such, e.g., beef tallow, hog lard, or distributed in the flesh, in milk, e.g., butter fat, or in liver, e.g., cod liver oil. In animal and vegetable oils and fats the predominating esters are glyceryl stearate, palmitate, oleate (an olefin acid), linoleate (a diolefin acid), linolenate (a tri-

field is the largest of all the outlets for oils, fats and waxes. The edible natural fats, butter, lard, tallow, have been supplemented in recent years by processed materials, such as oleomargarine, supplement to butter, and vegetable shortening, supplement to lard. Oleomargarine may be made by separation at a suitable temperature of liquid oleo oil from solid stearin of beef tallow, and then mixing oleo oil, coconut fat and cottonseed oil in the desired proportions and flavoring with butter fat or carefully ripened milk. Vegetable shortening is made from the liquid oils, usually cottonseed, peanut, corn, soy bean, by reaction with hydrogen in the presence of finely divided nickel which is later re-

VEGETABLE AND ANIMAL OILS, FATS AND WAXES—A SELECTED CLASSIFIED LIST

		% Oil and Fat in the Seed	Solidification Point, °C.	Specific Gravity	Index of Refraction n_D at °C.	Saponification Number	Iodine Number	Titer Test, °C.
A. Vegetable Oils:								
1. Drying......	1. Linseed........	40	−27	0.927–0.932	1.4725, 15°	188–195	170–190	19–21
	2. Chinawood * ...	55	− 3	0.936–0.943	1.51, 20°	190–195	150–175	37
	3. Perilla..........	30–40	....	0.932–0.937	1.483,	190–194	187–200	
	4. Walnut........	65	−27	0.92–0.93	1.4808, 20°	192–197	142–146	16
	5. Soy bean.......	20–25	− 8	0.920–0.926	1.4673, 40°	185–195	128–135	21–24
	6. Rapeseed.......	35–40	−2 − +10	0.913–0.916	1.475,	170–179	94–105	11–13
2. Semi-drying..	7. Cottonseed.....	25–35	....	0.922–0.927	1.4643, 40°	193–195	105	33–38
	8. Corn..........	10	−10 − −20	0.9105	1.4768, 15°	187	117	15–19
3. Non-drying...	9. Peanut (arachis, earthnut)....	45	0	0.916–0.925	1.4612, 40°	185–196	85	28–29
	10. Olive..........	40–65	−6 − +2	0.916–0.918	1.4698, 15°	185–196	79–90	17–21
	11. Olive kernel....	15	....	0.92	1.468, 25°	181–184	86–87	
	12. Castor........	50	−10 − −18	0.969	1.480, 15°	183–186	82–86	3
B. Vegetable Fats....	13. Coconut........	40–45	22	0.925	1.4488, 40°	233–253	7–10	20–22
	14. Palm..........	65–70	30–40	0.865–0.873	1.4503, 40°	196–205	11–83	42–45
	15. Palm kernel....	45–50	23	0.912	1.4431, 60°	242–250	10–18	20–24
	16. Cacao butter...	45	22–27	0.950–0.970	1.457, 40°	192–198	3.5	48–50
C. Animal Oils: Marine								
a. Fish.......	17. Salmon........		....	0.9258	1.478, 20°	182–188	160–190	
	18. Menhaden.....		−4	0.927–0.933	1.480, 15°	191–196	142–180	
	19. Herring........			0.92–0.93	1.48,	180–194	140	
b. Liver......	20. Cod-liver.....		0 − −10	0.92–0.93	1.48, 15°	187–197	150–180	17–18
c. Blubber....	21. Whale.........		....	0.871–0.878	1.46, 25°	122–144	80–93	22–24
	22. Seal...........		−2 − +3	0.924–0.926	1.474,	189–196	130–190	13–19
D. Animal Fats: Terrestrial Non-drying								
a. Body fats..	23. Beef tallow.....		27–35	0.943–0.952	1.451, 60°	193–200	35–46	43–45
	24. Hog lard.......		27–30	0 935	1.4539, 60°	195	52–77	36–39
b. Foot oils..	25. Neat's foot oil..		<0	0.92	1.469, 20°	194–199	58–70	20–26
c. Milk fats..	26. Butter fat.....		19–24	0.926–0.940	1.4650, 60°	216–240	26–38	36–38
E. Animal Waxes....	27. Spermaceti.....		42–47	0.945–0.960	....	123–135	4–7	
	28. Bee's wax......		60–63	0.96–0.97	1.4439, 75°	88–98	8–12	

* Also known as Tung Oil.

moved at the proper temperature. The gradual addition of hydrogen results in a product of less liquid and more solid, up to the limit of **hydrogenation,** when the product resembles closely beef tallow. At the desired stage the process is terminated, and the product used as shortening or for frying.

By reaction with water, acid or alkali, the oils and fats form glycerol plus fatty acids, which latter separate as salts when alkali is used. Addition of sodium chloride aids in the separation. These salts are soaps, and the ordinary hard soaps are sodium salts of stearic, palmitic, oleic acids. Potassium salts are jelly-like, and are known as soft soaps. Sodium and potassium soaps are soluble in water and yield a froth, known as lather or suds, when agitated in water. Calcium soaps are insoluble in water, and yield no froth upon agitation. When natural hard waters containing dissolved calcium or magnesium compounds, e.g., the bicarbonate, chloride or sulfate, are used with sodium soaps in washing and cleansing, calcium soaps separate as insoluble precipitates in proportion to the amount of calcium and magnesium contained in the water. This represents loss of soap for cleansing purposes as well as diminished froth and the accumulation of undesirable precipitate in the material being cleansed. Aluminum soaps are also insoluble but are used in certain lubricants and greases. Certain soaps of cobalt, manganese, lead, aluminum, are used as varnish dryers. Glycerol is recovered by distillation under diminished pressure (vacuum distillation) of the water

or sodium chloride solution portion, after separation of the sodium soaps.

When the sodium soaps are dissolved in water and dilute sulfuric or hydrochloric is added in slight excess, the free fatty acids of the soaps separate as an insoluble oily or fatty portion. These fatty acids may be separated into fractions as desired by fractional crystallization and filtration, and by fractional distillation under diminished pressure. The stearic acid portion is utilized in the manufacture of candles, when mixed with more or less paraffin wax to secure the desired consistency. The oleic acid portion may be hydrogenated to stearic acid, or treated with concentrated sulfuric acid and naphthalene to form a reagent, known as Twitchell's reagent, for converting oils and fats into glycerol plus fatty acids. While much of the soap which is made utilizes oils and fats that are in the edible class, inedible tallow and recovered greases are important sources. (See **Surface-Active Compounds; Esterification.**)

Other uses of oils and fats are for lubrication of machinery, ranging from light spindle oils to heavy greases, and in medicine, e.g., castor oil, chaulmoogra oil, cod-liver oil.

Oils and fats are recovered from seeds by pressing, cold or hot, or by extraction with a volatile solvent and then distilling off the solvent. Butter fat is recovered from milk by centrifuging. Tallow and lard are purified by heating to melting, the finer qualities being those recovered by less drastic heating and the lower qualities by more drastic heating.

When oils and fats are exposed to air and light, or contain enzymes, rancidity is developed. This is prevented by careful purification of the oil or fat and exclusion of air. Rancidity is hastened in the presence of free fatty acids in the oil or fat. (For further data on individual oils, see articles on **Cocoa, Castor, Coconut, Corn, Cotton, Olive, Palm, Peanut,** and **Soy Bean.**) (R.K.S.)

ESTRIN. Sex Hormones.

ETHANE. Ethane (C_2H_6 or $CH_3 \cdot CH_3$) is a colorless, odorless gas, boiling point —88° C., density 1.36 grams per liter at 0° C. and 760 mm. (specific gravity 1.05, air equal to 1.00), practically insoluble in water, moderately soluble in alcohol, burns when ignited in air with a pale faintly luminous flame, forms an explosive mixture with air over a moderate range, with excess air the products are **carbon dioxide** plus water, with deficiency of air carbon monoxide plus water. Ethane is among the chemically less reactive organic substances. It reacts, however, with **chlorine** (and with **bromine**) to form mixtures of chloro- (and bromo-) substitution compounds (one-half of the chlorine used forms hydrogen chloride). Ethane occurs, usually in small proportions, in natural gas. The fuel value of ethane is very high, 1730 B.T.U. per cu. ft. Ethane may be prepared by reaction of magnesium ethyl iodide in anhydrous ether (**Grignard's reagent**) with water or alcohols. Ethyl iodide (bromide, chloride) is preferably made by reaction of ethyl alcohol and **phosphorus** iodide (bromide, chloride). Important ethane derivatives, by successive oxidation, are (1) **ethyl alcohol** (C_2H_5OH or CH_3CH_2OH), (2) **acetaldehyde** (C_2H_5O or CH_3CHO), (3) **acetic acid** ($C_2H_4O_2$ or CH_3COOH). (R.K.S.)

ETHER. This term is used, with entirely different meanings, in chemistry and in physics. In chemistry it is used to designate a series of compounds which are discussed in this book under the heading **Alcohols and Ethers.** Chemists also use the term ether to designate a particular member of the series, that is, diethyl ether, ($C_2H_5)_2O$. This compound is also known as sulfuric ether, and as ethoxyethane. It is a colorless liquid, of characteristic odor, boiling point 35° C., inflammable and explosive with air when ignited by fire or electric spark. It is slightly soluble in water (1 volume ether in 10 volumes water) and slightly dissolves water (3 volumes water in 100 volumes ether), is miscible in all proportions with alcohol, dissolves **iodine** and many organic substances, e.g., **oils** and **fats**, no reaction with ordinary acids or alkalis, or with **sodium** or cold **phosphorus** pentachloride. Ether is prepared by treatment of **ethyl alcohol** with concentrated **sulfuric acid** in excess at about 140° C. For use as an **anaesthetic**, ether must be scrupulously pure. Ether is used (1) as an anesthetic by inhalation of the vapor, (2) as a **solvent** in the preparation of explosives and of collodion, and in the extraction of oils, fats, waxes, gums, resins, alkaloids.

In physics, the concept of the ether had its origin in the necessity for explaining the propagation of light and the existence of electric, magnetic, and gravitational fields of force. When it became evident that **light** is a wave phenomenon, the search for a medium became imperative. While sound does not traverse a vacuum, light does so perfectly; and electric, magnetic, and gravitational attractions are not interrupted by removing the intervening air. Yet all attempts to corral and examine portions of this supposed residual substance and to ascertain its properties by direct observation have totally failed. If the ether is actually a wave-propagating medium, the speed of the waves is such as to imply an extraordinarily great rigidity or a vanishingly small density, or both. (See **Vibrations and Waves.**)

One mode of attack has been to seek for evidence of an "ether drift," that is, a relative motion of matter with respect to the ether through which it moves; just as one traveling through still air experiences a wind. The results of such quests have proved ambiguous. The **Michelson-Morley experiment** gave a completely negative result. So did the Trouton-Noble experiment, which was an attempt to detect the electromagnetic effect of the motion of the electric charges in a **condenser**, as carried by the earth in its orbit. On the other hand, Airy's experiment on the **aberration of light** might be interpreted as indicating that when a transparent substance moves, the ether is dragged along, not with the full speed of the moving matter, but with a fraction of that speed expressed by Fresnel's so-called "coefficient of drag." This fraction is equal to $\dfrac{n^2 - 1}{n^2}$, in which n is the **refractive index** of the transparent medium; and its validity was further attested by the experiments of Fizeau on the propagation of light in rapidly moving water. The difficulties presented by such conflicting evidence have been avoided by Einstein, who is his theory of **relativity** dispenses with all assumptions as to the supposed stationary ether or motion with respect to it. (See **Anaesthesia.**) (R.K.S., L.D.W.)

ETHER DRIFT. Ether.

ETHYL ALCOHOL. Ethyl alcohol, ethanol, "grain alcohol" (C_2H_5OH or CH_3CH_2OH) is a colorless liquid, of pleasant odor, melting point —117° C., boiling point 78.5° C., constant minimum boiling point mixture—distillate—with water 78.15° C., 4.4% water, 95.6% ethyl alcohol, miscible in all proportions with water or ether, soluble in **sodium** hydroxide solution, when ignited burns in air with a pale blue, transparent flame producing water plus carbon dioxide, the vapor forms an explosive mixture with air and may be used in **internal combustion engines** under high compression. Ethyl alcohol reacts (1) with **sodium** metal forming sodium ethoxide, (C_2H_5ONa) plus hydrogen gas, (2) with **phosphorus** chloride, bromide, iodide, forming ethyl chloride, bromide, iodide, respectively, (3) with **sulfuric acid** concentrated, forming at 100° C. ethyl hydrogen sulfate ($C_2H_5OSO_2OH$), at 140° C. diethyl ether ($(C_2H_5)_2O$), at 200° C. ethylene ($CH_2:CH_2$), (4) with organic **acids**, warmed in the presence of sulfuric acid, forming **esters**, e.g., ethyl acetate ($CH_3COOC_2H_5$), ethyl benzoate ($C_2H_5COOC_2H_5$) (see various individual acids), (5) with magnesium methyl iodide in anhydrous ether (**Grignard's** solution) forming **methane** as in the case of primary alcohols, (6) with **calcium** chloride to form a solid addition compound ($4C_2H_5OH \cdot CaCl_2$), which is decomposed by water, (7) with **oxygen**, using **sodium** dichromate solution and sulfuric acid, to form **acetaldehyde** (and **acetic acid**), using air, in the presence of acetic bacteria, to form vinegar (dilute acetic acid along with the substances present in the alcohol used, e.g., wine, cider), (8) with **nitric acid** (a) concentrated, free from nitrogen tetroxide, to form ethyl nitrate, (b) dilute to form glycollic acid, (c) concentrated acid containing nitrogen tetroxide (fuming nitric acid) explosive reaction, (9) with **chlorine** (or **bromine**) to form chloral (CCl_3CHO) (or bromal).

The density of pure ethyl alcohol is 0.789, at 20° C., compared with water at 4° C. (the corresponding figure for methyl alcohol is 0.792) and the percentage of ethyl alcohol present in an ethyl alcohol-water solution may be determined from the density of the sample. Anhydrous ethyl alcohol, "absolute alcohol," may be obtained by removal of water by the reaction of the water with calcium oxide and then distilling the alcohol, or by distillation of ethyl alcohol-water with benzene.

Ethyl alcohol is made (1) by **fermentation** of many **carbohydrates;** directly from dextrose or levulose of fruit juices, e.g., grape, apple, and indirectly from sucrose, maltose, starch, cellulose after conversion of the latter group into the former. When starch (usually from corn or potatoes) is used, the starch is changed to maltose by means of diastase of "malt," which is produced by the germination of **barley.** Maltose from starch or sucrose of molasses is fermented by the addition of yeast which generates the **enzymes** maltase (converting maltose to glucose), invertase (converting sucrose to glucose plus levulose), zymase, effective in accomplishing the conversion of glucose to **ethyl alcohol** plus **carbon dioxide** ($\frac{1}{3}$ of the sugar is changed to carbon dioxide) in dilute solution (not over 18% alcohol). When starch or cellulose of wood is treated with dilute acid heated, dextrose is formed (about 2% of fermentable sugars is present in the waste liquor from the sulfite process of making wood into paper pulp) and this may be fermented by yeast, (2) by absorption of ethylene of coal gas or petroleum gas, and subsequent reaction with water, (3) by reduction of acetaldehyde in the presence of a catalyzer. From the solution, ethyl alcohol is separated, recovered and concentrated by fractional distillation. "Proof spirit" is defined by the United States Government as one containing 50% by volume of ethyl alcohol and 50% water, that is, 42.47% by weight of ethyl alcohol and of density 0.930 at 20° C. compared with water at 4° C.

Pharmacology and Medical Uses. The intoxicating agent of alcoholic beverages is ethyl alcohol. It is found in sherry, whiskey, brandy, wine, beer, ale, and all other alcoholic beverages. Considerable knowledge has been acquired about the various effects of ethyl alcohol upon man, and they will be discussed briefly in this article.

In the human body, alcohol is rapidly absorbed from the stomach and upper intestine. It is almost completely oxidized, and in the process energy is released and **carbon dioxide** and water are produced. The heat of combustion of pure alcohol is 7 calories per cc. About 2% of ingested alcohol usually escapes oxidation and is excreted, mainly through the kidneys and lungs.

Once absorbed, alcohol affects nearly every tissue of the body, but particularly the central **nervous system.** Contrary to lay opinion, it is not a stimulant, but a depressant. The feeling of well-being, the expansiveness of personality, impulsive speech and actions are all the result of loss of inhibitions, or removal of the normally restraining influences of the higher cerebral centers. Careful tests have shown that alcohol does not increase either the mental or physical abilities. It is the psychic effect of the drug which results in the individual's estimation of his own performance as greatly improved whereas, actually, his mental reaction time and manual skill are decreased.

On the **cardiovascular system** the chief effect is dilatation of the blood vessels, particularly those of the skin. This results in the commonly observed flushing of the skin following the ingestion of alcoholic beverages, and a feeling of warmth. Actually, however, peripheral vasodilatation and the increased sweating which goes with it result in heat loss, so that the temperature inside the body falls. This fact is particularly important in cold climates, where the irrational use of alcohol to "keep warm" may have serious results.

The **digestive system** is affected in several ways. Moderate concentrations stimulate the flow of saliva and gastric juices which favor digestive processes. This is particularly true of those who enjoy alcoholic beverages. Since the presence of food in the stomach decreases the concentration of alcohol, and therefore lessens any irritative effects, the most desirable way of taking it is with meals. High concentrations are irritative to the digestive-tract mucosa, cause hyperemia and in-

flammation and, in extreme instances, even massive hemorrhage.

Alcohol tends to increase the output of urine, not due to a direct diuretic effect, but because of the accompanying large amounts of fluid usually ingested. Contrary to popular belief, alcohol is not an aphrodisiac. The sexual behavior it provokes follows as a result of loss of inhibitions.

Drunkenness or alcoholic intoxication depends on the concentration of alcohol in the blood. Unlike other foods, it is absorbed unchanged from the gastrointestinal tract and into the circulation. The blood concentration varies greatly depending on the individual's degree of tolerance. Within 5 minutes after ingestion alcohol can be detected readily in the blood. In an abstainer, the concentration gradually reaches a peak and remains high for several hours. In an habitual drinker, the concentration reaches a lower peak more quickly and also falls off rapidly. The blood level which is indicative of intoxication is a matter of dispute. In general, a concentration of 0.20% in blood or urine means some degree of intoxication. Obvious intoxication is present in over 90% of individuals with a blood alcohol concentration of 0.30%, and the fatal range is between 0.50 and 0.80%. The mechanism of tolerance is not well understood. That there is a cross-tolerance to allied hydrocarbon anaesthetics is evidenced by the difficulty of inducing **anaesthesia** with chloroform, ether, or avertin in chronic alcoholics. Addiction to alcohol occurs, but it is unlike morphine addiction in that sudden abstinence need not provoke withdrawal symptoms. The tolerance of the alcohol addict has definite limits—usually only 2 to 4 times that of the occasional drinker.

Acute alcoholic intoxication is manifested by euphoria —a sense of great physical and mental power, noisiness, increased motor activity, progressing to clumsiness, incoordination, staggering gait, a feeling of remoteness, emotional instability, nausea, vomiting, poor control of urination, and finally loss of consciousness, stupor, and profound coma. Emergency treatment directed toward overcoming shock and stimulating respiration may be necessary. Death is rare unless coma persists for many hours, or unless some medical complication such as an acute infection supervenes.

Chronic alcoholism may result in a variety of syndromes, most of which are discussed under their respective headings. Depending on which organ systems are predominantly affected the following disorders may occur: alterations in personality, **delirium tremens,** acute and chronic alcoholic **psychoses** (acute alcoholic hallucinosis, Korsakoff's Syndrome, etc.). Failure to eat an adequate diet results in malnutrition, weight loss, and **deficiency diseases,** particularly **pellagra,** so-called alcoholic **polyneuritis,** and cirrhosis of the liver. The treatment of chronic alcoholism is a psychiatric as well as a medical problem. The chronic alcoholic is most often a mental defective or a psychopathic personality. In the latter case, long-term **psychotherapy** directed toward elimination of personal and social maladjustments is necessary for a "cure" to be effective and lasting. Hospitalization is often required, especially in the acute psychoses where a patient's suicidal and homicidal tendencies are a menace to himself and the community. General care, attention to diet, supplementary **vitamins,** and sedation are often valuable adjuvants to psychotherapy. Relapses are frequent, but the prognosis for recovery is better than in morphine addiction.

The therapeutic uses of alcohol are varied, particularly as interpreted by the laity. Externally, it is used as a rubifacient, to cool the skin and promote evaporation of moisture in fevers, to decrease sweating, and to keep the skin dry and resistant in bed-ridden patients.

Injection of alcohol around nerves or ganglia for the relief of pain is carried out in trigeminal neuralgia (tic douloureux), in certain cases of angina pectoris, and in some patients with inoperable cancer associated with intractable pain.

The ingestion of moderate amounts of alcoholic beverages by those who enjoy them is useful in improving the appetite and digestion, particularly in convalescents, and in aged or debilitated individuals. As a remedy for insomnia, a hot toddy at bedtime is a popular and effective remedy. Because of its vasodilatating effect on the coronary arteries of the heart, alcohol is excellent treatment for angina pectoris. The usefulness of alcohol in checking a "cold" perhaps rests more on the side results—for example, the relaxation and drowsiness induced are apt to keep a patient in bed—rather than on any actual pharmacological effects.

Besides its medical uses, ethyl alcohol is used (1) as an important solvent of chemicals and pharmaceuticals, in varnishes, tinctures, perfumes, (2) as an antiseptic in surgery, (3) for the preservation of hospital and museum physiological and pathological specimens, (4) in the production of ether, esters, and the source of the ethyl group (C_2H_5—) and the ethoxy group (C_2H_5O—), in organic chemical reactions, (5) for the production of vinegar, e.g., from wine, cider, (6) as a liquid fuel, clean and relatively safe, with possibility of extensive use in internal combustion engines, (7) in antifreeze solutions (20% by volume ethyl alcohol, 80% water, freezing point 18° F.; 36% alcohol, 64% water, —2° F.; 54% alcohol, 46% water, —29° F.). Stringent government regulations cover the manufacture, use and sale of ethyl alcohol in all forms. Alcohol not for beverage but for industrial purposes is tax-free. Tax-free industrial alcohol is either used under government supervision or, without such supervision, after being completely denatured by the addition of non-potable substances.

Ethyl alcohol may be detected by the formation of iodoform, a treatment with alkali, potassium iodide, and iodine solution.

Anhydrous ethyl alcohol is made from the constant boiling mixture with water (95.6% ethyl alcohol) (1) by heating with a substance such as calcium oxide, which reacts with water and not with alcohol, and then distilling, or (2) by distilling with a volatile liquid such as benzene (boiling point 79.6° C.), which forms a constant low boiling mixture with water and alcohol (boiling point 64.9° C.), so that water is removed from the main portion of the alcohol; after which alcohol plus benzene distills over (boiling point 68.3° C.), and finally anhydrous ethyl alcohol (boiling point 78.5° C.). Anhydrous ethyl alcohol is demanded for certain purposes as a solvent and reagent, among other applications is that of addition to gasoline or to benzene as motor fuel. In the presence of water, separation into two layers takes place, and the following data apply:

ALCOHOL-GASOLINE MIXTURES

% BY WEIGHT		WATER ADDED PER 100 GRAMS MIXTURE TO CAUSE SEPARATION	% ALCOHOL OF ALCOHOL-WATER PORTION IN PRODUCT
Alcohol	Gasoline		
10	90	0.46 g.	95.6
20	80	1.09	94.8
30	70	1.85	94.2
40	60	2.78	93.5
50	50	3.75	93.0

ALCOHOL-BENZENE MIXTURE

% BY WEIGHT		WATER ADDED PER 100 GRAMS MIXTURE TO CAUSE SEPARATION	% ALCOHOL OF ALCOHOL-WATER PORTION IN PRODUCT
Alcohol	Benzene		
10	90	1.0 g.	90.9
20	80	· 3.0	87.0
30	70	6.1	83.1
40	60	10.8	78.7
50	50	16.1	75.7

(R.S.M., D.M.H., R.K.S.)

ETHYL CHLORIDE. Chlorine; Anaesthesia.

ETHYLENE. Ethylene, ethene (C_2H_4 or $CH_2:CH_2$), is a colorless gas, of slight odor, and acts as an anesthetic when inhaled, boiling point —104° C., density 1.26 grams per liter at 0° C. and 760 mm. (specific gravity 0.97, air equal to 1.00), liquid at 10° C. and 50 atmospheres pressure, insoluble in water, burns when ignited in air with a luminous flame, its presence in coal gas is chiefly responsible for the luminosity of the latter, forms an explosive mixture with air, of high fuel value (1615 B.T.U. per cu. ft.). Ethylene reacts (1) with chlorine (bromine, iodine), to form ethylene dichloride ($C_2H_4Cl_2$ or $CH_2Cl·CH_2Cl$) (dibromide, diiodide), (2) with hypochlorous acid (hypobromous acid), to form ethylene chlorohydrin ($CH_2Cl·CH_2OH$), (bromohydrin), (3) with hydrogen iodide or bromide (not chloride) to form ethyl iodide (C_2H_5I) or ethyl bromide (C_2H_5Br), (4) with hydrogen in the presence of a catalyzer, e.g., finely divided nickel at 150° C., to form ethane (C_2H_6 or $CH_3·CH_3$), (5) with concentrated sulfuric acid at 160° C. to form ethyl hydrogen sulfate ($C_2H_5HSO_4$) from which ethyl alcohol is readily made, (6) with potassium permanganate, to form ethylene glycol ($CH_2OH·CH_2OH$), although glycol is made preferably from ethylene dichloride or chlorohydrin. Ethylene is made by the removal of water from ethyl alcohol, either by passing over heated bauxite, or by heating with concentrated sulfuric or phosphoric acid, and is recovered from gas mixtures in which it is present, e.g., oil-cracking gases where the olefin content may be as high as 35%. Ethylene is used (1) in the preparation of its derivatives, (2) as an anesthetic, (3) as a fuel with oxygen for high-temperature flames, (4) as a coloring and ripening agent for citrus fruits and tomatoes. Ethylene chlorohydrin is used as an agent for decreasing the dormant period of seeds. Fuming sulfuric acid is used as a reagent for the absorption and estimation of ethylene in a mixture of gases. (R.K.S.)

ETIOLATION. This is the effect of darkness on a living plant. It is a matter of common observation that plants grown in dark contain little or no chlorophyll and so are nearly white. Green plants placed in darkness lose their chlorophyll. Eventually, when the food reserves are exhausted, the plant dies.

Besides the lack of chlorophyll, plants grown in the dark have other characteristics. In dicotyledons, the internodes of the stem become excessively elongated and very slender. The leaves fail to expand normally. In monocotyledons the leaves become very long and usually very narrow, but the stem shows little change. Etiolated plants never bear flowers, unless the flower buds are well developed before the plants are darkened.

Internally the tissues are soft and lack strength, the cells being very large and having thin walls. Very little differentiation occurs, the conducting tissues being very much reduced. Leaves which form in darkness show very little of the structure characterizing a normal green leaf, but are almost entirely composed of loosely arranged **parenchymatous** cells.

Formative influence of light. Bean (Phaseolus) grown ten days in light (left), ten days in dark (right).

The cause of the conditions observed in etiolated plants is not at all understood. The cessation of **photosynthesis** and the consequent inability of the plant **to** manufacture foodstuffs does not explain the phenomenon, since no etiolation effects are observed in experiments in which photosynthesis is entirely stopped by withholding **carbon dioxide** from a well-lighted plant. This phenomenon is used to advantage in the growing of certain plants used for salads. French endives are grown in light until a food reserve is stored up, then held in storage, and later forced in darkened rooms. The bleaching of celery is a similar process, produced by covering the leaf petioles with earth or sheltering them with boards or paper. (R.M.W.)

ETIOLOGY. Knowledge of the cause of any disease or abnormal condition. (R.S.M.)

ETTINGSHAUSEN EFFECT. This phenomenon, discovered in 1887, is analogous to the **Hall effect** and appears to be closely related to it. If a strip of metal in which an electric current flows longitudinally is placed in a magnetic field with the plane of the strip perpendicular to the direction of the field, it is found that corresponding points on opposite edges come to different temperatures. If, to one looking along the strip in the direction of the current, and with the magnetic field downward, the decrease of temperature is toward the right, the effect is positive. This is the case with bismuth, in which the phenomenon was first observed by Ettingshausen. The same is true of antimony, nickel, and cobalt; but in iron the effect is negative. (See also **Nernst and Righi-Leduc Effects.**) (L.D.W.)

EUCALYPTUS OIL. Volatile Oils.

EUCLASE. The mineral euclase is a **silicate of beryllium** and **aluminum** corresponding to the formula $Be(AlOH)SiO_4$ which crystallizes in the **monoclinic** system. It has a perfect prismatic cleavage; hardness, 7.5; specific gravity, 3.1; luster, vitreous; is colorless to seagreen or blue. It has been used to a very slight extent for jewelry as its transparent crystals somewhat resemble the aquamarine. Euclase occurs in the Minas Geraes region, Brazil, associated with **topaz** and **beryl**, and also in the Ural Mountains, where it is found in gold-bearing sands. The name euclase is derived from the Greek meaning easiness and fracture, in reference to its easily cleaved crystals. (E.S.C.S.)

EUGENICS. A division of biological science closely related to sociology. It is concerned with the study of human **heredity** and with methods of improving the heritage of human beings.

Sir Francis Galton, a student of human heredity, was a pioneer in suggesting the possibility of securing a better heritage for mankind by the deliberate control of human reproduction. It is evident now that man cannot extend to himself, under his present social system, the degree of control that he applies to his domestic animals, but a few measures have seemed possible. Whether the possibility is more than theoretical is doubtful, but these measures are still advanced as the program of eugenics. They depend chiefly on the birth rate of different classes.

In the United States about 3.4 children per family is the average necessary for the maintenance of the population. This number is equaled only by the portion of the population at the lowest levels of mental development adequate for self-maintenance in society, while families of higher intelligence average far below it. Mental defectives who have to be confined in public institutions are not a serious source of concern, for many of them have no opportunity to reproduce, but, at the level of the low-grade moron, families average slightly more than four children. In the highest levels of the population, too, there is little need for concern. Families average slightly less than three children, but better opportunities probably offset this low figure. The great mass of respectable, self-maintaining, ambitious human beings between these extremes, however, average scarcely a child per family and so are far from maintaining their lines of descent.

The purpose of eugenics to correct this discrepancy demands for its realization some practical means of leveling the birth rate by increase in the upper brackets of society and by decrease in the lower. Sterilization has been suggested for the latter purpose and has been legalized by more than half of the states. It can be applied as a legal compulsion, however, only to institutionalized criminals and defectives, and this application does not reach the parts of society most in need of restriction. To increase the rate of reproduction in the higher middle classes is no easier. Since economic problems are an important cause of the limitation of families, it has been suggested that state subsidies for the care of children would be an aid. Unfortunately practical subsidization in the United States has been forthcoming only as charity and government aid, which have enabled the indigent to continue their overproduction at the expense of their more provident fellow citizens. Education in the serious import of the unbalanced birth rate and in the responsibility of the family to society have also been suggested, but unfortunately the foundations of human existence involve a stronger sense of responsibility to family than to society.

As a result of the lack of feasibility or of promise in the proposed measures of eugenics, the science is in the unenviable position of realizing the needs of mankind for the assurance of its future welfare and improvement, and of being impotent to accomplish any important gains toward the desirable end. Until the individual can be more completely subordinated to society it will undoubtedly remain at this impasse. (A.W.L.)

EUGLENOIDIDA. An order of 1-celled animals. Mastigophora. (A.W.L.)

EUHEDRAL. Used by mineralogists to describe a crystalline rock, usually igneous or metamorphic, in which certain of the mineral constituents, usually the **phenocrysts**, have well-defined crystal faces. (See also **Arkose, Subhedral** and **Auhedral**.) (R.M.F.)

EULAMELLIBRANCHIATA. In some classifications an order of **bivalve** mollusks containing the oysters. (A.W.L.)

EULERIAN COLUMN. Column.

EULER'S THEOREM ON HOMOGENEOUS FUNCTIONS. This is a statement of a mathematical result concerning a certain type of function.

If $u = f(x, y, z, \cdots)$ is a **homogeneous function** of two or more variables of the nth degree which has continuous first **partial derivatives**, then

$$x \frac{\partial u}{\partial x} + y \frac{\partial u}{\partial y} + z \frac{\partial u}{\partial z} + \cdots = nu.$$

(L.L.S.)

EULER'S THEOREM ON THE EXPONENTIAL FUNCTION. Euler's theorem is a remarkable mathematical result giving a relation between the **trigonometric functions** and the **exponential function.** It may be expressed by the formula

$$\cos x + i \sin x = e^{ix}.$$

(L.L.S.)

EUPHAUSIACEA. A small order of marine **crustaceans.** (A.W.L.)

EUPHORIA. This term describes a feeling of well-being which is not justified by the physical condition of the patient. It is seen in certain mental disturbances, manic depressive **psychoses,** paresis, in the terminal stages of **peritonitis,** and in tuberculosis. (D.M.H.)

EUROPEAN CORN BORER. Insecta, Lepidoptera. A small **moth,** *Pyrausta nubilalis,* whose **larva** bores in the stems of plants, especially Indian corn. The species is closely related to certain North American moths and was first noticed as a pest in the eastern part of the United States about 1920. Since then it has spread halfway across the continent. (A.W.L.)

EUROPIUM. Symbol: Eu. Atomic number: 63. Atomic weight: 152.0. (Isotopes: page 290.) Type of compound: Eu_2O_3. Color of salts: Rose. Discovered by Demarçay in 1906. A member of the **cerium** subgroup of the rare earth metals. (R.K.S.)

EURYALAE. Ophiuroidea.

EURYPTERID. Invertebrate Paleontology.

EUSTACHIAN TUBE. A slender canal between the **pharynx** and the middle **ear** of **vertebrates.** It permits the equalization of pressure on the two surfaces of the ear drum. (A.W.L.)

EUTAXIC. A term proposed by Keyes in 1901 for obviously stratified **sedimentary** ore deposits as contrasted with those which are unstratified. The latter he designated as **ataxic.** (R.M.F.)

EUTECTIC. This term is applied by **petrologists** to a discrete mixture of two or more minerals, in definite proportions, which have simultaneously crystallized from the mutual solution of their constituents. The eutectic point is the lowest temperature at any given pressure at which the above physical-chemical process may take place. The eutectic ratio is the ratio by weight of two minerals which originate by the above process.

The term eutectic is used in metallurgy in reference to that particular mixture, of a definite composition, of two or more given substances which has the lowest freezing point. The solid which separates at this temperature has the same composition as the liquid. (R.M.F., F.T.M.)

EUTECTOID. In binary alloy systems a eutectoid alloy is a mechanical mixture of two phases which form simultaneously from a solid solution when it cools through the eutectoid temperature. Alloys leaner or richer in one of the metals undergo transformation from the solid solution phase over a range of temperatures beginning above and ending at the eutectoid temperature. The structure of such alloys will consist of primary particles of one of the stable phases in addition to the eutectoid, for example ferrite and pearlite in low-carbon **steel.** (See **Metals and Alloys.**) (R.H.H.)

EUTHANASIA. Easy or painless death brought on to end a lingering, hopeless, painful disease. (R.S.M.)

EUTHERIA. Mammalia.

EVAPORATION. The evaporation of a liquid consists in the escape from the main body of the liquid of those **molecules** which, in their thermal agitation, are moving with a sufficient speed to break through the **surface tension;** that is, whose **kinetic energy** exceeds the **work function** of cohesion at the surface. Since only a small proportion of the molecules are at any instant located near enough to the surface and are moving in the proper direction to escape, the rate of the evaporation is limited. It is easy to see why it proceeds more rapidly with higher temperature, and why liquids of low surface tension are relatively volatile. Also, as the faster moving molecules emerge, those left behind have less average energy, and the temperature of the liquid is thereby lowered. If the evaporation takes place in a closed vessel, the escaping molecules accumulate as a **vapor** above the liquid. Many of them return to the liquid, such returns being more frequent, the greater the density and pressure of the vapor. Presently the processes of escape and return come to equilibrium; the vapor is then said to be "saturated," its density and pressure no longer increase, and the cooling effect ceases. Even a warm breeze cools the skin because it removes the evaporating perspiration and prevents saturation. (See **Hygrometer, Distillation,** and **Heat of Vaporization.**) (L.D.W.)

EVAPORATOR. Evaporators are used, (1) to concentrate a solution by volatilization of water it contains (as discussed in the article on **Distillation**), and (2) to produce pure water from sea water, or other impure source of supply.

An evaporator system may be single effect, in which the steam is produced from one evaporator, or multiple effect, in which the steam is produced from several evaporators in series. In a multiple effect system the vapor from one evaporator becomes the heating steam in the succeeding. Unusual conditions met in industrial or steam heating plants may require so large a fraction of make-up as to warrant double, triple, or quadruple effect evaporators. The central generating station ordinarily employs single effect and rarely requires more than a double effect system. The ratio $\frac{\text{vapor produced}}{\text{steam used}}$ is about 0.8 for the single effect, 1.5 for the double effect, and 2.5 for the triple effect system. Evaporator feed is sometimes preheated to increase evaporator capacity.

Evaporators are classed as film, flash, or submerged-tube types. The first and last are steam-tube types; in the former the raw water trickles over the hot tubes, in the latter the tubes are entirely surrounded by the water being evaporated. The flash type produces steam by dropping the pressure on water at the saturation temperature. The excess heat flashes part of the water into steam, then the remainder is drawn off, reheated, and again flashed. (F.T.M.)

EVERGREEN. Conifers.

EVOLUTE OF A PLANE CURVE. The evolute of a curve is the **locus** of the **center of curvature** of the given curve.

The **normal** to a curve at a point P is **tangent** to its evolute at a corresponding point Q, so that the **envelope** of the system of normals to a given curve is the evolute of that curve. (L.L.S.)

EVOLUTION. In astronomy, for the evolution of the planets, see **Solar System.** In mathematics, evolution is the operation of raising a number to a given positive integral **power** (i.e., of multiplying a number by itself a certain number of times). Its two **inverse operations** are **involution** and taking **logarithms.**

In mathematics, the operation of extracting a **root of a number** is sometimes called evolution. It is one of the inverse operations to **involution** (raising to positive integral powers).

In biology, evolution is a process of gradual transformation. In application to living things, it is the process by which their hereditary characteristics are modified through a series of generations, resulting ultimately in the production of new subspecific units and species differing in various degrees from the ancestral stock.

The idea of gradual change in living things as a normal part of their vital processes, and the idea of origin of species through these changes, were expressed in a crude form by the philosophers of ancient Greece, but it was not until 1859, when Charles Darwin published his famous book, the *Origin of Species,* that they were so firmly established as to be a permanent part of science. The idea of evolution now permeates all fields of science. In biology organic evolution, as opposed to special creation of species by a divine power, is regarded as an established principle, although it is still referred to as the theory of evolution.

The evidences of evolution are derived from two sharply contrasted fields, biology and **paleontology.** In the former the subsidiary anatomical sciences, **embryology** and **physiology,** reveal countless details of relationship among living things. All are related in their cellular and protoplasmic structure, and in taxonomic groups within the animal kingdom the resemblance becomes more and more detailed as the lower divisions are approached. Thus all animals resemble each other in metabolic processes, in contrast with plants. Within the kingdom a fairly sharp division into the 1-celled **Protozoa,** the loosely integrated multicellular **Parazoa** containing only the sponges, and the numerous phyla of **Metazoa,** is evident. Among the Metazoa **diploblastic** and **triploblastic** structure, **coelomate** and **acoelomate,** and metameric and unsegmented, mark other progressive divisions. The classification of the entire kingdom expresses all details of relationship. (See **anatomy; classification; tree of life.**)

The interpretation of these relations varies. To the person who is satisfied with the idea of special creation it is no more difficult to believe in the creation of related forms than of unrelated, but several types of evidence lead the biologist to entirely different conclusions. The existence of structurally similar organs (**homology**) adapted to widely different environmental conditions is among these evidences. An example is the group of appendages including the human arm, the flipper of the whale, and the wing of a bird. All are based on the **pentadactyl appendage** although they differ so widely that they are superficially unlike each other. Structures of similar uses but very different structure (**analogy**), such as the wings of birds and insects, indicate that animals may be adapted to a given environment in different ways, hence the many examples of homologous organs such as those mentioned above are strongly suggestive to the scientist that animals have been able gradually to become adjusted to environments different from those of their ancestors through structural modification. The vestigial organs recorded by anatomy, such as the human wisdom teeth, are additional evidence. Structures of no value to the individual, sometimes even harmful, can most satisfactorily be explained as persisting remnants of things once useful in an ancestral stage.

Only one biological science, embryology, reveals an actual transition in structure. This transition is the foundation of the recapitulation theory or biogenetic law which postulates that the individual during its development passes through steps representative of ancestral stages in its evolution. The idea must be used with caution, for no embryonic stage can be regarded as precisely like a preceding ancestral form and such conditions as **cenogenesis** are accompanied by correspondingly great modifications in individual development. In some structures, however, a transition occurs that is too closely like the sequence of structures in related groups of animals to be wholly insignificant. The circulatory system of the vertebrates is a particularly good illustration. In an embryonic bird or mammal it is first like that of a fish in many details, then like that of an amphibian, then it resembles the reptilian system, and at last it attains the distinctive development of its own class. This transition involves the formation and destruction of many parts and the remodeling of a structural plan fitted for a gill-breathing animal to form that of a lung-breathing type. Such preliminary stages cannot be explained as necessary to the ultimate pattern of the organic system in all cases. They are much more logically interpreted as remnants of ancestral stages, pointing to the origin of existing forms by evolution from ancestors in which they were definitive adult structures.

Paleontology adds to this evidence a fragmentary record of extinct inhabitants of the world in the form of fossils. These remains are in many cases associated with strata of sedimentary rocks that have enabled geologists to determine a sequence of past ages extending through many millions of years. In the chronology thus established the fossil remains appear also as a sequence, ascending from the simplest forms in the oldest fossil-bearing rocks to man in the most recent periods, with a dominance of various forms of gradually increasing complexity between. In this sequence only invertebrate remains appear in the earlier deposits. Later primitive fishes are found, and still later an age of fishes in which the group gained high development. The succeeding periods are an age of amphibians, an age of reptiles, and finally an age of mammals.

In many cases abundant fossils have made it possible to arrange series of forms leading by very gradual stages from some remote ancestor to an existing species. One of the most famous of these ancestral series is that of the horse. Beginning with a small browsing animal, *Eohippus,* with four toes on the front feet and three on the hind, it passes through the gradual loss of all but one toe on each foot, accompanied by increase in size and by the modification of the teeth for grazing. Sequences of this kind, and the many fossils of animals unlike the present fauna, leave no doubt that the population of the world has undergone tremendous changes, and no reasonable doubt that the changes have been based on the gradual modification of living things.

Accepting the occurrence of evolution, biological science is particularly interested in knowing how such changes have come about. Various theories have been proposed but exact knowledge of the nature of evolutionary change is still lacking.

Two schools of thought were emphasized in the earlier works on evolution following Darwin's contribution, one based on the ideas of the French biologist, Lamarck, and the other on those of Darwin. Lamarck emphasized the reaction of the individual to its environment through the use or disuse of inherited potentialities as a source of modifications. The results of such reaction in the individual body are commonly called acquired characteristics. Lamarck regarded them as the foundation of evolutionary changes through a succession of generations. Darwin's theory of natural selection, on the other hand, emphasized the great variation of heritable characteristics as the basis of evolution through the selective action of the environment. He pointed out that overproduction is a general tendency among living things. Assuming that variation within species results in some individuals being better equipped than others to meet environmental conditions, he concluded that these well-adapted individuals would have a better chance of surviving and perpetuating their inherited characteristics. He added that these characters would continue a progressive development in succeeding generations, to the ultimate production of new species. This theory is also known as the survival of the fittest.

Many objections have been proposed to both of these theories. Lamarck's idea that change occurs in the individual through reaction to the environment, or by use and disuse, is abundantly supported. We have no reason to believe that such changes are inherited, but whether or not they have some influence on the development of the following generation is less clear. Darwin erred chiefly in assuming the progressive development of selected characters. Apparently natural selection alone can result only in the splitting up of a species or in the contraction of its range of variation, but not in change beyond existing limits of variation. An even more serious objection is that the destruction of surplus individuals is largely accidental, offering no opportunity for the preservation of any particular hereditary characteristics.

One early result of this controversial approach was a futile attempt to establish some influence of body structures on reproductive tissues. The modern approach to this problem accepts the unity of the body and the close coordination of all of its parts, reproductive tissues included, but is still at a loss to prove that reactions of the body to environmental factors during the life of the individual have an effect on succeeding generations.

An important modern contribution is the mutation theory of De Vries, a Dutch botanist. Mutations are abrupt changes from the ancestral range of variation, which breed true, barring further change of the same kind. Mutations have been observed in many plants and animals, including the evening primroses with which De Vries worked, and the fruit fly, **Drosophila**, which has been so important in the study of **heredity**. The recognition of such changes provides a foundation for effective evolutionary change by natural selection, if only it can be demonstrated that some mutations are useful to the individual. Unfortunately most of those that have been observed are of no evident value. Even with the assumption that mutation plus selection may be an adequate explanation of evolution, we are faced with the difficulty of explaining the production of mutations, for they are only a recognized phenomenon of unknown source at present. They have been produced by the action of penetrating rays, such as x-rays and radium emanations, and in some cases by the effect of temperature on the parent animal. Here the problem returns to the interaction of the individual with its environment, and in the normal association of these factors the solution may yet be found.

Selective forces, including natural selection, **isolation** of limited stocks, and the deliberate control of reproduction by human interference (**selection**), may well bring about changes in the complex population of the world, and when associated with hybridization their potentialities are greatly increased. This type of evolutionary change does not, however, explain the initial appearance of diversity, as any adequate theory of evolution must.

Modern investigation of evolutionary processes is closely linked with the science of genetics. In this science many of the observed chromosomal irregularities associated with reproduction account for hereditary changes in organisms, to be classed with those mentioned above. Our ultimate explanation of evolutionary change may well come from studies of correlated environmental and hereditary changes in this field of study. (A.W.L., L.L.S.)

EXACT DIFFERENTIAL EQUATIONS. Ordinary Differential Equations of First Order and First Degree.

EXCESS AIR.
The theoretical quantity of air which should supply just that amount of **oxygen** required for chemical union with the molecules of a **fuel** is insufficient for combustion under actual conditions. This is because of the dynamic conditions attending the combustion of fuel in a furnace. In the relatively short time that the fuel and air are together in a region of high temperature, the imperfection of mixing methods, and the presence of the fuel in lumps and large particles, rather than separated molecules, combine to promote incomplete combustion unless there is an excess of oxygen over that which is actually needed. The excess air is ordinarily given as a percentage of the theoretical requirement, and the amount employed is determined by the degree of technological control of combustion practiced, the surface exposure of the fuel, i.e., whether the fuel is gaseous, atomized, powdered, or lump, and the perfection of design of the furnace and firing equipment. Excess air is shown by the presence of oxygen in the products of combustion. When no more than a trace of **carbon monoxide** is present in the flue gas, the excess air may be found by the following formula:

$$\text{Excess air} = \frac{20.9R}{CO_2(R+3)} - \frac{R+2.37}{R+3}.$$

In this formula the symbol R is employed to denote the "fuel ratio" of the coal, this being the ratio of carbon burned to hydrogen burned with oxygen from the air. The carbon is the total revealed by the ultimate analysis, less any fraction left unconsumed in the refuse. The hydrogen part of the ratio is ordinarily the free hydrogen, but if much sulfur is contained, the hydrogen should be corrected as follows:

$$H = \text{total hydrogen} - \frac{(\text{total oxygen} - \text{sulfur})}{8}.$$

R is generally from 50 to 25 for anthracite, 20 to 16 for semi-bituminous, and 16 to 12 for bituminous coals.

Fuels ranging from the hard and nearly smokeless anthracite through intervening ranks down to the brown and woody lignite have their combustible elements combined in many different ways but, strangely enough, the air required to produce a thousand heat units from any coal is very nearly a constant amount. This fact may be turned to good account in rapid determination of the air needed for combustion if the calorific value of the coal is known. Trial computations show that 1 lb. of combustion air is theoretically required for each 1340 B.T.U. produced in complete combustion, for any coal chosen.

The excess air requirements vary widely with the type of installation. Lump coal, fired by stokers, will usually be completely burned with the use of 50% of excess air, whereas in the hand-fired stokers, up to 100% may be necessary, and undoubtedly a great deal of domestic coal firing, unaided by technological information or experience, is carried out with from 150% excess air upwards. Pulverized coal and fuel oil may be satisfactorily fired with 5–15% excess air, and gaseous fuels fired with the aid of well-designed burners may be so effectively mixed with air that no excess air is needed. (F.T.M.)

EXCHANGE, TELEPHONE.
This designates the territory served by a group of central offices for which there is normally no extra charge for service. Thus the various offices of a given city constitute an exchange. (See **Telephony**.) (L.R.Q.)

EXCITATION. This term is used in several applications in electrical engineering. It is customary to speak of the field excitation of **dynamo** machines, meaning the current or voltage of the field circuit. In electron tube circuits the signal which excites a stage is commonly called the excitation. Thus in a radio **receiver**, the signal picked up by the **antenna** supplies the excitation for the first stage, the output of the first supplies the excitation for the next and so on. (L.R.Q.)

EXCITING CURRENT. This is the current which supplies the **core losses** and magnetizing current for a **transformer**. When the transformer is supplying a load this excitation current is one component of the total input, the other being the component which balances the load current. In an electrical distribution system having many customers' transformers (see **Electric Power Transmission**) which often have a low or zero load, the exciting current becomes quite important because of the line losses and poor power factor caused by it. (L.R.Q.)

EXCRETION. The removal of the waste products resulting from the chemical transformation of materials in the body.

The **oxidation** of materials derived from foods for the release of energy may produce carbon dioxide and water whether the compound oxidized is a **protein**, a **carbohydrate**, or a **fat**, but since proteins contain **nitrogen** and other elements in addition to **carbon**, **hydrogen**, and **oxygen**, they also give rise to more complex waste products. The chief nitrogenous wastes of animals are **urea** and **uric acid**. The elimination of all of these compounds and other substances of like derivation is excretion.

Many small animals, including both protozoans (**Protozoa**) and more complex forms, apparently discharge these wastes from the surface of the body generally, while in others a special **excretory system** occurs. Even in those forms which have a complex excretory system, any moist surface directly or indirectly exposed to the medium surrounding the animal is favorable for the diffusion of materials into or out of the body, and so may carry on excretion. The wastes passed out in this manner are largely carbon dioxide and water, although the discharge of water may also take out dissolved solids. Thus the lungs of a terrestrial vertebrate eliminate carbon dioxide and the sweat glands of the skin of some animals discharge water with other materials, including nitrogenous wastes, in solution. By far the greater part of the complex wastes is eliminated by the **excretory system**.

In complex animals other organs than those directly involved in the elimination of wastes may play an important intermediary role. The circulatory system of the **vertebrate**, for example, transports all wastes from the tissues where they are formed to organs which act upon them and finally to the centers which remove them from the body. The liver removes some substances, including complex organic compounds resulting from the destruction of old red blood cells, and discharges them in the **bile** by way of the intestine. It also transforms **ammonia** and **amino acids**, resulting from the oxidation of proteins, into urea which is returned to the blood to be removed by the **kidneys**.

Wastes are stored in the body during embryonic life and immature stages of some animals to be discarded with the tissues containing them at birth or transformation. The **allantois** of bird embryos serves as a reservoir for wastes and the **fat body** of insect larvae serves a similar purpose according to some observers. (A.W.L.)

EXCRETORY SYSTEM. An **organic** system whose principal or only function is the removal of complex **wastes** from the animal body.

Some animals lack an organized excretory system, discharging wastes from the surface of the body generally, but others, even among the 1-celled animals, have special excretory structures. The **contractile vacuoles** of protozoans are supposed to carry out this function.

In the flatworms a special excretory system based on the flame cell appears. Flame cells are large and hollow, with a group of cilia projecting into the cavity whose movement drives out the liquid discharged by the cell. The cavity of each flame cell joins a small duct and these ducts converge to form larger ducts which ultimately open at the surface of the body. Flame cells emptying by ducts into a vesicle connected with the caudal end of the alimentary tract also occur in rotifers.

Roundworms have two slender excretory canals along the sides of the body which unite to empty by a single pore near the anterior end.

In the segmented worms the body cavity becomes involved in excretion. Two forms of tubes, the coelomo-

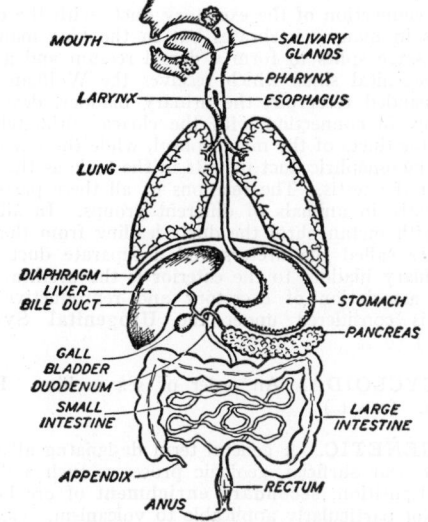

(From Carlson and Johnson's The Machinery of the Body, University of Chicago Press.)

ducts and nephridia, open from the **coelom** to the exterior in these worms. These organs are segmentally arranged ciliated tubes with a funnel-shaped inner end and a minute opening externally. They are variously associated in the excretory organs of different species and in some do not open into the coelom but are provided with cells much like flame cells which are called solenocytes.

Arthropods of different classes have special excretory structures, including the coxal glands of **scorpions**, said to be derived from coelomoducts, and the Malpighian tubules of insects. The latter are slender tubules opening into the alimentary tract at the caudal end of the stomach and blind at their other end.

The occurrence of **solenocytes** in the lancelets of the phylum **Chordata** is unusual in this phylum, since in the true vertebrates a pair of kidneys are the chief excretory structures. They are developed from intermediate mesoderm (**Embryology**.) The excretory unit in these organs is a minute tubule which has in its primitive form a ciliated funnel leading from the coelom. At their lateral ends the series of tubules unite to form a duct which grows back to empty into the **cloaca**. The tubule is associated with a knot of blood vessels near the coelomic opening (the nephrostome). In a more advanced stage of development excretory tubules lack the nephrostome and have the wall expanded to form Bowman's capsule, embracing the knot of blood vessels which is called a glomerulus. This unit, known as a renal corpuscle, is found in the kidneys of most vertebrates. The tubule leading from it is also specialized for the removal of wastes from the blood.

Three pairs of kidneys are found in different vertebrates and appear in succession in the embryos of the higher classes, the reptiles, birds, and mammals. In cyclostomes and embryos of fishes and amphibians the kidney are pronephroi, lying well forward in the body and made up of tubules of the primitive type. The pronephroi are vestigial in embryonic reptiles, birds and mammals. Functional kidneys in these embryos and in the adults of cyclostomes, fishes and amphibia are the mesonephroi, lying behind the pronephroi and made up of closed tubules. As they develop, these tubules connect with the duct formed by the pronephroi; this duct is then called the mesonephric or Wolffian duct. In adult reptiles, birds and mammals the mesonephroi are replaced by the metanephroi, lying still farther back. Their tubules develop in a mass of tissue surrounding a blind diverticulum of the mesonephric duct.

The connection of the excretory ducts with the cloaca persists in many vertebrates but in the true mammals this passage splits to form a dorsal rectum and a ventral urogenital sinus which receives the Wolffian duct. An expanded reservoir, the urinary bladder, developed ventrally in connection with the cloaca, ultimately receives the ducts of the metanephroi, while the remainder of the mesonephric duct persists in the male as the main duct of the testis. The relations of all these parts differ greatly in animals of different groups. In all animals with metanephroi the ducts leading from the kidneys are called the ureters and a separate duct from the urinary bladder to the exterior is the urethra.

The association of excretory and reproductive passages is considered under the **Urogenital System.** (A.W.L.)

EXOCYCLOIDA. An order of sea urchins. **Echinoidea.** (A.W.L.)

EXOGENETIC. A general term designatng all surficial, or near surficial, geologic processes such as: **erosion,** deposition, **secondary enrichment** of ore bodies, etc. Not particularly applicable to volcanism. (R.M.F.)

EXOPHTHALMIC GOITRE. Thyroid Gland.

EXOPHTHALMOS. An abnormal protrusion of the eyeball most often seen in **Graves' disease.** (See **Thyroid Gland.**) (D.M.H.)

EXOPODITE. Biramous Appendage.

EXOPTERYGOTA. Insects whose wings appear in the immature stages as external flaps. These insects have gradual or incomplete metamorphosis, hence the term embraces the **Paurometabola** and **Hemimetabola** of some writers. (A.W.L.)

EXOSKELETON. Skeletal System.

EXOTHERMAL CHANGE. Thermochemistry.

EXPANDER, VOLUME. This is the part of the communication circuit designed to expand the volume range back to the original value as it is normally compressed for transmission. (See **Compressor.**) (L.R.Q.)

EXPANDING UNIVERSE. Spirals.

EXPANSION. The term expansion refers commonly to a process in which a constant mass of a substance undergoes an increase in volume. In thermal expansion this is brought about by raising the temperature of the substance. Expansion processes are of great importance in engineering. Thus, the power derived from steam engines and turbines and internal combustion engines is produced as a result of expansion of gases and vapors. (See **Ratio of Expansion.**) Expansion of metals with increase of temperature is the operating principle of **thermometers, thermostats,** and many other useful devices. On the other hand the increase in size of pipes as they are heated creates a problem in engineering design. (See **Expansion Joint.**)

Each molecule of a body of matter in any state appears to monopolize a certain amount of space which, while it cannot be accurately called the volume of the molecule, does represent the contribution of that molecule to the volume of the whole body. The molecules are in a state of agitation; and it is to be expected that the space which any molecule monopolizes, or keeps clear for itself, will be larger, the greater the amplitude of its oscillations (just as one may make a posthole larger by working the post to and fro in different directions). We have here the fundamental reason for the expansion of bodies with rise of temperature. (See **Heat.**)

The rate of expansion of a substance with temperature has been expressed by several different "expansion coefficients." The one now usually employed, as regards change of volume, may be defined as the rate of change of the volume of a body of the substance with respect to temperature, divided by its volume at the zero of temperature:

$$\alpha_v = \frac{dv}{dt} / v_0. \tag{1}$$

Thus for iron at ordinary temperatures, α_v is about 0.000036 per degree centigrade. While not strictly constant, it is nearly enough so for ordinary purposes. By integration of (1) the volume at any temperature t is

$$v = v_0(1 + \alpha_v t). \tag{2}$$

These statements apply alike to solids, liquids, and gases. The fact that the coefficient α_v is nearly the same for all gases is expressed by **Charles's law.**

For solids we may also define a linear expansion coefficient, relating to the change in any one dimension l, thus:

$$\alpha_l = \frac{dl}{dt} / l_0, \tag{3}$$

from which the value of that dimension at any temperature t is

$$l = l_0(1 + \alpha_l t). \tag{4}$$

Since for an isotropic solid the volume at a given temperature is proportional to the cube of any dimension, it is easy to show that

$$\alpha_v = 3\left(\frac{l}{l_0}\right)^2 \alpha_l. \tag{5}$$

But for moderate temperatures l and l_0 are nearly equal, hence, practically, the volume coefficient is three times the linear. Thus the linear coefficient for iron is about 0.000012 per degree centigrade.

If from the known thermal capacity of a solid we deduct the calculated energy corresponding to change of thermal agitation in all the atomic degrees of freedom, and also the energy expended in expansion against the external pressure, the remainder may be taken as representing the work of expansion against cohesion, and hence used as a means of calculating the "internal pressure" arising therefrom. (See **Adhesion and Cohesion.**)

Much progress has been made in the preparation of substances having expansion coefficients of desired value. Thus "platinite" (54% iron, 46% nickel) has practically the same coefficient as glass and may therefore be used for lead-wires in vacuum tubes; while "invar" (64% iron, 36% nickel) has a linear coefficient of only 0.0000008 per degree centigrade, and is therefore suitable for clock pendulums. (L.D.W., F.T.M.)

EXPANSION JOINT. Metals constituting pipes have the property possessed by all materials of expanding with increase of temperature. Were they constrained to a fixed length, a reaction equivalent to the force required to compress the pipe through a deformation equal to the prevented expansion would be set up. For all but very short steam lines this force is too large to incorporate in the piping system. The same force would be present, theoretically, in the short line, but the supports would have enough elasticity to take the small expansion. In long lines the expansion is permitted by the use of suitable joints and bends.

Both packed and packless expansion joints are used for saturated steam at pressures up to 250 lbs. per sq. in. High temperature has a deteriorating effect on packing; however, packed joints have been designed for high temperature by protecting the packing by air-cooled sleeves. Expansion joints take up expansion at one point by allowing relative motion of the two sections of pipe connected by the joint. Usually one pipe end is anchored by a rigid connection to the body of the joint but occasionally the double slip joint in which both pipe ends are free to move in the joint is used.

When expansion is to be taken by the flexibility of the pipe itself various forms of pipe bends are used. This way of caring for expansion is free of the temperature-pressure limitations of the expansion joints and also of any maintenance work such as the repacking of joints. Consequently, it has been the standard for boiler and turbine leads and for long runs of high-pressure piping of all sorts. Its principal drawbacks are the added friction losses, the expense of fabrication (most bends are special jobs), and the space required. (F.T.M.)

EXPANSION OF DETERMINANTS BY MINORS. Determinants.

EXPANSION OF FUNCTIONS IN SERIES. In many mathematical investigations it is desirable to express a given function in a certain special form of representation known as an infinite series. One important use of such representation is for the calculation of numerical values of functions, as, for instance, in the preparation of mathematical tables.

Taylor's formula with a remainder for a **function** of one variable is:

$$f(a + h) = f(a) + hf'(a) + \frac{h^2}{2!}f''(a) + \frac{h^3}{3!}f'''(a) + \cdots$$
$$+ \frac{h^{n-1}}{(n-1)!}f^{(n-1)}(a) + R_n.$$

Various formulae have been given for R_n, some of which will be given presently.

If $f^{(n)}(x)$ is finite in an interval (a, b) for all values of n and if $R_n \to 0$ when $n \to \infty$, then the formula gives a **convergent infinite series (power series)** for the function $f(x)$.

Lagrange's form of the remainder is:

$$R_n = \frac{h^n}{n!}f^{(n)}(a + \theta h), \quad \text{where } 0 < \theta < 1.$$

Cauchy's form of the remainder is:

$$R_n = \frac{h^n(1 - \theta)^{n-1}}{(n-1)!}f^{(n)}(a + \theta h), \quad 0 < \theta < 1.$$

Let $|f^{(n)}(x)| \leq M_n$ in the interval to be considered; then

$$|R_n| \leq \frac{|h|^n}{n!} \cdot M_n.$$

When $a = 0$, Taylor's formula becomes Maclaurin's formula:

$$f(h) = f(0) + h \cdot f'(0) + \frac{h^2}{2!}f''(0) + \cdots$$
$$+ \frac{h^{n-1}}{(n-1)!}f^{(n-1)}(0) + R_n.$$

If $R_n \to 0$ as $n \to \infty$, we obtain a power series expansion for $f(h)$ in powers of h.

Some important expansions of elementary functions in power series are:

$$e^x = 1 + x + \frac{x^2}{2!} + \frac{x^3}{3!} + \frac{x^4}{4!} + \cdots,$$

convergent for all values of x;

$$\log_e(1 + x) = x - \frac{x^2}{2} + \frac{x^3}{3} - \frac{x^4}{4} + \cdots,$$

convergent for $-1 < x \leq 1$;

$$\sin x = x - \frac{x^3}{3!} + \frac{x^5}{5!} - \frac{x^7}{7!} + \cdots,$$

where x is in radian measure, convergent for all values of x;

$$\cos x = 1 - \frac{x^2}{2!} + \frac{x^4}{4!} - \frac{x^6}{6!} + \cdots,$$

(x in radian measure), convergent for all values of x;

$$(1 + x)^m = 1 + mx + \frac{m(m - 1)}{2!}x^2$$
$$+ \frac{m(m - 1)(m - 2)}{3!}x^3 + \cdots,$$

convergent for $-1 < x < 1$, when m is not a positive integer.

Taylor's formula with remainder for a function of two variables is:

$$f(a + h, b + k) = f(a, b) + hf_x(a, b) + kf_y(a, b)$$
$$+ \frac{1}{2!}[h^2 f_{xx}(a, b) + 2hk f_{xy}(a, b) + k^2 f_{yy}(a, b)] + \cdots + R_n,$$

or in symbolic form:

$$f(a + h, b + k) = f(a, b) + \left(h\frac{\partial}{\partial x} + k\frac{\partial}{\partial y}\right)f(a, b)$$
$$+ \frac{1}{2!}\left(h\frac{\partial}{\partial x} + k\frac{\partial}{\partial y}\right)^{(2)}f(a, b) + \frac{1}{3!}\left(h\frac{\partial}{\partial x} + k\frac{\partial}{\partial y}\right)^{(3)}f(a, b)$$
$$+ \cdots + \frac{1}{(n-1)!}\left(h\frac{\partial}{\partial x} + k\frac{\partial}{\partial y}\right)^{(n-1)}f(a, b) + R_n,$$

where

$$R_n = \frac{1}{n!}\left(h\frac{\partial}{\partial x} + k\frac{\partial}{\partial y}\right)^{(n)}f(a + \theta h, b + \theta k), \quad 0 < \theta < 1.$$
(L.L.S.)

EXPANSION REAMER. Reaming.

EXPANSION SHOE. Bearing.

EXPECTED VALUE. If $p_1, p_2, \cdots p_n$ are the **probabilities** that a random variable x should assume values $x_1, x_2, \cdots, x_{n-1}, x_n$, then the expected value of x, $E(x)$, is defined as

$$p_1 x_1 + p_2 x_2 + \cdots + p_n x_n$$

where $\sum_{i=1}^{n} p_i = 1$, and the x's are **discrete**. In case x is a **continuous** variable, then

$$E(x) = \int_{x=a}^{x=b} x P_x \, dx$$

where $P_x\,dx$ is the probability function of x, and the range of x is from a to b. The expected value should not be confused with the **mean**. Similarly we define the expected value of x^2, x^3 as

$$E(x^2) = \int_{x=a}^{x=b} x^2 P_x\,dx, \quad \text{and} \quad E(x^3) = \int_{x=a}^{x=b} x^3 P_x\,dx.$$

In the case of a discrete variable the integrals are replaced by summations. (L.A.A.)

EXPERIMENTAL ERROR. When experiments are repeated, the **statistics** calculated from each **sample** vary. If these variations are not attributable to any cause, they are called the experimental errors. This is particularly true in the **analysis of variance** where the experimental error is the portion of the variation remaining after all known causes of variation are removed. (L.A.A.)

EXPLICIT FUNCTION. If a **function** is defined by a relation between the **variables** giving an **equation** expressing one variable directly in terms of the other without the necessity of solving the equation, the function so defined is called an explicit function. (L.L.S.)

EXPLOSIVES. Explosives are substances that, either pure or with admixture of other substances, react rapidly with the production of local high temperature and the generation of large volumes of gases. The power from the expansion of the gases is utilized for propelling charges or for **blasting** objects, usually for military or industrial purposes. Gunpowder was the first explosive to be used. Not until 1865, when Abel perfected a process for washing nitrocellulose, "guncotton" (cellulose hexanitrate) thus making it safe to store and use, and in 1867, when Nobel discovered that nitroglycerine (glyceryl trinitrate) could be rendered safe by absorption in a porous material such as kieselguhr, were safe explosives available. In 1886, Nobel gelatinized these two explosives, nitrocellulose with nitroglycerine, and at the British government laboratory these were gelatinized with **acetone**. Such an explosive is in very general use as a propellant.

Black powder (gunpowder) consists of an intimate mixture of finely divided solids, 75% **potassium nitrate**, 15% **carbon**, 10% **sulfur**. Powders for sporting guns contain a slightly larger percentage of potassium nitrate (75 to 78%), smaller percentage of carbon (15 to 12%), and a variation in sulfur from 9 to 12%. Mining or blasting powders, where large volumes of gas are desired may have 14 to 21% carbon and 13 to 18% sulfur. When ignited, potassium nitrate supplies oxygen for the combustion of explosives, of carbon to carbon dioxide and of sulfur to sulfur dioxide. One gram of powder yields 250 to 300 milliliters of gas measured at 0° C. and 760 mm. pressure. The heat evolved per gram is 500 to 700 calories, and the temperature of the explosion is estimated at 2700° C.

In explosives of the smokeless powder type, that is, composed of nitrocellulose-nitroglycerine-trinitrotoluene, the material composing the explosive furnishes oxygen for its own combustion when exploded. One volume of nitroglycerin produces by explosion 1300 volumes of gas, measured under ordinary conditions or 10,000 volumes at the temperature of explosion. The speed of the explosive wave in various explosives is as follows:

EXPLOSIVE	SPEED OF EXPLOSIVE WAVE METERS PER SECOND
Nitroglycerin	1300
Dynamite	2700
Hydrogen plus oxygen	2800
Nitrocellulose	3800 to 5400

EXPLOSIVE	SPEED OF EXPLOSIVE WAVE METERS PER SECOND
Picric acid	6500
Ethyl nitrate	
(a) In rubber tube covered with cloth	1600
(b) In glass tube	2500

For comparison:

	SPEED OF SOUND WAVE
Hydrogen plus oxygen	515

Nitrocellulose is soluble in **acetone**, ethyl acetate (see **Esters**), **nitrobenzene** or **benzene**, insoluble in water, alcohol, ether, **acetic acid**, or nitroglycerin; when treated with **iodine** dissolved in **potassium** iodide solution followed by **sulfuric acid**, turns yellow (**cellulose** similarly treated turns blue); when treated with sodium **sulfide** solution or **ferrous** chloride solution, decomposes and is thus treated when it is desired to destroy the explosive. When carefully washed, nitrocellulose may be transported in the wet condition in wooden boxes placed inside zinc boxes, which are then hermetically sealed. Nitrocellulose may be ignited and burned in the open air without explosion. The ignition temperature of various explosives is as follows:

EXPLOSIVE	IGNITION TEMPERATURE
Nitrocellulose (not compressed)	220 to 250° C.
Nitroglycerine (explosive at 240° to 250°)	218
Black powder	288
Mercury fulminate	200

In order to produce an explosion it is usually necessary that a **detonator**, such as mercury fulminate, be used to set up the explosive wave. The explosive wave may be transmitted through a solid body with which portions of the explosive are in contact, as is shown by setting up a row of dynamite cartridges from 30 to 70 cm. apart. When the end cartridge is detonated the others are successively exploded by transmission of the explosive wave through the solid support.

Nitrocellulose is made by reaction of **cotton** with concentrated **nitric acid** in the presence of concentrated **sulfuric acid** (so-called "mixed" acid). The product is carefully washed free from acid with soda, and in composition is an **ester** (not a nitro-compound) cellulose hexanitrate, containing 13% of **nitrogen** and 85% insoluble in ether-alcohol mixture. Less highly nitrated celluloses, soluble in ether-alcohol mixture, are made and used for other purposes, for example, in plastics, collodion and photographic films.

Nitroglycerin is made by reaction of **glycerol** with nitric acid and sulfuric acid, in a process similar to that for nitrocellulose. The product is also an ester, glyceryl trinitrate, a colorless to yellow liquid, 23.4% nitrogen, freezing point 13° C., very sensitive to shock and dangerous to handle. Nitroglycerin is used as dynamite by absorbing in a porous material, such as kieselguhr, which is chemically inert, or wood pulp, which reacts in the explosion. Dynamite is safely handled and transported, and the explosive power of the contained nitroglycerin is practically unchanged upon detonation.

COMPOSITION OF VARIOUS DYNAMITES

1.	Nitroglycerin	72–75%
	Kieselguhr	24.5
	Sodium carbonate	0.5
2.	Nitroglycerin	40%
	Sodium nitrate	45
	Wood pulp	14
	Magnesium carbonate	1
3.	Nitroglycerin	20%
	Sodium nitrate	36

Ammonium nitrate........... 25 %
Roasted flour................ 18.5
Sodium carbonate........... 0.5

Safety explosives, used in coal mining, are designed to diminish the danger of igniting the mine gases. These are made of such materials that the rise in temperature upon explosion is relatively small, and the safety in use is thereby increased. Liquid carbon dioxide, upon fracture of a calibrated disk, allows safe expansion of gas with minimum breakage of coal.

COMPOSITION OF VARIOUS SAFETY EXPLOSIVES

1. Ammonium nitrate........... 37%
   Potassium nitrate............. 34
   Nitrobenzene................. 29

2. Ammonium nitrate............ 82%
   Dinitrobenzene............... 18

3. Ammonium nitrate........... 92%
   Trinitrotoluene............... 4
   Flour....................... 4

High explosives are used for blasting purposes and in bombs, shells, and mines, where great shattering effect is desired.

COMPOSITION OF VARIOUS HIGH EXPLOSIVES

1. Dynamite

2. Nitroglycerin................. 90.6%
   Nitrocellulose................ 8.8
   Calcium carbonate........... 0.6

3. Trinitrotoluene

4. Trinitrophenol (picric acid)

(R.K.S.)

EXPONENTIAL CURVE.

This is a very important type of mathematical curve, which is of use in many diverse kinds of problems, as in pure mathematics and in applied mathematical investigations in physics, chemistry, biology, finance, etc.

The exponential curve, whose equation is $y = e^x$, is shown in the accompanying figure. It crosses the Y-axis

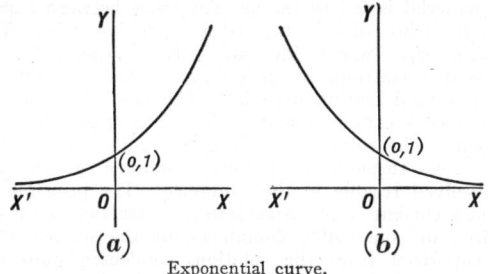

(a)   (b)

Exponential curve.

at the point (0, 1), is **asymptotic** to the X-axis on the left, and rises more and more steeply on the right. The graph of $y = a^x$ ($a > 1$) will be similar to that of $y = e^x$, except of different steepness.

The exponential curve, whose equation is $y = e^{-x}$, is also shown in the figure. It is the same as the preceding, except reversed in position.

The graph of $y = ae^{bx}$ or $y = a \cdot c^x$ will be similar to that of $y = e^x$, except that it will cross the X-axis at (0, a), and will have a different steepness (slope). (L.L.S.)

EXPONENTIAL EQUATIONS.

An equation in which the unknown is involved as an **exponent** is called an exponential equation.

Simple exponential equations can be solved by taking **logarithms** of both members and equating them, and solving the resulting **algebraic equation**. (L.L.S.)

EXPONENTIAL FUNCTION.

An exponential function is a **transcendental function** of the form $y = a \cdot b^x$, where a and b are **constants** ($b \neq 0$ or 1), and x is the **variable**; it has a constant base and a variable **exponent**.

It is often represented in the standard form $y = ae^{cx}$, where e is the Napierian base, and a and c are constants. (L.L.S.)

EXPONENTS. Powers and Exponents.

EXPOSURE.

Exposure is defined as the product of the illumination on the sensitive material and the time during which the material is exposed to this illumination or

$$E = it.$$

The unit of measurement is the candle-meter-second, which represents an exposure of 1 second to a source having a light intensity of 1 candle at a distance of 1 meter from the surface of the sensitive material.

Since the time is constant when a sensitive material is exposed in the camera, light reflected from the various parts of the subject produces exposures which are more or less proportional to their own brightnesses. The words "more or less" are used advisedly because the scatter of light within the lens prevents an exact proportionality between the exposures and the brightnesses of the corresponding parts of the subject.

The exposures on the sensitive material depend upon (1) the brightnesses of the subject, (2) the time, and (3) the amount of light transmitted by the lens. The brightnesses of the subject depend upon (1) the lighting, that is, the amount of light received by the different parts of the subject, and (2) by the reflectances of the subject. Thus the side of a house in shadow receives less light than the side in direct sunlight and the white house reflects more of the light than the dark shrubbery at its base. The time, of course, depends upon the timing of the shutter and its efficiency. (See **Shutters**.) The amount of light transmitted by the lens depends upon (1) the relative aperture ($f\#$) and (2) its construction, which determines the loss from absorption and scatter.

To produce a photograph in which the brightness differences of the original are properly rendered it is necessary that the exposures produced by these brightnesses result, upon development, in the density differences necessary for a good print. For satisfactory tone reproduction in the print, the minimum exposure for a given sensitive material is the exposure at that point on the D log E curve at which the gradient is 0.3 of the average gradient over a log E range of 1.5. (See **Sensitometry**.) The reciprocal of this exposure is a measure of the sensitiveness or "speed" of the sensitive material.

The time of exposure in making a photograph depends upon (1) the brightness of the optical image and (2) the sensitiveness (speed) of the sensitive material. However, since it is difficult to measure the brightness of the lens image, the time of exposure is frequently said to depend upon:

1. *The subject.* (a) Light, highly reflecting subjects, such as beach and snow scenes, distant landscapes, (b) dark objects, such as shrubbery, alleys, interiors, etc., (c) shaded objects, portraits in the shade, etc.

2. *The illumination.* (a) Time of day, (b) time of year, (c) bright, cloudy, dull.

3. *The f/number* of the lens used in making the picture.

4. *The sensitivity, or speed, of the film or plate,* i.e., the reciprocal of the exposure necessary to produce a negative from which a satisfactory print can be made. (C.B.N.)

EXPOSURE METER.

Instruments used in measuring light for the purpose of determining the proper exposure are known as exposure meters. Exposure meters designed for determining camera exposures may be divided into three classes:

1. Those which measure the *light reaching the subject* by the time required to darken a light-sensitive paper. These, formerly quite popular, are now obsolete.

2. Those which measure the *light reflected from the subject* visually. These are now nearly obsolete.

3. Those which measure the light reflected from the subject (or in some cases that falling on the subject) with light-sensitive photocells.

Meters of this last-mentioned type consist of a photocell which transforms the light falling on it into an electrical current and a microammeter which indicates on a dial the electrical current produced by the cell. The dial is graduated either in exposures or in foot-candles from which the exposure is calculated by movable scales.

Ordinarily the meter is held to face the subject so as to measure the total, or integrated, brightness of the entire scene. In this case, the exposure is determined principally by the highlights, since these reflect the most light, and care must be taken not to include too much of the sky, for example, or the rest of the subject will be underexposed. Exposures determined from the total amount of light reflected from the subject may lead to underexposure when the subject is generally light but contains dark objects, particularly if these are small and near to the camera. On the other hand, a subject which is generally dark but contains a small well-lighted area may be overexposed. In cases such as these it is best to measure the light reflected from that part of the subject which is of the greatest importance. This is done by proceeding to the area in question and measuring the light reflected from it alone. With most meters this means that the reading should be made from a distance no greater than the area whose brightness is to be measured.

Methods of determining the exposure from the light reading, or of dealing with exceptional subjects, vary with the different meters. For such information the literature of the manufacturer should be consulted. (C.B.N.)

EXPOSURE-DENSITY RELATIONSHIP. The response of a sensitive material, as shown by the relationship between density and exposure, is usually represented by a curve in which density is plotted against the *logarithm* of the exposure. This curve is known as (1) the *D log E curve;* (2) the *H & D curve,* after Hurter & Driffield, two English investigators who were the first to plot such curves; and (3) as the *characteristic curve* since it indicates the principal characteristics of a sensitive material insofar as the relationship between exposure, development and density is concerned.

The *D log E* curve may for convenience be divided into three portions (see figure), although with many

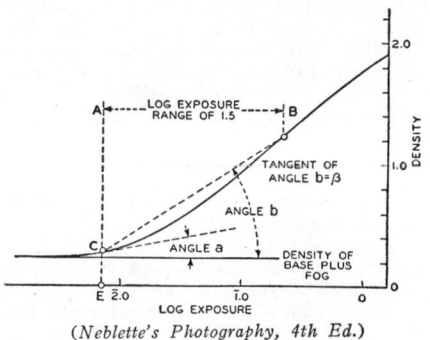

(*Neblette's Photography, 4th Ed.*)

negative materials the different portions are not readily discernible and the exposure range of the **sensitometers** now in general use is too short to include much, if any, of the upper portion. The lower portion, concave to the log *E* axis, is termed *period of increasing gradient*

(slope) or "toe" portion. Within this range of exposure the increase in density is greater than the increase in log exposure. This part of the curve is sometimes called the period of underexposure but this is true only in the sense that greater exposure is required for proportional representation in the *negative*. Parts of this portion of the curve may be used in general photography without producing the effect of underexposure in the *print*.

The central portion of the curve is approximately straight and is termed (1) the *straight-line portion,* (2) the *period of constant gradient* (constant slope) and (3) the *period of proportional representation*. Over this exposure range, density is proportional to log *E*. As density is, by definition, the logarithm of the opacity, it is clear that within this exposure range the opacities are proportional to the exposures which produced them. In other words, the opacities of the negative are proportional to the exposures produced by the different brightnesses of the subject; hence, proportional representation.

The upper portion of the curve is convex to the exposure axis, thus density increases less and less with increased exposure. This portion of the curve was termed by Hurter & Driffield the period of overexposure. This period terminates at the maximum density and is followed by the period of reversal in which increased exposure results in progressively lower density. The exposures required on negative materials for both the reversal period and for the maximum density are so great that these parts of the curve are seldom included in published *D log E* curves. This omission is of no consequence as this part of the curve is of no use in practice. (C.B.N.)

EXTENSION SPRING. Springs.

EXTRACTION. Extraction is the process of separation of a desired constituent from the other constituents of a mass. The term, as generally applied, refers to: 1. Mechanical extraction or expression, in such cases as the expression of vegetable oils from seeds by the application of high pressure. The material to be pressed is sometimes placed in layers between cloths which are folded at the edges to prevent expulsion of solid material during compression, and sometimes the material is fed to the annular space between a pair of interfitting slightly tapered compression rolls. The oil and press cake residue are collected separately. 2. **Solvent** extraction, in such cases as the recovery of oils from oil-bearing material. The material is placed in a porous container and subjected to treatment with solvent. The solvent containing some dissolved material passes through the porous membrane, leaving the undissolved residue in the container. The principle of counter-current (see **Dissolving**) extraction may be utilized in consecutive containers, or the solvent may be vaporized from the solution, condensed onto the material, and, by means of a syphon in the apparatus, withdrawn periodically to the solution compartment below, as in the Soxlet type of apparatus. When a third substance is of different solubility in two nonmiscible liquids, this substance may be separated from the solution of lower concentration by shaking with the more powerful solvent, and then separating the two liquid layers. The desired substance may be recovered from the solution by evaporation of the solvent. The effectiveness of separation is increased by the use of a given amount of extracting solvent in successive smaller portions rather than by a single extraction with the total amount.

Example. Upon shaking one volume of liquid *A* plus one volume of liquid *B*, suppose a concentration ratio of $\frac{1}{10}\frac{(\text{conc. in } B)}{(\text{conc. in } A)}$ of the third substance *C*.

CONCENTRATION RATIO $\frac{B}{A}$	VOLUME RATIO $\frac{B}{A}$	AMOUNT OF C IN		FRACTION OF C IN	
		B	A	B	A
10	1	$10\times1=10$	$1\times1=1$	$\frac{10}{11}=0.91$	$\frac{1}{11}=0.09$

Upon shaking one volume of liquid A plus one-half volume of liquid B, and, after separation, shaking one volume of liquid A (containing the residue of C) plus one-half volume of liquid B.

	CONCENTRATION RATIO $\frac{B}{A}$	VOLUME RATIO $\frac{B}{A}$	AMOUNT OF C IN		FRACTION OF C IN	
			B	A	B	A
First Extraction	10	0.5	10×0.5 $=5$	1×1 $=1$	$\frac{5}{6}=0.83$	$\frac{1}{6}=0.17$
Second Extraction	10	0.5	0.17×0.83	0.03	0.14	0.03
Combined..	....	....	....	......	0.97	0.03

A single equal-volume extraction would, therefore, remove 91% of C from A, whereas a double half-volume extraction would remove 97%.

$$\frac{\text{Concentration of solute in liquid } A}{\text{Concentration of solute in liquid } B} = \text{Approximately constant at a given temperature}$$

(See **Solvents**.)   (R.K.S.)

EXTRACTION CYCLE.

Extraction cycle refers to any arrangement whereby steam is bled from a turbine at one or more pressures for any purpose whatsoever; i.e., feedwater heating, process steam, heating steam, etc. The terms "bled steam" and "extracted steam" may be used synonymously, as may also "bleeder point" and "extraction point."

There are two types of extraction, i.e., extraction at constant steam pressure and extraction at whatever pressure exists in the **turbine** at the extraction point. Extraction at constant pressure requires that an extraction valve gear be provided to regulate casing pressure at the extraction point. This is necessary because, not only would the extraction pressure vary with different amounts of extracted steam demanded, but varying loads on the turbine would cause the casing pressure at the extraction nozzle to vary. The extraction valve gear is often complicated by the use of a control or pilot valve to operate the main extraction valve. Turbines equipped with extraction valve gear are naturally more expensive than the simpler forms which have no pressure governing on the extraction lines. Industrial use of extracted steam often requires that the pressure of the bled steam be kept constant. Also, industrial use of the extraction turbine differs from central station practice in that frequently a large portion of the total flow is extracted, whereas in the power plant only a small fraction of the total is used for feedwater heating. (F.T.M.)

EXTRACTS.

Extracts, or essences, are solutions of flavoring substances, mostly **volatile oils**, dissolved in alcohol or in water. (R.M.W.)

EXTRAEMBRYONIC MEMBRANES.

A series of structures developed in connection with the embryos of vertebrates but not as parts of the body itself. They relate the embryo to its environment in several ways. These membranes are the **allantois, amnion, chorion, serosa**, and **yolk sac**. (A.W.L.)

EXTRANEOUS ROOT OF AN EQUATION.

In the process of the **solution of equations**, they are often transformed into other derived equations, and one of the derived equations may be readily solved. It sometimes happens that one or more **roots** of the derived equation will satisfy the original equation and also that one or more roots of the derived equation will not satisfy the original equation.

An extraneous root of an equation is a value which satisfies a derived equation but does not satisfy the original equation.

Extraneous roots are liable to occur in solutions of **fractional equations** or **radical equations**, or in **trigonometric equations**. (L.L.S.)

EXTRATHECAL ZONE.

The projecting tissue about the base of a **coral polyp**, from which young polyps develop. (A.W.L.)

EXTRATROPICAL CYCLONE.

Wave Cyclones.

EXTRA-UTERINE PREGNANCY.

Pregnancy.

EXTRUSIVE ROCK.

Effusive.

EXUMBRELLA.

The upper or convex surface of the body of a **jellyfish** or other **medusa**. (A.W.L.)

EYE.

A sensory organ which is stimulated by light, particularly an organ whose stimulation results in the formation of a mental image of the objects from which the light is reflected or radiated.

Although most eyes enable the animal to form visual images, that is to see in the usual sense of the word, some light-sensitive organs are capable only of perceiving light and the direction from which it comes. Pigment spots in some of the 1-celled animals are supposed to be light-sensitive and in flatworms and a few insects the eyes are formed of a group of sensitive cells partially isolated by pigment. The structure of these eyes shows no possibility of their forming images.

Eyes of reasonable complexity are found in some of the segmented worms and three types of complex eyes occur in the phyla **Mollusca, Arthropoda**, and **Chordata**. These three forms of eyes have been extensively studied and are known in detail.

Arthropod eyes are of two kinds, simple and compound, of which one or both may occur in a single individual. The sensory end organ in both forms is the retinula, a group of visual cells surrounding a central optical rod or rhabdom. In many simple eyes a portion of the cuticula is thickened to form a biconvex lens opposite to a group of retinulae. Compound eyes are made up of many ommatidia, each consisting of a similar lens forming a facet of the cornea of the entire eye, and an underlying retinula, with intervening crystalline cells and, in some species, other structures.

Both mollusks and vertebrates, including, of course, man, have camera eyes, although their development and structure differ. All camera eyes have a lens suspended before a chamber lined with a sensory layer, the retina.

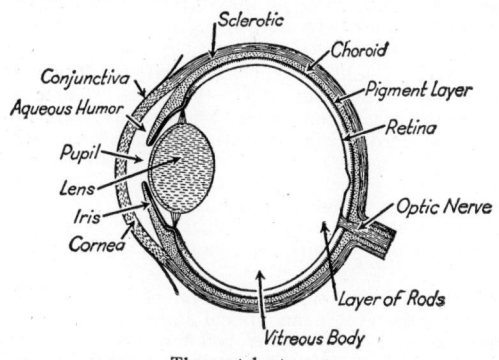

The vertebrate eye.

In front of the lens is another chamber and in front of that a transparent cornea which acts as a lens in terrestrial animals. The eye is insulated by a heavily pigmented layer which surrounds it except where the lens is suspended and extends in front of the lens as the iris. The iris, activated by muscles, controls the size of its central opening, the pupil, through which light enters the eye. Light passing through the lens is focused on the retina in a sharp image and the varied stimuli acting on nerve endings result in a definite mental picture. Such eyes are provided with muscles which direct them toward objects to be observed. They also have muscular focusing devices which move the lens in relation to the retina or vice versa, or control the curvature of the lens as in the human eye.

The action of the different kinds of eyes results in different kinds of **vision.** (A.W.L.)

EYE BAR. The eye bar is a heat-treated tension member formed from a single piece of steel. The finished eye bar consists of a body having a rectangular cross-section and two circular heads containing holes for **pins,** which are used to connect the eye bar when it forms part of a structure.

In the fabrication of an eye bar the ends of a steel plate, of the correct cross-sectional area and length, are heated and **upset** to form the heads. The heads are next rolled to remove any unevenness resulting from the upsetting operation. While the ends are still hot, holes are punched out which are smaller in diameter

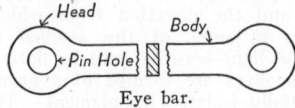
Eye bar.

than the finished pin holes. The bars are then subjected to special heat treatment which produces a high tensile strength. After cooling, the pin holes are bored to exact size simultaneously.

Eye bars make excellent tension members since the heat treatment enables them to carry higher tensile loads than the ordinary built-up steel members. As the eye bar is a very slender member it cannot be used where there is a possibility that it will have to carry compressive **stress.** These members are used in the cable anchorages of suspension bridges as well as for tension members of trusses. Eye bar chains have been used in preference to wire cable for suspension bridges. The tension members of cantilever **bridges** are often composed of eye bars. (C.W.C.)

EYE BOLT. Screw Fastenings.

EYE OF STORM. At the center of a hurricane there is usually a very small area in which winds are light and the sky nearly clear. This is known as the eye of the storm. A similar "eye of the storm" sometimes exists in temperate-zone cyclones. (P.E.K.)

EYELID. A fold of skin which can be drawn over the eye in **vertebrates** above the fishes. Three eyelids are the maximum. These are an upper and lower lid and a third eyelid or nictitating membrane which passes between the others and the eyeball from the inner to the outer margin of the eye. The eyelids contain glands whose secretion lubricates the apposed surfaces of the lids and eyeball, and in the mammals bears a row of stiff hairs, the cilia or eyelashes. (A.W.L.)

EYEPIECE. The lens, or system of lenses, which is closest to the eye in an optical instrument such as a **telescope** or a **microscope** is known as the eyepiece or ocular The eyepiece is usually a magnifying device

used for the purpose of detailed examination of the real image formed by the objective of the instrument. It is usually designed to act as a collimator to the light from the objective so that the light from each point of the image formed by the objective emerges in parallel or nearly parallel rays. Hence in using a telescope or microscope in proper adjustment, the eye should be focused as though looking at a distant object.

The simplest type of eyepiece is either a simple convex or concave lens, of relatively short focus, so placed as to serve as a magnifier for the image formed by the objective. The use of such simple eyepieces is shown in the diagrams in connection with the article on telescopes. Because of the **spherical** and **chromatic** aberrations of the simple lens of short focus, a combination of lenses is usually employed as an eyepiece. The two most common types of compound eyepieces are the

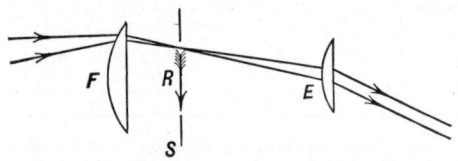
Fig. 1. Eyepiece.

Huygens (Fig. 1) and the Ramsden (Fig. 2). In these eyepieces the lens F is known as the field lens and the lens E as the eye lens. The Huygens eyepiece is placed slightly inside the focus of the objective and

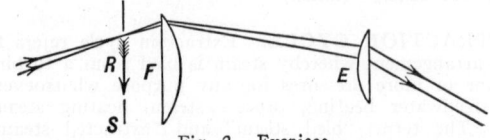
Fig. 2. Eyepiece.

the field lens of the eyepiece forms a real image R in the plane S from which the rays emerge parallel from E. In the Ramsden type the field and eye lenses combine to render the light from the real image R, formed in the plane S by the objective, parallel upon emergence from the eyepiece. Since the Huygens type eyepiece is placed inside of the principal focus of the objective, a **reticle** or **filar micrometer** cannot be used, although a reticle may be placed inside the eyepiece itself in the plane S. The Ramsden type, on the other hand, is focused directly upon the plane of the real image from the objective and a reticle or micrometer may be placed in this plane. Eyepieces of the Ramsden type, which are simple magnifiers focused upon the real image from the objective, are known as positive eyepieces; while eyepieces placed inside the principal focus of the objective, as in the case of the Huygens type, are known as negative eyepieces. There are many other types of positive and negative eyepieces which will be found discussed in treatises on optical instruments.

Both the positive and negative eyepieces give a view of the image from the objective in the same orientation as that image is formed. This means that the observer will see the image of a distant object inverted. While this is no disadvantage in microscopes and in astronomical telescopes, it is intolerable in a telescope or field glass to be used for observation of distant terrestrial objects. The simple concave lens, as used in the so-called Galilean telescope or opera glass, gives an erect image of a distant object. To avoid the aberrations of the simple concave lens various "erecting systems" are used in terrestrial telescopes. Some of these erecting systems employ prisms, as in the case of binoculars, or complicated systems of lenses. (W.K.G.)

EYRA. Cat.

F

F LAYER. Ionosphere.

F/SYSTEM. A method of designating the relative aperture of photographic lenses. The f number expresses the ratio of the diameter of the effective diaphragm aperture to the focal length of the lens. Thus, with a lens having an effective diaphragm of $1''$ and a focal length of $8''$, the ratio is $1:8$ which is written $f/8$. Following the recommendation of a committee of the Royal Photographic Society of Great Britain, at ratios greater than $1:4$ ($f/4$) lenses are marked so that the exposures increase as the power of 2. Thus:

f/number	4	5.6	8	11.3	16	22
Relative exposure	1	2	4	8	16	32

The difference in exposure for two f/values is the difference in the squares of the numbers. The difference in exposure, for example, in $f/4$ and $f/16$ is

$$\frac{16^2}{4^2} = \frac{256}{16} = 16$$

(C.B.N.)

FABRIC. In geology, the term proposed by Cross, Iddings, Pirsson and Washington, in 1902, for the shapes and arrangement of crystals in an igneous rock. (Compare with **texture** and **structure**.) Best defined as the arrangement of the constitute constituents in a rock, i.e., flow fabric in **lava, stratification** or **bedding** in **sedimentary** rocks, **foliation** in **metamorphic** rocks. (R.M.F.)

FABRICATION. Fabrication is the action of constructing or forming a **structure** composed of a number of separate elements which must be joined together in one way or another, according to a definite plan. The common methods of engineering fabrication include fusion methods, such as **welding; adhesion,** exemplified by gluing and **soldering;** and pinned connections, illustrated by bolting, riveting, and doweling. Fabrication will also include those operations necessary upon the several elements in order to fit them for assembly; also such trimming, polishing, or adjusting operations as will put the complete structure in its final shape. Included in this category are operations like **drilling, shearing, milling,** polishing, plating, etc. (F.T.M.)

FACE. The anterior and ventral part of the head, bearing the mouth, eyes, and in vertebrates the nose. (A.W.L.)

FACEPLATE. Lathe.

FACIES. In geology, the sum total of the inorganic and organic characteristics of a sedimentary formation. Obviously, different facies of a **sedimentary** formation (sedimentary time unit) are of the same age; but similar sedimentary facies may represent different formations. Fossils may be useful in determining the age of a facies, provided the types of organisms have not changed with the change in habitat, which, in the case of marine sediments (such as **limestones** or **shale**) they usually do. The term facies is also used to designate gradational types of **igneous** rocks which are supposed to have been differentiated from a parent **magma.** (R.M.F.)

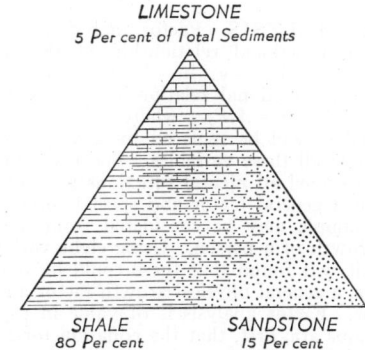

LIMESTONE
5 Per cent of Total Sediments

SHALE
80 Per cent

SANDSTONE
15 Per cent

Illustrating the relative abundance of the three principal types of sedimentary rocks and their intergradations, or facies. (*Field, Outline, Barnes & Noble.*)

FACING. Lathe.

FACSIMILE TRANSMISSION. This is the transmission by electrical means of any graphic material such as pictures, printed matter, maps, etc. There are three methods of transmission in use, transmission by radio, by land telephone lines and by submarine cable. Each of these introduces its own problems, but the fundamentals of the system are the same for all, the material must be broken into sequential elementary parts which may be transmitted by electrical means and then the parts converted back into a graphic presentation at the receiver. To be specific let us discuss the transmission of a picture although it should be kept in mind that exactly the same procedure applies to any graphic material. The picture is broken into the sequence of elemental parts by the process of scanning. This may be done by mounting the picture on a revolving drum and projecting a very small beam of light on or through it. The light is reflected (or transmitted) to a **phototube,** the light and hence the phototube output being proportional to the picture density. The light is moved along the picture (parallel to the axis of the drum) at such a rate that it displaces axially its own width for each revolution of the drum. Thus the spot of light progressively covers every spot on the picture. The output of the phototube is an electrical breakdown of the picture, and is then modified for transmission. For radio and land lines this means **modulation** upon a suitable **carrier.** There are several methods of doing this but these are merely details of the system. For submarine cable transmission the signal is amplified by d-c amplifiers and fed directly to the cable since these cables cannot handle high frequencies. At the receiving end the modulated signal is demodulated and fed to the recorder. This varies with different systems but two types are in wide use. One depends upon a variable light on a photographic paper or film. The light is varied in different ways, one being to use a specially constructed gas-filled lamp whose intensity varies with the signal, others use light valves, but in each case the result is a spot of light whose intensity varies with the picture focused on the paper. There are some direct-developing papers, but others require a development process after removal from the receiver. The other method of recording is to utilize a paper sensitized to electrical current passage and to pass the received signal current (after detection, of course) through it. Regardless of the method of recording it is necessary for the receiver to be synchronized with the transmitter. Sometimes this is accomplished by depending upon the stability of the power systems and the use of synchronous motor drives; other methods involve the transmission of synchronizing pulses periodically. Facsimile transmission has a wide field of application in the transmission of news pictures, legal documents, maps, etc. (L.R.Q.)

FACTOR ANALYSIS. Factor analysis is a technique by which a functional relationship of the form $z = a_1x_1 + a_2x_2 + \cdots + a_Nx_N$ in N variables may be replaced by a functional relationship of the form $z = A_1F_1 + A_2F_2 + \cdots + A_jF_j$ where $j < N$. Usually many variables will be replaced by a very few factors which include all these variables. Spearman in his first approach assumed all mental functions were explainable in terms of a general factor and special specific factors. Another example may be considered from the field of physical growth of a person. Using eight variables as a beginning it was found that these could be replaced by two factors, a general physical growth factor and a body type factor. Factor analysis is of value in psychology. Its limitations are, first, that the equation for z is linear in each variable or factor and, secondly, the factors are identified subjectively in the final analysis. (L.A.A.)

FACTOR OF SAFETY. The factor of safety is a number expressing the relation between the utmost endurance of a structural part, or of a complete structure, to the maximum actual demand that may be expected ever to be made upon it. But the factor of safety is not merely some ratio to allow for inaccuracies, lack of knowledge, or absence of confidence. Indeed, it has a very definite rational basis which becomes more apparent as the conditions which govern the factor of safety become known. Factor of safety has many different forms, a few of which will be given below. It represents a combination of the allowances necessary to be made in the use of practical data. The factor of safety will often include a factor which is not so much one of safety, but one of making due allowance for factors known to be present, but not definitely computable. If the engineer could definitely specify the usage and the care to which his product would be put, a large element of the so-called factor of safety would not be necessary. It must include allowance for unavoidable shocks or jars which might be expected during the working life of the structure. The material used may not be homogeneous in character, or uniform in all deliveries. Then there is always the desire to be well on the safe side when, due to failure of some part, life will be endangered. Usually, the factor of safety, at least for machine parts, is taken as the ratio of the **ultimate strength** claimed or accepted for the material, to the working stress used for design, and presumably reached under maximum design loading. When failure is measured by excessive deformation, the factor of safety might well be based on the elastic limit. Likewise, for elements subjected to repeated reversals of stress, the endurance limit should replace ultimate strength. In most old, well-organized fields of design, professional and standardizing societies have undertaken to specify working stresses for the commonly used materials, thus indirectly covering the factor of safety.

Factors of safety do not always bear this name; for example, the safety of a masonry **dam** against overturning is contained in a computed ratio of the overturning moment due to water pressure, divided by the stabilizing moment of the masonry weight. Safety in aircraft design is contained in a carefully and scientifically determined "load factor" made up in accordance

with certain rules promulgated by a governmental bureau. (See **Load Factor**.) (F.T.M.)

FACTOR THEOREM OF ALGEBRA. If $x - r$ is a factor of a **polynomial** $P(x)$, then $x = r$ is a **zero** of the **polynomial function** $P(x)$, and $x = r$ is a **root** of the **polynomial equation** $P(x) = 0$; conversely, if $x = r$ is a zero of the polynomial $P(x)$, or is a root of the equation $P(x) = 0$, then $x - r$ is a factor of $P(x)$. (L.L.S.)

FACTORIAL DEVELOPMENT. A method of determining the time of development, due to Alfred Watkins (1895). The time of appearance of the image is multiplied by a factor to obtain the time of development required for a given degree of development. The method is quite reliable with negatives which have been correctly exposed but leads to underdevelopment of overexposed negatives and overdevelopment in cases of underexposure. Factorial development has become practically obsolete as sufficient light cannot be used with panchromatic materials to enable the first appearance of the image to be seen. (C.B.N.)

FACTORIAL NOTATION. Binomial Formula.

FACTORING. To factor a **polynomial** means to find polynomials of lower degree than the given one whose **product** is the given polynomial.

By reversing the special **product formulae**, we obtain methods for factoring.

A polynomial with a common monomial factor may be factored by use of the formula

$$ax + bx + cx = x(a + b + c).$$

The difference of two squares may be factored by use of

$$x^2 - y^2 = (x + y)(x - y).$$

Trinomials which are perfect squares may be factored by use of

$$x^2 + 2xy + y^2 = (x + y)^2, \quad x^2 - 2xy + y^2 = (x - y)^2.$$

Trinomials of the form $x^2 + qx + r$ may be factored by use of

$$x^2 + (a + b)x + ab = (x + a)(x + b).$$

Trinomials of the form $px^2 + qx + r$ may be factored by use of

$$acx^2 + (ad + bc)x + bd = (ax + b)(cx + d).$$

The sum or difference of two cubes may be factored by use of

$$x^3 + y^3 = (x + y)(x^2 - xy + y^2),$$
$$x^3 - y^3 = (x - y)(x^2 + xy + y^2).$$

Polynomials may often be factored by grouping of terms and applying the preceding methods.

The process of **synthetic division** is often useful in factoring. (L.L.S.)

FACULAE. Sun.

FADING. The fading of radio signals is inherent in the transmission of such signals and at best can only be partially compensated for in the receiver by **avc** circuits, **diversity** reception, etc. The compensation may often be made entirely satisfactory if the fading is of the simplest type, but if it is selective, compensation is not always satisfactory. Radio waves going out from the transmitter travel along various paths to the **receiver**, some of the waves travel along the ground, others are reflected from the **ionosphere**. In the broadcast band fading is usually caused by signals which have been reflected from the ionosphere combining vectorially with signals which have traveled along the earth (these are called respectively **sky wave** and **ground wave**). The sky wave does not return to the earth near the transmitter so there is no fading in this region, and at great distances from the transmitter the ground wave has died out so again there is no fading due to this cause. In the intermediate region both waves may be present and if the phase of the two signals is such that they cancel fading results. Since the ionosphere is continually changing, the phase of the reflected sky wave may cause cancellation at one instant and addition of the signals at the next. Different frequencies travel somewhat different paths in the ionosphere so the time to reach the receiver is different for the different sideband frequencies. Thus one frequency may reach the receiver to add to the ground wave, while another may cancel. This produces what is known as selective fading and the output of the receiver is badly distorted. It should be realized that this is an effect of the transmission and not a characteristic of a given receiver. Both types of fading may be produced by two sky waves which have traveled different paths from the transmitter to the receiver. This is the cause of fading at the very high frequencies where the ground wave does not get far enough from the transmitter to cause any trouble. (L.R.Q.)

FAHLBAND. A Scandinavian term used by miners to describe **metamorphic** rocks containing richly disseminated ore minerals. (R.M.F.)

FAHRENHEIT SCALE. Temperature Scales.

FAILURE. The inability of a **structure** or a structural member to perform its proper function causes a condition known as failure. This condition may be the result of sudden fracture as in the case of brittle materials or the excessive **deformation** of ductile materials. Another cause of failure is a lack of equilibrium between the external **loads** and resisting forces such as exists in structures which fail by sliding or overturning. (C.W.C.)

FAIRED. An object is said to be faired if it is constructed to streamline shape, or has attached to it supplementary bodies which cause it to assume a shape of some degree of excellence of streamlining. The term has its major usefulness in aircraft nomenclature where many instances of fairing of parts in the exposed windstreams are present. Fairing of exposed struts, wheels, cabins, etc., has done much to reduce wind drag and increase performance. Sometimes the part is actually built in a streamline shape, and sometimes the streamline shape is obtained by enclosing the part in a streamlined case, or by attaching to it a shaped piece of some light material, such as balsa wood. The best faired object is one whose shape approaches that of a tear drop, having a ratio of length to width of approximately 3.5. (F.T.M.)

FALCON. Aves, Falconiformes. Large birds (**Aves**) of prey closely related to the hawks and eagles and like them in appearance. They are found throughout the world.

One species of the Old World is called the windhover, *Falco tinnunculus,* or, in common with other species, kestrel. Another is the merlin, *F. gesalon.* The peregrine falcon, *F. peregrinus,* has been widely used for catching game and other birds.

Several species of falcons and merlins occur in North America, among them the duck hawk, *F. anatum,* pigeon hawk, *F. columbanus,* and the little sparrow hawk, *F. sparverius.* (A.W.L.)

FALCONIFORMES. An order of birds of prey containing the **vultures, falcons, eagles,** and **hawks.** They have strong hooked beaks and, with the exception of the vultures, large curved claws used for grasping prey. The eyes are directed laterally, unlike those of the owls which have similar beaks and claws. (A.W.L.)

FALLFISH. Pisces, Teleostei. A **chub,** *Leucosomus corporalis,* of the eastern states. It lives in lakes and rapid streams and is a food and game fish of moderate worth. (A.W.L.)

FALLOPIAN TUBES (UTERINE TUBES, OVIDUCTS). The two fallopian tubes in the female are situated one on either side of the uterus, extending from its upper angle outward to the side of the pelvis, ending near the right and left ovaries. The end of each tube is surrounded by fringe-like processes called the fimbria. It is partly by means of these processes that the ovum, after discharge into the abdominal cavity, gains access into the tube through which it passes to reach the cavity of the uterus.

It is believed that fertilization by the male sperm cell takes place along the course of the tube.

The fallopian tubes are subject to infection, acute and chronic salpingitis, and abscess formation. They are the most common sites of extra-uterine or ectopic pregnancy. (For a general zoological discussion of this term, see **Reproductive System.**) (D.M.H.)

FALSE BEDDING. Cross-Bedding.

FALSE CIRRUS. Cirrus resulting from anvil heads of thunderstorms. (P.E.K.)

FALSE CLEAVAGE. This is also called strain-slip **cleavage** by the British geologists. It differs from the typical slaty cleavage in that it is obviously associated with incipient **foliation** of metamorphic rocks. (R.M.F.)

FALSE SCORPION. Pseudoscorpion. **Arachnida.** (A.W.L.)

FAMILY. 1. A group of animals consisting primarily of two parents and their offspring, sometimes with other related individuals. 2. A taxonomic subdivision of an order. In zoology family names are formed of the stem of the type genus with the ending -idae. Thus *Homo,* the genus to which man belongs, has the stem Homin- and forms the family name Hominidae. In botany the ending -aceae is used in a similar manner. (See **Taxonomy** and **Nomenclature.**) (A.W.L.)

FAN, CENTRIFUGAL. Centrifugal action whereby a gas is compressed has been described under the subject **Compressors.** When large volumes are to receive a small compression (pressure rises of 1 lb. per sq. in. or less) the device is called a fan. Centrifugal fans fall in this category and fulfill a variety of needs such as ventilating, heating, combustion draft, and drying service.

In contrast to the high-pressure centrifugal compressor, the fan has narrow blades and very little compression occurring in the blading. However, the blade action on the gas is to increase its speed, thus requiring a diffusion to gain pressure. This diffusion is accomplished in a scroll case surrounding the wheel and comprising the casing of the fan. Diffuser guide vanes may sometimes be inserted in the scroll case to improve efficiency by reducing turbulence. Simple radial blading is sometimes used on account of its cheapness and simplicity, but most fans have blading that is curved. Fig. 1 shows the appearance of wheels incorporating in the one case forwardly, and in the other, backwardly curved blading. Vector diagrams show that for the same relative velocity of the gas leaving the wheel and the same wheel speed, the absolute velocity is greater with forwardly curved blades. Achievement of efficient diffusion is therefore more important with forwardly curved blades and pressure increase is greater. Conversely for same pressure increase, forwardly curved blading may operate at lower rim speeds. However, backwardly curved

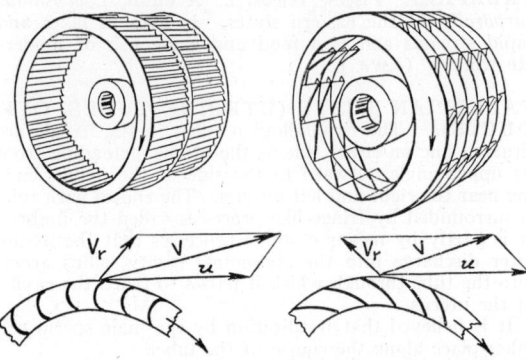

Forwardly curved blades.   Backwardly curved blades.

Fig. 1.

blading can be built so that somewhere near the best operating point (maximum efficiency) pressure rise diminishes more rapidly than volume increase and thereby induces a self-limiting feature in power consumption that is desirable in many applications. Power demand continues to increase with discharge in forwardly curved blading until well past the best operating point.

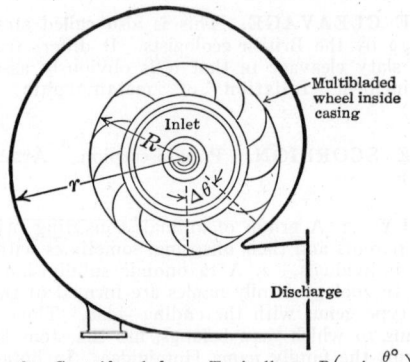

Fig. 2.   Centrifugal fan.   Typically $r = R\left(1 + \frac{\theta°}{300}\right)$

Since the pressure increments are small, an excellent approximation which simplifies the energy relation is to assume a constant volume flow through the fan. Given the weight of gas flowing per minute, W, and the draft, D (expressed in feet of air), the power imparted to the air is:

$$\text{Air horsepower} = \frac{WD}{33,000}.$$

Given the volume delivered per minute, V (in cu. ft.), and dynamic pressure, P (lb. per sq. ft.), power is also:

$$\text{Air horsepower} = \frac{PV}{33,000}.$$

The mechanical efficiency of a fan is the ratio of one of the above theoretical powers to the required drive power. Multivane centrifugal fans will usually exhibit an efficiency of from 70% to 80% at their optimum point, with radial plate type fans being somewhat poorer in performance. (F.T.M.)

FAN CHARACTERISTIC. A fan characteristic is a curve showing the relation between pressure and delivery. It is important because a fan operates entirely at the conditions depicted by the characteristic curve. Hence, it is a basis for fan selection. The characteristic

is determined by the shape of the blades. Blades curved forward in the direction of rotation have what is known as a rising characteristic—pressure increases with volume delivered. This characteristic is productive of low tip speed but fans having it can overload their drives if ignorantly handled. Backward curved blades have a drooping characteristic.

The centrifugal fan compresses the air or gas but slightly. In modern fan theory the work of compression is neglected and the action is assumed to be similar to a reversed **hydraulic turbine** or a **centrifugal pump**. (F.T.M.)

FAN, DRAFT. The mechanically-driven draft fan is often used to supplement or supersede **chimney** action in the production of **draft**. Without the use of fans it would not be possible to have the compact, high-capacity steam generator or the underfeed stoker. Draft fans are designated as plate (paddle wheel), multivane, or propeller type, the latter being but seldom used for this service. The plate fan is employed to some extent, but the multivane centrifugal fan is the most common type. (See **Fan, Centrifugal.**) Backwardly curved blade wheels are usually selected for forced draft service, because of the high speed, suitable for direct motor drive, the self-limiting power demand (a necessary feature when two or more fans are operated in parallel), and high static efficiency. Induced draft fans handle hot chimney gas. Forwardly curved blades which develop a given draft at lower speeds than those with backward curvature are frequently chosen for this service since the speeds and the centrifugal stresses in the wheels will be least. (F.T.M.)

FAN, PROPELLER. Propeller fans operate with flow of air parallel to the axis of rotation. They may be grouped into (1) fans with thin sheet blades of metal or composition material, and (2) fans with blades of airfoil section. Examples of the first class are the table fans (the common domestic "electric fan"), small ventilating fans, ceiling fans, unit heater fans, and radiator cooling fans. With the exception of certain low-speed ceiling fans, these are characterized by high rotative speed and low efficiency. The blades are often stamped in a single piece from a sheet of metal, then twisted slightly and mounted on the motor shaft. These fans are employed to move air but are not satisfactory if the air is to be forced against any pressure increment. The second class of propeller fans reflects a more scientific application of the axial flow principle. Whereas the first class moves the air largely by impulse against the face of the fan blades, the airfoil sections of the latter class perform in accordance with airfoil theory of lift. While more expensive, they are also more efficient, although the latter advantage is gained in sizes more suitable to industrial than domestic usage. The axial flow fan can move large volumes against light static pressures and is frequently more compact and more readily applied than the centrifugal type. (F.T.M.)

FAN, VENTILATING. Fans in this service are used to move air in predetermined directions for the purpose of changing the air in buildings, removing air charged with offensive odors, furnishing air to tunnels, etc. They are commonly fixed in position and connected to duct systems on the inlet or discharge sides, or both. Propeller fans have a limited use in this field, but ventilating fans are usually centrifugal because the application often requires overcoming of considerable static pressure. (F.T.M.)

FANGLOMERATE. The term proposed by Lawson in 1913 for a conglomerate composed of the coarser **clastic** sediments deposited at the head of **alluvial fans.** (R.M.F.)

FAN-TYPE MARKER. Radio Range.

FARAD. Capacitance; Electric and Magnetic Units.

FARADAY EFFECT. Magneto-optical Rotation.

FARADAY'S LAWS. The well-known Laws of Faraday, in electrochemistry, may be stated briefly as follows: 1. The amount of chemical action is proportional to the amount of electricity which has passed through the **electrolyte**. 2. Ions discharged by the same quantity of electricity are in the proportions of their chemical **equivalents**. (See **Electrochemistry**.) (R.K.S.)

FARSIGHTEDNESS. Hyperopia.

FASCIA. Layers of connective tissue composed largely of regularly arranged fibers. They cover muscles. Also marks in the form of bands. (A.W.L.)

FASTENING. Type of Fastening.

FAT BODY. 1. A large mass of fatty tissue found in **insects**. It serves for the storage of food during larval life, since it is much smaller in the adult, and is apparently a reservoir for nitrogenous wastes since it contains deposits of **uric acid**. 2. A mass of fatty tissue located near the **gonads** in Amphibia. (A.W.L.)

FATHEAD. Pisces, Teleostei. A small **minnow**, *Pimephales promelas*, found throughout the United States in sluggish streams. (A.W.L.)

FATHOMETER. Sounding.

FATIGUE. Failure of metal parts by progressive cracking caused by repeated application of stress. Most fatigue failures start at the surface where discontinuities in section such as square shoulders, screw threads, or even tool marks cause a high concentration of stress. Internal discontinuities may also start a fatigue crack, the most notable example being "transverse fissures" in rails which are believed to originate in areas within the rail section known as "flakes," a defect originating during the cooling period after hot rolling. Once a minute crack is started anywhere in the section, the root of the crack becomes the seat of high stress concentration upon subsequent applications of tensile stress, thus the crack will spread until the section is too weak to carry the load and the remaining portion will fracture suddenly. The portion of the section which failed progressively will be worn quite smooth due to the rubbing action of successive stress applications (e.g., alternate tension and compression in a rotating member loaded as a beam), while the suddenly fractured portion will have the usual crystalline appearance which is characteristic of fractures in heat-treated steels. For this or other reasons fatigue failures have wrongly been blamed on "crystallization" of the metal. All metals are crystalline and no alteration in the size or shape of the grains or crystals takes place in service during or before fatigue failure. (Exceptions might be made in the case of lead and other alloys which recrystallize when cold worked at room temperature. See **Recrystallization**.)

The fatigue strength, also called endurance limit, is the maximum stress which can be applied repeatedly without failure. In the case of steel, tests are run at a given maximum stress to 10,000,000 reversals or cycles of stress unless failure occurs earlier. It has been found that failures do not occur in steels after a successful run of this duration (4 days at 1700 rpm or less than 1 day

at 10,000 rpm). In the case of aluminum alloys and certain other non-ferrous metals, fatigue failures have occurred after much longer runs, hence tests are sometimes made to 500,000,000 cycles. The materials do not have a true endurance limit and the number of reversals of stress are stated in reporting the fatigue strength.

The most common test is a rotating beam type in which a carefully machined and polished sample is loaded as a beam while rotating in anti-friction bearings. As any point in the periphery rotates from top to bottom to top position the stress changes from maximum compression to maximum tension and back to maximum compression. From 4 to 8 or more individual tests at various maximum stress levels may be required to determine the endurance limit.

The endurance limit for smooth test specimens run in normal atmospheres at room temperature is an ideal or limiting value. In the case of steels not hardened, or heat treated to moderate hardnesses, the smooth specimen endurance limit is approximately one-half of the tensile strength. The endurance limit-tensile strength ratio is less than one-half for many other materials.

The presence of stress raisers, particularly at the surface, will lower the endurance limit. Notch sensitivity can be evaluated as the ratio of the endurance limit of a standardized notched specimen to that of a smooth specimen. Fatigue tests run in a corrosive medium, either gaseous or liquid, generally give much lower endurance limits than tests run in normal atmosphere. Corrosion-fatigue is responsible for many service failures of shafts and other stressed parts of pumps, engines, or processing equipment operating in corrosive media. Protective coatings are sometimes used to improve service life. **Nitriding** of alloy steel parts has proved very effective.

In order to guard against fatigue failures in critical parts such as connecting rods, they should be fabricated from high-quality steels using designs that avoid regions of stress concentration such as sharp fillets and engraved part numbers. They should be finished over all, avoiding tool or grinding marks. As a further aid in obtaining high fatigue strength such parts are being surface peened by a shot blasting process which work-hardens the surface and sets up compressive stresses in the surface layers. This raises the endurance limit, apparently by reducing the maximum tensile stresses which can be developed at the surface in normal operation. (R.H.H.)

FATIGUE LIMIT. Fatigue.

FATS. Esters.

FAUCES. 1. The exposed portion of the cavity of a **snail** shell. 2. The opening of the throat of a **vertebrate**, flanked by the tonsils. (A.W.L.)

FAULT. A fault is a great fracture in the crust of the earth along which movement has taken place with the result that the crustal blocks are displaced relative to one another. This movement is in most cases probably intermittent and the actual individual displacements may be very small, but by accumulation may reach tens, hundreds, or rarely thousands of feet. The fractures themselves may often be traced for many miles. The San Andreas fault, horizontal movement along a portion of which caused the earthquake that so severely damaged San Francisco, California, in 1906, has been traced for about 600 miles.

In describing faults, certain terms have been adopted. Those in most common use are given herewith: the fault plane is the plane of the fracture and may be vertical or at some other angle; the angle between the fault plane and the horizontal is called the angle of dip of the fault plane. The angle between the vertical and the fault plane is spoken of as the angle of hade or simply as the hade of the fault. The surfaces of the fault plane are called the walls of the fault. If the fault plane dips, the uppermost wall is called the hanging wall, the lower wall the foot wall. These terms are applied irrespective of whether the fault is normal,

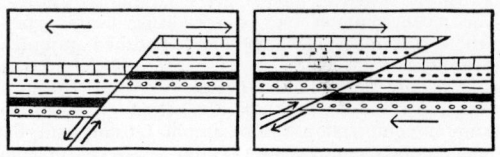

Fig. 1. Structure sections of a simple normal fault (on left) and of a simple thrust fault (on right).

with the hanging wall slipping down the dip of the fault plane, or whether it is a reverse fault with the hanging wall apparently pushed up the dip of the fault plane. A normal fault is sometimes spoken of as a gravity fault. The displacement measured along the fault plane is designated as the slip; the displacement measured vertically is called the throw; the displacement measured at right angles to the plane of the involved stratum is called the stratigraphic throw. The amount of horizontal displacement between the ends of a broken stratum measured at right angles to the direction of strike of the fault plane, is called the heave. The visible evidence of the trace of a fault plane at the earth's surface is called the fault trace. The block of the earth's crust which has moved downward, relatively speaking, to the other is called the downthrown block or referred to as the downthrow side of the fault. The other block is called the upthrown block or the upthrow side of the fault. If the strike of the fault plane is essentially at right angles to that of the bedding it is called a dip fault. A strike fault is one in which the

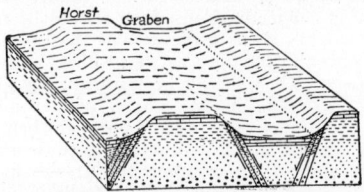

Fig. 2. Block diagram of one type of scenery produced by "normal" or block faulting. It may also be assumed that the graben has been pushed down, and the horst has been pushed up. Neither structure is necessarily entirely the result of tension. Note that the stratigraphic (vertical) order of the formations has not been duplicated or reversed. (*Field, Laboratory Manual, Princeton Univ. Press.*)

movement has been parallel to the strike of the strata involved. A compound fault involving several parallel displacements dipping in the same direction, resulting in a step-like arrangement, is referred to as a step fault. The term graben or trough fault refers to a downthrown area bounded on each side by two or perhaps more faults. A horst is an uplifted area bounded by two or more faults. (E.S.C.S.)

FAULT BRECCIA. Autoclastic.

FAULT SCARP. Scarp.

FAUNA. The animal population of a region. (F.T.M.)

FEATHER. The structure characteristic of the **vestiture** of birds (**Aves**). The development of a feather indicates that it is a modified scale, like those of reptiles and birds.

A feather consists of a central axis or rachis continuous with the hollow quill which is attached to the body. The rachis bears the flat vane of the feather which is made up of many slender barbs bearing barbules along each side. The barbules of adjacent barbs

interlock to form the continuous surface of flight feathers and the similar contour feathers of the body. Down feathers are of generally soft structure and lack barbules, and filoplumes are slender feathers with few barbs. (A.W.L.)

FEATHER, AERODYNAMIC SURFACE. Feathering.

FEATHER KEY. Key.

FEATHER STAR. Crinoidea.

FEATHER-BACK. Pisces, Teleostei. A fish (**Pisces**) of peculiar form found in Africa and the Oriental region. (A.W.L.)

FEATHERING. An object of flat plate shape in a fluid stream has maximum resistance to relative motion if its largest area be placed in an attitude perpendicular to the fluid stream, and minimum when the smallest area is so placed. Conditions occasionally arise when it is desirable to have a maximum resistance at one point of a cycle of events, and minimum at another. For example, during the cycle of a rowing stroke, the blade of the oar should have maximum resistance to motion in the water, whereas during the return stroke it should have minimum air resistance. Certain experimental types of lifting planes or **airfoils** have been built involving this same action. The act of first presenting the surface of maximum resistance, followed by the surface of minimum resistance on a return stroke, is known as feathering. In recent years controllable pitch propellers have been equipped with hub mechanisms and controls to permit the pilot to feather the blades when the engine was either throttled or inoperative. This action assists control and performance of multi-engined aircraft having one or more engines inoperative, also allows a single-engined airplane to reach higher terminal diving speeds. Feathering the propeller will keep a damaged engine from "windmilling" and increasing the damage. Feathering is also applied to the change in the angle of setting of the blades of a helicopter either to secure control or to equalize lift on the advancing and retreating sides. (F.T.M.)

FEED. Cutters, Cam.

FEED WATER. Feed water is water properly prepared for entering a **boiler** to take the place of that which is evaporated in the generation of steam. Although the instantaneous flow of feed water is not necessarily equal to the rate at which steam is boiled off, the total amount fed over a considerable period of time must equal the evaporation plus such loss as blowdown and steam released through the safety valve. The nearer a boiler design approaches the flash type, the nearer must the instantaneous flow of feed water be to the rate of evaporation. To be suitable for the service, feed water should be at a pressure enough above that of the boiler contents, so that it flows readily into the boiler when the feed valve is opened. It should have chemical purity to the required degree and be heated to a temperature as near that at which the boiler operates as is economically feasible. (F.T.M.)

FEEDBACK. Feedback, or the transfer of energy from a high level point to a low level point, in communications circuits may be either positive or negative. Positive feedback is that in which the signal fed back is in phase with the input signal; thus we might say that some of the output is sent back through the circuit to add to the regular signal present. For **amplifiers** this effect is normally very undesirable, although in some amplifiers it is used to increase the gain. If enough energy is fed back positively the input can be removed entirely and the signal kept going through the amplifier by the feedback. This is possible since a small signal at the input produces a much larger one at the output and only a small part of this needs to be returned to the input to keep the process going. When this condition is reached the amplifier oscillates or howls. While smaller amounts than this may be used, to increase the gain of an amplifier it is always at the expense of stability and fidelity. An oscillator is really an amplifier with a large amount of positive feedback. Negative feedback is that in which the signal fed back to the input is in phase opposition with the input. While this cancels part of the input it has such desirable effects that it is almost universally used for high-grade amplifiers. Negative, or inverse, feedback increases the stability, improves the frequency response, lowers distortion and noise and has other similar desirable results. The improvement along these lines is approximately proportional to the amount of feedback, but a compromise value which will not decrease the gain too much and yet give decided improvement in the other characteristics is used. Since many feedback circuits have phase and amplitude characteristics which vary with frequency an amplifier may have negative feedback at one frequency and positive at another. If the amplifier has sufficient gain in the positive feedback region it will oscillate. If the signal fed back is proportional to the output current it is termed current feedback and stabilizes the current, while if it is proportional to the output voltage it is voltage feedback and stabilizes the voltage. (L.R.Q.)

FEEDER. In electrical circuits this term is used to designate the lines running from a main **switchboard** to branch panels in an installation. In radio circuits it is used to designate the transmission line from the **transmitter** to the **antenna.** (L.R.Q.)

FEEDWATER HEATER. The purpose of heating feed water is threefold. First, if the water is heated by heat which would otherwise entirely or partially go to waste, the heating represents a saving of fuel.

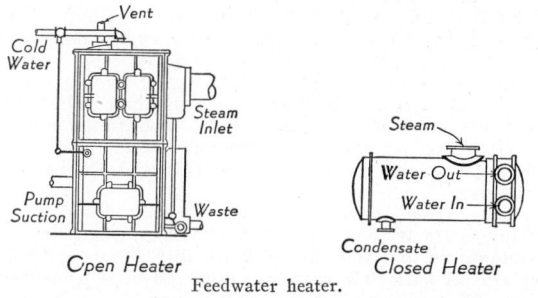

Open Heater   Closed Heater

Feedwater heater.

Second, the thermal stress induced in the **boiler** plate by contact with a stream of cold feed water is reduced, possibly eliminated. Third, the nearer the feed to saturation temperature the less the heat to be added in the boiler itself and the more the steam-raising capacity of each sq. ft. of heat-transfer surface. This becomes of increasing importance at higher boiler pressures.

Feedwater heaters are divided into two classes, the contact and the surface heaters. Economizer surface and a portion of the boiler surface are actual water heating surface; however, it is customary to refer only to equipment obtaining heat from steam as feedwater heaters.

The open heater is ordinarily built up in rectangular form, but heaters for other than near-atmospheric pressure are constructed in cylindrical form of cast iron or steel plate. The open heater is provided with tiers of trays, properly perforated and inclined to break up the

flow of water, delivered by gravity from a distributing trough, into a multitude of small cascading streams which present a large surface to the steam. It is possible to heat water to the temperature of saturated steam entering the heater if there are no non-condensable vapors. Heating is by direct conduction from steam to water so the effectiveness as a heater is not adversely affected by scale accumulation. A float-regulated valve admits the cold water required to supplement the returns in drips, condensate, and other uncontrolled feed supply.

The surface heaters are divided into steam-tube and water-tube types. Most heaters are of the water-tube type. These heaters can also be divided into straight tube and bent tube (U-tubes and steam coils), and into single or multi-pass. The surface heater is used when water is to be heated under pressure without direct contact with the steam. (F.T.M.)

FEEDWATER REGULATOR. Air, fuel, and water are the three variables entering into the production of steam "as wanted." The feedwater regulator is the governor of the **feed water** supplied to the boiler.

Soon after steam **boilers** came into use it was discovered that a disastrous explosion resulted if they boiled dry; also that the engine might be wrecked by water passing over with the steam. Naturally great care was taken to prevent the water level in the boiler from passing below or above the safe limits.

Under modern conditions it is necessary for feed water to flow into the boiler almost as rapidly as the steam flows out—and since boilers are approaching the flash type, it is plain that the feedwater regulation should be automatic, purely a machine function. It cannot be done successfully by hand.

There are several types of feedwater regulators, most of them either thermostatically or float operated. In

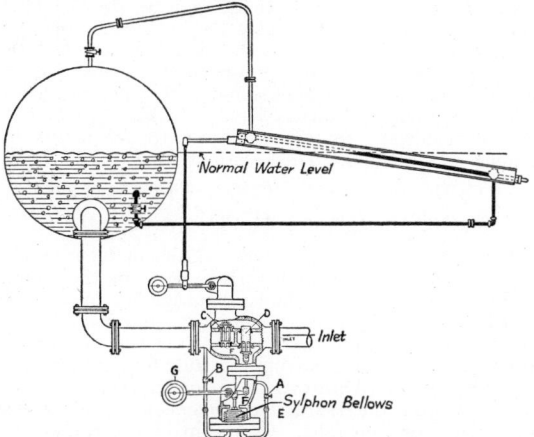

Feedwater regulator.

order to take advantage of the thermal storage contained in boiler drums between high and low water level, the regulator should gradually increase or decrease the feed on slowly increasing or decreasing steam demands, but it should decrease the feed during rapid increase of steam demand and increase it when the demand decreases rapidly, for in that way the fluctuation of boiler water level can offset the time lag of combustion response to changing steam requirements. (F.T.M.)

FEEDWATER TREATMENT. The purpose of a **feedwater** treatment system is to maintain the surfaces of the **boiler** in the same or approximately the same condition as when new. After entering the boiler, feed water is first heated to the saturation temperature, then evaporated in contact with the hot tube surface.

Unless the concentration of soluble salts and suspended particles in the boiler water is very great the steam will be free of all impurities the feed water might have contained except dissolved gases. The impurities are left behind in the boiler water whose concentration, as a result, increases. The point of evaporation being at the tube surface, there is every opportunity offered the impurities to deposit themselves on these surfaces as a scale. When untreated feed water produces enough scale on the boiler surfaces to interfere with heat transfer, or when it contains elements which either corrode or alter the strength of the boiler metal, feedwater treatment is necessary.

The higher the rate of heat transfer per square foot of surface, the more important it becomes to keep that surface scale-free, because the scale can both reduce the steaming capacity and cause overheating of the tubes. Natural waters usually contain dissolved salts and gases, also some organic and inorganic material in suspension. They rarely are neutral in reaction. The dissolved salts are chiefly the carbonates, sulfates, and chlorides of **calcium, sodium,** and **magnesium;** and occasionally some **iron, aluminum,** or **silica** salts. **Oxygen** and **carbon dioxide** are the gases. The suspended matter is usually alumina and silica in the form of mud and silt, or, if organic, sewage and industrial wastes.

The troubles caused by the feeding of water of undesirable quality are scaling, corrosion, foaming and priming, and **embrittlement.**

Boiler scale is due, mainly, to the cementing action of salts of calcium and magnesium. Calcium is the worst offender, particularly in the sulfate form. The formation of scale is caused by the dissolved salts in the boiler water reaching a concentration beyond which they are no longer soluble at the boiler temperature. At average boiler temperatures the sodium salts have solubilities running to thousands of grains per gallon. Magnesium sulfate is quite soluble in hot water and the chlorides are sufficiently soluble not to cause scale trouble. But calcium sulfate is less soluble in hot water than in cold. Its solubility varies between 150 grains per gallon in cold water to 5 grains in hot water, which explains why it is so troublesome a substance. Calcium and magnesium carbonates have solubilities so low at 212° F. that water containing them is treated merely by boiling and filtering. The reactions are:

$$Ca(HCO_3)_2 + heat \rightarrow CaCO_3 + CO_2 + H_2O$$
$$Mg(HCO_3)_2 + heat \rightarrow MgCO_3 + CO_2 + H_2O.$$

Scaling may take place in boiler drums or tubes, heater tubes, and feedwater piping. Its effect on the piping system is to choke the flow, requiring an increase of pressure to maintain water delivery. Its effect on heat transfer surfaces is to decrease the transfer. Treatment for scale consists of removing the scale-forming elements or replacing them with extremely soluble salts.

Corrosion is the destructive conversion of boiler metal into oxides or iron salts. Corrosion may occur at any place in the feedwater cycle, but it is found principally in boilers, heaters, and piping. It is due either to an acid condition of the water or to dissolved oxygen. From the standpoint of corrosion, scale is a protective agent. The corrosion may be a general loss of metal over the whole tube surface or a localized action. The latter is the more serious as it produces pitting. To prevent corrosion the boiler water is maintained alkaline and, if necessary, the feed water is deaerated to reduce the oxygen content to a suitable value.

Foaming refers to that condition of boiler operation where a stable foam is produced. It may or may not be accompanied by priming, which is the production of wet steam or, in the aggravated case, slugs of water. Priming can be produced by other causes than foaming, for instance, carrying too high a water level, insufficient disengagement area, or a pulsating steam demand that overtaxes the boiler steam storage. Foaming results also from saponification of the boiler water through mixture of oil or grease with the alkali. Floating organic matter is another source of foam. When foaming is due to concentration of salts in the water the condition is relieved by altering the treatment or by blowing down more of the concentrated water.

Chemical treatment is classed as external or internal, depending on whether the reactions are completed before the water enters the boiler or in the boiler. Internal treatment, if scientifically designed and controlled, is an effective method and is the best system where the proper treatment for the boiler might prove injurious to the steel tube economizers.

External water softeners are of two types, precipitation and base exchange. A precipitation softener embodies the principle of using calculated quantities of soluble reagents to react with the hardness in the raw water. Two treating tanks are used in the intermittent system, one supplying treated water to feed service, while the other is receiving its charge of chemicals and water or maintaining a quiescent condition so that the precipitate may settle out. Water flows continuously from inlet to outlet in the continuous type of softener. Reagents are added at the inlet and what precipitate does not settle out in the reaction tank is removed by filtration.

A base exchange softener removes the hardness by a simple filtration of the water through a bed of active material which exchanges its sodium base for the magnesium and calcium in the water. Natural and artificial **zeolites** are used as the active material. Natural zeolite is hard and dense; artificially prepared silicates such as permutit ($Na_2(Al_2Si_2O_8)$) are porous, thereby exposing a much greater surface to react with the water. (F.T.M.)

FEHLING'S SOLUTION. This is an alkaline solution of **copper** hydroxide and sodium or potassium **tartrate** in sodium hydroxide used either as a mild oxidizing agent or as a test for easily oxidizable groups such as **aldehyde** groups. The formation of cuprous oxide is a positive test, its color red but often yellow at first. (See **Carbohydrates.**) (R.K.S.)

FELDSPAR. Feldspar is the name of a group which includes the most important of the rock-forming minerals, making up perhaps as much as 60% of the earth's crust.

This group of minerals consists of three **silicates:** a **potassim-aluminum** silicate, a **sodium**-aluminum silicate, and a **calcium**-aluminum silicate ($KAlSi_3O_8$ $NaAlSi_3O_8$, and $CaAl_2Si_2O_8$) and their **isomorphous** mixtures.

The various members of the feldspar group show many characteristics in common. Crystallizing in the **monoclinic** and **triclinic** systems, they show similarity of crystal habit, cleavage and other physical properties as well as similar chemical relationships.

Orthoclase, $KAlSi_3O_8$, derives its name from the Greek words meaning right or straight, and fracture, because its two cleavages are at right angles to each other. It crystallizes in the monoclinic system and its crystals are usually prismatic; it occurs also in coarsely cleavable masses. Hardness, 6; specific gravity, 2.56–2.58; luster, vitreous to pearly; colorless to white, gray, yellow or red, rarely green. Twin crystals not uncommon.

Orthoclase is a common constituent of many igneous rocks and is often found in huge masses in pegmatite veins. Localities for orthoclase are so numerous as to prohibit a complete list. Adularia (from Adular) is essentially a pure potassium silicate; when pearly and opalescent it is called moonstone and frequently used for jewelry. These opalescent varieties are known to be an intergrowth of orthoclase and albite. A glassy kind of orthoclase, sanidine, is found in the trachytes of the Drachenfels, Germany. Beautiful moonstones come from Ceylon and Switzerland, in the United States from California and Virginia.

Orthoclase is found in the New England **pegmatites,** in New York, Pennsylvania, Virginia, North Carolina, Arkansas, Texas, Colorado, California and elsewhere. Its commercial use is in the manufacture of porcelain and as a constituent in scouring powders.

Microcline, $KAlSi_3O_8$, is chemically the same as orthoclase, but belongs to the **triclinic** system, the prism angle being slightly less than a right angle ($89° 30'$), hence the name microcline from the Greek meaning small, and to slope. Microcline is like orthoclase in all physical properties and can be distinguished from it surely only by optical examination. Under the polarizing microscope microcline displays a minute multiple twinning which results in a grating-like structure that is unmistakable. It is probable that much orthoclase would, upon proper examination, prove to be microcline. Amazon stone or amazonite is a beautiful green microcline occurring in the Ilmen Mountains in the Urals, Italy, Norway, Madagascar, and in the United States in the Pikes Peak region, Colorado, Virginia, North Carolina and sparingly in the pegmatites of New England.

The name amazon stone is derived from the application of this term to some green mineral found by the Spaniards among the aborigines of the Amazon Valley in South America. As no microcline is known to occur in the region there must have been some confusion with another green-colored substance.

A soda microcline, anorthoclase, is known, which is probably an isomorphous mixture of $KAlSi_3O_8$ and $NaAlSi_3O_8$, the sodium-aluminum silicate being in the greater proportion. The soda feldspar albite, $NaAlSi_3O_8$ and the calcium feldspar anorthite, $CaAl_2Si_2O_8$ form an isomorphous series from pure albite at one end to pure anorthite at the other, the two molecules appearing to be completely miscible one with the other. The members of this series are spoken of as the soda-lime (or lime-soda) feldspars, and as a group are called the plagioclase feldspars from the Greek meaning *oblique* and *fracture,* referring to the two cleavages at an angle that differs slightly from a right angle. Nearly always present are the striations, fine parallel lines, resulting from minute multiple **twinning,** which, never seen on orthoclase or microcline, are therefore an important diagnostic feature.

More or less arbitrarily, four intermediate **plagioclase** feldspars are recognized between albite and anorthite; these are listed below together with the approximate percentage of each molecule present.

	% of $NaAlSi_3O_8$	% of $CaAl_2Si_2O_8$
Albite	100 to 90	0 to 10
Oligoclase	90 to 70	10 to 30
Andesine	70 to 50	30 to 50
Labradorite	50 to 30	50 to 70
Bytownite	30 to 10	70 to 90
Anorthite	10 to 0	90 to 100

Albite is so called from the Latin, *albus,* in reference to its usual pure white color. It is a sodium aluminum silicate corresponding to the formula $NaAlSi_3O_8$. It crystallizes in the triclinic system commonly in tabular crystals. Twinning is very common, thin twinning lamellae producing a series of fine striations on certain crystal faces. There are two good cleavages at an angle of $86° 24'$ to each other. Hardness, 6; specific gravity, 2.62; luster, vitreous to pearly. It may be colorless to white or gray and transparent to opaque.

Albite is a relatively common and important rock-making mineral associated with the more acid rock types and in **pegmatite** dikes, often with rarer minerals like **tourmaline** and **beryl.** There are many famous localities in Europe in the Swiss and Austrian Alps, the Urals, the Harz Mountains, in Italy, France

and Norway. Brazil has yielded fine specimens. In the United States notable localities are Paris and Auburn, Maine; Chesterfield, Mass.; Haddam, Conn.; Amelia County, Va.; and the Pikes Peak region of Colorado. It is used in the ceramic industries and also in the manufacture of artificial teeth.

Anorthite was named by Rose in 1823 from the Greek meaning oblique, referring to its triclinic crystallization. The physical properties are essentially the same as for albite, except that the specific gravity of anorthite is somewhat greater, 2.74–2.76. Anorthite is characteristic of the basic igneous rocks such as gabbro and basalt. Anorthite is found in the lavas of Vesuvius and Monte Somma, Italy; in Finland, Japan, and in Sussex County, N. J.

The intermediate members of the plagioclase group are all very similar and with the exception of certain labradorites, cannot be distinguished from each other ordinarily save by optical means. Oligoclase is a common mineral in such rocks as **granites, syenites, diorites,** their extrusive equivalents and many **gneisses.** It is a frequent associate of orthoclase. The word oligoclase is derived from the Greek meaning little, and fracture, in reference to the fact that its cleavage angle differs slightly from 90°. Sunstone is mainly oligoclase (sometimes albite) spangled with flakes of hematite.

Andesine is a characteristic mineral of rocks such as diorites which contain a moderate amount of silica and related extrusives, such as andesites. Because of its occurrence in these latter andesine derives its name from them as well as from the Andes Mountains.

Labradorite is the characteristic feldspar of the more **basic** rock types like diorite, **gabbro, andesite** or **basalt** and it is usually associated with some one of the **pyroxenes** or **amphiboles.** Labradorite frequently shows a beautiful play of iridescent colors due to minute inclusions of another mineral. The classic locality for this mineral is of course Labrador, whence its name. It is a constituent there of the rock **anorthosite** and is found in the anorthosites of the Provinces of Quebec and Ontario and in the Adirondack region in New York State.

Bytownite, named from Bytown the former name for Ottawa, Canada, is a rare mineral occasionally found in the more basic rocks.

The feldspars crystallize from the **magma** in both extrusive and intrusive rocks; they occur as contact minerals, in veins and are developed in many sorts of **metamorphic** rocks, e.g., albite **schists.** They may also be found as mechanical deposits in various sedimentary rocks. (E.S.C.S.)

FELSITE. Felsites are defined by American geologists as dense, fine-grained, light-colored rocks rich in silica, hence classified with the **rhyolites,** from which some of them have been formed by **devitrification.** Felsites may occur as intrusive dikes but in general are found as **extrusive** rocks. They frequently occur interbedded with volcanic ash, tuff or breccia. According to American usage any light-colored lava whose ground mass or matrix is so fine-grained that the individual minerals cannot be distinguished by the naked eye (macroscopically) may be roughly classified as felsite, hence the prevalence of the term felsitic texture. When felsites show **phenocrysts** they are called felsite porphyries. The term felsite was first applied by Gerhard in 1814 to the fine ground mass (matrix) of **porphyries,** and is therefore one of the oldest, commonly used, petrological terms. (R.M.F.)

FEMALE. Sex.

FEMIC. This term is used by petrologists to designate the more common ferromagnesian (see **Iron; Magnesium**) minerals such as **pyroxene** and **olivine.**

Rocks which are relatively rich in femic minerals are said to be urafic. (R.M.F.)

FENCE. Woodworking.

FENNEC. Mammalia, Carnivora. Animals of two species which resemble foxes with enormous ears. One, *Vulpes zerda,* occurs in northern Africa, and the other, *V. famelicus,* in Syria and adjacent regions. (A.W.L.)

FENNEL. Carrot Family.

FERBERITE. Wolframite.

FER-DE-LANCE. Reptilia, Sauria. A large poisonous **snake,** *Lachesis lanceolatus,* of tropical America. It reaches a length of 7′ and is active at night. A **pit viper.** (A.W.L.)

FERGUSONITE. A mineral multiple oxide containing columbrium, tantalum and titanium. Essentially an oxide, or columtate-tantalate of yttrium. Crystallizes in the tetragonal system. Hardness, 5.5–6.5; specific gravity, 5.6–5.8; color, variable. Named after Robert Ferguson (1799–1865), Scotland. (R.M.F.)

FERMAT'S PRINCIPLE. This is a law of optics recognized nearly 300 years ago by Fermat. It states that when light proceeds by any path from a point A to another point B, the time required in its passage is either a minimum or a maximum as compared to other, arbitrarily chosen, adjacent paths. If the light is reflected from A to B by a plane surface, or is refracted at a plane surface on its way from A to B, the time is a minimum. For a curved reflecting surface, the time is a minimum if the surface has less curvature than the "aplanatic" surface osculating with it at the same point (i.e., the surface which gives rise to no **spherical aberration**); and this holds true also for a curved refracting surface. In these cases the law is known as the "principle of least time." But if the reflecting or refracting surface has greater curvature than the aplanatic surface at the same point, the time for the actual path is a maximum; that is, if the light could be made to follow a path through any other point of the curved surface than the one it actually passes through, it would do so in less time. For all points on a given aplanatic surface, the time is the same, and the light, if unobstructed, actually does follow paths through all of them. (L.D.W.)

FERMENTATION. While in recent literature there has been a tendency to extend the use of this term to include many **biochemical** reactions catalyzed by enzymes, this term applies strictly to the chemical conversion of **glucose** into **ethyl alcohol** under the influence of the enzyme **zymase.** (See **Carbohydrates.**) (R.K.S.)

FERNS. Pteridophytes. The ferns and their allies form the third division of the plant kingdom. This is a small division comprising only about 4500 species of plants, few of which are very large. But in past times, and especially in the **Carboniferous** period, plants of this division were more numerous and many of them of very large size.

At the present time the ferns are found in nearly all parts of the world, especially in regions where there is plenty of moisture. They are particularly numerous in the tropics. There also the largest of the ferns, called tree ferns, are found. These tree ferns have an erect usually unbranched trunk bearing at its top a crown of much dissected leaves of large size. Tree ferns are from 10–30′ in height; a few may reach a height of 50′.

In nearly all ferns the stem is a slender structure. In many species it grows underground as a creeping hori-

zontal **rhizome;** in other species it is shorter and erect, but seldom rises much above the surface of the ground. From the stem numerous fine wiry roots extend into

Tree ferns in African forest. (*American Museum of Natural History.*)

the ground. The leaves of ferns having creeping rhizomes are borne singly at the nodes; those of erect-stemmed ferns are borne in a group which forms a close crown. In many ferns the leaf, often called a frond, is pinnately compound, and the individual **pinnae** themselves compound. Other ferns have entire leaves. Young leaves are circinnately coiled, the tip of the leaf being in the center of a tight coil, which unrolls from the base upward. Often these young leaves are covered with a mass of dense brown hairs. On the lower surface of the frond reproductive bodies are formed. In many ferns these are found on all pinnae; in others they occur only on special pinnae which are commonly very much modified in size. These reproductive bodies are commonly borne in compact groups called sori, which are

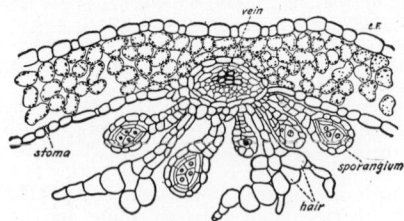

Section through leaf and sorus of *Polypodium* showing sporangia in various stages of development.

often covered by a protective structure, the indusium. The reproductive bodies are stalked sporangia containing **spores,** which are freed by the action of certain special cells of the **sporangium.** These cells have thick inner walls and in many ferns form a distinct row called the annulus. When the atmosphere is dry the cells of the

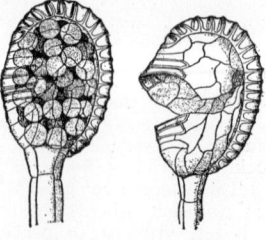

Fern sporangia. 1, Unopened sporangium filled with spores; 2, The empty sporangium after the annulus has returned to its first position.

annulus of a mature sporangium lose water and gradually contract. As a result, a considerable strain is exerted by the thin outer wall of these cells, so that the annulus is pulled backwards, a break occurring in a group of thin-walled cells known as the **stomium**. Finally the tension becomes too great and the annulus snaps back violently, catapulting the spores out of the sporangium.

These spores, carried by air currents to a region where moisture is sufficient, germinate. They do not, however, form a new fern plant. Instead they develop a small delicate green plant called the prothallium. In many ferns this is a heart-shaped body one cell thick. In others it is a branched object resembling certain species of algae, and in still other ferns it is a small tuberous body. From the lower surface of the prothallus numerous short **rhizoids** grow down and anchor it firmly in the substratum. On the lower surface also, reproductive bodies are formed. These consist of two kinds, commonly found on the same prothallus. One, usually in the basal portion of the prothallus, is the **antheridium**. An antheridium is a small multicellular organ in which are formed many small sperms. Fern sperms are spirally coiled cells, each having a group of **cilia** at the tip.

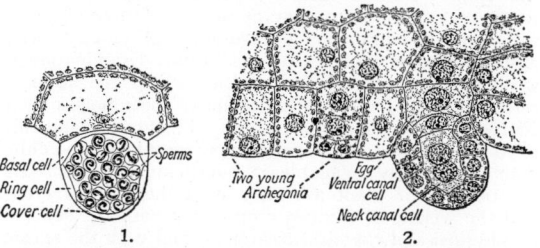

1, Mature antheridium of fern. 2, Two young archegonia of fern and one mature one. (*From Chamberlain's Elements of Plant Science, McGraw-Hill Co., Inc.*)

The female sex organ is the **archegonium**. These are commonly found near the notch of the prothallus. Each archegonium is a small organ consisting of a basal layer of cells surrounding the single large egg cell and a short tube surrounding a row of cells known as the neck cells. These break down, forming the neck canal, through which the sperm swim to unite with the egg. The fertilized egg or **zygote** immediately starts dividing and gives rise to a new fern plant. The prothallium of the fern is the **gametophyte** generation. It is green and

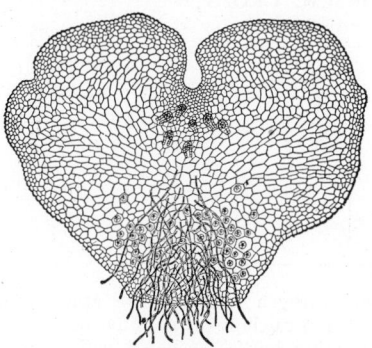

Fern prothallium. View of the under side showing archegonia near the apical notch and antheridia among the rhizoids near the base. (*From Sinnott's Principles and Problems, McGraw-Hill Co., Inc.*)

very much smaller than the sporophyte, of which it is entirely independent. Water is absolutely necessary for the prothallium to give rise to a new sporophyte.

The ferns are of very little importance to man. One species, the Christmas fern, *Polystichum acrostichoides,* has thick evergreen leaves which are often used for decorative purposes. This fern grows wild in open woods of the north temperate region. Another fern, the maidenhair, *Adiantum pedatum,* and related species,

is frequently grown as a decorative plant because of its delicate fronds. However, the fronds wilt too quickly to be of much use if cut from the plant. Species of

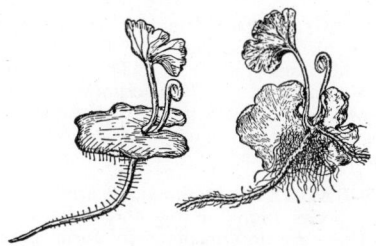

Fern embryo sporophyte still attached to its parent (the gametophyte) but differentiated into its parts and making its own food. Left, as seen from above; right, seen from below.

Osmunda, including the Cinnamon fern, the interrupted fern, and the royal fern, are often planted in shady places for ornament. From these ferns is obtained a

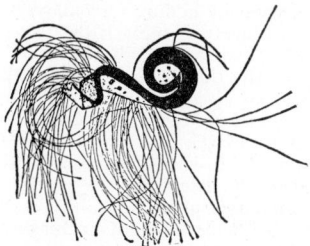

A single fern microgamete, killed and stained so that the parts of the cell are visible. (*After Steil.*)

coarse fiber used as a potting substance on which to grow epiphytic orchids.

Sometimes planted as a curiosity, *Camptosorus rhizophyllus,* the walking fern, gets its name because the tips of the fronds bend down to the ground and take root. New plants are formed at these points. This is a special

Plants of the walking fern, *Camptosorus rhizophyllus.*

form of vegetative reproduction. Several ferns propagate themselves by forming small bulbils on the surface of the fronds. Eventually these drop off and take root, giving rise to new plants. *Cystopteris bulbifera* is a fern propagating in this way. One of the commonest and best known ferns is the common **brake**, *Pteris aquilina,* which grows on dry hillsides and in open woods.

The Pteridophytes include not only the ferns or Filicales, but also the **Horsetails, Club-Mosses,** and the

Quillworts. These three groups are called the fern allies. (See also **Paleobotany**.) (R.M.W.)

FERRATES. Iron.

FERRET. Mammalia, Carnivora. 1. A plains species which preys chiefly on prairie dogs. Found from western Dakota to Montana and Texas. The black-footed ferret, *Mustela nigripes*. 2. The domestic ferret, a descendant of the European polecat, used to catch rats and mice and sometimes to pursue other animals into their burrows. (A.W.L.)

FERRET-BADGER. Mammalia, Carnivora. Animals of several species found in the forests of the Oriental region. They are intermediate in form between the slender carnivores of which the ferret is an example and the more stocky badgers. (A.W.L.)

FERRIC. Iron.

FERRITE. Iron; Steel.

FERROALLOYS. Most alloy additions to **cast iron** and **steel** are made by adding ferroalloys to the molten metal in the melting furnace or ladle. Ferroalloys have high contents of manganese, silicon, chromium, molybdenum, etc., with the balance principally iron. A standard grade of ferromanganese, for example, has 78–82% manganese, 15–19% iron, 6–8% carbon, under 1% silicon, under 0.35% phosphorus, and under 0.05% sulfur. In addition to the alloy content, the impurities or residual elements determine the grade and cost. A low carbon content is often desirable so that the alloy addition may be made without increasing the carbon content unduly.

Certain alloys are added to steel as pure metals rather than ferroalloys.. Included among these are nickel, copper, aluminum, and cobalt. (R.H.H.)

FERROMAGNETISM. Magnetism.

FERROUS. Iron.

FERRUM. Iron.

FERTILIZATION. The union of reproductive cells of the two sexes to form a new individual.

results. Since each germ **cell** has only one chromosome of a kind as a result of **meiosis**, the diploid number characteristic of the species is restored by this union.

The ovum reacts to the entry of the sperm head with peripheral changes which prevent the entrance of other male cells. A fertilization membrane separates from the surface of the egg and in some species a redistribution of cytoplasmic materials takes place.

Although some eggs develop normally without fertilization (**parthenogenesis**) and others may be stimulated artificially to do so, fertilization is the normal cause of development in most species. It is also important for the combination of hereditary qualities (see **Heredity**) of two individuals. (A.W.L.)

FERTILIZERS. In connection with the effective use of soils and the economic yield of crops, it is requisite that the exhaustion of soils by the removal of chemical elements of the crops be compensated by the addition of such elements as may not be spontaneously supplied by the soil. It has been found that the primary needs in this respect may be met by the addition of so-called fertilizers containing one or more of the three elements, **nitrogen, phosphorus, potassium**. Secondary needs require in some cases **sulfur, calcium**. Examination of a tabulated report of the constituents of the ashes of certain kinds of crops helps to make these points clear. Nitrogen does not appear in the ash, but must be supplied from the soil. **Silicon, magnesium, iron, chloride** are usually available in sufficient amounts. The constituents of plant ash vary with the kind of crops, and the weight of a given crop varies with the soil and its physical and chemical treatment, and with the season.

The content of each of the three chemical fertilizer elements is usually expressed in a characteristic manner, thus, for illustration, "4–8–2," which signifies that the percentage composition of the dry fertilizer is 4% **nitrogen** (N) element, 8% **phosphorus** pentoxide (P_2O_5), spoken of as "phosphoric acid," and 2% **potash** (K_2O). This expresses the composition of a "mixed" fertilizer, and on the same basis of expression "sulfate of ammonia" ((NH_4)$_2SO_4$) would be "21–0–0," "nitrate of soda" ($NaNO_3$) "16–0–0," "superphosphate of lime" ($Ca(H_2PO_4)_2 \cdot CaSO_4$) "0–35–0," "sulfate of potash" (K_2SO_4) "0–0–54," "muriate of potash" (KCl) "0–0–63."

CONSTITUENTS OF ASH OF NORMAL CROPS
POUNDS OF CONSTITUENTS PER ACRE OF GROUND

	Silica	Potash	Soda	Magnesia	Lime	Ferric Oxide	Chloride	Sulfate	Phosphate
Grain......	15	14	7	2	8	1	0	0	36
Straw......	233	33	1	28	12	6	4	13	11
Roots......	27	143	17	46	18	4	12	46	26
Tops.......	3	89	17	72	10	3	50	39	29
Hay........	78	38	12	45	7	1	4	9	15

In many species of animals germ cells of the male are transferred to the genital passages of the female in the process of **insemination** prior to fertilization, and in others both sexes discharge their **germ cells** into the water. In either case the **spermatozoa** swim about in the liquid surrounding them until they meet the ova.

When a spermatozoon meets an **ovum** either the entire male cell penetrates the surface and lodges in the cytoplasm of the egg or the tail alone is left outside. After entering, the head rotates so that the basal part is toward the interior of the ovum. At this point a mitotic figure develops. The nuclei of both eggs and sperm may enter the resting state as male and female pronuclei prior to their union. Whether this step occurs or not, the **chromosomes** derived from the two nuclei enter the mitotic figure and the first **cleavage** division

Common nitrogen fertilizers are **ammonium** sulfate, **sodium** nitrate, calcium **cyanamide**, organic substances, such as packing house recovered wastes, tankage, fish scrap, cotton seed meal, treated garbage and sewage, manure; **phosphorus** fertilizers are **calcium** dihydrogen phosphate, either as "superphosphate" or "treble superphosphate," **calcium** phosphate of "phosphate rock" or of bones; **potassium** fertilizers are potassium chloride or sulfate. Sulfur is supplied in three of the above, namely, ammonium sulfate, "superphosphate" and "sulfate of potash," or may be used as gypsum or sulfur; calcium is also supplied in "superphosphate" and "treble superphosphate," or may be used as gypsum or limestone.

The following are typical analyses of phosphate rocks and bones:

ANALYSES OF PHOSPHATE ROCK

Source	Phosphorus Pentoxide	Fluorine	Calcium Oxide	Carbon Dioxide	Aluminum Oxide	Ferric Oxide	Silicon Oxide
Florida.............	31–35%	3.6%	46–50%	2–3%	1%	1–2%	5–10%
Tennessee.........	32–37	3.6	48–50	2	1–2	1–4	2–12
South Carolina.....	26	3.4	42	4	1	1.5	13
Idaho.............	34	3.4	48	2	0.7	0.5	4
Montana..........	29	3.0	40	1	1.7	1.5	22
Wyoming..........	31	3.5	48	4	0.4	1	5
Quebec............	39	3.0	54	1	0.4	0.4	1
Morocco..........	35	4.0	53	4	0.3	0.3	1
Bone ash.........	40	0.1	54	1	0.0	0.2	0.5
Steamed bone meal.	35	0.1	48				

In making superphosphate about equal weights of Florida phosphate rock and sulfuric acid are used.   (R.K.S.)

FEVER. This term is used to describe elevation of the body temperature over the normal 37° C. or 98.6° F. Fever occurs most commonly in infections, but may accompany a variety of other diseases. The type of fever is often characteristic for the disease, and is described according to the constancy with which the elevated temperature is maintained. Thus, a continuous fever is one that is maintained at a fairly constant level for several days. When there are moderate fluctuations, the temperature is said to be remittent. If the temperature approaches or reaches normal during some part of the day but rises considerably at other times, the fever is said to be intermittent. Rises in temperature may be gradual, or very sudden following chilly sensations or shaking chills. Fever may terminate slowly by lysis or suddenly by crisis.

Conditions other than infections associated with fever are reactions to injections of foreign protein or certain chemical substances, excessive loss of water from the body, and hemorrhage or tumor in the region of the brain where the heat-regulating centers are located. Malignant tumors anywhere in the body, **leukemia,** cirrhosis of the liver, and other systemic diseases are frequently accompanied by fever.

The question is often debated as to whether fever is beneficial during an infectious disease, or should attempts be made to reduce it. Since its mode of production is not clearly understood, there is no ready answer. In general, fever of moderately high degree may be regarded as beneficial since it is known that at higher temperatures the body finds it easier to produce protective antibodies to fight infection. Furthermore, the increased temperature is harmful to micro-organisms, making it more difficult for them to grow and increase within the body. Excessively high fevers such as occur in sunstroke demand active measures toward their reduction; they indicate that the heat centers of the brain are probably temporarily overwhelmed and unable to cope with the emergency.   (D.M.H.)

FEVER BLISTERS. Herpes Simplex.

FIARD. Fiord.

FIBERS. Stem. Fibers are long, usually slender, thick-walled **sclerenchyma** cells. Usually they have pointed ends and lignified (**lignin**) walls. In most cases the central cavity of the fiber, the lumen, is very small, due to the thickness of the wall. Fibers are found in many parts of a plant, but are most frequent in the cortex and in the vascular cylinder. The name bast fibers is often given to fibers occurring in the pericycle, **cortex,** and in the **phloem** region. Plant anatomists prefer to designate all fibers according to the tissue to which they belong; xylem fibers, phloem fibers, pericyclic fibers and cortical fibers. Fibers may occur as single cells or in compact groups forming extensive

masses, giving to the part of the plant in which they occur considerable tensile strength. Masses of fibers of this sort give us **flax, hemp, jute, and ramie.** Jute is composed of fibers of the secondary phloem (inner bark). Flax, hemp, and ramie fibers are developed in the pericycle.

The single-celled outgrowths which surround the seeds of the **cotton**-plant are often called fibers. They are slender hollow cells of almost pure **cellulose,** and strictly are epidermal hairs, as are also the cells of **kapok.**   (R.M.W., B.S.M.)

FIBRIL. A minute thread-like structure **in a cell,** also the smaller components of the intercellular white fibers of connective tissue.

Fibrils occur near the surface of smooth muscle cells and connective tissue cells (border or myoglia fibrils and fibroglia fibrils respectively) and in fully differentiated muscle and nerve cells (myofibrils and neurofibrils respectively).   (A.W.L.)

FIBROID TUMOR. A benign **tumor** of fibrous or connective tissue. Fibroid tumors may occur in various parts of the body but are most often found in the uterus, where they may grow to great size, produce pressure symptoms, and abnormal bleeding. Uterine fibroids are the most common cause for removal of the uterus (**hysterectomy**).   (R.S.M.)

FIBROLITE. Sillimanite.

FICTITIOUS MEMBER. Graphical Statics.

FIDELITY. This is usually applied to an **amplifier** or communications circuit to indicate its ability to reproduce the original signal without **distortion.** A high-fidelity audio amplifier will amplify all audio frequencies equally while a poor-quality amplifier will badly attenuate certain frequencies, usually the low and high frequencies while passing the middle audio range. Many radio **receivers** have fidelity controls. These are in the **intermediate frequency** circuits and have the effect of broadening the pass band of such circuits. Most of the loss of the high frequencies in the radio is due to the various selective circuits cutting out the higher **sideband** frequencies, and, since these represent the higher audio frequencies, the highs are missing in the output. The use of fidelity controls allows the selective circuits to be broadened for receiving high-fidelity local programs where there is not likely to be adjacent station interference, yet allow the circuits to be returned to the selective condition for tuning in weaker stations which cannot override interference. (L.R.Q.)

FIELD. In electric **motors** and **generators** the field is the part of the machine which furnishes the magnetic flux which reacts with the armature to produce the desired machine action. In d-c machines the field may be shunt, i.e., connected in parallel with the armature, or series, i.e., connected in series with the armature. A given machine may have both types of fields, in which case it is a compound machine. For synchronous motors and generators the field is separately excited from a d-c source and is usually the rotating part of the machine. Field is also used in the sense of a magnetic field set up around a conductor or coil, or an electrostatic field between conductors at different potentials. These fields are represented by lines of magnetic or electrostatic flux. The term is also used to denote the energy radiated from an **antenna** system of a radio transmitter. This field is an electromagnetic type field consisting of both an electrostatic and a magnetic component. It is the cutting of the receiving antenna or circuit by one or both of these component fields which causes the signal voltage to be induced in the antenna and hence introduced into the main receiver circuit. (L.R.Q.)

FIELD INTENSITY. This is a measure of the strength of the magnetic or electric field at any point. In radio the field intensity at the receiving **antenna** determines the signal which will be induced in the antenna and hence is an important consideration in the design of the **transmitter.** By proper choice of the power and frequency of the transmitter and type of transmitting antenna considerable control may be exercised over the field intensity at the receiving point. (L.R.Q.)

FIELD SPLICE. Splice.

FIELDFARE. Aves, Passeriformes. A common thrush, *Turdus pilaris,* of northern Europe. (A.W.L.)

FIELDS OF FORCE. A field of force is commonly recognized by the fact that an appropriate test object placed therein gives evidence of a force acting upon it. Thus a stone in the earth's gravity field has weight, a charged pith ball near an electrified glass rod is urged toward or away from the rod, etc. The concept of such regions of space in which there is a condition of stress analogous to elastic tension or compression is due primarily to Michael Faraday. He initiated the "field theory" as a substitute for the older "action-at-a-distance" concept, and explained the observed behavior of appropriate objects in the neighborhood of magnets, electric charges, or gravitating masses as due to stresses along definite stress lines or "lines of force." Every student of physics is familiar with the experiment of

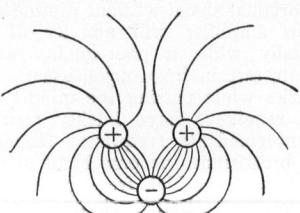

Typical electric field between charged conductors, mapped by lines of force.

mapping a **magnetic field** by means of iron filings sprinkled upon a card placed over a magnet. In this experiment, attention is first called to the direction of the lines as indicating the direction of the magnetic stress; later the student learns to interpret the closeness with which the lines are packed as corresponding to the intensity of the field. The lines are, of course, pure conventions; but they are nevertheless useful both qualitatively and quantitatively in discussions of field theory. Analogous concepts are applied in the case of

electric and gravitational fields, and are useful in the theory of the **electromagnetic field,** so ably developed by Maxwell. In recent years Einstein has undertaken to identify these different fields of force as components or aspects of a more general entity, and has developed his unified field theory as a feature of general **relativity.** (L.D.W.)

FIG. *Ficus carica.* Moraceae. Figs are the fruit of a shrub or small tree probably native to southwestern Asia. They have been cultivated since earliest recorded times, being widely used by the Hebrews, greatly improved by the Greeks, and highly valued by the Romans. The fig tree is now grown in cultivation in nearly all tropical countries and many subtropical regions; in the United States it is grown to some extent in the southwestern states, notably in California.

The plants have alternate leaves which are rather thick and rough surfaced above, but soft-hairy beneath. The leaves are deeply lobed in the cultivated varieties. The minute flowers are borne on the inside of the hollow receptacle, which develops into a pear-shaped body with a minute opening at its apex. The fruit developing from this is a synconium, composed of many small fruits inserted in the inner wall of a hollow fleshy receptacle. The narrow passage into this is partly closed by numerous small **bracts.**

There are four kinds of flowers in the fig. Staminate flowers, each having four pollen-bearing **stamens,** occur in the wild "caprifig." A few cultivated forms have staminate flowers. **Pollination** is brought about by using pollen from caprifigs, and is called caprification. Pistillate flowers, each having a single **pistil** which if pollinated produces a seed. These flowers are short-stalked. The third flower type is the gall-flower, so-called because a small wasp, *Blastophaga grossorum,* lays its eggs in them. The developing **larvae** causes the **ovaries** to become swollen **galls,** incapable of developing seeds. This type of flower occurs only in caprifigs, in the basal portion of the synconium. Lastly, in varieties of cultivated figs there are found sterile flowers, which will neither produce seeds nor become galls; these are called mule flowers. Caprifigs contain the first three types of flowers. If pollen is needed to insure fruit development, caprifigs must be planted, since they alone have pollenbearing flowers. So among Smyrna fig trees, the fruits of which fail to develop unless pollinated, caprifigs must be planted.

In Mediterranean countries, where figs are grown in abundance, three crops are produced each year. The first fruits, known as profichi, are formed in the spring. In the pistillate flowers of these the female wasp lays her eggs, so that galls are formed in the profichi. When the young wasps emerge, these profichi are gathered and hung among Smyrna figs. Wasps escaping have to crawl past the staminate flowers near the aperture of the synconium and so are dusted with pollen. The wasp then enters and pollinates a flower of the second crop, thus insuring fruit development. This second crop is known as mammoni. The third crop, the mammae, remain on the trees. It is in these that the wasp passes the winter.

Fig fruits are gathered and sometimes eaten fresh, but more frequently dried in the sun, then pressed together and shipped. Smyrna figs are often enlarged by pulling during drying. The fruits produce a mild laxative effect, and so are sometimes prescribed in cases of chronic constipation. From them a wine is sometimes made, and alcohol produced. The wood of the fig tree is occasionally used in cabinet work.

Propagation is usually by stem cuttings, but sometimes by budding or **grafting.**

Other species of *Ficus* are of some importance. In southern Asia the Sacred Fig, also called Peepul or Botree, occurs. This is a large tree with deltoid leaves, the apex of each leaf tapering into a long point. Any water falling during a tropical rain runs off rapidly from this point, so that the leaf surface does not long remain wet. The tree is held sacred by the Hindus.

There are many species of *Ficus* native in the American tropics, known as Strangling Figs. The seeds of these frequently germinate on the branches of other trees, sending aerial roots downward and developing slender stems which grow about the supporting plant. Eventually the roots enter the ground and become established there. The stems, enlarging and often anastomosing, grow tightly around the supporting plant, which is often strangled and killed.

All these plants have a milky juice which may be made into **rubber**. The juice of the India Rubber Tree, *Ficus elastica,* is used in this way.

The **Banyan** tree is another species of *Ficus*. (R.M.W.)

FIG INSECT. Chalcid Fly.

FIGWORT FAMILY. Scrophulariaceae. This family contains some 2500 species, most of which are herbs or shrubs. Its members are numerous in temperature regions, where many of them are common plants, as for example mullein, "butter-and-eggs," speedwell, and lousewort. Annuals, biennials and perennials are found in the family.

The flowers are zygomorphic, or bilaterally symmetrical, with the **calyx** and **corolla** both tubular and each composed of four or five lobes. In many plants of this family the **corolla** is distinctly 2-lipped, as in the Snapdragon. Usually there are four **stamens**, which are inserted on the corolla tube. The **ovary**, composed of two united **carpels**, becomes a dry capsule containing many small seeds. The flowers of this family are mostly pollinated by insects, such as bees, wasps, and flies, which seek the nectar secreted in a disk at the base of the ovary.

Many members of this family show a tendency towards parasitism. In these, the seed on germinating sends out a root which comes in contact with the roots of another plant, commonly a grass. On making this contact, the root sends into the grass root absorptive organs or haustoria, which take from the host certain materials in solution. The parasite grows into an ordinary green-leaved plant capable of carrying on **photosynthesis**, but with a root development insufficient for its own needs. Such plants are only partially parasitic.

In this family are found many plants grown by man as ornamentals. Some, such as Foxglove (*Digitalis*) and Veronica, are hardy biennials or perennials; others, like Snapdragon (*Antirrhinum*), are not hardy; while *Calceolaria,* a native of South America and Mexico, is a hot-house plant grown for its bizarre sac-like flowers of brilliant color. Drugs of medicinal value are also found in several plants of this family, the most important being **digitalis,** from species of foxglove. In early days many species were used as a source for home-made brews. Others are poisonous herbs. (R.M.W.)

FILAMENT. This is the direct heated type of **cathode** used in various electron **tubes**. It is also the heated element of the common electric **lamp**. (L.R.Q.)

FILAR MICROMETER. The filar micrometer is an instrument for measuring small distances in the field of an **eyepiece**. It consists fundamentally of two parallel wires, one of which is fixed and the other capable of motion in the direction perpendicular to its length by means of an accurately cut screw. The pitch of the screw is carefully determined for various temperature conditions and the head of the screw is graduated so that whole revolutions and fractions thereof may be read.

For astronomical purposes the filar micrometer has some modifications from the instrument as used for ordinary measuring purposes. In the accompanying figure we have *AB* a plate of brass carrying two wires *H* and *F* which are accurately perpendicular to each other. This

plate may be rotated and the index *I* sweeps over a circle graduated in degrees, with a vernier reading fractions of a degree. A second plate *CD* may be moved over the

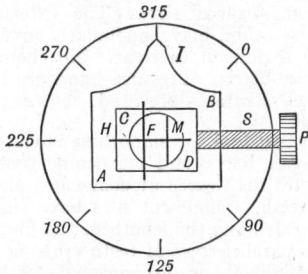

Diagram of filar micrometer.

surface of the first plate by means of the accurately calibrated screw *S* with the graduated head *P*. This plate carries on its lower surface, so that it will be practically in the same plane as *F*, a wire *M* which is set accurately parallel to *F*.

This instrument is so mounted that *F* and *M* are in the focal plane of the objective of a **telescope** with the optic axis of the instrument passing through the center of the opening in *AB*. When so mounted the filar micrometer is one of the most valuable instruments for measurement of small angular distance.

For the study of **double stars** the first adjustment of the instrument is to point the telescope at a star, preferably near the equator, and rotate the instrument until the star will move, due to the rotation of the earth, along the wire *H*. *H*, being now parallel to the equator, the reading (R_1) of the index is taken. Next the wires *M* and *F* are placed in coincidence and the head reading (D_0) taken. The telescope is then directed at the double star under investigation and rotated until *H* passes through both stars and the reading (R_2) of *I* is taken. Then holding one component of the pair of stars on *F* the screw is turned until *M* passes through the other star. The reading (D_1) of the head is taken, account being taken of the number of whole revolutions in the process. R_2-R_1 is defined as the **position angle** of the double star, and D_1-D_0 (when converted into angular units) is the distance.

The filar micrometer may also be used to determine the position of one astronomical body relative to another close object whose **spherical coordinates** are known. For this purpose the wire *H* is first held at setting (R_1), (i.e., parallel to the **celestial equator**), the wire *F* held on one of the two objects and the wire *M* set on the other. The distance thus measured is parallel to the equator (i.e., is proportional to difference in **right ascension** of the two objects). The instrument is then rotated through 90° and the difference in **declination** may be measured. (W.K.G.)

FILARIASIS. A parasitic disease of the tropics caused by various species of *Filaria*, a genus of **nematode** or thread worms. The parasites are commonly found in tropical countries but cases of infection have been reported in the Southern states. Both anopheline and culicine **mosquitoes** serve as intermediate hosts for the parasites. The adult forms develop in the human body after larval forms are transmitted by the bite of the infected mosquito. The incubation period is usually one year or less.

Symptoms depend on the degree of infestation and the parts of the body involved. The parasites show a predilection for the lymphatic structures. High fever, **lymphangitis** and swelling of the **lymph** nodes are common.

After repeated attacks permanent thickening of tissue occurs accompanied at times by great swelling of the affected parts. The swelling is caused by blockage of the lymph channels by the parasites obstructing the flow of lymph.

When the legs are involved the condition is called **elephantiasis**, as the legs become tremendously swollen with thickened, fissured skin. The external genitals, especially the scrotum, may be similarly involved.

The disease is difficult to treat. No chemotherapy is effective against filaria. Pressure bandages and surgical treatment of elephantiasis are used. (R.S.M., D.M.H.)

FILE. A cutting tool for smoothing surfaces, breaking corners, removing burrs, and sharpening tools. The cut of a file denotes the degree of coarseness and the character of the teeth. Single-cut files have single rows of parallel teeth extending the length of the file; double-cut files have two parallel rows of teeth crossing each other; rasps have individual, or disconnected, teeth, as shown in the figure.

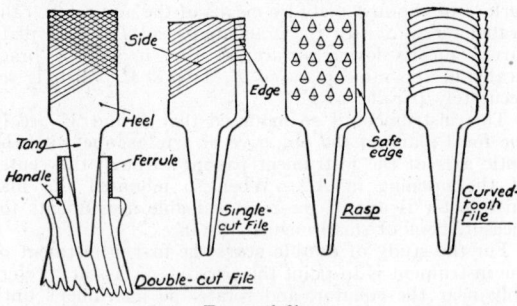

Types of file.

A mill file is rectangular in section, and is single-cut and tapered in both width and thickness. A flat file is like a mill file, but is double-cut. A hand file is double-cut, of parallel width and tapered thickness, and has a safe edge (one which is smooth and has no teeth). Half-round files are usually double-cut, with one flat and one curved surface. Round files are of circular, tapered section, and may be single or double-cut. Three-square and hand saw files are tapered and have a triangular cross-section, and are used for filing to sharp corners and for saw filing and sharpening. (H.C.H.)

FILE JIG. Jigs and Fixtures.

FILIBRANCHIATA. Lamellibranchiata.

FILICALES. Paleobotany.

FILLET. Fillet is the term employed to describe a concave section of a body which is used to reinforce a re-entrant angle formed by the intersection of two plane surfaces. It is purposely incorporated in patterns from which **castings** are to be made, since lines of stress radiating from sharp edges and corners are set up in the casting during cooling. These will be prevented, and the casting made measurably stronger, if the outside edges are rounded, and the inside filleted. Filleting is

also employed to improve the appearance of a corner, or hide a crack, and for the purpose of replacing an angularity with a smoothly rounded surface. An aircraft wing is faired into the fuselage by fillets. The junction of two parts, such as plates at right angles to each other, is often made with a weld of the fillet type, in which a small fillet of welding metal is laid down in the angle created by the intersection of the surfaces of the plates. (F.T.M.)

FILM AND FILM PROCESSING. Cameras, Motion-Picture.

FILM BASE. Photographic film base is made by combining a cellulose ester, or a mixture of cellulose esters, with a plasticizer, suitable solvents, thinners, and stablizers into a viscous liquid or *dope*. The film is cast, or formed, by flowing or spraying the dope in a thin layer onto a large, heated, slowly revolving chromium- or silver-plated drum. As the drum rotates, a portion of the solvent evaporates, causing the layer to set sufficiently, so the layer can be stripped from the drum before a complete revolution is made. Film base is dried and wound into rolls 50″ wide and 2000′ long. The thickness of the sheet varies according to the type of film and purpose for which it will be used. Roll film must be flexible and thin to allow it to be wound through the camera. Motion-picture film should be flexible enough to pass through the camera, or projector mechanism, yet sufficiently thick to prevent sprocket perforations from becoming torn during operation. Cut sheets have to be thick enough to lie flat in the filmholder and keep from curling during processing and storage. Representative thicknesses for film base are:

Roll film	85 microns
Motion picture film	140 "
Cut sheet	200 "

Since the solvents evaporated from the film base are inflammable and static electricity is generated during the casting operations, it is necessary to carry out the operation in sealed machines, in the absence of air or oxygen, in order to safeguard against the danger of explosion. The static charge is picked up by a series of small metal balls with short chains that are attached to metal rods which are grounded.

In modern film-casting machines the substratum layer, a bonding layer between film base and emulsion, and an antihalation backing are put on the base before it leaves the machine.

Two types of film base are manufactured: the cellulose nitrate used for professional motion pictures, and the safety base (noninflammable) used for professional cut sheet, amateur roll, and amateur motion-picture films.

Cellulose-nitrate base is made by treating cleaned cotton linters from upland cotton with a mixture of nitric acid, sulfuric acid and water at 40° C. with constant agitation for 20 minutes. The nitrated cotton is then washed, neutralized, bleached and dried. For safety, cellulose nitrate is moistened with alcohol and kept in metal drums.

During the operation, the cellulose molecule which contains several hundred glucose units, each unit of which has three replaceable hydroxy groups, is nitrated so that two and a fraction of the hydroxy groups are replaced by nitrate groups, or until the nitrogen per glucose unit ranges from 11.6–12.2%.

Cellulose nitrates mixed with camphor or triphenyl phosphates, as the plasticizers, and dissolved in solvents like acetone, alcohol, and toluene, forms a solution or dope which may be further thinned to proper coating viscosity.

Cellulose acetate is prepared by treating cellulose with an organic acid anhydride, an organic acid as diluent, and a catalyst. Cotton linters, treated with acetic anhydride, glacial acetic acid and sulfuric acid, are converted to cellulose acetate. The triacetate, first formed, is partially hydrolyzed until the product contains 35–40% acetyl, or two and a fraction acetyl groups, per glucose unit. The important steps in the manufacture of the acetate are: acetylation, hydrolysis, precipitation in water, neutralization, washing, centrifuging, and drying. Because cellulose acetate has a slow burning rate, it is used in the manufacture of safety film.

Safety film is made by mixing hydroxy cellulose esters with high-boiling organic plasticizers, as dibutyl phthalate, tri-phenyl, or tricresol phosphate, or dibutyl sebacate and solvents that are halogenated compounds—like propylene chloride, trichlor ethylene, and ethylene chloride, with a mixture of alcohols as methyl, ethyl, propyl, iso-propyl, -butyl and -amyl alcohols. Acetone and hydrocarbon thinners like toluene are also added.

Safety-base film can also be prepared by mixing cellulose esters of organic acids as cellulose-acetate and cellulose-propionate, or by using mixed esters as: cellulose acetate-propionate, or cellulose acetate-butyrate.

Cellulose ester solutions are concentrated and very viscous. They flow to a limited degree only. Actual thickness, during casting, is governed by a scraper.

Because of the low inflammability and the many improvements made in physical and mechanical properties during recent years, acetate and safety films have very largely replaced nitrate film. (S.M.T.)

FILM THEORY OR BOUNDARY LAYER THEORY.

The film theory as applied to heat or mass transfer has to do with the analysis of the physical way in which material or heat is transferred across a phase boundary where one or both of the phases may be flowing fluids. Examples of such a transfer would be heat flowing from a pipe to water moving inside, or water vapor passing from a wet surface into an air stream flowing over it. Any study of such processes must be concerned primarily with the major resistance to the flow. It is known that transfer by convection (heat) or mixing (mass) is very much faster than by conduction (heat) or diffusion (mass). If any part of the process involves conduction or diffusion, then this part will undoubtedly offer the greatest resistance to the transfer and hence will be the controlling variable.

Even though the fluid be moving past the surface in a turbulent manner, there will still be, next to the surface, a relatively stagnant *film* of the fluid. Through this film the heat or mass must pass, respectively, by conduction or by diffusion. Calculations involving transfer of heat or mass into a flowing stream are mainly concerned with the effective thickness and the properties of the film. Methods of increasing the rate of transfer are usually based on changes designed to reduce the film thickness or to change the properties of the fluid and hence increase the rate of transfer. Increasing the turbulence of the fluid tends to scrape off the film, thereby making it thinner; and raising the temperature (for liquids) lowers the viscosity of the film and makes it more easily rubbed away by the flowing fluid. Also, raising the temperature increases the rate of diffusion, and usually increases the heat conductance.

The actual film thickness is seldom known, but the way in which various factors (such as the viscosity, velocity, density, etc., of the fluid) affect it are often known, so that experimentally determined values for the rate of transfer under one set of conditions can usually be used to calculate the rate under a different set of conditions. (See **Heat Transfer.**) (D.E.M.)

FILTER AID. Filtration.

FILTER, CLARIFYING. Filtration.

FILTER, FREQUENCY.

A filter is a frequency discriminating network designed to select or pass certain bands of frequencies with low attenuation and cause very high attenuation to other frequencies. They are among the most important components of most of our communications systems. In the common radio **receiver** filters are used to smooth out the pulsating direct current from the **rectifier** to give a steady output of the **power supply.** In the **superheterodyne** filters in the **intermediate frequency** stages give the desired selectivity. Telephone **carrier** circuits are possible because

of filters which are used to separate the various channels and route the signals to the proper paths. While filters are used very widely, the telephone systems are probably the biggest users. The ordinary tuned or resonant circuit might be regarded as a special form of filter but is not commonly considered as such since it does not have a flat pass band.

Filters may be classified according to their characteristics (e.g., low pass, high pass, band pass) or according to their circuits (e.g., ladder, lattice, π and T). A low-pass filter passes all frequencies below a certain value, known as the **cut-off** frequency, with very little attenuation and then produces high attenuation for all above this value. A high-pass filter passes all above the cut-off frequency and a band-pass filter will pass a band or bands of frequencies and produce attenuation for all frequencies outside these pass regions. Filters are made, normally, of combinations of series and shunt elements which are as pure reactance as can be economically attained, the purer the reactance the better the filter. Fig. 1 shows the basic or prototype circuits for low-,

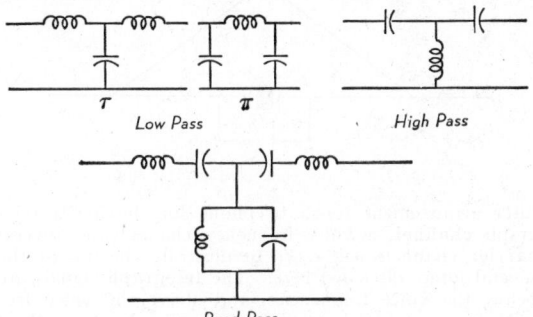

Fig. 1. Typical filter circuits.

high-, and band-pass ladder type filters. For more perfect action several sections may be connected in tandem, producing a ladder appearance. These sections may be connected as T or π circuits as shown. While these simple circuits are adequate for some purposes most filters are composed of several sections, each introducing a specific characteristic. A discussion of the low-pass T connected filter will serve to illustrate the usual practice. The prototype shown will cause little attenuation up to the cut-off frequency but neither will it cause a high attenuation just above the cut-off, the attenuation increasing very gradually. To overcome this difficulty a section known as a derived section is connected in series with the prototype. This derived section, shown in Fig. 2, gives a very high attenuation at the resonant

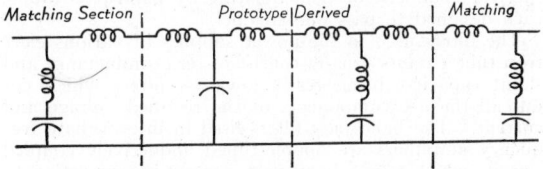

Fig. 2. Composite low-pass filter.

frequency of the shunt branch. If this resonant frequency is near the cut-off frequency the attenuation will rise rapidly at cut-off. Unfortunately this section does not keep a high attenuation as the frequency is raised so it must be used with the prototype to give high attenuation at all frequencies above the cut-off. Often several derived sections, each having a different resonant frequency for its shunt branch, are used. The input and output impedance characteristics of these sections are not very satisfactory so a matching section is connected on each end to correct this. All of these various sections are designed to match one another and to have the same cut-off so the resulting filter, called a composite filter, will have a low attenuation pass band with no irregularities and then a high attenuation for all frequencies above this band. Other types of filters are built up in a similar manner. Quartz crystals (see **Crystal**) are equivalent to **resistance, capacitance** and **inductance** networks having a very high **Q** so are ideal for filter components. Through the use of these crystals telephone carrier filters may be made to very close limits and thus many more carrier channels may be superimposed on the lines. These filters usually use a lattice type connection. (See Fig. 3.) A brief outline of the

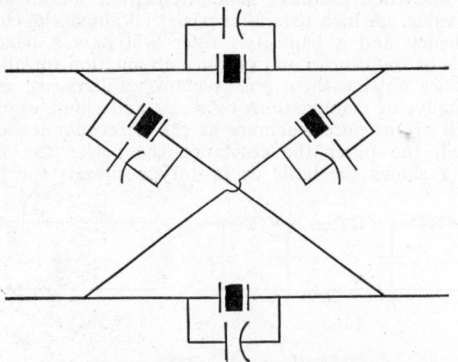

Fig. 3. Lattice-connected crystal filter.

filter arrangement for a telephone line having a telegraph **channel**, a voice frequency channel and several **carrier** channels will serve to illustrate the use of the several filters discussed here. The **telegraph** signals are below the voice frequencies, and above the voice frequencies are several **carrier** channels. Each of these bands represents a separate communication which must be routed along the proper path at the terminals, all going over the same pair of line wires between stations. A high-pass filter connected across the line with its output going to the carrier equipment will allow the carrier frequencies to pass to this equipment and will block the voice and telegraph frequencies which are below its cut-off. A low-pass filter connected across the line at the same point will pass these lower frequencies and block the carrier frequencies. Then if the output of this low-pass filter feeds a high-pass filter with cut-off at the lower voice frequency the voice currents may be passed on to the phone circuits and the telegraph signals blocked. Similarly a low-pass filter may be used to pass the telegraph and block the telephone currents. The carrier currents which were routed to the carrier circuits by the first high-pass filter are then routed to their respective channels for demodulating by band-pass filters, each designed to pass one channel.

The filters used in the power supplies of various electron tube circuits usually consist of series inductance and shunt capacity, being really low-pass filters which cut out all the a-c components of the rectified voltage and current. The band-pass filters used in intermediate frequency amplifiers are double-tuned, inductively coupled circuits which pass a band very narrow in proportion to the mid frequency. (See also **Decoupling Filter**.) (L.R.Q.)

FILTER, LIGHT. A light filter is a substance or device which changes the color or intensity of the transmitted light. It is widely used in the field of photography, spectroscopy, and microscopy. Light filters, once the adjunct of trained photographers only, are now in widespread use by amateur photographers who understand that atmospheric conditions, type of object to be photographed, and distance to be included, call for careful selection of light filter for best photographic results. (See **Color Correction**.)   (F.T.M.)

FILTER, MECHANICAL. Gas filters are used to separate solid or liquid particles from gases. An important application is the purification of air for human use. Although the physiological advantages secured in the breathing of clean, filtered air have long been known, only recently has the general public been able to experience the pleasure of this addition to the standard of living, through the increasing use of **air-conditioning** systems, all of which involve filtration. Especially are benefits to be gained where the air is abnormally polluted, as in passenger trains drawn by steam locomotives, in cities, or in the proximity of factories discharging a dusty waste. Many types of filters are used in the air conditioning of trains, homes, theaters, and stores, some of which are of the wet-spray types, others dry, containing mineral wool or felted pads through which the air is caused to pass, while others have oily surfaces to catch the dust. Filters are also to be found for cleaning the dust from air used in internal combustion engines, and for purifying dust or ash-laden gases from industries, and in the laboratory.

Liquid filters are used for separating solid particles from the liquid. They are used extensively in many chemical industries (see **Filtration**). In some cases the solids may be impurities; in others, they may be the product sought. It is often necessary to purify raw water by treating it chemically to precipitate the salts it may contain, or to coagulate impurities carried in suspension. Public water supply offers numerous examples of large-scale filters. A liquid filter may be made of paper, a porous membrane, or a layer of a porous material such as charcoal, coke, or sand. Two systems of large-scale filtration of raw water are practiced—one known as the slow sand filter, the other the rapid sand filter. The slow sand filter is simply a large water-tight basin containing graded sand to the depth of 3 or 4', overlaid with 3–5' of water. Water percolates downward through the sand bed, being removed by an underground system consisting of tile laid with open joints. After an extended period of time the effectiveness of the filter is reduced by accumulation of dirt on the filter bed, and cleaning is necessary. Then the filter is unwatered, and the top layer of sand and dirt removed for cleaning and renewal. A rapid sand filter employs a thinner filter bed, and is arranged for backwashing the bed to remove dirt. This cleaning action must be performed much more frequently than in the slow sand filter, because the rate of percolation through the bed is considerably faster. The usual arrangement of the rapid sand filter embodies a channel into which water from a sedimentation or coagulation basin flows. From this channel a number of lateral distributing gutters lead off at right angles. These distribute the water over the filter bed. Ordinarily, the filter bed will be about 3 or 4' thick, and will be submerged to a depth of 2 or 3'. The under drain system is much more complete and effective than in the slow sand filter, and consists of a series of parallel water channels which discharge into parallel pipes, all of which are connected to a common filtered water header. Some amount of suction, or negative head, on the drain system, is permissible.   (F.T.M.)

FILTER PRESS. Filtration.

FILTER, SAND. Filtration; Water.

FILTRATION, SEDIMENTATION, CENTRIFUGING, CLARIFICATION. These processes all involve the principles of separation of solids and liquids in the same system from each other. Filtration accomplishes this end by means of interposing a porous medium, usually of paper or cloth of the desired porosity, but sometimes of sand or woven metal in such a way that the solid or precipitate is retained by the membrane while the liquid filtrate passes through the pores of the membrane. In some cases both precipitate and filtrate are recovered, whereas in other cases either precipitate or filtrate is desired and recovered and the other discarded. The separation of precipitate and filtrate may be made more complete by washing the precipitate while on the filter or by repeated sedimentation before filtering. In sedimentation the solid material settles in the liquid at a rate depending upon the diameter of the particles (assuming these to be spherical), the difference in densities between the solid and liquid, and the viscosity of the liquid. This relation was expressed mathematically by Stokes, thus, where g is the acceleration due to gravity:

$$\text{Rate of fall of particle} = \frac{2g}{9} \times \frac{(\text{Radius of particle})^2}{\text{Coefficient of viscosity}}$$
$$\times \text{ Difference in densities of solid and liquid.}$$

It is evident, therefore, that, in a given case the larger particles of solid settle more rapidly than the smaller, and that in two parallel cases the greater the difference in densities of the solid and liquid, and the less the viscosity of the liquid, the more rapidly sedimentation takes place. In centrifuging, the machines in use are designed to subject the material which is treated to the action of centrifugal force. Sedimentation and centrifuging may be utilized to separate two immiscible liquids.

In the practice of filtration, two cases may be mentioned. First, in the case where the suspended solids to be recovered represent a *small* percentage of the total material to be treated. The filter cake is built up with a force and at a rate which depend upon the pressure with which the material is fed to the filter. It is desirable to start with a moderate to low pressure, and to

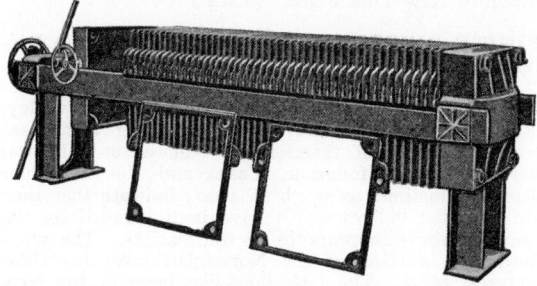

Plate and frame filter press (open), corner feed washing type.

increase the pressure gradually as the resistance of the filter cake increases. In this way a practically constant rate of flow of filtrate is maintained during the desired interval until removal of cake is demanded because of the slow rate of flow at the maximum pressure. Second, in the case where the suspended solids to be recovered represent a *large* percentage of the total material to be treated. The formation of the filter cake will depend largely upon the kind of solid particles composing the precipitate. When these are granular and of the same size throughout, the resistance builds up gradually and the pressure should be moderate to low and increased as the resistance increases. When the solid particles are gelatinous, or are of uneven size, the increase in resistance to filtration and the decrease in rate of flow of filtrate occur relatively quickly. High initial pressure is to be avoided on account of the increased compacting effect, due to the force with which the particles come in contact with the filtering medium. It is frequently arranged to build up a layer of the precipitate and to utilize this layer as filtering medium. In such cases, since the initial filtrate is cloudy, it is returned to the filter after the filtrate runs clear. Another device to increase the rate of filtration of fine particles is to build up a layer of inert granular material.

For filtering gelatinous materials, such as ferric or aluminum hydroxides, which tend to clog the filter, use is made of filter aids. These are usually finely divided

inert materials which are easy to filter and which have a large area per unit volume. Diatomaceous earth and

Conkey rotary vacuum filter in process of assembly for installation on sewage sludge at Detroit, Mich. (*Filtration Equipment Corporation.*)

paper pulp are examples. The filter aid is mixed with the solution to be filtered. The gelatinous or slimy

Conkey rotary vacuum filter in operation on sewage sludge at Elmira, N. Y. (*Filtration Equipment Corporation.*)

material tends to stick to the filter aid, thus keeping it from settling on and clogging the surface of the filter medium.

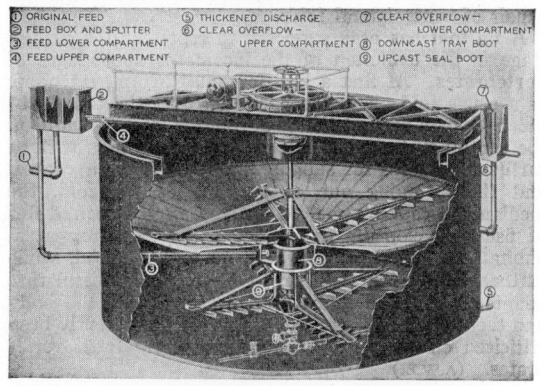

Dorr thickener. (*The Dorr Co.*)

The pressure utilized in filtering may be obtained by the use of negative pressure, that is, partial vacuum, or positive pressure, either hydrostatic head or pump. Filters are of the intermittent or batch and of the continuous types. In the latter type provision is made for the continuous removal of the filter cake as it is collected on a revolving filter.

Continuous separation of suspended solids and the liquid medium is also conducted, without the use of a filtering medium, by the application of the principles of sedimentation accompanied by the continuous separate withdrawal of sedimented layer and clarified supernatant liquid at the proper rates. This method of clarification is operated on an extensive scale. One modification of this method is utilized to remove the silt from the water of the Colorado River, immense settling basins in parallel arrangement being used. (R.K.S.)

FIMBRIATE. With a ruffled margin.

FIN. A broad thin appendage, primarily an organ of locomotion, and structures of similar origin adapted for other uses. True fins are found only in fishes (**Pisces**).

The two principal kinds of fins are the median and the paired. Median fins extend from the body along the median line of the dorsal surface, around the caudal end as the tail fin or tail, and forward as far as the vent. Paired fins include a pectoral pair attached to the pectoral girdle of the **skeleton** and the pelvic pair attached to the pelvic girdle. All fins are supported by bony or cartilaginous fin rays.

The median fin is broken up in most fishes to form separate dorsal fins and a ventral anal fin in addition to the caudal fin or tail. A caudal division of the dorsal portion is sometimes developed into the fleshy adipose fin. The tail fin may be heterocercal or homocercal. The former type has the end of the vertebral column extending into or toward its dorsal margin while the latter is evenly developed above and below the skeletal axis.

Paired fins are variously modified to form sensory lobes or supporting structures and the pelvic pair of some male sharks bear slender lobes, the claspers, which are thrust into the cloaca of the female during copulation. (A.W.L.)

FIN, AIRPLANE. A fixed or adjustable airfoil, attached to an aircraft approximately parallel to the plane of symmetry, to afford directional stability; for example, tail fin, skid fin, etc. (F.T.M.)

FIN FOLD. A projecting ridge of the body wall in the lancelets which extends the lateral surface of the body and aids in swimming. The ridge follows the same course as the fins of fishes, along the median line of the back, around the caudal end of the body, forward ventrally to the **atriopore** where it forks and continues forward as a pair of ventrolateral folds.

A fin fold has been supposed to represent the earliest stage in the evolution of the locomotor appendages of **vertebrates**. At first a keel for **equilibrium**, extending as a single fold along the back and over the tail, it divided to pass around the cloaca and extend forward on the ventral surface as a pair of folds. By the growth of cartilaginous rods for support and of muscle for control and the dropping out of portions of the fold, the median and caudal unpaired fins and also the paired fins of fishes developed. In the acquisition of a terrestrial habitat the unpaired fins were no longer useful, while further changes in the bony skeleton and musculature adapted the paired limbs first for supporting the body and then for propelling it. This theory has recently been criticized on the basis of studies of early fossil vertebrates. (A.W.L.)

FINCH. Aves, Passeriformes. Seed-eating birds (**Aves**) of many species, found chiefly in the northern hemisphere but to a limited extent in Africa and South America. They are small to moderately large and have a strong beak, usually conical and in some species very large.

In addition to the species whose names indicate their relation with the group, such as the greenfinches and the chaffinches, the finches include birds with distinctive names. The grosbeaks, **cardinals**, **brambling**, **siskins**, **linnets**, **redpolls**, **sparrows**, **canary**, and **crossbills** belong here. (A.W.L.)

FINFOOT. Aves, Gruiformes. Birds (**Aves**) of a few species found in Africa, South and Central America, and the Oriental region. They are similar to the cormorants in form and frequent the water. The toes are lobed. (A.W.L.)

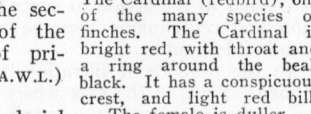

FINGER. Any of the second to fifth digits of the pectoral appendage of primates. (See **Hand**.) (A.W.L.)

The Cardinal (redbird), one of the many species of finches. The Cardinal is bright red, with throat and a ring around the beak black. It has a conspicuous crest, and light red bill. The female is duller.

FINGER LAKE. A glacial U-Valley or rock bowl which forms the basin for a fresh-water lake. Because of the character and origin of the basin, finger lakes are relatively long and narrow. Type locality, the finger lake region of New York State. (R.M.F.)

FINITE DIFFERENCES. Calculus of Finite Differences.

FIORD. A fiord is a glacially overdeepened **valley**, usually narrow and steep-sided, extending below sea level and occupied therefore by salt water. Typical fiords are to be found in Alaska and Norway; their depths, sometimes as much as 4000′, indicate that they are glaciated valleys which have been invaded by the sea after the disappearance of the glaciers. The word fiord is a variant of the Norwegian term for these features, *fjord*. The long fiord-like bays of the New England coast line are sometimes referred to as fiards. (R.M.F.)

FIR. Conifers.

FIRE. Combustible; Combustion.

FIRE BRAT. Insecta, Thysanura. A wingless **insect**, *Thermobia domestica*, clothed with silky scales and bearing long antennae and three slender filaments at the opposite end of the body. Related to the **silverfish** but found in warm places; economic importance and control similar. (A.W.L.)

FIRE DAMP. Methane.

FIRE EYE. Aves, Passeriformes. A common species of **ant bird** in Brazil. (A.W.L.)

FIREBALL. Fireball is a term used in astronomy to designate those **meteors** which are large enough to be apparently brighter than the **planet Jupiter**. They frequently leave a trail which may be visible for several minutes. Not infrequently a distinct sound is heard either during, or shortly after, the observation of a fireball. (W.K.G.)

FIRE-BRICK. Fire-brick is a type of brick capable of withstanding high temperatures and is used to line flues, stacks, furnaces, etc. Good resistance to heat flow is not to be secured simultaneously with refractoriness—indeed, the most refractory bricks generally have the highest thermal conductivities. Where necessary, insulation is added to minimize heat leaks. It is important for the refractory brick to be satisfactory on a number of points in addition to refractoriness, for resistance to melting is only one of several requirements to be met. Among these might be cited resistance to erosion by ash-laden gases, and to the fluxing action of molten slag. A good refractory should not spall badly under rapid temperature changes. The structural strength of fire-brick should hold up well as its temperature approaches the fusion temperature.

Modern installations often impose furnace conditions so severe that refractories other than fire-clay are needed. High aluminum and silicon carbide refractories are typical of these. The heat conductivities of the super-refractories are larger than those of fire-clay brick, and such construction should be backed up with high temperature insulation. Silicon carbide blocks are the most refractory and have the quality of resisting clinker adhesion better than ordinary fire-brick. Their fusion temperature is about 4000° F.

Clay fuses at from 2800° to 3200° F., the upper limit being for flint clay and the lower for the plastic form which, due to its cementing qualities, is especially valuable in fire-brick manufacture. Red brick is not suitable for refractory service, nor is insulating brick. There are several fire-clay furnace cements on the market that are adaptable to monolithic lining. The standard size of fire-brick and insulating brick is 9″ by 4½″ by 2½″. (F.T.M.)

FIRECLAY. This term is chiefly used by British geologists to designate the leached clays, rich in silica and alumina and low in alkalies and lime, which lie directly beneath coal beds. These clays are of economic importance because they are refractory, and do not melt when heated to high temperatures. (R.M.F.)

FIREFLY. Insecta, Coleoptera. Soft-bodied **beetles** with a luminous organ in the abdomen. The flashing of these insects apparently enables them to find mates. Also called lightning bugs. (A.W.L.)

FIREWALL. Industrial or commercial buildings, standing adjacent, with common division wall, may be required to have special attention given to this wall, as to apertures, thickness and material, so as to prevent ignition or transmission of conflagration from one building to another. Such a wall would be a firewall.

Aircraft having the **engine** mounted in the nose of the **fuselage** are required to have a firewall. This consists of a bulkhead of sheet steel, dividing the engine compartment from the remainder of the fuselage. (F.T.M.)

FIRN. Cirque.

FIRST ANGLE PROJECTION. Orthographic Projection.

FIRST DETECTOR. Superheterodyne.

FISCHER-TROPSCH PROCESS. Catalysis.

FISH. Pisces.

FISH FLY. Insecta, Neuroptera. Species related to the **corydalus.** The larvae are aquatic and the adults are found near water. (A.W.L.)

FISH LOUSE. Crustacea, Copepoda. Minute marine and fresh-water animals parasitic on fishes. (A.W.L.)

FISH MOTH. Silverfish.

FISHER. Marten.

FISHERIES. Pisciculture.

FISHER'S z DISTRIBUTION. Fisher's z distribution was derived by R. A. Fisher in 1924.

$$P_z dz = \frac{2 n_1^{\frac{n_1}{2}} n_2^{\frac{n_2}{2}}}{B\left(\frac{n_1}{2}, \frac{n_2}{2}\right)} \frac{e^{n_1 z}}{(n_1 e^{2z} + n_2)^{\frac{n_1 + n_2}{2}}} dz, \; -\infty < z < \infty$$

$z = \frac{1}{2} \log \frac{s_1^2}{s_2^2}$, $B\left(\frac{n_1}{2}, \frac{n_2}{2}\right)$ is the **Beta function,** s_1^2 is the estimated **variance** with n_1 degrees of freedom and s_2^2 is the estimated variance with n_2 degrees of freedom. The z distribution is used in the **analysis of variance.** The z distribution becomes the **normal curve** in case $n_1 = 1$, $n_2 = \infty$ after the transformation $z = \frac{1}{2} \log \chi^2$, $0 \le \chi^2 < \infty$; it becomes the χ^2 distribution after the transformation $e^{2z} = \chi^2$, $n_1 = n$, $n_2 = \infty$, $0 \le \chi^2 < \infty$; and becomes **Student's t distribution** after the transformation $z = \frac{1}{2} \log t^2$, $n_1 = 1$, $n_2 = n$, $0 \le t < \infty$. In the limit as n_1 and $n_2 \to \infty$ in any manner whatever, the z distribution approaches normality with asymptotic **mean** $\frac{1}{2}\left(\frac{1}{n_2} - \frac{1}{n_1}\right)$ and asymptotic variance $\frac{1}{2}\left(\frac{n_1 + 1}{n_1^2} + \frac{n_2 + 1}{n_2^2}\right)$. The z distribution is easily transformed into **Snedecor's F distribution.** Tables of the z distribution have been calculated. (L.A.A.)

FISSION. A process of **reproduction** by the subdivision of the parent body into two or more approximately equal parts which become independent individuals.

Fission is a common form of reproduction in the 1-celled animals. Division of the cell into two parts known as binary fission, is accomplished by **mitosis,** often of a modified form. In very simple species such as *Amoeba* reproduction is no more than cell division, but in species with constant body form each half of the cell differs from the other and reproduction is not complete until it has undergone a reorganization with the development of all structures characteristic of its kind.

Some protozoans (**Protozoa**), notably parasitic species, go through a process of subdivision into a number of parts simultaneously. This process is multiple fission or sporulation.

Fission occurs in multicellular animals of simple structure, including **polyps** and flatworms, by a gradual reorganization accompanied by the constriction and breaking of the body. (See **Radioactive Changes; Atomic Bomb.**) (A.W.L.)

FISSURE IN ANO. A break or crack in the skin just internal to the margin of the **anus.** It is an exceedingly common condition, one that usually complicates hemorrhoids and causes excruciating pain. The crack becomes infected, is raw and tender. Simple early cases can be treated locally with success. Many cases cannot be cured without operation. (D.M.H.)

FISTULA. A deep, chronically infected tract or artificial tube extending from the surface of the body deep into its tissues. Fistula in ano is a fistula which is located around the anal margin, which runs upward through the deep tissues around the **rectum,** usually opening into its lumen. Fistula in ano follows abscess formation around the lower rectum. Surgery is required for its cure. (D.M.H.)

FIT. The dimensional relationship between mating parts. In any machine or device there are certain relations between the dimensions of the component parts that are essential if the unit is to function properly. In a drilling machine, for example, the spindle should rotate freely but without any perceptible shake in the bushings. The bushings, on the other hand, should fit tightly in the frame. The difference in size between the bore of the bushing and the diameter of the shaft is referred to as clearance; the difference in size between the outer diameter of the bushing and hole in the frame is referred to as interference since the bushing is larger than the hole. Both of these intentional differences in the sizes of mating parts are termed allowances. Clearance is positive, interference negative, allowance.

The American Standards Association has adopted eight classifications of fits which are described in A.S.A. Bulletin B 4a-1925. A class 1 loose fit has a large allowance, provides for considerable freedom where accuracy is not essential, and is used in agricultural and mining machinery. A class 2 free fit has a liberal positive allowance, and is used as a running fit for speeds higher than 600 rpm. A class 3 fit has a medium positive allowance, and is used for the more accurate machine tool and automotive parts. A class 4 snug fit has a zero allowance, necessitates considerable precision, and is the closest fit that can be assembled by hand. A class 5 wringing fit has a zero to negative allowance, and gives practically metal-to-metal contact. A class 6 tight fit has slight negative allowance, requires light pressure to assemble, and is used for gears, pulleys, and extremely long contacts. A class 7 medium force fit has negative allowance, requires considerable pressure to assemble, and the parts are considered permanently assembled. A class 8 heavy force and shrink fit has considerable negative allowance, and is used for press fits in steel or for shrink fits where heavy force fits are impractical.

Four classes of screw thread fits are described in the A.S.A. Bulletin B 1.1 1935. A class 1 fit is quite loose, and is recommended where shake or play is not objectionable, and where rapid assembly is desired. A class 2 fit is used for the great bulk of commercial screw thread products, and represents a high quality of fit. Classes 3 and 4 represent extremely close fits, and are used only where freedom from vibration and play are absolutely essential. A class 4 screw is not yet adapted to quantity production, on account of the extremely small tolerances permitted. (H.C.H.)

FITCHEW. Polecat.

FITTIG REACTION. The formation of aromatic hydrocarbons from aryl or aryl and alkyl bromides by the use of sodium, e.g., bromobenzene plus ethyl bromide plus sodium forms ethylbenzene plus sodium bromide ($C_6H_5Br + C_2H_5Br + 2Na \rightarrow C_6H_5 \cdot C_2H_5 + 2NaBr$). (See **Würtz-Fittig Reaction.**) (R.K.S.)

FIX. In **navigation**, a position of a ship, as determined by the intersection of two or more **lines of position**, is known as a fix. Since lines of position are determined in all of the observational types of navigation, **pilotage**, **celestial**, or **radio**, it is obvious that we might define fix as any position of a ship determined by observational means.

Since all observational methods are subject to unavoidable errors of observation, a line of position is not strictly a line, but is a band of constant width, as in the case of a line determined by celestial navigation, or of width which increases from the point of observation, as in the case of pilotage or radio navigation. Hence, a fix is not strictly a point, but is a polygon whose area depends upon the accuracy of the observational material employed in determining the "line." With two lines of position, the area will be a quadrilateral and the area will be a minimum when the lines are at right angles to

each other. Three lines of position will seldom, in practice, intersect in a point, but will bound a triangle, and the standard symbol for indicating the position of a fix on a geographic plot, is a triangle with a dot at the center. This symbol is used no matter how many lines are used in determining the position, the point being the point of intersection of two lines, or the center of the polygon when more than two lines are used.

When the observations for the lines of position can be taken simultaneously, a fix can be immediately determined for the instant at which the lines were observed. Simultaneity of observation is seldom possible and the question of movement of the ship between observations must be carefully considered. It is safe to say that if the ship does not move more than 2 miles between observations, the fix may be considered as the intersection of the lines as directly plotted, with the time of the fix recorded as the average of the times of observation.

If the ship has moved appreciably (more than 2 miles) between observations, the method of running fix must be employed. To obtain a running fix, the time of the fix is usually selected as that of the time of observation of some one of the lines, and the other lines are advanced or retarded to that line by standard **dead-reckoning** methods. In certain particular cases, e.g., when a noon position is desired and the only observations available are one before noon and the other after noon, both lines are moved, one forward and the other back, to the desired time. Before considering a specific case it should be remembered that, by definition, a line of position is a line on which a ship is situated. For the purpose of moving the line, any conveniently located position may be selected, advanced or retarded by dead reckoning, and a line drawn parallel to the observed line through the DR point will be the advanced line. The use of parallel rulers is convenient, but not essential, for this purpose. In case the line of position is appreciably curved, due to the fact that the center of the circle of position is relatively close to the observer, the position of the center must be found, advanced by dead reckoning, and a new circle drawn, using the radius of the original circle.

The following situation illustrates the method for obtaining the running fix of a line of position obtained by celestial navigation and one obtained by pilotage:

At 0620 a ship is in L (latitude) = 42° 20'.8 N & Lo (longitude) = 68° 20'.1 W, and is proceeding at 16 knots on course 257°. At 0620 a line of position is obtained by celestial navigation, with the line running in the direction

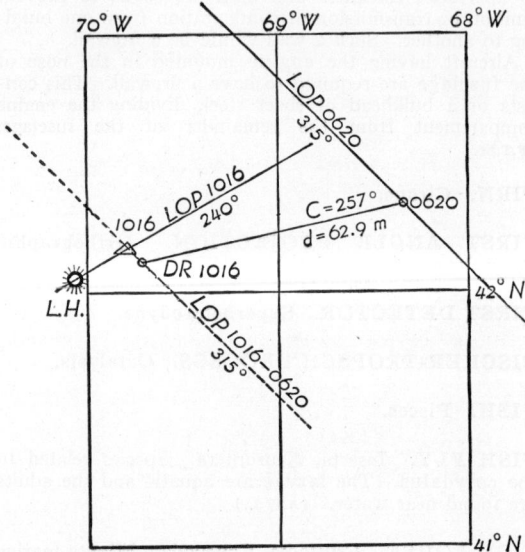

Fig. 1. Scale diagram on mercator plotting sheet.

$135°$–$315°$. At 1016 a lighthouse in L = N $42°$ $02'$.4 & Lo = $70°$ $03'$.7 W is sighted on bearing $240°$. The 1016 position of the ship is desired.

In Fig. 1 the problem is plotted on a **mercator plotting sheet** with all lines labelled in accordance with U. S. Navy procedure. Through the 0620 position the line of position (LOP) is drawn in the $135°$–$315°$. The course line is drawn in direction $257°$ and the 1016 DR position plotted on this line 62.9 miles (the distance run at 16 knots between 0620 and 1016) from the 0620 position. Through this DR position a line is drawn parallel to the 0620 LOP, thus giving the advanced line of position. Through the lighthouse a line is drawn in the $240°$–$060°$ direction and this is the 1016 LOP. The 1016 fix is determined to be at L = $42°$ $09'$.6 N & Lo = $69°$ $46'$.5 W. Other examples of obtaining and plotting lines of position and fixes will be found in the articles on **Celestial Navigation, Pilotage,** and **Radio Navigation.** (w.k.g.)

FIXATION. The process of removing silver halides remaining in photographic emulsions after development is termed "fixation." Since the developed emulsion contains metallic silver and undeveloped silver halides that are light-sensitive, it is necessary to remove the latter in order to stabilize the image. This is accomplished by converting the silver halides into complexes that are readily soluble in water. In addition, the emulsion is swollen, due to alkalinity of developer and imbibition of water, and because it contains partially oxidized developing agents that may produce stain, it is usually advisable to combine with the fixing agent, substances that stop development, prevent stain and harden or retard the swelling of gelatin.

Fixing Baths. According to their function, the components of fixing baths can be divided into the following heads:

1. Silver halide solvent. Many silver halide solvents have been suggested as fixing agents. Among these are:

Sodium thiosulfate	Sodium cyanide
(common hypo)	Sodium sulfite
Potassium thiosulfate	Ammonia
Ammonium thiosulfate	Sodium thiocyanate
Sodium chloride	Thiourea
(brine solution)	Potassium iodide
Potassium cyanide	Thiosinanine

Of these only the thiosulfates and cyanides are important. The others are too costly or possess properties that are objectionable. The cyanides are extremely toxic but are valuable as fixing agents for emulsions that contain a high concentration of silver iodide. For silver chloride and silver bromide, the most practical agent is sodium thiosulfate, $Na_2S_2O_3 \cdot 5H_2O$, which is known as hypo. Ammonium thiosulfate is a faster fixing agent but is more expensive.

The requirements for a good fixing agent are: (1) it should dissolve silver halides without affecting the silver image, (2) the complexes formed should be readily soluble in water and yet stable, so they will not decompose during washing, (3) the agent should not cause excessive swelling or softening of gelatin.

The exact reactions for the formation of thiosulfate complexes are still unknown. However, the thiosulfate ions must combine with silver ions and form soluble complexes as:

$$AgBr + 3Na_2S_2O_3 \rightarrow NaBr + Na_5Ag(S_2O_3)_3$$

$$2AgBr + 3Na_2S_2O_3 \rightarrow 2NaBr + Na_4Ag_2(S_2O_3)_3$$

$$AgBr + Na_2S_2O_3 \rightarrow NaBr + NaAgS_2O_3.$$

In used fixing baths, it is probable that many different reactions exist in equilibrium. The exact equations will depend on the relative concentrations of silver and thiosulfate. Experimental evidence has shown that $Ag_2S_2O_3$

is insoluble, and that $NaAgS_2O_3$ is slightly soluble, while $Na_5Ag_3(S_2O_3)_4$ and $Na_4Ag_2(S_2O_3)_3$ are readily soluble. The insoluble and slightly soluble complexes are unstable, and if left in the emulsion, will decompose in time, causing silver sulfide stain.

Slightly soluble and insoluble complexes can be avoided if emulsions fixed in partially exhausted baths are immersed, for a short interval, in a fresh fixing bath. The two-bath technique insures formation of soluble complexes and is, therefore, recommended.

2. The anti-staining agent. The function of an anti-staining agent is to neutralize the alkali of developer, stop development, and prevent stain due to oxidation of developing agents carried over with emulsion. While most acids are capable of neutralizing the developer, organic acids are preferred because these have a lower degree of dissociation and do not decompose or sulfurize hypo. Satisfactory anti-staining agents include:

Acetic acid	Maleic acid
Citric acid	Sulfuric acid
Tartaric acid	Sodium bisulfite
Boric acid	Sodium bisulfate
Lactic acid	Potassium metabisulfite

For acid hardening baths, acetic acid is undoubtedly the most satisfactory. Solid organic acids, like citric and tartaric, tend to form salts with alum and reduce hardening. The acid sulfites, sodium bisulfite and potassium metabisulfite, do not possess sufficient reserve acidity to prevent precipitation of aluminum sulfite.

The bisulfites are satisfactory, however, in acid fixing baths containing thiosulfate and sodium sulfite.

Sulfuric acid is preferred in chrom-alum fixing baths for tropical processing.

3. The preservative. The addition of acids to thiosulfate decomposes the thiosulfate into sulfurous acid and sulfur, destroying the fixing bath. Sodium sulfite, because it combines with sulfur and forms thiosulfate, is used as the preservative. Besides protecting the thiosulfates from the acid, sodium sulfite also acts as an antioxidant and prevents the partially oxidized developing agents from staining the gelatin.

4. Hardening agents. Salts of aluminum, chromium, and iron exert hardening effects on gelatin. The alums are most satisfactory for fixing baths. The colorless aluminum alum (potassium alum) can be used in both film and paper fixing baths. Chrom alum stains paper.

The hardening action of alum is due to hydrolysis of the salt into a hydrous oxide which is adsorbed by gelatin and forms a complex that is retained even when washed with hot water.

Sodium sulfate has been referred to as a temporary hardner, although it only prevents further swelling of the gelatin.

Formalin, tannin, quinone and quinoid bodies are hardening agents that function in alkaline or neutral solutions only.

Classification of Fixing Baths. Fixing baths are classified according to their composition.

1. Plain hypo—15–40% solutions are used in cases where no hardening is necessary and where the presence of acid will alter the silver grain or change the tone of prints.

2. Neutral fixing baths. These baths contain thiosulfate and sodium sulfite as a preservative. They have some stain-prevention properties and longer life than plain hypo.

3. Acid fixing baths. Acid fixing baths neutralize the developer, stop development and assist the sulfite in preventing stain. A typical bath is:

Sodium thiosulfate	240	grams
Sodium sulfite	10	grams
Sodium bisulfite	25	grams
Water to make	1.0	liter

Baths of this type are satisfactory when hardening is not desired, or if a hardening stop bath has been employed between developer and fixing baths.

4. Acid hardening baths. These baths normally contain four components—a silver halide solvent, a preservative, an anti-staining agent, and a hardener.

The four components may be contained in one solution or obtained by mixing two solutions, a hypo solution composed of sodium thiosulfate in water, with a hardner solution containing sodium sulfite, acetic acid and potassium alum. At times *buffers,* such as boric acid or borax, are added and anti-precipitants like sodium acetate and sodium citrate are included.

A general formula that is recommended for films, plates and papers is:

Hot water (52° C.–125° F.)....	600	cc.
Sodium thiosulfate..........	240	grams
Sodium sulfite, desiccated.....	15	grams
Acetic acid, 28%...........	48	cc.
Boric acid.................	7.5	grams
Potassium alum.............	15	grams
Cold water to make.........	1.0	liter

The above bath is buffered with boric acid and produces more satisfactory hardening throughout its useful life. It hardens satisfactorily over a pH range of 4.2–6.0 and does not sludge or sulfurize until a pH of 6.5 is reached.

The working life of the bath is from 15–25 8 x 10″ prints or films, depending on whether water- or acid-rinse baths are used.

Time of Fixation. The clearing time of an emulsion is an index to its time of fixation. Most authorities agree that the time for complete fixation is twice the clearing time. Factors that affect the time of fixation are:

1. Nature of emulsion.
2. Thiosulfate concentration.
3. Nature of thiosulfate ion.
4. Temperature.
5. Agitation.
6. Degree of exhaustion.

Test for Complete Fixation. Strips of unexposed photographic paper that have been fixed and washed with a batch of prints may be used for the test. The strips are partially immersed in a 0.2% sodium sulfide bath for one minute. If the strips show any yellowish-brown discoloration, they are improperly fixed. The prints should then be refixed in a fresh fixing bath. The test should not be conducted in a darkroom or near sensitive materials as the sulfide fumes produce chemical fog.

Test for Exhausted Fixing Baths. The test for exhaustion depends on the silver content of the fixing bath. The bath is considered exhausted if a 10 cc. sample treated with 1 cc. of 4% potassium iodide produces a yellow precipitate on standing. (s.m.t.)

FIXATION OF ATMOSPHERIC NITROGEN. Ammonia; Nitrogen.

FIXED ARCH. Arch.

FIXED BEAM.
A fixed or restrained beam is one having the ends so firmly connected to the supports that the tangent to the **elastic curve** at the ends remains fixed in direction under varying load conditions. Theoretically, this requires the support to be absolutely unyielding. In actual practice it is impossible to attain a perfectly rigid end support, and the design of a beam on the assumption of perfect restraint would be unduly optimistic. Therefore a design intermediate between perfect restraint and perfect freedom should be adopted

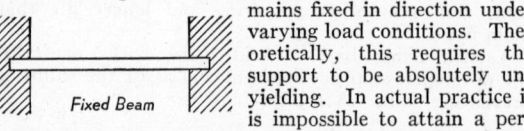

Fixed Beam

for built-in beams, the departure from the conditions of perfect fixity being based upon the rigidity of the end supports. Since a fixed beam under load receives an end moment where it is built in, it will be stiffer than a freely supported beam, and thus have a smaller deflection under the same load. (c.w.c.)

FIXED OILS.
These are fats, compounds of **glycerin** and various complex fatty acids. (See **Acids, Carboxylic; Esters;** and **Fats.**) They are often called non-volatile oils, in distinction to the essential or **volatile oils,** which are readily vaporized by heat. It is characteristic of them that they will leave a spot when dropped on paper. Many of them remain liquid at common room temperatures, others are solid at such temperatures. Solid forms are usually called fats, a purely arbitrary distinction since slight changes in temperature will cause many of them to change from liquid to solid or vice versa.

Fixed oils, especially those of economic importance, are largely obtained from the seeds of plants. They have a high energy value, so form a valuable food if they prove palatable.

Various methods are employed to obtain the oil from the vegetable tissues. Quite commonly the seeds containing the oil are subjected to great pressures in **hydraulic presses.** This may be done without heating, but is frequently facilitated by heating the seeds, the oil being then hot-pressed instead of cold-pressed. Hot-pressing usually increases the yield but causes the product to be less valuable. A third method of obtaining the oil is by means of solvents. Following expression of the oil from the plant tissues, various methods of refining, decolorizing and deodorizing are employed.

Fixed oils are usually classified into three groups, drying, semi-drying, and non-drying oils. Often a fourth group is made of those which are usually seen in solid form, the vegetable fats, although they differ but little otherwise from the other groups.

Drying oils are those which on exposure to air form a tough elastic film. Linseed oil from **flax** seeds is one of the most important and is largely used in making paints and varnishes. **Tung oil,** obtained from the fruits of *Aleurites cordata,* is a valuable oil much used in the manufacture of waterproof varnishes and quick-drying enamels. The tree, a native of China and Japan, has been introduced into Florida. Other drying oils are nut oil, from walnut seeds, poppy seed oil, **hemp** seed oil, and sunflower oil, the latter largely a product of Russia.

Non-drying oils are those which remain permanently greasy or sticky, becoming rancid after a time. Among these oils the most important are **olive** oil, castor oil from the seeds of the **castor** bean plant, rape seed oil, **peanut** oil, **almond** oil, used medicinally, and tea seed oil.

Semi-drying oils are intermediate in nature. The principal semi-drying oils are **cotton**-seed oil, **soybean** oil, **corn** or maize oil and sesame oil. The latter is obtained from the seeds of *Sesamun indicum,* a member of the Pedaliaceae, cultivated in India, China and Japan, where the oil is much used as a food oil and for cooking.

Non-drying oils which are ordinarily solid are palm and palm-kernel oil, coconut oil, and cocoa butter. Another interesting oil of this group is macassar oil, obtained from the seeds of *Schleichera trijuga,* one of the Sapindaceae, occurring in tropical Asia. The oil was formerly much used as a potential "hair restorer," necessitating the use of removable covers, or antimacassars, on the backs of upholstered chairs. The same tree also yields a useful timber. (r.m.w.)

FIXTURE. Jigs and Fixtures.

FLAGELLATES. This is a large group of organisms, of particular interest because of the position they occupy in the organic world. In this article they are treated as plants, but they are also treated as animals. (See **Mastigophora.**) They are usually 1-celled organisms, of extremely complex structure. Many of them have no cell wall of **cellulose,** lack the green pigment, **chlorophyll,** and are definitely animals. Others possess a distinct wall of cellulose and have **chloroplasts,** and are set off as plants. The separation of these two groups is not sharp, however, and some of the plant members are obviously very closely related to very similar animal forms. Therefore it is impossible to stress the differences which are used as a basis for classification. Characteristic of the Flagellates is the flagellum, a long lash-like extension from the **protoplast.** The vibrations of the flagellum propel the organism through the water, in which they usually occur.

The Flagellates are considered by many to represent the ancestral group from which higher organisms, both plant and animal, originated. Many of the animal flagellates occur as parasites in the bodies of higher animals. Some of these occur in the human body, commonly in the intestines. It has been suggested that certain species may cause severe **dysentery.** (R.M.W.)

FLAGELLUM. 1. A movable slender process arising from a **cell.** An organ of locomotion of certain one-celled animals (**Mastigophora**), occurring singly or in groups of two or four. Occurs in sponges, where the action of flagella of many **choanocytes** draws currents of water into the body.

In typical structure the flagellum has a delicate axial filament arising from a **blepharoplast** in the cell and surrounded by a sheath of protoplasm except at its tip. In many cases there is also a larger parabasal body connected by fibrils with the axial filament and with the nucleus. The flagellum is much larger than a cilium and moves in undulations.

2. The whip-like terminal portion of the antenna of a **crustacean.** (A.W.L.)

FLAME. Ignition.

FLAME CELL. Excretory System.

FLAME CUTTING. Cutting of ferrous metals by oxidation, using a stream of oxygen from a blowpipe or torch. The metal is preheated to a bright red, approximately 1500° F., by fuel gas jets in the cutting torch. The stream of oxygen is then applied through a central jet. Once oxidation of iron to Fe_3O_4 begins, the heat of the reaction plays a large part in the continuation of the process. Approximately 30% of the molten metal is removed without actual oxidation by the mechanical washing action of the stream of gas and burnt metal. The preheating gases are oxy-acetylene, hydrogen, natural gas, city gas, etc.

The oxygen lance is a form of flame cutting in which oxygen, supplied through an iron pipe, is the only fuel used. It is used for heavy-duty cutting.

Very heavy sections may be cut with the blow torch. Close dimensional tolerances can be maintained, using machine operated torches, thus flame cutting has become a production tool as well as a means of salvaging scrap. Underwater flame cutting is possible at depths of 135' or more using special practice. (R.H.H.)

FLAME HARDENING. Surface hardening of steel or cast iron by heating a thin surface layer to the hardening temperature with an oxy-acetylene flame, followed by rapid cooling. Depending on the nature of the part to be hardened, either the torch system or the work itself may be moved. Cylindrical parts are rotated before a stationary flame. An air jet or liquid spray following the torch is used to quench-harden the surface. The relatively cool metal in the interior hastens cooling of the surface by conduction. The depth of flame hardening may be less than $\frac{1}{16}''$ to about $\frac{1}{4}''$ depending on the thickness of the section and the service requirements. Distortion is generally less than in parts hardened by general heating and quenching.

Since no hardening agent such as carbon or nitrogen is added to the surface of the steel by this process, only steels having sufficient carbon to harden readily upon quenching are used for flame hardening. The most desirable range is 0.35–0.70% carbon. The hardening treatment is followed by a low-temperature tempering treatment to relieve quenching strains. Typical applications of flame hardening are gear teeth, cams, bearing surfaces, rail ends, crankshafts, and many other machine parts and tools. (R.H.H.)

FLAMINGO. Aves, Ciconiiformes. Large wading birds (**Aves**) of several species found in the warm regions of the world with the exception of Australia. They have very long legs and neck and a broad beak bent sharply downward at the middle. Red or rosy shades are characteristic in their plumage. (A.W.L.)

FLANGE. A flange is a rim or projection extending completely around the object which is flanged. Thus it is distinguished from an ear, which is a similar projection, but which extends only a small portion of the circumference. Flanges are employed for a great many different purposes, among which is the juncture of adjacent shafts by flanged **couplings,** the flanges providing area through which connecting bolts may be passed. Flanged wheels are commonly used to maintain the position of a wheel and axle group upon parallel rails; pipe flanges, for the connection of **pipes** which are not to be screwed together.

Low-pressure piping larger than 6-in. and high-pressure piping are, in the majority of cases, connected by companion flanges. Flanges are drilled to a standard templet and drawn tightly together by means of flange bolts. Alloy steel bolting to conform to A.S.T.M. specifications should be used when pressures exceed 160 lbs. per sq. in. or temperatures exceed 450° F., but below these limits commercial bolting should be used.

Flanges are made of cast iron, cast steel, forged and rolled steel. The face of the joint is always machined smooth. The Van Stone joint has proved to be a satisfactory connection and is widely used. For the most severe service the lap is made full pipe thickness with

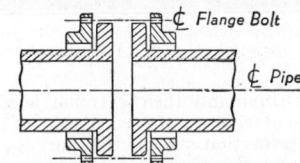

Loose flange (Van Stone) pipe joint.

a perfectly smooth finish to the laps. The flanges are loose on the pipe and the laps are clamped between them. The outside diameter and minimum thickness of pipe flanges are the same as those for fittings.

For the structural engineering significance of flange see **Bridge, Girder, I-Beam.** (F.T.M.)

FLAP. High Lift Devices.

FLASH. Forging.

FLASH POINT. The lowest temperature at which an oil will decompose to an inflammable gaseous mixture, demonstrable through its explosive quality, is its flash point. The flash point occurs at a temperature lower than the burning point, which is the lowest temperature at which the production of combustible gas occurs rapidly enough to support a steady flame. The flash point is an important characteristic of oils used for lubricating bearings, since any heating to the flash point would result in decomposition of the oil as a lubricant. On the other hand, the flash temperature is less important than the burning temperature in determining the fire risk of an oil. The flash point is tested experimentally by heating the oil under certain specified conditions in a cup. A thermometer is suspended in the oil so that the temperature may be read during the test. Periodically an open test flame is introduced through an opening in the cover to detect the slight explosive puff which follows when the flash point has been reached. (F.T.M.)

FLASH SPECTRUM.

At the instant of second or third contact during a total solar **eclipse** the edge of the moon is tangent to the **photosphere** of the sun as

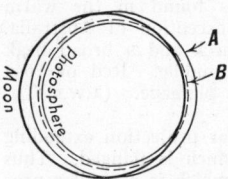

Fig. 1. Flash spectrum.

shown in Fig. 1. With the photosphere covered, the highly heated atmosphere of the sun, known as the **reversing layer** and the **chromosphere**, flashes into view. With the photosphere covered the continuous spectrum of the sun is cut off and the bright-line **spectrum** radiated by the atmosphere may be observed.

If a photograph is taken with an **objective prism** at the instant of second or third contact, a series of curved images of the atmosphere of the sun will be obtained, each in one particular radiation. As indicated by Fig. 1, radiations due to elements at the highest levels, as at *A*, will give longer curves than the radiations due to elements at lower levels such as *B*. Such a bright-line spectrum is referred to as a flash spectrum, since it is only visible for a few seconds at the instants of the contacts.

Fig. 2 shows a section of a flash-spectrum plate. The long curved lines are due to the so-called *H* and

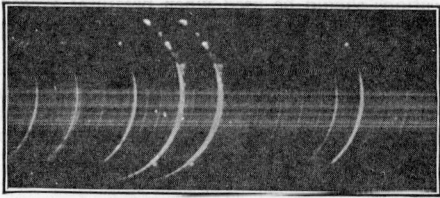

Fig. 2. The flash spectrum. Photographed by J. A. Anderson near the end of the total eclipse of January 24, 1925. (*Mount Wilson Observatory*.)

K lines of calcium and their extreme length indicates that this element rises to the very top of the chromosphere. The projections from the curves are due to prominences, and the breaks in the curves are caused by irregularities on the moon. Other lines, of shorter length, are due to elements in the lower regions of the solar atmosphere. A careful study of flash-spectrum plates gives the distribution of elements throughout the solar atmosphere. Such results are of great value for a variety of problems of solar research, such as the selection of radiations to be used with the **spectroheliograph** for obtaining photographs of the sun at different levels. (W.K.G.)

FLASHING, STRUCTURAL.

Flashing is a method of sealing joints on buildings, especially around roofs, chimneys, gutters, and valleys, in order to render them water-tight. The method consists of using strips, or shingles, of flashing material which are worked into the normal roof surface and turned over the joint. Where the flashing turns up, as along a brick wall, it is necessary to counter-flash, that is to let a strip of flashing

material into the brickwork and bend it down over the other flashing. A sloped shingle **roof** flashed against a brick wall requires flashing shingles which are worked under the top course of the regular shingles and turned up along the bricks. Corresponding counter-flashing let into the brick is bent down over these flashing shingles. A joint between the gutter and cornice is made weatherproof by flashing. Usually the flashing extends from under the lowermost course of shingles and is turned down over the edge of the gutter. The most common flashing materials are tin-coated sheet iron and copper. Lead and zinc are used to a limited extent. (F.T.M.)

FLASHING, THERMAL.

Liquids may exist with thermal stability at high temperatures provided they are subjected to sufficiently high pressure. Water, for example, may be heated to about 700° F. without boiling if under a pressure of 3200 lbs. per sq. in. It is true of liquids in general that the lower the pressure on them the lower the boiling temperature and the lower the heat contained in the "saturated" liquid. Thus high-temperature liquids when passed from a region of pressure sufficient for stability into a low-pressure region are not able to contain all the heat originally possessed as heat of fluid, and will be spontaneously partially evaporated by the surplus. This violent readjustment to thermal equilibrium is called "flashing," and is a common occurrence, having many uses (see **Deaerator** and **Evaporator**), and occasionally creating hazards. For example, the destructiveness of a boiler explosion arises mainly from the violence of flashing action since the water originally contained in a ruptured boiler drum at 600 lbs. per sq. in. pressure suffers an almost instantaneous *four hundred fold* expansion in volume. (F.T.M.)

FLASHLAMPS.

The photoflash lamp, developed about 1930, has completely replaced the older magnesium lamp and flashlight powder. The lamp consists of a globe of heat-resistant glass enclosing a low-temperature filament and finely drawn aluminum wire in an atmosphere of oxygen. When an electric current of 6 volts, or greater, is applied, the filament is heated to incandescence and ignites the wire, producing a brilliant flash. The duration and intensity of the flash vary with the size of the lamp, the amount and size of wire it contains, and the gas pressure.

The characteristics of a flashlamp are usually represented by time-intensity curves in which the time elapsing from the application of the electric current to the end of the flash is plotted in milliseconds (1/1000 second) against the light output of the lamp in lumens. Curves for two types of flashlamps are shown below.

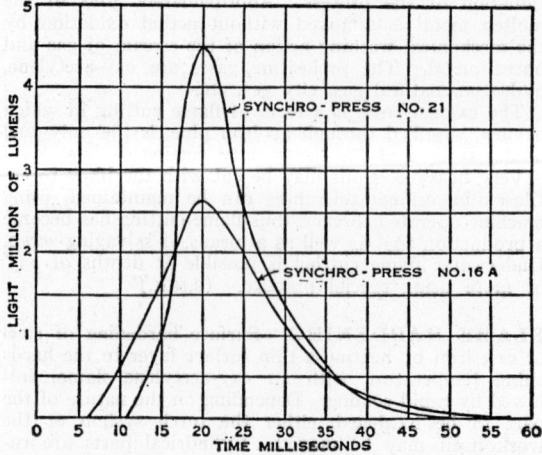

Time-light curve of Nos. 16A and 21 flashlamps. (*Neblette's Photography*.)

Flashlamps are ordinarily used with a synchronizer which is a device for synchronizing the flash of the lamp with the opening of the camera shutter. Since the blades of the modern lens shutter require approximately 5 milliseconds to open to full aperture, while most flashlamps require from 15–20 milliseconds to reach peak intensity, the synchronizer must delay the tripping of the shutter by 10–15 milliseconds after the electrical circuit of the lamp is closed. This allows the flash of the lamp to reach full intensity before the shutter opens for the exposure. This is accomplished in some synchronizers by a mechanical delaying gear and in electro-magnetic synchronizers by the use of a solenoid with an adjustable air gap in the shutter circuit. This air gap is adjusted so that from 10–15 milliseconds will elapse before the charge on the solenoid is sufficient to pull down the lever which trips the shutter. (C.B.N.)

FLAT HEAD. Pressure Vessels.

FLATFISH. Pisces, Teleostei. Fishes (**Pisces**) which lie on one side of the body when adult. The head is modified so that both eyes are on the upper side. Many species are valuable food fishes and the halibut is an important source of vitamin-bearing oil now widely used in place of cod-liver oil.

The flatfishes make up the family Pleuronectidae. Among the included species are the **halibut, plaice, flounder, turbot, brill, soles,** and **dabs.** (A.W.L.)

FLAT-PLATE AREA, EQUIVALENT. That area of an imaginary flat-plate, normal to an air stream, whose drag would be the same as that of an actual body at the same air speed is the equivalent flat-plate area of the body. The coefficient of drag of a flat-plate is known to vary with the **aspect ratio** and the area, but it is frequently assumed as 1.28. However, as the area is an imaginary one at all events, the drag coefficient may as well be assumed equal to 1. It is seen that equivalent flat-plate area might have two values, f and f', which are related thus: $f = 1.28 f'$.

If $D =$ drag, lbs., of a body in air stream of **dynamic pressure** q,

$$f = \frac{D}{q}$$

$$f' = \frac{D}{1.28q}$$

(F.T.M.)

FLATWORM. Platyhelminthes.

FLAX. *Linum usitatissimum.* Linaceae. Flax is the name given to the pericycle fibers of *Linum usitatissimum,* a plant native to Europe. (For a discussion of New Zealand flax, *Phormium tenax,* see the next article.) They were probably the first vegetable fibers to be used by man. Picture-writings at Thebes not only show the plant, but also give details of the processes used in making cloth from the fibers. Egyptians, Greeks, ancient Hebrews, and Romans knew the fiber and used it. Mummy-cloths are often of linen.

The flax plant is a slender annual attaining four feet in height and branching slightly. It has small **lanceolate** leaves and clear blue flowers. When mature it bears seed capsules containing ten seeds each and about 1/4" in diameter. Successful cultivation demands an abundance of **potassium** and **phosphorus** in the soil and plenty of moisture. The plant is cultivated not only for the fibers, but also for oil. The best fibers are obtained from plants grown in cool regions, while the best oil is derived from plants grown in tropical countries like India.

For preparing flax, the plants are pulled or cut before they are mature, and stripped of all leaves and seed capsules. The denuded stems are then tied in small bunches and immersed either in stagnant or slowly running soft water, where they are left for several days. During this time, the stems are attacked by **bacteria,** which bring about fermentation, causing a breaking down of the woody tissues and a partial separation of cells due to the action of the bacterial **enzymes** on the pectic substances binding the cells together. This process is called "retting." Sometimes the flax stems are spread on the ground in a thin layer and left exposed to the action of dew and sunshine for a few weeks. The same result is obtained, the process being now called "dew-retting." The retted stems are removed from the water, washed and cleaned of as much non-fibrous material as possible. This process is known a "Scutching." "Hackling" follows, and is a sort of combing which removes any remaining non-fibrous material. The fibers thus obtained are in reality bundles of cells which occurred as pericycle fibers in the stem. Good fibers vary from 12–36" in length, while many shorter ones are obtained. Short and tangled fibers are called "tow." The fibers vary in color from yellowish to dirty gray, largely depending on the attention paid to the retting process. They are soft and flexible, capable of division into smaller bundles of fewer cells. They are very strong, each cell possessing a uniformly thick wall which surrounds the very slender central cavity or lumen.

The principal use of the fibers is in the manufacture of thread requiring great strength, such as shoe thread, bookbinding thread, fish line and fish-net twine, and also in the making of fine cloth such as table-linen and handkerchief linen. All cloth made from flax fibers is called linen. Flax fibers conduct heat much more rapidly than cotton and so cloth made from them is cooler and much favored in tropical countries.

In addition to the fibers, flax plants yield a valuable oil, called linseed oil. (See **Fixed Oils; Esters.**) In making this the seeds are crushed by machinery, heated to 165° F. and treated with naphtha which extracts the oil. The seeds are about 40% oil. This oil, a drying oil, is used in the manufacture of paints, varnishes, and patent leather, as well as in making linoleum and oilcloth.

The oil cake left after the oil is pressed from the seeds is used directly as a cattle food or is ground up into oil meal and used for the same purpose. (R.M.W.)

FLAX, NEW ZEALAND. *Phormium tenax.* Liliaceae. (For a discussion of common flax, *Linum usitatissimum,* see preceding article.) The fibers of New Zealand flax occur as **sclerenchyma** sheaths surrounding the vascular bundles in the long, straight, rather stiff leaves of this plant. The plant, a native of New Zealand, has been introduced into Australia and other countries. In the United States, it is cultivated to some extent, often as an ornamental plant. The leaves, from 4–8' long and up to 8" wide, may be 20% fibrous material. To obtain the fiber the leaves are cut off and scraped to remove much of the non-fibrous material. After this the fibers are combed out and cleaned. They are very white, soft, flexible, lustrous and tough. They may be 5' long. The principal use of New Zealand flax is for binder twine, baling rope, and cordage, often in combination with sisal or other fibers. A fine cloth resembling linen duck can be woven from it. From 3000 to 5000 tons of this fiber are imported into the United States annually. (B.S.M.)

FLEA. Insecta, Siphonaptera. Small **insects** with transversely compressed bodies, sucking mouths, and no wings. They live as parasites in the adult stage on the bodies of mammals and more rarely on birds. The larvae have biting mouths and eat fragments of organic matter in the debris about the sleeping places of the hosts; they are never parasitic.

Pyrethrum powder is an effective deterrent for these pests. It can be dusted into the fur of dogs and cats

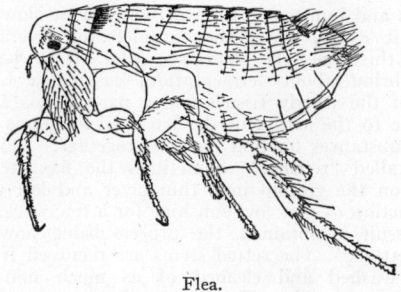

Flea.

and used about the house in regions where fleas are troublesome to man himself. (A.W.L.)

FLEXIBLE COUPLING. Coupling.

FLEXURE.
Flexure is a term which is used to denote the curved or bent state of a loaded beam. A horizontally located **beam,** transversely loaded with vertically directed load, offers an example of load-carrying ability derived through flexure. In flexure, an elastic structural material undergoes a deflection sufficient to set up in its material stresses which will support the load. Deflection under load is an essential and necessary part of the process of load carrying by a beam, for until the deflection has occurred, there are set up in the beam no resisting forces. Thus if an unloaded beam is perfectly straight and horizontal, it must assume a slightly curved position if any external loaded is supported by it. The only way in which a loaded beam could be straight would be to have had an initial deflection in a direction opposite to the loading.

The so-called flexure theory establishes a relation between the fiber stresses at any point in a beam and the bending moment causing these stresses. This theory is based on two fundamental assumptions whose validity, within ordinary working limits, has been established by experiment. The first assumption is that a cross-section which was a plane before bending remains a plane after bending. This implies that the unit deformations are proportional to the distance from the neutral axis. The second assumption is that the fiber stresses are proportional to the deformations resulting from these stresses. If a **tension** member is subjected to an axial load in a testing machine it will be found that, for stresses below the **proportional limit,** the ratio of the unit stress to the unit deformation is a constant called the modulus of elasticity. This would also be true if the test specimen were a short compression member. In order to reconcile the second assumption it must be further assumed that the fibers act similar to test specimens and that the modulus of elasticity is the same for tension and compression.

It will now be shown how deflection and load-carrying ability are interrelated in a beam. First, it will be assumed that the structural material is elastic, that is, within the elastic limit the stress is proportional to the strain inducing it, and that it is a homogeneous material. The results produced by materials not exactly meeting these specifications are usually in good accord with the theory based on these assumptions.

Assume that there is a beam of rectangular cross-section mounted horizontally between simple supports. If one were further to assume this material is weightless, the axis of the beam would be absolutely horizontal. Next a gravity load is placed on the beam, resulting in a certain deflection which sets up resisting couples within the beam, enabling it to carry the load. It must be evident that after a static condition is reached, the external **bending moment** thus imposed on the beam must, at any point, be balanced by an internal moment

arising out of the stresses in the material of the beam. (Next consider the enlarged section of the bent beam.) If the section is taken sufficiently small, it can be as-

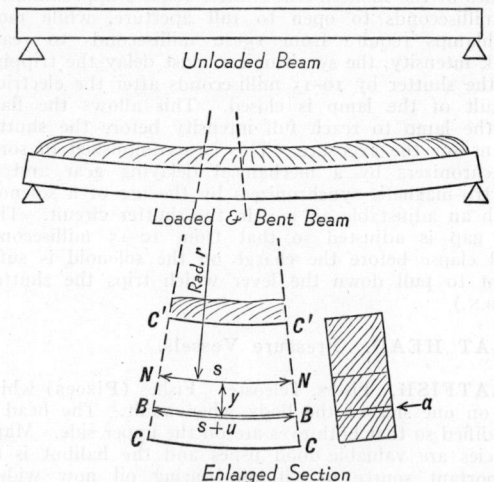

sumed to be bent in the arc of a circle whose radius is r. The upper fibers, i.e., $C'C'$, are naturally compressed or shortened in length, and the lower, CC, are stretched. At some intermediate plane, NN, there must exist an unstretched fiber whose length is the same as it possessed in the unloaded state. This axis of no strain is called the neutral axis. If its length is s, then the length of a typical fiber such as BB, located at a distance y from the neutral axis, is $s + u$, in which u represents the stretch. $\frac{u}{s}$ is the percentage stretch, or strain, of the material. From the geometry of the figure it is apparent that the strain $\frac{u}{s} = \frac{y}{r}$. Since stress is proportional to strain, the factor of proportionality being the modulus of elasticity E, it follows that the stress on $BB = p = E / \frac{y}{r}$. Referring to the cross-section of this small element of the beam, the end area of fiber BB is taken as a. The stress p acting on this area produces an elementary internal force ap. Above the neutral axis there are similarly produced forces, but oppositely directed. The sum of all these longitudinal forces is, of course, zero, since the beam is static; however, at any cross-section they produce, *in toto,* a torque or moment around the neutral axis which is exactly equal to the external bending moment at that section. For example, the moment of the force acting on the fiber BB is apy about the neutral axis. The total moment, then, is the Σapy about the neutral axis.

Substituting $\frac{Ey}{r}$ for p, the total moment equals

$$\frac{E}{r} \int ay^2 = \frac{E}{r} I.$$

The last step shows how the **moment of inertia** enters into the flexure formula. Since r is not a convenient quantity to work with, a substitution of $\frac{p}{y}$ is made for $\frac{E}{r}$ resulting in the common flexure formula:

$$M = \frac{pI}{y}.$$

In this formula, M is the bending moment at a section where the moment of inertia is I, and p is the unit stress at y distance from the neutral axis.

It is readily shown that the neutral axis is coincident with the center of gravity of the cross-section of the beam. From above, we extract the following equation:

$$ap = \frac{E}{r} \, ay$$

$$\int ap = \frac{E}{r} \int ay.$$

$\int ap$ is the total force within the beam parallel to the neutral axis, and is zero, as explained above, but this results also in $\int ay$ being equal to zero, which can be true only when the distance y is a moment arm around the center of gravity of the cross-sectional area.

The flexure formula is valid as long as the stresses are within the proportional limit. In the derivation of this formula it is assumed that the horizontal stresses are the only internal forces which resist the external bending. As a matter of fact, the true maximum tensile or compressive unit stress, called a principal stress, is the **resultant** of the bending and the shearing stress acting at the point. But, as has been previously stated, the stresses which are obtained by the flexure formula are reasonably correct for ordinary design purposes. (C.W.C., F.T.M.)

FLICKER. Aves, Piciformes. Moderately large **woodpeckers** of North America which differ from the other woodpeckers in feeding habits. Flickers eat many insects on the surface of the ground and are particularly fond of ants, which they catch on the long sticky tongue. Three species are recognized, the common or yellow-shafted flicker, *Colaptes auratus,* the red-shafted *C. cafer,* and the gilded flicker, *C. chrysoides,* of the southwest. The common flicker is sometimes called the high-hole.

Other species of the same genus occur in South America. (A.W.L.)

FLICKER PHOTOMETER. Bench Photometers.

FLIGHT. Locomotion through the air. Some animals travel through the air for short distances by gliding, supported by broad expansions of various parts of the body, but in true flight the animal is able also to support and propel itself by muscular activity. True flight occurs only in the insects, birds and bats. The pterodactyls, an extinct group of reptiles, were also able to fly.

The gliding animals include the **flying fishes,** an **amphibian** (the flying frog), a **reptile** (the flying dragon) ; and several species of mammals (**flying squirrels, flying lemur**). With the exception of the flying fishes all are provided with broadly expanded thin structures on which they coast down the air from one place to another. The supporting structures of the frog are enormous webbed feet, those of the dragon are broad membranes along the sides of the body supported by elongate ribs, and those of the flying lemur and flying squirrels are folds of skin along the sides of the body between the fore and hind legs. Flying fishes are supported during long leaps through the air by the greatly enlarged pectoral fins.

True flight also depends partly on the possession of broad light planes, the **wings.** These organs are moved by special muscles in such a way that their pressure against the surrounding air lifts and propels the body. Flight demands relatively great expenditure of energy, hence the flight muscles are highly developed. In the most rapid and powerful fliers this is especially true; the wings are smaller and the muscles larger in contrast with the broad wings and relatively smaller muscles of weaker fliers.

In the birds flight is accompanied by accessory adaptations for securing large amounts of oxygen, for storing energy in abundance, and to provide rigid skeletal support against the stresses of flight. The feathers are an insulating coat of high efficiency in the cold upper air. Birds are, in fact, primarily adapted for flight and in other respects conform to the limitations imposed by this adaptation. Only by such extensive adjustment to life in the air are they able to remain on the wing for many hours without rest, like the albatrosses during their everyday life and like many other birds during migration. The flight of birds, moreover, is the most rapid locomotion of living things, rivaling even the mechanical transportation achieved by man. Birds commonly fly 30 or 40 m.p.h. and speeds in excess of 100 m.p.h. have been recorded.

Some birds are adept at soaring, a type of flight in which the wings are held extended and the bird depends on air currents to carry it. It may ride ascending currents to high levels and may always glide to lower levels, moving only to maintain its equilibrium. The vultures very commonly fly in this way for long periods without flapping a wing.

Since the air is too light for an animal to float in it, all flying things must also have some other means of locomotion. The single known exception is a South American May fly. Although this insect walks beneath the water as a **larva,** during its adult life its legs are vestigial. When it once emerges from the water it must fly until the brief remainder of its life is ended, and when it once drops to the water or to the ground it never rises again. (A.W.L.)

FLIGHT, ARTIFICIAL AERIAL. The act or action of passing through the air is flight. The word is also used collectively to describe the like motion of multiple objects in flight. Thus, a *flight* of arrows. The act of flight is the exercise of ability to overcome the forces of gravitation and pass through the air, as by use of wings, while the action is descriptive of a journey through the air from one point to another. Natural flight is performed by the birds, mechanical flight by man-made devices—airplanes, helicopters, and ornithopters. Powered airships make flights; however, the flight of a free or captive balloon is better described as an *ascension*.

The science of flight and the machines employed to secure it are covered in many of the aeronautical articles of this volume. However, it may not be amiss to review, briefly, the history of flight. Mankind has apparently always been interested in flight. Probably from the first the natural flight of birds and the advantages of such travel inspired him with a desire to emulate this means of transportation. In ancient mythology there are many stories associated with the supposed flight of men in emulation of the birds. One of the better known of these legends is the legend of Icarus, the Greek, who essayed to escape from prison by fastening feathers to his body with wax. Unfortunately, he flew so near the sun that the wax was melted and Icarus plunged downward into the sea. Legendary horses have borne wings, and in the *Arabian Nights* Sinbad maneuvered some flights with the aid of the great roc. It took many centuries to take flight out of mythology into concrete proposals as to how it might actually be accomplished, and still more centuries went by before the mechanical flight of a man-carrying machine was a reality. For several centuries early in the Christian era there was little progress, but as scientists, inventors, and experimenters ventured farther along the road to artificial flight, progress accelerated. After 1910 the rate of progress was enormous, and men everywhere made great strides in pushing back the frontiers of flight in every direction.

With no thought of mentioning all the brave and gifted men who have given their efforts, and sometimes their lives, to advance man nearer to mechanical flight, a chronology of significant events and historical flights is presented.

1485 (approx.)..	Leonardo da Vinci	Research on the possibility of flight. Flying models built. Invention of the propeller.
1655..........	Robert Hook	Scientific evaluation of the possibility of flight.
1783..........	Joseph and Jacques Montgolfier	First hot-air balloon ascension.
	Pilâtre de Rozier, Marquis d'Arlandes	First free balloon ascension, man-carrying.
	Jacques Charles	Man-carrying ascension by hydrogen balloon.
1797..........	Andre Garnerin	The first successful parachute descent.
1810..........	George Cayley	Scientific analysis of the principles of flight. Built successful man-carrying glider.
1836..........	Charles Green	500-mile free balloon flight from London to Weilburg, Germany.
1842..........	William Henson	Designed a steam-powered monoplane of excellent possibilities. A full-scale machine was never built.
1852..........	Henri Giffard	Built and flew steam-engine powered, non-rigid airship capable of being navigated in light winds.
1872..........	Paul Haenlein	Built and flew semi-rigid airship with internal combustion engine power.
1889..........	Otto Lilienthal	Built several man-carrying gliders and made several hundred glider flights. Wrote a book summarizing his observations and experiments in aerodynamics.
1891..........	Octave Chanute	Conducted glider experiments, and wrote on the subject of aviation, in particular, stability.
1896–1899.....	Percy Pilcher	Made several hundred flights in a glider of his own design.
1896–1903.....	Samuel Langley	Scientific study of heavier-than-air flight, culminating in the building of powered models, about the success of which controversy raged for several years.
1898–1910.....	Alberto Santos Dumont	Built successful airships and airplanes which he himself flew.
1903..........	Wilbur Wright, Orville Wright	Built and flew an internal combustion engine powered biplane. The flight on December 17th is generally accepted as the first successful flight of a powered airplane.
1907..........	Walter Wellman	Pioneered in the use of dirigible for explorations.
	Gabriel Voisin, Charles Voisin	Foremost early builders of airplanes in France.
1909..........	Louis Bleriot	Flew the English Channel in a monoplane of his own design.
	Glenn Curtiss	Built and flew a biplane, winning the Gordon Bennett Trophy race at 46 m.p.h.
1910..........	Charles Hamilton	Won the New York *Times*-Philadelphia Public *Ledger* prize with a round-trip flight between New York and Philadelphia.
	Claude Graham-White	Piloting a Bleriot monoplane, won the Gordon Bennett race with an average speed of 61 m.p.h.
	Eugene Ely	Made the first take-off from a ship.
1910–1915.....	Lincoln Beachey	Pioneer acrobatic pilot.
1911..........	Earl Ovington	Piloted first official airmail plane in the United States.
	Harriet Quimby	First woman pilot in the United States.
1913..........	Igor Sikorsky	Pioneered in the construction of multi-engined planes.
1914..........	Count von Zeppelin	Constructed long-range dirigible airships.
1919..........		U. S. Navy NC-4 seaplanes cross the Atlantic.
	John Alcock, Arthur Whitten-Brown	First non-stop trans-Atlantic flight from Newfoundland to Ireland.
1920..........	Rudolph Schroeder	Piloted an airplane to over 30,000′ altitude.
1923–1933.....		Series of disasters to large dirigible airships diverted attention to heavier-than-air aircraft.
1923..........	John A. Macready, Oakley Kelley	Made the first non-stop East-West transcontinental flight over the United States.
	Juan de la Cierva	Developed the autogyro, first practical rotating-wing aircraft.
1925..........	Lincoln Ellsworth	Pioneered in Arctic and Antarctic explorations by air.
1926..........	Richard Byrd	Pioneered in Arctic exploration by air.
	Roald Amundsen	Arctic explorations by dirigible.
1927..........	Charles Lindbergh	Made famous non-stop flight from New York to Paris in the "Spirit of St. Louis," a Ryan monoplane.
	Clarence Chamberlin, Charles Levine	Made non-stop flight from New York to Eisleben, Germany, in Bellanca monoplane.
	Lester Maitland, Albert Hegenberger	First to complete a non-stop flight from the continental United States to Hawaii.
1928..........	Charles Kingsford-Smith, Charles Ulm	Flew the first Pacific crossing from the United States to Australia in a Fokker tri-motored monoplane.
	Charles Collyer, John Mears	First circumnavigated the globe by air, with a Fairchild monoplane.
1929..........	Ira Eaker, Carl Spaatz	Established refueling record of 150 hours, 40 minutes.
	James Doolittle	Pioneered in blind-flying technique; also, holder of several notable speed records.
	Richard Byrd	Pioneered in Antarctic exploration by air.
1931..........	Auguste Piccard, Charles Kipfer	Stratosphere balloon ascension to 52,000′.
1933..........	Wiley Post	Solo circumnavigation of the earth in less than 8 days.

1936..........	Hugo Eckener	Commanded the dirigible airship Hindenburg on several commercial trans-Atlantic flights. Radio range airway well established in the United States.
1937............	Igor Sikorsky	Built commercially successful helicopter.
1938..........	Howard Hughes	Circumnavigated the earth in less than 100 hours.
1939..		Russian pilots complete non-stop flight from Moscow to Vancouver across the North Pole; Russian pilots complete non-stop flight from Moscow to California, 6262 miles. Regularly scheduled trans-Atlantic airline flights begun; air-cooled engines reached 2000 hp. size.
1940..		Scheduled airline flights across the Pacific inaugurated; first all-blind instrument flight between termini; S. Campini, first jet-propelled airplane, a Caproni; high-altitude (24,000′) pressurized-cabin commercial airliner introduced; airlines of U. S. operate 12 months without fatality; U. S. Federal Airways System developed; popularity of private flying soars in U. S.; insurance rates drop; colleges adopt aviation courses.
1941............	F. Whittle	Developed successful jet-propelled plane; the RAF is credited with saving England from German invasion; Japanese air arm attacks U. S. base at Pearl Harbor; commercial stratoliner operations begin; air pick-up service developed; tailless aircraft again appear; mass production of military airplanes; commercial air flights on radio direction finders; airliners with 285 m.p.h. speed, capable of carrying 60 passengers 5000 miles appear.
1942..		Air force strength proves dominant military factor; Japanese aircraft sink British battleships "Repulse" and "Prince of Wales" off Singapore; pre-flight training introduced into U. S. secondary school curricula; four-engined Fortress in mass production; thousand-bomber air raids become a reality; battle of Midway fought and won by U. S. Naval air arm; jet-propulsion plane developed in the U. S.; first 70-ton flying boat—"Mars."
1943..		Four-engined Superfortress in mass production; allies attain air supremacy on nine war fronts; U. S. AAF grows to 44 times its 1940 strength; precision bombing introduced; air and seapower coordination stops Japanese; women pilots join men on dangerous missions; civil air patrol successful in anti-submarine operations.
1944..		West-east record flight across U. S. by "Constellation," non-stop, in less than 7 hours; jet-propelled fighters appear in combat; resonance ducts used to power flying bombs; long-range liquid-fuel rocket used as a war missile; external wing tanks used to extend range of fighter planes.
1945............	J. K. Newman, R. K. Smith	Mosquito bomber flight from England to India, 4700 miles in 12 hours, 25 minutes, average 378 m.p.h.
	E. W. Leach	American altitude record of over 9 miles in jet-propelled plane announced as having been attained in 1943 by Bell Aircraft Corp.
	J. R. H. Merfield	Mosquito bomber from Newfoundland to England, 2500 miles in 5 hours, 10 minutes, average 455 m.p.h.
	F. A. Armstrong	Four B-29s from Japan to Washington, D. C., 6544 miles lead plane's time, 27 hours, 29 minutes.
	H. J. Wilson	Jet plane 611 m.p.h. top, 606 m.p.h. average, off Herne, England.
	J. R. Holzopple	A-26 around the world, 24,859 miles, 96 hours, 50 minutes flying time; extensive prior use of radar in aerial navigation revealed; U. S. AAF operates on 5 air fronts; air transport system spans the world; fighter plane speeds increased to 500 m.p.h.; airpower credited with major role in defeat of Germany and Japan.
1946..		War-developed B-29 sets new record, carries 5000-kg. payload at 366 m.p.h. over a 2000-km. course; helicopter duration performance set at new high of 9 hours, 33 minutes, 27 seconds by a Sikorsky R-5; Lockheed P-80 sets speed record of 495 m.p.h. on a 62-mile low-level run; helicopter distance record set from Wright Field, Ohio, to Boston, Massachusetts, time 10 hours, 3 minutes; helicopter speed record brought from 76 m.p.h. to 110.5 m.p.h. by Sikorsky 5-A; airplanes flying 6 miles high used as experimental television relay points.

(F.T.M.)

FLINT. Flint is a rock composed essentially of a cryptocrystalline form of silica. It is very dense and tough, breaking with a **conchoidal** fracture; colors, usually dark grays, blues, or browns, often black. It occurs chiefly as nodules and masses in **chalks** and **limestones.** Flint is particularly interesting because it was used by primitive man for making instruments (artifacts) for thousands of years before he learned to use bone and metal. Flint still remained an essential mineral resource for making fire, including the flint locks on guns, until the close of the 18th century. From the dawn of civilization the best flint has come from Belgium and the coastal chalks of the British Channel and the Paris Basin. (R.M.F.)

FLINTY CRUSH-ROCK. A term proposed by the Scottish geologist Clough for the almost structureless, flinty portion of **mylonite,** an extreme product of dynamic **metamorphism.** (R.M.F.)

FLIPPER. An appendage of the aquatic **mammals** in which the digits are enveloped by continuous tissue so that the entire structure forms a flat paddle for swimming. Flippers occur in **seals, manatees, whales,** and related forms. (A.W.L.)

FLOAT, AIRPLANE. Seaplane.

FLOATING FOUNDATION. Foundation.

FLOCCULI. Sun.

FLOOD PLAIN. The relatively level surface, or surfaces, within a river valley, caused by the depositional work of the river, especially during flood or high-water. Compare and contrast with the **tapset** beds of **delta.** (R.M.F.)

FLOODING. Gas Absorption.

FLOOR BEAM. A **beam** which is the direct support of the floor **load** of a building and transfers this load to the adjacent girders or columns is a floor beam or floor joist. These beams may be of wood or steel.

The term floor beam is also used to designate the transverse beams of the floor system of a **bridge** which transmit load to the longitudinal **girders** or **trusses.** (C.W.C.)

FLORET. A small flower; for example, one of the flowers forming the head of the plants of the **Composite Family.** (R.M.W.)

FLORICAN. Bustard.

FLOTATION. Classification.

FLOUNDER. Flatfish.

FLOUR. Wheat.

FLOW. Fluid Flow.

FLOW, LAMINAR. A fluid is said to have laminar flow if the mass of flow may be considered as advancing in separate laminae with simple shear existing at the surface of contact of the laminae should there be any difference in mean speed of the separate lamina. If turbulence exists, its effect is confined to a lamina and there is no exchange of momentum between laminae. (F.T.M.)

FLOW LINES. Macrostructure.

FLOW METER. The flow of a fluid (see **Fluid Flow**) is one of the more difficult physical quantities to measure, especially if the fluid is a gas, and yet there often arises the necessity or desirability of measuring the flow of a vapor, liquid, or gas. Several meters have been developed for measuring these quantities. In distinguishing a flow meter from a quantity meter, flow is a quantity passing a given section of the conduit per unit of time: thus gal. per min. of water, cu. ft. per min. of gas, and lbs. per min. of steam, are flows. One of the most common principles employed in a flow meter is that of the interchangeability of pressure and velocity head of a fluid. Fundamental hydrodynamic theory indicates pressure, expressed as head of a fluid, and velocity to be interchangeable in the ratio as expressed by

$$v = \sqrt{2gh}.$$

The Venturi meter is used in the application of this principle to fluids. It is shown in Fig. 1. The instrument is

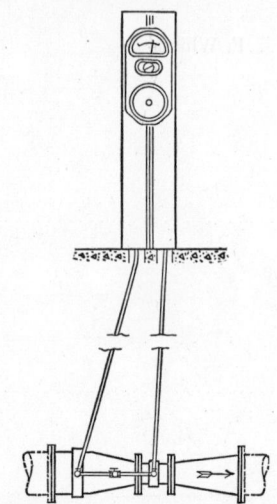

Fig. 1. (*Builders' Iron Foundry.*)

essentially a manometer device calibrated to read flow in cu. ft. per sec., gal. per min., etc. Recording and registering features may be added to the indicating meter. The Venturi meter principle is based on the reduction of pressure accompanying increase of velocity. (Bernoulli's principle.) The velocity is increased by inserting a fitting in the pipe line which converges to a minimum section and then diverges to the normal pipe size. Pressure leads are brought from the low-pressure region and from the normal-pressure region to the manometer of the instrument. The theory of the Venturi meter will be found in all standard works on hydraulics. The theory yields the equation

$$H = \frac{8Q^2}{\pi^2 g d^4}\left[\frac{d^4}{d_1^4} - 1\right],$$

in which H is the ft. of water of differential head produced by a flow of Q cu. ft. per sec. The internal diameter of the pipe is d, of the minimum section, d_1. The actual discharge Q will be from 0.96 to 0.99 of the theoretical Q. The above equation shows that when the dimensions of the Venturi throat fitting have been fixed, the flow can be measured proportional to the square root of the differential head H.

This does not exhaust the list of apparatus devised to measure flow of liquids. For example, **weirs,** weighing tanks (in conjunction with a chronometer), and displacement-type meters will measure flow.

Lighter fluids, such as air or steam, are measured in meters of the orifice type. This is explained in connection with Fig. 2. A constriction of flow produced by an orifice in a metal disk inserted between pipe flanges creates a differential head that is proportional to the square of the flow. This pressure head is taken off through two small pipe connections and transmitted to a mercury U-tube in one leg of which are a large number of graded-length contact rods connected to resistances.

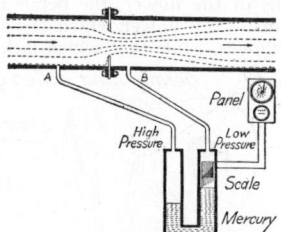

Fig. 2. Republic electric flow meter.

A flow of steam causes a certain pressure head on the manometer and a corresponding rise of the mercury level in the contact rod chamber. The rise of the mercury cuts out a definite amount of resistance, depending on the steam flow. The meter itself is an electrical instrument which measures the varying conductance in the instrument circuit, reading directly in units of flow. The flow meter is operated on ordinary a.c. and is unaffected by slight voltage variations. The number of rods in the contact chamber is so great that slight changes of flow are indicated on the meter. The indicating or recording instruments may be placed in any convenient location and connected to the manometer by a 2-wire circuit. (F.T.M.)

FLOW SHEET. Organization Chart.

FLOW, STREAMLINE AIR. Aerodynamics.

FLOW, TURBULENT AIR. Boundary Layer; Turbulence.

FLOWAGE (ROCK). This term, as used by geologists, signifies the internal movement of clays and rocks when stressed beyond the elastic limit. Important types of flowage are: **granulation** and **foliation**, although the latter is primarily a physical-chemical rather than a purely mechanical process, as in the other types of flowage. (R.M.F.)

FLOWER. The flower is that part of the plant which is concerned in the sexual reproductive process of angiosperms. The formation of the flower is preliminary to the production of fruit with its seeds.

Flowers may be borne at the tip of a stem or a branch, in which case they are said to be terminal flowers. Or they may be borne in the **axils** of leaf primordia and called axillary. In very many plants the structure which subtends the flower does not develop into a leaf like the other leaves of the plant. Instead, it may remain very small and inconspicuous, or it may grow larger, but be of shape quite unlike that of an ordinary leaf. These structures which subtend flowers are called bracts. In a few plants the bracts are large and brilliantly colored, so that at times they are much more showy than are the flowers. The scarlet bracts of the *Poinsettia* and the white or pink bracts of the Flowering Dogwood are examples. In many **monocotyledons**, as Palms and Arums, there is a single large bract which subtends and often more or less surrounds the flowers. Bracts of this kind are called spathes. The striped "pulpit" of the Jack-in-the-Pulpit and the white bract of the Calla Lily are well-known examples. The bract may surround and protect the flower in the bud.

Flowers may be borne singly or they may be associated in a cluster which is known as an inflorescence. A single flower is called a solitary flower, and the stalk which supports it is a peduncle. The stem which supports an inflorescence is also a peduncle, while the individual flowers of the inflorescence are supported on pedicels.

Anemonella, one of the buttercup family.

When there is a distinct axis extending through an inflorescence it is called a rachis. The arrangement of the flowers of an inflorescence varies in different groups. A common, and primitive, form of inflorescence is the raceme, in which the floral shoot grows at the apex and bears many pedicels, each ending in a single flower.

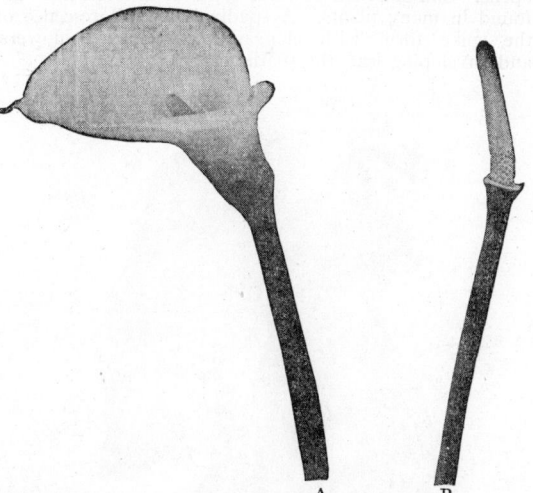

A   B

Spadix of Calla lily. This is not a flower but an inflorescence. In the center is a spike, the upper portion of which bears staminate flowers and the lower portion pistillate flowers. The showy white portion is a spathe—a special leaf surrounding the spadix. A, exterior view; B, with spathe removed to show spadix inside.

The first flowers to open are those at the base of the raceme. If, in an inflorescence of this sort, each branch is a raceme bearing several flowers, it is called a panicle, or a compound raceme. If the flowers of the raceme are borne directly on the main axis, the inflorescence becomes a spike. A secondary spike, common in grasses, is a spikelet. A catkin is a spike which droops. A corymb is a modified raceme in which the lower pedicels of the inflorescence grow faster than the upper ones, so forming a more or less flat-topped cluster. An umbel differs from a corymb in that it has no central rachis, all the pedicels of the inflorescence rising from a common point. More commonly umbels are compound, each of the main stalks of the inflorescence bearing an umbel at its tip. The inflorescence of the onion is an umbel, that of wild carrot a compound umbel. The inflorescence of the **composite family** is a head, which may be considered as an umbel in which the flowers are all sessile, without pedicels, on the apex of the stem. A cyme is an entirely different type of inflorescence. In the cyme the first flower to open is at the tip of the cluster. Below it on the stem are a number of bracts. From the axils of these bracts branches develop and also end in a

Upper portion of *Trillium* plant with whorled arrangement of leaves on the stem.

flower. This successive branching may be many times repeated, but always the flower terminates the stem and opens. Combinations of these types of inflorescences are found in many plants. A spadix is an inflorescence of the spike form with elongated axis, sessile flowers, and enveloping leaf, the spathe.

A flower of *Hippeastrum,* a member of the Amaryllis Family, showing three-parted stigma and the six stamens characteristic of the lily order, surrounded by perianth of six parts arranged in two whorls.

A flower consists of an axis, called the receptacle or torus, and, attached thereto, the pistils, the stamens, the petals, and the sepals. The pistils and stamens are the essential organs of the flower, the petals and sepals are accessory organs. Any flower which has all four organs is a complete flower, while that which lacks one or more

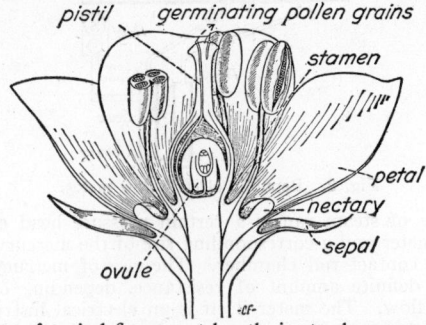

Diagram of typical flower, cut lengthwise to show arrangement of parts. Pollination has occurred. Pollen tube extends toward embryo sac within the ovary.

is incomplete. If the organs missing be either stamens or pistils, the flower is imperfect or unisexual. A perfect or bisexual flower has both sets of essential organs. If only the stamens are present the flower is staminate; if only pistils, it is pistillate. If the two kinds of flowers, staminate and pistillate, are found on the same plant, that species of plant is monoecious. When the two unisexual flowers are found on different plants, the species is dioecious. Infrequently flowers are borne which lack both stamens and pistils, and are sterile. When the flower lacks sepals and petals, it is a naked flower.

Flower of Saint John's-wort (polypetalous). The organs of this flower are unusually well exposed to view.

Flowers may also be distinguished as regular and irregular. Regular flowers are those in which all the members of any set of organs are alike, forming a flower which is radially symmetrical or actinomorphic. Often the organs of one or more sets are not alike, forming an irregular flower. An irregular flower may be bilaterally symmetrical or zygomorphic, one half being a mirror image of the other.

The accessory floral organs, the sepals and petals together, constitute the perianth. In the complete flower, the perianth is composed of a whorl of sepals which is called the calyx and inside the calyx a whorl of petals called the corolla. The sepals are usually green and small. The function of the calyx seems to be to protect the other parts of the flower before the flower bud opens. The petals are usually thin and bright colored or white. The term corolla is applied to all the petals together. The corolla appears in a wide variety of shapes and colors. Its function seems to be to attract animals, especially insects, to the flower and so bring about **pollination**. The number of sepals and petals is constant for each species of plant. In the flowers of monocotyledons there are usually three of each, while in dicotyledons it varies from four to many, with five a very common number. As a further attraction to insects, many flowers possess special glands called nectaries which

Flower of pea (irregular).

Rosa carolina, a wild species, sometimes cultivated. Notice the large number of stamens and single whorl of petals.

secrete a sweet fluid, nectar. Usually these nectaries are situated at the bases of the petals, though they may be found in many other places in the flower. Nectaries known as extrafloral nectaries are found on petioles of leaves or on the stipules.

The stamens, or microsporophylls, taken together constitute the androecium. Usually a stamen consists of

An inflorescence of the flowering dogwood, *Cornus florida.*

two parts, a stalk or filament and an anther. The filament may be short and stout, or more commonly long and slender, raising the anthers well above the base of

the flower. The anther when first formed is an undifferentiated mass of cells. As it develops, four groups of cells become set off from the surrounding cells. In these masses, which usually appear as linear strands certain cells undergo **meiosis** and become microspores. The sac which contains them is therefore a microsporangium. A microspore develops into a pollen grain. The anther sac, or sporangium, when mature usually opens by two longitudinal slits, or by special spores, and frees the pollen grains. The number of stamens in a flower varies from one to many. Often there are vestigial stamens, or staminodia, present in the flower; in some plants these are large and brightly colored, in others they are small and inconspicuous.

The pistil is the central organ of the flower. A single pistil or several pistils, which may be separate or partly, or even completely, united, is called a gynoecium. A pistil is composed of one or more modified leaves called carpels or megasporophylls. When there are two or more somewhat united carpels, the pistil is called compound. A pistil is composed of a basal ovary (ovulary), a terminal stigma, and usually an intermediate style, which is often long and slender. The stigma is a receptive organ, the surface of which is often either sticky or hairy. It is to the surface of the stigma that the pollen grains are carried when **pollination** takes place.

The style may be very much elongated to lift the stigma above the other parts of the flower and so increase the probability of pollination. The ovary has one or more cavities, or loculi. In these are located the ovules, which will become the seeds. The ovules are attached to the wall of the ovary or to a central column by a small stalk called the funiculus, through which the developing ovule receives nourishment. That region of the ovary to which the ovules are attached is called the placenta. The number of ovules in a single loculus varies from one to many.

Each ovule first appears as a minute rounded projection on the wall of the ovary or the columella. In the early period of its development this projection consists of an undifferentiated mass of cells known as nucellar tissue. One or two layers of cells, known as the integu-

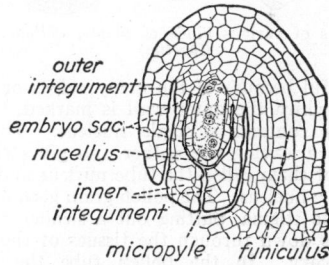

outer integument
embryo sac
nucellus
inner integument
micropyle   funiculus

Mature megagametophyte of lily within an ovule. (*Textbook of General Botany, Third Edition,* by Holman & Robbins, John Wiley & Sons, Inc.)

ments, rise from the base of the projection and finally almost completely surround the nucellar tissue. A minute opening through the integuments, called the micropyle, is left connecting the cavity of the ovary to the surface of the nucellar tissue. Within the nucellar tissue a very important series of **cell divisions** has taken place. While there are many variations of the process as it occurs in different species, the process is essentially as follows. Within the mass of nucellar tissue a single cell has become differentiated from all the others by its larger size and denser cytoplasmic content. This is the megaspore mother cell. It divides twice in rapid succession to form a row of four cells. These two rapidly succeeding cell divisions take place in such a way that the four cells have the haploid or reduced number of **chromosomes**. The process is known as **meiosis**. Three of the four cells disintegrate and are lost. The fourth, or megaspore, is usually the one nearest the micropyle. It enlarges greatly, while by three successive divisions its nucleus divides to form eight nuclei, all contained within the wall of the very much enlarged female gametophyte, commonly called the embryo sac. The arrangement of these eight nuclei is quite uniform. In most plants there are four of them at each end of the embryo sac. One from each end moves to the center of the embryo sac, where they form an intimate association and eventually fuse. These two are the polar nuclei. Of the three nuclei which remain at the micropylar end of the embryo sac, one becomes larger than the other two. This is the egg nucleus or megagamete; the other two form cells which are called the synergids. The three nuclei at the opposite end of the embryo sac form cells called the antipodals. The mature embryo sac is thus a seven-celled body, with seven nuclei.

The pollen grain is carried to the stigma by various agents. The pollen grains of different species of plants are very characteristically shaped, and are often strik-

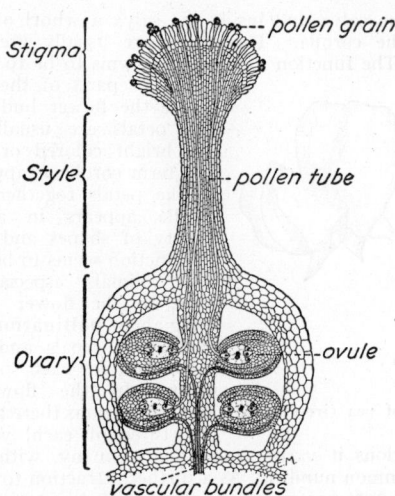

Stigma
Style
Ovary
pollen grain
pollen tube
ovule
vascular bundles

Longitudinal section of flower pistil showing pollen tubes growing through the style and entering the ovules. Diagrammatic.

Seeds with endosperm are known as albuminous seeds; those without endosperm as exalbuminous seeds.

The act of fertilization causes the immediate growth of the fertilized egg. In most plants a series of cell divisions takes place, forming a short filament of cells

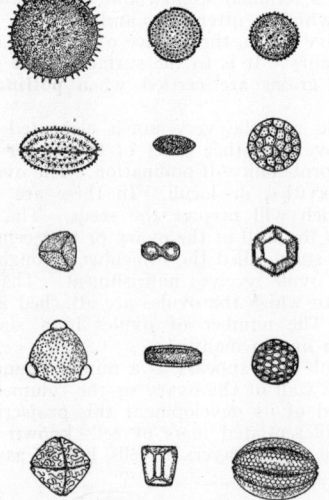

Pollen grains of various kinds of plants. (*Pope, Botanical Gazette.*)

ingly beautiful because of the many ridges or protuberances with which the outer wall is marked. At first a pollen grain contains a single nucleus. This nucleus divides before leaving the anther and gives rise to two nuclei, one of them called the tube nucleus and the other the generative nucleus. The pollen grain germinates when it reaches the stigma, putting out a slender pollen tube which grows down through the tissues of the style and into the ovary. In the pollen tube the generative nucleus divides to produce two sperm nuclei. There it grows towards an ovule, which it enters through the micropyle. The pollen tube continues to grow until its tip reaches the embryo sac. Into this the two sperm nuclei are discharged. One of them fuses with the egg nucleus of the embryo sac, while the other passes to the polar nuclei and fuses with them. The nucleus which is formed by the fusion of these three nuclei is called the primary endosperm nucleus; it contains three times the haploid chromosome complement. The act of fusion of the male nucleus with the egg nucleus is called fertilization. From the endosperm nucleus there is formed by repeated division and subsequent wall formation a mass of triploid tissue known as the endosperm, which surrounds the developing embryo. The endosperm nourishes the embryo during the early stages of its growth. In many plants such as the bean the endosperm is not formed at all or entirely absorbed by the developing embryo, while the seed is still immature; in others the endosperm forms a considerable part of the mature seed.

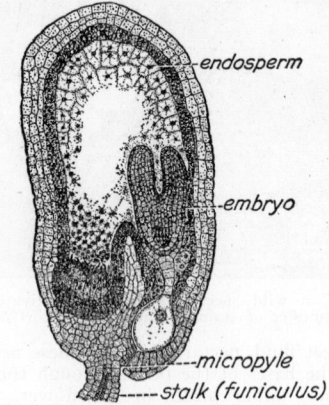

endosperm
embryo
micropyle
stalk (funiculus)

Ovule of shepherd's purse containing an embryo and endosperm. (*Chamberlain's Elements of Plant Science, McGraw-Hill Book Co., Inc.*)

which is called the proembryo. The appearance of the proembryo varies in different plants. The terminal cell of the proembryo becomes by repeated cell divisions a spherical mass of cells which is the beginning of the true embryo. The embryo grows rapidly and becomes differentiated into three regions, a primitive root, or radicle, a primitive shoot, and cotyledons. This embryo is surrounded by the tissues of the nucellus and the integuments, which have grown larger to form seed coats coincident with the growth of the embryo. The mature ovule becomes the seed, and the ovary which contains it becomes the fruit. (R.M.W.)

FLOWER PECKER. Aves, Passeriformes. Brightly colored birds (**Aves**) of the Oriental and Australian regions, related to the sun birds. (A.W.L.)

FLOWERING PLANTS. Angiosperms.

FLOW-STRUCTURE. A type of banding in effusive igneous rocks (lavas) due to the alignment of minerals, inclusions or gas cavities during the movement of the still molten but viscous material. It is not to be confused with foliation. (R.M.F.)

FLUE.

A flue is a channel for hot gases. The term is applied more to a channel for gases composing the products of combustion of a fuel than anything else. While frequently constructed of masonry or tile, or similar material, flues may also be sheet metal. Fire-tube boilers, in which the hot gases pass through the inside of instead of around the tubes, are said to have flues, since the tubes themselves are often so described. The chimneys which carry off the wastes of combustion from domestic fires are called flues, though flue refers more specifically to the space provided for passage of the gases, whereas chimney is inclusive, and one chimney may contain more than one flue, and may have architectural embellishments which are entirely independent of the flue. Boiler flues are generally seamless, welded, or riveted flues 5 to 18″ in diameter. Chimney flues should carry a lining of material which, by its smoothness, reduces friction loss, and by its refractoriness, is able to withstand continued heating and cooling without cracking or deteriorating in other ways, and which provides a continuous lining free from cracks which might cause a fire hazard by allowing hot gases to seep through the cracks in the masonry comprising a chimney. Flue-lining materials are fire-brick, plastic refractory cement, and hard burned terra cotta tile. (F.T.M.)

FLUE GAS.

Since the products of combustion of a fire are eventually led to the atmosphere through the flues of a chimney, the products of combustion are commonly referred to as flue gas. The fuel elements of all commonly used fuels are carbon and hydrogen. As air containing oxygen and nitrogen is supplied to make combustion possible, and as some excess air is generally needed, flue gas may be expected to be a composite gas composed of the following: Carbon dioxide, carbon monoxide, oxygen, nitrogen, and steam. There should be little or no carbon monoxide, for that is indicative of faulty or incomplete combustion. The oxygen content should be as low as it may be made consistent with maintenance of complete combustion. Since air is so largely composed of the inert gas nitrogen, the bulk of flue gas is also nitrogen. A typical analysis of dry flue gas by volume as produced by a coal fire, well tended, would be carbon dioxide, 12%; oxygen, 8%; nitrogen 80%. (F.T.M.)

FLUE GAS ANALYSIS.

While poor combustion of fuel is frequently detectable by the evolution of dense clouds of smoke, smokeless combustion may not necessarily be the best possible that may be obtained in the circumstances, since unnecessary quantities of excess air, or the product of incomplete combustion, carbon monoxide, may be present to render combustion inefficient, even though the flue gas is not badly colored. The flue-gas content is the fireman's best indication of actual combustion conditions. The combustion of coal or oil can be accurately gauged simply by the carbon dioxide content of the flue gas as long as the same fuel is employed. Consequently, the operating meter need record only carbon dioxide. However, for combustion test purposes, including performance runs for boiler efficiency, checking boiler operation, etc., complete analysis of the four principal gases in the products of combustion is required. For this analysis the Orsat apparatus is indispensable; indeed, the analysis is usually referred to as Orsat analysis. The apparatus conceived by that eminent French chemist analyzes a measured volume of a mixture of gases by the process of absorption. The volume remaining is measured, thus indicating by differences the gas absorbed. The remainder is then exposed to another reagent, which removes another gas. Upon remeasurement, the volume of that gas becomes known, and this process is continued a third time, so that the

percentages of carbon dioxide, carbon monoxide, and oxygen are determined. The remainder is assumed to be nitrogen, as it, in fact, is, for all practical purposes. The analysis obviously is one by volume, and on account of the construction and use of the apparatus, it is made at atmospheric pressure and temperature. (F.T.M.)

FLUID FLOW.

An ideal fluid is one which is not only continuous and homogeneous, but also incompressible. In addition, it should be inviscid, and should have physical properties which are unaffected by variations of temperature or pressure. A liquid resembles this ideal fluid more than does a gas, and consequently behaves with a great deal more regularity in fluid flow. A fluid flow may be steady or unsteady, depending on whether the flow passing any given cross-section of the conduit is constant. Furthermore, it may be uniform or non-uniform if, at different points along the flow, the velocity varies, due to changes of area or pressure. It also may be turbulent or laminar.

No fluid fulfills the requirements of the ideal fluid. All actual fluids have a certain amount of viscosity and compressibility. An essential of fluid flow is that continuity be satisfied by equal inflow and outflow from a certain region within a given period of time. A fluid flow is often thought of in the terms of streamlines, which are imaginary lines drawn in such a way that they are tangent to the direction of fluid motion. No quantity of the fluid may cross a streamline.

Elementary hydraulics always includes Bernoulli's law, and it is repeated here as being of great importance to the subject of fluid flow.

Bernoulli's law is:

$$\frac{p}{W} + \frac{v^2}{2g} + z = \text{a constant,}$$

and the symbols are defined as follows:

p = static pressure in lbs. per sq. ft.
W = specific weight of fluid in lbs. per cu. ft.
v = velocity in ft. per sec.
g = gravitational acceleration in ft. per sec. per sec.
z = potential, or "elevation," head in ft.

Bernoulli's law states that in steady flow, the total head is a constant at any point and equal to the sum of the pressure head $\left(\frac{p}{W}\right)$, the velocity head $\left(\frac{v^2}{2g}\right)$, and the potential head (z). Since there is actually a loss of head between any two points due to friction, the difference between the total heads at any two points must equal the friction head when the flow is steady.

In this equation, $\frac{v^2}{2g}$ represents the velocity head, a pressure which could be recovered by the efficient reduction of the velocity in a conduit of expanding cross-section. When this velocity head is multiplied by the specific weight of the fluid, it is reduced dimensionally to the unit of pressure, lbs. per sq. ft., and may be designated the dynamic pressure, in counter distinction to the static pressure p. In the case of an expansionable fluid, such as gas, the total energy at two points in the flow must be the same, that is, the heat energy plus the kinetic energy of motion must be constant. It is necessary to invoke this law of continuity of energy in dealing with the flow of gases or vapors through nozzles. (See Fluid Friction Orifice, Weir, Nozzle, Pitot Tube, Flow Meter, Aerodynamics, Boundary Layer.) (F.T.M.)

FLUID FRICTION.

The flow of any actual fluid must of necessity be attended by the presence of friction, due to the physical nature of fluids, none of which meets the requirements of the ideal fluid, as mentioned in fluid flow. A great deal of time and attention have been devoted to the study of the properties of a flowing fluid. The frictional effects present in the flow of liquids have been rationalized much more thoroughly than for vapors and gases. However, for all three the friction is found to depend upon the nature of the fluid itself, its **viscosity**, and upon the conduit which contains it. On account of the different molecular arrangement of liquids, vapors, and gases, the study of friction of fluids has become a specialized study of each of these three.

Fluid flow rarely follows the commonly accepted idea of **streamlines**, since the velocities necessary for viscous flow of this nature are almost always lower than those found expedient to employ. Most flows are turbulent in nature, and become turbulent at a definite velocity, the value of which was studied by Reynolds, and is incorporated in the well-known **Reynolds Criterion**. A general thermodynamic equation of energy of a fluid under flow conditions would be as follows:

Gain in kinetic energy+ gain in potential energy + net work received + energy liberated by any chemical change = change in heat content between two states.

In the case of a liquid, this equation can be considerably simplified:—in fact, it becomes **Bernoulli's** well-known equation—but in the case of compressible fluids, which may also undergo some change of form, such as condensation or compression, the longer equation applies. Most practical problems in fluid friction arise in connection with the flow of fluid through pipes, and for an extended discussion, including formulae, the reader is referred to **pipe friction**. (F.T.M.)

FLUID MECHANICS. Mechanics.

FLUID SEALS. Fluids are occasionally required to be held under pressure in enclosed regions through the walls of which one or more moving shafts must protrude. Unless a seal is provided the fluid will leak through the clearance, left for mechanical reasons between shaft and container, if its pressure exceeds that of the surroundings, or it will be diluted by inflow if at a lower pressure. Such seals are not difficult to provide if the shaft has no motion other than axial rotation or reciprocation. Common examples of such seals are (1) the "stuffing boxes" of double-acting reciprocating compressors, pumps, and engines where the piston rod passes through the cylinder, and (2) the stuffing boxes and "glands" of centrifugal pumps and steam turbines where rotating shafts protrude from the casing. **Piston rings**, although not described here, could be called fluid seals, as could also the method of sealing off fuel gas in large telescoping gas holders ("gas tanks") with a liquid. There are several applications of the floating inverted bell as a gas holder using a liquid seal.

A stuffing box is a recess in the wall surrounding the point of exit of the shaft, arranged to receive a soft and pliant packing such as treated hemp, or leather, which is compressed in the box and pressed firmly against the shaft by the pressure exerted against it from a *gland*. The gland is tightened against the packing by screw threads until leakage is reduced to a negligible amount but not enough to produce seizure or excessive frictional heating of the shaft.

Stuffing boxes are used primarily to seal against leakage around small shafts. Leakage glands are preferred for large diameter shafting—or whenever no mechanical contact is wanted in the seal. A typical example is the *labyrinth gland* of the steam turbine. Here the steam is allowed to flow through the clearance space but is given so many turns that the fluid friction and vortices resulting from numerous changes of direction account for the pressure difference across the gland with only a small flow of the fluid through it.

The diagram shows an outflow of leakage steam. When a vacuum exists inside the casing there would be an air inflow through this gland. To prevent this the gland

may be designed for admission of high-pressure steam or water to its midpoint, from which there is two-way leakage flow to the edges. Thus water or condensable steam replaces air inflow. (F.T.M.)

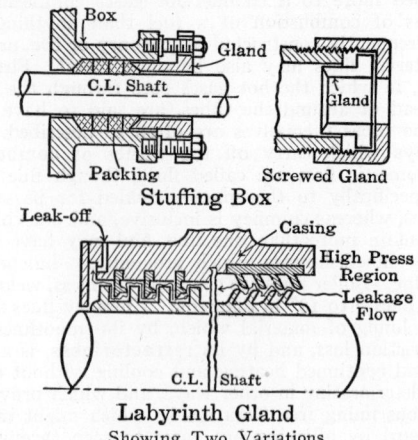

Stuffing Box

Labyrinth Gland
Showing Two Variations

FLUKE. 1. A parasitic flatworm. **Trematoda.** These worms attack many species of animals, passing through a complicated life cycle involving two or three hosts. The liver-fluke of the sheep, which spends part of its life in the body of a snail and becomes adult in the liver of the sheep, is a well-known example. A number of species attack man, particularly in the Orient and in warmer countries. They fall into four groups, the liver, lung, blood and intestinal flukes, depending on the part of the body in which they thrive. 2. The broad horizontal lobes of the whale's tail. 3. A fish, Paralichthys dentatus, belonging to the flatfish family. (A.W.L.)

FLUME. An open channel for conveying water for some special purpose, such as water power, washing, etc., is a flume. Flumes are frequently constructed of lumber having the boards placed in the direction parallel to the flow, these often being planed on the wetted side. However, flumes are also constructed of concrete, brick, etc. The flow in a flume is created by the slope of the bottom, releasing a certain amount of energy of position, which is converted into energy represented in the friction between the water and the flume, provided the flow in the flume is uniform. (F.T.M.)

FLUORSPAR. Fluorite.

FLUORESCEIN. Dyes.

FLUORESCENCE. Luminescence.

FLUORESCENT LAMP. Lamp, Electric.

FLUORINE. Symbol: F. Atomic number: 9. Atomic weight: 19.00. Density: 1.70 grams per liter, 0° C., 760 mm., or 1.31 when air equals 1.00. Melting point: −223° C. Boiling point: −187° C. No isotope, but of single atomic form: 19.

Fluorine is a pale yellow gas, poisonous, very reactive, combines with most other elements in the dark, except it does not combine with oxygen. Discovered by Scheele in 1771, but first isolated by Moissan in 1886, by **electrolysis** of fused potassium hydrogen fluoride in a **platinum** apparatus.

Flourine occurs as **calcium fluoride** (CaF_2) in the mineral **fluorite**, fluorspar, as sodium **aluminum** fluoride (Na_3AlF_6) in the mineral **cryolite** in Greenland, and with **calcium** phosphate as fluoride in the mineral **apatite**.

Acids: **hydrofluoric acid** (H_2F_2); hydrofluoboric acid (HBF_4) (see **Boron, Acids**); hydrofluosilicic acid (H_2SiF_6) (see **Silicon**).

Borofluoride: (See **Boron**.)

Bromide: Trifluorine bromide (F_3Br).

Fluoborate: (See **Boron**.)

Fluorides: **Sodium** fluoride (NaF), sodium hydrogen fluoride ($NaHF_2$), **potassium** fluoride (KF), potassium hydrogen fluoride (KHF_2), **ammonium** fluoride (NH_4F), ammonium hydrogen fluoride (NH_4HF_2), **silver** fluoride (AgF) are soluble fluorides; **calcium** fluoride (CaF_2), **strontium** fluoride (SrF_2), **barium** fluoride (BaF_2), **magnesium** fluoride (MgF_2) are insoluble fluorides. (See **Hydrofluoric Acid**.)

Fluosilicate: (See **Silicon**.)

Hydride: Hydrogen fluoride (H_2F_2), colorless gas, of marked odor, poisonous, melting point $-83°$ C., boiling point $19°$ C., very soluble in water yielding hydrofluoric acid. Formed by reaction of calcium fluoride and concentrated sulfuric acid upon heating. Hydrogen fluoride etches glass by reaction with silicate, forming volatile silicon tetrafluoride; forms hydrofluosilicic acid (see **Silicon**), hydrofluoboric acid (see **Boron**) and fluorides.

Iodide: Pentafluorine iodide (F_5I).

Silicofluoride: (See **Silicon**.)

Organic compounds: Organic fluorine compounds are made by reaction of the corresponding paraffin chlorocompounds with silver fluoride, mercurous fluoride, antimony trifluoride, titanium tetrafluoride, and the benzenoid fluoro-compounds by the diazo-reaction using hydrofluoric acid.

namental purposes; its softness, however, has been a bar to its general use. It is found also in Cumberland, England, Saxony, Bavaria, Baden, Austria, Czechoslovakia, Norway, Switzerland, and Italy. Colorless, transparent fluorite was formerly mined at Madoc, Province of Ontario, Canada; and in the United States has been found at Westmoreland and Chatham, N. H.; Trumbull, Conn.; Jefferson and St. Lawrence Counties, N. Y.; at Phoenixville, Pa.; Amelia Court House, Va.; and in commercial quantities in Kentucky and Illinois. It is used as a flux in the manufacture of **steel**, in making opalescent **glass**, **enamels** for cooking utensils, and for hydrofluoric acid. Its rare use for ornaments is due to its softness, above referred to. The name fluorite is derived from the Latin *fluo,* flow, in reference to its use as a flux. (E.S.C.S.)

FLUOROSCOPE. This device consists of a fluorescent screen mounted on one wall of a dark box having a hooded opening opposite to it into which an observer may look. When **x-rays** or other exciting radiations fall on the screen it glows brightly. The fluoroscope is generally used to observe x-ray shadows cast by parts of the human body or by other objects and, therefore, serves as a convenient means of making preliminary x-ray examinations. A material commonly used for the screen is barium platino-cyanide. (See **Luminescence**.) (L.D.W.)

FLUTTER. Oscillation of definite period but unstable character set up in any part of an aircraft by a momen-

REPRESENTATIVE ORGANIC COMPOUNDS OF FLUORINE

			BOILING POINTS
1.	Methyl fluoride (Fluoromethane)	CH_3F	$-78°$ C.
2.	Ethyl fluoride	C_2H_5F	-32
3.	Normal-propyl fluoride (1-fluoropropane)	$C_2H_5 \cdot CH_2F$	-3
4.	Iso-propyl fluoride (2-fluoropropane)	$CH_3 \cdot CHF \cdot CH_3$	-11
5.	Allyl fluoride	$CH_2:CH \cdot CH_2F$	
6.	Benzoyl fluoride	$C_6H_5 \cdot COF$	161 (745 mm.)
7.	Fluorobenzene (Phenyl fluoride)	$C_6H_5 \cdot F$	85
8.	Ortho-fluorobenzoic acid	$C_6H_4(F)(2)(COOH)(1)$	121
9.	Meta-fluorobenzoic acid	$C_6H_4(F)(3)(COOH)(1)$	124
10.	Para-fluorobenzoic acid	$C_6H_4(F)(4)(COOH)(1)$	183
11.	Fluorodichloromethane	$CHFCl_2$	14
12.	Fluorotrichloromethane	$CFCl_3$	25
13.	Difluorodichloromethane ("freon")	CF_2Cl_2	-29
14.	Para-fluorobromobenzene	$C_6H_4F(4)Br(1)$	
15.	Para-fluoroiodobenzene	$C_6H_4F(4)I(1)$	
16.	Fluoroform	CHF_3	20 (40 atm.)
17.	Benzotrifluoride	$C_6H_5 \cdot CF_3$	
18.	Carbon tetrafluoride	CF_4	-128

(R.K.S.)

FLUORITE. The mineral fluorite, **calcium** fluoride, CaF_2 is an isometric mineral with cubic habit, although **octahedrons** and **dodecahedrons** are not uncommon. Penetration twins are frequent. Fluorite may be also massive, granular to compact. It has a very perfect octahedral cleavage; brittle; hardness, 4; specific gravity, 3.01–3.25; luster, vitreous; color may be white or colorless, blue, blue-green, yellow, brownish-yellow or red. The blue kinds are often a delicate violet-blue, sometimes amythestine in tint. The streak is white; translucent to transparent. Certain specimens appear blue by reflected, green or yellow by transmitted light. Fluorite sometimes **phosphoresces** when heated or scratched, other varieties **fluoresce** beautifully under the influence of **x-rays** or **ultra-violet** light. Fluorite may occur as a vein deposit especially with the metallic minerals where it often forms a part of the gangue and may be associated with **barite, quartz** and **calcite**. It is a common mineral in the deposits of **pneumatolytic** origin and has been noted as a primary mineral in granites and similar rocks. One of the most famous of the older localities is Derbyshire, England, where under the name of Derbyshire "blue john" beautiful blue fluorite is used for or-

tary disturbance, and maintained by a combination of aerodynamic, inertial, and elastic characteristics of the member itself (example, buffeting). (See **Wing, Airplane**.) (F.T.M.)

FLUVIAL. Derived from the Latin, meaning river, and used by geologists and physiographers to denote a river as the agent, i.e., fluvial sediments. (R.M.F.)

FLUX. One use of the term flux is to designate a material which by its chemical action facilitates the **soldering** and **brazing** of metals. Such a flux applied to a metallic surface cleans it and renders it receptive to amalgamation with the solder or brazing metal. Some fluxes are **resin**, for soldering tin; muriatic acid, for galvanized **iron** and other **zinc** surfaces; and **borax**, for brazing.

A related use of the term flux is to designate the material added to the contents of a smelting furnace or a **cupola** for the purpose of purging the metal of impurities, and of rendering the slag more liquid. The flux most commonly used in iron and steel furnaces is limestone, which is charged in the proper proportions with the iron and fuel. The slag is a liquid mixture of ash, flux, and other impurities.

Magnetic flux is a term used in magnetism. A **magnetomotive force** will cause magnetic lines of force through a magnetic circuit. It is similar to a current flow created by voltage in an electric circuit. The magnetic flux is the number of lines of magnetic force set up in a magnetic substance, and thus flux becomes analogous to the current flowing in an electric circuit. The magnetic flux is numerically equal to the driving force called magnetomotive force, divided by the **reluctance** of the circuit, which is a quality analogous to the resistance of an electric circuit. The unit of the magnetic flux is the maxwell, and flux density is measured in units of maxwells per sq. cm., or gausses. (F.T.M.)

FLUX REFRACTION. This term refers to the fact that when a ferromagnetic body composed of two pieces of different magnetic **permeability** is placed in a magnetic field, or when a **dielectric** composed of two adjacent portions of different dielectric constant is placed in an electric field, the lines of magnetic induction in the former case, and the lines of electric displacement in the latter, if oblique to the interface, abruptly change their direction. The phenomenon is thus somewhat analogous to the **refraction** of light. But the law is different. Whereas, in the case of light the ratio of the sines of the angles of incidence and refraction is constant, in the case of flux refraction it is the ratio of the tangents of the angles that is constant.

For an electric current flowing across a boundary between two conductors of different electrical resistivity, there is a refraction of the lines of flow, likewise obeying the tangent law. (L.D.W.)

FLY. A 2-winged insect belonging to the order **Diptera.** Also commonly applied with some qualifying word to many flying insects with membranous wings, such as May fly, dragon fly, stone fly, and caddis fly. These four examples belong to as many different orders. (A.W.L.)

FLY CUTTER. Milling.

FLYCATCHER. Aves, Passeriformes. Insect-eating birds (**Aves**) of many species and world-wide distribution. The North American species belong to the family Tyrannidae and include the **kingbirds,** the wood **pewee,** and the **phoebes,** in addition to the several species named as flycatchers. (A.W.L.)

FLYING BOAT. Seaplane.

FLYING FOX. Bat.

FLYING LEMUR Mammalia, Insectivora. An arboreal Malayan animal, *Galeopterus temminckii,* of nocturnal habits. It glides from tree to tree by means of folds of skin stretching between the fore and hind legs and thence to the tail. A similar species, *Galeopithecus volans,* lives in the Philippine Islands. Both eat leaves and fruits. Not a true lemur, hence the names cobego and kaguan by which these animals are also known are less misleading. (A.W.L.)

FLYING SQUIRREL. Mammalia, Rodentia. Small arboreal animals of the northern hemisphere which glide

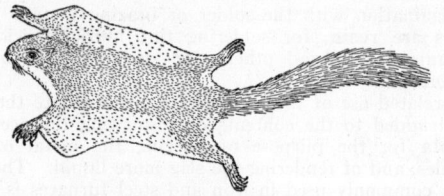

Flying squirrel.

by means of skin folds along the sides of the body from front to hind legs. They belong to the genus *Sciuropterus*. Their appearance is much like that of squirrels but they are as closely related to the ground squirrels and gophers. The African flying squirrels belong to a different family, the Aromaluridae. (A.W.L.)

FLYWHEEL EFFECT. Since the main purpose of applying a flywheel to any machine is to steady the speed, any similar action or arrangement of apparatus to produce such action might be called flywheel effect. The ability of a flywheel to steady speed lies in its capacity to absorb and release energy with small variations in speed. Now since the **kinetic energy** contained by a rotating flywheel is $\frac{1}{2}I\omega^2$, I being the **moment of inertia** of the **mass** about the center of rotation, ω being the angular velocity in radian units, energy will be absorbed when ω changes slightly only upon the condition that I be a large quantity. Consequently, flywheels are characterized by large moments of inertia. Mass alone is no criterion of flywheel effect, since moment of inertia involves as well the disposition of the mass; in other words, the shape of the body. Hence flywheels are not only massive, but the mass is placed as far as practicable from the center of rotation, as in a heavy rim. Flywheel effect, then, is to be secured wherever large masses having large amounts of inertia about the center of rotation are steadying rotational motion in the face of uneven power impulses by the absorption or release of kinetic energy by slight changes of angular rotation.

In the hydraulic turbine field, flywheel effect has a very definite and special meaning. The flywheel effect of a hydraulic turbine is the weight of its rotating element, multiplied by the square of the **radius of gyration** of the same. (F.T.M.)

FOAMING. Feed-Water Treatment.

FOCAL INFECTION. A localized infection. Formerly, it was thought that such infections might produce symptoms in other parts of the body by absorption of "toxins" liberated by bacteria. It is now believed that such occurrence is exceedingly rare. The most common sites of focal infection are the tonsils, teeth, paranasal sinuses, genito-urinary tract, appendix and gall bladder. (D.M.H.)

FOCAL LENGTH. Mirrors and Lenses.

FOCAL POWER. Mirrors and Lenses.

FOCI OF A CONIC. Conic Sections.

FOEHN WIND. On the lee side of mountains air flowing downhill dry-adiabatically with attendant heating. In western North America along the slopes of the Rockies, foehn winds are called **Chinooks.** (P.E.K.)

FOG. Condensation and consequent formation of water droplets (or ice crystals) in the air at the earth's surface will produce a fog. Fogs are classified in many ways. One of the simplest is the use of formation cause or process as a basis for differentiation among the various types.

1. Advection fogs are fogs that owe their existence to the flow of air from one type of surface to another. Surface temperature contrast between two adjacent regions is necessary in causing the formation of advection fogs.

a. The usual type of advection fog is formed when relatively warm and moist air drifts over much colder land or water surfaces. Examples of this type are found over land when moist air drifts over snow covered areas, or over water when moist warm air drifts over currents of very cold water. The latter happens with southerly or easterly winds blowing from the Gulf Stream over the Labrador Current.

b. Coastal and lake advection fog forms when warm and moist air flows offshore onto cold water (summer), or when warm moist air flows onshore over cold or snow covered land (winter).

c. Sea smoke, arctic fog, or steam fog forms in very cold air when it flows over warm water.

2. Radiation fog is a type that develops in nocturnally cooled air in contact with cool surface. Radiation fog forms over land and not over water because water surfaces do not appreciably change their temperature during hours of darkness.

3. Upslope fog is caused by dynamic cooling in air flowing uphill. Upslope fog will form only in air that is convectively stable, never in air that is unstable, because instability permits the formation of cumulus clouds and vertical currents.

4. Precipitation fog forms in layers of air which are cooler than the precipitation which is falling through them. The greater the temperature difference between relatively warm rain (or snow) and the colder air layer, the more rapidly will the fog develop. Fogs associated with fronts are largely precipitation fogs.

Visibility in fogs varies from a few feet up to a mile. Often in a fog blanket all ranges of visibility are present. (P.E.K.)

FOG DENSITY. The density of an unexposed portion of a sensitometric strip. The density due to fog is not the same in the exposed steps as in the unexposed portion of the sensitometric strip. The relation between the fog density and the total density is complicated, but in general it may be said that, as the total density increases, the fog density becomes less and less. This is due to the fact that the by-products released in the development of the exposed silver halide restrain the development of fog. Hence, in the higher densities, the development of fog is restrained to a greater degree than in the lower densities. (C.B.N.)

FOLIATION. Metamorphism.

FOLLICULIN. Sex Hormones.

FOLIUM OF DESCARTES. This is a plane curve represented by the **parametric equations**

$$x = \frac{3at}{1 + t^3}, \quad y = \frac{3at^2}{1 + t^3};$$

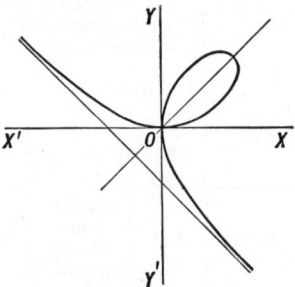

Folium of Descartes.

its equation in **rectangular coordinates** is

$$x^3 + y^3 - 3axy = 0.$$

It is shown in the accompanying figure. (L.L.S.)

FOLLICLE. In zoology, a glandular cavity or sac. The small spherical chambers in the thyroid gland, the fluid-filled cavity in the vertebrate ovary in which the ovum develops, and the chambers of the **arthropod** testis are examples of the application of the term. For the use of the term follicle in botany, see **Fruit**. For its detailed physiological discussion, see **Menstruation**. (A.W.L.)

FOLLICLE CELL. Cells of the ovarian **follicle** which nourish the ovum during its development. They occur in the Graafian follicles of vertebrates and in the open follicles of invertebrates. In the insects specialized cells of similar function contained within the follicle are termed nurse cells. (A.W.L.)

FOLLOWER. Cam; Milling; Lathe.

FOMALHAUT. Fomalhaut (Pisces Australis) is interesting as the most southern first-**magnitude** star visible from the latitude of New York. It rises almost simultaneously with **Capella** and is above the southern horizon during the evening in the autumn months. It was one of the Royal stars of **astrology**, ruling over the southern sky. The astrologers believe the star to portend eminence, power, and fortune to those born under its influence. (W.K.G.)

FONTANELLE. Any one of the places in an infant's skull that have not become bony. These places are usually situated at the junction of different bones forming the dome of the skull. At the age of 1 year they are normally filled in with bone. Fontanelles also occur in the skulls of other vertebrates, in their immature form. (R.S.M., A.W.L.)

FOOD. The raw material taken into the body of a living organism as a source of energy and material for the renewal of its own structure. The food of animals is chiefly organic, consisting of plant or animal tissue, either dead or alive, or waste products. The food must contain **protein** or another source of **nitrogen**. Animals can live on this substance alone but not on **fats** or **carbohydrates** without a nitrogenous compound to supplement them.

Food materials for human use should be (1) digestible, furnishing materials and energy adequate for body functions, (2) pure, not furnishing materials deleterious or capable of becoming deleterious to body functions, (3) palatable, of agreeable odor and taste, (4) such that a proper amount of indigestible roughage and bulk are furnished.

The first and fourth criteria, those of digestibility and indigestibility, are related directly to the composition of foods as such, and the changes undergone by foods in the body organs. The purity and palatability are largely affected by conditions outside the body, as by the methods used in gathering, storing and preserving, and in preparation for use. A satisfactory diet must supply sufficient amounts and kinds of various food materials to maintain health, and supply sufficient energy. It has been demonstrated that certain accessory materials such as the **vitamins** must be present in foods or must be supplied otherwise, in order to maintain health and prevent disease.

Flesh of terrestrial and marine animals, milk, fruits, and cereal grains seem to have been the principal diet of primitive men. One writer states that all civilizations have been built around some cereal grain as the principal source of food, and cites rice in Asia, wheat and rye in Europe, millet and sorghum in Africa, and Indian corn in America. In the United States, the average distribution of foods consumed has been estimated as meat and dairy products 37%, fish 2%, cereals 31%, Irish and sweet potatoes 13%, other vegetables 8%, miscellaneous foods 9%.

The non-skeletal portion—muscles and vital organs—of the human body are made up of about ¾ water and ¼ solids. These solids consist of about ¾ protein. Foods, therefore, must supply ample amounts and varieties of proteins. (See **Aminoacids, Polypeptides, and Proteins** for further data on proteins and food composition.) The sources are of animal or vegetable origin

The former include meats, poultry, fish, eggs, milk and cheese. The solid matter of all of these except fish and milk is approximately ½ protein and ½ fat (about ⅔–¾ of the total weight is water). In fish the ratio of protein to fat is frequently considerably greater than the above. Milk contains about 3.3% protein, 4% fat, 5% lactose and the remainder water. The milk product, cheese, contains protein, with fat varying from all in the milk to practically none, depending upon the quality of milk used.

FOODS

MEATS, POULTRY, FISH, AND EGGS

(See Aminoacids, Polypeptides, and Proteins; Esters (for fats))

Food—Edible Portion	Protein	Fat	Ash
Beef.	18%	13–22%	0.9%
Pork.	9	55	0.5
Lamb.	18	23	1.1
Fowls.	19	16	1.0
Salmon.	22	13	1.4
Halibut.	19	5	1.0
Eggs.	13	10	1.0

MILK AND CHEESE

Food	Protein	Fat	Sugar	Ash
Milk.	3.3% (3–6)	4% (3–4)	5%	0.7
Cheese.	18–30	27–37	. . . .	. . . .

NUTS

Food—Edible Portion	Pro-tein	Fat	Carbo-hydrates Non-cellulose	Cellu-lose (fiber)	Ash
Peanuts.	25.8%	38.6%	22%	2.5%	2.0%
Almonds.	21.0	54.9	15	2.0	2.0
Walnuts.	18.4	64.4	13	1.4	1.7
Brazil nuts.	17.0	66.8	7	. . . . . .	3.9
Filberts.	15.6	65.3	13	. . . . . .	2.4
Pecans.	9.6	70.5	15	. . . . . .	1.9
Chestnuts.	6.2	5.4	40	1.8	1.3
Coconuts.	5.7	50.6	28	. . . . . .	1.7

CHOCOLATE AND COCOA

Food	Protein	Fat	Carbo-hydrates	Ash
Chocolate.	13%	48%	30%	2.2%
Cocoa.	22	29	38	7.2

LEGUME

Food	Protein	Fat	Starch (Approx.)	Fiber	Ash
Beans, dried. .	25%	1–2%	55%	4%	4%

CEREALS

Food	Protein	Fat	Starch (Approx.)	Fiber	Ash
Wheat, flour, entire.	12%	2%	73%	. . . . . .	1%
Corn, kernel. .	10	4	72	2%	1.5
Oats, rolled. . .	17	7	66	1	2
Rye, kernel. . .	12	1.5	71	2	2
Barley, kernel	11	2.2	73	. . . . . .	2.5
Rice, cured. . .	8	2.0	76	1	1
Rice, polished	7	0.3	79	0.5	0.5

TUBERS

Food	Protein	Fat	Starch	Starch plus Sugar	Ash
Potatoes, Irish	2%	1%	15%	. . . . . .	1%
Potatoes, sweet	1.5	0.5	. . . .	22%	1

SUGARS AND STARCH

Food	Starch	Sugars	Dextrin
Sucrose (cane and beet sugar)	. . . . . .	100%	. . . .
Glucose (corn syrup).	. . . .	38.5	42.0%
Lactose.		100	. . . .
Maltose.		100	. . . .
Starch.	100%		. . . .

FATS AND OILS. (See Esters)

Food

Butter, average. . . . . Fat 84%.

Lard and suet, rendered. . . . . Fat 100%.

Oleomargarine
Shortenings, prepared } Composed of various mixtures of oleo oil of beef, neutral lard, cottonseed oil, cocoanut oil, peanut oil, hydrogenated vegetable oils. Fat 100%, except oleomargarine.

Olive oil
Corn oil } . . . . . Fat 100%
Cottonseed oil

FRUITS

Food—Edible Portion	Protein	Fat	Sugar plus Starch (approx.)	Fiber	Ash
Tomatoes. . .	0.9%	0.4%	33%	0.6%	0.5%
Oranges.	0.8	0.2	11	. . . . . .	0.5
Lemons.	1.0	0.7	7	1.1	0.5
Pineapples. . .	0.4	0.3	9	0.4	0.3
Bananas.	1.3	0.6	21	1.0	0.8
Apples.	0.4	0.5	13	1.2	0.3
Pears.	0.6	0.5	11	2.7	0.4
Peaches.	0.7	0.1	6	3.6	0.4
Plums.	1.0	. . . . . .	20	. . . . . .	0.5
Prunes.	0.9	. . . . . .	19	. . . . . .	0.6
Grapes.	1.3	1.6	15	4.3	0.5
Berries.	0.6–1.3	0.6–1.0	7–16	1.5–2.5	0.3–0.6
Squash.	1.4	0.5	8	1	1

Fruits also contain, in distinction to other foods, (1) organic acids, (2) tannins, both in varying percentages in different fruits and at various stages of ripeness for the individual fruit.

VEGETABLES

Food—Edible Portion	Protein	Fat	Sugar plus Starch (approx.)	Fiber	Ash
Carrots......	1.1%	0.5	8%	1%	1%
Beets.......	1.5	0.1	9	1	1
Parsnips.....	1.6	0.5	11	2	1
Turnips.....	1.3	0.2	7	1	1
Onions......	1.5	0.3	9	1	0.6
Cabbage.....	1.5	0.3	5	1	1
Cauliflower..	1.8	0.5	4	1	0.7
Lettuce......	1.2	0.3	2	1	1
Spinach.....	2.1	0.3	2	1	2

FOOD ADJUNCTS

Water (See **Water**)

Coffee ⎰ Contain caffeine, tannin and volatile oil (caffeol
Tea   ⎱   of coffee contributes the characteristic aroma)

Carbonated beverages

Alcoholic beverages

Salts

Spices

Vinegar

Flavoring extracts

Sugar, starch, water, it may be remarked, are the only pure individual chemicals of the 70-odd substances mentioned above. The others are mixtures.

Fats and carbohydrates furnish only three chemical elements, namely, carbon, hydrogen, oxygen, to the body, and proteins only these three and nitrogen and sulfur. The remaining elements required by the animal organism (see **Biochemistry**) are in the ash when food materials are burned in air. Study of the ash gives no information as to the *form* of chemical existence of the contained elements present in the food before burning, but simply proves their presence.

Calcium is so important for the growth of bones that special attention should be paid not only to its supply but to proper conditions for its assimilation. The proper assimilation of calcium and phosphorus are dependent upon the presence of vitamin D (see **Vitamins**) in the diet or upon sunlight.

Iodine is necessary for the proper functioning of the thyroid gland and iron for the haemoglobin of the blood.

ENERGY-PRODUCING VALUE OF VARIOUS FOODS

Carbohydrates:

Glucose	3.75 calories per gram
Sucrose	3.96
Starch	4.22
Glycogen	4.22

About 98% absorbed, so that the average physiological fuel value is about 4 calories per gram.

Fats:

Butter fat	9.2 calories per gram
Animal fats	9.4
Vegetable fats	9.4

About 95% absorbed, so that the average physiological fuel value is about 9 calories per gram.

Proteins:

Casein	5.85 calories per gram
Albumin	5.80
Gelatin	5.30
Gliadin	5.74
Legumin	5.62
Edestin	5.64

About 92% absorbed. Deducting the fuel value of the nitrogenous materials eliminated in the urine, e.g., urea 2.53 calories per gram (or 0.9 calorie per gram protein), creatinine, uric acid, the average net physiological fuel value is about 4 calories per gram.

The chemical analysis of a foodstuff does not throw light upon certain aspects of its dietary properties. The use of biological methods is necessary to ascertain the effectiveness of foods for the maintenance of growth in the young, health at all ages, and resistance to certain diseases. (See **Vitamins**.)

In milk, vitamins A, B, C, D occur in fairly well-balanced proportions. A quart of milk each day for every child and a pint for every adult, including that used in cookery, is the minimum amount recommended. Weight for weight, egg yolk contains 10 times as much vitamin A and 2 times as much vitamin B as milk. Nuts, meats and cereals are low in vitamins A and C. Fish liver oils are desirable to supply vitamin D, but sunlight, milk or egg yolk compensate.

A varied diet supplying energy foods—sugar, starch, fats, protein—supplemented by protective foods—milk, eggs, fresh fruits, green vegetables—appears to furnish all the dietary demands. The table on page 600 presents a summary of some important data regarding foods:

The treatment which foods undergo before being used is of great importance from the points of view of digestibility and palatability. Starch-containing foods are generally heated to render them edible. The treatments may be classified as follows:

1. Cooking
   Boiling, grilling, baking, frying
2. Preservation
   Cold storage, freezing, drying and evaporating, canning, preserving with sugar, vinegar, brine, smoking
3. Bleaching and Coloring
4. Flavoring

On account of the widespread adoption of refrigeration, data on the proper storage temperature of various foods is given.

PROPER STORAGE TEMPERATURE FOR VARIOUS FOODS

Meats, fresh	29–33° F.
Fowls	26–30
Fish, fresh	20–28
Canned meats	35–40
Eggs	31
Milk	35
Cheese	34
Butter	18–20
Lard	38
Oranges	34–45
Lemons	33–45
Bananas	34
Apples	32–36
Pears	34–36
Peaches	34–36
Canned fruits	35–40
Grapes	34–36
Cantaloupes	40
Berries	36
Potatoes	36–40
Carrots	34–35
Parsnips	34–35
Onions	36
Cabbage	34–35
Flour	36–40
Nuts, dried	35–40

For list of approved food colors, see Dyes. (R.K.S.)

SUMMARY OF SOME IMPORTANT DATA REGARDING FOODS

(√ MEANS AN EXCEL-LENT SOURCE)	PRO-TEINS	FATS	CARBO-HYDRATES	ASH	VITAMINS		
					A Anti-xerophthalmia	B Anti-neuritic	C Anti-scorbutic (Relative value)
Meats, poultry, fish	√	√	....	1%	Low	√	Negligible
Eggs.............	√	√		1	√	√	Negligible
Milk.............	√	√	(Lactose)	0.7	√	√	√
Cheese............	√	√			√	√	
Nuts.............	√	√		√			
Legumes..........	√		√(Starch)	4	Low	High	70 (Green, raw)
Cereals...........	√	0.3–7	√(Starch)	2	Negligible	Degermed—Neg. Germ—High	Negligible
Potatoes, Irish.....			√(Starch)		Low	√	20 (cooked)
Potatoes, sweet.....			√(Starch, sugar)		√		
Sucrose...........			√(Sugar, 100%)		None	None	None
Starch...........			√(Starch,100%)		None	None	None
Fats and oils:							
animal..........		√100%	....		Butter, cod-liver oil, high; others low		
vegetable........		√100%					
Fruits (contain also organic acids):							
citrus...........			(Sugar)	0.5		√	100
tomatoes........				0.5	√	√	100 (raw, canned)
pineapples.......			(Sugar)		0.3	....	70
bananas........			(Sugar, starch)	0.8		√	20 raw
apples.........			(Sugar)	0.3		√	20 raw
grapes........			(Sugar)	0.5			5
Vegetables:							
green...........			....	1–2			
white...........			√	1			
roots...........			2				
onions..........				1		√	100 (raw)
cabbage........						√	100 (raw) 5–30 (cooked)
turnip..........						√	70 (juice)
carrots..........					√	√	20 (raw)
spinach.........					√	√	20 (cooked)

FOOT. 1. The ventral protuberance of the body of a **mollusk**, usually an organ of locomotion. 2. The terminal portion of a jointed appendage which comes into contact with the supporting surface. In quadrupedal **vertebrates** the term is applied to both fore and hind appendages and in bipedal forms to the latter only. (A.W.L.)

FOOT CANDLE. A unit of illumination. The illumination in foot candles may be computed from

$$L = C/d^2$$

where L is the illumination in foot candles, C is the candle power of the source and d is the distance in feet from the source. (L.R.Q.)

FOOTINGS. A footing is a **foundation** used to distribute concentrated loads in walls or **columns** over a suitable area of soil or subsoil. Footings must be spread wider than the base of a column or wall for a number of reasons. They must give stability by reducing the unit pressure below that at which there might be local settlement along one side of the foundation. They must be thick enough to resist the punching shear of the column, and they must not flare so rapidly as to cause them to be weak in bending. Footing for walls, columns, etc. may be very conveniently and economically made of reinforced **concrete**. Except for very heavy loads, the footing slab is usually square or rectangular and of constant thickness. Design is usually made by semi-

rational rules and formulae since the distribution of stress in such a structural unit is very complex. For very heavy column loads the footing is usually made in the form

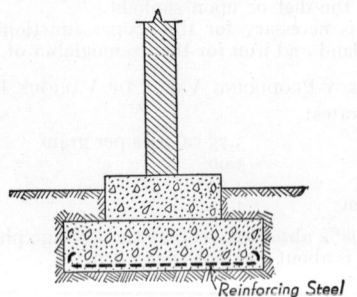

Section—Spread Footing

of a truncated pyramid or preferably in a series of steps giving the "stepped footing." The latter type, illustrated in the figure, predominates, due to the ease with which forms may be constructed. In case the soil is so weak as to require an extremely wide footing, a steel **I-beam** grillage is incorporated in the footing. (F.T.M., E.W.S.)

FOOT POUND. Work.

FOOTWALL. Fault.

FORAMEN. An opening, especially in bone, through which other structures pass, such as nerves and blood vessels. (A.W.L.)

FORAMINIFERA. Sarcodina; Invertebrate Paleontology.

FORBES LOG. Patent Log.

FORBIDDEN TRANSITIONS. Combination Principle.

FORCE. Our basic concept of force is that afforded by the muscular sense. We exert muscular effort upon objects and thereby produce certain effects; and when we observe similar effects produced by other means, we attribute them to the action of forces, whatever the ultimate nature of those agencies may be. Thus, if a strip of steel is clamped at one end and pulled sidewise at the other with the finger, it bends. A magnet placed near it will cause it to bend in the same way; and though we may know nothing of the nature of magnetic influence, we say that a force is acting upon the steel spring.

Force produces other effects than elastic deformation. **Newton's Laws of Dynamics** deal primarily with forces, and the first of his Laws points out that force, and force alone, can alter the motion of free bodies. He also recognized that the earth exerts forces upon all bodies near it, the measure of which is the weight of those bodies. Other examples are magnetic and electric interaction, cohesion, adhesion, and chemical affinity.

With so many objective manifestations, it becomes necessary to choose a suitable standard measure of force from among them. Elastic deformation and the sustaining of weight are actually used in familiar methods of force measurement, as with the spring balance and certain engineering testing machines. But these methods have the disadvantage of variability with time or place, since springs lose their elasticity and the intensity of gravity varies with latitude and altitude. A plan better adapted to accuracy is to accept Newton's second Law as the definition of our standard measure of force, and to agree that forces are to be considered proportional to the rates at which they cause free bodies to change their motion.

This "inertia measure" gives rise to a system of absolute units of force, recognized by Gauss and sometimes therefore called Gaussian units. Such units, for example, are the **dyne,** the **Newton,** and the less used poundal, each of which is defined as the force required to give some specified unit of mass a specified (linear) acceleration.

Expressed in these absolute units, the force required to give to the mass m an acceleration a is $f = m \cdot a$; this being the mathematical expression of the second Law in terms of these units. Thus, if a force of 1000 poundals were to be exerted upon a free mass of 500 lbs., the acceleration would be $a = f/m = 1000$ poundals/500 lbs. = 2 ft. per sec. per sec.

While the dyne and the poundal are thus conveniently related to problems of motion, and are invariable, they are awkward to apply in actual measurement. The units ordinarily used in practical engineering are gravitational, viz., simply the weights of suitable units of mass. Thus, tension in a truss rod is customarily expressed in pounds, the pound of force being the pull of gravity on a standard pound mass. While this pull differs by about one part in 160 between equator and poles, the variation with latitude is not important in engineering; and since gravitational units are so easily applied by simply balancing the unknown force against weights, they are in more general use than any others.

The **equilibrium of forces** is the subject-matter of **statics,** while the relation of forces to the motions of bodies is treated under **kinetics;** forces are also involved in the discussion of **work** and of **power.** (L.D.W.)

FORCE FIT. Fit.

FORCE PUMP. Air Pumps; Pumps.

FORCIPULATA. Asteroidea.

FORECAST OF WEATHER. Weather Forecasts.

FORE-SET BEDS. Delta.

FORESIGHT. Differential.

FORE-SKIN. The prepuce, a prolongation of the skin of the shaft of the **penis** or **clitoris,** covering the head of either of these organs. It is this fold of skin that is removed by **circumcision.** (R.S.M.)

FOREST. Wood.

FORESTER. Mammalia, Marsupialia. The great gray **kangaroo.** (A.W.L.)

FORGING. The shaping of metal by hammering or cogging. Only the malleable metals may be worked successfully. While most metals may be shaped in the cold state, the application of heat increases plasticity and permits greater deformation with any given expenditure of energy.

In usual forging practice, heated billets or stock are used. Extremely large billets or ingots are heated in soaking pits which are usually fired with gas or oil fuel and which often are designed to utilize regenerative heating principles. Billets or blooms of smaller size are heated in open muffle furnaces fired with gas or oil.

Stock forged by hand is generally small and this may be heated in a forge. A forge is a hearth made from fire-clay, firebrick, or tamped sand so arranged that a fire of coal, coke, or charcoal may be built upon it. Air is supplied from a blower through tuyeres, or openings, suitably placed. A hood above the forge removes the products of combustion from the forge room. In using a forge, the piece to be worked is buried in the hot coals of the thick carbonaceous fuel.

Care must be exercised in the heating of all metals for forging. Metals that are heated too long at too high a temperature oxidize rapidly, forming an excessive amount of scale which not only wastes metal but also prevents the production of smooth surfaces. Temperatures may be measured with thermocouples or with optical pyrometers and standard practices may be developed to give uniform results for any forgeable metal. Some metals are susceptible to carburization, decarburization, attack by sulfur gases, and intergranular oxidation. Furnace atmospheres during heating may be controlled to prevent any of the before-mentioned conditions. High sulfur fuel is a common source of sulfur contamination.

Many metals and alloys forge best in a given temperature range and those that may be worked only in a very narrow range of temperature are generally considered the most difficult to forge because this characteristic requires frequent reheating if the total deformation is great.

In addition to fabricating an identified shape, the operation of forging improves the quality of metals. The coarse crystals of metal resulting from solidification in an ingot mold are kneaded and refined. Blow holes and layers of slag are consolidated and usually welded together. This results in a more ductile and stronger product than cast metal with much greater resistance to shock and to fatigue stresses. Hammer forging imparts a higher degree of refinement on the surface of the work while pressed forgings exhibit better average quality throughout thick pieces.

Upsetting is the process of increasing the cross-section of stock at the expense of its length. Swaging or drawing-down operations increase length of stock at the expense of reduction in cross-section. Setting down is a localized swaging operation. Bending processes may be classified as angular or as curvilinear. Punching is an operation carried out to remove a slug of metal through shearing. Cutting-out is the process of cutting large holes by progressively using a hot chisel over a hole in the swage block. Saddening is a series of light shaping blows useful in preparing some metals for heavy forging operations.

Die blocks for power forging are made from carefully manufactured alloy steels. Such blocks must have the ability to withstand severe strains imposed by high pressures on hard metals. Long life with a minimum of impression wear is a prime requisite.

Power hammers are used for unit-production forging operations on work that cannot be feasibly hand-hammered, and also on comparatively small work where labor costs may be reduced by their use. There are two general types of hammers, steam or air hammers and trip and helve hammers. In the steam hammer, steam pressure is exerted on both faces of a vertically reciprocating piston, which is connected to the ram by a piston rod, to raise the ram and also to aid in striking the blow. The steam pressure on the downward stroke imparts additional velocity to the falling weight. The ram will therefore strike a blow whose full rating will be about double that of a gravity ram.

A trip hammer has a vertically reciprocating ram that is actuated by a toggle connection driven by a rotating shaft at the top of the hammer. The shaft is driven by a cone clutch which in turn is driven by a second shaft and pulley. The speed of the ram and the resultant effect of the blow are determined by the varying speed of the shaft. Trip hammers are built in sizes from 15 to 500 lbs.

Helve hammers are made in the same sizes as trip hammers and are used for similar classifications of work. The helve hammer consists of a horizontal wooden helve, pivoted at one end with a hammer at the other, and a cam or eccentric between the pivot and the hammer. The cam raises the hammer which falls and strikes a blow by the force of gravity. The action of a helve hammer is essentially that of a hand sledge since the wooden helve is somewhat elastic.

Light hammer blows are comparatively superficial in effect, while a heavy more slowly delivered blow penetrates and influences the material structure to a much greater extent. This effect is of particular importance in large forgings. Therefore, steam-actuated hydraulic presses are employed. Hydraulic presses for forging purposes range in size from 200 to 15,000 tons.

Drop forging is the process of shaping hot metal by forcing it into die cavities by the application of sudden blows. There are two principal forms of drop hammers that are used for drop-forging operations—the steam hammer and the board drop. The operation of the steam hammer has been described previously; the board drop has a hammer or ram fastened to maple boards whose upper ends pass between two rotating rolls. When the rolls are moved towards each other, they lift the board. At the top of the stroke the rolls automatically separate and release the board, permitting the ram to drop and strike the blow. Most board drop hammers are equipped with a device for changing the length of fall and consequently the force of the blow. Forging presses, in which the vertical ram is actuated by a pitman operated by an eccentric, are used for impact extrusion, hot forging, and hot and cold coining operations. These presses are available in capacities up to 2000 tons.

Forging rolls are often used for the breakdown operations preliminary to drop forging, and for reducing short thick sections to long slender sections. Their action is similar to that of rolling mills, but the forging rolls make use of only a portion of a revolution to reduce the stock; the remainder of the roll has a blank clearance space for the ends of the stock that are to be left full size. The operator stands at the back or emerging side of the rolls, and when the clearance space appears, he places the work into this space. When the reducing portion of the rolls comes in contact with the work, it reduces the stock and ejects it from the rolls towards the operator. At the next open portion of a revolution of the rolls, the stock is again inserted for a second reducing operation. By this method only as much of the length of stock is reduced in area as is required. Forging rolls contain a number of grooves, depending upon the number of passes required.

Machine forging, as distinguished from drop forging, is an upsetting or heading process applied to forgings made from bar stock. A forging machine consists essentially of three dies: a movable die opposed to a stationary die (the two are used for gripping the bar stock) and a third or header die which moves in a plane parallel to the parting surface between the clamping dies, and handles the major portion of the work of forging. The die operation is automatic and is controlled by a treadle-operated clutch. The operator inserts the bar, steps on the treadle and the movable die closes, gripping the bar. The header die moves forward to perform its operation and returns to starting position, and the movable die opens, completing the cycle. The operator then moves the bar to the next die cavity and again steps on the treadle to begin the next cycle. (H.C.H., M. E. CARRUTHERS, AMERICAN ROLLING MILLS CO.)

FORK-TAIL. Aves, Passeriformes. Indian birds related to the European chats. (A.W.L.)

FORM FACTOR. Form factor is a means for describing the shape of an alternating-current wave. The strength of a.c. constantly varies in magnitude and direction. The effective value of a.c. is equal to the d.c. which will produce the same heating effect as the a-c wave. The average value of an a.c. is that value which, multiplied by the length of time consumed in a cycle, would result in the same number of ampere-seconds as are included under the actual curve showing the wave form. The form factor is the ratio of the effective value of a current to its average value, and is smaller for a flat wave than for a peaked wave. (F.T.M., L.R.Q.)

FORMALDEHYDE. Formaldehyde (HCHO) is a colorless, soluble gas, boiling point $-21°$ C., commonly encountered as a 40% solution in water ("formalin"), which frequently contains methyl alcohol, up to about 20%, to increase the stability of the solution especially when exposed to winter temperatures. Both the gas and the solution have a pungent, irritating, characteristic odor, and the solution acts upon the skin to form a leather-like layer and frequently causes sores. Formaldehyde reacts with many chemicals in a marked manner, (1) with ammonio-silver nitrate (Tollen's solution), to form metallic silver, either as a black precipitate or as an adherent mirror film on glass, (2) with alkaline cupric solution (Fehling's solution), to form cuprous oxide, red to yellow precipitate, (3) with rosaniline (fuchsine, magenta) which has been decolorized by sulfurous acid (Schiff's solution), the pink color of rosaniline is restored, (4) with sodium hydroxide, yields methyl alcohol plus sodium formate, (5) with ammonium hydroxide, when evaporated, yields hexamethylene tetramine "urotropine" ($(CH_2)_6N_4$), white solid, melting point 263° C., (6) with sodium or hydrogen peroxide in sodium hydroxide, yields sodium formate, (7) with manganese dioxide and sulfuric acid, forms methylal, dimethoxymethane ($CH_2(OCH_3)_2$), colorless liquid, boiling point 42° C.

Formaldehyde gas, when cooled under certain conditions, yields trioxymethylene, metaformaldehyde ($(CH_2O)_3$); formaldehyde solution, when evaporated, upon standing, or upon being subjected to low temperatures, yields paraformaldehyde ($(CH_2O)_x$), white solid, from which formaldehyde is regenerated upon heating; dilute formaldehyde, in the presence of calcium hydroxide solution, yields a mixture of sugars called formose from which fructose ($C_6H_{12}O_6$) has been prepared, suggesting the intermediate formation in nature of formaldehyde in the photosynthetic process of the conversion of carbon dioxide to sugars. Formaldehyde stands chemically between methyl alcohol on the one hand—to which it can be reduced—and formic acid on the other hand—to which it can be oxidized. Formaldehyde is made by passing methyl alcohol vapor mixed with air over a catalyzer, e.g., smooth copper wire gauze, at a dull red heat, and collecting the resulting solution. Formaldehyde is commonly detected by the Schiff test (above), and confirmed by the formation of a dimethyl derivative with a melting point of 189° C. Formaldehyde is extensively used (1) as an antiseptic, disinfectant, and insecticide, in the preservation of glue, in the preparation of anatomical specimens, in embalming fluids, in treatment of pests in enclosed spaces and on seeds before planting, (2) as a constituent, along with a **phenol**, of synthetic **plastics** and **resins**, (3) as a hardener, stabilizer or preservative in casein, starch, and sugar preparations, (4) as a reducing agent in gold and silver metallurgy, and in silvering mirrors. (R.K.S.)

FORMALIN. Formaldehyde.

FORMATION. A distinct lithologic unit which may be used in geologic mapping. The term formation is usually confined to **bedded** or **stratified rocks**, including lava flows and volcanic **ejectamenta**. (R.M.F.)

FORMER. Forging.

FORMIC ACID AND FORMATES. Formic acid ($HCHO_2$ or $HCOOH$) is a colorless liquid, melting point 8.4° C., boiling point 100.5° C., miscible with water, alcohol or ether in all proportions. Formic acid solution reacts (1) with **hydroxides, oxides, carbonates,** to form formates, e.g., **sodium** formate, **calcium** formate, and with **alcohols** to form **esters,** (2) with silver of ammonio-**silver** nitrate to form metallic silver, (3) with **ferric** formate solution, upon heating, to form red precipitate of basic ferric formate, (4) with **mercuric** chloride solution to form mercurous chloride, white precipitate, (5) with **permanganate** (in the presence of dilute **sulfuric acid**) to form carbon dioxide and manganous salt solution. Formic acid causes painful wounds when it comes in contact with the skin. At 160° C., formic acid yields carbon dioxide plus hydrogen. When sodium formate is heated in vacuum at 300° C., hydrogen gas and sodium oxalate are formed. With concentrated sulfuric acid heated, sodium formate, or other formate, or formic acid, yields carbon monoxide gas plus water. Sodium formate is made by heating sodium hydroxide and carbon monoxide under pressure at 210° C. The following are representative esters of formic acid:

		BOILING POINT, °C.
Methyl formate	$HCOOCH_3$	32
Ethyl formate	$HCOOC_2H_5$	54
Glycol diformate	$HCOOCH_2CH_2OOCH$	174
Ethyl orthoformate	$HC(OC_2H_5)_3$	146

Formic acid may be obtained (1) from some natural products, e.g., the juice of the hairs of the stinging nettle plant, the juice of the giant nettle tree, the juice of ants and some caterpillars, (2) from sodium formate solution by addition of sodium hydrogen sulfate or dilute sulfuric acid, and then distilling, preferably in vacuum, (3) by reaction of **oxalic acid** and **glycerol** at 100° to 110° C., (4) from **lead** formate solid and **hydrogen sulfide** at 100° C. yielding anhydrous formic acid. Formic acid is used (1) in the textile and leather industries, (2) in reaction with glycerol at 220° C. to form allyl **alcohol,** (3) in the preparation of metallic formates, and **esters.** (R.K.S.)

FORMING. Fabrication of sheet and plate primarily by bending rather than by tension as in **drawing** or **stamping.** (R.H.H.)

FORMULAE, CHEMICAL. Chemical Composition.

FORWARD BEARING. Bearing.

FORWARD VOLTAGE. This is the voltage applied across an electron **tube** in a direction causing the **anode** to be positive and the **cathode** negative. (L.R.Q.)

FOSSA. 1. An anatomical term indicating a depression or furrow. 2. Civet. A species found in Madagascar. *Cryptoprocta ferox.* (A.W.L.)

FOSSIL AMPHIBIA. The first or earliest evidence of an **amphibian** is a primitive footprint in formations of upper **Devonian** *age.* The footprint is of particular significance because it suggests a transition stage in the development of the primitive limb structure of the first terrestrial **vertebrates** from the pectoral fins and limb girdle of the air-breathing ganoids, which also occur in the Devonian. The first known skeletons of amphibia occur in formations of Devonian age. These earliest known amphibia are called Stegocephalia, because their heads are covered or "roofed" with thick dermal bones. In some species both the back and stomach were covered with bony plates or scales. Thus the earliest known amphibia (Stegocephalia) were armored as compared with most of their modern descendants. Other particularly significant anatomical features are (1) a third, or pineal, eye, (2) presence of a ring of plates around the two lateral eyes, (3) the arrangement of the dermal plates in the head, (4) conical teeth, showing an infolding of the dentine and enamel. This type of tooth structure is particularly characteristic of one group called *Labyrinthodonts.* All of the aforementioned features

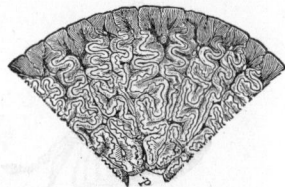

Transverse section of a labyrinthodont tooth. (*After Owen from Norton's Elements of Geology, Ginn and Company.*)

suggest ancestral relationship to the ganoid fishes. The fossil record of the amphibia is the poorest of all the terrestrial vertebrates. They are therefore not particularly valuable as index fossils, their place being taken by

A Pennsylvania amphibian (labyrinthodont), *Eryops.* This creature attained a length of 6 or 8 feet. (*American Museum of Natural History.*)

the **Mesozoic** reptiles and **Cenozoic** mammals. (See chart illustrating the geologic range of the vertebrates under the title **Vertebrate Paleontology.** (R.M.F.)

FOSSIL BIRDS. Probably one of the most famous fossils in the world is Archaeopteryx, or the flying reptile which had feathers. Fossil birds are rare, therefore it is all the more remarkable that this "missing link" between the reptiles and the birds should have been discovered as a natural lithograph in the fine-grained lithographic limestones of Upper Jurassic age at Solenhofen, Bavaria. Only two specimens of this genus are known. Archaeopteryx was about the size of a crow,

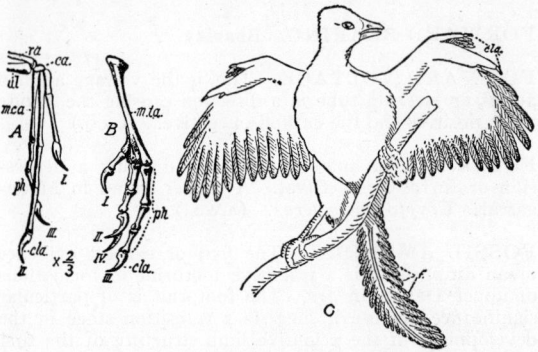

The earliest known bird, *Archeopteryx macrura*, from the Jurassic. *A*, right hand; *B*, right foot; *C*, restoration modified after Pycraft. (*Shimer's Introduction to the Study of Fossils, The Macmillan Co.*)

with the combined reptilian and bird-like claws upon each of the three fingers terminating the wings, and lateral arrangement of the feathers in the tail. There have been several theories as to the origin of Archaeopteryx. One is that it evolved from a small carnivorous bipedal **Triassic** dinosaur. Toward the close of the Mesozoic the birds evolved rapidly. Fossils from the **Cretaceous** show both flying and diving forms, such as *Ichthyornis*, shown in the accompanying illustration. The birds did

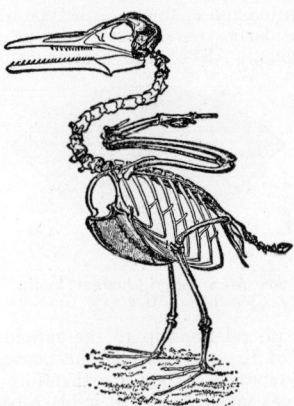

A Cretaceous toothed bird, *Ichthyornis victor*. Height, about 8 inches. (*After Marsh.*)

not lose their teeth until the **Tertiary**. It is also interesting to note that one of the largest known birds that ever lived, *Dinornis maximus* from New Zealand, stood 12′ high and has only recently become extinct. The nearest living relative of the ancestral bird (Archaeopteryx) appears to be the Hoactzin of the Amazon Valley. The embryonic history of this bird repeats many of the adult features of Archaeopteryx, and the young hoactzin still retains claws on its wings, but these claws disappear in the adult stage. As has been previously implied, birds do not make good index fossils primarily because of their rarity as fossils even in the later periods of the earth's history. (See chart illustrating the geologic range of the vertebrates listed under **Vertebrate Paleontology**.)   (R.M.F.)

FOSSIL FISH. The earliest fish may be represented by a "fish-scale" in the **Cambrian**. Problematical fish remains have also been found in the **Ordovician**, and small fin spines in the lower **Silurian**. A famous upper Silurian occurrence is the Ludlow bone bed of England. Probably none of the aforementioned bones and scales

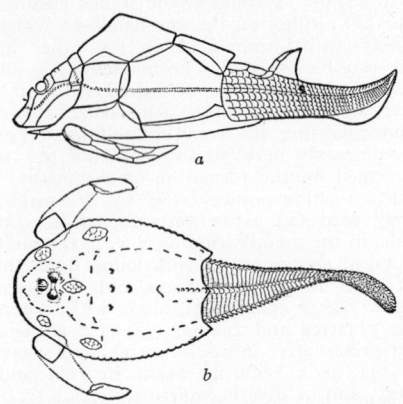

Devonian Ostracoderms; *a*, *Pterichthys testudinarius*, restored (*Dean after Woodward*); *b*, *Tremataspis*, restored (*after Patten*).

belonged to true fishes but to the so-called "bony-skinned fishes" or Ostracoderms, which are particularly characteristic of the **Devonian**. The Ostracoderms became extinct at the close of the Devonian, and there is no direct ancestral connection between them and either the sharks or the true fishes, although it is probable that

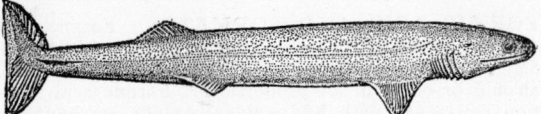

A Paleozoic (early Mississippian) Selachian or shark, *Cladoselache fyleri*. (*Restored by Dean.*)

all the types mentioned had a common ancestor. The Ostracoderms were armored with placoid bony plates or scales. They had no interior skeleton, and certain types such as Pteraspis and Cephalispis were probably adapted to bottom-feeding in relatively quiet marine waters. Other types such as Bothriolepis and Pterichthys prob-

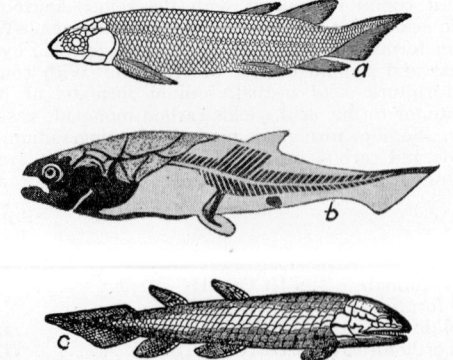

Devonian fishes: *a*, dipnoan. *Dipterus valenciennesi* (*restored by Traquair*); *b*, arthrodiran, *Coccosteus decipiens* (*restored by Woodward*); *c*, ganoid, *Osteolepis* (*restored by Nicholson*).

ably frequented fresh-water ponds and lakes. The earliest known sharks occur in the late Silurian or Devonian, and the true fishes (**Pisces**, or bony fishes) complete one phase of their development during the upper **Paleozoic** with a decline during the early **Mesozoic**. The **Cenozoic** has seen the gradual increase of a new phase with rapid and continuous expansion in diversity of form and adaptability during the late **Tertiary** and **Quaternary.** One of the most interesting groups of the fossil fishes are the **Dipnoi,** or lung fishes. The ganoids first appear in the Devonian and are probably closely related to the progenitors of the first terrestrial vertebrates. The structure of the pectoral fins foreshadows a primitive limb and foot. The arrangement of the bones in the head is somewhat similar to that in the earliest known **Amphibia** (Stegocephalians) also, a peculiar ring of bony plates around the eye, and the conical teeth of infolded dentine and enamel are similar to those of the earliest known Amphibians. (See chart on geologic range of the vertebrates under the title **Fossil Vertebrates.**) (R.M.F.)

FOSSIL MAMMALS.

The paleontological record suggests that the mammals have evolved from the reptiles through a common ancestor with the Theriodontia (beast-toothed reptile) of the **Permian.** The rise of the reptilian mammals began as early as the **Triassic,** as disclosed by small lower jaws and teeth. These animals have been classified as multituberculates, so named because the teeth have coned or tuberculated surfaces. Living descendants of this primitive group include the spiny ant-eater, **Echidna (duck bill)** and the duck-billed mole, *Ornythorynchus,* now living in Australia. Small mammalian jaws similar to those of insectivores have also been discovered in the **Cretaceous.** Most of the **Mesozoic** pro-mammals were probably egg-laying, relatively small in size and quite unable to usurp the place of the great host of **dinosaurs** which had become adapted to all types of environment. Following the extinction of the dinosaurs at the close of the Mesozoic, the mammals rapidly evolved soon after the beginning of the **Cenozoic,** and especially from the **Oligocene** on.

The reptile-like mammals of the Mesozoic, which may be termed pro-mammals, were followed by the archaic mammals of the **Paleocene.** Nearly all of the forms were small, primitive and generalized, with long and heavy tails, short limbs and five digits on each foot. The principal flesh-eaters were the creodonts, and of the modern orders of mammals only the **insectivores** and **marsupials** are present. The ancestors of the first mod-

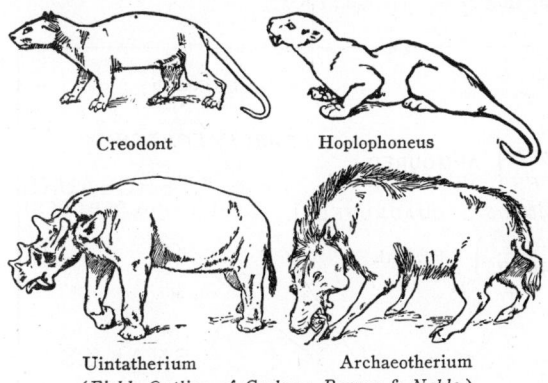

Creodont   Hoplophoneus

Uintatherium   Archaeotherium
(*Field, Outline of Geology, Barnes & Noble.*)

ern mammals such as Eohippus (a diminutive **horse**), cursorial **rhinoceroses,** the creodonts and other earlier archaic forms are also present but are rapidly being replaced by more active and intelligent types. One of the large-hoofed giant herbivores or hippopotamus-like forms is called uintatherium. The mammals first begin to take on a modern aspect in the Oligocene, including a large number of forms which grazed on the open plains. The true carnivores have replaced the creodonts, including Hoplophoneus, ancestor of the saber-toothed cats. The remainder of the Cenozoic saw the gradual evolution of the modern mammals. Among the many modern forms the paleontological record of the elephant has

been remarkably well worked out as shown by the accompanying charts. Because of the strongly marked climatic zones of the **Pleistocene,** as well as the glacial

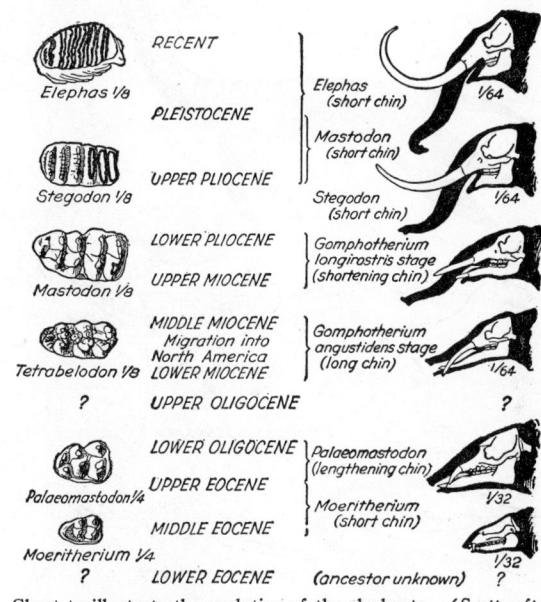

	RECENT	
Elephas 1/8	PLEISTOCENE	Elephas (short chin) 1/64
Stegodon 1/8	UPPER PLIOCENE	Mastodon (short chin)
	LOWER PLIOCENE	Stegodon (short chin) 1/64
Mastodon 1/8	UPPER MIOCENE	Gomphotherium longirostris stage (shortening chin)
Tetrabelodon 1/8	MIDDLE MIOCENE Migration into North America LOWER MIOCENE	Gomphotherium angustidens stage (long chin)
?	UPPER OLIGOCENE	? 1/64
	LOWER OLIGOCENE	Palaeomastodon (lengthening chin)
Palaeomastodon 1/4	UPPER EOCENE	Moeritherium (short chin) 1/32
Moeritherium 1/4	MIDDLE EOCENE	1/32
?	LOWER EOCENE	(ancestor unknown) ?

Chart to illustrate the evolution of the elephants. (*Scott, after Lull, modified by Sinclair. The Macmillan Co.*)

and interglacial cycles within the areas subjected to continental glaciation, the end of the Cenozoic is remarkable for the large number of mammals as well as the origin and evolution of man. (See **Paleontology of Man.**) (R.M.F.)

FOSSIL MAN. Paleontology of Man.

FOSSIL PLANTS. Paleobotany.

FOSSIL REPTILES.

The first or earliest evidence of reptiles is in the **Pennsylvanian.** Most of these reptiles were sluggish animals, probably not much more active than their amphibian progenitors, and equally adapted to both water and land. Some forms, however, such as the Pelycosaurs, or sail-backed lizards, had a great dorsal "fin," and thus appear to have been highly specialized. Most of the forms were distinctly lizard-like in appearance. The cotylosauria were the most primitive of the **Paleozoic** reptiles showing many of the ancestral **Stegocephalian** characters. Another and particularly interesting group of reptiles which lived during the **Permian** and **Triassic** periods are the Theriodontia. Fossil skeletons of these are found only in South Africa. Although these animals were reptilian, their teeth show a differentiation into incisors, canines, and molars, a decidedly mammalian characteristic. Because of this and certain other anatomical features the Theriodontia are considered by some paleontologists as the "missing link" between the reptiles and the mammals. The reptiles reached their all-time development during the **Mesozoic,** thus this era has been termed the age of reptiles. Among the Mesozoic reptiles the great group of the **Dinosaurs** command particular attention because they varied in size from the smallest bird-like creatures to huge forms, both carnivorous and herbivorous, which rival the mightiest animals that ever lived. The evolutionary vitality was particularly well expressed in the extreme adaptation of body forms to all types of habitat as outlined in the chart on page 605. The dinosaurs became extinct at the close of the Mesozoic, and the modern reptiles, such as the lizards, snakes, and turtles, are but a remnant of the great record of reptilian evolution. (R.M.F.)

FOUCAULT PENDULUM.

In 1851 Foucault performed his celebrated **pendulum** experiment at Paris, designed to give physical proof that the earth is in rotation about an axis. The pendulum, consisting of a very large iron ball suspended by a steel wire over 200' long, was suspended from the center of the dome of the Pantheon. Great care was exercised in the support for the wire so that no external forces should be effective at this point other than a vertical force to prevent the system from falling. On the floor, immediately under the pendulum, a layer of fine sand was placed so that the direction of the swing could be observed. The pendulum was started by drawing it to one side with a fine thread and, after the system was at rest, the thread was burned off, thus avoiding any lateral motion.

After such a pendulum is started swinging in one plane, it is soon observed that the plane of swing is apparently deviating slowly (in the clockwise direction in the northern hemisphere and the opposite in the southern). The rate at which the plane of swing deviates is equal to 15° per sidereal hour (c.f. time) multiplied by the **sine** of the **latitude**. Thus at the pole it would make a complete rotation in one sidereal day, while at the equator it would not rotate at all. At Paris (latitude 48° 50' N.) the rate of deviation is about 11° 18' per hour.

Foucault reasoned quite correctly that, in accordance with **Newton's Laws of Dynamics,** the direction in space of the plane of swing should not change unless the pendulum was acted upon by some external force other than that of gravitation and the counteracting force parallel to the direction of gravitation at the support. That the direction of the plane does apparently change can be accounted for only on the hypothesis that the earth is in rotation. Foucault's demonstration attracted wide scientific and popular attention and was accepted as a conclusive proof that the earth does rotate upon an axis, a fact which was not universally accepted at that time. (W.K.G., L.D.W.)

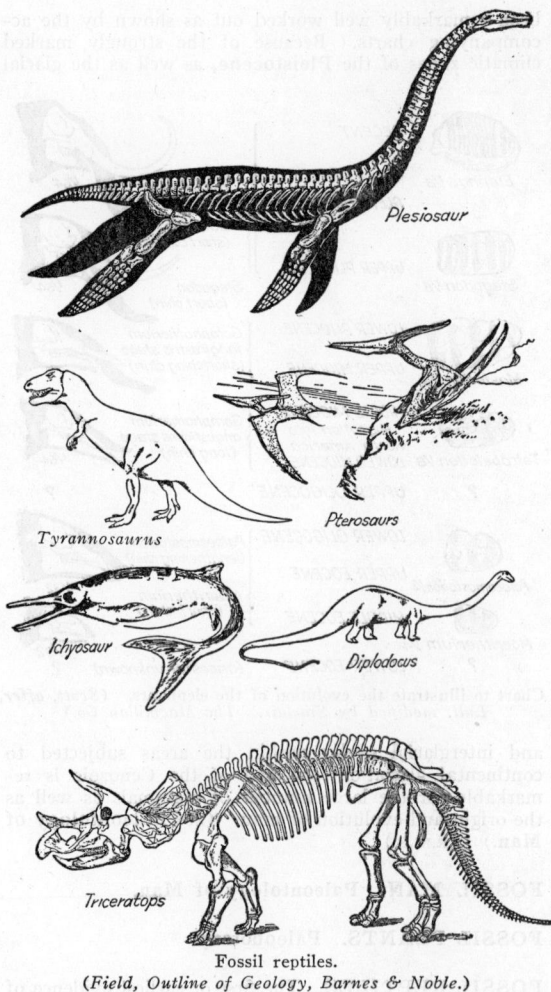

Plesiosaur

Tyrannosaurus

Pterosaurs

Ichyosaur

Diplodocus

Triceratops

Fossil reptiles.
(*Field, Outline of Geology, Barnes & Noble.*)

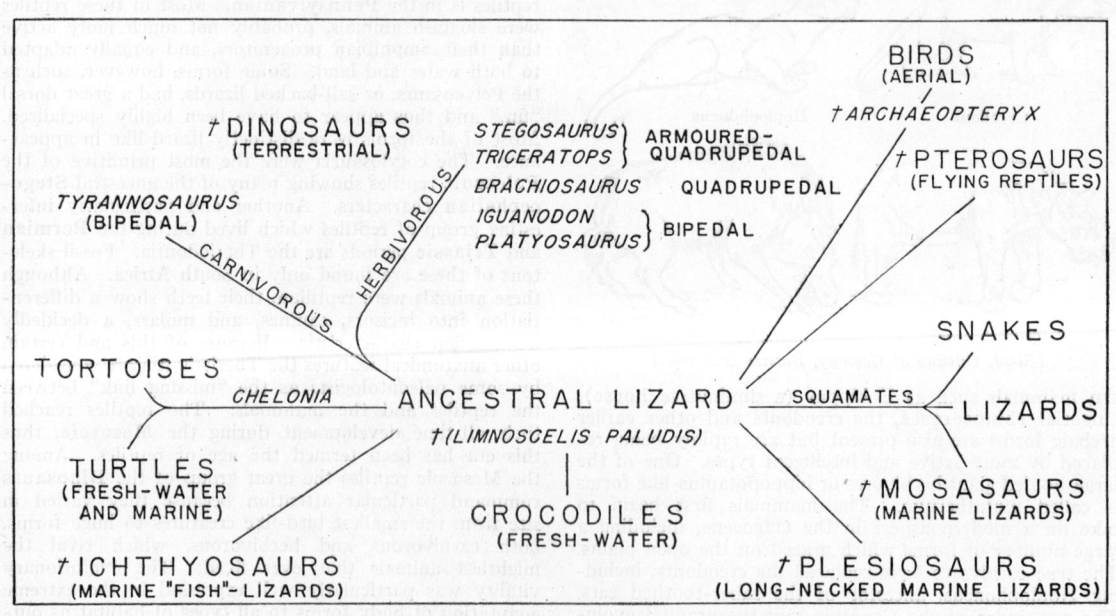

BIRDS
(AERIAL)

†*ARCHAEOPTERYX*

†DINOSAURS
(TERRESTRIAL)

STEGOSAURUS } ARMOURED-
TRICERATOPS } QUADRUPEDAL

†PTEROSAURS
(FLYING REPTILES)

TYRANNOSAURUS
(BIPEDAL)

BRACHIOSAURUS QUADRUPEDAL
IGUANODON
PLATYOSAURUS } BIPEDAL

CARNIVOROUS

HERBIVOROUS

SNAKES

TORTOISES

CHELONIA

ANCESTRAL LIZARD

SQUAMATES

LIZARDS

†(*LIMNOSCELIS PALUDIS*)

TURTLES
(FRESH-WATER
AND MARINE)

†MOSASAURS
(MARINE LIZARDS)

CROCODILES
(FRESH-WATER)

†ICHTHYOSAURS
(MARINE "FISH"-LIZARDS)

†PLESIOSAURS
(LONG-NECKED MARINE LIZARDS)

ADAPTIVE EVOLUTION OF THE REPTILES
† = extinct.
(*Field, Laboratory Manual, Princeton University Press.*)

FOUNDATIONS. The structural foundation is that part of a structure which transmits the loads to the supporting material. In the design of a foundation it is essential that the settlement shall be reduced to a minimum and that this settlement shall be uniform at all points. The first requirement may be fulfilled by providing a **bearing area** which is large enough to reduce the bearing pressure on the underlying material to a safe value. In the case of soils having low bearing values **pile** foundations can be used to reduce settlement. The second requirement may be secured by designing the foundations so that the **resultant** of the vertical **loads** passes through the **center of gravity** of the foundation. If the material under the foundation is structurally sound rock, having a bearing value within safe limits, there will be no appreciable settlement of the structures; but there is bound to be settlement in structures whose supporting medium is earth since it is a compressible material.

A floating foundation is a reinforced **concrete** mat which covers the entire area under the structure and transfers the loads to the supporting soil. This mat may be a thick slab or a system of beams and thinner slabs. Floating foundations are generally used for soils having very low bearing values.

For different types of foundation refer to **Caisson, Footings, Grillage, Pier** and **Pile.**

The machine foundation performs more than a simple bearing function. It must:

1. Distribute the weight of the machine, the machine bedplate, and its own weight over a safe subsoil area. If heavy unbalanced vertical kinetic forces (see **Mechanics**) are produced by the machine, they should be added to the dead weight and the bearing power of the soil must be well in excess of these vertical forces.

2. Provide sufficient **mass** to absorb machine vibration. Satisfactory foundation weight for this factor is not readily calculable. The accompanying table is given to provide an indication of minimum weights.

WEIGHT OF ENGINE FOUNDATIONS PER BRAKE HORSEPOWER

Prime Mover	Single Cylinder	Multi-Cylinder
Gas engine.....	2500 lbs.	1600 lbs.
Diesel engine...	2000 lbs.	1250 lbs.
Steam engine...	700 lbs.	500 lbs.
Steam turbine..	Not to exceed permissible deflection as stated by turbine manufacturer.	

3. Be rigid enough to prevent undue deflection of any part of the machine bedplate.

Machine foundations are usually made of concrete and unless an unyielding foundation soil is available, the concrete at the bottom of the foundation is subjected to tension. Since this is generally the case, the heavy foundation should have reinforcement near the bottom.

Large turbine (see **Steam Turbine**) foundations are not required to contain foundation masses comparable with those necessary for similar reciprocating units. Still, the necessity of providing space beneath the turbo-generator for condenser, pumps, generator, air cleaner, or cooler, and sometimes the throttle lead, materially complicates the design of the turbine foundation. This foundation does not carry the turbine upon a heavy bed-plate and hence a study of the foundation **deflections** is all important. No two foundations are alike. Reinforced concrete and structural steel foundations each have their advocates, but any installation should be figured upon a basis of comparable costs of the two types, because each is suited to a particular field of utility and economy. The concrete base gives more rigidity to the turbine, but it is claimed for steel that its flexibility is

an advantage in large units since it prevents distortive bowing of the shaft and attendant difficulties. The concrete foundation will require less maintenance; the steel type yields more available space below the unit.

When bringing a machine into alignment on its foundation, shims or sole plates are placed beneath the frame or bedplate and adjusted until the alignment is level in two directions. A spirit level can be used when aligning small machines, but the engineer's level is best for all classes of work. After alignment is secured, an earth dam is built around the top edge of the foundation. **Grout** is floated or forced under the bedplate until it is flush with the top or bottom surface. (C.W.C., F.T.M.)

FOUNDRY. Casting.

FOUR-CYCLE. An abbreviated expression for four-stroke cycle. The four-stroke cycle is one upon which either the **Otto** or **Diesel** types of **internal combustion engines** may operate, since it describes, not **thermodynamic** aspects of a cycle, but rather the sequence by which the cylinder is charged and exhausted. The four-stroke cycle is described as follows. Beginning with a suction or induction stroke, the **cylinder** is filled with a fresh charge by the outward motion of the **piston.** Next, on the return motion, this charge is trapped in the cylinder by closure of all **valves** leading to and from the cylinder, and is thereby compressed. On the next outward stroke, the power stroke, the fuel is burned or exploded to the accompaniment of energy liberation, a great deal of which is made usefully available on the power stroke. The final, or fourth, stroke is a return stroke, or exhaust stroke, during which the contents of the cylinder are exhausted through a port opened by an exhaust valve. Thus the four strokes are suction, compression, power, and exhaust. During the suction stroke, an inlet valve is open; during the exhaust stroke, an exhaust valve is open. The advantage of the four-cycle principle is that it gives a full stroke for induction of the fresh charge, and another full stroke for scavenging of the burned gas. In this way it promotes high **volumetric efficiency.** A disadvantage of the four-cycle principle is the intermittent delivery of power. This contributes to making the four-cycle engine bulky in comparison with the two-cycle, but by employing multi-cylindered engines a steady flow of power may be secured through overlapping of power strokes. (F.T.M.)

FOURIER'S LAW. Heat Transfer.

FOURIER SERIES. An infinite series of the type

$$\tfrac{1}{2}a_0 + a_1 \cos x + a_2 \cos 2x + a_3 \cos 3x + \cdots$$
$$+ b_1 \sin x + b_2 \sin 2x + b_3 \sin 3x + \cdots,$$

or, more briefly written,

$$\tfrac{1}{2}a_0 + \sum_{n=1}^{\infty} (a_n \cos nx + b_n \sin nx),$$

where the coefficients a_n and b_n are independent of x, is called a trigonometric series.

If we assume the possibility of expanding any arbitrary function $f(x)$ in a series of this form, and if we assume the uniform convergence of the series, the coefficients a_n and b_n are found to be given in terms of $f(x)$ by the formulae:

$$a_n = \frac{1}{\pi} \int_{-\pi}^{\pi} f(x) \cos nx \, dx, \quad b_n = \frac{1}{\pi} \int_{-\pi}^{\pi} f(x) \sin nx \, dx.$$

A trigonometric series whose coefficients are given by these formulae is often called the Fourier series of $f(x)$. The problem still remains of investigating the convergence of this series and the question as to whether the series represents the function.

A fundamental theorem concerning Fourier's series is the following Fourier's theorem:

Any single-valued function $f(x)$, which is continuous except possibly for a finite number of finite discontinuities in the interval $-\pi$ to π, and which has only a finite number of maxima and minima in that interval, may be represented by a Fourier series

$$f(x) = \tfrac{1}{2}a_0 + \sum_{n=1}^{\infty} (a_n \cos nx + b_n \sin nx),$$

where the coefficients a_n and b_n are given by the previous formulae in terms of integrals. (See **Harmonic Analysis; Vibration and Waves.**) (L.L.S.)

FOWL. Aves, Galliformes. Birds of the domesticated varieties. (A.W.L.)

FOWLERITE. Rhodonite.

FOX. Mammalia, Carnivora. **Vulpes.** Members of the **dog** family (Canidae) of moderate size and slender build, with a sharp muzzle, long bushy tail, and unusually large ears. Most species inhabit the northern hemisphere but a few are found in Africa with the closely related fennecs.

The fur of the fox is valuable. The excellent market for that of the silver fox has led to extensive breeding of these animals in captivity.

North America has several species of red foxes and several of gray. The silver fox is a variety of a red fox, and kit fox is a related species. (A.W.L.)

FOXGLOVE. Figwort.

FRACTIONAL EQUATIONS. An algebraic equation in which one or both sides are rational **fractional functions** is called a rational fractional equation.

Such an equation may in general be solved by clearing the equation of fractions by multiplying its terms by the **least common multiple** of the denominators of its fractions, thus reducing it to a **polynomial equation** (frequently of first or second degree).

In this method of solution, by clearing of fractions, **extraneous roots** may arise, so that the solution of each fractional equation should be checked by substituting in the original equation, and all values not satisfying this equation should be rejected. (L.L.S.)

FRACTIONAL FUNCTION. A rational fractional function is a **rational function** which is not a **polynomial function** (rational integral function); it involves the variable in a denominator of a **fraction.**

A rational fractional function can be expressed as a quotient of two polynomials. (L.L.S.)

FRACTIONS. A fraction is an indicated **quotient** of two numbers or two expressions. The dividend and divisor of the quotient are called respectively the numerator and denominator of the fraction.

A fundamental principle for operations with fractions is the following: The numerator and denominator of a fraction may both be multiplied or divided by the same number without changing the value of the fraction:

$$\frac{a}{b} = \frac{ma}{mb}.$$

A fraction is said to be in its lowest terms when its numerator and denominator have no common factor. A fraction can be reduced to its lowest terms by dividing the numerator and denominator by their **highest common factor.**

The signs of both numerator and denominator of a fraction may be changed without changing the value of the fraction. If the sign of either numerator or denominator is changed, the sign of the fraction is changed. Thus:

$$\frac{-a}{-b} = \frac{a}{b}, \quad \frac{-a}{b} = \frac{a}{-b} = -\frac{a}{b}.$$

Two or more fractions may be changed to equivalent fractions with a common denominator by multiplying the numerator and denominator of each fraction by the quotient obtained by dividing the **least common multiple** of all the denominators by the denominator of that fraction.

The sum (or difference) of two fractions with a common denominator is a fraction with the common denominator and whose numerator is the sum (or difference) of the given numerators:

$$\frac{a}{c} \pm \frac{b}{c} = \frac{a \pm b}{c}.$$

To add (or subtract) any two fractions (with different denominators), change each fraction to equivalent fractions with a common denominator (usually the least common denominator), and then combine them by the preceding rule:

$$\frac{a}{b} \pm \frac{c}{d} = \frac{ad}{bd} \pm \frac{bc}{bd} = \frac{ad \pm bc}{bd}.$$

The product of two fractions is a fraction whose numerator is the product of the numerators of the given fractions and whose denominator is the product of their denominators:

$$\frac{a}{b} \cdot \frac{c}{d} = \frac{a \cdot c}{b \cdot d}.$$

To divide one fraction by another, invert the divisor fraction and multiply the dividend by this inverted fraction:

$$\frac{a}{b} \div \frac{c}{d} = \frac{a}{b} \cdot \frac{d}{c} = \frac{a \cdot d}{b \cdot c}.$$

(L.L.S.)

FRACTURE. In medicine, a fracture is a breaking of any solid structure or organ of the body. Commonly, the term is used to describe a broken bone. Fractures are classified as to whether they are transverse or oblique; and comminuted, simple, compound, depressed, greenstick, or impacted. In a comminuted fracture, the bone is splintered into several fragments at the site of injury. In contrast to a simple one, a compound fracture is one in which the ends of the broken bone project through the skin; because of the possibility of bacterial infection in such an open wound, a compound fracture is always serious. A depressed fracture is a fracture of the skull in which a portion of the bone is pushed inward. The term greenstick is used when the break is in the shaft but does not extend all the way through the bone. Such a break is associated with bending and occurs only in children. In an impacted fracture one fragment is driven or jammed into the other.

Two of the commonest fractures are the so-called Colles' and Potts' fractures. A Colles' fracture is a break through the radius at the wrist; it occurs most frequently following a fall on the outstretched hand. In a Potts' fracture, the break is through the tibia and fibula at the ankle.

The treatment of fractures is directed toward the restoration of the normal anatomy and function of the part. In the immediate treatment of compound fractures, experiences in World War II have shown that debridement plus **penicillin** applied directly into the open wound is effective in preventing infection and reducing mortality. If there is displacement of bony fragments, they may be brought into alignment by closed reduction, i.e., manipulation or, if necessary, by open reduction, an operative procedure. Traction is used in certain instances to pull the bones into position and to maintain them so until healing takes place. Fixation may also be achieved by the application of a cast. Complications delayed union, and nonunion.

In mineralogy, fracture is the property of minerals to break with curved or uneven surface. When a mineral has three planes of cleavage it has no fracture. Some minerals, such as quartz, have no well-defined planes of cleavage, but may be distinguished by their fracture, such as conchoidal (shell-like), splintery, etc. (R.S.M., D.M.H., R.M.F.)

FRAME, TELEPHONE. In order to provide the necessary flexibility in the use of telephone lines, numbers, and switchboard equipment a system of cross connections is used. The lines terminate on one side of a rack structure, known as the main distributing frame, and the switchboard wires corresponding to the telephone numbers terminate on the other. Flexible cross connections which may be changed at will connect the two. Another frame, the intermediate distributing frame, is used to make the system still more flexible by having the wires from the answering jacks of the switchboard terminate on it. (L.R.Q.)

FRAMED STRUCTURE. A framed structure is a group of structural members (tension, compression or flexural members) which are joined together in the form of triangles or rectangles in order to support given loads and distribute them to the supports in a definite manner. Space frames, rigid frames and trusses are all framed structures. (C.W.C.)

FRANCOLIN. Aves, Galliformes. Birds (Aves) of Africa and the Oriental region, related to the partridges. (A.W.L.)

FRANGIPANGI. *Plumeria acutifolia.* Apocynaceae. This is a shrub or small tree native to Mexico and Central America, which is often cultivated for the beauty and fragrance of its flowers. The plant has but few, rather coarse, branches. Its deciduous leaves are thick, smooth, and oblong to ovate in shape; in length they vary from 5–15". The flowers, which often appear when the plants are leafless, are showy, and borne in terminal cymes. The flowers are red and extremely fragrant, and much used for decorative purposes. Several other species of *Plumeria* are known, some of which yield woods of some economic value. (R.M.W.)

FRANKLINITE. The mineral franklinite is a zinc-iron-manganese mineral whose formula may be written (Fe, Mn, Zn)(FeO$_2$)$_2$, but the composition varies considerably, in respect to the amounts of the several metals that may be present. Its isometric crystals have an octahedral habit; it may be coarse or finely granular or compact. It shows a parting parallel to the octahedron; fracture, uneven; brittle; hardness, 5.5–6.5; specific gravity, 5–5.2; luster, usually metallic, occasionally dull; color, black; streak, brown to black; opaque; may be slightly magnetic. Only in one place in the world does franklinite occur in quantity, at Franklin Furnace, New Jersey, from whence it was named. Here there are two bodies of this mineral, which is used as a zinc ore, about 3 miles distant from each other. The franklinite is found in pre-Cambrian limestones that are associated with gneisses believed to be of igneous origin and responsible for the mineralization. Associated minerals are willemite, zinc, silicate, and zincite, zinc oxide. Franklinite has been found at Eibach, Germany, in cubic crystals. (E.S.C.S.)

FRASCH PROCESS. Sulfur.

FRAUNHOFER LINES. The dark lines constituting the absorption spectrum exhibited by sunlight are frequently called the Fraunhofer lines. There are thousands of these lines, of which Fraunhofer, early in the 19th century, first observed the most prominent. To these particular lines he assigned letters for reference purposes. These lines, together with their origin and approximate wavelengths, are listed as follows:

A	Terrestrial oxygen	7594 A	(extreme red)
B	Terrestrial oxygen	6867 A	(red)
C	Hydrogen	6563 A	(red)
D$_1$	Sodium ⎫	5896 A	(yellow)
D$_2$	Sodium ⎭ doublet	5890 A	(yellow)
E	Iron	5270 A	(green)
F	Hydrogen	4861 A	(blue)
G	Iron and Calcium (group)	4308 A	(violet)
H	Calcium	3968 A	(extreme violet)

The lines of solar origin are due to absorption by gases and vapors in the solar atmosphere. (L.D.W.)

FREE FIT. Fit.

FREE GRID. This applies to the grid of an electron tube which has no electrical connection to it. Such a condition is highly undesirable, especially in vacuum tubes, since the grid will accumulate a charge which cannot leak off and which will alter the operation of the tube. If the charge remained fixed it would not be so serious, but the accumulated charge is very sensitive to external conductors, thus bringing the hand near a tube with a free grid will cause a redistribution of the electrons accumulated on the grid and cause the plate current to vary erratically. (L.R.Q.)

FREEMARTIN. A sterile cow, born as a twin of a bull. The sterility is due to the action of hormones produced by the male twin, which enter the circulation of the female in the extraembryonic membranes. (A.W.L.)

FREEZING POINT. The freezing point of a liquid is the temperature at which the solid form and the liquid form are in equilibrium. (See States of Matter; Melting Point.) (R.K.S.)

"FREON." A trademark for a group of halogenated hydrocarbons containing one or more fluorine atoms widely used as refrigerants and propellants. "Freon-12" is dichlorodifluoromethane (CCl$_2$F$_2$), used in the "insect bomb" (see DDT). (R.K.S.)

FREQUENCY. In electricity, frequency is the number of complete alternations per second of an alternating current. Sixty cycles per second is becoming the standard frequency for a-c generation in the United States. Elsewhere 25 and 50 cycles per second have some vogue. In an alternator, the number of alternations per second of the output is the speed, in revolutions per second, multiplied by the number of poles. The number of poles in alternators is usually 2 or 4 for steam turbine-driven alternators, or 24, 26, 28, 30, 36, 48, or 60 in the case of engine-driven alternators. The formula for frequency, cycles per second, in terms of the rotative speed, rpm, and the number of electrical poles, is:

$$\text{Frequency} = \frac{\text{rpm} \times \text{number of poles}}{120}$$

In acoustics, the frequency represents the number of sound waves passing any point of the sound field per second. (See Audio Frequency.) In the case of light or other electromagnetic radiation, frequency may be expressed in this same way but is usually so enormous (500 million per second for yellow light) that wavelengths are ordinarily used instead. Radio frequencies are commonly given in thousands of cycles (kilocycles) or millions of cycles (megacycles) per second. (F.T.M.)

FREQUENCY ALLOCATION. This is the assigned carrier frequency of a radio transmitting station. Since radio signals must all use the same medium of transmission, the so-called ether, it is necessary to select the particular signal desired by tuning the **receiver.** Signals whose frequencies are too close together cannot be separated, i.e., they cause interference. To minimize this effect the frequencies of stations are assigned by the controlling governmental agency, Federal Communications Commission in the United States, and are required to maintain this assigned frequency within very close tolerances. Various types of service are assigned frequencies in bands set aside for their use by international agreement. (L.R.Q.)

FREQUENCY BAND. This is usually applied to a group of closely related frequencies, but common usage applies it in two slightly different senses. The first of these is synonymous with **channel,** meaning the band of frequencies associated with a **carrier** under **modulation.** The second usage means a group of different carrier frequencies all designated for the same purpose. In this sense there may be many frequency channels within the given band. Thus the standard broadcast band contains many broadcast channels, each having its carrier spaced 10 kc. from the next. In some types of radio service the only restriction on the carrier value is that it be kept within the allocated band. (L.R.Q.)

FREQUENCY CHANGER. Wherever there is to be an interchange of electrical energy between two systems which operate at different **frequencies,** the tie together must be through the medium of a frequency changer, since it is impossible to have two **alternators** electrically connected to the same line operating at different frequencies. The one operating at higher frequency will tend to motor the other, and take unto itself enough of the load so that through the slowing down of it, and the speeding up of the slower, their frequencies will coincide. The simplest plan for connection of two such systems, and the one usually employed, is to couple, mechanically, two rotating machines of the **synchronous motor** or **generator** type, which have, respectively, the correct number of electrical poles to permit one to operate at the frequency of one system, while the other is at the frequency of the other system, both, of course, having the same rotative speed. This is relatively simple in the case of the drawing of power, say, from a 60-cycle system to supply energy for 25-cycle use, but has serious disadvantages as a means of interconnecting two power systems of different frequency, each having considerable amount of generating equipment. Any slight alteration in the relative frequencies of these two systems results in heavy fluctuations of the frequency changer load, and necessitates the employment of extremely large and bulky frequency changers. Nevertheless, this is the system usually employed. Fortunately, the necessity for such frequency changing is rare in this country at the present time. (F.T.M.)

FREQUENCY CHANNEL. Radio Communication; Carrier Frequency.

FREQUENCY DEVIATION. In amplitude-modulated or continuous-wave radio transmission the frequency deviation means the amount by which the **carrier** varies from its assigned value. This must be kept within close limits, for broadcast stations at the present time within ± 20 cycles. The deviation is the result of many factors, some of which are easily controlled while others are difficult. Since the **oscillator** is the frequency controlling part of the **transmitter** its frequency must be held to the proper tolerances. High Q, low-temperature effects, light load and stable voltages are among the things which must be realized to maintain a stable frequency. In frequency or phase **modulation**

the frequency deviation is the amount by which the frequency of the output varies under modulation. In these transmitters the carrier must be maintained very constant but under the influence of modulation the output frequency varies about the carrier as a mean value, the amount being determined by the type modulation and the audio-frequency signal. (L.R.Q.)

FREQUENCY DISTRIBUTION. A frequency distribution consists of the **variates** of a **serial distribution** (of more than 100 items) which have been grouped into mutually exclusive classes usually of equal width. From the frequency distribution any **statistics** may be calculated such as the **moments** or the **measures of central tendency.** The frequency distribution summarizes the data, making calculations and analysis much easier. The mathematical counterpart of the frequency distribution is the **probability function.** (L.A.A.)

FREQUENCY DIVIDER. Often in high-frequency measurements it is desirable to have available frequencies which are sub-multiples of some given frequency (i.e., the given frequency divided by a whole number). These sub-frequencies may be obtained by the use of a harmonic generator which, in this application, is called a frequency divider. **Oscillators** have a strong tendency to synchronize with an injected frequency which is not too different from their normal value. In certain types this tendency is so strong that a **harmonic** of the oscillator will synchronize with an injected frequency. It is this type which serves as a divider, the frequency to be divided is fed into the circuit of the dividing oscillator, causing the oscillator frequency to change so the proper harmonic is locked in with the injected signal. This, of course, means that the oscillator frequency is fixed at a sub-multiple value of the original frequency. The various harmonics of the synchronized oscillator then give a series of frequencies which might be said to be divisions of the original frequency. The **multi-vibrator** oscillator is widely used in this application. (L.R.Q.)

FREQUENCY DOUBLER. The oscillator of a radio **transmitter** is often operated at a lower **frequency** than the output frequency of the transmitter. This has several advantages; among them may be mentioned less reflected effect on the oscillator and hence more stable operation, elimination of **neutralization,** stronger **crystals** than if they were ground for the higher output frequency, and in the case of frequency **modulation** a proportionate increase in the degree of modulation. A frequency doubler is usually a class C **amplifier** with its grid **tank** tuned to the frequency of the preceding stage (oscillator or amplifier) and its output tuned to twice this frequency. While the output power is reduced by this operation there is still sufficient output to drive another amplifier, or even in some cases to serve as the transmitter output. The energy which might be fed back from the output to the input of such an amplifier cannot affect the input or a preceding stage since every other cycle (because of this double frequency) will be 180° out of phase. This reduces the effect on the oscillator and also eliminates the need for neutralization. Doubling is used extensively in frequency modulated transmitters since the **frequency deviation** is doubled each time the frequency is doubled. In some transmitters of this type there may be a dozen or more doubler stages for this purpose. (L.R.Q.)

FREQUENCY DRIFT. This is the gradual change of carrier frequency which a poorly designed radio **transmitter** may undergo with time. It is usually caused by temperature effects in the **oscillator.** (L.R.Q.)

FREQUENCY, FUNDAMENTAL. In many a-c generating devices there are several frequencies produced, all related in a definite manner. The lowest one of these frequencies is the fundamental frequency and the others are whole multiples of this. These multiples are the **harmonics**. In radio **transmitters** it is important to suppress all but the fundamental frequency of the output, since harmonics might cause interference with other signals. (L.R.Q.)

FREQUENCY HISTOGRAM. A frequency histogram is the graph of the **frequency distribution**. Rectangles are formed by using the **class interval** as the base and the frequency of the class as the height. Equal areas represent equal frequencies. (L.A.A.)

FREQUENCY METER. Most modern frequency meters for radio use are not **instruments** in the usual sense but are really several instruments and associated apparatus. The old absorption type wave-meter might be called a single instrument but it is not used except for rough measurements. It consists of a **coil** and **condenser** in series with some sort of indicator. The condenser may be varied to resonate the circuit (see **Resonance**) with the signal being measured. The indicator, which may be a simple lamp bulb or may be a vacuum-tube voltmeter, indicates resonance and the frequency is obtained from the condenser setting. Modern precision measurement of frequency utilizes a comparison process. The fundamental for this is the rotation of the earth. High-grade crystal **oscillators** are used to generate medium values of radio frequency, the output being compared with the rotation of the earth by synchronous clocks or by comparison with another oscillator which is so checked. Associated with this oscillator are various **multi-vibrator** oscillators serving as **frequency dividers** to give an extensive series of accurately known frequencies. The frequency to be measured is compared with the closest of these known frequencies by heterodyning and the resultant difference frequency (an **audio frequency**) determined in one of several ways. (See **Heterodyne**.) Very high frequencies are measured by exciting standing waves on an open or shorted transmission **line** and measuring the linear distance between nodes or anti-nodes of these waves. The frequency is then obtained by dividing the velocity (3×10^{10} cm./sec.) by the wavelengths in centimeters. For very accurate measurements several corrections must be applied to this method but it is sufficiently accurate for many purposes without the corrections. (See **Lecher Wires**.)

A common frequency-measuring **instrument** used for commercial power circuits is the vibrating reed instrument. This consists of many reeds, each tuned to a different frequency (commonly a half-cycle apart), which are set in vibration by the alternating flux of a **solenoid** connected across the circuit to be measured. Since the reeds are tuned, only those whose resonant frequency is very close to the circuit frequency will have appreciable amplitude, and the one closest to the circuit frequency will have the greatest amplitude. These reeds are mounted in a row or rows with the frequency scale on the adjacent face of the instrument. Another and more accurate frequency-measuring instrument for commercial circuits depends for its operation upon the effect on the moving element of the currents in two shunt circuits. One of these contains inductance and the other capacitance so the effect of a change in frequency is opposite in the two circuits. Since they are connected across the circuit to be measured their currents change with frequency and hence the moving element is deflected. (L.R.Q.)

FREQUENCY MODULATION. Modulation.

FREQUENCY MONITOR. Frequency monitor is used to give a continuous indication of any departure of a radio **transmitter's frequency** from its assigned value.

This is usually done by comparing the station frequency with that of a crystal controlled **oscillator** in the monitor. The monitor frequency and the transmitter frequency are **heterodyned** and the **beat frequency** measured. Some monitors are adjusted to give zero beat frequency when the station is on the correct value, others give a difference frequency of some convenient value such as 1000 cycles. The beat frequency is fed to a circuit feeding an indicating instrument which shows zero for correct transmitter frequency and deflects to the right or left for high or low transmitter frequency. The scale is calibrated in cycles and the instrument indicates cycles off frequency. (L.R.Q.)

FREQUENCY POLYGON. A graph of the frequency **distribution** formed by graphing the class frequencies as **ordinates** and the class marks as **abscissas** and then joining these points by straight lines. It is usually drawn only in the case of a **discrete variable**. (L.A.A.)

FREQUENCY RESPONSE. By the frequency response of a circuit is meant the range of frequencies which it will pass with an allowable loss. The expression is quite commonly applied to audio **amplifiers**, **microphones** and **loud-speakers**. The frequency response of an audio amplifier usually drops off markedly at the low end and at the high end, the exact point depending upon the circuit and the values of various constants used in it. The response of a high-fidelity amplifier (one which would reproduce without appreciable frequency discrimination) should cover most of the audio band, i.e., from about 20 to 15,000 cycles. However this is much better than necessary except for the most refined work and a range of 30 to 10,000 cycles is usually considered high fidelity. Modern microphones have responses which will cover this latter range while some will extend over the entire audio band. Loud-speakers require the use of dual speakers or special resonant chambers, horns, etc., to cover this range. (L.R.Q.)

FREQUENCY STANDARD. This is the standard of frequency used in frequency measurement (see **Frequency Meter**) and is usually an extremely stable crystal controlled **oscillator**. The fundamental standard is the rotation of the earth and other standards are checked against this. Standards which may be checked directly (by means of synchronous clocks) are called primary standards while those which are checked against primary standards are called secondary standards. (L.R.Q.)

FREQUENCY TRANSLATION. Often in communication circuits it is desirable to move a frequency **channel** bodily to another part of the spectrum. This process is frequency translation and is performed by beating the frequencies to be moved with another fixed frequency. The sum or difference of the original frequencies and the beating frequency then gives the same channel at another point in the spectrum and by the proper choice of the beating frequency this new location can be fixed as desired. The process is used extensively in carrier telephony. It is also used in the doubling process (see **Frequency Doubler**) of some frequency **modulation** systems since the necessary doubling would raise the final frequency too high if it were not translated back to a lower value. Translation differs from doubling in that it does not bear a harmonic relationship with the original frequency and does not change the absolute width of the channel. (L.R.Q.)

FRESNEL COEFFICIENT OF DRAG. Ether.

FRESNEL MIRRORS. Young's Interference Experiment.

FRICTION BETWEEN WIND AND THE EARTH.

Even though winds may blow with considerable velocity above the earth's surface, air cannot move rapidly in a very shallow surface layer where air molecules are caught in irregularities of the surface. This slowing of a wind at the earth's surface is the result of friction between moving air and the earth. The effect of friction extends to about 500 meters or 1500′. The first effect of reduced velocity is a reduction in Coriolis Force and therefore a pressure gradient which is not balanced. Because the pressure gradient is unbalanced, air flows across the isobars from high toward low pressure. Thus, unbalancing the pressure gradient causes air to converge toward low pressure and diverge from high pressure.

Over land the average deflection of air across the isobars at a height of 20–30′ is about 30°, and over water 20°. Reduction of wind velocity depends somewhat on the nature of the surface but, in general, above the friction level the wind velocity will be 1½ to 2 times that observed at the surface. (P.E.K.)

FRICTION BRAKE. Dynamometer.

FRICTION, FLUID. Fluid Friction, Pipe Friction, Drag.

FRICTION GEARING.

A non-positive form of power transmission for rotating shafts. In the usual forms, friction gears may consist of cylindrical wheels, for transmitting power between parallel-axis shafts, and beveled wheels, for transmitting power between shafts whose axes will intersect if produced.

If friction wheels are assumed to operate without slip, the surface speed of both wheels must be equal. The velocity ratio of a pair of wheels is therefore inversely proportional to their diameters, or, if a 4″ wheel rotating at 300 rpm drives a 5″ wheel, the larger wheel will rotate at 240 rpm. In friction wheel drives, the driven wheel should be made of the harder material for if slip occurs, the softer driving wheel will wear uniformly about its periphery, and will not cut grooves into the surface of the driven wheel. Driving wheels may be made of leather, paper, or fiber; driven wheels are generally made of cast iron or aluminum.

The power H that may be transmitted by cylindrical friction wheels is given by

$$H = fFV/33,000$$

where f is the coefficient of friction between the wheel surfaces, F the contact pressure, and V the peripheral velocity, in ft. per min. Restrictions imposed by the ability of the friction wheel material to withstand crushing, and by the very heavy loads that the contact pressures impose on the shaft bearings, limit the use of friction gearing to light load transmission. (H.C.H.)

FRICTION, MECHANICAL.

The chief causes of friction are the interlocking of the minute irregularities on the rubbing surfaces, adhesion between the surfaces, and the indentation of the softer by the harder body. Friction between solid bodies may be classified as sliding and rolling. The laws of sliding friction were investigated by Coulomb, who found that, approximately and within limits, (1) the friction between two surfaces is slightly greater just before motion begins than when the surfaces are in steady relative motion; (2) the friction is proportional to the force pressing the surfaces together; (3) it is independent of the area of contact, and (except at start) of the speed of relative motion. The constant ratio of the friction to the force pressing the surfaces together is called the coefficient of friction, some typical values of which are as follows:

Dry wood on dry wood	0.35
Leather on metal	0.55
Iron on stone	0.50
Wood on stone	0.40
Stone on stone or brick	0.65
Well-oiled metals	0.05

By means of such coefficients, it is possible to calculate what the friction will be between two bodies, as a wooden sill on a stone foundation, when the force pressing them together is given.

The angle at which a plane surface must be inclined for a solid block to slide steadily down it is the angle of friction; its tangent is the coefficient of friction between plane and block. Lubrication greatly reduces the coefficient by separating the solid surfaces. Rolling friction, due to the indentation of the surfaces in rolling contact, is much less than sliding friction, as illustrated by the use of ball-bearings. The viscosity of liquids and gases is sometimes called "internal friction." (L.D.W.)

FRICTIONAL ELECTRICITY.

This familiar phenomenon is technically known as tribo-electrification. When two dissimilar substances are rubbed together, they become oppositely electrified; and if either is an insulator, it retains a charge. For example, if glass is rubbed with silk, the glass becomes positive and the silk negative. Careful experiments make it appear probable that this is a type of contact potential difference, and that the friction serves only to bring about surface contact over a larger area. Accurately ground and polished disks of steel and glass, when pressed firmly together to ensure close contact and then separated, show the same effect, the glass again being positive. Various experimenters have arranged lists of substances in such order that when any two are pressed or rubbed together and then separated, the one higher in the list becomes positive with respect to the other; but the data have often been conflicting. Coehn concluded from his experiments that the potential difference of the charges developed by two contacting dielectrics is proportional to the difference between their dielectric constants, the one having the greater constant being positive. (L.D.W.)

FRIEDEL-CRAFTS REACTION.

Aluminum chloride anhydrous, introduced by Friedel and Crafts, is used as reagent, generally in carbon disulfide solution to avoid rise in temperature, for the preparation of (1) aryl-alkyl hydrocarbons, (2) di- and tri-phenylmethane and derivatives, and (3) aryl-alkyl and diaryl ketones.

1. Aryl-alkyl hydrocarbons. The reaction takes place between benzene or its homologues and the alkyl haloid, thus:

$$C_6H_5 \cdot H + Cl \cdot CH_3 \rightarrow C_6H_5 \cdot CH_3 + HCl$$

Benzene · Methyl chloride · Toluene · Hydrogen chloride gas evolved

$$C_6H_4\begin{matrix}H \\ H\end{matrix} + \begin{matrix}Cl \cdot CH_3 \\ Cl \cdot CH_3\end{matrix} \rightarrow C_6H_4(CH_3)_2 + HCl$$

Benzene · Methyl chloride · Xylene

2. Di- and tri-phenylmethane, derivatives. The reaction takes place between benzene and benzyl haloid or methylene haloid in the case of diphenylmethane, and between benzene and benzal haloid or chloroform in the case of triphenylmethane, thus:

$$C_6H_5CH_2Cl + HC_6H_5 \rightarrow C_6H_5CH_2C_6H_5 + HCl$$

Benzyl chloride · Benzene · Diphenylmethane

$$H_2CCl_2 + \begin{matrix}HC_6H_5 \\ HC_6H_5\end{matrix} \rightarrow C_6H_5CH_2C_6H_5 + \begin{matrix}HCl \\ HCl\end{matrix}$$

Methylene chloride · Benzene · Diphenylmethane

$$C_6H_5CHCl_2 + \begin{matrix}HC_6H_5 \\ HC_6H_5\end{matrix} \rightarrow C_6H_5CH\begin{matrix}C_6H_5 \\ C_6H_5\end{matrix} + \begin{matrix}HCl \\ HCl\end{matrix}$$

Benzal chloride · Benzene · Triphenylmethane

$$HCCl_3 + \begin{matrix}HC_6H_5 \\ HC_6H_5 \\ HC_6H_5\end{matrix} \rightarrow C_6H_5CH\begin{matrix}C_6H_5 \\ C_6H_5\end{matrix} + \begin{matrix}HCl \\ HCl \\ HCl\end{matrix}$$

Chloroform · Benzene · Triphenylmethane

3. Ketones. The reaction takes place between **benzene** and paraffin or benzenoid acyl haloid thus:

$$CH_3COCl + HC_6H_5 \rightarrow C_6H_5COCH_3 + HCl$$

Acetyl chloride   Benzene   Acetophenone

$$C_6H_5COCl + HC_6H_5 \rightarrow C_6H_5COC_6H_5 + HCl$$

Benzoyl chloride   Benzene   Benzophenone

The keto-group occupies the position para to alkyl already present. Two acyl groups have been placed in mesitylene to form diacetylmesitylene:

$$\begin{array}{c} CH_3 \\ H_3COC \diagup\!\diagdown COCH_3 \\ H_3C \diagdown\!\diagup CH_3 \end{array}$$

Summarizing: **Benzene** or its homologues plus paraffin-substituted haloid in the presence of **aluminum** chloride anhydrous react with the elimination of **hydrogen chloride**. In several cases an intermediate compound of the reactants with aluminum chloride has been identified. The reaction has been studied in detail by J. F. Norris and his co-workers.

Other reactions involving aluminum chloride anhydrous are:

a. **Xylene** plus **benzene** to yield **toluene,** and the reverse, namely, toluene to yield xylene plus benzene. Boiling temperature.

b. Benzene, toluene and homologues chlorinated by reaction with chlorine gas.

c. Benzene sulfinated by reaction with sulfur dioxide. Benzene sulfinic acid ($C_6H_5 \cdot SOOH$) formed. (R.K.S.)

FRIGATE BIRD. Aves, Pelecaniformes. *Fregata.* Marine birds (**Aves**) of slender build and powerful flight. They live chiefly on fish which they force other birds to give up to them. (A.W.L.)

FRITILLARY. Insecta, Lepidoptera. A **butterfly** of the genus *Argynnis.* Most species are red-brown with black markings and are spotted beneath with silver. Fritillary is also a genus of plants of the Lily Family, containing some 50 species growing in north temperate regions. They are mostly spring-blooming plants having bell-shaped flowers. The flowers are often spotted or mottled with dark purple and green spots. Most frequently grown are the snakes' head, *Fritillaria Meleagris* and the Crown Imperial, *F. imperialis.* Many of these plants have a strong fetid odor which prevents them from being too popular. (A.W.L., R.M.W.)

FROG. Amphibia, Anura. Tailless vertebrates with smooth moist skin and large hind legs, used for jumping and swimming. Closely related to the toads, which dif-

Three species of bullfrogs (*American Museum of Natural History.*)

fer in their more terrestrial habits and warty skin. The forms are not sharply separated; arboreal species of the family Hylidae are called both tree frogs and tree toads. (See **Railway Track.**) (A.W.L.)

FROG HOPPER. Insecta, Homoptera. Small jumping **insects** whose form faintly resembles that of the frogs. The immature insect sucks the sap of a plant and secretes about itself a protective frothy mass, hence they are also called spittle insects or spittle bugs. (A.W.L.)

FROG MOUTH. Aves, Caprimulgiformes. *Podargus.* Birds (**Aves**) of the Oriental and Australian regions with very short beaks and wide mouths. Related to the whip-poor-will and other goatsuckers. (A.W.L.)

FRONTAL APPENDAGE. An appendage of an extra pair between the second antennae of some **crustaceans.** (A.W.L.)

FRONTAL BONE. Bones of the **vertebrate** skull lying between and behind the eyes, usually paired but in most human skulls united to form the single large bone of the forehead. (A.W.L.)

FRONTAL ORGAN. An organ found on the front of the head of some **crustaceans.** Usually paired and probably sensory. (A.W.L.)

FRONTS, FRONTOGENESIS, AND FRONTOLYSIS. Surfaces of discontinuity between air masses are called frontal surfaces. The intersections of these frontal surfaces with the surface of the earth are called fronts. Frontal surfaces are also loosely spoken of as fronts. Actually the discontinuity usually extends through a varying zone of transition which is known as the frontal zone. By convention the exact frontal surface is taken as the boundary between the warm air and the frontal zone. Frontogenesis is the formation or intensification of a frontal surface. Frontolysis is the disintegration or weakening of a frontal surface. Practically all frontal zones have clouds associated with them. Some are accompanied by precipitation. A great percentage of temperate-zone precipitation and weather is directly the result of frontal activity.

The equilibrium slope of a front is given by the relation:

$$\tan \alpha = \frac{2\omega \sin \phi}{g} T_m \left[\frac{v_2 - v_1}{T_2 - T_1} \right],$$

where $\tan \alpha$ = Tangent of α, the angle of slope of the frontal surface to sea level.

ω = Angular velocity of the earth.

ϕ = Latitude angle.

T_m = Mean temperature of the air masses.

T_1, T_2 = Temperature of each air mass, respectively.

v_1, v_2 = Wind velocity parallel to the frontal surface in each air mass, respectively.

From this relation it is apparent that steep equilibrium frontal slopes result from high mean temperatures, high latitudes, large differences in the velocity components, and small differences in the temperatures between the two masses. The slope of a front is always such that cold air underlies warm.

There are two types of major frontal zones in the atmosphere resulting from the general circulation:

1. Between polar winds with an easterly component and sub-tropical winds with a westerly component surfaces of discontinuity develop known as polar fronts. These are the world's principal fronts and cause much of the storminess of the temperate zone.

2. Between the northeast and southeast trade winds a surface of discontinuity is maintained known as the equatorial or inter-tropical front. Tropical cyclones develop on the equatorial front, particularly when it is displaced far north and south.

When polar fronts have pushed far south in mid-winter discontinuities begin to develop between the polar air and colder arctic air to the north. These fronts are called Arctic Fronts.

When the cold air on one side of a front is replacing the warm air horizontally the front is a cold front. A relatively narrow band of cloudiness and precipitation normally accompanies a cold front. At the approach, passage, and recession of a cold front in the northern hemisphere the following phenomena commonly occur in order:

1. Sky and weather. Increasing cloudiness of alto-cumulus, cumulus, or stratocumulus types; overcast with heavy showers; clearing to scattered cumuli.

2. Temperature and wind. Wind from southerly or westerly direction with temperature steady or rising slightly; wind shifting, sometimes with violent gusts and high velocity, from southwesterly to northwesterly and temperature starting to drop; wind fresh from new direction and temperature dropping steadily.

3. Pressure. Barometer falling prior to wind shift, then rising rapidly after shift and continuing to rise.

When the warm air on one side of a front is replacing the cold air horizontally and overriding the cold air, the front is a warm front. Over the average warm frontal surface a wide band of clouds and precipitation are present. Precipitation fog is also likely to form over a considerable area. At the approach, passage, and recession of a warm front, the following usually occur in order:

1. Sky and weather. Increasing cloudiness, first cirrus, then cirrostratus, altostratus, rain, or snow; stratus, nimbostratus, scud, sometimes fog a considerable period of time; clearing to scattered or broken stratocumulus and cumulus type clouds.

2. Temperature and wind. Wind from some easterly quadrant and temperature steady and relatively low; wind shift at front passage from easterly quadrant to southerly quadrant and considerable temperature rise; more or less steady southerly winds with much higher temperature.

3. Pressure. Barometer falling until wind shift, then steady or falling very slightly.

Cold-front slopes are greater than, and warm-front slopes are less than, the equilibrium slope. Stationary fronts have approximately the equilibrium slope. Cold fronts move with relatively steady velocity although subject to some acceleration. Warm frontal movement is commonly erratic, but generally slower than cold fronts.

Warm and cold fronts are common to the temperate zone, particularly in winter when large tropical and polar air masses alternately surge northward and southward. The equatorial front is relatively stationary but moves north and south seasonally. (P.E.K.)

FROST. Frost, like snow, is the result of the **sublimation** of water vapor in saturated air. If there is excessive radiation from solid objects, as on a clear night in late fall, the air coming in contact with them may be chilled below the sublimation point, and spicules of ice grow out from the cold surfaces. The process is entirely similar to the formation of metallic crystals artificially from vaporized metals on the interior of a glass tube communicating with the vaporizing furnace. In either case the size of the crystals is a matter of time and the supply of saturated vapor. Frost is often observed around cracks in a wooden sidewalk, because of the damp air escaping from the ground below. The objects upon which frost forms most readily are those of low specific heat and high thermal emissivity, such as blackened metals; hence the marked accumulation of frost on the heads of rusty nails. The apparently erratic occurrence of frost in adjacent localities is due partly to differences of level, the lower areas becoming colder; but also largely to differences in absorptivity and specific heat of the ground, which, in the absence of wind, greatly influences the temperature attained by the superincumbent air. It should be understood that vegetation is not damaged by frost itself, but by cold air; the appearance of frost

merely indicates that the temperature has dropped below the freezing point. The formation of white frost on the indoor surface of window panes indicates low relative **humidity** of the indoor air; otherwise water would first condense in small drops and then freeze into clear ice. (L.D.W.)

FRUCTOSE. Carbohydrates.

FRUIT. While this word is often used to describe any product of the soil, to the botanist a fruit is the ripened ovary of the **flower.** Often, however, it is necessary to amend this definition in order to include certain other tissues which are a part (often a very large part) of a fruit. In the strawberry, for instance, the red pulp is not the ovary, but a very much enlarged and modified stem tip, the receptacle of the flower. Another illustration is found in the "fruit" of the pineapple, a large part of which is stem and not ovary. Often, too, it is difficult to distinguish a fruit from a seed, and many fruits are invariably called seeds. Sunflower and strawberry "seeds" are really fruits, as are the "seeds" of grasses.

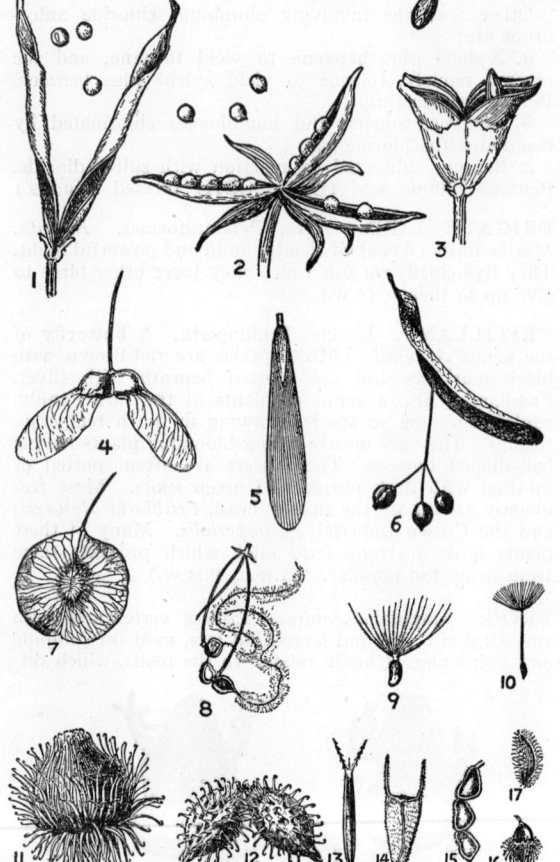

Dry fruits with devices for dispersal of their seeds. One to three, by sudden dehiscence. 1, wild bean; 2, violet; 3, witch hazel. Four to ten, by wind. 4, maple; 5, ash; 6, basswood; 7, elm; 8, *Clematis;* 9, thistle; 10, dandelion. Eleven to seventeen, by animals. 11, burdock; 12, cocklebur; 13, Spanish needle; 14, beggar's tick; 15, beggar's lice; 16, agrimony; 17, carrot.

There are many kinds of fruits. Usually they are separated into two classes, dry fruits and fleshy fruits. Dry fruits are again separated into dehiscent fruits, those which split open when ripe, and indehiscent fruits, which do not do so. Common dehiscent dry fruits are the legume, the follicle, and the capsule; dry indehiscent

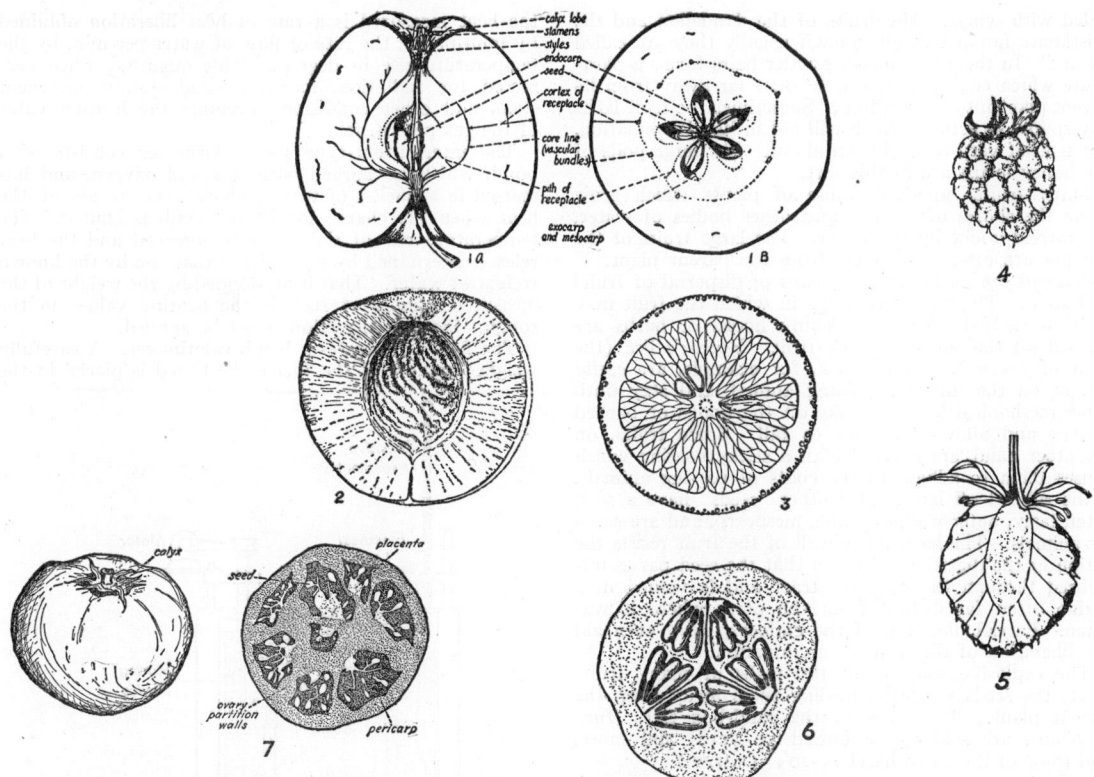

Fleshy fruits. 1a and 1b, apple, illustrating pome (*after Robbins*); 2, peach, illustrating drupe (*after Decaisne*); 3, orange, illustrating berry (*after Decaisne*); 4, raspberry, illustrating aggregate fruit (*after Chamberlain*); 5, strawberry, illustrating aggregate fruit (*after Brown*); 6, cucumber, illustrating pepo (*after Chamberlain*); 7, mature fruit of tomato: left, surface view; right, cross section. (*Redrawn.*)

fruits are the achene, the caryopsis or grain, the samara and the nut. A legume is a fruit which when ripe splits along both edges; it develops from a single **carpel**. The pods of peas and beans are legumes. A follicle is similar to a legume, but splits along one side only. Milkweed pods are follicles. The fruits of the columbine and larkspur are also follicles. A capsule is a dehiscent fruit which develops from a compound ovary. The fruit of a lily or an iris is a capsule. The achene is a single-seeded indehiscent fruit which when mature has the seed free from the ovary wall except at the point of attachment. The fruits of the buttercup are achenes; also the fruits of the strawberry, which are the small hard bodies borne on the surface of the "berry." The achene of the composite family differs from the others in having the calyx tube coalesced with the ovary wall. A caryopsis or grain is very similar to an achene, but has the seed coat fused with the pericarp so that the seed cannot be removed from the ovary wall. The fruits of all cereal grasses are caryopses. The samara is an indehiscent fruit which has a wing. The fruits of the maple and elm tree are samaras. A nut is a very hard-shelled fruit, usually 1-seeded. Walnuts, acorns, and beechnuts are examples.

Fleshy fruits include the pome, the drupe, the berry, and many special types often called aggregate and multiple fruits. A pome is a fleshy fruit developed from an epigynous flower having a compound ovary. The flesh of the pome consists of an enlarged and ripened receptacle together with the outer layers of pericarp. An apple is a pome. The tough paper-like part of the core is the endocarp. A drupe, or stone fruit, is developed entirely from the ovary. It is a 1-seeded fruit having a fleshy mesocarp and a hard endocarp. Peaches, cherries, prunes and olives are drupes. A berry is a fleshy fruit having all the pericarp fleshy, and con-

taining one or more seeds. The fruit of the tomato, the grape, the banana and the cranberry are true berries. The "skin" of the banana represents the receptacle of the flower, and is thus stem tissue. In many plants the fruit is very similar to a berry, but is partly composed of receptacle tissue which forms a hard rind. These fruits, such as squashes, cucumbers and melons, are called pepos. If the outer wall of a berry is leathery, as in the orange, the fruit is known as a hesperidium.

Aggregate fruits are those which are formed from a single flower which had many simple pistils. In these fruits the receptacle forms a considerable part of the whole. Raspberries and blackberries are aggregate. In some fruits of this class the separate fruits break from the receptacle, as in the raspberry, while in others they remain firmly attached. The individual fruit in the raspberry is a tiny drupe, called a drupelet. A multiple fruit is formed from the ovaries of many flowers growing in a compact mass. Mulberries and pineapples are fruits of this type.

As the fruit ripens, the ovary wall or pericarp grows. Three layers of cell tissue are usually recognizable as the ovary matures. The outermost layer or exocarp, which is usually a thin layer, often an epidermis only one cell thick. The innermost layer is the endocarp. Between these is the mesocarp, in which the vascular tissues ordinarily occur. The relative thickness and appearance of these layers vary greatly in different fruits.

Often the structure of the fruit is directly correlated with the dissemination of the seeds. The principal agencies for fruit dispersal are wind, water, and animals. Many fruits are provided with wings, thin blade-like structures which enable the fruits to drift slowly downward and generally away from the parent plants. The fruits of the maple, the ash, and the **ailanthus** are provided with wings. The fruits of the dandelion and the thistle are familiar to all, though usually they are called "seeds." In them a group of slender hairs forms a parachute which enables the fruit to drift far away from its parent plant into new regions. Sometimes the fruit lacks any special structures which will aid in its dissemination, but is itself very easily blown about. The large pods of the honey locust are of this sort.

Many fruits, especially those of plants which grow along the shores of streams and other bodies of water, are carried about by the water. The large fruits of the coconut are often carried far from the parent plant.

Animals are an important means of dispersal of fruits and seeds. There are two ways in which the fruit may be thus carried. Very commonly hooks or barbs are formed on the surface of the fruit. Beggar's lice (the fruit of *Bidens*), and burdock, for example, are easily caught on the fur of a passing animal, and stay until some mechanical injury breaks off the hooks or barbed bristles and allows the fruit to fall. Some fruits, on the other hand, are covered with a sticky coating which causes them to adhere to the coats of passing animals, to be rubbed off later. Still other "seeds" have a soft, often tasty, outer wall, or edible mesocarp, and are eaten by animals. The hard inner wall of the fruit resists the action of the digestive juices, so that the seed passes uninjured through the digestive tract and is voided in a region often far distant from the place where it was eaten. Partial digestion of the endocarp may even aid the liberation of the seed for germination.

The explosive splitting of the walls of many fruits ejects the seeds violently, hurling them away from the parent plant. The seeds of the sand-box tree, *Hura crepitans,* are said to be hurled 50 yards and more, and those of the witch hazel 15–20′. (R.M.W.)

FRUIT FLY. Drosophila.

FRUIT SUGAR. Carbohydrates.

FUCHSINE. Dyes.

FUCOXANTHIN. Pigments in Plants; Amino-acids and Proteins.

FUEL CALORIMETER. The heating value of a **fuel,** in terms of heat units such as the B.T.U., may be obtained experimentally with the use of a fuel calorimeter. As fuels exist in solid, liquid, and gaseous states, fuel calorimeters would, of necessity, be required to be able to measure the heat of either solids, liquids, or gases. The conditions of measurement of a gas call for an instrument different in many essential respects from that which would be satisfactory for a solid or liquid fuel, principally because it is impractical to measure out a gas in definite isolated quantities, as is so easily done with liquids and solids. Hence a gas calorimeter is a continuous-flow instrument, whereas the liquid and solid fuel calorimeter is an intermittent type, wherein a known weight is burned. Nevertheless, the principle underlying the measurement of heat in all fuel calorimeters is the absorption of heat by water, creating a temperature rise. Measurements of the quantity of water and temperature rise are used directly to determine the heat units, since a unit of heat raises the temperature of a unit weight of water one degree. In the English system, the B.T.U. is indicated when 1 lb. of water at about 60° is raised in temperature 1° F.

The Junker's gas calorimeter is a chamber wherein a known metered flow of gas is burned in an efficient-type burner, with liberation of heat which is absorbed in water as the products of combustion flow through the tubular passages of the calorimeter. The water which absorbs the heat is likewise flowing steadily through the calorimeter, being separated from the gas by the walls of the tubes. The temperature of the water entering and leaving is measured by thermometers, while the rate of flow of water is measured by catching and weighing it. Thus

the heat measured is a rate of heat liberation obtained by multiplying the rate of flow of water per min. by the temperature rise in degrees. This quantity, when corrected for radiation, moisture condensation, emergent stem, and other conditions, becomes the heating value of the gas per cu. ft.

The intermittent-type fuel calorimeter consists of a bomb which is charged with fuel and oxygen, and immersed in a bucket of water, which serves to absorb the heat when the charge inside the bomb is ignited. The temperature rise of the water is observed and the heat release determined by multiplying that rise by the known weight of water. That heat, divided by the weight of the measured sample of fuel, is the heating value—in the rough. Several corrections must be applied.

The figure illustrates a bomb calorimeter. A carefully weighed sample of the fuel to be tested is placed in the

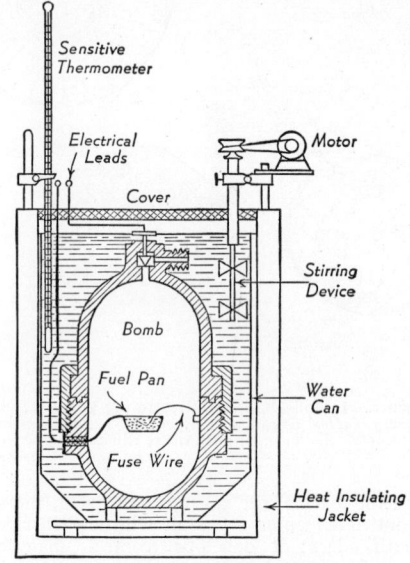

Emerson bomb calorimeter.

pan within the steel bomb. A calibrated length of fuse wire is drooped into the pan, then the bomb is screwed tightly together. It is next connected to a tank of oxygen and charged with that gas to a pressure of several atmospheres, after which it is removed and immersed in the water of the calorimeter. When the water and calorimeter have arrived at room temperature, the electric leads are connected to a source of voltage through a switch, a stirring device is started, and the circuit is closed. The rush of current fuses the calibrated wire, raising it to a temperature sufficient to ignite the fuel in the presence of oxygen. The fuel then burns rapidly —almost explosively—and the heat it liberates is absorbed by the water. Temperature rise is measured by a sensitive thermometer. A large amount of water is used, so that the temperature rise will be very small. Corrections must be made for radiation, emergent thermometer stem, heating value of the fuse wire, water equivalent of the calorimeter, and energy input of the stirring device. (F.T.M.)

FUEL INJECTION. Fuel injection is taken to mean the direct injection of fuel into the **combustion chambers** of an internal combustion engine by means of an injection valve located in the wall of the chamber. For *injection carburetor* see **Carburetor Types.** Although the spark ignition engine (Otto) has been built for fuel injection, it is nearly universal practice to carburet the spark ignition engine and use fuel injection for the compression ignition (Diesel) engine.

The injection of fuel into a Diesel cylinder is a precise,

split-second operation. A minute quantity of fuel oil must be injected as a spray into air compressed to 500 lbs. per sq. in. and over. It must become well mixed with the air and burn almost instantaneously, yet at a definite rate so that pressure rise can be controlled. In a 100 hp. cylinder (not engine) of an engine operating at the moderate speed of 750 rpm the amount of each injection is approximately two *drops* of oil. The injection must be started, completed, and the burning finished all in approximately .01 sec. This description is illustrative of quantities for one of the larger Diesels, at full load. Part load operation and smaller power capacities would each decrease the volume of fuel oil injected. To meter such minute volumes accurately under the action of speed responsive *governors* and to inject it with proper timing and atomization into a region of "hard" air (i.e., highly compressed) requires the very best of machine work in construction of the injection pumps, which are the heart of the system, and of the injection valves. Maximum penetration of the fuel into the combustion chamber requires a dense spray of small diameter, but good mixing of fuel and air needs a fine, well-atomized spray which, unfortunately, has poor penetration. In some systems mixing is greatly assisted by special designs of the **combustion chamber.**

The pumps are small diameter plunger type pumps which usually open a by-pass for cut-off of fuel rather than curtail the plunger stroke. The arrangements of injection on multi-cylinder engines could be classified as follows:

Individual Pump System. In this, the most popular system, an injection pump is provided for each engine cylinder. A spring-loaded injection valve is provided in each cylinder and the pump discharge is connected to it by high pressure capillary-type tubing.

Common Rail System. A single pump maintains injection pressure in a common header from which branches lead to the injection valves. But now, the pumps not providing metering of fuel, the injection valve must be mechanically operated (cams and rocker arms, etc.) to provide the timing and metering function.

Distributor System. This may also be a single-pump system, but with the injection valve drive relieved of the need for metering the fuel. A single injection pump meters the fuel for all cylinders at low pressure. The metered fuel is delivered through a rotating distributor valve which shunts the metered fuel to the various injection valves. The mechanically operated valve plungers then raise the oil pressure sufficiently for spraying. Thus the valve action times and pressurizes the fuel but does not meter it.

Unit Injector System. In this system the functions of the pump and injection valve are combined in a single mechanical unit attached to the cylinder in the usual location of the injection valve. It must be mechanically operated and the method of actuation must include the timing, metering, and pressurizing function. Oil is furnished to the unit injectors from a common header supplied by a low-pressure pump.

In the early development of Diesels the best results were secured with air injection wherein small quantities of highly compressed air were employed at the injection valve to drive the fuel into the combustion chamber in a finely atomized state. The necessity of a high-pressure air compressor (usually three-stage with intercoolers) was a disadvantage, so when builders were able to overcome the problems of airless, or "solid," injection the air injection principle was gradually abandoned. (F.T.M.)

FUEL OIL. The petroleum oil obtained from wells in various parts of the earth is rarely suitable as a fuel oil in its crude form. Fuel oils are the crude oil after some of the lighter and heavier fractions have been removed. (See **Fuels; Hydrocarbons; Petroleum Products.**)

The principal production fields in the United States are in Pennsylvania, Ohio, Texas, and California. Petroleum is composed of a series of hydrocarbons in various proportions. A typical analysis of crude oil would be 84% carbon, 13% hydrogen, 1% sulfur, 1% nitrogen, 1% oxygen. Practically all of the hydrogen is "free" (that is, it is in a state of combination, as with carbon, in which most or all of its heat of combustion is available), and since it is in larger proportions by weight than in coal, fuel oil might be expected to have a considerably higher heating value. The principal uses of fuel oil at the present time are in **Diesel** engines, for firing in the furnaces of steam **boilers,** and for use in domestic oil-burning furnaces. The value of a fuel oil may be gauged by the following properties:

Calorific Value. This, of course, is important because it shows the amount of energy available. The calorific value of fuel oils is from 18,000 to 20,000 B.T.U. per lb.

Viscosity. Viscosity is one of the most important qualities of a fuel oil. It determines the fluidity of the oil and is a fair indication of how readily the oil will atomize. A very viscous oil may prove troublesome to handle without heating to a point where viscosity is more satisfactory. Viscosity is generally stated in terms of seconds, Saybolt Universal, at some specified temperature, usually 100° F.

Ignition Quality. This characteristic measures the ability of fuel oil to ignite spontaneously in an engine cylinder. Good ignition quality is represented by high **cetane number.** Ignition quality is an index to possible detonation, ease of starting, and smoky exhaust.

Carbon Residue. The standard test for this characteristic is the **Conradson Test** which gives an indication of the amount of carbon that may be deposited as a scale in an engine combustion chamber. Although low-carbon residue is a desirable quality and will yield better engine operation, the lower cost of less completely distilled fuels makes them more economical in use.

Solid Impurities; Ash and Asphalt Content. The effect of these is to cause deposits in the combustion chamber and around the piston rings, necessitating increased maintenance. Ash is the most detrimental quality in fuel oil.

Water Content. Few oils are free from water, but a good oil will not contain more than 1% of water. Besides forming sludge, the water content decreases the calorific value of a fuel.

Sulfur and Acid. These components will cause corrosion of cylinders and valves under certain conditions.

Gravity. The Baumé scale is one of specific gravity.

$$\text{Degrees Baumé} = \frac{140}{\text{specific gravity}} - 130$$

(For liquids lighter than water)

The specific gravity is referred to water as 1. Suitable fuel oils range from 15–43° Baumé. The Baumé reading is useful in determining whether the oil is burdened with an undue amount of heavy residue which is difficult to burn successfully.

Flash and Boiling Points. These indicate the degree to which a fuel can be vaporized; also the inflammability and hence the fire risk of storage and handling. (F.T.M.)

FUELS. Fuels are those materials which when burned with air or **oxygen,** furnish heat energy. The rate at which a given fuel is burned determines the temperature at the point of burning. Since heat flows from bodies at a higher to those at a lower temperature in proportion to the temperature difference between them, the temperature of combustion is important when the rate of heat flow is a consideration.

The materials used as fuels fall into two grand divisions, (1) fuels—gases, liquids, solids—for ordinary combustion, (2) fuels for non-ordinary or **internal combustion engines,** motor and stationary. These are either gases or volatilizable liquids, usually **petroleum** or coal tar products, capable of rapid reaction (explosion) when mixed with air or oxygen and ignited. This *ignition* is usually accomplished by an electric spark, but high and rapid compression (that is, **adiabatic**, without heat loss, energy of compression being accounted for by temperature rise) is sufficient in certain cases, for example, in the **Diesel engine**.

The heat-evolving chemical elements in domestic and industrial fuels are **carbon** and **hydrogen**. Carbon as such is utilized in coke and charcoal, and hydrogen as such in hydrogen gas for high-temperature combustion, and in fuel gas mixtures, such as coal gas, water gas, producer gas and blast furnace gas. Carbon and hydrogen compounds are present (1) in the hydrocarbons of natural gas, petroleum and most of its products, and coal and most of its products, and (2) in the carbohydrates and lignins of wood and peat.

The naturally occurring fuel materials are (1) natural **gas**, (2) **petroleum**, (3) **coal** (a variety of forms), (4) **wood**, (5) **peat**, and (6) **lignite**. Artificial fuels are prepared (1) by the **destructive distillation** of coal (soft or bituminous coal) resulting in coal gas, coal tar and coke, (2) by the destructive distillation of wood resulting in charcoal, (3) by the fractional and destructive (cracking) distillation of petroleum, resulting in **gasoline**, and lower and higher boiling point distillates in "still" gases and in fuel oil residue, (4) by fuel gas reactions, resulting in water gas from coke and steam, in **producer gas** from coal or coke and air, oil-water gas from petroleum and steam. Other fuels are (1) blast furnace and other gases containing combustible material often with sensible heat at temperatures sufficiently high to be useful as sources of heat, (2) **alcohol**, (3) **benzene**, (4) **ethylene**, (5) **acetylene**.

Naturally Occurring Fuels: 1. *Natural gas.* Contains up to 95% of **methane**. Used as a fuel, and by incomplete combustion as a source of carbon pigment or lamp black. 2. *Petroleum.* (See **Hydrocarbons.**) Distilled to obtain gasoline, kerosene, lubricating oils, low boiling distillates, high boiling distillates, fuel oil, gas carbon. 3. *Coal.* *a.* Bituminous, of 12,000 to 15,000 B.T.U. per lb., of variable ash content as to amount of ash and as to the clinkering or fusing temperature of the ash. Used as fuel, or treated by destructive distillation to obtain coal gas, coal tar, coke, thence **benzene**, water gas, producer gas. Current and future developments are powdered coal, low-temperature distillation, hydrogenation. *b.* Sub-bituminous, of less than 12,000 B.T.U. per lb. Used as fuel. Important future reserve. *c.* Lignite and peat. Used as fuel. Important future reserve. *d.* Anthracite, of 14,000 B.T.U. per lb., more or less. Used as fuel, of diminishing importance. 4. *Wood.* Used as fuel, value as such markedly affected by the water content as heat from combustion is required to vaporize the water. Freshly cut wood usually contains 60% or more water, calculated on the weight of dry wood. Heating value of air dried woods approximately 21 (plus or minus 10%) million B.T.U. per cord, equivalent to approximately 0.8 short ton of coal of 13,000 B.T.U. per lb. Treated by destructive distillation to obtain methyl alcohol, woodtar, charcoal, and by burning to obtain carbon dioxide. 5. *Carboniferous shale.* Important future fuel reserve by treatment by destructive distillation to obtain hydrocarbon distillates. (See **Destructive Distillation.**)

Artificially Prepared Fuels. 6. *Coal gas.* Used as fuel. Treated to obtain benzene, toluene, ethylene. Oil-water gas is practically the same mixture as coal gas. Approximate composition: **methane** 30%, **hydrogen** 50%, **carbon monoxide** 6%, **ethylene** 2.5%. Fuel value: 550 (more or less) B.T.U. per cu. ft. 7. *Coal tar.* Treated by fractional distillation to obtain benzene, toluene (and other coal tar crudes). 8. *Coke.* Used as fuel and metallurgical reducing agent. Treated to obtain water gas, producer gas, carbon monoxide. 9.

Charcoal. Used as fuel and metallurgical reducing agent. 10. *Gasoline.* Used as motor fuel (and as solvent and diluent). (See **Hydrocarbons,** paraffin.) 11. *Low boiling naphtha.* Used as special motor fuel (and as solvent and diluent). 12. *High boiling kerosene and stove oil.* Used as fuel and illuminant. 13. *Fuel oil.* Distillation residue used as fuel in ordinary or internal combustion. Heating value approximately 19,000 B.T.U. per lb., which is more than 150,000 B.T.U. per gal. 14. *Water gas.* Made by reaction of coke and steam at furnace temperatures. Used as fuel, and source of **hydrogen** and **carbon monoxide.** Approximate composition: hydrogen 45%, carbon monoxide 45%. Approximate fuel value: 350 B.T.U. per cu. ft. 15. *Producer gas.* Made by reaction of coke or coal and air at furnace temperatures. Used as internal combustion fuel. Approximate composition: 25% carbon monoxide. Approximate fuel value: 100 B.T.U. per cu. ft. 16. *Blast furnace gas.* Used as fuel and for the sensible heat content. Approximate composition: 25% carbon monoxide. Approximate fuel value: 100 B.T.U. per cu. ft. 17. *Alcohol.* Future fuel possibility, either straight or blended with other liquid fuels. When 1.1% of water is present in a mixture containing 80% gasoline plus 20% alcohol separation into two layers takes place. (See **Ethyl Alcohol.**) 18. *Benzene.* Minor use as fuel. Future fuel possibility, either straight or blended with other liquid fuels. Excellent antiknock fuel for internal combustion, and miscible with gasoline. (See **Benzene.**) 19. *Ethylene.* Excellent fuel for local high temperatures, used in special burner with oxygen. (See **Ethylene.**) 20. *Acetylene.* Excellent fuel for local high temperatures, used in special burner with oxygen. (See **Acetylene.**) 21. *Hydrogen.* Excellent fuel for local high temperatures, used in special burner with oxygen. An important component of coal gas, water gas. (See **Hydrogen.**)

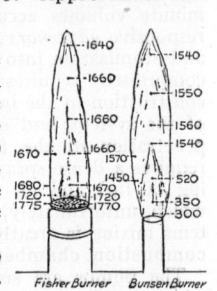

Fisher Burner   Bunsen Burner

Temperatures in degrees Centigrade of gas flames.

HEATING VALUE OF VARIOUS INDIVIDUAL FUEL SUBSTANCES

INDIVIDUAL FUEL GASES	HIGHER HEATING VALUE (Water Condensed as formed in Combustion)	
	In B.T.U. per cu. ft. at 60° F. and 30 in. (Hg) pressure	In kilogram-calories per gram-molecular weight
Methane............	995	211
Ethane.............	1730	368
Propane............	2465	526
Butane.............	3200	680
Ethylene...........	1615	332
Acetylene..........	1455	312
Hydrogen..........	319	68.4
Carbon monoxide.....	317	67.1

INDIVIDUAL FUEL LIQUIDS	HIGHER HEATING VALUE	
	Kilogram-calories per gram	Kilogram-calories per gram-molecular weight
Normal-pentane......	11.6	883
Normal-hexane.......	11.5	990
Normal-heptane......	11.5	1150
Normal-octane.......	11.4	1303
Methyl alcohol.......	5.34	171
Ethyl alcohol........	7.13	328
Normal-propyl alcohol	8.00	481
Benzene.............	10.0	782
Toluene.............	10.2	934

Individual Fuel Solids	Higher Heating Value Kilogram-calories per gram
Carbon, amorphous, to carbon dioxide	8.08 (97.0 Kg-Cal per gram molecular weight carbon dioxide)
Carbon, amorphous, to carbon monoxide	2.49 (29.9 Kg-Cal. per gram molecular weight carbon monoxide)
Cellulose...........	4.21

The examination of fuels is conducted with various purposes in mind. An "ultimate" analysis is designed to give the percentage of each element present, by burning carbon element to carbon dioxide, hydrogen element to water, sulfur element to sulfur dioxide or to sulfate, nitrogen element to nitrogen gas, phosphorus element to phosphate, inorganic residue as ash. The ash may be analyzed. The heating value of a fuel is determined by burning a definite weight or volume of a gas in a calorimeter. A "proximate" analysis is designed to give the percentage of water, of volatilizable material under given conditions of temperature and time, and of non-volatile combustible residue. A determination of the softening temperature of the ash may be made. A "rational" analysis is designed to give the percentage of each compound in the fuel. This is accomplished in the case of gases by measurements of volume before and after absorption in definite reagents or other treatment applied in a definite order. For complex mixtures of liquids and solids this type of analysis is difficult, and at present impossible in the case of some, for example coal, although some progress has been made. Special tests are sometimes demanded, as in studying the coking conditions of coal, and the explosive and power characteristics of internal combustion fuels. The velocity of explosion of certain gas mixtures has been measured by Dixon.

Gas Mixture Taken	Products Obtained	Observed Velocity of Explosion Meters per Sec.
$2H_2 + 2O_2$	$2H_2O + O_2$	2328
$8H_2 + O_2$	$2H_2O + 6H_2$	3532
$CH_4 + O_2$	$H_2O + COH_2$	2528
$C_2H_4 + 2O_2$	$2H_2O + 2CO$	2581
$C_2H_2 + O_2$	$2CO + H_2$	2961

Industries using fuels for the production of high temperatures are ceramics, Portland cement, glass, pig iron, steel, copper, zinc, lead, nickel, tin, destructive distillation of coal. The economies introduced in the use of coal since World War I are of such degree as to appreciably affect the demand for this fuel. In 1919, the fuel consumption per thousand gross ton miles by Class I steam railroads using coal-fired locomotives was 170 lbs., whereas in 1941 it was only 111 lbs.; and for public utility central station electric power plants using coal the fuel consumption in 1919 was 3.2 lbs. per kw-hr. and in 1941 the average was 1.3 lbs. (R.K.S.)

FULGURITE. A vertical, sometimes branching tube of fused quartzitic sand formed from the intense heat developed when the sand is struck by lightning. (R.M.F.)

FULLER'S EARTH. This is a fine-grained earthy substance similar to clay both in appearance and composition, but it lacks plasticity and is usually high in magnesia. It has the property of decolorizing oils and removing grease from raw wool. (E.S.C.S.)

FULMAR. Aves, Procellariiformes. Large marine birds (Aves) resembling the gulls. **Petrels.** (A.W.L.)

FULMINATES. Cyanic Acid.

FUMAROLE. Derived from the Latin *fumus,* smoke, the term fumarole is applied to openings in the earth's crust, often in the neighborhood of volcanoes, which emit steam and gases such as carbon dioxide, hydrochloric acid, and hydrogen sulfide. A special name, solfatara, from the Italian *solfo,* sulfur, is given to fumaroles that emit sulfurous exhalations. Perhaps the greatest area of fumarole activity is the famous Valley of Ten Thousand Smokes, adjacent to Katmai volcano, Alaska. (R.M.F.)

FUNCTIONAL. In medicine, functional means pertaining to or affecting the function of the body, and not its structure. A functional disorder is one in which there are no obvious anatomical changes. The term is used most commonly to denote disorders of psychogenic origin.

In mathematics, a functional is a function whose argument or independent variable is a curve or surface (or a corresponding function). It may also be described as a function of a function. (R.S.M., L.L.S.)

FUNCTIONS. If two variables are so related that to each value of one variable in a given range there correspond one or more values of the other variable, the second variable is called a function of the first variable.

The first variable is often called the independent variable and the function is sometimes called the dependent variable.

If one variable is so related to several variables that to each set of values of the last-mentioned variables there correspond one or more values of the first variable, the first variable is called a function of the other variables.

If a variable y is an explicit function of another variable x, the function is often denoted in general by such symbols as $f(x)$, $F(x)$, $\phi(x)$, etc., and we write $y = f(x)$, etc. The symbol $f(a)$ then denotes the value of the function $f(x)$ for the value $x = a$.

A function of two variables, as x and y, may be represented in general by such symbols as $f(x, y)$, $F(x, y)$, etc.; and similarly for functions of more than two variables.

A function which takes one value only, corresponding to any given value of the independent variable, is called a single-valued function.

A multiple-valued function is a function which takes more than one value corresponding to any given value of the independent variable.

Functions may be classified in many ways, as: explicit functions and implicit functions; continuous functions and discontinuous functions; algebraic functions and transcendental functions, and each of these classes into many sub-classes; periodic functions, etc. (L.L.S.)

FUNCTIONS OF A COMPLEX VARIABLE. The theory of functions of a complex variable may be described briefly as the differential and integral calculus of complex variables. (L.L.S.)

FUNCTIONS OF REAL VARIABLES. The theory of functions of real variables may be described briefly as a critical study of the fundamental concepts and processes of the differential and integral calculus. (L.L.S.)

FUNGI. Mushrooms, Toadstools, Smuts, Rusts, Molds and Mildews. The fungi are a group of plants of such diverse habit that they cannot easily be collectively described. Among the more than seventy thousand species described at the present time are many microscopic unicellular forms, as well as plants of elaborate structure and considerable size. Fungi grow in almost every habitat where organic substance exists and the external conditions are suitable. Many species are found in water, either fresh or salt. Others are adapted to life on land, or in the ground. Even in the Arctic regions fungi appear in numbers during the short summer. In the tropics fungi are particularly abundant, the hot, often humid climate greatly favoring their existence.

In one particular, however, fungi are all alike: they have no **chlorophyll,** and so are unable to synthesize food from simple substances. Instead, the fungi obtain their nourishment from various organic substances. Many attack living organisms, both plant and animal. These are parasites, often of great importance to man because of the damage caused to crops. Other fungi attack dead organisms, breaking down the organic compounds present into simpler forms and obtaining thereby their own sustenance—these are saprophytes.

Excepting the entire absence of chlorophyll, the structure of fungi resembles that of the **algae,** the other main division of the **Thallophytes.** The vegetative body of a fungus, except for the unicellular forms, is always composed of slender branching threads, or hyphae, making up what is known as the mycelium. Mycelia in many cases are colorless, but may contain pigments of every color, including green. Each hypha is ordinarily composed of a row of cells, each containing one (or more) minute nuclei; in many species cross walls are rarely formed, the hypha being coenocytic. Even the largest, most complex fungi are composed entirely of tangled masses of hyphae, which may be loosely aggregated or so densely packed as to form a hard body suggestive of woody structure, as for example in the Bracket Fungi.

In their reproductive processes fungi again remind one of the algae. Both asexual and sexual reproduction occur in the life histories of these plants, which also often show very distinctly an alternation of vegetative growth and reproductive activity. As may be expected, in so diversified a group of plants a considerable variety of reproductive processes occurs.

Among the lower forms, many of which occur in water, asexual reproduction is accomplished by means of zoöspores. The zoöspores are formed in sporangia from which they escape at maturity. After a period of motility, each zoöspore settles down, loses its cilia and at once gives rise to a new plant. In the non-aquatic fungi asexual reproduction ordinarily occurs by means of non-motile spores, called conidia, which have a rigid cell-wall. These conidia, often produced in immense numbers, are carried about by air-currents, sometimes to great distances, and on reaching a favorable habitat, germinate to form a new plant. The methods of sexual reproduction found in fungi are extremely varied, and can best be considered under the different groups. Sexual reproduction usually occurs in a distinct body, the sporophore, which forms in many fungi a very conspicuous part of the life cycle of a fungus. This is frequently the only part recognized by the ordinary observer. In each kind of fungus the sporophore assumes a very definite and distinct form. In the cup fungi the sporophore is frequently a saucer- or cup-shaped structure. Other types are found in the familiar mushroom; in the puff-balls; and in the Bird's-Nest Fungus, the sporophore here having many small somewhat spherical objects contained in an open cup-like body. In all of these, spores are formed, often in unbelievable numbers; a common puff-ball contains millions of them. The spores are borne about in the air currents, and germinate when brought to a favorable environment. It is obvious that many spores must fail to reach to a favorable spot, else the world would be overrun with fungi.

Fungi are separated into three classes, the **Phycomycetes,** in which the mycelium is non-septate and coenocytic; the **Ascomycetes,** characterized by having spores borne in special sacs or asci; and the **Basidiomycetes,** distinguished by the basidium, a spore-bearing cell which bears externally four spores, in some cases more or less. In addition to these three classes there is another group known as Imperfects, or Fungi Imperfecti, which con-

tains those forms of plants in which the sexual or perfect stage is not known, and which therefore cannot be assigned to one of three mentioned classes. Some botanists would include in the Fungi two other groups of plants: the **Bacteria** and **Slime-molds.**

Fungi are of prime importance to man, first because of the immense loss caused by saprophytic forms. These attack food-stuffs, causing complete spoilage; attack fabrics which they mildew and so ruin; and attack and cause the destruction of timbers. In spite of all this, it must be stated that such forms are very necessary; for the very processes of destruction they perform are necessary preparations for new growth. Were all rotting prevented the accumulation of dead matter would soon become so great as to hinder and stop life. Only the lower plants, notably fungi and bacteria, are able to break down complex matter to forms in which it is again available for higher organisms.

Somewhat less important to man are the parasitic fungi. They may attack his crops and cause immense damage, as in the case of Wheat **rust,** but usually remedies or preventatives are found to keep such pests under control, and to prevent them from becoming serious problems. Not many species attack man himself. Some, like the fungus which causes Athlete's Foot, are annoying. Since few are fatal, the fungi of this type have received little attention until recently.

Many edible forms of fungi are enjoyed by man. Of these many are species which grow wild. To distinguish those species which are edible from those which are harmful is an ever present problem. To style the edible species mushrooms and reject the others as toadstools does not solve the problem, since it first becomes necessary to define the terms "toadstool" and "mushroom." And there is no obvious distinction. · The only safe rule to follow is that of total abstinence from any doubtful species until one is absolutely certain that it is safe. It is thus only natural that man has turned to the cultivation of fungi. Of all the edible species known, only a few have been successfully cultivated. Of these only one is commercially important, *Agaricus* (*Psalliota*) *campestris.* The spawn of this fungus is planted on properly heated beds (usually underground) of well-rotted horse manure piled in long ridges. Spawn consists of bricks or flakes of prepared manure permeated with fungus mycelium. After planting, this mycelium rapidly grows through the beds, which are then covered with fine earth and left undisturbed for a time. After a month or so mushrooms begin to appear on the surface of the beds. They grow with astonishing rapidity, and must be cut soon after appearing, or they will reach maturity and disintegrate. The food value of mushrooms is probably much overrated, as most of them are actually over 90% water, with some nitrogenous matter and no fat. But they are much fancied for their palatability and fine flavor. It is interesting to note that one group of tropical ants feeds largely on fungus plants, which they grow in their nests in an advanced state of cultivation.

The poisonous nature of many fungi has received wide publicity, and probably accounts for the popular aversion to this group of plants. The toxic substances present in the fungus are products of its **metabolism,** not substances absorbed from without. It may be that these substances are some sort of waste products accumulating in the cells. Some people have suggested that they are a means of protection against animals which might otherwise eat the plant. However, many animals eat with impunity fungi which are violently toxic to man, so it seems difficult to maintain this explanation. The toxic substances present in fungi are various. Closely related species may contain quite different poisons; a single species may have more than one poison. The effect of the poison on the human body varies. One group of poisons is taken into the body some hours before its effects become evident. Then abdominal cramps and nausea develop; vomiting occurs; thirst arises and diarrhea. These symptoms continue for hours, usually (but not always) ending in death. This is the type of poisoning caused by *Amanita phalloides. Amanita muscaria* is less violent in its action. The poison of this fungus acts on the nerve centers, causing lack of coordination, illusions and delirium, as well as gastric disturbances. *Amanita muscaria* is only very rarely fatal. This fungus is used by native tribes of northeastern Siberia as a stimulant. In addition to the *Amanitas,* many other fungi are of poisonous nature. It also seems that the effect varies among different people. What one finds edible may be definitely toxic to another. This renders even more difficult the problem of satisfactorily determining harmful species. (R.M.W.)

FUNGUS GNAT.

Insecta, Diptera. Small two-winged flies whose larvae live on fungi and decaying vegetation. Family Mycetophilidae. (A.W.L.)

FUNICULAR POLYGONS AND CATENARIES.

If a closed loop of cord or rope is pulled at several points by forces in various directions, it forms a figure, plane or otherwise, known as a funicular polygon. The external forces acting on the loop at the vertices are, for **equilibrium,** subject to the same conditions as a set of non-concurrent **forces** acting on a rigid body; while the three forces concurrent at each vertex, including the tensions in the loop itself, may be represented by an equilibrium triangle, and the several triangles fitted together to form the equilibrium polygon for the external forces (Fig. 1). Fig. 2 gives the corresponding an-

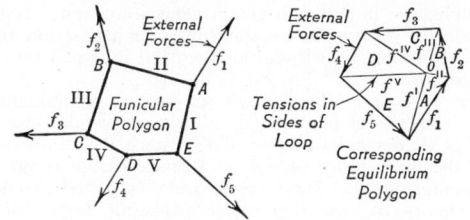

Fig. 1. Closed funicular polygon with diagram of forces.

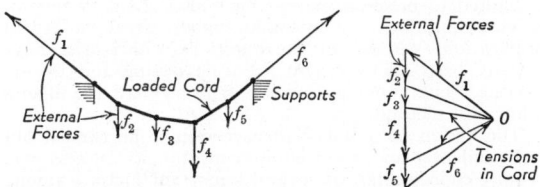

Fig. 2. Suspended cable with unequal loads.

alysis for an open cord supported at the ends and loaded by weights hung vertically from it. In Fig. 3 the weights are equal, have equal horizontal spacing, and

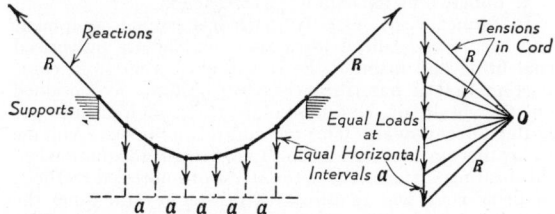

Fig. 3. Suspended cable with equal loads, as in a suspension bridge.

are hung close together. The form of the cord in this case approximates a parabola. This condition practically obtains with the cables of a suspension bridge. The point O in each figure is located by drawing from

the extremities of any side of the external-face polygon lines parallel to the sides adjacent to the corresponding vertex of the funicular polygon.

If a suspended cord is loaded uniformly along its length (not horizontally), as by its own weight, it assumes the form of a **catenary.** The equation of this curve may be written

$$y = \frac{a}{2}\left(e^{\frac{x}{a}} + e^{\frac{-x}{a}} \right) = a \cosh \frac{x}{a}.$$

(The hyperbolic cosine form is convenient for numerical computations.) a represents the Y-intercept of the curve. It is an interesting property of a catenary cable that if at any point it is hung over a pulley, and enough cable cut off to reach down to the X-axis (Fig. 4), the weight

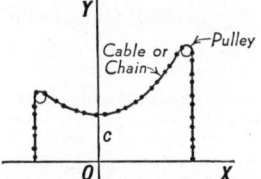

Fig. 4. Tension of chain in catenary balances weight of chain hanging down to X-axis.

of this portion will just sustain the tension on the other side of the pulley. The funicular-polygon theory is sometimes useful in solving equilibrium problems involving nonconcurrent forces. (See **Statics.**) (L.D.W.)

FUNICULUS.

A cord. Specifically: 1. A structure attaching the alimentary tract of the **bryozoans** to the body wall. 2. A slender segmented part of the antenna of some **insects,** just before its terminal segment. 3. A bundle of nerve fibers in its sheath. 4. Tracts of nerve fibers in the central nervous system of **vertebrates.** 5. The stalk which attaches the seed to the **placenta** of the fruit. (A.W.L.)

FUNNEL.

1. The tube leading out of the mantle cavity of the **squids** and related mollusks (**Mollusca**). The siphon. 2. The oral depression of **lampreys.** (A.W.L.)

FUR.

The fine soft hair of many mammals. Also the pelt, the skin of the animal bearing the fur, especially when made up to be worn as a scarf.

Fur shares with feathers the highest position as a protection against cold. It owes this property to the insulating value of the layer of air held among the fine hairs that compose it. In many animals the vesture consists of a thick woolly under layer interspersed with longer and heavier hairs which form a smooth surface.

The best furs are those of animals which live in the colder latitudes and especially the semi-aquatic species. The sea otter of the northern Pacific produces the most valuable fur, and mink and muskrat are examples of moderately valuable and low-priced furs. Fox and skunk are among the commercially important terrestrial species. All of these animals should be taken in the winter to furnish durable pelts. Summer fur is not only thinner but separates readily from the skin.

The preparation of fur for the market now involves so many processes, such as plucking, shearing, and dyeing, that only an expert can judge skins dependably. Trade names add to the confusion and give false dignity to many furs of very modest worth. Many of the seals on the market, for example, are clipped and dyed muskrat or rabbit fur. (A.W.L.)

FURANE AND RELATED COMPOUNDS.

Furane (C_4H_4O) contains a ring of 1 oxygen and 4 carbons, with 1 hydrogen attached to each carbon:

Beta, prime HC —— CH Beta

Alpha, prime HC₅ ₂CH Alpha

Furane is a colorless liquid, boiling point 32° C., insoluble in water, soluble in alcohol or ether. Furane vapor produces a green coloration on pine wood moistened with **hydrochloric acid**. Furane may be made from mucic acid (COOH(CHOH)₄COOH) by dry distillation into pyromucic acid (C₄H₃O·COOH) and then heating the latter under pressure at 270° C. Furane derivatives are known, namely, methyl, primary **alcohol, aldehyde, carboxylic acid,** in which the group attachment is at carbon number 2:

| Sylvane Alpha-methyl furane, boiling point 65° C. | Furfuryl alcohol Alpha-furyl carbinol boiling point 170° C. (750 mm.) | "Furfural," Alpha furfuraldehyde boiling point 160° C. (740 mm.) | Pyromucic acid, furoic acid, furane-alpha-carboxylic acid, Melting point 133° C. boiling point 230° C. |

(See **Furfuraldehyde**.)

Coumarone is benzo-furane (C₈H₆O or C₆H₄CH:CH·O

or , colorless liquid, boiling point 173° C.,

and diphenylene oxide is dibenzo-furane (C₁₂H₈O or

C₆H₄·C₆H₄·O or white solid, melting point

81° C., boiling point 288° C.

Gamma-pyrone (C₅H₄O:O(4)) is a gamma-ketone (4) containing a ring of 1 oxygen and 5 carbons with 1 hydrogen attached to each of 4 carbons, namely, 2,3,5,6,

C=O Gamma

Beta prime HC₅ ₃CH Beta

Alpha prime HC₆ ₂CH Alpha

Gamma-pyrone is a colorless liquid, melting point 32° C., boiling point 218° C.

Pyrone derivatives are known, e.g.,

| Alpha, Alpha prime dimethyl-gamma-pyrone | Chelidonic Acid Gamma-pyrone-alpha, alpha-prime-dicarboxylic acid |

Chromone is benzo-pyrone (C₉H₆O₂) or

white solid, melting point 59° C. and chromane is

colorless liquid, boiling point 214° C. (750 mm.)

Flavone is phenyl chromone: white solid

melting point 97° C.

Xanthone is dibenzo-pyrone (C₁₃H₈O₂ or C₆H₄ C₆H₄ or

white solid, melting point 174° C., boiling point

351° C., and xanthene is white solid, melting point

100° C., boiling point 315° C. From chromone and xanthone a number of yellow dyes are made, which dyes also occur in nature. Such dyes are chrysin, fisetin, buteolin, morin, quercetin, rhamnetin.

Where oxygen of furane is occupied by sulfur, thiophene is the compound, and of coumarone, benzothiophene; and where oxygen of furane is occupied by nitrogen (group —NH), pyrrole, and of coumarone, indole. (R.K.S.)

FURCA. A pair of divergent projections at the caudal end of the body of some **crustaceans**. (A.W.L.)

FURCULA. The jumping appendage of a spring tail. **Collembola.** (A.W.L.)

FURFURALDEHYDE. Furfuraldehyde (C₄H₃O·CHO(2)) is a colorless, odorous liquid, boiling point 162° C. When pentoses, e.g., arabinose, xylose (see **Carbohydrates**), are heated with dilute **hydrochloric acid**, furfuraldehyde is formed, recognizable by deep red coloration with phloroglucinol, or by the formation, with **phenylhydrazine** of furfuraldehyde phenylhydrazone (C₄H₃O·CH:NNHC₆H₅), solid, melting point 97° C. Furfuraldehyde is formed by the treatment of corn-cobs, bran, or wood, with **sulfuric acid**, and can be used in many instances where **formaldehyde** is utilized, as a disinfectant, insecticide, and for a raw material for the manufacture of synthetic plastics and resins; also as a solvent for organic substances, and to prepare furfuryl alcohol (C₄H₃O·CH₂OH(2)) and furoic acid (pyromucic acid, C₄H₃O·COOH(2)) for furoates. (R.K.S.)

FURNACE. A furnace may be said to be a chamber for **combustion**. In addition, it provides support and enclosure for the firing equipment and partial enclosure for equipment sometimes inbuilt into the furnace, as, for example, the boiler of a steam generating unit. It surrounds the region where the combustion reaction takes place, confining and isolating it so that it remains a controlled force.

The furnace is a relatively important component of any steam power plant. The success of a boiler installation is so dependent upon the furnace which serves it that the importance of correct furnace design cannot be over-emphasized. The furnace converts the latent chemical energy of raw fuel into a dynamic form (heat). There are many interesting, puzzling, and difficult problems in the field of furnace design. Even when this field is limited to boiler furnaces one finds a great variety of service conditions calling for an equally great variety of applications. Many of the questions which arise have already been answered by scientific testing and by experience, but some points are still met by the liberal allowance method.

The design of a boiler furnace cannot be carried out independently of other equipment, for its success will require coordination of several important factors, among which may be mentioned:

1. Type of combustion equipment.
2. Character of the fuel used, especially its ash content.
3. Draft equipment employed.
4. Air supply and degree of preheating.
5. Boiler, and its baffling arrangement.

In former years, especially when steaming equipment had been standardized for a time, and before pulverized coal firing had inspired the remarkable progress in combustion that it has, furnaces were customarily designed on the basis of certain volumetric requirements per **boiler horsepower**. This method related furnace volume to heating surface, and naturally became inadequate when this heating surface, under the stress of improved methods of firing coal, was required to transfer several times the heat it formerly did.

Recently, furnace volumes have been based on *heat liberation*, meaning the number of B.T.U. produced per hour per cubic foot of furnace volume. To fix upon furnace dimensions by this method requires only the division of the hourly heat liberation by some acceptable volumetric rate. This rate is not deductible from theory, but fortunately numerous data are available as to rates actually employed in successful designs.

There is much to be said for both theory and experience in boiler furnace design. The heat release method is hardly theoretical, but represents an enlightened approach compared with the older methods. When leaving long established practices behind, and in pioneering installations, experience may usefully be employed to season theory and computation.

Furnaces classified according to their wall construction are (1) solid refractory, (2) air-cooled refractory, (3) water-cooled, either totally or partially.

Recent progress in this field, stimulated by the introduction of high capacity firing, has greatly increased our knowledge of the part played by radiant energy in the furnace. Radiant energy reverberates back and forth in the furnace until gases are at their final flame temperature because flame is opaque to radiation. If cold black surfaces such as water walls are met in the process of reverberation large amounts of high-level energy will be absorbed, less will be reverberated to the flame whose temperature will, in consequence, be less. It is well established that a heat transfer of approximately 75,000 B.T.U. per sq. ft. per hour is realized. Not only water walls, but water screens and those portions of the boiler heating surface that "see" the furnace, absorb radiant energy. In many of these the fuels are burned in contact with the material being processed while in other, i.e., crucible and muffle furnaces, the material is separated from the combustion being placed in a separate chamber or crucible. For descriptions of some types of metallurgical furnaces see **Bessemer Process, Blast Furnace, Cupola, Electric Furnace, Open Hearth Process, Refractory.** (F.T.M.)

FURUNCLE. A boil. A localized infection of the skin and subcutaneous tissues occurring singly or multiply. The usual infecting agent is the *Staphylococcus*. (R.S.M.)

FUSAIN (FUSSIN). A term proposed by Stevenson in 1911 for what had previously been called "mineral charcoal" or "mother of coal." The highly oxidized cellulose and blackened woody fibers which form an important constituent of most true coals. (R.M.F.)

FUSE. A fuse is a common protective or circuit breaking device for low-voltage **electric circuits.** It is an over-current protector, and since the current must first heat the metal, there is a time delay in fuse "blowing" that is inversely proportional to the current. This characteristics is called "inverse time element." The ordinary fuse consists of a calibrated length of conductor whose resistance is so chosen that when a certain current flow through it is exceeded, it fails to lose by radiation enough of the resistance heat to keep its temperature below melting. The fuse is enclosed in a protective case which forms the contact points to connect it into its circuit. Sometimes the case contains a powder which helps to extinguish the arc which follows the blowing of the fuse element in high-capacity fuses. Sometimes, also, the case is provided with a glass cover, which allows one to discover at any time the integrity of the fuse. Fuses for 110-volt house circuits usually take the form of a plug which can be screwed into a socket. They are standardized at 5, 10, 15, 20, 25, and 30 amperes. Cartridge type fuses have been standardized with ferrule contacts up to 60 amperes, and knife contacts above that. There are a number of other types, such as expulsion fuses, thermal overload fuses, etc., that are not discussed here. The National Electric Code specifies that if fuses are used for motor protection, their capacity must not exceed 125% of the name-plate rating of the motor. The detonating device on a shell is a fuse as is the powder-filled cord used in setting off any charge. (F.T.M.)

FUSED SALT BATHS. Quenching, Heat Treating, Steel.

FUSEL OIL. Alcohols.

FUSELAGE, AIRPLANE. The fuselage is the body of an airplane. It must perform many functions such as providing adequate space for the power plant, fuel tanks, crew, passengers, and cargo, and necessary equipment and controls, and it must also transmit lifting, thrusting, balancing and landing loads from one part of the airplane to another. Probably because of the great number of possible combinations of the variables, fuselage designs are found in utmost variety of forms and constructions; nevertheless, all fuselages have much in common in that they will have similar outlines and arrangement of the primary components. Fuselages are smooth externally, with a minimum of projections into the airstream; they are slender, and generally have a fineness ratio (see **Parasite Drag**) greater than that which would yield the minimum drag. The attachments to the wing, and the space for crew and passengers, are usually in the forward 50–60% of the fuselage, and the long, tapering tail section carries the empennage for control and balance. As the fuselage accounts for one of the larger components of weight of the airplane, every effort is made to keep it as light as possible; however, there can be no compromise with necessary strength, as the fuselage supports parts vital to safety in flight, and any partial structural failure would invite catastrophe. In cross-section, fuselage outlines of various forms, i.e., oval, rectangular, elliptical, round, usually have one feature in common; that is, the depth exceeds the breadth. This is in deference to the primary nature of the fuselage acting as a cantilever beam supported by the wings and loaded at the tail with a balancing load.

Fuselages for single-engined airplanes are designed to bear an engine, mounted at the forward end. As the engine is a heavy concentrated load which in operation produces torque and thrust, this part of the fuselage structure must be made quite rigid, and therefore heavy. Extremely high loadings may occur on the forward end of this type of fuselage during hard landings on account of the sudden arresting of the downward momentum of the engine mass upon contact of the wheels with the ground. It is required that the engine compartment forward of the fuselage be separated from the fuselage by a firewall which is a thin partition of asbestos board or sheet metal. Multi-engined airplanes have the engines carried on the wing structure, in which case the forward end of the fuselage is neatly faired around in front of the control cabin.

In flight, the fuselage and contents consists of dead structural load, controls, instruments, crew, passengers, cargo, and miscellaneous other equipment, all of which is supported by reactions received from the wing structure, is pulled forward by thrust received through the engine mount, and is held in any desired attitude by balance and control forces set up on the tail surfaces. The operation of the rudder and ailerons will, in addition, impose side loads and twisting upon the fuselage, as will also the operation of a bi-motored airplane with one engine out of service. Unskillful landings are capable of producing extremely heavy impact loads at the juncture of the undercarriage with the fuselage, and will produce large bending moments in other parts of the fuselage. So, one sees that the fuselage must be able to resist torsion and bending in almost every direction. It must accommodate the mounting of a great deal of small equipment internally, and there are many cut-outs to be reckoned with, such as windows and doors, all of which complicate the problem of design.

Fuselages may be variously classified. On account of the many variations representing ideas of individual designers, the classification given here is not claimed to be all-inclusive.

   1. Fuselages classified according to seating arrangement.

a. Degree of exposure of the personnel.
  (1) Completely enclosed, as in cabin airplanes.
  (2) Open cockpits. Head and shoulders usually protrude above the fuselage, being partially protected from the airstream by forward windshields.
  (3) A modification of (b), wherein a glass or transparent plastic hatch, bubble, blister, or other form of cover, is used to enclose and protect the occupants. Whereas entry to a cabin airplane is obtained through side doors, this type is entered by having the cockpit cover hinged or sliding.
b. Relative position of the seats.
  (1) Tandem seating arrangement.
  (2) Side by side.
  (3) Both (a) and (b) apply mainly to the small two-place airplane. Three-, four-, and five-place airplanes usually incorporate a combination of side by side and tandem seating.
  (4) Large multi-engined planes designed to carry many passengers will typically exhibit a control compartment in the nose, where the pilot and co-pilot are seated side by side. Back of these there may be space provided for other crew members, such as flight engineer, navigator, etc., or there may be a galley and steward's compartment. Next aft would

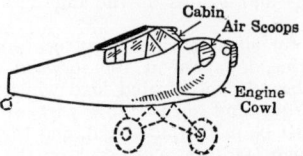

Fig. 1. Light plane type fuselage (trussed structural skeleton).

be the main passenger cabin, with seating arrangements similar to that of railway trains or motor buses, except that, in the smaller transports, there may be only a single row of seats on each side of a center aisle.

2. Structural types.
  a. Trussed frame structures.
    (1) Round steel tubes welded into a triangularly framed truss structure.
    (2) Round or square aluminum alloy tubes, riveted or bolted into a truss structure, sometimes including diagonal wires as tension members.
    (3) Trusses constructed by riveting members composed of drawn steel or aluminum alloy strip.
  b. Monocoque structures.
    (1) Full monocoque construction. This is rarely employed. Full monocoque construction relies on the strength of the skin or shell to carry all the load. Its

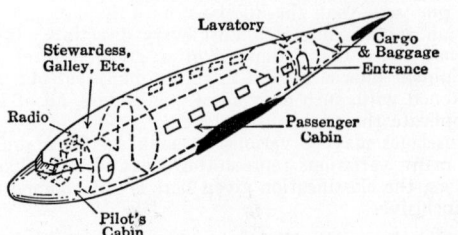

Fig. 2. Transport type fuselage (semi-monocoque construction).

only reinforcement consists of transverse bulkheads used to define the shape. Unless the curvature is considerable, the necessary skin thickness is uneconomically heavy; hence it is seldom used except for small diameter monocoque tubes such as booms.
    (2) Semi-monocoque construction. This is adaptable to large diameters, using thin skin, because in the semi-monocoque fuselage the skin is reinforced by an interior framework. This is the construction employed on most metal fuselages of the stressed skin type.
  c. Reinforced skin. A stressed skin covering is applied to a framework consisting of longitudinal stringers, transverse rings, and diagonals.
  d. "Geodetic" construction. A form of construction employing a diagonal basket-like weave of many closely spaced stringers to which is attached the surface covering.
3. Materials of construction.
  a. Wood. Although early airplanes were frequently constructed with wooden fuselages, these tended to become obsolete, but, latterly, the development of high-grade waterproof and flexible plywood, and the improvement in adhesives available, have made it possible to construct excellent stressed skin wood fuselages.
  b. Metal. The most common metal for welded steel tubular fuselage is seamless chrome-molybdenum steel. For stressed skin either 17 ST or 24 ST aluminum alloy is employed. Where the surface is not to be finished with paint, the alloy sheets are covered with a thin surface of a pure aluminum serving as a protecting coating. This sheeting is known by the trade name

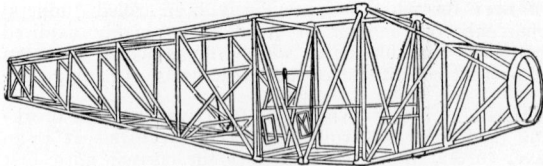

Fig. 3. Welded tubular steel fuselage.

"Alclad." Thin sheets of stainless steel also have been employed in fuselage construction.
4. Unconventional fuselage types.
  a. Flying wing. Here there is no separate fuselage, room to accommodate the normal fuselage functions being provided in a thickened midsection of the wing.
  b. Twin tail. When the empennage is carried by two booms extending back from the wing structure, the fuselage can be made in a better aerodynamic proportion, say with a length about three or four times its width.
  c. Canard. In the canard design the empennage is placed ahead of the fuselage.

The truss type of fuselage is built in the form of a rectangular beam with its flanges composed of tubes located at the four corners of its rectangular cross-section. These are the four longerons. These longerons, together with vertical web members, compose two side trusses which are usually built as an irregular Warren truss with welded joints. The trusses in the plane of the upper chord and lower chords give transverse rigidity, and diagonals are introduced to oppose torsion. The necessity of these diagonals, which pass through the interior of the fuselage, is a disadvantage of this construction, since they interfere with the full usage of the interior space for accommodation of cargo and personnel. Concentrated loads are comparatively easy to introduce into the trussed structure by applying them to the joints of the truss. Where redundancies occur, as around door openings, the corners must be heavily braced with reinforcement. As the best shape for the truss will not ordinarily be a good aerodynamic shape, a light-weight false work of formers is attached to the primary structure so that the covering, when attached, will be of a neatly faired, aerodynamic shape. The covering is ordinarily light-weight fabric attached securely to the formers, and made taut and smooth by repeated applications of airplane "dope."

Large fuselages of semi-monocoque construction represent the maximum strength with minimum weight type of construction, which is so desirable in aviation. Being a shell it has good torsional stiffness. This construction also leaves the interior comparatively clear of structural bracing. In spite of the recent importance of this construction it is not new, for as early as 1911 Handley-Page had built a wooden monocoque fuselage, and during the decade from 1910-1920, there were several airplanes built with metal monocoque fuselages. The cross-sectional shape is derived from bulkheads or rings, and the longitudinal profile is determined by stringers. The thin metal skin is riveted to these members (if plywood, the skin is glued and nailed to the members, which are then wooden) and combines with the underlying framework to carry the loads previously mentioned. The bulkheads are at much larger intervals than the stringers. In order to employ thin gauge skin without the stiffeners being at unreasonably close intervals, it is sometimes corrugated. Unlike the trussed fuselage, in the stressed skin design it is not easy to apply concentrated loads to the fuselage. At the points of reception of localized loads, the structure must be specially stiffened. While this might be done by employing especially stiff bulkheads, specially built-up assemblies which are designed to distribute loads around the shell are located at the points receiving concentrated load. In a semi-monocoque construction the stringers take the bending moment, and the shear flow is conducted by the skin. The calculations for design are de-

tailed and complex, but after the structure is built it is possible to check the design by applying strain gauges to the structure while it is being artificially loaded. Thus the adequacy of the design can be established before committing the structure to the air. Probably no one thing causes more difficulty in the arrangement of a monocoque structure to carry a load than the numerous cut-outs which have to be accommodated, such as window and door openings, cargo hatches, etc. However, in spite of the complications, successful and light-weight fuselages are constructed with stressed skin, and the advantages appear sufficiently important that it might be predicted that only the light airplanes of the future will employ trussed fuselages. (F.T.M.)

FUSION. A change from the solid to the liquid phase of matter. In crystalline bodies, and, we are beginning to understand, also in many other solids not exhibiting well-defined **crystal structure,** the atoms are held in positions of stable equilibrium by intermolecular forces. They of course move with thermal agitation, but their movements are oscillatory and do not carry them outside a limited range of distance from their equilibrium positions. Stable equilibrium may, however, become unstable when the system is disturbed beyond a certain limit. Thus if a solid body is sufficiently heated, the molecules break loose from their stable configuration and wander about or diffuse among each other. When this condition has become general, the body exhibits the characteristics of a liquid, and we say it has undergone fusion. In some cases, such as ice, the change is quite abrupt, the substance having a well-defined **melting point;** in others, like glass or pitch, it is gradual. The difference is probably due to the more uniform potential energy of the atoms in the former case, so that they all "break loose" at the same stage of thermal agitation; while in the latter case some atoms require more energy to dislodge them than others. In any case the process requires a supply of energy which is recognized as the **heat of fusion.** With most substances, fusion is accompanied by an increase in volume; but with some, like ice, the volume becomes definitely less. (L.D.W.)

GABBRO. Gabbro is a deep-seated and often very coarse-grained **igneous** rock composed of **plagioclase feldspar,** usually **labradorite** or **bytownite** and **monoclinic pyroxene** with occasionally as accessories **olivine** (when it is then called olivine gabbro), **biotite, magnetite, ilmenite** and **hornblende. Norite** is a variety of gabbro carrying **orthorhombic** pyroxene, usually **hypersthene** instead of the monoclinic sort. **Troctolite** is essentially olivine and plagioclase. **Quartz** gabbros are known and have probably been derived from **magmas** somewhat oversaturated with **silica.** On the other hand **essexites** represent gabbros whose parent magma doubtless had an insufficiency of silica resulting in the formation of **nephelite.** Gabbros are frequently rich in sulfides that may be of commercial value, a notable occurrence of which is at Sudbury, Canada. Here a norite carrying **chalcopyrite** and nickeliferous **pyrrhotite** forms the most important deposits of nickel known. Gold, silver and platinum are also recovered from this ore. (E.S.C.S.)

GADOLINIUM. Symbol: Gd. Atomic number: 64. Atomic weight: 157.3. (Isotopes: page 290.) Type of compound: Gd_2O_3. Color of salts: Colorless. Discovered by Marignac in 1886. A member of the **cerium** sub-group of the rare earth metals. (R.K.S.)

GADWALL. Aves, Anseriformes. A North American duck, *Anas strepera.* (A.W.L.)

GAHNITE—ZINC-SPINEL. The mineral gahnite is isometric with an **octahedral** habit but may appear as **dodecahedrons** or modified cubes. Chemically it is zinc aluminate corresponding to the formula $ZnAl_2O_4$. There is a tendency for cleavage parallel to the octahedron, fracture varies from conchoidal to uneven; brittle; hardness 7.5–8; specific gravity 4–4.6; luster vitreous; color ranges from dark green through various shades of greenish- or bluish-black, yellowish-black or grayish, subtransparent to almost opaque. Gahnite is found in association with other zinc minerals at several European localities, notably in Bavaria and Sweden. In the United States it is found at Franklin and Sterling Hill, New Jersey; at Rowe, Massachusetts and in Maryland, North Carolina, Georgia and Colorado. Gahnite was named in honor of the Swedish chemist, J. G. Gahn. (E.S.C.S.)

GAIN. Amplifier; Directional Antenna.

GALACTIC COORDINATES. As the modern theories regarding the structure of the sidereal **universe** became more and more firmly established, it became necessary to have a system of **spherical coordinates** for the representation of points relative to the plane of the **milky way,** or the **galactic plane.**

The galactic coordinate system is a system of spherical coordinates having as its fundamental plane the plane of the milky way (or galaxy). The adopted position of the pole of this plane is $12^h 40^m$ **right ascension** and 28° north **declination.** The plane of the milky way cuts the plane of the celestial **equator** at an angle of 62°. Galactic latitude is measured perpendicular to the plane of the galaxy along great circles drawn through the galactic poles and hence perpendicular to the galactic plane. Galactic longitude is measured in the plane of the galaxy from the point where this plane cuts the plane of the equator in right ascension $18^h 40^m$ to the point where

the great circle perpendicular to the galactic plane through the object intersects the galactic plane. (W.K.G.)

GALACTIC SYSTEM. Milky Way.

GALAGO. Mammalia, Primates. The African **lemurs.** The several species are long-tailed animals whose nearest relatives are the mouse lemurs of Madagascar. (A.W.L.)

GALAXY. Milky Way.

GALEA. Maxilla.

GALENA. The mineral galena, **lead** sulfide, PbS, crystallizes in the **isometric** system, usually in cubes or cube-octahedron combinations, less frequently in **octahedrons.** It is often found in cleavable masses, but may be granular or fibrous. The highly perfect cubic cleavage is an important characteristic of this mineral; it may, however, sometimes show an octahedral parting. Its hardness is 2.5; specific gravity, 7.3–7.6; luster, metallic; color, lead gray; streak, grayish-black; opaque. Galena is the most important ore of lead and in addition often carries values of **silver;** it is then known as argentiferous galena. It occasionally is actually mined as a silver ore. Sometimes galena contains small amounts of **zinc, cadmium, antimony, bismuth,** and **copper** as sulfides. Galena is a very common and widely spread mineral, it occurs in veins and beds in various rocks, both crystalline and sedimentary. Some of these deposits are doubtless replacements, others seem to show a close connection with intrusive igneous rocks. Of the many foreign localities might be mentioned the classic one at Freiberg, Saxony, and the silver mines of the Harz Mountains. This mineral has been found in the lavas of Vesuvius, in Italy, and fine specimens come from Cornwall and Cumberland, England. Australia, South America, Chile, and Peru produce galena. In the United States, Missouri, Illinois, Iowa, and Wisconsin contain large and important galena deposits. In Colorado and Idaho it has been mined for its silver content. Galena is usually associated with **sphalerite, smithsonite,** and at **Phoenixville,** Pennsylvania, with beautiful **pyromorphite** crystals. The name is derived from the Latin *galena,* a term which was applied both to the lead ore and slag from refining. (E.S.C.S.)

GALL. Gall is the secretion of the **liver** of vertebrates, also called bile. (A.W.L.)

GALL BLADDER. A pear-shaped organ (see figure) situated on the underside of the **liver** on the right side just below the lower ribs. It serves as a reservoir for the **bile** and by means of the cystic duct it communicates with the common duct through which the bile secreted by the liver passes to the **duodenum.** The gall bladder is about 3″ in length and 1–1¼″ in diameter. It holds about 1½ ounces of bile. When fatty substances are eaten, the normal gall bladder empties the stored, concentrated bile into the common duct. Upon passing into the duodenum the bile aids in the **digestion** of the **fat.**

The gall bladder under certain conditions is subject to acute and chronic infections (**cholecystitis**) and to stone formation (**cholelithiasis**). Under these conditions its function is lost and, since it is a source of infection to the body, and often the cause of continued distressing symptoms, it should be removed (cholecystectomy). No ill effects follow the removal of the gall bladder, as the common duct serves as a reservoir in its place. (R.S.M.)

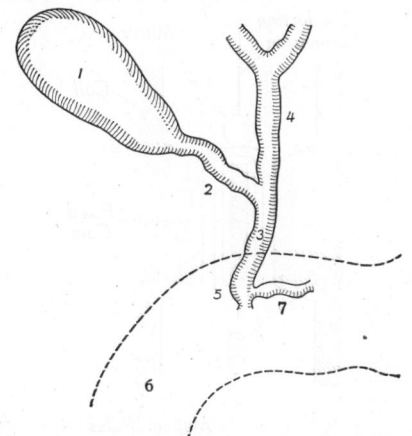

Diagram of biliary tract. 1. gall bladder; 2. cystic duct; 3. common bile duct; 4. hepatic duct; 5. opening of bile duct into the duodenum; 6. duodenum; 7. duct from the pancreas. (*Whipple, A. O., Surgery of the Biliary Tract, Nelson's Loose-Leaf Surgery, Volume 5.*)

GALL GNAT. Insecta, Diptera. Small 2-winged flies of many species constituting the family Cecidomyiidae. Most are plant feeders as larvae and produce **galls** on the plants that they attack. Others are predacious or scavengers. (A.W.L.)

GALL WASP. Insecta, Hymenoptera. A minute **insect** whose attack on plants produces **galls.** They are of many species, making up the subfamily Cynipinae. (A.W.L.)

GALLIFORMES. The order of birds (**Aves**) containing the more common domestic species and many game birds, such as the **quails, pheasants,** and **grouse.** They are predominantly ground birds although they may fly strongly for short distances and may roost in trees. (A.W.L.)

GALLINULE. Aves, Gruiformes. Wading birds related to the **coots** and **rails.** (A.W.L.)

GALLIUM. Symbol: Ga. Atomic number: 31. Atomic weight: 69.72. Density: 5.89. Melting point: 29.75° C. Boiling point: 1700° C. (Isotopes: page 290.)

Gallium is a white, tough metal, soft enough to be cut with a knife, fresh surface soon oxidized superficially to bluish-gray, burns in air upon being heated to 500° C. The temperature range at which gallium is a liquid permits its use as a substitute for mercury in high-temperature thermometers. Gallium metal is only slightly affected by water at room temperature, but reacts vigorously upon boiling; dissolves in **hydrochloric acid** and in **sodium** hydroxide; reacts vigorously with chlorine at room temperature. Discovered by Boisbaudran in 1875.

Gallium occurs in very small amount in **zinc blende, magnetite, pyrite, bauxite,** and **kaolin** of certain localities. A few parts per million is present in Oklahoma zinc ores. The recovery of gallium from zinc flue dust is effected by solution of the dust in excess of **hydrochloric acid,** addition of potassium chlorate, and distillation to remove **germanium.** When the residue is converted into sulfate, fractional electrolysis of the slightly acid solution removes **zinc,** and the gallium is obtained almost free from **indium.**

Hydroxide: Gallium hydroxide ($Ga(OH)_3$), white, gelatinous precipitate by reaction of **ammonium** hydroxide on gallium salt solutions; precipitate is soluble in **sodium** hydroxide to form sodium gallate.

Oxide: Gallium oxide (Ga_2O_3), white solid, infusible at red heat, obtained by ignition of gallium nitrate or hydroxide.

Salts: Gallium sulfate ($Ga_2(SO_4)_3$), nitrate (Ga ($NO_3)_3$), and chloride ($GaCl_3$, melting point 76° C.) are white, soluble in water, and upon boiling the solution a basic salt is precipitated. Gallium dichloride, gallous chloride ($GaCl_2$), melting point 164° C. (R.K.S.)

GALLOTANNINS. Tannins.

GALLS. Galls are abnormal outgrowths in plants caused by plant or animal **parasites,** which attack various parts of the plant. While no part of the plant is immune, galls most frequently occur in those regions composed of actively growing **cells,** such as the leaves, or the cortical tissues of the stem, or young roots. The irritation caused by the parasite may result in a tremendous **hypertrophy** of all cells affected or may cause numerous cell divisions which result in tremendous increase in the affected tissues.

The organisms which cause gall formation are many. **Nematode worms** often enter the roots of plants and cause the formation of irregular tumorous growth. These same organisms often infect the larger brown **algae** and cause hypertrophies, or at least gall-like malformations. Many parasitic **fungi** cause galls to form in the tissues which they attack. Galls occur in the leaves and stems of blueberry and cranberry bushes, due to fungus infection by *Exobasidium vaccinii,* a **basidiomycete.** The **hyphae** of the fungus penetrate the cells of the host, which enlarge tremendously in consequence. All **chlorophyll** in these enlarged cells is destroyed, and a red pigment forms, causing the galls to appear very conspicuous. Several species of *Taphrina,* a fungus of the **ascomycete** group, cause galls in the leaves of many plants. Those caused by *Taphrina aurea* in the leaves and fruits of poplar trees are especially common. Many **rusts** also cause gall formation.

However, probably the most striking and best known galls are caused by insects. (See **Gall-Wasps and Gall-Gnats.**) A gall-producing insect lays its eggs in the tissues of the plant. Apparently as a result of the irritations caused by the young **larvae,** the surrounding cells become greatly enlarged, and the gall is formed. The galls caused by each species of insect have a very characteristic shape. The leaves and stems of rose bushes, for example, are frequently infected. One insect causes a smoothly spherical gall to form; the gall produced by another is similarly shaped but studded with stiff spines; while a third causes the formation of a dense growth of matted, branched hairs, forming a structure an inch or more in diameter. Within, there may be a single insect larvae, or many, feeding on the loose **parenchymatous** inner tissues of the gall and protected from enemies by the firm outer layers. Often the young buds of willow twigs are parasitized, causing bud galls to form. As the bud grows older, the internodes enlarge tremendously in diameter but elongate very little, so that a gigantic bud is formed.

The leaves of oak trees are very commonly parasitized by gall-forming organisms, both fungus and insect. Considerable value attaches to these galls, because of the large accumulation of **tannin** occurring in the developing gall. In countries where cheap labor is available these galls are gathered in quantities. They are used in tanning leather, in the manufacture of ink, and in the preparation of an astringent ointment. (R.M.W., A.W.L.)

GALLSTONES. Cholelithiasis.

GALVANIZING. The application of a layer of zinc to the surface of iron and steel for protection from corrosion is known as galvanizing. Galvanizing is one of the most effective means of preventing rusting because of its low cost and ease of application compared with other metallic coatings.

In the hot dip process, the sheets or other articles to be coated must be free from scale, dirt, grease, etc., and are usually prepared by **pickling** and washing before immersion in molten zinc commercially known as spelter. Articles fabricated from iron and steel sheets and wire are hand-dipped. Sheets and wire are handled mechanically.

An increasing proportion of sheet-metal products is being coated as sheet or strip before fabrication. This requires a tightly adhering coating to prevent peeling during **stamping** or **forming** operations. In order to obtain good adherence in hot-dipped coatings special processing is necessary, especially with the heavier weights of coating which give longer protection. For lighter weight coatings a duplex bath consisting of a layer of molten lead under the molten zinc is often used. The steel sheet passes through the lead, which does not adhere, and up through the zinc. The time in which the steel is in contact with the spelter is greatly reduced and consequently less zinc is deposited. Another product which resists peeling is known as Galvannealed, a sheet which is annealed after dipping in order to form a coating consisting entirely of iron-zinc compounds. This coating is quite brittle and tends to powder when deformed but does not peel. A product known as Zincgrip is processed to minimize the formation of iron-zinc compounds and this sheet retains its coating intact even through extremely severe fabricating operations. Another method of applying a very ductile, adherent coating is by **electrodeposition**. This process is usually limited to rather light weights of coating because of the cost. (R.H.H.)

GALVANOLUMINESCENCE. Luminescence.

GALVANOMETERS.
A galvanometer is an instrument for measuring electric currents, usually by means of their magnetic effect. Observations are made by noting the deflection produced by the reactive **torque** exerted between an **electric circuit** and a magnet. Galvanometers may be divided broadly into two classes, according to whether the coil is stationary and the magnet turns, or vice versa.

Perhaps the most highly developed of the first type is the Kelvin astatic galvanometer. This has two magnets equally magnetized but antiparallel mounted on the same suspension, one above the other, and each magnet is surrounded by a coil. The two coils are joined in series and are oppositely wound, so that a current through them will turn their respective magnets in the same direction. The earth's uniform field has no effect upon such an astatic pair of magnets; but there is a large control magnet, placed above the pair, against whose field the current turns the suspended system. The movement is observed by the usual mirror-and-scale or **optical lever** device. Galvanometers of this type are now nearly obsolete.

Among galvanometers of the second type, that of d'Arsonval is best known. The magnet in this instrument is a fixed, permanent magnet of the horseshoe or double-horseshoe form, with a light, rectangular coil suspended in the strong field between its poles, the suspension carrying the feeble current. The current causes the coil to turn in the field. Often a fixed iron core is supported inside the movable coil to concentrate the field.

If these galvanometers are undamped, they will give a "throw" when a charge of electricity is sent through them, and the charge can be thereby measured. Such an instrument, with a heavy coil, called a ballistic galvanometer, is useful in capacitance measurements. The oscillations may be damped by shunting.

There are also string galvanometers, in which a straight, slender wire carrying the current is thrust to one side by a magnetic field; and vibration galvanome-

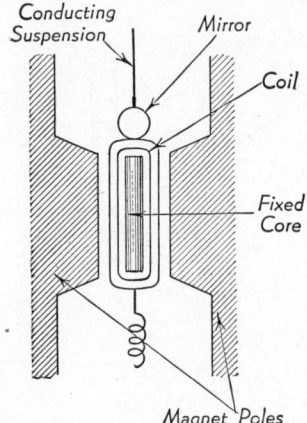

Essential parts of d'Arsonval galvanometer.

ters, in which the string vibrates in synchronism with the alternating current traversing it. (L.D.W.)

GAMBOGE. Garcinia. Guttiferae.

GAMETANGIA. Gamete.

GAMETE.
A sexual reproductive cell or **germ cell** which normally unites with another to produce a new individual.

The gametes of some primitive organisms are of one form; these are single cells which swim about in the water. Such organisms are said to be isogamous and the germ cells are called isogametes. In most species, however, only the male gametes retain the power of locomotion. The female gametes are larger inert cells and the organisms are called heterogamous. The male cells of these species are known as sperms or spermatozoa and the female cells as ova or eggs. Because of its smaller size the male gamete is also known as a microgamete, and the female gamete as a megagamete.

The union of two unlike gametes (heterogametes) is called heterogamy. The cell which is formed by the union of two gametes is called the zygote; from it a new plant or animal develops. The cells in which gametes are formed are called gametangia. In heterogamous plants, the gametangia-containing sperms are called antheridia; those containing eggs, either oögonia or archegonia. (In rare cases, a gamete does develop without two gametes having previously united.)

The development of two forms of gametes permits both the freedom of movement necessary to bring the two cells together for **fertilization** and the storage of the **protoplasm** and food necessary for the development of any body of reasonable size and complexity to a stage in which it can secure more materials for itself. By the delegation of one function to each kind of cell neither is subject to harmful restriction.

In most species of animals the sperm is a minute cell with a slender **flagellum** or tail whose undulating movements propel it through the water or the seminal fluid. The main part of the sperm is the head, which contains an apical body and the **nucleus**. Behind the head is a neck, or a middle piece of more complex structure, from which the sperm aster involved in the fertilization process sometimes develops. The sperms of some **worms** and **arthropods** lack the flagellum although many bear processes of other kinds. They are much less motile than flagellate sperms but they are said to move slowly by amoeboid action or by means of their processes.

Ova are more compact cells, often spherical in form. They contain abundant **cytoplasm** and in many species an enormous amount of food material (yolk, deutoplasm), as in the egg of a bird. Here the yolk is the egg cell or **ovum** proper but the living protoplasm is a

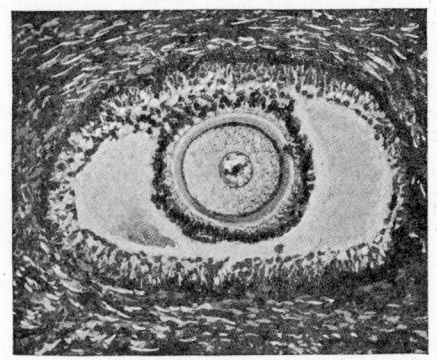

Graafian follicle and ovum of the cat, within the ovary.
(*Dr. B. F. Kingsbury.*)

tiny mass at some point on its periphery. Ova may also have special envelopes such as the albumen or white of the bird's egg, the shell membrane, and the shell. Insect eggs are enclosed by a shell called the **chorion** which is often beautifully sculptured and strangely shaped. Where such coverings occur, a minute opening, the micropyle, sometimes provides an entrance for the sperm. (A.W.L., R.M.W.)

GAMETOGENESIS. The formation of **gametes.** In animals, usually accompanied by **meiosis.** (A.W.L.)

GAMETOPHYTE. One of the two **generations** which **alternate** with each other in the life-history of many plants is called the gametophyte generation. It is the generation in which the gametes or sexual cells are formed, and so is frequently called the sexual generation. The cells of plants in this generation have the reduced or haploid number of **chromosomes,** which is half the number found in cells of the diploid sporophyte generation. (R.M.W.)

GAMMA. 1. The one-millionth part of a gram. 2. A unit of magnetic field intensity. 3. The tangent of the angle of the straight-line portion of the D log E curve and the log E axis is a measure of the degree of development and was termed by Hurter and Driffield the *development factor* and designated by the Greek letter γ (gamma). Gamma may also be defined as

$$\gamma = \frac{D_2 - D_1}{\log E_2 - \log E_1}$$

where D_2 and D_1 are densities on the straight-line portion of the D log E curve produced by log E_2 and log E_1, respectively.

Gamma is frequently regarded as expressing negative contrast, but the contrast of a negative depends upon several factors of which gamma is but one. Gamma, however, indicates the ratio of the negative contrast to that of the subject if the exposures lie entirely on the straight-line portion of the D log E curve, a condition not usually observed in the making of the negative.

The gamma of a D log E curve may be determined (1) by substituting values from the curve in the formula above, (2) by measuring the angle of the straight-line of the D log E curve and ascertaining the value of its tangent, (3) by constructing a right-angle triangle of which the log E axis is the base, the range in log E being equal to 1.0, and the straight-line portion of the curve the hypotenuse. The density at the point of intersection

of the perpendicular forming the right side of the triangle and the D log E curve is equal to gamma. (C.B.N.)

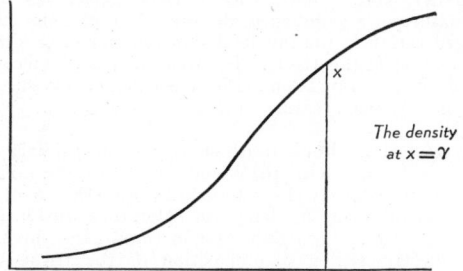

The density
at x = γ

GAMMA FUNCTION. The gamma function is a certain type of mathematical function which is met in advanced parts of **calculus;** it is a **transcendental function.** The infinite integral

$$\Gamma(\alpha) = \int_0^\infty x^{\alpha-1} e^{-x} dx$$

converges for all positive values of α, and defines **a function** of α which is called the gamma function.

A few of its important properties are:

$$\Gamma(\alpha + 1) = \alpha\Gamma(\alpha),$$

$$\Gamma(n)\Gamma(1 - n) = \frac{\pi}{\sin n\pi},$$

$$\Gamma(\tfrac{1}{2}) = \sqrt{\pi},$$

$$\Gamma(n + 1) = n!$$ when n is a positive integer.

Tables of values of the gamma function are published. (L.L.S.)

GAMMA GLOBULIN. The proteins of human blood serum consist of fibrinogen, albumin, and several globulins. The latter have been electrophoretically characterized as α_1, α_2, β, and γ globulins. The gamma globulin fraction has been found to contain the greatest concentration of **antibodies**—approximately 25-fold concentration of most of the antibodies present in normal human serum. This material is therefore an effective agent in providing passive **immunity** to certain diseases. When given intra-muscularly within 7 days after exposure, gamma globulin has the same effect as convalescent serum in preventing or modifying **measles;** because of its concentration it has the advantage of small dosage—$\tfrac{1}{10}$ to $\tfrac{1}{4}$ that of convalescent serum. If injected soon after exposure, gamma globulin is effective also in preventing **mumps** and infectious **hepatitis.**

GAMMA RAYS. Gamma rays are **x-rays** of short wavelengths emitted by **radioactive** substances. (R.K.S.)

GAMONT. An individual in the life cycle of certain one-celled animals whose subdivision produces **gametes.** (A.W.L.)

GANG PLANING. Planer.

GANGLION. In zoology, a ganglion is a small mass of nervous tissue isolated from the central system but containing cell bodies as well as fibers. Many ganglia bear special names. The brain of many **invertebrates,** for example, is also called the **cerebral ganglion,** and the more numerous centers of the molluscan nervous system bear names, such as the visceral ganglia and the pedal ganglia. The dorsal root of each nerve arising from the vertebrate **spinal cord** bears a spinal ganglion and the sympathetic system contains numerous ganglia.

In medicine, a ganglion is a tense globular cystic swelling, usually on the back of the wrist or the hand, communicating with one of the tendon sheaths or nearby joints. It is formed by a synovial membrane (see **Synovia**), and is filled with a thick gelatinous fluid. The nature of a ganglion is obscure. It represents either a degeneration in the involved synovial tissue, or simply a herniation of this tissue. The treatment is by mechanical rupturing, aspiration with a needle, or by surgical excision. (A.W.L., D.M.H.)

GANGRENE. Local death of tissue due primarily to total interference with the blood supply to the area or part involved. It is characterized by **anesthesia** of the part, loss of function, change in color to a dusky, cyanotic hue, lack of warmth, and in some cases, invasion by bacteria, causing decomposition of the tissue with spreading infection.

Many disease processes may terminate in gangrene. Thus in cases of injury, severe crushing of tissues may destroy their viability by interfering with the circulation. Again, inflammation, when intense, may shut off circulation by thrombosis or clotting, or by strangulation of blood vessels. This occurs in gangrenous appendicitis. Gangrene may also result from arrest of circulation, however produced, as is seen in various diseases causing obstruction of arteries or veins. Examples are severe hardening of the arteries (**arteriosclerosis**) common in the older age groups, particularly as a complication of **diabetes**, Raynaud's disease, and Buerger's disease. Chemical and physical agents, including **phenol**, or merely prolonged exposure to heat and cold, cause local death of tissue, just as does the prolonged constriction of a part. (D.M.H.)

GANGUE. The term gangue is used to refer to the non-valuable minerals associated with a metalliferous ore deposit. Commonly the gangue minerals are non-metallics such as **quartz** or **calcite**. (R.M.F.)

GANISTER ROCK. This term was originally applied to a **siliceous** underclay occurring in certain **coal** beds in the north of England. Now it is often applied to highly siliceous, fine-grained rocks used for refractory purposes or to a mixture of ground **quartz** and **fire-clay** used for furnace linings. (R.M.F.)

GANNET. Aves, Pelecaniformes. *Sula.* Large fishing birds (**Aves**) found on the seacoasts in the higher latitudes of both hemispheres. (A.W.L.)

GANOIDEI. A division of the fishes (Class **Pisces**) used in some classifications to include the orders **Chondrostei** and **Holostei.** (See also **Fossil Fishes.**) (A.W.L.)

GAP. In geology, a gap is an opening through a ridge connecting the valleys or lowlands on either side. Gaps may be formed by a river which earlier in the cycle of **erosion** was able to cut its way through the hard rocks now making up the ridge. If the stream is still flowing through this opening it is spoken of as a water gap, if the stream has disappeared because of its diversion or for other reasons it is then spoken of as a wind gap.

An electric gap is the distance separating two **electrodes** between which a **spark** or arc is caused to pass. Magnetically, a gap is the distance across an air gap separating two parts of a **magnetic circuit.** The clearance between pole pieces and rotor of **dynamo** machinery is such a gap.

In aviation a gap is a measurement used to describe the position of biplane wings. It is the distance between the upper and lower wings, measured from the mean aerodynamic **chord** of the upper wing to the mean **aerodynamic** chord of the lower wing. If the wings are staggered, the forward tri-section points of the mean aerodynamic chords are connected to determine the gap. The gap is frequently given in terms of the chord length,

thus a biplane wing cellule might be said to have a gap of 80%, which would mean that the gap was 80% of the wing chord. A gap of 100% is a relatively large one, though not considered a rarity. Small gaps, as low as 40%, have also been used. Larger gaps yield wing arrangements which are more efficient, aerodynamically, but smaller gaps lead to trimmer airplanes due to the compact arrangement of wings. (R.M.F., F.T.M.)

GAR, GARPIKE. Pisces, Holostei. Slender fishes (**Pisces**) with long, narrow jaws and sharp teeth. Three species in the rivers and lakes of North America and one

Garpike.

in Central America. The common garpike, *Lepisosteus osseus,* reaches a length of 4', the short-nosed gar, *L. platystomus,* about 3', and the alligator gar, *L. tristaechus,* as much as 18'. They are of little value as food fishes. (A.W.L.)

GARCINIA. Mangosteen; Gamboge. Guttiferae. A genus of trees and shrubs found wild in the tropics of the Old World. The leathery leaves are simple and opposite, and contain numerous oil glands. The fruit is a berry. Several species bear edible fruits. *Garcinia Mangostana* is the best known of these, the fruit being the famous mangosteen, considered by many to be the most delicious of all fruits. It has been introduced into tropical America.

Several species are tapped by cutting notches in the bark, through which **resin** exudes. This is the drug gamboge, a harsh cathartic, ordinarily mixed before use with other less violent drugs. (R.M.W.)

GARDENIA. Madder Family.

GAREFOWL. Aves, Charadriiformes. The great **auk**, *Pinguinus impennis,* now extinct. (A.W.L.)

GARFISH. Billfish.

GARGANEY. Aves, Anseriformes. A **duck**, the summer teal of Europe. *Anas guerguedula.* (A.W.L.)

GARIAL, GAVIAL. Reptilia, Crocodilia. A large fish-eating **crocodile** of India with an extremely long

Garial. (*N. Y. Zoological Society.*)

slender snout. The name gavial is said to have been an error for the vernacular name garial but it is much the commoner term and is perpetuated in the scientific names of the genus and family. (A.W.L.)

GARLIC. Allium.

GARNET. The name garnet is now applied to a group of very important minerals crystallizing in the isometric system and showing the same habit of **dodecahedrons** and **trapezohedrons.** They have the same general formulas, **orthosilicates,** and are to a limited degree **isomorphous.** Many different elements are included in the several varieties of garnet as **calcium, magnesium, aluminum,** ferrous or ferric **iron, chromium, manganese** and **titanium.** While garnets show no **cleavage** a dodecahedral **parting** is sometimes noted; fracture conchoidal to uneven; some varieties very tough and valuable for abrasive purposes. The hardness of garnet is 6.5–7.5; specific gravity, 3.1–4.3; luster, vitreous to resinous; colors, red, yellow, brown, black, green, or colorless; transparent to opaque. The word garnet is derived from the Latin *granatus,* a grain.

In general six varieties of garnet are recognized, based on their chemical composition: Grossularite (which is also called hessonite and cinnamon-stone), pyrope, almandine or carbuncle, spessarite, uvarovite, and andradite. Grossularite is a calcium-aluminum garnet which corresponds to the formula $Ca_3Al_2(SiO_4)_3$; the calcium may, however, be in part replaced by ferrous iron and the aluminum by ferric iron. The name grossularite is derived from the botanical name for the gooseberry, *grossularia,* in reference to the green garnet of this composition found in Siberia. Other shades are the well-known cinnamon brown, reds and yellows. Because of its inferior hardness to **zircon,** which mineral the yellow crystals resemble, they have been termed hessonite, from the Greek meaning inferior. Curiously enough, in the gem-bearing gravels of Ceylon both zircon and hessonite are found and indiscriminately called hyacinth, from the Greek, seemingly general term used by Pliny for the transparent varieties of corundum; later it was used for yellow zircons.

Grossularite is found in crystalline limestones with **vesuvianite, diopside, wollastonite** and **wernerite.** Among the many localities may be mentioned Siberia, the Urals, Italy, Switzerland, Mexico, and in the United States in Maine and New Hampshire.

Pyrope, sometimes called Cape ruby, is ruby-red in color and chemically a magnesium aluminum silicate with the formula $Mg_3Al_2(SiO_4)_3$; the magnesium may be replaced in part by calcium and ferrous iron. The color of pyrope varies from deep red to almost black. The transparent pyropes are used as gems, but some have a slight tinge of yellow. The name pyrope is derived from the Greek word meaning *fire-like.* A sub variety of pyrope from Macon County, North Carolina, is of a violet red shade and has been called rhodolite, from the Greek meaning *a rose.* In chemical composition it may be considered as essentially an **isomorphous** mixture of pyrope and almandite, in the proportion of two molecules of pyrope to one molecule of almandite. Pyrope is found at Teplitz and Aussig, Bohemia; in the Kimberley diamond mines in South Africa; in Australia and elsewhere. In the United States important localities are in Arizona, New Mexico, and Utah.

Almandite, sometimes called almandine, is the modern gem the carbuncle, although in Pliny's time this term was used for almost any red stone. The term carbuncle is derived from the Latin *carbunculus,* meaning a little spark. The name almandite or almandine is a corruption of Alabanda, a locality in Asia Minor where, in ancient times, these red stones were cut. Chemically, almandite is an iron-aluminum garnet corresponding to the formula $Fe_3Al_2(SiO_4)_3$. The deep red transparent stones are often called precious garnet and used for gems. Almandite occurs in metamorphic rocks like mica schists usually associated with typically metamorphic minerals

such as **staurolite, kyanite, andalusite,** etc. Good gem material comes from India and Brazil. Almandite is also found in Australia, Alaska, Africa, Norway, Sweden, Madagascar, and Japan. In the United States almandite is found in the **gneisses** of the Adirondack region of New York, sometimes of very large size, in New England and elsewhere.

Spessartite is manganese aluminum garnet, Mn_3Al_2 $(SiO_4)_3$. The name of this mineral is derived from Spessart in Bavaria, a well-known European locality. Spessartite of a beautiful orange-yellow comes from the Island of Madagascar. Violet-red spessartite has occurred in **rhyolites** in Colorado and Maine. Uvarovite is a calcium chromium silicate the formula being Ca_3Cr_2 $(SiO_4)_3$. It is a rather rare garnet, bright green in color, usually in small crystals associated with **chromite** in **serpentines,** sometimes in crystalline limestones or schists. Found in the Urals, South Africa, Canada, and in the United States in California and Pennsylvania. Andradite, calcium-iron garnet, $Ca_3Fe_2(SiO_4)_3$, is of variable composition and may be red, yellow, brown, green, or black, or of intermediate shades. The subvarieties topazolite, yellow or green, demantoid, green, and melanite, a black sort, are recognized. Andradite is found both in deep-seated igneous rocks like syenite as well as in serpentines, schists, and crystalline limetones. Demantoid has been called the "emerald of the Urals" from its occurrence there. Varieties of andradite are found in many localities in Europe; Italy, Switzerland, Norway, Saxony, etc., and in the United States at Franklin, New Jersey; Magnet Cove, Arkansas; and elsewhere. (E.S.C.S.)

GAS ABSORPTION. Gas absorption is a process whereby one gas or several gases are removed from a gaseous mixture by causing the desired gas or gases to be dissolved in some liquid, thus leaving undissolved the inert or undesired gases. An example is the removal of ammonia (NH_3) from an air-ammonia mixture by bringing the mixture into contact with water. The ammonia will dissolve into the water but the air will not.

The operation of absorption is ordinarily accomplished in an *absorption tower.* The tower is usually cylindrical and packed with coke, broken rock, or specially shaped forms. The most usual special packing is the *Raschig ring.* These rings are short tubular pieces with length about equal to the diameter. They are dumped into the tower and lie in a random fashion. Liquid is allowed to trickle down over the packing while the gas is pumped up through the tower where it swirls about the packing on its way up. In this way there is an intimate intermingling of the gas and the liquid. The tower must be tall enough so that there is sufficient time of contact between the gas and the liquid, and large enough in cross-sectional area so that the required volume of flow can be accommodated. If too much liquid is run down for the size of the tower, then the upward-moving gas will hold up the liquid, causing the tower to choke or *flood.*

The amount of gas which a liquid can dissolve at a given temperature is determined by Henry's Law, which states that the partial pressure of a gas in equilibrium with a solution is equal to a constant times its concentration in the solution, or

$$p = cx.$$

The constant, c, is different for each system and for each temperature, and it must be determined experimentally.

The concentration of the absorbable gas in the effluent from the tower must be somewhat less than the equilibrium concentration of the gas in the liquid. The difference between the two, i.e., the difference between the actual concentration and the equilibrium concentra-

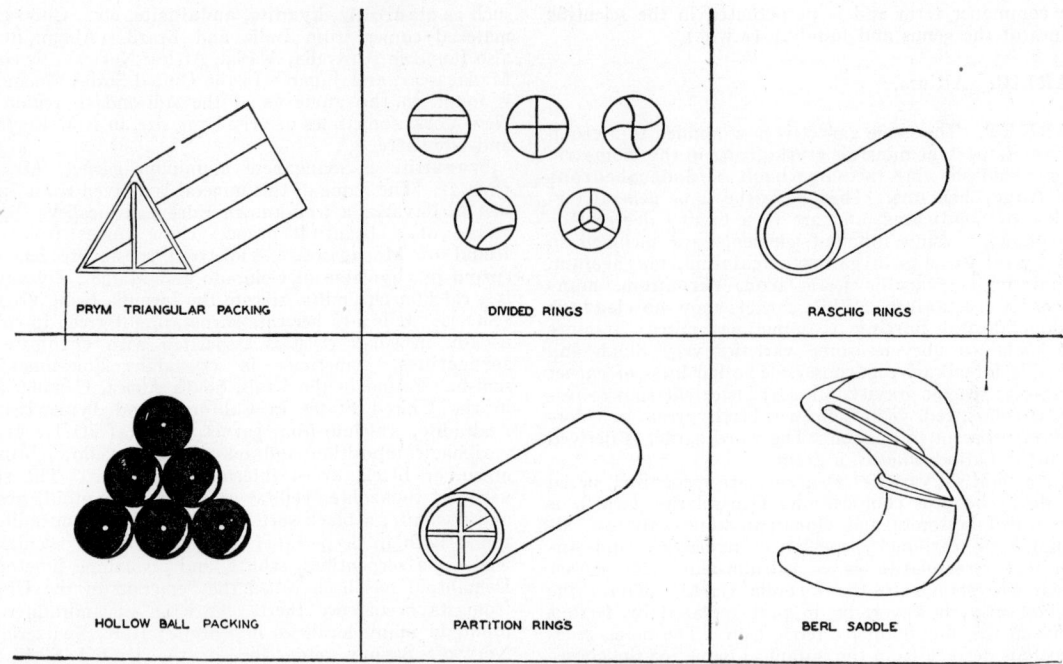

TYPES OF TOWER PACKING

tion, is necessary in order that there be a *driving force* to cause absorption to take place.

For each tower and set of operating conditions there is a specific absorption coefficient, K. K depends on the type and composition of the materials being operated on, the type of packing, the temperature, and the gas and liquid flow rates. The coefficient is defined as the amount of material absorbed per unit time per unit of contact area per unit of driving force. The area of contact through which the gas is being absorbed cannot be measured, therefore this unknown area is included with the coefficient, K, and determined experimentally as coefficient times the area, Ka. After the amount of material absorbed per unit time is determined for the whole tower by direct measurement, it is divided by the tower volume and by the driving force, giving the final form

$$Ka = \frac{\text{lb.}}{\text{(hr.)(cu. ft.)}(\Delta x)}$$

where Δx is the driving force. The Δx driving force is the difference between the concentration the liquid actually has and that which it would have if it were in equilibrium with the gas. Since Δx may vary throughout the tower, an average value must be used.

For any given installation the absorption coefficient depends on the liquid-flow rate, the gas-flow rate, the temperature and concentration of the liquid and the gas. Because the open cross-sectional area of the packed tower is not known, the gas and liquid rates are usually given in terms of superficial velocities. The superficial velocity is defined as the velocity the fluid would have if it were flowing through and completely filling the tower when empty of packing.

If slightly soluble gases are being absorbed (say oxygen into water) then turbulence in the liquid is essential but turbulence in the gas is not important. In this case the gas-flow rate has been found not to affect the absorption coefficient. When the gas is very soluble (say ammonia into water) the reverse is true, that is, the liquid rate is not important and the coefficient is affected greatly by the gas rate.

Typical examples of gas absorptions are: removal of SO_2 from flue gas, recovery of carbon dioxide from combustion gases, gasoline from natural gas, ammonia from coke oven gas, and the drying of air.

The reverse of absorption, where a dissolved gas is removed from a liquid by contacting the liquid with an inert gas, is called *stripping*.

Gas absorption need not necessarily be performed in a packed tower. It may be done in a bubble-cap tower (see **Distillation**), or in a tank where the gas is bubbled through a liquid. In order to obtain a good dispersion of the gas in the liquid it is common practice either to agitate the liquid violently with a stirrer (see **Mixing**) or to introduce the gas into the liquid through a porous plate. (D.E.M.)

GAS CALORIMETER. Fuel Calorimeter.

GAS ENGINE. Internal Combustion Engine.

GAS EQUATION FOR AIR. Equation of State for Air.

GAS GANGRENE. Infection of tissues around a wound by certain anaerobic bacteria which grow best deep in tissues away from the air. The infection is necrotic and rapidly spreading; it is accompanied by massive edema, gaseous infiltration and discoloration of the tissues. The organisms liberate a toxin which destroys tissue, particularly muscle, and they produce gas by fermenting muscle sugars.

Most of the information on gas gangrene was obtained during World War I. Many of the war wounds were infected with gas-producing organisms and from contamination either directly or indirectly with fecally contaminated soil. Gas gangrene is occasionally seen in civil medical practice in extensive wounds with considerable crushing of tissue, particularly in compound fractures.

The most common organism found in gas gangrene infections is *Bacillus Welchii*. It, as well as other of the gas-forming group, is a normal inhabitant of the human and animal intestinal tract and is present in soil with great constancy. In war and in some civilian injuries, organisms are carried into the depths of tissues by foreign bodies or bits of projectiles. In these types of injury variable amounts of tissue are destroyed or devitalized, thus furnishing the dead or necrotic material that gas organisms thrive upon. It is for this reason that debridement (excision of all devitalized or contaminated tissue) is immediately carried out as soon as the patient is hospitalized. This is one of the best methods of preventing this serious complication of wounds.

Good results have been obtained by injection of a polyvalent antitoxic serum against the common gas-gangrene organisms, both prophylactically and therapeutically. The **sulfanamides** and **penicillin**, used in conjunction with anti-serum and adequate surgery, are also effective in the treatment of gas-bacillus infection. (R.S.M.)

GAS OIL. Gas oil is the residual oil left after **gasoline** and kerosene are distilled from certain crude petroleums. (See **Fuels** and **Hydrocarbons, Paraffinic.**) (R.K.S.)

GAS THERMOMETER. When the standard measure of **temperature** was fixed upon as the pressure of a gas kept at constant volume, the constant-volume gas thermometer became the final arbiter of temperature measurement. (See **Thermometry.**) In this instrument the gas (preferably hydrogen or helium) is enclosed in a glass or fused quartz bulb that is connected to a mercury manometer, and facilities are provided to bring the gas always to the same volume and to indicate the gas

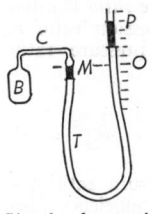

Sketch of essential parts of constant-volume gas thermometer.

pressure. The most common form, designed by Jolly (1874) is shown diagrammatically by the figure, in which B represents the bulb, T the flexible tube of the manometer, M the constant-volume mark at zero level, and P the level of the mercury indicating the pressure. The pressure is regulated by moving the right-hand mercury column up or down the scale. Slight corrections are necessary for the expansion of the bulb and for the difference of temperature between the gas in the bulb and that in the connecting tube C. The use of this thermometer depends upon the fact that changes of pressure are accurately proportional to the changes of temperature producing them. If a mercury thermometer is to be standardized between 0° and 100° C., for example, its bulb is placed along with the bulb B in a bath of adjustable temperature and the total pressure in B noted for each of these fixed points. The interval between these points is then divided into 100 parts, each of which corresponds to 0.01 of the whole change of pressure. Since the mercury at P is exposed to the atmosphere, the atmosphere pressure, taken from a good barometer, must be added to that measured by OP. The constant-pressure type of gas thermometer was perfected by Callendar and others, but proved somewhat impracticable and has fallen into disuse. (L.D.W.)

GAS TURBINE. The gas turbine is a means of producing mechanical work from heat energy, using a gas as the working medium. Its advantages and limitations were little understood prior to 1930. However, since that time the gas turbine has been subjected to intensive research and development and has become a competitor of the internal combustion engine. Many successful installations of gas turbine power plants have been completed since 1935, and the field has broadened to include mobile as well as stationary power.

In the gas turbine a stationary nozzle discharges a jet of gas (usually products of combustion) against the blades on the periphery of a turbine wheel, as shown in Fig. 1. The jet is thereby deflected and slowed while the blades receive an impulse force which is transmitted as a mechanical torque to the shaft. The prospective

jet speed is sometimes sufficiently high to warrant dividing the expansion into a series of *stages* with a set of nozzles and a row of blades in each stage, all blade

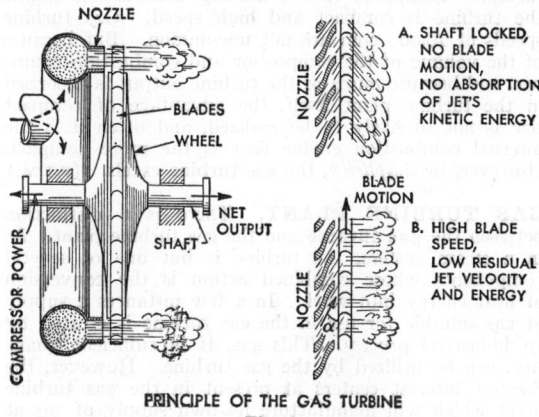

PRINCIPLE OF THE GAS TURBINE
Fig. 1.

wheels being mounted on the same shaft. By limiting the gas expansion per stage the blade speed and rpm of the shaft are suitably decreased. Were the blades themselves so shaped as to be virtual nozzles, some expansion would also take place in the gas as it went through the blading. The latter in consequence would receive a "reaction thrust" distinct from impulse action. Many gas turbine designs have employed the reaction principle.

A properly designed nozzle can produce almost an ideal (isentropic) adiabatic expansion of the gas. Any failure of the gas turbine to convert all the ideal

Fig. 2. Axial compressor unit for 23,000 cfm gas flow. Top half of casing removed. Turbine unit is in the foreground. (*Allis Chalmers Mfg. Co.*)

available energy into work at the shaft is mainly attributable to the blading—its clearance, friction, leakage, and residual gas velocity.

The term "engine efficiency" is frequently applied both to engine and turbine prime movers to denote perfection of thermodynamic design, using the ideal available energy as a standard. So, if e_i is the ideal amount of energy made available by the possibility of expanding a fluid between specified initial and final states, and e is the energy actually developed by a gas turbine working between the same limits, then

$$\eta_e = e/e_i.$$

η_e is the so-called *engine efficiency*, sometimes called internal efficiency.

It is quite essential to the success of a gas turbine plant that η_e be as high as possible. The range of η_e is 80% to 90%, a remarkable modern achievement. Had the efficiency been better in early experimental gas turbine plants, some might possibly have turned out successfully.

Gas turbines present no new difficult problems to the manufacturer. For the same inlet temperature the gas turbine is probably easier to design than the steam turbine. Compared to the internal combustion engine the turbine is compact and high speed. Gas turbine speeds of 10,000 rpm are not uncommon. But because of the volume of the compressor and combustion chamber, and because much of the turbine output is absorbed in the turbine plant itself, the advantage of compact size is not so likely to be realized, and often it is the internal combustion engine that is the more compact. However, in *simplicity,* the gas turbine excels. (F.T.M.)

GAS TURBINE PLANT. There is a distinction between the **gas turbine** and the gas turbine plant. As in a steam system, the turbine is but one of several components whose combined action is the conversion of heat energy into work. In a few instances a supply of gas suitable for use in the gas turbine is created by an industrial process. This gas, if in sufficient quantity, can be utilized by the gas turbine. However, the greatest interest centers at present in the gas turbine plant which will manufacture its own supply of gas at a pressure by compressing air and burning fuel in it. Such a plant will consist at least of an air compressor driven by the gas turbine, a combustion chamber wherein a liquid fuel is injected and burned at constant pressure equivalent to that of the compressor discharge. The resulting products of combustion form the working medium of the gas turbine. It is the expansion of volume of the working medium during the process of combustion that enables an ideal turbine to produce more work than an ideal compressor would consume. If this difference is sufficient to more than offset the internal imperfections of an actual compressor and turbine, some surplus work will become available as a net plant output. This net plant output (in B.T.U.), divided by the heating value of the fuel burned in the compressed air, is the over-all thermal efficiency of the gas turbine plant.

An elementary plant of this type is shown in Fig. 1. Since the working medium is changed in chemical com-

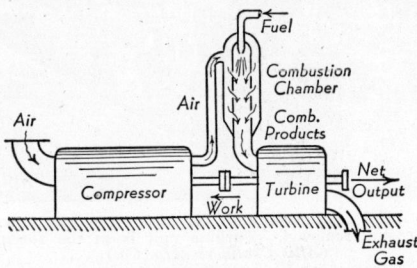

Fig. 1. Basic combustion gas turbine plant.

position by the combustion process, it is discharged from the system after yielding up some of its energy in the turbine and fresh air is continuously drawn into the compressor. The cycle, if there could be said to be one, is therefore an "open" one, since in a closed cycle the air passes around the cycle and receives heat by conduction instead of combustion. The open cycle of operations, shown in Fig. 2, is sometimes called the Brayton cycle. It has been used in internal combustion engines without significant success, but is well

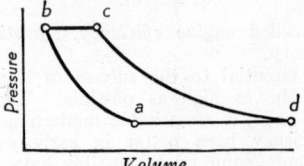

Fig. 2. Combustion gas turbine cycle.

adapted to a turbine cycle since the turbine can more readily produce complete expansion. Line *ab* represents compression of the air, *bc* the addition of heat at constant pressure. Expansion in the turbine is represented by *cd*, while *da* is imagined to be a constant pressure cooling of the air to ambient atmospheric temperature. It is this part of the cycle that is open in gas turbine plants.

This constant pressure combustion cycle is used in most gas turbine plants. Designating it the *combustion gas turbine cycle* is a means of distinguishing between it and the *explosion gas turbine cycle*. The latter type, which uses constant volume combustion, has been successfully constructed but is handicapped by the necessity of a mechanically operated valve system, like the internal combustion engine, and by operation with a fluctuating gas pressure.

Any liquid fuel which does not form corrosive products of combustion can be used in the open cycle. Since the products of combustion actually flow through the turbine, ash in them would be detrimental; consequently coal cannot be used as yet. Refined fuel oil and kerosene are used except where industry may have a gaseous fuel by-product to use.

The air standard efficiency of the Brayton cycle is the same as that of the Otto cycle, namely $1 - \frac{1}{r^{\gamma - 1}}$, in which r is the ratio of compression and γ the adiabatic polytropic gas exponent. In flow machines, such as the rotary compressor, the *pressure ratio* is often more useful than the volumetric compression ratio r. The theoretical cycle efficiency, using pressure ratio P_b/P_a as the variable, is

$$\eta_t = 1 - \frac{1}{\left(\dfrac{P_b}{P_a}\right)^{\frac{\nu-1}{\nu}}}.$$

This implies that an increase in pressure ratio increases cycle efficiency. The actual cycle efficiency also involves the internal machine efficiencies and a variable air-fuel ratio, since for a given temperature at point c in the cycle the higher the pressure ratio the greater the air-fuel ratio needed. For operation with a limiting maximum temperature in the combustion chamber (a practical criterion) the efficiency increases with pressure ratio up to a certain *optimum ratio,* then decreases with further increase of pressure ratio. This is shown

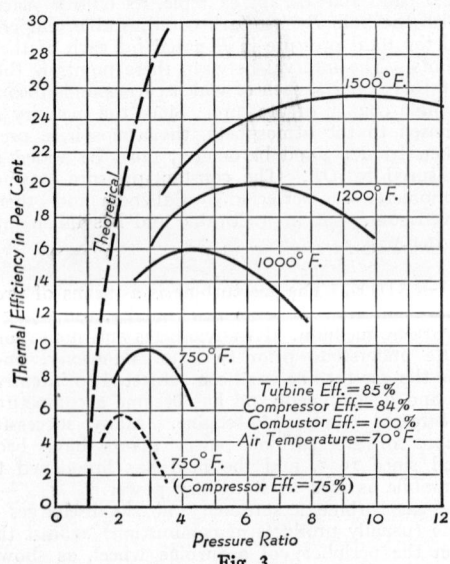

Fig. 3.

in Fig. 3. The actual efficiency of an open cycle combustion gas turbine plant is approximately

$$\eta = \frac{W_t \eta_t \eta_c - W_c}{\eta_c H_f}$$

W_c and W_t are the ideal amounts of work involved in processes ab and cd, respectively—in heat units. η_t and η_c are the internal efficiencies of turbine and compressor. H_f is the heating value of the fuel burned per cycle.

The two terms in the numerator of the above equation are of approximately the same magnitude, therefore small changes of η_t or η_c will cause η to vary widely. Since η is not particularly high anyway (on account of a limiting maximum temperature of about 1200° F. imposed by metallurgical considerations), this emphasizes the need for refinement of construction to secure the very highest actual efficiencies possible in the compressor and turbine, or else the W_c term equaling $W_t \eta_t \eta_c$ will leave zero efficiency and no net output for the plant. Illustrating the results possible in practice, η_t might, optimistically, be as high as 85%, η_c 80%. Then, using a pressure ratio of 4 and a maximum cycle temperature of 1000° F., the efficiency works out to be 13%, and somewhat more than 100 lbs. of air must be drawn into the compressor for each horsepower hour net work output. This is a contrast with the 9 lbs. used by a gasoline engine for the same output, but it should be remembered that the latter is a displacement machine whereas the gas turbine plant is a flow system, much better adapted to handle large volumes of the working medium. It should also be noted that the more of the turbine output the compressor requires the greater is the bulk of the equipment for a given *net* output. In its present stage of development, as much as ⅗ of the gross turbine output is required by the compressor.

The data quoted above might well dash the hope that the combustion gas turbine will actively compete with the steam and internal combustion prime movers. Although nearly perfect operation is assumed for the machines, the efficiency is so low that our gas turbine is surpassed by everything else except the non-condensing steam plant. However there are possibilities of improving the performance by metallurgical progress permitting higher maximum temperatures (1500° F. is being contemplated), and by thermodynamic schemes that reduce compressor demands and salvage part of the exhaust heat. Pushing up the maximum temperature limits will depend on developments in other fields of technology. But the thermodynamic modifications, called *intercooling, regeneration,* and *reheating,* may be applied at present. At the cost of complicating the flow pattern of a plant, whose virtue has always been simplicity, efficiencies may be raised until this prime mover becomes a practical competitor with the others, for over-all efficiencies of 20% to 30% seem to be possible, using today's maximum practical temperatures.

The intercooling principle consists of compounding the compression and intercooling between sections. This lowers the air temperature at constant pressure, reducing the volume and the compression work required in the following compressor stages. It requires a surface type intercooler with a coolant, air or water. Regeneration is effected by transferring heat energy from the exhaust gases to the compressed air after it leaves the compressor but before it enters the combustion chamber. This again requires a surface heat exchanger, one capable of standing up under high temperatures. The gain here is due to the smaller quantity of fuel which has to be injected to reach the same maximum gas temperature. Reheating is the process of burning the fuel in steps, the amount of fuel being proportioned to reach the maximum allowable temperature in each stage of combustion. For this the turbine must be compounded, with a combustion chamber inserted in the flow between the high and low pressure sections.

These principles are illustrated diagrammatically in Fig. 4. A marine gas turbine plant recently built, and incorporating intercooling, regeneration, and reheat, is

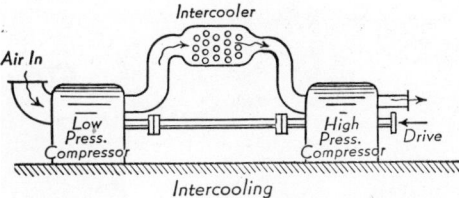

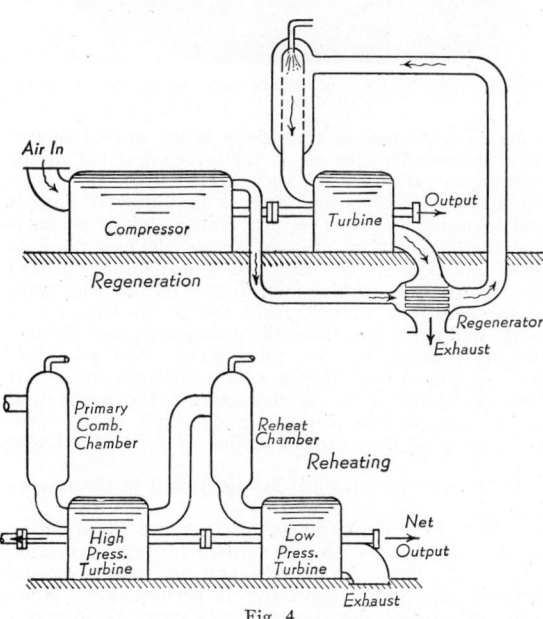

Fig. 4.

shown schematically in Fig. 5, with attendant temperatures noted. This plant is characterized by the use of a positive displacement rotary compressor rather

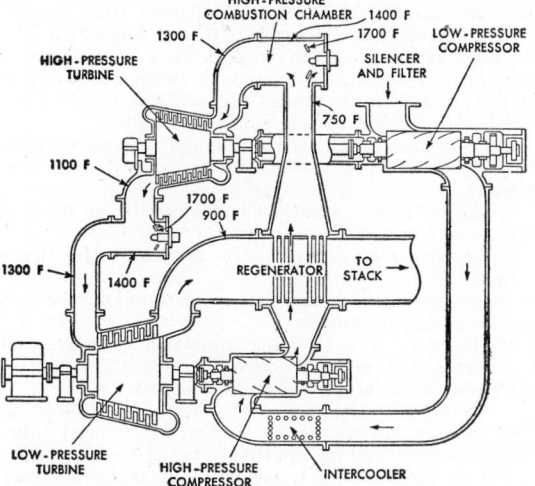

Fig. 5. An intercooled, regenerative-reheating gas turbine plant in schematic arrangement. (*Elliott Co.*)

than the axial flow type usually encountered. With a pressure ratio of 6.5 and a maximum throttle temperature of 1230° F., it is said to achieve an over-all efficiency of 29% when producing the rated net output of 2500 hp. The high-pressure section of the turbine drives the low-pressure section of the compressor, these

Fig. 6. View of the gas turbine plant for operation on the plan shown in Fig. 5.

being proportioned so that there is no surplus power. After intercooling, the air is further compressed in the high-pressure compressor, then heated by being passed through the regenerator. Fuel is next injected into it and burned. Plenty of oxygen remains in the products due to the great quantity of excess air used to hold down the temperature rise. After expansion through the high-pressure turbine more fuel is burned with some of the remaining oxygen, again raising the temperature to the limiting value. Next the products expand through the low-pressure turbine, after which they pass into the regenerator and thence to the exhaust stack. As this was designed for marine service, the net output of the low-pressure turbine is delivered to the propeller shaft at the output coupling of a speed-reducing gear box.

To date, the gas turbine has been used in the following fields:

1. Stationary power plant. Economical for stand-by service and possibly (in the future) for regular service.

2. Locomotive. An experimental unit was successful but not particularly impressive in performance. However, subsequent development may raise performance to competitive levels.

3. The Houdry process. In this gasoline cracking process compressed air must be provided for burning deposits from the catalyst. This process is accompanied by considerable increase in temperature and only a small drop of pressure. The gases discharged from process are put through a gas turbine, whose output supplies the necessary compression power. Here no net power is made, but as otherwise 2500 hp. would be required to drive the Houdry process compressors, the turbine-compressor combination is highly successful.

4. Supercharging of both spark and compressor ignition engines, using turbine-driven centrifugal compressors.

5. Marine power plant. Claimed to be as efficient and more compact than the steam plant.

6. Jet-driven airplanes. (See **Airplane, Jet Propelled.**)

Summarizing the comparative features of the gas turbine as a prime mover:

1. Mechanically it is simple compared to steam and I.C. plants, but in the endeavor to reach competitive efficiencies some of this advantage is lost.

2. An electric motor or I.C. engine is required to start the gas turbine plant. As the starter must bring the compressor well up toward operating speed, starting is not as simple as S.I. engines, but many compare favorably with C.I. engines and steam.

3. Like steam turbines, the gas turbine is not readily reversible. Steam engines and 2-cycle I.C. engines are best in this respect.

4. Turbine plants have less vibration than engine plants of similar size.

5. The gas turbine uses high temperatures. Even though the pressures are moderate, service life will be shortened by high temperatures.

6. With certain types of compressors, efficiency of the gas turbine plant is not as well maintained at part load as with steam or the I.C. engine. (F.T.M.)

GASES. States of Matter; Dynamics of Gases; Kinetic Theory; Ideal Gas Law; Characteristic Equations; Avogadro's Law; Boyle's Law; Charles' Law; Joule-Thomson Effect; Dalton's Law; Poisons.

GASKET. The gasket is a layer of packing material firmly held between contact surfaces on two pieces whose joint is to be sealed with the gasket. Gaskets are made in many different shapes to suit the shapes of the various mating pieces. The simplest type of gasket is that used to seal the joint between two circular flanges, as might be used in making a joint in piping. The gasket is made of a thin sheet of material, satisfactory for the service, having through it a hole corresponding to the internal diameter of the pipe. After being inserted between the flanges, flange bolts draw the latter tightly together, compressing the gasket material until it tightly seals the joint. Some gaskets are toroidal in shape, but most are flat. The services they perform range from that of sealing against leakage of the liquids in water lines to rendering gas-tight such high-temperature joints as those in engine exhaust manifolds. Naturally, a variety of materials would be required for these different conditions. In general, gasket material is rubber for water, corrugated copper for saturated steam, soft corrugated steel for superheated steam, asbestos for hot gas. (F.T.M.)

GASOLINE. A liquid hydrocarbon mixture obtained from four main sources: (1) natural gas or casing-head gasoline, by condensation under pressure or by adsorption of the liquefiable constituents of natural gas, (2) straight-run gasoline, by the fractional distillation of crude petroleum, (3) cracked gasoline by (a) thermal (heat-pressure) or (b) catalytic decomposition of crude petroleum or its distillate fractions, (4) synthetic gasoline by such reactions as (a) polymerization of hydrocarbons of lower, to those of higher, molecular weight, (b) hydrogenation of producer gas (Fischer-Tropsch), of petroleum or coal-tar distillate fractions, or of coal (Bergius). Cracked and synthetic gasolines show distinct and remarkable increases in production with resulting economy in the utilization of crude petroleum.

Catalytic cracking is conducted in the vapor phase, and therefore the distillates used must be those that

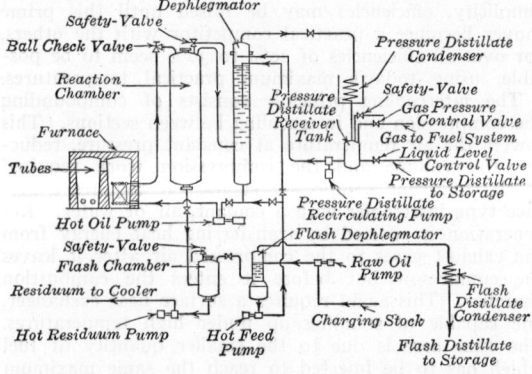

Dubbs thermal cracking process flow chart. (*Universal Oil Products Co.*)

can be vaporized at the operating temperature of the process. Steam is usually introduced simultaneously with the charging stock to facilitate the vaporization of the higher boiling portion. The chemical principles are similar in all three of the following processes, although the mechanical arrangements differ radically.

The fixed-bed catalytic process has been in commercial operation since 1936 by the Houdry Process Corporation, Socony-Vacuum Oil Company, and Sun Oil Company,

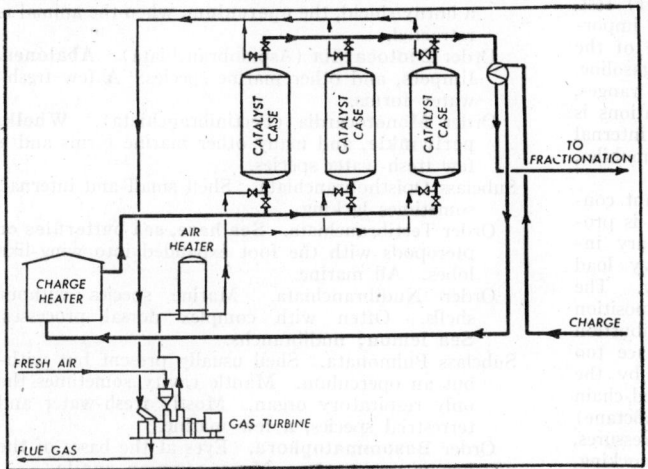

Houdry catalytic cracking process. ("*Refresher on Wartime Refining Technology,*" *National Petroleum Publishing Co.*)

producing high-quality motor fuel. In this process, alternate periods of cracking the distillate and regenerating the catalyst occur in each reaction compartment, these being arranged in a series so that the input and output of the system as a whole is continuous. The usual compartments are about 40′ in height and about 10′ in diameter. The feed stock is fed from below, and the resulting vapor removed at the top. There is considerable latitude in the choice of feed stock, and a corresponding flexibility in the range of product obtained. Early yields ranged around 45% gasoline of octane number 77 to 81.

The moving-bed catalytic process is one developed by Socony-Vacuum Oil Company—called "Thermofor"—from the type of clay-burning kiln originally used in regenerating the catalyst. Some 30 units of this type were constructed in the United States during the war. The process is distinguished by having the reaction and regeneration zones separated, although there is a continuous movement of the catalyst

through each zone. The counter-current principle is frequently employed, wherein the vaporized feed stock is admitted near the bottom of the reaction zone and the catalyst near the top, so that the fresh catalyst near the top meets first the more resistant portion of the oil vapor, and the partially spent catalyst near the bottom carries out the initial cracking. The concurrent principle is used in some installations, one advantage being that the operating pressure is less than in the counter-current type. The spent catalyst is purged of oil vapors by superheated steam and then moved in an elevator to the top of the regenerating kiln. Here the deposit of carbon on the catalyst is burned off by careful regulation of the air supply. The process yields 35 to 45% gasoline of octane number 85 to 93. A notable advance in the process has been the production of a specially active "bead" catalyst (1943) in the form of translucent spheres about 3.5 mm. in diameter which have superior crushing resistance and which increase the yields of product 15 to 30% compared to clay pellets.

Fluid catalyst cracking is also an outcome of World War II, although the Standard Oil Company of New Jersey was engaged in developing this process before the war. During the period 1941 to 1945 the number of fluid catalyst-cracking units operating in the United

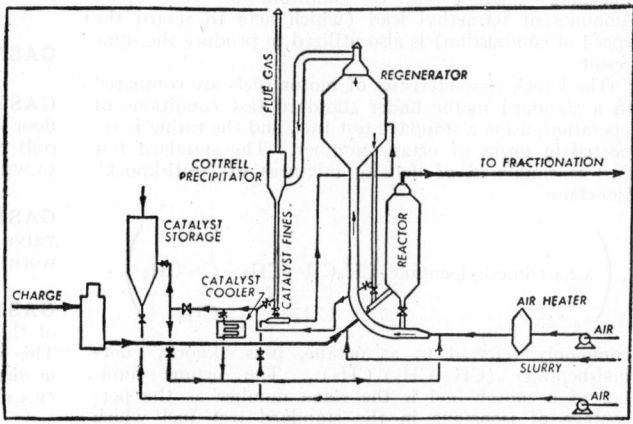

Fluid catalyst cracking process. ("*Refresher on Wartime Refining Technology,*" *National Petroleum Publishing Co.*)

States rose from none to 33. In this process the catalyst is in the form of a powder, a synthetic silica-alumina (SiO_2—Al_2O_3) material—ignited clay was first used and is still used in starting up units. By the use of the gas-lift principle and without any moving parts (changing the feed stock at the bottom and removing the vapor at the top) the catalyst is kept in a free-flowing condition in the reaction vessel, where the pressure is only about one atmosphere above the normal atmospheric pressure. The reaction vessel is a long cylindrical tank standing on end, say 50′ in vertical height and 25′ in diameter. The temperature in the reacting compartment is maintained between 425° C. (800° F.) and 525° C. (975° F.). On feed stock of 400° C. (750° F.) distillation dry-point the yields reported are about 43–46% of aviation gasoline having a distillation dry-point of 200° C. (400° F.) and octane rating of 91 to 94 without the addition of lead tetraethyl. The catalyst is recovered continuously from the bottom of the reaction compartment, and then the carbon, which is deposited on it in use, is carefully burned off by heated air, regulated so that the

Thermofor catalytic cracking process. ("*Refresher on Wartime Refining Technology,*" *National Petroleum Publishing Co.*)

temperature of burning is maintained between 540° C. (1000° F.) and 650° C. (1200° F.).

Specifications for various types of gasoline more or less commonly used in commerce are laid down involving color, odor, gums, distillation range, doctor test (for certain sulfur compounds), corrosion (of copper) test, acidity, and sulfur (total). Of these, the most important is the distillation test, which is a measure of the volatility and the range of volatility of the gasoline. Various grades, based upon different distillation ranges, are recognized. The necessity for such specifications is demanded by the extensive use of gasoline in internal combustion engines for motive power in automobiles, trucks, and airplanes.

The knocking power of gasoline is an important consideration. The knocking sound, or **detonation,** is produced in high-compression engines or in ordinary **internal combustion engines** when under heavy load ascending grades or when accelerating rapidly. The knocking property is dependent upon the composition of the **fuel.** Knocking indicates that the combustion of the fuel vapor in the cylinder is taking place too rapidly for complete utilization of the power by the cylinder of the engine. Benzene and branched-chain paraffins (such as 2,2,4-trimethylpentane or isoöctane) are non-knocking under prevailing engine pressures, straight-chain paraffins are most conducive to knocking, while olefins and cycloparaffins are intermediate in effect. Benzene is blended with gasoline to improve the anti-knocking quality, and the addition of very small amounts of tetraethyl lead (which acts to retard the speed of combustion) is also utilized to produce the same result.

The knock characteristics of motor fuels are compared in a standard motor under standard test conditions of operation, using a standard test fuel, and the rating is reported in terms of octane number. The standard test fuel is composed of definite mixtures of "anti-knock" isoöctane

$$\left(\text{2,2,4-trimethylpentane, } CH_3CH-CH_2-\overset{\displaystyle CH_3}{\underset{\displaystyle CH_3}{\overset{\displaystyle |}{\underset{\displaystyle |}{C}}}}-CH_3 \right),$$

commonly referred to as octane, plus "knock," normal-heptane $(CH_3(CH_2)_5CH_3)$. The octane number of a motor fuel is the same number as the percentage of isoöctane in the standard test fuel which matches the tested fuel in knock characteristics.

Federal Specification for United States Government Motor Gasoline includes *Distillation range.* A.S.T.M.[*2] method D86-30. When the thermometer reads 75° C. (167° F.) not less than 10% shall be evaporated. When the thermometer reads 200° C. (392° F.) not less than 90% shall be evaporated. The residue shall not exceed 2%. (R.K.S.)

GASTEROPODA, GASTROPODA. The **snails, slugs,** and allied forms, constituting a class of the phylum **Mollusca.** This group includes a large number of marine and fresh-water species and many that are terrestrial.

The chief structural characteristics of the gasteropods are these: 1. Most species have a dorsal visceral hump which is often spirally twisted. 2. A head is present, bearing eyes and tentacles. 3. The mouth is provided with a toothed organ called the radula. 4. The foot is usually a broad creeping organ. 5. Respiratory **ctenidia** lie in the mantle cavity of some species and in others the walls of the cavity are the respiratory organ. 6. In many species a shell, conical or spirally coiled, encloses the visceral hump.

Gasteropods are of relatively little economic importance. Snails are eaten in Europe and the abalones of the Pacific Coast are also used as food. The shell of the abalones furnishes beautifully iridescent mother-of-pearl for costume jewelry and there is an extensive traffic in the shells of many species among collectors.

The group is classified as follows:

Subclass **Streptoneura.** Usually with a shell closed by a horny shield, the **operculum,** when the animal is retracted.
Order **Diotocardia** (Aspidobranchiata). **Abalones, limpets,** and other marine species. A few fresh-water forms.
Order **Monotocardia** (Pectinibranchiata). **Whelk, periwinkle,** and many other marine forms and a few fresh-water species.
Subclass Opisthobranchiata. Shell small and internal, sometimes lacking.
Order Tectibranchiata. **Sea hare, sea butterflies** or pteropods with the foot expanded into wing-like lobes. All marine.
Order Nudibranchiata. Marine species without shells. Often with complex dorsal processes. **Sea lemon; nudibranchs.**
Subclass Pulmonata. Shell usually present but without an operculum. Mantle cavity sometimes the only respiratory organ. Mostly fresh-water and terrestrial species, a few marine.
Order **Basommatophora.** Eyes at the bases of the posterior tentacles. Many common **snails.**
Order Stylommatophora. Eyes at the tips of the posterior tentacles. Common snails and **slugs.** (See also **Invertebrate Paleontology.**) (A.W.L.)

GASTEROSTOMATA. Trematoda.

GASTRIC FILAMENT. Slender projections on the floor of the stomach of a **jellyfish,** containing stinging cells which kill living prey taken into the stomach. (A.W.L.)

GASTRIC SHIELD. A plate in the stomach of **bivalve** mollusks against which the **crystalline style** is worn away. (A.W.L.)

GASTRITIS. An inflammation of the lining membrane of the stomach occurring in an acute or chronic form. The common causes of gastritis are alcoholism, errors in diet, food poisoning, and irritant or corrosive poisons. (R.S.M.)

GASTRO-ENTERITIS. An inflammation or irritated condition of the gastrointestinal tract, due to errors in diet, infection or food poisoning, irritant poisons or alcohol. (R.S.M.)

GASTRO-ENTEROLOGY. The study of the diseases and functions of the **stomach** and **intestines.** A gastro-enterologist is a specialist in the diagnosis and treatment of the disorders of the stomach and intestines. (R.S.M.)

GASTROSTOMY. An artificial, and usually permanent, opening made through the abdominal wall into the stomach. This is done to prevent starvation when there is an obstruction in the **oesophagus** from a malignant growth, or scarring after drinking a corrosive chemical, etc., thus making it impossible for the normal passage of food into the stomach. (R.S.M.)

GASTROTRICHA. A group of minute animals found in fresh and salt water on the bottom and among the debris accumulated there. They move chiefly by means of **cilia** and have cement glands whose secretion attaches them temporarily to supports. They have a tubular alimentary tract and an excretory system consisting of two tubules with flame cells. The group is ranked by some writers as a class in the same phylum as the rotifers and by others as of uncertain relationship.

Two orders are recognized: **Macrodasyoidea,** made up of marine species with numerous cement glands, and **Chaetonotoidea,** made up of marine and fresh-water species with a single pair of cement glands at the caudal end of the body, or none. (A.W.L.)

GASTROVASCULAR SPACE. The central cavity (enteron) of coelenterates and ctenophores. (A.W.L.)

GASTROZOOID. A form of individual in hydrozoan colonies whose function is to digest food for the colony. (A.W.L.)

GASTRULA. The stage in embryonic development in which the initial differentiation of tissues is evident. The gastrula is typically a sac whose wall is composed of the two germ layers, an outer ectoderm and an inner endoderm. The cavity lined by the endoderm is the archenteron and the opening to the exterior is the blastopore.

The gastrula is formed from the **blastula** by the process of gastrulation. Typically the wall of the spherical blastula caves in on one side and the invagination progresses until this side is in contact with the opposite wall. In some **coelenterates,** however, the two germ layers appear as a solid mass of endoderm surrounded by a layer of ectoderm, and the archenteron forms by the splitting of the inner mass. In animals with abundant yolk, modifications also appear. In birds, for example, the stage approximating the blastula is a disk of cells on the surface of the yolk and the endoderm may be formed by the folding under of this layer at one point on the margin or by a more diffuse process of polyinvagination, recently discovered. Later the folded edge undergoes a concrescent growth until it doubles on itself and fuses to form the primitive streak, equivalent to a closed blastopore.

The mesodermal layer also appears in the gastrula of **triploblastic** animals. Its formation is extremely variable but it usually grows out from the indeterminate zone about the blastopore where ectoderm and endoderm join. (A.W.L.)

GATES. Sluices through a dam are conduits cast in the concrete and equipped with controls called sluice gates. The common types of sluice gates are the plain sliding, the gate, the butterfly, and the needle valve. The purpose of the sluice is to empty the reservoir if necessary, to control the head level, and to aid in passing floods. (See **Crest Gate.**)

In foundry practice, the pouring spout casting, which comes from the mold attached to the casting, is called a gate. (F.T.M.)

GAUGE. An instrument or device for measuring or comparing some physical characteristic, such as size, pressure, temperature, water level, force, etc. In the machine field, the term is usually applied to measuring tools (see **Measurement**). Electrical measuring devices are usually called instruments. (See **Pressure Gauge, Instruments, Electrical.**) (H.C.H.)

GAUGE BLOCK. Measurement.

GAUGE LINE. A gauge line marks the limits of any standard distance used repeatedly. Structural steel shapes are punched or drilled for rivets on lines called gauge lines as indicated in the accompanying figure. Structural handbooks contain tables giving the numerical value of these gauges for different sizes of structural shapes. The gauge lines may be varied to suit the details as long as the minimum required edge distance and clearance for punching, drilling, or riveting are maintained. In some fabricating shops the rivet holes in the webs of beams and channels are made with multiple punches or drills. (C.W.C., F.T.M.)

Standard rivet gauges.

GAUGE NUMBER. In the sheet metal industry, the term gauge is used to describe or define the thickness of sheet metal. Systems of gauge numbers are also in use for wire, rods, machine screws, and twist drills. A representative selection of gauge numbers is shown in the accompanying table. The United States Stand-

Number of Gage	U.S. Standard for Sheets and Plates	American or Brown & Sharpe	Birmingham Wire (BWG) or Stubs Iron Wire	Washburn and Moen	Music Wire	British Imperial Wire (SWG)	Twist Drills	Machine Screws
000	.375	.4096	.425	.3625	007	.372		
0	.3125	.3249	.340	.3065	.009	.3240		
2	.2656	.2576	.284	.2625	.011	.2760	.2210	.086
4	.2344	.2043	.238	.2253	.013	.2320	.2090	.112
6	.2031	.1620	.203	.1920	.016	.1920	.2040	.138
8	.1719	.1285	.165	.1620	.020	.1600	.1990	.164
10	.1406	.1019	.134	.1350	.024	.1280	.1935	.194

Gauge number. (*Hesse's Engineering Tools and Process, Van Nostrand.*)

ard Gauge is the recognized commercial standard for all uncoated sheet and plate iron and steel, and is the legal standard to be used in determining duties and taxes levied by the United States under act of Congress approved March 3, 1893. The American or Brown & Sharpe gauge is the recognized standard in the United States for wire and sheet metal of copper and other non-ferrous metals. The Washburn and Moen or American Steel and Wire Company gauge is the recognized standard for steel and iron wire, and is also called the U. S. Steel Wire Gauge. The Birmingham or Stubs Iron Wire gauge is nearly obsolete but is still used for specifying the thickness of brass and aluminum tubing, and by the Treasury Department of the United States in connection with the importation of wire. The Music Wire gauge designated by the American Steel and Wire Company has been adopted as the standard for piano and music wire upon the recommendation of the United States Bureau of Standards. The British Imperial Wire gauge is the official standard for Great Britain. The table of twist drill sizes is known as the Manufacturers' Standard, since all numbered sizes of twist drills are made in accordance with its specifications. The last column shows a few representative numbered machine screw diameters. (H.C.H.)

GAUGE, TRACK. Railway Track.

GAUR. Mammalia, Artiodactyla. A large wild **ox,** *Bibos gaurus,* of India, Burma, and the Malay Peninsula. It has been domesticated to a very limited extent. (A.W.L.)

GAUSS. The gauss, as now defined, is the practical c.g.s. unit of magnetic induction (see **Magnetism**). If a straight wire is passed across a region under magnetic influence so as to cut it with a speed of 1 cm. per sec. perpendicular to the direction of the induction, the value of the induction necessary to set up an electromotive force of 1 abvolt (0.00000001 volt) per cm. length of wire is 1 gauss. Or, if the induction is 100,000,000 gausses, the resulting electromotive force is 1 volt per cm. length of this moving wire. The induction in gausses is commonly represented by the number of "lines," or maxwells, per sq. cm. of normal cross-section.

For many years prior to 1932 the term gauss was used to designate that unit of magnetic field intensity which is now known as the **oersted**. This change in terminology was introduced to distinguish between magnetic induction and magnetic intensity as physical magnitudes. (L.D.W.)

GAUSS' THEOREM. Green's Theorem in Space.

GAUSSIAN DISTRIBUTION. Wherever statistical analysis has a part in physical theory or in the treatment of chance or accidental errors of measurement, frequency distributions and distribution functions are of importance. One of the most common types of distribution function is the so-called Gaussian distribution whose graphical representation gives rise to the "bell-shaped" figure so frequently met with in the treatment

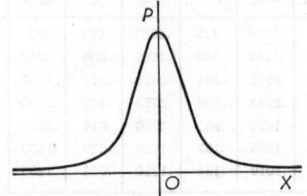

Typical Gaussian distribution curve.

of errors of measurement and the theory of least squares. It may be illustrated by a study of the distribution of a large number of shots on a target ruled with parallel, equidistant, vertical lines, the line bisecting the middle compartment being the "bull's eye" aimed at. The distribution of the shots may be tabulated by counting the shots and recording the percentages corresponding to the numbers found in the several vertical compartments numbered out both ways from the central (or zero) compartment. The outstanding characteristics of such a distribution are the tendency to a maximum at the center (zero mode), the progressive decrease of frequency with distance from the center, and the symmetry of the distribution with respect to this modal zero. In dealing with all such problems we are assuming that merely chance or accidental errors are involved, i.e., that the marksman honestly tried to hit the "bull's eye" and that his sights were properly adjusted, wind effects were properly compensated for, etc.

The mathematical analysis of the probability for distributions having the characteristics of pure chance or accident, carried out by Gauss and by Hagen along quite different lines of reasoning, result in the following form of distribution function:

$$p = \frac{k}{\sqrt{\pi}} e^{-k^2 x^2}.$$

In this expression p is the probability that any one of the statistical elements chosen at random (e.g., one of the shots on the target) will be found in the statistical interval numbered x from the modal zero (e.g., the xth vertical compartment from the center of the target). Or, it is the approximate percentage of all the elements which in the long run are found in the xth interval. k is a constant which depends upon the degree of con-

centration about the mode; its value being greater, the greater this concentration, as indicated by the fact that $\frac{k}{\sqrt{\pi}}$ is the probability of the element lying in the zero or modal interval (i.e., is the value of p for $x = 0$). In the target example k measures the excellence of marksmanship. In the language of statistics k is equal to about 0.95 of the reciprocal of the interquartile range, that is, the range of variation within which the central half of the elements lie.

Perhaps the most common example of the Gaussian distribution is that of the distribution of the accidental errors of varying size in measurements; in which case x is the size of the accidental error and k is the measure of precision. Again the components, parallel to any one axis, of the velocities of molecules in a pure gas (gravity and convection currents excluded) are distributed in like manner; though the actual speeds of the molecules follow the unsymmetrical and distinctly different **Maxwell distribution law**. Here k is dependent upon the temperature and molecular mass of the gas. Other examples are numerous in statistical physics. (L.D.W., W.K.G., L.L.S.)

GAUSSIAN FUNCTION. Normal Probability Function.

GAVIAL. Garial.

GAVIIFORMES. The loons. An order of large birds (Aves) with elongate bodies, short legs with webbed feet, and a long sharp beak. They are powerful swimmers and divers, and are named divers as well as loons. (A.W.L.)

GAYAL. Mammalia, Artiodactyla. A wild **ox** of northeastern India, related to the **gaur** but smaller and with less flattened horns. Domesticated to a limited extent. (A.W.L.)

GAY-LUSSAC LAW. Charles Law.

GAZELLE. Mammalia, Artiodactyla. **Antelopes** of moderate size which live in Africa and Asia. The numerous species are mostly inhabitants of deserts. They constitute the genus *Gazella* and with few exceptions are named gazelles. Among these exceptions are the **springbok**, the korin, and the aoul. (A.W.L.)

GEANTICLINE. Anticlinorium.

GEAR TOOTH FORMING. This is one of the most important processes in present-day engineering practice. Gears with cast teeth from sand molds, permanent molds, or metal dies are extensively used at the present time for comparatively slow-speed operation. Practically all watch and clock gearing is stamped from sheet metal and gives excellent results for the duty it must perform. Plastic molded gearing is also used to some extent. Gearing for precise operation, or for installation where high speeds and heavy loads prevail, is usually made with cut teeth which are often finished by various processes to attain high accuracy and comparatively noiseless operation.

Cutting gear teeth by using a form-type cutter is still an important and extensively used process. Mass-production spur gear form-cutting is generally performed on automatic gear-cutting machines. Two form-type cutters, one roughing and one finishing, are used and the process is analogous to cutting spur gears on the milling machine. The cutter slide advances to the work, feeds through for the cut and returns rapidly to starting position. The work then indexes and the cycle of operations is repeated. The roughing cutter has stepped teeth to break up the chips and removes most of the material, leaving just enough metal for the finishing cut.

Gears with teeth of almost any form can be used to generate conjugate teeth in a plastic blank by rolling the master gear and the blank together at the proper speed ratio equal to the ratio of their pitch diameters. This principle is used in producing hot-rolled gears, in which a master gear is rolled with a heated gear blank. (Good results are obtained but the process is not yet extensively used.) Involute teeth, however, will generate conjugate teeth of involute form in the application of this principle. Since a rack is an involute gear with a pitch diameter that approaches infinity as a limit, it will also generate involute teeth in a circular blank. This application is made use of in the gear hobbing process in which a **hob** whose tooth profile is of rack form is used for cutting spur and helical gears.

The conjugate generating principle is also employed in the widely used process of gear-shaping. A cutter is carried on a vertical reciprocating ram, and is similar to a spur pinion; it cuts on its upward stroke, backs away from the work, moves down to starting position, and then moves forward for a second vertical cut. Both the cutter and the work rotate slightly as the cutter descends, at a speed ratio inversely proportional to their pitch diameters. The gear shaping process has several advantages over other methods of gear tooth cutting. One cutter can be used for cutting all spur gears of the same pitch; the cutter has a very accurate profile, since it is possible to generate its tooth profiles after hardening by grinding; the cutter automatically corrects any tooth interference in the blank; and the finished gear has a generated profile. Another advantage of the gear-shaping process is that it can be used to cut internal gears.

Another type of machine which operates on an essentially similar principle is used for cutting herringbone gears with continuous teeth that have sharp apices.

Precision bevel gear tooth profiles may be cut by a single-point tool guided by a master template. A cutter that can pass through the small end of the tooth space is used, and moves along straight lines which intersect at a point corresponding to the apex of the gear. These cutter strokes are straight lines and the tool oscillates about the axis of the blank so that profiles of the correct form are generated. These two methods produce bevel gears with correct profiles over the entire face of the gear to the same degree of accuracy that spur gears are manufactured.

Spiral bevel and hypoid gear teeth are generated on special machines built for that purpose, by means of a rotating side-cutter tool that simulates the form of a mating gear. The rotating cutter machines one tooth space at a time and the work is then withdrawn and indexed for the next series of cuts.

Worm gear teeth are generated either by an infeed hob, or by a tapered hob that feeds tangentially to the gear to be cut. Both types of hobs are generally employed in conjunction with automatic gear-cutting machinery in which the gear blank rotates at a speed proportional to the speed of the hob.

Gear teeth may be finished after heat treatment by burnishing, shaving, lapping, and grinding.

Shaving processes are extensively employed for gear tooth finishing. A rotary cutting tool is provided with a series of serrations on the faces of the teeth, which act as cutting edges. The blank to be finished is mounted on an arbor or on a mandrel between centers. The cutting tool rotates at high speed and is at the same time rapidly reciprocated across the work.

Lapping is recognized as a highly efficient method of finishing gear teeth after hardening. One method uses a cast iron internal gear lap for finishing external spur gears by a reciprocating motion across the gear face, employing a rotary motion of both the lap and the blank to distribute the abrasive lapping compound. The tooth shape of the lap conforms to the tooth shape of the gear. Another method of lapping employs abrasive lapping wheels in which helical teeth are cut. (H.C.H.)

GEAR TRAIN. Two or more gears, transmitting motion from one shaft to another, constitute a gear train. If spur, bevel, or worm gears are used, the velocity ratio is inversely proportional to the numbers of teeth in the gears. A pair of spur gears, directly connected, result in a reversal of direction; if the driving gear drives an intermediate idler, which in turn drives the driven gear, the only effect of the idler is to cause the driven gear to rotate in the same direction as the driver. (The same effect can also be obtained by using an internal gear and a pinion.) If a two-gear idler, or compound gear, in which both idlers are fastened either to the idler shaft or to each other, is used, and where the driver engages one of the compound gears, and the driven gear the other compound gear, the velocity ratio is equal to the product of the two trains. The back gearing of a lathe or a milling machine is a familiar example of a compound gear train; a gear attached to the driving pulley drives a large gear mounted on the back gear shaft; a small gear on this shaft in turn drives the spindle gear. (See also **Differential, Epicyclic Train.** (H.C.H.)

GEARING. In the past, the term gear or gearing included mechanisms of all types (see **Valve Gear**), but modern terminology confines its usage to toothed wheels for transmitting rotary or reciprocating motion from one machine element to another. The principal forms of toothed gearing are **spur gearing, helical gearing, bevel gearing,** and **worm gearing.** (H.C.H.)

GEARMOTOR. Speed Reducers.

GEASTERS. Basidiomycetes.

GECKO Reptilia, Sauria. Small **lizards** of many species inhabiting the warmer regions of both hemispheres. Most geckos have adhesive disks on the toes. Several species are found in southern Florida and the southwestern states. (A.W.L.)

GEE NAVIGATION. Gee navigation is that type of **hyperbolic navigation** which was most effectively used by the Royal Air Force and the American Air Forces when operating against Europe from the British Isles.

For the transmission of the signals one master and two or more slave stations are used. The distance between stations is about 75 miles and the stations are located approximately on a circle with the master station between the slaves. The **frequencies** used are between 20 and 88 megacycles and the length of the pulses are of the order of magnitude of six microseconds.

The receiving equipment, frequently referred to by air navigators as the "Gee Box," contains a viewing **cathode ray tube** on which four fast sweeps can be exhibited simultaneously. These four sweeps appear as two pairs of parallel lines and on one of each pair a marker corresponding to the master pulse is shown. On each of the parallel sweeps a marker from one of the slave stations is shown, with its direction opposite to that of the master station. Fig. 1(a) indicates the appearance of the view scope showing the two recordings of the master station at M, and of two slave stations at S_1 and S_2, respectively. The operator next brings the slave "blips" into coincidence with those from the master and increases the velocity of the scope sweep to give more detail to the blips and hence closer setting. Fig. 1(b) shows the appearance of the scope when this adjustment has been made. Then time markers are thrown

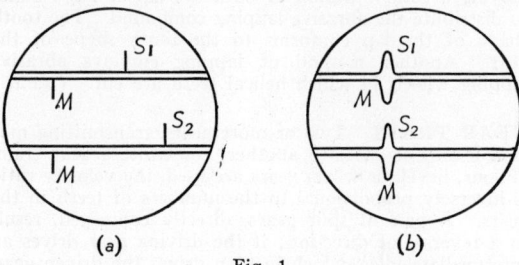

(a)   (b)

Fig. 1.

onto the scope and the differences of the time of arrival of each slave from the time of arrival of the master are read. The accuracy of the **lines of position** determined by Gee navigation varies with the square of the distance of the ship from the base lines between stations. On this line the accuracy is of the order of magnitude of 200 yds., when the navigator is on the base line, and about 1 mile at a distance of 400 miles from the base. Since time differences can be read simultaneously from the master and 2 slave stations, the fix can be determined by simultaneous observations without the necessity of using the running fix method. In practice between 2 and 3 min. is required to obtain the fix without any presetting of the instrument.

One of the most important uses of Gee hyperbolic navigation is for "homing" on a particular objective. A study of the chart will provide two lines of position, one from the master and one slave station and the other from the master and a second slave that intersect at the objective. The operator then sets the proper time differences on his Gee Box and proceeds until the markers for one line correspond. He then heads his plane so that this coincidence is maintained and knows that he is flying along one of the desired hyperbolic lines of position. This will be a true track, unaffected by drift, compass errors, or any other cause. A study of the rate of approach of the markers for the other line will give him the ground speed at which he is approaching his objective. When these two markers coincide he will know that he is at destination. If it is realized that the readings are independent of visibility—i.e., that they can be made in darkness as accurately as in daylight, or in thick fog as well as with unlimited visibility—the value of Gee navigation will be thoroughly appreciated. (w.k.g.)

GEGENSCHEIN. The gegenschein is a slight increase in intensity of the **zodiacal light** at a point on the **ecliptic** opposite to the position of the sun. The gegenschein appears as a soft glow against the sky, oval in shape, a few degrees wide and 10–15° in length. It is so faint that it cannot be observed on a night when there is any moon or when the patch falls in the vicinity of the **milky way**. (w.k.g.)

GEIGER-MÜLLER COUNTER. Counting Tube.

GEISSLER TUBE. Geissler manufactured a variety of gas discharge tubes at moderate exhaustion, which exhibited bright glow discharges and sometimes marked **fluorescence** effects. A form very useful in **spectroscopy** consists of two elongated bulbs, one containing the **cathode** and one the **anode,** and connected by a straight capillary section. The glow is most intense in the capillary, which is by its shape especially adapted to be placed in front of a spectroscope slit. Such tubes may be conveniently operated by an induction coil or a small transformer; though the latter is likely to overheat them and may even fuse the electrodes. (l.d.w.)

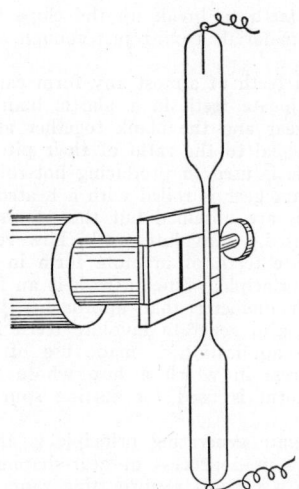

Common form of Geissler tube, in front of spectroscope slit.

GELADA BABOON. Mammalia, Primates. An Abyssinian **monkey,** *Theropithecus gelada,* of rather large size and chiefly black color, with a large mane. It resembles the true **baboons** in its elongate muzzle but differs in having the nostrils set well back from the tip. (a.w.l.)

GELATIN. Aminoacids, Polypeptides, and Proteins.

GEM, GEM STONES. A gem stone is a mineral substance which because of its beauty or rarity is in demand for ornamental purposes, chiefly personal adornment. The origin of such use for what we now call gem minerals is lost in the dim vistas of early human history. Ancient records describe the various gem stones, and archeologists find them in their investigations of bygone peoples. When we look at a collection of minerals with their bright colors and varying degrees of transparency or light-reflecting power, we cannot doubt that primitive man was much attracted by them and valued them greatly. We may imagine, too, that the occasionally found crystals with their regular geometric forms were more highly prized than broken fragments of the same minerals. Later they learned to polish them. Apparently the oldest form into which stones were shaped is that known as *en cabochon,* a French term derived from the Latin word for head and referring to its rounded shape. The forms were either hemispherical or hemiellipsoidal. The Emperor Nero is supposed to have had a large **emerald** cut en cabochon, and indeed for several centuries after his time this seems to have been the only sort of cutting employed. The supposedly accidental discovery in 1475 that diamonds would mutually scratch each other began the era of modern gem cutting. Previously it had been believed that diamonds were so hard that they could not be artificially shaped. At first, however, little progress was made in fashioning gems other than polishing a number of facets without any definite arrangement.

We owe to Vicenzio Peruzzi, a Venetian, the credit for devising the so-called "brilliant cut," the style of the modern **diamond** cutting, which, except for certain refinements due to a more thorough understanding of the behavior of minerals toward light, remains the same as in Peruzzi's day. At the present time transparent stones of all sorts are usually "brilliant cut," while translucent or opaque are cut en cabochon.

Since time immemorial, dealers in gems have used as the unit of weight the carat, undoubtedly introduced from the east. The word is derived from the Greek meaning a small horn, referring to the pods of the locust tree, *Ceratonia siliqua*, a common Mediterranean tree whose seeds were said to have been taken as the unit of weight in buying and selling gems. In the 19th century the actual weight of the carat differed slightly in different countries of Europe, from a little under to somewhat over ⅕ of a gram. The uniform use of the metric carat, exactly ⅕ of a gram, has been urged and to some extent adopted. (E.S.C.S.)

GEMINI. (The twins.) (Map, page 380.)

This constellation, which marks the third sign of the zodiac, has been recognized as a pair of twins from remote antiquity. The twins have not always been human, however, the Egyptians considering them as a pair of kids, and the Arabians as a pair of peacocks. By far the most familiar names for the two bright stars of this constellation are the names of the warrior brothers, Castor and Pollux, sons of Jupiter and Leda. Both of these stars are interesting objects in a 3-in. telescope, Castor being a fine **binary** and Pollux being a multiple star of at least six components. There is also a fine star **cluster** in this constellation which can easily be seen with a field glass and can be detected with the unaided eye on a clear moonless night. (W.K.G.)

GEMMULE.

Reproductive bodies of **sponges** enclosed in protective capsules which enable them to withstand severe conditions of temperature and drought. (A.W.L.)

GEMSBOK.

Mammalia, Artiodactyla. A South African **antelope** related to the **beisa**. *Oryx gazella.* (A.W.L.)

GENE.

A minute body or zone in a **chromosome** which governs the hereditary transmission and the development of a specific character of the individual.

The gene is regarded as a hypothetical unit of **heredity** and as such its existence has been judged by indirect evidence from the study of hereditary processes. In recent years the study of giant chromosomes of the salivary glands of the fruit flies has revealed fine details of structure which may be the genes. These chromosomes are differentiated longitudinally into many zones whose nature and arrangement appear to be constant, barring recognized types of change, and many of these transverse bands or zones have been identified as the points occupied by the hypothetical genes of other studies.

The action of genes is not definitely known, but they have been interpreted as self-perpetuating bodies that produce **enzymes** capable of influencing the action of other parts of the cell. (A.W.L.)

GENERAL CIRCULATION. Circulation of the Atmosphere.

GENERAL SOLUTION OF A DIFFERENTIAL EQUATION. Ordinary Differential Equations.

GENERATOR.

A generator is any equipment wherein originates a vital or chemical process. The two generators most frequently met are gas generators and electric generators. The method of generating the gas varies greatly, and there are several different kinds of gas generators. The acetylene generator, for example, has a chamber containing calcium carbide, into which water falls drop by drop, wetting the carbide, and reacting with it to produce acetylene.

Electric generators are built in all capacities, to suit the smallest and the largest installations. They can produce **alternating current** or **direct current,** depending on the design, but the origin of the electrical energy is the same, whether the final product be a.c. or d.c. To understand the action of the generator, examine Fig. 1,

which represents a soft iron **core** rotating between the poles of a permanent magnet, and having the slots on the surface, in which is embedded a coil of wire. It is ap-

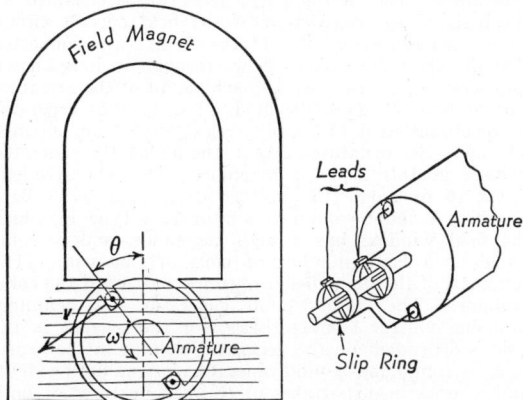

Fig. 1.   Elementary generator.

parent that as the coil rotates, carried by the soft-arm armature, it will cut across the **flux** lines extending from pole to pole. When this apparatus is connected to stationary leads through the medium of slip rings, and brushes resting thereon, it becomes an elementary **alternator.** If, instead of slip rings, a split segment, such as that shown in Fig. 2, is connected to the ends

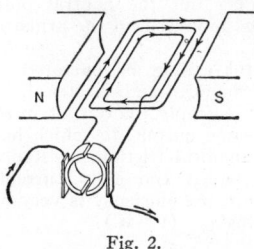

Fig. 2.

of the coil, the reverse of the current in the coil will occur when alternate segments of the slip ring (an elementary commutator) are opposite one of the **brushes.** This gives unidirectional current in the exterior leads, although it would be quite variable with only one coil in the **armature.** A uniform and unidirectional current is the result of many single-coil armatures so connected that the resultant current is the sum of several individual outputs. With sufficient overlap, the resulting current will be uniform and unidirectional. When the coil is revolving at a speed of ω radians per minute, at the position indicated, the speed of cutting vertically across flux lines is $v \cos \omega t$. The time is measured from the vertical position of the coil, and angle θ is ωt. When a wire cuts a magnetic field having a flux density represented by b and has a length and velocity represented by l and v, the voltage generated is $\dfrac{blv}{10^8}$ volts, thus the voltage generated any instant t, t being measured from the point of minimum generated voltage, is $\dfrac{blv}{10^8} \sin \omega t$.

The d-c generator is an ordinary dynamo machine having a multiple-coil winding, the ends of the coils of which are connected to a multiple-segment commutator. The armature is usually rotating, and the field stationary. The field sets up magnetic lines of force, which are cut by the conductors on the revolving armature, giving rise to a generated voltage, which is led through the commutator to a unidirectional external circuit. The iron core is built of laminations of iron insulated from each other by mill scale, or lacquer, or japanning, so that eddy currents which can be generated in the iron core, will be a minimum. The field windings are usually stationary, and the armature rotating; hence low voltage is the usual condition of use of d-c generators. The common d-c generator is classified on the basis of the connection of the field current circuit to the armature circuit. If the field is so connected that the armature current flows through it, it is known as a series field. In a series machine, all of the armature current flows through the field. This type of generator is sometimes used to supply series street lamp circuits. The more the armature current, the higher the generated voltage in this type of machine. This characteristic serves to overcome the voltage drop in a series light circuit. A shunt-wound generator is a type in which the field winding has a high resistance, and is composed of a large number of coils of fine wire. The terminals of the shunt field are connected across the commutator. The shunt-wound generator has definitely drooping voltage characteristics, for the current in the field is dependent on the generated voltage in the armature. A compound-wound generator, having both a shunt and a series field, partakes of the best features of both of the other types. When the shunt field is connected across the commutator, only, the compounding is called short shunt; when the shunt field is connected across the series field and the commutator, it is a long shunt generator.

The d-c generator may be adapted to the Edison three-wire system, in which there are 220 volts between outside wires and 110 volts between an outside and a neutral wire. This dual voltage arrangement is obtained by bringing out the neutral point of the coil through slip rings mounted on the armature shaft. (See Balance Coil.)

A d-c generator cannot be operated without certain losses, both mechanical and electrical. The efficiency of the generator is simply the output in electrical energy divided by the same output, to which have been added these losses: mechanical friction, resistance heating, core loss due to hysteresis and eddy currents. In a well designed generator, the efficiency is very high. It can be of the order of 95%. (F.T.M.)

GENET. Mammalia, Carnivora. *Genetta.* Animals related to the civets but with more slender bodies and shorter legs. The several species are found in Africa, Asia, and Europe. (A.W.L.)

GENETICS. Heredity.

GENITAL BURSA. A cavity lined with ciliated ectoderm at the base of each arm of the brittle stars (Asteroidea) into which the reproductive glands discharge. (A.W.L.)

GENITAL OPERCULUM. A small plate formed of the united vestiges of a pair of appendages, which covers the openings of the genital ducts of scorpions. (A.W.L.)

GENITAL PORE. An external opening on the ventral surface of a flatworm, communicating with the genital atrium into which the genital ducts discharge. (A.W.L.)

GENITAL RACHIS. A ring of tissue in the starfishes and sea urchins (Echinoidea) whose branches bear the reproductive organs. (A.W.L.)

GENITAL STOLON. A mass of cells associated with the axial organ of echinoderms. (See Echinodermata.) (A.W.L.)

GENITALIA. The organs of reproduction. (See Reproductive System.) (A.W.L.)

GENOTYPE. Heredity.

GENUS. A minor subdivision of the animal or plant kingdom composed of related species. Genera are subdivisions of families and subfamilies.

Genera are usually based on minor structural relations but in some cases superficial characters of color and pattern are made the basis of generic separation in groups of uniform structure. The distinctive characters of a genus are those of a type species (genotype) which should be stated in the original description of the genus and is determined in other cases by established rules of taxonomy. (A.W.L.)

GEOBENTHOS. All organisms that live on the surface of land masses. Aerial benthos. (See Distribution.) (A.W.L.)

GEOCENTRIC COORDINATES. Any system of coordinates on the celestial sphere which uses for its origin, or reference point, the center of the earth is known as a system of geocentric coordinates. Practically all cordinates published in an ephemeris or almanac are geocentric in character. (W.K.G.)

GEOCENTRIC PARALLAX. The origin of the apparent systems of spherical coordinates is a point on the surface of the earth, while the origin of the geocentric systems is at the center of the earth. For obvious reasons, all observations must be taken in the apparent system. For the solution of most problems, geocentric coordinates are desired. The transfer from one system to the other is made by applying a correction for geocentric parallax.

In the figure we have C the center of the earth of radius R, and O the position of an observer on the sur-

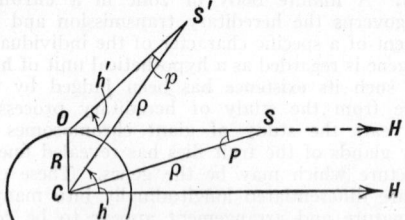

Geocentric parallax in altitude.

face. OC is the direction of gravity at O. OH is the direction of the astronomical horizon and CH is a parallel direction drawn through the center of the earth. S and S' represent two positions of an object at distance ρ from the center of the earth. S being the position when the object is on the horizon. At S' the object has an apparent altitude h' and a geocentric altitude h. P is defined as the horizontal parallax of the object and is the angle subtended at the object by the radius of the earth. For rigor the quantity usually defined is the mean equatorial horizontal parallax which is the angle subtended by an equatorial radius of the earth at the object, when the object is on the horizon, and at its mean, or average, distance from the earth. The equatorial horizontal parallax is tabulated in Ephemerides for all members of the solar system for selected dates. Inspection of the figure indicates that sin $P = R/\rho$.

The geocentric altitude h is greater than the apparent altitude h by the angle p which is defined as the geocentric parallax in altitude. In the oblique plane triangle COS' we have $\dfrac{R}{\rho} = \dfrac{\sin p}{\cos h'}$ but we have already seen that $R/\rho = \sin P$ whence sin P cos $h' = \sin p$. Now both P and p are such small angles that, without sensible errors for most problems, except those dealing with the moon, we have $p = P$ cos h' giving the geocentric parallax in altitude in terms of the equatorial horizontal parallax and the apparent altitude of the object. For objects outside of the solar system the value of P is far too small to be appreciable in even the most refined observations.

In case other spherical coordinates than altitude are to be used, the geocentric parallax in altitude may be transformed to the desired quantities by solution of the astronomical triangle or other triangles on the celestial sphere. (w.k.g.)

GEOCHEMISTRY. Geochemistry is that branch of the science of chemistry which deals with the **chemical composition** of and the past and present chemical changes in the earth. The subject matter is for convenience divided into three parts:

1. Atmosphere. (See **Air; Atmosphere.**)
2. Hydrosphere. (See **Water.**)
3. Lithosphere. (See **Lithosphere.**)

The chemical and physical weathering and transport of rock materials furnishes the inorganic portion of soils for crops. Local concentration of various minerals in the crust of the earth is a matter of great economic importance. (r.k.s.)

GEOCRONITE. A mineral sulfide of **lead, antimony** and arsenic, $Pb_5(Sb, AS)_2S_8$. Crystallizes in the orthorhombic system. Hardness, 2.5; specific gravity, 6.4±; color, gray to blue with metallic luster; opaque. (r.m.f.)

GEODE. A geode is a hollow concretion or nodule whose inside walls are lined with crystals, commonly of **quartz** or **calcite**. (r.m.f.)

GEODESY. Geophysics.

GEODETIC SURVEYING. Surveying.

GEODUCK. Mollusca, Lamellibranchiata. A giant clam found on the Pacific coast of North America. It attains a weight of more than 6 lbs. and is edible. (a.w.l.)

GEOGALE. Mammalia, Insectivora. *Geosale.* A small animal of Madagascar which resembles the mice (mouse). (a.w.l.)

GEOGRAPHICAL COORDINATES. Geographical coordinates provide a method for determining the position of a point on the surface of the earth by means of a system of **spherical coordinates.** Because of the fact that the earth is not a sphere but is in reality an oblate spheroid, technically the system of coordinates cannot be strictly spherical. The geographical method of representation of the position of points on a spherical earth by means of **latitude** and **longitude** was first applied by Ptolemy in the construction of his atlas of the world during the second century of the Christian era. (w.k.g.)

GEOGRAPHY. Literally, the study, description and mapping of the surface phenomena of the earth without, necessarily, a consideration of the origin of the phenomena. Compare and contrast with **geomorphology, geodesy** (cartography), geology, geophysics, meteorology, hydrology, oceanology (oceanography) and biology, of the geosciences; and history, economics and politics of the so-called social "sciences." (r.m.f.)

GEOLOGY. The study of the composition, structure, and history of the earth. The term is derived from the Latin, *geologia*, coined by Bishop Richard de Bury in 1473 to distinguish lawyers who study "earthy things" from theologians. First consistently used in its present sense in the latter part of the 17th century. The great mass of detail that constitutes geology is classified under a number of subdivisions which, in turn, depend upon the fundamental sciences, physics, chemistry and

biology. The principal subdivisions of geology are: **Mineralogy, Petrology, Structural Geology, Physiography** (Geomorphology), usually grouped under **Physical** or **Dynamical Geology**; and **Paleontology, Stratigraphy,** and **Paleogeography,** grouped under **Historical Geology.** The term **Economic Geology** usually refers to the study of valuable mineral (ore) deposits, including coal and oil. The economic aspects of geology are, however, much more embracive, including many subjects associated with Civil Engineering, Economic Geography and Conservation. Some of the more important of these subjects are: **Meteorology, Hydrology, Agriculture** and **Seismology.** Subjects which are also distinctly allied to geology are **Geophysics, Geochemistry** and **Cosmogony.** (r.m.f.)

GEOMETRIC MEAN. The geometric mean, G, of N positive **variates** is the Nth root of the product of the N variates.

$$G = (x_1 x_2 x_3 \cdots x_N)^{\frac{1}{N}}$$

It is found by use of logarithms.

$$\log G = \frac{\log x_1 + \log x_2 + \cdots + \log x_N}{N}.$$

The geometric mean is useful if the variates form an exponential or are approximately in a geometric progression. The geometric mean is occasionally useful in **index numbers.** It is another measure of **central tendency.** (See **Harmonic Mean; Arithmetic Mean.**) (l.a.a.)

GEOMETRIC MEANS. Geometric Progressions.

GEOMETRIC PROGRESSION. A geometric progression is a succession of numbers in which each term has a constant **ratio** to the preceding term; this ratio is called the common ratio. The terms "geometric progression" are often abbreviated by G.P.

There are two fundamental formulas for geometric progressions—one for the general term and one for the sum of any number of terms.

The general term or nth term of the geometric progression whose first term is a and whose common ratio is r is given by the formula

$$l = a \cdot r^{n-1}.$$

The sum of the first n terms of this progression is given by

$$s = \frac{a(1 - r^n)}{1 - r} = \frac{a(r^n - 1)}{r - 1} = \frac{rl - a}{r - 1} \quad (\text{if } r \neq 1).$$

The terms of a geometric progression between the first and last are called geometric means.

The geometric mean of two numbers is the middle term of a geometric progression whose first and last terms are the given numbers. It is given by the square root of the product of the given numbers.

If in a geometric progression $a, ar, ar^2, \cdots, ar^n, \cdots,$ the common ratio r is < 1 in absolute value, the sum s_n of the first n terms approaches $\frac{a}{1 - r}$ as a limit when n increases beyond all bounds ($n \to \infty$). In this case, we speak of the geometric progression as an **infinite** geometric series, and we say that its sum is $\frac{a}{1 - r}$. (l.l.s.)

GEOMETRIC SERIES. The infinite series $\sum\limits_{n=0}^{\infty} r^n$ is called a geometric series. It is **convergent** when $|r| < 1$ with sum $\frac{1}{1 - r}$; but is **divergent** for other values of r. (l.l.s.)

GEOMETRICAL OPTICS. This branch of physics treats light as if it were actually composed of "rays" diverging in various directions from the source and abruptly bent by **refraction** or turned back by **reflection** into paths determined by well-known laws. The idea that light travels in straight lines is here uppermost, while its wave character and other physical aspects are lost sight of. Thus the image of a point *A*, if "real," is simply another point *B* through which the rays diverging from *A* ultimately pass after the several reflections or refractions produced by the mirrors, lenses, etc., of the optical system. If the image *B* is "virtual," the rays appear to be diverging from it, but only because their direction has been so changed that if produced backward, the lines along which they now travel would intersect at *B*. A real image of a lamp may easily be formed by a reading glass; a virtual image, by a plane mirror.

The chief advantage of this mode of visualizing the behavior of light is the simplicity with which problems may be solved by geometrical constructions. The same formulae which are deduced by the methods of geometrical optics may be arrived at, but often with much more labor, by treating light as composed of waves and studying the changes of wave front. (See **Mirrors and Lenses, Optical Instruments, Eyepieces, Spherical Aberration,** etc.) (L.D.W.)

GEOMETRY. Geometry may be said to be a study of the properties of geometric figures. Elementary plane geometry deals with figures composed of points, straight lines and circles; elementary solid geometry deals with figures composed of points, planes, straight lines, and spherical, cylindrical and conical surfaces. **Analytic geometry** is an analytic (algebraic) treatment of geometric problems. There are other branches of geometry, such as **projective geometry,** differential geometry, etc. (L.L.S.)

GEOMORPHOLOGY. This term is gradually replacing the earlier term physiography to denote the full scientific interpretation of the origin of topographic features, or the purely physical attributes of scenery. This relatively distinct department of the earth sciences includes the study of the origin of all topographic features in terms of process, or processes of erosion and in their relation to geologic structure. (R.M.F.)

GEOPHYSICS. The term "geophysics," derived from the Greek, meaning the Physics of the Earth, was probably first used by the Germans. According to Dr. G. Angenheister, Director of the Geophysical Institute of the University of Göttingen, the word *"Geophysik"* appeared in Meyer's *"Conversationslexikon,"* published in 1853. The term must have been in use sometime before. Both A. Mühry (1863), and K. Löppritz, Sr., aided in procuring its general acceptance. (See A. Sieberg, *"Geologische Einführung in der Geophysik,"* 1927, p. 1.) Among the first comprehensive treatises in the English language on this subject is "Physics of the Earth's Crust," by Osmund Fisher, published in 1881 (O. T. Jones). The scope of geophysics is broad and difficult to define, especially in relation to geology. During recent years geophysical techniques have been highly developed, particularly in relation to prospecting for petroleum through the mapping of sub-surface geological structures, the traces of which may, or may not, be exposed at the surface of the earth. By 1925, the general field of geophysics is defined in the *"Zeitschrift für Geophysik"* as follows:

1. Motion and Constitution of the **Earth.**

a. Rotation; revolution; precession; mutation; fluctuation of the poles. (Subjects which are also directly related to astronomy and physical geography.)

b. Mass, weight, figure, density, and elasticity of the earth (frequently classified under the subject of **geodesy).**

c. Distribution of mass in the earth's interior; **isostasy.** (Intimately related to fundamental problems in geodesy, **geology,** and **cosmogony.**)

2. Deformations; drifts and vibrations.

a. Elevations and depressions of the crust, folding and mountain building, glaciation and glacier-movement; vulcanism. (Primarily geological problems.)

b. Tides of the **atmosphere,** of the oceans, and of the earth. (Problems in meteorology, physical oceanography and structural geology.)

c. Wave-motions and currents in air and water. (Meteorology and physical oceanography.)

d. Elastic deformation and seismic behavior of the earth. (Usually classified under seismology, and related to structural geology.)

3. **Electric and Magnetic Field** of the Earth.

a. The internal permanent magnetic field; its distribution and secular variation.

b. The external magnetic field and its periodic variation.

c. Earth-currents and **aurora.**

d. Atmospheric electricity; **radioactivity** of the earth, the oceans, and the atmosphere. (All of the above problems are included under the subject of **terrestrial magnetism,** a distinct and important branch of geophysics; it should be noted that the **compass** is the earliest type of geophysical apparatus.)

4. Cosmic Physics in its Relation to the Earth and its Atmosphere.

a. History of the earth; determination of the origin and age of the earth as a whole, and of its crust. (Usually considered under the subjects of cosmogony and historical geology.)

b. Solar constants; radiation of the earth and the atmosphere for light, heat, and wireless waves.

c. Relation of solar activity to terrestrial radiation and to the earth's electric and magnetic field.

d. Climatic changes (meteorology).

e. Cosmic radiation (discovered since 1925).

In terms of geophysics meteorology, hydrology, physical geography, physical oceanography, structural-inorganic-historical geology and cosmogony are allied subjects. In terms, however, of each of the "allied subjects," geophysical methods are included, among others, as contributary techniques.

The greatest field for the application of geophysical methods has been in geology, both from the point of view of fundamental science and economics. In the field of applied geophysics the many methods and techniques are classified as either gravitational, seismic, magnetic, electric or radioactive. (Mathematics and physical chemistry are directly allied to geophysics only through experimental physics.) In its purest sense, therefore, geophysics is a branch of experimental physics dealing, particularly, with the structure and, to a certain extent, the mass-composition of the earth, including its atmosphere and hydrosphere. Geophysics bridges the gap between physics and geology in its broadest sense. (R.M.F.)

GEOPOTENTIAL. If a unit mass is lifted above sea level to a given height, the geopotential of that level is given by the potential energy of the mass at that level. Unit geopotential is equal to the potential of unit mass lifted a unit distance in a force field of unit strength. Distance upward can be measured in terms of differences in geopotential. (P.E.K.)

GEOSCIENCE. A term proposed by R. M. Field in 1933 to include all the overlapping departments (divisions) of the physical and chemical earth sciences such as: geology, geophysics, geochemistry, geodesy, meteorology, hydrology, glaciology, oceanography, vulcanology (volcanology), seismology, geomagnetology (geomagnetism) and their numerous subdivisions and important interscientific and intertechnological relations. Later (1939) still further defined so as to include the relation of **natural resources** (their national and international geographic distribution and technological development) to human affairs. In contradistinction to geoeconomics and geopolitics which emphasize the "monetary" and "legal" rather than the scientific and technological environment of man. (R.M.F.)

GEOSTROPHIC WIND. The wind that is the result of a balanced pressure gradient and Coriolis Force. Geostrophic winds blow in straight or nearly straight lines. Low pressure is to the left of the wind direction in the northern hemisphere when the observer stands with back to the wind. (P.E.K.)

GEOSYNCLINE. Dana's definition of a geosyncline (1873) is a depression which has been produced by lateral compression and which is filled with **sediments**. Although Dana, in his original definition, suggested that subordinate ridges might be formed in the bottom of the geosyncline during its formation, it remained for Emile Haug, in his *"Traité de Geologie,"* to emphasize these ridges (geanticlines) in relation to the tectonics of the Alps. According to L. W. Collet, "A geosyncline is situated between two continental masses and is destined to be filled with sediments, some of which are derived from the **geanticlines** which develop in it." According to R. M. Field, "A geosyncline originates in a continental block as a great trough, the locus for the accumulation of marine and terrestrial sediments, which are derived from concomitant geanticlines formed in or on the margins of the geosyncline." The geophysical and geological study of the great island arcs, such as the East Indies and West Indies (1927-1937), strongly intimates that the foredeeps in front of the arcs represent geosynclines which have not been filled with sediments while they were being formed. The pronounced deficiency of gravity associated with these foredeeps suggests great down buckle of the crustal or continental type of rocks called Sial into the more basic subcrustal couch called **Sima**. (See also **Tectogene**.) (R.M.F.)

GEOTECTOCLINE. Term proposed by H. H. Hess and R. M. Field (1938) for the deformed prism of sediments in (of) the geosyncline. (See also **Synclinorium** as originally defined. See also relation of geotectocline to **Tectogene**.) (R.M.F.)

GEOTROPISM. Tropism, and also **Movement in Plants.**

GEPHYREA. Large marine worms. As adults they are not segmented but since the young show evidence of metameric segmentation they are placed with the segmented worms in the phylum **Annelida**. They have a large body cavity, one pair of **nephridia**, and in some species a few **setae**. The internal organs are not metamerically arranged.

This class is divided into three orders:

Order Echiurida. With a pair of setae near the anterior end. Body cylindrical with a slender anterior protuberance, the **prostomium.**
Order Sipunculida. No setae. Body slender, with a protrusible proboscis and a group of tentacles near the mouth.
Order Priapulida. No setae or tentacles. (A.W.L.)

GERANIOL. Alcohols and Ethers.

GERANIUM OIL. Volatile Oils.

GERBIL. Mammalia, Rodentia. Small burrowing animals of Asia and Africa. They resemble **rats** but have long hind legs and large eyes and move about by jumping. In these points they resemble the **jerboas** but they are less extreme. (A.W.L.)

GERENUK. Mammalia, Artiodactyla. An **antelope**, *Lithocranius walleri,* of eastern Africa with a very long neck and moderate spiral horns, turned forward sharply at the tips. Also called Waller's gazelle but not a true gazelle. (A.W.L.)

GERM. 1. Any microbe or **bacterium**. 2. A spore. 3. The substance from which the embryo develops. (R.S.M.)

GERM CELL. A sexual reproductive cell. **Gamete.** (A.W.L.)

GERM LAYER. The three tissues resulting from the first differentiation in the embryonic development of multicellular animals. They are formed from the presumptive areas of the blastula during gastrulation by a redistribution which results in three layers. The three are an outer ectoderm, an inner endoderm, and between the two the mesoderm.

Animals of the phyla **Porifera**, **Coelenterata**, and according to one interpretation the **Ctenophora**, develop only the first two germ layers and are said to be diploblastic. Multicellular forms of all other phyla have all three and are therefore **triploblastic**.

The chief parts of the body formed from the various germ layers are as follows (in these lists the terms are chosen to embrace both vertebrates and invertebrates and so do not all apply to the same animal):

Ectoderm: Outer cellular layers of the integument, their glandular derivatives, and the cuticula. Exoskeleton and exoskeletal structures such as setae, scales, feathers, hair, claws, hoofs and nails. Parts of sensory organs including the cornea and lenses of all types of eyes, external and internal ears of vertebrates. Lining of oral cavities and salivary glands, and in vertebrates the enamel of the teeth. Lining of the posterior part of the alimentary tract. The entire nervous system of most animals, including the nervous structures in the sense organs. A limited amount of muscular tissue. Organs of reproduction of some animals. Lining or covering of organs of respiration of many invertebrates.

Mesoderm: Lining of body cavity, circulatory system, water vascular system of echinoderms, and parts of excretory system. Muscular tissue. Bone. Teeth, except the enamel. The mesenchymal tissues such as cartilage, connective tissue, adipose tissue, and tendon. Blood. A limited part of the nervous system of starfishes. Reproductive organs.

Endoderm: Lining of the enteric cavity, including most of the alimentary tract of vertebrates and the limited mid-intestine of arthropods. Respiratory epithelium of vertebrates. Lining of parts of vertebrate excretory system. Reproductive organs and cells. (A.W.L.)

GERM PLASM. The essential reproductive tissue and the germ cells that it produces.

The concept of the germ plasm has been emphasized chiefly in the field of organic **evolution**. Since the germ cells of one generation produce both the body (soma, somatoplasm) and the germ plasm of the next, the continuity of this material is evident. It has been interpreted as the perpetual living substance, while the material of the body appears as an offshoot in each generation. In the 1-celled animals, however, there is no differentiation. (A.W.L.)

GERMAN MEASLES. Rubella.

GERMAN SILVER. Alloys; Brass and Bronze.

GERMANIUM. Symbol: Ge. Atomic number: 32. Atomic weight: 72.60. Density: 5.46. Melting point: 958.5° C. (Isotopes: page 290.)

Germanium is a silver-white, lustrous, hard, brittle metal; when heated in oxygen to 730° C. is partially oxidized to dioxide; unaffected by solutions of acids and bases but soluble in fused **sodium** hydroxide; in the form of powder is of dull gray color; combines with **chlorine** to form volatile tetrachloride. Discovered by Winkler in 1886, but predicted by Mendeléeff in 1871 as an element to be discovered with properties resembling **silicon.**

Germanium occurs in very small amount in many **sulfide** ores, such as American zinc ores (0.25% GeO_2), and the rare mineral **argyrodite** (silver germanium sulfide) of Saxony, and Bolivia. Separated from accompanying metals by fractional **distillation** of volatile germanium tetrachloride, then converted to dioxide, and then reduced to germanium metal by heating with **hydrogen, carbon,** or **aluminum.** Chemically related to **silicon** and tin.

Chlorides: Germanium dichloride ($GeCl_2$), white solid; germanium tetrachloride ($GeCl_4$), colorless liquid, boiling point 86° C., from the acid solution zinc metal precipitates germanium metal.

Hydroxide: Germanous hydroxide ($Ge(OH)_2$), yellow precipitate turning red on heating, formed by reaction of sodium hydroxide solution and germanous salt solutions, soluble in excess sodium hydroxide; germanic hydroxide not known.

Oxides: Germanous oxide (GeO), gray-black solid, soluble in hydrochloric acid, or sodium hydroxide solution; germanic oxide (GeO_2), white solid, by ignition of germanium metal or sulfide.

Sulfides: Germanous sulfide (GeS), brown to red precipitate; germanic sulfide (GeS_2), white precipitate; both by hydrogen sulfide with the respective salt solutions.

Germanomethane (GeH_4), melting point $-165°$ C., boiling point $-90°$ C.; germanoethane (Ge_2H_6), melting point $-109°$ C., boiling point 29° C.

Germanium tetramethyl ($Ge(CH_3)_4$), boiling point 43° C.; germanium tetraethyl ($Ge(C_2H_5)_4$), boiling point 163° C.

Germanochloroform ($GeHCl_3$), boiling point 75° C. (R.K.S.)

GERMARIUM. 1. A division of the ovary (see **Gonad**) of rotifers in which the eggs are formed. 2. A division of the follicles in the testes of insects in which the germ cells are formed but not completely differentiated. (A.W.L.)

GERMICIDE. Any substance or agent, physical or chemical, which is destructive to germs (**bacteria**). (R.S.M.)

GERMINATION. Seed.

GERSDORFFITE. A mineral related to cobaltite and ullmannite in the cobaltite group. A nickel, iron, cobalt, arsenic sulfide, (Ni, Fe, Co)AsS. Crystallizes in the isometric system. Hardness, 5.5; specific gravity, 5.9; color, white to gray with metallic luster; opaque. Named after von Gersdoffs (1842) at Schladming. (R.M.F.)

GESTATION. Pregnancy.

GEYSER. Derived from the Icelandic word *geysa,* meaning gush and descriptive of hot springs which at regular, or irregular, intervals throw a column of steam and hot water into the air. Geyser waters usually build up tubes or conduits of siliceous sinter. Geyser waters have been proved to be mainly **vadose** with approximately 10% of juvenile or magmatic water. Geyser action is the result of vadose water coming in contact

with steam arising from the solidifying **magma,** and periodically returning to the surface through the geyser tube, for the same reason that water is suddenly expelled from a test tube when heated too rapidly. The mechanics of geyser action are illustrated. The principal

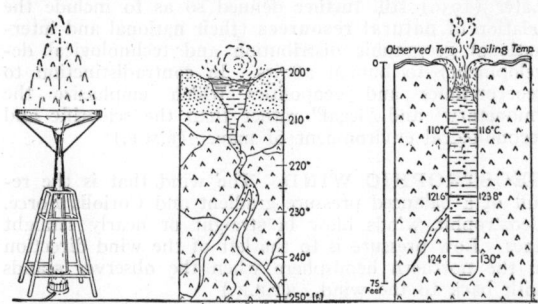

The mechanics of geyser action, as illustrated by laboratory experiment, and the hypothetical cross-sections of natural geysers. (*Field, Outline of Geology, Barnes & Noble.*)

geyser fields are in Wyoming (Yellowstone Park), New Zealand, and Iceland. (R.M.F.)

GEYSERITE. A loose or compact, sometimes concretionary, **siliceous** deposit, formed by geysers and hot springs from the material held in solution by the thermal waters. (R.M.F.)

GHOSTS. Diffraction Grating.

GIANT AND DWARF STARS. With the absolute **magnitudes** and **spectral classes** of a number of stars known, a diagram may be plotted showing absolute magnitudes as ordinates against spectral classes as abscissae. Such a diagram was first published by Russell of Princeton in 1913. He found a number of stars of approximately the same absolute magnitude running as a horizontal line across the top of the diagram and a series of stars of progressively decreasing absolute magnitude with spectral class changing from B to M. From the shape of the original diagram it became known as the "figure 7 diagram." In 1905, Hertzsprung noticed that there was a sharp distinction between M-type stars of high and low luminosity and he referred to them as

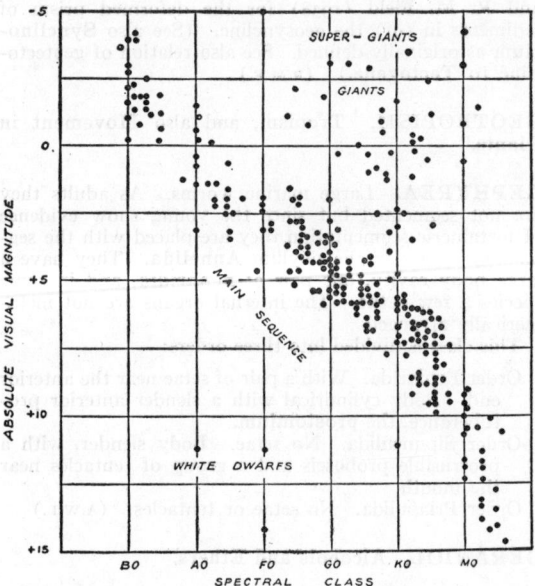

Fig. 1. Spectrum-luminosity diagram.

giant and dwarf stars. With the publication of the figure 7 diagram it became evident that the distinction between giant and dwarf classification could be extended to all spectral classes. However, the difference between the giant and dwarfs becomes less and less striking as we proceed from the M-type stars toward the B-types.

As the number of **parallax** determinations increased, and more and more stars were added to the figure 7 diagram, it became evident that the distinction between the giants and dwarfs was sharply indicated only for the M-types. Fig. 1 indicates such a diagram for a moderately large number of stars. It will be noted that the majority of the stars are found in a narrow region extending diagonally across the figure from the B to M spectral class. This is known as the main sequence.

The stars which appear on the diagram above the main sequence are known as the giant stars. It will be noted that the scattering of the points in the giant classification is far greater than for those along the main sequence and at present the term super-giant is applied to these giants with the greatest luminosity. Those stars which are found below the main sequence are known as the white dwarfs.

The **sun** is a main sequence star of G_0 spectral class and absolute magnitude about 4.9. Its luminousity is

giant and dwarf indicate size characteristics as well as brightness.

Calculations of the relative diameters of stars of differing spectral classes is a complicated process, but approximate values may be computed for all of the stars in terms of the known diameter of the sun. By means of the **interferometer** the actual diameters of a few of the stars can be measured and the agreement between the values derived on the basis of giant and dwarf classification and those obtained by observation provides a striking example of the validity of modern astrophysical theory.

Detailed studies of the individual stars, from purely dynamical considerations, indicate that the range of mass is probably not greater than 1000. Since we have seen that the ratio of diameter is of this same order of magnitude, the range of densities between the giant and dwarf stars may be as great as a million for the extreme cases. Such differences in density would produce important differences in the characteristics of the **absorption lines** observed in giants and dwarfs of the same spectral class. These differences have been observed and form the basis for the determination of **spectroscopic parallaxes,** and physical characteristics of the stars.

The table indicates approximate physical characteristics of stars of different classifications. (W.K.G.)

PHYSICAL CHARACTERISTICS OF TYPICAL STARS

STAR	SPECTRAL CLASS	TEMPER-ATURE IN °K.	DENSITY IN TERMS OF WATER	REFERRED TO SUN AS UNITY		
				Luminosity	Mass	Diameter
GIANTS						
Antares...................	M_0	3,100	0.0000003	3500	30	480
Aldeberan.................	K_5	3,300	0.00002	90	4	60
Arcturus..................	K_0	4,100	0.0003	100	8	30
Capella...................	G_0	5,500	0.002	150	4.2	12
MAIN SEQUENCE						
β Centauri................	B_1	21,000	0.02	3100	25	11
Vega.....................	A_0	11,200	0.1	50	3	2.4
Sirius A..................	A_0	11,200	0.4	26	2.4	1.8
Altair....................	A_5	8,600	0.6	9.2	2	1.4
Procyon..................	F_5	6,500	1.2	5.4	1.1	1.9
α Centauri A..............	G_0	6,000	1.1	1.12	1.1	1.0
The Sun..................	G_0	6,000	1.4	1	1	1.0
70 Ophiuchi A.............	K_0	5,100	0.9	0.42	0.9	1.0
61 Cygni A................	K_7	3,800	1.3	0.21	0.5	0.7
Krueger 60 A..............	M_3	3,300	9	0.002	0.3	0.3
WHITE DWARFS						
Sirius B..................	F	5,700	27,000	0.003	0.96	0.034
O_2 Eridani B..............	A_0	11,000	64,000	0.003	0.44	0.019

about 100 times greater than that of a main sequence M-type star, and about 100 times less than that of a B-type star in the same sequence.

Although the term giant was first used solely for indicating the relative brightness of these objects, the term has a far more striking meaning. The temperatures of all M-type stars are of the same order of magnitude; hence, the brightness per unit area must be approximately the same for both the giants and the dwarfs of the same class. Since the total luminosity of the giant M is about a million times that of the dwarf star, the areas of the two stars must be in that same ratio. This indicates that the diameter of a giant M star must be about 1000 times that of the dwarf and we see that the terms

GIB. Bearings.

GIB HEAD KEY. Key.

GIBBER. Glyptolith.

GIBBON. Mammalia, Primates. Several species of man-like **apes,** highly adapted for arboreal life. They live in the forests of southeastern Asia, especially in the region of the Malay Peninsula.

Gibbons are smaller and of more slender build than the other apes and are among the most agile of all primates in the trees. They have extremely long arms with which they swing long distances from bough to bough, and their movements are said to be very rapid. They differ from the other anthropoid apes in the presence of ischial callosities.

Three of the species are known as the siamang, *Symphalangus syndactylus,* hoolock, *Hylobates hulock,* and the wou-wou, *H. leuciscus.* (A.W.L.)

GIBRALTAR SKULL. Paleontology of Man.

GILA MONSTER. Reptilia, Sauria. A poisonous lizard, *Heloderma suspectum,* of the southwestern deserts. It attains a length of 18″ and is thick-bodied, with a stubby tail. The skin bears rounded tubercles instead of flat scales and is black or blackish with pink to yellow markings. (A.W.L.)

GILBERT. Magnetic Circuit.

GILL. A respiratory organ for the extraction of oxygen from the water and for the liberation of carbon dioxide.

Many small aquatic animals absorb oxygen through the surface of the body generally but the more complex forms have localized respiratory organs formed to present an adequate surface. They are usually thin plates of tissue or slender tufted processes and, with the exception of some aquatic insects, they contain blood or coelomic fluid which absorbs oxygen through their thin walls. In the insects a unique type of respiratory organ is the tracheal gill which contains air tubes. The oxygen of these tubes is renewed in the gills.

Gills are developed in starfishes and sea urchins (see **Echinoidea**) as thin protuberances on the surface of the body containing diverticula of the water vascular system. In the **crustaceans, mollusks,** and some **insects** they are tufted or plate-like structures at the surface of the body in which blood circulates. The gills of other insects are of the **tracheal** type and also include both thin plates and tufted structures, and in the larval dragon fly the wall of the caudal end of the alimentary tract (rectum) is richly supplied with tracheae as a rectal gill. Water pumped into and out of the rectum supplies oxygen to the closed tracheae.

Gills of vertebrates are developed in the walls of the **pharynx** along a series of gill slits opening to the exterior. Water taken into the mouth passes out of the slits, bathing the gills as it passes. Some fishes utilize the gills for the excretion of electrolytes. In some of the **amphibians** the gills occupy a similar position on the body but protrude as external tufts. (A.W.L.)

GILL BOOK. A series of many thin respiratory plates associated with the jointed appendages of the horseshoe crabs (see **Xiphosura**). (A.W.L.)

GILL CHAMBER. A partially enclosed space containing gills. In many invertebrates external gills project from the surface of the body. Such structures are very delicate and in many species are protected by folds of the body wall. The crayfish offers a good example, with the carapace extended down on each side of the body to form the outer wall of a chamber in which the gills lie. (A.W.L.)

GILL FILAMENT. A thread-like component of a gill. Also the ciliated ridges of the gills of **bivalve** mollusks. (A.W.L.)

GILL PLATE. The respiratory organ of some **bivalve** mollusks. It is formed of two thin plates or lamellae, each made up of united ctenidial filaments (ctenidium), and contains passages communicating with the mantle cavity and with the chamber above the gills. Water passes into these passages from the mantle cavity. (A.W.L.)

GILL RAKER. A comb-like structure along the inner margin of the gill arches of fishes (Pisces). These combs prevent the passage of food into the gill slits and direct it toward the esophagus. (A.W.L.)

GILL SLIT. A perforation of the body wall of vertebrates opening into the pharynx. In the fishes (Pisces) and amphibians the slits are associated with the gills but in terrestrial vertebrates they occur only in the embryo and in mammals they usually fail to open. The gill slits are paired, opening as a series on each side of the body. In the lampreys and most cartilaginous fishes (sharks, etc.) the openings are externally separate but in the bony fishes those of each side are covered by an operculum. (A.W.L.)

GILSONITE. Uintaite. The mineral Gilsonite, named for S. H. Gilson of Salt Lake City, is a variety of **asphaltum** that occurs in Uinta County, Utah. It is found in black lustrous masses which ignite easily. A less frequently used name for it is uintaite. (E.S.C.S.)

GILT-HEAD. Pisces, Teleostei. Marine fishes (Pisces) whose heads are marked with gold spots. (A.W.L.)

GINGER. *Zingiber officinale.* Zingiberaceae. Ginger is the dried **rhizome** of a perennial **monocotyledonous** plant, probably native to tropical Asia. The plant apparently no longer exists in the wild state, but is extensively cultivated in tropical countries and islands of both hemispheres. In China, the cultivation of ginger has been carried on since earliest times.

The plant has a fleshy, irregularly branched rhizome from which arise erect leafy stems 2–3′ in height. The leaves are grass-like; the flowers, borne on a separate stem, are yellow and of a distinctive shape resembling that of orchids. The inside tissues of the rhizome are white and richly spotted with resin dots. The plant is propagated by means of rhizome-cuttings, each cutting having an eye or bud which produces an erect stem.

When the leaves begin to turn yellow the plant is ready to harvest. The rhizomes are dug up and cleaned, then immersed in boiling water to kill the buds or eyes, and also to loosen the periderm or outer portion. The rhizomes are then peeled and dried.

Ginger is used principally as a condiment and as an aromatic stimulant. Several volatile oils are responsible for the characteristic odor, while a ketone, gingerol, or zingerone, produces the hot biting taste. Much ginger is used in preparing ginger ale. Formerly quantities of the rhizome were used in making Jamaica ginger. Ginger also appears on the market in another form, preserved ginger, a Chinese product made from uncured rhizomes. (R.M.W.)

GINGIVITIS. Inflammation of the gums due principally to dental pyorrhea, mouth infections (e.g. "trench mouth") and vitamin B deficiencies. (D.M.H.)

GINKGO. Paleobotany; Maidenhair Tree.

GINKGOALES. Paleobotany.

GINSENG. *Panax Ginseng* and *P. quinquelolium.* Araliaceae. The thick roots of the two species of ginseng, the first a native of Manchuria, the second of North America, are of curious interest because of the significance attached to their use. The plants are low-growing perennial herbs having compound leaves and compound **umbels** of small white flowers. They grow best in rich shady woods of hardwood trees. Even in cultivation they are usually grown in such localities. When mature, the thick fleshy roots are removed from the ground very carefully to avoid any damage. They are then dried and marketed. Their chief users are the Chinese, who attribute to the roots wonderful curative properties, which modern research has failed to verify. Especially valuable, from the primitive viewpoint, are roots having the general form of the human body, two lateral branch roots near the top simulating the arms, and a **dichotomy** in the lower half resembling two legs. Formerly such a root would be worth its weight in gold, but today the Chinese are becoming gradually educated by modern science away from such beliefs. (R.M.W.)

GIRAFFE. Mammalia, Artiodactyla. *Giraffa.* Large African animals with unusually long necks and legs. The head bears a pair of stubby horn-like structures, covered with skin. Giraffes are marked with large irregular brown spots on a lighter ground but this pattern is said to be inconspicuous in their natural habitat. (A.W.L.)

GIRDER. A girder is a large heavy beam capable of carrying both concentrated and uniformly distributed loads. Large rolled steel beams are frequently called girders although the name is generally applied to large beams which are made up of rolled steel sections connected by rivets or welding. In concrete construction the large beams which are used to support smaller beams are called girders. A girder, like a beam, resists transverse bending, and is loaded, ordinarily, by gravity load which is transferred by the girder to its supports. The common plate girder is a compound steel structure composed of plates and angles, bound together in one structure by the use of rivets or welding. **Plate girders** are used where strength requirements cannot be met by the largest available rolled steel sections. Due to their adaptability, plate girders are to be found in almost every form of construction embodying steel. **Bridges, cranes,** and buildings, show many examples of the plate girder.

The built-up plate girder roughly resembles an **I-beam** in shape. Its area may be thought of as subdivided into area of flanges and area of web. The flange sections are most useful in withstanding the bending, and the web resists most of the **shear** to which a girder is subjected. The arrangement of plates and angles in a plate girder is shown in the accompanying figure.

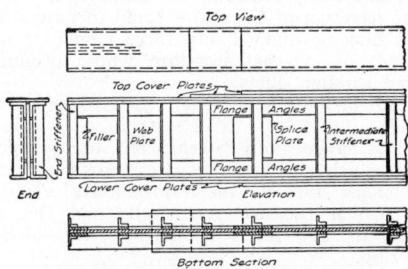

Principal parts of a plate girder.

As shown in the figure, the girder is built up of a web plate whose depth is nearly equal to the full depth of the girder, flange angles which are riveted near the top and bottom of the web plate and cover plates that are riveted to the flange angles. Since the flange chiefly resists bending, and bending moment is greatest at the center of a girder (for ordinary load conditions) the cover plate could be of a thickness increasing from minimum at the abutment to maximum at midspan. It is not practicable to specify a tapered plate, but the same effect is achieved by subdividing the total maximum required cover plate area into a number of plates in laminar arrangement, and achieving the taper effect by cutting off the plates where reduction of bending stress permits. Localized buckling

of the web must be resisted in order to permit the girder to develop its full strength. For this purpose stiffeners, consisting of angles arranged vertically, and riveted to the web and to the flange angles, are spaced periodically along the length of the girder. These are called stiffener angles, and may be smaller than the flange angles.

As the girder carries load by beam action, the **flexure** theory applies. The problem of design of plate girders begins with the computation of bending moment and shear. Generally, bending moment governs the design. A cross-section of the girder is then assumed, and the **moment of inertia** of the same computed. This moment of inertia must be such as to fit into the common flexure formula

$$\frac{M}{p} = \frac{I}{y}$$

and give a value of p which is within the safe working stress, but which is not uneconomically low. In the above equation, I is the moment of inertia of the section of the girder about its neutral axis, M is the bending moment, p is the working unit stress of the metal, and y is the distance from the neutral axis at the most stressed fiber of the girder, generally, at the extreme edge of the flange.

Most authorities require that the design of an important girder be carried through with an exact computation of the moment of inertia of some assumed section. If a determination of an economic section is made by trial and error, this moment of inertia method of design may become quite tedious. The number of trials can be greatly shortened if some approximation, which would guide the designer towards a correct selection of the proper structural shapes, could be employed. Such a method is outlined below. It is based on the assumption that a girder is made up of a simple rectangular web connecting rectangular flanges. Let the area of the web be w, and the area of each flange f, while the distance between the centers of gravity of the area of the flanges is h. The moment of inertia of this assumed area about the neutral axis which is taken to be on the axis of symmetry, is:

$$I = \frac{h^2}{2}(f + w/6).$$

If this expression be substituted in the flexure formula mentioned above, and note be taken of the fact that $y = \frac{h}{2}$, the flange area f is found to be given by the following equation:

$$f = \frac{M}{ph} - \frac{w}{6}.$$

As ordinarily given in structural texts, this formula represents the net flange area (area with rivet holes deducted). Consequently it has w divided by 8 instead of 6, the difference being accounted for by deduction of a certain amount of web area to account for rivet holes. If the flexure is obtained by some rapid estimating system, such as this flange area method, an arrangement of commercially procurable steel shapes can be set up, and the exact **moment of inertia** accurately established by the principles of **mechanics.**

The complete design of a steel plate girder includes also such problems as determining the riveting pitch in the flanges, the design of splices in the web plate, the spacing and riveting of stiffeners, and the strengthening of the ends by end stiffeners where the girder bears on its supports. (F.T.M.)

GIRDLE. 1. The part of the mantle of a **chiton** which surrounds the shell plates. 2. The skeletal structures of **vertebrates** by which the appendages are associated with the trunk. **Skeletal system.** (A.W.L.)

GIRT. Bent.

GIZZARD. A region of the alimentary tract (see **Digestive system**) with thick muscular walls and some adaptation for grinding food. The gizzards of birds are the best known examples. They have a tough lining and their grinding action depends on the movements of hard particles such as gravel contained in them. One of the fishes, the **gizzard shad**, has a stomach of similar nature. Many insects also have a gizzard but in this organ the supposed grinding structures are chitinous folds and teeth projecting into the cavity. The grinding action of the organ has been questioned by some observers. (A.W.L.)

GIZZARD SHAD. Pisces, Teleostei. A widely distributed North American fish (**Pisces**) whose stomach is developed like the **gizzard** of a bird. It occurs in both fresh and salt water. (A.W.L.)

GLACIAL DEPOSITS. DRIFT. The general term for glacial deposits, or sands, gravels, boulders, etc., which are the result of mountain or continental glaciation, is drift. Drift is classified as either stratified drift, the result of deposition by waters from the melting glacier, or, till (unstratified drift) which is apt to be coarsely graded sediments composed of clay, sand, gravel and boulders. Till may grade, in places, into stratified drift, but is principally transported and deposited by the ice. Both stratified drift and till also form distinctive topographic features, to such an extent that both mountain ranges and even broad continental areas which have been subjected to glaciation cannot be described as having been subjected to the normal cycle of **erosion**. When a glacier advances over old drift it may form cigar-shaped hills, called **drumlins**, whose longer axes are relatively parallel with the movement of the ice. **Till**, which is built up into long mounds and ridges at the frontal margin of the ice sheet, forms significant topographic features called moraines. The waters coming off from the front of a

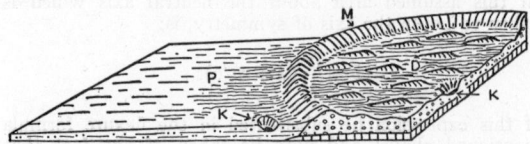

Block diagram showing a terminal moraine, *M*; an outwash plain, *P*; drumlins, *D*; and kettle holes, *K*. (*Modified after A. Penck.*)

melting ice-sheet deposit great sheets of stratified gravels, sands and clays. If ice-blocks have been covered by the outwash, when these ice-blocks finally melt they leave depressions in the outwash plain which fill with ground water to form ponds and lakes. These depressions are called kettle holes. (R.M.F.)

GLACIER. Wherever upon the earth's surface the temperature is sufficiently low and there is sufficient precipitation to produce a permanent snow field, there may glaciers be found. Other things being equal, perpetual snow is more likely to be found in high latitudes and high altitudes; as examples we have the extensive snow and ice field on Greenland and the Antarctic continent as well as valley glaciers of the Alps, of Alaska, the Rocky Mountains, the Andes, the Himalayas and elsewhere. Repeated thawing and freezing of the snow in perpetual snow fields permit the formation of coarse granular ice called nevé which passes into ice of the usual sort. On slopes, the accumulated ice will eventually begin to move, and as it fills a mountain valley, becoming literally a river of ice, it may be called a valley glacier. Even in the absence of great slopes ice will only accumulate to a limited thickness before it commences to spread out

in all directions from its place of accumulation. Such a mass of ice is called a continental ice sheet or continental glacier; Greenland is an example of such a sheet of continental ice. The exact mechanics of ice movement is not definitely known. (R.M.F.)

GLAND. An organ of epithelial structure which produces secretions necessary to the body, or which excretes waste materials from the system. Glands vary greatly in form and complexity and in the nature of their products.

The simplest glands are unicellular. In the glandular lining of the intestine (see **Digestive System**), for example, are isolated cells which secrete mucus. They are known as goblet cells because the mucus accumulates in a clear ovoid mass above the constricted base of the cell, approximating the form of a goblet.

Multicellular glands develop from the epithelial layers by local increase of cells and consequent expansion of

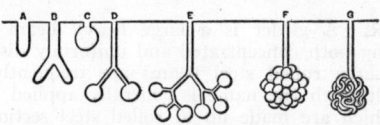

Diagram showing types of glands: *A*, simple tubular; *B*, branched tubular; *C*, simple acinous; *D* and *E*, branched acinous; *F*, compound acinous; *G*, compound tubular. (*From Kimber and Gray, Textbook of Anatomy and Physiology, Macmillan Co.*)

the layer into adjacent spaces or tissues. They include tubular, acinous, and alveolar structures. Tubular glands are slender tubes lined with glandular epithelium; acini are rounded groups of cells with a small central cavity; and alveoli are larger rounded chambers lined with glandular cells. Tubular glands may branch or coil. Many of the larger glands of the body, including the **pancreas** and **salivary glands**, are made up of great numbers of acini borne by complex branching ducts. These glands are said to be compound. In the most complex forms the secretion may leave the cells by minute canals, or similar canals between the cells may conduct it to the cavity of the acinus. This cavity empties into a short secretory duct lined with gland cells, and this in turn into the excretory duct. These smaller ducts join to form larger and larger passages, ultimately reaching the main duct which delivers the secretion of the entire gland to its destination.

Glands may be divided into three kinds:

1. Glands of external secretion whose products are discharged through ducts.

2. Glands of internal secretion, the endocrine or ductless glands.

3. Glands having both external and internal secretion.

Glands which produce cells are known as cytogenic glands. They include the reproductive glands (see **Gonad**) which produce germ cells and the **spleen**, **lymph** glands, and red bone marrow, in which blood cells develop.

Special glands are derived from all germ layers and are associated with all organic systems. They serve for hormone production, for lubrication, to prevent drying, for defense, in reproduction, and in many other ways. (See **Endocrine Glands**.) (A.W.L., D.M.H.)

GLAND, STEAM. Fluid Seals.

GLANDULAR FEVER. Infectious Mononucleosis.

GLASS. Glass is defined as "a liquid whose rigidity is great enough to enable it to be put to certain useful purposes" (Morey). The term "rigid liquid" includes the naturally occurring rock glasses, such as **obsidianite**, and the industrially formed metallurgical **slags**. Glasses, glazes of pottery, enamels on steel, obsidianite, fuzed quartz ware, and slags are all formed by rapidly cooling a previously molten mass without allowing sufficient time for crystallization to occur during the

boron, which functions somewhat as **silicon,** is available. Illustrative analysis of various glasses are the following:

ANALYSES OF VARIOUS GLASSES

Kind of Glass	Silicon Oxide	Sodium Oxide	Potassium Oxide	Calcium Oxide	Barium Oxide	Zinc Oxide	Lead Monoxide	Aluminum Oxide	Boron Oxide	Remainder
Common bottle—3 samples..	70–73	11–18	....	7–17	....	....	....	....	....	1.6–2.8
Window and plate—4 samples	71–73	12–14	....	11–16	....	....	....	....	....	0.1–1.3
Tableware and electric bulbs —2 samples..............	54–57	11–2	1–10	0.1–1.3	....	....	30–33	....	....	0.3–1.6
Optical instrument— $n_D = 1.6555$..........	20.0	....	....	....	....	....	79.9	....	....	0.1 ⎫ arsenic
1.5905..........	39.6	....	....	2.0	44.0	7.7	3.0	....	5.0	0.4 ⎬ trioxide
1.5179..........	68.5	12.0	5.0	....	9.7	1.0	....	....	3.5	0.2 ⎭
Chemical and heat resistant: Pyrex ($n_D = 1.47$).......	80.6	3.8	0.6	0.2	....	....	....	2.0	11.9	0.9
Jena....................	64.4	7.3	....	0.1	....	11.7	....	6.3	10.0	0.2

cooling process (undercooling). The viscosity characterizing liquids is absent, and the material appears to be solid.

Common glass is essentially a sodium calcium **silicate** in composition. In preparing the raw materials for the charge, **sodium** is supplied as soda ash (sodium carbonate, Na_2CO_3) or salt cake (sodium sulfate, Na_2SO_4), with **carbon** to reduce the **sulfate, calcium** as **limestone** (calcium carbonate, $CaCO_3$) or burnt lime (calcium oxide, CaO), and **silicon** as quartz, usually in the form of silica sand (silicon oxide, SiO_2), along with "cullet" or broken glass to assist in the fluxing and melting.

A representative charge for common glass is 100 parts by weight of silica sand, 35 of soda ash, 12 of lime, and 10 of niter (**sodium** nitrate, $NaNO_3$), which supplies some of the sodium and serves to oxidize ferrous to ferric. For plate glass one of the charges used is 100 parts by weight of silica sand, 32 of soda ash, 6.5 of salt cake, 0.3 of charcoal, and 32 of limestone. All materials of the charge that are not volatilized in the process remain in the glass produced. The green color due to the presence of ferrous iron, of much common glass is accounted for by the presence of **iron in** some of the materials of the charge. To produce colorless glass when iron is present, sodium nitrate or manganese dioxide is usually added. To produce colored glass small amounts of the desired coloring materials are added to the charge, for example, **cobalt** oxide for blue glass, **manganese** oxide for violet, **gold** or **selenium** for red, **uranium** oxide or **silver** for yellow, **ferric** oxide for brown. Addition of **calcium** fluoride, **arsenic** trioxide, **aluminum** oxide, **zinc** oxide, **calcium** phosphate, one or more, produces opalescent white glass, while black glass is produced by **iridium** oxide or mixtures of other oxides, such as those of cobalt, iron, nickel, manganese.

The raw materials are heated to a high temperature in a furnace either in individual pots or in hearth or tank furnaces. When the whole mass has been fused, time is allowed for the escape of gases, and then the liquid is withdrawn and worked. The working depends upon the unique property possessed by glass of being shaped while in a heated state, by blowing, molding or rolling, and retaining the shape upon being cooled. The range of temperature through which glass is thus workable and its viscosity in this range are important considerations.

For special purposes the composition of glass may be varied considerably. **Potassium** may be substituted for **sodium, lead** or **barium** for **calcium;** the ratio of these to silicon may be decreased markedly; and, finally,

Blast furnace slag has an approximate analysis somewhat as follows: calcium oxide 43%, **magnesium** oxide 2%, **aluminum** oxide 15%, silicon oxide 35%, residue—chiefly ferrous oxide plus **maganous** oxide plus **calcium** sulfide—5%.

Obsidianite has the following approximate analysis: calcium oxide 1%, **sodium** oxide 4%, **potassium** oxide 4%, aluminum oxide 14%, silicon oxide 75%, **ferric** oxide 1%, ferrous oxide 0.5%, iron disulfide 0.5%, water 0.5%.

When glass is cooled rapidly strains are set up in it but up to a certain point the resistance to shock, and to rapid, moderate changes of temperature is increased; but on scratching it is subject to shattering. To increase this effect, glass is sometimes cooled rapidly by immersion in oil, as for lamp chimneys. To decrease this effect, glass is usually cooled slowly or annealed as for lenses. Very slow cooling and aging cause devitrification of glass. Such glass crystallizes and is readily shattered.

Quartz glass, unlike ordinary glass, is transparent to **ultra-violet rays,** and is, therefore, used in the **mercury-arc lamp.** Sodium vapor lamps require a boro-silicate glass. Heat-absorbing glass, useful for skylights, is made so that it is effective in removing one-half of the infra-red or heat rays of sunlight. Safety or laminated glass is made by cementing together sheets of hardened glass by a thin layer of **cellulose** acetate or some other plastic. Such glass does not scatter when it fractures.

The system sodium (or potassium) oxide—calcium oxide—silicon oxide has been determined by Morey and coworkers. The lowest melting mixture, namely 5% sodium oxide, 20% calcium oxide, 75% silicon oxide, was found by them to have a melting temperature of 725° C. The system calcium oxide—aluminum oxide—silicon oxide, has been determined by Rankin and coworkers. (See **Cement.**)   (R.K.S.) ·

GLASS SNAKE. Reptilia, Sauria. A legless **lizard,** *Ophisaurus ventralis,* whose tail is exceptionally brittle.

Glass snake, *Ophisaurus ventralis.* (*N. Y. Zoological Society.*)

Although snake-like it may be recognized as a lizard by its small ventral scales and its eyelids. Central and southern states. (A.W.L.)

GLAUBER'S SALT. Sodium sulfate decahydrate.

GLAUCOMA. A common disease of the eye which usually occurs in middle and later life. The exact cause is unknown, but several predisposing factors have been observed. There is frequently an hereditary history. **Arteriosclerosis,** hyperopia, and abnormalities in the size of the eyeball and lens may play a rôle.

The symptoms of glaucoma are due to an increase of the pressure within the eyeball and the accompanying venous congestion. In the acute form there is diminished vision to blindness, pain in the eye and severe headache. The disease may be rapid or insidious in onset. Untreated, it terminates in blindness of both eyes. Treatment is with pupil-constricting drugs locally and, in advanced cases, surgery. (D.M.H.)

GLAUCONITE. Glauconite is a hydrous **silicate of potassium** and **iron** of somewhat variable composition. It has a dull green color and is often a constituent of marine deposits forming "green sand." It is believed to have been produced through the alteration of iron-bearing silicates chiefly **biotite** and possibly **augite** and **hornblende.** It occurs along the Coastal Plain of the Atlantic and Gulf States. Frequently found filling the interiors of the shells of globigerina, a common genus of the **foraminifera** (Protozoa). Since globigerina occurs as a deep sea deposit many European geologists have claimed that glauconite is only found in deep water. On the other hand typical "green sands" occur associated with sand and clays which are certainly of shallow marine origin. Glauconite derives its name from the Greek word meaning bluish-green. (R.M.F.)

GLAUCOPHANE. Glaucophane, essentially a complex **silicate** of **sodium,** and **iron** or **aluminum** $Na(Al, Fe)(SiO_3)_2$, is a rather rare mineral although it has been noted from widely separated occurrences. It is **monoclinic** and ordinarily is fibrous or granular. It is brittle; hardness, 6–6.5; specific gravity, 3–3.1; color, azure blue, blackish-blue or gray; luster, vitreous to pearly; translucent to opaque. Glaucophane is found only in the **metamorphic** rocks sometimes forming glaucophane schists. It is found in Switzerland, Italy, Siberia, Japan and in the United States chiefly in the rocks of the Coast Ranges in California and Oregon. The name glaucophane is derived from the Greek words meaning *bluish green,* and *appear.* (E.S.C.S.)

GLIDER. The glider is an **aircraft** obtaining its sustention from **aerodynamic** forces created by a wing of **airfoil** section. Unlike the airplane, the glider has no engine, and must depend either on gravitational energy obtainable by gliding down a slope, or upon a tow furnished by a surface vehicle or an airplane. The sailplane, or soaring glider, a refinement of the glider, is a more expensive, less rugged, and aerodynamically more efficient craft. It is so light and is built with so low a wing loading, (i.e., lbs. per sq. ft. wing area) that in the presence of upwardly directed air currents, such as are present along the windward slopes of mountain ranges, its sinking speed being lower than the upward component of air velocity, it is able not only to remain aloft, but actually to mount upwards considerable distances.

The glider has several spheres of usefulness. First, like yachting, it furnishes, to a high degree, an entertainment of a sporting nature. This is especially true of the sport of soaring. Second, since its controls are the same as those of the conventional airplane, it offers a means of preliminary pilot training at a minimum cost. It is said that one unable to coordinate the controls in a glider, cannot master a power plane. Third, it has many commercial possibilities as a towed air-freighter. Fourth, the glider is of proved tactical value in airborne offensive warfare.

There are three classes of gliders, primary, secondary, and cargo. The primary glider is relatively heavy, and crudely constructed. It is built around a keel or backbone rather than a fuselage, and there is no attempt to streamline the pilot's seat. Due to its steep gliding angle, flights in it are necessarily short. This type of glider has but little usefulness, and is not frequently built nowadays. The secondary glider, sometimes called dual-purpose or trainer glider, is more ruggedly built than the sailplane, and stands rougher usage, but will soar when subjected to the best wind conditions. Usually it is constructed as a single-place high-wing monoplane, with steel tube or wooden fuselage covered with fabric. The forward end is smoothly rounded and a detachable cockpit cover is slipped into place before soaring flights. The landing gear may consist either of a skid or a single wheel, aided by skids fore and aft. The controls are exactly the same as those of an airplane, but the light wing loading makes the glider much more sensitive to wind currents. Consequently, it is provided with extra large control surfaces in the form of aileron, elevator, and rudder.

The cargo glider is built primarily for towing by an airplane. Although it is large, the cargo glider is also efficient, aerodynamically, and a multi-motored airplane can pull two or three cargo gliders simultaneously. Airborne warfare has given great impetus to the construction of this type of glider since personnel and heavy armament may be safely landed on sites where airplane landings are impossible. (F.T.M.)

GLOBIGERINA OOZE. Oceanic Deposits.

GLOBULINS. Aminoacids, Polypeptides and Proteins.

GLOCHIDIUM. A larval form of fresh-water **bivalve** mollusks which lives as a parasite in the **gills** or skin of fishes. (A.W.L.)

GLOMERATE. The textural term, proposed by R. M. Field, for a **sedimentary rock** with a coarse and poorly graded texture, when the origin of the shape of the larger constituents has either been undetermined or is indeterminable. (R.M.F.)

GLOSSA. Labium.

GLOSSITIS. Inflammation of the tongue. It is seen commonly in pernicious **anemia** and **vitamin B** deficiencies. (D.M.H.)

GLOSSOPTERIS. Paleobotany.

GLOTTIS. The opening from the **pharynx** or throat of vertebrates into the **trachea.** (A.W.L.)

GLOW DISCHARGE. Ionized Gases.

GLOW PLUG. Hot Bulb.

GLOW WORM. Insecta, Coleoptera. Wingless females of certain **beetles.** They resemble larvae throughout life and are luminous. (A.W.L.)

GLUCINUM. Beryllium.

GLUCOSE. Carbohydrates.

GLUCOSIDES. Glucosides are substances that by reaction with water, either in presence of certain **enzymes** or of dilute acids or alkalis, yield a sugar, commonly

glucose, as one of the products, plus a principle characteristic of the individual glucoside. Most glucosides are soluble in cold or hot water, and in alcohol (95% C_2H_5OH), and insoluble or slightly soluble in ether (used to separate from alcohol solution). Most optically active glucosides are laevo-rotatory. The di- and polysaccharides are to be considered as glucosides. Glucosides occur in plants, especially in leaves, buds, young shoots where **metabolism** is active, and in the bark and seeds. **Anthocyanins,** the plant colors of flowers, are glucosides, as are also some **tannins.**

GLUTELINS. Proteins.

GLUTTON. Wolverine.

GLYCEROL. Glycerol, propantriol, glycyl alcohol, "glycerine" ($CH_2OH \cdot CHOH \cdot CH_2OH$) is a colorless, viscous liquid, of sweetish taste, odorless, boiling point 290° C., or at 12 mm. pressure 170° C., gradually solidifies at 0° C. to solid, melting point 18° C., miscible in all proportions with water or alcohol, insoluble in ether or chloroform, absorbs water on exposure to the

SELECTED REPRESENTATIVE GLUCOSIDES

GLUCOSIDE	FORMULA	MELTING POINT, °C.	HYDROLYSIS Sugar	HYDROLYSIS Principle
1. Aesculin........................ in horsechestnut bark	$C_{15}H_{16}O_9 \cdot 1\frac{1}{2}H_2O$	205	glucose	aesculetin
2. Amygdalin....................... in peach kernels, cherry laurel leaves, bitter almonds	$C_{12}H_{16}O_7 \cdot 3H_2O$	200 (anhyd.)	glucose glucose	mandelocyanides benzaldehyde+hydrocyanic acid
3. Arbutin........................ in bearberry leaves	$C_{12}H_{16}O_7 \cdot \frac{1}{2}H_2O$	165	glucose	hydroquinone
4. Coniferin...................... in sap of coniferous trees	$C_{16}H_{22}O_8$	185	glucose	coniferyl alcohol
5. Dhurrin........................ in sorghum seedlings, millet	$C_{14}H_{17}O_7N$	....	glucose	para-hydroxy-benzaldehyde + hydrocyanic acid
6. Digitalin...................... in digitalis	$C_{35}H_{56}O_{14}$	217	glucose	digitaligenin, digitalose
7. Digitonin...................... in digitalis	$C_{55}H_{90}O_{29}$	235 approx. decom.	glucose galactose	digitogenin
8. Digitoxin...................... in digitalis	$C_{34}H_{54}O_{11}$	240 (anhyd.)	digitoxose	digitoxigenin
9. Helleborein.................... 	$C_{37}H_{56}O_{18}$	200–230 decom.	glucose	helleboretin
10. Hesperidin..................... in unripe oranges	$C_{50}H_{60}O_{27}$	251	glucose rhamnose	hesperetin
11. Indican....................... in natural indigo	$C_{14}H_{17}O_6N \cdot 3H_2O$	100 (anhyd.)	glucose	indigo
12. Phloridzin.................... in bark of fruit trees	$C_{21}H_{24}O_{10} \cdot 2H_2O$	108 Remelts 170 decom.	glucose	phloretin
13. Quercitrin.................... 	$C_{21}H_{22}O_{12} \cdot 2H_2O$	168 decom. (anhyd.)	glucose rhamnose	quercitin
14. Salicin....................... in willow trees; used in medicine	$C_{13}H_{18}O_7$	201 Remelts 235 approx.	glucose	saligenin
15. Saponin....................... in soapwort root, forms foam with water, toxic to cold blooded animals	$C_{32}H_{52}O_{17}$	195 decom.	sugar	sapogenin
Tannins.......................... in nut galls	....	....	glucose	gallic acid
Anthocyanins...................... Red (with acids), violet (free), blue (with alkalis) pigments of flowers	....	....	....	anthocyanidins
Cyanin........................... 	$C_{15}H_{10}O_6$	....	glucose	cyanidin
Idaein........................... 	....	....	galactose	cyanidin
Pelargonin....................... 	$C_{15}H_{10}O_5$	....	....	pelargonidin
Delphinin........................ 	$C_{15}H_{10}O_7$	....	glucose	delphinidin + para-hydroxy-benzoic acid (R.K.S.)

GLUE. Glue is the product of **hydrolysis** of **proteins,** usually of animal origin. Waste products of slaughterhouses (i.e., hide clippings, hoofs, etc.) are the source of most commercial glue. (R.K.S.)

GLUME. A glume is an empty **bract,** often chafflike, which occurs at the base of the **spikelet** in grasses. Usually there are two to each spikelet. (R.M.W.)

atmosphere. Glycerol reacts (1) with **phosphorus** pentachloride to form glyceryl trichloride ($CH_2Cl \cdot CHCl \cdot CH_2Cl$), (2) with **acids** to form **esters,** e.g. glycerol monoacetate ($CH_2OH \cdot CHOH \cdot CH_2OOCCH_3$), glycerol diacetate ($C_3H_5(OH)(OCOCH_3)_2$), glycerol triacetate, triacetin ($CH_2OOCCH_3 \cdot CHOOCCH_3 \cdot CH_2-OOCCH_3$), glycerol mononitrates (alpha, $CH_2OH \cdot CHOH \cdot CH_2ONO_2$; beta, $CH_2OH \cdot CHONO_2 \cdot CH_2OH$), glycerol dinitrates (1,2, $CH_2OH \cdot CHONO_2 \cdot CH_2ONO_2$; 1,3, $CH_2ONO_2 \cdot CHOH \cdot CH_2ONO_2$), glyceryl trinitrate, "nitroglycerine" ($CH_2ONO_2 \cdot CHONO_2 \cdot CH_2ONO_2$), glyceryl tristearate, tristearin ($CH_2OOCC_{17}H_{35} \cdot CHOOCC_{17}H_{35} \cdot CH_2OOCC_{17}H_{35}$), indirectly, glycerol monophosphates (alpha, $CH_2OH \cdot CHOH \cdot CH_2OPO(OH)_2$, beta, $CH_2OH \cdot CHOPO(OH)_2 \cdot CH_2OH$), (3) with oxidizing agents, e.g., dilute **nitric acid,** to form glyceric acid ($CH_2OH \cdot CHOH \cdot COOH$), tartronic acid ($COOH \cdot CHOH \cdot COOH$), mesoxalic acid ($COOH \cdot CO \cdot COOH$), (4) with **phosphorus** plus **iodine,** to form allyl iodide ($CH_2 : CHCH_2I$), which with **hydrogen iodide** yields propylene ($CH_2 : CHCH_3$) and then iso-propyl iodide (CH_3CHICH_3), (5) with **sodium** or sodium hydroxide to form alcoholates, (6) with **sodium** hydrogen sulfate or **phosphorus** pentoxide heated, to form **acrolein** ($CH_2 : CHCHO$). Glycide alcohol ($CH_2OH \cdot CH \cdot CH_2$)

is obtained by treatment of glycerol alpha-monochlorohydrin ($CH_2OH \cdot CHOH \cdot CH_2Cl$), which is made by reaction of hypochlorous acid and allyl alcohol with barium hydroxide. With **hydrogen chloride,** glycide alcohol yields epichlorohydrin ($CH_2Cl \cdot CH \cdot CH_2$).

Glycerol is obtained (1) from vegetable and animal oils and fats, most of which are mainly glycerol esters of **stearic, palmitic** and **oleic acids** by treatment with alkali (**sodium hydroxide** commonly used), **acid** (sulfobenzone—or naphthalene—stearic acid, "Twitchell's reagent"), superheated steam, or an **enzyme** (lipase of castor beans). Glycerol is recovered from the water solution by **evaporation** under diminished pressure, and purified by treatment with decolorizing **carbon** followed by **filtration,** (2) by **fermentation** of glucose in the presence of **yeast** and **sodium** sulfite.

Glycerol may be detected by the characteristic odor of **acrolein,** found on heating with **potassium** bisulfate.

Glycerol is used (1) in the manufacture of high **explosives,** e.g., glyceryl trinitrate ("nitroglycerine"), which is the main component of dynamite, (2) in antifreeze solutions, especially for automobile radiators,

% GLYCEROL BY WEIGHT	SPECIFIC GRAVITY, 60° F.	FREEZING POINT, °F.
20	1.049	23
40	1.103	4
60	1.158	−28

(3) to maintain a moist condition in fruits and tobacco, (4) in cosmetics and skin preparations, (5) to prepare glycerol phosphoric acid, used in medicine, and "boroglyceride" used as a preservative. (R.K.S.)

GLYCINE. Aminoacids, Polypeptides, and Proteins.

GLYCOCOLL. Aminoacids, Polypeptides and Proteins.

GLYCOGEN. Carbohydrates.

GLYCOL. Glycol, ethylene glycol, ethandiol, ($CH_2OH \cdot CH_2OH$) is a colorless, viscous liquid, of sweetish taste, odorless, boiling point 197° C., miscible in all proportions with water or alcohol, slightly soluble in ether. Glycol reacts (1) with **sodium** to form sodium glycol ($CH_2OH \cdot CH_2ONa$) and disodium glycol ($CH_2ONa \cdot CH_2ONa$), (2) with **phosphorous** pentachloride to form ethylene dichloride ($CH_2Cl \cdot CH_2Cl$), (3) with **carboxy** acids to form mono- and di- substituted esters, e.g., glycol monoacetate ($CH_2OH \cdot CH_2OCOCH_3$), glycol diacetate ($CH_3COOCH_2 \cdot CH_2OOCCH_3$), (4)

with nitric acid (with sulfuric acid), glycol mononitrate ($CH_2OH \cdot CH_2ONO_2$), glycol dinitrate ($CH_2ONO_2 \cdot CH_2ONO_2$), (5) with **hydrogen chloride,** heated, to form glycol chlorohydrin (ethylene chlorohydrin, $CH_2OH \cdot CHCl$), (6) upon regulated oxidation to form glycollic aldehyde ($CH_2OH \cdot CHO$), glyoxal ($CHO \cdot CHO$), glycollic acid ($CH_2OH \cdot COOH$), glyoxalic acid ($CHO \cdot COOH$), **oxalic acid** ($COOH \cdot COOH$). In the preparation of glycol derivatives, important substances are (a) the sodium glycols, (b) ethylene dichloride (1,2-dichloroethane), best prepared by reaction of **ethylene** and **chlorine,** (c) ethylene chlorohydrin (1-hydroxy-2-chloroethane), best prepared by reaction of ethylene and **hypochlorous acid.** From these are readily made respectively, (a) **ethers,** e.g., glycol monoethyl ether ($CH_2OH \cdot CH_2OC_2H_5$), glycol diethyl ether ($CH_2OC_2H_5 \cdot CH_2OC_2H_5$), (b) ethylene diamine ($CH_2NH_2 \cdot CH_2NH_2$), ethylene dicyanide ($CH_2CN \cdot CH_2CN$), glycol itself, (c) hydroxyethylamine ($CH_2OH \cdot CH_2NH_2$), ethylene cyanhydrin ($CH_2OH \cdot CH_2CN$), ethylene oxide (by

sodium hydroxide $\begin{matrix} CH_2 \\ | \\ CH_2 \end{matrix}\!\!>\!\!O$). Glycol is made by reaction

of ethylene and chlorine or hypochlorous acid to form ethylene dichloride or ethylene chlorohydrin, respectively, followed by treatment of either of these with sodium carbonate solution heated under pressure. Glycol is also formed when ethylene is treated with **potassium** permanganate. Glycol is used (1) in antifreeze solutions, especially for automobile radiators, (2) in the prepara-

% GLYCOL BY VOLUME	SPECIFIC GRAVITY, 60° F.	FREEZING POINT, °F.
17	1.026	20
32.5	1.048	0
44	1.063	−20

tion of **ethers** and **esters,** especially nitrate for explosive, (3) as a solvent, substitute for glycerol. (R.K.S.)

GLYCOLLIC ALDEHYDE. Aldehydes, Ketones, and Related Compounds.

GLYCOSURIA. The presence of excess **sugar** (glucose) in the **urine.** By far the commonest cause is **diabetes,** where as much as 200 grams may be excreted daily. The glycosuria of diabetes is always accompanied by a high blood sugar. In some people extreme emotional stress or fright may produce an excess of sugar in the urine. Another form of glycosuria occurs with injuries or tumors of certain parts of the brain. Still another form occurs in some individuals in whom, regardless of the sugar intake, a certain amount of sugar is passed into the urine. Here the kidneys are said to have a low threshold for sugar. The condition is called renal glycosuria and is without pathological significance. Glycosuria is also seen in normal pregnancy, and in certain disorders of the thyroid and pituitary glands. (D.M.H.)

GLYPTOLITH. Dreikanter.

GNAT. Insecta, Diptera. Loosely applied to many small 2-winged flies. In such names as **buffalo gnat, gall gnat,** and **fungus gnat** it applies to specific groups. (A.W.L.)

GNATCATCHER. Aves, Passeriformes. Small birds related to the **kinglets.** One, the blue-gray gnatcatcher, *Polioptila caerula,* ranges over North America east of the Rockies and two other species occur in the southwestern states. (A.W.L.)

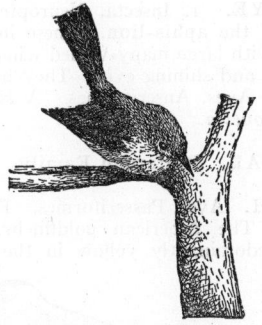

Blue gray gnatcatcher.
Polioptila caerulea. Bluish gray above, grayish white below. Outer tailfeathers white; inner ones black. Narrow black border on front and sides of head. Four and one-half inches long.

GNATHOBASE. The base of an appendage of the **arthropods**, formed for crushing food. **Spiders**, horseshoe crabs, and **scorpions** chew their food by such means. The gnathobases of a pair of appendages act together like the **mandibles** of other arthropods. (A.W.L.)

GNATHOCHILARIUM. The posterior element of the mouth parts of **millipedes**. It is formed of a pair of appendages and is similar to the **labium** of insects (A.W.L.)

GNATHOPOD. Appendages of crustaceans (**Amphipoda**) used for grasping food. (A.W.L.)

GNATHOSTOMATA. A term applied collectively to the members of the subphylum Vertebrata which have hinged jaws, in contrast with the funnel-shaped mouth of the Class Cyclostomata. It embraces the fishes, amphibians, reptiles, birds and mammals. (A.W.L.)

GNEISS. The gneisses are common and widely distributed rocks which have been derived by **metamorphic** processes from pre-existing formations that were originally either **igneous** or **sedimentary** rocks. Gneissic rocks are coarsely laminated and largely recrystallized but do not carry excessive quantities of the **micas**, chlorite or other platy minerals. Gneisses that are metamorphosed igneous rocks or their equivalent are termed **granite** gneisses, **diorite** gneisses, etc.; however depending upon their mineralogical composition, they may be called **garnet** gneiss, **biotite** gneiss, **albite** gneiss and so on. Orthogneiss designates a gneiss derived from an igneous rock; paragneiss, one from a sedimentary rock. The word gneiss is from an old Saxon mining term which seems to have meant decayed or rotten, or possibly worthless material. (R.M.F.)

GNETALES. Paleobotany.

GNOMONIC PROJECTION. This type of projection is used in producing, especially for use in navigation, what are frequently referred to as great-circle charts, because of the fact that great circles (geodesic lines) on the surface of the earth are projected as straight lines. In the gnomonic projection the chart is constructed by placing a plane tangent to the surface of the earth at some selected point and then projecting the surface features by extending radii from the center of the earth out until they meet the plane.

In the gnomonic projection the distortion, both of shape and of size, is very severe except for a very limited area immediately about the point of tangency with the earth. The great value of the charts lies in the fact that the shortest distance, even between very widely separated points, will be projected as a straight line. The govern-

ment issues a series of charts on this type of projection for all of the principal cruising areas of the world and they are of immense value to navigators for determining at a glance whether or not the following of the shortest course between two points (**great-circle course**) is practicable. (W.K.G.)

GNU. Mammalia, Artiodactyla. African **antelope**, *Connochaetes gnu,* of large size and ugly appearance. Also called wildebeests. These animals have a large head, strong curved horns, an erect bristly mane and a bristly muzzle. The withers are high and the tail hairy throughout its length. (A.W.L.)

GOAT. Mammalia, Artiodactyla. Hoofed animals belonging to the same family as the cattle, sheep, and antelopes, and not sharply distinct from the sheep. The males are usually bearded. The group includes numerous species of the Old World. One species occurs in North America, two in northern Africa, and the remainder in Europe and the more northern parts of Asia. They are primarily mountain animals.

In addition to goats named as such, the group includes the turs of the Caucasus, the Persian pasang, the **ibexes**, the **markhor**, and the **tahr**. The American species is found in the Rockies and Cascades, north to Alaska. It is called the mountain goat, *Oreamnos americanus.*

Wild goats are fine game animals and the domestic goat, *Capra hircus,* is valuable for its milk and hide, and as a source of mohair. (A.W.L.)

GOATSUCKER. Aves, Caprimulgiformes. All birds (**Aves**) of this order characterized by weak legs, a short weak beak and very wide mouth, and crepuscular habits. They catch insects while on the wing. The **frog mouths**, **oil bird**, and **nightjars**. Generally distributed except in the Australian region.

The North American goatsuckers are the chuck-will's-widow, **poor-will**, **whip-poor-will**, **nighthawk**, and parauque. (A.W.L.)

GOBLET CELL. Gland.

GOBY. Pisces, Teleostei. Shore fishes (**Pisces**) of the tropics. Many species of which a few enter fresh water. Family Gobiidae. (A.W.L.)

GODWIT. Aves, Charadriiformes. *Limosa.* Wading birds (**Aves**) allied to the sandpipers, with long legs and a long slender beak. The several species nest in the far north but migrate to the southern hemisphere in winter. (A.W.L.)

GOETHITE. The mineral goethite is a hydroxide of **iron** corresponding to the formula FeO(OH) crystallizing in the **orthorhombic** system. It occurs in prisms, but is often found in **foliated** or other massive forms. When observable it shows one good **cleavage** parallel to the **prism**; fracture, uneven; hardness, 5–5.5; specific gravity, 4.28; luster adamantine to dull; color, yellowish, reddish, brownish to nearly black; translucent to opaque. It is found associated with **hematite** and **limonite**, being perhaps in part an alteration product of the latter mineral. Goethite is used as an ore of iron. There are many European localities, including Bohemia, Saxony, Westphalia, and Cornwall. In the United States it is found in the **hematite** mines of the Lake Superior region and in Colorado. This mineral was named in honor of the German poet Johannes Wolfgang von Goethe. (E.S.C.S.)

GOITRE. Thyroid Gland.

GOLD. Symbol: Au (aurum). Atomic number: 79. Atomic weight: 197.2. Density: 19.3. Hardness: 2.5–3.

Melting point: 1063° C. Boiling point: 2600° C. No isotope, but of single atomic form: 197.

Gold is a yellow metal—the color is markedly affected by the presence of traces of other metals—very malleable and ductile—the ductility is diminished by the presence of other metals—soft—the softness is counteracted when desired by the addition of other metals—unattacked by air, water or hydrogen sulfide, not dissolved by hydrochloric or nitric acid, but soluble in aqua regia and converted into chloride by chlorine used either as gas or solution, soluble in solutions of cyanides in the presence of air. Very thin sheet gold is translucent and transmits greenish light. Discovery prehistoric.

Gold is one of the most ancient metals used in the arts, about ⅓ of the present output being so used, and the remaining ⅔ in coinage or in bars. The lettering of books and the decoration of porcelain utilize important amounts of gold. It occurs chiefly as native gold alloyed with silver, copper, lead, or other metals (60%–98% Au), in certain sands and quartz veins from which it is obtained (1) by mechanical methods, such as "washing," (2) by dissolving in sodium or potassium cyanide solution followed by precipitation of the gold by zinc metal, (3) by dissolving in mercury (quicksilver) and later distilling off the mercury.

Auricyanide: Potassium aurocyanide ($KAu(CN)_2$) and sodium aurocyanide ($NaAu(CN)_2$), by treatment of gold with potassium and sodium cyanide solutions, respectively, in the presence of air.

Bromide: Aurous bromide ($AuBr$) yellowish-gray solid, by heating auric bromide to a maximum temperature of 200° C.; auric bromide ($AuBr_3$) brownish-red soluble solid by reaction of gold with bromine solution.

Chloride: Aurous chloride ($AuCl$), yellow solid, by heating auric chloride to a maximum temperature of 175° C. for several days; auric chloride ($AuCl_3$), reddish-brown soluble solid, by reaction of gold with chlorine as gas or in solution, sublimes at 265° C. in a current of chlorine, is changed to aurous chloride as stated above, and when heated to temperature higher than 175° C. decomposes into gold metal and chlorine gas; chlorauric acid ($HAuCl_4 \cdot 3H_2O$), brown soluble solid, used in identifying certain organic bases.

Chloroaurates: Sodium chloroaurate, "sodio-gold chlorride" ($NaAuCl_4 \cdot 2H_2O$), yellowish-red crystals, soluble, as also lithium chloroaurate ($LiAuCl_4$) and potassium chloroaurate ($KAuCl_4$).

Hydroxide: Aurous hydroxide ($AuOH$), purple precipitate, by treating aurous bromide solution with sodium hydroxide solution in the cold; auric hydroxide ($Au(OH)_3$), brown precipitate, by treating auric chloride solution with sodium hydroxide solution.

Oxide: Aurous oxide (Au_2O), purple precipitate, by treating aurous bromide solution with a slight excess of sodium hydroxide and boiling the mixture; auric oxide (Au_2O_3), dark brown solid by heating auric hydroxide to 100° C.

Purple of cassius is precipitated when a solution of auric chloride is treated with a solution of stannous chloride of the proper concentration. The product is used in the preparation of ruby glass.

The occurrence of gold in sea water has attracted much attention. The various sources and reports show there is present from 5 to 250 parts by weight of gold per 100,000,000 of sea water. Although the *quantity* present is enormous, the cost of recovering the same has hitherto been greater than the value of the gold obtained. (R.K.S.)

GOLD-AMALGAM. A mineral alloy of gold and mercury, occurs as small grains or lumps. Color, white to yellow with metallic luster. (R.M.F.)

GOLDCREST. Aves, Passeriformes. A European kinglet. (A.W.L.)

GOLDEN-EYE. 1. Insecta, Neuroptera. The lacewing, adult of the **aphis-lion.** These insects are small and delicate, with large many-veined wings of yellowish- or green color and shining eyes. They have a disagreeable odor. 2. Aves, Anseriformes. A North American **duck.** *Glaucionetta.* (A.W.L.)

GOLDENSEAL. Buttercup Family.

GOLDFINCH. Aves, Passeriformes. **Finches** of several species. The American goldfinches, *Astragalinus tristis,* are predominantly yellow in the male sex and

Goldfinch. *Astragalinus tristis.* Bright yellow, with top of head, wings, and tail black. Two white bands on wings. Female duller. Male also much duller in winter. Length, five inches.

mostly olive in the female, while the European species is more brilliantly colored, with a red face. The latter species has been introduced into the United States but is apparently not established. (A.W.L.)

GOLDFISH. Pisces, Teleostei. A species, *Carassius auratus,* related to the carp, often bright reddish-golden in captivity. It is a native of Europe and Asia and is now established in some of the lakes and streams of the eastern half of the United States. As an aquarium fish it is available in many varieties of different form. (A.W.L.)

GOLGI APPARATUS. Cell.

GOMPHOTHERIUM. Fossil Mammals

GONAD. An organ in which sexual reproductive cells (**gametes**) are produced. In some of the simpler animals, gonads develop as temporary organs during the breeding season and in more complex forms they are permanent. They are ectodermal in the fresh-water polyps (*Hydra*), endodermal in the jellyfishes, and mesodermal in the higher forms. The gonads which produce male germ cells are called testes and those which produce egg cells are ovaries. In some snails the gonad produces both kinds of cells and is called an ovotestis. (A.W.L.)

GONANGIUM. A term applied both to the reproductive members of hydroid colonies and to the sheath which envelops them. Usually the reproductive **polyp** is called a blastostyle and the sheath a gonotheca. (A.W.L.)

GONAPOPHYSIS. Appendages of the insect **abdomen** which serve as accessory organs of reproduction. Copulatory organs of the male and egg-laying organs of the female. They are probably derived from jointed appendages. (A.W.L.)

GONDWANA LAND. Permian.

GONIATITE. Invertebrate Paleontology.

GONIOMETER. An instrument for measuring the angles between the reflecting surfaces of a crystal or a prism. Parallel rays from a **collimator**, impinging upon the polished surfaces, are reflected in different directions.

Two methods may be used. In one the crystal or prism is held stationary and the angle between the reflected beams from the two faces, received in succession by a telescope moving around a graduated circle, is measured on the circle; the angle between the two faces is then ½ of this (see figure). In the other method, the tele-

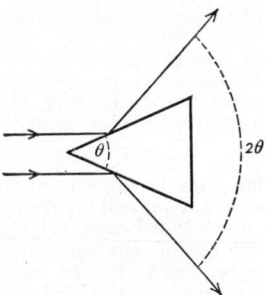

Angle between reflected rays is twice angle between prism faces.

scope is clamped in some convenient position and the crystal or prism is rotated so that first one and then the other face reflects light into it; the angle between the faces is the supplement of the angle through which the prism mounting is turned. An ordinary spectrometer may be used for the purpose (see **Spectroscope**). An instrument similar in geometrical principle, but employing **x-rays** instead of light and an ionization chamber instead of a telescope, is used for measuring angles between the atomic planes within crystals.

Sometimes in the use of a loop antenna for directional purposes (see **Directional Antenna**) it is inconvenient or impossible to rotate the loop. This is especially true for transmitting loops where the size becomes appreciable in order to improve the radiation efficiency. To overcome this difficulty the goniometer is used. As shown in the figure, the instrument consists of crossed sta-

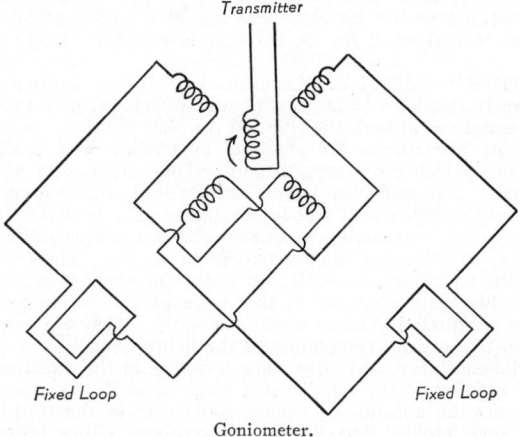

Goniometer.

tionary coils feeding fixed, crossed (90°) loops with the coupling to the moving coil proportional to the cosine of the angle of rotation. The movable coil is fed from the transmitter. The amount of energy transferred from the rotating coil to the fixed coils and hence to their antennae is determined by the position of the moving coil with respect to the others. The effect as far as the resultant field pattern of the antennae is concerned is exactly the same as if a single-loop antenna had been physically rotated. An extension of this principle is used with two sets of crossed loops in many radio range systems. A further advantage of the goniometer is that the antennae may be connected through a transmission line and thus it is not necessary that the operating posi-

tion be near the antenna. It may be used equally well for reception. (L.D.W., L.R.Q.)

GONOPOD. Modified jointed appendages of **centipedes,** used as accessory organs of reproduction. (A.W.L.)

GONORRHEA. An acute or chronic infectious disease, involving primarily the passages of the external genital organs and the **urethra.** Frequently, there is secondary involvement of more distant body structures, causing systemic manifestations.

Gonorrhea is the most common venereal disease. It is rarely a killing disease but due to its late complications, its obstinacy and its ability to cause permanent damage, it ranks next to syphilis as one of the most serious of infectious diseases.

Gonorrhea is as old as the human race. References to it are found in old Egyptian, Chinese and East Indian writings. It is mentioned in the Old Testament. Gonorrhea was not differentiated from syphilis until the early 19th century. Albert Neisser discovered the gonococcus as the causative organism in 1879. (See **Bacteria.**)

There are no figures as to the incidence of gonorrhea in the human race, as many infected never consult a physician and the disease is not ordinarily reported. At least 20% of those having gonorrhea suffer from complications.

The adult form of the disease results usually from sexual intercourse with an infected partner. Gonorrhea of the new-born results from infection of the eye at birth during passage through the infected vagina. It has always been the great cause of permanent blindness in the new-born, and for this reason prophylactic antiseptics are applied to the eyes of every new-born child. Accidental infection can occur through contaminated articles such as towels, etc., but is not especially common, although gonorrheal infection of the eyes in adults has occurred in this manner.

The primary infection involves the **mucous membrane** of the urethra in both male and female, and the mucous membrane of the **vagina** and cervix in the female. This lasts in its acute form for a relatively short time. It is marked by local swelling, burning, and pain of the regions involved. Urination is painful and frequent. There is a copious discharge of pus.

The serious complications of gonorrhea develop from direct spread of infection upward from the primarily infected local parts. In the male the upper portions of the urethra, the prostate, seminal vesicles and the testicles are commonly involved in the acute form. The infection in the prostate later becomes chronic and low-grade activity may remain for a considerable time. Later it may flare up and become acute. In the female, infection of the cervix, various local glands, uterus and ovary and fallopian tubes (salpingitis) are common. General peritonitis is a frequent complication of acute gonorrheal salpingitis. When the salpingitis becomes chronic operative procedures may be necessary. Salpingitis is frequently the cause of sterility.

Gonorrhea is also a systemic disease in some instances, the organism being carried to distant organs by the blood stream. Gonorrheal **septicemia** is rare and has a high mortality. Gonorrheal infection of the **heart** lining and valves (endocarditis) usually results fatally.

Gonorrheal arthritis is the most common complication, and may be a severe crippling disease. It sometimes appears long after the local disease seems to have been cured. Any or several joints may be involved. Fever is usually present with other constitutional symptoms. Pain is very acute. In a similar manner, tendon sheaths, muscles and bones may be involved. Neuritis and various neuralgia occur.

The treatment of gonorrhea involves local treatment of the primary disease and medical and surgical treatment for the various sequelae or complications. **Sulfonamides** are effective in a large per cent of cases.

Those organisms which are resistant to sulfonamides respond well to **penicillin**. (R.S.M.)

GONOSOME. The medusoid (see **Medusa**) of a hydrozoan (see **Hydrozoa**). This is the sexual stage in the alternation of sexual and asexual generations of these animals. (A.W.L.)

GONOTHECA. Gonangium.

GONOZOOID. 1. The attached medusoid (see **Medusa**) such as is borne by the **blastostyle** in some hydrozoan (see **Hydrozoa**) colonies. 2. Sexual individuals in the **salpian** colony. (A.W.L.)

GOOSANDER. Aves, Anseriformes. A merganser (**duck**) of the northern Hemisphere. *Mergus merganser.* (A.W.L.)

GOOSE. Aves, Anseriformes. Large swimming birds (**Aves**) with webbed feet and thick, strong beaks. They live chiefly on vegetation. Geese are found on all continents. They are strong fliers and some species migrate from their nesting grounds in the north to the southern hemisphere. Among the several North American species the Canada goose, *Branta canadensis,* is sometimes called brant and a related species is the black brant, *B. nigricans.*

Geese are excellent game birds and good food. The domestic goose is also valuable for its smaller feathers and down, which are preferred for filling pillows and cushions. (See **Die Casting.**) (A.W.L.)

GOOSEBERRY. Berry.

GOOSEBERRY STONE OR GROSSULARITE. Garnet.

GOPHER. Mammalia, Rodentia. 1. The pocket gophers. Stout-bodied burrowing animals of several species, found throughout the United States. They have fur-lined cheek pouches opening at the sides of the mouth. 2. Slender burrowing species of the central and western states, also called ground squirrels. The suslik or sisel is a European species.

Gophers are injurious to crops. The pocket gophers eat roots and the ground squirrels are more injurious to grain. In the prairie regions they are sometimes so abundant that they have to be destroyed by shooting or poison. (A.W.L.)

GORAL. Mammalia, Artiodactyla. *Urotragus.* Animals of several species related to the **goats**. They are slender and beardless and have small curved horns. They are distributed from the Himalayas to northern China. (A.W.L.)

GORDIACEA. Nematomorpha.

GORDIOIDEA. Nematomorpha.

GORILLA. Mammalia, Primates. The largest of the man-like **apes**. The gorillas are terrestrial animals with an enormous trunk, short bent legs, and long powerful arms. In walking they rest partly on the backs of the bent fingers. According to Akeley's observations they are poor climbers. The head of the gorilla is distinguished by the strong jaws and large teeth, the heavy ridges over the eye sockets, and the small ears. There are two species, the common gorilla, *Gorilla gorilla* and the mountain gorilla *G. beringei.*

Gorillas live in the forests of Africa and our knowledge of them has come mostly from field observations. They are apparently less intelligent than the chimpanzee.

Specimens formerly brought into captivity proved to be delicate and soon died. This difficulty has evidently been mastered by careful attention to diet and by protecting them from the respiratory diseases of man, to which they are very susceptible. The London Zoo, prior to the second World War, had a magnificent pair, Mok and Moina, which had been reared from infancy in captivity. Fine examples are owned also by various zoos in our own country.

The male gorilla was long regarded as a creature of the utmost ferocity. Since he may reach a height of more than 5′ and a weight of 400 lbs. he would, indeed, be a formidable opponent, but Akeley reported the gorillas of his experience to be very shy and difficult to approach so the older idea must be abandoned. (A.W.L.)

GOSSAN. This term is applied to the decomposed upper parts of mineral veins and ore deposits. It usually consists chiefly of hydrated **iron** oxide resulting from the weathering of **pyrite, chalcopyrite,** etc. Gossans have been important sources for the release of the relatively insoluble precious metals and gems which are washed away to form **placer** deposits. Many valuable gold ore bodies have been traced to their source by means of their derived placers. Also **secondarily enriched** sulfide ores of copper have been discovered beneath gossans which were originally prospected for the more precious metals. (R.M.F.)

GOUGE. Used by miners to designate soft or clay-like material between the sides of a mineral **vein** or **ore** deposit and the wall rock; also (structural geology), a layer of finely comminuted material between the walls of a **fault.** (See **Lathe, Woodworking.**) (R.M.F.)

GOUJON. Pisces, Teleostei. A large **catfish**, *Opladelus olivaris,* common in the larger streams of the Mississippi basin. Also called the mud cat. It sometimes attains a weight of 100 lbs. (A.W.L.)

GOURAMI. Pisces, Teleostei. *Osphromenus.* Brightly colored fishes (**Pisces**) of the fresh waters of the Old World tropics. The pelvic fins bear a long slender filament, apparently sensory. One species is valued as food and several small species are kept in aquaria. (A.W.L.)

GOURD FAMILY. Cucurbitaceae. A small family, largely restricted to tropical or warm climates, with representatives in both the Old and the New Worlds. Most of its 650 species are climbing or trailing herbaceous plants which grow very rapidly. They are mostly annuals. The stems are hollow and in most species abundantly supplied with stiff bristly hairs. The large leaves are borne alternately on the stem, have a distinct, often long, **petiole,** and show a variety of shapes. The **tendrils,** which are a conspicuous feature of many members of this family, appear in the **axils** of the leaves and are interpreted as stems modified greatly. They are very sensitive organs, responding to the lightest touch of any solid substance, and often show a change in the direction of twining in the middle of a single tendril. In many species the nutating or circling movement of the tendril is very rapid. The flowers are axillary, either borne singly or in various types of **inflorescence,** and are usually yellow or white. The plants are either **monoecious** or **dioecious.** The **calyx** is adnate to the inferior **ovary,** the corolla is 5-lobed and inserted on the calyx. The **stamens** are typically five, but show great variation in number through fusions. The inferior ovary is 1- to 3-celled and usually contains many flattened seeds. The latter lack **endosperm.** The fruit is a variety of berry called a **pepo,** differing from a berry in that the receptacle enters into the formation of the rind or outer wall. The germination of the seeds of the commonly grown members of this family exhibits one rather striking peculiarity. When the arched **hypocotyl** emerges from the seedcoats a small peg forms on its lower end. This peg prevents the seedcoats from sticking to the **cotyledons,** which are withdrawn and carried into the air by the straightening of the arched hypocotyl. Many members of this family are grown in cultivation, as for example squashes, pumpkins, and cucumbers, and certain ornamental species, like *Echinocystis, Momordica,* and some of the gourds.

Cucurbita, pumpkins and squashes. These are rather coarse annual vines having very rough bristly stems, large, long-stalked leaves and axillary (**axil**) flowers of two kinds. The staminate (**stamen**) flowers have long stalks while the pistillate flower stalks, or **peduncles,** are short. Staminate flowers have a rudimentary **ovary** while pistillate flowers have three staminodia, or vestigial stamens; in both kinds of flowers the 5-lobed yellow **corolla** is conspicuous. Insect-**pollination** is usual. It is probable that these plants are native to tropical America, where they have been long cultivated. Cultivation is now widespread in both the New World and the Old.

Pumpkins, *Cucurbita Pepo,* are of many varieties, sizes and shapes. Included here are the Field Pumpkin, Sugar Pumpkin, Pie Pumpkin and Mammoth Pumpkin, Fordhook, Scallop (Petty-pans), Crookneck Squashes, and Marrow Squashes. In this group the stems are prickly as a rule, and more or less 5-angled.

Squashes, *Cucurbita maxima,* are plants having cylindrical stems which are hairy rather than bristly. Here are found Hubbard squashes, turban squashes and mammoth squashes, the latter often of immense size and frequent occurrence, sometimes weighing over 100 lbs. *Cucurbita moschata* includes cushaw and cheese types of squashes.

Cucumis, Muskmelons, Cantaloupes and Cucumbers. The plants in this genus are considered to be natives of tropical Asia and Africa and the East Indian Islands, where they have been in cultivation for many centuries. In these plants the tendrils are unbranched, the staminate flowers are borne in small clusters in the axils of the leaves, while the pistillate flowers are solitary. Many more staminate flowers are formed than pistillate, to insure successful pollination, which is almost entirely by insects.

Cucumis Melo includes melons of various kinds, among them muskmelons and cantaloupes. Many varieties bear inedible fruits, some of which may be used in making preserves. Few of these are cultivated in American gardens. The fruits have a warted or ribbed skin, but never hairy or spiny. They are probably native to southern Asia.

Cucumis sativus is the cucumber. This is a native of the East Indies. There, and in Asia, cucumbers have been cultivated since earliest times. Many varieties have been developed. Certain varieties, grown under glass, are often seedless. Others are largely grown for pickling. The best pickling fruits are grown in regions having a cool climate. For pickling, the fruits are picked while still young and small. They are first salted in brine, after which they are bottled in vinegar, often with the addition of various spices or other flavorings, such as dill, or mustard, or peppers.

Gherkins, *Cucumis Anguria,* are native to the West Indies. Small cucumbers are also frequently called gherkins.

Citrullus. Watermelons, Citron and Colocynth. These are natives of Asia, Africa and southern Europe. They are coarse trailing vines with branched tendrils and lobed leaves.

Citrullus vulgaris includes the watermelon and citron as varieties. It is native to Africa, where it has been cultivated since the time of Egyptian supremacy. Watermelons are grown for the juicy tender flesh. In contrast to them, the flesh of the citron is firm and inedible when raw. It is grown largely for preserving or for pickling. Preserved citron is used in cakes and in the making of certain kinds of bread. Because the juice of the citron is rich in pectin, it is much used in making

jellies, especially with fruits naturally lacking in pectin and hence not capable of "jelling." Another kind of citron is made from the fruit of **Citrus media.**

Many members of this family are grown as ornamentals or for their curious and sometimes useful fruits. Many such are classed as gourds. One of these, *Lagenaria vulgaris,* a native of the Old World tropics, is known as the calabash gourd or bottle gourd, from the shape of its fruit. Excellent flasks are made from the woody pericarp.

Luffa cylindrica, the "bath sponge," is frequently seen in gardens. The **vascular** tissues of the **pericarp** form an intricate net which is sometimes used as a sponge. Many species of *Luffa* have edible fruits. Mostly they are natives of the Old World. Another gourd, a native of tropical America, is *Sechium edule,* grown for its edible fruit. Many other species of gourds have curiously ornamental fruits.

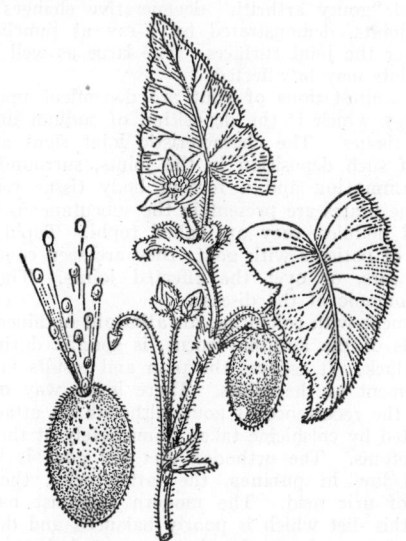

Squirting cucumber. Pressure develops inside the fruit, and as it is detached a hole is torn through which the seeds are violently discharged with the juice. (*Mottier's Textbook of Botany, P. Blakiston's Son & Co. After Jenkins.*)

Ecballium Elaterium, the squirting cucumber, found in Mediterranean regions, has a fruit which when mature is very turgid. When the fruit is broken, the seeds are forcefully ejected by the contraction of the pericarp. From the fruit is obtained a powerful purgative.

Citrullus Colycynthis is another Cucurbitaceous plant, the fruit of which yields a drug. In this species pulp of the fruit gives colocynth, used as a general tonic, an insecticide, a fungicide, a cathartic and as a treatment for certain forms of dropsy.

Echinocystis lobata, a native American plant, is frequently grown for ornament, and as a vine to cover unsightly places quickly. Its small white flowers are pleasantly fragrant. The staminate flowers are borne in long-stalked many-flowered inflorescences, while the pistillate are borne singly and are very short-stalked. The tendrils of this plant are especially sensitive to touch and move very rapidly. (R.M.W.)

GOUT. An hereditary constitutional disease characterized by recurrent attacks of acute **arthritis.** 95% of cases occur in males, and the disease usually comes on in middle life. Its cause is unknown, but a fundamental defect in **uric-acid metabolism** is involved. There is some difficulty in the disposal of uric acid. The blood level is high, but excretion by the kidneys is diminished. There is a popular conception that gout is more prevalent in men whose occupations are conducive to overindulgence in food and drink. This is not borne out by available statistics, nor is the idea that dietary and alcoholic excesses precipitate attacks in gouty individuals.

The onset of acute gout is dramatic. A healthy-appearing middle-aged man experiences sudden, severe pain usually in the metatarsal-phalangeal joint of the great toe. Within a few hours the affected joint is red, swollen and exquisitely tender. The pain is often excruciating, especially at night. Other joints—particularly in the foot—may be involved. During the acute attack, the blood uric-acid content is high. The attack subsides spontaneously usually within 10 days, with complete restoration of normal joint function. After the acute episode is over, the patient remains well until the disease recurs. The recurrences may be scattered over a period of years, and then cease altogether. In some instances the attacks recur frequently and persist for longer and longer periods. In these cases of chronic gout and "gouty arthritis" degenerative changes occur in the joints, demonstrated by x-ray as punched-out areas near the joint surfaces. The large as well as the small joints may be affected.

The manifestations of gout are dependent upon the **pathology** which is the deposition of sodium urate in various tissues. The characteristic joint signs are the result of such deposition in the joints, surrounded by an **inflammation** and a foreign body tissue reaction. When the urates are present in the subcutaneous tissues as small nodules, they are called tophi. Tophi occur in 50% of patients with gout; they are seen commonly on the ears, or over the affected joints. They are pathognomonic of the disease.

Treatment with colchicine, an **alkaloid** obtained from the seeds of the autumn crocus, is specific during the acute attack. It relieves the pain and results in rapid improvement of the joints. There is no way of preventing the recurrence of gout, although an attack can be aborted by colchicine taken immediately at the onset of symptoms. The orthodox diet for gout is low in fat and low in **purines**, the latter being the chief source of uric acid. The modern therapist has discarded this diet which is poorly balanced and deficient in the necessary foods; but it is suggested that patients avoid certain foods such as anchovies, sardines, sweetbreads, liver, kidney and brains, which are exceedingly high in purine content. (D.M.H.)

GOVERNOR. Ordinary governing consists of varying the power of a prime mover in accordance with the demands made upon it by the power user. By governor is generally understood the mechanical governor, used to effect a change of throttle position, spark advance, etc. The **centrifugal** force of rotating masses is the most common principle underlying governors. Fluid pressure produced by fans or centrifugal pumps, the rotors of which revolve with the prime mover shaft, have also been used, but greater dependence is laid upon the principle of rotating masses. A governor of this type is diagrammed in Fig. 1. The rotating fork R carries in its ends the pivots upon which are mounted the weights MM. By means of the bell-crank linkage and the sliding yoke S, the weight W tends to keep the flyballs in equilibrium in the position shown under the influence of the mass of the weight, transmitted through the connecting links, and the centrifugal force acting at the center of the gravity of the governor weight. Considering that the rotative speed of the governor is ω, the magnitude of this centrifugal force in weight units is:

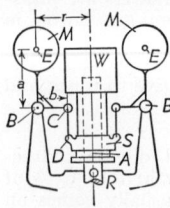

Fig. 1.
Fly-ball governor.

$$\frac{Wr\omega^2}{g}.$$

Heavier, slower speed governors are constructed with the heavier masses arranged as shown in Fig. 2, this type being known as the pendulum governor. In general, the higher the rotative speed of the prime mover being governed, the lighter will the governor weights be.

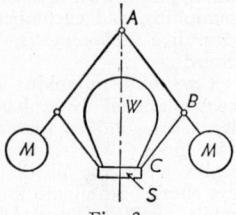

Fig. 2.
Pendulum governor.

Governor action on **internal combustion engines** for the purpose of regulating the speed to some normal value is undertaken in a number of ways. For example, the governor may hold the exhaust valve open, preventing any power stroke from being accomplished, or it may interrupt the ignition circuit, causing explosions to be missed by failure to ignite. Or the passage supplying the inflammable fuel air mixture may be throttled mechanically by a valve to vary the amount of fuel with which the cylinder is charged. In **Diesel engines,** governing is accomplished by varying the amount of oil pumped into a cylinder during that portion of the power stroke given over to the combustion of a fuel. This is done by a governor which either regulates the stroke of the fuel oil pump, or operates a variable by-pass in the discharge from the fuel oil pump. The steam engine may be governed with mass governors of two types, depending on the particular system of governing employed. The smaller, cheaper engines are governed by altering the throttle pressure through imposing an artificial pressure drop created by a governor valve between the boiler and the cylinder. This is known as a throttling governor, and is simplest, but least efficient. Governing of a **steam engine** consists of varying the amount of steam admitted to the cylinder per stroke. As just indicated, this can be done by throttling the steam pressure entering, but it can be accomplished, still allowing full pressure steam to enter, by varying the portion of the stroke during which the admission valve remains open. This is governing by changing the point of cut-off, and is known as cut-off governing. It is more efficient, but more complicated than the throttled governing. A throttling governor may be simply a flyball type, as already described, arranged so that the motion of the sleeve at a can be transmitted through a forked lever to the stem of the governor valve in the steam line. The automatic cut-off is usually built into the flywheel. It is a weight on the end of an arm which is pivoted to one of the flywheel spokes. Any motion of this arm in the direction of the centrifugal pull is opposed by a leaf or coil spring built into the governor. A mechanical link connects the governor with the eccentric, which drives the steam valve. Motion of the governor under varying centrifugal forces set up by variable power alters the position of the eccentric driving the valves, and thus alters the timing of the event of admission of steam to the cylinder. It can do this, for the various events of the steam engine cycle are under the control of the slide valve.

Governing of **steam turbines** is accomplished by three methods, viz.: (1) throttling at inlet, (2) varying number of inlet nozzles in action, (3) varying duration of full pressure puffs (blasts), of which there are several per second. In addition, some turbines are provided with hand-operated by-pass valves which, by admitting high-pressure steam to low-pressure stages, enable the turbine to carry more overload, though, of course, at reduced economy.

The basic actuation of governors will usually be found to employ a change brought about by change of centrifugal force during change of speed, the latter brought about by increasing or decreasing load. Usually the centrifugal force is opposed, mechanically, by governor springs which give the turbine a drooping speed characteristic. A single turbo-alternator not parallel with any other, and not expected to be paralleled with another, may have this curve as flat as is consistent with governor hunting—a fault induced by an oversensitive governor. The unit would operate back and forth over its range of load from $P = O$ to $P =$ full load with a speed regulation of $(s_1 - s_2)/s_2$. The speed regulation should, in no case, exceed 4%. But if the alternators are to be paralleled, their characteristics should have considerable droop, as too flat a characteristic would exaggerate the effect of slight variations in the slopes and make apportionment of the load a difficult matter. Shifting of tension on governor springs (usually done by remote control) has the effect of shifting the speed-load characteristic nearly parallel to itself.

To permit speed adjustment, the speed of a turbine at no load should be adjustable within 5% above or below the rated no-load speed. Within this range the emergency governor should be inoperative. Furthermore, the governing characteristic of a turbine should be such as to hold the speed rise upon instantaneous drop of full load within the operating limit of the emergency overspeed trip.

The gates of the **hydraulic turbine** require an operating force for their movement that is far greater than in the steam turbine, so the mechanism of a hydraulic turbine governor does not bear any resemblance to the delicate parts of the steam-driven machine. Indeed, the force is so large that the hydraulic governors are rated on the basis of the foot-pounds of energy they will produce in completing an operating stroke. A 30,000-foot-pound capacity is a small governor. Large ones may develop as much as 200,000 foot-pounds. Naturally, the force developed by sensitive flyballs operated by centrifugal force can be used only to control a relay which, in turn, will control the actual gate-shifting mechanism. An oil pressure governor consists of a governor proper and an oil-supply system. An oil pump, a pressure storage tank, a sump into which released oil can be drained and into which the suction of the pump dips, and suitable strainers to keep the oil clean, are the principal elements of the oil-supply system. The governor consists of the flyballs, the control valve, the servomotor, and such accessories as overspeed trips, and load-limit blocks.

The operation of a governor can be understood by reference to Fig. 3, which is a simplified diagram of the governor mechanism. The servomotor piston rod is the operating rod of the turbine gate-shifting ring. Oil is admitted to and drained from the servomotor by the control valve. This control valve is connected to the governing flyballs, and also indirectly to the servomotor rod. The operation is as follows: The change in speed of the turbine resulting from change of load is communicated to the governor through the drive belt. The flyballs seek a new position of equilibrium and move the control valve from closed position. Oil pressure is then admitted to the servomotor piston and it completes the motion required to adjust the gates to the new operating position. While completing this motion, the servomotor, through the medium of the bell-crank and connected rod, closes the control valve. Some large governors have a pilot valve to control the main control valve. Quick governor action is secured through using as large control valve lifts and oil passages as possible, and using heavy oil pressure. The natural tendency is then to cause the servomotor to override its correct position and leave the gates too far open or closed; consequently, there must be another movement of the whole system in search of the equilibrium point. Resonant conditions may be set up and a steady speed impossible to maintain. Stability of the system is unsatisfactory if the governors are too sensitive, yet there is need for close regulation so that the frequency will not suffer. Ordinary governors will complete a closing stroke, upon sudden release from full load, in from 2 sec. to 5 sec. (F.T.M.)

GRAB BUCKET. A grab bucket is an apparatus which is able to pick up a load of a bulk material by "biting" into the surface of the material. The particular usefulness of the grab bucket is that it may be lowered from the end of a **boom** onto the surface of the material to be moved, where it is operated to bite into this material, picking up a load, which can then be raised and deposited where wanted. The figure shows a grab bucket in open and closed positions. The pro-

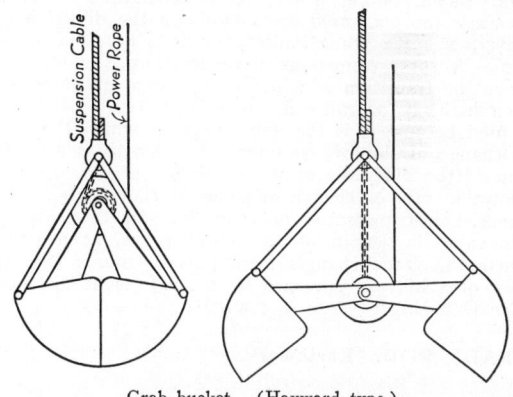

Grab bucket. (Hayward type.)

cedure by which this bucket is caused to close upon a load of material is based on the differential action of a two-step drum. A rope, to which power may be applied by winding it around a drum, passes to the bucket and has its end wrapped around the larger drum of the bucket. After power is applied to this rope, it unwraps from the drum on the bucket, turning that drum and wrapping the chain on the smaller diameter portion. The added leverage thus attained is sufficient to enable the drum to wind itself up on the chain, thus closing the bucket. The axis of the drum is pivoted to the center arms of the bucket. The digging power of a grab bucket of this kind is a function of the weight of the bucket, the sharpness of its cutting edges, the power applied to the operating rope, and the resistance of the material which it digs. Buckets of this type are not suitable for hard-packed material such as earth in origi-

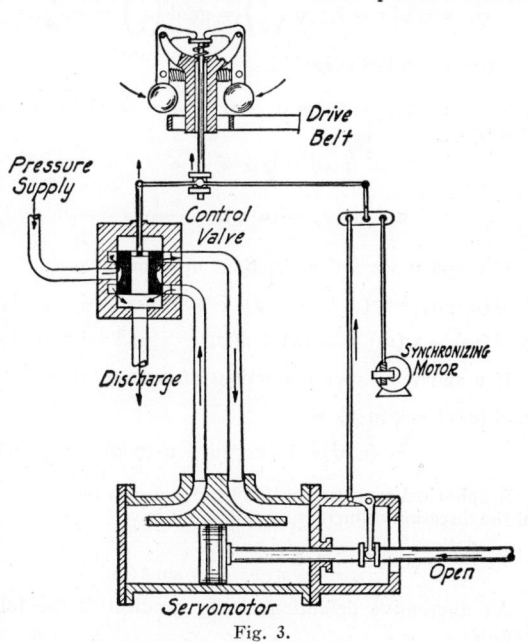

Fig. 3.

nal embankment, but are suited to handling grain, coal, ore, etc. They are built in capacities varying from ½ to 5 cu. yds. (F.T.M.)

GRABEN. Fault.

GRACKLE.
Aves, Passeriformes. In North America, several species of birds (**Aves**) with black plumage and iridescent metallic luster, related to the orioles and blackbirds. The great-tailed grackle, *Cassidix mexicanus*, which ranges from Texas into South America, is also called the jackdaw. It should not be confused with the European jackdaw. In India the hill mynas and related species are called grackles. (A.W.L.)

GRADE.
In geology, the term refers to the slope of the bed of a stream such that the water has just velocity enough to carry its load without either **erosion** or deposition. A stream valley is said to have become graded when its longitudinal profile is a smooth curve without waterfalls or rapids. The term grade is also used by students of **sedimentary rocks**, in a textural sense, to designate those grains of any sediment or sedimentary rock which are of the same size. The classification of grade-sizes is as follows:

NAME OF GRADE		RANGE OF DIAMETERS
Pebbles		Greater than 10 mm.
Gravel		10 mm. to 2 mm.
Sand	Very Coarse	2 mm. to 1 mm.
	Coarse	1 mm. to 0.5 mm.
	Medium	0.5 mm. to 0.25 mm.
	Fine	0.25 mm. to 0.1 mm.
Silt		0.1 mm. to 0.01 mm.
Clay		Less than 0.01 mm.

In highway, railway or municipal engineering the **slope of a line** is called the grade. Grades are usually expressed as percentages preceded by a plus or minus sign. As an example, a $+2\%$ grade indicates a rise of 2′ in every 100′ measured horizontally in the direction of travel; a -2% grade indicates a drop of 2′ in every 100′. A curve known as a vertical curve is used to make the transition at a point of change in the grade of a highway or railroad. A second-degree **parabola** is used because it is the only curve in which the rate of change of slope is constant. The length is a function of the difference of the connected grades and the allowable rate of change of slope of the parabola per hundred feet measured horizontally. In the case of highways the length of a vertical curve, at a point where the grade changes from plus to minus (at the crest of a hill), is governed by the safe sight distance. (See **Grinding**.) (R.M.F., C.W.C.)

GRADE ROD. Earthwork.

GRADED BEDDING.
A geological term denoting a type of bedding or stratification characterized by a cyclic or rhythmic deposition of coarse to fine sediments. A helpful criterion for determining the original position of the strata after they have been deformed. Graded bedding is generally supposed to be characteristic of off-shore rather than in-shore deposition. (R.M.F.)

GRADIENT.
In any field of atmospheric property, temperature for instance, lines of equal value can be drawn. Thus it is possible to draw isolines such as isotherms (lines of equal temperature). The gradient of the property is the maximum rate of change of the property with distance or space and is measured from the highest numerical values to the lowest.

In geology, this term is applied to streams to refer to the slope of their beds, as steep, gentle, or in terms of so many ft. per mile. (R.M.F., P.E.K.)

GRADIENT OF A SCALAR FUNCTION.
The gradient of a scalar function is a type of mathematical expression which occurs frequently in mathematical physics.

Let $\phi(x, y, z)$ be a **scalar function** of position. Form the **partial derivatives** $\frac{\partial \phi}{\partial x}$, $\frac{\partial \phi}{\partial y}$, $\frac{\partial \phi}{\partial z}$; they may be regarded as the rectangular components of a **vector**. This vector is defined as the gradient of ϕ, and is denoted by grad ϕ.

In terms of the unit vectors $\hat{\mathbf{i}}, \hat{\mathbf{j}}, \hat{\mathbf{k}}$, the gradient may be represented by

$$\operatorname{grad} \phi = \hat{\mathbf{i}} \frac{\partial \phi}{\partial x} + \hat{\mathbf{j}} \frac{\partial \phi}{\partial y} + \hat{\mathbf{k}} \frac{\partial \phi}{\partial z}.$$

This may be written symbolically:

$$\operatorname{grad} \phi = \left(\hat{\mathbf{i}} \frac{\partial}{\partial x} + \hat{\mathbf{j}} \frac{\partial}{\partial y} + \hat{\mathbf{k}} \frac{\partial}{\partial z} \right) \phi.$$

If we consider the expression $\hat{\mathbf{i}} \frac{\partial}{\partial x} + \hat{\mathbf{j}} \frac{\partial}{\partial y} + \hat{\mathbf{k}} \frac{\partial}{\partial z}$ as a symbolic vector, denoted by ∇, we may write

$$\operatorname{grad} \phi = \nabla \phi.$$

This symbolic vector ∇ is often called "del," (sometimes "nabla").

Grad ϕ represents both in magnitude and direction the greatest space rate of change of the function ϕ; or in other words, it has the direction and magnitude of the maximum **directional derivative** of ϕ.

The vector grad ϕ is normal to the ("level" or "equipotential") **surfaces** $\phi(x, y, z) = c$, where c is any constant.

Let $\hat{\mathbf{r}}$ be a unit vector with **direction cosines** l, m, n, so that $\hat{\mathbf{r}} = l\hat{\mathbf{i}} + m\hat{\mathbf{j}} + n\hat{\mathbf{k}}$. The component of grad$\phi$ (or $\nabla \phi$) in the direction of $\hat{\mathbf{r}}$ is

$$\hat{\mathbf{r}} \cdot \nabla \phi = l \frac{\partial \phi}{\partial x} + m \frac{\partial \phi}{\partial y} + n \frac{\partial \phi}{\partial z},$$

a scalar, which is the directional derivative of ϕ in the direction of $\hat{\mathbf{r}}$.

If $\mathbf{r} = x\hat{\mathbf{i}} + y\hat{\mathbf{j}} + z\hat{\mathbf{k}}$ and $\nabla = \hat{\mathbf{i}} \frac{\partial}{\partial x} + \hat{\mathbf{j}} \frac{\partial}{\partial y} + \hat{\mathbf{k}} \frac{\partial}{\partial z}$, we have

$$\nabla r = \operatorname{grad} r = \hat{\mathbf{r}}, \quad \nabla \left(\frac{1}{r} \right) = \operatorname{grad} \left(\frac{1}{r} \right) = -\frac{\hat{\mathbf{r}}}{r^2}.$$

$$\nabla r^n = \operatorname{grad} r^n = nr^{n-1}\hat{\mathbf{r}}.$$

If $\mathbf{r}$ is a variable vector, $\mathbf{r} = x\hat{\mathbf{i}} + y\hat{\mathbf{j}} + z\hat{\mathbf{k}}$, and if $\mathbf{a}$ is a constant vector, then

$$\operatorname{grad} (\mathbf{r} \cdot \mathbf{a}) = \nabla (\mathbf{r} \cdot \mathbf{a}) = \mathbf{a},$$

$$\mathbf{a} \cdot (\operatorname{grad} r) = (\mathbf{a} \cdot \nabla) r = \frac{\mathbf{a} \cdot \mathbf{r}}{r}.$$

If u and v are scalar functions of position, then

$$\operatorname{grad} (u + v) = \nabla(u + v) = \nabla u + \nabla v = \operatorname{grad} u + u \operatorname{grad} v,$$

$$\operatorname{grad} (uv) = \nabla(uv) = v(\nabla u) + u(\nabla v) = v \operatorname{grad} u + u \operatorname{grad} v.$$

If $\mathbf{u}$ and $\mathbf{v}$ are vector functions of position, then

$$\operatorname{grad} (\mathbf{u} \cdot \mathbf{v}) = \nabla(\mathbf{u} \cdot \mathbf{v}) =$$

$$(\mathbf{u} \cdot \nabla)\mathbf{v} + (\mathbf{v} \cdot \nabla)\mathbf{u} + \mathbf{u} \times (\nabla \times \mathbf{v}) + \mathbf{v} \times (\nabla \times \mathbf{u}).$$

In **spherical coordinates** if $\hat{\mathbf{r}}$, $\hat{\boldsymbol{\theta}}$ and $\hat{\boldsymbol{\phi}}$ denote unit vectors in the direction of increasing r, θ, ϕ, we have

$$\nabla = \hat{\mathbf{r}} \frac{\partial}{\partial r} + \hat{\boldsymbol{\theta}} \frac{1}{r} \frac{\partial}{\partial \theta} + \hat{\boldsymbol{\phi}} \frac{1}{r \sin \theta} \frac{\partial}{\partial \phi}.$$

An alternative definition of the gradient is the following:

Let ϕ be a scalar function of position, let δ be a small region of space and also its volume, surrounding a point P, and let ω be the bounding closed surface of δ, and let $d\sigma$ be an element on ω; let $\hat{n}$ be a unit normal to ω (outward drawn) at any point of ω. Then the gradient of ϕ at the point P may be defined by

$$\operatorname{grad} \phi = \lim_{\delta \to 0} \frac{1}{\delta} \int_\omega \hat{n}\phi d\sigma.$$

(L.L.S.)

GRADIENT WIND. The wind which blows when the pressure gradient, **Coriolis Force,** and centrifugal force are balanced, and when there is no acceleration. In clockwise systems in the northern hemisphere Coriolis Force balances both centrifugal force and the pressure gradient; in the counterclockwise system, the pressure gradient balances both centrifugal force and Coriolis Force. Gradient wind velocity depends on the latitude, the pressure gradient and the radius of curvature of the isobars. (P.E.K.)

GRAFTING AND BUDDING. Grafting is the process of inserting a part of one plant into another in such manner that the two unite and the inserted piece continues to grow. The part which is inserted is called the scion, the plant into which it is inserted is the stock. Budding is a similar process in which the part inserted consists of a bud with some of the bark adjoining it.

This process is possible because of the **cambium** cells. The successful union of the two pieces is caused by the formation of callus tissue by the cambium cells. Callus tissue is composed of a mass of **parenchyma** cells which fill in or grow over wounds, thus repairing the injury. In graft unions, the cells of the **callus** tissue soon begin maturing into cells of various types, as **xylem** and **phloem** cells, while others become typical cambium cells joining the cambium layer of stock and scion. In grafting, the cambium layers of the two parts are to be brought as closely together as is possible.

There are several methods of grafting. A very common method is known as cleft grafting. In this method a small twig having several buds is removed from the plant which is selected as desirable. The lower end of this twig is cut wedge-shaped. A branch of the plant used as stock is cut off, and a vertical cut made in the end. Into this cut the prepared scion is inserted in such position that its cambium layer and that of the stock come together. To prevent drying of the tissues the entire cut surface is covered with a prepared wax. Usually several scions are inserted in a branch of the stock. When union has taken place and the scion started to grow, all but one may be cut off.

Another method is whip grafting, which is used when the stock is too small for successful cleft grafting. In whip grafting, both stock and scion are cut in a long oblique cut. In the cut surface of each a vertical cut is made. They are then fitted together so that the parts of one slide into and against those of the other, with the cambium of one in contact with that of the other. The two parts are then bound firmly together and the whole covered with wax.

In budding a small bit of bark bearing a bud is removed from the selected plant. Usually, little wood is taken with this. In the stem of the stock a T-shaped cut is made in the bark and the flaps so formed loosened. The prepared bud is inserted under the flaps, which are then pressed down over it and bound tightly in place to insure contact between the two cambium layers. Wax is used here also to prevent loss of water.

In modern horticulture, grafting is a very important practice. Many plants, for instance, do not come true when grown from seed. It becomes necessary, therefore, to propagate such desirable plants vegetatively. This may be done in two ways. One is by means of cuttings, pieces of the plant which are rooted and grown into new plants. The other method is grafting, which is now done on an immense scale. Vegetative propagation must be used also in those plants which do not bear seed, as seedless oranges and seedless grapes.

Commonly the stock used in such cases is not a mature plant but a seedling. This is often chosen for its hardiness or its resistance to diseases and pests. The seedlings are allowed to grow until their roots are well established. The graft is then inserted at the base of the stem. As soon as union has taken place and the scion started to grow, the shoot of the stock is cut off, so that all substances absorbed by the root are sent into the scion. Grafting of this sort is used in producing nursery stock for rubber plantations, as well as nearly all common fruit trees.

Successful grafting can only take place between plants of the same kind or closely related. Others fail entirely to develop any union between the two parts. In nearly all cases the nature of the scion is constant after grafting, so that one can be sure of the product which will result. Because of this it is possible to graft several different scions on a single stock. Not infrequently one sees an apple tree bearing many different kinds of apples maturing at different times of the year. Dwarf apple and pear trees are produced by budding, using quince as stock. Grafting also hastens the time of fruiting, grafted plants coming into bearing earlier than those growing from seed.

Bridge grafting is done for a very different reason. Often trees are completely girdled at the surface of the ground by rodents, especially during the winter months. Damage of this sort is fatal to the tree unless quickly corrected. Correction is done by bridge grafting. This is done by trimming the edges of the girdled region and inserting small twigs across the gap in the bark in such a way that the cambium region of the strips is in contact with that of the tree in which it is inserted. Long sloping ends greatly increase the probability of such contact. These "bridges" unite with the damaged tissues and allow movement of materials to occur. Gradually the damaged and new tissues fill in the gap, and the damage is repaired. (R.M.W.)

GRAIN. Fruit; Wood; Abrasive.

GRAIN SIZE. The grain or crystal size of metals is determined by microscopic examination of a suitably prepared section. (See **Metallography.**) There are two principal standards of grain size in use in this country. Both are standards of the American Society for Testing Materials.

For most non-ferrous alloys, particularly **brass and bronze** and other alloys having homogeneous grain structures with **twin bands,** a set of ten **photomicrographs** having average grain diameters ranging from 0.010 to 0.200 mm. are used for direct comparison with microstructures at a magnification of 75 times.

The A.S.T.M. standard grain size chart for steels covers about the same range of average grain diameters but the comparison is made at 100 times magnification and the grain size is expressed by numbers from 1 to 8. In general, grain sizes 1 to 3 are considered coarse, 4 to 6 intermediate, and 7 to 8 fine. The grain size of steel can also be judged from a clean fracture if the steel can be fractured without appreciable plastic deformation. This is possible with most heat-treated machine steels and tool steels, but low-carbon steels are often too tough to break with a crystalline fracture. A series of standard fractures is available for direct visual comparison, and the numbering system for these standards coincides with that of the charts used for microscopic determination of grain size.

The grain size of metals is related to many important properties. In general, fine grain size is an indication of relatively high strength, **hardness,** and toughness while coarse grain indicates softness and **plasticity.** However, the **hardenability of steels** by heat treatment is highest for coarse grain steel. Coarse grain size is usually desirable for **creep** strength at elevated temperatures.

In the case of sheet and strip for drawing or stamping, coarse grain may give a rough surface. On the other hand, metal with too fine a grain size may lack plasticity and crack in the dies, therefore, a compromise must be reached.

The grain size of castings is generally much coarser than that of wrought products such as rod or sheet. In the case of steel castings the original coarse structure may be refined by heat treatment. This is not possible in the case of most non-ferrous alloys because they do not undergo a change in type of crystal structure on heating or cooling.

In the case of hot rolled or forged metals the finishing temperature has an important influence on grain size. A high finish-forging temperature, for example, will permit grain growth by recrystallization. In the case of metals finished by cold-working processes the final annealing temperature establishes the grain size. A high annealing temperature results in coarse grain size. (See **Recrystallization.**) (R.H.H.)

GRAININESS OF THE PHOTOGRAPHIC IMAGE. The term graininess is used to describe the mottled or grainy appearance of the image observed frequently in projected prints. It is due to the non-homogeneity of the silver deposit in the negative which is a direct consequence of the clumping of the silver halide grains and the scatter of light within the emulsion. Graininess is a matter of concern wherever a negative image must be greatly enlarged, as in miniature camera work and certain fields of applied photography.

For a given degree of enlargement, the graininess of the *positive* image depends upon:

1. *The Emulsion.* The graininess of an emulsion, other factors being constant, appears to increase with the average grain size and with the range of grain sizes; i.e., the size-frequency. There are other factors, such as the turbidity of the emulsion and the thickness of the coating, but these appear to be of less importance in producing the larger aggregates of silver which give rise to the degree of inhomogeneity associated with graininess. Manufacturers have succeeded in recent years in greatly reducing the graininess of emulsions and modern fine-grain emulsions represent a notable advance in this respect. Despite the great amount of attention given to fine-grain developers and processing, greater progress has been made by the manufacturers in reducing graininess through the improvement of emulsions. The first and most important step, therefore, in making fine-grain negatives is the use of a modern fine-grain film.

2. *The Density of the Image.* For a given emulsion and conditions of development, the graininess increases with the density up to a maximum which varies from 0.3 to 0.5. The graininess varies, therefore, with the tones, or brightness of the image, being more pronounced in the highlights of the positive and in areas of relatively uniform brightness, such as the sky. Factors which keep negative density to the minimum necessary for good printing quality, i.e., avoidance of overexposure and development to a relatively low **gamma**—assist in maintaining a fine-grain image.

3. *The Degree of Development or Gamma.* The graininess of the negative image, other factors being constant, increases with the degree of development or gamma. The increase in graininess with the gamma is more pronounced with emulsions tending towards graininess, than with fine-grain emulsions and is more marked with negative than with positive materials, such as motion picture positive film. For the finest grain, therefore, development should be to the lowest gamma which will produce negatives of good printing quality.

4. *The Composition of the Developer.* The only true fine-grain developing agent is paraphenylenediamine, and possibly orthophenylenediamine. With all other developing agents, the differences are slight if the negative is developed to the same gamma unless a solvent of silver halide is added to the developer. Developers, such as the familiar elon-hydroquinone-borax formula, employing an excess of sodium sulfite as a solvent of silver halide, produce a slight improvment in graininess and are popular where the degree of enlargement is not great. Developers containing more active silver halide solvents, such as potassium thiocyanate, produce images of finer grain but at the expense of emulsion speed. There is no known way of reducing graininess in development which does not necessitate a sacrifice in either emulsion speed, image contrast, or both. While much has been written on the subject of fine-grain developers, the fact remains that the composition of the developer is less important in producing fine-grain images than the emulsion.

5. *Conditions in Projection.* In general, all conditions affecting the contrast of the print have a like effect on its graininess. The use of a direct (non-diffused) light in the enlarger will result in prints of greater contrast and graininess than when a diffused light source is used. Increasing the diffusion of light through the use of ground glass, or opal glass, is effective, however, only up to a certain point. The use of a contrast grade of paper will result in an increase in graininess over the softer grades. Graininess will be reduced also by the use of matt or rough-surfaced paper, by diffusion of the image, or by the use of texture screens. (C.B.N.)

GRAM. A metric unit of mass or of weight, equal to about 1/28.35 of an avoirdupois ounce. Originally the gram was defined as the mass of 1 cc. of pure water at its maximum density (4° C.). But since the actual metric standard of mass is now the **kilogram** at Sèvres, the present gram is one-thousandth of this standard; this exceeds the original value in the approximate ratio 1.000028:1. One unfamiliar with metric weights may find it helpful to remember that a silver dime weighs about 2 grams. The avoirdupois pound is about 453.6 grams. (L.D.W.)

GRAM-MOLECULAR VOLUME. Chemical Composition.

GRAM-CHARLIER SERIES. The Gram-Charlier series consist of two types, Type A and Type B. Like the **Pearson system** of curves it may be derived from the hypergeometric probability function and may be regarded as based on **probability.** It may be used either as an empirical **frequency distribution** to explain actual frequency distributions or it may be used as an approximation to theoretical distributions (**probability functions**).

We define Type A (using 3 terms) by

$$P_t = \phi(t) - \frac{\alpha_{3:t}}{3!}\phi^{(3)}(t) + \frac{\alpha_{4:t} - 3}{4!}\phi^{(4)}(t),$$

where P_t are the ordinates of Type A, and

$$\phi(t) = \frac{e^{\frac{-t^2}{2}}}{\sqrt{2\pi}}, \quad t = \frac{x - \overline{X}}{\sigma_x}, \quad \alpha_{3:t} = \alpha_{3:x},$$

the measure of **skewness** of the function being approximated by Type A,

$$\alpha_{4:t} = \alpha_{4:x},$$

$$\phi^{(3)}(t) = \frac{d^3\phi(t)}{dt^3} = (-t^3 + 3t)\phi(t),$$

$$\phi^{(4)}(t) = \frac{d^4\phi(t)}{dt^4} = (t^4 - 6t^2 + 3)\phi(t).$$

The Type A series may be continued to more terms if desired.

$$P_t = A_0\phi(t) + A_1\phi^{(1)}(t) + A_2\phi^{(2)}(t) + \cdots + A_n\phi^{(n)}(t),$$

$$\phi^{(n)}(t) = \frac{d^n\phi(t)}{dt^n}.$$

In this case the coefficient of A_n is given by

$$A_n = \frac{(-1)^n}{n!} \int_{-\infty}^{\infty} H_n(t)P_t dt,$$

where $H_n(t)$ are the **Hermite polynomials**, remembering by

$$\int_{-\infty}^{\infty} t^n P_t dt = \alpha_n.$$

Tables of $\phi(t)$, and $\phi^{(n)}(t)$, n, 1 to 8 are available. To find the theoretical frequency f_t in the class whose limits are $t = t_1$ and $t = t_2$ in a frequency distribution where $\Sigma f_x = N$ we use

$$f_t = N \int_{t=t_1}^{t=t_2} P_t dt$$

$$= N \left\{ \int_{t=t_1}^{t=t_2} \phi(t)dt - \frac{\alpha_3}{3!} \left[\phi^{(2)}(t_2) - \phi^{(2)}(t_1) \right] \right.$$

$$\left. + \frac{\alpha_4 - 3}{4!} \left[\phi^{(3)}(t_2) - \phi^{(3)}(t_1) \right] \right\}.$$

The Gram-Charlier Type A may be used wherever the Pearson Type IV is indicated or whenever the frequency distribution is not too skewed. The Gram-Charlier Type A is an excellent approximation to the **Bernoulli probability function** when s in the Bernoulli distribution is rather large. The first term of the Gram-Charlier Type A is merely the **normal curve**. The Gram-Charlier distribution has been used by R. A. Fisher to approximate **Fisher's z distribution** when n_1 and n_2 are both more than 24, and in this case it is excellent. It has also been used to approximate **Student's t distribution** and the χ^2 **probability function** when the **degrees of freedom** in these distributions are more than 30.

The Type B series is defined in the case of a **discrete variate** by

$$P_x = c_0\psi(x) + c_1\Delta\psi(x) + c_2\Delta^2\psi(x) + c_3\Delta^3\psi(x) + c_4\Delta^4\psi(x),$$

where $\psi(x) = \dfrac{e^{-m}m^x}{X!}$, $m = \overline{X}$, the mean,

$$\Delta\psi(x) = \psi(x) - \psi(x-1) \text{ for } x = 0, 1, 2, \cdots \infty, \psi(-1) \equiv 0.$$

Then $c_0 = 1$, $c_1 = 0$, $c_2 = \frac{1}{2}(\mu_2 - m)$, $c_3 = -\dfrac{1}{3!}(\mu_3 - 3\mu_2 + 2m)$, $c_4 = \dfrac{1}{4!}[\mu_4 - 6\mu_3 + \mu_2(11 - 6m) + 3m(m - 2)]$.

It is generally used when the frequency distribution is skewed. It should be noted that the first term of the Type B series is the **Poisson distribution**. (L.A.A.)

GRAMPUS. Mammalia, Odontoceti. A large **dolphin**, *Orcinus orca*, of very vicious habits, also known as the killer. They hunt in groups and attack even large whales. (A.W.L.)

GRANITE. This name is applied to a common and widely occurring group of deep-seated igneous rocks consisting of **orthoclase, plagioclase, quartz, hornblende, biotite, muscovite** and minor accessories such as **magnetite, garnet, zircon** and **apatite**. Rarely a **pyroxene** is present. Ordinary granite always carries a small amount of plagioclase, but when this is absent the rock is then referred to as an alkali-granite. An increasing proportion of plagioclase feldspar causes granite to pass into granodiorite. A rock consisting of equal propor-

tions of orthoclase and plagioclase plus quartz may be considered a quartz **monzonite**. A granite containing both muscovite and biotite micas is called a binary granite.

The word granite comes from the Latin *granum*, a grain, in reference to the grained structure of such a crystalline rock.

Granite occurs as **stock**-like masses and as **batholiths** often associated with mountain ranges and frequently of great extent. Granite has been intruded into the crust of the earth during all geologic periods, except perhaps the most recent; much of it is of pre-Cambrian age. Granite is found extensively in Canada, New England, New York, the Appalachian region (especially the Piedmont), Wisconsin, Minnesota, South Dakota, Missouri, Oklahoma, Texas, the Rocky Mountains in general and the Pacific Coast States. There are innumerable foreign localities. (E.S.C.S.)

GRANITOID. A textural term derived from **granite** and signifying the relatively uniform and coarse-grain of **batholithic** rocks, such as granite, **syenite, anorthosite**, etc. In a typical granitoid rock each species of mineral occurs as a single generation; the **silicates** crystallizing first, and any surplus of free silica crystallizes last in the form of **quartz**, or is finally driven off with the surplus water to form quartz veins. (R.M.F.)

GRANODIORITE. Granite.

GRANULE. Cell.

GRANULITE (LEPTITE). This is a general term for a group of rocks that vary considerably in composition but for the most part seem to be derived by **metamorphic** processes from **quartz-feldspar** rocks. The classic locality for granulite is in Saxony, where there occurs a granular **gneiss** of quartz and feldspar plus such accessory minerals as **pyroxene** and **garnet**, with occasionally small quantities of **kyanite, spinel** and similar minerals. The Saxon granulites have a decided banded structure and seem to resemble **injection gneisses**. It appears reasonable to suppose that these and other granulites may have been derived from sedimentary formations severely altered by igneous processes. **Leptite** is a term used in the Scandinavian countries for fine-grained granulites that originally were rhyolitic tuffs and lavas.

Besides the Saxon and Scandinavian granulites these rocks are found in the northern highlands of Scotland, India, West Africa, and Canada. (E.S.C.S.)

GRAPE. *Vitis* sp. Vitaceae. Grapes are climbing plants, many of which have long been cultivated by man for their fruits and the various products obtained therefrom.

Climbing in grapes is made possible by tendrils, modified stems which coil tightly around any suitable support. These tendrils are usually interpreted as terminal portions of the stem which have been pushed to one side by the more rapid growth of an axillary (see **Axil**) bud. The leaves of grapes are simple, palmately lobed and alternate, with small stipules. The stems elongate rapidly and are of a coarse porous nature; the internodes of young stems are frequently hollow, the **nodes** solid. The flowers are borne in compact **panicles**. Each flower is small and inconspicuous. The **calyx** is a mere rim around the tip of the **pedicel**; the **corolla** five-parted and greenish. When the flower opens, the petals, united at their tips but free at the base, are forced away from the base of the flower and drop off. There are five stamens and a single pistil. The fruit is a 2-celled berry.

Commercial grapes are largely derived from three species, *Vitis vinifera*, the wine grape of Europe, a native of Asia, *Vitis Labrusca*, the northern fox grape of eastern North America, and *Vitis rotundifolia*, the southern fox grape. Many varieties and hybrids of these exist, as well as hybrids with other wild species. In commercial vineyards, grapevines are variously pruned to increase yield and improve quality. Pruning cuts are made through the nodes, to prevent the leaving of hollow internodes in which disease might gain entrance to the plant. Propagation of the grape is mainly by means of stem cuttings, a method which has been used in Europe for centuries.

Grapes are grown mainly for raisins and for wine. Both of these products come from the fruit of *Vitis vinifera*, which is much sweeter than other cultivated grapes. For the best raisins, selected ripe fruit is carefully sun-dried, after which the stems are removed, and the product packed. More commonly, grapes for raisin-making are first dipped in weak lye, then rinsed and dried either in the sun or, if necessary, artificially. After drying, the stems and seeds are removed and the product packed, entirely by machinery.

For wine-making the grapes are crushed and the juice allowed to ferment with yeast. If a dry wine is desired, fermentation is continued until all the sugar of the grape is changed to alcohol. Using both the skins and pulp of grapes with a colored flesh gives a red wine. Using colorless grapes, or those with colorless pulp and removing the skins, gives white wine. In sweet wine-making, fermentation is stopped before all the sugar is changed to alcohol, by adding alcohol, a procedure called fortifying the wine. Distilling wine produces brandy.

In addition to these two main uses, many grapes are eaten fresh. More are crushed to prepare fresh grape-juice, which is bottled without fermenting. Grapes are sometimes grown merely for ornamental purposes. (R.M.W.)

GRAPEFRUIT. Citrus Fruits.

GRAPE-LEAF FOLDER. Insecta, Lepidoptera. A moth, *Desmia funeralis,* whose larva eats the leaves of grape vines and lives in a fold fastened with silk. It is not an important pest. Arsenical sprays used for other insects destroy it. (A.W.L.)

GRAPE-LEAF SKELETONIZER. Insecta, Lepidoptera. A moth, *Harrisina americana,* whose larvae, working in groups, destroy the soft tissues of the grape leaf, leaving the network of veins. Rarely an important pest, and easily destroyed by arsenical sprays. (A.W.L.)

GRAPH OF A FUNCTION. The graph of a function $y = f(x)$ is the locus of the equation $y = f(x)$. (L.L.S.)

GRAPHICAL STATICS. The equilibrium of forces is often treated graphically in such practical problems

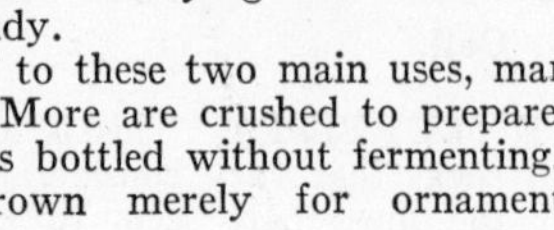

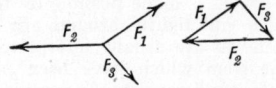

Fig. 1. Three forces in equilibrium.

as the stresses in the members of a framed structure. If three concurrent forces are in equilibrium, the three vectors drawn to a common scale to represent them may be made to form a closed triangle (Fig. 1); or if more than three, a closed polygon (Fig. 2). The principle may be extended and is much used in the calculation of the forces in a truss by means of the so-called stress diagram. A simple example is shown in Fig. 3, which represents a small roof-truss with equal loads resting on it at the joints $A, B, C, D, E,$ and supported by the upward reac-

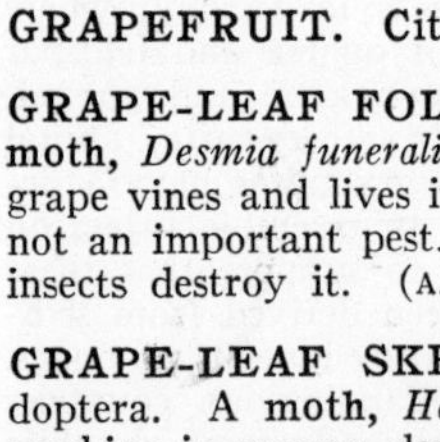

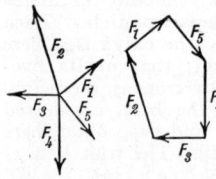

Fig. 2. Five forces in equilibrium.

tions of the walls at A and E. The several compartments of the figure are numbered, and both the external forces and the forces acting along the members

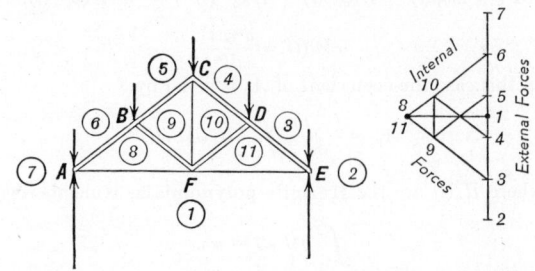

Fig. 3. Elevation of truss with corresponding stress diagram.

between these compartments are represented, both in magnitude and direction, by the lines joining the corresponding numbers in the stress diagram. For example, the compressive force in the strut BF is represented by the line 8–9, while the tension in the vertical rod CF is given by 9–10. The closed figure 5–4–10–9–5 in the stress diagram indicates the equilibrium of the forces acting at the joint C. This method of analysis is attributed to Maxwell (see **Statics** and **Bow's Notation.**)

In the graphical solution of some types of trusses it is found on reaching a particular joint that the arrangement of members is such that there are more than two unknowns. It is then necessary to replace the unknowns by a substitute system which reduces the number of unknowns at the joint to two. The substitute system consists of a single member inserted in such a way that the truss remains stable and determinate (see **Determinate Structure**). This member is called a substitute, fictitious or phantom member. When the solution reaches a joint where the stress in the members is unaffected by the substitution, the substitute arrangement is replaced by the original arrangement. The graphic procedure is reversed in direction to find the stress in the original members. (L.D.W., C.W.C.)

GRAPHIC GRANITE. A coarsely crystalline variety of **granite** or **pegmatite** composed almost entirely of **quartz** and **feldspar** which have intergrown in such a manner as to simulate semitic or cuneiform characters. (R.M.F.)

GRAPHIC REPRESENTATION OF EQUATIONS. An equation in two variables determines a **function** of one variable, and the **graph** of this function gives at the same time the graphic representation of the equation, which will be a **curve.**

An equation in three variables determines a function of two variables, which is represented by a **surface** in space. (L.L.S.)

GRAPHIC REPRESENTATION OF FUNCTIONS. A function of one **variable** $y = f(x)$ may be represented graphically by plotting as points in a plane pairs of corresponding values of the variable and the function as **rectangular coordinates** or **polar coordinates,** and drawing a smooth curve through these points. This graph of the function $y = f(x)$ is the same as the **locus** of the equation $y = f(x)$. (L.L.S.)

GRAPHIC SOLUTION OF EQUATIONS. The real **roots** of an **equation** with one unknown are represented graphically by the **abscissas** of the points where the **graph** of the equation or corresponding **function** cuts or touches the X-axis.

Sometimes the roots of an equation with one unknown can be found graphically by arranging the equation as an equality of two properly selected functions, plotting the graphs of these functions on the same diagram and finding their intersection points. This method can be used to great advantage frequently in **quadratic** and **cubic equations** and certain **transcendental equations.** (L.L.S.)

GRAPHITE (PLUMBAGO). Graphite is an allotropic form of **carbon.** It is produced artificially by heating **coal** or, more commonly, **coke** in the **electric** furnace. Graphite is also found in nature as a mineral. It crystallizes in the **hexagonal** system, often in the form of scales or plates, or in large foliated masses. It has a perfect **basal cleavage;** is soft; hardness 1–2, and feels greasy to the touch; specific gravity 2–2.2; luster metallic; color black to steel gray; **streak** black; opaque. Graphite is a rather widely distributed mineral and is found in a variety of rocks. It occurs in marbles, gneisses or schists; granites and other igneous rocks often carry graphite. It has been noted in pegmatites. It is likely that graphite has been formed by different processes, by magmatic separation of the graphite as an original constituent or as the result of assimilation of carbonaceous rocks, by pneumatolytic action, or by the metamorphism of sedimentary rocks that contained original carbonaceous matter. Well-known localities are in Siberia, on the Island of Ceylon, which is the chief producing district at present; England, Madagascar, Mexico, and Canada. In the United States it is found in the Adirondack region of New York State, in Massachusetts, Rhode Island, Pennsylvania, Alabama, New Mexico, and Montana. Graphite is important as material for crucibles, lubricants, paints, pencil "leads," etc. Graphite has been called black lead, but there is no connection between lead and graphite. The German mineralogist, A. G. Werner, devised the name graphite from the Greek meaning *to write,* with reference to its use in pencils.

In the metallurgy of steel and iron, graphite is an important constituent of **pig iron, cast iron,** and **malleable cast iron,** and of certain special die steels.

Graphite contributes to the wear-resistance of irons and steels by its lubricating action. When present in cast iron in excessive amounts or in the form of large interlocking flakes or films it greatly reduces the tensile strength. (E.S.C.S., R.H.H.)

GRAPHS. The graph of an equation is the **locus** (or totality) of all points whose **coordinates** satisfy the equation. The graph of a **function** of one variable is the locus of all points whose coordinates are the values of the variable and of the function. (L.L.S.)

GRAPHS OF TRIGONOMETRIC FUNCTIONS. Trigonometric Curves.

GRAPNEL. Broadly speaking, a grapnel is any device used to grapple with an object which is obscured to view, such as a submarine object. Grapnels generally take the form of grapnel hooks, which have several flukes, so that they will be certain to hook into any object with which they may come in contact. (F.T.M.)

GRAPTOLITES. Invertebrate Paleontology.

GRASS FAMILY. Gramineae. Of all plant families the Grass Family is the most important economically. Including many thousand species, the family is one of the largest in the plant kingdom. Members of the grass family were probably the first plants to be cultivated by the human race. Grasses are found everywhere plants can grow, from the coldest polar regions to the tropics, from the coasts to the upper limits of vegetation on mountains.

Most grasses are herbaceous plants of low stature. A few, notably the **Bamboos,** become woody plants of great height, and a small number are of clambering or trailing habit. The cereals, and many other grasses, are annuals, completing their growth in a single growing season; others are perennial plants. Some of the former are winter annuals, plants which start growth in one season, remain dormant over winter, and complete growth and fruit in the following season. Winter wheat is an example.

The root system of a grass plant is made up entirely of fine fibrous roots, which enlarge but little, remaining about the same diameter throughout their length. These roots are mainly **adventitious,** arising from the lowermost nodes of the stem. The roots of many grasses penetrate deep into the ground, so reaching supplies of moisture which enable the plant to live in dry regions where surface moisture is rare.

The stems of grasses, frequently called culms, are cylindrical and in most genera hollow except in the region of the nodes, where solid plugs occur. When young the stem is solid, but as growth continues the central portion fails to keep pace with the outer and gradually becomes hollow. Corn is an exception, the stems being permanently solid in that grass. In most grasses the stem grows erect, but frequently falls over during the growing season, due to climatic disturbances or to lack of suitable nutrient sources to give it strength. Such fallen stems do not remain so but gradually become erect through renewed growth in the nodal regions of the stem. The cause of such a growth is not definitely known. The upward bend, negative geotropism, may be produced by **auxin** which accumulates in the lower half of the node and stimulates overgrowth in that region. In many species of grass the lowermost nodes normally give rise to a number of buds which develop into lateral branches which give the plant a tufted appearance. Such basal branches are known as tillers, stools, and the habit of forming them as tillering or stooling. It is a valuable property of many cereals, and undesirable in others, for example, corn, where it causes a considerable reduction in yield. In a few grasses, the basal portion of the stem becomes enlarged by an accumulation of reserve food material, the plant being known as a bulbous grass. Many grasses develop underground stems known as rhizomes, from the nodes of which erect branch stems may develop, as well as numerous adventitious roots. These rhizomes may be short and the erect branches numerous, producing a tufted grass, or they may be long and wide spreading, as in the case of witch grass, *Agropyron repens,* also called quack or witch grass. Due to the readiness with which the joints of the rhizomes of the latter grass strike root and develop to erect stems, it becomes a pestiferous weed. Eradication by chopping up the rhizome with a hoe only serves to increase its numbers, each joint or node producing a new plant. Only by preventing the green tops from forming can the plant be controlled and eliminated, or of course by complete removal of the entire underground rhizome. In some grasses the stem grows out over the surface of the ground, being then known as a stolon. Rhizomes and stolons form an effective way of propagating the plant, and in many species insure considerable dispersal over a limited area.

The leaves of grasses are composed of two parts, a basal sheath which enwraps the stem and a flat elongate blade. The veins of the leaf are all parallel to one another, with few inconspicuous interconnecting veinlets. The blades of grasses grow from the bases, so that the apical portion is older and the cells of the basal portion retain for some time the ability to divide and increase. Because of this property grasses can be mowed by machines or cropped by animals, the upper portions of the blades being removed and the basal portion growing to renew the blade. Each node bears a single leaf, which is often reduced to a small scale, especially in the lowermost nodes, and in modified stems, such as rhizomes. At the junction of the sheath with the blade there occurs in many grasses a distinct structure called the ligule. This appears on the stem side of the leaf, and is a membranous or cartilagous fringe or ring.

The **inflorescence,** in grasses, is composed of large numbers of groups of flowers, called spikelets, attached to the main stem or rachis. These spikelets are variously arranged. If they grow directly from the main stem and the latter is unbranched, the inflorescence is said to be a spike. If the main stem produces many branches, which in turn branch, the resulting inflorescence is a panicle. The nature of the branches, whether long or short, spreading or appressed, determines the nature of the panicle. In other grasses the inflorescence is a raceme, the spikelets being borne on short unbranched lateral branches.

The individual spikelet of a grass is composed of a short axis called a rachilla from which arise a series of opposite overlapping **bracts.** The two lowermost bracts are called glumes; these are empty, that is, have no

A grass, red top (*Agrostis alba*). 1, panicle of flowers, ½ natural size; 2, single flower, consisting of three stamens and one pistil with two branching feathery styles, all enclosed by scales; ×15.

flowers formed in their axils. The next bract above the glumes is the lemma, in the axil of which is borne a flower. In many grasses each spikelet contains several lemmas, each with its associated flower. Opposite the lemma is the palea, which is not borne on the rachilla, but on a short pedicel, or flower-stalk. Opposite the palea and at the base of the **ovary** appear two minute scales, the lodicules. Three **stamens,** each with a long slender filament and a large **anther,** come next, while a single **pistil** grows at the apex of the pedicel. The pistil is composed of a 1-celled, 1-seeded ovary, two **styles** and two feathery stigmas. Many variations from the typical spikelet described occur in different species, the number of parts being increased, or parts being completely absent. In many species of grass, conspicuous prolongations on the glumes or the lemmas are noted—these are the awns.

Pollination in grasses is almost entirely by wind, the light dry pollen being scattered from the open anthers, often in conspicuous clouds. Grass pollen is a particularly common cause of hay-fever.

The fruit of grasses is 1-seeded, dry and indehiscent, that is, does not split open at maturity to liberate the seed. The ovary wall, or **pericarp,** is attached to the seedcoat. Within the latter is an abundant starchy

endosperm. Such a fruit is known as a **grain** or a karyopsis.

Considerable speculation has been advanced as to the probable origin of grasses, whether they are primitive **monocotyledonous** plants from which others such as lilies may have developed, or whether they are reduced plants. To many the available evidence indicates reduction from lily-like ancestors, a reduction in which two of the three pistil lobes of the ancestral form have been lost, also an entire whorl of stamens, and many of the **perianth** parts. The anatomy of the floral parts lends support to this conception; the vascular bundles suggesting that reduction has occurred. For example, in the pistil there are three vascular bundles, two passing to the styles, and the third bearing the ovule.

The vast importance of grasses to mankind has been previously noted. Grasses supply an important part of the food of the more valuable of man's domestic animals. Many of the forage grasses occur in the wild state, forming extensive ranges on which stock may be pastured. Others are cultivated, and used directly as forage or dried and stored as hay. The search for improved hay grasses has led to the introduction of many valuable species. Notable among them is Timothy, *Phleum pratense,* a grass particularly adapted for cool, moist climates. Another group of forage grasses is that of the millets, lush-growing annual grasses used not only as green forage and cut for hay, but also occasionally grown for their "seeds," used in poultry foods. Millets grow very rapidly and are drought-resistant, and are therefore valuable in regions where the rainfall may be deficient. Another source for bird seed is *Phalaris canariensis,* a grass which is not extensively grown in the United States, but widely in Argentina.

Vast in importance, but few in number, are the grasses grown for their fruit or grain; these are the cereal grasses. These are particularly important in regions of cooler climates, where the grain they produce during the short growing season is a valuable crop which can be stored and used as food during the remainder of the year. The principal cereals are **wheat, corn, oats, rye, barley,** and **rice.** (B.S.M.)

GRASSHOPPER. Insecta, Orthoptera. **Insects** of moderate to large size, usually with four wings, and with the hind legs long and strongly built for jumping.

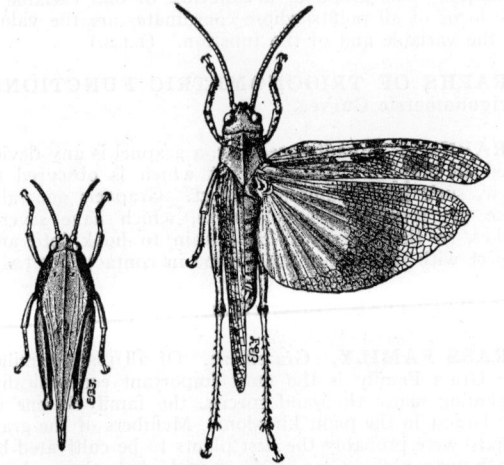

Grasshoppers. The figure at the left is a grouse locust, that at the right a common grasshopper.

The mouth parts are formed for biting and the front wings are thickened tegmina which conceal the hind wings when folded. The long-horned grasshoppers include the **katydids** and related species, which have very long many-jointed antennae. Short-horned grasshoppers have short antennae with a moderate number of joints and are the grasshoppers or **locusts** of common usage. The term locust applies properly to these insects, although it is also used for the cicadas.

The family Tettigoniidae, to which the long-horned grasshoppers belong, contains the katydids, **meadow grasshoppers, cone-headed grasshoppers,** and a number of wingless forms, including the cave or **camel crickets,** the **mormon cricket,** and the **sand cricket,** none of them true crickets.

The family Locustidae includes insects of more uniform appearance. The only members that differ strikingly from the grasshoppers of common experience are the little grouse locusts or **pigmy locusts.** They are less than an inch long and have the dorsal shield of the thorax prolonged over the entire body in an acute point.

In the plains and prairie states grasshoppers are sometimes a serious pest, destroying green crops completely. They are destroyed by the use of poison baits made of bran (25 lbs.), molasses (2 quarts), Paris green or arsenic (1 lb.), and six finely chopped lemons or oranges, mixed with 2–4 gal. of water, according to the prevailing humidity. The mixture should break apart slowly when squeezed into a ball. It is sown broadcast in the fields, preferably late in the day so that it may remain moist as long as possible.

In low crops, such as clover, a hopper trap has been used with great success. It is drawn through the field by horses. The hoppers are shoveled into sacks and when dead are spread out to dry in the sun before they are stored. They are an excellent winter food for poultry, hence this method of control is doubly profitable. (A.W.L.)

GRASSQUIT. Aves, Passeriformes. A Jamaican name for certain of the small birds more commonly called **buntings.** (A.W.L.)

GRATE. Bars which are arranged to support solid **fuel** for combustion are grates. The purpose of a grate is not only to support the fuel bed, but also to allow air to pass into it evenly. Service conditions of grates call for considerable structural strength, resistance to temperature and oxidation, and a shape serrated or rough edged, so as to leave air openings between adjacent grate bars. Furthermore, since ashes are generally dumped through the grate bar system, a means for increasing the spacing between bars so as to pass ash between them must be provided in such a way that motion of a shaking crank or lever will clear ash over the entire grate surface. Because of its cheapness and its resistance to oxidation, cast iron has been the material most used for the construction of grates. (F.T.M.)

GRATE EFFICIENCY. The efficiency of a grate is a measure of the effectiveness with which the grate or its equal prevents combustible from reaching the ashpit. Too frequent slicing of the fire-bed, a poorly maintained fuel bed, and a clinkering type of ash, are sources of combustible in the refuse. Analysis of ashpit contents shows that the combustible is practically all carbon, having a heating value of 14,540 B.T.U. per lb. If the ashpit contents analyze $Z\%$ combustible, the fraction of a pound of carbon per pound of coal thus represented is:

$$\frac{Z}{100 - Z} \times \frac{\text{(Per cent ash in the coal)}}{100}.$$

The grate efficiency is found by subtracting from 100% that per cent of the heating value of the coal represented by the combustible in the refuse at the rate of 14,540 B.T.U. per lb. (F.T.M.)

GRATICULE. The pattern of lines representing parallels of **latitude** and meridians of longitude on a map or chart is known as the graticule of the chart. A person familiar with the various types of **map projection** can usually tell by examination of the graticule the type of projection that was used in constructing the sheet. (W.K.G.)

GRATING. Diffraction Grating.

GRAVES' DISEASE. Thyroid Gland.

GRAVITATION AND GRAVITY. The distinction between these two terms is that between a universal property of matter and the special manifestation of that property exhibited in the vicinity of the **earth** or other celestial attracting mass and modified by the **centrifugal force** of planetary rotation.

Newton's conception of gravitation was expressed by his statement, to the effect that every particle of matter attracts every other particle with a force proportional to the product of the masses and to the inverse square of the distance. We are thus left to picture an infinitely complex network of attractions joining every two particles in the universe and tending to pull them together. Newton did not specify what the "particles" were assumed to be, whether atoms or otherwise. Faraday introduced a somewhat different picture in the form of a stressed medium, with its curved lines and tubes of force. The Einstein concept, again, envisages a space so warped by the presence of surrounding masses that a particle, which if projected into an empty space would follow a straight line, actually follows a more or less complicated curve, a geodesic line of this warped space, which represents physically the most direct path between any two points of the space. Furthermore, **Einstein's equivalence principle** makes no distinction between gravitation and centrifugal force.

The Newtonian law may be expressed by the equation $f = Gm_1m_2/r^2$, in which m_1 and m_2 are the masses of two particles, r the distance between them, and G the **gravitation constant.** For practical purposes the "particles" may be homogeneous spheres, r being the distance between their centers. Other bodies of finite size, such as cubes or cylinders, would not do, as they are not "centrobaric"; that is, there is no one point toward which their attraction is directed. The planets and stars being sensibly spherical, they may be treated approximately as particles. It was from the study of the **two-body problem** as applied to such objects that Newton deduced the conclusion expressed in his law.

At any point on the earth's surface, the earth's gravitational attraction is directed approximately toward its center. But since the earth rotates, the "weight" of a body is somewhat less than the earth's attraction for it, because of the centrifugal force, and is, furthermore, not in general directed toward the earth's center. A plumb line 10' long in the **latitude** of New York departs about ¼″ to the south from a line in the direction of the earth's geometrical center. This same influence accounts for the oblateness of the earth, and these two facts together, for the variation of gravity from equator to poles. Thus at the equator the weight of a gram mass is 977.99 **dynes,** while at the poles it exceeds 983 dynes. These facts are ascertained by observations with such instruments as **Kater's pendulum,** and the Eötvös balance (a kind of **torsion balance** used for gravity measurements), by means of which the acceleration that would be given to a freely falling body is indirectly determined. Various more or less complicated formulas expressing the intensity of gravity as a function of latitude and altitude have been proposed by Clairaut, Helmert, Hayford, and others, but these must always be modified to suit local conditions of topography and density of the crust. For purposes of standardization, the acceleration due to gravity adopted by international agreement is $g = 980.665$ cm./sec.². (L.D.W.)

GRAVITATION CONSTANT. The constant G in the equation expressing **Newton's** law of **gravitation,** which gives the attraction between two particles of masses m_1, m_2 at distance r as

$$f = G\frac{m_1 m_2}{r^2},$$

is called the Newtonian constant or the constant of gravitation. Newton himself was ignorant of its value. Not until 1798 did Cavendish utilize the **torsion balance** in the measurement of this important quantity. The obvious procedure is to place two known masses m_1, m_2 at a known distance r and measure the force of attraction f between them; from which $G = fr^2/m_1 m_2$. The constant might be interpreted as numerically equal to the force between unit masses at unit distance. In the actual experiment, large metal balls weighing several pounds each exert a torque on smaller balls mounted at the ends of the torsion balance beam. This torque, and hence the forces producing it, can be measured. The most recent determinations, made by P. R. Heyl at the Bureau of Standards in 1928, gave the value of G as 6.67×10^{-8} dyne cm.^{2}/g.^{2}. The attraction between two small spheres of one gram mass each with their centers 1 cm. apart would thus be 6.67×10^{-8} dynes. The exceedingly small magnitude of the quantity sought accounts for the difficulties in technique and the relatively low precision attained. (L.D.W.)

GRAVITY. Gravitation and Gravity.

GRAVITY FAULT. Fault; Normal Fault.

GRAVURE. Photomechanical Reproduction Processes.

GRAY SCALE. A term used in color photography to denote a series of achromatic tones ranging from black to white. A gray scale may be divided into three or more steps but 10 is a common number of divisions. A gray scale is generally included with the subject when making a color photograph so that measurements of its densities on the separation negatives or tripack will give the density range of that stage in the reproduction. A gray scale is helpful in controlling the processing stages in the analysis and synthesis of a color photograph. (H.C.C.)

GRAYLING. Pices, Teleostei. *Thymallus.* Fishes (**Pisces**) of the northern hemisphere, found in cold lakes and streams. Related to the salmon and trout. Two species in North America. (A.W.L.)

GRAYWACKE or GRAUWACKE. This term is of British origin and is not used extensively outside of western Europe. As originally defined graywacke designates hard, dark-colored, coarse sandstones and grits having an **argillaceous** matrix or cement and occurring among the lower **Paleozoic** formations of Wales, England. Many typical graywackes are similar to **basic arkoses,** the dark color being due to a preponderance of the **femic** minerals and **plagioclase feldspar.** (R.M.F.)

GREASE. A lubricating agent of higher viscosity than oils, consisting essentially of a calcium or sodium soap jelly emulsified with mineral oil. Greases are employed where heavy pressures exist, where oil drip from the bearings is undesirable, and where the motion of the contacting surfaces is discontinuous so that it is difficult to maintain a separating film in the bearing. Grease-lubricated bearings have greater frictional characteristics at the beginning of operation, causing a temperature rise which tends to melt the grease and give the effect of an oil-lubricated bearing. Calcium and sodium base greases are most commonly used; sodium base greases have higher melting point than calcium base greases but are not resistant to the action of water. Graphite, either by itself or mixed with grease, is also employed as a lubricant. Gear greases consist of rosin oil, thickened with lime and mixed with mineral oil, with some percentage of water. (See **Lubrication; Petroleum Products.**) (H.C.H.)

GREAT-CIRCLE CHART. Among navigators the **gnomonic projection** is commonly known as a great-circle chart, because of the fact that on this type of projection, great circles are projected as straight lines. (W.K.G.)

GREAT-CIRCLE COURSE. The shortest distance between any two points on the surface of a sphere is a great circle. For all practical purposes of navigation the earth may be considered as a sphere, and, hence, the shortest course which a vessel may follow, between any two ports is a great-circle course.

The great-circle course between two ports is frequently impractical for a ship to follow, because of the fact that it may lead across land or into dangerous waters. For example, the great-circle course between two points in the same **latitude,** but separated by 180° of **longitude,** will lead across the pole of the earth. Before deciding whether or not the great circle is practicable it is necessary to compute the course, computing a sufficient number of points so that the track may be plotted on a chart. Such computation is laborious and, to avoid the necessity of doing the computing, a **great-circle chart** may be used. On such a **chart** any great circle appears as a straight line and all that is necessary for the purpose of studying a great-circle course is to draw a straight line between the two points on the chart and examine it.

Even though the great-circle course does not lead the ship into danger, it is very difficult to follow such a course for it makes a different angle with each successive meridian and would require the helmsman to continually change his course. To avoid this difficulty, as well as to avoid dangers, and still approximate as closely as practicable to the shortest distance between the ports, the **composite course** is the type almost universally followed by vessels. (W.K.G.)

GREBE. Aves, Colybiformes. Swimming birds (**Aves**) with lobed toes, short legs and neck, and a sharp beak, in some species quite long. The grebes are found in temperate regions of both hemispheres and members of the same species may have a very wide range. The pied-billed grebe is also called the **dabchick.** (A.W.L.)

GREENBACK. Pisces, Teleostei. A species of **trout,** *Trutta smaragda,* reported from the headwaters of the Arkansas and South Platte rivers. (A.W.L.)

GREENOCKITE. The mineral greenockite is **cadmium** sulfide and is used as an ore of that metal. It is found rarely in **hexagonal** crystals, sometimes as earthy coatings on other minerals. Its hardness is 3–3.5; specific gravity, 4.9–5.0; luster, adamantine to earthy; color, yellow to yellowish-orange; subtransparent. It is found in Scotland, Bohemia, and France. Also in the United States at Franklin Furnace, New Jersey; and Marion County, Arkansas, where it occurs as a yellow coloring matter in **smithsonite;** and in Mono County, California. It was named for Lord Greenock. (E.S.C.S.)

GREENSAND. Glauconite.

GREENSHANK. Aves, Charadriiformes. A European **sandpiper,** *Tringa nebularia,* related to the **willets** of North America. It migrates into South Africa and Australia. (A.W.L.)

GREEN'S THEOREM IN SPACE. This is one of several mathematical results discovered by Green; it is of very great importance in all parts of mathematical physics. It gives a transformation of a surface integral in space into a triple (volume) integral, or vice versa.

Let $P(x, y, z)$, $Q(x, y, z)$, $R(x, y, z)$ be **continuous** together with their first **partial derivatives**, within and on the closed boundary **surface** S of a region V of space. Then

$$\iint_S \{P(x, y, z)dydz + Q(x, y, z)dzdx + R(x, y, z)dxdy\}$$

$$= \iiint_V \left(\frac{\partial P}{\partial x} + \frac{\partial Q}{\partial y} + \frac{\partial R}{\partial z}\right) dxdydz.$$

If dS denotes a surface element, and if α, β, γ are the **direction angles** of the outer normal to the surface, the **surface integral** above may be written

$$\iint_S (P \cos \alpha + Q \cos \beta + R \cos \gamma)dS.$$

This theorem is sometimes called Green's theorem in space, sometimes the **divergence** theorem, sometimes Gauss' theorem.

In **vector** language and vector notation, this theorem may be stated thus: The volume integral of the divergence of a vector function **F** of position taken over any region V of space is equal to the surface integral of **F** taken over the closed surface S bounding the region V:

$$\iiint_V \nabla \cdot \mathbf{F}dV = \iint_S \mathbf{F} \cdot \hat{n}dS.$$

Other important formulae derived from the preceding, and which are often referred to as Green's theorems also, are the following:

I. $$\iiint_V u \left(\frac{\partial^2 v}{\partial x^2} + \frac{\partial^2 v}{\partial y^2} + \frac{\partial^2 v}{\partial z^2}\right) dV$$

$$+ \iiint_V \left(\frac{\partial u}{\partial x}\frac{\partial v}{\partial x} + \frac{\partial u}{\partial y}\frac{\partial v}{\partial y} + \frac{\partial u}{\partial z}\frac{\partial v}{\partial z}\right) dV$$

$$= -\iint_S u \frac{\partial v}{\partial n} dS,$$

where u and v are functions of x, y, z, and where $\frac{\partial v}{\partial n}$ is the directional derivative of v along the inner normal to S; in **vector** form, this becomes:

$$\iiint_V u\nabla^2 vdV + \iiint_V \nabla u \cdot \nabla vdV = \iint_S u\nabla v \cdot \hat{n}dS.$$

II. $$\iiint_V (u\nabla^2 v - v\nabla^2 u)dV = -\iint_S \left(u \frac{\partial v}{\partial n} - v \frac{\partial u}{\partial n}\right) dS.$$

III. $$\iiint_V \nabla^2 udV = -\iint_S \frac{\partial u}{\partial n} dS.$$

IV. $$\iint_S \frac{\partial u}{\partial n} dS = 0 \quad \text{if } \nabla^2 u = 0,$$

i.e., if u is a solution of **Laplace's equation.**

V. $$\iiint_V \left[\left(\frac{\partial u}{\partial x}\right)^2 + \left(\frac{\partial u}{\partial y}\right)^2 + \left(\frac{\partial u}{\partial z}\right)^2\right] dV$$

$$= -\iint_S u \frac{\partial u}{\partial n} dS,$$

if u is a solution of Laplace's equation: $\nabla^2 u = 0$. (L.L.S.)

GREEN'S THEOREM IN THE PLANE. The English mathematician and physicist Green (1793–1841) discovered a number of mathematical results concerning transformations of various types of **integrals** into other useful types. Green's theorem in the plane, which is one of several so-called Green's theorems, expresses a **line integral** in the plane in terms of a **double integral** in the plane, or vice versa.

Let $P(x, y)$ and $Q(x, y)$ be two functions of x and y, which, together with their first **partial derivatives**, are **continuous** within and on the boundary C of a region S, then

$$\int_C (Pdx + Qdy) = -\iint_S \left(\frac{\partial P}{\partial y} - \frac{\partial Q}{\partial x}\right) dS,$$

where the right-hand side is a double integral of a function of x and y over a plane region S.

An immediate consequence of this theorem is: If P and Q are functions satisfying the preceding conditions, and if $\frac{\partial P}{\partial y} = \frac{\partial Q}{\partial x}$, then $\int_C (Pdx + Qdy) = 0$; and conversely, if $\int_C (Pdx + Qdy) = 0$, then $\frac{\partial P}{\partial y} = \frac{\partial Q}{\partial x}$. (L.L.S.)

GREENSTONE. Greenstone is an old field term for more or less altered **basalts** and **dolerites,** which because of the development of **chlorite** or perhaps **hornblende** or **epidote** develops a characteristic green color. Many **diabases** and **epidiorites** have been called greenstones. (R.M.F.)

GREGARINIDA. An order of 1-celled animals, parasitic in various invertebrates. (See **Sporozoa.**) (A.W.L.)

GREGARIOUSNESS. An association of animals of the same species which may be of benefit to the individual but is not essential. The incidental grouping of animals, as in the swarms of maggots in a dead body, is not an association of this type, but the grouping of caterpillars of certain moths, even though the group originates in a like manner by the deposition of eggs in a mass, must be regarded as a gregarious association because the maintenance of the group is due to the behavior of the individuals. They are free to scatter but do not.

Herds of grazing animals cooperate for the common defense and such animals as the killer whale and the wolves are able to attack large animals by hunting in groups, but in all such cases the individual is able to subsist without the assistance of his fellows. (A.W.L.)

GREISEN. An old German **petrological** term originally proposed by Werner for an **igneous** rock of granitic or aplitic texture composed principally of **quartz,** alkali feldspar, the fluorine-rich **micas,** and sometimes containing **topaz.** Greisens are pneumatolytically altered granites which are closely associated with the development of the tin ore mineral, **cassiterite.** (R.M.F.)

GREYHEN. Aves, Galliformes. The female of the Eurasian black **grouse.** (A.W.L.)

GRIBBLE. Crustacea, Isopoda. A small marine **crustacean,** *Limnoria lignorum,* which bores into submerged timbers. A source of serious damage to docks and piling. (A.W.L.)

GRID. The grid is the control electrode inserted in the path of the electrons in an electron **tube.** Its primary function is to exercise some sort of control on the tube without collecting any more **electrons** itself than necessary. This is distinguished from the operation of the **anode** which, in a triode, is intended to collect all of the electrons. In the usual vacuum tube the control grid is an open mesh or grid-like structure near the **cathode.** Voltages impressed on this grid exert considerable influence upon the passage of electrons across the tube space. The screen grid is an additional open type element inserted between the control grid and the anode to reduce the **interelectrode capacitance** between them. The suppressor grid is inserted between the screen and anode to reduce secondary emission. Some special-purpose vacuum tubes, e.g., the mixer or first detector of a **superheterodyne,** have other grids, each however controlling in some way the electron stream which is eventually collected by the anode. In many gas-filled tubes and in the **cathode ray tube** the grid is a solid cap-like element with a single hole for the passage of the electrons. (L.R.Q.)

GRID-CONTROLLED RECTIFIER. Thyratron.

GRID-GLOW TUBE. Thyratron.

GRID LEAK. This term is frequently applied to the resistance in the grid circuit of a vacuum **tube.** (L.R.Q.)

GRIGNARD REACTION. **Magnesium** metal plus haloid (see **Chlorine, Bromine, Iodine**) hydrocarbon $\left(Mg\!<^R_X\right)$ in anhydrous **ether,** introduced by Grignard, is used as reagent for the preparation of:

1. Secondary alcohols by treatment with **aldehydes** or **formic acid** esters.
2. Tertiary alcohols by treatment with **ketones or esters** (not formic acid esters).
3. Primary **alcohols** by treatment with **formaldehyde.**
4. Alcohols by treatment with **oxygen** gas or **hydrogen peroxide.**
5. **Hydrocarbons** by treatment with water, alcohols, **phenols, amines.** With R X as ethyl bromide (C_2H_5Br) the volume of ethane gas (C_2H_6) evolved may be utilized as a measure of hydroxyl (OH—) or amino (NH₂—) radical. One mol of ethane for each equivalent of either.
6. **Carboxylic acids** (group —COOH) by treatment with **carbon dioxide.**
7. **Sulfinic acids** (group —SOOH) by treatment with **sulfur dioxide.**
8. Organic **phosphines, arsines, mercury** di-, **tin** tetra-, **lead** tetra- compounds.

An intermediate compound, identified in numerous cases, but generally not isolated, is formed, and this is decomposed subsequently by addition of water or acid with the formation of magnesium hydroxyhalide (magnesium halide solution when acid is used) plus the main product. The main product is usually extracted with ether followed by recovery upon evaporation of the ether. "Within recent years no single group of compounds has proved of such value in synthetic chemistry as these [Grignard] reagents." (Bernthsen-Sudborough, 1930.)

Illustrative examples follow:

1. Secondary alcohols

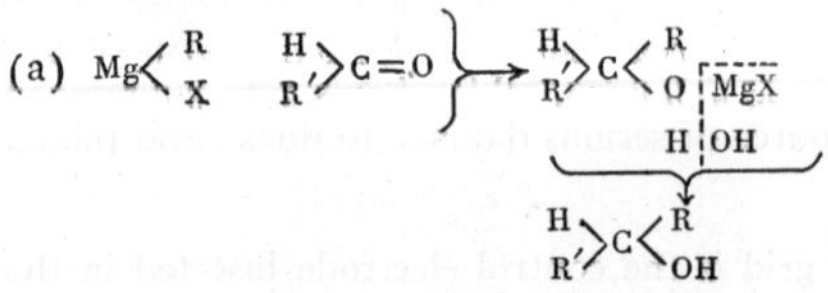

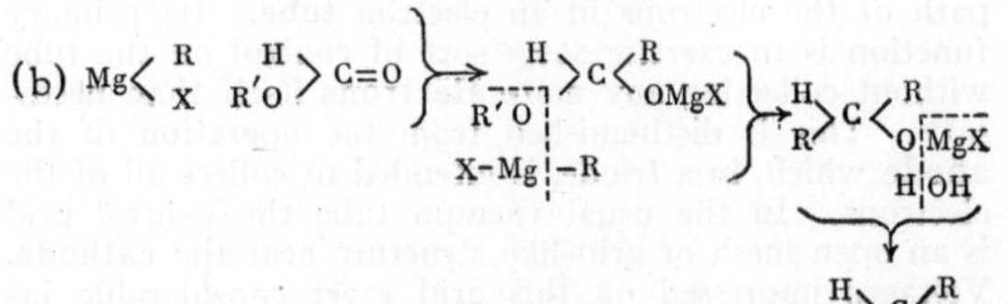

(a)

(b)

2. Tertiary alcohols

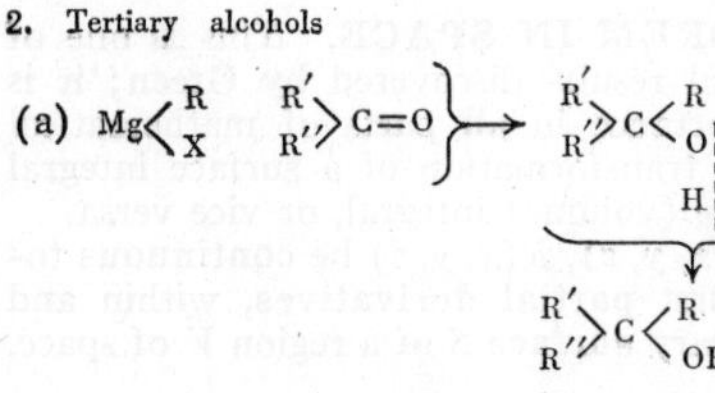

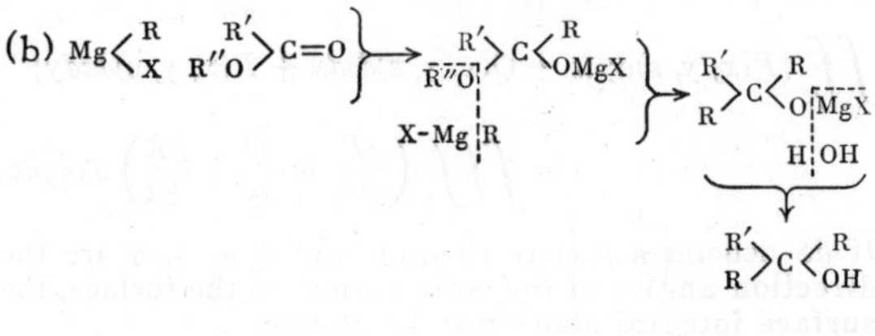

(a)

(b)

3. Primary alcohols

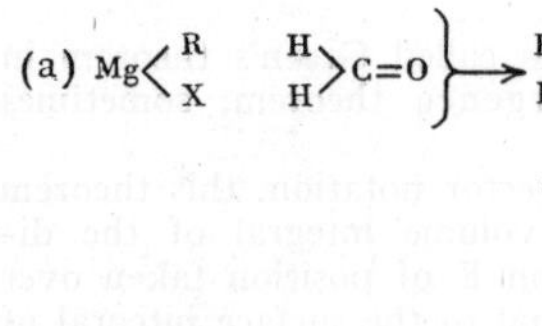

(a)

4. Alcohols

(a)

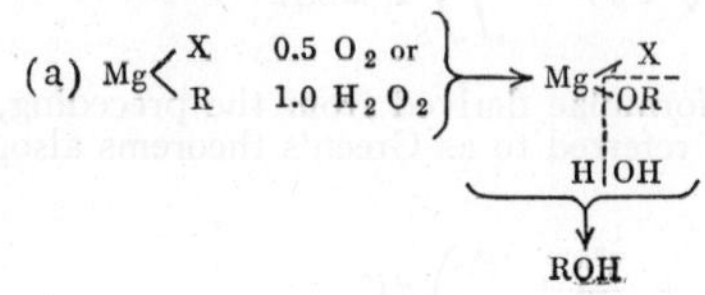

5. Hydrocarbons

(a)

(b)

(c)

(d)

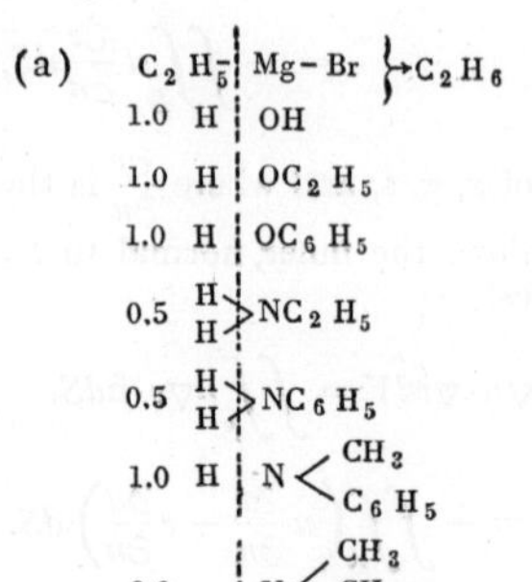

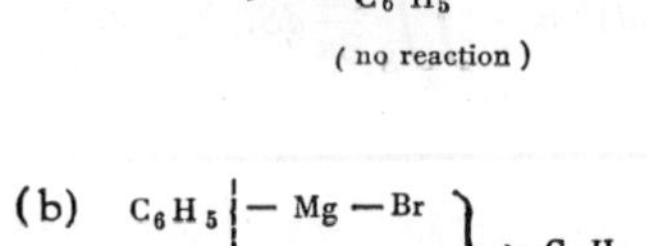

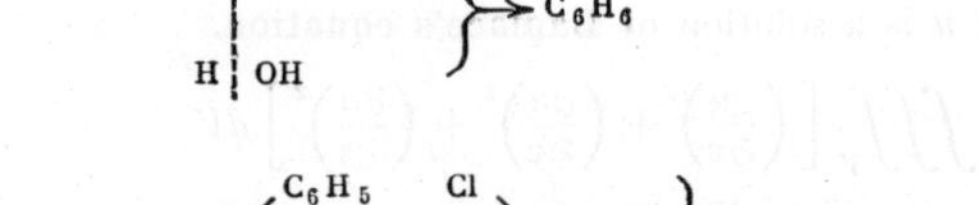

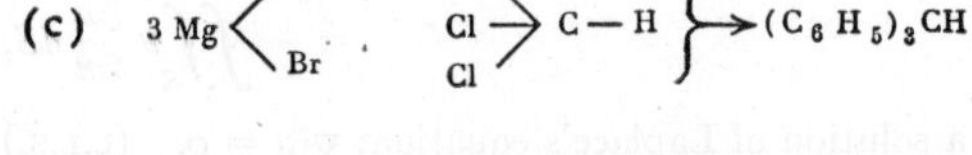

6. Carboxylic acids

(a) $Mg \Big\langle {R \atop X}$ $C \Big\lessgtr {O \atop O} \Big\} \longrightarrow R-C \Big\lessgtr {O \atop OMgX}$

$HO \mid H$

$\downarrow$

$R-C \Big\lessgtr {O \atop OH}$

7. Sulfinic acids

(a) $Mg \Big\langle {R \atop X}$ $S \Big\lessgtr {O \atop O} \Big\} \longrightarrow R-S \Big\lessgtr {O \atop OMgX}$

$HO \mid H$

$\downarrow$

$R-S \Big\lessgtr {O \atop OH}$

8.(a) Triphenylphosphine

$3 Mg \Big\langle {C_6H_5 \atop Br}$ $\begin{matrix} Cl \\ Cl \\ Cl \end{matrix} P \Big\} \longrightarrow (C_6H_5)_3 P$

(b) Triphenylarsine

$3 Mg \Big\langle {C_6H_5 \atop Br}$ $\begin{matrix} Cl \\ Cl \\ Cl \end{matrix} As \Big\} \longrightarrow (C_6H_5)_3 As$

(c) Mercury diphenyl

$2 Mg \Big\langle {C_6H_5 \atop Br}$ $\begin{matrix} Cl \\ Cl \end{matrix} Hg \Big\} \longrightarrow Hg(C_6H_5)_2$

(d) Tin tetraphenyl

$4 Mg \Big\langle {C_6H_5 \atop Br}$ $Cl_4 Sn \Big\} \longrightarrow Sn(C_6H_5)_4$

(e) Lead tetraphenyl

$4 Mg \Big\langle {C_6H_5 \atop Br}$ $2 Cl_2 Pb \Big\} \longrightarrow Pb(C_6H_5)_4$

9. Beta-beta-Diphenylpropionic acid from cinnamic acid

$Mg \Big\langle {C_6H_5 \atop Br}$ $\begin{matrix} CH-C_6H_5 \\ \| \\ CH-COOH \end{matrix} \Big\} \begin{matrix} C_6H_5\cdot CH \cdot C_6H_5 \\ \mid \\ H-CH-COOH \end{matrix}$

HOH

The choice of haloid hydrocarbon is in some cases important in determining the yield of the desired substance. (R.K.S.)

GRILLAGE. A grillage is a system of timber or steel beams which is used under columns to spread the loads over a comparatively large area. Timber grillages, consisting of layers of wooden beams, laid at right angles to each other, are generally used for temporary construction, although there are instances in which they have been inclosed in concrete for permanent construction. If this grillage is used for permanent foundation it should be either entirely submerged or creosoted to withstand deterioration.

The steel grillage consists of one or more layers or tiers of beams which are encased in concrete. If there are two or more tiers the beams in one tier are laid at right angles to those in the next tier. The individual beams in each tier are held in place by rods and pipe separators, cast iron separators or steel diaphragms. Since the concrete-encased steel grillage has more resistance to bending than the ordinary reinforced concrete spread footing it can be used to distribute heavy column loads over large areas. (C.W.C.)

GRINDING. The process of removing solid particles by means of solid or sectional abrasive wheels rotating at comparatively high speeds (6000 ft. per min. or greater). Originally employed for sharpening tools, it has beecome a useful and accurate production process for both hardened and unhardened metal parts.

Grinding machines are quite varied in type and function. Bench and floor grinders usually consist of a motor, with a two-wheel spindle replacing the motor shaft, and are used for tool sharpening and general off-hand grinding. Plain and universal grinding machines are essentially similar to engine lathes, with the carriage and tool post replaced by a wheel stand carrying a grinding wheel and spindle, and are used for grinding cylindrical and conical work. Internal grinders are used for finishing cylindrical and conical holes. Surface grinders employ either a rotary or a reciprocatory table for holding parts on which plane surfaces are required. Centerless grinding machines are used for mass production finishing of cylindrical and conical parts. The part to be ground is supported between the grinding wheel, a work rest, and a suitable regulating or feed wheel which imparts axial motion to the part. Internal centerless grinders, for finishing roller bearings races and bushings, are also used. Disk grinders are used for plane surfacing operations, are similar to bench and floor grinders, but are equipped with metal disks to which abrasive disks are cemented. Grinding wheels consist of abrasive grains held together by some bond such as clay, shellac, or rubber. The hardness of the abrasive, the shape and form of the grain fracture, and the tenacity of the bond are each important in grinding operations. The grade of a grinding wheel denotes its hardness, which cannot be accurately determined by the bond mixture or the method of manufacture. Wheel grade is often indicated by letters, running from E, soft, to Z, extremely hard. Grinding wheels are dressed or trued by metal "star" wheels or by mounted diamonds, to remove metal particles or dull grains of abrasive, and to restore their original shape and accuracy. (See Abrasive.) (H.C.H.)

GRIP. Rivet.

GRISON. Mammalia, Carnivora. *Galictis.* A small South American animal with long slender body and tail and short legs. Related to the weasels. (A.W.L.)

GRIT. An old term for coarse-grained sandstones whose components are angular or "gritty." There is a tendency to use it for any coarse-grained sandstone without regard to the angularity of the fragments. (R.M.F.)

GRIVET. Guenon.

GROSBEAK. Finch.

GROSSULARITE. Garnet.

GROUND. In electrical terminology, a ground is a conductor connected to earth, or a large conductor whose potential is taken as zero (e.g., the steel frame of a car). A ground may be an undesirable, inadvertent, or accidental path taken by an electrical current in its effort to reach ground potential; or it may be the deliberate provision of conductors well connected to the ground by means of plates buried therein, or similar device.

There is always the possibility that, during the life of an insulated conductor, the insulation may be punctured or broken down and a ground occur. Usually, a ground develops rapidly into a low-resistance path through which currents of damaging magnitude may flow. **Insulation** may be damaged in many ways—by the effect of moisture, or chemical vapors, by age, heat, abrasion, breaking, or crushing. Two-wire d-c systems are permanently grounded on one side of the line, three-wire d-c systems permanently grounded on the neutral wire. The same applies to two- or three-wire single-phase a-c systems. The common grounding point of station three-phase lines is the generator neutral.

The grounding system of the a-c generating station fulfills two distinct functions. The first is the grounding of non-current-carrying parts, the second is the furnishing of a ground connection for generator or transformer neutral to provide for the operation of a ground protection system. A common ground **bus** is employed, to which are connected the frames of all electric machines, the cases of instruments, transformers, circuit breakers, the secondaries of current and potential transformers, the switchboard ground bus, conduits, insulator bases, building structural steel, etc. Thus, if the grounding system is effective, a zero, or earth, potential will be established on all metal parts which might otherwise be dangerous in case a ground developed. To the common ground bus is also connected the fault bus, when used.

Grounds should be detected as soon as possible after they occur and the defective section immediately taken out of service. If the ground persists, in a short time an otherwise small repair job may become a large one. Lamp types ground detectors are used to a considerable extent on low-voltage circuits because they are reliable and cheap. It is important that the low-voltage control circuits be kept as free of grounds as the main circuits. They may be applied to high-voltage lines through the interposition, between lamps and line, of potential transformers. Also the vacuum tube type and the electrostatic type of ground detector are available for direct connection to the high-voltage lines.

In the terminology of building construction, a ground is a strip of wood about 2″ wide and as thick as the plaster, which is applied to the framing when nailing room is needed. Grounds are used around windows and doors, at baseboards, cornice, etc. (F.T.M.)

GROUND EFFECT. Persons of an inquisitive turn of mind may occasionally have wondered what action replaces the concentrated reaction of the wheels of an airplane against the ground after the airplane is airborne. The answer is that there is a slight increase of atmospheric pressure upon the surface of the earth, which is maximum directly below the airplane but which dwindles to zero only at an infinite distance. However, nearly all the airplane weight is accounted for by the pressure increase over a circular area of a diameter of 20 times

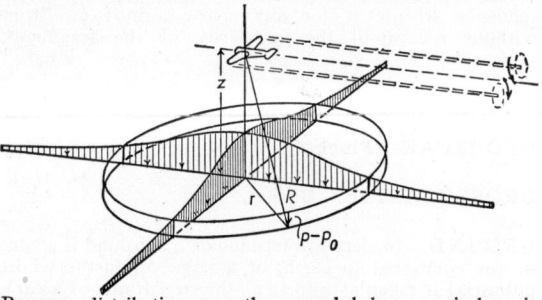

Pressure distribution over the ground below an airplane in flight. (*N.A.C.A.*)

the height of the airplane over it. The figure shows the calculated distribution of pressure from an airplane flying at altitude z over the earth's surface. By use of the circulation theory and replacing the wing by a simple vortex, an equation for the increase of atmospheric pressure $p - p_0$ due to the support of the airplane at radius r may be derived.

When this is multiplied by the area on which it presses, i.e., $2\pi r dr$ and integrated to infinity, the result is no more or less than the weight of the airplane. Of course the pressures are ordinarily very small, since the area carrying the weight is so large. For example, the increase of pressure under an airplane of 4000 lbs. weight flying at 100′ directly overhead is only about .0005 lb. per sq. in.

However, there is one flight condition where this increase of atmospheric pressure and density is of real moment. It is when an airplane is flying just a few feet off the surface of the ground, especially a low-wing monoplane, where only 4 or 5′ may separate ground and wing. Here the strength of the bound vortex is sufficient to build up a substantial change in atmospheric properties, with effects on airplane flying qualities usually called *ground effect*. Induced drag is reduced and the stall occurs at a smaller apparent angle of attack. The airplane may seem, to the pilot attempting a landing, to "float" instead of immediately stalling to a normal landing. While, in general, this is not hazardous to landings once the pilot becomes familiar with an airplane's floating tendencies, it does offer a peculiar and distinct hazard at the take-off of overloaded airplanes, since the ground effect might aid the airplane to become airborne, but would practically disappear as soon as a climb to altitude was started. (F.T.M.)

GROUND ICE. Ice will occasionally form in stream beds, especially in the lee of rocks or other obstructions which tend to reduce the velocity of the water and allow freezing to take place. Such ice formations are known as ground ice or anchor ice. (R.M.F.)

GROUND LOOP. An uncontrollable violent turn of an airplane while taxiing, or during the landing or take-off run. (F.T.M.)

GROUND MORAINE. When a valley glacier melts completely away the debris carried on or within it is dropped upon the valley floor, forming a deposit called ground moraine. The ground moraine from the melting of the great **Pleistocene** ice sheets is usually spoken of as till. (R.M.F.)

GROUND PEARL. Insecta, Homoptera. The iridescent covering secreted by some of the scale insects which live on the roots of plants. Used as ornaments. (A.W.L.)

GROUND ROD. Earthwork.

GROUND WATER. At varying depths below the surface of the earth, depending upon wet or dry seasons, underground structures, and other natural and unnatural factors, is a zone which is saturated with water most of which comes from rain which has penetrated the ground. The upper surface of this saturated zone is called the water table, and the water itself, the ground water or the sub-surface water. The region above the upper surface of the water table is called the zone of aeration or vadose zone.

There is a lower limit to the saturated zone as well as an upper limit. Little ground water exists at depths below 2000 or 3000′. Deep down in the earth's crust the pressure must be so great that all pores in the rocks are completely closed; thus at depths of several miles below the surface there could exist no zone of saturation.

The ground water moves through the rocks and unconsolidated materials of the earth near the surface, constantly seeping into streams and lakes to maintain these bodies of water between rains. If this seepage is sufficiently strong on hillsides or elsewhere springs may result. A well is simply an opening dug deep enough to encounter the zone of saturation.

In certain cases the ground water will flow through porous tilted beds called aquifers from higher to lower localities, establishing a "head" which is sometimes suf-

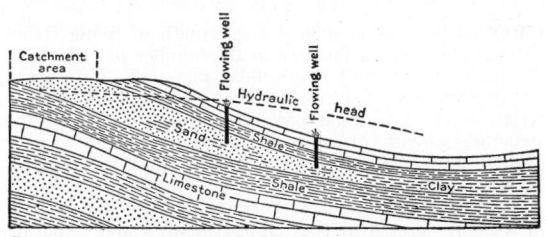

Structure section illustrating flowing artesian wells in a monocline. (*After U. S. Geological Survey.*)

ficiently great to cause the water to flow out under pressure and rise above the surface of the ground, when the aquifer is penetrated by a drill. Such a source of water is called an artesian well from Artois, France, a classic locality for such waters. Artesian conditions exist along much of the Atlantic Coastal Plain of the United States and in North and South Dakota, Nebraska, Kansas, Illinois, Indiana, Missouri and Arkansas. Since the supply of underground water is largely dependent upon structure, the geology of water supply is one of the most important economic phases of the earth sciences. From the point of view of their origin, ground waters are classified as juvenile, connate and meteoric. Juvenile waters are of volcanic or magmatic origin, hence original. Connate waters are those in which the sediments were originally deposited. Meteoric waters are those of atmospheric origin.

All pure water, and most of all of the underground waters are of meteoric or surface-water origin. (R.M.F.)

GROUND WAVE. The energy which reaches the radio receiving **antenna** from the **transmitter** by travel along the surface of the earth rather than by reflection from the **ionosphere** is called the ground wave. The ground wave is unaffected by seasonal or diurnal variations and is consequently very reliable for communication. However, it is attenuated by absorption of the earth and gradually becomes too weak to furnish a reliable signal. This attenuation depends in a complicated way upon the frequency, the soil conductivity and dielectric constant, but increases markedly with frequency. Thus, while it is suitable for communication over several thousand miles at the lower radio frequencies, over a hundred or two in the broadcast band, it becomes almost useless at the high frequencies. See **fading** for its effect on the total received signal. (L.R.Q.)

GROUNDHOG. Woodchuck.

GROUP. For the significance of this term in Geology, see **Era.**

GROUP DRIVE. There is some controversy as to the respective merits of individual and group drives. Individual motor drives are used on many modern machine tools, are very convenient, and are extremely efficient when the power capacity of a machine is definitely known and when it is generally operated at full capacity. In many machine tools operated at high rates of production, two or more motors are often used on the same machine for more efficient operation of its separate elements.

Group drive by means of a number of countershafts and a lineshaft driven by a large motor is less expensive in initial cost than individual motor drives. Individual motor-driven machines must each have motors

that will, on occasion, deliver the full power requirement of that machine even though the machine may usually be operated at half its power capacity. A large group drive motor of a capacity equal to the aggregate capacity of the individual motors is not only less expensive than the small separate units, but it is frequently possible to use a group drive motor with a capacity of three-fourths or one-half that of the individual aggregate. To illustrate: suppose a shop has ten lathes, each of which has a maximum power capacity of 5 hp.; individual drives for these machines will aggregate 50 hp. even though the machines may operate at half-capacity most of the time. For a group drive, however, a 30 hp. motor would probably furnish sufficient power capacity to drive all the machines, even if several were to operate at full load at one time. The ratio of total anticipated power required to total machine capacity is termed the *diversity factor;* in the foregoing example, the diversity factor is .60, or 60%. (H.C.H.)

GROUP MODULATION. In some telephone carrier circuits and in radio links of telephone circuits several carrier **channels** are treated as a single group and the whole modulated upon a new **carrier.** This is known as group modulation. In the reception of such a system, of course, the received signal must first be demodulated into the various channels and then in another step each channel must be demodulated to obtain the original voice currents. (L.R.Q.)

GROUP VELOCITY. The velocity of propagation of an **interference** pattern between two or more wave trains traveling in the same direction with different speeds. It may be quite different from the velocity of any one of the component wave trains. If there are more than two components, the character (wave form) of the resultant wave changes as the "group" progresses, so that the group velocity becomes ambiguous. For two components the analysis is fairly simple.

To illustrate, first suppose for the moment that the wave train A of shorter wavelength λ is standing still, and the other, B, of wavelength $\lambda + \Delta\lambda$ is moving past it in the positive direction (see figure). For example,

$$A \ |\ |\ |\ |\ |\ |\ |\lambda|\ |\ |\ |\longrightarrow v$$
$$B \ |\ |\ |\ |\ |\ |\ |\ |\ |\ |\ |\ |\longrightarrow v+\Delta v$$
$$\underset{X}{\overset{\uparrow}{}} \quad \lambda+\Delta\lambda$$

Two sets of waves traveling with different velocities. Resultant maximum is at X.

let $\lambda = 1$ cm. and $\lambda + \Delta\lambda = 1.1$ cm., and let the velocity Δv of the train B relative to the (stationary) train A be $+3$ cm. per sec. As often as B moves forward 0.1 cm., the coincidence or **beat** maximum X moves backward 1 cm.; consequently X moves with respect to A with the velocity -30 cm. per sec., which is -10 times, or in general $\dfrac{\lambda}{\Delta\lambda}$ times, the velocity Δv with which B moves. (The analogy to a **vernier** should be quite apparent.) Now suppose that an additional velocity v is imposed upon both wave trains, so that now A moves with velocity v and B with velocity $v + \Delta v$. If $v = +100$ cm. per sec., A moves with this velocity, B moves 103 cm. per sec., but X moves only $100 - 30 = 70$ cm. per sec. That is, the velocity of the interference maximum X is $u = v - \lambda \cdot \Delta v / \Delta\lambda$. This is the group velocity; usually written

$$u = v - \lambda \frac{dv}{d\lambda}.$$

In the case of media in which there is **dispersion**, v is a function of λ; where there is no dispersion, $u = v$, since $\dfrac{dv}{d\lambda}$ is then zero.

Take the case of sodium light traveling through carbon bisulfide. This light has two close components with respective wavelengths 5890 angstroms and 5896 angstroms (in air). The refractive index for the 5890-angstrom component being about 1.64, the velocity v of this component in CS_2 is about 1.83×10^{10} cm. per sec. Now the dispersion of CS_2 in this part of the spectrum is such that $\frac{dv}{d\lambda}$ is readily computed to be 3.81×10^{13} cm. per sec. per cm., while the wavelength λ in CS_2 is 3590 angstroms or 3.59×10^{-5} cm. Hence the group velocity u is 1.83×10^{10} cm. per sec. $- 3.59 \times 10^{-5}$ cm. $\times 3.81 \times 10^{13}$ cm. per sec. per cm. $= 1.69 \times 10^{10}$ cm. per sec.

Michelson, using the same revolving-mirror method as for the **electromagnetic constant**, actually obtained this velocity in carbon bisulfide, showing that it is the group velocity which this method really measures. (L.D.W.)

GROUPER. Pisces, Teleostei. A marine game fish (**Pisces**) of the West Indies and the Atlantic Coast. Related to the sea bass. (A.W.L.)

GROUPS. A set of elements G is said to form a group when they satisfy the following conditions:

1. There is an operation by which to each ordered pair of elements A and B of G there is associated an element C of G, denoted by $C = AB$.

2. For this operation the associative law holds: $(AB)C = A(BC) = ABC$ for any three elements A, B, C of G.

3. There exists: (a) a unit-element E in G such that $EA = A$ for each element A of G, and (b) to each element A of G there exists a reciprocal (or inverse) element A^{-1} of G such that $A^{-1}A = E$.

If the commutative law $AB = BA$ holds for any elements A and B of G, the group G is called an Abelian group. (L.L.S.)

GROUSE. Aves, Galliformes. Game birds with compact rounded bodies and legs feathered to the feet. The closely related ptarmigans have both legs and feet feathered.

Grouse are birds of the northern hemisphere. The ptarmigans, including the red grouse of the British Isles and the willow grouse, are found at high altitudes and in the north. Most of these birds have white plumage in the winter. Grouse vary in habits, some frequenting woodlands and others open ground.

Among the best known North American species are the ruffed grouse, *Bonasa umbellus,* the prairie chicken, *Tympanuchus,* and the sage grouse, *Centrocercus urophasianus.* The heath hen, *Tympanuchus cupido,* an eastern species resembling the prairie chicken, has recently become extinct. Two western species, the Franklin grouse, *Canachites franklini,* and dusky grouse, *Dendagapus obscurus,* are locally called the fool hen. (A.W.L.)

GROUSE LOCUST. Grasshopper.

GROUT. Grout is a mixture of cement and water or cement, sand and water of such consistency that it will flow, or may be forced by pressure, into small confined spaces.

It is widely used at present to seal geological faults, cracks, crevices, or other cavities in the rock **foundations** of dams. Grout is also excellent for use in connection with **column footings** and machine foundations which require level bearing surfaces. Since a column footing can never be poured to an exact elevation it is usually built up to within about an inch of its final elevation. Steel shims (fillers) are placed on the top of the footing so as to provide the correct elevation for the bottom of the base plate of the column.

The column base which is an integral part of the column is then set on the shims. After the anchor bolts, which are used to fasten the column to the footing, have been tightened, the space between the top of the footing and the bottom of the base plate is filled with grout. (C.W.C., E.W.S.)

GROWTH. Increase in size. Growth of living structures depends upon increase in the number of cells or in the bulk of cells and intercellular material. It is based on the process of intussusception through which materials received as food become an integral part of the structures already present. Accretional growth is of very limited occurrence in living things and is not independent of intussusception.

Most animals exhibit determinate growth; that is, they increase in size until they approximate a limit characteristic of their kind. A few mature within rather wide limits according to the amount of food available. In the adult body the capacity of various tissues to continue their growth varies, but in all cases tissues which are worn away in the course of normal life have the power of renewal and some, such as the bone-producing cells of vertebrates, are capable of becoming active for the restoration of damaged structures. These aspects of growth are closely associated with **regeneration**.

The rate of growth in different parts of the body also varies, as also does the rate of total growth at different periods of life. Most mammals increase in size rapidly during early life and gradually slow down as maturity is approached, while man grows rapidly during infancy, slowly during childhood, rapidly again during youth, and more slowly toward the completion of his size. In his body the nervous system most rapidly approaches its maximum size and the reproductive system lags until the onset of maturity. Some of the glandular tissues increase rapidly before maturity and then decrease in bulk. The balance of all of these processes when normal food is available results in the gradual process of general growth, and the attainment of stability in adult life is a result of their correlation with external factors. Although no one factor is wholly responsible for growth, **hormones** of the pituitary and thyroid glands are of great importance in its regulation in vertebrates. Deficiency of either gland may result in dwarfing, and pituitary excess sometimes causes human beings to attain unusual height. Persons taller than $7'$ are probably due in all cases to such abnormality.

Plant growth is indeterminate. In the higher plants primary growth is confined to the tips of stems and roots, secondary growth to **cambium** layers which produce wood and bark. The cambiums and the undifferentiated tissues at the tips of stems and roots are called meristems. Meristem cells divide rapidly and some of them finally become the mature cells of the plant. Each cell starts to grow, like an animal cell, by adding more protoplasm but finally increases tremendously in size by taking up a quantity of water to form a large central vacuole. Tissues are differentiated by the accumulation of excess food (cellulose, lignin, suberin) on the outside of each cell in the form of a cell wall. Certain columns of cells thicken their side walls, digest their end walls, and then die, leaving long tubes (vessels) which conduct water. Other cells die from an excess accumulation of impervious wall material and become **fibers** or **cork** cells. Others remain alive for a season or two and manufacture, transport, or store food, much more food than the plant can ever use. Some few cells become concerned with the isolation of meristems in reproductive organs (ovules, seeds). These isolated meristems produce the cells of new plants. The life of a plant need never terminate. There is no adult stage as in animals. **Propagation** may serve to keep a single set of meristems in action continuously. Stem tips and annual plants die, but only because the terminal meristems are captured in the developing ovary of the flower. (See also **Cambium, Roots, Stem,** and **Wood.**) (A.W.L., P.A.W.)

GROWTH SUBSTANCE. Auxins.

GRUB. The soft white larvae of certain beetles. They have few legs, limited to the anterior part of the body. Other animals of similar appearance are also called grubs. (A.W.L.)

GRUIFORMES. The **rails, coots, cranes** and related species. An order of wading and swimming birds (**Aves**) of varied form with lobed toes or with neither lobes nor webs. Feet never fully webbed. (A.W.L.)

GRUNERITE. Cummingtonite.

GRUNT. Pisces, Teleostei. A term applied to food fishes (**Pisces**) of several marine species on the Atlantic Coast. (A.W.L.)

GUACHARO. A native name for the oil bird. (A.W.L.)

GUAN. Aves, Galliformes. Game birds (**Aves**) of Central and South America, related to the curassows. One species, the chachalaca, *Ortalis vetula,* enters southern Texas. (A.W.L.)

GUANACO, HUANACO. Mammalia, Artiodactyla. A wild representative, *Lama huanacos,* of the group commonly called **llamas,** the New World representatives of the camel family. The guanaco ranges from high altitudes in Ecuador to Tierra del Fuego and is larger than the vicuna, the one other wild species. (A.W.L.)

GUANIDINES. Amines and Amides.

GUANINE. Alkaloids.

GUARD CELLS. Stoma.

GUARD RAIL. Railway Track.

GUARD STAKE. Station.

GUAVA. *Psidium Guajava,* and other species. Myrtaceae. Guava, a shrub or small tree indigenous to tropical America, has oblong, short-petioled leaves and white flowers. The fruits have a curious penetrating odor and vary greatly in appearance. Their slightly acid, seedy pulp is principally used in making guava jelly. The plant has been widely introduced into tropical countries throughout the world; in the United States it is grown in Florida and southern California. (R.M.W.)

GUAYULE. *Parthenium argentatum.* Compositae. A perennial shrubby plant native to the semi-desert regions of north central Mexico and southwestern Texas which has long been a minor commercial source of **rubber.** The plants are seldom more than 2′ high, and bear leaves with a gray, silvery appearance. Formerly wild plants were harvested on a considerable scale for rubber but most guayule rubber at the present time comes from cultivated plants. Guayule is grown as a crop plant principally in the vicinity of Salinas, California. The plants are grown from seed which must, however, be subjected to a special washing treatment followed by treatment with sodium hypochlorite before a good percentage of germination can be obtained. In practice the seeds are usually sown in the late winter or early spring and the plants transplanted from the nursery beds the following January. The plants are usually not harvested until they are 5 years old. The harvested plants are cut into small pieces and the rubber

separated by mechanical maceration of these pieces in a pebble mill. In the best selected strains of guayule about 20% of the dry weight of the plant is rubber at the end of 5 years. The yield per acre is about 1600 lbs. of rubber for 5-year-old plants and 2200 lbs. per acre for 6-year-old plants. The rubber obtained from guayule is of good quality, but contains a considerable percentage of resins which must be extracted before it can be used for many purposes. (B.S.M.)

GUDGEON. Pisces, Teleostei. *Gobio.* Small freshwater fishes (**Pisces**) of Europe. (A.W.L.)

GUEMAL. Deer.

GUENON. Mammalia, Primates. A group of African **monkeys** of the genus *Cercopithecus.* All species are slender and of moderate or small size. The genus includes the vervet, grivet, mona monkeys, patas, nisnas, ludio, hocheur, and several other species named as monkeys. (A.W.L.)

GUEREZA. Mammalia, Primates. An African thumbless **monkey** of the genus *Colobus.* (A.W.L.)

GUILLEMOT. Aves, Charadriiformes. *Uria.* Birds (**Aves**) with short legs, webbed feet, and upright posture, related to the auks. The several species are found chiefly about the northern oceans. (A.W.L.)

GUINEA FOWL. Aves, Galliformes. African birds (**Aves**) of several species, related to the pheasants. All have some dark plumage with light spots, and brightly colored bare skin about the head and neck. The common Guinea fowl, *Numida meleagris,* is among the common domesticated species. (A.W.L.)

GUINEA PIG. Cavy. These little rodents are widely kept as pets and are useful to medical science as laboratory animals. Because of their high rate of reproduction they have also been bred for the study of heredity. (A.W.L.)

GUINEA WORM. Nemathelminthes, Nematoda. A large roundworm, *Filaria medinensis,* parasitic in man. It sometimes reaches a length of more than a yard. The worm lives in the superficial tissues, forming an abscess open to the surface, and can be removed by gradual traction on the end of the body exposed in this opening. The eggs are dropped in the water and the young develop in the bodies of water fleas. The species occurs in tropical Asia and Africa. (A.W.L.)

GULDINUS. Theorems of Pappus and Guldinus. (See **Pappus' Theorems.**)

GULL. Aves, Charadriiformes. Water birds (**Aves**) with webbed feet and long narrow wings. Their flight is powerful and easy, and they are more often seen in the air than on the water. They are common along the seashore and on larger bodies of fresh water, but Franklin gulls are often seen far from water, even following the plow to pick up insects. With the exception of the sabine gull they may be distinguished from the closely related terns by the square end of the tail. The kittiwake is a related species of the North Pacific and Bering Sea. (A.W.L.)

GUM-BICHROMATE PROCESS. A process of photographic printing once quite popular among salon exhibitors in France, where it originated, and in America but now practically obsolete. It is based upon the fact that a colloid such as gelatin, gum arabic, etc., is rendered insoluble upon exposure to light in the presence of a bichromate such as ammonium or potassium bichromate. An outline of the process follows: Heavy drawing paper is first coated with a mixture of gum arabic, potassium bichromate and a ground pigment of the color desired. When dry, the paper is exposed in contact with the negative to daylight, or a strong source of artificial light. No image is visible as a result of this exposure, but on placing the paper in water, the portions not exposed to light, i.e., the highlights of the image, dissolve slowly leaving an image consisting of pigment in the hardened gum. As the scale of tones is short, it is generally necessary to resort to multiple printing. Usually, at least three printings are necessary; one for the shadows, one for the halftones and a third for the highlights. The second and third printings are made by coating the paper a second and third time with coating mixtures of different consistencies, exposing and washing out the unattached pigment as with the first coating. Obviously, registration marks on the paper and the negative are necessary in order to superimpose the three printings.

The gum bichromate process produces prints with deep, rich shadows and affords considerable control over local tone values. (C.B.N.)

GUMBO. Till.

GUMMA. A soft spongy inflammatory tumor which occurs in late untreated syphilis. The common sites for gummata are the skin, liver, brain, and lung. Treatment with anti-syphilitic drugs causes them to disappear. (D.M.H.)

GUMS. These are carbohydrates of complex structures which are formed as decomposition products in many plants. They are particularly common in plants growing in very dry regions. They have neither taste nor odor and are insoluble in alcohol and ether. Many gums are soluble in water, while others unite readily with water to form a mucilaginous product, or swell greatly in water.

Gums are employed in making adhesives, in calicoprinting, in sizing fabrics, both silk and cotton, and in confectionery. Medicinally they are widely used because of their soothing properties, and as a vehicle for insoluble substances.

While a great many plants form gums, only a few of these are of any importance to man. One of these, gum arabic, is a product of an acacia tree of tropical Africa. This tree, *Acacia Senegal,* grows in very dry regions. It has bipinnately compound leaves which have a gray color, and small yellow flowers borne in axillary racemes. The gum exudes from cracks which are formed in the bark of the stem and branches at the beginning of the dry season when very rapid drying of the plant occurs. Gum arabic is entirely soluble in water. Another gum is obtained from a small shrub growing in desert regions of southwestern Asia. This shrub, *Astragalus gummifera,* a legume, has pinnate (see Pinna) leaves. The leaflets fall off, leaving the leaf axis as a stiff thorn. The plant has small yellow flowers and small 1-seeded pods. The gum, known as gum tragacanth, is only partially soluble in water. In America several small shrubs of the dry plains yield a gum called mesquite. One of these shrubs is *Prosopis glandulosa.* Many species of *Prunus,* including the common cherries and plum trees, yield a dark-colored gum which is entirely insoluble in water. (R.M.W.)

GUN COTTON. Explosives.

GUNDI. Mammalia, Rodentia. A nocturnal burrowing animal living in rocky country near the Sahara desert. (A.W.L.)

GUPPY. Pisces, Teleostei. *Lebistes.* A small species of the killifish family, popular for tropical aquaria. The males are brightly colored and variable, the females larger and duller. They bear living young and are easily reared if protected from the cannibalistic parents. (A.W.L.)

GURNARD. Pisces, Teleostei. *Trigla.* Marine fishes (Pisces) with high angular heads and finger-like appendages on the pectoral fins. These processes are tactile and also aid in supporting the fish on the bottom. Some are called the sea robins and one British species is known as the piper. The flying gurnards and beaked gurnards belong to a related family. They are widely distributed marine fishes whose bodies are strangely formed and armored with bony plates. (A.W.L.)

GUSSET PLATE. A gusset plate is a flat plate connecting two or more structural members where they meet at a joint. Stress is transferred between the members through the gusset plate by riveted, bolted, or welded connections. A gusset plate should be of a shape giving a minimum waste of material, which can

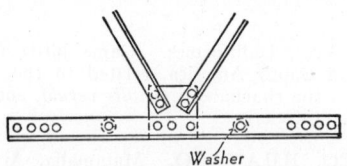

Washer

be fabricated in the shop with minimum amount of labor. For this reason it should be cut with straight edges. The thickness of a gusset plate should be sufficient to give *bearing* value, so that the material or the rivet will not be crushed. Minimum thicknesses of gusset plates are usually ¼″ for inside protected structures and ⅜″ for outside exposed structures. The area between rivet holes should be great enough to transmit the stress from one member to another. Examples of gusset plates are to be found in all types of welded and riveted steel structures, and in gussets which strengthen and make the joints in the rib structure of an airplane wing. (F.T.M.)

GUSTS. Transient but rapid fluctuations of wind velocity. Gusts are the result of turbulent air flow. Gusty winds usually vary radically in direction. (P.E.K.)

GUTTA PERCHA. *Palaquium Gutta,* and related species. Sapotaceae. Gutta percha is prepared from the latex found in the stem and leaves of certain trees native in Malaysia and various South Sea Islands. To obtain the latex, which does not flow readily from living trees, the tree may be felled and a series of rings cut in the bark. From these the latex oozes and may be gathered. Such a method is naturally very destructive to continued production. A more desirable method is practiced in plantations of today. Fresh leaves are gathered and chopped up and crushed. The crushed mass is then boiled in water and the gum removed and pressed into blocks.

In South America a related tree, *Mimusops Balata* (Sapotaceae) yields a similar gum of somewhat inferior quality. This tree is usually tapped by cutting a row of zigzag gashes which connect one with another. Down these the latex flows, to be gathered in a cup at the bottom, and later coagulated in trays.

Gutta percha is a yellowish or brownish somewhat leathery solid containing up to 90% of a hydrocarbon, gutta. On heating, it becomes plastic and is very resistant to water. It is therefore much used as insulation for marine cable and electrical equipment, as well as in the manufacture of golf balls and in dentistry. The cheaper balata is also used in the manufacture of beltings. (R.M.W.)

GUTTATION. The loss of liquid water from intact plants is called guttation. This process should not be confused with **transpiration** which is the loss of water vapor. Guttation occurs most commonly from the leaves, the exuded drops of water appearing at the tips or margins of the leaves. The water is not pure but contains traces of sugars and other solutes. Guttation occurs through distinctive structures called hydathodes or water stomates. In external structure a hydathode resembles an enlarged **stomate**. In temperate regions guttation can most often be observed on cool, late spring mornings following a warm day. Exuded drops of water can be observed at the margins or tips of many, but by no means all, kinds of herbaceous plants at this season. The exudation of water is believed to result from a root pressure (see **Ascent of Sap**) which is imposed on the sap in the xylem ducts. The drops of water exuded in this process are often erroneously considered to be dew. The quantities of water lost by most species of plants in guttation are negligible compared with the quantities lost in transpiration. (B.S.M.)

GWYNIAD. Pisces, Teleostei. *Coregonus.* A whitefish (**Pisces**) of the English lakes. (A.W.L.)

GYMNOLAEMATA. Bryozoa.

GYMNOPHIONA. A class of **amphibians** made up of legless burrowing species which resemble worms or small snakes. They live in the tropics of South America and the Old World. About 40 species are known. (A.W.L.)

GYMNOSPERMS. The Gymnosperms form one of the two main divisions of the seed plants or **Spermatophytes**. The characteristic feature of the Gymnosperms is the occurrence of the **ovule** on the surface of the scale which bears it, and not surrounded by the **ovary** wall. In most Gymnosperms the reproductive bodies are borne in cones. The overlapping scales of the ovulate cones protect the developing **embryo** and cover the seed until it is mature.

The Gymnosperms are the most primitive of seed plants. Arising early in geological time, these plants became abundant and widespread in the **Carboniferous** period. From that period to the present gymnosperms have decreased in numbers, many groups becoming entirely extinct. Today there are about 500 species of Gymnosperms, occurring in nearly all sections of the world, but attaining their greatest development in the temperate zones. There they often form a dominant forest tree, with a comparatively small number of species, but many individuals. Other species occur in small isolated groups of few individuals occupying a very limited area.

Gymnosperms are woody plants. The majority of them are trees, which often attain immense size, being among the largest known plants. A few are low shrubby plants, and a very small number of vine-like species exist. Nearly all Gymnosperms are plants of xerophytic habit, that is, fitted to survive in regions in which water is not abundant. Some, like *Welwitschia* of the arid deserts of southwestern Africa, live in regions where the annual rainfall is but a fraction of an inch.

The living orders of Gymnosperms are the **Cycads**, Ginkgo, the **Conifers**, and the Gnetales. The last order, composed of three genera of widely different habit, has no single distinguishing feature, except the presence of vessels in the secondary wood. The three genera are: 1. *Gnetum,* a genus of tropical Africa, most of whose species are lianas, growing over other plants. A few are trees or shrubs. 2. *Welwitschia mirabilis,* a most striking odd plant growing in deserts of southwestern Africa. The plant has a large turnip-shaped stem which tapers into a long slender taproot which goes deep into the ground to reach the water table. It has two thick leathery leaves which last throughout the life of the plant, growing at the base as they wear away at the tip. 3. *Ephedra,* comprising low much-branched shrubs found in desert regions of Africa and America. *Ephedra* yields the drug ephedrine, a crystalline alkaloid which is used to give relief to sufferers from asthma and hay fever. (See also **Paleobotany**.) (R.M.W.)

GYMNOSPORE. Asexual reproductive cells which are capable of active locomotion by amoeboid movement or by **cilia** or **flagella**. (A.W.L.)

GYMNURA. Mammalia, Insectivora. Small animals of Burma, the Malay Peninsula, and adjacent islands. They resemble the shrews in appearance, with a sharp nose, long tail like that of a rat, and short legs. The name is that of a genus but is also used as a common name. (A.W.L.)

GYNANDROMORPH. An abnormal individual whose body shows the characteristics of the two sexes in different parts. Not synonymous with **hermaphrodite** although this term is sometimes applied to these abnormalities.

Gynandromorphs are fairly common among the insects, where they are often of the bilateral type. Such individuals have one side of the body male and the other female, with a sharp boundary in the median line. Mosaic gynandromorphs present an irregular distribution of the sexual characters. (A.W.L.)

GYNECOLOGY. That branch of medicine and surgery which covers diseases of women related to the genital organs and associated structures. (R.S.M.)

GYPSUM. The mineral gypsum is hydrous **calcium** sulfate, $CaSO_4 \cdot 2H_2O$. It occurs as flattened **monoclinic** crystals, often twinned, transparent cleavable masses, called selenite, or silky and fibrous, called satin spar; it may also be granular or quite compact. It is a soft mineral, hardness 1.5–2; has two good cleavages which yield rhombic plates whose angles are 66° and 114°. Its specific gravity is 2.31–2.33; luster vitreous to silky or pearly; color, colorless to white and gray, may be tinted red, yellow, blue, brown, etc., by impurities; transparent to opaque. A very fine-grained white or lightly tinted variety of gypsum is called alabaster, and prized for ornamental work of various sorts. Gypsum is a very common mineral, thick and extensive beds of which are associated with **sedimentary rocks**. The largest deposits known occur in strata of **Permian** age. Besides being a result of deposition in sea and lake waters, gypsum has been deposited by hot springs, from volcanic vapors, and by sulfate solutions in veins. Notable foreign localities for gypsum are in Greece, Czechoslovakia, Austria, Saxony, Bavaria, Italy, France, Spain, England and Mexico. In the United States well known localities are at Lockport, New York; the Mammoth Cave, Kentucky; Ellsworth, Ohio; Grand Rapids, Michigan; Hermosa, South Dakota; Wayne County, Utah; and San Bernardino County, California. In Canada the Provinces of New Brunswick and Nova Scotia have large gypsum deposits. The word gypsum is derived from the Greek meaning to cook, in reference to the burnt or calcined mineral. Because the gypsum from the quarries of the Montmartre district of Paris has long furnished burnt gypsum used for various purposes this material has been called plaster of Paris. (E.S.C.S.)

GYPSY MOTH. Insecta, Lepidoptera. A **moth**, *Porthetria dispar*, introduced from Europe and now a serious pest in the northeastern United States. The caterpillars are able to defoliate shade and forest trees and also attack apple and sometimes the conifers. The damage and control are the same as in the case of the brown-tail moth. (A.W.L.)

GYRO COMPASS. Compass.

GYRO PILOT. Compass.

GYROCOTYLIDEA. Cestoda.

GYROSCOPE. A gyroscope is a heavy wheel or disk which may be set into rapid rotation and, because of the conservation of its angular momentum, serves to illustrate in various ways the tendency of rotating bodies to maintain a fixed axis of rotation or to exhibit precession. In its usual form the "gyro," or rotating disk, is mounted in a ring so that it can be handled while spinning; and this is frequently hung in a second ring or frame to form gimbals. In this case the outer ring or frame may be moved or turned in any manner, without bringing to bear a torque tending to change the direction of the axis of rotation.

The complete theory of the gyroscope is far too complex to be included in a work of this character and reference must be made to advanced texts on mechanics. (See **Instruments, Aircraft**.) (L.D.W., W.K.G.)

H

H SECTION. In many circuits, particularly in the telephone field, it is necessary to have both sides of a circuit balanced to **ground** to reduce the effect of interfering influences (see **Inductive Interference**). This is accomplished by dividing any series impedance into two equal parts and inserting one in each line, rather than putting the total amount in one line. (See figure.) (L.R.Q.)

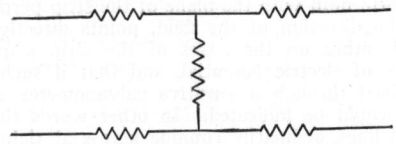

HABIT. As used by the mineralogist, this term denotes the sum of the external characteristics of a mineral. It is also, but more rarely, applied to rocks. (R.M.F.)

HACHURE. A short line drawn parallel to the slope as a means of illustrating topography on a map. (R.M.F.)

HACKLY. Used by mineralogists to describe a jagged **fracture.** (R.M.F.)

HACKSAW. Saw.

HADDOCK. Pisces, Teleostei. An important marine food fish, *Gadus aeglefinus,* taken on both sides of the Atlantic. It is closely related to the cod. Usually dried and smoked for the market. (A.W.L.)

HADE. Fault.

HAEMATOXYLIN or LOGWOOD. *Haematoxylon campechianum.* Leguminosae. Logwood is obtained from a small tree which is native in Central America, but has been extensively planted in the West Indies and South America. The tree, seldom more than 25' tall, has **pinnately** compound leaves with smooth obovate leaflets, and fragrant yellow flowers in terminal **racemes.** The fruit is a dry 2-seeded pod.

The wood is very hard and yellow; on exposure to air it turns red. It has a rather pleasant odor. To obtain the dye the sapwood is cut away and the heart wood cut up into chips. From these chips the dyestuff is extracted. Logwood dye is used to color cottons, woolens, silks and leathers. To make the dye **mordants** must be added, in this case the salt of some metal, usually iron. Haematoxylin is also used in the making of **inks,** and as a histological stain in the preparation of organic tissues for microscopic examination. Following ferric ammonium sulfate, as a mordant, it is an excellent dye for chromosomes. Small quantities are used medicinally, in the form of extracts of the wood, in cases of chronic diarrhea. It is also a mild astringent. The active principle in this case is tannin, not the dye. (R.M.W., B.S.M.)

HAEMOCOELE. A cavity in which blood or haemolymph circulates. Well developed in the **arthropods** and **mollusks** where it superficially resembles a true body cavity. It is associated with some tubular blood vessels whose contractions propel the blood in it, and these movements are supplemented by the shifting of its contents as a result of body movements. (A.W.L.)

HAEMOCYANIN. A **protein** containing copper which combines readily but unstably with oxygen and serves as a respiratory medium. It occurs in the blood of **arthropods** and **mollusks.** When oxidized it is bluish in color. (A.W.L.)

HAEMOGLOBIN. Hemoglobin.

HAEMOLYMPH. The **blood** of higher invertebrates, consisting of a clear plasma and white cells but without red cells. Respiratory pigments are dissolved in the plasma. It contains a lower percentage of water than the blood of more primitive forms. (A.W.L.)

HAEMOSPORIDIA. Sporozoa.

HAFNIUM. Symbol: Hf. Atomic number: 72. Atomic weight: 178.6. Melting point: 1700° C. (Isotopes: page 290.)

Hafnium is similar in chemical properties to zirconium, but separable from the latter by repeated fractional crystallization of the double **potassium** fluoride (K_2HfF_4), which collects in the mother liquor, from potassium zirconium fluoride which collects in the crystals. Discovered by Coster and Hevesy in 1922 in the mineral **zircon** of Norway by the Moseley **x-ray** spectrographic method of analysis, and later found to be present in almost all **zirconium** minerals and chemicals (most of these containing about 5% Hf in the zirconium).

The use of hafnium metal in lamp filaments and radio tubes has been proposed.

Oxide: Hafnium oxide (HfO_2), white. (R.K.S.)

HAGFISH. Cyclostoma.

HAIL. A precipitation is often called hail if it is in the form of lumps or pellets of ice. Strictly, the term should be applied only to the globose or sometimes irregular masses which fall at the beginning of a thunderstorm in hot weather. The most typical hailstones are somewhat oblately spheroidal and of the size of a hazelnut, though very rarely they attain a diameter of 2 or 3″ or larger. When broken, they reveal a structure of concentric alternate layers of clear and opaque white ice.

The fact that true hail never occurs except during a violent thunderstorm indicates that wind plays an essential rôle in its formation. It seems probable that hailstones have their origin in those vortices or whirlwinds in whose interior masses of cloud-laden air rush upward with great speed and then circle downward outside, only to be drawn inward and upward again, like a gigantic smoke-ring. The lower part of such a thundercloud may be composed of raindrops and the upper part of snow crystals. A raindrop, blown upward into the colder region, freezes and collects a layer of snow, then descends to have more water frozen upon it, then upward for more snow, and so on, until it finally becomes too heavy to circulate further, and falls. Often several hailstones freeze together into a shapeless, nodular mass. Large hailstones, naturally, are somewhat dangerous and often very destructive. (See **Hydrometeors.**) (L.D.W.)

HAIRSTREAK. Insecta, Lepidoptera. Small **butterflies,** those of the temperate zone dull-colored and those of the tropics often brilliant. The hind wings of most species bear hair-like tails. With the coppers and blues they make up the family Lycaenidae. (A.W.L.)

HAIR-TAIL. Pisces, Teleostei. Predacious fishes, (**Pisces**), chiefly tropical, whose bodies taper at the posterior end to a slender point. They are long, slender, scaleless, and have the fins reduced excepting a long dorsal and the pectorals. (A.W.L.)

HAIRWORM. Nemathelminthes, Nematoda. Long slender roundworms of small size which live as parasites in the bodies of **invertebrates**, chiefly insects. (A.W.L.)

HAKE. Pisces, Teleostei. Inferior marine food fishes (**Pisces**) related to the cod. Both sides of the Atlantic and colder waters of southern hemisphere. The squirrel hake and white hake are also called codling and the silver hake, *Merluccius bilinearis*, is known as the whiting or stockfish. (A.W.L.)

HALATION. The halo which is sometimes observed in photographic images of bright objects, or sources of light, is known as halation when due to light reflected back into the emulsion from the rear of the support. Ordinarily, the halo produced by halation is indistinguishable from that due to the spreading of light within the emulsion which is known as irradiation. The size and density of the halo, when due to halation, increase with the exposure and the degree of development and decrease with an increase in the turbidity and thickness of the emulsion.

Halation can be almost entirely prevented by placing a light-absorbing layer either on the back of the support or between the emulsion and the support. Light-absorbing layers of this type are termed *backings*. Dyes which are decolorized by the sulfite in the developer have replaced the old mixtures of lamp black and similar materials which had to be removed by washing in water and discolored the developer so that it could no longer be used.

The tendency of negative materials to show halation may be lessened by incorporating a gray dye in the film support, or by double coating, first with a slow- and then with the regular high-speed emulsion. The first of these methods is used for certain motion-picture negative materials and the latter, in conjunction with a backing, for professional sheet film. (C.B.N.)

HALF-VALVE PERIOD. Decay Coefficient.

HALF-WAVE. Rectifier Circuits.

HALIBUT. Pisces, Teleostei. A large flatfish, *Hippoglossus*, reaching a length of 6' and a weight of 400 lbs. Found in all of the northern seas.

The flesh of the halibut is inferior to that of some of the smaller flatfishes but the species is valuable as a source of vitamins. Halibut-liver oil is reported to be about 10–30 times as rich in vitamin D and 80 to 200 times as rich in vitamin A as cod-liver oil. (A.W.L.)

HALITE. Rock Salt. The mineral halite is naturally occurring **sodium** chloride, NaCl, common salt. It is isometric with cubic habit and cleavage. It is brittle; hardness, 2.5; specific gravity, 2.4–2.6; luster, vitreous; colorless when pure, but usually white, yellow, red or blue. It is soluble. Halite occurs interbedded with sedimentary rocks in all parts of the world and in all but the very oldest rocks. It frequently occurs in association with **anhydrite** and **gypsum**. In the United States this type of "salt beds" have been exploited in Michigan, New York, Ohio and Pennsylvania. Louisiana produces salt from great sub-surface dome-shaped

masses often 2000–4000' thick. The salt domes of the Gulf Coastal Plain are particularly important as subsurface structures on the flanks of which are apt to occur large and important pools of **petroleum**. Poland, Saxony, Austria and France possess well known deposits of salt as well as Russia, England, Algeria, India and China. Salt is chiefly used in cooking and as a preservative; in the manufacture of soda ash for the glass industry, and as a source of many **sodium** compounds. It derives its name from the halogen group of elements to which **chlorine** belongs. (R.M.F.)

HALL EFFECT. In 1879 Hall, at Johns Hopkins University, discovered that if a strip of gold-leaf, carrying an **electric current** longitudinally was placed in a **magnetic field** with the plane of the strip perpendicular to the direction of the field, points directly opposite each other on the edges of the strip acquired a difference of electric potential, and that if such points were joined through a sensitive galvanometer a feeble current would be indicated. In other words the equipotential lines, ordinarily running across at right angles to the edges, were skewed into an oblique position, and the electric lines of flow in the plate were deflected to one side.

If one looks along the strip in the direction of the current, with the magnetic field directed downward, then with strips of antimony, cobalt, zinc, or iron the electric potential drop is toward the right and the effect is said to be positive; while with gold, silver, platinum, nickel, bismuth, copper, and aluminum, it is toward the left and the effect is called negative. The transverse electric potential gradient per unit magnetic field intensity per unit current density is called the "Hall coefficient" for the metal in question. A special case known as the "Corbino effect" occurs when a circular disk carrying a current radially is placed at right angles to a magnetic field; the result being a current component around the disk.

The Hall effect as first observed, in gold, agrees in direction with the lateral thrust on current-carrying armature conductors in a motor, and suggests the identity of these phenomena. But this explanation is upset by the reverse action in some metals, and is further complicated by many other experimental facts. At present the Hall effect cannot be said to be fully understood. (See Ettingshausen, Nernst, and Righi-Leduc Effects.) (L.D.W.)

HÄLLEFLINTA. A Swedish term for hard, dense **metamorphic** rocks composed chiefly of microscopic crystals of quartz and feldspar with occasional **phenocrysts**. Accessory minerals may be **hornblende**, **chlorite**, **hematite** or **magnetite**. The texture and composition of hälleflinta suggests that it is the metamorphosed equivalent of **acid** lava flows or **tuffs**. (R.M.F.)

HALLER'S ORGAN. Sense organs of uncertain function borne on the anterior legs of some **ticks**. (A.W.L.)

HALLEY'S COMET. This is probably the most famous of all the **comets** and is deserving of special mention. The general subject of comets is discussed elsewhere, and this article, therefore, will confine itself to Halley's comet alone.

Halley's comet is of special interest because of the fact that it was the first comet whose return was predicted. When Halley computed the **orbit** of the great comet observed in 1682, he found the elements to be almost identical with those of prominent comets observed and studied by Kepler in 1607 and by Apian in 1531. He noticed that the interval between 1531 and 1607 was not exactly equal to the interval between 1607 and 1682, but suspected that the difference might be due to attractions by other planets. He was unable to predict just what effects the attractions of Jupiter might produce on the next return, but suspected that they would retard it and predicted the return for the early part of 1759. In the meantime, mathematical astronomy had developed to such a point that April, 1759, was predicted by Clairaut. The comet actually came to **perihelion** on March 13th of that year. At the next return in 1835, the predicted date of perihelion differed from the observed date by only 2 days, and for the return in 1910 the agreement was practically perfect.

Extensive calculations made by Crommelin and Cowell after the observations in 1910 carried the dates of perihelion passages back through the centuries. Examination of ancient records prove that the comet was observed and recorded at every perihelion passage back to 87 B.C. In some cases, the descriptions are complete enough to prove, both that the comet has always been a striking object and also that the ancient records were surprisingly accurate as to position of the comet.

Due to **perturbations** of the planets the periods between successive perihelion passages have varied considerably. The average is about 77 years. Historians have, at times, attempted to use observations of comets as a means for fixing dates, employing the hypothesis that any bright comet appearing at about the correct date was Halley's. This is a very dangerous practice for there have been many others of as equally striking appearance as Halley's and their appearances were always noted.

At the return in 1910 the comet was first picked up by Wolf at Heidelberg on September 11, 1909, when it was 310,000,000 miles from the sun, and it was followed photographically until July 1, 1911, when it was 520,000,000 miles from the sun. When close to perihelion it was a magnificent object, particularly during the early part of May, when it was observed in the morning sky. On May 19th, about a month after perihelion passage, the comet passed directly between the earth and the sun, but no change in the brightness of the sun could be observed even with the most delicate instruments. At one time the tail of the comet extended across the sky for a distance of nearly 120°, appearing as a broad bright band much like the milky way. On May 21st the earth certainly grazed the tail of the comet and probably passed through it.

In 1949 the comet will reach its greatest distance from the sun, being at that time more distant than the planet **Neptune**. It should return to perihelion on April 29, 1986. (W.K.G.)

HALLUCINATION. A mental state in which sensory impressions are not based on reality. The sensory impressions may be of sight, hearing or smell. Hallucinations are most frequently seen in some forms of acute **alcoholism, opium,** and **cocaine** habituation, delirium from high fever and in certain forms of insanity.

An hallucination should not be confused with (1) a delusion, which differs in being a false belief, not correctable by reasoning or evidence of the senses, or (2), an illusion—a false interpretation of a sensory image. (R.S.M.)

HALO. This term applies to a class of phenomena observed in the sky in connection with the **sun** (or sometimes the **moon**) and due to particles of frost suspended in the air. The minute spicules of ice, in falling, take some definite attitude determined by their shape. Some are needle-like and assume a horizontal position, some are flat disks or stars and fall with their planes horizontal, while others, made up of both disks and rods, behave like a parachute. The sunlight is refracted by each type in a characteristic manner and dispersed into colors; it is also reflected from their external surfaces without dispersion. The most commonly observed effects are: the colored parhelia (commonly called "sun dogs") on a level with the sun and 22° each side of it; a 22° halo passing through the parhelia; a 46° halo, usually faint; and certain other appended arcs or partial halos due to refraction and reflection by the variously oriented crystals.

In addition to these phenomena may be mentioned the colored rings, called coronae, often surrounding the sun or moon. These are due to **diffraction** by particles of fog or mist. The diameter of the **corona** is greater, the smaller the fog particles. (L.D.W.)

HALOGENS. The chemical elements **fluorine, chlorine, bromine,** and **iodine.** (R.K.S.)

HALTERE. The vestigial hind wing of a 2-winged fly (**Insecta, Diptera**). In this order the 2-winged condition is due to the loss of the hind wings as organs of flight. They persist, however, as small knobbed organs, in some species large enough to be seen with the naked eye just behind the bases of the functional wings. They contain sense organs, and both static and chordotonal functions have been attributed to them. The evidence seems inconclusive. (A.W.L.)

HALYS. Reptilia, Sauria. Asiatic vipers of several species, related to the copperhead and water moccasin of North America. One small species of Ceylon and India is called the carawila. (A.W.L.)

HAMADRYAD. Cobera.

HAMMER. Sheet Metal Processes; Forging; Press Working.

HAMMERHEAD. Pisces, Plagiostomi. *Sphyrna.* Large **sharks** of the tropical seas. The head is greatly widened to form lateral projections bearing the eyes. (A.W.L.)

HAMSTER. Mammalia, Rodentia. Burrowing animals of Europe and Asia. The common hamster, *Cricetus,* attains a length of 1'. It is sometimes extremely numerous and is then a serious pest to the farmer. It damages crops of all kinds. The flesh is eaten and the fur is useful, although not of high quality. (A.W.L.)

HAND. The terminal portion of the pectoral appendage of **mammals,** developed for grasping and in some species largely freed from locomotor uses. True hands appear only in the **primates.**

The skeletal structure of the hand includes the series of five bones, the metacarpals, which attach it to the wrist, and the five divergent series of phalanges located in the digits. Of the digits, one, the thumb, is placed and articulated so that it can be opposed to the other four, which are fingers. As a result the appendage can be used for grasping like a forceps and also as a prehensile organ by folding the fingers back against the palm. In some of the monkeys the prehensile method of grasping is more important in moving through the trees and the thumb has shifted and become smaller so that it can no longer be opposed.

The human hand is the most versatile grasping organ in the animal kingdom and has been of primary importance in the manufacture of tools and machines, and distinguishes man from all other living things. (A.W.L.)

HAND SET. This is the part of the modern telephone which contains the transmitter and receiver, i.e., the part which the user holds when talking. While this placing of transmitter and receiver on the same handle seems a rather insignificant accomplishment, the necessity of avoiding feedback from the receiver into the transmitter makes very careful acoustical design necessary. In addition the electrical and mechanical construction of the transmitter must be such that it can be operated in almost any position as the user will hold the instrument in innumerable positions. (L.R.Q.)

HANGER. Bearings.

HANGING VALLEY. Under normal conditions a tributary stream enters the main stream at grade, that is, at the same level. Under certain circumstances the tributary valley may be at a greater elevation than the main valley into which the tributary stream will plunge, forming a waterfall. In such cases the tributary valley is called a hanging valley, and the stream in it is said to be out of adjustment with the main stream.

Hanging valleys originate in the following ways: by glacial action, the main glacier cutting down its valley faster than a tributary glacier; by river action, the main stream eroding its bed faster than the tributary stream; by faulting, the tributary stream flowing off the upthrown block. A fourth type of hanging valley, much less common, may result from a stream plunging over wave-cut cliffs or other escarpments into a lake or ocean basin. (R.M.F.)

HANGING WALL. The use of this term in geology is discussed in the article on **Fault.** (R.M.F.)

HANGNEST. Aves, Passeriformes. A group of birds (Aves) whose nests are woven of vegetable fiber, grass, and hair, and are suspended from small branches. The Baltimore oriole is a common North American species. Others occur from the southwestern states to Brazil. (A.W.L.)

HANUMAN. Langur.

HAPALOPS. Miocene.

HAPLOSPORIDIA. Sporozoa.

HARD COAL. Coal.

HARD FACING. Deposition of a hard wear resistant alloy on a metal surface. The material to be deposited is generally in the form of a welding rod and may be applied by gas or arc welding. Such surfaces are usually finished by grinding.

While hard facing or hard surfacing is usually a maintenance operation, it is also used in new production. The surfacing material may be cemented carbides, nonferrous Stellite-type alloys, or iron-base alloys with alloying additions such as chromium, tungsten, manganese, silicon, nickel, and carbon. While hard facing is most often applied to steel, cast iron and some of the non-ferrous alloys such as Monel metal can also be coated. Typical applications are metal-working dies, oil well drilling tools, excavating equipment, shafting, and rolling mill rolls. (R.H.H.)

HARDENABILITY OF STEEL. The hardenability of steel refers to the ease with which it can be hardened rather than the maximum hardness value attainable. For example, a 1-in. diameter bar of a certain 0.20% carbon alloy steel can be hardened to 50 Rockwell "C" in the center by quenching in oil. A similar bar of plain carbon steel requires a drastic quench in brine to attain the same hardness, and therefore, has a lower hardenability. Neither bar can be quenched to a greater hardness because 50 Rockwell "C" is the maximum attainable for a 0.20% carbon steel. A 0.40% carbon steel can be hardened to a maximum of about

60 Rockwell "C" and the maximum for high-carbon steel is about 65 Rockwell "C."

Of the several methods for determining the relative hardenability of steels, the Jominy test is the most widely used. A cylindrical specimen 1" in diameter and about 3" long is heated to the hardening temperature and quenched in a special fixture which holds the specimen in a vertical position and directs a stream of water on the bottom surface. The stream takes an "umbrella" shape and does not wet the sides. Cooling occurs progressively from the bottom to the top of the cylinder and the cooling rate at any distance from the bottom is known and reproducible from one sample to another. The hardness along the length of a quenched Jominy bar decreases from bottom to top. The distance from the bottom, expressed in sixteenths of an inch, to the point where the hardness is 50 Rockwell "C" is one method of reporting the hardenability.

One of the principal functions of alloying elements in steel, such as manganese, chromium, nickel, molybdenum, etc., is to increase the hardenability. Whereas prodigious amounts of expensive alloys were formerly used to insure full hardening, especially in medium and heavy sections, wartime shortages have focused attention on the use of as little alloy as possible within the hardenability requirements. A large number of National Emergency Steels have been developed containing relatively small additions of a number of elements, thus utilizing the residual alloy contents normally present in much of the steel scrap now available. (R.H.H.)

HARDENING OF METALS. There are three principal methods of hardening metals and alloys, cold working by plastic deformation, precipitation hardening, and heat treating as applied to steel. The last two methods involve heating and cooling operations. Alloying itself is a hardening process since an alloy addition generally hardens a pure metal. (R.H.H.)

HARDENING OF THE ARTERIES. Arteriosclerosis.

HARDIE. Forging.

HARDNESS. The significance of this term as applied to solids has various interpretations. Commonly, it refers to the resistance of the substance to surface abrasion, so that of two solids, the one that will scratch the other, as diamond scratches glass, is the harder. Again, it may denote rigidity, or lack of plasticity, or even strength; in some cases a combination of several such properties. The Mohs' Scale of Hardness, used by mineralogists, is as follows: 1. Talc (softest). 2. Gypsum. 3. Calcite. 4. Fluorite. 5. Apatite. 6. Orthoclase. 7. Quartz. 8. Topaz. 9. Corundum. 10. Diamond (hardest). Example: If a mineral can be scratched by orthoclase and will not scratch apatite its hardness is between 5 and 6.

In metallurgy and engineering hardness is determined by methods based on resistance to penetration by an indenter of greater hardness than the material being tested. Aluminum, copper, lead, magnesium, tin, and their alloys as well as plastics are generally indented by hardened steel balls ranging in size in the various tests from $\frac{1}{16}$" to 10 mm. in diameter. The same methods may be used for soft steels and irons, but for heat-treated steels and all other alloys which develop high hardness special diamond indenters, or in some cases sintered tungsten carbide balls, are used. In all of the technological tests the indenters are impressed into the test material under carefully regulated loads; thus the relative size of the resulting indentation becomes a measure of hardness. The operating principles of the instruments most widely used in this country follow:

Brinell. The indenter is a 10-mm. diameter hardened steel ball. A sintered tungsten-carbide ball is also coming into use, especially for testing hard metals. The load applied is generally 500 kg. for soft metals and 3000 kg. for steels and hard metals. Brinell hardness is equal to the load (kg.) divided by the surface area (sq. mm.) of the impression made in the test material. Tables are available for direct conversion to hardness from the diameter of the indentation as measured with a calibrated magnifier after removal of the piece from the testing machine.

Rockwell. Indenter is $\frac{1}{16}''$-, $\frac{1}{8}''$-, or $\frac{1}{4}''$-diameter steel ball or a conical diamond having an apex angle of 120° and a slightly rounded point. The various scales used are designated by letters. Rockwell "B," for example, indicates a 100-kg. load on a $\frac{1}{16}''$-diameter ball. Rockwell "C" indicates a 150-kg. load on the diamond indenter. Rockwell "30T" designates a load of 30 kg. on a $\frac{1}{16}''$-diameter ball. (An instrument of higher sensitivity known as the Rockwell Superficial Tester is used for loads of 15, 30, and 45 kg.) The size of the indentation is measured by a **dial gauge** as the final depth minus a small preliminary penetration produced by a minor preload of 10 kg. The Rockwell hardness values are arbitrary numbers having an inverse relationship to the depth of the indentation.

Vickers. Also known as Diamond Pyramid Hardness. Indenter is a square-based diamond pyramid with included angle between faces of 136°. Loads may vary from 1 to 120 kg. with 10, 30, and 50 kg. in common use. Hardness is equal to load (kg.) divided by surface area (sq. mm.) of the permanent indentation. It is determined directly from optical measurements of the diagonals of the indentation which appears square at the surface of the metal.

Tukon. A recently developed instrument for determining hardness under very light loads down to 25 grams. The small indentations are measured at high magnifications up to 1000 times. The indenter is a diamond pyramid that makes an elongated impression, one diagonal being 7 times the other in length.

Eberbach. Also used for very light loads. Consists of a spring-loaded, Vickers-type diamond pyramid indenter arranged for use on a metallurgical microscope.

Scleroscope. Depends on the height of rebound of a diamond-tipped body falling under the force of gravity from a fixed height. The instrument is relatively small and is portable. One type reads directly on a graduated dial.

While there is overlapping in the field of useful application of the various hardness tests, each has certain special qualifications. The Brinell test makes a large indentation, giving an average hardness value for several grains even in rather coarse-grained metals; however, it cannot be used on small or thin specimens. The various Rockwell tests are widely used, especially for rapid production inspection of parts. The Vickers test, which originated in England, is less rapid than the Rockwell but has the advantage of a single scale covering the hardness of all metals from lead to the hardest tool materials. The Tukon test makes it possible to determine the hardness of very thin sheets and of thin metallic coatings such as chromium plate, or zinc on galvanized steel. The Scleroscope test is used principally on heavy forgings or castings which cannot be placed in an indentation-type instrument, or for field tests where a portable instrument is required. It is less accurate than the indentation-type instruments. An instrument known as the King Brinell, employing the standard Brinell principle, is also available in portable form for shop or field tests.

It is often desirable to be able to convert from one hardness scale to another. Unfortunately, no single set of conversions is accurate for all materials.

TYPICAL HARDNESS VALUES

MATERIAL	BRINELL 500 kg.	BRINELL 3000 kg.	ROCKWELL	VICKERS 50 kg.
Aluminum, annealed......	23		H 45	25
Magnesium alloy..........	63		B 21	63
Armco iron..............	66	73	B 31	71
Yellow brass, annealed....	72	82	B 40	77
Copper, cold rolled........	99	83	B 55	110
Mild steel, annealed.......	107	117	B 70	123
Aluminum alloy, 24st......	130	144	B 78	146
Stainless steel, annealed....	121	145	B 80	153
Yellow brass, cold rolled....	174	178	B 91	189
Ni-Moly steel, quenched in water, tempered at 1200° F.		241	C 23	255
Same, 1000° F...........		293	C 31	310
Same, 800° F............		363	C 38	380
High speed tool steel.......		684	C 62	740

(L.D.W., R.M.F., R.H.H.)

HARD-PAN. The term which prospectors and miners give to the sub-surface or basal layers of placer deposits in which the gold-bearing gravels have been cemented and hardened. The same term is also used to designate till or boulder clay which has been cemented by **limonite**. (R.M.F.)

HARE. Mammalia, Rodentia. Rodents (**Rodentia**) with long ears, large hind legs and small front legs, and a very short tail. Adapted for speed and for jumping. The Old World species burrow and those of the New World often occupy holes in the ground although they do not burrow. The rabbits and hares belong to the same genus and the two names are not sharply distinct. The North American species are more commonly called rabbits but the white or snowshoe rabbit of the north is also called the varying hare, and the white-tailed jack rabbit is known as the prairie hare. Hares are indigenous to all continents except Australia.

Among the domesticated species of hares and rabbits many breeds have been developed. These animals are kept for pets and as a source of meat. They are also valuable laboratory animals in medical science and have been used in the study of heredity. The fur is fine and soft but the pelts are relatively weak. They are valuable for linings and when sheared and dyed are used in inexpensive garments. Rabbit fur is also used in making felt. (A.W.L.)

HARELIP. A congenital deformity in which there is a failure of fusion of the maxillary and globular processes resulting in a cleft in the upper or lower lip. It is more common in the upper lip and may be associated with a cleft palate. It may be double, in which case there is a division on either side of the mid line of the lip. The treatment is surgical. (D.M.H.)

HARMONIC ANALYSIS. Not only is it possible to combine two or more simple **harmonic motions** of different period, amplitude, and phase to form a complex motion, but there are also means of analyzing the resultant motion, when the latter is given, to find its component harmonics. For example, if the wave form of such a complex tone as that produced by a bell or a saxophone is accurately graphed by means of a phonodeik (see **Vibrations and Waves**, and **Musical Sounds**), the equation of the vibratory motion can be deduced in such form as to show the separate components. Fourier showed that the same analysis is possible for any periodic motion, however complicated. The equation, called **Fourier's series**, may be written

$$y = a \sin 2\pi nt + b \cos 2\pi nt + c \sin 4\pi nt + d \cos 4\pi nt$$
$$+ e \sin 6\pi nt + f \cos 6\pi nt + \cdots,$$

in which y is the displacement of the vibrating particle and t is the time. The fundamental frequency n and the constants a, b, c, d, etc., must be calculated from the given wave form or the data from which it is plotted.

There is a type of instrument, called a "harmonic analyzer," which automatically computes the coefficients; or it may be done mathematically, though the process

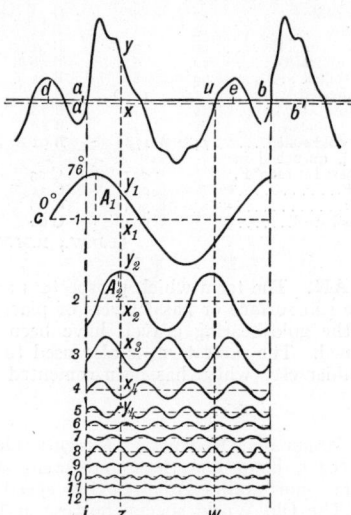

Records of a complex sound and twelve of its components.
(*D. C. Miller.*)

is very laborious. The accompanying figure shows the wave form and the twelve components of a complex tone, analyzed by Professor D. C. Miller. (L.D.W.)

HARMONIC CURVE.
By a harmonic curve is usually meant a curve whose equation is of the form $y = a \sin nx$ or $y = a \cos nx$. (L.L.S.)

HARMONIC DIVISION OF A LINE-SEGMENT.
A line-segment is said to be divided harmonically by two points if these two points divide the segment internally and externally in the same ratio. The two points of division are called harmonic conjugates. (L.L.S.)

HARMONIC FUNCTIONS.
A harmonic function is any real **function** u which satisfies **Laplace's equation** $\dfrac{\partial^2 u}{\partial x^2} + \dfrac{\partial^2 u}{\partial y^2} + \dfrac{\partial^2 u}{\partial z^2} = 0$, and which, together with its first and second **partial derivatives**, is **continuous** and single-valued throughout a certain region. (See also **Laplace's Equation.**) (L.L.S.)

HARMONIC LAW.
Keplerian Laws.

HARMONIC MEAN.
Given the N positive **variates** $x_1, x_2, x_3, \cdots, x_N$, the harmonic mean, H, is defined as the reciprocal of the **arithmetic mean** of the reciprocals,

$$\frac{1}{H} = \frac{\dfrac{1}{x_1} + \dfrac{1}{x_2} + \cdots + \dfrac{1}{x_N}}{N}.$$

The arithmetic mean of N unequal positive variates is greater than the **geometric mean** which in turn is greater than the harmonic mean. The harmonic mean is used in time and rate problems. (L.A.A.)

HARMONIC MOTION.
A distinct type of periodic motion, or vibration, characteristic of elastic bodies; illustrated by a bird-cage bobbing up and down at the end of a spiral spring, or (approximately) by the piston of a steam engine. It may be either simple, with only one **frequency** and amplitude, or made up of two or more simple components and consequently of more complex character. The essential feature of simple har-

monic motion is that, with its range extending to equal distances on both sides of an equilibrium position or origin, the **acceleration** is always toward the origin and directly proportional to the distance from it. With elastic vibrations this is easily seen to follow from Hooke's law, since the force tending to restore the deformed body to equilibrium is proportional to the deformation. (See **Elasticity.**) The motion is called "harmonic" undoubtedly because the vibrations of bodies emitting musical sounds are of this character. Any simple harmonic motion may be represented by the equation

$$y = a \cos (2\pi nt + \varphi),$$

in which y is the distance at time t, a is the amplitude, n is the frequency or number of vibrations per unit time, and φ is the phase constant, such that when $t = 0$, $y = a \cos \varphi$.
It is interesting to note the relationship between harmonic and circular motion. If a peg is inserted in the face of a circular disk or wheel and the latter uniformly rotated, the motion of the peg, as viewed with the wheel seen edgewise, is simple harmonic. In fact, uniform circular motion is made up of two simple harmonic components of the same period and amplitude at right angles, one being a quarter-period ahead of the other in phase. If the two harmonic components have a phase difference other than a quarter-period, the resultant in general is motion in an ellipse; while if they have unequal periods, the path is one of a class of more or less complicated loci called "Lissajous' curves." (L.D.W.)

HARMONIC PROGRESSION.
A harmonic progression is a succession of numbers whose **reciprocals** form an **arithmetic progression**.
The harmonic mean between two numbers is the middle term of a harmonic progression whose first and last terms are the given numbers. The harmonic mean between a and b is given by

$$H = \frac{2ab}{a + b}.$$

If A, G and H are respectively the arithmetic, geometric and harmonic mean of two numbers, then $G^2 = A \cdot H$. (L.L.S.)

HARMONIC RINGING.
This is a system of individual ringing used on party telephone lines. The bell or ringer armature is tuned mechanically to a certain frequency. Then if the coils are excited by currents of this same frequency the armature response is great due to the resonant effect while at other frequencies where the resonant gain is not present the response is too small to give any sound. When the bells of the several phones on the party line are each tuned to a different frequency selective ringing (i.e., the ringing of only the called party) is accomplished by using the proper frequency ringing current. Such a system is not needed on a two-party line but is increasing in use on multi-party lines. (L.R.Q.)

HARMONIC SERIES.
The infinite series $\sum\limits_{1}^{\infty} \dfrac{1}{n}$ is called the harmonic series. It is **divergent**. (L.L.S.)

HARMONICS, ELECTRIC.
In a great many types of a-c generators there are present various overtones or multiple frequencies, all related by integral multipliers to the fundamental **frequency**. These higher frequencies are called harmonics of the lower or fundamental frequency. In power systems the only harmonics present are normally the lower odd-numbered ones, i.e., third, fifth, etc. Of these the third is by far the most common and hence the cause of the most trouble. Because of the non-linear relation between flux and mag-

netomotive force in iron most **transformers** draw third harmonic currents which in turn produce third harmonic voltage drops and losses. In a three-phase circuit these currents can flow out over the line wires and return over the neutral. If there is no neutral there can be no third harmonic current in the line wires and unless the transformer banks are connected with at least one set of windings in delta the induced voltages will not be sinusoidal, but will contain third harmonics. There are usually some harmonics present in the output of alternators because of non-sinusoidal flux distribution, tooth and slot effects, etc. Much of this can be reduced by proper coil layout and connection. These higher harmonics also cause appreciable telephone interference unless precautions are taken (see **Transposition**). Unless properly tuned many radio circuits will generate harmonics. In **frequency doublers,** etc., this characteristic is utilized to advantage, but normally these harmonics must be suppressed. In any event, it is undesirable to radiate any harmonic of the carrier from the **antenna.** Various circuit devices are used to suppress these unwanted frequencies but even then there is often some harmonic radiation which accounts for the reception of stations at double, triple, etc., their normal frequency. (L.R.Q.)

HARMOTOME. The mineral harmotome is a **zeolite,** composition approximately $(K_2Ba)Al_2Si_5O_{14} \cdot 5H_2O$; it is monoclinic but often forms double twins giving the effect of a square prism. It is a brittle mineral; hardness 4.5; specific gravity 2.44–2.50; luster vitreous; color, white to gray or perhaps yellow, red or brown; white streak; translucent. Harmotome like other zeolites is found in cavities in basalts and similar rocks, sometimes in **trachytes** or in **gneisses,** occasionally as a gangue mineral in veins of metallic minerals. Some well-known localities are in Bavaria, in the Harz Mts., Norway and Scotland. Harmotome occurs in the United States with stilbite, near Port Arthur, Lake Superior. The name harmotome comes from the Greek meaning joint and to cut, referring to the division of the pyramid formed by the prismatic faces of the mineral when in the twinned position. (E.S.C.S.)

HARPY. Aves, Falconiformes. *Thrasaëtus.* Large crested **eagles** related to the true buzzards. The several species range from Mexico to Patagonia. (A.W.L.)

HARRIER. Aves, Falconiformes. Moderately large **eagles** of several species, and a group of **hawks.** The harrier eagles are mostly limited to Africa but one species extends to Asia and Europe. The hawks are found in all continents. They are slender birds with long wings, and in general are useful as destroyers of reptiles and rodents. The marsh hawk, *Circus hudsonius,* is a North American harrier. (A.W.L.)

HARTEBEEST. Mammalia, Artiodactyla. *Bubalis.* An African **antelope** with large ringed horns, irregularly spiraled with the tips pointing back. There are several species. (A.W.L.)

HARVESTMAN. Arachnida, Phalangida. Spiderlike animals, most species with small oval bodies and extremely long slender legs. Those with shorter legs are more easily confused with the true spiders but all may be recognized by the segmented abdomen. Daddy longlegs. (A.W.L.)

HASHISH. Hemp.

HAUSEN. Pisces, Chondrostei. The giant **sturgeon,** *Acipenser nuso,* of the Mediterranean and the large rivers and inland seas of Europe and western Asia, also found on both sides of the Atlantic. Specimens weighing over 3000 lbs. are said to have been taken in the Volga River, although half that weight is large for the species. (A.W.L.)

HAÜY'S LAW. Crystallography.

HAWK. Aves, Falconiformes. Birds (**Aves**) of prey with hooked beaks and large curved claws, closely related to the eagles, falcons, harriers, and others and

Cooper's Hawk. (*American Museum of Natural History.*)

not sharply distinguished as a group. Hawks are found on all continents. North America has many species, including **buzzards, harriers, goshawks** and other forms. Most of them are beneficial as destroyers of vermin but the sharp-shinned (*Accipiter velox*), and Cooper (*A. cooperi*) hawks destroy too many birds, including poultry, to be regarded as friends. (A.W.L.)

HAWK MOTH. Insecta, Lepidoptera. Large **moths** composing the family Sphingidae, one of the largest of the order. These moths have a long, rather stout body projecting beyond the narrow wings. The front wings are much longer than the hinder pair, and because of their limited surface they are vibrated rapidly in flight. The moths have long tongues and visit deep-throated flowers. From their habit of hovering as they probe the flower for nectar they are also called hummingbird moths. Another common name is sphinx moth. (A.W.L.)

HAWKSBILL. Turtle.

HAWTHORN. Rose Family.

HAY. Grass Family.

HAY FEVER. (Allergic Coryza.) This condition is one manifestation **allergy** seen in subjects who are hypersensitive to **pollens,** other airborne substances, and certain **foods.** It is characterized by an intense irritation of the membranes of the upper respiratory tract and the eyes. It is seen in two forms: (1) the seasonal type which is due to pollens (2) the non-seasonal type which occurs in attacks throughout the year and is due to animal emanations, vegetable or seed powders, or ingested substances.

A common form of hay fever is seasonal hay fever, sometimes known as Rose Cold. Any pollen may cause it in a sensitive individual. The spring type lasts from March to the beginning of June and is due usually to pollens of trees—elm, birch, maple, oak, and hickory. The summer type occurring through June and part of July is due to pollen of **grasses** such as red-top, timothy, June and orchard grass, sweet vernal, and plantain. The fall type begins in August and lasts until frost, and is usually due to the pollens of the ragweeds. Over 50% of hay fever subjects are sensitive to more than one pollen. Those that are sensitive to only one pollen are relieved of the hay fever as soon as pollenization of the particular plant stops. During the following years attacks begin at the same time of year.

The symptoms of hay fever are local and consist of attacks of severe itching, congestion, and weeping of the eyes. Sneezing is apt to be violent due to the irritative reaction within the nose. Swelling of the nose may cause obstruction to nasal breathing. An irritating discharge from the nose is usually present. The entire attack closely resembles a severe head cold.

The most frequent complication of hay fever is sinusitis and this occurs in a large percentage of cases. Asthma is present in 15–20% of hay fever cases.

Non-seasonal hay fever (perennial hay fever, vasomotor rhinitis) may be continuous or occur in paroxysmal attacks throughout the year. The substances causing this form may be inhaled or ingested. They are animal dust and vegetable or seed powders. House dust is one of the most common exciting agents. The specific substance in house dust has not been identified. The dust of hay and straw causing hay fever is not related to the pollens of these substances that cause the seasonal type.

The determination of the specific offending substances is accomplished by means of skin tests, scratch tests, or the intracutaneous injection of dilutions of various test substances. A positive test is said to occur when a wheal of irregular outline, surrounded by a zone of irritation, is produced. The wheal may vary from ¼ to several inches in diameter. Marked itching may be present at the site. Sensitivity tests may also be carried out by instilling dilutions of test substances into the conjunctival sac of the eye.

The treatment of the acute attack of hay fever consists of the local application to the nose of ephedrine-like drugs—"privine," "neosynephrin," etc., which shrink the mucous membrane, diminish the discharge, and control the itching. Similar drugs taken orally ("propadrine") achieve the same result.

Prophylactic treatment against hay fever is successful in many cases, especially those of the seasonal type. Subcutaneous injections of a solution of the offending substance are begun 10 weeks in advance of the expected attack. The dose injected is very dilute at first—the strength being gradually increased. The effects of the treatment are not permanent and the desensitization must be repeated each year. Results are quite satisfactory even in those cases where asthma complicates the picture.

Some forms of hay fever, especially of the non-seasonal type, can only be prevented by avoidance of the exciting agent. (R.S.M., D.M.H.)

HAZE. Very small particles of salt and dust in the air reduce visibility and cause the atmosphere to appear off-color. Against a dark background the veil appears bluish, and against a bright background it seems yellowish or orange. This is known as dry haze. Very small water droplets cause a haze more grayish than dry haze known as damp haze or mist. (P.E.K.)

HEAD. In zoology, the head is the region of a bilaterally symmetrical animal body lying at the front end in relation to the ordinary direction of locomotion, or, in bipedal vertebrates like man and some of the birds, at the highest level.

The development of a head is indicated in animals which are without sharply separated body regions, such as the flatworms. This process of cephalization is closely correlated with bilateral symmetry. The portion of a bilateral animal which goes first inevitably is the first to encounter new sources of stimuli, and shows some concentration of sense organs. Usually the chief nerve center, a cerebral ganglion or brain, also develops here. The concentration of sense organs and nervous control in the head remains characteristic of the region throughout the animal kingdom and in most groups is accompanied by the location of the mouth in the head, together with associated structures for securing food. (See **Hydrostatic Pressure.**) (A.W.L.)

HEADACHE. Headache is not a disease but one of the most common symptoms met in medical practice. It may be the first symptom of a number of grave organic diseases. Before treating persistent headache lightly, the cause should be determined. The explanation of the mechanism of headache is difficult. The pain is often a pressure phenomenon within the skull. This may be due to increased or decreased pressure within the blood vessels of the brain or in the brain substance itself. The causes of these pressure disturbance are exceedingly numerous.

Headache varies in location, character, and severity of pain and time of occurrence. Often pain in the face and outside the skull is indistinguishable from headache; such pain may accompany toothache, sinusitis, eyestrain, and neuralgia of facial nerves.

The causes of headache fall into three classifications:

1. *Organic causes.* These include injury to the brain, either from simple compression or fracture of the skull, syphilis, abscess and tumor of the brain or its coverings, cerebral accidents (stroke, hemorrhage, or embolism) meningitis, arteriosclerosis, middle-ear disease, diseases and disorders of the eye, high or low blood pressure, etc.

2. *Toxic Causes.* The inhalation of poisonous gases, especially carbon monoxide, and over-indulgence in alcohol, tobacco, etc. Headaches accompanying any severe infection or illness, kidney disease (nephritis), constipation, digestive disorders, allergic diseases, and menstrual disorders are sometimes loosely termed "toxic," although the mechanism of their production is obscure.

3. *Functional Causes.* These include mental or emotional strain, fatigue, and the emotional maladjustments of the *psychoneuroses.* Migraine is discussed separately. (D.M.H.)

HEADER. Any pipe, conduit, duct, or channel, which acts as a central point of distribution of a fluid flow to several branch lines, is a header. An example is the steam header, which is usually a large steam pipe well anchored, to which several boilers supply steam through boiler lead pipes, and from which steam is taken for such uses as the individual case presents. Header ducts are used where a fan is to supply air to several sources. Cases often arise, also, where a liquid, such as water, oil, etc., is to be distributed from a header for several uses. (F.T.M.)

HEADING. The direction of the forward end of the keel of a ship (either air or seaborne) is known as the heading of the ship. Unless a qualifying adjective is used with the term, heading is the direction referred to true north. Compass heading, or magnetic heading, may be converted to heading by applying the compass corrections. (W.K.G.)

HEADSTOCK. Lathe, Milling.

HEART. Circulatory System.

HEART FAILURE. Heart failure is a loose term which covers several types of disordered heart function. Congestive heart failure denotes passive engorgement of the vascular system, due to weakness of the heart so that it cannot pump forth the blood which fills its chambers in diastole. Back pressure, high venous pressure and stasis result. The patient has dyspnoea, congested swollen organs, and often edema of the dependent portions. Digitalis is the most important drug in the treatment of congestive failure. Hypertension, arteriosclerosis, rheumatic fever, and syphilis are the common causes of heart disease resulting in congestive failure.

The heart failure which results from sudden severe damage to the heart, such as occurs in coronary occlusion, usually produces acute circulatory failure, and "collapse" or "shock." Weakness, feeble, rapid pulse, and decrease in blood pressure are the characteristic features. These phenomena are the results of diminished output of blood by the heart, occurring suddenly. (D.M.H.)

HEARTBURN. A burning sensation in the esophagus and stomach caused either by some irritating substance which has been eaten, or by increased acid production in the stomach. It occurs frequently in peptic ulcer with hyperacidity of the gastric juice. (R.S.M.)

HEARTWOOD. Wood.

HEAT. That heat is a form of **kinetic energy** has been known only since the work of Rumford and Davy in the first decade of the nineteenth century. They succeeded in boiling water and melting ice by heat generated mechanically. Prior to their work, heat had been believed to be a substance, a sort of gas called "caloric." This was no doubt suggested by the shimmering appearance in the air surrounding a hot object. No one appears, at that time, to have tested the question as to whether a heated body weighs more than a cold one, or to have attempted the isolation and analysis of the supposed caloric as it escaped from the body on cooling. This well illustrates the very imperfect state of scientific thought a century ago.

The chaotic agitation of **molecules** which we now associate with heat, and the violence of which determines the temperature, is strikingly exhibited, though on a much altered scale, by the **Brownian movement.** When a substance is heated, its molecules receive impulses which result in the acceleration of their motions of translation, of rotation, and sometimes of internal vibration. With most gases composed of diatomic molecules, a simple calculation based upon the known **specific heat** and upon the **kinetic theory** shows that 60% of the energy, at ordinary temperatures, goes into the translational molecular motion and the other 40% to rotational motion; the internal vibrations apparently do not begin until higher temperatures are reached. This apportionment is in accord with the principle of **equipartition of energy** and the **quantum theory.**

Although we now recognize that heat is energy, it is still customary to express quantity of heat in the old water-temperature measure, by means of **British thermal units** or of **calories**; and whenever heat quantities so expressed are used in thermodynamic calculations, it is necessary to use the **mechanical equivalent of heat** as a conversion factor between these and the ordinary dynamic units of energy (foot-pounds or **ergs**). (See **Temperature, Calorimetry, Thermal Convection, Thermal Conduction, Thermal Radiation, Thermodynamics,** etc.) (L.D.W.)

HEAT BALANCE. A **heat** balance **is** a method of accounting for all heat units in a process or change during which heat is transferred. Examples of cases where heat balances might be undertaken are: 1. Determining the nature and the magnitude of the various losses which occur when coal is burned in a steam boiler **furnace.** 2. Accounting for all heat units during the operation of a prime mover such as a **Diesel engine** or a **steam turbine.** 3. Determining the distribution of heat in a static heating device such as a water heater supplied with steam.

Heat balance work is based upon the first law of **thermodynamics,** a statement of which is: Energy may not be created or destroyed, but may be converted from one form to another. The significance of this law applied to the heat balance is that the total energy may be accounted for by straight addition, hence striking a heat balance resembles bookkeeping, with heat supplied on the credit side of the ledger, and various heats usefully employed on the debit side. One way of showing a heat balance is a tabular form, another shows the heat as a stream, properly branched and subdivided to indicate the distribution of heat. Briefly, a heat balance might be said to be the bookkeeping by which heat supplied is shown to be equal to the sum of heat utilized and lost. (F.T.M.)

HEAT BALANCE IN THE ATMOSPHERE. Heat received from the sun is the primary source of energy for the earth. Some slight amount of heat is received from the earth's interior by virtue of radioactive rocks but this need not be considered in view of its comparable smallness. Total heat received from the sun, directly below the sun, at the outer limits of the atmosphere (the amount that would be received at the earth's surface if passage were unaffected by the atmosphere and clouds) is very nearly 1.94 gram-calories per sq. cm. per min. This great quantity of heat is distributed in such a way that the maximum is received directly below the sun with a decreasing amount received as the distance from the heat equator increases. Tropical areas, for this reason, are warm and polar regions cold.

Not all the sun's radiation is received at the earth's surface. Clouds and snow reflect about 75% of solar radiation falling upon them; land surfaces reflect an average of 10–30% of that falling on them; water reflects varying percentages of solar radiation from 70% when the sun is only 5° high to less than 2% when the sun is over 50° above the horizon. Some solar radiation is absorbed by the atmospheric gases and some by water vapor in the air. Another part is lost to the earth by scattering in the atmosphere. Altogether solar radiation is distributed as follows:

1. Approximately 42% is sent back into space by reflection.

2. 15% is absorbed by the atmosphere and its impurities and cloud particles.

3. 43% is received and absorbed by the earth's surface.

Clouded days prevent considerably more solar radiation from reaching the earth than is received on a clear day. Loss on a clear day is approximately 17% of the total amount, but on a clouded day the loss is about 78%. Average cloudiness is about 52%. Deserts are conspicuously clear and therefore receive a much larger percentage of the incoming solar heat than do continental west coasts which have considerable cloud cover. Snow-covered regions lose a large percentage of their incoming solar heat when compared with forest and vegetation-covered lands. Water surfaces, averaged the world over, do not reflect a large percentage but water is capable of absorbing large quantities of heat with only a small temperature change. Influence of local terrain on solar radiation plays a considerable role in determining the daily and seasonal temperatures of that area.

The earth receives its heat from a number of direct sources.

1. About 17% is direct solar radiation.

2. 10% is sky radiation (from scattered solar radiation).

3. 70% is long-wave radiation received from the atmosphere which surrounds the earth.

4. 3% is received by contact with warm surface air currents.

It should be realized, however, that all this energy regardless of its immediate source, was received initially from the sun.

There is no accumulation of heat on the earth which indicates a radiative heat balance. Radiation received by the earth goes into several categories.

1. 7% goes to space by radiation through transparent bands in the atmosphere (transparent to radiation from a black body at 300° abs.).

2. 78% goes to the atmosphere by radiation where it is absorbed and redistributed.

3. 15% is used for evaporation and is carried to the atmosphere where it adds to the store of atmospheric heat. Water vapor is the principal absorber of earth

radiation as it passes through the atmosphere. Carbon dioxide and ozone also have some strong absorption bands. (P.E.K.)

HEAT COIL. This is a device used in a telephone circuit to serve somewhat the same function as a **fuse.** However, instead of opening the circuit, it grounds it. The heat coil consists of a coil through which the current flows and a plunger held back by a fusable link and tending to be displaced by a spring. The normal current through the coil produces no effect but a sufficiently high current generates enough heat to melt the holding link and release the plunger which grounds the circuit. (L.R.Q.)

HEAT CONTENT. Heat may be absorbed by a substance in several ways. It may be absorbed in the form of external work done when the heated substance expands against a pressure. It may be absorbed by increasing internal energy associated with the motion of the molecules (commonly measured by temperature). Again, heat may be absorbed by change of state of the substance (see **Heat of Fusion**, and **Heat of Vaporization**), examples of which are the vaporizing of a liquid, the melting of a solid, etc. The total thermal energy possessed by a substance includes that present in these various forms. Heat content at the absolute zero temperature (minus 460° F., or minus 273° C.) is zero. However, an arbitrarily taken datum for heat content often proves more valuable than with the absolute heat content; thus, for example, steam tables which display the heat content of steam do so upon the assumed basis of zero heat content at 32° F. (See **Enthalpy.**) (F.T.M.)

HEAT OF COMBUSTION. Fuels; Thermochemistry; Fuel Calorimeter.

HEAT OF FORMATION. Thermochemistry.

HEAT OF FUSION. Very simple experiments show that the **fusion** of a given mass of any crystalline substance requires a definite quantity of **heat.** The quantity required per unit mass, without any change of temperature, is called the heat of fusion of the substance. It may be measured by means of a **calorimeter.** The fused substance is introduced into the calorimeter at a temperature somewhat above its melting point and allowed to cool, the heat evolved being measured. At the melting point it ceases to cool for a time, but continues to give out heat as it solidifies; and when all congealed, it begins to cool again. At this stage the process is terminated; and the total heat evolved, with corrections for the cooling before and after solidification calculated from the known specific heats, gives the heat of fusion. For ice the value is about 79.71 calories per gram.

The freezing of 10 lbs. of water gives out 1440 B.T.U., which is equivalent to the burning of 0.1 lb. of coal or about 2 cu. in. of kerosene, or which would raise the temperature of 1 gallon of water 171° F. This fact is sometimes utilized to prevent vegetables from freezing in unheated basements, by setting tubs of water near them; or likewise to protect the battery and radiator of a car on very cold nights. (L.D.W.)

HEAT OF REACTION. Thermochemistry.

HEAT OF SUBLIMATION. Sublimation.

HEAT OF VAPORIZATION. The **evaporation of a** given mass of any liquid requires a definite quantity of **heat,** dependent upon the liquid and upon the temperature at which it evaporates. The quantity required per unit mass at a fixed temperature is called the heat of vaporization of the substance at that temperature. It may be measured by allowing the vapor to condense in a suitable **calorimeter,** the heat thus evolved, corrected for fall of temperature before and after condensation,

being observed. (The heat evolved in condensing is equal to that absorbed when the liquid evaporates.) The result is often surprising. For example, the evaporation of water at the boiling point requires about 540 calories per gram, or more than five times the heat required to raise its temperature from freezing to boiling. The explanation is the large amount of energy necessary to separate the molecules against their cohesion, and the much smaller amount (about 7.4% of the whole) which is used in expanding the vapor against atmospheric pressure. At lower temperatures the value is still greater, because the cohesion is then more effective; with water, for each degree below the normal boiling point about 0.6 calorie per gram must be added to the heat of vaporization. Trouton found that the heat of vaporization per mole for different liquids bears a nearly constant ratio to the absolute temperature of the boiling point. (L.D.W.)

HEAT TRANSFER. Heat can be transferred by three different methods. by conduction, where the heat must diffuse through solid materials or through stagnant fluids; by convection, where the heat is carried from one point to another by actual movement of the hot material (common in fluids); and by radiation, where heat is transferred by means of radiant wave energy.

The amount of heat transferred by conduction can be determined by using Fourier's law:

$$Q = \frac{KA\Delta T}{L}$$

where Q = Heat transferred per unit time.
A = Area perpendicular to the heat flow through which it is passing.
L = Thickness of the body of matter through which the heat is passing.
K = A conductivity constant dependent on the nature of the material and its temperature.
ΔT = The temperature difference between the hot and cold sides of the substance through which the heat is being transferred.

This law of conduction may also be stated

$$Q = UA\Delta T$$

in which U is the *conductance*. The reciprocal of conductance is resistance. It is the resistance that is additive when several conducting layers lie between the hot and cool regions. In a multiple layer partition having an overall conductance, U, and layer conductances U_1, U_2, etc.,

$$\frac{1}{U} = \frac{1}{U_1} + \frac{1}{U_2} + \frac{1}{U_3} \cdots \text{etc.}$$

Thus, when dealing with several different materials in layers, such as might be found in a furnace wall, the form:

$$Q = \frac{A\Delta T}{\dfrac{L_1}{K_1} + \dfrac{L_2}{K_2} + \dfrac{L_3}{K_3} \cdots \text{etc.}}$$

may be used, although this must be modified if the walls are curved. The subscripts 1, 2, 3, etc. refer to the various layers.

When a solid partition is conducting heat from one fluid to another, as in the accompanying figure, a thin layer of fluid remains stagnant against each face of the partition and becomes, in effect, another conducting layer outside of which the free stream fluid temperature exists. Although this film is extremely thin, its relative resistance to heat flow may be high compared to the solid partition. This is especially true in heat exchangers where the solid wall material is metal of good conducting properties. Although **film conductance** is difficult to study or evaluate, being under the control of

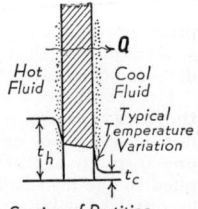

Section of Partition
Bathed with Hot
and Cool Fluids

several variables (i.e., fluid velocity, viscosity, density, turbulence), its importance to conductive heat transfer is considerable.

Typical conductivities of some common materials are cited in the following table.

MATERIAL	TEMPERATURE °F.	CONDUCTIVITY B.T.U. PER HR. PER SQ. FT. PER DEG. PER FT. THICKNESS
Air	50	.014
Aluminum	100	122
Asbestos, loose	400	105
Brick, fire	2000	.9
Brick, red	100	.27
Concrete, stone	60	.6
Copper	100	220
Corkboard	80	.025
Insulating wallboard	60	.03
Magnesia, powdered	400	.06
Rock wool	90	.023
Rubber, soft	60	.10
Steel	100	25
Water	100	.35
Wood, pine	60	.07

Transfer of heat by *convection* implies a carrying medium traveling from hot to cool region. While this might be a solid (conveying, each trip, heat in the amount of weight $\times$ specific heat $\times \Delta T$), the important cases of convection arise in connection with fluids in steady motion as the carrying media. Dependent on whether the fluid is under forced, controlled motion or is moved by natural density differences caused by the heat-

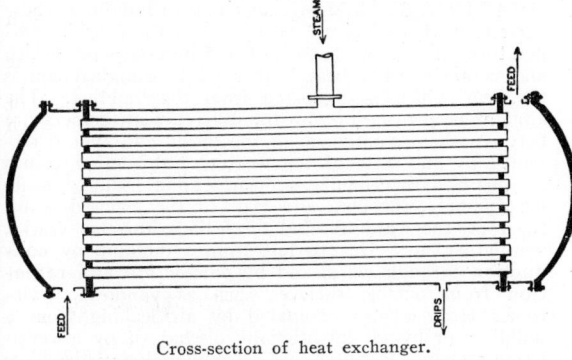

Cross-section of heat exchanger.

ing itself, the convective process is termed "forced" or "free." Heat transfer in forced convection is more readily analyzed and predicted than in free convection, although the latter is important in the liberation of heat from steam radiators, hot walls, pipe lines, and other static hot surfaces about which circulate "free" fluid currents. For transfer of heat by convection, use is made of the concept of the *heat transfer coefficient "h"* which is defined as: $h =$ heat transferred per unit time per unit area per unit temperature difference.

The coefficient h for forced convection in conduits can be studied experimentally by using the theoretical relationship:

$$\frac{hD}{k} = K \left(\frac{Du\rho}{\mu}\right)^a \left(\frac{C\mu}{k}\right)^b$$

where $\frac{hD}{k} = Nusselt\ group =$ heat ransfer coefficient $\times$ diameter of flowing stream $\div$ conductivity of fluid.

$K =$ An experimentally determined constant.

$\frac{Du\rho}{\mu} = Reynolds\ group =$ Diameter of flowing stream $\times$ velocity of stream $\times$ density of fluid $\div$ viscosity of fluid.

$a, b =$ Constants experimentally determined.

$\frac{C\mu}{k} = Prandtl\ group =$ specific heat of fluid $\times$ viscosity of fluid $\div$ conductivity of fluid.

After K, a, and b have been determined for one system, h can be evaluated for any other geometrically similar system.

Other theoretical equations can be used for other situations, such as flow outside of pipes, condensing vapors (steam), and natural convection in tanks. Many cases have been studied so that information is available for almost every practical problem.

When h has been determined it is put into the equation $Q = hA\Delta T$, where Q is the heat transferred per unit time, A is the area through which it is being transferred and ΔT is the temperature difference across the material through which the heat is flowing. For design purposes, the known facts are usually "Q" and the properties of the fluid, both dynamic and physical. By proper manipulation of the known facts, together with the various formulae, the term "A" (the area) can be determined and from it the equipment size needed can be calculated.

The amount of heat transferred by radiation can be determined by use of the *Stefan-Boltzmann law*:

$$Q = bA(T_1^4 - T_2^4)$$

where Q is the amount of heat transferred per unit time, b is a constant, A is the area of the radiating surface, T_1 is the absolute temperature of the radiating body and T_2 is the absolute temperature of the receiving body. Various correction factors are introduced into the formula to account for the shape of the bodies, their thermal radiation characteristics and the properties of the media through which the radiant rays must pass while traveling from radiator to absorber. The thermal radiation characteristics are its emissivity, a measure of its ability to radiate at a given temperature, its absorptivity, a measure of its ability to absorb heat and its reflectivity, which measures its ability to reflect without absorbing.

Radiant energy travels in a straight line. Therefore to transmit it to an object out of sight of the radiator requires a reflector, such as a furnace wall, to deflect the rays to their objective.

It is possible to set up controlled laboratory radiation between simple plane surfaces and determine therefrom accurate coefficients to incorporate into radiation equations. However, the radiation of heat from furnace gases, consisting of non-luminous gases, luminous carbon particles in flame, ash globules, etc., to the walls and tubes of a steam generator in commercial operation at variable load, is another matter. Here, empirical data which are gathered and interpreted from field tests on similar equipment, must still be resorted to however great the designer's urge to go back to basic laws of heat transfer.

Radiant heat transfer in furnaces is roughly proportioned to the difference in the fourth power of the absolute temperatures of the radiating and receiving surfaces. The water wall surface is approximately at boiler saturation temperature, while the superheater surface varies from this to somewhat above the temperature of the steam at the superheater outlet. However, the mean radiating temperature of the furnace gases is usually over 2200° F. The fourth power of the receiving surface temperature is thus seen to be small compared to the fourth power of the transmitting surface temperature; consequently the latter controls the transmittance, and boiler tube temperature does not need to be considered a variable to be accounted for.

The accompanying figure shows some of the arrangements in which radiant heat-absorbing surface is disposed. It may be used to illustrate another of the diffi-

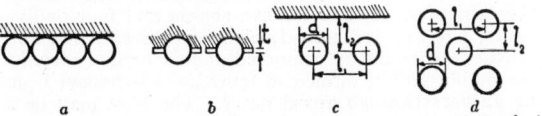

a   b   c   d
(n rows deep)

(Morse's Power Plant Engineering and Design.)

culties which beset the designer in following a rational or semi-rational form of radiation analysis. Projected radiant surface is one thing; actual radiant energy receiving surface may be quite a different area. For example, suppose the tubes of case (a) to be separated and spaced l_1 inches on centers. The *projected* areas of cases (a) and (c) would then be the same, but it seems obvious that re-radiation from the wall causes more of a (c) tube to receive radiant energy than is the case with an (a) tube. Also, if δ is a factor correcting projected area to *equivalent* absorbing surface, what value should be assigned to it in the case of a bank of tubes which may receive by re-radiation some radiant energy deep in the tube bank? Here δ has a minimum value of 1, but some investigators have derived expressions which indicate that δ may have a magnitude of 3 or more.

Some of the more common cases of industrial heat transfer are:

1. Radiation from fuel beds and luminous gases to absorptive surfaces such as boilers, cylinder walls, etc.

2. Radiation from heat generators such as drying lamps.

3. Convection of heat out of combustion regions.

4. Convection of heat from hot surfaces under either free or forced convection.

5. Conduction of heat through the tubes of boilers, heaters, heat exchangers, condensers, etc.

6. Conduction in walls, pipe covering, and other so-called "heat insulators." (D.E.M., F.T.M.)

HEAT TRANSFER COEFFICIENT. Heat Transfer.

HEAT TREATING. Heating and cooling of metals to effect changes in properties. **Annealing** and normalizing are generally for the purpose of softening or improving the grain structure. Patenting is also a softening process in which cold drawn carbon-steel wire is heated above its critical temperature range followed by cooling to below this range in a molten lead or molten salt bath, with subsequent cooling to room temperature.

While heat treating includes the softening treatments, it most often implies hardening and strengthening. In the case of steels this requires heating to above the critical temperature range followed by rapid cooling (**quenching**) in oil, water, or brine, except in the case of special grades which harden on cooling in air. This is followed by **tempering,** a low-temperature reheating treatment which reduces the internal stresses caused by the hardening treatment. Tempering may be carried to a high enough temperature to reduce somewhat the extreme hardness of the as-quenched steel and increase the toughness and ductility, depending on the requirements of the part. (See **Steel.**)

Another important form of heat treatment for hardening is **precipitation hardening.** (See also **Carburizing, Case Hardening,** and **nitriding.**) (R.H.H.)

HEATER. Cathode; Tube, Electronic.

HEATH FAMILY. Ericaceae. This is a small family composed mainly of woody plants. Nearly all plants of this family grow in acid peaty soil, often covering extensive areas. Most of the heaths have a pronounced **xerophytic** habit. The leaves are usually entire, leathery and have a thick cutinized upper surface. Many are evergreen. The flowers may be solitary, but are more frequently in **racemes.** Usually they are regular. There are generally four or five **sepals** and four or five **petals,** more or less united; eight or ten **stamens** and a single inferior **ovary** composed of four or five **carpels.** The **pollen** grains, formed in fours, are discharged from the **anthers** through apical pores. The fruit may be a **capsule,** a **drupe** or a **berry.** Pollination is mainly by bees.

Many plants of this family are used by man. Rhododendrons and Azaleas are frequently cultivated because of their beautiful bright-colored flowers, and many hybrids have been formed. There are over 200 species. Eastern Asia is especially rich in plants of this genus. *Epigaea repens,* the trailing arbutus or mayflower of eastern North America, is another plant much sought for its delicately fragrant flowers. It is not easily grown in cultivation. Species of *Gaultheria,* including the red-berried *G. procumbens,* called checkerberry, or wintergreen, contain an oil which is sometimes distilled from the plant. The fruits of blueberries, huckleberries and cranberries, members of this family, are widely used. In recent years considerable attention has been given to the blueberries and several new varieties having very large berries have been developed. Species of *Erica* cover vast areas of moor in England. One species, *Erica arborea,* grows in southern Europe and northern Africa. It is a stout bush several feet high, with a yellow hard close-grained wood. From the rootstocks of this plant briarwood is obtained. This is used in making briarwood pipes. (R.M.W.)

HEATH HEN. Aves, Galliformes. A grouse of the eastern United States, resembling the prairie chicken of the central states. The last of these birds were rigidly protected on Martha's Vineyard and increased to a large number by 1916. Later, disease introduced with domestic turkeys affected the birds, and other destructive agents further reduced their numbers until the last individual disappeared in 1931. (A.W.L.)

HEATING, BUILDING. Portions of buildings where persons work or live are heated when the outside temperature falls below that considered necessary for health and comfort. The heat that must be supplied equals the heat which is dissipated from the building. The amount needed varies with the difference in temperature between that maintained on the inside and that determined by the state of the weather. But other things also have a determining influence upon building heating; namely, the area and type of the exposed walls, the roof, the windows, and the leakage through cracks, ventilators, etc. Heat is lost from a building by **conduction** through walls, and by **convection** and **radiation** from outside surfaces such as windows, walls, roof. Heat is also dissipated by air leaking from a building, or by air intentionally discharged by a ventilating system. Research and accumulated statistics have provided the data from which the amount of heat lost through walls, windows, roof, etc., can be determined, given the composition of the walls.

The heat transfer coefficient is stated as the number of B.T.U.'s lost per sq. ft. of surface area per hour per degree difference of temperature. By multiplying the coefficient by the exposed surface and the difference between outside and inside temperature, part of the heat loss is determined. This, however, does not take care of heat loss by leakage or infiltration. To allow for that a certain number of air changes per hour are included in the needed heating. For the ordinary room having windows, it is assumed that to allow for leakage, the heating system must supply heat to take care of a complete change of air each hour. (Halls, stores, factories, may need two or three times this allowance.) By computing the heat needed for each heated room, and totalling these for the building, the required heat output of a heating system may be determined.

A person gives off some 400–600 B.T.U. per hour. It will be necessary to reckon with this source of heat in any heating analysis of a public building, where large numbers of people may gather in one room. The building heating system and ventilation are jointly considered in a large building, especially a public one. With any attempt to ensure adequate heating, the modern building owner usually will find it economical to spend money for heat insulation, since by proper coordination of investment in heat insulation, heating system, and **fuel,** a minimum annual heating cost can be secured.

Heat insulation is available commercially in a number of forms. The air space between studding, if blocked at the top floor levels, becomes a dead air space, and is effective in insulating against heat loss. Plaster and brick are fairly good conductors; wood, building paper, fiber wall board, are heat insulators. Rock wool, either granular, fluffed, or in bats, is widely used where the maximum of heat insulation is attempted.

Central heating means the supply of heating service to a group of surrounding buildings from a central heating plant. The heat-carrying medium is sometimes steam, sometimes hot water. District heating is similar to central heating, but a distinction can be drawn as follows: A central system can be thought of as that which supplies a group of buildings which are under common building superintendence, or having common aim, as, for example, buildings of an educational institution, or a manufacturing plant. District heating, on the other hand, is a public utility service and applies to the heating of buildings in densely occupied city sections, from a public utility heating plant which sells heating service.

The instances where electric heating is of use are almost numberless in this day, since one may add to the large assortment of domestic electric devices, such as irons, corn poppers, percolators, etc., an equally great variety of commercial devices, like annealing furnaces, bakers' ovens, glue pots, vulcanizers, etc. However, the use of electricity for building heating, though apparently successful in some experimental installations, is at a great disadvantage because of the high cost of electric heat compared to that obtained directly from the fuel. (F.T.M.)

HEATING SYSTEMS. While the words "heating system" generally convey the idea of one of the indirect systems employed at the present time to heat homes or buildings, direct radiation is, itself, a system of heating. A stove located in the room which it is heating proves to be a very efficient means of getting heat from a fuel into the air of the room. As a system, it suffers from the following disadvantages:

1. Unsightliness.
2. Multiplicity of heating units required for a building containing a number of rooms.
3. Unequal distribution of heating.
4. Fire hazard.

While most of the heat supplied by a stove is black body radiation, there is some considerable amount of convection resulting from air currents sweeping up over the heated portions.

A fireplace delivers no heat by convection; rather, it withdraws heat from a room by leakage up the flue. The heating effect of a fireplace comes entirely from radiation. As a heating system, a fireplace is inefficient and wasteful of fuel. The adequate heating of buildings seems to be obtained best by the use of indirect heating, that is, generation of the heat at a central point, as a **furnace,** then loading that heat onto the medium which conveys it to the various rooms in the required amount. The different heating systems in use today can be classified on the basis of the heat-conveying medium, i.e., warm air, hot water, steam. Each of these systems is briefly described below.

In a warm-air heating system the furnace is enclosed by a casing, leaving an air space between casing and furnace. From the casing ducts lead out to the different rooms. Air is supplied to the casing, either directly from the furnace room, from outside the building, or from return ducts which withdraw the air from the rooms. The air in the casing being in contact with the heated furnace, expands, becomes lighter than normal, and rises in the ducts until it is discharged in heated condition into the rooms through registers located in the floor or side wall. The older warm-air systems had a circulation that was maintained entirely by the levity of the heated air leaving the furnace, but had certain disadvantages. For example, the effect of wind blowing against one side of the building caused infiltration which opposed duct air pressure in the windward rooms, an action which tended to give unequal distribution of heat, leaving the rooms on the windward side underheated, and those on the leeward side, overheated. Also, the furnace had to be centrally located, the basement was encumbered with large numbers of ducts, and homes which were not roughly cubical in shape were not well adapted to this style of heating.

Recent developments in the field of warm-air heating have led to the employment of forced circulation of the hot air by a motor-driven blower. Having this, it is not necessary to be so careful about friction losses, and smaller ducts, trunk-line systems, and filters are possible. Since the available pressure is of much higher order than that obtained from gravity alone, there is no difficulty about forcing air equally into all rooms. This system of heating, furthermore, is admirably adapted to **air-conditioning** requirements. The operation of a forced circulation warm-air heating system is controlled by a thermostat located in a representative room. The **thermostat** is connected electrically to a small draft-regulating motor, and in effect becomes a switch for that motor. A certain differential of a few degrees is allowed in a room. When the temperature sinks to a predetermined point, the thermostat operates the draft-controlling motor, which in turn closes the check damper and opens the draft so that the rate of combustion is increased. The temperature of the air in the bonnet of the casing rises until a thermostatically operated switch located there starts the blower. The warm air is then circulated in the rooms until the temperature rises to a predetermined limit, upon which the room thermostat regulates the draft. Soon the slower rate of combustion results in a cooler furnace bonnet, and the blower is thermostatically stopped. This cycle is repeated as frequently as needed, but the room and bonnet thermostatic switching do not necessarily stay in synchronism.

Hot-water heating systems have been very popular in recent years. When the reason for this is sought, it will be found that hot water, as compared to steam or gravity-circulated warm air, offers a more uniform and steady heat. It is free from the unequal heat distribution and bulky duct work of the warm-air system, and from the condensate troubles of the steam or vapor system. Like the steam system, it has no summertime usefulness, and offers no possibility of air conditioning. The radiators and piping, furthermore, are larger than those for an equivalent steam system. In a hot-water system the water is heated in a heater similar to a boiler except that it is completely filled with water, and no boiling takes place. Most hot-water systems operate at atmospheric pressure, which limits the temperature to which the water may be raised to slightly over 200° F. However, for heating homes, this temperature is entirely adequate for the coldest weather. Over much of the heating season, temperatures of 140–180° F. are sufficient. From the heater the water passes to the radiators located in the rooms, where the heat is transferred to the air. Delivery of heat to the air cools the water which, as a result, becomes more dense than that in the supply mains, and it sinks to the heater; thus circulation is maintained by gravity and is due to the difference in density between the columns of hot and cool water between the radiator and the heater. Central hot-water heating systems, and occasionally domestic systems, work on forced circulation created by a motor-driven circulating pump. This gives more flexibility to the system and permits the use of smaller pipes. The control of a hot-water heating system is vested in a thermostat which operates the check damper and draft. The expansion elements of the thermostat are immersed in, or are subjected to the temperature of, the hot water in the heater. (The thermostat tends to maintain uniform water temperature by adjusting the rate of combustion.) The temperature of the rooms is controlled by altering the temperature of the water leaving the heater through opposing the action of the thermostat with adjustable weights or springs provided for that purpose, or by a variable rate of by-pass of the water around, instead of through, the heater.

Steam heating, although lacking flexibility, and sometimes having difficulties of clearing condensate from the system, is, nevertheless, a convenient medium for indirect heating because a considerable heating effect is obtained from a relatively small volume of the circulating fluid. Because the heat is released by condensing, latent heat of evaporation is made available, whereas in hot-water systems, only a portion of the heat of the liquid is available. Since latent heat is several times the amount of heat of the liquid, the size of a steam heating system compares favorably with any other of equivalent heating capacity. It is widely employed for large buildings, and to a more limited extent, for residences. As the circulation depends on the difference in density between steam and water instead of between hot water and cool water, it is quite positive. In laying out a steam heating system, it is very important to pitch the return pipes so that condensate will drain properly. In some cases the return and supply pipes are the same, that is, the condensate flows against the steam, but this is suitable only for small inexpensive installations. Two-pipe heating systems are found in many different arrangements involving differences of points of admission of steam to radiators, methods of expelling air, and returning of condensate to the boiler.

Although experimental at present, the "heat pump cycle" may possibly, in the future, receive favorable attention as a building heating system. The heat pump is a reversed refrigeration process. In the normal refrigeration system the refrigerant is compressed, the resulting heat is removed by the radiation of the condenser and then the refrigerant is allowed to expand to cool the stored material, i.e., it absorbs heat from this material. The heat pump uses this same principle, but utilizes a different part of the system for its useful function. It draws air or water from some outside source and cools it by having the refrigerant (if we may still call it this) absorb its heat. This refrigerant is then compressed and the heat removed at the condenser and utilized for heating purposes. This system has the advantage that by proper utilization of the apparatus it may be used for summertime cooling and the disadvantage of requiring a large motor for driving the compressor. (F.T.M.)

HEATING VALUE. The heating, or calorific, value of a **fuel** is the quantity of heat produced by the combustion, under specified conditions, of unit weight or volume of the fuel. The heating value of a fuel may be calculated by formula which may be derived for any fuel by multiplying the percentage of each chemical element present by its heating value per unit weight, and adding the products for all combustible elements in the fuel. Thus for coal, whose combustible elements consist of carbon, hydrogen, and sulfur, the heating value is:

$$14{,}540C + 62{,}000H + 4000S \text{ (B.T.U. per lb. of coal)}.$$

The numbers in the above formula are the heating values per lb. respectively, of carbon, hydrogen, and sulfur. In the use of this formula, it is essential that only that portion of the element that is actually free to burn be employed. For example, all coal contains some moisture. Now the hydrogen present in this water is not free to

burn (i.e., it is already combined with oxygen). Therefore the figure used for H in the foregoing formula should not include the hydrogen present as water.

Heating value by formula will not necessarily be the same as that obtained experimentally with the fuel calorimeter. The difference lies not in the accuracy of the experiment, nor of the calculation, but in the possible endothermic or exothermic reactions which take place when a compound fuel, such as a hydrocarbon, is burned. The volatile matter of coal must be broken down into the elements of carbon and hydrogen by heat-absorbing action before they may reunite with the oxygen during combustion. For this reason, experimentally determined heating values are less than those which are computed by formula, which make no reference to endothermic or exothermic reactions. Approximate heating values of some of the common fuels are: coal, 13,000 B.T.U. per lb.; natural gas, 1000 B.T.U. per cu. ft.; artificial gas, 300 B.T.U. per cu. ft.; gasoline, 19,000 B.T.U. per lb.; wood, 5000 B.T.U. per lb.

Many thermodynamic analyses require the use of a "lower heating value" which may be obtained from the above values by subtracting an allowance for the latent heat of evaporation of the steam found in the products of combustion. Lower heating value might be thought of as "sensible heating value." (F.T.M.)

HEAVE. Fault.

HEAVISIDE LAYER. Ionosphere.

HEAVY MINERALS. Sedimentary Minerals.

HECTOCOTYLUS ARM. One of the arms or tentacles of the male of certain species of **cephalopod** mollusks, specialized for the insemination of the female. It may serve as an intromittent organ, introducing **spermatophores** into the mantle cavity of the female, and in the paper nautilus, *Argonauta,* it breaks away from the body and enters the mantle cavity of the female, remaining inside for some time. (A.W.L.)

HEDGEHOG. Mammalia, Insectivora. Small compact animals with short legs and tail and a sharp nose. The entire upper surface is covered with sharp spines which protect the entire body when it is curled up. The numerous species live in Africa, Europe, India, and the remainder of Asia north of the Himalayas. They are nocturnal animals. The European hedgehog, *Erinaceus europaeus,* is also called the urchin. (A.W.L.)

HEEL. The prominence at the posterior end of the foot. It is based on the projection of one bone, the calcaneum, behind the articulation of the bones of the lower leg. In the long-footed mammals, both the hoofed species and the clawed forms which walk on the toes, the heel is well above the ground at the apex of the angular joint known as the hock or hough. In plantigrade species it rests on the ground. (A.W.L.)

HEIGHT OF INSTRUMENT. Earthwork.

HELIATHIN. Dyes.

HELICAL GEARING. For high pitch-line velocities and heavy loads, some form of "twisted tooth" gear is generally used. Two important types are helical gears and double-helical or herringbone gears. Both helical and herringbone gears are essentially spur gears with teeth twisted across the face in the form of a helix about the axis of rotation.

When spur gear teeth engage, the contact extends across the entire tooth on a line parallel to the axis of rotation, and may result in noise and shock at high speeds. In helical gear engagement, contact begins at one end of the entering tooth and gradually extends along a diagonal line across the tooth face as the gears rotate. The nature of the contact is such that with sufficient face width, two or more teeth are in contact

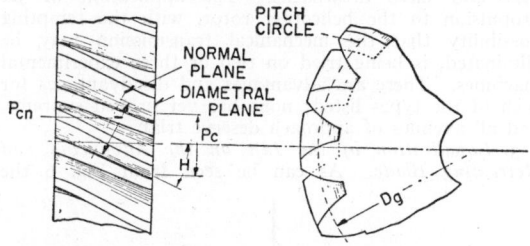

and are carrying the load at all times. Helical gears are therefore used for transmission ratios as high as 10:1, and at pitch-line velocities up to 2000 ft. per min. for commercially-cut units. Herringbone gear sets of special design have been successfully operated at pitch-line speeds of 12,000 ft. per min.

Tooth elements of helical gears are similar to those of spur gears. The *helix angle H* of the tooth is measured between the line tangent to the tooth helix at the pitch circle and the shaft axis. In any pair, the gears have teeth with mating right-hand and left-hand helices. The usual method of tooth measurement is by diametral pitch P_d, which corresponds to circular pitch P_c in the diametral plane, perpendicular to the axis of rotation. By using standard pitches, the pitch diameters (and therefore the center distance of helical gear sets) can be given in commonly used fractions or integers; consequently, a spur gear set of a certain size can be replaced directly by a similar helical gear set. Actual tooth thickness depends upon the pitch and the size of the helix angle; if the circular pitch P_c be held constant, the actual tooth thickness measured perpendicular to its elements will decrease as the helix angle H is increased. A different cutter is required for every change in helix angle, although the pitch may remain constant. To eliminate an extensive variety of cutters, commercially available helical gears are made in several standard helix angles, among which are 7° 30′, 15°, and 23°.

By using a standard pitch in a plane normal to the tooth helix, the pitch diameter of a helical gear can be varied to suit a particular center distance by changing the helix angle. In this method of tooth measurement, the normal diameter pitch P_n corresponds to a normal circular pitch P_{cn} in the normal plane. Helical gear teeth designed with normal diametral pitches may be cut with standard spur gear cutters or hobs.

The pitch diameter D_g of a helical gear, based upon normal pitch P_n, is given by

$$D_g = N_g / P_n \cos H$$

where N_g is the number of teeth in the gear. The power transmitting capacity of helical and herringbone gears may be found by methods analogous to those used for spur gearing.

End thrust inherent in single helical gears can be eliminated by the use of herringbone gears that consist virtually of two integral single helical gears of opposite hand, which absorb the axial thrust within the gear. Herringbone gears are extensively used for hoisting and mining machinery, rolling mills, sugar mill and lumber machinery, turbine and compressor drives—in fact for nearly all heavy duty, high transmission ratio applications. (See **Spiral Gearing.**) (H.C.H.)

HELICOPTER. The conception of the helicopter appears first in the Chinese top with two airscrews made of feathers turning in opposite directions. The origin of the Chinese top is lost in antiquity and it even antedates the sketches made in the 16th century by Leonardo da Vinci. It is not certain that the Italian artist-inventor actually flew any helicopter models, but his writings showed good understanding of the lifting screw, and he made use of the Greek word *helix* meaning spiral or twist in connection with the idea of flight. Much later, a combination of *helix* with the word *pteron,* meaning wing, led to the term helicopter whose correct pronunciation is hěl′ĭ-kŏp′tĕr. Centuries intervened, however, between understanding and realization and there was no truly successful helicopter until the Focke-Wulf FW-61 appeared in Germany in 1936, and Igor Sikorsky built and flew the VS-300 in 1937.

Contrast with the airplane and the autogiro serves to define the helicopter. In the airplane a fixed wing provides lift or sustention; in the autogiro the blades revolving freely like those of a windmill have the same function. In both airplane and autogiro forward thrust or propulsion is given by a propeller which absorbs the whole power of the engine. But in the helicopter the power of the engine is used to drive the blades of the rotor which provides both sustention, at all times, and propulsion (when its axis of rotation is tilted forward)

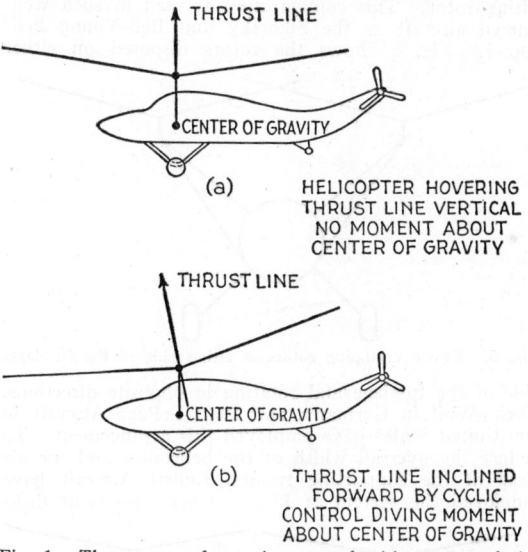

Fig. 1. The process of securing control with a rotor of the flapping-blade type.

(Fig. 1). It is because of this dual function of its rotor that the helicopter is superior in efficiency to the autogiro.

The **autogiro,** invented by the Spaniard, Juan de la Cierva, and first flown successfully in 1921, has been superseded by the helicopter for a reason other than superior performance. The helicopter affords the true vertical flight of which the autogiro is incapable. The helicopter can hover in the air, rise vertically and come down vertically and thus carries man a step further in his conquest of the air. Early inventors thought that to secure direct lift or vertical flight was the sole requirement of a practical aircraft. They were wrong; other objectives must be attained:

1. *Ability of the Blades to Turn or Autorotate in Case of Power Failure.* Thus the helicopter rotor is effectively converted into an autogiro rotor and can come down without power, safely on an inclined path, and somewhat hazardously in vertical descent. Autorotation after engine failure is readily secured by decrease of the blade pitch or setting. The helicopter is then transformed into an autogiro.

2. *Ability to Take up the Torque or Turning Moment of the Lifting Rotor.* To secure a thrust or lift of at least 10 lbs. per hp., the rotor must have a much greater diameter for a given horsepower than the conventional airplane propeller. It also revolves more slowly and

the engine is frequently geared down, 10 to 1 or even more. But gearing down the engine 10 to 1 also means that the torque or turning moment in the shaft is increased in the same ratio. The torque of the rotor ceases to be the subsidiary factor that it is in the airplane propeller and counteracting this torque becomes a prime element in the design of the helicopter. Fig. 1

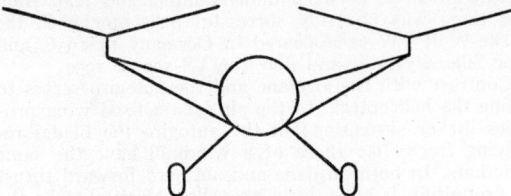

Fig. 2.   Two rotors at either side of the fuselage.

shows the arrangement which has become almost conventional; the thrust of an auxiliary tail rotor, also driven by the engine, counteracts the torque of the main lifting rotor. This configuration is used in such well-known aircraft as the Sikorsky and Bell-Young helicopter. Fig. 2 shows the rotors disposed on either

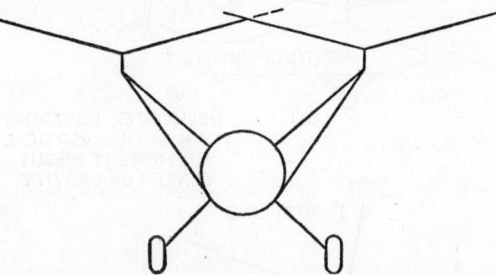

Fig. 3.   Two overlapping rotors at either side of the fuselage.

side of the fuselage and rotating in opposite directions. Focke-Wulf in Germany, and Platt-LePage Aircraft in the United States have employed this arrangement. To reduce the over-all width of the helicopter and the air drag of the supporting trusses, Kellett Aircraft have employed the system of Fig. 3, where the rotor disks

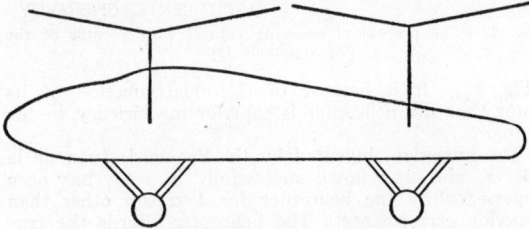

Fig. 4.   Two rotors in tandem.

overlap to some extent. The arrangement of Fig. 4 is utilized in the PV Engineering Forum and Rotorcraft machines, and has the advantage of a large central space for the passenger. Bendix and Hiller among others have resorted to coaxial, contrary turning rotors, placed one above the other (Fig. 5). Such coaxial

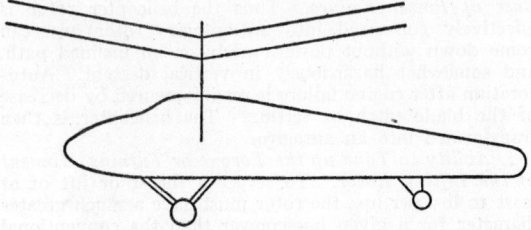

Fig. 5.   Two superimposed coaxial rotors.

machines eliminate the loss of power, the complication of a long transmission, and the hazard to the public of the tail rotor. They are also compact, but are complex and have drawbacks. The application of jet propulsion to the helicopter rotor, with the tempting possibility that the mechanical transmission may be eliminated, is being tried on two or three experimental machines. There are advantages and disadvantages for each of the types listed; none has yet proved supreme, and all avenues of approach deserve trial.

3. *Equalization of the Lift on the Advancing and Retreating Blade.* As can be seen from Fig. 6 the

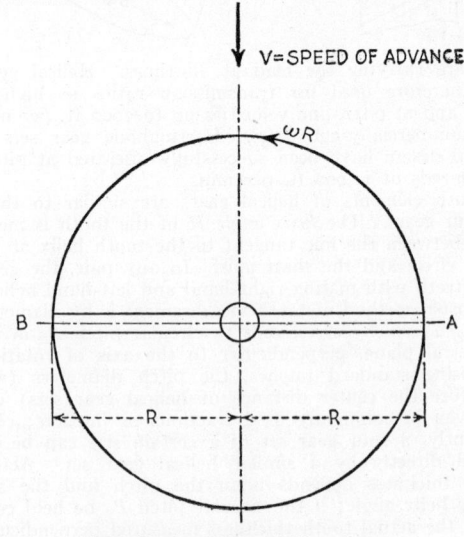

RELATIVE SPEED AT A = $\omega R + V$
RELATIVE SPEED AT B = $\omega R - V$

Fig. 6.   The advancing blade in forward flight meets the air at greater resultant speed than the retreating blade.

advancing blade encounters a more rapid airflow than the blade retreating from the wind of forward flight. Therefore, the advancing blade also experiences a greater thrust or lift. If the blades are rigidly mounted to the rotating shaft the single-rotor helicopter suffers from a powerful rolling moment. Equalization of thrust on the advancing and retreating blades and elimination of the rolling moment create just as fundamental a problem as taking up the torque of the main rotor. One method is to use superimposed coaxial rotors, but while rolling moments are thus avoided there is no true equalization of lift, so that vibration effects are present. The most generally adopted method is that of the flapping hinge, one of the valuable conceptions of the the de la Cierva type which is shown diagrammatically in Fig. 7.

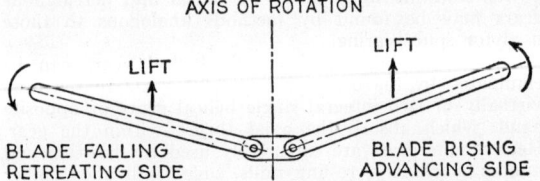

Fig. 7.   Flapping to equalize lift in forward flight.

With the free-flapping hinge, no moments can be transmitted to the remainder of the helicopter. The blade on the advancing side rises and receives an additional flow from above. In so doing it decreases its angle of attack and loses lift so that substantial, although not complete, equalization of lift is secured. The flapping motion of the blades is such that it describes the surface of a shallow cone. Flapping introduces difficulties, such as varying loads in the plane of rotation which have to be met by the introduction of a vertical or lag hinge as shown in Fig. 8. Instead of the flapping, other de-

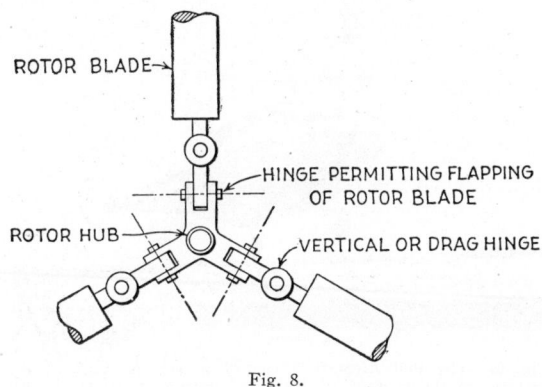

ROTOR BLADE→

HINGE PERMITTING FLAPPING OF ROTOR BLADE

ROTOR HUB→

VERTICAL OR DRAG HINGE

Fig. 8.

signers have sought equalization of lift by using two blades continuous across the axis of rotation and oscillating together about a "see-saw" hinge. Still others have advocated feathering or change in the angle of attack with the angle on the advancing side decreasing. Feathering and flapping are aerodynamically equivalent, since both tend to equalize lift on the advancing and retreating sides.

4. *Control.* In hovering flight, stabilizing surfaces such as are used in the airplane become ineffective, and the helicopter is unstable. Fortunately the time required for its unstable oscillation is long, so that good control by the pilot can make up for lack of stability. The control of the rotor is achieved by cyclical variation of the blade angle or pitch of the blade. In the airplane, control is secured in direct fashion by displacing a control surface such as the elevator and thereby producing a pitching moment, either up or down. In the rotor with flapping blade, the application of cyclic feathering cannot directly produce a pitching or rolling moment since the blade is freely pivoted at its root. Cyclic feathering displaces the tip path plane or the plane in which move the tips of the blade. The direction of the thrust is changed correspondingly as illustrated in Fig. 1. It is only when the direction of the thrust line is altered that a moment about the center of gravity of the aircraft is created. Thus, helicopter control is indirect, and while it is powerful at all speeds, it is accompanied by a certain lag to which airplane pilots must accustom themselves. The mechanism of the control, leading from the pilot's cockpit to the blade, is also more complex than in the airplane since there must be transmission of motion to revolving blades. Tilting plates or swash plates become necessary, with a stationary outer part connected to the cockpit, and an inner revolving part, with a ball race between the stationary and the revolving components of the swash plate. Another difficulty besides hovering stability lies in the fact that air-stream conditions for the blades vary as they revolve, and these variations of force and moment can produce "shaking" or continual motion of the control stick, to the discomfort and fatigue of the pilot.

5. *General Requirements.* The foregoing requirements are those which are special to the helicopter. There are others which are similar to those an airplane has to meet. In the early development stages of the helicopter, the mere ability to achieve vertical flight was so astounding that other requirements were neglected. As the art progresses, greater speed is called for, with a top speed of more than 100 m.p.h. Climb on an inclined path of over 1000′ per min. is deemed necessary. While vertical climb is an essential element of the helicopter, it should be noted that both climb

and ceiling are far greater when the helicopter flies on an inclined path. More useful load, a weak spot of the helicopter, is also demanded. Perfect vision forward and down unimpeded by the instrument board, freedom from mechanical difficulties, passenger accommodation and comfort are also sought. The evolution from mere flight to comfortable transportation proceeds rapidly.

Piloting Problems. No discussion of the helicopter is complete without reference to piloting. From some points of view piloting a helicopter is easier than flying an airplane because in landing or in selecting a landing spot the factor of haste is removed. The pilot can hover carefully and, provided the power of his engine holds out, alight safely in the most difficult and restricted terrain. Also, it has been shown again and again that helicopters can operate in snow and fog which ground all other aircraft. In other respects the helicopter is more difficult to fly than the airplane. Because of its instability it has to be "flown" all the time, so that the pilot enjoys neither physical nor mental relaxation. A flight of a couple of hours becomes fatiguing. Excessive vision forward through a hemispherical expanse of transparent plastic may produce vertigo. On the whole, it is easy to learn how to fly a helicopter, perhaps easier for a man who has never flown at all, than for the experienced airplane pilot who mistakenly believes that he can "check out" on a helicopter on a first attempt, without specific instruction. There are many differences in piloting a rotary and a fixed-wing aircraft. One of these differences is the effect produced when the control stick is pulled back sharply. In the airplane this results in increasing the angle of attack and lifting capacity of the airfoil, so that the speed of flight is decreased. Pulling back the stick in the airplane is thus a natural prelude to landing. In the helicopter pulling back sharply on the stick can produce rearward flight which may be quite disconcerting to the airplane pilot. As stated previously, helicopter control is achieved by changing the direction of the thrust line in indirect fashion. Thus, if a helicopter pilot moves his stick to the right, the helicopter first of all moves to the right and only later does it roll to the right. The indirectness of the process may annoy the airplane pilot. Rudder control in the cockpit is as in the airplane, a matter of foot pedals. In a single-rotor helicopter, rudder control is obtained by changing the blade pitch of the tail rotor and hence its thrust. The unbalanced thrust has to be compensated for by a rolling inclination of the helicopter, which is troublesome. In the case of two coaxial rotors, the blade pitches of the two rotors are changed differentially—one rotor receiving an increase in pitch and the other a decrease. The torques of the two rotors are then no longer equal, so that the fuselage receives a net turning moment.

Another difficulty in piloting the helicopter arises from the fact that to the four classical controls of the airplane—rudder, ailerons, elevator, and throttle—there is added a fifth, namely pitch control; and pitch and throttle must be carefully coordinated to give altitude control. Imagine the helicopter on the ground. The pilot selects a low pitch while revving up the rotor. Then, when he wishes to leave the ground, he must increase pitch and throttle simultaneously in the correct proportions. When the helicopter has climbed high enough to have cleared trees or other local obstructions, the pilot may decide to fly horizontally. He pushes his stick forward and flies forward. But because a helicopter is basically more efficient in forward flight than in vertical flight, he now needs less pitch and throttle than in the vertical climb. Therefore, both pitch and throttle must be decreased in the correct proportions, otherwise the helicopter will climb sharply on an inclined path. This problem of an additional control and correlation of pitch and throttle are being met by a synchronization of pitch and throttle by the use of a constant speed governor acting automatically to give either correct pitch or correct throttle setting. The technique of making a dead-stick landing and flaring out at the end of the glide requires skill. We conclude that learning to fly a helicopter is not difficult, but the art has to be given attention even by an experienced airplane pilot.

Mechanism. In comparing the cost of a helicopter with the cost of an airplane, we have to remember that the helicopter needs all the mechanism of the airplane and much more besides. The full power of the engine cannot suddenly be applied to the rotor without shock to the blades, so that a clutch must be interposed between the engine and the transmission box. This clutch should preferably be automatic and controlled by centrifugal force. The transmission box has to give a speed reduction of 10 to 1, yet be a compact and light device. Long shafting revolving at slow speed and carrying high torque has to be provided. To secure autorotation in case of engine failure, the pitch of the rotor blades must be decreased and simultaneously the rotor must be completely disengaged from the engine. Hence a free-wheeling or over-running device must be interposed between the rotor and the transmission box. The mechanism for changing blade pitch either differentially or simultaneously is also likely to be complex. A brake for the rotor has to be provided when the helicopter is on the ground.

Maintenance. Just as the mechanism of the helicopter is more complex than the mechanism of the airplane, so its maintenance presents additional tasks. Blades must be carefully balanced, and the pitch of all blades kept to exactly the same value, if vibrations are to be avoided. Fabric-covered rotor blades have proved particularly troublesome and are being replaced by plywood or metal covering.

Safety. If a blade of a helicopter should suddenly let go, the ensuing vibration might be so violent as to prevent the pilot from jumping. Engine failure would be most serious if it occurred 50' or so from the ground so that forward flight with autorotation could not be established in time. But these two hazards aside, the consensus is that the helicopter, by virtue of its hovering characteristics, is safer than the airplane, and capable of operating under weather conditions which ground other types of aircraft.

Immediate Possibilities. Sikorsky helicopters gave the most useful military and naval service in several theaters of war in rescue work at sea; in ambulance work; in landing supplies to men stranded in the jungle; in detailed reconnaissance; in observation work from the decks of small merchant ships. The helicopter is also likely to be useful to the armed forces in acting as a flying crane, and in close artillery observation. Its military future is thus completely secure. In civil use, because of high initial cost, the need of careful maintenance, and the lack of stability, the helicopter is not likely to serve as a personal aircraft for several years to come. On the other hand, it has immediate and assured applications in industry. These applications are numerous and include: surveying; oil pipe laying; delivery of drugs and first aid in cases of disasters; coast patrol and coastal rescue work; forest patrol and fire fighting, etc.

Two Commercially Licensed Helicopters. On March 8, 1946, Bell Aircraft Corporation received the first commercial license ever granted by the Civil Aeronautics Administration on its Basic Model 47 with the designation NC-1H. Fig. 9 shows one of the Model 47 to be put into service as a flying relief ship by United-Rexall Drug Company. This Model 47 is a 2-place helicopter with side-by-side seating and good vision, equipped with a Franklin Air-cooled 6-cylinder engine of 175 hp. The 2-bladed rotor is controlled by a gyroscopic device mounted under the main rotor and helping to keep the aircraft on an even keel. The gross weight is 2100 lbs.

of which 607 lbs. is useful load. It has a top speed of over 100 m.p.h., an operating speed of 80 m.p.h., and a range of approximately 250 m.p.h.

Fig. 9. The Bell Aircraft Model 47 illustrates many of the features of the modern helicopter. Vision forward is secured over almost the entire hemisphere by liberal use of Plexiglas, yet the occupants have a definite point of horizon reference and a sense of protection from enclosures at the sides. In addition to the two main wheels, two nose wheels are provided which are completely swiveling. The scarcely visible blades of the tail rotor are protected by a long skid projecting from the rear end of the fuselage. Engine and transmission are placed in back of the cabin. The vertical shaft on which the two blades are mounted is placed back of the cabin and passes through the center of gravity of the whole helicopter. The rotor is disposed at some distance above the top of the fuselage.

The other helicopter for which CAA license has also been received is the Sikorsky S-51 (Fig. 10) which is a comfortable 4-place model with three passengers

Fig. 10. The four-seater Sikorsky S-51 in which three passengers are seated behind the pilot.

seated side by side, just aft of the pilot. The S-51 has a heating and ventilating system, cabin air control and defrosting of the pilot's windows, indicative of the desire to give passenger comfort. The gross weight is 4900 lbs. of which 1250 lbs. is useful load. The three-bladed rotor has a diameter of 48' 0". Equipped with a 450 hp. Pratt & Whitney Wasp, the top speed is reported at 103 m.p.h. and cruising at 80 miles. (It is understood, however, that in both the Bell Model 47 and the Sikorsky S-51, the CAA has not granted license for the highest attainable speed, but has reduced the placarded speed by perhaps 20 miles below top speed—on the ground that higher speeds produce stall effects at the tip of the retreating blade, vibration and possible decrease in control.) (Dr. Alexander Klemin.)

HELICOSPORIDIA. Sporozoa.

HELIOTHERAPY.
The treatment of disease by exposure of the body to the rays of the sun or to ultraviolet rays from an artificial source. The method consists either of general exposure of the body as a tonic measure or a stronger local exposure for use in various skin diseases and in indolent ulcers or slow-healing infected wounds. In tuberculosis, heliotherapy has proved of little value in the pulmonary form, but in

other forms, such as tuberculosis of bone, **lymph glands, spine, abdomen,** it may be of some value. (R.S.M., D.M.H.)

HELIOZOA. Sarcodina.

HELIUM. Symbol: He. Atomic number: 2. Atomic weight: 4.003. Density: 0.1785 gram per liter, 0° C., 760 mm., or 0.138 when air equals 1.000. Melting point: $-272.2°$ C. Boiling point: $-269.0°$ C. No isotope but of single atomic form: 4.

Helium is a colorless, odorless gas, of negative chemical properties with ordinary materials, except under the influence of electric glow discharge or **electron** bombardment helium forms compounds with **tungsten, iodine, sulfur, phosphorus.** In a **vacuum** electric discharge **tube** shows green to canary-yellow glow. Discovered first in the vapors surrounding the sun by Lockyer in 1868, through the yellow spectral line near the two yellow lines of sodium, then by Ramsay in 1895 in the mineral clevite.

Helium occurs (1) in minerals of **uranium** and **thorium,** such as **clevites, pitchblende, carnotite, monazite,** and also in **beryl,** (2) in mineral waters (1 part He per thousand of water, in some Iceland waters), (3) in volcanic gases, (4) especially in certain natural gases of the United States. The first discovery of this kind was made in Kansas. The richest helium wells are in Utah. In northeastern Texas, four wells have produced 55 million cu. ft. of helium. The fields in which are located the wells having the greatest percentage (1.3–8.0%) of helium are now held as government reserves. The primary use for helium is in inflating airships. Helium is a very light gas. Its lifting power is 92.64% that of hydrogen (of course, the comparative net buoyancy in **airships** is much less than this), and it is non-inflammable, (5) in ordinary air, about 1 part in 200,000.

The most striking properties of helium are its radioactivity, when emitted as the positively charged ($+2$) alpha particles in **radioactive changes,** its formation in radioactive change by uranium-**radium** and thorium-containing substances, emitting alpha particles, later losing the charge to become helium, and recently, its production artificially by bombardment with **lithium** or **boron** with high-velocity **protons** or alpha rays.

The liquefaction of helium was accomplished by Onnes in 1908 in Leiden, and Keosom in 1926 succeeded in solidifying helium in the same laboratory. Helium is the most difficult of all the gases to liquefy. An astonishing property of certain metals is that at the temperature of liquid helium they possess electrical **super-conductivity,** e.g., mercury, lead, and tin.

Helium-oxygen atmospheres are utilized in high-pressure breathing work, such as diving suits, caisson work, because helium is inert, less soluble in the blood stream than nitrogen, and diffuses 2.5 times more rapidly than nitrogen. The time for de-gassing is materially reduced, as also the hazard resulting from the gas collecting in the joints. (R.K.S., R.M.F.)

HELIX. A helix is a **skew curve** or twisted curve in space which is defined as the **locus of a point** which moves on a right circular **cylinder** in such a way that the distance it moves parallel to the axis of the cylinder varies directly as the angle it turns through around the axis.

Parametric equations of the helix are:

$$x = a\cos\theta, \quad y = a\sin\theta, \quad z = b\theta,$$

where θ is a parameter. (L.L.S.)

HELIX ANGLE. Helical Gearing.

HELLBENDER. Amphibia, Urodela. A large aquatic **salamander,** *Cryptobranchus alleghaniensis,* of the

Mississippi river system. It reaches a length of 18" and has a flattened head and body, short legs, and a compressed tail. The **gills** are concealed, but otherwise it resembles the mudpuppy. (A.W.L.)

HELLGRAMMITE. Insecta, Neuroptera. The large aquatic larva of the dobson fly, *Corydalus.* It lives in running water and is an excellent bait for bass. (A.W.L.)

HELMINTHOLOGY. A biological science dealing with the **worms.** Since many worms are parasitic, the term parasitology is more commonly used. The study of roundworms is important in agriculture and has resulted in the science of nematology (see **Nematoda**) which is properly a subsidiary of helminthology. (A.W.L.)

HELVE HAMMER. Forging.

HEMATEMESIS. The vomiting of blood which arises from hemorrhage high in the gastro-intestinal tract, usually in the stomach or duodenum. The hemorrhage may be massive. Benign peptic ulcer is the commonest cause. Other causes are alcoholic gastritis, **cancer** of the stomach, and ruptured esophageal varices in **cirrhosis** of the liver. Vomited blood is dark brown in color and is described as "coffee ground" material. (D.M.H.)

HEMATITE. The mineral hematite, **ferric oxide,** Fe_2O_3, occurs as thick or thin tabular **rhombohedral** forms, sometimes in pyramids but rarely in **hexagonal** prisms. It also assumes **botryoidal,** columnar and lamellar shapes, and may be granular or compact. Its hardness is 5.5–6.5; specific gravity varies from as low as 4.2 to as high as 5.25; luster, metallic to earthy or dull; color, dark gray to black; earthy forms may be different shades of red; streak, red to red-brown; translucent (in very thin flakes) to opaque. Hematite with a metallic luster is called specular iron. It is a widely distributed and common mineral, found in **igneous, sedimentary** and **metamorphic** rocks as beds and veins, having probably been formed in many different ways under very different conditions. Beautifully crystallized hematite has been found in the Urals of Russia; Rumania; Switzerland; the Island of Elba; Alsace, France; Cumberland, England; and Brazil. Perhaps the greatest hematite region in the world lies along the southern and northwestern sides of Lake Superior in Michigan, Wisconsin and Minnesota where this mineral has long been mined. Extensive beds of hematite are found throughout the Appalachian region from New York to Alabama, being principally mined near Birmingham in the latter state. Hematite occurs in quantity in Nova Scotia and Newfoundland. It is the most important ore of iron, and has other industrial uses in paint manufacture, polishing compounds, etc. The name hematite is derived from the Greek word meaning blood. (E.S.C.S.)

HEMATOLOGY. That branch of medicine having to do with the study of the **blood,** the blood-forming tissues and the diseases of the blood. (R.S.M.)

HEMATOMA. An accumulation of free blood in the body tissues forming a localized mass. This usually follows an injury in which rupture of blood vessels takes place. (R.S.M.)

HEMATURIA. The presence of **blood** in the **urine.** This condition is found in certain forms of **nephritis** and with injury, tumors, stones, or calculi in the urinary tract. It is also seen in **scurvy** and in some cases of severe **sepsis.** (R.S.M.)

HEMICHORDATA. A subphylum of the phylum Chordata containing only a few primitive marine animals without common names. The genus *Balanoglossus* has lent its name to the forms most commonly seen, although some belong to other genera. They are worm-like animals which live in mud and sand at the bottom of the ocean. The central nervous system is dorsal in this group but it remains partly or wholly at the surface. The **notochord** is limited to the anterior part of the body and is sometimes connected with the alimentary tract. Gill slits vary from one to many pairs. The group is also commonly named Enteropneusta and rarely Adelochorda.

There are two orders:

Order Balanoglossida. Worm-like animals with many gill slits and with a fleshy proboscis before the mouth. *Balanoglossus* and related forms.
Order Pterobranchia (Cephalodisca). Sessile animals, some solitary and some colonial. One pair of gill slits. A proboscis and branching tentacles lie before the mouth and the intestine is U-shaped. *Cephalodiscus* and *Rhabdopleura*. (A.W.L.)

HEMIGALE. Civet.

HEMIMETABOLA. A division of the **insects** characterized by incomplete metamorphosis. The immature insect differs conspicuously from the adult in form and is adapted to an entirely different mode of life; in this the group resembles the **Holometabola**. The young have compound eyes, however, and the wings develop externally as in the **Paurometabola**. The group includes the three orders, **Plecoptera, Ephemerida,** and **Odonata,** all with aquatic larvae which are called naiads. (A.W.L.)

HEMIMORPHITE. Calamine.

HEMIPLEGIA. Apoplexy.

HEMIPODE. Aves, Golliformes. *Turnix.* The **bustard-quails** of the Australian, Oriental, and Ethiopian regions. They are unusual in the larger size and brighter colors of the female and in the fact that the male incubates the eggs and cares for the young. (A.W.L.)

HEMIPTERA. The true **bugs,** an order of insects containing about 21,000 species, many of economic importance. They have a piercing and sucking mouth and live on the blood or juices of animals or the sap of plants. The wings, when present, are usually distinctive. The basal half is thicker than the terminal, and the tips overlap partially so that the margins of the wings form an X on the back. The chinch bug and bedbug are species of economic importance.

Bugs of several families are aquatic and some forms live on the surface of the water, supported by the surface film. The swimming forms are the **water boatmen, back swimmers,** and giant water bugs and the **water striders** skate on the surface. One of the last, *Halobates,* is the only marine insect known. Shore forms include the **toad bugs.** On dry land the order is represented in almost every possible habitat. (A.W.L.)

HEMITROPIC. Used by mineralogists to describe a crystal that appears to be composed of two halves of the same crystal turned partly around. (R.M.F.)

HEMLOCK, POISON. Carrot Family.

HEMOGLOBIN. A chemical compound which is the coloring matter of red **blood** cells. Its composition is complex. One portion of the molecule is a protein known as globin; the other portion is hematin, and contains iron. (See **Aminoacids, Polypeptides, and Proteins.**)

Hemoglobin, having the power of combining with oxygen, is the means of supplying the body cells with oxygen. When oxidized, it is oxy-hemoglobin. In the circulation when the oxygen is liberated, the hemoglobin is known as reduced hemoglobin.

The amount of hemoglobin may easily be estimated from a fresh specimen of the blood. Normally it varies from 80 to 100% or 12.0 to 14.5 grams. In **anemia** the percentage is reduced. (R.S.M.)

HEMOLYTIC JAUNDICE (Congenital jaundice, familial jaundice). A chronic disease due primarily to increased fragility of the red **blood** cells accompanied by destruction of these cells, with resultant **anemia** and **jaundice.** The jaundice is due to the presence of excessive amounts of the blood pigment, bilirubin, in the circulation. The congenital or familial form is hereditary; it is treated successfully by removal of the **spleen.**

Acute hemolytic anemia and jaundice occur as toxic reactions to certain drugs, the **sulfonamides,** lead, etc., and occasionally in severe infections. (D.M.H.)

HEMOPHILIA. An hereditary disease marked by excessive and prolonged **hemorrhage** after trauma. Often the injury is so minor that it would normally not cause bleeding; occasionally the hemorrhage seems to start spontaneously. Hemophilia is a disease of males, transmitted from generation to generation only through females. It behaves as a sex-linked Mendelian recessive (see **Heredity**). A female capable of transmitting the disease does so to about $\frac{2}{3}$ of her male children, while $\frac{2}{3}$ of her female offspring are conductors of the disease.

The cause of hemophilia is unknown. The characteristic pathologic finding in the blood is failure of the **coagulation** mechanism. The clotting time is therefore much prolonged, but once formed, the clot retracts normally.

The treatment of hemorrhage in a hemophiliac is rest, **morphine,** and the local application of thrombic material such as thromboplastin (see **Coagulation**) derived from rabbit lung. **Transfusion** of small amounts of blood will decrease the clotting time and control the hemorrhage. Since the patient is subject to bleeding throughout his life, he must always take special care to avoid trauma. It is wise for him to know the whereabouts of a suitable blood donor for emergencies. (D.M.H.)

HEMOPTYSIS. The expectoration of blood which arises from hemorrhage in the **lungs.** The blood is usually bright red and frothy. **Tuberculosis** and lung **tumors** are the commonest causes of hemoptysis. (D.M.H.)

HEMORRHAGE. Marked bleeding in any part of the body. The common types of hemorrhage, excluding hemorrhage following accident and injury or occurring operatively or post-operatively are: 1. Pulmonary. Pulmonary hemorrhage varies markedly in degree. The blood is usually bright red and frothy due to mixture with air. Tuberculosis and cancer are common causes of pulmonary hemorrhage and **hemoptysis.** 2. Gastric hemorrhage occurring in **peptic ulcer** when a blood vessel is eroded in the ulcer bed. The blood is vomited (**hematemesis**) and is usually dark red or brown due to admixture with the gastric juices. 3. Uterine hemorrhage, due to premature separation of the **placenta,** or occurring after delivery. It may also be due to malignant or benign **tumors** of the **uterus,** various glandular disorders causing excessive bleeding with the menses, and abortion. 4. Rectal bleeding usually due to acute infection (**dysentery, typhoid**) or to ulceration or malignant growths in the intestinal tract. In this type of hemorrhage the expelled blood may be black or tarry in color. Occasionally bleeding from **hemorrhoids** is severe enough to be called a hemorrhage, although slight blood loss is more characteristic. 5. Nasal hemorrhage, the common nose bleed which is not usually serious. 6. Cerebral hemorrhage, hemorrhage from an **artery** in the **brain,** causing the so-called "stroke" or **apoplexy.** (R.S.M.)

HEMORRHOIDS. Hemorrhoids are small **tumors,** made up of dilated varicose veins, occurring at or near the rectal outlet. They are popularly called "piles." Hemorrhoids are divided into three groups: external, internal and a combination of the two. External hemorrhoids occur at the external margin of the rectum beneath the skin. Internal hemorrhoids are situated just inside the rectum and are covered by mucous membrane which lines the rectum. Long-standing internal hemorrhoids, however, may protrude through the anal opening. In cases of long duration, both internal and external hemorrhoids may be present.

The chief factor in the development of hemorrhoids is **infection** of the veins in this region. The main predisposing factor is chronic **constipation.**

The chief symptoms of hemorrhoids are bleeding, pain, and spasm of the anal muscles. The diagnosis can be made only on careful proctoscopic examination of the rectum and anus.

The treatment of choice of all varieties of hemorrhoids is surgical excision, which when expertly done is a simple procedure causing little or no post-operative pain. If internal hemorrhoids alone are present, and there is no acute infection at the site, they can be treated by the multiple-injection method, consisting of the injection of a sclerosing substance into the veins, causing them to be obliterated by scar tissue. This can be done in the surgeon's office and when done by experts in the technique, fairly satisfactory results can be obtained. If the causative factors are not corrected, hemorrhoids will usually recur no matter what treatment is used. (D.M.H.)

HEMP. *Cannabis sativa.* Cannabinaceae. Hemp is obtained from the stem pericycle of a tall hollow-stemmed annual which is a native of central and western Asia. In cultivation the slight branching which characterizes the plant is considerably reduced by planting thickly. The plants grow from 5–16′ in height. They have digitately compound dark green leaves and small inconspicuous flowers which are of two kinds, occurring on different plants. The staminate (see **Stamen**) flowers appear in small axillary (see **Axil**) clusters on male plants, and the pistillate (see **Pistil**) flowers are borne in leafy spikes on female plants. The fruit, an **achene,** is a hard ovoid structure, often called hemp seed. Cultivation of hemp has been carried on in China for many centuries. From that country its culture has spread to many countries, Europe taking it up long before the Christian era dawned. It is widely grown in the United States, although extensively in only a few states, such as Wisconsin and Kentucky. Hemp grows best in regions having a warm humid growing season of about five months; the plants grow rapidly, soon shading the ground so effectively as to suppress all other plants present, for which reason its culture is sometimes recommended as a way to eradicate obnoxious weeds. When the staminate flowers are mature, the plants are ready for harvest; to delay after that is not desirable, since the male plants die soon after flowering; furthermore, after flowering the fibers become coarser. Harvesting and the treatment of the plants after harvesting are very similar to those of **flax** plants. The hemp plants are cut off or pulled up, denuded of leaves, roots and tops, and tied in bunches and left to dry for about 2 weeks. They are then immersed in water to ret. In retting the intercellular substance of the stems is acted upon by **bacteria** and softened so that the fibers are readily cleaned of surrounding tissues. Scutching removes the woody tissue, after which the rough hemp fibers are

hackled, or drawn over coarse combs which pull out the fibers.

Hemp fibers are coarse and rather harsh, and much less pliable than flax fibers. Furthermore, they are dark colored and not easily bleached without injury. So the principal use made of them is in the making of rope and coarse twine, sail cloth, the warp of carpet and belt and upholstery webbing, all products where strength and durability are the principal aim, and appearance of little consequence. Short fibers of hemp, called tow, are used in packing joints in iron pipes, and as pump packing, also as a stuffing for upholstery. The woody waste from hemp fiber production has been used in paper making.

From hemp seeds, a valuable oil is pressed out; it is used in the making of soft soaps, and also in mixing paints and varnishes. The seeds themselves are used as bird food.

From the hemp plant also a drug is obtained. This substance is located in the glandular hairs of the leaves and stem. From the pistillate flowers and fruits is obtained a resinous substance, which is smoked in the Orient under the name of hashish or bhang. As a medicine the drug may have a quieting effect on the nervous system, but large doses are dangerous. The Mexican preparation, marihuana, is very similar to hashish in its ill effects. (See also **Manila Hemp.**) (R.M.W.)

HENNA. *Lawsonia inermis.* Lythraceae. This plant, a native shrub of Africa and Asia, is widely cultivated in tropical countries. The small flowers are inconspicuous, but very fragrant. The leaves are powdered and made into a paste which applied to the hair or beard gives it a bright red color. It stains the skin yellow and is used by some Oriental people to color the hands and feet. From the fragrant flowers, a rich perfume is obtained, which is used in oils and ointments. (R.M.W.)

HENRY. Inductance; Electric and Magnetic Units.

HENRY'S LAW. Gas Absorption; Solutions.

HEPATICS. Bryophytes.

HEPATITIS. Inflammation of the liver, due to infection, chemical poisons or obstruction of the bile channels.

Acute infectious hepatitis (infectious jaundice) is an epidemic disease which is common during war time and a serious problem on military fronts because of its high incidence. The etiological agent is a **virus** whose characteristics are not yet completely known. Fever, malaise, gastro-intestinal symptoms and jaundice constitute the clinical picture. The signs and symptoms persist for about 6 weeks; the mortality is low. Fatal cases show extensive destruction of liver cells. Infectious hepatitis is probably transmitted by the oral route; the virus is excreted in the stools during the acute phase of illness. The disease can be prevented by the intramuscular injection of gamma globulin. The incubation period is long—averaging 30 days, and the progress of an epidemic can be halted by the early administration of gamma globulin to susceptible individuals who have been exposed, i.e., children and young adults.

Acute catarrhal jaundice is thought to represent the **endemic** form of infectious hepatitis. It is a similar but milder disease, usually occurring sporadically. Syphilitic hepatitis is rare.

Homologous serum jaundice is a form of jaundice produced by a virus similar to that responsible for infectious hepatitis. It has a longer incubation period (60 to 90 days) and is often a more severe disease, but patients have the same type of symptoms and signs in the two conditions. Homologous serum jaundice does not occur naturally but only on the injection of human blood products into the body. Since the virus circulates in the blood for several weeks before the clinical disease becomes manifest, blood drawn from an apparently healthy donor who is actually in the incubation period will contain the virus. When such blood is given in the form of a transfusion, convalescent serum, or a vaccine, to a susceptible recipient, the latter will develop jaundice 2 to 3 months or more later. It is not yet known whether gamma globulin will prevent homologous serum jaundice.

Acute yellow atrophy is an inflammatory disease of the liver, which may occur without known cause, as in toxemia of pregnancy, or as a reaction to toxic chemicals such as arsphenamine, trinitrotoluene, etc. The liver undergoes rapid necrosis and shrinks to half its normal size. The patient is acutely ill, with headache, delirium and deep jaundice. Death in coma usually follows within a few days. In the subacute form, recovery sometimes occurs.

Obstruction to the biliary system from stones or infection may lead to inflammation of the liver cells and secondary hepatitis. (D.M.H.)

HEPATOPANCREAS. A digestive gland of the mollusks (Mollusca) and arthropods which discharges into the stomach. (A.W.L.)

HERCULES. (Map, page 380.) A large and important constellation between Lyra and Corona Borealis. The constellation contains no strikingly bright stars and hence is somewhat difficult to locate. Once found, however, it is a fertile field for a small telescope. In 1934 Hercules received considerable notice because of the brilliant Nova that appeared in it just before Christmas. Perhaps the most interesting object in the constellation is the remarkable star cluster which was first noted by Halley in 1714. While the cluster can be distinguished as such in a telescope of only 2-in. aperture, it requires a telescope of larger than 6-in. to really appreciate the magnificence of the object. In addition to the star cluster there are several double stars to be observed with small telscopes, many of them having components of different colors. (W.K.G.)

HERCYNIAN REVOLUTION. Permian.

HEREDITY. The transmission of developmental potentialities from one generation of living things to the next through the process of reproduction. The materials of the parent bodies from which a new individual develops are its actual heritage. During its own embryonic (embryology) development the potentialities of this heritage are expressed in the structural characteristics of the new body, normally like those of the parents or those of a more remote generation of ancestors. This fact leads to the common statement that the organism inherits certain characters; while not precisely true, the interpretation is permissible for all ordinary purposes of description.

The fact of inheritance is obvious. It has been expressed for ages in unscientific observations. Attempts to determine the scientific foundations of inheritance are relatively recent, however, and the establishment of a sufficient body of facts relating to heredity to constitute a science has occurred only during the 20th century. This science is called genetics.

The first steps in genetics were taken by plant hybridizers of the 18th and 19th centuries, chiefly in Europe, and culminated in the experiments of Gregor Johann Mendel, a monk at Brno, Czechoslovakia, then Brünn in Austria. Mendel's results were published in 1866 and lay almost unnoticed until 1900, when they were corroborated by three scientists in the birth of modern genetics. The published report of Mendel's work repeated the significant observations of his predecessors and added a simple mathematical analysis that had not previously been expressed. As a result of the far-reaching importance of this work the term Mendelian heredity is now commonly applied to the established fundamentals with which all subsequent discoveries have been correlated.

Mendelian heredity depends on three fundamental concepts: 1. The organism is a mosaic of unit characters capable of separate hereditary transmission. 2. A unit character may mask a related unit character completely when the potentialities for the development of both are present in the same individual. This principle is called dominance, and the masked character is said to be recessive. 3. Unit characters may be segregated during reproduction, regardless of the combinations in which they have been associated.

To these concepts modern genetics has added that the association of different related unit characters in one individual may result in the development of both in different parts of the body, in a mosaic inheritance, or in an intermediate condition through blending inheritance.

Some characters, particularly of a quantitative nature, are not amenable to these rules unless through a very complex association of underlying hereditary unit characters. Such characters must be studied by statistical methods. They were the foundation of another attempt to formulate laws of inheritance made by Sir Francis Galton, from which we retain the law of ancestral inheritance and the law of filial regression. The former indicates that each parent contributes one-quarter of the total heritage of the individual, each grandparent one-sixteenth, and so on in a rapidly diminishing percentage. The law indicates the great reduction of the possibility of a hereditary character reappearing after a lapse of generations. Filial regression is the tendency of extreme parents to produce offspring less extreme than themselves. Thus tall parents beget tall children, but usually shorter than themselves. Galton studied human inheritance and in addition to his mathematical analyses, so necessary in this field, took the initial steps in proposing deliberate control, which led to the modern science of eugenics.

Modern science has also added to early discoveries the definite recognition that hereditary potentialities are resident in the chromosomes of body cells and that definitely located genes within these chromosomes are the determiners through which specific unit characters are brought to expression. The behavior of chromosomes is strictly in harmony with the transmission of characters by Mendelian heredity (Meiosis; Fertilization). Since nothing was known of chromosomes during Mendel's life, this correlation had to await further advances in cytology.

Mendel's chief contributions were derived from the study of garden peas, in which he observed seven pairs of unit characters, all similar in behavior. He noted, for example, that seed colors included two unit characters, yellow and green. When he crossed parent plants of the two strains the resulting hybrid seeds were entirely yellow, indicating the dominance of this color over green. He then inbred the hybrids, and in their offspring both yellow and green seeds appeared in the ratio of three yellow to one green. Related unit characters of this kind are said to be alleles or allelomorphs. It is now known that their genes occupy the same position in the paired chromosomes of the cell, while only one can be represented in the single chromosome of a germ cell. Since each parent contributes one chromosome to each pair in its offspring, it may also contribute one gene of an allelic pair. The one parent plant contributed a gene for yellow, the other for green, and through dominance the offspring were yellow. Segregation, however, enabled these hybrids to transmit either yellow or green during their reproduction, and through random fertilization all possible combinations of these determiners were established. The characters are commonly represented by symbols, using a capital letter for the dominant and a small letter for the related recessive, as Y and y for yellow and green respectively. For the pair of characters mentioned, the following diagram is representative:

Parental generation (P):	YY	yy
Germ cells:	Y	y

Hybrids of first filial generation (F₁): Yy

Gametes of F₁ generation

and their combinations

in the F₂ generation,

in a Punnett square:

	Y	y
Y	YY	Yy
y	Yy	yy

The YY and yy individuals in this diagram are homozygous, while the Yy individuals are heterozygous. Since all YY and Yy individuals look alike, due to the dominance of Y, they belong to the same phenotype, but since their hereditary potentialities are different they belong to different genotypes. The yy individuals from hybrid parents are known as extracted recessives. There are twice as many heterozygotes as homozygotes of either kind in this 3:1 ratio because similar individuals in this category result from reciprocal combinations of genes, half of the individuals receiving the dominant from one parent and half from the other. Examples of this kind, involving only one pair of allelic characters, are known as monohybrids.

Additional complexity arises in dihybrids, trihybrids, and polyhybrids of still more characters through the free reassortment of the unrelated pairs of alleles. Thus peas from smooth yellow seeds crossed with others from wrinkled green seeds, a dihybrid combination, produce only yellow smooth seeds in the F₁ generation, but when inbred these plants give rise in the F₂ generation to the four possible combinations: smooth yellow, smooth green, wrinkled yellow, and wrinkled green, in the ratio 9:3:3:1. The reason is evident in the following diagram:

	SY	Sy	sY	sy
SY	SY SY	Sy SY	sY SY	sy SY
Sy	SY Sy	Sy Sy	sY Sy	sy Sy
sY	SY sY	Sy sY	sY sY	sy sY
sy	SY sy	Sy sy	sY sy	sy sy

In this diagram each pair of symbols above and at the left side represents the contribution of one parent in one of its germ cells, and in the small squares the possible combinations from the two parents are shown. Dominance prevails as in the monohybrid.

In a trihybrid free reassortment results in an F₂ ratio of 27:9:9:9:3:3:3:1. The number of phenotypes is always a power of two indicated by the number of pairs of alleles under consideration.

The study of heredity in animals has shown that these principles are applicable in that kingdom as well as in plants, but relatively few animals are sufficiently prolific to demonstrate complex ratios. The fruit fly, *Drosophila melanogaster*, has been the most productive of all genetic subjects, while man and the domestic animals yield very limited Mendelian data.

Modern genetics, largely from studies of the fruit fly, has disclosed many principles as corollaries of simple Mendelian heredity. The more important are as follows:

Multiple alleles: More than two unit characters may be related to each other as alleles. In such cases only two of the series may be present in any one individual, and dominance is in a graded series, as may be determined by experimental results.

Multiple factors: More than one gene may be necessary for the production of a single unit character. If two genes are essential for its appearance and either alone is incapable of expression, they are said to be **complementary.** If one expresses itself alone, a gene that modifies this expression is **supplementary.** If two are capable of producing the same effect whether present singly or in combination, so that the resulting character is absent only from homozygous recessives, they are said to be **duplicate** genes. In all cases recombination of the genes during reproduction follows the same course as in simple Mendelian heredity but the resulting phenotypic ratios differ because fewer unit characters are involved.

Lethal genes: Some genes completely inhibit development or modify it in such a way that the individual dies. They also modify the usual ratios of associated characters.

Linkage: Some characters, although not allelic, are inherited in definite groups; they are said to be linked. Modern genetics shows that linkage is due to the presence of genes for the linked characters in the same chromosomes.

Crossing over: Linkage relations are sometimes interrupted in a limited number of individuals, permitting some reassortment of normally grouped characters. This change is due to the breaking of paired chromosomes in **synapsis** and the reunion of their fragments in new combinations to form similar chromosomes, sometimes with new combinations of genes.

Translocation: This change is a shifting of the relations of genes in the chromosomes, due to looping, fusion, and rupture, or to the attachment of fragments to other chromosomes. It may result in the duplication of genes within a chromosome or in a change in the serial arrangement of the included genes.

The inheritance of sex has also been shown in many cases to depend on a simple chromosomal mechanism. Males of many species have an x chromosome without a synaptic mate or with a y chromosome mate that is evidently abortive. The females of such species have two x chromosomes. In the formation of germ cells all eggs receive an x chromosome while half of the sperm cells receive an x chromosome and half a y or none. Random combination of these cells restores the xx combination in one-half and x or xy in the other, thus producing half females and half males. Other investigations have shown that the quantitative balance between the sex and other chromosomes is the active factor in conditioning the differentiation of the sexes.

This disclosure also explains the phenomenon of sex linkage. Genes lying in the sex chromosomes, mostly in the x chromosomes but a few in the y, are inevitably transmitted and expressed in some definite relation with sex, hence they are said to be sex linked. Such characters need have no active sexual role.

The findings of genetics have been of great practical value in plant and animal breeding. Although the improvement of cultivated plants and domestic animals by **selection** preceded by many years the formulation of scientific principles of heredity, the discovery of these principles has made possible much more precise and efficient procedure in the establishment of useful strains. **Hybridization** and selection together are the chief means of improvement. Applied by scientists they have brought about many modifications of living things and have disclosed many facts concerning heredity. Corn has been studied in detail and subjected to many experiments, both practical and purely scientific. Tomatoes, radishes, various cereals, and flowers of many species have also commanded attention. More has been done with plants than with animals because the domestic animals are less amenable to experiment. From the practical point of view plants are more satisfactory subjects because desirable hybrid strains may often be propagated by cuttings, grafting, and other asexual methods which avoid the segregation that is inevitable in sexual processes. Only rigid selection can establish desired hybrid combinations in plants or animals that must be produced sexually.

The study of human heredity depends entirely on observation of the family, since controlled mating is impossible. Genealogical records have furnished a large amount of valuable material and the records of public institutions such as prisons and asylums have been equally useful to the geneticist. Such records are not to be compared with scientifically assembled experimental data, but they leave no doubt that the principles of heredity worked out in the study of other organisms are also applicable to man. In a few cases they have also disclosed adequate evidence of a specific type of Mendelian inheritance of human characters.

The clearest evidences of human heredity are found in the behavior of simple structural defects, such as the appearance of extra digits (polydactylism), the fusion of bones in the digits (symphalangism), and shortness of the fingers (brachydactylism). These defects are transmitted as Mendelian unit characters allelic to normal structure. Red-green color-blindness (vision) is one of the most striking examples of inheritance in man. It is a sex-linked recessive allele of normal vision. Both x chromosomes of the female must carry the gene for the defect if she is to be color-blind, whereas the male may be color-blind if he receives such a gene in his one x chromosome. Females may be heterozygous carriers of the defect, with normal vision; males are either strictly normal or defective. In this type of inheritance the male always receives the genes for his characters from his mother, therefore a carrier mother may have some color-blind sons. A color-blind man and a genotypically normal woman cannot produce color-blind children, but all of their daughters are carriers. On the other hand, a color-blind woman and a normal man will produce carrier daughters and color-blind sons. Hemophilia is inherited in a like manner, except that the recessive genes for hemophilia are lethal in the homozygous condition in the female.

Since man is concerned chiefly with his behavior, in the broad sense of the term, the inheritance of ability is much more important. It is, however, extremely complex, depending upon many simpler heritable unit characters that are only slightly understood. Nevertheless the inheritance of specific types of ability, and particularly of various degrees of ability, are well established by our records. Deficient families usually produce deficient children, and superior families normally maintain their superiority. There are occasional exceptions in both cases but in the long run superiority, mediocrity, and mental and social inadequacy tend to persist generation after generation in family lines. These general terms are not to be taken to indicate hereditary traits in the strict sense of scientific genetics, but the heritage of individual organization whose expression in the behavior of the individual is quite evidently amenable to the principles so readily appreciated in the study of simpler structural unit characters. The possibility of practical application of this knowledge is the field of eugenics.

The study of all aspects of heredity is now pursued extensively and is the foundation of a voluminous and constantly increasing literature. (A.W.L.)

HERMAPHRODITE. An animal with functional reproductive organs of both sexes. The condition is common among the flatworms and segmented **worms** and occurs in a few species of **echinoderms** and mollusks (**Mollusca**). Among the vertebrates the occurrence of both sexes in one individual is rare and the sexes appear at different periods. Animals in which such a transition is possible are sometimes influenced by external conditions and during the transformation may be functionally hermaphrodite. This is true of some fishes and **amphibians.** (A.W.L.)

HERMAPHRODITE DUCT. The duct of the ovotestis (**gonad**) of certain **snails.** (A.W.L.)

HERMAPHRODITISM. Only about twelve cases of supposedly true hermaphroditism in the human race have been reported. The term signifies the presence of all of the functioning genital organs of both sexes in one individual. The cases mentioned above were supposed to have both testicles and ovaries present. The ability to impregnate as well as to conceive has never been reported in the one individual.

Many cases of pseudo-hermaphroditism have been seen. In this condition the genital organs, internal or external, do not conform either totally or in part with the sexual glands (testicles or ovaries) present. In the male hermaphrodite, testicles are present but may be abdominal in position. The penis is small and more nearly resembles a large **clitoris;** the **scrotum** is divided by a cleft resembling the female labia with a small short **vagina. Uterus** and tubes are not present.

The female hermaphrodite has a large clitoris more like a small penis, rudimentary vagina, a uterus and ovaries. Various in-between stages may be present, giving a very bizarre picture where the sex can only be determined by operation or in some instances by determination of the amount of sex **hormone** in the blood and urine.

In general an hermaphrodite is best raised as a female. The bodily characteristics may not be sharply differentiated in regard to sex. (R.S.M.)

HERMITE POLYNOMIALS. The Hermite polynomials $H_n(t)$ are defined by

$$H_n(t) = \frac{(-1)^n}{e^{-\frac{t^2}{2}}} \frac{d^n e^{-\frac{t^2}{2}}}{dt^n}.$$

We list the first few $H_0(t) = 1$, $H_1(t) = t$, $H_2(t) = t^2 - 1$, $H_3(t) = t^3 - 3t$, $H_4(t) = t^4 - 6t^2 + 3$, $H_5(t) = t^5 - 10t^3 + 15t$, etc. They satisfy the recurrence relationship $H_{n+1}(t) = tH_n(t) - nH_{n-1}(t)$. When n is odd $H_n(t)$ is a polynomial of $\frac{n+1}{2}$ terms and contains only odd powers beginning with t^n. When n is even $(n > 0)$ $H_n(t)$ is a polynomial of $\frac{n}{2} + 1$ terms in even powers starting with t^n. The Hermite polynomials are orthogonal with weight function $\frac{e^{-\frac{t^2}{2}}}{\sqrt{2\pi}}$ that is,

$$\int_{-\infty}^{\infty} H_m(t) H_n(t) \frac{e^{-\frac{t^2}{2}}}{\sqrt{2\pi}} dt = 0, \quad m \neq n$$

$$\int_{-\infty}^{\infty} H_m(t) H_n(t) \frac{e^{-\frac{t^2}{2}}}{\sqrt{2\pi}} dt = n!, \quad m = n.$$

The Hermite polynomials may also be found by expansion of the generating function in powers of x.

$$e^{xt - \frac{x^2}{2}} = H_0(t) + H_1(t)x + H_2(t)\frac{x^2}{2!}$$
$$+ H_3(t)\frac{x^3}{3!} + H_4(t)\frac{x^4}{4!} + \cdots$$

The Hermite polynomials find applications in **statistics** in the **Gram-Charlier** Type A series and are also useful in certain applications in physics to problems of heat and quantum mechanics. (L.A.A.)

HERNIA (Rupture). The protrusion of a portion of tissue or organ through a weakness or abnormal opening in any part of the body. Thus it is possible to have a herniation of the brain through an opening in the skull, or herniation of muscle if an opening occurs in the covering membranes of the muscle.

Common types of hernia such as femoral, inguinal, umbilical (navel) occur as protrusions of a sac-like process of the lining membrane of the abdominal cavity—the **peritoneum**—through a weakness at certain sites in the abdomen. These sites are usually where structures such as blood vessels pass out from the abdominal cavity, or did during uterine life. These sites are frequently congenitally weak, so that if any strain causing increased abdominal pressure occurs, this weakness is accentuated and a hernia results. Chronic cough, constipation, heavy lifting, violent games, etc., are all precipitating causes of hernia since they increase the pressure within the abdomen. As continued pressure increases the size of the sac-like process, there may be forced into it a loop or small portion of intestine or some easily movable organ in the abdomen. Sometimes, when the opening through which the rupture occurs is small, the intestinal organ forced into the sac cannot be pushed back into the abdomen. Such a condition is known as an incarcerated hernia. This produces acute symptoms with local pain, often vomiting, and an emergency operation is necessary. If this is not performed, the blood supply may be shut off, resulting in a strangulated hernia. **Gangrene** may develop in the strangulated portion.

Common types of hernia are:

1. Inguinal Hernia. Develops along the course of or through the wall of the inguinal canal, where, in the male, the ducts leading from the testicle pass upward along the lower portion of the abdomen above the groin into the abdominal cavity. In the female the corresponding structure, the round ligament of the **uterus**, runs through the inguinal canal. Inguinal herniae are of several varieties; they are more common in men than in women. Some herniae of long standing will push their way downward into the **scrotum**.

2. Femoral Hernia. Occurs in the upper anterior portion of the thigh just below the groin. The weakness here results from imperfect closure around the large femoral blood vessels which pass from the abdomen into the thigh. Femoral herniae are more common in women.

3. Umbilical Hernia. This type of hernia occurs at the navel, where, during fetal life, the blood vessels connecting the fetal and maternal circulations passed. Imperfect closure after birth produces a weak navel through which a hernia may develop.

4. Ventral Hernia. May occur in any portion of the abdominal wall where a weakness develops or occurs around an operative wound. Post-operative herniae are apt to occur in fat or weak abdominal walls where there has been an infected or drained wound, or some post-operative complication which has increased the abdominal pressure. Prolonged coughing, vomiting or distention of the abdomen before complete healing has occurred may thus produce herniation.

There are other less common sites for hernia, such as the diaphragm, where the hernial sac protrudes through the diaphragm into the chest. (R.S.M.)

HEROIN (Di-acetyl morphine). This drug is rarely used at present in medicine because it is more toxic than **morphine,** is more markedly habit forming, and its powers to allay pain and promote sleep are decidedly less than those of morphine or **codeine.** It is much used by drug addicts. More than any other narcotic, it

destroys the emotional values of a subject, obliterates all traces of remorse or responsibility and gives a sense of exhilaration and inflation of the ego. It is taken as snuff, as pills, or hypodermically. (R.S.M.)

HERON. Aves, Ciconiiformes. Long-legged wading birds (**Aves**) with a sharp slender beak and when adult with plumes or a crest. They live chiefly on fish.

Herons are found throughout the world. The most widely known North American species are the great blue heron, *Ardea herodias,* the green heron, *Butorides virescens,* and the egret, *Egretta.* The last is a white bird which bears beautiful plumes known as aigrettes during the breeding season. It was once threatened with extinction through the use of these plumes as ornaments for hats, but the remaining birds are adequately protected. (A.W.L.)

HEROTA. Mammalia, Artiodactyla. An African **antelope,** also known as Hunter's hartebeest. (A.W.L.)

HEROULT FURNACE. Electric Furnace.

HERPES SIMPLEX (Fever blisters). An acute eruption of tiny blisterlike lesions occurring usually in groups on the lips, face, or genitalia, due to infection with a specific **virus,** herpes virus. This agent is related to the virus causing herpes zoster, but is distinct from it. The lesions are self-limiting, but tend to be recurrent.

Herpes simplex occurs following exposure to wind and sun, in association with the common cold, and gastrointestinal disturbances, and during the course of certain febrile diseases. Pneumonia, malaria, and meningitis are frequently accompanied by herpetic lesions. (D.M.H.)

HERPES ZOSTER OR SHINGLES. A disease due to infection and inflammation of sensory nerves, caused by a **virus,** which appears to be closely allied to the virus of chicken-pox. In some instances it occurs in epidemics. The first symptom is usually severe pain over the course of a nerve. Constitutional symptoms characteristic of an acute infection may appear—that is, malaise, fever, chills, etc. Within a few days a rash breaks out in that portion supplied by the irritated nerve. Small blisters appear which may ulcerate, healing in 3–6 weeks. Scarring often results. The disease is self-limited, although injections of **pituitary** extract or **sodium** iodide may shorten the course of illness and lessen the pain. In older people the neuralgic pains are often particularly severe and may persist for months or years after the acute disease has disappeared. (R.S.M.)

HERRING. Pisces, Teleostei. Food fishes (**Pisces**) of numerous species, found in both fresh and salt water. The common herring, *Clupea harengus,* is among the

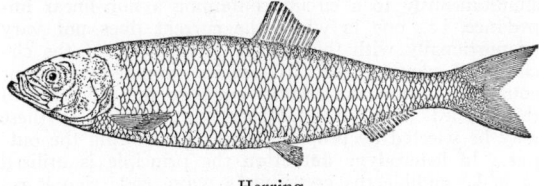

Herring.

most important of the smaller food fishes on the New England coast and northward, and several fresh-water species of the genus *Leucichthys* are equally important in the fisheries of the Great Lakes. The smaller sizes are canned as "sardines," especially in the United States and Norway. (A.W.L.)

HERRING OIL. Esters.

HERRINGBONE GEARING. Helical Gearing.

HERSCHEL EFFECT. A photographic effect observed first by F. W. Herschel in 1839. In that year Herschel observed that an image on printing-out-paper (silver chloride) is destroyed (bleached out) upon exposure to red light. This is now known as the visual Herschel effect. The destruction of a latent image in a gelatin emulsion (which must not be dye-sensitized) is now known as the latent Herschel effect. Non-color sensitive materials may be made sensitive to the infrared by exposing to red light after exposure. (See Clark, *Infra-Red Photography*.) (C.B.N.)

HERTZIAN RADIATION. Electric Oscillations and Electric Waves; Electromagnetic Radiation.

HESPERIDIUM. Fruit.

HESSIAN FLY. Insecta, Diptera. One of the worst pests of wheat. It is a small 2-winged fly, *Phytophaga destructor,* a member of the gall-gnat family, which was introduced into the United States in the Revolutionary period. The larva lives between the base of a leaf and the stem of the wheat plant and either kills or weakens the plant so that no grain develops. Other cereals are attacked to some extent.

Fall plowing and burning stubble aid in destroying many insects. The most effective means of avoiding damage to winter wheat is to sow late enough to avoid the attack of most of the adults. They live no more than ten days and the date of emergence is known for various regions, hence late planting subjects the crop only to the light infestation due to the eggs deposited by the relatively few flies which emerge late. (A.W.L.)

HESSITE. A mineral telluride of silver, Hg_2Te, with some gold. Crystalline form not obvious at normal temperatures. Hardness, 2–3; specific gravity, 8.24–8.25; color, gray with metallic luster; opaque. Named after H. G. Hess (1802–50) of St. Petersburg. (R.M.F.)

HESSONITE. Garnet.

HETEROCYCLIC COMPOUNDS. This is a class of organic compounds containing one or more rings in which there are elements other than carbon present. Examples are furane, pyrrole, and pyridine. (See Carbon for index to articles on heterocyclic compounds.) (R.K.S.)

HETERODYNE. This term is used in communications terminology as an adjective or a verb, but in either case it concerns the beating together in an electrical circuit of two frequencies to produce new frequencies which are the sum or difference of the original ones. When two voltages of different frequencies are applied simultaneously to a circuit containing a non-linear impedance, i.e., one in which the current does not vary proportionally with the voltage, the output of the circuit will contain new frequencies, among them one equal to the sum and another equal to the difference of the applied frequencies. Either one or both of these may be selected by properly tuning or filtering the output. In heterodyne detection the principle is utilized to make audible the continuous wave code signals received. By applying to the input circuit of the vacuum tube detector the incoming signal and another locally generated signal which differs by some audio amount, say 1000 cycles, the output of the detector will contain a signal whose frequency is the difference, 1000 cycles. This 1000-cycle signal will go on and off with the keying of the original signal. A similar system is used in the first detector of the superheterodyne receiver where the difference frequency is made high (in the radio-frequency range) and this new radio frequency is further amplified. In both radio and telephone applications the principle of heterodyning is used to shift the location of a single frequency or a frequency channel in the spectrum. (See Beats.) (L.R.Q.)

HETEROGAMY. Gamete. In addition to the widely accepted application of the word to the occurrence of male and female gametes of different form, it has been applied in one work to the peculiar life cycle of the plant lice, in which generations with different characteristics succeed each other. The latter use is not to be recommended. (A.W.L.)

HETEROMORPHOSIS. Deviation from normal form. Malformation or deformity and also less extreme departures incidental to slightly different conditions in the animal or its environment. (A.W.L.)

HETERONEMERTEA. Nemertea.

HETEROTHALLISM. Phycomycetes.

HETEROTRICHIDA. Ciliata.

HEULANDITE. The mineral heulandite is a monoclinic zeolite whose crystals are often quite suggestive of orthorhombic forms. Its chemical composition is probably $(Ca,Na_2)O \cdot Al_2O_3 \cdot 6SiO_2 \cdot 5H_2O$; strontium may be present. Heulandite has one good cleavage; is brittle with a conchoidal fracture; hardness 3.5–4; specific gravity, 2.18–2.22; luster, vitreous to pearly; color, white to gray, red or brown; streak white; transparent to translucent. Occurs chiefly in cavities in basaltic rocks with other zeolites, but may be found in granites, pegmatites, gneisses, and schists. Famous localities are in Iceland, India, the Harz Mountains, Italy, Switzerland, Scotland, Nova Scotia; and in the United States at Bergen Hill and West Paterson, New Jersey. This mineral was named for the English mineralogist Heuland. (E.C.L.S.)

HEXACTINELLIDA. The glass sponges, constituting a class of the phylum Porifera. The spicules of the skeleton are silicious (see Silicon) and of 6-rayed form. Many of the species have a large central cavity, resulting in a tubular or vase-like form, and when freed of organic matter appear to be made of spun glass. These sponges are found in deep water in the ocean. Venus' flower basket, *Euplectella,* and the glass-rope sponge, *Hyalonema,* are the most common examples. (A.W.L.)

HEXAGONAL SYSTEM. Crystallography.

HEXAGONITE. Tremolite.

HEXAMETHYLENE TETRAMINE. A compound of formaldehyde and ammonia. (See Formaldehyde.) (R.K.S.)

HEXAPODA. Synonymous with Insecta. (A.W.L.)

HIBERNATION. The passing of the winter in a state of torpor. The condition is unavoidable to the cold-blooded animals when the surrounding temperature falls low enough to slow the chemical processes of their metabolism below the level necessary for normal activity. Under such conditions they live through the winter if the body is able to endure the lowest temperatures to which it is subjected. Even in such cases preparation may be made for the winter by entering a more resistant form. Some rotifers and plant lice, for example, produce specially formed winter eggs and many insects pass the winter in the inert pupal stage.

Hibernation is more striking in warm-blooded animals, since their bodily processes are not necessarily slowed by low surrounding temperatures. The species that hibernate, like some of the bears, store up reserve energy through the warm season and hide away for the winter. During this period their activity is reduced to the very slow respiration and reduced circulation. By spring they have used up the fat stored during the previous year.

Many animals hide away in nests or burrows for the winter without hibernating in the strict sense. Their activities are, of course, lessened, but they seek food or live on stores accumulated previously, like the squirrels and beavers. The honey-bee is an exceptional example of this kind among cold-blooded animals. The colony maintains its warmth by the activity of some of the bees, and all individuals are intermittently active throughout the winter. (A.W.L.)

HICCOUGH OR HICCUP. An intermittent sudden contraction of the **diaphragm**. The condition is due to a great variety of causes which may irritate either the nervous pathways leading to the motor centers controlling the diaphragm, the motor centers themselves, or the pathways from the centers to the diaphragm. The condition occurs at any age.

Hiccough commonly follows swallowing very hot or irritating substances, or occurs with disorders of the esophagus or stomach, such as **gastritis** or dilation of the stomach. Hiccough may occasionally occur after operations and at times may prove very severe. It occurs quite often in **peritonitis** or in any severe infection such as **typhoid**. In severe toxic conditions, especially **uremia** and alcoholism, hiccough may be severe and exhausting. An inflammation or tumor near the centers controlling the diaphragm will cause hiccough. An epidemic variety of hiccough has been described, the disease lasting about a week. Many cases of prolonged hiccough are of psychogenic origin. Attacks in such people may last for weeks, but the disorder is not present while eating. Many of these attacks may be aborted by sudden emotion such as fear, anger, etc.

To stop hiccough, the well-known measures such as holding the breath, drinking cold water, sudden fright, etc., are usually effective in mild cases. In severe ones, the insertion of a stomach tube, the injection of apomorphine or hyoscine, or the inhalation of 5–10% carbon dioxide in oxygen may be necessary. (R.S.M.)

HIDDENITE. Spodumene.

HIGH. A region over which the atmospheric pressure is greater than the surrounding area; an abbreviation for region of high pressure. Anticyclonic winds blow about a high. (P.E.K.)

HIGH BLOOD PRESSURE. Hypertension.

HIGH CONTACT. A probability function (or a frequency distribution) is said to possess high contact, if the function and all its derivatives vanish at the upper and lower limits of the variable x. (L.A.A.)

HIGHER DERIVATIVES. The derivative of the derivative of a function $y = f(x)$ is called the second derivative of $f(x)$, and is denoted by $D_x^2 y$ or $\dfrac{d^2 y}{dx^2}$ or $f''(x)$ or y''. Similarly, the derivative of the second derivative is called the third derivative, and is denoted by $D_x^3 y$ or $\dfrac{d^3 y}{dx^3}$ or $f'''(x)$ or y'''. In general, the nth derivative is the derivative of the $(n-1)^{st}$ derivative, and is denoted by $D_x^{(n)} y$ or $\dfrac{d^n y}{dx^n}$ or $f^{(n)}(x)$ or $y^{(n)}$.

The curves $y = f'(x)$, $y = f''(x)$, $\cdots$ are called the first, second, $\cdots$, derived curves corresponding to the curve $y = f(x)$.

The relations between the derived curves are best brought out by drawing the curves one below the other. The **ordinate** of the first derived curve is the **slope** of the original curve at corresponding points. Correspond-

ing to points on the original curve which are **maxima and minima,** the first derived curve crosses the X-axis; corresponding to **points of inflection** of the original curve, the first derived curve has maxima and minima. Corresponding to points of inflection of the original curve, the second derived curve crosses the X-axis; corresponding to a maximum point on the original curve, the second derived curve has a negative ordinate, and for a minimum point, a positive ordinate. (L.L.S.)

HIGHER MOMENT. The higher moments about any origin a are $\mu'_{3:x-a}, \mu'_{4:x-a}, \mu'_{5:x-a}$, where

$$\mu'_{n:x-a} = \frac{\Sigma(x-a)^n f_x}{\Sigma f_x}, \quad n > 2.$$

The higher central moments are defined by

$$\mu_{n:x} = \frac{\Sigma(x-\overline{X})^n f_x}{\Sigma f_x}, \text{ where } \overline{X} \text{ is the mean. The higher}$$

product moments in two **variables** x and y are defined by

$$\mu'_{n,m:x-a, y-b} = \frac{\Sigma(x-a)^n(y-b)^m f_{xy}}{\Sigma f_{xy}}.$$

n and m either or both greater than one. Similarly

$$\mu_{n,m:x, y} = \frac{\Sigma(x-\overline{X})^n(y-\overline{Y})^m f_{xy}}{\Sigma f_{xy}}$$

where $\overline{X}$ is the mean of the x's and $\overline{Y}$ is the mean of the y's. The results for the higher moments are in terms of integrals for a **probability function** of a **continuous variate.** (L.A.A.)

HIGHER PLANE CURVES. By the term "higher plane curve" is usually meant any plane **curve** which is not a **straight line** or a **conic section.**

Among the most important and most interesting higher plane curves are the following: **bipartite cubic, cardioid, Cartesian oval, catenary, cissoid, conchoid, cubical parabola, cycloid, epicycloid, folium of Descartes, hyperbolic spiral, hypocycloid, lemniscate, limacon, lituus, logarithmic spiral, ovals of Cassini, parabolic spiral, rose curves, semi-cubical parabola, serpentine, spiral of Archimedes, strophoid, trisectrix of Maclaurin, trochoid, witch of Agnesi.** (L.L.S.)

HIGHER PRODUCT MOMENT. The higher product moment for a bivariate distribution in x and y is given by

$$\mu'_{n,m:x-a, y-b} = \frac{\Sigma(x-a)^n(y-b)^m f_{xy}}{\Sigma f_{xy}}$$

where n and m are both greater than one. Similarly the higher central product moments are defined by

$$\mu_{n,m:x, y} = \frac{\Sigma(x-\overline{X})^n(y-\overline{Y})^m f_{xy}}{\Sigma f_{xy}}, \quad n > 1, \quad m > 1.$$

In the case of the **probability function** in two **variables** $P_{xy}\, dx\, dy$ of two continuous variates the summation is replaced by a double integral over the **range** of x and of y. (L.A.A.)

HIGHEST COMMON FACTOR. The highest common factor of several **polynomials** is the polynomial of highest degree that is a factor of each of them. It is usually found by factoring the given polynomials separately, and picking out the highest common factor by inspection.

In cases where the polynomials are not readily factored, the following method, known as Euclid's algorithm, may be used.

Let P and P' denote two polynomials and suppose that the degree of P is greater than or equal to the degree of P'. Divide P by P' until a remainder R_1 is obtained whose degree is less than the degree of P'. Divide P' by R_1 and obtain a remainder R_2 whose degree is less than that of R_1. Continue this process until the remainder $R_k = 0$; then R_{k-1} (the preceding remainder) is the highest common factor of P and P'. (L.L.S.)

HIGH-FREQUENCY PHOTOGRAPHY.

High-frequency photography includes both still and motion-picture photography, the latter being much the more important. In still photography an ordinary camera is used, the lens being left open, and a series of exposures made at the desired frequency on the same film by means of an intermittently flashing light source. Photographs of this kind are practical only if the time interval is relatively short, as, for example, a golfer's "swing" or a batter's "cut" at a pitched ball. Even then their usefulness for analytical purposes is, in some cases, seriously affected by the overlapping of parts of the image.

In high-frequency (slow-motion) cinematography the exposing rate may range from 32 to several thousand frames per sec. When the film is projected at the usual rate, the time interval recorded on the film is expanded accordingly. If, for example, the exposing rate is 128 pictures per sec. and the projection rate 16 per sec., the time interval recorded on the film is expanded 8 times; i.e., $128/16 = 8$. Thus motion photographed in one second requires 8 seconds in projection. With higher taking speeds, the expansion of time is correspondingly greater. At 3200 exposures per sec., for example, the time interval is expanded—or magnified—200 times, assuming projection at the rate of 16 images per sec.

The usual form of motion-picture camera with intermittent film movement may be employed if the exposing rate is not much above 200 per sec. To reach higher exposing speeds, the intermittent movement must be discarded for a continuously moving film. Sharp images can be obtained on a continuously moving film by one of two methods: (1) by employing an intermittently flashing light such as the stroboscope, the duration of each flash being made sufficiently short to produce a sharp image on the moving film, and (2) an optical system consisting of rotating (a) lenses, (b) mirrors, or (c) prisms which cause the image to move during the exposure at the same rate as the film and thus produce a sharp image.

High-frequency cinematography with the stroboscope has been brought to a high degree of perfection by Edgerton and his coworkers at the Massachusetts Institute of Technology, who have developed equipment capable of making 10,000 exposures per sec., or even greater if necessary. High voltages, however, are required for such high speeds and the equipment is quite bulky. (C.B.N.)

HIGH-HOLE. Flicker.

HIGH LIFT DEVICES.

The lifting ability of an **airfoil** is improved by increasing its camber (over limited range), also by delaying the separation of the air flow from the lifting surface as **angle of attack** is increased. Most airfoils employed as airplane wings are of low camber because their drag coefficients at low attack angles (high speed) are more favorable than those of the thick highly cambered airfoils.

It is known that the maximum lift coefficient of a highly cambered airfoil is greater than that of a flat, thin one. Safety in landing an airplane demands low landing speed, which is obtainable by using wings with high maximum lift coefficients. On the other hand, high maximum speed is only to be obtained with the thin, low-drag type airfoil. Many manufacturers of aircraft have adopted the **flap** as a means for securing low landing speeds on aircraft fitted with inherently high-speed wings. The flap, as illustrated in the accompanying figure, is akin to a flexible trailing edge, and by depressing the flap, the effective camber of the airfoil may be increased. The **drag** is also increased,

giving a steeper gliding angle with flap down than without the use of the flap. Apart from any disadvantage, this factor may actually be beneficial, since with flap position variable and under control of the pilot, the angle of normal gliding is alterable, permitting steep approaches to small obstructed fields. Among the flaps having received a great deal of attention, have been the simple flap, the split flap, Zap flap, and the Fowler flap. The split flap is simplest to apply, and has been extensively used. The operation is simply that of a rotation of a panel on the lower side of the trailing edge about its forward edge. In the Zap flap, the rotation is accompanied by a backward movement of the pivoting point. In the Fowler flap, the motion

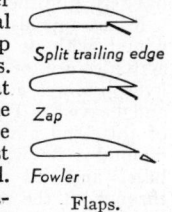
Split trailing edge

Zap

Fowler
Flaps.

is not one of simple rotation, and the effect is one of increase of airfoil area as well as change of camber. Most flap designs are patented.

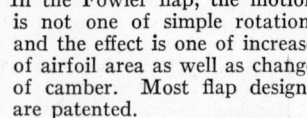

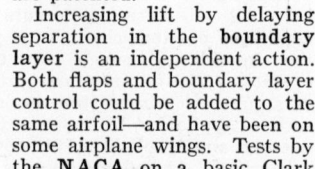
Air flow at high angle of attack with and without slot.

Increasing lift by delaying separation in the **boundary layer** is an independent action. Both flaps and boundary layer control could be added to the same airfoil—and have been on some airplane wings. Tests by the **NACA** on a basic Clark Y airfoil equipped with slot and flap revealed that the high lift devices were capable of doubling the maximum lift coefficient. Delay of air-flow separation by sucking off the incipient vortices around the **stagnation point**

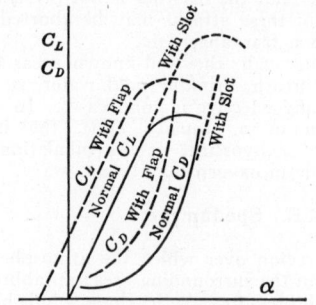
Effect of flaps and slots on lift and drag.

produced amazing lift coefficients but has not been exploited on account of the compressor energy needed. Also very small high-speed jets to impart energy to the boundary layer, and rotating cylindrical nose sections have been given tests with excellent results. However, boundary layer control by leading edge *slots* is definitely valuable for lifting airfoils and lacks the structural and mechanical complications attending the other boundary layer control proposals. Hence slots and flaps have been the only high lift additions commercially employed to date. In the light of probable future aeronautic development, the other possibilities should not be discounted. Some attempts have also been made to intensify the action of the flap by using multiple-hinged flaps, although such flaps do not appear thus far to have gone beyond the wind tunnel stage. (F.T.M.)

HIGH LINE TELEPHONE.

This is a **carrier** system of telephone communication superimposed on the regular electric power transmission lines. It is used quite extensively by power companies for interplant communication. Basically the system does not differ in principle from the ordinary carrier systems. The signals are applied to and taken from the lines by coupling capacitors which allow the high frequencies to pass but stop the 60-cycle power currents to a large extent. To reduce the effect of various stray 60-cycle signals which leak through the system, filters to take out the interference are often provided. (L.R.Q.)

HIGH-PRESSURE PHENOMENA.
The earlier researches in this field were associated with the study of the liquefaction and the critical states of gases; for example, the work of Andrews (1861). The critical pressure of water, for example, is something over 2000 kilograms per sq. cm. The hydrostatic pressure at the greatest ocean depths must be about 1000 kilograms per sq. cm. But these would now hardly be considered "high" pressures, since with modern technique it is possible to attain pressures as great as 30,000 kilograms per sq. cm. The usual means of attaining high pressures is the "intensifier," which is merely a double free piston, that is, a straight rod with a large piston on one end and a small one on the other, each in its own cylinder. Any pressure applied to the larger piston is multiplied in the smaller cylinder by the ratio of the two areas. The chief problem is that of packing to prevent leaks, and this has been met by special devices perfected by Bridgman, Poulter, and others. (See Pressure Gauges.)

Substances often exhibit unfamiliar properties at high pressure. For example, the minimum volume of water, at 4° C. under normal pressure, occurs at lower and lower temperatures as the pressure is increased; and finally, at about 2500 kilograms per sq. cm., a minimum no longer exists. Solid bismuth kept at 250° C. melts at a pressure of 5600 kilograms per sq. cm.; but liquid sodium at 150° C. solidifies at 7200 kilograms per sq. cm. Some oils behave like sodium, so that they cannot be used as the media in high-pressure apparatus. The thermal expansion of liquids under great pressure decreases with temperature instead of increasing as it normally does. When liquids are subjected to 12,000 kilograms per sq. cm., the work of compression causes them to become almost boiling hot. Many other properties have been studied in detail, such as density, electrical resistance, thermal conductivity, viscosity, dielectric constant, and polymorphic transitions.

It is interesting, but at present futile, to speculate as to the state of matter subjected to such pressures as must exist in the far interior of the earth or of the sun—millions of times greater than anything artificially attainable—and especially at the inconceivable temperatures also prevailing there. It is not improbable that states of aggregation quite unknown to us are produced under these conditions. (L.D.W.)

HIGH-SPEED PHOTOGRAPHY.
The term high-speed photography is applied both to single photographs made with extremely short exposures and to high-frequency photography which consists of a number of high-speed photographs made in rapid succession. (See High-Frequency Photography.)

Since the shortest exposure practical with mechanical shutters is not much over 1/1000 of a second, high-speed photography has employed the electric spark and, in recent years, the gaseous discharge lamp, the development of which for photographic purposes is due largely to the work of Edgerton and his coworkers at the Massachusetts Institute of Technology. With either of these a highly intense source of illumination is obtained the duration of which may be as short as 1/1,000,000 of a second, although an exposure of 1/25,000 of a second is sufficient for all except extremely high-speed subjects.

The photographs are made (1) in a darkened room using an ordinary camera with the uncovered lens focused on the subject and the exposure made by discharging the condenser across the spark gap, or through the discharge tube, or (2) in daylight using a synchronizer which opens the shutter immediately before the discharge and closes it directly thereafter. When the photograph is made in daylight with a synchronizer the time which the shutter is open must be sufficiently short to prevent the exposure of the film from the pervailing daylight. In either case, the duration of the actual exposure is determined by the duration of the flash.

Spark photographs of bullets in flight were made as early as 1865. Since the development of the gaseous discharge lamp, high-speed photography has been employed extensively in the study of machinery moving at high speed, chemical reactions which occur at high speed such as the detonation of explosives, etc., as well as for the analysis of athletic form, acrobatic dancing, the ballet, etc. (C.B.N.)

HIGHWAYS.
Of the historical importance of highways as aids in the progress of civilization, and the closer association of peoples of the earth, little can be said in this article. The highly interesting way in which development in road building is interwoven with the affairs of man as he stumbled forward from a barbaric state has, however, been adequately treated in literature. Suffice to say here that the modern highway reaches a degree of perfection attained only through many years of attempts to suit the road type to the vehicular traffic carried by it with due regard to existing economic factors. Usually perfection of types lags behind progress in the invention and building of vehicles which travel over them. It seems necessary for vehicular improvement to precede roads satisfactory for the operation of the vehicle.

A road system plays a very important part in the economic affairs of a country such as the United States, particularly in these days of widespread automobile ownership. Nowhere else in the world are there to be found such conditions of congested, high-speed, or long-distance traffic. The large sums collectible through automotive taxation have made possible a network of hard-surfaced highways, the main arteries of which are representative of great progress in the accommodation of large amounts of high-speed traffic. The special intersections where it is unnecessary for any cross traffic to exist, but in which the full choice of direction of travel of an ordinary crossroads is still available to the driver; aerial highways, in which the roadway is carried over congested districts, and entirely removed from local traffic; highway tunnels driven under rivers; and many other gigantic engineering enterprises, are part of the highway picture of the present time. Roads and highways are comparatively expensive, seldom costing less than $2000 per mile for the most elementary form of earth road, ranging from that upward to $30,000 per mile for an ordinary two-lane concrete road, and into the hundreds of thousands of dollars per mile for super highways. These gigantic sums are raised through taxation which reaches nearly every person in the United States. The rates, however, direct and indirect, are not exceptionally high in that there are few persons indeed who do not make considerable use of highways.

The simplest form of road is that which is constructed by leveling a right-of-way about 10' wide by excavating, filling, blasting, felling trees, grubbing roots, and providing ditches for drainage. Such roads are satisfactory only for the lightest form of horsedrawn traffic. If the type of road just described is adequately drained by deepening the ditches on the side and lowering the water table, then provided with an earth surface which is more of a road material than the original soil, a road is produced which is serviceable for light traffic the year round.

The most important element of a highway is its foundation. This should be firmly compacted, and have sufficient bearing power to keep a tire or rim of a vehicle from cutting through, and it should also be sufficiently porous so that water will readily drain from it. Deep side ditches are provided to drain the water from this foundation, keep it firm and hard, and reduce frost heavage. Earth as a road foundation is not altogether satisfactory unless it happens to exist in proportions of sand and binder which are suitable. The sand is necessary for bearing power, while clay or silt is necessary for binding the particles of sand together.

A sand-clay road is made by forming the sub-grade, properly drained, and spreading on top of that from 8 to 12" of sand-clay mixture in proportions of about 40% clay and silt to 60% sand. If the natural soil is to be used, and is deficient in either clay or sand, the deficiency may be made up by spreading the other material in proper amounts, after which it is mixed with harrows or plows. The sand-clay road must be crowned so as to shed water directly and at low velocity to the ditches, otherwise it will wash badly under heavy rains. Maintenance consists of scraping from ditches into the road and crowning up the surface, also adding more topsoil as needed.

The gravel road is suitable for heavier traffic than the sand-clay or earth, and is more expensive to construct. A comparatively wide roadway is built up in which a trench 6 to 10" deep and of the desired width is left to be filled with a mixture of gravel and sand-clay. This type of road tends to deteriorate to a washboard-like surface under the wear of high-speed automotive traffic, particularly in the hollows, and must be scraped when damp.

The macadam road is one of the oldest types, and one of the first really improved roads to be invented. A macadam road is one whose foundation is built of crushed stone in graded sizes, the larger stones being placed at the bottom, the smaller at the top. A waterbound macadam surface has stone screenings or other suitable fillers spread on the top surface, watered and rolled. This type of surface has not been able to withstand modern traffic conditions, due to the raveling of the surface layer, so road engineers have turned to different fillers. **Asphalt, tar,** and heavy asphaltic oil are materials that have been used for filler binders on modern macadam surfaces. Hard-surfaced roads may be constructed of some bituminous substance in conjunction with crushed stone, or of concrete with or without a bituminous surface. Practically all hard-surface mileage is of this type, and surfaces like brick, cobble stone, or blocks are but little used except for special traffic conditions in and around cities.

Bituminous hard-surfaced roads will be briefly described. Bituminous macadam roads have a sub-base of

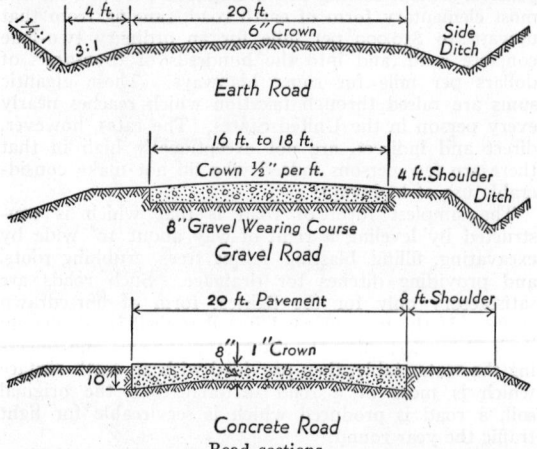

Earth Road

Gravel Road

Concrete Road
Road sections.

gravel or stone. An old waterbound macadam road can readily be converted, using the existing waterbound macadam as a foundation or base. Upon this, when suitably leveled and smoothed, may be placed a binder course of about 2½" of ½" to 1½" stone. Upon this layer is sprayed the bituminous binder from a pressure-distributing truck, following which stone chips are spread and

brushed into the interstices. A penetration of at least 1" should be secured. After lightly rolling the stone chips a second application of bituminous binder is made and the surface rolled with a heavy roller to shape. Various other methods may be employed dependent upon the size and void space of the binder course. This type is known as penetration macadam. A bituminous concrete road is built upon a well-compacted base of stone or gravel, or upon a **concrete** slab. It is essentially a premixed, bituminous concrete, consisting of asphalt, sand, and finely crushed rock. It is laid hot, and immediately rolled to about a 2" wearing course. Most bituminous concrete mixtures are patented.

The concrete road is made up of Portland cement concrete from 6 to 10" thick, poured in slabs. Unless the foundation is exceptionally well compacted, the concrete slab should have metal reinforcement. Adjacent slabs should be separated by a small space to allow for expansion. The spaces should be filled with asphalt to keep moisture from the sub-grade. A typical concrete section is shown in the accompanying figure. After the slab is poured it must be kept moist for several days in order properly to cure the concrete and develop its full value as a road surface. While the concrete mixture employed is not standardized, a 1:2:3 mixture is about what is ordinarily used. The coarse aggregate may be either gravel or crushed rock, and the fine aggregate washed sand.

A standard 2-lane highway should have a pavement surface from 18 to 20' wide, with 4 to 6' earth shoulders on the side of that. Where traffic conditions are exceptionally severe, it has been necessary to increase this width and add more lanes for traffic. The 3-lane highway should not be used, if at all avoidable, since it does not increase the traffic capacity of the road in proportion to the added expense, and statistics show such roads to be extremely productive of traffic accidents. The 4-lane highway is far superior in that cars going the same direction may pass without invading an area set aside for traffic moving in the opposite direction. In spite of this fact, however, many head-on collisions have occurred on 4-lane highways, and are particularly disastrous because of the high average speed of traffic on these highways. It appears that it may eventually be necessary to separate traffic moving in opposite directions by a central strip which is unpaved, creating, in effect, two parallel roadways. (F.T.M., E.W.S.)

HILL MYNA. Aves, Passeriformes. The **grackles** of India and adjacent regions. They are glossy black birds with brightly colored wattles and in some species other bright marks. When kept in captivity they learn to talk readily and imitate the inflections of the human voice much more faithfully than the parrots. (A.W.L.)

HIND. The female **deer,** especially of the European red deer. (A.W.L.)

HINGE LIGAMENT. A tough elastic connection between the dorsal margins of the two valves of the shell of **bivalve** mollusks. (A.W.L.)

HINNY. The hybrid offspring of a stallion and a jennet, reciprocal of the cross of jack and mare by which mules are produced. The difference between mules and hinnys is a disputed point. Some writers claim that hinnys are smaller and lacking in the qualities desired in mules, while others say that both forms fall within the range of variation to be expected in the hybrid. (A.W.L.)

HIP. The joint at the attachment of the human thigh to the body. Also the adjacent portion of the thigh where it merges with the buttocks and less commonly the corresponding part of the leg in other animals. (A.W.L.)

HIPERNIK. Nickel Alloys; Magnetic Materials.

HIPPOPOTAMUS. Mammalia, Artiodactyla. A large African animal with a bulky body, short strong legs, and a very broad muzzle. Hippopotami have been known to attain a length of 12' and a weight of 4 tons. They are largely aquatic in habits and live entirely on vegetation, including both water plants and terrestrial species. They sometimes do great damage to crops.

A second species, the pigmy hippopotamus of western Africa, reaches only half the length of the larger animal and is found in the forests and swampy lands. (A.W.L.)

HIRUDINEA. The leeches, a class of segmented worms (Phylum **Annelida**), well known for their habit of sucking blood. Marine and fresh-water species are known, and in the moist tropical forests terrestrial species occur. They often attach themselves to bathers.

The members of this class are distinguished from other annelids by the following characters: 1. The body is relatively short, usually with thirty-two segments. 2. The external segments are annuli numbering from two to fourteen to each metamere. 3. Each end of the body bears a sucker. 4. The mouth is usually provided with three toothed plates or jaws. 5. The alimentary tract is provided with an enormous pouched crop in which blood is stored prior to digestion. 6. The anus opens dorsal to the posterior sucker. 7. The coelom is partially obliterated by a peculiar **mesenchymal tissue.**

8. At the anterior end of the ventral nerve cord several ganglia **(ganglion)** are fused to form a large mass.

Leeches were once extensively used in medicine for letting blood and are still of minor importance for this purpose. Otherwise they are of no importance to man save as an occasional annoyance. They eat small aquatic animals as well as the blood of vertebrates and some species are entirely predacious.

Two orders are recognized:

Order Rhynchobdellida. With a protrusible proboscis, colorless blood, and no jaws. Marine and fresh-water.

Order Gnathobdellida. With jaws and red blood. No proboscis. Fresh-water and terrestrial. The medicinal leech belongs to this order. It is native to Europe but is naturalized in ponds and streams of the eastern United States. (A.W.L.)

HISTIDINE. Aminoacids; Polypeptides; Proteins.

HISTOBLAST. A group of cells in the immature stages of **insects** from which some organ of the adult is developed. The forerunners of the wings in insects with complete metamorphosis, for example, are small thickened layers of cells in the **larva.** Also called imaginal disks. (A.W.L.)

HISTOLOGY. The science which deals with the minute structure of living things. The study of the struc-

---

GEOLOGICAL TIME-SCALE (See **Historical Geology.**)

ERA = time GROUP = rocks	PERIOD = time SYSTEM = rocks	LIFE RECORD (FOSSILS) BOTH ANIMALS AND PLANTS
CENOZOIC Age of mammals and modern flora	QUARTERNARY	Periodic glaciation and origin of man (Pleistocene). The transformation of the ape-like ancestor into man may have begun in the Pliocene. Culmination of mammals (Miocene). Rise of higher mammals (Oligocene). Vanishing of archaic mammals (Eocene).
	TERTIARY upper lower	
MESOZOIC Age of reptiles	CRETACEOUS	Rise of the archaic mammals in the interval between the Mesozoic and the Tertiary. This ERA is remarkable for the great development of the ammonites which became extinct at the end of the Cretaceous. The mollusks are more highly developed in this ERA than in the preceding one. Culmination and extinction of most reptiles (Cretaceous). Rise of flowering plants (Comanchean); birds and flying reptiles (Jurassic); dinosaurs (Triassic).
	JURASSIC	
	TRIASSIC	
Upper PALEOZOIC Age of amphibians and lycopods	PERMIAN	Periodic glaciation and extinction of many Paleozoic groups during and after the Permian. Rise of modern insects, land vertebrates and ammonites (Permian); primitive reptiles and insects (Pennsylvanian); ancient sharks and echinoderms (Mississippian).
	CARBONIFEROUS	
Middle PALEOZOIC Age of fishes	DEVONIAN	First known land floras (Devonian) not very different from those of the Carboniferous. Earliest evidence of a terrestrial vertebrate in the form of a single footprint from the Devonian of Pennsylvania. Rise of lung-fishes and scorpions (first terrestrial air-breathers) in the Silurian.
	SILURIAN	
Lower PALEOZOIC Age of higher (shelly) invertebrates	ORDOVICIAN	Rise of nautiloids, armored fishes, land plants and corals. Also the first evidence of colonial life (Ordovician). First known marine faunas; dominance of trilobites; rise of animals with hard shells or exo-skeletons (Cambrian).
	CAMBRIAN	
PROTEROZOIC Primordial life	PRE-CAMBRIAN	Fossils almost unknown except for a few problematical forms in the Proterozoic. Fossils unknown in the Archeozoic.
ARCHEOZOIC Most ancient life		

NOTE. Geological time-tables are so constructed as to show the oldest periods at the bottom and the youngest periods at the top. *To get the proper order and sequence of events always read from the bottom to the top.* (Field, *Geology Manual,* Princeton University Press.)

ture and functions of cells is the special province of **cytology**, leaving the study of special forms of **cells** and their association in **tissues** and **organs** as the field of histology, but histology necessarily includes much cytological matter.

The science is made up of two subordinate fields, general and special histology. In the former are considered the specialization of cells in the multicellular body and the characteristics and classification of the tissues in which they are grouped. The details of minute structure of the organs and organ systems are the materials of the latter. This field of histology is necessarily extensive and detailed, even in the study of a single species.

Histology recognizes five principal kinds of tissues, epithelium, nervous tissue, mesenchymal (connective and supporting) tissues, muscular tissue, and vascular tissue. All organ are made up of these components and further detail of histology are included under the various organs and organ systems and under the topics **epithelium, connective tissue, nervous tissue, mesenchyme, cartilage, bone, muscular tissue, blood,** etc. (A.W.L.)

HISTONES. Aminoacids; Polypeptides; Proteins.

HISTORICAL GEOLOGY. The study and description of the origin and evolution of the earth and its inhabitants (animals and plants). The technical methods employed are included under the general term, **stratigraphy**. The outstanding events or important "chapters" in the history of the earth are outlined in the form of a geologic Time-Scale or Time-Table; see page 713. The study of the extinct forms of life, from the earliest known **fossils** up to, but not including, the Recent epoch, is called **Paleontology**. The history of the earth from its astral stage up to the oldest known rocks is included under the general term **cosmogony**. The numerous methods for determining the age of the earth as well as the relative ages of geological events are included under the term **chronology**, which in turn may be considered as included under the more general term Stratigraphy. The fundamental data of geologic history are local sequences of formations, and the chronologic equivalences of formations in different regions. Through correlation all formations are referred to a general time scale, of which the (fundamental) units are periods. The formations made during a period are collectively designated a system. The fundamental criteria used in delimiting geologic periods are (1) **Unconformities**, (2) Cycles of Sedimentation, (3) Index Fossils. The ultimate aim of the historical geologist is paleogeography, or the reconstruction of the consecutive geographies of the past. (R.M.F.)

HISTORY AND EVOLUTION OF CHEMISTRY.

Pure Chemistry

1755	Black. Studies on carbonic acid and carbonates.
1766	Cavendish. Discovered inflammable air (named hydrogen by Lavoisier in 1783) as a distinct substance.
1772	Rutherford. Isolated nitrogen.
1774	Priestley. Discovery of oxygen.
	Scheele. Discovery of oxygen.
1776	Scheele. Discovery of oxalic acid and uric acid.
1777	Lavoisier. Fundamental studies of combustion and of respiration.
	Wenzel. Law of Mass Action.
1778	Scheele. Discovery of chlorine.
1779	Scheele. Discovery of glycerol.
1780	Scheele. Discovery of lactic acid.
1781	Cavendish. Showed hydrogen burns to form water.
1782	Scheele. Discovery of hydrogen cyanide.

1782	Lavoisier. First quantitative synthesis of water. Fermentation of sugar yields ethyl alcohol and carbon dioxide.
1784	Charles. Law of expansion of gases. Pressure varies directly as the absolute temperature.
1785	Scheele. Discovery of malic acid.
1787	Lavoisier. Classification of compounds.
1791	Richter. Law of neutralization of acids and bases.
1796	Lampadius. Discovery of carbon disulfide.
1799	Walter. Discovery of picric acid.
1800	Nicholson and Carlisle. Quantitative electrolysis of water.
1801	Hare. Oxyhydrogen blowpipe.
1802–06	Proust-Berthollet controversy on constancy of chemical composition, Proust winning by proving the constancy.
1805	Northmore. Liquefaction of chlorine.
1806	Proust. Law of constant chemical composition.
1807	Dalton. Law of multiple combining proportions.
	Dalton. Law of partial pressures of gases in mixtures.
	Dalton. Atomic theory.
	Davy. Isolated sodium and potassium.
	Davy. Electrochemical theory.
1808	Gay-Lussac. Law of simple combining volumes of gases.
	Gay-Lussac. Law of expansion of gases. Volume varies directly with absolute temperature.
	Davy. Isolated magnesium, calcium and barium.
1810	Berzelius. Isolated silicon.
	Davy. Elementary nature of chlorine.
1811	Avogadro. Molecular hypothesis.
	Berzelius. System of chemical nomenclature.
	Berzelius. Dualistic theory.
1814	Frauenhofer. Solar spectral lines.
1815–22	Gay-Lussac. Work on cyanogen.
	Prout. Hypothesis that other elements are composed of hydrogen.
	Biot. Optical activity of sugar solutions.
1817	Arfvedson. Discovery of lithium.
	Berzelius. Exact ratios of atomic weights. A revised table in 1826.
1819	Mitscherlich. Isomorphism of crystals.
	Garden. Isolated naphthalene.
1821	DuLong and Petit. Law of atomic heat.
	Cagniard de la Tour. Critical phenomena of gases and liquids.
1823	Faraday. Liquefied chlorine, sulfur dioxide, hydrogen sulfide, carbon dioxide, ammonia, nitrous oxide, and cyanogen.
1824–32	Gay-Lussac. Methods of volumetric analysis.
	Carnot. Studies in themodynamics.
1825	Faraday. Discovery of benzene and butylene.
1826	Unverdorben. Discovery of aniline.
	Balard. Discovery of bromine.
1827	Dumas. Molecular weights by vapor density methods.
	Wöhler. Isolated aluminum.
1828	Wöhler. Formation of urea from ammonium cyanate.
1829	Döbereiner. Triads of elements.
1830	Robiquet. Discovery of the enzyme emulsin in almonds and in the seeds of Rosaceae.
	Liebig. Modern combustion furnace and method for carbon and hydrogen determinations in organic substances.
1831	Dumas. Method for nitrogen determination in organic compounds.
	Liebig. Discovery of chloroform.
	Leuchs. Discovery of the enzyme ptyalin of saliva.
1832	Dumas and Laurent. Isolated anthracene.
	Liebig. First theory of radicals.

1832	Liebig and Wöhler. Studies on the radical benzoyl; benzoin condensation.
1832	Serullas. Discovery of iodoform.
1833	Faraday. Laws of electrolysis, ions carry simple or multiple charge.
1833	Mitscherlich. Discovery of benzene sulfonic acid.
	Graham. Studies on basicity of such acids as phosphoric.
1834	Dumas. Discovery of substitution in organic compounds.
	Mitscherlich. Preparation of benzene and of nitrobenzene from benzoic acid.
	Payen and Persoz. Discovery of the enzyme diastase from malt.
1835	Thilorier. Solidified carbon dioxide.
	Schwann. Discovery of the enzyme pepsin of gastric juice.
1835	Dumas. Quantitative synthesis of water by passing hydrogen over heated copper oxide.
1836	E. Davy. Discovery of acetylene.
	Laurent. Preparation of phthalic acid from naphthalene.
1837	Bunsen. Studies on the radical cacodyl.
1840	Hess. Law of constant heat summation.
1840–70	Stas. Researches on atomic weights.
1843	Laurent. Definition of atomic and molecular equivalents.
	Pasteur. Dextro and laevo tartrates.
1844	Gerhardt and Laurent. Classification of organic compounds.
1847	Dumas. Preparation of acetamide.
1849	Fehling. Reduction of cupric salt in alkaline medium to cuprous oxide as test for sugars.
	Frankland. Ethyl iodide plus zinc at 150° C. yields butane.
	Kolbe. Electrolysis of acetates yielded ethane.
1850	Wilhelmy. Studies on the rate of inversion of sucrose solutions.
	Hofmann. Prepared alkyl amines.
1851	Graham. Diffusion, dialysis, and osmosis in solutions.
1852	Frankland. Principle of valency.
1852–62	Joule and Thomson. Observations of change of temperature by emergence of compressed gas through small orifice.
1853	Hittorf. Migration of ions.
1853–1908	Thomsen. Thermochemical investigations.
1854	Wurtz. Alkyl iodide plus sodium heated to yield paraffin hydrocarbon.
	Deville. Thermal dissociation of substances.
1857	Bunsen. Methods of gas analysis.
1858	Cannizzaro. Rational symbols and formulas.
	Gerlach. Introduced biological stains.
	Kekulé and Couper. Quadrivalent carbon atom.
1859	Kirchhoff. Black body radiation.
	Griess. Discovered diazo compounds.
1860	Bunsen and Kirchhoff. Invention of the spectroscope; discovery of cesium.
1861	Bunsen and Kirchhoff. Discovery of rubidium.
1862	Graham. Crystalloids and colloids.
1863	Reich and Richter. Discovery of indium.
	Fittig. Aryl haloid plus sodium heated yields benzenoid hydrocarbon.
1864	Guldberg and Waage. Law of Mass Action.
1867	Kekulé. Formula of benzene; oscillation of double bond, 1872.
1867–70	Meyer, L. Molecular volumes and the periodic system.
1868	Hofmann. Determination of molecular weights by vapor displacement.
	Janssen and Lockyer. Discovery of helium in the sun.
1869	Andrews. Variation of volume of carbon dioxide with change of pressure.
	Mendeléeff. Periodicity of the elements.

1869–85	Berthelot. Thermochemical investigations.
1873	Meyer, V. Determination of molecular weights by vapor displacement.
	van der Waals. Equation of state of gas and vapor.
	Maxwell. Theory of electricity and magnetism.
1874	Boisbaudran. Discovery of gallium.
	Volhard. Volumetric titration of silver as thiocyanate.
	Le Bel and van't Hoff. Stereoisomerism of tetrahedral carbon atom.
1875	Gibbs, Willard. The phase rule.
	Stoney. The electron.
1876	Kohlrausch. Electrical conductivity of solutions.
1877	Cailletet. Liquefied methane, ethylene, acetylene, nitric oxide.
	Friedel and Crafts. Organic reactions with anhydrous aluminum chloride.
1878	van't Hoff. Stereochemistry of the nitrogen atom.
	Raoult. Properties of solutions.
1878–87	Ostwald. Electrical conductivity of organic acids.
1879	Crookes. Suggested the nature of cathode rays.
1880	Crookes. Cathode rays.
1881	van der Waals. Gas equation.
1883	Kjeldahl. Estimation of nitrogen as ammonia in organic compounds.
	Wroblewski and Olszewski. Liquefied nitrogen.
1884	Dewar. Vacuum-walled flask for liquefied gases at low temperature.
	Le Chatelier and van't Hoff. Principles of shifting of chemical equilibrium.
1885	Le Chatelier. Thermocouple for measuring temperature perfected.
	Callender. Resistance thermometer introduced.
1886	Goldstein. Discovery of anode rays.
	Moissan. Isolation of fluorine.
1887	Arrhenius. Ionic theory of electrolytic dissociation.
1888	Nernst. Diffusion theory of electromotive force of solutions.
1892	Le Chatelier. Optical pyrometer invented.
1893	Amagat. Variation of the product of pressure and volume with change of pressure for gases.
	Ramsay. Work on the rare gases.
	Cleve. Discovery of helium in minerals.
	Roentgen. Discovery of x-rays.
	Linde and Hampson. Oxygen and nitrogen by fractionation of liquefied air.
1893–1908	Landolt. Conservation of matter in chemical reactions re-proved.
1894	Rayleigh and Ramsay. Discovery of argon.
1895	Olszewski. Liquefaction of hydrogen.
	Morley. Composition of water by direct weighing of hydrogen, oxygen, and water.
1896	Becquerel. Radioactivity of uranium.
1897	Thomson. Nature of cathode rays as streams of electrons.
	Sabatier. Catalytic hydrogenation of unsaturated organic compounds.
1898	Wien. Nature of anode rays.
	Curie, P. and Marie. Discovery of radium.
1899	Walden. Investigation of active malic and chlorosuccinic acids.
1900	Guye. Atomic weights from gas densities.
	Grignard. Magnesium—ether—organic halide as organic reagent.
	Gomberg. Triphenylmethyl free radical (at least 2 aryl groups to each C).
1902	Rutherford and Soddy. Spontaneous decomposition hypothesis of radioactive changes.
	Morse. Optical pyrometer improved by use of glow lamp filament.
1904	Barkla. X-rays as (1) ether impulses, (2) characteristic rays.

Pure Chemistry (Continued)

1905   Dewar. Charcoal as gas absorbent at low temperature.
1906   Miers. Refractive index of crystallizing solution.
1907   Willstätter. Structure of chlorophyll.
1908   Onnes. Liquefaction of helium; approach to absolute zero.
Rutherford and Geiger. Detection and counting of single alpha particles.
Millikan. Precision measurements of the charge on an electron.
1911   Rutherford. Hypothesis of nuclear atom.
Soddy. Mesothorium isotopes.
1912   Laue and W. H. and W. L. Bragg. Structure of crystals by x-rays.
1913   Thomson. Spectrum of neon in anode-ray tube.
Fajens, et al. Alpha and beta particle shift in periodic table.
1914   Moseley. Atomic numbers by x-ray spectra of the elements.
1917   Hull. Structure of crystals by x-rays.
Debye and Scherer. Structure of crystals by x-rays.
1918   Kendall. Thyroxine isolated.
1919   Aston. Mass-spectrograph for ascertaining the atomic weight of isotopes.
Rutherford. Artificial disintegration of elements.
1921   Harkins and Hayes. Fractionation of chlorine isotopes.
Bronsted and Hevesy. Fractionation of mercury isotopes.
Bragg, W. H. Structure of naphthalene by x-rays.
1922   Banting and collaborators. Insulin isolated.
1923   Coster and Hevesy. Discovery of hafnium.
Debye and Hückel. Strong electrolytes ionized completely. Quantitative treatment of same.
1925   Noddack, W. and Ida. Discovery of rhenium.
Niewland. Synthetic chemicals from acetylene, namely, monovinylacetylene and divinylacetylene.
1929   Paneth and Hofeditz. Methyl free radical of half life 0.006 second.
Fleming. Penicillin discovered; 1939, Florey extended investigation; 1941, Florey and Heatley visited U. S. A. with the result that the U. S. Northern Regional Research Laboratory, Peoria, Illinois, undertook a study of production to which A. J. Moyer contributed largely.
1930   Bothe and Becker. Discovery of neutron.
Mauss and Mietsch. Synthesis of the antimalarial drug "Atabrine."
1932   Urey. Discovery of deuterium.
Anderson. Discovery of positron.
1934   Joliot, F., and Joliot-Curie, Mme. Artificial radioactivity.
1937   Protactinium recognized by the International Committe on Atomic Weights.
1939   Meitner, Hahn, Frisch, Fermi, Bohr, Dunning. Fission of uranium.
1944   Woodward and Doering. Synthesis of quinine.

Applied Chemistry

1620   Coal used for smelting iron ore.
1635   Winthrop. Survey of American chemical resources.
1650   Glauber. Hydrochloric acid from sodium chloride and sulfuric acid.
1746   Roebuck. Chamber process for sulfuric acid.
1747   Marggraf. Discovered sugar in beet juice.
1755   Nordhaussen. Fuming sulfuric acid.
1769   Coke manufactured.
Watt. Steam engine patent.
1778   Baumé. Hydrometer scale.

1779   Achard. Manufacture of beet sugar.
1784   Cort. Puddling process for wrought iron.
1788   Le Blanc. Soda process for conversion of sodium chloride into carbonate via sulfate and sulfide.
1792   Murdoch. Manufacture of coal gas.
1793   Harrison. Manufacture of sulfuric acid in U.S.A.
1800   Volta. First electric battery.
1801   duPont Company founded.
1806   Coal gas used for lighting in U.S.A.
1823   Braconnot. Discovered guncotton.
1827   Gay-Lussac. Tower for absorbing nitrogen oxide gases from sulfuric acid chamber process at Chauncy, France.
1828   Neilson. Introduced use of hot air for iron blast furnace.
1830   Kuhlmann. Discovered platinum as catalyzer for ammonia oxidation.
1831   Ure. Fulminates as detonators.
Phillips and Peregrine. Patented essential details of contact process for sulfuric acid.
1834   Runge. Discovered phenol and pyrrole in coal tar.
1836   Daniell. Electric battery (copper-zinc).
de la Rive. Introduced electroplating.
1839   Goodyear. Vulcanization of rubber.
1841   Bunsen. Electric battery (carbon-zinc).
1842   Kingsford. Production of corn starch.
1844   Burt. Discovered Lake Superior iron ore deposits; 1877, first shipment of ore.
1845   Hofmann. Discovered aniline in coal tar.
Petroleum discovered in Pennsylvania.
1847   Maynard. Collodion made in U.S.A.
1850   First fertilizer plant in U.S.A.
1853   Watt and Burgess. Soda process (sodium hydroxide) for paper pulp from wood.
1856   Perkin. Discovered mauveine.
Bessemer. Steel process using air blast converter.
1858   Hofmann. Discovered aniline in coal tar.
Hofmann. Prepared rosaniline.
1859   Drake drilled first oil well (Titusville, Pa.). First petroleum refinery in Pennsylvania.
Glover. Tower for supplying nitrogen oxide gases to sulfuric acid chamber process.
Plante. Lead-lead dioxide storage battery.
1861   Solvay. Soda process for conversion of sodium chloride into carbonate by ammonium hydrogen carbonate forming sodium hydrogen carbonate.
1862   Nobel. Introduced dynamite.
1867   Caro. Prepared Bismarck brown.
Tilghman. Sulfite process (calcium hydrogen sulfite plus sulfurous acid) for paper pulp from wood.
1868   Weldon. Process for chlorine by hydrochloric acid and manganese dioxide with recovery of manganese.
Pasteur. Studies in acetic acid fermentation.
Leclanché. Dry cell electric battery.
1869   Gräbe and Liebermann. Alizarin from anthracene.
Hyatt, J. W. and I. S. Manufacture of celluloid at Albany, N. Y.
1872   Baeyer. Introduced fluorescein.
Electrolytic refining begun in U.S.A.
Saylor. Portland cement manufactured in U.S.A.
1873   Linde. Ammonia compression refrigeration.
1875   Winkler. Contact process for sulfuric acid.
1878   Fischer, O. Introduced malachite green.
Behr. Glucose from corn.
1880   Deacon. Process for chlorine by hydrogen chloride-air mixture over cupric salt catalyzer.
1881   Brin. Process for oxygen, using barium peroxide

1882 Baeyer. Synthesis of indigo.
Edison. Introduced incandescent carbon electric lamp.
1884 De Laval and Parsons. Steam turbine.
First Solvay soda plant in U.S.A. at Syracuse, N. Y.
1885 Welsbach. Incandescent gas mantle.
Bradley. Heating of ores by electricity.
Castner and Kellner. Electrolytic process for sodium hydroxide from sodium chloride, using mercury cathode.
Gayley. Dry air blast used in the pig iron blast furnace; 1894, perfected.
1886 Hall (U.S.A.); Héroult (France). Electrolytic process for aluminum from oxide; 1888, commercial production at New Kensington, Pa.
Mond. Nickel carbonyl for volatilizing and refining nickel.
1888 Florida phosphate deposits discovered.
Willson. Manufacture of calcium carbide in electric furnace.
1889 Introduction of cyanide in gold metallurgy.
1890 Castner. Electrolytic process for sodium from fused hydroxide.
Acheson. Manufacture of silicon carbide in electric furnace.
1891 Chardonnet. Manufacture of artificial silk at Besançon, France.
Frasch. Mining of sulfur by superheated water in Louisiana.
1892 Le Sueur. Diaphragm type of electrolytic cell.
1893 Moissan. Electric arc furnace.
Moissan. Silicon carbide.
1894 Moissan. Calcium carbide.
Cross and Bevan. Viscose for artificial fiber.
Power plants started at Niagara Falls, N. Y.
1895 Hampson-Linde. Liquefaction of air and fractional distillation of same.
1896 Acheson. Graphite from coal in electric furnace.
1897 Large-scale manufacture of indigo in Germany.
1898 Sabatier. Catalytic hydrogenation of unsaturated organic compounds; 1902, Norman. Commercial hydrogenation of liquid fatty oils to semisolid or solid fats.
Goldschmidt. Thermite reaction of aluminum powder and iron oxide for production of high temperature.
Knietsch. Contact process for sulfuric acid patented.
Dow. Electrolytic chlorine for bleaching powder.
1900 Héroult. Industrial smelting of steel in electric furnace.
Edison. Nickel-nickel dioxide storage battery.
Ostwald. Patented oxidation of ammonia by air, using platinum catalyzer.
Wesson. Vacuum deodorization of fatty oils.
Taylor. Carbon disulfide manufactured in electric furnace at Torrey, N. Y.
1901 First contact process sulfuric acid plant in U.S.A.
1902 Readman and Parker. Manufacture of phosphorus in electric furnace.
1903 Birkeland and Eyde. Electric fixation of atmospheric nitrogen in Norway.
1904 Betts. Electrolytic refining of lead, using fluosilicate.
1905 Frank and Caro. Cyanamide process from nitrogen and calcium carbide; 1907 commercial production at Niagara Falls, N. Y.
Becket. Pure ferroalloys by reduction of oxides by silicon.
1906 De Forest. Invention of radio tube detector.
Hybinette. Electrolytic process for refining of nickel.

1906 U. S. Food and Drugs Act.
Cottrell. Electrostatic precipitation of suspended dust and mist particles from air.
1908 Baekeland. Phenol-formaldehyde resins and plastics.
1909 Bradley and Lovejoy. Electric fixation of atmospheric nitrogen at Niagara Falls, N. Y.
1910 Langmuir. Gas-filled electric lamp.
Coolidge. Ductile tungsten filament.
1911 Dreyfus. Cellulose acetate for artificial fiber.
1912 Brearley. Stainless or corrosion resistant steels.
Burton. First cracking still for petroleum.
1913 Haber. Perfected process for synthesis of ammonia from nitrogen and hydrogen.
Coolidge. Filament x-ray tube.
1914 Sullivan and Taylor. Pyrex, a hard borosilicate glass.
1915 McAfee. Introduced aluminum chloride anhydrous in petroleum refining.
Weizmann. Fermentation of starch to produce acetone and normal butyl alcohol.
Activated sludge process of sewage recovery.
Gas carbon black in rubber for tires.
High chromium iron alloys introduced.
1916 Commercial production of artificial leathers.
Mustard gas used in warfare.
1917 Pyrethrum extract in kerosene as insect spray.
1918 Dubbs. Cracking of petroleum by continuous process.
Gibbs. Phthalic anhydride by oxidation of naphthalene vapor by air, using vanadium pentoxide catalyst.
1919 Sperry. Electrolytic process for white lead.
1920 Phosphoric acid by electric furnace smelting.
Downs. Sodium metal by electrolysis of fused sodium chloride.
John. Urea-formaldehyde resin; 1929, commercial production in U.S.A.
1921 Potash beds discovered in Texas and New Mexico; 1931, developed.
1922 Fink. Chromium plating.
Sheppard and Eberlein. Electrolytic deposition of rubber.
Continuous cracking in petroleum refining.
1923 Midgley. Tetraethyl lead as anti-knock compound in motor fuel.
Quick-drying, low-viscosity nitrocellulose lacquer by du Pont Company, using spraying method.
1924 High alumina early strength cement.
Amyl compounds from pentanes in natural gas.
1925 Ethylene glycol and related solvents made from petroleum still gases by Carbide and Carbon Chemicals Corp.
Langmuir. Welding by use of atomic hydrogen.
Antioxidants for rubber.
Steenbock and Hess. Ultra-violet irradiation of sterols, producing vitamin D.
1926 Ethylene ripening of citrus fruits.
Quick freezing of foods.
Regenerated cellulose films and tubes.
1927 Vanadium pentoxide as catalyzer for sulfur trioxide from sulfur dioxide plus air.
1928 Cellulose acetate rayon produced in U.S.A.
Detergents composed of sodium salts of sulfuric acid esters of higher aliphatic alcohols (C_{10} to C_{18}) produced in Germany; 1934, in U.S.A.
1929 Egloff. Gum inhibitors for gasoline.
1930 Vinylite resins from acetylene.
Commercial production of synthetic ethyl alcohol from ethylene by Carbide and Carbon Chemicals Corp.
Carothers and Collins. Neoprene synthetic rubber (chloroprene).

Applied Chemistry (Continued)

1930 Iodine from California oil well brines.

Hydrogenation of petroleum at Bayway, N. J.

Midgley. Dichlorodifluoromethane (Freon) as refrigerant.

1931 Hill. Methylmethacrylate resin; 1937, produced commercially in U.S.A.

1932 The alkene polysulfide rubber "Thiokol" produced in U.S.A.

1933 Synthetic camphor from turpentine.

Ipatieff. Polymerization of olefin hydrocarbons, and alkylation of aromatic hydrocarbons with olefins, using calcined phosphoric acid catalyst.

Ascorbic acid (vitamn C) synthesized.

1934 Bromine from sea water at Cape Fear, N. C.

Isooctane as aviation fuel; 1936, first tank carload produced in U.S.A.

1935 Synthetic rubber, Buna type, in U.S.A.; 1940, butyl type; 1941, large-scale production of Buna-S type started in U.S.A.; 1943, GR-S type plant at Institute, W. Va., including both styrene from benzene plus ethylene followed by dehydrogenation, and butadiene from ethyl alcohol.

Sulfanilamide recognized as a medicinal chemical; 1939, licensed for sale in U. S. interstate commerce; 1938, sulfapyridine; 1939, sulfathiazole; 1940, sulfadiazine and sulfaguanidine.

Kodachrome motion-picture process for amateurs by Eastman Kodak Company.

Thiamine chloride (vitamin B_1) synthesized.

1937 Ethylcellulose produced in U.S.A.

Houdry. Fixed bed catalytic cracking of petroleum.

Grosse. Butenes to butadiene, using aluminum oxide plus chromium oxide catalyst.

Hickman. High-vacuum distillation of fish oils; 1941, centrifugal still for the fractionation of oils.

1938 Synthetic fiber nylon produced at Seaford, Del., by du Pont Company.

1938 Vinylidene chloride resin produced in U.S.A.

Butadiene from ethyl alcohol produced commercially.

1938 Linn and Grosse. Alkylation of hydrocarbons, using anhydrous hydrogen fluoride; 1942, commercial production in U.S.A.

1939 Heavy metal non-silicate glasses, having high refractive index for a given dispersion, introduced by Eastman Kodak Company.

Polyvinyl acetal resin.

Nicotinic acid (vitamin PP) and riboflavin (vitamin B_2) synthesized.

1940 Dichlorodiphenyltrichloroethane (DDT) antilouse agent; 1943, commercial production in U.S.A.

Nitroparaffins (C_1 to C_3) produced commercially by Commercial Solvents Corporation.

1941 Toluene produced from petroleum in U.S.A.

Extraction of magnesium from sea water at Freeport, Tex., by Dow Chemical Company.

1942 High-speed continuous tin electroplating.

1943 Penicillin produced commercially in U.S.A.

Fluid catalytic (catalyst falling through the reaction zone) cracking of petroleum.

Polyethylene resin produced commercially in U.S.A.

1944 Methylolurea resin treatment of wood.

Organic silicon resins (Silicones) produced commercially in U.S.A.

1945 Uranium atomic bomb.

AWARDS, MEMORIALS, AND LECTURES. Nobel Prize Awards, Perkin Medal Awards, Memorial Lectures, Faraday Lectures.

Nobel Prize Awards. The Nobel prize awards have been made annually beginning in 1901, five years after the death of the donor, Alfred Bernhard Nobel. The amount of each prize varies with the income from the fund, but is usually about $40,000. The awards in chemistry and physics are made by the Swedish Academy of Science, Stockholm, and in physiology or medicine by the Caroline Medical Institute, Stockholm. The following awards have been made:

Year	Chemistry (Most important discovery or improvement)	Physics (Most important discovery or improvement)	Physiology or Medicine (Most important discovery)
1901	J. H. van't Hoff	W. C. Roentgen	E. A. von Behring
1902	Emil Fischer	H. A. Lorenz and P. Zeeman	Sir Ronald Ross
1903	Svante Arrhenius	H. Becquerel, P. and Mme. Curie	N. R. Finsen
1904	Sir Wm. Ramsay	Lord Rayleigh	I. P. Pawlow
1905	A. von Baeyer	Ph. Lenard	R. Koch
1906	H. Moissan	J. J. Thomson	C. Golgi and S. Ramon
1907	E. Buchner	A. A. Michelson	C. L. A. Laveran
1908	E. Rutherford	G. Lippmann	P. Ehrich and E. Metchnikoff
1909	W. Ostwald	G. Marconi and F. Braun	Th. Kocher
1910	O. Wallach	J. D. van der Waals	A. Kossel
1911	Marie Curie	W. Wien	A. Gullstrand
1912	V. Grignard and P. Sabatier	Gustaf Dalen	A. Carrel
1913	A. Werner	H. K. Onnes	C. Richet
1914	T. W. Richards	M. von Laue	R. Barany
1915	R. Willstätter	W. H. Bragg and W. L. Bragg	Not awarded
1916	Not awarded	Not awarded	Not awarded
1917	Not awarded	G. Barka	Not awarded
1918	F. Haber	M. Planck	Not awarded
1919	Not awarded	J. Stark	Jules Bordet
1920	Walther Nernst	C. E. Guillaume	A. Krogh
1921	Frederick Soddy	Albert Einstein	Not awarded
1922	F. W. Aston	Niels Bohr	A. V. Hill and O. Meyerhoff
1923	Fritz Pregl	R. A. Millikan	F. G. Banting and J. J. R. McLeod
1924	Not awarded	K. M. G. Siegbahn	W. Einthoven
1925	Richard Zsigmondy	Jas. Franck and Gust. Hertz	Not awarded
1926	T. Svedberg	Jean B. Perrin	Johan Fibiger
1927	Henrich Wieland	A. H. Compton and C. T. R. Wilson	J. Wagner Jauregg

Year	CHEMISTRY (Most important discovery or improvement)	PHYSICS (Most important discovery or improvement)	PHYSIOLOGY OR MEDICINE (Most important discovery)
1928	Adolf Windaus	O. W. Richardson	Ch. Nicolle
1929	A. Harden and H. von Euler-Chelpin	Duc de Broglie	F. G. Hopkins and C. Eijkmann
1930	Hans Fischer	Raman	Karl Landsteiner
1931	Carl Bosch and Fred. Bergius	Not awarded	Otto Warburg
1932	Irving Langmuir	Werner Heisenberg	Charles Sherrington and D. Adrian
1933	Not awarded	P. A. M. Dirac and Erwin Schroedinger	Thomas H. Morgan
1934	H. C. Urey	Not awarded	G. R. Minot, W. F. Murphy and G. H. Whipple
1935	F. Joliot and Mme. Joliot-Curie	James Chadwick	Hans Spemann
1936	P. J. W. Debye	Carl D. Anderson and V. G. Hess	Sir Henry Dale and Otto Loewi
1937	Walter N. Haworth and Paul Karrer	Clinton J. Davisson and George P. Thomson	Albert von Szent-Györgyi
1938	Richard Kuhn (declined)	Enrico Fermi	Corneille Heymans
1939	Adolph F. Butenandt (declined) Leopold Ruzicka	E. O. Lawrence	Gerhard Domagk (declined)
1940–42	Award suspended	Award suspended	Award suspended
1943	Georg Hevesy	Otto Stern	Henrik Dam E. A. Doisy
1944	No announcement	I. I. Rabi	Joseph Erlanger H. S. Gasser

(We are indebted to the American-Scandinavian Foundation for assistance in preparing this list.)

Perkin Medal Awards. The Perkin medal awards have been made annually, beginning in 1906, fifty years after the discovery of the first synthetic dyestuff, mauveine, by William Henry Perkin. The award is made by the Society of Chemical Industry, American Section, to a chemist residing in the U.S.A., for the most valuable work in applied chemistry. The work may have been done at any time during his career. The selection is made by a committee representing the Society of Chemical Industry, American Chemical Society, Electrochemical Society, American Institute of Chemical Engineers, and Société de Chimie Industrielle.

The following awards have been made:

1906	Perkin, W. H.	1928	Langmuir, Irving
1908	Herreshoff, J. B. F.	1929	Sullivan, E. C.
1909	Behr, Arno	1930	Dow, Herbert H.
1910	Acheson, E. G.	1931	Little, A. D.
1911	Hall, Charles M.	1932	Burgess, Charles F.
1912	Frasch, Herman	1933	Onslager, George
1913	Gayley, James	1934	Fink, Colin G.
1914	Hyatt, John W.	1935	Curme, G. O., Jr.
1915	Weston, Edward	1936	Lewis, W. K.
1916	Baekeland, L. H.	1937	Midgley, Thomas, Jr.
1917	Twitchell, Ernest	1938	Tone, F. J.
1918	Rossi, A. J.	1939	Landis, Walter S.
1919	Cottrell, F. G.	1940	Stine, C. M. A.
1920	Chandler, Chas. F.	1941	Dorr, J. V. N.
1921	Whitney, W. R.	1942	Ittner, M. H.
1922	Burton, W. M.	1943	Wilson, Robert E.
1923	Whittaker, M. C.	1944	Du Bois, Gaston F.
1924	Becket, F. M.	1945	Bolton, E. K.
1925	Moore, H. K.	1946	Frary, F. C.
1926	Moore, R. B.		
1927	Teeple, John E.		

Chemical Society, London, See Memorial Lectures; Faraday Lectures.

Memorial Lectures. The memorial lectures are delivered, since 1893, before the Chemical Society, London, following the death of an eminent foreign chemist. These lectures furnish a valuable source of information concerning the life and work of these eminent non-British chemists. The following lectures have been delivered:

YEAR	MEMORIAL LECTURE	YEAR	MEMORIAL LECTURE
1893	Stas	1896	Hofmann
1893	Kopp	1896	Helmholtz
1895	Marignac	1896	L. Meyer
1897	Pasteur	1912	Becquerel
1898	Kekulé	1913	van't Hoff
1900	V. Meyer	1913	Ladenburg
1900	Bunsen	1920	E. Fischer
1900	Friedel	1923	Baeyer
1901	Nilson	1923	van der Waals
1901	Rammelsberg	1927	Onnes
1902	Raoult	1928	Arrhenius
1905	Wislicenus	1930	Richards
1906	Cleve	1932	Wallach
1909	Wolcott Gibbs	1933	Ostwald
1909	Mendeléeff	1935	Mme. Curie
1910	Thomsen	1935	Brauner
1911	Berthelot	1936	Hantzsch
1912	Moissan	1937	Le Chatelier
1912	Cannizzaro	1938	Franklin

Faraday Lectures. The Faraday lectures are delivered, since 1869, before the Chemical Society, London, upon invitation, by an eminent foreign chemist. These lectures furnish a valuable source of information concerning the work of these eminent non-British chemists. The following lectures have been delivered:

YEAR	FARADAY LECTURER	TITLE
1869	Dumas	Eulogy of Faraday
1872	Cannizzaro	Some Points on the Theoretical Teaching of Chemistry
1875	Hofmann	Liebig
1879	Wurtz	On the Constitution of Matter in the Gaseous State
1881	Helmholtz	Modern Development of Faraday's Theory of Electricity
1889	Mendeléeff	Periodic Law of the Chemical Elements
1895		Presentation of the Faraday Medal to J. W. Strutt by Lord Rayleigh
1904	Ostwald	Elements and Compounds
1907	Emil Fischer	Synthetic Chemistry in Relation to Biology
1911	Richards	Fundamental Properties of the Elements
1914	Arrhenius	Theory of Electrolytic Dissociation
1924	Millikan	Atomism in Modern Physics
1927	Willstätter	Problems and Methods in Enzyme Research

Year	Faraday Lecturer	Title
1930	Bohr	Chemistry and the Quantum Theory of Atomic Constitution
1933	Debye	Relation Between Stereochemistry and Physics
1937	Haber	
1938	Pictet	

(R.K.S.)

HISTORY OF GEOLOGY.

B.C.

400 "If one is sufficiently lavish with time, everything possible happens" (Herodotus).

A.D.

200 The first appreciation of changes in sea level.

1500 Fossils generally recognized by students of geology as the evidence of extinct organisms.

1517 Formations now well above sea level recognized as having been originally deposited beneath the sea.

1570 Description of the stratigraphy of the coal beds of Great Britain (Owen).

1564 The earth thought to be 4004 years old (Lightfoot).

1740 Recognition of difference between igneous and sedimentary rocks (Guettard).

1760 Earth considered to be about 75,000 years old.

1795 The first treatise approaching the modern concept of dynamical geology (Hutton).

1809 The first geological map of a wide region (Maclure).

1822 The first extensive and careful use of fossils for the purpose of correlation (William Smith).

1830 The emphasis of the important doctrine of "Uniformitarianism" and the development of the modern systemic division of the Cenozoic (Lyell).

1840 The development of the modern systemic division of the Paleozoic (Sedgwick and Murchison) and the importance of facies (Murchison and Lonsdale).

1841 The recognition of continental glaciation (Agassiz).

1842 The appreciation of the true structure of a folded mountain range and the birth of the idea of the geosyncline (Rogers Brothers and Dana).

1846 The appreciation of the permanency of ocean basins and continents (Dana).

1855 The recognition of isostasy (Airy).

1863 First paleogeographic map of a specific period (Dana).

1870 The recognition of Permian glaciation (Fedden).

1872 The earth thought to be 20 million years old (Kelvin).

1873 The recognition of the importance of cycles of sedimentation (Newberry). The recognition of the geosyncline as fundamental to structural and stratigraphical geology.

1878 The introduction of the theory of nappes (Heim).

1884 The recognition of the full effect of low-angle overthrusts (Peach and Horne).

1899 The earth thought to be 89,222,900 years old as determined by the age of the oceans (Jolly).

1907 The recognition of pre-Cambrian glaciation (Coleman).

1911 The European concept of the geosyncline (Haug).

1929 The lithic history of the earth thought to be over 1 billion, 500 million years, as determined by the lead-uranium method.

1934 Discovery of lineal distribution of negative gravity anomalies in relation to island arcs and the consequent development of tectonophysics (Meinesz).

(R.M.F.)

HITCH. Pisces, Teleostei. A fresh-water fish of the minnow family found in the Coast Range. (A.W.L.)

HIVES. Urticaria.

HOACTZIN. Fossil Birds.

HOARHOUND. Mint Family.

HOATZIN, HOACTZIN. Aves, Galliformes. A peculiar South American bird (**Aves**) of doubtful relationship. It lives along streams of the Amazon valley and eats fruit and other vegetation. The young have a clawed digit on the margin of the wings which they use to grasp boughs in climbing. (A.W.L.)

HOB. A milling cutter with form-type teeth of helicoidal shape and with profiles such that conjugate surfaces on cylindrical parts may be machined by rotating the work and the hob at a constant velocity ratio. Hobs are extensively used for cutting spur gears, and hobbing is the only really precise method of cutting heavy-duty worm wheels. (See **Worm Gearing.**) Two types of gear hobs are commonly used; the radial or infeed type, and the tapered or tangential feed hob. The latter is superior, particularly for hobbing worm gears with high helix angles and high pressure angle. Hobbing processes are also used for spline cutting, and for generating ratchet teeth.

The term hobbing is also used to designate a method of die sinking, in which a hardened master punch, a duplicate of the part to be formed, is pressed into an unheated die blank so that the shape of the hob is reproduced in the die impression. This method of producing die cavities is simpler than die sinking by cutting away the material, since it is considerably easier to machine the surface of the hob than to machine the die cavity. It is also advantageous in the production of multiple die cavities, since a single hob can be used for a series of duplicate dies. The process is also referred to as "hubbing." (H.C.H.)

HOBBY. Aves, Falconiformes. A falcon of moderate size. It ranges over Europe and Asia and migrates into Africa. (A.W.L.)

HOCHEUR. Guenon.

HOCK, HOUGH. The joint at the attachment of the foot and the leg in animals which walk on the toes (digitigrade or unguligrade), commonly applied to domestic animals. It corresponds to the ankle joint of other species. Also the back of the human knee. (A.W.L.)

HODGKIN'S DISEASE. A disease of unknown origin characterized by non-painful enlargement of the **lymph nodes** accompanied by fever, anemia, and in the later stages, wasting.

The disease seems to be an infection with neoplastic manifestations. At one time it was thought to be a kind of tuberculous infection. As yet nothing specific is known either as to cause or treatment. Pathologically, the lymph nodes and spleen show increased numbers of lymphoid and endothelial cells, eosinophiles, and giant cells. Necrosis and fibrosis of varying degree are present.

The non-painful swellings of the lymph glands in various parts of the body—particularly the neck and within the chest—at first may not be accompanied by symptoms. Soon, however, weakness, wasting, and anemia become marked. The disease usually results in death after 2 or 3 years either from secondary infection or through pressure on the trachea from enlarged glands.

The disease may occur in an acute form, death developing in a few weeks.

In some cases, the abdominal organs may be involved primarily, with little or no involvement of the peripheral lymph nodes. In the generalized form, lymph nodes and internal organs are all involved.

Palliative treatment is obtained by the use of **radium** or **x-ray** over the enlarged masses. Transfusions control the anemia and give symptomatic relief. There is nothing known, to date, that will arrest or cure the disease. (R.S.M., D.M.H.)

HODOSCOPE. Cosmic Rays.

HOFMANN REACTION. Rearrangements.

HOGBACK. Ridge-like topographic features, the result of the differential **erosion** of highly tilted hard and soft strata. The steeper, or dip-slope, side is developed on the harder or less soluble formation, while the gentler slope is developed on the opposite side, on the softer rocks. (Compare with **Escarpment; Scarp; Cuesta**.) (R.M.F.)

HOG SUCKER. Pisces, Teleostei. A fish (**Pisces**) of moderate size found in clear streams in the northeastern quarter of the United States. (A.W.L.)

HOHLRAUM. Black Body.

HOIST. Any device for lifting materials, weights, articles, etc., may be called a hoist. Hoists often compose a part of other apparatus whose purpose may extend to movement of material other than vertically. For example, the bridge **crane** incorporates within it a hoist for vertical lift. The energy required for lifting is derived ultimately from a number of various sources. For example, in the hoisting field one finds such varied power sources as compressed air, internal combustion engines, hydraulic power, steam and electric power. The pneumatic drives may be either a direct lift supplied by air acting on a **piston** connected directly to the load, or it may be employed in compressed air engines, whose crankshaft is geared to the hoisting apparatus. In the **internal combustion engine** type hoist, the gasoline engine is generally used for the light-capacity hoist, and the **Diesel engine** for heavier hoists. It has the advantage over other drives for portable service, such as locomotive cranes, power shovels, etc. Hydraulic drives of hoisting machinery take the form of water pressure applied to a piston which slides in a long cylinder. The hydraulic drive usually requires some other source of power for driving the pump required to force the water into the cylinder. Hydraulic drives were once quite popular for elevators in buildings of moderate height, but the type has lost favor because of the speed of electrically driven hoists.

The hoist which is driven by an electric motor is probably the type most used today. It will be found in a variety of sizes ranging from the high-speed passenger elevator of the tall office building to the small electrically driven chain hoist. Steam as a drive for hoisting machinery is used principally for stationary hoists moving heavy loads such as mine lifts, and is used to a certain extent for portable hoisting. An example in the latter class is the steam shovel.

The essential parts of a hoist are a rope or chain which is wrapped around a drum or drive sheave. A hook, grapnel magnet, or other device for handling the load is attached to the free end. The rotation of the drum winds up the rope, thus shortening the distance between the drum and the load. If the drum is fixed in position over the load, naturally the load must be hoisted. To drive the drum, one of the power supplies just mentioned is connected with the drum through a suitable speed-reducing, torque-increasing mechanism. A gear train is often used. These component parts when supplied with a brake controlling the speed during lowering of weights, are the essential elements of all hoists except the direct acting. The rope employed in hoists is either good quality Manila rope, or twisted steel rope. Manila is used principally for small or relatively unimportant hoists; multiple strand, greased flexible wire rope for others.

Various small hand-operated lifting devices may also be truly classified as hoists. Among these might be mentioned the winch, the **capstan**, the screw jack, hydraulic jack, and the **chain block**. (See **Elevators**.) (F.T.M.)

HOLLYHOCK. Mallow Family.

HOLMIUM. Symbol: Ho. Atomic number: 67. Atomic weight: 164.94. No isotope, but of single atomic form: 165. Type of compound: Ho_2O_3. Color of salts: Yellow. Discovered by Cleve in 1879. A member of the **yttrium** sub-group of the rare earth metals. (R.K.S.)

HOLOCRYSTALLINE. The term applied by **petrologists** to **igneous** rocks composed entirely of **crystals**; in contradistinction to igneous rocks which are partly or entirely composed of natural glass, such as **obsidian**. (R.M.F.)

HOLOGAMY. A type of **fertilization** of 1-celled organisms in which interchange of material takes place between two cells indistinguishable from ordinary individuals. Conjugation. Also called macrogamy. (A.W.L.)

HOLOGONIA. Nematoda.

HOLOMETABOLA. A division of the insects characterized by complete **metamorphosis**. The insect as it hatches from the egg is a **larva** which differs conspicuously from subsequent stages and is adapted for a different mode of life in many species. It has no compound eyes and the wings develop internally. Caterpillars, grubs, and maggots are larvae of this type. When fully grown the larva transforms into a **pupa**. This stage is relatively inert and in some species is incapable of movement. It is often hidden in a subterranean cell or in a cocoon prepared by the larva before its transformation. The adult or imago emerges from the pupa.

The insects included here make up the orders **Coleoptera, Strepsiptera, Neuroptera, Mecoptera, Trichoptera, Lepidoptera, Hymenoptera, Suctoria**, and **Diptera**. (A.W.L.)

HOLOSTEI. The garpikes and bowfins, a small order of the class **Pisces**. (A.W.L.)

HOLOTHUROIDEA. The sea cucumbers, a class of the phylum **Echinodermata**.

These animals differ from other echinoderms in several particulars: 1. The principal axis is elongated and the animal rests on its side. 2. The body wall is soft because of the reduction of the calcareous ossicles. 3. A branching respiratory tree extends from the alimentary tract into the body cavity.

Sea cucumbers are used as food in the Oriental region. They are dried for the market and in this form are called trepang or bêche-de-mer.

The class includes five orders:

Order Aspidochirota. Tropical species with shield-shaped tentacles. In shallow water.
Order Elasipoda. Benthonic species of deep water.
Order Dendrochirota. Shallow water species with branching tentacles.
Order Molpadonia. Burrowing species. Tentacles unbranched or slightly branched.
Order Apoda. (Synaptida, Paractinopoda). Burrowing species without respiratory trees. (A.W.L.)

HOLOTRICHIDA. Ciliata.

HOLOTYPE. As used by biologists and paleontologists to mean the specimen to which all others should be ultimately referred to determine the species. The holotype does not necessarily have to be the originally described species (type) and frequently is not. (R.M.F.)

HOMEOTYPE. As used by biologists and paleontologists for a specimen which has been identified by an authority by comparing it with the type. (R.M.F.)

HOMOATROPINE. Alkaloids.

HOMOCLINE. Group of strata which dip in one and the same direction. Never a complete structure and usually representing the limb of an **anticline** or syncline. (R.M.F.)

HOMOGENEOUS COORDINATES. If the ratios of **coordinates** (one more than necessary) are used instead of the coordinates themselves, we have homogeneous coordinates. Their use renders equations involving them homogeneous. (L.L.S.)

HOMOGENEOUS DIFFERENTIAL EQUATIONS. Ordinary Differential Equations of First Order and First Degree, also Linear Differential Equations.

HOMOGENEOUS FUNCTIONS. A homogeneous function is a type of mathematical expression, defined as follows:

A **function** $f(x, y)$ is called homogeneous in x and y if $f(\lambda x, \lambda y) = \lambda^n f(x, y)$; the **exponent** n is then called the degree or order of the function. (L.L.S.)

HOMOGENEOUS SYSTEMS OF LINEAR ALGEBRAIC EQUATIONS. Linear Algebraic Equations, Systems of.

HOMOIOTHERMY. Warm-bloodedness. The maintenance of a body temperature above that of the environment is common among animals, hence the usual terms warm- and cold-blooded are inaccurate. Cold-blooded forms are those whose body temperature fluctuates with that of the surrounding air or water, so that the animal's activity is directly conditioned by external temperatures. They are more accurately described as poikilothermal. In contrast, homoiothermal animals tend to maintain a constant body temperature in spite of external fluctuations. Fluctuations are normal, although the human body usually maintains a constant temperature.

Only birds and mammals are homoiothermal. Both regulate the body temperature by producing excess heat and by regulating its radiation from the surface. Regulation is accomplished by nervous control of the blood vessels near the surface, by insulating vestiture, and by the evaporation of water from the body. When the surrounding air is warm the blood flows more freely near the surface of the body and more heat is radiated but when the air is cold less blood reaches the surface and the heat is conserved. In air too warm to permit adequate radiation the animal reduces its activity, exposes as much surface as possible, and either sweats or pants. The evaporation of water either from the mouth or from the sweat glands absorbs heat from the underlying tissues. Vestiture plays a passive role as an insulating coat but it is capable of some regulation, especially in the birds. The erection of the feathers provides a thicker and looser covering of high insulating value and their depression results in less interference with radiation.

Homoiothermy is one of the highest adaptations of living things, since it provides for the maintenance of optimum conditions for the vital processes of the body. Through it the animal becomes virtually independent of one of the most important of the fluctuating environmental conditions. (A.W.L.)

HOMOLOGOUS SERIES. Two organic compounds are said to be homologous if their molecular formulae differ by CH_2, or a multiple of CH_2. (R.K.S.)

HOMOLOGY. Fundamental structural relationship based on similarity of embryological development and evolutionary history. The antithesis of analogy, which is superficial likeness based on adaptation for similar uses.

The anterior appendages of terrestrial vertebrates, for example, are regarded as fundamentally similar structures, derived from the **pentadactyl appendage**, yet they include the wings of birds, flippers of aquatic mammals, and a great variety of less extreme adaptations, including the legs of animals and the arms of man. In contrast, the wings of birds and of insects are broad thin structures used for flight but in structure and origin they show no resemblance beyond this point and so are analogous. (A.W.L.)

HOMOPLASY. More commonly designated as analogy. Homology. (A.W.L.)

HOMOPTERA. The cicadas, leaf hoppers, plant lice, scale insects, and numerous other forms, constituting a large order of insects. They have sucking mouths which differ from those of most bugs in that the slender proboscis arises from the hind margin of the head and extends back between the legs. The wings, when present, are membraneous. The order includes about 16,000 species.

Many members of this order, particularly the plant lice, scale insects, and **phylloxerans**, are economically important. (A.W.L.)

HOMOSEXUALITY. Perverted sex practices among those of the same sex. (R.S.M.)

HOMOTHALLISM. Phycomycetes.

HONEY. A thick sweet liquid formed by **bees** from the nectar of flowers and to a limited extent from the juice of fruits and honey-dew. Honey contains a large percentage of simple sugars, **essential oils** of the flowers from which it is derived, and a minute amount of **formic acid** which acts as a preservative. Its flavor depends on the flowers from which it comes.

The honey produced by **honey-bees** is marketed in several forms and various grades. The lighter grades are more widely demanded for table use and the darker grades are sold for baking, candy making, and the compounding of medicines. In the United States, Californian white-sage honey ranks as the finest of the white honeys, with honey from orange blossoms and white and alsike clover next. Immense quantities of honey are produced from alfalfa in irrigated regions and from sweet clover in the Middle West. Both are very light in color but of poorer flavor. Dark honey comes chiefly from buckwheat, fall flowers such as goldenrod and asters, and from many plants in the southern states. Usually only the better grades of honey are produced in the small sections commonly called comb honey, but these grades and inferior honeys as well are removed from the comb and sold in jars or pails as extracted honey. A third form is chunk honey, consisting of pieces of comb honey in containers filled with extracted honey. (A.W.L.)

HONEY BUZZARD. Aves, Falconiformes. A bird (**Aves**) related to the eagles, named from its habit of robbing the nests of bees and wasps and eating the larvae. One species lives in Europe and Asia and others in the Oriental region. (A.W.L.)

HONEY CREEPER. Aves, Passeriformes. Small birds (**Aves**) of tropical South America and the West

Indies. Related to the warblers. They visit flowers like the hummingbirds but are incapable of hovering flight. One species is called the banana-quit. (A.W.L.)

HONEY DEW. A sweet secretion produced by **plant lice.** When these insects are abundant on trees it sometimes spots the leaves and anything below the tree like a heavy dew. It is freely sought by ants and is sometimes gathered by bees, but it makes a very inferior honey. (A.W.L.)

HONEY EATER. Aves, Passeriformes. Birds of the Australian region. They have long tongues with which they secure nectar from flowers. The group includes the parson bird, the stitch bird, and several species called white eyes. (A.W.L.)

HONEY GUIDE. Aves, Piciformes. Birds (**Aves**) of several African and Oriental species. They lay their eggs in the nests of other birds and are named from their reputed habit of leading the way to nests of bees. (A.W.L.)

HONEY PECKER. Aves, Passeriformes. Small brilliantly colored birds (**Aves**) of the Oriental and Australian regions, related to the sun birds. One Australian species is called the diamond bird. (A.W.L.)

HONEY-BEE. Insecta, Hymenoptera. Insects of several species which live in colonies, build combs of wax secreted by the body, and in the cells of these combs raise their young and store honey and pollen. They make up the family Apidae.

A single species, *Apis mellifica,* is kept for the production of honey and wax in Europe and North America. This species was introduced into North America from

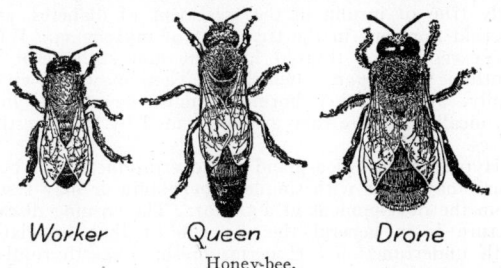

Worker   Queen   Drone

Honey-bee.

Europe and is the chief species of economic importance in all parts of the world. It occurs in several varieties, including the dark German or black bee, the golden to leather-colored Italian, and the gray Carniolan. Italian bees predominate in the United States; the better strains are prolific, good-tempered, hardy, and otherwise desirable.

The honey-bee colony normally contains a single queen which may be active for 3 years. During the active summer season males or drones are produced in considerable numbers, but they are not tolerated in the winter or in times of scarcity of food. A strong colony for honey production also contains from 60,000 to 100,000 or more workers. The queen is the only normal female, although workers are also of this sex.

Owing to the economic importance of the honey-bee it has been observed in detail and an extensive literature has accumulated on the habits of bees, their structure, the organization of the colony, and all phases of the management of apiaries and the production and marketing of honey. (A.W.L.)

HOODOO. In geology, a columnar or pillar-like erosional remnant which has been carved and sculptured from relatively horizontal formations by differential erosion. The form and subsidiary features of Hoodoos

may be partly governed by joint planes and the differential hardness of the stratified sediments. The term applies particularly to eccentric and peculiar forms which are especially noticeable because of their fancied resemblance to animals and artifacts. (R.M.F.)

HOOK GAUGE. Weir.

HOOKE'S LAW. Elasticity.

HOOKWORM. Nemathelminthes, Nematoda. A minute worm which develops in the soil and lives as an adult in the human intestine, causing a disease which is also called hookworm (Uncinariasis, Ankylostomiasis). This parasitic disease is quite widespread and has existed for thousands of years. The **parasite,** however, was not discovered until 1838 by Dubrin. It is found throughout the world and is most prevalent near the equator. In many countries of the world nearly 100% of the population is infected. In the southern United States it exists for the most part in the poorer rural districts. There it is a serious problem. Medical treatment for the elimination of the worms is available but to prevent reinfection in such climates is a difficult matter, even though the method is simply the protection of the hands and feet from contact with earth in which the worms are likely to occur.

The disease is spread by hookworm ova in the feces. A single female hookworm living in man's intestinal tract will pass several hundred eggs daily. Under favorable conditions, **embryos** hatch out in the soil in 1–3 days. Five days later in the larval stage the parasites are able to again infect humans. When the skin—usually the feet—comes in contact with mud the parasites are able to infect an individual, penetrating with ease through a break in the skin, and are then carried throughout the body by the bloodstream. The larvae reach the lungs, are carried by the bronchial mucus to the pharynx and are swallowed, thus reaching the intestinal tract. Six to eight weeks after an individual is infected, eggs are passed in the feces.

Since these worms feed upon the blood it is to be expected that a severe **anemia** would develop with infection of large numbers of the parasites. Practically all of the symptoms of this disease result from the profound anemia. Heavy infection in children results in stunted growth, apathy, lack of energy, and retarded mental development.

The diagnosis is made with ease by microscopic examination of the stools and identification of the ova.

The prognosis in hookworm infestation is excellent unless the disease has progressed too far. Rapid recovery ensues if treatment with thymol (see **Phenol**), oil of chenopodium, or other similar drug is used, and all worms expelled from the gastro-intestinal tract. Reinfection must be prevented. (R.S.M., A.W.L.)

HOOLOCK. Gibbon.

HOOPOE. Aves, Piciformes. Birds (**Aves**) of the Old World with a very high crest and a long sharp beak. One species, *Upupa epops,* lives in Europe and others in the Oriental region, Africa, and Madagascar. (A.W.L.)

HOPLOCARIDA. The mantis shrimps, **Crustacea.**

HOPLONEMERTEA. Nemertea.

HOPLOPHONEUS. Fossil Mammals; Oligocene.

HOPS. Mulberry Family.

HORIZON. The visible horizon is the line where "earth and sky meet." In astronomy the term horizon is used to describe the great circle cut out on the **celestial sphere** by a plane perpendicular to the direction of **gravity.** In case this plane is tangent to the surface of the earth, the horizon so described is the apparent horizon; if the plane passes through the center of the earth, we have the geocentric horizon.

The difference in direction between the visible and the apparent astronomic horizon is known as the dip of the horizon. In the figure, $O'H'$ represents the direction of the visible horizon from an observer at a station O' elevated above the surface of the earth by an amount h. OH represents the direction of the astronomic horizon as defined above for the observer on the surface of the earth at O. The angle HAH' is the dip of the horizon and may be shown to be very

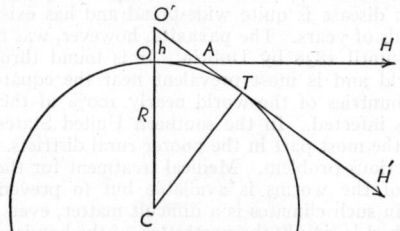

Visible and astronomical horizons.

approximately given by the relation: the dip of the horizon (expressed in minutes of arc) is equal to the square root of the height of the observer above the surface of the earth (expressed in ft.). The distance $O'T$ from the observer to the visible horizon is approximately given by: The distance of the visible horizon (expressed in miles) is given by the square root of 3/2 the height of the observer above the surface of the earth (expressed in ft.). This distance is frequently very much increased by an effect known as looming of the horizon, produced by **refraction** of light in heated (or cooled) layers of the air near the surface. Both the expressions for the dip and distance of the horizon are applicable only when the point of observation of the visible horizon (the point T) is actually on the surface of the earth, e.g., the sea horizon. (w.k.g.)

HORIZONTAL CONTROL. Station.

HORIZONTAL COORDINATE SYSTEM. The horizontal coordinate system is a system of **spherical coordinates** on the **celestial sphere** which uses the **horizon** as a fundamental plane. Planes perpendicular to the horizon cut out great circles on the celestial sphere known as **vertical circles.** The fundamental direction selected in the fundamental plane is true south. The **azimuth** of a point on the celestial sphere is the angular distance, measured in the plane of the horizon, from the true south direction to the point of intersection of the vertical circle through the object with the horizon. There are several different methods for expressing azimuth, but the astronomical method is to measure azimuth from the south through the west through 360°. The **altitude** of a point on the celestial sphere is the angular distance, measured along the vertical circle through the point, from the plane of the horizon to the point.

The horizontal system of spherical coordinates is frequently referred to as the altazimuth system. (w.k.g.)

HORMONE. A hormone is a distinct chemical substance formed by one organ and acting in a specific manner on the function of another organ (or organs) of the body. Hormones are produced mainly by the glands of internal secretion—the ductless or endocrine glands. They may also be liberated by glands having an external as well as an internal secretion, for example, the **pancreas, testicles,** and probably the **liver.** The

duodenum produces a hormone called secretin which stimulates the flow of pancreatic juice into the intestine.

The endocrine glands and the hormones produced by them are listed below. A discussion of the function of each hormone will be found under the heading of the various glands.

GLAND	HORMONE
Adrenal	
Medulla.......	Epinephrine
Cortex........	Cortin
Thyroid.........	Thyroglobulin
Pituitary	
Anterior lobe...	Growth hormone
	Thyrotropic (thyroid stimulating)
	Adrenotropic (adrenal stimulating)
	Gonadotropic (gonad stimulating)
	Lactogenic (milk gland stimulating)
Posterior lobe..	Pituitrin, composed of
	Pitocin (uterine stimulating)
	Pitressin (pressor substance)
Pancreas........	Insulin
Parathyroid......	Parathyroid hormone
Gonads	
Ovaries........	Estrin (follicular hormone)
	Progestin (corpus luteum hormone)
Testes.........	Testosterone (androsterone)

There are a number of diseases associated with abnormalities of hormone production—both with increased and decreased function of the particular gland. The common ones occurring with underfunction are **myxedema** (hypothyroidism), **diabetes,** and the syndromes resulting from failure of the **pituitary** gland and the **gonads.** In these diseases, replacement therapy with commercial preparations—desiccated gland substance or synthetic hormones—may alleviate the signs and symptoms, or even control them completely. This is particularly true of insulin in the treatment of diabetes, and thyroid hormone in the treatment of myxedema. With the exception of thyroid hormone and stilbesterol (a synthetic **estrogen**), which are active when given orally, all the other hormones must be given hypodermically, because they are inactivated by the digestive juices.

Hyperfunction of a gland and overproduction of hormone may occur with simple **hyperplasia** or may result from the development of a **tumor.** The ensuing disease picture is in general the opposite of that associated with underfunction. Hyperinsulinism, hyperthyroidism (Graves' disease), and hyperpituitarism (**acromegaly**) are the common diseases of this type. Hyperinsulinism and hyperthyroidism are treated by surgical removal of a part of the gland. This usually results in complete cure of the disease.

Hormones are also known to the botanist, occurring in minute quantities in various parts of the plant. Their presence, usually not where they are formed, produces very definite results. Growth substance, formed in the tips of plants, and especially in young seedlings, seems to be the cause of the great sensitiveness the plant has to light (see **Movements in Plants**). It seems to stimulate cells on the unlighted side of the plant to elongate greatly and so causes the plant to bend towards the light. Other hormones, called wound hormones, seem to be formed wherever the tissues of a plant are injured. Because of the presence of these hormones, cells in the vicinity of the wound are stimulated to rapid division, which causes tissue to form over the wound and close it. (R.S.M., R.M.W., D.M.H.)

HORN. A hard translucent material formed by the development of epidermal cells containing a substance known as **keratin.** The outer layers of the skin are keratinized and the nails, claws and hoofs of mammals are formed of similar material. Horn is also developed in large amounts in the appendages of the head which go by the same name. Horns may be bony cores sheathed in horn or solid bony growths. The former occur in cattle and the latter in deer. The median horn or horns borne on the head of rhinoceroses are quite unlike true horns. They are formed of aggregated hair-like components firmly based on roughened areas of the underlying bones. (A.W.L.)

HORN, ELECTROMAGNETIC.

Horn radiators are used to obtain directional radiation characteristics which could not be obtained as conveniently with simple antennae (see **Directional Arrays**). As such directors they are used both with conventional antennae and with wave guides, but in either case they serve to direct the radiation in a pattern from the open end of the horn in a manner determined by the dimensions of the horn. The important dimensions are the horn opening (in terms of wavelength of the radiation) and the flare angle. While theoretically an infinitely long horn will give a radiation pattern whose angle conforms to that of the horn, those of practical length do not confine the beam to quite this degree. For example a horn with an angle of 15° may give a radiation pattern which spreads 23°. The common horns may be divided into three classes, sectoral, pyramidal and biconical. The sectoral

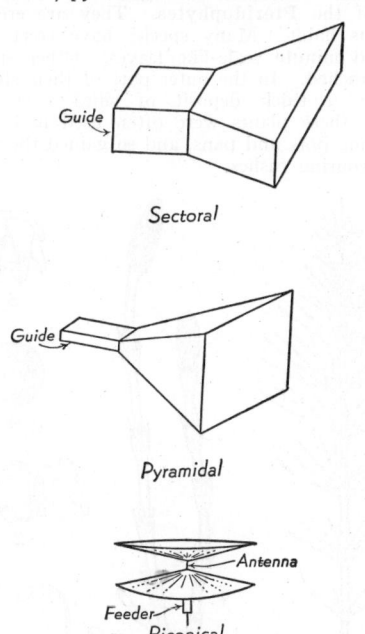

Sectoral

Pyramidal

Biconical

horn has two sides which are parallel and the other two flared. The pyramidal horn has all sides flared. The conical horn is really a pyramidal horn with a circular cross-section. The biconical horn consists of two cones with their vertices coinciding or adjacent to one another. The first two types are used where a singly directed beam of radiation is desired, the exact pattern in both vertical and horizontal planes being determined by the dimensions. The biconical horn gives a uniform pattern in a plane perpendicular to the axis and highly directional in any plane containing the axis.

Sectoral and pyramidal horns may be excited by more or less conventional antennae or by wave guides. In the former case, a short section of wave guide is attached to the end of the horn and this is excited by the antennae. In the latter case, the horn is really a flared extension of the guide and may be looked upon as a impedance-transforming section for matching the impedance of the guide to that of free space. Biconical horns are excited by a variety of antenna arrangements in the space between vertices of the horns. Because of space con-

siderations horns are not feasible except at ultra high frequencies, but in the microwave region they are widely used as radiators. (L.R.Q.)

HORNBILL.
Aves, Coraciiformes. Large birds (**Aves**) of the African and Oriental regions, characterized by the very large beak, many species with a large prominence above the base of the upper mandible extending back onto the head. (A.W.L.)

HORNBLENDE.
The mineral hornblende is a complex **silicate** which is probably an **isomorphous** mixture of three molecules, a **calcium-iron-magnesium** silicate, an **aluminum**-iron-magnesium silicate and an iron-magnesium silicate. Manganese and alkalis are sometimes present as is also **titanium**. It is **monoclinic**, with prismatic crystals, often pseudo-**hexagonal**. Bladed, fibrous, columnar, granular and compact massive varieties also are common. It has a perfect prismatic **cleavage**; hardness, 5–6; specific gravity, 2.9–3.4; color, green, greenish-brown, brown and black; luster, vitreous to silky; transparent to opaque. Hornblende is a common constituent of many of the **igneous** rocks such as **granite, syenite, diorite,** or **gabbro,** of **gneisses and schists** and is the principal mineral of the **amphibolites**. Hornblende alters easily to **chlorite** and **epidote**. A variety of hornblende that contains little (less than 5%) of iron oxides is gray to white in color and named edenite, from its locality in Edenville, N. Y. Very dark brown to black hornblendes which contain titanium ordinarily are called basaltic hornblende from the fact that they are usually a constituent of basalts and similar rocks. Well-known localities for hornblende are in Czechoslovakia, Mt. Vesuvius, Italy; Norway, Sweden and in the United States in Massachusetts, New Hampshire and New York. Black hornblende is found in Renfrew County, Canada. The word hornblende is derived from the German *horn,* and *blende,* to blind or dazzle. The term blende was often used to refer to a brilliant non-metallic luster, for example, zincblende. (E.S.C.S.)

HORNBLENDITE.
A coarse-grained rock related to **gabbro** which consists almost wholly of **hornblende**. **Olivine** being present, this rock may grade into a hornblende-peridotite (cortlandtite). Hornblendite is a rare rock type and of relatively little importance. (E.S.C.S.)

HORNED TOAD.
Reptilia, Sauria. *Phrynosoma.* Small spiny **lizards** of the southwestern states and Mexico. They have short broad bodies and short tails,

Horned toads. (*New York Zoological Society.*)

hence the confusion of terms in the common name. Horned lizard is a better term.

Horned toads are desert animals and are capable of living for incredibly long periods without food or water. They cannot, however, survive for the long periods of years which have sometimes been claimed. (A.W.L.)

HORNER'S METHOD.
Horner's method is a method of successive approximations for finding the approxi-

mate value of an irrational root of a **polynomial equation** to any desired degree of accuracy.

It may be summarized in general terms as follows: Locate the root between successive integers; the smaller integer is the integral part of the root. Now **transform** the given equation $P(x) = 0$ into another equation $P_1(x) = 0$, whose roots are those of $P(x) = 0$ diminished by the integral part of the root, so that $P_1(x) = 0$ has a root between 0 and 1. Locate this root between successive tenths; the smaller tenth is the tenths part of the root. Next, transform the equation $P_1(x) = 0$ into a new equation $P_2(x) = 0$, whose roots are those of $P_1(x) = 0$ diminished by the tenths part of the root, so that $P_2(x) = 0$ has a root between 0 and 0.1. Locate this root between successive hundredths; the smaller hundredth is the hundredths part of the root. Continue this process as far as necessary to obtain the desired degree of accuracy. Special devices may be used in the location of the root between consecutive tenths or hundredths, etc. **Synthetic division** should be used for the various transformations. (L.L.S.)

HORNET. Insecta, Hymenoptera. A name loosely applied to many of the larger **wasps,** particularly the species which build paper nests. (A.W.L.)

HORNFELS. A more or less general term applied to fine-grained, massive, and frequently speckled rock, the result of contact **metamorphism** developed in **slates** by **granitic** intrusions. (R.M.F.)

HORNSTONE. Old English synonym for **flint** and **chert.** (R.M.F.)

HOROSCOPE. Astrology.

HORN-TAIL. Insecta, Hymenoptera. Large **sawflies** whose **larvae** bore in the trunks of trees. The adults have a cylindrical body and in the female sex a short strong ovipositor which is the source of the name horn-tail. With this organ holes are drilled into the wood of the tree for the deposition of the eggs. (A.W.L.)

HORSE. Mammalia, Perissodactyla. Hoofed animals with a single toe on each foot, encased in a massive hoof. The teeth are very high-crowned grinding structures. The term applies properly not only to the domestic horse, *Equus caballus,* but also to any member of the family Equidae though many of the wild species are commonly known by other names. Of these species the best known are the **zebras,** a number of species of animals marked with conspicuous stripes. The **quagga** or couagga is striped only on the anterior half of the body. Both zebras and quagga live in Africa. Wild asses occur in northeastern Africa and in the deserts of Asia, among them the **kiang** or kulan of Mongolia and Tibet and the **onager** or ghorkhar of western India and adjacent areas. These animals are among the fleetest and hardiest known. Two wild species of central Asia, the **tarpan** and Prejevalski's horse, *Equus przewalskii,* are most closely related to the domestic horse and probably represent the original stock from which it was derived.

Although much of the evolutionary history of the horses is known from North American fossils, no horses existed on this continent when it was dicovered by Europeans and the wild horses of the West are entirely feral.

Under domestication many varieties of horses have been developed for riding, driving, draft animals and other uses. They have also been crossed with the domestic ass to produce mules for various purposes, and have been hybridized experimentally with other species. (A.W.L.)

HORSEHAIR WORM. Nematomorpha.

HORSEPOWER. The survival of this old unit of **power** recalls the crude beginnings of the **English system** of measures. James Watt, in seeking for a means of expressing the power of steam engines and water wheels at the dawn of the industrial revolution brought about by the extensive use of machinery, turned to the horse as a familiar source of power and one in terms of which power values would be easily comprehended. It is said that he actually experimented with horses, using the best draft animals available, and that it was as a result of his observations that the horsepower has now become standardized as 550 **foot-pounds** of work per second. This is equivalent to about 746 **watts,** so that a kilowatt is approximately $1\frac{1}{3}$ hp. (L.D.W.)

HORSES. A term used by miners to describe fragments of the wall rock included in the ore-bearing vein. (R.M.F.)

HORSESHOE CRAB. Xiphosura.

HORSETAILS, OR SCOURING RUSHES. Equisetales. The horsetails, or Equisetales, form a small section of the **Pteridophytes.** They are erect plants of various habit. Many species have erect columnar stems and minute scale-like leaves. Other species are much branched. In the outer part of their stems there is usually a thick deposit of silica (see **Silicon**). Therefore these plants were often used in early times for scouring pots and pans, and so gained their appellation of scouring rushes.

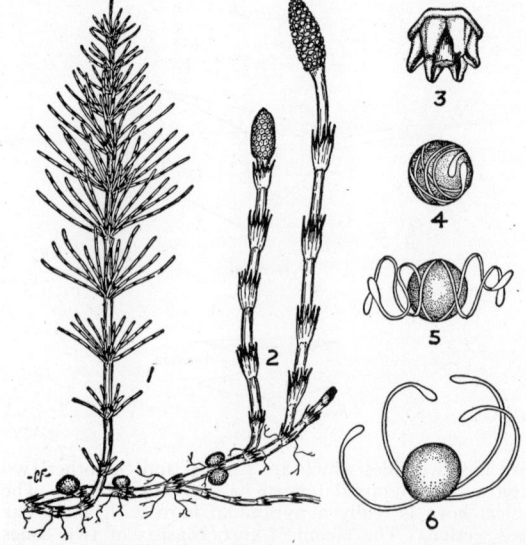

One of the horsetails, *Equisetum arvense.* 1, a vegetative shoot; 2, spore-bearing shoots, with a portion of the rhizome and a few roots; each shoot is terminated by a strobilus; 3, a portion of the strobilus much enlarged; the small white sacs are sporangia borne on scale-like sporophylls; the powdery dark masses beneath these are composed of spores; 4, 5 and 6, spores much enlarged; in 4 the elaters are closely coiled around the spore; in 5 and 6 the elaters are partially uncoiled.

At the tips of the ordinary stems or of special reproductive stems the **sporangia** are borne. These sporangia occur beneath the edges of small umbrella-shaped stalks which are aggregated into small cones. The **spores** are all alike. Each has four long flat appendages, called elaters, which curl and uncurl with changes of humidity, causing tangled masses of spores to be liberated. The spore develops into a small, much branched **prothallus** or gametophyte. On the upper surface of the prothalli **antheridia** containing **sperms** or **archegonia** containing eggs are borne. The sperms, multiciliate spiral cells, swim to the archegonia and unite with the egg to form a **zygote**. This at once develops into a new horsetail. (See also **Paleobotany**.) (R.M.W.)

HORST. Fault.

HOST. An animal which is used as a source of food by a parasite. The parasite may live on the surface of the body or within it and may be harmless or harmful, but in all cases the host is the source of its food. (A.W.L.)

HOT WORK. Forging, rolling, pressing, extruding, swaging, drawing, or forming of metals at temperatures above their **recrystallization** temperature. (R.H.H.)

HOT-BULB. The semi-Diesel **engine** is not a true compression ignition engine in that the **piston** instroke does not compress the air sufficiently so that the resulting temperature rise is high enough to cause ignition in a cold engine. The semi-Diesel engine, however, does not use electrical **ignition** in the manner of the gasoline engine. It has a hot-bulb, which is a certain mass of metal incorporated in the cylinder head in such a way that a portion of it projects slightly in the combustion space. Before starting the semi-Diesel, the hot-bulb is thoroughly heated by applying a blow torch to its exterior surface. It thus provides a focal center of high temperature which produces ignition during the starting of this type of engine. A similar service is performed by "glow plugs" in engines so equipped. The glow plug resembles a common spark plug except that the insulated electrode instead of terminating in a gap connects to a loop of resistance wire. When a moderate voltage is impressed on the plug the resistance wire glows and provides a focal high-temperature point. (F.T.M.)

HOT-WATER HEATING. Heating Systems.

HOTWELL. A hotwell is a tank or container in which heated liquid collects. An example is the hotwell attached to and made part of a steam **condenser** of the surface type. As the steam is condensed, the condensate drops to the bottom of the condenser shell and flows into the hotwell, from which it is pumped. (F.T.M.)

HOUR ANGLE. The hour angle of a celestial object is the **spherical coordinate**, in the equatorial system of coordinates, which is measured in the plane of the celestial **equator** from the local **meridian**, in the direction of apparent rotation of the celestial sphere, to the intersection of the hour circle through the object with the equator. Since **time** and **hour angle** are practically synonymous (e.g., the hour angle of the **mean sun** is local mean time) the determination of hour angle is vitally necessary for the determination of local time and hence **longitude**.

At sea, hour angle is determined by measuring the **altitude** of the object by means of the **sextant**, reducing the observed altitude to true **geocentric**, and solving the **astronomical triangle**. For the solution of the triangle both the **declination** of the object and the **latitude** of the observer must be known. The declination may be immediately obtained from the tabulated **coordinates** of the object, but the latitude can be obtained only by some previous observation. In case the ship is in motion the latitude must be obtained by **dead reckoning** from the previously determined position. (W.K.G.)

HOUR CIRCLE. Equatorial Coordinates.

HOUSE FLY. Insecta, Diptera. A true **fly**, *Musca domestica*, well known for its habit of frequenting

houses and alighting on all kinds of food. Since it also visits filth of any kind it is an important carrier of disease, especially typhoid fever, and has been the object of public health crusades for many years. With the improvement of sanitation the danger has been lessened, although it has not been entirely eliminated.

The house fly breeds in horse manure and in various kinds of decaying organic matter. Proper disposal of such wastes is an important measure in the control of the insect. (A.W.L.)

HOWLER. Monkey.

HUBARA. Bustard.

HÜBERNITE. Wolframite.

HUCHO. Pisces, Teleostei. A European fish (**Pisces**), *Hucho hucho,* related to the trout and salmon. (A.W.L.)

HUCKLEBERRY. Heath Family.

HUIA. Aves, Passeriformes. A New Zealand bird (**Aves**) related to the starlings. The male has a short straight beak while that of the female is long and curved. (A.W.L.)

HULL, AIRPLANE. Seaplane.

HUM. This is the annoying 60- or 120-cycle tone sometimes heard in the output of communication equipment. It may be introduced in the circuit in a number of ways, inductive or capacitive coupling with adjacent circuits carrying 60-cycle current, by the use of a.c. for heating the **filaments** of vacuum **tubes** or by induced effects from the **heaters** of indirectly heated tubes, by improper filtering of the output of the **rectifiers** supplying the d-c voltages for the operation of the system, etc. This tone is very annoying and sometimes difficult to eliminate, but fortunately careful design and layout of the equipment usually keep the hum level below the value which becomes noticeable. The use of shielded components in high gain **amplifiers**, indirectly heated **cathodes** in vacuum **tubes**, careful filtering, and the use of push-pull amplifiers are means commonly employed to reduce hum. (L.R.Q.)

HUMIDIFIER. Air Conditioning and Humidity.

HUMIDITY. The most obvious way of expressing the humidity or moisture content of the **air** is to state the percentage, by weight, of water vapor in its composition. This is the absolute humidity, and can be determined by passing a measured quantity of air through a tube containing an absorbing substance which removes all the vapor and may be weighed before and after the absorption. For many purposes, however, a more useful quantity is the relative humidity, which expresses the vapor content as a fraction or percentage of the concentration necessary to render the vapor saturated at the given temperature. (See **Vapors**.) Specifically, the relative humidity of the air at any temperature is the ratio of the actual vapor pressure of the water vapor contained therein to the maximum or saturated vapor pressure of water vapor at the same temperature. At the **dew point**, the relative humidity is 100%. A rise of temperature without the addition of more vapor reduces the relative humidity (but not the absolute humidity), while a fall of temperature increases it and may bring about saturation. Various forms of **hygrometer** have been devised to measure relative humidity. (L.D.W.)

HUMMINGBIRD. Aves, Micropodiformes. Small birds (**Aves**) of the New World whose wings are moved so rapidly in flight that they produce a low-pitched sound. They are capable of hovering in one spot in the air, and habitually poise before the flowers which they visit for nectar and insects. Most species of hummingbirds inhabit tropical America but they range to Patagonia and several species enter the United States. Only one, the ruby-throated hummingbird, enters the eastern states. (A.W.L.)

Ruby-throated humming-bird, *Archilochus colubris*. Shiny green above, dusky underparts, with ruby-red throat in male but not in female. Length, three and three-quarter inches.

HUMUS. Soil.

HUNTING. The tendency of rotating mechanism which normally should operate at constant speed to pulsate in speed above and below the normal point, is known as hunting. It may occur in prime movers controlled by **governors** which are too isosynchronous, or in electric apparatus where rotating and stationary parts are electrically coupled. The nature of such **coupling** is essentially elastic, and may, under certain circumstances, lead to hunting action on the part of the rotor. (See **Damper Winding**.) Governors which hunt must be corrected by the use of **dash pots** or other damping devices, and the introduction to the governor characteristic of a slight amount of speed regulation. (F.T.M., L.R.Q.)

HURRICANES. Over all oceans, near the equator, with the exception of the South Atlantic there develop occasionally tropical cyclones which are intense vortices covering relatively large areas. They are particularly frequent over the western part of the Atlantic and Pacific, over the Indian Ocean, the Caribbean Sea and the Gulf of Mexico. Such storms are known as hurricanes over the ocean waters bordering on southeastern U.S.A. All are the same type of storm. Surface pressure in a hurricane is very low at the center or eye of the storm but rises rapidly outward toward the periphery. Because of the large pressure gradient, winds are of high velocity, blowing counterclockwise in the northern hemisphere and clockwise south of the equator. Velocities of 100 m.p.h. are relatively common near the storm's center,

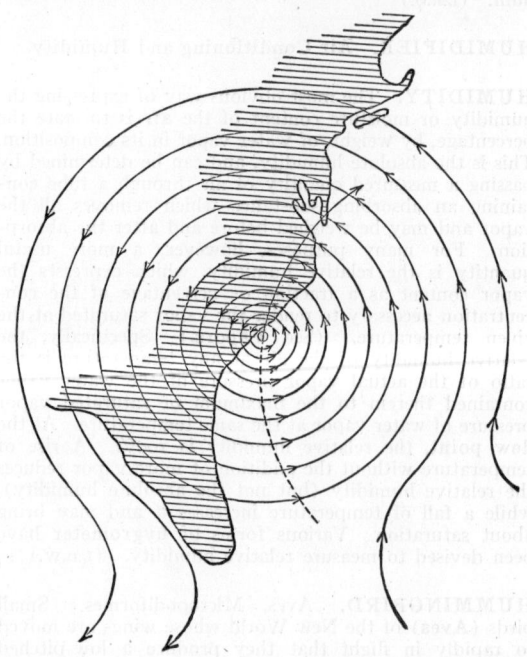

and 50–75 m.p.h. over a wide area can be expected. It is probable that velocities of 175 m.p.h. and perhaps more have occurred in more than one such storm. Mountainous waves and confused seas result from such winds. That quadrant of the storm where the velocity of wind and the velocity of the storm are additive is the quadrant where the greatest wind velocity is present; this is usually the north or east quadrant in the northern and the south or east quadrant in the southern hemisphere. Only in the center of the storm, the "eye of the storm," is there little or no wind. The sea remains confused, however. A hurricane moving toward a coast line drives a rising sea before it in the dangerous quadrant. This apparent mound of water is known as the storm wave, or hurricane wave. This hurricane wave, driven by very strong on-shore winds, often lifts ocean water to great heights along coastal areas. Great depth of clouds and torrential rain normally occur over a fairly wide area about the storm center. Rainfall locally amounts to as much as 40–50"; a 20-in. rainfall during the passage of a single storm is common.

North American hurricanes originate near the equator over the Atlantic, the Caribbean, or Gulf of Mexico. From the point of origin they move in a westerly direction for several days, then begin to curve northward. After they pass the 30th latitude north, they usually assume a path somewhat east of north and begin to accelerate considerably. Hurricane season is that period from June to November but an occasional one may occur out of season. (P.E.K.)

HUTIA. Mammalia, Rodentia. Large arboreal rodents of the West Indies. They resemble rats but have a blunt muzzle and a moderately long tail. Related to the coypu. (A.W.L.)

HUYGENS' PRINCIPLE. A well-known method of analysis applied to problems of wave propagation. It recognizes that each point of an advancing wave front is in fact the center of a fresh disturbance, and the source of a new train of waves; and that the advancing wave as a whole may be regarded as the resultant of the secondary waves arising from points in the medium already traversed. This view of wave propagation facilitates the study of various phenomena, such as **diffraction**. For example, if two rooms are connected by an open doorway, and sounds are produced in a remote corner of one room, a person in any part of the other will hear the sounds as proceeding from the doorway, which is indeed the case. So far as the second room is concerned, the vibrating air in the doorway is the source, and from it sound waves enlarge in all directions through the second room. The same is true of light reaching a slit or passing the edge of an obstacle, though this is not quite so easily observed because of the short wavelength. The **interference** of light from variously distant areas of the moving wave front accounts for the maxima and minima observable as diffraction fringes. (L.D.W.)

HYACINTH. Lily Family; Zircon.

HYADES. The Hyades is an open V-shaped cluster of stars in the **constellation** of Taurus. References to this group are to be found in all of the ancient literatures, Virgil referring to them as the "rainy Hyades." The group is exceedingly rich in **double stars**, and even with a small telescope and low **magnifying power**, they present a beautiful appearance.

The Hyades form one of the best known of the so-called moving star **clusters**. The brightest star of the Hyades, **Aldebaran**, is not a member of the cluster, but has an independent motion through space and just happens to be in its present position at this time. (W.K.G.)

HYALITE. Opal.

HYBRID. An organism produced by parents belonging to different species or to different strains of the same species. A hybrid combines characteristics derived from the two parent stocks and in some cases is more desirable than either. Beauty of flowers, productivity of various plants, and appearance and hardiness of animals have been enhanced by controlled **hybridization.**

When a hybrid is once secured its propagation is hampered by the fact that the diverse hereditary characters are reassorted in hereditary transmission by sexual reproduction. Hybrids are often infertile but even when they are capable of producing offspring they rarely breed true. The **mule** is the only animal hybrid of great value and it is produced always by parents of the two species, horse and ass. Plant hybrids are not subject to this limitation for they can usually be propagated by **bulbs, cuttings,** or **grafts.** (A.W.L.)

HYBRID COIL. The bridging transformer used in coupling a two-way telephone circuit to the repeater station or for coupling two one-way circuits to a two-way circuit. The coil is so wound that when the line is properly balanced by a balancing network there is no reaction between the output and input connections of the transformer. (See **Telephony.**) (L.R.Q.)

HYBRIDIZATION. The cross breeding of different strains or species of organisms. The crossing of species is interspecific hybridization and the interbreeding of strains of the same species is intraspecific. Since the differences between species are greater than variation within a species, intraspecific hybridization is rarely impossible but crosses between species are usually possible only if the species are closely related.

The difficulty of making an interspecific cross is sometimes due to anatomical differences or mental reactions which prevent mating and sometimes to the failure of **germ cells** of the two species to unite normally in **fertilization.** If neither of these obstacles prevents, a hybrid individual results. Usually it shows some of the characteristics of each parent species. A familiar example is the mule. Crosses of this kind sometimes occur between closely related species in nature but most of the recorded examples have been produced under experimental conditions.

Intraspecific crosses between breeds of animals and varieties of plants are constantly being carried out by animal and plant breeders for the production of new forms. Among plants especially, the new varieties offered each year are usually produced in this way. (A.W.L.)

HYDATID. The bladder worm or cysticercus of a species of **tapeworm,** *Echinococcus granulosus,* which forms a large fluid-filled cyst in the liver or other organs of the hoofed animals and man. It may grow to a diameter of 6″ and contains many scolices. The adult lives in the dog. (A.W.L.)

HYDATOGENESIS. A term used by **petrologists** to designate the process by which rocks are formed from highly aqueous solutions. Some petrologists limit the use of the term to rocks which have been deposited from water-rich **magmatic** solutions. (R.M.F.)

HYDNOCERAS. Invertebrate Paleontology.

HYDRANTH. A form of individual which receives and digests food in colonies of **hydrozoan** coelenterates. It is a **polyp** attached to the remainder of the colony at its base and with a mouth surrounded by a circlet of tentacles at the free end. (A.W.L.)

HYDRARGYRUM. Mercury.

HYDRARIAE. Hydrozoa.

HYDRASTINE. Alkaloids.

HYDRASTININE. Alkaloids.

HYDRATES. Water.

HYDRATUBA. The attached form, resembling a polyp, which develops from the first larval stage of some **jellyfishes.** It may bud off other similar individuals and at certain seasons is subdivided by constrictions to form **ephyrae** which become jellyfishes. (A.W.L.)

HYDRAULIC FILL. An embankment or other fill in which the materials are deposited in place by a flowing stream of water, with the deposition being selective, is termed hydraulic fill. Gravity, coupled with velocity control, is used to effect the selected deposition of the material.

Where **borrow pits** containing suitable material are accessible at an elevation such that, after being washed from the bank by a powerful stream from a large, high-pressure **nozzle,** the earth can be sluiced to the fill, hydraulic fill is likely to be the most economical construction. Even where the elevation is not realized, the material can be washed into pools, then elevated to the **sluice** with a dredge pump. In the construction of a hydraulic fill **dam,** the edges of the dam are defined by low embankments or dykes which are carried upwards as the fill proceeds. The sluices are carried parallel to and just inside these dykes. The sluices discharge their water-earth mixture at intervals, the water then fanning out and flowing towards the central pool which is maintained at the desired level by discharge control. While flowing from the sluices, the coarse material is first deposited, then, as the central pool is reached, the water velocity is diminished and the fine materials are deposited to form an impervious central section. The water flow must be well controlled at all times, otherwise the central section may be bridged by tongues of coarse material which would allow the water a free passage through the dam. (F.T.M.)

HYDRAULIC FRICTION. Fluid Friction.

HYDRAULIC GRADE LINE. Hydrokinetics.

HYDRAULIC MACHINE TOOLS. Hydraulically actuated machine tools are coming into extensive use in the metal-working industry. They offer great flexibility of speed and feed control, elimination of shock, and possess the ability to stall against obstruction, thus protecting parts or tools from breakage. Hydraulic actuation also permits slip, or slowing-up motion, when the cutting tool is overloaded. If a hydraulic drive were used to actuate the carriage of a lathe, accurate thread cutting would not be possible because the slip would not be compensated for at the end of the movement. In mechanically actuated feeding mechanisms, any slip that may result from slight deformations of the mechanism must be made up, and may therefore result in an increased rate of motion during some portion of the total movement.

One form of hydraulic circuit employs a constant-delivery pump which may move the piston at a rate corresponding to only a fraction of the displacement of the pump. The excess displacement escapes through a relief valve in the feed line into the oil reservoir. The rate of motion of the piston is therefore controlled by the setting of the relief valve.

Another form of hydraulic circuit receives its energy from the displacement of a variable-delivery pump, where the piston speed is changed by varying the amount of liquid delivered by the pump. Hydraulic circuits for intermittent or varying rates of feed and for rapid traversing motions in combination with feeding motion are in extensive use on various types of machine tools. Two or more cylinders, with varying rates of motion, may be driven from the same pump by proper application of valves and control equipment. Rotary motion may be obtained by using a rotary hydraulic motor instead of a feed cylinder. (H.C.H.)

HYDRAULIC PRESS. Hydrostatics; Press Working.

HYDRAULIC RADIUS. The theory of hydraulics indicates that the ratio of the frictional area to the volume of the fluid stream is an important dimension governing the friction loss. The hydraulic radius, which expresses this fact, is the cross-sectional area of flow divided by the wetted perimeter of a cross-section of the conduit. The hydraulic radius of a circular pipe flowing full of water is one-fourth of the diameter. The hydraulic radius of an open canal is the cross-sectional area of the stream divided by the wetted perimeter of the cross-section excluding the length in contact with the air. (F.T.M.)

HYDRAULIC TURBINE. The fundamentals of the turbine were incorporated into the wheels built before the turn of the nineteenth century, but its principal development has occurred since that time. Beginning with Fourenyon and his outward flow turbine, Jonval, Boyden, Swain, and Francis rapidly brought the reaction turbine to an advanced stage of development. By 1875 the inward flow turbine, as perfected by Francis, and which now bears his name, had established itself in the lead, a position which it maintained until about 1900 when the impulse, or Pelton, type of wheel had progressed to the point of dominating the high head field.

The inherent slow speed of the Francis type runner on low heads was a fault that the propeller type runner was designed to cure. During the decade 1910-1920 progress was made with this type of wheel and by 1920

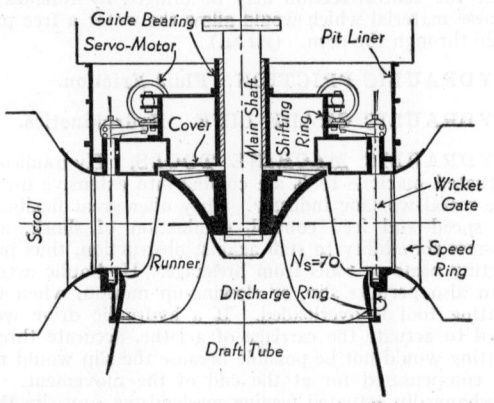

Cross-section showing component parts of a Francis turbine.

the propeller type runner, often called the Nagler runner, was definitely established in the hydroelectric field. Later it was arranged so that the blades could be adjusted and set at different angles to accommodate changes in elevation of the forebay level without undue loss of efficiency. The success of the propeller type turbine encouraged American adoption of the Kaplan turbine, on which the blade adjustment is performed automatically, being under the same control as the turbine gates.

Hydraulic turbines have been built for very high heads as well as large capacities. Heads as high as 5000' have been utilized (3000' in the U. S.) and runners have been constructed to develop as much as 70,000 hp. At present though, a 30,000-hp. turbine is a large one, and ordinary sizes range from 1000 to 10,000 hp.

As between impulse and reaction types, the action in the impulse turbine is easiest to understand. There is no difficulty in visualizing the transformation of pressure head into velocity head at the nozzle, nor of understanding the push, or impulse, that is given to the buckets by the stream of water. The jet is directed upon the rotor tangentially, and hence this type is also called the tangential turbine. The velocity of the jet of water is only slightly less than the free spouting velocity under the effective head h. Impulse buckets are divided into two halves by a "splitter" and the axial thrusts which would otherwise have to be borne by special bearings are equalized.

The essential difference between the impulse and reaction types is that, in the former the entire energy received by the wheel is in the velocity form, while in the latter it may be partially in the velocity form, but is also, in a large measure, still in the pressure form. The reaction of conversion of residual pressure into velocity in the runner is the source of much of the torque delivered to the reaction turbine. If the turbine were blocked stationary and had its gates opened, the water would issue from the turbine as from a nozzle. Now, by removing the blocking, let these nozzles begin to rotate and the absolute velocity of water leaving them is found to be diminishing, the energy having been absorbed by the runner. At the best speed the final velocity will be just sufficient to enable the water to clear the runner. At this time the wheel may be absorbing from 90-95% of the energy that the water had in the pressure form just before reaching the turbine gates.

In a Francis turbine the water flows inward, then downward and into the draft tube.

The various types of hydraulic turbines are classified according to the kind of runner each employs. This is logical because the runner is the most important part of the turbine. Turbine runners are made of bronze, plate steel, cast steel, and cast iron.

The Pelton wheel is either a solid or open disk, to the rim of which are attached buckets upon which a jet of water is played from a stationary nozzle. A horizontal shaft is the usual arrangement but vertical shaft units have also been put into operation. The advantage of using the vertical arrangement is that more than one jet can be played on the buckets; this is obtained, however, at the expense of some loss of efficiency. The Pelton wheel is overhung on the bearing and often, for additional capacity, two wheels are overhung on the same generator. Variable power demand is met by decreasing the amount of water in the jet, by deflecting the jet from the buckets, or both. Some turbines of this type have a relief jet which opens as the main jet closes. Afterwards, a **dash pot** slowly closes the relief jet—slowly enough to prevent a large pressure rise in the penstock. The same thing is also accomplished by deflecting the jet from the wheel upon loss of load, then slowly closing the valve controlling the jet.

The Francis turbine is rarely a horizontal shaft machine, except in small sizes and where it is desired to avoid the expense of excavation for a vertical setting. The standard runner consists of two crowns between which the buckets or blades are placed. It is best adapted to vertical setting. In order to pass the large discharges possible in a high **specific speed** wheel, the buckets are curved downward. Some axial flow action is present in runners of high specific speed. Water is admitted to the runner through guide vanes and gates.

Loss of efficiency at part load is sometimes a serious fault as, for instance, where only one or two units are installed in an isolated plant. The feature of the Kaplan turbine is that the blade angles and gates are adjusted simultaneously by the governor mechanism so that the blades are always in the position best suited for full utilization of the flow, through the reduction of eddying and shock losses. The result is that the efficiency at part load holds up remarkably well.

Conveying the water from the **penstock** and directing the proper amount of it correctly against the runner requires first, a scroll case; second, a speed ring; and third, turbine gates.

The scroll case for medium and high head development is circular in form. In plan, it leads from the penstock and wraps, in spiral form, around the speed ring. The cross-section of the spiral at any point should be such that the water flows with uniform velocity. This leads to the spiral form because the water is being delivered to the turbine uniformly around the entire circumference.

The speed ring is that part of the turbine which joins the discharge ring with the turbine cover and pit liner. The ribs between the top and bottom portions must be strong enough to support the dead weight above the casing, consisting of concrete, generator, and turbine rotative parts; hence the speed ring is a very important part of the turbine.

Inside the speed ring and rigidly bolted to it is the inlet gate mechanism. The mechanism is operated by the governor which, by opening or closing the gates, can maintain a control of speed under variable load. Gates are of the guide vane type and, while various types of gates have been used, the wicket gate is in general use at the present time. Its principal advantage is its efficiency. Shock losses at part gate opening are reduced to a minimum in the wicket gate. It is not particularly tight and has many wearing parts, most of which are bronze bushed and grease lubricated. (F.T.M.)

HYDRAULICS. Hydraulics is the dynamics of liquids (hydrodynamics), especially applied to the practical problems of engineering. Although this general definition is entirely correct, in common usage hydraulics is the study of water at rest or in motion. This conception of hydraulics is used in this article. A basic proposition of hydraulics is that water is incompressible. While this condition is not completely met in fact, the compressibility of water is so small as to be negligible for practically all propositions of hydraulics. The viscosity of water varies with the temperature and is one reason for change of conditions of water flow in pipes with changing temperature. The unit weight of fresh water is usually taken as 62.4 lbs. per cu. ft.

The science of hydraulics is divisible into **hydrostatics** and **hydrokinetics.** Hydrostatics is the hydrodynamics of liquids considered apart from their motion: hydrokinetics is the hydrodynamics of moving, especially flowing liquids. Among the subjects included in any study of hydrostatics are the following: (1) the pressure on a submerged area of any shape or inclination, (2) the measurement of pressure on water at rest by manometers or pressure gauges, (3) buoyancy and flotation. Practical application of (1) is to be found in problems associated with water gates, large valves, pressure against dams, tanks, hydraulic presses, etc.

Hydrokinetics includes a great many different phases of hydraulics. Most of these will be found treated in specialized articles, references to which are given below. The flow of fluids supplies many cases of the application of hydraulic science. Flows of steady, uniform, unsteady and non-uniform types, and the friction losses occasioned thereby, in closed or open conduits; the measurement of flows and the discharges under given conditions, are part of this phase of hydraulics; also, there is to be considered the flow of water through openings, such as **orifices, nozzles,** and **weirs.** The flow of water in pipe lines offers a great many problems in addition to friction: the discharge through different sections of branching and looping pipes, siphons, fittings, valves, etc., is included. Measurement of discharge of large amounts of water, as in stream and river flow, offers problems different from those met in closed conduits. Furthermore, the forces occasioned by deviated flows of water, as met in hydraulic turbines, the pump, and other hydraulic machinery, are fit subjects to be included in any study of hydromechanics. (See **Fluid Flow, Fluid Friction, Dams, Hydrostatics, Hydrokinetics, Flow Meter, Orifice, Weir, Hydraulic Turbine, Head.**) (F.T.M.)

HYDRAZINES, HYDRAZONES AND OSAZONES. Hydrazine ($H_2N \cdot NH_2$) is a colorless, fuming liquid, melting point 1° C., boiling point 113° C., when heated to about 350° C. decomposes into nitrogen and

DIAGRAM SHOWING RELATIONSHIP OF HYDRAZINES

	PRIMARY	SECONDARY	TERTIARY	QUATERNARY
1. Hydrazine M.P. 1° C. B.P. 113° C.	2. Phenylhydrazine M.P. 20° C. B.P. 243° C.	3. 1,1-diphenylhydrazine M.P. 34° C.	5. Triphenylhydrazine M.P. 142° C.	6. Tetraphenylhydrazine M.P. 148° C.
		4. 1,2-diphenylhydrazine (Hydrazobenzene) M.P. 131° C.		
	7. Methylhydrazine B.P. 87° C. (745 mm.)	8. 1,1-dimethylhydrazine B.P. 81° C. (747 mm.)		
		9. 1,2-dimethylhydrazine B.P. 62° C. (717 mm.)		

TABLE SHOWING OTHER HYDRAZINES

Other Hydrazines	Formula	Melting Point, °C.	Boiling Point, °C.
10. Ethylhydrazine	$C_2H_5NHNH_2$		100 (710 mm.)
11. Diethylhydrazine	$(C_2H_5)_2NNH_2$		97
12. 1,2-Diethylhydrazine	$C_2H_5NH \cdot NHC_2H_5$		
13. 1,1-Ethylphenylhydrazine	$\begin{matrix}C_2H_5\\C_6H_5\end{matrix}\!\!>\!\!NNH_2$		237
14. 1,2-Ethylphenylhydrazine	$C_2H_5NH \cdot NHC_6H_5$		237–240
15. Benzylhydrazine	$C_6H_5CH_2NHNH_2$	26	135 (30 mm.)

ammonia, burns when ignited in air with violet-colored flame, forms hydrate with water ($H_2N \cdot NH_2 \cdot H_2O$), which hydrate yields hydrazine and water in a vacuum or upon heating moderately, and at about 180° C. nitrogen and ammonia. Hydrazine is a base, slightly weaker than ammonium hydroxide, and forms two series of salts (1) hydrazine monohydrochloride ($H_2N \cdot NH_2 \cdot HCl$), mononitrate ($H_2N \cdot NH_2 \cdot HNO_3$), hemisulfate ($H_2N \cdot NH_2 \cdot \frac{1}{2}H_2SO_4$), (2) hydrazine dihydrochloride ($H_2N \cdot NH_2 \cdot 2HCl$), dinitrate ($H_2N \cdot NH_2 \cdot 2HNO_3$), monosulfate ($H_2N \cdot NH_2 \cdot H_2SO_4$), all soluble in water except the last. Hydrazine azide ($H_2N \cdot NH_2 \cdot HN_3$) is a soluble solid, melting point 65° C. Hydrazine hydrate attacks glass, rubber, cork, and therefore silver vessels are preferred in its manipulation.

Hydrazine may be made by converting one-half of a given amount of ammonia into chloramine (NH_2Cl) by sodium hypochlorite solution in the presence of a colloid such as glue or gelatin, by heating. The remaining one-half of ammonia reacts with chloramine to form hydrazine. The product is then cooled to 0° C. and sulfuric acid added in amount to react with the hydrazine to form hydrazine monosulfate ($H_2N \cdot NH_2 \cdot H_2SO_4$), insoluble solid, melting point 254° C. ($H_2N \cdot NH_2 \cdot \frac{1}{2}H_2SO_4$ is soluble in water).

Hydrazine is a powerful reducing agent, for example, cupric salt solutions changed to cuprous chloride (in HCl) or to copper (in NaOH), silver salt solutions to silver, mercuric salt solutions to mercury, sulfur trioxide to sulfur sesquioxide (S_2O_3), and may be determined by titration with potassium permanganate solution.

Phenylhydrazine is a colorless liquid, slightly soluble in water, miscible in all proportions with alcohol or ether, forms salts with acids, e.g., phenylhydrazine hydrochloride ($C_6H_5NH \cdot NH_2 \cdot HCl$), melting point 241° C. appr., is a powerful reducing agent, with alkaline cupric salt solution (Fehling's solution) yields cuprous oxide precipitate, reacts with carbonyl group ($=CO$) of aldehydes or ketones yielding phenylhydrazones, white solids, of definite melting point and utilized in identifi-

Phenylhydrazone of	Melting Point, °C.	Boiling Point, °C.
Acetaldehyde	99	236 (20 mm.)
Acetone	27	163 (50 mm.)
Acetophenone	105	
Benzaldehyde	155	
Benzophenone	137	
Benzoin, alpha	156 appr.	
Benzoin, beta	106	
Cinnamaldehyde	168	
Cyclohexanone	77	
Glyoxal	180	
Salicylaldehyde	142	
d-Glucose, alpha	159	
d-Glucose, beta	140	
l-Arabinose	152	

cation of aldehydes and ketones, e.g., acetaldehyde phenylhydrazone ($CH_3CH:NNHC_6H_5$).

Phenylhydrazines, as hydrochloride solution plus sodium acetate react with polyhydroxy aldehydes or ketones yielding osazones or diphenylhydrazones, yellow solids, of definite melting point and utilized in identification of sugars, e.g., phenyl-d-glucosazone ($CH_2OH(CHOH)_3C:(NNHC_6H_5)CH:(NNHC_6H_5)$) plus aniline ($C_6H_5NH_2$) plus ammonia. Glucose and fructose yield identical osazones, melting point 205° C. decom. The specific rotatory power of the osazone is also an important consideration.

Osazone of	Melting Point, °C.
Glucose	205
Fructose	205
Mannose	205
Galactose	214
Maltose	205
Lactose	200
Arabinose	167
Xylose	115–158
Sucrose (none)	
Raffinose (none)	

Attention should be directed to the difference between osazones and osones. An osone is formed by reaction of an osazone with hydrochloric acid, e.g., glucosone ($CH_2OH(CHOH)_3CO \cdot CHO$).

Phenylhydrazine is made by reduction of benzene diazonium chloride ($C_6H_5N_2Cl$) by stannous chloride ($SnCl_2$) plus hydrochloric acid, the diazonium chloride being formed by reaction of aniline and nitrous acid cold.

1,1-Diphenylhydrazine is a white solid, slightly soluble in water, soluble in alcohol or ether, forms salts with acids, is a reducing agent, reacts with aldehydes or ketones yielding 1,1-diphenylhydrazones, and with polyhydroxy aldehydes or ketones yielding osazones, of definite melting point. 1,1-Diphenylhydrazine is made by reduction of diphenylnitrosamine ($(C_6H_5)_2N \cdot NO$) by zinc plus acetic acid, the nitrosamine being formed by reaction of diphenylamine ($(C_6H_5)_2NH$) and nitrous acid.

1,2-Diphenylhydrazine (hydrazobenzene) is a white to pale yellow solid, very slightly soluble in water, soluble in alcohol or ether, upon standing or upon heating changes into azobenzene, red to orange, solid, melting point 68° C., the hydrogens of the amino groups ($=NH$) are replaceable by acetyl ($CH_3CO—$) or nitroso ($NO—$) groups as for secondary amines. 1,2-Diphenylhydrazine reacts (1) with oxidizing agents, e.g., ferric chloride or air to form azobenzene, (2) with strong reducing agents, e.g., sodium amalgam, to form aniline, (3) with strong acids, to form benzidine hydrochloride ($(4')H_2N \cdot C_6H_4 \cdot C_6H_4 \cdot NH_2(4) \cdot HCl$). 1,2-Diphenylhydrazine is formed by reduction of nitrobenzene by zinc plus sodium hydroxide solution.

Tetraphenylhydrazine is a white solid, soluble in chloroform, acetone, benzene, or toluene, and upon standing is changed into triphenylamine plus azobenzene. In solution, tetraphenylhydrazine dissociates into nitrogen diphenyl $((C_6H_5)_2N\cdot)$, free **radical**, which in toluene at 90° C. reacts with **nitric** oxide (NO). Tetraphenylhydrazine is formed by oxidation of diphenylamine $((C_6H_5)_2NH)$ by lead dioxide. (R.K.S.)

HYDRAZOATES. Hydrazoic Acid and Azides.

HYDRAZO-COMPOUNDS. Aniline and Azo-, Diazo-, and Related Compounds; Hydrazines.

HYDRAZOIC ACID AND AZIDES. Hydrazoic acid (HN_3) is a colorless, odorous, poisonous liquid, boiling point 37° C., soluble in water, volatile in steam, made anhydrous by removal of water from 91% hydrazoic acid by **calcium** chloride, explosive. Hydrazoic acid reacts (1) with metals, e.g., magnesium, aluminum, zinc, iron, to form azides or hydrazoates (or trinitrides), (2) with heavy metal salt solutions to form insoluble azides, e.g., silver azide (AgN_3), mercurous azide (HgN_3), lead azide (PbN_6). Silver, mercuric, **cuprous** azides decompose in the light to form nitrogen gas plus the metal. Lead azide is used as a detonator for the **explosive** trinitrotoluene (T.N.T.), (3) with **ammonium** hydroxide, to form ammonium azide $(NH_4\cdot N_3)$, (4) with **hydrazine**, to form hydrazine azide $(N_2H_4\cdot HN_3)$, (5) with **sodium** hypochlorite plus **acetic acid**, to form chlorazide (ClN_3), explosive, (6) with sodium amalgam (sodium dissolved in mercury), to form **ammonia** mainly (and some hydrazine), (7) with **potassium** permanganate, to form **nitrogen** gas plus water.

Hydrazoic acid is formed (1) by reaction of ethyl or amyl **nitrite** in sodium hydroxide solution (sodium azide formed), then acidifying with dilute **sulfuric acid** and distilling. Hydrazoic acid is recovered mainly in the early portion of the condensate, (2) by reaction of ammonia gas and sodium heated to about 300° C. (sodamide formed), and then treating the residue with dry nitrous oxide gas at about 200° C. The product is dissolved in water, then acidified and distilled as above.

The azide group $(-N_3)$ resembles the halogen groups $(-Cl, -Br, -I)$ in several reactions, and in the properties of several compounds.

Soluble azides react with **ferric** salt solutions to produce a red color, similar to that of ferric **thiocyanate**. Sodium azide is not explosive, even on percussion, and nitrogen may be evolved upon heating. With **iodine** dissolved in ether cold, silver azide forms iodine azide (IN_3), yellow explosive solid.

Ester: methyl azide (CH_3N_3), boiling point 20° C., explosive when heated to about 500° C. (R.K.S.)

HYDRAZONES. Hydrazines, Hydrozones, and Osazones; Azo-, Diazo-, and Related Compounds.

HYDRIDES. Hydrides are binary compounds of **hydrogen** and some other element. Sodium hydride (NaH) and calcium hydride (CaH_2) react with water to yield hydrogen gas and the respective hydroxides. When lithium hydride (LiH) is electrolyzed hydrogen is evolved at the *anode* (hydrogen in this case an anion) and lithium metal at the cathode. (R.K.S.)

HYDRIODIC ACID AND IODIDES. Hydriodic acid (HI) is a colorless solution formed when hydrogen iodide gas is dissolved in water, commercially of strength 10% HI, frequently colored brown by **iodine**. There

is a maximum constant boiling point 127° C. (774 mm.) at 57% HI (distillate) for mixtures of hydriodic acid and water. Hydriodic acid is used in the preparation of iodides, and as an important reagent in organic chemistry.

All metallic iodides except **silver** iodide, **mercurous** iodide, mercuric iodide, **lead** iodide, **cuprous** iodide, **thallium** iodide, and **palladium** iodide, are soluble. The iodides of **antimony, bismuth, tin** require a little free acid to keep them in solution.

Dilute hydriodic acid reacts with hydroxides, oxides, carbonates, sulfides, metals in a manner chemically analogous to dilute **hydrochloric acid**; with solutions of some salts, e.g., silver nitrate, to yield the corresponding iodide, e.g., silver iodide, precipitate. Higher strengths of hydriodic acid react with oxygen of the air upon standing to yield free iodine, which imparts a brown color to the solution, thus indicating the reducing character of the acid.

Hydriodic acid is made by the reaction (1) of iodine and **hydrosulfuric acid** (or **sulfurous acid**), (2) of **phosphorus** plus iodine plus water, with subsequent distillation in all cases.

Two common tests for iodides are as follows:

1. Silver nitrate produces a yellow precipitate insoluble in **nitric acid**, slightly soluble in **ammonia** and soluble in **potassium** cyanide and in **sodium** thiosulfate.

2. On treatment with **chlorine** water and shaking with **carbon disulfide** a violet color due to free iodine is produced in the carbon disulfide layer. (R.K.S.)

HYDROBIOLOGY. Limnology.

HYDROBROMIC ACID AND BROMIDES. Hydrobromic acid (HBr) is a colorless solution formed when hydrogen bromide gas is dissolved in water, commercially of strength 48% HBr. Sometimes colored yellow to red by **bromine**. There is a maximum constant boiling point 126° C. (760 mm.) at 48% HBr (distillate) for mixtures of hydrobromic acid and water. Hydrobromic acid is used in the preparation of bromides, and as an important reagent in organic chemistry.

Dilute hydrobromic acid reacts with hydroxides, oxides, carbonates, sulfides, metals in a manner chemically analogous to dilute **hydrochloric acid**; with solutions of some salts, e.g., silver nitrate, to yield the corresponding bromide, e.g., **silver** bromide, precipitate. Higher strengths of hydrobromic acid react with oxygen of the air upon standing to yield free bromine, which imparts a yellow to red color to the solution, thus indicating the reducing character of the acid.

Hydrobromic acid is made by the reaction (1) of **phosphorus** tribromide and water, (2) of bromine and **sulfurous acid** (or hydrosulfuric acid), with subsequent distillation in all cases.

Metallic bromides except silver bromide and mercurous bromide are soluble in water, but **lead** bromide, **cuprous bromide** and thallium bromide are only slightly soluble.

Two common tests for bromides are as follows:

1. Silver nitrate produces a yellow precipitate insoluble in nitric acid and soluble in **ammonia,** in **potassium** cyanide, and in **sodium** thiosulfate.

2. On treatment with **chlorine** water and shaking with **carbon disulfide** a brown coloration due to free bromine appears in the **carbon disulfide** layer. (R.K.S.)

HYDROCARBONS. Hydrocarbons are compounds of **carbon** and **hydrogen** only. (See **Gasoline, Lubrication, Petroleum Products.**) Hydrocarbons possess, as a whole, a considerable range of properties. This is illustrated by the following selections:

SELECTED REPRESENTATIVE HYDROCARBONS

Hydrocarbon	Physical State	Density	Melting Point, °C.	Boiling Point, °C.	Solubility in		
					Water	Alcohol	Ether
1. Methane.........	Gas	Air = 1.00 0.56	−184	−161	Slight	Solub.	Solub.
2. Ethane..........	Gas	1.36	−172	−88	Slight	Solub.	Solub.
3. Ethylene........	Gas	0.98	−169	−104	Mod. Slight	Solub.	Solub.
4. Acetylene.......	Gas	0.91	−82	−84	Very Slight	Solub.	Solub.
5. Normal-hexane....	Liquid	Water = 1.000 0.661	−94	69	Insol.	Solub.	Infin.
6. Cyclohexane......	Liquid	0.779	6.5	81	Insol.	Infin.	Infin.
7. Benzene.........	Liquid	0.878	5.5	79.6	Insol.	Infin.	Infin.
8. Toluene.........	Liquid	0.866	−95	110.5	Insol.	Infin.	Infin.
9. Cymene.........	Liquid	0.857	−73.5	176	Insol.	Solub.	Solub.
10. Limonene........	Liquid	0.842	−97	177	Insol.	Infin.	Infin.
11. Pinene.........	Liquid	0.878	−55	154	Very Slight	Infin.	Infin.
12. Biphenyl........	Solid	1.041	69	255	Insol.	Mod. Solub.	Solub.
13. Diphenylmethane.	Solid	1.006	27	262	Very Slight	Solub.	Solub.
14. Naphthalene......	Solid	1.145	80	218	Insol.	Slight	Solub.
15. Anthracene.......	Solid	1.25	218	342	Insol.	Insol.	Very Slight

These particular hydrocarbons (in the foregoing table) are characterized as to:

1. Density: The four gases, from 0.56 (methane) to 1.36 (ethane) (air = 1.00) and other hydrocarbon gases above 1.36, e.g., normal-butane 2.05.
   The seven liquids, from 0.661 (normal-hexane) to 0.878 (benzene; pinene) (water = 1.000).
   The four solids from 1.006 (diphenylmethane) to 1.25 (anthracene) (water = 1.000).
2. Melting point: From −184° (methane) to 218° C. (anthracene).
3. Boiling point: From −161° (methane) to 342° C. (anthracene).
4. Solubility in water: Practically insoluble, except ethylene.
5. Solubility in alcohol: Soluble, except naphthalene, anthracene.
6. Solubility in ether: Soluble, except anthracene.

Further remarks on solubility: In acetone, acetylene is very soluble. In chloroform, cymene is very soluble. Other solvents of note are carbon tetrachloride, carbon disulfide. Hydrocarbons, speaking generally, are soluble in each other.

BEHAVIOR ON TREATMENT WITH BROMINE (IN CCl₄)

Hydrocarbon	No Gas Evolved But Red Color Disappears	Hydrogen Bromide Evolved and Red Color Disappears (White Fog When Breathed Upon)	No Change
1. Methane			√
2. Ethane			√
3. Ethylene	√		
4. Acetylene	√		
5. Normal-hexane			√
6. Cyclohexane			√
7. Benzene		√	
8. Toluene		√	
9. Cymene		√	
10. Limonene	√		
11. Pinene	√		
12. Biphenyl		√	
13. Diphenylmethane		√	
14. Naphthalene		√	
15. Anthracene		√	

In general, on treatment with bromine in carbon tetrachloride, and warming:

1. Unsaturated hydrocarbons cause disappearance of the red color and no gas evolved. Reaction is addition, e.g., ethylene forms ethylene dibromide, acetylene forms acetylene tetrabromide.
2. Benzenoid hydrocarbons cause disappearance of the red color and white fog, when breathed upon, of

hydrogen bromide. Reaction is substitution—bromination, e.g., toluene forms chlorotoluene (ortho or para, or both).
3. Saturated hydrocarbons, chain or cyclic, cause no change, e.g., normal-hexane, cyclohexane.
4. Chlorine behaves similarly to bromine.

BEHAVIOR ON TREATMENT WITH FUMING SULFURIC ACID (AT ROOM TEMPERATURE)

Hydrocarbon	Soluble, Heat Evolved	No Change
1. Methane		√
2. Ethane		√
3. Ethylene	√	
4. Acetylene	√	
5. Normal-hexane		√
6. Cyclohexane		√
7. Benzene	√	
8. Toluene	√	
9. Cymene	√	
10. Limonene	√	
11. Pinene	√	
12. Biphenyl	√	
13. Diphenylmethane	√	
14. Naphthalene	√	
15. Anthracene	√	

With sulfuric acid fuming:

1. Unsaturated hydrocarbons react. Reaction is addition, e.g., ethylene forms ethyl hydrogen sulfate; or reaction is polymerization, e.g., pinene.
2. Benzenoid hydrocarbons react. Reaction is substitution—sulfonation, e.g., naphthalene forms naph-

thalene sulfonic acid (alpha or beta, or both, or disulfonic acid).

3. Saturated hydrocarbons, chain or cyclic, cause no change, e.g., normal-hexane, cyclohexane.

stitution—nitration, e.g., toluene forms nitrotoluene (mono-, 1,2 or 1,4; or di- 1,2,4 or 1,2,6; or tri- 1,2,4,6); **oxidation** frequently occurs, e.g., **anthracene** forms anthraquinone.

BEHAVIOR ON TREATMENT WITH FUMING NITRIC ACID (AT ROOM TEMPERATURE)

HYDROCARBON	SOLUBLE, HEAT EVOLVED	NO CHANGE	HYDROCARBON	SOLUBLE HEAT EVOLVED	NO CHANGE
1. Methane		√	9. Cymene	√	
2. Ethane		√	10. Limonene	√	
3. Ethylene	√		11. Pinene	√	
4. Acetylene	√		12. Biphenyl	√	
5. Normal-hexane		√	13. Diphenylmethane	√	
6. Cyclohexane		√	14. Naphthalene	√	
7. Benzene	√		15. Anthracene	√	
8. Toluene	√				

With **nitric acid** fuming:

1. Unsaturated hydrocarbons react. Reaction is addition, e.g., pinene and limonene to form nitrosites and nitrosates; cleavage frequently occurs at the unsaturated bond accompanied by violent reaction.

2. Benzenoid hydrocarbons react. Reaction is sub-

3. Saturated hydrocarbons, chain or cyclic, cause no change, e.g., normal-hexane, cyclohexane; secondary and tertiary hydrocarbons moderately easily oxidized to **ketones** and tertiary **alcohols,** respectively.

The most important hydrocarbons are classified in accordance with their composition as follows:

CLASSIFICATION OF HYDROCARBONS
(Upon Basis of Composition)

CLASS	EXAMPLES	INDIVIDUAL FORMULA	STRUCTURAL FORMULA	CLASS FORMULA
1. Paraffin...............	Methane	CH_4	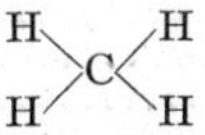	C_nH_{2n+2}
	Ethane	C_2H_6	$H_3C \cdot CH_3$	
	Normal-hexane	C_6H_{14}	$CH_3(CH_2)_4CH_3$	
2. Cycloparaffin...........	Cyclohexane	C_6H_{12}	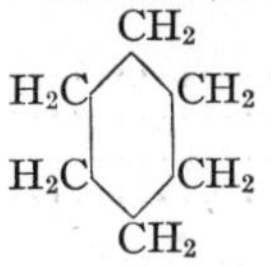	C_nH_{2n}
3. Olefin (one double bond)..	Ethylene	C_2H_4	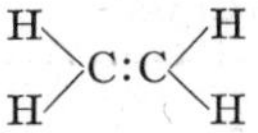	C_nH_{2n}
4. Cyclo-diolefin..........	Limonene	$C_{10}H_{16}$	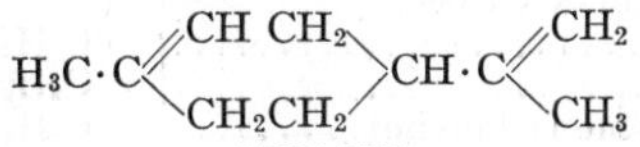	C_nH_{2n-4}
5. Acetylene (one triple bond)	Acetylene	C_2H_2	$HC:CH$	C_nH_{2n-2}
6. Benzenoid..............	Benzene	C_6H_6	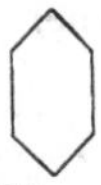	C_nH_{2n-6}
(a) one ring............	Toluene	C_7H_8	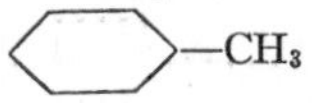	
(b) two rings or more, not doubly adjacently interlocked..........	Cymene	$C_{10}H_{14}$	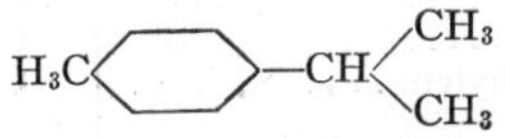	
	Biphenyl	$C_{12}H_{10}$	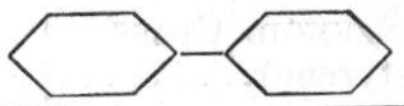	C_nH_{2n-14}
(c) two rings or more, doubly adjacently interlocked..........	Diphenyl methane	$C_{13}H_{12}$	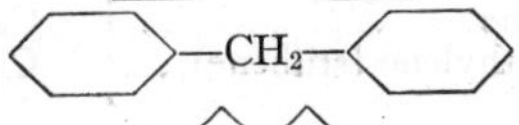	
	Naphthalene	$C_{10}H_8$	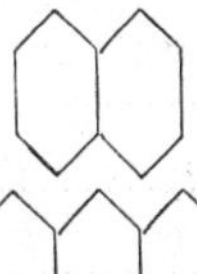	C_nH_{2n-12}
	Anthracene	$C_{14}H_{10}$	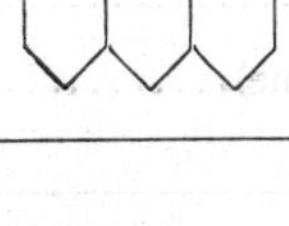	C_nH_{2n-18}

The most fundamental classification on the basis of reactivity appears to be into benzenoid and non-benzenoid (and mixed, e.g., toluene, which in many reactions behaves as a benzenoid, but in some others as a non-benzenoid hydrocarbon.)

CLASSIFICATION OF HYDROCARBONS
(Upon Basis of Reactivity)

			Melting Point, °C.	Boiling Point, °C.
Paraffin Class				
1. Methane	CH_4	(structure)		−161
2. Ethane	C_2H_6	$H_3C \cdot CH_3$		−88
3. Propane	C_3H_8	$CH_3 \cdot CH_2 \cdot CH_3$		−45
4. Normal-butane	C_4H_{10}	$CH_3(CH_2)_2CH_3$	−135	0.6
5. Iso-butane	C_4H_{10}	$(CH_3)_3CH$	−145	−10
(2-Methylpropane) (Trimethylmethane)				
6. Normal-pentane	C_5H_{12}	$CH_3(CH_2)_3CH_3$		36
7. Tetramethylmethane	C_5H_{12}	$(CH_3)_4C$		9.5
(2,2-Dimethylpropane)				
8. Normal-hexane	C_6H_{14}	$CH_3(CH_2)_4CH_3$		69
9. Normal-heptane	C_7H_{16}	$CH_3(CH_2)_5CH_3$		98
10. Normal-octane	C_8H_{18}	$CH_3(CH_2)_6CH_3$		125
11. Normal-nonane	C_9H_{20}	$CH_3(CH_2)_7CH_3$		150
12. Normal-decane	$C_{10}H_{22}$	$CH_3(CH_2)_8CH_3$		174
13. Hexacontane	$C_{60}H_{122}$	$CH_3(CH_2)_{58}CH_3$	101	
Mixed Paraffin-Benzenoid Class				
14. Toluene and successors (see benzenoid)				
15. Bibenzyl	$C_{14}H_{14}$	$C_6H_5CH_2CH_2C_6H_5$	52	284
16. Diphenyl methane	$C_{13}H_{12}$	$(C_6H_5)_2CH_2$	27	262
17. Triphenyl methane	$C_{19}H_{16}$	$(C_6H_5)_3CH$	92	359
18. Tetraphenyl methane	$C_{25}H_{20}$	$(C_6H_5)_4C$	282	431
19. Tetraphenyl ethane	$C_{26}H_{22}$	$(C_6H_5)_2CH \cdot CH(C_6H_5)_2$	209	383
Cycloparaffin Class				
20. Cyclopropane	C_3H_6	$(CH_2)_3$		−34 (750 mm.)
(Trimethylene)				
21. Cyclobutane	C_4H_8	$(CH_2)_4$		12 (725 mm.)
22. Cyclopentane	C_5H_{10}	$(CH_2)_5$		50
23. Cyclohexane	C_6H_{12}	$(CH_2)_6$		81
24. Cycloheptane	C_7H_{14}	$(CH_2)_7$		118
Dicycloparaffin Class				
25. Camphane	$C_{10}H_{18}$	$CH_3 \cdot C_7H_9(CH_3)_2$	152	160
Olefin Class				
26. Ethylene	C_2H_4	$H_2C:CH_2$		−104
27. Propylene	C_3H_6	$CH_3CH:CH_2$		−48
28. Alpha-butylene (1-butene)	C_4H_8	$CH_3CH_2CH:CH_2$		−5
29. Beta-butylene (2-butene)	C_4H_8	$CH_3CH:CHCH_3$		1
30. Iso-butylene (Methylpropene)	C_4H_8	$(CH_3)_2C:CH_2$		−6
31. Normal-amylene (alpha)	C_5H_{10}	$CH_3(CH_2)_2CH:CH_2$		40
32. Normal-amylene (beta)	C_5H_{10}	$CH_3CH:CHC_2H_5$		36
33. Iso-amylene (alpha)	C_5H_{10}	$(CH_3)_2CHCH:CH_2$		25
34. Iso-amylene (beta)	C_5H_{10}	$(CH_3)_2C:CHCH_3$		38
35. 1-Hexene	C_6H_{12}	$CH_3(CH_2)_3CH:CH_2$		64
36. 1-Heptene	C_7H_{14}	$CH_3(CH_2)_4CH:CH_2$		99
37. Melene	$C_{30}H_{60}$			380
38. Tetramethyl ethylene	C_6H_{12}	$(CH_3)_2C:C(CH_3)_2$		73
Mixed Olefin Benzenoid Class				
39. Phenyl ethylene (styrene)	C_8H_8	$C_6H_5CH:CH_2$		146
40. Propenyl benzene	C_9H_{10}	$C_6H_5CH:CHCH_3$		175
41. Sym-diphenyl ethylene (stilbene)	$C_{14}H_{12}$	$C_6H_5CH:CHC_6H_5$	124	
Cyclo-olefin Class				
42. Cyclobutene	C_4H_6	$\overline{CH_2CH_2CH:CH}$		3 3
43. Cyclopentene	C_5H_8	$\overline{CH_2CH_2CH_2CH:CH}$		45
44. Cyclohexene	C_6H_{10}	$\overline{CH_2CH_2CH_2CH_2CH:CH}$		83
45. Cycloheptene (suberene)	C_7H_{12}	$\overline{CH_2CH_2CH_2CH_2CH_2CH:CH}$		114

CLASSIFICATION OF HYDROCARBONS—*Continued*

(Upon Basis of Reactivity)

			MELTING POINT, °C.	BOILING POINT, °C.
DIOLEFIN CLASS				
46. Allene (propadiene).............	C_3H_4	$CH_2:C:CH_2$	-32	-3
47. Bivinyl........................	C_4H_6	$CH_2:CH\cdot CH:CH_2$	-109	-4
(1,3-butadiene) (biethylene)				
48. Biallyl (1,5-hexadiene)...........	C_6H_{10}	$CH_2:CHCH_2CH_2CH:CH_2$		60
49. Isoprene (3-methyl-1,3-butadiene)...	C_5H_8	$CH_2:CHC(CH_3):CH_2$		34
CYCLO-DIOLEFIN CLASS				
50. Cyclohexadiene (1,2).............	C_6H_8			78.5
51. Cyclohexadiene (1,3).............	C_6H_8			80.5
52. Cyclohexadiene (1,4).............	C_6H_8			85.5
53. Limonene......................	$C_{10}H_{16}$	$(1)CH_3\cdot C_6H_8\cdot C_3H_5(4)$		177
(1-methyl-8(9)-isopropylenecyclo-hexene-1)				
DICYCLO-OLEFIN CLASS				
54. Alpha-pinene...................	$C_{10}H_{16}$	$(2)CH_3\cdot C_7H_7(CH_3)_2(7,7)$		154
(2,7,7-trimethylbicyclo[3.1.1] heptene-2)				
55. Camphene......................	$C_{10}H_{16}$	$(2)CH_2:C_7H_8(CH_3)_2(1,1)$	50	160
(1,1-dimethyl-2-methylene-bicyclo[2.2.1]heptane)				
56. Terpenes (discussed below along with cycloölefin hydrocarbons)				
ACETYLENE CLASS				
57. Acetylene.....................	C_2H_2	$CH:CH$		-84
58. Methyl acetylene (allylene).......	C_3H_4	$CH_3C:CH$		-27
59. Ethyl acetylene.................	C_4H_6	$CH_3CH_2C:CH$		19
MIXED ACETYLENE BENZENOID CLASS				
60. Sym-dimethyl acetylene..........	C_4H_6	$CH_3C:CCH_3$		29
(crotonylene)				
61. Phenyl acetylene...............	C_8H_6	$C_6H_5C:CH$		142
62. Sym-diphenyl acetylene (tolane)....	$C_{14}H_{10}$	$C_6H_5C:CC_6H_5$	60	300
DIACETYLENE CLASS				
63. Biacetylene (Butadiyne)..........	C_4H_2	$CH:CC:CH$		
64. Bipropargyl (1,5-hexadiyne).......	C_6H_6	$CH:CCH_2CH_2C:CH$		85
BENZENOID CLASS				
(a) one ring				
65. Benzene.......................	C_6H_6	$(CH)_6$	5.5	79.6
66. Toluene.......................	C_7H_8	$C_6H_5\cdot CH_3$	-95	110.5
67. Ortho-xylene...................	C_8H_{10}	$(1)CH_3\cdot C_6H_4\cdot CH_3(2)$	-27	144
68. Meta-xylene....................	C_8H_{10}	$(1)CH_3\cdot C_6H_4CH_3(3)$	-54	139
69. Para-xylene....................	C_8H_{10}	$(1)CH_3C_6H_4\cdot CH_3(4)$	-15	138
70. Ethyl benzene..................	C_8H_{10}	$C_6H_5\cdot CH_2CH_3$		136
71. Isopropylbenzene (cumene)........	C_9H_{12}	$C_6H_5\cdot CH(CH_3)_2$		153
72. Para-methylisopropyl-benzene......	$C_{10}H_{14}$	$(1)CH_3\cdot C_6H_4\cdot CH(CH_3)_2(4)$		175
(cymene)				
(b) Two or more rings, not doubly adjacently interlocked				
73. Biphenyl......................	$C_{12}H_{10}$	$C_6H_5\cdot C_6H_5$	69	255
74. Diphenyl methane...............	$C_{13}H_{12}$	$(C_6H_5)_2CH_2$	27	262
75. Sym-diphenyl ethane (bibenzyl)....	$C_{14}H_{14}$	$C_6H_5CH_2\cdot CH_2C_6H_5$	51	284
76. Triphenyl methane..............	$C_{19}H_{16}$	$(C_6H_5)_3CH$	93	359
77. Tetraphenyl methane............	$C_{25}H_{20}$	$(C_6H_5)_4C$	285	431
78. Sym-tetraphenylethane...........	$C_{26}H_{22}$	$(C_6H_5)_2CH\cdot CH(C_6H_5)_2$	207	380
79. Unsym-tetraphenyl ethane........	$C_{26}H_{22}$	$(C_6H_5)_3C\cdot CH_2(C_6H_5)$	144	
80. Dibenzylmethane...............	$C_{15}H_{16}$	$C_6H_5(CH_2)_3C_6H_5$		299
(alpha-gamma-diphenyl propane)				
81. Dibenzylethane.................	$C_{16}H_{18}$	$C_6H_5(CH_2)_4C_6H_5$	52	
(alpha-delta-diphenyl butane)				
82. Para-diphenylbenzene............	$C_{18}H_{14}$	$(1)C_6H_5\cdot C_6H_4\cdot C_6H_5(4)$	205	427
(diphenylphenylene)				
83. Diphenylene methane (fluorene)....	$C_{13}H_{10}$	$C_6H_4\cdot CH_2\cdot C_6H_4$	116	295

(*Continued on next page*)

CLASSIFICATION OF HYDROCARBONS—*Continued*

(Upon Basis of Reactivity)

			MELTING POINT, °C.	BOILING POINT, °C.
BENZENOID CLASS—*Continued* (c) Two or more rings, doubly adjacently interlocked				
84. Naphthalene....................	$C_{10}H_8$		80	218
85. Anthracene....................	$C_{14}H_{10}$		218	342
86. Phenanthrene..................	$C_{14}H_{10}$		100	340
87. Naphthalene ethylene............ (acenaphthene)	$C_{12}H_{10}$		95	278
NOT BENZENOID				
88. Dihydronaphthalene(1,4).........	$C_{10}H_{10}$		15	212
89. Tetrahydronaphthalene (1,2,3,4)....	$C_{10}H_{12}$			207
90. Decahydronaphthalene............	$C_{10}H_{18}$			190 (approx.)
91. Dihydroanthracene (9,10).........	$C_{14}H_{12}$		108	313
92. Hexahydroanthracene............. (1,2,3,4,9,10)	$C_{14}H_{16}$		63	290
MISCELLANEOUS				
93. Carotene (yellow)................ (Dicyclic and olefin linkages)	$C_{40}H_{56}$			

CLASS OF HYDROCARBONS			CHARACTERISTICS OF THE CARBON LINKAGES
Paraffin.............	No ring	Saturated	All single linkages
Cycloparaffin........	One closed ring	Saturated	All single linkages
Olefin...............	No ring	Unsaturated	One or more double linkage
Cycloölefin..........	One or more closed ring	Unsaturated	One or more double linkage
Acetylene...........	No ring	Unsaturated	One or more triple linkage
Benzenoid...........	One or more closed ring	Unsaturated	Oscillating double linkages in groups of three

In hydrocarbons of no ring, the chain of carbons may be single or branched, the limit of branching of the chain being two branch carbons to any one chain carbon.

Formula............	$\begin{matrix} H \\ H \end{matrix}\!>\!C\!<\!\begin{matrix} H \\ H \end{matrix}$	$\begin{matrix} H \\ H \end{matrix}\!>\!C\!<\!\begin{matrix} H \\ CH_3 \end{matrix}$	$\begin{matrix} H \\ H \end{matrix}\!>\!C\!<\!\begin{matrix} CH_3 \\ CH_3 \end{matrix}$	$\begin{matrix} H \\ CH_3 \end{matrix}\!>\!C\!<\!\begin{matrix} CH_3 \\ CH_3 \end{matrix}$	$\begin{matrix} CH_3 \\ CH_3 \end{matrix}\!>\!C\!<\!\begin{matrix} CH_3 \\ CH_3 \end{matrix}$
Name..............	Methane	Methyl methane Ethane	Dimethyl methane Propane	Trimethyl methane	Tetramethyl methane
Boiling Point, °C......	−161	−88	−45	−10	9.5
Designation.........		Primary Normal	Secondary Iso	Tertiary	Quaternary

In hydrocarbons having more than one closed ring, the attachments may be of various types.

Single ring   Two rings or more, not doubly adjacently interlocked

Formula

Name Benzene Biphenyl Diphenyl methane Triphenyl methane Tetraphenyl methane

Toluene   Para-diphenyl benzene   Fluorene

Alpha-pinene
(2,7,7,-trimethylbicyclo
[3.1.1] heptene-2)

Limonene
(1-methyl-8(9) isopropylene
cyclohexene-2)

Two rings or more, doubly adjacently interlocked

Naphthalene   Anthracene

Acenaphthene   Phenanthrene

All hydrocarbons are combustible, and many of them have preferred uses as fuels, e.g., methane, ethylene, acetylene, paraffins, benzene, and as solvents, e.g., benzene, toluene, xylenes, paraffins, terpenes. These and others have special importance individually.

PARAFFIN HYDROCARBONS. Paraffin hydrocarbons are characterized in general by lack of marked chemical reactivity, and by having the lowest specific gravity of the liquid hydrocarbons (paraffin range about 0.62 to 0.77). Paraffin hydrocarbons occur in natural gas and petroleum, and are the predominant constituents in the hydrocarbon portion when coal is destructively distilled at low temperatures. By the fractional distillation of petroleum there is obtained a wide range of mixtures of paraffins, illustrated by the following table.

Distillate	Typical Yields by Distillation of Crude Petroleum from		
	Pennsylvania	Texas	California
Gasoline.......	28%	22%	22%
Naphtha.......	10	9	9
Kerosene......	8	12	12
Gas or stove oil	7	42	42

From the remaining portion other distillates are obtained by further distillation at atmospheric pressure and by the use of vacuum.

By means of the comparatively recent perfection of processes, any type of petroleum oil, kerosene distillate, gas oil, fuel oil residue, or crude oil may be converted into gasoline with yields of 50% to 75%. These so-called cracking processes, which are of both thermal and catalytic types, result in the decomposition, upon heating under high pressure, without or with catalysts of the raw material into gases, gasoline distillate, and carbon.

Gasoline is a liquid hydrocarbon mixture obtained from three main sources: 1. Straight-run gasoline, by the fractional distillation of crude petroleum. 2. Cracked gasoline, by heat-pressure decomposition without or with catalysts of crude petroleum or any distillate fraction. 3. Natural gas or casing-head gasoline, by condensation under pressure or by absorption of the liquefiable constituents of natural gas. The production of each in the United States is of the order of 55%, 35%, 10%, respectively, and the percentage of cracked gasoline shows rapid increase, with accompanying economy in the utilization of crude petroleum for gasoline production.

Federal Specification for United States Government Lubricating Oil, class D, for Internal Combustion Engines other than Aircraft and Diesel.

Viscosity. A.S.T.M.[2] method D88–30. The Saybolt universal viscosities of the six grades (S.A.E.[3] numbers 20 to 70) shall lie within the limits:

S.A.E.[3] NUMBER	SECONDS AT 130° F.		SECONDS AT 210° F.	
	Not Less Than	Less Than	Not Less Than	Less Than
20	120	185		
30	185	255		
40	255	...	...	75
50	...	...	75	105
60	...	...	105	125
70	...	...	125	150

Flash point. F.S.B.[1] method 110.32. The flash point shall not be lower than the minimum for each grade:

S.A.E.[3] NUMBER	FLASH POINT MINIMUM FOR PASSING
20	340° F.
30	350
40	370
50	395
60	425
70	460

Pour point. F.S.B.[1] method 20.12. The pour point shall not be higher than 40° F. In the case of S.A.E.[3] 20 and S.A.E.[3] 30 oils, the Government reserves the right to require a pour point of not higher than 15° F., or under adverse climatic conditions, a pour point of not higher than 0° F.

Color. F.S.B.[1] method 10.2. The color shall not be darker than the darkest for each grade:

S.A.E.[3]	DARKEST A.S.T.M.[2] COLOR NUMBER
20	7.5
30	8
40	6 dil
50	7 dil
60	8 dil
70	8 dil

Carbon residue. A.S.T.M.[2] method D189–30. The carbon residue shall not be higher than the maximum for each grade:

S.A.E.[3]	MAXIMUM A.S.T.M.[2] CARBON RESIDUE
20	0.60%
30	0.80
40	1.00
50	1.70
60	2.00

Oxidation number. F.S.B.[1] method 340.1. The Sligh oxidation number shall not be higher than 50. The oxidation number is the number of milligrams of precipitate from 10.0 grams of oil after subjection to the specified treatment with oxygen.

Neutralization number. F.S.B.[1] method 510.31. The neutralization number shall not be higher than 0.30. The

[1] Federal Specifications Board.
[2] American Society for Testing Materials.
[3] Society of Automotive Engineers.

neutralization number is the number of milligrams of potassium hydroxide required for neutralization of 1 gram of oil.

Corrosion. F.S.B.*[1] method 530.31. A clean, freshly polished copper strip shall not show more than extremely slight discoloration when submerged in the oil for three hours at 212° F.

CYCLOPARAFFIN HYDROCARBONS. Cycloparaffin hydrocarbons are characterized in general by lack of marked chemical reactivity, and by moderately low **specific gravity** of the liquid hydrocarbons (cyclohexane 0.78). Cycloparaffins resemble paraffins, except that in the former the ratio of **carbon** to **hydrogen** is slightly less than in the latter, and there is one or more closed rings of carbon atoms in the former and none in the latter. Cycloparaffin hydrocarbons occur in certain petroleum fields, notably Baku, Russia. When the **calcium** salt of a higher dibasic **acid** in the oxalic acid series is subjected to dry **distillation,** a cyclic **ketone** is formed by the separation of calcium carbonate, and this ketone can be reduced to the corresponding cycloparaffin hydrocarbon. By the action of **sodium** on dichloro (or dibromo) paraffins (chlorine or bromine not attached to the same or adjacent carbon atoms), sodium chloride separates upon heating and cycloparaffin hydrocarbon is also formed.

OLEFIN HYDROCARBONS. Olefin hydrocarbons are characterized by marked chemical reactivity, as shown by the behavior of such reagents as (1) **bromine in carbon tetrachloride.** The red-colored solution is decolorized by olefins by addition of bromine at the olefin linkage ($>C:C<$), (2) **sulfuric acid,** fuming. Olefins react, generating heat, by addition or polymerization, (3) **nitric acid,** fuming. Olefins react violently, (4) **potassium** permanganate in sodium carbonate solution. The purple-colored solution is decolorized by olefins, (5) **hydrogen,** in the presence of finely divided nickel at 150° C. Olefins form paraffins by addition of hydrogen at the olefin linkage, finally, (6) olefins react, by addition at the olefin linkage, with **hydrogen iodide, hydrogen bromide, hypochlorous acid, hypobromous acid, ozone.**

The olefin hydrocarbon ethylene is a constituent of coal gas (up to 8%), and in larger percentage (up to 30%) in the gases from petroleum subjected to the cracking process. Cracking process gas contains propylene (up to 20%), and butylenes (up to 9%).

Olefin hydrocarbons are formed (1) by the **destructive distillation** of coal, and the cracking of petroleum, (2) by the removal of water from **alcohols,** by passing the alcohol vapor over heated **bauxite,** or by heating with concentrated sulfuric or **phosphoric acid,** (3) by heating paraffin chlorides, bromides, or iodides with **sodium** hydroxide in alcoholic solution, (4) by heating dihalogen saturated hydrocarbons with **zinc,** (5) by **electrolysis** of the **sodium** salts of dibasic acids, which in the case of **succinic acid** ($COOH \cdot CH_2 \cdot CH_2 \cdot COOH$) yields **ethylene** plus **carbon dioxide** (2 volumes) and **hydrogen** (1 volume).

CYCLOÖLEFIN AND TERPENE HYDROCARBONS. Cycloölefin hydrocarbons are characterized by marked chemical reactivity, as shown by the behavior of such reagents as (1) **bromine in carbon tetrachloride.** The red-colored solution is decolorized by cycloölefins by addition of bromine at the olefin linkage ($>C:C<$), (2) **sulfuric acid,** fuming. Cycloölefins polymerize and dissolve in the reagent, (3) **nitric acid,** fuming. Cycloölefins react violently, sometimes with explosive violence, (4) **potassium** permanganate in **sodium** carbonate solution. The purple-colored solution is decolorized by cycloölefins, (5) hydrogen, in the presence of finely divided nickel at 150° C. Cycloölefins form cycloparaffins by addition of hydrogen at the olefin linkage.

Cycloölefin hydrocarbons predominate in the hydrocarbon secretions of plants known as essential or **volatile**

*[1] Federal Specifications Board.

oils, the terpenes ($C_{10}H_{16}$). These are found in **coniferous** trees, especially pine, from which **turpentine** is obtained, and in the skin of **citrus fruits,** which yield limonene. Terpenes, especially turpentine, are desirable solvents of many organic materials such as **resins** and **gums** and find application as such in preparing **paints, varnishes, lacquers.**

Cymene, para-methylisopropyl benzene ($C_{10}H_{14}$) or (1) $CH_3 \cdot C_6H_4 \cdot C_3H_7(4)$ contains the carbon skeleton of the terpenes, limonene ($C_{10}H_{16}$, which is a monocyclo-1(2),8(9)-diene and alpha-phellandrene ($C_{10}H_{16}$), which is a monocyclo-1(2),5(6)-diene. Completely hydrogenated cymene is known as menthane ($C_{10}H_{20}$).

The terpene, alpha-pinene ($C_{10}H_{16}$) is a dicyclo-monoene compound having three methyl groups (2,7,7) attached to two (2,7) of its seven carbon atoms. The three bridges of the dicyclo heptene nucleus (between the two carbon atoms common to each ring) contain respectively 3,1,1 carbon atoms (expressed [3.1.1]).

ACETYLENE HYDROCARBONS. Acetylene hydrocarbons are characterized by marked chemical reactivity, as shown by the behavior of such reagents as (1) **bromine in carbon tetrachloride.** The red-colored solution is decolorized by acetylenes by addition of bromine at the acetylene linkage ($-C:C-$), (2) **sulfuric acid,** concentrated. Acetylenes dissolve in the reagent, (3) **nitric acid,** fuming. Acetylenes react violently, (4) **hydrogen** in the presence of finely divided nickel at 150° C. Acetylenes form olefins or paraffins by addition of hydrogen at the acetylene linkage, (5) ammonio-**cuprous** (or **silver**) salt solution. Acetylene having the group $-C:CH$ yield precipitates, explosive when dry.

Acetylene hydrocarbons are formed (1) by reaction of metallic acetylides or certain metallic carbides with acids or water. **Calcium** carbide and water yields acetylene plus calcium hydroxide, (2) by heating dihalogen saturated hydrocarbons with sodium ethoxide, (3) by **electrolysis** of the sodium salts of the fumaric acid type, which in the case of fumaric acid ($COOHCH:CHCOOH$) yields acetylene plus **carbon dioxide** (2 volumes) and **hydrogen** (1 volume).

BENZENOID HYDROCARBONS. Benzenoid hydrocarbons are characterized by moderate reactivity, as shown by the behavior of such reagents as (1) **bromine in carbon tetrachloride,** usually on warming. The red-colored solution is decolorized by benzenoids by substitution of bromine for **hydrogen** of the benzenoid with the simultaneous separation of **hydrogen bromide,** which is evolved as a colorless gas and recognized by the fog formed when breathed upon, (2) **sulfuric acid,** fuming, usually on warming. Benzenoids react by substitution of the sulfonic acid group ($-SO_3H$) for hydrogen of the benzenoid with the simultaneous formation of water (not visible), (3) **nitric acid,** fuming, frequently on warming. Benzenoids react by substitution of the nitro group ($-NO_2$) for hydrogen of the benzenoid with the simultaneous formation of water (not visible), (4) with methyl (or other) chloride, bromide, iodide, in the presence of **aluminum** chloride anhydrous. Benzenoids react by substitution of the methyl (or other) group for hydrogen of the benzenoid with the simultaneous separation of hydrogen chloride, bromide, iodide, (5) with **acetyl** (or other) **chloride,** bromide, in the presence of **aluminum** chloride anhydrous. Benzenoids react by substitution of the acetyl (or other) group for hydrogen of the benzenoid with the formation of a ketone and the simultaneous separation of hydrogen chloride, bromide.

Benzenoid hydrocarbons are the predominant constituents in the hydrocarbon portion when **coal** is **destructively distilled** at high temperatures. This sources furnishes **benzene, toluene, xylenes, naphthalene, anthracene** of commerce. Since 1941, however, more toluene and xylenes have been produced in this country from petroleum sources by catalytic cyclization of paraffins and dehydrogenation of cycloparaffins than from all the coal distillation sources. (R.K.S.)

HYDROCAULUS. The common stalk bearing the individuals of a colony in the **hydrozoan** coelenterates. (A.W.L.)

HYDROCELE. A localized collection of fluid about the **testicle** within the membrane (*tunica vaginalis*) enveloping the testicle. It is cured by surgery. (R.S.M.)

HYDROCEPHALUS. A disorder of the normal circulation and absorption of fluid within the cavities (ventricles) of the **brain.** There are blocking and increased accumulation of cerebrospinal fluid in the ventricles. This may be caused by tumors, congenital deformities, infection, or injury.

Hydrocephalus usually occurs early in childhood. It is characterized by progressive and extreme enlargement of the skull, atrophy of the brain through pressure, mental impairment, and convulsions. Unless the process is arrested, imbecility, blindness, and eventual death from malnutrition or some intercurrent infection results.

The treatment is unsatisfactory and the outlook is usually hopeless. If the obstruction is caused by infection, recovery may take place. Surgical procedures may relieve the block, but the child will still be a mental defective if pressure atrophy of the brain has already begun. (D.M.H.)

HYDROCHLORIC ACID AND CHLORIDES. Hydrochloric acid, "muriatic acid" (HCl), is a colorless solution formed when hydrogen chloride gas is dissolved in water, commercially of strength 18° Baumé (specific gravity at 60° F., water at 60° F., 1.1417, 27-92% HCl); 20° Baumé (specific gravity at 60° F., water at 60° F., 1.1600, 31.45% HCl); specific gravity 1.19 (approximately 36% HCl). Sometimes colored yellow by ferric iron. There is a maximum constant boiling point 108.58° C. (760 mm.), at 20.22% HCl (distillate) for mixtures of hydrochloric acid and water.

A commonly used strength for dilute hydrochloric acid is 18.25 grams HCl per 100 milliliters of solution (5 normal). Dilute hydrochloric acid reacts (1) with many hydroxides, e.g., **sodium** hydroxide, to yield the corresponding chloride, e.g., sodium chloride, solution, (2) with many ordinary oxides, e.g., **magnesium** oxide, to yield the corresponding chloride, e.g., magnesium chloride, solution, (3) with many carbonates, e.g., **calcium** carbonate, to yield the corresponding chloride, e.g., calcium chloride solution plus carbon dioxide gas, (4) with many sulfides, e.g., **ferrous** sulfide, to yield the corresponding chloride, e.g., ferrous chloride, solution plus **hydrogen** sulfide gas, (5) with many metals, e.g., **zinc** (but not copper) to yield the corresponding chloride, e.g., zinc chloride, solution plus hydrogen gas, (6) with some special oxides, e.g., lead or manganese dioxide, to yield **lead** or **manganese** chloride plus chlorine gas, (7) with solution of some salts, e.g., silver nitrate, to yield the corresponding chloride, **silver** chloride, precipitate. Higher strengths of hydrochloric acid usually react similarly to the dilute. Hydrochloric acid sometimes reacts as a reducing acid, e.g., (6) above.

The uses of hydrochloric acid have been suggested by the chemical reactions previously cited, and the largest quantities are used for reaction with metals, e.g., zinc recovery from galvanized iron scrap; in the production of chlorides; and, previous to the electrolytic production of **chlorine**, for chlorine.

All metallic chlorides, except **silver** chloride and mercurous chloride, are soluble in water, but **lead** chloride, **cuprous** chloride and **thallium** chloride are only slightly soluble. Metallic chlorides when heated melt, and volatilize or decompose, e.g., **sodium** chloride, melting point 804° C.; **calcium, strontium, barium** chloride volatilize at red heat; **magnesium** chloride crystals yield magnesium oxide residue and hydrogen chloride; **cupric** chloride yields cuprous chloride and chlorine.

Two common tests for chlorides are as follows:
1. An aqueous solution gives a precipitate with **silver** nitrate which is insoluble in **nitric acid** but soluble in **ammonia, sodium** thiosulfate, or **potassium** cyanide.
2. On treatment with **chlorine** water and shaking with **carbon disulfide** no dark coloration is produced as is the case for the bromides and the iodides. (R.K.S.)

HYDROCLADIUM. Small branches bearing sessile polyps in **hydrozoan** coelenterates of the family Plumulariidae. (A.W.L.)

HYDROCORALLINAE. Hydrozoa.

HYDROCYANIC ACID AND CYANIDES. Hydrocyanic acid, prussic acid (HCN), is a solution of hydrogen cyanide gas in water of characteristic odor and very poisonous, is soluble in all proportions in water, alcohol, or ether, and anhydrous liquid hydrogen cyanide is a solvent for many salts and organic substances. Hydrogen cyanide reacts with **hydrogen** at 140° C. in the presence of a **catalyzer**, e.g., **platinum** black, to form methyl **amine** (CH$_3$NH$_2$); when burned in air, produces a pale violet flame; when heated with dilute **sulfuric acid** forms formamide (HCONH$_2$) and ammonium formate (HCOONH$_4$); when exposed to sunlight with **chlorine** forms cyanogen chloride (CNCl), plus **hydrogen chloride.** An important reaction of hydrogen cyanide is that with **aldehydes** or **ketones**, whereby cyanhydrins are formed, e.g., acetaldehyde cyanhydrin (CH$_3$CHOH·CH), and the resulting cyanhydrins are readily converted into alpha-hydroxy acids, e.g., alpha-hydroxypropionic acid (CH$_3$·CHOH·COOH).

Metallic cyanides are (1) soluble, e.g., **sodium** cyanide (NaCN), **potassium** cyanide (KCN), **calcium** cyanide (Ca(CN)$_2$), **mercuric** cyanide (Hg(CN)$_2$), **aurous** cyanide (AuCN), (2) insoluble, e.g., **silver** cyanide (AgCN), **cuprous** cyanide (CuCN), (3) complex, (a) decomposed by dilute **sulfuric acid** and not affected by dilute **sodium** hydroxide, e.g., sodium silver cyanide (NaAg(CN)$_2$) solution, sodium cuprous cyanide (NaCu(CN)$_2$) colorless solution, (b) changed only to the acid by dilute sulfuric acid and reactive with dilute sodium hydroxide, e.g., **potassium** ferrocyanide (K$_4$Fe(CN)$_6$) yields, with dilute sulfuric acid, **hydroferrocyanic acid,** cupric ferrocyanide (Cu$_2$(Fe(CN)$_6$)) yields, with dilute sodium hydroxide, cupric hydroxide.

Sodium cyanide solution dissolves certain metals (1) with absorption of oxygen, e.g., gold, silver, mercury, lead, (2) with evolution of hydrogen, e.g., copper, nickel, iron, zinc, aluminum, magnesium; and solid sodium cyanide, when heated with certain oxides, e.g., **lead** monoxide (PbO), **stannic** oxide (SnO$_2$), yields the metal of the oxide, e.g., lead, tin, respectively, and sodium **cyanate** (NaCNO). Two classes of **esters** are known, cyanides or nitriles, and iso-cyanides, iso-nitriles or carbylamines, the latter being very poisonous and of marked nauseating odor.

Methyl cyanide (CH$_3$CN), boiling point 82° C., formed by reaction of (1) methyl iodide and **potassium** cyanide, (2) acetamide and phosphorus pentoxide. Methyl iso-cyanide (CH$_3$NC), boiling point 60° C., formed by reaction (1) of methyl iodide and **silver** cyanide, (2) of methyl **amine, chloroform** and **sodium** hydroxide solution warmed. Ethyl iso-cyanide (C$_2$H$_5$NC), boiling point 78° C. Phenyl iso-cyanide (C$_6$H$_5$NC), boiling point 78° C. at 40 mm. pressure.

Hydrogen cyanide is made by reaction of **sodium** cyanide and dilute **sulfuric acid.** Its uses are for fumigation of ships, cars, rooms, and fruit trees. Sodium cyanide is used for recovering gold and silver from ores (cyanide process). The detection of hydrogen cyanide is accomplished (1) by evaporation to dryness with yellow **ammonium** sulfide, whereby **thiocyanate** is formed, which yields red-colored solution upon addition of **ferric** salt solution. It is said that 1 part of cyanide may thus be detected in 4,000,000 parts of solution, (2) by boiling with ferrous plus ferric salt solution to which sodium hydroxide has been added. Upon addition of excess dilute sulfuric acid to the product, a blue precipitate forms, if cyanide was present, or a blue solution if the amount was small. Cyanides are also detected by the formation, with silver nitrate solution, of a white precipitate that is soluble in excess of cyanide. On treatment of cyanides with dilute sulfuric acid, the characteristic almond-like smell of hydrogen cyanide (poisonous) is produced. (R.K.S.)

HYDRODYNAMICS. Hydraulics.

HYDROELECTRIC POWER.

This is electric power derived from generators, driven by hydraulic turbines or water wheels. The principal elements of this kind of power plant are a site to provide flow and head of water, a turbine-generator machine, and a discharge channel.

The desirability of a hydro site rests not only upon its topography and stream flow, but also upon the cost of fuel, the trend of fuel cost, and the load factor to be expected. When coal prices are rising rapidly more attention will be given to the hydro projects than when coal prices are on the decline. The length and cost of transmission lines from plant to load also present problems of some moment in any consideration of hydroelectric development. On the other hand, it can be said for hydro developments that many of them serve more than one purpose and sometimes the power development is of secondary importance. Construction of impounding areas for flood control or for irrigation frequently offers the possibility of producing electrical energy as a by-product. Another feature commending hydro development is the ability to store energy, in the hydraulic form, over long periods of time in a reservoir; also the adjustment, through correct impounded volumes, of a variable stream flow to meet a variable load demand.

Hydroelectric plants can be classified as follows:

1. Extent of impounded volume.
   a. Storage plants.
   b. Run-of-river plants.
2. Status in the power system.
   a. Peak load plant.
   b. Base load plant.
   c. Isolated plant.
3. Head.
   a. High-head development.
   b. Medium-head development.
   c. Low-head development.

The low-head plant has a characteristic design differing in all essentials from the high-head plant. The medium-head plant may partake of the characteristics of either the high- or low-head plants as its working head approaches either the high- or low-head range. There is no definite line of demarcation between high, medium, and low heads; however, a head of more than 500′ can be considered a high-head development, and one lower than 50′ a low-head development. Briefly, the characteristics of the low-head plant are: vertical, reaction type, runners using large volumes of water and requiring large water passages. Substructure is both extensive and expensive, and intake works are large and complicated. Large diameter generators are made necessary by the low rotational speeds. Characteristics of the high-head plant are: horizontal impulse turbines, small volumes of water at high pressures, plant at some distance from the dam. The advantage of smaller and simpler substructure is offset by the presence of a long water conduit, or penstock, between dam and plant. The turbines are high-speed and allow smaller generator diameter. The high speed is accounted for by the high

heads used. Inherently, the impulse turbine has a low characteristic speed.

The possible hydroelectric development sites along the flow of a stream are of two types, namely, those suitable for run-of-the-river plants and those offering natural impounding basins for storage plants. In general, the run-of-the-river plant is cheaper than the storage plant of equal capacity, but it suffers seasonal variation of output more or less proportional to the variation of stream flow.

Storage plants give a greater proportion of firm power which can be delivered day by day on a regular schedule. This firm power is in more or less direct ratio to the degree of regulation of the flow of the stream and this in turn is a function of the impounded volume. Complete regulation of stream flow is rarely possible or practical although 80–90% regulation is not infrequent.

In any storage plant the theoretical energy or power available over and beyond the firm power developed is known as flash power or flood peak power. Firm power commands, commercially, a considerably higher rate than flash power.

When integrated with steam generating plants, hydroplants are frequently used to give peak power outputs to take care of peak load conditions and thus avoid the expensive standby service of additional steam generating equipment. Such service, of course, may still permit the delivery of a certain amount of firm power.

If all the run-of-river plants were located upstream from the storage plants they would be operated continuously on a base load plan, because, were they idle, their small reservoirs would quickly overflow and water would be wasted over the crest gates. If, however, they are located between storage plants, the run of the river, as far as they are concerned, is just what the storage plants are passing on to them. So, located downstream from a storage plant, a run-of-river plant will produce an increase in output when the storage plant increases its output.

In the hydroelectric plant the turbines and generators are the main items of equipment. The hydroelectric superstructure, as usually laid out, has one large building housing the main units and an electrical bay, or wing, of one or more stories in which are located the switching equipment, offices, storerooms, and most of the auxiliary equipment.

Hydro sites that are developed to use but part of the normal stream flow are exceptions to the general rule. Only rarely is a development made where conservation of the water and its use in the most efficient manner are not paramount features of operation. Failure to give due cognizance to this feature may wipe out the net operating profit; hence a continuous, watchful scrutiny of all natural factors which can affect the station operation is a duty of the operating personnel.

A hydraulic turbine suffers loss of efficiency at heads above or below the designed value because of shock losses. At the correct head there will be one point of best efficiency, somewhere between 80–95% of full load. When a number of units are installed in a plant, and when steam reserve is available, it is generally possible to operate the units near the point of best efficiency. There are four faults of operation and maintenance which can reduce the maximum energy production of a plant. They are:

1. Waste of water over spillways.
2. Improper distribution of the load between the station units.
3. Water leakage through valves, gates, dam or flow line.
4. Wear on moving parts, especially corrosion or erosion of the runner.

The relative simplicity of hydroelectric equipment makes hydraulic efficiency of the turbine the principal consideration. (F.T.M.)

HYDROFERRICYANIC ACID AND FERRICYANIDES.

Hydroferricyanic acid ($H_3Fe(CN)_6$) is a brownish-green non-volatile solid, soluble in water to form a strongly acidic brown-colored solution, which decomposes in the light. **Potassium** ferricyanide ("red prussiate of potash," ($K_3Fe(CN)_6$), red soluble solid, the ordinary source of ferricyanide, is made by reaction of potassium ferrocyanide ($K_4Fe(CN)_6$), obtained as a by-product of coal gas works, and **chlorine** in solution, and then crystallization. **Potassium** ferricyanide, unlike potassium ferrocyanide, is soluble in alcohol. From potassium ferricyanide concentrated solution, by treatment with **sulfuric acid**, solid hydroferricyanic acid may be obtained, and then dried in vacuum. When potassium ferricyanide and dilute sulfuric acid are warmed, **hydrogen cyanide** is evolved, but, with concentrated sulfuric acid warmed, **carbon monoxide** is evolved. When potassium, sodium, or ammonium ferricyanide solution is added to certain metallic salt solutions, characteristic results are obtained:

With **silver** nitrate, silver ferricyanide ($Ag_3Fe(CN)_6$), brick red precipitate; with **cadmium** nitrate, cadmium ferricyanide ($Cd_3(Fe(CN)_6)_2$), orange precipitate; with **cupric** sulfate, cupric ferricyanide ($Cu_3(Fe(CN)_6)_2$), yellowish-green precipitate; with **ferrous** sulfate, ferrous ferricyanide ("Turnbull's blue," $Fe_3(Fe(CN)_6)_2$), deep blue precipitate or colloidal solution (this is a common test for ferricyanides); with **ferric** sulfate, no precipitate but brown to green solution. With **nitric acid** or acidified solution of **sodium** nitrite, potassium or sodium ferricyanide solution is changed to nitroferricyanide ("Nitroprusside," $Na_2Fe(NO)(CN)_5$), recoverable by crystallization. (R.K.S.)

HYDROFERROCYANIC ACID AND FERROCYANIDES.

Hydroferrocyanic acid ($H_4Fe(CN)_6$) is a white, non-volatile solid. **Sodium** or **potassium** ferrocyanide ("yellow prussiate of potash," $Na_4Fe(CN)_6$) is obtained as a by-product of coal gas works by fixing the **cyanogen** compounds of the gas by reaction with **ferrous** compounds and then boiling with **calcium** hydroxide suspension. Soluble calcium ferrocyanide is converted to soluble sodium or potassium ferrocyanide by reaction with sodium or potassium carbonate, and then filtering off calcium carbonate precipitate and evaporating the filtrate. From sodium ferrocyanide solution, by treatment with **sulfuric acid**, hydroferrocyanic acid may be obtained by extraction with ether, and then evaporating the ether solution. When sodium ferrocyanide and dilute sulfuric acid are warmed, **hydrogen cyanide** is evolved, but, with concentrated sulfuric acid warmed, **carbon monoxide** is evolved. When sodium, potassium, or ammonium ferrocyanide solution is added to certain metallic salt solutions, characteristic results are obtained: with **silver** nitrate, silver ferrocyanide ($Ag_4Fe(CN)_6$), white precipitate slowly turning blue; with **cadmium** nitrate, cadmium ferrocyanide ($Cd_2Fe(CN)_6$), white precipitate; with **cupric** fulfate ($Cu_2Fe(CN)_6$), reddish-brown precipitate; with **ferrous** sulfate, ferrous ferrocyanide ($Fe_2Fe(CN)_6$), white precipitate when pure ferrous but usually pale blue precipitate; with **ferric** sulfate, ferric ferrocyanide ($Fe_4(Fe(CN)_6)_3$) or $Fe_3(Fe(CN)_6)_2$), deep blue precipitate ("Prussian blue") (this is a common test for ferrocyanides). Sodium and other soluble ferrocyanides are oxidized to ferricyanides, when in acid solution, by **chlorine, nitric acid, nitrous acid, hydrogen peroxide,** dichromate, permanganate, **lead** dioxide, **manganese** dioxide. (R.K.S.)

HYDROFLUORIC ACID AND FLUORIDES.

Hydrofluoric acid (HF or H_2F_2) is a colorless solution formed when hydrogen fluoride gas is dissolved in water, commercially of strengths 30% HF, and 60% HF. There is a maximum constant boiling point 111° C. (750 mm.) at 43% HF (distillate) for mixtures of hydrofluoric acid and water. Since hydrofluoric acid attacks glass, the container must be resistant, e.g., of lead or ceresine.

Hydrofluoric acid is used (1) to etch and frost glass, (2) to volatilize silicon oxide, (3) in the preparation of fluorides, fluosilicates, fluoborates. (See **Boron**.)

Metallic fluorides are soluble as follows: **sodium, potassium, ammonium, silver, mercurous, mercuric, thalium;** the remaining fluorides are insoluble.

A common test for fluorides is treatment (of the dry sample) with hot concentrated sulfuric acid. This produces hydrogen fluoride, which can be identified by its property of etching glass. (R.K.S.)

HYDROGEN.

Symbol: H. Atomic number: 1. Atomic weight: 1.0080. Density: 0.0899 gram per liter. 0° C., 760 mm., or 0.070 when air equals 1.000. Formula of hydrogen gas: H_2. Melting point: $-259.1°$ C. Boiling point: $-252.7°$ C. Critical temperature: $-239.9°$ C. Critical pressure: 12.8 atmospheres.

Hydrogen is a colorless, odorless, tasteless gas, suffocating but not toxic. **Palladium**, finely divided, adsorbs 1000 to 3000 times its own volume of hydrogen, and most of the gas is retained at a temperature as high as 100° C. Also, when finely divided as by reduction of the oxide powder, the following metals adsorb appreciable quantities of hydrogen: **cobalt, gold, nickel, iron.** Hydrogen was recognized as a distinct substance by Cavendish in 1766, who called it "inflammable air." Isotopes (1) (99.98%), (2) (0.02%) (deuterium), (3) (small percentage) (tritium). Used (1) in the oxyhydrogen and atomic hydrogen flames for high-temperature **welding**, cutting and melting of metals, (2) in the fixation of **nitrogen** as ammonia, (3) in the **hydrogenation** of fatty oils, unsaturated hydrocarbons, coal tar, coal, (4) in the formation of methyl **alcohol** by reaction with **carbon monoxide** in the presence of a **catalyzer**. In **fuel** gas mixtures, such as coal gas, water gas, and producer gas, hydrogen is a fuel constituent of low fuel value, namely, 320 B.T.U. per cu. ft.

Hydrogen occurs chiefly combined with **oxygen** in water, with **carbon** in hydrocarbons, with carbon and oxygen (without and with nitrogen) in a vast variety of organic substances, and with other elements in acids and bases.

When ignited, hydrogen burns in air with a pale blue to colorless, non-luminous flame; when mixed with air or oxygen and ignited, is explosive; combines violently with **chlorine** in sunlight or magnesium light to form **hydrogen chloride** (HCl); when heated with **sodium, calcium,** or related metals, yields the corresponding hydride; reacts with **nitrogen** to form **ammonia** (NH_3) in the presence of a **catalyzer**; with **sulfur**, to form **hydrogen sulfide** (H_2S) upon heating; with **cupric** oxide, cuprous oxide, ferric oxide, ferroferric oxide, ferrous oxide to form copper metal in the cases of the first two and iron metal in the cases of the last three plus water in all cases. Similar reactions take place with oxides of **nickel, tin, lead,** but not with oxides of zinc, aluminum, magnesium; with unsaturated **organic** compounds, to form corresponding saturated compounds, e.g., **oleic** acid ($C_{17}H_{33}COOH$) to form **stearic** acid ($C_{17}H_{35}COOH$), in the presence of a catalyzer.

Hydrogen is prepared (1) by the **electrolysis** of water solutions of salts, e.g., sodium chloride, of bases, e.g., sodium hydroxide, of acids, e.g., sulfuric acid, (2) by the reaction of dilute hydrochloric or sulfuric acid and a metal, such as **zinc** or iron, but not copper or mercury, (3) by the reaction of heated steam and a metal, such as zinc or iron, forming zinc oxide and ferroferric oxide, respectively, but not copper or mercury, (4) by reaction of silicon or aluminum with sodium hydroxide solution heated, forming sodium silicate and sodium aluminate, respectively, (5) from water gas, by separation from **carbon monoxide** by fractional liquefaction, (6) in the cracking of natural gas for carbon black, (7) from water gas and steam by catalytic conversion, (8) from methane and steam on catalysts at high temperatures, in the last instance generally accompanied by nitrogen for ammonia synthesis.

"Heavy hydrogen," **deuterium,** hydrogen isotope 2 (0.02%), symbol D, atomic weight 2.01363, is ultimately obtainable by prolonged electrolysis of the residual water of electrolytic cells from which ordinary hydrogen (isotope 1, 99.98%) has been largely removed. Approximately one part by weight of isotope 2 is present in ordinary water along with 5000 parts by weight of isotope 1. The discovery of deuterium made possible in **electron** bombardment studies the use of a projectile of atomic mass 2, intermediate between hydrogen of atomic mass 1 and helium of atomic mass 4. By means of its compounds, deuterium makes possible the extended study of chemical structure and reactions.

Below −220° C. the specific heat of hydrogen is that of a monatomic gas like **helium** (He). Practically pure para-hydrogen may be obtained by adsorption of ordinary hydrogen, which is three-fourths ortho and one-fourth para, on charcoal at about −225° C. The melting point of para-hydrogen is 0.13° C. lower (ortho-hydrogen 0.04° C. higher) than ordinary hydrogen, and the boiling point at 60 mm. pressure is 0.13° C. lower (ortho-hydrogen 0.04° C. higher) than ordinary hydrogen. Para-hydrogen reverts slowly to ordinary hydrogen, but immediately in the presence of platinized asbestos.

At high temperatures the loss of heat from a glowing wire in hydrogen is larger than expected on regular assumptions. This is believed to be due to dissociation of ordinary hydrogen into atomic hydrogen (H).

Temperature, °C	Dissociation of Hydrogen At 760 mm.	At 1 mm.
1730	0.33%	8.7%
2230	3.1	57.5
2730	34	99.3

When hydrogen is passed through an electric arc between tungsten poles, a considerable transformation into atomic hydrogen occurs, and when a stream of this gas strikes a surface a large evolution of heat takes place through recombination to ordinary hydrogen. This atomic hydrogen flame is of temperature sufficiently high to melt **tungsten** (melting point 3370° C.). The half-life period of the hydrogen atom is one-third second at 0.5 mm. pressure.

Acids: Hydrogen is present in all solutions of acids. (See **Acids,** general.)

Hydrocarbons: (See **Hydrocarbons.**)

Hydrogenation: (See **Hydrogenation.**)

Hydrides: Of oxygen—H_2O, water; H_2O_2, hydrogen peroxide.

Of Non-metals:		Of Metals:	
Boron	B_2H_6 and others	Lithium	LiH
Carbon	CH_4 and others	Sodium	NaH
Silicon	SiH_4 and others	Potassium	KH
Nitrogen	NH_3 and others	Rubidium	RbH
Phosphorus	PH_3	Cesium	CsH
Arsenic	AsH_3	Calcium	CaH_2
Sulfur	H_2S	Strontium	SrH_2
Selenium	H_2Se	Barium	BaH_2
Tellurium	H_2Te	Cuprous	Cu_2H_2
Fluorine	H_2F_2		and others
Chlorine	HCl	Germanium	GeH_4
Bromine	HBr		and others
Iodine	HI	Cerium	CeH_3
		Antimony	SbH_3
		Columbium	CbH_3

Oxides: (See **Water;** and **Hydrogen Peroxide.**) (R.K.S.)

HYDROGEN CHLORIDE. Hydrochloric Acid.

HYDROGEN CYANIDE. Hydrocyanic Acid and Cyanides.

HYDROGEN EMBRITTLEMENT. Embrittlement.

HYDROGEN IONS. Reactions Involving Recombination of Ions.

HYDROGEN PEROXIDE. Hydrogen peroxide, hydrogen dioxide (H_2O_2), is a colorless (blue in thick layers), odorless liquid, melting point −2° C., boiling point 152° C., decomposition 84° C. at 68 mm. pressure (68° C. at 26 mm.), soluble in water in all proportions, usually encountered as a dilute solution (3% H_2O_2, "10-volume solution," that is, one volume of solution yields 10 volumes of oxygen) although available up to 30% strength. Acetanilide is frequently added in small amount (0.002%) to decrease the decomposition on storage. Hydrogen peroxide is formed in 90 to 100% concentration when atomic hydrogen reacts with oxygen gas at ordinary temperature.

Hydrogen peroxide is used (1) as a bleaching agent as stated, (2) as an antiseptic and disinfectant, (3) as an oxidizing agent.

Hydrogen peroxide reacts (1) with **alkalis** to form peroxides, (2) with **potassium** iodide solution, in presence of ferrous sulfate, to liberate **iodine.** This reaction serves to indicate the presence of as small an amount as 1 part by weight of hydrogen peroxide in 25,000,000 parts of water, (3) with **lead** sulfide (PbS), brown solid, to form lead sulfate ($PbSO_4$), white solid, and sometimes used to brighten the lead pigment of darkened oil paintings, (4) with lead dioxide to form lead oxide, (5) with **sulfites,** especially in alkaline solution, to form sulfates, (6) with **nitrites** to form **nitrates,** (7) with **arsenites** to form arsenates, (8) with **ferrous** compounds to form ferric, (9) with chromic compounds to form chromates (see **Chromium**), (10) with permanganates (see **Manganese**), in acid solution to form manganous plus oxygen gas of twice the volume available from the hydrogen peroxide used, (11) with dichromates in acid solution cold to form perchromic acid, blue solution, more soluble in ether than in acid, (12) with **titanic** salt solutions to form pertitanic acid, yellow solution (see **Titanium**), (13) with colored organic materials, e.g., litmus, indigo, to destroy the color, and thus used for **bleaching** hair, silk, feathers, straw, ivory, teeth, bones, gelatin, flour. When hydrogen peroxide solution is treated with finely divided **platinum** or other substances, or comes in contact with rough surfaces, e.g., ground glass, **oxygen** is evolved (water also formed).

Hydrogen peroxide is prepared from **barium** peroxide by treatment with ice-cold dilute acid; when **sulfuric acid** is used barium sulfate insoluble may be separated by filtration. Other peroxides, e.g., sodium peroxide, react similarly with acids to form hydrogen peroxide plus the salt corresponding to the peroxide and acid used. Hydrogen peroxide is formed when ether is exposed to sunlight, when a hydrogen-oxygen flame impinges on ice, and when water in a quartz vessel is exposed to **ultra-violet light.** (R.K.S.)

HYDROGEN SCALE. Thermometry.

HYDROGEN SULFIDE AND SULFIDES. Hydrosulfuric Acid and Sulfides.

HYDROGENATION. Nickel, prepared in finely divided form by reduction of nickel oxide in a stream of hydrogen gas at about 300° C., was introduced by Sabatier (1897) as a **catalyzer** for the reaction of **hydrogen** with unsaturated organic substances to be conducted at about 175° C. Nickel, while not the only catalyst, has proved one of the most successful in such reactions. **Platinum** black or **palladium** black is sometimes used at lower temperatures, and finely divided copper metal or nickel oxide at higher temperatures. The unsaturated organic substances that are hydrogenated are usually those containing the olefin linkage (C:C), but those containing a triple bond (—C —) and (—C:N) may also be hydrogenated.

Ethylene (or acetylene) with hydrogen—all properly purified, as the catalyzer is easily poisoned—reacts in the presence of properly prepared nickel at the properly controlled temperature to form ethane; benzene (C_6H_6) to form cyclohexane (C_6H_{12}); styrolene ($C_6H_5CH:CH_2$) or phenylacetylene ($C_6H_5C:CH$) to form ethyl cyclohexane ($C_6H_{11}\cdot CH_2CH_3$); phenol (C_6H_5OH) and di- and trihydroxylphenols to form the respective hydroxycyclohexanes with change of properties from those of phenols to those of tertiary alcohols; naphhthalene ($C_{10}H_8$) to form tetra- or decahydronaphthalene; anthracene ($C_{14}H_{10}$) to form di- or hexahydroanthracene; aniline ($C_6H_5NH_2$) to form cyclohexylamine ($C_6H_{11}NH_2$) a strong base; methyl cyanide (CH_3CN) to form ethyl amine ($CH_3CH_2NH_2$); oximes to corresponding amines; nitrobenzene ($C_6H_5NO_2$) to aniline ($C_6H_5NH_2$).

The reaction is applied on a large scale to the hydrogenation of oleate ester fats under the influence of pressure, advantage thereby being taken of the decrease in volume upon the fixation of hydrogen by the fat. (See **Equilibrium.**) Oleate ester fats are hydrogenated to stearate ester fats of lower melting point and the product may be obtained of intermediate hydrogenation—and consequently of intermediate melting point—as desired. Artificial shortenings are thus prepared from liquid fatty oils such as cottonseed, peanut, corn oil. Oleic acid behaves similarly.

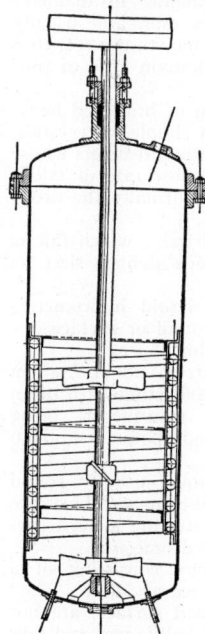

Apparatus used in the hydrogenation of oils under pressure.

The catalyzer may be supported on such a material as kieselguhr. The method of preparation of the catalyzer and the avoidance of impurities are the two greatest difficulties of the process. Nickel is more active when reduced at 320–350° C. unsupported and at 500° C. on a support than when higher temperatures are used; nickel reduced at 700° C. is almost inactive.

Great interest has attached to the extension of the hydrogenation reaction to the technical field of unsaturated liquid hydrocarbons such as petroleum and coal tar fractions, and also to the hydrogenation of coal. (See **Catalysis.**) Temperatures of the order of 450° C. and pressures of the order of a hundred atmospheres are utilized. A British report states that 100 tons of dry ash-free coal will yield 62 tons of "petrol," 28 tons of gas, and a residue containing ash and 6 tons of solid carbonaceous matter, and then an additional equal weight of coal is requisite to operate the process. (See **Methyl Alcohol** and **Ammonia.**) (R.K.S.)

HYDROGRAPH. By graphing the discharge of a stream as ordinate against time sequence as the abscissa, a hydrograph of the stream flow is obtained. The hydrograph proves to be an important source of information in **hydroelectric power** design. The reliability of the information it contains increases as the period of time over which the hydrograph extends is lengthened. Hydrographs extending over periods of less than 10 years are liable to be deceptive in the information they convey regarding maximum and minimum flows. The United States Geological Survey water supply papers

form a valuable and important reference source for data upon which hydrographs are constructed. (F.T.M.)

HYDROGRAPHIC MAP. Map.

HYDROGRAPHIC SURVEYING. Surveying.

HYDROID. One of the two forms of individuals in the **coelenterates.** The **polyp.** This form is a tubular or sac-like individual whose body wall is composed of two cellular layers separated by a thin mesogloea. The latter contains some cells derived from the other layers but is not developed as a third cellular layer. The hydroid is usually attached to the stalk of a colony or directly to a supporting surface. Its cavity opens at the free end of the body, and the mouth is surrounded by a circlet of slender tentacles except in specialized individuals found in some colonial species. The other form of coelenterate individual is the **medusa.** (A.W.L.)

HYDROKINETICS. The flowing of liquids is due to three principal causes: pressure difference, gravity, and inertia. **Bernoulli's law** expresses an ideal condition fulfilled by the three components of "head" corresponding to these three causes. The value of this head (whether constant or not) is, at a given point (x, y, z) of the liquid,

$$e + \frac{p}{\rho g} + \frac{v^2}{2g} = F(x, y, z). \qquad (1)$$

The terms of this expression represent lengths, usually given in cm. or ft. The assumption of constant density requires that the product of the speed of flow by the cross-section of any conserved portion of the stream shall be constant and that the streamlines (paths of the moving particles) therefore converge as the speed increases. If one could assume that the function F is really constant, or if it were possible to obtain F as a known function of the coordinates of the moving particle, then all hydrokinetic problems could be solved by applying suitable mathematics to the equation which would thus develop from (1).

Various attempts have been made to do this. Useful formulae result from assuming F constant (Bernoulli's law) and applying the equation to special cases. But when such formulae are tested, the calculated results are found to be in error, in every case indicating that appreciable energy has been lost in friction. While some improvement is obtained by introducing a friction factor, it has on the whole been found more satisfactory to employ empirical formulae adapted to each type of problem. Thus we have the Darcy formula

$$v = D\sqrt{\frac{d(F_1 - F_2)}{l}}, \qquad (2)$$

for the speed of flow in a pipe of length l and diameter d, running full, and with a difference of total head $F_1 - F_2$ at the two ends; D being a constant to be determined by experiment. Also the Chezy formula

$$v = C\sqrt{\frac{as}{u}}, \qquad (3)$$

giving the speed of flow in an inclined channel, like a ditch or a sewer; a being the cross-section of the flow, u the length of channel perimeter covered by the liquid, s the fall per unit length, and C an experimental constant.

The "hydraulic grade line" is a convenient concept in connection with flow through pipes. This is an imaginary line so drawn that each point of it lies vertically above (or below) the pipe at a distance equal to the pressure head $p/\rho g$ at the corresponding point of the pipe. In the case of a **siphon**, part, at least, of the conduit rises above this line, which means that the pressure in this portion is less than atmospheric.

Among the more difficult problems are those of vortex motion (like a whirlpool) and turbulent flow; and the general treatment of flow through a cavity of given shape under given boundary conditions, which presents some analogies to the electric current and the conduction of heat. (L.D.W.)

HYDROLITE. Sinter.

HYDROLOGIC CYCLE. Hydrology.

HYDROLOGY. The science, or study, of water, especially in relation to its occurrence in streams, lakes, underground structures, and as snow. The study of glaciers, their origin and geological effects is usually included under the heading of Glaciology. The term hydrology is derived from the Greek meaning water, and reason, hence the science of water, including its discovery, uses, control and conservation. Since water ranks first of all the natural resources, the science of hydrology is of great practical importance. The basis of hydrology is the hydrologic cycle. All terrestrial (fresh) waters are derived from the great oceanic reservoirs through evaporation and precipitation as rain or snow, hail or sleet.

The basis of rainfall data is the rainfall records collected by the United States Weather Bureau from its 106 climatological divisions. Rainfall is expressed in inches per year, and the volume that has fallen in a given time is expressed in acre feet, meaning the volume which would cover an acre to the depth of one foot. Most of the moisture that falls on the ground eventually returns to the sea by means of rivers. This is called the run-off. Stream flow past a stated point is measured in cu. ft. per sec., a term that is universally shortened to "second-feet." The difference between precipitation and run-off represents the amount of rain water which is, (1) temporarily caught in local depressions, such as lakes and swamps, (2) added to the **ground water** reservoir, (3) returned to the atmosphere by evaporation, or the transpiration of plants.

A great variation in rainfall in different parts of the country is to be noted, as is also the variation of precipitation at one point from year to year. This might be expected on the basis of the physical factors which cause rainfall.

The data of most importance to the study of water supply are the average annual precipitation, the minimum annual precipitation, and the frequency of occurrence of dry years. There exists only an indirect, and sometimes unrecognizable, connection between rainfall and run-off. Much depends on the topography of the watershed, its extent, and its geology. Of the volume precipitated in a rainfall, a large amount is evaporated from the watershed surfaces, more is absorbed by plants and trees, some goes into natural ground storage, and that which remains after these demands of nature have been satisfied will appear as run-off in the streams draining the watershed. The effect of a rainfall on stream flow must be interpreted in the light of the following conditions:

1. The character of the rainfall. If it is alternated with sunny periods, the evaporation will be large.

2. Temperature. The warmer the atmosphere, the more evaporated moisture it is able to contain.

3. Porosity of the soil. This governs the rate of percolation of the water from the surface to the depths.

4. Inclination, depth, and character of the underlying rock strata. This affects the amount of ground storage and its rate of flow to run-off.

It has been estimated that nearly half of the people of the United States obtain their water from wells (underground water). Within recent years the shallow, hand-dug wells have been supplemented by deeper, drilled wells, each one of which penetrates one or more water-bearing strata called aquifers. When the water in a drilled well rises to the surface under its own pressure, it is called an **artesian well.** Due to the acts of Man, as well as Nature, water contributes to such catastrophes as floods and epidemics. A deficiency of water results in droughts. It should be fairly obvious that a careful and continuous study of the hydrology of a country is highly essential to the general welfare of its people. (R.M.F., F.T.M.)

HYDROLYMPH. The watery body fluid or blood of lower invertebrates. (A.W.L.)

HYDROLYSIS. Reactions Involving Water.

HYDROMETEORS. Condensation products of atmospheric processes often appear as hydrometeors or bodies of falling liquid and solid water:

1. Rain is liquid water drops ranging in diameter from 0.5 mm. to approximately 5.0 mm. and usually falls with a velocity ranging from 3 meters per sec. to 8 meters per sec. Rain is the most common type of precipitation.

2. Snow is solid water in the form of branched hexagonal crystals sometimes mixed with simple ice crystals.

3. Drizzle is numerous very small liquid droplets whose diameter is less than 0.5 mm. and whose rate of fall is usually less than 3 meters per sec. Normally the drops seem to float downward.

4. Sleet is frozen rain drops (or drizzle) which fall as particles of ice. (International usage defines sleet as a mixture of rain and snow.)

5. Freezing rain and drizzle are liquid hydrometers that freeze upon impact with the ground or surface objects, glazing such surfaces with a film of ice.

6. Hail is stones of ice ranging from about $\frac{1}{16}''$ in diameter to as much as 4 or 5". Hail stones are often transparent, but more frequently translucent, being formed of alternate layers of clear and opaque ice. Hail usually falls from thunderstorms.

7. Snow pellets are whitish, opaque, usually round pellets with a structure something like snow. They are compressible and often burst when striking a hard surface. They are a shower type of hydrometeor.

8. Small hail is small ice particles consisting of a nucleus of soft hail with a shell of ice about it. They do not rebound when striking a hard surface, are not compressible or crisp. Usually small hail is wet and falls at a temperature near or slightly above 32° F. (P.E.K.)

HYDROMETER. A well-known device for making quick, approximate measurements of the **densities** or **specific gravities** of liquids. It consists essentially of a long, slender glass float weighted at the lower end, and provided with a scale so graduated that the depth to which the instrument sinks in the liquid indicates the specific gravity by direct reading on the scale. The numbering of the scale necessarily increases from the top downward. The instrument is sometimes so proportioned that the numbering begins with unity at the top, being applicable only to liquids of the density of water or heavier; in others it increases from the top to unity at the lower end and is thus for use with light liquids only; in some cases it has unity at the middle and applies to both light and heavy liquids. To be very sensitive, the stem carrying the scale must be slender. Obviously the scale intervals corresponding to equal increments of density cannot be equal if the stem is of uniform diameter; in fact, they are inversely proportional to the square of the density, being much smaller at the lower than at the upper end of the scale. To avoid this, some hydrometers are graduated with an arbitrary scale having uniform spacing, like that of Baumé, the readings of which may be converted into true density by means of a table. Nicholson devised a hydrometer for measuring the densities of small solids, the specimen being placed on the hydrometer first above and then below the

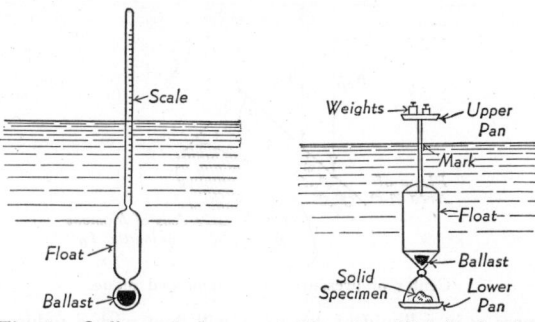

Fig. 1. Ordinary hydrometer for liquids.

Fig. 2. Nicholson's hydrometer for small solids.

surface of the water in which the instrument floats, and its volume deduced from the resulting change in buoyant force. (L.D.W.)

HYDROPHYTES OR WATER PLANTS.
These plants, which can grow only where there is an abundance of water, form an interesting and very distinct group of plants. Those hydrophytes which are members of the **Spermatophyta** are probably all plants which have reverted to an aquatic habitat, since it is assumed that all land plants originally evolved from plants growing in water. The reverting land plants may first have become marsh plants and then gradually developed into definite hydrophytes.

An aqueous environment presents conditions far more constant than an aerial one does. In the tropics such conditions permit the plants to grow throughout the year. In colder regions there is a definite winter period when growth must cease. Many hydrophytes of temperate regions merely sink to the bottom and remain dormant during the winter. Others accumulate food reserves in **rhizomes,** which remain rooted in the bottom and renew growth in the spring. Still others form winter buds, consisting of large apical buds surrounded by many closely packed leaves containing much reserve food material. A few hydrophytes form small tubers.

The stems of hydrophytes contain a very small amount of **vascular** tissue, since support is largely afforded by the water and conduction is not a great problem. In many of these plants the stem is very porous so that the plant floats in the water. The leaves of hydrophytes are of two types. Submerged leaves are thin and of various shapes; some, like **eel grass** leaves, are long and ribbon-like; others, like **bladderworts,** are finely dissected; while others are reduced to awl-shaped structures of small size. Floating leaves are usually large, undivided, and with **stomata** on the upper surface. Many hydrophytes show interesting leaf changes as the plant grows; leaves formed under water are finely dissected; when the stem emerges from the water it bears entire leaves. By changing the water level as the stem grows it is possible to cause repeated alternation of dissected and entire leaves. Once formed, the nature of the leaf cannot be changed.

The submerged surface of most hydrophytes is slimy with a secretion of a mucilaginous substance which probably protects the plant against excessive diffusion of substances present in the cell-sap and also against external enemies.

Reproduction in hydrophytes occurs both asexually and sexually. The flowers of nearly all hydrophytes are wind and insect pollinated, apparently a hangover from the time when they lived on land. A few have become modified to such an extent that **pollination** takes place on the surface of the water, the pollen floating about thereon and eventually reaching the stigma. A small number of hydrophytes are pollinated under water.

Nearly all **Algae** and many **Fungi** are hydrophytes, so also are several **Bryophytes** and a few **Pteridophytes.** In the flowering plants there are many water plants, such as the water lilies, bladderworts, eel grass and pond-weeds. Many are very interesting plants; several are aquarium plants serving to oxygenate the water; few are of any economic value. (R.M.W.)

HYDROPONICS.
The term hydroponics has been suggested as a technical designation for the soilless culture of plants. The possibilities of such a method of growing plants have received considerable attention in recent years. The practice of the soilless culture of plants is an outgrowth of the laboratory techniques of sand cultures and solution cultures, long used by plant physiologists. In these techniques plants are grown with their roots immersed in a solution containing certain necessary mineral salts or rooted in a sand medium which is kept moistened with such a solution. The soilless culture of plants is similar in principle but larger in scale. In one version of the method the plants are supported in a matrix of peat, excelsior or some similar material on a wire screen with their roots dipping into the solution below. Aeration of the solution must also be provided if the best results are to be obtained. In another method the plants are rooted in a medium of sand, gravel, or some similar material contained in a shallow tank into which the solution is automatically pumped at suitable intervals. Between pumpings the solution gradually drains back into a reservoir tank.

The elements known to be necessary in chemically detectable amounts for the development of plants are carbon, oxygen, hydrogen, nitrogen, phosphorus, sulfur, potassium, magnesium, calcium, iron, manganese, boron, copper, zinc, and perhaps molybdenum. The first three of these elements are obtained by the plant from atmospheric gases or from water absorbed from the soil. The others are all absorbed in the form of mineral salts from the soil. Of the elements absorbed as salts the iron, manganese, boron, copper, zinc, and molybdenum are required in relatively minute quantities and are often called micronutrient elements. The principal elements which must be provided in the form of dissolved salts in hydroponic techniques, therefore, are nitrogen, phosphorus, sulfur, potassium, calcium, and magnesium.

Numerous solutions have been devised for use in the solution or sand culture of plants on both large and small scales. One solution which has been widely and successfully used for such purposes is made as follows: To each liter of water (preferably distilled or rain) add 1 molar solution of the following salts as indicated: 1 cc. KH_2PO_4, 5 cc. KNO_3, 5 cc. $Ca(NO_3)_2$, and 2 cc. $MgSO_4$. To this solution then add 1 cc. per liter of a solution of micronutrients made as follows: 2.5 g. H_3BO_3, 1.8 g. $MnCl_2 \cdot 4H_2O$, 0.1 g. $ZnCl_2$, 0.05 g. $CuCl_2 \cdot 2H_2O$, and 0.075 g. MoO_3 per liter of distilled water. Also add to each liter of the solution made as described above, 1 cc. of a 0.5% solution of iron tartrate. The solution must be replaced with a fresh one at suitable intervals, and it is often necessary to add more of the iron solution between replacements.

Crop yields of at least some kinds of plants fully equal to those obtained on fertile soils can be obtained by hydroponic methods. The raising of crops by this method, however, probably will prove to be economically sound only for certain intensive types of agriculture or under certain special conditions. Some greenhouse floricultural and horticultural crops are now being grown successfully by this method. In regions where there is no soil, or the soil is extremely infertile, but in which the climate is suitable to the development of plants, it seems likely that hydroponic techniques may prove useful. They have been used with some success, for example, on some of the coral islands of the Pacific Ocean. (B.S.M.)

HYDROQUININE. Alkaloids.

HYDRORHIZAE. The root-like processes by which colonies of **hydrozoan** coelenterates are attached to a supporting surface. This term is in the plural. The singular is hydrorhiza. (A.W.L.)

HYDROSPHERE. The discontinuous envelope of **water,** both fresh and salt, which covers a major portion of the lithosphere. The bulk of the hydrosphere is contained within the deeper depressions of the surface of the earth. These depressions are termed ocean basins, and the water within them, oceans. Since the ocean basins are not large enough to hold the entire hydrosphere, seas are formed by the overflow of the oceanic waters on the continents. Geologists classify seas as epicontinental (epeiric) or relict. Technically, lakes, rivers and underground waters are also part of the hydrosphere. In general therefore, the term hydrosphere is used mainly to distinguish the watery covering of the earth from the lithosphere on which, and in which, in part, it rests. (R.M.F.)

HYDROSTATIC EQUATION. Atmospheric Structure.

HYDROSTATIC PRESSURE. The pressure created by a superimposed layer of a liquid is hydrostatic pressure. As the science of **hydraulics** is commonly understood, the fluid is water, and hydrostatic pressure is considered as pressure due to the existence above the point of measurement of the pressure of a head of water. The intensity of hydrostatic pressure is commonly expressed as lbs. per sq. in. A head of $2.31'$ of fresh water creates a hydrostatic pressure of 1 lb. per sq. in. At a given depth of immersion in water, the pressure acts with equal intensity in all directions, that is, hydrostatic pressure is not directional in effect. Hydrostatic pressures are measured by means of **pressure gauges** of the Bourden tube type, or **manometers** of the U-tube type. (F.T.M.)

HYDROSTATICS. This branch of physics has to do with the **equilibrium** of liquids and the laws relating to liquid **pressure.** A study of these laws makes it clear that the components of pressure in a liquid fall naturally into two classes, according to the way in which they are produced; namely, (1) pressures due to forces applied externally, as by the atmosphere or by the piston of a pump, and (2) those due to causes operating throughout the body of liquid, such as gravity or inertia.

Pascal's law applies only to the first class, and states that any pressure in an enclosed liquid, originating in forces applied at its boundary, is communicated with unaltered intensity to all parts of the liquid. A familiar illustration of this fundamental law is the hydraulic press, which consists of two communicating cylinders, usually of different diameter, fitted with pistons, the force acting upon one piston and the force exerted by the other being in proportion to their areas.

The pressure in an enclosed liquid due to its own weight, on the other hand, increases uniformly with the depth below its highest point, and is equal to the product of the depth by the weight per unit volume. For fresh water, the pressure at depth h ft. is $62.4 h$ lbs. per sq. ft. The pressure of water against a submerged plane area is equal to the average intensity of pressure times that area. (See **Center of Pressure.**)

$$h_{cp} = \frac{I_o}{A h_{cg}}.$$

Problems of flotation, draft, and buoyant stability always involve the density of the liquid and the volume and shape of the floating object. A floating body of

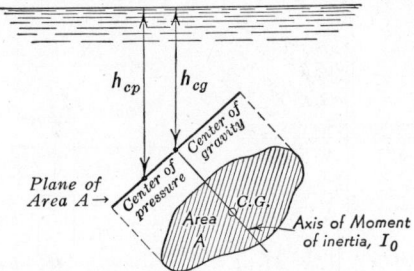

Center of pressure on a submerged plane.

mass m in a liquid of density ρ, will float with a volume v submerged, v determined by the relationship $v = \frac{m}{\rho}$. The buoyancy may be said to be the force which is equivalent to the weight of the liquid displaced by the submerged portion of the floating object. Buoyancy and weight do not, in general, act in the same vertical line. The weight acts at the **center of gravity** of the floating object, the buoyancy at the center of gravity of the displaced liquid, called the center of buoyancy. The relative positions of the buoyancy and the weight when a floating object is disturbed from an upright floating position, determines whether it is stable or unstable flotation. If the vertical drawn through the center of buoyancy passes above the center of gravity of the body, there is a righting moment, and the body is stable, whereas if it passes below the center of gravity, it is unstable in that the buoyancy tends to tip the object still further. The intersection of the line of buoyancy with the axis of symmetry of the floating body is the metacenter, and the distance from the metacenter to the center of gravity is the metacentric height. The latter is used to measure the stability of a hull. Another case of flotation is illustrated by the balance of the **hydrometer.** It is apparent from the above that with a given weight, the volume of immersion varies inversely with the density of the liquid. In other words, a floating body rides higher in a denser liquid. This fact is put to use in the hydrometer, which has a given weight and which is immersed in fluids to measure their density. The hydrometer is calibrated to read the volume submerged directly in terms of density of the liquid.

An important general principle of hydrostatics is that which determines the free liquid surface in equilibrium. The direction of the surface at any point is perpendicular to the resultant of all forces acting upon a particle at that point. Thus, if only gravity is acting, the surface is horizontal or "level"; but if there are capillary forces, or if the external pressure is not uniform, the surface is inclined. An interesting case is that of a liquid rotating uniformly in a cylindrical tub; the surface then assumes the form of a paraboloid of revolution, symmetrical about the vertical axis. (See **Capillarity, Barometer, Pumps, Pressure Gauges,** etc.). (F.T.M.)

HYDROSULFURIC ACID AND SULFIDES. Hydrosulfuric acid (H_2S) is a colorless solution formed when hydrogen sulfide is dissolved in water, decomposes slowly in the presence of air forming sulfur. Hydrogen sulfide is removed completely from solution by boiling. A strong reducing agent, usually with the separation of **sulfur,** e.g., with **nitric acid** (nitric oxide formed), with concentrated **sulfuric acid** (**sulfur dioxide** formed), with **chlorine** (**hydrochloric acid** formed), with permanganate (**manganous** formed in the presence of acid), dichromate (**chromic** formed in the presence of acid).

Sulfides. (See **Sulfur.**) (R.K.S.)

HYDROSULFUROUS ACID AND HYDROSUL-FITES. Commercial names far hyposulfurous acid and hyposulfites. Not to be confused with thiosulfate, e.g., sodium thiosulfate, $Na_2S_2O_3$, photographers' "hypo." (R.K.S.)

HYDROTHERMAL METAMORPHISM. Metamorphism.

HYDROTROPISM. Movement in Plants.

HYDROXYAZO COMPOUNDS. Azo-, Diazo-, and Related Compounds.

HYDROXYL IONS. Reactions Involving Recombinations of Ions.

HYDROXYLAMINES AND OXIMES. Hydroxylamine (H_2NOH) is a white, odorless solid, melting point 33° C., boiling point 56° C. at 22 mm., explosive, soluble in all proportions in water or alcohol. Hydroxylamine is: 1. A weak **base** forming with **acids** soluble salts that decompose more or less violently when heated, e.g., hydroxylamine hydrochloride (hydroxylammonium chloride, $H_2NOH \cdot HCl$), melting point 151° C., nitrate ($H_2NOH \cdot HNO_3$), hemisulfate $H_2NOH \cdot \frac{1}{2}H_2SO_4$. Dihydroxylamine oxalate and trihydroxylamine phosphate are insoluble in water. Hydroxylamine hydrochloride is

OXIMES OF	MELTING POINT, °C.	BOILING POINT, °C.
Benzophenone	143	
(diphenylketoxime)		
Alpha-Benzil monoxime	134	
Beta-Benzil monoxime	113	
Alpha-Benzil (syn)	236	
(benzyldioxime)		
Beta-Benzil (anti)	206 dec.	
(benzildioxime)		
Gamma-Benzil (anti)	165 (anhyd.)	
(benzildioxime)		
Phloroglucinol	155 expl.	
(phloroglucinol-1,3,5-trioxime)		
Quinonemonoxime	125	144 dec.
Quinone-1,4-dioxime	240 appr. dec.	
Anthraquinone-9-oxime	224 dec.	

Hydroxylamine hydrochloride may be made by reduction of **nitric oxide** gas when the latter is passed into tin plus hydrochloric acid heated, then evaporating and dissolving hydroxylamine in alcohol, which may be distilled off. All the oxides and oxygen acids of **nitrogen,** except nitrous oxide, may be reduced to yield detectable amounts of hydroxylamine, ethyl nitrate (CH_3ONO_2) being used to advantage. Better yields of hydroxylamine are reported as having been obtained by electrolytic reduction of nitrites or nitric acid.

	BETA, OR N-	BETA, BETA, OR N, N-	ALPHA, OR O-
1. H $\quad\quad$>NOH H Hydroxylamine M.P. 33° C.	2. C_6H_5 $\quad\quad$>NOH H Beta-phenylhydroxylamine M.P. 81° C.	3. C_6H_5 $\quad\quad$>NOH C_6H_5 Beta-beta-diphenylhydroxylamine M.P. 60° C. dec.	4. H $\quad\quad$>NOC₆H₅ H Alpha-phenylhydroxylamine

OTHER HYDROXYLAMINES	FORMULA	MELTING POINT °C.	BOILING POINT °C.
5. Beta-Methylhydroxylamine	CH_3NHOH	42	62 (15 mm.)
6. Alpha-Methylhydroxylamine	H_2NOCH_3		
7. Beta-ethylhydroxylamine	C_2H_5NHOH	59 dec.	
8. Alpha-ethylhydroxylamine	$H_2NOC_2H_5$		68
9. Beta-benzylhydroxylamine	$C_6H_5CH_2NHOH$	57	123 (50 mm.)
10. Alpha-benzylhydroxylamine	$H_2NOCH_2C_6H_5$		118 (30 mm.)

soluble in alcohol. 2. A weak acid forming with bases soluble salts, e.g., sodium hydroxylamite (H_2NONa). Hydroxylamine salt solution is a powerful reducing agent, more especially in alkaline than in acid solution, for example, **cupric** salt solutions changed to cuprous oxide, **silver** salt solutions to silver, **mercuric** chloride solution to mercurous chloride, ferric salt solutions (in acid) to ferrous. **Ferrous** hydroxide in sodium hydroxide is, however, oxidized by hydroxylamine to ferric hydroxide plus ammonia.

Hydroxylamine reacts with carbonyl group (=CO) of **aldehydes,** ketones or quinones, yielding oximes, white solids, of definite melting point and used in identification of aldehydes and ketones, e.g., acetaldehyde oxime ($CH_3CH:NOH$).

OXIMES OF	MELTING POINT, °C.	BOILING POINT, °C.
Acetaldehyde	47	114
(acetaldoxime)		
Acetone	60	136
(dimethylketoxime)		
Acetophenone	59	dec.
(methylphenylketox-ime)		
Alpha-Benzaldehyde	35	118 (15 mm.)
(anti) (benzaldoxime)		
Beta-Benzaldehyde (syn)	129	

Beta-phenylhydroxylamine, N-phenylhydroxylamine, is a white solid, slightly soluble in water, very soluble in alcohol or ether, forms salts with acids, e.g., beta-phenylhydroxylamine hydrochloride ($C_6H_5CHNOH \cdot HCl$), upon exposure to air the water solution forms azobenzene ($C_6H_5N:NC_6H_5$). Beta-phenylhydroxylamine reacts (1) with oxidizing agents, such as **chromic** acid or ferric chloride, to form nitrosobenzene (C_6H_5NO), (2) with reducing agents, such as tin plus hydrochloric acid, to form **aniline** ($C_6H_5NH_2$), (3) with alkaline **cupric** salt solution (Fehling's solution) at room temperature to form cuprous oxide, (4) with ammonio-**silver** salt solution (Tollen's solution) at room temperature to form silver, (5) in the presence of **hydrochloric** acid to form para-aminophenol ($HO \cdot C_6H_4 \cdot NH_2(1,4)$). (See **Azo-, Diazo-, and Related Compounds.**)

Beta-phenylhydroxlamine is formed by reduction of nitrobenzene (1) by zinc and calcium chloride or ammonium chloride solution, (2) by electrolysis in acetic acid plus sodium acetate solution.

Diphenylhydroxylamine is prepared by reaction of nitrosobenzene and phenylmagnesium bromide in anhydrous ether, followed by treatment with water (magnesium hydroxybromide also formed.) (R.K.S.)

HYDROZOA. A class of the phylum **Coelenterata** composed chiefly of small animals without common names. Many species are colonial and the colonies of a few, such as the Portuguese man-of-war, are quite large.

The class differs from the other coelenterates in the occurrence of both **hydroid polyps** and **medusae** in the same species, usually in alternating generations. In many colonies additional specialization occurs among the polyps for the performance of different functions; the gonozooids, gastrozooids, and **dactylozooids** of the siphonophores are such individuals. Hydrozoan medusae differ from the jellyfishes and are called **medusoids.** In some cases they are specialized forms which remain attached to the colony and show no resemblance to medusae, but the free-swimming forms differ from medusae only in details of structure. The medusoids are sexual reproductive individuals.

Relatively few species of hydrozoans live in fresh water. **Hydra,** the most widely known genus, includes a number of species without a medusa stage. The polyps are solitary and carry on both asexual and sexual reproduction. Several fresh-water medusae for which no polyp stage has been discovered are also known from lakes in Europe, Africa, and the Americas. The marine species are numerous.

The following orders are recognized:

Order Hydrariae. Small cylindrical polyps. Solitary. Mostly in fresh water.

Order hydrocorallinae. Marine colonial forms with a calcareous covering which sometimes resembles coral.

Order tubulariae. Mostly colonial forms. Polyps without a protective sheath (hydrotheca). Medusae free or sessile, often ovoid or bell-shaped.

Order Campanulariae. Colonial species with polyps of two kinds, **hydranths** and **blastostyles.** A sheath called the perisarc invests the colony and extends about each individual polyp.

Order Trachomedusae. Polyps minute where known, in some species never discovered. Medusae free-swimming. Marine and fresh-water.

Order Narcomedusae. Marine species without known polyps. Medusae with lobed margin.

Order Siphonophora. Free-swimming colonial species. Polyps of several specialized types, in many species borne by a stalk which is expanded at one end to form a hollow float, the **pneumatophore.** Portuguese man-of-war and others. (A.W.L.)

HYENA. Mammalia, Carnivora. *Hyaena.* Large animals slightly resembling wolves but more closely related to the civets. They live in Africa, India, and the eastern Mediterranean countries. They are heavily built, with massive heads and disproportionately long front legs. Hyenas are said to be cowardly and skulking but at times vicious. (A.W.L.)

HYGROGRAPH. Meteorological Instruments.

HYGROMETERS. Any apparatus for measuring atmospheric **humidity,** either absolute or relative, is called a hygrometer. The most common relative humidity instrument is the so-called wet-and-dry-bulb thermometer, or psychrometer. Two exactly similar mercury thermometers are mounted side by side. The bulb of one is wrapped in a wick dipping into water, and is thereby kept wet. This bulb is cooled by the evaporation, the rate of which, and hence the resultant cooling, depends in a definite manner upon the relative humidity of the air. A table accompanies the instrument, giving the humidity in terms of the actual air temperature as indicated by the dry-bulb thermometer, and the difference of temperature between the wet bulb and the dry bulb.

The dew point hygrometer is an apparatus for indicating the **dew point,** from which the relative humidity can be calculated when the air temperature is known. It usually consists of a small metallic cup or thimble, the outside of which is polished like a mirror. The cup contains a small quantity of ether, into which dips a thermometer. Air is bubbled through the ether, whose rapid evaporation lowers the temperature of the cup; when it reaches the dew point of the surrounding air, a film of moisture suddenly appears upon its surface.

Various hygrometers have been designed which depend for their operation upon the longitudinal shrinkage of organic fibers or hairs in damp air, being arranged so that the shrinkage causes a pointer to move over a dial. These instruments, called absorption hygrometers, have not, however, proved reliable and do not give consistent readings. (L.D.W.)

HYGROSCOPE. Meteorological Instruments.

HYMEN. A layer of membranous tissue which almost completely closes the outer opening of the **vagina** during virginity. It may be quite tough or very delicate in structure. In rare cases it completely closes the opening. This latter condition requires surgical interference when menstruation begins. (R.S.M.)

HYMENOPTERA. One of the large orders of **insects,** including **ants, bees, wasps, saw flies,** and many species without common names. The mouth is formed for biting or for biting and sucking and the wings, when present, are four in number and membranous. Metamorphosis is complete. The order includes plant-eating, parasitic, and predacious species, and in the ants and bees displays some of the finest examples of social organization. The order includes about 70,000 species.

Owing to its extent and diversity this division of the insects includes many species of economic importance. Some of the **saw flies** and **gall wasps** are harmful to plants and on the other hand the fig insects are beneficial and the galls produced by some gall wasps are of commercial value. Many parasitic species are of undoubted value in holding in check important insect pests. Ants are sometimes very troublesome and the large carpenter bee sometimes damages wood in construction. The most important single species is the honey-bee, which is of great value as a producer of honey and wax and in the cross pollination of fruit trees. (A.W.L.)

HYOCYAMUS. Potato Family.

HYOSCINE. Alkaloids.

HYOSCYAMINE. Alkaloids.

HYPABYSSAL. A general term sometimes used by structural geologists and petrologists to designate those igneous rocks such as **sills** and **dikes** which have congealed under less pressure than the **plutonic** or deep-seated rocks, but under greater pressure than the effusive rocks (lavas). (R.M.F.)

HYPERBOLA. The hyperbola is one of the conic sections, and may be cut from a right circular cone by a plane which cuts both nappes of the cone.

The hyperbola is the **locus** of a point which moves so that the difference of its distances from two fixed points is constant.

The fixed points are called the foci. The curve consists of two open branches. It is symmetric about the line through the foci, and about a line through the center perpendicular to the preceding line; the center is the point midway between the foci. It is also symmetric about the center. (Fig. 1.)

The segment (or length) between the vertices is called the transverse axis; the vertices are the points where the curve is cut by the line through the foci. The constant in the definition of the hyperbola is equal to the transverse axis length. If this transverse axis is $2a$, and the distance between the foci is $2c$, then if $b^2 = c^2 - a^2$, $2b$ is called the conjugate axis (or sometimes the segment of this length symmetric about the center, and along the line through the center perpendicular to the line through the foci).

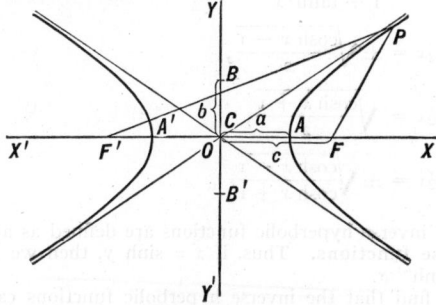

Fig. 1. Hyperbola.

The latus rectum (or focal width) is the chord through the focus perpendicular to the transverse axis.

The eccentricity of the hyperbola is the ratio c/a; it is > 1.

If the origin of a system of **rectangular coordinates** is taken at the center of a hyperbola, and the X-axis along the transverse axis, the equation of the curve is

$$\frac{x^2}{a^2} - \frac{y^2}{b^2} = 1,$$

where a and b are the semi-transverse and semi-conjugate axes.

The equation $\frac{x^2}{b^2} - \frac{y^2}{a^2} = 1$ $(a > b)$ represents a hyperbola with foci on the Y-axis.

The lines $x = \pm\frac{a}{e}$ are called the directrices of the hyperbola $\frac{x^2}{a^2} - \frac{y^2}{b^2} = 1$. They have the property that the ratio of the distance of any point of the hyperbola from a focus to the perpendicular distance of the point from the corresponding directrix is equal to the eccentricity.

The equation $\frac{(x-h)^2}{a^2} - \frac{(y-k)^2}{b^2} = 1$ represents a hyperbola with center at (h, k) and axes of symmetry $x = h$, $y = k$.

The hyperbola $\frac{x^2}{a^2} - \frac{y^2}{b^2} = 1$ has the lines $\frac{x}{a} + \frac{y}{b} = 0$ and $\frac{x}{a} - \frac{y}{b} = 0$ as **asymptotes**. (Fig. 2.)

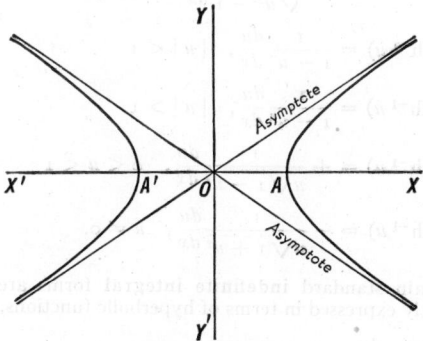

Fig. 2. Hyperbola.

The branches of the hyperbola approach indefinitely near its asymptotes as the tracing point recedes to infinity.

The asymptotes serve as a convenient guide in drawing the hyperbola.

Two hyperbolas are called conjugate hyperbolas if the transverse and conjugate axes of one are, respectively, the conjugate and transverse axes of the other. If $\frac{x^2}{a^2} - \frac{y^2}{b^2} = 1$ is one hyperbola, the conjugate is $\frac{y^2}{b^2} - \frac{x^2}{a^2} = 1$.

Two conjugate hyperbolas have the same asymptotes.

A construction (by continuous motion) based on the definition is the following (Fig. 3): Fasten thumb

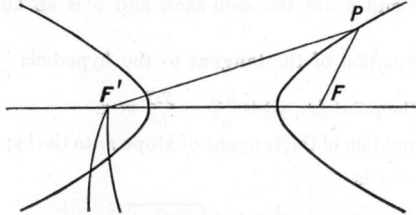

Fig. 3. Construction of hyperbola.

tacks at the foci. Pass over F' and around F a string whose ends are held together. If a pencil is tied to the string at P, and both strings are pulled in or let out the same length, the point P will describe a hyperbola. If the transverse axis is to be $2a$, the strings must be adjusted at the start so that the difference between PF' and PF is $2a$.

An equilateral or rectangular hyperbola is one in which the transverse and conjugate axes are equal. If its center is at the origin and its transverse axis is along the X-axis, its equation is $x^2 - y^2 = a^2$, where a equals the semi-transverse or conjugate axis.

Its asymptotes bisect the quadrants and are therefore perpendicular.

If the hyperbola be referred to its asymptotes as rectangular coordinate axes, its equation becomes $2xy = a^2$. (Fig. 4.)

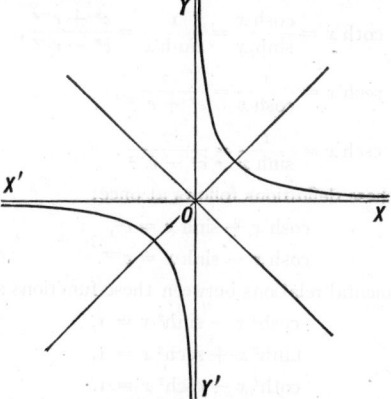

Fig. 4. Hyperbola.

A construction often used for an equilateral hyperbola when the asymptotes and one point A on the curve are given is as follows (Fig. 5):

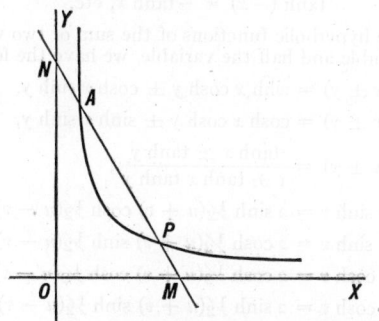

Fig. 5. Construction of hyperbola.

Let OX and OY be the asymptotes and A the given point. Draw any line through A to meet OX at M and OY at N. Lay off $MP = AN$. Then P is a point on the required hyperbola.

The simplest **parametric equations** of a hyperbola are:

$$x = a \sec\phi, \quad y = b \tan\phi,$$

where a and b are the semi-axes, and ϕ is an auxiliary angle.

The equation of the **tangent** to the hyperbola $\dfrac{x^2}{a^2} - \dfrac{y^2}{b^2}$

$= 1$ at the point (x_1, y_1) is $\dfrac{x_1 x}{a^2} - \dfrac{y_1 y}{b^2} = 1$.

The equation of the **tangent** of **slope** m to the hyperbola $\dfrac{x^2}{a^2} - \dfrac{y^2}{b^2} = 1$ is

$$y = mx \pm \sqrt{a^2 m^2 - b^2}.$$

Applications of the hyperbola are: Boyle's law in physics, and sound ranging. (L.L.S.)

HYPERBOLIC FUNCTIONS. In many applications the exponential functions e^x and e^{-x} occur in certain combinations which have many analogies to the trigonometric functions. Their geometric representation is related to a **rectangular hyperbola** in a way similar to that in which the trigonometric functions are related to a **circle**. These combinations are therefore called hyperbolic functions.

The hyperbolic functions are defined as follows:

$$\sinh x = \tfrac{1}{2}(e^x - e^{-x}),$$
$$\cosh x = \tfrac{1}{2}(e^x + e^{-x}),$$
$$\tanh x = \frac{\sinh x}{\cosh x} = \frac{e^x - e^{-x}}{e^x + e^{-x}},$$
$$\coth x = \frac{\cosh x}{\sinh x} = \frac{1}{\tanh x} = \frac{e^x + e^{-x}}{e^x - e^{-x}},$$
$$\operatorname{sech} x = \frac{1}{\cosh x} = \frac{2}{e^x + e^{-x}},$$
$$\operatorname{csch} x = \frac{1}{\sinh x} = \frac{2}{e^x - e^{-x}}.$$

From these definitions follows at once:

$$\cosh x + \sinh x = e^x,$$
$$\cosh x - \sinh x = e^{-x}.$$

Fundamental relations between these functions are:

$$\cosh^2 x - \sinh^2 x = 1,$$
$$\tanh^2 x + \operatorname{sech}^2 x = 1,$$
$$\coth^2 x - \operatorname{csch}^2 x = 1.$$

For the negative of the variable we have:

$$\sinh(-x) = -\sinh x,$$
$$\cosh(-x) = \cosh x,$$
$$\tanh(-x) = -\tanh x, \text{ etc.}$$

For the hyperbolic functions of the sum of two variables and of double and half the variable, we have the formulae:

$$\sinh(x \pm y) = \sinh x \cosh y \pm \cosh x \sinh y,$$
$$\cosh(x \pm y) = \cosh x \cosh y \pm \sinh x \sinh y,$$
$$\tanh(x \pm y) = \frac{\tanh x \pm \tanh y}{1 \pm \tanh x \tanh y};$$
$$\sinh u + \sinh v = 2 \sinh \tfrac{1}{2}(u + v) \cosh \tfrac{1}{2}(u - v),$$
$$\sinh u - \sinh v = 2 \cosh \tfrac{1}{2}(u + v) \sinh \tfrac{1}{2}(u - v),$$
$$\cosh u + \cosh v = 2 \cosh \tfrac{1}{2}(u + v) \cosh \tfrac{1}{2}(u - v),$$
$$\cosh u - \cosh v = 2 \sinh \tfrac{1}{2}(u + v) \sinh \tfrac{1}{2}(u - v);$$
$$\sinh 2x = 2 \sinh x \cosh x,$$

$$\cosh 2x = \cosh^2 x + \sinh^2 x = 2 \cosh^2 x - 1$$
$$= 2 \sinh^2 x + 1.$$
$$\tanh 2x = \frac{2 \tanh x}{1 + \tanh^2 x};$$
$$\sinh \tfrac{1}{2}x = \pm \sqrt{\frac{\cosh x - 1}{2}},$$
$$\cosh \tfrac{1}{2}x = \sqrt{\frac{\cosh x + 1}{2}},$$
$$\tanh \tfrac{1}{2}x = \pm \sqrt{\frac{\cosh x - 1}{\cosh x + 1}}.$$

The inverse hyperbolic functions are defined as are all **inverse functions.** Thus, if $x = \sinh y$, then we write $y = \sinh^{-1} x$.

We find that the inverse hyperbolic functions can be expressed in terms of **natural logarithms**, as follows:

$$\sinh^{-1} x = \log(x + \sqrt{1 + x^2}),$$
$$\cosh^{-1} x = \log(x \pm \sqrt{x^2 - 1}), \quad x \geqq 1,$$
$$\tanh^{-1} x = \tfrac{1}{2} \log \frac{1 + x}{1 - x}, \quad -1 < x < 1,$$
$$\coth^{-1} x = \tfrac{1}{2} \log \frac{x + 1}{x - 1}, \quad x < -1, \quad x > 1,$$
$$\operatorname{sech}^{-1} x = \log \frac{1 \pm \sqrt{1 - x^2}}{x}, \quad 0 < x < 1,$$
$$\operatorname{csch}^{-1} x = \log \frac{1 + \sqrt{1 + x^2}}{x}.$$

The **derivatives** of the hyperbolic functions and of the inverse hyperbolic functions are:

$$\frac{d}{dx}(\sinh u) = \cosh u \frac{du}{dx}, \quad \frac{d}{dx}(\cosh u) = \sinh u \frac{du}{dx},$$
$$\frac{d}{dx}(\tanh u) = \operatorname{sech}^2 u \frac{du}{dx}, \quad \frac{d}{dx}(\coth u) = -\operatorname{csch}^2 u \frac{du}{dx},$$
$$\frac{d}{dx}(\operatorname{sech} u) = -\operatorname{sech} u \tanh u \frac{du}{dx},$$
$$\frac{d}{dx}(\operatorname{csch} u) = -\operatorname{csch} u \coth u \frac{du}{dx};$$
$$\frac{d}{dx}(\sinh^{-1} u) = \frac{1}{\sqrt{u^2 + 1}} \frac{du}{dx},$$
$$\frac{d}{dx}(\cosh^{-1} u) = \pm \frac{1}{\sqrt{u^2 - 1}} \frac{du}{dx}, \quad |u| > 1$$
$$\frac{d}{dx}(\tanh^{-1} u) = \frac{1}{1 - u^2} \frac{du}{dx}, \quad |u| < 1$$
$$\frac{d}{dx}(\coth^{-1} u) = \frac{1}{1 - u^2} \frac{du}{dx}, \quad |u| > 1$$
$$\frac{d}{dx}(\operatorname{sech}^{-1} u) = \pm \frac{1}{u\sqrt{1 - u^2}} \frac{du}{dx}, \quad 0 < u < 1$$
$$\frac{d}{dx}(\operatorname{csch}^{-1} u) = -\frac{1}{u\sqrt{1 + u^2}} \frac{du}{dx}, \quad u \neq 0.$$

Certain standard **indefinite integral** forms are conveniently expressed in terms of hyperbolic functions, thus:

$$\int \frac{du}{\sqrt{u^2 + a^2}} = \sinh^{-1} \frac{u}{a} + C, \quad a > 0$$
$$\int \frac{du}{\sqrt{u^2 - a^2}} = \pm \cosh^{-1} \frac{u}{a} + C, \quad u > a > 0$$
$$\int \frac{du}{a^2 - u^2} = \frac{1}{a} \tanh^{-1} \frac{u}{a} + C, \quad |u| < a, \quad a > 0$$

$$\int \frac{du}{a^2 - u^2} = \frac{1}{a} \coth^{-1} \frac{u}{a} + C, \quad |u| > a.$$

Expansion of the hyperbolic functions into **series** gives:

$$\sinh x = x + \frac{x^3}{3!} + \frac{x^5}{5!} + \frac{x^7}{7!} + \cdots,$$

$$\cosh x = 1 + \frac{x^2}{2!} + \frac{x^4}{4!} + \frac{x^6}{6!} + \cdots$$

The hyperbolic functions are related to the trigonometric functions by the following formulae, in which $i = \sqrt{-1}$;

$$\sin (ix) = i \sinh x, \quad \cos (ix) = \cosh x,$$

$$\sinh (ix) = i \sin x, \quad \cosh (ix) = \cos x.$$

The graphs of the hyperbolic functions are shown in Figs. 1–3.

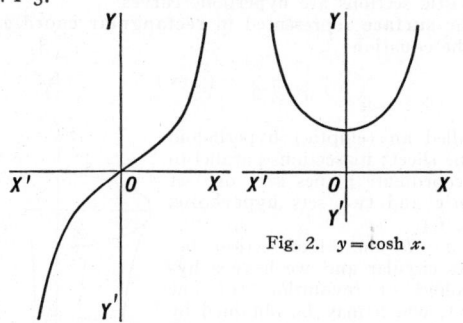

Fig. 1. $y = \sinh x$.

Fig. 2. $y = \cosh x$.

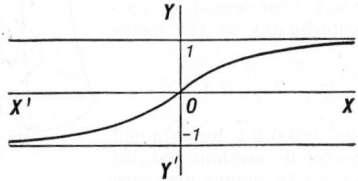

Fig. 3. $y = \tanh x$.

The hyperbolic functions are related to the hyperbola as follows (Fig. 4): Take the rectangular hyperbola

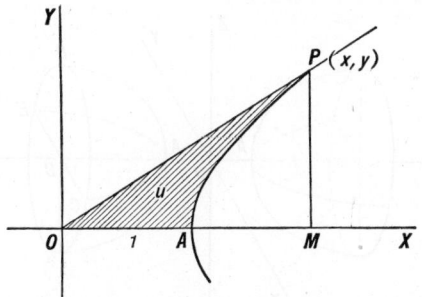

Fig. 4. Hyperbolic sector.

$x^2 - y^2 = 1$, let $P(x, y)$ be any point on the hyperbola, and let u be the area of the sector OAP between the curve, the X-axis and OP. Then

$$x = \cosh 2u, \quad y = \sinh 2u.$$

The function $\tan^{-1} (\sinh x)$ is called the gudermannian of x and is denoted by gd x. Then

$$\frac{d}{dx} (\text{gd } x) = \text{sech } x, \quad \int \text{sech } x \, dx = \text{gd } x.$$

(L.L.S.)

HYPERBOLIC GEOMETRY. Non-Euclidean Geometry.

HYPERBOLIC LOGARITHMS. Logarithms.

HYPERBOLIC NAVIGATION. The term hyperbolic **navigation** is used to describe the general method for determining a **line of position** by measuring the difference in distance of the navigator from 2 stations of known position.

The difference in distance is determined by measuring the difference in time of arrival of signals transmitted from the 2 stations. Although a great variety of signaling methods are theoretically possible, at the present time only **radio** waves are used in hyperbolic navigation. One system, which uses continuous-wave signals, is known as **Decca**. **Loran** and **GEE** are systems using signals transmitted as pulses.

Since **electro-magnetic** waves travel with a speed of about 186,000 miles per sec., the difference of arrival times of the signals will be very small. The unit of time, which is used in all systems at present, is the microsecond (0.000001 sec.); and a difference of arrival of one of these units indicates a distance difference of 0.186 miles from the 2 transmitters.

From the very nature of the problem the transmission must be made from 2 stations, separated by distances ranging from 75 to 1200 miles, and the synchronization of the two signals is of fundamental importance. One station of each pair is known as the master, and the other as the slave. In some systems, e.g., GEE, 2 or more slave stations may be used with a single master. The cycle of transmission always begins at the master station and the signal travels out in all directions. The arrival of the master signal at the slave "triggers off" the slave. This signal in turn travels out in all directions. Operators are continually on watch at both stations, each monitering the signals for the other to detect the slightest variations of frequency of carrier waves, interval between pulse, characteristics of the signals, etc. At the slave station adjustable delay circuits are available so that, once the cycle has been started, the slave station may transmit simultaneously with the master, or be delayed by any desired amount.

Points of constant difference of time of arrival of the 2 signals will fall on spherical hyperbolas with the transmitters at the foci. For navigational purposes we need only to consider the lines of intersection of these surfaces with the surface of the earth. The total number of distinguishable lines in any system is equal to the time required for the signal to travel from the master to the slave and back again, divided by the smallest time interval that can be measured by the receiving equipment. This number is different for the various systems in use, being about a thousand for GEE and eight or ten thousand for Loran and Decca. Certain fundamental lines, representing integral multiples of distance or time difference, are superimposed on regular navigational charts. Such a family of curves is shown in Fig. 1. Tables are also published which contain the data for determining the fundamental lines. By graphical interpolation on the chart, or mathematical interpolation from the tables, the navigator can determine a hyperbolic line of position, using the observed difference in time of arrival of the signals, and the particular stations being used. The accuracy of this line varies from about 200 yards up to 2 miles, depending upon the distance of the observer from the base line between stations, and upon the type of equipment and system being used.

Although the navigator's equipment differs in details for GEE, Decca, and Loran, nevertheless the fundamental characteristics of all are the same. For all systems a simple single wire antenna is sufficient. The receiver is a good quality radio receiver which contains amplifiers of high fidelity. The indicating system must contain a timing system employing a crystal oscillator and this must be capable of maintaining stability to within a few parts in a million for periods of a few minutes. Manual controls are provided for adjusting the timing equipment

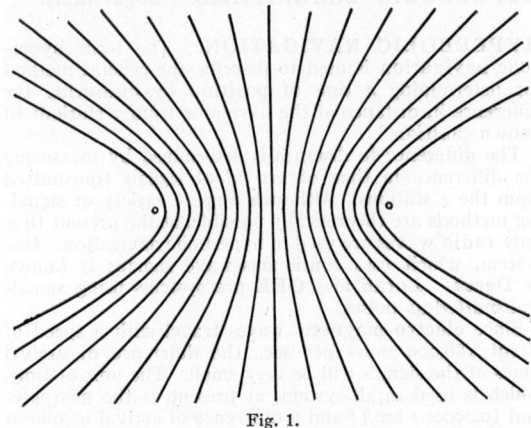

Fig. 1.

into exact step with the received signals. The signals themselves are "seen" on the viewing plate of a **cathode ray tube,** known as an oscilloscope or "view scope." Sweep generators, controlled by the oscillator, send traces across the face of the cathode tube. The time required for these traces to cross the viewing disk may be adjusted to various specified values for various stages in the process of measuring the difference of arrival time of the signals. The signals themselves appear as "pips" on the sweep traces. At low sweep speed these "pips" appear as short lines standing above or below the horizontal sweep line. With high sweep speed the pips are drawn out to indicate the actual characteristics of the received signals. Time indicators, governed by the oscillator, may be set up along the sweep lines and the time differences seen directly. For more accurate readings calibrated delay circuits may be used to bring the pips, magnified by high-speed sweeping, into coincidence. More detailed discussion of the methods of time measurement will be found in the articles dealing with the individual systems (see GEE, Decca, Loran).

Fig. 2 indicates a fix as determined by a hyperbolic navigation system employing 1 master station, M, and

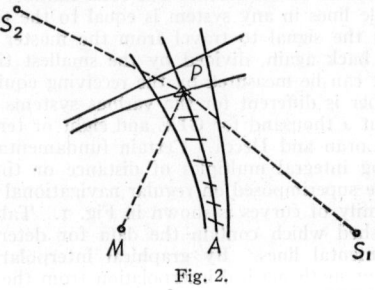

Fig. 2.

2 slaves, S_1 and S_2, as in GEE navigation. The diagram also indicates the value of hyperbolic navigation for "homing" on a point A. The navigator obtains a fix at P and then sets his indicating equipment so that the pips from M and S_1 are in coincidence on the view scope. He then heads his plane or ship so that these pips remain in coincidence and he must be following the hyperbola PA. By taking observations on the M S_2 pair at intervals he can determine the rate at which he is approaching his objective. (W.K.G.)

HYPERBOLIC PARABOLOID. Paraboloids.

HYPERBOLIC SPIRAL. The plane curve whose polar coordinate equation is $r\theta = a$ is called a hyperbolic spiral. It is shown in the accompanying figure. (L.L.S.)

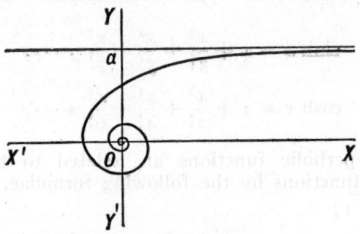

Hyperbolic spiral.

HYPERBOLOIDS. The hyperboloids are certain types of mathematical surfaces in which certain characteristic sections are hyperbolic curves.

The **surface** represented in **rectangular coordinates** by the equation

$$\frac{x^2}{a^2} + \frac{y^2}{b^2} - \frac{z^2}{c^2} = 1$$

is called an (elliptic) hyperboloid of one sheet; its sections parallel to the coordinate planes are: one set **elliptic** and two sets **hyperbolic.** (Fig. 1.)

If $a = b$, one set of sections becomes circular and we have a hyperboloid of revolution (of one sheet), which may be obtained by revolving a **hyperbola** about its conjugate axis.

The surface represented in rectangular coordinates by the equation.

$$\frac{x^2}{a^2} - \frac{v^2}{b^2} - \frac{z^2}{c^2} = 1$$

is called an (elliptic) hyperboloid of two sheets; its sections parallel to the coordinate planes are: one set elliptic and two sets hyperbolic. (Fig. 2.)

If $b = c$, one set of sections becomes circular, and we have an hyperboloid of revolution (of two sheets), which

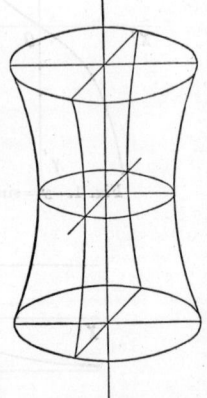

Fig. 1. Hyperboloid of one sheet.

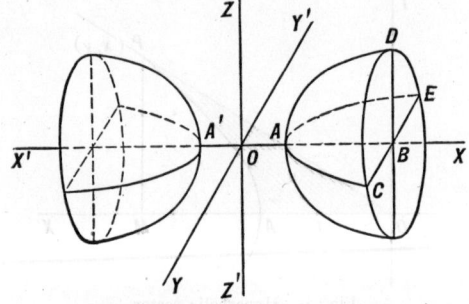

Fig. 2. Hyperboloid of two sheets.

may be generated by revolving an hyperbola about its transverse axis. (L.L.S.)

HYPERBOLOIDS OF REVOLUTION. Hyperboloids.

HYPERCOMPLEX NUMBERS. A hypercomplex number is a number represented by a form $x = x_1 e_1 + x_2 e_2 + \cdots + x_n e_n$, in which $x_1, x_2, \cdots, x_n$ are ordinary real or complex numbers, and $e_1, e_2, \cdots, e_n$ are n independent units. (L.L.S.)

HYPERFINE STRUCTURE. Many spectrum lines, when examined under high resolution, turn out to be "multiplets," i.e., to be composed of two, three, four, or more, closely packed, fine lines. (See **Atomic Spectra.**) This multiplicity is attributed to a quantization (successive, finite differences) in the energy of rotation of the atomic nucleus. Abrupt changes in this energy take place simultaneously with the much larger electronic transitions which determine the location of the multiplet as a whole. The rotational quantization thus results in slight differences in the total energy of the emitted quanta, and hence in a group of different frequencies. Atomic spectra often have series of multiplets. For example, the two D lines of sodium belong to a series of doublets. Hartley (1883) found that all the multiplets of a given series have the same frequency separations between their components. Kossel and Sommerfeld pointed out that the arc spectra of elements of even **atomic number** have odd multiplets and vice versa. Thus sodium (atomic number 11) exhibits doublets, and zinc (30), triplets. (See also **Zeeman Effect** and **Stark Effect.**) (L.D.W.)

HYPERGEOMETRIC PROBABILITY FUNCTION. Let

$$P_x = \frac{{}_{Tp}C_x \cdot {}_{Tq}C_{N-x}}{{}_{T}C_N},$$

$x = 0, 1, 2, \cdots N$ or T_p whichever is smaller, N is the number of **variates** in the **sample** chosen from a **population** of T variates consisting of Tp variates of one kind termed "success" and Tq items of another kind termed "failure." We may interpret P_x as the probability of obtaining x successes when sampling from such a population as described above. The mean of the hypergeometric probability function is Np, the variance is $Npq\left(\dfrac{T-N}{T-1}\right)$. The hypergeometric function reduces to the **Bernoulli probability function** when T approaches infinity, and when $N = 1$. In case T is large and p is small, the **Poisson probability function** may be used as an approximation to the hypergeometric probability function. The hypergeometric function has applications in **industrial statistics**, in sampling from a finite population, and has served as a basis of the **Pearson system** of **probability functions.** (L.A.A.)

HYPERHARMONIC SERIES. The infinite series $\sum_1^\infty 1/n^p$ is called the hyperharmonic series; it is **convergent** when $p > 1$ and is **divergent** when $p \le 1$. (L.L.S.)

HYPERMASTIGIDA. Mastigophora.

HYPERMETROPIA. Vision.

HYPERNASTY. Movement in Plants.

HYPEROARTIA. Cyclostoma.

HYPEROPIA (Farsightedness). A decrease in the refractive power of the eye so that rays of light coming from nearby objects cannot be focused on the retina, but those coming from distant objects can be properly refracted. As a result, close images are blurred while distant ones are clear. The cause may be too shallow an eyeball, or not a great enough convexity of the lens (see **Eye, Vision**). The defect is corrected by means of glasses having convex lenses. (D.M.H.)

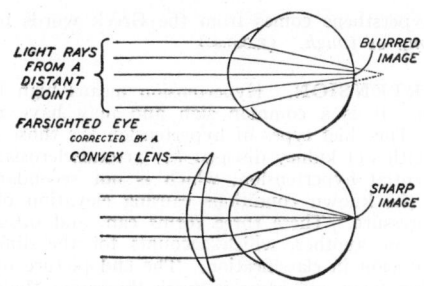

(*Carlson and Johnson's The Machinery of the Body, University of Chicago Press.*)

HYPEROTRETA. Cyclostoma.

HYPERSENSITIZING OF EMULSIONS. The sensitiveness of photographic emulsions may be increased, i.e., the time of exposure reduced, (1) by bathing in ammonia or hydrogen peroxide before exposure, (2) by prolonged exposure to a weak source of light, (3) by exposure, either before or after exposure in the camera, to mercury, or to acetic acid. The increase in sensitivity in all cases varies greatly with different emulsions, some showing no appreciable increase.

Hypersensitizing with ammonia is generally unsatisfactory because of the increased fog and the necessity of using the material immediately after treatment to avoid heavy fog and loss in speed. Hydrogen peroxide is too unreliable for practical use as a hypersensitizer.

The effective speed of some medium-speed panchromatic materials is increased from 2–4 times when exposed from 30–60 minutes to light of such intensity as to produce a **fog density** of about 0.2. The increase in speed is obtained only if the general exposure is given *after* exposure in the camera.

Exposure of either the exposed or unexposed emulsion in a sealed container to a small quantity of mercury for 36–72 hours at room temperature increases the speed by as much as 100% in some cases. While convenient, the utility of the method is limited as high-speed films show an appreciable increase in fog.

Muller and Bates of the Ansco Research Laboratories have discovered that many modern high-speed emulsions are increased in speed if exposed to 1% acetic acid and formic acid in equal quantities for 1–2 hours directly before development. The degree of hypersensitizing increases with the time up to a maximum which varies with the emulsion and the temperature. Prolonged exposure results in softening of the gelatin and an actual loss in speed.

The increase in speed is confined principally to the lower densities, becoming progressively less as the density increases. (C.B.N.)

HYPERSTHENE. The mineral hypersthene is an **orthorhombic pyroxene**, chemically a ferro-magnesian silicate, differing from enstatite in that the **iron** content is considerable (FeO being greater than 15%). It is usually found as a massive mineral, whose crystals tend to be prismatic or tabular in habit. It has a distinct prismatic **cleavage**; fracture, uneven; brittle; hardness, 5–6; specific gravity, 3.4–3.5; luster, pearly to somewhat metallic; color, brownish-green, brown, greenish-black to grayish-black; streak, grayish-brown; translucent to opaque. Hypersthene is often associated with **labradorite in gabbro** and **norite** and in extrusive rocks like **andesite**. It is occasionally encountered in **meteorites.** Hypersthene is associated with **pyrrhotite** in Bavaria, with labradorite on the Isle St. Paul, Labrador. It is also found in Montmorency County, Quebec; and in the United States in the rocks of the Cortlandt series in the Hudson River Valley, and the **andesites** of Colorado and northern California. The word hypersthene comes from the Greek words meaning *strong* or *tough*. (E.S.C.S.)

HYPERTENSION. Hypertension means high blood pressure. It is a common sign and may have many causes. The chief types of hypertension are those associated with (1) kidney disease; (2) arteriosclerosis; and (3) essential hypertension, which is not secondary to any of the known conditions causing elevation of the blood pressure. These three forms can, and often do, overlap one another, which accounts for the difficulty and confusion in classification. The end picture of hypertension from any cause is much the same: the heart is enlarged and the vascular system almost invariably shows some arteriosclerosis; the kidneys are always involved, with damage to the glomeruli, and degenerative changes in the kidney vessels.

1. Kidney disease, inflammatory, obstructive and vascular, is accompanied by hypertension. It is a transient feature of acute nephritis and a constant one in chronic nephritis. Chronic **pyelonephritis** of bacterial origin eventually leads to hypertension. The vascular type of nephritis, nephrosclerosis, is characterized pathologically by degenerative changes in the small vessels of the kidney, the arterioles, and clinically by the various manifestations of hypertension.

2. Arteriosclerosis, as is seen almost uniformly in old age, is often accompanied by mild elevation of the blood pressure, usually unassociated with symptoms.

3. Essential hypertension is a clinical term which covers a variety of types of cases not included in (1) and (2). Benign essential hypertension is a chronic, slowly progressing disease, while the so-called malignant form may occur early in life and run an extremely rapid and severe course. The benign form after a number of years of mild severity may suddenly change and run the course of malignant hypertension. Both forms eventually lead to death from heart disease in 60% of cases, from cerebral hemorrhage or thrombosis (stroke) in 20%, and from uremia in 10%. Nevertheless, long life, even in the face of very high blood pressure, is not uncommon.

The signs and symptoms of hypertension are the same regardless of cause, and vary with the degree of elevation of pressure. Headaches, dizziness, and loss of a sense of well being are early symptoms. Cardiac symptoms eventually occur as enlargement of the heart and strain on its capacities develop: **dyspnoea,** cough, paroxysms of shortness of breath, and finally congestive heart failure appear. Renal failure and uremia are often terminal. In malignant hypertension, particularly, hemorrhages into the retina of the eye and white spots or scars in the eye grounds are common.

The treatment of hypertension is symptomatic. If the patient is a hard-driving, overactive individual, helping him adjust to a slower pace is one of the most important aspects of therapy. Rest, sedation and reduction in weight if the patient is obese are useful. When congestive failure develops, **digitalis** is given, as well as **diuretics,** and other measures to regulate fluid balance. (D.M.H.)

HYPERTHYROIDISM. Thyroid Gland.

HYPERTROPHY. This is the excessive growth or development of any part of an organ or organism. (R.M.W.)

HYPHA. Basidiomycetes.

HYPIDIUM. Pleistocene.

HYPNOSIS. The induction by psychical means of a state resembling sleep. The subject to be hypnotized usually lies in the recumbent position and is directed to fix his attention on some object. The psychiatrist repeats over and over in a monotonous manner that the individual is growing tired, drowsy, sleepy. By this means a trancelike state of varying depth is induced; the conscious mind of the individual is in abeyance, and he is in complete obedience to the commands of the hypnotist. The value of this state therapeutically is that it is the most effective way of employing suggestion. The patient is highly receptive; since his critical faculties are not operating he cannot reject the statements and suggestions made by the hypnotist.

Hypnosis is particularly useful in the treatment of hysteria (see **psychoneurosis**). An hysterical paralysis, or hysterical blindness, often clears completely under the influence of suggestion while the patient is under hypnosis; this is accomplished by the therapist's repeating over and over that the limb is well, or the eyes see. The connotation of mysticism and an evil spell that shrouded the practice of hypnosis in the early days is unjustified. Hypnotism is now recognized as a useful psychotherapeutic measure although its application is limited to the treatment of very few conditions. (D.M.H.)

HYPNOTIC. Any remedy or **drug** used to induce or maintain sleep. Drugs of this order produce their effect by depressing the higher centers of the **brain,** the excitability of which often prevents sleep. Because of the nervous factors involved in insomnia the taking of hypnotic drugs may lead to a drug habit. The commonly employed hypnotics are chloral hydrate, paraldehyde, the bromides, barbital and its derivatives. (D.M.H.)

HYPO. Sodium Thiosulfate.

HYPO ELIMINATORS. Hypo is removed readily from films and plates by washing in water and, consequently, there is little need for a hypo eliminator. Where water is scarce, or the time of washing must be short, a 3% solution of ammonium hydroxide may be useful in removing the last traces of hypo. After washing for 10 minutes, the negatives are placed in the hypo eliminator for 3 minutes and then washed for 3 minutes.

The investigations of Crabtree, Eaton and Muehler have shown that the last traces of hypo cannot be removed from photographic papers by washing in water and the use of a hypo eliminator is necessary if the highest degree of permanency is desired.

The hypo eliminator recommended is as follows:

Water......................	400 cc.	16 oz.
Hydrogen peroxide (3% sol)...	125 cc.	4 oz.
Ammonia (3% sol)...........	100 cc.	3¼ oz.
Water to make.............	1 liter	32 oz.

One gallon (4 liters) of this solution, *which should be used but once,* is sufficient for 50 8 x 10″-prints or their equivalent.

The prints should first be washed for 30 minutes, then placed in the hypo eliminator for 6 minutes and washed again for 10 to 15 minutes.

The slight yellowing of the highlights which occurs on some papers may be removed by placing the prints in a 1% solution of sodium sulfite upon their removal from the eliminator. (C.B.N.)

HYPOBROMITE. Hypobromous Acid and Hypobromites.

HYPOBROMOUS ACID AND HYPOBROMITES. Hypobromous acid (HOBr) is a yellow solution, of characteristic odor. It is unstable and when distilled under reduced pressure solutions containing only less than 1% can be obtained; decomposes into **bromine** and **bromic acid,** completely at 60° C. into bromine (and water).

Prepared by reaction (1) of bromine and mercuric oxide suspension in water, mercuric bromide being simultaneously formed, (2) of sodium hypobromite and an acid, excess acid yielding bromine.

Sodium hydroxide solution reacts with bromine to form bromide and hypobromite.

Sodium hypobromite solution reacts with ammonium salts or urea yielding nitrogen gas quantitatively. (R.K.S.)

HYPOCHLORITE. Hypochlorous Acid and Hypochlorites.

HYPOCHLOROUS ACID AND HYPOCHLORITES.
Hypochlorous acid (HOCl) is a yellow solution of characteristic odor. It decomposes upon standing, the rate depending upon the concentration; the exposure to light; upon the presence of a catalyst, such as cobaltous hydroxide (see Cobalt), which promotes the evolution of oxygen; and upon the acidity or alkalinity. A powerful oxidizing agent, e.g., manganous chloride solution changed to manganese dioxide, insoluble, by calcium or sodium hypochlorite, and prolonged boiling yields green manganate solution or pink permanganate solution; and bleaching agent for many organic colors.

Prepared by the reaction (1) of chlorine monoxide (Cl₂O) and water, (2) of sodium hypochlorite and an acid, excess acid yielding chlorine and oxygen, (3) of chlorine and mercuric (see Mercury) oxide suspension in water, mercuric chloride being simultaneously formed.

All hypochlorites are soluble in water, and the solutions are decomposed at 100° C. to form the corresponding chlorate and chloride. The most important hypochlorites are those of sodium, calcium, and silver, which last substance decomposes quickly to form silver chloride, white precipitate, and silver chlorate in solution. Sodium hydroxide solution reacts with chlorine to form chloride and hypochlorite. Sodium hypochlorite solution reacts with ammonium salts or urea yielding nitrogen gas quantitatively. A common means of detecting hypochlorites is the production of a blue color (caused by free iodine) with starch iodide paper by hypochlorites in weakly alkaline solution. Again, silver nitrate solution precipitates part of the hypochlorite as white silver chloride. (R.K.S.)

HYPOCHONDRIA.
Undue anxiety about one's own health, often accompanied by simulated symptoms of disease. Hypochondria is common in the psychoneurotic. (D.M.H.)

HYPOCOTYL. Seed.

HYPOCYCLOID.
A hypocycloid is a type of mathematical plane curve defined as follows:

If a circle of radius b rolls upon the interior of a fixed circle of radius a, a point on the first circle describes a hypocycloid (Fig. 1). Its parametric equations are:

$$x = (a - b) \cos \theta + b \cos \left(\frac{a - b}{b} \theta \right),$$

$$y = (a - b) \sin \theta - b \sin \left(\frac{a - b}{b} \theta \right).$$

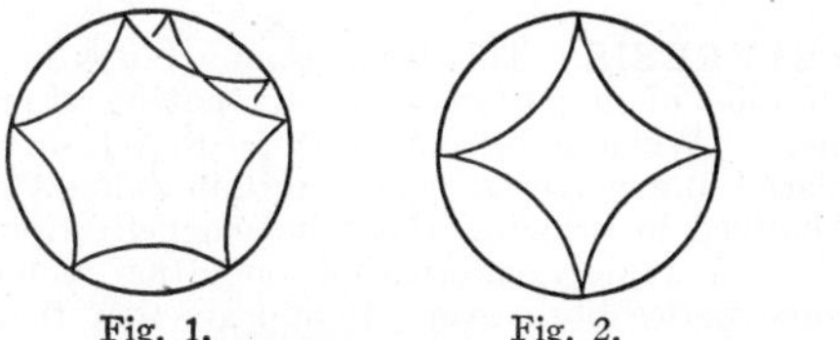

Fig. 1.   Fig. 2.
Hypocycloids.

When $a = 4b$, we get the hypocycloid of four cusps, also called the astroid (Fig. 2). Its equation in rectangular coordinates is

$$x^{2/3} + y^{2/3} = a^{2/3}.$$

It has parametric equations of the form $x = a \cos^3 \phi$, $y = a \sin^3 \phi$. (L.L.S.)

HYPODERMIS.
The cellular layer of the integument (integumentary system) in the invertebrates, which secretes the outer cuticula. (A.W.L.)

HYPOGENE.
Originated by the geologist Charles Lyell for all igneous rocks which assumed their form, fabric and texture at great depths beneath the surface of the lithosphere. (See also Plutonic. Contrast with Hypabyssal.) (R.M.F.)

HYPOGYNY. Fruit.

HYPOID GEARING. Bevel Gearing.

HYPOIODITE. Hypoiodous Acid and Hypoiodites.

HYPOIODOUS ACID AND HYPOIODITES.
Hypoiodous acid (HOI) is a greenish yellow solution, of characteristic odor. It is unstable, and cannot be distilled unchanged.

Prepared by reaction (1) of iodine and mercuric oxide (see Mercury) suspension in water, mercuric iodide being simultaneously formed, (2) of sodium hypoiodite and an acid, excess acid yielding iodine.

Sodium hydroxide solution reacts with iodine to form iodide and hypoiodite, the latter decomposing in a few hours at ordinary temperatures to form iodide and iodate. (R.K.S.)

HYPONITROUS ACID AND HYPONITRITES.
Hyponitrous acid (H₂N₂O₂) is a white solid, explosive even at as low a temperature as 0° C., soluble in water, more soluble in ether, can thus be extracted from water solution by ether and the latter evaporated, water solution decomposes quickly into nitrous oxide plus water. Hyponitrous acid is non-reactive with hydriodic acid (a strong reducing agent), but reactive with permanganic (see Manganese) acid (a strong oxidizing agent) to form nitrous or nitric acid.

Prepared (1) by reaction of silver hyponitrite (Ag₂N₂O₂) and hydrogen chloride in anhydrous ether, and evaporation of the resulting solution, (2) by reaction of hydroxylamine (H₂NOH) plus nitrous acid (HONO).

Sodium hyponitrite (Na₂N₂O₂) is formed (1) by reaction of sodium nitrate or nitrite solution with sodium amalgam (sodium dissolved in mercury), after which acetic acid is added to neutralize the alkali. Sodium stannite (see Tin), ferrous hydroxide, or electrolytic reduction with mercury cathode may also be utilized, (2) by reaction of hydroxylamine sulfonic acid and sodium hydroxide. Silver hyponitrite is formed by reaction of silver nitrate solution and sodium hyponitrite. (R.K.S.)

HYPOPHARYNX.
A protuberance on the floor of the mouth of insects. It is somewhat similar to the tongue of vertebrates and is sometimes called the lingua. (A.W.L.)

HYPOPHOSPHORIC ACID AND HYPOPHOSPHATES.
Hypophosphoric acid (H₂PO₃ or H₄P₂O₆) is a solid, melting point 55° C., decomposing in solution to form phosphorus plus phosphoric acids. Hypophosphoric acid is used in solution and is a reducing agent, but only with strong oxidizing agents, such as potassium permanganate; and the acid is unaffected by zinc and dilute sulfuric acid (distinction from phosphorous acid). Dehydration of hypophosphoric acid does not yield phosphorous tetroxide; hydration of

phosphorous tetroxide does not yield hypophosphoric acid but phosphorous plus phosphoric acids.

Hypophosphoric acid is formed by reaction (1) of yellow phosphorus and potassium permanganate in sodium hydroxide medium, (2) of red phosphorus and calcium hypochlorite solution, (3) also one of the products of slow oxidation at ordinary temperatures of phosphorus in moist air.

There are recorded the following sodium hypophosphates: Na_2PO_3 (or $Na_4P_2O_6$), $NaHPO_3$ (or $Na_2H_2P_2O_6$), $Na_3H(PO_3)_2$ (or $Na_3HP_2O_6$), $NaH_3(PO_3)_2$ (or $NaH_3P_2O_6$); and the dimethyl ester of hypophosphoric acid: $(CH_3O)_2PO$. There is evidence in support of each of the formulas H_2PO_3, $H_4P_2O_6$ for hypophosphoric acid.

Ester: Dimethyl hypophosphate $((CH_3)_2PO_3$ or $(CH_3O)_2PO)$. (R.K.S.)

HYPOPHOSPHOROUS ACID AND HYPOPHOSPHITES.

Hypophosphorous acid (H_3PO_2, or $H \cdot PO_2H_2$) is a colorless liquid, melting point 26.5° C., density 1.493.

Hypophosphorous acid is miscible with water in all proportions and a commercial strength is 30% H_3PO_2. Hypophosphites are used in medicine.

Hypophosphorous acid is a powerful reducing agent, e.g., with copper sulfate forms cuprous hydride (Cu_2H_2), brown precipitate, which evolves hydrogen gas and leaves copper on warming; with silver nitrate yields finely divided silver; with sulfurous acid yields sulfur and some hydrogen sulfide; with sulfuric acid yields sulfurous acid, which reacts as above; forms manganous immediately with permanganate.

Hypophosphorous acid is formed by reaction of barium hypophosphite and sulfuric acid, and filtering off barium sulfate. By evaporation of the solution in vacuum at 80° C., and then cooling to 0° C., hypophosphorous acid crystallizes.

Sodium hypophosphite ($NaPO_2H_2$), the only sodium hypophosphite, is formed (1) by reaction of yellow phosphorus and sodium hydroxide solution (phosphine simultaneously formed), (2) by reaction of hypophosphorous acid and sodium hydroxide, and evaporating. Sodium hypophosphite, upon heating, yields sodium phosphate and sodium phosphide. Common tests for the hypophosphites are as follows:

1. Zinc reduces dilute sulfuric acid solution of hypophosphites to phosphine recognizable by odor (difference from phosphates).

2. Barium chloride produces no precipitate (difference from phosphites). (R.K.S.)

HYPOPHYSIS.
An endocrine gland attached to the ventral surface of the brain, also called the pituitary gland. (A.W.L.)

HYPOPUS.
A larval form of certain mites. It has eight legs but no mouth. Ventral suckers enable it to attach itself to another animal for transportation. (A.W.L.)

HYPOPYGIUM.
The protruding male genital organs of some flies. In some species they form a conspicuous appendage at the tip of the abdomen, much like an additional segment. (A.W.L.)

HYPOSTOME.
A projection at the free end of the body of a hydroid polyp in which the mouth opens. (A.W.L.)

HYPOSULFUROUS ACID AND HYPOSULFITES.
Hyposulfurous acid ($H_2S_2O_4$) is a yellow solution rapidly oxidized in air to sulfurous acid and then to sulfuric acid. Commercially known as hydrosulfurous acid and its salts as hydrosulfites (but not to be confused with "hypo" which is sodium thiosulfate).

Hyposulfurous acid is a powerful reducing agent, e.g., with copper sulfate forms cuprous hydride (Cu_2H_2), brown precipitate, which evolves hydrogen gas and leaves copper on warming, with silver nitrate yields finely divided silver, with permanganate yields manganous. Hyposulfurous acid is formed by reaction of sodium hyposulfite and an acid.

Sodium hyposulfite, sodium hydrosulfite ($Na_2S_2O_4 \cdot 2H_2O$) is formed (1) by reaction of zinc and sulfurous acid (or sodium hydrogen sulfite), yielding zinc hyposulfite and then converted by sodium chloride into sodium hyposulfite, (2) by electrolysis of sodium hydrogen sulfite and then addition of sodium chloride.

Sodium hyposulfite is used to bleach sugar, indigo, wood pulp. With moist hydrogen sulfide, sulfur is precipitated and sodium thiosulfate simultaneously formed. (R.K.S.)

HYPOTHYROIDISM. Thyroid Gland.

HYPOTRICHIDA. Ciliata.

HYPOTYPE.
A species identified in print as belonging to the species in question. (R.M.F.)

HYPOXANTHINE. Alkaloids.

HYRACODON. Oligocene.

HYRACOIDEA.
The hyraces, less properly called the coneys. They are small animals, superficially resembling rodents. The fore feet have four toes and the hind feet three, part of them with nails instead of claws. The several species live throughout Africa and some extend north to Syria. An order of mammals. (A.W.L.)

HYRACOTHERIUM (Eophippus). Fossil Mammals.

HYRAX.
Mammalia, Hyracoidea. Any of the small animals of this order. There are two genera, *Procavia*, which includes the cony, and *Dendrohyrax*. (A.W.L.)

Hyrax. (*N. Y. Zool. Soc.*)

HYSTERECTOMY.
The operation of removing the uterus. This is usually done by a lower abdominal incision, although in certain instances it has been removed by the vaginal route. Hysterectomy is usually performed for cancer or fibroid tumors of the uterus. (R.S.M.)

HYSTERESIS.
This term usually refers to magnetic hysteresis, of importance in alternating-current machinery. When a ferromagnetic material such as iron is placed in a magnetic field, a certain amount of energy is involved in bringing about its magnetization. If the field is a rapidly alternating one, the material may become noticeably warm. It appears that the repeated changes of orientation in whatever it is within the substance that responds to the reversals of field are opposed by something like viscous friction.

A quantitative study of the process indicates that, as the field intensity H increases, the magnetic induction B

also increases in a manner characteristic of the substance. This is conveniently represented by a graph, such as OS (see figure). Upon reducing the intensity H, the induc-

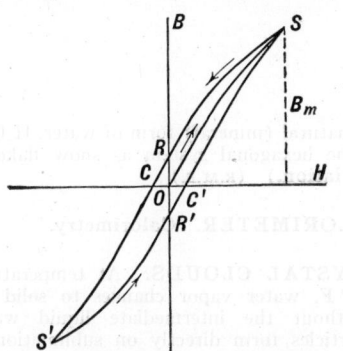

Typical B-H hysteresis curve for a ferromagnetic material.

tion B does not fall off as it was built up, but follows a different course, SRC, so that there is some residual induction OR even when H has fallen to zero. A reverse intensity OC, called the coercive force, must be applied to demagnetize the material completely. From this point the cycle proceeds over the path $CS'R'C'S$, thus completing a closed curve called a hysteresis loop. The initial graph OS is not retraced. The amount of energy converted into heat during the cycle is proportional to

the area of the loop. Steinmetz found that this energy is represented by the formula $aB_m^{1.6}$ in which B_m is the maximum induction during the cycle, and a is a constant for the given material, called the Steinmetz coefficient. An instrument known as a hysteresigraph has been devised to trace hysteresis loops automatically for any specimen of magnetic material under test.

Electric hysteresis is a somewhat analogous phenomenon exhibited by **dielectrics** in the electric field and gives rise to heating in a-c condensers.

Some solids exhibit what is called elastic hysteresis, in which the variables corresponding to H and B in the magnetic case are the stress and the strain or deformation. Elastic bodies such as metals operating at stresses below the **proportional limit** also undergo hysteresis. If B represents positive or tensile stress and H represents positive strain or elongation, the action of an elastic metal under cyclic or reversed stress can be represented by $OSC S' CS$. By definition of the proportional limit the line OS should be straight; however, highly sensitive measurements show a slight curvature, leading to the development of a loop of appreciable thickness. The area within the loop is proportional to the loss of energy in a complete cycle of reversed stress. If it were not for this loss in energy even in the most perfect elastic bodies, **perpetual motion** might be possible in a tuning fork or spring vibrating in a vacuum. (See also **Damping**.) (L.D.W., R.H.H.)

HYSTERIA. Psychoneurosis.

I-BEAM. An I-beam is a structural shape whose cross-section resembles the capital letter I. The I-beam is rolled from **steel** for ordinary use. **Aluminum** is sometimes used where the structure must be exceptionally light as in the case of bridge floors. Brass and other materials are also used. The size is designated by its over-all depth from the top of the top **flange** to the bottom of the lower flange, and by its weight per ft. of length. A beam of this shape has a disposition of material such as to allow the material to be stressed most efficiently, thus reducing waste due to inclusion of

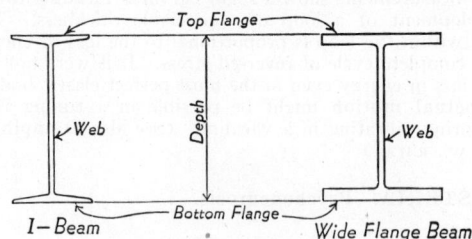

slightly stressed material. A beam, carrying **load**, deflects slightly, which puts part of the beam in compression and part in tension. The farther a given amount of beam material is from the **neutral axis**, the more effective it is in resisting load. (See **Flexure** theory.) The I-beam with its two flanges, connected by a thin vertical web, provides the proper resistance to load with a minimum of material. When the proportions of width to depth are nearly equal the beam is known as an H beam. Another type of I-shape beam is called the wide flange beam. This is a popular shape for building columns and floor beams. (C.W.C., F.T.M.)

IBEX. Mammalia, Artiodactyla. **Goats** of several species, all with very large horns in the male sex. The species found in the Alps is also called the steinbok or

Asiatic ibex. (*Field Museum of Natural History.*)

bouquetin, *Capra ibex,* and the Arabian species is called the beden. Others live in Egypt and the Himalayas. (A.W.L.)

IBIS. Aves, Ciconiiformes. *Ibis.* Long-legged wading birds (**Aves**) with long curved beaks. Related to the storks. The numerous species occur in all continents. (A.W.L.)

ICE. A natural (mineral) form of water, H_2O. Crystallizes in the hexagonal system as snow flakes, or hail. (See **Glaciation.**) (R.M.F.)

ICE CALORIMETER. Calorimetry.

ICE CRYSTAL CLOUDS. At temperatures below about 15° F. water vapor changes to solid water directly without the intermediate liquid water stage. Cloud particles form directly on sublimation nuclei as ice crystals, and such clouds are then composed of ice crystal particles. Cirro-form clouds are of the ice crystal group. (P.E.K.)

ICE FORMATION ON AIRCRAFT. Liquid droplets can remain liquid at temperatures well below 32° F. but, when such supercooled droplets are disturbed, they freeze almost instantly. Both cloud droplets and rain drops appear in the atmosphere at temperatures below 32° F. and both present considerable hazard to aircraft in flight because the drops and droplets impinging on the craft freeze to it and form a layer of glaze which not only weights the plane but alters the leading edge of the wings to destroy **lift.** Very small cloud droplets

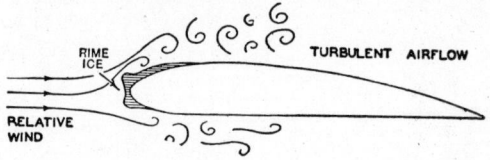

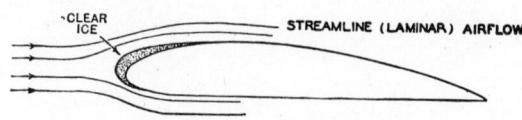

Ice formation on an airfoil.

tend to flow around the surface of an aircraft with the **airstream** and therefore present no great hazard. Small cloud droplets are found primarily in stratus clouds and in stable air.

Intermediate-size drops penetrate the friction layer and impinge on the aircraft surfaces but may not break upon impact. They freeze where they strike and form a rough, irregular, opaque rime ice coating which usually is not too difficult to break with de-icer boots.

On initial impact, large drops break and freeze but, in freezing, release enough heat of fusion to warm remaining parts of the drops somewhat. At temperatures usually encountered, the warmed part of a drop actually reaches a temperature of 32° F. and then tends to flow along the surface of the craft in a thin film which freezes rapidly due to exposure to the air, as well as evaporation. This conversion process takes place quite quickly but it does permit the formation of two different types of ice. Ice formed on initial impact is predominately non-crystalline and opaque rime ice; ice formed from the remainder of the drop is usually crystalline and clear. The temperature of the droplets or drops, therefore, determines primarily the type of ice that will be predominant, although both types form over a wide range of temperatures. At temperatures of 32° F. or less, down to about 28° F., clear ice is predominant; at temperatures below 15° F. rime is predominant.

Water droplets grow largest in rising currents of air. Cumulus clouds which grow in rising currents, orographically lifted clouds, and frontal clouds, therefore, are the most serious cloud icing hazards, although icing does occur in other types of clouds. Icing in supercooled rain drops is always a serious hazard because the total water (which forms ice) intercepted by the craft is considerable and all that impinging on the craft's surfaces adheres as ice in one form or other. Cold air regions beneath warm frontal surfaces are the most serious icing regions of freezing rain. (P.E.K.)

ICELAND MOSS. Lichens.

ICELAND SPAR. Calcite.

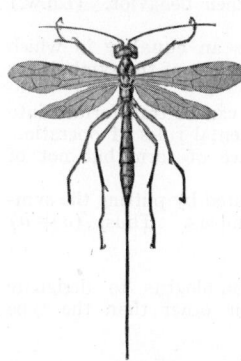

Ichneumon fly.

ICHNEUMON. 1. Insecta, Hymenoptera. Any **insect of** a large number of species which live as parasites on other insects in the larval stage. They resemble wasps in appearance. 2. Mammalia, Carnivora. The Egyptian **mongoose,** *Herpestes ichneumon.* (A.W.L.)

ICHTHYOLOGY. The science that deals with facts pertaining to fishes (**Pisces**). The species of fishes are so numerous that the classification of the group is an important part of this science, and their value as food is so great that the regulation of fisheries and measures for maintaining the supply of food and game fishes make up an important practical field. (A.W.L.)

ICHTHYOPSIDA. Cyclostomes, fishes, and amphibians. A division of vertebrates characterized by moist skin and by the presence of gills during part or all of the life. The eggs are usually deposited in the water and the embryo develops without an **amnion,** hence the name Anamnia is also applied to the group. (A.W.L.)

ICHTHYORNIS. Fossil Birds.

ICHTHYOSAUR. Fossil Reptiles.

ICING, AIRCRAFT. For many years the formation of ice on airplanes in flight set considerable limitation on their usefulness. When faced with weather bearing the threat of an icing condition, pilots either had to turn back, or risk the hazard and perhaps have to "bail out." When airlines first began to carry passengers, flights were made only under favorable weather conditions. Many travelers were reluctant to consider air transportation during winter months. To maintain scheduled operation, improved techniques, including methods of combating icing, were developed. Even today, when icing conditions are known to exist, the airlines frequently change their flight plans, even though this may inconvenience passengers. However, this may not always be possible, and, of course, in military operations, the aircraft should be able to complete their missions regardless of the weather.

Meteorological conditions which are conducive to the formation of ice on aircraft are the presence of visible moisture and an atmospheric temperature lower than 34° F. These conditions may result in ice formed on wings, control and tail surfaces, propellers, and fuselage. On account of the adiabatic expansion in the Venturi

of carburetors, ice may form there when atmospheric temperatures are considerably above 34° F. Visible moisture in the air may take the form of sleet, snow, or water droplets, supercooled below freezing. Water can remain liquid below 32° F. in the atmosphere for several reasons. Sometimes water vapor condenses on microscopic sea-salt particles, thus lowering the freezing temperature of the droplets. Or the droplets may be so small that surface tension prevents freezing. Or again, there may not be sufficient agitation of the droplets to prevent them becoming supercooled water. When these supercooled droplets of water strike the leading surfaces of an airplane, part of each droplet instantly changes into ice, with the rest of it quickly freezing by evaporative cooling. Three forms of ice may be produced, viz., glaze ice, rime ice, and rime frost. Glaze ice is formed from freezing rain near the freezing temperature. It is quite solid and smooth, but covers the most area. Rime ice is formed from droplets of supercooled water and from wet snow or sleet mixed with glaze ice, being porous and opaque, and of irregular rough shapes on the leading edges of the airfoils. It is likely seriously to affect the lifting properties of a wing. It is the most frequently encountered form of icing. Rime frost is formed at very low temperatures from very small droplets. Its chief effect is to alter the landing characteristics of an airplane.

As ice may hold the potential threat of interruptions of flight, making scheduled flights hazardous and landings difficult, great efforts have been made to overcome it. These efforts have been, first, accurately and scientifically to chart the weather, with the purpose of routing flights to avoid icing conditions, and second, to provide devices for freeing the airplane in flight of accumulations of ice. The airplane has high sustained speed. For this reason, ice forms only on its frontal area. The smallness of this area simplifies the problem of keeping the plane free of ice. The problem has been attacked by three general methods—chemical pastes or liquids, heat, or by methods which remove the ice mechanically. Chemical pastes are difficult to apply smoothly. Abrasion and water erosion tend to roughen the surface and make frequent application necessary. Antifreeze materials and heat applied to the leading edge only permit melted ice to run back and re-freeze. In the case of heat, the entire wing must be heated. Consequently, the method of most promise is that of removing the ice mechanically.

Dr. W. C. Geer of Ithaca, New York, was commissioned by the Daniel Guggenheim Foundation to study the ice problem. As the result of considerable experimentation, he found that the most practical method is to allow a little ice to form, and then break it mechanically so the air stream can remove it. Since Dr. Geer's studies, the mechanical **de-icer** has been developed for application to practically all exposed parts of the airplane which have a tendency to pick up ice. In the case of propellers and windshields, however, antifreeze liquids are employed to de-ice. (F.T.M.)

ICONOSCOPE. This is a form of electron **tube** used as a camera tube in **television.** The scene to be transmitted is focused on the mosaic consisting of many very minute photo-electric cells. These are formed by treating a mica sheet with silver oxide and reducing the oxide to metallic silver in such a way that many little globules of silver rather than a continuous sheet result. These silver particles are then photosensitized by treating with caesium as in the conventional **phototube.** The mica is backed by a conducting coating so the silver units form little **condensers** with the backing. The scene focused on this mosaic causes the photosensitive particles to emit electrons proportional to the light falling on each. These emitted electrons are removed by the circuit connections to the tube. Thus the various midget condensers formed by the particles and the backing of the mica are charged, the amount of charge depending upon the light at the corresponding part of the picture and upon the time during which the charge is allowed to build up. These condensers are periodically discharged by a scanning beam of electrons which is swept back and forth across the picture until every particle has been touched in sequence. This electron beam restores the negative charge which the photo-electric action had removed. This sudden restoration of charge gives a pulse of current in the circuit connected to the other plate of the little condenser, i.e., the backing conductor. This pulse magnitude depends upon how many electrons had to be restored and hence upon the brilliance of the picture at that point. Since the charges are restored in sequence there will be a sequence of pulses

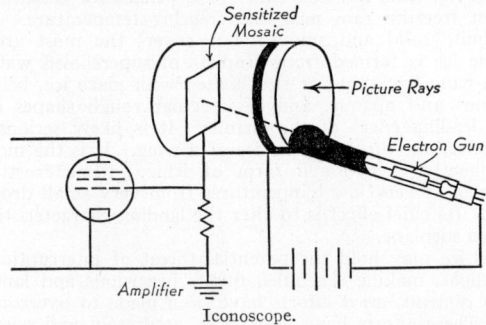

Sensitized Mosaic
Picture Rays
Electron Gun
Amplifier
Iconoscope.

in the output circuit which represents the orderly dissection of the picture into minute parts for transmission. (See figure.) (L.R.Q.)

ICTERUS. Jaundice.

ICW. Interrupted Continuous Wave.

IDE. Pisces, Teleostei. A European fish (**Pisces**), *Idus idus,* related to the roach and dace. (A.W.L.)

IDEAL GAS LAW. An "ideal gas" would, if kept at a constant temperature, behave as respects volume and pressure in strict accord with **Boyle's law.** If now the temperature is also allowed to vary, we must combine the **law of Charles** (or of Gay Lussac) with Boyle's law, yielding the **Boyle-Charles law:**

$$pv = p_0v_0(1 + at), \qquad (1)$$

in which p_0v_0 is the value of the pressure-volume product pv when the temperature t is zero, a is the coefficient of expansion of the gas, practically the same for all gases, and in the ideal case equal to the reciprocal of the absolute temperature of the scale zero. If the centigrade scale is used, the value of a is approximately $1/273.2$ per degree. Substituting this, Eq. (1) may be written

$$pv = \frac{p_0v_0}{273.2°} (t + 273.2°), \qquad (2)$$

which is one expression for the ideal gas law.

The factor $t + 273.2°$ will be recognized as the **absolute temperature** T of the gas. And since the ideal gas obeys Boyle's law, the product p_0v_0 is constant however p_0 and v_0 may vary between themselves. We may thus denote the coefficient $p_0v_0/273.2°$ by a single constant symbol, say R, and the ideal gas equation then takes the usual form

$$pv = RT. \qquad (3)$$

The value of R depends, of course, upon the quantity of gas used, since at any pressure p_0 it is proportional to the volume v_0. For 1 gram of air, R equals about $2,868,000 \frac{\text{g cm.}^2}{\text{sec.}^2 \text{deg.}}$. At the zero of temperature and at

any given pressure p_0, the gram molecular weights, or moles, of all pure gases have equal volumes. (This follows from **Avogadro's law.**) Hence if one mole of any pure gas is used, R will always have the same value, in c.g.s. units about $8.314 \times 10^7 \frac{\text{g. cm.}^2}{\text{sec.}^2 \text{deg.}}$; which is called the "ideal gas constant." Many physical formulae involve a quantity which may be regarded as the ideal gas constant per molecule, that is, the above molar gas constant divided by the number of molecules in a mol, 6.064×10^{23}, giving $1.3805 \times 10^{-16} \frac{\text{g. cm.}^2}{\text{sec.}^2 \text{deg.}}$. This is the "Boltzmann constant."

Since actual gases, even those with the smallest molecules, hydrogen and helium, do not obey the ideal gas law exactly, various empirical **characteristic equations** have been devised to represent their behavior. (L.D.W.)

IDENTITIES. An identity is an **equality** in which both members are equal for all values of the symbols for which the members are defined.

In an identity, either member can be transformed into the other by use of the fundamental rules of operation. An identity involves a difference of form but not of value.

An identity is frequently indicated by putting the symbol $\equiv$ between the two members. Thus, $(a + b)(a - b) \equiv a^2 - b^2$. (L.L.S.)

IDEOTYPE. Used by paleontologists to designate a specimen collected from some other than the type locality. (R.M.F.)

IDIOBLAST. A term proposed by Becke in 1903 for pseudo-**idiomorphic** crystals occurring in the **meta-morphic** rocks. (R.M.F.)

IDIOCY. Complete congenital imbecility. (See **Mental Deficiency.**) (R.S.M.)

IDIOMORPHIC. The term proposed by Rosenbusch for those minerals in **igneous** rocks which are well crystallized and therefore display their **crystal** form with a high degree of perfection. (See also **Euhedral** and **Automorphic.**) (R.M.F.)

IDIOSYNCRASY. 1. Individual hypersensitivity to the effects of any drug, food, or physical or chemical agent. 2. A habit or temperament peculiar to any individual. (D.M.H.)

IDLER. Gear Train.

IDOCRASE. Vesuvianite.

IGNEOUS ROCK. Igneous rocks are rocks which have solidified (congealed) with, or without, **crystallization** from hot natural solutions such as **magma** or **lava.** Igneous Rocks are classified by their **texture,** fabric, chemical (mineral) composition, and their field relation-

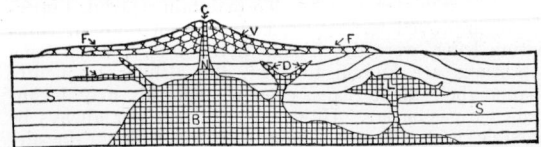

Diagrammatic structure section illustrating modes of occurrence of igneous rocks. S, strata; B, batholith of plutonic rock; L, laccolith; D, dikes; I, intrusive sheet or sill; V, volcano; N, neck of volcano; F, lava flow; C, crater.

ship or mode of occurrence. Under mode of occurrence igneous rocks are classified as intrusive (**plutonic**) or extrusive (**effusive**). The intrusive rocks are classified according to the shape and size of the intrusive body and its relation to the other formations which it intrudes. Typical intrusive are **batholiths, laccoliths, sills** and **dikes.** The extrusive types are called **lavas.** Over 700 species of igneous rocks have been described, the bulk of which are intrusives. (R.M.F.)

IGNITER. Ignitron.

IGNITION. Ignition is the initiation of **combustion.** It is necessary for the inception of any flame and is accomplished by raising a mechanical mixture of the combining substances, such as carbon and oxygen, to the "ignition temperature." This temperature may be defined as the lowest temperature which will cause flame to start and spread through a combustible mixture. The ignition temperature varies with the substance. Some typical values are:

	°F.
Carbon	750
Carbon monoxide	1250
Hydrogen	1090
Sulfur	470
Isoöctane	980
Heptane	500
Phosphorus, yellow	93

The most common source of high-temperature energy used for exciting molecular activity to the point of ignition is the ordinary match. On account of its low ignition temperature, yellow phosphorus was at one time employed as the igniting element of matches; however, its use was hazardous, and in modern match manufacture the tip is composed of a mixture of various materials, including phosphorus, glue, ground glass, chlorate, and potash. The common match is composed of a paraffined splint dipped into a combustible compound which, upon hardening, forms the bulb of the match head. The head is redipped, to form an eye, in a compound which will ignite spontaneously by friction against a rough surface. In safety-match manufacture, the phosphorus compound is usually affixed to the side of the box and the head contains the oxidizing agent.

Rays of the sun, properly focused through a lens, are capable of heating some substances to their ignition points. Mechanical (flint and steel) and electrical (spark) energy can be used for the same purpose, as can also compression (of air), and fluid friction (meteors).

In certain apparatus, combustion is not continuous, and ignition must be intermittently and precisely accomplished. The most familiar example is the **internal combustion engine.** Two common variants of this engine are the spark ignition (Otto) and the compression ignition (Diesel) engines. In either case the fuel and air must be mixed and raised to the ignition temperature, or above. Each of these actions is time-consuming, with the result that the crankpin describes a finite arc between the moment of inception of the ignition action and the moment of actual ignition.

An explosive mixture is compressed in the spark ignition engine, but although mixing is reasonably thorough, there is some small but finite elapsed time after appearance of the spark at the **spark plug** of the ignition system and the completion of combustion, which leads to the timing of the spark somewhat in advance of dead-center position. Some delay in pressure rise after ignition is desirable to reduce **detonation.**

An engine which employs compression ignition provides the necessary ignition temperature by compression of air to a compression ratio sufficiently high that the polytropic temperature rise will cause the air to equal or exceed the ignition temperature of the fuel employed. In the Diesel engine a liquid fuel is employed, generally a petroleum derivative. It is sprayed into the hot air, mixed, and ignited. Both the mixing process and the rise of temperature of the fuel are time-consuming, and an ignition delay occurs which must be offset by advancing the moment of fuel ignition before dead-center position. The number of degrees of crankpin travel between beginning of injection and the action of ignition is termed the **delay angle.** It is desired to have the delay angle as small as possible so that the cylinder will not be overloaded with fuel when ignition occurs, otherwise detonation may result. Note that ignition delay promotes detonation in the compression ignition engine, and opposes it in the spark ignition engine. (F.T.M.)

IGNITION QUALITY. This characteristic measures the ability of fuel oil to ignite spontaneously in the engine cylinder. It is a quality which assumes increasing importance the higher the rotative speed of the engine. It is believed that combustion knock in a Diesel engine is caused by delay in ignition during the first part of the injection, causing accumulation of fuel which burns simultaneously with the fuel injected during the latter part of the injection. Good ignition quality, as represented by high Cetane Number, minimizes the delay period and the tendency of the engine to knock. Ease of starting and occurrence of a smoky exhaust at light loads are also dependent, to a degree, on ignition quality. The Cetane Number scale is derived from the standard practice of using a test fuel composed of cetane and alpha-methol-napthalene. Of these two, cetane has very good ignition qualities, whereas alpha-methol-naphthalene is very poor. The Cetane Number of a fuel is the percentage of cetane, in a mixture of the two, which will give the same ignition quality as the fuel oil under test. The average high-speed Diesel engine requires a fuel having ignition quality of better than 45 Cetane Number, and will scarcely run at all on fuels rated lower than 25 Cetane. Cracked oils show a considerably lower Cetane Number than straight run oils. (F.T.M.)

IGNITION SYSTEM. The ignition system here described is that used in engines operating on the **Otto cycle.** In this cycle, an inflammable mixture is compressed just before each power impulse. This mixture cannot be compressed until it reaches the spontaneous ignition temperature (as is the case in the **Diesel cycle**) because the presence in the cylinder of the combustible mixture would cause ignition to be uncontrolled and irregular. This engine must, therefore, depend upon some system of ignition which is under control, and which is automatically synchronized with the action of the cycle. Although early engines were operated on hot tubes and open flames, in these days ignition systems are universally electrical in nature. In a **gasoline** engine the gasoline-air mixture has an **ignition temperature** varying somewhere between 650 and 850° F. The ignition system must produce a focal point of heat sufficient to exceed this temperature and to provide it at exactly the right instant. The average time allowable in a gasoline engine operating at about 2000 rpm for ignition and explosion of the mixture following ignition, is only about one one-hundredth of a second. Furthermore, it has been found necessary to vary the timing of ignition with respect to the cycle of operation in such a way that ignition occurs relatively earlier in the cycle the higher the rotative speed. These conditions are well met by the electrical spark system of ignition, in which a high-voltage spark jumps across a stationary spark gap. Due to the high speed of electrical impulses in wires, there is no difficulty in accurately timing the ignition in an electrical system. However, the high-voltage jump spark method of ignition is not the only electrical form in use, and both the low-tension and high-tension ignition systems will be briefly described.

A low-tension ignition system is one in which the voltage at the ignition points is insufficient to cause a spark to jump a static gap. The **electrodes** of the gap are brought together, and then rapidly separated. In this way, a low-voltage supply is able to cause an electric arc to follow the separation of electrodes. **Fig. 1** shows a system of this nature. It consists of a source of **voltage** (a battery), an induction coil, and ignition points. The

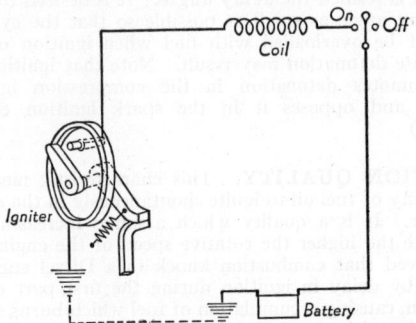

Fig. 1. Low tension system for single cylinder stationary engine.

ignition points and their operating mechanism constitute what is known as the igniter. The igniter is mounted in the **combustion** chamber, and is composed of one stationary, and one moving, electrode. The moving member is mechanically operated from the crankshaft by means of a shaft extending through the cylinder wall, and in this way the igniter is synchronized with the cycle. When the igniter points are pushed together, making contact, an electric current flows through the coil, igniter points, and battery. If, then, the current is suddenly interrupted through rapid opening of the igniter points, the battery voltage, aided by self-induced voltage of the induction coil, will cause an arc to follow the opening of the points. This arc is the source of ignition. This low tension system is only occasionally used on engines, usually on relatively slow-moving, fixed, or semi-portable engines.

The high-tension system, shown by **Fig. 2**, is the system used on most gasoline and gas engines, and is

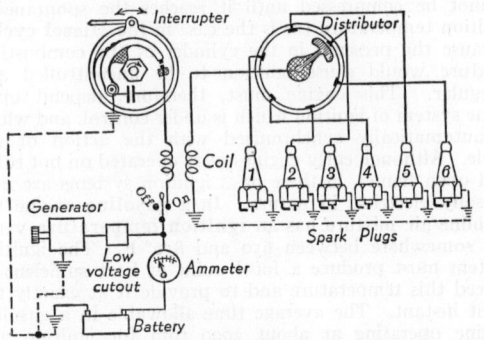

Fig. 2. Typical high tension ignition system for a 6-cylinder automotive engine.

familiar to many persons through its use as the ignition system of the automobile engine. It is well adapted to multi-cylinder engines, and is especially suitable where other electrical services are required; for example, lighting and electric cranking. Two systems of high-tension ignition should be mentioned. One, the battery and coil system, is most popular today on automobiles; the other, the **magneto** system, is the more widely used on aeronautical and fixed or semi-portable engines. The magneto system of ignition operates very much in the same manner as the coil and battery, except that, having all of the separate elements consolidated in one piece of apparatus, it is a much more compact ignition system. On the other hand, an automobile, with its need of a battery for purposes other than ignition, is best serviced with the coil and battery system.

Coil and battery electrical ignition will be explained in connection with the figure. A source of low voltage, i.e., the battery at 6 volts, is connected to the primary of an induction coil through an automatic switch which is opened at regular intervals by the engine itself. This automatic switch is called the interrupter or timer. Whenever the interrupter opens the circuit, the induction coil generates a high voltage which is sufficient to cause a spark momentarily to jump the static spark points which are in the combustion space of the cylinder. At the moment of ignition, the cylinder is filled with a compressed gas, some of which occupies the space between the spark points. The gap extends between .010 and .020″, so that it is apparent that the induction coil must generate a very high voltage in order for the secondary current to be able to overcome the resistance of the compressed gas between the points. These points are made part of a spark plug which is screwed into the cylinder, thus providing a way of attaching the points, one of which must be insulated. The use of a single coil to supply high-voltage current to the spark plugs of a multi-cylinder engine, involves the use of another mechanically operated device, the **distributor**. The distributor picks up the high-voltage impulse as it comes from the induction coil, and shunts it to the cylinder ready to receive ignition action. Depending upon the sequence with which high-voltage leads are attached between spark plug and distributor, the engine will operate with a definite sequence of power strokes derived from the different cylinders. There are certain standard arrangements of this sequence for the common multiples of cylinders employed in gasoline engines. These arrangements are called the firing orders. The firing order must be well chosen to prevent rocking or galloping of the engine.

In the common **four-cycle** engine, one ignition impulse serves a cylinder for two complete revolutions. Because of this fact, the distributor rotates at one-half engine speed. Now the interrupter is usually driven by a **cam** which does not necessarily have to rotate at one-half crankshaft speed, but which can conveniently be made to do so, since a one-half speed shaft must be provided for the distributor. If the interrupter is operated by a cam revolving at one-half crankshaft speed, the cam must have as many lobes on it as there are cylinders in the engine. The adjustment of the instant of ignition in a cycle with due regard to the demands of variable speed

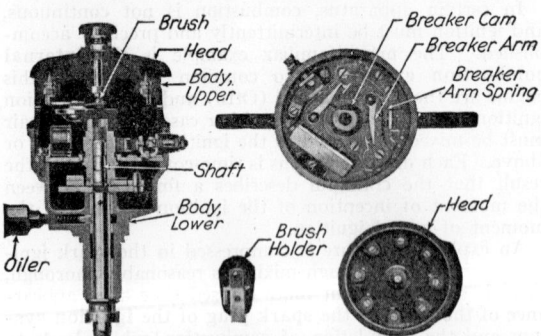

Fig. 3. North-East ignition unit (Packard).

operation, can be done through a slight rotative displacement of the contact-carrying case of the interrupter. Formerly this action was accomplished manually, but at present progress in electrical ignition systems has enabled manufacturers to develop automatic spark advances, where centrifugal weights opposed by springs adjust the interrupter to a position of spark advance which will be best for the speed of the engine. A **condenser** paralleled across the interrupter points absorbs energy which would otherwise be manifested as a small arc following the opening of the points. It is convenient to mount the condenser, interrupter, and distributor as a unit enclosed in a single case, and driven by a one-half speed shaft. A unit of this type is shown in Fig. 3. (F.T.M.)

IGNITRON. This is an electron tube of the mercury-arc type having a special starting principle. The tube consists of a mercury pool, to serve as **cathode**, and an **anode** for the main part of the circuit and an auxiliary electrode, the igniter, which dips into the mercury pool. Since all mercury pool tubes are essentially cold cathode devices until the arc is started some means must be provided for initiating the arc. For rectification of a.c. where no provision is made for keeping the arc alive from cycle to cycle or for control purposes where the tube may be alternately turned on and off under the influence of auxiliary equipment, the arc must be restarted at intervals. In the ignitron this is accomplished by the igniter, a rough-surfaced material which will not be "wet" by the mercury. The resultant points of contact between the igniter rod and the mercury will carry very high current densities if a pulse having only a fair current value is passed through it. This high current density, possibly coupled with high fields where the mercury and rod are not actually in contact, cause the creation of a minute cathode spot at the junction of the mercury surface and the rod. This gives the electron emission necessary to start conduction to the main anode if the latter is positive. The current between the spot and the anode then will develop the spot to its normal size. This tube has the advantages of the ordinary mercury-arc tube plus the great advantage of an easily controlled starting mechanism not involving any moving parts. It has rapidly come into extensive use for applications requiring ease of control and high current capacities; especially capacities of short-time duration, such as motor control, welding control, rectifiers in electrochemical processes, etc. (L.R.Q.)

IGUANA. Reptilia, Sauria. Large **lizards** of South America and the West Indies. They eat vegetation and insects and are largely **arboreal**, although they are at home on land or in the **water**. The eggs and flesh of some species are eaten.

The name is applied to other lizards than the true iguanas, including members of the same family and others less closely related. Among these species are a land lizard of the Galapagos Islands and a marine species of the same archipelago which eats seaweed, and the unrelated monitors of the Old World. The family Iguanidae also includes the **basilisks**, anolis lizards, and **horned toads.** (A.W.L.)

IGUANODON. (Duck-billed dinosaur.) **Fossil Reptiles.**

IJOLITE. Ijolite is a wholly crystalline, granular igneous rock consisting essentially of **nephelite and pyroxene.** (E.C.E.S.)

ILLINIUM. Discovery announced by Harris, Yntema, and Hopkins in 1926 and presumably a member of the **cerium** sub-group of the rare earth metals of atomic number 61. See **Chemical Composition, I. Elements.** (R.K.S.)

ILLUMINATION. The illumination of a surface is the luminous flux which it receives per unit area. The three most common units are: the foot-candle, or 1 lumen per sq. ft.; the phot, or 1 lumen per sq. cm.; and the lux, or 1 lumen per sq. meter. (See **Photometry.**) The phot is thus equal to 10,000 luxes and to about 929 foot-candles, and is appropriate to only very intense illuminations; the milliphot (0.001 phot) is usually more convenient.

Illumination obeys a cosine law, like the radiation from a surface and for similar reasons (see **Cosine Emission**

Law). That is, if the illumination is I_0 for zero angle of incidence, then for any other angle θ it is $I = I_0 \cos \theta$. Since for perpendicular incidence the illumination from a concentrated source of luminous intensity L at distance r is $I_0 = kL/r^2$, the illumination at incidence angle θ is

$$I = k \frac{L}{r^2} \cos \theta.$$

The value of the constant k depends on the units of I, L, and r; for example, if I is in foot-candles, L in candles, and r in ft., $k = 1$ (lumen per candle).

The illumination desirable at any point varies of course with the circumstances. For reading tables or office desks a typical night-time illumination should be from 6 to 8 foot-candles. Actual illuminations are measured by means of an illumination photometer or **illuminometer.**

The term illumination is also applied to the science of providing and directing light for a specific purpose associated with the activities of mankind. The light itself is radiant energy whose wavelength lies within a band to which the human eye is sensitive. It is produced from an incandescent source such as that furnished by the sun or artificially by lamps. The sources of incandescence employed most frequently are open flames and closed flames, electric arcs, incandescent filaments, and luminescent gas. In point of historical precedence, the flames of open fires were undoubtedly the first artificial illumination employed by man. This method was next improved on by selecting certain sticks of wood which contained resinous parts which provided torches that would remain alight while carried from point to point. Then animal and vegetable fats and waxes were variously employed in candles and lamps, yielding a more portable type of illumination which could be employed for illumination only, and without the necessity of receiving large amounts of heat as well. With the discovery and refining of **petroleum,** the kerosene lamp, which was so much more effective as a source of light than anything previous, came for a short period of time into almost universal use. The kerosene lamp is still used in isolated homes removed from the conveniences of the more efficient kinds of lighting, and where lowest cost lighting is wanted. Defects associated with odor, fire hazard, nuisance of maintenance, and inferiority of illumination, have discouraged the continued use of the kerosene lamp.

A fuel gas may be employed as a means of securing illumination. Among such gases might be mentioned ordinary artificial **illuminating gas, acetylene gas,** air gas (gasoline-air mixture), and natural gas. The use of natural **or** illuminating gas is confined largely to cities and larger towns wherein is installed a gas distribution system. For isolated homes, acetylene gas, which is generated from the action of water on **calcium** carbide, and air gas lamp units are available. As a gas flame is practically transparent when the gas is completely burned by a suitable burner, illumination from gas can be obtained only by heating to incandescence finely divided particles or objects placed in the flame. An ordinary slit type gas burner having an open flame produces a certain amount of carbon particles which, when heated in the flame, emit a yellowish light. This yields illumination of low intensity which is often unsteady, and is certainly inefficient. The addition to plain open-flame gas burners of a mantle consisting of a mineral which, when subjected to the heat of the burning gas, becomes white hot and emits a strong steady light, greatly improved gas lighting. The standard Welsbach mantle is composed chiefly of thorium oxide supported on a cotton or artificial cellulose base which burns out when the mantle is first lit. The luminous shell of mineral left must be very gently handled in order to prevent its destruction. The fragility of the gas mantle has been its chief defect, as it is an efficient source of illumination, quiet and odorless. Invention of the incandescent lamp, and the building up of electric service distribution networks which reach a large portion of the population, have made gas lighting an inferior and less desirable form of illumination. To-day in the United States most homes that are not il-luminated by electricity have retained the kerosene lamp in more or less improved forms. The above statement holds true even though a great many rural homes will be found enjoying acetylene gas illumination. The rela-tively high illuminating efficiency of electric lighting may be demonstrated by quoting the energy consump-tion per unit of illumination of a few illuminants. The unit ordinarily employed in this field is the lumen per watt of input. Open-flame gas burners and kerosene lamps yield about $\frac{1}{4}$ of a lumen per watt of power. The mantle-type gas lamp may provide as much as $1\frac{1}{2}$ lumens per watt, but the modern incandescent lamp can deliver 8 lumens per watt, and the mercury-arc lamps 30 lumens per watt, or over 100 times as much illumina-tion per unit of energy used. (See **Electric Lighting**.)

The merits of an illumination system may be judged by the freedom from fluctuation of light, by color, by intensity of lighting on the working plane, and by diffu-sion which prevents glare. Glare is due to excessive illumination of surfaces, the light reflected from which reaches the eye. Glare is reduced or eliminated by diffu-sion, which can be obtained by special designs of lighting units or by special treatment of the walls of an interior. There are three methods of supplying artificial illumination to the region in which the illumination is desired. With specific reference to interior illumination, the region in which the illumination is most desired is the lower part of a room, in which persons live and move. Also, the source of light is usually above, as it has been customary to place lamps near the ceiling, with sufficient clearance so that they will not interfere with freedom of movement of the individual about the room. The direct downward supply of light from the lamp is known as direct lighting. Direct lighting is efficient as measured by **candles** of luminous intensity per watt of energy, but is likely to have some glare, and to illuminate dif-ferent regions of the interior unequally. Indirect lighting, the second lighting system to be mentioned, avoids glare, and incidentally loses considerable illuminating power by concealing the source of light from view and directing it upon a reflecting surface, usually the ceiling, which re-flects it in a diffused state throughout the interior being illuminated. Shadows are softened, the general effect is more pleasing than in direct lighting, but freedom of choice of ceiling surface is somewhat restricted, the in-stallation is more expensive, and high-powered lamps must be provided or portable lamps provided for direct illumination where high lighting intensity is needed. The third system of lighting, known as the semi-indirect, is intermediate between direct and indirect. Semi-indirect lighting is obtained by enclosing the lamp in a translucent bowl having a surface which diffuses the light as it is transmitted.

The **candle power** of a lamp is what primarily deter-mines brightness of the illumination provided by it. The standard candle is the reference source for illumination, and is maintained in most of the national standardizing laboratories. Since a lamp in space may provide il-lumination in all directions, it can be considered a spher-ical source of illumination. The average candle power, taken in all directions from the lamp, is known as the mean spherical candle power. Often the shape of a lamp, or of its accessories, interferes with transmission of light equally in all directions. The candle power in a hori-zontal direction is frequently employed. The average horizontal candle power is expressed in candles, output is expressed in lumens. The luminous flux from a source of one spherical candle power is 12.57 lumens (4π). Recommended intensities of illumination for a number of different interiors are given in the accompanying table.

INTENSITIES OF ILLUMINATION

	FOOT-CANDLES
Auditoriums, churches...................	2–4
Armories, exhibition halls................	3–6
Corridors, stairways.....................	1–2
Drafting rooms.........................	10–15
Foundries, forge shops, boiler rooms.......	3–6
Industrial yards.........................	0.2–0.5
Instrument shops.......................	6–15
Laundries, machine shops................	5–10
Offices................................	6–10
Schools, libraries.......................	6–10
Show windows..........................	10–70
Stores.................................	4–10
Tool rooms, press rooms, textile mills......	8–16

The lumens which are actually received upon a work-ing area are fewer in number than those which are emitted by the sources of illumination, because of the absorption by walls, ceiling, drapes, etc. The ratio of the lumens reaching the working area to those emitted from the lamp or lamps is the utilization factor. This factor is determined chiefly by the character of the walls and ceiling, the number of windows and hangings, and the type of reflectors or lighting fixtures in which the lamps are mounted. The utilization factor affords a means for connecting the desired intensity of illumination with the lighting installation. (L.D.W., F.T.M.)

ILLUMINOMETER OR ILLUMINATION PHO-TOMETER. An instrument for measuring **illumina-tion.** The older, standard forms of this instrument employ a **photometer** of ordinary type to measure the luminous intensity due to light reflected from a white matt surface or diffused by a white translucent screen exposed to the illumination to be measured. The balance may be secured as usual by the **bench photometer** method, by screening down the comparison lamp, or by turning the diffusing surface through a known angle and relying upon the cosine law (see **Illumination**). A dif-ferent type, known as a foot-candle meter, utilizes the Bunsen screen principle, there being however a long row of translucent spots lighted from behind by a lamp at one end, hence unequally along the row. At some point in the row this balances the illumination from in front, which is to be measured, and a scale indicates the foot-candles directly. In a still more recent instrument the illuminated element is a copper oxide photovoltaic cell (see **Photovoltaic Effects**) connected to a sensitive cur-rent meter reading directly in foot-candles. All of these instruments must, of course, be portable to be of prac-tical use. (L.D.W.)

ILLUSION. A false interpretation of a sensory image. (See **Hallucination.**) (D.M.H.)

ILLUVIATION. A special term in soil nomenclature implying the process by which clay is added to the sub-soil horizon. (R.M.F.)

ILMENITE. A mineral in the hematite group. Com-position, $FeTiO_3$. (R.M.F.)

IMAGE DISSECTOR. This is an electron tube serv-ing as a camera tube for a television system. The pic-ture to be transmitted is focused on a photosensitive surface, causing electrons to be emitted from each part of the surface in proportion to the amount of light in that particular part of the picture. These electrons are drawn down the tube by a positive anode but are kept focused in an electron reproduction of the picture. This focusing is done by magnetic fields. The focusing fields or auxiliary fields are varied periodically so the electron picture is swept back and forth, up and down, as it moves down the tube. Placed in the path of this electron picture is an aperture opening into an **electron multi-**

plier. Each part of the electron stream is thus swept across this aperture and so the output of the electron multiplier will vary with the various parts of the picture. This output, then, represents the electrical unravelling of the picture into an orderly sequence of parts to be transmitted. (L.R.Q.)

IMAGE FREQUENCY. In the **superheterodyne** receiver the **intermediate frequency** is obtained by beating the incoming radio signal with the locally generated signal and selecting the difference (see **Heterodyne**). Since the intermediate frequency stages are tuned to this particular value of frequency they will respond to it regardless of how it may be generated. If the local oscillator is designed always to be higher in frequency than the incoming signal the normal intermediate frequency signal is obtained by subtracting the incoming frequency from the local one. Exactly the same result is obtained in heterodyning if the oscillator is below the incoming signal by the same amount. Thus, if, for example, a set is tuned to 550 kc. and the intermediate frequency is 450 kc., the local oscillator will be tuned to 1000 kc. (giving a difference between the incoming and local of 450). If at the same time a signal is received from a station at 1450 and this signal gets to the **grid** of the first detector it will give a difference frequency of 450 also and hence will go through the i.f. stages along with the desired one. Once the signals have reached the i.f. stages in this manner there is no way of separating them. Both the station at 550 and the one at 1450 will contribute to the loud-speaker output. The frequency of 1450 is said to be the image frequency of the one of 550. (L.R.Q.)

IMAGES. Geometrical Optics.

IMAGINAL DISK. Histoblast.

IMAGINARY NUMBERS. Numbers, also **Complex Numbers.**

IMAGO. An insect in the adult stage. (A.W.L.)

IMBIBITION. In color photography, the transfer of a dye from a matrix, such as a gelatin-relief image, onto a prepared support. Both surfaces are moist and the dye diffuses from the gelatin relief into the support, which is generally coated with a suitably prepared layer of gelatin. (H.C.C.)

IMBRICATE STRUCTURE. The type of compound, low-angle (almost horizontal) thrust faults which produce mechanical piles similar in arrangement to slates on a roof. This type of compound faulting consists of the lower thrust plane or sole, the intervening multiple thrusts (imbricate) and the overlying thrust plane or thrust proper. Imbricate structure is particularly characteristic of the North West Highlands of Scotland. (R.M.F.)

IMHOFF TANK. The Imhoff tank is a chamber of special design suitable for reception and purification of **sewage.** It may be used for the clarification of sewage by plain sedimentation with digestion of the sludge thereby collected. It consists of an upper chamber in which sedimentation takes place and from which the collected solids slide down inclined bottom slopes to an entrance into a lower chamber in which the collection of sludge and digestion take place. The chambers are otherwise completely separated with normal flow of sewage through the upper sedimentation chamber and no flow of sewage in the lower or digestion chamber. The latter must be provided with separate gas vents and sludge discharge lines for withdrawing the sludge after thorough digestion which takes place progressively for a time interval of 6–9 months. The Imhoff Tank is in

effect a 2-story septic tank which by means of this separation retains the advantage of the simplicity of the septic tank without its many disadvantages, most of which accrue from the intimate mixing of fresh sewage and septic sludge in the same chamber.

Imhoff Tanks are being superseded by plain sedimentation tanks with mechanical devices for continuously collecting the sludge which is conveyed to separate sludge digestion tanks. Such an arrangement gives improved sedimentation results and permits, by temperature control, a much more rapid, complete and satisfactory digestion of the sludge in the separate tanks. (E.W.S.)

IMIDES. Amines and Amides.

IMINO-COMPOUNDS. Imino-compounds are organic compounds containing the imino-group ($<$NH), e.g,. dimethylamine ((CH_3)$_2$NH), dibenzamide ((C_6H_5-CO)$_2$NH), succinimide $\left(\begin{array}{c}CH_2CO\\ \\CH_2CO\end{array}\right.NH\Big)$, pyrrole (($CH_4$-NH), uric acid $\left(CO\begin{array}{c}NH-CO-C-NH\\ \quad\quad\quad\quad\parallel\\NH\text{———}C-NH\end{array}CO\right)$. (R.K.S.)

IMMUNITY. The power of the body to resist or overcome a specific disease: the state of being immune.

Natural immunity is immunity developed by an individual without having the disease or inoculations against it. Congenital immunity is immunity that is present at birth. Acquired immunity is immunity that is acquired during life. Active immunity is immunity produced by the stimulation of antibody formation by one of the following means: 1. Recovery from a specific disease. 2. Cumulative exposures to infection without actually having suffered a noticeable attack of the disease. 3. Treatment with a **vaccine.** 4. Injection of toxins of bacteria. Passive immunity results from the transfer of antibodies from an immune to a non-immune subject by **transfusion of blood** or serum. (D.M.H.)

IMPACT. Impact is the action of two bodies in collision, whereby the velocity of one or both bodies is changed. In the case of direct impact, the velocity of the moving bodies is in the direction of the normal (perpendicular) to the bodies at the point of contact. Otherwise the impact is oblique. The impact is central when the centers of gravity of the two bodies lie on the line of impact (normal to the bodies at the point of contact). The momentum of a body is its mass multiplied by its velocity. A law of impact is that the sum of the momentums of the two masses before and after impact is the same, provided the bodies are perfectly elastic, and no energy is absorbed in permanent plastic deformation.

The impact coefficient (coefficient of restitution) is the ratio between the differences of velocities of the two bodies after impact to the same differences before impact. This coefficient would be unity for impact of perfectly elastic bodies, and zero for fully inelastic bodies. To find the energy lost in an imperfect impact (one in which the impact coefficient is some number less than one), the masses of the two bodies, M_1 and M_2, may be substituted into the following formula. This formula has in it also the differences of velocity of the bodies after collision. Let f be the impact coefficient.

$$\text{Energy lost} = \frac{M_1 M_2(1 - f^2)(v_1 - v_2)^2}{2(M_1 + M_2)}.$$

(F.T.M.)

IMPACT LOAD. Load.

IMPACT STRESS. Load.

IMPACT TOUGHNESS. In the design of metal structures and machines the static stresses can ordinarily be calculated and a material of suitable strength selected.

The impact loading requirements are usually less definitely known, and the translation of these requirements into a material specification is very difficult.

In testing rails, car wheels, and certain structural parts such as railway draft gears, a simple drop test is used. The weight, height of fall, and number of blows to failure give a relative measure of impact resistance.

In determining the impact toughness of metals as materials rather than as finished structures or machine parts, tests are made in a pendulum-type testing machine which breaks a standard beam-type sample with a single blow. The height of swing of the weighted pendulum past the anvil after fracturing the sample is related inversely to the energy absorbed in breaking the sample. For example, a tough steel will absorb a large proportion of the total energy of the pendulum so that the swing past the anvil will be small. The results are expressed in foot lbs.

While some tests on relatively brittle materials are made on square-sectioned beam samples, in most cases a notch is machined opposite the striking position to localize fracture as in the Izod and Charpy tests. This introduces stress concentration and a state of combined stresses in the vicinity of the notch, which tends to restrict normal plastic deformation and to produce a brittle fracture quite independent of the velocity effect. In fact, some prefer to call this the notched-bar test rather than an impact test because the energy required to break a notched specimen differs little from that required to fracture a similar specimen by slow bending methods. Furthermore, the velocity developed by a free falling pendulum of reasonable length is relatively small compared to the velocities encountered in moving vehicles, in certain machine parts, or in ballistics, therefore the so-called impact test is more nearly a static than a dynamic test.

High-velocity impact testers which break a test specimen in tension have been developed. The interpretation of test results of this type is still somewhat uncertain.

It is known from experience that impact failures are more likely to occur at low temperatures than at normal room temperatures. Of the ferrous alloys, only austenitic **stainless steels** and certain other alloy grades containing nickel retain a large proportion of their normal room temperature toughness at low temperatures. The aluminum, copper, and nickel base non-ferrous alloys do not develop low-temperature brittleness. The Charpy notched-bar impact test is generally used to determine impact toughness at various temperatures. Other tests such as the tension test fail to detect low-temperature brittleness. (R.H.H.)

IMPALA, PALA. Mammalia, Artiodactyla. *Aepyceros.* Moderately large African **antelopes** of several species. They have long slender horns, slightly spiraled and ringed through most of their length. (A.W.L.)

IMPEDANCE. Alternating-Current Circuits.

IMPEDANCE, CHARACTERISTIC. Characteristic Impedance.

IMPEDANCE MATCHING. Impedance matching is the process of making equal the **impedance** looking both ways from a junction point of two parts of a circuit. This serves two important functions; it gives a condition for maximum power transfer from one circuit to another and also prevents reflection of voltage and current waves. Impedances are usually matched by using the **transformer** which multiplies the value of an impedance connected to its secondary terminals by a factor equal to the square of the primary to secondary turns ratio. Other methods include networks of resistances, networks of reactances, tuned circuits, quarter-wave and half-wave transmission lines, open- and short-

circuited transmission line segments in parallel with the circuits at proper matching points, and numerous others. Matching is very important in long smooth transmission lines to prevent reflection from the load. The latter must be equal to the **characteristic impedance** of the line and, if it is not so, must be transformed by one of the methods previously described. (L.R.Q.)

IMPELLER. The impeller is the rotating member of a **centrifugal pump.** It has backwardly curved vanes mounted on a hub attached to the pump shaft. When these vanes are cased with disks on either side, they are known as shrouded impellers. The water is admitted to the pump in such a way that it comes in contact with the impeller first near the central or hub portion. The impeller rotates in its cases with close clearances, and the water is constrained to remain in the impeller, which, by virtue of its rotation and the action of centrifugal force, drives the water from the periphery at high velocity. (F.T.M.)

IMPETIGO (Impetigo contagiosa). An acute infectious skin disease characterized by the occurrence of discrete, thin-walled vesicles and pustules which rupture, form yellow crusts, and heal without scarring. The lesions may be single or multiple; once started, they often develop and spread very rapidly. The etiologic bacteria are **streptococci**; **staphylococci** cause a similar but deeper type of skin lesion, which is loosely referred to as impetigo also. Ammoniated mercury, gentian violet and more recently **sulfonamide** ointments are used successfully in the treatment of the disease. (D.M.H.)

IMPLICIT FUNCTION. If a **function** is defined by a relation between the **variables** given by an **equation** which must be solved in order to express one variable in terms of the other, the function so defined is called an implicit function. (L.L.S.)

IMPOTENCE. A disturbance of sexual function in the male which precludes satisfactory **coitus.** It varies from premature ejaculation to total loss of erection. Impotence should not be confused with **sterility,** which, in the male, means an absence of normal **spermatozoa** and therefore failure of reproduction. An impotent man may be fertile in that his testicles produce spermatazoa; and a sterile individual may be potent.

The causes of impotence are psychic and organic. The psychic are by far the more frequent. Some form of **psychoneurosis** is at the base of most cases. Organic causes are local lesions of the genitalia interfering with the transmission and delivery of semen, endocrine diseases of the testes and other ductless glands, and organic lesions of the nervous system. (D.M.H.)

IMPOUNDING RESERVOIR. When the required flow of water from a reservoir or from a stream is larger than the minimum rate of flow of that stream in a dry season, a reservoir of the impounding type is needed. Ordinarily, storage is relied upon to a greater or lesser extent to regulate flows. For a study of storage conditions the mass curve is indispensable. The mass curve can best be explained by reference to the accompanying figure. The point of origin of the mass curve is a month when the reservoir is known to be full, as after spring floods. The ordinates of the curve represent the total flow, or mass of water, from the date of origin. River gauging records have mass expressed in acre-feet but, for the purpose of the mass curve, acre-feet are converted into "day-second-feet" so that the slope of a straight line drawn on it will represent a rate of flow in second-feet. A day-second-foot is the volume of water equivalent to a flow of one cu. ft. per sec. for one day. One day-second-foot = 1.98 acre-feet. The line AD represents a uniform flow of 1000 second-feet. Its ordinate is $1000 \times 30 = 30,000$ day-second-feet per month. The mass curve of the non-uniform stream flow is $ABDC$. If lines parallel to AD are drawn tangent to the mass curve, the significance of the points of tangency A, B, C is that, on their dates, the stream flow was equal to the steady flow

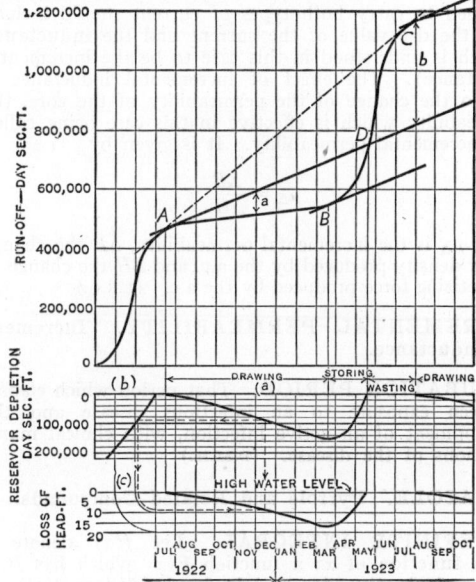

Mass curve and its derivatives.

represented by the slope of AD. At A and C the stream flow passes from greater than to less than the required regulated flow, while at B it changes from less than to greater than the required flow. Having passed A and C the stream discharge becomes less than the required regulated flow, and the difference is made up by drawing on the storage in the reservoir. At the end of November, 1922, the reservoir would have supplied the number of day-second-feet indicated by the ordinate a. Passing B the stream flow has increased and is now in excess of the flow requirements, consequently the storage is regained, and regained completely at D. From D to C there is wastage over the crest gates amounting to a maximum as indicated by the ordinate b. The slope of the line AC is the maximum regulated flow that could be realized without wastage over the crest gates or ultimate depletion of the reservoir. (F.T.M.)

IMPROPER INTEGRALS AND INFINITE INTEGRALS. If $f(x)$ is continuous for $x \geqq a$ and if the definite integral $\int_a^t f(x)dx$ approaches a limit as $t \to \infty$, we denote this limit by $\int_a^\infty f(x)dx$ and call it an infinite integral:

$$\int_a^\infty f(x)dx = \lim_{t \to \infty} \int_a^t f(x)dx.$$

The infinite integral is then called convergent.

We define $\int_{-\infty}^a f(x)dx$ in a similar way.

The infinite integral $\int_{-\infty}^{+\infty} f(x)dx$ is defined by:

$$\int_{-\infty}^{+\infty} f(x)dx = \int_{-\infty}^a f(x)dx + \int_a^\infty f(x)dx.$$

If $f(x) \to \infty$ as $x \to a$, the integral $\int_a^b f(x)dx$ is defined by:

$$\int_a^b f(x)dx = \lim_{\delta \to 0} \int_{a+\delta}^b f(x)dx \quad (\delta > \mathrm{o});$$

this is called an improper integral, and it is said to be convergent.

Similarly, if $f(x) \to \infty$ as $x \to b$, we define:

$$\int_a^b f(x)dx = \lim_{\delta \to 0} \int_a^{b-\delta} f(x)dx \quad (\delta > \mathrm{o}).$$

If $f(x) \to \infty$ as $x \to c$, where c is between a and b, we define:

$$\int_a^b f(x)dx = \lim_{\delta \to 0} \int_a^{c-\delta} f(x)dx + \lim_{\delta' \to 0} \int_{c+\delta'}^b f(x)dx.$$

These improper integrals are also called convergent. (L.L.S.)

IMPULSE. Momentum.

INBREEDING. The mating of closely related animals. As practiced in stock breeding, brothers and sisters are often mated, or parents and their offspring. In most human societies the mating of individuals more closely related than first cousins is not approved, and even cousin marriages are regarded as close inbreeding.

Inbreeding is popularly regarded as a weakening process but this is not entirely true. Closely related individuals are more likely to have a similar complex of hereditary potentialities than those from different lines of descent, hence hereditary characters which may be masked in them have a better opportunity of appearing in their offspring according to the processes of Mendelian heredity. These characters may, however, be either desirable or undesirable. In either case if the character in question can be concealed by some other it may appear more often in an inbred line. (A.W.L.)

INCISOR. A sharp-edged cutting tooth. The incisors are located at the front of the jaws of **mammals**, between the canines. (A.W.L.)

INCLINATION. Inclination is that **element** of the **orbit** of a celestial object which indicates the angle between the plane containing the orbit of the object in question and some reference plane. In the case of orbits of members of the solar system the reference plane is the ecliptic, while in orbits of **binary stars** inclination refers to the angle between the plane of the orbit of the stars and the plane perpendicular to the line of sight. (W.K.G.)

INCLINED PLANE. Machines.

INCLINOMETER. Clinometer.

INCLUSION. This term is used chiefly to connote a fragment of a foreign rock or mineral in an igneous rock. It may also be used to refer to gas or liquid enclosed in a mineral crystal.

In metallurgy, inclusions are small particles of non-metallic compounds embedded in iron or steel. The principal types are oxides, sulfides, and silicates, all of which are more or less hard and brittle. They may originate as slag particles or pieces of refractory from the furnace or ladle which become entrapped upon solidification of the molten metal. More commonly they are the result of reactions within the metal itself during the finishing or deoxidation period and during pouring and solidification. The term "sonim," from solid non-metallic inclusion, is sometimes used.

In high-quality heat-treated machine and tool steels subject to impact and repeated stresses, inclusions are considered objectionable. They can be objectionable in any steel if present in large amounts, particularly in segregated areas forming continuous or semi-continuous stringers or plates known as **laminations**. In normal amounts and when well distributed their effect on strength, ductility, and other properties is negligible. (E.S.C.S., R.H.H.)

INCOMPETENT. Used by structural geologists to designate those formations which when subjected to compressional deformative processes tend to be deformed by multiple fracture or crumpling of the relatively weaker formations. Typical incompetent formations are **shale** and highly carbonaceous sediments especially when interstratified with such **competent** formations as **quartzite, sandstone, limestone** or **dolomite**. (R.M.F.)

INCOMPLETE BETA FUNCTION. The incomplete Beta function $B_x(p, q)$ is defined

$$B_x(p, q) = \int_0^x x^{p-1} (1-x)^{q-1} dx,$$

$$0 \le x \le 1, \quad p > 0, \quad q > 0.$$

Generally, however, we desire

$$I_x(p, q) = \frac{B_x(p, q)}{B(p, q)}.$$

Tables of $B_x(p, q)$ and $I_x(p, q)$ have been provided by K. Pearson. The two functions are helpful in finding the area under a Pearson Type I function, or the theoretical class frequency. They are also useful in finding the levels of significance of Snedecor's "F" and Fisher's z. Tables of the levels of significance of the Beta function (Pearson Type I) have been constructed. These may be used to find the **confidence limits** in the **Bernoulli probability function.** (L.A.A.)

INCOMPLETE GAMMA FUNCTION. The incomplete Gamma function is defined

$$\Gamma_x(n) = \int_0^x e^{-x} x^{n-1} dx, \quad n > 0, \quad 0 \le x < \infty.$$

Tables of the incomplete Gamma function have been prepared by Karl Pearson, using the notation $\Gamma_u(p + 1) = \int_0^{u\sqrt{p+1}} e^{-v} v^p dv$. Let $p = n - 1$ and we see that this corresponds to $x = u\sqrt{n}$. More generally we need

$$I_x(n) = \frac{\Gamma_x(n)}{\Gamma(n)} \quad \text{or} \quad I(u, p) = \frac{\Gamma_u(p + 1)}{\Gamma(p)}$$

in Pearson's notation. The incomplete Gamma function is used to find the area or the frequency in a class in **Pearson's Type III function.** However, it is much easier to use Salvosa's tables of the Type III function than it is to use Pearson's tables. The range covered by Pearson's tables is larger while the subdivisions of the argument are finer in Salvosa's tables. (L.A.A.)

INCONSISTENT STATISTIC. A statistic which is not consistent is said to be inconsistent. Such statistics should never be employed. (See **Consistent Statistic.**) (L.A.A.)

INCONSISTENT SYSTEMS OF LINEAR ALGEBRAIC EQUATIONS. Systems of Linear Algebraic Equations.

INCREMENTAL INDUCTANCE. This is the inductance which an iron-cored coil will offer to a.c. when it is superimposed on d.c. through the coil. This condition occurs very frequently in communication and electronic circuits since many of these involve direct currents for establishing an operating point and then superimpose the a-c signal. The d-c produces a certain amount of saturation in the core so the flux conditions presented to the a.c. are not the same as if no d.c. were present. When the core is in this partially saturated condition the flux changes produced by the a.c. are not as great as they would be otherwise and hence the back e.m.f. of the coil or its inductive effect is reduced. Since the actual inductance presented depends upon the degree of saturation the rating of a coil which is designed to carry both types of current should include both the d-c value of the current and the **inductance** (which is understood in this case to be the incremental inductance). The effect of incremental inductance is due to the change of the permeability of the core, the permeability which is effective in this case being called the incremental permeability. It is given by

$$\mu_\Delta = \frac{\Delta B}{\Delta H}$$

where μ_Δ is the incremental permeability, ΔB the change in flux density produced by the a.c. and ΔH the change in magnetizing force produced by the a.c. (L.R.Q.)

INCREMENTAL PERMEABILITY. Incremental Inductance.

INCUBATION PERIOD. That period which elapses between exposure to an infectious disease and the development of an active infection, with clinical manifestations of the disease. (D.M.H.)

INDAZOLE. Pyrrole and Related Compounds.

INDEFINITE INTEGRAL. Let $f(x)$ denote a given **function** of x; a function $F(x)$ which has $f(x)$ for its **derivative** (or $f(x)dx$ for its **differential**), is called an integral of $f(x)$ (or of $f(x)dx$).

If $f(x)$ is **continuous** and has an integral $F(x)$ in the interval (a, b), then every integral of $f(x)$ in (a, b) is of the form $F(x) + C$.

Hence, if $F(x)$ be any particular integral of $f(x)$, the general integral of $f(x)$ is $F(x) + C$, where C is an **arbitrary constant.** A general integral is also called an indefinite integral. The constant C is called the constant of integration.

The symbol for the general integral or indefinite integral of $f(x)$ is

$$\int f(x) dx.$$

To integrate $f(x)$ is to express $\int f(x)dx$ in terms of known functions. We may interpret the symbol $\int$ as a symbol for this operation of integration and call $f(x)$ the integrand.

By definition

$$\frac{d}{dx} \int f(x)dx = f(x), \quad \text{or} \quad d \int f(x)dx = f(x)dx.$$

Hence, integration is the **inverse operation** of **differentiation,** that is, it is the operation which differentiation undoes.

No general process of integration exists. The process depends ultimately on recognizing a function as a derivative of a known function. It consists, therefore, of reversing differentiation rules. (L.L.S.)

INDEPENDENT CHUCK. Chuck.

INDEPENDENT VARIABLE. Functions.

INDETERMINATE EQUATIONS. An indeterminate equation is an **equation** with more than one unknown or **variable.** Indeterminate equations are sometimes called Diophantine equations, after the Greek mathematician Diophantus, who studied them.

An indeterminate equation in two variables is frequently studied to determine positive integral values of the variables which satisfy it; similarly for more unknowns. (L.L.S.)

INDETERMINATE FORMS. In various mathematical investigations, when limiting processes are applied to certain types of expressions involving special combinations of **functions**, an indeterminate form arises, which is one having no meaning in itself. To assign useful meanings to these forms, methods (which belong mostly to the **calculus**) have been devised, as follows:

The form $0/0$:

If the fraction $f(x)/\phi(x)$ takes the form $0/0$ when $x = a$, it is often possible to discover a factor common to $f(x)$ and $\phi(x)$ which vanishes when $x = a$, and after the removal of this factor, to find $\lim_{x \to a} f(x)/\phi(x)$ directly.

A more general method is: If $f(a) = 0$ and $\phi(a) = 0$, and if $\lim_{x \to a} \dfrac{f'(x)}{\phi'(x)}$ exists, then

$$\lim_{x \to a} \frac{f(x)}{\phi(x)} = \lim_{x \to a} \frac{f'(x)}{\phi'(x)}.$$

If $f'(a) = 0$ and $\phi'(a) = 0$, but if $\lim_{x \to a} \dfrac{f''(x)}{\phi''(x)}$ exists, then

$$\lim_{x \to a} \frac{f(x)}{\phi(x)} = \lim_{x \to a} \frac{f''(x)}{\phi''(x)};$$

and similarly for higher derivatives.

The form ∞/∞:

If the fraction $f(x)/\phi(x)$ takes the form ∞/∞ when $x \to a$, but if $\lim_{x \to a} \dfrac{f'(x)}{\phi'(x)}$, exists, then

$$\lim_{x \to a} \frac{f(x)}{\phi(x)} = \lim_{x \to a} \frac{f'(x)}{\phi'(x)};$$

and similarly with higher derivatives.

The forms $0 \cdot \infty$ and $\infty - \infty$:

These forms can frequently be reduced to the form $0/0$ or ∞/∞ and treated as in the preceding.

The forms 1^∞, 0^0, ∞^0:

A function of the type $[f(x)]^{\phi(x)}$ may take one of these forms. If we put $u = [f(x)]^{\phi(x)}$, take logarithms and get $\log u = \phi(x) \log f(x)$, which will then take the form $0 \cdot \infty$, which can be evaluated by the preceding methods. (L.L.S.)

INDETERMINATE STRUCTURE. A statically indeterminate structure is one which cannot be solved by the equations for static equilibrium. These equations state that the components of the forces acting on a body, taken in any two directions, must be equal to zero and that the sum of the moments of these same forces, taken around any moment center, must equal zero. If the axial stresses in the members of a structure are changed by altering the length of one of the members a very small amount the structure is classified as indeterminate.

When there are more **reactions** than equations for static equilibrium the structure is externally indeterminate. After these reactions have been calculated the stresses in the members of the structure become statically determinate unless internally indeterminate, that is, contain redundant members (see **Redundancy**).

In general, statically indeterminate structures can be analyzed by methods such as the energy theory or the deflection theory, although some are so complicated that the analyst must resort to experimental methods. The analysis of any indeterminate structure requires a knowledge of the size, shape and elastic properties of the individual members.

Examples of indeterminate structures are triangular frameworks containing redundant members, rigid frames having **loads** transmitted by the rigidity of the joints, fixed or 2-hinged **arches**, suspension **bridges** with stiffening **trusses**, building frames under the action of **lateral** forces, beams with built-in end supports, beams fixed at one end and simply supported at the other end and beams continuous over a number of supports. (See **Least Work.**) (C.W.C., F.T.M.)

INDEX FOSSIL. Paleontology; Stratigraphy.

INDEX NUMBER. An index number is usually the ratio of 2 sums. It is a statistical device to measure changes in business, wages, prices, cost of living, production and consumption. Formulas are of two types, aggregative or relative and may be weighted or unweighted. There is always a base period to which other periods are compared. A typical aggregative formula weighted by quantities in the base year is

$$I = \frac{\Sigma p_1 q_0}{\Sigma p_0 q_0} = \frac{p_1{}^{(1)}q_0{}^{(1)} + p_1{}^{(2)}q_0{}^{(2)} + \cdots + p_1{}^{(N)}q_0{}^{(N)}}{p_0{}^{(1)}q_0{}^{(1)} + p_0{}^{(2)}q_0{}^{(2)} + \cdots + p_0{}^{(N)}q_0{}^{(N)}}$$

an index number of price where $p_1{}^{(N)}$ is the price of the Nth item in the period compared to the base period, $p_0{}^{(N)}$ is the price of the Nth item in the base period, $q_0{}^{(N)}$ and $q_1{}^{(N)}$ referring, respectively, to the quantities of the Nth item in the base period and the period being compared to the base period. (L.A.A.)

INDEX OF CORRELATION. The index of correlation, ρ_{yx} is defined as

$$\rho_{yx} = \sqrt{1 - \frac{\sigma_e{}^2}{\sigma_y{}^2}},$$

where $\sigma_e{}^2$ is the square of the **standard error of estimate**,

$$\sigma_e{}^2 = \frac{\Sigma(y - y^1)^2}{N},$$

where y is an actual value, y' is the theoretical or estimated value found from a curvilinear function $y' = f(x)$. The **variance** of y is $\sigma_y{}^2$. The main distinction between the **coefficient of correlation** and the index of correlation is that r_{xy} depends on linear **regression**. To test the significance of ρ_{yx}, the **analysis of variance** is used. (L.A.A.)

INDEX OF DETERMINATION. The index of determination is $\rho^2{}_{yx}$, the square of the **index of correlation**. (L.A.A.)

INDEX OF DISPERSION. The purpose of the index of dispersion for the **Bernoulli probability function** and the **Poisson distribution** is to test homogeneity in certain types of data. The index of dispersion for the Bernoulli probability function is defined as

$$D = \frac{\Sigma(x - \overline{X})^2}{npq} = \frac{\Sigma(x - \overline{X})^2}{\overline{X}\left(1 - \dfrac{\overline{X}}{n}\right)},$$

where the Σ extends over the number of **samples** N, $\overline{X}$ is the **mean** of the number of successes in each sample, n is the number of specimens examined in each sample. Then D is distributed approximately as χ^2 with $N - 1$ **degrees of freedom**. If $np \geq 5$, this approximation is highly satisfactory. For $np < 5$ it should be handled with care.

For the **Poisson distribution**, the index of dispersion D is defined as

$$D = \frac{\Sigma(x - \overline{X})^2}{\overline{X}},$$

where x is the value from each sample and $\overline{X}$ is the mean of the N samples. D is again distributed approximately as χ^2 with $N - 1$ degrees of freedom. This approximation is quite satisfactory for $m \geq 2$, where m is the mean in the Poisson distribution. (L.A.A.)

INDEX OF MULTIPLE CORRELATION. The index of multiple correlation P is defined as

$$P = 1 - \frac{\sigma_e{}^2}{\sigma_y{}^2},$$

where

$$\sigma_e{}^2 = \frac{\Sigma(y - y^1)^2}{N},$$

y is the actual value of an observation, y' is the value obtained by a curvilinear function involving more than one independent **variable**, σ_y^2 is the **variance** of the variable y. The index is essentially the same as the **coefficient of multiple correlation**. The main difference is that in the case of multiple correlation the theoretical or estimated value y' is found by use of the independent variables in the first degree only. (L.A.A.)

INDEX OF MULTIPLE DETERMINATION.
The index of multiple determination is the square of the **index of multiple correlation** P^2. (L.A.A.)

INDEX OF PRECISION.
The index of precision, h, is defined as

$$h = \frac{1}{\sqrt{2}\,\sigma_x}$$

when x is **normally distributed**. As the σ_x decreases, the index of precision increases. It is infrequently used. (L.A.A.)

INDEX OF REFRACTION.
Refractive Index.

INDEX OF ROOT OF NUMBER.
Roots of Numbers.

INDEXING.
Milling.

INDICATED HORSEPOWER.
The horsepower developed in the cylinder of any **piston** and **cylinder engine** is called indicated horsepower because it can be measured by the **indicator** mechanism. This power is the result of a gas or vapor pressure pushing against a piston, and is larger than the crankshaft horsepower by the amount of friction and windage losses of the engine.

If, during a certain power stroke in a cylinder, P is the average effective pressure acting against each unit of piston area, the piston will receive and pass on to the piston rod a push of $P \times A$ pounds. In the above, A is the piston area. When the piston moves through a stroke measured as L feet, the work represented by the motion of the push of PA pounds is PAL. In the ordinary double-acting steam engine, the number of these power strokes per minute is double the number of revolutions per minute, since there is an action of steam against the piston on both the out and in stroke. Thus in one minute there are $2PLAN$ foot-pounds of work done in the cylinder. The number of foot-pounds per minute in a horsepower being 33,000, the above explains the origin of the common steam engine horsepower formula,

$$IHP = \frac{2PLAN}{33,000} \text{ per cylinder.}$$

The actual use of this formula requires knowing the bore and stroke of the engine from which A and L are determined, the operating speed in revolutions per minute, and the mean effective pressure P. This mean effective pressure can be obtained experimentally with the use of the indicator, which is an autographic device by means of which an engine draws its cycle on a PV plane. The average height of the figure so drawn is the mean effective pressure.

In the case of multi-cylinder internal combustion engines which may be two- or four-stroke cycle, single- or double-acting, the $2N$ of the above formula should be replaced by the number of power strokes made per minute by the engine. The result would be engine indicated horsepower. (F.T.M.)

INDICATOR.
Chemical indicators are discussed in the article on **Reactions Involving Recombinations of Ions**.

In engineering the indicator is an **engine** instrument the purpose of which is to give an autographic record of the pressure-volume relationship of the working medium within the cylinder as the engine goes through its **cycle** of operations. The cross-section of a **steam engine** indicator is shown in the accompanying drawing. The rec-

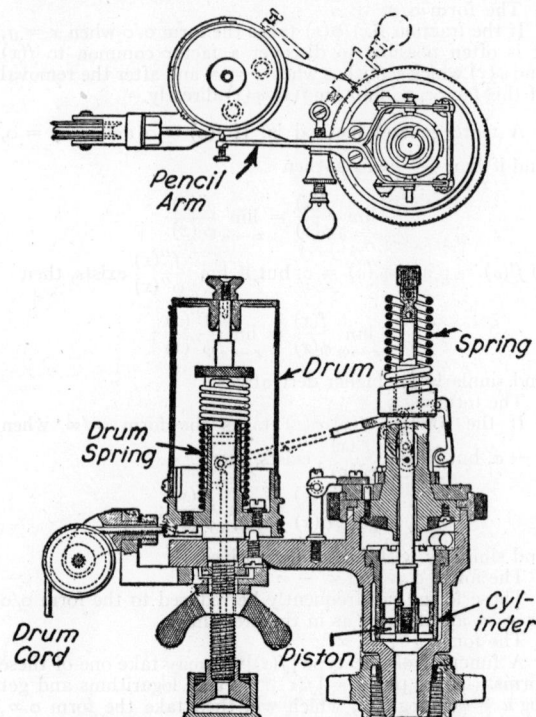

Sectional view of Maihak standard indicator. (Backarach Industrial Instrument Co.)

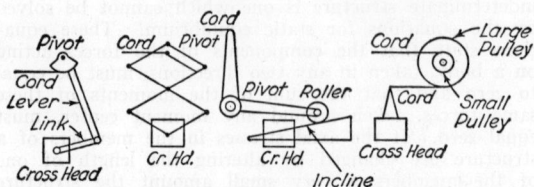

Indicator reducing motion.

ord is traced on a paper of oblong shape about 3″ by 6″, which is wound around and fastened to the drum. The lower portion of the drum is a pulley around which is wound a cord. The end of the cord is made fast to the drum, and the other end is attached either directly to the crosshead or through a **reducing mechanism**. The rotation of the drum under the influence of a pull on the cord is always opposed by a coil spring. In action, the cord transmits to the drum an oscillatory motion which is synchronized with the motion of the piston in the cylinder. This places the paper wound around the drum in a position (with respect to a fixed pencil point) which is proportional to the stroke or to the volume of piston displacement at any instant. The other element of the indicator, the pressure element, consists of a piston moving in a cylinder and opposed by a calibrated spring. The indicator cylinder is connected to the engine cylinder by short interconnecting piping through which the instantaneous pressures in the cylinder are transmitted to the indicator piston. The latter rises against the spring pressure until the point of equilibrium is reached. The position of the piston in the indicator cylinder is proportional to the pressure in the engine cylinder. This position is transferred to the record as a corresponding vertical motion of the pencil arm. At the extremity of this arm, the pencil writes on the drum. It draws a record which, by virtue of the features pointed out above, is a diagram of the cycle of pressure vs. the volume in the engine cylinder. Engines with high rotative speed (over 750 rpm) may not readily be tested with the above described indicator because the inertia of the indicator parts prevents an accurate tracing of the rapidly changing state of the working medium in the engine cylinder. Special high-speed indicators, electrical in nature, have been developed which trace a pressure-time graph. Using data from this curve and the known bore, and rpm of the engine, a pressure volume diagram may be constructed. (F.T.M.)

INDICOLITE. Tourmaline.

INDIGESTION. Impairment of digestion and the symptoms resulting therefrom. The term is a loose one and covers digestive disturbances which may occur as part of the symptom complex of a number of diseases. The symptoms complained of vary, but usually include discomfort and a feeling of fullness in the abdomen, belching, sour eructations and "heartburn." (D.M.H.)

INDIGO. Indigofera tinctoria. Leguminoseae. The indigo-yielding plant is a shrub 4–6′ tall, growing wild in southern Asia. It has **pinnately** compound leaves, which are downy beneath, and bears reddish-yellow flowers.

For thousands of years the plant has been grown in India, where it was first discovered to yield a dye which would give cotton or woolen cloth a deep blue color of great permanence. Curiously enough, no indication of this color is found in the plants. They contain, however, a glucosidal (see **Carbohydrates**) substance which can be extracted from the shoots with water, and which yields indigo blue. This substance is insoluble in water. Before it can be used to color cloth, the dye-stuff must be dissolved in lime-water (or any **alkaline** solution), and some fermenting substance, such as molasses. Then a new compound, indigo white, is formed. When the cloth to be dyed is soaked in this solution and then exposed to the air to dry, it soon acquires the deep blue color characteristic of indigo. (For the chemistry of Indigo, see **Pyrrole and Related Compounds.** See also **Dyes.**)

The development of synthetic indigo has caused the use of the natural product to decline considerably. (R.M.W.)

INDIUM. Symbol: In. Atomic number: 49. Atomic weight: 114.76. Density: 7.28. Hardness: 1.2. Melting point: 155° C. Boiling point: 1450° C. (Isotopes: page 290.)

Indium is a silver-white metal, softer than **lead**, malleable, ductile and crystalline: stable in dry air: on heating in air burns with a blue flame to form oxide; does not decompose water at 100° C.; dissolves in **hydrochloric, sulfuric** or **nitric acids** but not in **sodium** hydroxide; combines with **chlorine** and with **sulfur.** Discovered by Reich and Richter in 1863 by means of spectroscope.

Indium occurs in very small amount in **zinc blende, tungsten, tin** and **iron** ores of certain localities. The recovery of indium from zinc flue dust (sometimes, 1 part per thousand) is effected by treating with a slight deficiency of hydrochloric acid and allowing to stand. The residue is subjected to a series of treatments until finally pure indium sulfate is obtained, a solution of which when electrolyzed yields compact indium metal. A thin surface layer of indium is used on some bearings.

Chlorides: Indium monochloride (InCl), dark red crystals; indium dichloride (InCl$_2$), colorless crystals; indium trichloride (InCl$_3$), white crystals. Volatile indium salts, such as the chlorides, color the Bunsen flame blue-violet, which accounts for the derivation of the name indium (indigo blue).

Hydroxide: Indium hydroxide (In(OH)$_3$), white, gelatinous precipitate by sodium hydroxide with indium salt solutions, soluble in excess sodium hydroxide, but reprecipitated on boiling.

Oxides: Indium monoxide (InO), black, pyrophoric powder, by heating indium sesquioxide in hydrogen at 300° C.; indium sesquioxide (In$_2$O$_3$), light yellow powder, by burning indium in air, or by ignition of indium nitrate below 850° C., above 850° triindium tetroxide (In$_3$O$_4$) is formed. (R.K.S.)

INDOLE. Pyrrole and Related Compounds.

INDRI. Mammalia, Primates. A lemur, *Indris brevicaudatus*, of Madagascar, with a slender body, disproportionately large arms and legs, and a small head resembling that of a dog. (A.W.L.)

INDUCED DRAG. A large portion of the air drag on an airfoil or wing is derived from a source which is independent of surface friction, turbulence, impact, etc. This *induced* drag is theoretically zero if the span is infinitely long. If the Æ were 100 or more, the induced part of the drag would, for all practical purposes, have disappeared. With constant angle of attack, the coefficient of induced drag depends only on Æ, hence activities like smoothing and waxing wing surfaces, fairing projections, etc., would not affect it. Aircraft needing reduction of drag to least possible values (such as sailplanes) must perforce not only have smooth, aerodynamically clean surfaces, but also wings of high Æ. Long tapered wings with Æ's as high as 20 have been used. This might be contrasted with the powered airplane where 5 to 8 is found to be the usual range. (See **Airfoil.**) (F.T.M.)

INDUCED RADIOACTIVITY. Artificial Disintegration of Elements.

INDUCTANCE. The inductance of a circuit (such as a coil) is the rate of increase in magnetic linkage with increase of current. If we have a coil of several turns, carrying a steady current, a certain magnetic flux will, as a result, be linked with the coil, depending upon the size and shape of the coil, the number of turns, and the material occupying the surrounding space. If the current is now slightly increased, the resulting increase in flux may or may not be proportional to the change in current; if not, we shall have to consider a very small increase in each. The "linkage" is the product of the flux through the coil by the number of turns. Since magnetic flux is ordinarily expressed in lines or maxwells (see **Magnetic Field**), the linkage may be expressed in maxwell-turns. The inductance unit called the henry corresponds to a rate of linkage increase of 10^8 maxwell-turns per ampere of current. This is a rather large unit, hence the millihenry and microhenry are commonly used.

Since an increase of 10^8 maxwell-turns per sec. induces 1 **volt** of electromotive force in the circuit, a coil may be said to have 1 henry of inductance if an increase of 1 ampere per sec. results in the self-induction of 1 volt; this is an alternative definition of the henry.

Mutual inductance may be explained in the same way and expressed in the same units; except that the current in one circuit causes a flux linkage, or induces an electromotive force, in another, neighboring or coupled circuit. An **induction coil** or a **transformer** involves this kind of inductance, as well as the self-inductance of each circuit. A coil arranged to have its inductance adjustable and provided with a scale to indicate its inductance in millihenrys is called an inductometer. Inductances of circuits are usually measured in terms of the impedances which they offer to **alternating currents.** For very high frequencies the observed impedance of an inductive circuit presents anomalies due to the "skin effect," as a result of which the inductance apparently changes slightly with frequency in the high-frequency range. (L.D.W.)

INDUCTION COIL. A device for obtaining a high, intermittent voltage from a source of low, steady voltage, such as a battery. This is accomplished by electromagnetic induction. A coil of relatively few turns, called the primary, and provided with a soft iron core, is connected in series with the battery and with some form of interrupter which renders the current intermittent. The rapid variations of magnetic flux thus produced give rise to correspondingly high electromotive forces in the secondary, a coil of many turns

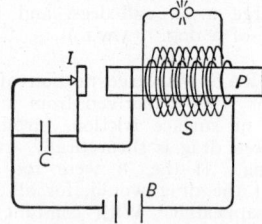

Diagram of an induction coil, showing battery *B*, interrupter *I*, and condenser *C*.

of fine wire wound on the same core and thus effectively coupled with the primary. The secondary electromotive force of course reverses with each make and break of the primary circuit. The performance is improved by placing a **condenser** across the terminals of the interrupter. This takes up the surge of current that would otherwise cause destructive sparking at the interrupter gap, while the discharge of the condenser causes a higher peak electromotive force in the secondary. The change of flux during the charging and discharging of the condenser while the interrupter is open is much more rapid than that while the primary circuit is closed. This gives a partially unidirectional character to the secondary voltage and makes the induction coil suitable for operating Geissler and small x-ray tubes. The largest use of the induction coil at present is for ignition in the operation of internal combustion engines, where it is commonly called a spark coil. Devices of somewhat similar construction, known as vibrators, are commonly used to derive the voltages necessary for operating radio apparatus from low-voltage batteries. (L.D.W.)

INDUCTION FIELD. This is the magnetic field set up around a conductor by the current in the conductor and which causes a back electromotive force to be induced in the conductor. The magnetic fields present around d-c and low-frequency circuits are considered to be of this sort since the error in the assumption is entirely negligible. All the energy stored in this field is returned to the circuit when the current flow is stopped. The induction field is contrasted with the radiation field which has a large value at radio frequencies and whose energy is not returned to the circuit but is radiated outward giving electromagnetic waves in space (the waves by which we communicate in radio). Many short-distance wireless communication systems utilize the induction field since it does not extend very far and hence does not interfere with regular radio reception at any great distance. (L.R.Q.)

INDUCTION FURNACE.   Electric Furnace.

INDUCTION GENERATOR. When an induction **motor** is driven by mechanical means above synchronous speed its **slip** becomes negative and the machine will act as a **generator** if it has the proper excitation. The induction generator draws its exciting current from the line which puts a definite limit on the usefulness of the machine. For the line current to produce the correct field the load must take a leading current from the generator. This requirement for a **capacitive load** may be met by static **condensers** (almost never used) or synchronous machines on the line. The speed of the induction generator does not determine the **frequency** which is fixed by synchronous machines on the same system. The load which the induction machine can supply does, however, vary with the speed. Since induction generators cannot supply lagging power factor loads and require an appreciable leading current for excitation they are not widely used to give increased generating capacity to a system. The generating characteristic of over-driven induction motors is important in certain applications, notably railway work, where it is used for dynamic braking. (L.R.Q.)

INDUCTION HEATING.   Electric Furnace.

INDUCTION, MATHEMATICAL. The method of mathematical induction is a general method of proof of theorems in which a positive integral variable is involved. It consists of two main parts: (1) direct verification of the theorem for the smallest admissible value of the positive integer involved, (2) the algebraic proof that if the theorem is true for any value of the integer, it is true for the next greater value. In conclusion, the theorem is proved by combining the above two parts. (L.L.S.)

INDUCTION MOTOR.   Motor, Electric.

INDUCTION REGULATOR. This is a specially constructed **transformer** designed for regulating the line voltage of a-c systems. The single-phase regulator is essentially a transformer with one winding arranged so it may be rotated with respect to the other. When the axes of the two windings coincide the induced voltage in the secondary will be a maximum while when the axes are 90° apart the induced voltage will be a minimum. The primary winding is connected across the line and the secondary in series as shown in the figure. This causes the voltage of the secondary to add vectorially to the original line voltage. By rotating the movable winding the magnitude of the secondary voltage may be changed and it may also be made to add to or subtract from the primary voltage. The resultant output voltage can, then, be varied by plus or minus the secondary voltage. The capacity of the regulator needs to be only a small fraction of the capacity of the line since only a small part of the power is transformed, the remainder going through by direct conduction. The operation of the regulator is usually made automatic by voltage-operated relays controlling driving motors on the rotor. To prevent the series winding offering high impedance when its axis is 90° from the primary a shorted winding is placed on the primary core 90° from the primary winding. Thus this winding serves as a shorted winding coupled to the secondary when it is in its zero position. If it were not for this auxiliary winding the current through the main secondary would be limited to a very low value when the secondary is in its minimum voltage position. Any three-phase wound-rotor **induction motor** with blocked rotor may be used as a three-phase induction regulator or the machine may be built especially for this application, the principle of operation being the same in either case. The secondary voltages will always be of the same magnitude regardless of the position of the rotor but the phase relation of the secondary and primary voltages may have any value from 0° to 360° so the vector addition will give a variable or adjustable output. (See figure for connections.)   (L.R.Q.)

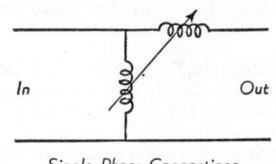

Single-Phase Connections

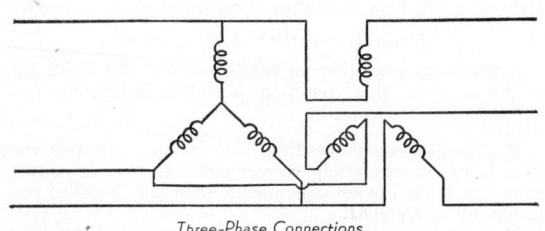

Three-Phase Connections

INDUCTIVE COORDINATION. This is the co-ordinated system of line **transpositions** carried out by the telephone and power system operators. (See **Inductive Interference.**)   (L.R.Q.)

INDUCTIVE INTERFERENCE. When a telephone line is paralleled by a power line or even another telephone line there is almost certain to be induced interference in the telephone line. This interference is due to voltages and corresponding currents induced in the line by voltages or currents in the paralleling line. If a.c. flows in a line it causes an alternating flux to be set up around the wires. This flux extends outward for a considerable distance and may link another line inducing voltages in series with this second line. Because of their extremely small magnitude the effects produced by telephone lines in power lines are of no consequence but those produced by the power line in the telephone line are very serious since, although they may be small, they are comparable to the normal signal voltages. Sixty-cycle currents are below the transmission limits of most telephone equipment but may cause trouble in telegraph circuits. However, harmonics of the power circuit and especially high-frequency transients induced by switching and lightning cause objectionable interference. Another type of induced voltage is caused by the electrostatic flux from the high-voltage transmission line. The telephone line in this field will assume a potential corresponding to its capacitance with respect to the ground and the power line. Noise currents in the terminal equipment of the phone circuits may be eliminated or materially reduced by accurate balancing of the various line and equipment impedances with respect to ground in the telephone system and by a co-ordinated system of line **transpositions.** By transposing the lines (both telephone and power) very nearly equal voltages will be induced in both wires of the telephone line and hence there will be no net voltage across them or in series with them to cause the flow

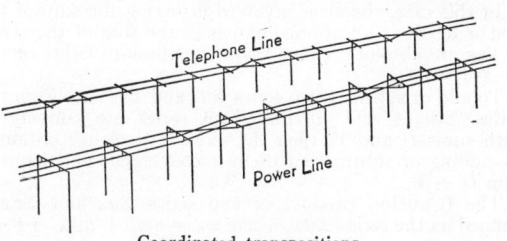

Coordinated transpositions.

of noise currents. Telephone lines are also transposed to avoid **crosstalk** between them. These transpositions must be carefully worked out for all lines concerned if full benefit is to be realized. Inductive interference from power lines is largely eliminated by the shielding by the sheath of a **cable** and is completely eliminated by the use of the **coaxial cable.** An elementary coordinated system of transpositions is shown. (L.R.Q.)

INDUCTIVE LOAD. An electrical load consisting of resistance and inductance, that is to say, of **coils, solenoids,** windings, etc., is said to be inductive since the induction in such windings delays the current to the point where variations of current in an a-c circuit lag a certain degree behind variations of voltage. (See **Alternating Currents.**)   (F.T.M., L.R.Q.)

INDUCTOMETER. Inductance.

INDUCTOR. An inductor is a conductor or bundle of conductors on an electrical machine in which the voltage is induced by the cutting of lines of flux. The main conductors along the surface or in the slots of a generator **armature** are inductors.   (L.R.Q.)

INDURATED. This term as used by petrologists signifies rocks which have been hardened by the action of heat.   (R.M.F.)

INDURATION. The state of increased resistance or hardness in any tissue or organ. This may be an indication of inflammation, abscess or tumor.   (R.S.M.)

INDUSIUM. Ferns.

INDUSTRIAL STATISTICS. Industrial statistics consists of the applications of statistical theory to the problems of industry and engineering. The theory includes **descriptive statistics, probability functions,** the **Bernoulli, Poisson,** and **normal probability functions,** analysis of variance, Student's t distribution, Snedecor's F distribution, consumer protection, production control, the control of defectives in the manufacturing process and **quality control** in general.   (L.A.A.)

INEFFICIENT STATISTIC. An inefficient statistic is one which is not efficient. Roughly speaking, it does not make a complete use of all the **variates** in a sample. (See **Efficient Statistic.**)   (L.A.A.)

INEQUALITIES. The notation $a < b$ means: a is less than b, and the notation $a > b$ means: a is greater than b. The notation $a \leqq b$ means: a is either less than or equal to b; similarly for $a \geqq b$.

The rules for operating with inequalities are expressed by:

$$\text{if } a < b, \text{ and } b < c, \text{ then } a < c,$$
$$\text{if } a < b, \text{ then } a + c < b + c,$$
$$\text{if } a < b \text{ and } c > 0, \text{ then } a \cdot c < b \cdot c,$$
$$\text{if } a < b \text{ and } c < 0, \text{ then } a \cdot c > b \cdot c.$$

If the sense of an inequality is the same for all values of the symbols for which its members are defined, the inequality is called an absolute or unconditional inequality.

If the sense of an inequality holds only for certain values of the symbols involved, but is reversed or destroyed for other values of the symbols, the inequality is called a conditional inequality.

The sense of an inequality is not changed if both members are increased or decreased by the same number.

The sense of an inequality is not changed if both members are multiplied, or divided, by the same positive number. The sense of an inequality is reversed if both members are multiplied, or divided, by the same negative number. (L.L.S.)

INERT GASES. These are the elements which are found in the first group of the periodic table and consist of helium, neon, argon, krypton, xenon and radon. (R.K.S.)

INERTIA. Inertia is one of the few properties manifested by all kinds of matter. While its name infers inaction (no doubt from the fact that bodies do not set themselves into motion), inertia is an actual opposition to any alteration of motion. That is, the acceleration of a body in any sense, either as to speed or direction of motion, requires the application of force proportional to the mass of the body and to the amount of acceleration. The reaction to this force, acting always through the **center of mass**, is a measure of the inertia of the body.

It is interesting to speculate as to wherein the inertia of a body resides. Is it inherent in the body itself, or does it in some way concern also its physical environment? The discovery of electromagnetic mass and electromagnetic inertia, and the identification of the latter as a property of the **electromagnetic field** surrounding a moving charge, suggests that the ordinary inertia of neutral masses may be similarly traceable to some reaction in the neighboring space. And when we consider that all matter is, in the last analysis, probably electrical in character, the idea that inertia may actually be an electromagnetic reaction appears not unreasonable. (See **Moment of Inertia.**) (L.D.W.)

INFANTILE PARALYSIS. Poliomyelitis.

INFECTIOUS MONONUCLEOSIS (Glandular fever). An acute disease characterized by fever, sore throat, enlargement of the spleen and lymph nodes, and leukocytosis with an increase in large mononuclear cells up to 50 to 85% of the total count. These cells are abnormal lymphocytes which are typical for a number of virus diseases.

The cause of infectious mononucleosis is unknown. The disease seems to be contagious . It occurs chiefly in the age group 15 to 30 years. The clinical picture is extremely variable, but the course is usually mild. The average duration is 2 to 4 weeks, and treatment is symptomatic. The diagnosis is established by the characteristic blood findings plus a serological test. The serum of patients with infectious mononucleosis contains so-called hetrophile **antibodies** which agglutinate sheep red cells. This test is positive the second week, often in very high titer. (D.M.H.)

INFINITE GEOMETRIC PROGRESSION. Geometric Progressions.

INFINITE SEQUENCES. An infinite sequence is an endless set of elements such that we can establish a one-to-one correspondence between these elements and the natural numbers $1, 2, 3, \cdots, n, \cdots$. An infinite sequence may be represented in general by $u_1, u_2, u_3, \cdots, u_n, \cdots$, or more briefly by $\{u_n\}$.

A sequence $\{u_n\}$ is called convergent when it has a limit; that is, if a number u exists such that for any arbitrary number $\epsilon > 0$ there exists a number N such that $|u - u_n| < \epsilon$ for every $n > N$, then the sequence $\{u_n\}$ is convergent and has the limit u. We write $u = \lim_{n \to \infty} u_n$.

If a sequence is not convergent, it is frequently called divergent. (L.L.S.)

INFINITE SERIES. Let $u_1, u_2, u_3, \cdots, u_n, \cdots$ be an **infinite sequence**; form the partial sums: $s_1 = u_1$, $s_2 = u_1 + u_2$, $s_3 = u_1 + u_2 + u_3, \cdots, s_n = u_1 + u_2 + \cdots + u_n, \cdots$. The infinite sequence $s_1, s_2, s_3, \cdots, s_n, \cdots$

is called the infinite series whose terms are $u_1, u_2, \cdots, u_n, \cdots$; it is denoted by

$$u_1 + u_2 + u_3 + \cdots + u_n + \cdots, \text{ or by } \Sigma u_n, \text{ or } \sum_{n=1}^{\infty} u_n.$$

If the infinite sequence $\{s_n\}$ converges and has a limit S, then the infinite series Σu_n is said to converge and to have the sum S; otherwise the series is usually said to be divergent.

The general principle of convergence for series is: A necessary and sufficient condition for the convergence of a series Σu_n is that, for any $\epsilon > 0$, an index N exists such that for every $n > N$ and for every value of p, we have

$$|u_{n+1} + u_{n+2} + \cdots + u_{n+p}| < \epsilon.$$

A necessary condition for convergence of the series Σu_n is $\lim_{n \to \infty} u_n = 0$; this condition is not sufficient for convergency.

If $\Sigma|u_n|$ is convergent, so also is Σu_n. In this case Σu_n is called absolutely convergent. If Σu_n is convergent but $\Sigma|u_n|$ is not convergent, then Σu_n is called conditionally convergent.

Many tests of convergence and divergence of series are known. We can give only a few of the most useful of these here.

Comparison test: If a positive term series Σc_n is known to be convergent, and if Σa_n is a positive term series to be tested and if $a_n \leq c_n$ for all values of n (or at least for all values of n beyond a certain point), then Σa_n is convergent. If Σd_n is a positive term series known to be divergent, and if $a_n \geq d_n$ for all values of n (or for all values of n beyond a certain point), then Σa_n is divergent.

Ratio test: If Σa_n is a positive term series to be tested, and if $\dfrac{a_{n+1}}{a_n} < r < 1$ for all values of n (or for all values of n beyond a certain point), then Σa_n is convergent; if $\dfrac{a_{n+1}}{a_n} \geq 1$ for all values of n (or all beyond a certain point), then Σa_n is divergent.

Also, if $\lim_{n \to \infty} \dfrac{a_{n+1}}{a_n} = l$ and $l < 1$, then Σa_n is convergent, and if $l > 1$, then Σa_n is divergent, but if $l = 1$, there is no test.

Radical test: If Σa_n is a positive term series to be tested, and if $\sqrt[n]{a_n} < r < 1$ for all values of n (or for all beyond a certain point), then Σa_n is convergent, and if $\sqrt[n]{a_n} \geq 1$ for all n (or for all beyond a certain point), then Σa_n is divergent.

Also, if $\lim_{n \to \infty} \sqrt[n]{a_n} = l'$, then Σa_n is convergent if $l' < 1$, and is divergent if $l' > 1$, but if $l' = 1$ there is no test.

Integral test: If $f(x)$ is a positive monotonic decreasing function which approaches 0 as $x \to \infty$, then the series $\Sigma f(n)$ is convergent or divergent according as the definite integral $\displaystyle\int_a^{\infty} f(x)dx$ is convergent or divergent (that is, according as $\lim_{\lambda \to \infty} \displaystyle\int_a^{\lambda} f(x)dx$ exists (finite) or not).

Alternating series test: If a series Σu_n has its terms alternately positive and negative, and steadily decreasing, so that each term is numerically less than the preceding, and if $u_n \to 0$ as $n \to \infty$, then Σu_n is convergent.

In this case, the error involved in taking the sum of the first n terms as an approximation to the sum of the series is less in absolute value than the absolute value of the $(n + 1)$st term.

To add or subtract two series Σu_n and Σv_n is to form the series $\Sigma(u_n \pm v_n)$. If the given series are convergent with sums U and V, then the series $\Sigma(u_n \pm v_n)$ obtained by adding or subtracting them is convergent and has the sum $U \pm V$.

The (Cauchy) product of two series Σu_n and Σv_n is defined as the series Σw_n, where $w_n = u_1v_n + u_2v_{n-1} + \cdots + u_nv_1$.

If Σu_n and Σv_n are both absolutely convergent, then the product series Σw_n is absolutely convergent, and if their sums are U and V, then the sum W of the product series is $W = U \cdot V$. (Cauchy's theorem.)

If Σu_n and Σv_n are convergent with sums U and V, and if at least one of them is absolutely convergent, then the product series Σw_n is convergent and its sum is $W = U \cdot V$. (Mertens' theorem.)

If Σu_n and Σv_n are convergent with sums U and V, and if the product series Σw_n is convergent with sum W, then $W = U \cdot V$. (Abel's theorem.)

If two series are only conditionally convergent, their product series may be convergent or may be divergent. (L.L.S.)

INFINITESIMALS. Limits.

INFLAMMATION. The response of the body to infection or irritation. It is characterized by redness, heat, swelling, and pain. The redness and heat are due to the increased blood supply to the involved area. The blood vessels are dilated and engorged, and there is a loss of plasma fluid from them into the tissue spaces. This results in **edema** or swelling. The swelling distends the tissues, compresses nerve endings, and thus causes pain. The white blood cells or leucocytes take an important part in inflammation. They crowd the tissue spaces, carry on their work as **phagocytes,** picking up bacteria and cellular debris. They aid in walling off an infection and preventing its spread. As the inflammatory reaction subsides, repair of the damaged tissue takes place. If the tissue is one capable of complete regeneration, new cells of the same type may completely replace the old ones. This phenomenon is seen in minor inflammations of the skin. In other tissues, such as nervous tissue, regeneration may be very limited or absent; the damaged cells will then be replaced by fibrous scar tissue. This latter form of repair occurs with all inflammations of great size which cause marked cellular destruction. (D.M.H.)

INFLECTION, POINT OF. Concavity and Convexity of Plane Curves.

INFLORESCENCE. Flower.

INFLUENCE LINE. An influence line is a graphical way of representing the effect of a certain variable circumstance upon a given condition. In particular, the influence line as applied in structural engineering represents the variable effect of a single moving unit concentrated load upon the shear, bending moment, reaction, or any other function of a structure such as a beam, **truss,** or **bridge.** The influence line is plotted in reference to a

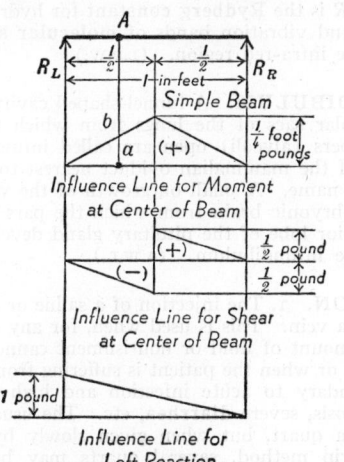

A

R_L R_R

Simple Beam

Influence Line for Moment at Center of Beam

Influence Line for Shear at Center of Beam

Influence Line for Left Reaction

base or zero line. Positive or tensile effects are represented above the line and negative or compressive effects below. The ordinate of the influence line is the ratio of the effect to the concentrated load producing it. If the load is in lbs. or tons the effect is in lbs. or tons. It is very useful for locating the position of the load which will produce maximum effect. For instance, the influence line for **bending moment** at the center of the beam shown above indicates that the maximum moment for this point will occur when the moving load which may be taken as unity is directly over the point. Any other ordinate such as *ab* represents the bending moment at the center due to a load of unity at point *A*. The maximum moment at the center of this beam due to a uniform load of *w* lbs. per linear ft., covering the entire length, may be computed by multiplying the area of the influence line by *w*.

$$\text{Area} = \frac{l}{4} \times l \times \frac{1}{2} = \frac{l^2}{8}$$

$$\text{Maximum moment at center} = \frac{wl^2}{8}.$$

The results obtained from influence lines may also be found by analytical methods but the graphical solution is usually much more efficient from the standpoint of time. This is particularly true when computing the stresses in the members of a large bridge due to its own dead weight and moving loads. (C.W.C., F.T.M.)

INFLUENT. In hydrology designates that portion of a stream which contributes water to, rather than derives water from, the ground-water zone. (R.M.F.)

INFLUENZA. A highly contagious disease which usually occurs in epidemic or pandemic form, spreading with great rapidity. The last great pandemic occurred in 1918–20. Previous epidemics occurred in 1889–92, 1847–48, 1836–37, 1830–33, 1780–82, 1729–32.

Influenza in its uncomplicated form is rarely serious, but it may become so because it predisposes to a particularly fatal secondary infection of the lungs, an extremely severe **pneumonia.**

The disease is caused by a filterable **virus.** Influenza virus is now known to occur in many different strains which are immunologically different. At least two, designated influenza A virus and influenza B virus have been identified in epidemics. In a typical case, the principal symptoms are sudden onset with fever, prostration, severe aching pains in the back and limbs, blood-shot eyes, headache and progressive **inflammation** of the respiratory **mucous membranes.** The disease is self-limited, and runs its course in 2 to 4 days. Complications such as **bronchitis, sinusitis,** and middle-ear infection (**otitis media**) are not uncommon in the simple form. The fatal **pneumonia** form usually occurs in pandemics only. (D.M.H.)

INFRA-RED PHOTOGRAPHY. The first photographs by infra-red radiation appear to have been made about 1880 by Sir William Abney using a specially prepared collodion emulsion. Abney is reported to have photographed a boiling tea-kettle but efforts by others to repeat his work were not particularly successful. In 1903 the first real infra-red sensitizer, Dicyanine, was discovered. While the sensitizing action extended to a wavelength of 960 millimicrons in the infra-red, the exposure was too long for general photography and Dicyanine-sensitized plates were used chiefly in infra-red spectroscopy. The discovery of more efficient sensitizers in the early twenties, beginning with Kryptocyanine and Neocyanine and followed by the penta- and tetra-carbocyanines, has made it possible to prepare films and plates whose sensitivity in the infra-red is such that they can be used for general photography, including aerial and motion-picture photography.

Infra-red-sensitive films and plates may be divided into two classes: (1) materials of relatively high speed to the extreme red and infra-red, i.e., from approximately 700 to 900 millimicrons, and (2) materials sensitive to much longer wavelengths but of lower sensitiveness. The former are used for general photography, for aerial photography and cinematography; the latter for spectroscopy in the infra-red and other scientific applications requiring sensitivity to wavelengths longer than about 900 mμ. All infra-red-sensitive materials are sensitive to violet and blue and to the extreme visible red as well. Photographs made without a filter resemble those made on an ordinary blue-sensitive material. For most purposes it is sufficient to use an orange or light-red filter which will absorb blue and violet light. In this case, the picture is made partly by infra-red and partly by the extreme red. The result, however, is generally only slightly different from that obtained with infra-red radiation alone. For true infra-red photographs, a visually opaque filter transmitting the infra-red only must be used. No filter, however, is required when photographing hot bodies such as an electric flatiron, hot castings, or high-pressure boilers, provided these show no visible glow.

All infra-red sensitive materials must be loaded and developed in total darkness, as safelight screens, even those for panchromatic films and plates, transmit the infra-red freely.

Certain precautions are necessary when making pictures with infra-red-sensitive materials. The bellows and shutter blades of some cameras, although perfectly safe for ordinary photographic films and plates, transmit the infra-red and fog infra-red sensitive films. The slides of some film and plate holders transmit the infrared sufficiently to cause fog. While some modern lenses are corrected for the infra-red, with most the focal distances for the visible and for the infra-red are different. Usually it is sufficient to extend the lens a distance equal to 2% of the focal length beyond the visual focus. Even this may often be ignored when using a lens of short focal length or a small diaphragm.

Infra-red photographs of landscapes are quite different from those made in the usual way. Green foliage is reproduced light and blue sky and water almost black. The shadow portions of the subject are dark and without detail. The general effect is that of a photograph made by moonlight, particularly if the print is made rather dark. As a matter of fact, most night scenes in motion pictures are really infra-red photographs.

Since infra-red radiation is not scattered by atmospheric haze, as is light, distant objects are rendered sharper and more distinctly in infra-red photographs. Objects invisible to the eye because of the intervening haze are often reproduced sharply in an infra-red photograph. In fact one of the most important applications of infra-red photography is in photographing distant objects, whether from the ground or the air. Infra-red photographs, however, cannot be made through dense fog.

Infra-red portraits are unusual in appearance but not generally pleasing. The lips and eyes are quite dark, almost black, while the face is light and translucent.

The scientific applications of infra-red photography are numerous and important.

1. *In Medical Photography.* To study diseases and conditions affecting the venous pattern not revealed by light. To study the progress of healing beneath certain scabs. Photographs of the eye to determine atrophy. Photographs of histological specimens to reveal structures below the surface and invisible to the eye.

2. *In Industry.* To study irregularities in the dyeing and weaving of textile fibers, the interior of furnaces, the detection of carbon in lubricating oils, infra-red spectroscopy of metals and alloys.

3. *In Astronomy.* To detect nebulae and stars otherwise invisible because of astronomical haze or because their radiation lies chiefly in the infra-red. Infra-red spectroscopy for the determination of the composition, the temperature, and the movement of stars and nebulae.

4. *In Criminology.* Deciphering writing or printing which has been crossed out with other inks to render it illegible. To obtain copies of charred documents, detect erasures, reveal finger prints, identify blood and other stains, uncover secret writings, etc. (C.B.N.)

INFRA-RED RADIATION. That range of radiation which extends from the limit of the visible red to the ultra-short **Hertzian radiation,** and therefore comprises wavelengths from approximately 0.00077 mm. to 0.4 mm. or longer. This radiation is observed and measured by means of very sensitive thermal detectors such as the **bolometer,** the **radiomicrometer,** and different types of **radiometer;** while its **spectrum** is studied by the "spectrobolometer," and its **spectral energy distribution** by the "spectroradiometer." Much of the region may also be photographed, as is the visible spectrum, but special plates are necessary. Pictures may even be taken by infra-red radiation in total visual darkness. Also, since these longer waves penetrate haze with less absorption, landscapes obscured by haze may be photographed by using an infra-red filter with suitable plates. Glass and water are nearly opaque to infra-red radiation, while rock salt is relatively transparent to it. (See **Thermal Radiation.**)

The strong absorption of infra-red by many substances renders it a useful means of applying heat energy. We utilize this principle in baking or toasting bread. Many modern industries use large banks of infra-red lamps for drying paints or lacquers on their products, or for similar purposes.

A convenient laboratory source of infra-red radiation is the "silica pencil," a heated rod of silicon dioxide. When thermal radiation is allowed to fall on quartz reflectors, narrow bands of infra-red **residual radiation** may be isolated for special study. The spectra of elements exhibit lines in the infra-red (see **Atomic Spectra**); for example, hydrogen gives the Paschen series, the Brackett series, and the Pfund series. The wave-number formulae for these are, respectively,

$$w = R\left[\frac{1}{3^2} - \frac{1}{(n+3)^2}\right],$$

$$w = R\left[\frac{1}{4^2} - \frac{1}{(n+4)^2}\right],$$

$$w = R\left[\frac{1}{5^2} - \frac{1}{(n+5)^2}\right];$$

in which R is the **Rydberg constant** for hydrogen. The rotation and vibration bands of **molecular spectra** are also in the infra-red region. (L.D.W.)

INFUNDIBULUM. A funnel-shaped cavity or organ. The alveolar sacs of the lungs from which the minute air chambers (alveoli) open are called infundibula and the end of the mammalian oviduct nearest to the ovary bears this name. A small outgrowth of the ventral wall of the embryonic brain from which the pars nervosa of the posterior lobe of the pituitary gland develops is also named the infundibulum. (A.W.L.)

INFUSION. 1. The injection of a saline or sugar solution into a vein. This is used when, for any reason, the normal amount of fluid or nourishment cannot be given by mouth or when the patient is suffering from dehydration secondary to acute infection and high fever, diabetic acidosis, severe **diarrhea,** etc. The usual quantity is about a quart, but when given slowly by the continuous-drip method, several quarts may be given in

24 hours. 2. The product obtained by steeping a **drug** for the extraction of its medicinal principles. (R.S.M.)

INFUSORIA. A subdivision of the phylum **Protozoa**, no longer included by some taxonomists but used by others in place of the subphylum **Ciliophora** or the class **Ciliata.** (A.W.L.)

INFUSORIAL EARTH. Diatomaceous earth. Diatoms.

INGESTION. The reception of food into the body. The simplest type of ingestion is found in some of the 1-celled animals, whose **protoplasm** merely flows around the food particle and engulfs it. In all groups some means of bringing food to the body or of catching it or cutting it from the parent plant or animal occur. It must then be taken into the body for digestion by an engulfing action of single cells like that of one-celled animals, by ciliary action, or by a muscular process of swallowing, all of which are forms of ingestion. (A.W.L.)

INGOT. A casting designed for reduction by **hot working** to a semi-finished product such as a billet or to a finished product such as a bar, plate, or sheet. Steel ingots are cast in massive cast-iron ingot molds which extract the heat faster than a sand mold and facilitate both the casting and handling of the ingots. (R.H.H.)

INHIBITORS. Corrosion.

INIA. Dolphin.

INITIAL CONDENSATION. The ordinary **steam engine** is double-acting and has at each end of the **cylinder** a port which serves both for the admission and exhaust of steam. The use of one port both for admission and exhaust causes a loss known as initial condensation. This term refers to the condensation of some amount of the incoming steam on the walls of the port and cylinder head. The reason for this condensation is to be found in the cooling of the same surfaces by the outflow of the exhaust steam. Thus the ports have a cycle of heating and cooling which is in unison with the engine cycle. Unfortunately, the heat given up by the port during the cooling cycle is released to steam which is on its way out of the cylinder. Thus any initial condensation, though it may be re-evaporated, is an entire loss in so far as availability of the heat for work is concerned. This is no minor loss; in fact, it is one of the major sources of thermal loss of the dual-flow steam engine. The uniflow engine, which represents the most significant advance in steam-engine practice in recent years, derives its advantage chiefly from the elimination of this loss. Steam flows from the ports towards the center of the cylinder in the uniflow engine. The cool exhaust steam does not backflow through the entrance ports. It escapes through ports located in the center of the cylinder, uncovered by the piston at the end of its stroke. This 1-way flow prevents cooling of the admission ports, and is the essential difference between the uniflow and conventional engine. The Corliss engine, with its valves in the end of the cylinder, but having separate exhaust and admission valves, has much less initial condensation than the single-ported engine. (F.T.M.)

INITIAL STRESS. Stress.

INITIAL TENSION. Stress.

INJECTION. This term is commonly used to designate the act of placing or forcing a fluid, **drug, serum,** or **vaccine** into a vein, tissue (such as a muscle), or a portion of the body (such as the rectum). A hypodermic injection is one that is made with a syringe and needle into the tissues beneath the skin. An intramuscular injection is similar to a hypodermic injection, except that a longer needle is used so that the material injected may be introduced into the muscle. This is used for drugs which would be painful if injected subcutaneously and also because faster and more thorough absorption is so obtained. In an intravenous injection the drug is introduced directly into the blood stream by inserting a needle into a vein. The most rapid absorption is obtained in this way.

The term injection is also used in another sense to denote the state of being congested, inflamed, or reddened, as in beginning **inflammation** when the blood vessels of the affected part are greatly engorged. (R.S.M.)

INJECTION, FUEL. Fuel Injection.

INJECTION GNEISS. A gneiss whose banding or **foliation** is wholly or partly due to interlaminar injection of **granitic** magma into already **metamorphosed** and foliated rocks. (R.M.F.)

INK. Inks are aqueous solutions containing organic ferrous compounds, as, for example, the tannate. They are used for writing. (See **Tannins, Iron.**) (R.K.S.)

INK SAC. A glandular sac found in the **squids** and related species (**cephalopod** mollusks) which secretes a dark fluid. The duct of the gland opens into the intestine near the anus. When the animal is aroused the ink is discharged into the mantle cavity and thence through the siphon into the surrounding water. The ink of the **cuttlefish** is the source of the pigment, sepia. (A.W.L.)

INLIER. Used by geologists to designate a structural block, the sides (fault planes) of which separate it from younger formations. Sometimes incorrectly used to designate the partially buried hills or **monadnocks** of an old erosion surface. (R.M.F.)

INORGANIC CHEMISTRY. Chemistry.

INSANITY. A legal term used to describe an individual who, because of mental illness, must be denied the right of personal liberty, as his behavior is such that he is dangerous to himself and society. Popularly the term is used to cover **psychoses, either schizophenia,** or the many depressive psychoses. (D.M.H.)

INSECT PESTS. Animals whose ways of living come into competition with those of man may become pests under some conditions. While some animals, like the poisonous snakes, are gradually eliminated as the land becomes more heavily settled and more extensively cultivated, others are tenacious enough to persist against the advance of civilization. Thus mice and rats continue to be a problem to the city dweller, but only on farms are wolves and gophers and woodchucks likely to give trouble. Some of our pests are obscure invertebrate parasites. Against their attacks we need expert medical assistance. Against such pests as rats and mice, or even larger mammals, the familiar methods of poisoning, trapping and shooting are adequate. These facts leave a considerable number of pests that trouble man in any surroundings and require special methods of control. They are chiefly insects (**Insecta**) and other arthropods (**Arthropoda**), and because they are very likely to be small and obscure and ubiquitous, this discussion of general methods of dealing with them seems of practical value.

These little creatures are so numerous and so varied that they come into conflict with human enterprises in many ways. While entomologists have discovered special means of dealing effectively with various species which are commercially important on a large scale, such as the **codling moth** and the **Hessian fly,** the layman may still master a limited amount of basic information through which he may approach the control of any insect pest intelligently. A work of this kind cannot embrace a multitude of special methods. The purpose of this article is, therefore, to summarize the basic approach to the control of insect pests and the underlying reasons for that approach.

According to their specific contacts with human economy, pests may be divided into those which attack man himself and those which damage things of value to him.

The insects that attack man to a troublesome degree in the temperate zone are comparatively few. The lice (**Louse**) are important in some strata of society and occasionally to human beings whose normal standards of cleanliness are a protection in themselves. If, under special conditions, one must rid his person of lice he may resort to mercurial ointment, which should be used under the direction of a physician, or he may be able to resort to a newer method in the use of DDT (dichloro-diphenyltrichloroethane) as a dusting powder. This insecticide is evidently somewhat dangerous in ointment form. Good housekeeping and high standards of personal hygiene usually prevent trouble with **bedbugs.** Bedbugs may require only the treatment of a single bed, in which case kerosene or fly spray, painted or sprayed into crevices where the bugs may hide, is normally effective, or they may infest an entire dwelling. In the latter case fumigation of the structure is the most effective procedure. It should be carried out by a professional pest-eradicator, since cyanide fumigation is a very dangerous process. The mosquitoes and flies (**Fly**) which "bite" us are sometimes distressing but rarely harmful in themselves. Against such pests as flies and mosquitoes the screening of our houses is a partial defense. When we must be exposed to them in the open in unusually large numbers, repellent compounds may be applied to exposed portions of the skin. The "bite" of all of these insects is really a puncture made by sharp piercing mouth parts which enable the insect to reach the blood on which it feeds. No true biting insect attacks man to an important degree.

Far more serious than the actual damage done to the human body is the possibility that an insect may transmit the germs of a disease, but this danger is far greater in tropical and subtropical regions. Malaria and yellow fever, transmitted by mosquitoes, are classic examples of insect-borne diseases. African sleeping sickness and typhus fever, transmitted by the tsetse fly and by the body louse respectively, are also well known. Protection against these insects where such risks must be run is, of course, an important corollary to the treatment of the diseases themselves.

In the tropics **bot flies** and the **chigoe** flea are also important to man. Since they live as internal parasites the removal of the insect itself is necessary for cure.

Such related pests as the harvest mite, the jigger of temperate North America, and the itch mite, *Sarcoptes scabiei,* come under the same category. The former causes temporary annoyance, but the latter is a troublesome parasite requiring medical attention.

Our domestic animals and birds are subject to similar attacks. Bot flies, horse flies, other biting flies, lice, parasitic flies such as the **sheep tick, flea, bird louse, mite,** and true **tick** may all be serious pests and in some cases may transmit diseases. For most of them repellent powders such as pyrethrum, rotenone, and the newer DDT, and commercial dips, are effective.

In addition to the insects attacking animals we are troubled by those which destroy animal products, notably the **clothes moths** and dermestid beetles (carpet beetles). These forms destroy fur, feathers, and woolens. Fortunately, mothproofing treatments have been devised for carpets and woolen clothing and are readily available on the market. Fur and feathers must still be safe-guarded by proper storage when they are not in use. Serious infestations of these pests may often be traced to a breeding place in some forgotten part of the home. An old woolen garment in the attic, stored feathers or a discarded fur may produce clothes moths over a period of years to spread into other garments and into our rugs. The elimination of such breeding places does away with much painstaking work in the protection of our possessions.

The pests attacking plants are legion. Some are biting forms which consume the tissues of the plant while others merely suck the juices. The former may live on flowers or on specific parts of flowers, they may eat leaves or fruits or roots, and they may bore into various parts of the plant as they eat. If they attack a part that is valuable to man, the damage is necessarily greater than if they merely weaken the plant, hence leaf-eaters are usually less serious than borers in stems and fruits. The latter may kill the plant or render its products useless, while the former often do relatively little damage. Such species as the squash and peach borers, the codling moth and the **boll weevil** are good examples of individually serious pests. The **tomato worm** and the tent caterpillars are among the pests which become serious only when excessive numbers of them destroy a considerable part of the leaves of the plant. In the home garden in some parts of the United States the Colorado potato beetle and the Mexican bean beetle are often destructive pests. For any region, state entomologists and county agents can furnish specific information on local pests and their control.

The economically important sucking insects, although most of them are very small, often develop enormous numbers of individuals, hence plant lice (**Aphid**) and scale insects (**Scale**) often seriously injure crop plants and orchard trees. Many species occur, some attacking only a single species of plant while others may be found on various kinds.

Pests of this kind may also be important in transmitting diseases, as in the case of the common melon aphis which carries mosaic disease to numerous species of garden plants.

From the practical point of view, the recognition of biting and sucking insects does not require a detailed knowledge nor observation of insect anatomy. If portions of the plant tissue are cut away, the insect doing the work is a biting form. If the plant remains intact but shrivels, turns yellow, or wilts, it is very likely to be losing more sap than it can spare to some sucking insect or borer. Plant diseases may be responsible for such changes.

Obviously a biting insect will consume anything on the surface of the tissues that it eats, hence the arsenical poisons sprayed or dusted on the plant are effective means of control. Lead arsenate is commonly used against such insects. Sucking insects, however, penetrate the thin layers of poison applied to the surface and suck the unpoisoned juices below. Against them, poisons that kill on contact must be used. Here nicotine sulfate and lime-sulfur sprays are important. Any such spray must touch the insect in order to kill it, hence spraying must be carried out in such a way as to reach the part of the plant on which the insects are lodged. Special difficulty arises when insects lodge in inaccessible places, but these problems are not likely to be important in the ordinary house or garden.

Even after plant products are harvested they are still subject to attack by such insects as the grain **weevils,** the Mediterranean flour moth, and the voracious wood-eating termites (**Isoptera**).

Fumigation of storage spaces is effective against most of these pests and, indeed, against any insect whose occurrence makes the use of the method possible. Volatile poisonous substances such as carbon disulfide and hydrocyanic acid gas are extremely effective but dangerous to use, the former because of its poisonous and explosive properties and the latter because it is a fatal poison. They should be used only in extreme cases and by experts. In the home the elimination of forgotten breeding places is an important measure. An old box of flour or prepared cereal left on an unused shelf may be a source of starch-eating insects for months.

Methods of dealing with insect pests have reached a high degree of complexity. They may be summarized as follows:

1. *Mechanical Methods.* Destruction of pests by hand-picking or by jarring them from plants into a container holding a little water covered with a film of kerosene is often adequate for small-scale infestations such as the tomato worms and potato beetles in the home garden. It is also important in the well-kept home, where the occasional clothes moth or beetle may be killed by hand. Now that succulents and cacti are raised in the home in large numbers, this method is of particular importance. Some of these plants are badly troubled by mealy bugs, which can be destroyed by using, a sable pencil or any similar small brush moistened with alcohol. Even crops raised on a large scale may sometimes be protected by mechanical means, as in the use of the hopper-dozer for the destruction of grasshoppers among plants of moderate height like the clovers.

2. *Repellents.* Volatile materials such as naphthalene (moth balls) and paradichlorobenzene or creosote, when sufficiently concentrated, provide adequate protection. The last, used for the treatment of wood in construction, is rather more than a repellent, however, since it renders the wood unfit for consumption by termites.

3. *Poisons.* The majority of chemical protective measures are designed for the actual destruction of the pests, hence the substances used may be classed as poisons. The division into contact and stomach poisons has already been mentioned, including nicotine sulfate and lime-sulfur in the former category and the arsenates in the latter. Substances which enter the respiratory system of the insect form a border category since they include violent poisons like hydrocyanic (prussic) acid, oils such as kerosene, which evidently cause suffocation, and whale-oil soap whose action is evidently similar.

The application of true poisons may be conditioned by the mode of life of the pest. The ants which enter our homes, for example, come from colonies whose organization guarantees a constant renewal of their numbers no matter how many are poisoned in the course of their wanderings. Sirups including a small percentage of a fairly slow poison such as sodium arsenite are usually effective against these pests since they are carried to the nest and distributed to the individuals of the reproductive caste thus destroying the entire colony. Such ant poisons are commercially available.

The successful use of poisons against insect pests requires an appreciation of the tolerance of the host plant or animal for the same material and an understanding of the effectiveness of distribution attained by the method of application used. Formerly most poisons were used as sprays. Some of the most effective, e.g., kerosene, were very harmful to plants, but combined with soap in an emulsion this insecticide could be used safely. Nicotine sulfate must also be used with care, since it damages foliage if too concentrated. Insecticides are applied at present in four ways: as sprays, by dusting in a fine dry powder, as fumigants, and as aerosols. In the last form they are combined with a highly volatile vehicle. A quantity effective over a considerable area can be confined within a small container in liquid form and vaporized as needed.

One of the most promising of the newer insecticides is DDT, but too little is yet known about its effects upon man and other animals to recommend its general use. The substance is insoluble in water but soluble in organic materials such as oils. Its volatility is low. It is used against many types of insects in sprays and dusts. During the present war its effectiveness against body lice

has been given considerable publicity. Unfortunately its use on the human body is attended by some risk since its toxicity is high, its action cumulative, and it may be absorbed through the skin. It is safest as a emulsion or a dry powder and most dangerous in solutions and ointments. For ordinary uses the arsenates, nicotine sulfate, pyrethrum and rotenone powders, naphthalene, paradichlorobenzene, and commercial mothproofing compounds will be found adequate when used according to directions.

4. *Biological Control.* Most of the important insect enemies of crops have been introduced from other lands and have increased rapidly because of favorable conditions or the absence of natural enemies. One goal of the economic entomologist is to restore natural checks so that the pest can no longer persist in damaging abundance. The milky Typhia parasite of the Japanese beetle is one example. Many years ago the San José scale was brought under control in the West by the introduction of a natural enemy from its home in the Orient. This enemy is a predacious beetle of the family commonly called ladybirds. Several species are now known to feed on the scale insects. Since such methods of control are self-operative once they are established, their importance is obvious.

No insect has yet been made extinct through deliberate human effort, so we shall probably always have such pests. They are not likely to be of grave importance in the temperate zone but occasionally they demand attention and usually we can cope with them by simple and readily available means. In some cases the services of a professional pest eradicator may be needed to handle extensive infestations. (A.W.L.)

INSECTA. The insects, a class of the phylum **Arthropoda.** This is by far the largest taxonomic division of the animal kingdom, including approximately three and one-half times as many species as there are of all other animals. About 450,000 species of insects have been described.

Insects are characterized by the hard exoskeleton and jointed appendages of the phylum. Among other arthropods they are distinguished by several characteristics: 1. The body is divided into head, **thorax**, and abdomen. 2. Both compound and simple eyes may be present. 3. They have one pair of antennae. 4. The thorax bears a maximum of three pairs of legs. 5. Wings are found in the adults of many species. Two pairs occur in most orders. 6. The abdomen is usually without jointed appendages although a few primitive forms bear modified derivatives of these structures. 7. Respiration is accomplished by means of tracheae, which are air tubes opening from the exterior and branching among the tissues. 8. Metamorphosis is complex in some orders although lacking in the most primitive forms.

The immense numbers of insects indicate remarkable biological success. They have invaded all possible habitats save the ocean; here only one form, *Halobates*, a genus of marine water striders, is known. On land and in fresh water insects are found in almost every imaginable habitat. They eat plants, animals, and dead organic matter of all kinds and many are parasitic. They walk, run, jump, fly, swim, and burrow. Moreover, their small size permits them to live in very limited habitats, hence the plant feeders include species which are confined to flowers, leaves, stems, fruits, or roots, and parasites are not limited to larger animals but find adequate hosts in other insects; even insect eggs are attacked by parasitic insects. Predacious species are, of course, confined to smaller prey such as other insects and the small members of other groups. This extreme diversity makes it impossible for human beings to avoid experience with insects. The blood-sucking mosquitoes and flies and the scavenger clothes moths and beetles force themselves upon us.

The contacts of insects and man are often of great economic importance, both beneficial and harmful. Among useful insects the honey-bee and silkworm are the outstanding examples, but among harmful species it is more difficult to select, for many species have caused appreciable losses in various fields of human activity. In agriculture the plant-feeding species such as the Colorado potato beetle and the **codling moth, the gipsy moth** and the **Japanese beetle** are important pests, and the **bot flies** have at times been serious parasites of domestic animals. Man suffers direct injury from many parasites such as the lice (**louse**), **fleas, black flies,** and **mosquitoes,** and in the tropics from the **chigger,** bot flies and other species, but the most serious damage done by insects in his life is through their role as carriers of disease. Malaria and yellow fever are transmitted by mosquitoes, typhus by the body louse, and bubonic plague by fleas. (See **Insect Pests.**)

The classification of insects is necessarily complex and it is impossible for any individual to have detailed knowledge of more than limited subdivisions of the class. Two subclasses, **Apterygota** and **Pterygota,** are commonly recognized. The former includes only primitive wingless insects, the latter the winged species and those which are secondarily wingless. In these subclasses 19–33 orders are recognized by various authorities. The following tabulation summarizes a classification which is widely used:

Subclass **Apterygota.**

Order **Protura.** Rare and primitive insects of very small size.

Order **Thysanura.** The **silverfish** or fish moth and allied species.

Order **Collembola.** The spring-tails.

Subclass **Pterygota.**

Order **Orthoptera. Grasshoppers, crickets, cockroaches, katydids, mantises, walking-sticks** and their allies.

Order **Isoptera.** The white ants or **termites.**

Order **Plecoptera.** The **stone flies.**

Order **Ephemerida. May flies,** locally called shad flies, salmon flies, **June bugs** and Canadian soldiers.

Order **Odonata. Dragon flies** and **damsel flies.**

Order **Zoraptera.** Rare insects without a common name.

Order **Corrodentia.** The **book lice** and psocids.

Order **Mallophaga. Bird lice** or biting lice.

Order **Anoplura.** True or sucking lice.

Order **Embiidina.** Rare insects with no common name.

Order **Thysanoptera.** The **thrips.**

Order **Hemiptera.** The true **bugs.**

Order **Homoptera. Cicadas, leaf hoppers, plant lice, scale insects,** etc.

Order **Dermaptera. Earwigs.**

Order **Coleoptera.** The **beetles.**

Order **Strepsiptera.** The stylopids.

Order **Neuroptera. Lacewings, dobson flies, hellgrammites, ant lions, alder flies,** and others.

Order **Mecoptera.** The **scorpion flies.**

Order **Trichoptera.** The **caddis flies.**

Order **Lepidoptera. Butterflies, skippers,** and **moths.**

Order **Hymenoptera. Saw flies, ants, bees, wasps,** and many parasitic forms.

Order **Suctoria.** The **fleas.**

Order **Diptera.** The 2-winged flies, including true flies, mosquitoes, midges, gnats, and others.

(A.W.L.)

INSECTIVORA. An order of mammals made up of small animals, mostly of nocturnal habits, which live on insects and other small invertebrates and in a few cases on vegetation. **Moles, shrews, hedgehogs** and related species. (A.W.L.)

INSECTIVOROUS PLANTS. These are plants which are able to obtain a part of their **nitrogen** supply from the bodies of small insects and other animals which are trapped by the plants in various ways. They are also frequently called carnivorous plants. All of them are green and capable of living without this animal nitrogen, but many seem to thrive better if they have it. They are plants which grow in marshy or boggy places, where the supply of available nitrogen may be very slight. Some of them are water plants.

Insectivorous plants may be divided into three distinct groups, distinguished by the manner in which the plant captures the insects. In one group, the insects are attracted to the plant's leaves by a glandular secretion which is sticky and holds them fast. In some species in this group, the glandular hairs fold inward over the prey to hold it firmly and to aid in digesting it. The most common plants of this group are the sundews, species of the genus **Drosera.** They are small bog plants of fairly common occurrence. In some species the leaves are linear, in others round and long-**petioled.** In all, the upper surface of the leaf is covered with long tentacle-like hairs with swollen tips. This tip secretes a copious quantity of a colorless sticky substance which glistens in the sunlight and attracts many small insects. When the insect alights on the leaf or on one of the hairs, it stimulates the latter to fold inward, gradually carrying the insect towards the center of the leaf. The stimulus is transmitted to other nearby hairs, which fold in likewise, until the insect is carried to the leaf center and pressed firmly against the surface. A digestive **enzyme** is there secreted, which acts on the **proteins** of the animal body, changing them to a soluble form which can be absorbed by the leaf. When digestion is completed, the glandular hairs unfold, and the leaf is ready for another victim.

Another common plant of this group is the Butterwort, *Pinguicola vulgaris.* In this plant the surface of the leaf is shiny with the sticky secretion from the glands, while the edges of the leaf are rolled inward. Insects attracted by the sticky surface are gradually caught under the enrolled margin and there digested. Other plants in this group are tall herbs or low shrubby plants; in them no movement of the glands occurs, the prey being held by the sticky secretions alone.

The second group of insectivorous plants comprises all those in which the leaf is variously modified to form a pitcher in which the prey is entrapped. Plants of this group are often very striking objects. They occur in widely scattered regions. In the eastern part of North America are found species of the genus *Sarracenia,* which are commonly called pitcher-plants. They are found in open marshes where plenty of light will reach them. The leaves occur in basal rosettes, and are green, often deeply mottled with red. Each leaf has the form of an open pitcher with a distinct lip or flange at the top, and a green wing down one side of the pitcher. Around the mouth of the pitcher, on the inner side, are numerous glands which secrete a fluid which attracts insects. Lining the inside of the pitcher are numerous stiff pointed teeth or bristles which project sharply downward, so that it is easy for the insect to crawl down into the pitcher, but practically impossible to crawl up. The lower part of the pitcher is usually full of water, into which the unfortunate insect eventually falls and is drowned. Either due to the action of **bacteria** or that of secretions from glands in the surface of the leaf, the proteins of the animal body become assimilable by the leaf, which thus obtains nitrogenous matter. In California is found *Darlingtonia californica,* the leaves of which are even more remarkable. The same pitcher-like structure occurs, but the opening of the latter is covered by the overgrowing of the upper part of the leaf, which ends in a brightly colored flap. In the arching part of the pitcher which covers the opening there is a clear translucent space, against which insects persistently move until exhausted, then fall into the pitcher below. Even more remarkable pitchers are found in the leaves of the genus *Nepenthes,* which is native in the East Indies, Malaya, and Madagascar. These plants are herbs or shrubby plants which grow in bogs. At first they have leaves like other plants, but soon very much modified leaves appear and aid the plant in climbing upward to a position having a favorable light supply. The mature leaves of these plants are remarkable objects. In each there is a flat green blade which is prolonged at its tip into a long slender stem-like structure which often tightly twines about any supporting object. Beyond this portion of the leaf the pitcher is found. In size the pitcher varies according to species, from a small object an inch or so long, to bodies capable of holding a quart or more of liquid. The rim of the pitcher is rolled inward and provided with an abundance of glands which secrete a substance attractive to insects. The lower part of the inner surface of the pitcher is slippery, offering no foothold to insects seeking to escape. Often the pitcher is rather brightly colored. Beyond the pitcher the leaf is again prolonged into a slender stalk which ends in an expanded flap which frequently stands over the opening of the pitcher. Because of their curious habit many species of *Nepenthes* are cultivated as hot-house plants.

Other species of pitcher plants are found in South America and in Australia.

The third group of insectivorous plants is composed of those plants which capture their prey by some sort of movable trap. In this group are found Venus' fly-trap, the Bladderworts, and Aldrovanda.

Venus' fly-trap, *Dionaea muscipula,* is a small plant found in the Carolinas. The leaves form a basal rosette close to the ground. Each leaf has a broad blade-like **petiole** which abruptly narrows at its tip and bears a remarkable blade. The latter is formed of two halves which are joined by a movable hinge down the center. The edges of each half are fringed by long stiff bristles, while on the upper surface of each are borne three long slender trigger hairs which are sensitive to contact with any solid object. Each trigger hair is jointed at its base. The upper surface of the blade is abundantly supplied with small glands. When any insect alights on the leaf and comes in contact with one of the trigger hairs, a stimulus is given which causes the two halves of the blade to fold together with the bristles around their edges interlocking and so trapping the insect securely. Once caught the insect is slowly digested by the leaf.

Aldrovanda vesiculosa is a rootless plant found in quiet waters in Europe and Asia. Its leaves are stalked, and have a hinged blade with trigger hairs, and digestive glands very similar to those of Venus' fly-trap. Small aquatic animals are trapped in the leaves of this plant.

The Bladderworts belong in the genus *Utricularia,* which contains some 200 species, widely distributed in tropical and temperate regions. One of the commonest species, *Utricularia vulgaris,* occurs in ponds and slow streams in both Europe and North America. This plant never has any roots, even when it first develops from the seeds. The leaves of the plant are finely dissected and bear small bladders which give the plant its name. The bladders have a narrow opening which is surrounded by a group of radiating hairs which form a funnel-like approach to the opening. This opening is provided with a "trap-door." The inside of the bladder bears many 4-parted hairs. Should a small aquatic creature chance to venture into the funnel of hairs and touch certain sensitive hairs, the bladder suddenly expands, creating a current of water which sweeps the unfortunate creature into the bladder. The "trap-door" closes behind him and he is held prisoner until he dies. Other species of bladderworts have ordinary undissected leaves. In the tropics certain species are found only

in the waters contained in the pitchers formed by the wide leaves of certain members of the **Pineapple** family. (R.M.W.)

INSEMINATION. The introduction of the seminal fluid, bearing the reproductive cells of the male, into the genital passages of the female. **Copulation. Mating.** (A.W.L.)

INSEQUENT STREAM. A stream which starts to cut its valley at the edge of a cliff or on the crest of a ridge. (R.M.F.)

INSERT. Plastic Molding.

INSERTION LOSS. The insertion loss of a piece of apparatus, usually expressed in **db.**, is the loss introduced in an electrical circuit by the insertion of the apparatus. Thus in many communication circuits the connecting of essential components of the system may introduce an insertion loss which must often be compensated for by additional **amplifier** gain. (L.R.Q.)

INSOLATION. One of the processes of weathering. Extreme diurnal (daily) changes in temperature such as occur on high plateaus, and especially high deserts, causing relatively rapid expansion and contraction of the rocks so that they crack and disintegrate. Differences in temperature between day and night have been registered as high as 120° F. Insolation is only effective on bare rock faces, as a slight blanket of soil or debris serves to protect the rock from relatively rapid heating and chilling. (See **Heat Balance in the Atmosphere.**) (R.M.F.)

INSTABILITY. Stability and Instability.

INSTANTANEOUS CENTER. The instantaneous center is the imaginary point about which a body having general motion may be considered to be rotating for the instant. The instantaneous center is not necessarily on the body; in fact, it can be, in the case of rectilinear motion, infinitely distant. (See **Centrode.**) (F.T.M.)

INSTAR. A single period of growth of an arthropod larva, ending with a moult or **ecdysis.** The term is applied chiefly to insects. In this group normal growth is completed in a fixed number of instars, the last one terminating with pupation. Under abnormal conditions, however, the **larva** may grow slowly and moult more than the usual number of times. It is evident from such cases that the length of instars and their number are conditioned by more factors than growth of the animal to a size which demands a larger integument. (A.W.L.)

INSTRUMENT. The term instrument describes a wide range of objects, but from a technical standpoint, an instrument is a tool or mechanism for scientific and professional service. As far as tools are concerned, instrument might be taken to designate such tools as are delicately, accurately, and scientifically constructed, and whose use requires above average dexterity and technique.

Nearly every technical field exhibits a great number of specialized scientific mechanisms which might be called, with reason, instruments. Obviously, the scope of mechanisms included under so general a heading is too extensive for detailing here. However, a great many measuring devices, meters, gauges, and the like, in their construction, will satisfy the definition of instrument given above. A great many meters and gauges are called instruments, although in some cases there could be some reasonable doubt as to whether they could rightfully be so designated. For further details, see **Instruments, Electrical; Instruments, Mechanical; Meters.** (F.T.M.)

INSTRUMENT TRANSFORMER. Transformer.

INSTRUMENTS, AIRCRAFT. Flight by airplane now involves varied instruments which are read by the pilot or crew member to yield information on the current condition of the engine and other parts of the airplane, on the attitude of the airplane, and on its position and direction of travel. Early aviators flew with few instruments. So may also the modern aviator if he uses small airplanes flown at low altitudes in visual contact with a fixed base. Even so, such flights are accompanied by improved peace of mind of the pilot if instruments indicate continuously that the engine is functioning normally. A tachometer, an engine oil-pressure gauge, and an engine oil-temperature gauge are considered to be the minimum complement of "condition" instruments for any internal combustion engined airplane, however small it may be. At the other extreme, the instrumentation of the large multi-engine transport or military airplane is complex. There are from 40–60 instruments (not including warning lights, switches, etc.) on the board of a modern airliner.

Condition instruments are to indicate the operational state and integrity of crucial parts of the airplane. This group includes engine instruments, fuel and lubricating oil instruments, flap, landing gear, and cowling position indicators, fire warnings, etc. It would also include instruments designed to indicate the readiness-to-function of the several auxiliary systems—electric, pneumatic, or hydraulic.

Attitude instruments show the angular position of the airplane with reference to a horizontal plane; also the rate of change of attitude. They include bank, pitch, turn and bank, rate of climb, and gyro horizon instruments.

Position instruments serve to fix in the pilot's mind the position of his airplane in space and the direction of travel. This classification includes all navigational instruments and communication devices. The minimum group of these instruments required for cross-country contact (visual) flying consists of a magnetic **compass,** an altimeter, and an air-speed indicator. But in more complete instrumentation, "position" instruments also include radio compass, drift indicator, terrain clearance (radio) indicator, and chronometer. (See also **Automatic Pilot.**)

These instruments are either self-actuating or remotely actuated by hydraulic, electric, mechanical, or pneumatic systems. The instruments themselves might be classified by principle of operation as follows:

1. By pressure of liquid or gas on a diaphragm, Bourdon tube, or collapsible capsule. Examples: engine oil pressure, hydraulic brake system pressure, altimeter, air speed.

2. By thermal action, creating pressure, electric resistance, or voltage. Examples: oil temperature, cylinder-head temperature.

3. By mechanical action. Examples: centrifugal tachometer, mechanical position indicators.

4. By gyroscopic action. (Gyroscope electrically or pneumatically driven.) Examples: turn and bank, gyro horizon.

5. By direct electrical action. Examples: remote position indicators, electric tachometer.

An explanation of the construction and action of all aircraft instruments is beyond the scope of the present writing. A few selected instruments will be explained in principle.

Altimeter. The altimeter is used to determine the altitude of the aircraft. The instrument is simply an aneroid **barometer** calibrated to read in feet of altitude instead of barometric pressure. It thus reads

pressure, not density, altitude. The heart of this instrument is a circular evacuated capsule, one face of which is fastened to the case, while the other is attached through a system of multiplying levers (terminating in a chain wrapping around the spindle) to a spindle carrying a pointer which moves over a scale. A U-shaped spring holds the capsule distended and atmospheric pressure tends to collapse it.

When mounted in an airplane which is climbing to higher altitudes, the capsule is subjected to weakening atmospheric pressure, allowing the spring to distend the capsule and move the pointer. A small hairspring keeps the mechanism taut and the pointer will move upon the slightest movement of the capsule face in either direction.

The instrument may be made to indicate altitude above any given point, e.g., the airport runway, by rotating the dial containing the scale until the pointer

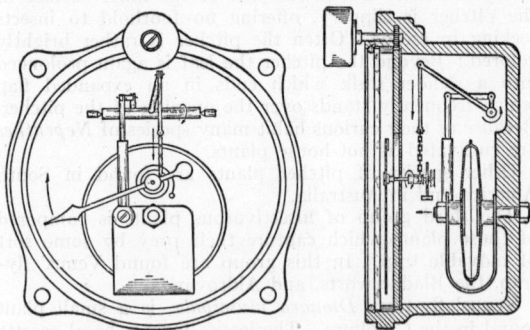

Fig. 1. Simple altimeter. (*Pioneer Instr. Div. of Bendix Aviation Corp.*)

registers with zero altitude, this being done when the airplane is on the runway. A knob having a pinion engaging a gear on the periphery of the dial makes this adjustment. The mechanism of an air-speed indicator is similar except that the case is air-tight and subjected to static-tube pressure, while pitot-tube pressure is communicated to the interior of the capsule. (See **Pitot-Static.**) The U-spring is omitted, but not the hairspring. The distention of the capsule is thus a function of the difference between static and dynamic pressure, i.e., the velocity head. Thus the pointer may travel over a dial calibrated in m.p.h. The dial is usually calibrated for air of sea-level density, and a correction must be applied to indicated air speed to obtain true air speed at an altitude.

Gyro Horizon. This is an instrument designed to simulate the attitudes of an airplane. A miniature airplane on the face of the instrument appears to bank or climb and does so in imitation of the real airplane's attitude. The pilot therefore is relieved of any necessity of interpreting numerical instrument readings in terms of attitude. This instrument makes use of the gyroscope—its rigidity in space and its property of precession.

A spinning gyro rotor with vertical axis is borne by a gyro housing which, in turn, is supported in lateral horizontal bearings in a gimbal ring. The latter is supported from the instrument case by being pivoted on a fore-and-aft axis. The spinning axis will tend to remain vertical while the airplane (and also the instrument case which is attached to it) is banked or pitched. A horizon bar controlled by the gyro remains horizontal while a small airplane silhouette attached to the case is tilted with the plane. Not only will the horizon bar remain horizontal, but also it will move up or down to indicate diving or climbing flight.

To the pilot riding the airplane through non-level flight, the real horizon is seen "tilted" during banked flight; also the pilot sees the nose of the airplane rise above the horizon while climbing, and sink below it when gliding. Since he is accustomed to seeing the real horizon tilted, and above or below his center of vision he knows just what the relative position of the horizon bar on this instrument means in terms of his actual flight attitude; consequently, when flying through rain or overcast, or under any condition where a horizon of land, water, or cloud layer cannot be seen, the pilot readily substitutes the impressions gained from the gyro horizon for those he derives normally from the real horizon.

The gyro is air or electrically driven. Air-driven gyros are small turbine wheels driven by an air jet induced by a vacuum applied to the instrument exhaust. This vacuum may be maintained by a Venturi tube with a suction line at the throat, or by a vacuum pump, engine-driven. The spent air is exhausted from the gyro housing through four equally spaced ports partly covered by pendulous vanes. Any tendency of the gyro axis to depart from vertical is followed by a motion of one or more of these vanes, upsetting the normal balance of the four exhaust jets. The resulting horizontal air reaction on the gyro housing results in a precession which precesses the gyro

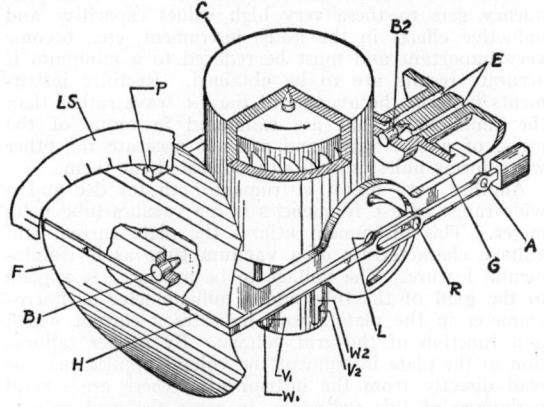

Fig. 2. Air-driven gyro housing during precession. (*Sperry Gyroscope Co., Inc.*)

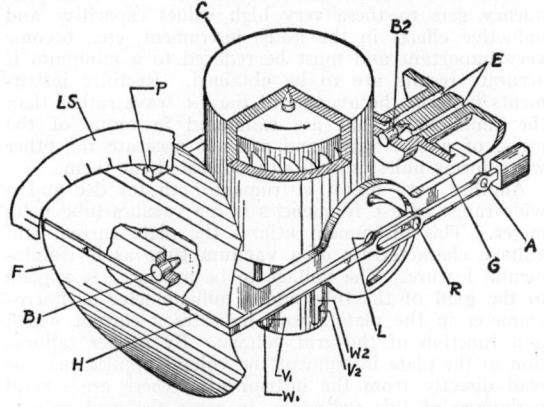

Fig. 3. Principle of the gyro horizon. (*NACA.*)

housing until it has resumed the erect position in which the exhaust jets are again in balance.

For high-altitude use, electric-driven gyros have been developed since their drive is not affected by diminished air density. The erecting action produced by the vanes in the air-driven model is replaced by a liquid level device in which the liquid level controls torque motors on the roll and pitch axis. (See also, **Blind Flight, Blind Landing, Pressure Gauges.** (F.T.M.)

INSTRUMENTS, ELECTRICAL. The most common electrical instruments are the various meters which, when properly applied and connected, indicate or record the characteristics of flow in an electrical circuit, characteristics which are, by nature, indeterminable by visual inspection of the conductors. The principal electrical instruments are: the **ammeter**, to show flow of current; the **voltmeter**, to show electrical pressure on a circuit; the **wattmeter**, for power; the watt-hour meter, for electrical energy; the synchroscope, which is used to effect parallel operation of current-carrying machinery; the power-factor meter; the volt-ampere meter or **volt-ammeter**; the **frequency meter**; and the ground detector. Indicating instruments are used chiefly for

operating guidance, recording instruments, for operating supervision and calculation of performance, and integrating types for cost allocations.

The electrical instrument is commonly a comparatively rugged instrument, enclosed in an iron case having a glass front through which the register of a pointer on a scale may be read. Few electrical instruments for d-c service are applicable to an a-c circuit, and vice versa.

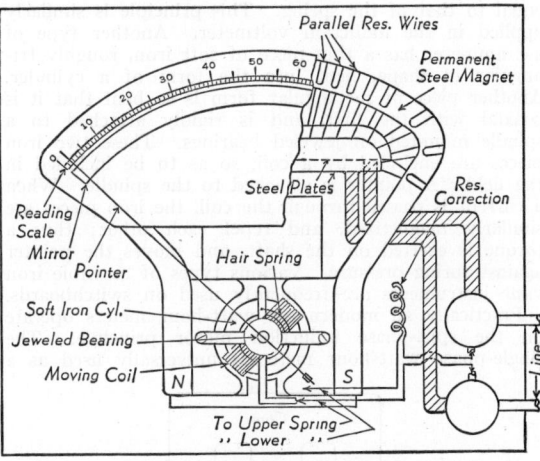

Direct-current ammeter.

The operating principles of instruments include electromagnetic induction, magnetic attraction, and resistance heating. A typical d-c ammeter is the d'Arsonval type consisting of a light coil mounted in the field of a permanent magnet. When the coil is energized by passing the current through it, a torque (see **Statics**) is set up which tends to set the coil to include a maximum of flux lines. A pointer is attached to the moving coil, and the motion of this system is opposed by a coil spring. The spring controls the throw of the pointer, which otherwise would run completely off the scale for all currents. As it is impossible for the delicate moving coil of this instrument to carry large currents, a shunt is employed. The current through the shunt bears a definite ratio to the current through the ammeter, so that the ammeter can be calibrated to read directly the current which is being metered. The shunt is often incorporated within the case of fixed-range instruments.

The voltmeter is similar in construction to the ammeter. One difference arises from the fact that ammeters are used to measure currents, are connected in series in the circuit, and must embody a low-resistance shunt in parallel with the moving coil, while voltmeters are used to measure potential difference, are connected across the circuit, and so must have high-resistance coils in series with the instrument coil.

In a-c practice, induction type instruments are very common. An induction type meter depends on the interaction of an inducing and an induced current. In the type shown in the figure just below, a laminated

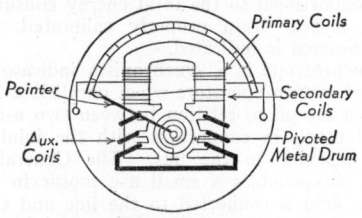

Induction Ammeter

iron core is surrounded by one or more coils of wire. When current flows through the coils, an alternating magnetic flux is set up in the air gap of the core. The effect of a rotating field may be secured through the action of more than one group of coils, in which the currents differ in **phase**. A drum or disk, usually of aluminum, is supported by pivots in the air gap of the core, and a torque is produced by the rotating field reacting on currents induced in this movable element. A hair-spring produces counter-torque and brings the pointer to rest when the torque produced by the coil is equal to that of the spring. This principle is similarly applied in the induction voltmeter. Another type of a-c ammeter has a thin piece of soft iron, roughly triangular in shape, bent into the form of a cylinder. Another piece of rectangular form is so bent that it is coaxial with the first, and is rigidly attached to a spindle mounted on jeweled bearings. These two iron pieces are encircled by a coil, so as to be included in the field. A pointer is attached to the spindle. When a current is passed through the coil, the iron pieces are similarly magnetized, and repel each other; thus a torque is exerted on the shaft, and moves the pointer against spring pressure. Various types of movable iron vane instruments are frequently used on switchboards.

Practically all modern a-c watt-hour meters operate on the split-phase induction motor principle. The single-phase watt-hour meter is universally used as a

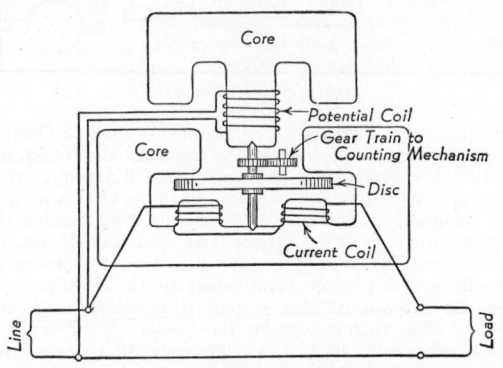

Induction watt-hour meter.

meter for the kilowatt-hour consumption of domestic customers. The essential parts of the watt-hour meter are a current coil, a potential coil, and a rotating disk. The current coil is connected in the line, and the potential coil is across the line. Both coils are mounted on iron cores, and placed in close proximity to the rotating disk. The alternating flux produced by these windings sets up eddy currents in the disk, and these eddy currents, in turn, set up fields which react on the field of the opposite coil. The combined action of the two coils sets up a torque which, at any instant, is proportional to the instantaneous values of the voltage and the current which is in phase with it; hence the meter acts independently of the power factor of the circuit, and measures watt hours, not volt-ampere hours. The speed at which the disk rotates is proportional to power being used; therefore, the number of revolutions the disk turns is proportional to the total energy consumed. By means of a gear train properly calibrated, the total energy consumed is indicated.

The synchroscope is a synchronism indicator for the guidance of station operators when paralleling machines. It indicates the phase relation between two a-c circuits. The usual use is in connection with the joining of an incoming machine to the bus. The General Electric synchroscope resembles a small a-c motor in construction. The field is connected to the line and the armature to the incoming machine. When the frequency of the machine is different from that of the line the resultant field in the synchroscope constantly changes its position, thus making the armature revolve in one direc-

tion. When the frequencies are the same the field is stationary in space and the pointer comes to rest in a position to show the phase difference between the two voltages. When the phase difference is zero the pointer will come to rest on the mark indicating perfect synchronism and the switch can be closed.

The growth of high-frequency applications in recent years has given impetus to the development of instruments capable of accurately measuring electrical quantities at frequencies ranging from ordinary **audio frequencies** to very high **radio frequencies**. Among the types developed as a result of this demand are the thermocouple and the rectifier instruments. The first of these utilizes the thermoelectric effect (see **Thermoelectric Phenomena**). The current to be measured is passed through a heater wire which is in contact with or adjacent to a thermocouple. A voltage is induced in the couple when current passes through the heater wire so an instrument connected to the couple can be calibrated to indicate the current in the heater. Since the heat is proportional to the square of the current, the scale is an expanded or square law type. These instruments can be constructed to give accurate results to a hundred or more megacycles so are almost universally used for measuring radio frequency currents in this range. The rectifier meter uses some type of **rectifier** to convert the a-c to d-c and then utilizes a regular d'Arsonval type d-c instrument. At low frequencies (audio values) the rectifier is commonly a small bridge type copper oxide unit. At **ultra high frequencies** (radio) some type of rectifying **crystal** such as silicon is used with a sensitive milliammeter. When the frequency gets to these very high values capacitive and inductive effects in the leads, instrument, etc., become very important and must be reduced to a minimum if accurate results are to be obtained. Rectifier instruments indicate the average of the a-c wave rather than the r.m.s. value but are calibrated in terms of the r.m.s. of a sine wave and are not accurate for other wave forms unless specifically calibrated for them.

Another very useful instrument both for d-c and a wide range of a-c frequencies is the vacuum-tube voltmeter. This instrument utilizes the plate current-grid voltage characteristic of a **vacuum tube** as its fundamental feature. The voltage to be measured is applied to the **grid** of the tube and a milliammeter or microammeter in the plate circuit indicates a current which is a function of the grid voltage. By proper calibration of the plate instrument the voltage applied may be read directly from the instrument. There are several variations of this voltmeter, in some the grid voltage being measured is balanced by an internal voltage of the instrument (so-called slide-back voltmeter), in others the voltage applied to the grid is limited and multipliers must be used for larger values. In the better types of this latter group the normal plate current is balanced out in the milliammeter so it reads only the part due to the voltage being measured. These instruments are extremely useful since the input impedance may be made that of the grid of the vacuum tube (many megohms resistance and only a few micromicrofarads of capacity) and the frequency range is limited only by the frequency limit of the vacuum tube. (See also **Frequency Meters**.) (F.T.M., L.R.Q.)

INSTRUMENTS, MECHANICAL. With the broad definition of the term **instrument, mechanical instrument** could apply descriptively to an almost infinite variety of devices used in modern science and technology. By restricting the meaning to measuring instruments in common use for the measuration of physical quantities the following classification was devised. This classification is not of mechanically-operated instruments which are used to measure mechanical quantities. (For measuration of electrical quantities, see **Instruments, Electrical**.)

1. **Temperature** measurement. **Thermometry.**
   a. Glass tube mercury thermometers.
   b. Gas-filled bulb and tube thermometers.
   c. Vapor pressure thermometer.
   d. Electrical resistance thermometer.
   e. Thermocouple thermometer or pyrometer.
2. **Pressure** measurement.
   a. Standard, Bourdon tube type, **pressure gauge.**
   b. Helical tube or diaphragm type low-pressure gauge.
   c. Vacuum gauges. **Manometers.**
   d. **Draft gauges** (inclined glass tube, diaphragm, and liquid-sealed bell types).
3. **Flow** measurement.
   a. Steam **flow meters.**
   b. Water flow meters. Measure condensate, feed water, process flows, etc.
   c. Air flow meters. When these are used, they are generally in the form of a differential draft gauge.
   d. Gas flow meters. Positive displacement and orifice types.
4. **Fuel** measurement.
   a. Coal. Coal is usually weighed in batches, although belt conveyor weighers and some pulverized coal weighers are continuous.
   b. Gas meters.
   c. Oil meters. Positive displacement type.
5. **Carbon dioxide** measurement for combustion. The types of carbon dioxide meters in present use employ one of the following principles:
   Chemical—Modifications of the Orsat apparatus.
   Electrical—Based on measurement of the conductivity of flue gas.
   Mechanical—Flue gas density balanced against air.
6. **Speed** measurement.
   a. Vibrating reed tachometer (local reading).
   b. Electrical tachometer (remote reading).
   c. Clock type tachometer (local reading).
   d. Revolution counter.
7. Level recorders.
   Liquid level in boilers, tanks, canals, etc.
   Pulverized material in bins or silos.
8. Gong alarms.
   Gong alarms, with or without annunciators, are used to give warning of high temperatures, of high water, or low water in tanks.
9. Atmospheric measurements.
   Barometer, hygrometer, thermometer.
                                                    (F.T.M.)

INSULATION, HEAT. Bare surfaces at temperatures considerably above atmospheric lose much heat to the atmosphere. The B.T.U. per hour loss from bare pipe may not, on first thought, seem to amount to much, but if it be remembered that this loss is nearly steady 8760 hours per year (unless the pipe is out of service part of the time), and that, in the case of main steam lines, the B.T.U. so lost are high potential heat and therefore more valuable than the average B.T.U. in a pound of steam, it will be understood why practically every hot pipe in the modern plant or factory is covered. Cold pipes are also insulated to keep heat out, and insulation for this service is common in refrigeration plants. By keeping heat in hot lines not only is there a conservation of B.T.U. which have, at considerable expense and trouble, been transferred to the fluid, but also there is the avoidance of an uncomfortably overheated atmosphere in the vicinity of the pipe. Besides the pipe itself, fittings, valves, ducts, boiler drums, tanks and heaters are insulated.

In order to prevent great heat loss, a furnace wall must have an insulating layer, or else be built so thick that, in addition to being costly, it is inelastic and has large thermal storage capacity. In most cases the temperature of the inside furnace wall is near enough to that of furnace gases to be taken the same. The outside wall temperature will be enough higher than the surrounding atmosphere to discharge to it a heat flow sufficient to cause the temperature drop between the furnace and outside wall. Heat is discharged from the outside wall to the atmosphere by convection and radiation. It has been found that the total heat so transferred to still air is about 50% higher than the radiation component.

Other common uses of insulation are in the walls of cold rooms (cork board), coverings for hot and cold pipes, and the walk and ceilings of dwellings and other buildings. For the latter purpose a variety of materials are available, viz.: rock wool, insulating panels of ground wood, etc.

A good pipe covering should, of course, be non-conducting. As perfect non-conductors are not yet available, those materials whose conductivities are the lowest are best. Insulation should be able permanently to withstand the temperature to which it will be subjected; that is, it should be stable and resist deterioration over the working life of the pipe. It should be easily molded and applied and have the requisite mechanical strength. No insulation commercially procurable will overload the pipe by its dead weight, for density is not one of the attributes of a good insulator. In fact, the non-conducting properties seem chiefly to be derived from the presence of large numbers of air cells. The materials most commonly used are asbestos, "magnesia" (magnesium carbonate), cork, hair felt, wool felt, rock wool, and diatomaceous earths. Most commercial insulations are either built up from corrugated asbestos paper, or laminated asbestos paper artificially roughened to produce air spaces, or are molded, or felted with asbestos, etc. A very common and effective insulation for temperatures up to 600° F. is the molded "85% magnesia," so called because it is 85% carbonate of magnesium and 15% binder. Pipe insulation for higher temperatures should have an inner layer of some special high-temperature insulation, since 85% magnesia alone will deteriorate. Painting with aluminum or bronze paint will greatly decrease the radiation losses.

As in the case of furnace wall design, calculations of heat transfer through pipe covering take the form of an addition of resistances of a series circuit.

Manufacturers publish insulation efficiency tables for the various standard thicknesses of their different grades of insulation. The "efficiency" of an insulating material is expressed as the per cent heat saved by using the insulation, compared to what would have been lost had the surface been left bare. Like many other design problems, the amount of insulation to apply is, basically, an economic problem. The cost of the covering must be weighed against the saving of heat. (See **Thermal Conduction.**) (F.T.M.)

INSULATOR, ELECTRIC. Electric Power Transmission.

INSULIN. Insulin is a **hormone** which is vitally concerned in the **metabolism** of **carbohydrates.** It is secreted by the islet tissue of the **pancreas** and discharged directly into the blood stream. Although its presence was surmised, and it was already named "insuline" as early as 1909, it was not until 1922 that Banting and Best succeeded in isolating the hormone from pancreatic tissue.

Insulin is a complex **protein,** which must be injected to be effective, since gastric juices destroy its activity. Its mode of action is not completely understood, but it is known to promote the **oxidation** of carbohydrate, and to act as an essential in the intermediary metabolism of carbohydrate. An insufficient supply of insulin results in the clinical picture of **diabetes,** with low blood sugar, and all of the characteristic signs and symptoms of the disease. Diabetes is successfully controlled by insulin in varying doses, depending on the severity of the disease. Insulin derived from hog or beef pancreas, or the crystalline product may be used. With these forms, usually repeated injections must be made throughout the day—2 to 4 times. Protamine-zinc-insulin has been developed to overcome this disadvantage. The protamine form acts slowly over a 24-hour period, and only one daily injection is therefore required.

Hyperinsulinism results from an excessive supply of insulin, and produces a picture roughly opposite to that found in diabetes. If hyperinsulinism follows an accidental overdose of insulin given to a diabetic, the blood sugar may fall to an extremely low level, and the condition is called insulin shock. The clinical picture varies from simple hunger to irritability, pallor, weakness, excessive perspiration, double vision and coma. Insulin shock is now induced purposely in the treatment of certain **psychoses**. Improvement of the symptoms of insulin shock by giving carbohydrate by mouth as orange juice, or glucose intravenously, is sudden and dramatic. **Tumors** of the islet tissue of the pancreas also may cause hyperinsulinism.

The relationship of insulin to other **endocrine glands** is complex. The **pancreas, adrenal, pituitary, thyroid** and **liver** are all interrelated in their action on carbohydrate metabolism. (D.M.H.)

INTEGRAL EQUATIONS. An integral equation may be described in a general way as an equation in which an unknown function appears under a **definite integral**.

A linear integral equation of the first kind is an equation of the form

$$\int_a^b K(x, t)u(t)dt = f(x),$$

where $u(t)$ is an unknown function to be found, $K(x, t)$ and $f(x)$ are known functions, and the limits a and b are known. The function $K(x, t)$ is called the kernel of the equation.

A linear integral equation of the second kind (or Fredholm integral equation) is an equation of the form

$$u(x) = f(x) + \int_a^b K(x, t)u(t)dt,$$

where $u(t)$ is an unknown function to be found, $K(x, t)$ and $f(x)$ are known functions, and a and b are given. The function $K(x, t)$ is called the kernel of the equation.

If $f(x) \equiv 0$, the resulting equation

$$u(x) = \int_a^b K(x, t)u(t)dt$$

is called a homogeneous linear integral equation of the second kind.

The equation

$$u(x) = f(x) + \int_a^x K(x, t)u(t)dt$$

is called Volterra's linear integral equation of the second kind.

Non-linear integral equations have also been investigated.

Three types of solutions of linear integral equations of the second kind have been developed:

1. The first method, which may be called the method of successive substitutions, is due to Neumann, Liouville, and Volterra. It gives the unknown function $u(x)$ in the equation

$$u(x) = f(x) + \lambda \int_a^b K(x, t)u(t)dt$$

as a power series in λ, whose coefficients are functions of x; this series has in general a finite radius of convergence.

2. The second method, due to Fredholm, gives $u(x)$ as a ratio of two power series in λ, each series converging for all values of λ. In the numerator the coefficients of the powers of λ are functions of x, while the denominator is independent of x. The solution is obtained by considering the integral equation as the limiting form of a system of n linear algebraic equations in n variables as n becomes infinite.

3. The third method, which was developed by Hilbert and Schmidt, gives the unknown function $u(x)$ in terms of a set of so-called fundamental functions. These functions are in the ordinary case the solutions of the corresponding homogeneous equation

$$u(x) = \lambda \int_a^b K(x, t)u(t)dt.$$

This equation is in general satisfied only by $u(x) \equiv 0$, but there exists a sequence of numbers $\lambda_1, \lambda_2, \cdots, \lambda_n, \cdots$, called characteristic constants (or fundamental numbers) for each of which the equation has a finite solution: $u_1(x), u_2(x), \cdots, u_n(x), \cdots$. These functions are the fundamental functions. The solution of the given integral equation is then obtained in the form

$$u(x) = \sum_{n=1}^{\infty} a_n u_n(x),$$

where the a_n are arbitrary constants. (L.L.S.)

INTEGRAL FUNCTION. An integral function is a **function** (of a real or complex variable) defined by a **power series** which converges for all finite values of the variable; this includes **polynomial functions** as special cases. An integral function may also be defined as a function of a complex variable which is **analytic** at all finite points of the complex plane. Simple examples of integral functions (besides the polynomials) are the **exponential function**, and the sine and cosine (**trigonometric) functions.** (L.L.S.)

INTEGRAPH. An integraph is an instrument which draws the integral curve of a given curve mechanically; from it areas may be read off as ordinates. (L.L.S.)

INTEGRATING METERS. The ordinary electric service meter measures the total of electric energy used over a period of time. It is in principle much like a **wattmeter**, except that the movable coil is replaced by a motor armature rotating against a magnetic damping arrangement. The speed of revolution is proportional to the torque, which, in turn, is proportional to the product of current and electromotive force. The total revolution of the armature is recorded on dials by pointers suitably geared to the armature shaft. Since angle = revolution speed × time, the reading of the dial is proportional to current × electromotive force × time, that is, to amperes × volts × hours or watt-hours. The dial may thus be graduated directly in watt-hours or in convenient multiples or sub-multiples thereof. (See **Power** and **Watt.**) (L.D.W.)

INTEGRATING PHOTOMETERS. The usual types of **photometer** give the luminous intensity of a source as viewed from one direction only. If the lamp under examination is turned around its vertical axis and the measurement thus obtained from various horizontal directions averaged, the result is the mean horizontal **candle power**. Usually a still more significant quantity is the mean spherical candle power, that is, the average luminous intensity from all directions. To obtain such averages, some type of integrating photometer is used.

One plan is to reflect the light emitted in various directions toward one spot by means of a system of mirrors appropriately placed, and to measure the resulting illumination at that point. The most common device at present, however, is the "sphere" or "globe" photometer. This has a large hollow globe painted white inside with barium sulfate paint. The lamp is mounted at any convenient point inside, and the resulting diffuse illumination of a transparent screen covering an opening in one side is measured by a photometer. (The screen must be protected from the direct rays of the lamp.) It was shown by Sumptner (1892) that under these conditions, the illumination on the screen is proportional to the mean spherical candle power and is independent of the orientation or the position of the lamp. Comparison may thus be made with a standard lamp of known spherical candle power by the substitution method. A photovoltaic cell may be substituted for the translucent screen and the readings thus made electrically. (See Illuminometer.) (L.D.W.)

INTEGRATION. Indefinite Integral, also Definite Integral.

INTEGRATION AS A PROCESS OF SUMMATION. Definite Integral.

INTEGRATION, TECHNIQUE OF. Standard integrals:

$$\int u^n\, du = \frac{u^{n+1}}{n+1} + C \quad (n \neq -1),$$

$$\int \frac{du}{u} = \log_e u + C, \quad \int e^u\, du = e^u + C,$$

$$\int \sin u\, du = -\cos u + C, \quad \int \cos u\, du = \sin u + C,$$

$$\int \sec^2 u\, du = \tan u + C, \quad \int \csc^2 u\, du = -\cot u + C,$$

$$\int \sec u \tan u\, du = \sec u + C,$$

$$\int \csc u \cot u\, du = -\csc u + C,$$

$$\int \tan u\, du = -\log \cos u + C = \log \sec u + C,$$

$$\int \cot u\, du = \log \sin u + C,$$

$$\int \sec u\, du = \log (\sec u + \tan u) + C,$$

$$\int \csc u\, du = \log (\csc u - \cot u) + C,$$

$$\int \frac{du}{u^2 + a^2} = \frac{1}{a} \tan^{-1} \frac{u}{a} + C,$$

$$\int \frac{du}{u^2 - a^2} = \frac{1}{2a} \log \frac{u-a}{u+a} + C, \quad (|u| > |a|),$$

$$\int \frac{du}{a^2 - u^2} = \frac{1}{a} \tanh^{-1} \frac{u}{a} + C, \quad (|u| < |a|),$$

$$\int \frac{du}{a^2 - u^2} = \frac{1}{a} \coth^{-1} \frac{u}{a}, \quad (|u| > |a|),$$

$$\int \frac{du}{\sqrt{a^2 - u^2}} = \sin^{-1} \frac{u}{a} + C, \quad (|u| < |a|),$$

$$\int \frac{du}{\sqrt{u^2 \pm a^2}} = \log (u + \sqrt{u^2 \pm a^2}) + C,$$

$$\int \frac{du}{\sqrt{u^2 + a^2}} = \sinh^{-1} \frac{u}{a} + C,$$

$$\int \frac{du}{\sqrt{u^2 - a^2}} = \cosh^{-1} \frac{u}{a} + C, \quad |u| > |a|,$$

$$\int \frac{du}{u\sqrt{u^2 - a^2}} = \frac{1}{a} \sec^{-1} \frac{u}{a} + C, \quad (|u| > |a|).$$

Integration by substitution:

An integral $\int f(x)dx$ may often be reduced to a standard form by expressing x as a function of another variable t. If $x = \phi(t)$, the substitution $x = \phi(t)$, $dx = \phi'(t)dt$ will transform $\int f(x)dx$ into an equivalent integral in t.

Integration by parts:
This is based on the formula

$$\int u\, dv = uv - \int v\, du.$$

Integration of rational fractions:
When the function to be integrated has the form of a rational fraction, it should be decomposed into a sum of **partial fractions**; each of these partial fractions can then be integrated separately by standard forms.
Trigonometric forms:

1. Type $\int \sin^m x \cos^n x\, dx$, where either m or n is an odd integer.
Suppose n is an odd integer. Write the integral $\int \sin^m x \cos^{n-1} x(\cos x\, dx)$; by use of $\cos^2 x = 1 - \sin^2 x$, we can express $\cos^{n-1} x$ rationally in terms of powers of $\sin x$ since $n - 1$ is an even integer, and since $\cos x\, dx = d \sin x$, we obtain an expression in powers of $\sin x$ each multiplied by $d \sin x$, so that they can each be integrated by the power law

$$\int u^n\, du = \frac{u^{n+1}}{n+1} + C.$$

2. Type $\int \begin{Bmatrix} \sin \\ \cos \end{Bmatrix} mx \cdot \begin{Bmatrix} \sin \\ \cos \end{Bmatrix} nx\, dx$:

To evaluate $\int \sin mx \cos nx\, dx$, we use the trigonometric identity $\sin mx \cos nx = \frac{1}{2}[\sin (m + n)x + \sin (m - n)x]$. To evaluate $\int \sin mx \sin nx\, dx$, we use the identity $\sin mx \sin nx = \frac{1}{2}[\cos (m - n)x - \cos (m + n)x]$. To evaluate $\int \cos mx \cos nx\, dx$, we use the identity $\cos mx \cos nx = \frac{1}{2}[\cos (m + n)x + \cos (m - n)x]$.

3. Type $\int \sin^m x \cos^n x\, dx$, where m and n are both even integers.
Transform the integrand function into an integrable form in terms of sines and cosines of multiples of x by use of the trigonometric identities:

$$\sin^2 x = \frac{1}{2}(1 - \cos 2x), \quad \cos^2 x = \frac{1}{2}(1 + \cos 2x),$$

$$\sin x \cos x = \frac{1}{2} \sin 2x.$$

4. Type $\int \tan^n x\, dx$ or $\cot^n x\, dx$, where n is any integer.
These forms may be reduced to standard forms by use of $\tan^2 x = \sec^2 x - 1$, or $\cot^2 x = \csc^2 x - 1$, and $\sec^2 x\, dx = d (\tan x)$ and $\csc^2 x\, dx = -d(\cot x)$.

5. Type $\int \tan^m x \sec^n x\, dx$ or $\int \cot^m x \csc^n x\, dx$, where n is an even integer.
In this case we use the relations $\tan^2 x = \sec^2 x - 1$, $\sec^2 x = 1 + \tan^2 x$, $d \tan x = \sec^2 x$, and similar forms in terms of cot x and csc x, and reduce to use of the power law

$$\int u^n \, du = \frac{u^{n+1}}{n+1} + C.$$

Integration of linear radical forms:

If the integrand contains a linear radical $\sqrt[n]{ax+b}$, the substitution $a + bx = z^n$ will reduce it to a rational form.

Integration by trigonometric substitutions:

Expressions involving $\sqrt{a^2 - x^2}$ or $\sqrt{x^2 \pm a^2}$ are often most easily integrated by aid of one of the following substitutions:

when $\sqrt{a^2 - x^2}$ occurs,

> put $x = a \sin \theta$, $\sqrt{a^2 - x^2} = a \cos \theta$,

when $\sqrt{x^2 + a^2}$ occurs,

> put $x = a \tan \theta$, $\sqrt{x^2 + a^2} = a \sec \theta$,

when $\sqrt{x^2 - a^2}$ occurs,

> put $x = a \sec \theta$, $\sqrt{x^2 - a^2} = a \tan \theta$.

Integrals with quadratic radicals:

If the integrand contains $\sqrt{x^2 + ax + b}$, substitute $\sqrt{x^2 + ax + b} = z - x$, and we obtain a rational integrand in terms of z.

If the integrand contains $\sqrt{-x^2 + ax + b}$, substitute $\sqrt{-x^2 + ax + b} = (\alpha - x)z$ or $(\beta + x)z$, where $\alpha - x$ and $\beta + x$ are the factors of $-x^2 + ax + b$; the integrand becomes rational in z.

An integral in which the integrand is a rational function of $\sin x$ and $\cos x$ is transformed into one having a rational algebraic integrand by the substitution $\tan \frac{1}{2}x = z$, then

$$x = 2 \tan^{-1} z, \quad dx = \frac{2dz}{1 + z^2},$$

$$\sin x = \frac{2z}{1 + z^2}, \cos x = \frac{1 - z^2}{1 + z^2}.$$

Tables of integrals are published, giving more or less extensive classified lists of integrals. (L.L.S.)

INTEGUMENT, INTEGUMENTARY SYSTEM.

For the use of this term in botany, see **Seed.** The body of every multicellular animal is covered with a layer of tissue adapted to meet the external conditions that prevail in the normal environment of the species. This covering is the integument and together with all of the specialized structures derived from the cellular layers it constitutes the integumentary system.

The general functions of the integumentary system are protection against mechanical damage and desiccation, the transmission of materials which must pass into or out of the body, the conservation of heat, and the reception of stimuli. Among the invertebrates such rigid supporting structures as the animal may possess are often developed in the integument, so that it becomes a **skeletal system** or exoskeleton as well as a covering for the body.

The integument is always at least partly ectodermal in origin. In the simpler animals such as the **coelenterates** it is an epithelial layer containing cells specialized for the reception of stimuli, defensive cells, and simple cells bearing contractile basal processes. It bears **cilia** in some of the flatworms, in the **rotifers** and **bryozoans**, and in some of the mollusks (**Mollusca**), and so aids in bringing food to the animal and in locomotion. In parasitic flatworms it degenerates into a **syncytium** and produces a non-cellular **cuticle**, and in roundworms, segmented **worms**, and **arthropods** it also secretes an external cuticle although it remains cellular. It gives rise to the **setae** of the segmented worms and arthropods and to the external portions of sensory organs, and contains glands of various kinds. In the arthropods the cuticula is highly developed as an exoskeleton. The integument also lays down the hard deposits of **corals** and secretes the shells of **brachiopods** and mollusks.

The integument of **vertebrates** is a **skin** composed of two layers, an inner corium or dermis derived from mesoderm and an outer cuticle or epidermis which is ectodermal. It produces various hard structures including the scales of fishes, scales of reptiles, birds and mammals, feathers, hair, horns, claws, hoofs, and nails. In addition it contains glands of various kinds, including the mucus glands of fishes and amphibians and sweat glands of mammals and forms parts of the sensory organs. (A.W.L.)

INTELLIGENCE TESTS. Mental Deficiency.

INTENSIFICATION.

A process for increasing the density and contrast of a photographic image by depositing thereon silver, or a compound of mercury, copper, lead, uranium, or other metal. Intensification may also be effected by mordanting dyes to the image, but this method is not in general use.

Intensification with silver is accomplished by means of an acid solution of silver nitrate and a developer, usually metol, which results in the silver being precipitated preferentially on the silver deposits of the image. The following formula (Crabtree & Muehler) is recommended:

Solution 1

Silver nitrate, crystals........	60 grams	2 oz.
Distilled water to make......	1 liter	32 oz.

Solution 2

Sodium sulfite, desiccated.....	60 grams	2 oz.
Water to make..............	1 liter	32 oz.

Solution 3

Sodium thiosulfate (Hypo)....	105 grams	3½ oz.
Water to make..............	1 liter	32 oz.

Solution 4

Sodium sulfite, desiccated.....	15 grams	½ oz.
Elon......................	24 grams	350 gr.
Water to make..............	3 liters	96 oz.

For use, add 1 part #2 to 1 part #1, shaking or stirring thoroughly. The white precipitate which forms is dissolved by the addition of 1 part of #3. Allow to stand until clear, then add with constant stirring 3 parts of #4. Use at once. The time of intensification should not exceed 20 minutes or the negative may be stained. After intensification, fix in a 30% solution of hypo and wash thoroughly.

One of the simplest and most reliable methods of intensification consists in bleaching the image in a solution of potassium or ammonium bichromate and hydrochloric acid—or chromic acid and sodium chloride—followed by redevelopment in an ordinary developer. The chemistry of the process is not definitely known but the increase in density is due to the formation on the image of a reduction product of the chromate whose composition is assumed to be either CrO_2 or Cr_2Cl_2.

The following forms a satisfactory bleaching solution:

Potassium bichromate.....	90 grams	3 oz.
Hydrochloric acid, C.P....	64 cc.	2 oz.
Water to make..........	1 liter	32 oz.

For use dilute 1 part of the above with 10 of water.

After bleaching the image, wash in running water until every trace of the bichromate has been removed and develop in an ordinary metol-hydroquinone developer. Borax and other metol-hydroquinone developers containing a high concentration of sulfite, however, should not be used.

After redevelopment, wash for 15–20 minutes in running water and dry.

The process may be repeated if the density and contrast are insufficient after the first application.

There are a number of methods of intensification with mercury. Probably the most popular consists in bleaching the negative in a solution of mercuric bromide and blackening in (1) an ordinary developer, (2) a 10% solution of sodium sulfite, (3) strong ammonia, (4) a 2% solution of sodium sulfide, (5) a 10% solution of sodium sulfantinioniate. When the image is redeveloped the process may be repeated if necessary; little or no increase in density or contrast is obtained if the other blackeners are used. The permanency of images intensified with mercury is always questionable and these processes offer no advantage over the chromium intensifier just described.

If the bleached image is blackened in silver cyanide, however, an extremely high degree of intensification is obtained. This process is known as Monkhoven's intensifier and is especially suitable for the intensification of negatives of drawings, printed matter and other black and white subjects. The silver cyanide solution is prepared as follows:

Potassium cyanide	15 grams	2 oz.
Silver nitrate	22.5 grams	3 oz.
Water to make	1 liter	1 gal.

Each is dissolved separately in water and the solutions mixed. It is then allowed to stand for an hour and filtered.

NOTE: The cyanides are deadly poisons and should be handled with care.

A useful single solution intensifier is compounded as follows:

Mercuric iodide	2 grams	90 gr.
Potassium iodide	2 grams	90 gr.
Hypo	2 grams	90 gr.
Water to make	100 cc.	10 oz.

This solution may be used until exhausted. While the intensifier contains hypo, negatives should be washed before intensification. The action of the intensifier is progressive and may be stopped when the required density and contrast have been reached by washing under the tap, followed by 15–20 minutes washing in a tray or washing tank.

The intensified image is not permanent. (C.B.N.)

INTENSITY SCALES. Systems of rating equal intensity by various earthquake effects on persons and things. The Rossi Forel scale is simple, has ten grades, and has been long in use. The Mercalli scale is more accurate and has twelve grades. A modified form called the Wood-Neumann scale is used in the United States, and is coming into more general use, both in complete and abridged form. (R.M.F.)

INTERCEPT. Celestial Navigation.

INTERCHANGEABLE MANUFACTURE. The principle of interchangeability requires manufacture to such specification that component parts of a device may be selected at random and assembled to fit and operate satisfactorily. This necessitates parts made to definite limits of error, fitting gauges instead of mating parts. Interchangeability does not necessarily involve a high degree of precision; stove lids, for example, are interchangeable, but not particularly accurate.

Interchangeable manufacturing procedure requires specification for the permissible error in each dimension, as well as for the size of the part, since it is practically impossible to manufacture to an exact dimension, and is therefore based upon the limit system of dimensioning. In limit dimensioning, nominal size is a designation given to a dimension which has no specified limits of accuracy, but is a close approximation to a standard size. Basic size is an exact theoretical size from which all variations are made. Allowance is an intentional difference in the size of mating parts. Tolerance is the permissible variation in the size of a part. Limits are the extreme permissible dimensions of a part.

Two types of tolerance are used in limit dimensioning. Unilateral tolerance is related to a basic size in one direction only. In a running fit, for example, the hole size may vary between limits of 1.2500″–1.2514″ where the tolerance is .0014″, above the basic size of 1.2500″. A bilateral tolerance is one in which the variation from the basic size is in both directions. In a similar example, the hole size may vary between 1.2493″–1.2507″, giving a tolerance of +.0007″ and −.0007″ from the basic size.

The use of tolerances makes it impossible to maintain exact theoretical allowances for fits; where more precise allowances are desired, selective assembly is used. Selective assembly consists of trial selection of mating parts to obtain a desired precision of fit. An example of this process is in the assembly of ball bearings. All ⅜″ diameter balls are not alike, since there is a tolerance of about .0001″ in manufacture. The balls are sorted into three groups: .3751″, .37505″, and .3750″ in diameter, and assembled in similarly selected races. The assembled ball bearing is not strictly interchangeable, since a replacement ball may be slightly under or over the size of the rest of the balls in the assembly.

Fractional dimensions indicate that ordinary scale measurements are sufficiently precise, and thereby imply limits, in general, of plus or minus .010″ on each dimension. (H.C.H.)

INTERCLASS CORRELATION. When two variables x, y are correlated and each variable refers to a different group, then we have an interclass correlation as distinct from the **correlation** among items in the same group. If we find the correlation between the height and weight of people we are determining an interclass correlation. If we find the correlation among the weights of adults in the same family we are finding an **intraclass correlation.** (L.A.A.)

INTERCONNECTION. Each year witnesses interconnections and mergers which are slowly uniting the power and light companies of the United States into one vast interconnected network of transmission lines. The justification for interconnection can be presented on a financial basis, but the ability to render intersystem assistance during local trouble and in that way to prevent interruption of service to the customer should be the basic reason for interconnection, even though no definite financial expression may be attached to it. Of more tangible value is the use of the more efficient plants as base load stations. Off-peak power may be exchanged on some prearranged basis when the load peaks on one system do not occur simultaneously with those on another. This has the effect of delaying, for a time, the purchase of new equipment to care for increasing peaks. Also, it renders economical the installation of additional capacity beyond that justified by the gain in the individual system; that is, a new station or unit causes temporary surplus capacity and consequent annual charges for idle equipment which may be reduced by adjacent systems installing their new equipment alternately. A decrease in the total of emergency standby capacity is possible when two or more systems operate interconnected. Systems predominantly hydroelectric, but with steam standby, realize the maximum economy in interconnected operation because of the diversity of stream flow in different localities.

Interconnection relieves the necessity of splitting up the plant capacity into a number of small units for the sake of uninterrupted service.

A superpower system is a vast interconnected system which has for its basis the maximum exploitable water power of the country, relying on excess flows at one point to counteract low water at another, and having steam plants suitably located to care for deficiencies in water power. While it would be possible to transmit over great distances, actually transfers of energy between groups constituting the superpower system would rarely occur over distances greater than 200 miles. (F.T.M.)

INTERELECTRODE CAPACITANCE. This is the capacitance which is present between the various electrodes of an electron **tube**. Thus the **grid** and **plate** of a **vacuum tube** are normally not connected directly to one another but because they are conductors separated by a dielectric (the vacuum space in the tube) they have a certain capacitance. The same situation exists between the other electrodes of the tube. These various electrode capacitance offer paths through which alternating currents may flow between the electrodes. Since these currents are not under the control of the signal they are very undesirable in most cases. In an **oscillator** this interelectrode capacitance is utilized to transfer energy from the plate to the grid circuit and thus make the circuit oscillate. However, in an **amplifier** such action is normally highly undesirable and various means such as neutralizing and screen grids are employed to reduce it. While the capacitance in a receiving tube, for example, may be only a few micromicrofarads appreciable current may flow through it at high frequencies. (L.R.Q.)

INTERFACIAL TENSION. Surface-Active Compounds.

INTERFERENCE. This is the undesirable mixing of two signals and is commonly used to designate the mixing of one or more undesired signals with a desired communications signal. Various forms of interference are power-line interference, i.e., noise introduced by a power circuit in a communication circuit (see **Inductive Interference**), **Crosstalk**, **Cross-Fire**, heterodyning of desired and undesired stations in radio reception and many others. (See **Fit; Gearing.**) (L.R.Q.)

INTERFEROMETERS. The term interferometer may be applied to any arrangement whereby a beam of light from a large, luminous area (as a sodium flame) is separated into two or more parts by partial reflections, the parts being subsequently reunited after traversing different optical paths. The two components then produce **interference**. The best known instrument is that of Michelson, shown diagrammatically in the accompanying figure. The original beam a is separated at

Diagram of Michelson interferometer.

the surface AM of a glass plate, part of it ($1,1'$) going to a mirror M_1 and part ($2,2'$) going on through a second, exactly similar plate B to the mirror M_2. They reunite at AM and are observed together at E. One of the mirrors, M_1, is mounted on a micrometer screw so

that its distance from AM can be varied, the phase difference of the reunited beams thereupon passing through a series of cycles. If M_1 or M_2 is not quite perpendicular to the beam reflected by it, the field at E is crossed by interference fringes, which move across the field as the mirror M_1 is moved. Each complete cycle corresponds to a displacement of M_1 equal to a half-wavelength. The Fabry and Perot interferometer is somewhat simpler in design, but utilizes multiple reflection and produces very sharp fringes (high resolving power).

Interferometers are used for precise measurements of wavelength, for the measurement of very small distances and thicknesses by using known wavelengths, for the detailed study of the **hyperfine structure** of spectrum lines, for the precise determination of refractive indices, and, in astrophysics, for the measurement of **double-star** separations and the diameters of very large stars. (L.D.W.)

INTERFLUVE. The area or watershed between two neighboring streams or rivers. (R.M.F.)

INTERGRANULAR CORROSION. Corrosion.

INTERLACING. Scanning.

INTERLOBATE MORAINE. If large glaciers and continental ice sheets advance irregularly so that their margins are lobate, when the margins retreat by melting the resulting **terminal moraines** of boulders, clay, and sand simulate the original interlobate shape of the glacier or glaciers, and therefore such moraines are called interlobate. (R.M.F.)

INTERMEDIATE FREQUENCY. Superheterodyne.

INTERMEDIATES. Coal Tar Products and Intermediates.

INTERMETALLIC COMPOUND. A solid compound of two or more metals united in definite atomic proportions as in the case of chemical compounds but not following the usual rules of chemical valence. Examples are $CuAl_2$, Mg_2Si, $CuSn$, Bi_2Te_3. Compounds between metals and **metalloids** such as Fe_3C and MnS are generally classed as intermetallic compounds also. In alloys these compounds are of great importance. They may form directly from the molten metal upon solidification or from **solid solutions** upon cooling to lower temperatures. They generally are present in the **microstructure** as hard particles embedded in a more plastic matrix. They are often soluble to a limited degree in the matrix and thus make it possible to harden and strengthen the alloy by a **precipitation-hardening** heat treatment. (R.H.H.)

INTERNAL COMBUSTION ENGINE. The internal combustion engine is one in which **combustion** of a **fuel** takes place within the **cylinder**, and the products of combustion form the working medium during the power stroke. As it is a self-contained power supply unit, the internal combustion engine assumes a position of great importance in the power field, especially for that class of service where portability, light weight, and compactness are important. Witness the majority of self-propelled vehicles powered by internal combustion engines. The principal cycles of internal combustion engines in use at present are the **Otto** and **Diesel** cycles. Other cycles are of some historical importance, but these two have supplanted the others in the course of time by possessing certain points of superiority such as economy, reliability, compactness, etc.

An internal combustion engine consists primarily of a cylinder, almost always stationary, and a **piston**, generally single acting, which, together, form a **combustion chamber** of variable volume. Both of these parts are constructed of metal, and as the temperatures attained during combustion are well above the ability of uncooled metals to withstand, the cylinder of an internal combustion engine must be adequately cooled by transferring through the cylinder wall a certain amount of the heat contained in the gases of combustion. This is accomplished in a practical way by surrounding the cylinder with a **jacket** of cooling water, or by providing it with an extended outer surface of fins so that air can absorb enough heat to keep the metal cool. The required motion is given to the piston by a crank and connecting rod mechanism which also serves to take from the piston the power developed by gas pressure.

An engine operating with flaming gas in its cylinder would not last long with simple metal to metal contact of the moving parts, therefore a lubricating system, embodying oil as the lubricant, is an important feature of every internal combustion engine. The most difficult job is that of lubricating the piston in the cylinder. During a portion of the stroke, at least, the lubricated wall is exposed to incandescent gases which tend to burn off the film of lubricating oil. The cooling system must be adequate to maintain the metal surfaces cool enough to save the lubricating film.

The events of the cycle upon which an internal combustion engine works are controlled chiefly by the operation of **valves** located in ports leading to and from the cylinder. Generally, an admission or inlet valve, and an exhaust valve, are provided in each cylinder. The operation of these valves is derived mechanically from the crankshaft through the **valve gear** system.

The combustion of fuel in an internal combustion engine is not a continuous affair, but a series of individual explosions, each one requiring a metered amount of fuel to be individually ignited. For this reason, every internal combustion engine must incorporate an **ignition system**, whose function it is to supply in proper time the ignition temperature required for combustion. The internal combustion engine is of a type tending to deliver its power cyclically, and in a fashion which would be very fluctuating unless balanced by the use of heavy **flywheel**, or by overlapping of power impulses through multi-cylindered arrangements. It is usual, in fact, to build internal combustion engines with more than one cylinder so that the delivery of power will be more uniform, and flywheel proportions will not be excessive. The supply of fuel to multi-cylindered engines from a common source, and the conduction of exhaust from them, leads to another service feature for the internal combustion engine, namely, the inlet and exhaust **manifolds**.

The production of power by this type of engine represents a thermodynamic conversion of a portion of the heat energy developed into mechanical energy. The heat energy enters the engine latently in the form of fuel. Mechanical energy appears as power available at the crankshaft. Unavailable or rejected heat is found in exhaust, cooling, and friction. The conversion of the energy of the fuel into useful power takes place about as follows: Air is brought into the cylinder and, either after, before, or during compression, depending on the cycle, fuel is introduced into the air and mixed with it. Upon ignition of this fuel, the heat developed raises the pressure of the products of combustion, or, at least, maintains the pressure during some motion of the piston. The fact that the piston has, against one face of it, a gas pressure greatly exceeding that on the other, inevitably results in the transmission of energy through the train of mechanism consisting of moving piston, wrist pin, connecting rod, and crankshaft. During the motion of the piston, the gases of combustion expand and are cooled somewhat. It has not been found economical to build an engine sufficiently bulky to expand the gases until they reach ordinary atmospheric temperature, and

there is always considerable heat loss in the exhaust. In spite of the losses the internal combustion engine of the present is the most efficient prime mover that has been devised. (See **Diesel Engine, Otto Engine, Two-cycle, Four-cycle, Aeronautical Engine, Engine, Ignition System, Carburetion.**) (F.T.M.)

INTERNAL FRICTION. Viscosity.

INTERNAL GEAR. Spur Gearing.

INTERNAL PRESSURE. Adhesion and Cohesion.

INTERNAL SAC. An organ of larval **bryozoans** by which the animal attaches itself to a supporting surface before transforming into the adult. (A.W.L.)

INTERNATIONAL DATE LINE. In accordance with the fundamental definition of civil time the date changes when the mean sun crosses the **meridian** at lower culmination, i.e., at midnight. It would be possible for a traveler to travel around the earth on a sufficiently high parallel of **latitude** in 24 hours. Say such an observer starts out at one o'clock in the afternoon, i.e., with the sun about 1 hour west of his meridian, and travels westward at just the proper rate to maintain the sun in that position. For such a traveler the date would not change and he would return to his starting point on the same date, as he would reckon it, as that on which he set forth on his journey. If the traveler had taken 100 days to complete his journey instead of only 1, he would still return to his starting point a day earlier, on his own reckoning, than the date of those who had remained at home. To avoid confusion of this sort the so-called international date line has been established.

This date line is approximately 180° west of Greenwich in **longitude** but is adjusted so that, so far as is possible, the Pacific insular possessions of the different countries shall carry the same date as the home nations, and also so that the line shall not cross any land. Ships crossing the date line from east to west skip one day. That is, if it is Monday when the ship arrives at the line from the east it immediately becomes the same hour on Tuesday after crossing the line and the day so dropped is omitted from the log book. On crossing the line from west to east a day is repeated. (W.K.G.)

INTERNATIONAL TEMPERATURE SCALE. Thermometry.

INTERNODE. Stem.

INTERPHONE. This is a small private telephone communication system installed in a building for use in communicating within the building only. Such systems may be miniature replicas of the regular telephone systems or they may consist only of two or three stations operating as the simplest type of telephone system. In such systems the switching may be done by pushbuttons or switches at each phone, or even in some operated without any switching on a sort of party line, or they may utilize regular dial or automatic switching equipment. Often in the more extensive systems provision is made for connecting all or certain selected phones with the public telephone lines. (L.R.Q.)

INTERPOLATION. Interpolation is a process by which an appropriate value is placed between tabulated values of a function.

Simple interpolation, by first differences, is based on a principle of proportional parts: when the variable (or argument) changes by a small amount, the change in the tabulated function is very nearly proportional to the change in the variable.

For more accurate work, interpolation by second and higher differences is needed, and more complicated interpolation formulae are then used. (L.L.S.)

INTERPOLE. This is an auxiliary pole placed between the main poles of a d-c motor or generator to give a flux which will aid in commutation. When a machine is operating under load the armature current sets up a flux which combines with the field flux, giving a resultant distortion of the main flux. When a coil is being commutated it is, of course, short-circuited by the brushes so any voltage present in the coil will cause a circulating current around the coil by way of the brush or brushes. (See Commutation.) Such currents are very undesirable as they introduce extra losses and cause unnecessary burning of the brushes and commutator. If the coil being commutated has its sides lying in a position of zero flux while undergoing commutation there will be no field induced voltage. However, the current in the coil is reversed during commutation (by very definition of commutation) and as a reversal of current causes an inductive or reactance voltage to be induced the coil still has this e.m.f. to cause currents. The distortion of the field flux by armature reaction and this reactance voltage make it necessary to provide interpoles or shift the brushes from the mechanical neutral. When brushes are shifted they must be shifted forward in the direction of rotation for generators and backwards for motors, hence must be changed if the rotation is reversed. In addition this is a not too satisfactory method for curing the trouble since the amount of shift is proportional to load. Introducing interpoles between the main poles and connecting the interpole windings in series with the armature gives an auxiliary flux which will tend to counterbalance the distortion of the flux by the armature current and at the same time give an auxiliary flux which will induce a voltage in the armature coils to counteract the reactance voltage. Thus much improved commutation results and shifting of brushes is eliminated. (L.R.Q.)

INTERQUARTILE RANGE. The interquartile range is defined as $Q_3 - Q_1$ where Q_3 and Q_1 are the third and first quartiles in a distribution. It is a measure of variability and is not recommended. In its place the standard deviation should be used. (L.A.A.)

INTERRAY. A division of the body of a starfish (Asteroidea) between the axes of two rays or arms, or a corresponding division of the sea urchins which lack radiating arms but have radii indicated in the structure of the compact body. (A.W.L.)

INTERRUPTED CONTINUOUS WAVE. An interrupted continuous wave is the result of chopping up or interrupting a normal continuous wave radio signal at an audio rate. This interruption is distinct from the normal keying and is at a rate high enough to give several interruptions even for a dot of the International code. The result is an audio output from the receiver without the use of some method of heterodyning. The interruption of the wave, however, causes the radiation of many more sideband frequencies and hence the need of a wider channel. For this reason such transmissions are not widely used. (L.R.Q.)

INTERSTITIAL CELL. 1. Small undifferentiated cells in the outer layer of polyps such as *Hydra* which develop into the cnidoblasts and reproductive cells. 2. Cells between the reproductive structures of the ovaries and testes of vertebrates. They have been interpreted as glands of internal secretion but the point is disputed. (A.W.L.)

INTERTENTACULAR ORGAN. A temporary duct leading from the coelom to the exterior within the tentacle-bearing ridge (lophophore) of bryozoans. Germ cells escape through it. (A.W.L.)

INTESTINAL OBSTRUCTION. An organic or functional block interfering with the continuity of the bowel. This is a serious condition which is compatible with life for only a short period. It is one of the grave surgical emergencies.

The causes of intestinal obstruction are many. The organic ones include all lesions which mechanically obstruct the bowel, such as cancer in its walls, adhesions constricting the gut, twists in the intestines, intussusception, etc. A hernia which becomes incarcerated and strangulated is one of the commonest causes. Functional obstruction of the bowel results in paralytic ileus in which the peristaltic movement of the intestinal-wall muscles is inhibited, and the bowel dilates. This occurs as a postoperative complication in abdominal surgery, and is part of the picture of peritonitis.

The symptoms of mechanical obstruction are intermittent abdominal pain, vomiting and constipation. The abdomen may be distended and tender. With the aid of a stethoscope, great roaring peristaltic rushes can be heard. In paralytic ileus, the abdomen is quiet since there is complete absence of peristaltic movement.

The treatment of mechanical obstructions is surgical. In ileus, suction by means of a tube inserted through the mouth, esophagus and stomach to the bowel, relieves the gaseous distention; drugs are used to help restore tone and movement to the intestinal muscles. (D.M.H.)

INTESTINE. Digestive System; Alimentary Tract.

INTRACLASS CORRELATION. Often we are interested in finding the correlation among items within a group such as the correlation in the heights of brothers. In this case, both heights are entered as an x value and a y value since there is no way of differentiating as to which variable is x and which is y. Let $p =$ the number of groups or families, $k =$ the number of variates in each family, $\bar{x}_i =$ mean of the ith group, $\bar{x} =$ the mean of all kp variates, $s^2 = \dfrac{\displaystyle\sum_{j=1}^{k}\sum_{i=1}^{p}(x_j - \bar{x}_i)^2}{p(k-1)}$, then r the intraclass correlation coefficient is calculated by the formula

$$ r = \frac{1}{k-1}\left\{ \frac{k\displaystyle\sum_{i=1}^{p}(x_i - \bar{x})^2}{ps^2} - 1 \right\}. $$

The intraclass coefficient of correlation cannot have a value less than $\dfrac{-1}{k-1}$, and if k is very large as it may be in most cases, then r will be positive. The intraclass coefficient of correlation may be tested for significance by the analysis of variance,

$$ F = \frac{k\displaystyle\sum_{i=1}^{p}(\bar{x}_i - \bar{x})^2/p - 1}{\displaystyle\sum_{j=1}^{k}\sum_{i=1}^{p}(x_j - \bar{x}_i)^2/p(k-1)} $$

with $n_1 = p - 1$ degrees of freedom, and $n_2 = p(k-1)$ degrees of freedom. (L.A.A.)

INTRAFORMATIONAL. This term is used by stratigraphers to denote textures and structures developed in sedimentary rocks at the time of the deposition of the sediments and thus previous to their consolidation through lithification. While normal structures, such as different types of bedding and stratification, may be considered as intraformational, the term is usually confined to such phenomena as mud-cracks (desiccation fractures), raindrop imprints, and intraformational or

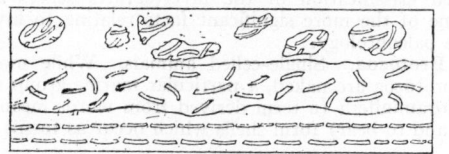

Intraformational conglomerates, in cross-section. (*Field, Outline, Barnes & Noble.*)

"edgewise" conglomerates whose "pebbles" or **phenoplasts** were formed from mud while it was still plastic. (R.M.F.)

INTROMITTENT ORGAN. A protuberance or protrusible organ of the male for the introduction of the seminal fluid into the genital passages of the female in the act of **copulation.** (A.W.L.)

INTROVERT. In zoology, a portion of the body which can be thrust out or retracted by a process of involution. Applied to the proboscis of sipunculids (**Annelida**) and sometimes to the eversible portion of the body of the **bryozoans.** (A.W.L.)

INTRUSIVE ROCK. Igneous Rock.

INTUBATION. The insertion of a tube into the **larynx** through the throat to permit breathing when the larynx becomes closed through swelling such as occurs in severe laryngeal diphtheria. This procedure was formerly used extensively, but since the widespread use of **diphtheria** immunization and antitoxin there is less occasion for it at present. It was first introduced by Dr. Joseph O'Dyer of New York City in 1885. (R.S.M.)

INTUSSUSCEPTION. The telescoping of one section of the intestine into the part below, causing **intestinal obstruction.** Intussusception may occur in either the small or large intestine; it is generally caused by the pulling up of one section of bowel over another of marked spasm just above. Other causes are **polyps, cancer,** and foreign bodies. Intussusception occurs most often in infants and children. The clinical picture is striking: the onset is sudden and marked by severe abdominal pain and vomiting occurring in paroxysms. Collapse and **shock** soon follow with pallor, feeble pulse and subnormal temperature. Without surgery **peritonitis** develops; the temperature rapidly rises, and death soon occurs. Bloody stools are almost a constant accompaniment of this disease, with diarrhea in which only blood and mucus are passed. Death often results unless surgery is undertaken, and the mortality rises in direct proportion to the time lost before operation.

For the use of this term in zoology, see **Growth.** (D.M.H.)

INULIN. Carbohydrates.

INVAR. Metals and Alloys; Expansion; Nickel Alloys.

INVARIANTS AND COVARIANTS. An invariant of a **quantic** is a **function** of the coefficients of the quantic which is transformed into the same function of the new coefficients, multiplied by a power of the modulus of the transformation, when the variables of the quantic are subjected to a **linear transformation.**

A covariant of a quantic is a function of both coefficients and variables of the quantic which retains its form, multiplied by a power of the modulus of the transformation, when the variables of the quantic are subjected to a linear transformation. (L.L.S.)

INVERSE FEEDBACK. Feedback.

INVERSE FUNCTIONS. If $y = f(x)$ is a given **function** of x, then x regarded as a function of y is called the inverse function of the given function.

Examples of pairs of inverse functions are: the square of a variable and the square-root function, the **exponential function** and the **logarithmic function,** the **trigonometric functions** and the **inverse trigonometric functions.** (L.L.S.)

INVERSE HYPERBOLIC FUNCTIONS. Hyperbolic Functions.

INVERSE OPERATIONS. Two operations are said to be inverse operations when each one counteracts the effect of the other.

Inverse operations occur in many places in mathematics. In elementary algebra, the operations of **addition** and **subtraction,** the operations of **multiplication** and **division,** the operations of **evolution** and **involution,** the operations of **involution** and taking **logarithms** are examples of inverse operations. In calculus, the fundamental operations of **differentiation** and **integration** are inverse operations. (L.L.S.)

INVERSE PEAK VOLTAGE. This is the peak voltage which may be safely impressed across an electron **tube** in the inverse direction, i.e., with the **anode** negative and the **cathode** positive. If this voltage rating of the tube is exceeded there is danger of the tube conducting in the inverse direction (see **Arc Back**) and thus ceasing to act as a **rectifier.** In certain circuits this type conduction will result in currents of destructive magnitude. (L.R.Q.)

INVERSE TRIGONOMETRIC FUNCTIONS. The function inverse to the sine function is defined as follows: if $y = \sin x$, then x is the angle whose sine is y, denoted by $x = \sin^{-1} y$ or $x = \arcsin y$; similar notation is used for the other inverse trigonometric functions.

If $y = \sin x$, there are infinitely many angles x (all coterminal) satisfying this relation, hence $x = \arcsin y$ is an infinitely many-valued function, and similarly for the other inverse trigonometric functions.

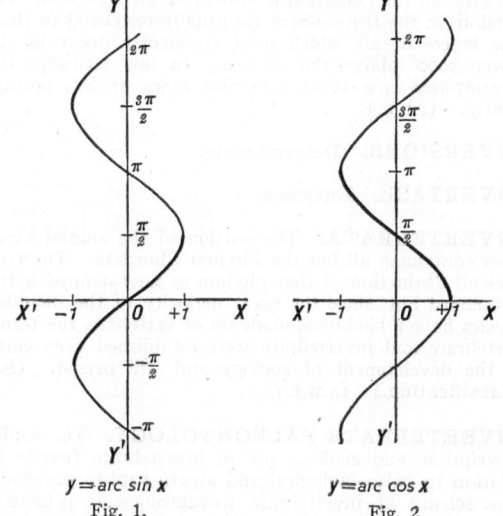

$y = \arcsin x$ $y = \arccos x$
Fig. 1.   Fig. 2.

The principal value of the inverse sine, $\arcsin y$, is the value which lies between $+90°$ and $-90°$ (or $+\dfrac{\pi}{2}$ and $-\dfrac{\pi}{2}$). It is often denoted by $\operatorname{Arc}\sin y$ or $\operatorname{Sin}^{-1} y$.

Usage is not uniform in regard to the definition of principal values.

Sometimes, Arc cos y is taken as the value between $0°$ and $180°$ (or 0 and π); sometimes Arc tan y is taken as the value between $+\dfrac{\pi}{2}$ and $-\dfrac{\pi}{2}$.

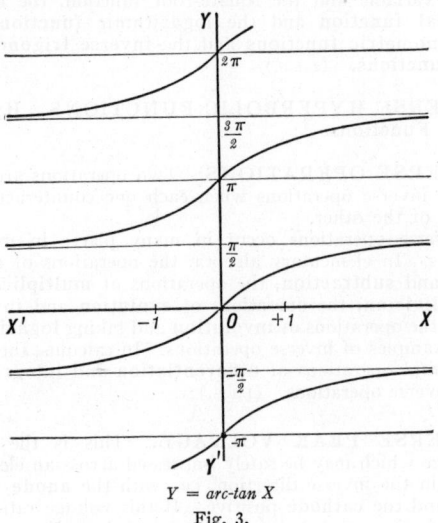

$$Y = \text{arc-tan } X$$
Fig. 3.

General values of inverse trigonometric functions are given by:

$$\text{arc sin } x = n\pi + (-1)^n\alpha, \text{ where } \alpha = \text{Arc sin } x,$$

$$\text{arc cos } x = 2n\pi \pm \alpha, \qquad \text{where } \alpha = \text{Arc cos } x,$$

$$\text{arc tan } x = n\pi + \alpha, \qquad \text{where } \alpha = \text{Arc tan } x,$$

where n is any positive or negative integer or 0.

The graphs of the inverse trigonometric functions are shown in Figs. 1–3. (L.L.S.)

INVERSION. The temperature below the stratosphere normally decreases with altitude. When temperature increases with altitude, normal conditions are inverted, and the condition is said to be an inversion. Inversions in the troposphere are usually restricted to shallow layers of air which most frequently occur in the lower 5000' above the surface. In low latitudes the stratosphere is a slight inversion more or less permanently. (P.E.K.)

INVERSIONS. Determinants.

INVERTASE. Enzymes.

INVERTEBRATA. The portion of the animal kingdom containing all but the Phylum Chordata. The true skeletal distinction of that phylum as now defined is the notochord but, since the great majority of the included species have a backbone made up of vertebrae, the terms vertebrate and invertebrate were established very early in the development of zoology and still persist. (See **Classification.**) (A.W.L.)

INVERTEBRATE PALEONTOLOGY. The study, description, and geologic use of invertebrate **fossils** in relation to paleo-biological and **stratigraphic** problems. The science of invertebrate paleontology is primarily founded upon invertebrate zoology. Since thousands of invertebrate fossils, ranging in age from the **Cambrian** to the **Pleistocene,** have been figured and described, it is not possible to list them all in a general science encyclopedia. Also there is no single reference work in existence which covers the entire subject. For detailed

information the student must consult special bibliographies which list the references in a large number of special papers and monographs. The following condensed classification of the invertebrates serves as an outline of the more significant facts relating to invertebrate paleontology.

I. Protozoa. Single-celled animals. While most of the protozoa are naked, a particular marine group called the foraminifera (a term derived from words meaning a hole and to bear) form shells which occur as fossils from

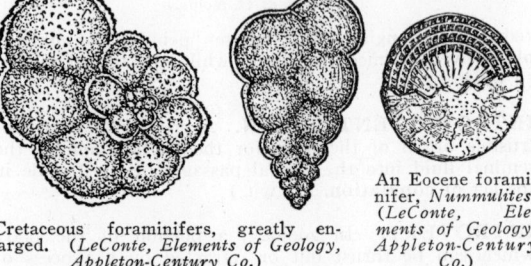

Cretaceous foraminifers, greatly enlarged. (*LeConte, Elements of Geology, Appleton-Century Co.*)

An Eocene foraminifer, *Nummulites.* (*LeConte, Elements of Geology, Appleton-Century Co.*)

the Cambrian to the present. Foraminifera are an important constituent of chalk. Due to the large number of distinctive and rapidly evolving species foraminifera are useful index fossils, especially in the **Mesozoic** and **Cenozoic** periods.

II. Porifera or Sponges. The name porifera is derived from words meaning a pore and to bear. Fossil sponges occur from the Cambrian to the Pleistocene. Except for a few species they are not particularly valuable index fossils. A particularly interesting genus is *Hydnoceras,* a delicate glass sponge which has left its imprint in the fine-grained muds of **Devonian Age.**

III. Graptolites. This term is derived from words meaning written and stoned, because the fossils look like pencil marks on slate. Graptolites were colonial marine animals with chitinous skeletons. Their stratigraphic range is from the **Ordovician** to the **Silurian,** inclusive. Because the graptolites evolved rapidly and have a world-wide distribution they are excellent index fossils.

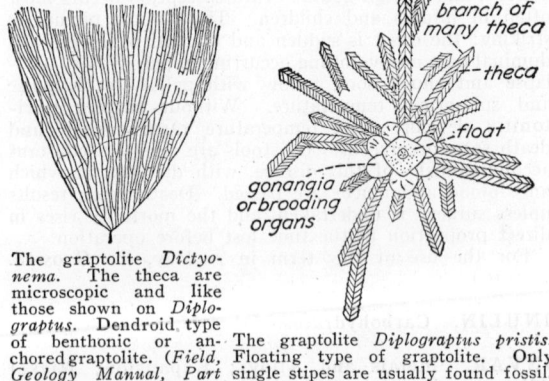

The graptolite *Dictyonema.* The theca are microscopic and like those shown on *Diplograptus.* Dendroid type of benthonic or anchored graptolite. (*Field, Geology Manual, Part II, Princeton University Press.*)

The graptolite *Diplograptus pristis.* Floating type of graptolite. Only single stipes are usually found fossil. (*Field, Geology Manual, Part II, Princeton University Press.*)

The accompanying sketches illustrate two of the principal types of Graptolites.

IV. Corals. The geologic record of fossil **corals** is from the Ordovician to the present day. Important reef builders in the tropical waters of the present oceans and seas. The earliest forms are both colonial and single. The accompanying illustration is that of a single or rugose coral of Ordovician age. Corals are important and useful index fossils in the strata of certain geologic periods.

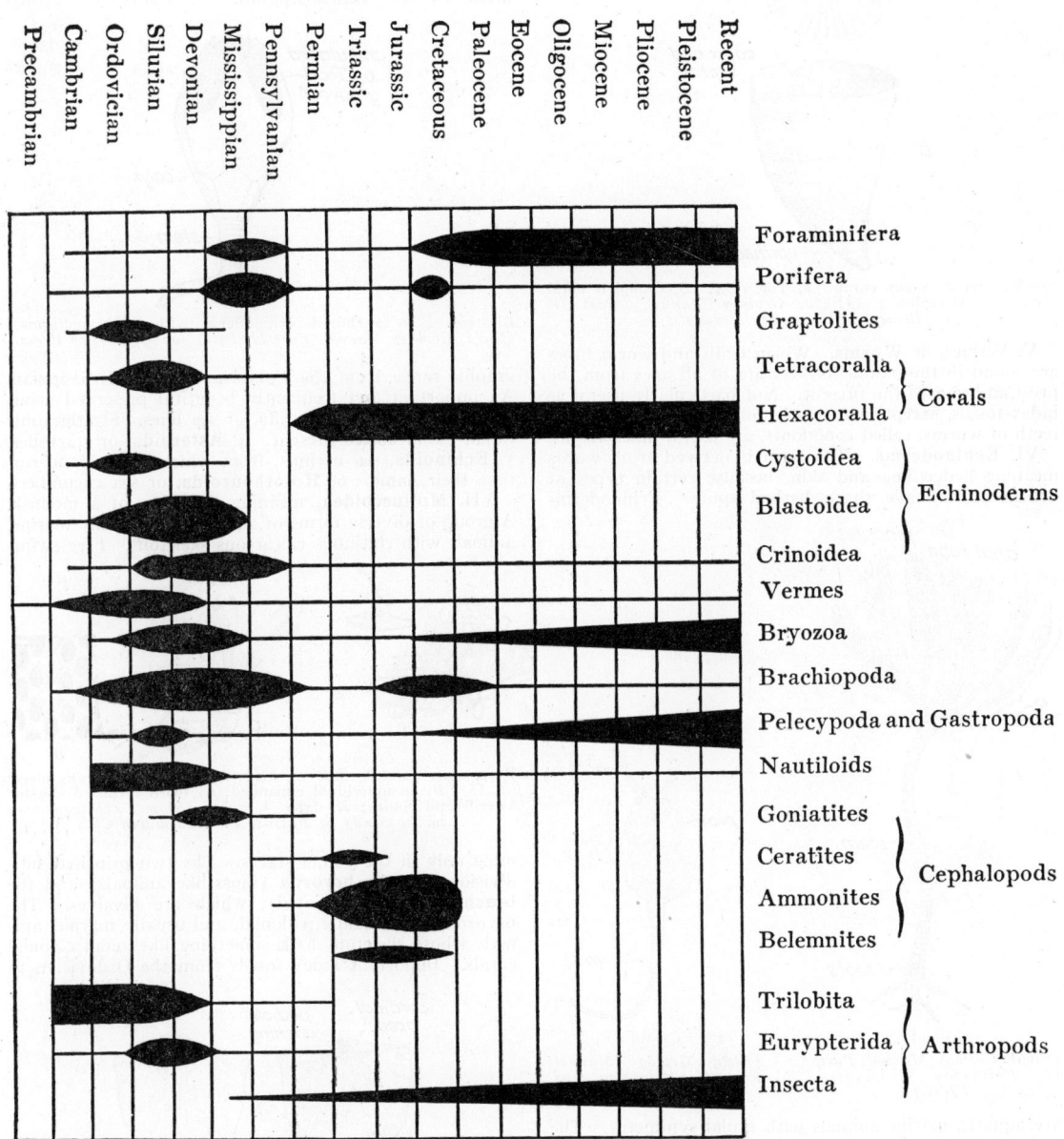

GEOLOGIC RANGE OF THE INVERTEBRATES

NOTE: The "lifelines" are swelled during the periods in which there were the greatest number of genera and species, or when the particular class of organisms is most important as "index fossils."

(Field, Laboratory Manual, Princeton University Press.)

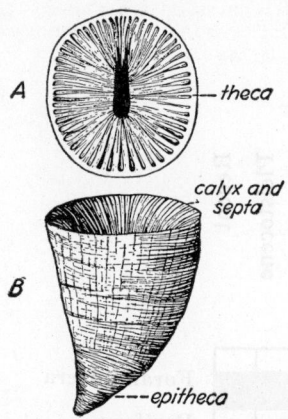

Single, rugose or cup coral. *A*, Top view. *B*, Complete coral skeleton. (Corallite.) (*Field, Geology Manual. Part II, Princeton University Press.*)

V. Vermes, or Worms. Worm trails and worm tubes are found in the sedimentary strata of all ages from the pre-Cambrian to the present. Not particularly useful as index fossils, except in the early Silurian. The jaws and teeth of worms, called conodonts, are useful index fossils.

VI. Echinoderms. The term is derived from words meaning hedge hog and skin, because certain types of **echinoderms** have sharp barbed spurs. Echinoderms

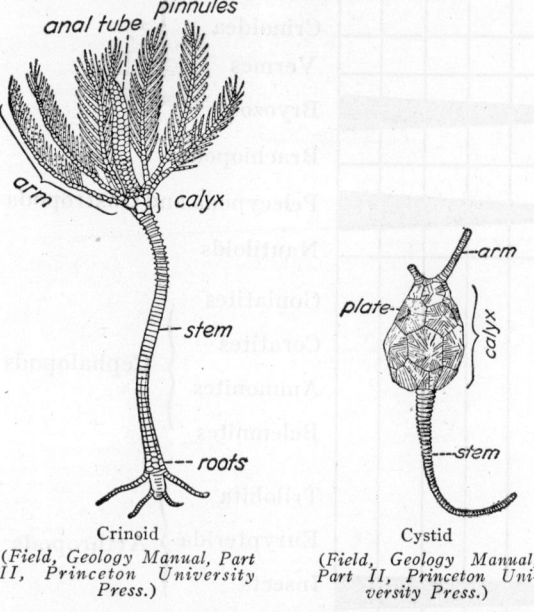

Crinoid
(*Field, Geology Manual, Part II, Princeton University Press.*)

Cystid
(*Field, Geology Manual, Part II, Princeton University Press.*)

are aquatic, marine animals with radial symmetry. The test or skeleton is composed of plates of **calcium** carbonate, with or without a chitinous covering (when living), and usually in the shape of a cup or "calyx"

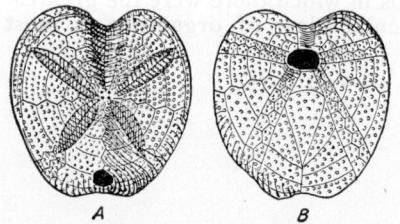

"Sea Urchin," Echinoid. *A*, Dorsal view. *B*, Ventral view. (*Field, Geology Manual, Part II, Princeton University Press.*)

with or without "arms." Some forms are attached to the sea bottom by means of "stems" and "roots"; others are floating or free swimming. The echinoderms are subdivided into the following characteristic groups. 1. Cystoids (Cystids). Stratigraphic range from the Cambrian to the **Mississippian.** 2. Blastoids. Strati-

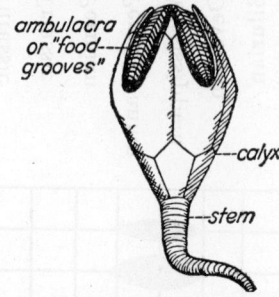

Blastoid. An anchored echinoderm without free "arms." (*Field, Geology Manual, Part II, Princeton University Press.*)

graphic range from the Cambrian to the Mississippian. An important and frequently beautiful preserved genus is *Pentramites.* 3. **Crinoids,** or sea lilies. Stratigraphic range Cambrian to present. 4. **Asteroids,** or starfishes. 5. **Echinoids,** sea urchins, from which the echinoderms take their name. 6. **Holothouroids,** or sea cucumbers.

VII. Molluscoidea, meaning the form of a mollusk. A group of diverse forms of aquatic, and usually marine, animals with chitinous calcareous skeletons. Free swim-

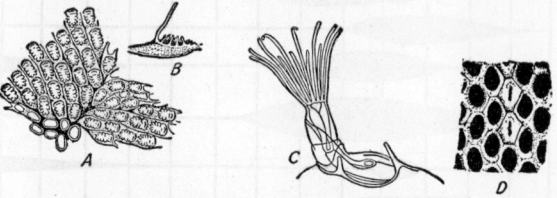

Bryozoans. *A*, Portion of modern colony seen from above (× 15); *B*, an individual expanded; *C*, fossil form. A-C after Verrill and Smith; *D*, from Ulrich. (*Shimer's Introduction to the Study of Fossils, The Macmillan Co.*)

ming only in the young stages. The two principal subdivisions are the **bryozoa** (moss-like animals) and the **brachiopoda** (arm-footed), which are bivalves. The bryozoans are aquatic, colonial, and usually marine, animals whose skeletons look something like small colonial corals. Important index fossils from the Ordovician to

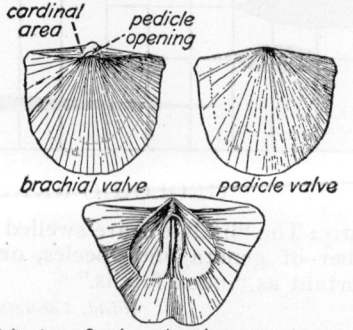

Rafinesquina. A calcareous brachiopod having no interior skeleton. (*Field, Geology Manual, Part II, Princeton University Press.*)

the present. Important reef builders, especially in the **Paleozoic.** The skeleton of the brachiopod is composed of two parts or valves which are formed of either

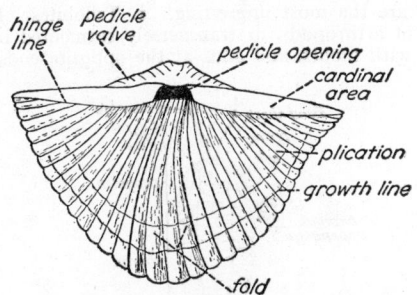

Front view of spirifer, showing full view of brachial valve. (*Field, Geology Manual, Part II, Princeton University Press.*)

"horn" or calcium carbonate. The brachiopod skeletons are principally distinguished from the pelecypods (clams, etc.) by a different type of bilateral symmetry.

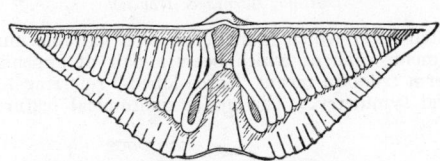

Interior of brachial valve of spirifer showing brachidia, or spires. (*Field, Geology Manual, Part II, Princeton University Press.*)

The higher forms have internal skeletons. Brachiopods are excellent index fossils, especially in the Paleozoic, where they share their importance with the graptolites

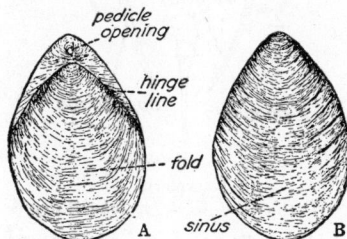

Brachiopod, "Lamp Shell." A, Front view. B, Rear view (Pedicle valve.) (*Field, Geology Manual, Part II, Princeton University Press.*)

and the trilobites. The accompanying figures illustrate some of the principal types of brachiopods.

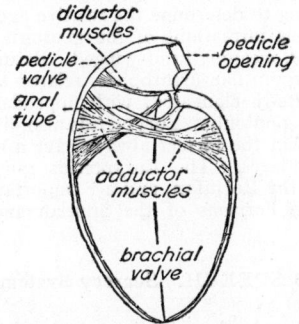

Cross-section of brachiopod (lamp shell) showing musculature. (*Field, Geology Manual, Part II, Princeton University Press.*)

VIII. **Mollusca,** meaning soft-bodied animals. This class includes the **Pelecypoda,** meaning axe-footed bivalves whose shells are formed of calcium carbonate with a "horny" covering. No interior skeleton, even in the higher types. The skeleton is distinguished from that of the brachiopods by a different type of bilateral symmetry. A few species, such as the oyster, have no

symmetry. All pelecypods are aquatic, but may be either fresh-water or marine. Most species are attached to the bottom in the adult stage, but a few are free-

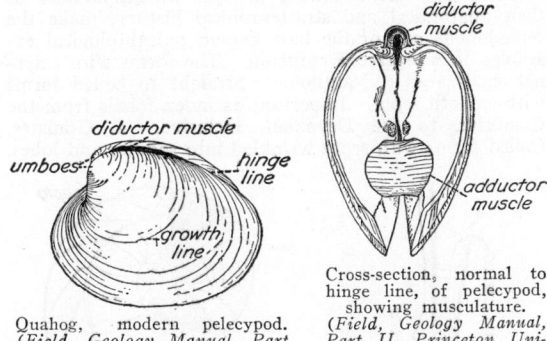

Quahog, modern pelecypod. (*Field, Geology Manual, Part II, Princeton University Press.*)

Cross-section, normal to hinge line, of pelecypod, showing musculature. (*Field, Geology Manual, Part II, Princeton University Press.*)

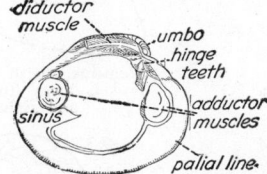

Interior view of left valve of pelecypod. Showing muscle scars. (*Field, Geology Manual, Part II, Princeton University Press.*)

swimming. Pelecypods are not particularly good index fossils except at certain horizons in the Mesozoic and Cenozoic.

IX. **Gastropoda,** or snails, meaning stomach-footed. Mollusca with single unchambered shells composed of calcium carbonate. All shapes and types of ornamentation, frequently well-preserved. Gastropods are only important as index fossils in the **Canadian,** and at certain horizons in the Mesozoic and Cenozoic.

X. **Cephalopoda.** Meaning head-footed. Existing forms are **nautilus, cuttlefish, octopus,** etc. Shells composed either of calcium carbonate or horn, and either external or internal in relation to the living animal. The shell differs from that of the gastropods because the in-

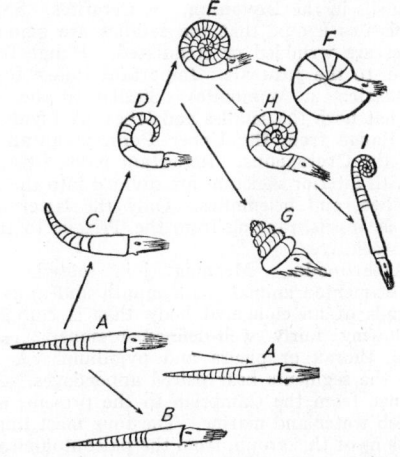

Adaptative evolution of the Nautiloids. (*After Dunbar, Organic Adaptation to Environment, Yale University Press.*) *A,* Straight form, poor swimmers; *B,* first adaptation to bottom habitat, straight, slightly flattened form; *C,* slightly coiled; *D,* partly coiled; *E,* fully coiled; *F,* coiled and involuted; *G,* second adaptation to bottom habitat. Coiled and twisted; *H-I,* decadent (gerontic) stages. Partly uncoiled. NOTE: The ammonoids also passed through a somewhat similar adaptive evolution.

terior is divided into a number of compartments by means of platforms or septa, the animal living only in the outer compartment. The cephalopods are naturally divisible into the following groups, which, because of their anatomical and stratigraphical history, make the cephalopods one of the best known paleontological examples of **Adaptive Evolution**. The forms with external shells are: 1. Nautiloids. Straight to coiled forms with smooth septa. Important as index fossils from the Cambrian to the Devonian, inclusive. 2. Goniates. Coiled forms with septa wrinkled into saddles and lobes.

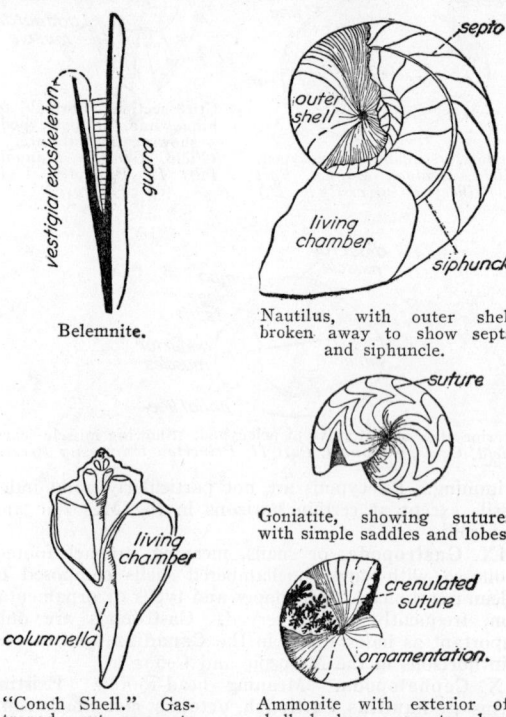

Belemnite.

Nautilus, with outer shell broken away to show septa and siphuncle.

Goniatite, showing sutures with simple saddles and lobes.

"Conch Shell." Gastropod, cut away to show interior whorls.

Ammonite with exterior of shell broken away to show dendritic form of sutures.

(*Field, Geology Manual, Part II, Princeton University Press.*)

Range from the **Silurian** to the **Permian**. Important index fossils in the **Devonian**. 3. Ceratites. Similar to the goniatites except that the saddles are smooth and the lobes are wrinkled or crenulated. Range from the Devonian to the **Jurassic**. Important **index fossils** in the **Triassic**. 4. Ammonites. Similar to the ceratites except that both the saddles and lobes are highly crenulated. Range from the Upper **Pennsylvanian** to the close of the **Cretaceous**. Important index fossils. The forms with interior skeletons are divided into the **squids**, **cuttlefishes**, and belemnites. Only the latter are important as fossils, ranging from the Triassic to the Cretaceous, inclusive.

XI. Arthropoda. Meaning joint-footed. Transversely segmented animals with mouth and anus at opposite ends of an elongated body that is composed of the following fairly well-defined regions, "head" or cephalon, thorax or pleura and pygidium. A few or most of the segments bear paired appendages. Arthropods range from the Cambrian to the present, and are both fresh water and marine. The four most important subdivisions of this group, from the paleontological point of view, are the Trilobites, Ostracods, and Eurypterids. The first insects appear as fossils in the Carboniferous and several of the ancestral types of the older order appear in the Permian. The fossil insects of the **Tertiary** are particularly interesting as they prove that social life in the insect world began as long ago as the **Oligocene**. From the Paleontological point of view however the tri-

lobites are the most interesting. 1. Trilobites. Extinct group of arthropods, or transversely segmented invertebrates with mouth and anus at the opposite ends of an

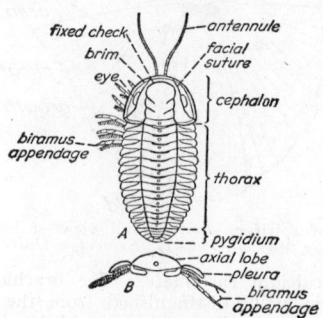

Trilobite. *A*, dorsal view; *B*, cross-section of thorax. (*Field, Outline, Barnes & Noble.*)

elongated body which is made up of a variable number of segments each of which bears a pair of appendages. The term trilobite means "three-lobed," referring to the bilateral symmetry. The major anatomical features of

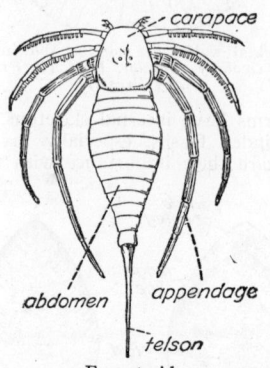

Eurypterid.
(*Schubert, Textbook of Geology, John Wiley & Sons.*)

the external skeleton are shown in the accompanying figure. Although one of the highest orders of the invertebrates, trilobites occur among the oldest known fossils of the Cambrian period, becoming extinct at the close of the Paleozoic Era. Trilobites are important as index fossils, especially in the Cambrian, Ordovician, Silurian and Devonian periods, and are of great aid to the geologist in helping to determine the relative ages of the oldest fossiliferous formations of the **geologic time-scale**. 2. Eurypterids, meaning head-winged. Extinct marine or estuarine scorpion-like arthropods, related to the horseshoe crab. Body elongated, with appendages attached to the head region only. Eurypterids also differ from the trilobites in that the former always have a regular number of appendages. The Eurypterids range from the Cambrian to the **Permian,** and are important index fossils in certain horizons of the Silurian and Devonian. (R.M.F.)

INVERTED SPEECH. Secrecy Systems.

INVERTER. An inverter is a device for converting d.c. into a.c. but the term is commonly employed to designate a gas-filled electron-tube circuit for performing this function. In recent years the development of reliable **thyratrons** and **ignitrons** and of inverter circuits utilizing these tubes has opened up the possibility of high voltage d-c transmission. One of the major deciding factors in the selection of a.c. for the standard electrical supply was the ease with which it could be transformed in voltage for transmission and distribution. (See **Elec-**

tric Power Transmission.) D-c transmission at high voltage has certain advantages over a.c. but heretofore there has been no suitable means for transforming volt-

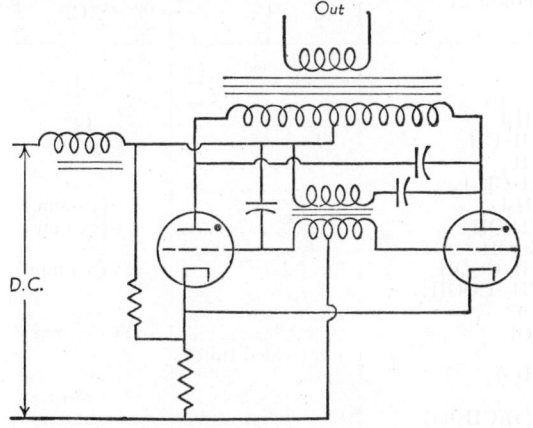

Self-excited parallel type inverter

ages. The inverter offers the possibility of generating power as a.c., then stepping it up to the desired transmission voltage, rectifying it with high-efficiency gas-filled **rectifiers**, transmitting as high-voltage d.c. with attendant advantages, inverting it to **a.c.** at the receiving end and stepping down to the normal distribution voltage by using **transformers**. While the system sounds complicated, the greatly decreased losses of d-c transmission and the high efficiency of gas-filled tubes indicate that this type of transmission may offer serious competition to the established a-c system. Certain experimental systems are already in use.

A basic self-excited inverter circuit is shown. This circuit is so set up that the tubes conduct alternately, causing the d.c. to switch back and forth through the primary of the output transformer. This will induce a-c voltage in the secondary. There are numerous variations of inverter circuits but all use the tubes to switch the d-c supply back and forth to produce an a-c output. (L.R.Q.)

INVOLUTE. The locus of a point on a curve rolled along another curve termed the directrix of the involute. In gearing and in gear-tooth formation, the involute is the locus of a fixed point on a straight line rolled on a circular directrix termed the base circle of the involute. (H.C.H.)

INVOLUTION. The operation of raising numbers to powers. (L.L.S.)

INYOITE. Colemanite.

IODIC ACID AND IODATES. Iodic acid (HIO_3) is a white soluble solid. Upon heating to $110°$ C. forms iodic anhydride, **iodine pentoxide** (I_2O_5). A solution of iodic acid can be concentrated by evaporation up to 70% HIO_3. Iodic acid is an oxidizing agent in hot solution, smoothly converting **sulfur** to **sulfuric acid, phosphorus** to **phosphoric acid, carbon** to **carbon dioxide,** silicon to silicic acid.

Prepared by reaction (1) of iodine and **chloric acid** (25% $HClO_3$), (2) of iodine and **nitric acid** ("fuming") —the yield is low, but the product, upon evaporation, is pure, (3) of iodine and excess **chlorine** in water.

Potassium iodate is formed (1) by electrolysis of potassium iodide solution, with stirring, heated, and over some hours, (2) by reaction of iodine and potassium chlorate solution, (3) by allowing potassium **hypoiodite** solution to stand a few hours at ordinary temperature.

Metallic iodates are solids, moderately soluble in water, except that silver, lead, barium, thallous iodates are insoluble. Potassium hydrogen iodate ($KH(IO_3)_2$) recalls potassium hydrogen fluoride (KHF_2), an acid salt. Iodates, when heated, evolve oxygen and leave the iodide (or oxide) as residue.

Test: Upon addition of hydriodic acid (or iodide plus dilute sulfuric acid), iodate liberates iodic acid, which reacts with hydiodic acid to form iodine. (R.K.S.)

IODIDE. Iodine.

IODINE. Symbol: I. Atomic number: 53. Atomic weight: 126.92. Density: 4.94. Melting point: $113.5°$ C. Boiling point: $184.4°$ C. No isotope, but of single atomic form: 127.

Iodine is a violet to black solid, of characteristic odor, color of vapor violet, easily purified by **sublimation,** insoluble in water, soluble in alcohols, ether, **carbon disulfide, carbon tetrachloride.** Discovered by Courtois in 1812. Used (1) in **photography,** (2) in certain chemicals and **dyes,** (3) as an antiseptic solution in ethyl alcohol (3% to 7%), (4) as a reagent for testing starch, and in estimating such substances as **sulfites,** (5) in medicine, various salts of iodine are given internally in the treatment of such diseases as arteriosclerosis, hypertension, syphilis, actinomycosis, emphysema, chronic bronchitis, some forms of arthritis, and in certain thyroid disorders. When overdosage occurs, or when an idiosyncrasy is present in the patient, acne and other skin disorders may result.

Iodine occurs in the ashes of sea plants, e.g., kelp, especially California and Bay of Biscay; in the petroleum oil well brines of California; and in small percentages in sodium nitrate of Chile, which source for a long time furnished the world's supply.

Iodine is soluble (1) in carbon disulfide, in chloroform, and in carbon tetrachloride, yielding a violet solution, and upon evaporation leaves a residue of iodine crystals, (2) in ether, in ethyl alcohol, and in potassium iodide solution, yielding a brown solution. Iodine reacts (1) with **sodium** thiosulfate solution to form dithionate, (2) with **sulfurous acid** to form **sulfric acid,** (3) with **hydrosulfuric acid** to form **sulfur,** (4) with starch to form a dark blue solid.

Acids: Hydriodic acid (HI), hypoiodous acid (HOI), iodic acid (HIO_3), periodic acid (H_5IO_6 or HIO_4). (See each acid.)

Bromide: Iodine bromide (IBr).

Chlorides: Iodine chloride (ICl); iodine trichloride (ICl_3).

Fluorides: Iodine pentafluoride (IF_5).

Hydride: Hydrogen iodide (HI), colorless gas when pure (frequently contains free iodine as violet to brown gas), melting point $-51°$ C., boiling point $-35°$ C., very soluble in water, yielding **hydriodic acid.** Formed by reaction of red phosphorus, iodine, and water under the proper conditions.

Hypoiodite. (See **Hypoiodous Acid.**)

Iodate. (See **Iodic Acid.**)

Iodides: **Sodium** iodide (NaI), **potassium** iodide (KI), **ammonium** iodide (NH_4I) are soluble iodides; **silver** iodide (AgI), **mercurous** iodide (HgI), **mercuric** iodide (HgI_2), **lead** iodide (PbI_2) are insoluble iodides; **nitrogen** iodide ($N_2H_3I_3$). (See **Hydriodic Acid.**)

Oxides: Iodine dioxide (IO_2 or I_2O_4), pale yellow solid, formed by grinding iodine and cold fuming **nitric acid;** iodine pentoxide (I_2O_5), white solid, decomposes upon heating to $300°$ C. into iodine and oxygen, formed by heating iodic acid to $170°$ C. Reactive with, and used to detect **carbon monoxide,** with accompanying formation of iodine and carbon dioxide.

Periodate. (See **Periodic Acid.**)

Organic Iodo-Compounds:

Hydrocarbons are generally without reaction with iodine or hypoiodous acid, except under special condi-

SELECTED REPRESENTATIVE ORGANIC COMPOUNDS OF IODINE

Name	Formula	Melting Point, °C.	Boiling Point, °C.
1. Methyl iodide	CH_3I		45
2. Ethyl iodide	C_2H_5I		72
3. Normal-propyl iodide (1-iodopropane)	$C_2H_5 \cdot CH_2I$		102
4. Iso-Propyl iodide (2-iodopropane)	$CH_3 \cdot CHI \cdot CH_3$		89
5. Vinyl iodide (iodoethylene)	$CH_2:CHI$		56
6. Allyl iodide	$CH_2:CH \cdot CH_2I$		101
7. Benzyl iodide	$C_6H_5 \cdot CH_2I$	24	93 (10 mm.)
8. Glycol iodohydrin (ethylene iodohydrin)	$CH_2I \cdot CH_2OH$		85 (25 mm.)
9. (Mono)iodoacetic acid	$CH_2I \cdot COOH$	82	
10. Alpha-iodopropionic acid	$CH_3 \cdot CHI \cdot COOH$	45	105 (0.3 mm.)
11. Beta-iodopropionic acid	$CH_2I \cdot CH_2 \cdot COOH$	82	
12. Acetyl iodide	$CH_3 \cdot COI$		108
13. Benzoyl iodide	$C_6H_5 \cdot COI$		135 (25 mm.)
14. Cyanogen iodide	$CN \cdot I$	146 (sealed tube)	
15. Iodofurane	$C_4H_3O \cdot I(2)$		
16. Iodobenzene (phenyl iodide)	C_6H_5I		189
17. Ortho-iodotoluene (1,2-)	$C_6H_4(I)(2)(CH_3)(1)$		211
18. Meta-iodotoluene (1,3-)	$C_6H_4(I)(3)(CH_3)(1)$		204
19. Para-iodotoluene (1,4-)	$C_6H_4(I)(4)(CH_3)(1)$	35	211
20. Alpha-iodonaphthalene	$C_{10}H_7 \cdot I(1)$		305
21. Beta-iodonaphthalene	$C_{10}H_7 \cdot I(2)$	54	309
22. Ortho-iodophenol (1,2-)	$C_6H_4(I)(2)(OH)(1)$	40	186 (160 mm.)
23. Meta-iodophenol (1,3-)	$C_6H_4(I)(3)(OH)(1)$	40	decom.
24. Para-iodophenol (1,4-)	$C_6H_4(I)(4)(OH)(1)$	93	decom.
25. Ortho-iodoaniline (1,2-)	$C_6H_4(I)(2)(NH_2)(1)$	60	
26. Meta-iodoaniline (1,3-)	$C_6H_4(I)(3)(NH_2)(1)$	33	
27. Para-iodoaniline (1,4-)	$C_6H_4(I)(4)(NH_2)(1)$	67	
28. Ortho-iodonitrobenzene (1,2-)	$C_6H_4(I)(2)(NO_2)(1)$	53	
29. Meta-iodonitrobenzene (1,3-)	$C_6H_4(I)(3)(NO_2)(1)$	37	
30. Para-iodonitrobenzene (1,4-)	$C_6H_4(I)(4)(NO_2)(1)$	171	
31. Ortho-iodobenzoic acid (1,2-)	$C_6H_4(I)(2)(COOH)(1)$	162	
32. Meta-iodobenzoic acid (1,3-)	$C_6H_4(I)(3)(COOH)(2)$	187	
33. Para-iodobenzoic acid (1,4-)	$C_6H_4(I)(3)(COOH)(3)$	269	
34. Iodoso benzene	C_6H_5IO	210 (expl.)	
35. Iodoxy benzene	$C_6H_5IO_2$	236 (expl.)	
36. Methylene iodide	CH_2I_2		180
37. Ethylene iodide (1,2-diiodoethane)	$CH_2I \cdot CH_2I$	81	
38. Ethylidene iodide (1,1-diiodoethane)	$CH_3 \cdot CHI_2$		178
39. Diiodoacetic acid	$CHI_2 \cdot COOH$	110	
40. Ortho-diiodobenzene (1,2-)	$C_6H_4I_2(1,2)$	27	286
41. Meta-diiodobenzene (1,3-)	$C_6H_4I_2(1,3)$	40	285
42. Para-diiodobenzene (1,4-)	$C_6H_4I_2(1,4)$	129	285
43. Diphenyl iodonium iodide	$(C_6H_5)_2I \cdot I$	182	
44. Iodoform	CHI_3	119 (subl.)	
45. Methyl iodoform (1,1,1-triiodoethane)	$CH_3 \cdot CI_3$	93 (decom.)	
46. Benzotriiodide (phenyl iodoform)	$C_6H_5 \cdot CI_3$		
47. Triiodoacetic acid	$CI_3 \cdot COOH$	150 (decom.)	
48. Carbon tetraiodide	CI_4	decom.	
49. Tetraiodoethylene	$CI_2:CI_2$	190	
50. Hexaiodoethane (ϵ)	$CI_3 \cdot CI_3$		
51. Hexaiodobenzene	C_6I_6	340 (decom.)	

tions, but olefins react readily with hydrogen iodide by addition, to form, for example, ethyl iodide (CH_3CH_2I) from ethylene.

Oxygen-function compounds, e.g., **ethyl alcohol, acetaldehyde, acetic acid,** react with **phosphorus** iodide to form corresponding oxygen-function iodides, e.g., ethyl iodide (C_2H_5I), ethylidene diiodide (CH_3CHI_2), acetyl iodide (CH_3COI).

Iodoform is made by reaction of **acetone or ethyl alcohol** with sodium **hypoiodite.**

Use is made of the diazo-reaction (see **Azo- and Related Compounds**) to introduce iodine into benzenoid compounds. Iodine-substituted carboxylic acids are made by heating the selected chloro or bromo acid, with potassium iodide, e.g., monochloroacetic acid yields monoiodoacetic acid (potassium chloride is formed). Many of the iodo-compounds are used as reagents or as inter-mediate compounds in organic chemistry. When paraffin iodo-compounds are treated with sodium hydroxide dissolved in alcohol, hydrogen iodide is removed, e.g., ethyl iodide (CH_3CH_2I) yields ethylene ($CH_2:CH_2$). (R.K.S.)

IODINE NUMBER. Esters.

IODOFORM. A yellow, insoluble, crystalline powder (for its chemical properties, see **Iodine**) with a very penetrating and disagreeable odor. It is antiseptic in action, but only when in direct contact with raw surfaces. It is used in infected cavities and abscesses which are packed with gauze saturated with the powder. In certain susceptible subjects toxic symptoms may result. (R.S.M.)

IOLITE. Cordierite.

ION. Reactions Involving Recombinations of Ions; Ionized Gases.

ION COUNTER. Counting Tube.

IONIC THEORY. Reactions Involving Recombination of Ions.

IONIUM. Symbol: Io. A radioactive element of the uranium-radium series. (See **Radioactive Changes.**) (R.K.S.)

IONIZATION CONSTANT. Reactions Involving Recombination of Ions; Acids, Carboxylic; Amines.

IONIZATION CHAMBER. The term applies to a variety of enclosures used in the study of **ionized gases** or of ionizing agencies. The essential features are a closed vessel containing a gas at normal or altered pressure, and furnished with two electrodes kept at different potentials. These may be in the form of parallel plates or of coaxial cylinders, or one of them may be the vessel itself with the other inside and insulated from it. When the gas between the electrodes is ionized by any means, as by **x-rays** or **radioactive** emission, the ions move to the electrodes of opposite sign, thus creating an ionization current which may be measured by a **galvanometer** or an **electrometer.** If, when ions are being produced at a fixed rate, the potential difference between the electrodes is gradually increased, a point is reached at which further increase of voltage causes no increase in current, because the ions are removed as fast as they are formed. This limiting current is called the saturation current, and its value may be used as a measure of the rate of ionization and hence of the intensity of the ionizing radiation. (See also **Counting Tube.**) (L.D.W.)

IONIZED GASES. Various agencies, such as fast-moving **electrons, alpha particles,** various forms of **radiation,** and high temperature, are capable of dislodging electrons from **atoms** or **molecules** of a gas and thereby leaving them positively charged. Some of the dislodged electrons may attach themselves to other molecules and render them negatively charged. In some cases two or more electrons may be removed from the same molecule, or a molecule with a double positive charge may unite with a singly charged negative molecule, forming a singly charged complex, etc. Such charged atoms, molecules or molecular groups are called ions, and their production from neutral molecules is called ionization. The complete separation of an electron from a molecule or an atom requires a definite amount of energy. This may be expressed in **ergs,** but is more commonly given in **electron volts** (1.59×10^{-12} erg), its value being the "ionizing potential." A less amount of energy may excite the atom or molecule to emit radiation, but will not ionize it.

If an ionized gas is left to itself, the ions soon recombine and become neutral. But if it is subjected to an electric field, as in an **ionization chamber,** the ions pass to the electrodes, such a migration being an "ionization current." Such currents, commonly called electric discharges, are attended by diverse phenomena and vary widely in character from the silent glow discharge to the **lightning** stroke.

At ordinary pressures, discharges may be classified into four types. (1) If the voltage between two electrodes in open air is gradually increased, the electrodes become surrounded with a luminosity. This "glow" or "corona" gives way, at the negative electrode first, to (2) a "brush," composed of hair-like branches. (3) Finally the disruptive **spark** passes. (4) Under other conditions an **arc** may be formed. If, however, the electrodes are enclosed in a tube and the pressure reduced, a point is reached at which the tube becomes filled with a beautiful luminosity. Close examination shows this to have structure. Very close to and surrounding the cathode is a thin, luminous layer c, the cathode glow (see figure);

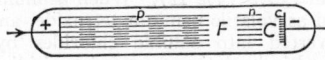

and outside this, the Crookes dark space C. Next, extending toward the anode, is the short negative glow n, then the Faraday dark space F. From this to the anode extends the long positive column p, with its regular, transverse striations. As the pressure is further reduced, the cathode dark space enlarges and the other features dwindle toward the electrodes until they finally disappear at about 0.001 mm. pressure. From this point on, the **cathode rays** are the predominant feature.

Upon exploring the discharge in a Crookes tube with suitable probes, it is found that in certain regions the positive and negative ions are so nearly equal in number as to neutralize each other's effect. Such a region is called a "plasma." The plasma may be surrounded by a "sheath" of ionized gas in which ions of one sign greatly predominate, the effect being that of a **space charge.** (See also **Anode Rays, Canal Rays, Cloud Chamber, Cosmic Rays, Counting Tube, Crookes Tube, Electrodeless Discharge, Geissler Tube, Vacuum Tube.**) (L.D.W.)

IONIZING POTENTIAL. Ionized Gases.

IONOSPHERE. A layer of ionized air high above the earth's surface, the existence of which was surmised by Heaviside and verified later by Kennelly. It is also called the Kennelly-Heaviside layer. Its importance in the transmission of radio signals is now well recognized. Waves from a transmitting station T, proceeding obliquely upward and encountering this layer, are deflected (reflected or refracted) downward, so that a distant receiving station R may receive waves from T either by the direct path along the surface or over the longer route via the ionosphere. Were it not for this layer, much more of the energy emitted from the transmitter would escape into space.

The height of the ionosphere has been ascertained by measuring the time interval between an emitted signal and its "echo." It appears to be composed of at least three layers, E, F_1, F_2, at heights varying from 40–50 to 175 or 200 miles. The two higher layers together constitute the Appleton or F layer, the name Kennelly-Heaviside being now usually applied to the lowest, or E layer. There is reason to believe that the ionization is produced largely by sunlight, especially by the absorbed ultraviolet radiation. The ionosphere generally recedes at night, sometimes apparently to the region outside the earth's shadow, and approaches the earth during the daylight hours. (See **Atmosphere.**)

Interference between waves reaching the receiver by different routes, that is, after different numbers of reflections by ionosphere and ground, gives rise to the troublesome phenomenon of radio "fading." (L.D.W.)

IPECAC. (Ipecacuanha) the root of *Cephaëles ipecacuanha* from Brazil. (See **Madder Family.**) It is an irritant and promotes secretion, especially of the skin and respiratory tract. It is used as an expectorant in respiratory infections. In larger doses it is an **emetic.** (R.S.M.)

IRIDIUM. Symbol: Ir. Atomic number: 77. Atomic weight: 193.1. Density: 22.42. Hardness: 6.0–6.5. Melting point: 2350° C. **Isotopes:** page 290.

Compact iridium is a white, very hard metal, and is not attacked by acids. Discovered by Tennant in 1804. Iridium metal is used chiefly as an alloy with **platinum,** which latter metal is thereby hardened. Standard measures and weights are made of such an alloy (10% Ir), also pen points, and parts of scientific apparatus and surgical tools.

Iridium occurs native with platinum, and with **osmium** as osmiridium (50%–75% Ir). When osmium is present it is removed as volatile osmium tetroxide, and the residue is converted into soluble chlorides by chlorine. From the resulting solution ammonium iridium chloride ($(NH_4)_2IrCl_6$) is precipitated, and then ignited and fused to obtain iridium metal.

Chloroiridate: Ammonium chloroiridate ($(NH_4)_2IrCl_6$) is insoluble in alcohol.

Hydroxide: Iridium hydroxide ($Ir(OH)_4$), dark blue precipitate, by excess of **sodium** hydroxide solution, and boiling. (R.K.S.)

IRIDOSMINE. A mineral alloy of iridium and osmium in the platinum group. (R.M.F.)

IRIS. The pigmented structure in front of the lens of the **eye.** It is contracted and expanded by muscular action so that the opening in its center, known as the pupil, varies in size according to the brightness of the light. By this means the quantity of light that reaches the sensitive retina is regulated. An iris occurs in **molluscan** and **vertebrate** eyes. In the **insect** eye a group of iris cells shut out light except from the direction of the lenticular cornea. Also a plant. (A.W.L.)

IRON. Symbol: Fe (ferrum). Atomic number: 26. Atomic weight: 55.85. Density: 7.86. Hardness: 4–5 (iron), 5–8.5 (steel). Melting point: 1535° C. Boiling point: 3000° C. (**Isotopes:** page 290.)

Iron is a silver-white metal, capable of taking a high polish, ductile, malleable; can be welded when hot; pure iron is attracted by a magnet but does not retain the **magnetism** (silicon steel is preferred for **electromagnets** because it retains magnetism even less than pure iron). Discovery prehistoric.

Iron is seldom encountered as a "chemically pure" metal, but usually as steel (ordinary, with **carbon** as the essential alloying element; special, with various elements); iron is the most largely used of the metals, and considerable scrap metal is recovered. Steel is used in construction of machines and apparatus where workability is required, and definite resistance to corrosion is supplied by the metal. The metal is at times protected (1) by depositing a surface layer of metal, e.g., **tin, zinc, lead, chromium, nickel, aluminum,** (2) by formation of a chemical surface layer, such as the black oxide of iron, or the phosphate, (3) by painting the surface, (4) by applying oil or grease. Commercially pure iron is satisfactory for galvanized sheet, for porcelain enameling sheet, and for direct current electromagnets.

Iron occurs abundantly as oxide, **magnetite** (ferroferric oxide, triiron tetroxide, Fe_3O_4), black, **hematite** (ferric oxide, Fe_2O_3), red, **limonite** (hydrated ferric oxide) yellow to brown; as carbonate, **siderite** (ferrous carbonate, $FeCO_3$), colorless to pale brown; as sulfides, **pyrite** (iron disulfide, FeS_2), brass colored, **arsenopyrite** (iron sulfoarsenide, FeAsS), and as **aluminosilicates** in most rocks, evidenced by green coloration of ferrous, by yellow to red to brown of ferric, and by black of ferroferric. The element iron is fourth in abundance of the elements of the earth's crust (5.1% of the solid crust) and is responsible for practically all of the coloration of ordinary rocks. Native iron is rarely found except in **meteorites.** In the United States about 85% of the iron ore mined comes from the Lake Superior region of Minnesota and Michigan, and, in the south, Alabama is a producing state. The ore is at present commercially usable when somewhat above 50% Fe content. The industrial nations (Britain, France, Germany, U.S.S.R., Japan, the United States) are the great consumers of iron and steel. The oxide is smelted in a **blast furnace** with **carbon** (coke) and **flux** (limestone) at a high temperature. In operation, all products are withdrawn as liquid or gas, the heavier liquid iron settles beneath the lighter liquid **slag,** and each is drawn off separately. The iron is cast into cast iron bars or treated immediately in a furnace for the removal of impurities, such as carbon, **silicon, sulfur, phosphorus,** then the desired additions, such as carbon, **manganese,** are made, and the resulting steel is fabricated.

Four allotropic forms of iron are known, namely, (1) alpha-iron below 769° C., (2) beta-iron between 769° C. and 906° C., (3) gamma-iron between 906° C. and 1404° C., (4) delta-iron between 1404° C. and 1535° C.; the properties of iron are markedly affected by the addition of other elements, notably carbon, silicon, sulfur, phosphorus, manganese in steel, and **chromium, manganese, nickel, molybdenum, tungsten, vanadium** in special steels. Iron is scarcely attacked in dry air, but is rapidly corroded in moist air at ordinary temperatures, forming iron rust, hydrated ferric oxide; burns when heated in air to form ferroferric oxide; reacts with steam at a red heat to form ferroferric oxide plus hydrogen gas; dissolves in **hydrochloric** or dilute **sulfuric acid,** forming ferrous salt solution and **hydrogen gas,** dissolves in hot concentrated sulfuric or in cold dilute **nitric acid,** forming ferric salt solution; no reaction (passive) with cold concentrated nitric; no reaction with alkalis, except hot solutions of high concentration.

Acetate: Ferrous acetate "iron liquor" ($Fe(C_2H_3O_2)_2 \cdot 4H_2O$), prepared as an industrial chemical—a black solution—by reaction of scrap iron and pyroligneous acid (see **acetic acid**), and then evaporation. Used as a mordant in dyeing; ferric acetate ($Fe(C_2H_3O_2)_3$), formed by the reaction of iron and acetic acid and oxygen of the air, and then **crystallization.** Used in textile and leather dyeing, as a wood preservative, and in medicine; basic ferric acetate formed as a precipitate upon boiling ferric acetate solution.

Bromides: Ferrous bromide ($FeBr_2$), green solid, soluble, formed by reaction of iron and **hydrobromic acid,** and then evaporating out of contact with air; ferric bromide ($FeBr_3 \cdot 6H_2O$), red solid, soluble, formed by reaction of ferrous bromide solution and **bromide,** and then evaporation; ferric bromide anhydrous ($FeBr_3$), red solid, formed by reaction of iron and dry bromine upon heating.

Carbide: Iron carbide, cementite (Fe_3C), black hard solid, melting point 1837° C., formed by reaction of iron and carbon at high temperature, an important component of steels.

Carbonate: Ferrous carbonate ($FeCO_3$), white precipitate, when pure, but soon changes to green in air, then to black ferroferric hydroxide, and then to red ferric hydroxide, formed by reaction of soluble ferrous salt solution and **sodium** carbonate solution.

Carbonyl: Iron tetracarbonyl ($Fe(CO)_4$), gas, formed by reaction of iron and **carbon monoxide** at 80° C., burns with a yellow flame; iron pentacarbonyl ($Fe(CO)_5$), yellow liquid, boiling point 103° C., formed by reaction of finely divided iron and carbon monoxide at ordinary temperatures, is decomposed by light.

Chloride: Ferrous chloride ($FeCl_2 \cdot 4H_2O$), greenish solid, soluble, formed by reaction of iron and **hydrochloric acid,** and then evaporation out of contact with air; ferric chloride ($FeCl_3 \cdot 6H_2O$), reddish-yellow solid, soluble, formed by reaction of ferrous chloride solution and **chlorine,** and then evaporation, used in photography and as an oxidizing agent; ferric chloride anhydrous ($FeCl_3$), red solid, formed by reaction of iron and dry chlorine upon heating. Used as a reagent in organic chemistry; ferroferric chloride ($FeCl_2 \cdot 2FeCl_3 \cdot 18H_2O$), yellow solid, soluble; ferric ammonuium chloride ($FeCl_3 \cdot NH_4Cl$), orange solid, soluble.

Citrate: Ferric citrate ($Fe(C_6H_5O_7)_3 \cdot 3H_2O$), reddish-brown solid, soluble, formed by reaction of ferric hydroxide and **citric acid,** and then evaporation. Used in preparing **blue print** paper; ferric ammonium citrate, brown solid, soluble. Used in preparing blue print paper.

Ferrate: Sodium ferrate (Na_2FeO_3), purple solution, formed (1) by fusing **sodium** nitrate and finely divided iron, and then extracting with water, (2) by reaction of ferric hydroxide suspension in concentrated **sodium** hydroxide and addition of **chlorine** in excess.

Ferrites: By fusion of ferrous oxide with the appropriate basic oxide, hydroxide or carbonate, e.g., calcium ferrite ($Ca(FeO_2)_2$), barium ferrite ($Ba(FeO_2)_2$), magnesium ferrite ($Mg(FeO_2)_2$), ferrous ferrite ($Fe(FeO_2)_2$), sodium ferrite ($NaFeO_2$).

Ferricyanides: Potassium ferricyanide, red prussiate of potash ($K_3Fe(CN)_6$, red crystals, soluble to reddish-brown solution, formed by reaction of potassium **ferrocyanide** solution and an oxidizing agent such as chlorine, and then crystallizing; ferrous ferricyanide, "Turnbull's blue" ($Fe_3(Fe(CN)_6)_2$) blue precipitate by reaction of ferrous salt solution and potassium **ferricyanide** solution, unattacked by acids, but blue color destroyed by alkalis. Used as a pigment.

Ferrocyanides: Sodium ferrocyanide, yellow prussiate of soda ($Na_4Fe(CN)_6 \cdot 10H_2O$) and potassium ferrocyanide, yellow prussiate of potash ($K_4Fe(CN)_6 \cdot 3H_2O$) are yellow solids, soluble, formed by treating "spent oxide" of coal gas works with **calcium** hydroxide to extract the ferrous cyanide as soluble calcium ferrocyanide, and converting, with **sodium** or **potassium** carbonate, into the respective ferrocyanide. Used (1) in the preparation of other ferrocyanides (e.g., ferric ferrocyanide), and of ferricyanides, (2) in blue print paper, (3) in tanning; ferric ferrocyanide, "Prussian blue" $Fe_4(Fe(CN)_6)_3$ blue precipitate formed by reaction of ferric salt solution and sodium ferrocyanide solution, unattacked by acids, but blue color destroyed by alkalis. Used as a pigment.

Hydroxides: Ferrous hydroxide ($Fe(OH)_2$), white precipitate, when pure, but soon changes in air to ferroferric hydroxide and then to ferric hydroxide, formed by reaction of soluble ferrous salt solution and **sodium** or **ammonium** hydroxide solution, soluble in acids, insoluble in alkalis; ferroferric hydroxide ($Fe(OH)_2 \cdot 2Fe(OH)_3$), black precipitate, soon changes in air to ferric hydroxide, formed by reaction of mixture of soluble ferrous and ferric salt solution and **sodium** or **ammonium** hydroxide solution; ferric hydroxide ($Fe(OH)_3$), reddish-brown gelatinous precipitate, formed by reaction of soluble ferric salt solution and sodium or ammonium hydroxide solution, soluble in acids, insoluble in alkalis.

Iodide: Ferrous iodide ($FeI_2 \cdot 4H_2O$), grayish-black solid, soluble, formed (1) in solution by reaction of soluble ferric salt solution and **potassium** iodide solution, with accompanying separation of iodine, (2) by reaction of **iron** and **iodine.**

Nitrate: Ferrous nitrate ($Fe(NO_3)_2 \cdot 6H_2O$), greenish solid, unstable, formed (1) by reaction of ferrous hydroxide or carbonate and dilute **nitric acid,** (2) by **barium** nitrate and ferrous sulfate solutions, filtration, and then evaporation at low temperature out of contact with air; ferric nitrate ($Fe(NO_3)_3$), violet solid, soluble, formed by reaction of iron metal, or ferric oxide, and dilute nitric acid, and then crystallization. Solution is colorless when freshly prepared in excess of nitric acid.

Oxalate: Ferrous oxalate ($FeC_2O_4 \cdot 2H_2O$), yellow precipitate, formed by reaction of soluble ferrous salt solution and **ammonium** oxalate solution, yields ferrous oxide on heating at 160° C. out of contact with air; ferric oxalate ($Fe_2(C_2O_4)_3$), greenish solid, soluble, formed by reaction of ferric hydroxide and **oxalic acid,** and then crystallization; ferrous potassium oxalate ($Fe(C_2O_4)_2 \cdot K_2C_2O_4 \cdot 2H_2O$), yellow solid, soluble; ferric potassium oxalate ($K_3Fe(C_2O_4)_3 \cdot 3H_2O$); ferric ammonium oxalate (($NH_4)_3Fe(C_2O_4)_3 \cdot 3H_2O$), green solid, soluble.

Oxides: Ferrous oxide (FeO), black solid, insoluble, formed (1) by heating ferrous oxalate at 160° C. out of contact with air, (2) by heating ferric oxide or ferro-ferric oxide and **hydrogen** gas at 300° C., (3) by heating iron metal and steam above 570° C., soluble in hydrochloric or dilute **sulfuric acid** to ferrous salt; ferroferric oxide, black oxide of iron, **magnetite,** "iron scale," (Fe_3O_4), black solid, insoluble, formed (1) by heating iron metal, ferrous or ferric oxide in air, (2) by heating iron and steam below 570° C., believed to be ferrous ferrite ($Fe(FeO_2)_2$); ferric oxide, red oxide of iron, **hematite** (Fe_2O_3), red solid, insoluble, formed by heating ferric hydroxide or nitrate to a high temperature. The mineral is the important source of iron, and when pulverized is used as a paint pigment, various naturally occurring hydrated forms serve as brown or yellow pigments—umbers and siennas from **limonite.**

Perferrates: Sodium perferrate (Na_2FeO_4), formed (1) by fusing **sodium** peroxide and ferric hydroxide, (2) by heating finely divided iron, ferrous oxide, or ferric oxide in **sodium** nitrate to a red heat; barium perferrate ($BaFeO_4$) precipitate, formed by reaction of sodium perferrate solution and barium chloride solution.

Sulfates: Ferrous sulfate, green vitriol, "copperas" ($FeSO_4 \cdot 7H_2O$), green solid, soluble, formed by reaction of iron and dilute **sulfuric acid** and then crystallizing. Obtained as a by-product in the cleaning or "pickling" of steel. Used widely (1) as a source of ferrous compounds, and of ferric compounds by easy oxidation. (2) as a disinfectant, water purifier, wood preservative, weed exterminator, (3) in the preparation of inks, pigments, medicines, (4) in the purification of coal gas, (5) in the textile and leather industries; ferrous ammonium sulfate, "Mohr's salt" ($FeSO_4 \cdot (NH_4)_2SO_4 \cdot 6H_2O$), green solid, soluble, more easily purified, more stable than ferrous sulfate; ferric sulfate ($Fe_2(SO_4)_3$), brownish solid, soluble, formed (1) by reaction of ferric hydroxide and sulfuric acid, and then evaporation, (2) by reaction of iron and hot concentrated sulfuric acid, (3) in solution by reaction of **chlorine** and ferrous sulfate in acid solutions. Used (1) to prepare iron alum, (2) in water purification as basic ferric sulfate, ferric "persulfate" or "subsulfate," yellow solid, soluble, (3) in medicine; ferric ammonium sulfate "iron alum" ($Fe_2(SO_4)_3 \cdot (NH_4)_2SO_4 \cdot 24H_2O$), violet solid, soluble, more easily purified than ferric sulfate.

Sulfides: Ferrous sulfide (FeS), black precipitate, formed by reaction of soluble ferrous salt solution and **sodium** or **ammonium** sulfide, soluble in acids, also formed by heating finely divided **iron** and **sulfur,** reacts with dilute **hydrochloric** or **sulfuric acid** to yield ferrous salt and hydrogen sulfide; iron disulfide, pyrite (FeS_2), brass-colored mineral, when heated in air yields ferric oxide plus **sulfur dioxide** gas; ferric sulfide (Fe_2S_3), black precipitate, formed by reaction of soluble ferric salt solution and sodium or ammonium sulfide.

Sulfite: Ferric sulfite ($Fe_2(SO_3)_3$), red solution, by reaction of soluble ferric salt solution and **sodium** hydrogen sulfite solution, is unstable, quickly forming ferrous sulfite and dithionate.

Thiocyanate: Ferric thiocyanate ($Fe(CNS)_3$), red solution, by reaction of soluble ferric salt solution and ammonium **thiocyanate** solution. Used as a delicate test for ferric (1 part ferric in solution in 1,500,000 parts of water may be detected), more soluble in ether than in water and easily concentrated in the ether layer by mixing.

Ferrous salts are generally green, sometimes yellow to brown by slight oxidation, and in solution are generally green, but are quickly oxidized in part by air to ferric, brownish. Ferrous is chemically related to zinc, cobaltous, nickelous, manganous, magnesium. Ferroferric compounds are black. Ferric salts are generally violet to brown, sometimes orange or greenish, and in solution generally red and on boiling yield red precipitate of basic salt or ferric hydroxide; ferric hydroxide, in various partially dehydrated forms in nature, is brown to yellow. Ferric is chemically related to chromic, aluminum. (R.K.S.)

IRON AGE. Paleontology of Man.

IRRADIATION. 1. Exposure to **x-rays, radium,** or other forms of radioactivity. 2. Exposure to **ultraviolet** rays to increase the **vitamin D** content of a substance. (D.M.H.)

IRRATIONAL EQUATIONS. Radical Equations.

IRRATIONAL FUNCTION. An irrational function is an **algebraic function** which is not a **rational function;** it therefore involves one or more root-extractions involving the variable. (L.L.S.)

IRRATIONAL NUMBERS. Number.

IRRATIONAL ROOTS OF POLYNOMIAL EQUATIONS. Polynomial Equations.

IRREGULAR VARIABLES. There are many **variable stars** which vary in brightness in such non-systematic manner as to be best designated as irregular variables. In some cases their manner of variation and spectral class is so similar to that of the **long period variables** that they are classed as members of this class of stars by some authorities. Notable among stars of this type are the bright stars α **Orionis** and α **Herculis,** both of which are **giant red stars.**

There are certain groups of irregular variables which have enough characteristics in common that they may be considered as groups. Such groups are usually designated by some typical star of the group. The RV Tauri group resemble to some extent the **Cepheids** although their periods, of the order of magnitude of 75 days, is far too long to permit their inclusion in the true Cepheid class. These stars are of G and K **spectral classes** and their light curves are characterized by shallow minima coming between two deep ones. A few stars of the R Coronae Borealis type remain, often for several years, practically constant in brightness and then drop suddenly one or two magnitudes, returning to normal brightness after a few oscillations.

The problem of the irregular variables is at present the subject of a great deal of research and observation, but to date there is no adequate explanation of their variability. (W.K.G.)

IRRITABILITY. A fundamental property of living matter which enables it to be stimulated by external factors. Although the simplest forms of living things are capable of receiving stimuli from various environmental factors such as light, mechanical contacts, and chemical compounds, the property is most highly developed in animals with a complex nervous system. The **sense organs** of such animals show a high degree of specialization and diversification of this property. (A.W.L.)

IRROTATIONAL VECTOR. If the **curl** of a **vector function** of position vanishes everywhere in a certain region, the function is said to be an irrotational vector (or a lamellar vector) in this region.

If a given vector function **v** is the **gradient** of a scalar function ϕ, then **v** is irrotational. (L.L.S.)

ISACOUSTS. Lines of equal sound of an **earthquake.** (R.M.F.)

ISALLOBARS AND ISALLOBARIC FIELDS. Atmospheric pressure changes at every point in the atmosphere from time to time. It is possible to measure such changes with considerable accuracy. The unit used to express pressure change (or pressure tendency) is conventionally the total net change occurring in a 3-hour interval. It is customary to indicate the nature of the change because the pressure change character may have varied during the selected time interval.

Three-hourly pressure changes are plotted on a synoptic chart (weather map) and lines drawn to join points of equal pressure change. Care is exercised, however,

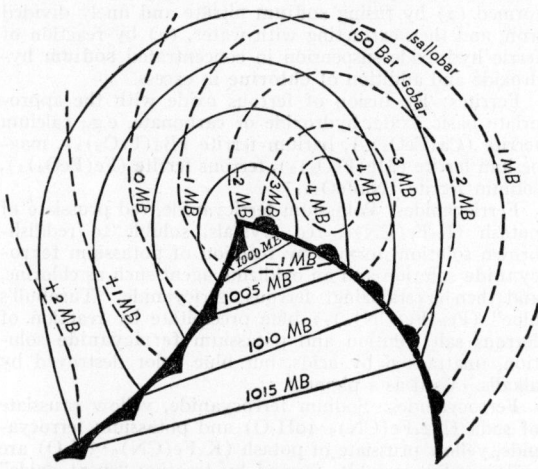

Isallobars in isallobaric field of wave cyclone.

in noting the character of the change in judging the real value of Δp. Lines joining points of equal pressure change are isallobars. Isallobars taken together constitute an isallobaric field. Where there is present an isallobaric field superimposed on a pressure field, a component of the actual wind blows along the isallobaric gradient which is directed perpendicular to the isallobars toward regions of greatest pressure fall or least pressure rise, as the case may be. Normally the isallobaric wind component is small unless the isallobaric field is pronounced. When there is no isallobaric field, the wind is defined approximately by the orientation and spacing of isobars themselves. Isallobaric fields also are used to compute the movement of pressure centers, wedges, troughs, cols, fronts, and isobars. Centers of low pressure will tend to move toward the region of greatest pressure fall, whereas centers of high pressure tend to move toward regions of maximum pressure rise. Both cases have modifying factors and the direction is not always exactly as would be expected from a casual glance. In all cases of computation of movement, it is necessary to know the isallobaric field, the pressure field, and the orientation of the system whose movement is to be computed.

In establishing an isallobaric field it is imperative that the diurnal pressure change be accounted for and a correction applied. Diurnal pressure changes do affect movement of pressure systems, but the effect is transient and therefore must be neglected in any study extending for more than 6 hours into the future.

In general, isallobars and isallobaric fields are as useful in weather forecasting as the pressure field itself. (P.E.K.)

ISINGLASS. Mica.

ISCHIOPODITE. Biramous **appendage.**

ISCHIUM. Skeletal System.

ISOBARS. Lines joining points of equal pressure. The field of isobars is known as an isobaric field. Isobaric fields are of prime importance in determining winds and weather. (P.E.K.)

ISOBASE. In geology, a **contour** line which intimates the former equal elevation of a series of **points,** or level surface, now deformed. (R.M.F.)

ISOBATH. Meaning equal depth. 1. A line on a land surface all points on which are the same vertical distance above the ground water table. 2. A structural contour or isopach. (R.M.F.)

ISOCLINE. Vertical duplication of formations by close folding, as illustrated in the accompanying diagram. (R.M.F.)

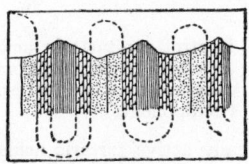

Isoclinal folds.

ISOCYANATES. Cyanic Acid and Cyanates.

ISOCYANIDES. Hydrocyanic Acid and Cyanides.

ISOELECTRIC POINT. In solutions of proteins and related compounds, the **hydrogen ion** concentration at which the dipolar **ions** are at a maximum is the isoelectric point. At this point the solution shows minimum **conductivity**, osmotic pressure, and viscosity. At this pH the protein shows the least swelling with water and does not undergo cataphoresis. That is, the colloidal particles move toward neither electrode. Proteins coagulate best and contain the least amount of inorganic matter at their isoelectric points. (See also **Aminoacids, Polypeptides, and Proteins.**)

ISOELECTRIC POINTS

SUBSTANCE	ISOELECTRIC POINT (pH)
Proteins:	
Glutenin	4.5
Gelatin	4.7
Egg albumin	4.8
Serum albumin	5.4
Edestin	5.7
Oxyhemoglobin	6.8
Gliadin	9.2
Clupeine	12.1
Dipeptides:	
Glycylglycine	5.5
Alanyl glycine	5.5
Leucyl glycine	5.7
Aminoacids:	
Glycine	6.6
Alanine	6.7
Leucine	6.5

(R.K.S.)

ISOGAMY. A type of sexual reproduction in which the male and female germ cells are similar in form. (A.W.L.)

ISOGEOTHERM. Depths of equal temperature in the earth. (R.M.F.)

ISOGONIC CHART. A chart showing lines of equal magnetic declination, or lines on which the variation of the magnetic needle from true north is the same. (See also **Agonic Line.**) (R.M.F.)

ISOGONIC LINE. Compass Corrections.

ISOLATION. A principle of organic **evolution.** Animals separated from the remainder of their kind are supposed to have a characteristic heritage which determines the nature of their descendants. This foundation may differ from that of the rest of the species and so **may** give rise in time to a different variety or species **as** a result of the isolation, independent of the adaptive **value** of the heritage.

Isolation may be geographic or biological. The Galapagos Islands offer famous examples of geographic isolation. They are regarded as the remnants of a once continuous land mass and are now inhabited by distinct varieties of birds of several species. These varieties are characteristic of certain islands and do not pass from one to the other, hence they are interpreted as the descendants of birds which remained on the islands when they were first formed.

Biological isolation may be due to cytological incompatibility, physical differences or to time of reproductive activity. Some varieties of dogs, for example, are physically incapable of interbreeding although they belong to the same species. Time is a decisive factor in interbreeding among the insects, especially in species whose adult life is short. If some individuals emerge early and others late in the normal season, those of the two groups have little or no chance to interbreed. The result is that any distinctive characters of the two are likely to be perpetuated in different strains. (A.W.L.)

ISOLEUCINE. Aminoacids, Polypeptides, and Proteins.

ISOMAGNETIC. Lines of equal magnetic force, but not necessarily **isogonic.** Isomagnetic lines may represent local magnetic anomalies such as caused by magnetic ore bodies, magnetic minerals in sediments, or the vertical rather than the horizontal deviation of the compass or magnetic needle. (R.M.F.)

ISOMERISM AND STEREOISOMERISM. Two substances are said to be isomeric when they possess the same ordinary molecular **formula** and different properties. A simple instance is that of C_3H_8O, which represents two propyl **alcohols**, one (A) of boiling point 98° C. and density 0.804, the other (B) of boiling point 82° C. and density 0.789. These two alcohols differ in other properties including their chemical behavior. It is believed that the difference is due to the arrangement of **atoms** in each **molecule,** thus:

$$
\begin{array}{c}
\text{H} \quad \text{H} \quad \text{H} \\
\text{(A)} \;\; \text{HC}-\text{C}-\text{COH, propanol 1, ethylcarbinol, normal-propyl} \\
\text{H} \quad \text{H} \quad \text{H} \qquad\qquad \text{alcohol, } C_2H_5 \cdot CH_2OH
\end{array}
$$

$$
\begin{array}{c}
\text{H} \quad \text{H} \quad \text{H} \\
\text{(B)} \;\; \text{CH}-\text{C}-\text{CH, propanol 2, dimethylcarbinol, iso-propyl} \\
\text{H} \quad \text{OH} \quad \text{H} \qquad\qquad \text{alcohol, } (CH_3)_2CHOH
\end{array}
$$

Therefore, the ordinary molecular formula, C_3H_8O, represents two different isomeric substances; $C_2H_5 \cdot CH_2OH$ and $(CH_3)_2CHOH$ are isomers; and the phenomenon displayed by these is known as isomerism.

The ordinary molecular formula, C_4H_{10}, corresponds to two butanes, namely, normal-butane, $CH_3CH_2CH_2CH_3$, (diethyl) of boiling point 0.6° C. and density 0.60, and iso-butane $(CH_3)_3CH$ (2-methyl propane, trimethyl methane) of boiling point −10° C. and density 0.60.

In any series of compounds, as alcohols and **hydrocarbons** above, the number of isomers increases rapidly with increase in the molecular formula, thus, there are four isomeric alcohols of C_4H_9OH, and eight of $C_5H_{11}OH$, and three isomeric hydrocarbons of C_5H_{12}, and five of C_6H_{14}.

The hydroxyacids represented by the ordinary formula $C_3H_6O_3$ show other forms than the two plain isomers:

$$
\begin{array}{l}
\text{CH}_3 \\
| \\
\text{HCOH} \qquad \text{alpha-hydroxypropionic acid,} \\
| \qquad\qquad\qquad \text{lactic acid (3 forms)} \\
\text{COOH} \\
\text{H}_2\text{COH} \\
| \\
\text{HCH} \qquad \text{beta-hydroxypropionic acid,} \\
| \qquad\qquad\qquad \text{hydracrylic acid (1 form)} \\
\text{COOH}
\end{array}
$$

Plain isomerism is not sufficient to account for these three known forms of **lactic acid**. The explanation that is accepted was offered by van't Hoff and LeBel, independently (1874). This explanation demands that the four valencies of the **carbon** atom be disposed *equally in space*. The carbon atom may be pictured as occupying the center of a regular tetrahedron. The four **valencies** are disposed from the carbon atom at the center towards the four corners of the tetrahedron. When *four different groups* are attached (at the corners of the tetrahedron) to the carbon atom (at the center of the tetrahedron) it is possible to arrange the resulting compound in *two different ways*. Upon projection these two forms appear thus:

$$\begin{array}{ccc}
CH_3 & & CH_3 \\
| & & | \\
H-C-OH & \text{and} & HO-C-H. \\
| & & | \\
COOH & & COOH
\end{array}$$

These are called space-isomers or stereoisomers, and the phenomenon displayed is known as stereoisomerism. One of these forms rotates the plane of **polarized light** to the right (dextro-lactic acid, d-lactic acid or sarcolactic acid) and the other to the left (laevo-lactic acid or l-lactic acid). A mixture of the two forms in equal amounts is ordinary lactic acid (dextrolaevo-lactic acid, dl-lactic acid, racemic-lactic acid) from sour milk, which does not rotate the plane of polarized light. These are the three forms of lactic acid, namely, dextro, laevo, and dextrolaevo or externally compensated. From dl-lactic acid, the other two forms may be obtained in the following manner, (1) d-lactic acid, by treatment with the mold, *Penicillium glaucum,* whereby l-lactic acid is destroyed, (2) l-lactic acid, by treatment with an optically active base, **strychnine,** followed by fractional crystallization, and subsequent treatment with an acid.

These two methods are of general application to racemic or externally compensated stereoisomers, the first, **biochemical,** in which one-half of the racemic substance is destroyed by the organism used, and the second, resolution, which can be applied in the case of acidic or basic stereoisomers. A third method, involving plain **crystallization,** is described below (Pasteur's separation of tartrates).

The simplest alcohol displaying stereoisomerism is amyl **alcohol,** which is known in the three forms dextro, laevo, dextrolaevo or racemic.

$$\begin{array}{c}
C_2H_5 \\
| \\
CH_3-C-H \\
| \\
CH_2OH
\end{array}$$

The carbon atom, to which are attached the four different groups to produce stereoisomerism or optical isomerism, is known as *asymmetric,* and, when it is desired to identify this carbon atom as such, it is written or printed more prominently than non-asymmetric carbon atoms. The projection formulas of the four forms of **tartaric acids** are as follows:

$$\begin{array}{c}
COOH \\
| \\
H-C-OH \\
| \\
H-C-OH \\
| \\
COOH
\end{array}$$ inactive or meso-tartaric acid (internally compensated; possesses a plane of symmetry; optically inactive).

$$\begin{array}{c}
COOH \\
| \\
H-C-OH \\
| \\
HO-C-H \\
| \\
COOH
\end{array}$$ dextro-tartaric acid (arrangement of groups around each asymmetric carbon atom is cumulative; optically active; dextrorotatory).

$$\begin{array}{c}
COOH \\
| \\
HO-C-H \\
| \\
H-C-OH \\
| \\
COOH
\end{array}$$ laevo-tartaric acid (arrangement of groups around each asymmetric carbon atom is cumulative; optically active; laevo-rotatory).

$\left\{\begin{array}{l} \text{d-tartaric acid} \\ \text{l-tartaric acid} \end{array}\right\}$ Dextrolaevo-tartaric acid, racemic tartaric acid (externally compensated; optically inactive; can be resolved into d and l components).

The two optically active tartaric acids when crystallized differ in the arrangement of the faces—one is the mirror image of the other. Pasteur (1848), observing this difference, was able to separate the two optically active forms of ammonium sodium tartrate crystals made from racemic tartaric acid.

When stereoisomeric substances are synthesized using *inactive* materials, the product is optically inactive, since, it is believed, the chances of dextro and laevo forms being produced are equal. Pope and Read (1914) prepared optically active forms of chloroiodomethanesulfonic acid

$$\begin{array}{c}
H \\
| \\
Cl-C-I \\
| \\
SO_2OH
\end{array}$$

The sugars, $C_6H_{12}O_{16}$, have attracted much attention in this field. (See **Carbohydrates.**) Many substances in nature, especially the alkaloids and glucosides, are optically active.

Ethylene Stereoisomerism. While it might appear that succinic acid, $C_4H_6O_4$, could demonstrate the phenomenon of stereoisomerism, the same has not been detected. It is assumed that there is free rotation about the single bond joining the two central carbon atoms and the position finally assumed is one of equilibrium, one form only being known:

$$\begin{array}{cc}
\begin{array}{c}
H \\
| \\
H-C-COOH \\
| \\
H-C-COOH \\
| \\
H
\end{array} &
\begin{array}{c}
H \\
| \\
H-C-COOH \\
| \\
HOOC-C-H \\
| \\
H
\end{array}
\end{array}$$ (only one form of **succinic acid** known)

But in the case of olefin linkage between the central carbon atoms in an analogous compound, stereoisomerism is detected. Thus, two acids, $C_4H_4O_4$, are known, namely, **maleic** and **fumaric:**

$$\begin{array}{cc}
\begin{array}{c}
H-C-COOH \\
\| \\
H-C-COOH
\end{array} &
\begin{array}{c}
H-C-COOH \\
\| \\
HOOC-C-H
\end{array}
\end{array}$$

Maleic acid	Fumaric acid
Cis-butenedioic acid	Trans-butenedioic acid
or	or
Cis-1,2-ethenedicarboxylic acid	Trans-1,2-ethenedicarboxylic acid

Each of these acids has distinctive properties. Maleic acid, melting point 130° C., decomposes above 135° C., with partial transformation into maleic anhydride.

$$\begin{array}{c}
H-C-C=O \\
\quad\quad\quad\searrow O, \\
H-C-C=O
\end{array}$$ solubility 79 grams per 100 grams water

at 25° C., optically inactive; fumaric acid, melting point 287° C., boiling point 290° C., no anhydride, 0.7 gram per 100 grams water at 17° C., optically inactive.

Oleic and elaidic acids display this type of isomerism.

Carbon-Nitrogen (group $> C = N -$) *Stereoisomerism.* Analogous to ethylene isomerism is that of carbon-nitrogen in such compounds as the **oximes.**

Benzaldehyde with **hydroxylamine** yields two oximes. Stereoisomerism is utilized to explain the difference in structure.

C_6H_5—C—H	C_6H_5—C—H	C_6H_5—C
‖	‖	‖
HON	NOH	N
Benzantialdoxime	Benzsynaldoxime	Phenyl cyanide
Melting point 35° C.	Melting point 128° C.	Boiling point 191° C.
Optically inactive	Optically inactive	(For comparison)

Benzsynaldoxime is transformed into benzantialdoxime by treatment with acids, and is more readily converted by loss of water into phenyl cyanide (this is a debated point).

The **ketone** para-chlorobenzophenone with hydroxylamine yields two stereoisomeric oximes:

H_5C_6—C—$C_6H_4Cl(4)$ H_5C_6—C—$C_6H_4Cl(4)$
‖   ‖
HON   NOH

Cyclic Cis-Trans Stereoisomerism in non-benzenoid compounds. This phenomenon is illustrated by the two forms of hexahydroorthophthalic acid ($C_6H_{10}(COOH)_2$ (1,2)):

Cis-Hexahydro-1,2 phthalic acid
Melting point 192 C.
Soluble in water
Plane of symmetry (the dotted line), therefore, optically inactive

Trans-Hexahydro-1,2 phthalic acid
Melting point 221° C.
Soluble in water
No plane in symmetry, therefore, two optically active forms

This type of isomerism is believed to explain the existence of two forms of glucose, namely, alpha and beta (not dextro and laevo, which is of the lactic acid and tartaric acid non-cyclic type). (See **Carbohydrates**.) (R.K.S.)

ISOMETRIC REPRESENTATION. Pictorial Representation.

ISOMETRIC SYSTEM. Crystallography.

ISOMORPHOUS. The term applied to two or more minerals whose molecules can form intimate crystalline mixtures. Such mixtures are called solid **solutions,** in which the physical properties vary with the chemical proportions. The bulk of the silicate minerals are isomorphous, especially **albite** and **anorthite,** varying mixtures of which produce different species of **plagioclase** feldspar. (R.M.F.)

ISONITRILES. Amines and Amides.

ISOPACHOUS LINE. ISOPACH. A line, on a geological (stratigraphical or structural) map, all points on which show equal thickness of a formation. (R.M.F.)

ISOPHANE. A line plotted on a map through regions presenting uniform association of biological and climatic factors. Isophanes of the United States curve upward from the southeast toward the northwest. Phenomeridians crossing the isophanes at intervals in degrees of longitude corresponding to the spacing of the isophanes complete the division of the map into quadrangles. In each quadrangle the same conditions prevail within certain limits of variation. This system has been used for the computation of safe dates for sowing wheat to avoid the attack of the **Hessian fly** in various parts of the United States. (A.W.L.)

ISOPLETH. Alignment Chart.

ISOPODA. Crustacea.

ISOPORS, ISOPORIC FOCI. Lines or contours joining points of equal geomagnetic (compass direction) change are called *isopors*. The geographic coordinates of rapid geomagnetic changes are called *isoporic foci*. These foci are concentrated in the large continents and the sub-Atlantic lithosphere. (R.M.F.)

ISOPRENE. Hydrocarbons.

ISOPTERA. The white ants or termites. An order of insects of great economic importance and biological interest, made up of social species which eat wood and other vegetable matter. Most termite species are tropical or subtropical but a few live in temperate regions.

Termites have biting mouth parts and are moderate to small soft-bodied insects. They live in dark nests and tunnels except when the winged sexual individuals emerge to leave the parent colony. The bodies of these flying individuals are dark but the termites that remain in the nest are whitish with dark heads. They do not resemble ants in form, hence their similar habits are probably responsible for the name white ant. The temporary wings of termites are long and slender and the two pairs are similar in form. In most species the veins near the anterior margin are strong and the rest are faintly marked. The wings are shed after the swarming termites find a new nesting place.

The termite colony contains workers, soldiers, and reproductive individuals of both sexes. The workers are developed in subordinate castes in several species. Soldiers have large heads and strong jaws. The queen in some colonies becomes relatively enormous through the expansion of the abdomen as the eggs develop and is quite helpless. The workers feed and groom her and carry away her eggs.

It has been shown conclusively that termites depend on **protozoans** in the intestine for the digestion of the wood that they eat. Very few animals can digest cellulose, which is the chief compound in wood, but the protozoans do so and the termites utilize the products developed by the protozoans. This relationship is one of the finest examples of **symbiosis** among animals.

Because of their wood-eating habits termites sometimes do great damage to buildings. Their habit of building tunnels wherever they go and of remaining concealed in the wood where they work often results in their presence being unsuspected until the honeycombed timbers give way. When they once enter buildings they are not restricted to wood but damage papers, books, clothing, carpets and many other things. In regions where they are plentiful no timber in construction should be left in contact with the ground. Even a small contact may be a point of entry. Where timber must be exposed to attack it can be protected by impregnation with creosote, but the most effective type of construction demands masonry wherever contact with the ground must be made. Even in such structures termites may traverse several feet of masonry, building tunnels as they go, and may work through small cracks into the wooden parts of the building. Where termites have already entered, blocking their entrance and destroying the colony both inside the building and out with creosote or fumigants are usually effective methods of control. Special equipment and methods are available commercially for this work. (A.W.L.)

ISOSEISMALS. Lines of equal earthquake intensity according to one of the intensity scales. (R.M.F.)

ISOSEISMIC LINE. A line connecting all points on the surface of the earth where an earthquake has been determined to be of equal intensity. (See **Earthquakes.**) (R.M.F.)

ISOSTASY. A term proposed by C. E. Dutton in 1889 for the theory of **gravitational** balance between relatively broad, contiguous areas of different average altitudes or topographic relief. The term is derived from the Greek meaning equal standing (balance), and has been applied particularly to the isostatic equilibrium of the major topographical features of the earth—continents and ocean basins. Thus the continents are assumed to "stand high," relative to the ocean basins, because they are lighter (or less dense) material. Whenever any large segment of the earth's crust is out of balance with its surroundings, adjustment is assumed to take place by means of solid flow in a deep subcrustal zone. Since the surfaces of the continents are constantly being worn down by erosion, if it were not for isostasy, early in the geologic history of the earth all land areas would have been reduced to ocean level. The theory of isostasy was originally proposed to explain major problems in structural geology, and has been amply justified by **geophysical** and **geodetic** surveys. In the application of the theory there has been considerable disagreement between geophysicists and geologists as to the minimum areas of relief (topographic features) which may be subject to isostatic balance, especially in mountainous regions which are characterized by low angle **overthrusts** and consequent, horizontal translations of great thicknesses of the **lithosphere.** The general principle of isostasy has, however, been accepted by an increasing majority of geologists, geodesists, and geophysicists, and probably no other geophysical-geological theory is so firmly entrenched in geodetic, geophysical and geological literature. The most recent outstanding exponent of Isostasy was William Bowie (died, 1940). (R.M.F.)

ISOTHERMAL. Having the same temperature.

ISOTHERMAL ATMOSPHERE. Any layer of air in which the temperature does not vary with altitude is known as an isothermal layer; if the whole atmosphere were of constant temperature in the vertical it would be an isothermal atmosphere. (P.E.K.)

ISOTHERMS. Lines joining points of equal temperature. Isotherms can be drawn for any surface or cross-section. (P.E.K.)

ISOTOPES. Chemical Composition.

ITABRITE. A variety of **quartzite** rich in iron minerals. It is named from the type locality at Itabira, Brazil. (E.S.C.S.)

ITACOLUMITE. This is a **sandstone** with a peculiar flexibility which is due to the interlocking of its constituent grains, permitting a limited amount of distortion without fracture. The name is derived from *Itacolumi,* a mountain in Brazil. (E.S.C.S.)

ITERATED INTEGRALS. Double Integrals and Triple Integrals.

ITERATIVE IMPEDANCE. Characteristic Impedance.

IVORY. The dentine of teeth, a material similar to bone but harder and of different minute structure. It is deposited outside of the layer of cells that produce it in the form of small tubules extending toward the outside of the tooth.

The chief sources of ivory for commercial purposes are the tusks of various animals. Elephants' tusks have only a little enamel at the tip and are solid ivory except where the pulp cavity invades the base. The tusks of walruses have also been an important source of ivory, although they are inferior to elephant ivory. The material is used extensively for carved ornaments. (A.W.L.)

IZARD. Mammalia, Artiodactyla. A name applied to the **chamois** in the Pyrenees. (A.W.L.)

J

JABIRU. Aves, Ciconiiformes. The giant storks of South America, Africa, and Australia. (A.W.L.)

JACAMAR. Aves, Piciformes. South American birds (**Aves**) related to the woodpeckers. (A.W.L.)

JACANA. Aves, Charadriiformes. Shore birds (**Aves**) of South America, Africa, and the Oriental region. They have long legs and tail and very long toes. (A.W.L.)

JACARE. Caiman.

JACK. A jack is a portable device for lifting heavy loads through short distances by hand power. Power is applied manually to a lever or bar. The mechanism of the jack allows the operator to exert a great **mechanical advantage** and lift a weight many times that which would be possible unaided by the jack. The mechanical advantage is secured by a great reduction in motion, and consequently jacks are slow in action.

The requirement of being readily carried by hand from place to place necessarily limits the capacity which may be built into a jack. However, it may be built to lift several tons. The mechanical advantage of a jack may be secured by screw, lever, or hydraulic action. The screw jack has a threaded screw fitting to a nut which is a part of the base of the jack. The threaded portion is revolved by bars inserted in it, and is topped by a bearing plate, which presses against the load being lifted, and which rubs on the top of the jack as the latter is turned. In its usual form, the lever jack has a rack and pawl, the pawl being mounted on the end of a lever whose fulcrum point is very close to the pawl, thus giving a big mechanical advantage to any force exerted at the other end of the long lever. The hydraulic jack, which is even more powerful than the other two types, has a very small piston which is actuated by hand, and which forces oil into a cylinder where it can act against a much larger piston. This larger piston is connected to the lifting portion of the jack, and the mechanical advantage which is, roughly, the ratio of the areas of the large and small pistons, may be made as large as wanted. (F.T.M.)

JACKAL. Mammalia, Carnivora. An animal resembling the wolf and of similar habits. The common jackal, *Canis aureus,* ranges from southeastern Europe to Ceylon and into northern Africa, and the remaining species are African. Jackals are predators and scavengers. They also eat some fruits and occasionally damage sugar cane. (A.W.L.)

JACKDAW. Aves, Passeriformes. A small European crow, *Corvus monedula,* typically with a gray collar but also occurring in a wholly black form. Another species of **daw** occurs in Asia and the name jackdaw is applied to the great-tailed **grackle** in the southwestern United States. (A.W.L.)

JACKET. A jacket is a covering whose function is associated with retention or extraction of heat from that which it covers. The **cylinder** of a **steam engine** is sometimes covered with a wooden jacket whose purpose is to prevent heat loss from the cylinder. The jackets of **internal combustion engines** remove the heat from the cylinder. While some internal combustion engines are air-cooled, the majority are liquid-cooled, usually with water which is circulated in jackets which surround the cylinder head and sometimes extend the full length of the cylinder. These jackets may be cast on the cylinders when the latter are poured, or they may be made of thin sheet metal welded to the cylinder. The latter practice produces a lighter weight engine, but is much more expensive than the cast type water-jacketed cylinder. Small engines can have the water led into the jacket and conducted away from it by simple openings, but a more definitely controlled flow of cooling water is necessary for a large cylinder. In addition, other parts such as exhaust **manifolds,** and pistons, may be water-jacketed. About one-third of the heat of the fuel supplied to an internal combustion engine finds its way into the contents of the cooling jacket. (F.T.M.)

JACK-IN-THE-PULPIT. Aroids.

JACOBIAN. A Jacobian is a type of mathematical expression, named after the great German mathematician Karl Gustav Jakob Jacobi (1804–1851).

Let F_1 and F_2 be two **functions** of u and v. The **determinant**

$$J = \begin{vmatrix} \dfrac{\partial F_1}{\partial u} & \dfrac{\partial F_1}{\partial v} \\ \dfrac{\partial F_2}{\partial u} & \dfrac{\partial F_2}{\partial v} \end{vmatrix} = \dfrac{\partial F_1}{\partial u}\dfrac{\partial F_2}{\partial v} - \dfrac{\partial F_1}{\partial v}\dfrac{\partial F_2}{\partial u}$$

is called the functional determinant or Jacobian of F_1 and F_2 with respect to u and v. It is frequently denoted by $\dfrac{\partial(F_1, F_2)}{\partial(u, v)}$.

In general, if F_1, F_2, $\cdots$, F_n are functions of u_1, u_2, $\cdots$, u_n, then the functional determinant or Jacobian of F_1, F_2, $\cdots$, F_n with respect to $u_1, u_2, \cdots, u_n$ is defined to be the determinant

$$J = \begin{vmatrix} \dfrac{\partial F_1}{\partial u_1} & \dfrac{\partial F_1}{\partial u_2} & \cdots & \dfrac{\partial F_1}{\partial u_n} \\ \dfrac{\partial F_2}{\partial u_1} & \dfrac{\partial F_2}{\partial u_2} & \cdots & \dfrac{\partial F_2}{\partial u_n} \\ \cdot & \cdot & \cdots & \cdot \\ \dfrac{\partial F_n}{\partial u_1} & \dfrac{\partial F_n}{\partial u_2} & \cdots & \dfrac{\partial F_n}{\partial u_n} \end{vmatrix},$$

and is often denoted by $\dfrac{\partial(F_1, F_2, \cdots, F_n)}{\partial(u_1, u_2, \cdots, u_n)}$. (L.L.S.)

JADE. Jade is a general term for a compact green mineral substance much prized for ornamental purposes in China and Japan and to a less extent in the Western World. Jade is, properly speaking, either a compact **actinolite** called **nephrite,** a variety of **amphibole,** or jadeite, a **monoclinic pyroxene.** It is easily worked and many prehistoric implements have been found of this material in Mexico, Switzerland, France, Greece, and Egypt. The word jade is derived from the Spanish *pietra di hijada,* kidney stone, because it was supposed to be beneficial to diseases of the kidneys. Nephrite is derived from the Greek word for kidney, the allusion being the same as in the case of jade. (E.S.C.S.)

JADEITE. The mineral jadeite, essentially **sodium-aluminum silicate,** $NaAl(SiO_3)_2$, is a monoclinic **pyroxene** usually appearing in crystalline masses, or may be granular, fibrous, or compact. It has a prismatic **cleavage;** splintery **fracture;** hardness, 6.5–7; specific gravity, 3.3–3.5; luster, vitreous to pearly; color, various shades of green, bluish-green, greenish-white or almost white;

translucent to opaque. The processes that have acted to form this mineral are little understood both because of the confusion that exists between jadeite and **nephrite**, and the fact that the localities are not well known. Jadeite is found in Burma and China and has been reported from Mexico. It has probably resulted from the **metamorphism**, at great depths, of rocks rich in soda and aluminum, such as **nephelite syenites**. Its association in Burma with **serpentine** suggests its origin in more **basic igneous** rocks.

Jadeite is a tough and yet rather easily worked substance and has long been used for ornamental purposes. Evidence has been found in Europe, Mexico, Egypt, and elsewhere that it was used in prehistoric times for both ornaments and implements. The word jadeite has been formed by adding -ite to jade, the general term used for all green-colored tough compact stones that have been used as indicated above. (E.S.C.S.)

JAEGER. Aves, Charadriiformes. Birds closely related to the **gulls**, from which they differ in the elongate middle tail feathers. Four species nest in the Arctic regions and migrate into Europe, the southern United States, and even the southern hemisphere. One is antarctic and one lives on the west coast of South America. Also called skuas. (A.W.L.)

JAGUAR. Mammalia, Carnivora. A large South American cat, *Felis onca*, found chiefly in the jungles but also in open country. It is tan, marked with rings

Jaguar. (*N. Y. Zool. Soc.*)

and dots of black, resembling the leopard. The jaguar is larger than the leopard and differs in details of structure and markings. (A.W.L.)

JAGUARONDI. Mammalia, Carnivora. A small **cat**, *Felis jaguarondi*, of uniform brownish or blackish color. It ranges from northern Mexico to Brazil and Paraguay. (A.W.L.)

JAPANESE BEETLE. Insecta, Coleoptera. A small bronze-green **beetle**, *Pompillia japonica*, about three-eighths of an inch long, related to the May beetles. It is native to Japan and was first noticed as a pest in the United States about 1916. From New Jersey, where it first appeared, it has now spread about 500 miles westward. It attacks fruits, shrubs, forest and shade trees, and many crop plants, and the larvae damage the roots of grasses. The control of the pest is still a serious problem since sprays strong enough to kill the beetles injure many plants and trees. In cultivated lands the working of the ground kills the grubs, and strong sprays of **lead** arsenate (8 lbs. per 100 gal. of water) are effective against the adults on some fruits, shade trees, and shrubs.

In 1944 a new type of control became commercially available. It consists of a mixture of the spores of a milky-disease organism, *Bacillus popilliae,* and powdered talc. By spreading this material at 10-foot intervals in infested areas the disease, which kills the larvae of the beetles, is established much more promptly and

effectively than through its natural spread. It is a very effective means of reducing the beetle population. (A.W.L.)

JARARACA. Reptilia, Sauria. A poisonous **snake**, *Lachesis jararaca*, which ranges from eastern Brazil to Ecuador and Peru. A **pit viper**. (A.W.L.)

JARGOON or JARGON. Zircon.

JARNO TAPER. Taper.

JAROSITE. The mineral jarosite is a **basic** hydrous sulfate of **potassium** and **iron** corresponding to the formula $K_2Fe_6(OH)_{12}(SO_4)_4$. It is formed in the outcrops of ore deposits during oxidation of **iron** sulfides. It is a **hexagonal** mineral with basal **cleavage**; is brittle; hardness 2.5–3.5; specific gravity 3.15–3.26; luster vitreous to dull; color, dark yellow to yellowish-brown; shining yellow **streak**; translucent to opaque. Jarosite was originally reported from and named for Barranco Jaroso in the Sierra Almagrera, Spain. It has been found in Bohemia, France, the Island of Elba, Siberia and Bolivia. In the United States it is found in Arizona, Colorado, Texas, New Mexico, Utah, Nevada, and South Dakota. (E.S.C.S.)

JASMINE OIL. Volatile Oils.

JASPER. This mineral normally occurs as a **red chalcedony**, the coloring matter being **hematite** or **limonite**. Other varieties of jasper are yellow, green, or dull blue. Jasper forms an important constituent of the hard iron ores of the Lake Superior iron ranges. (R.M.F.)

JASPILITE. A rock made up of alternating layers of siliceous material such as **quartz** or **chalcedony**, and red **jasper** or **hematite**. Jaspilites are often contorted and **brecciated** to a considerable degree, and are sometimes polished for ornamental use. (E.S.C.S.)

JAUNDICE (ICTERUS). A condition in which the bile pigments stain the skin, mucous membranes, tissues, and body fluids. It occurs commonly in many disorders and diseases.

Jaundice may be divided into three types: 1. Obstructive jaundice. This form accounts for about 85% of all cases and is seen when there is any obstruction of the bile ducts leading from the liver to the duodenum. (See diagram page 627.) The obstruction may be caused by gallstones within the ducts, inflammation of the duct wall, shutting off the flow of bile by swelling (cholangitis), tumor within the duct walls or causing pressure from without, scar tissue formation, or kinking, or spasm of the bile ducts. 2. Hemolytic jaundice due to excessive destruction of red blood cells with liberation of their pigments. (See **Hemolytic Jaundice**.) 3. Toxic and infectious jaundice. This occurs when the liver cells have been damaged by toxic substances, either chemical, virus, bacterial, or spirochaetal in nature. It is seen in poison from chloroform, phosphorus, and arsenic compounds and may occur with syphilis and several acute infections. It is also seen in Weil's disease, acute infectious **hepatitis**, and catarrhal jaundice.

Jaundice usually appears first as a yellow discoloration of the eyeballs. Gradually, the skin of the entire body becomes yellow, and may itch violently, especially in chronic jaundice. The urine varies from yellow to dark brown. The blood serum is bile-stained and in severe cases the clotting time of the blood is greatly prolonged so that bleeding is hard to control. The cerebrospinal fluid is usually not colored. The stools, when there is complete obstruction and absence of bile in the intestine, are clay-colored. With jaundice there are usually mental depression, headache, and drowsiness. In severe cases there may be stupor or coma.

Other symptoms and signs depend on the disease causing the jaundice. (R.S.M.)

JAW. The assemblage of bones before and behind or above and below the mouth of a **vertebrate.** The lower jaw is movably articulated with the skull to work against the upper jaw for biting and chewing and both bear teeth in most species.

The earliest vertebrate jaws, as exemplified by the **sharks,** consisted entirely of cartilage and were not rigidly attached to the brain case. With the acquisition of an outer covering layer of dermal bones the upper jaw was firmly attached to the cranium and the primitive jaw **cartilages** receded in importance, although they, or the replacement bone superseding them, still served as the jaw joint. In the evolution of the mammals from the mammal-like reptiles a new jaw joint was formed, this time between dermal bones, and the old jaw joint became incorporated in the middle ear where it still functions as the joint between the malleus and incus. (A.W.L.)

JAY. Aves, Passeriformes. Birds **(Aves)** of several species related to the magpies and crows. Most jays live in the northern hemisphere but a few species occur in the Oriental region and northern Africa. They are noisy birds, often brightly colored, and sometimes interesting in habits, but some are too destructive of the eggs and young of other species and too quarrelsome to be desirable.

The blue jay, *Cyanocitta cristata,* is the most widely known of the North American species. More than a

Blue jay, *Cyanocitta cristata.* Blue above, with white bands on the wings. White below. A black band around the neck. Narrow black bars on wings and tail. A crest on the head.

dozen others occur in the western states where the Oregon jay, *Perisoreus obscurus,* and the Canada jay or whiskey jack, *P. canadensis,* are friends of campers. These species visit camps and human habitations freely in search of food. (A.W.L.)

JEJUNUM. That portion of the small intestine which extends from the end of the **duodenum** to the beginning of the **ileum.** The jejunum is about 8' long and continues into the ileum without any line of demarcation. It occupies the upper and left part of the abdomen lying below the stomach and spleen on the left side. Its coils are freely movable and are connected with the posterior abdominal wall by wide folds of membrane **(peritoneum)** called the mesentery, through which run the blood and lymph vessels and nerves supplying it. (R.S.M.)

JELLYFISH. Scyphozoa. Coelenterata. Medusa.

JENNET. Mammalia, Artiodactyla. The female **ass.** (A.W.L.)

JENNY HANIVER. A curiosity resembling a misshapen human figure, made by variously distorting and

decorating a **ray.** Made in the past by New England fishermen and sometimes found among collections of curios. (A.W.L.)

JERBOA. Mammalia, Rodentia. Small jumping animals of Asia and northern Africa. They resemble mice with long tufted tails and very long hind legs. The

Egyptian jerboa. (*N. Y. Zool. Soc.*)

small fore legs are not used in locomotion. The Asiatic jerboas have five toes on the hind feet and the African three. The most common Asiatic species is also called the alagdaga, *Allactaga indica.* (A.W.L.)

JET. The geological material, jet, is not a mineral in the true sense of the word. It is a hard and compact variety of lignite, coal-black in color. It can be easily polished and has been used for the manufacture of cheap ornaments. Jet is found in England, France, Spain, Germany, and the United States. (See also **Jet Propulsion.**) (E.S.C.S.)

JET ENGINE. Technically this term might include any device for propulsion which uses the reaction force obtained from the acceleration of a fluid stream (see **Jet Propulsion**). However, it is proposed to restrict the present treatment to combustion jets, wherein the reaction of a powerful jet of products of combustion expelled rearward from the engine is the source of propulsion. Present-day jet engines meeting this specification are classified as follows:

1. Gas turbine systems or turbo-jets.
2. Resonance ducts.
3. Aero-thermodynamic ducts (Athodyd).

All of these draw in air for combustion from the surrounding atmosphere, and it is in this respect that they differ from rocket motors.

Propulsion by jet action never becomes efficient unless the propelled body is moving at high speed. Airplanes furnish this condition, in some measure, as do also the so-called "flying bombs." In general, surface vehicles do not. Jet engine development therefore was nurtured in the aviation field, receiving great impetus from wartime demands for planes with lighter, more powerful engines, especially those well adapted to the high-speed range.

Gas Turbine Systems. One method of producing jet reaction is to impart a high velocity to the fluid by means of adiabatic expansion. But this necessitates a pressure drop. Since the discharge is back to the same atmosphere from whence the air was initially drawn, a means of compression has to be provided. Briefly, this system has a gas turbine driven compressor. Both centrifugal and turbine type (axial flow) compressors have been used. Between discharge from the compressor and entrance to the turbine, the air receives heat by the direct combustion in it of a suitable quantity of fuel. This increases the volume and temperature at constant pressure. The energy available is then sufficiently high so that the work of compression can be taken out of the gases by the turbine, with enough energy remaining to impart the high final kinetic energy and speed to the jet.

Fig. 1 shows the basic features of the turbo-jet engine. The single rotating shaft mounts the impeller of a centrifugal compressor and the wheel of a single stage impulse gas turbine. Air, sometimes at subzero temperatures, is rammed into the engine cowling by the forward motion of the whole engine. It speedily finds the air in-

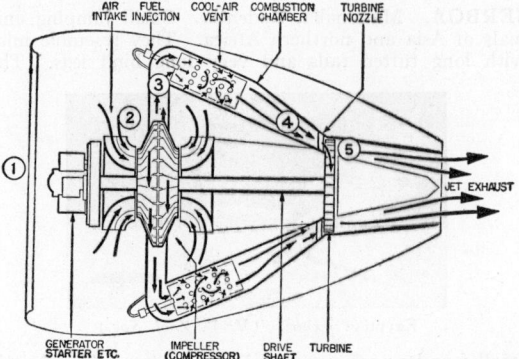

Fig. 1. Principle of the combustion gas turbine jet engine. (*General Electric Co.*)

take of the compressor and quickly is discharged in a compressed condition into the combustion chamber. Here fuel (kerosene) is burned in it in sufficient quantity to raise the temperature of the gases to say 1500° F. at the turbine nozzle. To obtain as complete combustion as possible, and to cool the chamber walls, the air-fuel ratio is temporarily reduced by feeding only part of the air through the burner. The remainder enters the combustion chamber through cooling vents. After partial expansion in the turbine nozzles the gas flows at high speed through the turbine blading, spinning the wheel at many thousand revolutions per minute. All the turbine output is used to drive the compressor; consequently only a portion of the available energy is consumed in this way. The gases emerge from the turbine wheel with considerable velocity and some positive pressure above final exhaust pressure. This is employed by the exhaust nozzle to raise the gases to the final jet speed by adiabatic expansion. It is seen that the compression and combustion process is devoted to preparing a flow of gas, composed of products of combustion and excess air, for adiabatic expansion through a nozzle. The gas turbine is merely incidental, a means of driving the compressor.

Theoretical considerations of energy and of momentum yield equations for thrust and efficiency based on engine speed u, jet speed v, and mass flow m. (See **Jet Propulsion**.) This approach has the virtue of simplicity but fails to include the factors which are responsible for whatever numerical value v may possess. With reference once again to Fig. 1, note that 5 stations along the line of flow of the fluid are selected as State Points in a description of the engine process from the thermodynamic viewpoint. State 1 is the free atmosphere at temperature T_1. The ram of air into the engine cowl is of the nature of a reversed and imperfect expansion of a gas jet. Borrowing an equation from **Jet, Gas**, with suitable subscripts for the case under consideration, the ideal ram pressure

$$P_2 = P_1 \sqrt[z]{\frac{u^2 z}{64 R T_1} + 1}.$$

The relation $T_2/T_1 = (P_2/P_1)^z$, whose use was necessary in formation of the equation for P_2 from the gas jet equation, will also serve to determine the compressor intake temperature, for the actual ram pressure P_2 will be 80% to 90% of the ideal. Using symbols as follows,

$$W_c = \text{ideal compressor work} = \frac{R T_2}{z}\left[\left(\frac{P_3}{P_2}\right) z - 1\right].$$

η_c, η_t = internal efficiencies of compressor, turbine.
R = 53.4 for air, nearly the same for gas turbine products.

$z = (c_p - c_v)/c_p.$
c_p, c_v = specific heats, constant pressure and constant volume.
h = enthalpy.
J = mechanical equivalent of heat.

The remaining steps in determining the magnitude of the final velocity, which call v, for a limiting T_4, are, based on a pound of air:

$$h_3 - h_2 = \frac{W_c}{J \eta_c} = c_p(T_3 - T_2).$$

This defines the temperature T_3 to be expected from a pressure ratio $\dfrac{P_3}{P_2}$. The combustion process which raises the temperature from T_3 to T_4 is essentially constant pressure, making the necessary heat addition $c_p(T_4 - T_3)$, although some adjustment might be made for a drop of pressure between points 3 and 4. Since all the turbine output goes to the compressor, the enthalpy at point 5 can be determined by $h_4 - h_5 = \dfrac{W_c}{J \eta_c} + $ energy required by the auxiliaries. Considering $T_{5'}$, the temperature at exit from turbine nozzle, and T_5, the temperature at exit from turbine blades,

$$h_4 - h_{5'} = \left[\frac{W_c}{J \eta_c} + \text{auxiliary energy}\right] \times \frac{1}{\eta_t} = T_4 c_p - T_{5'} c_p.$$

Thus $T_{5'}$ is fixed, and also the pressure $P_{5'}(= P_5)$ from the polytropic relation $[P_4/P_{5'}]^z = T_4/T_{5'}$. The actual temperature drop from 4 to 5 is $\eta_t(T_4 - T_{5'})$. With P_5 and T_5 the final jet velocity is given by the jet equation (see **Jet, Gas**)

$$v = 8\sqrt{\frac{R T_5}{z}\left[1 - \left(\frac{P_5}{P_4}\right)^z\right]}.$$

To obtain good thermal efficiency the temperature T_4 needs to be as high as conditions permit. The cooling of these engines is therefore a critical feature of their design. Cooling air bled off the compressor is used to air-jacket many of the high-temperature regions. Air circulation through hollow turbine blades and other ingenious features has characterized these designs. Although not impressive in thermal efficiency, these pure jet engines are capable of powering airplanes of higher speed than the propeller type engines whose propeller blades are the first part of the airplane to be affected by compressibility.

A new type of power plant has been developed, having both propeller and jet output. Designed primarily to drive large high-speed military transports and bombers, this gas turbine with propeller is designed for installation in the wings of multi-engined aircraft or in the nose of a single-engine plane. A diagram of it shows that the air rams into the nose of the engine through ducts

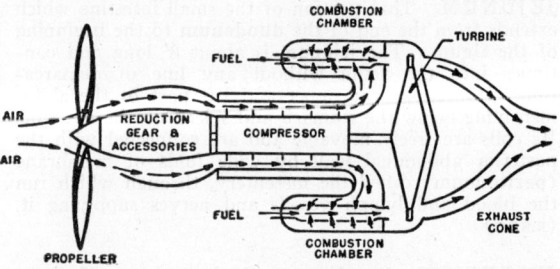

Fig. 2. G-E propeller drive gas turbine. (*General Electric Co.*)

opening forward. This air is compressed by axial flow units in the forward part of the engine and then is forced into combustion chambers. There fuel is injected and burns intensely. This raises the temperature and velocity of the gases, which then, with great energy, strike the buckets of the turbine wheel. The turbine, spinning more than 10,000 times a minute at a temperature over 1500° F., absorbs the *major part* of the available energy in the gases.

The turbine powers the compressor and through reduction gears drives the propeller. Reactive thrust created by the energy remaining in the gases passing through the turbine wheel and discharging rearward is utilized in jet propulsion.

The power generated by these new units already is great and engineers see no basic difficulties in increasing the propulsive output of this type of gas turbine to almost any force visualized as necessary to drive the projected mammoth planes of the future.

Fig. 3. The propeller-jet engine. (*General Electric Co.*)

The range of planes powered by gas turbines of this type will be extensive. At slow speeds and low altitudes, the gas turbine uses more fuel than a reciprocating engine cruising under similar conditions, but when operating at full power the turbine uses less fuel than a conventional engine operating at maximum power. The

turbine functions most efficiently at high altitudes where the air is colder than at lower altitudes.

The speed limits on planes powered by the propeller-jet turbine are unquestionably the compressibility barriers that are reached by all propeller driven planes at speeds slightly exceeding 500 m.p.h.

Planes powered by these gas turbines may not reach the extreme high speeds attained by the pure jet propelled airplanes. However, while this type of gas turbine may not power the fastest planes, it will give ranges greater than may be possible with pure jet propulsion.

No complicated ignition or carburetion or timed injection system is needed on the combustion jet engine, but starting is more of a task as the compressor must be set spinning at high speed in order to initiate the cycle.

Resonance Duct Engine. In 1944 a new type of jet engine was introduced on a war weapon. Without touching on its value as a weapon, the technical features of its propulsion system are to be described. The "buzz bomb," or V-1, to mention two of the names ascribed to this flying bomb, was an unmanned aircraft carrying an explosive charge which was detonated when the bomb crashed at the end of its 100 to 200 mile flight. Gyro stabilizing and other controls which could be pre-set guided it to the target. By far the most interesting thing about this weapon was the jet engine that propelled it. Although possessing poor thermal efficiency, it was extremely light weight, simple, and inexpensive to construct, just the features needed for the expendable nature of its use. The principle of the "intermittent duct engine" is explained by Fig. 4. The duct itself is a long

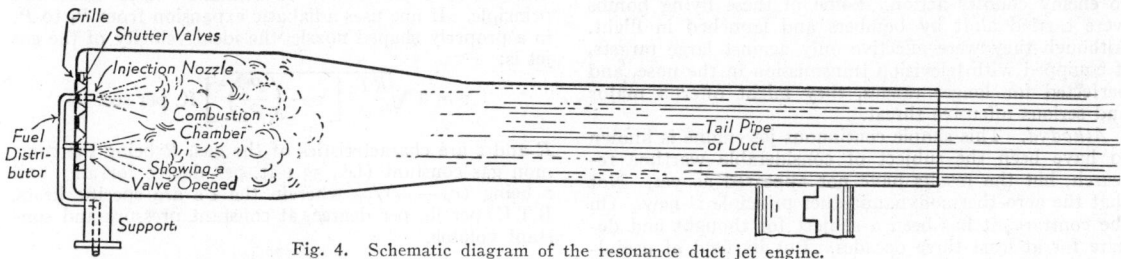

Fig. 4. Schematic diagram of the resonance duct jet engine.

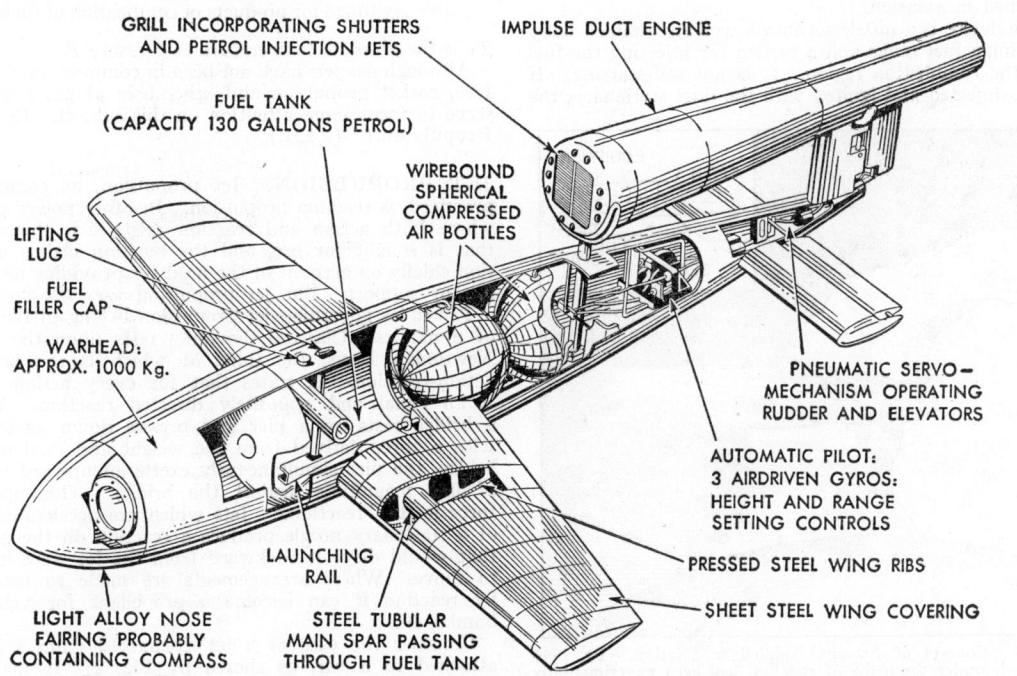

Fig. 5. Flying bomb with jet propulsion. Resonance duct type. (*British Information Service.*)

tube (11 ft.) enlarged at the forward end to serve for the combustion chamber. Across the front is a grill equipped with large numbers of thin spring steel flap valves which will open inward to allow air to enter, but will close when fuel burning in that air raises the combustion chamber pressure. Following each explosion a sound wave and column of gases travel down the duct and a partial vacuum is induced behind the grill. So more air enters, mixes with fuel that is sprayed into the chamber continuously, and explodes. These explosions follow each other in rapid sequence. Rates of 40 cycles per sec. have been mentioned. The ignition is by electric spark. However, after the combustion chamber is hot, ignition is continued by residual flame. The duct must be of the correct length to harmonize the explosion impulse waves with the natural period of oscillation of the grill valves. The source of thrust is, of course, the change in momentum of the gases shooting rearward out of the duct.

The flying bomb, which was able to make something like 300 m.p.h., is shown in Fig. 5, on page 815. Because of the low thermal efficiency, about 1 gal. of gasoline for each mile to be covered is required. This is carried in a fuel tank from which it is forced by compressed air pressure into fuel jets located in the front grill. A big disadvantage of this engine, and one which ultimately prevented its becoming a decisive weapon, was the trouble of getting it started. It had to be projected at about 250 m.p.h. from a long launching track before the resonance action could begin to function. This rendered the launching sites large targets vulnerable to enemy counter action. Some of these flying bombs were carried aloft by bombers and launched in flight. Although they were effective only against large targets, if equipped with television transmission in the nose, and perfected for longer ranges, they might offer a major and serious offensive threat.

Athodyd. This simple continuous jet engine is known to have been the subject of considerable wartime research, but the results have not appeared as yet. Not that the aero-thermodynamic-duct principle is new. On the contrary, it has been a subject for thought and debate for at least three decades. But its field of usefulness seems to lie in speed ranges only now being approached in aviation.

The device is a nozzle set into a well streamlined body containing fuel tanks and a system for injecting the fuel into the combustion region. It is not self-starting. If fuel is injected and ignited with the duct stationary, the exhaust will blow out of the duct in both directions, for there is nothing there to cause it to do differently. But when the duct is in motion at high speed, the air ram into the front end will block the possibility of exhaust there unless it is attempted to burn the fuel too rapidly. The thrust is theoretically

$$F = \frac{\rho u^2}{2} (A_m - A_1)$$

where ρ = mass density of the air.
u = duct speed.
A_m = mouth area.
A_1 = entrance area.

The $(A_m - A_1)$ term must be limited in magnitude because of certain thermodynamic and aerodynamic considerations, and large thrusts must necessarily imply high speeds, which enlarge the magnitude of dynamic pressure $\rho u^2/2$. Power is proportional to the product of thrust and velocity, which makes it proportional to u^3. Thus we find Athodyd to be a device suitable only for high speeds, but its power, insignificant at conventional speeds, rises mightily as those high speeds (thought to be near the speed of sound) are approached. The ideal efficiency is 100%, but whether that can be approached in an actual case depends on whether the duct design permits a decent amount of combustion to occur before the jet carries out of the duct. (F.T.M.)

JET, GAS. As gas is a compressible fluid, the velocity attained in a jet cannot be evaluated from Bernouli's principle. If one uses adiabatic expansion from P_1 to P_2 in a properly shaped nozzle, the ideal velocity of the gas jet is:

$$v = 8 \sqrt{\frac{RT_1}{z} \left[1 - \left(\frac{P_2}{P_1} \right)^z \right]} \text{ ft. per sec.}$$

R and z are characteristics of the gas, R being the common gas constant (i.e., 53.4 ft. per degree for air), and z being $(c_p - c_v)/c_p$ wherein the c's are specific heats, B.T.U. per lb. per degree, at constant pressure and constant volume.

$z = .286$ for air under 1000° F.
$z = .23$ to .28 for products of combustion of fuels.

T_1 is the absolute temperature at pressure P_1.

Although gas jets have not been in common use heretofore, rocket propulsion and other uses of gas jets may serve to focus more attention on this subject. (See **Jet Propulsion**.) (F.T.M.)

JET PROPULSION. Jet propulsion, as commonly practiced, is reaction propulsion. In other power plants where both action and reaction exist, it is the action that is sought for use, and the reaction claims attention chiefly on account of the need for providing anchorage or support. But jet propulsion *uses the reaction.*

Principles governing jet propulsion belong in the field of dynamics. Classical dynamics rests upon the three Newtonian Laws, the third of which is the law of action-reaction. It states that for every action there is an equal and oppositely directed reaction. When a bridge rests on a pier and presses down against it with a force derived from the weight and load of the bridge, the pier simultaneously exerts an upward thrust of like magnitude against the bridge. This upward force is the "reaction." Jets which are accelerated out of a stationary nozzle produce a reaction on the nozzle that would drive it backward from the jet were it free to move. Where arrangements are made to use this jet reaction it can become a propellant for vehicles, bombs, etc.

Consider the case of a jet discharged from a body attached to a car, as shown by Fig. 1. It may be imagined that the source of this jet is a large sky-

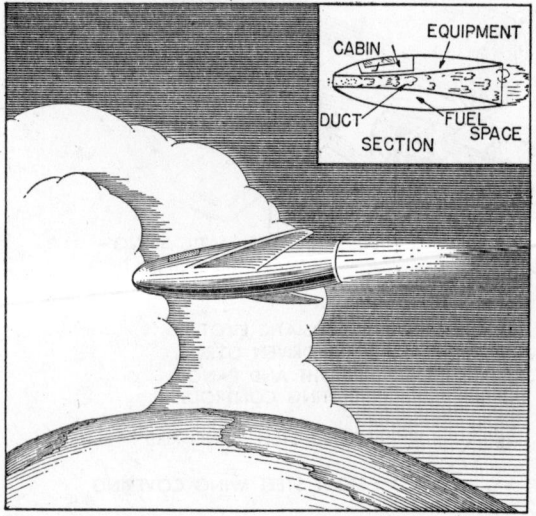

Fig. 6. Concept of Athodyd applied to aviation. By 1946 Athodyd, under the name of ram jet, had been experimentally applied to radio-guided missiles.

rocket. If the car is light weight and equipped with anti-friction bearings, one can readily imagine that it would be accelerated forward speedily upon ignition of

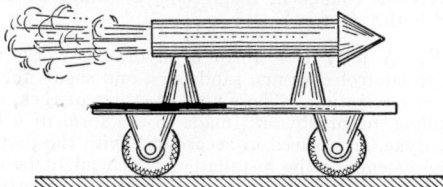

Fig. 1. Simple reaction propulsion.

the rocket. Now many may possess a feeling that this, in some way, is because the jet from the rocket pushes against the air; something in the manner of gas in an engine cylinder pushing against the piston. They may be surprised to learn that the rocket would use its fuel even more efficiently if operated in the absence of an atmosphere—as in the outer regions of space.

When the rocket contents burn, the products are expelled at high velocity and a forward thrust is imparted to the rocket case. If a constant rate of discharge of m slugs per sec. is assumed, and v is the velocity of discharge, relative to the rocket, which takes place through the nozzle mouth at section aa, Fig. 2, then a reaction

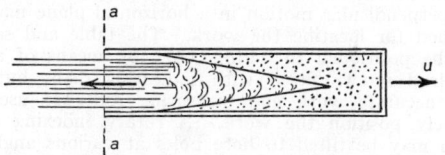

Fig. 2. Thrusting rocket.

force is set up which is caused by the momentum change suffered by m. This force is mv. Due to the rocket's symmetrical shape, the air pressure acting externally on it is balanced save for the area at aa and an equal parallel area at the closed end of the rocket body. Let P_m be the gas pressure at aa, A the area there, and P_a atmospheric pressure. The net static pressure is $A(P_m - P_a)$ acting with mv.

$$F = mv + A(P_m - P_a) \text{ gross rocket thrust.}$$

The net thrust would require air drag to be deducted from this force, also a gravitational component in some cases.

If the nozzle expands the jet to atmospheric pressure $P_m = P_a$ then $F = mv$. This is often cited as the thrust equation of true rockets. The same can be obtained from an energy analysis. The original kinetic energy of mass mt, t being time, and the realized heating value Q that creates the velocity v in that mass are, together, equivalent to the thrust work Fut plus the residual energy existing at the final absolute speed of $v - u$. Written symbolically,

$$\tfrac{1}{2}mtu^2 + \tfrac{1}{2}mtv^2 = Fut + \tfrac{1}{2}m(v - u)^2$$

this becomes, as before,

$$F = mv.$$

The efficiency of propulsion for a fixed jet velocity, v, and variable rocket speed, u, is given by the equation

$$\eta = \frac{2uv}{v^2 + u^2}.$$

Hence it appears that, theoretically at least, the thrust remains constant, but efficiency of use of the energy developed by combustion increases with u until it becomes 100% when $u = v$, thereafter falling off as u is increased above v.

In a jet engine the fuel is carried in the vehicle which the engine propels, but oxygen is secured by taking in air from the surrounding atmosphere. As very high

air-fuel ratios are employed in combustion jet engines, a good approximation, with simplifying features, is to neglect the mass of the fuel and consider that heat only is added to the air passing through the engine.

A much simplified diagram of the engine in Fig. 3 is illustrative of jet propulsion in any fluid which may

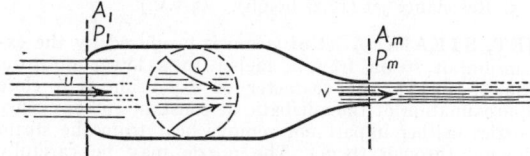

Fig. 3. Jet engine inversion.

somehow or other be accelerated as it passes through the engine. Although, in reality, the engine moves forward through a stationary fluid with a velocity of u, it is sometimes convenient to "invert" the relative velocities and consider the engine at rest in a fluid stream moving at velocity u past it. The region enclosed by the dotted boundary contains whatever means may be employed to accelerate the fluid flow from u to v. For example, this would consist of the compressor, burner, and exhaust nozzle of a turbo-jet airplane engine. Or it would be the fuel injection and combustion apparatus of the Athodyd which is a simple but intriguing jet engine using only air ram for compression.

Jet propulsion is secured as follows from the combustion jet engine. Let m be the mass per sec. of the air involved. As seen in Fig. 3, air rammed into the entrance scoop comes to rest, producing a ram pressure at the entrance of P_1. After fuel heat Q is added to it there is an expansion producing velocity and causing a jet to be expelled from the mouth of the nozzle at velocity v relative to the engine. Let P_m be the gas pressure at exit, A_m the area of the nozzle there, $A_1 =$ area of entrance, and $P_a =$ atmospheric pressure; then the thrust $F = mv - mu + (P_m A_m - P_1 A_1) + P_a(A_1 - A_m)$.

When the engine is in motion through still air the energy it receives in t time is Fut. The residual, or unused, energy is $\tfrac{1}{2}mt(v - u)^2$. Efficiency being useful energy divided by useful energy plus losses,

$$\eta = \frac{Fut}{Fut + \tfrac{1}{2}mt(v - u)}.$$

Monographs by different authors are found to interpret this equation differently, depending on their assumptions regarding the PA terms of the expression (above) for thrust F. Assuming that $P_m = P_1 = P_a$, ideally,

$$\eta = \frac{2(uv - u^2)}{v^2 - u^2}.$$

In the case of Athodyd the engine action is principally an expansion of gases at constant (ram) pressure, making $v = u$ and causing A_m to exceed A_1.

No mechanical means of compression is furnished. If Athodyd has a speed of u, a dynamic pressure increment ΔP can be built up in the Athodyd body by air ram. Fuel is burned so that a velocity v equal to u will exist at the mouth. Since $v = u$ no momentum change is involved and thrust is derived from static pressure. The equation given above for F becomes $F = \Delta P(A_m - A_1)$ when $P_1 = P_a + \Delta P = P_m$ and $v = u$. As residual velocity $(v - u)$ is zero, the efficiency of Athodyd, operating under the assumed conditions, is 100%.

The thrust of the compression jet engine is mv at standstill, and decreases by the term m as velocity is increased. Athodyd has zero thrust until some dynamic pressure $\dfrac{\rho u^2}{2}$ exists.

It has been proposed that marine vessels, in particular submarines, be propelled by the reaction obtained by imparting a high velocity to a quantity of water taken in through a lengthwise duct, then discharged rearward.

Some cases illustrating reaction propulsion are:

1. The skyrocket.
2. War rockets (Bazooka, etc.).
3. Liquid fuel rockets (V-2).
4. Jet engines for airplanes.
5. Resonance jet (Buzz bomb). (F.T.M.)

JET, STEAM. A jet of steam is produced by the expansion of steam from a higher to a lower pressure. Almost always, the character of expansion is a close approximation of the adiabatic at constant entropy since nozzles neither impart nor remove heat from the steam flowing through them. The nozzle may be carefully shaped based on the theory of compressible flows through nozzles, or it may be a crude nozzle such as a small bored hole in a plate. The less perfect the **nozzle**, the lower the steam jet velocity will be and the more uncertain the jet direction. Steam is so light in weight per unit volume and its jet velocity usually so large that the effect of gravity on the path of the jet is negligible, so the axis of the nozzle determines the position of the jet for a considerable distance from the nozzle mouth.

The ideal velocity of a steam jet resulting from an adiabatic expansion may be calculated from the principle of conservation of energy. If 1 lb. of steam in expanding through a nozzle (which it approached at a velocity of V_1) suffers a change of enthalpy of Δh B.T.U., and if conditions of expansion are ideal, namely, no friction, turbulence, over-or-under-expansion, or supersaturation (see **Supersaturated vapor**), the jet velocity $V_2{}^2$ will be:

$$V_2{}^2 = 224\sqrt{\Delta h} + V_1{}^2.$$

This is derived by equating the loss of enthalpy to a gain in kinetic energy. Actual jet velocity approaches closely the ideal.

Although of low density, steam jets may represent great power because of extremely high velocities which may be produced with moderate pressure drops. The power of steam jets is put to use in numerous ways. **Steam turbines** are machines for producing jet power, then absorbing it on blades which, in turn, produce power as a rotating torque at a shaft to which the blades are indirectly attached. Certain refrigerating apparatus uses steam jets for compression, as do also ejectors for evacuating air from regions of high vacuum. Steam jets are the source of power for **injector** pumps and are often used for blowing and cleaning operations. (F.T.M.)

JET, WATER. A jet is a rapidly moving fluid stream launched into space from a **nozzle, orifice,** or other mouthpiece. The velocity of the jet is obtained from a pressure drop in the fluid as it passes the orifice or nozzle. The water jet will attain a theoretical velocity of $\sqrt{2gh}$ ft. per sec. (g = acceleration of gravity) when it is produced by an effective pressure head of h ft. of water. The actual jet velocity is 1 or 2% less than the theoretical.

A jet of water discharged horizontally follows a downwardly curved path on account of the attraction of gravity. The path taken by a freely spouting jet, initially discharged in a horizontal direction under head h is closely approximated by the parabola $x^2 = 4hy$, in which y is the drop of the jet below the level of the nozzle at a distance x horizontally out from the nozzle.

A water jet of A sq. ft. cross-sectional area moving at v ft. per sec. contains energy at a rate equivalent to $566\ Av^3$ horsepower. To obtain this power from the jet a machine would have to bring the water to rest without friction or turbulence. Impulse **hydraulic turbines** recover about 70% of the available power by deflecting the jet with blades or buckets mounted on the rim of a rotating disk. The force acting on a blade

which deviates a jet of water through a velocity change of Δv is $\dfrac{62.4\Delta v}{g}$ lbs. per cu. ft. of water per sec. flowing. Δv may be a change in magnitude, a change in direction, or both. (F.T.M.)

JETTY. A jetty is a form of hydraulic works employed to control currents, sand bars, and shoals for the benefit of navigation. These are structures of rock, concrete, piling, or brushwork, made in the form of a long narrow dyke, and placed in accordance with the particular requirements of the installation which might be control of silt through maintenance of current, or control of waves or currents which might tend to produce sand bars. Sometimes jetties are in pairs, one on each side of the current, to confine and direct it, though single jetties are also useful in certain locations. The most famous jetties of the United States are those of the lower Mississippi, which serve to maintain the ship channel through the Mississippi Delta. (F.T.M.)

JEWFISH. Pisces, Teleostei. The giant **sea bass,** *Stereolepis gigas*, a game fish of the California coast which reaches a weight of 500 lbs. The name is also applied to other fish. (A.W.L.)

JIG BORER. A vertical-spindle boring machine for accurately boring and locating holes. It is similar in principle to an upright drill press, but is equipped with a table and a saddle similar to a milling machine so that perpendicular motion in a horizontal plane may be obtained for locating the work. The table and saddle may be positioned to within .001" by means of accurate lead screws fitted with indicating dials, but for more accurate work end measuring rods are used to precisely position the work. A rotary indexing table which may be tilted to bore holes at various angles is also available. (See **Boring.**) (H.C.H.)

JIGGER. Chigger.

JIGGING. Classification.

JIGS AND FIXTURES. Jigs and fixtures are as important in modern manufacturing and mass production processes as cutting tools and machine tools. A jig is a device used in the manufacture of interchangeable parts to hold and locate the work and to guide the cutting tool. A fixture is a similar device, but is used only for holding and locating the work.

The figure shows several views of a jig for holding and locating a bushing W while drilling holes A and Z.

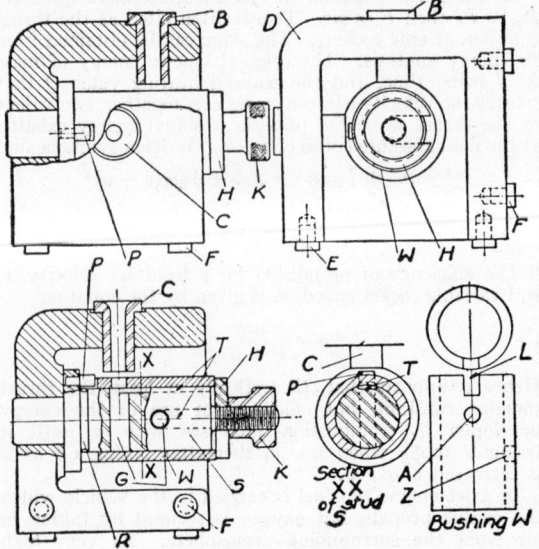

The jig has a cast iron frame D into which a hardened steel stud S, for supporting the work, is pressed. S has a large integral collar, with a chip recess R at its juncture with the body of the stud, into which a locating pin P is pressed. P fits the slot L in the work and serves to locate L with respect to A and Z. The work is held in place during the drilling operation by a horseshoe washer H and a knurled nut K. K is slightly smaller than the bore of W so that the work may be removed by giving K a fraction of a turn, taking off H, and sliding the work over K. The drills are guided by drill bushings B and C which are pressed into D. Drill bushings are made of tool or high-carbon steel, glasshard, and lapped for a snug fit with the drill. They have a wide bell-mouth at the top so as to facilitate entry of the drill. The jig is supported on four feet E and F on each of two sides; the feet are made of casehardened common steel and are pressed into holes in D.

In operation, the operator places a blank W on S, slips H in place and clamps W with K. He then places the jig on the drill press table resting on feet F and brings the drill down. If the jig is not perfectly aligned with the drill, the contact of the drill with the bell-mouthed hole in bushing C will move the jig sufficiently to permit entry. Hole A is drilled, the drill is withdrawn, and the jig is turned over so that it rests on feet E, and the drilling procedure is repeated for hole Z. The operator then removes the work and the cycle is complete. The stud S is provided with two grooves T, which permit easy removal of W even though the drilled holes have burrs resulting from the operation. The clearance holes G in the stud permit the drill to go completely through the wall of the bushing and provide a place for chips.

It is often more convenient and economical to purchase commercially available parts such as drill bushings, assembled clamping devices, skeleton drill jig frames, jig feet, and similar parts, and fabricate drill jigs by assembling these parts, rather than to completely build the entire tool. Several commercial jigs are available at present.

File jigs are used as templates for filing parts to shape and size. In its usual form, a file jig consists of a hardened steel plate with a beveled gauging edge, with some means for attaching the jig to the work. The operator locates the jig on the piece to be filed, clamps both in a vise, and files the part to the gauging edge. File jigs are employed for parts difficult to mill, and also in cases where the quantity of pieces required does not warrant the expense of a special milling cutter.

Milling fixtures are used for locating and holding work and are similar to drill jigs, but have no media for guiding the cutting tool. Fixtures are usually fastened rigidly to the table of a milling or boring machine; adjustment for position is generally made by moving the machine table.

Comparatively inexpensive milling fixtures can be made by fitting a milling machine vise with false or substitute jaw plates for locating and holding the work. The jaw plates supplied with the vise are removed, the false jaws substituted, and held in place with the jaw screws. When more than one part is to be held at the same time, equalizing or self-adjusting clamps are fitted to the false jaws. (H.C.H.)

JIMSON WEED. Potato Family.

JOINT, ANATOMICAL.
In anatomy, a joint is a connection between rigid units of the **skeleton**. Joints are of two types, movable and fixed. The latter (synarthroses) provide for the formation of firm skeletal structures where more than one rigid part is involved, and movable joints (diarthroses) enable the separate divisions of the body and jointed appendages to act together as systems of levers, moved by muscles.

In the **arthropods** separate hard parts of the exoskeleton are known as sclerites. Since they are merely plates developed in a continuous tissue the flexible regions of the integument between them serve as connections. These joints are known as sutures and vary from well-marked and freely movable to completely obliterated unions. Where the principal sclerites of a segment unite along the sides of the body and between the sclerites of adjacent segments the unions are called conjunctivae. At the attachment of the segments of legs to each other and the articulation of legs and wings with the body the same principle prevails. The union of an appendage with the body is sometimes supplemented by small articular sclerites.

In the fixed joints of **vertebrates** the bones are united by tough **connective tissue**, including many white fibers which continue into the substance of the bone, or by a zone of cartilage (synchondrosis).

Movable joints of vertebrates develop in continuous tissue between the ends of the bones. The tissue splits to form a closed joint cavity filled with synovial fluid and surrounded by a connective tissue capsule. In this capsule a fibrous layer continues into the periosteum covering the bones. It is thickened in places to form the strong ligaments that bind the ends of the bones together. Folds or plates of fibrous tissue project into the cavity, forming articular disks and menisci, as in the knee joint. The entire cavity is lined with a looser layer of the capsule except at the ends of the bones, where articular cartilages are exposed to the **synovial fluid**. (See **Joint, Geological.**) (A.W.L.)

JOINT EFFICIENCY.
This term refers to a strength ratio in connection with either riveted or welded joints. The joint efficiency is defined as the ratio of the strength of a section of the joint to the strength of an analogous section of solid plate. The joint efficiency, in pressure vessel design, usually requires an increase in the theoretical thickness of the plate which would otherwise be required for a seamless vessel. To illustrate, if the theoretical thickness of a seamless vessel, for a given set of conditions, is $1''$, the same vessel with a welded joint having an efficiency of 80% would require a wall thickness of $1\frac{1}{4}''$. Welded joint efficiencies vary from 55 to 95%; riveted joint efficiencies from 45 to 95%. (H.C.H.)

JOINT, GEOLOGICAL.
In geology, a joint is a fracture in a rock with no apparent relative displacement as in the case of a **fault**. Fractures are exceedingly common in all types of rocks and are frequently arranged in definite relation to each other, such as to produce joint systems more or less constant over considerable areas. Depending on the supposed cause of a given joint system the fractures may be described as tension joints, or compressional joints. In **igneous rocks** a set of joints may be developed by contraction during their period of cooling and solidification within the earth's crust. Igneous rocks, after complete solidification, may later have sets of tension or compression joints superimposed on the first set by the relief of pressure, due to the **erosion** of the overlying formation, or to insolation or frost action after the removal of the overlying formations. Joints are of great importance in quarrying and all operations which require the removal of bed rock. Well-jointed rocks are relatively easily taken out and split into smaller blocks for various constructional purposes. (See **Joint, Anatomical.**) (R.M.F.)

JOINTER. Woodworking.

JOINTWORM.
Insecta, Hymenoptera. Small **insects** whose larvae work in the stems of grains and grasses, sometimes causing swelling at the joints or the forma-

tion of growths. Their work may weaken the stem and cause it to break. (A.W.L.)

JOIST. Floor Beam.

JOLY SCREEN. Bench Photometers.

JOMINY TEST. Hardenability of Steel.

JOULE. The M.K.S. absolute unit of **work** and **energy**, equal to one meter-newton, or to 10 ergs. It is roughly ¾ of a **foot-pound.** The "international joule" is defined in terms of the international **ampere** and the international **ohm.** (See **M.K.S. System; Watt.**) (L.D.W.)

JOULE'S LAW. The law commonly referred to by this name expresses the quantity of heat generated by a steady **electric current** as proportional to the **resistance** of the conductor in which the heat is generated, to the square of the current, and to the time of its duration: $H = KRI^2t$. If the resistance is in ohms, the current in amperes, the time in seconds, and the heat in calories, the constant K has the value $0.2388 \frac{calories}{joule}$. For example, an electric grill of 22 ohms resistance, operated at 110 volts, carries 5 amperes. In 10 minutes, therefore, the quantity of heat generated is $0.2388 \frac{calories}{joule}$ $\times$ 22 ohms $\times$ 25 amperes^2 $\times$ 600 sec. $= 78,800$ calories; this amount would heat a gallon of water about $37°$ F. In electric power distribution systems much energy is wasted through heating of conductors, which, because of the form of the above equation, is often called the "RI^2 loss."

Another, although less familiar, law due to Joule states that in an ideal or perfect gas (see **Kinetic Theory** and **Ideal Gas Law**), the "internal energy" does not change during any process such as compression or expansion unless there is a change in temperature. That this statement is not strictly true of actual gases was demonstrated by Joule himself (see **Joule-Thomson Effect**). (L.D.W.)

JOULE-THOMSON EFFECT. A result observed in the so-called "porous plug experiment," performed by Joule and Thomson (Lord Kelvin) in 1852. Some years earlier (1844) Joule had tried liberating a compressed gas into a vacuum. He sought for a possible cooling of the gas as a whole as the result of work against intermolecular forces, but found none.

This was probably because his thermometer was not sufficiently sensitive. With Kelvin he later arranged an apparatus to pump gas slowly and steadily through a plug of cotton, carefully insulated against thermal conduction. The gas, before entering the plug, was maintained at strictly constant pressure and temperature, the energy being furnished entirely by the pump and not at the expense of the internal energy of the gas. Hence, unless there were other energy transformations than that involved in working against the constant (atmospheric) pressure beyond the plug, the gas merely served as a means of conveying energy from the pump to the outer atmosphere and should not change temperature at all, even though it expanded during its passage through the plug.

But in most cases there was a slight fall of temperature, proportional to the pressure difference on the two sides of the plug; with hydrogen at ordinary temperatures there was a slight rise of temperature. The cooling is taken to mean that expansion of a gas usually involves work against intermolecular attraction. Why hydrogen shows the opposite effect is not so clear; it is possibly due to an intermolecular repulsion. At higher temperatures, hydrogen behaves like other gases, while at suffi-

ciently low temperatures and high pressures other gases behave like hydrogen. The data obtained from these experiments have served to give the exact relation between the standard gas thermometer scale and the Kelvin thermodynamic scale. (See **Temperature Scales.**) (L.D.W.)

JOURNAL BEARING. Bearings.

JULIAN DAY. In making calculations involving long intervals of time the use of the **calendar** date and hours, minutes, and seconds introduces both confusion and ambiguity. To avoid these difficulties astronomers have adopted a system of chronological reckoning using merely the mean solar day as a unit. According to a system first proposed in 1582, at the time of the adoption of the Gregorian Calendar, a date is expressed as the number of days elapsed since the beginning of an arbitrary "Julian Era," January 1, 4713 B.C. On this system of reckoning the use of hours, minutes, and seconds is eliminated by expressing the instant of occurrence of an event as a decimal part of a day, the day being assumed to begin when the mean sun is at upper culmination (mean noon) at Greenwich, England. On this system of reckoning an event which occurred at 2 hr. 43 min. 34.6 sec. P.M., eastern standard time, on January 24, 1937, would be recorded as occurring at J. D. (Julian Day) 2,428,558.32193. The Nautical Almanacs of the various governments give tables for converting any date to the proper Julian day number. (W.K.G.)

JUMPING HARE. Mammalia, Rodentia. A burrowing animal, *Pedetes caffer*, of the South African deserts. Its fore quarters are much like those of rabbits and the ears are long, but the tail is long and bushy like that of a squirrel and the hind legs are much larger proportionately than those of rabbits. It runs on the hind legs alone. (A.W.L.)

JUMPING MOUSE. Mammalia, Rodentia. Small mouse-like **rodents** with very large hind legs and long tail. They differ from the kangaroo mice in the absence of cheek pouches. Several genera, one Asiatic. Related to the jerboas of the Old World. (A.W.L.)

JUMPING PLANT LOUSE. Insecta, Homoptera. A small sucking **insect** which lives on plants. Family Chermidae. The numerous species include several of economic importance, among them the pear-tree psylla. Spraying with kerosene emulsion or nicotine sulfate is recommended as a check for this insect and the adults may be destroyed in winter by scraping off and burning the rough outer bark of the trees. The cylindrical galls on the leaves of hackberry trees are caused by another species. (A.W.L.)

JUMPING SHREW. Mammalia, Insectivora. Small African animals of several species, all with prolonged snouts and large hind legs. (A.W.L.)

JUNCO. Aves, Passeriformes. *Junco.* Small North American birds (**Aves**) of several species, mostly colored

Junco. *Junco hyemalis.* Slate gray above and on the throat and breast. Belly white. Outer tail feathers white.

in shades of gray and white. They breed chiefly in colder regions and in higher altitudes, migrating to the valleys and to more southern latitudes, although they thrive even in cold winters. (A.W.L.)

JUNE BUG. Insecta. A common name for the **May beetles,** large brown scarabaeid beetles whose larvae are the common white grubs of the garden. Applied along Lake Erie to the May flies. (A.W.L.)

JUNGLE FOWL. Aves, Galliformes. Wild game birds (**Aves**) of India and other parts of the Oriental region. One species, the red jungle fowl, is ancestral to the domestic game cock and resembles it in color. The jungle fowls are supposed also to be remotely ancestral to all domestic fowls. (A.W.L.)

JUPITER. (See tables of planetary data, page 1109.) Jupiter, the "giant planet," is the fifth of the major **planets** in order of distance from the sun. Whether we consider bulk or mass, Jupiter is truly a giant planet, for it is larger than all of the other planets combined in both of these characteristics, having a volume more than 1300 times that of the earth and a mass nearly 317 times the earth.

In spite of its great distance from the sun, the huge surface area and high **albedo** reflect a large amount of sunlight, with the result that the planet appears as one of the brightest objects in the night sky, being exceeded only by the **moon** and **Venus.** In a telescope Jupiter is a most interesting object, showing a distinct flattening at the poles of rotation, and reddish bands parallel to the planet's **equator.** With a large **telescope** and moderately high **magnifying power** a wealth of detail may be seen on the surface. Most of these markings are only semipermanent and drift about over the surface to a considerable extent.

The rotation period of the planet has been determined from these surface features and found to be the most rapid of all of the major planets, completing one rotation in slightly less than 10 hours. The rotation period varies with **latitude** on the planet, being the most rapid at the equator and diminishing toward the poles.

The semipermanent character of the surface features, the variation in rotation period in different latitudes, the high reflecting power, and the low mean density (0.242 that of the earth) indicate that the planet probably has a relatively small solid core, surrounded by a very thick layer of **atmosphere.** The characteristics of this atmosphere have been the subject of a large amount of research, particularly within recent years. The modern results all indicate that the atmosphere consists largely of ammonia and methane. The temperature of the surface of the planet is too low to be accurately measured, but theoretical studies, based on the distance of the planet from the sun and its reflecting power, indicate that the temperature is in the vicinity of 133° K. (−207° F.). At temperatures of this value and under the calculated pressure of the atmosphere the ammonia gas would be in unstable condition and would be continually condensing and evaporating. This leads to the conclusion that the semipermanent markings seen on Jupiter may well be clouds of droplets of ammonia.

Jupiter has 11 known **satellites,** four of which are easily visible with a pair of field glasses. The watching of the constantly changing apparent positions of these objects as they revolve about Jupiter forms an interesting program for those possessing moderate telescopic equipment. These four satellites were probably the first celestial objects ever "discovored," having been found by Galileo shortly after he completed his first telescope and applied it to astronomical observations in 1610. In size and mass the so-called Galilean satellites are comparable with our own moon, the inner two having diameters very close to that of our own moon while the outer two are half as large again. The planes of the orbits of these inner satellites are so nearly in the plane of the orbit of Jupiter that they pass within the shadow of the planet or cast a shadow on the surface of the planet at practically every revolution. Observations of the times of **eclipse or occultation** of these Galilean satellites led to the first determination of the finite velocity of light. The other seven satellites of Jupiter are too small and faint to be observed except with the largest telescopes, and even with such instruments the outer four can only be observed photographically. Their motions are subject to tremendous **perturbations** by the inner satellites and the analysis of their motions forms a complicated problem in celestial mechanics which has never been completely solved. The three outer satellites revolve about the planet in **retrograde** sense. (W.K.G.)

JUPURA. Kinkajou.

JURARA. Reptilia, Testudinata. The giant **tortoise,** *Podochemis expansa,* of the Amazon River basin. (A.W.L.)

JURASSIC PERIOD. A major subdivision of the **Mesozoic Era** of the geologic time-scale. Type locality, the Jura Mountains, Switzerland. The formations of this system were first studied in the south of England by William Smith, the father of **stratigraphy,** and the period was named by A. Brongniart in 1829. The Jurassic period began approximately 150,000,000 years ago and lasted for 40,000,000 years. In North America the formations of this system are best exposed on the Pacific Coast, where they occur both in the Rocky Mountain and Pacific **geosynclines.** No upper Jurassic strata are known to occur in eastern North America. Aeolian "Red Beds" of Early and Middle Jurassic age occur in the western interior and are especially well exposed in the Colorado Plateau. Marine Jurassic formations occur in the Arctic, also Africa, South America, Australia, New Zealand, Asia, the Himalayas, and Japan. The principal economic products of Jurassic age are: **gold** (Sierra Nevada), **coal,** and lithographic **limestone** from Solenhofen, Bavaria. Plant life during this period was essentially like that of the Triassic. Among the marine invertebrates the **pelecypods** and **cephalopods** are the most important fossils. Sharks and the modern type of fishes were abundant. The complete adaptive or radial evolution of the reptiles (Saurians) during this period is proved by the fossil skeletons of Ichthyosaurs (fish lizards), Plesiosaurs (marine lizards), Teleosaurus (ancestral crocodile), Pterosaurs (flying reptiles) (see **Fossil Reptiles**); and a number of terrestrial herbivorous and carnivorous **dinosaurs,** such as Diplodocus, Stegosaurus, Ceratosaurus, and Allosaurus. Perhaps the most famous fossil of this or any other geologic period is Archaeopteryx, the "missing link" between the reptiles and the birds. Severe crustal deformations occurred near the close of the Jurassic, especially in the Cordilleran region, with the birth of the Sierra Nevada and the Cascade Mountains. These mountain-building movements, which typify the close of the Jurassic Period in North America, are referred to as the Sierra Nevada Revolution. (R.M.F.)

JUTE. Jute is a fiber obtained from *Corchorus capsularis* and *Corchorus olitorius* (Tiliaceae), largely grown in India. The plants are woody, sparsely branched annuals growing 10–12′ in height, with ovate **leaves** and yellowish-white flowers. The phloem fibers are developed in the inner bark. In India the plant is grown mainly in the rich soil of river valleys, and cultivated by human labor. Harvesting, also by hand, is done 4 or 5 months after planting. The plants are pulled up, the roots and tops removed, and the stems tied in bunches and immersed in water for 2 or 3 weeks, during which time "retting" occurs, so that the fibers separate more or less easily from the remaining tissues. The stalks, still in water, are then pounded with mallets, rinsed thoroughly, and wrung until the non-fibrous material is removed. After this the fibers are hung up to dry. The fibers are yellowish-white and lustrous, soft and without great strength. It is the cheapest textile fiber in use today and is used in immense quantities, cotton alone outranking it. The chief use is in the manufacture of burlap, from which are made sacks used in packing potatoes, sugar, grain, etc. Burlap is also used as a backing for linoleum. Cheap brown twine is often made from jute. The principal drawbacks to an even wider use of jute are its brittle nature, lack of durability, and lack of resistance to moisture. Nevertheless, about ¾ of a billion pounds are used annually in the United States. (R.M.W.)

JUVENILE WATER. Geyser.

K

K SERIES. X-Ray Spectra.

KAGU. Aves, Gruiformes. A bird (**Aves**), *Rhinochetus jubatus,* of the island of New Caledonia, related to the sun bittern. It is about the size of a domestic fowl with longer legs and beak and a long drooping crest. (A.W.L.)

KAGUAN. Flying Lemur.

KAKA. Parrot.

KAKAPO. Parrot.

KAKAR. Muntjac.

KALIUM. Potassium.

KALONG. Mammalia, Chiroptera. The Malayan fox bat, *Pteropus,* largest of all bats, with a wing spread of 5′. (A.W.L.)

KAME. Irregularly shaped mounds and depressions associated with **terminal moraines.** A kame topography is usually the result of a rather mixed set of glacial conditions, including both stratified and unstratified **drift,** and frequent **kettle holes.** One peculiarity of the term is that it is never used for a single mound or depression but rather to designate the character and origin of the general kame type of topography found only in glaciated regions. (R.M.F.)

KANGAROO. Mammalia, Marsupialia. Pouched mammals of Australia and adjacent islands. The typical kangaroos, *Macropus,* have large hind legs and a strong tail. They sit upright and move by springy leaps, not

Kangaroo.

touching the front feet to the ground. The largest kangaroos reach a height of more than 6′ and a weight of 200 lbs.

Moderate and small-sized members of the group which resemble the true kangaroos in form are called wallabies. The rock wallabies, *Petrogale* and *Peradoreas,* differ in the slender tufted tail and the hare wallabies, *Lagorchestes,* are small animals with an evenly furred tail.

Tree kangaroos, *Dendrolagus,* have the hind legs only moderately large and the tail long and evenly furred. They are arboreal, as the name suggests.

The Australian rat kangaroos and the Tasmanian jerboa kangaroo resemble the animals for which they

are named, although one of the rat kangaroos has a bushy tail.

All of these animals eat vegetation and some cause severe damage to crops. (A.W.L.)

KANGAROO RAT. Mammalia, Rodentia. Burrowing **rodents** of the western United States with long tufted tail and long hind legs. Related to the pocket mice and kangaroo mice. Several genera. (A.W.L.)

KANKA. A local Indian term for stone, which has been applied by some geologists to **concretions** of **calcium** carbonate that occur in otherwise unconsolidated sediments or **alluvium.** (R.M.F.)

KAOLINITE, KAOLIN. Kaolinite is the most common of a group of hydrous **silicates** of **aluminum** which result from the breaking down by weathering of mineral aluminum silicates such as the **feldspars, nephelite,** etc. Kaolinite when pure corresponds to the formula $H_4Al_2Si_2O_9$, and occurs in white, clay-like masses. Impurities may cause various colors or tints. Microscopic study shows kaolinite to be **crystalline** and **monoclinic;** it is also found, very rarely, in **hexagonal** scales. It has a perfect basal **cleavage;** is flexible but not elastic; hardness, 2–2.5; specific gravity, 2.6–2.63; luster, pearly to dull; color, white when pure, as described above, but may be yellow, red, blue, or brown; translucent to opaque. Kaolinite is a mineral of widespread occurrence, well distributed throughout the world. The finest kaolinite locality in Europe is said to be in France, from whence the clay is obtained for porcelain ware. Cornwall and Devonshire in England supply large quantities of this mineral. In the United States Pennsylvania, Virginia, Colorado, Georgia, and South Carolina contain deposits of kaolinite. The word kaolin or kaolinite is said to be a corruption of a Chinese word *kauling,* the name of a locality where this mineral is found. Kaolinite is very important commercially in the manufacture of china and pottery. (E.S.C.S.)

KAPOK. This is a downy substance obtained from the inside of the seed-pods of several species of trees known as silk-cotton trees. Its principal use is in pillows and mattresses and some types of life preservers. Java produces a large part of the marketed crop. (R.M.W.)

KARAKUL. The short, tightly curled wool of lambs of certain Asiatic breeds of domestic sheep. The lambs are killed very young to secure the most curly wool. The pelts are known as astrakhan or karakul, the latter term applying especially to those of better quality. (A.W.L.)

KÁRMÁN VORTICES. Within a certain range of **Reynolds number** the vortices in the wake of a **bluff body** are arranged in a definite pattern, as shown in the figure. This pattern would be visible when the body

was moved through fluid at rest. If the fluid moves past a stationary body, the observer must move with the fluid in order to see these vortices. The vortices persist some distance downstream of the body, being finally dissipated by internal friction. At higher Reynolds numbers they disappear and the flow becomes completely irregular.

Although others had previously observed and studied the vortices, it remained for Th. Von Kármán to explain them as the only stable arrangement the vortices could assume within a limited range of Reynolds number. (F.T.M.)

KARYOLYMPH. Cell.

KARYOPSIS. Grass Family.

KARYOSOME. Cell.

KATABATIC WINDS. Cold-air drainage downhill toward lower terrain. In desert ravines katabatic winds locally reach high velocities. (P.E.K.)

KATABOLISM. Metabolism.

KATACLASTIC. Cataclastic.

KATAMORPHISM. Anamorphism.

KATER'S PENDULUM. Kater devised a number of rigid **pendulums** for comparing the accelerations of gravity at different places on the earth. Any pendulum of fixed length, carried from place to place, will swing in periods inversely proportional to the square root of the value of this acceleration, g, at the respective stations. Therefore if the pendulum has been timed at a station at which the value of g is accurately known, its period at any field station gives at once the value of g at that station.

Kater's best known pendulum is reversible, being provided with two knife edges, facing each other, and carefully adjusted to be at conjugate points with respect to each other. It follows that when such a pendulum is accurately timed, and the distance between the two knife edges accurately measured, the value of g can be calculated by using the formula for the period T of an ideal simple pendulum: $T = 2\pi\sqrt{\dfrac{l}{g}}$, or $g = \dfrac{4\pi^2 l}{T^2}$. When this is applied to the Kater's pendulum, l is taken as the distance between the knife edges. Such a pendulum can thus be used for absolute gravity measurements. (See **Gravitation and Gravity.**) (L.D.W.)

KATYDID. Insecta, Orthoptera. Large winged **insects** with long hind legs formed for jumping and very long slender antennae. They belong to the family of long-horned **grasshoppers**. The true katydid is found throughout the United States east of the Rockies and sings normally in three syllables which have been interpreted as ka-ty-did. Other insects of a different subfamily are commonly called katydids because of their similar appearance and habits. All of these insects are bright green and have leaf-like wings. A pink form of the katydid occasionally appears. (A.W.L.)

KAURI GUM. Resins.

KEA. Parrot.

KEBER'S ORGAN. Paired glandular organs of **bivalve** mollusks which serve as accessory organs of excretion. They are associated with the wall of the **pericardial cavity** or of the **auricles**. (A.W.L.)

KEEP-ALIVE. Mercury-Arc Tube.

KELP. Algae.

KELVIN SCALE. Thermometry.

KENNELLY-HEAVISIDE LAYER. Ionosphere.

KENOTRON. This is a high-vacuum diode rectifier **tube.** The tube contains a hot **cathode**, either filament or indirectly heated type, and an **anode.** Since electrons are emitted by the cathode and may be drawn to the anode when the latter is positive but the reverse process cannot take place, the tube will pass current in only one direction, hence serves as a rectifier. These tubes are extensively used for rectifying high-voltage alternating currents to produce high-voltage direct currents. They are used for the rectifiers in radio **receivers** as the noise introduced by gas-filled rectifiers would be objectionable. They are also used at high voltages where gas-filled tubes are not suitable because of **arc-back** difficulties. At intermediate voltages (up to a few thousand) the gas-filled rectifier is more generally used as it is more efficient. (L.R.Q.)

KEPLERIAN LAWS OF PLANETARY MOTION. After years of labor in attempting to develop a theory of planetary motion which would satisfy the accumulation of planetary positions determined by Tycho Brahe, Johann **Kepler** decided to abandon the superstition that the circle is the only perfect curve and hence must be followed by the planets. In the early part of the 17th century he announced three fundamental laws of planetary motion which satisfied Tycho's observations. These laws may be stated as follows:

1. Each planet moves in an ellipse with the sun at one focus.

2. The radius vector of each planet passes over equal areas in equal intervals of time. (The law of areas.)

3. The square of the period of revolution of a planet about the sun is proportional to the cube of the mean distance of the planet from the sun.

When first published, the laws were without theoretical foundation, being empirically derived from observational data. They created a tremendous stir and were declared heretical since they abandoned the circle as the only possible path for a planet. About 50 years later **Newton** was able to show that the laws are a direct consequence of motion under the action of the **law of universal gravitation.** It has also been shown that all types of **orbital motion** are characterized by these Keplerian laws.

The harmonic law as first stated by Kepler was independent of the masses of the bodies involved, but Newton was able to show that the masses are actually involved and obtained a more rigorous expression for it. Assume M_1 to be the mass of one body, M_2 to be the mass of the other, R to be the mean distance between the two bodies, P the sidereal period of revolution of one about the other, and K a numerical constant involving the constant of universal gravitation and we have:

$$K(M_1 + M_2) = \frac{R^3}{P^2}.$$ The masses of the planets are all inappreciable in comparison with the mass of the sun, and accordingly the expression $K(M_1 + M_2)$ is virtually a constant when considering the motions of the planets about the sun. In cases such as **binary star** orbits, and motions of **satellites** about primaries, the rigorous expression of the harmonic law is of immense value in the determination of masses of celestial objects.

While the development of the theory of relativity has necessitated slight modifications in the original laws of planetary motion, nevertheless, together with the law of universal gravitation, they form the basis upon which the entire structure of celestial mechanics and related fields rests. (W.K.G.)

KERATIN. A chemically complex material of which horns, nails, claws, hoofs, and the scales of reptiles, birds, and mammals are formed. Hair and feathers also contain much keratin. It is present in the external layers of the skin, where it develops by the transformation of clear granules of keratohyalin of lower layers. (A.W.L.)

KERATOPHYRE. The name given by Gümbel, in 1874, to felsitic and porphyritic rocks resembling **hornfels**, and which occur in the Bavarian Alps. Later the use of the term was restricted to **porphyries** and **porphyrites** but differing from them by the abundance of **anorthoclase** instead of either **orthoclase** or the soda-lime feldspars. The term has also been restricted to rocks of this type which are pre-**Tertiary** in age. (R.M.F.)

KERATOSA. Demospongiae.

KERNITE. Colemanite.

KEROGEN. Oil Shale.

KEROSENE. Hydrocarbons.

KERR ELECTRO-OPTICAL EFFECT. Electric and Magnetic Double Refraction.

KERR MAGNETO-OPTICAL EFFECT. Magneto-Optical Rotation.

KESTREL. Aves, Falconiformes. A **falcon** of Europe and Asia. The common species is also called the windhover and a smaller related species is known as the lesser kestrel. (A.W.L.)

KETENES. Aldehydes, Ketones, and Related Compounds.

KETO-FORM. Tautomerism.

KETONES. Aldehydes, Ketones, and Related Compounds.

KETTLE HOLE. During the melting stages of the continental glaciers of the **Pleistocene Period** large masses of ice were imbedded in the stratified **drift.** The melting of these huge ice blocks resulted in the slumping of the loose material to form well-defined and steep-sided depressions. These depressions may be as much as 100′ deep and a mile or more in diameter, often containing small lakes, whose surfaces represent the level of the ground water in the surrounding glacial sediments. (R.M.F.)

KEUPER. Copper-bearing shales. (See **Triassic.**)

KEY. A machine element for preventing relative rotation. The sunk key is the commonest form, and must be carefully fitted to prevent rotation in the keyway, and consequent crushing or deformation. The Pratt and Whitney, or drop-seat key is applicable to conditions where the hub must move on the shaft and key, but the keyway must be end-milled. The Woodruff key fits in a semi-circular recess in the shaft, and automatically adjusts itself in instances where both the hub and shaft are tapered. The gibhead taper key will prevent both axial and rotative motion, since it is driven into place against a tapered keyway in the hub. It is usually equipped with a head, as shown, to permit withdrawal when required. The feather key is employed to prevent relative rotation but permits axial motion. It may be demonstrated that the axial force necessary to move a hub along a keyed shaft is twice as great when one key is used as when two or more keys are used. For this reason, and on account of strength considerations, multi-splined shafts are employed where shafts or hubs must

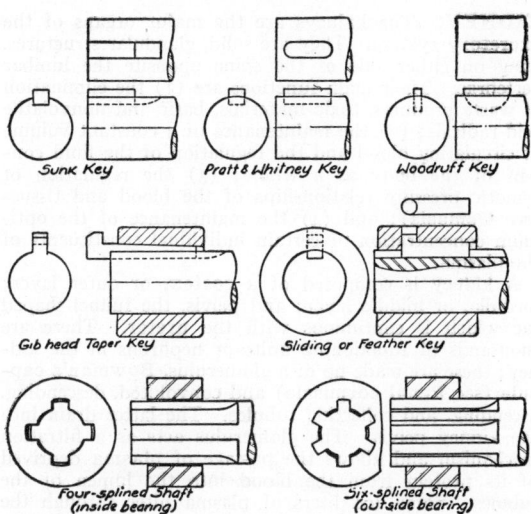

Sunk Key   *Pratt & Whitney Key*   *Woodruff Key*

Gib head Taper Key   *Sliding or Feather Key*

Four-splined Shaft (inside bearing)   *Six-splined Shaft (outside bearing)*

move axially under load. Modern production methods have made it possible to machine both shaft and hole with comparative ease. (H.C.H.)

KEYING. This is the process of breaking up the telegraph currents or radio signals into the dots and dashes of the particular code used for transmission of intelligence. The keying may be done automatically by means of a prepared tape or other mechanism, or it may be done manually. There are two principal types of hand-operated keys, the conventional telegraph key consisting of a lever moved down by the hand and up by a spring and thus making and breaking the contacts, and the semi-automatic key or "bug." This latter is a sidewise operated device which is thrown to one side to transmit dashes, the length of the dash being determined by the time it is held in the dash position, and thrown to the other side to transmit the dots, these being transmitted automatically by the key mechanism and the number of dots being determined by the time the key is held to the dot position. This semi-automatic key is capable of faster operation, is less tiring to the operator and in the hands of a skilled operator transmits better formed characters than the conventional key. In telegraph circuits the keying consists of opening and closing a d-c circuit or of changing the polarity of the d-c circuit (see **Telegraphy**). In radio telegraph transmission keying is accomplished by interrupting the radio-frequency signal at some point in the **transmitter**, preferably at an intermediate or **buffer amplifier.** For break-in keying, i.e., where the receiving operator may operate his transmitter and interrupt the sending operator when desired, oscillator keying is used. This gives rise to some frequency shift unless the circuit is carefully designed. In any type of radio keying care must be exercised to prevent key clicks or slurring of the code characters. Key clicks are interfering signals in nearby receivers and are caused by spurious frequencies generated when the key makes and breaks the circuit too sharply. This may be avoided by introducing lag circuits (resistance, capacitance and inductance in various combinations) in the keying circuit. These lag circuits will destroy the sharpness of the code if overdone, however, so they must be adjusted for a compromise between key clicks and slurred characters. (L.R.Q.)

KEYWAY. Key.

KIANG. Mammalia, Perissodactyla. A **wild ass,** *Equus hemionus,* of Tibet and Mongolia, also called the kulan. (A.W.L.)

KIDNEY. The kidneys are the major organs of the **excretory system.** They are solid, glandular structures, lying on either side of the spine opposite the lumbar vertebrae. Their main functions are (1) the elimination of waste products, toxic materials, basic and nonvolatile acid radicals; (2) the maintenance of a constant volume of circulating blood and the regulation of the fluid content of the body as a whole; (3) the regulation of osmotic pressure relationships of the blood and tissues (see **Osmosis**); and (4) the maintenance of the optimum concentration of certain individual constituents of the plasma.

A kidney is composed of a **cortex**, or outer layer; medulla, or middle layer; and pelvis, the funnel-shaped sac which is continuous with the **ureter.** There are thousands of functioning units or nephrons in the kidney; these are made up of a glomerulus, **Bowman's capsule** (see **Renal corpuscle**) and convoluted, descending, ascending, and collecting tubules. The latter drain into the kidney pelvis. The glomerulus acts as a filtration mechanism and allows the passage of **plasma** deprived of its protein from the blood into the lumen of the tubules. Some 100 liters of plasma filter through the glomeruli each 24 hours. Since only 1.5 liters of urine are excreted daily, 98.5 liters of fluid must be reabsorbed. The tubules exert a selective action on the various materials present in the filtrate, reabsorbing almost completely substances like sugar, chloride, sodium, and bicarbonate, which are necessary to the body economy, and excreting waste substances like urea, creatinine and sulfate. Whether or not the tubules in addition secrete certain substances has not yet been established.

The kidneys are subject to a variety of disease conditions. Inflammation (**Nephritis**), infection (**pyelitis** and **pyelonephritis**), congenital defects, **tumors,** and stones are the more common. (D.M.H.)

KILLDEER. Aves, Charadriiformes. A North American **plover**, *Oxyechus vociferus,* named for its call. Breeds throughout temperature North America and winters as far south as northern South America. (A.W.L.)

KILLED STEEL. Steel deoxidized by aluminum, ferrosilicon, ferrotitanium, etc., so that there is little or no evolution of dissolved gases during cooling in the mold. Ingots of killed steel are relatively free from **blowholes** but have a large central shrinkage cavity or **pipe** unless a chamber of heat-insulating brick, called a hot top, is used to keep the top of the ingot molten and to feed the shrinkage cavity with molten metal.

Compared with **rimming** and semi-killed steels, killed steels are generally finer grained and tougher. In the case of low-carbon **drawing** grade sheets, killed steels are "non-aging"; that is, after **temper rolling** the tendency to form **stretcher strains** is permanently eliminated. (R.H.H.)

KILLIFISH. Pisces, Teleostei. Small fishes (**Pisces**) of fresh and salt water. The family Cyprinodontidae is known as the killifish family, and in it the genus *Fundulus* contains the typical killifishes. The common killifish is also called the mummichog. (A.W.L.)

KILOGRAM. Originally the kilogram was dependent upon the **gram** as the standard unit of **mass** in the **metric system.** The kilogram was simply 1000 grams, and the gram, in turn, was defined in terms of the centimeter and the density of water. Now, however, the kilogram is the standard, and the gram is defined as one-thousandth part of it. The primary standard kilogram is the mass of a cylinder of platinum kept in the archives of the French government at Sèvres. It was constructed with the object of duplicating and perpetuating the original gram in a form practicable for

use; unfortunately, it was made too large by about 28 parts in a million. (See **Liter.**) (L.D.W.)

KILOWATT. A unit of **power**—defined as 1000 watts —which ordinarily serves as the commercial measure for electrical power. Electrical power of one kilowatt used steadily for one hour involves an energy consumption of one kilowatt-hour. A kilowatt is equal to about 1.34 hp. (F.T.M.)

KIMBERLITE. The name applied to a mica **peridotite** which occurs at Kimberley and other places in South Africa, the source of rich deposits of **diamonds.** These valuable **gem stones** were originally found in the decomposed kimberlite which, being colored yellow by limonite, was termed "yellow ground." Deeper workings disclosed the less altered rock, kimberlite, which the miners call "blue ground." (E.S.C.S.)

KINEMATICS. The physics of abstract motion, without regard to forces or bodies of matter, which are treated under **Mechanics.** Two aspects need special consideration: the motion of points and the motion of rigid figures. Whatever system of space coordinates is found simplest may be used, and may be transformed from one system to another as desired.

The motion of a point is completely specified by giving each of its three rectangular coordinates (for example) as a function of the time; that is, by writing its "equations of motion":

$$\left. \begin{aligned} x &= f_1(t), \\ y &= f_2(t), \\ z &= f_3(t). \end{aligned} \right\} \qquad (1)$$

In many cases the information represented by these equations is supplied only indirectly in the statement of a problem. For example, an expression may be given for the component of the linear **velocity** or the linear **acceleration**, as a function of either the time or the distance. In such case a differential equation is first obtained. Thus if it is specified that the X-component of the acceleration is constant and equal to a, the differential equation is

$$\frac{d^2x}{dt^2} = a; \qquad (2)$$

and the corresponding equation of (uniformly accelerated) motion is its solution,

$$x = \tfrac{1}{2}at^2 + v_0t + x_0, \qquad (3)$$

in which the integration constants x_0, v_0, stand respectively for the initial distance (value of x when $t = 0$) and the initial velocity. By eliminating t from the equations of motion (1) three geometric equations may be arrived at, one in x, y, one in x, z, and one in y, z, any two of which determine the path of the moving point in space.

Again, if the Y-component of acceleration is proportional to the coordinate y and of opposite sign,

$$\frac{d^2y}{dt^2} = -k^2y, \qquad (4)$$

may be written in which k^2 is a positive constant. The solution of this is

$$y = A \sin kt + B \cos kt, \qquad (5)$$

which is a general equation of **simple harmonic motion.**

A rigid 3-dimensional figure has six **degrees of freedom.** These require six equations of motion, which may, for example, express the three components of linear motion of the centroid of the figure with respect to the three rectangular axes, and the three components of angular motion or rotation about the same three axes. In such a case it is often convenient to use the notation of vector analysis. It may be shown that any motion of a rigid figure is at any given instant equivalent to a linear motion of its centroid in some definite direction, plus a rotation of the figure about some definite axis through the centroid.

The laws and equations of kinematics are of constant service in problems of **kinetics**. (L.D.W.)

KINESCOPE. This is a **cathode ray tube** used as the picture tube in a **television** receiver. The signal representing the picture intensity is fed to the **grid** of the electron gun so the intensity of the beam varies with the intensity of the original scene. This variable beam is swept back and forth, up and down by deflecting voltages impressed on the deflecting plates and synchronized with the scanning of the camera tube (see **Iconoscope** and **Image Dissector**). The beam of electrons then hits the fluorescent end screen of the tube and reproduces the scene being televised. (L.R.Q.)

KINETIC ENERGY. The most obvious way in which a body can manifest **energy** is to be in motion. Experience impels us to get out of the way when we see a rapidly moving, massive object approaching. We know that the more massive it is, and the faster it moves, the more work (and the more damage) it will do when it strikes. A simple course of reasoning based on **Newton's laws of dynamics** leads to the formula $E = \frac{1}{2}mv^2$ for the kinetic energy (in absolute units) of a mass m moving with a speed v. Thus a mass 100 grams moving with a speed of 1000 cm. per sec. has $\frac{1}{2} \times$ 100 g. $\times$ 1000^2 cm.^{2}/sec.^{2} = 50,000,000 g. cm.^{2}/sec.^{2} (or ergs) of kinetic energy. When the moving body is brought to rest by a force of average value f, and continues to move a distance, d, after this force is applied, it does an amount of work, fd, equal to its kinetic energy $\frac{1}{2}mv^2$. If d is very small, f must be large. Thus, if the stone in the above example were stopped in a space of 0.01 cm. (as in striking a hard obstacle), we should have 0.01 cm. $\times f$ = 50,000,000 g. cm.^{2}/sec.^{2}, or f = 5,000,000,000 g. cm./sec.^{2} (or dynes), which is equivalent to 5100 kilograms or more than 5.6 tons.

The kinetic energy of a body rotating with angular speed ω (radians per sec.) about an axis for which its moment of inertia is I, is $E = \frac{1}{2}I\omega^2$; or $2\pi^2 In^2$, where n is the number of rotations per sec. The theory of **relativity** gives slightly higher values for the kinetic energy, but the differences are negligible except for very great speeds. (L.D.W.)

KINETIC THEORY. Many phenomena formerly attributed to unknown or hypothetical forces are now known to be due to the invisible motions of small particles, molecules, atoms, or electrons. Notable among these are gas and **vapor pressure, evaporation**, and **diffusion of fluids**.

Gases expand indefinitely when released, not because of repulsion between the molecules as formerly supposed (though the **Joule-Thomson** effect under certain conditions may involve this), but because the molecules are in rapid motion and do not stop unless they collide with something. Air is not "forced" out through a tire puncture; only those air molecules pass out which, in their aimless wanderings, happen to encounter the opening. Molecules also pass in from the outside; but since there are several times as many per unit volume inside as outside, many more pass out than in. This continues until, a statistical equilibrium being reached, the air inside is no more dense than that outside, and the tire is "flat." The rapidity with which this takes place emphasizes the speed of the molecular motion and the relative insignificance of the "internal friction" opposing it.

What appears to be a steady pressure is due to the incessant impacts of the gas molecules on any surface exposed to them. If n molecules of equal mass m are released in an enclosure of volume v, and if their speeds

are $u_1, u_2, \cdots, u_n$, it is easy to show that the average pressure set up by these impacts, neglecting the effects of collisions and gravity, is

$$p = \frac{m}{3v}(u_1^2 + u_2^2 + \cdots + u_n^2). \qquad (1)$$

This may be written

$$p = \frac{1}{3}\frac{nm}{v}\frac{\Sigma(u^2)}{n};$$

or since nm/v is the gas density ρ, and $\Sigma(u^2)/n$ is the mean square molecular speed $\overline{u^2}$,

$$p = \frac{1}{3}\rho\overline{u^2}. \qquad (2)$$

This relation gives the mean square speed as $\overline{u^2} = 3p/\rho$; which is the square of the effective speed, corresponding to average kinetic energy. From this it may be shown that the average speed is

$$\bar{u} = \sqrt{\frac{8p}{\pi\rho}}, \qquad (3)$$

easily determined since p and ρ are measurable. (See **Maxwell Distribution Law**.) Again (1) may be written

$$pv = \frac{2}{3}(\frac{1}{2}mu_1^2 + \frac{1}{2}mu_2^2 + \cdots + \frac{1}{2}mu_n^2) = \frac{2}{3}E; \qquad (4)$$

in which E is the total kinetic energy of linear motion of the molecules. From this it follows that the absolute temperature T of the gas bears a constant ratio to this total kinetic energy, and hence to the average translational kinetic energy of the molecules. (See **Ideal Gas Law**.)

Further analysis shows that when gravity is considered, the pressure in an undisturbed pure gas of uniform temperature, at an elevation h, is given by

$$p = p_0 e^{-3gh/\overline{u^2}} \qquad (5)$$

in which p_0 is the pressure at the zero of elevation. This is a form of Laplace's "law of atmospheres," useful in barometric altitude determinations.

A quantity much used in kinetic theory is the "mean free path," which is the average distance traversed by a molecule between collisions. There are ways of calculating this and also the effective diameters of molecules, and these data lead to many conclusions as to frequency of collisions, rate of diffusion, etc. (See **Brownian Movement**.) (L.D.W.)

KINETICS. This branch of physics deals with the effects of forces or of torques upon the motions of material bodies. There are separate articles on **Kinematics, Dynamics of Rotation, Moment of Inertia, Power, Centrifugal Force, Precession, Orbital Motion, Pendulums**, etc. The basis of this subject, so far as the classical theory is concerned, consists in **Newton's laws of dynamics**, which may be extended to include **d'Alembert's principle** of kinetic equilibrium.

The motions of a particle or of a rigid body may be either "free" or "constrained"; that is, it may be at liberty to move in any manner in obedience to the applied forces or torques, or there may be present material barriers which limit its linear motion to a certain path or surface, or its rotation to a certain axis. Thus a projectile or a planet is free to travel and to rotate as it will; but a railway train must follow a fixed track, and a grindstone must revolve, or a **pendulum** must swing, about a fixed axis. These latter cases may be dealt with by considering that the barriers offer opposing forces or torques in equilibrium with those components tending to cause motion against them, and by treating as "effective" only those components acting in directions in which motion is permissible.

A simple problem of this type is that of a block of mass ma sliding without friction down an inclined plane inclined at an angle α (see figure). Its weight mg (in which g is the acceleration due to **gravity**) is resolved into components f and p, along and perpendicular to the plane, only the former of which is effective. This force $f = mg \sin \alpha$, acting on the mass m, gives it the constant

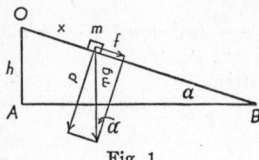

Fig. 1.

acceleration $a = f/m = g \sin \alpha$. Therefore the motion of the block along OB is expressed by

$$x = \tfrac{1}{2} g \sin \alpha \cdot t^2 + v_0 t + x_0, \qquad (1)$$

in which $x = Om$, x_0 is the initial value of x, and v_0 is the initial velocity (see **Kinematics**). An interesting problem of constrained motion is to determine in what path from O to B the block would slide down in the least time; it turns out to be not a straight line, but a **cycloid**.

Many kinetic problems are greatly simplified by applying the conservation of **energy** principle. Thus for any path, the final speed of the block starting at O and sliding to B is $\sqrt{2gh + v_0^2}$, where h is the vertical height of O above B. This may be shown by using (1), as an equation of pure kinematics, with $x_0 = 0$ and $x = OB = h/\sin \alpha$. It is necessary first to solve the equation, which now takes the form

$$\frac{h}{\sin \alpha} = \tfrac{1}{2} g \sin \alpha \cdot t^2 + v_0 t,$$

for t, obtaining as the (positive) root

$$t = \frac{\sqrt{v_0^2 + 2gh} - v_0}{g \sin \alpha}. \qquad (2)$$

Then the final speed is $V = at + v_0 = g \sin \alpha \cdot t + v_0$, which, using the value of t from (2), gives

$$V = \sqrt{2gh + v_0^2}. \qquad (3)$$

But let us now use instead the fact that the kinetic energy at B is equal to that at O plus the loss of potential energy in descent through height h, which is mgh. The kinetic energy at O and at B are, respectively,

$$E_O = \tfrac{1}{2} m v_0^2, \quad E_B = \tfrac{1}{2} m V^2;$$

so that

$$\tfrac{1}{2} m V^2 - \tfrac{1}{2} m v_0^2 = mgh. \qquad (4)$$

Solution of this gives at once

$$V = \sqrt{2gh + v_0^2},$$

which is the same as (3). The most significant feature of this is that the solution is here reached without considering the nature of the motion at all. Indeed, if a simple pendulum is pulled aside until the ball is at an elevation h above its lowest level and then given a push with initial speed v_0, it will attain the speed given by (3) as it descends to that lowest level; though no algebraic equation corresponding to (1) can be derived to describe its motion on the way down. The problem has thus become one of energetics rather than of kinematics.

Problems involving impacts or collisions are best handled by means of the principles of conservation of momentum and conservation of energy. In this way, for example, it may be shown that if an elastic sphere A strikes centrally a stationary elastic sphere B of greater mass, B will take part of the energy and momentum and A will rebound; if A and B are of equal mass, B will take all the energy and momentum and A will stop; while if A is more massive than B, B will take part of the energy and momentum and A will continue more slowly in its original direction.

By the use of d'Alembert's principle, problems of kinetics become equivalent to problems of statics. In case a body or a system is in frictionless, uniform rectilinear motion, in which there are no accelerations, classical dynamics introduces no element of the problem due to the motion; the situation is exactly as if the system were at rest. But with accelerations there develop what may be called "inertia forces," namely, the reactions opposing the forces causing the accelerations. If there is appreciable friction, that factor enters also as another opposing force.

Suppose a heavy boat is at rest in the water, in static equilibrium. Its mass is $m = 5000$ kilograms. A tow line begins pulling on the boat. When, at the end of 30 sec., the boat has attained a speed of 0.6 meters per sec., the force of water friction on the boat is 3 kilograms. It is desired to find the tension in the tow line at that moment.

We take the direction of motion as positive and let the required tension be f. The reaction force of inertia is negative and in gravitational units is equal to ma/g, where a is the acceleration. We then have the equation of equilibrium,

$$f + \text{inertia force} + \text{friction force} = 0$$

or

$$f - \frac{5000 \text{ kilograms} \times 0.02 \text{ meters/sec.}^2}{9.8 \text{ meters/sec.}^2} - 3 \text{ kilograms} = 0,$$

from which $f = 13.2$ kilograms. The last two force components enter just as if the boat were at rest and two other ropes were pulling backward on it, one with a force of 10.2, the other with 3 kilograms. The same principle applies, using vector addition, when the forces concerned are not collinear. (L.D.W.)

KINETOGENESIS. An evolutionary theory of E. D. Cope which assumes that animal structure is shaped by mechanical stresses due to movement. In relation to length of limb, stresses along the axes of the bones are supposed to have lengthened them whether these stresses were due to repeated impact against the ground as in the hoofed animals or to stretching as in the arms of the primates. (A.W.L.)

KING CRAB. Xiphosura.

KINGBIRD. Aves, Passeriformes. A large North American flycatcher. The most common species is found from the Atlantic to beyond the Rockies and is called the **bee martin** and tyrant flycatcher as well as the kingbird. This species and the several others found in the West and South are noisy and quarrelsome birds. (A.W.L.)

Kingbird, *Tyrannus tyrannus.* Dark slaty-gray above, with a concealed orange spot on the top of head. White below. Tail black, with a white band across the tip.

KINGFISHER. Aves, Coraciiformes. Birds (**Aves**) of many species, mostly with strong sharp beaks used for catching fish. Some species eat insects and reptiles more than fish and have broader beaks, hooked at the tip. One of the species, *Dacelo gigas*, with a broad beak lives in Australia and the Papuan Islands and has been named from its peculiar call the laughing jackass. Kingfishers are found on every continent. (A.W.L.)

KINGLET. Aves, Passeriformes. Very small birds (**Aves**) related to the gnatcatchers. The two North

American species are the golden-crowned, *Regulus satrapa,* and the ruby-crowned, *Corthylio calendula.* Both are grayish and olive with a bright crown patch. In the golden-crowned kinglet this patch is orange in the male and yellow in the female, and is concealed. In the ruby-crowned it is bright red and is exposed. It occurs only in the male. The goldcrest and firecrest of Europe are related species. (A.W.L.)

KINGSBURY THRUST BEARING. This is a special type of **babbitted** bearing for supporting large rotating masses, such, for example, as the principal bearing of a vertical shaft hydroelectric turbine. Very small frictional losses with large bearing pressure and high velocities are made possible by dividing the bearing surface into a number of pivoted segments which tilt in such a way as to scrape a wedge of oil between them and the rotating collar against which the bearing is mounted. (F.T.M.)

KINKAJOU. Mammalia, Carnivora. An animal, *Cercoleptes caudivolvulus,* related to the raccoons, which resembles the common cat slightly in appearance and has a long prehensile tail. It is yellowish-brown in color. The species ranges from central Mexico to Brazil. Also called jupura. (A.W.L.)

KINORHYNCHA. Minute worms, not exceeding 1 mm. in length, which live in sand and mud on the ocean floor. The body is made up of rings but the internal organs are unsegmented. They progress by thrusting out and retracting the head and sometimes adjacent segments.

The relationships of these animals are uncertain. They are sometimes included in the phylum **Rotifera** as a separate class. One order, Echinodera, is recognized. (A.W.L.)

KIPPER. A European name for the breeding male of the Altantic **salmon.** A cured (Yarmouth) herring. (A.W.L.)

KIRCHHOFF'S LAWS OF NETWORKS. Two laws relating to electric networks carrying steady currents. The general case is that of *n* points or junctions, each one of which is connected with each of the *n* − 1 remaining points by a conductor containing a source of electromotive force. Kirchhoff's two statements are as follows:

1. If conductors forming part of a network carrying a steady current meet at one point, the sum of the currents flowing toward the point is equal to the sum of those flowing away from it; or the algebraic sum of all the currents in these conductors is zero.

2. Starting at any one of the junctions of such a network and following any succession of the conductors which form a closed path, around either way to the starting point, the algebraic sum of the products formed by multiplying the resistance of each conductor by the current through it is equal to the algebraic sum of the electromotive forces encountered on the journey. (In this reckoning, we call all currents moving with us positive, and all electromotive forces tending to cause such currents positive.)

Maxwell has set forth a general method of calculating the currents and the relative potentials of the junctions when the resistances and electromotive forces in the several branches of a network are given. For a network of *n* points, this method involves the solution of *n* − 1 simultaneous, first-degree equations. The work is often simplified, however, by the circumstance that some of the conductors or some of the electromotive forces are absent. (L.D.W.)

KIRCHHOFF'S LAW OF RADIATION. Thermal Radiation.

KIROUMBO. Aves, Coraciiformes. A peculiar bird **(Aves)** of Madagascar and several similar species of other Pacific islands. Related to the rollers. (A.W.L.)

KISSING BUG. Insecta, Hemiptera. A name applied late in the 19th century to a supposedly deadly **bug** which was said to bite human beings about the face. It probably applies to some of the large predacious bugs of the family Reduviidae, commonly called **assassin bugs.** One species known as the big bedbug enters houses and sucks human blood, inflicting a painful but by no means dangerous wound. (A.W.L.)

KITCHEN MIDDEN. Primitive refuse pile of shells of edible **mollusks,** bones and implements, dating from the **Paleolithic** to the present. (R.M.F.)

KITE. Aves, Falconiformes. Large birds **(Aves)** of prey related to the eagles. They are found on all continents. Three species, the swallow-tailed kite, *Elanoides forficatus,* white-tailed kite, *Elanus leucurus,* and Mississippi kite, *Ictinia misisipiensis,* occur in North America. (A.W.L.)

KITTIWAKE. Aves, Charadriiformes. *Rissa.* Marine birds **(Aves)** of the far north, resembling gulls. One species lives in the north Pacific and a second in all polar seas of the north. (A.W.L.)

KIWI. Aves, Apterygiformes. *Apteryx.* Stoutly built flightless birds **(Aves)** of New Zealand. The several species are all about as large as domestic fowls. They have strong legs, a long curved beak, slender hairlike feathers, and rudimentary wings which are concealed by the plumage of the body. (A.W.L.)

KLIPSPRINGER. Mammalia, Artiodactyla. A very small African **antelope,** *Oreotragus oreotragus.* It is an active and agile animal which inhabits rocky and mountainous country. The males have very short horns. (A.W.L.)

KLYSTRON. This is an electron **tube** of the velocity-modulated type used in **ultra high frequency** circuits. At these extremely high frequencies (measured in terms of hundreds or thousands of megacycles) the conventional **vacuum tubes** become useless because of lead and electrode inductance and capacitance, and transit-time effects, so a radically different approach is necessary. The klystron is one solution of the problem. This vacuum tube consists of a **cathode, grid** and perforated **anode** somewhat like the electron gun of the

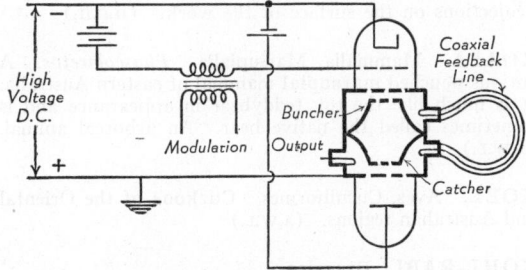

Modulated klystron oscillator.

cathode ray tube (however, here the electrodes are grid-like structures rather than the cap with a single hole) followed by two **cavity resonators** separated a calculated distance and finally a collector. All electrodes except the collector have grid-like surfaces so the electrons can pass on through them. The beam of random-velocity electrons passing through the grid is accelerated by the positive potential applied to the first resonant cavity structure, causing this structure to serve as an anode. These electrons pass through the grids into the cavity which is called the buncher. Here the standing waves in the cavity act on the electrons and cause them to change speed so they arrive at the second cavity, called the catcher, in bunches (having passed out of the first into a field free space and then into the second through the grids in the sides of the cavities). Here the energy of the electrons is absorbed by the field and contributes to the useful output and normally supplies also the driving energy for the buncher. The electrons then pass on to the collector and return to the cathode. By proper adjustment of the voltages and spacings of the cavities the circuit may be made to oscillate or amplify as desired. A circuit of a modulated klystron oscillator is shown. (L.R.Q.)

KNEE BRACE. Bent.

KNEE TOOL. Turret Lathe.

KNOT. In **navigation,** the term knot is used as an abbreviation for the phrase "**nautical miles per hour.**" The use of this term started during the period when the **log** chip and line were used to determine the speed of a ship through the water.

It should be clearly understood that the word itself indicates a speed and, if a ship is moving through the water at a speed of 15 nautical miles per hour, the speed of the ship relative to the water is 15 knots. This should never be expressed as "15 knots per hour" since such a phrase means that the ship has an **acceleration** of 15 nautical miles per hour per hour.

The term knot is used for speed by the U. S. Navy and Merchant marine for both air and seaborne ships. It is also used by the ocean-going vessels of nearly all other nations. Because of the fact that the nautical mile is, by definition, closely related to angular measure on the surface of the earth, the knot is a far more convenient speed unit for use in **dead-reckoning** problems than is the statute **mile** per hour. (W.K.G.)

KNURLING. The process of forming a series of fine ridges upon the periphery of a circular part such as a screw head or knob to facilitate its rotation by hand, although knurling is often done for a purely ornamental effect. Knurling may consist of small, closely spaced ridges which are either at an angle to the axis of the surface of revolution or parallel with it like the milled edges of a coin. A knurling tool generally has a pair of rolls with diagonal teeth inclining in opposite directions, which are pressed against the unhardened work and rotate with it, thus forming small diamond-shaped projections on the surface of the work. (H.C.H.)

KOALA. Mammalia, Marsupialia. *Phascolarctos.* A curious pouched **marsupial** mammal of eastern Australia. It is much like the toy teddybear in appearance and is sometimes called the native bear. An arboreal animal. (A.W.L.)

KOEL. Aves, Cuculiformes. **Cuckoos** of the Oriental and Australian regions. (A.W.L.)

KOHL-RABI. Brassica.

KOLM. The term applied to highly bituminous "coallike" lenses which occur in the lower **Paleozoic,** alumshales of Sweden. The ash of this "coal" contains a relatively high content of the **radioactive minerals** which have been used to determine the age of the formations in which they occur. (R.M.F.)

KONZI. Mammalia, Artiodactyla. An African **antelope** belonging to the hartebeest group. (A.W.L.)

KORIGUM. Mammalia, Artiodactyla. An **antelope** of central Africa, also called the Senegal antelope. One of the hartebeest group. (A.W.L.)

KORIN. Mammalia, Artiodactyla. A central African **antelope,** one of the gazelles. (A.W.L.)

KORSAKOFF'S PSYCHOSIS. Ethyl Alcohol.

KOSTINSKY EFFECT. A photographic effect discovered by the Russian, Kostinsky (1906). It was found that if two areas of high density, e.g., star images, are close together the restraining by-products formed in the process of development retard the development of the adjacent portions of the images so that the distance separating the two images is greater than is actually the case. The effect assumes considerable importance in astronomical photography. It is more marked with developers such as pyro which are greatly restrained by the oxidation products formed in development and less so by metol-hydroquinone, diaminophenol, and paraminophenol. (C.B.N.)

KRA. Mammalia, Primates. The crab-eating **macaque,** *Macacus cynomolgus,* a monkey of the Oriental region. (A.W.L.)

KRAIT. Crait.

KRENNERITE. Sylvanite.

KRYPTON. Symbol: Kr. Atomic number: 36. Atomic weight: 83.7. Density: 3.708 grams per liter, 0° C., 760 mm., or 2.87 when air is 1.00. Melting point: —157° C. Boiling point: —152.9° C. (**Isotopes:** page 290.)

Krypton is a colorless, odorless gas, of negative chemical properties with ordinary materials. Discovered by Ramsay and Travers in 1898, in ordinary air to the extent of 1 part krypton in about 1,000,000 air. (R.K.S.)

KUDU. Mammalia, Artiodactyla. African **antelopes** of two species, the common, *Strepsiceros strepsiceros,* and the lesser kudu, *S. imberbis.* The males have long spiral horns. (A.W.L.)

KUMQUAT. Citrus Fruits.

KUNDT CONSTANT. Faraday Effect.

KUNZITE. Spodumene.

KURTOSIS. Kurtosis is a property of a distribution or **probability function** depending on its height or peakedness at the **mean** combined with the property of having a long or short tail at the extremities of the distribution. A measure of kurtosis is $\alpha_{4:x} = \frac{\mu_{4:x}}{\sigma_x^4}$, where $\mu_{4:x}$ is the fourth central **moment,** σ_x is the **standard deviation** of the distribution. If $\alpha_{4:x} = 3$, the distribution is said to be mesokurtic, if $\alpha_{4:x} > 3$ leptokurtic, if $\alpha_{4:x} < 3$ platykurtic. If a probability function is represented by a Type A **Gram-Charlier series,** consisting of terms up to and including $\phi^{(4)}(t)$, and if $\alpha_{4:x} > 3$, then such a function has a higher ordinate at the mean than the **normal curve.** Under the same conditions if $\alpha_{4:x} < 3$, the distribution has a lower ordinate at the mean than the normal curve. If the distribution function is not representable by a Gram-Charlier Type A series, then $\alpha_{4:x} > 3$ does not guarantee that the distribution will have a higher peak than the normal curve. In the normal curve $\alpha_{4:x} = 3$. In Fig. 1 we have $\alpha_{4:x} > 3$, and $\alpha_{4:x} = 3$. In Fig. 2 we have $\alpha_{4:x} < 3$, $\alpha_{4:x} = 3$. (L.A.A.)

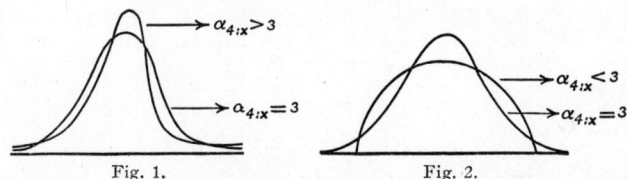

Fig. 1.   Fig. 2.

KYANITE. The mineral kyanite is an **aluminum silicate,** corresponding to the formula Al_2SiO_5. It is triclinic, and has a good **cleavage** parallel to the **macropinacoid.** Its hardness varies considerably, depending on the crystallographic direction from 5 to 7.25; specific gravity, 3.56–3.67; luster, vitreous to pearly; color, commonly blue to white, but sometimes gray to green or nearly black; transparent to translucent. Usually found in long-bladed crystals or columnar to fibrous structures. Kyanite is found in such **metamorphic** rocks as **gneisses** or mica **schists.** Of the many European localities for fine specimens might be mentioned the Ural Mts., Russia; Czechoslovakia; Austria; Trentino, Italy; the St. Gotthard region, Switzerland; and France. In the United States: Chesterfield, Massachusetts; Litchfield, Connecticut; and Gaston County, North Carolina, have furnished fine specimens. Kyanite derives its name from the Greek work meaning *blue,* in reference to the delicate blue of the inner portions of the bladed crystals. (E.S.C.S.)

L

L SECTION. This refers to an elementary section of a network such as a **filter** where the components are connected in the form of an L, i.e., one component in series with one side and the other in shunt across the two sides of the circuit. (L.R.Q.)

L SERIES. X-Ray Spectra.

LABARIA. Reptilia, Sauria. A poisonous South American **snake** belonging to the pit vipers. It ranges from eastern Brazil north into the Guianas. Related to the jararaca. (A.W.L.)

LABIUM. 1. A lip or lip-shaped structure. 2. The posterior element of the insect's mouth parts, derived from a pair of jointed appendages.

The labium consists typically of three segments with a pair of appendages called labial palpi. The basal segment is the submentum, the next segment the mentum, and the terminal segment the ligula. The ligula is sometimes complex, consisting of a median glossa, sometimes very long, and two basal lobes known as paraglossae of very variable size. The labial palpi arise from the base of the ligula and consist of four segments or less. In some species the palpus is attached to a distinct part called the palpiger. (A.W.L.)

LABOR (CHILDBIRTH). The process of separation of the mature or nearly mature **fetus** and **placenta** from the interior of the **uterus** and its expulsion from the mother.

The first stage of labor begins near the end of the tenth lunar month after conception. The onset of labor is marked by the appearance of cramp-like pains in the lower abdomen, which gradually increase in frequency and severity. The pains are due to muscular contractions of the uterus which brings about dilation of the **cervical** opening of the uterus in the **vagina.** At this time there may be a little blood with the vaginal discharge—"the show." After the pains have continued for a variable time—usually 12–15 hours—there is a gush of considerable fluid from the vagina due to rupture of the membranes, "bag of waters," with escape of the amniotic fluid.

A short time after the rupture of the membranes the second stage of labor begins. The labor pains increase in severity, frequency, and duration. During this stage the child passes from the uterus through the dilated cervix and vagina to the outside world. The **umbilical** cord is severed at this time. The second stage of labor usually lasts about 2 hours in the first pregnancy. In succeeding pregnancies the time is usually shorter.

After the birth of the child the third stage of labor begins. At first the labor pains cease, the uterus contracts and becomes much smaller. After a few minutes contractions begin again and the placenta (afterbirth) is expelled. This last procedure may be completed by the obstetrician. Considerable blood may be lost during this stage.

Normal labor varies in duration for the three stages, depending on the size of the child, size of the mother's pelvis, and the character of muscular contractions. The three stages usually take about 18 hours—16 for the first, 2 hours for the second, and 15–30 minutes for the third stage. The birth of the first child usually takes 6 hours longer for the three stages.

Labor is sometimes prolonged in primiparous women over 35 years of age because of the loss of elasticity and distensibility of the cervical tissue.

The complications of labor are obstruction due to disproportion between the size of the fetus and the size of the pelvis, and poor muscular contractions of the uterus. The use of obstetrical forceps may be necessary to aid the progress of the child. Other complications are hemorrhage, premature separation of the placenta, **eclampsia,** and rupture of the uterus. (R.S.M., D.M.H.)

LABRADORITE. Feldspar.

LABRUM. The anterior or upper element of the insect's mouth parts. It is not derived from jointed appendages like the other mouth parts and is usually a simple flap, movably attached to the front of the head. (A.W.L.)

LABYRINTH. Stuffing Box.

LABYRINTHODONTS. Fossil Amphibia.

LACCOLITH. An intrusive, **igneous rock** mass which has been injected along the bedding plains of **sedimentary** formations in such a way as to dome the overlying strata. Typical laccoliths have been fed by **dikes** and are mushroom-shaped with symmetrical domes and flat floors. Because of the mechanics involved in their origin, laccoliths are formed in the upper portion of the **lithosphere** where the static pressure of the overlying formations is not too great to prohibit their local uplift by the intruded **magma.** (R.M.F.)

LACERTILIA. Reptilia.

LACEWING. Golden-Eye.

LACING BAR. Compression Member.

LACINIA. Maxilla. Also a movable appendage of the **mandible** of some isopod **crustaceans,** the lacinia mobilis. (A.W.L.)

LACQUER. Pigments; Paints; Varnishes; Resins; Nitrocellulose; and Paints.

LACTAMS. Aminoacids, Polypeptides, and Proteins.

LACTIC ACID AND LACTATES. Lactic acid alpha-hydroxypropionic acid ($H \cdot C_3H_5O_3$ or $CH_3 \cdot CHOH \cdot COOH$) is a colorless liquid, melting point 18° C., boiling point 122° C. at 15 mm. pressure, miscible with water, alcohol or ether in all proportions. **Calcium** lactate, on account of its solubility characteristics (10 grams per 100 grams cold water and very soluble in warm water) is of importance in the separation and recovery of lactic acid. Calcium lactate plus dilute **sulfuric** acid yields lactic acid plus calcium sulfate, and the latter may be separated by **filtration.** Lactic acid may be obtained by **evaporation** of the filtrate in **vacuum.** Lactic acid, when heated at atmospheric pressure, changes to lactide; when oxidized cautiously, changes to pyruvic acid; when heated with dilute sulfuric acid yields **acetaldehyde** plus **formic acid.** Lactic acid may be obtained (1) by the fermentation of **lactose,** "milk sugar" (thus accounting for the presence of lactic acid in sour milk), **sucrose,** molasses, **glucose,** starch (corn starch, potato starch) in the presence of **calcium** carbonate forming calcium lactate, from which lactic acid is obtainable, (2) by reaction of glucose and sodium hydroxide solution under proper conditions, (3) by reaction of sodium hydroxide solution and alpha-chloropropionic acid, (4) by reaction of acetaldehydecyanhydrin and water. The physiological formation of lactic acid in the animal organism is responsible for the fatigue of muscles.

The following are esters of lactic acid:

Methyl lactate ($CH_3CHOHCOOCH_3$), boiling point 145° C.

Ethyl lactate ($CH_3CHOHCOOC_2H_5$), boiling point 154° C.

Lactic acid is used in the textile and farming industries, and in certain food industries, e.g., pickles, essences. (See Isomerism and Stereoisomerism.) (R.K.S.)

LACTIDES. Acids, Carboxylic and Related Compounds.

LACTIMS. Aminoacids, Polypeptides, and Proteins.

LACTONES. Acids, Carboxylic, and Related Compounds.

LACTOSE. Carbohydrates.

LACUNAR TISSUE. A tissue peculiar to echinoderms. It occurs in strands containing associated spaces which serve as circulatory channels. Unlike true blood vessels these spaces have no epithelial lining. (A.W.L.)

LADLE. Foundry Practice.

LADY BUG. Insecta, Coleoptera. Small oval beetles, strongly convex and with relatively small legs. The common name applies chiefly to the more common red species, marked with black and white, but the family Coccinellidae to which they belong contains many others. (A.W.L.)

LADY'S-SLIPPER. Orchid Family.

LAGOON. Barrier Beach.

LAKE HERRING. Pisces, Teleostei. Fresh-water herrings of several species, all important food fishes. Also called ciscoes. These fishes occur in the Great Lakes and many smaller bodies of water. They include the most important commercial species of the Great Lakes fisheries. (A.W.L.)

LAKES. Dyes.

LAMARCKIAN THEORY. A theory of organic evolution first expressed in 1809 by the French scientist, Jean Baptiste Lamarck. Lamarck formulated two principles, first that use strengthens and develops an organ and that disuse weakens it and causes it to become atrophied, and second that the results of use and disuse are inherited and so influence the development of a species through a succession of generations. The first of these principles is abundantly proved, while the second is not true.

In discussions of evolutionary theory since Darwin's work was published the term acquired characters has become synonymous with the organic changes which Lamarck emphasized and the inheritance of acquired characters, or Lamarckian evolution, has been generally discredited. Closer perusal of Lamarck's work, however, indicates that he emphasized the reaction of the organism to environmental conditions as a source of change. The results of this reaction are not inherited as such, but whether the reaction in one generation adds to the capacity of the next generation for similar reaction is neither proved nor disproved, hence the Lamarckian theory still has a valid place in organic evolution. (A.W.L.)

LAMBERT. Brightness.

LAMBERT PROJECTION. The Lambert modified conformal conic projection (commonly known as the Lambert, and sometimes as the Gauss conformal) has been used for many years in the construction of maps. Within the past decade this projection gained rapidly in favor for use in constructing charts for air navigators.

The projection is actually a mathematical type, but may be quite accurately described as a conical projection, differing from the simple conical in that the cone is not tangent to the earth's surface, but cuts it on two latitude parallels known as standard parallels. This type of projection is particularly valuable for portraying large longitudinal areas, e.g., the entire United States. The graticule of the Lambert chart shows parallels of latitude as concentric circles centered at the nearer pole, and the meridians of longitude as straight lines converging on this pole. From simple geometric considerations it is obvious that the meridians and parallels must be perpendicular. The angle of convergence of the meridians, and therefore the radii of the parallels, depends upon the distance of the nearer pole from the center of the area.

The great advantage of the Lambert projection is that the scale of distance is uniform, for all practical purposes, all over the chart. To indicate the accuracy of this statement consider the Lambert projection on which the series of aeronautical charts of the United States are constructed by the Coast and Geodetic Survey. The standard parallels for these charts are N 45° and N 33°. The scale of distance, if considered as unity on the standard parallels, is 0.994 at the central parallel, expands to 1.010 at the extreme north boundary of the United States, and to 1.023 at the tip of Florida. Comparing the Lambert with the Mercator distance scales we find that if we consider the Mercator scale as unity at N 39°, the scale at the northern limits of the United States is 1.154 while at the tip of Florida it is 0.846.

The non-orthogonal graticule of the Lambert chart is a distinct disadvantage for general navigational problems since neither the rhumb line nor the great circle is straight. However, the uniformity of scale is of such great advantage that air navigators have begun to use this type of chart even in preference to the Mercator, particularly when navigating in good visibility over land or where good radio aids are available. A straight line between two points on a Lambert chart is referred to as a Lambert line. Since the meridians on the graticule are convergent this will not be a rhumb line. However, the distance measured along this line will be less than that along the rhumb line and only slightly greater than the great-circle distance between the two points. This line is frequently used by aviators during conditions of good visibility. The standard procedure in using the Lambert line is to draw a straight line on the chart and pick out conspicuous landmarks separated by about 25 miles. Then follow the rhumb line indicated by measurement of the angle between the Lambert line and the meridian nearest the starting point. This will lead you to a point some distance from the first landmark but within visibility. From this first landmark a new heading is adopted for the second, and so on to destination. In case a rhumb line is desired for the entire route, the Lambert line is drawn as before and the course measured from the meridian half-way between the point left and point to be arrived at. This rhumb line will appear as a curve on the Lambert chart, but can be laid down by any one of a number of standard methods with sufficient accuracy for the selection of landmarks.

The convergence of the meridians prevents the use of the Lambert chart for graphical solution of dead-reckoning problems, and introduces difficulties in plotting lines of position obtained either by radio bearings or by celestial observation. The ease with which such problems can be solved on the Mercator chart seems to cast doubt on the statement to the effect that "the Lambert Chart will completely supersede the Mercator for all navigational purposes." (W.K.G.)

LAMELLA. In botany the **middle lamella** is the compound layer composed of the primary walls and the cement-like intercellular substance which occurs between the primary walls of two cells. From this layer is obtained **pectin** which is added to concentrated fruit juices in the preparation of jellies. The term lamella is also applied to each of the concentric growth layers in large starch grains.

In zoology the term lamella is used with two common meanings. It may be: 1. A thin leaf or plate, such as a lamella of bone. 2. A flat plate formed by the fusion of **ctenidial filaments** in the **bivalve** mollusks. Two lamellae united by bridges of tissue form a gill through which water circulates under the influence of **ciliary** action in the persisting open spaces. This form of gill is the source of the name **Lamellibranchiata** applied to the class containing these animals. (R.M.W., A.W.L.)

LAMELLAR VECTOR. Irrotational Vector.

LAMELLIBRANCHIATA. The bivalve mollusks, a class of the phylum **Mollusca** including the **clams, mussels, oysters, scallops** and related species. Many of these animals are valuable for food, and pearls and mother-of-pearl are produced by them. The class is also named Pelecypoda.

Bivalve mollusks differ from other members of the phylum in the following characters: 1. The body is transversely compressed. 2. The **mantle** forms two lobes extending down along the sides of the body. In most species these lobes unite at the posterior end to form two passages, an upper excurrent and a lower incurrent siphon. Currents of water carry food and oxygen into the mantle cavity through the lower opening and a current bearing wastes passes out of the upper. 3. Each mantle fold secretes a valve of the shell formed of calcareous matter covered outside by a horny periostracum and inside by nacre, commonly called mother-of-pearl. The two valves of the shell are joined by a hinge ligament and the articulation is strengthened in some species by interlocking teeth. 4. The gills are thin plates on each side of the body in most species. They are formed of united ctenidial filaments. 5. The foot is a muscular wedge-shaped protuberance at the anterior end of the body.

All bivalves are aquatic. Most species creep slowly by thrusting the foot into the muddy or sandy bottom but some propel themselves by jets of water squirted from the siphons or forced from the mantle cavity by rapidly closing the valves. The species vary from fresh water forms about $\frac{1}{8}''$ long to giant marine shells more than a yard long.

The class is divided into four orders:

Order Protobranchiata. Gills in the form of small leaflets, two rows on each side of the body. Marine species.

Order Filibranchiata. Marine **mussels, scallops,** etc. Gills composed of filaments united only by ciliary junctions.

Order Eulamellibranchiata. Gill filaments united to form continuous plates. Fresh-water **clams** or mussels, marine clams, **oysters, shipworms,** etc.

Order Septibranchiata. Gills replaced by a horizontal partition between the upper and lower divisions of the mantle chamber. A few marine species.

(See also **Clam, Mussel, Mother-of-Pearl, Pearl,** and **Oyster.**) (A.W.L.)

LAMINAR FLOW. Turbulence.

LAMINARIA. Algae.

LAMINATION. A discontinuity in a plane parallel to the surface of a sheet, strip or plate of any metal. Laminations may vary considerably in size and distribution. Common causes of laminations are shrinkage cavities resulting from the solidification of metal **ingots** and less frequently gross segregations of nonmetallic **inclusions** that may be trapped in the ingot when the metal solidifies.

In most electrical machinery there are certain iron core parts which are subjected to an alternating flux, hence they will have alternating voltages induced in them. These voltages give rise to circulating currents which represent a loss and consequent heating of the machine. These currents are known as eddy currents and the loss is commonly called eddy current loss. To reduce this loss to a minimum the core parts which have alternating flux are laminated, i.e., composed of thin sheets of magnetic material, each insulated from the other. This breaks the circuit for the eddy currents into many very small ones so the induced voltages will be too small to produce appreciable currents. Often the laminations are varnished for insulation but frequently the natural oxide scale is depended upon to perform this function. As the frequency goes up the laminations must be made thinner if the loss is to be kept down. Thus at audio frequencies the laminations are much thinner than at commercial power frequencies and at **radio frequencies** (iron cores can only be used at the lower radio frequencies) the magnetic material is powdered and mixed with a binder to give the effect of extremely thin laminations. (See **Metamorphism** and **Lava.**) (R. S. BURNS, L.R.Q.)

LAMMERGEIER. Vulture.

LAMP, ELECTRIC. The principal electric lamps are the **arc lamp,** vapor lamp, the incandescent resistance lamp and the fluorescent lamp. The common vapor lamps in use now are the mercury-vapor, giving a bluish-green light, and the sodium lamp, giving a yellow or orange light. Other gases are used for special purpose lamps such as advertising displays where a wider range of colors is desirable. The Cooper-Hewitt lamp is the oldest commercial form of vapor lamp and makes use of a mercury pool **cathode.** As with all types of vapor lamps the light is produced principally by the recombination of the vapor ions and electrons, the gas being ionized by the passage of an electric current through it. (See **Discharge.**) The mercury liquid is contained in the vapor lamp in a state of high vacuum. When the tube is cold it is necessary to strike an arc by tilting the tube until an electrode dips in the pool of mercury. The accompanying figure shows the principle of the electrical connections of a single-phase, a-c Cooper-Hewitt lamp.

A recent development is the high-pressure mercury-vapor lamp. These lamps, in conjunction with a high reactance **transformer,** are self-starting without any tilting or moving mechanism and have a higher efficiency than the low-pressure Cooper-Hewitt lamps.

The quality of light from the mercury-vapor lamp is bluish-green in color, and has a tendency to distort the colors of objects viewed in it. However, this light is good for industrial purposes, as it permits small details to be more easily seen by workmen than does white light. On account of the intense light delivered by the mercury vapor lamp, fewer units are necessary for adequate lighting of large manufacturing areas.

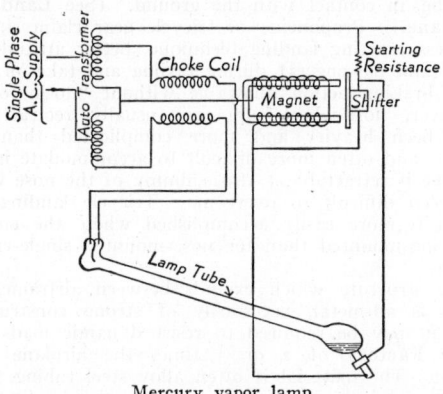

Mercury vapor lamp.

In the ordinary incandescent lamp, light is emitted from a highly heated resistance wire having a high melting point so that it may remain in a state of incandescence many hours without fusing or breaking. The resistor or filament, is hermetically sealed in a glass bulb, which is evacuated, or which is filled with inert gas, such as nitrogen. Most modern incandescent lamps have filaments of drawn tungsten wire, and go by the name of Mazda lamps. The incandescent lamp is rated on the basis of the power input required. Measured in watts, the standard sizes are the 25-, 30-, 40-, 60-, 100-, 200-, 500-, 750-, and 1000-watt lamps. Lamps both smaller and larger than these generally used sizes are commercially available. The lumens per watt of power consumed varies from about 7 for the smaller to 20 for the 1000-watt modern incandescent lamp. The average life has been raised until it is now about 1000 hours per lamp.

The most recent development in the lamp field is the fluorescent lamp which is a type of vapor lamp, but converts the radiation from the ionized vapor into a more usable type by means of a fluorescent coating on the inside of the lamp. These lamps are tubular in form, contain a filament-type electrode at each end, have the inner wall of the tube coated with some phosphor (a material which will fluoresce under ultra-violet excitation) and are filled with mercury vapor at low pressure. The accompanying figure illustrates the method of operation. When first connected to the line

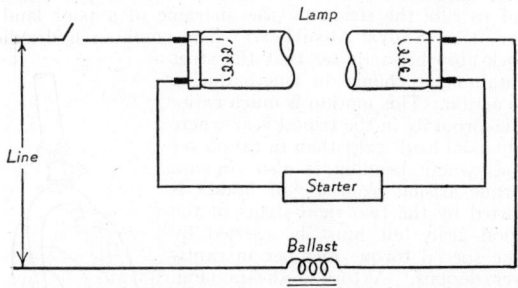

Fluorescent lamp connections.

the filaments heat and emit electrons. After a short time interval the starter opens and high voltage appears across the electrodes, causing the vapor to break down into a self-maintaining **discharge**. This discharge is rich in ultra-violet radiation which excites a visible radiation in the fluorescent coating of the tube. The ballast limits the current to a safe value since the discharge is not inherently self-limiting. The color of the light may be controlled by the type phosphor used for

coating the lamp. The efficiency is much higher than incandescent lamps, for example, a 40-watt fluorescent lamp will produce about the same number of lumens as a 150-watt incandescent lamp. The better control of color of light and the better efficiency are causing a widespread change-over to this type lighting in many applications.

One of the most striking developments of the present century has been the growth of electric lighting. The reason for this growth is to be found in the multiple advantages of electric lighting over other methods of illumination. Principal advantages might be summarized as follows:

1. Convenience in use.
2. Adequate illumination, eliminating physical handicaps resulting from eye-strain.
3. Safety from fire hazard of oil and gas lights.
4. Artistic lighting units made possible by the use of electricity. (F.T.M., L.R.Q.)

LAMP SHELL. Tongue Shell; and Invertebrate Paleontology.

LAMPREY. Cyclostomata.

LAMPROPHYRE. An old group term originally proposed by Gümbel and later redefined by Rosenbusch to include **basic** dike rocks of **porphyritic** texture whose phenocrysts are **femic** minerals such as **augite, hornblende** and **biotite**. The term is derived from the Greek meaning glistening, and referring particularly to the abundant biotite which occurs in the particular variety of Lamprophyre called **Minette**. (R.M.F.)

LANCELET. Cephalochordata.

LANCEOLATE. Shaped like a lance; that is, narrow and tapering to a point, as in lanceolate leaves. (R.M.W.)

LAND BREEZE. At night air in contact with land cools, and that in contact with water does not. When sufficient difference in temperature arises between the land and the sea air, there springs up a breeze flowing off-shore from cold land onto warm water. This wind is known as the land breeze. Its basic cause lies in the fact that constant pressure and density surfaces intersect and form a solenoidal field; the resulting circulation tends to bring the density surfaces parallel with the pressure surfaces. (P.E.K.)

LAND SUBDIVISION. The standard system of subdivision of lands for legal and other purposes is that employed in the United States land subdivision system, which involves the location of subdivisions with regard to a set of principal axes. In this system the units of subdivision are the township and the section. The origin is the intersection of a true parallel of **latitude** called the base line, and the true **meridian** called the principal meridian. Some 35 of these origins have been established and wholly, or in part, govern surveys in most of the states of the Union, outside the original 13 colonies. Secondary axes are established at intervals of 24 miles north and south of the base line, and east and west of the principal meridian. These divide the area into quadrangles bounded by meridians and parallels. The meridians are each 24 miles long, but while the southern bounadry of each quadrangle is 24 miles in length, the northern boundary, in the northern hemisphere, is less than 24 miles, due to the convergence of meridians. These 24-mile boundaries are termed standard parallels and guide meridians. The parallels are, of course, continuous, but the meridians, with the exception of the principal meridian, are broken at the parallels. On the other hand, the parallels are curves, and the meridian lines are straight.

Each of these quadrangles is divided into 16 parts by north and south range lines at 6-mile intervals, and by east and west township lines at 6-mile intervals. The area enclosed by range lines and township lines is slightly less than 36 square miles. It would be 36 square miles except for the convergence of the range lines which are meridional in nature. Beginning with the eastern side of the township, the sections are laid off by section lines north and south and east and west, at 1-mile intervals. Thus there are 36 sections in a township, each of which is 1 mile square except on the westerly edge, where, due to the convergence of the meridional section lines, the sections will have a latitudinal dimension of less than 1 mile with the deficiency being greater in the northern sections than in the southern.

The base line and principal meridian are used for legal descriptions of all sections whose survey relates to their origin. Any given township may be located by giving it a range number east or west from the principal meridian, and a township number north or south from the base line; for example, range 6 East, township 7 South. The section lines divide each township into 36 sections, which are numbered consecutively from east to west in the first tier, west to east in the second, and so on, beginning with number 1 in the northeast corner, and ending with number 36 in the southeast corner. (F.T.M.)

LANDING, AIRCRAFT SYSTEMS. Blind Landing.

LANDING GEAR, AIRPLANE.

The structure which supports an airplane in contact with land is called its landing gear. As here interpreted, the term is applicable only to airplanes intended to operate normally from land surfaces, although others have considered "landing gear" to include the understructure of float-type seaplanes.

The components of all landing gear are (1) wheels (alternately skis for snow), (2) structural connection between wheels and airframe, and (3) some means of reducing the shock of landing. Modifications sometimes found include brakes, retraction, fairing, steering.

The method of airplane support by the landing gear is often classified as *conventional* or *tricycle*. The "conventional" designation is used to indicate a type, formerly in general use, at a time when usage of the tricycle type is rapidly increasing and it threatens to become the conventional type of the future. Both types support the airplane at three points, two of which are furnished by main wheels located close to the airplane center of gravity, while the third (the auxiliary wheel) is at the tail in the conventional type, and at the nose in the tricycle type. The wheels are always located so that the resultant of airplane mass acts inside an imaginary triangle joining the three wheels. Both types are shown in Fig. 1 in normal rest position. In a range of normal landing attitudes the main wheels touch

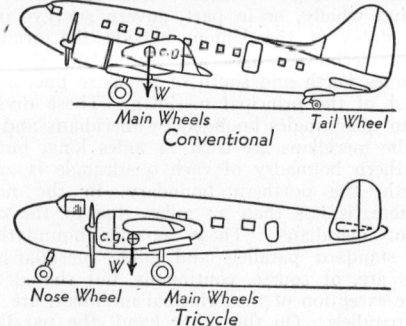

Fig. 1. Method of support.

first, followed by a rotation (caused by weight and inertia forces acting through the center of gravity (C.G.) bringing the tail or nose wheel, as the case may be, in contact with the ground. (See **Landings, Airplane.**) Proponents of tricycle gear claim simpler and less exacting landing technique, better attitude for pilot (and passengers) during taxiing and take-off, and more braking action allowable without "nosing-over." However, nose-wheel supports (usually rectractable) have been heavier and more complicated than tail wheels, and often more difficult to accommodate in the fuselage if retractable. Also, shimmy of the nose wheel has been difficult to overcome. Tricycle landing-gear design is more easily accomplished where the engines are wing-mounted than for nose-mounted single-engine designs.

The structure which extends between airframe and wheel is all-metal, necessarily of strong construction since it may be required to resist dynamic loads (see **Load Factor**) of 2 or 3 times the airplane dead weight. The material is often alloy steel tubing, steel, and magnesium castings. Loads on the landing-gear structure arise not only from the normal reaction between wheel and ground, but also from horizontal sidewise forces caused by landing with side drift, and torque created by use of brakes.

These loads are carried by a variety of structures in different airplanes; however, all may be classified as (1) tripod, (2) full cantilever, or (3) semi-cantilever. The tripod type has the greatest rigidity, but the most aerodynamic drag. Also, it is not very well suited for retraction designs. The full cantilever types shows better aerodynamic performance, but tends to be heavier than the truss type for the same rigidity. Large bending moments are imposed on the fuselage-landing gear joint. Semi-cantilever types have been widely adopted for retractable gear for the reason that the retraction arm may also constitute a brace for the main load-carrying strut. All three types are diagrammed in Fig. 2, with shock absorption and re-

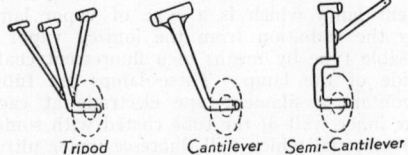

Tripod Cantilever Semi-Cantilever
Fig. 2. Structural types (without shock absorption).

traction omitted. Shock absorption is needed to make good landings smooth and comfortable for passengers, and to ease the strain on the airframe of a poor landing. (See **Shock Absorber.**) The common hydraulic shock-absorber requires that the strut containing it change in length during the action. This motion is much easier to incorporate in the tripod gear where only axial loads exist than in cantilever types where bending is also present. Torque arising from use of brakes is resisted by the two rigid struts of the tripod gear, but must be carried by some special torque member in cantilever designs. A torque shears (Fig. 3) is often used to maintain wheel alignment.

A further classification of landing gear may be: (1) fixed, (2) retractable. For many years all land planes had fixed landing gear, usually of the tripod type. These had the advantages of simplicity, ruggedness, and reliability, but did expose a parasitic area to air action, resulting in an aerodynamic drag. At low airplane speeds (say below 100 m.p.h.) the landing-gear drag

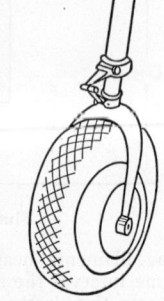

Fig. 3. Torque shears for cantilevered gear.

is not of great importance and fixed landing gear imposes no great penalty on performance. Parasitic drag increases as the square of air velocity, so the effect mounts rapidly with increase of air speed once it begins to be noticeable. To achieve high speeds (200–600 m.p.h.) retraction of all parasitic surfaces into a streamline fuselage or nacelle is a practical necessity. At moderate speeds considerable reduction of required engine hp. will partly compensate for the trouble of including retractable landing gear. Engine power varies about as the cube of air speed, so to raise a flight speed from 110 m.p.h. to 130 m.p.h. would require 65% more power unless drag can be reduced by retraction and other means. All high-speed airplanes now have retractable gear, while at the other extreme, several small airplanes for private use have been able to make speeds of 120–140 m.p.h. on 80 to 100 engine hp. with retractable gear, whereas formerly barely more than 105 m.p.h. would be expected on this power.

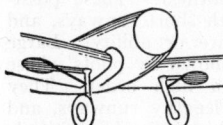

Fig. 4. Typical retraction single-engined airplane.

On single-engine airplanes, main wheels are retracted into the fuselage or into the wings, generally the latter. Retraction into the engine nacelle is common on multimotored designs. In any event, the problem of retraction requires mechanical ingenuity for successful solution, since the necessary space is always hard to find, and certain interior structural members will need to be cleared. Some of the motions employed are:

1. Sideways into the wing or wing center section. When extended, the retraction member serves as a lateral brace.

2. Backwards, with a 90° rotation of the landing-gear strut, so that the plane of the wheel will be horizontal and it will retract into the lower surface of a wing. Where this is actuated by a brace, the brace is longitudinal when the gear is extended.

3. Backwards or forwards into nacelle with a jack-knife action on the main strut. This is necessary where (as on bombers) the strut is long and simple rotation would land it outside the available retraction region. Some of these mechanisms are quite complicated. (Fig. 5.)

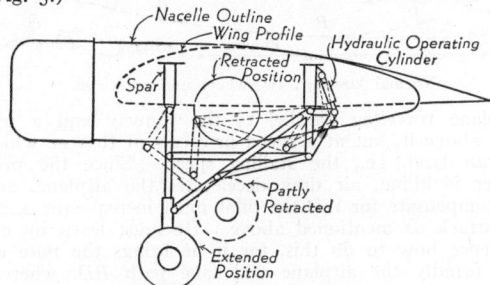

Fig. 5. Retraction mechanism.

Operation of retraction mechanism by the pilot is by remote control since ordinarily the gear is accessible to neither pilot nor crew member. Systems employed are hydraulic, electric, and mechanical. Hydraulic systems operate from engine oil pressure through control valves accessible to the pilot. The ultimate power motion is the stroke of a plunger in or out of a double-acting oil cylinder. Electric systems are, of course, controlled by switches and generate power either by motor or solenoid. Light-weight motors operating nut and thread actuators are common types of final drives.

Pure mechanical systems are seldom used for actuating landing gear as they are slow in action (hand-cranking by pilot, etc.), but they are sometimes installed as an emergency measure.

Retractable landing-gear installation is further complicated by the need of the following components:

1. Indicating mechanism to assure the pilot that the wheels are in correct "down" position before he lands. The gear is rarely visible from the control cabin.

2. Locking mechanism to lock the jointed landing gear in its extended position.

3. Limit stops on power supply to prevent heavy stresses acting on the structural members after being extended or retracted.

4. Automatic landing-gear well doors to open when gear is actuated for lowering, and close after gear is retracted. These doors are mainly to "fair" over the opening for aerodynamic reasons.

The wheels are generally cast from aluminum or magnesium alloy, and consist of little other than the

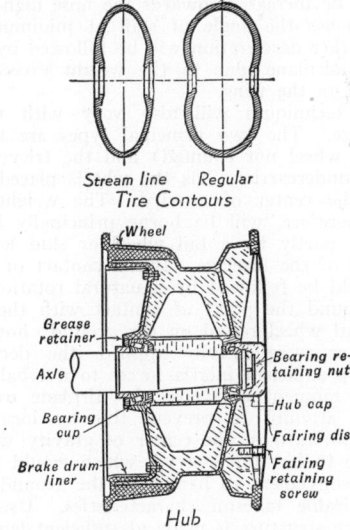

Stream line   Regular
Tire Contours

Fig. 6. Airplane wheel.

hub, on which the tire and tube are mounted, and inside of which are located wheel bearings and brake drum. Several different designs of brake shoe have been used. Small airplanes frequently have brakes actuated by cable from the brake pedals. This is not satisfactory for large brakes nor for retractable wheels, these being generally hydraulic in action. A typical hydraulic braking system would consist of a reservoir to hold the brake fluid and compensate for any slight leaks in the connecting lines, a master cylinder to build up the system pressure, and a wheel cylinder whose piston actuates the brake shoe. The master cylinder piston can be actuated directly from the brake pedals or (as on some of the large airplanes) by hydraulic or electric servo-motor. Wheels are individually braked to aid in controlling ground loops and to facilitate taxiing downwind.

Tires for airplanes are of the pneumatic tube and casing type similar to automobile tires, except that the ratio of contour diameter to rim diameter is much greater. Wheels are constructed for both drop-center and removable-flange methods of mounting tires. Streamline tires are made for use on fixed landing gear where careful external streamlining is attempted. (F.T.M.)

LANDINGS, AIRPLANE. Obviously, every airplane flight must be terminated by a landing. The landing maneuver has as its object the transfer of the airplane weight from wings to wheels in such a way as to produce the minimum practical impact and deceleration. As this may be done in a variety of ways, both expertly and inexpertly, producing variations in **load**

factor and requiring variable techniques, various types of landings will be described individually.

The difference between the landing of a helicopter (or airship) and an airplane is that the airplane is always required to exceed a certain minimum forward speed in order to obtain the necessary sustension. This is called the stalling speed, and is an important factor to be taken into account in landing maneuvers. Until the time when the wheels are to be put in contact with the ground, this speed must be exceeded, otherwise the airplane will begin to fall in an uncontrolled way. At the stalling point the **angle of attack** has its maximum value, and the attitude of the airplane for horizontal flight is nose high, whereas, progressively, at speeds higher than the stall the nose lies closer to the horizontal. It will be seen, then, that for horizontal flight at decreasing velocity the attitude of the airplane must be increased towards the nose high position until it reaches the angle of stall at minimum flying speed. Further deceleration will be followed by a stalling of the airplane, due to the weight exceeding the existing lift on the wing.

Landing techniques will also vary with types of undercarriage. The two principal types are the two-wheel (tail wheel not counted) and the tricycle. The two-wheel undercarriage has the wheels placed slightly ahead of the center of gravity. The weight of the airplane, therefore, will be borne principally by these wheels, but partly by a tail wheel or skid located at the aft end of the fuselage. Initial contact of the two wheels would be followed by a natural rotation of the airplane around the point of contact with the ground until the tail wheel rested on the ground; however, if wheel brakes were heavily applied, the deceleration might set up sufficient inertia forces to overbalance the deadweight moment and nose the airplane over in a crash. An attempt to prevent this by locating the wheels far forward of the center of gravity will cause the airplane to be "tail heavy," which would result in its being inconvenient to handle on the ground, and to have undesirable take-off characteristics. Usually the undercarriage structure is made of sufficient length that the attitude of the grounded airplane is approximately that for the stalling angle for horizontal flight.

Although for many years the two-wheel undercarriage was standard, latterly three-wheel, or tricycle, undercarriages have received general approval. The center of gravity is slightly ahead of the two main wheels in this arrangement, hence their initial contact with the ground rotates the airplane into nose-down position, bringing the nose wheel into contact with the ground. Brakes may then be applied heavily without danger of nosing over. Other advantages of this system are that the angle of attack and the lift are automatically decreased upon contact with the ground, thus eliminating a tendency to regain an airborne condition if the landing speed is unduly high, and the attitude of the plane is more conducive to good visibility and control during taxiing.

The nature of an airplane landing is such that a *runway* of 1000' or more is required for a normal landing. As the safest landings are made *into* the ground wind, airports will usually have a number of runways so that the pilot of the approaching plane may select a direction into or nearly into the wind for a landing. An airplane with a landing speed of 50 m.p.h. in still air would land at 30 m.p.h. into a 20 m.p.h. wind, whereas the landing speed would become 70 m.p.h. downwind. Furthermore, control of the airplane during the landing run is far simpler for upwind landings; in fact, there is practically no control downwind unless the airplane is equipped with individual wheel brakes. Hence downwind landings are generally taken to be the mark of ignorant, inexperienced, or rash pilots, except in light ground winds with airplanes of high landing speed.

Prior to the beginning of the landing maneuver, the airplane must be put into an approach to the runway. This maneuver varies not only with the size and type of airplane, but also with the individual policies and preliminary training of the pilot. Some prefer to make a long (1 mile or more) straight-away approach during which the airplane is put in its normal glide attitude and corrections are made for side drifting; however, pilots of light airplanes usually prefer a short right-angle approach with the expectation of making their final turn in the direction of the runway not much more than ¼ mile from its beginning. Arguments in favor of the short approach are that position relative to the landing area may be more accurately gauged, and more readily corrected if defective, thus decreasing the likelihood of gliding past the runway before achieving a landing (over-shooting), or failing to reach the runway (under-shooting). These possibilities loom large on airports with short runways, and with pilots of moderate experience or ability. Large airplanes, such as transports, are flown by pilots whose experience and abilities are of a high order. They are operated from airports with lengthy runways, and are a type of airplane for which the low-altitude maneuvers necessary in a short approach are not recommended. Hence long approaches are the rule in their operation.

Normal Landing—Two-wheel Undercarriages. The approach maneuvers place the airplane a few hundred feet from the end of the runway in a normal glide the horizontal projection of which coincides with the axis of the runway. The *normal glide* is unaccelerated flight at a speed of 5–15 m.p.h. above the stalling speed. The angle of glide depends on the drag of the airplane (**parasite drag** and **aspect ratio**), and the pilot's use of flaps or other drag-increasing action such as fish-tailing, slipping, etc. The angle also depends on the ground wind, being steeper the higher the wind. As the airplane approaches to within 30–50' of the ground in this glide, the pilot will use his controls to deflect the **elevator**, thereby leveling the direction of flight, as shown by the arc *AB*, Fig. 1. This will leave the

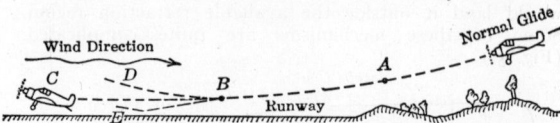

Normal landing; two-wheel undercarriage.

airplane traveling parallel to the runway and a few feet above it, but at a speed in excess of that at which it can land, i.e., the stalling speed. Since the propeller is idling, air drag decelerates the airplane, and to compensate for this the pilot must increase the angle of attack as mentioned above. He must learn by experience how to do this, for if he brings the nose up too rapidly the airplane will take path *BD*, whereas if he makes insufficient allowance, the path *DE* will result. With the proper control manipulation, the airplane will arrive at the point *C* at the stalling attitude and the stalling speed, whereupon it will settle. In a normal landing the wheels are within 2–3' of the ground when this **stall** occurs, and the airplane settles smoothly and naturally into a "three-point landing." If the pilot has misjudged his altitude, and the wheels are perhaps 10–20' from the ground when the stall begins, the airplane can achieve considerable vertical velocity before contact with the ground, with shock which will usually cause structural damage. This is called a *pancake landing,* and is a sign of inept piloting, except that an expert might elect to pancake an airplane when making an emergency landing into a forest.

Normal Landing—Tricycle Undercarriage. Up to the point *B* of Fig. 1, the tricycle-equipped airplane would be flown in a manner similar to that of one with a two-wheel undercarriage. From B on, the maneuver would be different. Instead of holding the airplane off the ground while it was losing speed, it would be flown into gentle contact with the ground along a path something like BE. As the main wheels touch the runway surface the weight of the airplane would cause the nose to drop until the nose wheel also touched the surface. This will cause a decrease in angle of attack and wing lift so that there will not be a tendency for the airplane to become airborne again even though it has a speed in excess of stalling. With this undercarriage the airplane will be brought to a fast stop by heavy use of the brakes.

Cross-Wind Landing. When airplanes must be landed on a runway where the wind is not parallel to the axis of the runway, the cross component must be allowed for, or the airplane will land with some side drift which may overstrain the undercarriage, cause a wing tip to touch the ground surface, blow out a tire, or one of several other mishaps. The lighter the wing loading of the airplane, the more vulnerable it is to cross winds during landing. There are two methods for making a smooth cross-wind landing. One is to lower the windward wing slightly when approaching the runway, thus slipping the airplane sidewise into the wind just enough so that the path over the ground is maintained in a straight line in the same direction that the airplane is headed. Just before contact the wings are leveled with the ailerons, although in case of a strong cross wind it may be desirable to land with the windward wing low, i.e., on one wheel. The other method is to head into the wind slightly, but with wings horizontal. The airplane can then be caused to maintain a straight line down the runway, but it will not be headed in its direction of motion. Just before contact the airplane is ruddered rapidly so as to head it in the direction of its motion, this being necessary because normally the main wheels are not of a castering type. There are other variations, and often a pilot will use a combination of methods.

Belly Landing. Emergency landings of airplanes equipped with retractable undercarriage are sometimes, by the choice of the pilot, made with the wheels retracted. This is usually done where the emergency landing area looks soft or irregular, or where the distance available for landing run is quite limited. The airplane is landed on the under side of the fuselage. The center of gravity is low, and the frictional drag with the ground high. These are both of advantage in preventing a serious crash, and belly landings often can be accomplished with but minor damage to the fuselage.

Power Landing. Power landings are made with throttle partly open, and the propeller delivering an appreciable thrust. Two types of power landings will be described, viz., the two-wheel landing and the power stall landing. When landing in a high wind, it is inadvisable to throttle the engine. The approach to the landing is not made with a normal glide, but with the use of some amount of the engine power. This improves the control and allows the pilot to combat unexpected gusts better. The airplane is flown into contact with the runway in the position for level flight, and, if possible, taxied into the lee of a windbreak or to where airport personnel may help with the ground handling. During such taxiing the tail should not be allowed to come down, that is, the angle of attack should be kept low. The power stall landing is used in order to land with minimum forward speed, when landing on muddy or swampy ground, or in snow. It is also a customary method of "landing" seaplanes at night. In a power stall landing the airplane is headed into the wind with the throttle partly closed and speed reduced by raising the nose slightly until settling begins. The rate of settling is controlled by the use of the throttle, which should not be reduced to the point where the rudder and elevator begin to feel in-

effective. In the night landing of a flying boat, this attitude is maintained until the boat strikes the water. The crew has knowledge of the progress of the landing operation from a sensitive altimeter which will note the height from the water, and from the rate of climb meter, which will show also the rate of descent. Airplanes with wheel undercarriages when brought to a power stall landing in daylight are generally landed with the throttle closed a few seconds before actual contact is made. Pilots must always be cautioned on the manner of terminating a power stall approach if it is desired to resume a normal glide. Full power must be employed, and the angle of attack reduced very gradually, otherwise there may be a decrease of angle of attack disproportionate to the speed. (See **Airplane Speeds; Airplane; Controls, Airplane Flight.**) (F.T.M.)

LANDSCAPE MARBLE. A British popular and trade term for argillaceous limestones of Liassic (lower Jurassic) age quarried near Bristol, England, and characterized by a tree-like pattern. (R.M.F.)

LANGUR. Mammalia, Primates. *Semnopithecus.* A group of Old World monkeys characterized by slender build and by extremely long tails. The legs are longer than the arms. They eat principally the leaves and young shoots of trees. Langurs live only in Asia and some of the East Indian islands.

Among the species of langurs that bear distinctive names are the hanuman, lutong, douc, negro monkey, leaf monkeys, bear monkey, white monkey, and purple-faced monkey. (A.W.L.)

LANIER EFFECT. A photographic effect observed by A. Lanier (1896) involving the acceleration of development by the addition of potassium iodide to the developer, or by bathing the plate or film in a solution of potassium iodide before development. The effect is more marked with developers of low reduction potential, particularly hydroquinone, and is hardly apparent with developers of high reduction potential. (C.B.N.)

LANTERN FLY. Insecta, Homoptera. Any member of the family Fulgoridae, which differs from the related leaf hoppers and other families in having the antennae inserted at the sides of the head. The North American species are small but one giant Brazilian species has a wing spread of six inches. A large prominence on the head of this species was once said to be luminous, hence the name lantern fly has persisted although none of these insects is actually luminous. (A.W.L.)

LANTHANUM. Symbol: La. Atomic number: 57. Atomic weight: 138.92. Density: 6.15. Melting point: 826° C. No isotope, but of simple atomic form: 139. Type of compound: La_2O_3. Color of salts: Colorless. Discovered by Mosander in 1837. A member of the cerium sub-group of the rare earth metals. (R.K.S.)

LAP JOINT. Riveted Fastenings.

LAPILLI. During volcanic eruptions there may be thrown into the air quantities of melted rock which cools into solid fragments before it falls back on the earth. This material is classified roughly according to size. Fragments 2 or 3″ or more in diameter are called bombs, those of smaller size down to the dimensions of bird shot are called lapilli, still finer particles are called volcanic ash or dust. Derived from the Latin *lapillus,* a little stone. (E.S.C.S.)

LAPIS LAZULI. Lazurite.

LAPLACE'S EQUATION. Laplace's equation is a type of partial differential equation of the second order in three independent variables. In most of its many applications, these three variables represent 3-dimensional space coordinates, while the one dependent variable represents one of a number of physical point-functions fulfilling the condition expressed by the equation.

In **rectangular coordinates** (x, y, z), Laplace's equation takes the form

$$\frac{\partial^2 u}{\partial x^2} + \frac{\partial^2 u}{\partial y^2} + \frac{\partial^2 u}{\partial z^2} = 0 \qquad (1)$$

where u is the dependent variable. The operator $\frac{\partial^2}{\partial x^2} + \frac{\partial^2}{\partial y^2} + \frac{\partial^2}{\partial z^2}$, often abbreviated ∇^2, is called the Laplacian.

In **cylindrical coordinates** (r, θ, z), the equation corresponding to (1) is

$$\frac{\partial^2 u}{\partial r^2} + \frac{1}{r}\frac{\partial u}{\partial r} + \frac{1}{r^2}\frac{\partial^2 u}{\partial \theta^2} + \frac{\partial^2 u}{\partial z^2} = 0; \qquad (2)$$

while in **spherical coordinates** (ρ, α, β), one of several forms is

$$\frac{1}{\rho^2 \cos \beta}\left[\cos \beta \frac{\partial}{\partial \rho}\left(\rho^2 \frac{\partial u}{\partial \rho}\right) + \frac{1}{\cos \beta}\frac{\partial^2 u}{\partial x^2} + \right.$$
$$\left. \frac{\partial}{\partial \beta}\left(\cos \beta \frac{\partial u}{\partial \beta}\right)\right] = 0. \qquad (3)$$

Each of the three equivalent forms has its advantages in certain situations, but the rectangular form (1) is the one usually referred to when Laplace's equation is mentioned.

We shall here cite two familiar physical examples to which Laplace's equation applies; there are many others.

1. Consider a region of space in which there is an electric field (due to electric charges in the vicinity) but no free electricity, that is, no such **space charge** as exists in a vacuum tube in operation. At any point in this space there is an electric **potential**, which varies with the position of the point and is therefore a function of its coordinates. It is shown in electrostatic theory that if the potential is designated by u, it satisfies Laplace's equation in any of the forms (1), (2), or (3), whatever be the distribution of electric charges responsible for it. This fact makes it possible to trace the lines of force and equipotential surfaces in the region, by means of special solutions of Laplace's equation, called harmonic functions, which satisfy the "boundary conditions" imposed by the arrangement of the neighboring charges. As to which form of the equation is to be used, this depends upon which system of coordinates is most conveniently adapted to the shape and arrangement of the charged bodies.

2. If specified parts of the surface of a solid thermal conductor (such as a block of copper) are kept at different specified constant temperatures, heat will flow from the warmer toward the colder boundaries within the conductor. When this flow has become steady, the temperature of the conductor takes on a definite constant value at each point, dependent upon the location of the point. If the temperature is designated by u, it now satisfies Laplace's equation in each of its three forms, and the lines of flow and isothermal surfaces may be determined accordingly. A similar procedure is applied to potential in the steady flow of electricity through a metallic conductor. (L.D.W.)

LAPPING. The process of producing an extremely accurate, highly finished surface by means of a lap, which is a block charged with **abrasive**. Lapping reduces the possibility of wear on close-fitting running parts or on the surfaces of measuring equipment, by reducing the minute ridges and serrations left by machining and grinding operations to a more uniform bearing surface. Lapping may be done by hand or by machine. If a part is to be hand lapped to a final accurate dimension, a lap or mating form is made from a metal somewhat softer than the part to be finished. The surface of the lap is charged with a fine abrasive, or a small amount of abrasive mixed with grease, oil, or alcohol. A flat lap has a carefully-trued surface with a series of grooves in it. The lapping compound is smeared on the face of the lap and the work is rubbed over the face along an ever-changing path. The grooves in the face of the lap act as channels for any excess abrasive and oil. Very little pressure is used in order to eliminate the danger of scoring the work or stripping the lap. Hand lapping requires skill and time. The amount of material removed by lapping should not exceed .0002" to .0005".

A squaring lap is one in which the edge of the work is lapped at right angles to the bottom surface. A cylindrical lap is usually made of brass and is split and has a set screw so that the lap can be adjusted for wear. A helical groove of irregular lead in the surface of the lap serves as a channel for the abrasive and oil.

Mechanical lapping is analogous to hand lapping and is often the final stock-removing operation for sizing and finishing gauges and other commercial precision parts. Three types of lapping media are ordinarily used: bonded abrasives for the usual run of commercial precision work; metal laps and loose abrasive mixed with a lubricant for gauge manufacture; and abrasive paper or cloth, for crankshaft lapping.

For satisfactory precision lapping, the lap surfaces must be true planes. The process of generating true planes is very similar to the process of originating straight edges. Three laps are hand-scraped to a surface plate and one of the laps is then fastened to the table of a drill press. The second lap is placed on the first with a film of abrasive and oil between the contact surfaces. By means of a short-throw crank attached to the drill spindle, the upper lap is made to move in a circular eccentric path across the lower lap, which also causes the upper lap to rotate slowly about its own axis. This motion of the upper lap reduces the faces of both laps until they match. The first lap is then worked with the third, and then the second with the third, continuing until a true surface has been attained. Such laps may be used without any additional abrasive for finishing operations since the lap surface retains enough abrasive for this purpose. (See also **Gear Tooth Forming, Superfinishing.**) (H.C.H.)

LAPSE RATE. The rate at which temperature decreases or lapses with altitude. It is the vertical temperature gradient. Since temperature normally decreases with altitude in the troposphere, it is convenient to assign positive values to the rate of temperature change with altitude: Lapse rate, therefore, is defined as the rate of change of temperature with altitude and is positive when the temperature decreases. (P.E.K.)

LAPWING. Aves, Charadriiformes. *Vanellus.* Birds (**Aves**) of several species resembling the plovers in appearance and habits. They occur in Europe, Asia, Africa and the Americas. (A.W.L.)

LARAMIDE REVOLUTION. Cretaceous.

LARD. Esters.

LARK. Aves, Passeriformes. Song birds (**Aves**) of many species, confined to the northern hemisphere. The skylarks of Europe and Asia are the most famous members of the group because of the quality of their song. The common European species, *Alauda arrensis*, is established in Oregon and North America also has a native species, the horned lark, *Otocoris alpestris*. The **mea-dowlark** is more closely related to the blackbirds and orioles. (A.W.L.)

LARKSPUR. Buttercup Family.

LARVA. An immature form of animals that undergo **metamorphosis** between emergence from the egg and the attainment of adult life. The larva is often very different from the adult.

The larvae of many invertebrates of **sessile** habit, such as the **sponges** and some **coelenterates**, are ciliated (**cilia**) organisms which swim about for a time before attaching themselves to the permanent support where they are to develop. In the two **phyla** mentioned the larva is little more than a ciliated **gastrula**. It is filled with solid endoderm in the coelenterates and is called a planula.

The **flukes** also begin life as a ciliated larva known as a miracidium. This form gives rise to more complex larvae called rediae and these in turn produce tailed larvae called cercariae. The cercaria is transformed into the adult. In the same phylum the tapeworms hatch as 6-spined hexacanth larvae and pass through a bladderworm stage before becoming adults.

Roundworms of many species pass through one or more larval stages in which they are worm-like but differ in habits and in some structural details from the adults.

Some of the segmented worms, **mollusks, echinoderms,** and **chordates,** hatch as complex larvae with localized zones of cilia for locomotion. The trochophore or trochosphere larva of **annelids** and mollusks has a ciliated alimentary tract with mouth and anus 90 degrees apart and a belt of cilia around the middle of the body. Larvae of echinoderms also have a bent alimentary tract but the cilia are arranged in one or more bands, sometimes of intricate form. These larvae bear various names: bipinnaria, auricularia, pluteus, of the starfishes, sea cucumbers and sea urchins and brittle stars, respectively, or collectively the dipleurula. The bipinnaria resembles the tornaria larva of **Balanoglossus.**

Among the **arthropods** the high development of metamorphosis is accompanied by great diversity of larval forms. In the class **Crustacea** these forms seem to represent previous evolutionary stages and are named after groups of the class whose adults they resemble. Among them are the Nauplius, Cypris, and Cyclops larvae and many others. A single individual may pass through several of these stages in the course of its development. Insects present an entirely different type of metamorphosis in which the larval characteristics appear to have been acquired later in the course of evolution than those of the adult, as an adaptation to special conditions (**Cenogensis**). In species with complete metamorphosis only the first immature stage is called the larva. In this stage butterflies and moths are called caterpillars, some of the beetles grubs, and many flies maggots. Species with less complex metamorphosis are called nymphs or naiads during development (in gradual and incomplete metamorphosis respectively).

(See also **Actinula, Ephyra, Glochidium, Hydratuba, Hypopus, Pilidium, Scyphistoma,** and **Veliger.**) (A.W.L.)

LARVACEA. The appendicularians. A class of minute marine animals belonging to the subphylum **Tunicata.** These forms have a trunk and tail and resemble the larvae of the tunicates. (A.W.L.)

LARYNGITIS. Inflammation or infection of the **larynx** occurring commonly as part of a general upper respiratory infection or as a complication of the common cold. Pulmonary **tuberculosis** is not infrequently complicated by tuberculous laryngitis. **Syphilitic** laryngitis is rare. **Croup** is a form of laryngitis. (D.M.H.)

LARYNX. The larynx or voice box is in the upper anterior part of the neck in front of the lower part of the **pharynx** and just above the **trachea.** It is composed of a series of cartilages adapted for phonation. The cavity of the larynx is divided into three compartments by two paired folds of **mucous membrane** stretching across it. These folds are the vocal cords. The two lower ones are the true vocal cords. They are set into vibration by currents of air which pass over them during breathing. In speaking, laughing, etc., the air currents are altered to produce different sounds. The pitch, volume of sound, and tautness of the vocal cords are under voluntary control, and are modified by voluntarily changing the force of the air current.

The larynx is subject to inflammation and infection (**laryngitis**), tumors (**cancer**) and trauma. Laryngeal **diphtheria** is a particularly fatal form of the disease. (See **Sound Production.**) (D.M.H.)

LASSO CELL. Colloblast.

LATENT PHOTOGRAPHIC IMAGE. The invisible, but developable, image formed on exposure of a light-sensitive emulsion to light is termed the *latent* image. Formerly thought (1) to consist of a subhalide, i.e., a silver halide containing less halide than the normal, or (2) a physical modification of the silver halide crystal, it is now generally considered to be composed of metallic silver adsorbed to silver bromide. This assumption is without direct experimental confirmation but is supported by much indirect evidence. It is well known, for example, that if the exposure producing a latent image is greatly increased a visible image is obtained. This suggests, although it does not necessarily follow, that the visible "print-out" image is a continuation of the same process which produces the latent image. Since the visible image, formed on prolonged exposure, is known to consist of metallic silver, it is evident that the latent image may also be regarded as consisting of silver. Furthermore, the latent image is destroyed by solvents of silver, such as nitric acid, or by strong oxidizing agents such as potassium permanganate or persulfate. Substances which reduce silver halide to silver produce a latent image; i.e., render an emulsion developable without exposure to light. Finally, no other explanation of the process of development fits the known facts so well as that which regards the latent image as consisting of a center, or nucleus, of silver on the grain of silver halide which serves as a center for the reduction of the entire grain to metallic silver by the developing agent.

While the latent image is generally regarded as consisting of silver, the mechanism by which silver is formed when light quanta are absorbed by the silver halide crystals of a photographic emulsion is not entirely clear.

According to one theory (Gurney-Mott) the quanta absorbed by the silver halide crystal produce free electrons which are trapped—or held—by the sensitivity centers. These sensitivity centers are small centers, or nuclei, of silver sulfide on the silver halide crystal and are formed during the digestion of the emulsion (see **Photographic Emulsions**) with heat or ammonia, as a result of a reaction taking place between substances normally present in the gelatin and the silver halide. As the electrons thus trapped are negatively charged, the sensitivity center acquires a negative charge and thus attracts positive charges. Therefore, silver ions, which bear a positive charge are drawn to the sensitivity center until its negative charge is neutralized. The sensitivity center then holds metallic silver bearing no free charge. The sensitivity center thus forms the nucleus on which the free silver constituting the latent image is built up.

Not all students of the subject, however, agree that the latent image consists of silver. Huggins, for example, suggests that the effect of exposure is to convert the cubic crystal, typical of the silver halides, into a crystal of a different type, in fact of the type represented by zinc sulfide. His calculations suggest that the energy (light quanta) necessary for such a transformation is more nearly in accord with that required for the formation of a latent image than is the case if the result of exposure is to form metallic silver. This theory is of much more recent date (1943) than the other and less fully developed. (C.B.N.)

LATERAL. A force which acts on a structure or a structural member in a transverse direction is sometimes called a lateral load. The wind blowing upon the exposed surface of a **bridge** or building at right angles to its length or upon the stationary or moving traffic using the bridge constitutes one type of lateral load. The sway of a moving train on a bridge or the centrifugal force transmitted if the bridge is on a curve is a type of lateral loading. A moving crane supported on girders exerts a side thrust on the girders which may also be included in this classification.

Trusses and **girders**, which constitute the main load-carrying members of bridges, are not ordinarily designed to carry side loads of this nature, and consequently have very little strength in that direction. For this reason, the trusses or girders of a bridge are joined together in a horizontal plane by a system of lateral bracing composed of **struts** and diagonals. The diagonal members are known as laterals. This lateral bracing stiffens the whole bridge and opposes any sidewise **deflection** or vibration.

The term lateral is also used in connection with **sewerage** systems. Any sewer which serves the abutting property owners and in which each owner has an equal right is a common sewer. A lateral sewer is one which has no other common sewer flowing into it. (C.W.C.)

LATERAL BRACING. Bridge.

LATERAL LINE ORGANS. Sensory organs located on the head and in a line along the side of the body in the fishes (**Pisces**). They are unlike the sensory organs of terrestrial vertebrates but are supposed to perceive vibrations of low frequency. (A.W.L.)

LATERAL MORAINE. The **talus** or other material from the sides of a glacial valley accumulated upon the glacier and are carried away by it. Thus the mass of debris distributed along the lateral edges of the glacier is called a lateral **moraine**. In the case of valley glaciers which have disappeared, their former existence may often be proved by the traces of lateral moraines left along the sides of the valley. (R.M.F.)

LATERAL SEWER. Lateral; Sewerage System.

LATERITE. The sub-aerial decay of rocks in tropical regions, having a distinctly moist or rainy climate, results in the development of a residual, reddish, and usually sticky soil frequently containing **concretions.** The principal products of lateritization are the hydrated oxides of **aluminum** and **iron** either in the crystalline or amorphous form. If the concentration of iron oxide is sufficiently high the laterite may be valuable as an iron ore. If, on the other hand, the concentration of alumina is high the laterite may be valuable as an ore of that metal. (R.M.F.)

LATEX. Latex is a milky substance found in many plants. It is a complex emulsion in which such substances as **proteins, alkaloids, starches, sugars, oils, tannins, resins,** and **gums** are found. In most plants the latex is white; but in some it is yellow; in others, orange or scarlet.

The cells or vessels in which latex is found make up the laticiferous system. There are two very different ways in which this system may be formed. In many plants the laticiferous system is formed from cells laid down in the **meristematic** region of the stem or root. Rows of these cells are formed. The cell walls separating them are dissolved, so that continuous tubes, called latex vessels, are formed. This method of formation is found in the **poppy family;** in the rubber plant, *Hevea brasiliensis;* and in the Cichorieae, a section of the **composite family** distinguished by the presence of latex in its members. Dandelion, lettuce, hawkweed, and salsify are members of the Cichorieae.

In the **milkweed** and **spurge** families, on the other hand, the laticiferous system is formed in a very different way. Early in the development of the seedling latex cells are differentiated. As the seedling grows these latex cells grow, forming a branching system which extends throughout the plant. So in the mature plant the entire latex system results from the growth of a single cell or group of cells which were present in the **embryo.**

The laticiferous system is found in all parts of the mature plant, including roots, stems and leaves, and sometimes the fruits. It is particularly noticeable in the cortical (see **Cortex**) tissues.

Many functions have been attributed to latex. Some regard it as a form of stored food, while others consider it a sort of excretory substance in which waste products of the plant are deposited. To still others it is a substance which protects the plant in case of injuries, the latex exuding and drying to form a covering which prevents the entrance of harmful bacteria and fungi. Similarly it may be a protection against browsing animals, since in some plants it is very bitter, and in others, poisonous.

Many products useful and valuable to man are obtained from latex. First among these is **rubber,** obtained from the latex of many plants. **Chicle,** widely used as the base of chewing gum, is another latex product. So also is **opium,** with its many derivatives. (R.M.W.)

LATHE. A lathe is essentially a machine tool for producing and finishing surfaces of revolution. The machine is designed to hold and revolve work about an axis of rotation so that it may be subjected to the action of a cutting tool moving in a horizontal plane through the axis of the work. When the cutting tool moves in a longitudinal direction or parallel to the axis, the operation is known as turning; when it moves in a transverse direction, it is known as facing. In hand lathes for woodwork and some metal work, the cutting tool is guided by hand; in engine lathes the cutting tools are generally guided by the machine tool itself.

Figure 1 shows the front view of a wood-turning lathe. The rotating spindle carries a live or spur center

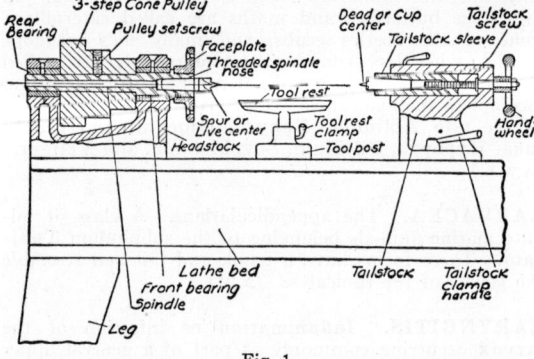

Fig. 1.

on which the piece to be turned is placed. The other end of the work is supported by and rotates on the nonrotating dead or cup center.

Many modern lathes have a motor built into the headstock with the spindle serving as the motor shaft. The lathe shown, however, is belt driven and has a three-step cone pulley which is driven from a countershaft cone pulley. The spur center fits in a tapered hole in the spindle which is hollow, so that the center may be removed by inserting a rod from the rear. The headstock is fastened to the lathe bed and carries the bearings in which the spindle rotates.

The tailstock may be moved anywhere along the lathe bed and can be clamped in place at any point. The tailstock is keyed to the bed, however, so that the headstock and tailstock centers will always be in alignment. The tailstock carries a non-rotating sleeve, keyed to the tailstock, which may be advanced or retracted by means of the tailstock sleeve screw operated by the handwheel.

The dead center fits in a Morse Taper hole in the sleeve and may be removed by retracting the sleeve, thereby bringing the end of the tailstock screw against the rear of center and forcing it out. Major adjustment for work length is made by moving the tailstock and clamping it in position; subsequent fine adjustment to provide the proper bearing for the work is made by turning the handwheel. The sleeve clamp handle is used to clamp the sleeve after the center is adjusted to the work.

The tool post may be placed anywhere on the lathe bed and clamped in position; the tool rest is then adjusted for height and set either parallel to the lathe axis or in an angular position, and clamped by the tool rest clamp. In wood turning, the stock is centered and driven on the live center with a mallet. The dead center is brought up to approximate position by moving up the tailstock and clamping it. The tailstock sleeve is then adjusted and clamped so that the dead center is imbedded in the work (a small quantity of mineral oil having been previously applied to the center). The tool post is placed in position and clamped, and the tool rest is brought to the proper height and clamped.

The spurs on the live center cause the work to rotate. Cutting is performed by moving the tool towards the headstock so that the bearing will take the cutting thrust. An outside bevel gouge is generally used for converting the stock from its rough original shape to a cylindrical form. The skew or shaving chisel is used to obtain a smooth finish after roughing with the gouge. Scraping chisels are employed for both finishing cuts and for grooving and angular cuts. The cut-off or parting chisel is generally used for grooving work at various points to definite diameters to serve as a guide in turning, thus eliminating frequent measurement. Turning chisels are equipped with wooden handles. The handle is held in the left hand, and the right hand guides the cutting edge. The parting chisel, however, is held in the right hand while a caliper to check the groove diameter is held in the left hand.

The *engine lathe* is used for metal turning, and is essentially similar to the wood-turning lathe, except that it is fitted with a power-actuated carriage and cross-slide for clamping and holding the cutting tool. The headstock and tailstock of an engine lathe are similar to those on a wood-turning lathe, but the engine lathe tailstock rests on a saddle to which it is cross-keyed, so as to permit the tailstock center to be adjusted for alignment or misalignment with respect to the spindle center. The centers have 60° conical points to fit center holes in the work. Their shanks are fitted to self-holding tapers in the headstock spindle and tailstock sleeve. The live center is often left soft, but the dead center is hardened since the work rotates on it.

The carriage is supported on the lathe bed ways and can move in a direction parallel to the spindle axis. The front wall of the carriage is termed the apron which provides a support for the operating hand wheel and control levers. The cross-slide is mounted on the carriage, and can move at right angles to the spindle axis. It is operated by the cross-slide screw which turns in a nut fixed to the carriage. On the cross-slide is mounted a saddle in which the compound rest moves. The compound rest is similar to the cross-slide, except that it can be swung around at an angle. The compound rest slide is actuated by a screw which rotates in a nut fixed to the saddle. The tool post fits in a tee slot in the compound rest, and the tool holder is adjusted and supported by a rocker fitting in a concave ring, and clamped by the tool post screw.

For turning and facing, the carriage is driven from the headstock spindle by gearing or belting, through a feed shaft. For thread cutting, where a definite amount of carriage movement is required for every spindle rotation, a lead screw, geared to the spindle, is used for the motion of the carriage.

The size of a lathe is determined by the diameter and length of work that may be swung between centers. A $14'' \times 54''$ lathe, for example, will handle parts that are $14''$ in diameter and $54''$ long. Lathes of comparatively small swing are generally termed bench lathes. Gap lathes are large engine lathes with a space or gap in the bed at the headstock which will permit work to be handled, particularly in facing operations, that would exceed the capacity of the ordinary engine lathe of similar proportions. Very large lathes, particularly those employed in ordnance manufacture, are often made without legs with a bed mounted directly on a concrete base.

There are three important methods of holding and rotating work in engine lathes, which may be referred to as turning between centers; chuck work; and faceplate work. In turning between centers, the work is supported by the 60° conical points of the live and dead centers, turns with the live center, and on the dead center. The work must therefore have 60° center holes in each end, machined by using a pilot drill and a 60° countersink, or a center drill. (The function of the drilled hole is to insure contact on the sides of the conical hole and not at the extreme point.) The work is rotated by a lathe dog, illustrated in Fig. 2, which is clamped to the work by means of a set screw. The tail of the dog fits loosely in a slot in the lathe faceplate and turns with it. In many lathes, the faceplate F screws on the spindle nose. In modern heavy-duty lathes, the faceplate is seated and keyed on a standard tapered spindle nose, and is drawn on and held in place by a "pull-on" nut engaging external threads on the hub of the faceplate. Lathe chucks are held in the same manner.

Draw-in chucks and collets are used for bar work, and are designed to fit on the heavy-duty spindle nose if the live center is removed. The collets, which are of various sizes, fit in the chuck and are clamped with a removable key or wrench. Drawing in the spring collet forces its outer surface against the taper on the inside of the chuck; releasing the collet causes its jaws to open by their spring action. Bars of any length may be held in the chuck, extending entirely through the hole in the spindle if necessary. Collets for all standard sizes of circular rod are available as well as collets for hexagonal bar stock and cylindrical metric sizes.

Fig. 2 illustrates the essential operations in turning cylindrical work on a lathe. The work shown is a cylindrical bar of cast iron. The casting is about $\frac{3}{16}''$ larger in diameter, and has been previously cut to a length of $\frac{1}{16}''$ greater than the finished part. The first operation is not illustrated, and consists of centering each end of the casting and drilling center holes with a center drill. A bent-tail lathe dog is clamped to one end of the bar which is placed between the live and dead centers of the lathe. The lathe dog tail fits loosely in the slot in the faceplate and serves to drive the work. The dead center is lubricated with some substance such as white-lead and oil, and adjusted to the work by turning the tailstock sleeve screw handwheel so that the work will rotate on the dead center without any perceptible looseness or "shake." The tailstock is then clamped. A straight-shank turning tool with an inserted tool bit is placed in the tool post and adjusted by means of the supporting rocker so that the tool point or cutting edge will be very slightly above center. The tool post screw is tightened so as to clamp the tool holder in place.

Fig. 2 shows the initial rough-turning operation about half completed. A cut sufficiently deep to get under the

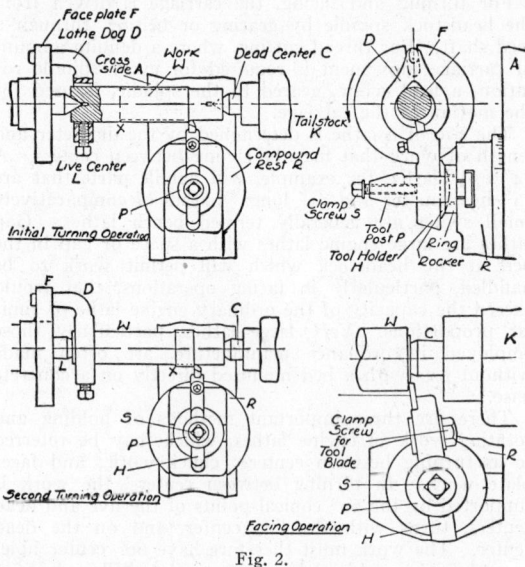

Fig. 2.

casting scale is taken, but at least .015″ to .025″ must remain for finish turning. The depth of cut is set by adjusting either the cross-slide or the compound rest, and the operator starts the longitudinal feed of the carriage by turning the carriage hand wheel. As soon as the operator observes that the feed and cut are satisfactory, he engages the power feed for the carriage, and the turning continues until point X is reached. At this point the lathe spindle rotation and the feed are stopped, the dead center is withdrawn, the work is removed from the centers, and the carriage is returned by hand to its position. The lathe dog is taken off the end of the bar and placed on the other end of the work which is then replaced between the centers, thus reversing the work end for end. Without disturbing the tool setting, the balance of the bar is then rough-turned as shown in the second turning operation. The second roughing cut is run slightly past point X to be certain that no ridge remains.

The same procedure is followed for the finishing cut. The tool is generally resharpened or stoned for this operation. A trial finishing cut is taken; this cut is just long enough so that the bar diameter may be measured, which may be done by using outside calipers, vernier calipers, or a micrometer, depending upon the degree of precision required. If the diameter is correct, the longitudinal power feed of the carriage is engaged, and a finishing cut is taken for about two-thirds of the length of the bar. The work is then reversed end for end, and the remainder of the bar length is finish-turned with the same tool setting. The finished surface on the second finishing cut is generally protected by inserting a piece of copper or brass between the work and the point of the lathe dog set screw.

After the cylindrical surface of the work is finished, the turning tool is removed, and a right-hand offset side tool is substituted. The carriage is set in the approximate position required for the end facing cut and

clamped. The compound rest is set around at an angle and used to provide adjustment for the depth of cut. The cross-slide is employed to feed the tool towards the axis of the work, and is generally hand-operated because the distance to be traversed is so short. After one end is squared, the work is reversed between centers, the other end squared, and the bar brought to correct length. The facing or squaring operation generally leaves a burr at the center holes, which must be removed by filing at the bench. In some instances it is necessary to obtain a smoother finish than that provided by turning. In such cases finishing by filing and polishing with emery cloth may be resorted to. An ordinary single-cut mill file is employed for lathe work. When a filed finish is required, the operator generally turns the bar .0005″ to .001″ oversize for filing allowance. Too much filing may destroy the precision of the bar; too little will not permit the removal of the turned ridges.

In machining long bars of small diameter, the bar has a tendency to spring away from the cutting tool so that a truly cylindrical surface cannot be procured. For such work some form of auxiliary support other than the use of the centers is required. The follower rest is a support that is bolted to the carriage with bearing blocks that can be adjusted to fit the turned surface. Whenever the bar has several diameters so that a follower rest cannot be employed, a steadyrest may be used. In this case it is necessary to turn a small portion of the bar to size, which can generally be done by taking light cuts. The steadyrest is clamped to the lathe bed, and its bearing blocks are adjusted to the turned surface or "seat" on the bar. In this application the steadyrest acts as a center bearing, but is not as effective as the follower rest which provides a support immediately to the right of the turning tool, or practically at the point where the cutting force is exerted.

Tapered or conical work may be turned in the lathe by three methods. The first method necessitates setting-over the tailstock an amount equal to one-half the taper, based upon the actual length of the stock. The set-over method is not applicable to large tapers since the centers will not fit properly in the center holes. The taper attachment method is accurate but requires a taper-turning attachment on the lathe. In this method the clamping bolt for the cross-slide feed screw nut is removed so that the motion of the cross-slide is no longer controlled by its screw, but is guided by a block moving in the taper attachment guide. The tool adjustment and setting is effected by the compound rest. Another method of turning tapers, particularly large tapers, is to set the compound rest at the required angle, clamp the carriage in place, adjust and set the tool by using the cross-slide, and cut the taper by hand-feeding the compound rest.

Hollow cylindrical work is usually machined in a lathe by drilling and boring the hole while the part is held in a chuck; the work is then removed and driven or pressed on a mandrel, which is swung between centers to turn the exterior periphery and face the ends. The hole in the work may be bored by a boring tool held in the tool post, or drilled and reamed by taper-shank twist drills and reamers held in the dead center socket in the tailstock sleeve.

In thread cutting, it is necessary that the carriage advance a definite amount for every revolution of the spindle, which is accomplished by connecting the lead screw to the stud, an auxiliary shaft rotating at the same speed as the spindle, by spur gears. Engine lathes are equipped either with a change gear box or with loose gears, called change gears, to provide suitable speed ratios for the stud and the lead screw. If, for example, the lead screw has 6 threads per inch, and a 6-pitch screw is to be cut, the gears on the stud and screw should have the same number of teeth; if a 12-pitch thread is to be cut, the gear on the stud should have one-half as many teeth as the gear on the screw.

Since lathes are equipped with a limited number of change gears, it is often necessary to compound the gearing to cut special thread pitches.

Lathes are also used for boring brackets, frames, etc., that cannot be held in a chuck or on a faceplate; this may be accomplished by removing the compound rest, mounting the work on the cross-slide, and using a boring bar held between the lathe centers. Springs may be wound on a lathe by using a mandrel of suitable size, and using the lead screw to provide a constant lead for the spring. A special tool or wire holder to furnish the necessary tension on the spring wire is required. (H.C.H.)

LATIN SQUARE.

The Latin square is a type of design used in the **analysis of variance.** Consider a table of r rows, r columns arranged in such a manner that each treatment occurs once and only once in every row and every column. The total sum of squares equals then the sum of squares among column **means,** among row means, and among treatment means. We are interested in testing the hypothesis that treatment means are equal, assuming the **variates** are **normally distributed** with common **variance** σ^2. The analysis of variance proceeds as follows:

Mean square among column means

$$r \frac{\Sigma(\overline{X}_c - \overline{X})^2}{r - 1} = V_c$$

Mean square among row means

$$r \frac{\Sigma(\overline{X}_r - \overline{X})^2}{r - 1} = V_r$$

Mean square among treatment means

$$r \frac{\Sigma(\overline{X}_t - \overline{X})^2}{r - 1} = V_t$$

Mean square due to error

$$\frac{1}{r - 2} \{(r + 1)V_T - V_c - V_r - V_t\} = V_e$$

Total mean square

$$\frac{\sum_{j=1}^{r} \sum_{i=1}^{r} (X_{ij} - \overline{X})^2}{r^2 - 1} = V_T$$

To test the hypothesis of equality of treatment means we compute **Snedecor's** $F = \dfrac{V_t}{V_e}$ with $n_1 = r - 1$, the **degrees of freedom** of the numerator and $n_2 = (r - 1)(r - 2)$, the degrees of freedom of the denominator. In this analysis $\overline{X}_c$ = mean of a column, $\overline{X}_r$ = mean of a row, $\overline{X}_t$ = mean of a particular treatment, Xij is the variate in the ith row and jth column, and $\overline{X}$ = mean of all the variates. The Latin square is an effective way of finding the causes of variation when there are 3 sources of variation and it is desired to eliminate the variation due to 2 of the sources of variation, the third source of variation generally called a treatment.

Latin squares are effective designs in agriculture and industry, particularly if the square consists of less than 10 rows and 10 columns. (See **Randomized Blocks.**) (L.A.A.)

LATITUDE.

The celestial latitude of a point on the **celestial sphere** is the spherical coordinate measured from the plane of the **ecliptic** along a great circle passing through the object and the poles of the ecliptic.

Because of the fact that the earth is not a perfect sphere there are several different sorts of terrestrial latitude in use. In Fig. 1 we have $PEP'E'$ an ellipse representing a section of the earth in the plane of a

meridian. C is the geometric center of the earth and the line COZ' is the line to the geocentric **zenith** of the point O. The angle ECO (φ') is the geocentric latitude of the point O.

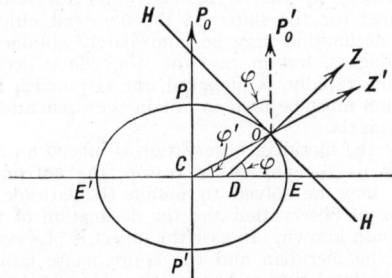

Fig. 1. Astronomic and geocentric latitudes.

The line DOZ represents the direction of **gravity** at the point O and extends to the astronomic zenith of O. The angle EDZ (φ) is the astronomic latitude of the point O. The difference between the astronomic and geocentric latitude of a point, the angle $COD = \varphi - \varphi'$, is defined as the reduction of latitude for the point O.

Because of local influences, such as massive mountains in the vicinity, the direction of the plumb line may not be strictly perpendicular to the surface of the earth. The geographical latitude of a point is the angle, measured in the plane of the local meridian, between the equator and a line drawn perpendicular to the theoretical **geoid** (surface of the earth) through the point in question. The difference between astronomic and geographic latitude is always relatively small, but by no means an inappreciable angle, and is known as station error. Station error is commonly between 4 and 6 seconds of arc, but occasionally amounts to 30 or 40 seconds.

In Fig. 1 CP represents the axis of rotation of the earth, which, if extended, will pierce the celestial sphere in its pole of rotation. The parallel line OP'_0 is the line from the observer at O to the pole of rotation of the celestial sphere and the line HOH represents the plane of the astronomic **horizon** at O. HOP'_0 is the **altitude** of the pole of rotation at O and inspection of the figure will indicate that this is equivalent to the angle EDZ. This gives rise to the common definition that the astronomic latitude of a point is the altitude of the pole of rotation of the celestial sphere at the point.

Astronomic latitude may be determined in a variety of ways by observation of the celestial objects. The most direct method is to observe the altitude of some object on the meridian whose **declination** is known. In Fig. 2 we have a representation of the celestial sphere drawn in the plane of the local meridian of the point O. In the figure, HOH' represents the plane of the horizon; $HPZQH'$ represents the local meridian; OP the direction of the pole of rotation; OQ the direction of the equator. $HOP = \varphi$ (the astronomic

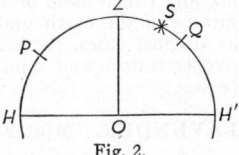

Fig. 2.

latitude of O), and $H'OQ = 90 - \varphi$. Since $H'S$ represents the altitude of a celestial object S which is on the meridian, and QS represents the **declination,** δ, of the object, we have at once the relationship that $\varphi = \delta + 90° - $ altitude. This is the method of determination of latitude most commonly used at sea and presents two fundamental difficulties to the navigator. The instant that the object is on the meridian must be accurately known and also the declination of the object observed. In case both the Greenwich time and the **longitude** are known, the instant that the object should reach the meridian may be calculated in advance from the **right ascension** of the object; and the observation of altitude is taken at the predetermined instant. Before chronometers came into use it was necessary to watch the object very carefully and record the maximum altitude attained by the object. In case the object was the sun, the time that the maximum altitude was obtained was the local apparent noon and was used by the navigating officer for setting the watch time for the ship. If the observed object is a star the declination may be immediately obtained from star catalogues, but in case the sun, whose declination is changing rapidly, is observed, the Greenwich time of observation must be used to obtain the declination from the **ephemeris.**

In case the meridian observation is missed on account of clouds or for any other reason, the **astronomical triangle** may be solved to obtain the latitude if the local time of observation and the declination of the object are both known. In case the object is observed very close to the meridian and the approximate latitude as well as the local time is known, the observation may be "reduced to the meridian" by tables published in a variety of places such as "Bowditch American Practical Navigator."

Modern navigational methods for determining latitude are discussed elsewhere. (See **Celestial Navigation, Dead Reckoning, Fix, Radio Navigation.**)

A meridian altitude of an object is always effected by the correction for **astronomical refraction** which is always subject to error unless the object observed is close to the zenith. For accurate determination of latitude for purposes of geodetic surveying the **zenith telescope** is used and the method will be described under the description of this instrument. (W.K.G.)

LATITUDE AND DEPARTURE. Bearing.

LATTICE. Crystal Structure.

LATTICE BAR. Compression Member.

LAUDANINE. Alkaloids.

LAUE PATTERN. Crystal Structure; X-rays.

LAUGHING GAS. Nitrous oxide. (See **Nitrogen.**)

LAVA. Molten material which has poured out on the surface of the earth and, due to relief of pressure, may have lost much of its original gas and water content during its relatively rapid consolidation. The term lava is used for both the liquid and the consolidated state of the igneous material. Lava may be erupted either by volcanoes or from fissures. The most extensive lava flows are fissure eruptions, such as the Columbia Plateau **basalts** in Oregon or the plateau basalts of the Deccan, India, which are derived from basic **magma.** Had this magma, either basic or acid, cooled slowly beneath the surface of the earth under great pressure and with all its original gases, the resulting rock would have had a coarser texture and somewhat different mineral content. (R.M.F.)

LAVENDER. Mint Family.

LAVENDER OIL. Volatile Oils.

LAW OF AVERAGES. The distribution of **sample means** possesses less **variability,** that is, it has a smaller **standard deviation** than the distribution of the **variates** themselves. This is readily seen since the standard deviation of sample means is $\frac{\sigma_x}{\sqrt{N}}$ (N is the size of the sample), while the standard deviation is σ_x. (See **Law of Large Numbers.**) (L.A.A.)

LAW OF LARGE NUMBERS. There are various laws of large numbers but the essential idea is exactly the same in each case. If the size of a **sample** is in-

creased indefinitely or becomes very large, sample estimates of **population parameters** will ordinarily be very accurate. **Bernoulli's theorem** is the first illustration of a law of large numbers.

In a second law of large numbers we state a corresponding theorem for **sample means** $\overline{X}$ as compared with the population mean m. In the limit as the size, N, of a sample increases, the **probability** of obtaining a discrepancy between a sample mean and a population mean of more than $\epsilon > 0$ in absolute value approaches 0, provided that the **variable** possesses a **standard deviation.**

$$\lim_{n \to \infty} P(|\overline{X} - m| > \epsilon) = 0.$$

There are other "laws of large numbers" somewhat more general. (L.A.A.)

LAW OF SMALL NUMBERS. If a **variable** is distributed in a **Poisson distribution,** then the variable is said to be distributed according to the law of small numbers or rare events, provided m, the mean is small. Bortkiewicz gives as an example the **frequency distribution** of the number of deaths per army corps per year in the Prussian army caused by horse kicks. Another example occurs in **industrial statistics** when the **universe** is essentially infinite, such as the number of defects on an article, since this number might possibly be very large. If the mathematical conditions for a Poisson probability function are not closely observed, the law of small numbers cannot be assumed arbitrarily just because the event is a rare event. (L.A.A.)

LAW OF SUPERPOSITION. Superposition.

LAWYER. Pisces, Teleostei. A fresh-water fish (Pisces) related to the cods. Also known as the **burbot** and ling. Most common in northern North America. (A.W.L.)

LAZURITE. The mineral lazurite or lapis lazuli has been used since ancient times for jewelry and other ornamental purposes. Ground to powder it forms the pigment ultramarine, now, however, largely superseded by artificial preparations. Lapis lazuli is a mixture of minerals, lazurite being the chief component. This mineral is **isometric,** and chemically a **sodium, calcium, aluminum sulfo-chloro-silicate.** Lapis lazuli has a hardness of 5-5.5; specific gravity 2.4; color, various shades of blue; luster, vitreous to greasy; translucent to opaque. Localities are Afghanistan, Siberia, Chile, and California. (E.S.C.S.)

LEAD. Symbol: Pb (plumbum). Atomic number: 82. Atomic weight: 207.22. Density: 11.34. Hardness: 1.5. Melting point: 327.5° C. Boiling point: 1620° C. (Isotopes: page 290.)

Lead is a white to bluish-gray metal, soft, malleable, and slightly ductile; tarnishes in air, forming a film of oxide, forms oxide scum upon heating the molten metal in air; soluble in dilute **nitric acid; hydrochloric or sulfuric acid** attack lead only slightly, the extent depending markedly upon the concentration and the temperature; slowly dissolves in water and consequently the use of lead constitutes a health hazard due to its toxic effect; attacked by solutions of organic **acids** or **sodium** hydroxide. Lead is the end-product of the **uranium-radium** and **thorium** series. Discovery prehistoric. Lead is one of the four most largely produced and utilized metals, and considerable scrap metal is recovered. Used (1) in construction and apparatus where workability is demanded, and definite resistance to corrosion is supplied by the metal, (2) as a constituent of various alloys, especially solder, type metal, pewter, and fusible alloys, (3) for storage battery plates, (4) for shot and bullets, (5) as a protective coating for iron and steel.

Lead occurs principally as sulfide ore (**galenite**, galena PbS) in Missouri, Idaho, and Utah of the United States, Broken Hill of New South Wales, Mexico, and Spain. The sulfide ore is roasted in air, resulting in a variable mixture of sulfide, sulfate, and oxide. The temperature is then raised with the exclusion of air, whereupon the sulfide reacts both with sulfate and with oxide to form **sulfur dioxide** gas and molten lead, which latter is drawn off and cast into blocks. Another method of obtaining lead is by fusion of the sulfide with scrap iron metal, thus yielding molten **ferrous** sulfide **slag** and molten lead separable by difference in densities. The refining of lead is conducted (1) by passing air over the surface of molten lead and removing the oxidized layer or "scum," (2) by **electrolysis** of lead fluosilicate solution in **hydrofluosilicic acid.**

Lead is commercially produced to standards of very high purity. The minimum lead content permitted by the A.S.T.M. Specifications for Pig Lead (7 classifications) is 99.73% and fully refined metal averaging 99.99% lead is obtainable. Large quantities are used for the production of paint pigments and lead tetraethyl for gasoline. Lead is soft and ductile and is readily worked by most of the popular methods, particularly by rolling and extrusion. It is easily formed and readily joined by welding (burning) or by soldering, and can be bonded to steel or be used as a liner for steel, wood, concrete, etc., as required. Lead is widely used because of its excellent resistance to atmospheric corrosion, soil corrosion, and attack by certain of the mineral acids, notably sulfuric and phosphoric. It generally does not resist the action of the organic acids nor the oxidizing mineral acids, such as nitric, and is attacked by alkalis and distilled water.

Acetates: Lead acetate, "sugar of lead" $(Pb(C_2H_3O_2)_2 \cdot 3H_2O)$, white crystals, soluble, formed by reaction of lead oxide and **acetic acid,** and then crystallization. Used (1) to furnish a soluble lead salt, (2) as a mordant in dyeing and printing textiles, (3) as a paint and varnish drier; basic lead acetate, white crystals, soluble, formed by reaction of lead acetate solution and lead oxide, and then crystallization. Used (1) as a coagulating, clarifying, and de-acidifying agent for many organic solutions, (2) in weighting silk.

Arsenate: Lead arsenate, arsenate of lead $(Pb_3(AsO_4)_2)$, white precipitate, formed by reaction of soluble lead salt solution and **sodium** arsenate solution. Used as an insecticide.

Azide: Lead azide (PbN_6), white precipitate, formed by reaction of soluble lead salt solution and **sodium** azide solution (white solid, formed by reaction of sodamide $(NaNH_2)$ upon heating in nitrous oxide N_2O) gas). Used as a detonator.

Borate: Lead borate $(Pb(BO_2)_2)$, white crystals, insoluble, by reaction of lead oxide and **boric acid** solution. Used as a paint and varnish drier, and in preparing special types of **glass.**

Bromide: Lead bromide $(PbBr_2)$, white precipitate, formed by reaction of soluble lead salt solution and **potassium** bromide solution, melting point 373° C.

Carbonates: Lead carbonate $(PbCO_3)$, white precipitate, formed by reaction of soluble lead salt solution and **sodium** carbonate solution in the cold; basic lead carbonate, formed by reaction of (1) soluble lead salt solution and hot **sodium** carbonate solution, (2) lead sheets, **carbon dioxide** and **acetic acid,** for "white lead" paint pigment, the quality depending largely upon the conditions of the reaction.

Chlorides: Lead chloride $(PbCl_2)$, white precipitate, formed by reaction of soluble lead salt solution and **hydrochloric acid** or **sodium** chloride solution in the cold, markedly soluble in hot water, melting point of lead chloride 501° C.; lead tetrachloride $(PbCl_4)$, yellow liquid, formed by reaction of lead dioxide or chloride and concentrated **hydrochloric acid** in the cold, is explosive on warming.

Chromates: Lead chromate, "chrome yellow" $(PbCrO_4)$, yellow precipitate, by reaction of soluble lead salt solution and **sodium** dichromate or chromate solution, melting point of lead chromate 844° C. Used as a pigment; basic lead chromate, red solid, insoluble, formed by heating lead chromate and sodium hydroxide solution.

Fluoride: Lead fluoride (PbF_2), white precipitate, formed by reaction of soluble lead salt solution and **sodium** fluoride solution, melting point of lead fluoride 855° C.

Hydroxide: Lead hydroxide $(Pb(OH)_2)$, (probably a basic hydroxide, formed), white precipitate, formed by reaction of soluble lead salt solution and **sodium** or **ammonium** hydroxide solution, soluble in **nitric acid** or excess sodium hydroxide, insoluble in ammonium hydroxide, yields lead oxide and water at 130° C.

Iodide: Lead iodide (PbI_2), yellow precipitate, formed by reaction of soluble lead salt and **potassium** iodide solution, melting point of lead iodide 402° C.

Nitrates: Lead nitrate $(Pb(NO_3)_2)$, white crystals, soluble, formed by reaction of lead oxide and **nitric acid,** and then crystallization, decomposes on heating leaving lead oxide residue. Used to furnish a soluble lead salt; basic lead nitrate, formed by reaction of lead nitrate solution and lead oxide.

Oxalate: Lead oxalate (PbC_2O_4), white precipitate, formed by reaction of soluble lead salt solution and **ammonium** oxalate solution, yields lead suboxide on heating at 300° C. out of contact with air.

Oxides: Lead suboxide (Pb_2O), black solid, formed by heating lead oxalate at 300° C. out of contact with the air; lead, oxide, lead monoxide, "litharge," "massicot" (yellowish), "yellow lead oxide" (PbO), yellow solid, insoluble, formed by heating in air any of the following: lead metal, lead dioxide, trilead tetroxide, or lead nitrate. Melting point of lead oxide 888° C., easily reduced to lead metal by heating with dry reducing agents. Used (1) for making many lead compounds, (2) in the manufacture of storage **battery** plates, (3) in compounding **rubber,** (4) in certain **glasses** and enamels for ceramic ware; trilead tetroxide, plumbo-plumbic oxide, "minium," "red lead" (Pb_3O_4), red solid, formed by heating lead oxide at 400° C. for some time, at higher temperatures decomposes to form lead oxide and oxygen, with **hydrochloric acid** yields lead chloride and chlorine gas, with **nitric acid,** lead nitrate and dioxide. Used (1) as a paint pigment especially on iron and steel, (2) in certain glasses and enamels for ceramic ware, (3) in packing metal pipe joints, (4) in compounding rubber; plumbic oxide, lead sesquioxide (Pb_2O_3), reddish-yellow solid, by reaction of **sodium** hypochlorite solution and lead hydroxide; lead dioxide, "brown lead oxide" (PbO_2), brown solid, formed (1) by reaction of trilead tetroxide and **nitric acid,** and then separating the dioxide precipitate from the solution of lead nitrate, (2) by **electrolysis** (at the anode) of lead oxide in the grids of the storage battery (lead metal simultaneously formed at the cathode electrolyte sulfuric acid). Used (1) as an oxidizing agent, e.g., in matches, (2) as one electrode of the lead storage battery.

Phosphate: Lead phosphate $(Pb_3(PO_4)_2)$, white precipitate, by reaction of soluble lead salt solution and **sodium** phosphate solution.

Sulfates: Lead sulfate $(PbSO_4)$, white precipitate, formed by reaction of soluble lead salt solution and **sulfuric acid** or **sodium** sulfate solution; basic lead sulfate, "sublimed white lead," white solid, formed (1) by reaction of lead sulfate and lead hydroxide in water (slow reaction), (2) by roasting galenite in a current of air. Used as a paint pigment called "sublimed white lead."

Sulfide: Lead sulfide (PbS), brownish-black precipitate, formed by reaction of soluble lead salt solution and **hydrogen sulfide** or **sodium** or **ammonium** sulfide, soluble in dilute nitric acid.

Tetraethyl lead ($(C_2H_5)_4Pb$), colorless liquid, decomposes at 125° C., formed by reaction of ethyl chloride and lead **sodium amalgam.** Used as an "anti-knock" in small concentrations in **gasoline** motor fuel. (See **Worm Gearing.**) (R.K.S., J. S. SMART, JR.)

LEAD AND TIN ALLOYS. Due to its softness, pure lead will very gradually flow or "creep," particularly at elevated temperatures, under low sustained stresses such as the oil pressure in lead-covered power conducting cable, or due to its weight in the case of a deep tank lined with sheet lead. For such purposes lead containing .06% copper ("Chemical" or "Acid" lead) is preferred, due to its superior resistance to "creep."

The addition of antimony in amounts up to 12% greatly improves the casting properties and increases the hardness very materially. These properties make possible the casting of intricately shaped antimonial lead storage-battery grids which, including the weight of the lead oxide paste applied to them, constitute the largest single use for the metal.

Tin and lead in various proportions form a highly useful series of alloys generally known as the soft solders which are used for joining copper, iron, nickel, lead,

ing points of which can be varied to suit a wide range of requirements. The type metals of the printing industry are lead-tin-antimony alloys having the requisite hardness and good casting properties needed for high-fidelity reproduction.

The **Babbitts,** or white-metal bearing alloys, are generally classified as either tin-base or lead-base. The true tin-base Babbitts contain only tin, antimony and copper, and have been used for many years. The practice of adding up to 25% lead to the tin Babbitts to reduce their cost is to be avoided since the net result is an expensive series of alloys with inferior properties to the inexpensive lead-base Babbitts. The lead-base bearing alloys of the older type usually contain lead, antimony and tin, and while not considered the equal of the tin-base alloys for severe service have been widely employed due to their low cost. In the past few years the lead-base alloy containing arsenic has found very extensive use and has come to the fore of this group since it has successfully met many automotive and other severe service requirements. All of these alloys render their most efficient service when used in the form of a thin lining bonded to a bronze or steel shell. (J. S. SMART, JR.)

REPRESENTATIVE LEAD AND TIN ALLOYS

NAME	Pb	Sn	Sb	Cu	Bi	Ag	Cd	TYPICAL APPLICATIONS
Lead Alloys								
Chemical or acid lead.	99.9			.06				Tank linings, coils, etc., power cable sheath.
Cable sheath	98.9		1.0					Telephone cable sheath.
Hard lead	96–92		4–8					Cast shapes, wrought sheet and pipe.
Battery grid metal	92–88	.25	8–12					Cast battery grids.
Solders								
Soft solder	50	50						General purposes, most popular solder.
Wiping solder	{ 60 { 60	40 37.5	2.5					For wiping joints in cables, lead pipes, etc.
"Fine solder"	40	60						For making joints at low temperature.
Solder	95–97.5					5–2.5		High temperature solder.
Fusible Alloys								
Wood's metal	25	12.5			50		12.5	Melts in hot water at 154° F. Wets glass. Wide range of melting points possible with changes in composition for automatic sprinkler systems and other safety devices.
Matrix metal	28.5	14.5	9		48			For anchoring punches, etc., in jigs and fixtures. Expands on freezing.
Bending alloy	26.5	13.5			50		10	Filler for tubes, etc., during bending. Melts out in hot water.
Type Metals								
Electrotype	93	3	4					
Linotype	84	4	12					
Stereotype	80.5	5.75	13.75					
Monotype	76	8	16					Single type.
Tin Base Babbitts								
ASTM #2		89	7.5	3.5				General usage.
ASTM #3		83.3	8.3	8.3				Hard Babbitt.
Lead Base Babbitts								
ASTM #15	82.5	1.0	15	.5	1.0 As			General usage.
ASTM #8	80	5	15					General usage.
ASTM #7	75	10	15					General usage.

zinc and even glass. The solders can be applied by means of a soldering tool, by wiping, by hot-dipping or by special machines as in the tin-can industry. Numerous compositions are used, the most popular of which are listed in the accompanying table.

Further additions of bismuth, cadmium, and antimony to the tin-lead alloys result in the low melting or "fusible" alloys widely used as safety devices, the melt-

LEAD, ENGINE. This is a steam engine term and means the distances the valve has uncovered the inlet port when the piston has reached dead center position. Lead is employed to decrease **wire drawing** of the steam engine cycle. (F.T.M.)

LEAD POISONING. This form of metal poisoning is often seen as an occupational disease. There are about 150 occupations which may expose the worker to lead poisoning, either from inhaling lead dust or absorbing it through the skin. These occupations include the manufacture of white and red **lead,** rubber, printing materials, storage batteries, pottery, etc. It is much less common now than 20 years ago due to better factory conditions. Poisoning may also occur from the use of hair dyes, cosmetics, and food or water contaminated with lead.

Lead is a cumulative poison, the metal being stored in the solid portion of the body skeleton. For this reason, even after exposure has ceased, its excretion from the body may continue for years.

The symptoms of chronic poisoning are great weakness, constipation, severe abdominal cramps ("lead colic"), marked anemia, palsy, and, at times, psychic manifestations and paralyses.

The red **blood** cells of a patient with lead poisoning show a peculiar stippling that is characteristic of this disease. (Figure 2, Plate B, facing page 180.) The deposit of lead sulfide about the blood vessels near the margins of the gums sometimes causes a visible "lead line."

The prognosis is good in most cases of lead poisoning, although symptoms may persist for many years. Paralyses and mental symptoms in advanced cases may be permanent.

Treatment is directed towards relief of symptoms, removal of the cause, and increased excretion of lead from the body. Calcium gluconate intravenously, potassium iodide, and saline cathartics are used. (D.M.H.)

LEAF. The leaf is a most important organ of the plant, since it manufactures food (see **Photosynthesis**). Typically leaves consist of a broad thin lamina borne on a slender stalk and green in color.

The leaf originates as a small protuberance from the surface of the growing tip of the stem. Numerous divisions of the cells of this protuberance produce a structure from five to eight cells thick. Many of these leaf primordia are borne together on the stem tip, and, together with any protecting scales which may cover them, form the **buds** of the stem. At first all the cells of these leaf primordia are alike. Very early in their existence, however, certain cells become distinct by their somewhat elongated shape. These cells are the beginnings of the **vascular** elements. Cell divisions continue in these small bodies until there are present in the bud recognizable but very small leaves which are folded in various ways. In woody plants this development takes place in the year previous to that in which the leaf will unfold. With the advent of the new growing season, growth of the many minute cells of these tiny leaves is very rapid, so that within a few days' time the leaf has unfolded and grown to its mature size. During this enlargement many changes have taken place in the cells of the leaf.

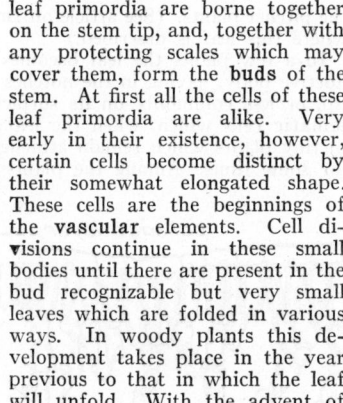

Leaf of apple, illustrating all parts—blade, petiole, and stipules.

~Blade
~Petiole
~Stipule

The mature leaf is commonly composed of two distinct parts, the broadly expanded, thin green blade, and the petiole or stalk which supports it and connects it with the stem. In many plants there is formed at the base of the petioles a pair of outgrowths called stipules, which in some plants may take the form of a complete sheath. This sheath is well developed in members of the **Carrot Family.** Sometimes the petiole is completely lacking, the blade being attached directly to the stem; leaves of this kind are called sessile leaves. Less frequently the blade of the leaf is lacking, the petiole being expanded into a flattened object looking much like a

blade. Certain Australian trees, species of *Acacia* and *Eucalyptus,* exhibit this peculiarity. Leaves of such plants often show progressive changes from those having well-developed blades to those in which the blade is completely lacking, showing clearly that the flattened portion present is a modified petiole. Such flattened petioles are not uncommon, but usually the blade is present, as is the case in the lemon tree. Leaves may be deciduous, falling off at the end of a single growing season, or evergreen and persistent through several seasons. In nearly all cases the leaf fall is brought about by the development of a definite **abscission layer.** In many plants such a layer is formed not only at the base of the petiole but also at the point where the petiole joins the blade.

The shape of the blade is extremely varied, ranging from very slender linear leaves to those which are broader than they are long. The margin of the leaf may be entire, that is, without indentations of any sort, or toothed or lobed in various ways, until some are incised nearly to the midrib. If the leaf is completely divided into separate segments it is said to be a compound leaf, in contrast with the undivided leaves, which are simple leaves, no matter how deeply they may be lobed. If the sections of a compound leaf all come from a common point, the leaf is said to be palmately compound; if they are borne along a central axis, the leaf is pinnately compound. While such infinite variations do exist, the leaves of any single species of plant are recognizably constant in shape.

The blade of the leaf is supported by a framework of veins which are also very characteristically arranged. In many leaves, especially in **dicotyledons,** one vein, usually extending through the center of the blade, is more prominent than the others. This is called the main vein or midrib. The others are lateral veins. In most dicotyledons the veins branch abundantly to form an intricately anastomosing network, which reaches all parts of the leaf. In most **monocotyledons** the midrib and lateral veins extend in parallel lines from base to apex of the leaf. Between these many minute veinlets exist, too small to be readily seen, reaching all parts of the leaf.

The cellular structure in leaves is very constant. (see next page for illustrations.) Covering the entire surface of the leaf is the epidermis, a layer of tabular cells. On the upper surface of the leaf the epidermal cells are frequently covered with a layer of cutin, a waxy substance which is impervious to water and so greatly reduces the loss of water by evaporation from the leaf surface. Epidermal cells contain a scant peripheral **cytoplasm,** and a large central **vacuole** full of cell-sap. Usually there are no **chloroplasts** present in the epidermal cells. The cells of the epidermis of the lower surface are similar to those of the upper, but with a less evident cuticle. In the epidermis of the leaf, particularly that of the lower surface, there are many minute openings, called **stomata,** which permit a ready exchange of gases between the interior of the leaf and the external air. Each stoma is surrounded by a pair of guard cells containing chloroplasts. These cells close the stoma by collapsing and open it by expanding. All cells occurring between the upper and lower epidermal layers are called mesophyll cells. Beneath the upper epidermis the mesophyll cells form a very distinct layer, called the palisade mesophyll. These are elongated cells with their long axis perpendicular to the surface of the leaf. They contain large numbers of chloroplasts. In them, furthermore, active **photosynthesis** takes place. Occupying all the rest of the leaf is a loose tissue composed of irregularly arranged rounded cells known as the spongy mesophyll. Numerous intercellular spaces separate these cells from one another. Ramifying through the leaf just below the palisade cells are the veins. Each vein is composed of three types of cells. Some of them are thick-walled **xylem** cells which carry water and dissolved mineral matter to all parts of the leaf. Others are **phloem** cells which carry food away from the green cells of the leaf where they are elaborated. The xylem cells are towards the top of the leaf, the phloem cells towards the bottom. Outside these and often forming a conspicuous tissue are masses of **fibers,** or collenchyma, thick-walled cells which give support to the leaf.

Leaves are often greatly modified. (Various types of leaves are illustrated on the following page.) In many have leaves modified into tendrils, slender thread-like objects which twine tightly around any suitable object with which they may come in contact. Sometimes only the tip of the blade functions in this way, and sometimes only the stipules are thus modified, as in the Carrion flower, *Smilax herbacea.* Many plants of the legume family have pinnately compound leaves, some of the segments of which are changed into tendrils. Weirdest of all are the leaves of species of *Nepenthes,* one of

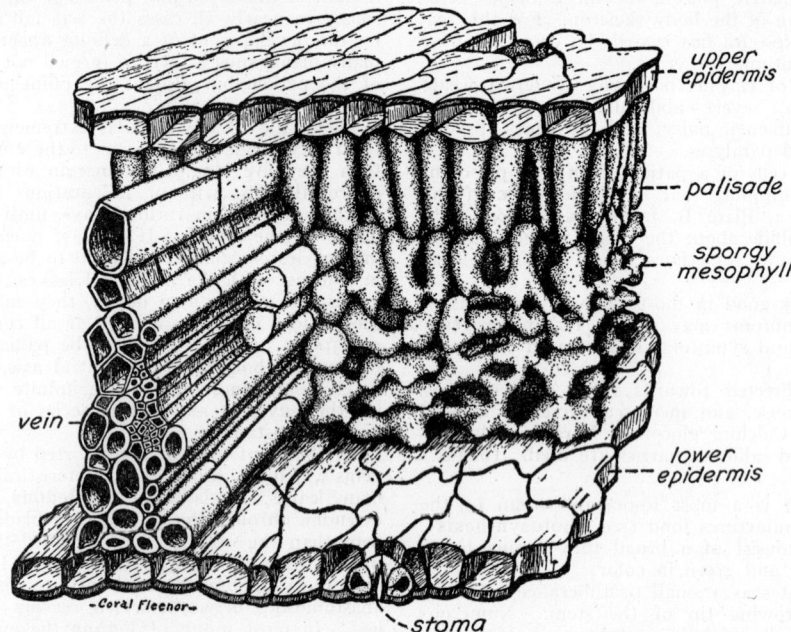

A portion of the blade of a leaf cut so as to show the internal structure. The cell contents are not represented.

plants they become greatly enlarged and fleshy, and serve as organs of storage of water and food. Many rock garden plants, such as species of *Sedum,* have leaves of this type. Of similar nature are the scale-like leaves which form the greater part of many bulbs, such as those of many lilies. The common onion is composed of the closely enwrapped bases of leaves, swollen with food material. In other plants modification of the leaves the pitcher plants. (See article on **Insectivorous Plants,** where this leaf is described.)

In a few plants the leaf becomes a vegetative reproductive body, having in the notches of its margin, at its tip, or less commonly on its surface, groups of **meristematic** cells which, when the leaf is mature, give rise to tiny plants which remain attached to the parent leaf for some time. Among the plants in which reproduc-

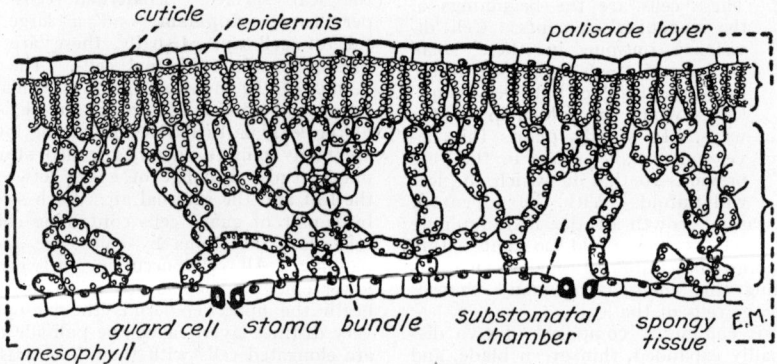

Cross-section of an apple leaf.

becomes extreme as, for example, in the **Pitcher plants** and **bladderworts.**

In other plants, such as the common barberry, the leaf is reduced to sharp-pointed branched spines; in many cases all gradations between these spines and typical leaves may be found on a single branch. In some plants, as the Locust, *Robinia Pseudacacia,* only the stipules are modified to short sharp spines. Many plants tion of this type occurs are species of *Bryophyllum* and *Kalanchoë.*

The principal function of the leaf is to carry on **photosynthesis.** To do this the leaf must receive adequate light. Leaves are not distributed haphazardly on the stem, but in a very definite way which assures them the maximum of light. In many plants the leaves are in pairs on opposite sides of the stem. Each successive pair usually grows out at right angles to the pair beneath it, thus preventing overshadowing. Leaves may occur in whorls, in which case there will be three or more leaves growing from each node of the stem.

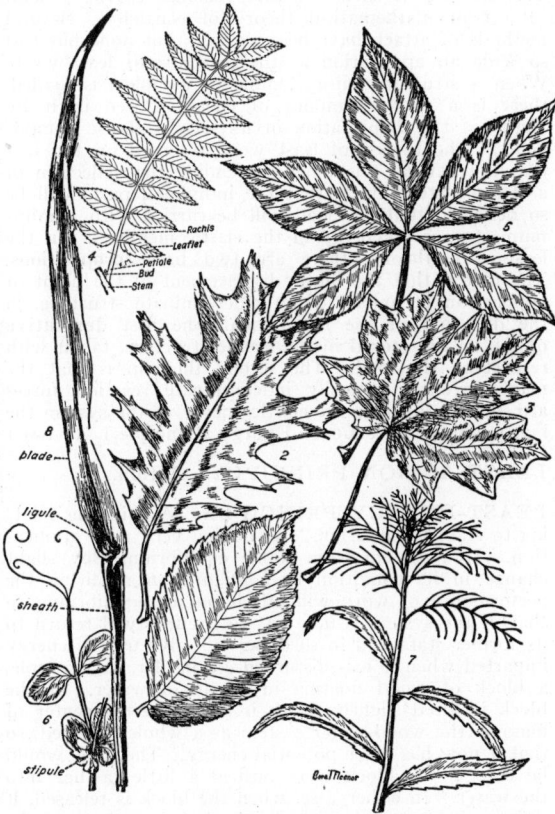

Types of leaves. 1, 2, elm leaf and oak leaf, both pinnately netted veined; 3, maple leaf, palmately netted veined; 4, black walnut leaf, pinnately compound; 5, buckeye leaf, palmately compound; 6, a pea leaf, with stipules, tendrils, and two unmodified leaflets; 7, portion of a plant of the water mermaid, *Proserpinaca,* with upper leaves modified by immersion in water; 8, grass leaf.

In many plants the leaves are alternate, each node bearing a single leaf. In every case alternate leaves arise from the stem in such a way that a line passing around the stem and through the junction of the petiole with the stem forms a regular spiral. Examination of this spiral shows that the leaves are distributed on it in a very exact mathematical arrangement. In the simplest case the leaves are in two longitudinal rows along the stem, every third leaf being directly above the first; in the next arrangement there are three longitudinal rows, the fourth leaf of the spiral being above the first. In another, and very common, arrangement, there are five rows of leaves, with the sixth leaf above the first. Other more complicated arrangements are found. A common way of indicating the arrangement of leaves on a stem is by means of common fractions. The fraction may be determined by starting at any one leaf and passing by the shortest way around the stem to the next higher leaf and so on until the leaf vertically above the first is reached; the number of turns about the stem from the first to the last leaf gives the numerator of the fraction, and the number of leaves passed gives the denominator. The fractions obtained form a very exact mathematical series; they are ½, ⅓, ⅖, ⅜, ⁵⁄₁₃, ⁸⁄₂₁, ¹³⁄₃₄, etc. The arrangement of leaves on a stem is called phyllotaxy.

Sometimes the exact arrangement is more or less obscured by twisting of the stem during growth. The leaves themselves turn considerably during their growth, petioles twisting to one side or the other, or elongating unequally, in such a way as to bring the blade into a position to receive the most favorable light. (R.M.W.)

LEAF HOPPER. Insecta, Homoptera. Any **insect** of the large family Cicadellidae or Jassidae. They are small to moderate jumping insects which often come freely to light.

Many species are of economic importance and, since they have sucking mouths, they must be attacked with contact poisons such as nicotine sulfate or kerosene emulsion. These sprays are effective against the tender immature insects but the adults cannot readily be killed. (A.W.L.)

LEAF INSECT. Insecta, Orthoptera. Large **insects** of the Old World tropics related to the walking-stick insects. They have leaf-like wings and in some species the body and legs are extended in flat processes which also resemble leaves. (A.W.L.)

LEAF MINER. Larval **insects** which work in the soft tissue of leaves between the upper and lower epidermis. They are necessarily small and are sometimes able to complete their development on a very small part of the food available in a single leaf. The burrow or mine shows as a brownish or transparent patch in the leaf and its form is characteristic of the insect making it. The larvae of many of the smallest moths and of some sawflies are leaf miners. (A.W.L.)

LEAF SPRING. Springs.

LEAKAGE CURRENT. This is the current which flows or "leaks" along the surface or through the body of an insulator. Except under abnormal conditions such as dirty or moist surfaces or in electronic circuits having very minute currents the leakage is usually negligible. (L.R.Q.)

LEAKAGE REACTANCE. This is the inductive reactance caused by the flux which links only one coil of a **transformer.** The useful flux, of course, links both windings and is the medium of transfer of energy between them. Leakage reactance is one of the major internal impedance components of the transformer. (L.R.Q.)

LEAST COMMON MULTIPLE. The least (or lowest) common multiple of several **polynomials** is the polynomial of lowest degree which contains each of them as a factor. It is usually found by factoring the polynomials separately, and then finding the least common multiple by inspection. (L.L.S.)

LEAST SQUARES. Principle of Least Squares.

LEAST SQUARES, METHOD OF. The method of least squares is a statistical method for obtaining the most probable value of a quantity from a set of **physical measurements,** and for obtaining a quantity which is an indication of the precision of the most probable value.

In the development of the theory of the method of least squares two lines of attack have been employed: the observational and the mathematical. In each of these certain fundamental assumptions are necessary. In the observational method the assumption is made that the arithmetic mean of a series of equally reliable observed values of a given quantity, each observation being freed from all systematic **errors of measurement,** is the most probable value of the quantity. To approach the theory from the purely mathematical point of view it is assumed that the accidental errors of observation follow the **Gaussian distribution.** These two assumptions are.

in fact, interdependent. If they are justifiable for the given set of observations, then it may be proved that the sum of the squares of the residuals will be a minimum. If from an individual observation we subtract the most probable value of the quantity, a residual is obtained. It is the purpose of the method of least squares to determine a quantity from a series of observations, whether directly observed, indirectly observed, or computed from observational material, such that the sum of the squares of the residuals shall be a minimum.

In the discussion of the Gaussian distribution it is found that the probability of occurrence of a residual of magnitude x may be expressed by

$$P = \frac{h}{\sqrt{\pi}} e^{-h^2 x^2}.$$

In this expression h is a measure of the precision (or relative agreement) of the individual observations of the measured quantity. In practice it is customary to express the precision of a measured quantity, as obtained from a number of independent observations from which all systematic errors have been removed, by certain functions of h rather than by h itself. In practically every case the function is inverse in character, i.e., the smaller the value of the precision indicator the greater is the value of h. The standard method for indicating precision adopted by most American and English workers in the physical sciences is the so-called probable error, which is symbolically expressed by the double sign $\pm$. If the probable error of a most probable quantity A is $\pm R$, then the chance that the true value of the quantity is between $A + R$ and $A - R$ is equal to the chance that the quantity has any value whatsoever outside of these limits. For example, if the length of a certain rod is published as 356.25 cm. $\pm$ 0.15 cm., then the chance that the true length of the rod is between 356.40 cm. and 356.10 cm. is equal to the chance that the length is anything outside of these two limits. Other methods for expressing the precision of a set of observations are in use, such as, for example, mean error, standard deviation, radical mean square error, etc. The meanings of these and other terms for expressing precision will be found in treatises on statistical analysis. The symbol $\pm$ should be reserved for probable error as defined above and should not be used for other terms for expressing precision, unless adequate notice is given to the reader. The methods for actually computing the precision measures of directly observed quantities, indirectly observed quantities, and quantities obtained by computation from observed values, involve too much detail to be included in a work of this character.

It should be carefully noted that the expression of the precision of a set of observational data is based upon certain fundamental assumptions and that if these assumptions are not valid for the particular set of observations, then the expression of precision is meaningless. It has been found that in most cases the accidental observational errors in the fields of astronomy, geodesy, and physics satisfy the Gaussian distribution law and hence least-square adjustments of the results are justifiable. This is not the case in the fields of biology and psychology, where one-sided variations in the observational material may occur to produce unsymmetrical distribution of the observational errors. In such problems great care must be exercised in expressing the precision of the results. A least-square discussion can never improve the quality of the observations, neither can it remove the effects of systematic errors, although a careful discussion of the residuals may indicate the presence of systematic errors. It must never be assumed that the method of least squares is some magical process which can be indiscriminately applied to all types of observational or statistical data. (W.K.G.)

LEAST TIME PRINCIPLE. Fermat's Principle.

LEAST WORK. The three equations of static **equilibrium** (see **Statics**) are insufficient to determine the analysis of certain types of structures which, in consequence, are called statically indeterminate structures. The analysis of such structures belongs to the general subject of mathematical theory of elasticity. Several methods of attack have been devised, but none has had so wide an application as the principle of least work. When a structure, either simple or complex, is loaded, there is a certain amount of energy stored in it by virtue of the deformation of its several elastic components. The theory of least work is based on the fact that the natural phenomena attending the deflection of a stressed structure is that the individual parts will be so deflected that the load will be carried with a minimum storage of energy in the elastic members. In the least-work theory, there are two basic propositions. The first is that the linear displacement of the point of application of a load of an indeterminate structure, in the direction of the load, equals the first **derivative** of the energy stored in the elastic structure, taken with respect to the load. The second theorem is that the magnitudes of statically indeterminate reaction forces are such as to make the elastic energy of the system the best possible. (See **Least-Energy Principle.**) (F.T.M.)

LEAST-ACTION PRINCIPLE. Action.

LEAST-ENERGY PRINCIPLE. A principle relating to stable equilibrium, and having very wide application. If a system is in stable equilibrium, any slight change in its condition or configuration requiring the performance of work will put it out of equilibrium, so that, if the system is now left to itself, it will return to its former state and in so doing will give up the energy imparted when it was disturbed. Consider, for example, a block of wood floating in a pail of water. If the block is lifted slightly, work is done and the center of mass of the wood-water system as a whole is raised, so that it now has more potential energy. The same would be true if the block were pushed a little farther into the water. In either case, when the block is released, it resumes its former level and the potential energy of the system diminishes to its former minimum value. This illustrates the general principle, which is that a system is in stable equilibrium only under those conditions for which its potential energy is at a minimum.

The principle of least energy is one aspect of the principle of virtual work (see **Equilibrium of Forces.**) (L.D.W.)

LEATHER. Tannins.

LEATHER-JACKET. Insecta, Diptera. The tough-skinned **larvae** of some species of **crane flies.** They live in the ground in pastures, hay fields, and grain fields and are sometimes serious pests. Since they come to the surface at night they can be destroyed by the use of poison baits. (A.W.L.)

LE CHÂTELIER LAW. Equilibrium.

LECHER OSCILLATOR. Electric Oscillations; Electric Waves.

LECHER WIRES. This is a special type of transmission line used primarily for the measurement of frequencies in the high radio-frequency range (see **Frequency Meter**). Basically they consist of two parallel wires a few wavelengths (in terms of the frequency to be measured) long which are adjustable in length. While usually adjusted by sliding a shorting bar along the wires they are sometimes fixed in physical length, either open or short-circuited at the remote end, and varied electrically by tuning a series condenser. In any event, if the system is coupled to a source of high frequency the induced waves will be reflected from the end and if the line is electrically an integral number of half-wavelengths long standing waves will result. The nodes or anti-nodes of these may be detected with a suitable detector and the spacing of corresponding points will be a half-wavelength apart, thus allowing the frequency to be calculated. (L.R.Q.)

LEDEBURITE. The eutectic alloy of the iron-iron carbide alloy system. It contains 95.7% iron, 4.3% carbon and forms at 2066° F. from liquid metal on cooling. (R.H.H.)

LEDLOY. Steels containing very small additions of lead to improve machineability. (R.H.H.)

LEECH. Hirudinea.

LEEK. Allium.

LEEWAY. The difference between the actual direction in which a ship is moving relative to the surface of the water and the direction in which the keel of the ship is pointing is known as leeway. Leeway is usually produced by the pressure of the wind against the side of the vessel and is much more pronounced in the case of sailing vessels than in internally powered ships. The amount of leeway can best be determined by observing the angle between the wake of the vessel and the line of the keel.

In determining the true course of the vessel the leeway is treated in the same manner as a compass correction. In case the wind is blowing against the left side of the vessel the vessel is said to be on the port tack and the true course will be to the right of the course indicated by the keel. Hence for a ship on the port tack leeway has the same effect as an east or positive compass correction, on the starboard tack leeway is applied as a west or negative correction. (W.K.G.)

LEG. A jointed appendage, used for locomotion on a solid supporting surface by walking, running, or jumping. The jointed appendages of arthropods and of vertebrates are termed legs. The term is applied to similar structures used for other purposes, such as swimming or burrowing, but those which have undergone great modification have also received special names. The anterior appendages of birds and primates are distinguished as wings and arms, respectively, but in quadrupedal vertebrates they are still called legs. In human anatomy, the term leg is restricted to the region between the knee and the ankle. (A.W.L.)

LEGENDRE FUNCTIONS. The differential equation

$$(1 - x^2)\frac{d^2y}{dx^2} - 2x\frac{dy}{dx} + n(n + 1)y = 0$$

is called Legendre's equation. It cannot be solved in terms of elementary functions. It defines a new class of functions, called the Legendre functions; they are also sometimes called surface zonal harmonics. Let n be a positive integer.

Legendre functions of the first kind (also called Legendre polynomials or Legendre coefficients) may be defined by

$$P_n(x) = \frac{(2n - 1)(2n - 3)\cdots 3.1}{n!}\left[x^n - \frac{n(n - 1)}{2(2n - 1)}x^{n-2}\right.$$
$$+ \frac{n(n - 1)(n - 2)(n - 3)}{2.4(2n - 1)(2n - 3)}x^{n-4}$$
$$\left. - \frac{n(n - 1)(n - 2)(n - 3)(n - 4)(n - 5)}{2.4.6(2n - 1)(2n - 3)(2n - 5)}x^{n-6} + \cdots\right].$$

$P_n(x)$ is a particular solution of Legendre's equation.

Legendre functions of the second kind may be defined by the series

$$Q_n(x) = \frac{n!}{(2n + 1)(2n - 1)\cdots 3.1}$$
$$\left[\frac{1}{x^{n+1}} + \frac{(n + 1)(n + 2)}{2(2n + 3)} \cdot \frac{1}{x^{n+3}}\right.$$
$$\left. + \frac{(n + 1)(n + 2)(n + 3)(n + 4)}{2.4(2n + 3)(2n + 5)} \cdot \frac{1}{x^{n+5}} + \cdots\right],$$

when the series is convergent. $Q_n(x)$ is another particular solution of Legendre's equation. Then the general solution of Legendre's equation is $y = c_1P_n(x) + c_2Q_n(x)$, where c_1 and c_2 are arbitrary constants. (L.L.S.)

LEGUME. Fruit.

LEIBNITZ' RULE FOR SUCCESSIVE DERIVATIVES OF A PRODUCT. This is a formula for the nth derivative of the product of two functions in terms of the successive derivatives of the factors. It is:

$$\frac{d^n}{dx^n}(uv) = \frac{d^nu}{dx^n} \cdot v + \binom{n}{1}\frac{d^{n-1}u}{dx^{n-1}} \cdot \frac{dv}{dx} + \binom{n}{2}\frac{d^{n-2}u}{dx^{n-2}} \cdot \frac{d^2v}{dx^2}$$
$$+ \cdots + \binom{n}{r}\frac{d^{n-r+1}u}{dx^{n-r+1}} \cdot \frac{d^rv}{dx^r}$$
$$+ \cdots + \binom{n}{1}\frac{du}{dx} \cdot \frac{d^{n-1}v}{dx^{n-1}} + u \cdot \frac{d^nv}{dx^n}.$$

The coefficients are binomial coefficients. (L.L.S.)

LEMBERG'S SOLUTION. A solution of logwood (see Tannins) digested in aqueous aluminum chloride which is used as a stain and reagent for distinguishing between calcite or aragonite and dolomite. Both calcite and aragonite are stained violet by the reaction while dolomite remains unchanged. (R.M.F.)

LEMMA. The lower one of the two bracts which immediately subtend the floret in the spikelet of grasses. (R.M.W.)

LEMMING. Mammalia, Rodentia. Small animals of northern latitudes. The European lemmings resemble the woodchuck in form but are much smaller, and the American species, *Synaptomys*, are like short-tailed mice. The common lemming, *Lemmus lemmus*, of northern Europe is noted for its occasional migrations. Many thousands of the animals take part in these migrations, crossing mountains, fording streams, and always pushing straight on until they enter the sea and are drowned. (A.W.L.)

LEMNISCATE. The lemniscate is a plane curve which may be defined as the locus of a point which

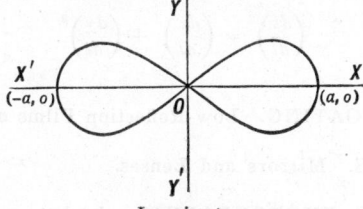

Lemniscate.

moves so that the product of its distances from the points $(-a, 0)$ and $(a, 0)$ is equal to a^2.

Its equation in rectangular coordinates is

$$(x^2 + y^2)^2 = 2a^2(x^2 - y^2),$$

and in polar coordinates is

$$r^2 = 2a^2 \cos 2\theta.$$

(L.L.S.)

LEMON. Citrus Fruits.

LEMON GRASS OIL. Volatile Oils.

LEMUR. Mammalia, Primates. The most primitive animals of the order containing man and the apes and monkeys. Lemurs have a well-developed thumb and great toe, like the other **primates,** but the second toe bears a sharp claw instead of a nail. The body is generally more like that of a squirrel than like the apes and monkeys and the face in many species is peculiarly expressionless, with large staring eyes.

Lemurs constitute a family Lemuridae, containing, in addition to the species named as lemurs, the indri, the sifakas or propitheques, the galagos, the awantibo, the pottos, the lorises or slow lemurs, and the avahi. They center in Madagascar but a few species occur in eastern Africa, southern India, and islands of the Oriental region. The tarsiers and the aye-aye are closely related to the true lemurs. (A.W.L.)

LEMUROIDS. Paleocene.

LENARD RAYS. Cathode Rays.

LENGTH OF PLANE CURVE ARC. The length of a curve arc is defined as the **limit** of the length of a broken line inscribed in the arc, as the number of pieces increases indefinitely and each piece approaches zero.

Let s represent arc length from a fixed point on a plane curve whose **equation in rectangular coordinates** is $y = f(x)$. Then

$$ds^2 = dx^2 + dy^2,$$

and

$$\frac{ds}{dx} = \sqrt{1 + \left(\frac{dy}{dx}\right)^2}, \quad \frac{ds}{dy} = \sqrt{1 + \left(\frac{dx}{dy}\right)^2},$$

and the arc length is

$$s = \int_a^b \sqrt{1 + \left(\frac{dy}{dx}\right)^2} \cdot dx = \int_c^d \sqrt{1 + \left(\frac{dx}{dy}\right)^2} \cdot dy.$$

If the equation of the curve in **polar coordinates** is $r = f(\theta)$, then

$$ds^2 = dr^2 + r^2 d\theta^2,$$

and

$$\frac{ds}{d\theta} = \sqrt{r^2 + \left(\frac{dr}{d\theta}\right)^2},$$

and arc length is given by

$$s = \int_\alpha^\beta \sqrt{r^2 + \left(\frac{dr}{d\theta}\right)^2} \cdot d\theta.$$

If the curve is given by **parametric equations:** $x = f(t)$, $y = g(t)$, then

$$\left(\frac{ds}{dt}\right)^2 = \left(\frac{dx}{dt}\right)^2 + \left(\frac{dy}{dt}\right)^2.$$

(L.L.S.)

LENS COATING. Low-Reflection Films on Glass.

LENSES. Mirrors and Lenses.

LENSES, PHOTOGRAPHIC. A photographic lens must form a sharp image on a flat surface extending over an angle of from 15 to 45° from the axis (depending upon the type of lens) at as large an aperture as possible. The problem facing the lens designer is a difficult one; to obtain good definition a high degree of correction is required but in a photographic lens, unlike most optical instruments, the same high degree of correction must be made to include the oblique pencils, and the image must be corrected for a flat field of

greater area in proportion to focal length and effective aperture. Progress in the fulfillment of these somewhat contradictory requirements has been hampered to some extent by the materials available. For nearly half a century, until the introduction of the so-called Jena glass in 1886, the glasses necessary for the correction simultaneously of astigmatism and spherical aberration were unavailable. It was only after the introduction of the newer glasses by Schott of Jena that the lens designer was able to produce a large-aperture lens which would at the same time provide good definition over a large field. Attempts in recent years to increase the effective aperture of the photographic lens still further without sacrificing the definition or the area have resulted in objectives of complex construction. The very recent introduction of newer types of glass by the Eastman Kodak Company should have an important bearing on future development in the field of photographic optics.

In the account which follows no attempt will be made to discuss the photographic lens from the historical point of view, nor can every make and type of lens be included. The principal types of lenses will be discussed and the purposes for which generally used.

1. Single lenses. Single meniscus lenses of the type shown in (1) of the accompanying group are used on

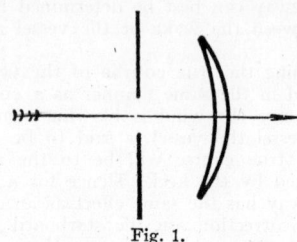

Fig. 1.

only the very cheapest of box cameras. The *single achromatic* lens shown in (2) is much more common even on inexpensive box cameras. Lenses of this type

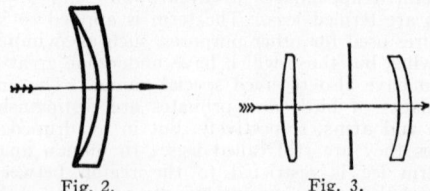

Fig. 2.   Fig. 3.

are generally made to work at from f/12.5 to f/16 but the definition is not sufficiently good to stand a high degree of enlargement. In some fixed-focus cameras a single lens is placed in front, as shown at (3). If this front lens is positive, it is moved to one side when making pictures of *distant* objects; if a negative lens is used then the front lens is removed for nearby objects.

2. The rapid rectilinear or aplanat. This lens (4), consisting essentially of two single achromatic lenses

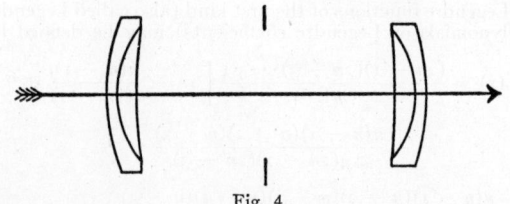

Fig. 4.

with a diaphragm between, was formerly in general use by both amateur and professional photographers. It is now obsolete having been replaced by other designs which are no more difficult to manufacture and provide superior definition at as large, or larger, aperture. The older lens with a relative aperture of about f/8 will be found on many hand cameras made 10 or 15 years ago.

3. The portrait lens. Many kinds of lenses are used for portrait work but the term portrait lens is applied almost exclusively to a type of lens designed by Josef Petzval in 1841 and shown in (5). This lens was the

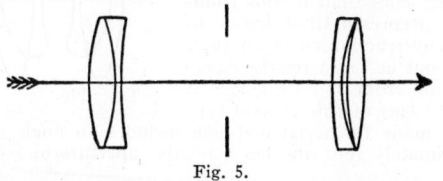

Fig. 5.

first large-aperture lens having a maximum aperture of f/6 when first introduced which was increased by various manufacturers over the next 20 years to f/2.2 in the shorter focal lengths. The definition on and close to the axis is exceptionally good for a lens of such large aperture but falls off rapidly. The field is sharply curved and the illumination diminishes rapidly from the center outwards. These defects are not so noticeable when a lens of long focal length is used for portraits of the head and shoulders and lenses of this type are still favored by many portrait photographers. However, even for portraiture, lenses of the Petzval type are being displaced more and more by other types.

4. Convertible lenses. The term convertible is applied to double lenses, the front or rear components of which may be used separately to obtain a different focal length. Lenses of this type are popular with commercial photographers, particularly for outdoor photography. In use, the single component should always be placed to the rear so that the diaphragm is in front. The single component usually requires about twice the exposure of the complete lens, the increase depending upon the largest aperture at which satisfactory definition is obtained. In most cases, the diaphragm plate attached to the shutter is marked separately for the lens as a whole and for each of the two components, if these differ in focal length.

While the rapid rectilinear, or aplanat, should be considered as a convertible lens since the front and rear halves may be used alone, the modern photographer is concerned chiefly with the convertible anastigmatic lenses. The *Dagor* of Goerz (6), the *Protar* of Zeiss (7), and the Turner-Reich (8) are probably the best known in this country. The maximum aperture of these lenses varies

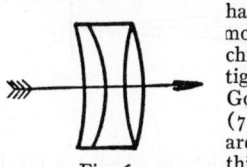

Fig. 6.

from f/6.8 to f/7.7 and the single components at from f/11 to f/16. Lenses of this type cover a large field in proportion to their focal length and, particularly at

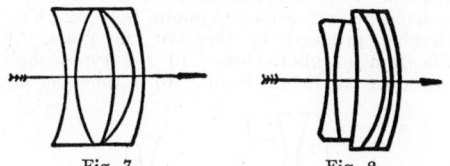

Fig, 7.   Fig. 8.

the smaller apertures, may be used as medium wide-angle lenses covering an angle of from 30–40° from the axis. This type of lens has received little attention from lens designers in recent years probably because of its relatively small aperture.

5. Triplet lenses. The first really successful anastigmat lens with a large relative aperture was patented by H. D. Taylor in 1895 and placed on the market by Taylor, Taylor and Hobson of Leicester, England, as

the *Cooke lens* (9). This lens is still popular, being used on hand cameras where a simple and relatively inexpensive lens is required and also as a portrait

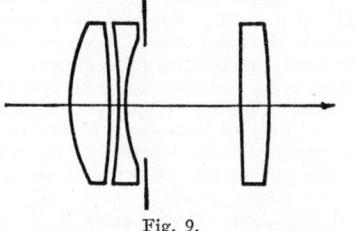

Fig. 9.

lens. It gives good definition at apertures up to f/4.5 and f/3.5 over a field of from 25–30 degrees. The *Heliar* of Voightlander (10) is a modification of the

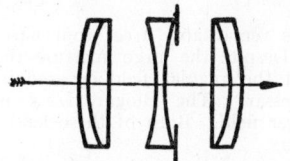

Fig. 10.

simple triplet, the outer elements being composed of two cemented lenses. This results in improved definition over a larger field. A lens of similar design by Dallmeyer of London, known as the *Pentac,* has a relative aperture of f/2.9. The Biotessar of Zeiss (Jena) (11)

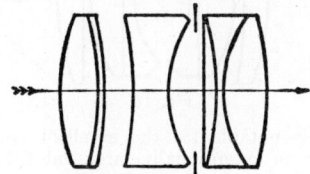

Fig. 11.

with a relative aperture of f/2.7 is similar except that the rear element consists of three cemented lenses.

One of the most popular lenses ever developed is the *Tessar* of Zeiss (12) which may be regarded as a triplet in which the rear lens is composed of two cemented lenses. This lens was designed by Paul Rudolph in 1903 and, since the patents have expired, has been widely copied by other lens manufacturers. The *Velostigmat,* Series of Wollensak, the *Paragon* of Ilex, the *Ektars* (f/3.5, f/4.5 and f/6.3) of Eastman Kodak Co., the *Serrac* of Dallmeyer, the *Skopar* of Voightlander and the *Xenar* (f/2.8, f/3.5, f/4.5) of Schneider are of this type.

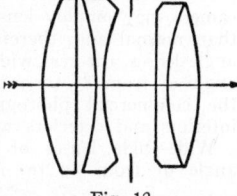

Fig. 12.

The *Aviar* of Taylor, Taylor and Hobson (Leicester, England) with a relative aperture of f/4.5 is a development of the original triplet lens with the central lens divided into two separate lenses of lower power (13). This permits the use of shallower curves and results in improved definition over a larger field than is obtained with the triplet. Lenses of this type are used by the press photographer on reflex cameras and in the longer focal lengths by commercial and portrait photographers except where an aperture larger than f/3.5 or an angular field exceeding 40° is required.

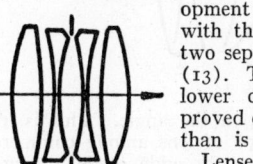

Fig. 13.

6. Lenses for miniature cameras. Lenses for the precision miniature camera require a higher degree of correction than lenses to be used on larger cameras because of the small image size and the necessity for subsequent enlargement. For lenses the maximum aperture of which does not exceed f/4.5 or, at the most, f/3.5 satisfactory results may be obtained with lenses of the triplet type, or better yet with one of the modified forms, and in particular the Tessar type.

The *Sonnar* (14) of Zeiss, with a relative aperture of f/1.5, is apparently a development of the earlier Tessar. Although the curves are deeper and the design more

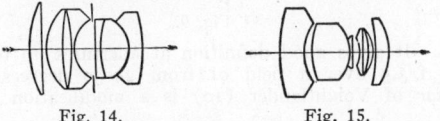

Fig. 14.   Fig. 15.

complex it is remarkably free from flare and internal reflections. Despite the large aperture the definition is excellent, but the angular field covered is smaller than with the Tessar. The Biogon f/2.5 of Zeiss (15) covers a larger field. Both of these lenses are designed for the *Contax*.

The *Ektar* f/2 of the Eastman Kodak Company, the *Biotar* f/1.4 of Zeiss and the *Xenon* f/2 of Schneider are based upon the so-called Gaussian design (16).

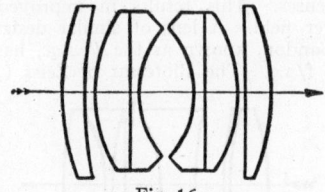

Fig. 16.

This design is notable for the excellent correction obtainable over an exceptionally large flat field at a large aperture.

7. Wide-angle lenses. In general, any lens which includes a wider angle than that normally used is termed a wide-angle lens. For the purposes of this discussion, however, any lens which is designed for use on a plate or film whose diagonal is greater than the focal length will be regarded as a wide-angle lens. While both the press photographer and the user of the miniature camera may employ lenses which include a wider angle than normal and, therefore, may be regarded as wide-angle lenses, the real wide-angle lens (which includes an angle of from 85 to 90°, or higher) is used chiefly by the commercial photographer in making pictures of interiors and exteriors when working in close quarters.

Wide-angle lenses of the type in (17) include an angle of from 85 to 95° at apertures varying from

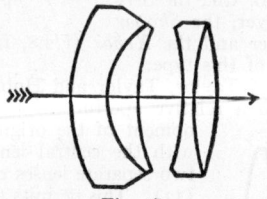

Fig. 17.

f/16 to f/18, depending upon the angle. This is, perhaps, the most popular type of lens among commercial photographers for an extremely wide angle, owing to its excellent definition over a wide angle and to its freedom from flare when windows, lighted lamps, etc., are included in the picture.

The construction shown in (18) does not, in general, include quite as wide an angle but has the advantage that

it may be made to work at a larger aperture. Lenses of this design may have a maximum relative aperture of from f/6.8 to f/8. Usually a smaller diaphragm must be used if the widest possible angle is to be included.

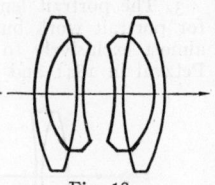

Fig. 18.

The design shown in (19) is more popular in Europe and the British Isles than in this country. Representative lenses of this construction cover an angle of about 90° at a relative aperture of from f/9 to f/11. A special lens of this general type (20) made for aerial mapping includes an angle of approximately 100° and has a relative aperture of f/6.3.

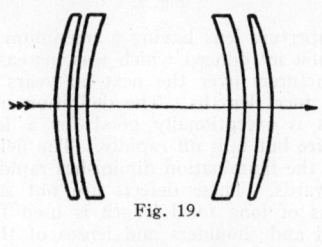

Fig. 19.

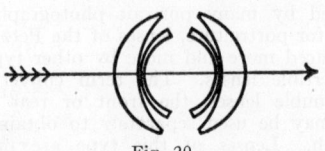

Fig. 20.

8. Telephoto lenses. The term *telephoto* is applied to lenses in which the front element is positive and the rear a negative—or diverging—lens which increases the size of the image formed by the positive element. This type of lens produces a larger image for the same lens-to-image distance (bellows extension) than a lens of the usual type. In other words, the focal length of the lens is greater than the lens to image distance. (Fig. 21.)

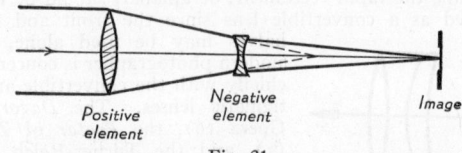

Positive element   Negative element   Image

Fig. 21.

The older *compound telephoto* lens in which the focal length can be varied by altering the distance between the positive and negative elements has the disadvantage that good definition is obtainable only at relatively small apertures. This so limits the usefulness of this type of lens that it is now practically obsolete.

The fixed-magnification telephoto, however, has made considerable progress in the last 20 years. Fixed-magnification telephoto lenses of the types shown in (22) produce an image from 2 to 3 times as large as

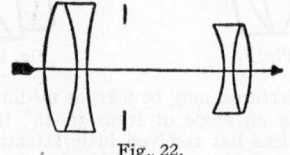

Fig. 22.

the lenses usually fitted to the cameras for which they are designed and provide good definition at apertures ranging from f/4.5 to f/6.3. The definition, generally speaking, is not equal to that of the best anastigmat lenses and there is usually some distortion, neither of which is sufficient to be of any concern to the average photographer. Telephoto lenses with an equivalent focal length of 40″, for example, have been widely used in aerial photography for reconnaissance photographs.

It should be noted that some lenses for miniature cameras and amateur cinematography, which are described as telephoto lenses or as "producing telephoto results" are, in reality, simply lenses of longer focal length mounted in long metal tubes which provide the increased extension necessary. (C.B.N.)

LENTICELS.

The young stems of plants are covered with a single layer of cells known as the epidermis. As the stem grows older, this epidermis is lost and replaced by a thicker protective tissue known as cork. The cork cells have walls which are suberized and impervious to gases. The living cells within the stem require an exchange of gases with the outside atmosphere. This exchange occurs through lenticels. A lenticel is a mass of thin-walled **parenchyma** cells loosely arranged so that air spaces are numerous. Through the lenticel gases pass readily. Lenticels appear on the surface of the stem as rough masses, usually

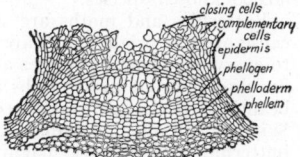

Stems of apple tree showing external view of lenticels.

Section through lenticel of cherry. (*Eames and MacDaniels' Introduction to Plant Anatomy.*)

protruding somewhat, and either circular or somewhat elongate in shape. They are very irregularly distributed. (R.M.W.)

LENTICULAR CLOUDS. Clouds.

LENTICULAR PROCESS.

A process of color photography in which minute semi-cylindrical lenses (lenticules) on the film base, in conjunction with a banded color filter on camera and projector lenses, act as an optical color-separation system for both analysis and synthesis. The lenticular process is thus essentially a screen method in which the screen is formed optically on the film during exposure. The lenticules generally take the form of small ridges embossed on the base side of the film, several hundred per in. Such a film is loaded in the camera with the emulsion away from the lens so that the light must pass through the lenticules and film base before exposing the film. The shape of the lenticules is such that they act as minute lenses and the camera lens aperture is thus imaged onto the emulsion. If the camera lens is covered with a banded color filter, the light transmitted by any area of the aperture will depend on the color of the light reflected from the subject and the filter-band color covering that portion of the aperture. The density of the silver image developed in the film will thus be determined.

If a lenticular film is exposed and processed by reversal development and then put into a projector with a similar optical system, the proper quantity of light will be projected through each band of the color filter and the original scene will be reproduced in color by means of an additive synthesis on the screen. Lenticular processes appear at first sight to be a relatively simple solution to the problems of color analysis and synthesis but physical and optical problems as well as projection difficulties have limited any widespread commercial usage. The lenticular system is applicable to two color processes and may also be used as an analysis

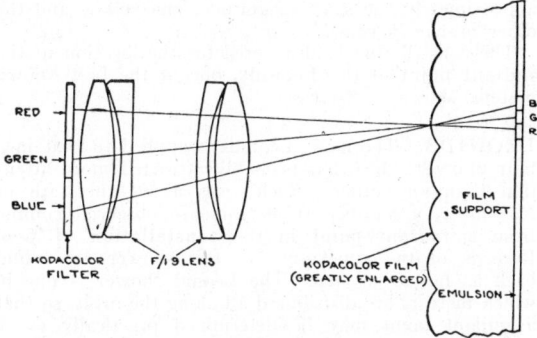

General optical system of Kodacolor lenticular process.

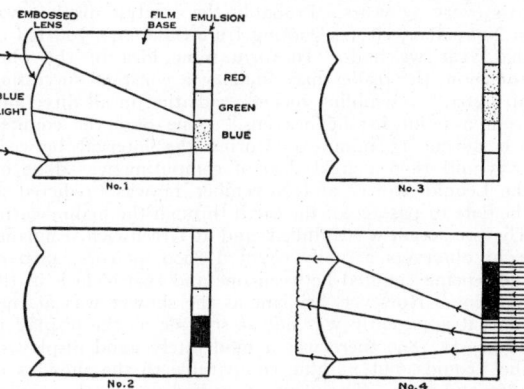

Enlarged sections of films show image formation in four stages: (1) exposure, (2) development, (3) bleaching and re-exposure, (4) reversal development and final projection. (*Neblette's Photography.*)

system for a process in which the color synthesis may be carried out by other than additive projection. (H.C.C.)

LENZ'S LAW.

A general law of **electromagnetic induction**, stated by H. F. E. Lenz in 1833. It points out that the electromotive force induced by the variation of magnetic flux, with reference to a conductor, in the manner discovered by Faraday, is always in such direction that, if it produces a current, the magnetic effect of that current opposes the flux variation responsible for both electromotive force and current. An outstanding illustration is the drag on a generator armature; if the armature circuit is closed, the rotation is opposed by a torque arising from the reaction between the field and the current in the armature conductors. Power must therefore be applied to drive the machine; and the greater the armature current, the more power is required. The effect known as magnetic damping depends upon this principle. A copper disk, when spun between the poles of a strong magnet, quickly comes to rest because of the opposing torque. This arrangement serves as a speed regulator in watt-hour meters. The oscillations of **galvanometer** coils are often damped in a similar manner. (L.D.W.)

LEO

(The lion) (Map page 380). The **constellation** of Leo is one of the most easily distinguished of all of the **zodiacal** constellations. The "sickle" of this, the fifth sign of the zodiac, is known to all watchers of the spring and early summer skies. The brightest star in the group, Regulus, is a **double** star but cannot be resolved with telescopes smaller than 3-in. aperture because of the fact that in small instruments the bright star masks the fainter one. Gamma Leonis is one of the finest of the doubles with its two components of approximately the same magnitude, one yellow and the other orange in color.

The constellation is also noted for the location of the **radiant point** of the **Leonids,** one of the best known **meteor** showers. (W.K.G.)

LEONIDS. The name Leonids is applied to that **meteor shower** which has probably attracted more attention than any other. Each year, about the 12th of November, a number of meteors are observed coming from a **radiant point** in the **constellation** of **Leo.** Records of the appearance of this shower are found back as far as 585 A.D. The Leonid shower is one in which meteors are distributed all along the orbit, so that a radiant point may be determined practically every year, and also there is a very strong condensation of the meteors into a swarm through which the earth used to pass every 33 years. Probably the greatest display was in November, 1833. Quoting from Silliman's Journal of that year we find: "To form some idea of the phenomenon, the reader may imagine a constant succession of fireballs, resembling rockets, radiating in all directions from a point in the heavens." One observer counted 650 during 15 minutes. During the interval between 1833 and 1866 a great deal of computing was done on the Leonid shower and November 13 was predicted as the date of passage of the earth through the main swarm. The prediction was fulfilled, and at Greenwich, England, eight observers actually counted 8000 meteors, 4860 of them being counted between one and two o'clock in the morning. However, brilliant as the shower was at that time, it apparently was not as striking as the display in 1833. In 1899 there was a moderately good display of the Leonids, but nothing comparable to the showers of 1866 and 1833. The newspapers had promised so much to the general public that the failure of the shower to come up to the expectations proved a rather serious blow to astronomy. The explanation for the failure of the shower to live up to its prediction is to be found in the fact that Jupiter passed very close to the swarm during 1899 and deflected it from the earth's orbit. Further **perturbations** have so deflected the orbit that in 1932, 1933, and 1934 no real shower was observed, although enough meteors were seen during each November to permit of a determination of the radiant point.

It is not possible to make any definite predictions for the future. It should be pointed out, however, that in the centuries preceding 1833 there are several instances where we find no record of striking displays on the basis of the 33-year period. It is very possible that perturbations may again bring the main swarm into such a position that the earth will again pass through it, giving rise to showers comparable with that of 1833. (W.K.G.)

LEOPARD. Mammalia, Carnivora. A large cat of the Oriental region and Africa. Its fur is tawny, marked with black rings and spots, although a black variety occurs in which the spots are faintly traceable. A species known as the snow leopard, *Felis uncia,* lives at high altitudes in central Asia, and in southeastern Asia a short-legged species called the clouded leopard, *F. nebulosa,* is found. The former is spotted and the latter blotched and striped. The cheetah, *Cynaelurus jubatus,* of India and Africa, is a slim long-legged animal marked with small black spots on a tawny to reddish ground. It has been trained for use in hunting and is also called the hunting leopard. (A.W.L.)

LEPIDODENDRON. Paleobotany.

LEPIDOLITE. This member of the **mica** group of minerals is a **silicate** of **potassium, lithium** and **aluminum,** sometimes with **sodium, florine,** or rarely **rubidium.** Crystals of lepidolite are **monoclinic** but often **pseudo-hexagonal; cleavage,** basal and perfect, being susceptible of splitting into thin laminae; hardness, 2.5-4; specific gravity, 2.8-3.3; luster, pearly; color, reddish to violet, grayish-blue, gray to white. A variety carrying rubidium is yellowish-gray; translucent. It usually is found as granular to scaly masses, in short stocky prisms or less often in eaily cleavable sheets. Lepidolite is characteristic of **pegmatite** veins, frequently being associated with other lithium-bearing minerals such as **tourmaline, spodumene, amblygonite,** and others. It occurs in the Ural Mountains, Czechoslovakia, the Island of Elba, and Madagascar, where it is often found in large sheets. It is found in the pegmatites of New England, California, South Dakota, and New Mexico. The name lepidolite is derived from the Greek meaning scale. (E.S.C.S.)

LEPIDOPTERA. The **butterflies, skippers, and moths.** An order of insects characterized by sucking mouths in the adult stage, complete metamorphosis, a **larva** with biting mouth parts, and two pairs of wings covered at least in part with a vesiture of flattened scales. The second order of insects in size, with about 90,000 known species.

Butterflies and moths are widely distributed and because of their bright colors are among the animals known to everyone. The butterflies are diurnal and so are readily observed, but most moths are nocturnal, hence many beautiful species are rarely seen unless they are sought. Skippers are an intermediate group more nearly like the butterflies. The most magnificent species of all three forms are tropical but representatives are found even in the Arctic regions.

The adults visit flowers for nectar or take no food. In the larval stage most species are plant feeders but a few carnivorous forms are known and some are scavengers. The order includes many economic species, among them the clothes moths, the bee moth, and the cut worms. (A.W.L.)

LEPROSY. A chronic infectious disease caused by *Mycobacterium leprae.* The disease presents a great variety of signs and symptoms, depending on what tissue or organ of the body is involved.

Leprosy is a disease of antiquity, and there is evidence that it has existed at least since 2000 B.C. References to the disease are found in the Old Testament. While it has been common in the Orient for several thousands of years, it appeared as a scourge in Europe in the 11th and 12th centuries, and did not subside until the 16th century when segregation of the victims was carried out on a large scale. At present the disease occurs endemically and sporadically, chiefly in the Orient, Australia, Asia, on the Mediterranean, and in Central and South America. There are various other foci of sporadic cases such as some parts of northern and central Europe, the West Indies, Louisiana, Minnesota, and South Carolina in the United States, and several in Canada. Occasional cases are encountered in the larger seaports, both Atlantic and Pacific.

The mode of infection is not definitely known. The nasal mucosa, the gastrointestinal tract, and the skin have been considered as possible portals of entry. The disease is definitely contagious, but years of exposure and contact seem to be necessary for its transmission.

The organism causing leprosy is similar to the tubercle bacillus in appearance and staining characteristics. It is found in great numbers in the nodules occurring under the skin, in discharges from the nose and throat, and in discharges from ulcers. It was first discovered in 1873 by Hansen.

The period of incubation is variable. It may be from a few months to 20 or 30 years. The symptoms of the disease depend largely on the tissue attacked. There are two general types of the disease: (1) nodular leprosy in which the skin is primarily involved; and (2) maculo-anaesthetic leprosy, in which there is an involvement of nervous tissue. A mixed form also occurs showing symptoms of both forms.

In the first stages of nodular leprosy, brownish-red spots appear on the skin, usually on the limbs and face, covering large and small areas of the skin. Later nodular thickenings appear at these sites. The face may show the so-called leonine appearance, due to the thickening of the skin in the region of the forehead, eyes, lobes of ears and around the nose and mouth. The entire skin assumes an unhealthy, dusky appearance. Some of the thickened areas ulcerate and fingers and toes may rot off. Ulceration also appears in the nose and throat and the voice becomes hoarse. The eyes are affected similarly, and blindness may result. This form of the disease may last 10, 20 years, or longer without treatment. Many of the patients die of complicating disorders such as pneumonia, nephritis, tuberculosis and malnutrition.

Maculo-anaesthetic leprosy is characterized by flat, red to brown lesions on the skin, distributed symmetrically. These lesions gradually become insensitive to pain (anaesthetic); trauma, even burns, may occur without the patients feeling pain. Ulceration of the area and contractions will produce deformities.

The outlook for recovery in leprosy has been somewhat improved by modern treatment with chaulmoogra oil, or preparations derived from this oil, used either by subcutaneous or intravenous injections. This has actually arrested cases of leprosy, when the disease was not too far advanced. (R.S.M., D.M.H.)

LEPTITE. Granulite.

LEPTOCARDIA. Cephalochordata.

LEPTOSTRACA. Crustacea.

LESBIANISM. Perverted sexual practices between women. Homosexuality. (R.S.M.)

LESION. Any wound, injury, diseased area or area of local degeneration. (R.S.M.)

LETTER SIZE DRILL. Drilling.

LETTUCE. *Lactuca sativa.* **Composite Family.**

LEUCINE. Aminoacids, Polypeptides, and Proteins.

LEUCITE. The mineral leucite is a **metasilicate** of **potassium** and **aluminum** corresponding to the formula $KAl(SiO_3)_2$. It is **isometric** at a temperature of about 600° C. and psuedo-isometric at lower temperatures, at which the mineral may possibly be **monoclinic** or even **triclinic.** The external forms remain isometric. It has a conchoidal fracture; is brittle; hardness, 5.5–6; specific gravity, 2.45–2.50; luster, vitreous; color, white or some shade of gray; translucent to opaque. It is commonly found in the more recent lavas of high alkali content. Leucite is seldom reported from **plutonic** rock types. It is a relatively rare mineral. It is found plentifully at Vesuvius and Monte Somma and elsewhere in Italy, and in Germany in the Tertiary volcanic district of the Eifel. In the United States leucite has been found in the Leucite Hills of Wyoming, the Highwood Mts. of Montana and as pseudomorphs in New Jersey and Arkansas. Its name is derived from the Greek word, referring to its white color. (E.S.C.S.)

LEUCO BASE. Dyes.

LEUCOCYTE. Blood.

LEUCON. Synonymous with rhagon. **Porifera.** (A.W.L.)

LEUKEMIA. A disease of the blood-forming organs in which abnormal numbers of mature and immature **leucocytes** or white blood cells appear in the blood stream. When the bone marrow, from which granular leucocytes originate, is hyperplastic or overgrown, the leukemia is of the myeloid type, and the peripheral blood contains great numbers of immature granulocytes. (See Fig. 6, Plate B, facing page 180). When there is overgrowth of the lymph nodes, the leukemia is of the lymphoid type (lymphatic leukemia), and immature lymphocytes crowd the circulation. Monocytic leukemia is the third type and is believed to be a disease of the reticulo-endothelial system, which consists of specialized monocytic cells found in the liver, spleen, and lymph nodes.

The cause of leukemia is unknown. In some respects the disease resembles an infection, and in other ways it simulates a malignant **tumor.** The various types may be observed in either sex at any age, although most cases occur between 35 and 55 years. The acute form is commonest in early childhood. The disease occurs in various animals as well as in human subjects.

The onset of chronic leukemia is gradual. **Anemia** occurs as a constant feature and its characteristic symptoms, weakness, pallor, shortness of breath, may be the first to appear. The spleen becomes greatly enlarged and its weight alone may cause a dragging sensation in the left abdomen. Basal **metabolism** is increased and fever of some degree is usually present. In the lymphatic form there is general enlargement of the lymph nodes. The circulating blood may contain several hundred thousand white blood cells. Because of the reduction of platelets (see **Blood**) there is a marked bleeding tendency.

Death occurs in chronic leukemia from exhaustion due to progressive anemia, from intercurrent infection, or from **hemorrhage.** With x-ray therapy, remissions lasting for varying periods of time may occur. Patients may carry on a normal life for several years, but death is the eventual outcome in all cases.

Acute leukemia differs from the chronic form in that its onset is abrupt; and its course is rapid and fatal, usually after not more than 8 weeks. The circulating blood is crowded with great numbers of young cells, frequently so immature that an accurate diagnosis as to cell type is impossible. A severe anemia always develops. Moderate enlargement of lymph nodes and spleen, hemorrhage into the skin and mucous membranes, and **stomatitis** with infection and swelling of the gums occurs. There is no effective treatment; x-ray irradiation usually makes the acute disease worse. (D.M.H.)

LEUKOPENIA. A decrease in the normal number of leucocytes in the **blood** stream. This is a usual accompaniment of certain stages of some infectious diseases, while in other infections it is of grave prognostic significance, and indicates a failure of one of the lines of defence of the body. In **agranulocytosis** leukopenia is the chief manifestation of the disease. (R.S.M.)

LEUKORRHEA. A whitish mucoid discharge from the vagina. This may occur normally in slight degree at the end of a menstrual period, but where marked or persistent, is a sign of infection or disease in the **uterus, cervix, or vagina.** (R.S.M.)

LEVEL. The term level, as an adjective, applies to any direction perpendicular to the force of gravity, and is synonymous with horizontal. The familiar device commonly known as the spirit level finds place not only in the tool kits of bricklayers and carpenters, but is an essential feature of many delicate physical, astronomical, and engineering instruments. It depends upon the simple principle that an air bubble seeks the highest point of the container holding the liquid in which it is formed. The glass tube of a level is either slightly curved, like a sausage, with convex side upward, or is ground with a

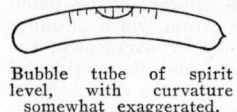

Bubble tube of spirit level, with curvature somewhat exaggerated.

curved inner surface. If such a tube is supported on a rigid base, the bubble contained therein always comes to equilibrium at the same point whenever the base has the same given inclination to the horizontal. If the instrument, with its base, be now reversed end to end, the bubble will move to a new position unless the base is exactly horizontal in both positions. By not moving, the bubble indicates that the surface is horizontal along that direction. If the bubble does move, horizontality can be obtained by adjusting until the bubble occupies a point midway between the two positions; then a second reversal should produce no change. If the tube is provided with a scale, the level may be made a very sensitive instrument for measuring angular changes of inclination. The larger the radius of curvature of the tube, the more sensitive is the level.

A surveyor's level is an instrument for determining differences of elevation directly. It is of use in obtaining comparative levels of two points, or in defining the profile of a certain path, such as a roadway, drainage ditch, etc. A level of this type has an accurately made bubble level which is attached to and made exactly parallel with a telescope. In the wye level this telescope rests in Y-shaped supports. These supports are held in turn by the instrument base, which may be adjusted by hand, and which is attached, usually by screwing, to the top of a tripod, upon which the instrument rests when in use. The bubble tube being parallel to the center of the telescope, the latter will automatically be leveled, ready for a horizontal sight, when the bubble tube is level. The dumpy level has the telescope and supports cast in one piece or rigidly connected. Since there are fewer movable parts than in the wye level it can be adjusted more easily and remains in adjustment for a longer period of time. In use, a level is set up and the base is leveled up by means of leveling screws, until the bubble stays in the center of the bubble tube however the telescope is rotated. The telescope barrel contains a ring having cross-hairs, so that when an object is viewed through the eyepiece, the cross-hairs will center on the center of the object being sighted. The imaginary line which joins the optical center of the objective and the cross-hairs in the tube is known as the line of sight. This line should coincide with the line of collimation. The latter is usually the geometric axis of the telescope. The telescope is slightly different from the ordinary field glass, which has no cross-hairs to be focused on the eyepiece. Some levels are erecting, that is, provide an upright image, while others of a more precise or simplified type, produce an inverted view of the object and are known as inverting levels. The dumpy level is of the latter type. (See **Level Rod.**) (L.D.W., F.T.M.)

LEVEL ROD. The length of sight which may be taken with a surveying instrument may be increased—also the facility and accuracy with which the reading is taken—by the employment of a special rod upon which to take a sight. A level rod, then, is a measuring stick graduated and plainly marked in feet, tenths of feet, and hundredths of a foot with the subdivisions and numerals receiving a special treatment to render them easily distinguishable. Size and coloring of the graduations are of great aid to this end. The rod is usually of wood, metal bound, having metal fittings. Telescopic features permit a 12-ft. rod, for example, to be compressed to 6′

in length for ease in carrying. A large movable circle with cross lines, known as a target, aids in centering the cross-hairs of the instrument upon the rod. (Some-

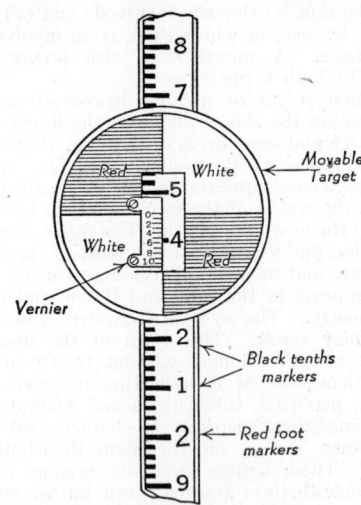

Level Rod. (Vernier target unnecessary except for levelling requiring highest degree of precision.)

times this target is equipped with a vernier making it possible to read elevations to thousandths of a foot. The levelman sets the center of the target which is read by the rodman.) (F.T.M.)

LEVER. Machines.

LEVULOSE. Carbohydrates.

LEWIS EQUATION. Spur Gearing.

LEXIS DISTRIBUTION. Coefficient of Dispersion.

LEXIS RATIO. Coefficient of Dispersion.

LEYDEN JAR. The earliest form of electrical condenser; attributed to two or three investigators, but generally credited to Muschenbroeck of the University of Leyden (1745). It is said that the discovery arose from an attempt to fix or imprison electricity by making a solution of it in water. A glass bottle contained the water, into which dipped a nail driven through the cork. The bottle was held in the hand while the water inside was charged from a static machine by means of the nail. When the operator, with his hand still around the bottle, touched the connecting wire, he received a severe shock. In the usual form of Leyden jar, the hand and the water are replaced by outer and inner coatings of tin or aluminum foil, the inner coating being provided with a rod and chain for convenience in charging. These condensers have low capacitance, but are rugged enough to withstand considerable voltages and are useful for many electrostatic experiments. In one interesting form the coatings are removable, leaving the glass jar itself charged. A highly charged Leyden jar may, incidentally, be somewhat dangerous to the operator, unless carefully handled. (L.D.W.)

LIANAS. In the tropics climbing plants often have woody stems of considerable size, which are called lianas. Many families of plants contain lianas. Some lianas are of economic importance; species of *Landolphia*, native in the forests of Africa, yield **rubber.** The means by which lianas hold themselves in place are many. Some have tendrils which twine tightly around any available support. Others have **adventitious roots** which attach them to the stems of sturdier plants; still others are provided with hooks of different kinds, which prevent them from sliding from their support. The stems of many lianas twine tightly around each other, and other plants. Lianas are often a very conspicuous feature of tropical forests, hanging in long festoons and trailing irregularly in tangled masses over the ground. (R.M.W.)

LIBIDO. The normal sexual desire. (R.S.M.)

LIBRA. (The scales or the balances) (Map, page 380.) Libra is a small constellation best known because it is the seventh sign of the **zodiac** and its symbol is taken for the autumnal **equinox** (i.e., the point where the sun apparently crosses the celestial **equator** from north to south). The brightest star in the constellation is a wide **double** which can easily be resolved by a field glass. This star carries the Arabic name Zuben el Genubi, the southern scale. (W.K.G.)

LIBRATIONS. The term libration is applied in astronomy to many periodic oscillations. In particular it is applied to slight apparent oscillations of the **moon,** whereby observers on the earth are enabled to observe somewhat more than 50% of the moon's surface. There are three principal librations of the moon: a libration in lunar **latitude,** a libration in lunar **longitude,** and a diurnal (or daily) libration.

The libration in lunar latitude arises from the fact that the **orbit** plane of the moon is inclined at about 6.5° to the plane of the moon's equator. This produces an effect relative to the earth similar to the terrestrial effects relative to the sun which produce the **seasons.** During one half of the **month** the north lunar pole is directed slightly toward the earth, while during the remainder of the month south lunar pole is toward the earth.

The libration in lunar longitude is due to the fact that while the rotation of the moon is quite uniform with a period equal to the period of revolution about the earth, nevertheless the orbital motion is not uniform but is in accordance with the **Keplerian law of areas.** Consider that a certain point on the surface of the moon is directly toward the earth on a date when the moon is in **perigee.** Since the moon is moving more rapidly in its ellipse at perigee than at any other part of its orbit, by the time that the **elongation** has increased to 90° the selected point has not completed one quarter of a rotation and is not directly toward the earth. The result is that the observer will see a slightly different hemisphere of the moon than he did at perigee. By the time the moon has reached apogee the selected point will again be toward the observer and he will have the same view of the moon as at perigee. At this point the moon's orbital motion is a minimum and the selected point will complete a quarter rotation before a quarter revolution is completed and another slightly different hemisphere of the moon will be visible.

The diurnal libration is due to the fact that when the moon is rising the observer sees slightly "over the top" of the moon and at setting slightly **under** the bottom. This is really a libration of the observer rather than a libration of the moon but is classed with the latter.

The combined result of the librations is that about 41% of the moon is always visible from the earth, or would be if the sun were shining upon it, 41% is never visible, and the remaining 18% is either visible or invisible, depending upon the particular position of the moon relative to the earth.

The term libration is also applied to certain periodic **perturbations** in the orbits of members of the **solar system.** (W.K.G.)

LICHENS. Lichens are perennial plants which are of very great interest because of their unique nature. For while lichens are treated as plants, and separated into genera and species just as other plants are, in reality they are rather a combination of two plants growing together in an association so intimate that they appear as one. Indeed, either of the component plants alone possesses none of the characteristics shown by the two in combination. Such an association of two organisms, living together and apparently of mutual benefit to each other, is called **symbiosis;** lichens are often cited as outstanding examples.

The components of a lichen are always an alga and a **fungus.** The algal constituent is usually one of the simple green **algae,** or, more rarely, a blue green one. The alga can live perfectly well by itself, and is often found growing free on rocks or tree trunks in regions where the lichen would exist. The fungal component is usually a member of the **ascomycetes.** In lichens growing in cooler regions it is always one of this group. There are certain lichens found in the tropics, however, in which the fungal component is a **basidiomycete.** In the lichen, the fungus alone is capable of fruiting, although the algal cells do divide and so increase in number. Lichens having an ascomycete fungus are usually called ascolichens, while those with a basidiomycete are called basidiolichens or hymenolichens. It is very difficult to see how such an association came about, and to determine whether it is really a case of symbiosis or whether it is not parasitism, one of the plants living on the other. Unquestionably the fungus benefits from the presence of the alga, since the latter carries on **photosynthesis,** making food materials which are used by the fungus. The latter, lacking **chlorophyll,** cannot manufacture its own food. It is possible that the alga benefits by the added moisture gathered by the fungus, that the latter protects the alga against desiccation. Certainly the association is well established, and seemingly has been so for a long period of time. Lichens have been "made" artificially, that is, the two components have been grown separately in pure cultures, and when brought together have produced a lichen. So lichens can be formed anew, always with a very constant appearance characterizing the particular form considered.

Lichens are found nearly everywhere where civilization has not killed them. They are found on the surface of rocks and soil: they occur on the bark of trees: in the tropics they may be found on the surface of thick evergreen leaves of trees: a few species even grow on rocks submerged by the tides. They are found on mountain tops which are not perpetually covered by snow. They occur from the tropic regions to the polar regions which are not permanently ice-covered. But, being slow-growing organisms and affected adversely by various gases, they are not usually found in the vicinity of large cities.

The shape of the lichen body or thallus is very diverse. Some species are flat crusts growing on or even in the surface of the substratum, whether the latter be trunk of tree or barren rock. Lichens of this type are called crustose. In others the thallus is split up into many radiating divisions, and is called a foliose lichen. Many others have an erect, often much-branched thallus and are called fruticose lichens. The color of the thallus may be yellow, orange, brown, gray or black.

The greater part of the lichen thallus is composed of **fungus** hyphae, which form a compactly tangled mass. The algal cells, usually called gonidia because of the early conception that they were the reproductive cells of the plant, occur in an irregular loosely arranged layer near the outer surface of the thallus. Short irregular branches from the fungus hyphae grow tightly around each algal cell, sending into it short absorbing structures called haustoria. The surface of the lichen is composed of enlarged thick-walled fungus cells which form a compact layer over the more loosely arranged central portions. **In** many of the crustose and foliose lichens there are many **rhizoids** which anchor the plant firmly.

One of the ways in which a lichen reproduces is by means of soredia. These are minute bits of lichen, formed on the surface, and composed of one or more of the gonidia together with a small mass of closely associated hyphae. Often these soredia are so numerous as to give to the lichen a powdery appearance. Either through disintegration of the lichen body, or because the continuity of the hyphae breaks down, soredia become free from the thallus. They are then easily spread by wind or by water. They may even be carried about unintentionally by the many small insects and other animals which feed on lichens. Lichens also reproduce by means of **spores;** that is, the fungus component forms special reproductive structures very similar to those formed by similar fungi not forming lichen thalli. In the ascolichens these reproductive structures are open cups or mounds, called apothecia. Commonly these apothecia are of a different color from that of the thallus. Inside each apothecium is a layer of asci, containing ascospores which are freed onto the surface of the ascus-containing layer and disseminated by wind or other agencies. An ascospore from a lichen apothecium can become a lichen only when it chances to reach an algal cell and can form an association with it. Since this will happen only rarely, reproduction of lichens is mainly accomplished by asexual or vegetative means. The formation of spores by the fungus would appear to be a persistence of a habit which was necessary when the fungus lived independently, but is of no value in its present condition.

A lichen; reindeer moss, *Cladonia rangiferina.* (*Smith, Lichens, Cambridge University Press.*)

It has been mentioned that lichens are a foodstuff for many of the lower animals. But some of them are also eaten by higher animals. Reindeer moss, *Cladonia rangiferina,* is the principal food of the reindeer, and may also be used as fodder for other animals. Reindeer moss, not a moss at all, is an erect much-branded lichen which grows abundantly over wide stretches of barren soil. The dense grayish-green tufts grow continuously at the tops, becoming 6–10" tall and attaining great age. Another lichen, *Cetraria islandica,* or Iceland moss, may be used as stock food. In habit it resembles reindeer moss but is coarser and less branched. The latter species may also be eaten by man. Like all lichens it contains an abundance of acid which gives a bitter astringent taste and causes violent digestive disturbances if eaten in any quantity. To get rid of this, the "moss" is gathered when moist, it being soft and pliable then, thoroughly washed to remove the bitter substances and dried. In this condition it may be stored, or it may be powdered first. To use this powdered "moss," it is soaked in water to remove the cetraric acid present, and boiled. It forms a tasteless jelly which is used in making soups, bread and porridges. Another lichen, known as rock tripe (*Gyrophora hyperborea*), likewise bitter and nauseating, if boiled and treated like Iceland moss, becomes

edible, if not palatable, and has been eaten by Arctic travellers in times of need. The manna of the Bible probably refers to another lichen, *Lecanora esculenta,* which occurs in desert lands.

In former times many lichens were used medicinally, as treatment for lung troubles, as purgatives, and as tonics.

Another and more important use of lichens was as a source of dyes. To obtain the dye the soaked lichen was treated with an alkaline substance, which acted on the acid of the lichen. The purple dye, orchil, was obtained from *Rocella tinctoria,* a lichen of the seaside, in this manner. Only animal fibers can be dyed by it. Another blue dye, litmus, is obtained from *Lecanora tartarea.* Litmus is prepared as a dark brown powder which is soluble in water. Litmus paper is paper which has been soaked in litmus solution. Acids cause litmus to turn red; while alkalis, even in small quantities, will make it blue. Holland produces nearly all the litmus used. Many other lichens yield dyes, some of them blue, others red or crimson, brown and yellow. Today, however, they are little used, modern synthetic dyes having supplanted them.

In those days when powdered wigs were in fashion, another and important use was made of lichens. This was as a hair powder, because of the fact that powdered lichens would retain for a considerable time any fragrance that was given them. So the lichens were packed with sweet-smelling flowers, or other fragrant substances, then dried and powdered. Often other substances, such as musk, were used with the powdered lichen. (R.M.W.)

LICHI. Mammalia, Artiodactyla. An African **antelope.** The species inhabits swamps in the south central part of the continent. (A.W.L.)

LICORICE. *Glycyrrhiza glabra.* Leguminosae. Licorice is obtained from an herb of southern European countries. The plant grows 4 or 5′ tall and has many pinnately-compound, pale green leaves and purplish flowers resembling those of the perennial pea. The fruit is a smooth pod containing several seeds. Licorice is a yellowish substance extracted from the rootstock. In water it swells to a jelly-like mass. It is used in medicines to hide the taste of unpleasant-tasting substances and in the treatment of head colds. But by far the greater part of the licorice root imported is used as a flavoring in chewing tobacco. (R.M.W.)

LIFE ZONE. A geographical area characterized by approximate uniformity of temperature conditions. The life zones are conditioned both by latitude and by altitude, hence a given zone need not be continuous but may, for example, appear at intervals along a mountain range. These zones are of limited value in the study of animal distribution.

In North America three principal zones are recognized, the Boreal, the Austral, and the Tropical. The first is subdivided into the Arctic, the Hudsonian, and the Canadian, ranging from the conditions of the far north to those of Canada and some parts of the northern United States and of more southern mountain regions. The second includes a Transition zone and the Upper and Lower Austral, and the last is not subdivided. (A.W.L.)

LIGAMENT. 1. The connection between the two parts of the shell of **bivalve** mollusks. 2. A strong fibrous band spanning the joint between the ends of two bones in the vertebrates. **Connective tissue. Joint.** (A.W.L.)

LIGATURE. A thread-like material or wire used for tying off blood vessels or other structures of the body during surgical operations. The material may be absorbable (catgut) or non-absorbable (silk or linen) or metal wire. Ligatures are made in various grades of thickness and tensile strength. (R.S.M.)

LIGHT. This term properly refers to the range of **electromagnetic radiation** frequencies associated with **vision**; though the physicist is apt to think of more objective manifestations, such as **photovoltaic effects**, and sometimes even oversteps the limits of the visible range by calling **infra-red** or **ultra-violet** radiation, and even **x-rays**, "light." The wavelengths of visible light extend approximately from 4000 angstroms (extreme violet) to 7700 angstroms (extreme red). Compared with radiation as a whole, this is an extremely limited range. It appears to be an inevitable limitation, however, on account of the strong absorption of most substances for radiation on both sides of it. The **quantum theory** of radiation applies of course to light, the energy quanta of which are called photons. Some of the optical phenomena so readily interpreted on the wave theory, such as **reflection, refraction, interference, diffraction,** and polarization of light, offer difficulties when studied in terms of quanta; the laws of **photoelectric phenomena, photoconductivity,** the **spectrum,** etc., become, on the other hand, much more intelligible. **Geometrical optics** is easily expressed in terms of either. The well known **Huygens principle** and **Fermat principle** apply to any radiation, including light, as does the **electromagnetic constant** representing the speed of radiation in a vacuum. Discussions of **photometry** and of **color** are, however, usually in terms of visual sensation and hence are confined to light proper, though the physical ideas involved are not subject to such limitation. (L.D.W.)

LIGHT, AERONAUTICAL.

Anchor. Light or group of clear lights carried on an aircraft to indicate its position at night while at anchor.

Boundary. Any one of the lights designed to indicate limits of landing area of airport or landing field.

Course. Light projected along course of an airway so as to be visible chiefly from points on or near the airway.

Identification. Group of lights, clear and colored, carried on rear part of an airplane for identification at night.

Landing. Light carried by an aircraft to illuminate the ground while landing.

Obstruction. Red light designed to indicate position and height of an object hazardous to operation of aircraft.

Position. For U. S. commercial aircraft, these consist of a red light on left wing tip, a green light on right wing tip, and a white light at the tail. The wing-tip lights must be visible from dead ahead and from the side. (F.T.M.)

LIGHT ALLOYS. The **aluminum alloys** which have densities approximately $\frac{1}{3}$ that of steel and the **magnesium alloys** with densities approximately $\frac{1}{4}$ that of steel are known as the light alloys. Other possible light alloys are those having lithium or beryllium as the base metal but no commercially important compositions have been developed. (R.H.H.)

LIGHT CURVE. In the study of **variable stars** and in kindred problems in astronomical research, it is desirable to graphically represent the variation of radiation intensity with time. A diagram in which light intensity, on any convenient scale, is plotted as ordinates against time as abscissae is known as a light curve. As the number of observations increases it is frequently possible to detect a periodic variation in the light intensity. After a provisional period has been determined, some convenient epoch is selected and all of the observations are reduced to the cycle of variation embracing the selected epoch by the use of the provisional period. In order that the resulting points may fall on a regular curve, it is frequently necessary to apply a number of corrections to the provisional period. The curve drawn through the plotted points, all reduced to the selected epoch by means of the repeatedly corrected period, is known as the mean light curve. Examples of light curves will be found in the articles on **long period variables** and on **Cepheids.** (W.K.G.)

LIGHT FILTERS. A homogeneous optical medium that is characterized by its absorption in certain regions of the spectrum. A filter is used to change or control the total or relative energy distribution of a beam of light. In photography a filter may be placed over the light source or in some part of the optical path traversed by the light in reaching the photographic emulsion, generally over the camera lens.

A photographic light filter may consist of a glass cell containing a liquid of the proper absorption, a piece of glass or gelatin containing suitable dyes or colorants, or some similar material having the desired absorption as well as satisfactory physical and optical properties.

Photographic filters are generally named in a manner to indicate their use. A color filter, therefore, means a filter showing selective absorption in the visible spectrum, while a neutral density filter is characterized by non-selective absorption in the visible spectrum. Infra-red and ultra-violet filters transmit those spectral regions. The largest group of light filters is the color filters, which may be divided into two main classes, the correction and contrast filters. This division is somewhat arbitrary and there are filters that are not readily so classified.

The correction, or compensating filter, is characterized by a relatively gradual change from absorption to transmission in any given region of the spectrum. Such filters, therefore, afford redistribution of the energy in the light source. Photometric filters, light yellows and many pale green filters, fall into this group. They are used in photography to alter the relative brightness with which differently colored objects are reproduced.

Contrast filters are distinguished by their rather abrupt change, over a short wavelength region, from complete absorption to high transmission. Such filters are used to isolate regions of the visible spectrum and to pass the chosen region quite freely while absorbing all other visible light. Typical contrast filters are the highly saturated red, green, blue and deep yellow photographic filters. Light filters of this type are used extensively in commercial contrast photography, color photography and microscopy.

Polar screens are often used as photographic light filters and exhibit non-selective absorption throughout the visible spectrum, but do absorb **polarized light.** Since the light received at the camera is often polarized, such a filter allows the brightness of various parts of the subject to be altered without changing the relative color balance of the light received by the photographic emulsion.

Graduated light filters are occasionally useful in photography and consist of light-absorbing media, the densities of which are not constant over the surface. Such filters are used to equalize the brightness differences between sky and foreground in landscape photography and to equalize the illumination in color cameras. The sky filter to control brightness in landscape photography is a deeper yellow on the upper portion and pale yellow or clear below. When placed over the lens properly the light from the bright sky is filtered more effectively than light reaching the film from the foreground. Graduated filters for the control of illumination in color cameras may be neutral or colored and are placed in front of the film plane inside the camera. (H.C.C.)

LIGHT YEAR. The light year is a popular method of expressing large distances. It is the distance that light will travel in the course of 1 year. The **velocity of light** is approximately 186,000 miles per sec. or 300,000 kilometers per sec. and there are approximately 31,560,000 sec. in a **mean solar year.** Accordingly a light year represents a distance of approximately 5.88×10^{12} miles (nearly six million-million miles) or 9.461×10^{12} kilometers.

For purpose of comparison with other astronomical units of distance we find that a star at a distance of one **parsec** (parallax $1''$) is at a distance of 3.258 light years, and that a star at a distance of 1 light year is over 63,000 **astronomical units** from the earth. Hence, the astronomical unit bears about the same relation to the light year as the inch does to the mile. (W.K.G.)

LIGHT-FILTER FACTOR.

LIGHT-FILTER FACTOR. Light filters absorb a part of the incident light and, therefore, transmit less than was present before filtration. In photography the quantity of light present determines the exposure given to the emulsion. Every filter, therefore, has a filter factor that applies to conditions of use and is the increase in exposure necessary to compensate for the light absorbed by the filter. The filter factor is generally determined by the exposure increase necessary with the filter when photographing a neutral gray subject, such as a **gray scale**.

The numerical value of the filter factor, or multiplying factor as it is often called, is not constant since it depends not only on the color of the filter itself, but also on the energy distribution or color of the source of light, and on the relative color sensitivity of the photographic emulsion being used. With a given red filter and panchromatic film, for example, the filter factor would be somewhat greater in daylight than in average tungsten illumination which contains a relatively greater proportion of red light compared to other portions of the spectrum. Likewise, the filter factor for a given yellow filter in daylight would be considerably greater with an orthochromatic emulsion (sensitive only to blue and green) than with a panchromatic emulsion that was sensitive to all colors. (H.C.C.)

LIGHTNING.

LIGHTNING. The electrical condition of the earth's surface and of the atmosphere is quite different in stormy weather from its normal, fair-weather state. Over a level stretch of country in fine weather, there is distributed a negative surface charge estimated at about 0.00027 electrostatic unit per sq. cm. or 0.0014 coulomb per sq. mile. Above this, the electric **potential** of the atmosphere increases with elevation at the rate of about 100 volts per meter, the upper atmosphere being, apparently, positively charged. The earth, the atmosphere, and the **ionosphere** thus form a vast **condenser**, through the dielectric of which there is constant leakage because of ionization. What maintains the charges against this leakage is not well understood.

In a rapidly developing rainstorm clouds become charged, positively at the top and negatively below. According to C. T. R. Wilson, this is brought about by the differential falling rate of large and of very small drops, the former becoming for some reason negatively and the latter positively charged. The cloud thus acts as a huge **static machine** with drops as carriers, which operates until the electric stress becomes so great as to cause a discharge of lightning between the charged surfaces of the same cloud, or between two clouds, or between a cloud and the induced charge on the earth under it. These activities of course greatly modify the distribution of charge and potential in the surrounding area, changes which can be detected by electrometers suitably placed.

It has been estimated that over the entire earth the frequency of lightning averages about 100 flashes every second and that this rate of discharge represents something like 4,000,000,000 kilowatts of continuous power. The flashes are often very long, sometimes several miles, and have been estimated to be from $4-6''$ in diameter. Often several flashes, each of very short duration, follow in quick succession over nearly but not quite the same path, thus producing the illusion of forking. Because light travels at the rate of 186,000 miles per sec. and sound at about 1100 ft. per sec., lightning flashes are seen and vanish long before the rumble of thunder is audible unless discharge occurs near the observation point. "Sheet lightning," so called, is merely the reflection or scattering of light from distant flashes by clouds. "Ball lightning" is a very rare manifestation not at all understood. Observers describe it as a small incandescent, hissing globe, moving slowly along in the air, sometimes indoors, now and then touching obstacles and scorching them, sometimes exploding. (L.D.W.)

LIGHTNING ARRESTER.

LIGHTNING ARRESTER. Lightning arresters are applied to electric lines which are exposed to direct or induced **lightning** disturbances, in order safely to conduct the high-frequency or high-voltage lightning disturbance to ground. These may be divided into two types, those used on power circuits, and those used on communication circuits. The former presents the more difficult problem, since a power **arc** tends to follow the **ionized** path created by lightning.

A lightning arrester on a power circuit should normally have so high a resistance to current that at the normal line voltage there will be only a negligible leakage of current through the arrester. Upon reception of abnormal voltages, this arrester should provide a direct and positive path to the ground, and should interrupt any power arc that may tend to follow the high-voltage surge. At present several arrangements satisfy these requirements. The simplest form of lightning arrester is a horn gap, in which two horns, one on the line and one connected to the ground, are separated by a small, set gap above which they flare to a wider separation. When excessive voltage causes breakdown of the resistance of this gap, and a discharge to ground, the power arc which follows will usually be extinguished by the electromagnetic and thermal effects of the arc, which cause it to rise higher on the horns until extinguished by their divergence, creating an ever longer arc path. On the high-voltage transmission lines employed today this extinguishing action is not always reliable, and various types of film and valve arresters have been devised. The first of these was the aluminum **cell**, or electrolytic arrester. It was necessary to connect this arrester to the line through a horn gap, and since its continued operation required periodic recharging, it necessitated some attendance, which is unnecessary in the more recently perfected valve-type arresters. Typical of the latter is the oxide film arrester, which consists of a number of cells in series. A cell is made of a **lead** oxide compound held between metal plates. The assembled cell is covered with lacquer insulation, and a certain number of them is built up into a stack which can be connected from line to ground. The cell would be a good conductor but for the lacquer. A discharge of lightning punctures the latter and causes local heating, which changes the nature of the oxide from conductor to insulator. This cuts short the current which the line voltage forces to follow the lightning discharge. These dry type arresters do not need daily attendance.

The protection of communication circuits and antennae is simpler due to the lack of high voltage on these circuits. A common type consists of two carbon blocks held at a definite spacing in porcelain. These are connected from the circuit to ground so that an abnormal voltage will break down the gap and ground itself. Since there is no power voltage behind the arc, the gap clears itself as soon as the disturbance has passed. (F.T.M.)

LIGHTNING BUG.

LIGHTNING BUG. Insecta, Coleoptera. Common **beetles** of the family Lampyridae which have a luminous organ located on the under surface of the last few segments of the abdomen. They are especially active in warm damp places early in the night, flashing at intervals of a few seconds. Occasionally large numbers of these insects flash synchronously, a phenomenon that has not been satisfactorily explained. The light apparently serves them in finding mates. (A.W.L.)

LIGNIN. Lignin and **cellulose** are the chief constituents of wood. Lignins are complex substances of unknown composition believed to be polysaccharides. In the manufacture of paper from **wood** it is necessary to remove the lignin, and this is often accomplished by treatment of the wood fibers with such agents as **sulfur dioxide**—**calcium** bisulfite, or sodium sulfate—**sodium sulfide**, or **sodium** hydroxide. For the details of these various methods, see **Paper**. (R.K.S.)

LIGNITE. Coal.

LIGNUM VITAE. *Guaiacum officinale* and *G. sanctum.* This is the heartwood of a tree growing native in the West Indies. It is a valuable, tough resinous wood and very heavy, being the heaviest of all commercial woods. A cu. ft. weighs 76 lbs. Its principal use is in the making of bowling balls, pulley sheaves and mallet heads. (R.M.W.)

LIKELIHOOD. The method of maximum likelihood for the estimation of **population parameters** was introduced by R. A. Fisher. It is extremely important since under certain limitations it affords an easy way of finding **consistent, efficient,** and **sufficient statistics.** Let $P(x, \theta)$ be a **probability function** dependent on a single parameter θ. Then the likelihood of obtaining the **sample** $x_1, x_2, \cdots, x_N$ is defined as the product of the N values of $P(x_i, \theta)$.

$$L = P(x_1, \theta) \cdot P(x_2, \theta) \cdots P(x_N, \theta).$$

We maximize L by considering θ the **variable** or, what is the same thing, maximizing the log L, and solve the resulting equation for θ. In the case of the **normal probability function** the maximum likelihood estimate for the population **mean** is the sample mean, and for the population **variance** the maximum likelihood estimate is the sample variance. The method of maximum likelihood is very useful in the fitting of the **Pearson system,** although the resulting equations are rather difficult to solve. It is also very helpful in finding the asymptotic variance of efficient statistics. (L.A.A.)

LILY FAMILY. Liliaceae. The Lily Family has representatives in all parts of the world, and more especially in the drier regions of the temperate zone. Several members of the family are important vegetables, notably asparagus and onions, while a great many more are cultivated for ornament. Among the latter are the true lilies (the genus *Lilium*), tulips, and hyacinths.

Most members of the Lily Family are herbaceous plants with a shallow fibrous root system. A few species of *Aloe* and *Dracaena* are shrubby or even small trees. Characteristic of the family are underground rhizomes or bulbs, storage organs which enable the plant to survive in regions where protracted dry seasons occur. As a rule, these plants have linear undivided leaves which do not show division into **petiole** and **blade.** The **inflorescences** of the family are very diverse. In some genera the flowers are solitary, in others they occur in **racemes,** while **umbels** occur in still others. The **perianth** of the flower has six separate members in two whorls of three, which are very much alike in size, shape, and color. The **stamens** have conspicuous **anthers.** The ovary is superior, 3-celled, and bears a single **style** with a 3-lobed **stigma.** The fruit is a capsule or a berry.

Members of the genus **Allium** are extensively cultivated for food; less widely grown, but forming an important crop in western countries, is **Asparagus.** In the Eastern World, bulbs of certain species of *Lilium* are

used as foods. In the Western World members of this genus are used entirely for ornament. White-flowered species are extensively grown indoors to flower at Easter time. (R.M.W.)

LIMA BEAN. *Phaseolus lunatus.* **Bean.**

LIMAÇON. The limaçon is a plane curve which may be defined geometrically as follows: Draw a circle of radius b passing through the origin of a set of rectangular axes and with its center on the X-axis. Draw a secant line OS cutting the circle at B, and extend OB to P so that $BP = 2a$. As the secant rotates about O, the point P describes the limaçon, as shown in Figs. 1–3.

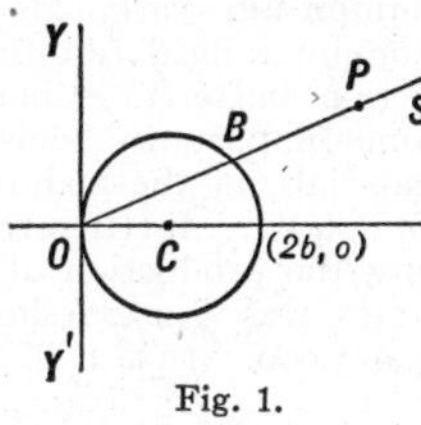

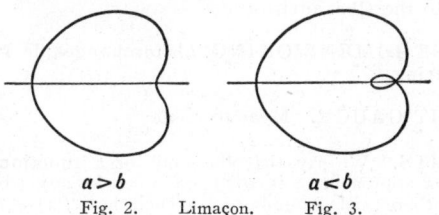

Fig. 1.

Fig. 2.   Limaçon.   Fig. 3.

$a > b$ $a < b$

In **polar coordinates,** the equation of the limaçon is

$$r = a - b \cos \theta.$$

If $b = a$, the curve becomes the **cardioid.** (L.L.S.)

LIMB. A jointed appendage, especially of the **vertebrates.** Applied usually to **primates** as a collective term to indicate both arms and legs. (A.W.L.)

LIMBURGITE. An igneous rock with a glassy base, which displays **phenocrysts** of **olivine** and **pyroxene.** (E.S.C.S.)

LIME. For the fruit of this name, see **Citrus Fruits.** In chemistry, lime is a term applied to **calcium** compounds. Limestone, calcium carbonate; caustic lime, calcium oxide; lime water, calcium hydroxide solution; slaked lime, calcium hydroxide solid. Percentage of lime expressed in analyses of chemicals is for calcium oxide (CaO). (R.K.S.)

LIMESTONE. A general term for the great, variable group of **carbonate** rocks whose chief constituent is **calcium** carbonate, usually in the form of the mineral **calcite.** Although limestones are quite common among the **sedimentary rocks,** occurring as true limestones or marbles in all periods from the **Archean** to the present, most of them contain large amounts of impurities, of which the more common are, **magnesia, silica, iron oxide,** iron hydroxide, **clay** and organic matter. Carbonate rocks relatively rich in magnesia are called magnesian limestones. When the magnesia is in the form of the mineral **dolomite** the rock is called either dolomitic limestone or dolomite, according to the abundance of that mineral. Limestones which are composed chiefly of shells and shell fragments are called **coquina.** Limestones which contain **argillaceous** material are called either shaly limestone or calcareous **shales,** according to the proportions of calcite to clay. Limestones which have a high **bituminous** content may be called stink stein because they emit an unpleasant sulfurous odor when struck with a hammer. It has not yet been definitely determined whether fine-grained relatively unfossiliferous limestones are entirely the product of ground-up calcareous shells, or straight chemical precipitates, either of organic or inorganic origin. Probably, however, the bulk of the limestones, especially those formed during the earliest periods of the earth's history, are composed of **drewite**, which in turn is a bacterial-algal precipitate of fresh or brackish water origin but relatively insoluble in sea water, and therefore preserved when transported and deposited in the sea. At the present time calcareous muds are being precipitated from sea water through the reaction of **ammonium** carbonate on **calcium** sulfate, the ammonium radical resulting from the action of the putrifaction bacteria. Limestones are of considerable economic importance being used: as building and decorative materials, in the manufacture of portland **cement**, lime for plaster, **fertilizers**, **flux**, chemical lime, etc. In 1934 the production of building limestone in the United States was 539,300 short tons valued at approximately $3,655,000. (R.M.F.)

LIMICOLAE. An old order of birds, partially equivalent to the **Charadriiformes**. (A.W.L.)

LIMIT DIMENSIONING. Interchangeable Manufacturing.

LIMIT GAUGE. Measurement.

LIMITS. We say that the limit of a **function** $f(x)$ when x approaches a is equal to l, when for any arbitrary $\epsilon > 0$ there exists a number $\delta > 0$ such that $|f(x) - l| < \epsilon$ whenever $|x - a| < \delta$. We write $\lim\limits_{x \to a} f(x) = l$, or $\lim_{x \to a} f(x) = l$, or $f(x) \to l$ when $x \to a$.

If a variable continually increases but remains less than some constant c, it approaches a limit; and this limit is either c or some lesser number.

If a variable continually decreases but remains greater than some constant c, it approaches a limit; and this limit is either c, or some greater number.

The limit of the sum of two variables which approach limits is the sum of those limits.

The limit of the product of two variables which approach limits is the product of those limits.

The limit of the quotient of two variables which approach limits is the quotient of those limits, unless the limit of the divisor is o.

A variable which approaches a limit o is called an infinitesimal.

When a variable v increases beyond all bounds, it is said to become infinite, and we write $v \to \infty$.

The symbol ∞ does not denote a number; it merely denotes a mode of variation. (L.L.S.)

LIMNOBENTHOS. Aquatic benthos. **Distribution.** (A.W.L.)

LIMNOLOGY. A division of biology that deals with the living organisms, both plant and animal, in fresh water. It includes the classification and biology of these organisms and their ecological relations. (A.W.L.)

LIMNOSCELIS PALUDIS. (Ancestral lizard.) **Fossil Reptiles.**

LIMONITE. The mineral limonite, hydrated oxide of iron, corresponds to the formula $Fe_2(OH)_6Fe_2O_3$, but is often very impure due to the admixture of sand and clay. It is not found crystallized but grades from loose porous material to compact masses. Its hardness is variable but pure material is 5–5.5; specific gravity 3.6–4; usual luster, dull to earthy but may be silky to submetallic; color, various shades of yellowish-brown, sometimes nearly black; streak, yellowish-brown; opaque. Limonite is a secondary mineral from the alteration of various other iron-bearing ores or minerals, it is of widespread occurrence and used both as an ore of iron and as a **pigment**. Limonite has been formed

in marshy and boggy areas and is frequently called bog iron ore. Limonite is an important ore of iron in Lorraine, Luxemburg, Bavaria and Sweden. It is found in Saxony, Austria, and England. In the United States limonite is found particularly in Connecticut, Massachusetts, Pennsylvania, New York, Virginia, Tennessee, Georgia and Alabama, but these deposits are of little economic importance at the present time. (E.S.C.S.)

LIMPET. Mollusca, Gasteropoda. Marine and freshwater animals related to the snails, with a low conical shell, not spirally twisted. In the common limpets the shell is solid and in the keyhole limpets it is either notched in front or perforated between that point and the apex. Mollusks of the family Capulidae, more closely related to some of the species with coiled shells than to the true limpets, also have shells which are not spiral and are called limpets. One form, *Crucibulum*, is called the cup and saucer limpet and another, *Crepidula*, the boat limpet or slipper shell. (A.W.L.)

LIMPKIN. Aves, Gruiformes. Birds (**Aves**) of two species resembling the rails but larger. One, *Avamus vociferans*, ranges from Florida through the Antilles and Central America and the other, *A. scolopaceus*, lives in tropical South America. Also called courlans. (A.W.L.)

LINE. This term is used in two senses in electrical circuits, the first being as the conducting system for electrical energy between two remote points of the system and the second in terms of the scanning in **facsimile** and **television** systems. For a discussion of the power transmission line see **Electric Power Transmission.** In communications circuits the line takes on added properties, or perhaps we should say that certain properties become much more important. This is due to the much shorter wavelengths used in such circuits so the lines used will frequently be several wavelengths long. Thus at 60 cycles (commercial power **frequency**) a line would have to be a little over 3000 miles long to be a wavelength while in the middle of the audio range (5000 cycles) it would only need to be about 37 miles and at the ultra high radio frequencies only a few inches. If a line has a length of the order of a wavelength or more and is not matched (see **Matching**) reflection and resultant standing waves will result. Often this is a serious drawback to its useful function but for high radio frequencies this property is used to great advantage. A line having standing waves appears at its input terminals to be a resonant circuit (alternately as a series and a parallel type) when adjusted to multiples of a quarter-wave in length. Since at the very high radio frequencies a line quarter-wavelength long is only a few inches such lines are often used to serve as resonant circuits for **tanks in oscillators, amplifiers,** etc. Used in this connection they are superior to ordinary lumped inductance and capacitance circuits as they have a much higher **Q** and when properly constructed will cause much lower radiation losses. If these lines are adjusted in length to something other than quarter-wave multiples they act as inductance or capacitance, depending upon the exact length. Such lines are used for a variety of functions, e.g., **tanks,** radio-frequency **chokes,** tuning elements, coupling elements, etc. They are widely used at frequencies of a hundred or so megacycles and at still higher frequencies are the only satisfactory means for tuning and coupling circuits. In these applications the **co-axial line** or **wave guide** are ordinarily used since they do not radiate but at the lower frequencies parallel lines are sometimes used for convenience. Various short- and open-circuited lines are useful for impedance **matching** at **radio frequencies.** In power circuits the losses of the line, while important from an efficiency standpoint are not too serious but in telephone circuits these losses may be an appreciable part of the total power and consequently are a serious problem. Also a line which is long enough to be comparable to a wavelength will produce appreciable phase shift and this may cause trouble in communication circuits. Important constants of lines may be summed up in the following equations:

$$\gamma = \sqrt{zy}$$
$$|I| = |I_s|e^{-\alpha l}$$
$$\gamma = \alpha + j\beta$$
$$Z_0 = \sqrt{z/y} \qquad I = I_s e^{-\gamma l}$$

where z is the series **impedance** of the line per unit length, y the shunt **admittance** per unit length, γ the **propagation constant**, α the **attenuation constant**, β the **wavelength** or phase constant, Z_0 the **characteristic impedance**, $|I|$ the magnitude, I the vector current at a distance l from the sending end, I_s the sending end current, e the base of the natural logarithms. The phase shift along the line may be found from

$$\text{Phase} = \beta l.$$
(L.R.Q.)

LINE DIAGRAM. Skeleton Diagram.

LINE FINDER. This is one of the automatic switching units used in a machine-switching telephone office and serves to locate the line initiating a call when the receiver is removed from the hook by the calling party. (See **Telephony.**)   (L.R.Q.)

LINE INTEGRAL. A line integral is an important mathematical concept which is a natural extension of the idea of the **definite integral** of a function of one variable.

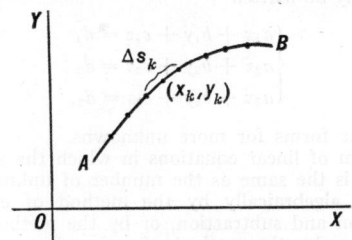

Let C be an arc of a plane curve, joining points A and B and let $f(x, y)$ be a **continuous function** of x and y in a region containing C. Divide C into n sub-arcs by points on C and denote the kth sub-arc by Δs_k. Let (x_k, y_k) be any point on Δs_k, and form the sum

$$\sum_{k=1}^{n} f(x_k, y_k)\Delta s_k = f(x_1, y_1)\Delta s_1 + f(x_2, y_2)\Delta s_2 + \cdots$$
$$+ f(x_n, y_n)\Delta s_n.$$

The **limit** of this sum (when the limit exists), as each $\Delta s_k \to 0$ (and $n \to \infty$) is called a line-integral of $f(x, y)$ along the path C, and is denoted by

$$\int_C f(x, y)ds.$$

Now let $P(x, y)$ and $Q(x, y)$ be continuous functions of x and y in a region, and let C be a given path (arc of a plane curve). As before, divide C into n sub-arcs and let Δs_k denote the kth sub-arc, and let Δx_k and Δy_k be the **projections** of Δs_k on the X-axis and Y-axis respectively. Form the sums

$$\sum_{k=1}^{n} P(x_k, y_k)\Delta x_k \quad \text{and} \quad \sum_{k=1}^{n} Q(x_k, y_k)\Delta y_k;$$

if these sums approach limits as each $\Delta s_k \to 0$ (and $n \to \infty$), these limits are defined as line-integrals of P and Q, respectively, along the path C, and are denoted by

$$\int_C P(x, y)dx \quad \text{and} \quad \int_C Q(x, y)dy.$$

The combination

$$\int_C P(x, y)dx + \int_C Q(x, y)dy,$$

which is generally written

$$\int_C (Pdx + Qdy)$$

is the usual way in which line integrals occur.

The line integral $\int_C f(x, y)ds$, defined first, can be expressed in the form $\int_C Pdx + Qdy$.

Line integrals are sometimes called curvilinear integrals. If the path C is given by an equation of the form $y = \phi(x)$, we may replace y by $\phi(x)$ and dy by $\phi'(x)dx$ in the integral $\int_C Pdx + Qdy$, and obtain an ordinary **definite integral** in one variable of the form $\int_{x_1}^{x_2} F(x)dx$.

Similarly, if path C is given by an equation of the form $x = \psi(y)$.

If the path C is given by **parametric equations** $x = \phi(t)$, $y = \psi(t)$, we may express P and Q in terms of t and dx and dy by $\phi'(t)dt$ and $\psi'(t)dt$, respectively, and obtain an ordinary integral of the form $\int_{t_1}^{t_2} F(t)dt$.

A line integral

$$\int_C (Pdx + Qdy + Rdz)$$

along a path C in space may be defined in a similar manner to that in a plane; in this case P, Q, R are functions of three variables x, y, z.

By use of **Green's theorem in the plane,** it can be shown that, if P and Q are two functions of x, y which, together with their first **partial derivatives**, are **continuous** within and on the boundary of a certain region, then the necessary and sufficient condition that the line integral in the plane $\int_C (Pdx + Qdy)$ may be independent of the path C in this region is that $\dfrac{\partial P}{\partial y} = \dfrac{\partial Q}{\partial x}$.

If the preceding line integral $\int_C (Pdx + Qdy)$, in which one end point of path C is a variable point (x, y), be denoted by u, then $\dfrac{\partial u}{\partial x} = P$, $\quad \dfrac{\partial u}{\partial y} = Q$.

In order that the line integral $\int_C (Pdx + Qdy + Rdz)$ in space have the same value for all paths C joining the same two points, it is necessary and sufficient that

$$\frac{\partial P}{\partial y} = \frac{\partial Q}{\partial x}, \quad \frac{\partial Q}{\partial z} = \frac{\partial R}{\partial y}, \quad \frac{\partial R}{\partial x} = \frac{\partial P}{\partial z},$$

provided P, Q, R and their partial derivatives are continuous in a region surrounding the paths C.

If this line integral be denoted by u, then

$$\frac{\partial u}{\partial x} = P, \quad \frac{\partial u}{\partial y} = Q, \quad \frac{\partial u}{\partial z} = R,$$

provided one end point of C is a variable point (x, y, z).
(L.L.S.)

LINE INTEGRAL OF A VECTOR FUNCTION.

A line integral of a vector function is a mathematical notion of importance in mathematical physics.

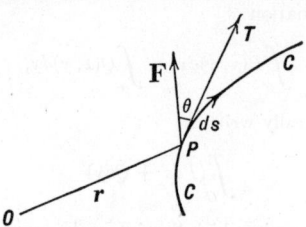

Let $F(r)$ be a **vector function** of the position vector r. Then the line integral $\int_C F \cos \theta \, ds$ along a path C, where θ is the angle between F and the element of arc ds, is called the line integral of the vector function $F(r)$, and is denoted by $\int_C F \cdot dr$ in vector notation. (L.L.S.)

LINE OF APSIDES.

A line which contains the major axis of an **ellipse** is known as the line of apsides of the ellipse. In astronomy the term is used to indicate the line joining **perihelion** and **aphelion** points in an **orbit** and extending to infinity to cut the **celestial sphere**. (W.K.G.)

LINE OF NODES.

The line of nodes is the astronomical term applied to line of intersection of any two fundamental planes. The line of nodes for the **moon** is the line of intersection of the plane containing the moon's orbit with the plane of the **ecliptic**. The line of nodes for any member of the **solar system**, other than **satellites**, is the line of intersection of the plane of the orbit of the object with the plane of the ecliptic. The line of nodes for the earth is the line of intersection of the plane of the earth's **equator** with the plane of the ecliptic. (W.K.G.)

LINE OF POSITION.

In **navigation** any line on the surface of the earth upon which a ship is known to be located, is called a **line of position**. If two or more lines of position are known, the ship must be at their point of intersection. A position determined by the intersection of lines of position is known as a **fix**. Lines of position are usually circles, either great or small. In most cases, the distance of the center is so great in comparison with the length of the plotted line that the curvature is not apparent in the plot. In some cases, however, we may find both points of intersection of two circles of position appearing on the plot. In these cases a dead-reckoning position will indicate which one is the desired fix. Lines of position are obtained by methods of **pilotage, celestial navigation,** or **radio navigation**. (W.K.G.)

LINE SQUALL.

Extremely turbulent roll-type squall cloud usually found at the leading edge of squall lines associated with rapidly moving cold fronts. (P.E.K.)

LINE SPECTRA.

Atomic Spectra.

LINE-OF-SIGHT TRANSMISSION.

This term is commonly applied to the range of transmission of very high frequency radio stations such as **television** and frequency-modulated stations. At these high frequencies the radio waves are not reflected from the **ionosphere** so no **sky wave** returns to earth, thus limiting the range to that covered by the **ground wave**. This is refracted around the surface of the earth to only a slight extent, so while line-of-sight is not strictly accurate it is fairly descriptive. It is also inaccurate in that radio waves, even at these frequencies, will pass through objects which are opaque to visible light. (L.R.Q.)

LINEAR ALGEBRAIC EQUATIONS.

A linear algebraic equation in one unknown is a **polynomial equation** of the first degree, and may be written in the form $ax + b = 0$. It may be solved by transposition and division.

A linear algebraic equation in two variables is represented by the general form $ax + by + c = 0$. It may be satisfied in general by an unlimited number of pairs of values of x and y. The graphic representation of such an equation gives a straight line in the plane.

A linear algebraic equation in three variables is represented by the form $ax + by + cz + d = 0$. It is represented graphically by a plane. (L.L.S.)

LINEAR ALGEBRAIC EQUATIONS, SYSTEMS OF.

A system of linear equations in two or more unknowns is a set of linear equations which are considered together with the object of determining whether they are satisfied by one or more sets of values of the unknowns and of finding these values if such exist.

A solution of such a system of equations is any set of corresponding values of the unknowns which satisfy the equations of the system.

The general form of a system of two linear equations in two unknowns may be written

$$\begin{cases} a_1x + b_1y = c_1 \\ a_2x + b_2y = c_2, \end{cases}$$

that of a system of three linear equations in three unknowns may be written

$$\begin{cases} a_1x + b_1y + c_1z = d_1 \\ a_2x + b_2y + c_2z = d_2 \\ a_3x + b_3y + c_3z = d_3, \end{cases}$$

and similar forms for more unknowns.

A system of linear equations in which the number of equations is the same as the number of unknowns may be solved algebraically by the method of elimination by addition and subtraction, or by the method of substitution, or by the method of comparison, or by **determinants**. A system of two equations with two unknowns may also be solved graphically.

If a system of linear equations with the number of equations the same as the number of unknowns has only one solution, the equations are said to be **independent and consistent**. If the system has no solution, the equations are called **inconsistent**. If the equations of the system are satisfied by an unlimited number of values of the unknowns, the equations are called **dependent**.

An inconsistent pair of linear equations in two unknowns is represented graphically by a pair of parallel straight lines. A dependent pair of linear equations in two unknowns is represented graphically by one straight line used twice.

A system of two linear equations in two unknowns may be solved by the method of elimination by addition and subtraction, by multiplying both equations by such multipliers that the coefficients of one of the unknowns are the same in both equations, then adding or subtracting these equations to eliminate the one unknown, obtaining a linear equation in one unknown, which may be immediately solved; substitution of this value of the unknown in one of the original equations again gives one linear equation with one unknown. Or, the second unknown may be found by eliminating the first unknown as in the preceding.

In the method of substitution, one of the given equations is solved for one unknown in terms of the other and this is substituted in the other equation, obtaining one linear equation in one unknown.

In the method of comparison, both of the given equations are solved for the same unknown in terms of the other unknown, these are equated, and the resulting linear equation in one unknown solved.

For a system of three equations in three unknowns, the algebraic procedure for solution would be: Eliminate one of the unknowns from one pair of equations and also from another pair; solve the resulting two equations for the two unknowns in them, as just described; substitute the resulting values in one of the given equations to obtain the value of the third unknown. An exactly similar procedure applies to systems of more equations with more unknowns, the number of equations being the same as the number of unknowns.

A system of two linear equations with two unknowns may be solved graphically by constructing the **graphs** (straight lines) of the two equations (in **rectangular coordinates**) and finding the **coordinates** of the intersection point.

By making use of **determinants**, the solution of systems of linear equations may be expressed in convenient and compact form as follows:

The solution of the system of equations

$$\begin{cases} a_1x + b_1y = c_1 \\ a_2x + b_2y = c_2 \end{cases}$$

is given by

$$x = \frac{\begin{vmatrix} c_1 & b_1 \\ c_2 & b_2 \end{vmatrix}}{\begin{vmatrix} a_1 & b_1 \\ a_2 & b_2 \end{vmatrix}}, \qquad y = \frac{\begin{vmatrix} a_1 & c_1 \\ a_2 & c_2 \end{vmatrix}}{\begin{vmatrix} a_1 & b_1 \\ a_2 & b_2 \end{vmatrix}},$$

provided the denominator $\begin{vmatrix} a_1 & b_1 \\ a_2 & b_2 \end{vmatrix} \neq 0$. If this denominator is zero and if the numerators are not both zero, the equations have no solution and are inconsistent; if all three determinants are zero, the equations are dependent.

The solution of the system of equations

$$\begin{cases} a_1x + b_1y + c_1z = d_1 \\ a_2x + b_2y + c_2z = d_2 \\ a_3x + b_3y + c_3z = d_3 \end{cases}$$

is given by

$$x = \frac{\begin{vmatrix} d_1 & b_1 & c_1 \\ d_2 & b_2 & c_2 \\ d_3 & b_3 & c_3 \end{vmatrix}}{D}, \quad y = \frac{\begin{vmatrix} a_1 & d_1 & c_1 \\ a_2 & d_2 & c_2 \\ a_3 & d_3 & c_3 \end{vmatrix}}{D}, \quad z = \frac{\begin{vmatrix} a_1 & b_1 & d_1 \\ a_2 & b_2 & d_2 \\ a_3 & b_3 & d_3 \end{vmatrix}}{D},$$

where $D = \begin{vmatrix} a_1 & b_1 & c_1 \\ a_2 & b_2 & c_2 \\ a_3 & b_3 & c_3 \end{vmatrix}$, provided $D \neq 0$. If $D = 0$, the equations are inconsistent or dependent.

A system of n linear equations in n unknowns has a single solution if the determinant of the coefficients is not zero; the value of any unknown can be expressed as a fraction whose denominator is the determinant of the coefficients and whose numerator is the determinant obtained from the denominator determinant by replacing the column of coefficients of this unknown by the column of constant terms (when these are on the right hand side of the equations). This is known as Cramer's rule, and may be expressed symbolically thus:

The solution of the system of equations

$$\begin{cases} a_{11}x_1 + a_{12}x_2 + \cdots + a_{1n}x_n = c_1 \\ a_{21}x_1 + a_{22}x_2 + \cdots + a_{2n}x_n = c_2 \\ \cdots\cdots\cdots\cdots\cdots\cdots\cdots \\ a_{n1}x_1 + a_{n2}x_2 + \cdots + a_{nn}x_n = c_n \end{cases}$$

is given by

$$Dx_1 = C_1, \quad Dx_2 = C_2, \quad \cdots, \quad Dx_n = C_n,$$

where

$$D = \begin{vmatrix} a_{11} & a_{12} & \cdots & a_{1n} \\ a_{21} & a_{22} & \cdots & a_{2n} \\ \cdots & \cdots & \cdots & \cdots \\ a_{n1} & a_{n2} & \cdots & a_{nn} \end{vmatrix} \neq 0,$$

and C_k $(k = 1, 2, \cdots, n)$ is what D becomes when the elements of its kth column are replaced by $c_1, c_2, \cdots, c_n$ respectively.

The system of equations is inconsistent if the denominator determinant $D = 0$ and if any one or more of the numerator determinants $C_1, C_2, \cdots, C_n$ is $\neq 0$.

If the constant terms $c_1, c_2, \cdots, c_n$ of a system of linear equations are all zero, the equations are called homogeneous. Such a system is therefore of the form

$$\begin{cases} a_{11}x_1 + a_{12}x_2 + \cdots + a_{1n}x_n = 0, \\ a_{21}x_1 + a_{22}x_2 + \cdots + a_{2n}x_n = 0, \\ \cdots\cdots\cdots\cdots\cdots\cdots\cdots \\ a_{n1}x_1 + a_{n2}x_2 + \cdots + a_{nn}x_n = 0. \end{cases}$$

If a system of n homogeneous equations in n unknowns has a solution other than the trivial one where each unknown is 0, then the determinant of the coefficients $D = 0$. Conversely, if in such a system the determinant $D = 0$, then the system has infinitely many non-trivial solutions.

In general, a system of n equations with less than n unknowns will have no common solution, except under certain special conditions; a system of n equations with more than n unknowns will have an unlimited number of sets of values of the unknowns satisfying it. (L.L.S.)

LINEAR CORRELATION. If 2 **variables** x and y are connected by a linear relationship $y = a_1x + b_1$ and $x = a_2y + b_2$, then x and y are linearly correlated. The two equations connecting x and y are called **regression** lines. These are readily found as $y - \overline{Y} = \frac{r\sigma_y}{\sigma_x}(x - \overline{X})$ and $x - \overline{X} = \frac{r\sigma_x}{\sigma_y}(y - \overline{Y})$ where r is the **coefficient of correlation**, $\overline{X}$, $\overline{Y}$ are the respective **means**, σ_x and σ_y are the respective **standard deviations**. The **standard error of estimate** of y is given by

$$\sigma_e{}^2 = \frac{\Sigma(y - y')^2}{N}$$

for N observations where y is the actual value, and y' is the value given by the regression line. The values of a_1 and b_1 may also be found by solving the normal equations

$$\Sigma y = a_1\Sigma x + Nb_1$$

$$\Sigma xy = a_1\Sigma x^2 + b_1\Sigma x$$

for a_1 and b_1. This is the method of moments which, in this instance, is the same as the method of **least squares** for evaluating a_1 and b_1. Similarly a_2 and b_2 may be found. We may test a regression line for significance by the **analysis of variance**, and also test whether the regression is linear or curvilinear.

In a **correlation** table, the regression line of y on x, $y = a_1x + b_1$, is the line of best fit in the sense of least squares for the means of the columns, and the regression of x on y, the line of best fit for means of rows. (L.A.A.)

LINEAR DEPENDENCE AND INDEPENDENCE OF FUNCTIONS. If n functions $f_1(x), f_2(x), \cdots, f_n(x)$ are connected by an identical relation of the form

$$c_1f_1(x) + c_2f_2(x) + \cdots + c_nf_n(x) = 0,$$

where the c's are not all 0, the functions are said to be linearly dependent. If no such relation exists, they are called linearly independent.

For example, $\sin x$, $\cos x$ and $\sin (x + a)$ are linearly dependent; but x, e^x and $\sin x$ are linearly independent. (L.L.S.)

LINEAR DIFFERENTIAL EQUATIONS. The ordinary differential equation

$$\frac{d^n y}{dx^n} + P_1 \frac{d^{n-1} y}{dx^{n-1}} + P_2 \frac{d^{n-2} y}{dx^{n-2}} + \cdots + P_{n-1} \frac{dy}{dx} + P_n y = R,$$

in which the coefficients P_1, P_2, $\cdots$, P_n are given **functions** of x which do not depend upon y, and R is a given function of x, is called a linear differential equation of order n.

If $R \equiv 0$ (identically), the equation is called a homogeneous linear differential equation, otherwise it is called non-homogeneous.

Fundamental properties of linear differential equations are the following:

If y_1 is a **solution** of a homogeneous differential equation, then cy_1 is also a solution, where c is an arbitrary constant.

If $y_1 \, y_2$, $\cdots$, y_n are n **linearly independent** solutions of a homogeneous differential equation of the nth order, then $y = c_1 y_1 + c_2 y_2 + \cdots + c_n y_n$ is also a solution, and is the **general solution.**

If y_p is a **particular solution** of a non-homogeneous differential equation (without any particular constants of integration), and if y_1, y_2, $\cdots$, y_n are n linearly independent solutions of the corresponding homogeneous equation (of order n), and if c_1, c_2, $\cdots$, c_n are n arbitrary constants, then the general solution of the non-homogeneous equation is $y = y_p + c_1 y_1 + c_2 y_2 + \cdots + c_n y_n.$

Let us now consider the case of linear differential equations with constant coefficients, i.e., the case where the coefficients P_1, P_2, $\cdots$, P_n are constants.

Denote $\frac{dy}{dx}$ by Dy, $\frac{d^2 y}{dx^2}$ by $D^2 y$, etc., so that D denotes the symbolic operator $\frac{d}{dx}$. Then the differential equation can be written in the form

$$(D^n + A_1 D^{n-1} + A_2 D^{n-2} + \cdots + A_{n-1} D + A_n) y = R,$$

where the A's are constants and R is a function of x. The expression in the parentheses may be considered as a symbolic polynomial in D, and denoted by $f(D)$ where f denotes a polynomial function. The non-homogeneous equation is then $f(D)y = R$, and the corresponding homogeneous equation is $f(D)y = 0$.

According to one of the fundamental theorems stated above, the solution of a non-homogeneous equation depends on the general solution of the corresponding homogeneous equation; so we consider first the homogeneous equation.

For a given homogeneous equation $f(D)y = 0$, the function $y = e^{mx}$ will be a solution of the equation if m satisfies the **polynomial equation** $f(m) = 0$, which is called the characteristic equation or auxiliary algebraic equation.

Let m_1, m_2, $\cdots$, m_n be the roots of this equation $f(m) = 0$ (assuming the differential equation to be of the nth order). Then there are three cases to consider.

(1) If the roots m, m_2, $\cdots$, m_n are real and distinct, the general solution of the differential equation is

$$y = c_1 e^{m_1 x} + c_2 e^{m_2 x} + \cdots + c_n e^{m_n x}.$$

(2) If two or more of the roots m_1, $\cdots$, m_n are equal, the solution in (1) must be modified: if r of the roots are equal to a, the group of corresponding terms in (1) is to be replaced by

$$e^{ax}(c_0 + c_1 x + c_2 x^2 + \cdots + c_{r-1} x^{r-1}).$$

(3) If **complex** values occur among the roots m_1, $\cdots$, m_n, the solution in (1) may be put in a more convenient form, by use of **Euler's theorem,** as follows: corresponding to a pair of conjugate complex roots $a + i\beta$ and $a - i\beta$, the solution will contain the terms

$$e^{\alpha x}(A \cos \beta x + B \sin \beta x),$$

where A and B are arbitrary constants; this can also be written

$$A' \cdot e^{\alpha x} \cos (\beta x + B'),$$

where A' and B' are arbitrary constants.

Let us now consider the non-homogeneous linear differential equation with constant coefficients.

The general solution of a homogeneous equation $f(D)y = 0$ is called the complementary function of the corresponding non-homogeneous equation $f(D)y = R$; we shall denote it by y_c. Then if y_p denotes any particular solution of $f(D)y = R$, the general solution of the non-homogeneous equation will be $y = y_c + y_p$.

To find the particular solution y_p, various methods are available; we shall mention only a few.

1. Method of undetermined coefficients: In this method we assume a form which the particular solution will take, using undetermined coefficients, which are then determined by substitution in the differential equation. For the most common cases, we take the following assumed forms:

a. If R is a polynomial in x, say

$$R = a_0 x^n + a_1 x^{n-1} + \cdots + a_n,$$

assume, in general,

$$y_p = A_0 x^n + A_1 x^{n-1} + \cdots + A_n,$$

where A_0, A_1, $\cdots$, A_n are undetermined coefficients; but if D^m is a factor (of highest degree) of $f(D)$, then assume

$$y_p = x^m (A_0 x^n + A_1 x^{n-1} + \cdots + A_n).$$

b. If $R = ce^{ax}$, assume, in general,

$$y_p = Ae^{ax},$$

where A is an undetermined constant; but if $(D - a)^m$ is a factor of $f(D)$, assume

$$y_p = x^m \cdot Ae^{ax}.$$

c. If $R = c_1 \sin ax$ or $c_2 \cos ax$ or $c_1 \sin ax + c_2 \cos ax$, assume, in general,

$$y_p = A \sin ax + B \cos ax,$$

where A and B are undetermined constants; but if $(D^2 + a^2)^m$ is a factor of $f(D)$, assume

$$y_p = x^m (A \cos ax + B \sin ax).$$

d. If $R = e^{ax} \phi(x)$, where $\phi(x)$ is a function of the type in a or c, put $y = e^{ax} \cdot z$, divide out e^{ax} and we obtain an equation of the type in a or c.

2. Short-cut rules:

a. If $R = e^{ax}$, then $y_p = \dfrac{1}{f(a)} e^{ax}$.

b. If $R = \sin ax$ (or $\cos ax$), then

$$\frac{1}{f(D^2)} = \frac{1}{f(-a^2)} \cdot \sin ax \text{ (or } \cos ax).$$

3. Method of successive differentiation of the given differential equation: Differentiate successively the given equation and obtain, either directly or by elimination, a homogeneous linear differential equation (with right-hand member 0). Solve this homogeneous equation for its general solution, which will consist of the complementary function y_c of the original equation plus additional terms involving constants of integration. Substitute these additional terms in the original differential equation, equate coefficients of like terms and determine the coefficients.

4. A factoring method: If $f(D)$ can be factored into linear factors $(D - m_1)(D - m_2) \cdots (D - m_n)$, we may

reduce the given differential equation to a system of n first order linear equations thus: For definiteness, suppose $n = 3$, so that $f(D) = (D - m_1)(D - m_2)(D - m_3)$. Put

$$(D - m_1)y = u, \quad (D - m_2)u = v, \quad (D - m_3)v = R.$$

We may solve these first order linear equations successively, starting with the last, and finally obtain y as a function of x.

There are also other methods, such as the method of variation of parameters, a method based on a decomposition of $1/f(D)$ into partial fractions, etc., which cannot be described here.

The solutions of a few important special forms of linear differential equations (with constant coefficients) are listed here:

1. The equation $\dfrac{d^2x}{dt^2} - k^2x = 0$ has the solution:

$$x = c_1 e^{kt} + c_2 e^{-kt}.$$

2. The equation $\dfrac{d^2x}{dt^2} + k^2x = 0$ has the solution:

$$x = c_1 \cos kt + c_2 \sin kt = A \cos (kt + \alpha) = B \sin (kt + \beta).$$

3. The equation $\dfrac{d^2x}{dt^2} + k^2x = a$ has the solution:

$$x = c_1 \cos kt + c_2 \sin kt + \frac{a}{k^2} = A \cos (kt + \alpha) + \frac{a}{k^2}$$
$$= B \sin (kt + \beta) + \frac{a}{k^2}.$$

4. The equation $\dfrac{d^2x}{dt^2} + k^2x = a \cos nt + b \sin nt$ $(n \neq k)$ has the solution:

$$x = c_1 \cos kt + c_2 \sin kt + \frac{1}{k^2 - n^2}(a \cos nt + b \sin nt).$$

5. The equation $\dfrac{d^2x}{dt^2} + k^2x = a \cos kt + b \sin kt$ has the solution:

$$x = c_1 \cos kt + c_2 \sin kt + \frac{1}{2k} \cdot t(a \sin kt - b \cos kt).$$

6. The equation $\dfrac{d^2x}{dt^2} + 2l\dfrac{dx}{dt} + k^2x = 0$ has the solution:

(a) if $l^2 = k^2$: $x = e^{-lt}(c_1 + c_2 t)$,

(b) if $l^2 > k^2$: $x = e^{-lt}[c_1 e^{\sqrt{l^2 - k^2} \cdot t} + c_2 e^{-\sqrt{l^2 - k^2} \cdot t}]$,

(c) if $l^2 > k^2$: $x = e^{-lt}[c_1 \cos \sqrt{k^2 - l^2} \cdot t$
$$+ c_2 \sin \sqrt{k^2 - l^2} \cdot t]$$

or

$$x = A e^{-lt} \cos (\sqrt{k^2 - l^2} \cdot t + \alpha)$$
$$= B e^{-lt} \sin (\sqrt{k^2 - l^2} \cdot t + \beta).$$

A linear differential equation of the type

$$x^n \frac{d^n y}{dx^n} + A_1 x^{n-1} \frac{d^{n-1} y}{dx^{n-1}} + \cdots + A_{n-1} x \frac{dy}{dx} + A_n y = R,$$

where the A's are constants and R is a function of x, may be reduced to a linear differential equation with constant coefficients by the substitution $x = e^z$ or $z = \log x$. Such an equation is sometimes called a homogeneous linear equation, although this conflicts with another use of the term.

Differential equations may occur in systems of linear differential equations in several unknown functions. The general method for solving such equations is to combine the equations so as to give an equation in one unknown function, i.e., eliminate all but one unknown function from the equations, and then use the previous methods.   (L.L.S.)

LINEAR DIFFERENTIAL EQUATION OF FIRST ORDER. Ordinary Differential Equations of First Order and First Degree.

LINEAR FUNCTION. A linear function is a **polynomial function** of the first degree, and is therefore a **function** of the form $ax + b$, where a and b are constants and x is the **variable**.

The graphic representation in rectangular coordinates of any linear function $y = ax + b$ is a straight line with slope a and cutting off on the Y-axis an intercept b.   (L.L.S.)

LINEAR TRANSFORMATION. A linear (homogeneous) transformation of a set of variables x_1, x_2, $\cdots$, x_n is given by a system of n **linear equations**

$$\begin{cases} a_{11}x_1 + a_{12}x_2 + \cdots + a_{1n}x_n = y_1, \\ a_{21}x_1 + a_{22}x_2 + \cdots + a_{2n}x_n = y_2, \\ \cdots \cdots \cdots \cdots \cdots \cdots \cdots \cdots \\ a_{n1}x_1 + a_{n2}x_2 + \cdots + a_{nn}x_n = y_n \end{cases}$$

with given coefficients a_{ij}.   (L.L.S.)

LINEAR VECTOR FUNCTION. A linear vector function is a type form of mathematical expression which finds its principal application in certain parts of mathematical physics.

Let $\mathbf{v}$ be a **vector variable** expressed in terms of a system of non-coplanar vectors $\mathbf{a}_1$, $\mathbf{a}_2$, $\mathbf{a}_3$ by $\mathbf{v} = v_1\mathbf{a}_1 + v_2\mathbf{a}_2 + v_3\mathbf{a}_3$. Then if

$$\mathbf{f} \equiv f(\mathbf{v}) = v_1 f(\mathbf{a}_1) + v_2 f(\mathbf{a}_2) + v_3 f(\mathbf{a}_3),$$

and if $f(\mathbf{v})$ is continuous, then $\mathbf{f} \equiv f(\mathbf{v})$ is called a linear vector function.

If $f(\mathbf{v})$ is a linear vector function, then

$$f(k\mathbf{v}) = kf(\mathbf{v}), \quad f(\mathbf{u} + \mathbf{v}) = f(\mathbf{u}) + f(\mathbf{v}).$$
(L.L.S.)

LINEN. Flax.

LINES OF FORCE. Fields of Force.

LINESHAFT. Shaft.

LING. Burbot.

LINGUATULIDA. Pentastomida.

LINGULA. Invertebrate Paleontology.

LININ. Cell.

LINK COUPLING. This is a rather common system of inductive coupling used in radio transmitter circuits. The diagram is almost self-explanatory, the primary

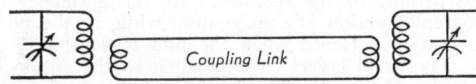

Link-coupled tank circuits.

circuit has a coupling winding of a few turns coupled to it, this being connected by a short transmission line to a similar coil inductively coupled to the second circuit. (See also **Coupled Circuits**.)   (L.R.Q.)

LINKAGE. Electromagnetic Induction; Inductance; Coupling; Heredity.

LINNET. Finch.

LINOLEUM. Flax; Jute.

LINSANG. Mammalia, Carnivora. *Linsanga*. Slender predacious animals with short legs and very long tails. Related to the civets. Several species are Oriental and one African.   (A.W.L.)

LINSEED OIL. Flax; Esters.

LINTEL. A lintel is a **beam** which carries the wall **load** of a building over an opening such as a window or a doorway. It is supported by the walls on either side of the opening. If these walls extend far enough on both sides of the opening to be able to resist a horizontal thrust, **arch** action can be developed in the wall above the opening. In this case the loading is usually assumed to be triangular with a maximum intensity at the center of the **span.** (C.W.C.)

LION. Mammalia, Carnivora. *Felis leo.* One of the best known species of the **cat** family from its long eminence as the "king of beasts." Like the tiger, the lion reaches a length of 10′ from tip to tip and a weight of 500 lbs. It is uniformly tawny as a rule but varies from much lighter yellowish shades to very dark brown. The male usually has a full mane but this also is a variable character; males with no mane have been found.

Lions are nocturnal in habit and are said to be generally shy, although they will attack man under some conditions. They sometimes hunt in groups, sharing the animals that they kill, and sometimes eat the carcasses of animals that they have not killed, even when badly decomposed.

The lion ranges over all of Africa and into Asia as far as Mesopotamia and northwestern India. Many of the specimens seen in menageries and with circuses were born in captivity. (A.W.L.)

LIP. A fleshy fold at the external orifice of a cavity, especially those which bound the mouth of the vertebrates. (A.W.L.)

LIPARITE. The term liparite is synonymous with **rhyolite,** but used chiefly by European geologists. The word was derived from the Lipari Islands where liparites are quite common. (E.S.C.S.)

LIPOMA. A benign **tumor** made up of fat cells. Lipomas occur commonly in the subcutaneous tissues about the head and neck. They cause no symptoms. (D.M.H.)

LIPPMANN FRINGES. The Lippmann **interference** fringes or laminae constitute an interesting photographic demonstration of stationary waves of **light.** A film of special fine-grained photographic emulsion is backed by mercury, which serves as a reflector. Monochromatic light falling normally upon the film and reflected by the mercury gives rise to interference in stationary waves whose nodes and antinodes are in planes parallel to the reflector. At the antinodes the photographic action is a maximum, while at the nodes there is none. Hence when the film is developed the silver deposits in layers corresponding to the antinodes, ½ wavelength apart. The laminar structure therefore depends upon the color of the light. Neuhauss has succeeded in making photomicrographs of the cross-section of the film, resembling a cut jelly-cake.

The silver layers may be made to act as reflecting planes whose spacing is just right to cause reinforcement by interference when the reflected light is of the same wavelength as that to which the plate was exposed. White light is thus selectively reflected, so that if part of the plate had been exposed to green light, that part would reflect green, etc. Lippmann adapted this principle to the production of photographs in natural colors; but the technique of the process is very difficult and it has not come into practical use. (L.D.W.)

LIQUATION. A process of magnetic differentiation believed to take place as a result of the separation of two immiscible liquids from the parent **magma.** (E.S.C.S.)

LIQUEFACTION OF GASES. Consideration of this subject at once emphasizes a distinction which must be made between gases, in the technical sense of that term, and vapors. Properly speaking, a gas cannot be liquefied at all until it is first reduced to a vapor by lowering its temperature below the critical point. (See **Critical State.**) The problem thus resolves itself into two parts, one of **refrigeration** and one of compression; and further, from a practical standpoint, one of storage.

To liquefy carbon dioxide (as for making carbonated drinks) requires only a compressor and a bath of ice-water to keep the temperature down during compression; and when liquefied, the CO_2 can be stored indefinitely in sealed steel drums capable of withstanding 1200 or 1500 lbs. per sq. in. of pressure. The reason is that the critical temperature of CO_2 is about 31° C. Atmospheric temperature exceeds this amount only in very warm weather. Ammonia (NH_3), with a critical temperature of 132° C., is still easier to liquefy. But nitrogen (and hence **air**) must first be cooled below −147° C. and kept there during the compression, a requirement far beyond the reach of ordinary refrigeration methods.

In 1857 the process of "regenerative cooling," also known as the "cascade" method, was devised by Siemens, to be perfected later by Linde, Hampson, and others. The essential features of this technique are as follows: The gas to be liquefied is first pre-cooled by ordinary means, such as ammonia or "dry ice," or even cold water, then highly compressed while kept cool by removal of the heat of compression. The container is surrounded by a jacket, and the highly compressed gas is released through a small escape valve into this jacket. The escaping gas falls in temperature, both on account of doing external work against the pressure in the jacket and, with most gases, also on account of the **Joule-Thomson** effect. The jacket communicates with the compression pump, and is so designed that the cool gas, to reach the outlet, must circulate around the chamber containing the warmer, high-pressure gas, thus acting as a "heat exchanger." It should be clear that as each unit mass of the gas traverses again and again the cycle composed of compressor, cooling coil, escape valve and jacket, it must become progressively colder; until finally its critical temperature is reached. Further operation now results in the formation of liquid drops, which accumulate in the lower part of the jacketed high-pressure chamber. from which the liquid can be drained off.

Hydrogen and helium are much more difficult to liquefy than other gases, because of their very low critical temperatures (−240° C. and −268° C., respectively), and because in the case of hydrogen the Joule-Thomson effect is a rise instead of a fall of temperature until very low temperatures are reached. Hydrogen must therefore be pre-cooled by evaporating liquid air or otherwise, while helium requires evaporating liquid hydrogen as a pre-cooling agent. Hydrogen was first liquefied by Dewar (1898) and helium by Onnes (1908).

Liquid air, hydrogen, etc., cannot be stored in closed containers, as can ammonia and carbon dioxide, but must be kept at atmospheric pressure in a Dewar flask, or thermos bottle, the low temperature being maintained by the evaporation of the liquid itself, as long as it lasts. (L.D.W.)

LIQUID CRYSTALS. Cybotaxis.

LIQUID STRUCTURE. Cybotaxis.

LIQUID-EXPANSION THERMOMETER. This familiar instrument for measuring **temperature,** developed by Fahrenheit and others, makes use of the relative expansion of a liquid and its transparent container. The use of mercury as a thermometric substance is recommended by its high boiling point

(+357° C.) and low freezing point (−39° C.), and by the constancy of its expansion coefficient. Alcohol is often substituted, because of its much lower freezing point (−114° C.), and because its expansion coefficient is more than six times that of mercury; it is also lighter and cheaper. Its boiling point, however, is so low (78° C.) that it cannot be used for high temperatures; and it must be stained to be easily visible. In some mercury thermometers an inert gas is introduced above the mercury, the pressure of which, as the mercury expands, raises the boiling point of the mercury and hence increases the range. The Beckmann mercury thermometer has a very large bulb and a very fine bore with a storage reservoir for mercury at the top. It is used only for differential temperature measurements, with a range of only a few degrees, and is graduated to hundredths of a degree.

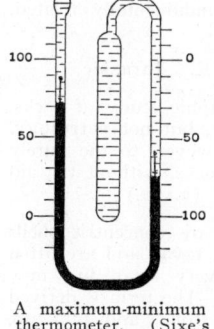

A maximum-minimum thermometer. (Sixe's form.)

Some thermometers are designed to indicate the maximum or the minimum temperature attained during a given period. The best maximum thermometers employ mercury, with a constriction just above the bulb at which the mercury thread separates when the temperature starts to fall, leaving the top of the column to mark the highest temperature. Minimum thermometers employ alcohol, with a light, solid index or marker just inside the free surface. As the temperature falls, this marker is pushed down by the surface tension, and remains at the lowest point attained. In a thermometer devised by Sixe, both maximum and minimum temperatures are similarly indicated by a small iron marker, which can be adjusted by means of a magnet.

Liquid thermometers are subject to certain inherent errors, among which are those due to the unequal temperature of bulb and stem and to the imperfect recovery of the volume of the bulb after heating. The latter effect may accumulate over a long period, requiring recalibration from time to time, especially if the instrument is used over wide ranges of temperature. (L.D.W.)

LIQUIDS. States of Matter; Hydrostatics; Hydrokinetics.

LIROCONITE. The mineral liroconite is a hydrous **arsenate** of **copper** and **aluminum** of uncertain formula. It is a **monoclinic** mineral usually in very small crystals; hardness 2–2.5; specific gravity 2.9; vitreous luster; color various shades of blue, blue-green and green. It appears to be a secondary mineral associated with other copper compounds as **malachite.** It is found in Czechoslovakia and Cornwall, England. (E.S.C.S.)

LISSAJOUS CURVES. Harmonic Motion.

LISSAJOUS FIGURES. One of the many uses of the cathode ray oscilloscope is its use as a frequency comparison instrument so oscillators may be calibrated or unknown frequencies determined by visual comparison with a known frequency on the tube screen.

When two sine waves, varying about axes at right angles, are combined the resultant figure is no longer a sine wave but varies with the relative time phase of the waves and with their relative frequency. For example, if the waves have the same frequency the resultant is a straight line when they are in time phase (or 180° out of phase) and is an ellipse for all other values of phase position. For equal amplitudes of the original waves and 90° phase the ellipse is the

special case of the circle. If the frequencies of the two waves are not the same the resultant becomes more complicated but gives a definite pattern whenever the frequencies are in the ratio of whole numbers to one another. Fig. 1 shows a graphical construction for a frequency ratio of 1:2.

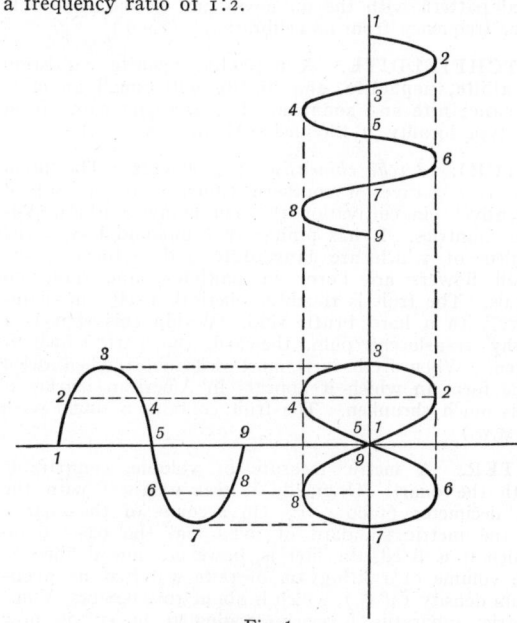

Fig. 1.

If sine waves of voltage are applied to the vertical and horizontal deflecting plates of an oscilloscope tube it does the job of combining them and gives the resultant on the screen. Thus two waves of equal frequency applied to the two sets of plates give an ellipse (the straight line and circle may be considered special cases of the ellipse), two waves with frequencies in the ratio of 1:2 give a figure similar to Fig. 1, and so on. If the frequency ratio of the waves is not a ratio of whole numbers the oscilloscope screen pattern will not be stationary and may even move so rapidly that it appears as a blur. However, whenever two waves whose frequency ratio is that of whole numbers the pattern, which is a Lissajous figure, will appear stationary. The frequency ratio of the applied waves will be equal to the ratio of the number of horizontal to the number of vertical tangencies (if the figure is imagined as inscribed in a rectangle). Thus in Fig. 2, there are two horizontal tangencies and three vertical ones so the fre-

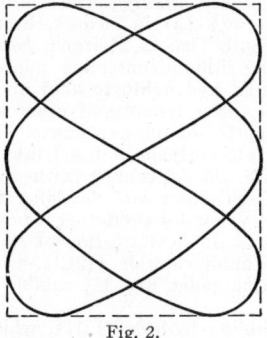

Fig. 2.

quency ratio of the waves applied to the horizontal and vertical plates is 2:3. The extension of this method to the calibration of an oscillator is obvious, a known frequency is applied to one set of plates and the oscillator output to the other, then as the frequency of the oscillator is varied a known ratio of frequencies will be obtained each time a stationary figure is obtained. A variation is to use a calibrated variable-frequency oscillator to determine a given unknown frequency by varying the oscillator until its output gives a fixed pattern with the unknown and noting the oscillator frequency from its calibration. (L.R.Q.)

LITCHFIELDITE. A **nephelite syenite** consisting of **albite, nephelite,** and **biotite** with small amounts of **cancrinite** and **sodalite.** It derives its name from the type locality at Litchfield, Maine. (E.S.C.S.)

LITCHI. *Litchi chinensis.* Sapindaceae. The litchi tree is a native of southern China, and has spread extensively in cultivation through many southern Asiatic countries. It has **pinnately** compound leaves, the leaflets of which are **lanceolate** and leathery. The small flowers are borne in **panicles,** and have no petals. The fruit is roughly spherical, $1-1\frac{1}{2}''$ in diameter, with a hard brittle rind. Within this rind is a fleshy translucent pulp, the aril, the part which is eaten. When fresh it is most delectable; when dried (the form in which it appears in American markets), it is much shrunken. The fruit contains a single seed. (R.M.W.)

LITER. A metric measure of volume, comparable with the quart. Originally it was identical with the cu. decimeter (1000 cc.). On account of the change in the metric standard of mass and the basis upon which it is fixed, the liter is, however, now defined as the volume of 1 **kilogram** of pure water at its maximum density (4° C.), which is about 1000.028 cc. Volumetric apparatus formerly graduated in cc. is now commonly rated in milliliters. (L.D.W.)

LITHARGE. Lead.

LITHIFICATION. **Lithify.** Literally to turn to stone. A term commonly applied to the consolidation and hardening of sediments so as to form a **sedimentary rock.** (R.M.F.)

LITHIUM. Symbol: Li. Atomic number: 3. Atomic weight: 6.940. Density: 0.534 at 20°. Melting point: 186° C. Boiling point $>$ 1200° C. Isotopes 6 (7.9%), 7 (92.1%).

Lithium is a silver-white metal; harder than **sodium** but softer than lead; tough and may be drawn into wire or rolled into sheets; tarnishes rapidly in air, preserved under naphtha; reacts with water forming lithium hydroxide solution and **hydrogen** gas. Discovered by Arfvedson in 1817.

Lithium occurs in **lepidolite** (lithium mica, lithium aluminosilicate, 1%–3% Li) **spodumene** (lithium aluminosilicate, 4% Li), **amblygonite** (lithium aluminum fluorophosphate, 8% Li) in Saxony, France, Manitoba and Quebec, South Dakota, Arizona, New Mexico and California. The lithium-containing mineral is digested with concentrated **hydrochloric acid** and the resulting soluble chloride, after removal of other metals, is converted into slightly soluble **carbonate.** Lithium metal is obtained by **electrolysis** of fused lithium **potassium** chloride mixture out of contact with air. Lithium is used (1) as a deoxidizer and degasifier for nonferrous castings, (2) as vapor for producing a protective atmosphere in furnaces for heat treating of steel.

Chloride: Lithium chloride (LiCl), white solid, deliquescent, melting point 614° C., soluble in water and in alcohol.

Hydride: Lithium hydride (LiH), white crystals, by heating lithium metal in hydrogen.

Hydroxide: Lithium hydroxide (LiOH), white solid, melting point 450° C., soluble.

Oxide: Lithium oxide (Li_2O), white solid, by heating lithium metal in oxygen or dry air, reactive with water to form soluble lithium hydroxide; lithium peroxide (Li_2O_2), white solid, by reaction of **hydrogen peroxide** and lithium hydroxide solution in alcohol, and later dehydration of the precipitate.

Other soluble salts: Lithium sulfate (Li_2SO_4); lithium nitrate ($LiNO_3$); lithium perchlorate ($LiClO_4$) (soluble in alcohol).

Slightly soluble salts: Lithium carbonate (Li_2CO_3); lithium phosphate (Li_3PO_4); lithium fluoride (LiF).

Lithium carbonate and citrate are used in medicine as "lithia water," a **diuretic** of doubtful value.

Volatile lithium salts, such as the chloride, color the bunsen flame carmine red. (R.K.S.)

LITHOCYST. A small hollow organ containing a solid particle, found at the base of a tentacle on the margin of certain **hydrozoan** medusae. It is also called an otocyst or statocyst. Superficially like the tentaculocysts of jellyfishes but not fundamentally related. (A.W.L.)

LITHOGRAPHIC LIMESTONE. Jurassic.

LITHOLOGY. Literally the graphic study of rocks, hence a synonym for **petrography,** but not **petrology.** This term is usually restricted, however, to the purely descriptive macroscopic study of rocks, without the aid of the **petrographic microscope.** (R.M.F.)

LITHOPHYSAE. Lithophysae are concentric shells of crystalline material occurring in **lavas** and are often very fragile. They are usually very small but may reach a diameter of several inches. The term is derived from the Greek words meaning stone, and puff up. (E.S.C.S.)

LITHOPONE. Pigments, Paints, Varnishes; Barium; Zinc.

LITHOSPHERE. The term lithosphere, from the Greek meaning a stone, and a sphere, refers to the solid, rocky outer portion of the earth as distinguished from the barysphere, a term also derived from the Greek meaning heavy, and sphere, and designating the unknown interior which is supposed to consist of matter heavier than the surface materials. The true lithosphere, which is formed only of the types of rocks which are now observable to the geologist, is assumed to be only approximately 60 miles thick. A common but inaccurate synonym of lithosphere is "crust."

About 99.5% of the solid outer crust of the earth is made up of eleven elements, combined in various ways, forming mainly **silicates** and **oxides,** and less commonly **carbonates** and **phosphates.** Clarke and Washington (1925) designate as petrogenic or rock-forming elements the following 22 arranged in the table in order of their abundance in the 10-mile crust of igneous and sedimentary rocks. They designate as metallogenic elements **copper, zinc** and all elements of higher atomic weight. (See **Chemical Composition.**)

ELEMENTS IN ORDER OF THEIR ABUNDANCE IN THE 10-MILE CRUST OF IGNEOUS AND SEDIMENTARY ROCKS OF THE EARTH

Element	%	Element	%
1. Oxygen	46.71	12. Carbon	0.094
2. Silicon	27.69	13. Manganese	0.090
3. Aluminum	8.07	14. Sulfur	0.052
4. Iron	5.05	15. Barium	0.050
5. Calcium	3.65	16. Chlorine	0.045
6. Sodium	2.75	17. Chromium	0.035
7. Potassium	2.58	18. Fluorine	0.029
8. Magnesium	2.08	19. Zirconium	0.025
9. Titanium	0.62	20. Nickel	0.019
10. Hydrogen	0.14	21. Strontium	0.018
11. Phosphorus	0.13	22. Vanadium	0.016
		Remainder	0.057
	99.47		0.530

Most rocks are, accordingly, silicates or aluminosilicates of five elements, namely, iron, calcium, sodium, potassium, magnesium. Excluding the rare gases, helium, neon, argon, only six of the elements (namely, lithium, beryllium, boron, nitrogen, scandium, cobalt) up to and including nickel (number 28 in the periodic classification) are not contained in the above list. Beyond nickel only three elements (namely, strontium, barium, zirconium) appear.

Among the considerations of particular significance are (1) the high percentage of silicon, (2) the *relatively* high percentage of elements similar to silicon, namely, aluminum (3rd), titanium (9th), zirconium (19th), (3) the unique position of iron (4th) above, but number 26 in the periodic classification, (4) the positions of the alkali metals (sodium 6th, potassium 7th) and alkali earth metals (magnesium 8th, calcium 5th, strontium 22nd, barium 16th), (5) the low percentages of the familiar elements, hydrogen, carbon, nitrogen, sulfur, phosphorus, chlorine, (6) in most cases the even-numbered elements are relatively more abundant than the adjoining odd-numbered elements. Rocks in the 10-mile crust of the Earth: Igneous rocks, 95% (75% of this is $SiO_2 + Al_2O_3$); shale, 4 (74% of this is $SiO_2 + Al_2O_3$); sandstone, 0.75 (78% of this is SiO_2); limestone, 0.25 (84% of this is $CaCO_3$); total, 100.00 (of the total 60% is SiO_2, and 15% Al_2O_3).

Barnett (1924) has estimated that the interior of the earth—77.5% of the whole—is an irregular core of metallic substances (iron 90%; nickel, cobalt, copper 7%) in a fused state, probably more or less mixed with silicates, whilst an irregular sheet of silicates as slag, more or less mixed with metal, constitutes the remaining 22.5%. (R.K.S., R.M.F.)

LITMUS. Lichens; Indicators.

LIT-PAR-LIT. A term derived from the French, meaning bed by bed, which is used to define banded **gneisses** produced by the injection of igneous material along planes of **foliation**, usually with an accompanying alteration by **contact metamorphism.** (R.M.F.)

LITTORAL. Inhabiting the shore line of the ocean in shallow waters and in the tidal zone which is periodically exposed to the air. (A.W.L.)

LITUUS. The lituus is a spiral curve represented by the equation $r^2\theta = a$ in **polar coordinates.** (L.L.S.)

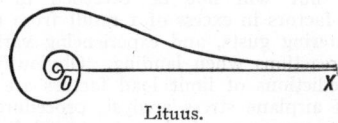

Lituus.

LIVE CENTER. In machine tools, the center which rotates with the work. (See **Dead Center; Lathe.**) (H.C.H.)

LIVE LOAD. Load.

LIVER. A large gland associated with the **digestive system** of vertebrates. It secretes the bile which is discharged into the intestine, absorbs from the blood the products of **carbohydrate** digestion and stores them as glycogen, acts on nitrogenous wastes and returns them to the **blood** in the form of **urea** and related compounds, and destroys worn-out red corpuscles. Other equally important functions are deamination, detoxication, production of fibrinogen and prothrombin and storage of the anti-anemia principle.

The bile discharged through the intestine plays an uncertain role in the digestion of fat and carries with it some of the more complex waste products of the body.

In structure the **vertebrate** liver is very complex. It develops as a hollow outgrowth of the embryonic gut just behind the stomach which forks to produce the gall-bladder and the liver. The connection with the gut persists as the common bile duct. In the adult the liver cells are arranged in cords, separated by blood channels with incomplete lining known as sinusoids. Within the cords minute bile capillaries between the cells converge to larger and larger ducts which ulti-

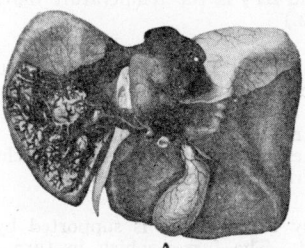

A

Liver. View is from undersurface, showing the gall-bladder (A).

mately form the main hepatic duct. The gland receives blood from an arterial supply and also from the portal vein. The latter drains blood from the capillaries of the intestine and breaks up into sinusoids in the liver. These small passages are drained by the hepatic vein.

The name liver has also been applied to large digestive glands in the phyla **Mollusca, Arthropoda,** and **Echinodermata.** (A.W.L.)

LIVERWORTS. Bryophytes.

LIZARD. Reptilia, Sauria. Mostly small animals, some species attaining a length of several feet, usually elongate with short legs and a long tail. They are closely related to the snakes but differ in having eyelids and in having the ventral surface of the body as well as the upper covered with small scales.

The classification and nomenclature of these animals is confused. The lizards are sometimes grouped with the snakes but some authorities regard them as a separate order of **Reptilia.** (See also **Fossil Reptiles.**)

There are many species of lizards but with the exception of the poisonous **Gila monster** and the edible **iguanas** they are of no economic importance. A few of the smaller species are eaten to a limited extent.

Among the lizards whose names do not indicate the association with this group are the geckos, iguanas, swifts, chameleons, chuck-walla, mountain boomer, horned toads, glass snakes, Gila monster, race runner, skinks, agamas, molochs, anolises, basilisks, scheltopusik, blind worms, monitors, teju, and **amphisbaena.** (A.W.L.)

LLAMA. Mammalia, Artiodactyla. A domestic animal, *Lama huanacus glama,* found in the high altitudes of western South America and on lower ground in the southern part of the continent. It has moderately long legs and a long neck, and is one of the New World representatives of the camel family. The animal is a source of wool, hides, meat and milk and is used as a beast of burden.

The name llama is sometimes applied to the entire group including the wild **vicunia** and **guanaco** and the domestic **alpaca** as well as the true llama. Both of the domestic species are supposed to have been derived from the guanaco. The llama is the larger animal but does not produce such fine wool. (A.W.L.)

LLOYD'S MIRROR. Young's Interference Experiment.

L.M.T.D. Log mean temperature difference. An average temperature difference between the hot and cold side of a piece of equipment, e.g., a heat exchanger, for use where the temperature difference may vary along the equipment. Its form is

$$\frac{\Delta T_1 - \Delta T_2}{2.3 \, \log \dfrac{\Delta T_1}{\Delta T_2}}$$

where ΔT_1 is the temperature difference at one end of the equipment and ΔT_2 is the temperature difference at the other. (D.E.M.)

LOACH. Pisces, Teleostei. Small bottom-feeding fishes (**Pisces**) of Europe and Asia. They have a long slender body and a group of barbels near the mouth. The European species are eaten.

A few fishes of similar form but not closely related are called African loaches. (A.W.L.)

LOAD. Any **force** which is supported by a body is called a load. The forces which in turn support the given body are called reactions. A concentrated load is a theoretical force having a contact area infinitely small compared with the area of the surface of the body upon which the force acts. A distributed load is one whose area of contact covers, wholly or partially, the area of the supporting surface of the body. Distributed loads are uniform if the intensity is the same for each unit of area covered by the load. When this intensity varies, the distributed load is non-uniform.

Loads may be classified as central, torsional or bending, depending upon the effect on the body. A central load is a concentrated load whose line of action passes through the **center of gravity** of the surface under consideration. A distributed load is a **central** load if the line of action of the **resultant** acts through the center of gravity of the surface. If the central force is at right angles to the surface it is called a direct or axial load. Tensile loads are axial loads which cause an increase in the length of a member. Compressive loads are axial loads which decrease the length of a body. Central forces which act in the plane of the surface under consideration are shearing loads. A torsional load is a force which causes a body such as the **shaft** of a machine to twist about its longitudinal axis. Loads which tend to change the **radius of curvature** of a body are called bending loads. Forces which bend a beam are known as transverse loads. An eccentric load, which is a type of bending load, is a force acting normal to a surface but not passing through the center of gravity of the surface.

A classification of loads which is important to the structural designer, is that which distinguishes between dead and live loads. A permanent load acting on a structure, such as its own weight, is called a dead load. Variable or moving loads are classed as live loads. Snow, wind and merchandise constitute variable loads. Moving loads are made up of the weight of people, vehicular traffic, railroad trains or street cars. The dead weight of traveling **cranes** and their loads must often be considered as live loads when designing the supporting structure.

Structural and machine elements are frequently subjected to moving loads. That part of the total kinetic energy of the moving body which is absorbed by the resisting element itself is known as an energy load. If the moving body and the supports of the resisting element are assumed to be inelastic, all of the kinetic energy is absorbed by the resisting body.

In structural engineering the kinetic energy absorbed by a structural element is usually small because the velocity is relatively low. Consequently it is common practice to use a force which is a certain percentage of the weight of the moving body and assume that its action on the element is equivalent to the dynamic effect. These percentages are obtained from empirical formulae and the equivalent load is called an impact load.

A load which is applied to a machine or structural element many times during the useful life of the element is called a repeated load. The force on a **piston** rod of a steam engine and the live load on bridges are typical examples. (C.W.C., F.T.M.)

LOAD, AIRCRAFT.
Basic Load (*Stress Analysis*). The load on a structural member or part in any condition of static equilibrium of an airplane. When a specific basic load is meant, the particular condition of equilibrium must be indicated in the context.

Designed Load (*Stress Analysis*). A specified load below which a structural member or part should not fail. It is the probable maximum applied load multiplied by the factor of safety. Also, in many cases, an appropriate basic load multiplied by a design load factor.

Full Load. Weight empty plus useful load; also called gross weight.

Normal Load (*Stress Analysis*). The load on that part of a wing assumed to be unaffected by tip losses or similar corrections. In any given case, it may be a basic, design, gross, net, or ultimate load, depending on the context.

Pay Load. That part of the useful load from which revenue is derived, viz., passengers and freight.

Ultimate Load (*Stress Analysis*). The load that causes destructive failure in a member during a strength test, or the load that, according to computations, should cause destructive failure in the member.

Useful Load. The crew and passengers, oil and fuel, ballast other than emergency, ordnance, and portable equipment. (F.T.M.)

LOAD FACTOR, AERODYNAMIC. The ratio of dynamic to static load on a structure is sometimes referred to as the load factor. Thus load factor becomes an allowance for acceleration and impact. This term is in common use in the field of aeronautics to specify design loads and to report flight-maneuvering loads. The following discussion will relate to this field.

Load factor in aviation is the ratio of the external load, applied or expected to be applied to an airplane, to the design weight. The limit factor describes the load which it is assumed or known may be safely experienced but will not be exceeded in operation. Flight load factors in excess of 1 result from maneuvering, encountering gusts, and experiencing various kinds of ground reactions when landing. Obviously assumptions or predictions of limit load factors are an essential part of airplane stress analysis procedure. Whenever a pilot causes an airplane to travel in a curved flight path, as in a loop, a turn, a zoom, he imposes a load factor in excess of 1 on the airplane. Fighter aircraft must have inbuilt strength to resist exceptionally high acceleration loadings resulting from combat maneuvers. But to build equivalent safety into a commercial airplane in order to safeguard the few pilots who might wish to engage in violent acrobatic maneuvers is a penalty on the speed, economy and pay load which is not warranted by the general conditions of usage. Consequently, government specifications for **airworthiness** require only strength to meet all emergencies of normal reasonable use of the airplane. A placarded maximum diving speed is displayed in the pilot's compartment and the designer has based his structural strength on load factors predicted to arise from encountering gusts, or maneuvering, at speeds not to exceed the placard speed. Thus it would be entirely possible for a foolhardy pilot to damage structurally, or even crash an airplane certified by the government as being airworthy if he were to dive it to a speed considerably in excess of the placarded, and then do an abrupt pull-up. The **CAA** states that at any speed much greater than about twice the stalling speed, the pilot has it in his power to pull the wings off almost any (civilian) airplane.

Some typical maneuvering load factors are:

60° banked turn	2.0
80° banked turn	5.7
Loop	3.0
Pull-ups	4.0
Spin	2.5
Landing (normal)	2.5

Gust load factors vary between 3 and 4.5 for most commercial designs. (F.T.M.)

LOAD FACTOR, ELECTRIC. The ideal electric load, from the standpoint of equipment needed and operating routine, would be one of constant magnitude and steady duration. Such an ideal load is shown in Fig. 1(a). The cost to produce an elementary area of this load curve (i.e., one kilowatt-hour) could be from ½ to ¾ of that to produce the same unit under the more frequently realized condition illustrated in Fig. 1(b). Hence the problem of variable load is a

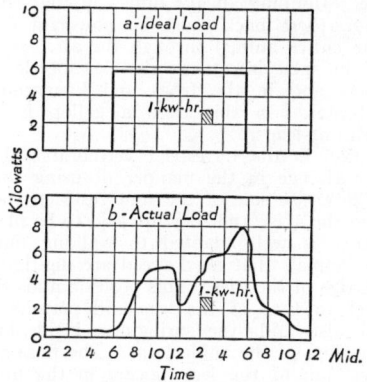

Fig. 1.

vital one, for, from the industrial viewpoint, the cost of manufactured articles includes an energy charge as an element of no inconsiderable proportion, while from the utility viewpoint, the chief concern is to put each kilowatt-hour on the transmission line at as low a production cost as possible.

The general conclusion is that industrial processes and domestic uses impose highly variable demands upon the capacity of a plant. The exceptions do not disprove this as a basic operating condition of most generating equipment. Even though the characteristics of the demand made by any one user will hardly be understood until his conditions of use are fully investigated, one might suppose, for purposes of illustration, that this has been accomplished. Then Fig. 2 might represent the domestic demands of two adjacent resi-

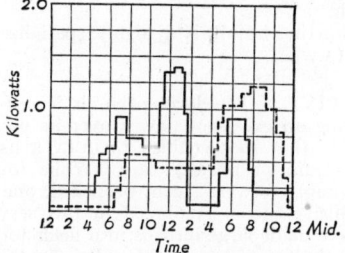

Fig. 2. Individual customers' load curves.

dences. There is, apparently, no great similarity between them, nor would one expect a similarity unless he knew the family life of the two sets of occupants to be similar. Furthermore, the next residence might have a still different form of load curve, and so on for still other residences. This is all the result of the natural fact of the individual differences of persons. However, as the number of customers increases, the effect of individual differences is submerged to the general use conditions of the community, and the resulting load curve is typical.

The variable load problem has injected into the language a number of terms. The basic information is the operating data of demanded load plotted against time sequence. This is commonly referred to as a load curve and usually appears with kilowatts as the power unit and hours as the time unit, the sequence being the 24 hours beginning with midnight. The most important variations of this are the monthly and annual load curves, each of which is the average of the daily load curves over the period named.

Some information other than the kilowatt-hour magnitude of energy produced is needed to describe an operating condition. Evidently the relation of the peak load to the average in some measure satisfies this requirement. This relationship is expressed in the load factor. The daily, monthly, or annual load factor is the average load over the time specified divided by the maximum peak. The latter is seldom taken as the maximum instantaneous peak but rather as the maximum 15-minute, half-hour, or even hour-long peak. In general, the length of time over which the peak load shall be measured is increased as the period of time in which it might occur is increased from the daily to the monthly or the annual period. Load factor should not be confused with power factor, with which it has nothing in common. (F.T.M.)

LOAD LINE. This is the graph of current against voltage of the load in the plate circuit of a **vacuum tube** and is superposed on the plate characteristic of the vacuum tube. Fig. 1 shows a typical family of plate characteristic curves with a load line plotted on them.

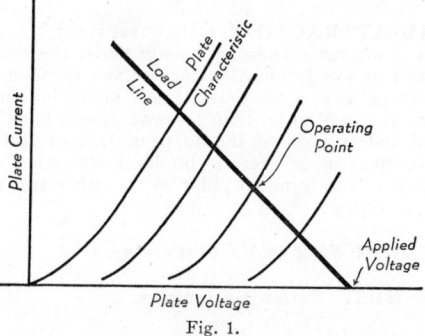

Fig. 1.

In this case the load is assumed to be resistance. Since the vacuum tube does not follow any simple law of voltage and current variation and the current must satisfy both the load conditions and the tube characteristics, a combined plot of this type is the only simple method of solving the circuit for the plate current. The load line starts at the point on the horizontal axis corresponding to the applied voltage and has a negative slope equal $1/R$. Where the load line crosses any given tube curve indicates the plate current for that value of grid voltage. The abscissa of this point gives the drop across the tube and the applied voltage minus this value gives the load voltage. By assuming various values of grid voltage a complete picture of the plate current for any assumed variation of grid voltage may be obtained. Other types of loads may be handled in a similar manner, the load line representing the current-voltage characteristic of the load. (L.R.Q.)

LOADING. Telephony.

LOADING COILS. Telephony.

LOADING POTS. These are the metal tanks in which loading coils are placed for protection from the elements. They are usually mounted on pole-supported racks and have the cables entering through weatherproof seals. (See **Telephony.**) (L.R.Q.)

LOADING, WING. Performance Parameter; Airplane.

LOAM. Soil.

LOBATA. Ctenophora.

LOBECTOMY. Surgical removal of a lobe of a gland or organ, such as the lung. (R.S.M.)

LOBSTER. Crustacea, Decapoda. A large marine animal (Decapoda) resembling the crayfishes in form. The American lobster, *Homarus americanus,* found on the Atlantic Coast from Labrador to North Carolina, attains a length of over 20″ and a weight of almost 30 lbs., but as a result of large numbers being caught for food they are now rarely taken above 1′ long. The Norway lobster, *Nephrops norvegicus,* a smaller species with more slender pinchers, is among the marine species known as crayfishes, as also are the spiny lobsters of the warmer latitudes in both Atlantic and Pacific oceans. The spiny lobsters, *Palinurus vulgaris,* differ conspicuously from the true lobsters in the absence of the large pinchers. Both are important as food.

The American lobster is taken in traps (lobster pots) baited with dead animal matter. The catch has declined from a peak of 100,000,000 lobsters annually to a small fraction of that figure but measures of protection and propagation are expected to aid in restoring partly their former abundance. The similar European lobster has followed the same course. (A.W.L.)

LOCAL ATTRACTION. In surveying, the deviation of a magnetic compass needle from the magnetic meridian as the result of a local magnetic field is referred to as local attraction. Local attraction may be detected at a station by taking forward and back bearings of one or both of the adjacent transit lines. If these bearings agree there is no local attraction at the station or at the terminal point of the other transit line or lines. (C.W.C.)

LOCATION SURVEY. Surveying.

LOCK NUT. Screw Fastenings.

LOCKJAW. Tetanus.

LOCOMOTION. The process of moving from place to place, a characteristic power of most animals and a lesser distinction between them and the majority of plants.

Locomotion is necessary to animals because their food is organic and in most environments does not reach the animal through external forces. Even the sessile animals, which may or may not be capable of some locomotion, often accomplish the same end by bringing food within reach through their own activities.

In the water the weight of the surrounding medium is so great that the animal may float, and the resistance offered to its body is sufficient to be utilized for propulsion. The body is so shaped that resistance is little in the direction of locomotion but great where propulsive effort is expended. Projections from the surface which beat against the water like oars or push or pull by undulating are common organs of locomotion here. They include cilia and flagella in one-celled and small multicellular forms, specialized jointed appendages of arthropods, and fins and flippers of vertebrates. Undulation of the body itself is a sufficient means of propulsion in some animals.

Some aquatic forms rest on the bottom and the terrestrial animals are forced to rest on some solid support at least intermittently because the air is too light to float them. In many of these forms the friction of contact with a solid is utilized by the development of movable supporting appendages which are shifted alternately to change the animal's position. This means of locomotion is known as walking. Other animals, notably the worms, creep through the action of muscles in the body wall. The body is progressively elongated and shortened, parts being thrust ahead and then drawn up to the maximum point of advance. In this type of locomotion they are aided by suckers or setae in some cases to grip the supporting surface.

Running may involve no other difference from walking than more rapid movement or it may also involve a change in the order of movement of the appendages and in their position when used, as in the various gaits of a horse. Jumping always differs in that the appendages set farthest back must be powerful enough to project the entire animal through the air. It is highly developed in such insects as the flea beetles and the grasshoppers and in the frogs and kangaroos among the vertebrates. In this class a gallop is no more than a series of leaps.

Locomotion in the terrestrial vertebrates also shows progressive change in the manner of using the appendages. The entire sole of the foot rests on the ground in man and the apes, and they are said to be plantigrade. This posture is well adapted to walking but not to running. Animals that need speed are digitigrade, resting on the tips of the toes. This position adds the length of the feet to that of the legs and permits a longer stride. It also adds the springiness incidental to the greater freedom of the ankle joint. The final expression of this position of the leg appears in the unguligrade (**Ungulata**) hoofed animals where only the hoof comes into contact with the ground. Man is plantigrade in walking and at rest but rises to his toes when he runs.

The locomotion of snakes is a highly specialized creeping process in which the ribs serve as the movable appendages and the grip of the body on the ground is provided by the broad scales of the ventral surface which project backward.

Climbing animals may merely run along branches, aided by sharp claws to provide a secure grip. The sloths, however, have the claws developed as great hooks which suspend them in an inverted position. They walk as well as hang upside down. The primates show the most extreme specialization for a form of locomotion in the trees called brachiation. Their pectoral appendages are arms, adapted for grasping and suspension, and the pelvic appendages are supporting legs. They move by swinging from branch to branch or by shifting from one hold to another, as human beings climb.

Locomotion in the air is a highly specialized process of flight. (A.W.L.)

LOCOMOTIVE. The locomotive is a self-propelled vehicle having an excess of power over its own propulsive needs, so that it is enabled to draw a useful load, generally a train of railway cars. While formerly all railway locomotives were steam propelled, one now sees other propulsive media employed in this service, some of which have made large inroads into fields formerly the province of the steam locomotive. Briefly, the types of locomotives encountered at present are: first, steam driven; second, **electric motor** driven (this reference is principally to the electric locomotive obtaining its power from third rail or elevated trolley. While the storage battery locomotive has a definite place in industrial haulage and switching, it is not suited to line service. Many of the light-weight streamlined passenger trains powered by Diesel engines have electric drive); third, **Diesel,** and other **internal combustion engine** driven locomotives. Due to the high **load factor** and uniform operating conditions of rail locomotives, the Diesel engine is particularly well suited for use where internal combustion drive is employed. There is at present a growing tendency on the part of the railroads, stimulated by public interest, to employ specially built Diesel powered train units for high-speed passenger runs. Other locomotive types are the **steam turbine** and the **compressed air** locomotives; the former, having enjoyed

the locomotive, and the products of **combustion** pass forward through the tubes to the smoke box, from whence they are discharged upwards to the atmosphere through a short stack. This type of boiler, in large sizes, has tubes of sufficient size so that **superheater** elements may be installed in them, and the use of superheated steam has greatly improved the performance of the steam locomotive. Due to the use of stay-bolt construction to hold the inner and outer shells of the water leg around the furnace, the boiler pressures carried are limited to approximately 250 lbs. per sq. in. The combustion system includes oil firing, hand firing of lump coal on grates, and **stoker** firing of coal. **Draft** is obtained by aspirator action of the exhaust discharging into the stack. As the locomotive is non-condensing in operation, sufficient supply of water for a run must be carried in the tender tank. From the tender tank it is pumped and

Locomotive.

some measure of success in Sweden, is now being tried in the United States. The latter is a common mine type. There is some experimentation with gas turbine propelled locomotives.

The main function of a locomotive is to exert a draw bar pull at the coupler. This it accomplishes by applying a rotative effort to driving wheels which rest on the track. Since the draw bar pull cannot possibly exceed the weight on the drivers multiplied by coefficient of friction between driver and track, it is seen that there is not the same urge for light-weight construction in a locomotive as in the power plant, for example, of the transport airplane. The elements of the locomotive consist, in part, of a frame and running gear, the latter consisting, in part, of a number of wheels called drivers, to which the propulsion is transmitted in the form of a rotation. The reactions developed at the wheel bearing are transmitted through the frame to the coupler. Another element is the source of power. In the common steam locomotive this is an expansion steam engine receiving steam from a **boiler,** and exhausting against atmospheric pressure. By means of a throttle **valve** located on the steam line leading to the engine, and by **valve gear** adjustments, the power developed in the engine cylinder is varied to suit the needs. In the electric locomotive, two or more motors with special windings are so arranged with controllers that their power output may be varied in several steps. They draw their power from a third rail, or overhead trolley. In a Diesel locomotive the engine is built in multicylindered form for smoothness of operation, and in order to obtain a high speed design. The engine is **governor** controlled, and operates at constant speed, driving an **electric generator.** This generator is connected through the control arrangement with the driving motors, which are mounted on the locomotive trucks.

Typical of the Diesel drive are the specifications of a recent twelve-car streamlined train. The total locomotive **horsepower** is 3600, this being divided between four 900-horsepower Diesel engines. These engines are 12 cylinder V-type, two-cycle full Diesels, of 8-in. bore and 10-in. stroke, operating at 750 rpm. The engines are water-cooled, and connected directly to d-c generators.

In the steam locomotive, another essential element is the **boiler** in which the steam is raised. Through many years of evolution, the boiler has become standardized on a horizontal fire tube type, with completely water-cooled **furnace.** The furnace end is placed at the rear of

heated by an **injector,** which is the most common means used to supply feed water to a locomotive boiler.

The principal structural material of the steam locomotive is steel. There is a great deal of fabrication of plate steel and steel shapes, which are assembled both by welding and riveting. In recent years, the heavier portions, including the frames and cylinders, have been made of steel castings. Wheels are cast iron, cast steel, or wrought steel, and revolve in plain wick-lubricated **bearings,** or, as is the case on many recent locomotives, in roller bearings.

Locomotives are frequently classified by their wheel arrangements, the method of classification being based

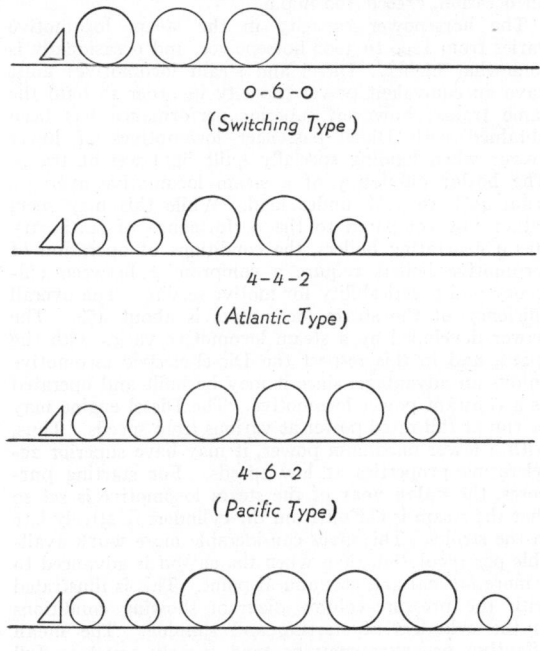

0-6-0
(*Switching Type*)

4-4-2
(*Atlantic Type*)

4-6-2
(*Pacific Type*)

4-8-2
(*Mountain Type*)

Typical steam-locomotive wheel arrangements.

upon the position and number of wheels of the leading truck, the drivers, and the trailing truck. Some typical wheel arrangements, with corresponding classifications, are shown in the accompanying figure. The **cylinders** are mounted with axis longitudinal, and in line with the center line of the drivers. Steam locomotives are two-, three-, or four-cylindered, the most common arrangement being two cylinders. Where three cylinders are used, the third cylinder is mounted under the boiler, and between the other two. Although compound-expansion locomotives have been built, the common locomotive is a single-expansion type. The **connecting rod** which extends from **crosshead** to one of the drivers is enabled to transmit propulsion equally to all drivers by use of the **side rod,** to which the **crank pins** of all drivers are connected. The trailing truck under the cab is frequently fitted with a booster engine, which is used solely to help get a train started, after which it is disconnected. On some locomotives, the booster drives on the leading tender truck.

Typical dimensions of a modern steam freight locomotive are: boiler pressure, 245 lbs. per sq. in.; cylinders, 25-in. bore by 34-in. stroke; total weight, 400,000 lbs.; weight on drivers, 260,000 lbs.; diameter of drivers, 69 in.; draw bar pull, 64,000 lbs.

In both freight and passenger service, train schedules have been stepped up in speed. In freight service, higher speeds mean less time in transit for perishable goods, and savings such as reducing amount of care needed for livestock in transit, reducing amount of ice needed to refrigerate, etc. In passenger service, high-speed trains with modern coaches redirected public attention to the railroads, and have succeeded in reestablishing the latter as a personal transportation medium competing with other modern high-speed systems. As an outgrowth of this trend, several special streamlined locomotives have been built and placed in service. While the first of these were Diesel engine powered, several streamlined steam locomotives have now been built. The beneficial effects of streamlining do not begin until speeds in excess of 50 m.p.h. are made. However, the streamlined trains are given schedules calling for high speeds, speeds which may, on occasion, exceed 100 m.p.h.

The horsepower capacity in the steam locomotive varies from 1200 to 3000 horsepower, and occasionally is somewhat higher. Diesel and steam locomotives must have an equivalent power capacity in order to haul the same trains; however, superior performance has been obtained with Diesel passenger locomotives of lower power when hauling specially built light-weight trains. The **boiler efficiency** of a steam locomotive averages from 40% to 50% under load. While this may seem rather low compared to the performance of stationary steam generating boilers, the conditions of operation of locomotive boilers require a compromise between efficiency and practicability for motive service. The overall efficiency of the steam locomotive is about 4%. The power developed by a steam locomotive varies with the speed, and in this respect the Diesel-electric locomotive enjoys an advantage, since it may be built and operated as a constant power locomotive. The Diesel engine may be run at full rated power at various train speeds. Thus, with a lower maximum power, it may have superior accelerating properties at low speeds. For starting purposes, the **valve gear** of the steam locomotive is set so that the steam is cut off from the cylinders relatively late in the stroke. This gives considerable more **work** available per revolution than when the cut-off is advanced to a more normal and economical point. This is illustrated with the pressure-volume diagram showing conditions in the cylinders at starting and running. The **mean effective pressure** may be made nearly equal to full boiler pressure with very late cut-off; however, this operation is so wasteful of steam that it is employed only for starting heavy loads.

The complete locomotive includes, in addition to the elements mentioned above, an enclosed cab containing the principal train and locomotive controls conveniently

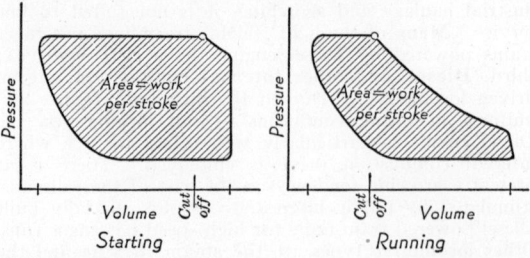

Cylinder pressure-volume diagrams for two load conditions.

arranged for operation by the engineman. Shelter and good vision, important qualities which the cab should offer, are apparently more easily obtained in the Diesel and electric designs than in the steam locomotive having a cab at the rear of the boiler. The locomotive must have air brake equipment and control, as well as lighting and various safety devices.

The steam locomotive has been evolved and perfected through approximately a century of mechanical progress. The design has become standardized, with the result that in spite of size, weight, and power, the steam locomotive is relatively cheap at present. With the exception of water-borne carriers, there is no way of transporting goods which costs so little per ton mile carried as the locomotive drawn railway freight. (F.T.M.)

LOCUS OF AN EQUATION. The locus of a point subject to one or more conditions is the set (or totality) of all points satisfying the given condition or conditions.

The locus of an **equation** in two **variables** is the set of all points in a plane whose **coordinates** satisfy the given equation and only such points. Such a locus is in general a **plane curve.**

To simplify the plotting of the graph (or locus) of a given equation:

1. The intercepts on the axes should be found.
2. The symmetry of the locus should be investigated.
3. The extent of the locus should be investigated.
4. The existence of horizontal and vertical **asymptotes** should be investigated.

Suppose a plane locus is given by its equation in **rectangular coordinates.**

To find the X- and Y-intercepts of the locus, put $y = 0$ in the equation and solve for x, and put $x = 0$ and solve for y, respectively.

Two points are said to be symmetrical with respect to an axis if this axis is the perpendicular bisector of the segment joining the given points.

Two points are said to be symmetrical with respect to a center if this center is the mid-point of the segment joining the given points.

A figure or curve is said to be symmetrical with respect to an axis or center if every point on the figure or curve has a symmetrical point (with respect to the axis or center) which also lies on the figure or curve.

To test the equation for symmetry:

If an equation is unaffected by replacing y by $-y$, the locus is symmetric with respect to the X-axis.

If an equation is unaffected by replacing x by $-x$, the locus is symmetric with respect to the Y-axis.

If an equation is unaffected by changing both x and y into $-x$ and $-y$, respectively, the locus is symmetric with respect to the origin.

For an algebraic curve (whose equation is an **algebraic equation** in x and y): the locus is symmetric with respect to the X-axis if no odd powers of y occur in the equation; it is symmetric with respect to the Y-axis if no odd powers of x occur in the equation; it is symmetric with respect to the origin if every term in the equation is of even degree or is of odd degree.

The extent of a locus is determined by finding the values of one variable for which the corresponding value or values of the other variable are imaginary or complex. Complex coordinates cannot be plotted.

An asymptote of a curve is a straight line which the curve approaches closer and closer so that the distance between the curve and the line may be made as small as we please by going out sufficiently far along the curve.

To find the horizontal and vertical asymptotes of an algebraic curve, solve the equation of the locus for each variable in terms of the other, let the one variable increase beyond all bounds numerically and find the corresponding limiting value of the other.

The locus of an equation in three variables is the set of all points in space whose coordinates satisfy the given equation and only such points. Such a locus is in general a surface.

The locus of two equations in three variables is in general a curve in space. It may be regarded as the intersection curve of the two surfaces represented by the two given equations taken separately.

In the discussion of the equation of a **surface,** we should examine the following items:

The intercepts of the surface on the coordinate axes should be found by putting the coordinates equal to zero in pairs in the equation of the surface and solving for the remaining coordinate.

The traces of the surface on the coordinate planes (intersection curves of the surface with these planes) are found by putting in turn each coordinate equal to zero in the equation; this gives the equations of the curves.

Then the curves of intersection of the surface by planes parallel to the coordinate planes are found by putting each coordinate in turn equal to various constant values.

The symmetry of the surface may be investigated by methods analogous to the case of plane curves. (L.L.S.)

LOCUST. Grasshopper. Cicada. The term is more properly applied to the short-horned **grasshoppers.** These include the migratory locusts which sometimes appear in Africa, Asia, and the plains of North America as serious crop pests. (A.W.L.)

LOESS. Loess is a buff-colored, wind-blown deposit of fine silt or marl, usually unstratified, which is often exposed in bluffs with steep to vertical faces. Loess is found in the United States in the Mississippi valley from Louisiana to Iowa, and along the course of the Missouri. The average thickness of the loess here is about 20′, but may range to 50–100′. Loess also occurs in central Europe, Mongolia and China where it is said to attain a thickness as great as 300′. The loess of the United States and Europe is believed to be the finer materials first transported and deposited by the waters of the melting ice sheets of the glacial period, and later blown to considerable distances and sometimes deposited in lakes. The Asiatic loess seems to be wholly wind transported, the source of the dust being, perhaps, the great deserts of central Asia. In the latter case the accumulation of such thick deposits is attributed to the binding power of successive generations of grasses whose former existence is suggested by a network of narrow tubes. (R.M.F.)

LODE. Ore Body.

LODESTONE. Magnetite; Magnetism.

LOG. In **navigation** the term log is used with two different meanings: a speed-measuring device, and a record book. Prior to the middle of the 19th century the speed of a ship relative to the surface of the water was measured by the log chip and line. "Heaving the log" was a duty that was performed every time the ship's bell was struck, i.e., every half-hour, and the speeds determined were entered in a book, which came to be known as the log book. Since this log book was always available to the watch officer, it became customary to enter all important incidents relating to the operations of the ship, behavior of members of the crew, conditions of the weather and sea, and, in fact, anything that the watch officer might think worthy of recording. This "deck log" was turned over to the Captain and he abstracted all important material and made up the ship's log. At present, practically every department of the ship, deck, engine room, ordnance, steward, etc., keeps its individual rough log and from these the ship's log is made up under the supervision of the Captain.

The oldest and perhaps even now the most accurate and reliable method for determining the speed of a ship relative to the water is by use of the log chip and line. The log chip is a wooden quadrant, loaded along its circular edge so that it will float upright in the water. It is attached to the log line by a 3-legged bridle, the upper leg of which is attached to the apex of the log chip in such a manner that a sharp jerk will free it. Then the log chip may be easily hauled back to the ship. When the log chip is thrown overboard it floats nearly submerged, with the flat surface perpendicular to the motion of the ship, and the line runs out over the stern with the speed at which the ship is moving through the water. For measuring the speed at which the line runs out, a sand glass was usually used. For speeds under 4 **knots** a 28-second glass was employed, and for higher speeds a 14-second glass was available. The line was divided into lengths by threading pieces of fish line through it at specified distances. In the pieces of fish line, 1, 2, 3, etc., knots were tied to indicate the number of lengths that had run out in the given time. The distance between markers was determined by the ratio between 28 seconds to the number of seconds in an hour:

$$\frac{distance}{6080 \text{ ft.}} = \frac{28 \text{ sec.}}{3600 \text{ sec.}} \text{ hence distance} = 47' \, 3.5''.$$

In order that the log chip may be well clear of the turbulence in the wake, a certain amount of stray line is provided, with a red marker indicating the beginning of measurement.

To "heave the log" two men are necessary, one to tend line and the other to operate the sand glass. The first operator throws the chip overboard and gives the word "Tip" when the red marker passes through his fingers. When the last grain of sand runs through the glass the timekeeper calls "check" and the line tender grabs the line. This gives the sudden jerk necessary to free the bridle. He then notes the number of "knots" and approximate fractions thereof that have passed through his fingers and reports that number to the officer of the deck, the number of **knots** being equal to the speed of the ship in **nautical miles** per hour.

The use of the log chip and line gives the instantaneous speed of the ship, and the values obtained over a day must be averaged to obtain the total distance run. Many different types of **patent logs** have been devised to give both speed and distance run. They are all subject to unexpected errors and should be checked frequently by comparison with the log chip and line. On the high-speed ships of the modern navy, the old-fashioned line is impracticable. To check patent logs the number of seconds required for a given length of line to run out is determined by means of a stop watch and the speed calculated. (See **Patent Log** and **Airspeed Meter.**)   (W.K.G.)

LOGARITHMIC CHART. A method of representing change by means of a curve drawn on chart paper having right-angled, logarithmically spaced coordinates. Semi-logarithmic charts have one rectilinear and one logarithmic scale. Logarithmic scales are often used because many functions and formulae will take a straight-line form on this type of chart, in contrast to a curved form on rectilinear paper. (H.C.H.)

LOGARITHMIC CURVE. Many of the properties of **logarithms** are exhibited graphically by the logarithmic curve. The logarithmic curve, whose equation is $y = \log_e x$, is shown in the accompanying figure. It is

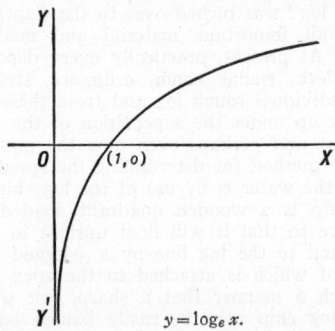

$y = \log_e x$.

the same as the **exponential curve**, whose equation is $y = e^x$, except in different position. It crosses the X-axis at the point $(1, 0)$, is **asymptotic** to the Y-axis as $x \to 0$, and to the right of the point $(1, 0)$ it rises more and more slowly.

The graph of $y = \log_a x$ (for example, $y = \log_{10} x$) is of the same form as the preceding, except of different **slope**. (L.L.S.)

LOGARITHMIC DIFFERENTIATION. Sometimes a **derivative** is found most easily by taking **logarithms** on both sides of the defining functional equation and then differentiating. This is called logarithmic differentiation. (L.L.S.)

LOGARITHMIC EQUATIONS. A logarithmic **equation** is an **equation** in which the unknown occurs in a **logarithm**. Such an equation may often be solved by using the definition of logarithms to reduce the equation to an **exponential** form. (L.L.S.)

LOGARITHMIC FUNCTION. If x and y are two **variables** related by the **exponential functional** relation $x = B^y$, where $B > 0, \neq 1$, then y is called a logarithmic function of x, and we write $y = \log_B x$. The logarithmic function is a **transcendental function**. It is the **inverse function** to the exponential function. (L.L.S.)

LOGARITHMIC MEAN TEMPERATURE DIFFERENCE. Mean Temperature Difference.

LOGARITHMIC PAPER. Logarithmic paper is paper ruled from scales on the two perpendicular axes, in which at least one of the scales is logarithmic. In semi-logarithmic paper, the one scale is a uniform scale while the other scale is a **logarithmic scale**. In log-log paper, both scales are logarithmic scales. (L.L.S.)

LOGARITHMIC SCALE. A logarithmic scale is one in which the distance from the origin to any scale mark is proportional to the **logarithm** of the number attached to that mark.

Thus, the accompanying figure shows a (common) logarithmic scale with the numbers 1, 2, 3, $\cdots$, 10, $\cdots$ attached to the division marks; if we take OI as the unit of length, then the distances OA, OB, OC,

etc., are represented by log 2, log 3, log 4, etc., so that $OI = \log 10 = 1$. In going from left to right, the scale marks will become closer and closer together.

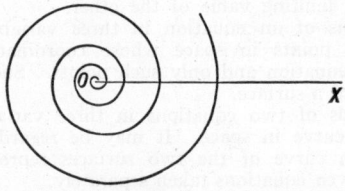

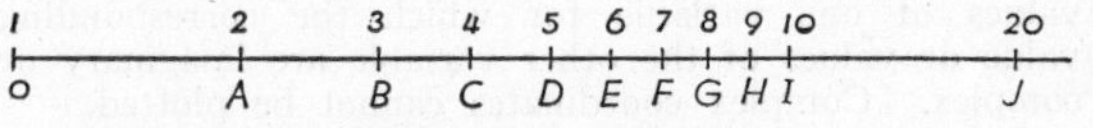

Logarithmic scale.

The logarithmic scale is applied in the **slide-rule** and in **logarithmic paper**. (L.L.S.)

LOGARITHMIC SPIRAL. This is the spiral curve represented by the equation in **polar coordinates**: $r = e^{a\theta}$ or $\log r = a\theta$. It is also sometimes called the

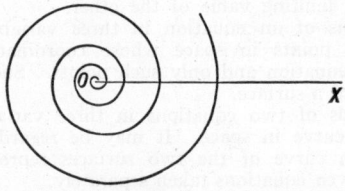

Logarithmic spiral.

equiangular spiral, since it cuts all radii from the origin at a constant angle. (L.L.S.)

LOGARITHMS. Logarithms are **exponents**, relative to a given base. Calculations involving **multiplication**, **division**, raising to **powers** and extraction of **roots** can usually be carried out more quickly and with less labor by the use of logarithms than by direct arithmetic calculation. Certain **equations** require the use of logarithms for their solution.

The logarithm of a number to a given base is the **exponent** of the **power** of the **base** which equals the given number; if $N = B^L$, then $L = \log_B N$. The logarithm of a number N to a base B is denoted by $\log_B N$.

For any positive base $B \neq 1$, we have:

$$\log_B 1 = 0, \quad \log_B B = 1.$$

The fundamental laws of logarithms are:

1. The logarithm of a product of two or more positive factors is equal to the sum of the logarithms of the factors:

$$\log (MN) = \log M + \log N.$$

2. The logarithm of the quotient of two positive numbers is equal to the logarithm of the dividend minus the logarithm of the divisor:

$$\log (M/N) = \log M - \log N.$$

3. The logarithm of a power of a positive number is equal to the logarithm of the number multiplied by the exponent of the power:

$$\log M^n = n \log M.$$

4. The logarithm of a real positive root of a positive number is equal to the logarithm of the number divided by the index of the root:

$$\log \sqrt[n]{M} = \frac{1}{n} \log M.$$

Any positive number not equal to 1 can be used as a base of a system of logarithms. However, only two particular numbers are in general use as bases, namely 10 and a number denoted by e.

The system of logarithms which uses 10 as a base is called the common system (or Briggs' system) and logarithms to the base 10 are called common logarithms. This system is especially adapted to the application to numerical calculation and is used principally for such work.

The number e is an **irrational** and **transcendental number** defined by the **infinite series**

$$1 + \frac{1}{1!} + \frac{1}{2!} + \frac{1}{3!} + \cdots,$$

or by the limit

$$\lim_{n \to \infty} \left(1 + \frac{1}{n}\right)^n,$$

and its approximate value is 2.71828.

Logarithms to the base e are called natural logarithms (or sometimes Napierian logarithms, or hyperbolic logarithms). They are particularly adapted to analytical work, but are not of much direct service as a mere aid to computation.

The symbol $\ln N$ is frequently used for natural logarithms instead of $\log_e N$.

General properties of common logarithms are:

A common logarithm in general consists of two parts, an integral part called the characteristic, and a non-integral part, less than 1, usually expressed as a decimal, called the mantissa.

The mantissa of the common logarithm of a number is independent of the position of the decimal point in the number, and depends only on the succession of digits that make up the number.

The characteristic of the common logarithm of a number is independent of the digits that make up the number but depends entirely on the position of the decimal point in the number.

The characteristic of the common logarithm of a given positive number may be found by the following rules:

The characteristic of the common logarithm of any positive number greater than 1 is positive and is 1 less than the number of digits in the integral part of the number. The characteristic of the common logarithm of a positive number less than 1 is negative and is numerically 1 more than the number of ciphers between the decimal point and the first significant figure in the given number.

So-called tables of logarithms contain the mantissas of logarithms of numbers, to a specified number of decimal places. Four-place, five-place, six-place, seven-place, eight-place and even fourteen- or more place tables are published and in use. A graphic table has also been published which has many features to recommend it.

In the use of a table of logarithms, **interpolation** based on a principle of proportional parts is generally required. This principle is: a small change in the number produces a change in its logarithm which is very nearly proportional to the change in the number.

The anti-logarithm of a given logarithm is the number whose logarithm is the given value.

The co-logarithm of a number is the logarithm of the reciprocal of the given number and is equal to minus the logarithm:

$$\operatorname{colog} N = \log (1/N) = -\log N.$$

The co-logarithm is frequently used to simplify logarithmic calculation when quotients are involved.

Formulae for change of base of logarithms are the following:

$$\log_b N = \log_a N \cdot \log_b a,$$

$$\log_b a \cdot \log_a b = 1,$$

$$\log_e N = \log_{10} N \cdot \log_e 10,$$

$$\log_{10} N = \log_e N \cdot \log_{10} e,$$

$$\log_e 10 \cdot \log_{10} e = 1,$$

$$\log_{10} e \approx 0.434294, \; \log_e 10 \approx 2.302585,$$

$$\log_e N \approx 2.302585 \log_{10} N,$$

$$\log_{10} N \approx 0.434294 \log_e N.$$

The number $\log_{10} e \approx 0.434294$ is called the modulus of the common system of logarithms with respect to the natural system; and the number $\log_e 10 \approx 2.302585$ is called the modulus of the natural system with respect to the common system.

The calculation of logarithmic tables is usually based on the use of logarithmic series. The basic series is:

$$\log_e (1 + x) = x - \frac{x^2}{2} + \frac{x^3}{3} - \frac{x^4}{4} + \cdots,$$

which is **convergent** when $|x| < 1$, but it is too slowly convergent for practical computational purposes. It can be transformed into the **series**

$$\log_e (N + 1) = \log_e N + 2 \left[\frac{1}{2N + 1} + \frac{1}{3} \frac{1}{(2N + 1)^3} \right.$$

$$\left. + \frac{1}{5} \frac{1}{(2N + 1)^5} + \cdots \right],$$

which is convergent for all values of N. By taking $N = 1, 2, 3, \cdots$ successively, the logarithms of integers can be calculated from this series. It is only necessary to calculate the logarithms of prime numbers in this way, as logarithms of composite numbers can be found by combination of these.

A graphical representation of many of the properties of logarithms is furnished by the **logarithmic curve**. (L.L.S.)

LOGISTIC CURVE. The logistic curve is a growth curve used to describe functions which continually increase, gradually at first, more rapidly in the middle growth period, and slowly again, reaching a maximum at the end of the growth. We write its equation

$$y = \frac{k'}{1 + e^{a + bx}}, \quad b < 0,$$

the symmetrical logistic. A more general form is

$$y = k_1 + \frac{k_2}{1 + e^{a + bx + cx^2}}, \quad c < 0,$$

the asymmetrical or skew logistic. The values of k'^2, k_1, k_2, a, b, and c are found by the method of averaging, the method of **least squares**, the method of selected points, or by a transformation $1/y$ using moments. The logistic curve was discovered by Verhulst and has been applied extensively by Pearl and Reed to the growth of the population of various regions and countries. (L.A.A.)

LOGWOOD. Haematoxylin. *Haematoxylon campechranun.* Leguminosae.

LOIN. The lower or posterior part of the back, near the hips. Usually in the plural. (A.W.L.)

LONGERON. A principal longitudinal member of the framing of an airplane fuselage or nacelle, usually continuous across a number of points of support. (F.T.M.)

LONGITUDE. The longitude of a point on the surface of the earth is the angular distance measured along the earth's **equator** from the **meridian** through Greenwich, England, to the point where the local meridian of the point cuts the equator. Longitude may be expressed either in units of time (hours, minutes and seconds) or angle (degrees, minutes and seconds). It is measured east or west from Greenwich through 12 hours or 180°. For convenience in **navigation**, west longitude is marked plus (+) and east longitude minus (−).

Since time is defined as **hour angle**, the difference between Greenwich time and local time must be equal to the local longitude. In case Greenwich time is greater than local time the longitude will be west, if smaller the longitude is east. Both the Greenwich and local time must be expressed in the same system, i.e., the hour angle of the same object must be considered.

The longitude of a celestial object is the coordinate in the ecliptic system of **spherical coordinates** measured along the **ecliptic** from the **vernal equinox** in the direction of the sun's annual motion to the great circle through the pole of the ecliptic and the object.

Determination of Terrestrial Longitude. The problem of determination of terrestrial longitude resolves itself into two parts: the determination of local time and the determination of the Greenwich time at the same instant.

The most accurate method for determination of local time is to observe the instant of passage of a star of known right ascension across the local meridian. The **meridian circle** is used for this purpose and is essentially an **altazimuth** instrument accurately fixed in the azimuth of the local meridian. By using a large number of stars with accurately known right ascension it is possible to obtain local time with an accuracy of about one thousandth of a second.

The meridian circle is a fixed instrument and not available for ordinary field work or for navigational purposes. For such purposes several different methods for determining local time are available. The method most commonly employed is to observe the **altitude** of some celestial object of known **declination** either with an **altazimuth** instrument or with a **sextant.** If the **latitude** of the observer is known the **astronomical triangle** is then solved to obtain the hour angle of the observed object. If the object observed is a star the **right ascension** of the star added to the computed hour angle will be the sidereal time at the instant of observation. In case the altitude of the sun has been measured, the computed hour angle will be the local apparent time.

In spite of the fact that methods for measuring the altitude of celestial objects have been known for 2000 years and the solution of the astronomical triangle understood for nearly half that period, it was not until the 18th century that any method for determining Greenwich time at sea was available for the solution of the longitude problem. The first practical method was the employment of the **moon's** motion through the stars as a clock hand. The angular distance of the moon from certain bright stars was tabulated in **Ephemerides** for Greenwich time throughout the year. The measurement of **lunar distance** to find Greenwich time was far from accurate and it was not until late in the 18th century that the perfection of the **chronometer** made it possible for navigators to carry Greenwich time with them. Formerly, the chronometer rate had to be determined while the ship was in port and then the chronometer was treated with great care in order that this rate should remain constant. At present time signals are sent out by radio by observatories of the various nations so that navigators may determine the Greenwich time and hence their chronometer corrections with great accuracy. In fact the errors introduced by measurement of altitude and determination of dead-reckoning latitude, which must be used in the solution of the astronomical triangle, are larger than the errors in the Greenwich time.

Modern navigational methods for determining longitude are discussed elsewhere. (See **Celestial Navigation, Dead Reckoning, Fix, Radio Navigation.**) (W.K.G.)

LONG-JAW. Pisces, Teleostei. One of the freshwater **herrings,** found in deep water in the Great Lakes. (A.W.L.)

LONG-PERIOD VARIABLES. Examination of the curve appearing in the article on **variable stars,** indicates that there is a considerable group of variables with periods greater than 100 days. This group of objects is known as the long-period variables. More

than half of the long-period variables have periods between 250 and 400 days. The accompanying figure (Light curve of the long period variable χ Cygni)

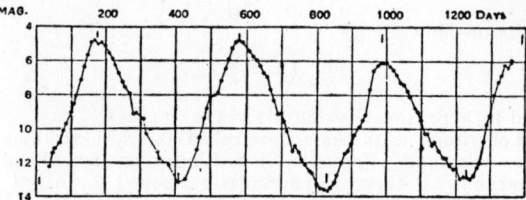

Light curve of the long period variable χ Cygni. (*From observations, during the years 1922 to 1925, by the American Association of Variable Star Observers.*)

shows the **light curve** of a long period variable. Examination of the diagram indicates at once that the characteristics of variation do not repeat themselves exactly from cycle to cycle. This irregularity in the shape of the light curve from cycle to cycle is characteristic of all long-period variables. All of the long-period variables are red stars, most of them being M **spectral class,** with a few in the N, R, and S classes.

There is no adequate explanation for the variability of the long-period variables. The pulsation theory, as discussed for the **Cepheid** variables has many attractive features, but on the basis of this theory we should expect the stars to be hottest when they are of smallest diameter and this is exactly contrary to the observed diameters of **Mira.** The similarity between the forms of the light curves of certain long-period variables and the **sun-spot** curve has frequently been commented upon, and the statement is sometimes made that our sun is a long-period variable with period of about 11 years. However, the period is much longer than any known long-period variable and, furthermore, the sun is a G-type star, so it seems that variability in the sun and in long-period variables must be totally different phenomena. (W.K.G.)

LONG-RANGE FORECASTING. Weather Forecasting.

LONG WAVES IN THE PREVAILING WESTERLIES. There develop in the westerlies, particularly during the cold months of the year, certain perturbations which cause the westerlies to blow alternately northward and southward in a sinusoidal wave pattern but always with a component of velocity directed from west to east. Wave crests are associated with anticyclones at ground and near ground levels, whereas troughs are associated with cyclones. There is, therefore, a definite relation between the sinusoidal perturbations of the westerlies and large-scale surface weather phenomena. Progression eastward, or retrogression westward, of the crests and troughs is usually about the same as surface anticyclones and cyclones. Sinusoidal perturbations can, therefore, be used for prognostic purposes computing their velocities. These velocities are given by the formula:

$$C = U - \frac{bL^2}{4\pi^2}$$

in which C = The velocity of the wave.
U = The west to east component of air motion.
L = The wave length of the waves.
b = The rate of change in the coriolis parameter northward.

Wave velocities may be positive, zero, or negative; positive indicating easterly movement, negative, westerly movement.

Isotherms aloft also assume partial or complete sinusoidal form under some conditions. The following rela-

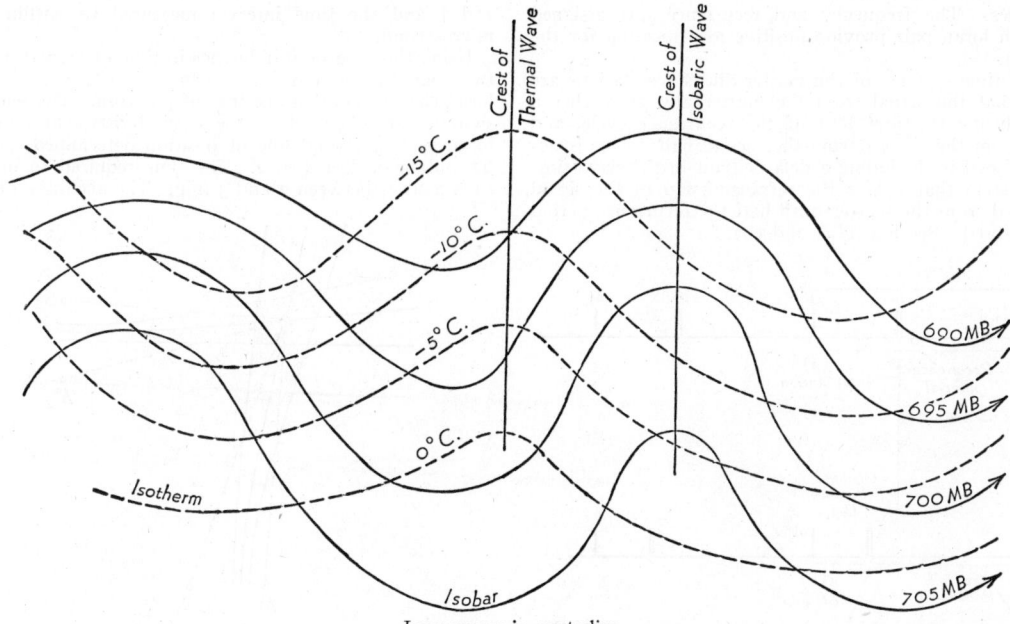

Long waves in westerlies.

tions between amplitudes of the thermal and the streamline wave apply:

$$C = U\left(1 - \frac{A_s}{A_t}\right)$$

in which A_s = Streamline amplitude.
A_t = Thermal amplitude.

From this the following conclusions are possible:

1. If the amplitude of the thermal wave is greater than the streamline wave and the two are in phase, cyclones and anticyclones will have a slow eastward component of movement.

2. If the thermal wave is 180° out of phase with the streamline wave, cyclones and anticyclones move eastward rather rapidly. The smaller the amplitude of the thermal wave in relation to the streamline wave, the more rapid will be the movement of surface systems.

3. If the amplitude of the thermal wave is less than the streamline wave and the two are in phase, cyclones and anticyclones will move with a slow westward component (retrograde motion).

4. If the amplitudes of the thermal and streamline waves are the same and the waves are in phase, the surface systems will have no east-west component of movement.

The first two of these conditions are common: the latter two occur but are not usual. (P.E.K.)

LOOMING. Mirage.

LOON. Aves, Gaviiformes. *Gavia.* Large birds (**Aves**) with short legs, webbed feet, and strong sharp beak. They are powerful swimmers and divers and are called divers as well as loons. (A.W.L.)

LOOP ANTENNA. A loop antenna, when used with radio transmitters or receivers, possesses valuable directional properties. (See **Directional Antenna.**)

On ships and airplanes, a loop antenna equipped with a **compass card,** whose 000°–180° line is parallel to the longitudinal axis of the ship, may be used to obtain relative **bearings** of radio stations. In using the loop, the position of minimum intensity is sought for. In other words, the indicating needle on the compass card is in a direction perpendicular to the plane of the loop, and the observer rotates the antenna until the position of minimum signal intensity is found. The reading of the dial will then give a **line of position** through the ship and the radio station. With the simple loop there is no method for telling on which side of the instrument the station is located. To overcome this difficulty, the loop antenna is combined with a non-directional antenna as described under **radio compass.** (See **Radio Bearing, Radio Beam, Radio Navigation.**) (W.K.G.)

LOPHOPHORE. A ridge surrounding the mouth or closely related to it in the phyla **Bryozoa, Brachiopoda,** and **Phoronidea.** It bears the tentacles. (A.W.L.)

LOPOLITH. The term proposed by Grout in 1918 for large concordant **lenticular, igneous** intrusives which differ from **laccoliths** in that the center is depressed so that its upper surface, together with the overlying strata, form a basin instead of a dome. (R.M.F.)

LORAN. The word loran is derived from the phrase long range navigation and is used to indicate the system of **hyperbolic navigation** which has the longest range of all systems thus far developed. In common with GEE navigation, loran signals are of the pulse type and, in the progress of the developmental research, many different combinations of **radio frequencies** and pulse rates were used. One of the systems developed is known as Standard Loran and has been recommended by the U. S. Coast Guard for use in commercial air and sea **navigation.** Standard loran has a range over water of about 750 nautical miles in daylight and 1400 during the night. The range over land is very much less than this. Since loran is a hyperbolic system, the stations must operate in pairs as "master and slave." The slave station is essentially a relay station. The plans of the Coast Guard provide for a system of about 70 stations which will effectively cover the sea lanes along both coasts of the United States, the **great circle tracks** from this continent to the British Isles and, by using some of the conquered islands in the Pacific, the various important trade routes over that ocean.

Standard loran operates on radio frequencies of 1950, 1850, and 1750 kc. Fourteen recurrence rates, i.e., intervals between signals, are in use on each of the frequencies. The frequency and recurrence rate assigned to each loran pair provide positive identification for the station.

The time systems of the master and slave stations are such that the signal from the master always reaches a ship during the first half of the recurrence cycle, and that from the slave during the second half. This is accomplished by including a delay circuit in the slave timing system that delays the retransmission of the signal received from the master until half the recurrence period has elapsed. See Fig. 1(a) and (b).

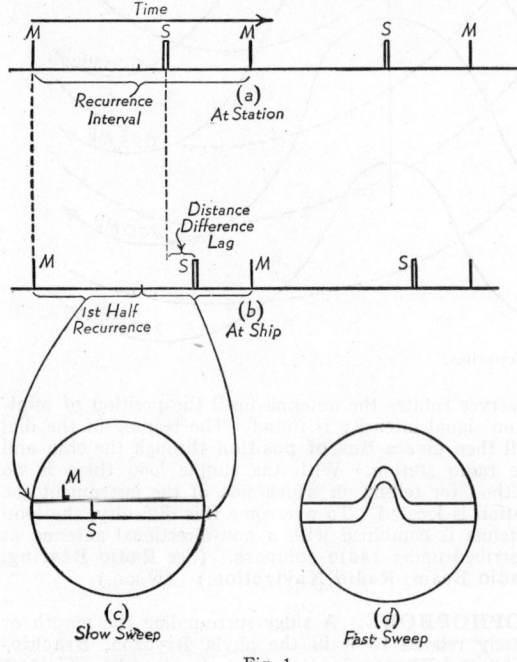

Fig. 1.

Standard receiving equipment has been designed for ships and planes in which both the receiving and timing units are present, with selector switches permitting the operator to set on the frequency and recurrence rate assigned to any loran pair he wishes to use. Differential amplifiers, synchronized by the timing circuit to the recurrence rate of the station, act on both the master and slave signals to deliver them at equal strength to the indicator unit.

The slow sweep of the oscilloscope ("viewing scope") appears as two parallel lines, one covering the first half of the recurrence cycle and the other the second half. Hence the signals received from the master appear on one line and from the slave on the adjacent parallel line. See Fig. 1(c). An adjustment is provided to allow for correction of slight variations in the crystal control of the timing unit and, when this is properly set, the desired signals remain stationary on the scope. The signals from other stations, which may be within range and operating on the same frequency, will drift along the line since their recurrence interval will be different from that of the pair being used.

When the signals are properly "set up" on the scope, a set of time markers is thrown on the screen and a determination is made of the time interval between reception of signals from the master and slave. This will include the delay interval at the slave station, but, since this is standard for each recurrence rate, it may be allowed for. A delay circuit is now introduced and the signals brought into approximate coincidence. With this adjustment made, a fast sweep spreads out the signals so that close coincidence may be established [see Fig.

1(d)] and the time interval measured to within one microsecond.

Using this measured difference in time of arrival of the two signals, the navigator then uses either tables or loran charts to obtain one line of position. The selector switches are then set to the characteristics of another loran pair, a second line of position determined, and a fix obtained. See Fig. 2. The time required to obtain such a fix is between 2 and 3 min. The accuracy of the

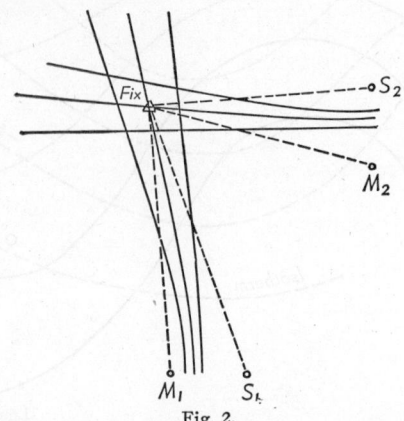

Fig. 2.

determined fix is of the same order of magnitude as that obtained from good celestial navigation and, of course, is independent of the state of visibility.

Standard loran may be used for homing purposes in the manner described in the articles on hyperbolic and GEE navigation. Among the many advantages of loran over the various "old-fashioned" systems are the facts that the determined fix is independent of any navigational instruments (such as compass, chronometer, radio equipment, etc.) and that the ship can proceed along a loran line with nothing operating other than the loran receiver. Furthermore if anything is wrong with the timing sequence of a particular station the monitors on shore will pick it up and the station will be shut down. This means that any loran fix obtained is good, since nothing will be received if the station is at fault.

It should be realized that loran depends for its effectiveness upon the transmission of electromagnetic waves. The characteristics of the waves received may be influenced by a variety of causes between the transmitter and the receiver. Among other troubles is found the interference between ground and sky waves, combined with the fact that the height of the reflecting layer of the sky waves varies.

In the attempt to increase the range of loran beyond that at present attainable with standard, several other systems have been developed with some successes for each. In the so-called SS (Skywave-Synchronized) loran the distance between stations is increased to 1200 or 1300 miles so that the effects of the ground wave are negligible. Several sky waves arrive at the receiver with the equipment designed to bring these into synchronism. Unfortunately there is no way of avoiding the errors introduced by the variability in the height of the reflecting layer. The minimum errors of about nine tenths of a mile occur when the lines cross at right angles and, to bring this condition to certain desired regions of the earth, the stations may be arranged at the corners of a large square, with pairs of stations at the opposite corners of the "SS Quadrilateral." This system was used with some success by the RAF in operations against Germany.

Still another type of loran takes advantage of the long-distance transmission of low-frequency radiation. The pulse lengths must be much longer in this LF loran than in standard, and there is a corresponding decrease in the accuracy of time difference measurement. However, both the daylight range and the range over land appear to be as great as that for nocturnal conditions with the Standard, and at extreme distances the accuracy of fix determination seems to be nearly as good as the Standard. One great advantage of LF loran is that the waves never penetrate the reflecting layer, and the signal shapes at long distances are much less confused than with the Standard system. (W.K.G.)

LORENTZ-FITZGERALD CONTRACTION. A hypothesis proposed by Fitzgerald and extended by Lorentz to yield an explanation of the negative result of the **Michelson-Morley experiment.** Fitzgerald suggested that when a body moves through space, it experiences a compression or shrinkage in the direction of the motion. Lorentz showed how such an effect might be expected on the basis of the electromagnetic theory

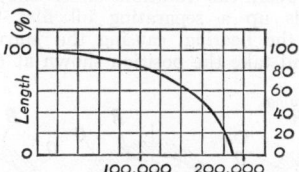

Speed (miles per second)

Graph showing falling off of length with speed. The percentages are of the "static length."

and the electrical constitution of matter. That is, he deduced that when a body moves through space, its dimension parallel to the line of motion should become less by an amount dependent upon its speed. If the speed of the body is v, and that of light is c, then the contraction is in the ratio $\sqrt{1 - \frac{v^2}{c^2}} : 1$. For the earth moving in its orbit (about 18.5 miles per sec.), this contraction amounts to about one part in 200,000,000, which, on the diameter of the earth, would be about 2.5". Small as this is, it accounts exactly for Michelson and Morley's result by making the source of light and the mirror draw closer together when the system is moving lengthwise. It is the only satisfactory explanation of that result if we regard light as traveling through a stationary **ether.** Viewed from a different angle and expressed in different terms, this hypothesis is now embodied as one of the basic principles of **relativity.** (L.D.W.)

LORICA. A structure enclosing the body in many species of rotifers. Formed of thickened integument. Also a shell or test surrounding the body in certain species of **Protozoa.** (A.W.L.)

LORICATA. 1. A group of ciliate (cilia) protozoans protected by a test. 2. A group of rotifers in which the body is protected by a shell. 3. The Crocodilia, a division of the reptiles. 4. A suborder of edentate mammals including the **armadillos** and **pangolins.**

A similar name, Loricati, is used for a group of fishes.

The term is now obsolete, in spite of its former wide application, and is rarely met in modern treatises on classification. (A.W.L.)

LORIQUET. Parrot.

LORIS. Mammalia, Primates. Animals of the warmer part of southeastern Asia and the East Indian islands. **Lemurs.** They have large staring eyes and from their very slow movements have also been named slow lemurs. The lorises are forest animals of nocturnal habits. The slender loris is known as *Loris gracilis,* the slow loris is *Nycticebus tardigradus.* (A.W.L.)

LORY. Parrot.

LOSCHMIDT NUMBER. Avogadro's Law.

LOTUS. Water-Lilies. Nymphaeaceae.

LOUD-SPEAKER. Speaker.

LOURI. Plantain Eater.

LOUSE. Insecta. An external parasitic insect found on warm-blooded animals, both birds and mammals. Two kinds of lice occur, the bird or biting lice and the

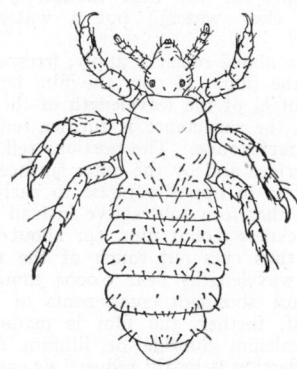

The head louse of man.

true or sucking lice. The former make up the order **Mallophaga** and the latter the order **Anoplura.**

As the names suggest, the bird lice have biting mouth parts and the sucking lice have piercing and sucking mouths. Bird lice eat bits of feathers and hair and cuticular scales while true lice suck blood. The true lice and a few species of bird lice live on mammals.

Three species of true lice, the head louse, *Pediculus humanus capitis,* crab louse, *Phthirus pubis,* and body louse, *Pediculus humanus corporis,* are parasitic on man. Head lice can usually be removed with a fine-toothed comb and the other species yield to mercurial ointments. The body louse is sometimes dangerous as a carrier of typhus fever.

Many other species of lice infest domestic animals and wild species. (A.W.L.)

LOUSE FLY. Insecta, Diptera. True flies specialized to live as external parasites on birds and mammals. They include the **sheep** tick (*Melophagus ovinus*), **bat ticks,** and **bee lice** as well as certain winged species which are recognizable as flies. Found chiefly on owls and other birds of prey. (A.W.L.)

LOVE BIRD. Parrot.

LOW. A region over which atmospheric pressure is lower than the surrounding area; i.e., an abbreviation for region of low pressure. Cyclonic winds blow around a low. (P.E.K.)

LOW-REFLECTION FILMS ON GLASS. The problem of troublesome reflections from glass surfaces, such as those from windshields, show windows, clock faces, and lenses, has been recently solved by the superposition of transparent films of suitable material and thickness. The action of these films depends jointly upon two different optical principles.

1. Reflection of light occurs only at an interface between two different media. The ratio of the quantity of the light reflected by any reflector to that of the light incident upon it is called the reflection factor, or the reflectance, of that reflector. In the case of a glass

surface covered with a transparent film, the reflection takes place partly at the outer surface of the film and partly at the interface between film and glass. It may be shown that the total reflectance of such a combination is a minimum when the absolute **refractive index** of the film is equal to the square root of that of the glass. Therefore it is desirable to find a film material which fulfills this condition as nearly as possible and at the same time makes a durable coating adhering firmly to the glass. The materials which at present appear most satisfactory in these respects are calcium fluoride (CaF_2) and lithium fluoride (LiF). Water would be excellent: wet glass reflects very little light (which accounts for the near invisibility of a glass object under clear water); but a water film soon evaporates.

2. All of the above remarks apply, irrespective of the thickness of the film. If now the film be made of a thickness about ¼ of the wavelength of the (normally) incident light, the reflectance is further reduced almost to zero by **interference**. The portions reflected by the film-glass interface and by the outer surface of the film are in such case in opposite phase to each other, and their effect is therefore subtractive instead of additive. A film of thickness 0.0001 mm. (or about 4 millionths of an inch) thus cuts out much of the reflection of light having wavelengths near 0.0004 mm., which includes the most abundant components of visible daylight. And if, further, the film is made of such a material as calcium fluoride or lithium fluoride, the amount of reflection is vastly reduced as compared with that from the untreated glass surface. Details of the chemical methods of applying low-reflection films to glass will be found in numerous technical papers.

It also follows that, since less light is reflected from glass provided with such films, more light is transmitted. This feature has proved of value in the treatment of lenses in optical instruments. (L.D.W.)

LUBBER'S LINE. Compass.

LUBRICATION.
Lubricants are employed in engineering practice for two reasons: (1) to diminish friction between the moving surfaces of machine parts; and (2) to decrease friction between a cutting tool and the material to be cut, and at the same time serve to dissipate the heat developed in the machining operation. When the heat-dissipating function is of primary importance, the lubricant is usually referred to as a coolant.

Lubricants may be classified either by state, as liquid, semi-solid, or solid, or by origin as mineral, vegetable, or animal. Animal lubricants are obtained from the fat of common animals and can be classified as hard fats (stearin) and soft fats (lard) or naturally occurring combinations. They can be separated into further subdivisions of hardness and softness. Vegetable lubricants include rape seed oil, sperm oil (whale oil), cottonseed oil, soybean oil, castor oil and linseed oil. They vary in properties from solid to liquid, depending on their origin. Because animal and vegetable lubricants are more complex, chemically speaking, they are also more unstable, and the simpler mineral oil lubricants are usually preferred for machine applications.

Mineral oils are refined from crude oil by separating and purifying the various heavy components of the crude. Some crude oils (Pennsylvania) contain less gum-forming and asphaltic components than do others (Texas). For this reason the methods of treatment vary widely. The various fractions of the crude oil are usually separated by distillation, and undesirable components are removed by special processing such as acid treatment. After refining to obtain improved properties various special chemicals may be added, such as anti-oxidants, anti-gumming compounds and substances to change the "oiliness" characteristic. In the latter class are compounds which tend to concentrate at the surface rather than in the main bulk of the oil so that a small amount of the added constituent may have an effect quite disproportionate to its actual volume.

There are a few special cases where contamination by the oil may be detrimental to a product and the product itself may be used as the lubricant. Such cases include the manufacture of edible oils and some chemicals of an oily nature such as concentrated sulfuric acid (oil of vitriol).

The mechanism of lubrication in a bearing is shown diagrammatically in the figure. At A, the clearance space between the journal and bearing is filled with oil, and the journal rests on the bearing surface in metal-to-metal contact. (The load on the journal acts vertically downward.) As the journal begins to rotate in the direction shown by the arrow, the journal rolls upward to the right, as shown at B. Since oil adheres to the surface, however, the rotation under suitable circumstances builds up a separating oil film between the journal and the bearing, causing the journal to move to the left and take the position shown at C.

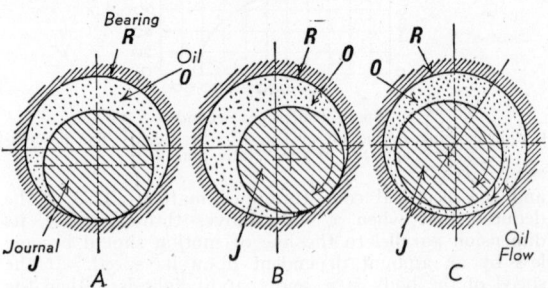

If the oil film shown at C separates the journal and bearing surfaces sufficiently so that no metal-to-metal contact (induced by the minute irregularities in the surfaces) can occur, the induced friction in the bearing is dependent largely upon the force necessary to shear the oil film, and thick-film or perfect lubrication is said to prevail. When metal-to-metal contact exists for all or part of the time, thin-film or imperfect lubrication is present, and the character of the contacting surfaces and the nature of the materials used for journal and bearing are important. The latter characteristics are important in thick-film lubricated bearings, however, since metal-to-metal contact may prevail during starting and stopping conditions. Thick-film lubrication is difficult to attain in oscillating or reciprocating bearings, since the oil film is continually undergoing changes in character and size because of the reversal of the contact surface motions.

The coefficient of friction f in thick-film lubricated bearings may be estimated from the following:

$$f = .002 + \frac{473}{10^{10}}\left(\frac{Zn}{p}\right)\left(\frac{D}{c}\right)$$

where Zn/p is the *bearing modulus*, D is the diameter of the journal or bearing in inches, and c is the total clearance in inches, or difference in diameter between the bearing and the journal. This expression is valid for small- and medium-sized bearings, up to about 3 or 4″ in diameter, with length-diameter ratios varying from .75 to 2.8. For thin-film lubricated bearings, the coefficient of friction f may be estimated from the following:

$$f = K\sqrt[4]{p/V}$$

where p is the unit pressure on the bearing, in pounds per square inch of projected area (length times diameter), V is the surface velocity of the journal, in feet per minute, and K is a constant, varying from .006 for a well-lubricated, carefully attended bearing, to .045 for reciprocating bearings in unfavorable locations, given only casual or indifferent maintenance, with a restricted or intermittent supply of lubricant.

Bearings may be lubricated in many ways. Intermittent lubrication is accomplished by using grease or oil and the lubricant is usually applied by an operator or through an oil hole, oil cup, or grease cup. Limited lubrication insures a continuous supply of a limited quantity of the lubricant, and is effected by a drop feed oil cup which permits a constant supply of oil through an adjustable needle valve or by a pad or wick which presses against the journal as it rotates, and which permits the oil to feed by capillary action to the surfaces in contact. Continuous lubrication insures an adequate supply of oil to the bearing surfaces and is effected in numerous ways. Ring- and chain-oiled bearings have a loose ring or chain resting on the journal which brings oil from a reservoir in the bearing housing to the top of the journal as it rotates. In bath lubrication, the journal is partly or wholly submerged in a pool of oil. Splash lubrication is used on reciprocating mechanisms, as in internal combustion engines where the shaft is inclosed and the reciprocating member can dip into a reservoir of oil at each stroke. Pressure lubrication employs a circulating system where the oil is pumped from a reservoir to the bearing and returns by gravity to the reservoir. (See **Hydrocarbons**.) (H.C.H., D.E.M.)

LUDER LINES. Temper Rolling.

LUDIO. Guenon.

LUDWIG'S ANGINA. Stomatitis.

LUES. Syphilis.

LUMBAGO. Acute pain in the back due to local infection or **sprain** of the back muscles. (R.S.M.)

LUMBAR PUNCTURE. Cerebrospinal Fluid.

LUMBER. Lumber is timber which has been sawed into boards and planks (see also **Wood**). If the logs are cut into quarters, and then sawed across the annual rings, the resulting product is known as quarter-sawed lumber. Boards which are sawed tangential to the annual rings are flat-sawed, and most lumber is so produced. Lumber is known as rough lumber, surfaced lumber, or worked lumber, depending on whether it is rough as it comes from the mill, dressed by a planer, or specially cut to make matched, lapped, or patterned lumber. On the basis of quality, lumber is graded into two main divisions, select and common. The former class is used mainly for special interior trim, and is more expensive than common lumber, since in select lumber the number, size, and type of defects permissible are restricted. There are four grades of select lumber. Grade A must be free from defects; grade B allows a few defects or blemishes; grade C allows that small number of defects and blemishes which may be covered by paint; and grade D allows that number of defects and blemishes which do not detract from the finished appearance. Grades A and B are suitable for natural finishes, and C and D for painted finishes. Grade A lumber is very difficult to obtain. Usually it is wanted in small quantities, which can be obtained by cutting out suitable portions of Grade B lumber. There are few lumber requirements but which may be satisfactorily executed in Grade B lumber, so when one wishes to have the best it is customary to specify, not A, but "B or better." Common lumber is graded as No. 1 common, No. 2 common, etc. The grading and selection of lumber cannot be exact, and hence there can be no definite demarcation between grades. Nos. 1 and 2 common may generally be used for house framing and other purposes with little or no waste, and this only in the extent the defects permit. Most wooden construction is with No. 2 or No. 3 common lumber.

The common unit of measurement of lumber is the board foot, which is the volume of wood in a board 12″ wide, 1′ long, 1″ thick. It is $\frac{1}{12}$ of a cu. ft. However, the purchaser of a board foot will get $\frac{1}{12}$ of a cu. ft. only in rough lumber, since it is customary to charge for finished lumber upon the basis of the rough board which was used. Thus a piece of 1″ × 8″ dressed lumber will usually measure about $\frac{7}{8}″ \times 7\frac{3}{4}″$, but if one contracted for 100 such boards, each 10′ long, he would be billed with the board feet in the rough lumber before planing. In the example quoted, the board feet are 100 × 10 × $\frac{2}{3}$ = 667 (1 ft. of 1 × 8 board has $\frac{2}{3}$ the volume of 1 ft. of 1 × 12 board). Standard commercial practice in this country is to market lumber in even lengths, as in 6, 8, 10, 12, 14, and 16 ft. Generally the cost per unit is the same for all lengths, but premium may be required for extra long lengths. (F.T.M.)

LUMEN. The internal cavity of a hollow organ. Usually applied only to tubular organs such as the blood vessels. The lumen is also a unit of measurement in photometry, being defined as the quantity of light or luminous flux received upon a unit surface, all points of which are at a unit distance from a concentrated source of one spherical candle intensity. (A.W.L.)

LUMINESCENCE. Broadly this term refers to the emission of light due to any other cause than high temperature. A firefly, for example, illustrates bioluminescence; certain chemical reactions give out light (chemiluminescence); and some electrolytic rectifiers are the source of galvanoluminescence. Triboluminescence is observed upon vigorously grinding certain solids, notably ordinary sugar.

A large variety of substances become luminescent when stimulated or "excited" by suitable radiation or by emissions such as cathode or beta rays. This phenomenon is apparently quite complex and is exhibited in various aspects. In some cases the light is emitted only so long as the exciting emission is maintained; it is called fluorescence. The screen of a fluoroscope thus responds to x-rays. In other cases the luminescence persists after the excitation is removed and it is then called phosphorescence. Thus zinc sulfide, under certain conditions, glows brightly for a time after exposure to daylight or lamplight, but the luminosity decays rapidly and disappears, usually within a few minutes. Some materials exhibit thermoluminescence; that is, they become luminescent, after exposure to excitation, upon being raised to a sufficiently high temperature. **Resonance radiation** may be regarded as a type of fluorescence in certain gases.

Stokes pointed out that when luminescence is excited by radiation, the frequency of the luminescence is usually less than that of the incident radiation. This is of course always true when visible luminescence is excited by ultra-violet, x-rays, or gamma rays.

Numerous theories regarding luminescence have been proposed, by Lenard, Kowalski, Perrin, Baly, and others. They agree mostly in assuming that the emission of luminescence is due to the removal of electrons from molecules by the energy of the exciting rays and the release of part or all of the energy upon their return. The **quantum theory** of radiation and electronic processes has done much in recent years to clarify certain aspects of the phenomena. (L.D.W.)

LUMINOSITY OF A STAR. The intrinsic brightness of a star may be expressed either in terms of its **absolute magnitude** or in terms of the sun's brightness as unity. The luminosity of a star is defined as its intrinsic brightness in terms of the brightness of the sun as unity. That is to say if our sun were replaced by a star of luminosity 100 the light received by the earth would be 100 times as great. (W.K.G.)

LUMINOUS EFFICIENCY. Photometry.

LUMINOUS INTENSITY. Photometry; Candle Power.

LUMMER-BRODHUN SCREEN. Bench Photometers.

LUMP SUCKER. Pisces, Teleostei. *Cyclopterus.* Coastal fishes (**Pisces**) of the colder seas of the northern hemisphere. They have an adhesive organ úsed to attach them to rocks. The body is short and stout. (A.W.L.)

LUNAR DISTANCE. For a determination of terrestrial **longitude** the Greenwich **time** must be known. Prior to the invention of chronometers and the subsequent development of radio broadcasting of time, the determination of Greenwich time at sea was very complicated. For many centuries it was realized that the position of the **moon** relative to the stars might be considered as a clock hand.

Up to 1912 the American **Ephemeris,** and corresponding publications of other nations, published the distance of the center of the moon from the center of the sun, the four brighter **planets,** and certain bright stars, for every three hours of Greenwich time. Only those objects were used which were within convenient distance of the moon at the given time. All that was necessary for a navigator to determinue his Greenwich time at any instant was to measure with his **sextant** the distance between the bright limb of the moon and a tabulated "lunar distance star." Then, after applying necessary corrections, by interpolation from the tables he could find the Greenwich time corresponding to the instant of observation.

While this problem is fundamentally very simple in theory, in practice the method will not yield very accurate results. In the first place, the published lunar distances were **geocentric** in character and referred to the center of the moon. The problem of correcting the observations for **parallax, refraction,** semi-diameter of the moon, etc. (known as "clearing the observation") was quite complicated and subject to numerous errors. Furthermore, the motions of the moon could not be predicted with accuracy greater than several seconds of arc. Finally, it must be realized that the moon is moving, on the average, with a motion of only about $11''$ in 20 sec. of time. Eleven seconds of arc is about the limit of accuracy of a single sextant observation, and hence 20 sec. of time is about the limit of accuracy of a determination of Greenwich time by the method of lunar distance. An error of this magnitude would introduce an error of about 5 miles in a longitude determination for a ship on the **equator.** (W.K.G.)

LUNAR INTERVAL. Tides.

LUNG. Respiratory System.

LUNG BOOK. A respiratory organ of **spiders** and some other **arachnids.** It consists of many thin plates or lcaves in a small chamber on the abdomen which opens by a narrow aperture. Air circulating between these leaves supplies oxygen to the blood inside them. (A.W.L.)

LUNG FISH. A fish with an air bladder opening from the **pharynx** which can be filled with air gulped through the mouth and serving as a lung. Although some members of the other orders are able to breathe air, the term applies specifically to species found only in Australia (*Neoceratodus*), Africa (*Protopterus*), and South America (*Lepidosiren*), constituting the order or

subclass Dipnoi. The dipnoids live in transient streams and swamps. Some species pass the dry season in cells which they form in the muddy bottom as the water dries up. They resemble the amphibians in some details of structure. (See also **Fossil Fishes.**) (A.W.L.)

LUNULE. 1. A crescentic mark in the pattern of any animal. 2. A depression in front of the umbo on the shell of some **bivalve** mollusks. (A.W.L.)

LUPANINE. Alkaloids.

LUSTER. This term is used by mineralogists to describe the appearance of the surface of a mineral, usually a crystal face, in reflected light. The principal types of luster are: metallic, adamantine, vitreous, resinous, greasy, pearly. The degrees of luster may be defined as: splendid, shining, glistening, or dull. Schillerization is a peculiar form of sub-metallic luster observed in different directions in certain minerals such as schillerspar, **diallage, hypersthene,** etc. (R.M.F.)

LUTECIUM. Symbol: Lu. Atomic number: 71. Atomic weight: 174.99. No isotope, but of single atomic form: 175. Type of compound: Lu_2O_3. Color of salts: Colorless. Discovered by Urbain in 1907. A member of the **yttrium** sub-group of the rare earth metals. (R.K.S.)

LUTH. Reptilia, Testudinata. A large marine **turtle,** *Dermachelys coriacea,* reaching a length of six feet. Its **carapace** is formed of bony plates connected together but not joined to the spinal column or ribs. Also called the leathery turtle. Its flesh is not palatable. (A.W.L.)

LYCOPHORE. A larval form of tapeworm with ten hooks for locomotion. It lodges in **annelids** and **mollusks. Cestoda,** subclass Cestodaria. (A.W.L.)

LYCOPIN. Pigments in Plants, Aminoacids and Proteins.

LYCOPODIALES. Club mosses. The order Lycopodiales contains about 500 species, most of which are included in two genera, *Lycopodium* and *Selaginella.* The species of *Lycopodium* are trailing plants often called ground pines, or ground hemlock, as well as club mosses. Many of them are common plants of dry open places in the temperate zone. The plants have long creeping stems growing on the surface of the ground or several inches beneath it. From this prostrate stem short **dichotomously** branched roots extend down into the ground, and erect branches grow upward. The stems are covered with many small, pointed, dark green leaves. In the more primitive species the reproductive structures or **sporangia** are found in the **axils** of ordinary leaves. In other species the sporangia are borne in the axils of modified leaves which are aggregated at the tip of an erect branch, forming a slender cone or strobilus. The many **spores** borne within the sporangia are all alike and for this reason *Lycopodium* species are said to be homosporous. The spores are disseminated by the wind, and in time develop into gametophytes. The **gametophytes** of *Lycopodium* species are extremely small tuberous bodies which grow slowly and reach maturity only if they are invaded by an endophytic **fungus.** Generally the gametophyte or prothallus develops underground. In the upper surface of the prothallus both **antheridia** and **archegonia** are found. Each antheridium contains many straight, biciliate **sperms.** These swim to the egg, with which one unites, forming a **zygote.** From this the new **sporophyte** develops. At first the sporophyte depends on the gametophyte for its food substances. Thus it obtains nutriment by means of a special absorbing structure called a suspensor which grow into the tissue of the gametophyte.

Lycopodium plants are widely used as material from which to make Christmas wreaths. For this purpose the entire plant is often ripped from the ground. The spores of *Lycopodium* are also gathered and sold under the name of Lycopodium powder. Formerly these spores were used in making explosive mixtures and for flash lights.

The genus *Selaginella*, containing some 400 species, is most abundant in the tropics. A few species of small plants are found in temperate regions. In the tropics there are both terrestrial and epiphytic species. The general habit of the plant is much like that of *Lycopodium*. The sporangia are formed in the axils of leaves at the tips of the branches, forming terminal strobili or cones. At the base of the **sporophylls** there is also a small scale, called a ligule, of unknown function. The sporangia are of two kinds, one, a **megasporangium**, containing four large spores, called megaspores; the other a **microsporangium** containing many small spores or microspores. Species of *Selaginella* are therefore heterosporous, a character which distinguishes them from *Lycopodium* species. The microspores, while still within the sporangial wall, start to develop and become multicellular bodies or microgametophytes. In each microgametophyte are formed many biciliate sperms. The megaspore also develops without leaving the sporangium. Its contents divide and form a mass of cells so great that the wall of the megaspore is ruptured. Archegonia are formed in this mass of cells protruding from the cracks in the megaspore wall. The eggs within the archegonia are fertilized by the sperms. At once development begins and a new sporophyte is formed. First formed is the suspensor, a small absorbing organ which pushes into the gametophyte tissue. Stem and roots form later. Often these structures are well developed while still within the sporangium wall. From an evolutionary standpoint this life cycle is very significant. The microspores correspond to the pollen grains in the Spermatophytes. The embryo, surrounded by gametophyte tissue, gives a clue to the possible origin of the seed.

Several species of *Selaginella* are grown in cultivation because of their delicate appearance. Usually they require so much moisture around them that it is difficult to keep them alive in ordinary rooms. One xerophytic species, *Selaginella lepidophylla*, is sometimes seen in cultivation. The plant, a native of the drier regions of America, when dry forms a tight little ball. On receiving sufficient moisture this ball unfolds, forming a flat plant whose much-branched stems spread out on the surface of the ground. Large numbers of these plants are gathered and sold under the name of resurrection plants. (See also **Paleobotany.**)   (R.M.W.)

LYCOPSIDA. Lycopsida are **vascular** plants with the following characteristics: The **stele** is a solid cylinder; the leaves are small and spirally arranged on the stem; no breaks, or leaf gaps, occur in the central cylinder; and the **sporangia** are borne on the upper surface of the leaves. Lycopsida are considered the more primitive types of vascular plants. The members of the order Lycopodiales are Lycopsida.   (R.M.W.)

LYDITE. Basanite.

LYMPH. A clear fluid which circulates in the tissue spaces of vertebrates and passes into the venous system by way of a tubular **lymphatic system.** It is derived from the liquid plasma of the **blood** but is more watery and contains no red corpuscles. It serves as an intermediary between the blood itself and the **tissues of** the body.

Lymph is found lying free in the serous sac cavities of the body, i.e., the **peritoneum, pleura,** and the spaces in the **brain** filled with cerebro**spinal fluid,** which may also be classified as lymph, although it differs in composition from the fluid found in the lymph vessels.

Lymph is derived from the plasma of the blood either by filtration, diffusion, or **osmosis** through the capillary walls or by active secretion of endothelial cells making up capillary walls. Its composition is very similar to the blood plasma.

The function of lymph is to bring nourishment to the tissue cells and to return waste matter and other toxic material to the blood stream by way of the lymphatic vessels, or directly into the blood stream through the capillary walls. Lymph has been compared with the body fluids of invertebrates, commonly designated as **hydrolymph** and **haemolymph.**   (A.W.L., R.S.M.)

LYMPH GLAND, LYMPH NODES. Lymph.

LYMPHATIC SYSTEM. A tubular system supplementing the **blood** vascular system of vertebrates. It collects fluid, chiefly lymph, from the tissue spaces and returns it to the venous circulation.

The smaller tubules of the system resemble **capillaries** and the larger ducts, called lymphatics, are similar to **veins,** although of more delicate structure in relation to their size. Like veins, they have valves which aid in promoting flow through the movements of the surrounding muscles. They are irregular in diameter, forming reservoirs at some points and dilating in the amphibians, reptiles, and birds to form lymph hearts whose pulsations propel the lymph toward the heart. The smaller vessels converge like blood vessels to form larger trunks. In man the chief vessels are the thoracic duct, and the right lymphatic duct, which empty into the large veins at the sides of the neck. Along the course of the lymph vessels are groups of lymph nodes. They serve as filters which localize and retard the spread of toxic and infective elements that are being returned to the blood stream. The lymph nodes also serve as centers for formation of **lymphocytes** which form one of the main divisions of blood cells.

Besides returning to the general circulation waste products from the cells the digestive lymphatics absorb nourishment from the products of **digestion** in the intestine, returning these food products to the blood stream.   (A.W.L., R.S.M.)

LYMPHOCYTE. A variety of white **blood** cell or leucocyte which is manufactured by **lymph nodes.** Lymphocytes are classified as large and small and together they make up about 25–30% of the white blood cells under normal conditions. This relationship is disturbed **in** disease.   (R.S.M.)

LYMPHOGRANULOMA INGUINALE (Lymphopathia venereum). A specific venereal disease characterized by a transient primary sore followed by **adenitis** of the lymph nodes in the groin, inguinal **buboes, suppuration,** and **fistulae.** The disease is transmitted by sexual intercourse. The etiological agent is a filterable **virus.**   (D.M.H.)

LYMPHOSARCOMA. A malignant **tumor** arising in lymph glands, or in an organ which contains lymph tissue. The gastro-intestinal tract, and the spleen, bone marrow, the entire lymphatic system may be involved. The disease is always fatal, but may run a relatively long course. X-ray treatment gives temporary relief.   (D.M.H.)

LYNX. Mammalia, Carnivora. Moderately large **cats** of the northern hemisphere, characterized by conspicuous ear tufts and a fringe of long fur about the throat. Among the several North American species most are called wildcats or bobcats, and the Canada lynx, *Lynx canadensis,* alone is known as the lynx. Several species of lynx also occur in Europe and Asia.   (A.W.L.)

LYRA. (The harp.) (Map, page 380.) This **constel-lation**, while small in size, contains a number of most interesting objects for an observer with a telescope, whether the instrument be large or small. The constellation is most easily distinguished by the equilateral triangle with the star **Vega** at one of its apexes.

Vega is an interesting star for a variety of reasons. In the first place, it is the brightest star in the northern celestial hemisphere. It lies almost in the direction in which the **sun** and all of the **planets** are moving due to solar motion. At present, this star is a long ways from the pole of rotation of the **celestial sphere,** but, due to **precession** this star will be the pole star about 12,000 years hence.

The star Epsilon Lyrae is one of the most famous multiple stars in the entire sky. A field glass will show this star to be double, and a four-inch telescope on a good night will show that each component is also double.

Several other interesting doubles are to be found in the constellation. Also in this constellation is the famous ring **nebula,** which, while not impressive in a small telescope, is a very interesting object in an instrument larger than a 6″.

In zoology, the term lyra designates a sound-producing or stridulating apparatus of certain **spiders.** It consists of a group of spines projecting over a concavity in the surface to which they are attached. Various other forms of stridulating organs are found associated with the pedipalps and chelicerae of different species of spiders; all consist of structures formed for the production of vibrations. (W.K.G., A.W.L.)

LYRE BIRD. Aves, Passeriformes. *Menura.* A large bird (**Aves**) with an ornamental tail in the male sex which simulates the form of a lyre. The several species are Australian. (A.W.L.)

LYRIDS. The Lyrids are **meteor showers** which are observed about April 20 of each year. The orbit of the radiant point was definitely associated with the orbit of comet 1861 I. by Weiss. Records of showers from this radiant are found back as far as 687 B.C. The report written by the Chinese in 15 B.C. indicates that during the Lyrids shower of that year "after the middle of the night, stars fell like rain." Several other accounts of striking showers during April are on record, in particular we find many newspaper accounts of the Lyrid shower of 1803, which was observed over the United States from North Carolina to New Hampshire. The Richmond, Va., *Gazette* of April 23, 1803, gives a long and vivid account of the shower occurring on the morning of April 20, stating that "from one until three those starry meteors seemed to fall from every point of the heavens, in such numbers as to resemble a shower of sky rockets."

A few scattered members of this shower are observed coming from the radiant point in the **constellation** of **Lyra** every year, but there has not been any very striking display since 1803. Since there have been striking showers in the past, the assumption is made that there is a large swarm of meteors at some undetermined point along the orbit, and that we may be treated to another brilliant display during some April in the future. (W.K.G.)

LYRIFORM ORGAN. A sensory organ, probably an organ of smell, found in **spiders.** These organs occur on most segments of the legs and on the under surface of the body. They consist of a series of slits and are associated with special nerve endings. (A.W.L.)

LYSINE. Aminoacids, Polypeptides, and Proteins.

LYSIS. 1. The gradual decline of the symptoms of disease, referring especially to the gradual abatement of fever. 2. The dissolution of cells, e.g., bacteria, or red blood cells, changing them from solids to liquids. (D.M.H.)

M

M SERIES. X-Ray Spectra.

M TYPE MARKER. Radio Range.

MAAR. A volcanic explosion crater without a prominent volcanic cone. (R.M.F.)

MACADAM ROAD. Highways.

MACAQUE. Mammalia, Primates. **Monkeys** of several Asiatic and one African species. They are related to the mangabeys but are stouter and have a slightly longer muzzle. The tail varies from long to rudimentary. These monkeys make up the genus *Macacus*. Among the included species are the bonnet monkey, lion-tailed monkey, pig-tailed monkey, and magot. (A.W.L.)

MACARONI. Wheat.

MACASSAR OIL. Fixed oils.

MACAW. Parrot.

MACE. Nutmeg. *Myristica fragrans.* Myristicaceae.

MACHINE FORGING. Forging.

MACHINE SCREW. Screw Fastenings.

MACHINE TOOL. As based upon common usage, a machine tool is any power-driven non-portable machine which is designed primarily for shaping and sizing metal parts, either by chip removal or abrasion, from raw materials such as castings, forgings, bar and tube stock, plates and pressed parts. Machines for producing these raw materials, such as forging machines, rolling mills, or die-casting machines, are not classified as machine tools. Metal-cutting machines such as punches and shears, and metal-bending machinery, are also excluded from the machine tool classification. Several other definitions of the term machine tool are in use by the Department of Commerce and the Treasury Department of the United States Government, which are somewhat broader than the preceding explanation. The most important machine tools are those used for boring, broaching, drilling, gear tooth formation, grinding, hobbing, milling, planing, turning and tapping, and will be described under these heads. (See **Lathes, Turret Lathes,** and **Screw Machines.**) (H.C.H.)

MACHINES. The term machine applies traditionally in physics to any one of those simple devices—lever, inclined plane, etc.—which might with greater propriety be called "elements of mechanism." These elementary machines fall into two general classes, those dependent upon the vector resolution of forces (inclined plane, wedge, screw, toggle joint), and those in which there is an equilibrium of torques (lever, pulley, wheel-and-axle). Their detailed explanation is to be found in any textbook of high school physics. For each of the several types, there is a factor known as the "mechanical advantage," which theoretically represents the ratio of the force exerted by the device to the force acting upon it; though on account of friction and elasticity, the actual ratio of the force may differ somewhat from this theoretical value. For a lever, it is equal to the ratio of the "arms"; for an inclined plane, with the force acting parallel to the plane, it is the cosecant of the angle of inclination; etc. Ordinarily a more practicable measure of the mechanical advantage is the ratio of a small displacement produced by the operator of the machine to the resulting displacement of the load by the machine. Thus, if the handle of an automobile jack is moved 1″, in lifting the car 0.002″, the mechanical advantage is 500.

The principles of **efficiency** (ratio of output energy to input energy) are well illustrated by the action of machines. No mechanism can operate without loss of energy through friction or otherwise; hence the efficiency of a machine is always less than unity. Otherwise, it would be possible to realize **perpetual motion.** (L.D.W.)

MACKEREL. Pisces, Teleostei. Marine fishes (**Pisces**) with compressed spindle-shaped bodies and a widely flaring forked tail. The large family Scombridae which they make up includes the important food fishes of moderate size called **mackerels,** *Scomber scombrus,* and a number of moderate to large game and food fishes. The latter are the **king fish,** sierra or pintado, *Scomberomorus regalis,* **bonitos, albacore,** and tunny, more commonly called the **tuna,** *Thunnus thynnus.* One of the smaller genera, *Pneumatophorus,* is called the thimble eye or chub mackerel. (A.W.L.)

MACKEREL SKY. Both altocumulus and cirrocumulus clouds often appear arranged in uniform bands similar in appearance to a mackerel's back. The reference pertains particularly to altocumulus. (P.E.K.)

MACLAURIN'S FORMULA. Expansion of Functions in Series.

MACRODASYOIDEA. Gastrotricha.

MACRONUCLEUS. A large nucleus in the bodies of many 1-celled animals, accompanied by one or more small nuclei known as micronuclei. The macronucleus is apparently the controlling center for the ordinary activities of the animal. It develops under certain conditions from a subdivision of the micronucleus. (See **Cell.**) (A.W.L.)

MACROSCOPIC. Derived from the Greek, meaning broad and visual. Used by **petrographers** to describe the features of a rock which are large enough to be seen by the naked eye, hence the antithesis of microscopic. A better synonym is megascopic, from the Greek, meaning large and visual. (R.M.F.)

MACROSTRUCTURE. The internal structure of metals as observed in a ground or polished section with the naked eye or at low magnification. The surface is usually deeply etched to develop flow lines, **dendritic** structure, weld structure, segregation, porosity, **laminations,** laps, seams, and other structural discontinuities.

The macrostructure of castings is characterized by dendrites, while forgings or rolled products generally have flow lines, which are indicative of the plastic deformation which the part underwent during its production. In steel the flow lines result from alignment of non-metallic inclusions as minute stringers in the direction of elongation of the part. The flow lines developed in forgings are useful in checking die design and forging practice. The quality of tool steel and other high-grade bar stock is often checked by etching cross-sections and examining the macrostructure. (See **Microstructure**.) (R.H.H.)

MACULOSE. A term proposed by Holmes in 1919 to denote a group of spotted **contact-metamorphosed** rocks, including mottled **slates, schists,** and **hornfels,** as distinguished from those exhibiting well-defined **foliated** structure. (R.M.F.)

MAD TOM. Pisces, Teleostei. **Catfishes** of small size belonging to several species of the genus *Schilbeodes.* They occur throughout the Mississippi River system, the Great Lakes region, and the eastern states generally. (A.W.L.)

MADDER FAMILY. Rubiaceae. A family comprising some 4500 species, particularly abundant in tropical regions. Some species are found in temperate regions, and a few in Arctic climates. The family includes trees, shrubs, and herbs having opposite entire or sometimes toothed leaves with **stipules,** the latter often large and conspicuous. The flowers are perfect, regular and **epigynous,** and 4- or 5-parted. Few members of the family are important. Species of *Gardenia,* natives of tropical Old World regions, are frequently grown for their fragrant showy flowers. *Rubia tinctorium,* the madder plant, was formerly a very important source of the dye madder, also called alizarin. Now, however, the dye is prepared synthetically. Gambier, *Uncaria Gambii,* a climbing plant native in tropical Asia and the Oceanic Islands, yields quantities of pyrogallol (see **Phenol**) **tannin,** extracted with boiling water from the leaves and young shoots. This is used in tanning leathers, often mixed with other **tannins.** It is also used as an astringent in medicines.
Ipecac, *Cephaelis Ipecacuanha,* a native of South American tropics, is a shrubby plant, the roots of which are six millimeters thick. From the dried roots and the lower part of the stem the drug ipecac is obtained. Used in small doses, ipecac is a stimulant; in large doses it is an eliminant, causing vomiting, sweating, and elimination through the kidneys and bowels. It is a very efficient means for clearing an overloaded stomach. **Quinine** and **coffee** are two other important products from members of this family. (R.M.W.)

MADREPORITE. A perforated plate on the surface of the body of **echinoderms.** The perforations lead into the water vascular system. (A.W.L.)

MAFIC. Femic.

MAGDALENIAN. Paleontology of Man.

MAGENTA. Dye.

MAGGOT. The soft-bodied **larva** of many species of two-winged **flies.** They are often white but some species are brightly colored. No organs of locomotion are present. Most maggots hatch from the egg in the midst of an abundant food supply of decaying organic matter or living plant or animal tissues and are able to move about sufficiently for their needs by wriggling the body. (A.W.L.)

MAGIC EYE TUBE. Tuning Indicators.

MAGMA. The term for molten material. A natural, complex, liquid, high-temperature, **silicate** solution ancestral to all **igneous rocks,** both intrusive and effusive. The locus of a magma is within the **lithosphere** (crust) under great pressure and an impenetrable cover which helps the magma to retain its original gases and water vapor in solution. The origin of magma is not known but it is generally assumed that separate magma chambers may exist within the lithosphere. (R.M.F.)

MAGMATIC STOPING. A term proposed by R. A. Daly in 1906 for a process by which large intrusive rock bodies, such as **batholiths,** might be able to take the space which must have been previously occupied by other rocks. This process assumes that the pre-existing rocks are shattered, or stoped, at the roof of the **magma** chamber and sink to lower levels in the magma, where they are melted and assimilated. The outer margins of the magma are assumed to be kept molten by the aid of 2-phase convection currents as well as by the relatively rapid foundering of the wall rocks and their assimilation in depth. When the magma congeals at the top of the chamber so as to show the angular fragments which have been stoped from the walls of the chamber, these fragments are called **xenoliths.** (R.M.F.)

MAGMATIC WATER. Water and water vapor given off by volcanoes proves that water is an important constituent of all **magmas.** (R.M.F.)

MAGNAFLUX. Magnetic Particle Inspection Test.

MAGNESITE. The mineral magnesite is **carbonate** of **magnesium,** $MgCO_3$. It is a hexagonal mineral, but usually found massive. It has a **rhombohedral** cleavage; conchoidal fracture; brittle; hardness, 3.5–4.5; specific gravity, 3; luster, vitreous to silky; color, white, gray, yellow, or brown; transparent to opaque. Most magnesite is believed to have been derived from the action of carbonated waters upon rocks rich in magnesium. Magnesium-bearing waters, on the other hand, may have in some cases acted upon **calcite** or **dolomite.** Magnesite deposits are known in Greece, Austria, Norway, India, Australia, and South Africa. In the United States magnesite is found in California and Nevada, some of which deposits seem to be of original sedimentary character. Magnesite is in demand for the manufacture of refractories and various compounds of magnesium used in medicine or for other purposes. (E.S.C.S.)

MAGNESIUM. Symbol: Mg. Atomic number: 12. Atomic weight: 24.32. Density: 1.74. Hardness: 2. Melting point: 651° C. Boiling point: 1110° C. Isotopes: 24 (77.4%), 25 (11.5%), 26 (11.1%).
Magnesium is a silver-white metal, malleable and ductile when heated; unattacked by dry **oxygen,** by water or **alkalis** at room temperature; when heated to about 800° C. reacts in air or steam and emits a brilliant white light of high actinic power; reactive with acids including **carbonic** at room temperature; reactive upon heating with **nitrogen, phosphorus, arsenic, sulfur, chlorine,** in some cases with such vigor as to constitute a hazard. Recognized by Black in 1755, and isolated by Davy in 1808.
Magnesium is used in increasing amounts (1) as an alloy with **aluminum** in light, strong construction, e.g., **airplanes,** and also used (2) in photographic flashlight powders, incendiary bombs, signal flares, (3) as a deoxidizer in the casting of metals, (4) as a "getter" in radio tubes, (5) as a reagent in organic chemistry, e.g., **Grignard reaction.**
Magnesium occurs generally in rocks, especially **limestone** (average 8% MgO) and **igneous rocks** (average 3.5% MgO); as the important mineral **dolomite** (magnesium calcium carbonate mixtures), notably in the Alps, and as **magnesite** (magnesium carbonate, $MgCO_3$), obtained commercially in Austria, Greece, Russia, and the states of California and Washington; in ocean water as the metal second in abundance to **sodium** (about 1 part of magnesium to 10 parts of sodium), in salt deposits, lakes, and brines, notably Stassfurt, Germany, and the state of Michigan, mainly as chloride or sulfate; present in natural hard waters of the earth's surface; present in chlorophyll (see **Pyrrole and Related Compounds**). Magnesium metal is obtained by **electrolysis** of fused magnesium chloride.

Acetate: Magnesium acetate $(Mg(C_2H_3O_2)_2 \cdot 4H_2O)$, white solid, soluble, formed by reaction of magnesium carbonate and **acetic acid**, and then crystallizing.

Arsenates: Magnesium ammonium arsenate $(MgNH_4 AsO_4)$, white precipitate, by reaction of soluble magnesium salt solution and **sodium** arsenate in the presence of excess **ammonium** hydroxide, upon igniting yields magnesium pyroarsenate $(Mg_2As_2O_7)$, white solid.

Borate: Magnesium borate $(Mg_3(BO_3)_2$ or $Mg (BO_2)_2)$, white precipitate, by reaction of soluble magnesium salt solution and **sodium** borate, used as a preservative.

Boride: Magnesium boride (Mg_3B_2), brown solid, by reaction of **boron** oxide and magnesium powder ignited.

Bromide: Magnesium bromide $(MgBr_2 \cdot 6H_2O)$, white solid, soluble, formed by reaction of magnesium carbonate and **hydrobromic acid**, and then crystallizing. Is present in many salt brines, and in traces in ocean water, constituting the source of bromine.

Carbonates: Magnesium carbonate, magnesia alba $(MgCO_3)$, white solid, insoluble, formed by reaction of soluble magnesium salt solution and **sodium** carbonate or bicarbonate solution. Present in carbonate minerals and rocks, **magnesite** (more or less pure magnesium carbonate), **dolomite** (magnesium calcium carbonate mixtures), dolomitic **limestone**. When ignited yields magnesium oxide and **carbon dioxide**; when treated with acids yields the corresponding magnesium salt and carbon dioxide, but with **carbonic acid** yields soluble magnesium bicarbonate. Used as a heat refractory and insulating material, and as a source of magnesium compounds; magnesium bicarbonate $(Mg(HCO_3)_2)$, colorless solution, by reaction of magnesium carbonate and carbonic acid, yields, upon boiling, magnesium carbonate white solid and carbon dioxide; magnesium ammonium carbonate $(MgCO_3 \cdot (NH_4)_2CO_3 \cdot 4H_2O)$, white precipitate (soluble in ammonium chloride solution) by reaction of soluble magnesium salt solution and excess ammonium carbonate.

Chlorides: Magnesium chloride $(MgCl_2 \cdot 6H_2O)$, white solid, soluble, formed by reaction of magnesium carbonate (or hydroxide, or oxide, or metal) and **hydrochloric acid**, and then crystallizing, loses hydrogen chloride when heated, yielding magnesium oxychloride. Used as a dressing and filler for cotton and woolen fabrics, in paper manufacture, in cements, refrigerating brines, in ceramics, and as a source of magnesium for its insoluble magnesium compounds and magnesium metal; anhydrous magnesium chloride $(MgCl_2)$, white solid, soluble, formed (1) by heating magnesium chloride crystals in a current of dry **hydrogen chloride**, (2) by heating magnesium ammonium chloride. Melting point $712°$ C. Used as the **electrolyte** in manufacture of magnesium metal; magnesium ammonium chloride $(MgCl_2 \cdot NH_4Cl \cdot 6H_2O)$, white solid, soluble, when heated yields anhydrous magnesium chloride residue; magnesium potassium chloride, **carnallite** $(MgCl_2 \cdot KCl \cdot 6H_2O)$, white solid, soluble, when heated fuses to anhydrous mixture magnesium potassium chloride; magnesium oxychloride, white solid, insoluble, formed (1) by heating magnesium chloride crystals, (2) by mixing magnesium chloride solution and magnesium oxide. Used as a cement when mixed with fillers, such as wood flour, sand powder, marble powder, cork, talc, for flooring.

Chromate: Magnesium chromate $(MgCrO_4 \cdot 7H_2O)$, yellow solid, soluble, formed by reaction of magnesium carbonate and **chromic** acid solution and then evaporating.

Citrate: Magnesium citrate, citrate of magnesia $(Mg_3 (C_6H_5O_7)_2 \cdot 4H_2O)$, white solid, soluble, formed by reaction of magnesium carbonate and **citric acid**, and then evaporating. Used in medicine and in effervescent beverages.

Fluoride: Magnesium fluoride (MgF_2), white precipitate, by reaction of soluble magnesium salt solution and **sodium** fluoride solution. Used in ceramics.

Fluosilicate: Magnesium fluosilicate $(MgSiF_6)$, white solid, soluble, formed by reaction of magnesium carbonate and **hydrofluosilicic acid**, and then evaporating. Used in hardeners for concrete, and in ceramics.

Hydroxide: Magnesium hydroxide, "milk of magnesia" $(Mg(OH)_2)$, white precipitate, by reaction of soluble magnesium salt solution and **sodium** hydroxide solution. Used in medicine as a suspension in water ("milk of magnesia"), and in sugar refining.

Hypophosphite: Magnesium hypophosphite $(Mg(H_2 PO_2)_2 \cdot 6H_2O)$, white solid, soluble, formed by reaction of magnesium carbonate and **hypophosphorus** acid, and then evaporating. Used in medicine.

Iodide: Magnesium iodide $(MgI_2 \cdot 8H_2O)$, white solid, soluble, formed (1) by reaction of magnesium carbonate and **hydriodic acid**, and then evaporating, (2) by heating magnesium metal and iodine. Used in medicine.

Lactate: Magnesium lactate $(Mg(C_3H_5O_3)_2 \cdot 3H_2O)$, white solid, soluble, formed by reaction of magnesium carbonate and **lactic acid**, and then evaporating. Used in medicine.

Nitrate: Magnesium nitrate $(Mg(NO_3)_2 \cdot 6H_2O)$, white solid, soluble, formed by reaction of magnesium carbonate and **nitric acid**, and then evaporating. Used in pyrotechnics.

Nitride: Magnesium nitride (Mg_3N_2), yellow solid, with moist air or water yields ammonia and magnesium hydroxide, formed by heating magnesium to a high temperature in **nitrogen** or **ammonia** (**hydrogen** gas evolved), or air (when a quantity of magnesium powder is ignited and then left undisturbed until cool, magnesium nitride is formed in the inner part of the mass).

Oleate: Magnesium oleate $(Mg(C_{18}H_{33}O_2)_2)$, yellow solid, insoluble, formed by reaction of soluble magnesium salt solution and **sodium** oleate. Used in varnish driers.

Oxalate: Magnesium oxalate $(MgC_2O_4 \cdot 2H_2O)$, white solid, insoluble, formed by reaction of soluble magnesium salt solution and **ammonium** oxalate solution.

Oxide: Magnesium oxide, magnesia, "burnt magnesia" (MgO), white solid, reacts slowly with water to form magnesium hydroxide, absorbs carbon dioxide from the air to form magnesium carbonate, is readily soluble in acids, insoluble in alkalis; formed (1) by heating magnesium carbonate to high temperature (**carbon dioxide** gas evolved), (2) by heating magnesium hydroxide, nitrate, sulfate, or oxalate, (3) by burning magnesium metal in air or **oxygen** (emission of brilliant white light of high actinic power). Used as a heat refractory and insulating material, in compounding rubber, in cosmetics, as a filler for paper; magnesium peroxide (MgO_2), white solid, insoluble, formed by reaction of soluble magnesium salt solution and sodium peroxide. Used in bleaching woolen and silk fabrics, and as antiseptic.

Phosphate: Magnesium ammonium phosphate $(MgNH_4 PO_4)$, white precipitate, by reaction of soluble magnesium salt solution and **sodium** phosphate in the presence of excess ammonium hydroxide, upon igniting yields magnesium pyrophosphate $(Mg_2P_2O_7)$, white solid.

Salicylate: Magnesium salicylate $(Mg(C_7H_5O_3)_2 \cdot 4H_2O)$, white solid, soluble, formed by reaction of magnesium carbonate and **salicylic acid** in water. Used in medicine.

Sulfate: Magnesium sulfate, "Epsom Salt" $(MgSO_4 \cdot 7H_2O)$, white solid, soluble, formed by reaction of magnesium carbonate and **sulfuric acid**, and then evaporating, removable from natural brines or salt deposits. Used as a source of magnesium for its insoluble compounds, as a dressing and filler for cotton and silk goods, and in dyeing and printing and fireproofing cotton, as sizing for paper, in cosmetics and lotions, and in medicine.

Sulfide: Magnesium sulfide (MgS), formed (1) by heating magnesium sulfate and **carbon** to a red heat, (2) by heating magnesium and **sulfur** to ignition (Hazard!).

Tungstate: Magnesium tungstate ($MgWO_4$), white precipitate, by the reaction of soluble magnesium salt solution and ammonium **tungstate**. Used as a luminescent paint, and fluorescent x-ray screen. (R.K.S.)

MAGNESIUM ALLOYS. Magnesium was first made and some of its mechanical properties determined in 1830, but until about 1910 its only use was in the form of flashlight powder, ribbon and sheets and for the deoxidation of certain non-ferrous alloys, especially those containing nickel. The American development dates from 1915. Because of their lightness, about ¼ the weight of steel and ⅔ that of the aluminum alloys, and also because of their easy machinability the magnesium alloys are being used to an increasing extent.

Commercial magnesium and many of its alloys are resistant to ordinary types of atmospheric corrosion but may be seriously corroded in bad industrial atmospheres or along the sea coast, especially if in direct contact with salt water. Many methods of chemical treatment to increase the corrosion-resistance have been developed, of which a typical one is a mixture of sodium dichromate and nitric acid. For severe conditions, the chemically treated alloys are protected by paint, varnish or enamel.

For most working and operating conditions, the fire risk is unimportant as the heat will be dissipated because of the relatively high thermal conductivity before the burning point is reached. A fire hazard exists if a machining operation produces very fine chips or powder and is always present in grinding operations. This hazard is eliminated by grinding under oil or by adequate removal of the fine particles by suitable ventilation.

Designation *	Dowmetal C	Dowmetal H	Dowmetal O-1	Dowmetal FS-1
Condition	Heat treated and aged	Heat treated	Heat treated and aged	Hard rolled
Form	Castings	Castings	Extruded shapes	Sheet
Approximate chemical composition	Al 9.0 Zn 2.0 Mn 0.2	Al 6.0 Zn 3.0 Mn 0.2	Al 8.5 Zn 0.5 Mn 0.2	Al 3.0 Zn 1.0 Mn 0.2
Typical physical properties				
Tensile strength, lbs. per sq. in	40,000	40,000	53,000	43,000
Yield strength, lbs. per sq. in	23,000	14,000	35,000	33,000
Elongation, % in 2 in	3	12	9	11
Modulus of elasticity, lbs. per sq. in. $\times 10^6$	6.5	6.5	6.5	6.5
Brinell hardness number	84	55	85	73
Rockwell hardness number	E90	E66	E90	E83
Melting point, °C	599	613	610	637
Melting point, °F	1,110	1,135	1,130	1,160
Specific gravity	1.82	1.83	1.77	1.80
Electrical resistivity				
Micro-ohms per cc. (room temp.)	13.9	14.9	11.9	8.8
Thermal conductivity, cal./sec./sq. cm./°C./cm	0.131	0.123	0.149	0.194
Coefficient of linear expansion per degree cengrade $\times 10^{-6}$	14.0	14.0	14.0	14.0

* Courtesy of The Dow Chemical Company.

(R. S. WILLIAMS, M.I.T.)

With one or two exceptions, all magnesium alloys contain aluminum in amounts from 2.5–12%. All contain manganese in amounts from 0.1–0.3%. Many carry zinc in amounts from 0.5–3% and a very few have small percentages of cadmium or silicon. Intensive studies of the effects of other alloying elements are in progress. The alloys are available in the form of ingot, castings, forgings, sheet, extruded shapes, tubes and wire. Sand castings vary in tensile strength from 14,000–30,000 lbs. per sq. in. while die casting produces alloys with strengths up to 33,000 lbs. per sq. in. Castings may be heat-treated by annealing at 330–400° C. (626–752° F.) followed by air-cooling. The tensile strength increases to 29,000–34,000 lbs. per sq. in. with an increase in ductility and toughness. The heat-treated castings may be further strengthened by aging (heating at about 175° C. (340° F.) to 39,000 lbs. per sq. in., but the ductility and shock-resistance are decreased. Forging or extrusion will increase the tensile strength to 32,000–47,000 lbs. per sq. in. depending on the composition and treatment and with an increase in ductility. The elastic modulus of the alloys is rather low, 6,500,000 as compared to about 30,000,000 for steel, so that rigid structures must be built with somewhat heavier sections than would be indicated by calculations based on tensile strength alone.

MAGNET. In ancient times it was known that stones containing magnetite (see **Iron**) had qualities which were not the property of other stones. It was found that they would attract iron, and when freely supported would turn so that their axis would take a north-south direction. These lodestones were the earliest magnets, but now are natural curiosities only, since the magnets

Graphical magnetic field of a bar magnet.

used in compasses, instruments, magnetos, and all the various array of equipment which embodies a magnetic field produced by a magnet, are artificially constructed of hardened steel, magnetized by a strong magnetic flux. This type of magnet may be said to be a permanent magnet, in contrast to the **electromagnet.**

A magnet is a body possessing the property of attracting magnetic substances. The so-called permanent magnet should be used where a constant magnetic field is to be produced, since a well-made permanent magnet loses its magnetism very slowly, and then only up to a certain point, after which it is said to be aged. Thereafter it maintains a constant degree of magnetism unless subjected to strong de-magnetizing effects. Manufacturers of permanent magnets for precision instruments artificially age them during the process of manufacture. A bar which has been magnetized is found to have poles. These are centers where magnetic attraction is strongest. If the magnet is free to turn, the pole which points northerly is called a north pole, and the other a south pole. Like poles repel each other with a magnetic force, and unlike poles attract each other. The earth is a large magnet having magnetic poles somewhere near, but not coincident with, the geographic poles. Since unlike poles attract each other, and the north pole of a magnet is taken as that which points northerly, it is the earth's south magnetic pole which lies near the north geographic pole. A magnet, delicately made, and freely suspended, becomes a **compass**. (F.T.M.)

MAGNETIC ANALYSIS. Mass Spectograph.

MAGNETIC BEARING. Bearing.

MAGNETIC CIRCUIT.

A magnet, or a coil of wire carrying a current, is the seat of an influence which extends outward from it and is called a **magnetic field**. The flux from a bar magnet or from a straight electromagnet issues from one end of the magnet or coil, bends around, and re-enters at the other end. This can be exhibited by exploring the region with a compass needle. If there is provided an iron frame or ring extending from one pole of the magnet or coil around to the other, and in case of the coil, running clear through it, the magnetic flux is not only concentrated largely in the iron but is much greater in total amount than if the induction is entirely in the air. Even a short air gap in the iron reduces the flux considerably.

The analogy of such a magnetic path to an electric circuit is easily seen. The magnetic flux corresponds to a current. The magnet or coil corresponds to a battery, and provides magnetomotive force just as a battery supplies electromotive force. The amount of flux produced by a given magnetomotive force depends upon the dimensions and material of the "magnetic circuit," e.g., the length and cross-section of the iron ring followed by the flux and the permeability of the iron; just as the dimensions and material of the electric conductor determine its resistance. This attribute of the magnetic circuit (corresponding to resistance) is called its reluctance.

These ideas are expressed quantitatively for the purpose of practical calculations. The magnetomotive force $\mathfrak{M}$ is commonly given in "ampere-turns." Thus, a coil of 50 turns carrying a current of 4 amperes has a magnetomotive force of 200 ampere-turns. Another unit of magnetomotive force sometimes used is the "gilbert," equal to $5/2\pi$ or 0.794 ampere-turn. The flux ϕ is expressed in lines or "maxwells." Just as the resistance of an electric circuit is defined as the ratio of the electromotive force to the current, so the measure of the reluctance $\mathfrak{R}$ of a magnetic circuit is the ratio of the magnetomotive force to the flux. We then have the relation

$$\phi = \frac{\mathfrak{M}}{\mathfrak{R}},$$

a sort of magnetic Ohm's law, known as Bosanquet's law.

There are approximate formulae, used in electrical engineering, for calculating $\mathfrak{M}$ and $\mathfrak{R}$, and hence ϕ, from the specifications of a coil, its core, air gaps, etc. Such formulae are extensively used in designing transformers, generators, motors, etc. (See **Electromagnet.**) (L.D.W.)

MAGNETIC DAMPING. Lenz's Law.

MAGNETIC DEFLECTION. Cathode-Ray Tube.

MAGNETIC COMPASS. Compass.

MAGNETIC FIELD.

The region surrounding a **magnet** or an **electric current** is endowed with peculiar properties, the most familiar manifestation of which is the **torque** experienced by a small magnet, such as a compass needle, when placed in such a region. For any point of the field, there is only one direction in which the small magnet will come into stable equilibrium; and when this takes place, the direction in which the north pole of the magnet points is called the direction of the field. Upon exploring a magnetic field by moving the small magnet about in it, it is found that the field direction in general follows curved lines of force. If the field is due to the current in a conductor, these lines form completely closed curves enclosing the conductor; if due to a magnet, they apparently enter the iron at the south pole and emerge at the north pole, inferring that they complete themselves through the iron as they do through a coreless, current-carrying helix.

In any magnetic field there is at each point a certain magnetic intensity, a vector whose direction is that of the field. If one pole of a very long, slender bar magnet of unit pole strength is inserted into the field, that pole is acted upon by a force which, expressed in dynes, is a measure of the magnetic intensity. The unit magnetic intensity is the **oersted** (formerly the **gauss**). An instrument used for measuring the intensity is called a **magnetometer**. (See **Magnetic Flux**.)

Just as there is an **electric potential** at every point of an electric field, a **magnetic potential** exists at every point of a magnetic field. The difference in the magnetic potential at two points is measured by the work necessary to move a unit magnetic pole against the field from one point to the other. This difference is sometimes called a magnetomotive force, in analogy to **electromotive force**.

The lines of magnetic intensity around a current-carrying wire are circular, having the plane of the circle perpendicular to the axis of the wire. The direction of the whorls of force is determined by the "right-hand rule." When the wire is grasped by the right hand, the fingers encircling the wire, and the thumb pointing along the wire in the direction of the current, the fingers encircle the wire in the direction of the lines of force. This rule is used to determine in which direction the north pole would lie in a helix or solenoid of wire, for, instead of having a straight wire encircled by magnetic whorls, the wire itself is bent into a circular form by being wound in a solonoid or helix. The lines of force will then produce an axial magnetic field, so that one end of the solenoid is equivalent to a north pole, the other to a south pole. (L.D.W., F.T.M.)

MAGNETIC FLUX.

The magnetic flux through any closed figure, such as a circle, a rectangle, or a loop of wire, is the product of the area of the figure by the average component of magnetic induction (see **Magnetism**) normal to that area. Thus, if a rectangle 5 cm. $\times$ 8 cm. is placed in a region where there is a uniform magnetic induction of 2500 **gauss**, and at an angle of 30° with the lines of induction, the magnetic flux through it is 2500 gauss $\times$ 40 cm.^{2} $\times$ sin 30° = 50,000 gauss-cm.^{2} or "maxwells." The magnitude of this quantity is often conventionally represented by imagining the lines of induction to be so spaced that the number of them through a given area is equal to the number of gauss-cm.^{2} or maxwells of flux through that area. The flux in the above example would be commonly expressed as 50,000 "lines." When a coil has several (n) turns and each turn has approximately the same flux (ϕ) through it, the effect is the same as for a single loop with the flux $n\phi$ through it. This product, which is called the "linkage," is expressed in "maxwell-turns" or "line-turns."

The magnetic flux or the linkage through a loop or a coil may be measured by putting into the circuit a ballistic (undamped) galvanometer and then suddenly removing the flux (or the coil). If the resistance of the whole circuit and the constant of the galvanometer are known, the flux may be calculated from the "throw" of the galvanometer. (L.D.W.)

MAGNETIC INDUCTION. Magnetism; Gauss.

MAGNETIC INTENSITY. Magnetic Field; Oersted.

MAGNETIC MATERIALS. The electrical industry is dependent on four basic types of materials, good con-

(maximum ratio of magnetic induction to magnetizing force) is surpassed by other materials. When used as a core in a field induced by a.c. the so-called "iron losses" are relatively high. Iron losses consist of energy losses related to the area of the hysteresis loop, and of eddy-current losses caused by induced current circuits within the metal. High values of residual induction (*OR* on the typical hysteresis curve on page 759) and coercive force (*OC*) are indications of high hysteresis loss. A low value of electrical resistivity results in high eddy current losses. However, other factors such as the thickness of the sheets in laminated cores, and to a leser degree grain size also affect eddy current losses. Values for commercial quality magnetic iron are compared with other representative soft magnetic materials in the following tabulation. By special heat treatments and the use of specially purified iron much higher values of permeability and lower hysteresis losses can be obtained.

SOFT MAGNETIC ALLOYS	SATURATION INTRINSIC INDUCTION, B GAUSSES	MAXIMUM PERMEABILITY B/H	RESISTIVITY MICROHM CM.	HYSTERESIS LOSS * ERGS/CC/ CYCLE	RESIDUAL * INDUCTION, BR GAUSSES	COERCIVE FORCE HC * OERSTEDS	MAXIMUM EXTERNAL ENERGY
Ingot Iron.	21,600	5,000	11	2,600	8,000	0.9	....
4% Silicon Steel.	19,600	8,000	57	1,200	....	....	....
50% Ni Hipernik	15,200	55,000	48	200	7,000	0.07	....
78.5% Ni Permalloy	10,700	100,000	16	200	6,000	0.05	....
3.8% Mo-78.5% Ni Permalloy.	8,500	100,000	55	200	5,000	0.50	....

PERMANENT MAGNET ALLOYS							
Carbon Steel (0.9 C, 0.5-0.8 Mn).					8,600	48	170,000
Chromium Steel (0.9-1.0 C, 2.0 Cr).					10,000	50	220,000
Tungsten Steel (0.7-0.8 C, 6 W).					10,000	70	280,000
Cobalt-Chromium Steel (0.9 C, 16 Co, 9 Cr, 1.0-1.5 Mo).					8,000	190	500,000
Cobalt Magnet Steel (0.9 C, 36 Co, 5 Cr, 4 W).					9,600	240	860,000
Comal (12 Co, 17 Mo).					10,300	245	1,100,000
Cast Alnico I (12 Al, 20 Ni, 5 Co, Balance Fe).					7,300	440	1,400,000
Cast Alnico II (10 Al, 17 Ni, 12.5 Co, 6 Cu, Balance Fe).					7,350	560	1,600,000
Sintered Alnico II (10 Al, 17 Ni, 12.5 Co, 6 Cu, Balance Fe).					6,900	520	1,430,000
Cast Alnico IV (12 Al, 28 Ni, 5 Co, Balance Fe).					5,300	730	1,300,000
Cast Alnico V (8 Al, 14 Ni, 24 Co, 3 Cu, Balance Fe).					12,500	550	4,500,000

* In the case of magnetically soft materials, these values are for a magnetizing force, H, of 10 kilogausses.

ductors of electricity, high-resistivity conductors capable of withstanding high temperatures, insulators, and magnetic materials. In recent years the greatest developments have been in insulators and magnetic materials with resultant improvements in motors, generators, transformers, and instruments of all kinds.

The ferromagnetic elements are iron, nickel, and cobalt. Of these iron is the only one which has important magnetic applications in pure, or commercially pure form. All three elements, and many others which themselves are non-magnetic are used in special alloys in which certain magnetic characteristics are developed to a high degree. It is even possible, by alloying, to develop magnetic material from certain nonmagnetic elements.

There are two distinct groups of ferromagnetic materials, those which are easily demagnetized, and those which are not. These are often designated as soft and hard magnetic materials. In many respects iron is an excellent soft magnetic material. It attains the highest saturation value of magnetic induction found in any material with the exception of certain cobalt alloys. (See **Magnetism.**) However, its maximum permeability

With a good grade of silicon electrical steel the hysteresis loss can be cut in half and the eddy-current loss reduced even more because of the high resistivity compared with iron. This results in much more efficient operation of **alternating-current** equipment. A special type of silicon steel known as Hipersil having marked directional properties has been developed recently. A reduction of 25% in the weight of a transformer may be made without reduction in performance provided the design takes advantage of these directional properties.

Certain high-nickel alloys of the Permalloy type have very high permeabilities at low and moderate inductions and give low hysteresis losses. For many special applications in instruments and communications equipment these higher cost alloys are economically justified. Power transformers, motors and generators, etc., are designed for silicon electrical steels.

In the use of any of these materials for a-c applications, it is desirable to reduce the thickness of individual laminations to reduce eddy-current losses. For special applications at higher than usual power frequencies, extremely thin strip is used, approaching 1/1000″ thick. The most common thickness at 60-cycle power frequencies is about 14/1000″. In the communications field where the use of very high frequency current greatly aggravates the problem of eddy-current losses, alloys of the Permalloy type are used. They are produced in powdered form, given a very thin insulating film, and compacted under high pressure into a suitable core shape. While this is a product of **powder metallurgy,** it is unusual in that no sintering is required. (A final heating is applied to relieve the stresses caused by pressing and thus greatly improve the magnetic properties.)

In the production of magnetically soft materials it is important that elements such as carbon and sulfur which disturb the continuity of the base-metal crystal structure be reduced to low residual amounts, and that the material be used in a strain-free condition. High annealing temperatures resulting in coarse grain structures are used. These conditions are, for the most part, reversed when magnetically hard materials are sought. These materials must be capable of reaching high inductions, but upon removal of the direct magnetizing force the magnetic flux remaining in the circuit should be high, therefore a hysteresis loop of large area is desired. A good over-all measure of the quality of a permanent magnet is the "maximum external energy." This is the maximum value which can be obtained for the product of the co-ordinates of a point on the curve between B_r and H_o (line *RC* on page 759).

In order to obtain a high external energy value, high carbon and alloy contents are used and the steels are quenched to a high hardness. Processing and heat-treating temperatures are kept as low as possible and no attempt is made to relieve the severe internal stresses set up by formation of **martensite.**

Magnet of fair quality can be made of ordinary high-carbon steels, however, the advantages of the chromium and tungsten types of alloy steels are apparent. There are many special grades containing cobalt, three of which are included in the table. Still greater improvement in coercive force, which is a measure of resistance to de-magnetization, is attained in the Alnico compositions which are **precipitation hardening** type alloys. These compositions are hard and brittle as cast and must be finished by grinding. Small-size magnets can be produced by **powder metallurgy** to high dimensional accuracy. The magnetic properties are equivalent to those of the cast alloys and the mechanical strength is greater, thus the principal limitations are size and cost. (R.H.H.)

MAGNETIC MOMENT. Magnetism.

MAGNETIC PARTICLE INSPECTION TEST (MAGNAFLUX). A rapid non-destructive test for fatigue cracks, grinding checks, weld defects, stress-cor-rosion cracks, seams, and other surface and subsurface defects in steel and other magnetic alloys.

The piece to be inspected is magnetized by means of a solenoid or by-passing a heavy current through the piece itself. At cracks or other discontinuities, local flux-leakage fields are formed. A magnetic powder applied to the magnetized piece will be attracted strongly to the areas of flux leakage, forming a pattern which re-veals the location and extent of the defects. The powder may be applied dry by dusting on the work from a porous bag or shaker (dry method), or it may be sus-pended in a light oil which may be flushed over the work or used as an immersion bath (wet method).

Magnaflux inspection can be used to advantage on steel bars, shapes, plates, or castings before fabrication, on finished parts, and on parts which have been in service. Steels of "aircraft quality" must meet Magnaflux in-spection specifications.

Non-magnetic metals such as austenitic stainless steels and the non-ferrous alloys can be examined for defects by a related process known as "Zyglo." The part is dipped into a penetrating fluorescent liquid, rinsed, dried, and powdered. Under the capillary action of the pow-der the liquid marks the surface at cracks, seams, or other defects. The piece is then viewed under "black" (ultra-violet) light which excites fluorescence. (R.H.H.)

MAGNETIC PENDULUM. A bar magnet poised on a pivot or suspended by a thread in a magnetic field having a horizontal component will, if disturbed and released, oscillate in a horizontal plane as a magnetic pendulum. This arrangement is commonly employed in the **magnetometer** method of measuring the earth's mag-netic field intensity, or in comparing field intensities in different localities. The period of oscillation of such a pendulum. This arrangement is commonly employed in magnet, the horizontal component H of the magnetic field intensity, the **moment of inertia** I of magnet and supports with respect to the oscillation axis, and (if a suspending fiber is used) the torsion coefficient K of the fiber. For small amplitudes the period is

$$T = 2\pi \sqrt{\frac{I}{MH + K}}.$$

With a pivoted magnet, $K = 0$; while if the bar is unmagnetized, $M = 0$, and the formula degenerates into that for a **torsion pendulum.** The moment of inertia I may be determined experimentally in the same manner as with a torsion pendulum, but using $MH + K$ in place of K. In order to find K alone, the magnet is replaced by a non-magnetic bar having the same moment of inertia, and the torsion pendulum method applied. (L.D.W.)

MAGNETIC POLE. Magnet; Magnetism.

MAGNETIC POTENTIAL. Potential; Magnetic Field.

MAGNETIC RECORDING. Sound Recording.

MAGNETIC RESONANCE ACCELERATOR. Cy-clotron.

MAGNETIC STORMS. Terrestrial Magnetism.

MAGNETIC VECTOR. Radiation Field.

MAGNETISM. Centuries ago, minerals were discov-ered which exhibited strong attraction for iron and served as compass needles. These "lodestones" are in fact iron ores, usually Fe_3O_4. It was found that a steel bar stroked with such a natural magnet becomes itself a magnet with poles dependent upon the direction of stroking. With care the poles may be strongly localized in limited regions, almost points, and the forces between poles are then found to obey **Coulomb's law** of inverse squares. Two unit poles at a distance of 1 cm. in a vacuum attract or repel each other with a force of 1 dyne. When placed in a **magnetic field,** a bipolar magnet experiences a torque proportional to the field intensity and to the sine of the angle between the axis of the magnet (line joining poles) and the field direc-tion. The maximum torque per unit field intensity (which obtains when the magnet and the field are at right angles) is the magnetic moment of the magnet.

When a substance having magnetic properties is placed in a magnetic field, the field intensity gives rise to a magnetizing force H within the body, and the substance acquires a certain magnetization I, depending upon H and upon the susceptibility I/H of the substance. The resulting magnetic induction, $B = H + 4\pi I$, is a quantity much used in magnetic theory. If the field intensity is gradually built up from zero, the induction may be represented by a curve with H and B as coordi-nates. As magnetization proceeds, the curve approaches a straight line inclined at 45° (if B and H are on same scale); which means that I approaches a steady value, a condition called saturation. The ratio B/H is at any stage the magnetic permeability of the substance. If now the field is diminished, it is found that a certain reverse intensity OC, called the coercive force, is usually required to reduce the induction to zero. This is due to **hysteresis.** Some alloys of iron show extraordinary peculiarities; for example, permalloy (22% iron, 78% nickel) has enormous initial permeability, while perminvar (iron, nickel, and cobalt) has almost constant permeability for different fields, so that the graph OS is practically straight. (See **Magnetic Materials,** and typical hysteresis curve on page 759.)

Substances may exhibit magnetic properties not only in different degrees, but in radically different ways. There are three fairly distinct classifications: paramagnetic, ferromagnetic, and diamagnetic substances. The many paramagnetic substances, and the small group of ferromagnetic substances (iron, nickel, cobalt, gadolinium, and some alloys), both have positive susceptibility (I increases with H), but their magnetization curves are different and the phenomena are attributed to quite different causes. Any ferromagnetic substance becomes paramagnetic at a sufficiently high temperature (see **Curie-Weiss Law**). With diamagnetic substances, such as copper, silver, and bismuth, the susceptibility is negative and the permeability slightly less than 1; so that their magnetization curves lie in the fourth quadrant.

The physical nature of magnetism has been the subject of much speculation and research. It appears now that paramagnetism is probably due to the orbital motion of electrons in the atom, there being an excess of electrons revolving one way over those revolving the other; that in the atoms of ferromagnetic substances the number of electrons spinning one way on their axis is unequal to the number spinning the other way, while relatively large groups called "domains," made up of atoms having parallel magnetic moments, act together when the substance is magnetized and somehow add enormously to the induction; and that diamagnetism arises from the **precession,** in the applied field, of atoms which are not even paramagnetic, but which thus give rise to a feeble magnetization opposed to the field. The magnetic moments of paramagnetic atoms appear to be made up of very small units called magnetons, and the processes causing changes in magnetization take place in accordance with **quantum** conditions. (See also **Barkhausen Effect, Barnett Effect, Magnetic Circuit, Magnetic Flux, Magnetostriction, Electromagnetism.**) (L.D.W.)

MAGNETITE or MAGNETIC IRON-ORE (LODESTONE). The mineral magnetite, ferroferric oxide, Fe_3O_4, is **isometric,** commonly occurring in **octahedrons, dodecahedrons,** and massive, granular, laminated, etc. It is brittle with an uneven fracture; cleavage is not distinct, but with pressure an octahedral **parting** may develop; hardness, 5.5–6.5; specific gravity, 5.18; luster, metallic to dull; color, **iron black; streak,** black. It is opaque and strongly magnetic; when possessing polarity is known as lodestone. Magnetite is a common mineral in the **igneous rocks,** especially those of the ferromagnesian varieties, and is found in many **metamorphic** types. It is associated with **corundum** in emery. In northern Sweden are what may be the largest magnetite deposits in the world, believed to have been formed by segregation in the **magma.** Magnetite is also found in Norway, in the Urals, Italy, Switzerland, Australia and Brazil. In the United States the Pre-Cambrian rocks of the Adirondacks contain large beds of magnetite. This mineral is found also in New Jersey, Arkansas, Utah, and in Canada in Quebec and Ontario. The lodestone or natural magnet is found in Siberia, the Harz Mountains, the Island of Elba, and at Magnet Cove, Arkansas. The name magnetite is said to be derived from the district of Magnesia, near Macedonia. There is, however, a fable that it was named for a shepherd, Magnes, whose

iron-bound staff and shoes with iron nails stuck to the ground in which magnetite was present.

This mineral is an important ore of iron, containing 72% of metallic iron. (E.S.C.S.)

MAGNETO. The magneto is a device for producing alternating currents of high voltage properly synchronized for distribution to the spark plugs of an engine of the spark ignition (Otto cycle) type. The same function may generally be admirably performed by a coil-and-battery system, but the compactness and light weight of the magneto, and its reliability, have caused it to be widely used for *ignition*. The magneto is but seldom seen in that most familiar example of spark ignition, the automobile engine, because various electrical requirements of the automotive vehicle besides ignition have promoted the use of systems embodying a substantial storage battery which can also supply electric current to an ignition system. Such are termed coil-and-battery systems. (See **Ignition System.**)

The magneto has a single function—ignition. Except for spark plugs, wiring harness, and switch, it is a complete ignition system. Nowadays it is to be found on stationary engines, tractor engines, boat engines, and others where the ignition system is to be independent of other electrical services; and on the aircraft engine, where its proved reliability is highly respected.

In reality the magneto is an electric generator, induction coil, interrupter, and distributor, all consolidated in one small compact unit. Three different systems of generation of the low voltage electrical pulses are found. These, which are illustrated in Fig. 1, may be called (1) the armature type, (2) the inductor type, and (3) the

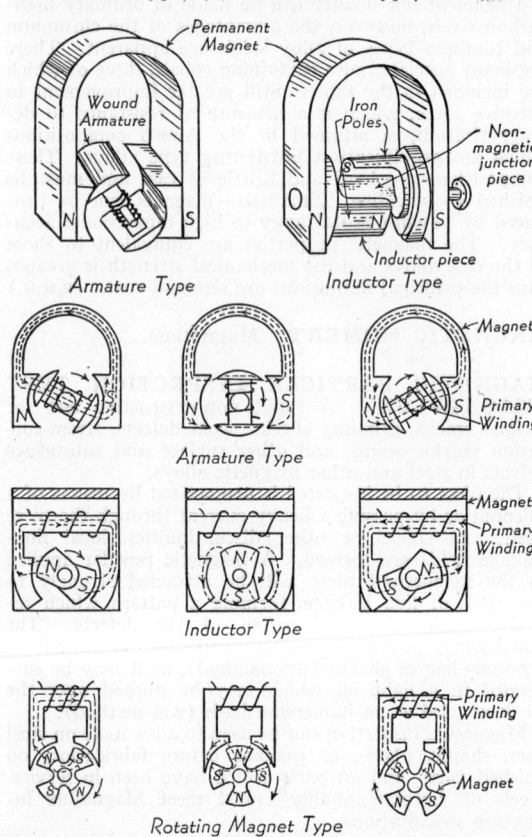

Fig. 1. Diagram of flux flow and magnetic field variation.

rotating magnet type. The armature and inductor types have stationary permanent magnets. In one a wire-wound armature is rotated between pole shoes, as in a dynamo machine, and in the other a soft iron "inductor," or flux changer, is rotated between the pole shoes.

The magnetizing influence which a magnet can exert on a magnetic circuit is called magnetomotive force. It is the "voltage" of a magnetic circuit. Flux in a magnetic circuit is equivalent to current in an electrical circuit. It may be represented by imaginary lines of force, as in Fig. 1. The stronger the magnetomotive force of a circuit, the greater will be the flux. The material of the magnetic circuit may be composed of anything through which the magnet can force flux, even air, but is mainly ferrous because of the excellent magnetic "conductivity" of this material, especially soft iron. Only the smallest of air gaps can be permitted because of the high "reluctance" of such to convey flux. If the amount of flux associated with a coil is varied by any process whatsoever, an electromotive force is induced in the coil. In all magnetos motion of the rotor induces a change of flux linkage through the primary coil winding. The method of varying flux linkage is shown in Fig. 1. Flux lines are shown dotted, with assumed direction of flux being from a north to a south pole. The diagrams are self-explanatory, although possibly some comment will be helpful in tracing the magnetic circuit of the inductor type. In the figure it is seen that the rotating soft iron inductor has N and S pole pieces which remain magnetically in contact with the poles of the permanent magnet, although continuously rotating. The N and S inductors are magnetically separated by the middle section of the inductor being non-magnetic (brass). Consequently the flux lines leaving the N pole of the magnet pass through the N end of the inductor, thence to the pole shoe and through the soft iron core of the primary winding to the other pole shoe, then across the air gap into the S end of the inductor, back to the S pole of the permanent magnet and through the magnet to the starting point.

The voltage induced in the primary winding is insufficient to jump a spark across the gaps of the **spark plugs,** but if a circuit breaker (called interrupter or breaker) is connected in the coil circuit to open the previously short-circuited coil at the instant the flux change reaches maximum (not coincident with maximum flux density), the resulting surges of primary coil current apply extra magnetomotive force to the magnetic circuit. Then, if a secondary coil of a great many turns is wound around the primary coil, extremely high voltages will be induced in it. It is these voltages which provide the "pressure" necessary for a jump spark. The secondary coil lead could be connected directly to the spark plug of a single cylinder engine. A 4-cylinder, 4-cycle engine needs 4 precisely timed sparks every 2 revolutions of the crankshaft. Consequently, the magneto rotor speed, the number of poles, and the interrupter design must all be correlated so as to produce the 4 high voltage surges. There must, in addition, be a rotary switch to receive these surges from the secondary winding and direct them in proper succession to the 4 spark plugs. This is called the distributor and is usually an integral part of the magneto. The distributor must be geared to rotate at one-half crankshaft speed. The distributor rotor (finger, brush) makes its connection with the magneto secondary through the drive shaft. The distributor case is molded of high-grade insulation with conducting inserts to which are connected the wires which carry the voltage to the plugs. In some distributors the rotor "brushes" the contact points as it passes them, whereas in others the current must jump a small gap between the moving finger and the stationary contact point.

The primary circuit must be broken as many times per single revolution of the rotor as there are permanent or induced poles on the rotor. The break is mechanically accomplished by a cam. Usually the cam is mounted directly on the rotor shaft. In a 4-cycle engine the number of firings required per revolution is one half the number of cylinders. The required magneto shaft-crankshaft speed ratio is the number of cylinders divided by twice the number of poles on the rotor. Fig. 2 shows typical breaker points. The movable point is mounted on a spring. Points are made of platinum

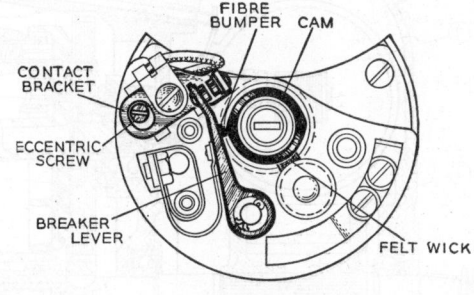

BREAKER
Fig. 2. (*American Bosch Corp.*)

iridium alloy, and have a condenser connected across them to eliminate arcing and burning from self-induction in the primary circuit, and also to assist in bringing primary current to zero rapidly, thereby increasing the voltage produced by the secondary winding.

Illustrating a standard type magneto for 6-cylinder engine ignition, Fig. 3 shows an external appearance and a section through a rotating magneto-type of magneto. As the auxiliary view consisting of a circuit diagram shows, the 2-pole inductor rotor rotates between pole shoes A and B in order to create reversal of magnetic flux through the coil. The inductor rotor shaft which is mounted on ball bearings (4) rotates at $1\frac{1}{2}$ times crankshaft speed, producing 2 sparks for every 360° of inductor rotor travel. The condenser (6), coil (2), and breaker assembly are stationary, although the latter may be shifted by arm (11) so as to advance or retard the timing of the spark relative to piston position. One end of the primary coil is connected to the breaker, and a lead from the secondary coil is carried to the distributor plate (16) for connection to a contacting brush which establishes contact with a circular, pick-up conducting ring in the distributor rotor. The distributor rotor shaft is mounted on ball bearings (14) and is driven at half crankshaft speed by the 1:3 ratio between the drive gear (8) and the distributor gear (13). A connection is made internally in the molded insulating material of the distributor rotor from the pick-up ring to a small inset segment in the distributor rotor. This segment is located at a greater radial distance from the distributor shaft than is the pick-up ring. Six brushes extending from sockets in the interior of the distributor plate bear against the face of the distributor rotor which, as it rotates, permits the small segment to make contact with one brush after another until all 6 brushes have received sparks. The brushes are connected to sockets on the exterior of the distributor plate. The high-tension cable to the spark plugs is inserted into these sockets.

Magnetos are employed regularly in modern aircraft engine ignition. That field of use poses operating problems not encountered elsewhere, some of which are worthy of description. Since inductor and rotating magnet types have no high voltage lead-off from the rotor, they are favored over the armature (shuttle) type. A typical magneto system is pictured in Fig. 4. Two magnetos are provided for the engine (9 cylinder, radial, indicated by the wiring harness) so that two independent sources of ignition will exist. Each cylinder has two spark plugs, each plug being connected separately to a different magneto. The magnetos are mounted on the engine crankcase and gear driven by it. The ignition switch, which is at the pilot's control, is turned to short circuit the breakers in order to stop

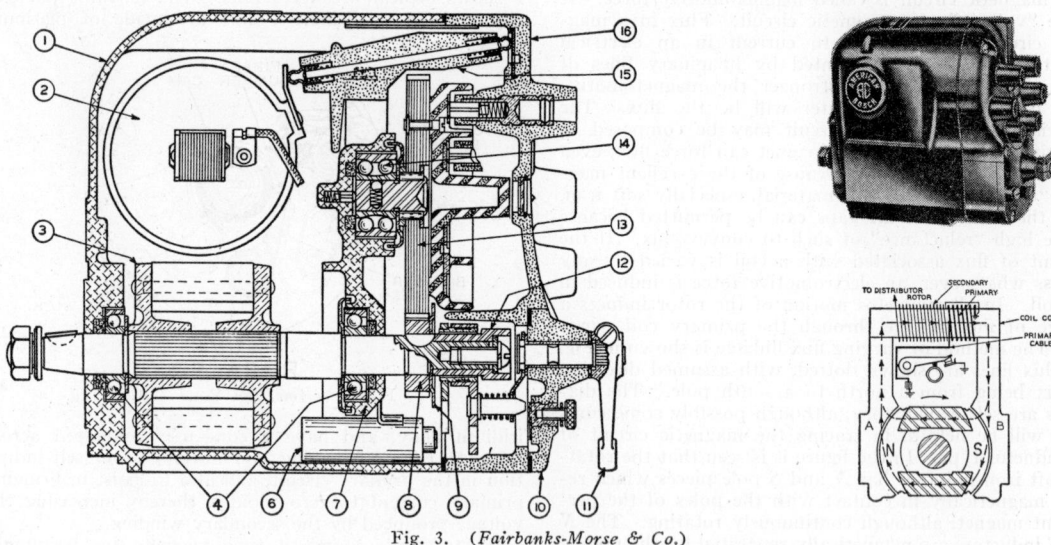

Fig. 3. (*Fairbanks-Morse & Co.*)

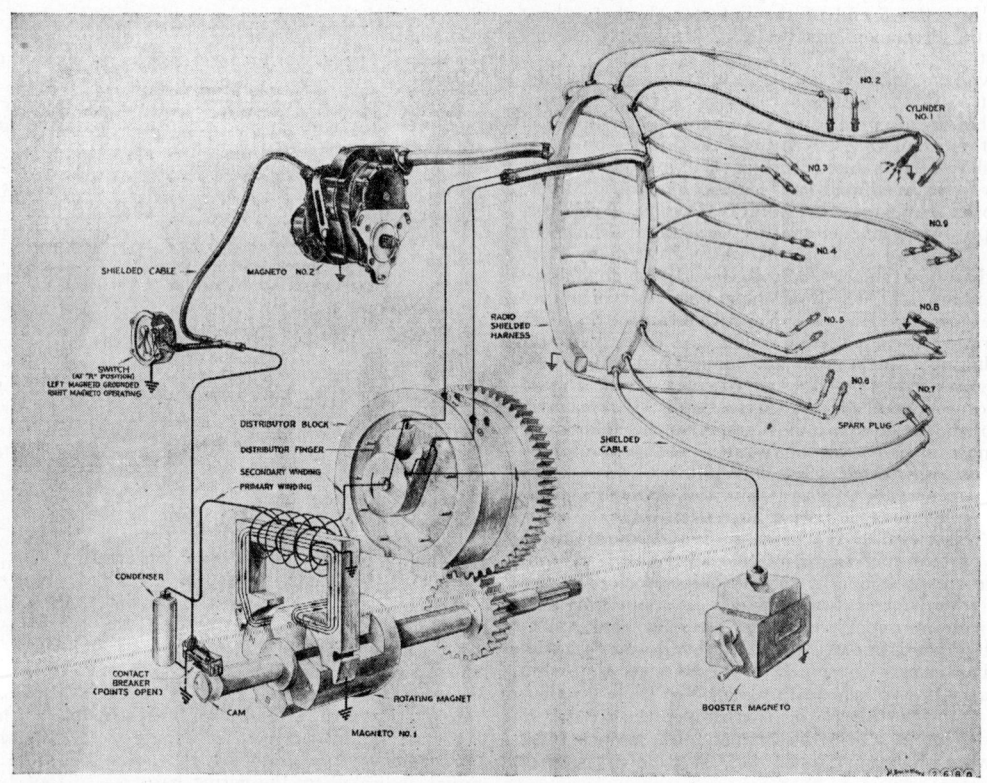

Fig. 4. Dual ignition system, nine-cylinder aviation engine. (*Scintilla Magneto Div., Bendix Aviation Corp.*)

the engine. In some cases, during cranking of the engine the magneto rotor speed is too slow to produce a satisfactory spark. A supplementary high voltage source known as a booster may be used during cranking, or an impulse coupling of the magneto rotor to the engine may be supplied. The booster magneto is usually hand-cranked and is connected to the spark plug through a distributor finger which trails the normal finger so as to retard the spark enough to prevent engine "kickback" during cranking. The impulse coupling is a spring-like mechanical linkage between the engine and magneto shafts which will wind up and let go at just the right moment to spin the magneto shaft fast enough to generate the necessary spark voltage. After the engine has started and reached a predetermined speed where normal magneto action is adequate, a centrifugal weight-actuated lock renders the impulse device inoperative. The ignition lag necessary to reduce kickback of an engine which normally operates on about 20° advance of spark before top dead center can be built into the distributor finger design used with a booster magneto (or vibrating coil and battery), or into the impulse coupling. The case on which the breaker is mounted, if permitted some small angular adjustment, can become a means of retarding the spark, for if the breaker shaft were rotating at crankshaft speed, a 5° rotation of the case in the direction of rotation of the cam would delay the spark 5° of crankshaft travel. Some magnetos have this spark-retarding feature, as do the breakers on most all coil and battery systems, but it is uncommon in aircraft ignition.

A single-unit dual magneto consists of a magneto unit containing 1 rotating magnet, 2 coils, and a double breaker assembly driven from an accessory drive shaft. There are 2 separate distributor assemblies, which are often driven by gearing from the rotor shaft.

As shown in Fig. 5, the condensers and coils are located in the breaker housing. There is some saving in weight

sheathing. This sometimes necessitates modifications of the magneto which would not otherwise be practical. High altitudes promote corona discharge from ignition systems over the high-tension (high-voltage) components. This not only leaks off energy needed in the spark plugs, but also, through the formation of ozone, accelerates the deterioration of wiring harness insulation. Flashovers may take place. Since these difficulties are aggravated as higher altitude flights become common, means have been sought for their elimination. One method is to supercharge the ignition system—that is, enclose the troublesome parts in an airtight casing and supply an air pump to maintain in it dry air at pressures that suppress **corona** difficulties. This is particularly suitable for modifying existing engines and their magnetos. Other methods involve relocation of magnetos so as to shorten the high-tension lines. Special high-altitude magneto systems have been devised, of which one will be described.

Corona can be suppressed if high-tension circuits are eliminated. But the high tension is necessary at the spark plugs. This difficulty is overcome in the high-altitude system by delaying the creation of high tension until the timed ignition pulse has reached the vicinity of the spark plug. There, the high-tension pulse needed by the spark plug is produced by a small transformer coil. The magneto is low tension, as are the distributor and wiring harness. There is a simplification and lightening of these parts, but a transformer coil is needed for each spark plug. This high-altitude, low-tension magneto ignition system is diagrammed in Fig. 6.

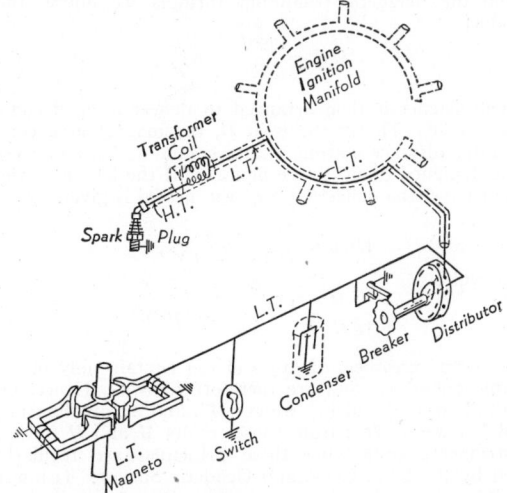

Fig. 6. Diagram of magneto ignition suitable for high altitudes (9 cyl. radial engine).

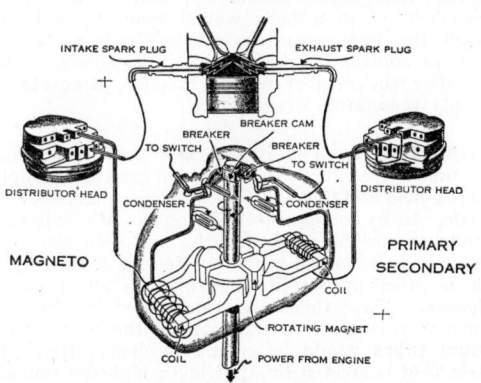

Fig. 5. Single-unit dual magneto.

over 2 complete magnetos, and the reliability of ignition appears to be nearly equivalent to the double magnetos.

The ignition system of aviation engines could be much like that for any other service, except for the following service conditions:

1. The growing importance of radio communication to navigation and piloting requires special attention to prevent broadcasting of interference static from the ignition system.

2. Increasing demands for high altitude equipment have accentuated the effect of corona loss, ionization flashover, etc.

3. The need in high-power engines for a system that will fire successfully even though the normal insulation throughout the ignition system is greatly impaired.

Radio communication is relieved of interference by shielding the ignition equipment, especially the wiring and spark plugs, by enclosing it in metallic tubing or

In modern aircraft engines of high-power rating there is a tendency for an electrical shunting of spark plug gaps by formation of lead compounds over the insulator surface, carbon tracking, and moisture on the terminals. From 8000 to 10,000 volts is normally required at the gap, but leakage paths may so accumulate as to cause irregular firing, especially where many hours' service between spark plug changes are expected. Magneto ignition systems using a high frequency condenser discharge to activate the plugs are being installed experimentally on large aircraft engines. The system is made low tension as far as the spark plug. Special plug designs are used, having the transformer coil integral in the spark plug bushing. From a magneto a charging pulse, created by the magneto breaker, is passed through a rectifier to charge a large condenser. A synchronized contactor (like a breaker, but acting by closing the circuit) and distributor unit discharges high-frequency, low-voltage transient current through the spark plug primary coil. A high-frequency, gap-jumping voltage is induced in the plug coil secondary, The high frequency discharge is not so readily shunted by leakage paths, and it is claimed that the high frequency system will continue to fire plugs that would have shorted out in an ordinary high tension system. (F.T.M.)

MAGNETOMETER. An apparatus used for measuring moderate magnetic field intensities, or sometimes the magnetic moments of magnets; frequently both. The standard laboratory equipment is set up in two ways, giving two results from which may be calculated both the magnetic field intensity (or rather, its horizontal component H) and the magnetic moment M of a short bar magnet forming part of the equipment. The terrestrial magnetic intensity is usually measured in this way, and Gauss employed this method in verifying the magnetic inverse-square law.

In the first arrangement, the magnet is used as a **magnetic pendulum**, whose **moment of inertia** I is known and whose period of oscillation T is measured. It is

Diagram of magnetometer; δ is the deflection of the needle at O, due to magnet in the "end-on" position at distance a.

usually suspended by a fiber of known torsion coefficient K; in some cases K is small enough to be neglected. From the magnetic pendulum formula we obtain the product

$$MH = \frac{4\pi^2 I}{T^2} - K. \qquad (1)$$

The magnet is then arranged to deflect a small compass needle. The needle is at O, the magnet at a considerable distance a from it, and commonly east or west of and aligned upon it (see figure). If the length of the magnet is l, the deflection δ of the needle is given by

$$\tan \delta = \frac{M}{H} \frac{32a}{(4a^2 - l^2)^2},$$

from which

$$\frac{M}{H} = \frac{(4a^2 - l^2)^2}{32a} \tan \delta. \qquad 2)$$

The second members of (1) and (2) contain only measurable quantities, and are therefore known. Hence, by multiplying (1) and (2) we get M^2, and by dividing (1) and (2) we get H^2; from those results M and H follow. Instruments embodying these principles are regularly used by the U. S. Coast and Geodetic Survey. (L.D.W.)

MAGNETOMOTIVE FORCE; Magnetic Circuit.

MAGNETON. Magnetism.

MAGNETO-OPTICAL ROTATION. Some substances are in themselves optically active, that is, they rotate the polarization plane of **polarized light** passed through them. In 1845 Faraday discovered that glass and other substances devoid of this property acquire it when placed in a strong magnetic field. This is the so-called Faraday effect. The light must traverse the substance along the lines of force. The direction of the rotation is reversed if the field is reversed, but is the same with respect to the observer whether the light is going or coming, so that a beam passing one way and reflected back has its rotation thereby doubled (which is not the case with natural activity). Some substances produce right-handed, some left-handed rotation in light traveling in the direction of the field. In any case, except for ferromagnetic substances, the rotation per unit thickness is for a given wavelength proportional to the

field intensity, its value per cm. per oersted intensity being called the Verdet constant. Some typical values of the constant, with sodium light, are: flint glass, 0.061'; water, 0.013'; air, 0.000007'. For ferromagnetic substances the rotation is proportional to the magnetization (see **Magnetism**) rather than to the intensity, as shown by Kundt. The rotation increases with diminishing wavelength.

Another type of magnetic rotation was observed by Kerr in 1877. A beam of plane-polarized light vibrating either in or perpendicular to the plane of incidence has its polarization plane rotated upon reflection from the polished pole-piece of a strong magnet. The polarization is at the same time rendered slightly elliptical. (L.D.W.)

MAGNETOSTRICTION. The term literally implies magnetic contraction, but is generally understood to include a number of closely allied phenomena relating to ferromagnetic substances under magnetic influence.

1. When an iron rod is subjected to a gradually increasing longitudinal magnetic field, it at first increases slightly in length (Joule effect) and later the length diminishes; when the magnetic intensity has reached about 250 oersteds, the rod has returned to its original length, and further increase of intensity causes it to contract (Villari reversal point). Nickel contracts rapidly at first and then remains nearly constant, while some iron-nickel alloys lengthen without reversal. The following phenomena are also recognized. 2. The Guillemin effect: the tendency of a bent ferromagnetic rod to straighten in a longitudinal field. 3. The Wiedemann effect: the twisting of a rod carrying an electric current when subjected to a magnetic field. 4. The Villari effect: a change of magnetic induction within an iron rod under longitudinal stress (inverse Joule effect).

Magnetostriction has been put to practical use in the magnetostrictive resonator, essentially an iron rod maintained in longitudinal elastic vibration by a high-frequency current in a helix wound upon it, and used, through the joint operation of the Joule and Villari effects, to control the frequency of the current, somewhat after the manner of the familiar **piezo-electric** (crystal) **resonator.** (L.D.W.)

MAGNETRON. This is a vacuum tube which functions under the joint action of an externally applied magnetic field and the electric field between its anode and cathode. In its common form it consists of a cylindrical **cathode** (a straight wire filament) and a coaxial anode structure. In some tubes the **anode** is a single cylinder while in others it is split lengthwise for all or part of its length. These tubes are widely used in ultra high frequency oscillator circuits where the conventional **vacuum tubes** would be completely inoperative. The electric field is created by applying a high d-c potential between the filament and anode structure while the magnetic field is applied longitudinally by external permanent or electro-magnets. When the tube is properly connected to a resonant **line** it can be made to operate as an **oscillator** for certain values of the applied fields. A magnetron of suitable dimensions may be made to generate frequencies measured in thousands of megacycles (wavelengths of a few centimeters). (L.R.Q.)

MAGNIFYING POWER. Crudely defined, the magnifying power of an optical instrument is the ratio of the apparent size of an object as seen through the instrument to the apparent size of the same object as seen without the instrument. To define "apparent size of an object" in terms which will permit of rigorous discussion requires certain assumptions. We may define the apparent size of an object as the angle which the object subtends at the eye, or as the size of the image which the lens of the eye forms of the object on the retina of the eye (retinal image). Using this definition we shall find that the apparent size of the rising moon is actually less than that of the moon when up on the meridian in spite of the fact that the rising moon always "looks" much larger than when it is up away from objects with which direct comparison can be made.

Even with the assumption of apparent size defined as above, the significance of the term magnifying power depends upon the type of instrument which is under consideration. For a **telescope**, used to view distant objects, the magnifying power may be directly defined as the ratio of the size of the retinal image obtained with the instrument to the size of the retinal image obtained without any optical aid. When a positive **eyepiece** is used the magnifying power may be shown to be directly equal to the ratio of the focal length of the object glass of the telescope to the focal length of the eyepiece. For example, a telescope with an object glass of 10 feet focal length will have a magnifying power of 60 when used with a positive eyepiece of 2" focal length. It should be noted that changing the eyepiece will change the magnifying power of the telescope; e.g., if an eyepiece with one-half inch focal length were used with the above telescope, the magnifying power would be 240. Hence, the question regarding the magnifying power of a telescope with interchangeable eyepieces is meaningless unless the particular eyepiece is specified for the particular telescope. In telescopes, binoculars, field glasses, etc., purchased from dealers in optical supplies, usually the eyepieces are not interchangeable and the magnifying power specified by the maker depends upon the lenses supplied with the instrument. In case negative eyepieces are used with a telescope a so-called "effective focal length" must be known for the eyepiece before the magnifying power can be calculated.

In the case of a **microscope**, either simple or compound, the magnifying power is usually defined as the ratio of the retinal image obtained for a given object while using the microscope to the retinal image that would be obtained of the same object without the instrument with the object at a standard distance from the eye, usually 250 mm. For a compound microscope the magnifying power is approximately equal to 250 times the length of the tube, divided by the product of the focal lengths of the objective and eyepiece, the three lengths all expressed in millimeters. The magnification effected by a projection instrument, such as an enlarging camera or a stereopticon, is logically expressed as the ratio of the diameter of the real image formed on the screen to the diameter of the object itself. If the focal length of the projecting lens is f and its distance from the screen is d, this magnetification is approximately equal to $\dfrac{(d-f)}{f}$. (W.K.G., L.D.W.)

MAGNITUDE. Stellar Magnitude.

MAGPIE. Aves, Passeriformes. Moderately large long-tailed birds (**Aves**) related to the crows. The common North American species, *Pica pica*, ranges from Alaska to Arizona and eastward into Iowa. It is a black and white bird. The yellow-billed magpie, *P. nuttalli*, flies only in California. The European and Asiatic species of magpies are more brightly colored.

The magpies build very large untidy nests and are noted for their curiosity, adaptability, and noisiness. (A.W.L.)

MAHOGANY. *Sweitenia Mahagoni* and other species. Meliaceae. The name mahogany belongs properly to a group of trees which are found in Central America and the West Indies, and to a few related species growing in tropical Africa. These, species of the genus *Sweitenia*, are large trees with **pinnately** compound leaves like those of ash trees and small flowers in **panicles** in the leaf **axils**. The term mahogany is also frequently and

mistakenly applied to many dark red woods, not mahogany.

The trees grow in a variety of habitats, often in most inaccessible places, so that it is very difficult to get the cut logs to the market. Mahogany is usually classified according to the region from which it comes, as Cuban mahogany, Honduras mahogany, etc.

The first mahogany to appear in Europe was brought in as ballast in a ship, the heavy logs being very suitable for that purpose. In port it was necessary to remove these in order to get in more cargo. So it was offered to English woodworkers. At first it was rejected as too hard to work and of little use, being heavy and dark-colored. Gradually it found favor, until it became a highly prized wood for fine furniture making. The first mahogany used in cabinet work was Spanish mahogany, *Sweitenia Mahogoni*. For a long time mahogany from San Domingo held first rank and was eagerly sought after. It was a very hard dark wood which could be given a very high polish, and was very durable. The supply is now nearly exhausted. Cuban mahogany is another variety which gives a dark red wood, and which finishes with a very fine glossy surface. In this, as in some other varieties, the wood is frequently marked with very small white pores of chalk-like substance. With age the wood of these varieties gradually darkens; it does not, however, lose its beautiful smooth finish. Other species of *Sweitenia* are shipped from Panama, from Mexico, and from South American countries.

The heavy logs of mahogany are removed from their native forests and shipped to American or foreign markets. In some varieties the logs are short and thick; often they are not perfect, having cracks or imperfections caused by branching. Some of the larger logs may be hewn square before removing from the forest, making them lighter and easier to handle. Specially valuable are those logs which when cut show a wavy grain or other irregularities. Even more valuable are blocks of mahogany which come from a large forking of the stem; from these the beautifully grained crotch mahogany is obtained. This is usually cut into thin **veneers**. In drying, mahogany shrinks very little, and once dry, is very durable, twisting or warping very little.

African mahogany is becoming increasingly valuable. It is obtained from large trees of the genera *Khaya* and *Entandrophragma*, both members of the same family as mahogany, and yielding woods very similar to it. Often appearing as mahogany are certain species of the genus *Cedrela*, which includes Spanish **cedar**, native trees of Central and South America. Many entirely unrelated woods, such as birch, maple, and even softwoods like whitewood, are frequently stained to simulate mahogany. True mahogany need not be stained. (R.M.W.)

MAHSEER, MAHASIR. Pisces, Teleostei. A large fresh-water fish (**Pisces**), *Barbus mosa*, of the Oriental region, related to the carps. It attains a length of 6'. (A.W.L.)

MAIDENHAIR TREE. *Ginkgo biloba*. Ginkgoales. The maidenhair tree is the sole surviving member of an order of **Gymnosperms** which in **Mesozoic** times was very abundant and widely distributed. Doubt exists whether the tree grows wild in any region today, though many people state that it does on the mountain slopes of China. It has been cultivated in the temple gardens of China and Japan for centuries. In late years it has been widely planted in the New World.

The tree often has a tall slender pyramidal shape when young; others, and especially older specimens, are widespreading. The branches are of two kinds: a long shoot which grows rapidly in length and which is composed mostly of woody tissue; and short shoots or spurs which elongate very slowly. These short shoots have a large pith, a thick **cortex** and very little wood. A short shoot may sometimes (especially in the case of injury to the long shoot) become a long shoot. The leaves of the ginkgo tree are somewhat variable in shape. Those on the long shoot are wedge-shaped and deeply notched, those of the short shoot broadly wedge-shaped and little or not at all notched. The veins are furcate, forked, as in the ferns. It is to the leaves that the tree owes its common name, Maidenhair tree, since their shape suggests that of the maidenhair fern. The trees are **dioecious,** the two types of flowers being borne on different trees. The male **strobilus** is composed of many **sporophylls,** each with two **sporangia.** The female flowers are also numerous. Each consists of a long slender **peduncle** or stalk bearing two **ovules.** In most cases one of those aborts early. The pollen grains, which consist of three small disk-shaped cells and one relatively large one, are carried to the ovule by the wind. There it forms a pollen tube which digests its way through the tissue surrounding the **gametophyte.** A pollen tube which has nearly reached the gametophyte contains two large **sperms,** each of which has a spiral coil of cilia at its anterior end. One of these sperms passes to one of the large eggs contained in the female gametophyte, and joins with it. The nucleus of the sperm unites with that of the egg, which is then said to be fertilized. The ovule containing the fertilized egg enlarges. When mature it is about an inch in diameter and green. Its outer covering is fleshy and has a curious rancid odor which is very noticeable when the fruit is crushed. Within this fleshy coat is a dry covering surrounding the gametophyte and the embryo plant. The seeds germinate readily, forming a long tap root and a short erect shoot. Young plants are rather susceptible to low temperatures, requiring some protection in the northern states. (R.M.W.)

MAIZE. Corn.

MALACHITE. The mineral malachite is a **basic carbonate** of **copper** corresponding to the formula $CuCO_3 \cdot Cu(OH)_2$. It is **monoclinic,** crystals tending to be acicular, but usually found massive. It is a brittle mineral; hardness, 3.5–4; specific gravity, 3.9–4.03; luster, vitreous to silky or dull; color, green; streak, green; translucent to opaque. Malachite is an alternation product found associated with other copper-bearing minerals. It is a rather common mineral and is found quite widely distributed. Large quantities have been found in the Ural Mts.; it is also found in Germany, France, England, the Belgian Congo, Rhodesia, and Australia. In the United States beautiful radiated masses of fibrous crystals have been found in Berks County, Pennsylvania. It has been found in Tennessee at Ducktown, and in Arizona, Nevada and Utah. Malachite besides being an ore of copper has been used for various ornamental purposes. The word malachite is derived from the Greek meaning, a *mallow,* because of its green color. (E.S.C.S.)

MALACOLITE. Diopside.

MALACOSTRACA. Crustacea.

MALARIA. A disease transmitted by the bite of infected anopheline mosquitoes, characterized clinically by recurrent paroxysms of chills, fever and sweating. In the mosquito the parasite develops on the stomach wall and in the salivary glands, while in man it grows in the red blood cells. Man, monkeys, birds, fish and cattle are subject to the disease.

Malaria is widespread in tropical and subtropical countries. Probably no other disease has disabled and killed as many people throughout the world. The combined deaths in all parts of the world where the disease occurs is probably 2,000,000 each year. In the United States, where malaria is prevalent in the southern states only, there are nearly a million cases and 5000 deaths; in India, 100,000,000 cases and 1,000,000 deaths occur annually. On the battle fronts of World War II, malaria has been one of the major problems.

In man, malaria is produced by four specific parasites which are not infectious for the lower animals. *Plasmodium vivax, P. malariae, P. falciparum* and *P. ovaie.* They are transmitted by various species of anopheline mosquitoes, and undergo definite development in the mosquito necessary for their survival.

Vivax or *tertian malaria* is by far the most common type. The incubation period is about 6 days, and paroxysms of chills and fever appear on the 14th day after the bite of an infected mosquito. They continue to recur every other day, as the parasite completes its 48-hour cycle of development. During the paroxysm, the patient first goes through a "cold stage" during which he has chilly sensations, his skin is blue, his teeth chatter and there is violent shaking. After an hour, the "hot stage" is ushered in, with a rapid rise in temperature to as high as 107° F.; the skin is hot and dry and the patient complains of severe headache. The fever lasts about 2 hours, and is followed by the "sweating stage," during which there is profuse perspiration, the temperature falls to normal, the headache disappears, and although weak and drowsy, the patient feels well.

P. ovale produces a disease practically identical with tertian malaria.

Quartan malaria, produced by *P. malariae,* has an incubation period of 18 to 40 days. The paroxysms occur every 72 hours, and are longer and somewhat more severe than those accompanying tertian malaria.

Falciparum or *estivo-autumnal malaria* is a more severe disease than the other types, and is the cause of practically all deaths due to malaria. The paroxysms occur irregularly after a 12-day incubation period. They are severe, and accompanied by high temperatures. The so-called cerebral, algid, hemorrhagic and pernicious types of malaria represent forms of falciparum malaria with different localizations of the parasite. In the cerebral type, the onset is rapid with delirium and coma, and death may occur in several hours without return to consciousness. "Black-water fever" or hemorrhagic malaria is a type in which hemolysis or dissolution of the red cells occurs, and dark urine due to the presence of hemoglobin is an outstanding feature. In the algid form, there are vomiting, diarrhea, and subnormal temperature.

The treatment of malaria with **quinine** is one of the oldest examples of specific chemotherapy in medicine. Quinine has been used as a preventive in countries where malaria is endemic, and is given orally or hypodermically in the acute disease. Because of the shortage of quinine, a synthetic drug, **atabrin,** is now used, both as a preventive, and alone or in conjunction with quinine as a curative drug during the acute attack. A course of treatment with atabrin and quinine extends over one week. Although these drugs control the paroxysms of malaria, the complete eradication of the parasites is extremely difficult to accomplish. The disease therefore tends to recur again and again, even though adequate treatment be given. A malarial infection confers a low-grade very brief immunity after the acute attack has subsided.

The control of malaria is a matter of the eradication of the mosquito vector. Wherever anopheline mosquitoes breed, they must be diligently sought out and killed. This involves drainage of swamps, and clearing out of all possible breeding places. In a malaria area houses must be adequately screened, and people should not be out of doors in the evenings when anopheline mosquitoes do most of their biting. The application of mosquito-repellents to exposed areas of skin is also effective. (D.M.H.)

MALEIC ACID. Acids, Carboxylic.

MALEO. Aves, Galliformes. A peculiar bird **(Aves)** of Celebes and neighboring East Indian islands. The head and neck are covered with naked red skin and the crown bears a black prominence resembling a helmet. The plumage is mostly black but that of the breast and belly is salmon colored. The large eggs are buried in hot sand. (A.W.L.)

MALIC ACID AND MALATES. Malic acid, hydroxysuccinic acid ($H_2 \cdot C_4H_4O_5$ or $COOH \cdot CH_2 \cdot CHOH \cdot COOH$) is a white solid, melting point $133°$ C., decomposes at $150°$ C., soluble in water, alcohol, or ether. Calcium malate, on account of its slight solubility, is of importance in the separation and recovery of malic acid. Calcium malate plus dilute **sulfuric acid** yields malic acid plus calcium sulfate and the latter may be separated by **filtration.** Malic acid may be obtained by evaporation of the filtrate. Malic acid may be obtained (1) from some natural products, e.g., the free acid in the juice of unripe apples and gooseberries, often in conjunction with **citric** or **tartaric acid**; potassium hydrogen malate is present in rhubarb and currants; calcium malate in sugar "sand" by evaporation of the sap of the sugar maple, (2) by synthesis. Malic acid is a dibasic acid, that is, two series of **salts** and **esters** are known. On heating under pressure to $200°$ C., it is converted into fumaric acid. Malic acid is used in medicine and in the production of various salts and esters.

Ester: Ethyl malate $\begin{array}{l} CH_2 \cdot COOC_2H_5 \\ | \\ CHOH \cdot COOC_2H_5 \end{array}$ boiling point $253°$ C. (R.K.S.)

MALIGNANT. A term used to describe (1) an infection or disease which is especially virulent and rapidly grows worse, (2) **tumors**—cancer and sarcoma—which are non-encapsulated, spread to distant structures, and usually result fatally. (D.M.H.)

MALIGNITE. This rock is best described as a **nephelite syenite** with larger proportions of **iron** and **magnesia** than is usual in **alkali** rocks. It has received its name from the type locality on the Maligne River, Province of Ontario, Canada. (E.S.C.S.)

MALINGERER. One who pretends sickness deliberately. (R.S.M.)

MALKOHA. Aves, Cuculiformes. A **cuckoo** of Ceylon. (A.W.L.)

MALLEABILITY. Metals, Physical Properties of.

MALLEABLE CAST IRON. Malleability is understood, in most cases, to be that property which makes it possible to hammer, press, roll, etc., a metal to some finished shape. However, in the case of malleable cast iron, the malleable product denotes a casting which is much stronger and possesses shock resisting qualities to a much greater extent than gray cast iron. A malleable iron casting is produced from an iron casting which in the untreated state is hard and brittle. This casting is caused to have as large a silicon content as possible, since it has been found that this element promotes the change from the hard casting to the malleable form during the heat treatment.

The malleabilizing of iron castings is the conversion of cementite **carbon** into a finely dispersed graphite form, although a part of malleabilizing may result from burning out of the surface layer of the carbon by heating the casting in the presence of the oxidizing agent. The hard castings are annealed to the malleable form by packing them in an annealing box surrounded by an oxidizing agent. Crushed iron oxide and mill scale have been used for this packing, but the malleable product may be produced even if an inert packing, such as sand, be used. However, since some increase in strength is obtained by packing in an active material, castings are so treated. After packing, the boxes are placed in an annealing furnace, and held at a temperature of about $800°$ C. for 3–5 days, after which they are slowly cooled, and are malleable. There are two types of product:—white-heart and black-heart malleable iron. The latter is made by conducting the annealing operation at a higher temperature and shorter period of time than for the white heart, and it is more satisfactory when large, thick castings are to be annealed. (F.T.M.)

MALLOPHAGA. The **bird lice** or biting lice, constituting a small order of insects. They have flattened bodies with many spines directed backward, short legs, and biting mouths. Wings are lacking. Most species live as external parasites on birds but a few are found on mammals. (A.W.L.)

MALLOW FAMILY. Malvaceae. The plants of this family include herbs, shrubs and trees (the latter tropical), and are rich in mucilaginous substance. The leaves are alternate, and in most cases **palmately** lobed and veined, with small deciduous **stipules.** The flowers are regular and perfect, often large and showy, and variously borne. They have five (or rarely, fewer) more or less united **sepals,** five petals, and numerous **stamens,** which characterized the family by having their filaments joined to form a tube which surrounds the **styles.**

The most important member of this family is the **cotton** plant, whose fibers outrank in commercial importance all others. Okra or Gumbo, *Hibiscus esculentus,* a native of tropical Africa, is another member of importance. It is a coarse annual plant with large veiny leaves and showy axillary flowers. The slender 5-ribbed pods are used in soups, or when young, are cooked and used in salads. Okra has also been used as a source of fibers for paper manufacture. Another member of some importance is *Althaea officinalis,* the Marsh Mallow. The underground rootstock of this plant is not only rich in mucilage, but is also used medicinally in ground form, the bark being removed before grinding.

Many members of the family are grown as ornamental plants, among them being the Hollyhock, *Althaea rosea,* the so-called flowering maples, *Abutilon* sp., species of *Malva,* the true Mallows, and the Rose of Sharon, *Hibiscus syriacus,* which becomes a large bush or even a small tree, with showy pink or white flowers. (R.M.W.)

MALONIC ACID. Alcohols and Ethers.

MALPHIGIAN TUBULES. Small tubules closed at one end and connected with the alimentary tract at the other. They occur in the spiders and other **arachnids, centipedes** and millipedes, and **insects,** and function as excretory organs. The tubules vary greatly in form and number in different species. (A.W.L.)

MALT. Barley.

MALTA FEVER. Undulant Fever.

MALTOSE. Carbohydrates.

MALUS'S LAW. A law applying to the intensity of **polarized light** as affected by the polarizing apparatus. If a beam of plane-polarized light is passed through a **Nicol prism,** for example, the intensity (flux density) of the emergent beam falls off, as the prism is rotated, from a maximum value when the transmission plane of the prism coincides with the plane of vibration of the light, to zero when it is at right angles to that direction. The intensity varies as the square of the cosine of the angle through which the prism has been thus rotated. The same law applies to the effect of a glass reflector, reflecting always at the polarizing angle, as the plane of reflection is rotated around the stationary, polarized incident beam. (L.D.W.)

MAMMALIA. Animals characterized by warm blood and hairy vestiture. The young of most species develop in the body of the mother and all are nourished by milk secreted by special glands. The teeth are of four kinds, incisors, canines, premolars and molars (**Dentition**). A class of the phylum **Chordata**, including the creatures popularly called animals without further qualifications.

The mammals are one of the dominant forms of animal life. The most important domestic species belong in this class and man himself is a mammal, hence their economic importance cannot be overemphasized. They are the source of most of the animal products used by man and include also some harmful species of predators and rodents.

The classification of the mammals is briefly as follows:

Subclass Prototheria. Egg-laying mammals.
  Order **Monotremata.** The duck-bill (**Platypus**) and spiny ant-eater (**Echidna**) of the Australian region.
Subclass Eutheria. Young developed at least partly in the body of the mother.
Division Didelphia. Pouched mammals. Young born in an early stage of development and carried in a pouch on the abdomen of the female until able to walk.
  Order **Marsupialia. Kangaroos** and **wallabies, opossum, koala,** and others.
Division Monodelphia. Placental mammals. Young connected with the body of the mother throughout embryonic development by a special structure called the **placenta.**
  Section **Unguiculata.** Clawed animals.
  Order **Insectivora.** Small furry animals, mostly nocturnal. They eat insects and smaller animals and some vegetable matter. **Shrews, moles,** etc.
  Order **Dermoptera.** The flying **lemurs** of the Oriental region.
  Order **Chiroptera.** Winged mammals. The **bats.**
  Order **Carnivora.** The flesh-eating species, with large canine teeth and sharp-edged molars. **Lion** and **tiger, weasels, bears, skunks, seals, walruses,** etc.
  Order **Rodentia.** Gnawing animals. Incisor teeth like chisels. **Rabbits, squirrels, rats, mice, porcupines,** etc.
  Order **Edentata.** Toothless or with imperfect teeth. The **sloths, armadillos** and **ant-bears.**
  Order **Pholidota.** Scaly ant-eaters or **pangolins.**
  Order **Tubulidentata.** The **aard-varks.**
  Section **Primates.** With nails instead of claws or hoofs. Appendages often differentiated as arms and legs. Thumb and great toe opposable to the other digits or anatomically similar to such appendages.
  Order **Primates.** The **monkeys, baboons, apes,** man, and other related species.
  Section **Ungulata.** The hoofed animals and related forms with heavy nails on the appendages.
  Order **Artiodactyla.** The even-toed ungulates. The axis of the foot is through two digits. **Swine, cattle, sheep, deer, camels** and other species.
  Order **Perissodactyla.** Odd-toed ungulates. Axis through the middle digit of the foot. **Horses, rhinoceroses, tapirs** and related species.
  Order **Proboscidea. Elephants.** Nose and upper lip prolonged to form a trunk or proboscis.
  Order **Sirenia.** Aquatic species. **Dugongs** and **manatees.**

Order **Hyracoidea.** Small animals resembling rodents. The hyraces or **conies.**
Section **Cetacea.** Highly specialized marine animals.
  Order **Odontoceti.** The toothed **whales.** Sperm whales, **porpoises, dolphins,** and related forms.
  Order **Mystacoceti.** Species with baleen or whalebone in place of teeth. Right whales, rorquals, and baleen whales. (A.W.L.)

MAMMARY GLAND. A large gland which secretes milk for the nourishment of the young of **mammals.** Mammary glands are structurally related to the sweat glands. They are normally functional only in the female although they develop as rudiments in the male sex. In most species the ducts of the glands open on a prominence, the teat or nipple, from which the young suck the milk, but in egg-laying mammals the milk is merely licked from the surface upon which it exudes and in some of the pouched mammals the young are temporarily attached to the teats within the pouch by an oral sucker and the milk is discharged into their mouths. (A.W.L.)

MAMMARY ORGAN. An organ developed in some of the **bryozoans** which nourishes the embryos during their attachment to the parent. (A.W.L.)

MAMMATO-CUMULUS. Clouds.

MAN. Paleontology of Man.

MANAKIN. Aves, Passeriformes. Brightly colored birds (**Aves**) of Central and tropical South America. **Chatterers.** (A.W.L.)

MANATEE, MANATI. Mammalia, Sirenia. Completely aquatic animals with a horizontally flattened oval tail, no hind limbs, and fore limbs developed as flippers. They are superficially similar to the whales but differ in many details of structure and are apparently derived from different ancestral stock. They live only in shallow coastal waters and estuaries and eat aquatic plants. Also called sea cows.

The several species of the genus *Manatus* are distributed on both shores of the Atlantic and in the Oriental and Australian regions. *M. latirostris* is found in Florida. (A.W.L.)

MANDIBLE. A structure associated with the mouth and used for biting, or if used differently, evolved from a biting organ. The term is applied to the pair of biting organs derived from jointed appendages in the **arthropods,** and, in the sucking mouths of some **insects,** to the slender piercing structures which have been shown to have the same origin. It is also used for the lower jaw of **vertebrates** and for a prominent bone of that jaw which makes up the entire structure in the mammals but is associated with other bones in the jaws of fishes (**Pisces**), **amphibians** and reptiles. The two parts of the beak in birds are known as the upper and lower mandibles. (A.W.L.)

MANDIBULAR GROOVE. A groove in the **carapace** of some **crustaceans** just behind the **mandible.** It may accompany the **cervical groove,** in which case it is located farther forward. (A.W.L.)

MANDREL. An arbor with a taper of about .001″ per in., upon which hollow cylinders are held for turning and facing the exterior surfaces. (H.C.H.)

MANDRILL. Mammalia, Primates. A peculiarly ugly species of African **baboon,** *Papio maimon.* (A.W.L.)

MANEUVERABILITY. That quality in an aircraft which determines the rate at which its altitude and direction of flight can be changed. (F.T.M.)

MANGABEY. Mammalia, Primates. *Cercocebus.* An African **monkey** of a small group of species also called the white-eyelid monkeys. They are slender animals with a fairly long muzzle and long tail. Related to the macaques. (A.W.L.)

MANGANESE. Symbol: Mn. Atomic number: 25. Atomic weight: 54.93. Density: 7.2. Hardness: 5. Melting point: 1260° C. Boiling point: 1900° C. No isotope, but of single atomic form: 55.

Manganese is a silver-white metal, not notably hard (becomes hard on alloying with **carbon**), brittle, capable of taking a brilliant polish but readily oxidized upon heating, reacts with water upon boiling, soluble in dilute acids. Discovered by Scheele in 1774.

Manganese occurs chiefly as **pyrolusite** (manganese dioxide (MnO_2)) in the Caucasus Mountains of Russia, India, Brazil, the Gold Coast of Africa, and the United States. 1. Treating pyrolusite in a **blast furnace** with **iron** ores and **carbon** furnishes ferro-manganese (78% Mn) and speigeleisen (12–33% Mn), both used in the production of steel. 2. Pure manganese may be obtained by ignition of the oxide with aluminum powder, or by electrolytic method from the sulfate. Small percentages of manganese are added to steel as a deoxidizer, and large percentages, say 12%, produce a very tough steel. An alloy of 20% manganese, 20% nickel, 60% copper possesses good resistance to corrosion, and by heat treatment develops greater hardness and strength than any other copper alloy. 3. When pyrolusite is heated with sodium carbonate and nitrate, sodium manganate is formed, which is then extracted with water. This is the substance from which manganese compounds are commonly obtained.

Acetate: Manganese acetate ($Mn(C_2H_3O_2)_2 4H_2O$), pale pink crystals, soluble.

Chlorides: Manganese chloride, manganous chloride, manganese dichloride ($MnCl_2 \cdot 4H_2O$), rose-red crystals, soluble to practically colorless solution by reaction of **hydrochloric acid** on manganese oxides, hydroxide, carbonate, sulfide and subsequent crystallization; manganese tetrachloride, manganese perchloride ($MnCl_4$), green solid.

Dithionate: Manganese dithionate (MnS_2O_6), by reaction of manganese dioxide suspension with **sulfurous acid.** With **sodium** carbonate yields sodium dithionate solution and manganese carbonate precipitate.

Hydroxide: Manganese hydroxide, manganous hydroxide ($Mn(OH)_2$), white precipitate but rapidly turning brown in the air by oxidation to manganic hydroxide, formed by reaction of manganese salt solution and **sodium** hydroxide solution; manganic hydroxide ($Mn(OH)_3$), brown solid.

Manganates: Sodium manganate (Na_2MnO_4) and potassium manganate (K_2MnO_4), green soluble solids, formed (1) by oxidation of other forms of manganese by heating with **sodium** carbonate and nitrate, and (2) regulated reduction of permanganate. Manganate solution is readily oxidized to permanganate in the presence of acid.

Nitrate: Manganese nitrate, manganous nitrate ($Mn(NO_3)_2 \cdot 6H_2O$), rose-red crystals, soluble to practically colorless solution.

Oxides: Manganese monoxide, manganous oxide (MnO), grayish-green solid, formed (1) by heating of manganese **oxalate**, hydroxide or carbonate in the absence of air, (2) by reduction of higher manganese oxides by heating with **hydrogen;** manganese sesquioxide, manganic oxide (Mn_2O_3), brownish-black solid, by heating manganese dioxide at 700° C.; trimanganese tetroxide, mangano-manganic oxide (Mn_3O_4), brownish-black solid, by heating any oxide of manganese at 1000° C.; manganese dioxide, manganese "peroxide" (MnO_2), black solid, by heating manganese nitrate to 200° C., used as pyrolusite in dry cell batteries, and to color glass and ceramic ware, also in the preparation of a dryer for paint and varnish oils, and in the chemical laboratory for the preparation of **chlorine** from **hydrochloric acid;** manganese heptoxide (Mn_2O_7), red oil, dangerously explosive, formed by reaction of potassium permanganate and concentrated sulfuric acid.

Permanganates: Sodium permanganate ($NaMnO_4$) and potassium permanganate ($KMnO_4$), purple-black soluble solids, the latter readily crystallized, formed by oxidation of acidified manganate solution. Permanganate solution, purple, is readily reduced to manganous in the presence of acid, e.g., by the reaction of ferrous salt.

Sulfate: Manganese sulfate, manganous sulfate ($MnSO_4 \cdot 4H_2O$), pink solid, soluble to practically colorless solution.

Sulfide: Manganese sulfide, manganous sulfide (MnS), pink precipitate by reaction of manganous salt solution and **ammonium** sulfide solution. (R.K.S.)

MANGANITE. The mineral manganite is a hydrous **oxide** of **manganese** corresponding to the formula $MnO(OH)$, it occurs in prismatic **orthorhombic** crystals, sometimes in massive columnar forms. It is a brittle mineral; hardness 4; specific gravity 4.2–4.4; luster, submetallic; color, steel gray to iron black; streak, red-brown to almost black; opaque. Manganite is of secondary origin and it may itself alter to **pyrolusite.** It is usually associated with other manganese minerals. It is found in the Harz Mts., Germany; Sweden; Cornwall and Cumberland, England; and in the United States in Michigan. It is an ore of manganese. (E.S.C.S.)

MANGEL-WURZEL. Beet.

MANGO. Mangifera indica. Anacardiaceae. The mango is a long-lived tree, often developing a massive trunk and widely spreading branches. Its **lanceolate** leaves are evergreen and about 4″ long. The flowers are numerous, small, pink and borne in **racemes.** The ovoid fruits, 1–5″ in diameter, are 1-seeded berries having a thick rough greenish rind and a pleasantly aromatic orange-colored flesh esteemed by many. This fruit is eaten fresh or in salads. Unfortunately, fruits frequently occur which are not at all agreeable, because of the fibrous nature of the flesh and the unpleasant sour taste.

Reproduction is either by seedlings, which do not always come true, or by grafting. The tree is extensively cultivated in tropical regions, and is now grown in Florida, southern California, and tropical America. For successful growth hot moist weather is necessary, followed by a short dry period for successful ripening of the fruit. (R.M.W.)

MANGOSTEEN. Garcinia. Guttiferae.

MANGROVE. *Rhizophora Mangle.* Rhizophoraceae. The mangrove is a moderate-sized tree which grows on low, often submerged, coastal lands. It is found, for instance, in all tropical American coasts. The leaves of the plant are opposite, entire, dark green, and rather tough. The flowers are borne in small clusters and are perfect, with four **sepals,** four pale yellow linear petals, four to twelve **stamens** and single 2-celled inferior **ovary.** Only one **ovule** develops. The seed usually germinates while the fruit is still attached to the tree. A long thick **hypocotyl** grows from the fruit, and attains a length of 5–10″ and a diameter less than ¾″. Eventually an **abscission layer** develops, so that the fruit, in which the young seedling is well advanced in germination, falls to the soft muddy ground, in which the new tree will grow. Because the lower part of the hypocotyl is heaviest, it strikes the ground first and so sinks into a position most favorable for further growth. Should it fail to penetrate into the mud, the upper and lighter portion of the germinated fruit causes it to float until a favorable environment is reached. In a favorable location the hypocotyl puts out many roots, which anchor the young plant; then the epicotyl quickly grows. It is characteristic of the mangrove that from the stem and branches there grow out arching prop roots which soon form an intricate mass in which is deposited silt and all sorts of debris floating in the water. Because of this the mangrove causes a gradual building up of the land around it, until eventually the black slimy mud in which it grows gives place to a low coastal land which gradually becomes usable by man.

In addition to its land-forming function, the mangrove has other uses. The wood is dark red or reddish-brown, fine-grained, and hard; it is used in charcoal making. The bark contains tannin and so is employed in tanning hides. From the young shoots a reddish dye may be obtained, which, however, is of little value. (R.M.W.)

MAN-HOLE. A man-hole is a means of ingress and egress to a region, ordinarily and normally not occupied by humans, but occasionally necessitating inspection or repair by them. Sewers, tanks, boiler drums, and many other structures are equipped with one or more man-holes which are often only of sufficient size to permit a man to crawl through with considerable difficulty. Normally these openings are closed by man-hole covers. (F.T.M.)

MANIA. A state of extreme excitement or exaltation. Mania is characterized by violence, hallucinations, delusions, and often requires restraint to protect the patient as well as others. Mania occurs as an acute toxic reaction to drugs and certain infections, or as a part of the picture in psychoses, particularly those of the manic-depressive type. (D.M.H.)

MANIC-DEPRESSIVE PSYCHOSIS. A severe mental illness characterized by alternating phases of excitement with elation and profound depression. Between these episodes the individual is normal in a typical case. Occasionally patients experience only one attack; in others, the alternating episodes become continuous.

The cause of this psychosis is unknown. There is a definite hereditary tendency and women are more often affected than men. The manic-depressive is almost always an extrovert and often is of the short, stocky type of body build. This disease is little understood, but is thought to be related to a disturbance in the emotional life, usually based on fears, uncertainties and insecurity. It is a more complete and serious personality disorder, and unlike a simple psychoneurosis, results in abnormal behavior.

The manic phase is usually preceded by a mild depression for a few days. Then, gradually, excitement begins. At this stage the patient is buoyant, full of self-confidence, liberal with his money, aggressive and boastful. As his excitement increases he becomes talkative, argumentative, full of schemes and pranks. His good humor is sustained until he is crossed in his wishes, when he becomes angry and vociferous in abuse. He quickly recovers his good spirits, and is able to be extremely active for days with little sleep, and declares he never felt better in his life. This period is classified as one of hypomania. It passes imperceptibly into the period of mania, when the mounting excitement reaches a pitch of joyous exaltation, accompanied by rapid uninterrupted speech, with continuous flight from one subject to another, singing, dancing, aggressive sexual behavior, and sometimes violent destructive motor activity. Throughout the episode, the patient remains oriented as to his environment. Hallucinations rather in the nature of transitory illusions occasionally occur. Delusions usually of the grandiose wish-fulfillment type may occur, and ideas of persecution are not rare. The average duration of manic attacks is 6 months.

The depressive phase in a large per cent of individuals appears after several manic episodes a few years apart. It begins with a mild depression of spirits. The patient lacks confidence in himself and feels inadequate. Gradually he slips into a severe depression. He sits with his head hung down, his body stooped; his attitude is one of hopelessness, despondency and profound unhappiness. Thought and speech are slow; ideas of guilt, remorse and hypochondriasis predominate. Ideas of persecution may be present, but hallucinations are uncommon. In extreme forms, there is complete absence of motor activity, and stupor may develop. Sometimes threats and attempts at suicide are carried out. In some patients the depressive state is combined with mania, and instead of joy coloring the manic phase, loud grief and woe predominate.

Both the manic and the depressive stages of this psychosis often represent defense reactions on the part of the patient—unconscious defences against his fears, or insecurity. The elation is an over-compensation, an effort to bluff himself into believing he is not afraid. This mechanism in mild form is used frequently by normal people, but in the manic it becomes a fixed pattern which he cannot escape. The depressive phase, also, is an unconscious defense in the form of a withdrawal from the active life which is full of disappointment and insecurity.

The prognosis for a single manic or depressive episode is good, but recurrences are common—Treatment during these episodes is symptomatic—restraining and attempting to quiet the manic, and caring for the depressed individual, protecting him from suicide, giving him an adequate diet and tube-feeding if necessary. During the normal intervals between the psychotic episodes, psychotherapy directed toward discovering the patient's emotional problems and helping him to adjust himself to them, offers the most hopeful method of preventing future attacks. (D.M.H.)

MANIFOLD, ENGINE. With particular attention to the internal combustion engine in the multi-cylindered form, the manifold is that part which distributes a common fuel-air mixture uniformly to each of the several cylinders, or which gathers up the exhaust gases after they issue from the cylinders and combines them into one exhaust stream. Since the great majority of internal combustion engines are multi-cylindered, and further, since the purpose of multi-cylinders is to provide more uniform flow of power, the manifolds must be well chosen and correctly patterned, or else they will tend to offset the advantage of the multi-cylinder arrangement.

Considering problems of manifolding from the standpoint of gasoline engines, in which they are most difficult, the condition which the manifold must meet is that of equal distribution of fuel and air to the different cylinders without permitting any segregation of the heavier fuel particles, and without creating much pressure drop by friction. The inlet manifold is attached at its discharge end to the inlet ports of the cylinders, and at its other end to the outlet of the carburetor. A carburetor does not perfectly vaporize and mix a fuel with air so that an equal distribution of the volume of the fluid passing through the manifold is not necessarily an equal distribution of the fuel to the separate cylinders because of the inertia of the heavier fuel particles; therefore bends have to be carefully laid out, and any factors which might tend to deflect the stream unevenly into one limb of the manifold should be eliminated. Inlet manifolds may be either updraft or downdraft, depending on the relative location of carburetor and cylinders. It is much simpler to maintain uniform mixture in a downward moving stream than in the upward. Many manifolds are provided with what is known as a hot spot, which is a portion of the manifold upon which the heavier gasoline particles will strike, and which by the heat it contains will immediately vaporize these heavier particles. In an ordinary T-shaped manifold, the hot spot should be placed at the top of the vertical stem, where it will form a target for the ascending droplets of gasoline. While shape of the inlet manifold is a matter of great importance, the material of it is not, since the inlet manifold is kept cool by the inflowing mixture. It should be tight and smooth internally.

The exhaust manifold conveys a stream of heated gas which may, at times, cause it to reach red-hot temperatures, and it must be constructed of a material capable of resisting such conditions, and be provided with some means of expanding and contracting with variation of temperature. Some large engine exhaust manifolds are water-jacketed. The ideal exhaust manifold has a separate lead from each cylinder, each lead being streamlined into a common stack or exhaust pipe at some distance from the cylinder. If this is not possible, the manifold should be divided into sections so that two cylinders will not exhaust into the same section simultaneously. Where the noise of the exhaust is not objectionable, or is less objectionable than the back pressure created by the exhaust manifold, exhaust may either be freed directly from the ports, or from short, straight exhaust stacks discharging directly to the atmosphere. (F.T.M.)

MANILA HEMP. *Musa textilis.* Musaceae. The Manila Hemp plant is a perennial herb having an underground **rhizome** from which the very large leaves rise directly. The apparent stem, which may be as much as 20′ in height, is composed of the tightly overlapping, broad leafstalks. The flower stem grows up through the center of the column formed by the leafbases and bears inconspicuous flowers covered by reddish bracts. The fruit is green, banana-like and filled with numerous seeds. The native home of the plant is the Philippine Islands.

To obtain the fibers the so-called stem of leaf-bases is cut down and at once cut in long narrow strips. By drawing these over a dull knife the fibers are cleaned of most of the non-fibrous substance surrounding them, leaving uniform strands 6–10′ or more long. The fibers are white to buff-colored and lustrous, coarse and very strong. They are extremely durable, little affected by water, and hence much used for making cables, hawsers and marine cordage. The finer strands may be spun into thread and from that **coarse** cloths woven. Manila waste and worn-out manila products are used in making a very tough paper known as manila paper. The plant is extensively cultivated, especially in the Philippine Islands. The cultivated plant yields a better grade of fiber. **Another** name given to the fiber is **abaca.** (R.M.W.)

MANOMETER. Pressures which are so small that the Bourdon tube **pressure gauge** is not accurate, are conveniently measured by a liquid manometer. This instrument is one which balances fluid heads in a glass tube so that readings may be taken by comparing the registry of meniscii on a scale mounted alongside the tube. In its simple form, it consists of a U-tube, one end of which is open to the atmosphere, and the other to the region where the pressure is to be measured. If the pressure is different from atmospheric, the liquid with which the manometer is partially filled will stand higher in one leg of the tube than the other. As shown in the accompanying illustration, the manometer is connected to a

vacuum, the balancing head on the low-pressure leg. Since the other leg is open, the pressure head y is either the vacuum or the gauge pressure. The magnitude is yw

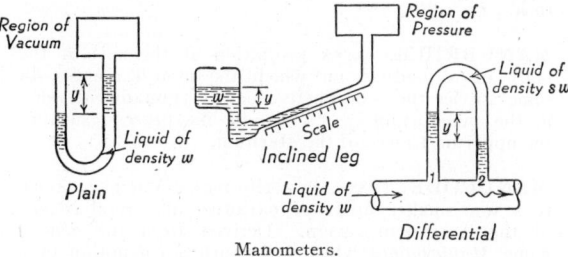

Manometers.

lbs. per sq. ft. when y is measured in ft. and w in lbs. per cu. ft. To make the instrument more sensitive, one leg may be inclined at a large angle to the other so that a given vertical displacement of the meniscus will travel a considerably larger distance along the scale. Of course in this type the other leg must be of enlarged cross-section so that its level will not vary appreciably as the meniscus travels along the scale. The scale can be calibrated to read any desired pressure units.

Very small pressure differences are measured by differential manometers, a simple form of which is shown. Suppose that a liquid of specific weight w is flowing along a pipe and it is desired to measure the friction loss between points 1 and 2. An inverted U-tube is connected, having its upper portion filled with a non-miscible liquid of specific weight sw. The difference of pressures at points 1 and 2 will then be $wy (1—s)$. Very small differences of pressure may thus be indicated by a readable displacement y if s approaches unity; that is, if the two liquids are nearly of the same density. (F.T.M.)

MANTIS. Insecta. Orthoptera. A large **insect** of predacious habits. The body is moderately broad and bears four wings. The first segment of the thorax is long and rather slender, adding to the reach of the powerful raptorial front legs. The head is prominent and has large eyes.

From the fancied suppliant air of these voracious insects as they await their prey with the forelegs uplifted they are called praying mantises, and their owlish expression has given them the rarer name of soothsayers. The common species is *Mantis religiosa.* (A.W.L.)

MANTIS FLY. Insecta, Neuroptera. Small predacious insects superficially like the **mantis** in form. They make up the family Mantispidae and are also called mantispas. (A.W.L.)

MANTIS SHRIMP. Crustacea, Hoplocarida. Moderately large marine **crustaceans** with a pair of powerful grasping appendages formed like those of the mantis. They are named for their superficial resemblance to the shrimps on one hand and to the mantis on the other. (A.W.L.)

MANTISSA OF A COMMON LOGARITHM. Logarithms.

MANTLE. 1. A fold of the dorsal body wall of mollusks. It secretes the shell, forms a respiratory chamber, and in some species encloses the body. 2. The wall of the body of **tunicates,** beneath the enclosing tunic. 3. Folds of tissue forming a dorsal and a ventral flap which secrete the shell in the **brachiopods.** (A.W.L.)

MANTLE ROCK (REGOLITH). The term mantle rock is applied to all of the loose unconsolidated material which is found at the surface of the earth and covers, more or less completely, the solid rock beneath.

Regolith, derived from the Greek word meaning a cover and stone, is essentially synonymous with mantle rock. (R.M.F.)

MANUBRIUM. 1. A projection at the axis of the body of a medusa on which the mouth opens. In some species the reproductive bodies (gonads) develop in the manubrium. 2. A handle-like process, such as the uppermost piece of the sternum. (A.W.L.)

MANUCODE. Aves, Passeriformes. A name applied to a few smaller birds of paradise of several islands of the Australian region. Derived from the generic name *Manucodiata* which is in turn a corruption of a Malay name. (A.W.L.)

MANUL. Mammalia, Carnivora. A wild cat, *Felis manul*, of Siberia, Mongolia and Tibet. It is about the size of the domestic cat and varies from buff to silver-gray in color. Also called Pallas' cat. (A.W.L.)

MAP. A drawing which describes portions of the earth's surface is called a map. Since there are many different purposes for which maps are used, different kinds of maps are made. For example, there are the topographic maps, land subdivision maps, general geographic maps, geologic maps, etc. The features which are expressed on a map will vary with the degree of specialization of use of the map, even to the point of excluding all features except those which are to be emphasized. For example, railway maps show little other than the line of the railway and the towns along its route. Since the surface of the earth is in three dimensions, it is necessary to use some special method of showing the altitude of the surface. Otherwise the drawing would be capable of showing only breadth and length of an area. A topographic map is one on which the altitude is shown by some conventional means, usually contour lines or hachures. Other features of the landscape, such as bluffs, woodland, streams, marshes, may be included on the topographic map by generally accepted symbols. The standard symbols of the U. S. Geological Survey are more often used than any others. (See Contour.)

A planimetric map is one which shows natural and artificial features of the earth's surface. Contours are not drawn on this map. A hydrographic map is a map showing the shore line and the configuration of the bottom of a body of water (see Hydrographic Surveying).

A map may be oriented or interlocked with adjacent plats by giving the direction of either true or magnetic north on the map, or by placing thereon some of the meridians and parallels of the earth. The former method is suitable to maps of small areas, the latter for depicting the larger areas, states, and countries. Maps of large areas must show the surface of a sphere, as that is the shape of the earth. It is not possible to represent accurately a spherical surface on a flat surface, therefore some approximation, or some method of representation which is correct in some, but not all respects, must be employed in mapping large areas so that the effect of curvature is not neglected. (See Map Projections and Charts.)

Most maps are of no value in navigation and some of them are actually dangerous to use for this purpose since they emphasize some features in such a manner as to be misleading. As an extreme case we have Mark Twain's balloon pilot, who lost his way because a state boundary was crossed without the change in color of the land that was indicated on the map!! Charts should always be used for navigational purposes.

The scale of a map is the fixed ratio between corresponding distances on the map and on the territory which the map portrays. Scales are numerical or graphical. A numerical scale may be given in terms of one inch on the map equals a certain whole number of feet or miles on the ground; or it may be stated as one unit on the map equals a certain number of the same units on the ground. When the latter method is used, the scale is in fractional form having a numerator equal to unity. This is called the representative fraction. A graphical scale is a subdivided reference line drawn on the map to the same scale as that used in plotting the map. The subdivisions are marked in terms of the units of measurement of the numerical scale. A graphical scale should always be given if distances are to be scaled because the materials on which maps are made are subjected to changes in linear dimensions due to climatic conditions. A graphical scale is also necessary for use in connection with scaling distances from a photographic reproduction of an original map when the size has been changed. (F.T.M., W.K.G., C.W.C.)

MAP PROJECTIONS. The numerous methods for representing the surface of the earth on a plane are known as map projections. The term originated from the geometric-projection methods of drawing lines from some point out through the surface of the earth to a developable surface. The surfaces most commonly used in the geometric projections were the plane, the cylinder, and the cone. In most cases, the distortions introduced by the strictly geometric projections are intolerable for modern purposes, and the term map projection has been expanded to include all sorts of purely mathematical methods for representing the surface of the earth on a plane. The pattern of lines on the map or chart representing meridians of longitude and parallels of latitude form what is known as the graticule of the projection.

The growth of geopolitics and the increase of air travel have provided powerful stimuli to the general subject of cartography and many ingenious map projections have been devised. It is impossible to include anything like a complete list of modern maps and charts, but particular attention is directed to those which are of value in modern navigation such as: Mercator, gnomonic, polyconic, cylindric, Lambert and stereographic projections. (W.K.G.)

MAPLE SUGAR. Sugar.

MARAL. Mammalia, Artiodactyla. A Persian deer, *Cervus elephas maral*, of the red deer group. (A.W.L.)

MARBLE. Technically speaking this term should only be applied to metamorphosed, recrystallized limestones and dolomites. It is generally used as a trade term, however, for any crystalline calcium carbonate rock of pleasing pattern and color when cut and polished. Some marbles are almost pure white, as in the case of the best statuary marble, but various impurities produce marbles of all colors, shades and patterns. In the northern Appalachians there is a great belt of marbles which, in Vermont, have been developed into one of the greatest marble-producing regions in the world. Here are quarried and mined a variety of marbles which compete with the Italian travertines and breccias. Including the more common stone, which is bluish gray, are such decorative types as red, verd antique, black, green and pink. Other producing states are Alabama, Massachusetts, North Carolina, Maryland and Virginia. The amount of marble produced in the United States in 1933 was equivalent in value to $6,236,508. (R.M.F.)

MARCASITE. The mineral marcasite, sometimes called white iron pyrites, is, like ordinary pyrites, disulfide of iron corresponding to the same formula, FeS_2. Marcasite, however, crystallizes in the orthorhombic system often yielding serrate, spear-shaped twins, hence the name "cock's comb pyrites." It is a brittle mineral; hardness, 6–6.5; specific gravity, 4.85–4.90; luster, metallic; color, light bronze-yellow; streak, grayish-black; opaque. Marcasite alters very easily and may disintegrate with the formation of sulfuric acid and iron sulfate. Fossils replaced by marcasite are therefore often destroyed after being placed in collections. Marcasite is found in numerous places in Europe, Czechoslovakia, France, England, etc.; in Mexico, and in the United States in the lead districts of Illinois, Wisconsin ·and Missouri. The name marcasite is believed to be of Arabic origin and formerly was applied to common pyrite. (E.S.C.S.)

MARE. The female of the **horse** and related species with the exception of the asses. . Females of these animals are called jennets. (A.W.L.)

MARE'S TAIL. Detached fragments of fibrous cirrus appearing in streaks with feather-like tufts at one end. (P.E.K.)

MARGAY. Mammalia, Carnivora. A wild **cat** of moderated size. It is reddish marked with black spots. Mexico to Paraguay. (A.W.L.)

MARIALITE. Wernerite.

MARIHUANA. Hemp.

MARINE BIOLOGY. The study of all living things found in the ocean and their interrelations.

The masses of water in the oceans are so great that all human knowledge of the living things within them must be a very small part of their story. Most investigations have been conducted in the shallower waters near the shores. Specially equipped ships have occasionally conducted extensive towing operations in deep as well as shallow waters but this work is hampered by the great expense involved and discloses only such forms of deep-sea animals as can be taken in relatively small nets. The recent use of the bathysphere by Beebe and Barton for direct observation down to a half-mile depth has shown that large fishes also occur in the abysses.

Many marine laboratories have been established primarily for the study of marine biology. Among them the station· at Naples is famous. In the United States, stations are located at Woods Hole in Massachusetts, Cold Spring Harbor on Long Island, Friday Harbor on Puget Sound, and in other favorable places. The Bermuda Biological Station for Research, formally opened in 1932, is one of the most favorably located institutions for all oceanographic work. (A.W.L.)

MARINER'S COMPASS. Compass.

MARJORAM. Mint Family.

MARKHOR. Mammalia, Artiodactyla. A wild *goat, Capra falconeri,* of the Himalayas. It is a large and variable species with long horns, spirally twisted although in variable forms. (A.W.L.)

MARL. Marl is a loose earthy deposit of **calcium** or **magnesium** carbonate mixed with clay in varying proportions. It is usually gray but the color may be affected by the presence of other substances, such as **iron** oxides. Marls are believed to have accumulated in freshwater basins as they are frequently found carrying shells of fresh-water **mollusks,** then called shell marl. If sand is present in any quantity it is then called sandy marl. There are considerable deposits of marl in the **Tertiary** formations of the Atlantic and Gulf States. (E.S.C.S.)

MARLIN. 1. Pisces, Teleostei. Large marine game fishes (**Pisces**) taken in the warmer waters of the Atlantic. Related to the swordfishes and sailfishes. 2. Aves, Charadriiformes. The marbled **godwit.** (A.W.L.)

MARMOSET. Mammalia, Primates. Small **monkeys** of Central and South America. They have only thirty-two teeth, four less than the other American monkeys, and the thumb is not opposable to the fingers. All digits but the great toe bear claws instead of nails. They constitute the family Hapalidae.

Most of these monkeys are called marmosets but the common Brazilian species, *Hapale jacchus,* is also known as the ouistiti and the group of long-tusked marmosets, *Mystax,* are called tamarins. One species, *Midas aedipus,* of the Isthmus of Panama, bears the French name pinché. (A.W.L.)

MARMOT. Mammalia, Rodentia. Stout-bodied burrowing animals of moderately large size. The common **woodchuck** or groundhog, *Marmota monax,* is a familiar North American species, and in the western states others occur, one of them, *M. caligata,* known as the whistler. In Europe and Asia the bobac, *M. bobac,* alpine marmot, *M. marmota,* golden marmot, and several other species occur. The fur is sparse and rather coarse but it is used to a limited extent. The flesh is edible. (A.W.L.)

MARQ SAINT HILAIRE. Celestial Navigation.

MARS. (See planetary tables, page 1109.) Mars, the "ruddy planet," is the fourth **planet** in order of distance from the sun. It has been observed from remote antiquity since its ruddy color and relatively rapid motion among the stars make it a very conspicuous object. Within the past 50 years there has been a great deal of speculation relative to the possibility of there being intelligent life on this planet, and, for this reason, there has probably been more printers' ink expended in pseudo-scientific articles about conditions on Mars than on any other astronomical object, with the possible exceptions of the sun and moon.

Since the **orbit** of Mars lies entirely outside of the orbit of the earth, the planet can never be seen in the crescent **phase.** However, at **quadrature,** Mars does present a distinctly gibbous-phase condition as seen with a telescope.

Mars is best observed at **opposition,** for during this **configuration** it is on the **meridian** at midnight. The distance from the earth at an average opposition is about 78,200,000 kilometers (48,600,000 miles), but at a "favorable opposition" (i.e., with the earth at aphelion and Mars at **perihelion**) this distance is reduced to 55,700,000 kilometers (34,600,000 miles). These favorable oppositions occur at intervals of from 15 to 17 years.

The question regarding the possible habitability of Mars, or of any other planet, may be said to hinge upon two questions, not mutually exclusive: the **temperature** of the surface, and the existence and character of an **atmosphere.** The temperature of the surface of Mars has been a vexing one for a long time, but has apparently been definitely settled by the brilliant research both at the Lowell and Mt. Wilson observatories. Using independent methods, both agree that the temperature of the planet's surface is approximately 283° K. (50° F.) at the equatorial regions at noon—a condition well adapted for life as we know it on the earth. The question regarding the atmosphere of Mars cannot be said to be as definitely settled as the surface temperature. The **surface gravity** of the planet is such that the planet could retain an atmosphere of oxygen, nitrogen, and other heavy gases, and probably water vapor as well, but could not hold the light gases such as hydrogen and helium. The clarity with which the surface features of the planet can be observed and the absence of any definite cloud effects seem to indicate conclusively that there is very much less atmosphere than we have on the earth. At the Mt. Wilson Observatory, using the most powerful instruments and most delicate tests, it has been shown that the amount of oxygen over the planet's surface cannot exceed a thousandth part of that over the surface of the earth, and that the planet's atmosphere and surface must be exceedingly dry.

Through even a moderately large telescope, Mars is a very beautiful object, with a number of large and permanent surface markings clearly visible. From the motion of these surface markings the rotation period of the planet has been definitely determined as about 24.5 hours. It has also been determined that the planet's equator is inclined to the plane of the orbit of the planet at about 24°. Both of these figures indicate that the day and night conditions and **seasonal** changes on Mars are comparable with those on the earth, with the exception, however, that the Martian year is about twice as long as the corresponding period on the earth.

The greater portion of the surface of Mars is reddish in appearance, but there are large and well-defined greenish-gray areas, and brilliant white caps at the polar regions. The reddish areas do not show any seasonal changes, and are believed to be desert areas, possibly similar in character to regions on the earth such as Sahara. The dark areas, unfortunately known as seas, show definite seasonal variation in color, having a more strikingly greenish tint during the summer than in the winter. The seasonal changes of these dark areas have been ascribed by many reliable authorities as due to some form of vegetation and there seems to be no need for rejecting this possibility, in spite of the low oxygen content of the atmosphere and scarcity of water. The changes in appearance of the polar caps with season is most striking. During the winter season in the northern Martian hemisphere the polar cap is extremely prominent, while no cap whatsoever appears at the south pole. As spring approaches in the north the north cap shrinks and the south cap grows until the north cap entirely disappears. It has long been believed that the polar caps are composed of snow, but whether this is actually frozen water or of some other form of frozen gas is not definitely known.

The question regarding the fine details on the surface of Mars, in particular the so-called canals, is still a puzzling one. In 1877 and again in 1879 Schiaparelli announced the discovery of a number of fine dark lines crossing the surface of the planet. During the next 50 years a great many observers searched for the canals with varying degrees of success. Schiaparelli, Lowell, E. C. Pickering, and a number of other observers saw the canals and were able to construct detailed maps of them. On the other hand, Barnard and a number of other excellent observers were never able to see any trace of the canals. All of the observers who have been able to see the canals agree that they show great variations in visibility and also that they show seasonal variations. Even taking this into account, it is difficult to understand why Barnard, one of the most skillful of observers, and working with the best instrumental equipment in the world, should never be able to see the canals at all. To date the canals have never been positively photographed, and until their reality is proved beyond serious doubt, it is futile to speculate regarding their origin or purpose.

Mars has two **satellites,** both discovered by Hall in Washington in 1877. They are both very small and relatively close to the surface of the planet. The inner one is so close that it revolves about the planet in about 7.5 hours, about ⅓ of the rotation period of the planet, and hence rises slowly in the west and sets in the east, contrary to all other celestial objects observed from Mars. This is the only case on record in the solar system where a satellite has a period of revolution shorter than the rotation period of the planet. (W.K.G.)

MARSH GAS. Methane.

MARSH TEST. Arsenic.

MARSH MALLOW. Mallow Family.

MARSUPIAL MOLE. Mammalia, Marsupialia. A pouched burrowing animal, *Notoryctes typhlops,* found in a very limited part of Australia. It is about the size of the true moles and resembles them in the soft fur, the rudimentary eyes, the enormously developed claws, and the lack of external ears. Also called the pouched mole. (A.W.L.)

MARSUPIALIA. The pouched mammals, including the kangaroos, opossums, and many less familiar forms. The order is characterized by the presence of a marsupium or pouch on the abdomen of the female in which the young complete their development. They are born in a very early stage. With the exception of the opossums of the Americas, all marsupials occur in the Australian region.

The principal forms in this order are the **kangaroos** and **wallabies, phalangers, wombats, bandicoots, dasyures, pouched mole,** and **opossums.** (A.W.L.)

MARSUPIUM. A pouch formed of a fold of skin on the abdomen of the female. The teats open into this pouch and the young, born in a very early stage of development, are transferred to it. They adhere to the teats by a temporary sucker and the milk is discharged into their mouths. (A.W.L.)

MARTEMPERING. Steel.

MARTEN. Mammalia, Carnivora. Slender animals with short legs and a moderately long bushy tail. Related to the weasels but larger. The several species live in the northern hemisphere.

Some of the most valuable furs marketed are those of martens, and all species bear fur of fine quality. The sable of northern Asia, *Martes zibellina,* and the marten of northern North America, *M. americana,* produce the most valuable fur, although that of the fisher is also good. The fisher, *M. pennanti,* formerly ranged over most of North America, but is now found only in the wilder northern areas. It is also called the pekan and the American marten is sometimes known as the American sable. (A.W.L.)

MARTENSITE. Steel.

MARTIN. Aves, Passeriformes. Any of several species of **swallows,** represented on all continents except Australia. They have the short beak and wide mouth of the swallows and either a forked or a square tail. They nest about houses, on cliffs, and in burrows and hollow trees. North America has one species, the purple martin, *Progne subis.* The male is glossy blue-black and the female somewhat duller. These birds commonly nest in the cornices of buildings or in bird houses where they live in colonies. (A.W.L.)

MASK. A peculiar hinged appendage associated with the mouth of the immature **dragon fly.** It can be extended to catch the animal's prey. A modified **labium.** (A.W.L.)

MASONRY. Masonry is a term applied to structures composed of individual units laid in and bound together by mortar. The common materials of masonry construction are brick, stone, and tile. Masonry is one of the most durable and permanent types of construction, since the materials which enter into it are but little affected by the elements. However, the quality of masonry construction depends on the mortar and caliber of workmanship employed in its construction. Masonry is used chiefly in the walls of buildings, retaining walls, **piers**, buttresses, **arches**, **foundations**, and monuments. Brick masonry is used more frequently than any other type. Brick walls may be either solid or veneered. Veneered construction has strength imparted by a framework of wood or a rough masonry wall composed of concrete or tile blocks, over which is placed a layer of bricks which give weatherproofing and a finished appearance to the wall. The veneered wall is often superior to a solid brick wall. Solid brick masonry is

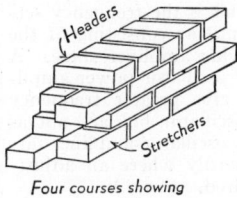

Four courses showing
Bonding and Jointing

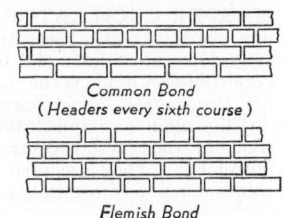

Common Bond
(Headers every sixth course)

Flemish Bond

BRICK

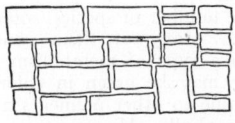

Ashlar (Uncoursed)

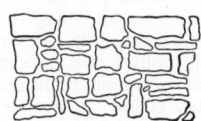

Rubble

STONE

Masonry.

made of two or more layers of brick with the bricks running longitudinally, bound together with bricks running transverse to the wall. These brick arrangements are known respectively as stretchers and headers. The method of alternating stretchers and headers gives rise to different bonds, such as the common bond, the English bond, and the Flemish bond. There is not a great deal of difference between these bonds from the utilitarian standpoint, but the appearance of the finished wall is somewhat different, and these bonds have been developed to suit different tastes.

Blocks of cinder concrete, or of ordinary concrete, and blocks of hollow tile are known as building blocks. They are much larger in size than the ordinary bricks, and lay up much faster in the wall. Furthermore, cinder and tile masonry have but little water absorption compared to solid brick masonry. While these blocks are used alone for commercial walls such as factories, garages, etc., their use in the best grade of construction is limited to backing up brick veneer and to interior partitions and cellar walls.

Stone which is worked into the masonry may be dressed or rough. Stone masonry wherein the stones are dressed to flat surfaces is known as ashlar masonry. Stone masonry with irregularly shaped stones is rubble masonry. Both ashlar and rubble masonry may, by the selection of stones, be laid in course, but a great deal of stone masonry is uncoursed.

The strength of masonry walls is dependent upon the bond between the building material and the mortar, but not entirely so, since the complete filling of the space between adjacent bricks or stones with a mortar which hardens so interlocks the units that a strong masonry wall would be obtained even with no adhesion between mortar and brick. The irregularities of stone surfaces, and the artificially made grooves or recesses of building blocks, aid this interlocking action. (F.T.M.)

MASS. Few terms are used in physics with greater frequency and assurance than "mass of a body," and few are more difficult to define. Mass is often confused with weight, a mistake not helped by the use of the same names for the units of mass and of weight (e.g., gram). What appears to be the only inseparable attribute of mass is **inertia;** and, indeed, for the purposes of dynamics, inertia may be taken as the measure of mass. One body has twice as much mass as another body if it offers twice as much force in opposition to the same acceleration, no matter where the two bodies may be in the universe. The **relativity** theory teaches, however, that the inertia of a body, and hence its mass, is dependent upon its motion. If in any state we consider the body at rest, and call its mass in that state m_0 (the "rest mass"), then when the body is given a velocity v with respect to that state, the mass becomes $m = m_0(1 - v^2/c^2)^{-\frac{1}{2}}$, in which c is the velocity of light; so that at a speed of about 160,000 miles per sec., the mass of a body would be doubled. The concept of the "electromagnetic mass" of electric charges was developed by Maxwell (see **Electromagnetic Field**). (L.D.W.)

MASS ACTION, LAW OF. Equilibrium.

MASS CURVE. Impounding Reservoir.

MASS DIAGRAM. Earthwork.

MASS, LAW OF CONSERVATION OF. Matter cannot be created nor destroyed. This law holds within the experimental error of the most precise **chemical reactions.** However, matter is believed to be converted into energy according to the Einstein equation (energy equals mass times the velocity of light squared) in many of the atomic disintegrations. (See **Radioactivity**.) (R.K.S.)

MASS POTENTIAL. Potential.

MASS PRODUCTION. Measurement.

MASS RATIO. Rockets.

MASS SPECTROGRAPH. Any type of apparatus for sorting streams of electrified particles in accordance with their different **masses** by means of deflecting fields. If a particle of mass m (grams) carrying a charge E (e.m.u.) and moving with the **kinetic energy** EV corresponding to its passage through an accelerating potential drop V (volts), enters a uniform transverse **magnetic field** of intensity H (oersteds), it will follow a circular path whose radius is

$$r = \frac{1}{H}\sqrt{\frac{2mV}{E}}.$$

H and V may be given known values, and r may be observed; then from the above formula, the mass

$$m = \frac{H^2 r^2}{2V}E$$

may be calculated. If the particles are ionized molecules or atoms (e.g., positive rays), E will be some multiple of the electronic charge $e = 1.602 \times 10^{-20}$ (e.m.u.), so that particles of any one mass will move on different radii according as they happen to be singly, doubly, ... , multiply ionized; while all singly ionized particles moving on different radii will be known to have masses in proportion to r^2. The figure shows, in purely schematic form, one type of apparatus arranged for this purpose. (In the Aston spectrograph the particles pass through both an electric and a magnetic field, the separation produced by the one being offset by a convergence due to the other; in which case V need not be known.) By means of such "magnetic analysis," much has been learned concerning the actual masses of atoms and other particles, and the **isotopes** of elements. (See **Chemical Composition**.) (L.D.W.)

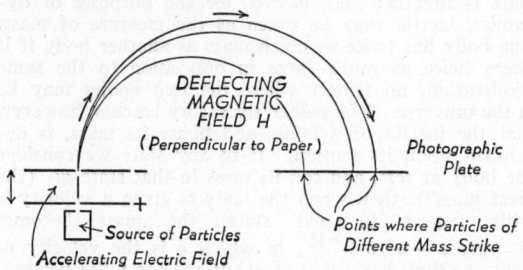

DEFLECTING MAGNETIC FIELD H
(Perpendicular to Paper)

Photographic Plate

Source of Particles
Accelerating Electric Field

Points where Particles of Different Mass Strike

The particles follow semicircular paths and make spots where they strike the plate.

MASSASAUGA. Reptilia, Sauria. A small **rattlesnake**, *Sistrurus catenatus,* found throughout the eastern half of the United States and south into Texas and Mexico in low swampy ground. It does not attain a length of 3′ and is consequently less dangerous than the larger rattlers, but like all others of the group it is poisonous. (A.W.L.)

MASSIF. A prominent structural and topographical unit in a mountain range of the Alpine type. Frequently referred to as a central massif because it represents the **granitic** or **metamorphic** core of a deeply **eroded** and highly deformed mountain range with characteristic **nappes.** (R.M.F.)

MASS-LUMINOSITY RELATION. From purely theoretical reasoning, based upon the hypothesis that the material of which a **star** is constructed follows the gas laws, Eddington was able to show that there should be a relationship between the mass and the total radiation from a star. The details of the theory are far too complex to be included here, but it should be emphasized that the conclusions are based upon purely theoretical reasoning and not upon a statistical study of previously determined masses and **absolute magnitudes.** The results obtained from the theory are graphically represented by the curve shown in the accompanying figure. The curve was calculated from pure theory and

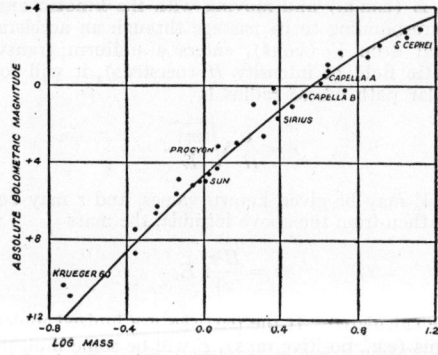

Mass-luminosity curve. (*From a diagram by Eddington.*)

the plotted points represent stars with observationally determined masses and luminosities. The agreement between theory and observation is remarkably close for **giant and dwarf stars** of all **spectral classes.** The only class of stars for which the theory completely fails to agree with observational results is the abnormal group known as the **white dwarfs.** (W.K.G.)

MASTAX. The chewing or grinding apparatus of the alimentary tract of **rotifers.** It is an expanded chamber with muscular walls in which chitinous jaws or trophi work together to break up the food. (A.W.L.)

MASTECTOMY. Surgical removal of the breast. The operation is usually performed for the removal of tumors, benign or cancerous. Simple mastectomy is removal of the breast alone. Radical mastectomy is often carried out if the tumor is a **cancer.** It consists of removal of the breast, the underlying muscles and the lymph glands in the axilla, in an effort to remove all tumor cells and thus prevent recurrence. (R.S.M.)

MASTER OSCILLATOR. This is the frequency setting **oscillator** of a radio **transmitter** consisting of the oscillator and at least one following amplifier stage. A very common usage is the master oscillator-power amplifier transmitter. Such circuits give greater frequency stability than a simple loaded oscillator but are not as good as the crystal controlled oscillator-amplifier arrangement. They are used primarily where an adjustable frequency transmitter is desired. (L.R.Q.)

MASTIC. Resins.

MASTIGOPHORA. One-celled animals whose bodies bear one or more **flagella** as organs of locomotion. A class of the phylum **Protozoa.**

This group contains a large number of species, some of great economic importance. Many of them contain chlorophyll and so carry on a type of metabolism like that of green plants. They may be green in color or blue, yellow, brown or red, due to other pigments associated with the green chlorophyll. Many species are colonial and some show a division of labor within the colony which foreshadows true multicellular organization. A few of the colonial species have collared cells like the sponges.

The **trypanosomes** which cause African **sleeping sickness** and various diseases of animals and the Leishmanias which cause kala azar and Oriental sore are the most important species economically.

The classification of the group is as follows:

Subclass **Phytomastigina.** Usually with chlorophyll, living as plants.
Order **Chrysomonadida.** Very small. One or two flagella. Color yellow to brown, rarely green or bluish. Sometimes form pseudopodia.
Order **Cryptomonadida.** No pseudopodia. Color varied.
Order **Dinoflagellida.** Marine and some freshwater forms. Body enclosed in a cellulose sheath, often beautifully sculptured.
Order **Phytomonadida.** Small rounded green species. Many colonial.
Order **Euglenoidida.** Usually elongate, with one or more flagella in a pit at the anterior end. With or without green color. Sometimes with a pigment spot (stigma) near the anterior end.
Order **Chloromonadida.** A few rare green species without a stigma.
Subclass **Zoomastigina.** Without pigment bodies. Free-living and parasitic species.
Order **Pantostomatida.** With pseudopodia and flagella.
Order **Protomonadida.** One of three flagella. Sometimes form pseudopodia. Mostly parasitic. **Trypanosomes** and others.
Order **Polymastigida.** Three to eight flagella. Mostly minute, living in the alimentary tract of animals.
Order **Hypermastigida.** With numerous flagella. Living in the alimentary tract of insects. (**Isoptera**). (A.W.L.)

MASTITIS. Chronic or acute inflammation or infection of the breast. (D.M.H.)

MASTODON. Fossil Mammals.

MASTOIDITIS. An acute infection of the air cells of the mastoid process of bone behind the ear. This usually occurs as a complication of middle-ear infection (**otitis media**) which reaches the mastoid by direct extension. The symptoms are fever, pain, tenderness and swelling over the mastoid process, occurring during the course of an ear infection. Early treatment with **sulfonamides** or **penicillin** may control the infection completely, but if the process is extensive and associated with abscess formation and bone destruction, surgery will be necesary in addition to chemotherapy. Mastoiditis is dangerous because of the proximity of the infection to the brain and its coverings and the danger of spread to these structures. (D.M.H.)

MASTURBATION. Producing an orgasm in oneself by the hand or some other mechanical friction of the genital organs. It is more common than supposed, occurs in both sexes and often is quite common in childhood. The habit does not produce physical disorders or injury. It may have an injurious mental effect produced by the sense of guilt, the conflicts, and fears which may lead to the development of a **psychoneurosis.** (R.S.M.)

MASURIUM. Discovery announced by Noddack, Tacke and Berg in 1925 and presumably of atomic number 43. (R.K.S.)

MATCH. Ignition.

MATCHING. Impedance Matching.

MATÉ. *Ilex paraguayensis.* Aquifoliaceae. Maté yerba, or Paraguay tea, is a drink made from the leaves of a small evergreen tree found in the forests of Paraguay and southern Brazil. The branches of the tree are cut off and dried, after which the leaves are broken off, dried more thoroughly and ground to a powder. The drink is prepared by pouring boiling water over the powder, and has a pleasant mild odor and taste. As yet it has not gained great popularity as a beverage in the United States. It contains caffeine. South American natives consume great quantities, using a curious instrument for the purpose—a tube at the lower end of which is a flattened spoon-shaped part, perforated to permit the liquid to pass. Through this instrument, called a bombilla, maté is sucked up. (R.M.W.)

MATERIAL BALANCE. A method of using the law of conservation of matter. All the material entering a process must equal all the material leaving it. It provides a basis for evaluating unknown amounts of material. (D.E.M.)

MATERIALS HANDLING. Conveyor, Elevator, Hoist, Dredge, Crane, Cableway.

MATHEMATICAL STATISTICS. Mathematical statistics may be considered as that part of **statistics** which is concerned with the mathematical proofs and theorems employed in statistics. Topics in mathematical statistics are **sampling**, the theory of statistical inference, **correlation** theory, **statistical probability**, the Neyman-Pearson theory of testing hypotheses, theory of **distribution functions**, theory of characteristic functions, the **analysis of variance,** and the design of experiments. (L.A.A.)

MATING. The process of securing a consort of the opposite sex for the purpose of reproduction.

Among some animals sexual union seems to follow the meeting of the sexes without preliminaries while among others a period of more or less elaborate courtship precedes the actual choice of mates and sexual union. Under the simpler conditions the sexes must be together at the time when their reproductive cells are ripe, but this results usually from the normal course of development, aided in some cases by special behavior such as the swarming of marine **annelids.** When sexually mature these worms respond to the same conditions by vigorous activity which brings them, in some cases, to the surface waters in great numbers at twilight. Here their mating takes place.

Birds and mammals offer many examples of special courtship. As a rule the male is active in the process and the female receives his attentions, which may involve no more than close personal interest and may include a great variety of display. Birds often sing at their best during this period and often go through strange evolutions to display the beauties of their plumage. Displays such as that of the **prairie chicken** combine vocal efforts, such as they are, with display of plumage and mimic combat with other males, and in many species both of birds and of mammals combat between males is a prominent feature of the mating period.

The acceptance of a mate by the female is followed by their physical union, or **copulation,** and in many cases by mutual preparation for the care of the young. Some species, however, including the **wrens,** include nest building in mating activities. The male provides one or more nests as a practical accompaniment for his abundant song while he seeks a wife. (A.W.L.)

MATRICES. A matrix is an ordered set of mn elements a_{ij} arranged in a rectangular array, having m rows and n columns. The notation commonly used is

$$A \equiv \begin{Vmatrix} a_{11} & a_{12} \cdots a_{1n} \\ a_{21} & a_{22} \cdots a_{2n} \\ \cdot & \cdot \\ a_{m1} & a_{m2} \cdots a_{mn} \end{Vmatrix} \text{ or } \begin{pmatrix} a_{11} & a_{12} \cdots a_{1n} \\ a_{21} & a_{22} \cdots a_{2n} \\ \cdot & \cdot \\ a_{m1} & a_{m2} \cdots a_{mn} \end{pmatrix}.$$

If $m = n$, the matrix is called a square matrix of order n.

Any two matrices A and B of the same number of rows and of columns are said to be equal if and only if $a_{ij} = b_{ij}$ for every i and j.

If the matrix is square, from the elements of this matrix a **determinant** can be formed, which is called the determinant of the matrix. From the elements of a rectangular matrix, by striking out rows or columns or both, determinants can be formed.

A matrix A is said to be of rank r if there exists at least one determinant of A of order r which is not zero, while all determinants of A of order higher than r are zero.

The idea of the rank of a matrix is of importance in stating the conditions under which a **system of linear equations** has a solution. (L.L.S.)

MATRIX. Printing film used in color photography for the transfer of an image to another support. Dyed relief films or differentially hardened films are generally called matrices when used in an **imbibition** process. (H.C.C.)

MATRIX MECHANICS. Quantum Mechanics.

MATTERHORN. Cirque.

MAVIS. Aves, Passeriformes. The European song thrush, *Turdus philomelus,* a bird that resembles the wood thrush of North America. (A.W.L.)

MAW WORM. Nemathelminthes, Nematoda. A large roundworm parasitic in the intestine of the horse. When present in large numbers its attack is serious. A member of the genus *Ascaris* which contains the **eelworm** and other species of similar habits. (A.W.L.)

MAXILLA. 1. A bone of complex shape that forms the greater part of the upper jaw in the **vertebrates.** 2. One of a pair of jointed appendages developed as mouth parts in the **arthropods.** They lie behind the **mandibles.** The typical maxilla consists of a basal segment, the cardo, bearing a second segment, the stipes, from which two parts arise. On the outside the **palpifer** serves as the attachment for a sensory palpus consisting of several joints. Mesially the stipes bears the subgalea from which arises the galea, of one or two segments, and a cutting or chewing part called the lacinia. The entire structure may be greatly modified. In the **butterflies,** for example, the maxillae are developed into long flexible structures which fit together to form a tubular proboscis. (A.W.L.)

MAXILLARY GLAND. An excretory organ of **crustaceans.** Its duct opens at the base of the **maxilla.** (A.W.L.)

MAXILLIPED. A jointed appendage of the body region of **arthropods,** modified to serve as an accessory mouth part. The term is applied to one to three pairs of the thoracic appendages of some **crustaceans** and to the poison claws of the **centipedes.** (A.W.L.)

MAXIMA AND MINIMA. Maxima and minima of **functions** of one variable:

If, as x increases and passes through a value x_1, a function $f(x)$ ceases to increase and begins to decrease, then $f(x)$ is said to have a maximum at $x = x_1$ and $f(x_1)$ is called a maximum value of the function; if $f(x)$ ceases to decrease and begins to increase, then $f(x)$ is said to have a minimum at $x = x_1$ and $f(x_1)$ is called a minimum value of the function.

A maximum value of a function as just defined is not necessarily the greatest value of the function, nor is a minimum the least value. Maxima and minima should properly be called relative maxima and minima. Together they are often called extremes of the function.

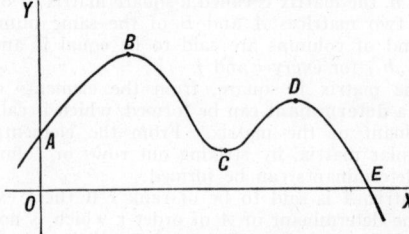

For the graph of the function, the points corresponding to the maximum and minimum values are called turning points of the curve. Thus, in the accompanying figure, the points B and D are maximum points and C is a minimum point of the graph.

In the case of both maxima and minima, the value of the first derivative at $x = x_1$, $f'(x_1)$, is either 0 or ∞.

If, as x passes through x_1, increasing, $f'(x)$ changes sign from $+$ to $-$, then $f(x_1)$ is a maximum; if from $-$ to $+$, $f(x_1)$ is a minimum. If $f'(x)$ does not change sign in passing through x_1, then $f(x_1)$ is neither a maximum nor a minimum.

If at $x = x_1$ we have $f'(x_1) = 0$ and if also $f''(x_1)$ is negative, then $f(x_1)$ is a maximum, but if $f''(x_1)$ is positive, then $f(x_1)$ is a minimum.

Hence, to determine maxima and minima of a function $f(x)$:

First find all the real values x_1 of x for which $f'(x)$ is 0 or ∞; these are called critical values. To test the critical values, to find which give maxima and which minima, we may use either of the following two methods:

(1) If $f'(x)$ changes sign from $+$ to $-$ as x passes through a critical value x_1, increasing, then $f(x_1)$ is a maximum, but if $f'(x)$ changes from $-$ to $+$, $f(x_1)$ is a minimum.

(2) If $f''(x_1)$ is negative, then $f(x_1)$ is a maximum, but if $f''(x_1)$ is positive, then $f(x_1)$ is a minimum.

At a maximum or minimum, $x = x_1$, we must have $f'(x_1) = 0$ (or ∞), but it may happen that $f''(x_1)$ is also 0. In this case the preceding tests fail.

If $f'(x_1) = 0$ and $f''(x_1) = 0$, and if also $f'''(x_1) = 0$ and $f^{(4)}(x_1) < 0$, then $f(x_1)$ is a maximum, but if $f'''(x_1) = 0$ and $f^{(4)}(x_1) > 0$, then $f(x_1)$ is a minimum.

For the general case:

Suppose that $f'(x_1) = 0$, $f''(x_1) = 0$, $f'''(x_1) = 0$, $\cdots$, $f^{(k-1)}(x_1) = 0$, but $f^{(k)}(x_1) \neq 0$, then if k is odd, $f(x_1)$ is neither a maximum nor a minimum, but if k is even, then $f(x_1)$ is a maximum if $f^{(k)}(x_1) < 0$, and a minimum if $f^{(k)}(x_1) > 0$.

Maxima and minima of functions of several variables:

A function $f(x, y)$ is said to have a maximum value at a point (a, b) if $f(a, b)$ is its greatest value in some region about (a, b), i.e., if $f(a + h, b + k) - f(a, b) < 0$ for all values of h, k (except 0, 0) which are numerically less than some positive number d. In like manner, $f(x, y)$ is said to have a minimum value at (a, b) if $f(a + h, b + k) - f(a, b) > 0$ for all values of h, k (except 0, 0) such that $|h| < d$, $|k| < d$.

A function $f(x, y)$ can have a maximum or minimum only at points where $\dfrac{\partial f}{\partial x} = 0$ and $\dfrac{\partial f}{\partial y} = 0$.

If at a point (a, b), we have $\dfrac{\partial f}{\partial x} = 0$ and $\dfrac{\partial f}{\partial y} = 0$, and if also $\left(\dfrac{\partial^2 f}{\partial x \partial y}\right)^2 - \dfrac{\partial^2 f}{\partial x^2} \cdot \dfrac{\partial^2 f}{\partial y^2} < 0$, then $f(a, b)$ will be a maximum if $\dfrac{\partial^2 f}{\partial x^2} < 0$ and $\dfrac{\partial^2 f}{\partial y^2} < 0$, and will be a minimum if $\dfrac{\partial^2 f}{\partial x^2} > 0$ and $\dfrac{\partial^2 f}{\partial y^2} > 0$.

For maxima and minima of functions of three variables, there are similar conditions. (L.L.S.)

MAXWELL-BOLTZMANN LAW. Equipartition of Energy.

MAXWELL COLOR TRIANGLE. Color.

MAXWELL DISTRIBUTION LAW. A law expressing the relative numbers of **molecules** in a gas which have various given speeds, or various given kinetic energies, of thermal agitation at any instant. In its usual forms, it is limited by certain simplifying conditions, viz., uniformity of temperature, absence of turbulence or convection currents, negligible effect of gravity, and purity of the gas (molecules all of equal mass). The

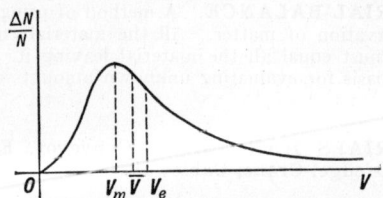

Graph of the Maxwell distribution law.

law may be expressed in various terms. For example, if N is the total number of molecules, the proportion of them having speeds confined to the interval Δv between $v - \frac{1}{2}\Delta v$ and $v + \frac{1}{2}\Delta v$ (and hence having v cm. per sec. as their representative speed), is

$$\frac{\Delta N}{N} = \frac{4h^3(\Delta v)}{\sqrt{\pi}} v^2 \varepsilon^{-h^2 v^2}. \tag{1}$$

h is a constant which may be shown to be equal to

$$h = 6.034 \times 10^7 \sqrt{\frac{m}{T}}, \tag{2}$$

in which T is the absolute temperature and m is the mass of one molecule in grams. From (1) it is easily deduced that the modal (most frequent) speed of the molecules is

$$v_m = \frac{1}{h} ; \qquad (3)$$

the mean speed is

$$\bar{v} = \frac{2}{\sqrt{\pi} h} ; \qquad (4)$$

and the effective speed (corresponding to average kinetic energy) is

$$v_e = \sqrt{\frac{3}{2}} \cdot \frac{1}{h}. \qquad (5)$$

This last quantity, and hence h, can be calculated from the density and the pressure of the gas. (See **Kinetic Theory.**) (L.D.W.)

MAXWELL'S EQUATIONS. A set of four classic formulae of the electromagnetic theory. They deal with certain **vector** quantities pertaining to any point of a region under varying electric and magnetic influence. If the point is in empty space, the equations are somewhat simplified; in general, provision must be made for the presence of dielectrics, conductors, or magnetizable bodies. In these equations, H is magnetic intensity, B is magnetic induction, E is electric intensity, D is electric displacement, ρ is electric space density, u is conduction current density, t is time, c is the electromagnetic **constant.** The "curl" and the "divergence" of a function are well-known operators of **vector analysis.** The equations, in this notation, are

$$\operatorname{curl} H = \frac{1}{c} \frac{\partial D}{\partial t} + \frac{4\pi u}{c}, \qquad (1)$$

$$\operatorname{div} B = 0, \qquad (2)$$

$$\operatorname{curl} E = -\frac{1}{c} \frac{\partial B}{\partial t}, \qquad (3)$$

$$\operatorname{div} D = 4\pi \rho. \qquad (4)$$

(L.D.W.)

MAXWELL'S LAW OF RECIPROCAL DEFLECTIONS. Deflection.

MAY FLY. Ephemerida.

MAYFISH. Pisces, Teleostei. A species of **killifish,** *Fundulus majalis,* found in shallow bays from Massachusetts to Florida. (A.W.L.)

MAYFLOWER. Heath Family.

McLEOD GAUGE. Pressure Gauges.

McQUAID-EHN TEST. A method of determining **grain size** of steels, particularly carburizing grade steels. The sample is **carburized** at 1700° F. by the pack method and cooled slowly. The **microstructure** of the case then contains **pearlite** grains outlined by **cementite** in the grain boundaries, thus the McQuaid-Ehn grain size can readily be determined by comparison with standard charts.

When the pearlite and cementite tend to agglomerate, leaving patches of ferrite, the structure is said to be "abnormal." So-called "abnormal" steels tend to be fine-grained and to have lower **hardenability** and less tendency to distort in the hardening treatment than "normal" steels. The abnormal steels are also tougher after heat treatment. Determination of this characteristic of the microstructure is a part of the McQuaid-Ehn test. (R.H.H.)

MEADOW-BROWN. Insecta, Lepidoptera. **Butterflies** of dull gray-brown color marked with eye-like spots, in some species set in a yellow patch on the fore wings. Family Satyridae. (A.W.L.)

MEADOWLARK. Aves, Passeriformes. A common North American bird (**Aves**) more closely related to the blackbirds and orioles than to the true larks. The eastern meadowlark, *Sturnella magna,* is less attractive than the western, *S. neglecta,* which has a brief but glorious song. Both have the characteristic yellow breast with a black chevron at the throat. (A.W.L.)

MEAGRE. Pisces, Teleostei. Fishes (**Pisces**) of numerous species, widely distrbuted in the oceans and to a limited extent in fresh water. Most are of moderate size but one species of the Old World reaches a length of six feet. Excellent food fishes. (A.W.L.)

MEALY BUG. Insecta, Homoptera. Small active scale **insects.** The body is covered with a granular white secretion.

Mealy bugs are important greenhouse pests. Spraying with whale-oil soap (1 lb. per gal. of warm water) is recommended for resistant plants and with fir-tree oil (1 part to 20 parts water) for ferns and orchids. (A.W.L.)

MEAN. Arithmetic Mean.

MEAN EFFECTIVE PRESSURE. Effective pressure acting against a **piston** face and transmitted into a thrust in the piston rod is at any instant the difference between the pressures on the two sides of the piston. In the **steam engine, gasoline engine, air compressor,** and other piston and cylinder mechanisms, this effective pressure varies throughout the stroke. The average of the effective pressure is known as the mean effective pressure. It may be predicted by means of equations and empirical constants, and also found experimentally by the use of the **indicator.** The mean effective pressure is represented by the average height of the diagram drawn by an indicator. (F.T.M.)

MEAN FREE PATH. Kinetic Theory.

MEAN SEA LEVEL PRESSURE. The value assigned to surface pressure of the Standard Atmosphere: 29.92″ of mercury, or 1013.2 millibars. (P.E.K.)

MEAN SUN. The mean, or average, sun is a purely fictitious object used as a reference point for measuring mean or civil **time** by the rotation of the earth. The term mean sun comes from the fact that the day as determined by use of it, i.e., the mean solar day, is very nearly equivalent in length to the average of the lengths of the different apparent solar days throughout the year.

The apparent motion of the true sun through the stars is produced by the actual motion of the earth in an elliptical **orbit** about the sun, the plane of which is inclined to the plane of the **equator.** To remove the irregularity in the apparent motion of the sun due to the **elliptical motion** of the earth a fictitious object is assumed to move in the plane of the **ecliptic** with constant angular velocity which passes through **perihelion** coincident with the true sun. The mean sun is a fictitious object assumed to be moving in the plane of the equator with constant angular velocity passing through the **vernal equinox** coincident with the fictitious object just defined. (W.K.G.)

MEAN TEMPERATURE DIFFERENCE. Heat transfer surfaces, for which one or both of the media on opposite sides of the surface exhibit variable temperatures, obviously are subject to a variable temperature difference. Since the temperature difference is of primary importance to heat transfer through a conducting surface, its mean, or average, value should be carefully investigated. Consider a case of heat transfer surface, illustrated by a tube surrounded by a medium at t', and through which flows another medium which rises from t_i at inlet to t_o at outlet of the tube. The arithmetical mean temperature difference is the simple average of the terminal temperature differences; arithmetical m.t.d.

$$= t' - \frac{t_i + t_o}{2}.$$

The arithmetical m.t.d. is only an approximation of the true thermal m.t.d. When the heat received by the flow w of the medium in the tube past a unit of the surface dS ($= wcdt$ in which c is specific heat) is equated to heat passed through that area under the temperature difference $t' - t$ existing across $dS (= KdS(t' - t)$ in which K is the conductivity assignable to the existing conditions of heat transfer), the integration for true over-all m.t.d. may be simplified to:

$$\text{Thermal m.t.d.} = \frac{t_o - t_i}{\log_e \dfrac{t' - t_i}{t' - t_o}}.$$

If the first medium changed in temperature from t'_i to t'_o instead of remaining constant, then the

$$\text{Thermal m.t.d.} = \frac{(t'_i - t_o) - (t'_o - t_i)}{\log_e \dfrac{t'_i - t_o}{t'_o - t_i}}.$$

(F.T.M.)

MEAN VALUES OF FUNCTIONS.

The mean value of a function $f(x)$ over an interval (a, b) is defined by

$$\frac{\displaystyle\int_a^b f(x)dx}{b - a}.$$

The mean value of a function $f(x, y)$ over an area S is defined by

$$\frac{\displaystyle\iint_S f(x, y)dS}{S}.$$

The mean value of a function $f(x, y, z)$ over a region V of space is defined by

$$\frac{\displaystyle\iiint_V f(x, y, z)dV}{V}.$$

(L.L.S.)

MEAN VALUE THEOREM FOR DERIVATIVES.

Let $f(x)$ be a function which has a finite derivative at all points of the interval (a, b). Then there exists a value ξ of x between a and b such that

$$f(b) - f(a) = f'(\xi)(b - a).$$

(L.L.S.)

MEAN VALUE THEOREMS FOR INTEGRALS.

The first law of the mean for integrals is:

$$\int_a^b f(x)dx = (b - a)f(\xi),$$

where $a \leqq \xi \leqq b$.

The second law of the mean for integrals may be written:

$$\int_a^b f(x)\phi(x)dx = \phi(a)\int_a^\xi f(x)dx,$$

where $a \leqq \xi \leqq b$, provided $f(x)$ is **continuous,** and $\phi(x)$ is continuous and is also a positive monotonic decreasing function in the interval (a, b), and

$$\int_a^b f(x)\phi(x)dx = \phi(a)\int_a^\xi f(x)dx + \phi(b)\int_\xi^b f(x)dx,$$

where $a \leqq \xi \leqq b$, provided $f(x)$ and $\phi(x)$ are continuous functions, and $\phi(x)$ is a monotonic decreasing function in the interval (a, b) without being always positive.

There are similar formulae for the case when $\phi(x)$ is an increasing function. (L.L.S.)

MEANDER LINE. Traverse.

MEASLES.

A common and highly contagious disease of childhood which tends to recur in epidemics every 2 to 4 years. It is world wide in distribution. The etiologic agent is a filterable **virus.**

The incubation period of measles is 8 to 11 days. The disease is ushered in as an upper respiratory infection. During the period of invasion the signs and symptoms are slight fever, malaise, congestion of the **mucous membranes** of the nose and throat, and **conjunctivitis.** After 36 hours pathognomonic lesions appear in the mouth. These are Koplick spots, small bluish-white specks which are seen inside the lower lip or opposite the molar teeth. The rash begins 3 to 4 days after the onset of fever, and lasts 4 to 5 days. Characteristically it starts at the hair line and spreads down over face, neck, trunk and eventually over the entire body. The lesions start as tiny flat red spots; they enlarge and spread to become confluent in many areas. At the height of the disease the temperature may be as high as 105°, and the patient suffers from itching and burning of the skin. Marked sensitivity of the eyes to light, and cough, are present at this stage; the patient looks and feels miserable. After the eruption reaches its peak it begins to fade, in the order of its appearance, and at the same time the patient improves dramatically. The fever and constitutional symptoms resolve rapidly.

The complications of measles are serious. **Encephalitis,** probably due to the virus itself, is fortunately rare. It has a high mortality and, in those patients who survive, permanent sequellae such as muscular weakness, spasticity and mental deterioration are usual. Secondary infections—**otitis media,** cervical **adenitis,** and bronchopneumonia—are common complications. Bronchopneumonia is the most frequent cause of death in measles.

The treatment of measles is symptomatic once the disease is full blown. If a susceptible child is seen soon after exposure, the disease can be prevented, or at least modified, by the injection of **convalescent serum,** gamma globulin, or placental extract. (D.M.H.)

MEASURE OF CENTRAL TENDENCY.

A measure of central tendency is an **average.** It is a typical value of the set of **variates.** The common measures of central tendency are the **mean, median, mode, harmonic mean** and the **geometric mean.** (L.A.A.)

MEASURE OF POSITION.

A measure of position is a **statistic** or a **population parameter** determined by the position of the **variate** in the **frequency distribution** or in the **probability function.** Typical measures of position are the **median,** the **quartiles, quintiles,** and **percentiles.** (L.A.A.)

MEASUREMENT.

A fundamental process in all production systems, generally involving comparison with some accepted standard or a mating part. The significance of a measurement is determined by the degree of accuracy to which elements may be measured. Two types of accuracy are important: accuracy of form and accuracy of size. The former involves not only the duplication of irregular profiles, but also the accuracy of form embodied in straight-edges, squares, true cylinders and cones, etc. Accuracy of size implies comparison with some accepted standard; in linear measurements, the fundamental unit is the yard, which is the

distance at a temperature of 62° F. between two fine
lines on gold plugs in a bronze bar at Westminster, England, and the meter. In 1866, the United States Congress established the following relation between the
yard and the meter:

$$\frac{1 \text{ yard}}{1 \text{ meter}} = \frac{3600}{3937}.$$

The simple ratio of one inch equal 25.4 mm. has been
approved by the American Standards Association for
United States industrial practice, since it involves only a
negligible error as far as ordinary industrial practice is
concerned.

Methods of measurement are largely controlled by
methods of manufacture, which may be roughly classified as unit production and mass production systems of
manufacture. In the former, articles are produced, usually in limited quantities, by skilled artisans using
standard tools and machines, often from comparatively
simple drawings. In the mass production system, parts
are produced by comparatively unskilled operators, usually in large quantities, using highly specialized tools, machines, and measuring equipment. Such processes involve the transfer of skills employed by the designer,
the tool-maker, the tool-setter, and others to the unskilled operator, making it possible for him to produce
without extensive knowledge of the function or application of the parts manufactured.

Standard measuring equipment includes the scale,
Fig. 1, which is usually graduated in thirty-seconds and

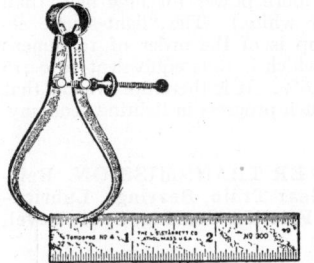

L. S. Starrett Co.   L. S Starrett Co.

Fig. 1.

sixty-fourths of an inch. A capable mechanic can estimate within a third of a division, or to an accuracy of
.006″. Some scales are provided with a hook attached
to one end, to facilitate taking measurements over
rounded corners, or from an edge. Calipers, Fig. 1, are
used for transferring scale measurements. Scribing or
hermaphrodite calipers have one caliper and one pointed
leg, and are used for scribing parallel lines and arcs on
bosses or the ends of bars to locate centers. Dividers
are calipers with two pointed legs for layout work,
scribing circles, or spacing lines.

For more precise work, vernier calipers are used. The
vernier caliper, Fig. 2, consists of a scale with a fixed
head; a movable head may be adjusted along the scale,
and the reading between the jaws of the instrument
read on the scale by means of a reference point and a
vernier. The vernier caliper can be used for measurements accurate to within .001″. The micrometer or
micrometer caliper employs the principle of a screw
moving in a fixed nut; the screw has forty threads per
in., and the thimble of the micrometer is divided into
twenty-five equal parts, permitting direct reading to
.001″. By means of an appended vernier, measurements
to .0001″ may be taken. Inside micrometers and screw
thread micrometers are also available.

In many machine shops, precision gauge blocks are
used for measurement and reference purposes, and consist of small hardened steel blocks of a definite thickness,
held to a tolerance of a few millionths of an inch. Precision gauge blocks may be used singly or in combination

to give an almost unlimited variety of sizes within their
range. Two or more blocks may be assembled by
"wringing" their surfaces together, since these are so

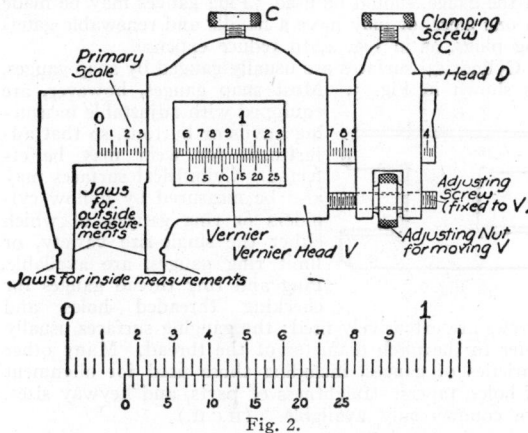

Fig. 2.

nearly perfect planes that they will resist separation
in a direction perpendicular to their contact faces with a
force considerably greater than the atmospheric pressure. Gauge blocks may be used for setting gauges, for
measuring between pins and plugs, and for height measurements, in which instance they are often used in conjunction with surface gauges. A surface gauge consists
of a base fitted with an adjustable arm which carries
either a scriber point of some form of dial test indicator.

For hole measurements, a plug may be fitted to the
hole, and the diameter of the plug measured with calipers or a micrometer. In some instances, a telescoping
gauge, which is a tee-shaped device having a sliding
part in the head of the tee, is used. The sliding leg may
be locked after insertion in the hole, and a measurement
taken by a micrometer over the head. The combination
set, Fig. 3, may be used as a center head, for locating

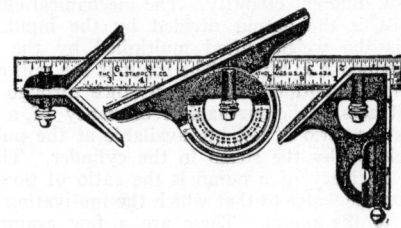

L. S. Starrett Co.

Fig. 3.

the centers of cylinders, as a bevel protractor, or as a
square, in conjunction with the scale on which these
devices are carried.

In the mass production system, single-size or special
gauges are usually employed. A fixed gauge is a
measuring tool that conforms to an established dimension and will show whether a certain part is correct or
not, but will not indicate the degree of error. Fixed
gauges are not necessarily non-adjustable, since many
have measuring surfaces that can be adjusted for wear or
for a limited size range. Limit gauges are those which
indicate the permissible variation in the size of a part;
a progressive limit plug gauge, for gauging the size of a
hole which has a basic size of .188″, with a tolerance of
+.001″, is shown in Fig. 4. Progressive gauges enable

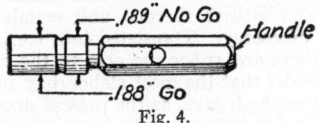

Fig. 4.

the gauging operation to be performed in one operation, but are not applicable to blind holes, for which double-end gauges, with the high and low limits on either end of the gauge, should be used. Plug gauges may be made in one piece, or may have a handle and renewable gauging plugs, as in Fig. 4, to reduce expense.

Cylindrical surfaces are usually gauged by snap gauges, as shown in Fig. 5. Most snap gauges, however, are equipped with adjustable measuring anvils or buttons, so that adjustment for wear may be effected. Cylindrical surfaces may also be measured by hollow cylinders or ring gauges, of which either the single-size variety, or limit ring gauges, are available. Plug and ring thread gauges for checking threaded holes and screws are extensively used; the gauging surfaces usually refer to the pitch diameter of the thread. Many other varieties of gauges, including those used for alignment of holes, tapers, straightness of parts, and keyway sizes, are commercially available. (H.C.H.)

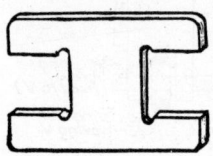
Fig. 5.

MEASURING WORM. Insecta, Lepidoptera. A **caterpillar** which loops the body by drawing the hind legs up close to the front legs as it crawls. The movement is associated with the lack of most of the legs near the middle of the body. Caterpillars of the family Geometridae and a few species of Noctuidae are of this type. (A.W.L.)

MECHANICAL ADVANTAGE. Machines.

MECHANICAL EFFICIENCY. An efficiency, including only the effect of losses arising from mechanical sources, mainly friction, would be mechanical efficiency. A **machine** such as a **hoist** has an efficiency expressible only as the mechanical efficiency. A **steam engine** has not only a mechanical efficiency which measures the friction in windage losses, but a thermal efficiency which is an entirely different quantity. The mechanical efficiency of a hoist is the output divided by the input. The former is the weight lifted multiplied by the height through which it is lifted. The input is the pull exerted on the hoist, multiplied by the distance through which that pull acts. The mechanical efficiency of a steam engine is the ratio of net hp. available at the pulley to hp. developed by the steam in the cylinder. The mechanical efficiency of a pump is the ratio of power expended on the water to that which the motivating source supplies to the pump. These are a few examples of mechanical efficiency, and might be multiplied endlessly since any moving machine incurs friction losses, and consequently has a mechanical efficiency of less than 100%. (F.T.M.)

MECHANICAL EQUIVALENT OF HEAT. It was in 1840 that Dr. James P. Joule began his classic researches on the quantitative relationship between **heat** and **work.** Rumford and Davy had established the fact that heat is a form of energy; it remained to determine how many **foot-pounds** are equivalent to one British **thermal unit.** In the arrangement adopted by Joule, a heavy weight, in descending, was caused to drive a mechanism for "churning" water in a calorimeter. The energy furnished by the weight in its descent, with correction for friction losses, was then equated to the heat indicated by the rise in temperature of the agitated water, with the usual calorimeter corrections; the result indicated that 1 British thermal unit equals about 774 foot-pounds of energy. Translated into c.g.s. units, this meant that there are 41,600,000 ergs to the calorie; and when we consider that the most elaborately precise modern electrical methods give, as the present accepted value of "J," 41,855,000 ergs per calorie or about 778 foot-

pounds per British thermal unit, it is clear that Joule's experimental skill must have been of a high order.

It has been customary, in writing thermodynamic equations involving both thermal and mechanical energy terms, to include the factor J in the former in order to reduce them to the same denomination; but the tendency at present is to express quantities of heat directly in ergs or joules, thereby avoiding the necessity of this factor. (See **Thermodynamics.**) (L.D.W.)

MECHANICAL EQUIVALENT OF LIGHT. Since any radiation is a flux of energy, it should be possible to express the emission of **light** in equivalent power units. Great precision in such measurements is hardly possible, because of the difficulty of separating the visible sharply from the invisible components of radiant energy. One procedure is to enclose a lamp, whose total power input (and hence its total output) is known, in a jacket of transparent material which strongly absorbs throughout both the infra-red and the ultra-violet, so that only the visible radiation comes through. The absorbed energy is deduced calorimetrically from the rise in temperature of the jacket, while the luminous intensity is obtained photometrically in the usual way. The dynamical power corresponding to the light is the difference between the total power output of the lamp and the power absorbed by the jacket. Measurements of this type carried out by Ives in 1926 with lamps giving white light gave an equivalent to 0.0016 watt per lumen. (Since the visibility of light varies greatly with the color, being a maximum near the middle of the visible spectrum, the lumen would represent more power for blue light than for average light or for white.) The "light-source efficiency" of a good lamp is of the order of 10 lumens per watt input power, which is thus equivalent to 0.016 watt per watt or only 1.6%. It is therefore evident that there is still room for much progress in lighting economy. (L.D.W.)

MECHANICAL POWER TRANSMISSION. Belting, Chain, Gearing, Gear Train, Bearings, Lubrication, Speed Reducers, Friction Gearing, Spur, Bevel, Worm Gearing.

MECHANICAL PROPERTIES OF METALS. Metal, Physical Properties of.

MECHANICAL RATING. Speed Reducers.

MECHANICS. Mechanics is that science which deals with the effects of forces upon bodies at rest or in motion. The laws and phenomena of gases and liquids and solid bodies have a part in this subject, and it is one of the basic studies of engineering, physics and astronomy. It is customary to subdivide mechanics into the study of liquids (**hydraulics, hydrodynamics** and **hydrostatics**), the study of the action of gases (**pneumatics**), and the study of rigid or elastic particles or bodies of solid materials. It is to the latter field that the term mechanics is frequently restricted. For convenience it is further subdivided into **statics, kinematics,** and **kinetics,** each of which is treated in this book. Statics deals with bodies at rest, in equilibrium under the action of forces or of torques; kinematics deals with abstract motion and kinetics treats of the effect of forces or of torques upon the motions of material bodies. Modern usage favors the term **dynamics,** reserving mechanics for the more practical phases of the field (machinery, building, etc.).

Fluid mechanics is that branch of mechanics which deals with those fundamental laws which apply to all fluids (liquids or gases) at rest or in motion. (F.T.M., C.W.C.)

MECOPTERA. Insects with four narrow wings with numerous veins. The head is prolonged downward in a beak bearing biting mouth parts at the tip. The relatively few species inhabit moist woods and are not commonly known.

This order includes two chief forms, the **scorpion flies** and a group of slender long-legged insects usually known by their generic name, *Bittacus*. In the former the tip of the abdomen is modified so that it resembles that of a scorpion slightly. *Bittacus* has the general appearance of the crane flies but for its four wings, and is chiefly remarkable for the grasping joints at the tips of the legs. (A.W.L.)

MEDIAL MORAINE. If one or more tributary glaciers coalesce with the main glacier the **lateral moraines** unite to form trains of debris on the surface of the glacier at or near its center, these are called medial moraines. (R.M.F.)

MEDIAN. The median, a **measure of central tendency,** is the value of that **variate** which divides a distribution into 2 equal parts. In a **serial distribution,** after the variates have been arranged in order of magnitude, the median is the value of variates $\frac{N+1}{2}$ if N is odd; if N is even, the median is the **arithmetic mean** of variates $\frac{N}{2}$ and $\frac{N}{2}+1$ where N is the number of variates. The median then is clearly a **measure of position.** The median in a **frequency distribution** is found by determining the value of variate $\frac{\Sigma f_x}{2}$ by linear interpolation in the **cumulative frequency distribution.** The chief virtue of the median is its simplicity and ease of calculation. It always exists and in some instances it is superior to the mean as a measure of central tendency. It is commonly used if the distribution of variates is considerably skewed. The sum of the **absolute value** of the deviations about the median is a minimum. (L.A.A.)

MEDIAN LINE (of an airfoil profile). **Airfoil.**

MEDIASTINUM. The space in the **chest** between the **lungs,** around and above the heart, bounded in front by the anterior wall chest and in back by the posterior chest and spine. The space contains many blood vessels, nerves, and lymph glands. Infections and tumors of the various structures are the chief pathological disturbances. The mediastinum is difficult of access because the greater part of the space is filled with large blood vessels and the heart. (D.M.H.)

MEDINA WORM. Guinea Worm.

MEDIOSILICIC. A term proposed by Clarke, in 1911, to describe **igneous rocks** whose **silica** content is between 52 and 66%. (R.M.F.)

MEDULLA. 1. An axial structure found in some hairs. 2. The central portion of the **adrenal glands.** 3. The medulla oblongata, the posterior region of the vertebrate **brain.** (A.W.L.)

MEDUSA, HYDROMEDUSA. A form of **coelenterate** in which the body is shortened on its principal axis and broadened, sometimes greatly, in contrast with the **hydroid** or **polyp.** It varies from bell-shaped to a thin disk, scarcely convex above and only slightly concave below. The upper or aboral surface is called the exumbrella and the lower surface, on which the mouth opens, the subumbrella. The latter may be partly closed by a membrane extending inward from the margin. This structure is called the velum. The digestive cavity consists of a central chamber, the stomach, and radiating canals which extend toward the margin. These canals may be simple or branching and few or many. The margin of the disk bears tentacles and sensory organs.

In the class Hydrozoa medusae are the sexual individuals of many species, alternating in the life cycle with asexual **polyps,** but in the Scyphozoa or **jellyfishes** proper the medusa alone is well developed. (A.W.L.)

MEDUSOID. The **medusa** of certain **coelenterates** of the class **Hydrozoa.** A hydromedusa. Medusoids

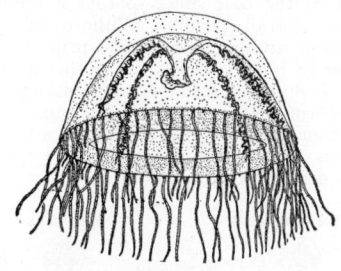

Medusoid.

differ from the free-swimming jellyfishes to which the term medusa is applied in the usual presence of the velum and in the simpler digestive cavity. The term is also applied in some cases to the young medusae budded from the larval **polyp** stage of the **jellyfishes.** (A.W.L.)

MEERKAT. Mammalia, Carnivora. A South African animal of the **civet** family, related to the **mongooses.** The name is sometimes applied to the penciled mongoose. (A.W.L.)

MEERSCHAUM. Sepiolite.

MEGAGAMETE. Gamete.

MEGALONYX. Pleistocene.

MEGANUCLEUS. Macronucleus.

MEGAPODE. Aves, Galliformes. Dull-colored birds **(Aves)** of the Pacific islands, from the Philippines to Australia. They have strong legs and feet and resemble turkeys slightly. The eggs are deposited in mounds of decaying vegetation which generates the heat necessary for incubation. The family includes the **brush turkeys** of the Australian region and the **maleo** in addition to the true megapodes. (A.W.L.)

MEGASCLERES. The large spicules in **sponges.** They form the chief support of the body wall. (A.W.L.)

MEGASCOPIC. Macroscopic.

MEGATHERIA. Pleistocene.

MEIGEN'S REACTION. The use of a solution of **cobalt** nitrate as a stain and reagent for determining the difference between **aragonite** and **calcite.** Aragonite, when boiled in the solution, is tinted lilac, which remains visible in thin section. Calcite and **dolomite** may be colored blue, but the color does not show in thin section. (R.M.F.)

MEIONITE. Wernerite.

MEIOSIS. A process of **cell division** accompanied by the reduction of the chromosomes from the diploid condition, in which two of a kind occur in the same cell, to the haploid, with only one of a kind. In a great majority of animals the **cells** of the body are diploid and the reduction occurs in the process of maturation, which culminates in the formation of the sexual reproductive cells (Gametes).

The germ cells in the gonads of the male (spermatogonia) subdivide repeatedly by mitosis and the resulting cells may undergo meiotic division. The foundation of the process is the same as that of mitosis but the behavior of the chromosomes differs. The similar chromosomes pair in a process of synapsis as the primary spermatocyte develops from the spermatogonium, and each chromosome of the pair splits so that a group of four parts, called a tetrad, results. Without further change two maturation divisions ensue, the primary spermatocyte splitting to form two secondary spermatocytes and each of these dividing into two spermatids which develop into spermatozoa. Each of the four cells contains one of the four members of each tetrad of the original cell, and thus bears representative halves of only one-half of the original chromosomes.

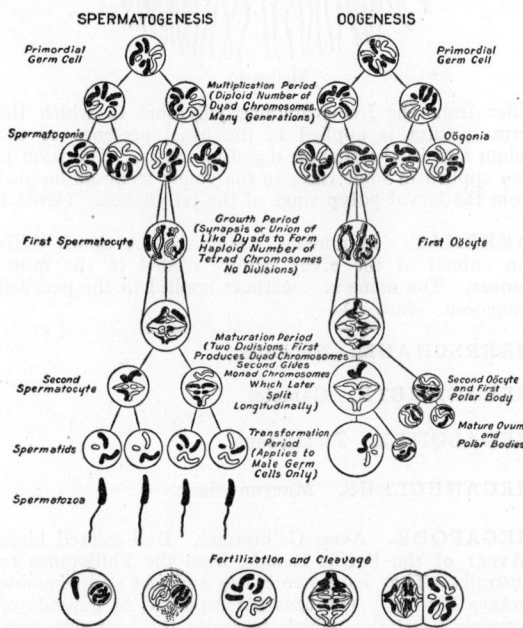

Diagrams of spermatogenesis, oogenesis, and fertilization. Chromosomes derived from female are shown in solid color, those from male in outline. Sex chromosomes are shown with roughened contours. The diploid complex consists of a double series, both as to size and shape, with the exception that the male has only one sex chromosome. Note that it is always derived from his mother. (E. Carothers, 1931.)

In the formation of egg cells the chromosomes behave in the same way, but in the first maturation division the primary oocyte gives rise to a large secondary oocyte containing most of the cytoplasm and a small polocyte or polar body containing almost none. The second maturation division of the secondary oocyte is similar, and in some cases the first polar body also divides equally to form two. The net result is then one ovum and three polar bodies from each oogonium. The polar bodies disintegrate.

The union of the two kinds of gametes in the process of fertilization restores the diploid number of chromosomes.

In plants, meiosis never accompanies the formation of gametes. In Spirogyra and related algae meiosis reduces the diploid chromosome number of the zygospore to the haploid number characteristic of these algae. In the Basidiomycetes it accompanies basidiospore formation. In the Bryophytes, Pteridophytes, and Spermatophytes, it reduces the diploid chromosome number of the spore-producing (sporophyte) generation to the

haploid number of the gamete-producing (gametophyte) generation. Haploid spores grow into haploid plants which produce gametes by equational mitosis. These gametes fuse in pairs (egg with sperm) to produce diploid plants which produce haploid spores by meiosis.

Meiosis is best defined as a series of two rapidly succeeding mitoses in which each chromosome divides but once. This reduces the chromosome number by half because in any other two succeeding mitoses each chromosome would divide twice and all of the original chromosomes would be represented in each of the four resulting cells.

It should be understood that the term diploid (2n) refers to the existence in an individual of two distinct sets of chromosomes one from the mother, one from the father. These two sets are identical in size, shape, and form. They may differ in the genes which they contain. Each chromosome has a mate, a homologous chromosome. Furthermore the chromosome is not *formed* during mitosis; it maintains its individuality throughout the life of the cell. In the nucleus it is in the form of a long thread, a string of genes. In the early stage of the first meiotic division each long thread-like chromosome pairs with its homologue. Then each chromosome contracts and divides lengthwise into two parts, the chromatids, each gene reproducing itself in the process. Each pair of chromosomes now consists of four chromatids forming the tetrad. During the process of contraction the chromatids twine about each other and may break between genes and exchange whole portions. If one chromatid carries the dominant genes ABCDE and the homologous chromatid the recessive genes abcde, such an exchange might produce the two new combinations ABcde and abCDE. This would be known to students of heredity as crossingover. Unequal crossingover may produce the combinations ABCcde and abDE, the C-locus being duplicated in the first chromatid and lost in the second. During the early stages of meiosis other types of chromosomal interchange or gene rearrangement may occur. Meiosis therefore provides, not only for the accurate separation of the duplicate sets of chromosomes in a diploid individual so that each gamete or spore contains one complete set, but also for profound changes in the hereditary constitution of a species. The abnormal effects are enhanced by the use of x-rays applied to organs in which meiosis is taking place. Colchicine, applied to cells during meiosis, prevents the chromatids from separating and produces, eventually, the tetraploid (4n) condition. Tetraploid plants are larger, often more valuable, than diploid ones. (A.W.L., P.A.W.)

MELANCHOLIA. A form of mental illness characterized by great depression, anxiety and fears. *Involutional melancholia* is the common form which comes on at the menopause, or in males in their late 50's. At this period there is a conscious recognition that the peak of life has passed, early dreams and desires cannot be fulfilled, ambition and life's forces are waning. Often a disturbing experience, such as breaking up of the home, loss of position, or the death of one upon whom the individual felt dependent precipitates the disease. Frustration, disappointment and fears of a lonely future activate old conflicts and culminate in profound depression. Remorse, feelings of guilt, delusions, and hallucinations may occur.

About 40% of patients with involutional melancholia recover, although the convalescence may last several years. Shock treatment with metrazol, a powerful central-nervous-system stimulant, insulin, or electric shock which produces convulsions, have been most successful in treating this type of psychosis. (D.M.H.)

MELANISM. A condition of unusually dark pigmentation. Melanism is conspicuously shown in occasional aberrant individuals of certain butterflies whose wings are usually colored yellow or red-brown over most of their area. Melanic (melanistic, melanotic) individuals have the normal black markings extended into the lighter areas or in extreme cases have these areas entirely obscured. The black leopard is a melanic phase of the spotted leopard and the black fox is related in the same way to the red fox. (A.W.L.)

MELANOCRATIC. The term proposed by W. C. Brögger for eruptive igneous rocks containing more than 50% of **femic** minerals. Derived from the Greek, meaning black. (R.M.F.)

MELAPHYRE. An old general term proposed by Brongiart, in 1813, for altered **amygdaloidal andesites** and **basalts**. (R.M.F.)

MELON. *Cucumis Melo.* **Gourd Family.**

MELTING POINT. With the exception of certain glassy or resinous substances devoid of crystal structure and some that are chemically unstable, every pure substance has a more or less well-defined melting point (or freezing point) under any given pressure. By this is meant a temperature at which the solid and the liquid phases of the substance may exist together in equilibrium, without either phase changing into the other. Thus, at atmospheric pressure, a mixture of ice and water kept at 0° C. will not change in relative proportions unless heat is applied or withdrawn; and the temperature of the mixture will remain constant so long as both phases are present. Applying heat will merely melt some of the ice, withdrawing it will freeze some of the water. The melting point is in general affected by pressure; for some substances it is lowered by increased pressure, for some it is raised, according to whether the liquid or the solid phase has the greater density. The melting points of substances in an impure state (mixture or solution) are often profoundly modified by the presence of the impurity; thus solder, a mixture of lead and tin, has a much lower melting point than either pure metal. The **freezing-point law** and the properties of **solutions** give further information on this aspect of the subject.

MELTING POINTS OF SOME ELEMENTS

Aluminum	660° C.	Platinum	1774° C.
Copper	1083° C.	Silver	960° C.
Gold	1063° C.	Sulfur	113° C.
Iron	1535° C.	Tin	232° C.
Lead	327° C.	Zinc	419° C.
Mercury	−39° C.		

(L.D.W.)

MEMBRANELLE. A thin projection adjacent to the mouth of most ciliate **protozoans**, formed of fused **cilia**. (A.W.L.)

MENDELISM. Heredity.

MENEGHINITE. A mineral lead-antimony sulfide, $Pb_{13}Sb_7S_{23}$. Crystallizes in the orthorhombic system. Hardness, 2.2; specific gravity, 6.36; color, gray with metallic luster; opaque. Named after Meneghini (1811–1889) of Pisa. (R.M.F.)

MENHADEN. Pisces, Teleostei. An abundant marine fish (**Pisces**), *Brevoortia tyrannus,* related to the herrings. Massachusetts to Florida. Edible and also used for oil and fertilizer. (See **Esters.**) (Also called mossbunker, bug fish and fat back.) (A.W.L.)

MENINGITIS. Inflammation of the meninges, the membranes covering the brain and spinal cord. Meningitis is most often caused by infection with meningococci, but may be due to streptococci, staphylococci,

pneumococci, influenza bacilli or tubercle bacilli. (See **Bacteria.**) Ocasionally sterile meningitis due to an irritant such as blood in the **cerebrospinal fluid** occurs.

Meningococcus meningitis (cerebrospinal fever, spotted fever) is an epidemic disease, although sporadic cases occur. Outbreaks are common among soldiers in wartime. The causative organism may be carried for weeks or months in the nose and throat without producing symptoms. When the meningococci invade the blood stream, chills, fever, and a characteristic petechial or hemorrhagic rash appear; and when they localize in the meninges, the characteristic signs of meningitis develop: severe headache, irritability, stiff neck, vomiting and sometimes delirium and coma. The organism can usually be found in the spinal fluid, or cultured from it in the acute stage of the disease.

Until the advent of modern chemotherapy, the mortality rate for meningococcus meningitis was high—about 35%. Anti-meningococcus **serum** reduced this figure slightly, but the **sulfonamides** have replaced serum completely. Sulfadiazine is the drug of choice, and if it is given early in an uncomplicated case, the prognosis is excellent. Complications include ear infections (**otitis media**), eighth (auditory) nerve involvement with deafness, purulent **conjunctivitis**, **arthritis**, which is relatively common, and **endocarditis**, which is rare. A rare and extremely virulent form of the disease is accompanied by hemorrhages into the vital organs, particularly the **adrenal** glands, and death invariably ensues within a few hours after onset.

Meningitis caused by organisms other than the meningococcus has a similar clinical picture but a higher mortality, in general. **Sulfonamides** and **penicillin** are used in the treatment of the coccal infections, and antiserum when the agent is the influenza bacillus. There is no effective treatment for tuberculous meningitis; this disease is almost uniformly fatal, although **Streptomycin** appears to have some therapeutic value. (D.M.H.)

MENISCUS. In anatomy, this term denotes a crescent-shaped piece of cartilage in a joint. The medial and lateral meniscus are two flat pieces of cartilage on either side of the knee joint between the femur and the tibia. They serve as cushions between the long surfaces. At times they become injured or torn loose from their attachments. This causes considerable pain and interference with function, with accumulation of fluid in the knee joint. In such a case, operation is performed and the injured meniscus is removed. No interference with function follows such a procedure. (D.M.H.)

MENOPAUSE. That period of life at which permanent cessation of the menses occurs. The average age for the onset of the menopause is 47 years. The menstrual periods may cease abruptly, or they may be irregular over a period of several years, gradually diminishing, and finally stopping. The cause of the menopause is a sudden decrease in ovarian hormone production. Although the period may be completely asymptomatic, various nervous and **circulatory system** manifestations are common. When they are severe, the menopause may be stormy both from the psychological and the physiological point of view. The symptoms include hot flashes, increased sweating, irritability, depression, fatigue, and emotional instability. Transient **hypertension** occurs not infrequently, and **obesity** often appears at the time of the menopause. The symptoms may be controlled by the administration of stilbesterol, a synthetic hormone preparation. (D.M.H.)

MENORRHAGIA. Menstruation.

MENSES. Menstruation.

MENSTRUATION. The monthly discharge of blood from the uterus which begins with puberty and continues until the menopause. It is absent during pregnancy, after removal of the uterus or both ovaries, and after exposure of the ovaries to sufficient radiation from x-ray or radium. It may cease during severe illnesses.

The primitive explanation of menstruation (and this explanation is still believed by many uninformed people) was that of periodic purging of the woman's body of poison. The Greek word for menstruation was the same as that for catharsis.

The menstrual cycle is dependent upon the activity of the anterior pituitary and ovarian hormones. In brief, two hormones from the ovary, estrogen and luteinizing hormone, acting in sequence, are directly responsible for the changes that take place in the uterus during the 28-day cycle. The production of these ovarian hormones is controlled by hormones secreted by the pituitary gland, the follicle-stimulating and luteinizing hormones.

Immediately following menstruation, follicles mature in the ovary under the stimulation of the pituitary follicle-stimulating hormone. Follicles consist of egg cells with their protective and nutrient covering. (See illustration page 629.) During their growth, a hormone is secreted by them into the blood stream. This follicular hormone is called by various names—estrogenic hormone, female sex hormone, theelin, folliculin, estrin, etc. Estrogenic hormone is responsible for growth and regeneration of the lining of the uterus, helping to prepare it for reception of the ovum if pregnancy is to take place.

Ovulation or escape of the ripe ovum from the ovary by rupture of the follicle occurs between the 10th and 17th day of the menstrual cycle (the cycle beginning on the first day of menstrual flow, extending to the beginning of the next period—an average duration of 28 days). Following ovulation a new development begins in the collapsed follicle which then becomes the corpus luteum. During development of the corpus luteum the follicular hormone is still produced, but, in addition, a second hormone is secreted under stimulation of the luteinizing hormone of the pituitary. This is called progesterone (progestin, corpus luteal hormone) and it causes the lining of the uterus to change from a vascular to a secreting type of mucosa, further preparing the uterus for implantation of the fertilized ovum. If the ovum is not fertilized (if pregnancy does not take place) the corpus luteum degenerates and the production of the follicular and luteal hormone ceases. The withdrawal of these hormones is believed to cause the onset of menstrual bleeding and the sloughing away of the lining of the uterus.

The periodicity of the menstrual cycle is characteristic; 80% of normal women have a cycle of 28 days between periods. The cycle may be longer or shorter. Under normal conditions great variation may be seen.

The average duration of flow is from 3 to 5 days and the amount of blood lost varies from 2 to 8 ounces. The menstrual discharge consists of blood, mucus, cellular debris and bacteria. Menstrual blood does not clot. It is normal for lassitude, a feeling of heaviness in the pelvis, mild headache, irritability, depression, frequency of urination, painful swelling of the breasts to accompany the menstrual period.

The average age of onset of menstruation is 13.9 years; it is influenced by physical health, temperament, and environment and to a less extent by race and climate. Menstruation ceases with the menopause or change of life which occurs in the majority of women at 47 years.

Abnormalities of the menstrual cycle include (1) amenorrhea, (2) dysmenorrhea, (3) menorrhagia, and (4) metrorrhagia.

Amenorrhea is the absence of the normal period. It may be a physiological suppression of the menses as before **puberty,** during **pregnancy** and **lactation,** or after the **menopause.** It occurs pathologically in structural abnormalities such as failure of development of the genital tract, endocrine disorders of the ovary, **pituitary** or **thyroid** glands; at times in severe infections; **tuberculosis;** prolonged wasting diseases; **diabetes;** and **psychoses.**

Dysmenorrhea is painful menstruation. It is most often functional, caused by congestion of the pelvic organs. Psychogenic factors may play a significant role. Organic causes are endocrine disturbances, obstruction and infections in the genital tract, particularly gonorrheal infection of the **fallopian** tubes (**salpingitis**).

Menorrhagia is excessive flow of blood during the menstrual period. The common causes are **fibroid tumors, cancer, polyps** and other tumors of the uterus; endocrine disturbances in the pituitary and ovary resulting in overgrowth and abnormal vascularity of the lining mucosa of the uterus during each cycle; infections of the uterus and tubes, particularly gonorrheal salpingitis.

Metrorrhagia is irregular bleeding from the uterus, independent of menstruation. It is due to tumors of the uterus (cancer or a fibroid), pelvic infections, and **abortion,** or ruptured ectopic **pregnancy** (see **Pregnancy**). (D.M.H.)

MENTAL DEFICIENCY. Arrested or incomplete development of the mind due to arrested or incomplete development of the brain. Mental defectives are classed as (1) idiots, those whose mental development never gets beyond that of a normal child of 3 years; (2) imbeciles, who reach the mental age of 7 years; (3) feebleminded individuals, who reach the mental age of 10 years, and (4) morons, whose mental age is 12 years. The morons are often intelligent enough to provide for themselves and lead a simple normal social life.

The causes of mental deficiency are numerous. In *primary* mental deficiency heredity is a prominent factor. Mental deficiency is a complex of hereditable characteristics and does not behave as a simple Mendelian recessive. Yet it has been established that if two mental defectives marry, all their children will be mental defectives; if a mental defective marries a normal, there is a strong likelihood that mental deficiency will appear in an early subsequent generation. *Secondary* mental defect results most commonly from **encephalitis,** either of the epidemic variety, or as a complication of such diseases as measles. Deficiency of the hormone produced by the **thyroid gland** is responsible for cretinism and mental defect. Degenerative processes and mental deterioration early in life are occasionally associated with **epilepsy** and juvenile **paresis.** In children who are blind and deaf and are thus deprived of stimuli which are necessary to normal mental development, a degree of mental deficiency commonly exists.

The diagnosis of mental deficiency rests on the presence of abnormalities of development and behavior, plus the score in certain standard intelligence tests. Failure of normal development may be indicated by the late appearance of bladder and rectum control, and of walking and talking. In the older child, stunted growth, impaired muscle power, extreme backwardness at school, senseless or asocial behavior, inability to fix the attention, all may point to mental defect. However, since these characteristics may be based on environmental factors and personality difficulties as well as on mental deficiency, caution must be exercised in making the diagnosis. In some types of mental deficiency, bodily characteristics are diagnostic. This is true of cretinism (see **thyroid gland**) and Mongolian idiocy. In the latter the child at birth has certain stigmata: a small round head, oblique eyes with narrow lid slits, squat nose, short stubby fingers and hypermobility of the joints. Mongolism occurs as an unaccountable aberration in families of normal mental heredity.

Intelligence tests have been devised and standardized to test the mental capacities of an individual. They do assume a certain amount of education, however, and a completely illiterate person would fail in such a test though he be of normal intelligence. The tests are merely aids and are useful in the hands of experienced psychologists only. They supplement the real criterion of mental deficiency which is social failure; they must be interpreted in relation to reports from the home, or school, or employer, and by observation of the patient himself. The common standard tests are known as the Merrill-Palmer for children under the mental age of six, and the Stanford-Binet for all over six. These consist of simple exercises in logic, reasoning ability, and powers of observation. The Porteus performance tests indicate manipulative skill. (D.M.H.)

MENTHOL. Alcohols and Ethers.

MENTHONE. Aldehydes, Ketones, and Related Compounds.

MENTUM. 1. Labium. 2. The chin. (A.W.L.)

MERCAPTANS. Thioalcohols and Related Compounds.

MERCATOR PROJECTION. The Mercator method of projecting the surface of the earth on a plane was invented by Gerard Mercator, who lived in Flanders during the latter part of the 17th century. At the present time, 90% of the chart work of the deep-sea navigator is done on the Mercator Chart. To realize the general outline of the method, say that we have a terrestrial globe and wish to peel off the surface. One method of procedure would be to make cuts through the surface along **meridians** of **longitude** and remove the sectors of the surface thus obtained. If the material of which the surface was constructed was of sufficient elasticity, these segments could be placed on a plane with a small amount of stretching. The segments would be tangent along the equator, but would separate as the poles were approached, thus forming a discontinuous map. To make the map continuous requires stretching the segments along parallels of **latitude,** the stretching increasing as higher latitudes are reached. This east-west stretching will introduce distortion in the shape of objects, e.g., a circular object would be distorted into an ellipse. To maintain the shape of objects the same amount of stretching must be introduced in the north-south direction in any latitude, as is necessary in the east-west direction in that latitude to make the segments tangent. The final result is that objects retain their true shape, but there is considerable alteration in the relative sizes of surface features in different latitudes.

On the completed Mercator Chart meridians of longitude and parallels of latitude appear as perpendicular straight lines. The meridians are equally spaced, but the distance between successive parallels becomes greater and greater as we proceed away from the equator. This distance becomes so great that Mercator projection is not used beyond 70° latitude. The actual computation of the distances of the successive parallels from the equator, taking into account the ellipticity of the earth, is a complicated mathematical task. The results of the computation have been tabulated in a number of places (e.g., Bowditch—"American Practical Navigator," and similar publications) being generally known as the meridional parts.

The great advantages of the Mercator Chart over all others are that meridians and parallels are perpendicular straight lines, and also **rhumb lines** are straight lines.

Problems in **dead reckoning** may be solved graphically on the Mercator Chart using straight lines to indicate the courses of the vessel. To facilitate such problems and avoid the marking-up of actual charts Mercator plotting sheets are published by the different governments, which show meridians and parallels drawn on the proper relative Mercator scale. (See **Small Area Plotting Sheet.**) (W.K.G.)

MERCATOR SAILING. In case the distance run by a ship does not exceed 300 **nautical miles** the various problems of **dead reckoning** and the **sailings** may be solved without sensible error by methods which assume that the surface of the earth is a plane. The distances run by modern steamships and aircraft during a single day frequently exceed this limit and the true shape of the earth's surface must be taken into consideration. The method of solving the various problems connected with **course** and distance in such cases is usually one which depends for its theory upon the **Mercator projection** and is known as Mercator sailing.

In the accompanying figure we have the representation of the general problem, drawn on a section of a

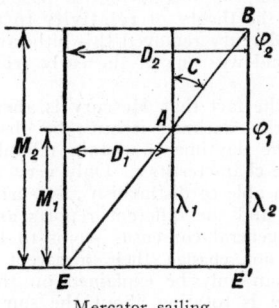

Mercator sailing.

Mercator Chart. The vessel is proceeding from A in **latitude** φ_1 and **longitude** λ_1, to B, in latitude φ_2 and longitude λ_2. The course is extended to cross the **equator,** EE' in the point E. The angle C is approximately the **rhumb-line** course between the two points. M_1 and M_2 represent the stretched distances of the parallels φ_1 and φ_2 from the equator, the values being taken from tables of meridional parts. D_1 and D_2 represent the difference of longitude of the points A and B from the point of equator crossing of the extended course.

Now call $M_2 - M_1 = m$ and $D_2 - D_1 = \delta\lambda$ (the longitude difference between A and B) and we have at once from the figure $\delta\lambda = m \tan C$. The value of C thus computed, taking into account the Mercator stretching, is frequently referred to as the Mercator course between A and B. The distance along AB is a stretched distance and, if calculated from the above figure, or obtained graphically, would be greater than the distance actually traveled between A and B. To avoid this discrepancy the distance is computed by the methods of **plane sailing,** solving the plane triangle using the Mercator course as computed above for the vertex angle and the difference of latitude as one leg. The so-called Mercator distance is obtained as $d = (\varphi_1 - \varphi_2)$ secant C, the value of $(\varphi_2 - \varphi_1)$, the difference of latitude, being expressed in minutes of arc which is practically identical with nautical miles. (W.K.G.)

MERCURY. Mercury is the name of a **planet,** and of a chemical **element.** They will be discussed in that order in this article.

The planet Mercury (see planetary tables, page 1109) is the closest planet to the sun of those thus far discovered. It has been known from remote antiquity and there are recorded observations of it as far back as the 3rd century B.C. Mercury is so close to the sun that at its maximum **elongation** it is less than 30° away, as seen from the earth, with the result that it will never rise more than two hours before the sun, nor be above the horizon in the evening more than two hours after sunset in the latitudes of the United States. The early astronomers failed to recognize the planet as the same object when seen east and west of the sun; it being known as Apollo when seen west of the sun in the early morning, and as Mercury when seen east of the sun in the evening.

As well as being the closest planet to the sun the orbit of Mercury is the most eccentric of all planetary orbits. This high eccentricity, coupled with the proximity of the sun, give to the planet a velocity of more than 36 miles per sec. when at perihelion, a value more than twice as great as that for the earth. After orbits had been computed on the basis of the Newtonian and Keplerian laws and all perturbations applied for the gravitational effects of all known planets, there still remained a progressive motion of the longitude of perihelion. For many years this unexplained perturbation was attributed to the attractions of an unknown planet between Mercury and the sun. The search for this intra-mercurial planet, provisionally called Vulcan, formed a part of many eclipse programs during the past century. The application of the theory of relativity to the orbit calculations for Mercury removed this hitherto unexplained perturbation and was one of the early triumphs of the theory.

Because of the fact that Mercury is always very low in the sky during darkness it must be observed with a telescope in the daytime to obtain reliable results regarding surface characteristics. Only very few observers have ever been able to distinguish any surface markings on the planet and the different reports are quite conflicting. The general consensus seems to be that these markings do not change their apparent positions, a result which can only be explained on the hypothesis that the planet is rotating about the sun in the same period as it revolves, as is the case with the moon relative to the earth.

The physical conditions of the surface of Mercury are quite comparable with those on the moon. The surface gravity and reflecting power of both the moon and Mercury are very comparable and these results have led to the conclusion that Mercury is without atmosphere and hence barren of life as we know it on the earth. Observations of the temperature of the sunlit surface of Mercury give a value of over 600° K. (621° F.), a temperature comparable with that of molten lead.

When inferior conjunction occurs with the sun close to one of the nodes of the apparent path of the planet, the planet will pass between the earth and the sun. This phenomenon, known as a transit of Mercury, occurs about 13 times in each century. Transits of Mercury formerly were used as a method for determination of solar parallax, but this method for determining the value of this important constant has been superseded by other methods. At the time of a transit of Mercury the planet appears as a very small dot visible only with a telescope, and moving slowly across the disk of the sun.

The chemical element mercury: Symbol: Hg (hydrargyrum). Atomic number: 80. Atomic weight: 200.61. Density: 13.546 at 20° C. Melting point: −38.87° C. Boiling point: 356.90° C. Isotopes (page 290).

Mercury or quicksilver is a silver-white liquid metal—the only metal that is liquid at ordinary temperatures (the melting point of gallium is 29.75° C.); forms alloys, called amalgams, with most metals, but not with iron or platinum; does not wet glass but forms a convex surface when in a glass container; is slightly volatile at ordinary temperatures and a health hazard due to its poisonous effect; slowly tarnishes in moist air; upon heating in air or oxygen, somewhat below its boiling temperature of 357° C., forms mercuric oxide slowly, as in the classical experiment by Lavoisier on the composition of air; may be purified by distillation and condensation (health hazard); unattacked by dilute hydrochloric or sulfuric acid, but dissolved by dilute or concentrated nitric acid with the formation of mercurous and mercuric nitrates, respectively, and by hot concentrated sulfuric acid with the formation of both mercurous and mercuric sulfates; unattacked by alkalis. Discovery ancient.

Mercury is used (1) in scientific apparatus on account of its remarkable range of properties, (2) in the preparation of dental and other amalgams, (3) recently, in some industrial boilers instead of steam, (4) in the manufacture of fulminate of mercury explosive, vermilion (mercuric sulfide) pigment, and other mercury compounds, (5) in the amalgamation process for the recovery of gold and silver, (6) in medicine. In the treatment of syphilis, it has been employed for centuries. Mercury bichloride is used as an antiseptic. Mercurial diuretics are effective in treating edema due to heart disease. Mercury occurs infrequently as free metal, but chiefly as the red colored sulfide (cinnabar, HgS) in Spain (the Amaden mine has been continuously worked for over 2500 years and furnishes all of Spain's production), Italy, California, Oregon. The ore is roasted with air and the exit gases passed through a condensing system where the mercury is collected.

Bromides: Mercurous bromide (HgBr), pale yellow precipitate, formed by reaction of mercurous salt solution and potassium bromide solution; sublimes at 345° C.; mercuric bromide (HgBr₂), white, sparingly soluble precipitate, formed by reaction of mercuric salt solution and potassium bromide solution. Melting point of mercuric bromide 237° C.

Chlorides: Mercurous chloride, "calomel" (HgCl), white precipitate, formed by reaction of mercurous salt solution and sodium, potassium or ammonium chloride solutions or hydrochloric acid. Melting point of mercurous chloride 302° C. Turned black by ammonium hydroxide; mercuric chloride, "corrosive sublimate" (HgCl₂), formed by heating mercuric sulfate and sodium chloride whereupon mercuric chloride, melting point 277° C., sublimes. Used in medicine; mercurammonium chloride, "infusible white precipitate" (NH₂HgCl), by adding ammonium hydroxide to mercuric chloride solution; mercurodiammonium chloride, "fusible white precipitate" (HgCl₂·2NH₃), formed by adding mercuric chloride slowly to a hot mixture of ammonium hydroxide and ammonium chloride.

Chromates: Mercurous chromate (Hg₂CrO₄), red precipitate, formed by reaction of mercurous salt solution and potassium chromate solution; mercuric chromate (HgCrO₄), red precipitate, formed by reaction of mercuric salt solution and potassium chromate solution.

Cyanide: Mercuric cyanide (Hg(CN)₂), white solid, soluble—the only soluble cyanide of the heavy metals—slightly ionized and giving no precipitate with sodium hydroxide. Used in the manufacture of cyanogen gas.

Fulminate: Mercuric fulminate (Hg(CNO)₂), dark brown powder, formed by the reaction of mercury, alcohol and concentrated nitric acid. Used to detonate explosives of various types in military, industrial, and sporting practice.

Iodides: Mercurous iodide (HgI), greenish-yellow precipitate, formed by reaction of mercurous salt solution and potassium iodide solution, melting point of mercurous iodide 290° C. with decomposition; mercuric iodide (HgI₂), pale yellow turning to red precipitate, formed by reaction of mercuric salt solution and potassium iodide solution, melting point of mercuric iodide 259° C.; potassium mercuriiodide (K₂HgI₄), yellow crystals, formed by addition of excess of potassium iodide to mercuric iodide and crystallization of the mixture. Used in the presence of sodium hydroxide as an important reagent (Nessler's) for ammonia, with which it forms the "iodide of Millon's base" (HgO(NH₂)HgI), yellow to brown precipitate.

Nitrates: Mercurous nitrate ($HgNO_3 \cdot H_2O$), white crystals, formed by reaction of mercury metal in excess and dilute nitric acid, and crystallization; mercuric nitrate ($Hg(NO_3)_2 \cdot \frac{1}{2}H_2O$), white crystals, formed by reaction of mercury metal with excess of concentrated nitric acid, and crystallization.

Oxides: Mercurous oxide (Hg_2O), black precipitate, formed by reaction of mercurous salt solution and sodium hydroxide solution; mercuric oxide (HgO), (1) yellow precipitate, formed by reaction of mercuric salt solution and sodium hydroxide solution, (2) red solid, by ignition of mercuric nitrate, or by heating mercury somewhat below its boiling point in an enclosed volume of air or oxygen.

Sulfates: Mercurous sulfate (Hg_2SO_4), white precipitate, formed by reaction of mercurous salt solution and sodium sulfate solution; mercuric sulfate ($HgSO_4$), white solid, soluble, formed by reaction of mercury metal and excess hot concentrated sulfuric acid; basic mercuric sulfate "turpeth mineral" ($HgSO_4 \cdot 2HgO$), lemon-yellow solid, formed by reaction of mercuric sulfate with water, but soluble in sulfuric acid.

Sulfides: Mercuric sulfide (HgS), (1) black precipitate, as final result of reaction of mercurous or mercuric salt solution with hydrogen sulfide. With mercurous, free sulfur is also formed and with mercuric, the color change is first yellow, then red, brown, black successively. Insoluble in dilute nitric acid, but soluble in aqua regia, slightly soluble in ammonium sulfide, but soluble in sodium sulfide and hydroxide mixture. Sublimes at 446° C., (2) red solid ("vermilion," "artificial cinnabar"), by sublimation of black mercuric sulfide, or by grinding mercury metal and sulfur under slight pressure. Sublimes at 580° C. Used as a pigment in paints, rubber, and plastics.

Thiocyanates: Mercurous thiocyanate (HgCNS), grayish-white precipitate, by reaction of mercuric salt solution and potassium thiocyanate solution; mercuric thiocyanate ($Hg(CNS)_2$), white precipitate, by reaction of mercuric nitrate or sulfate solution and potassium thiocyanate solution. An explosive, which upon burning, produces "Pharaoh's serpents" ash.

Numerous organic compounds of mercury have been prepared. "Mercurochrome" is the disodium salt of 2,7-dibromo-4-hydroxymercurifluorescein, used in medicine.

All mercury-containing substances, when dry and mixed with dry sodium carbonate and heated in a glass tube, yield a metallic mercury mirror in the colder part of the tube.

All solutions of mercury salts deposit mercury metal on a strip of copper placed in the solution. (W.K.G., R.K.S., D.S.H.)

MERCURY ARC. The electric discharge through mercury vapor, between electrodes either of mercury or of some solid metal, is among the richest sources of ultraviolet radiation and has long been used as such. In the more common forms now in use at least one electrode is of mercury, deposited in a suitable reservoir at the end of a quartz tube. As these tubes are operated on moderate voltage, it is necessary to start or "strike" the arc by temporarily running a small stream of mercury through the tube from one electrode to the other. This makes a mercury conductor which quickly grows hot and fills the tube with mercury vapor, after which the mercury stream is broken and the arc is self-sustaining. The temperature is not nearly so high as in solid-electrode arcs, and these lamps are quite efficient. If the mercury vapor becomes too dense, the conductivity falls off; therefore in some forms, as the Cooper-Hewitt lamp, arrangements are provided for condensing the vapor to a fixed density. The bulb must be made of quartz for ultra-violet because glass is highly opaque

to that radiation. In some forms, a fluorite window is inserted, instead, in a glass bulb. Mercury arcs with glass tubes have proved useful to some extent for illumination and especially in photography, the light being a highly actinic blue-green.

The mercury-arc tube is used primarily as a power rectifier but occasionally has grids inserted and is used for power control purposes. Since the voltage drop across the tube is small and is constant the tube losses may be made relatively small by using fairly high circuit voltages. At a few hundred volts it is the most efficient converter for a-c to d-c transformation. In the larger power applications the tubes are operated on polyphase circuits, going even as high as 24 or more phases, with a common cathode and multiple anodes. This gives a d-c output which has very little ripple and hence needs little if any filtering. (L.D.W., L.R.Q.)

MERCURY-ARC TUBE. Mercury Arc.

MERCURY VAPOR CYCLE. Although water vapor has been the standard working medium of the commercial power plant employing an external combustion cycle, it was recognized long ago that water vapor had physical properties not altogether desirable at either the high temperature or low temperature end of the cycle, but at no time has any vapor been discovered which would be more satisfactory at both extremes of the expansion range than steam. The use of two vapors in series makes it possible to keep an extensive temperature range and eliminate several of the disadvantages attending the use of a single vapor. The mercury-steam is the only binary vapor cycle operated on a commercial scale at present. This cycle was pioneered by Emmet. After experimenting with small-size units, the problems peculiar to mercury vapor were solved, and a few large mercury vapor installations have recently been built.

The advantages of mercury as a binary cycle vapor can be briefly stated.

1. At high temperatures its vapor pressure is moderate.

2. The liquid is of sufficient density to be returned to the boiler by gravity, thus eliminating a boiler feed pump.

3. Its high vapor density results in moderate spouting velocities and simple turbines can be used. Exhaust passages can be small, and the size of the heat-exchanging mercury condenser is not excessive.

4. No tube scaling or other difficulties attending poor feed water are possible. Feed treatment is eliminated as mercury is a stable liquid. (F.T.M.)

MERGANSER. Duck.

MERIDIAN. On the celestial sphere a meridian is a great circle passing through the poles of rotation and, hence, perpendicular to the celestial equator. The local meridian is the great circle passing through both the poles of rotation and also through the zenith. It is both a vertical circle and also an hour circle. However, the local meridian differs from the ordinary circles on the sphere in that it apparently remains fixed as the celestial sphere apparently rotates once each day. An object on the local meridian is said to be in culmination and if it is on that section of the meridian between the horizon and above the pole it is said to be in upper culmination, if below the horizon, or below the pole and still above the horizon, it is said to be in lower culmination. The spherical coordinates both of hour angle and astronomical azimuth are measured from the local meridian in the direction of apparent rotation of the celestial sphere.

A terrestrial meridian is a curve cut on the surface of the earth by a plane containing the earth's axis of rotation. The local meridian of a point on the surface of the earth is the meridian passing through that point. The plane of the local terrestrial meridian coincides with the plane of the local celestial meridian. Terrestrial **longitude** is measured from the local meridian of Greenwich, England.

See **Bearing** for the meaning of meridian as used in **plane surveying**. (W.K.G.)

MERIDIAN CIRCLE. A meridian circle is a telescope adjusted so that the **collimation plane** of the instrument is in the plane of the local meridian, and the telescope may be rotated about a horizontal axis. The instrument is usually fitted with a circle, accurately graduated in degrees, minutes, and seconds, which is perpendicular to the axis of rotation and hence is in the plane of the meridian. In case the instrument does not carry the circle in the meridian, the instrument is known as a transit circle.

At the principal **focus** of the telescope is placed a **reticle** with an odd number of vertical wires, the middle one of which is in the collimation plane. One horizontal wire through the optic axis of the telescope is usually present.

The instrument is used to determine the **equatorial coordinates** of the stars, when the local sidereal **time** and terrestrial **latitude** are known; or, conversely, to determine accurate local sidereal time by observation of stars of known right ascension. The local sidereal time of the instant of passage of a star across the middle wire of the reticle must be the **right ascension** of the star. The declination is obtained from the readings of the graduated circles when the star passes through the field of view along the horizontal wire. The instrument is also used for the accurate determination of terrestrial **longitude** by determining the local sidereal time and knowing the corresponding Greenwich time. (W.K.G.)

MERISTEMATIC. Parenchyma.

MERLE. Aves, Passeriformes. The common **blackbird**, *Turdus merula*, of southern Europe. A French name. (A.W.L.)

MERLIN. Falcon.

MEROGONY. Development of an egg deprived of its nucleus, or of a cytoplasmic fragment of an egg, after **fertilization** by a normal sperm. The process is biologically interesting as proof that the nucleus of the male germ cell is capable of bringing about embryonic development although the cell itself is unable to develop because its cytoplasm is chiefly in the form of specialized structures. (A.W.L.)

MEROPODITE. Biramous Appendage.

MEROSOMA. The broad anterior part of the **abdomen** of scorpions. Preabdomen. (A.W.L.)

MESA. A flat-topped, steep-sided, table-like mountain capped with a formation or stratum which is relatively horizontal and resistant to **erosion**. When such a topographic feature is less than 1 sq. mi. in area it is usually called a butte. (R.M.F.)

MESCAL. Cactus.

MESENCHYME. A diffuse tissue formed chiefly from the middle germ **layer** (mesoderm). It appears in the embryo as a mass of scattered angular or stellate cells with long processes.

The mesenchyme of vertebrates forms the **connective tissues**, **bone**, **cartilage**, and other special tissues. (A.W.L.)

MESENTERY. A thin **tissue** which attaches the intestine to the wall of the body and in some cases holds it suspended in the body cavity. It is formed in animals with a **coelom** by the splitting of the mesoderm (**germ layer**) around the alimentary tract to form this cavity. Separate splits occur on the two sides of the body. In some parts they meet in the median line and become confluent, and elsewhere a thin mesodermal partition persists between them, continuous with the lining of the cavity and with a mesodermal covering of any organ around which the advance of the cavity has taken place. This median partition in connection with the intestine is the mesentery.

The mesentery of vertebrates persists only above the alimentary tract as a dorsal mesentery save in very limited regions. The liver grows into the ventral mesentery, which persists as the gastro-hepatic ligament between liver and stomach and as the falciform ligament between liver and body wall. The regions of the dorsal mesentery are named for the regions of the tract with which they are connected. That of the stomach is the mesogastrium, that of the duodenum the mesoduodenum, and that of the large intestine the mesocolon. In mammals the dorsal mesogastrium forms a saccular extension called the great omentum.

In the sea anemones and related **coelenterates** the tubular stomodaeum is held in place by radiating septa from the body wall formed of a central **mesogloea** covered with endodermal tissue. These structures are also called mesenteries. Similar thin septa extending inward from the body wall but not reaching the stomodaeum bear the same name. Those which connect with the stomodaeum are primary mesenteries, the next longest are secondary mesenteries or metacnemes, and still shorter septa are tertiary mesenteries. (A.W.L.)

MESH CONNECTION. This is a polyphase circuit connection in which the phase elements are connected end to end as opposed to the star connection where all phase elements have one end in common. The **delta connection** is the three-phase mesh connection and is by far the most common. (L.R.Q.)

MESOCARP. Fruit.

MESODERM. Germ Layers.

MESOGLOEA. A layer of material between the ectoderm and endoderm in **diploblastic** animals, including **sponges**, **coelenterates**, and possibly **ctenophores**. The layer does not form like the other **germ layers** as a compact mass of cells but is of unorganized material containing specialized cells derived from the formed layers. In the sponges these cells form the spicules and the reproductive cells and in the coelenterates they include many nerve cells.

The jellyfishes have a thick jelly-like mesogloea which is responsible for their common name, and in the ctenophores this layer is similar but contains all of the muscular tissue of the body. Some observers regard it as a true mesoderm in the latter group. It is, in any case, at the border between the diploblastic and triploblastic forms. (A.W.L.)

MESOHIPPUS. Fossil Mammals; Oligocene.

MESON. Nuclear Particles.

MESOPHYLL. Leaf.

MESOPHYTES. Plants which are adapted for growth in soil with a plentiful supply of both air and water, as distinguished from **Hydrophytes** adapted to an excess and **Xerophytes** adapted to a deficiency of water. All common garden plants and the large majority of wild plants are mesophytes. Certain mesophytic trees such as the oak may adapt themselves during growth to root submersion by overgrowth of the base of the trunk. (P.A.W.)

MESOTHORIUM. Symbol: MsTh₁, and MsTh₂. Two radioactive elements of the thorium series. (See **Radioactive Changes**.)

MESOTRON. A **nuclear particle** which appears to originate as a **secondary emission** from the incidence of **cosmic rays** on matter. It carries a single elementary charge, either positive or negative, and has a mass estimated at from 100 to 200 times that of the **electron**. The term "mesotron" indicates that this mass is intermediate between that of the electron and any other known mass, such as the **proton**. Some writers prefer the shorter term "meson," and several other synonyms have been used.

The mesotron was first recognized by Dr. Carl D. Anderson in 1936, by means of **cloud chamber** tracks whose properties could not be attributed to any particle then known. The extraordinarily high ionization and penetrating power, together with the small curvature of the tracks in a magnetic field, indicated a mass much greater than that of the electron or the **positron**. Mesotrons are of such rare occurrence that insufficient data are yet available to yield a precise determination of the mass, but its magnitude is unquestionably of the order indicated above.

The numerical equality of the mesotronic charge to that of the electron makes it doubtful that the mesotron is an independent entity. Indeed there is indirect evidence that mesotrons are radioactive, which means that they have a complex structure. This evidence is based largely on the fact that mesotrons have a habit of disappearing whenever they traverse any considerable distance, and the number so disappearing seems to be proportional to the distance, and not to the density of the matter as in ordinary absorption. The inference is that a certain percentage suffer radioactive disintegration, with products not so far recognized as such. The very short half-life (decay coefficient) deduced from the observations would preclude any possibility of mesotrons having come from interstellar space; hence they must be regarded as secondaries to the original cosmic radiation. (L.D.W.)

MESOZOIC. A major subdivision of the geologic time-scale. The **era** of "Middle" life, or the age of reptiles. Subdivided from the base up, into the following periods: **Triassic, Jurassic, Comanchean, Cretaceous.** The Era was characterized by: the rise of **dinosaurs** (Triassic); rise of birds and flying reptiles (Jurassic); rise of flowering plants (Comanchean); great development of **ammonites** which became extinct at the end of the Cretaceous; culmination and extinction of most reptiles, and rise of the **archaic mammals** between the Cretaceous and the **Tertiary.** The Mesozoic era began 200 million years ago and lasted 140 million years. (R.M.F.)

METABASIPODITE. A subdivision of the **basipodite** of the **biramous appendages** of some crustaceans, also called the preischiopodite. (A.W.L.)

METABENTONITE. Injurated or lithified **bentonite** in which the "clay" mineral is beidellite. Important horizon markers, of volcanic ash, in the **Ordovician** stratigraphy of the Appalachian Mountains and the upper Mississippi Valley. (R.M.F.)

METABOLISM. The interchange of materials between living organisms and the environment, by which the body is built up and energy for its vital processes is secured.

Within the body of the individual both constructive and destructive processes take place. The incorporation of materials is known as anabolism and the breaking down of these materials for the release of the energy contained in them is katabolism.

Among different kinds of living things three principal types of metabolism are recognized, that carried on by green plants, that of plants which have no chlorophyll, including bacteria, yeasts, molds, etc., and that of animals.

The metabolism of green plants, based on the process of photosynthesis, is also carried on by some 1-celled creatures classed with the **protozoans**. They are not strictly plants or animals but are intermediate forms with some characteristics of each. Briefly this process consists of the utilization of water, carbon dioxide, and mineral salts as raw materials for the formation of complex organic compounds rich in energy. The green substance, chlorophyll, is the active agent in the process and the sun is the source of energy. **Carbohydrates** are formed by the union of carbon dioxide and water, with the release of oxygen. By the reorganization of these compounds and the addition of nitrogen, sulfur, and other elements from inorganic salts the **fats** and **proteins** are made.

Such plants as the bacteria and yeasts secure energy by chemical transformations of many kinds. Some act on carbohydrates and produce carbon dioxide and alcohol. Others oxidize ammonia and produce nitrites, and still others form nitrates from atmospheric nitrogen. The net result of their activities is that all compounds discarded by either green plants or animals, or left in their bodies at death, are transformed into substances which can be used again by one or another form of organism. Thus the three types of metabolism are related in a cycle. Any one alone would gradually transform the available foods into unavailable wastes, and would automatically come to an end. Together they carry on an endless circulation of materials.

Animals are utterly dependent on the complex compounds as foods and so must rely directly or indirectly on the green plants. They break the compounds down partially in digestion and resynthesize similar compounds from the resulting products. Both plants and animals utilize compounds of the three groups as sources of energy, releasing the energy by oxidation in the process of respiration. The resulting simplified wastes are then disposed of by excretion.

From the chemical changes included in metabolism, life is maintained, and heat, mechanical energy and electric currents are produced within the body. Metabolism may be measured as heat produced. To produce the changes of metabolism, oxidation of food must take place. Oxidation in the body is a slow process, but the same amount of heat is produced in this way as occurs when the food products are burned outside of the body.

The unit of measure used for metabolic heat production is the kilogram calorie. This unit represents the heat necessary to raise 1 kilogram of pure water 1° C.; 1 gram of carbohydrate yields 4.1 kilogram calories; 1 gram of fat yields 9.3 kilogram calories; 1 gram of protein yields 4.1 kilogram calories; 1 gram alcohol yields 7 kilogram calories.

Heat production, and therefore metabolism, may be measured directly by the **calorimeter** (by means of a compartment so constructed that all heat given off by the body may be measured); and indirectly by the respiration calorimeter (by measuring the oxygen absorbed by the lungs and the carbon dioxide given off). In medicine the respiration calorimeter is used exclusively for measuring the basal metabolism. The heat produced by a fasting individual at rest 12–15 hours after the last meal is called basal metabolism. Thus basal metabolism forms a standard for comparison of metabolism under varying conditions of health and disease. Particularly does it indicate the activity of the thyroid gland. (A.W.L., R.S.M.)

METACENTER. Buoyancy; Hydrostatics.

METACENTER HEIGHT. Buoyancy.

METACHROSIS. The change of colors in animals. Changes occur relatively slowly in fishes, fairly rapidly in some reptiles, and as a rapid play of different shades in the octopus and related mollusks. (A.W.L.)

METACNEME. A secondary **mesentery** of sea anemones and related **coelenterates.** (A.W.L.)

METACRYST or **METACRYSTAL.** A term proposed by A. C. Lane, in 1902, for large pseudo-porphyritic crystals (**porphyroblasts**) which occur in the **metamorphic rocks.** (R.M.F.)

METAGENESIS. Alternation of Generations.

METAL SPRAYING. A method of coating metal or other surfaces by spraying with molten metal from a portable spray gun. The coating metal or alloy is generally fed into the gun as wire or in powdered form. It is melted in an oxyacetylene or oxyhydrogen flame and blown out in finely divided form by an air blast. The spray consists of semi-molten particles which impinge on the base material and generally form an adherent coating of somewhat lower density than normally cast metal. Metals having low melting temperatures such as zinc, aluminum, and tin are the easiest to spray. However, copper, brass, nickel, iron, stainless steel, and many other metals can be applied. Zinc-sprayed coatings are used to protect steel from corrosion. Aluminum coatings on steel provide heat resistance for mufflers and exhaust pipes, combustion chambers, etc. In addition to the application of protective coatings, metal spraying is widely used in building up worn bearing surfaces of shafts, cylinders, rolls, and other machine parts. (R.H.H.)

METALLIC TONING PROCESS. Many subtractive processes of color photography form their colorant images by the conversion of the black silver images into metallic compounds having suitable hues. Reds are obtained by conversion with uranium or copper, magenta with nickel salts, blues and cyans with iron salts and yellows with cadmium or lead salts.

These metallic-toning procedures have been widely practiced in the production of 2-color motion pictures and to a lesser extent in making 3-color prints on paper. Such processes are relatively simple and yield reproductions of good photographic quality although the colors obtained are not so brilliant as those obtained with some other methods of color printing. (H.C.C.)

METALLOGENY. The genetic study of ore deposits (R.M.F.)

METALLOGRAPHY. Study of the structure and properties of **metals and alloys,** principally by microscopic and x-ray diffraction methods. The term is also used in a broader sense to include the processing of metals by mechanical and heat treatments and the fabrication and testing of finished products. In this usage it is synonymous with **physical metallurgy.**

Metallographic samples are prepared for microscopic examination by grinding and polishing a flat surface to a mirror finish, finishing with a very fine abrasive such as rouge or levigated alumina. Intermediate abrasives may be a series of successively finer emery papers but the final abrasive is used on a wet polishing cloth mounted on a horizontal rotating disk. The small sample

may be held by hand or in an automatic polishing device. It is possible to polish most metals so that no scratches are visible when observed at a magnification of 1000 times. Electropolishing can also be used to obtain a smooth scratch-free surface.

Unless there are non-metallic impurities present, such as **inclusions,** or other particles whose appearance and mechanical properties are markedly different from those of the base metal, such as **graphite** flakes in cast iron, the polished metallographic specimen does not yield any useful information. Etching of the polished surface, usually in dilute acids, develops the grain structure of the metal by preferential attack of the grain boundaries or of one or more of the constituents present in alloys having duplex structures. The metallurgical microscope is equipped with a camera so that the microstructure may be photographed. (See **Dendrite, Cast Iron,** and **Steel** for typical examples of photomicrographs.)

The practical limit of magnification for any microscope using visible light is about 2000 times. Ultra-violet microscopes have been used up to about 10,000 times. The new **electron microscope** is capable of much higher magnification. In examining opaque objects such as metals with the electron microscope it is necessary to make a replica of the surface with a transparent material. The thin film of variable thickness which is stripped from the prepared metal surface is examined by transmission methods in the electron microscope. The resolving power of this instrument far surpasses that of light microscopes and many important metallographic applications are being found. (R.H.H.)

METALLOID. An element which behaves chemically as both a metal and a non-metal, such as arsenic, antimony, or tellurium. In steel metallurgy, however, the term metalloid is applied to the elements carbon, manganese, silicon, phosphorus, and sulfur when present in small amounts in iron and steel. (R.H.H.)

METALLURGICAL CHEMISTRY. See each metal.

METAL-MARK. Insecta, Lepidoptera. **Butterflies** of small or moderate size, in many species marked with metallic spots and dashes. They constitute the family Rhiodinidae (Erycinidae). Relatively few species occur in temperate climates but in the tropics they are numerous and varied. (A.W.L.)

METALS AND ALLOYS. The pure metals are described under their own titles, and various alloy groups such as **Brass and Bronze, Aluminum Alloys, Magnesium Alloys, Nickel Alloys, Lead Alloys, Steel, Cast Iron,** and **Stainless Steel** are also discussed.

Process metallurgy is the science of extraction of metals from their ores. The study of metals and alloys from the standpoint of their production or alloying from metallic base materials, their mechanical and heat treatment, fabrication, and testing are known as physical metallurgy. **Metallography** is generally restricted to the study of the structure of metals and alloys by means of the microscope and x-ray. There is some conflict, however, in the use of these terms, especially physical metallurgy and metallography.

Most metals are soluble in one another in the liquid state and alloying procedures usually involve melting; however, alloying by treatment in the solid state without melting is accomplished by the methods of **powder metallurgy.** When molten alloys solidify they may remain soluble in one another or may separate into intimate mechanical mixtures of the pure constituent metals. More often there is partial solubility in the solid state and the structure consists of a mixture of the saturated **solid solutions.** Another important type of solid phase is the **intermetallic compound** which is characterized by hardness and brittleness and usually has only limited solid solubility with the other phases present. Many possibilities exist and these can be portrayed by diagrams showing the relationships between the solid and liquid phases at various temperatures under equilibrium conditions.

It is a well-known fact that many important alloy combinations have properties which could not be predicted on the basis of the properties of the constituent metals. For example, copper and nickel, both having good electrical conductivity, form solid-solution type alloys having very low conductivity, or high resistivity, making them useful as electrical resistance wires. In some cases very small amounts of an alloying element produce remarkable changes in properties, as in **steel** containing less than 1% carbon with the balance principally iron. Steels and the **age-hardening** alloys depend on **heat treatment** to develop special properties such as great strength and **hardness**. Other properties which can be developed to a much higher degree in alloys than in pure metals include **corrosion-resistance**, oxidation-resistance at elevated temperatures, abrasion- or wear-resistance, good **bearing** characteristics, **creep** strength at elevated temperatures, and **impact toughness**. (R.H.H.)

METALS, PHYSICAL PROPERTIES OF. Most substances that are chemically classified as metals have certain characteristic and almost unique physical properties. Among these are: high electrical and thermal conductivity, attributed to free electrons; great opacity and high reflectivity for light, due to the same cause, and responsible for the "luster" commonly associated with metals; malleability—a sort of plasticity by virtue of which a metal may be cold-worked and rolled into thin sheets; ductility—a combination of malleability and toughness which permits a metal to be drawn into wire. Metals in their normal, pure state are crystalline. They exhibit a characteristic effect upon **polarized light** reflected from their surfaces. A highly useful property is their ability to mix with each other to form an almost endless variety of alloys, having properties often subject to wide control. For example, an alloy of 60% copper with 40% nickel, called constantan, developed for use in resistance coils, has almost invariable electrical resistivity under varying temperatures. Some substances such as selenium, not chemically metals, exhibit one or more of these metallic physical properties. Mercury is the only metal with a melting point below room temperature.

In engineering and metallurgical usage a distinction is made between physical and mechanical properties. **Hardness, tensile strength, impact toughness,** and many other properties vary with the mechanical and heat treatments which a metal has undergone. These are known as mechanical properties. On the other hand, the **modulus of elasticity,** coefficient of thermal **expansion, specific heat,** etc. are essentially constant for a given composition and are designated as **physical properties.** (L.D.W., R.H.H.)

METAMERE. A division of the animal body occurring as one of a series along the principal axis. Segments of this type are well developed in the earthworms. As a general rule each segment of such a body contains similar internal organs but a certain amount of specialization of the segments is evident both internally and externally in different parts of the body. **Arthropods** and **vertebrates** are also metameric but in these animals the specialization of segments is more advanced and has given rise to body regions. The arthropods have a head, thorax, and abdomen, with metameres clearly evident only in the abdomen in many species. **Vertebrates** have a head, trunk, and tail, and as a rule show metameric segmentation only internally. (A.W.L.)

METAMORPHIC ROCKS. Metamorphism of rock material, whatever its original nature and origin, may produce such profound changes that the resulting mass has distinctive characters which warrant a new classification. The chief agents of metamorphism are pressure and heat, and circulating liquids and gases, and usually a long time interval is required for their operation. Metamorphic changes affect mineral composition, texture, and structure of rocks. Minerals that compose most rocks are definite chemical combinations, which commonly are stable only under definite conditions. If these minerals are subjected to radically new conditions, they will tend to change slowly into new chemical combinations, stable under the new conditions. For example, **bituminous coal** that has been subjected to long-continued pressure and increased temperature in a belt of deformation becomes changed into **anthracite** coal through loss of volatile constituents and concentration of fixed carbon.

Metamorphic rocks have acquired distinctive physical as well as chemical characteristics. Commonly they have taken on a definite banding or foliation, partly by mechanical elongation and flattening of mineral particles already present, partly through growth of new mineral grains with an orientation imposed by stresses. A metamorphic rock that has coarse and rather crude banding is called a **gneiss** (pronounced nīce); one that has close foliation, and splits readily because of parallel arrangement of flat or elongate mineral grains, is known as a **schist. Slate** may be regarded as an incipient schist in which the individual mineral particles are too small to be distinguishable. Two common types of metamorphic rock commonly have no distinctive banding, although they retain in some degree their original primary sedimentary stratification. One of these, **quartzite,** results from recrystallization and thorough cementation of the grains in quartzose sandstone; the other, **marble,** is formed by recrystallization in limestone or dolomite. (See also **Rocks.**) (C.R.L.)

METAMORPHISM. Derived from the Greek, meaning change of form. Used by geologists to designate rocks derived from pre-existing rocks by mineralogical textural and structural changes within the original mass. All metamorphic rocks are either **crystalline** or **cryptocrystalline,** the crystallization or recrystallization usually producing alignment and segregation of the minerals into bands. This type of banding is called **foliation.** In the case of the fine-grained metamorphic rocks, such as **slate,** the foliation results in a high degree of fissility or cleavage. **Schists** usually display both banding and cleavage. The processes of metamorphism may be roughly classified as follows: 1. Contact metamorphism. The result of igneous intrusions into sedimentary, **igneous** or **metamorphic rocks,** producing at the contact bands or **aureoles** of metamorphism. This type of metamorphism causes alteration of the intruded rock by the heat and solutions of the **magma.** Sometimes also the margins of the intrusive body are chemically affected by impregnations of solutions from the rock which is intruded. A number of varieties of valuable **ore** deposits occur in zones of contact metamorphism.

Regional metamorphism is from a quantitative point of view, a type of greater importance, and originates well below the surface of the earth (zone of metamorphism) very slowly, often the result of differential pressures; but high temperature, hydrostatic pressure, and the effects of solutions may produce certain types of regionally metamorphosed rocks. Sedimentary rocks may be metamorphosed with or without apparent crystallization.

Hydrothermal metamorphism is the result of hot, aqueous solutions circulating through fractures, and usually causing the alteration of the adjacent rocks. (R.M.F.)

METAMORPHOSIS. Development of the individual after birth or hatching, involving marked change in form as well as growth and differentiation.

Metamorphosis usually accompanies change of habitat or of habits. In some forms, however, it is merely development through a succession of forms which probably represent ancestral stages in the evolution of the species. The first type is illustrated by many **insects** and by **amphibians.** Immature **dragon flies** are aquatic although the adults are flying insects, and **frogs** undergo a transition from the aquatic tadpole to the air-breathing, if not entirely terrestrial, adult. Change of habits is illustrated by the transformation of free-swimming young of many aquatic invertebrates into **sessile** adults, and by the development of adult **butterflies** and **moths** with sucking mouths from caterpillars which eat solid food. The **crustaceans** afford many examples of transformation through several immature forms without conspicuous change of habits or of habitat.

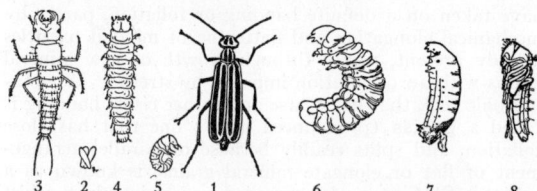

3   2   4   5   1   6   7   8

Hypermetamorphosis of a blister beetle. Number shows stage in development. (*After Riley and Chittenden.*)

The immature stages are wholly or partly designated by the term **larva.** In the complex metamorphosis of **insects,** however, only the first stage is called the larva and even it sometimes bears a different name. The distinction depends on the nature of the metamorphosis.

Some insects hatch from the egg with the general form of the adult and the attainment of the adult stage is marked chiefly by the completion of the wings. This type of metamorphosis is said to be gradual and in its early stages the insect is called a **nymph.** The orders that develop in this way are grouped as the **Paurometabola.** A few orders are aquatic in early life and are then called naiads. They transform directly into the terrestrial adult and are known as the **Hemimetabola,** or insects with incomplete metamorphosis. The **Holometabola,** with complete metamorphosis, pass through a larval stage, then enter an inactive stage known as the **pupa,** and finally become the conspicuously different adult. A few beetles undergo hypermetamorphosis with a sequence of different larval forms preceding pupation. (A.W.L.)

METANILIC ACID ($NH_2C_6H_4SO_3H$). This is a **dye** intermediate and the meta isomer of **sulfanilic acid.** (R.K.S.)

METAPHOSPHATE. Phosphoric Acid.

METAPLASM. Inactive or lifeless matter included in living **protoplasm.** (A.W.L.)

METAPODIUM. The posterior part of the foot of mollusks. (A.W.L.)

METASOMA. The slender posterior part of the abdomen of **scorpions.** Postabdomen. It bears the sting at its tip. (A.W.L.)

METASOMATISM. This term was proposed by Naumann to designate that type of mineral alteration whereby one mineral is dissolved and removed by the solution which introduces and deposits a new mineral in its place. Metasomatism is therefore a process of replacement with or without the formation of **pseudomorphs,** and is frequently a very important process in the formation and enrichment of ore deposits. (R.M.F.)

METASTABLE STATE. A peculiar state of pseudo-equilibrium, in which the system has acquired energy beyond that for its most stable state, yet has not been rendered unstable.

By using great care, water at 760 mm. pressure may be heated several degrees above its normal boiling point, say to 105° C., yet not boil. In this condition it has received heat energy beyond that normally required for liquid-vapor equilibrium, energy which it might be expected to release by spontaneously exploding into steam; and only a slight disturbance will precipitate that change, but the disturbance must come from some external source. This situation may properly be called a metastable, that is, beyond the stable, state.

The term is most commonly applied in physics to that state of an atom which is described as "excited" (as by the application of an electric field), but which for some reason does not release its excitation energy as radiation unless disturbed by some additional influence. The energy of **resonance radiation** is spontaneously released by an atom returning from some "resonance state" to a more stable, lower-energy level; but an atom in a metastable state must be prodded into action, so to speak, by some such disturbance as a **collision.** The electron voltage required to transform an atom in its most stable condition to a metastable state has been called the "transformation potential" corresponding to that state. (L.D.W.)

METASTASIS. A secondary implantation of **tumor** cells at a distant site from the primary tumor. In **cancer,** metastasis may occur by the transportation of detached cells through blood or lymph channels. The term is used also to describe **abscesses** which arise as a result of widespread distribution of **bacteria,** usually through the blood, from a primary focus of infection. (D.M.H.)

METASTOMA. A structure behind the **mandibles** in some **crustaceans.** Formed of the fused **paragnatha.** (A.W.L.)

METASTOMIUM. The posterior part of the head of segmented **worms.** (A.W.L.)

METATHERIA. Mammalia. Synonymous with Didelphia.

METATYPE. A **topotype** which has been approved by the author of the species. (R.M.F.)

METAZOA. All animals except the 1-celled species and the sponges. Characterized by multicellular structure and by the organization of the body in compact tissues and organs coordinated by a nervous system. In most metazoans the food is digested in a special cavity, although some may be taken into cells as in the sponges and protozoans. (A.W.L.)

METEOR SHOWER. This term is applied to indicate a number of **meteors** coming from the same general part of the sky known as a **radiant point.** (W.K.G.)

METEORIC WATER. Ground Water.

METEOROLOGICAL INSTRUMENTS. Some atmospheric properties can be measured but others must be judged or estimated. Properties, like cloud types, total sky cover, visibility, generally are estimated by a competent observer. Those atmospheric properties that are measured include: water-vapor content, temperature, pressure, wind direction and velocity at the surface and aloft, cloud-base height and movement, rain and snow fall, evaporation, upper level pressure, temperature, and humidity, duration of sunshine and amount of heat received.

1. Water-vapor measuring devices.

a. A psychrometer is a combination of wet- and dry-bulb thermometers. A conventional sling psychrometer consists of two thermometers mounted parallel, with the wet bulb slightly lower than the dry. This instrument is whirled at approximately 3 meters per sec. for 3 or more minutes until the wet-bulb reading reaches a steady reading. The difference between dry- and wet-bulb reading is known as the wet-bulb depression and is correlated to other humidity criteria in tabular or graph form for instant use. Ventilation is required for most psychrometers.

b. Hair hygrometers are instruments in which strands of human hair mounted under tension expand with increasing relative humidity and contract with decreasing relative humidity. One end of a set of strands is fixed and the other end operates a set of levers whose initial movement is magnified mechanically to cause a pointer or other indicating device to ride over a calibrated scale. Hair hygrometers, after they are calibrated, indicate relative humidity which, in conjunction with tables of specific humidity and temperature, can be converted to other humidity criteria. Hygrographs are hygrometers in which a pen records relative humidity continuously on a rotating drum. The drum is operated by a clock mechanism.

c. Dew-point hygrometers are instruments which measure the temperature at the time of formation and evaporation of dew. A mean of the temperature at time of formation and evaporation is approximately the actual dew-point temperature. Some instruments require visual observation of dew drops on the instrument and the disappearance of the same drops, but others use photoelectric cells to determine the time of dew formation and evaporation. Dew-point hygrometers must contain both heating and cooling mechanisms, a polished surface upon which the dew forms, and a means of measuring the temperature of the surface or the temperature of the air in contact with the surface.

d. Absorption hygrometers, diffusion hygrometers, and optical hygrometers are other instruments in limited use. Some absorption instruments are constructed to alter the passage of electrical current in a thin film of the absorbing material in relation to the relative humidity of the air surrounding the film. Other absorption instruments require the passing of a limited weight of air through a drying agent which is weighed before and after the passage of air.

2. Thermometers.

a. Ordinary mercury or alcohol thermometers are used for surface temperature measurements, including wet-bulb temperature measurements.

b. Maximum thermometers are so constructed that the column of mercury will break as soon as the temperature begins to fall. A small segment of the mercury column, therefore, remains in place until the instrument is read, at which time it can be shaken down.

c. Minimum thermometers have within the bore a small index which moves downward toward the bulb as the column of fluid contracts but remains in its nearest-to-the-bulb position as the fluid again expands. Thus the index can be read and then readjusted to the top of the fluid column. Alcohol is the usual fluid used.

d. Bimetal thermometers are so constructed that they have a bending coefficient in relation to the temperature. They are made of two metal strips with different coefficients of thermal expansion welded together, usually Invar and brass or Invar and steel. One end of the strips is fixed and the other end operates a series of magnifying levers which in turn operates an indicating device. Thermographs are usually bimetal instruments in which a pen records temperature continuously on a rotating drum.

e. The Bourdon thermometer is a flat hollow tube filled with an organic liquid. As the temperature of the liquid increases, its expansion causes the tube to straighten. It is not used to any great extent in meteorology.

f. Electrical thermometers depend on the thermal coefficients, or on thermoelectrical properties, of dissimilar metals. Any metal that has a more or less linear thermal coefficient of resistance over the necessary temperature range can be used to measure temperature by its resistance to the passage of electricity. A **Wheatstone bridge** circuit is usually employed, one branch of which is the thermometer coil. Thermocouples are the other type of electrical thermometers, potentiometers being used to measure the output of the thermocouple. Electrical thermometers have the advantage of the practicability of remote installations with the recording dials placed where easily accessible.

3. Pressure indicators.

a. Mercury barometers are the standard instruments for measuring atmospheric pressure. Generally speaking, they are simply a tube erected vertically with the bottom end submerged in a pool of mercury. The top of the tube is sealed and evacuated of all air and other gases except mercury vapor. Mercury rises in the sealed tube above the level in the pool in direct relation to the air pressure on the mercury pool. Many refinements must be built into a mercury barometer before it is sufficiently accurate for precise pressure measurements.

b. An aneroid barometer consists of one or more disk-like cells inside which a partial vacuum is maintained. When one side of the cell is fixed, the other side will move and operate a system of levers which in turn cause an indicator to move over a scale. If atmospheric pressure falls, the disk expands; if atmospheric pressure increases, the disk contracts. Aneroid barometers must be calibrated against a mercury barometer and checked frequently. A barograph is an aneroid barometer which operates a pen over a rotating drum and thus records pressure continuously.

4. Wind-measuring devices.

a. Surface wind direction is measured by use of a wind vane which is nothing more than an arrow with considerable tail surface. Air flowing past the tail surface keeps the arrow pointed in the direction from which the wind is coming. The vane is usually mounted some 20–30' above the ground or water level.

b. Wind velocity is measured generally by use of a cup anemometer, although there are other available instruments. A cup anemometer consists of a set of three or four semispherical or conical cups mounted on a wheel whose axle is vertical. When the wind blows it forces the cups and the wheel to rotate; the wind velocity is related to the total number of rotations over a short period of time. Various types of counters are used to determine the rotations per min. or per sec.

Plate anemometers are the oldest in terms of use, having been used since about 1660. They consist of a simple plate mounted in such a manner that the wind blows against it, deflecting it in the direction toward which the wind is blowing. A pointer or scale mounted along the plate measures the deflection of the plate which can be correlated to the wind velocity.

Pressure-tube anemometers employ the difference in static and dynamic pressure in a wind blowing across a tube mounted vertically, and into a tube mounted with its opening directly into the wind. This type of anemometer is not in great use in meteorology but is used almost exclusively in aircraft as an air speed indicator.

c. Winds above the surface of the earth are measured by use of pilot balloons which have a sensibly constant ascension rate and drift with the wind as they rise. A theodolite is used to follow the balloon as it drifts. Azimuth and altitude angles observed each minute permit the path of the balloon to be plotted. Wind direction and velocity at any altitude are determined from the plotted track of the balloon. At night the balloon may be fitted with a light.

5. Cloud observations.

a. Cloud types and amount of sky cover are estimated by an observer.

b. Cloud heights are often measured by use of ceiling balloons and ceiling lights. A small balloon with a relatively constant ascension rate is released, and the time from release until it penetrates the cloud base is a measure of the heights of the cloud base. At night a spotlight mounted vertically forms a spot on the cloud base, and a clinometer is used to measure the altitude angle of the spot.

c. Direction and relative speed of cloud movement are measured with a **nephoscope** which is a grid or a mirror with reference points enabling an observer to note details of cloud motion. Neither type is in extensive use.

6. Rain- and snow-fall measuring devices.

a. Rain gauges measure, by collection in a cylindrical tube, the total fall of rain during one or more rain storms, or over a period of time.

b. Snow fall is conveniently measured with a ruler and the total fall correlated to rain by a 10 to 1 reduction. More accurate measurement is achieved by lowering a cylinder vertically into the snow and melting the depth of snow collected into its rain equivalent.

7. Evaporation is generally measured by exposure of a water surface to the atmosphere. Loss of water from a pan or vat is a measure of evaporation. Other specialized devices are in use for particular problems such as transpiration from plants.

8. Sunshine measurement.

a. Duration of sunshine is measured by making use of the sun's rays to burn a record on suitable sensitized or photographic paper.

b. Pyrheliometers measure the intensity of solar radiation. They absorb radiation with a consequent rise in temperature in the absorbing medium of the instrument which in turn is converted into calories.

9. Upper-air temperature, humidity, and pressure measurement.

a. A meteorograph is a clock-operated instrument in which pens trace on a rotating drum a record of temperature, relative humidity, and pressure. The instrument can be carried aloft by the use of a balloon, airplane, or kite. A bimetal element operates the temperature pen; human hair strands operate the humidity pen; and an aneroid cell operates the pressure pen. Frequent calibration is necessary.

b. Radio-meteorographs utilize small radio transmitters to send modulated signals which have equivalents in temperature, humidity, and pressure. A balloon carries the measuring devices and the transmitter aloft to great altitudes. Four different methods of sending the signals are used, (1) time-spaced signals, (2) audio-frequency change, (3) radio-frequency change, and (4) codes. Measurements of temperature, relative humidity, and pressure are introduced into the transmitter which transmits the measurements in terms of a radio signal. An electrical thermometer is generally used. The humidity element is usually a special human-hair instrument or an absorption instrument which utilizes the electrical resistance of lithium chloride in a thin film on a tube. Pressure elements are generally aneroid cells. (P.E.K.)

METEOROLOGY. Science of the atmosphere including all its aspects. In a more restricted sense in wider use it is the science of weather, being particularly concerned with the physics of the elements which make the weather. (P.E.K.)

METEORS (OR METEORITES). Meteors, or "shooting stars," are visitors from space which come into the atmosphere of the earth with high velocities. They are made visible by being raised to incandescence by resistance offered by the atmosphere of the earth.

The term meteor comes from the fact that since the late 18th century they have been associated with the atmosphere of the earth. The terms **fireballs, bolides,** and **meteorites** or aerolites are also applied to meteors. The difference between the various objects depends merely upon their observed characteristics and there is considerable evidence in support of the hypothesis that they are all made up of the same fundamental materials. The term meteor is sometimes restricted to that class which does not attain a brightness greater than zero stellar **magnitude** (i.e., is not brighter than the **planet Jupiter**). Objects which attain a brightness greater than this, and which leave behind them trails which may persist for several minutes, are known as fireballs. Frequently, the direction of flight of a fireball is observed to change suddenly, and not infrequently a distinct sound is heard either during or shortly after the passage of a fireball. A fireball which is observed to explode into several fragments is known as a bolide, and frequently a loud detonation is heard following the explosion. Meteors, fireballs, and bolides, in the true sense of the terms, are so completely disintegrated during their passage through the earth's atmosphere that they fall to the surface of the earth as dust. Any one of these classes of objects which is large enough to penetrate the atmosphere and arrive at the surface of the earth as a solid body is known as a meteorite, or aerolite.

Meteors are observed most frequently by pure chance, but at present there are a number of systematic photographic surveys in progress. The most effective method for determining accurate data regarding the heights, velocities, and other characteristics of a meteor is to have two cameras situated about 20 miles apart, with the distance between them accurately known. If the same meteor is photographed by both cameras, it will be found to be projected against different backgrounds of stars on the two plates. From the difference in position relative to the stellar background, the angle subtended by the meteor at the length of the distance between the cameras may be determined, the triangle solved, and the height of the meteor obtained. If one of the cameras is equipped with a rotating shutter, which will produce breaks in the meteor trail at accurately timed intervals, the lengths of the segments of the trail may be measured and the velocity at which the object is moving be determined after the height has been found.

The results of the recent and accurate surveys are just beginning to appear, but from these results thus far obtained it appears that more than half the meteors enter the earth's atmosphere with velocities less than 42 kilometers per sec. (26 miles per sec.), while the remainder have velocities greater than that. Furthermore, those with velocities less than 42 kilometers per sec. are found to be members of meteoric showers emanating from **radiant points,** and hence are actually members of the **solar system.** It may be shown theoretically that any true member of the solar system when situated at a distance from the sun equal to that of the earth, cannot have a velocity greater than 42 kilometers per sec. This seems to indicate that half of the observed meteors are actually visitors to the solar system from interstellar space. How many such objects there may be in interstellar space we have no means for determining at present. A survey which has been in progress by

Whipple at Harvard fails to show any meteor with velocity greater than 42 kilometers per sec.

Another result of the recent surveys indicates that the average height of the middle of a meteor trail is about 89 kilometers (55 miles). The fact that this is the same as the height of one of the layers of the earth's atmosphere which reflects radio waves has led to interesting speculation as to the constitution of this region of the atmosphere.

The determination of the actual number of meteors which enter the atmosphere of the earth in a given period of time is a difficult task. The most recent surveys indicate that there may be as many as a billion each day. The size of the average meteor may be determined from studies of the brightness of the object and is found to average about the order of magnitude as the size of a drop of water. A hundred million objects of this size, and of the density of the average meteorite, would have a total mass of about 100 tons. Such an increase in the mass of the earth would slightly increase the rotation period and would increase the radius by about ¼″ in the 3000 million years that the earth has had its present form.

The composition of meteors may be determined only from meteorites or from fragments of meteoric dust which have been collected. A more complete discussion of the actual composition of meteorites will be found elsewhere. There has never been anything found in meteorites which has not previously been found on the earth, and the average percentages of the different chemical elements found in meteorites is comparable with the distribution found elsewhere in the solar system. The wholly metallic meteorites are chiefly an iron-nickel alloy and are called siderites. **Sideriolites** are meteorites composed of both metallic and silicate materials. Meteorites made up almost entirely of silicates are called **aerolites.**

Nothing is known regarding the origin of those meteors which enter the earth's atmosphere with velocities greater than 42 kilometers per sec., and hence come from outside of the solar system. Of the slower objects we know that some of the radiant points are following orbits similar or identical with orbits of comets. The cases are too frequent and the agreement too close to be assigned to mere chance and hence we know that many meteors are definitely connected with comets. As soon as the question as to how many of the slow-moving meteors are members of meteoric showers is answered, we shall be able to make more definite statements regarding the origin of these terrestrial visitors. Such objects, however, must be actual members of the solar system.

Since by far the greater percentage of meteors fall to the surface of the earth as fine dust, there is no danger to the earth or its inhabitants from these objects. The meteorites, however, are a potential source of danger. The air wave and earthquake which would follow the impact on the surface of the earth of a mass of, perhaps, several hundred tons, moving with a velocity of many miles per sec., are almost beyond imagination. The results of such impacts are found in the huge meteor crater in Arizona and in the more recent fall in northern Siberia. Fortunately, large meteors are very rare and they seem to have the happy faculty of selecting uninhabited portions of the earth for their descent. The fall of such an object on a city would completely destroy it. (w.k.g.)

METEPIPODITE. A broad thin lobe on the outer side of the appendage and at its upper end in crustaceans which have the flattened form of **biramous appendage** called a **phyllopodium.** An epipodite. Also called the branchia. (a.w.l.)

METER. Standard Meter.

METERS. Meter and instrument have frequently been used interchangeably, but now there is a tendency to restrict the use of *meter* to instruments which record and/or integrate as well as indicate. These devices in various forms are used extensively in research and instruction, also in commercial installations such as factories and power plants.

Instruments are used in industry for a number of reasons. For efficient operation, the personnel must know, not guess at, the conditions of pressure, temperature, flow, etc., of the process. Indicating instruments are satisfactory for operating guidance but recording or integrating meters serve better the calculation of performance, the allocation of cost, and supervision.

Unless a record is needed for operating supervision or for plant calculations, the initial expense and maintenance of recording instruments plus the fact that index instruments are more easily read precludes the use of recording instruments as operating guides alone. However, many applications will be found for which recording instruments will be selected. There are two types of recording instruments: those using circular charts and those using strip charts. The circular charts must be replaced each day; the strip charts last several days. Circular charts are: (1) less expensive than strip charts, (2) more easily planimetered, (3) rugged and easily filed, (4) of a form to expose a full day's record at all times, (5) accurately held in position by the centering point. Considering points in favor of the strip chart: (1) they are more suitable than circular charts when many records are to be centralized or when multiple records of draft, temperature, etc., are to be put on one chart; (2) being electrically operated, strip chart meters are, as a rule, more accurate than circular chart meters; (3) the chart speed may be changed. (f.t.m.)

METES AND BOUNDS. A method of describing, in surveying terms, tracts of land for the purpose of establishing legal titles is known as metes and bounds. In early days the boundaries of a piece of land were described in a haphazard manner, but as property became more valuable it was necessary to define these boundaries in terms of definite quantities. Legal titles called deeds now commonly contain the length and **bearing** (direction) of the individual sides, a description of the **corners,** the names of the adjoining property owners, and the calculated area. Thus, if one of the original corners is known, the boundaries of a tract of land may be relocated from this description by metes and bounds. (c.w.c.)

METHANE. Methane, "marsh gas," "fire damp" (CH_4), is a colorless, odorless gas, boiling point $-161°$ C., density 0.72 gram per liter at 0° C., 760 mm. (specific gravity 0.55, air equal to 1.00), practically insoluble in water, moderately soluble in alcohol, burns when ignited in air with a pale, faintly luminous flame, forms an explosive mixture with air, when between 5% and 13% methane, with excess air the products are **carbon dioxide** plus water; with deficiency of air, **carbon monoxide** plus water, when mixed with air and passed over **magnesium** oxide at 350° C. and 500° C. **formaldehyde** is found among the products. Methane is among the chemically less reactive organic substances. It reacts, however, with **chlorine** (and similarly with **bromine**) to form mixtures of methyl chloride (CH_3Cl), methylene chloride (CH_2Cl_2), chloroform ($CHCl_3$), **carbon tetrachloride** (CCl_4), depending upon the conditions (one-half of the chlorine used forms **hydrogen chloride**), or may form free carbon with explosive violence, as is the case when one volume of methane plus one volume of chlorine is exposed to direct sunlight. Methane occurs in natural gas, of which it is the main constituent; forms in coal mines and swamps, where vegetable matter is decaying under water (anaerobic fermentation); and is a constituent of coal gas. The fuel value of methane (995 B.T.U. per cu. ft.) is high (hydrogen and carbon monoxide about 320 units each). Carbon monoxide and hydrogen react to form methane in the presence of nickel catalyst. Methane is also formed by reaction of magnesium methyl iodide in anhydrous ether (**Grignard's reagent**) with substances containing hydroxyl (—OH) group. Methyl iodide (bromide, chloride) is preferably made by reaction of methyl alcohol and **phosphorus** iodide (bromide, chloride).

Important methane derivatives, by replacement of each hydrogen atom, step by step, by hydroxyl are (1) **methyl alcohol** (CH_3OH or $H \cdot CH_2OH$), (2) **formaldehyde** ($CH_2(OH)_2$ by loss of water to CH_2O or $H \cdot CHO$), (3) **formic acid** ($CH(OH)_3$ by loss of water to $CHO(OH)$ or $H \cdot COOH$), (4) **carbonic acid** ($C(OH)_4$ by loss of water to $CO(OH)_2$ or $HO \cdot COOH$ or H_2CO_3). Methyl alcohol is, therefore, the simplest **alcohol**, formaldehyde the simplest **aldehyde**, and formic acid the simplest **carboxylic acid**. (R.K.S.)

METHYL ALCOHOL. Methyl alcohol, methanol, "wood alcohol" (CH_3OH or $H \cdot CH_2OH$), is a colorless liquid of pleasant odor, boiling point 64.5° C., miscible in all proportions with water, alcohol, or ether, when ignited burns in air with a pale blue transparent flame, producing water plus **carbon dioxide**, the vapor forms an explosive mixture with air. Methyl alcohol reacts (1) with **sodium** metal, forming sodium methylate, sodium methoxide (CH_3ONa) plus **hydrogen** gas, (2) with **phosphorus** chloride, bromide, iodide, forming methyl chloride, bromide, iodide, respectively, (3) with **sulfuric acid** concentrated, forming dimethyl **ether** ((CH_3)$_2O$), (4) with organic acids, warmed in the presence of **sulfuric acid**, forming **esters**, e.g., methyl acetate (CH_3COOCH_3), methyl salicylate ($HO(2)C_6H_4COO$ CH_3), possessing characteristic odors (see the various individual acids), (5) with magnesium methyl iodide in anhydrous ether (**Grignard's solution**), forming **methane** as in the case of primary alcohols, (6) with **calcium** chloride, forming a solid addition compound ($4CH_3OH \cdot CaCl_2$), which is decomposed by water, (7) with **oxygen**, in the presence of heated smooth copper or silver forming **formaldehyde**. The density of pure methyl alcohol is 0.792 at 20° C. compared with water at 4° C. (the corresponding figure for ethyl alcohol is 0.789), and the percentage of methyl alcohol present in a methyl alcohol-water solution may be determined from the density of the sample. Anhydrous methyl alcohol may be obtained by fractional **distillation** (anhydrous ethyl alcohol must be obtained by special treatment). Methyl alcohol is obtained (1) by the destructive distillation of hardwoods at about 350° C. along with **acetic acid** and small percentages of **acetone** in the water condensate. After neutralization of the acetic acid, methyl alcohol is recovered by distillation, usually accompanied by acetone and **acetaldehyde**, and then purified by further distillation, (2) by reaction of **carbon monoxide** plus **hydrogen** in the presence of a **catalyzer**, such as zinc oxide or basic zinc chromite, at 400° C. and 150 to 250 atmospheres pressure, (3) by regulated reduction of formaldehyde or formic acid.

A common test for methyl alcohol is by its **oxidation** in air with a hot copper wire to form **formaldehyde**.

Methyl alcohol is used (1) as the source of the methyl group (CH_3—) in organic chemistry, using methyl alcohol or methyl chloride, bromide, or iodide, and of the methoxy group (CH_3O—), using methyl alcohol or sodium methoxide, (2) in the manufacture of formaldehyde, (3) as a solvent for many organic substances, especially in the preparation of lacquers, varnishes, (4) as a denaturant in the form of crude wood spirit for ethyl alcohol. (R.K.S.)

METOL. Developing Agents.

METRIC SYSTEM. The metric system had its origin in the need, imposed by the development of scientific thought, for immutable, and at the same time conveniently related, units of physical measure. Prior to the end of the 18th century, uniformity and consistency in this matter were notably absent. The French Revolution revolutionized many things besides government. Even the system of number notation was brought into question, but fortunately was not disturbed. One outgrowth of the upheaval was the metric system. In 1791 a committee of the French Academy, including the well known Lagrange and Laplace, made a report to the National Assembly which, after much delay, was finally put into effect. Their report fixed the standard of length as the **meter,** defined as one ten-millionth of the earth's meridian quadrant at sea level, and determined from elaborate geodetic surveys between Barcelona, Spain, and Dunkirk, France. The unit of mass then became a secondary standard, viz., the **gram,** based on the centimeter and the density of water; but this was soon displaced by the present standard **kilogram,** which departs rather seriously from the original ideal. All subdivisions and multiples in the system are decimal, with terminology in accord with the following scheme:

deci- = one-tenth (decimeter, decigram)
centi- = one-hundredth (centimeter, centigram)
milli- = one-thousandth (millimeter, milligram)
deka- = ten times (dekameter, dekagram)
hecto- = one-hundred times (hectometer, hectogram)
kilo- = one-thousand times (kilometer, kilogram)

While the metric system has come into general use throughout continental Europe and most of the civilized world, it has yet to gain general recognition in Great Britain and the United States, for other than scientific purposes. (See **C.G.S. System; M.K.S. System.**) (L.D.W.)

MEXICAN ONYX. Travertine.

MIASKITE. An old term proposed by Rose, in 1839, for an **oligoclase-nephelite-syenite** in which the chief **mafic** mineral is **biotite.** The type locality for this **igneous** rock is at Miask, in the Ural Mountains. (R.M.F.)

MICA. The mica group of minerals includes several closely related species, having a highly perfect basal **cleavage;** all are **monoclinic** with a tendency toward pseudo-**hexagonal** crystals, and are closely similar in chemical composition. The highly perfect cleavage, the most prominent characteristic of the group, is explained on a basis of **x-ray** studies of these minerals, which seems to show a sheet-like arrangement of the atomic structure and a hexagonal grouping of **atoms** which apparently explains the pseudo-hexagonal crystals above mentioned. The word mica is believed to have been derived from the Latin *micare,* meaning to shine, in reference to the brilliant appearance of this mineral, especially when in small scales.

The following members of the mica group are treated under their respective headings: **Biotite, Muscovite, Lepidolite,** and **Phlogopite.** (E.S.C.S.)

MICA-SCHIST. Schist.

MICHELSON-MORLEY EXPERIMENT. In 1881 two American physicists, Michelson and Morley, first carried out an experiment which was destined not only to raise doubts regarding the behavior of the supposed **ether,** but ultimately to revolutionize a large part of our thinking about physical phenomena. The details of the experiment are somewhat complicated, involving an adaptation of the **interferometer,** but its essential principle may be set forth as follows:

A source of light S and a mirror M, facing it, are mounted on a rigid base at a distance l apart (see figure).

Source of light S and mirror M at opposite ends of a rigid bar which carries them with it when it moves.

Light travels from S to M and back in a certain time, expressed by $t_1 = 2l/c$, where c is the speed of light. Suppose the whole apparatus is set moving lengthwise with a speed v. If this is actually the relative motion of apparatus and ether, so that the ether is "blowing past" the source and mirror with the speed v, then a very simple calculation shows that the time now required for the journey $SM + MS$ should be greater, namely, $t_2 = 2l/c \left(1 - \dfrac{v^2}{c^2}\right)$. It is true that the term v^2/c^2 is necessarily small; but by utilizing the earth's orbital speed of about 18.5 miles per sec. as the value of v, the change to be expected was brought well within the reach of observation. This was accomplished by turning the whole apparatus, first across, then parallel to, the earth's orbit.

No difference whatever could be detected between observations in the two positions; the result indicated that $t_1 = t_2$. And no one has yet been able to obtain even an approximate approach to the effect predicted.

Naturally this discrepancy aroused much speculation. The only satisfactory explanation in classical terms was that offered by Lorentz, following a suggestion of Fitzgerald, and known as the **Lorentz-Fitzgerald contraction**, which has now been incorporated in the foundation of the **relativity theory**. (L.D.W.)

MICROBE. Bacteria.

MICROCLIMATE. The aggregate of physical conditions in limited habitats, corresponding to the climate of large geographical areas. (A.W.L.)

MICROCLINE. Feldspar.

MICROCRYSTALLINE. The term applied to the ground mass of an **igneous rock** in which the individual crystals can only be made out with the aid of a microscope. Compare with **crypto-crystalline**. (R.M.F.)

MICRO-FELSITE. The term for the ground mass of an **igneous rock** which is not glass but which is still so finely grained that it is extremely difficult, or impossible, to determine the constituent minerals by **polarized light**. Many such textures are probably the result of devitrification. (R.M.F.)

MICROGAMETE. Gamete.

MICROLITES. The general term for microscopic tabular or prismatic **crystals**. Microlites differ from **crystallites** in that the former may be distinguished by the petrographic microscope under **polarized light**. (R.M.F.)

MICROMETER. The micrometer represents a general principle of **physical measurement**, used on various instruments such as **comparators, spherometers, compensators, interferometers**, etc. It is essentially a screw of accurately known, uniform pitch (commonly 1 mm. or 0.5 mm.), provided with a large head whose periphery is divided into equal parts, forming a scale. Turning the screw through a given number of these parts causes the shaft to travel through a distance which is a proportionate fraction of the pitch. For example, if the pitch is 0.5 mm. and the head is divided into fiftieths, each scale division corresponds to a travel of 0.01 mm. A familiar application is the micrometer caliper, an instrument resembling an ordinary screw clamp. The screw is, however, of the micrometer design, with its scale reading zero when the caliper is closed. When unscrewed and closed again upon a thin plate or wire, the scale reading of the caliper shows the thickness of that object, in hundredths of a millimeter or thousandths of an inch. Measuring **microscopes** and astronomical **telescopes** are frequently equipped with a filar micrometer. (See **Measurement**.) (L.D.W.)

MICRON. The 1/1000 of a millimeter. A unit of measure used for microscopic objects. (L.D.W.)

MICRONUCLEUS. A small nucleus found in many one-celled animals associated with a large **macronucleus**. It is active in **cell division**. Most species have only one micronucleus but in some two or more occur. (A.W.L.)

MICROPALEONTOLOGY. Paleontology.

MICROPEGMATITE. A microscopic intergrowth of **quartz** and **feldspar** similar to the coarser-textured **graphic granite**. A form of **pegmatite**. (R.M.F.)

MICROPHONE. A microphone is a device for converting sound waves into corresponding electrical variations. Since the development of the first crude telephone **transmitter** or microphone by Bell in 1875 many types of microphones have been invented but most have been discarded and now only the single-button carbon, double-button carbon, crystal, velocity and dynamic types are used to any extent. In a microphone two things are important, the sensitivity, i.e., the degree of electrical variation for a given intensity of sound wave, and the fidelity or ability to reproduce the audio frequencies in their proper relative magnitudes without generating other frequencies in the process. In **telephony** the first requirement is the more important as only a limited range of frequencies is used and it is not essential that these be reproduced in exactly the correct relative magnitude. Thus the single-button carbon microphone, or transmitter as the telephone engineer calls it, is universally used in this application. Fig. 1a shows the basic principle of this microphone. The sound wave striking the diaphragm causes the pressure on carefully selected carbon granules in the cup to be varied, thereby varying the resistance of the carbon between the cup and the diaphragm. This causes the current in the circuit to vary and the speech may then be transmitted electrically. The single-button microphone has the highest sensitivity, but is limited in its frequency response, has appreciable background noise and requires a battery for operation. The double-button microphone operates on the same principle but uses two carbon-containing cups, back to back with the diaphragm between them. This produces a higher fidelity but less sensitive instrument and is used for some public address and special service radio work but not for regular broadcast service. The crystal microphone is of medium sensitivity and gives good fidelity. It utilizes the piezo-electric effect of Rochelle salt, the sound waves striking the **crystal** or a diaphragm connected with it produce mechanical distortions with resultant voltage variations across the crystal. This microphone is a high impedance device which may be connected to the **grid** of an amplifying tube without an impedance-matching **transformer**. It is widely used in public address work and radio other than broadcasting. The velocity microphone, shown in Fig. 1b, consists of a thin metallic ribbon suspended between the poles of strong permanent magnets. Sound waves striking this ribbon produce pressure differences on the two sides, causing it to move with the waves and the magnetic field then induces a voltage in the ribbon. This is a very low impedance microphone and is connected to an amplifier tube through a transformer. The dynamic microphone is similar to the dynamic **speaker**, having a coil attached to a diaphragm so the speech waves will cause it to move in the field of the magnets and thus generate a voltage in the coil. This is connected to an amplifier through a transformer. Both the velocity and dynamic microphones are of low sensitivity, requiring about one more stage of vacuum-tube amplification than the other types, but they have uniform frequency response over a very wide range so are almost universally used for high quality sound and radio work.

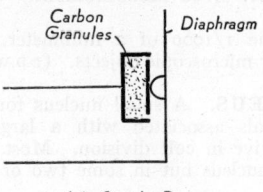

(a) Single Button

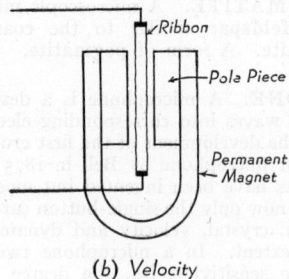

(b) Velocity

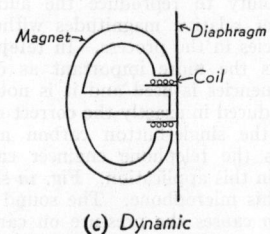

(c) Dynamic

Fig. 1. Common microphones.

form frequency response over a very wide range so are almost universally used for high quality sound and radio work. Combination microphones having both velocity and dynamic elements have been developed. Since the polarities of the output voltages are different in the two for different directions of sound travel, the combination can be utilized to give a directional unit which is dead from the back, thus greatly facilitating the use of the microphone in noisy locations or where the reproducing speaker may feed back to the microphone. (L.R.Q.)

MICROPHONIC NOISE. This is the high-pitched note which comes from the **speaker** of an **amplifier** or radio **receiver** when the set is subjected to a mechanical shock. It is caused by mechanical vibration of the vacuum-tube electrodes altering the electrical response of the **tube.** (L.R.Q.)

MICROPHOTOGRAPHY. The processes employed in making photographs of microscopic size; i.e., photographs requiring enlargement to be visible to the unaided eye. Developed about 1860, chiefly in France, microphotography for a long time was utilized chiefly in the making of novelties. These contained a tiny photographic reproduction of a passage of scripture, a famous painting, or a familiar scene, and a small magnifying glass for viewing. Micrometer rulings for optical instruments are another application of microphotography.

The chief application of microphotography today is in the reproduction of books, bank checks, newspapers and other printed matter on a greatly reduced scale for filing purposes. This field is sometimes known as *documentary reproduction.* It has grown rapidly in recent years. There are special cameras and high resolution films for making such copies on 35- or 16-mm. motion picture film, and readers for convenient reading of the small film records. The amount of printed material which can be purchased in the form of copies on film is growing steadily and now includes the daily editions of several of the larger newspapers. These film copies will be available when the originals have disintegrated. Furthermore, new copies can be made at any time quickly and at low cost. Other important applications are the reproduction of rare manuscripts for the use of scholars throughout the world, or to make source material available to the interested public to whom the originals cannot be entrusted; the publication of books or technical papers where the number of copies does not justify the cost of printing; the making of duplicate records of material for storage at different places as a safety measure and the reduction of bulky material for storage. (C.B.N.)

MICROPODIFORMES. The swifts, hummingbirds, and related species. An order of birds characterized by strong flight. Although the swifts resemble the swallows superficially more than they do the hummingbirds, these two forms are anatomically related. (A.W.L.)

MICROPOIKILITIC. A term used by mineralogists and petrologists to describe minerals which contain microscopic inclusions of other minerals which have no chemical relationship to each other or to the host mineral. (R.M.F.)

MICROSCLERES. The smaller **spicules** in the tissues of **sponges.** (A.W.L.)

MICROSCOPE. The optical instrument that bears this name consists essentially of two parts. 1. The objective is a lens combination, usually of small aperture and short focal length, which forms a real, inverted, and much enlarged image of the object at a point high up in the microscope tube, very much as a stereopticon objective throws an enlarged picture upon a distant screen. The objective-lens system is composed of several positive lenses, the first of which is hemispherical with its plane surface facing the object. Following this is a larger convexo-concave or "meniscus" lens, and then two still larger plano-convex achromatic lenses. 2. The **eyepiece** or ocular is placed beyond, with its focal plane coinciding with this image, and acts as a collimator, so that one looking into it sees a virtual image, apparently at an infinite distance, and subtending a wide angle. The instrument is focused by varying the distance between objective and object. The **magnifying power** of the microscope depends upon the relative focal length of objective and eyepiece; its resolving power depends upon the dominant wavelength of the light used (see **Diffraction**). In the **ultramicroscope**, the object has a special type of illumination. Sometimes a drop of transparent oil is placed between the anterior surface of the objective and the slide cover in order to cut down the effect of refractive spherical aberration and the loss of light by reflection from these surfaces (oil-immersion objective). In measuring microscopes, a transparent scale or a **filar micrometer** may be introduced into the ocular focal plane; or a pair of cross-hairs may be used and the whole instrument moved laterally by a micrometer screw, thus forming a comparator. See also **Petrographic Microscope.** (L.D.W.)

MICROSEISM. Earthquakes.

MICROSPORANGIUM. Flower.

MICROSPORIDIA. Sporozoa.

MICROSTRUCTURE. Metallography.

MICROWAVE. These are extremely short radio waves, only a few centimeters or less in wavelength. (L.R.Q.)

MIDBRAIN. Mesencephalon. **Brain.** (A.W.L.)

MIDDLE LAMELLA. The middle lamella consists of the primary walls which surround the cells of plants and the intercellular layer between these walls. This composite layer is usually made up of pectic materials (colloids which have a great affinity for water), one of which is calcium pectate. The function of the middle lamella is to hold adjoining cells together. Sometimes, particularly in mature fruits, the middle lamella substance breaks down. As a result the cells of the fruit separate easily, giving to the fruit a meal-like character. (R.M.W.)

MIDDLE-LATITUDE SAILING. The term middle-latitude sailing is applied to that modification of the **plane-sailing** problem in which the earth is assumed to be a sphere. The plane-sailing problem is illustrated in the figure, in which (L_1, Lo_1) are the terrestrial coordinates (**latitude** and **longitude**) of a point

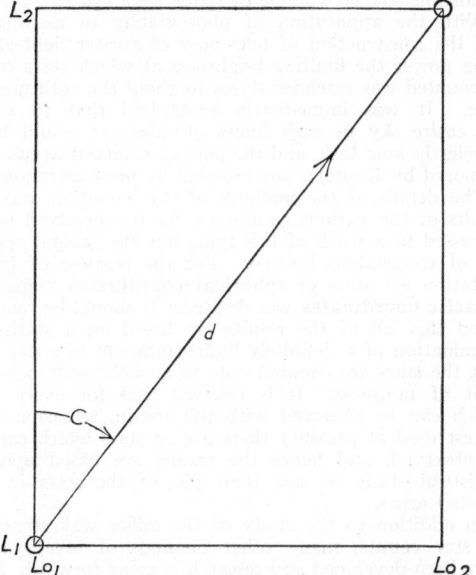

which a ship left, C is the **course** or **track** followed, d is the distance in nautical miles along the track, and (L_2, Lo_2) are the terrestrial coordinates of the point arrived in.

By definition, the difference in latitude $L_2 - L_1$ can be expressed directly in **nautical** miles by calling one minute of arc a nautical mile. However, the difference of longitude $Lo_2 - Lo_1$ cannot be expressed in nautical miles of **departure** (easting or westing) without making use of the relation departure $= (Lo_2 - Lo_1)' \cos L$, in which $Lo_2 - Lo_1$ is expressed in minutes of arc and L is the latitude of operation. From the figure it is evident that the value of the computed departure would differ, depending upon what value of L between L_1 and L_2 is used. For distances under 600 miles and in latitudes lower than 60° a sufficiently close approximation to the departure distance can be obtained by using for

L the value $\dfrac{L_1 + L_2}{2}$ or the "middle latitude." Outside

the limits stated for latitude and distance, other methods, such as **Mercator sailing** will yield more accurate results. Graphical solutions of the middle-latitude problem may be made on **small-area plotting sheets.** (W.K.G.)

MIDGE. Insecta, Diptera. **Insects** resembling mosquitoes but with strong veins only near the anterior margin of the wing and without scales along the veins. Only a few can bite. Family Chironomidae. Insects of the family Dixidae, called dixa midges, also resemble mosquitoes but differ in the lack of scales on the wings and in the venation. (A.W.L.)

MIDRIB. Leaf.

MIGMATITE. The term proposed by Sederholm, in 1907, for composite gneiss, the result of **lit-par-lit** injection of granitic **magma** between the thin, closely spaced **foliation** planes of **schistose** rocks. Migmatites are closely related to **injection gneisses** and probably represent a form of **contact metamorphism** in the wall rock, or roof, of a granitic **batholith.** (R.M.F.)

MIGRAINE. Severe paroxysmal headache frequently associated with vomiting and visual disturbances.

The cause of this disorder is unknown. **Heredity** is a strong factor in many cases. Neurotic and mentally active individuals are more apt to be affected than phlegmatic and dull ones. Women are 2–3 times more commonly affected than are men.

Many theories are advanced for its causation: 1. Transient spasm of the cerebral arteries with or without pressure disturbance of the cerebrospinal fluid in the cavities of the brain. 2. Disturbances of the sympathetic nervous system brought on by toxic, psychic, emotional, or fatigue states. These disturbances of the sympathetic nervous system could cause the vascular disturbances cited above in (1). 3. Various allergic states with sensitivity to food or other allergens. 4. Disturbances in the function of the endocrine glands, particularly the pituitary and the ovaries.

Migraine attacks occur as paroxysms which come at more or less regular intervals. Most often they begin during or after puberty and in some cases disappear in both sexes at the menopause or climacteric. In women, the attacks often occur with the menstrual periods. The attack may last one, two, or three days and utterly prostrate the patient. Many patients can tell when the attack is coming on. The headache is usually on one side, and is of a boring, intense type difficult to bear. Most often it is located around the eyeball, temporal region, or forehead, later spreading to involve a larger area. Flashes of light, or scintillating zigzag lights before the eyes are common. Jarring movements, bright lights and noise all accentuate the headache. Dizziness, nausea, and vomiting are frequent.

The treatment of migraine is varied, and different measures are effective in different patients. Gynergen (ergotamine tartrate) is the most effective drug now in use. The inhalation of 100% oxygen, sedation with **barbital** compounds, and desensitization with histamine are also employed. (D.M.H.)

MIGRATION. Movement from one region to another over greater distances than are covered by the incidental wanderings of the individual.

The most familiar example is the seasonal migration of birds from their nesting grounds at higher latitudes to regions nearer the equator and back. Some species cover thousands of miles in these semiannual flights, while others winter only a few hundreds of miles from their summer homes.

The underlying causes of bird migration have been the subject of too extensive discussion to be reviewed here. They are undoubtedly closely correlated with the ability to migrate, which results from specialization for flight. Birds are able to see long distances from the altitudes at which they fly, sometimes thousands of feet above the ground, and are able to proceed far more rapidly than any running animal. The advantages of wintering in a region of plenty are obvious, and the release of pressure during the breeding season by removal to less crowded regions is scarcely less so. At that time not only hungry adults must be fed but voracious young as well. Recent experimental work has emphasized the role of the endocrine glands and of seasonal changes in the environment in stimulating them as factors partially explaining the mechanism of bird migration.

Migrations of other animals are less common, but the movements of the great herds of bison on the western plains were a similar adjustment to seasonal food supply. The migrations of the European **lemming**, ending in the destruction of great numbers, are less easily understood. It has been suggested that they originate in a scarcity of food.

Many species of fish also migrate, either from one part of the ocean to another or between fresh and salt water. Two of the most noteworthy migrations are those of the salmon and the eel. Eels grow up in fresh water and migrate to the Atlantic Ocean in the neighborhood of Bermuda to spawn. Their young gradually work their way back to the rivers to develop. Salmon, on the other hand, are chiefly ocean fish, but during the breeding season they migrate far up the coastal rivers to spawn in fresh water.

Migrations of invertebrates have also been noted, but as a result of the small size and slow locomotion of most species they are usually mass movement over relatively short distances. Flying insects, however, often cover hundreds of miles, and in the case of some species of butterflies great hordes of individuals may take part in the migration. The southward flights of the North American milkweed butterfly are the most common example. Occasionally these butterflies move in such hordes that they cover large trees when they come to rest at night. The northward flight of the cotton moth in the summer is also noteworthy. This species is abundant in Ohio, far beyond the nearest cotton, which is its sole food plant. (A.W.L.)

MILDEW. Fungi.

MILE. The English or statute mile of 5280′ is that most commonly used by English-speaking people. This distance is supposed to be equivalent to 1000 paces of the Roman Legions stationed in Britain. Navigators and geographers prefer to use the nautical mile and the geographic mile respectively. (See **Nautical Mile.**) (W.K.G.)

MILK FACTOR. Cancer.

MILKWEEDS. Asclepiadaceae. A family of some 325 genera with over 1700 species of shrubs, woody vines, and perennial herbs. All contain a milky juice from which **rubber** may be made. The family is particularly abundant in the tropics, especially in Africa. The forms are extreme: many are **lianas** of great length; others have leaves modified into **pitcher**-like forms; some are **epiphytes;** while many, especially in Africa, are very much like **Cacti** in appearance. Indeed, many species are sold under the name of cactus. The uses made of various milkweeds are many and varied. The young shoots of many species are eaten as greens; other species yield dyes. The juices of several are violent poisons, as *Gonolobus,* from which an arrow-poison is obtained, and *Cynanchum,* which is used to stupefy fish. Many are grown for their weird shape, or because of their beauty, as the Wax-plant, *Hoya carnosa.* (R.M.W.)

MILKY WAY. The milky way, or galaxy, appears to the naked eye as a broad irregular band of misty light which encircles the sky. The telescope shows that it is in reality made up of myriads of faint stars interspersed with gaseous **nebulae.** Its light is due to these faint stars, and if all of the stars which can be distinguished as individuals with the unaided eye were blotted out, the milky way would still remain undiminished in intensity against the otherwise dark sky. A mere casual examination of the milky way on a dark night will reveal that its intensity varies from point to point, which is an indication that the faint stars are not uniformly distributed.

For many years the milky way itself, and the clustering of faint stars in the general direction of its plane, have been taken to indicate, not that the stars were actually closer together in that direction, but rather that the volume of space which contains all of the stars is more extensive in that direction. The first line of attack on the general problem of the distribution of matter in space was made by Sir William Herschel, in 1784. His method, which is essentially that of the modern methods of star counting, consisted in counting the number of stars visible in his telescope when it was directed to various directions in the sky. As a result of his observations he announced that all of the stars were contained in a volume of space shaped like a "grindstone," with the plane of the grindstone in the direction of the plane of the milky way. The relative dimensions of this grindstone-shaped volume of space were 850 to 155.

With the application of photography to astronomy and the construction of telescopes of greater light-gathering power the limiting brightness at which stars could be counted was extended down to about the 19th magnitude. It was immediately recognized that to cover the entire sky to such limits of faintness would be a hopelessly long task, and the plan of **selected areas** was proposed by Kapteyn and adopted by most astronomers.

The details of the methods of star counting and the results of the various counts are far too involved to be discussed in a work of this type, but the general results are of tremendous interest. For the purpose of interpretation a system of spherical coordinates known as **galactic coordinates** was devised. It should be remembered that all of the results are based on a statistical examination of a definitely limited amount of space and that the stars are counted only to an arbitrarily selected limit of faintness. It is believed that for every star which can be observed with the 100-in. telescope (the largest used at present) there are 29 stars which cannot be observed, and hence the results are based upon a statistical study of less than 5% of the stars in the selected areas.

In addition to the study of the milky way structure by star counts, many other methods of investigation have been developed and research is going forward. The existence of absorbing material in "empty space" has been confirmed from a variety of observations. We have even been able to make approximate determinations of the character and density of this "cosmic dust" or "cosmic haze," to mention merely two of the various descriptive terms applied to it. We are confident that it is not uniformly distributed and that there exist large, and relatively dense, clouds of dark material known as dark **nebulae.** Until the effects of this material on star counts and other data, in all directions and distances from our sun, have been established, we cannot do much more than guess as to the size and detailed structure of the milky way. All that we are justified in stating is that the original hypothesis of Herschel still remains and we believe that our milky way, or galaxy, occupies a more or less lens-shaped volume of space with the maximum diameter of the order of magnitude of one hundred thousand **light years.** The sun is not far from the central plane of the lens, and is of the order of magnitude of thirty thousand light years from the center. The total mass of this galactic system is probably as high as two hundred billion (2×10^{11}) times the mass of our sun.

The problems relating to the dynamics and evolution of this galactic system are far from definitive solution. We have an abundance of evidence, from an analysis of the observed stellar motions, that the milky way is in general rotation with a period of the order of two hundred million years. Questions as to whether or not this motion is uniform, the location of the axis of rotation, etc., are far from being definitely answered.

It is interesting to speculate upon whether or not there are other galactic systems in space, and, if so, how they compare in size with our own. It is confidently believed that the extra-galactic spirals represent other galactic systems. When we attempt to compare the masses and dimensions of these outer objects with those of our own galaxy we are faced with two practically insurmountable problems: the dimensions of our own are very imperfectly determined, and the dimensions for the others are equally uncertain. At one time we believed that our own system was by far the largest of all. That statement must be modified now, and the best that we are entitled to state is that our own system probably has much in common with the Great Spiral in Andromeda. (w.k.g.)

MILLER INDEX. Crystallography.

MILLERITE (CAPILLARY PYRITES). The mineral millerite is nickel sulfide, NiS, whose slender hexagonal interwoven crystals so suggestive of hairs has led to the application of the name "capillary pyrites." It occurs also as radiated masses and coatings. It is brittle; hardness, 3–3.5; specific gravity, 5.3–5.6; luster, metallic; color, brass-yellow, often with an iridescent tarnish. Millerite is found in association with other nickel-bearing minerals and other sulfides. European localities are Bohemia, Westphalia, Wales, etc.; and in the United States at Antwerp, New York; with pyrrhotite in Lancaster County, Pennsylvania; at St. Louis, Missouri; Keokuk, Iowa; and Milwaukee, Wisconsin. In Canada millerite occurs in Oxford, Quebec, and in the famous Sudbury District, Ontario. It is used as an ore of nickel. Millerite was named for the English mineralogist, W. H. Miller. (e.s.c.s.)

MILLER'S THUMB. Pisces, Teleostei. A small fish (Pisces), *Cottus bairdi*, with very large pectoral fins and the head and anterior part of the body large and somewhat flattened. Common in the northeast quarter of the United States and southward in the mountains. In clear streams and lakes. One of the fresh-water sculpins. (a.w.l.)

MILLET. Gramineae. Cereal and forage grasses of several different genera are included in the term millet. All have a fibrous root system, ample foliage, and rather small grains. They are grown extensively in Occidental countries as forage crops and in many Oriental countries for human food as well as for forage.

Pearl millet, or *Pennisetum glaucum*, is an erect grass with solid stems 3–8' tall. The inflorescence is a dense cylindrical spike 6"–1' long. The spikelets are 2-flowered, the lower floret being staminate (see **Stamen**), the upper pistillate (see **Pistil**). Cross-pollination normally occurs. This plant is very variable in the structure of its inflorescence and has been given a number of names, such as African millet, Japanese millet, and Indian millet.

Panicum miliaceum is an erect grass 2–3' in height and with frequent branching from the basal portion of the stem. The inflorescence is a panicle from 4–12" long and variable in appearance in different varieties.

Foxtail millet, *Chaetochloa italica*, is an erect grass 2–5' tall, with occasional branches from the base of the

stem. The leaves are long, rather broad and tapering to a sharp point. The inflorescence is a narrow spike 4–8" long. The elliptical spikelets are subtended by a group of long bristles and are 2-flowered. Cross-pollination occurs in this grass, as is also the case in the preceding species. Hungarian, German, Siberian, and Common Millet are varieties of this plant, an Old World native which is now widely distributed in North America.

Barnyard millet *Echinochloa Crus-galli*, is a coarse branching annual from 2–4' tall. The inflorescence is a branched panicle with 2-flowered spikelets densely crowded on one side of the branches. This grass is another European native which has become a widely distributed weed in both cultivated and waste land in North America.

Millets as a group are drought-resistant plants widely distributed as forage or pasture grasses and for hay. The seeds are a valuable poultry food. A valuable feature of the group is their very rapid growth, which allows them to be used at times when other crops have failed; about 6 weeks after planting the seed, the plant is ready to cut for hay. (r.m.w.)

MILLIBAR. The unit of pressure used in meteorology is the millibar, which is 1/1000 part of a bar. A bar is 1,000,000 dynes per sq. cm. A millibar, therefore, is 1000 dynes per sq. cm. "Bar" is often used by physicists and acoustic engineers to denote 1 dyne per sq. cm. Also spelled barye. (p.e.k., l.d.w.)

MILLING. The process of removing material by multitoothed rotating cutters. Milling machines may be roughly classified into four groups: column and knee type, bed type, planer type, and rotary type.

Column and knee type milling machines are made in three styles: universal, plain, and vertical spindle. They are used for both toolroom and manufacturing work because of their adaptability and because of the ease with which they may be handled. Figure 1 shows the essential features of a modern motor-driven or

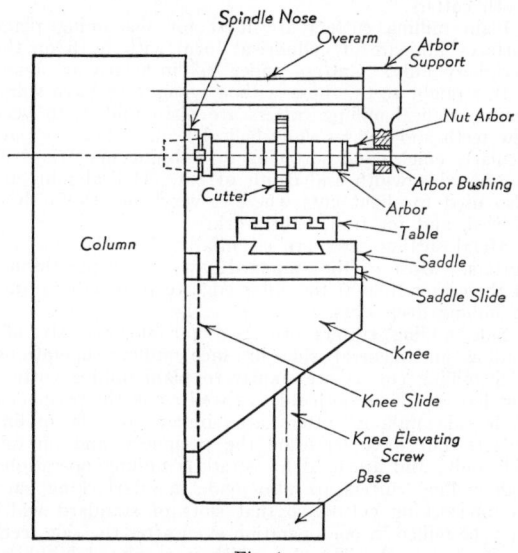

Fig. 1.

constant-speed plain milling machine. The machine has a horizontal spindle rotating in anti-friction bearings in the column. The table is mounted on and slides in dovetail guides on the saddle. The saddle is mounted on and slides in dovetail guides on the knee, which is free to move up and down on the face of the column, and is supported by a screw within a telescoping cover. In operation, the milling cutters are either attached to the spindle nose or carried on an arbor or shaft which is driven from it. All three elements, table, saddle, and knee, may be either power- or hand-fed by screws turning in fixed nuts.

Universal milling machines are similar to plain milling machines, but the saddle is mounted on and swivels on a clamp bed which in turn slides on the knee, thus permitting the saddle to swing at an angle, and permitting table motion at other angles than 90° to the spindle axis. Vertical milling machines are similar to plain milling machines, but the spindle is vertical.

Modern heavy-duty milling machines are equipped with a standardized spindle end which has a locating taper hole in the spindle. The arbor is seated by turning a draw-in-bolt which extends through a hole in the spindle, and screws into a threaded hole in the arbor. The arbor is driven by a driving key on the spindle nose which fits into slots in the arbor shoulder. The arbor support provides a cylindrical bearing for the pilot of the arbor, and in many instances, an intermediate arbor support, serving as a bearing for an oversize collar, is employed. The arbor support is often connected to the knee by overarm braces for additional rigidity.

There are three types of milling cutters: hole type cutters which are mounted on the milling machine arbor; shank type cutters which have an integral arbor or shank and are mounted in the hole in the spindle nose, or in an adapter or collet that fits the spindle nose; and face type cutters which are fastened directly to the spindle nose by screws in the body of the cutter. Milling cutters may have teeth on the periphery only, on the ends only, or on the periphery and ends. Peripheral teeth may be straight or parallel to the cutter axis, or they may be helical, in which case they are usually, although incorrectly, referred to as spiral teeth. Spiral teeth give the cut a shearing action, which reduces the stress on each individual tooth and prevents the shock that occurs as each tooth of a straight tooth cutter meets the work. Spiral cutters produce much smoother surfaces and require less power to operate than straight-tooth cutters.

Plain milling cutters are used for machining plane surfaces, and are of cylindrical form with teeth on the periphery only. Cutters under ¾″ in width are made with straight teeth, those with a wider face have spiral teeth. Helical milling cutters are plain mills with very few teeth and a very short helical lead. They are particularly efficient on surfacing or slabbing operations of considerable width and depth of cut. Helical mills are also used for light cuts where a very smooth finish is desired, and for frail, light work.

Metal slitting saws are essentially thin plain milling cutters. Most of the standard saws are made thinner at the center than at the outer edge to provide clearance in milling deep slots.

Side milling cutters are used for slotting, straddle milling and general side or face milling operations. Side milling cutters are similar to plain milling cutters, but have teeth on both sides as well as on the periphery. Half side milling cutters are similar to side milling cutters, but have teeth on the periphery and on one side only, and are used for straddle milling operations. Side milling cutters are also made in interlocking pairs as interlocking cutters so that slots of standard width may be milled in one operation even after the side teeth are resharpened. The slot width is regulated by shims or washers placed between the hubs of the two parts of the cutter.

Shank type cutters with an integral shank may be held directly in the hole in a taper nose spindle, or held in an adapter or collet which fits in the spindle nose hole in standardized spindle ends. Fig. 2 illustrates three extensively used shank type cutters which have Brown and Sharpe taper shanks with tang ends. Cir-

cular ends with a threaded hole for a milling machine draw-in-bolt are also obtainable. Slotting or two-lipped end mills have two cutting lips and are somewhat similar

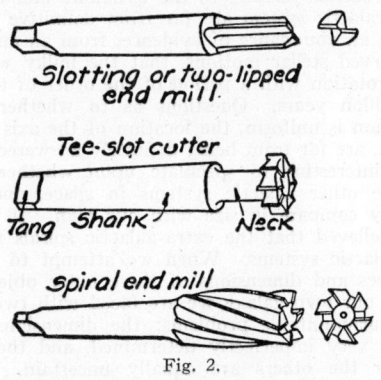

Fig. 2.

in action to twist drills. They may be used for milling slots with semicircular ends by first feeding to depth and then moving laterally. The depth of cut in solid stock is generally limited to one-half the diameter of the cutters, but deep slots can be produced by successive cuts.

Spiral end mills are similar to plain milling cutters but have teeth on the end face as well as on the periphery. End mills are used for a variety of surfacing operations, particularly for surfaces that cannot be conveniently reached with hole type cutters. The cutter is made with a deep center hole to permit many resharpenings, but cannot be used for originating blind end slots from the solid metal since the teeth do not extend to the center. In machining slots by end milling, it is necessary to drill a starting hole whose diameter is at least equal to the mill diameter. Spiral end mills with straight shanks are extensively used for die-sinking and other similar operations. Straight shank mills are either held in an arbor with a straight hole by a set screw, or held in a spring collet. Tee-slot cutters are used for milling the base groove or portion of a tee-slot. The body slot must be cut before the tee-slot cutter can be used, and the neck of this cutter must be smaller than the body slot width.

Shell end mills are used in conjunction with an arbor and are more economical than solid end mills of the same size, since one arbor will serve for several cutters and need not be discarded when a cutter is worn out or broken. Face milling cutters are essentially shell end mills with inserted teeth. Each tooth is held by a taper bushing and a screw which wedges it tightly in place, but will permit adjustment for sharpening. Face type cutters are used for surfacing operations, and consist of a body of cylindrical form with inserted teeth that have cutting edges on the periphery and on one face. The back of the cutter has a cylindrical recess that fits over the spindle nose to locate the cutter. The cutter is held to the spindle nose by four fillister head screws passing through holes in the body of the cutter and screwing into tapped holes in the spindle nose. The cutter has a key slot in the back that fits the driving keys on the spindle. Fly cutters are single-toothed form cutters. Since one tooth does all the cutting, it reproduces its shape very accurately in the work, but a very fine feed is required. The cutter can be easily ground from tool bit stock, and often serves as a means of milling intricate shapes that will not warrant the expense of special formed cutters.

Cutting speeds on the milling machine depend upon the character of the work, the type of cutter, the condition of the machine and, in many instances, upon the experience and ability of the machine operator. A reasonable basis for surface cutting speeds, in ft. per min., for high-speed steel cutters, is as follows:

annealed tool steel, 60 to 80; machinery steel and cast iron, 80 to 100; and brass, 150 to 200. Carbon steel cutters should operate at approximately half these speeds.

Feed rates in milling are expressed in two ways: in. per min., or thousandths of an inch per revolution of the spindle. Delicate or fragile work requiring an accurate finish will need fine feeds, while heavy work, from which a considerable amount of metal is to be removed, can be subjected to coarse feeds. The type of cutter employed also has a definite bearing on the feeds.

A good commercial finish can usually be obtained by using a feed rate of from .030″ to .050″ per revolution of the cutter. Finer feeds, such as .015″ per revolution, will result in an excellent finish.

Milling machine work is held in a vise whenever possible, since work may be set up more quickly than by using clamps or bolting the work to the table. Hole type cutters are used in preference to end milling cutters, although face mills will afford production rates that compare favorably with those attained by peripheral cutters. Milling machine spindle attachments are employed for milling, drilling and boring operations on column and knee type milling machines. The attachment may be used for machining angles, undercuts, and other surfaces that are difficult or impossible to reach with standard equipment. It is also used in place of a vertical milling machine on small and medium sized work. A rotary milling attachment consists of a circular table which is supported by and rotates on a base which is bolted to the milling machine table. The rotary table is driven by a worm gear and worm, which is in turn rotated by a handwheel and crank in hand attachments, or by a universal shaft drive from the table in power attachments. The rotary attachment is used for circular milling and for indexing, for which graduations are provided at the edge of the table.

Indexing is one of the most important applications of the milling machine. A set of universal index centers consists of a headstock and a footstock. The headstock has a hollow head in which a spindle carrying a 40-tooth worm wheel is mounted. A single-threaded worm, which may be rotated by a crank, is in mesh with the worm wheel. The nose of the spindle is threaded to carry a chuck if desired, and has a taper hole for a center or an arbor. It is possible to swing the head in its bearings so that the nose of the spindle can be set to any angle from 10° below the horizontal to 5° beyond the perpendicular.

The work to be indexed can be held on a mandrel or arbor in the spindle nose of the headstock, or it can be held in a chuck which is screwed to the spindle. Many parts, however, are held between the headstock and footstock center or on a mandrel between these centers. The footstock is bolted to the milling machine table, and carries a sleeve which holds the footstock center. The sleeve may be adjusted axially by a crank. The center can be adjusted vertically, and can be set at an angle out of parallelism with the base when it is desired to mill tapered work.

Plain indexing makes use of index plates mounted on the bearing for the worm shaft and locked in place by a pin. Each index plate has six series of holes, and the crank arm, fastened to the worm shaft, is adjustable so that the locking pin in the crank handle can fit any of the series of holes in a particular plate.

To illustrate plain indexing, suppose it is desired to cut 60 slots in a cylindrical bar that is mounted between the headstock and footstock centers. The plate with a 15-hole circle is used and the crank is set to fit this circle. If the crank is turned from one hole on this circle to the next hole, the crank will have made $\frac{1}{15}$ of a revolution. Since the worm and worm gear ratio is 40:1, the spindle will therefore make $\frac{1}{15} \times \frac{1}{40}$ of a turn, or $\frac{1}{600}$ of a turn. In order for the spindle to index $\frac{1}{60}$ of a turn, it is therefore necessary to move the crank 10 spaces at

a time. The crank will be moved from hole No. 1 to hole No. 11, to hole No. 21, etc. With the three index plates usually supplied, plain indexing can be used for all divisions up to 50; even numbers, except 96, up to 100; and many others. Two methods of indexing, compound and differential, are used for divisions that cannot be obtained by plain indexing. In compound indexing, the crank is turned as in plain indexing and locked in place in the plate, but the plate locking pin is then withdrawn and the plate rotated a definite amount to give a combination of motions that will afford the desired division. The method is tedious and time-consuming, and is rarely employed except when differential indexing equipment is not available.

In differential indexing, the index plate is not fixed, but is geared to the crank in such a manner that it moves a small fraction of a turn for every revolution of the crank. In this way, all indexing between 2 and 382 divisions may be handled; data for plate selection and necessary gearing may be found in shop handbooks.

Spiral milling is the process of cutting helices on a universal milling machine by connecting a universal dividing head by change gears to the table lead screw, so that the dividing head spindle rotates in conjunction with the movement of the table.

The profiler is a vertical milling machine with two spindles, one of which carries a tracer or guide point, and the other an end mill whose cutting end is a duplicate of the tracer end. The operator follows a master template with the tracer and the end mill cuts a similar contour in the work. (H.C.H.)

MILLIPEDE. Diplopoda.

MILLON'S REAGENT. Aminoacids, Polypeptides, and Proteins.

MIMETITE. The mineral mimetite is a **chloro-arsenate** of **lead** corresponding to the formula (PbCl)Pb$_4$ (AsO$_4$)$_3$. It is **hexagonal;** brittle; hardness 3.5; specific gravity, 7.0–7.25; luster, resinous; color, usually yellow to brown but may be colorless or white; translucent. Mimetite is a rather rare secondary mineral occurring in altered lead deposits. Found in Bohemia; Saxony; Cornwall and Cumberland, England; South West Africa; Mexico; and in the United States in Pennsylvania and Utah. The name mimetite is derived from the Greek word meaning imitator, because of the similarity of mimetite and **pyromorphite.** (R.M.F.)

MIMICRY. The resemblance of an animal to some other living thing or inanimate object. Mimicry may involve both color and form, hence it is closely related to **coloration.** It is supposed to benefit the mimic either by concealing it from its enemies, by causing them to mistake it for something undesirable, or by enabling it to approach its prey without giving alarm.

Some cases of mimicry are probably incidental, though it is easy to imagine value in all of them. The brightly colored king snakes, for example, may be avoided because of their resemblance to the poisonous coral snake, but it is difficult to know that this is true. On the other hand, some butterflies are distasteful to monkeys, and by actual experiment it has been shown that after one unpleasant taste the monkey avoids both the unpalatable species and others that resemble it closely. Here there can be no question of the value of the mimicry.

Mimicry of inanimate or inmovable objects is shown by many insects. Tree hoppers may look like stout thorns, some caterpillars resemble stubby dead twigs, leaf insects and leaf butterflies are much like leaves, and some moths that rest on the trunks of trees are scarcely distinguishable from bark. One of the most remarkable examples is that of a little North American moth which resembles a bird-dropping. Its front wings are chalky white with mottled gray markings. When at rest they are folded close about the body and hind wings and the tip of one overlaps the other, thus concealing the symmetry of the insect. Most moths found near the ground in the woods perch among dry leaves or on the under side of leaves or on stems of low plants, but this one remains in full view on the upper surface of a leaf, where a bird-dropping would naturally fall.

Aggressive mimicry occurs in the **ambush bugs** and some of the **robber flies**. The ambush bugs hide in flowers, where their color and markings make them nearly invisible. They pounce on insects which visit the flowers for nectar. Some robber flies capture bees for food, and in one genus differ greatly from the usual form and resemble bumblebees very closely. (A.W.L.)

MINE SURVEYING. Surveying.

MINERAL.
A natural inorganic substance having a characteristic range of chemical composition, usually a definite **crystal form**, and exhibiting other specific, physical characteristics such as **cleavage, fracture, hardness, color, luster,** and **heft** (specific gravity). The study of minerals is called mineralogy. One who studies minerals is called a mineralogist. (R.M.F.)

MINERAL OIL. Petroleum.

MINERALIZERS.
The term applied by petrologists to magmatic gases such as water vapor, **hydrogen,** and compounds of the rarer elements and volatile substances which are retained in solution under varying degrees of pressure and temperature, but which tend to escape from the effusive **lavas** under atmospheric pressure alone. These original constituents of **magmas** are essential to the origin of all types of intrusive and relatively coarsely crystalline **igneous rocks,** as they tend to:

1. Lower **viscosity** and thus lower the temperature at which certain minerals crystallize.

2. Act as **catalyzers** in aiding the formation of compounds and crystals.

3. Combine with other elements and compounds to form minerals which require mineralizers.

4. Concentrate metallic and other compounds which otherwise would tend to remain dispersed. (R.M.F.)

MINERALOGIST. Mineral.

MINERALOGY. Mineral.

MINETTE.
The term proposed by Voltz, in 1822, for "ironstones" of **Jurassic** age in the Briey basin and Lorraine, France. A **lamprophyric dike** rock whose principal minerals are **orthoclase** and **biotite.** (R.M.F.)

MINIVET.
Aves, Passeriformes. A brightly colored **shrike.** The several species inhabit eastern Asia and India. (A.W.L.)

MINK.
Mammalia, Carnivora. A slender semiaquatic animal with short legs, partially webbed toes, and a short, moderately bushy tail. Related to the polecats. One species, *Mustela sibiricus,* occurs in Siberia, one, *M. lutreola,* in eastern Europe, and one, *M. vison,* throughout North America. The fur varies from light to very dark brown and is thick and soft, mixed with long glossy hairs. It is among the more valuable furs commercially. (A.W.L.)

MINNOW.
A small fish (**Pisces**), especially certain species related to the carp, dace, and chubs, which never exceed a few inches in length. The members of the genus *Hybopsis* are the typical minnows but the name is not limited to this genus. (A.W.L.)

MINOR OF A DETERMINANT. Determinants.

MINT FAMILY.
Labiatae. The greater number of the 3000 species of this family are herbaceous plants, widely distributed in the temperate regions. The family also includes a few small trees and shrubs, found mostly in the American tropics. The characters distinguishing these plants are so outstanding as to make identification easy. Commonly the stem is 4-angled, appearing square in cross-section, with the leaves opposite, simple, and without **stipules.** The flowers are borne in **racemes** or more frequently in dense axillary **cymes.** The flowers are irregular, with the **calyx** composed of five **sepals** united to form a tube, and the **corolla** composed of fused petals which form a 2-lipped tube, the upper lip of three, the lower of two lobes. The **stamens,** either two or four in number, are inserted on the corolla tube. The **pistil** is composed of a bicarpellate **ovary,** each of whose **carpels** becomes constricted early in development to form a 4-parted ovary, from which a 2-lobed style arises. The fruit is commonly composed of four

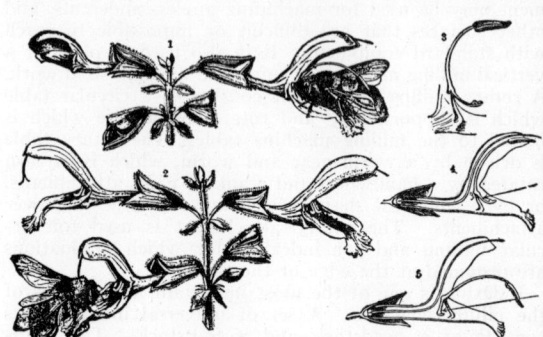

Pollination by insects of sage, *Salvia glutinosa.* The flower is visited by a bee which pushes the stamen in such a way that its pollen-covered anther strikes the insect on the back (1). The anther (3) is hinged and has a projecting appendage which the bee touches with its head, thus tilting the anther over (see 4 and 5). When the bee visits another and older flower (2) in which the style and stigma project, cross pollination is effected. (*Kerner, Natural History of Plants, Henry Holt & Co.*)

achenes or nutlets. The flowers of this family are mainly cross-pollinated by insects: in some species the corolla tube is short enough to allow bees to obtain the nectar located in a disk at the base of the ovary. Pushing into the corolla tube to reach this nectar causes the pollen in the anthers to be shaken onto the insect's back, where it will be carried to the stigmas of another flower. Other species having longer corolla tubes are cross-pollinated by butterflies.

Many members of this family have **volatile oils** located in epidermal glands on the leaves, giving to the plants characteristic odors. Because of these volatile oils many of them are useful to man, some as condiments, some as **perfumes,** and some as **drugs.** Food products are rare in this family, the genus *Stachys* having species which form tubers, eaten in some European countries. *Salvia* species are widely grown because of the showy scarlet flowers, often accompanied by brightly colored bracts. The leaves of the garden sage, *Salvia officinalis,* are used as flavoring for poultry dressing. The stamens of *Salvia* species are interestingly modified to insure pollination. The filaments are long and arching under the upper lip of the corolla, and so attached that a downward projecting portion stands across the corolla tube. Any insect pushing against this projection in order to reach the nectar at the base of the flower causes the filaments to swing down, bringing the anthers against his back, on which the pollen is dusted. After discharge of the pollen from the anther, the style elongates so that the forked stigma is in a position to touch the back of a pollen-laden insect and thus receive pollen, insuring cross-pollination. This stamen structure of *Salvia* is almost unique. Several species of *Monarda*, native plants of North America, are cultivated for their showy red or pink flowers. These plants are variously known as Bee Balm, Horse Balm, Oswego Tea. The leaves of some of them are used medicinally.

The leaves of *Ocimum Basilicum* (basil), *Thymus vulgaris* (thyme), and *Origanum vulgare* (marjoram), as well as *Salvia,* are used for flavoring. *Rosamarinus officinalis* gives rosemary oil; *Lavandula vera* and other species, oil of lavender; and *Pogostemon Patchouly,* oil of Patchouli—all these oils being used in making perfumes.

Many species of *Mentha* yield valuable **volatile oils,** used in medicine and as flavoring. *Mentha Pulegium* is pennyroyal. American pennyroyal is another mint, *Hedeoma pulegioides,* yielding a similar oil used as a stimulant and an emmenogogue. *Scutellaria lateriflora* contains a volatile oil used as a tonic and anti-spasmodic. *Marrubium vulgare,* or horehound, contains a volatile oil used as a treatment for catarrh and chronic infections of the lungs.

Certain species of Mints, commonly called nettles, are troublesome weeds, often with strong rank odors. Of these two species of *Nepeta* are worthy of note. One is *Nepeta Glechoma,* the Gill-over-the-ground, or Ground-ivy, a prostrate trailing vine, freely rooting at the nodes and bearing small deep blue flowers. It is often a persistent weed in lawns and gardens. The other is *Nepeta Cataria,* Catnip or Catmint, an erect, branching plant with pale bluish flowers and hairy leaves. It has a stimulating effect on cats; the dried leaves and stems, sometimes put up in cloth bags, are frequently sold to cat owners. (R.M.W.)

MIOCENE. Next to the last **period** in the **Tertiary** subdivision of the **geologic time-scale.** Type locality, the Paris Basin. The term Miocene was first proposed by Lyell in 1832. Miocene sedimentation started approximately 20,000,000 years ago and continued for about 12,000,000 years. In the United States the marine formations are best developed on the Pacific Coast, where they reach a maximum thickness of some 21,000'. There is no distinct evidence of an unconformity between the **Oligocene** and Miocene, but the **paleontological** record shows that the plants and animals of the Miocene are distinctly modern, especially those occurring in the terrestrial deposits which cover a wide area of the Great Plains. There was considerable volcanic activity during this period, as disclosed by beds of volcanic ashes and **andesitic** and **basaltic** lava flows in the John Day Basin and Columbia River Plateau. Volcanic **agglomerates** also occur interbedded with ashes and conglomerates in the Yellowstone Park Region. The Miocene is also an important period of mountain building, especially in the Alps, Apennines, and Himalayas. The plants of this period were similar to modern types, including the grasses, pines, and hard woods (Sumach, Elm, Oak). Most of the **fossil** marine **invertebrates** belong to the same genera living at the present time. During the Miocene there was a great development of modern mammals, especially in North and South America. The principal and peculiar South American types are: Cladosictis (carnivorous **marsupial**), Stegotherium (giant **armadillo**), Propalaeohoplophorus (glyptodont) Hapalops (giant ground **sloth**). The principal North American forms are the primitive types of dogs, horses, camels, antelope, elephants, and rodents. The development of grass-covered plains formed a suitable environment for the grazing instead of browsing herbivores. Mammalian adaptation is shown in both the tooth and limb structure, as well as in the increased size of the brain. For a description of the mineral resources of this period see the **Tertiary.** (R.M.F.)

MIRA. Mira (omicron Ceti) has the distinction of being the first **variable** star ever announced as such. In 1596 the Dutch astronomer Fabricius noticed a star of about the third **magnitude** which had not previously been recorded. The star faded within a few weeks, but was again seen and recorded by Bayer in 1603. In 1638 Holwarda, another Dutch astronomer, again observed the star and found that after disappearing it returned to visibility about 11 months later. At maximum brightness the star is easily visible to the naked eye, having a magnitude of about 3.5, but at minimum it can be seen only with a telescope of aperture greater than 1", for its magnitude is only about 9.

Many determinations of the period of variability have been made, the time from maximum to maximum averaging about 330 days, with variations between times of successive maxima amounting to as much as one month. The diameter of the star is of the order of magnitude of 260,000,000 miles, or large enough to contain the sun and all the members of the **solar system** in their **orbits** out to beyond the planet **Mars.** Coupled with the variation in brightness there is also a variation in diameter amounting to about 32,000,000 miles, together with a temperature variation of roughly 500° K. (900° F.). (W.K.G.)

MIRACIDIUM. Larva.

MIRAGE. A curious atmospheric phenomenon caused by the **total reflection** of light at a layer of rarefied air. The most familiar manifestation is observed in warm weather on paved highways. The air next the pavement becomes heated and rarefied in comparison with that above it, so that at a sufficient angle of incidence, objects beyond the area are mirrored as if by polished silver, giving the almost irresistible impression that one is looking at a layer of water. Travelers in hot desert regions are sometimes thus deceived. Much more rarely the appearance appears in the air at a higher level than the observer. In either case the images are inverted; and because of the irregular contour of the air layer, they are usually distorted. A somewhat different effect, known as "looming," is produced by the refraction of light passing from rarefied air to a lower and denser layer. This results in distortion, making distant objects appear grotesquely elongated vertically, or in lifting into view objects beyond the horizon. It is most frequently observed at sea. (L.D.W.)

MIRRORS AND LENSES. We shall here be chiefly concerned with spherical or plane reflecting and refracting surfaces. A spherical mirror may be treated as a 1-base zone of a spherical surface, the axis of which is the straight line passing through the center of curvature and the pole of the zone. When the diameter of the mirror is small compared with its **radius of curvature,** and when the rays make only small angles with the axis, so that **spherical aberration** may be neglected, such a mirror produces fairly sharp images which are easily calculated from the laws of **reflection.** Taking the pole O as origin and the mirror axis as X-axis (Fig. 1), and representing the radius OC by r, the image of a point $P_1(x_1, y_1)$ in the plane XY is the

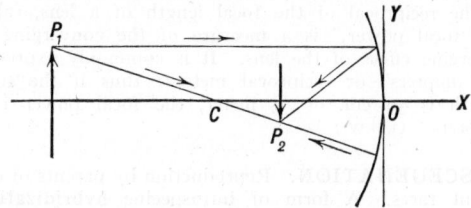

Fig. 1. Formation of real image by concave mirror.

point P_2 whose coordinates, for either a concave or a convex mirror, are

$$\left.\begin{array}{l} x_2 = \dfrac{rx_1}{2x_1 - r}, \\[2mm] y_2 = \dfrac{-ry_1}{2x_1 - r}. \end{array}\right\} \quad (1)$$

r is $+$ or $-$ according as the mirror is convex or concave. If x_2 turns out positive, it means that the image is virtual. In Figure 1, r, x_1, and x_2 are all *negative*, as is y_2, so that the image is real and inverted. If the incident rays are parallel to the axis, the focus of the reflected rays, called the "focal point" of the mirror, is on the axis at $x = r/2$. For a plane mirror, $r = \infty$ and $x_2 = -x_1$, $y_2 = y_1$.

Spherical lenses have various combinations of convex, concave, or plane surfaces. There is always a point on the axis, called the "optical center," such that if a ray in traversing the lens is in line with this point, the entering and emerging parts of the ray are parallel. For a very thin lens this point may be considered at the center of the lens and is a suitable origin (O, Fig. 2). If

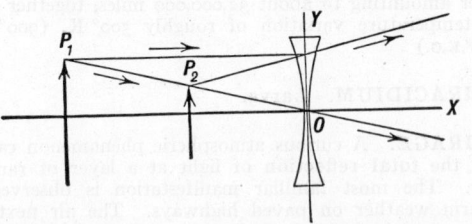

Fig. 2. Formation of virtual image by "negative" lens.

the radii of curvature r_1, r_2 are large compared with the diameter of the lens, and if the refractive index of the lens is n, the equations giving the image of the point x_1, y_1 made by a thin lens are

$$\left.\begin{array}{l} x_2 = \dfrac{r_1 r_2 x_1}{r_1 r_2 + (n - 1)(r_2 - r_1)x_1}, \\[3mm] y_2 = \dfrac{r_1 r_2 y_1}{r_1 r_2 + (n - 1)(r_2 - r_1)x_1}. \end{array}\right\} \quad (2)$$

r_1 and r_2 refer to the left and right surfaces, respectively, which is the order in which the light encounters them. P_1 and P_2 in either figure are called conjugate points of the system. The "focal length" is obtained by letting $x_1 = \infty$ and calculating x_2 from (2), which gives

$$f = \dfrac{r_1 r_2}{(n - 1)(r_2 - r_1)}. \quad (3)$$

This enables us to write (2) in simpler form:

$$\left.\begin{array}{l} x_2 = \dfrac{fx_1}{f + x_1}, \\[2mm] y_2 = \dfrac{fy_1}{f + x_1}. \end{array}\right\} \quad (4)$$

If x_2 turns out negative, the image is virtual; if y_2 is negative, it is inverted. For lenses of appreciable thickness or of strong curvature, the calculations are not so simple, and in general there are two unequal focal lengths, depending upon which way the rays pass through the lens.

The reciprocal of the focal length of a lens, called its "focal power," is a measure of the converging or diverging effect of the lens. It is commonly expressed in "diopters" or reciprocal meters; thus if the focal length is 50 cm. or ½ meter, the focal power is 2 diopters. (L.D.W.)

MISCEGENATION. Reproduction by parents of different races. A form of intraspecific **hybridization**. (A.W.L.)

MISFIT STREAM. U-Valley.

MISPICKEL. Arsenopyrite.

"MISSING LINK." Paleontology of Man.

MISSISSIPPIAN PERIOD. A geologic period in the **Paleozoic** Era. Term first proposed by H. S. Williams in 1891. Type locality, Mississippi Valley. The period began about 280 million years ago and lasted for about 25 million years. The term Mississippian is roughly equivalent to the more general term, Lower **Carboniferous**. In Britain, the formations of this system are grouped under the terms **Culm** and **Mountain Limestones**, which, as in the United States, immediately succeed the **Devonian** and are followed by the upper Carboniferous, or **Pennsylvania** System (U. S.), and Coal Measures (Britain). The formations of this system are chiefly sandstones and shales in the **Appalachian Geosyncline**, representing **delta** and estuarine deposits of considerable thickness which pass Westward into thinner facies of marine **shales** and **limestones**. In the Rocky Mountain region occur a great thickness of marine Mississippian called, locally, the **Madison Limestone**. The marine life of the Mississippian is chiefly characterized by **echinoderms** and **foraminifera**. Petroleum occurs in the Mississippian formations of Southeastern Ohio, West Virginia, Southeastern Pennsylvania and Eastern Kentucky. (R.M.F.)

MISSOURITE. The term proposed by Weed and Pirsson, in 1896, for a **basic, intrusive igneous** rock of **granitic texture** containing an abundance of **pyroxene**, **olivine** and **leucite**. This rock is the deep-seated equivalent of leucite **basalt**. Type locality, Highwood Mts., Montana. (R.M.F.)

MIST. Suspended, very small water droplets with a grayish appearance are internationally recognized as mist. This type of phenomena is known as damp haze in the United States. (P.E.K.)

MISTLETOE. Parasitic Plants.

MITE. Arachnida, Acarina. Minute animals related to the spiders. They have a compact body without well marked regions. The many species include scavengers and external parasites, many of them serious pests of domestic animals and man.

The group includes the harvest mites, also called the **chigger** or jigger in the United States, the red **spider**, and numerous species known as mites. Among the latter are the scab mites, mange mites, and itch mites which attack animals and man, and the gall mites which live on plants.

The forms called fresh-water and salt-water mites are ticks. (A.W.L.)

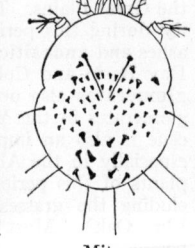

Mite.

MITER GEARING. Bevel Gearing.

MITOCHONDRIA. Cell.

MITOSIS. Cell Division.

MIXED ACID. Mixtures of **nitric acid** and **sulfuric acid** used for nitrating organic substances, such as glycerol, toluene, phenol, in the manufacture of **explosives** and **plastics**. A standard acid consists of 36% nitric acid, 61% sulfuric acid and 3% water. Different strengths are used for various purposes, for example,

Nitration of	Mixed Acid Used		
	H₂SO₄	HNO₃	H₂O
Cellulose, for military nitro-cellulose................	74	18	8
Glycerol, for nitroglycerin...	50	50	—
Toluene, for mononitrotol-uene................	58	24	18
Dinitrotoluene, for T.N.T...	68	24	8

The price of mixed acid is based on the "N unit" (the price of 1 lb. of contained HNO₃) and on the "S unit" (the price of 1 lb. of contained H₂SO₄) in the quantity of acid involved. (R.K.S.)

MIXED CYCLE. High-speed internal combustion engines perform their thermodynamic cycles of operation so rapidly that, whether originally intended to operate on the **Diesel** or the **Otto** principle, the actual cycle has a combustion that exhibits both constant volume and constant pressure phases. Spark ignition engines must receive the timed spark in advance of piston dead center position. Compression ignition engines must have the injector timed to begin the fuel spray also ahead of dead center position. In either case, on account of high rotative engine speed the combustion is only partially completed before the piston begins the expansion stroke. The remaining fuel is burned approximately at constant

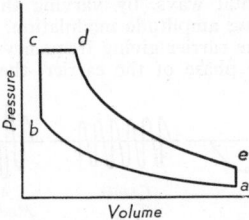

pressure. A conventionalized cycle is shown in Fig. 1. Combustion extends from b to d. The air standard efficiency expression is:

$$E = 1 - \frac{1}{r^{\gamma-1}}\left[\frac{ZR^\gamma - 1}{(Z-1)+Z\gamma(R-1)}\right]$$

where r = Ratio of compression along ab.
γ = Adiabatic gas constant (specific heat ratio).
R = Cutoff ratio V_d/V_c.
Z = Combustion pressure ratio P_c/P_b.

All factors which influence either Otto or Diesel cycle efficiency appear here, and in addition a factor introducing characteristics of the constant volume phase of combustion. (F.T.M.)

MIXER. Superheterodyne.

MIXING. Mixers for agitating mobile liquids can be divided into two general types, namely (1) propeller, which gives a flow parallel to the axis of rotation, and (2) radial, which gives a flow perpendicular to the axis of rotation. The most common radial types are paddles and turbines.

The power required to drive an impeller can be found by use of the following formula:

$$P = KN^{3-a}D^{5-2a}\rho^{1-a}\mu^a$$

where P = Power input to the shaft.
K = A constant characteristic of the tank geometry.
N = Shaft speed.

D = Impeller diameter.
ρ = Density of the fluid.
μ = Viscosity of the fluid.
a = A number varying from one to zero and dependent on the value of the modified Reynolds number, $\dfrac{ND^2\rho}{\mu}$, "a" has a value of one at low values of the Reynolds number.

Once the constant K and the variation of "a" with the Reynolds number have been established in a laboratory scale tank the equation can be used to "scale up" to other geometrically similar systems. Both tank and impeller must be kept similar.

Mixers may be used wherever agitation of a liquid is desired. Such cases may include blending of liquids, solution of solids, crystallization, gas absorption, suspension of a solid, e.g., a catalyst, or for **heat transfer** from tank walls or from a coil in the tank. The problem of anticipating the size, shape and speed of an impeller to do a specific job is still for the most part unsolved and recourse must be made to trial and error testing. Among the various criteria used as a measure of agitation intensity are: power input to the stirrer, periferal velocity, shaft speed, and Reynolds number. No one criterion has been found to hold for all cases.

The mixing impeller is usually mounted on a vertical shaft concentric with the tank, although side entering shafts or tilted shafts are used, the shaft usually having no bottom bearing because of lubrication and corrosion difficulties. The drive may be either direct or by pulley.

To gain increased turbulence and to break up the vortex caused by the swirling liquid, *baffles* are frequently put into the tank. These are usually placed vertically at or near the outside wall of the tank, and causing the liquid swirling around the edge of the tank to be deflected into the center.

Thin liquids are usually stirred by small high-speed mixers, and thick liquids by large slow-speed types. When the mix becomes semisolid or doughy, a *kneading machine* is required. These machines consist of two Z-shaped knives which rotate slowly in a trough, and which lift, press, cut, and mix the materials. (D.E.M.)

MIXING OF AIR. Condensation in the Atmosphere.

MIZAR. Mizar (ζ **Ursae Majoris**) is, perhaps, the most interesting star in the "big dipper." It is probably the first **double star** ever observed. The fourth **magnitude** star Alcor forms with it a naked-eye double, and Mizar itself has a close companion which is telescopically visible. It was the first star observed as double by Riccioli in 1650. Tradition says that observation of the pair Mizar-Alcor was considered a good test of eyesight among the American Indians. If this was a difficult pair for them to separate, their eyesight could not have compared very favorably with that of modern times, for it is an easy double for most people.

As well as being the first visual double star to be discovered, Mizar also has the distinction of being the first **spectroscopic binary** discovered. In 1889 E. C. Pickering discovered that the **spectral lines** of this star were alternately double and single, a phenomenon which can only be adequately explained if the star is a close binary. In 1908 the fainter companion of Mizar and also the more distant, bright companion Alcor were both found to be spectroscopic binaries. (W.K.G.)

M.K.S. SYSTEM. That view of the **metric system**, in which it is thought of as based directly upon the present-day metric fundamental standards, the meter, the kilogram, and the mean solar second; whereas the **c.g.s. system** is based on the centimeter, gram and second. For example, the m.k.s. unit of work or energy, the joule, is equal to 1 kg. meter²/sec.² or 1 meter-newton, while the corresponding c.g.s. unit, the erg, is 1 gram cm.²/sec.² or 1 centimeter-dyne.

Many writers prefer the m.k.s. system because the units are in general more convenient for practical purposes. The whole intricate array of electrical and magnetic units used by engineers, for example, is on the m.k.s. basis. While the demonstration of this fact would in many cases involve quite technical applications of the theory of physical dimensions, a single, comparatively simple example will serve to illustrate it.

In the article on **Electromagnetism** is mentioned the lateral force or thrust experienced by a conductor carrying a current across a magnetic field, as in an electric motor. If the current is denoted by I, the magnetic field intensity by H, and the length of the current's path, at right angles to the field, by l, then this force f is proportional to the product IHl. Electrical and magnetic units may be so defined that the force is numerically equal to that product:

$$f = IHl. \tag{1}$$

For our present purpose, the relation is more conveniently written

$$I = \frac{f}{Hl}. \tag{2}$$

In the c.g.s system the unit of force is the dyne, that of length is the centimeter, and that of magnetic intensity is the oersted, or maxwell per sq. cm. The above equation (2) now serves to define the c.g.s. absolute electromagnetic unit current. This unit, called the abampere, is that current which, flowing across a magnetic field of 1 oersted magnetic intensity, would experience a lateral force of 1 dyne per cm. length of wire. Thus, using Eq. 2,

$$I = 1 \text{ abampere} = \frac{1 \text{ dyne}}{1 \frac{\text{maxwell}}{\text{cm.}^2} \times 1 \text{ cm.}} = 1 \frac{\text{dyne cm.}}{\text{maxwell}}. \tag{3}$$

Now let us use m.k.s. units and take $f = 1$ newton $= 100,000$ dynes, $l = 1$ meter $= 100$ cm., and $H = 1$ weber/meter² $= 10^8$ maxwells/10,000 cm.² $= 10,000$ $\frac{\text{maxwells}}{\text{cm.}^2}$. Using these values in Eq. 2, we have

$$I = \frac{100,000 \text{ dynes}}{10,000 \frac{\text{maxwell}}{\text{cm.}^2} \times 100 \text{ cm.}} = \frac{1}{10} \frac{\text{dyne cm.}}{\text{maxwell}}$$

$$= \frac{1}{10} \text{ abampere.} \tag{4}$$

But this current, $\frac{1}{10}$ abampere, is called the ampere. The ampere might thus be defined as that current which, flowing across a magnetic field of intensity 1 weber/meter², would experience a lateral force of 1 newton per meter length of wire. Clearly, therefore, the ampere is the m.k.s. absolute electromagnetic unit current.

Similarly, the absolute coulomb, volt, ohm, henry, farad, etc., that is, the so-called "practical" units of electrical engineering, all belong to the m.k.s. absolute joule and watt in the solution of the practical problems of electromechanics. (L.D.W.)

MOA. Aves, Apterygiformes. Giant flightless birds (**Aves**) of New Zealand. Now extinct, although they probably existed until about 500 years ago. They are known from skeletons, feathers, dried remains of soft parts, and egg shells. The largest moas, *Dinornis maximus,* were about 12′ tall. (A.W.L.)

MOCCASIN. Reptilia, Sauria. A name applied to two poisonous snakes of North America, the **copperhead** and the **cottonmouth**. (A.W.L.)

MODAL CLASS. The modal class in a **frequency distribution** is the class which has the greatest frequency. (L.A.A.)

MODE. The mode, one of the common **measures of central tendency**, is the value of that **variate** which occurs most frequently in a distribution. In a **serial distribution** it is found by observation. In a **frequency distribution** an empirical formula for the mode is $l + c \dfrac{\Delta_1}{\Delta_1 + \Delta_2}$ where l is the lower closed **class limit** of the **modal class**, c is the class interval, Δ_1 is the **absolute value** of the difference of the frequency of the modal class and the class preceding the modal class, and Δ_2 is the absolute value of the difference between the frequency of the modal class and the class succeeding the modal class. In a **probability function**, the mode is the value of x which determines the maximum value of the function. Usually distributions are unimodal.

A term used in Iddings and Washington's classification of the **igneous rocks** (1902) for the actual mineral composition expressed quantitatively in percentages by weight. As opposed to the chemical composition, expressed in the standard mineral molecules in terms of **oxides,** and called the **norm**. (L.A.A., R.M.F.)

MODEL ANALYSIS. Similitude.

MODULATION. Modulation is the process of impressing the intelligence to be transmitted upon the higher frequency radio or telephone carrier wave. The carrier may be altered in accordance with the intelligence (speech, music, television picture signal, etc.) in three fundamental ways, by varying the amplitude of the carrier giving amplitude modulation, by varying the frequency of the carrier giving frequency modulation, or by varying the phase of the carrier, thereby producing

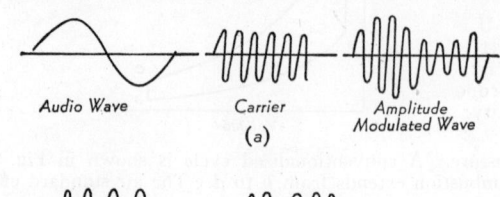

Audio Wave   Carrier   Amplitude Modulated Wave

(a)

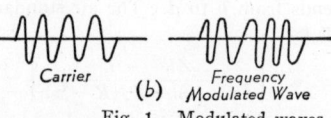

Carrier   (b)   Frequency Modulated Wave

Fig. 1. Modulated waves.

phase modulation. Until fairly recently amplitude modulation was used almost exclusively for radio and is still used exclusively for telephone carrier work. Reference to Fig. 1a will indicate the principal characteristics of an amplitude-modulated wave. The amplitude of the carrier wave is caused to vary in accordance with the audio-frequency wave, i.e., the envelope of the carrier is identical with the audio wave. While the figure shows the carrier modulated with a single tone, the usual case involves a very complex sound wave, but even here the envelope is the same as the sound wave. Such a modulated wave may be analyzed and is found to be really several constant amplitude waves of different frequency. These frequencies are the original carrier and the original carrier frequency plus and minus every audio frequency. Thus for ordinary speech which is a continuous band of frequencies the modulated wave will consist of a band of frequencies from the carrier plus the highest audio to the carrier minus the highest audio. These extra, or side, frequencies constitute the sidebands. This is the reason a definite channel width is required to transmit a modulated signal and why the receiver must pass a band of frequencies in order to reproduce the original sound. The degree of modulation is known as the modulation factor and for amplitude modulation is the increase (or decrease) of amplitude under modulation divided by the original amplitude. It is thus limited to 1 or 100%, since the wave cannot be reduced below zero. Frequency modulation, which has undergone very rapid development in the last few years, is obtained by varying the frequency of the carrier in accordance with the audio wave. The degree of modulation is often expressed by the deviation ratio which is the frequency modulation factor divided by the audio frequency. Fig. 1*b* indicates this type of modulation. It will be noted that while the carrier wave crosses the axis at equal intervals the modulated wave crosses at varying intervals as its frequency is changed. This wave is the combination of a much wider band of frequency components than the amplitude-modulated wave. Primarily for this reason frequency modulation is used only on very high frequencies. One of the major advantages of a system using this type modulation is its freedom from noise and interference. This inherent freedom from noise makes it possible to transmit a wider range of audio frequencies than is ordinarily possible with amplitude modulation. FM has thus become known as the high-fidelity system although the real distinction is the difference in noise, since, if noise could be eliminated, amplitude modulation could give just as wide response. The use of very high frequencies limits the coverage range of the frequency modulation stations, but since they have low interference characteristics the stations may be located closer together and thus produce the desired complete coverage. Phase modulation is used to a limited extent only. Modulation may be produced by several methods but **vacuum tubes** are used in all modern radio installations while vacuum tubes and oxide rectifier units are widely used in **telephony.** (L.R.Q.)

MODULUS. A ratio; see **Bearing Modulus, Section Modulus, Modulus of Elasticity.**

MODULUS OF ELASTICITY. The ratio of the unit **stress** to the unit **deformation** of a structural material is a constant, as long as the unit stress is below the **proportional limit,** and is called the modulus of elasticity. The shearing modulus of elasticity is frequently called the modulus of ridigity. (See **Ultimate Strength, Proportional Limit, Tension Test.**) (C.W.C.)

MODULUS OF RIGIDITY. Modulus of Elasticity.

MODULUS OF RUPTURE. The modulus of rupture in bending of a material is found by testing a transversely loaded beam of constant cross-section to failure and substituting the maximum **bending moment, moment of inertia** of the cross-section and the distance from the **neutral axis** to the extreme fiber in the **flexure** formula:

$$p = \frac{My}{I}.$$

The torsional modulus of rupture is obtained by testing a shaft of constant, circular cross-section to failure and then substituting the maximum **torque,** polar moment of inertia of the cross-section and the radius in the torsion formula:

$$s_s = \frac{Tc}{J}. \quad \text{(See \textbf{Torsional Stress.})}$$

The bending or torsional modulus of rupture may be used to predict the maximum bending or torsional moment which a member can resist. (C.W.C.)

MODULUS OF RESILIENCE. Resilience.

MOERITHERIUM. Fossil Mammals.

MOHS SCALE. Hardness.

MOISTURE CONTENT OF THE AIR. Humidity.

MOLAR. The broad grinding teeth of mammals, located at the back of each jaw. (A.W.L.)

MOLAR SOLUTION. Concentration.

MOLASSES. Sugar.

MOLD. Casting; Phycomycetes.

MOLE. Mammalia. Burrowing animals of small size, highly specialized for life underground. They have short legs but the front pair are powerful with broad feet and strong claws. The eyes are rudimentary and there are no external ears.

The true moles of the Palaearctic region belong to the order Insectivora, which also contains the closely related shrew moles and web-footed moles. All of these forms are related to the shrews. The golden moles of Africa, *Chrysochloris,* are animals of similar form and habits, related to the **tenrecs.**

In the order Marsupialia a single rare Australian animal adapted for subterranean life is called the pouched mole or **marsupial mole.**

The name appears also in the mole voles, **mole rat,** and Cape mole **rats,** all burrowing animals of the Old World belonging to the order Rodentia. They are much less extremely adapted than the moles.

Mole (or mol) is also a common name for gram molecule. (A.W.L.)

MOLE CRICKET. Insecta, Orthoptera. Burrowing crickets whose large fore legs give them a superficial resemblance to moles. (A.W.L.)

MOLECULAR EXCITATION. In physics, excitation usually refers to the process of putting an atom or a molecule into a condition in which the total energy of its interior mechanism is greater than it is in the normal or "ground" state. (This does not refer to energy of translational motion of the particle as a whole.) According to the **quantum theory,** the energy required to accomplish such a change must be supplied in certain definite amounts or quanta. A less amount than this "excitation limit" apparently cannot be received by the atom or molecule. The necessary energy may be supplied by a **collision** with another atom or with an electron (as a cathode particle), or by the advent of a radiation quantum. In the former case the impinging atom or electron must have at least a certain speed (see **Critical Potential**); in the latter, the frequency of the radiation must be sufficient to give its quanta the necessary energy (see **Planck's Law**).

Upon returning to the normal state, or to a state of lower excitation, the atom or molecule yields up this extra energy; ordinarily by emitting a radiation quantum of this amount and of corresponding frequency; as in the emission of **light** or **x-rays,** by imparting the energy, in a "second-class" collision, to another atom or molecule, or by hurling off particles of itself, as in induced radioactivity. Familiar examples of excitation are the production of incandescence by heating, and the glowing of phosphorescent substances after exposure to light or x-rays. (L.D.W.)

MOLECULAR PUMP. Air Pumps.

MOLECULAR RAYS. If a narrow opening or slit is made in the wall of an enclosure containing a gas or a vapor, surrounded by a vacuum, those molecules which chance to encounter the opening pass out through it—a process known as effusion. The escaped molecules move in a wide variety of directions. But if, instead of a simple slit, we have a succession of similar, parallel slits in plates set one in front of the other, the only molecules finally emerging will be those whose directions of motion are nearly parallel with the common axis of the slits. Such an emergent stream is a beam of "molecular rays." Much study has been devoted in recent years to such beams. By heating the enclosure, the vapors of metals may be studied in this way. To detect a beam of metallic molecular rays, a "target" of very cold glass or porcelain may be interposed for it to condense upon in the form of a visible spot. The motions of the molecules are unidirectional (few collisions) and their speeds should be in accord with the **Maxwell distribution law**; a point which has been ingeniously verified by Eldridge. Many studies have been made possible regarding the properties and behavior of individual molecules; the **Stern-Gerlach experiment** is a notable example. The importance of this field is just beginning to be realized. (L.D.W.)

MOLECULAR SPECTRA.

The spectra of substances in the molecular state, like **atomic spectra**, are really made up of lines, though they are much more complicated. The transitions in a molecule which release the most energy (largest quanta) are due to electron changes, as in atoms, and the results of these changes are observed as lines in the **ultra-violet** region. But there are other ways in which a molecule can release or absorb energy. Thus the component atoms oscillate with reference to each other within the molecule, and this motion apparently is "quantized," i.e., changes abruptly from one state to another of different energy. (See **Quantum Theory.**) But these "vibrational" energy changes are much less than the electronic, so that the resulting quanta and spectrum lines are of much lower frequency, and appear in the extreme red or near **infra-red**. Again, the molecule rotates, and the quantization of its rotational energy results in the emission of quanta of still lower frequency, appearing as lines in the far infra-red.

Atomic spectra (due to electronic transitions) are characterized by series of lines progressively crowded together toward a "series limit." This is due to a variety of possible transitions of successively greater energy. The same is true of the changes in molecular rotational energy, but here the differences between the successive quantum energies are so very small and the lines are thus crowded so close together that a whole series of them appears merely as a "band," coming to a sharply defined edge on the low-frequency side and fading away gradually on the other side, or vice versa. Not only this, but the vibrational and electronic transitions are accompanied by rotational transitions, giving combined spectra which, on account of the close-grained character contributed by the rotational component, is composed of bands like the rotational bands themselves.

Thus a molecular spectrum appears as an array of bands instead of distinct lines, but arranged, like lines, in groups and series. The study of these bands and their groupings has furnished a surprising amount of information as to the structure and internal mechanism of molecules. (L.D.W.)

MOLECULAR WEIGHTS. Chemical Composition.

MOLECULES. Chemical Composition.

MOLLIER DIAGRAM.

The properties of a vapor, as recorded in vapor tables, may be displayed graphically in a number of ways, among which the most used, and probably the most valuable, is the charting upon a plane whose coordinates are enthalpy or total heat and **entropy**. Generally, the total heat is made the ordinate, and entropy the abscissa. This chart of the properties of vapor is named the Mollier Diagram, and is of considerable use in tracing both theoretical and actual **expansions** of vapor. A throttled expansion on the Mollier Diagram is parallel to the constant heat lines, and **adiabatic** expansion is parallel to the constant entropy lines. Pressure, quality or superheat, and total temperature are shown on the Mollier Diagram as series of lines curved and inclined to the axes. Thus all characteristics of a vapor except volume may be displayed on the Mollier Diagram. (F.T.M.)

MOLLUSCA.

A major division of the animal kingdom containing the **snails, oysters, clams, mussels, squids, octopus, nautilus** and related forms. **Mollusks** are the most highly developed of the unsegmented invertebrates and are both diverse in form and numerous in species.

The phylum is characterized by the following structures: 1. The body is unsegmented. 2. A well-developed head is found in most species. 3. The body bears a ventral muscular protuberance, the foot. 4. A fold extends in most species from the dorsal wall, enclosing a cavity associated with respiration. The fold is the **mantle** and the cavity, the mantle cavity. 5. In many species the mantle secretes a shell. 6. The **circulatory system** consists of tubular vessels and open spaces, with a **heart** made up of a ventricle and two auricles.

Some mollusks are important as food. Clams, oysters, and scallops are the most familiar of the edible species but others are eaten. Pearls and mother-of-pearl are also molluscan products.

The phylum is divided into the following classes:

Class **Amphineura**. Without a distinct head. Shell absent or composed of a series of plates. **Chitons.**
Class **Gasteropoda**. With a distinct head. Shell absent, conical, or spiral. **Snails** and related forms.
Class **Scaphopoda**. Head indistinct. Shell cylindrical.
Class **Lamellibranchiata** (Pelecypoda). Head indistinct. Shell of two lateral parts (bivalve). **Clams, mussels, oysters, etc.**
Class **Cephalopoda**. Head distinct, with long tentacles. **Squids, octopus, nautilus, etc.**

See also **Invertebrate Paleontology**. (A.W.L.)

MOLLUSCOIDEA.

A name originally applied to a major group of animals including the **Bryozoa** and **Tunicata** and now used by some biologists to include the Bryozoa and the **Brachiopoda**. (See also **Invertebrate Paleontology**.) (A.W.L.)

MOLOCH.

Reptilia, Sauria. An Australian **lizard** found in arid country. It is about 8" long and resembles the horned toads of the southwestern United States in form, but is covered with stout spines which give it a forbidding appearance. Like the horned toads it does not eat readily in captivity. (A.W.L.)

MOLPADONIA. Holothuroidea.

MOLYBDENITE.

The mineral molybdenite is **sulfide** of **molybdenum**, MoS_2. Its **hexagonal** crystals are usually tabular to short prismatic, but if in massive form it may be **foliated** or granular. Has a perfect basal cleavage; is sectile; hardness, 1–1.5; specific gravity, 4.7–4.8; luster, metallic; color, very slightly bluish, lead gray; streak, greenish-gray; opaque. Molybdenite is one of the few minerals soft enough to give a distinctly greasy feel. Molybdenite is found as a contact mineral with **cassiterite** and **wolframite**, in granite **pegmatites** and sometimes in granites, **syenites**, or **gneisses**. It is found associated with tin ore in Saxony and Bohemia; in Norway, England, Australia; and in the United States in Washington County and Oxford County, Maine; in New Hampshire, Connecticut, Pennsylvania, and Washington. Its name is derived from the Greek meaning lead, was formerly applied to minerals containing lead, to graphite and to molybdenite as well. Later the term was restricted to the latter mineral. It is an ore of molybdenum, and the chief commercial source in the United States is the Climax mine, Colorado. (E.S.C.S.)

MOLYBDENUM. Symbol: Mo. Atomic number: 42. Atomic weight: 95.5. Density: 10.2. Melting point: 2622° C. (Isotopes, page 290.)

Molybdenum is a silver-white, tough, malleable metal softer than glass, not oxidized by air at ordinary temperatures but above 600° C. burns to form white molybdenum oxide; dissolved by dilute **nitric acid,** and by **aqua regia;** made passive by concentrated nitric acid, and attacked by fused **alkalis.** Chemically related to **chromium, tungsten** and **uranium** elements. Discovered by Scheele in 1778.

The principal use of molybdenum metal is in the production of alloy steels, 0.15 to 0.35% molybdenum in steel for heat-treated machine parts and for some structural grades, up to 1.5% for *tool steel* and heat-resistant grades, up to 4% for *stainless steel* (especially corrosion-resistant), and up to 8.5% for high-speed tool steel. A certain molybdenum high-speed cutting steel contains 8.0 to 9.5% Mo, 3.5 to 4.0% chromium, 1.3 to 1.8% tungsten, and 0.9 to 1.3% vanadium. Molybdenum steel is extensively used in the automotive, airplane, and petroleum refining industries.

Molybednum occurs as **molybdenite** (molybednum sulfide, MoS_2), and **wulfenite** (lead molybdate, $PbMoO_4$). Roasting of molybdenite in a current of air yields white crystalline sublimate of trioxide, from which the metal is obtained by reduction with **aluminum** or **carbon** at high temperatures. Fusion of the ore with carbon and **sodium** carbonate in a blast furnace yields a sodium molybdate matte from which ferro-molybdenum is made in an electric furnace.

Hydroxide: Molybdenum trihydroxide ($Mo(OH)_3$), brownish black solid: molybdenum tetrahydroxide ($Mo(OH)_4$, possible $MoO(OH)_2$).

Oxides: Molybdenum dioxide (MoO_2), brown and sometimes blue solid by reduction of the trioxide or ignition of ammonium molybdate ($NH_4)_2MoO_4$; molybdenum sesquioxide (Mo_2O_3), black solid by the reduction of the trioxide with zinc metal; molybdenum trioxide (MoO_3), white, somewhat volatile, solid, yellow when hot, greenish-yellow after ignition, soluble in alkalis to form molybdates, which form complex compounds with excess trioxide. The trioxide is the source to most molybdenum compounds.

Many reducing agents, e.g., **hydrogen, zinc** metal, **tin** metal, **ferrous** salts, **sulfurous acid,** sucrose react with molybdates usually forming a mixture known as "Molybdenum blue." Lower valence forms are readily oxidized to molybdate by ignition or by nitric acid.

Ammonium phosphomolybdate, yellow precipitate, of variable composition, and ammonium arsenomolybdate, similar, are important compounds in the identification of **phosphates** and **arsenates** respectively. A common test for molybdenum is as follows: **ammonium** phosphate added to a molybdate solution strongly acidified with **nitric acid,** produces a yellow crystalline precipitate. Arsenates give the same type of precipitate but arsenic can be eliminated by other tests. (See **Arsenic.**) (R.K.S.)

MOMENT. Moment consists of the product of a quantity and a distance to some significant point connected with that quantity. The principal moments are moments of forces, moments of lines, moments of areas, and moments of masses. Two types of moments are statical moment and the **moment of inertia.** Unless specifically stated to be otherwise (see **Statics**), the word moment would be taken to mean statical moment. A physical picture of moment may be obtained by con-

sidering the moment of a **force** (called torque). It is the magnitude of the force multiplied by the moment arm which is a perpendicular dropped from the moment center to the line of action of the force. This moment is the turning effect on a body against which the force is applied. The moment of an area is the magnitude of the area multiplied by the perpendicular distance from the centroid of the area to the axis of moments. Similarly the moment of a solid is its weight multiplied by the distance from its center of mass to the axis of moments.

The summation of moments of a force enters into one of the three equations of statical equilibrium (see **Statics**). In a case of equilibrium, this summation must always equal zero about any chosen moment center. The moment of an area about a line is of use in finding the centroid of an area, since the magnitude of the area multiplied by the distance to a parallel line through the centroid must equal the summation of all the incremental areas multiplied by the perpendicular distances from the centers of these areas to this reference line. Thus in the irregular figure shown the distance from the Y-reference axis (any convenient line) to a

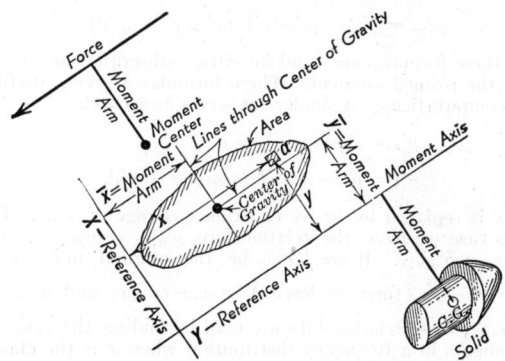

Moments of force, area and mass.

parallel line through the centroid of the figure is determined by summing the elementary moments, i.e., ay, and equating the summation to the product of the total area, A, and the centroidal distance $\bar{y}$.

$$\bar{y} = \frac{\text{Summation of } ay}{A}.$$

In a similar manner the location of a line, through the center of gravity, may be obtained in relation to the X-reference axis (any other convenient line not parallel to the Y-reference line). That is

$$\bar{x} = \frac{\text{Summation of } ax}{A}.$$

The intersection of these two lines, whose location is given by $\bar{x}$ and $\bar{y}$, is the center of gravity of the area. The center of gravity of a solid may be found in a similar manner by the use of three reference axes.

The use of the statical moment in finding centroid of areas or masses extends to areas and shapes which are made up of a number of elementary areas or masses whose centroids are common knowledge. Thus a trapezoid could be considered as made up of two triangles; a rivet, of a hemisphere and a cylinder. This method is also useful in computing centroids when the outline of the figure is expressed by a mathematical equation, since then the summation of ax and ay may be obtained as definite integrals. Cases of irregular figures which would not be analyzable by either of these methods can be treated by graphical means.

In the case of *statistical* moment, let us consider a variate x with either a **probability function** P_x, or a frequency distribution f_x; then we define the nth moment about any point a as

$$\mu'_{n:x-a} = \frac{\Sigma(x-a)^n P_x}{\Sigma P_x} \quad \text{or} \quad \frac{\Sigma(x-a)^n f_x}{\Sigma f_x}$$

and for the case of a continuous probability function by

$$\mu'_{n:x-a} = \int_{x=x_1}^{x=x_2} (x-a)^n P_x dx$$

where $\int_{x=x_1}^{x=x_2} P_x dx = 1$.

We note $\mu'_{1:x} = \overline{X}$, and $\mu'_{1:x-a} = \mu'_{1:x} - a$. If $a = \overline{X}$, the **mean**, we obtain in all cases the central moments $\mu_{n:x} = \mu'_{n:x} - \overline{x}$. Some formulas for obtaining the central moments are

$$\mu_{0:x} = 1, \quad \mu_{1:x} = 0,$$

$$\mu_{2:x} = \text{the variance} = \sigma_x{}^2,$$

$$\mu_{2:x} = \mu'_2 - \mu'_1{}^2,$$

$$\mu_{3:x} = \mu'_3 - 3\mu'_2\mu'_1 + 2\mu'_1{}^3,$$

$$\mu_{4:x} = \mu'_4 - 4\mu'_3\mu'_1 + 6\mu'_2\mu'_1{}^2 - 3\mu'_1{}^4;$$

all these formulas are valid for either subscript x or $x - a$ for the primed moment. These formulas are very useful in computations. A similar set serves as a check

$$\mu'_3 = \mu_3 + 3\mu_2\mu'_1 + \mu'_1{}^3,$$

$$\mu'_4 = \mu_4 + 4\mu_3\mu'_1 + 6\mu_2\mu'_1{}^2 + \mu'_1{}^4.$$

If x is replaced by bx we call this a change of scale. In this case we have the relationships $\mu'_{n:bx} = b^n\mu'_{n:x}$, and $\mu_{n:bx} = b^n\mu_{n:x}$. If we let x be the original units and $d' = \dfrac{x-a}{c}$, then we have $\overline{X} = a + c\mu'_{1:d'}$, and $\mu_{n:x} = c^n\mu_{n:d'}$; these relationships are used in finding the central moments in a frequency distribution where c is the **class interval**, x is the class mark in original units, and d' is the class mark in **class interval** units about an arbitrary origin a. The **higher moments** $\mu_{3:x}$ and $\mu_{4:x}$ are used to define $\alpha_{3:x} = \dfrac{\mu_{3:x}}{\sigma_x{}^3}$, a measure of **skewness**, and $\alpha_{4:x} = \mu_{4:x}/\sigma_x{}^4$ a measure of **kurtosis**. The notion of moments may be extended to any number of **variables**. (L.A.A., F.T.M.)

MOMENT COEFFICIENT. The reaction of air upon an **airfoil** varies with the **angle of attack** both in magnitude and position of its result. The moment of the air reaction about some moment center located on the chord and the airfoil is, to a fair degree of approximation, at least, equal to the lift multiplied by the distance to the center of pressure. A coefficient which will bear the same relationship to moments as do coefficients of lift and drag to the lift and drag forces, is obtained by dividing the moment by $\dfrac{\rho}{2}SV^2c$. ρ is the mass density of air, S wing area, V air speed, c wing chord. The moment coefficient so defined is nearly a constant if the moment center used coincides with the aerodynamic center of the airfoil. On most airfoils, this is about one-quarter of the chord back of the leading edge. It is this quality of the aerodynamic center that causes its frequent use as a reference point in aerodynamic analyses of airplanes. (F.T.M.)

MOMENT OF INERTIA. The moment of inertia of a body with respect to any straight line as an axis has reference to the opposition which the body would offer, by reason of its **inertia**, to being set rotating about that axis if the body were rigid and free to rotate. It represents the **torque** or force moment required, per unit angular acceleration about the axis (in radian measure, e.g., radians per sec. per sec.). Thus if the moment of inertia is I and the angular acceleration is α, the required torque is $T = I\alpha$, which equation corresponds to the familiar $f = ma$ in the dynamics of linear motion.

If the body is divided into elements of mass dm, each at its respective distance r from the axis, it is easy to show that the moment of inertia is expressed by the summation or integral

$$I = \int r^2 dm. \tag{1}$$

Following are formulae for the moments of inertia of certain homogeneous solids with respect to the axes specified (M is the total mass of the body in each case):

Particle distant r from axis.............. Mr^2

Sphere of radius R, with respect to any diameter........................ $\frac{2}{5}MR^2$

Cube of edge L, with respect to axis through center parallel to edge................ $\frac{1}{6}ML^2$

Rectangular plate, dimensions $A \times B$, with respect to axis perpendicular to it at center............................ $\dfrac{M}{12}(A^2 + B^2)$

Cylinder of length L and radius R, with respect to axis perpendicular to its length at center..................... $M\left(\dfrac{L^2}{12} + \dfrac{R^2}{4}\right)$

Cylinder of radius R, with respect to its own longitudinal axis.................... $\frac{1}{2}MR^2$

Any body with respect to any axis distant r from the center of mass, the value for a parallel axis through that point being I_0...................... $I_0 + Mr^2$

Experimental methods of obtaining moments of inertia by the use of a **torsion pendulum** are explained in any laboratory manual of elementary dynamics.

For certain purposes it may be desirable to know at what one distance from the axis all the particles of the body of mass M would have to be placed to give it the same moment of inertia I that it actually has. This distance is the radius of gyration, and is expressed by the formula

$$R = \sqrt{\frac{I}{M}}. \tag{2}$$

The "principal axes" of a body through a given point are axes of maximum or minimum moment of inertia. (See **Dynamics of Rotation**.)

By reason of its mathematical analogy to Eq. 1, the quantity expressed by

$$I = \int r^2 da, \tag{3}$$

in reference to any plane figure, is called the **areal moment of inertia** of the figure with respect to a given straight line in its plane. The figure is divided into elements of area da, each element multiplied by the square of its distance r from the axis, and the products summed as indicated in (3) to get the areal moment of inertia. This quantity is purely geometric and has of course no actual connection with inertia or mass. One of its important applications is in the theory of **flexure** of elastic rods or beams. If E is the (Young's) elastic modulus of the material and I the areal moment of inertia of the cross-section with respect to the **neutral axis**, the bending moment or flexural torque required to bend the rod to a **curvature** C is given by

$$T = EIC. \tag{4}$$

(L.D.W.)

MOMENT-RESISTING JOINT. Rigid Frame.

MOMENTUM. The momentum of a body is the product of its mass and linear velocity, while the moment of momentum (or angular momentum) of a body is the product of its moment of inertia and angular velocity. Thus linear momentum is MV, and angular momentum is $I\omega$. Both linear and angular momentum are vector quantities.

$$M = \text{mass.}$$
$$I = \text{moment of inertia.}$$
$$\omega = \text{angular velocity.}$$

Because of the relation of momentum to force as set forth in the second of Newton's laws, this is a fundamental concept of dynamics.

A body tends to continue unchanged in momentum unless acted upon by external forces such as applied working forces, resistance, friction or air drag. The force F acting on a body for t seconds alters its momentum by Ft. The product Ft is known as the impulse, with a free body equal to change of momentum. In rotation, the corresponding quantity is the moment of impulse, that is, the product of the time by the applied torque or force moment.

One of the consequences of Newtonian dynamics is the principle of conservation of momentum. This states, in effect, that no operation of forces between the bodies in a system can change the momentum of the system as a whole, which is the vector sum of the momenta of its particles. As a simple example, take the collision of two balls. They may have very different velocities before and after the encounter, but their resultant momentum is the same; and this is true whether they collide "head on," or one overtakes the other, or one is at rest and is struck by the other; also whether they are moving in the same or in different lines. In the event of a collision or other suddenly applied force, the change of momentum resulting is the measure of the impulse, defined above. This is well illustrated by the impulse given to a croquet ball when another ball in contact with it is held by the foot and struck with the mallet. (F.T.M., L.D.W.)

MONADNOCK. Peneplain.

MONAL. Aves, Galliformes. Brightly colored **pheasants** of several species found in the higher forests of the mountains of Asia. (A.W.L.)

MONAZITE. The mineral monazite is essentially a **phosphate** of the rare-earth metal **cerium,** $CePO_4$, but other rare-earth metals are usually present. So constant is the presence of **thorium** that monazite is the chief source of thorium dioxide. It is **monoclinic,** but found ordinarily as translucent yellow to brown grains with a resinous luster, often as sand. Its hardness is 5.–5.5; specific gravity, 4.9–5.3. Monazite is found in **granites, pegmatites** and similar rocks, but rarely in any concentration. The commercial deposits are residual sands. The Ilmen Mts. in Russia, Norway, India, Madagascar, South Africa and Brazil are well known for their monazite deposits. In the United States monazite is known from Connecticut, New York, Virginia, North Carolina and Idaho. Monazite derives its name from the Greek word meaning solitary, in reference to the relative rarity of this mineral. (E.S.C.S.)

MONCHIQUITE. The term proposed by Rosenbuch and Hunter, in 1890, for a **basic microcrystalline** or **porphyritic dike** rock composed chiefly of **femic** minerals with little or no **feldspar,** in a matrix of **analcite.** Accessory minerals may be **olivine, nepheline** and **leucite.** (R.M.F.)

MONEL METAL. Nickel Alloys.

MONGOLIAN IDIOT. Mental Deficiency.

MONGOOSE, MUNGOOSE. Mammalia, Carnivora. Slender animals with short legs and a long tail, related to the civets but without scent glands. They live in Africa and the Oriental region and have been introduced successfully into the West Indies and other regions. Mongooses kill many small animals and are valuable for destroying rats and other vermin but will attack useful animals when the vermin supply is low. They are particularly noted for their ability to kill snakes, including the poisonous species. The India mongoose, *Herpestes mungo,* is readily tamed and is often kept to free premises of undesirable pests.

The Egyptian mongoose, *H. ichneumon,* is also called the ichneumon. (A.W.L.)

MONKEY. Mammalia, Primates. A name applied to most members of the order excepting the man-like **apes** and the **lemurs.** It is not as accurately defined as the names of the various kinds of monkeys but, in general, monkeys differ from the other forms in the possession of nails on all fingers and toes and in their long tails. Apes, some **baboons,** and a few monkeys have short tails or none, the **marmosets** have claws except on the great toe, and the lemurs and related forms have claws on at least one digit.

The monkeys of the New World form a well-marked group constituting the family Cebidae, characterized by the widely separated nostrils, the non-opposable **thumb,** and the usually long and prehensile **tail.** All of them inhabit the tropical forests of South and Central America.

The family includes monkeys of several types. These are the woolly monkeys, woolly spider monkeys, spider monkeys, and squirrel monkeys, and under distinctive names the sapajous or capuchin monkeys, the dourou-colis, titis, sakis, uakaris, and howlers.

The Old World monkeys, including the baboons, make up the family Cercopithecidae, characterized by the thin septum of the nose, the usually opposable thumb, and the lack of a prehensile tail. They occur in Africa and the Oriental region and one species, the Barbary ape, is established at Gibraltar.

This family includes, in addition to the baboons, the **langurs, guenons, mangabeys, macaques,** the peculiar proboscis monkey, the black ape and gelada baboon, and the African thumbless monkeys. (A.W.L.)

MONOCHROMATIC ILLUMINATOR. An instrument used to supply a beam of light having some desired, narrow range of wavelengths; sometimes called "monochromator." The common form resembles a prism spectroscope. White light, entering the fixed **collimator** as usual through a narrow slit, is dispersed by a special, 4-sided prism after one internal total reflection (see figure). The image of the resulting **spec-**

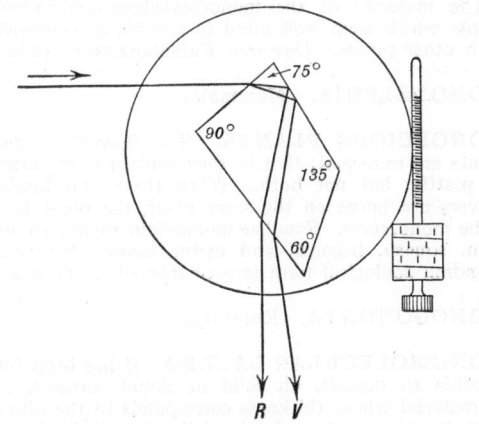

Diagram of dispersing prism and micrometer tangent screw of monochromator.

trum falls on a metal plate, and a second fixed, narrow slit in this plate allows light of approximately only a single wavelength to emerge. The purity of this beam depends upon the narrowness of the two slits, also upon **diffraction** at the slits and **scattering** by the prism. The prism is mounted so that it can be rotated by a tangent screw, thus bringing different parts of the spectrum to the second slit as desired. The tangent screw may be provided with a graduated head reading directly in wavelengths; but too much reliance must not be placed on the indications, since the adjustment is affected by fluctuations of temperature. For **ultra-violet** or **infra-red** radiations the prism and lenses must be of materials other than glass, or mirrors may be used instead of lenses. (L.D.W.)

MONOCLINE. If horizontal or slightly inclined beds or stratified formations change their **dip** by increasing their steepness of inclination and then flatten out or resume their normal gentle dip, such a structure is called a monocline. Monoclinical folds may pass into faults. (R.M.F.)

MONOCLINIC SYSTEM. Crystallography.

MONOCOQUE. Fuselage.

MONOCOTYLEDONS. One of the two subclasses of the **angiosperms**, containing about 25,000 species, many of which are of the greatest value to man. All the **cereal grains** and other grasses belong in this subclass. **Bananas, pineapples** and **palms** are tropical monocotyledons. Many monocotyledons are widely cultivated for their beautiful flowers. Among these are lilies, cannas, irises, and **orchids**.

Certain features are characteristic of plants of this subclass, distinguishing them from the **dicotyledons**. The stems show secondary growth only in a very few monocotyledons. The vascular bundles are usually scattered irregularly in the ground tissue of the stem. The leaves are typically parallel-veined, that is, the several main veins run parallel to one another from the base of the leaf to its tip. The parts of the flowers are generally in multiples of three; that is, there are three **sepals**, three **petals**, three (or six) **stamens**, and a compound **pistil** composed of three **carpels**. The **embryo** of the **seed** is quite unlike that of dicotyledons, having a single seed leaf or cotyledon. In many species this cotyledon is greatly modified, becoming an absorbing organ which never emerges from the seed during germination.

Almost all the monocotyledons are herbaceous plants of small size. A few, such as certain species of *Yucca*, **bamboo**, and the **palms**, become tree-like. Some of the palms have stems 50–100′ tall.

The majority of the monocotyledons are perennial plants which seem well fitted to survive in competition with other plants. (See also **Paleobotany.**) (R.M.W.)

MONODELPHIA. Mammalia.

MONOECIOUS PLANTS. The flowers of many plants are unisexual; that is, they contain only **stamens** or **pistils**, but not both. When these two kinds of flowers are borne on the same plant, the plant is said to be monoecious. Familiar monoecious plants are oaks, corn, squash, begonia, and castor beans. The corresponding zoological term is hermaphrodite. (R.M.W.)

MONOGONONTA. Rotatoria.

MONOMOLECULAR LAYERS. It has been found possible to deposit, on solid or liquid surfaces, films of material whose thickness corresponds to the effective diameter of a single molecule of the deposited substance. Darrow, for example, in 1925 succeeded in laying down on glass successive layers of certain fatty acids, each of which resulted in the same increase in thickness, as shown by the diffraction effect on **x-rays**. This increase was found to depend upon the number of carbon atoms in the molecule and to increase by the same amount with each additional pair of carbon atoms. Any such layer, one molecule thick, is called a monomolecular layer, or monolayer.

Monolayers deposited on liquid surfaces have been found to affect the rate of evaporation. A vapor may be adsorbed on a glass surface in successive monolayers, the number of which in the completed film (at equilibrium with the vapor) being greater, the greater the density and pressure of the vapor. Frazer observed this with water and with methyl alcohol.

The deposition of molecules by adsorption on metal surfaces differs from that on glass or on liquids, in that the arrangement of the adsorbed molecules depends upon the atomic spacing of the metal in its crystal lattice. This fact, discovered by Langmuir, might be illustrated by the marbles on a Chinese checkerboard, which fit into definite pockets of predetermined spacing. To complete the analogy, Emslie found that if the metal surface is only partially covered, so that there are vacant "pockets," the molecules jump spontaneously from one place to another, remaining at the same site for perhaps several minutes. (L.D.W.)

MONOPISTHOCOTYLINEA. Trematoda.

MONOPISTHODISCINEA. Trematoda.

MONOPLANE. An airplane with but one main supporting surface, sometimes divided into two parts by the fuselage:

High-Wing Monoplane. A monoplane in which the wing is located at, or near, the top of the fuselage.

Low-Wing Monoplane. A monoplane in which the wing is located at, or near, the bottom of the fuselage.

Midwing Monoplane. A monoplane in which the wing is located approximately midway between the top and bottom of the fuselage.

Parasol Monoplane. A monoplane in which the wing is above the fuselage. (F.T.M.)

MONOSACCHARIDES. Carbohydrates.

MONOTOCARDIA. Gasteropoda.

MONOTREMATA. The egg-laying mammals. A primitive order containing only the duck-billed **platypus** and the **spiny anteaters** of the Australian region. (A.W.L.)

MONSTER. An abnormal individual. A sport. Departure from the normal range of variation of the species which does not breed true.

Monsters occur in a small percentage of animals as a result of accidents during embryonic development. In man and other mammals they include 2-headed individuals, 1-eyed individuals, and various degrees of duplication in the body up to Siamese twinning. Some of them are presumably due to hereditary factors although the conditions surrounding the embryo may be responsible. The latter cause is especially evident among the invertebrates, where experimental modification of the environment during development may cause monstrosities to appear. (A.W.L.)

MONTH. Originally the term month was used to indicate the period of time required for the **moon** to pass from some particular phase (i.e., full moon) back to the same phase again. Astronomically the term is used to indicate the period of revolution of the moon from any reference point back to that reference point again. The sidereal month is the time it takes the moon to make one revolution from a given star back to the same star again as seen from the center of the earth. It averages 27^d.32166 (mean solar days) but it varies approximately 7 hours on account of **perturbations**. From the purely mechanical point of view this is the true revolution period of the moon. The synodic month is the period described above (i.e., the period between successive full moons) and averages 29^d.53059 varying by more than 13 hours principally because of eccentricities in the moon's **orbit**. Other types of month which are in use with their average lengths in days are: Tropical (from one celestial **longitude** back to the same again) 27^d.32156, Nodical (from one **node** back to the same) 27^d.21222, Anomolistic (from some point in orbit back to same) 27^d.55460, and Solar ($\frac{1}{12}$ of a tropical **year**) 30^d.43685. (w.k.g.)

MONUMENT (Geologic Term). **Cirque; Corner.**

MONZONITE. A deep-seated granular **igneous** rock composed of equal amounts of **orthoclase** and **plagioclase feldspars** together with **hornblende, augite** or biotite as accessories. The name is derived from Monzoni in Tyrol. It may be considered as a variety of **diorite.** (e.s.c.s.)

MOON. (See tables of satellite data, page 1272.) The moon, the **satellite** of the earth, is, next to the sun, the most conspicuous of all of the astronomical objects. In spite of its apparent brightness and the fact that it has been the subject of countless legends and superstitions throughout the existence of mankind, the moon is in reality a small and unimportant member of the vast universe of celestial objects.

The physical and orbital data regarding the moon will be found in the tables of satellites of the solar system (page 1272), and we shall limit ourselves to the peculiar features of this, our closest neighbor in space. All observational evidence and theoretical calculations point to the fact that the moon is devoid of any **atmosphere.** Its distance from the sun averages, over a period of time, the same as the distance of the earth from the sun with the result that the moon will receive the same amount of heat as does the earth. Knowing the reflecting power of the moon we can calculate the surface temperature of the moon. It is found that at noon on the moon the temperature is approximately 400° K. (261° F.), while at midnight, the temperature falls to about 120° K. (−243° F.). These calculated values have been verified observationally. Such extremes of temperature coupled with the lack of atmosphere preclude the possibility of there being on the moon any form of life such as we know it on the earth.

Probably the most strikingly interesting characteristic of the moon, as seen by the average observer without a telescope, is the so-called **phase change.** The moon shines solely by reflected sunlight and the phase changes can be best understood by reference to the accompanying figure on which the names of the various phases are given. It should be noted that at the time of new

moon the dark side of the moon is toward the earth and the moon is invisible, in fact it is a safe statement to make that no one has ever seen a strictly "new" moon, except possibly at the time of a solar **eclipse.** It should further be noted that the illuminated side of the moon is always toward the sun so that, against a dark sky, the horns of the crescent always point away from the horizon. The various weather proverbs connected with "the moon pouring water," etc., are entirely without foundation. The angle which the line connecting the horns of the crescent makes with the horizon depends entirely upon the relative positions of the earth, sun and moon and has no connection whatever with the weather.

At the time when the moon is slightly beyond new phase, and appears as a thin crescent in the western sky, the earth, as seen from the moon, is practically in full phase. That part of the moon which is not illuminated by direct sunlight is lighted by "earthlight" and is visible as "the old moon in the new moon's arms." The apparent difference in size between the sunlit crescent and the earthlit moon is an optical illusion due to the fact that a brighter object always appears larger than a fainter one.

The moon rotates on its axis in the same direction that the earth rotates, and in the same direction that the moon revolves about the earth. This means that the moon keeps approximately the same face toward the earth at all times. Hence, at periods of full moon we always see the face of the "man in the moon" and never the back of his head. However, due to effects known as **librations,** the moon apparently rocks slightly in all directions. As a result of these librations, about 41% of the moon's surface is always visible from the earth, about 41% has never been observed from the earth, and the remaining 18% is invisible or visible, depending upon the particular time of observation and position of the observer on the surface of the earth.

The study of the surface geography of the moon, or selenography, has been carried on ever since the invention of the telescope and that portion of the surface which is visible to the earth has been very carefully mapped. From the reflecting power of the surface and the characteristics of the reflected light we find that the surface is composed of a brownish-yellow rock. Much of the lunar nomenclature was developed 2 or 3 centuries ago and the descriptive terms are unfortunate. For example, the so-called seas, which are the most conspicuous features in a small telescope, and go to make up the imaginative figures of the "man," the "lady," the "crab's claw," etc., to the naked eye, are in reality broad flat plains. In addition to these plains we find mountain ranges with peaks comparable in height to the highest peaks on the earth; rills, or cracks in the surface as much as ½ mile wide and of undetermined depth; rays, or bright streaks radiating out from certain points on the surface; and the so-called lunar craters.

Certainly the most remarkable of all of the surface features are the so-called craters. In accordance with a system of lunar nomenclature first introduced in the middle of the 17th century by Riccioli these craters bear the names of distinguished scientists and philosophers. The craters have a generally circular form, with high mountain ranges rising abruptly from the surface of the moon and frequently with an isolated peak near the center. There are over 30,000 of these craters ranging in size from great walled plains with diameters of nearly 150 miles down to small craterlets 1000′ or less in diameter. In some cases the floors of the craters are depressed below the surrounding surface of the moon, while in others the floors are elevated. The surface of the plain inside the walls is very rough in some cases and very smooth in others.

There is no complete theory to account for all of the lunar surface features. It is possible that the plains and large mountain ranges may be accounted for on the

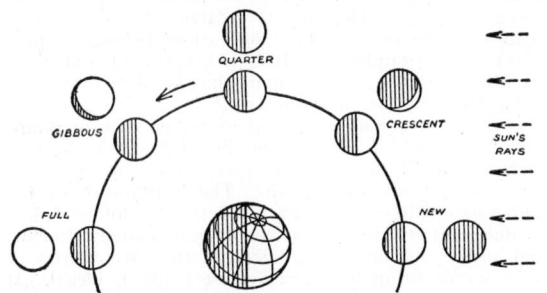

The phases of the moon. The outer figures show the phases as seen from the earth.

same theories as those which explain the major geologic features of the earth; but the craters have not thus far been explained. There are two main theories for the origin of the lunar craters: the volcanic and the meteoric. Of these the volcanic is the older and is the theory which gave rise to the name "crater" for these features. The chief objective to this theory is the great difference in size between the explosive volcanoes on the earth, such as Vesuvius, which do not exceed a few miles in diameter, and the great walled plains on the surface of the moon. The meteoric theory was advanced during the latter half of the 19th century and was revived about 25 years ago. Airplane photographs of craters produced on the surface of the earth by bombs bear striking resemblance to the appearance of the lunar formations. Also, there are several meteoric craters on the surface of the earth, notably the great crater near Winslow, Arizona. However, the tremendous size of the lunar craters as compared with similar features on the earth appears as an almost insurmountable objection.

Throughout the ages much has been written and said regarding the influences of the moon on mankind and upon the weather. Practically all of these influences are purely imaginative and the various proverbs relative to the proper time for planting corn, etc., may be relegated to the field of pure superstition. The theory that the first frost of the fall always occurs during the "bright of the moon" may be readily explained by the fact that the first frost will occur on a night when the sky is clear, with a correspondingly high rate of radiation from the surface of the earth, and on clear nights the moon frequently is bright. The amount of energy which the earth receives from the full moon is only about 1/500,000 part as much as is received from the sun, and hence it is obvious that any lunar effects on climate must be negligible. The moon is, however, the dominant factor in the production of tides and through this medium exerts a powerful influence on commerce, and may, indirectly, produce some effects on weather conditions near the coast. There is also some slight, but distinct relationship between the changes in distance of the moon from the earth and terrestrial magnetism. (W.K.G.)

MOONEYE. Pisces, Teleostei. Moderately large fishes (**Pisces**) of several species found in rivers and lakes of eastern and central North America. One species is also called the silver bass, *Hiodon tergisus*. They are not valuable as food. (A.W.L.)

MOONSTONE. Feldspar.

MOORHEN. Aves, Gruiformes. An English name loosely applied to **gallinules** and related species. (A.W.L.)

MOOSE. Mammalia, Artiodactyla. A large deer with a broad muzzle, prehensile upper lip, and high shoulders. The male has broad palmate antlers. Two species occur in North America, one, *Alces americana*, ranging over the northern United States and Canada and the other in Alaska. The European species, closely related to the common species of North America, is called the elk. (A.W.L.)

MORAINE. The general term for debris of all sorts originally transported by glaciers or ice sheet long since melted away. The following are commonly recognized types of moraines: Lateral moraines, medial moraines, terminal moraines, recessional moraines, interlobate moraines and ground moraines. (R.M.F.)

MORBIDITY. 1. The state of being sick or diseased. 2. The sickness rate or ratio of the number of individuals in a community attacked by a given disease to the num-

ber remaining well. Morbidity cases are usually expressed as cases per hundred thousand of population. (D.M.H.)

MORDANT. A substance which unites with a dyestuff to form an insoluble compound, called a lake, which colors materials permanently. (See **Dyes**.) (R.M.W.)

MOREPORK. Aves. A name applied in New Zealand to a species of **owl** and in Tasmania to a **nightjar.** (A.W.L.)

MORGANITE. Beryl.

MORIBUND. The state bordering on death. (R.S.M.)

MORMON CRICKET. Insecta, Orthoptera. A large wingless long-horned **grasshopper** of the western United States. It varies in color from pale green or yellow to black. It eats vegetation of all kinds and is a cannibal and scavenger, even eating dead animals. When abundant it is a serious crop pest and one which is difficult to combat. (A.W.L.)

MORMYR. Pisces, Teleostei. Peculiar African freshwater fishes (**Aves**). They vary greatly in form but many have the jaws prolonged into a snout. Family Mormyridae. (A.W.L.)

MORNING GLORY. Sweet Potato.

MORNING SICKNESS. Nausea and vomiting which commonly occur in the first few months of **pregnancy.** It is called morning sickness because it occurs upon rising. (R.S.M.)

MORPHINE. This is probably the most valuable drug in medicine. It is the chief and most powerful of the alkaloids of **opium.** It is usually given as morphine sulfate, hypodermically or occasionally orally. Morphine is most frequently used for the relief of pain of all types. It also induces sleep, depresses respirations, and is a general nervous system depressant. The morphine derivatives have similar actions in varying degrees. Dilaudid has a greater analgesic effect but a minimal hypnotic one. Dilaudid and Pantapon, a mixture of purified opium alkaloids, are sometimes used successfully in individuals who develop toxic reactions to morphine. In addition to relief of pain due to trauma, morphine is used in treating heart pain, particularly that of **coronary occlusion.** It is effective in biliary and renal **colic**, and intractable pain associated with **cancer** and other malignant diseases. It is also an important drug in treating congestive **heart failure** and severe **hemorrhage.**

Habituation to morphine and its derivatives is easily established. For this reason the drugs are used with great caution, and never in chronic diseases unless the pain is associated with a hopeless situation such as inoperable cancer. Heroin has greater powers of addiction than the other opium derivatives because it produces greater **euphoria.** It is a favorite of addicts in the United States. It, like morphine, is taken by addicts hypodermically or intravenously.

Contrary to popular belief, there are no signs or symptoms of morphine intoxication in addicts. As long as such individuals have an adequate supply of the drug, they are in excellent health. The appearance of the addict as a shifty eyed, emaciated, poorly clothed fellow is due entirely to the psychological maladjustments, and social and financial difficulties that are a part of obsessive devotion to a drug whose traffic is illegal. If the supply is cut off, withdrawal symptoms appear— lacrimation, yawning, sneezing, tremor, and sweating being the first. The desire for the drug becomes an overpowering obsession; the individual is restless, demanding, and displays mounting excitement. Weakness and psychic depression are marked. Vomiting, colic, and diarrhea are common. Twitching, headache, double vision, and occasionally mania occur. Heart failure and death may result in extreme cases if morphine is not administered.

The treatment of addiction is psychiatric and medical. With the help and cooperation of a competent psychiatrist, a plan of withdrawal of the drug is carried out. The permanence of a cure often depends on the nature of the addiction. In the rare instance where it occurs in a normally adjusted person through injudicious medication, the prognosis is good. If, however, the patient has deep psychoneurotic conflicts, the problem is more difficult, and permanent success is proportional to the effectiveness of psychotherapy in adjusting the disturbed personality. (D.M.H.)

MORPHOLOGY. The division of biological science which deals with structure. Due to the ease and accuracy with which structural differences can be recognized they have always served as the prime means of recognizing both animals and plants and have consequently served as the most useful basis for **taxonomy**. In other fields of biology, such as embryology, histology, cytology and anatomy the first studies were morphological descriptions, followed later by investigations of function. See also **Geomorphology**. (A.W.L.)

MORSE TAPER. Taper.

MORSE THUMP. Inductive Interference.

MORTALITY. The total deaths in a population. Usually expressed as a mortality rate of so many deaths per thousand of population. It is a fundamental factor in studies of population development. Maintenance of a normal population depends upon the birth rate being sufficient to balance the mortality or death rate, and all factors tending to influence the reproduction of animals or their destruction influence this balance. (A.W.L.)

MORTAR. Mortar is the material which is used to bind together the stone or brick of **masonry** construction. Mortar is mixed wet to a plastic state, spread upon the brick to a thickness of from ¼–½″, and the next course of brick pressed into it. The material of mortar hardens in a short time, firmly locking the bricks or stone together in a solid masonry structure. Ordinary mortar consists of a cementing agent and a filler. Lime, sand, and cement are the materials from which mortar is usually made. The mortar plays no small part in masonry construction, since 1 cu. yd. of brick masonry, requiring approximately 500 bricks, will take from ¼–⅓ cu. yd. of mortar. (See **Calcium; Cement**.)

A test comparison of lime and cement mortars shows the latter to be much stronger. However, cement and sand mortar does not work easily under a trowel, and lime paste, made by slaking lime with water, is added to render the mortar more easily worked. This does not materially impair the strength. In earlier days, a great deal of masonry was laid in mortar consisting entirely of lime and sand. The proportions of Portland cement, lime, and sand used in brick construction nowadays is approximately in the volumetric ratio of 1:1:3.

The use of mortar is not confined to masonry, for the mixture used to plaster interior walls or to stucco exterior walls is called mortar when in the state ready for application by workmen. Plaster mortar consists of lime and sand, with fiber or hair as a binder. Such mortars are usually applied in two or more layers, the "scratch" coat being a sand and lime mixture applied directly to lath or other base, for the purpose of keying in the plaster and providing a base for the second coat of the same material. After the second coat the plaster may be finished with a thin coat of lime paste to which plaster of Paris may be added for hardness. This gives a surface known as "hard finish." Alternatively, the wall may receive a lime-sand finish, leaving it somewhat rough with sand grains showing, in which case the work is known as "sand finish." Pigments may be added to the sand finish coating in order to give desired coloration. The thickness of the finish coat is very thin compared to the scratch coat. Stucco is applied to brick, stone, wood or metal lath, and is made of cement, lime, and sand. It is applied in 2 or 3 coats, the exterior coat being colored with mineral colors not affected by the cement, lime, or weather. Colored sands usually give better results than the addition of pigments. It is customary for the surface coat of stucco to be left rough, as the appearance is much superior to a troweled finish. (F.T.M.)

MORTISE. Dado.

MOSAIC. Iconoscope; Photogrammetry.

MOSAIC VISION. A theoretical interpretation of the action of compound **eyes** of **arthropods**. Each visual unit (ommatidium) of such an eye forms an image of part of the object toward which it is directed and the total image seen by the animal is supposed to be composed of many such partial images formed by the many units in the eye. This type of vision is supposed to be inferior to that of other kinds of eyes in the sharpness of the image perceived but to be extremely sensitive to motion. (A.W.L.)

MOSASAUR. Fossil Reptiles.

MOSELEY'S LAW. X-Ray Spectra.

MOSQUITO. Insecta, Diptera. A small 2-winged fly with slender body, long legs, and narrow wings bearing scales along the veins. The **larvae** are aquatic. Male mosquitoes feed on plant juices and only the fe-

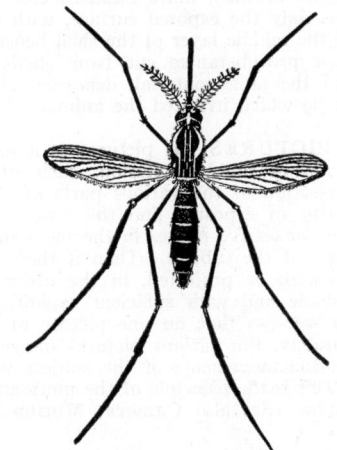

Yellow-fever mosquito, *Stegomyic fasciata*. (*F. B. Howard, U. S. Dept. of Agriculture.*)

males suck blood, but in some species neither sex sucks blood.

Mosquitoes are well known as a nuisance and in warmer climates they are also dangerous because some species transmit disease. **Malarial** parasites are carried by mosquitoes of the genus *Anopheles* and **yellow fever** by species of *Aedes*. In regions were these diseases occur the destruction of mosquitoes by draining swampy areas where they may breed, and by applying oil to water that cannot be drained, is important. The protection of patients from the attack of mosquitoes so that the disease cannot be carried to others is accomplished by adequate screening, a measure which is valuable both for comfort and safety in all dwellings. (A.W.L.)

MOSS AGATE. Agate.

MOSSES. Bryophytes.

MOTH. Insecta, Lepidoptera.

Insects with the four wings at least partly scaly, the mouth parts formed for sucking, and the antennae rarely clubbed near the tips. No one character serves to distinguish all moths from the butterflies and skippers. Most moths are nocturnal but many are diurnal or **crepuscular.** Most of them have the antennae slender and tapering or broadened by **setae** or by processes from the segments, forming a comblike (pectinate) structure, but a few have an expansion just before the tip like that found in most skippers. Most of the caterpillars of moths pupate in a cocoon or a subterranean cell or in the tissues of plants but a few form a brightly colored naked **pupa** like those of the butterflies. The term does not apply to a principal division of the order but includes many families making up one entire suborder and most of the second. (A.W.L.)

MOTHER. The female parent.

Also applied to any unit which gives rise to others, as mother cell, mother colony, etc. (A.W.L.)

MOTHER LIQUOR.

The liquid left behind after a solid has been crystallized from a solution. It usually contains the impurities which were present in the solution. (D.E.M.)

MOTHER-OF-PEARL.

The smooth lining of the shells of **mollusks,** consisting of calcareous and organic matter, usually iridescent and sometimes brightly colored. Also called nacre.

A large amount of mother-of-pearl is used commercially for pearl buttons, knife handles, etc. For such uses it forms only the exposed surface, with a variable thickness of the middle layer of the shell beneath. Blister pearls are protuberances cut from shells and true pearls are of the same material, deposited about some foreign particle which irritated the animal. (A.W.L.)

MOTION PICTURES.

If pictures of a moving object are made in rapid succession, each picture will record the position of the various parts of the subject at the moment of exposure and the series of pictures will represent successive phases in the movement of the different parts of the subject. Then if these same pictures are viewed, or projected, in the order in which they were made and with sufficient rapidity—not less than 16 per sec.—so that no one picture of the series is seen separately, the various pictures merge into one another and the movements of the subject will be reproduced. This is the principle of the motion picture or cinematograph. (See also **Camera, Motion Picture.**) (C.B.N.)

MOTMOT. Aves, Coraciiformes. *Momotus.*

Birds (**Aves**) with a long tail and serrated beak, found in Mexico, Central and South America. Several species are reported to nest in burrows in the banks of streams. (A.W.L.)

MOTOR BOATING.

This is the effect produced by positive **feedback** producing low-frequency oscillations in an audio **amplifier** so the output sounds like the putput of a motor boat. (L.R.Q.)

MOTOR, ELECTRIC.

An electric motor is a machine which, receiving electrical energy, converts it into mechanical energy. Since there are so many different types of electric motors, it seems logical to begin a discussion of them with some attempt at classification of the principal types. Such a classification is given below, and let the reader note that the motors here mentioned are diagrammed in the same order in the illustrations.

A. Direct-current types.
   1. Shunt.
   2. Series. } Straight or Interpole
   3. Compound.]
B. Alternating-current types.
   1. Synchronous.
   2. Induction.
      a. Polyphase.
         (1) Squirrel-cage rotor.
         (2) Wound rotor.
            (a) Slip ring.
            (b) Brush shifting.
      b. Single-phase.
         (1) Split-phase.
         (2) Repulsion-induction.
         (3) Universal.
         (4) Condenser.
         (5) Series.

Electric motors are built in a range varying from outputs of $\frac{1}{100}$ of a **horsepower** up to well over 1000 hp. A 50-hp. motor is considered a large one, and the majority of electric motors now in use ranges between $\frac{1}{4}$ and 10 hp. Standard motor sizes above the small fractional sizes are $\frac{1}{4}$, $\frac{1}{3}$, $\frac{1}{2}$, $\frac{3}{4}$, 1, $1\frac{1}{2}$, 2, 3, 5, $7\frac{1}{2}$, 10, 15, 20, 25, 30, 40, and 50-hp. Sixty-cycle synchronous speeds are 3600, 1800, 1200, 900, 720, 600, 514, and 450 rpm. Full-load induction motor speeds are 2–5% less than these. The efficiency of the electric motor ranges from 75–95%. It is higher in large motors than in small. Induction motors are more efficient the higher the rated speed, but d-c motor efficiency is little affected by speed. Efficiency is often secondary to reliability; nevertheless, it is a factor to be considered, particularly if the drive is heavy and the motor is well loaded over a considerable part of the time. **Direct-current** motors are much less frequently employed than **alternating-current,** because of the preponderance of a-c over d-c systems. However, speed control and starting torque are so excellent with d.c. that it is frequently used in a-c territory where these characteristics are important. D-c power for motors is commonly obtained from the a-c supply through motor-generators, **mercury-arc tubes,** or grid-controlled **rectifiers,** with voltages ranging from 110 to 600. The extra expense of the converter installation lays some handicap upon the employment of d-c motors, and a number of methods have been devised to vary the speed of a-c types, but, in the main, the latter are constant speed.

The losses sustained by a motor in converting electrical to mechanical power arise chiefly through the electrical and magnetic characteristics, but also, in some degree, through bearing friction and windage. The losses are, then, the **resistance** losses occasioned by current flowing through the conductors of the **armature,** the **field,** and the **controller,** the core losses of **hysteresis** and eddy currents, friction and windage. The **cores** of all motors must be built up of laminations insulated from one another by lacquer or enamel, otherwise the core loss becomes excessive.

The motor is so compact that in large sizes, although the efficiencies are high, the heat liberation per unit volume becomes sufficient to need the positive ventilation secured by fans and impellers. In the small open-frame motor the windage of the motor itself is generally sufficient for cooling. Larger sizes are cooled by air forced through the windings by impellers mounted on the motor shaft, or by external fans. In sheltered locations a simple open frame is employed, but motors may be had completely enclosed so that they may be installed where they will be exposed to the weather. Intermediate between these are such frames as the drip-proof types which are protected from overhead drips, the enclosed and ventilated, and the totally enclosed, the latter being cooled by ventilating air brought to it in ducts. The totally enclosed type would be necessary, for example, in an atmosphere wherein a spark or incipient fire could create an explosion.

The relation between input, output, and efficiency are expressed by the following equations, wherein

$$P = \text{horsepower output}$$
$$\eta = \text{efficiency}$$
$$I = \text{line current}$$
$$V = \text{line voltage}$$
$$p.f. = \text{power factor}$$

For all d-c motors

$$P = \frac{\eta I V}{746}.$$

For single-phase a-c motors

$$P = \frac{\eta I V p.f.}{746}.$$

For three-phase a-c motors

$$P = \frac{\sqrt{3}\eta I V p.f.}{746}.$$

The frames of motors are frequently of cast iron, although welded construction has been employed successfully. The **magnetic circuit**, of course, is of ferrous material, and the conductors are copper. Insulating material is rubber, cambric, varnish, lacquer, enamel, asbestos, mica, wood, fiber-glass and fiber. **Brushes** are usually carbon. The usual position for which a motor is desired is with horizontal shaft, in which case the bearings are held by suitable recesses in the main frame casting, and lubrication achieved by one of the methods outlined under **bearings**. Vertical shaft motors are built having thrust bearings to carry the weight of the motor and any externally attached weight. The end thrust for horizontal shaft motors connected to belts or gears can readily be carried by plain collar thrust bearings. But if the drive itself sets up an additional end thrust, special thrust bearings may be necessary.

Important individual characteristics of electric motors include the following: starting torque, normal speed, **speed regulation,** reversibility, efficiency, and cost. Motors are selected upon the basis of the voltage available, the peak load, the necessary reliability, the desired speed range, and the **load factor.** It is common practice to use polyphase motors for all motors larger than 3 hp., although single-phase motors larger than this will occasionally be found. In order to secure the most economical size, 2300 volts is employed where possible for motors of output exceeding 100 hp.; 440 volts is suitable for medium-sized motors, and 220 volts for small motors, except that fractional hp. motors are almost always 110 volts. The size of wire leading from the supply to the motor is made adequate to deliver the rated current at not more than 2 or 3% voltage drop. A feeder supplying one motor should, according to the electrical code, have conductors of a current capacity not less than 110% of the rated motor current. However, much larger conductors may be required to take care of the starting current. The capacity of wires carrying the starting current of an a-c motor started by a **compensator** should be 200% of the rated current if the rated current is above 30 amperes or 250% if the rated current is below 30 amperes. In the case of across-the-line starting, 300% of the rated current should be used. D-c motors started by a starting box should have leads with current-carrying capacity about 125% that of rated current.

All motors should receive some sort of protection, either **fuses,** thermal **cutouts,** or **circuit breakers** actuated by relays.

The circuits of the principal types of electrical motors will now be explained briefly. The reader should consult the illustrations for amplification of the text. The shunt motor has a wound **armature,** the ends of the windings of which are brought to a **commutator,** upon which rest **brushes.** The incoming leads are connected to these brushes so that line voltage is impressed across the windings of the armature. The stationary **field coils** are connected across the brushes in shunt arrangement so that they receive a constant voltage. In the illustration, only one coil is shown, but any practical machine would be multipolar. When the motor is running, the coils of the armature cut the lines of force of the magnetic **field,** and so generate an internal voltage known as the counter-electromotive force. The sum of this counter-electromotive force and the resistance drop through the armature must equal the impressed voltage. Consequently the current taken is much larger when the motor is revolving slowly than when it is up to speed. This also explains why weakening the shunt-field current (and thereby the lines of force) causes the armature to increase its speed, since the weakened field causes less counter-voltage, hence allows more current to flow in the armature. Then more torque is produced, which increases the speed until the back voltage allows just the right current to flow to carry the existing load. The torque of the motor is produced by the magnetic reaction existing between the magnetism of the stationary field and the electromagnetic field surrounding the armature conductor.

Unlike the shunt motor, whose field current is practically constant at all speeds, the series motor produces a field which is maximum during starting and decreases as the motor comes up to speed. For this reason, the series motor has a powerful starting torque, and is used for hoists, traction motors, and the like. The shunt motor is essentially a constant-speed type; the series, a variable speed type. A motor having better speed regulation and starting torque can be obtained by a compound winding having both shunt and series fields. However, the simplicity of the shunt-field motor, coupled with the possibility of effecting a reasonable variation in speed by a variable resistance in the field circuit, has caused it to be widely used. It has been found, though, that any considerable weakening of the field is accompanied by sparking at the commutator, due to the demagnetizing **armature reaction.** Small poles, located between the shunt-field poles, and wound with series coils, will compensate for the distortion of the field flux, and such motors are known as interpole motors. (See **Interpole.**)

In the a-c field, the above classification shows a primary division into synchronous and induction types. Of these, the **induction motor** is by far the more important; but the strictly constant speed feature of the synchronous motor has caused its selection in certain cases. The **synchronous motor** is practically an **alternator** operated inverted. It has a polyphase stator winding which carries the main line current. The field is wound on the rotor and is excited by d.c. brought to it by brushes resting on slip rings. The synchronous motor is stable only when operating at a synchronous speed corresponding to the frequency of the system, and if it is loaded to where it lags ever so slightly behind this synchronous speed, it quickly "falls out of step" and comes to rest. The disadvantages of the synchronous motor are principally two: 1. Its constant speed. 2. It requires d-c excitation. Modern construction of polyphase synchronous motors results in good starting torque. The single-phase synchronous motor has no starting torque, but with three-phase the motor may be made self-starting if copper bars similar to the

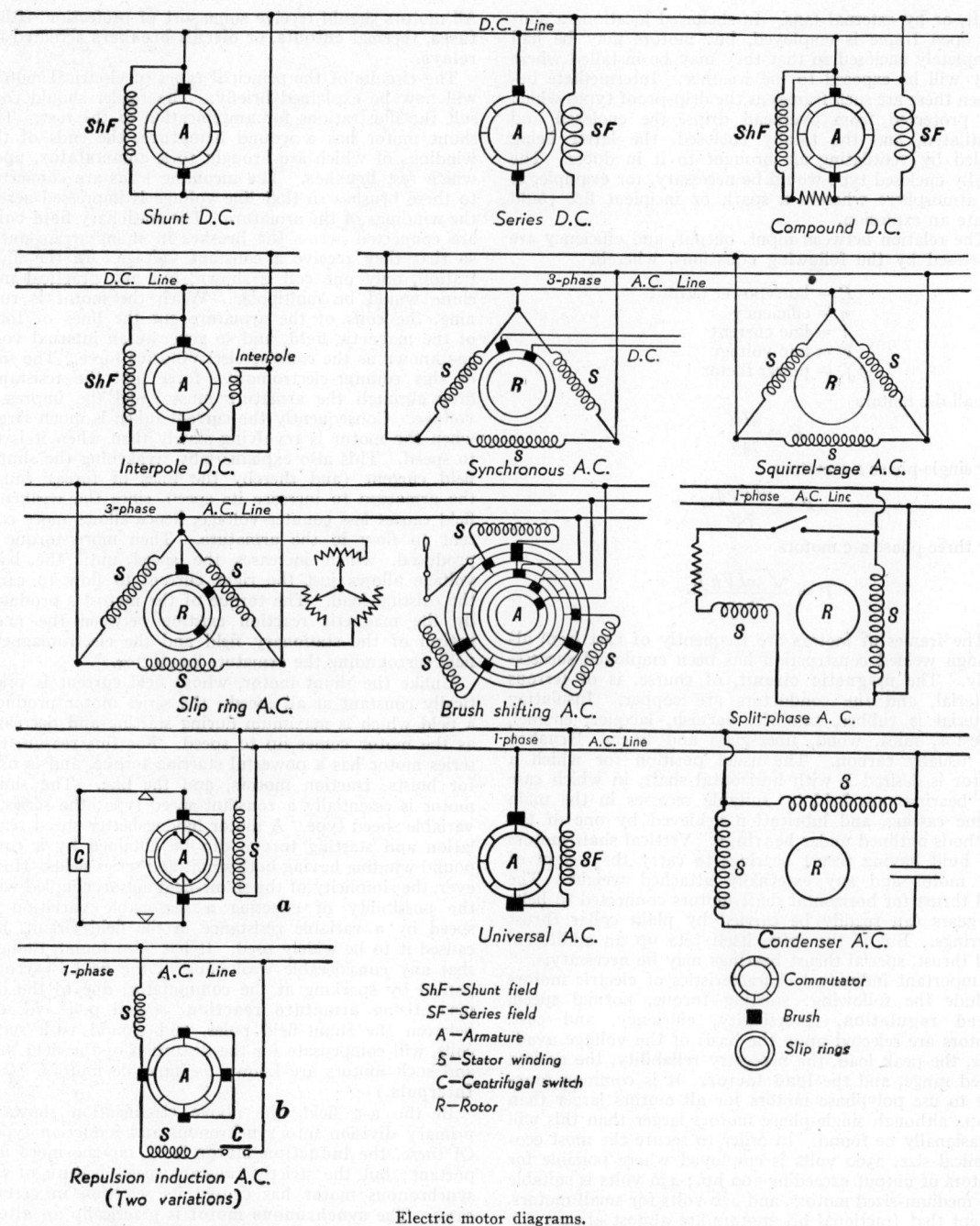

ShF—Shunt field
SF—Series field
A—Armature
S—Stator winding
C—Centrifugal switch
R—Rotor

Commutator
Brush
Slip rings

Electric motor diagrams.

rotor of a squirrel-cage motor are embedded in the rotating field and connected to end rings. To start the motor, the d-c field is open-circuited and the stator windings connected to the line. The motor will then come up to speed, operating as a squirrel-cage induction motor, after which the field current may be applied, upon which the rotor will lock itself into step with the frequency of the system. A synchronous motor is generally used only in large sizes where it has the advantage of providing some power factor correction, since one of the characteristics of this motor is that a leading current will be drawn if the d-c field is over-excited, and this can be adjusted to neutralize the lagging current drawn by induction type motors.

The three-phase squirrel-cage motor is the simplest and most reliable electric motor made. It has a powerful starting torque, and good efficiency, and would probably replace all other types were it not for the following reasons: It is essentially a constant speed motor, it draws a lagging current, and it is not built single-phase. The stationary windings are connected either in **Delta** or **Y**, as may suit the individual design, and are so arranged as to produce a rotating field in the space occupied by the rotor. The rotor is a shaft upon which is built up a laminated steel core carrying embedded in its surface copper or aluminum bars which are parallel to the shaft. The inductive action of the field on this "cage" (if the core were removed, the bars would resemble the familiar squirrel exercising cage) sets up in the latter induced currents whose magnetic field reacts against the rotating field set up by the stator winding, producing a torque. If the rotor were turning in synchronism with the rotating field there would be no induction and no rotor currents. Therefore it is seen that the rotor cannot possibly operate it at full synchronous speed, even though idling. The difference in speeds is expressed by slip, i.e., difference in speeds divided by synchronous speed, and a certain amount of slip is necessary to secure inductive action. As mentioned before, this varies from 2–5% of synchronous speed. A squirrel-cage motor with rotor blocked acts like a transformer with short-circuited secondary, thus explaining the high starting torque.

There are installations where a polyphase motor is wanted, having some degree of speed control, and which may be brought up to speed more slowly than is customary with the squirrel-cage motor. For this service the more expensive wound-rotor and brush-shifting types may be used on three-phase circuits. The wound-rotor principle is employed chiefly on large motors. As its name implies, the wound rotor has polar windings in the rotor, the ends of which are joined either in Y or Delta, and brought to three slip rings. The currents induced in the rotor are brought out through these slip rings to an external three-phase resistance, which may be varied at will from zero to maximum. The operation is much like that of a squirrel-cage motor, except that, for starting, the rotor current is decreased by inserting the maximum of resistance in the external circuit. This is gradually decreased as the motor comes up to speed, until all of the resistance is short-circuited and the motor is operating inductively with a normal slip. Given constant torque, this motor may be varied in speed by varying the external resistance, but it is somewhat less efficient than the brush-shifting type, because of the energy consumed in the resistance. The brush-shifting motor is used where considerable speed variation is desired at good efficiency, as, for instance, when driving fans or pumps of large size. The brush-shifting motor has the primary winding on the rotating armature, similar to d-c practice. This winding is connected to the three-phase line through slip rings. Another winding, called an adjusting winding, is also placed on the rotor; in fact, in the same slots, but is connected with a commutator, which is made fairly wide. The three-phase stator secondary windings are brought out individually to six brushes which bear on the commutator, and are connected as shown in the diagram. Each set of three brushes is joined by a yoke, so that they may be moved simultaneously, and each pair is placed on opposite ends of the commutator. When these yokes are moved with respect to one another, they cause to be included a certain number of commutator segments in each secondary coil. When each pair of brushes is on a common commutator bar, the motor runs as a straight induction motor at slip frequency. By moving the brushes apart by rotating a yoke, the voltage induced in the commutator coil is added to that in the secondary, and the motor speeds up. These voltages may be subtracted by moving the brush in the opposite direction, resulting in slowing down the motor. Since the forces needed to move the yoke are very small, it may be readily operated by the light pressures produced in an automatic control system.

Turning next to the single-phase motor, it is entirely possible for a single-phase motor to operate inductively like the squirrel-cage motor, provided it can be brought up to speed, but a single-phase squirrel-cage motor has no starting torque, so there have been developed numerous ways of doing this for single-phase motors, most of which are of small size. In the split-phase motor, an inductance and resistance are used to displace the voltage at the mid-point so as to get an arrangement resembling

a two-phase impressed voltage. Of course the starting torque obtainable is inferior to that of a polyphase motor, but is sufficient to start a motor attached to a drive requiring low starting torque. A fan illustrates this service. For heavy starting duty the starting torque of a single-phase motor is created by repulsion, which shifts over to induction as the motor comes up to speed. Several systems have been invented, two of which are illustrated. At a, the armature windings are brought out to a commutator, upon which rest two brushes connected externally by a low resistance conductor. A stator winding is connected across the line. The short-circuited armature has induced in it the large current necessary to secure starting torque. As the motor comes up beyond a certain speed, a centrifugally operated switch lifts the brushes from the commutator and applies to it a ring which short-circuits all the segments. When this is done the motor operates as a straight induction motor. This principle, known as repulsion-induction—i.e., repulsion starting and induction running—is employed in most small motors which are to produce large starting torques on single-phase supply. At b is another repulsion-induction principle, less complicated mechanically. Here the switch operates during starting, and is closed for induction operation.

A universal motor is a series motor which may be operated on either d.c. or a.c. It is usually employed in small sizes only, there being a compensating coil to prevent armature sparking and to improve the power factor.

The condenser motor is a split-phase motor, having the phase displaced by capacitance rather than inductance. It is superior to the former in starting torque, efficiency, and power factor, but more expensive and slightly more bulky. The starting torque of the modern condenser motor compares favorably with the repulsion-start motors, and, since the cost is less, the condenser motor is replacing the repulsion type in many applications. In some of these motors the condenser is disconnected by a centrifugal switch after starting; in others it is left in the circuit to improve the operating characteristics and the power factor of the motor. (F.T.M., L.R.Q.)

MOTOR VEHICLE. By motor vehicle is designated that class of road vehicles, employed generally on the public highways, which serve some useful purpose, such as the carrying of passengers or freight, and which are propelled by a **motor** which is an in-built part of the vehicle. A survey of the motor vehicles in operation at present reveals that there are principally three types, the passenger car, or automobile; the motor coach, or **bus**; and the motor truck. Almost without exception, these are propelled by **internal combustion engines,** of which the bulk are of the **Otto cycle** type, the remainder operating on the **Diesel** cycle. The application of the engine to propulsion of road vehicles was experimented with during the latter part of the 19th century, and the industry might be said to have been born commercially in the 10 years centered around the turn of the century. In the United States, Duryea and Haynes introduced the gasoline-engined automobile to the American public. By 1905, several manufacturers were regularly producing automobiles, and 5 years later they were being turned out by factories which embodied the essentials of the present quantity production methods. The American system of automotive building tended to crowd out the many small independent builders, and the decade from 1915 to 1925 saw consolidations and eliminations which greatly reduced the number of bona fide automobile manufacturers. This period also saw the change from an automobile which was principally an assembly of independent manufacturers' parts into a machine which was largely manufactured, fabricated, and assembled under one management.

The principal parts of an automobile are:

1. Frame.
   > Steel frame of channel, I-beam, or tubular members, mainly located in horizontal plane and providing rigidity by beam action principally. The other elements are assembled on this frame.
2. Running gear.
   > Wheels.
   > Axles.
   > Springs.
   > Brakes.
   > Steering device.
3. Propulsion.
   > Engine and clutch.
   > Change speed gears.
   > Drive shaft.
4. Body.
   > Including upholstery, glazing, doors, and interior fitting.
5. Accessories and auxiliaries.
   > Radiator.
   > Fuel tank.
   > Defroster, etc.

Until very recently, there had been no questioning of the suitability of the method of construction involving the building of a rigid underframe to which the other elements of the motor vehicle were attached by bolting or riveting. Nevertheless, with design of the motor vehicle, principally the automobile, emerging from an earlier formative stage into one where more emphasis can be placed on scientific design and research, the desirability of an underframe construction for the high-speed automobile is open to debate. It has been established that a truss construction, wherein the truss members are located in the side walls of the body, is productive of a greater degree of the right sort of rigidity than is an underframe, and at the same time, is lighter. In motor coach construction, especially where there is more emphasis upon reduction of dead weight, and where the service conditions are particularly severe, the so-called chassisless construction, in which the trussing is incorporated in the body, has found favor with designers.

The running gear, which is assembled to the frame, is commonly of a 4-wheel type, the wheels having typically a tread of about 54″, and a wheelbase of 110–130″. The over-all length of an automobile, bumper to bumper, may often exceed the wheelbase by as much as 30% of the latter. Almost from the inception of the automobile, the pneumatic tire, having an outer casing and an inner air-tight tube, both being founded on rubber as the material, has been standard. These have, however, been built in a wide variety of sizes, many of which were unnecessary and tended to increase the tire cost to the user. There is, however, some standardization among the larger producers of modern and low-priced cars, upon the 16-in. tire, that is, the tire which will be mounted on a rim 16″ in diameter. **Brakes** are applied by internal expanding shoes faced with suitable material, which bear upon drums rigidly fixed to the wheel. These brakes are actuated either mechanically by cables and jointed rods, or hydraulically by oil pressure acting against the piston, whose movement is carried mechanically to the brake-shoe cams. Early automobiles were often steered by a tiller, but with increasing road speeds, it was soon found that the most suitable steering mechanism was a wheel, arranged in a position comfortable to the driver. In the United States, this position is founded on the driver's occupying the forward and left-hand seat in the automobile. The principal requirements of the steering device are, first of all, reliability; secondly, freedom from road shocks; and thirdly, ease of manipulation. The automobile builders soon found that the bolster and kingpin steering,

which was suitable with horse-drawn vehicles, had to be abandoned in favor of steering knuckles, by means of which the wheels are fastened to yokes at the ends of a fixed axle. The pneumatic tire, backed up by spring shock absorbers in the steering mechanism and an irreversible gear connection between the hand wheel and steering mechanism, combine to take the road shock from the steering wheels. At the same time this incorporates a reduction ratio so that the full swing of the road wheels is accomplished only through several turns of the steering wheel.

The automobile engine is fairly well standardized today on 6- and 8-cylinder gasoline engines, arranged either in line or as a V. These engines are characterized by **poppet valves**, pressure oil **lubrication**, water cooling, **carburetion**, and high-tension spark plug **ignition**. The engine and clutch are usually built as a unit, to which the change-speed gear box is attached so intimately that in effect it becomes a unit with the engine. From the gear box a drive shaft extends to the rear axle and transmits the power as a **torque**. Although each new year sees improvements in automobile engines, these are chiefly an aggregate of small refinements. New designs endeavor to increase efficiency or output, to reduce size or weight, to promote quieter, smoother operation, and to increase life or decrease maintenance. The automobile industry is viewing, and has for some time been cognizant of, the possibility that the **Diesel** engine may prove as suitable or even more desirable than the gasoline engine for the propulsion of motor vehicles. It seems entirely possible that these two engines may both prove suitable for the automotive field, and that certain combinations of circumstances will point to the selection of one of them where other circumstances might indicate the alternative engine. The over-all operating thermal efficiency of the automobile is very low. It is lower than the efficiencies one customarily associates with the **Otto cycle**, because public demand for performance has led to the creation of automobiles whose engines customarily operate at part load, under which conditions the Otto engine efficiency is penalized. Also, in addition to mechanical friction of the drive and running gear, the automobile engine must furnish an ever-increasing amount of electrical power for the operation of the electrical accessories, i.e., lights, horn, defrosters, cigarette lighters, radio, etc. The power required to propel has increased with the demands for higher speed operation. While these might have been offset by proper streamlining, the public mind changes very slowly, and since true streamlining involves radical changes in the conventionally styled car, designers have had to produce the higher speed by engines of greater power. These conditions have led to this situation: Although refinements of engine design have increased their efficiency, the demands for higher peak powers and the auxiliary services have so offset these engine refinements that the mileage achieved per gal. of fuel has remained substantially constant for many years.

Engineers could in a very short time produce a design which would, to a mind unprejudiced by many years of conventional car shapes, appear eminently suitable as a road vehicle. Furthermore, this vehicle would not be any slower, although it would consume only about half as much fuel per mile traveled as the current models. It would be more comfortable and just as safe. But, unfortunately, it would require altogether different styling, and such radical changes as engine in rear, chassisless construction, and others. While these developments are expected to be consummated with time, the fact that the motor vehicle is the personal possession of a large part of the American public, coupled with the slowness of a large population to change its preconceived ideas, prevents any immediate attainment of the technically proven and technically desirable. (See also **Axle, Brakes, Bus, Carburetion, Clutch, Detonation, Diesel**

Engine, Differential, Highways, Ignition System, Internal Combustion Engine, Otto Engine.) (F.T.M.)

MOUFLON. Mammalia, Artiodactyla. A European wild **sheep**, *Ovis musimon,* found on the islands of Sardinia and Corsica. The rams have very large horns. (A.W.L.)

MOULT. 1. Ecdysis. 2. The shedding of old feathers by birds preparatory to the development of new. Most birds moult at least once a year, beginning just after the breeding season. Most flying birds shed the large flight feathers in pairs but a few shed them all together and temporarily lose the power of flight. The rest of the plumage is also shed and renewed little by little. Moulting is accompanied in many species by the seasonal changes in plumage which make some species so different in summer and winter. 3. The shedding of the outer layer of the skin by reptiles. (A.W.L.)

MOUNTAIN BOOMER. Reptilia, Sauria. The collared **lizard**, *Crotaphytus collaris*, a small species ranging from central California to Missouri and southward. In limited areas as far north as Idaho and Oregon. (A.W.L.)

MOUNTAIN BREEZE. Katabatic Wind.

MOUNTAIN LION. Puma.

MOUNTAINS. Mountains may be classified in three chief groups: 1. Mountains of accumulation (volcanoes). 2. Mountains formed by crustal movements. 3.

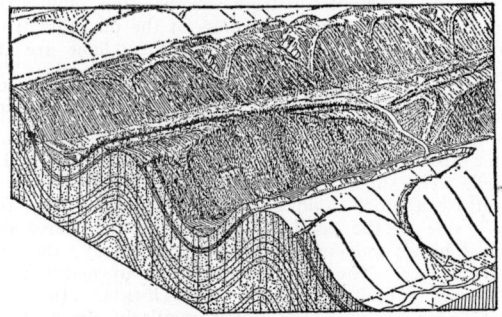

Block diagram showing slightly eroded mountain folds. Jura Mountains, Switzerland. (*After W. M. Davis.*)

Residual mountains (erosional remnants). Structural mountains are those whose form and relief have not, as yet, been particularly modified by erosion. The ridges are still **anticlinal** and the valleys **synclinal**. Later, in the cycle of erosion, the synclines may become mountains and the anticlines, valleys. If the folded mountainous region is ultimately reduced to a **peneplain,** and the peneplain is then lifted without further folding, a new cycle of erosion operating on the same, but base-leveled, structure will develop the type of topography now seen in the Appalachian mountains. (See **Antecedent Stream.**) (R.M.F.)

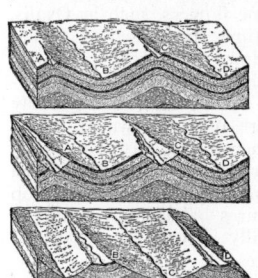

Block diagrams showing how erosion may develop anticlinal valleys and synclinal mountains. (*Tarr, New Physical Geography, Macmillan Co.*)

MOURNING CLOAK. Insecta, Lepidoptera. A large butterfly, *Euvanessa antiopa*, of Europe and North America. The wings are very deep maroon above, bordered with yellowish-white, and are slightly angular.

The species hibernates in the adult stage and is often in flight on the first warm days of spring. (A.W.L.)

MOUSE. Mammalia, Rodentia. A name loosely applied to many species of small burrowing and gnawing **rodents** with slender bodies, long tails, and either the front legs or both pairs short. They are related to the hamsters, jerboas, lemmings, rats, and voles.

Most of the North American species are included in the family Muridae, of which the house mouse is a typical species. The family also contains the grasshopper or scorpion mice, harvest mice, the pine mouse, red-back mice, meadow or field mice, also called voles, and a group variously named wood, deer, **vesper,** and white-footed mice. The family Heteromyidae contains the pocket mice and kangaroo mice and the family Zapodidae the jumping mice. In addition to these details of classification, a number of species in various groups bear special names. Mice of one or another group are found on every continent, although they predominate in the northern hemisphere. The house mouse is *Mus musculus.* (A.W.L.)

MOUSTERIAN. Paleontology of Man.

MOUTH. The external orifice leading into the alimentary tract. Commonly but less correctly applied to designate the terminal cavity of the tract which is the buccal or oral cavity in the vertebrates. (A.W.L.)

MOVABLE BRIDGE. Bridge.

MOVEMENT IN PLANTS. Movement of plants is largely confined to movements of the individual organs as, except in a few of the simplest forms, entire plants cannot move from place to place. Even the movements of stems, roots, flower parts and other organs of plants often escape notice because of the slowness with which most of them take place. The vigor and reality of the autonomous movements of such plant organs can easily be demonstrated by "time lapse" photography in which the plant is photographed at regular intervals over a period of time with a motion picture camera and the resulting film projected at such a speed that a period of several weeks may be represented on the screen in the course of a few minutes.

Most of the movements which occur in the organs of the higher plants may be classified as: growth movements, turgor movements, and hydration movements. Growth movements of plant organs result from enlargement of cells or increase in the number of cells or both. When increase in the size or number of the cells is not uniform throughout an organ curvatures or other growth movements result. Growth movements are usually subdivided into three groups: 1. Tropic movements or tropisms. 2. Nastic movements or nasties. 3. Nutations. Tropic movements are those which occur under the influence of an environmental factor which operates with greater intensity from one direction than from others, the direction of the resulting movement often bearing a direct relation to the direction from which the factor operates. A tropism is said to be *positive* when the organ bends towards the direction from which the factor is acting, and *negative* when it bends in the opposite direction.

Phototropism is undoubtedly the most familiar of the tropic movements of plants. When a plant, such as a potted plant on a window sill, is so placed as to be more strongly illuminated from one direction than others, the stem bends towards the source of strongest illumination. In young seedlings this reaction may occur within a few minutes. In part this reaction results from a direct retarding influence of light on the enlargement of cells. The cells on the more strongly illuminated side of the stem do not elongate as much as those on the weakly illuminated side and the stem thus becomes bent towards the stronger source of illumination. In many plants this may be the main or only mechanism of phototropic reaction. In other plants, however, a more complex mechanism operates in addition to the one just described. This mechanism of phototropism has been studied principally in the **coleoptiles** of grasses. It has been known for many years that if the extreme tip of a coleoptile is shaded with a cap of opaque material or if the tip is cut off, that bending, which takes place in the basal portion, fails to occur. Further investigations have shown that **auxin,** which is apparently made in the coleoptile tip from a precursor translocated up from the grain, is displaced laterally away from the side of the more intense illumination under the influence of such illumination. There may also be some actual destruction or inactivation of the auxin on the side exposed to the higher intensity of light. Downward translocation of auxin in a coleoptile is almost strictly longitudinal, hence the concentration of auxin reaching the elongating cells near the base of the coleoptile on the side away from the stronger light is greater than that reaching the cells on the other side. Since, within limits, cell elongation occurs in proportion to auxin concentration, the cells on the weakly illuminated side of the coleoptile elongate more than those on the opposite side and the coleoptile bends towards the light.

If a young potted plant be placed in a horizontal position it can be observed within a few days that the growing stem has turned in an upward direction and the growing primary root in a downward direction. In

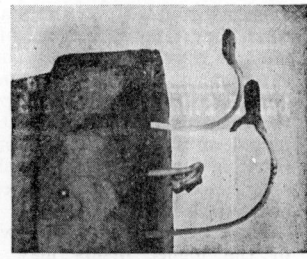

Sunflower seedlings showing a negative reaction to gravity when the pot in which it was growing turned on its side.

Sunflower seedlings growing in a normal position.

other words the stem exhibits the reaction of *negative geotropism* while the root exhibits that of *positive geotropism.* If such a horizontally placed plant is slowly rotated about the stem as an axis such tropic movements do not occur, showing that they are induced in stationary plants by the unilateral effect of the factor of gravity. Auxin apparently also plays an important role in geotropic curvatures of plants. Gravity, like light, apparently has an influence on the distribution of auxin in a plant organ. More than half of the auxin from the tip of a horizontally placed coleoptile, for example, is found in the lower half of the tip. Upward curvatures of horizontally oriented coleoptiles and stems are therefore a result of a greater concentration of auxin reaching the elongating cells on the lower side of the organ than on the upper side. The positive geotropism of roots may also be interpreted in terms of an auxin mechanism. In explaining this effect, however, it is necessary to recall the principle (see **Auxins**) that in a range of concentrations above the optimum that auxins have a retarding effect on the elongation of cells and that the optimum favorable effect on cell elongation varies according to the tissue involved. The optimum concentration for elongation of root cells is

known to be considerably less than that for stem or coleoptile cells. Hence the greater concentration of auxin in the lower half of a horizontally placed root as a result of the influence of gravity apparently falls in the range which checks cell elongation rather than favors it. The root cells therefore elongate more on the upper side than on the lower and the root turns down.

Hydrotropism is the term applied to the turning of roots from drier to moister regions of the soil. This reaction of roots is probably not as frequent as is sometimes supposed, but does occur in some species. Hydrotropic curvatures result from differences in the rate of elongation of the cells on the opposite sides of the root but at the present time there is no explanation of the exact mechanism responsible for this differential growth rate.

Many plant organs react to a contact with solid objects, a phenomenon to which the name *thigmotropism* is applied. Thigmotropic reactions are exhibited most clearly by tendrils, although they also may be shown by stems, petioles or other plant organs. A young tendril is a slender, cylindrical organ which makes slow circular movements in space (*nutation*—see later) during its elongation. As soon as the tendril comes in contact with a solid object rapid growth reactions are initiated. The growing tendril tip rapidly winds around the support, often within a period of a few minutes. As a result of further growth reactions the lower straight part of the tendril may become coiled thus drawing the stem to which it is attached closer to the support.

Nastic movements are those which occur in plant organs when the initiating factor affects all parts of the growing organ uniformly or when the factor, although acting largely or entirely from one direction, evokes the same kind of a reaction regardless of the direction from which it acts. Temperature and diffuse light are the most important factors evoking nastic movements. The movements of bud scales, young leaves, and flower petals are examples of nastic movements. In a young flower, for example, the lower sides of the petals grow faster than the upper, and the flower bud remains closed. The subsequent opening of the bud is brought about by a more rapid growth of the upper than the lower side of the petals. The first of the above described nastic movements is called *hyponasty;* the second *epinasty.* The leaves and flowers of many kinds of plants exhibit *photonastic* movements. In some species, for example, the leaves regularly droop from the horizontal position every night; in other species the leaves regularly assume a more nearly vertical position at night than during the daytime. Many such leaf movements, however, are turgor movements rather than nastic movements. Photonastic movements of flowers are more familiar. Some flowers, such as Oxalis, are regularly open in the daytime and closed at night; others such as the Evening Primrose are regularly open at night and closed in the light. *Thermonastic* movements also occur in both leaves and flowers. Many early spring flowers, for example, are open on warm bright days but remain closed on other bright days on which the temperature is lower. Thermonasty of leaves is well shown by the rhododendron in which the leaves droop and roll markedly at temperatures below approximately the freezing point, but normally assume an expanded condition at higher temperatures.

Nutation is the term applied to the irregular spiral pathway which a growing stem tip usually traces as it grows through space. This type of growth movement results from unequal rates of growth in different vertical segments around the stem axis and seems to occur independently of any effect of environmental factors.

The most spectacular examples of nutation are shown in twining plants. The stem tip of such a plant is usually long and slender and devoid of leaves and often swings through a wide angle as the stem grows ahead through space.

Turgor movements of plant organs are those which are caused by reversible changes in the volume of plant cells. Innumerable examples of such movements in plants can be cited but many of the most pronounced turgor movements occur in species in which pulvini are located at the base of the leaf blade or petiole or both. Such structures occur most commonly in species of the pea family. Externally a *pulvinus* appears as a short, swollen portion of the petiole. Structurally a pulvinus is composed of a compact mass of thin-walled cells which surround a central vascular strand. Movements of leaves or leaflets result from sudden decreases in the turgor of the cells on one side of the pulvinus relative to the cells on the other, the petiole or leaf blade bending towards the side of lower turgor. Loss of turgor of the cells on one side of a pulvinus apparently results from outward movement of water from

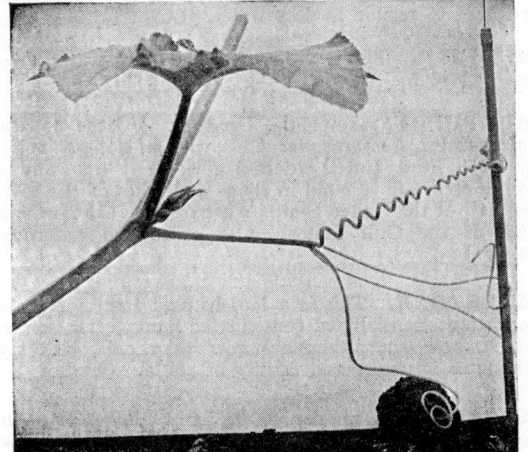

Tendril of squash attached to a slender rod. Note that a portion of the tendril twines clockwise and a portion counter clockwise. Slow-motion pictures show that this is accomplished by the central portion turning in a circle as the tendril contracts.

the cells into the surrounding intercellular spaces but the details of the mechanism are not well understood. Turgor movements originating in pulvini may be brought about under the influence of a number of different environmental factors. In the sensitive plant, *Mimosa pudica*, which affords the most striking and probably most familiar example of turgor movements in plants,

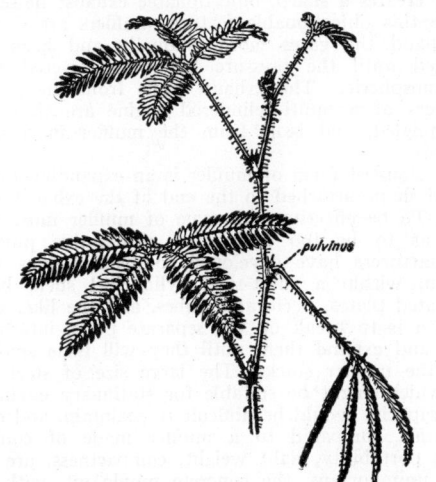

Sensitive plant. One leaf is reacting to the stimulus of touch.
(*Pfeffer's Physiology of Plants, Clarendon Press, Oxford.*)

leaf movements may be induced by physical contact, exposure of the plant to various gases, electrical shock, heat, inadequate water supply, change from light to darkness and in other ways. In this species a pulvinus is present at the base of the petiole of each bipinnately compound leaf and also at the base of each leaflet and at the base of each part of the compound leaf. Merely touching one of the terminal leaflets sharply is usually sufficient to cause the entire leaf to close up. It is evident that the "stimulus" applied to the terminal leaflet has been transmitted to other parts of the leaf, resulting in turgor movements originating in many pulvini. Detectable turgor movements may occur in the leaves of a sensitive plant within a fraction of a second after the "stimulus" has been applied, but the mechanism by which such a rapid transmission of the effect occurs is unknown. Recovery of turgor occurs much more slowly, requiring from several to many minutes.

Hydration movements occur in many non-living parts of plants as a result of changes in the degree of hydration of the cell walls. Such movements are usually inconspicuous and mostly of little biological significance. The splitting of dry seed pods and capsules and the movements of the awns of some grasses are examples of hydration movements. (B.S.M.)

MOVING CLUSTERS.

Statistical studies of the various stellar motions, such as **proper motion, radial velocity,** and **space velocity,** have proved conclusively that there are a number of groups of stars that are moving through space together. These motions can be considered under two main headings: moving clusters and star streams.

A relatively small number of stars which are known to be moving through space together in parallel paths constitute what is known as a moving cluster. Due to a perspective effect, similar to that discussed in connection with meteoric **radiant points,** the proper motions of the members of the moving cluster will appear to be converging upon, or diverging from, a point on the celestial sphere. The apparent convergence of proper motions of a moving cluster is illustrated by the figure which shows a group of stars in the **constellation** of **Taurus** which form part of the so-called Taurus move-

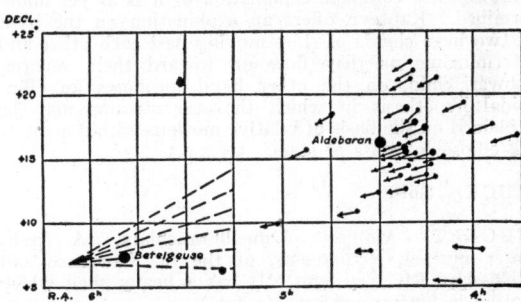

Convergence of stars in Taurus cluster. (*Adapted from a diagram by Lewis Boss.*)

ing cluster. About 100 stars have been found which belong to this cluster. A number of moving clusters have been found in various parts of the sky and they frequently carry the name of a constellation in which a number of the members of the cluster are found, e.g., the Ursa Major group, the Scorpio-Centaurus group, the Orion group, etc. It must not be understood that all members of a particular moving cluster are to be found in the constellation for which the group is named, for such is rarely the case. For example, members of the Ursa Major cluster are to be found not only in the constellation of **Ursa Major,** but also in **Canis Major, Corona Borealis, Auriga,** and others.

A study of the geometry of a moving cluster will indicate that the angular distance of any member of the cluster from the convergent or divergent point is the angle β as defined in the discussion of **space velocity**. It may be further shown that if the radial velocity of only one member of the cluster can be determined, not only the distance of this star, but also the distance of every member of the cluster, may be determined. With these data at hand, it is possible to find the space distribution of all of the members of the cluster. An analysis of the Taurus cluster indicates that all members of this group are contained within a roughly globular volume of space about 35 **light years** in diameter with its center about 130 light years from the sun. The velocity of the cluster indicates that 800,000 years ago the cluster was only 65 light years from the sun, while 65,000,000 years hence it will have receded to such a distance that it will appear as a sparse globular star **cluster** not more than 20' in diameter. At present the sun is within the Ursa Major cluster, with the members moving by on both sides. This cluster is somewhat larger than the Taurus group, having a diameter of 500 light years. So far as is known, the sun is not a member of any moving cluster.

The common motions of so many different groups of stars cannot be accounted for on any chance arrangement and at once suggests a common origin for all members of a moving cluster. This hypothesis receives support from the fact that in many cases the various members of a cluster are of similar **spectral class**. Jeans has shown that the attractive forces of stars other than members of the cluster will eventually produce disintegration of the group and on this basis he calculates ages of the order of magnitude of 10^{12} years for the group.

In the early part of the present century Kapteyn was able to show that the space velocities of the stars were not at random, but had two fundamental preferential directions. It is as though there were two great rivers or streams of stars, one moving toward the constellation of **Orion** and the other toward the constellation of Scutum (in the southern hemisphere). The points toward which the streams are moving are known as the vertices of preferential motion, and both are close to the plane of the **milky way**.

While the fact of star streaming is quite definitely established, a complete explanation of it is as yet undetermined. Kapteyn offers an explanation on the basis of two huge clouds of stars moving past each other and intermingling as they flow on toward their vertices. Schwarzschild, on the other hand, proposes an ellipsoidal hypothesis in which the star streams may be explained on the basis of relative motions within a rotating ellipsoidal mass of stars. (W.K.G.)

MUCK. Soil.

MUCKET. Mollusca, Lamellibranchiata. A fresh-water **mussel**, *Actonomais*, of the St. Lawrence and Mississippi River systems. It has a heavy shell which is used in button-making. (A.W.L.)

MUCOUS GLAND. A form of gland whose cells produce a secretion rich in **mucus**. Glands may be of this type or may produce watery secretions, in which case they are said to be serous. Some contain secretory cells of both kinds and are said to be mixed. (A.W.L.)

MUCOUS MEMBRANE. A membrane composed of epithelial cells (see **Epithelium**). Mucous membranes line those canals, cavities, and tracts which communicate with the external air, such as the nose and throat, **respiratory** tract, **generative** and **urinary** passages, and the **digestive system**. (D.M.H.)

MUCUS. A clear slimy secretion secreted by animals where surfaces must be lubricated or moistened. It is produced at the surface of the body by fishes (**Pisces**)

and **amphibians**. Terrestrial **vertebrates** secrete it in the linings of the **respiratory** and **digestive systems** in abundance. (A.W.L.)

MUD DAUBER. Insecta, Hymenoptera. Any of several species of **wasps** which make their nests of mud. (A.W.L.)

MUD EEL. Amphibia, Urodela. A long slender aquatic species, *Siren lacertina*, with external gills, small front legs but no hind legs, and three pairs of gill slits. The **eel** is found in swamps throughout the southern half of the United States, east of Texas. (A.W.L.)

MUD SKIPPER. Pisces, Teleostei. Coastal fishes (**Pisces**) of Africa and the Oriental region, including *Periopthalmus*. They have prominent eyes and strong pectoral fins which they use in moving about on the muddy shores and in climbing to a limited extent. They are said to be distinctly mud fishes, incapable of thriving if forced to remain in deep water. (A.W.L.)

MUDFISH. Pisces, Holostei. The fresh-water dog-fish or **bowfin**. (A.W.L.)

MUDPUPPY. Amphibia, Urodela. A large aquatic **salamander**, *Necturus maculatus*, with an elongate body, flattened head, tufted external gills, short legs, and a compressed tail. Found in rivers and lakes of the eastern half of the United States, reaching only the northern part of the Gulf States. Also called the waterdog. (A.W.L.)

MUDSTONE. This is a loosely used term for rocks consisting essentially of consolidated muds, often sandy, which, however, although harder than clay, lack the laminated structure and fissility of **shale**. (R.M.F.)

MUFFLER. A muffler, as its name implies, is some device for silencing or muffling an objectionable noise. Release of the gases from the cylinders of **internal combustion engines** on the exhaust stroke at pressures considerably above atmospheric, has created a condition which has had to be met by the use of mufflers. All things considered, it has been found best to operate the internal combustion engine with incomplete expansion, i.e., release above atmospheric pressure. When the exhaust valve opens, the pressure in the cylinder exceeds that in the atmosphere by 20–50 lbs. per sq. in. The result of the sudden rush of gases out of the cylinder under this driving force is an explosive expansion which creates a sharp, objectionable exhaust noise. To reduce this objectionable feature, mufflers are designed to expand the gases more gradually and keep them confined until the pressure has been lowered nearly to atmospheric. The exhaust puffs from the separate cylinders of a multi-cylindered engine are thoroughly intermingled, and issue from the muffler in a steady stream.

The simplest form of muffler is an expanding tube of conical shape attached to the end of the exhaust manifold. To be effective, this type of muffler must be so large as to be too bulky for automotive purposes. Manufacturers have developed compact designs which contain, within a sheet-steel cylindrical shell, baffles, perforated plates, perforated tubes, and the like, whose function is to break up the separate puffs, intermingle them, and expand them until they will issue smoothly from the muffler stack. The large size of steel mufflers which might be suitable for stationary engines of large capacity would be difficult to maintain, and costly to install, compared to a muffler made of concrete. Where portability, light weight, compactness, are relatively unimportant, the concrete muffle pit, with considerable internal volume, supersedes the automotive-type muffler. However, because of the large-scale production of the latter types, they are applied to many engines which might otherwise have been served with muffle pits. The chief drawback of a muffler is that the more effective it is in reducing noise, the more it handicaps the engine performance by creating a back pressure upon the engine manifold. One may find examples where the attractiveness of silent operation has been a factor of sufficient importance to warrant a high degree of muffling, and other instances, where, although quietness of operation would be highly desirable, mufflers have been omitted because they reduced horsepower, added weight, and constituted hazards, due to their high operating temperature. (F.T.M.)

MUGGER. Reptilia, Crocodilia. The Indian crocodile, *Crocodylus palustris,* distributed from Beluchistan to Burma and south to Ceylon and other Oriental islands. From the native name, magar. (A.W.L.)

MUGWORT. Artemisia.

MULBERRY FAMILY. Moraceae. Widely scattered in all but the coldest regions of the world are the more than nine hundred species of the Mulberry Family, including trees, shrubs, and herbs. Many are of great economic importance. In Asiatic tropics, for instance, *Ficus elastica* has been used as a source for rubber, while *Ficus carica,* a related plant, is the cultivated fig. Another tropical member, *Artocarpus incisa,* is the breadfruit. This tree has large glossy incised leaves and bears large fruits which are roasted and eaten. Hemp, *Cannabis sativa,* is an important fiber plant. *Humulus Lupulus* yields hops, while various species of *Morus* are important as food for silk worms.

All members of the Mulberry Family contain a milky juice. The flowers are borne in axillary spikes or heads. Many members have dioecious flowers, that is, the staminate (see **Stamen**) and pistillate (see **Pistil**) flowers are borne on different plants, while others are monoecious. The staminate flower has a variously 3- to 6-parted **calyx**, no **petals**, and one to four stamens with filiform filaments. The pistillate flower has a calyx of three to five, more or less united sepals and a single 1- to 2-celled superior ovary. The fruit varies greatly in different members of the family.

The true mulberries, trees or shrubs of the genus *Morus* are widely distributed plants of temperate regions. The leaves are alternate. On a single tree one may observe interesting variations of the leaves; on one shoot they may all be entire, while on a nearby shoot they are variously and irregularly lobed or divided. The flowers develop early in the growing season. The plants are either monoecious or dioecious. The staminate flowers are borne in long catkins and soon fall from the tree. The calyx is divided into four lobes and there are four stamens inserted at its base. The pistillate inflorescence is a short dense catkin. The flowers have a 4-lobed calyx and a single 1-celled ovary. After pollination the calyx lobes become greatly swollen and fleshy, the individual fruits pressing together tightly to form a multiple fruit. These fruits may be white or pink in the White Mulberry, red in the Red Mulberry, and black in the Black Mulberry. The white mulberry, *Morus alba,* is grown in the Orient largely to supply food for silk worms. The roots of the tree yield a yellow dye, and the wood is used for various purposes. The black mulberry, *Morus nigra,* is grown largely for the fruits, which are greedily eaten by birds, including domestic poultry. The wood is also valuable. Red mulberry, *Morus rubra,* furnishes wood used in making shoe lasts, and for other purposes. All species of mulberry are frequently planted as decorative trees.

Another member of the Mulberry Family which is very valuable is a perennial climbing plant, *Humulus Lupulus,* or the Common Hop. This plant has an extensive underground stem, or rhizome, from which rise the annual climbing stems, which twine in a clockwise direction around any supporting object. The hollow stem is ridged, with downward pointing hairs along each of the ridges. The opposite leaves are large and **palmately** veined. Hops are usually dioecious plants. The staminate inflorescence is a loose panicle; the pistillate, spike-like, with conspicuous bract-like structures subtending each branch. The staminate flowers have a 5-parted calyx and five stamens; there is no corolla. The pistillate flower is a single ovary, partially surrounded by a small **bract** and having two long hairy **stigmas;** around the ovary is a cup-like **perianth.** Hops are wind-pollinated. After fertilization the bracts enlarge greatly. On their outer surface, and also on the surface of the perianth and the subtending bract, yellow grains develop. These grains are called hop-meal and are multicellular cup-shaped bodies developing from single epidermal cells. The cells of these bodies secrete a yellow substance which fills the cup-shaped hollow and which contains the substances which make hops valuable—an essential oil, resins, **tannin,** and a bitter substance, probably **alkaloidal** in nature. The resins are bitter and germicidal.

The principal use of hops is in the brewing of beer. Hops are prepared for the brewing process by drying and bleaching. Prepared hops are boiled with the sweet beer wort, which extracts the bitter principle and imparts to the beer an aroma due to the essential oil of the hops. (R.M.W.)

MULE. Mammalia, Perissodactyla. A hybrid between the domestic horse and ass, produced by mating a mare and a jack. The reciprocal cross of stallion and jennet is called a hinny.

Mules have the large ears, small hoofs, and tufted tail of the ass and the stature of the horse. They are strong and hardy, resistant to disease and adverse conditions. Although they are often of uncertain temper they are such valuable work animals that the mule-breeding industry of the United States has reached a value of $500,000,000 per year at its peak. Mules are bred for various purposes, including riding and driving as well as work of heavier nature. Since they are infertile, they are always bred by crossing the two species. (A.W.L.)

MULITA. Mammalia, Edentata. A South American armadillo. (A.W.L.)

MULLERIAN DUCT. A duct of a pair developed in vertebrates of both sexes, from which the oviducts and, in the mammals, the uterus and vagina of the female are formed. The Mullerian ducts of the male become vestigal structures in the adult. (A.W.L.)

MULLET. Pisces, Tetelostei. Fish (Pisces) of several families. Those of the family Mugilidae are the gray mullets of European terminology and those of the family Mullidae are the red mullets. Several species of the family Catostomidae are called mullets in North America, or mullet suckers because of their close relationship with the fishes of the latter name. One of these species, the common red horse or white sucker, is confusingly called the red mullet.

Most of these fishes are coastal forms which ascend estuaries and even the wholly fresh parts of streams. The red mullets are chiefly tropical. Some of the Mullidae are good food fishes. The Mugilidae are less valued and the Catostomidae are also of moderate worth. (A.W.L.)

MULTIPARA. A woman who has given birth to several children. (D.M.H.)

MULTIPLE CORRELATION. We shall discuss multiple curvilinear correlation. For the theory of ordinary multiple correlation, see the **coefficient of multiple correlation.** In the case of curvilinear correlation, we have the **regression** curve in four variables as the sum of polynomials in x_i of degree n,

$$x_1' = a_1 + a_{21}x_2 + a_{22}x_2^2 + \cdots + a_{2n}x_2^n$$
$$+ a_{31}x_3 + a_{32}x_3^2 + \cdots + a_{3n}x_3^n$$
$$+ a_{41}x_4 + a_{42}x_4^2 + \cdots + a_{4n}x_4^n.$$

The constants are readily determined by the method of **least squares.** Then we define

$$\rho^2_{1\cdot234} = 1 - \frac{\Sigma(x_1 - x_1')^2}{\Sigma(x_1 - \bar{x})^2}$$

and $\rho_{1\cdot234}$ is the multiple curvilinear index of correlation of the **variable** 1 on 2, 3, 4, remembering x_1 is the actual value of the **variate** and x_1' the value found from the regression curve. Similarly we may define $\rho_{2\cdot134}$, etc. The theory of multiple curvilinear correlation has not been extensively developed. The results have been stated for four variables. Similarly the theory may be extended to m variables.   (L.A.A.)

MULTIPLE DIE. Die Casting.

MULTIPLE EFFECT. Evaporator.

MULTIPLE INTEGRALS. Double Integrals and Triple Integrals.

MULTIPLE PROPORTIONS, LAW OF. If two chemical **elements** unite to form more than two compounds, the different amounts of one which unite with a fixed amount of the other stand in the ratio of small whole numbers, e.g., **hydrogen** and **oxygen** can form either **water** or **hydrogen peroxide.** Analysis shows that 2 grams of hydrogen combine with 16 grams of oxygen to form water; on the other hand, 2 grams of hydrogen combine with 32 grams of oxygen to form hydrogen peroxide. The ratio of oxygen which combines with a fixed amount of hydrogen (in this case 2 grams) is 16 to 32 or 1 to 2. (See **Chemical Composition.**)   (R.K.S.)

MULTIPLE ROOTS OF AN ALGEBRAIC EQUATION. Polynomial Equations.

MULTIPLE SCLEROSIS (Insular Sclerosis, Disseminated Sclerosis). A disease of the central nervous system characterized by degeneration of nervous tissue in multiple areas causing many varied symptoms, mostly of the motor system. It is one of the common organic diseases of the nervous system and is more prevalent in Europe than in America.

The cause of the disease is not known. It occurs in young adult life—usually in the 20's and early 30's.

The symptoms are extremely varied. Some of the more common ones are disturbances of speech, tremor, disturbances of gait, nystagmus, spastic paralysis of the extremities, paralyses of the eye muscles, and atrophy of the optic nerves.

The disease tends to have periods of remission but eventually leads to a fatal result. No proven case has recovered, although a patient may live for many years. There is no efficient treatment.   (R.S.M., D.M.H.)

MULTIPLE-SLIDE MACHINE. Press Working.

MULTIPLETS. Atomic Spectra; Hyperfine Structure.

MULTIPLEX CIRCUIT. This is a system of automatic or teletypewriter telegraph service in which several messages may be transmitted simultaneously over the same **line.**   (L.R.Q.)

MULTIPLICATION. Multiplication is one of the fundamental operations with **numbers,** by which two or more numbers are combined to give a number; the result is called the product.

This operation is subject to several fundamental rules or laws: the commutative and associative laws, and in combination with addition, the distributive law.

The commutative law for multiplication of **numbers** is expressed by the formula

$$a \cdot b = b \cdot a$$

for any two numbers a and b; in words, the product of any two numbers is the same in whatever order they are multiplied.

The associative law for multiplication of numbers is expressed by

$$(ab)c = a(bc)$$

for any three numbers a, b, c; in words, the product of any three numbers is the same in whatever manner they are grouped.

The distributive law for addition and multiplication of numbers is expressed by the formula

$$a(b + c) = ab + ac$$

for any numbers a, b, c; in words, the product of a factor by a sum of two terms is equal to the sum of the products of the factor by each of the terms of the sum. A similar statement applies when the sum contains more than two terms.

The product of two numbers of the same sign is positive, that of two numbers of unlike signs is negative, the **absolute value** being the product of the absolute values of the numbers.

To find the product of two **polynomials,** we multiply each term of one polynomial by each term of the other, and add these results.

The following special product formulae are of frequent use:

$$a(b + c + d) = ab + ac + ad,$$
$$(a + b)(c + d) = ac + ad + bc + bd,$$
$$(x + a)(x + b) = x^2 + (a + b)x + ab,$$
$$(a \pm b)^2 = a^2 \pm 2ab + b^2,$$
$$(a + b)(a - b) = a^2 - b^2,$$
$$(a \pm b)^3 = a^3 \pm 3a^2b + 3ab^2 \pm b^3,$$
$$(a \pm b)(a^2 \mp ab + b^2) = a^3 \pm b^3,$$
$$a^n - b^n = (a - b)(a^{n-1} + a^{n-2}b + \cdots + ab^{n-2} + b^{n-1}),$$
$$a^n - b^n = (a + b)(a^{n-1} - a^{n-2}b + \cdots - b^{n-1}) \text{ if } n \text{ is an even integer,}$$
$$a^n + b^n = (a + b)(a^{n-1} - a^{n-2}b + \cdots + b^{n-1}) \text{ if } n \text{ is an odd integer.}$$

From these one may obtain at once certain useful special quotients.

By reversing the above product formulae, we obtain formulae for **factoring.**   (L.L.S.)

MULTISTIGMATEA. Thaliacea.

MULTITUBERCULATES. Fossil Mammals.

MULTIVIBRATOR. Relaxation Oscillator.

MUMMICHOG. Pisces, Teleostei. The common **killifish,** *Fundulus heteroclitus*, also called the **mud fish.** A small shore fish, common from Maine to Mexico. (A.W.L.)

MUMPS (Epidemic Parotitis). A generalized contagious disease with marked local involvement of the **parotid glands,** and occasional involvement of other glandular structures, **especially** the **testicles.**

Mumps occurs most frequently between 5 and 15 years of age. The disease is caused by a filtrable virus and one attack usually gives immunity through life.

Mumps may be transmitted before signs of glandular swelling appear, and contagion usually persists while this swelling lasts. The incubation period varies from 8 to 30 days, usually averaging from 15 to 18 days. The first symptom is usually swelling and pain of one or both of the parotid glands. At times the other salivary glands are involved—the sublingual and submaxillary glands. In rare cases these glands alone may be involved instead of the parotid.

The most frequent complication is inflammation of the testicles, or orchitis, occurring in 15–30% of cases in adults. Very rarely does it occur before puberty. Orchitis is accompanied by pain and swelling of the testicle, malaise, and fever. It is a serious complication because it may result in testicular atrophy and sterility. Meningo-encephalitis is the next most frequent complication. It has a good prognosis. Inflammation of the ovaries is less common. Other complications which are uncommon are pancreatitis, pneumonia, and bronchitis. Deafness may occur and be permanent.

Convalescent serum and gamma globulin are of use prophylactically after exposure. There is no specific treatment. (D.M.H.)

MUNGOOSE. Mongoose.

MUNIA. Aves, Passeriformes. Any bird of numerous species of weaver finches constituting the genus *Munia*. They are native to Africa and the Oriental region. The commonest species is the rice bird, paddy bird, or Java sparrow, which is regarded as a pest in the rice fields and has been kept extensively as a cage bird in Europe. (A.W.L.)

MUNTJAC. Mammalia, Artiodactyla. Small Asiatic deer of several species. They stand only about two feet high and have small two-tined antlers somewhat like those of the American prong horn. The common Indian species, *Cervulus muntjac,* is called the kakar. (A.W.L.)

MUNTZ METAL. Brass and Bronze.

MURAENA. Eel.

MURIATE. Term applied to chlorides. Muriate of potash, potassium chloride; muriatic acid, hydrochloric acid. (R.K.S.)

MURINE. Opossum.

MURMUR. An abnormal sound heard over the heart area, either by the ear placed against the chest wall, or with the aid of a stethoscope. Murmurs are classed as organic or functional; and loud, soft, rough, blowing, high-pitched, low-pitched. Organic murmurs are associated with structural deformities of the valves, shrinking and constriction, or dilatation, such as occur with rheumatic fever, bacterial endocarditis, arteriosclerosis and syphilis. The deformity may be part of the acute lesion in rheumatic fever, or it may represent the healing and residual scarring of the valve in rheumatic heart disease. The flow of blood over such deformed valves is responsible for the abnormal sounds. Functional murmurs are frequently heard in normal hearts, in normal healthy individuals; the mechanism of their production is a matter for debate. In profound anemia a different type of functional murmur is heard. It is due to dilatation of the heart and its valves, and disappears when the anemia is overcome. (D.M.H.)

MURRAY COD. Pisces, Teleostei. A fish belonging to the large sea-bass family. (A.W.L.)

MURRAY LOOP. This is a bridge type measurement made on communication lines to locate accidental grounds. As may be seen in the figure, it is a Wheat-

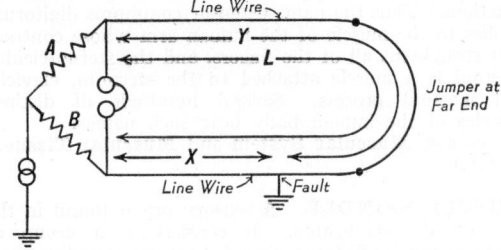

stone bridge in which the line wires form two arms. The balance point gives

$$X = \frac{BL(r_1 + r_2)}{2r_2(A + B)}$$

where r_1 is the resistance per ft. of the upper wire and r_2 of the bottom wire. Other quantities are shown on the diagram. (L.R.Q.)

MURRE. Aves, Charadriiformes. Moderately large marine birds (**Aves**) related to the auks. The name is, to some extent, synonymous with **guillemot** but different species have been named as one or the other by ornithologists. They are powerful swimmers and divers. (A.W.L.)

MURRELET. Aves, Charadriiformes. Small species of marine diving birds (**Aves**) of several genera related to the auks and murres. (A.W.L.)

MUSCHELKALK LIMESTONES. Triassic.

MUSCLE. An organ formed of a bundle of contractile fibers attached to parts of the body which are moved in relation to each other when it shortens. Among the invertebrates the fibers of a muscle are loosely associated but in the vertebrates they are bound together and enveloped by special tissues.

The typical vertebrate muscle is surrounded by a connective tissue sheath, the external perimysium, which continues into it a series of septa (**septum**) called the

Cross-section of muscular tissue, enlarged, showing bundles of muscular tissue.

internal perimysium. Between the septa lie bundles of muscle fibers, each surrounded by a delicate continuation of the connective tissue closely joined to the surface of the fiber. Blood vessels and nerves supplying the muscle course through the perimysium.

Muscles differ in form according to the relations of their fibers to the tendons with which they connect. If the tendon is terminal, the fibers are longitudinal. In penniform muscles the tendon runs along one side and the fibers are oblique to the axis of the muscle, resulting in a feather-like appearance. If the tendon runs down the middle the result is a bipenniform muscle, and even more complex patterns of this type occur. Digastric muscles have a tendon at each end and another in the middle connecting two contractile portions.

Individual muscles throughout the body have received special names according to their anatomical relations or functions. Thus the name extensor communis digitorum applies to the muscle of the human arm whose contraction straightens all of the fingers, and the sterno-cleido-mastoid is a muscle attached to the sternum, clavicle, and mastoid process. Several hundreds of distinct muscles of the human body bear such names.

(See also **Muscular System** and **Muscular Tissue.**) (A.W.L.)

MUSCLE SPINDLE. A sensory organ found in the muscles of **vertebrates.** It consists of a group of modified muscle fibers surrounded by nerve endings and is supposed to be stimulated by the variations of pressure resulting from the contraction of the muscle. (A.W.L.)

MUSCOVITE. POTASH MICA. The mineral muscovite is an **orthosilicate** of **alminum** and **potassium** which crystallizes in the **monoclinic** system although frequently found in pseudo-**hexagonal** forms. Usually tabular in **habit,** the most prominent characteristic is the highly perfect basal **cleavage** yielding remarkably thin laminae which are often highly elastic. Hardness, 2–2.25; specific gravity, 2.76–3; luster, vitreous to pearly; color, colorless through grays, browns, greens, yellows, and rarely violet or red; transparent to translucent.

Muscovite is the commonest mica, being **found in granites, pegmatites, gneisses,** and **schists,** and as a **contact metamorphic** mineral, or as a secondary mineral resulting from the alteration of **topaz, feldspar, kyanite,** etc. In pegmatites it is often found in immense sheets which are commercially valuable. A complete list of occurrences of muscovite would be impossible. In the United States excellent specimens are found in the pegmatites of New England, where they are associated with rarer minerals like **tourmaline, beryl,** etc. Pennsylvania, Maryland, Virginia, North Carolina, Georgia, South Dakota, and New Mexico also furnish large and fine examples of this mineral.

The name muscovite comes from Muscovy-glass, a name formerly much used for this mineral because of its use in Russia for windows. It is in much demand for the manufacture of insulating and fireproofing materials and to some extent as a lubricant. (E.S.C.S.)

MUSCULAR SYSTEM. The entire assemblage of contractile structures by which movement is produced in complex animals. **Muscular tissue** consists of several types of specialized cells, variously disposed. Some are scattered in the walls of hollow organs or among other tissues and some are bound together to form separate **muscles.** The latter are more conspicuous since they are separate organs, but all contractile structures properly belong to the muscular system.

In the simplest animals with special muscular tissues, the **worms,** muscle cells pass across the loose tissues within the body and form layers in the body wall. Their disposal is often in a circular layer, with fibers running around the body, and a longitudinal layer with fibers parallel to the main axis. The contractions of these layers lengthen and shorten the body and so carry on the creeping movements characteristic of worms. Special groups of muscle fibers also govern such special structures as the setae of earthworms. These muscles have an attachment to the body wall called the origin, and an insertion in the tissue surrounding the part to be moved.

In animals with a rigid skeleton an arrangement of muscle fibers like that of the body wall of worms persists in hollow organs like the alimentary tract, and in organs like the heart and urinary bladder a less regular

arrangement provides for the uniform contraction of all parts of the wall. Locomotion and similar movements, however, are due to the action of muscles on skeletal

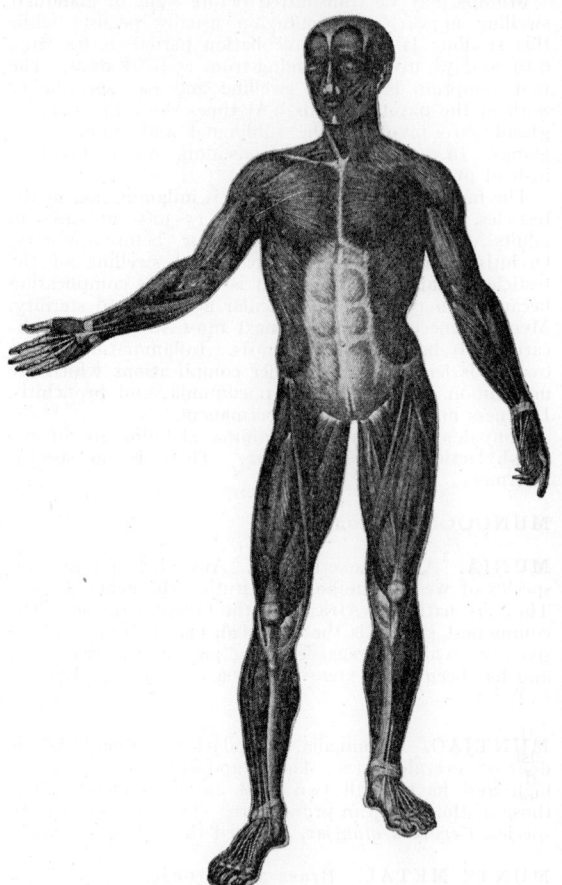

Muscular system.

supports which serve as systems of levers. Usually the shortening of a single muscle regulates the movement of one part in relation to another; it is said to have its insertion in the part moved and its origin at the other point of attachment. When a muscle is used for slowing a movement, it is stretched; and when it is used for equilibrium, as in standing, the length does not change. These activities are all included under the term contraction for historical reasons.

The nature of movements varies greatly. Extension of jointed appendages, flexion, retraction, rotation, and other movements are carried out by opposed systems of muscles. Any movement is positive, due to the contraction of a specified muscle, and the return of the part to its former position results from the contraction of an opposed muscle, sometimes aided by the action of gravity. The anatomy of the muscular system and the relations of specific muscles have been worked out in great detail in the human body, in other vertebrates, and in some of the **arthropods.** (A.W.L.)

MUSCULAR TISSUE. Muscular tissue is made up of long slender cells specialized for contraction. It is usually mesodermal in origin. There are three kinds of muscular tissue, smooth, cardiac, and striated or skeletal. Among the invertebrates all muscle may be smooth, as in many worms, or striated, as in most arthropods. All three kinds are found in **vertebrates** and cardiac muscle is characteristic of the vertebrate **heart.** Muscle **cells** may contain a few scattered myofibrils, or a large number, occupying most of the cytoplasm. The myofibrils consist fundamentally of protein chains which are responsible for the contractile properties of muscle.

Among the invertebrates the contractile cells vary greatly. Those of **nematode** worms, for example, consist of a contractile and a noncontractile portion and some muscle fibers of **mollusks** show incipient striation in the form of oblique refractory lines.

The smooth muscle tissue of vertebrates consists of slender tapering cells with a nucleus placed centrally. These cells are involuntary in action and are found in the walls of the alimentary tract, excretory system, blood vessels, and other organs.

Striated muscle is so named because the many fibrils in its cytoplasm are made up of alternating zones of different refractive quality which cause the fiber to show light and dark transverse bands. The fiber is long and cylindrical and contains many nuclei located at the periphery. It is surrounded by a delicate membrane, the sarcolemma. Most striated muscle is the foundation of voluntary movements and from its extensive association with the skeleton it receives the name skeletal muscle.

Cardiac muscle, like skeletal, is striated, although the striations are much finer. It differs in the general branching of its fibers and in its centrally placed nuclei. Cardiac muscle is rhythmically contractile, initiating the heart beat of vertebrates. The nerves which inervate the vertebrate heart serve to modify but not to initiate the contraction of the heart muscle. (A.W.L.)

MUSCULO-EPITHELIAL CELL. A cell forming part of the layers in **coelenterates** such as *Hydra*, provided with basal contractile processes which serve as muscles. (A.W.L.)

MUSHROOM. Agarics and Basidiomycetes.

MUSHROOM POISONING. Eating certain types of mushrooms, notably the *Amanita phalloides, A. muscaria* and *A. verna*, produces severe poisoning. Muscarine and other toxins are responsible. Clinically one or two types of symptoms occur, depending on which variety of toxin is at work. One is chiefly a gastroenteritis, with nausea, vomiting, diarrhea and abdominal pain. With this type the prognosis is fairly good. In the other form, the toxin attacks the nervous system primarily, and headache, muscular cramps, blindness and coma occur. The mortality in such cases is high. (D.M.H.)

MUSICAL SOUNDS. The chief characteristics whereby musical sounds are distinguished from each other and from non-musical sounds are pitch and quality.

Pitch is that which the normal ear recognizes as corresponding to position in the musical scale. It is dependent primarily upon the frequency of the predominant sound waves; for example, middle *c* on the piano has a fundamental frequency of about 264 per sec. The frequency of a musical tone may be measured by means of a sonometer in comparison with a standard **tuning fork.** The sonometer string is tuned to unison with the unknown frequency, and then with the fork of known frequency, by varying its length at constant tension by means of a movable fret; the frequencies are then in the inverse ratio of the lengths. A "siren" may also be used for this purpose; it has a disk with equally spaced holes through which air is blown as the disk is rotated. This gives the frequency by direct count.

Quality is an attribute due to the complexity of a tone. Besides the lowest (and usually the loudest) frequency, called the fundamental, there may be a series of overtones, higher frequencies arising from the vibration of the source in parts or segments. Thus, a string may vibrate as a whole, giving the fundamental, or in halves, thirds, fourths, etc. The exact quality of a tone depends upon the relative intensity of the several components, which may be ascertained by means of a device called a "phonodeik" for amplifying the wave form and by subsequent **harmonic analysis.** Harmony or discord is dependent upon the loudness and frequency of the **beats** between the principal components of the tones concerned. (See **Combination Tones.**)

The musical scale at present used in occidental countries is a geometrical progression of frequencies with ratios or "intervals" each equal to $^{12}\sqrt{2}$ or 1.05946. An octave comprises twelve such intervals and has the frequency ratio 2. Various harmonious combinations of the notes on this scale form the chords of ordinary music. In addition to pitch and quality, music utilizes differences in intensity, duration, and spacing (rhythm) of musical sounds, all of which contribute to that subtle attribute known as "expression." (L.D.W.)

MUSK DEER. Mammalia, Artiodactyla. A small deer, *Moschus moschiferus,* of the Himalayas. It stands twenty inches high. Antlers are lacking and the upper jaw bears a pair of tusks which may project several inches in the males. In northwestern China a second related species, *M. sifanicus,* occurs. (A.W.L.)

MUSK HOG. Peccary.

MUSK OX. Mammalia, Artiodactyla. **An animal,** *Ovibos moschatus,* of Arctic America related to the sheep but resembling oxen. It is about ⅔ as large as the American bison and is clothed with long hair. The horns are broad at the base but become rapidly narrower as they curve downward from the forehead over the sides of the head. The more slender tips turn abruptly upward.

Musk oxen are hunted by the Eskimos for their hides and flesh. (A.W.L.)

MUSKALLUNGE. Pisces, Teleostei. A large freshwater game fish (**Pisces**), *Esox nobilior,* of North America, found chiefly in colder waters of the northern states and Canada, but occasionally as far south as North Carolina. It is a slender species related to the pikes and pickerel and attains a length of 8' and a weight of 100 lbs. Specimens over 40 lbs. are, however, rarely taken.

The name is among the most abused of all fishes. It is also spelled muscalonge, muskellunge, maskinongy, masquinonge, and occasionally in other ways. (A.W.L.)

MUSKEG. Swamp.

MUSKMELON. Gourd Family.

MUSKRAT. Mammalia, Rodentia. Moderately large amphibious **rodents** of North America. They are stoutly built, short-legged with partially webbed hind toes, and have a compressed tail. They inhabit swamps and streams, making houses of sticks and also burrowing in the banks. Two species are recognized, one a dark brown animal, *Ondatra rivalicia,* of the coastal part of Louisiana and the other an extremely variable species, *O. zibethica,* found over most of the continent.

Muskrat fur consists of a fine woolly undercoat and long glossy hairs. It is the best of the inexpensive furs and because of the ability of the animal to withstand extensive trapping is now the most important fur commercially. Approximately 15 million pelts are taken annually in North America, many of them from fur farms. The natural colors of the fur are beautiful and many pelts are made up in this state, but many also reach the market as Hudson seal after being plucked, clipped and dyed black.

The muskrat has also been introduced successfully into Europe. (A.W.L.)

MUSQUASH. The Muskrat.

MUSSEL. Mollusca, Lamellibranchiata. A name applied to many **bivalve** mollusks, both marine and fresh-water species. Used to some extent as a synonym of clam, especially in reference to the fresh-water species. Most of the marine forms are normally called either mussels or clams, the species of the family Mytilidae bearing the former name. They include the edible mussel of both American coasts and Europe.

Fresh-water mussels are eaten but they are not an important source of food. They are taken in large numbers from the larger rivers of the country for their shells, which are used in making buttons, and for the pearls which they contain. As many as 50,000 tons of shells have been marketed in a single year, chiefly from the Mississippi River. (A.W.L.)

MUSTANG. Mammalia, Perissodactyla. Western American **horses** descended from stock introduced by the Spanish conquerors. Also known as Indian ponies and bronchos. They are noted for their endurance, agility and spirit. (A.W.L.)

MUSTARD. Brassica.

MUSTARD GAS. Poison Gases.

MUTAROTATION. Carbohydrates.

MUTATION. A characteristic different from any in the ancestry of the individual displaying it, which breeds true. Mutations are like sports in showing abrupt departure from the hereditary characters of the species to which they belong but they differ in being transmissible.

Changes of this type have been important in the study of **heredity**, since they afford contrasts with existing hereditary characters. They have usually been regarded as fortuitous changes issuing from the heritage but it has been found possible to increase their appearance by treating breeding stock with x-rays, radium, heat, and other environmental factors. They are a possible basis of evolutionary change. (A.W.L.)

MUTUAL CONDUCTANCE. This is one of the common vacuum-tube coefficients and indicates the effect on the plate current of the grid voltage, other quantities being held constant. It is also known as the transconductance. Mathematically it is

$$G_m = [di_p/de_g]_{E_p \text{ constant}}$$

where di_p is an infinitesimal change of plate current and de_g is an infinitesimal change of grid voltage, while E_p is the plate voltage. (L.R.Q.)

MUTUAL INDUCTANCE. Coupled Circuits and Inductance.

MUTUAL INDUCTION. Electromagnetic Induction.

MYALGIA. Pain in a muscle or muscles usually due to injury or inflammation. (R.S.M.)

MYCELIUM. Ascomycetes, Basidomycetes, Phycomycetes, and **Fungi.**

MYCETOZOA. Sarcodina. An interesting group of organisms formerly classed with the plants are **slime molds,** related to the fungi, and now included with the 1-celled animals in the phylum **Protozoa.** They consist of a creeping mass of **protoplasm** containing many nuclei. This plasmodium occasionally produces fixed reproductive bodies (sporangia) in which their resemblance to plants is at its height. (A.W.L.)

MYCORHIZAE. Mycorhizae are, strictly speaking, **hyphal** threads of various **fungus** plants, **growing** in intimate association with some part of the higher plant, usually the root. Two different kinds of mycorhizae are known. In one kind, known as ectotrophic mycorhiza, the fungus hyphae form a close welt over the surface of the root with hyphal threads penetrating into the root between the **cortical** cells. This kind is particularly common on the tips of tree-roots, and often causes swelling of the infected roots. The fungus seems to take the place of root hairs, which are not developed in its presence. Apparently the general welfare of the tree is not absolutely dependent on the presence of mycorhizae, though they may materially benefit the tree. The other kind of mycorhiza is the endotrophic form, in which the fungus hyphae are found within the cells of the higher plants. **Orchids and Heaths** are well known examples in which endotrophic mycorhizae are found. The fungus enters the plant through the epidermis of the root, passes into the cells of the cortical **parenchyma** and lives there. From these cortical cells the fungus may extend into all the tissues of the higher plant, even reaching the **ovary.** Plants supporting endotrophic mycorhizae seem much more dependent on their presence than those with ectotrophic forms. Indeed, the successful growing of orchids from seed may depend on the presence of the fungus associate which enters the orchid embryo.

Plants growing in bogs, regions in which there is usually a lack of sufficient nitrogen, almost always have mycorhizae. Apparently the mycorhizae obtain the **nitrogen** necessary for the welfare of the higher plant.

This association of mycorhizae with higher plants is often described as a case of symbiosis, a living together of two organisms for mutual benefit. The higher plant receives a greater supply of nitrogen and so benefits; the **fungus** obtains food from the higher plant. (R.M.W.)

MYDRIATIC. Any drug that causes dilation of the pupil of the eye. Such drugs are **atropine, belladonna, cocaine,** etc. (R.S.M.)

MYELITIS. This term refers to general or local involvement of the spinal cord by inflammation or softening, resulting from infection, injury, or certain poisons or toxins. Infective myelitis may occur as a complication in many diseases, of which the most common are **measles, chicken pox, mumps, syphilis, diphtheria,** and the various forms of **meningitis.** Toxic myelitis may be due to poisoning by **carbon monoxide, gasoline, chloroform** and other chemicals. Traumatic myelitis may be due to direct or indirect injury about the spine. The disease may be acute or chronic.

Symptoms occurring in myelitis depend on its cause and on the extent of the spinal cord involvement. The common ones are weakness, paralysis, pain and sensory changes. The disease is serious and the mortality is high, varying according to the extent and degree of involvement of the spinal cord. (D.M.H.)

MYLONITE. The term proposed by Charles Lapworth in 1885 for a massive **chert**-like rock occurring at the contact of the shear planes associated with the low-angle **overthrusts** of the North West Highlands of Scotland. Mylonite is therefore developed through the process of **dynamic metamorphism** and derives its name from the Greek, meaning to crush. (R.M.F.)

MYNA. Aves, Passeriformes. Indian birds (**Aves**) closely related to the starlings. (A.W.L.)

MYOCARDITIS. An acute inflammation of the muscular wall of the **heart.** Acute myocarditis is almost always secondary to any of the acute infections such as acute rheumatic fever, diphtheria, malaria, scarlet fever, erysipelas, septicemia, etc. It may result from direct invasion of the heart muscle by the infective organism or, on the other hand, from the toxins liberated by bacteria and carried to the heart by the blood stream. It is most common in childhood and early life. Often it complicates an acute and infectious disease which masks the symptoms and makes the diagnosis difficult or impossible. Acute myocarditis may be very mild and permit an early and complete recovery, or it may be so severe that death results within a few days.

Chronic myocarditis is a term no longer used. (R.S.M., D.M.H.)

MYOCARDIUM. The muscular layer which makes up the greater part of the wall of the vertebrate **heart.** It is lined with the thin endocardium and covered with a thin epicardium, continuous with the lining of the pericardial cavity. (A.W.L.)

MYONEME. Fine contractile filaments in the cytoplasm of 1-celled animals. (A.W.L.)

MYOPIA. Short sight or **near-sightedness,** due to an increase in the refractive power of the **eye** so that images

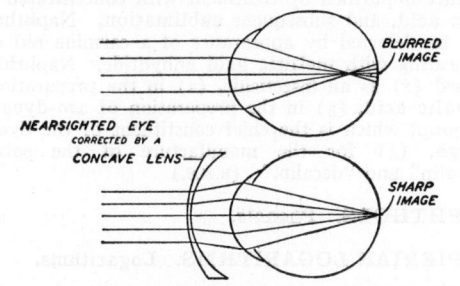

(Carlson and Johnson's The Machinery of the Body, University of Chicago Press.)

from distant objects are focused before reaching the retina and the image is blurred. The fault may be in too great a convexity of the lens, or an eyeball that is too deep. A concave lens is used to correct this condition. (See **Vision.**) (D.M.H.)

MYRIAPODA. A class of the phylum **Arthropoda,** now obsolete. It has been subdivided to form the four classes, **Chilopoda, Diplopoda, Pauropoda** and **Symphyla.** (A.W.L.)

MYRIENTOMATA. Protura.

MYRMECOPHILE. An animal which makes its home in the nests of ants. Among the insects that have become adapted to this mode of life are **crickets, beetles,** and larval flies. In most cases the myrmecophiles seem to feed at the expense of the ants, which may in turn eat secretions produced by their guests. (A.W.L.)

MYRRH. *Commiphora Myrrha.* Burseraceae. Myrrh is obtained from a spiny tree of small size, which grows wild in northeastern Africa and Asia Minor. It has trifoliate leaves, in which the lateral leaflets are very much smaller than the terminal, and small red flowers borne in the axillary buds of the leaves of the previous year. These flowers are **dioecious.** The fruit is a small **drupe.** A **resin** exudes from the stem of the tree, collecting in small lumps. This is myrrh, long known for its fragrance. It is used in perfumes and incense, and in medicine, and was one of the substances used by the Egyptians in embalming their dead. Several other species of the genus yield similar resins, also called myrrh. (R.M.W.)

MYSIDACEA. Crustacea.

MYSTACOCETI. The whalebone **whales,** an order of marine mammals including the largest living animals of this class. They are characterized by the presence in the mouth of plates of baleen or whalebone whose fringed ends serve as a sieve through which water is strained to remove the small animals contained in it. The order includes the fin whales or rorquals, the humpback whale, the pigmy whale of the southern hemisphere, the gray whale, the sulfur-bottom whale, and the right whales. (A.W.L.)

MYXEDEMA. Thyroid Gland.

MYXINOIDEA. Cyclostomata. The hag fishes.

MYXOMYCETES. Slime Molds.

MYXOSPONGIDA. Demospongiae.

MYXOSPORIDIA. Sporozoa.

N

N SERIES. X-Ray Spectra.

N.A.C.A. Abbreviation for National Advisory Committee for Aeronautics. A large and influential research agency of the United States Federal Government with headquarters in Washington, D. C., and laboratory facilities at Langley Field, Virginia, Cleveland, Ohio, and Moffet Field, California. Its publications, viz., *Technical Reports, Technical Notes,* and *Technical Memoranda,* have been issued since 1918, and form an important and frequently consulted body of aeronautical information. (F.T.M.)

N.A.C.A. FIVE-DIGIT AIRFOIL FAMILY. Airfoil Classification.

N.A.C.A. FOUR-DIGIT AIRFOIL FAMILY. Airfoil Classification.

NACELLE. An enclosed shelter for personnel or for a power plant. An airplane nacelle usually is shorter than the fuselage, and does not carry the tail unit. (F.T.M.)

NACRE. Mother-of-pearl.

NADIR. Horizontal Coordinate System.

NAGYAGITE. A mineral compound of uncertain chemical formula in which the essential elements are lead, gold, tellurium and sulfur. Crystallizes in the monoclinic system. Hardness, 1–1.5; specific gravity, 7.41±; color, dark gray with metallic luster. (R.M.F.)

NAIAD. An aquatic insect larva. **Hemimetabola. Larva.** (A.W.L.)

NAKONG. Mammalia, Artiodactyla. One of the larger harnessed **antelopes** of Africa. This species differs from most harnessed antelopes in its uniform grayish-brown color. The horns are long and spiral. Also called the sititunga. (A.W.L.)

NANNOPLANKTON. The portion of the floating and drifting aquatic animals (**plankton**), including minute species, which pass through ordinary nets and must be secured by centrifuging. (A.W.L.)

NAPE. The back of the neck. (A.W.L.)

NAPHTHA. Hydrocarbons.

NAPHTHALENE. Naphthalene $\left(C_{10}H_8 \text{ or } \right)$ is a colorless, odorous (odor of "moth balls"), solid, melting point 80° C., boiling point 218° C., sublimes, insoluble in water, slightly soluble in cold alcohol, soluble in hot alcohol or in ether. Naphthalene reacts (1) with oxidizing agents, e.g., **sodium** dichromate, plus **sulfuric acid,** to form 1,4-naphthaquinone, (2) with oxygen of the air, in the presence of **vanadium** pentoxide as a **catalyzer** at 425° C., orthophthalic acid is formed, with **chlorine in chloroform,** to form naphthalene tetrachloride ($C_{10}H_8Cl_4(1,2,3,4)$), white solid, melting point 182° C., (3) with concentrated **sulfuric acid,** to form various naphthalene sulfonic acids. This reaction is important on account of the ease of transforming the sulfonic acid group (—SO_2OH) into the hydroxyl group

(—OH) by heating with **sodium** hydroxide, and because of the solubility in water of the sulfonic acids (see **Thioalcohols and Related Compounds**). With **mercuric** salt present as a catalyzer sulfuric acid forms ortho-phthalic acid, (4) with concentrated **nitric acid,** to form alpha-nitronaphthalene ($C_{10}H_7NO_2(1)$), yellow solid, melting point 59° C., (5) with **picric acid** ((1)HO·C_6H_2·(NO_2)$_3$($2,4,6$)), to form yellow crystalline naphthalene picrate, melting point 149° C., (6) with hydrogen to form "tetralin" ($C_{10}H_{12}$) and then "decalin" ($C_{10}H_{18}$).

Naphthalene is obtained from coal tar in the fraction distilling between 180° C. and 200° C., from which the crystals are separated by **filtration** after cooling. The product is purified by treatment with concentrated **sulfuric acid,** and subsequent **sublimation.** Naphthalene may be detected by appearance of a carmine red color on heating with **mellitic acid** anhydride. Naphthalene is used (1) as an insecticide, (2) in the preparation of **phthalic acid,** (3) in the preparation of azo-**dyes** and indigotin, which is the chief constituent of the dyestuff **indigo,** (4) for the manufacture of the solvents "tetralin" and "decalin." (R.K.S.)

NAPHTHOLS. Phenols.

NAPIERIAN LOGARITHMS. Logarithms.

NAPOLEONITE. Corsite.

NAPPE. A great overfold of the "crust" of the earth including a thick **stratigraphic** series of formations frequently representing one or more geologic periods of the earth's history. Under great continued compressional stress the overfold may be faulted and ride forward so as to produce a profound and relatively horizontal translocation of the earth's "crust." Some thrust masses of this type are supposed to have traveled as much as 200–300 miles from the region in which the rocks were originally formed. The type locality of nappes is the Swiss Alps, where they form the principal structural features of this highly deformed and structurally complex mountain system. (R.M.F.)

NARCOMEDUSAE. Hydrozoa.

NARCOTINE. Alkaloids.

NARES. Terminal orifice of the nasal passages. The external openings are the external or anterior nares and the openings at the communication of these passages with the **pharynx** in air-breathing **vertebrates** are the posterior or internal nares. (A.W.L.)

NARWHAL. Mammalia, Odontoceti. A moderately large animal, *Monodon monoceros,* of the **dolphin** family, characterized by the single long straight tusk, spirally twisted, which extends forward from the upper jaw. The tusk is a single tooth, usually that of the left side. Its mate in the male and both of these teeth in the female are rudimentary. The narwhal has no other teeth except a few irregular rudiments. The tusk attains a length of 8′ and the animal reaches a length of 12–16′. (A.W.L.)

NASAL BONE. One of a pair of small bones forming the bridge of the human **nose** and the corresponding portion of the skull in other vertebrates. (A.W.L.)

NASO-LABIAL GROOVE. A groove extending from the nostril to the mouth in some of the fishes (**Pisces**) and **amphibians**. (A.W.L.)

NASTIES. Movement in Plants.

NATRIUM. Sodium.

NATROLITE. The mineral natrolite, one of the **zeolites**, is a **sodium aluminum silicate** corresponding to the formula $Na_2Al(AlO)(SiO_3)_3 \cdot 2H_2O$. It is **orthorhombic**, crystallizing in slender prisms of nearly square cross-section which are terminated by flat pyramids. There are also fibrous to compact varieties. Natrolite is a brittle mineral; hardness, 5–5.5; specific gravity, 2.2; luster, vitreous; color, red, yellow, white, or colorless; transparent to opaque. Natrolite is found with other zeolites in fissures and cavities in basaltic and related rocks. Czechoslovakia, France, Italy, Norway, Scotland, Ireland, Iceland, Greenland, and South Africa contain well-known localities for natrolite. In the United States it is found in the Triassic **traps** of New Jersey. Its name natrolite refers to its soda content. (E.S.C.S.)

NATURAL FREQUENCY. This term is commonly applied to **coils** and **antennae** in communication circuits. In the former it designates the frequency at which the inductance of the coil and its **distributed capacity** produce resonance. Referred to the antenna it means the lowest frequency at which there will be a standing wave on the antenna. (L.R.Q.)

NATURAL GAS. Petroleum.

NATURAL LOGARITHMS. Logarithms.

NATURAL NUMBERS. Number.

NAUPLIUS. A form of **larva** occurring in some of the **crustaceans**. It is the first stage to hatch from the egg and is a minute ovoid creature, unsegmented, with three pairs of appendages. (A.W.L.)

NAUTICAL ASTRONOMY. The term nautical astronomy is applied to those problems in the transformation of **spherical coordinates** and in the solution of the **astronomical triangle** which are of particular importance to the navigator. Among such problems may be cited the determination of **latitude, longitude,** and **azimuth** from a single **altitude** of a celestial object as taken with a **sextant**. (See **Celestial Navigation**.) (W.K.G.)

NAUTICAL MILE. This is the fundamental unit of distance used in navigation and, for purposes of convenience, has been defined as 6080′. Rigorously the nautical mile was defined as the length of 1 minute of arc on a great circle drawn on the surface of a sphere with the same area as the earth. In accordance with this rigorous definition the length of the nautical mile is 6080.27′.

The nautical mile is frequently confused with the geographical mile, which is defined as the length of 1 minute of arc on the earth's **equator.** The geographical mile has a length of 6087.15′.

Owing to the fact that the earth is an oblate spheroid and flattened at the poles the length of 1 minute of arc measured along a **meridian** varies in different latitudes. It is shortest at the poles and longest at the equator, having an average length of 6076.82′.

With sufficient accuracy for navigational purposes the minute of arc on the meridian, the minute of arc on the equator, and the nautical mile may all be considered as 6080′. It will be noted that such an assumption will never introduce an error greater than 0.8%. (W.K.G.)

NAUTILUS. Mollusca, Cephalopoda. The name of a genus of marine **mollusks,** also anglicized as a common name. The animal lives in the last and largest chamber of a shell which is coiled in a flat spiral. The three or four existing species are found in shallow to moderately deep water in the tropical oceans. They are eaten by the natives of some of the Pacific islands.

The paper nautilus, *Argonauta,* is not a true nautilus, but is more closely related to the octopus. (See also **Invertebrate Paleontology**.) (A.W.L.)

NAVEL. The scar on the abdomen of mammals where the **umbilical cord** was attached during prenatal life. The **umbilicus.** (A.W.L.)

NAVEL ORANGE. Parthenocarpy; Citrus Fruits.

NAVIGATION. A literal translation of the Latin word *navigare* is "to conduct a ship," and navigation is sometimes defined as the art or science of conducting a ship from place to place on the surface of the **earth.** However, the actual conducting of a ship more properly belongs to seamanship or airmanship than to navigation.

The subject matter of navigation, in the present usage of the term, can be considered under one of three main headings: 1. The **sailings.** To find the **direction**, or directions, in which to **head** a ship in order that it may proceed on the most practical **course** from one place to another. 2. **Dead reckoning.** To find the position of a ship at any particular instant, provided that its position at some previous time is given and that, since leaving this position, the headings, distances run on each heading, and the effects of environment and motions of the medium supporting the ship are known. 3. **Piloting, contact flying, celestial navigation.** To find the position of a ship at any instant by means of observations, not necessarily visual, of one or more objects on the surface of the earth or on the **celestial sphere.**

The number of individual methods and instruments employed in modern navigation is far too great to even list in an article of this character. The more important instruments and methods will be discussed under their individual titles throughout this volume. (W.K.G.)

NAVIGATIONAL COMPUTER. A navigational computer is fundamentally a circular **slide rule**, with the slide and scale carrying material of particular importance to navigators. On most navigation computers the slide is marked off in units of time, either minutes or hours and minutes, and the scale is graduated in miles. By setting 60 minutes on the slide to the speed of the ship (distance traveled in 1 hour) any other number of the scale will be the distance traveled in the corresponding time indicated on the slide. On some computers different colors are used on the scale and slide to indicate other frequently used calculations, e.g., rate of fuel consumption as a function of air speed.

By using "cut-outs" and auxiliary scales on the computers, the conversion from calibrated **air-speed meter** reading to true air speed, the conversion from **altimeter** reading to true altitude, and other such important conversions, may be readily made.

There is no standard type of computer and many navigators have gotten into serious difficulty by using a new type of instrument without carefully studying the description issued with it. (W.K.G.)

NAVIGATORS' STARS. A list of 55 stars has been designated as the "navigators' stars." The list was selected to cover the entire **celestial sphere** in such a manner that a navigator, no matter at what season or in what part of the earth he may be operating, will have two or three navigators' stars available for observation. The names and positions of these stars are listed in the

Air **Almanac,** published by the U. S. Naval Observatory, and in a number of other publications. (w.k.g.)

NEANDERTHALOIDS. Paleontology of Man.

NEAR-SIGHTEDNESS. Myopia.

NEATSFOOT OIL. Esters.

NEBULA. The term nebula (stellar nebulosa) was originally used by astronomers to describe any luminous spot which remained fixed relative to the **stars.** Before the application of the **telescope** to astronomy probably the only objects to which the term applied were those which we now refer to as **star clusters,** although there is a tenth-century reference to the great **spiral** in Andromeda. Following the application of the telescope to the study of the heavens, many more nebulous objects were discovered. Originally these objects were grouped into three main classifications: the diffuse nebulae, the planetary nebulae, and the spiral nebulae. Further research indicated that the spiral nebulae were very different from the other two. The application of the large telescopes of the present century has proved that they are in reality groups of stars, the term nebula has been dropped, and they are now referred to simply as spirals.

The diffuse nebulae are divided into two main groups, the dark and the bright. For many years the study of photographs of the Milky Way have shown "black holes" and other configurations that gave distinct evidence that clouds of dark material were obscuring stars that were beyond. Detailed investigation of the spectra of the spectroscopic binaries has shown that, although the great majority of the absorption lines from these objects shows the periodic displacement due to the motions of the components, nevertheless there are a few lines which remain stationary. The only plausible explanation is that these "stationary lines" are due to absorption by gaseous material in what had hitherto been referred to as empty space. When the color characteristics of stars in any one of the **spectral classes** are carefully studied, it is found that the more distant ones show a distinct reddening. The reddening can be well explained by the assumption of interstellar dust, which produces a reddening effect in much the same way that our atmosphere produces that reddening of sunlight which is more evident when we see that object through a thick layer (e.g., at sunset).

The study of this interstellar material forms one of the most interesting problems of modern astrophysical research, but the conclusions at the present time are not positive. Suffice it to state that interstellar material exists and that it contains gases, in atomic or molecular state, and also some form of dust-like material. The temperature of the interstellar region is of the order of magnitude of absolute zero and we are unable to state just what sort of "dust" might be formed under these conditions.

There is considerable evidence that the interstellar material is present in all parts of space. However, there is a definite tendency for the mixture of gas and dust to collect in clouds and condensations. When one of these dark clouds is between the earth and a region where the stars are apparently most numerous, e.g., in the plane of the **milky way,** we can observe the darkness of the cloud against the background of stars. Such obscuring clouds of interstellar material are known as dark nebulae.

When one of these clouds is in the vicinity of a bright star, the intense radiation from the star will illuminate the cloud and we observe what is known as a bright diffuse nebulae. Studies of the spectra of these objects show that the light is made up both of reflected starlight and also of radiation from the interstellar material. The character of the reflected light is similar to that from the nearby stars. The radiation from the nebulous material itself, however, is of quite a different character from star light. This nebular spectrum is of the bright-line type and is produced by the absorption of radiation from the star by the gas atoms, and then reradiation of this energy in frequencies characteristic of the gas atoms and their states of **ionization.** This hypothesis receives almost positive confirmation from the fact that the character of the nebular spectrum depends upon the spectral type of the nearby stars. If the star is hot (B-type) the nebular spectrum is rich with bright lines, but in the vicinity of a relatively cool star (M-type) the nebular spectrum is almost entirely that of reflected star light.

When the spectra of the bright nebulae were first studied, and before the modern hypothesis as to their general structure was even guessed, 2 bright lines were observed to be present which could not be identified with any known terrestrial or solar element. It was supposed that these lines were due to some material that existed only in the nebulae, and the element was called Nebulium. We now know that the lines are due to the oxygen atom, which is in such abnormal condition that the particular lines could never appear in any laboratory spectrum.

In addition to the diffuse nebulae, there are others that are known as planetaries, because of the fact that they have quite definite shapes and look more or less like a planet when observed with a telescope. The general appearance of these objects, on detailed photographs, is that of a shell of gaseous material. They are generally elliptical in form, and may have the appearance of being made up of several elliptical shells with their axes at various angles to each other. Frequently a very blue, and hence probably very hot, star is observed at the center of the shell. Even when the star itself is not seen, astronomers are of the opinion that it is actually there, but that the shell is thick enough in the line of sight to prevent its being seen from the earth.

The spectra of the planetaries is, in general, the same as that of the diffuse nebulae found in the vicinity of B-type stars. In some cases, where the central star can be studied, variations in its light have been found and the nebula radiation is found to vary with that of the star. It is evident that the planetary nebulae are actually stars with very extended and attenuated atmospheres. There is no hypothesis to completely explain why some stars are in this condition, whereas the vast majority are not. Careful studies of the spectral lines from the planetaries indicate that the shell may be expanding. Expanding shells of gas have been observed around some **novae,** but the rate of expansion is far in excess of that found in the planetaries.

NEBULIUM. Nebula.

NECK. In zoology, a slender region connecting two other parts of a body. Particularly the region between the head and trunk of many vertebrates. The vertebrate neck contains a series of cervical vertebrae, respiratory and food passages, and part of the central nervous system, but it is not invaded by the body cavity. It varies greatly in relative length, a long neck usually compensating for long legs to enable the animal to reach the ground.

In geology, a **volcanic conduit** of **plug-**like character consisting of lava and other volcanic products, which has been exposed by **erosion.** (a.w.l., r.m.f.)

NECROPSY. An autopsy examination of the body to discover or verify the cause of death, and to study the pathology of disease. (d.m.h.)

NECROSIS. The local death of cells results in changes in the tissue known as necrosis. These consist of disintegration of the cellular structure with destruction of the nucleus and coagulation or liquefaction of the cytoplasm (see **cell**). The causes of necrosis include interference with the blood supply of a tissue (see **gangrene**), physical injury, and deleterious actions by bacteria or their toxins. (D.M.H.)

NECTARY. Flower.

NECTOCALYX. A modified **medusa** occurring in coelenterates of the order **Siphonophora. Hydrozoa.** (A.W.L.)

NECTONEMATOIDEA. Nematomorpha.

NEEDLE BEAM. Underpinning.

NEEDLE BEARING. Bearings.

NEEDLING. Underpinning.

NEGATIVE FEEDBACK. Feedback.

NEGRO BUG. Insecta, Hemiptera. Small shining black **bugs** with a smooth convex upper surface. They resemble beetles superficially. Most of the abdomen is covered by a greatly enlarged sclerite of the thorax which also conceals most of the wings. (A.W.L.)

NEKTON. The portion of a population made up of animals which are capable of directive locomotion through a fluid medium. Usually applied only to aquatic animals, including the fishes, although flying creatures constitute a similar part of the terrestrial fauna and may be called an aerial nekton. (See also **Plankton** and **Invertebrate Paleontology.**) (A.W.L.)

NEMATHELMINTHES. The roundworms, a major division of the animal kingdom containing both free-living and parasitic species, many living in the human body. The most widely known species are the intestinal worms of man and the domestic animals, the hookworms, the spiny-headed worms, and the horsehair worms.
  The phylum is characterized as follows: 1. The body is not segmented and is usually cylindrical, tapering at the ends. 2. The ectodermal covering is noncellular and secretes an elastic cuticle. 3. The body wall includes an inner layer of muscle cells divided into contractile and noncontractile parts. 4. The alimentary tract is a slender tube. In some species it is tubular for only part of its length and in some it is completely lacking. 5. The excretory system consists of two lateral tubes. 6. The nervous system includes a ring around the esophagus and longitudinal cords extending through the body. A dorsal and a ventral cord are the chief nerve tracts. 7. The space between the body wall and the gut is filled with large cells containing many **vacuoles** which join extensively and form an apparent body cavity.
  The phylum is divided into three classes which differ enough to be segregated by some biologists:

  Class **Nematoda.** Without a spiny proboscis. Intestine present. The typical roundworms. The **hookworms, eelworm,** and **Guinea worm** are among the species parasitic in man. Some species attack plants and some live in earth or water.
  Class **Nematomorpha** (Gordiacea). The **hairworms** or horsehair worms. No spiny proboscis. Body cavity lined with epithelium. Alimentary tract present. Adults free in water, larvae parasitic in insects or crustaceans.
  Class **Acanthocephala.** With a spiny proboscis. No gut. Parasitic in the intestine of vertebrates. (A.W.L.)

NEMATOBLAST. A term proposed by Becke, in 1903, for pseudo-**porphyritic** minerals of fibrous habit developed in **metamorphic** rocks. (R.M.F.)

NEMATOCYST. A structure discharged by the stinging cells of coelenterates (**cnidoblast**). It consists of a flask-shaped body bearing barbs and a long slender filament. (A.W.L.)

NEMATODA. The threadworms. A class of the phylum **Nemathelminthes** containing worms of many habits, some parasitic in man and the domestic animals. The body lacks a spiny proboscis and is marked by slender longitudinal lines along the sides. These lateral lines follow the excretory tubes.

  Two orders are recognized:
  Order Hologonia. Genital tract an unbranched tube. Both large and small species, parasitic in man and domestic animals. Trichinella (**Trichina**), a serious parasite sometimes taken into the human body in insufficiently cooked pork, is a member of the order.
  Order Telogonia. Genital tract of female with two or more branches. Most nematodes belong in this order, including free-living, plant-feeding, and parasitic forms. The **Guinea worm, eelworm,** and **hookworm** belong here. (A.W.L.)

NEMATOMORPHA. The **hairworms** or **horsehair worms,** a class of the phylum Nemathelminthes. These worms receive their popular name from the old belief that horsehairs soaked in water turn into them. Their abrupt appearance in watering troughs and small pools is, in fact, due to their living in the bodies of insects and crustaceans as larvae and emerging when they take on the adult form to live in the water.
  They differ from the true **nematodes** in the absence of lateral lines and in the presence during adult life of a limited body cavity lined with epithelium. In appearance they are very long and slender, bluntly rounded at the anterior end and blunt or forked at the posterior end.
  They have also been called Gordian worms from their occasional massing in tangled clusters reminiscent of the famous Gordian knot.

  The class includes two orders:
  Order Gordioidea. Fresh-water and terrestrial species.
  Order Nectonematoidea. Marine species only. (A.W.L.)

NEMERTEA. Marine worms with a flattened body, often long and ribbonlike in form. They are unsegmented and have no body cavity, hence they are sometimes included with the flatworms as a class of the phylum **Platyhelminthes.** More often they are made a separate phylum because the alimentary tract is a tube opening with an anterior mouth and a posterior anus. Like the free-living flatworms, they have ciliated (**cilia**) integument. They are also provided with an eversible proboscis associated with but not derived from the alimentary tract. No common name is available save nemertean or nemertine worms.
  These worms live among seaweed or at the bottom of the ocean and prey on living animals or eat dead ones. They are not economically important. A few fresh-water species and a few parasitic forms are known.

  The phylum is divided into four orders:
  Order Paleonemertea. Long and slender. Mouth behind brain. Outer muscles of body wall circular. Marine species.
  Order Heteronemertea. Long and slender. Mouth behind brain. Outer muscles of body wall longitudinal. Marine.
  Order Hoplonemertea. Either long and slender or short and thick. Mouth in front of brain. Marine, fresh-water, terrestrial, and parasitic species, the last living on crabs.
  Order Bdellonemertea. Short, flat, and broad, with a sucker at the posterior end. Three species which

live in the gill chamber of marine and fresh-water mollusks. (A.W.L.)

NEODYMIUM. Symbol: Nd. Atomic number: 60. Atomic weight: 144.27. Density: 6.96. Melting point: 840° C. (Isotopes: page 290.) Type of compound: Nd_2O_3, blue. Color of salts: rose-red. Discovered by Welsbach in 1885. A member of the **cerium** sub-group of the rare earth metals. (R.K.S.)

NEOLITHIC. Paleontology of Man.

NEON. Symbol: Ne. Atomic number: 10. Atomic weight: 20.183. Density: 0.9004 gram per liter, 0° C., 760 mm., or 0.696 when air equals 1.000. Melting point: $-248.7°$ C. Boiing point: $-245.9°$ C. Isotopes: 20 (90.0%), 21 (0.27%), 22 (9.73%), have been separated by, **diffusion;** identified and quantitatively estimated by the mass **spectrograph.**

Neon is a colorless, odorless gas, of negative chemical properties with ordinary materials. Discovered by Ramsay and Travers in 1898.

Neon is present in ordinary **air** to the extent of 1 part neon in about 65,000 air. In a vacuum electric discharge tube neon shows a crimson glow, and is widely used in advertising illumination. One of the early elements to be discovered having the property of **isotopism.** (R.K.S.)

NEOPITHECUS. Paleontology of Man.

NEOPLASM. Any new or abnormal overgrowth of cellular tissue. A neoplasm is a cellular **tumor** and may be either benign or malignant. (R.S.M.)

NEOPRENE. The generic name for synthetic rubber made by the polymerization of 2-chloro-1,3-butadiene, which is itself made by the action of hydrogen chloride on monovinylacetylene. Vulcanized neoprenes are outstanding in their resistance to oils, greases, many chemicals, as well as to heat, sunlight, and ozone. (R.K.S.)

NEOSALVARSAN. Arsphenamine.

NEOTENY. The retention of characteristics of immature stages during adult life. Neoteny and **paedogenesis** are difficult to distinguish in some cases. The latter is the attainment of sexual maturity by structurally immature animals. Both are illustrated by distinct cases among the insects, where immature and adult life are marked by distinct stages. The **larvae** of some **gall gnats** produce young, sometimes for a series of several generations, but the fact that some individuals become winged adults marks the condition as paedogenesis. On the other hand, the females of some **beetles** (Phengodidae), when they have attained the stage corresponding to the winged adult males, still retain the form of the larva and are called glow worms. This is a distinct case of the persistence of larval characters in adult life, or neoteny. Among the vertebrates the phenomenon of neoteny has been most extensively studied in the **axolotl.** (A.W.L.)

NEOTYPE. A specimen selected to represent the type when the type has been lost. (R.M.F.)

NEOZOIC. In historical geology the time from the end of the Mesozoic to the present. (Compare with Cenozoic.) (R.M.F.)

NEPENTHES. Insectivorous Plants.

NEPHELINITE. A dense, sometimes **porphyritic** rock made up almost wholly of **nephelite** and **augite.** If olivine is present the rock is then classified as a nephelite **basalt.** (E.S.C.S.)

NEPHELITE-SYENITE. A coarse crystalline igneous rock composed chiefly of alkali-**feldspars**, **nepheline**, and **femic** minerals, such as the soda-**pyroxenes** and **amphiboles.** Accessory minerals are other soda-**feldspathoids,** zircon, apatite, and **sphene.** Nepheline syenites occur in Canada in the provinces of Quebec, Ontario, and British Columbia. The principal foreign localities are Norway, Greenland, Sweden, the Ural Mountains, the Pyrenees, Italy, Brazil, China, and the Transvaal region. (R.M.F.)

NEPHOSCOPE. Meteorological Instruments.

NEPHRIDIUM. The excretory tubule in **annelid** worms. It consists typically of a convoluted tubule with a ciliated (**cilia**) opening called the nephrostome at the inner end, communicating with the body cavity. At the outer end it empties by a minute pore on the surface of the body. True nephridia are associated with the **coelomoducts** in the excretory systems of annelids. (A.W.L.)

NEPHRITIS, BRIGHT'S DISEASE. Inflammation or injury to the **kidney,** involving in varying degrees the cellular, intercellular, and blood-vessel tissue of the kidney. The term does not include the pus-producing infections which may involve the kidney.

Nephritis was first accurately described by Richard Bright in 1836. The term now covers a variety of disease states of unknown etiology which may be acute or chronic, all of which are characterized by the presence in the urine of **albumin** (albuminuria) and at times blood, cells and casts. **Edema, hypertension** and **uremia** are also characteristic features. The main types of nephritis are (1) glomerulonephritis, acute or chronic, (2) arteriolar nephrosclerosis, (3) nephrosis, (4) miscellaneous nephritides.

Acute glomerulonephritis, like **rheumatic fever,** seems definitely to be related to infection of the upper respiratory tract with hemolytic streptococci. The exact mode of pathogenesis is not understood, but it is a well-recognized fact that virtually all cases of the disease follow **streptococcus tonsillitis, pharyngitis,** or **sinusitis,** etc. At times the respiratory infection may be so mild as to pass unnoticed, but immunological techniques will establish the recent presence of such infection. Predisposing factors are largely dependent upon conditions conducive to spread of respiratory infections.

Acute glomerulonephritis usually begins with hematuria (blood in the urine), headache, puffiness about the face, and decreased urinary output. The urine contains albumin, red blood cells, and casts of the kidney tubules. Transient elevation of the blood pressure is characteristic. In the severe or fulminating variety, fever, generalized edema, marked hypertension, visual disturbances, varying degrees of uremia, nausea, vomiting, delirium, convulsions, coma and death may ensue. Commonly, one or more features, such as edema and backache, dominate the picture, and the course is relatively mild. There is no specific treatment: bed rest and restriction of sodium chloride in the dict to control the edema are the most important measures. The use of sulfonamides has not yet been clearly evaluated. The prognosis in acute cases is good, the immediate mortality being 4%. Most patients recover completely, but a few go on to latent or chronic disease.

Chronic glomerulonephritis does not have such a clear-cut relation to hemolytic streptococci as the acute form, and its etiology is even more obscure. The onset is usually gradual, with weakness, lassitude, edema, headache and hypertension. The urine is of low specific gravity, contains albumin, casts, but few or no red cells. **Anemia** is a common finding. The disease usually runs a long course over a period of years, with variations in signs and symptoms. At times, edema may be massive

Eventually death occurs from hypertension and **heart failure**, or kidney failure and uremia. The treatment of the various stages is symptomatic, and guided by the symptoms present. Bed rest, salt restriction, **transfusions, digitalis**, etc., may be indicated.

The **pathology** of glomerulonephritis is characterized by various degrees of inflammatory and degenerative change in the glomeruli, the blood vessels, and the interstitial tissue of the kidney. In the acute form, edema, infiltration with inflammatory cells, and small areas of hemorrhage appear. In the chronic form the picture is one of scarring and obliteration of the glomeruli, with destruction of the majority of functional units.

Arteriolar nephrosclerosis is the renal lesion occurring as part of the picture of hypertensive cardiovascular disease. About 10% of patients with this disease die of kidney failure and uremia. The clinical picture is the same as with chronic glomerulonephritis with emphasis on the hypertensive feature, but the kidneys at autopsy show characteristic vascular lesions. The arterioles, or small arteries, have markedly thickened walls with necrotic inflammatory changes. The glomeruli and tubules show varying degrees of degeneration and obliteration and there is fibrosis of the interstitial tissue. Sometimes, however, the pathological picture is indistinguishable from that of chronic glomerulonephritis.

Nephrosis is a term used to describe the edematous stage of chronic glomerulonephritis, and to describe a distinct disease, separate from glomerulonephritis. This latter, or true lipoid nephrosis, occurs as a chronic disease of children, characterized by massive edema and albuminuria without hypertension or hematuria. Pathologically it is primarily a disease of the kidney tubules. The clinical course is over a period of weeks up to several years, but the patient eventually recovers and remains well. Death from nephrosis is rare, and when it occurs is commonly due, for no apparent reason, to pneumococcus **peritonitis**. Therapy consists of a high protein diet, salt restriction, and **diuretic** drugs.

Miscellaneous nephritides comprise a diverse group including transfusion nephritis, which follows **transfusion** with incompatible blood; syphilitic nephritis, a rare manifestation of tertiary **syphilis;** focal nephritis occurring in the course of an infection when bacterial emboli lodge in the kidneys; and arteriosclerotic nephritis, a mild disease due to sclerosis of the large kidney vessels. (D.M.H.)

NEPHROMIXIUM. An excretory organ of segmented worms formed of the united **nephridia** and **coelomoducts.** (A.W.L.)

NEPHROSTOME. The ciliated (cilia) opening of the excretory tubules of segmented worms and vertebrates which communicates with the **coelom.** It is characteristic of the primitive form of tubule in both groups but is lacking in many species. The connection with the body cavity is then supplanted by an association of the excretory organs with the **circulatory system.** (A.W.L.)

NEPTUNE. (See tables of planetary data, page 1109). Possibly the most interesting feature of the **planet** Neptune is the circumstance which led to its discovery. After Uranus was discovered an orbit was computed, and, by reckoning backwards, it was found that the planet had been observed as a star on several occasions many years previous to its announcement as a planet. The **orbital** elements computed from the observations made shortly after the discovery did not satisfy the old observations accurately, and, what was worse, the planet soon began to deviate from the computed path. Even the introduction of **perturbations** by all of the then-known planets failed to completely remove the deviations, and after a prodigious amount of computing, Leverrier, a young French astronomer, and Adams, an Eng-

lishman, both predicted that a new planet would be found in the **constellation** of **Aquarius.** Search was made in this vicinity by the German astronomer Galle on the basis of Leverrier's prediction, and by Challis at Cambridge, England, using Adams' prediction. Challis observed the new planet first, but failed to recognize it until after Galle had announced the discovery on September 23, 1846. The discovery of Neptune can safely be regarded as one of the greatest successes of the Newtonian gravitational theory and the methods of orbit computation.

Neptune itself is a planet slightly less than 4 times the diameter of the earth, and a mass slightly greater than 17 times that of the earth. Its mean density is only 0.29 times that of the earth, and this low value, coupled with the high value of the **albedo** (0.52), indicates that the planet probably has a thick layer of atmosphere.

Neptune is invisible to the naked eye, having a stellar **magnitude** of about 7.7, but can be readily observed in any telescope with aperture greater than 1″. In small instruments it can only be distinguished from the stars by observing the change in position from night to night, but with larger instruments it appears as a small greenish disk. Observations indicate that the disk is apparently circular and that there are no distinguishable surface markings. Hence, telescopic observations tell nothing regarding the rotation period. In 1928 Moore and Menzel, at the Lick Observatory, found from spectroscopic observations, employing the **Doppler** principle, that the planet has a rotation period of slightly less than 16 hours, and that the planet is rotating in the same directional sense as the majority of the other members of the solar system.

Neptune has one **satellite**, discovered by Lassell very shortly after the discovery of the planet. Telescopically the object is very faint and can only be observed with large instruments. From the brightness it is estimated that the satellite is similar to our own **moon** in size. The orbit of the satellite is inclined at about 20° to the plane of the planet's equator and the satellite revolves about the planet in the direction opposite to that of the majority of the members of the solar system. (W.K.G.)

NEPTUNIUM. Chemical Composition, I. Elements.

NERNST EFFECT. If heat is flowing through a strip of metal and the strip is placed in a magnetic field perpendicular to its plane, a difference of electric potential develops between the opposite edges. This phenomenon, discovered by Nernst in 1886, is analogous to the **Hall effect,** but with a longitudinal flow of heat replacing the longitudinal electric current. If, to one looking along the strip in the direction of the heat flow, and with the magnetic field directed downward, the transverse potential drop is toward the right, the Nernst effect is said to be positive. (But Nernst at first used the opposite convention.) Bismuth, in which the phenomenon was first observed, shows the positive effect, iron the negative. (See also **Ettingshausen** and **Righi-Leduc** effects.) (L.D.W.)

NERNST HEAT THEOREM. Entropy.

NEROLI OIL. Volatile Oils.

NERVE. A slender cord made up of nerve fibers (**neuron**). In the **vertebrates** the nerve is surrounded by loose connective tissue called the epineurium. Each small bundle of fibers within the epineurium is surrounded by a thin compact perineurium, and from this layer thin septa, the endoneurium, run between the irregular groups of fibers within the bundle.

Nerves form the communicating paths between the central **nervous system** and the various parts of the body, as well as the connections between ganglia. They have been given special names according to their anatomical distribution, and a few functional properties have been indicated by descriptive terms. Thus sensory or afferent nerves carry impulses from sense organs to the central nervous system and motor or efferent nerves lead out to muscles and other effectors. Most nerves of the body contain fibers of each kind and so are mixed nerves. Nerves arising from the brain in the vertebrates are called cranial nerves and those connected with the spinal cord are spinal nerves. Typically all of these main nerves of the vertebrate are supposed to be based on the form of the spinal nerves, which connect with the cord by two roots. The dorsal root bears a spinal ganglion and carries all the sensory fibers and the ventral root lacks a ganglion and is motor. All sensory nerves bear a ganglion or arise from a sensory layer such as the retina of the eye and no purely motor nerves have ganglia.

The motor components of these nerves grow out from cells in the central nervous system and the sensory components grow into the central system from ganglia or from sensory cells and also out from the ganglia toward the periphery of the body. (A.W.L.)

NERVE IMPULSE.

A progressive transfer of a condition of excitation along a **nerve** fiber, initiated by a stimulus acting upon a sensory organ, by a cell within the nervous system, or experimentally by direct stimulation of the nerve fiber.

The nerve impulse involves measurable chemical changes and energy consumption accompanied by changes in electrical potential associated with membrane depolarization. The electrical changes have been extensively used in the investigation of nerve physiology and of the brain. The passage of the nerve impulse over the synapse is accompanied by the action of acetylcholine in cases which have been studied, while terminal nerve fibers may form either acetylcholine or sympathin, the latter being closely related to adrenalin.

The impulse activates some organ of the body or enters the nervous system, where it is relayed to other parts. (A.W.L.)

NERVOUS COORDINATION.

The attainment of harmonious action of all component parts of the animal and the adjustment of its behavior to environmental conditions by communication and regulation through the **nervous system.** All coordinative processes are of this type or of the type carried on by the secretion and distribution of **hormones.** Nervous coordination has the advantage of rapidity, since the **nerve impulse** travels much more rapidly than materials can be transported. It is responsible for the immediate adjustment of the animal to fluctuating conditions, while chemical coordination is extensively involved in the maintenance of the normal organic processes which are a more uniform part of the animal's activity. Coordination in general is, however, due to close interaction of the two types; neither is wholly independent of the other.

The process of coordination through the nervous system is based on the general sensitiveness of **protoplasm** known as irritability and on the property of conductivity by which some result of stimulation passes rapidly through the adjacent substance. All living cells are irritable to some extent. They respond to some stimulus with characteristic activity. Sensory organs possess this property to a high degree and in addition are specialized to receive a certain kind of stimulus. **Nervous tissue** is specialized for ready activation and for the rapid conduction of the impulse generated within it.

The organs which receive stimuli as a special function for the benefit of the animal, whether relatively simple cells or extremely complex organs like the human **eye,** are known as **receptors.** They have special nerve endings, often associated with other structures, and are con-

nected by afferent or sensory nerves with other parts of the nervous system or, in very simple animals, with some organ capable of acting in response to the condition from which the stimulus arose. Organs of the latter category are muscles, glands, electric organs, light organs, and some pigment cells, and are known collectively as effectors. Usually one or more cells of the nervous system intervene between the sensory cell associated with the receptor and the motor cell which communicates directly with the effector. The more complex this nervous chain, the more intricately may nerve impulses be relayed through different paths in the body, but in all cases the net result is the same. All adjustments are due to the reception of stimuli by sense organs, both internal and external, followed by appropriate reaction of other parts of the body. (A.W.L.)

NERVOUS SYSTEM.

An organic system specialized for the ready reception of stimuli and the rapid transmission of a resulting change in its substance called the **nerve impulse** to other parts of the body. The function of the nervous system is the regulation and coordination of the various parts of the body known as **nervous coordination.**

All animals except **protozoans** and **sponges** have a nervous system derived from the ectoderm. In the **coelenterates** it consists of scattered cells associated in a network which extends through the body without marked centralization. In the flatworms the network persists, but in it nerve cells are massed near the anterior end of the body to form a cerebral ganglion or **brain,** and definite longitudinal paths can be traced from this center back through the body. This centralization is associated with the change from radial to bilateral **symmetry** and persists in all bilaterally symmetrical animals. With the attainment of radial symmetry by the echinoderms it is again lost. The departure from the nerve net is marked by the attainment of synaptic organization. In the nerve net impulses pass in either direction over nerve fibers and in the synaptic system they pass over the synapse in only one direction. The ends of the fibers of different cells are closely associated in the **synapse** through which the impulse in one fiber acts as a stimulus to the others.

In all of the more complex nervous systems a major center, the **brain,** is present as a part of the central

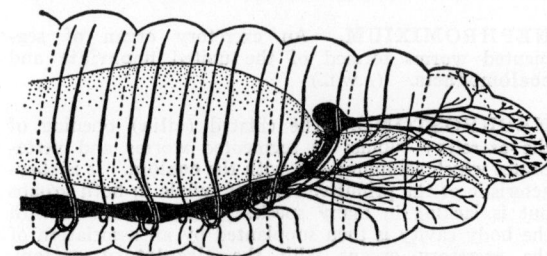

Nervous system of the earthworm. Drawing showing a lateral view of the arrangement of the larger nerve trunks in the left half of the anterior segments of the earthworm.

nervous system, and supplementary centers called ganglia, each containing a small number of nerve cells, are scattered through the outlying parts of the system. All of these parts are connected by nerve cords or **nerves** and are joined in the same way with other structures of the body.

The most common type of nervous system among invertebrates consists of a small brain lying above the alimentary tract near the anterior end of the body. It is connected by cords passing down around the sides of the gut with a ventral nerve cord. In the **annelid** worms this cord is a chain of ganglia, one lying in each segment. Conclusive evidence from **embryology** and minute anatomy shows that the primitive cord was a paired structure and that each segment contained a pair of ganglia connected by transverse nerves as well. In most existent species the adult shows no visible evidence of the paired condition.

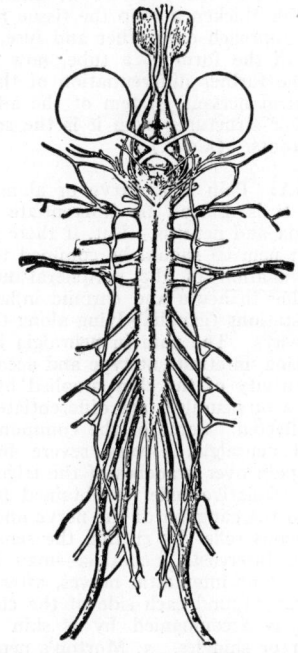

Ventral view of the nervous system of frog. (*After Woodruff. Foundations of Biology, from Ecker, The Macmillan Co.*)

In the **arthropods** a concentration of ganglia in some species has resulted in one large ganglionic mass near the **anterior** end of the ventral cord.

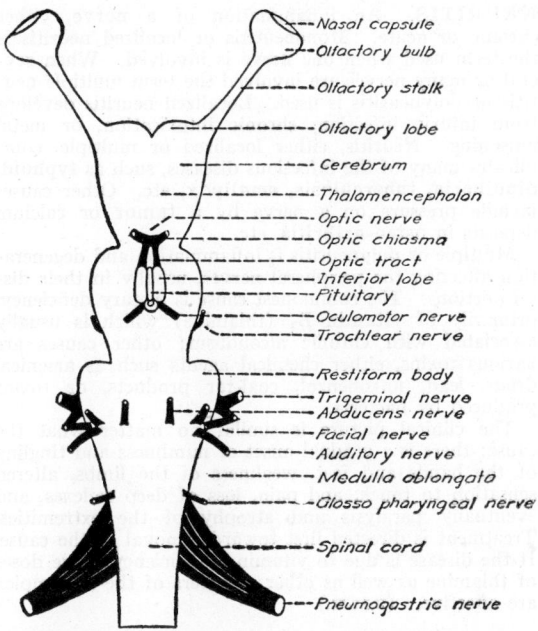

- - - Nasal capsule
- - - Olfactory bulb
- - - Olfactory stalk
- - - Olfactory lobe
- - - Cerebrum
- - - Thalamencepholon
- - - Optic nerve
- - - Optic chiasma
- - - Optic tract
- - - Inferior lobe
- - - Pituitary
- - - Oculomotor nerve
- - - Trochlear nerve
- - - Restiform body
- - - Trigeminal nerve
- - - Abducens nerve
- - - Facial nerve
- - - Auditory nerve
- - - Medulla oblongata
- - - Glosso pharyngeal nerve
- - - Spinal cord
- - - Pneumogastric nerve

Ventral view of dogfish brain, showing cerebral nerves. (*Drawn by W. J. Moore.*)

Many **mollusks** have a number of pairs of ganglia of about equal size as nerve centers, although the cephalopods have a concentrated brain equal to any other invertebrates.

In the **vertebrates** the entire central nervous system lies above the alimentary tract. The brain is large and complex in most classes. From it a **spinal cord** runs back along the axis of the body. Both brain and spinal cord bear nerves which extend throughout the animal.

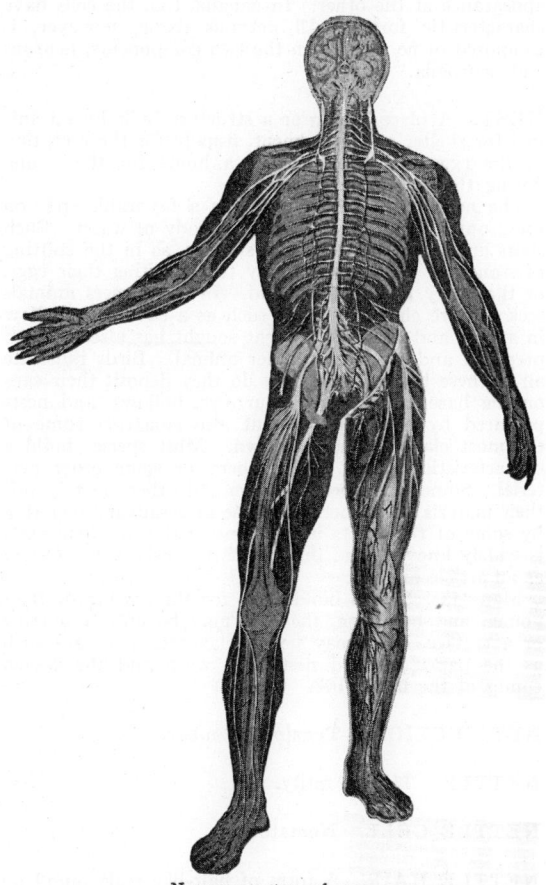

Nervous system of man.

Below the spinal cord two chains of ganglia connected by slender nerves with each other and with the spinal nerves constitute the sympathetic chains of the autonomic nervous system. Nerves of this system supply the viscera and are widely distributed elsewhere in the body. In cooperation with the parasympathetic fibers of the cranial and sacral nerves, they control the functions which cannot be deliberately regulated by individual desire, such as the movements of the alimentary tract. (A.W.L.)

NERVOUS TISSUE. Nervous tissue proper is composed of cells specialized to a high degree in the properties of irritability and conductivity. These cells are called **neurons.** With the exception of a portion of the nervous system of echinoderms which is said to be mesodermal, all nervous tissue originates from the ectoderm. More complex nervous systems such as those of the vertebrates also contain a supporting tissue called **neuroglia**, chiefly from the same source.

The bodies of nerve cells in most animals are concentrated in the central nervous system and in ganglia, while **nerves** are composed wholly of fibers and accessory sheaths. Even in the central system the tissue is **not** uniform, for some parts are fiber tracts and others are **nuclei** containing cell bodies or more extensive cellular layers like the cerebral cortex. Various regions, moreover, contain cells of special forms. Cells of the cortex, for example, include several forms with rounded to pyramidal bodies and a moderate number of processes, while in the cerebellum (**brain**) one form of cell is small, giving the tissue a granular appearance, and another, known as a Purkinje cell, bears a single process at one pole and numerous branching processes of mossy appearance at the other. In ganglia, too, the cells have characteristic forms. All nervous tissue, however, is composed of no more than the two components, neurons and neuroglia. (A.W.L.)

NEST. A place chosen or a structure built by an animal for shelter or concealment, usually for the reception of the eggs or young and as a home for the young during their early development.

The nest may be no more than a favorable spot on the ground or on the bottom of a body of water. Such spots are sometimes modified slightly, as in the shifting of stones by fishes preparatory to depositing their eggs, or they may be used as found. In other cases animals seek a more elaborate home such as a cave or a hollow in a tree, and often the retreat sought has already been prepared and used by another animal. Birds illustrate all of these habits. Not only do they deposit their eggs on the bare ground or in burrows, hollows, and nests prepared by other species, but also construct some of the most elaborate nests known. Most species build a characteristic nest of sticks, fibers, or some other material. Some are expert weavers and others merely pile their materials together. The use of a salivary secretion by some of the swifts in the construction of their nests is widely known from the use of the nests in the Orient as an article of food.

Many insects also build nests for the rearing of their young, and here, too, the nest may be only a burrow or a crevice, or it may be an elaborate structure such as the paper or mud nest of a wasp and the waxen combs of the honey-bee. (A.W.L.)

NET SECTION. Tension Member.

NETTLE. Mint Family.

NETTLE CELL. Nematocyst.

NETTLE HAIR. A form of hair-like scale found on some **caterpillars** which causes an irritation of human skin resembling that of nettles. Caterpillars of the Io and the brown-tail moths have larger poisonous spines of similar properties. The irritation caused by some species is very mild but others are much more severe. (A.W.L.)

NETWORKS. Kirchhoff's Laws.

NEUMANN BANDS. Mechanically formed **twin bands** of microscopic dimensions in iron or steel, believed to be formed by spontaneous reorientation of a portion of a crystal when subjected to sudden deformation or impact. (R.H.H.)

NEURAL CREST. A strip of tissue formed from the ectoderm between the outer layer of the **neural fold** and the **neural tube.** The two strips later fuse and then separate and divide into longitudinal series of masses, metamerically arranged, from which are formed the ganglia of the dorsal roots of the spinal **nerves**, the sympathetic ganglia, nerve sheath cells, pigment cells of the skin, and a part of the mesenchyme. (A.W.L.)

NEURAL FOLD. A longitudinal fold of tissue in the vertebrate embryo which contributes to the formation of the **neural tube.** A pair of these folds arises along the dorsal surface, flanking the median line, very early in development. (A.W.L.)

NEURAL TUBE. The embryonic structure from which the entire central **nervous system** of the vertebrates and much of the peripheral system are developed.

Early in embryonic life the ectoderm thickens along the upper surface of the body in the middle line. On each side of this thickened strip the tissue rises in folds which finally approach each other and fuse, leaving the median strip in the form of a tube, now lying inside the body. The further differentiation of the tube produces the central nervous system of the adult and the growth of other structures from it is the source of the motor nerve roots. (A.W.L.)

NEURALGIA. Pain in a **nerve** or along the course of a nerve. It is difficult to differentiate sharply between neuralgia and neuritis. But, if there is a distinction, the term neuritis should be confined to acute and chronic inflammation of the peripheral nerves, while neuralgia is due to acute and chronic inflammation of the pathway stations (ganglia) lying along the course of nervous pathways. The pain in neuralgia is usually of a sharp, shooting, intermittent type and accompanied by increased sensitivity of the skin supplied by the nerve. Many varieties of neuralgia are differentiated according to the part affected. Some of the common forms are: 1. Trigeminal neuralgia, a very severe form marked by agonizing pain over branches of the trigeminal nerve in the face. Palliative relief is obtained by injections of alcohol into the ganglion of the nerve and permanent and instantaneous relief by cutting the sensory root of the nerve. 2. Intercostal neuralgia—may be mistaken for **pleurisy** as the intercostal nerves, after leaving the spinal cord, run around each side of the chest between the ribs. It is accompanied by a skin eruption in **herpes zoster** or shingles. 3. Morton's neuralgia—pain in the joint of the third and fourth toe caused by pinching of a nerve in this region. 4. Sciatic neuralgia. (See Sciatica.) (R.S.M.)

NEURASTHENIA. Psychoneurosis.

NEURITIS. An inflammation of a **nerve**, either chronic or acute. Mononeuritis or localized neuritis is the term used when one nerve is involved. When several or many nerves are involved the term multiple neuritis or polyneuritis is used. Localized neuritis develops from injury, infection, chronic intoxication, or metal poisoning. Neuritis, either localized or multiple, complicates many of the infectious diseases, such as **typhoid, diphtheria, tuberculosis, smallpox,** etc. Other causes include pressure on a nerve by a **tumor** or calcium deposits in **osteo-arthritis,** etc.

Multiple or polyneuritis is inflammation and degeneration affecting the peripheral nerves, usually in their distal portions. The commonest cause is dietary deficiency, primarily of **vitamin B₁** (thiamine) which is usually associated with chronic alcoholism; other causes are various toxins, either chemical agents such as arsenical drugs, lead, nitrobenzol, coal-tar products, or toxins produced by bacteria.

The clinical picture is similar no matter what the cause: there is a gradual onset of numbness and tingling of the hands and feet, weakness of the limbs, altered sensation to touch, and pain, loss of deep reflexes, and eventually paralysis and atrophy of the extremities. Treatment is directed first toward removal of the cause. If the disease is due to vitamin B deficiency, large doses of thiamine as well as other members of the B complex are effective. (D.M.H.)

NEUROGLIA. Supporting **cells** of the central **nervous system** derived, with the possible exception of microglia, like the nervous tissue from the ectoderm. The cells are of various forms. A common form in the vertebrate nervous system is the astrocyte, which has many fine processes radiating in all directions from the cell body. In some astrocytes the processes are fibrous. They form a network among the nerve cells and fibers. (A.W.L.)

NEURON. A nerve cell with all of its processes. The processes include the dendrites which carry impulses to the cell body and the axon which carries an impulse from the cell body to some other nerve cell or to an organ such as a muscle. At the end the processes branch. This part of the axon is called the terminal arborization. The axon may also bear branches nearer to the cell body, called collaterals.

Many nerve fibers in the central system and in peripheral nerves are surrounded by a sheath consisting of a fatty material called myelin, and in the nerves a cellular neurolemma or sheath of Schwann invests the whole. The latter sheath ends before the terminal arborization, and the myelin some distance before that. (A.W.L.)

NEUROPODIUM. The ventral part of the pseudopodium of segmented worms. (A.W.L.)

NEUROPTERA. Insects of varied form and habits, including the dobson fly, the golden eyes or lacewings, the alder flies, and the ant lions. The order is characterized by the four membranous wings, usually with a large number of branching veins which are united in some species to form a network. The mouth is formed for biting but in some larvae the mandibles and maxillae fit together to form a piercing and sucking organ. The insects are predacious.

Neuroptera are of little economic importance save as food for fishes. The larvae of lacewings prey on aphids, hence they are of some assistance in holding these pests in check.

The order contains about 1200 species. (A.W.L.)

NEUROPTERIS. Paleobotany.

NEUROSIS. Psychoneurosis.

NEUROSYPHILIS. Syphilis involving principally the central nervous system. The common forms are locomotor ataxia, syphilis of the spinal cord, paresis, syphilis of the brain and meningo-vascular syphilis. Neurosyphilis is one of the late stages of untreated syphilis occurring many years after the primary infection. (R.S.M.)

NEUTRAL AXIS. The line of intersection of the neutral plane and any normal cross-section of a structural member is called the neutral axis. When a straight bar of homogeneous material (rectangular for sake of illustration) is bent the fibers which are nearer the axis of flexure will shorten and those on the opposite side will lengthen. There is one plane within the bar, parallel to its edges, in which there is no deformation of the fibers; this plane is known as the neutral surface. The flexure theory shows that the neutral axis passes through the center of gravity of the cross-sectional area of any member subject to transverse bending only. (C.W.C.)

NEUTRAL SURFACE. Neutral Axis.

NEUTRAL WIRE. Three-Wire System.

NEUTRALIZATION. In certain vacuum-tube amplifier circuits energy may be fed back from the plate to the grid circuit through the interelectrode capacity and thus cause undesirable oscillations. The introduction of the screen grid in the tube is one means of preventing this feedback. Another widely used remedy is to balance out the feedback with externally produced opposite feedback. This is called neutralization. When the energy for neutralization is fed from the plate to the grid circuit it is known as plate neutralization, while if the feedback energy is obtained from the grid and fed to the plate circuit it is plate neutralization. Both are widely used for radio-frequency amplifiers. Neutralization is not ordinarily needed in audio circuits because of the lesser effect of feedback at the lower frequencies. (See also **Acid**; **Bases**; and **Salts**.) (L.R.Q.)

NEUTRINO. Certain radioactive changes result in a distribution of energy among the particles involved, apparently in conflict with the laws of conservation of energy as well as of momentum; unless it be assumed that some particle, as yet not directly observed, partakes in the process in such a way as to carry off the missing energy and fill out the apparently unbalanced momentum. Since the hypothetical particle is not retained within the walls of any ordinary container, it must have great penetrating power and is therefore presumably neutral. Once called the "neutret," it is now commonly designated as the neutrino. Pending further experimental evidence, all assumptions regarding the properties of the neutrino, if it exists, must be regarded as tentative. (See also **Nuclear Particles**.) (L.D.W.)

NEUTRON. In the course of experiments on the "atom-smashing" effects of alpha rays from thorium upon substances containing the lighter atoms, such as hydrogen, beryllium, and carbon, early experimenters detected a secondary emission of extraordinary penetrating power which, however, produced no visible tracks in a cloud chamber, though it could be detected by a Geiger counting tube. Naturally, this was interpreted as evidence of a very hard gamma ray arising from artificial disintegration of the light atom. But when these secondary rays were sent through a cloud chamber, there would sometimes appear a single cloud track originating mysteriously in mid-air and proceeding with every evidence of great speed and energy. Now, gamma rays and other photons not only do not produce cloud tracks themselves, but, since their mass is vanishingly small, their collisions with atomic particles cannot give the latter any considerable "recoil" momentum. (See **Compton Effect**.) Thus physicists were confronted with the dilemma of conceding an exception to the conservation laws of energy and momentum in the case of photon collision, or of recognizing the existence of a new nuclear particle having curious properties.

It was Chadwick who, in 1932, first showed that the second of these alternatives would meet all the known facts, provided it be assumed that the new particle is in the nature of a proton bereft of its positive charge and therefore electrically neutral. This particle, whose reality is now universally admitted, is the neutron. Having protonic mass, it readily imparts momentum to other particles upon collision; while the fact that it is neutral enables it to pass through the open structure of atoms without interference except on the rare occasion of an actual mechanical collision with an electron or a nucleus. The one property accounts for the effect observed in the cloud chamber, the other for the ability of neutron rays to penetrate even several cm. of lead without marked absorption. Subsequent research has shown that the neutron is not a neutral proton, but has a mass about 1% greater, and a magnetic moment differing both in magnitude and in sign, from that of the proton. It is now believed that the nucleus of deuterium or "heavy hydrogen," called the deuteron, is composed of one proton and one neutron. (L.D.W.)

NÉVÉ. Glacier.

NEWARK SERIES. This is a geological term, discussed under **Triassic.**

NEWT. Amphibia, Urodela. Air-breathing **salamanders** which are at least partially aquatic in habits. The many species are found chiefly in the northern hemisphere. (A.W.L.)

NEWTON. The m.k.s. absolute unit of force; viz., that force which, if applied to a free body having a mass of 1 kilogram, would give it an acceleration of 1 meter per sec. per sec. It is equal to 100,000 **dynes,** and approximately equal to 0.102 of the weight of 1 kilogram mass under standard gravity or to about 3.6 ounces of force. The joule, m.k.s. unit of work or energy, is equal to 1 meter-newton. The newton is named for Sir Isaac Newton, whose second dynamic law is the basis of the absolute measure of force. (See **M.K.S. System; Newton's Laws of Dynamics.**) (L.D.W.)

NEWTONIAN POTENTIAL FUNCTION. Potential.

NEWTON'S LAW OF COOLING. From experiments upon the cooling of bodies, Newton concluded that, over moderate temperature ranges, the rate of cooling is proportional to the difference between the temperature of the cooling body and that of the surrounding medium. This may be expressed as a differential equation, the solution of which for the temperature T as a function of the time t is

$$T = T_m + (T_o - T_m)e^{-At}$$

in which T_m is the air temperature and T_o is the value of T at the beginning of the time interval t. A is a constant depending upon the size, shape, material, and surface of the body. While the law is only roughly approximate, it is nevertheless very useful for calorimeter corrections, etc. (See **Thermal Radiation** and **Stefan-Boltzmann Law.**) (L.D.W.)

NEWTON'S LAWS OF DYNAMICS. The classical or Newtonian **dynamics** rests upon certain propositions first enunciated in systematic form by Sir Isaac Newton, which he set forth as three "Laws" of force and motion.

The first Law states, in effect, that bodies of matter do not alter their motions in any way except as the result of **forces** applied to them. A body at rest remains at rest, or if in motion it continues to move in the same direction with the same speed, unless a force is impressed upon it. It is quite conceivable that Newton's interpretation of "force" was the primitive concept which we all have, based on muscular effort, and that he regarded this statement as the expression of a natural law connecting force with motion. On the other hand he may have recognized in this first Law, as we now do, an objective definition of force, namely, that which is capable of altering bodily motions in the face of an opposition called **inertia** whose nature is even now not fully understood.

The second Law is made up of two distinct parts: 1. When different forces are allowed to act upon free bodies, the rates at which the **momentum** changes are proportional to the forces applied. 2. The direction of the change in momentum caused by a force is that of the line of action of the force. Part 1 may now be regarded, from the more rigorous viewpoint of **dimensional analysis,** as a definition of the standard measure of force. Two forces are judged equal if they produce change of momentum at equal rates; one force is twice as great as another if it changes the momentum at twice the rate; etc. The absolute or Gaussian force units, such as the **dyne,** so much used in dynamic theory, are based upon this system of measure. Part 2 emphasizes the **vector** character of force, and points out that no single force, acting alone, can cause a change of

motion in any direction save that of its own line of action. If the effect is apparently in some other direction, as when a string operates over a pulley, the force is always combined with one or more auxiliary forces, the **resultant** of all of them being in the direction of the observed change of motion.

The third Law asserts the equality of "action and reaction." In the case of forces acting on bodies at rest, the principle is easily illustrated. When a steel truss rests on a pier and presses downward upon it with a force of 100,000 lbs., the pier exerts an upward thrust or "reaction" against the truss, also of 100,000 lbs., and this thrust, tending to bend the truss upward, is a most important factor in computing the stresses in the truss members. The Law also applies to forces acting upon bodies free to yield and to receive acceleration; a fact not explicitly stated by Newton, and discussed more fully elsewhere as **d'Alembert's principle.**

These propositions constituted the unquestioned foundation of dynamics until about the beginning of the 20th century. So far as practical operations with bodies of ordinary size are concerned, they still answer every purpose. It is only when we consider motions with velocities comparable to that of light, or attempt to analyze the mechanics of bodies of the atomic and electronic order of magnitude, that the Newtonian dynamics breaks down and must be replaced by a system founded upon the postulates of **relativity** and the concepts of the **quantum theory.** (L.D.W.)

NEWTON'S METHOD FOR SOLUTION OF EQUATIONS. Let $x = a$ be a first approximation to a real **root** of an **equation** $f(x) = 0$, found by trial or by graph or otherwise. Then a second approximation to the root is given by the formula

$$a' = a - \frac{f(a)}{f'(a)}.$$

A new approximation may be obtained by substituting a' for a in this formula again, and so on until the desired degree of accuracy is attained. The formula is based upon finding the point where the tangent at $x = a$ to the curve cuts the X-axis. (L.L.S.)

NEWTON'S RINGS. An interference phenomenon, easily observed by laying a slightly convex lens upon a flat glass plate. When the lens and plate are arranged so that monochromatic light is reflected at a suitable angle to the observer's eye, the point of contact is seen to be surrounded by a series of concentric, alternately bright and dark rings, which become closer together with increasing radius. The rings are due to the interference of light at the film of air between the glass surfaces, which film increases in thickness with increasing distance from the contact point. If the radius of curvature of the convex surface is R, and if, counting the central contact-spot as the zero ring, we number the rings in order, both bright and dark, from the center out, the radius of the Nth ring in monochromatic light of wavelength λ is approximately

$$a = \sqrt{\frac{NR\lambda}{2}}.$$

With white light, the bright rings become colored spectra, the overlapping of which at larger values of N causes the system to become indistinct and disappear. (L.D.W.)

NEW ZEALAND FLAX. Flax.

NEYMAN-PEARSON THEORY OF TESTING HYPOTHESES. In testing a statistical hypothesis we may make two types of **errors,** Type I and Type II. If we reject a true hypothesis we commit an error of Type

I. If we accept a false hypothesis, we commit an error of Type *II*. Naturally no error is committed in calling a true hypothesis true and a false hypothesis false. Now in the Neyman-Pearson theory, the probability of making a Type *I* error is controlled in advance as .05 or .01 and then that test of the statistical hypothesis is preferable which minimizes the **probability** of Type *II* errors while Type *I* errors are controlled. If a test exists which controls Type *I* errors and minimizes the probability of a Type *II* error for any hypothesis which is false, then we say such a test is uniformly most powerful. When uniformly most powerful tests exist, they are preferred above all others. When a uniformly most powerful test does not exist, other tests have been developed. In the Neyman-Pearson theory it is important to consider what the alternative hypotheses are in relation to the hypothesis being tested. The theory is a recent development originated by J. Neyman and E. Pearson and has been greatly extended by them and by A. Wald. (L.A.A.)

NICCOLITE. A nickel arsenide mineral, NiAs, crystallizes in the **hexagonal** system but is usually found massive. Color, light copper-red; hardness, 5.0–5.5; specific gravity, 7.33–7.67; luster, metallic; opaque. Found in several European localities and in the Province of Ontario, Canada; in the United States at Franklin, New Jersey, and Silver Cliff, Colorado. It is used as an ore of nickel. (E.S.C.S.)

NICHOL'S RADIOMETER. Radiometer.

NICKEL. Symbol: Ni. Atomic number: 28. Atomic weight: 58.69. Density: 8.9. Melting point: 1452° C. Boiling point: 2900° C. (**Isotopes:** page 290.)

Nickel is a silver-white metal, harder than **iron,** capable of taking a brilliant polish, malleable and ductile, magnetic below 345° C. Compact nickel is not oxidized on exposure to air at ordinary temperatures; soluble in **nitric acid;** does not react with **alkalis;** becomes passive in concentrated nitric acid. Finely divided nickel dissolves 17 times its own volume of **hydrogen,** and is extensively used as a **catalyzer** in the **hydrogenation** of oils. Discovered by Cronstedt in 1751.

Nickel is used (1) in **electroplating,** as a protective and ornamental coating for less resistant metals, especially iron and steel, (2) in coins of small denominations (25% Ni, 75% **copper**), (3) in nickel steel "invar" (36% Ni, 64% iron) of low coefficient of thermal expansion for standards of length, "permalloy" (80% Ni, 20% iron) for sheathing electric **cables,** 45 to 80% nickel and 55 to 20% iron for electromagnetic applications requiring high permeability and low hysteresis loss, (4) in monel metal (about 62% Ni and 31% copper), nichrome or chromel wire (50%–80% Ni, 11%–25% **chromium,** remainder iron) of high electrical **resistance** for heating units; german or nickel silver (10%–30% Ni, 50%–65% copper, remainder **zinc**), (5) in the alkaline (Edison) storage battery (see **Accumulator**), (6) extensively in corrosion resistant apparatus, (7) as an alloying element in steel, generally 1 to 5% nickel but larger percentages in some stainless steels.

Nickel occurs as sulfide (**pentlandite**) (NiS·2FeS) in Sudbury, Ontario, the ore averaging 3% nickel and 1.5% copper along with iron and some precious metals, and as silicate carrying 4%–8% Ni and no copper in New Caledonia. The Sudbury ore is roasted, and then treated in a **blast furnace** to obtain nickel and copper sulfides (25% metals), then in a **converter,** the product being composed of about 56% nickel, 24% copper and 20% **sulfur.** This is roasted to form nickel and copper oxides, and the copper oxide removed by dilute **sulfuric acid.** The residue is reduced to metallic nickel by heating with water gas (**hydrogen** and **carbon monoxide**), and the metal treated at 50° C. with **carbon monoxide**

(Mond process), to form volatile nickel carbonyl, a very poisonous gas, which is then decomposed at 200° C. on nickel shot, nickel depositing on the shot and carbon monoxide being regenerated and used again.

Acetate: Nickel acetate, nickelous acetate (Ni(C$_2$H$_5$O$_2$)$_2$), green solid, soluble.

Carbonyl: Nickel carbonyl (Ni(CO)$_4$), liquid, melting point −25° C., boiling point about 43° C., by the reaction of carbon monoxide and nickel metal at about 50° C., and decomposed at 200° C. to nickel and carbon monoxide.

Chloride: Nickel chloride, nickelous chloride (NiCl$_2$·6H$_2$O), green crystals, soluble.

Cyanide: Nickel cyanide (Ni(CN)$_2$), greenish-yellow precipitate by nickel salt solution and **potassim** cyanide solution; nickel potassium cyanide (K$_2$Ni(CN)$_4$), soluble, by excess of potassium cyanide with nickel salt solutions. Unchanged in boiling solution, and recovered by crystallization as reddish-yellow solid.

Hydroxides: Nickel hydroxide, nickelous hydroxide (Ni(OH)$_2$), light green precipitate by reaction of nickel salt solution and **sodium** hydroxide solution; nickelic hydroxide (Ni(OH)$_3$).

Nitrate: Nickel nitrate, nickelous nitrate (Ni(NO$_3$)$_2$·6H$_2$O), green crystals, soluble.

Oxides: Nickel monoxide, nickelous oxide (NiO), green to grayish green (yellow when hot) solid, by heating nickelous hydroxide or carbonate; nickel sesquioxide (Ni$_2$O$_3$), black solid, by heating nickelous nitrate, but the identity of this oxide is questioned; trinickel tetroxide (Ni$_3$O$_4$), gray solid, by heating the sesquioxide in **hydrogen** at 190° C.; nickel dioxide or its hydrate (NiO$_2$·nH$_2$O), black solid, by reaction of nickel salt solution and **sodium** hypochlorite solution, but the identity of this hydrate is questioned.

Sulfate: Nickel sulfate, nickelous sulfate (NiSO$_4$·6H$_2$O), green crystals, soluble.

Sulfide: Nickel sulfide, nickelous sulfide (NiS), black precipitate by reaction of nickel salt solution with **ammonium** sulfide solution, relatively insoluble (after precipitation) in **hydrochloric acid.**

Dimethylglyoxime or diacetylglyoxime yields a characteristic red precipitate with nickel salts, of delicacy sufficient to detect one part of nickel in 400,000 parts of solution.

Nickel salts are green in solid or solution. (R.K.S.)

NICKEL ALLOYS. Although not produced in large tonnages, there are many nickel alloys that are well-known because of their special properties. The normal compositions of several of these alloys are listed. In most cases the balance consists of relatively small amounts of silicon, manganese, carbon, and sulfur. In nearly all cases there are many modifications of the trade-named alloys.

Commercially pure wrought nickel in the form of sheets, wire, tubing, etc., has many applications because of its corrosion-resistance. These include utensils and food-processing equipment, marine construction, coinage, and chemical equipment. Nickel also has many radio and other electrical applications such as spark-plug wires. Electroplated nickel has been widely used as a protective coating on steel.

The 80% Ni-20% Cr alloy and other compositions with 45-65% Ni (e.g., Nichrome) are used as resistance wires for heating elements. Constantan has a high electrical resistivity and a very low temperature-coefficient of resistivity. It is widely used with copper as a thermocouple element for temperature measurement. Inconel is both a heat- and corrosion-resistant alloy generally used in sheet and other wrought forms for applications such as exhaust manifolds, dairy equipment, and chemical apparatus.

Permalloy, Hipernik, and Perminvar are representative of a large group of high nickel **magnetic** alloys.

Monel metal is a high-strength, corrosion-resisting alloy available in many wrought and cast forms for use in processing equipment, marine applications, household appliances, etc. One variety, "K" Monel, can be heat treated by **precipitation hardening** to about twice the strength of annealed Monel.

The Hastelloy-type alloys have excellent resistance to hydrochloric, sulfuric, and other acids.

K42-B is one of the newer high-temperature alloys. It has exceptionally high **creep** strength and oxidation-resistance at temperatures as high as 1500° F.

Invar has a very low temperature-coefficient of thermal expansion. It is used for measuring tapes, instruments, and bimetallic thermostats. Elinvar has a low temperature-coefficient of elasticity which makes it useful for springs in watches and precision instruments.

The 35% Ni and 25% Ni heat-resistant alloys are used principally as castings for furnace parts. Lower nickel—higher chromium compositions are generally classified as **stainless steels.**

The 30% Ni-copper alloy and the nickel-silver alloys are listed under **brass and bronze.**

REPRESENTATIVE NICKEL ALLOYS

Alloy	Ni	Cu	Fe	Cr	Mo	Co	Others
Nickel	99	..	...	...	..	..	..
Resistance alloy	80	..	...	20	..	..	..
Inconel	79	..	7	13.5	..	..	..
Permalloy	78.5	..	11.5	...	..	..	..
Monel	67	30	1.5	...	..	..	..
"K" Monel	66	29	1	...	..	..	3 Al
Nichrome	62	..	22	15	..	..	..
Hastelloy B	63	..	5	...	30	..	..
Hastelloy A	58	..	18	...	18	..	2 Mn
Hipernik	50	..	50	...	..	..	..
K42-B	46	..	7	20	..	25	2 Ti
Perminvar	45	..	30	...	..	25	..
Constantan	45	55	...	...	..	..	..
Invar	36	..	64	...	..	..	..
Heat resistant alloy	35	..	49	15	..	..	..
Elinvar	34	..	57	4	..	..	2 W
Cu-Ni	30	70	...	...	..	..	..
Heat resistant alloy	25	..	55	20	..	..	..
Nickel coin	25	75	...	...	..	..	..

(R.H.H.)

NICKEL IRON. Iron.

NICKEL SILVER. Brass and Bronze.

NICOL PRISM. One of the best known devices for producing plane-**polarized light.** It consists of two

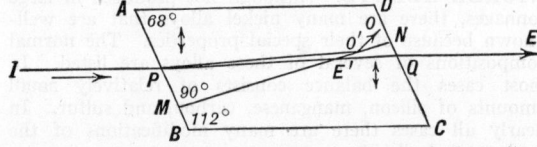

Diagram of Nicol prism. Double-headed arrows indicate direction of optic axis.

pieces of Iceland spar (pure calcium carbonate) cut as shown in the figure. The optic axis of each is approximately indicated by the double arrow, and they are cemented together with colorless Canada balsam along the plane MN. If the incident beam IP is unpolarized, it suffers **double refraction** at P, dividing into an ordinary component PO' and an extraordinary component PE'. The refractive index of Iceland spar for the ordinary ray (sodium light) is 1.658 and for the extraordinary it is 1.486, while that of Canada balsam for both is 1.53. The ordinary ray therefore encounters at O' a less refractive medium and, the incidence being at an angle larger than the critical angle, is totally reflected $(O'O)$; while the extraordinary ray incident at E' encounters a more refractive medium, therefore cannot suffer total reflection and most of it passes on along $E'Q$, emerging along QE completely plane-polarized with its vibration plane in the plane of the paper. Modifications of this prism, having different shapes and using

other cements, have been designed for special purposes. The use of such polarizers is somewhat limited by the scarcity and costliness of large Iceland spar crystals of suitable quality. (See also **Petrographic Microscope.**) (L.D.W.)

NICOTINE. Alkaloids.

NICTITATING MEMBRANE. The third **eyelid** of vertebrates. (A.W.L.)

NIDAMENTAL GLAND. Glands of the female reproductive system of **cephalopod** mollusks which secrete an elastic envelope about the egg. (A.W.L.)

NIGHTHAWK. Aves, Caprimulgiformes. **Nightjars** or **goatsuckers** of the New World. They have a very short beak and wide mouth, adapted for taking insects in flight, and are on the wing late in the day. Their

Nighthawk. *Chordeiles virginianus.* Mottled-blackish brown and reddish; lighter below, with wavy brown bars. A large white patch on each wing; the male has white on throat and tail also.

flight is easy and powerful and their long dives, terminating in a peculiar hollow boom, are a memorable exhibition.

The common nighthawk, *Chordeiles virginianus,* is widely distributed in North America and winters far into South America. A second species, the Texan nighthawk, *C. acutipennis,* enters the southwestern United States. (A.W.L.)

NIGHTINGALE. Aves, Passeriformes. A **warbler,** *Luscinia megarhyncha,* of western Europe, noted for its song. Farther east two other species, the eastern, *L. pheilomella,* and Persian, *L. hafizi,* nightingales, are found. (A.W.L.)

NIGHTJAR. Aves, Caprimulgiformes. Birds (**Aves**) with mottled plumage, a wide mouth and a short beak. They are insect-eaters, flying chiefly at twilight. Every continent has some of the numerous species except Australia.

In North America the **nighthawk** and **whip-poor-will** are the most widely known representatives of the group, with the **poor-will,** chuck-will's widow, *Antrostomus carolinensis,* and Merrill parauque, *Nyctidromus albicollus,* as less widely distributed species.

Also called goatsuckers. (A.W.L.)

NIGHTSHADE. Potato Family.

NILGAI. Mammalia, Artiodactyla. An Indian **antelope,** *Boselephas tragocamelus,* of moderate size. The male has small horns, only slightly curved. (A.W.L.)

NIMBOSTRATUS. Clouds.

NIMBUS. Clouds.

NIOBIUM. Columbium.

NITRATES. See each element for inorganic and organic nitrates; **Nitric Acid.**

NITRATION. Nitration is the process of the addition of the NO_2 group to a carbon atom, usually replacing hydrogen attached to carbon. The addition may be accomplished by means of (1) strong nitric acid, (2) mixed acid, nitric and sulfuric, (3) a nitrate plus sulfuric acid, (4) nitrogen pentoxide, (5) a nitrate plus acetic acid. More than one hydrogen atom can be replaced in a compound, but each succeeding one is replaced with more difficulty than the preceding.

Paraffin hydrocarbons may be nitrated by nitric acid in the vapor phase, giving rise to the nitroparaffins, which are starting compounds for other aliphatic compounds.

Certain aliphatic alchols such as glycerine form esters with nitric acid such as nitroglycerine. Those are misnomers, since they are not nitro compounds. Nitroglycerine, for instance, is glyceryl trinitrate.

Aromatic compounds are usually nitrated in the liquid phase. Various rules of addition govern the position of the entering nitro group, depending upon the conditions. For example, in the benzene series, the second nitro group can enter in the ortho, meta or para position, but the presence of some other group usually fixes the position of the entering nitro group. For example, it enters meta to a nitro, sulfonic, or carbonyl group, and ortho and para to a chloro, bromo, or hydroxy group. Various other rules govern other conditions in other aromatic series.

One of the great uses of nitration is to break into a pure hydrocarbon, which is usually more difficult to do by other means. The nitro group may then be changed and another group take its place. Typical examples are nitration of ethane to form nitroethane, and of benzene to form nitrobenzene, which is easily changed to aniline.

An important economic consideration in any nitration process is the recovery of the spent acid. Since the nitration reaction forms water, the reagents gradually become diluted to a point where they will not react any more. The water may be taken up during the reaction by removing it with oleum or acetic anhydride, a practice which still leaves large amounts of the reagents at the end of the process.

The nitric acid is usually concentrated by distilling it from sulfuric acid solution which retains the water.

Since the reaction used concentrated sulfuric acid, ordinary iron vessels can be used, but the neutralization process must be carried out in lead-lined tanks. Good agitation and adequate cooling facilities are necessary to avoid any local overheating and the formation of higher nitrated compounds. (D.E.M.)

NITRE. Nitre or saltpeter is a naturally occurring mineral form of **potassium** nitrate, KNO_3. It is found only in small amounts as **orthorhombic** crystals or crystalline masses or crusts in limestone caverns or in soils. It is not an important mineral. (R.M.F.)

NITRIC ACID AND NITRATES. Nitric acid (HNO_3) is a colorless solution, commercially of strength 36° Baumé (specific gravity at 60° F., water at 60° F., 1.330, 52.30% HNO_3); specific gravity 1.42 (approximately 69% HNO_3). Sometimes colored yellow to brown by nitrogen tetroxide. Higher strengths of nitric acid than 69% HNO_3 are used as "fuming nitric acid" (85% to 95% HNO_3). "Mixed acid" is a mixture of nitric and **sulfuric acids.** There is a maximum constant boiling point 120.5° C. (760 mm.) at 68% HNO_3 (distillate) for mixtures of nitric acid and water. A commonly used strength for dilute nitric acid is 31.5 grams HNO_3 per 100 milliliters of solution (5 normal).

Dilute nitric acid reacts (1) with many hydroxides, e.g., **sodium** hydroxide, to yield the corresponding nitrate, e.g., sodium nitrate, solution (2) with many ordinary oxides, e.g., **magnesium** oxide, to yield the corresponding nitrate, e.g., magnesium nitrate, solution (3) with many carbonates, e.g., **calcium** carbonate, to yield the corresponding nitrate, e.g., calcium nitrate, solution plus **carbon dioxide** gas (4) with some sulfides, e.g., **copper** sulfide, to yield the corresponding nitrate, e.g., copper nitrate plus **hydrogen sulfide gas,** (5) with many metals, e.g., **copper,** to yield the corresponding nitrate, e.g., copper nitrate, solution plus **nitric oxide** gas, (6) with cold freshly prepared solution of **ferrous** salt, when carefully stratified by concentrated sulfuric acid beneath, to yield a brown boundary layer of solution (nitrates react similarly).

REPRESENTATIVE ESTERS OF NITRIC ACID

Methyl nitrate.......................	(CH_3ONO_2) explodes at 65° C.
Ethyl nitrate........................	$(C_2H_5ONO_2)$, boiling point 88° C.
Glycol dinitrate.....................	$(C_2H_4(ONO_2)_2)$, explodes at 114° C.
Glycerol alpha-mononitrate...........	$(CH_2OHCHOHCH_2ONO_2)$, melting point 58° C.
Glycerol beta-mononitrate............	$(CH_2OH \cdot CHONO_2 \cdot CH_2OH)$, melting point 54° C.
Glycerol 1,2-dinitrate...............	$(CH_2OH \cdot CHONO_2 \cdot CH_2ONO_2)$, explosive.
Glycerol 1,3-dinitrate...............	$(CH_2ONO_2 \cdot CHOH \cdot CH_2ONO_2)$, boiling point 148° C. at 15 mm. pressure.
Glyceryl trinitrate (nitroglycerine).......	$(CH_2ONO_2 \cdot CHONO_2 \cdot CH_2ONO_2)$, melting point 13° C., explosive at 260° C.
Cellulose trinitrate.................	$(C_{12}H_{10}(OH)_7(ONO_2)_3)$
Cellulose tetranitrate...............	$(C_{12}H_{10}(OH)_6(ONO_2)_4)$
Cellulose pentanitrate...............	$(C_{12}H_{10}(OH)_5(ONO_2)_5)$
Cellulose hexanitrate (guncotton).......	$(C_{12}H_{10}(OH)_4(ONO_2)_6)$

After the nitric acid has been driven off, the temperature is raised and the water driven off the sulfuric acid, thereby concentrating the latter.

As an example of nitration, let us consider the preparation of nitrobenzene. Mixed acid consisting of strong sulfuric plus nitric acid is slowly added to benzene in a closed iron vessel provided with stirrer and reflux condenser. The acid must be added to the benzene. If it were done the other way the benzene which was added first would be quickly nitrated all the way to a trinitrobenzene. The temperature is maintained from 45–55° C. After the nitration is finished, the nitro compound is separated from the acid by decantation, since the nitrobenzene is lighter and does not mix with the acid. The nitrobenzene is washed with water and with dilute caustic or sodium carbonate solution and then again with water to give a neutral product. To obtain dinitrobenzene the reaction would be run with stronger acid and at a temperature of about 100° C.

Higher strengths of nitric acid react similarly in kind in the cases of (1), (2), (3), (6) above, but not, in general, in the remaining cases, (4) **sulfides** react to yield the corresponding nitrates, but accompanied by **sulfate** or **sulfur,** (5) reactions with metals depends upon the metal and the strength of nitric acid. **Copper** and concentrated nitric acid yield copper nitrate and **nitrogen tetroxide. Iron** is made passive by concentrated nitric acid, so that when dipped into copper sulfate solution copper metal is not deposited (on nonpassive iron a deposit of copper metal forms). **Tin** and **antimony** yield stannic and antimonic oxides respectively.

Concentrated nitric acid is thus (7) a powerful oxidizing agent and further examples of its action are the oxidation of **sulfur** to **sulfuric acid,** of **hydrogen sulfide** to sulfur, of **hydriodic acid** to **iodine,** and of cane sugar (sucrose) to **oxalic acid,** (8) a powerful nitrating agent, thus **phenol** is nitrated to nitrophenols (mono- ortho- or para-, di- 2,4-nitro, tri-2,4,6-nitro), (9) an esterification agent, e.g., **glycerol** esterified to glyceryl nitrates (mono, di, tri-nitrate).

In order to obtain nitric acid, (1) sodium nitrate is heated with the proper amount of sulfuric acid, whereupon nitric acid is **distilled** and condensed, with some accompanying loss due to formation of nitrogen oxides, (2) free nitrogen of the air is "fixed" by combination with hydrogen to form ammonia (see graph, page 79) which may then be mixed with air or oxygen at a suitable temperature and passed over a **catalyzer**, which is usually smooth compact **platinum** in the form of very fine wire gauze in order to expose a maximum surface for the weight used. The resulting nitrogen oxides—nitric oxide and nitrogen tetroxide—are absorbed in water, and the dilute acid obtained is subsequently concentrated by distillation. Nitrogen oxides are also obtained by passing air (oxygen and nitrogen) through an electric arc. This is the "arc" process for fixing nitrogen.

The uses of nitric acid have been suggested by the chemical reactions previously cited, and the largest quantities are used in nitration and esterification of organic substances for explosives, plastics and dyes, and as an acid to prepare nitrates.

All nitrates are soluble in water. A few are decomposed by water with the formation of insoluble basic nitrates, e.g., **bismuth** nitrate ($Bi(NO_3)_3$) to bismuth oxynitrate ($BiONO_3$), which are dissolved by excess nitric acid. Most nitrates (distinction from basic nitrates) are less soluble in nitric acid than in water, e.g., **lead** nitrate. Metallic nitrates, when heated, behave in an individually characteristic manner, e.g., **potassium** nitrate, melting point 333° C., evolves oxygen gas at high temperatures, leaves a residue of nitrite; **ammonium** nitrate, melting point 170° C., evolves nitrous oxide gas at somewhat higher temperatures; **copper** nitrate melts upon heating and at higher temperature evolves nitrogen dioxide and oxygen gases, leaving a residue of cupric oxide; **silver** nitrate, melting point 212° C., at 320° C. decomposition begins into oxygen and silver nitrite, at high temperatures the residue is silver.

A common test for nitrates is as follows: An aqueous solution of the nitrate is carefully mixed with an equal volume of concentrated **sulfuric acid**, cooled and a layer of **ferrous** sulfate is carefully poured down the side of the test tube to form two layers in the tube. A brown ring at the interface indicates presence of nitrate. (R.K.S.)

NITRIC OXIDE. Nitrogen.

NITRIDING.
Surface hardening of alloy steels by heating at 900–1200° F. in an atmosphere of partially dissociated ammonia. As in **cyaniding**, hardening results from the formation of nitrides of iron and of certain alloying elements that may be present in the steel. Much longer heating time is required than in **carburizing** practice, and while the depth of penetration is generally less, the maximum hardness at the surface is higher, 900–1100 D.P.H. (*Vickers* Brinell) compared to 800–900 D.P.H. for an average carburized case. Nitriding also differs from carburizing in that the parts are fully heat-treated to develop the required core properties before the nitriding treatment. Because of the comparatively low temperature of the process, distortion and dimensional changes are at a minimum. Nitrided steels have good corrosion-resistance when used for valves, pump parts, shafting, and bearing surfaces operating in steam, crude oil, gasolines, and gaseous products of combustion. The fatigue strength is also improved by nitriding.

Other typical applications are piston pins, crankshafts, cylinder liners, timing gears, gauges, and ball and roller bearing parts. (R.H.H.)

NITRILES. Hydrocyanic Acid and Cyanides.

NITRITES.
See each element; see also **Nitrous Acid**, for inorganic and organic nitrites.

NITRO- AND NITROSO-COMPOUNDS.
Nitro-compounds contain the nitro-group (—NO_2) attached directly to **carbon** atom; nitroso-compounds contain the nitroso-group (—NO) similarly attached. A very important member of this group is nitrobenzene, which upon reduction yields a variety of products, important in the synthesis of drugs and dyes.

ALKYLNITRO-COMPOUNDS:

Primary	Secondary	Tertiary
$CH_3CH_2 \cdot NO_2$	$(CH_3)_2CH \cdot NO_2$	$(CH_3)_3C \cdot NO_2$
Nitroethane	Nitrodimethylmethane (2-nitropropane)	Nitrotrimethylmethane

ISOMERIC NITRITES:

$CH_3CH_2 \cdot ONO$	$(CH_3)_2CH \cdot ONO$	$(CH_3)_3C \cdot ONO$
Ethyl nitrite	Isopropyl nitrite	1,1-dimethylethyl nitrite

ALKYLNITROSO-COMPOUNDS:

$(CH_3)_3C \cdot NO$
Nitrosotrimethyl-methane

NITRATES:

$CH_3CH_2 \cdot ONO_2$	$(CH_3)_2CH \cdot ONO_2$	$(CH_3)_3C \cdot ONO_2$
Ethyl nitrate	Isopropylnitrate	1,1-dimethylethylnitrate

NITROSAMINE:

$(C_2H_5)_2N:NO$
Diethylnitrosamine

Upon reduction, nitro- and nitroso-compounds form the corresponding **amine**; nitrosamines form the corresponding **hydrazine**.

Upon oxidation, nitroso-compounds form the corresponding nitro-compounds.

Upon treatment with **sodium** hydroxide solution, nitrites and nitrates form the corresponding alcohol plus **sodium** nitrite and nitrate, respectively. Primary and secondary nitro-compounds, with **sodium** methylate ($NaOCH_3$) in alcohol form salts of isonitro-compounds, $CH_3CH_2NO_2$ yielding $CH_3CH:NO(ONa)$, and $(CH_3)_2CHNO_2$ yielding $(CH_3)_2C:NO(ONa)$. These salts are derived from an acid form $\left(-CH:N\diagdown^O_{OH}\right)$ of the pseudo-acid $\left(\text{true nitro-compounds } \left(-CH_2 \cdot N\diagdown^O_O\right)\right)$.

Upon treatment with **nitrous acid**, primary nitro-compounds form nitrolic acids, e.g., nitroethane ($CH_3CH_2 \cdot NO_2$) yields ethylnitrolic acid $\left(CH_3C\diagup^{NOH}_{NO_2}\right)$, which dissolves in **sodium** hydroxide to form $\left(CH_3C\diagup^{NONa}_{NO_2}\right)$ red color; secondary nitro-compounds form pseudo-nitrols, e.g., 2-nitropropane ($(CH_3)_2CHNO_2$) yields 2,2-nitrosonitropropane $\left((CH_3)_2C\diagup^{NO}_{NO_2}\right)$, colorless, solid but on fusion or in solution changes to blue color; tertiary nitro-compounds are unaffected.

Upon treatment with **sodium** hypobromite (or hypochlorite) primary and secondary nitro-compounds form bromo- (or chloro-) nitro-compounds, thus, nitroethane $CH_3CH_2 \cdot NO_2$) yields 1-bromo-1-nitroethane $(CH_3CHBr \cdot NO_2)$, and 1,1-dibromo-2-nitroethane $(CHBr_2 \cdot CH_2NO_2)$; 2-nitropropane $((CH_3)_2CH \cdot NO_2)$ yields 2-bromo-2-nitropropane $((CH_3)_2CBr \cdot NO_2)$; tertiary nitro compounds are unaffected.

SELECTED REPRESENTATIVE NITRO-COMPOUNDS

Nitro-Compounds	Formula	Melting Point, °C.	Boiling Point, °C.
1. Nitrobenzene..........................	$C_6H_5 \cdot NO_2$	6	211
2. 1,2-Dinitrobenzene....................	$C_6H_4(NO_2)_2$ (1,2)	116	319
3. 1,3-Dinitrobenzene....................	$C_6H_4(NO_2)_2$ (1,3)	90	302
4. 1,4-Dinitrobenzene....................	$C_6H_4(NO_2)_2$ (1,4)	172	299
5. 1,2,3-Trinitrobenzene.................	$C_6H_3(NO_2)_3$ (1,2,3)	127	
6. 1,2,4-Trinitrobenzene.................	$C_6H_3(NO_2)_3$ (1,2,4)	61	
7. 1,3,5-Trinitrobenzene.................	$C_6H_3(NO_2)_3$ (1,3,5)	121	decom.
8. 2-Nitrotoluene........................	$CH_3C_6H_4(NO_2)$ (2)	−11	222
9. 3-Nitrotoluene........................	$CH_3C_6H_4(NO_2)$ (3)	15	231
10. 4-Nitrotoluene........................	$CH_3C_6H_4(NO_2)$ (4)	51	238
11. 2,4-Dinitrotoluene....................	$CH_3C_6H_3(NO_2)_2$ (2,4)	70	300
12. 2,6-Dinitrotoluene....................	$CH_3C_6H_3(NO_2)_2$ (2,6)	61	
(Five others)			
13. Trinitrotoluene ("T.N.T.")..........	$CH_3C_6H_2(NO_2)_3$ (2,4,6)	81	240 expl.
(Three others)			
14. 2-Nitrophenol........................	$HOC_6H_4 \cdot NO_2$ (2)	44	214
15. 3-Nitrophenol........................	$HOC_6H_4 \cdot NO_2$ (3)	96	194 (70 mm.)
16. 4-Nitrophenol........................	$HOC_6H_4 \cdot NO_2$ (4)	113	
17. 2,4-Dinitrophenol....................	$HOC_6H_3(NO_2)_2$ (2,4)	114	
(Five others)			
18. 2,4,6-Trinitrophenol (picric acid)...	$HOC_6H_2(NO_2)_3$ (2,4,6)	122 expl.	>300
(Three others)			
19. 2-Nitrobenzaldehyde.................	$C_6H_4(CHO)(NO_2)$ (1,2)	41	153 (23 mm.)
20. 3-Nitrobenzaldehyde.................	$C_6H_4(CHO)(NO_2)$ (1,3)	38	
21. 4-Nitrobenzaldehyde.................	$C_6H_4(CHO)(NO_2)$ (1,4)	58	164 (23 mm.)
22. 2-Nitrobenzoic acid..................	$C_6H_4(COOH)(NO_2)$ (1,2)	147	
23. 3-Nitrobenzoic acid..................	$C_6H_4(COOH)(NO_2)$ (1,3)	140	
24. 4-Nitrobenzoic acid..................	$C_6H_4(COOH)(NO_2)$ (1,4)	240	subl.
25. 2-Nitrobenzyl alcohol...............	$C_6H_4(CH_2OH)(NO_2)$ (1,2)	74	270
26. 3-Nitrobenzyl alcohol...............	$C_6H_4(CH_2OH)(NO_2)$ (1,3)	27	175 (3 mm.)
27. 4-Nitrobenzyl alcohol...............	$C_6H_4(CH_2OH)(NO_2)$ (1,4)	93	185 (12 mm.)
28. 1-Nitronaphthalene..................	$C_{10}H_7(NO_2)$ (1)	59	304
29. 2-Nitronaphthalene..................	$C_{10}H_7(NO_2)$ (2)	79	165 (15 mm.)
30. 2-Nitro-1-naphthol..................	$C_{10}H_6(OH)(NO_2)$ (1,2)	128	
31. 4-Nitro-1-naphthol..................	$C_{10}H_6(OH)(NO_2)$ (1,4)	164	
32. 1-Nitro-2-naphthol..................	$C_{10}H_6(NO_2)(OH)$ (1,2)	103	
33. 9-Nitroanthracene (nitrosoanthron).	$C_{14}H_9 \cdot NO_2$ (9)	146	>360
34. 1-Nitroanthraquinone...............	$C_6H_4(CO)_2C_6H_3(NO_2)$ (1)	230	subl.
35. 2-Nitroanthraquinone...............	$C_6H_4(CO)_2C_6H_3(NO_2)$ (2)	185	270 (7 mm.)
36. Nitromethane........................	CH_3NO_2	−28	101
37. Dinitromethane......................	$CH_2(NO_2)_2$		
38. Trinitromethane (nitroform)........	$CH(NO_2)_3$	15	126
39. Trichloronitromethane (chloropicrin)........	CCl_3NO_2		112
40. Tetranitromethane...................	$C(NO_2)_4$	13	
41. Nitroethane.........................	$CH_3CH_2NO_2$	<−50	115
42. Dinitroethane.......................	$CH_3CH(NO_2)_2$		
43. 1-Nitropropane......................	$C_3H_7NO_2$		131
44. 2-Nitropropane......................	$(CH_3)_2CHNO_2$	−93	120
45. 1-Nitrobutane.......................	$C_4H_9NO_2$		151
46. 1-Nitropentane......................	$C_5H_{11}NO_2$		172
47. 1-Nitrohexane.......................	$C_6H_{13}NO_2$		193
48. Nitroethyl alcohol...................	$CH_2OHCH_2NO_2$	<−80	194
49. Nitrobromoform (bromopicrin)......	NO_2CBr_3	10	Expl.
50. Nitrochloroform (chloropicrin)......	NO_2CCl_3	−64	112
51. Nitrofurane.........................	$C_4H_3O \cdot NO_2$	28	
52. Nitrourea...........................	$OC{<}^{NH_2}_{NHNO_2}$	155 dec.	
53. Nitroguanidine......................	$HNC{<}^{NH_2}_{NHNO_2}$	246	
54. 1,2-Nitroaniline.....................	$C_6H_4(NO_2)(NH_2)$ (1,2)	72	
55. 1,3-Nitroaniline.....................	$C_6H_4(NO_2)(NH_2)$ (1,3)	114	>285
56. 1,4-Nitroaniline.....................	$C_6H_4(NO_2)(NH_2)$ (1,4)	146	

SELECTED REPRESENTATIVE NITROSO-COMPOUNDS

Nitroso Compounds	Formula	Melting Point, °C.	Boiling Point, °C.
1. Nitrosobenzene...........................	C_6H_4NO	68	58 (18 mm.)
2. 2-Nitrosotoluene.........................	$C_6H_4(CH_3)(NO)$ (1,2)	72	
3. 3-Nitrosotoluene.........................	$C_6H_4(CH_3)(NO)$ (1,3)	53	
4. 4-Nitrosotoluene.........................	$C_6H_4(CH_3)(NO)$ (1,4)	48	
5. 4-Nitrosophenol (4-quinoneoxime)...........	$C_6H_4(OH)(NO)$ (1,4)	125	144 dec.
6. 4-Nitrosonaphthol-1 (4-naphthaquinoneoxime).	$C_{10}H_6(OH)(NO)$ (1,4) or $C_{10}H_6(O)(NOH)$ (1,4)	193	
7. 2-Nitrosonaphthol-1......................	$C_{10}H_6(OH)(NO)$ (1,2)	163 dec.	
8. 1-Nitrosonaphthol-2......................	$C_{10}H_6(OH)(NO)$ (2,1)	109	
9. 4-Nitrosoaniline.........................	$C_6H_4(NH_2)(NO)$ (1,4)	173	
(quinoneimideoxime).....................	$(C_6H_4(NH)(NOH)$ (1,4))		
10. N-Nitrosomethylaniline..................	$C_6H_5N{<}^{CH_3}_{NO}$	13 an~~	128 (20 mm.)
11. 4-Nitrosophenylaniline..................	$C_6H_5NH \cdot C_6H_4NO$	145	
12. 1-Nitrosonaphthylamine-2................	$C_{10}H_6(NH_2)(NO)$ (2,1)	151	
13. Diphenylnitrosamine....................	$(C_6H_5)_2N \cdot NO$	66	
14. Dimethylnitrosamine....................	$(CH_3)_2N \cdot NO$		153 (774 mm.)
15. Diethylnitrosoamine....................	$(C_2H_5)_2N \cdot NO$		177

(R.K.S.)

Alkylnitro-compounds are made (1) by reaction of the alkyl iodide (see **Iodine**) and **silver nitrite**. Higher alkyl members yield increasing proportions of nitrite along with the nitro-compound, but these frequently may be separated by fractional **distillation**. Tertiary alkyl iodides do not behave in this manner; (2) by reaction of alpha-substituted halogen acids and **sodium** nitrite, followed by loss of **carbon dioxide**, e.g., chloroacetic acid ($CH_2Cl \cdot COOH$) yields nitroacetic acid ($CH_2NO_2 \cdot COOH$) and then nitromethane plus **carbon dioxide**, (3) by reaction of the **hydrocarbons** with **nitric acid**.

BENZENOID NITRO- AND NITROSO-COMPOUNDS:

Mononitro-compound Dinitro-compound Trinitro-compound

Nitrobenzene   1,3-Dinitrobenzene   1,3,5-Trinitrobenzene

Nitroso-compounds

Nitroso benzene   Diphenyl nitrosamine

Under the proper conditions of concentration of **nitric acid** and of temperature, **benzene** forms mainly nitrobenzene, nitrobenzene forms mainly 1,3-dinitrobenzene, and 1,3-dinitrobenzene, mainly 1,3,5-trinitrobenzene.

When nitrobenzene is treated (1) with **zinc and calcium** chloride or ammonium chloride solution, beta-phenylhydroxylamine (C_6H_5NHOH) is formed, and from this by treatment with **chromic acid** or **ferric chloride** nitrosobenzene is formed, (2) with **tin** or **iron** and **hydrochloric** acid, aniline ($C_6H_5NH_2$), is formed and from this by treatment with **nitrous acid** followed by treatment with **stannous** chloride plus **hydrochloric acid** phenylhydrazine ($C_6H_5NH \cdot NH_2$) is formed. (For other reactions of nitrobenzene see **Azo-, Diazo- and Related Compounds**.)

Mono- or poly- substituted nitro-compounds are changed in whole or in part to the corresponding amino-compounds by proper choice of reducing agent and temperature, e.g., in acid medium 1,3-dinitrobenzene yields 1,3-phenylendiamine ($C_6H_4(NH_2)_2(1,3)$), and with ammonium sulfide yields 3-nitraniline (($1)H_2NC_6H_4NO_2(3)$). When diphenylnitrosamine is reduced, 1,1-diphenylhydrazine (($C_6H_5)_2N \cdot NH_2$) is formed.

NITROBENZENE. Nitro and Nitroso Compounds.

NITROCELLULOSE. Explosives.

NITROGEN. Symbol: N. Atomic number: 7. Atomic weight: 14.008. Density: 1.2505 grams per liter, 0° C., 760 mm. or 0.967 when air equals 1.000. Formula of nitrogen gas: N_2. Melting point: $-209.86°$ C. Boiling point: $-195.8°$ C. Critical temperature: $-147.1°$ C. Critical pressure: 33.5 atmospheres. Isotopes: 14 (99.62%), 15 (0.38%).

Nitrogen is a colorless, odorless, tasteless, non-toxic gas, found free in the **atmosphere** (78.03% by weight nitrogen) mixed with **oxygen, argon, carbon dioxide**, and water vapor. Nitrogen was recognized as a simple gas by Lavoisier about 1776, although previously isolated by Rutherford in 1772. The free gas **nitrogen** is among the least reactive substances chemically, but many of its compounds display marked reactivity, e.g., **nitric acid, glyceryl nitrate, cellulose** nitrates, nitrotoluene, **picric acid,** the last four being important explosives. Nitrogen occurs combined locally in Chile as **sodium** nitrate, and as a constituent of many plant and animal proteins (see **Aminoacids**). In the atmosphere nitrogen serves as a diluent for the oxygen in the processes of burning and respiration. Nitrogen is an important element in plant nutrition, e.g., in the form of nitrates, in animal nutrition, e.g., in the form of **proteins**, and many of its compounds are important **explosives, dyes**, and **drugs**. The fixation of atmospheric nitrogen is accomplished in nature by certain **bacteria** of the soil. Several processes have been exploited for the commercial fixation of nitrogen of the atmosphere. Of these, one of the first was the combination of nitrogen and oxygen into nitric oxide by means of the electric spark, developed in Norway, another was the formation of **aluminum** nitride and thence **ammonia**, but the process which has survived practically to the displacement of all others is the combination of nitrogen and **hydrogen** gases in the **catalytic** reaction to form **ammonia**. (See **Chemical Changes**, also **Nitric Acid**.)

Nitrogen, mixed with about 1% **argon**, may be

obtained from the air by passing the latter over heated **copper** or **iron** to remove the **oxygen**, or pure by fractional **distillation** of liquid air whereby the nitrogen distills off before the oxygen. Pure nitrogen may also be obtained by heating such compounds as **ammonium** nitrite, **ammonium** dichromate, and collecting the gas. Mixed with **carbon monoxide** in producer gas, nitrogen may be utilized without separation by first making methyl alcohol from carbon monoxide and hydrogen and then using hydrogen and nitrogen for ammonia. When nitrogen at low pressure is subjected to the silent electric discharge, activated nitrogen is produced. Activated nitrogen displays a golden yellow afterglow upon cessation of the current, increased by cooling and decreased by heating. This form of nitrogen is very active with **phosphorus**, with alkali metals (forming **azides**), with the vapor of **zinc, mercury, cadmium,** arsenic (forming nitrides), with many metallic chlorides (forming a green fluorescence), and with **hydrocarbons** (forming **hydrocyanic acid** and cyanides). The transformation of nitrogen to activated nitrogen is partial, and its return to ordinary nitrogen takes place rapidly, in about one minute.

Acids: Nitrogen is a constituent of several acids. In increasing order of oxidation the following are those without carbon: **hydrazoic acid** (HN_3); **hydroxylamine** (H_2NOH); **hyponitrous acid** ($H_2N_2O_2$); **nitrous acid** (HNO_2); **nitric acid** (HNO_3); those which contain **carbon:** **hydrocyanic acid** (HCN); **cyanic acid** ($HCNO$); **fulminic acid** ($HONC$); **uric acid** ($C_5H_4N_4O_3$).

Bases: Nitrogen is a constituent of several bases. In increasing order of oxidation the following are those which do not contain carbon: **ammonia** (NH_3); **hydrazine** (N_2H_4); **hydroxylamine** (H_2NOH); those which contain carbon: **amines,** such as methyl amines, phenyl amines; **hydrazines,** such as phenyl hydrazine; **pyridine; ureas; semicarbazides; guanidines.**

Chlorides: Nitrogen chloride (NCl_3), yellow volatile oil, odor of chlorine, very explosive in light, on heating, or with various substances, such as turpentine. With water yields **ammonia** and **hypochlorous acid;** formed by reaction of **ammonium** chloride concentrated solution with excess **chlorine;** nitrosyl chloride ($NOCl$), orange yellow gas, boiling point $-5°$ C., formed by distilling **sodium** nitrite plus **phosphorus** pentachloride.

Hydrides: **Ammonia** (NH_3); **hydrazine** (N_2H_4); **hydrazoic acid** (HN_3).

Iodide: Nitrogen iodide (NI_3), reddish solid, very explosive by slight mechanical shock, in light, or on warming.

Nitrides: **Magnesium** nitride (Mg_3N_2), yellow solid; boron nitride (BN), white solid; trisilicon tetranitride (Si_3N_4). Formed by reaction of certain elements (or

SCHEME SHOWING THE INTERRELATIONSHIPS OF NITROGEN-CONTAINING SUBSTANCES

Ammonia	NITROGEN In the atmosphere		Nitric oxide	Nitrogen trioxide	Nitrogen dioxide	Nitrogen pentoxide
Ammonium compounds		Nitrous oxide			Nitrogen tetroxide	
Metallic ammines		Hyponitrous acid	Hydronitrous acid	Nitrous acid		Nitric acid
		Metallic hyponitrites		Metallic nitrites		Metallic nitrates
						Saltpeter in nature
Organic amines				Organic nitrites		Organic nitrates
Hydrazine	Hydroxylamine					
	Metallic hydroxylamates					
Organic hydrazines	Organic hydroxylamines					
Hydrazoic acid Metallic azides Organic azides						
Cyanogen Hydrocyanic acid Metallic cyanides		Cyanic acid				
		Metallic cyanates				
Organic cyanides		Organic cyanates Urea				
	Proteins In Plant and Animal Substances					
Pyrrole	Tetrapyrroles Chlorophyll In green leaf chloroplasts. Haemoglobin In red blood corpuscles.					
Pyridine						

their oxide plus carbon) and nitrogen (or ammonia) upon heating.

Oxides: Nitrous oxide (N_2O), colorless gas, boiling point $-90°$ C., of pleasant odor, of sweetish taste. Used as an anaesthetic in minor operations—when used as such extreme caution must be exercised to avoid the presence of poisonous oxides of nitrogen. Made by heating ammonium nitrate, and collecting and purifying the gas; nitric oxide (NO), colorless gas, boiling point $-152°$ C., reactive with air or oxygen forming brown nitrogen tetroxide, and in early times used thus to test "the goodness of air." Made (1) by the reaction of dilute **nitric acid** and **copper,** (2) by the catalytic oxidation of **ammonia** and air, (3) in the electric arc process of nitrogen-oxygen (air) fixation; nitrogen trioxide (N_2O_3), brown gas, behaving at ordinary temperatures as if a mixture of nitric oxide and nitrogen tetroxide, boiling point $-27°$ C., the anhydride of nitrous acid, and with water forms nitrous acid; nitrogen tetroxide, nitrogen dioxide, nitrogen peroxide (N_2O_4), brown gas, color deepens with increase of temperature to $140°$ C. when there is present nitrogen dioxide (NO_2) only. Made by reaction of concentrated nitric acid and **copper**; nitrogen pentoxide (N_2O_5), white to yellowish solid, melting point $29.5°$ C., the anhydride of nitric acid, and with water forms nitric acid. Formed by dehydrating nitric acid with **phosphorus** pentoxide, and recovered from the mixture by volatilizing in a current of dry air.

Other compounds of nitrogen are discussed as follows:

Acridines. (See **Pyridine and Related Compounds.**)
Alkaloids.
Amines.
Aminoacids.
Aminoazo-compounds. (See **Azo- and Related Compounds.**)
Aminoguanidines. (See **Amines and Amides.**)
Ammines.
Ammonia.

SCHEME SHOWING INTERRELATIONSHIPS OF NITROGEN-FUNCTION ORGANIC COMPOUNDS

Related to	$H—N\begin{smallmatrix}H\\\\H\end{smallmatrix}$ Ammonia	$H\begin{smallmatrix}\\\\H\end{smallmatrix}N—N\begin{smallmatrix}H\\\\H\end{smallmatrix}$ Hydrazine	$H\begin{smallmatrix}\\\\H\end{smallmatrix}N—OH$ Hydroxylamine	$HO—N=O$ Nitrous acid	$HO—N\begin{smallmatrix}O\\\\O\end{smallmatrix}$ Nitric acid
$>$CH of benzene	Pyrrole and derivative Pyridine and derivative				
—CH_2OH $>$CHOH $>$COH of alcohols, and phenols	Amines • Primary Secondary Tertiary Quaternary ammonium compounds	Hydrazines Mono Unsym. di Sym. di (Hydrazo comp.[1])	Hydroxylamines Mono N Di N	Nitrites Nitroso compounds (Nitrosamines) (Diazo and Azo compounds [1]) Nitroparaffin compounds	Nitrates Nitroaromatic compounds
—CHO $>$CO of aldehydes, ketones, and quinones		Hydrazones (Osazones)	Oximes		
H—COOH Carboxylic acids	Cyanides Isocyanides Amides (Anilides) Aminoacids (Polypeptides) (Proteins)				
HO—COOH Carbonic acid	Cyanates Isocyanates Fulminates Cyanamides Carbamates [2] Ureas [2] (Ureides) [3] Semicarbazides [2] Semicarbazones [4] Guanidines [2] Aminoguanidines [2]				

[1] Hydrazo, Azo and Azoxy comp. form a series.
[2] See **Amines and Amides.**
[3] Includes Purine comp.
[4] From aldehydes and ketones, similar to hydrazones and oximes.

$H—N\!\!=\!\!N\!\!\equiv\!\!N$

Hydrazoic acid
Metallic azides
Organic azides

(R.K.S.)

Ammonium compounds.
Anilides. (See **Amines and Amides.**)
Azides. (See **Hydrazoic Acid and Azides.**)
Azines. (See **Pyridine and Related Compounds.**)
Azo-compounds.
Azoles. (See **Pyrrole and Related Compounds.**)
Azoxy-compounds. (See **Azo- and Related Compounds.**)
Biuret. (See **Amines and Amides.**)
Carbamic acid. (See **Amines and Amides.**)
Carbazoles. (See **Pyrrole and Related Compounds.**)
Chlorophyll. (See **Pyrrole and Related Compounds.**)
Cyanamides.
Cyanates. (See **Cyanic Acid and Cyanates.**)
Cyanic acid.
Cyanides. (See **Hydrocyanic Acid and Cyanides.**)
Cyanogen.
Cyanuric acid. (See **Cyanic Acid and Cyanates.**)
Diazines. (See **Pyridine and Related Compounds.**)
Diazoles. (See **Pyrrole and Related Compounds.**)
Diazoamino compounds. (See **Azo-, Diazo- and Related Compounds.**)
Diazo-compounds. (See **Azo-, Diazo-, and Related Compounds.**)
Esters. (See corresponding acids.)
Fulminates. (See **Cyanic Acid and Cyanates.**)
Fulminic acid. (See **Cyanic Acid and Cyanates.**)
Guanidines. (See **Amines and Amides.**)
Hydrazines.
Hydrazo-compounds. (See **Azo- and Related Compounds, and Hydrazines.**)
Hydrazoic acid.
Hydrazones. (See **Hydrazines, Hydrazones and Osazones.**)
Hydrazoates. (See **Hydrazoic Acid, Hydrazoates, and Azides.**)
Hydrocyanic acid.
Hydroferricyanic acid.
Hydroferricyanides. (See **Hydroferricyanic acid and Hydroferricyanides.**)
Hydroferrocyanic acid.
Hydroferrocyanides. (See **Hydroferrocyanic acid and Hydroferrocyanides.**)
Hyponitrites. (See **Hyponitrous acid and Hyponitrites.**)
Hyponitrous acid.
Hyroxyazo-compounds. (See **Azo-, Diazo-, and Related Compounds.**)
Hydroxylamines.
Imides. (See **Amines and Amides.**)
Imines. (See **Amines and Amides.**)
Indazoles. (See **Pyrrole and Related Compounds.**)
Indigo. (See **Pyrrole and Related Compounds; Dyes.**)
Indoles. (See **Pyrrole and Related Compounds.**)
Isocyanates. (See **Cyanic acid and Cyanates.**)
Isocyanic acid. (See **Cyanic acid and Cyanates.**)
Isocyanides. (See **Hydrocyanic acid and Cyanides.**)
Isonitriles. (See **Hydrocyanic acid and Cyanides.**)
Isoquinolines. (See **Pyridine and Related Compounds.**)
Lactams. (See **Aminoacids, Polypeptides, and Proteins.**)
Lactims. (See **Aminoacids, Polypeptides, and Proteins.**)
Nitrates. (See **Nitric acid and Nitrates.**)
Nitric acid.
Nitriles. (See **Hydrocyanic acid and Cyanides.**)
Nitrites. (See **Nitric acid and Nitrates.**)
Nitro-compounds.
Nitrolic acids. (See **Nitro- and Nitroso-Compounds.**)
Nitroso-compounds. (See **Nitro- and Nitroso-Compounds.**)
Nitrous acid.
Nucleic acids. (See **Proteins.**)

Osazones. (See **Hydrazine, Hydrazones, and Osazones; Carbohydrates.**)
Oxazoles. (See **Pyrrole and Related Compounds.**)
Oximes. (See **Hydroxylamines.**)
Phenylhydrazones. (See **Hydrazines and Hydrazones.**)
Piperidines. (See **Pyridine and Related Compounds.**)
Polypeptides. (See **Aminoacids, Polypeptides and Proteins.**)
Porphyrans. (See **Pyrrole and Related Compounds.**)
Porphyrins. (See **Pyrrole and Related Compounds.**)
Proteins. (See **Aminoacids, Polypeptides, and Proteins.**)
Purines.
Pyrazoles. (See **Pyrrole and Related Compounds.**)
Pyridines.
Pyrroles.
Pyrrolidines. (See **Pyrrole and Related Compounds.**)
Quinolines. (See **Pyridine and Related Compounds.**)
Semicarbazides. (See **Amines and Amides.**)
Semicarbazones. (See **Amines and Amides.**)
Tetrazoles. (See **Pyrrole and Related Compounds.**)
Triazoles. (See **Pyrrole and Related Compounds.**)
Ureas. (See **Amines and Amides.**)
Ureides. (See **Purine and Uric Acid Compounds.**)
Urethanes. (See **Amines and Amides.**)
Uric Acid. (See **Purine and Uric Acid Compounds.**)
Note: Sulfur-nitrogen organic compounds are listed under **Sulfur.**

NITROGEN CYCLE. Bacteria.

NITROGLYCERIN. Explosives.

NITROLIC ACIDS. Nitro- and Nitroso-Compounds.

NITROSYL CHLORIDE. Nitrogen.

NITROUS ACID AND NITRITES. Nitrous acid (HNO_2) is a blue solution, unstable, the blue color soon disappears, especially on warming, with the formation of brown nitrogen tetroxide (see **Nitrogen**) and nitric acid in solution. Nitrous acid reacts in some cases as an **oxidizing** agent, in other cases as a **reducing** agent, and is unusually interesting on that account. As an oxidizing agent, nitric oxide is usually formed, e.g., with **ferrous** salt solution changed to ferric, with **hydrosulfuric acid** to **sulfur**, **sulfurous acid** to **sulfuric acid**, **hydriodic acid** to **iodine**; nitrogen is formed with **urea**, with **ammonia** and with **formaldehyde**; ammonia with aluminum in **sodium** hydroxide medium. As a reducing agent, nitric acid is formed, e.g., with **potassium** permanganate changed to manganous, with dichromate to **chromic**, with chlorate to **chlorine**, with bromate to **bromine**, with iodate to **iodine**, with **hypochlorite** in **sodium** hydrogen carbonate medium to **chloride.**
Prepared by reaction (1) of **nitrogen** trioxide (or nitric oxide plus **nitrogen** dioxide) and water, (2) **barium** nitrite solution and **sulfuric acid,** and filtering off barium sulfate, (3) sodium nitrite solution and an acid.
Sodium nitrite is formed (1) by heating sodium nitrate solid and **lead,** with stirring, preferably in an iron dish, and upon cooling, dissolving the nitrite and filtering off lead oxide, (2) by reaction of nitric oxide plus nitrogen dioxide with **sodium** hydroxide or carbonate solution.
Silver nitrite is insoluble, and formed as a precipitate by reaction of sodium nitrite solution and silver nitrate solution.
With numerous organic substances characteristic colors are developed. Nitrous acid is (1) used in organic chemistry as an important reagent, e.g., with **amines**, and (2) the color developed in dilute solutions is used as a

method of quantitatively estimating nitrous acid, e.g., **sulfanilic acid** plus alpha-naphthylamine develops a red color. A complex nitrite, potassium cobaltinitrite, yellow insoluble, is important in the detection of **potassium** in salt solution, soluble sodium cobaltinitrite is used as the reagent.

Esters: methyl nitrite (CH_3ONO), boiling point $-12°$ C.; ethyl nitrite (nitrous ether) (C_2H_5ONO), boiling point $17°$ C.

Indication of the presence of a nitrite is given by the appearance of brown fumes on treatment with dilute **sulfuric acid** in the cold. (R.K.S.)

NITROUS OXIDE. Anaesthesia; Nitrogen.

NOBEL PRIZE AWARDS. In chemistry, physics, and physiology or medicine. (See **History and Evolution of Chemistry.**)

NOCTILUCA. Protozoa, Mastigophora. A genus of **protozoans** with one phosphorescent species. This minute form is sometimes so abundant in the ocean that the water appears luminous at night. (A.W.L.)

NOCTULE. Mammalia, Chiroptera. A **bat** found in Europe, Asia, the Oriental region, and northern Africa. It has a wing spread of more than a foot. (A.W.L.)

NOCTURIA. Increased frequency of urination at night. It occurs with infections of the urinary tract, obstruction (e.g., benign prostatic hypertrophy), and in advanced kidney disease (e.g., nephritis) in which there is loss of the normal power of the kidney to concentrate urine. (D.M.H.)

NODDY. Aves, Charadriiformes. A group of birds (**Aves**) related to the terns. They are chiefly tropical. (A.W.L.)

NOISE AND VIBRATION. The two chief characteristics of vibration are **frequency** and **amplitude**. A mechanical vibration which is felt as a shake or tremor differs from one which manifests itself to the senses as noise or sound only in the value of its frequency. When the frequency is low, say less than 15 oscillations per sec., vibration is felt as a shake or tremor, and then is only perceptible to the senses if its amplitude amounts to .016″ or more. When the frequency lies between 15 and 30 oscillations per sec., the shake or tremor is accompanied by noises such as clicks and rumbles. For example, all major earthquakes are accompanied by a rumble, so deep that observers usually differ as to whether the sensation it produces is felt or heard. When the frequency lies about 30 oscillations per sec., vibration passes into the range of sound. The limits of frequency for musical sounds are usually given as 40 to 15,000 oscillations per sec. The upper limit of audibility extends to about 15,000 oscillations per sec., depending upon the age of the person. For a vibration of given frequency, the intensity is proportional to the square of the amplitude, or linear magnitude, of the disturbance. Thus in the case of sound, the amplitude of vibration determines the loudness while the frequency determines the pitch. For vibrations of low frequency which are felt as tremor or shake, various experiments have been made to determine the limiting amplitude and frequency at which they become apparent to the senses.

The use of **insulation** for localizing vibration in structures is only one phase of the complicated problem of vibration control. The general problem, whether considered in relation to structures or machinery, is a dynamic rather than a static one. Probably it is for this reason that greater progress toward a solution has been made in mechanical than in structural lines. The general problem of vibration control may be classified under several heads. In the design of permanent machinery installations such as lighting and power units in office buildings and apartments; batteries of looms, printing presses, or machine tools in shops and factories; marine installations; automotive and aircraft power plants, etc., the first step consists in eliminating the unbalanced effects in rotating parts by means of dynamic balancing. In general, however, this by itself is never completely effective. In reciprocating machinery, where the power fluctuates, it is of utmost consequence to design parts affected by vibration so as to avoid critical speeds. In buildings and other structures, critical speeds are those speeds of operation at which synchronous vibration is set up in one or more members of the structure. See **Resonance.** If the design of the structure cannot be altered, synchronism may be destroyed by changing the speed of the machine causing vibration. When vibration is due to outside sources such as street traffic, some other remedy, such as insulation, must be adopted.

Mechanical insulators fall into two general classes: Those that are elastic and resilient, such as rubber, felt, and cork; and those which are almost entirely lacking in these qualities, such as sand, gravel, lead, and asbestos. Properly speaking, there is no such thing as a vibration absorbent, since this implies dissipation of energy through internal work in the form of molecular friction, heat losses, etc. Resilient materials store up energy like a spring and give this up again when released without appreciable loss, and their insulating properties depend on this feature; whereas non-resilient materials like sand and gravel serve mainly to destroy the rigidity, or interrupt the continuity, of otherwise rigid structures.

The most general principle of vibration control is that vibration should be damped out as near as possible to its source. When the source of vibration is under control, as in the case of a stationary machine in a building, this principle implies that the machine shall be of correct design, properly balanced, and with adequate foundation, thereby preventing so far as possible the occurrence of vibration, or reducing it to a minimum, before any attempt is made to introduce resilient supports or vibration dampers for short-circuiting residual vibration.

When the source of vibration lies entirely outside control, as in the case of vibration in a building due to street traffic, the use of resilient dampers or isolators often becomes the chief remedy. Also when the problem is to prevent vibration from affecting some particular object, such as a piece of delicate scientific apparatus, local insulation by the use of resilient supports is in general the most effective means of securing results.

The sensation of **sound** is produced, in general, by compressional waves transmitted through the air from some vibrating source. Sound may also be communicated to the ear through the medium of a solid body which transmits vibration directly to the bones of the skull, or by means of a metal plate held firmly between the teeth. When the sound wave is periodic, a musical sound is produced, the pitch of which depends on the frequency of vibration. When the sound wave is nonperiodic, the resulting sensation is what we call noise.

The transmission of sound implies a transmision of wave energy. The numerical amount of such wave energy measures what is called the **intensity** of the sound. Sound intensity is thus a definite physical quantity, and is defined as the amount of energy transmitted per unit of time through a unit area of a plane, normal to the direction of travel of the sound wave.

When a sound is transmitted through the air, it proceeds outward from the source in spherical waves, alternating between compression and rarefaction with the same frequency as its source. When such a sound wave in air impinges on a surface, such as the wall of a room, part of its energy is reflected, part absorbed, and part transmitted through the wall, in proportionate amounts depending on the construction of the wall and the nature of its surface.

Experiments on soundproofing structures, such as those carried out at the U. S. Bureau of Standards for a period of years on standard building construction and more recently on aircraft cabins, as well as somewhat similar but independent developments in connection with the insulation of sound stages of talking picture studios, have disclosed certain basic facts with regard to practical methods for preventing sound transmission.

For ordinary building construction it has been found that in the case of solid walls such as those of brick or masonry, whatever tends to make the wall stiffer or heavier, tends to improve its sound-insulating properties.

However mechanical may be the age in which we live, the foundations of this or any nation's wealth is based on human resources. The nervous and muscular system which directs and controls human effort, whether mental or manual, is, in fact, a very highly organized machine, and from a purely mechanical standpoint its efficiency depends very largely on the prevention of fatigue and the restoration of used tissues through sleep and relaxed muscular and nervous tension. With all forms of mechanical waste carefully guarded against by scientific management, authorities on industrial economics assure us that the greatest waste in industry today is that caused by nervous fatigue induced by excessive and incessant noise.

During the past few years tremendous advances have been made in the solution of problems in vibrating systems by the use of dynamical analogies. By establishing analogies between electrical, mechanical and acoustical systems, the solution of vibrating problems in mechanical and acoustical systems is reduced to the solution of an electrical network. In this way the widespread knowledge of electrical networks is used to solve problems in complex dynamical systems—that are so difficult by classical methods—in the most simple and logical way. These methods are applied to problems in electricity, dynamics, acoustics and mechanics. They include the development and design of sound systems, electromechanical transducers, machines and mechanisms, and the control of noise, vibrations and oscillations. These methods may also be applied to transient phenomena, such as impulsive forces in engines and machines. Specifically, dynamical analogies is an indispensable scientific tool in the analysis, study and design of mufflers, dampers, vibration isolators, of microphones, loud speakers, pickups, galvanometers, and other electromechanical transducers as well as filters and corrective networks. (See *Dynamical Analogies*, by H. F. Olson, D. Van Nostrand Co., Inc.)

NOISE SUPPRESSOR (Radio). This is a vacuum-tube circuit designed to suppress undesirable noise in a receiver. Suppressors are of two types, interchannel noise suppressors and large signal suppressors. The first acts as a sort of **automatic volume control** to cut off the audio **amplifier** when no **carrier** is being received. The second is a peak limiting circuit to suppress all signals above a certain amplitude. (L.R.Q.)

NOMENCLATURE. For **chemical nomenclature**, see article under that heading. In the biological sciences, nomenclature is the names of species and the groups in which they are classified, and the procedure governing the selection of such names. A part of **taxonomy.**

All zoological nomenclature is dated from Linnaeus' *Systema Naturae*, Tenth Edition, 1758. The same scientist is credited with the establishment of the system of binomial nomenclature which applies to every species a name consisting of the name of the genus to which it belongs, followed by the name of the species. The name of the author of the species or an abbreviation is also appended, and it is permissible, if the species is later removed to another genus, to place this author's name in parentheses and follow it with that of the author of the new combination.

In modern zoology a series of International Rules of Zoological Nomenclature prepared by the International Commission on Zoological Nomenclature is regarded as the authoritative code for naming taxonomic divisions and for the regulation of certain other procedure of taxonomy. Disputed points are submitted to the Commission and opinions on such points are published from time to time.

A general provision is that all scientific names must be published in such form as to satisfy the Commission's definition of publication. In addition a priority rule establishes the first valid application of a name, all names subsequently published to apply to the same unit becoming synonyms.

The selection of names for species is practically unrestricted. They are Latinized in their application to scientific uses but they may be derived from any language and may be descriptive or not. Custom imposes a limitation in some groups, as in a family of moths with specific names ending in -*ana* and another with the ending -*ella*. Names of species are written with a small initial letter and are italicized in text.

The names of genera are chosen with equal freedom, but it is provided that no name shall be used a second time in the same kingdom. If this is inadvertently done, the second application is said to be preoccupied and is replaced by a new name when the duplication is discovered. Names of genera are single words, written with an initial capital and italicized in text.

Subfamily names are formed of the stem of the name of the type genus plus the ending -inae, family names of this stem with the ending -idae, and superfamily names with the ending -oidea. These and all names of higher divisions are capitalized but are not placed in italics.

Procedure in botanical taxonomy differs in detail. Here the International Rules of Botanical Nomenclature, adopted and later revised by the International Botanical Congresses of various years, correspond as an authoritative guide to the zoological rules. Their provisions are in general harmonious. A few important points of difference follow: 1. The starting point of botanical nomenclature for most groups is Linnaeus' *Species Plantarum,* Edition I, 1753, but for certain mosses and fungi other pioneer works have been adopted. 2. A list of generic *nomina conservanda* has been adopted to render certain established usages immune from strict application of the rules. 3. The rules specify the ending -ales for orders, -ineae for suborders, -aceae for families, -oideae for subfamilies, -eae for tribes and -inae for subtribes. 4. The specifications for the establishment of types of genera are much less detailed and precise.

In modern botany there is no single set of rules constituting a code for the selection of names for the various taxonomic categories. Two distinct codes are in use in the United States, one of which, the American Code, corresponds closely to the International Rules of Zoological Nomenclature. Each International Botanical Congress has made progress in the formulation of acceptable rules governing the selection of appropriate names, priority rights, adequate publication, and other matters, but general agreement has not yet been reached.

All botanists agree on the use of the system of binomial nomenclature for species. Each species is given a name consisting of the name of the genus followed by the specific name and the official abbreviation of the name of the author of the combination. In common practice the latter item is usually omitted.

The central problem of nomenclature lies in the selection of the generic name. Each must be different and different from any name used for a genus of animals. It is always a Latin or Latinized noun in the nominative singular. It may be masculine, feminine, or neuter; the most practical method of determining the gender is that of consulting a manual for the gender of an adjective which designates a species in the genus. The generic names of trees and of genera named for persons are always feminine. The name is always printed in italics with an initial capital.

Species names are usually Latin or Latinized adjectives which must agree in case and gender with the generic name. Such adjectives are italicized but never capitalized. If the specific name is an appositive noun or a proper noun in the genitive case it is usually capitalized. In rare cases the name is a common noun in the genitive case, singular or plural, and is not printed with a capital. These rules for capitalization lead to confusion and there is a growing tendency among botanists, especially horticulturalists, to follow the lead of the zoologists by omitting the capital from all species names.

An apparent exception to the system of binomial nomenclature is the use by botanists of a third name for varieties within species. The practice is especially common in the naming of marketable new varieties of cultivated plants. Varietal names follow the rules for specific names. They may be written after the specific name to form a trinomial or, more commonly, may be separated from the specific name by the abbreviation for the word variety (var.). The abbreviation is not italicized.

Family names are usually made from the root of the name of the principal genus within the family by adding the ending -aceae. The names of six large families constitute an exception to this rule. The names of these families follow with, in parentheses, the new names which have been proposed for them: Compositae (Asteraceae), Cruciferae (Brassicaceae), Gramineae (Poaceae), Labitae (Menthaceae), Leguminosae (Fabaceae), and Umbelliferae (Apiaceae).

Related families are collected into orders. The ordinal name has the ending -ales. Names of sub-families have the ending -eae. Names of families, sub-families (tribes), and orders are never italicized. (See **Taxonomy.**) (A.W.L., P.A.W.)

NOMOGRAM. Alignment Chart.

NONCALCAREA. A division of the **sponges** (Porifera) used by some zoologists to include the species without calcareous spicules. Equivalent to **Hexactinellida** and **Demospongiae** of this work. (A.W.L.)

NONCONFORMITY. Unconformity.

NON-EUCLIDEAN GEOMETRY. A non-Euclidean geometry is a geometry in which not all the axioms and postulates of Euclid are assumed. In particular, the classical non-Euclidean geometries are obtained by replacing the parallel postulate of Euclidean geometry by other assumptions.

In the hyperbolic geometry, usually credited to the Russian mathematician Lobachevski, all of Euclid's axioms are accepted with the exception of the parallel postulate, which is replaced by the assumption that through any point there are two or more lines which do not intersect a given line in the plane. In hyperbolic geometry, many theorems are the same as in Euclidean geometry, but many are different; for example, in hyperbolic geometry, the sum of the three angles of a triangle is less than two right angles.

In elliptic geometry, the parallel postulate of Euclid is replaced by the assumption that through a given point there are no lines which do not intersect a given line in the plane. (L.L.S.)

NON-FERROUS METALS. Metals and alloys in which iron is absent or present only as an impurity or alloying addition. (See **Brasses and Bronzes, Light Alloys, Solder, Nickel Alloys, Babbitt,** etc.) (R.H.H.)

NON-INDUCTIVE RESISTANCE. A resistance constructed in such a manner that the inductive effects are reduced to a minimum. Such construction is often required for high-frequency work. (L.R.Q.)

NONPAREIL. Aves, Passeriformes. The painted bunting, *Passerina ciris,* of the southern United States, a small bird brightly colored with red, blue, and green. (A.W.L.)

NON-SINUSOIDAL WAVE. This is any periodic wave which is not a pure sine wave. Such waves may, however, be analyzed into numerous sine components and often circuits may be analyzed by considering these one at a time. (L.R.Q.)

NORDMARKITE. The term proposed by Brögger, in 1890, for an **alkali syenite** containing free **silica** in the form of **quartz,** the principal **femic** minerals being **biotite** and **aegirine.** This type locality is at Nordmaken, Norway. (R.M.F.)

NORITE. Gabbro.

NORM. Used by **petrologists** to describe **igneous** rocks according to their standard, or normative **mineral molecules** as calculated in terms of **oxides.** (Contrast with **Mode.**) (R.M.F.)

NORMAL. A perpendicular; in descriptive geometry, a normal view of a line is one in which the direction of sight is perpendicular to the line, and shows the true length of the line. (H.C.H.)

NORMAL CURVE. The normal curve is given by

$$y_x = \frac{1}{\sqrt{2\pi}\,\sigma_x} \exp. - \frac{(x-m)^2}{2\sigma_x{}^2}.$$

m = **population mean,** σ_x = **standard deviation.** It is often given in standard units $y_t = \dfrac{e^{-\frac{t^2}{2}}}{\sqrt{2\pi}}$. (See **Normal Probability Function; De Moivre-Laplace Theorem.**) (L.A.A.)

NORMAL DISPERSION. A series is said to have normal dispersion if D, the **coefficient of dispersion,** is 1. The Bernoulli series of trials, a series of trials in which the trials are independent and the **probability** of success is constant from trial to trial, has $D = 1$. (See **Bernoulli Probability Function.**) (L.A.A.)

NORMAL FAULT. Fault.

NORMAL LINE TO A PLANE CURVE. Tangents and Normals to Plane Curves.

NORMAL LINE TO A SURFACE. Tangent Plane to a Surface.

NORMAL PROBABILITY FUNCTION. The equation of the normal probability function is in **standard units**

$$P_t dt = \frac{1}{\sqrt{2\pi}}\, e^{-\frac{t^2}{2}} dt, \quad t = \frac{x-m}{\sigma_x},$$

m = **mean of the population,** σ_x = **standard deviation** of the population. It occurs frequently in **statistics** particularly as a limiting distribution. The distribution of sample means and sample variances approach the normal probability function as the size of the sample increases, under very general conditions. **Snedecor's F distribution,** the χ^2 distribution, **Fisher's z distribution** and **Student's t distribution** are others which approach a normal distribution as the **degrees of freedom** approach infinity. The distribution function

$$H(t) = \int_{-\infty}^{t} P_t dt, \; P_t, \text{ and } \int_{0}^{t} P_t dt$$

have been tabulated. The ordinates of the normal probability function, P_t, form a symmetrical curve with mean zero and variance 1. It has 2 points of inflection at $t = \pm 1$. Approximately 68% of the area under the curve lies between $t = \pm 1$, 95% between $t = \pm 1.96$, and 99.7% for $t = \pm 3$. Sometimes the normal probability function is called the Gaussian function or the **normal curve**. (L.A.A.)

NORMAL SOLUTION. Concentration.

NORMALIZING. Annealing.

NORMALLY DISTRIBUTED. A variate is said to be normally distributed if it is distributed in a **normal curve** or a **normal probability function**. (L.A.A.)

NORTH. For many centuries, north has been the fundamental direction used by navigators and surveyors. During this long period a loose usage of the term has become prevalent. It seems desirable to set down certain standard meanings accepted by the majority of modern navigators and astronomers.

True north (unless a qualifying adjective is used with north, true north is to be assumed) is the direction along the geographical **meridian** of the observer, in the plane of the observer's **horizon,** toward the north pole of rotation of the earth. Compass north is the direction in the plane of the horizon toward which the north-seeking end of the **compass** points. Unless otherwise stated, compass north refers to north defined by the magnetic compass, if another type of compass is used it should be clearly indicated, e.g., **gyro-compass** north, etc. Magnetic north is the direction in the plane of the observer's horizon toward the north magnetic pole of terrestrial **magnetism.** For methods of conversion from any one of these three "norths" to any other see **Compass Corrections.** (W.K.G.)

NORTH POLAR SEQUENCE. Stellar Magnitude.

NORTH STAR. Polaris.

NORTHFIELDITE. The term proposed by Emerson in 1915 for an exceedingly quartz-rich **granite** containing 83% of **quartz** and 13% of **soda—orthoclase** feldspar. The type locality is at Northfield, Mass. (R.M.F.)

NOSE. A protuberance on the face of air-breathing **vertebrates** through which the respiratory passages lead. It is also associated in these passages with the sense of smell through the presence of the sensory olfactory epithelium.

A term used by **structural geologists** to define the end of a pitching **anticline.** Usually topographically expressed as a small open fold. (A.W.L., R.M.F.)

NOSE HEAVINESS, AIRPLANE. The condition of an airplane in which the nose tends to sink when the longitudinal control is released in any given attitude of normal flight. (See tail-heavy.) (F.T.M.)

NOSE LEAF. Complicated folds of skin, sometimes very extensive, on the snout of some **bats.** They are supposed to be provided with delicate sensory organs akin to the organs of touch. (A.W.L.)

NO-SEE-'EM. Black Fly.

NOSTRIL. The external orifice of a respiratory passage in the air-breathing **vertebrates.** There are usually two nostrils, but in some of the whales the respiratory system opens by a single aperture, called the nostril or blow-hole. External **naris.** (A.W.L.)

NOTCH SENSITIVITY. Fatigue.

NOTOCHORD. A longitudinal stiffening rod found in all **embryonic** chordates and in the adults of some of the lower members of this phylum (**Chordata**). It lies between the central **nervous system** and the alimentary tract (**digestive system**) and is the axis around which the spinal column develops. As the bony structure forms, the notochord is almost crowded out of existence. In the human body small remnants of it form the nuclei pulposi in the intervertebral disks. (A.W.L.)

NOTOPODIUM. The dorsal portion of the **parapodium** of **annelid** worms. (A.W.L.)

NOVA. There is probably no class of stars that attracts so much popular attention as the novae or "new stars." These objects suddenly appear in the sky at unpredicted time and place, in some cases becoming the brightest object in the sky for a few days. They then fade away and disappear from naked-eye observation, but may be followed telescopically for indefinite lengths of time. Many novae which do not attained naked-eye brilliancy are discovered and studied telescopically. It is impossible to give a definite estimate of the total number of novae which appear each year, for undoubtedly many escape detection. Bailey has estimated that 10 or more reach a brightness of the ninth **stellar magnitude** or greater each year. During the first 35 years of the present century there were 5 novae which reached conspicuous brightness.

In the accompanying figure we have represented the **light curves** of three of the bright novae of the present century. The ordinate scale of brightness is expressed in

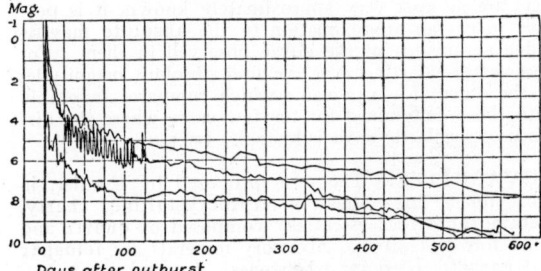

Mag.

Days after outburst

Light curves of Nova Aquilae, 1918; Nova Persei, 1901; and Nova Germinorum, 1912. They are designated in order of decreasing height. (*Harvard College Observatory Annals.*)

stellar magnitude, and, since magnitude 6 is the limit of naked-eye visibility, the length of time that each was visible to the naked eye may be determined from the time scale at the bottom. These curves are characteristic of most novae, with the very rapid rise to maximum and then the relatively slow and irregular decline. Examination of photographic records indicates that novae are not actually "new stars" at all, but rather are faint stars which, for some unexplained reason, suddenly increase in intensity. An increase of 10 magnitudes is by no means uncommon, and this represents an increase of light intensity amounting to 10,000 fold.

Coupled with the increase in light intensity of a nova there is a correspondingly remarkable change in **spectral characteristics.** While the spectral changes in different novae vary to a considerable extent, nevertheless, there are certain stages of development which are more or less characteristic of them all. During the period of rise, in the few cases where increasing novae have been detected in time for observation, the star is of the hot, blue A-type, with the absorption lines displaced very strongly to the violet. As the star starts to decline the color changes from white to yellow, and bright lines, particularly of hydrogen and ionized iron, appear. The bright lines then broaden out to bands of irregular structure which soon completely mask the continuous spectrum of the star. A few days later dark lines again make their appearance, and these are displaced far to the violet. Soon bright lines again appear, frequently of the type which are characteristic of the gaseous **nebulae,** except that they are broad. As the brightness of the star further decreases it eventually settles down to a peculiar O-type spectrum with bright lines superimposed on a continuous and dark line absorption spectrum.

There has been a great deal of research on the novae during the present century. The best theory is that, for some completely unexplained reason, a relatively faint star explodes and blows off an outer shell of gaseous material. Since this shell would rapidly expand after leaving the surface of the star, the displacement of the spectral lines to the violet may be explained as a **radial velocity** shift. There is further confirmation of the possibility of the expanding shell of gas in the fact that a nebulous envelope has actually been observed in certain novae. While the explosion and expanding shell hypothesis is by far the best of any thus far proposed, there is no explanation available as to what causes the terrific explosion necessary to drive the material out of the star against the enormous gravitational attraction. Other theories for the formation of novae from the collisions or very close approach of two stars have been advanced. In view of the approximately known distribution of stars in space and the average **space velocities** of the stars, the probability of a collision can be computed and is found to be far too small to account for the large number of novae which are observed.

Novae have been observed telescopically in some of the **extra-galactic** nebulae such as the great spiral in Andromeda. Since the distances of some of these objects are at least very approximately known, it is possible to get an approximation to the **absolute magnitudes** at maxima of the novae observed in them. For these extra-galactic novae we find absolute magnitudes of the order of -4, a value which compares favorably with those determined for the few cases where the distance of a galactic nova is known. (W.K.G.)

NOVACULITE. The term proposed by Cordier, in 1868, for a fine-grained or **crytocrystalline, cherty, metamorphic** rock essentially composed of **quartz** and other forms of **silica.** Accessory minerals are **feldspar** and garnet. Used for whetstones. (R.M.F.)

NOVOCAINE. Anaesthesia; Cocaine.

NOZZLES. A nozzle is a converging or converging-diverging tube attached to the outlet of a pipe, hose, or pressure chamber, the purpose of which is to convert efficiently the pressure existing in a fluid into velocity. Examples may be cited of nozzles for liquids, gases, and vapors. It may or may not be necessary to take compressibility of the fluid into consideration. Water in a hose or pipe at a pressure measured by a head of h ft. should, theoretically, spout freely from the pipe with a velocity equal to $\sqrt{2gh}$, in which $g = $ **acceleration** due to gravity. However, unless a nozzle were applied, there would be a tendency of the water to begin to increase in velocity far enough back in the pipe so that a great deal of the head would be consumed in overcoming friction. A nozzle throttles the discharge down to a smaller stream, and allows a pressure to be carried in the pipe or hose adjacent to the nozzle. The percentage of ideal jet velocity actually obtained in a nozzle is known as the velocity coefficient. This coefficient ranges between .85 and .98 for nozzles of different shapes and designs. The discharge of water from a nozzle needs to be further corrected by allowing for the velocity of approach. As the water moves through the converging section of the nozzle towards the tip, it is increased in velocity, resulting in some tendency to offset reduction by velocity coefficient. Multiplying the computed velocity by

$$\sqrt{\dfrac{1}{1 - \left(\dfrac{\text{tip diameter}}{\text{root diameter}}\right)^4}}$$

allows for this velocity of approach and gives the actual velocity at the tip of the nozzle.

The flow of air through nozzles may or may not involve compression. For example, compressed air jets used for cleaning have considerable expansion in the nozzles, from which high velocity is derived. Another case of an air nozzle is the entrance cone of a wind tunnel, in which an air stream is restricted in cross-sectional area, and increased in velocity. Here compressibility is not generally taken into account.

Steam nozzles are very definitely based upon expansion of steam. The nozzles of a steam turbine have to change the heat contained in steam into work, and direct the course of the steam onto the blades. The liberation of heat during the adiabatic expansion which occurs in steam nozzles increases the velocity of flow. Heat is liberated by a drop of pressure made possible by proper design of nozzle areas. The velocity attained by steam through **adiabatic** expansion can be found by application of the law of conservation of energy. The sum of the heat energy and kinetic energy in steam approaching a nozzle is equated to that leaving the nozzle, since there is no work done on the nozzle itself. This yields the formula:

$$V_2{}^2 - V_1{}^2 = 50{,}200H.$$

Here the V's are the velocity of approach and emergence from the nozzle, and H is the heat released by adiabatic expansion. When v cu. ft. of steam are flowing with a velocity of V ft. per sec., they require a flow area of $\dfrac{v}{V}$, so at any point of a nozzle the area is determined by the density of the steam and its velocity, both of these being determined by the pressure. The physical phenomena attending the variation in velocity and volume with change of pressure create a condition of a converging nozzle, that is, diminishing area, until the pressure becomes 58% (55% for superheated steam) of the initial pressure, after which the nozzle diverges. The final velocity, that is, the velocity at the mouth of the steam nozzle, is slightly less than that given by the above equation, because (1) reversible adiabatic expansions are not obtained, (2) there are frictional losses on the nozzle walls. (See **Jet; Pressure Vessels.**) (F.T.M.)

NUCELLUS. Flower.

NUCLEAR PARTICLES. Particles of atomic or sub-atomic mass, supposed to enter into the structure of the nuclei of atoms. Listed among these at present are the **alpha particle,** the **electron,** the **mesotron,** the **neutrino,** the **neutron,** the **positron,** and the **proton.** Some writers go so far as to include the **photon,** or quantum of **radiation,** under this head, since in some circumstances photons emerge from atomic nuclei and behave in certain respects like particles. Photons differ, however, from the other entities mentioned in that they have various masses (because they have various quantum energies), but always the same speed in a vacuum, whereas each of the others always has the same mass but may move with various speeds. (L.D.W.)

NUCLEIC ACIDS. Aminoacids, Polypeptides, and Proteins.

NUCLEOLUS. These are small rounded bodies found in the **nucleus** of the **cell.** Generally each nucleus contains a single nucleolus, but in many cases there are several. The function of the nucleolus is unknown. It seems to be a mass of accumulated material which is used in the **metabolic** processes going on in the nucleus. (R.M.W.)

NUCLEUS. 1. Cell. 2. A group of cell bodies in the central **nervous system** of vertebrates.

Examples are the red nucleus in the mid-brain, through which impulses are routed for the control of subconscious muscular movements, and Deiter's nucleus, lying at the junction of the medulla with the **hind-brain.** Through this center impulses pass for muscular action involved in the maintenance of equilibrium. (A.W.L.)

NUDA. Comb jellies (ctenophores) of large size and ovate or conical form, without tentacles. They constitute a class of this name in the phylum **Ctenophora.** It contains the single family Beroidae. (A.W.L.)

NUDIBRANCHIATA. Gasteropoda.

NULL HYPOTHESIS. The null hypothesis is the hypothesis which is being tested under the assumption that it is true. (L.A.A.)

NULL METHOD. Physical Measurement.

NUMBER. Historically, the various types of numbers of algebra were introduced gradually, step by step, as the need for them arose.

In the early stages of the development of man, the process of counting gave rise to the natural numbers or positive integers, 1, 2, 3, 4, 5, etc. These numbers serve to answer the question "How many?"

The process of measurement of quantities led naturally to the introduction of the positive rational fractions, examples of which are $\frac{1}{2}$, $\frac{2}{3}$, $\frac{5}{7}$, etc. These numbers are needed to assist in answering the question "How much?" From an algebraic viewpoint, the positive rational numbers were introduced in order to make division always possible. Positive rational fractions can be expressed as quotients of positive integers.

The positive irrational numbers were introduced through the incommensurable quantities which occur frequently in geometry; for example, the diagonal of a square is incommensurable with its side. Examples of such numbers are: $\sqrt{2}$, $\sqrt[3]{2}$, $5 + \sqrt{7}$, π, etc. From an algebraic viewpoint, the positive irrational numbers were introduced in order to help to make the extraction of roots always possible.

Quantities that could be measured in two opposite senses suggested the idea of the negative numbers; for example, temperature is measured above and below zero, and latitude is measured north and south of the equator. Algebraically, the negative numbers were introduced in order to make subtraction always possible. The negative numbers consist of several types: the negative integers -1, -2, -3, -4, etc., the negative rational fractions, such as $-\frac{3}{4}$, $-\frac{8}{5}$, $-\frac{1}{7}$, etc., and the negative irrational numbers, such as $-\sqrt{3}$, $-\sqrt[5]{6}$, $-1 - \sqrt{2}$, etc.

The number zero, 0, is often used to indicate absence of quantity. Algebraically, zero is introduced to indicate the result of subtracting a number from itself.

The positive integers, the negative integers, zero, the positive rational fractions, and the negative rational fractions are grouped together to form the system of rational numbers.

The numbers of the rational number system together with the positive and negative irrational numbers are grouped to form the real number system.

Comparatively late in the historical development of algebra, the pure imaginary numbers and the complex numbers were introduced in order to make the extraction of roots of negative numbers and the solution of certain types of equations always possible. Examples of pure imaginary numbers are $\sqrt{-1}$, $\sqrt{-5}$, $-3\sqrt{-1}$, etc.; examples of complex numbers are $2 + 3\sqrt{-1}$, $-1 - \sqrt{-2}$, $\frac{1}{2} - \frac{1}{2}\sqrt{-3}$, etc.

The square of any real number is always a positive real number. But the pure imaginary numbers have the property that their squares are negative real numbers; thus, the square of the imaginary number $\sqrt{-3}$ is -3. Any complex number may be expressed as an indicated sum of a real and a pure imaginary number, as for example $2 + \sqrt{-3}$. The term "imaginary number" is sometimes used to mean complex number.

An irrational number cannot be expressed as a quotient of positive integers nor as a quotient of rational fractions. It may, however, be represented approximately by rational fractions, to any degree of approximation desired. Thus, the irrational number $\sqrt{2}$ may be represented approximately by the rational numbers 1.4, or 1.41, or 1.414, or 1.4142, etc. In fact, any irrational number can always be inclosed between pairs of rational numbers whose differences are as small as we please. Thus, $\sqrt{2}$ lies between 1.4 and 1.5, between 1.41 and 1.42, between 1.414 and 1.415, etc.

The absolute value (or numerical value) of a positive number is the number itself, while the absolute value of a negative number is the number with its sign changed. The absolute value of a number a is denoted by the symbol $|a|$. Thus, $|5| = 5$, $|-8| = 8$.

Real numbers may be represented graphically by points on a straight line, whose distances from some arbitrarily selected origin are the corresponding real numbers.

An algebraic number is a number which satisfies a **polynomial equation** in one variable with integral coefficients. A transcendental number is a number which is not algebraic, i.e., is not a root of a polynomial equation in one variable with integral coefficients.

Irrational numbers are either algebraic, as $\sqrt{2}$, $\sqrt[3]{5}$, $1 - \sqrt{3}$, etc., or transcendental, as π, e, etc.

Any type of number may be regarded as contained in the complex number system. If in the complex number $a + bi$, we have $b = 0$, the number is a real number a; if $a = 0$, the number is purely imaginary. (L.L.S.)

NUMBER SIZE DRILL. Drilling; Gauge Number.

NUMERICAL INTEGRATION. Approximate Integration.

NUNATAK. A topographic or physiographic term of Eskimo origin and meaning the top of a hill or mountain which projects above an ice sheet or continental glacier. (R.M.F.)

NUSSELT GROUP. Heat Transfer.

NUT. Screw Fastenings; Fruit.

NUTATION. For the use of this term in botany, see **Movements in Plants.**

Nutation in astronomy is a short periodic change in **precession.** Precession is caused by the attracting force of the sun and the moon tending to pull the equatorial bulge of the earth into the plane of the **ecliptic.** The amount of this force is changing throughout the year as the **declinations** of the sun and the moon change. For example, twice during each year both the sun and the moon are on the equator, and at those times their precessional forces are zero. The principal nutation is due to the periodic change in the plane of the moon's orbit, and has a period of about 19 years. Most of the nutation effects are periodic in character, but a complete account of them is beyond the scope of this work. (W.K.G.)

NUTCRACKER. Aves, Passeriformes. Birds (**Aves**) related to the crows and jays. The few species are confined to the northern parts of the northern hemisphere. The Clarke nutcracker, *Nucifraga columbiana*, lives principally at higher altitudes in the mountains of western North America. It is associated with coniferous forests. (A.W.L.)

NUTHATCH. Aves, Passeriformes. Small climbing birds (**Aves**) which cling in any position to the bark of trees as they search for food. They eat both insects

Nuthatch.

and seeds. The most common North American species is the white-breasted nuthatch, *Sita carolinensis,* found throughout the United States east of the Rockies. The red-breasted nuthatch, *S. canadensis,* is more common in Canada and the mountains but migrates in winter as far as the southern states. The pigmy nuthatch, *S. pygmaea,* is a western species. Nuthatches are found on all other continents except South America, although in Africa they are confined to the north. (A.W.L.)

NUTMEG. *Myristica fragrans.* Myristicaceae. Nutmegs are the seeds of a tree native to the Molucca Islands. The tree grows 60′ tall, and has pointed **lanceolate** leaves. The trees are **dioecious**, pistillate (see **Pistil**) and staminate (see **Stamen**) flowers being borne on separate trees. Since the trees are frequently grown in cultivation, especially in favorable localities, it is necessary to plant some of both sexes to insure cross-pollination and seed formation. The flowers are pale yellow, the fruit a dark orange-colored berry containing a single large brown seed. Surrounding the seed is a branched deep red **aril**, which on drying becomes pale brown. The seed is the nutmeg of commerce, the aril is mace. Both the seed and the aril contain an aromatic oil, but only poor quality seeds are used for the oil. Both nutmeg and mace are well-known spices used in great quantities in the United States. (R.M.W.)

NUTRITION. The process of incorporating in the body the substances necessary for growth, repair, and energy. Constructive **metabolism.** Anabolism.

In the usual sense the term designates certain aspects of the entire process of securing food and incorporating it in the body, hence words describing the manner of nutrition indicate fundamental relations of the animals involved. All organisms which ingest animals or plants as food, as is true of most animals, carry on holozoic nutrition. They are also described as zootrophic or

heterotrophic organisms. Those which have chlorophyll and are able to synthesize food from inorganic materials like the green plants and some one-celled animals, have holophytic nutrition, or are called autotrophic or phytotrophic organisms. Animals which absorb dissolved organic materials have saprozoic nutrition and those which live in another organism and use its materials carry on parasitic nutrition. No exact line can be drawn between parasitic and saprozoic animals for some parasites merely absorb food from the contents of the alimentary tract in which they live and so are, in the strict sense, saprozoic. The bodily association and dependence of parasitism also enter into the definition of this condition. (A.W.L.)

NUX VOMICA. Strychnine.

NYALA. Mammalia, Artiodactyla. An African harnessed **antelope**, *Tragelaphus angasi,* related to the bongo and nakong. It is blue-gray with faint white stripes. Found in swampy jungles of eastern Africa. (A.W.L.)

NYLON. This is a generic term applied to "any long-chain synthetic polymeric amide which has recurring amide groups as an integral part of the main polymer chain, and which is capable of being formed into a filament in which the structural elements are oriented in the direction of the axis." A typical reaction mixture is 14.8 parts of pentamethyleneamine, 29.3 parts of sebacic acid, and 44 parts of mixed xylenols. The filament nylon is silky in appearance, relatively insensitive to moisture and mildew, and of superior elastic recovery (better than silk) and tensile strength (better than silk and rayon). Nylon is specially adapted for hosiery, knitted and woven goods, shoe laces, bristles for brushes, and transparent wrapping film; also for parachute fabrics, glider tow-ropes, and as fabric for airplane tires. (R.K.S.)

NYMPH. An immature insect of the forms which have gradual metamorphosis (**Paurometabola**). Nymphs resemble adults but have wings in the developmental state. (A.W.L.)

NYMPHAEACEAE. Water Lilies.

NYMPHOMANIA. Abnormal degree of sexual desire in the female. (R.S.M.)

NYSTAGMUS. An involuntary rapid movement of the eyeball which may be from side to side, up and down, circular, or a combination of these movements. It may be due to incoordination of the eye muscles, disturbances of the nerves of the eye, or to disturbances of the vestibular canals in the ear. (R.S.M.)

O

O SERIES. X-ray Spectra.

OAKS. *Quercus* sp. Fagaceae. The oaks are trees and shrubs of the north temperate region. All the more northern species are **deciduous** plants. Many of those in the southern part of the range have evergreen leaves, and are often called live oaks. In Asia and the Pacific coast of North America, oaks are found in regions approaching tropical conditions.

The 300 species of this genus have simple alternate leaves. The flowers are of two kinds, borne on the same tree. The pistillate (see **Pistil**) flowers are borne singly and are surrounded by an **involucre** of many scales beyond which the **stigmas** protrude. The staminate (see **Stamen**) flowers are borne in long slender pendant **catkins.** Pollination is by wind. The fruit is an acorn, a **nut** of characteristic cylindrical shape, capped by the small persistent style-base, and seated in the scaly involucre, which forms a cup partially or almost wholly surrounding the nut.

Many of the oaks are valuable trees, yielding woods which have a variety of uses. In early times, before the day of the sawmill, oaks were much used in the construction of buildings. Often the oaks used for this purpose were split into thin planks, a method of preparation which served well to bring out the attractive grain of the **wood.** This grain is due partly to the numerous large vessels which are formed periodically every spring and appear as very evident dark lines or streaks in the wood, and partly to the large **vascular** rays which appear as irregular flakes, especially when the wood is split in a radial plane. In modern construction oak is often used as paneling or flooring. To obtain the best grain, the wood is quarter-sawed, that is, cut in such a way that the flat surfaces shall be as nearly radial as is possible. Because of its beauty and also its durability, oak wood is also much used in furniture making. In America the principal species used for wood is white oak, *Quercus alba.* In Europe several species are used, among them the British oak, *Quercus Robur.* Often these European oaks are trees of remarkable size, and are preserved because of their rugged beauty. The wood is very strong and durable, and finds considerable use in ship construction. Formerly much more was used for this purpose.

Accidents sometimes cause the formation of oak wood of special properties and value. The trunks of fallen trees may lie buried for long periods of time in bogs or elsewhere. Sometimes, when removed, these logs are found to be perfectly sound and to have developed a rich dark brown or nearly black color, which makes them especially sought after for furniture making. Such oak is known as bog oak. Living trees frequently develop large irregular growths, known as burls, in which a very irregular much-contorted grain is found. The custom of cutting back the top of the tree, causing the development of numerous **adventitious buds,** a practice known as pollarding, causes a similar irregular grain. These burls are used for making **veneers.** Another species of oak, *Quercus Suber,* yields **cork.**

Oaks are valuable sources of **tannin.** In many species the bark is the source of the tannin, but in *Quercus Aegilops,* a native of eastern Europe and Asia, a tannin known as valonia is obtained from the cup and the young acorns.

African oak, a strong, heavy wood, comes from African trees of other genera than *Quercus.* This wood is rarely used, due to the difficulty of removing the heavy wood from its native forest. (R.M.W.)

OATS. *Avena sativa.* Gramineae. Oats are annual cereal grasses native in temperate regions of the Old World. The several species of oat plants are characterized by their closed leaf-sheath, a wide-branched **panicle,** and by a special type of **inflorescence.** In the **florets** of wild oat plants the **lemma** has the midrib prolonged as a prominent **awn,** the basal portion of which is spirally twisted. In many cultivated forms this awn has been eliminated. The grain of oats is not easily separated from the surrounding husk, composed of the lemma and **palea,** which are neither palatable nor digestible.

Oats are principally adapted to growing in a climate having cool summers and abundant moisture. However, the plants are very hardy and tolerant of adverse conditions. Oats are used principally for grain. The straw is used either for bedding livestock or as a rough cattle food. The grain is an important article in horse feeding. A comparatively small part of the total oat harvest is used for human consumption in the form of rolled oats and oatmeal. In manufacturing rolled oats the grain is first carefully cleaned and graded according to size, then passed between two millstones so placed that by the revolution of one against the other the husks are removed and withdrawn by suction. The cleaned grains are steamed, and in the case of "Quick Oats," partially cooked, and then passed between heavy rollers and dried, after which they are ready for the market. (R.M.W.)

OBESITY (Adiposity). A state of the body in which excessive fat is stored in the tissues. This is one of the most common disorders to which the human race is subject. The prevention and treatment of obesity are important, not only for the sake of appearance, but also because obesity decreases human efficiency, shortens life and predisposes the subject to many disorders.

The fundamental cause of obesity is the absorption of a greater number of calories than is necessary for the total amount of energy expended by the body. It is also true that, while two persons may eat the same amount of food, one may gain and the other lose weight. This difference is due to factors regulating the amount of energy expended. One who gains weight and who does not apparently eat to excess (such a person usually eats more than he realizes or admits) is usually phlegmatic, reacts slowly to outside stimuli, is not easily worried, dislikes exercise or expenditure of energy in any form, sleeps longer and more soundly, and relaxes more completely than either the thin person or the one of normal weight. A few people who fall in the overweight group have disturbances of the endocrine glands, especially the **thyroid** or **pituitary** or both. But most obesity is due to the one factor of simply liking to eat large quantities of fattening food—carbohydrates, starches and fats.

Marked obesity can cause shortness of breath and fatigue on slight exertion. It also seems to predispose to certain diseases or accentuate their symptoms. An individual with heart disease may develop heart failure sooner than a thin person with the same degree of cardiac disease. Arthritics having pain in the legs or back notice considerable improvement in their symptoms, as a rule, with reduction in weight. Obesity is a factor in **diabetes.** Obese patients seem to be unable to cope with such diseases as **pneumonia** as well as the normal individual.

Prevention of obesity is simpler than its treatment. Treatment lies along two lines. First, the amount of food must be reduced and at the same time a balanced diet, adequate in minerals and vitamins, achieved. The rapid loss of weight by radical diets is harmful. Second, increase in exercise, which unfortunately increases the appetite. (D.M.H.)

OBJECTIVE. Optical Instruments; Telescope; Microscope; Camera.

OBJECTIVE PRISM. In the ordinary laboratory spectrograph a collimating lens is necessary in order that the light from the narrow slit may be sent through the dispersing agent in a parallel beam. At the principal focus of the camera lens the images of the slit in the different radiations are commonly known as the spectral lines.

The stars are at such tremendous distances that they subtend infinitesimal angles and the light from them reaches the earth in parallel beams. Hence the slit and collimating lens of the laboratory spectrograph are unnecessary and the dispersing agent, usually a prism, may be placed directly before the objective lens of the telescope or astrographic camera. With such an instrument, commonly known as an objective prism, instead of single-point images of the stars, there appear on the photographic plate a series of dots, each of them in some particular radiation from the stars, or, in other words, the spectra of the stars. In order that the spectra may have appreciable width or that the spectral lines may have length instead of appearing as mere dots, the refracting edge of the prism is set parallel to the spherical coordinate of right ascension and the telescope driven a bit too fast or too slow, slightly "trailing" the images.

The objective prism has certain important advantages over the slit spectrograph for astronomical purposes. On each plate taken with the instrument we obtain spectra of a great number of different stars. Furthermore, the loss of light at the narrow slit is obviated and the exposure time necessary to obtain a good spectrogram will be greatly diminished. In contrast to the advantages in speed of the objective prism there is the serious disadvantage that it is impossible to obtain good comparison spectra with such an instrument, and the determination of accurate wavelengths of the different radiations is practically impossible.

The objective prism is used primarily for the determination of the spectral classes of the stars and to determine the relative intensities of the different lines and regions of the spectra. Within recent years attempts have been made to determine radial velocities of the stars with this instrument and results obtained are of sufficient accuracy for statistical purposes. The flash spectrum is often obtained by use of the objective prism. (W.K.G.)

OBLIQUE COORDINATES IN A PLANE. Let any two non-perpendicular intersecting lines in a plane be chosen as reference axes; let P be any point in the plane, and through P draw lines parallel to the axes. The distances, parallel to the axes, from the axes to the point P are called oblique coordinates of the point. (L.L.S.)

OBLIQUE REPRESENTATION. Orthographic Projection.

OBLIQUE TRIANGLES. Triangles.

OBLIQUES (PAIRED). Photogrammetry.

OBSEQUENT STREAMS. Consequent Streams.

OBSIDIAN VOLCANIC GLASS. Highly acidic lavas (those containing a preponderance of silica), when chilled very rapidly and congealed, without appreciable crystallization, into a rigid liquid solution.

Such a solid is called a glass to distinguish it from a crystalline substance. Pure obsidian is hard, with conchoidal fracture and vitreous luster. Thin chips appear smoky-gray or red in transmitted light. The exceptionally dark color of obsidian, as compared with chemically related crystalline rocks (rhyolite and granite) is due to the fact that a small amount of coloring matter is much more effective in a solution than when segregated in a few dispersed dark-colored minerals. (R.M.F.)

OBSTETRICS. The branch of medicine and surgery which has to do with the management of pregnancy, labor and the complications and disorders arising from them. (R.S.M.)

OCCIPITAL BONE. A bone of the vertebrate skull located at the posterior end of the dorsal wall. It is a median unpaired bone. (A.W.L.)

OCCLUSION. Adsorption; Wave Cyclones; Minerals, Rocks.

OCCULT MINERALS. A term proposed by Iddings, in 1913, for mineral compounds which, because of the chemical composition of an igneous rock, should be either potentially or actually present, but which cannot be detected by the petrographic microscope. (R.M.F.)

OCCULTATION. When the moon passes between a star and the earth, the light of the star is cut off and the star is said to be occulted. The term eclipse, which is technically correct for this phenomenon, is reserved for circumstances involving the sun, earth, and moon, and for such things as eclipsing binaries, satellites of Jupiter, etc. As the moon passes between the earth and a star in its revolution about the earth the star disappears behind the eastern edge of the moon and reappears again at the western edge. The disappearance and reappearance are practically instantaneous; proving both that the moon has no sensible atmosphere and that the star appears sensibly as a point of light. The interval between disappearance and reappearance depends fundamentally upon how closely the center of the moon appears to pass across the star.

Since the disappearance and reappearance are practically instantaneous, the times of the phenomena can be determined with great precision. Observations of the instants of occultation of stars by the moon may be used to determine the position of the moon with great precision. A large number of such occultations are observed both by professional and amateur astronomers, and the results used to verify and correct the theory regarding complicated motions of the moon. The differences in the time of occultation of a given star, as observed by widely separated observers, may be used both to determine the distance of the moon and also to determine the difference in terrestrial longitude of the two observers. This method of determination of difference in longitude was the most accurate available before the development of modern methods for distribution of Greenwich Time. (W.K.G.)

OCEANIC DEPOSITS. The map on the facing page shows the distribution of the different types of deep-sea or oceanic sediments. (R.M.F.)

OCELLUS. 1. The simple eye of arthropods. Usually very small. 2. An eye-like spot. The wings of many butterflies and moths bear such markings. (A.W.L.)

OCELOT. Mammalia, Carnivora. A moderately large South American cat, *Felis pardalis,* which ranges north to the Rio Grande. It is tawny or reddish, marked with black spots and blotches.

This species has been recorded from southern Texas. (A.W.L.)

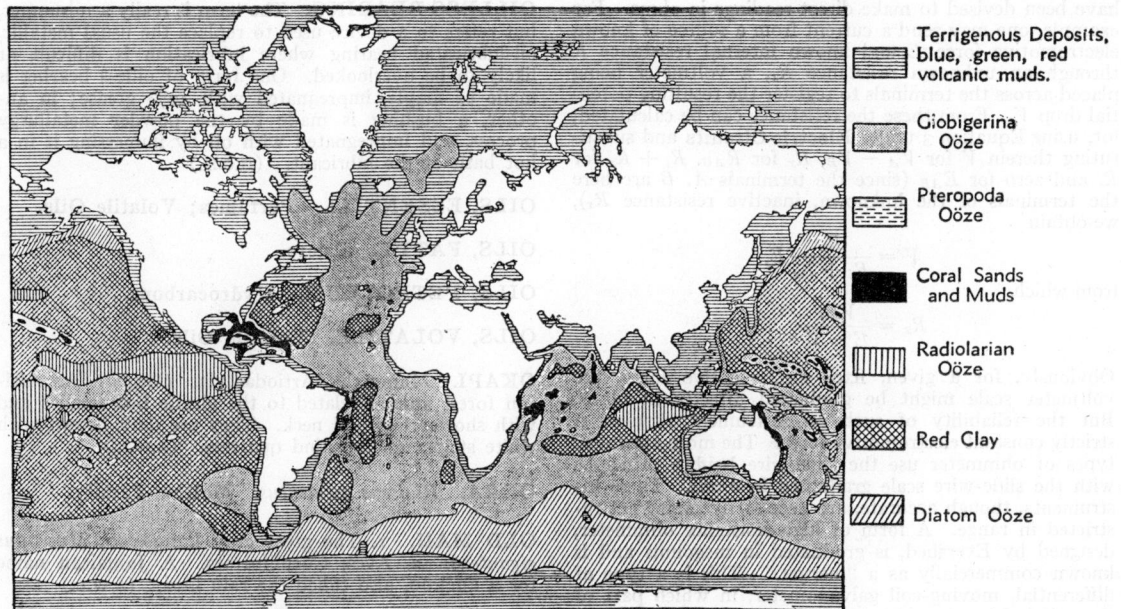

Distribution of marine-continental and oceanic deposits. (*After Scott. After Collet.*)

Legend:
- Terrigenous Deposits, blue, green, red volcanic muds.
- Globigerina Ooze
- Pteropoda Ooze
- Coral Sands and Muds
- Radiolarian Ooze
- Red Clay
- Diatom Ooze

OCTAHEDRITE. The mineral octahedrite, **titanium** dioxide, TiO_2, is a rare mineral crystallizing in the **tetragonal** system. It has the same chemical composition as **rutile** and **brookite.** It commonly occurs in **octahedral** crystals, either acute or obtuse, hence the name. (See **Rutile.**) (E.S.C.S.)

OCTANE NUMBER. Detonation.

OCTANE RATING. Gasoline.

OCTOPUS. Mollusca, Cephalopoda. A genus of marine **mollusks** commonly called **devil fishes** but also known by the anglicized generic name. They have a large head bearing eight similar arms or tentacles, each with many suckers. The arms are webbed at the base, and in a related genus the webs unite them almost to the tips. (See also **Invertebrate Paleontology.**) (A.W.L.)

ODONATA. The dragon flies and damsel flies, an order of insects containing moderate to large species with slender bodies and four narrow net-veined wings. Their flight is powerful and they are well adapted to take their insect prey on the wing. The mouth is formed for biting. In the immature stages these insects are aquatic. Commonly called devil's darning needles. The order contains about 2700 species and is represented in all parts of the world. (A.W.L.)

ODONITE. The term proposed by Chelius, in 1892, for a basic **igneous rock** with a **porphyritic** texture whose **phenocrysts** are **labradorite** and **augite** in a matrix composed of **feldspar** and **hornblende.** (R.M.F.)

ODONTOCETI. The toothed whales, an order of marine mammals containing the **dolphins, narwhal, porpoises,** etc. A few species live in the large tropical rivers, including the Ganges and the Amazon. (A.W.L.)

ODONTOLITE. Turquoise.

OENOCYTE. A kind of large **cell** found in clusters associated with the **tracheae** and fat bodies of insects. Larval oenocytes have been interpreted as **endocrine** glands and those of adults are apparently for the storage of waste products. (A.W.L.)

OERSTED. The practical unit of magnetic intensity. The term oersted, prior to 1932, was used to designate the practical unit of magnetic reluctance (see **Magnetism**). By international agreement, however, it now replaces the term **gauss,** which formerly denoted a unit of magnetic field intensity; while the unit of magnetic reluctance is left without a name. The value of the oersted of magnetic intensity may be expressed in the same way as that used in defining the gauss of magnetic induction; but it is simpler to state that if a unit north magnetic pole is placed in a vacuum traversed by a magnetic field of intensity 1 oersted, the pole is urged in the direction of the field with a force of 1 **dyne.** (L.D.W.)

OERSTED EXPERIMENT. Electromagnetism.

OGIVE. The graph of the **cumulative frequency distribution,** or the graph of the **distribution function** in the case of a **continuous variate.** (L.A.A.)

OHM. The practical unit of electrical **resistance.** It is defined in two ways. The absolute ohm is that resistance which causes a potential drop of 1 absolute **volt** when a steady current of 1 absolute **ampere** flows through it; it is equal to 10^9 abohms (see **Electrical and Magnetic Units**). The international ohm is based upon a specified conductor; viz., it is the resistance of a uniform thread of mercury in a capillary tube of such diameter that the thread is 106.3 cm. long and weighs 14.4521 grams, its temperature being 0° C. (This gives the thread a cross section of almost exactly 1 sq. mm.) Unfortunately the two units are not quite equal, the international ohm being larger by about 46 parts in 100,000. In more familiar terms: a No. 16 copper wire (0.05″ in diameter), having a resistance of 1 ohm, would be about 264′ long. The resistance of a 110-volt, 50-watt lamp in full operation is 242 ohms. Occasional use is made of the megohm (1,000,000 ohms) and the microhm (0.000001 ohm). (L.D.W.)

OHMMETERS. The accurate measurement of resistance is somewhat tedious, and various instruments have been devised to make direct readings in ohms. For example, one may send a current from a source of known electromotive force E and known internal resistance R through an unknown resistance R_x, a voltmeter being placed across the terminals to register the resulting potential drop V. From these the resistance can be calculated; for, using Equation 3 under **Electric Circuits** and substituting therein V for $V_A - V_B$, R_x for R_{AB}, $R_i + R_x$ for R, and zero for E_{AB} (since the terminals A, B are here the terminals of the unknown, inactive resistance R_x), we obtain

$$V = \frac{R_x}{R_i + R_x} E,$$

from which

$$R_x = \frac{V R_i}{E - V}.$$

Obviously, for a given, fixed electromotive force, the voltmeter scale might be graduated directly in ohms. But the reliability of such an instrument requires a strictly constant electromotive force. The more approved types of ohmmeter use the slide-wire **bridge** principle, with the slide-wire scale graduated in ohms. These instruments, though accurate, are necessarily somewhat restricted in range. A form of high-resistance ohmmeter, designed by Evershed, is graduated in megohms and is known commercially as a "megger." This is a type of differential, moving-coil galvanometer, in which part of the coil is in series with the unknown resistance, while another part, carrying current from the same source, is independent of that resistance. The galvanometer reading depends upon the relative currents in the two parts, and hence upon the unknown resistance. (L.D.W.)

OHM'S LAW. This very familiar law of **electric conduction**, stated by George Simon Ohm in 1827, is expressible in various forms, of which the following is typical: The steady electric current in a metallic circuit is proportional to the constant total electromotive force operating in the circuit: $I = KE$. The constant K, known as the "conductance" of the circuit, is the reciprocal of the **resistance** R; so that the equation may be written in the more usual form

$$I = \frac{E}{R}.$$

Emphasis must be placed on the constancy of the electromotive force and the current. For, if the current varies, the effects of **inductance** and **capacitance** set up extra electromotive forces, positive or negative, which render the law expressible in general only by a differential equation. (See **Transients, Alternating Currents,** and **Electric Circuits.**) Also there are certain kinds of conduction for which the law is not valid; notably that of **ionized gases, thermionic vacuum tubes** and **photoelectric cells.** (L.D.W.)

OIL. Volatile Oils; Fixed Oils; Paints; Hydrocarbons; Lubrication; Lubricants.

OIL HOLE DRILL. Drilling.

OIL POOLS. Petroleum.

OIL SANDS. Petroleum.

OIL SHALE. A general term applied to a group of fine black to dark brown **shales** rich enough in **bituminous** material (called kerogen) to yield **petroleum** upon appropriate distillation. Oil shales are of only of economic importance in those countries which are notably deficient in petroleum. The United States has vast reserves of oil shales. (R.M.F.)

OIL WELLS. Petroleum.

OILLESS BEARING. The term is really a misnomer, but refers to a device used to replace the usual metallic, oil-lubricated bearing where lubrication is difficult or likely to be overlooked. One form of oilless bearing is made of maple, impregnated with 40% grease; in another, a bushing is made by the powder metallurgy process, and impregnated with oil by immersing it in a hot bath of the lubricant. (H.C.H.)

OILS, ESSENTIAL. Perfumes; Volatile Oils.

OILS, FATTY. Esters.

OILS, PETROLEUM. Hydrocarbons.

OILS, VOLATILE. Volatile Oils.

OKAPI. Mammalia, Artiodactyla. *Okapia.* An African forest animal related to the giraffe but smaller and with shorter legs and neck. They are dark brown with white stripes on the hind quarters. (A.W.L.)

OKRA. *Hibiscus esculentus.* Mallow Family.

OLD MAN. 1. The great gray **kangaroo,** *Macropus giganteus.* 2. Aves, Cuculiformes. A Jamaican name for a **cuckoo,** also called the rain bird. (A.W.L.)

"OLD RED" SANDSTONE. Devonian.

OLD SQUAW, OLD WIFE. Aves, Anseriformes. A small **duck,** *Clangula hymenalis,* of the northern latitudes of the northern hemisphere, migrating into the United States in winter. It is chiefly black and white and the male has a long slender tail. Also called the long-tailed duck. (A.W.L.)

OLDHAMS COUPLING. Coupling.

OLEFIN. Hydrocarbons.

OLEIC ACID AND OLEATES. Oleic acid ($H \cdot C_{18} H_{33} O_2$ or $C_{17} H_{33} \cdot COOH$ or $CH_3 (CH_2)_7 CH : CH (CH_2)_7 \cdot COOH$), is a colorless liquid, melting point $14°$ C., boiling point $286°$ C. at 100 mm. pressure, insoluble in water, miscible with alcohol or ether in all proportions. Oleic acid differs from **stearic acid** chemically by possessing 33 instead of 35 hydrogen atoms in the radical ($C_{17} H_{33} \cdot COOH$ (oleic acid), $C_{17} H_{35} COOH$ (stearic acid)). It is possible to convert oleic acid and oleate esters into stearic acid and stearate esters by treatment with **hydrogen** gas in the presence of finely divided **nickel** as a **catalyzer** at $250°$ C. under pressure as in the **hydrogenation** of oils and fats. Either by careful oxidation, or by addition of ozone (see **Oxygen**) and splitting, oleic acid yields products of 9 carbon atoms, thus leading to the conclusion that the double bond is in the center of the carbon chain. Oleic acid adds **bromine** or **iodine** in definite amounts to confirm the conclusion that one double bond is contained. **Nitric acid** converts oleic acid into elaidic acid ($C_{17} H_{33} COOH$), melting point $51°$ C. (oleic and elaidic acids are related, cis- and trans-, as maleic and fumaric acids).

Oleic acid may be obtained from glycerol trioleate, present in many liquid vegetable and animal non-drying oils, such as olive, cottonseed, lard, by **hydrolysis.** The crude oleic acid after separation of the water solution of **glycerol** is cooled to fractionally crystallize the stearic and **palmitic acids,** which are then separated by filtration, and fractional distillation under diminished pressure. Oleic acid reacts with **lead** oxide to form lead oleate, which is soluble in ether, whereas lead stearate or palmitate is insoluble. From lead oleate oleic acid may be obtained by treatment with **hydrogen sulfide** (lead sulfide, insoluble solid, formed). With sodium oleate, a soap is formed. Most soaps are mixtures of sodium stearate, palmitate, and oleate.

Representative esters of oleic acid are: methyl oleate ($C_{17}H_{33}COOCH_3$) boiling point 190° C. at 10 mm. pressure; ethyl oleate ($C_{17}H_{33}COOC_2H_5$) boiling point 205° C. at 10 mm. pressure; glyceryl trioleate (triolein) (($C_3H_5(COOC_{17}H_{33})_3$) boiling point 240° C. at 18 mm. pressure.

Oleic acid is used in the preparation of metallic oleates, such as aluminum oleate for thickening lubricating oils, for water-proofing materials, and for varnish driers. As the glyceryl ester oleic acid is one of the constituents of many vegetable and animal oils and fats. (R.K.S.)

OLEO SHOCK ABSORBER. Shock Absorber.

OLFACTORY ORGAN. Organs of the sense of smell. These organs are stimulated by minute quantities of material such as diffuse through the air in the form of gases or vapors. In a broad sense they sample materials in the animal's environment in the form of gases or vapors, although these substances probably dissolve in the body fluids before they act on the sensory ending. Like the organs of taste, olfactory organs are chemoreceptors, reacting to chemical properties, and in some animals it is difficult to distinguish accurately between the two senses. The fact that they are activated by minute concentrations, sometimes as low as one part in one million of air, gives them a wider range of perception. Whereas taste demands actual contact with a solid or liquid material, the volatile products may diffuse rapidly and over long distances from such a source and are readily carried by air currents.

Sense organs of many lower animals which live in water are more logically interpreted as organs of taste or as more primitive chemoreceptors, since materials must reach them in solution. Some aquatic insects and fishes, however, have olfactory organs enclosed in cavities which open to the exterior. Whether reached by water or not, they are classed as olfactory organs from their resemblance to such organs in related terrestrial forms.

The olfactory organs of insects are most abundant on the **antennae.** They consist of blunt processes or flat plates, associated with sensory nerve endings. In some cases they are grouped in the lining of depressions. As many as 39,000 have been reported on a single antenna and five or six thousand are frequently present.

In the **vertebrates** the olfactory cells lie in the epithelium lining a pair of olfactory pits which form in the embryo as depressions at the anterior end of the head. These cells are connected with the brain by the fibers of the olfactory nerves, the most anterior pair of cranial nerves of known function. In the air-breathing vertebrates the olfactory epithelium becomes part of the lining of the nasal passages.

The keenness of smell and its varied uses are well illustrated by dogs. Man has very nearly abandoned the sense through depending more on vision for distant perception, although he is usually more aware of odors when he cannot see. The sense is keener in man than we usually realize, but it is much less keen than in other animals. (A.W.L.)

OLIGOCENE. A geologic period of the **Tertiary,** of the Lower **Cenozoic** era of the **geologic time-scale.** The term was proposed by Beyrich in 1854. Type locality near Paris, France. Maximum thickness of strata in Italy. This period began approximately 36,000,000 years ago and lasted for about 16,000,000 years. In the United States marine sediments overlap the Cretaceous and earlier Tertiary sediments of the Atlantic border of South Carolina and the Gulf of Mexico. Marine sediments also occur on the Pacific Coast. Terrestrial sediments are well developed in the easterly Great Plains and Oregon (John Day Basin). The "Bad Lands" of South Dakota, eastern Wyoming, and North

Dakota (Black Hills) are important collecting localities for the fossil mammals of this period. In the Paris Basin occur fresh and brackish water deposits, which contain numerous **fossil vertebrates, invertebrates,** and plants (see **Paleobotany**). The Oligocene formations are also well developed in Germany, and in the Alps. The marine invertebrates and fishes of the Oligocene are similar to those in the Eocene. Among the mammals the true carnivores have replaced the **creodonts.** The principal types of mammals are the Archaeotherium (giant pig), Poebrotherium (ancestor of the camels), Mesohippus (early horse), Hyracodon (cursorial rhinoceros), and Hoplophoneus (progenitor of the sabertoothed cats). (See also **Fossil Mammals.**) For mineral resources of this period see the **Tertiary.** (R.M.F.)

OLIGOCHAETA. Chaetopoda.

OLIGOCLASE. Feldspar.

OLIGOTRICHIDA. Ciliophora.

OLIVE. *Olea europaea.* Oleaceae. The olive is a small tree indigenous to the eastern Mediterranean region. It has **lanceolate** evergreen leaves, small inconspicuous flowers, and a purplish **drupe,** the flesh of which is very bitter in the natural state. Cultivation of the tree has continued through many centuries, gradually spreading not only to all Mediterranean countries, but abroad to suitable regions both in the Old and the New Worlds. In the United States, olive growing is largely restricted to California, and even there it is not a major crop. The wood of the tree is used to a limited extent.

The principal product is the oil, which is expressed from the flesh of the fruit. To obtain this oil the fruit is picked when ripe and usually allowed to dry a bit to remove some of the water contained in the flesh. Pressing freshly picked fruits yields a much higher grade of oil. Pressing is often done in a rather primitive machine. The fruit is first crushed and then firmly pressed. The best grade of oil is known as virgin oil.

Olive oil is widely used as a food or in the preparation of food, owing to its lack of objectionable taste. Cheaper grades are used in soap-making. Sardine packers require large quantities for packing their product. In recent years olive oil has been much adulterated with cottonseed, sesame, peanut, and other oils. (See **Esters.**) (R.M.W.)

OLIVENITE. The mineral olivenite is a complex basic **arsenate** of **copper** with the formula $Cu_3As_2O_8 \cdot Cu(OH)_2$. It appears as prismatic, sometimes acicular **orthorhombic** crystals or may be fibrous to granular and earthy. It is brittle; hardness, 3.; specific gravity, 4.1–4.4; luster, vitreous; color, usually some shade of green but may be brownish, less frequently yellowish or grayish; translucent to opaque. Olivenite is a rare secondary mineral found associated with other copper minerals in various localities in France, England, South West Africa, and in the United States in the Eureka and Tintic mining districts of Utah. The name olivenite is derived from its olive-green color. (E.S.C.S.)

OLIVINE. The mineral olivine is an **orthosilicate** of **magnesium** corresponding to the formula $(Mg,Fe)_2SiO_4$, in which the ratio of magnesium to iron is found to vary considerably. Olivine crystallizes in the **orthorhombic** system in somewhat flattened forms but may occur massive or granular. It has a conchoidal **fracture** and is rather brittle; the hardness is 6.5–7; specific gravity, 3.27–3.37; luster, vitreous; color, olive-green; may be reddish from the oxidation of the iron. It is transparent to translucent. Olivine occurs both in **igneous rocks** as a primary mineral as well as in certain rocks of **metamorphic** origin. It has also been discovered in **meteorites.**

Olivine crystallizes from **magmas** that are rich in magnesia and low in silica which form such rocks as **gabbros, norites, peridotites** and **basalts.** The metamorphism of impure **dolomites** or other sediments in which the magnesia content is high and silica low seems to produce olivine.

Transparent olivines of good color are sometimes used as a gem, often called *peridot,* the French word for olivine; it is also called chrysolite from the Greek meaning gold, and stone. Olivine occurs in the lavas of Vesuvius and Monte Somma and in the Eifel district of Germany. Gem material comes from St. John's Island in the Red Sea, Upper Burma, and from Minas Geraes, Brazil. In the United States olivine localities are Orange County, Vermont; Webster and Jackson Counties, North Carolina. Arizona and New Mexico have also furnished some gem material. (E.S.C.S.)

OLM. Amphibia, Urodela. A European **salamander** with a long snake-like body, very small legs, and tufted external gills. It is found in subterranean waters, and in correlation with this habitat has rudimentary eyes. (A.W.L.)

OMMATIDIUM. The structural unit of the compound eye of **arthropods.** (A.W.L.)

ONAGER. Mammalia, Perissodactyla. A wild **ass,** *Equus onager,* of western India. (A.W.L.)

ONCHOSPHERE. An early embryo or **larva** of the tapeworm (**Cestoda**). It is a small rounded organism with six hooks which hatches from the egg after it is taken into the alimentary tract of a host. After hatching the onchosphere lodges in various tissues, depending on the species to which it belongs, and develops into the bladder worm or another intermediate stage. (A.W.L.)

ONION. Allium.

ONYCHOPHORA. Small, soft-bodied, creeping animals, slightly like caterpillars in appearance. They are of limited distribution in warm countries and have no common name. The name of one genus, *Peripatus,* is sometimes applied indiscriminately to all members of the group.

These animals form a class of the phylum **Arthropoda** characterized as follows: 1. The body is metameric but segments are not distinctly marked externally. 2. The head consists of three segments but is not distinctly separated from the body. 3. The first segment bears a pair of **antennae.** 4. The mouth is provided with a pair of jaws and a pair of oral **papillae.** 5. Each segment of the body bears a pair of legs which are fleshy protuberances ending with a pair of claws. 6. Each leg contains an excretory tubule. 7. Respiration is carried on by air tubes or **tracheae** whose external orifices are irregularly placed. 8. Unlike other arthropods, these animals have **cilia** in the alimentary tract and reproductive system.

Peripatus.

Onychophora are regarded as the most primitive of the terrestrial arthropods. Their structure suggests the ancestral form of the insects. (A.W.L.)

ONYX. Agate.

ONYX MARBLE. A trade term for a variegated, crystalline **limestone** which resembles the cryptocrystalline variety of **quartz** called **onyx.** Not a true **marble** since it is not a **metamorphosed** limestone. Onyx is usually formed as encrustations around springs which

are particularly rich in **calcium** carbonate. Much used for interior ornamental purposes. (R.M.F.)

OOECIUM. A brood pouch in which the **embryo** develops in **bryozoans.** Nourishment is supplied during this period by specialized cells in the walls of the ooecium. Regarded as a modified **zooecium** and also called the ovicell. (A.W.L.)

OÖGONIUM. Gamete.

OÖLITE. The term is from the Greek meaning egg and stone. Oölites are well-rounded sand-like particles, originally formed of calcite but sometimes subsequently altered to either dolomite, or entirely silicified. Structure typically concentric about a nucleus, and often with radial lines. In the accompanying figure, A and B (typical oölites) and C are probably the excrement of worms. Oölites are relatively common constituents of

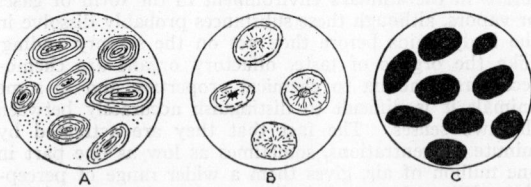

Different types of oölites. *A* and *B* typical oölites, *C* probably the excrement of marine worms. (*Field, Outline of Geology, Princeton University Press.*)

limestones, often forming distinct beds. Oölites are now forming on the shores of Great Salt Lake, but no authentic cause is known of marine oölites being formed at the present time. Coarse-grained oölites, in which the particles are about the size of peas, are called pisolites, from the Greek, meaning pea and stone. (R.M.F.)

OÖLOGY. A division of biology which deals with the eggs of birds. It is scarcely a science, but is rather a part of **ornithology.** (A.W.L.)

OÖSPORE. An oöspore is a thick-walled plant **cell** formed by the union of two unlike **gametes.** It is therefore a fertilized egg surrounded by a special protective coating. Certain **algae** such as Vaucheria have oöspores. (R.M.W., B.S.M.)

OÖTYPE. A region of the genital passages of flatworms (**Platyhelminthes**) at the junction of the oviducts and vitelline ducts. At this point the walls are thickened and a special secretion aids in forming the shell of the egg. (A.W.L.)

OOZE. Ooze is a general term used to designate the mud found on the ocean bottoms at abyssal depths and composed largely of the calcareous and silicified shells of minute surface living marine organisms, called **plankton.** (R.M.F.)

OPAH. Pisces, Teleostei. A beautifully colored Mediterranean and Atlantic fish (**Pisces**), *Lampris luna,* also called the sunfish. It reaches a length of 4'. (A.W.L.)

OPAL. The mineral opal is amorphous hydrated **silica,** the water content being sometimes as much as 20%. It has, of course, no crystal form, occurring as irregular **veins** and masses. It has a conchoidal fracture; hardness 5.5–6.5; specific gravity 2.1–2.3; luster, vitreous or greasy to dull; color, very variable, colorless, white, milky-blue, gray, red, yellow, green, brown and black. Often a beautiful play of colors may be observed in the gem varieties. These interference colors result from minute cracks in the opal which are filled with secondary silica. It has also been suggested that this effect is due to the physical arrangement of thin lamellae in the opal during its solidification. Besides the gem varieties which show the delicate play of colors, there are other sorts of common opal such as the milk opal, a milky bluish to greenish kind; resin opal, which is honey-yellow with a resinous luster; wood opal, resulting from the replacement of the organic matter of wood by opal, and **hyalite,** a colorless glass-clear opal sometimes called Muller's Glass. Opal is a mineral gel, which is deposited at relatively low temperatures and may occur in the fissures of almost any type of rock. Hungary, Australia, Honduras, Mexico and in the United States, Nevada and Idaho, have been the sources of gem opals. Hyalite comes from Czechoslovakia, Mexico, Japan and British Columbia. Other common varieties of opal are widespread in their occurrence. The word opal is derived from the Latin *opallus*. (R.M.F.)

OPEN DELTA.

This is a three-phase transformer connection using two single-phase **transformers** connected to form a V or open **delta** across the three lines. Such a connection has about 58% of the capacity of full delta using transformers of the same rating. It is often used for temporary work anticipating a later completion of the delta or for emergency service when one transformer of the complete delta requires servicing. (L.R.Q.)

OPEN HEARTH PROCESS.

The open hearth process developed by Sir William Siemens is the principal method for making **steel** in the United States. This process is briefly described as follows: A rectangular covered furnace is provided with a hearth in the shape of a large shallow dish in which is charged **pig iron,** which may be either cold "pigs" or molten pig iron direct from a blast furnace, scrap, and iron ore, together with limestone. The charge is heated and melted at a very high temperature. The source of this high temperature is the combustion of fuel which may be oil, powdered coal, natural gas, coke-oven gas, or producer gas, for which the air supply has been highly preheated. The high temperature in the furnace, coupled with the oxidizing ore, and the presence of considerable excess air in the furnace, brings about the oxidation of part of the carbon and of other elements which are to be removed from the iron. In this respect the action of the open hearth furnace is seen to be the same as the **Bessemer** process, which is one of oxidation. The chemical reactions are under better control in the open hearth process, and a greater variety of steels can be produced than in the Bessemer furnace. The loss of iron is lower, and the product is of higher quality. For these and other reasons, the open hearth process is the leading method of steel-making in the United States.

The secret of obtaining the high temperatures necessary for the successful operation of an open hearth furnace lies in the regenerative heating of the air for combustion. The furnace is equipped with ports at both ends, through which the preheated air enters the melting chamber. The fuel, either gas or oil, is introduced through high-pressure burners located in the ends of the furnace. The furnace is fired from one end at a time, and the heated gases pass out of the ports at the other end into heat exchangers, which are large chambers in which is set up a checker work of brick. An open hearth furnace is fired alternately from either end at about 15-min. intervals, and there are two sets of the checker-work chambers, in one of which air is being heated and brick being cooled, while in the other bricks are being heated by the products of combustion of the fuel.

There are two modifications of the open hearth process which differ in the kind of materials used for refractories, the composition of the charge, and the slag produced.

The basic open hearth process uses raw materials high in sulfur and phosphorus which are removed by a "basic" slag composed mostly of lime, iron oxide, and silica, the lime being usually at least twice the silica content. The refractories in contact with the molten bath and slag are predominately magnesia. In the acid open hearth process, the raw materials are low in sulfur and phosphorus. The "acid" slag formed is low in lime and high in silica. The furnace refractories are predominately silica.

The furnace itself is made of brick having steel reinforcement on the sides and ends. The hearth is made of basic or acid materials laid on a concrete foundation, and may be either stationary or tilting. The open hearth furnace may have a capacity of 15–350 tons of metal, a common size being 200 tons. It is about 80′ long by 20′ wide, and the pool of metal is generally about 3′ deep. The side walls are laid up with first quality silica brick, as is also the arched roof. In addition to the ports previously mentioned, the walls are pierced by several charging doors, with slag holes, and by a tapping hole. The slag holes are placed at the normal upper level of the bath, and the tap holes at the bottom so that all of the iron may be drained.

The manufacture of steel by this process requires, in addition to the furnace, a large amount of other equipment. There must be calcining cupolas to produce the calcined dolomite used for lining ladles and hearths; storage bins for limestone, ore, pig iron, scrap; conveyor systems; machines for dumping measured amounts of materials into the furnaces; ladles and cranes; ingot molds; ingot stripping machines; hot metal mixers; blast furnaces for producing molten pig iron; hot metal mixers in which molten pig iron may be held until needed; and frequently, gas producer plants.

The actual making of steel by the open hearth process can be described as follows. When the furnace is ready for the heat, the limestone is charged, then the scrap, ore, and cold pig iron. The combustion is stepped up to its maximum rate, and the melting down of the charge begins. If the heat is one in which some of the charge is molten iron which has been obtained from an auxiliary cupola, blast furnace or mixer, it is not added until the melting is nearly finished. After a few hours the charge lies molten on the hearth, and the purification of the iron is carried out by oxidation. Due to the presence of the iron oxides, the limestone, and the excess oxygen in the combustion region, the action of the ore definitely precedes that of the limestone, and the reaction in the furnace has two definite stages, the one known as the ore boil, the other as the lime boil. During the ore boil, usually occuring shortly after the addition of hot metal from the blast furnace, considerable slag is "flushed" from the furnace by the violence of the reaction of the ore with carbon to form carbon monoxide. When the carbon is nearly oxidized out of the bath, the furnace men take samples of the metal, which are cast in small molds. These test bars are broken, and the experienced melter can tell from the appearance of the fracture what the approximate carbon content was at the time of sampling. More accurate control of the carbon content is obtained by rapid chemical analysis or by special magnetic tests on a rod-shaped sample.

There are two methods of making steel. One is to estimate the rate of elimination of the carbon and to tap the heat when the carbon content has dropped to that which is desired in the steel. The other is to reduce the carbon content to about .10% and recarburize in the ladle with Spiegel, coal, or graphite. When the iron approaches the final state desired, the temperature of the pool of metal must be raised to that which is proper for pouring. This should be possible by controlling the furnace carefully during the period just preceding the tapping, but sometimes it is necessary to add pig iron, whose carbon content will give some slight boil which will increase the temperature to that suitable for pouring (about 2900° F.). The heat (molten metal) is then tapped into a large crane ladle, and additions of manganese, carbon, and other alloys are made in the ladle, depending on the type of steel desired. Exceptions to this procedure are to be noted in the case of nickel and copper, both of which are added in the furnace. After thus finishing the steel, the ladle is carried over the mold track, and a train of cars, each carrying an ingot mold, is brought up under it. The molten steel is then teemed into the ingots one by one, after which the ingot train moves off to the stripper, which removes the molds from the red hot ingots. (F.T.M., R.H.H.)

OPEN TRAVERSE. Traverse.

OPEN-BILL. Aves, Ciconiiformes. *Anastomus.* Birds (**Aves**) of Indian and African species related to the storks. The upper **mandible** is straight and the lower curved, so that complete closure is impossible. (A.W.L.)

OPERA GLASS. Telescope; Binocular.

OPERATIONAL DIVISIBILITY. The segregation or division of simultaneous operations on an automatic or semi-automatic chucking machine to obtain approximately equal processing time on all work or tool stations. To illustrate, suppose a part machined on an automatic turret lathe involves facing, drilling, and boring operations requiring one minute each, and a turning operation that requires two minutes. Since the total time per piece on a machine of this character is based upon the time of the longest operation, the turning operations would probably be handled at two adjacent work stations in two stages of one minute each, thus making the actual production time one minute per piece, instead of the two minutes that the turning operation would necessitate. (See **Screw Machine.**) (H.C.H.)

OPERCULUM. A flap covering an opening. 1. An enlarged branch of a tentacle found in some of the marine **annelid** worms. It closes the mouth of the tube when the animal is retracted. 2. The margin of the **zooecium** of **bryozoans** which closes in when the body is retracted. 3. A horny plate which fits into and closes the shell aperture of certain **gasteropod** and cephaloped mollusks when the body is retracted. 4. A plate formed of a pair of rudimentary appendages which covers the external apertures of the genital ducts of **scorpions.** 5. A hinged flap on the side of the head in most fishes (**Pisces**). It covers the entire series of gill slits. Also called the opercle. 6. The lid of the capsule in the mosses. (A.W.L.)

OPHICALCITE. An old, but still consistently used term, proposed by Brongiart in 1813, for a variety of crystalline **calcite** and **serpentine.** Frequently used as a decorative **marble.** (R.M.F.)

OPHITIC TEXTURE. A term proposed by Michel-Levy, in 1877, for the characteristic texture of **dolerites,** in which the **pyroxene** crystals are penetrated by laths of **plagioclase** feldspar. This type of texture differs from poikilitic in that in the latter type of texture the pyroxene crystals entirely enclose a **number** of laths of plagioclase. (R.M.F.)

OPHIURAE. Ophiuroidea.

OPHIUROIDEA. The brittle stars, a class of **echino-**derms resembling starfishes with a well-marked disk and slender arms. The class is distinguished chiefly by this sharp demarcation of disk and arms, which accompanies the restriction of visceral organs to the disk. In addition the tube feet are without suckers and the madreporite lies on the oral surface. (A.W.L.)

OPHTHALMIA. Conjunctivitis. Acute infection of the **eye** or the membrane around the eye. Ophthalmia of the newborn is a very severe form caused by **gonorrhea.** Until measures were taken for applying **antiseptics** to the eyes of all infants directly after birth, it was one of the great causes of blindness. (R.S.M.)

OPHTHALMOSCOPE. An instrument made up of lenses, and producing a beam of light by means of which the retina of the eye can be seen. It is one of the valuable diagnostic instruments as in many diseases significant findings are present in the eye grounds. (R.S.M.)

OPIATE. Any **drug** derived from **opium** or, as the term is commonly used, any drug which produces sleep. (R.S.M.)

OPISTHOBRANCHIATA. Gasteropoda.

OPISTHOSOMA. The abdomen of arthropods. (A.W.L.)

OPIUM. The dried milk-juice obtained from incising the unripe **poppy.** The principal opium-producing countries are in Asia Minor, India and China.

Opium contains a series of closely related **alkaloids,** but only three of these are of medical importance. These are (1) **morphine** and its related compounds marketed as dilaudid, dionine, heroin, pantapon, (2) **codeine,** and (3) papaverine. Morphine and codeine are central nervous system depressants. Papaverine relaxes smooth muscle, particularly that of blood vessels.

Opium is used as a powder, a tincture (laudanum), and a camphorated tincture (paregoric). Its main actions are those of morphine. It has a marked constipating effect which is made use of in the treatment of **diarrhea.** The smoking of opium is common in the Far East. This is the least pernicious form of opium addiction. (See **Poppy.**) (D.M.H.)

OPOSSUM. Mammalia, Marsupialia. Pouched **mammals** of moderate size, with long scaly tails and sharp noses. With the exception of the water opossum they are arboreal animals which eat insects or a general diet. The pouch is rudimentary in some species.

The common opossum, *Didelphys virginiana,* ranges from the Great Lakes to the Gulf and westward to Oklahoma. Other species range from Mexico to southern Brazil. The water opossum or yapock, *Chironectes,* of this region is a swimming animal which lives on fishes and other aquatic animals. (A.W.L.)

OPPOSITION. Planetary Motion.

OPTIC AXIS. Double Refraction.

OPTICAL ACTIVITY. Isomerism; Polarized Light.

OPTICAL INSTRUMENTS. Optical instruments may be divided into two general classes: (1) those used for optical projection, such as the stereopticon and the **camera;** and (2) those used as an aid to natural **vision,** such as the **telescope** and the **microscope.** In the first class, a real image of the object to be represented is formed on a screen or photographic plate by means of a lens system or a mirror. In the second, the eye of the observer is placed so as to view a virtual image formed by the optical system as a whole, which may or may not involve the formation of a real image in the interior of the instrument. In either type there is an objective lens or mirror, and in the second class the final virtual image is formed by an **eyepiece** or ocular. There are also usually one or more "stops," which are opaque screens with circular openings serving to limit sharply the field of view and incidentally to intercept stray light. Some cameras and projectors have stops of adjustable diameter. (See **Geometrical Optics, Magnifying Power,** and **Vision.**) (L.D.W.)

OPTICAL LEVER. A common device for amplifying and measuring small rotations. The object rotated carries a small mirror, which, reflecting a beam of light, deflects it through twice the angle of rotation to be measured. Light from a lamp is thus reflected as a bright spot moving along a scale, or the image of a fixed scale is viewed in the mirror by means of a reading telescope. The most common applications are to **galvanometers, electrometers,** etc., using a torsion suspension; but the principle is also often adapted to devices such as that used for measuring Young's modulus, and in situations where a **micrometer** might otherwise be used. (L.D.W.)

OPTICAL PYROMETER. Several **pyrometers** have been devised, by means of which the temperature of a very hot surface is determined from its incandescent **brightness.** One commercial type is illustrated in the figure. The hot body is viewed through a sort of tele-

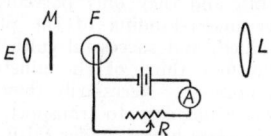

Arrangement of parts of optical pyrometer. (Diagrammatic.)

scope, whose objective L produces at F a real image of the glowing surface. At this point F is placed a lamp filament, which is thus viewed through the eyepiece E against the hot surface as a background. A monochromatic filter M is interposed before both, so that their brightness is compared in one spectral region only. The current in the filament is so adjusted by means of the rheostat R that the filament becomes invisible against the bright background. The ammeter A then gives the current, from which the temperature may be deduced; or the ammeter scale may be graduated to read temperatures directly. In another type the balance is secured by keeping the current constant and introducing an absorbing wedge between the filament and the objective, as in a **wedge photometer.** In still others, the temperature is determined, not by the total brightness, but by the relative brightness at two selected wavelengths. (L.D.W.)

OPTICAL ROTATION. Polarized Light.

ORAL CAVITY. The cavity usually called the mouth. It is formed in the **vertebrates** of an embryonic depression, the stomodaeum, which forms in the ectoderm of the under side of the head and unites with the embryonic gut just behind its anterior end. The depression is deepened by the growth of processes from the body wall at the level of the pharynx which form the upper and lower jaws. Later the olfactory pits break through to join it and in this stage, which persists in the **amphibians,** the cavity is common to the **respiratory** and **digestive systems.** In a more advanced stage shelf-like partitions grow out from the lateral walls and join to form the palate, which divides the cavity into respiratory and oral portions, as in man. Also called the buccal cavity. (A.W.L.)

ORAL FUNNEL. The depression leading to the mouth in the lampreys (**Cyclostomata**). (A.W.L.)

ORAL GROOVE. A shallow depression in the surface of the one-celled animals resembling *Paramecium.* It is

lined with **cilia** which create a current along it toward the cytopharynx at the rear, where food particles are gathered together to be taken into the body. (A.W.L.)

ORANGE. Citrus Fruits.

ORANGITE. Thorite.

ORANG-UTAN. Mammalia, Primates. A large man-like **ape** found in Borneo and Sumatra. It reaches a height of less than 5′ and a weight of about 150 lbs. The sharply depressed bridge of the nose accentuates the prominence of the rounded muzzle. This feature, together with the weak legs, long arms, and prominent abdomen, gives the species a grotesque appearance.

Orangs have been kept in captivity more successfully than gorillas but they also are delicate in the severe climates of Europe and America. They have been found very intelligent, but they are less desirable subjects for study than chimpanzees because of their uncertain temper. This is especially true of older males.

The name is Malayan for man of the woods. (A.W.L.)

ORBICULAR STRUCTURE. The term proposed by Delesse, in 1849, for concentric shells of different minerals, frequently formed around a **xenotithic** nucleus, in **granites, diorites** and other granitoid, intrusive igneous rocks. Synonyms for orbicular (in the above sense) are spheroidal and nodular. (R.M.F.)

ORBIT. In anatomy the orbit is the bony depression in the **vertebrate** skull in which the **eye** is seated.

The path which a celestial object follows in its motions through space, relative to some selected point, is known as the orbit of the object. The solution of the **two body problem** indicates that, in the case of two objects moving under the influence of their mutual gravitational attractions, the relative orbit of one to the other will be a **conic section.** The character of the conic will depend upon initial conditions. In the case of members of the **solar system,** because of the relative very large mass of the sun in comparison with any of the other members, the orbits of the members may be conveniently represented as ellipses with the sun at one focus. In the case of the **satellites** of the various members the orbits of the satellites may be represented as ellipses with the primary object at one focus. Any deviations from the two body problem may be treated as departures of **perturbations** from the simple conic.

To define completely the orbit of an object at any particular instant and to permit the determination of

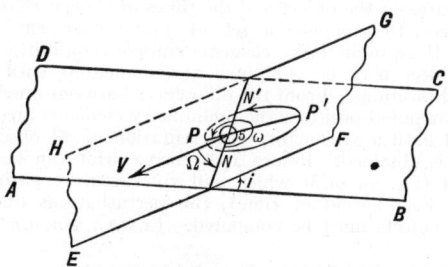

Diagram of orbit of member of solar system.

the position of the object in this orbit at any future time, six quantities are necessary, known as the seven elements of the orbit. In the accompanying figure we have a diagram of a planetary orbit about the sun. The plane $ABCD$ represents the plane of the ecliptic, and the plane $EFGH$ represents the plane containing the orbit. These two planes must intersect in a line, NN', which must pass through the sun since both the earth (the plane of the ecliptic being the plane of the earth's orbit) and the planet are revolving about the sun. The elliptical orbit of the planet is represented by $PNP'N'$ with the direction of motion of the planet as indicated, PP' representing the major axis of the ellipse, and P the point where the object is closest to the sun (the **perihelion** point). The line SV represents, in the plane of the ecliptic, the direction to the vernal equinox. We will assume that we are looking down on the plane of the ecliptic from the direction of its north pole. The points N and N', where the planet is on the ecliptic, are known as the nodes; the point N, where the planet is passing from south to north of the ecliptic, being the ascending node and the point N' the descending node. To define the orbit plane the orbital element, i, the inclination, gives the angle between the orbit plane and the plane of the ecliptic; while the element, Ω (the angle VSN), gives the celestial **longitude** of the ascending node. The size and shape of the orbit in its plane are given by the orbital elements, a, semi-major axis (usually expressed in **astronomical units**) and, e, the eccentricity of the **conic**. To locate the position of the conic in the orbit plane we use either ω (the angle NSP measured in the direction of motion of the planet), or, more commonly, $\bar{\omega}$, the longitude of **perihelion** which is the sum of the two angles, Ω and ω. It should be noted that Ω is measured in the plane of the ecliptic while ω is measured in the orbit plane so that $\bar{\omega}$ is not strictly a longitude. To locate the position of the object in its orbit the element, T, which is some epoch or date when the object is at perihelion and the rate at which the object is moving in the orbit are necessary. Occasionally we find the rate of motion of the object given as an orbital element, but it is strictly not an independent element, since it may be derived from a. Listing the six elements as defined above we have, i, Ω, a, e, $\bar{\omega}$, and T. Unfortunately, several different systems of symbolic notation have been used by different authors and great care must be exercised in interpreting any symbolic description of an orbit.

In describing the orbits of satellites the plane frequently used is the plane of the planet's orbit about the sun and the inclination, i, of the satellite orbit is referred to it. Also, instead of locating the position of the orbit in the plane by perihelion, the point where the satellite is closest to the planet is used (in the case of satellites of Jupiter the point is known as perijove). In the case of **binary star** orbits, i gives the inclination of the orbit plane to the plane perpendicular to the line of sight, and instead of perihelion the point where the smaller star is closest to the primary, the periastral point, is used.

The problem of orbit computation is far too complex to be included in a work of this character. In general three observations of an object, giving the **spherical coordinates** of the object and the times of observation, are sufficient to compute a set of preliminary elements. From these preliminary elements subsequent positions of the object may be computed and compared with observed positions. From the differences between observed and computed positions the preliminary elements are corrected until a satisfactory representation of all observations is obtained. Before a definitive orbit can be obtained (i.e., an orbit which will give accurate positions for a long period of time), the perturbations due to other objects must be computed. (A.W.L., W.K.G.)

ORBITAL MOTION. Two Body Problem.

ORCHID FAMILY. Orchidaceae. The Orchids, which form the second largest family of plants, are the highest development of the **Monocotyledons**, just as the **Composite Family** marks the highest point of evolution reached by the **Dicotyledons**. But between the two families a most striking contrast exists. The Composites are most beautifully formed to enjoy a more abundant life; the numerous small flowers are massed in compact heads, so pollination is almost certain of accomplishment. The reduction of the fruit to a single

ovule and the presence of various barbs or scales or bristles, called the pappus, greatly favor the probability of successful continuance of the species. In the 9000-10,000 species of Orchids, the individual flowers are usually very conspicuous objects often of bizarre form and rare beauty. But they depend entirely on insects for **pollination** and often exhibit elaborate modifications to insure successful insect pollination. The seeds are minute and borne in tremendous numbers, few of which germinate and grow to maturity. As a result, orchid plants are relatively rare, a fact which has contributed not a little to the zeal with which collectors have sought these plants.

Originally orchids were known only from the reports brought back by travelers from the tropics who spoke of the brilliant colors, the curious forms and the delightful fragrance and also of the mystery and folk-lore which often attached to orchids. Later botanists gained a wider knowledge of the family from dried specimens, in which color, fragrance, and, to a considerable extent, form, were lacking. In time, however, living plants were obtained by collectors and carried to western lands. There they were grown, usually without much success, since it was assumed that they could grow only in a very hot humid atmosphere. Only the most tolerant species stood this, and they only partially. Gradually, however, better understanding of the plants' requirements was obtained, and successful culture followed, so that during the first third of the nineteenth century orchids became popular. Necessarily they are expensive to collect, they are not easy to transport, and they can be grown only in glass houses under fairly uniform conditions of temperature and humidity.

Orchids are primarily plants of the tropical rain forests, where they grow in greatest abundance. In these forests they are found mostly as **epiphytes**, plants growing on other plants. Such plants grow attached to the branches of large trees, or massed in a crotch of the tree, or even attached to the trunk. Often they occur high up on the topmost branches of lofty trees, where they are inaccessible except when the supporting tree falls, bringing all its attached plants to earth. Other species of orchids, including nearly all those found in extra-tropical regions, are terrestrial. All orchids, wherever they grow, are herbaceous perennials. A few species are **saprophytes**, lacking **chlorophyll** entirely, and obtaining their food by absorption from the soil of complex organic substances.

In terrestrial orchids the roots are rather coarse and sparsely branching. In many species one of these roots becomes greatly swollen with food, as the growing season advances, forming a tuberous body which will be used to promote rapid growth in the following season. n epiphytic orchids two kinds of roots are found. One of these grows tightly appressed to the supporting plant, is not affected by gravity but is negatively **phototropic**, growing into crevices in the supporting plant. The other roots are the aerial roots, coarse branching objects which hang down, often in conspicuous masses, from the base of the plant. The epidermis and the outer portion of the **cortex**, which is called the velamen, are composed of dead cells with perforated walls which readily absorb any water which may come to them and retain it tenaciously. The inner tissues of the cortex are green and capable of carrying on **photosynthesis**. In some orchids slender absorbing branches grow out from the base of the other roots and penetrate the mass of debris which frequently collects at the base of the plant.

The stems of terrestrial orchids are erect and leafy, and terminated by the **inflorescence**. In many epiphytic forms the leaves are dropped at the end of the growing season, the bare stem remaining. The internodes of such stems are often conspicuously swollen, forming pseudobulbs in which are stored water and food reserves. In other epiphytic species the leaves are fleshy and serve for storage.

The most characteristic and in many species the most conspicuous part of the plant is the flower. In all orchids the flowers are irregular but formed on a very uniform pattern. In all there is a 6-parted **perianth** composed of two groups of three members each. One of the inner three differs from all the others and is designated the lip or labellum. This becomes variously fringed in some orchids, broadly expanded in others, and even saccate (slipper-like), as in the Lady's Slipper. It is commonly much more brilliantly colored than the other five parts and gives to the flower its showiness. It serves as a landing place for insects, and is often a definite factor in bringing about pollination. Often various outgrowths such as spurs add further complexity to the flower. The essential organs of the flower, the **stamens** and **pistil**, are united into a single body, the column, which is a characteristic feature of the orchid flower. Its structure varies somewhat in the different groups of the family. In orchids there are only one or in some species, two, **anthers** present. The greater number of species have a single anther, which has two lobes, each filled with a mass of pollen grains held together by fine elastic threads. Such a mass is called a pollinium. The threads all unite to form a slender stalk which ends in a sticky mass resulting from the breaking down of certain cells in the rostellum. The latter is a special organ which represents one of the **stigmas** of the flower.

An insect comes to the orchid flower seeking the nectar which it contains. In getting this nectar the insect comes in contact with the sticky mass of cells which become firmly cemented to some part of the body, often the eyes, or the **antennae**. When the insect leaves the flower, it drags the pollinium out. The latter may be in a position so that when the insect enters the next flower, the pollen comes directly in contact with the stigma and pollination is accomplished. In many cases, however, a striking change occurs. The stalk of the pollinium, due to changes in its water content, bends through an angle of 90°, and so brings the pollen mass into a position which will insure its reaching the surface of the stigma of the next flower visited by the insect. There are many, often complex, variations in this process of pollination in orchids, all on this general method. In the second group of orchids the pollen is not combined into masses.

The **ovary** of the orchid flower is inferior; that is, all the floral organs are borne at the apex of the ovary, which contains an immense number of ovules attached to its walls. After fertilization, the ovules develop into very minute, light seeds, which are easily blown about by air currents.

Raising of orchids from seed is a very difficult problem, in which the ordinary grower is seldom successful. Development from seeds is very slow, several years being required to bring the plants to maturity. Propagation is usually by means of cuttings. Orchids are commonly grown for the beauty of the flowers, or for the weird shapes and striking colors which the latter often show. Only one of them, **Vanilla**, is of any great importance commercially. In times past many orchids were reputed to have medicinal value.

Common orchids found in the United States are the *Cypripediums* or Lady's Slippers, the *Habenarias* or Fringed Orchids, *Spiranthes* or Lady's Tresses, *Epipactis*, the Rattlesnake Plantain, and *Pogonia*, the Grass Pink. (R.M.W.)

ORCHITIS. Inflammation of the testicle. This may occur in **mumps, gonorrhea, syphilis** or **tuberculosis** or it may result from injury. The disease is marked by pain, a sense of weight or fullness, and swelling of the scrotum. (D.M.H.)

ORDER. A taxonomic division of the class. (See **Classification** and **Nomenclature.**) In this work the orders of the animal kingdom are listed with brief notes or for reference to the larger groups under which they are discussed.

Orders are characterized by important details of structural development and are the least of the major taxonomic divisions. The families into which they are divided are in many cases determined by minor evolutionary trends. (A.W.L.)

ORDINARY DIFFERENTIAL EQUATIONS. An **equation** which connects a **function** y of a single **independent variable** x with its **derivatives** of the first n orders, of the form:

$$F\left(x, y, \frac{dy}{dx}, \frac{d^2y}{dx^2}, \cdots, \frac{d^ny}{dx^n}\right) = 0,$$

where F is a polynomial function, is called an ordinary differential equation and its order is defined to be n.

If several functions $y, z, \cdots$ are connected with one another and their derivatives by as many equations as there are functions, we have a system of ordinary differential equations; as for example,

$$\frac{dy}{dx} = f(x, y, z) \quad \text{and} \quad \frac{dz}{dx} = \phi(x, y, z).$$

By a solution or a primitive or an integral of an ordinary differential equation in x and y is meant a function $y = f(x)$ which satisfies the equation.

The most general function which satisfies a differential equation of order n contains n arbitrary constants. A solution containing n arbitrary constants is called the general solution. If special values of the constants are used in the general solution, we obtain particular solutions. An equation may sometimes have a singular solution, which is a solution of the equation which cannot be obtained from the general solution for any particular values of the constants; a singular solution contains no arbitrary constant.

In applied problems, we are in most cases concerned with particular solutions. The determination of the general solution is usually a necessary preliminary step, after which the required particular solution is found by determining the arbitrary constants by using given initial conditions.

The order of a differential equation is the order of the highest derivative that occurs in it.

When an ordinary differential equation is rational and integral with respect to all the derivatives that occur in it, its degree with respect to the derivative of highest order is called the degree of the equation.

The geometric interpretation of an ordinary differential equation of the first order is as follows: Any differential equation $F(x, y, y') = 0$, of the first order and n^{th} degree, represents an infinite set of curves, n of which (real or imaginary) will pass through any assigned point (x_1, y_1), and will there have the n slopes obtained by solving the differential equation algebraically for y'. (L.L.S.)

ORDINARY DIFFERENTIAL EQUATIONS OF FIRST ORDER AND FIRST DEGREE. A differential equation of the first order and degree $\frac{dy}{dx} = f(x, y)$ may be written in the differential form $M\,dx + N\,dy = 0$, where M and N are functions of x and y.

The simplest type of equation here is the separable case. If it is possible to separate the variables and write the equation in the form $f(x)\,dx = g(y)\,dy$, it may be integrated at once.

If in the differential equation $M\,dx + N\,dy = 0$, the functions M and N are **homogeneous functions** of x and y, the equation may be solved by making the substitution $y = vx$, for the resulting equation in v and x is then separable.

To solve an equation of the form

$$(ax + by + c)\, dx + (\alpha x + \beta y + \gamma)\, dy = 0,$$

make the substitution $x = x' + h$, $y = y' + k$, choose h and k so that $ah + bk + c = 0$, $\alpha h + \beta k + \gamma = 0$, then the equation becomes homogeneous, unless $a/b = \alpha/\beta$, in which case the substitution $z = ax + by$, or $z = \alpha x + \beta y$ will give rise to an equation in which the variables are separable.

A linear differential equation of the first order is an equation of the form

$$\frac{dy}{dx} + Py = Q,$$

where P and Q are functions of x (but not of y). Such an equation may be integrated by multiplying through by $e^{\int P dx}$ (an integrating factor). The solution may be written

$$y e^{\int P dx} = \int Q e^{\int P dx}\, dx + c.$$

The equation $\frac{dy}{dx} + Py = Qy^n$ may be reduced to a linear equation by the substitution $z = y^{1-n}$; it becomes

$$\frac{dz}{dx} - (n-1)Pz = -(n-1)Q.$$

A differential equation $M\, dx + N\, dy = 0$ is called an exact differential equation if the left member is an exact differential of some function of x and y.

The necessary and sufficient condition that $M\, dx + N\, dy = 0$ be exact is that $\frac{\partial M}{\partial y} = \frac{\partial N}{\partial x}$.

If the differential equation is not exact, it may be possible to find a function, in general involving x and y, such that when both sides are multiplied by it, the equation becomes exact; such a factor is called an integrating factor. (L.L.S.)

ORDINARY DIFFERENTIAL EQUATIONS OF FIRST ORDER AND HIGHER DEGREE THAN THE FIRST.

In dealing with these equations, it is customary to abbreviate the derivative by the letter p.

We may denote the given equation, of the first order, by $F(x, y, p) = 0$.

There are three cases that may be considered separately, depending on whether the equation $F(x, y, p) = 0$ can be solved algebraically for p in terms of x and y, or for y in terms of x and p, or for x in terms of y and p.

If $F(x, y, p) = 0$ can be solved for p in terms of x and y, we obtain thus several **differential equations of the first order and first degree**, which may each be solved for y in terms of x. If these solutions are $f_1(x, y, C) = 0, f_2(x, y, C) = 0, \cdots$, the general solution of $F(x, y, p) = 0$ may be written

$$f_1(x, y, C) \cdot f_2(x, y, C) \cdots = 0.$$

If $F(x, y, p) = 0$ can be solved for y in terms of x and p, we obtain one or more equations of the type $y = \psi(x, p)$. If we differentiate each such equation $y = \psi(x, p)$ with respect to x, we obtain a differential equation of the first order in p which does not involve y, and may be solved for p in terms of x, say $\phi(x, p) = C$. For any particular value of C, the equations $y = \psi(x, p)$ and $\phi(x, p) = C$ may be interpreted as **parametric equations** of a curve in terms of the parameter p; the set of all such curves represents the solution of the equation $y = \psi(x, p)$.

In a similar manner we may treat the case when $F(x, y, p) = 0$ can be solved for x in terms of y and p. Having found x in terms of y and p, we differentiate

with respect to y and replace $\frac{dx}{dy}$ by $1/p$. We may then proceed as before.

In differential equations of the first order and higher degree than the first, singular solutions may occur. Let $F(x, y, p) = 0$ be such an equation, and let $f(x, y, C) = 0$ denote its general solution. The equation $f(x, y, C) = 0$ represents a family of curves, which may have an **envelope**, i.e., a curve $\phi(x, y) = 0$ which is touched at each of its points by one of the curves of the preceding family. Then $\phi(x, y) = 0$ will be a solution of the differential equation, but it will not generally be included among the particular solutions obtained by assigning particular values to C in the general solution $f(x, y, C) = 0$. Such a solution $\phi(x, y) = 0$ is called a singular solution of the given differential equation.

A differential equation of the form $y = px + f(p)$ is called a Clairaut equation. Its general solution is $y = Cx + f(C)$. Another solution is $x = -f'(t)$, $y = tx + f(t)$, which is a singular solution. (L.L.S.)

ORDINARY DIFFERENTIAL EQUATIONS OF THE SECOND ORDER.

An ordinary differential equation of the second order may in general be written in the form

$$\frac{d^2 y}{dx^2} = F\left(x, y, \frac{dy}{dx}\right).$$

We shall here consider only very briefly several important special types, in which one of the variables is absent from the equation.

If the function y is lacking in the differential equation, put $\frac{dy}{dx} = p$, then $\frac{d^2 y}{dx^2} = \frac{dp}{dx}$, and the differential equation takes the form $\frac{dp}{dx} = \phi(x, p)$, which is an **ordinary differential equation of the first order** in p as a function of x. After finding p as a function of x, we may replace p by $\frac{dy}{dx}$, and again solve a first order differential equation to express y as a function of x.

If the independent variable x is lacking in the differential equation but y is present, put $\frac{dy}{dx} = p$, so that $F\left(x, y, \frac{dy}{dx}\right)$ becomes a function $\psi(y, p)$; then since $\frac{d^2 y}{dx^2} = \frac{dp}{dx} = \frac{dp}{dy} \cdot \frac{dy}{dx} = p\frac{dp}{dy}$, the given differential equation becomes $p\frac{dp}{dy} = \psi(y, p)$, which is a first order equation with p a function of y. After finding p as a function of y, we replace p by $\frac{dy}{dx}$ and solve another first order differential equation for y as a function of x.

A special device for differential equations of the type: $\frac{d^2 y}{dx^2} = f(y)$ is the following: Put $\frac{dy}{dx} = p$, then the equation becomes

$$\frac{dp}{dx} = f(y).$$

Multiply both sides by $2p$, then since $2p\frac{dp}{dx} = \frac{d}{dx}(p^2)$, and $2pf(y) = 2f(y)\frac{dy}{dx}$, we have $\frac{d}{dx}(p^2) = 2f(y)\frac{dy}{dx}$. Integrating, we get $p^2 = 2\int f(y)dy + c = \phi(y)$, say. Extract the square root, replace p by $\frac{dy}{dx}$, and integrate again by separation of variables, obtaining finally y as a function of x. (L.L.S.)

ORDINATE OF A POINT. Rectangular Coordinates in a Plane.

ORDOVICIAN. A period of the **Paleozoic Era.** Type locality, border of Wales and Shropshire, England. The formations of this system were first studied and described by Charles Lapworth in 1879. The Ordovician period began 440 million years ago and lasted for 60 million years. In 1874 J. D. Dana had already proposed a post-**Cambrian** system, the **Canadian.** This system has only been recently recognized in Britain (Northwest Highlands of Scotland) where it is still considered as upper Cambrian by some British geologists (1937). In 1911, E. O. Ulrich proposed a new system (period) called the Ozarkian, to include some of the upper Cambrian and some of the lower Canadian. There is still considerable doubt as to the systemic importance of the Ozarkian, especially as it appears to be absent in Great Britain. Ordovician formations are well exposed in North America due to the fact that a large part of the continent was submerged during this period. The marine sediments are of two principal types, limestones and shales, the former contain "shelly **fossils**" and the latter, principally **graptolites.** Thus the American Ordovician covers an enormous area and has an endless variety of local developments, for during the vast lapse of time which is included in the period, the shallow **epeiric seas** were continually shifting their positions, outlines and depths. The United States Ordovician contains no **igneous rocks** with the exception of volcanic ashes, called bentonites. The formations of this system are well exposed in Portugal, Switzerland, Bohemia, Austria, Hungary, Ireland, eastern Asia, China, Manchuria, Siberia, Himalayas, Burma, Morocco, Australia, New Zealand, northern Argentina, Bolivia, and eastern Peru. The maximum thickness of Ordovician strata, 40,000 feet, occurs in Australia. All classes of marine invertebrates occur as fossils, a number of which classes are now extinct. The principal types are **graptolites, corals, echinoderms, bryozoans, pelecypods, nautiloids, trilobites** and **ostracods.** (See also **Invertebrate Paleontology**). The strata of eastern North America are an important source of petroleum. Because of the wide surficial distribution of the Ordovician limestones they are much quarried for foundation structures, road metal, flux for the reduction of iron ores, and especially for the lime used in mortar, whitewash, fertilizers and Portland cement. In Vermont and Tennessee occur valuable deposits of **marble.** Lead and zinc ores in rocks of Ordovician age are mined in Iowa, southern Wisconsin and northern Illinois. Phosphates of the same age are mined in central Tennessee. In eastern North America the Ordovician closed with a period of mountain building named the **Taconic Revolution** by J. D. Dana, in 1895. (R.M.F.)

ORE. A mineral aggregate in which the valuable metalliferous minerals are sufficiently abundant to make the aggregate worth mining. Types and origins of ore deposits are illustrated at the top of the page. (R.M.F.)

ORFE, GOLDEN ORFE. Pisces, Teleostei. A domesticated variety of the ide, a European fish of the **carp** family. This form is golden-yellow. Introduced into England during the 19th century. (A.W.L.)

ORGAN. A multicellular structure made up of various **tissues** for the performance of some complex function. The stomach, for example, contains in its walls tissues of all of the five principal divisions. Its function is the digestion of food and this end is gained by the cooperative exercise of the simpler functions of all of the tissues composing it. (A.W.L.)

ORGANELLE. A specialized structure of the single **cell** composing the body of the **protozoans.** Organelles are comparable with the **organs** of multicellular animals

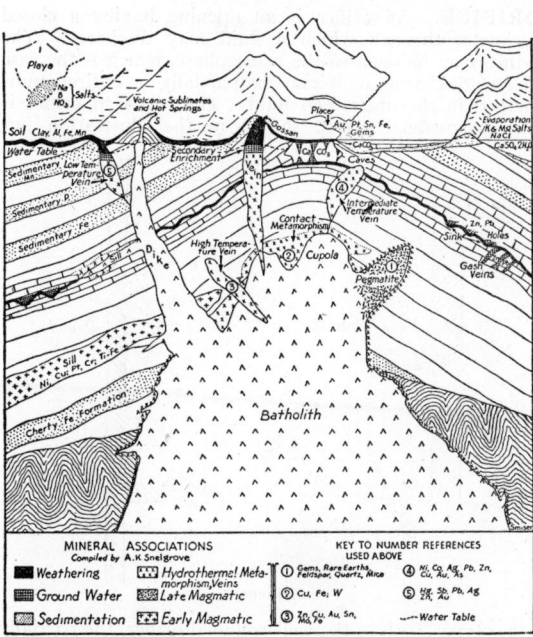

MINERAL ASSOCIATIONS
Compiled by A.K Snelgrove

KEY TO NUMBER REFERENCES USED ABOVE

Diagrammatic illustration of the origin of the ore deposits. (*Field, Outline of Geology, Princeton University Press.*)

in their relation to individual life but differ in their simplicity of structure. Also called cell organs. (A.W.L.)

ORGANIC CHEMISTRY. Chemistry.

ORGANIC DISEASE. In contrast to a functional disease, an organic disease is one in which definite structural changes take place in some organ or tissue. These changes may be temporary or permanent in character. (D.M.H.)

ORGANIZATION. The association of specialized parts in the body of an organism for the coordinated performance of various functions for the welfare of the entire individual. All living things, no matter how simple, show some degree of organization, hence it is regarded as a distinctive property of living matter. Some of the details of animal organization are taken up under **anatomy** and others under the various organ systems. (A.W.L.)

ORGANIZATION CHART. A graphical representation of the coordination, rank and function of industrial organization. An organization chart usually shows the varied personnel of an industry or plant by means of rectangular blocks, arranged in order of rank, with "leaders" or connecting lines or links to show their functional relationship. Flow sheets are similar in appearance to organization charts, but are usually employed to show the relationships and continuity of the various steps of industrial processes. (H.C.H.)

ORGANOGENY. The formation of **organs** in the embryo. After the formation of the **germ layers** and their initial stages of differentiation, each layer or its subordinate parts gives rise to certain of the organs characteristic of the adult by processes described under **embryology.** (A.W.L.)

ORIBI. Mammalia, Artiodactyla. *Ourebia.* South African **antelopes** of dainty build. They are less than 2′ high, with sharply pointed horns 4 or 5″ in length. The color is tawny above and white below. (A.W.L.)

ORIENTATION. Metallography.

ORIFICE. An orifice is an opening having a closed perimeter through which a fluid may discharge. The orifice may be open to the atmosphere, which is the case of free discharge, or it may be partially or entirely submerged in the discharged fluid. The standard orifice is the sharp-edged orifice shown in the illustration but

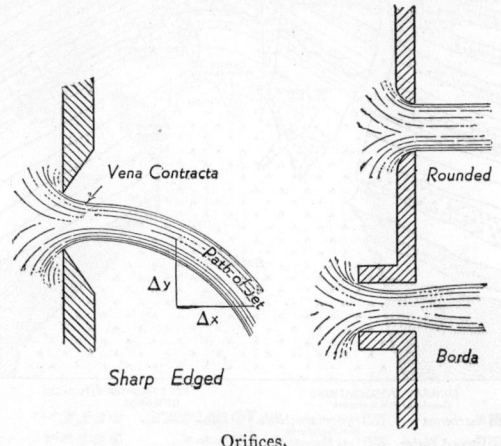

Orifices.

other types, such as the well-rounded orifice, the partially rounded orifice, the Borda mouthpiece, and the short tube orifice, have their special uses. An orifice may be very small, as in the case of those used for leak ports or for calibration, or large, as illustrated by sluice gates in a dam. The head of water on the orifice is measured from the water level surface to the center line of the orifice. Should the head above the orifice be so small as to be less than approximately the vertical dimension of the orifice, the following remarks will not apply, as this would come under the special case of large orifices under low heads.

The streamlines in water approaching a sharp-edged orifice converge on the orifice from all directions, and so continue to converge for approximately one-half of the orifice diameter downstream. The jet contracts to a section somewhat smaller in diameter than the orifice, after which it increases in size. The contracted section is known as the *vena contracta*. The ratio of the cross-section of the jet at the vena contracta to the area of the orifice is known as the contraction coefficient. Friction in the orifice slows the velocity to somewhat lower value than the ideal free spouting velocity which is $\sqrt{2gh}$. The ratio of actual to spouting velocity is the velocity coefficient. Since the discharge is the product of velocity and area, the discharge coefficient is the product of velocity and contraction coefficients. It has a numerical value of 61% for the average sharp-edged orifice. The discharge from an orifice of area a is

$$Q = .61a\sqrt{2gh}.$$

The path taken by a jet discharging freely horizontally under head of h is parabolic in shape due to the pull of gravity acting on a particle having, originally, horizontal motion only. The equation

$$x^2 = 4K_v^2hy$$

gives a curve of the path of the center of the path of the jet. K is the velocity coefficient which averages 98% for sharp-edged orifices. Suppression of the contraction of a jet increases the discharge from an orifice. An orifice on the side wall of a tank near the bottom has a higher coefficient of contraction than one which is located farther away from the bottom. Similarly an orifice with the upstream edges rounded has a higher coefficient of contraction than one with sharp edges. The discharge may be as much as 30% greater for well-rounded

orifices. Orifices which are submerged, orifices which are squared instead of circular, and orifices in which the water approaches with a high velocity, cannot be treated by the equation given above without corrections being made for these special conditions.

The above discussion relates to the flow of water through an orifice. Orifices are much in use for measuring flows of vapors and gases. The method employed is to place an orifice of some type in the pipe or duct carrying the fluid. By means of a **manometer,** or pressure gauge, the upstream and downstream pressures are measured, and the discharge can be determined from those readings coupled with the known area of the orifice. The flow of a gas through an orifice depends on the area of the orifice, the upstream pressure, the temperature, and a factor which involves gas constants, such as the ratio of the specific heats at constant pressure and constant volume, and the ratio of the upstream and downstream pressures. The formula for the weight of gas flowing is $\dfrac{C_1C_2aP}{\sqrt{T}}$ (pounds per second). C_1 is the constant just mentioned, C_2 the velocity of approach correction, P the upstream pressure, a the area of the orifice, and T the absolute temperature. Many steam flow meters are based on the principle of the orifice. A sharp-edged, or thin-plate, orifice is clamped between the flanges at some joint of a flanged steam line. Pressure leads are taken from upstream and downstream sections to an instrument which is a pressure-measuring device, but which may be calibrated to read steam flow. (See **Flow Meter.**) (F.T.M.)

ORIGINATION. In mechanical terminology, origination is the process of developing or producing a part without reference to any previous or existing standard. By reference to the definition of a straight-edge, it may be seen that three straight-edges may be originated by cutting and scraping; once the standard is established, however, any number of straight-edges may be made by machining and finishing them by reproduction, or comparison with the existing standard. (See **Lapping.**) (H.C.H.)

ORIOLE. Aves, Passeriformes. Brightly colored birds (**Aves**) of the family Oriolidae, found in all parts of the Old World. The North American birds to which this name is applied belong to the family Icteridae and are more closely related to the blackbirds than to the true orioles. The orchard oriole, *Icterus spurius*, and the Baltimore oriole, *I. galbula*, are the most widely known species of the latter group. (A.W.L.)

ORION. (Map, page 380.) This constellation is, on the whole, the richest and most impressive of all of the constellations. The "belt and sword" of Orion are frequently referred to in both ancient and modern literature and even found recognition as the shoulder insignia of the 27th division of the United States Army during both World Wars, probably because of the fact that the division was commanded by General O'Ryan. Despite the wide area covered, the physical characteristics of many of the stars are so similar that there is considerable evidence in support of the theory that they have a common origin.

The star Betelgeuse must be considered as an exception to this class for it is quite different from the other bright members of the constellation. Its color is a distinct yellow-orange as contrasted with the blue-white tint of the typical "Orion star." It is a **giant star,** its diameter, as measured with the interferometer, being about 300 times that of the **sun.**

The middle "star" of the sword is not a star at all, but is a huge gaseous **nebula.** This is one of the very few nebulae that can be seen with any satisfaction in a small instrument. Of course the larger the instrument the finer the view.

Several of the bright stars in the constellation are multiple objects, and the possessor of a 4-in. telescope will find the stars of Orion worth more than a passing glance. (W.K.G.)

ORNITHOLOGY. The biological science which deals with facts relating to birds (**Aves**). The large number of species and the great range of habits of birds have furnished abundant material for purely scientific studies, and the beauty and general interest of the group have made bird study a hobby of many persons. Economically birds are important as destroyers of insect pests, as game, and less frequently as predators. In these roles they have commanded the attention of the Biological Survey and of the many conservation boards and commissions of fish and game of the various states. (A.W.L.)

ORNITHOPTER. Ornithopters are the natural result of man's attempts to gratify his desire to fly. An ornithopter is a flying machine which, like a bird, flies by flapping its wings. It is quite logical that efforts should have been made to copy bird flight. The first attempts to fly were with ornithopters. One is recorded as having been built by Leonardo da Vinci in 1490, but, like the many ornithopters which have been built since that day, it was not a success. The ornithopter is too complicated mechanically, and there is not a close enough resemblance between the flight of birds and that with which man must be content to permit much success in the field of the flapping wing machine, even though flying ornithopter models have been built. (F.T.M.)

ORNITHORYNCHUS. Fossil Mammals.

OROGENY. Epeirogeny.

OROGRAPHICAL WEATHER PHENOMENA. Any weather phenomena caused by the flow of air over prominent features of the terrain. Air flowing uphill undergoes certain changes because of this orographical lifting.

1. As long as the air is unsaturated, it cools at the dry-adiabatic rate of approximately 5.5° F. per 1000'.

2. As soon as the air is saturated, it cools at the pseudo-adiabatic rate which depends on the amount of water being condensed by cooling, but in any event less than the dry-adiabatic rate.

3. Clouds, rain, or snow are often formed in this orographically lifted air. Thunderstorms occur if the air is unstable.

Down-slope warming such as occurs on the eastern slopes of the U. S. Rockies is thus of orographic origin. The Indian monsoon rains and the rains of the U. S. and European northwest coasts are large-scale orographic phenomena. (P.E.K.)

ORPIMENT. This mineral, like realgar its frequent associate, is an **arsenic** sulfide. Orpiment, however, is the trisulfide corresponding to the formula As_2S_3. It is **monoclinic** with a resinous to somewhat pearly luster; color, various shades of lemon yellow; translucent to nearly opaque. Orpiment is found in association with **realgar**, although a somewhat rarer mineral. It is believed to be formed from the alteration of other arsenic-bearing minerals. It occurs in Czechoslovakia, Rumania, Macedonia, Japan, and in the United States in Utah, Nevada, and Wyoming. The name orpiment is derived from a corruption of the Latin *auripigmentum,* meaning golden paint, because of its color as well as the belief that it contained gold. It was formerly much used as a dye, but its place has been largely taken by an artificial product of similar composition. (E.S.C.S.)

ORRIS ROOT. Volatile oils.

ORSAT ANALYSIS. Flue Gas Analysis.

ORTHITE. Allanite.

ORTHOCHROMATIC REPRODUCTION. The word orthochromatic is derived from the Greek roots *ortho* (correct) and *chroma* (color) and orthochromatic is loosely defined as the proper (or correct) reproduction of color in monochrome. However, since only one characteristic of color, i.e., brilliance, can be reproduced in monochrome, orthochromatic photography is more properly described as the correct reproduction of the brilliance characteristic of color into monochrome. For orthochromatic reproduction, the variation in the sensitivity of the photographic material to color should correspond with that of the eye. In other words, the distribution of sensitivity with wavelength of the film or plate should be the same as that of the eye, which, of course, is represented by the color-brilliance curve. It is obvious that, except for a limited color range, orthochromatic reproduction can be obtained only with a film or plate which is sensitive throughout the visible region, or from a wavelength of approximately 400 mμ to 700 mμ. Such materials are termed panchromatic: pan (all) color. Therefore, "orthochromatic films," which are widely used in photography, do not rightfully deserve this designation because, as they are insensitive to red, they do not reproduce red color values correctly. While considerable progress has been made in such materials within the last decade, it is not possible as yet to produce a film or plate whose variation in sensitivity with wavelength corresponds at all closely to that of the eye. The principal difficulty lies in the predominant sensitivity of the silver halides in the short-wave region, but many of the most useful dye sensitizers sensitize too powerfully in the long-wave region to produce the desired result. It is necessary, therefore, to employ light filters if orthochromatic reproduction is to be obtained. Orthochromatic filters are generally yellow since the principal requirement is to absorb violet and blue, but some sensitive materials and light sources necessitate green filters. In all cases in which orthochromatic reproduction is required, the filter recommended by the manufacturer of the sensitive material should be used.

It should be noted that the filters recommended do not in fact result in true orthochromatic reproduction as this would involve greatly increased exposures and the danger of insufficient detail in colors at both ends of the spectrum, namely the violet and deep red, but only in what may be regarded as a satisfactory approach to it.

Orthochromatic reproduction is not always the end in view. In many cases the visual effect of color is reproduced more accurately if orthochromatic reproduction is departed from and one or more of the colors rendered relatively lighter, or darker, than its color brilliance would justify. Color contrasts, for example, which depend upon hue rather than the brilliance characteristic must be distorted to produce a difference in tone which will reproduce the visual contrast adequately in monochromate. (See **Light Filters.**) (C.B.N.)

ORTHOCLASE. Feldspar.

ORTHOGENESIS. A theory of organic **evolution** which assumes the occurrence of progressive change in living things along definite lines of development.

In **fossil** series it is always possible to see orderly progression or retrogression from form to form. According to some evolutionists this type of development has taken place under the impulse of unknown controlling factors as a directive tendency. Others point out that one result of evolution is certain to be an ancestral series which, if isolated, appears to be orthogenetic.

Since every heritage must have certain potentialities, the·evolution of any group must proceed along the lines made possible by these potentialities. To this extent orthogenesis is an established principle. No comprehensive theory of evolution, however, fails to admit the existence of other factors, at least in a directive capacity, and so orthogenesis is not to be regarded as an explanation of organic evolution. (A.W.L.)

ORTHOGNEISS. Gneiss.

ORTHOGONAL CIRCLES. Circles.

ORTHOGRAPHIC PROJECTION. A method of representing solid objects on a plane surface, using parallel rays or projectors, in contrast to a cone of rays as used in **perspective.** The rays are perpendicular to the plane of representation, in contrast to oblique representation or projection, in which the rays are parallel, but at an angle to the plane of representation. In auxiliary projection, the plane of representation is inclined with respect to the principal axis of the object, although the rays are perpendicular to the plane of representation, and usually perpendicular to an inclined face of the object.

Six principal orthographic views are possible, although usually only four, the top view or plan, the front view or front elevation, and the left-side and right-side views or side elevations are employed. The usual arrangement of views in American practice is shown in the figure; in

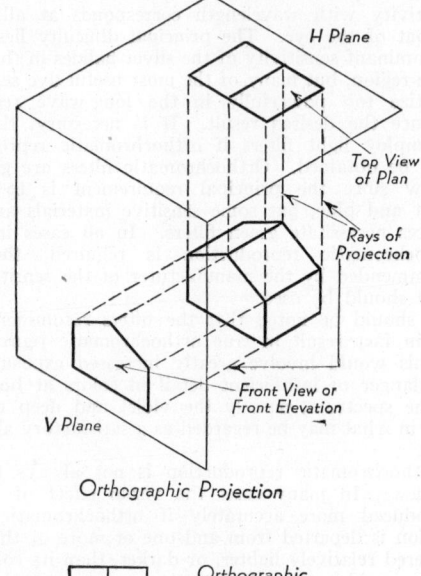

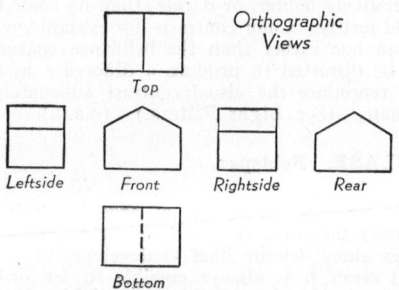

Orthographic Projection

Orthographic Views

Europe, a view arrangement in which the top view is below the front view, and the so-called right-side view is to the left of the front or top view, referred to as first angle projection, is extensively used. (H.C.H.)

ORTHOPEDICS. A division of surgery having to do with the diagnosis and treatment of the diseases of bones and joints. (D.M.H.)

ORTHOPTERA. Insects of many forms with many common names, constituting one of the larger orders. The principal members of the group are the **grasshoppers, crickets, mantises, stick insects,** and **cockroaches.** Locusts are grasshoppers. Some authorities place these major forms in separate orders, but the general tendency is to group them together.

The order is characterized by biting mouth parts and gradual metamorphosis (**Paurometabola**). When wings are present there are two pairs, the front wings somewhat thickened as tegmina and the hind pair broad and membranous, folding beneath the tegmina when at rest.

These insects have a wide range of habits and occupy many terrestrial habitats. The locusts are important crop pests in some areas and cockroaches are common household pests. (A.W.L.)

ORTHORHOMBIC SYSTEM. Crystallography.

ORTHOSILICATE. Silicon; Garnet.

ORTHOSITE. The term proposed by Turner, in 1900, for a rock of granitic texture but composed almost entirely of the feldspar, **orthoclase.** (R.M.F.)

ORTOLAN. Aves, Passeriformes. A European bird, *Emberiza hortulana,* whose flesh is regarded as an unusual delicacy. Many of the birds are caught in nets and fattened for the market. One of the **buntings.** (A.W.L.)

ORYX. Mammalia, Artiodactyla. A genus of **antelopes** found in the desert regions of Africa and thence to Syria. The name is also anglicized as a common name of some species. All species have very long horns, straight or slightly recurved. The **gemsbok, beisa,** leucoryx and beatrix antelope are oryxes. (A.W.L.)

OSAZONES. Hydrazines, Hydrazones and Osazones.

OSCILLATION. An oscillation is to and fro, or vibratory motion, of an object, a wave, electrons, etc. Mechanical oscillation is exemplified in the **pendulum,** by its regular swing back and forth. The center of oscillation of a pendulum is one of its conjugate points. (See **Kinetics.**) Electrical oscillation occurs in a circuit in which electricity surges back and forth. Electromagnetic waves are oscillatory in nature, and can be produced by **oscillators.** The number of oscillations per second is known as the **frequency** of oscillation, and its reciprocal is the period. (F.T.M.)

OSCILLATOR. The oscillator is the heart of the radio **transmitter** since it generates the high-frequency carrier signal essential for this type communication. The early oscillators used in radio consisted of inductance-capacitance circuits to which a surge of electrical energy was applied by the breakdown of a spark gap or an arc. This energy then surged back and forth between the inductance and capacitance until it was all dissipated as radiation and as circuit losses. This type oscillator produced damped oscillations and is no longer used. Modern oscillators use **vacuum tubes** in various circuit arrangements. Since the ordinary vacuum tube is inherently an amplifying device, it can be used as an oscillator by feeding back some of the output energy to the **grid** so the tube effectively drives itself. The various oscillator circuits in common use are utilizations of different means of doing this. For stable frequency characteristics crystal-controlled oscillators are used in most stations. In these a quartz **crystal** is connected in the grid circuit of the vacuum tube so the voltages produced by the crystal when it vibrates mechanically control the grid and hence the plate output of the tube.

Enough energy is fed back, usually through the grid-plate capacitance of the tube, to keep the crystal in a steady state of vibration. Because of the high frequency stability of the mechanical vibrations, the output frequency of the circuit is remarkably stable, in some cases where extreme care is exercised amounting to 1 part in 20,000,000. In modern broadcast stations the crystals maintain the frequency accurately to 2 or 3 cycles. Where continuously adjustable frequency output is needed some type of self-controlled oscillator is needed. The Hartley, shown in the figure, is one of

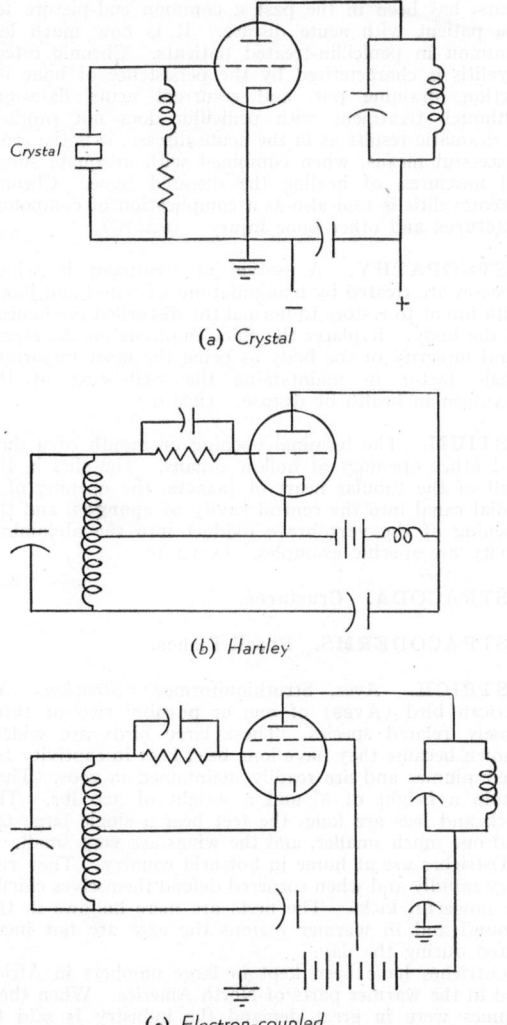

Crystal

(a) Crystal

(b) Hartley

(c) Electron-coupled

the simplest. The energy is fed back from the plate to the grid circuit through the inductive coupling of the two sections of the coil. The frequency is determined by the inductance and capacitance values in the tuned circuit. A somewhat more complicated but more stable variable circuit is the electron-coupled oscillator shown at (c). This uses a tetrode or pentode tube, utilizing the screen as the plate of an oscillator whose output is coupled by the electron stream in the tube to the main plate circuit. The one shown employs the Hartley circuit. At very high frequencies the oscillator frequency is often controlled by a resonant transmission line. At ultra high frequencies special tubes and types of circuits must be used. In these the frequency is largely determined by the transit time of the electrons in the tube. (L.R.Q.)

OSCILLOGRAPH. This name applies to a variety of instruments, including certain vibration **galvanometers**, which automatically trace, or at least render visible, curves representing variable currents, electromotive forces, or other electrical quantities. For example, the electric oscillations in a circuit may cause the electromagnetic vibration of a filament bearing a small mirror which, reflecting a beam of light, leaves a trace on a moving photographic film; or a diaphragm may be vibrated through electrostatic forces controlled by varying potential. The most generally adaptable forms are of the **cathode-ray** type, originated by Braun. In these, a slender stream of cathode particles, usually from a hot-filament cathode, issues from a small anode opening (like negative **canal rays**), and is deflected by means of a rapidly varying magnetic or electric field or both, so that its far end oscillates in faithful reproduction of the currents or voltages responsible for the fields. The stream may fall on a fluorescent screen where its vibrations become visible (in which case the instrument is often called an **oscilloscope**), or may leave its trace on a photographic plate or moving film. In one form, the motion along the time axis, called the "sweep," is secured by causing the stream to deflect under the influence of an increasing electric field controlled by a slowly charging **condenser**. (L.D.W.)

OSCILLOSCOPE. Oscillograph.

OSCULATING ORBITS. Orbit.

OSCULUM. The opening of the central cavity (paragaster, gastral or paragastric cavity) of **sponges**. Water passes into this cavity through the small pores in the body wall and flows out of it by the osculum. (A.W.L.)

OSE. Esker.

OSMETERIUM. A scent gland found in some insects. A common example is the forked eversible organ just behind the head of the caterpillars of swallowtail butterflies. This organ is thrust out when the insect is alarmed. It is brightly colored and gives off a powerful odor, usually reminiscent of the animal's food plant but too strong to be agreeable to the human nose. (A.W.L.)

OSMIUM. Symbol: Os. Atomic number: 76. Atomic weight: 190.2. Density: 22.48. Hardness: 7. Melting point: 2700° C. (Isotopes, page 290.)

Compact osmium is a bluish-white metal, and is not attacked by acids. Discovered by Tennant in 1804.

Finely divided osmium is a **catalyzer** for the reactions, (1) **hydrogen** plus **oxygen** at 50° C. forming water, (2) hydrogen plus **nitrogen** at high pressures and 400–500° C. forming **ammonia**.

Osmium occurs native with **platinum**, and with **iridium** (osmiridium, 20–40% Os). By boiling finely divided osmium-containing material with **aqua regia**, volatile, poisonous osmium tetroxide distills over, and is absorbed in **sodium** hydroxide solution. The resultant red solution, containing sodium osmate (Na_2OsO_4), is later acidified, and reduced to osmium metal by zinc metal.

Osmium tetroxide is used in preparing microscopic slides, where it stains selective portions of tissues black.

Chlorides: Osmium dichloride ($OsCl_2$); osmium trichloride ($OsCl_3$); osmium tetrachloride ($OsCl_4$).

Hydroxide: Osmium hydroxide ($Os(OH)_4$), brown precipitate, by solutions of hydroxides, e.g., **sodium** hydroxide.

Oxides: Osmium monoxide (OsO); osmium sesquioxide (Os_2O_3); osmium dioxide (OsO_2); osmium tetroxide, or osmic acid (OsO_4). The last compound has melting point 40° C., boiling point 100° C.

Sulfides: Osmium disulfide (OsS_2), brown precipitate, in acid solution by **hydrogen sulfide;** osmium tetrasulfide (OsS_4), in neutral solution.

Osmium tetroxide is reduced by **ferrous** sulfate solution to osmium dioxide. Osmium compounds are reduced to osmium metal by ignition in hydrogen. (R.K.S.)

OSMOSIS. The passage of liquids or gases from solution through membranes. When solutions of different degrees of concentration are separated by a membrane (a common form of membrane for experiment is a very thin sheet of collodion) substances diffuse through from the solution in which they are more concentrated to that in which they are less so, provided that the membrane is permeable to them. The term osmosis is restricted to such movements of the solvent, which in organisms is usually water. Dissolved substances may dialyze if the membrane is permeable to them. (See **Diffusion of Fluids.**) (A.W.L.)

OSPHRADIUM. A branching sensory organ of **mollusks,** supposed to be olfactory in function. It sometimes resembles a **ctenidium.** (A.W.L.)

OSPREY. Aves, Falconiformes. A large bird of prey of almost worldwide distribution. It is a skillful fisher and is known in North America as the fish hawk, *Pandion haliaëtus.* (A.W.L.)

OSSICLE. 1. The calcareous plates in the body wall of **echinoderms.** In some species, notably the sea urchins, they are connected to form a shell and in others, especially the sea cucumbers, they are small and scattered. 2. A small bone, such as the chain of three auditory ossicles (hammer, anvil and stirrup) in the middle **ear** of mammals. (A.W.L.)

OSSIFICATION. The formation of bone, or the conversion of tissue into bone or a bone-like substance. (R.S.M.)

OSTEITIS. A low-grade inflammation of bone marked by tenderness, pain and enlargement of the bone. (R.S.M.)

OSTEITIS FIBROSA CYSTICA. Parathyroid Gland.

OSTEO-ARTHRITIS. Arthritis.

OSTEOLEPIS. Fossil Fishes.

OSTEOLOGY. A division of vertebrate anatomy which deals with the **skeletal system.** The study of bones. Comparative osteology is the study and comparison of the bones of different races and different species of organisms. (A.W.L., R.S.M.)

OSTEOMYELITIS. An acute or chronic infection of bone, most commonly due to hemolytic *Staphylococcus aureus* but occasionally to the *streptococcus, pneumococcus,* or to *B. typhosus.* The organisms often reach the blood from an insignificant **furuncle** or other infection on the skin, and are carried in the blood stream to the bone. Previous injury at the site of infection seems to play a part in lowering the resistance of the bone and allowing the organisms to gain a foothold.

Osteomyelitis is a disease of childhood, most of the cases occurring between the ages of 2 and 12 years. It is commoner in boys than in girls. The bones most frequently involved are the femur (thigh), tibia (leg), and humerus (upper arm). When fully developed, the infection may involve the covering of the bone, the bone itself, and the marrow cavity. Widespread destruction and **necrosis** may occur.

Acute osteomyelitis is usually sudden in onset with a chill and high fever, pain, tenderness, redness and swelling of the involved bone. The child is extremely sick, dehydrated, and may be delirious or even comatose. Staphylococci can often be found in the blood stream by **culture.** The treatment of acute osteomyelitis has been improved enormously since the introduction of the **sulfonamides** and **penicillin.** Penicillin, administered early, will sterilize the blood stream and control the bone infection with remarkable rapidity. If the infection has gone on to the stage of pus and **abscess** formation before the drug is begun, surgical drainage may be necessary in addition to drug therapy.

Chronic osteomyelitis, which may go on for many years, has been in the past a common end-picture for the patient with acute disease. It is now much less common in penicillin-treated patients. Chronic osteomyelitis is characterized by the persistence of bone infection, draining pus, and recurrent acute flare-ups. Although treatment with penicillin does not produce as dramatic results as in the acute disease, it is the most successful means, when combined with adequate surgical measures, of healing the diseased bone. Chronic osteomyelitis is seen also as a complication of compound **fractures** and other bone injury. (D.M.H.)

OSTEOPATHY. A system of treatment in which diseases are treated by manipulations of bones and joints with intent to restore to normal the disturbed mechanism of the body. It places the chief emphasis on the structural integrity of the body as being the most important single factor in maintaining the well-being of the organism in health or disease. (D.M.H.)

OSTIUM. The terminal opening or mouth of a duct and other openings of hollow organs. The slits in the wall of the tubular heart of **insects,** the opening of a radial canal into the central cavity of **sponges,** and the opening of the **vertebrate** oviduct into the abdominal cavity are specific examples. (A.W.L.)

OSTRACODA. Crustacea.

OSTRACODERMS. Fossil Fishes.

OSTRICH. Aves, Struthioniformes. *Struthio.* An African bird (**Aves**) of one or possibly two or three closely related species. These large birds are widely known because they have long been kept in captivity for their plumes and are readily maintained in zoos. They attain a height of 8' and a weight of 300 lbs. The neck and legs are long, the feet bear a single large toe and one much smaller, and the wings are very small.

Ostriches are at home in hot arid country. They run very rapidly and when cornered defend themselves chiefly by powerful kicks. The nests are mere hollows in the ground and in warmer regions the eggs are not incubated during the day.

Ostriches have been kept in large numbers in Africa and in the warmer parts of North America. When their plumes were in great demand the industry is said to have attained a value of about $10,000,000 annually.

The **rheas** are called American ostriches. (A.W.L.)

OTITIS MEDIA. Infection of the middle ear commonly due to staphylococci and streptococci. This condition is very common, and is most frequently seen in children. It often led to **mastoiditis** in the days before **sulfonamides** were in use. The infection reaches the middle ear through the **Eustachian canal** from the **pharynx.** It usually occurs secondary to an acute upper respiratory infection such as **coryza, pharyngitis,** or acute **tonsillitis.** Children with adenoids and large tonsils are more subject to ear infections. Coughing, hard blowing of the nose, and the use of nasal douches or sprays often spread the infection from the throat to the ear. As soon as there is sufficient pus in the cavity of the middle ear to cause pressure, the pain becomes excruciating. Rise in temperature and loss of hearing in the affected ear are present.

The treatment in certain cases is with **sulfonamides**. Later, when there is pressure on the ear drum from the accumulated pus, incision of the drum is necessary as well as chemotherapy. Most cases recover shortly and heal within 2 weeks with little or no impairment of hearing after 2 months. When the symptoms and discharge continue for a longer period, mastoid involvement is usually present. (R.S.M., D.M.H.)

OTOCYST. Lithocyst.

OTOPORPA. A protuberance extending upward from the base of a **lithocyst** in some of the jellyfishes. (A.W.L.)

OTTER. Mammalia, Carnivora. Elongate animals with short legs, broad head, and webbed toes. They are excellent swimmers, even catching fish apparently for the pleasure of the pursuit.

Otters are found on every continent except Australia. Different species vary from two to three feet in length, exclusive of the long tail, but there is relatively little difference in their general appearance. *Lutra canadensis* is found in North America.

Otter fur is thick, soft and glossy, and is among the most valuable commercially. The fur of the sea otter, *Latax lutris,* one of the largest species, is regarded as the most valuable of all furs. The species inhabits the northern Pacific, in both Asiatic and American waters, where it feeds on marine invertebrates. It was once abundant but through indiscriminate trapping it has become very rare. (A.W.L.)

OTTO CYCLE. The possibilities of an **internal combustion engine** which would operate on the **four-stroke cycle** were analyzed in the nineteenth century by de Rochas. The action of an engine, which was at that time described only, is in all fundamental respects the same as that used today. Shortly after this internal combustion engine analysis was made, the German Otto built an engine which operated on the four-stroke cycle, and which has come, since, to be called the Otto cycle. The Otto cycle is carried out by a series of operations, namely, and in order:

1. Suction during outward stroke of the piston.
2. Compression during inward stroke of the piston.
3. Expansion during outward stroke of the piston following ignition at inward dead center.
4. Exhaust during inward stroke of the piston.

Although the first engines built by the Otto works were four-stroke cycle (**four-cycle**) engines, the Otto engine can be adapted to a two-stroke cycle (**two-cycle**). An engine which operates on the Otto cycle has considerable clearance volume (when the piston is on dead center) into which the air and fuel drawn in on the suction stroke are compressed. As the piston pauses instantaneously on dead center position, the charge is **ignited** and **combustion** of the charge occurs at constant volume. The heat added by this combustion greatly increases the pressure. During the expansion stroke, this pressure is lowered, and a considerable amount of the heat is taken from the gas to be converted into work. At the end of expansion, the pressure is dropped to the exhaust point, and the exhaust gases are pushed out of the cylinder, which is thus cleared for the next suction stroke. In the figure which illustrates the Otto cycle in theory and in practice, the cycle is shown on the pressure-volume plane. The ideal cycle is shown solid, and departures from it, made by actual cycles, are shown dotted. The ideal four-stroke cycle is made up of two isovolumetric changes, one representing the combustion of fuel at constant volume, the other the rejection of heat to the exhaust at constant volume, and two **adiabatic** changes, one representing the com-

pression of the charge, the other the power stroke. Finally, there are two coincident constant pressure changes, representing the ideal exhaust and suction

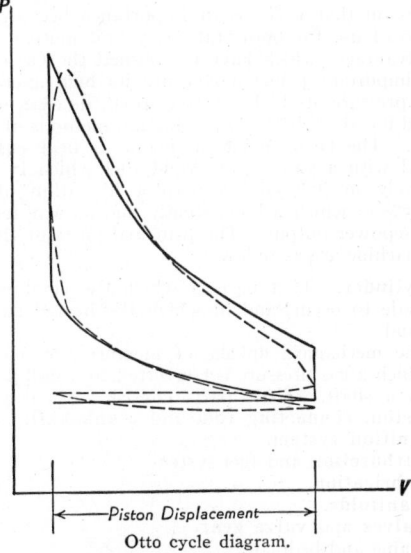

Otto cycle diagram.

strokes. An actual working cycle will not have quite the same processes. Combustion of a fuel is not instantaneous in nature. That occurring in the gasoline engine occupies an appreciable length of time compared to piston motion, so that isovolumetric pressure rise is not possible. The point of ignition must be advanced before inward dead center, or there will be considerable loss of work by delayed combustion. An exchange of heat between cylinder cooling water jackets and cylinder contents tends to destroy the adiabatic relationship assumed for the ideal cycle. The exhaust cannot be at constant pressure because of the inertia of gases leaving the cylinder and the definite rate of valve action of mechanically driven valves. Furthermore, the exhaust and suction lines cannot be coincident, since exhaust must be above atmospheric pressure by the amount of friction losses through the ports and manifolds, while suction must be below atmospheric pressure for the same reason. The cylinder horsepower of an engine operating on this cycle is $\frac{PLAN}{33,000}$, wherein P is the mean effective pressure, and N is the number of power strokes per minute. The N of a four-stroke cycle, single-acting engine is one-half the number of revolutions per minute multiplied by the number of cylinders per engine. L and A are the stroke in feet and piston area in units corresponding to those of P.

The ideal efficiency of this cycle is expressed by the equation

$$E = 1 - \frac{1}{r^{y-1}}$$

r is the compression ratio, and y the ratio of the specific heats at constant pressure and constant volume. This air standard efficiency is higher than the actual efficiency because it neglects cooling losses and friction losses. The actual efficiency of the Otto cycle is usually between 20 and 25%. The compression ratios employed have been steadily raised by manufacturers in order to increase the efficiency of their product, for, as is seen in the above equation, an increase of r will increase the efficiency of the Otto engine. Modern motor cars are built having r as large as 7, but only by the use of aluminum cylinder heads, which conduct the heat from the cylinder more rapidly. An average value of r for the Otto cycle would be 5. (F.T.M.)

OTTO ENGINE.

The Otto cycle is described in another section. The **internal combustion engine,** which operates on that cycle, is of importance because of its widespread use for both stationary and motive service. The advantages which have established the Otto engine as an important prime mover are its high average effective pressure, its high rotative speed, its ease of starting, and its adaptability to production methods of manufacture. The Otto engine, acting as a heat engine, is supplied with a gaseous or liquid fuel which is burned explosively in its cylinder with a liberation of heat, some 25% of which will eventually find its way into useful horsepower output. The principal parts of the Otto cycle machine are as follows:

1. **Cylinder.** That part in which the combustion is made to occur, and in which the heated gases expand.
2. The mechanical linkage of members, by means of which a gas pressure is converted to a useful torque on a shaft. These parts consist, ordinarily, of a **piston, connecting rod, and crankshaft.**
3. **Ignition system.**
4. **Carburetion** and fuel system.
5. **Lubrication.**
6. **Manifolds.**
7. **Valves** and **valve gear.**
8. Frame and bedplate.
9. Combustion chamber cooling system.

The piston, cylinder, connecting rod, and crankshaft form the principal parts. These are integrated and aligned by being fastened to a frame or bedplate, which is often of the fully enclosed type, so that spray lubrication may be used. Valves operated by the valve gear provide the function of admitting and releasing the working medium to and from the cylinder, while manifolds distribute or collect the gases in the multi-cylindered engines. The two auxiliary systems of carburetion and ignition are essential in the gasoline engine, the carburetion to prepare the fuel, and ignition to ignite it. An average heat balance for this type engine would be as follows:

Output	25%
Cooling	36%
Exhaust	34%
Friction	5%
Total	100%

A typical multi-cylindered engine is arranged with cylinders in line or in banks, or radially. The cylinders are aligned so that the connecting rods may bear on the same crankshaft. This crankshaft drives the necessary auxiliaries, and delivers the remaining power as useful output at a coupling, pulley, or clutch. Each cylinder has at least one inlet and one exhaust valve mechanically operated and synchronized with the cycle. The ignition apparatus is located in the combustion chamber either as an igniter head or a spark plug. The openings to the cylinder, which are opened and closed by the valves, lead to ports, to which manifolds are connected. Auxiliaries which must be driven by the engine, and which consume part of its gross output, are the aforementioned valves, the lubricating oil pumps, cooling water circulating pumps, the **generator** for current to the ignition system, the igniters or timers of the ignition system, and sometimes an air fan. The cylinders are mounted on a **crankcase,** which provides alignment and also supports the crankshaft bearings.

The operation of an Otto engine may be described beginning with the suction stroke. At the beginning of the suction stroke, the inlet valve will be open, and will remain so until the outward stroke is completed, drawing in through the carburetor and manifolds a mixture of gasoline and air in explosive proportions. The quantity drawn in per stroke depends on the **volumetric efficiency** created by the induction system. On the succeeding inward stroke of the piston the inlet valve remains open for possibly 10 to 15 degrees of crankshaft travel because there is a certain inertia of the gas column moving in the manifolds and ports, and the flow will not immediately reverse upon reversal of the piston travel. The inlet ports should remain open as long as any gas will flow into the cylinder, so that the volumetric efficiency will be maximum. After closure of the inlet valve, the gas is compressed into the clearance space at the extremity of the cylinder. This clearance space is the combustion space, and its shape has been given a great deal of study in connection with **detonation.** In this space is located the **spark plug** which receives the high-voltage impulse from the ignition system when the crank pin is still 10 to 15 degrees (sometimes even more) before dead center. The advance of this spark before dead center is variable, so that it may be advanced more at high rotative speeds, thus aiding completion of the explosion before expansion begins. On the outward stroke, which is the power stroke, both valves remain closed until the crank is about 30 degrees before outward dead center, when the exhaust valve begins to open. The valve is given this lead on the piston position in order that it may be fully open at the dead center position, so that on the inward exhaust stroke a complete scavenging of burned gases from the cylinder can be a possibility. In fact, the exhaust valve remains open until after the piston is slightly past inward dead center position.

Power is nearly maximum at rated speed. The power strokes per minute in a four-stroke cycle engine of a single-acting type is one-half the number of revolutions per minute times the number of cylinders. In a two-stroke single-acting engine it is the number of revolutions per minute times the number of cylinders. In the double-acting engine, it is always twice that of the corresponding single-acting engine. Thus a two-cycle, double-acting engine would have four times as many power strokes for the same rotative speed as a four-cycle single-acting engine. The amount of power developed by the Otto engine may be varied in several ways, among which are the following:

Quantity Governing. In quantity governing, the amount of air mixed with a unit of fuel is kept the same, but the quantity fed per cycle is decreased for decreased power, and vice versa. This system is the type ordinarily employed on the automobile engine, where the quantity is governed by interposing an artificial resistance between the carburetor and the cylinder in the form of a valve, which may be opened or closed by the driver. The disadvantage of quantity governing is that it reduces the efficiency because the full compression pressure is not reached when the engine is operating throttled.

Quality Governing. With this the full compression pressure is reached at the end of the compression stroke because the amount of air charged per cycle remains constant. Power is varied by altering the mixture of fuel and air. This system of governing is faulty in that unexplosive mixtures are produced at very light loads, causing a hit and miss type governing. It is suitable to gas, but not gasoline, engines.

Hit and Miss Governing. An inexpensive simple form of constant speed governing used on light portable and stationary engines, is known as hit and miss governing. It is governing by means of varying the number of power strokes per minute by causing some normal power strokes to be missed. This can be done by one of three ways:—by opening the switch in the ignition circuit, causing failure to ignite; by holding the exhaust valve open during the suction stroke; and by leaving the inlet valve closed during the suction stroke.

The combustion of gasoline in the Otto engine is as follows: (Gasoline is a mixture of many hydrocarbons, for which the formula C_8H_{18} is taken as a fair average composition.)

$$C_8H_{18} + 12.5O_2 = 8CO_2 + 9H_2O.$$

Calculations based on this equation show that 15.2 lbs. of air are required for the complete combustion of each pound of gasoline. A deficiency of air results in the formation of **carbon monoxide** and **hydrogen**, but if there is only a small deficiency, the hydrogen may be assumed as being completely burned. Theoretically, **oxygen** and carbon monoxide should not be present in the combustion of a perfect mixture, but actually oxygen will be found in the combustion of lean mixtures, and carbon monoxide will persist in the products of combustion of rich mixtures. The limits of explosive mixtures of gasoline and air are between 12 and 19 lbs. of air per pound of gasoline. Maximum power is derived at about 13 lbs. of air, the ideal conditions call for 15.2 lbs., and maximum economy is obtained with even leaner mixtures.

There is no more critical part of the Otto cycle than the valves, for these are required to remain pressure-tight under the severe condition of high operating temperature. The exhaust valves of heavily loaded engines operate continuously in an atmosphere of flame, and become red hot. The valves, especially the exhaust valve, which has not the benefit of the cooling derived from an incoming charge, are sources of trouble in the Otto engine. By the use of special cooling and of alloy steels, these failures have been greatly reduced in number. The poppet valve is used on nearly all Otto engines. The only other type to be employed at all is the sleeve valve, in which a sleeve is placed between the piston and the cylinder. The piston slides smoothly in the sleeve, which has some slight motion relative to the cylinder walls. The sleeve is driven by eccentrics or cams, so that a slot in it registers with a fixed port on the cylinder wall when the cylinder is to be opened to the manifolds. Two concentric sleeves are used in the Knight sleeve valve design. Of much more importance from the standpoint of usage is the poppet valve. This valve has a disk-like head attached to a stem. The stem reciprocates in a valve guide under the action of a cam which bears against the end of the stem, or which operates a tappet which, in turn, bears against the valve stem. The head has a face which is a portion of a cone, and which sets in a conical seat in the cylinder, or combustion chamber. There is no sliding action between the valve and its seat. Mushroom and tulip type poppet valves are shown in the accompanying diagram. The mushroom type is the

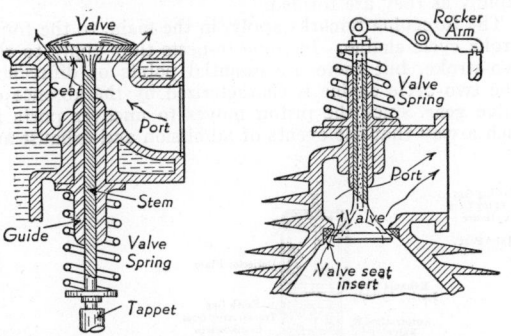

Mushroom type inlet valve — water - cooled cylinder.

Tulip type exhaust valve— air-cooled cylinder.

Fig. 1.   Poppet valves.

simplest, and is suitable for ordinary work. When valves run continuously at very high temperatures, and are made fairly large, the mushroom head has a tendency

to break or warp, because its strength is derived by cantilever action, and there is a point of weakness at the attachment of the stem. The tulip valve is in tension instead of bending. It is less liable to leak and gives a better streamlining in the port. This type of valve, when constructed with a hollow stem, partially filled with metallic sodium, which assists in the transfer of heat from the head to the rest of the stem, has been able to perform satisfactorily in the high output aeronautical type engines. The materials which are used must be alloy steel to withstand the corrosion, abrasion, scaling, burning, and high stress met in service. While carbon steel may be used for inlet valves, chrome nickel or chrome silicon exhaust valves are required. The valve seats are often cut and reamed into the cast iron which comprises the cylinder block of small and medium duty engines. Large valved engines, and engines having aluminum cylinder heads, require inserts of aluminum bronze, stellite, or special alloys, to form the valve seats.

Valves are actuated by an operating gear deriving its motion ultimately from the crankshaft. Since the motion of a poppet valve is a reciprocation of short throw, the **cam** is the most practical intermediary between the rotative motion of the crankshaft and the reciprocating motion of the valve. Due to the fact that the valve needs to reciprocate only once during the two revolutions of a four-stroke cycle, the cam shaft must be a half-speed shaft driven from the crankshaft by gears or chain. The cam bears against a follower, which may be of roller type, or flat face. This follower may be the end of the valve stem, or it may be a tappet, push rod, or rocker arm between cam and valve stem, depending on the relative location of the cam shaft and valves. Any discussion of valve operating gear would necessarily be based on some knowledge of common cylinder head arrangements (Fig. 2). These might be classified as T, L, I, and F arrangements, these letters being symbolic of the shape of the

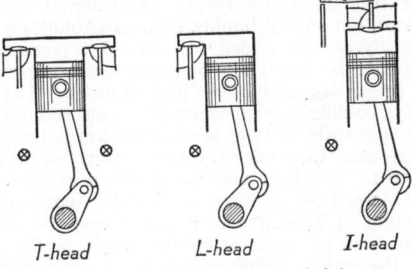

T-head   L-head   I-head

⊗—Camshaft location

Fig. 2.   Common cylinder shapes as governed by valve position.

cylinder head. In the T head, the combustion chamber has pockets on either side, in one of which is located the exhaust valve, in the other the inlet valve. Earlier engines were so built, but the construction has been abandoned because two cam shafts were necessary. The L head arrangement is the most common today. The valves are both located in the same pocket on the same side of the cylinder. In-line multi-cyclinder engines, then, have all the valves in a single line, so that they may be driven by cams, all of which are machined on a common cam shaft. When the valves are located overhead in the top of the cylinder instead of in a special valve pocket, the arrangement is known as the I head. The valves open downward into the cylinder in the I head, whereas in the T and L they open upward. I head arrangement has advantages of more direct gas flow to the cylinder, thus promoting higher volumetric efficiency; but has the disadvantage of a more complicated valve gear. In the F head, which has been but little used, the exhaust and inlet valves are mounted one over the other in a valve pocket, one being similar to an L head valve, the other to an I head valve. (See **Valve Gear.**)

Valves in L head engines are set directly in line with the cam shaft, so that the cam either bears directly on the end of the valve stem, or indirectly, through a relatively short follower. As the cam provides positive motion in one direction only, the valve is opened against a spring pressure. The valve spring holds the cam follower against the cam. On I head engines, the cam shaft must either be placed above the cylinder, a somewhat awkward position, or motion must be transmitted from its crankcase location to the valve by push rod and rocker arm.

The high cylinder temperatures of internal combustion engines would quickly destroy any lubricating film if positive cooling of some sort were not adopted. By positively abstracting about one-third of the heat of combustion to a cooling system, a temperature gradient is maintained through the cylinder wall, and the moving parts can be adequately lubricated. The heat so abstracted is loaded on either air or water. In a great many cases air is also the final cooling medium for the water. The water is circulated through a **radiator**, where it is cooled by a current of air drawn through the radiator. Direct air cooling requires extending the outer surface of the cylinder, because computations show that a plain cylinder would have a temperature exceeding 1800° F. Cylinders are finned and flanged so that the external surface will be sufficient to dissipate the required amount of heat to a stream of air. A positive circulation of air over the fins is also necessary. This is accomplished by fans and by baffles which direct the air over the finned surface. Air cooling is most successful where the cylinders are individual, whereas liquid cooling is better suited to en-bloc construction. The liquid is contained in a thin layer held around the cylinder head, or completely around the cylinder, by the cooling **jacket**. This may be a metallic cover welded or otherwise fastened around the cylinder, leaving a water space between it and the cylinder, or the water space may be an opening in the casting made by cores. Most liquid cooling is with water, but special liquids, such as alcohol, glycerine, ethylene glycol, are added for several reasons, among which are to be noted low freezing temperatures, anti-rusting properties, and better heat transfer. The water used in a cooling system must be clean and relatively pure, so that scale will not accumulate inside the jackets. It can be used in a once-through system, or it can be recirculated. The heat may be absorbed as heat of the liquid, obtained by an increase of water temperature to a maximum of around 200° F., or it may be absorbed as latent heat in evaporating the water in the jacket. Usually the water-cooling system is of the recirculation type, the heat being ultimately given to the air by means of radiator, cooling tower, or spray pond.

The lubrication of this type of engine is of the utmost importance because of high speed, temperature, and pressure, coupled with the interdependence of all parts upon the proper functioning of the remainder. Parts needing lubrication are the piston and cylinder, the valve gear, the connecting rod bearings, the crankshaft, and camshaft bearings.

Auxiliaries such as fans, generators, starters, also have their points of lubrication. Three systems of lubrication most used are the splash, the semi-force feed, and the force feed, all of these relating to the enclosed crankcase type engine. In the splash system oil is pumped into troughs located so that the ends of the connecting rods dip into a pool of oil and splash it about inside the crankcase. The semi-force feed system has lubrication of the crankshaft bearings and the camshaft bearings by oil which is conveyed to them under pressure through tubes. The connecting rod bearings and cylinders are lubricated by splash. There is neither trough nor splash in the full force feed. Oil from the pump is carried to the crankshaft bearings through tubes, and to the connecting rod bearings through drilled passages in the webs of the crank. Centrifugal force throws it out of the connecting rod bearing as the latter revolves, so that the crankcase is filled with an oil mist which lubricates the piston and wrist pin. Occasionally one will find the connecting rod drilled in its entire length so that the piston end of the connecting rod also receives oil under pressure. The oil so delivered to the bearings drains into the crankcase, from which it is either taken as needed by the oil pump, or is immediately removed and placed in an external tank by a scavenging oil pump. These two systems are known as the dry and the wet sump systems.

During the operation of the engine there is a steady loss of oil. Some is burned in the combustion chamber, some leaks through gaskets and around shafts, and some is blown out of the crankcase breather as a mist. That which is burned in the combustion chamber accounts largely for the accumulation of carbon so often found there. The incomplete combustion of fuel, and dust and dirt from the air, contribute somewhat to combustion chamber deposits, but to a much lesser degree than does oil. Some oil in the combustion chamber is necessary if the piston is to be properly lubricated, but wear or sticking of the piston rings may cause the piston to assume some pumping action, and deliver to the combustion chamber a great deal more oil than would be necessary for cylinder lubrication, certainly more than is desirable from the standpoint of the condition of the combustion chamber. In addition to creating a carbon deposit, an excessive amount of oil may foul the sparking plug points, and cause sticking of the valves. Carbon in the cylinder head provides a black surface which initially aids heat transfer, but, as it thickens, greatly retards heat transfer, due to the low heat transfer coefficient of carbon. Then hot spots remain in the cylinder and pre-ignition and detonation are liable to occur. The thicker the accumulation of carbon, the hotter the combustion chamber becomes, and carbon reaches a maximum thickness beyond which accumulations are burned off as rapidly as they are formed.

The foregoing remarks apply, in the main, to the four-stroke cycle engine. In some respects they apply to the two-stroke, but there are essential points of difference. The two-cycle engine is characterized by the absence of valve gear, since the piston moves to uncover ports in such a way that the events of admission and exhaust are

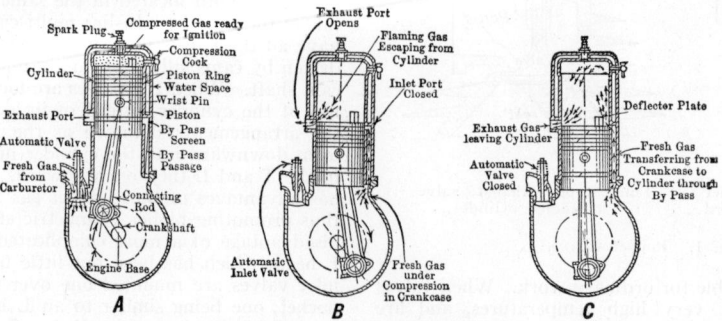

Fig. 3. Two port two-cycle engine.

accomplished without the use of valves. Fig. 3A represents the two-cycle engine nearing the inward dead center, with the compressed gas ready for ignition. Ignition occurs, and the resulting combustion raises the pressure and drives the piston outward on a power stroke. Near the end of the outward stroke the piston uncovers an exhaust port in the lower part of the cylinder wall, and the residual pressure drives the exhaust gas quickly out through this port (Fig. 3B). Further outward motion to the dead center position then uncovers a port leading from the crankcase to the cylinder. The crankcase is filled with a slightly compressed mixture of gasoline and air, which is rapidly passed through this port (Fig. 3C). Entering the cylinder, being deflected upwards, it tends to drive the remaining burned gas out through the exhaust port, which is still open. As the piston moves on its inward stroke, it closes these ports and compresses the gas into the clearance space. The same motion tends to create a vacuum in the airtight crankcase, with the result that a fresh charge is introduced into the crankcase to be compressed on the down stroke. The comparatively shorter length of time available in this engine for charging and scavenging the cylinder makes it impossible to attain volumetric efficiencies comparable with the four-stroke cycle, except for relatively small, slow-moving engines. (F.T.M.)

OTTRELITE. Chloritoid.

OUNCE. Mammalia, Carnivora. The snow leopard of Asia, *Felis uncia,* a moderately large cat found at high altitudes in the central part of the continent. The fur is long and thick, grayish or slightly tawny above with darker black-ringed spots. (A.W.L.)

OUTCROP. Everywhere beneath the **soil** at greater or less depths there exists the solid continuous masses of rock which make up the earth's crust. Wherever rock appears protruding through the soil cover it is spoken of as an outcrop. (E.S.C.S.)

OUTLIER. A term used by **structural geologists** to define an area of younger rock strata which has been geographically separated by **erosion** from the principal area within the same region. Not to be confused with **klippe.** (See also **Inlier.**) (R.M.F.)

OUTPUT METER. This is an **instrument** used for measuring the output of audio **amplifiers.** It consists of a resistor coupled to the amplifier through an impedance-matching **transformer** and a rectifier voltmeter connected across the resistor. The voltmeter is usually calibrated so it indicates the watts dissipated in the resistor either directly in watts or in **decibels** above some reference value. (L.R.Q.)

OUZEL. Aves, Passeriformes. 1. The ring ouzel, *Turdus torquatus,* a mountain bird (**Aves**) of central and northern Europe. A thrush. 2. The **blackbird** of Europe and eastern Asia, also a **thrush.** 3. The water ouzel, or dipper. The name water ouzel applies to several species of the northern hemisphere which haunt rapid streams. Although not a swimming bird it wades into deep water, where it is said to make its way by using both feet and wings. The North American species, *Cinclus mexicanus,* is characteristically a western mountain bird. Its dull gray color is more than offset by its interesting habits and by a glorious song, heard all too rarely.
　Sometimes spelled ousel. (A.W.L.)

OVALS OF CASSINI. This is the name given to the plane curve defined as the **locus** of a point that

moves so that the product of its distances from two fixed points $(-a, 0)$ and $(a, 0)$ is equal to a constant k^2. (Figs. 1 and 2.)

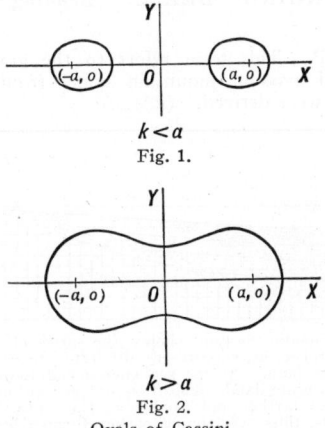

$k < a$
Fig. 1.

$k > a$
Fig. 2.
Ovals of Cassini.

Its equation in **polar coordinates** is
$$(r^2 + a^2)^2 = 4a^2 r^2 \cos^2 \theta + k^4.$$
If $k = a$, we get the **lemniscate.** (L.L.S.)

OVARIOLE. One of the tubular components of the insect ovary. It consists of regions in which the egg cells are formed and in which they grow and mature. (A.W.L.)

OVARY. The ovaries are the sex glands of the female. In humans they are two in number, 3–4 cm. in diameter, almond-shaped, and situated on either side of the **pelvis** in the folds of the broad ligament which supports the **uterus.** With the onset of puberty, cyclic changes take place in the ovaries; after the **menopause** they diminish in size and activity.
　The function of the ovaries is the production of **ova,** and the elaboration of several hormones concerned in reproductive mechanisms. The ovaries are therefore glands of internal secretion, or **endocrine glands.** Two types of hormones are produced: the estrogenic group of which estrin is the most important, and corpus luteum hormone or progesterone. These substances are concerned with the changes characteristic of **menstruation,** those following impregnation which are necessary for the development of the fertilized ovum, and the changes in the mammary glands occurring during **pregnancy.** Further, the ovary produces **hormones** that cause development of the female genital organs and the secondary sex characteristics. The ovary is directly stimulated or inhibited by hormones from other endocrine glands, notably the **pituitary** and in a lesser degree by the **thyroid** and **adrenal glands.**
　For the use of the term ovary in botany, see **Flower.** For its general use in zoology, see **Gonad.** (D.M.H.)

OVEN-BIRD. Aves, Passeriformes. 1. The European willow **wren,** *Phylloscopus trochilus,* and other birds which build domed nests. 2. A North American Warbler, *Seiurus aurocapillus,* also called the golden-crowned thrush. 3. South American birds of several species which build mud nests resembling old-fashioned ovens. Genus *Furnarius.* (A.W.L.)

OVERCAST. When the sky is more than $\frac{9}{10}$ covered with clouds. Viewed from on top (aircraft view) it is said to be an undercrest. (P.E.K.)

OVERFOLD. A term used by **structural geologists** to define an overturned **anticline** resulting in a **recum-**

bent fold often simulating a normal **stratigraphic** pile of strata for which it may be mistaken. (R.M.F.)

OVERHANGING BEAM. Bending Moment; Shear.

OVERLAP. This term refers to the gradual burial of the land mass or mountain slopes from which the **sediments** were derived. (R.M.F.)

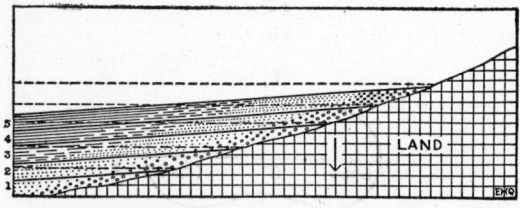

This diagrammatic section shows the principle of overlap. Horizontal broken lines represent different stages of sea level relative to the land. As the sea encroached toward the right upon the subsiding land, deposition of the sediments 1, 2, 3, 4, 5, extended farther and farther to the right, later formed beds, thus overlapping earlier formed beds.

OVERSATURATED. As used by **petrologists,** this term, proposed by Shand in 1915, signifies **igneous** rocks containing free **silica of magmatic origin, and in** the form of **quartz** or **tridymite.** (R.M.F.)

OVERTHRUST. A low-angle **fault** of the thrust or compressional type frequently resulting in the translocation of a mass of rocks several miles. Because of

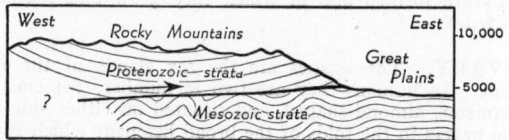

Diagrammatic structure section showing how the great body of Proterozoic strata has been thrust-faulted from the west for miles over upon Late Mesozoic strata in Glacier National Park, Montana. The Mesozoic beds were folded by action of the over-riding block. Length of section, about 25 miles. Vertical scale, much exaggerated.

the low angle of the thrust as well as the possible inversion of great piles of formations this type of faulting or failure of the **lithosphere** has led, and still leads, to considerable technical difficulties in determining the **stratigraphic** history in regions where low angle faulting predominates. (R.M.F.)

OVERTONES. Vibrations and Waves; Musical Sounds.

OVICELL. Ooecium.

OVIPOSITOR. An organ used by some of the **arthropods** to deposit their eggs. It consists of a maximum of three pairs of appendages formed to transmit the egg, to prepare a place for it, and to place it properly. In some of the insects the organ is used merely to attach the egg to some surface, but in many parasitic species (Hymenoptera) it is a piercing organ as well. It is used by the **grasshoppers** to force a burrow in the earth to receive the eggs and by **cicadas** to pierce the wood of twigs for a similar purpose. Both long-horned grasshoppers and sawflies cut the tissues of plants by means of the ovipositor. None of these examples is quite as remarkable as the **ichneumon flies** (parasitic Hymenoptera) which have a slender ovipositor several inches long, used to drill into the wood of tree trunks. These species are parasitic in the larval stage on the larvae of wood-boring insects, hence the egg must be deposited in the burrow of the host.

The sting of **wasps** and **bees** is also an **ovipositor,** in this case highly modified and associated with poison glands. (A.W.L.)

OVO-TESTIS. Gonad.

OVULE. Flower.

OVUM. Gamete.

OWL. Aves, Strigiformes. Nocturnal birds (**Aves**) of prey with hooked beaks and strong curved talons. They differ from the other birds of prey in having the eyes directed forward and surrounded by a more or less evident disk of small radiating feathers. Owls have soft thick plumage and are characteristically silent in flight as a result.

The North American species range in size from the 6-in. Acadian owl to the 2-ft. great horned owl, *Bubo virginianus,* and a similar range of size is noted in the rest of the world from the pigmy owls of the Old World to the great hawk-owl of Australia. The group is represented on all continents.

Although owls are characteristically predacious they do not disdain insects as food. The smaller species catch mice and other small animals, and larger owls prey upon animals as large as rabbits. The great horned owl is said to be an undesirable marauder in poultry yards at times. Certainly it catches birds, but owls in general probably do more good than harm in their destruction of small rodents. (A.W.L.)

OWLET MOTH. Insecta, Lepidoptera. A common name for the **moths** of the great family Noctuidae. Used chiefly in Europe. (A.W.L.)

OX, OXEN. Mammalia, Artiodactyla. Hoofed animals with two functional toes on each foot, with hollow horns in both sexes, and with a long tail, usually tufted at the end. The horns are never spiral as in sheep, goats, and some antelopes, but are usually curved. They are mostly of massive build, with a short head, broad muzzle, and short thick neck.

Most of the species of this group are indigenous to the Old World. North America has the **musk ox** and the **bison,** and South America and Australia are entirely without representatives. A number of species have been domesticated and are used in all favorable parts of the world as a source of meat, milk, and hides, and to a lesser extent as draft animals. The common ox is *Bos taurus.*

The group includes the **aurochs,** humped cattle or **zebu,** the **gaur, gayal, banting, yak, bisons, buffalos** and the **musk ox,** which is related to the sheep. The European bison is also called the wisent or zubr, and the Indian buffalo is sometimes known as the arna. In the Philippines the **tamarao** and **carabao** are the buffalos used for draft purposes. The **anoa** is a small species related to the buffalos and antelopes; it is found in Celebes. (A.W.L.)

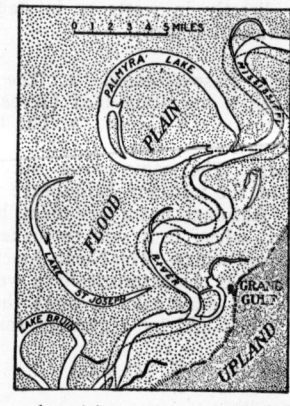

Meanders and ox bow lakes of part of Mississippi River flood plain in 1883 (heavy lines) and 1896 (dotted lines). (*By William Davis, based upon Government Surveys.*)

OX BOW LAKE. The type of lake developed on the flood plains, especially in the delta regions, of large rivers. Ox bow lakes represent the detached or truncated former meanders of the main stream which, from time to time, straightens its course by forming cut-offs or short cuts. (R.M.F.)

OXALIC ACID AND OXALATES. Oxalic acid ($H_2 \cdot C_2O_4 \cdot 2H_2O$ or $COOH \cdot COOH \cdot 2H_2O$) is a white solid, melting point of crystals 101° C. (of anhydrous 180° C.), sublimes at 150° C., soluble in water, alcohol, or ether. **Calcium** oxalate, on account of its solubility characteristics (0.007 gram per liter of water), is of importance in the separation and recovery of oxalic acid. Calcium oxalate plus dilute **sulfuric acid** yields oxalic acid plus calcium sulfate, and the latter may be separated by filtration. Oxalic acid may be obtained by evaporation of the filtrate. With concentrated **sulfuric acid** heated, calcium oxalate, or other oxalate or oxalic acid, yields equal volumes of **carbon dioxide** and **carbon monoxide** gases. Oxalic acid (or oxalates), after being dissolved in dilute sulfuric acid, is readily oxidized to carbon dioxide by **permanganate**.

Representative esters of oxalic acid are: Methyl oxalic acid ($COOH \cdot COOCH_3$), melting point 37° C., boiling point 108° C. at 12 mm. pressure; dimethyl oxalate (($COOCH_3)_2$), melting point 54° C., boiling point 163° C.; diethyl oxalate (($COOC_2H_5)_2$), melting point −41° C., boiling point 186° C.

Oxalic acid may be obtained (1) from some natural products, e.g., wood sorrel, and other members of the oxalis family, as **potassium** hydrogen oxalate, the bark of certain species of **eucalyptus** (sometimes containing 20% calcium oxalate in the cells and cell walls), (2) by reaction of acid with an oxalate, e.g., calcium oxalate plus sulfuric acid as above. Sodium oxalate is made (1) by heating **sodium** formate at 300° C. in vacuum, with the evolution of **hydrogen** gas, (2) by reaction of **carbon dioxide** plus metallic **sodium** at 360° C., (3) by heating wood powder ("sawdust"), particularly of **coniferous** woods, with **sodium** hydroxide (addition of potassium hydroxide enables one to use a moderate temperature, say 220° C.), (4) by oxidation of sucrose or starch with nitric acid. Oxalic acid is a dibasic acid, that is, two series of salts are known, a third series is also known, thus, sodium oxalate ($Na_2C_2O_4$), sodium binoxalate ($NaHC_2O_4$), sodium tetroxalate ($NaH_3(C_2O_4)_2$).

Oxalic acid is used (1) in the preparation of oxalates, e.g., **titanium** potassium oxalate, and esters, (2) in the purification of certain chemicals, e.g., **glycerol**, stearates, (3) in bleaching straw, (4) in ink and rust remover, (5) in the leather and textile industries, (6) in the manufacture of dyes, (7) in the preparation of glyoxalic and glycollic acids by regulated reduction.

A common test for oxalic acid is as follows: Heat solution with resorcinol (see **Phenol**) in a test tube and on cooling add carefully a layer of sulfuric acid. A blue ring at the junction of the two layers indicates oxalic acid. (R.K.S.)

OXAZOLE. Pyrrole and Related Compounds.

OXIDATION. Reactions Involving Oxidation-Reduction.

OXIDES. Oxygen; and individual elements.

OXIMES. Hydroxylamines and Oximes.

OX-PECKER. Aves, Passeriformes. African birds, (**Aves**) of a group related to the starlings. The common name refers to their habit of climbing about the bodies of domestic cattle in search of ticks and other external parasites. They also visit wild animals for the same purpose. (A.W.L.)

OXYACETYLENE WELDING. Welding.

OXYGEN. Symbol: O. Atomic number: 8. Atomic weight: 16.0000. Density: 1.42904 grams per liter, 0° C., 760 mm., or 1.105 when air equals 1.000. Formula of oxygen gas: O_2 (formula of ozone: O_3). Melting point of oxygen: −218.8° C. Boiling point: −183° C. Critical temperature: −118.8° C. Critical pressure: 49.7 atmospheres. Isotopes: 16 (99.76%), 17 (0.04%), 18 (0.20%).

Oxygen is a colorless, odorless, tasteless, non-toxic gas, found free in the atmosphere (23.15% by weight of oxygen in dry air, 20.98% by volume) mixed with **nitrogen, argon,** the rare gases, **carbon dioxide,** and water vapor. Liquid oxygen is strongly magnetic. Molten **silver** dissolves about ten volumes of oxygen, giving it up upon cooling, with an effect known as "spitting." Oxygen was discovered by Priestley in 1774 by heating **mercuric** oxide, and independently in the same year by Scheele. Oxygen is necessary for the burning of substances. The temperature of ignition of each combustible substance is more or less characteristic, for example, **phosphorus** in air is ignited at 34° C., **ether** in air 340° C., **ethyl alcohol** in air 560° C., **kerosene** in air about 300° C., **hydrogen** in air about 600° C.

Oxygen occurs combined with **silicon, aluminum, iron** and other metals in all rocks (average of the solid crust of the earth 46.7% oxygen), as a constituent of practically all plant and animal substances, except **hydrocarbons,** as the most abundant element in the ocean (85.8% oxygen). The process of respiration of animals involves the reaction of free oxygen with the animal organism at the temperature of its surroundings. The burning of fuels when heated to the **ignition** temperature is a process of combination with free oxygen, as also the corrosion of iron, which is an important reaction at ordinary temperature.

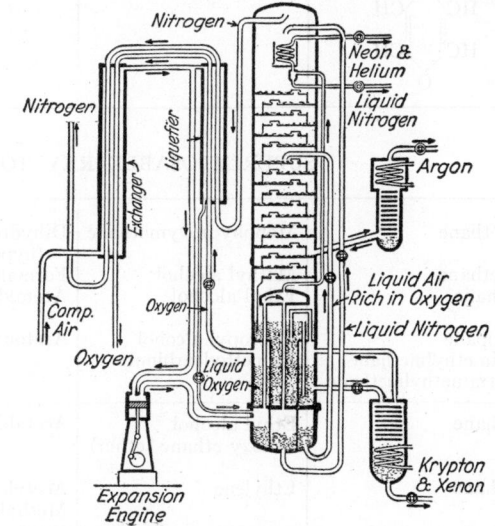

Diagrammatic representation of apparatus for rectification and fractionation of liquid air.

Oxygen is prepared (1) by the fractional distillation of liquid air, (2) by the **electrolysis** of water containing an **electrolyte** such as **potassium** hydroxide, (3) by heating **potassium** perchlorate, which is preferable to potassium chlorate on account of the diminished hazard, and by heating **barium** peroxide or **lead** dioxide, (4) by the reaction of sodium peroxide and water, forming also sodium hydroxide.

When oxygen is subjected to the silent electric discharge, activated atomic oxygen is produced. Atomic oxygen displays an afterglow upon cessation of the cur-

SCHEME SHOWING INTER-RELATIONSHIPS OF OXYGEN-FUNCTION NON-BENZENOID ORGANIC COMPOUNDS

{ CH_4
  $H \cdot CH_3$
  $CH_3 \cdot CH_3$
  $(CH_3)_2CH_2$
  $(CH_3)_3CH$
  $(CH_3)_4C$
	CH_3OH
	$H \cdot CH_2OH$
	CH_3CH_2OH
	$(CH_3)_2CHOH$
	$(CH_3)_3COH$
		{ $CH_2(OH)_2 - H_2O$
		  $H \cdot CHO$
		  $CH_3 \cdot CHO$
		  $(CH_3)_2CO$
			{ $CH(OH)_3 - H_2O$
			  $H \cdot COOH$
			  $CH_3 \cdot COOH$
				{ $C(OH)_4 - H_2O$
				  $HO \cdot COOH$
				  $CO(OCH_3)_2$

$CH_3 \cdot CH_3$ $\begin{matrix} CH_3 \\ | \\ CH_3 \end{matrix}$	$CH_3 \cdot CH_2OH$ $\begin{matrix} CH_3CH_2 \\ CH_3CH_2 \end{matrix} \!\!> O$ $\begin{matrix} CH_2 \\ \| \\ CH_2 \end{matrix}$	$CH_3 \cdot CHO$ $\begin{matrix} CH \\ \| \\ CH \end{matrix}$ $CH_3CO \cdot OCH_3$	$CH_3 \cdot COOH$ $\begin{matrix} CH_3CO \\ CH_3CO \end{matrix} \!\!> O$ $\begin{matrix} CH_2 \\ \| \\ CO \end{matrix}$	
		$CH_3CH(OC_2H_5)_2$	$CH_3CH(OC \cdot CH_3)_2$	

HC═CH
HC CH
  O
 	
C═O
HC CH
HC CH
  O
	$\begin{matrix} CH_2OH \\ | \\ CH_2OH \\ | \\ CH_2OH \\ | \\ CHO \\ | \\ CH_2OC_2H_5 \\ | \\ CH_2OH \end{matrix}$
	
	$\begin{matrix} CH_2 \\ | \\ CH_2 \end{matrix} \!\!> O$
		$\begin{matrix} CHO \\ | \\ CHO \end{matrix}$
		
		$\begin{matrix} CH_2OH \\ | \\ COOH \end{matrix}$ $\begin{matrix} CHO \\ | \\ COOH \end{matrix}$
			$\begin{matrix} COOH \\ | \\ COOH \end{matrix}$	

SUPERIMPOSABLE KEY TO NAMES OF ABOVE COMPOUNDS

Methane Methane Ethane Propane Trimethylmethane Tetramethylmethane	Monohydroxymethane Methyl alcohol Ethyl alcohol Isopropyl alcohol Trimethylcarbinol	Dihydroxymethane (hypothetical) Formaldehyde Acetaldehyde Acetone (ketone)	Trihydroxymethane (hypothetical) Formic acid Acetic acid	Tetrahydroxymethane (hypothetical) Carbonic acid Dimethylcarbonate (ester)
Ethane Ethane	Ethyl alcohol Ethoxy ethane (ether) Ethylene	Acetaldehyde Acetylene Methyl acetate (ester)	Acetic acid Acetic anhydride (acid anhydride) Ketene	
		Diethoxy acetal	Diacetyl acetal	

Furane
Pyrone
	Ethylene glycol
	Glycollic
	  aldehyde
	Ethoxyethyl alcohol
	(ether)
	Ethylene oxide
		Glyoxal
		Glyoxyllic Glyoxalic
		  acid acid
			Oxalic acid	

SUPPLEMENTARY SCHEME FOR OXYGEN-FUNCTION BENZENOID COMPOUNDS

C_6H_6 $C_6H_5 \cdot CH_3$ $(C_6H_5)_2CH_2$ $(C_6H_5)_3CH$ $(C_6H_5)_4C$ $C_6H_5 \cdot H$	$C_6H_5CH_2OH$ $(C_6H_5)_2CHOH$ $(C_6H_5)_3COH$ $C_6H_5 \cdot OH$ (1) HOC_6H_4OH (4)	C_6H_5CHO $(C_6H_5)_2CO$ (1) $O{=}C_6H_4{=}O$ (4)	C_6H_5COOH

SUPERIMPOSABLE KEY TO NAMES OF ABOVE COMPOUNDS

Benzene Toluene Diphenylmethane Triphenylmethane Tetraphenylmethane Benzene	Benzyl alcohol Diphenylcarbinol Triphenylcarbinol Phenol 1,4-dihydroxyphenol (hydroquinone)	Benzaldehyde Benzophenone (ketone) 1,4-benzoquinone (quinone)	Benzoic acid

(R.K.S.)

rent, and the oxygen is notably active with **hydrogen bromide** forming water and bromine, with **hydrogen sulfide** forming sulfur, sulfur dioxide, sulfur trioxide, and sulfuric acid, with **carbon disulfide** forming carbon monoxide, carbon dioxide, and sulfur dioxide and, strangely, reduces **molybdenum** trioxide to a white oxide not reducible with **hydrogen.** The concentration of atomic oxygen obtainable by the silent electric discharge through oxygen is estimated at 20%.

Ozone (O_3) is a blue gas, of characteristic odor, formed when ordinary oxygen is subjected to electrostatic discharge. Density: 1.5 times that of oxygen gas. Melting

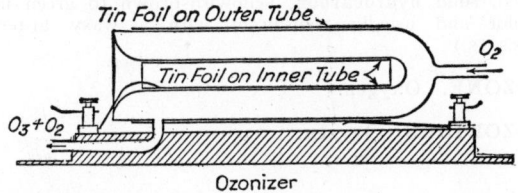

Ozonizer

point: $-251.4°$ C. Boiling point: $-111.5°$ C. Explosive by percussion, or under variations of pressure. Ozone reacts (1) with **potassium** iodide, to liberate **iodine,** (2) with colored organic materials, e.g., litmus, indigo, to destroy the color, (3) with **mercury,** to form a thin skin of mercurous oxide causing the mercury to cling to the containing vessel, (4) with **silver** film, to form silver peroxide (Ag_2O_2) black, produced most readily at about 250° C., (5) with tetramethyldiaminodiphenylmethane ($(CH_3)_2N \cdot C_6H_4 \cdot CH_2 \cdot C_6H_4 \cdot N(CH_3)_2$ in alcohol solution with trace of **acetic** acid to form violet color (**hydrogen peroxide,** colorless; **chlorine** or **bromine,** blue; **nitrogen tetroxide,** yellow). In contrast to **hydrogen peroxide** ozone does not react with dichromate, permanganate, or titanic salt solutions. Ozone reacts with olefin compounds to form ozonide addition compounds. Ozonides are readily split at the olefin-ozone position upon warming alone, or upon warming their solutions in glacial acetic acid, with the formation of aldehyde and acid compounds which can be readily identified, thus serving to locate the olefin position in oleic acid ($C_{17}H_{33} \cdot COOH$) as midway in the chain $CH_3(CH_2)_7CH{:}CH(CH_2)_7COOH$. Ozone is used (1) as a bleaching agent, e.g., for fatty oils, (2) as a disinfectant for air and water, (3) as an oxidizing agent.

Oxides are known of many elements. Metallic oxides are (1) reactive with water, e.g., **calcium, strontium,**

barium oxides (CaO, SrO, BaO), (2) insoluble, e.g., ferric oxide (Fe_2O_3), **ferrous** oxide (FeO), ferroferric oxide (Fe_3O_4), **cupric** oxide (CuO), **cuprous** oxide (Cu_2O); and may be otherwise subdivided, (a) reactive with **hydrochloric** acid to yield **chlorine,** e.g., **lead** dioxide (PbO_2), plumboplumbic oxide (Pb_3O_4), **manganese** dioxide (MnO_2), (b) reactive with water or dilute acid to yield **hydrogen peroxide,** e.g., **sodium** peroxide (Na_2O_2), **barium** peroxide (BaO_2). Non-metallic oxides are (1) reactive with water, e.g., **carbon** **dioxide** (CO_2), **sulfur dioxide** (SO_2), **sulfur trioxide** (SO_3), **phosphorous** trioxide (P_2O_3), **phosphorus** pentoxide (P_2O_5), boron oxide (B_2O_3) forming the corresponding acids (H_2CO_3, H_2SO_3, H_2SO_4, H_3PO_3, H_3PO_4, H_3BO_3); reactive with water to form two acids, e.g., nitrogen trioxide (N_2O_3), yielding **nitrous** plus **nitric** acids (HNO_2 and HNO_3), **phosphorus** tetroxide (P_2O_4) yielding **phosphorus** plus **phosphoric** acids (H_3PO_3 and H_3PO_4), (2) insoluble or non-reactive with water, e.g., **carbon** monoxide (CO), nitric oxide (NO), nitrous oxide (N_2O), silicon oxide (SiO_2). Amphoteric oxides are reactive both with acids and with bases to form salts, e.g., **aluminum** oxide (Al_2O_3) with **sulfuric** acid yielding aluminum sulfate ($Al_2(SO_4)_3$), with **sodium** hydroxide yielding sodium aluminate ($NaAlO_2$), **lead** monoxide (PbO) with **nitric** acid yielding lead nitrate ($Pb(NO_3)_2$) with sodium hydroxide yielding sodium plumbite (Na_2PbO_2), zinc oxide (ZnO) with hydrochloric acid yielding zinc chloride ($ZnCl_2$), with sodium hydroxide yielding sodium zincate (Na_2ZnO_2).

Organic compounds of carbon, hydrogen, oxygen only are discussed as follows:

Acetals. (See **Aldehydes and Related Compounds.**)
Acids. (See **Acids, Carboxylic, and Related Compounds.**)
Acid anhydrides. (See **Acids, Carboxylic, and Related Compounds.**)
Alcohols.
Aldehyde acids. (See **Acids, Carboxylic, and Related Compounds.**)
Aldehydes.
Anhydride, acid. (See **Acids, Carboxylic, and Related Compounds.**
Anhydride, alcohol. (See **Alcohols and Related Compounds** (ethers.)
Anthocyanins. (See **Glucosides.**)
Benzoin, Benzil, and Related Compounds.
Carbohydrates.
Cellulose. (See **Carbohydrates.**)
Coumarone. (See **Furane and Related Compounds.**)

Esters. (See also individual acids.)

Ethers. (See Alcohols and Related Compounds.)

Fats. (See Esters, including Oils, Fats, and Waxes.

Furane.

Glucosides.

Hydroxyacids. (See Acids, Carboxylic, and Related Compounds.

Hydroxyaldehydes. (See Aldehydes and Related Compounds.

Hydroxyketones. (See Aldehydes and Related Compounds.)

Ketenes. (See Aldehydes and Related Compounds.

Ketone acids. (See Acids, Carboxylic, and Related Compounds.)

Ketones. (See Aldehydes and Related Compounds.)

Lactones. (See Acids, Carboxylic, and Related Compounds.)

Lactides. (See Acids, Carboxylic, and Related Compounds.)

Oils, fatty. (See Esters, including Oils, Fats, and Waxes.)

Oxides. (See Alcohols and Related Compounds.)

Phenols.

Phthaleins. (See Phthalic Acid, Phthalates, and Phthaleins.)

Pyrones. (See Furane and Related Compounds.)

Quinones. (See Phenols and Quinones.)

Saccharides. (See Carbohydrates.)

Starch. (See Carbohydrates.)

Sterols. (See Alcohols and Related Compounds.)

Sugars. (See Carbohydrates.)

Tannins.

Waxes. (See Esters, including Oils, Fats, and Waxes.)

OXYPYROLINE. Aminoacids, Polypeptides, and Proteins.

OYSTER. Mollusca, Lamellibranchiata. Bivalve mollusks of the family Ostreidae. Pearl oysters are species of different families found only in warmer oceans, while the edible oysters include a European species, *Ostrea edulis,* a species of the Atlantic coast of North America, *O. virginica,* and the Pacific coast oyster, *O. lurida.* The common oyster of commerce has been transplanted successfully to the Pacific coast. All are marine.

Oysters are the most important of the edible invertebrates. They have been cultivated since the days of the Roman Empire and are now gathered from the oyster beds of the United States alone at an estimated rate of 25,000,000 bushels per year. Fortunately they are extremely prolific, a single oyster of the common American species producing as many as 20–60 million eggs in a breeding season. The young oysters develop in the gills of the parent into ciliated (cilia) larvae called spat which swim freely for a few days and then settle to the bottom and become attached. From that time they are permanently fixed. In cultivated beds they may be moved to favorable places after once attaching themselves, and after 3–5 years they attain market size. The beds require protection against starfishes and other natural enemies, and for the consumer's sake they must be guarded against pollution by human wastes. Oysters that have lain near sewage outlets are dangerous as possible carrier of typhoid germs, although when they are cooked before eating they are likely to be sterilized. (A.W.L.)

OYSTER PLANT, SALSIFY. *Tragopogon porrifolius.* Composite Family.

OYSTER-CATCHER. Aves, Charadriiformes. Birds (Aves) of few species but worldwide distribution. They are related to the avocets, from which they differ in being more stoutly built. They frequent the coasts and eat bivalve mollusks, among a much greater variety of food. The American species is *Haematopus palliatus.* (A.W.L.)

OZALID. Drawing Reproduction.

OZARKIAN. Ordovician.

OZOKERITE. A mineral wax which occurs as a natural, solid **hydrocarbon,** yellowish-brown to green in color and usually translucent with a waxy luster. (E.S.C.S.)

OZONE. Oxygen.

OZONIDES. Oxygen.

P

P SERIES. X-Ray Spectra.

PA. Public Address.

PACA. Mammalia, Rodentia. A stoutly built **rodent,** *Agouti paca,* about 2′ long, marked with rows of light spots on a fawn to blackish ground color. It occurs through most of South America east of the Andes. Related to the agoutis. (A.W.L.)

PACHYDERMATA. An obsolete name for a group of mammals, including the **elephants.** It is reflected in the name pachyderm sometimes applied to these animals. The word means thick-skinned. (A.W.L.)

PACIFIC SUITE. A term proposed by A. Harker, in 1896, for the chemically, structurally, and geographically related **igneous rocks** of the Pacific coast line. Chemically the rocks of this suite are described as calc-alkali, and are represented by such types as **andesites, granodiorites** and their relatives, as compared with the alkali igneous rocks of the **Atlantic Suite.** (R.M.F.)

PAD. Attenuator.

PADDLE FISH. Pisces, Chondrostei. Large fishes (**Pisces**) related to the sturgeons. The snout is prolonged into a spatulate protuberance. Only two species are known, one in the large rivers of China and the other in the Mississippi and its larger tributaries. The latter, *Polyodon spathula,* is also called the spoon-bill, spoon-beaked sturgeon, and spoon-billed catfish. (A.W.L.)

PADDY BIRD. Aves, Passeriformes. The Java sparrow, *Ardeola grayii,* a **munia.** (A.W.L.)

PAEDOGAMY. Autogamy.

by enlargement and deformity of the skull, spine and long bones.

The disease develops after 60, the first sign being an increase in the diameter of the head and a bowing of the legs, causing shortness of height. Some cases are marked by severe pain over the involved bones.

Treatment is unsatisfactory. **Calcium, vitamin D, ultra-violet** light, and **x-ray** are used. (R.S.M.)

PAHOEHOE. An Hawaiian term introduced to geological nomenclature by C. E. Dutton, in 1883, and signifying a lava flow with a smooth, ropy surface. Contrast with **aa.** (R.M.F.)

PAINT. Pigments, Paints, Varnishes; Protective Coatings; Resins.

PAINTER. Puma.

PALA. Impala.

PALATE. The partition between the oral cavity and the nasal passages. The anterior portion contains flat bones and is called the hard palate, while the posterior soft palate is made up entirely of soft tissues. (A.W.L.)

PALEA. A small, usually thin **bract** or scale borne on the **axis** of a grass flower, just above the **lemma.** (R.M.W.)

PALEOBOTANY. This is the study of ancient plants which are known today only through **fossil** remains. These are of two general types: prints and petrifactions. Most of our knowledge of fossil plants has been derived from the study of petrifactions in which the original structure of the plant has been replaced by mineral material.

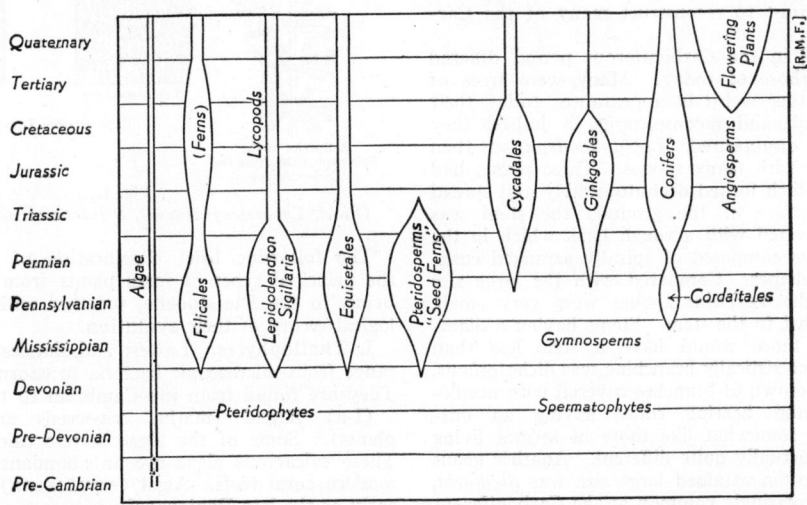

Geologic range of plants. (*Field, Geology Manual, Part II, Princeton University Press.*)

PAEDOGENESIS. The attainment of sexual maturity by animals during an immature stage of development. Closely related to **neoteny.** (A.W.L.)

PAGET'S DISEASE. *Osteitis deformans.* A chronic disorder of the bones of unknown origin characterized

Originally the study of fossil material was a laborious process, for sections of the material were necessary and sections of such fossils were not easily made. At first these sections were obtained by cutting the fossil in two, grinding the cut surface smooth and examining it, or fastening it firmly to a glass slide and making a second cut close to the slide, thus getting a thin section which was then ground as thin as possible. Naturally the preparation of fossil material was not only a tedious process, but also one that wasted much of the material.

In recent years an entirely different and much more satisfactory method has been developed. One surface of the fossil is first ground flat. This flat surface is then etched with **hydrochloric acid** or **hydrofluoric acid** according to the material composing the fossil. When the acid has etched deeply enough, it is washed off with water and the surface dried. The dried surface is flooded with a celloidin solution and the latter allowed to dry. The dried celloidin film is then carefully peeled from the surface of the fossil, removing with it a thin film of fossil material. The peeled surface is reground and the process repeated as long as desired, or until the fossil is completely "sectioned." The thin peels may be either mounted on glass slides, like sections of living material, or they may be dried under pressure, so that they remain flat. The advantages of fossil peels for scientific investigations are many. They are so thin that they can easily be examined with high magnifications; they can be rapidly made; they permit serial sections of the material. In this process little of the material is wasted. Finally, fossil peels permit examination of material which can be sectioned only with the greatest difficulty by any other method. This method of preparation has greatly advanced the study of fossil life.

Paleobotany yields much interesting information about the plants of prehistoric times, notably of the **Carboniferous** and **Devonian** periods. Probably the best known fossil plants are those which grew during the Carboniferous period of the **Palaeozoic** era. The many types of coal are geochemical compounds and mixtures of plants including their spore cases, oxidized cellulose (mother of coal), lignite and bacterial by-products The nature of the processes by which coal has been formed is such that coal is often not the best material to study to get a picture of the fossil plants composing it. But found associated with the coal are mineral masses called coal balls. Within these, plant remains are often preserved with beautiful detail. Microscopic studies of peels and thin sections of coal balls have made it possible for the paleobotanist to reconstruct many of the Carboniferous plants.

Plants growing in the Carboniferous period differed strikingly from those of today. Many were trees of large size but rather weird in appearance; today their relatives are small and inconspicuous, if indeed they have not entirely disappeared. *Lepidodendron* was then a common genus with many species. These plants had radiating roots which forked **dichotomously** and spread out near the surface of the ground; the stem was columnar and covered with a rough bark which in the younger parts was composed of spirally arranged cushions of various shapes. Compared with the large size of these stems the **vascular** tissues were very small, giving little support to the stem. Stems having a diameter of 10″ or more would have a stele less than 3″ across. Characteristically branching was dichotomous, forming an open crown of branches covered with needle-shaped leaves, and bearing cones having an outward appearance somewhat like those of several living **conifers**, but structually quite different. Another genus whose members often attained large size was *Sigillaria*, which became increasingly common as the Carboniferous period continued. These plants also had tall thick stems each bearing at its top a crown of branches covered with needle-like leaves. Giant **horse-tails**, known as *Calamites*, were also abundant.

All these plants and many others have a structure which suggests that they grew in a very uniform climate. From the Proterazoic to the late **Silurian** or Devonian, the only fossil plants are calcareous **algae**, usually marine.

In the Devonian have been found the first terrestrial fossil plants, which are the progenitors of the Carboniferous flora. The plants were without roots or leaves, but had a prostrate **rhizome** from which arose erect sparingly branched stems, many of which bore **sporangia** at their tips. These stems had a well developed vascular system. Other Devonian plants were larger, some being 10–12′ high, and much branched. As the Devonian period continued, the plants became larger and more like those in the Carboniferous.

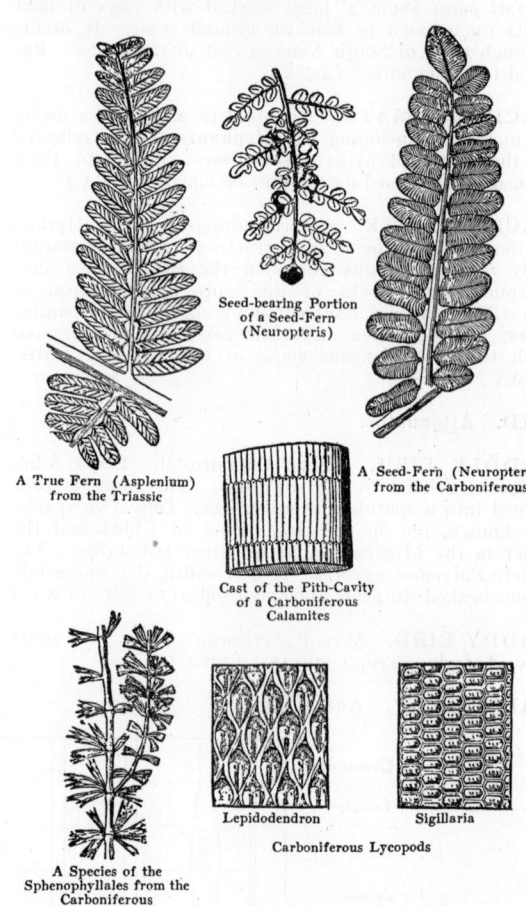

Seed-bearing Portion of a Seed-Fern (Neuropteris)

A True Fern (Asplenium) from the Triassic

A Seed-Fern (Neuropteris) from the Carboniferous

Cast of the Pith-Cavity of a Carboniferous Calamites

Lepidodendron   Sigillaria

Carboniferous Lycopods

A Species of the Sphenophyllales from the Carboniferous

Fig. 1.

(*Field, Laboratory Manual, Princeton University Press.*)

The following brief classification of plants includes the principal types of fossil plants from the **pre-Cambrian** to the **Pleistocene**, together with the paleontological record of their **evolution**.

I. Thallophytes. Lowest and simplest plants. They range from microscopic bacteria to enormous sea-weeds. They are found from pre-Cambrian to the present.

(I-a) **Algae** (marine sea-weeds and fresh-water plants). Some of the algae secrete calcium carbonate. These calcareous algae are an abundant constituent of modern coral reefs. Algal (**cryptozoan**) reefs occur as early as the late **Proterozoic**.

(I-b) **Fungi.** These plants depend upon other living plants or organic debris for food. Include familiar forms such as mildew, yeast, wheat rust, and mushrooms. Not important as fossils.

II. Bryophytes (moss plants). These plants possess organs which resemble true stems and leaves, though no true roots or woody tissue. Small and delicate forms, living largely in moist habits. Occur from **Carboniferous** to present. Not important as fossils.

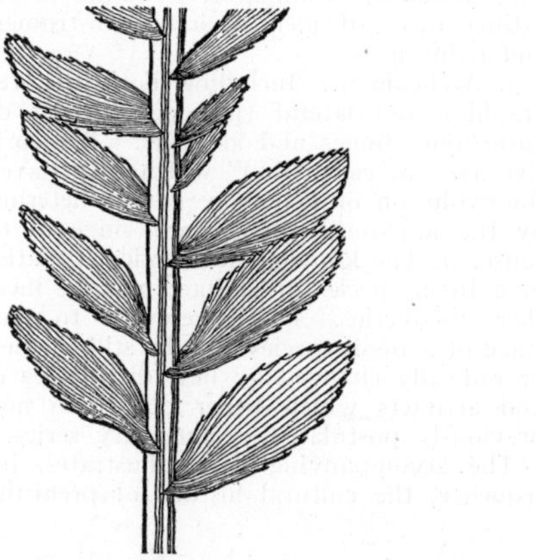

Fig. 2.
A Cycad (*Plagiozamites*) from the Carboniferous. (*Field, Laboratory Manual, Princeton University Press.*)

III. **Pteridophytes** (fern plants). These plants have acquired roots and woody tissue in stems, and larger and more complex plant body. **Devonian** to present.

(III-a) **Filicales** (true **ferns**). A dominant Carboniferous type, illustrated on page 1032. 4000 living species, ranging from small temperate ferns to large tropical tree-ferns.

(III-b) **Lycopodiales** (club mosses). Large tree forms in Carboniferous. 500 living species, all small and herbaceous. *Lepidodendron* and *Sigillaria* are best known Paleozoic types, illustrated on page 1032.

(III-c) Equisetales (**Horsetails**). Tall tree types (*Calamites*) in Carboniferous. 25 species now living, all small and herbaceous, illustrated on page 1032.

(III-d) Sphenophyllales. Small, thin-stemmed plants (illustrated on page 1032) of late Paleozoic. Now extinct. Closely related to Equisetales.

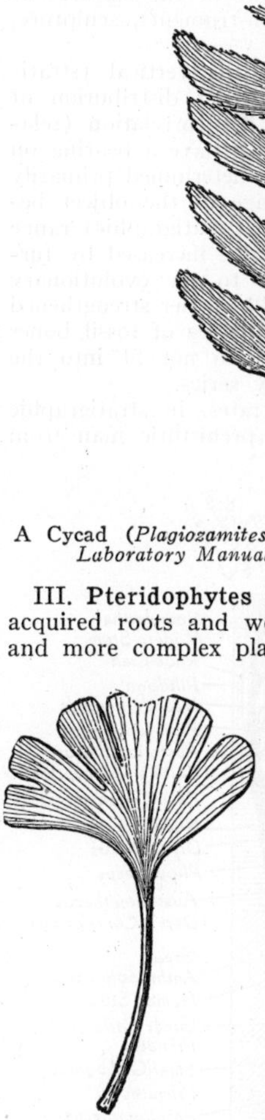

Fig. 3.
A Jurassic Ginkgo. (*Field, Laboratory Manual, Princeton University Press.*)

IV. **Spermatophytes** (seed plants). Dominant terrestrial forms of modern flora. Most highly evolved plants. Not dependent upon moisture for fertilization. Three important groups:

(IV-a) Cycadolfilicales or Pteridosperms (seed-ferns). Important Paleozoic group; now extinct, illustrated on page 1032. Possessed fern-like foliage but also true though simple seeds. Perhaps represent transitional form between ferns and modern seed-plants. An important fossil genus is *Glossopteris* which is found in the **Permian** formations of both South America and Africa.

(IV-b) **Gymnosperms** (naked seeds). Seeds not protected by a covering, but borne on scales, usually in cones. Trees mainly evergreen, with scales or needle-shaped foliage. Five groups:

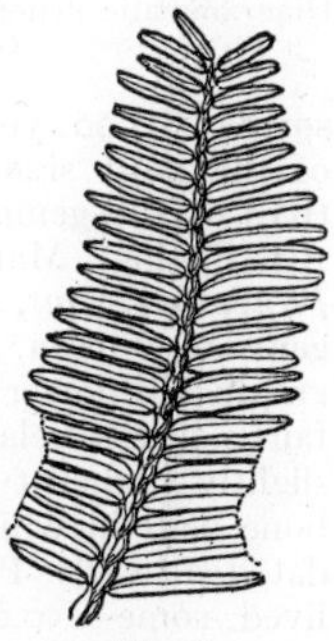

Fig. 4.
A Common Tertiary Conifer (*Sequoia*). (*Field, Laboratory Manual, Princeton University Press.*)

(IV-b-1) Cycadales. Tropical **cycads** or sago-palms. Dominant in **Mesozoic** in 3 families, only one of which has survived. (Figure 2 above.)

(IV-b-2) Cordaitales. Tall Paleozoic tree form. Now extinct. Supposed ancestor of modern conifers.

(IV-b-3) Ginkgoales (Figure 3 above). Peculiar tree, dominant in Mesozoic. Only 1 species in Western China now living.

(IV-b-4) Coniferales. Modern evergreen **conifers.** Mesozoic to present (Figure 4). 350 species now living, mostly temperate forms, such as the pines, spruces, hemlocks and cedars.

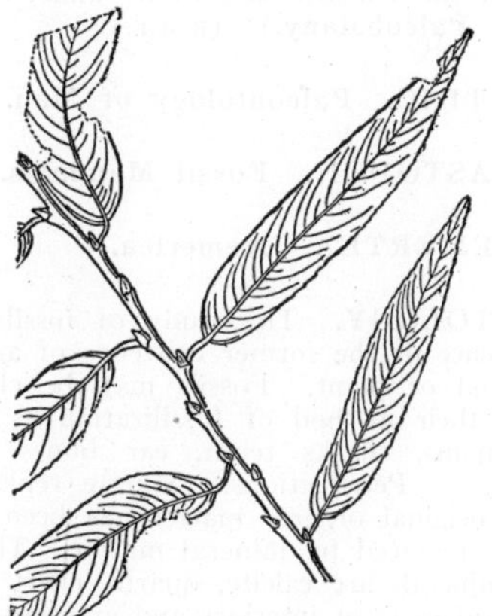

Fig. 5.
A Tertiary Willow (*Salix*). A common Dicotyledon. (*Field, Laboratory Manual, Princeton University Press.*)

(IV-b-5) Gnetales. Desert type. Peculiar and unimportant group.

(IV-c) **Angiosperms** (enclosed seed). Seed protected by seed coats and enclosed in a ripe ovary, the fruit. Only group which produces real "flowers." Leaves usually broad and **deciduous.** Represent highest type and most numerous of modern plants. Upper **Cretaceous** to present. Two groups:

(IV-c-1) **Monocotyledons.** Plant begins growth with single seed leaf or **cotyledon.** Leaves parallel-veined. No growth rings in the stem. Parts of the flowers arranged in threes or multiples of three. Mostly herbaceous. Supposedly derived from primitive dicotyledons. 30,000 species now living; grasses, sedges, etc.

(IV-c-2) **Dicotyledons.** Plant begins growth with two cotyledons. Leaves net-veined. Annual growth rings present in the stem. Floral parts arranged in multiples of four or five. 100,000 species now living; oaks, willows, maples, chestnuts. (Figure 5.) R.M.F., R.M.W.)

PALEOCENE. The earliest period in the **Cenozoic** Era of the geologic time-scale. (See **Historical Geology.**) It is also spoken of as basal **Tertiary.** The term was first proposed by Schimper in 1874. The greatest thickness of the formations of this system occur in Wyoming. The Paleocene began approximately 60 million years ago and lasted for about 10 million years. In the United States the sediments are mainly of terrestrial origin, occurring as intermontane deposits of sands, gravels, and clays. The Paleocene is chiefly remarkable in that the **dinosaurs** have been replaced by the archaic or earliest types of mammals, including survivors of the **Mesozoic** Multituberculates. Principal types are creodonts (archaic flesh-eaters), amblypods and **condylarths** (primitive hoofed animals). Also several species of primitive **insectivores,** lemuroids (ancestral monkeys) and **carnivores.** (R.M.F.)

PALEOCHRONOLOGY. Stratigraphy, Paleontology, Evolution, Historical Geology.

PALEOCLIMATOLOGY. The study of ancient climates which obtained during the previous periods of the earth's history, as determined by the criteria derived from **sedimentary** rocks and the included **fossils.** Certain types of fossils such as **corals** are assumed to have lived in tropical seas. **Limestones** are assumed to have been formed only in warm seas. **Aeolian** sandstones, red sediments, **playa** deposits, salt, **gypsum,** and desiccation products in general, are usually evidence of aridity. **Tillites** are evidence of glacial climates. Fossil plants are particularly good evidence as to the climatic conditions under which they lived. (See **Paleobotany.**) (R.M.F.)

PALEOLITHIC. Paleontology of Man.

PALEOMASTODON. Fossil Mammals.

PALEONEMERTEA. Nemertea.

PALEONTOLOGY. The study of fossils. A fossil is the evidence of the former existence of an organism, either animal or plant. Fossils may be classified according to their method of fossilization as follows: 1. Actual remains, sharks teeth, ear bones of whales, **chitin,** etc. 2. Petrifactions. Minute replacements in which the original organic matter has been completely or partially replaced by mineral matter. The principal replacing minerals are calcite, quartz, chert and pyrite. 3. Molds and casts of interiors and exteriors. 4. Prints of leaves, jellyfish, etc., sometimes showing carbonized traces of organic matter, as in the case of some fossil plants. 5. Coprolites, fossil excrement. 6. Tracks, trails and burrows. The geologist is principally interested in fossils, not from the point of view of paleobiology, but rather as valuable aids to **stratigraphy** in helping to determine the relative ages of the **sedimentary rocks** in which the fossils occur. Unfortunately the fossil record is so incomplete that the general principle of evolution is of little or no value for determining the age of the smaller divisions of the geological time scale. Because of the complexity of the fossil record and the host of fossil types, including whole classes of organisms, which are now extinct, Paleontology has been subdivided into the following natural divisions: 1. Micropaleontology, or the study of microscopic fossils, especially the **foraminifera.** 2. Invertebrate Paleontology. 3. Vertebrate Paleontology. 4. Paleobotany. 5. Stratigraphic Paleontology. (R.M.F.)

PALEONTOLOGY OF MAN. Since the time of Charles Darwin's publication "On the Origin of Species" (1859), certain formations from the **Pleistocene** have been found to contain fossil remains which are anatomically different from any existing race of men, but which are also more "human" than any other existing type of anthropoid. Within the last few years the study of fossil man, including the study of his increasing mentality, as disclosed by his creative art (artifacts), has led to a special branch of geological history called the Paleontology of Man. The accompanying diagrammatic presentation of the fossil and stratigraphic record of man is in the form of a family tree, buried in the stratigraphic record of successive **sedimentary** formations containing numerous fossil animals and plants, which help to date, or correlate, the ancestral stock and divergent branches of *Homo Sapiens.* The generic and specific names represent specific types, certain of which are supposed to be "milestones" in man's total evolutionary journey from a lower form of anthropoid to his present state. The geological proofs of the evolution of man depend upon the following studies, both in the field and in the laboratory:

1. Paleontology. The comparative anatomy of the fossil bones, principally the bones of the skull, teeth, and, to a certain extent, the limbs. The comparison of analogous bones of different fossil types. The compari-

son of the analogous bones of fossil types with those of living anthropoids, including man.

2. Artifacts. Comparative studies of the cultures of extinct races of men, including instruments, sculpture, and painting.

3. Archeology. Including both the vertical (stratigraphic) and lateral (paleogeographic) distribution of anthropoid bones and artifacts. The correlation (relative age) of each "find" which may have a bearing on the evolution of man may not be determined primarily by the apparent evolutionary stage of the object because: *a.* The known total vertical (stratigraphic) range of a fossil species or artifact may be increased by further discoveries. *b.* Concepts as to the evolutionary stage of a fossil species may be still further strengthened or radically changed by new discoveries of fossil bones and artifacts which either do or do not fit into the previously postulated evolutionary series.

The accompanying chart illustrates, in stratigraphic sequence, the cultural history of prehistoric man from

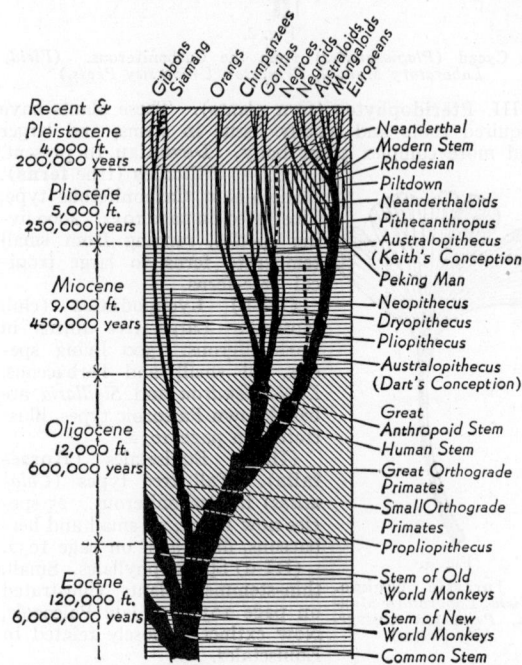

Diagrammatic synopsis of human evolution. (Keith.) (*Field, Outline, Barnes & Noble.*)

some 350,000 years ago to the present day. A few of the most significant progenitors, or prehistoric relatives of the genus *Homo* are:

1. Peking Man (*Sinanthropus*). This type, and *Pithecanthropus*, are the oldest known primitive progenitors of man. While *Sinanthropus* may occur stratigraphically lower than *Pithecanthropus,* it appears to be fairly closely related to *Pithecanthropus,* and possibly a slightly higher type (see chart of cultural ages). The bone beds in which the Peking Man was discovered are dated as early Pleistocene, and it is estimated that he lived some 250,000 years ago. The discoveries were made in a limestone cave near Peking. The undisturbed cave deposits contain 110' of richly fossiliferous deposits which date the "human" remains.

2. Java Man (*Pithecanthropus*). Has been thought, and still may be, the nearest known approach to the "missing link." Total evidence intimates an animal having a brain of 900.94 cc., or approximately that of the lowest limit amongst the primitive living races of mankind (aboriginal Australian woman). The discovery was made in the same stratigraphic horizon as a

number of **fossil mammals** of **Pliocene** age, near Trinil, Java, in 1891. The restoration of the type depends upon a skull cap, some teeth and a femur which may or may not have belonged to the same individual, or to the same species.

3. **Piltdown Man** (*Eoanthropus*). The most ancient fossil man so far discovered in England. Cranial capacity some 300–400 cc. greater than that of the Peking Man. While not in the direct evolutionary line of the existing races of man, it probably represents a higher western race which was probably contemporaneous with a lower eastern type. The skull was found at Piltdown, Sussex, England, in river gravels. The associated artifacts have been determined as probably Pre-Chellean in age, and may be older than the skull. The exact age of the formation is still under dispute, but it appears to be Pliocene or early **Pleistocene.**

4. **Neanderthal** (*Homo neanderthalensis*). A well-established race with progenitors probably well down in the Pliocene, and probably contemporaneous with the earlier races previously cited. Cranial capacity on the order of 1600 cc. meters, or possibly somewhat larger than that of modern man. This race appears to have died out during the late Pleistocene and is not directly ancestral to the living races of man. The Neanderthal race had a wide geographic distribution, including eastern and western Europe and Africa. The first discovery was the Gibraltar skull in 1848, but the type locality is a cave in the Neanderthal gorge of the valley of the Düssel, a German tributary of the Rhine. Numerous complete skeletons are known. The jaw, flatness of the cranium, pronounced eyebrow (supra-orbital) ridges and thickness of bone in the skull are decidedly simian. The development of retouch, and the flake for special types of flint instruments have tended to supersede the more primitive general utility tool called a hand stone or coup de poing of the earlier Chellean and Acheulian cultures.

5. **Crô-Magnon** (*Homo Sapiens*). The highest type of the extinct races of men, both anatomically and mentally. Not directly ancestral to the modern races

Sketches of stone implements of various cultural ages used by Paleolithic man. Upper left, Chellean; upper right, Mousterian; and lower, Solutrian. (*The Nature of the World and Man, University of Chicago Press.*)

of men, because of distinct differences in the skull. Abundant fossil skeletons have been found, including children and adults of both sexes. Skull large, and brain size exceeding that of the average modern man. On the other hand the skull retains primitive characters. It is long and narrow (dolichocephalic) with relatively broad face, large eye sockets and pronounced supra-orbital ridges. Complete skeletons show that the males were broad-chested and fairly erect, with an average stature of 6′ 1½″. The culture ranges from Aurignacian to the close of the Paleolithic (Azilian). The Aurignacian culture includes flint, bone and horn instruments, including the burin, or engraving tool. Also the first evidence of prehistoric art in the form of sculptures, tinted engravings on the walls of limestone caverns, and art moblier, or figurines. During the Solutrian occurred the culmination of the retouch in the making of **flint** instruments, as exemplified in the **pointes en feuille de laurier,** or laurel-leaf points. The follow-

ing Magdalenian period represents the culmination of the Paleolithic or prehistoric culture. With the decadence of flint, and the increasing use of bone instruments also occurs the climax of Paleolithic art, including engraving, sculpture and painting. The Azilian shows a complete revolution in culture with a degeneration of the retouch in making flint instruments and the foreshadowing of the polished stone instruments of the Neolithic. (R.M.F.)

PALEOSTRACA. Xiphosura.

PALEOZOIC. The **Era** of "ancient" life or the age of invertebrates. Subdivided from the base up, into the following **Periods: Cambrian, Ozarkian, Canadian, Ordovician** (lower Paleozoic), **Silurian, Devonian** (middle Paleozoic), **Mississippian, Pennsylvanian, Permian** (upper Paleozoic). The lower Paleozoic was characterized by: the first known marine faunas; domi-

A highly generalized map of North America showing a rather typical arrangement of the principal Paleozoic positive and negative areas, and also the Appalachian and Cordilleran geosynclines, from Middle Cambrian to Middle Pennsylvanian time.

nance of **trilobites;** rise of animals with hard shells (Cambrian); rise of **nautiloids,** armored **fishes** and **corals;** also first evidence of colonial life (Ordovician). The middle Paleozoic was characterized by: the rise of **lung-fishes** and **scorpions** (Silurian); first known land floras (Devonian) not very different from those of the Pennsylvanian; earliest evidence of a terrestrial vertebrate in the form of a single footprint from the Devonian of Pennsylvania. The upper Paleozoic was characterized by ancient sharks, and **echinoderms** (Mississippian); primitive reptiles and insects (Pennsylvanian); periodic glaciation and extinction of many Paleozoic groups during and after the Permian. Rise of modern insects, land vertebrates and **ammonites.** (Permian.) The Paleozoic Era began 500 million years ago and lasted for 300 million years. (R.M.F.)

PALINGENESIS. A term proposed by Sederholm in 1907 for the reversal of **petrological** processes, such as the melting of **batholithic** and **metamorphic rocks** *in situ.* Theoretically the completion of a cycle of rock genesis. (R.M.F.)

PALISADES DISTURBANCE. Triassic.

PALLADIUM. Symbol: Pd. Atomic number: 46. Atomic weight: 106.7. Density: 12.16. Hardness: 4.8. Melting point: 1553° C. (Isotopes: page 290.)

Compact palladium is a white metal, and due to its property of softening at temperatures below the melting point can be easily welded. Heated in air to redness palladium becomes coated with oxide. The molten metal absorbs **oxygen**, and on cooling "spits" like **silver**. Finely divided palladium adsorbs 1000 to 3000 times its volume of hydrogen gas, and retains most of this gas when heated to 100° C. **Acetylene** is adsorbed similarly. Palladium is slowly dissolved by boiling **hydrochloric** or **sulfuric acid** and by cold **nitric acid**; is readily dissolved by **aqua regia**; upon heating reacts with **chlorine** and with **sulfur**; upon fusion with **potassium** hydrogen sulfate forms palladous sulfate, soluble in water; reacts with fused **sodium** hydroxide. Discovered by Wollaston in 1803.

Palladium metal is used in dentistry, in non-magnetic springs for watches and clocks, as a permanent mirror when deposited on glass, in jewelry as white gold when alloyed with **gold**, being superior in whitening power to **platinum** in this respect.

Palladium occurs native in platinum ores, sometimes to the extent of 2%, and in sufficient amount in the nickel ores of Canada to be recovered in the process of their manufacture. When **osmium** and **ruthenium** are present, they are removed as volatile oxides; when platinum and **iridium**, by precipitation of the **ammonium** chloro-compounds, insoluble in alcohol, upon addition of **ammonium** chloride; and precipitation of palladium as the ammonium chloro-compound with ammonium chloride and **chlorine** leaves rhodium in the solution. The ammonium chloropalladate is ignited and reduced by heating with hydrogen to yield palladium metal.

Chlorides: Palladous chloride ($PdCl_2$), absorbs **carbon monoxide** gas, a reaction which is used for estimating this gas in mixtures; palladium tetrachloride ($PdCl_4$), more commonly as chloropalladic acid (H_2PdCl_6).

Chloropalladate: Ammonium chloropalladate ((NH_4)$_2$ $PdCl_6$), slightly soluble in water, and insoluble in alcohol.

Oxides: Palladium monoxide (PdO), black; palladium dioxide (PdO_2), black. (R.K.S.)

PALLET, PALETTE. A structure formed of horny and calcareous materials, associated with the siphons of **shipworms**. A pair of pallets lie near the end of the siphons and serve as a plug for the burrow in which the worm lives. (A.W.L.)

PALLIAL GILL. Folds of the **mantle** which serve as respiratory organs in some of the **bivalve mollusks**. (A.W.L.)

PALLIAL LINE. The scar on the inner surface of the shell of a **bivalve mollusk** which marks the marginal attachment of the **mantle** folds. (A.W.L.)

PALLIUM. A mantle-like structure. 1. The cortex of the cerebrum (**brain**) with its underlying white substance. 2. The **mantle** of mollusks. (A.W.L.)

PALM FAMILY. Palmaceae. This family of plants contains some 1100 species, most of which are tropical. Many are of large size. A tall woody stem bearing at its top a crown of large compound leaves characterizes most of them, although some are short and bushy, while a few are vine-like. Only rarely does branching occur, although many do develop numerous basal offshoots. The leaves are either **pinnately** or **palmately** compound. The **inflorescence** is either a simple or a compound **spike**, surrounded by a **spathe** which may become extremely large. The flowers are regular, with their parts in threes or multiples of three. Wind **pollination** generally occurs in palms. The **fruit**

is a berry, a drupe, or a nut which in many species has a fibrous mesocarp.

Many species of palms are planted extensively, because of their ornamental habit, in tropical and subtropical countries. Large numbers are also grown as greenhouse or conservatory plants. A few species are of great economic value. Chief among these is the **Coconut** palm. Somewhat less valuable than the Coconut is the Date Palm. This plant, *Phoenix dactylifera*, is a tree whose mature trunk often attains a height of 75–100'. From the base of this trunk numerous **adventitious** roots extend into the soil to a depth of 20' or more. From the base of the stem many basal offshoots develop. The pinnate leaves are from 10–20' long, with the individual pinnae sharp-pointed, linear and from 1–3' long. These leaves remain attached for an indefinite time to the trunk, giving it a very ragged appearance. In cultivation the old leaves are removed. The inflorescence is a large branched spike enclosed in a large tough spathe. In a mature tree a single flower cluster may be composed of several thousands of flowers. The flowers are of two different sexes, borne on different plants, and are small, wax-white and of firm texture. The pistillate (see **Pistil**) flowers have three **carpels**, each with a short curved **stigma**. The carpels are almost surrounded by the **perianth**, composed of three united **sepals** and three **petals**. The staminate (see **Stamen**) flowers, bearing six stamens, are of larger size than the pistillate and more showy. The fruit is a drupe. After fertilization only one of the three carpels develops, the other two being suppressed. At first the carpel is wax-white, but some time after fertilization it becomes green and remains so during growth. As the drupe becomes mature, the color changes to yellow or red or a blending of these colors, according to the variety of date. When ripe the color of the fruit varies from pale straw color through deep amber to a deep purple. The seed of the date contains an abundance of **endosperm** and a rather small **embryo**.

Many varieties of dates are grown in cultivation. Since cross-pollination is usually necessary to produce seeds, propagation by seeds cannot be used to increase the number of plants. For such plants would not remain true to the seed-bearing parent in type but would be affected by the pollen from the staminate plant; such crossing would cause the formation of new varieties with the possibility that less desirable forms might arise. Furthermore, about half the seedlings would be staminate, producing no fruit. To perpetuate desirable varieties propagation is largely by means of the basal offshoots. These shoots are cut from the parent and planted, ensuring new plants identical with the parent.

Cultivation of the date has been carried on in Arabia and adjoining countries since prehistoric times. It is probable that the plant is native to this region, even though it is unknown in the wild state. Today the date palm is grown widely in regions having a sufficiently hot climate. The Mesopotamian valley is the chief producing region of the world. In the United States the date can be grown successfully only in a very limited region in the southern parts of California, Nevada and Arizona.

The principal product of the plant is the edible fruit. In the Orient the fruit is often fermented, yielding a very potent alcoholic drink, and also vinegar. Various materials used in building houses, and making baskets, ropes, and household articles are also obtained from the tree in regions where other sources of such materials may be lacking. (See **Esters**.)

Another useful palm is the Oil Palm, *Elaeis guineensis*, a native plant occurring in large numbers in the forests of the west coast of Africa. The plant is also extensively cultivated elsewhere in Africa and in the East Indies, particularly in Sumatra. From the fruit of this palm two valuable oils are obtained. The orange-yellow **pericarp** yields palm oil, much used in the making of soaps. The **endosperm** yields a white pleasantly flavored oil known as palm-kernel oil, also used in making soaps. In Africa the fruits are eaten by the natives.

In India, *Phoenix sylvestris* is widely cultivated as a plant from which sugar is obtained. Many other palms are used as sources for sugar in regions where they grow. (R.M.W.)

PALMATE. Leaf.

PALMITIC ACID AND PALMITATES. Palmitic acid ($H \cdot C_{16}H_{31}O_2$ or $C_{15}H_{31} \cdot COOH$ or $CH_3(CH_2)_{14} \cdot COOH$) is a white solid, melting point $64°$ C., boiling point $272°$ C. at 100 mm. pressure, insoluble in water, moderately soluble in alcohol, soluble in ether. Palmitic acid is present as cetyl ester in spermaceti from which, by **hydrolysis**, the acid may be obtained; it is present in bee's wax as the melissic ester; and in most vegetable and animal oils and fats, in greater or less amounts, as glyceryl tripalmitate or as mixed **esters**, along with **stearic** and **oleic acids**. Palmitic acid is separated from stearic and oleic acids by fractional vacuum **distillation**, and by fractional **crystallization**. With **sodium** hydroxide, palmitic acid forms sodium palmitate, a soap. Most soaps are mixtures of sodium stearate, palmitate and oleate.

Representative esters of palmitic acid are: methyl palmitate ($C_{15}H_{31}COOCH_3$) melting point $30°$ C., boiling point $195°$ C. at 15 mm. pressure; ethyl palmitate ($C_{15}H_{31}COOC_2H_5$), melting point $24°$ C., boiling point $185°$ C. at 10 mm. pressure; cetyl palmitate ($C_{15}H_{31}COOC_{16}H_{33}$), melting point $54°$ C.; glyceryl tripalmitate (tripalmitin) ($C_3H_5(COOC_{15}H_{31})_3$), melting point $65°$ C., boiling point $310°$ C., approximately.

As the glyceryl ester, palmitic acid is one of the constituents of many vegetable and animal oils and fats. (See **Esters**.) (R.K.S.)

PALOLO WORM. Annelida, Polychaeta. Marine worms of the genus *Leodice* which live in burrows in coral rock. They have two body regions. In the posterior part the reproductive organs develop, and when they are mature this part breaks away from the anterior region and swims to the surface of the sea. The attainment of sexual maturity occurs at a specific time, and during this period the water swarms with the reproductive portions of the worms. The anterior part remains in the burrow and during the succeeding year produces another reproductive region. One species lives in the southern Pacific Ocean and another in the West Indies. (A.W.L.)

PALP, PALPUS. An appendage of sensory functions. 1. In some of the segmented worms thick fleshy protuberances on the head are called palpi. 2. **Bivalve mollusks** bear two pairs of flap-like appendages near the anterior end of the body which are called palps. 3. The **maxillae** and **labium** of insects each bear a pair of segmented appendages called palpi. (A.W.L.)

PALPIFER. Maxilla.

PALPIGER. Labium.

PALPITATION. The consciousness of temporarily abnormal action of the **heart**, characterized by increased force, rate or irregularity of the beat. Palpitation is a symptom present in many conditions; it is not a disease in itself. It is commonly seen in emotional, nervous or excitable subjects, in acute fevers, following the drinking of tea, coffee or alcohol in certain individuals, and often as a symptom of organic heart disease. (R.S.M.)

PALSY. Paralysis agitans.

PAMPEAN FAUNA. Pleistocene.

PANAMA HAT PALM. *Carludovica palmata.* Cyclanthaceae. This plant is very abundant in tropical forests of America, where it forms large clumps of long-**petioled** leaves. In contrast to true palms, the Panama hat palm is stemless. It is frequently planted as an ornamental plant. For making Panama hats, the young leaves are gathered, cut into narrow strips, bleached, and woven into the hats. These hats are actually manufactured only in a section of Equador, and not in Panama. (R.M.W.)

PANCREAS. A large digestive gland whose duct empties into the alimentary tract (**Digestive System**). The name is sometimes applied to a mass of tissue surrounding the duct of the principal digestive gland of the **squid** but it belongs characteristically to the **vertebrates**. The vertebrate pancreas develops as a group of outgrowths from the embryonic (**embryo**) gut just behind the stomach. In the higher vertebrates one outgrowth is dorsal and one or two ventral, but all normally unite to form one mass which usually discharges by one of the original ducts, although in some species both persist. The ventral duct often joins the duct of the **liver**. In man this duct becomes the pancreatic duct, joining the hepatic duct to discharge into the intestine by the common bile duct. The pancreas is a compound acinous gland. (See **Glands**.) The human pancreas is a large, elongated gland situated on the posterial abdominal wall behind the stomach above the level of the navel, lying between the spleen on the left and the duodenum on the right. The pancreas has both an external and an internal secretion.

The external secretion of the pancreas empties into the duodenum. This digestive fluid contains at least three important **enzymes**: 1. Trypsin, which interacts with an enzyme secreted by certain cells of the small intestine. It causes **protein** food substances to be broken down into simpler forms so that they may be absorbed and utilized by the body. 2. Amylase, which assists in the digestion of **starch**. 3. Lipase, which breaks down **fatty** foods into glycerine and fatty acids.

The internal secretion of the pancreas is formed by small cell groups scattered throughout the gland substance. They are called Islands of Langerhans and their function is to secrete **insulin** into the blood stream. Insulin is necessary for the oxidation of **sugar** by the body. When insufficient insulin is manufactured by the gland **diabetes** occurs. When too much insulin is discharged into the blood stream hyperinsulinism develops and is characterized by a low blood sugar accompanied by weakness and fainting spells.

The pancreas is stimulated and pours out its secretion due to the action of secretin, a **hormone** originating in the walls of the duodenum. This hormone is activated by the passage of acid food from the stomach into the duodenum. The hormone is then absorbed in the blood stream where it acts upon the pancreas.

The pancreas is subject to acute and chronic **infections** and **tumor** and **cyst** formation. (A.W.L., R.S.M.)

PANDA. Mammalia, Carnivora. A peculiar animal, *Aelurus fulgens,* of the southeastern Himalayas. Its body is about 2' long, exclusive of the long furry tail, and the legs are strong and only moderately long. The head is shaped much like that of the raccoons, which seem to be fairly closely related. Also called the red cat-bear from its slight resemblance to both forms. (A.W.L.)

PANEL POINT. Bridge; Truss.

PANGOLIN. Mammalia, Edentata. The scaly ant-eaters, a group of peculiar Old World animals which are covered with overlapping horny scales of large size. They are slender animals with a long tail and short legs bearing powerful claws. Like other anteaters they have a sharp snout and long sticky tongue and live chiefly on termites.

The several species are confined to Africa and the southeastern part of Asia, where the most common one is *Manis pentadactyla*. (A.W.L.)

PANICLE. Flower.

PAN-IDIOMORPHIC. A term proposed by Rosen-busch signifying an **igneous rock** whose texture is **idiomorphic.** (R.M.F.)

PANTHER. Puma.

PANTOGRAPH. Parallel Mechanism.

PANTOGRAPH TROLLEY. This is a type trolley often used on electric locomotives to connect with the overhead trolley wire. It is so named because of its resemblance to the drafting instrument of the same name. (L.R.Q.)

PANTOPODA. Synonym of **Pycnogonida.**

PANTOSTOMATIDA. Mastigophora.

PAPAL MITRE. Mollusca, Gasteropoda. A marine shell of the Indian Ocean. It is white with red spots. The name is a translation of the scientific name *Mitra papalis*. (A.W.L.)

PAPAVERINE. Alkaloids.

PAPAYA. Papaw. *Carica Papaya*. Caricaceae. This is a tropical American tree with a straight rarely branching trunk from 6–25′ tall. On the upper portion of the stem is borne a crown of large compound, long-petioled leaves. The pale yellow, fragrant flowers are of two kinds, pistillate (see **Pistil**) and staminate (see **Stamen**), borne on different plants. Papayas are therefore **dioecious plants.** The smooth-skinned fruits vary considerably in shape and size, those of wild plants being not much larger than eggs, while cultivated fruits are much larger. The orange-colored flesh is sweet and juicy, and surrounds a central cavity which contains the numerous seeds. Usually the fruit is eaten raw, but may be used in salads or cooked. It is not a good shipping fruit, so rarely appears in United States markets.

From the fruit and sap of the plant is obtained an **enzyme,** papain, which is used as a digestive aid, its action being much like that of pepsin. The leaves, cooked with meat, are said to tenderize meat.

The plant is widely introduced in many tropical lands. It is extensively grown in the Hawaiian Islands and to some extent in Florida and California. Propagation is mainly by seed.

The name papaw, sometimes given to this fruit, is also applied to a North American tree, *Asimina triloba*, with which it should not be confused. (R.M.W.)

PAPER. One of the most important factors in the progress of civilization has been paper, a thin flat tissue composed of closely matted fibers obtained almost entirely from plant sources. In modern life paper finds a variety of uses, for writing, for containers, wrappers, wall covering, and—perhaps most important—in all the forms of printing: newspapers, magazines, books.

The art of making paper seems to have been discovered first by the Chinese, who were making paper as early as the beginning of the Christian era. From China the process was carried to Arabia and thence to Europe. Paper was not an important article at first and, since it is not a very durable substance under ordinary conditions, could not compete with parchment or vellum as a medium for the written word. In the fifteenth century writing became more general and the demand for a cheaper material increased. Paper became an important product. At this time paper was made largely from vegetable fibers reclaimed from cloth (especially linen), as had been done since the invention of paper in China. This paper was made entirely by hand, as is done even today in the manufacture of certain expensive types of paper. In making hand-made paper, a pulp is formed by soaking the vegetable fibers in water in a vat. From this vat the pulp is dipped out in a mold, the bottom of which is a fine screen. By a deft motion of this mold the soft pulp is spread over the screen in a thin layer of matted fibers. The water in the pulp drains off, leaving a rather firm mass which is turned out on a piece of felt. More pieces of half-dried pulp spread on felt are added. The whole pile is then pressed to squeeze out more of the water, press the fibers closer together and form a firm sheet. These are then removed from between the felts, pressed again, and dried. During the final treatment surface sizing is added to render a surface more suitable to receive ink. Sheets of hand-made paper are naturally of limited size and expensive.

To meet the great demand for paper, machine methods were developed. This increased demand for paper also led to the utilization of material which could be obtained in quantities much greater than rags. Out of this developed the vast pulp industry which today converts vegetable material, mostly soft woods such as spruce and fir, as well as poplar, into a white felt-like mass of fibrous substance, known as pulp. Several methods are in use to obtain the fibers from the wood. In one, the logs are barked and then ground by pushing against large grindstones. By this process the separate fibers of the wood are dissociated, their ends being more or less frayed in the process. Water flowing over the grinding surfaces keeps the temperature resulting from friction from rising unduly, and also removes the fibers. These are then washed and drained, undesirable substances such as bits of bark and other materials removed, and then pressed into sheets. These sheets of pulp are folded into bundles, which are then shipped to the paper mills. Pulp mills are commonly located near the source of wood supplies. Ground pulp is largely made from spruce wood and is principally used in making newsprint paper. Usually it is mixed with some chemically prepared pulp so that the paper may have greater strength to resist tearing in handling.

Chemical pulp is prepared by several processes. In these the wood is first barked, and then further cleaned by hand or other mechanical equipment to remove bark, pitch, and other foreign material as completely as possible. It is then cut up into chips which are screened to remove the fine sawdust and dirt. The chips are cooked or digested in very large vertical tanks or digesters. Digestion is accomplished by means of various chemicals. In the sulfite process, **sulfur dioxide** dissolved in **calcium** bisulfite or **magnesium** bisulfite is used. In the soda pulp process, a solution of **sodium** hydroxide in water is used.

In the sulfate or "kraft" process sodium sulfate (Na_2SO_4) is reduced by heating with carbonaceous matter in a furnace—usually a rotary kiln—to form sodium sulfide (Na_2S), which is then used in water solution with sodium hydroxide (NaOH) as the cooking liquor. Digesting is done under pressure and at high temperatures. The sulfite process is used with coniferous woods such a spruce, fir and hemlock; the soda process with poplar and other deciduous woods, and the sulfate process for coniferous woods. After digesting, the fiber mass is washed to free it of chemicals and pressed into sheets of pulp.

To convert the pulp into paper, it is first put into large tub-like containers with water. There it is thoroughly beaten by constantly passing between a roll and stationary plate, the surfaces of both being made up of narrow, closely spaced steel bars. During the beating a certain amount of blue dye is added to neutralize the yellow tint otherwise present. Later alum, sizing and bleaching materials are added. In the case of rag pulps the bleaching is done in the washer beaters, but in the case of wood pulps it must be carried out in large tanks designed specifically for the purpose. The chief bleaching agents are chlorine and calcium hypochlorite. When the mass is thoroughly beaten and mixed, it is poured on a very fine copper screen which is in the form of an endless belt several feet wide. Vibration of the screen spreads the material in a thin uniform layer. As it is carried along much water in the soft mass drains out. To prevent the material from running over the edge thick rubber belts are placed on either side. Mechanical suction removes much of the water left after draining, after which the felted mat passes between rollers on a thick woolen felt which is carried between a series of heavy rollers. Passing between these the paper is pressed into a firm thin sheet, which then has sufficient strength to pass without support of screen or felt through a long series of heated rolls. From these it may go directly to the calender machines, stacks of heavy rolls which give to it a smooth firm surface. If a better surface is required, paper then goes to the coating machines, where it receives a coating of a clay or other mineral pigments and casein, which imparts a very smooth glossy surface suitable for fine reproduction of photographs. Many variations are found in the details of the various processes leading to the production of finished paper.

The principal woods used in the making of paper are spruce, hemlock, southern pine, poplar and fir, with smaller quantities of several other woods. Attempts have been made at various times to use other plant materials, such as cotton and corn stalks, and various straws. So far these have given only slight promise because of the difficulty and expense of obtaining the fibers. Large quantities of cereal straw, namely, oat, wheat, rye, and barley, are used in making brown, coarse straw pulp which is principally used for making paper boards for corrugated paper. From waste paper, pasteboard and cheap grades of paper are made. Large quantities of the better quality waste paper are deinked and used in the manufacture of various grades of printing paper. Unbleached waste papers and lower grade waste which may be badly soiled are disintegrated without deinking and used for the manufacture of container board, boxboard, and some of the low quality grades of paper. (See **Carbohydrates.**) (R.M.W., R.K.S.)

PAPER LOCATION. Surveying.

PAPER WASP. Insecta, Hymenoptera. The hornets and yellow-jackets. **Wasps** which build their nests of coarse paper made by chewing fragments from weathered or partially decayed wood. Some build subterranean nests and others suspend the nest from the eaves of buildings or from the boughs of trees. (A.W.L.)

PAPPUS. Composite Family.

PAPPUS' THEOREMS. These are two mathematical results obtained by Pappus (c. 300 A.D.) concerning areas and volumes of solids generated by rotation of a plane figure about an axis.

If an arc of a curve in a plane is rotated about an axis not cutting it, the area generated by the arc equals the length of the arc times the perimeter of the circle described by its **centroid.**

If a closed region in a plane is rotated about an axis in the plane not cutting it, the volume generated by the region equals the area of the region times the perimeter of the circle described by the centroid of the region. (L.L.S.)

PAPULA. Thin-walled projections from the surface of starfishes. They increase the surface available for respiratory and excretory interchange. (A.W.L.)

PAPYRUS. *Cyperus Papyrus.* Cyperaceae. In early times writing was done on sheets of material known as papyri. These were prepared from the stems of a huge species of reed growing on river banks and in marshes in Egypt and other eastern countries. This plant, *Cyperus Papyrus,* has a thick **rhizome** which often grows several feet in length. The pith of the rhizome is edible, and frequently is used as food in eastern countries. From the rhizomes arise the erect stems of the plant, which are from 3–12′ tall and bear at the top a radiating mass of **flower** spikes. In making papyri, the stems were first cut into thin strips. These were placed side by side until the desired width, varying from 4–12″, or more, was obtained. Across this layer another was placed, with the strips at right angles to the first. It is possible that the strips were interwoven. This material so arranged was then immersed in water and left to soak for some time. After soaking the layer was beaten flat and into a coherent sheet which was then dried in the sun. Separate sheets of papyrus were pasted together so that rolls of considerable length were obtained. So prepared, papyrus forms a thin flexible sheet of great durability, though some of the larger sheets tended to tear and break apart with use. Papyrus making was principally carried on in Egypt, but the plant was introduced into other eastern countries in early times, and attempts at papyrus making undertaken. The product was often inferior to the Egyptian papyri. (R.M.W.)

PARABASAL BODY. A structure associated with the **nucleus** of some of the flagellate protozoans (**Mastigophora**). Its functions are unknown. (A.W.L.)

PARABOLA. The parabola is one of the conic sections, and may be obtained by cutting a right circular cone by a plane which is parallel to an element of the cone.

A parabola is the **locus** of a point which moves so that its distances from a fixed line and a fixed point are equal. (Fig. 1.) The fixed line is called the directrix,

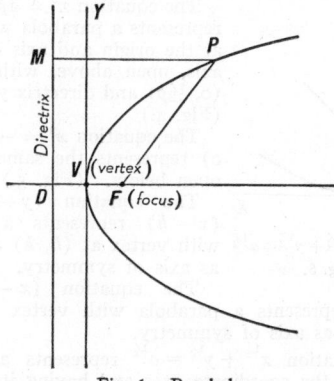

Fig. 1. Parabola.

and the fixed point the focus. The line through the focus perpendicular to the directrix is called the axis of the parabola; it is an axis of symmetry of the curve. The point midway between the focus and the directrix, which is the intersection of the curve and the axis, is called the vertex of the curve. The chord of the curve through the focus and perpendicular to the axis is called the latus rectum (or focal width) of the parabola. The line joining any point of the curve to the focus is called a focal radius.

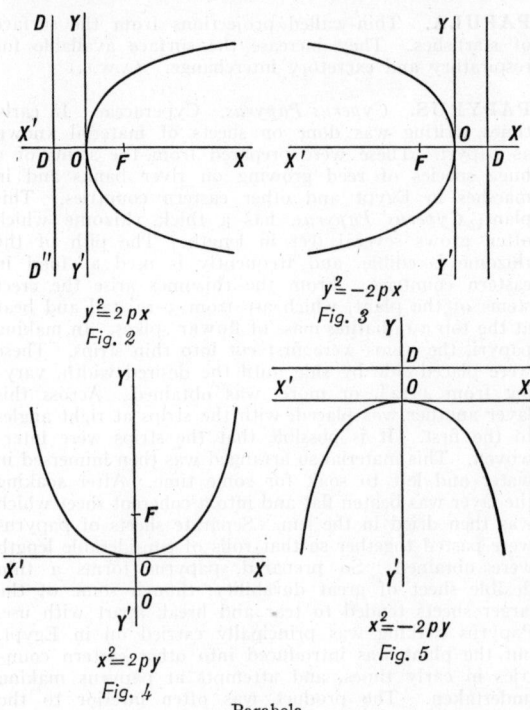

$y^2 = 2px$
Fig. 2

$y^2 = -2px$
Fig. 3

$x^2 = 2py$
Fig. 4

$x^2 = -2py$
Fig. 5

Parabola.

If the origin is taken at the vertex and the X-axis along the axis of the parabola, and if p is the distance of the focus from the directrix, then the equation of the parabola is

$$y^2 = 2px, \text{ (Fig. 2),}$$

if **rectangular coordinates** are used, and if the curve is open to the right. The focus is at $(\frac{1}{2}p, 0)$ and the equation of the directrix is $x = -\frac{1}{2}p$.

The length of the latus rectum is $2p$.

The equation $y^2 = -2px$ $(p > 0)$ represents the same parabola as the preceding, but open to the left, with focus at $(-\frac{1}{2}p, 0)$ and directrix $x = \frac{1}{2}p$. (Fig. 3.)

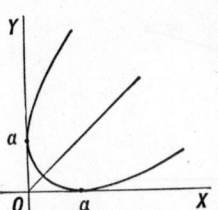

Parabola $x^{1/2} + y^{1/2} = a^{1/2}$
Fig. 6.

The equation $x^2 = 2py$ $(p > 0)$ represents a parabola with vertex at the origin and axis on the Y-axis, open above, with focus at $(0, \frac{1}{2}p)$ and directrix $y = -\frac{1}{2}p$. (Fig. 4.)

The equation $x^2 = -2py$ $(p > 0)$ represents the same parabola open below. (Fig. 5.)

The equation $(y - k)^2 = 2p (x - h)$ represents a parabola with vertex at (h, k) and $y = k$ as axis of symmetry.

The equation $(x - h)^2 = 2p (y - k)$ represents a parabola with vertex at (h, k) and $x = h$ as axis of symmetry.

The equation $x^{1/2} + y^{1/2} = a^{1/2}$ represents a parabola tangent to the coordinate axes and having its principal axis bisecting the first (and third) quadrants. (Fig. 6.)

The parabola may be constructed geometrically in several different ways.

Given the focus and directrix of a parabola, it may be constructed point by point as follows: (Fig. 7.) Draw the axis MX, construct the vertex V as the mid-

point of MF. Through any point A on the axis to the right of V draw a line AB parallel to the directrix. From F as center with radius MA strike arcs to intersect AB at P and Q. Then P and Q are points on the curve. By changing the position of A we may construct as many points on the curve as desired.

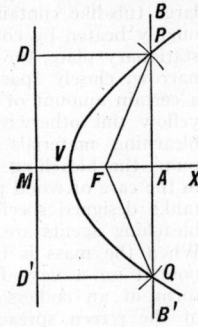

Fig. 7. Construction of parabola.

A construction by continuous motion is the following: (Fig. 8.) An arc of a parabola can readily be drawn in the following way: At the vertex A of a draughtsman's triangle fasten one end of a string of length AB. Fasten the other end at the focus F of the required parabola and place the other leg BC of the triangle along the directrix. Hold the string taut by pressing it against the side of the triangle with the point of a pencil at P. If the side BC of the triangle is now made to slide along the directrix, the point P will describe an arc of a parabola.

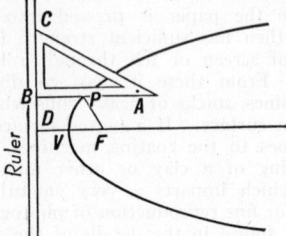

Fig. 8. Construction of parabola.

When the span AB and height OH of a parabolic arch are given, points on the arch may be constructed as follows: (Fig. 9.) Draw the rectangle $ABCD$. Divide

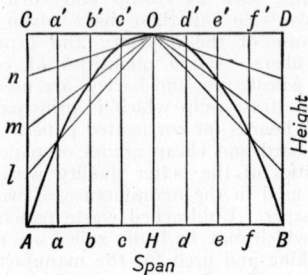

Fig. 9. Construction of parabolic arch.

AH and AC into the same number of equal parts. Starting from A, let the successive points of division be: on AH: a, b, c and on AC: l, m, n. Draw ua' perpendicular to AB and draw Ol, mark point of intersection of aa' and Ol, and do likewise for the points b and m, and c and n. The points of intersection are points on the parabola required.

The equation of the tangent to the parabola $y^2 = 2px$ at the point (x_1, y_1) is $y_1 y = p(x + x_1)$.

The equation of the tangent with **slope** m to the parabola $y^2 = 2px$ is $y = mx + \dfrac{p}{2m}$.

Some applications of the parabola are:

The path of a projectile near the surface of the earth (air resistance being neglected) is a parabola.

A cable of a suspension bridge, if the load is uniformly distributed along the bridge, assumes a parabolic form.

A parabolic mirror is one whose reflecting surface may be generated by revolving a parabola about its axis. (Searchlights, locomotive headlight, reflecting tele-scope.)

If a pan of water is rotated about a vertical axis, the surface of the water assumes a parabolic shape. (L.L.S.)

PARABOLIC SPIRAL. The parabolic spiral is a geometric curve of spiral form. It is the graph of the equation in polar coordinates $r^2 = a^2\theta$. (L.L.S.)

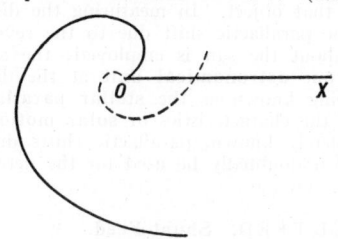

Parabolic spiral.

PARABOLOID OF REVOLUTION. Paraboloids.

PARABOLOIDS. The paraboloids are geometrical surfaces belonging to the class known as **quadric surfaces.**

The **surface** represented in **rectangular coordinates** by the equation

$$\frac{x^2}{a^2} + \frac{y^2}{b^2} = 2cz$$

is called an elliptic paraboloid, since its sections parallel to the coordinate planes are one **elliptic** set and two **parabolic** sets. (Fig. 1.)

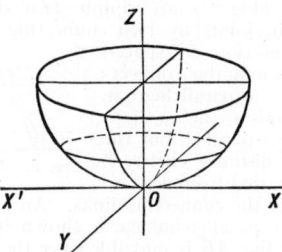

Fig. 1. Elliptic paraboloid.

If $a = b$, the elliptic set of sections becomes circular, and the equation represents a paraboloid of revolution, which may be obtained by revolving a **parabola** about its axis of symmetry.

The surface represented in rectangular coordinates by the equation

$$\frac{x^2}{a^2} - \frac{y^2}{b^2} = 2cz$$

is called a hyperbolic paraboloid (Fig. 2) since its sections parallel to the coordinate planes are: one set of **hyperbolas** and two sets of parabolas. (L.L.S.)

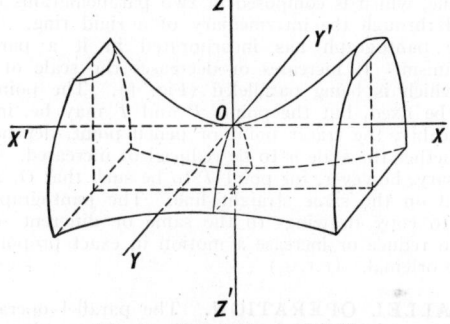

Fig. 2. Hyperbolic paraboloid.

PARACHUTE. The parachute is an aerial life-saving device, and is worn by pilots and passengers of aircraft as a safety measure. The safety parachute is not used except in emergencies, when its wearer is faced with the necessity of quitting the aircraft in midair. This discussion omits consideration of the parachute combination employed in exhibition jumps. The parachute is made up of a circular silk (pongee in cheaper grades) or nylon canopy about 24′ in diameter. At intervals around the edge of this canopy are attached shroud lines which lead to risers. These are straps of webbing joining the ring, to which are gathered the shroud lines, with the harness, which is worn by the user, and in which he is suspended when the parachute is in use. At the center of the canopy is a circular vent edged with rubber bands which tend to keep the vent closed, but which will stretch and allow it to expand to ease the shock when the canopy is opened. The parachute is worn as a back pack, a seat pack, or lap pack. The seat pack is the most convenient type for the average person since the back may rest comfortably against the seat and the pack will act as a cushion to sit on. The lap pack is used chiefly for quick detachable chutes and for exhibition jumps. A parachute is packed into a canvas cover, shroud lines and all. The pack is held closed by a flexible wire which passes through eyelets in the pack cover. Two things cause the mainsail to be thrown out of the pack when the flexible wire, which is called the rip cord, is pulled to release the parachute. Rubber bands attached to the canvas cover pull it away from the parachute itself and spring ribs throw a small pilot parachute into the airstream, where it forms an anchor which drags out the mainsail.

The recommended practice when leaving an airplane is to dive headlong, pulling the rip cord when clear of the aircraft so that there may be no danger of the aircraft fouling some part of the chute. A variation of this method of releasing the canopy is employed by paratroopers (air-borne troops landing by parachute). Requirements of rapidly timed jumps of a squad from a troop carrier, often at low altitude for tactical reasons, are better met by automatic than by individual release. The release wire is pulled by a strap attached to the airplane, this strap being long enough to allow the jumper to clear the airplane before the slack is absorbed and the release pulled. A "Jumpmaster" times the jumping intervals and the system allows all the paratroopers in a troop transport to be landed in a small area without danger of fouling parachutes. Its success depends on straight, level flight of the airplane. Emergency jumps from aircraft in distress are more safely made if the jumper has control over the moment of release of the canopy. About three-fifths of a second after the rip cord is pulled, the mainsail is inflated and the velocity of descent is checked. While successful parachute landings are made from jumps below 500′, any casual parachute jump from below that altitude may be considered extremely hazardous. The standard rate of descent of a 24-ft. parachute is 17 ft. per sec., which is said to be equivalent to a jump from a 10-ft. wall.

In addition to the parachuting of humans, the parachute has been used to drop circulars, food, supplies, and consigned cargo to points inaccessible to any other transportation, and to points not possessing a landing field. So important is the requirement of correct and careful packing of the parachute that the Civil Aeronautics Administration undertakes to license parachute-packing ability. A parachute which has been correctly packed within three months of the time of use and which is in good physical condition, is certain to open when released. (F.T.M.)

PARACONIINE. Alkaloids.

PARADOXURE. Mammalia, Carnivora. An anglicized form of the generic name *Paradoxurus*, applying to the civets. (A.W.L.)

PARAESOPHAGEAL OR PERIESOPHAGEAL CONNECTIVES. The slender nerve cords that connect the dorsal brain of invertebrates with the ventral nerve cord. (A.W.L.)

PARAFFIN. Petroleum; Petroleum Products.

PARAFORMALDEHYDE. Formaldehyde.

PARAGASTER. The large central cavity of **sponges,** also called the gastral, paragastral, or paragastric cavity. It is superficially like the enteric cavity of the gastrula of higher animals but it is formed by an involution of the layer of cells which cover the external surface of the **larval** sponge. Water passes into it through the canal system of the body wall and flows out through the osculum. It is not a digestive cavity. (A.W.L.)

PARAGASTRIC CAVITY. Paragaster.

PARAGENESIS. The term applied by **petrographers** to the succession, or order of development, of the minerals in an **igneous** or **metamorphic** rock. More particularly applied to the order of crystallization of related minerals in a vein or ore-body, including processes of alteration. (R.M.F.)

PARAGLOSSA. Paragnatha.

PARAGNATHA. 1. The lobes into which the lower lip or metastoma of the **crustacean** mouth is sometimes divided. 2. Lobes or appendages borne by the **hypopharynx** of the insect mouth. They lie between the mandibles and **maxillae** and have also been called paraglossae, superlinguae and maxillulae. They are homologous with the paragnatha of crustaceans. (A.W.L.)

PARAGNEISS. Gneiss.

PARAGUAY TEA. Maté.

PARALDEHYDE. A powerful sleep-producing drug which may be given by mouth, by vein, or by rectum. It is a volatile liquid with a penetrating burning taste. It is rapidly absorbed and produces sleep very quickly. After paraldehyde is given, the disagreeable odor of the drug is very noticeable as it is excreted in the breath. (See **Acetaldehyde.**) (D.M.H.)

PARALLAX. As an observer moves about, the relative positions of distant objects seem to change. This apparent change of position of distant objects, due to the actual change of position of the observer, is technically known as parallax; and the amount of apparent shift is known as the parallactic shift. The amount of parallactic shift is inversely proportional to the distance of the object.

In taking readings of instruments employing scales and pointers, care must be taken that the eye of the observer and the pointer are both in a line perpendicular to the plane of the scale. To facilitate this, many instruments have a mirror in the plane of the scale so that the observer may eliminate parallax errors by bringing the reflected image of his eye coincident with the pointer and its reflected image.

In astronomy parallax effects play a very important part. In transferring from one system of coordinates to another parallax effects must always be applied when the location of the origin changes. For example, observations are taken from the surface of the earth and **geocentric parallax** must be applied when transferring the observations to the center of the earth. Distances of many celestial objects are expressed in terms of parallactic angles. The angle subtended by an equatorial radius of the earth at the distance of the sun is known as **solar parallax,** while the angle subtended by the radius of the earth at the distance of any other member of the solar system is known as the horizontal parallax of that object. In measuring the distances of the stars the parallactic shift due to the revolution of the earth about the sun is employed; the angle subtended by one **astronomical unit** at the distance of the star being known as the **stellar parallax** of the object. As the characteristics of **solar motion** become more completely known, parallactic shifts due to this motion will undoubtedly be used for the determination of distances. (W.K.G.)

PARALLEL FEED. Shunt Feed.

PARALLEL MECHANISM. Parallel mechanisms are those in which the links which compose the mech-

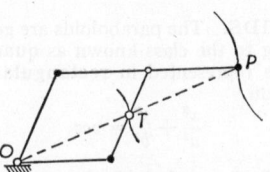

Fig. 1. Pantograph.

anism are pinned together in a **parallelogram.** The parallel rulers (Fig. 2) are simply two straight edges joined with pin joints by two connecting links having points of connection so spaced that the rulers and the connecting links form a parallelogram. In this mechanism the straight edges are always parallel, but the perpendicular distance between them may be varied from zero up

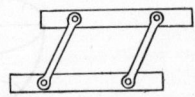

Fig. 2. Parallel rulers.

to the length of the connecting links. An interesting application of the parallel linkage is shown in Fig. 3. In this device the line *AB* is movable over the plane of the

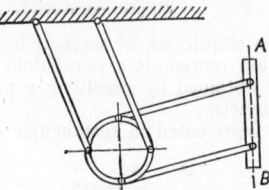

Fig. 3. Drafting machine.

mechanism, within limits. In all positions, however, the line will be parallel to the position shown in the illustration. This linkage is the basis of the universal drafting machine, which is composed of two parallelograms connected through the intermediary of a rigid ring.

The pantograph has incorporated in it a parallel mechanism. It increases or decreases the scale of the line which is being paralleled (Fig. 1). The point *O* must be fixed, but the points *P* and *T* may be, interchangeably, the tracer point or pencil point, depending on whether the scale is to be reduced or increased. It is necessary, however, for point *T* to be such that *O, T, P* are all on the same straight line. The pantograph is used to copy drawings to the same or different scale, and to reduce or increase a motion in exact proportion to the original. (F.T.M.)

PARALLEL OPERATION. The parallel operation of equipment is taken in contrast to series or "booster" connections. Parallel operation, of course, implies at least two units, although the individuals do not necessarily have to be of the same capacity. Generally speaking, in the parallel operation the quantities of output of the units are additive, whereas in series operation certain characteristics, such as pressure, voltage, and the like are additive, and the quantity output is the same as for each machine. Machines operated in parallel discharge to a common collective device such as a **header** or a **bus** or a **conveyor**. Among the more common instances of parallel operation are found the operation of two or more power plants feeding electrical energy into the same distribution network, the parallel operation of electric generating equipment within a plant, the parallel operation of prime movers such as **steam** or **hydro-turbines**, and the parallel operation of **transformers**. Industry offers numerous examples of parallel operation of production machines, the output of which is collected by some mechanical conveyor system.

On account of the interesting technical problems associated with the parallel operation of power equipment, particularly electric generating equipment, this field will be described in some detail. D-c **generators** can be paralleled on a bus, and the load divided between them at will by adjustment of their shunt fields. Shunt-wound generators of like voltage may be operated in parallel merely by being connected to the buses of the same polarity. In the case of compound-wound generators, and most d-c machines are of this type, an equalizer bus must be connected between the series fields of the two **armatures**. Should the machines be paralleled without the equalizer, they might run satisfactorily until some slight unbalance caused one of them to supply a little more terminal voltage. This would lead to a cumulative action due to the presence of series coils, and there would be a heavy surge of load to that machine, which would cause the other to operate as a motor accompanied by excessive circulation of current. By using an equalizer bus, there is a stabilizing action due to the series field currents being determined by the total load current and the series field resistances of the different machines. Any tendency to increase the armature current of one machine will have the effect of strengthening the field of the other, and increasing the load carried by the other. As stated before, shifts of load between d-c generators in parallel are accomplished by altering the shunt-field strength, and no manipulation need be given to the prime mover governors.

When a-c generators are connected in parallel, no such simple method of dividing load is possible, since variation of excitation only causes more wattless current to circulate between the machines. A division of load in this field is accomplished by prime mover governor action, and the principles involved are explained under **governor**. As pointed out there, division of load is accomplished by adjustment of the prime mover governor. Reverting to the a-c generator itself, for parallel operation two alternators must have the same **phase** sequence and **frequency**; their voltages must be equal and in phase. The phase sequence, once

correctly established, need not be considered upon subsequent paralleling operations unless the generator or its leads has been altered in the meantime. Visual indication that these electrical quantities are correct is obtained in routine operation by the voltmeter and synchroscope. A diagram of typical synchronizing connections is shown herewith.

When transformers are to be connected in parallel, they should have the following similar characteristics stated in order of importance: (1) equal voltage ratings, (2) the same ratios of transformation, (3) equivalent **impedances** which are inversely proportional to their current ratings, (4) equal ratios of equivalent resistance to equivalent **reactance**. (F.T.M., L.R.Q.)

PARALLEL PERSPECTIVE. Perspective.

PARALLEL SAILING. The term parallel sailing is frequently used to designate the problem of converting the distance traversed by a ship along a parallel of **latitude** into difference of **longitude**. This problem is solved under the subject of **departure**. (W.K.G.)

PARALLELS. Steel bars of rectangular section generally used in pairs for supporting work parts and for measurement. They are finished to a tolerance of one-half thousandth of their nominal size, and the surfaces are parallel or perpendicular within this limit. In addition to solid parallels, large hollow bars, known as box parallels, and adjustable parallels with a range of heights, are used by machinists and toolmakers. (H.C.H.)

PARALYSIS. Loss of motor activity in some part of the body. Paralysis occurs when the nerve impulse has met with interference. This may occur in the brain centers, the spinal-cord centers or pathways, nerve trunks or individual nerves, or the nerve endings in the muscles. It may be temporary or permanent, depending on the cause. The cause may be due to a specific disease, to chemical or bacterial poisons, toxins or injury to nervous tissue. (R.S.M.)

PARALYSIS AGITANS (Shaking Palsy, Parkinson's Disease). A chronic organic disorder of the central nervous system characterized by a spontaneous tremor, weakness and rigidity of muscles.

The cause is unknown. Degenerative changes take place in cells and blood vessels in the basal ganglia of the **brain**. One form of paralysis agitans occurs after epidemic **encephalitis**.

The onset of the disease is gradual and the symptoms progress slowly. The small muscles are most frequently involved; the "pill-rolling" tremor of the fingers is characteristic. Eventually weakness and rigidity appear. Rigidity of the face muscles gives a mask-like appearance to the face which does not change with any emotion.

Patients with this disorder may live for years. Death is usually not due directly to the disease, but to complications or intercurrent infections. (D.M.H.)

PARAMAGNETISM. Magnetism.

PARAMETRIC EQUATIONS. A plane curve is usually represented by a single equation in two **variables** representing **rectangular coordinates** or **polar coordinates**. Sometimes it is preferable to represent the curve by two equations expressing the coordinates separately in terms of a third variable called a parameter; these equations are then called parametric equations.

We may also have parametric equations of surfaces and of curves in space. (L.L.S.)

PARAMORPHISM. A term used by mineralogists and petrologists to denote the passage of one mineral into

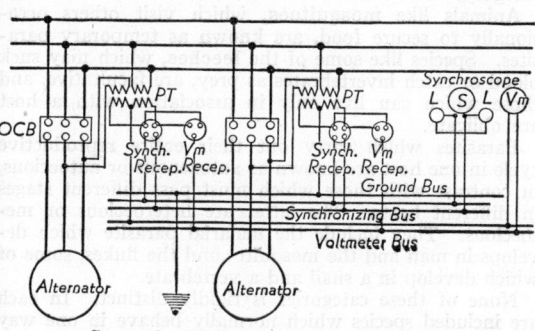

Connection for synchronizing alternators,

another without any fundamental change of chemical elements. A specific example is uralite, secondary hornblende, paramorphic after augite. Paramorphism has also been used to describe the process of metamorphism by which a rock is completely changed. (R.M.F.)

PARAPODIUM. A lobed appendage formed as a protuberance of the lateral body wall in some of the segmented worms (chiefly **Polychaeta**). The typical parapodium consists of two principal divisions, a dorsal notopodium and a ventral neuropodium. Each may be divided into subordinate lobes. A slender basal process, the dorsal cirrus, extends upward and outward from the edge of the notopodium and a similar ventral cirrus is borne by the lower edge of the neuropodium. Each of these main divisions also bears a cluster of **setae** and contains a strong supporting rod, the **aciculum.**

In various species of marine **annelids** the parapodia are modified in many ways to serve as respiratory organs, protective structures, and organs for the creation of water currents through the tubes occupied by the worms.

It is possible that parapodia were the forerunners of the **biramous appendages** of **arthropods.** (A.W.L.)

PARAQUE. Aves, Caprimulgiformes. A large bird of Mexico and Texas which resembles the poor-will. One of the goatsuckers or **nightjars.** (A.W.L.)

PARASITE AREA, EQUIVALENT. Flat-Plate Area; Equivalent.

PARASITE DRAG. Aerodynamic drag which is unaccompanied by useful lift is parasitic. In everyday parlance this interpretation is not always followed. The drag of airplane tail surfaces, although accompanied by useful reactions of a balancing or maneuvering nature, is usually included in the parasite drag. Some writers have wished also to refer to profile drag of wings as parasitic, leaving only induced drag nonparasitic. However, in the generally practiced nomenclature of airplanes, no part of wing drag is parasitic, and all other drag is parasitic.

All true parasitic drag may be subdivided into (1) skin friction, (2) turbulence. The smoother the surface in contact with the air, the lower the surface frictional drag, but parasitic drag is apt to consist largely of the effects of a turbulent wake formed by the "bluffness" of the body. To reduce turbulence by streamlining or "fairing" a body so as to avoid all turbulence in the wake may not be desirable since it may unduly increase skin friction because of the larger surface area. Also the boundary layer may reach the separation point farther forward on the drag body. Experience with total drags of shapes with varying degrees of streamlining bears out these conclusions. If by "fineness ratio" (equal to length/breadth ratio) one describes degree of streamlining, a ratio of approximately 3.5:1 is found to have less drag than any other. This is the shape assumed by a freely falling drop of water, and is the shape striven for in faired airplane parasitic bodies such as struts and braces. Often other considerations render it impractical to build the optimum fineness ratio into parasitic components of aircraft. The fuselage must usually have a much greater fineness ratio in order to fulfill its other requirements. Streamline wire must be shaped so its drag is as low as possible in either 0° or 180° position of its longitudinal axis; thus it is made oval instead of teardrop in section.

The usual methods of expression for this type of drag are:

1. Coefficient of drag, C_D, in the basic drag equation. S is some significant surface or projected area of the object, while ρ and V have the usual meaning in **drag.**

2. **Equivalent flat plate area.**

3. Disk ratio. Ratio of the area of a disk of equal drag to the maximum area of the object, projected on a plane normal to the air stream. This is sometimes used to compare drags of fuselages, a streamlined fuselage having a disk ratio of ⅙ as against a possible ⅓ for a poor shape.

4. Actual drag of a unit length of the object at 100 m.p.h. This is sometimes employed in the case of struts, wires, and the like, normal to the air stream. Thus if a certain wire had a drag of 0.15 lb. per ft., the drag of a 12-ft. wire at 150 m.p.h. would be $12 \times .15 \times (\frac{100}{150})^2$ (+ an allowance for drag of the two terminal fittings).

The principal parasitic components of an airplane are listed here, with some mention of devices or methods employed to reduce said drag to the minimum in the effort to improve speed.

1. *Fuselage.* Reduction or streamlining of proturberances such as cabin enclosures, cockpit covers, etc. Adoption of streamline shapes of optimum fineness ratio. Maximum smoothness of external surface, attained with flush riveting, waxing, etc.

2. *Engine and Cowling, Nacelles, etc.* Engine-fuselage combination for single-engine airplane, engine-nacelle combinations for multi-engine airplanes. Parasitic drag, in this case, is involved with the **cowling and cooling** of the engine and as improved cooling is perfected, parasitic drag is decreased.

3. *Landing Gear, Consisting of Wheels, Struts, Floats, etc.* Fixed undercarriages are partially streamlined by fairing the struts, employing streamline-shaped tires, and partially covering wheels with streamline "pants." But drag can be completely eliminated by mechanically retracting the landing gear, wheels and all, into fuselage, wings, or nacelles. The additional expense, complexity, and vulnerability to mechanism defects are considered to be justified by the increase of speed if the cruising speed is over 150 m.p.h., but is questionable in the case of cruising speeds of 100 m.p.h. or less.

4. *Bracing Struts and Wires.* Streamline strut and wire sections, replacing round types, reduce drag materially. High-speed airplanes are constructed with internally braced wings and empennage for the elimination of this form of drag.

5. *Empennage.* The drag of stabilizer, elevator, fin, rudder may be diminished by employing smooth, low-drag profiles having as small an area and large an aspect ratio as other conditions permit. (See also **Airfoil, Drag, Induced Drag.**) (F.T.M.)

PARASITIC ANIMALS. In general, parasitism is an association of living things in which one lives on or in the body of the other and consumes materials available in that body without rendering any service in return. The organism which provides maintenance is the host and the one that lives at its expense is a parasite.

Parasites are classified in various ways. Those which remain on the surface of the body, such as **lice** and **ticks,** are called ectoparasites, and those which live in the alimentary tract or in tissues of the host are endoparasites. One-celled animals, **flukes,** roundworms, the larvae of **bot flies,** and other forms belong to the latter category.

Animals like **mosquitoes,** which visit others occasionally to secure food, are known as temporary parasites. Species like some of the **leeches,** which may suck blood or catch invertebrates as prey, are facultative, and those which can live only in association with a host are obligate.

Parasites which carry out their entire reproductive cycle in one host are known as autonomous or autoecious, in contrast with those which must pass different stages in different hosts. The latter are heteroecious or metoecious. They include the malarial parasite which develops in man and the mosquito, and the flukes, some of which develop in a snail and a vertebrate.

None of these categories is rigidly distinct. In each are included species which normally behave in one way but may adjust their mode of life to meet unusual circumstances, or species whose normal habits are not limited to one or the other type.

No phylum of animals is made up entirely of parasitic species, but many members of the phyla **Protozoa, Platyhelminthes, Nemathelminthes,** and **Arthropoda** are parasites, and entire orders of these phyla are composed of parasitic species. In many other groups a few parasitic species are known.

The habit of depositing eggs in the nests of other animals, practiced by the European **cuckoo** and the **cowbird,** and by some of the **wasps** and **bees,** is also referred to as parasitism. (A.W.L.)

PARASITIC ANTENNA. Directional Antenna.

PARASITIC OSCILLATION.
This is an undesirable oscillation occurring in radio-frequency **amplifiers** or **oscillators** usually caused by circuit elements becoming ineffective at frequencies very remote from their normal operating frequencies. Thus very high frequency parasitic oscillations may occur because the normal tuning condensers act essentially as short-circuits and the normal tuning inductances are effectively open-circuits while the interelectrode and wiring capacitances and lead inductances become the **tank circuits.** Low-frequency parasitics may occur because the normal tuning inductances are ineffective and the radio-frequency **chokes** become parasitic tank elements. Parallel-operated vacuum **tubes** are especially susceptible to parasitic oscillations, becoming push-pull oscillators at the parasitic frequencies. Adjustment of leads, insertion of chokes or resistors, etc., are expedients for eliminating these oscillations. (L.R.Q.)

PARASITIC PLANTS.
Parasitic plants obtain part or all of their food from other living organisms, called hosts. Except for the lowest groups of plants, the host of a parasitic plant is another plant. The greatest number of parasites are numbered among the **bacteria** and **fungi,** many of which cause diseases of great economic consequence.

Among flowering plants there are relatively few parasites. Of these there are two different types, differing very greatly in appearance. One type contains those plants which are only partially parasitic, the other those which are total parasites, depending on the host entirely for their nourishment.

In the first group the parasites are green and appear very much like ordinary plants. Indeed, some of them, such as the eyebright, a species of *Euphrasia* of the Figwort Family, can live independently, but usually attach their roots to those of other plants and obtain a part of their food requirements from the host. Another member of the same family, the common cow wheat, *Melampyrum americanum,* of open woods is less independent. Without a host, its growth is stunted. To come to maturity and fruit it must find a host. Since, like all the other members of this family, it grows where host plants (commonly grasses) are abundant, little difficulty in meeting a host occurs. The roots of the parasite fasten to those of the host and send sucking organs, called haustoria, into the host roots. From these the parasite draws the water and mineral salts it needs. It seems to affect the host plant very little. Some members of this family do cause certain changes in the host plant, whose roots are stimulated by the parasite haustoria to become much enlarged, producing knob-like growths. In this group of parasites, the principal difference from ordinary non-parasitic plants is a deficiency of roots.

Another type of partial parasites, including the various mistletoes, do not grow on the ground, but attach themselves to the branches of trees. In these parasites the seeds are usually very sticky and attractive to birds. The latter carry them from one tree to another. On germinating, the seed forms a structure called an adpressorium, which attaches itself firmly to the branch. Then

a hard, penetrating, root-like growth pushes into the wood of the host until it reaches the water-conducting cells. There it may form numerous peg-like growths or branches, which greatly increase the absorbing surface. Meanwhile the seed has sent out a normal stem which bears green leaves, and which branches abundantly. This plant carries on **photosynthesis** *normally*, but gains all its water supply and mineral salts from a host plant. Often such parasites are mistaken for branches of the host, and again for **epiphytes.** In these plants, no normal root system ever develops.

Complete or total parasites are strikingly different. They are usually entirely lacking in **chlorophyll,** and so cannot carry on photosynthesis. They must obtain from the host plant their food supply, in the form of **carbohydrates,** fats, and proteins already synthesized by the host. Parasites of this group have the leaves reduced to minute scales or frequently entirely lacking. Roots also are missing, being replaced by haustoria, the structures which enter the host and absorb nutrients from it.

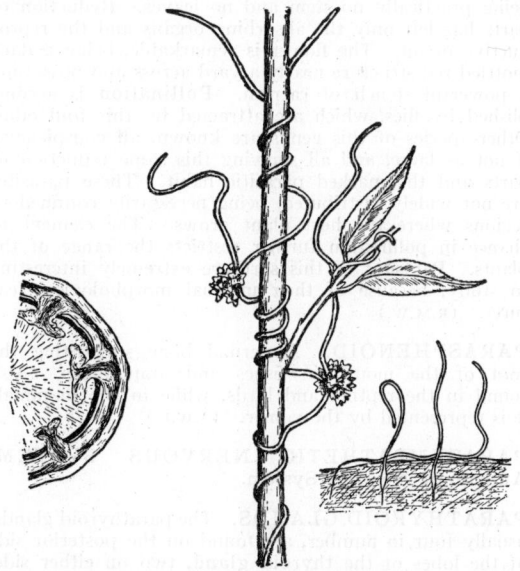

Dodder, a parasite. Middle, habit sketch; right, seedlings of dodder; left, section through stem of host and parasite showing haustoria of dodder reaching the fibro-vascular bundles of host. (*After Strasburger, Noll, Schenck and Schimper, Lehrbuch der Botanik, Gustav Fischer, Jena.*)

A common parasite, and one widely distributed and often very destructive, is the dodder (species of the genus *Cuscuta*), which parasitizes many crop plants, as well as wild hosts. Dodders are twining vines in the same family as the morning glory, to which they are closely related. The seed of the dodder germinates and forms a short root which attaches itself firmly to the soil, but does nothing more. Then a long slender stem is formed. This swings about in a wide arc until it makes contact with some green plant. Around this the stem of the dodder winds tightly. Soon that part between the host and the ground withers and dries, leaving the parasite to grow entirely at the expense of the host. This it does very rapidly, pushing a series of haustoria into the stem of the host and absorbing from it all necessary food. The mature dodder is a long, slender, much-branched vine. The stem is regularly divided into nodes, at which are borne small scale leaves and, when mature, compact clusters of waxy white flowers. The entire plant is yellowish or pinkish and has very little chlorophyll. Under favorable conditions a dodder plant will cover an area of many sq. ft. with a tangle of slender branching stems which pass from one host plant to another. Common species will attach to many different species of plants as hosts.

The broomrapes (species of *Orobanche*) form another group of parasites. These attach themselves to the roots of various host plants. The parasite is a low-growing plant, lacking chlorophyll, with greatly reduced leaves and small but attractive flowers. It is of little importance.

In the tropics many species of parasitic flowering plants are found, some of them very striking. In many South American parasitic species the reduction of the plant body has continued until there remains but a short thick stem rising from the host plant's roots, and a mass of flowers, the latter often brilliantly colored. (See **Orchid Family.**)

The extreme of reduction of parts occurs in certain parasites of the East Indies and Malayan jungles. One species, *Rafflesia Arnoldii*, is worthy of note. It grows attached to the spreading roots of species of *Cissus,* forming in them tumorous enlargements. From these the flower of the parasite bursts out directly, there being practically no stem and no leaves. Reduction of parts has left only the absorbing organs and the reproductive organ. The flower is remarkable, being a dark mottled red structure nearly a yard across and possessing a powerful stench of carrion. **Pollination** is accomplished by flies, which are attracted by this foul odor. Other species of this genus are known, all conspicuous, if not as large, and all showing this same reduction of parts and the marked parasitic habit. These parasites are not widely distributed, being necessarily confined to regions where the host plant grows. The element of chance in pollination further restricts the range of the plants. Parasites of this sort are extremely interesting to study, because of their unusual morphological features. (R.M.W.)

PARASPHENOID. A dermal bone supporting the roof of the mouth in fishes and amphibians. Also found in the reptiles and birds, while in the mammals it is represented by the vomer. (A.W.L.)

PARASYMPATHETIC NERVOUS SYSTEM. Autonomic Nervous System.

PARATHYROID GLANDS. The parathyroid glands, usually four in number, are found on the posterior side of the lobes of the **thyroid gland,** two on either side. Accessory glands may be present.

These glands of internal secretion control the calcium and phosphorus **metabolism** of the body. Complete removal results in death with painful spasms, twitching, and convulsions. The symptoms are due to overexcitability of the nervous system due to the low concentration of **calcium** in the blood. Without proper function of the parathyroids calcium cannot be utilized by the body. **Vitamin D** enters into this process, acting in some manner with parathyroid **hormone.** Collip has described and prepared an extract of these glands which relieves symptoms from hypofunction.

Overactivity of the parathyroids causes a condition known as **Osteitis fibrosa cystica** (Von Recklinghausen's bone disease). In this condition there is often a tumor of the parathyroid with overproduction of its hormone and withdrawal of calcium from the bones into the blood. This produces cystic areas in the bones. Often **kidney** stones are formed from the excess of calcium excreted in the **urine.** Surgical removal of the involved gland results in clinical cure.

Deficiency of the parathyroid hormone results in the condition known as tetany, characterized by painful spasm of the muscles due to deficiency of calcium in the blood. (D.M.H.)

PARATYPE. Any specimen of animal or plant, living or **fossil,** considered by the describer to be similar to the type or **holotype.** (R.M.F.)

PARATYPHOID FEVER. Paratyphoid and **typhoid** fevers are distinct and separate diseases. Paratyphoid fever is an acute general infection caused by the paratyphoid bacillus A or B, having the same clinical signs and symptoms and pathological changes as are seen in typhoid fever except in milder degree. The organism is in an intermediate position between *Bacillus typhosus* and *Bacillus Coli.*

An attack of, or immunization against, typhoid fever does not give immunity to paratyphoid fever. Until World War I, paratyphoid was considered a rare disease. It was found at that time that not only did the inoculations against typhoid fail to give protection against paratyphoid, but that the disease was far more prevalent than previously believed. At present inoculations are given using a triple vaccine which protects against typhoid and paratyphoid A and B.

The sources of infection are the same as of typhoid, as is the manner of contagion. The incubation period varies from 3 to 15 days. Paratyphoid cannot be distinguished from typhoid by clinical signs or symptoms, but only by means of laboratory tests. It is a milder infection than typhoid as a rule, of shorter duration, and with a lower mortality.

The complications of paratyphoid are similar to those of typhoid, and the chief causes of death are pneumonia, hemorrhage, and perforation of the intestines. (R.S.M.)

PARAZOA. A major division of the animal kingdom containing only the **sponges** (Porifera). In contrast with the 1-celled Protozoa on the one hand and the many-celled Metazoa on the other, these animals are made up of many cells, but the cells are organized in tissues only to a limited extent and are not closely coordinated by a nervous system. (A.W.L.)

PARENCHYMA. In zoology, this term is commonly used in two ways. 1. It designates the essential or functional elements of an organ in contrast to the connective elements or framework. 2. Parenchyma also designates a loosely compacted tissue of **mesodermal** origin which fills the space between the viscera and the body wall in flatworms and some roundworms.

In botany, parenchyma tissue is the fundamental tissue found in all parts of the plant. From parenchyma cells all other kinds of cells are formed. A typical parenchyma **cell** is thin-walled and more or less rounded in shape. Mutual pressure of many parenchyma cells against one another causes them to become angular. They are generally isodiametric.

Parenchyma cells contain living **cytoplasm** and are potentially capable of dividing to form new cells at any time. In certain parts of the plant parenchyma cells have definite functions. For instance, in the growing tips of both stems and roots there are groups of cells which divide rapidly, adding to the length of the stem or root. These groups of actively dividing cells make up the meristematic region. Many of the parenchyma cells, especially in the outer cortical tissues of stems and in leaves, contain **chloroplastids** and are green. Cells of this kind are called chlorenchyma cells. In them **photosynthesis** is carried on. In other parenchyma cells food reserves are stored. **Carbohydrates** may move through the plant in the parenchyma cells. Pith, rays, most of the cortex, and leaves (excepting the veins) are composed of parenchyma cells. (A.W.L., R.M.W.)

PARENTHESES. In algebraic expressions it is sometimes desirable to group several parts together to indicate that they are to form a single unit of investigation. For this purpose, parentheses, ( ), brackets, [ ], braces, { } and occasionally a vinculum, —, are used.

Rules for the removal of parentheses in algebraic expressions are: If a parenthesis is preceded by a plus sign, the parentheses may be removed without any change in the terms within it; if a parenthesis is preceded by a minus sign, the parentheses may be removed if all the signs of the terms within it are changed.

By reversal of these rules, parentheses may be inserted in expressions. (L.L.S.)

PARESIS, GENERAL (Paretic Neurosyphilis, General Paralysis, General Paralysis of the Insane, Dementia Paralytica).

General paresis is a chronic infection and inflammation of the brain and its coverings occurring as a manifestation of late or tertiary **syphilis**. It is characterized by progressive dementia and a generalized paralysis which eventually terminates fatally.

About 3% of syphilitics develop paresis.

Pathologically, the changes in the brain are characterized by degeneration of the brain cells, especially in the frontal lobes which control the higher mental processes. Cerebral atrophy is usually marked, and the brain coverings are thick and adherent.

The symptoms may assume any form. Since the onset may be sudden or insidious, paresis may resemble and be mistaken for any kind of mental disease. The diagnosis is easily established by finding a positive spinal fluid Wassermann test.

In the early stages, changes in personality, lack of judgment, carelessness, may lead to financial and moral difficulties. These early symptoms may last weeks, months, or years. Visual, auditory, and speech defects are common.

During the fully developed stage of the disease, characteristic symptoms are euphoria with grandiose ideas, delusions of wealth and power. In others, depression may be marked, with anxiety and fear. There is also a form characterized by delusions of persecution. Convulsive seizures followed by temporary paralysis are common. Progressive dementia is present with all the above syndromes.

In the late stages the paralysis is marked, involving all muscles, and the patient is bedridden until death occurs either by intercurrent infection or by respiratory paralysis.

The prognosis is more hopeful at present since more cases receive early treatment. Remission of the disease for varying periods has been obtained by means of fever therapy together with tryparsamide, which has more effect on neurosyphilis than the common arsenical compounds.

Juvenile paresis is much more common than juvenile tabes and is a form of congenital syphilis. It is characterized by a rapidly progressing dementia or feeble-mindedness.

Often central nervous system syphilis appears as a combination of paresis and **tabes dorsalis**, known as tabo-paresis. (D.M.H.)

PARESTHESIA.

Any abnormal sensation on the surface of the body. It may be described by the patient as burning, tickling, itching, pricking, etc. Paresthesias occur in nervous-system diseases involving the spinal cord, in pernicious **anemia**, and polyneuritis. (R.S.M., D.M.H.)

PARIETAL.

One of the large bones of a pair which form the sides and roof of the human skull, and the smaller bones of similar relations in the skulls of other vertebrates. The word is used in a similar descriptive sense to refer to the walls of various organs. (A.W.L.)

PARKERIZING.

A protective iron phosphate coating applied by immersing iron or steel in a hot solution of manganese dihydrogen phosphate for 30–60 minutes. Bonderized coatings are much thinner iron phosphate coatings applied by spraying or dipping in hot phosphate solutions containing accelerators. Similar processes use nitrate salts, oxalic acid, and chromic acid as the reactive agents. Bonderizing and similar processes are often applied to galvanized steel to provide a suitable base for painting. (R.H.H.)

PARKINSON'S DISEASE. Paralysis Agitans.

PARONYCHIA.

Infection of the tissues surrounding the finger nail. Paronychia commonly follows injury, especially from a "hang nail." Treatment is surgical. (R.S.M.)

PAROTID GLAND. Salivary Glands.

PAROTITIS.

Infection of the **parotid glands**. Parotitis may occur as a complication of any prolonged illness, especially after operations or illnesses in the aged and debilitated. **Mumps** is a form of parotitis. (D.M.H.)

PARRAQUET. Parrot.

Also spelled parakeet, parrakeet, and in various other ways. (A.W.L.)

PARROT.

Aves, Psittaciformes. Birds (**Aves**) with a strong hooked beak and with two toes of the foot directed forward and two back. They are represented on all continents except Europe but are confined chiefly to tropical and subtropical areas. They eat nuts and other fruits and seeds, although the kea parrot, *Nestor notabilis,* of New Zealand is said to have acquired a taste for mutton.

The brilliant colors of many parrots and their ready imitation of human speech have led to their being kept as cage birds for many centuries, hence they are familiar even beyond their natural range.

There are many subsidiary forms of parrots. The nestor parrots of New Zealand and the neighboring islands are dark-colored birds, including the kea and the kaka. The bad reputation of the former for destroying sheep has recently been denied, hence it is impossible to give accurate information on this point. It is a mountain bird. Cockatoos are crested species with the hook of the beak transversely ridged below; an Australian species is called the cockatiel. The lories and loriquets are small species of the Australian region. Macaws are large brilliantly colored birds of the American tropics with a very large beak and long tail. They are sometimes seen in captivity. Conures are small species, including the only North American representative of the order, commonly called the Carolina paraquet, *Conuropsis carolinensis.* The term paraquet, spelled in various ways, designates many small species of typical parrot of the Oriental and Australian regions. Among them are the broadtail, turquoisine, and budgerigar, as well as other species with less striking names. They are closely related to the small parrots so often seen in cages under the name love-birds.

A curious member of the order is the owl-parrot or kakapo of New Zealand, which constitutes a distinct family. It is a flightless bird of owl-like appearance, barring its more brilliant colors, and is largely nocturnal in habits. (A.W.L.)

PARROT-FISH, PARROT-WRASSE.

Pisces, Teleostei. Brightly colored marine fishes (**Pisces**) with a prominent beak formed by the partial coalescence of the teeth. They occur in tropical waters, chiefly about coral reefs. (A.W.L.)

PARSEC.

The parsec is a unit of distance used for expressing distances between stars and other members of the **sidereal universe**. Technically an object is at a distance of one parsec when it has a **stellar parallax** of 1″ (one second of arc) or, in other words, one **astronomical unit** would subtend an angle of one second at the distance of one parsec.

Expressed in other units of distance:

1 parsec = 3.26 **light years.**
= 206265 astronomical units.
= 1.02×10^{13} miles or about 20 millions of millions of miles.
= 3.08×10^{13} kilometers.

Within recent years in the discussion of distances between **extra-galactic** objects the parsec is not large enough to be convenient and the terms kiloparsec (1000 parsecs) and even megaparsec (1,000,000 parsecs) have been proposed. (W.K.G.)

PARSLEY. Carrot Family.

PARSNIP. Carrot Family.

PARTHENOCARPY. The development of a fruit without fertile seeds. In most flowering plants fruit production depends upon **pollination**, fertilization, and the subsequent growth of embryos in the seeds. The size and shape of fruits are largely determined by the number and location of the fertile seeds. Cucumbers, for instance, sometimes are pollinated in such a way that seeds develop only in the stem end, resulting in an undesirable, constricted blossom end.

Certain fruits, such as the banana and the "seedless" or navel orange, are normally parthenocarpic. It is now possible to induce parthenocarpy in plants which normally fail to produce any fruits at all without pollination by spraying the open flowers with auxin. "Seedless" tomatoes, apples, and possibly other fruits may be produced commercially in this manner. The dilute auxin solutions produce parthenocarpic fruits whether applied to the stigma or to the surface of the ovary from which the style and stigma have been removed. (P.A.W.)

PARTHENOGENESIS. The development of eggs without **fertilization.** In some groups of animals eggs normally develop in this way, either for a series of generations interrupted occasionally by a normal fertilization or as a special part of the reproductive process associated with the development of some young from fertilized eggs. The **rotifers** are extensively parthenogenetic. In some cases males are not known and it is possible that parthenogenesis has entirely superseded normal sexual reproduction. Among the insects plant lice reproduce parthenogenetically for many generations, but in the temperate zones the onset of cold weather is accompanied by the appearance of a sexual generation. The honey-bee and other related species carry on parthenogenesis to a limited extent. Male or drone honey-bees are produced from unfertilized eggs and females, both queens and workers, from eggs which have been fertilized.

Artificial parthenogenesis has been induced experimentally by subjecting eggs to varied chemical and mechanical stimuli. The eggs of marine invertebrates, such as sea urchins, have been especially favorable subjects, but Loeb, in one of the most famous experiments, produced several frogs from unfertilized eggs by the mechanical stimulus of pricking the egg with a needle. (A.W.L.)

PARTIAL CORRELATION. We often desire to find the dependence of two variables x_1 and x_2 when other variables x_3, x_4, $\cdots$, xN are kept constant. This is readily accomplished by partial linear correlation and we refer to the **coefficient of partial correlation** for a complete discussion. The more general theorem of partial curvilinear correlation is in very many respects similar but somewhat more complicated. (L.A.A.)

PARTIAL DERIVATIVES. Let $u = f(x, y, z, \cdots)$ be a **function** of two or more variables. If all the variables except one, say x, are held **constant** and x is given an increment Δx, and if Δu is the corresponding increment of u, and if we form the ratio $\Delta u/\Delta x$ and take the **limit** when $\Delta x \to 0$, then the limit

$$\lim_{\Delta \to 0} \left(\frac{\Delta u}{\Delta x} \right)$$ is called the partial derivative of u with

respect to x, and is denoted by $\frac{\partial u}{\partial x}$ or u_x. Similarly, we

define the partial derivatives $\frac{\partial u}{\partial y}$, $\frac{\partial u}{\partial z}$, etc.

Partial derivatives of a function of two variables may be given a geometric interpretation as follows: If $z = f(x, y)$

represents a **surface**, then $\frac{\partial z}{\partial x}$ represents the **slope** of the

curve of intersection of the surface and a plane perpendicular to the Y-axis; and similarly for $\frac{\partial z}{\partial y}$.

A change of variable in partial derivatives may be effected as follows: Let u be a function of a first set of independent variables x, y, z, $\cdots$; let us introduce a second set of variables r, s, t, $\cdots$, related to the first set of variables x, y, z, $\cdots$ by explicit or implicit functional relations. Then

$$\begin{cases} \dfrac{\partial u}{\partial r} = \dfrac{\partial u}{\partial x}\dfrac{\partial x}{\partial r} + \dfrac{\partial u}{\partial y}\dfrac{\partial y}{\partial r} + \dfrac{\partial u}{\partial z}\dfrac{\partial z}{\partial r} + \cdots, \\[2mm] \dfrac{\partial u}{\partial s} = \dfrac{\partial u}{\partial x}\dfrac{\partial x}{\partial s} + \dfrac{\partial u}{\partial y}\dfrac{\partial y}{\partial s} + \dfrac{\partial u}{\partial z}\dfrac{\partial z}{\partial s} + \cdots, \\[2mm] \dfrac{\partial u}{\partial t} = \dfrac{\partial u}{\partial x}\dfrac{\partial x}{\partial t} + \dfrac{\partial u}{\partial y}\dfrac{\partial y}{\partial t} + \dfrac{\partial u}{\partial z}\dfrac{\partial z}{\partial t} + \cdots, \\[2mm] \cdots \cdots \cdots \cdots \cdots \cdots \cdots \cdots \end{cases}$$

the number of equations being the same as the number of variables in the second set and the number of terms on the right in each equation being the same as the number of variables in the first set. Also

$$\begin{cases} \dfrac{\partial u}{\partial x} = \dfrac{\partial u}{\partial r}\dfrac{\partial r}{\partial x} + \dfrac{\partial u}{\partial s}\dfrac{\partial s}{\partial x} + \dfrac{\partial u}{\partial t}\dfrac{\partial t}{\partial x} + \cdots, \\[2mm] \dfrac{\partial u}{\partial y} = \dfrac{\partial u}{\partial r}\dfrac{\partial r}{\partial y} + \dfrac{\partial u}{\partial s}\dfrac{\partial s}{\partial y} + \dfrac{\partial u}{\partial t}\dfrac{\partial t}{\partial y} + \cdots, \end{cases}$$

Let $u = f(x, y, z, \cdots)$ be a function of two or more variables, and let x, y, z, $\cdots$ be functions of a single independent variable t. Then

$$\frac{du}{dt} = \frac{\partial u}{\partial x}\frac{dx}{dt} + \frac{\partial u}{\partial y}\frac{dy}{dt} + \frac{\partial u}{\partial z}\frac{dz}{dt} + \cdots$$

is called the total derivative of u with respect to t.

Partial derivatives of higher order are defined thus: In

general, the partial derivatives $\frac{\partial u}{\partial x}$, $\frac{\partial u}{\partial y}$ of a function

$u = f(x, y)$ are themselves functions either of x or of y or of both x and y. They may then be differentiated partially with respect to x or y. Thus, we get

$$\frac{\partial^2 u}{\partial x^2} = \frac{\partial}{\partial x}\left(\frac{\partial u}{\partial x}\right), \qquad \frac{\partial^2 u}{\partial y^2} = \frac{\partial}{\partial y}\left(\frac{\partial u}{\partial y}\right),$$

$$\frac{\partial^2 u}{\partial x \partial y} = \frac{\partial}{\partial x}\left(\frac{\partial u}{\partial y}\right), \qquad \frac{\partial^2 u}{\partial y \partial x} = \frac{\partial}{\partial y}\left(\frac{\partial u}{\partial x}\right),$$

$$\frac{\partial^3 u}{\partial x^3} = \frac{\partial}{\partial x}\left(\frac{\partial^2 u}{\partial x^2}\right), \qquad \frac{\partial^3 u}{\partial x \partial y^2} = \frac{\partial}{\partial x}\left(\frac{\partial^2 u}{\partial y^2}\right), \text{ etc.}$$

Frequently the notation u_{xx} is used for $\frac{\partial^2 u}{\partial x^2}$, u_{xy} for $\frac{\partial^2 u}{\partial x \partial y}$,

etc.

For functions having continuous first partial derivatives,

$$\frac{\partial^2 u}{\partial x \partial y} = \frac{\partial^2 u}{\partial y \partial x} \quad \text{or} \quad u_{xy} = u_{yx}.$$

(L.L.S.)

PARTIAL DIFFERENTIAL EQUATIONS. A partial differential equation is an **equation** which involves **partial derivatives;** it therefore involves an unknown function of two or more independent variables. A solution of such an equation is a relation among the variables, dependent and independent, that satisfies the equation.

In general the solution of a partial differential equation involves arbitrary functions, just as the solution of an ordinary differential equation involves arbitrary constants.

The problem of solving partial differential equations is inherently more difficult than that of solving ordinary differential equations, so no further indication of methods of solution can be given here. (L.L.S.)

PARTIAL DIFFERENTIATION. Partial Derivatives.

PARTIAL FRACTIONS. When a given rational fraction is resolved into a sum of simpler fractions, usually with linear and quadratic denominators, the resulting fractions are often called partial fractions.

The following fundamental theorem lies at the basis of the method of partial fractions:

Any rational proper fraction with real coefficients may be resolved into a set of partial fractions, of the following types:

1. To any linear factor, as $ax + b$, occurring once in the denominator of the given fraction, there corresponds a single partial fraction of the form $\dfrac{A}{ax + b}$, where A is a constant;

2. To any linear factor, as $ax + b$, occurring r times in the denominator, there corresponds a set of r partial fractions of the form

$$\frac{A_1}{ax + b} + \frac{A_2}{(ax + b)^2} + \cdots + \frac{A_r}{(ax + b)^r},$$

where $A_1, \cdots, A_r$ are constants;

3. To any quadratic factor, as $ax^2 + bx + c$, occurring once in the denominator, there corresponds a single partial fraction of the form $\dfrac{Ax + B}{ax^2 + bx + c}$, where A and B are constants;

4. To any quadratic factor, as $ax^2 + bx + c$, occurring r times in the denominator, there corresponds a set of r partial fractions of the form

$$\frac{A_1x + B_1}{ax^2 + bx + c} + \frac{A_2x + B_2}{(ax^2 + bx + c)^2} + \cdots + \frac{A_rx + B_r}{(ax^2 + bx + c)^r},$$

where $A_1, B_1, \cdots, A_r, B_r$ are constants.

The undetermined coefficients in these assumed forms of partial fractions may be found by clearing of fractions and equating coefficients of like powers of x on both sides of the equality, which is an identity in x; or by other special devices, such as substituting special values of x. (L.L.S.)

PARTICULAR SOLUTIONS OF A DIFFERENTIAL EQUATION. Ordinary Differential Equations.

PARTING. In geology, a small joint in a sedimentary rock, especially coal. In mineralogy, the tendency of crystals to separate along certain planes that are not true cleavage planes. (R.M.F.)

PARTRIDGE. Aves, Galliformes. Game birds (**Aves**) of numerous species, related to the pheasants and turkeys. The francolins of Asia and Africa are included here. True partridges are similar to the **quails.** The latter, although more generally known in North America by the name quail, are members of the group. They are found over Europe, Asia, Africa, and North America.

Aside from the common quail or bob-white of North America, the names quail and partridge are both applied to the several western species, and in the southern states even the bob-white becomes the partridge. To confuse the term still further the ruffed **grouse** is often called a partridge in the northern states. (A.W.L.)

PARTURITION. The process of **labor,** or the act of giving birth to a child. (R.S.M.)

PASCAL'S LAW. Hydrostatics.

PASCAL'S TRIANGLE. Binomial Formula.

PASCHEN-BACK EFFECT. Zeeman Effect.

PASSERES, PASSERIFORMES. The largest order of birds (**Aves**), containing many families and several thousand species. All of the common songbirds belong here, including such familiar forms as **sparrows, warblers, wrens, thrushes,** and similar forms. (A.W.L.)

PASSION FLOWERS. *Passiflora* sp. Passifloraceae. The passion flowers are mostly tropical American plants which climb by means of axillary **tendrils.** The leaves are of various shapes, sometimes varying greatly on a single plant. The flowers are borne singly or in small cymes in the **axils** of the leaves, and are of a variety of colors, some species having white flowers, others blue, and still others scarlet. The five **sepals** are borne on the margin of a cup-like receptacle, as are the five **petals** and the corona, an outgrowth from the receptacle, which in many species is cut into numerous slender segments colored like petals. Five **stamens** are borne at the base of the ovary; these bend outward over the corona. The fruit is a **berry,** in which the seeds are surrounded by a fleshy **aril.** Many species of passion flower are extensively grown for their curiously beautiful flowers. In tropical America the edible fruits appear on the market. (R.M.W.)

PASSIVITY. When **iron** is immersed in **nitric acid** concentrated, there is no visible reaction (Keir, 1790), although nitric acid dilute results in a marked reaction with iron. Upon removal of the iron from the nitric acid concentrated and immersion in **copper** sulfate solution, the iron is not plated by copper, although this occurs with ordinary iron. Iron in such a condition is described as passive iron and the phenomenon is known as passivity.

Passivity of iron is produced by other means than the use of nitric acid concentrated, for example, (1) by immersion in solutions of **chromic acid, iodic acid, arsenic acid, potassium** permanganate, potassium dichromate, **lead** nitrate, **hydrogen peroxide,** (2) by anodic oxidation in sulfuric acid **electrolyte.** Other metals than iron are subject to passivity by anodic oxidation, e.g., **cobalt, nickel, aluminum.** In sulfuric acid dilute aluminum as **anode** withstands 25 volts, and in ammonium borate solution 500 volts. Reversal of the current permits the current to pass, thus aluminum may serve as a rectifier of a.c. in a suitably constructed cell.

Formation of a thin film of oxide on the surface of the metal was proposed as the explanation for the behavior (Schonbein, 1836). Passivity is readily removed by percussion, by immersion in a non-oxidizing acid such as **hydrochloric acid,** or by heating in an atmosphere of **hydrogen.** Passive iron shows the **photoelectric** effect less than ordinary iron (Allen, 1914), thus indicating a change of surface in the former. (R.K.S.)

PATCHOULI. Mint Family.

PATELLA. 1. A short segment of the jointed leg of **spiders** and related forms, between the femur and tibia.

2. The kneecap of man and the corresponding bone of other vertebrates. (A.W.L.)

PATENT. A United States patent grant gives the inventor the right to exclude all others from making, using, or selling his invention for the term of 17 years, but it does not give the patentee the right to make, use, and sell his own invention if it is an improvement on some unexpired patent whose claims are infringed thereby. The Patent Office in its investigation preceding the issue of a patent does not consider whether the invention infringes prior patents.

A patent is granted only upon a regularly filed application, complete in all respects, upon payment of the fees, and only after a determination of utility and completeness of disclosure of the invention, and a search to determine its novelty.

No patent is granted upon a mere idea or a suggestion.

There must be a complete description of the invention and it must be accompanied by drawings suitably illustrating the same, if it is of a machine or other device that can be illustrated. If the device is not operative and not so clearly set forth as to make it capable of manufacture from the description, no patent can issue.

An application for patent must be made by the inventor *only*, and no person who has not actually created a portion of the invention is entitled to be considered a joint inventor. A patent issued to more than one inventor where only one has actually invented the device is invalid. A person who makes a financial contribution merely is not a joint inventor, but the invention may be assigned to him.

Patents are not granted for useless devices, for printed matter, for methods of doing business, for improvements in devices which are the result of mere mechanical skill, nor for machines that will not operate, particularly for alleged perpetual-motion machines.

A patent is not granted for a new composition of matter unless the component parts thereof, as well as the manner of making and using the same, are fully disclosed in the application when filed.

No protection is afforded by the Patent Law prior to the actual issue of a patent. The terms "Patent applied for" and "Patent pending" have no effect in law but give information that an application has been filed.

Once a patent has been issued it is out of the jurisdiction of the Patent Office, and, therefore, the office is not concerned with questions of infringement, the scope of a patent, or any other questions that arise out of the grant. These matters are within the jurisdiction of United States District Courts.

Protection of the Patent Law extends throughout continental United States, Alaska, Hawaii, and the Canal Zone, and, upon compliance with certain regulations, to Porto Rico, the Philippine Islands, the Virgin Isles, and Guam. (F.T.M.)

PATENT LOG. The term patent log is applied to any one of a large group of instruments for recording the speed of a ship through the water, and also the distance run through the water in a given interval of time.

The screw of the ship itself is, in a sense, a patent log, for the speed of the ship through the water is proportional to the rpm of the screw, and the distance run is proportional to the total number revolutions in a given time. However, the distance that the ship will move for a single revolution depends upon a number of variable factors such as the trim of the vessel, the speed of the vessel, the state of the sea, etc.

The earliest, simplest, and, perhaps, the most reliable of the various types of patent logs is the so-called taffrail log. This instrument consists of a spinner which is towed astern of the ship, well beyond the turbulence produced by the screw. The revolutions of the spinner are transmitted to a recording mechanism which was originally at the stern, or taffrail, of the ship. In modern installations the recording dials may be located on the bridge or wherever they will be of most use to the navigating staff. The dials show the speed of the ship at any instant, and also the distance run since the indicator was set to zero. The instrument must be continually watched to see that it is not fouled by seaweed or debris thrown overboard from the ship. Furthermore, it must be frequently checked by the log chip and line, or some other method, to be certain that the blades have not been bent by striking objects floating in the water.

The spinner in the Forbes type of patent log projects below the keel of the ship, at the turning center of the vessel, and its revolutions operate dials similar to those of the taffrail log.

The principle of the **Pitot tube** is used in another type of patent log. The tube itself is below the ship, at the turning center, and operates dials similar to those with the other instruments. (W.K.G.)

PATENTING. Heat Treating.

PATHOGENIC. A term used to describe the agents producing disease. Most commonly, bacteria are so described. (D.M.H.)

PATHOLOGY. That branch of medicine which is concerned with the structural changes caused by disease. Pathologists perform necropsies, examine the organs and tissues grossly, and prepare sections of these for microscopic examination. (D.M.H.)

PATINA. The geochemically altered surface of any discrete object such as a mineral, pebble or rock. A film, usually green, formed on copper and bronze after long atmospheric exposure. Particularly used by archeologists to describe the altered surface of artifacts. (R.M.F.)

PATTERN. Sheet Metal Processes; Casting.

PAUROMETABOLA. A division of the insects characterized by gradual metamorphosis. The insect hatches from the egg in a form resembling the adult and adapted for the same mode of life. It is called a nymph. As growth proceeds the nymphs of winged species acquire rudimentary wings in the form of small external flaps, which increase in size with each moult and become fully formed wings at the last transformation.

The cicadas are in an order of this group but they differ from most of the other forms in having nymphs adapted for life underground and consequently less like the adults than is usual.

The group includes the orders Orthoptera, Isoptera, Corrodentia, Dermaptera, Thysanoptera, Hemiptera, and Homoptera. (A.W.L.)

PAUROPODA. Minute animals related to the centipedes and millipedes. Usually regarded as a distinct class of the phylum Arthropoda. They resemble millipedes in having the segments of the body fused in pairs above but the legs are not grouped two pairs together. They have a distinct head and body, branched antennae, and no eyes. The body consists of twelve segments.

Pauropods live in damp places under debris on the surface of the ground. They have been recorded from Europe and the Americas. (A.W.L.)

PAVEMENT. Highways.

PAX. Private automatic exchange. This is a small machine switching telephone system used for interphone service within the confines of a privately owned building or group such as a factory. It may or may not be connected by trunk lines with the public telephone system. (L.R.Q.)

PAXILLA. A modified spine of starfishes. A thick process bearing small spines at the end. (A.W.L.)

PBX. This is a private branch exchange, a small manual telephone system used for interphone service with connecting trunks to the outside or public telephone system. Frequently such branch exchanges are supplied operating current from the main telephone office. (L.R.Q.)

PEA FAMILY. Leguminosae. The Pea Family is second only to the **Composite Family** among the **dicotyledons,** with respect to the number of species it includes. Of its more than 10,000 species in nearly 500 genera, many are trees or shrubs, especially those in tropical regions. Herbaceous species are numerous in temperate regions. Many climbing plants, also, are found in the family. Leguminous plants are found in all sorts of environments and climates.

Nearly all the plants in this family have **pinnately** compound leaves. The **stipules** present in the leaves are sometimes modified to persistent **spines.** The flowers are either regular or irregular. When regular, the flowers have five **sepals,** commonly more or less united, five **petals,** a varying number of **stamens** and a single **pistil.** Irregular flowers are of the type known as papilionaceous, a name given because of the fancied resemblance of the flower to a butterfly. In flowers of this type the **calyx** has five unequal, more or less united sepals, which frequently persist during development of the fruit, five separate petals, showing very constant difference in form. The upper one, called the standard, is large and showy; the two lateral to this, called wings, are smaller in size; and the two lower ones are more or less united into one unit, called the keel or carina. Within this keel are the ten **stamens,** which may be separate but in many genera are united in groups, the nine lower ones having their filaments more or less completely joined, while the tenth stamen remains free. The pistil has a somewhat flattened **ovary,** a long **style,** and a terminal **stigma.** The ovary contains several **ovules.** The mature fruit is called a pod or legume, which when mature often splits open with sufficient force to eject the seeds to considerable distances. The seeds in most cases have large food reserves stored in the thick cotyledons.

There are three subfamilies in the Leguminoseae. The Mimosoideae have regular flowers and valvate **corolla.** The Caesalpinioideae have irregular (zygomorphic) flowers. These two subfamilies are essentially tropical. The Papilionoideae have irregular (papilionaceous) flowers, and include most of the important cultivated forms.

Many members of this family supply man with important foodstuffs, such as **beans,** peas, and **peanuts,** while others are important forage crops for domestic animals. The high **protein** content of the plant is the principal reason for its importance as a food source. Clovers and **alfalfa** are not only valuable as forage plants, but furnish an excellent hay. Legumes are also of immense value because of their association with nodule-forming **bacteria,** resulting in a considerable accumulation of nitrogenous substance, which is later liberated into the soil, greatly enriching it. Other members of the family yield valuable **dyes, gums** and **resins,** and **oils;** many are sources of timber.

The leaves of many genera of legumes are interesting because of their ability to move. In many of them the leaflets fold together at night, so that the blade of the leaflet is vertical. Of particular interest in this connection is the Sensitive plant, *Mimosa pudica,* the leaves of which respond very quickly to external stimuli. A light blow will cause the many leaflets to fold together,

and the whole section of the compound leaf to bend down. A sudden breeze or change of temperature will produce the same result. Recovery from the shock is gradual. When stimulated by a series of successive shocks, the plant recovers more and more slowly each time. (R.M.W.)

PEA WEEVIL. Insecta, Coleoptera. A **beetle** which attacks growing peas in the pod. The most effective control measures are to avoid planting infested seed or to fumigate it with **carbon disulfide** before planting. (A.W.L.)

PEACH. Rose Family.

PEACH-TREE BORER. Insecta, Lepidoptera. *Synanthedon.* Any of three species of **moths** whose **larvae** burrow in the sapwood and inner bark of peach trees, sometimes killing younger trees. The moths belong to the family Aegeriidae, characterized by the long body and narrow wings, more or less free from scales. All three species attack other fruits as well as peaches.

The larvae of all species are destroyed by digging them out of their burrows and by burning badly infested boughs or entire trees. The common species is also treated by placing a ring of paradichlorobenzene (see **Chlorine**) on the surface of the ground about two inches from the trunk and covering it with earth. It should be left 4–6 weeks and should be applied on dates ranging from early September in New York to the middle of October in Georgia. The treatment is not recommended for trees younger than 3 years, and for those up to 6 years not more than ¾ of an ounce of the material should be used. (A.W.L.)

PEACH-TWIG BORER. Insecta, Lepidoptera. The **larva** of a small **moth,** *Anarsia lineatella,* which bores in the young twigs and fruit of peaches and other fruit trees. It is an important pest, chiefly in the western states, where various spraying programs have been found effective in controlling it. (A.W.L.)

PEAFOWL, PEACOCK. Aves, Galliformes. *Pavo.* Large birds (**Aves**) of the Oriental region, related to the pheasants and Guinea fowls. They are beautifully colored and the males have gorgeous tail feathers. They are kept to a limited extent as ornamental birds for the garden but are scarcely to be regarded as domesticated. Several species are known. (A.W.L.)

PEAK CURRENT. Rectifier.

PEAK VOLTAGE. Rectifier.

PEANUT. *Arachis hypogaea.* Leguminosae. A native legume of South America which is now widely cultivated in warm climates throughout the world. The plant is a bushy annual with pinnately compound leaves and rather showy yellow flowers. After fertilization, the flower stalk elongates greatly and bends downward so that the ovary is pushed into the ground. There it develops into the familiar peanut with its two or more seeds. The total peanut crop harvested in the United States in 1932 was over a billion lbs. Much of the crop is roasted in the shell and so marketed. Large quantities are ground for peanut butter, or crushed for peanut oil. (See **Esters.**) (R.M.W.)

PEAR. Rose Family.

PEARL. A gem formed by **bivalve mollusks,** particularly by several marine species known as pearl oysters. Pearls are formed as a protection against the irritation caused by foreign objects, either parasites or bits of gravel, which lodge inside the shell. A fold of soft tissue envelops the foreign particle and deposits layer after layer of nacre on it, similar to the mother-of-pearl lining the shell.

Pearl oysters occur in all of the tropical seas, but the ancient center of pearl fishing is at Ceylon. Pearls of considerable value are also taken from the fresh-water mussels caught for the button industry in the Mississippi River system. A large majority of fresh-water pearls, however, are poorly shaped or of undesirable color, and many are too small to be of much value. (A.W.L.)

PEARLITE. Steel.

PEARSON SYSTEM OF PROBABILITY FUNCTIONS.

Karl Pearson developed his system of probability and frequency functions from 1895 to 1916. We follow C. C. Craig's simplified exposition. The Pearson system is useful in fitting an actual **frequency distribution** by means of the **sample moments**. The sample moments in conjunction with the criterion, δ, tell which Pearson type will explain the sample results. This Pearson type is then considered as a hypothesis which is tested against the actual frequencies by the **chi-square test of goodness of fit**.

The second use of the system is to approximate a theoretical **probability function** by means of the first 4 moments of the theoretical probability function. This use is regarded as empirical and a convenience until more exact results are obtained.

In fitting an actual frequency distribution either the method of **moments** or the method of maximum **likelihood** is available. While the method of maximum likelihood is preferable on theoretical grounds, it is somewhat arduous in practice. We shall explain the method of moments.

The Pearson system is based on the differential equation,

$$\frac{dy}{ydt} = \frac{a-t}{b_0 + b_1 t + b_2 t^2},$$

where $t = \dfrac{x - \overline{X}}{\sigma_x}$, **standard units.**

This differential equation may be derived from the **hypergeometric probability function** and consequently the Pearson system may be considered as having a basis in **probability.** The system consists of 3 main types, 9 transitional types, and the **normal curve.** The types cover a very wide range of bell-shaped, J-shaped, and U-shaped functions. These will be briefly explained in terms of the criterion,

$$\delta = \frac{2\alpha_4 - 3\alpha_3^2 - 6}{\alpha_4 + 3}.$$

If δ is negative, we have main *Type I;* if δ is positive and the roots of $b_0 + b_1 t + b_2 t^2$ are complex, we have main *Type IV;* if δ is positive and the roots of $b_0 + b_1 t + b_2 t^2$ are real, we have main *Type VI.* The types occurring most frequently are *I, III, IV,* and *VII.* The values of a, b_0, b_1, and b_2 are

$$a = \frac{-\alpha_3}{2(1 + 2\delta)},$$

$$b_0 = \frac{2 + \delta}{2(1 + 2\delta)},$$

$$b_1 = \frac{\alpha_3}{2(1 + 2\delta)},$$

$$b_2 = \frac{\delta}{2(1 + 2\delta)}.$$

We assume hereafter that $\alpha_3 \geqq 0$. If $\alpha_3 < 0$, we can make it positive by changing the signs of all **variates.**

TYPE I, $\delta < 0$, $y = C(t - r_1)^{m_1}(r_2 - t)^{m_2}$, $r_1 \leq t \leq r_2$

$$r_1 = \frac{-\alpha_3 + \sqrt{D}}{2\delta}, \quad r_2 = \frac{-\alpha_3 - \sqrt{D}}{2\delta},$$

$$D = \alpha_3^2 - 4\delta(\delta + 2), \quad m_1 = \frac{1 + \delta}{\delta}\frac{\alpha_3}{\sqrt{D}} - \frac{1 + 2\delta}{\delta},$$

$$m_2 = -\frac{1 + \delta}{\delta}\frac{\alpha_3}{\sqrt{D}} - \frac{1 + 2\delta}{\delta},$$

$$C = \frac{\Gamma(m_1 + m_2 + 2)}{\Gamma(m_1 + 1)\Gamma(m_2 + 1)(r_2 - r_1)^{m_1 + m_2 + 1}}.$$

The *Type I* function may be J-shaped, U-shaped, or bell-shaped.

TYPE IV, $\alpha_3 \neq 0$, $\delta > 0$, $\alpha_3^2 < 4\delta(\delta + 2)$.

$$y = Ce^{\frac{v\pi}{2}}[(t + r)^2 + s^2]^{-m}e^{-v\tan^{-1}\frac{t+r}{s}},$$

$$-\infty < t < \infty, \quad \frac{v}{2} = -\frac{1 + \delta}{\delta}\frac{\alpha_3}{\sqrt{-D}}, \quad m = \frac{1 + 2\delta}{\delta},$$

$$r = \frac{\alpha_3}{2\delta}, \quad s = \frac{\sqrt{-D}}{2\delta}, \quad D = \alpha_3^2 - 4\delta(\delta + 2),$$

$$C = \frac{s^{2m-1}}{G(2m - 2, v)},$$

$$G(2m - 2, v) = \int_0^\pi \sin^{2m-2}\varphi\, e^{v\varphi}\, d\varphi.$$

The *Type IV* function is always bell-shaped.

TYPE VI, $\alpha_3 > 0$, $\delta > 0$.

$$y = Cz^{m_2}(z - \alpha)^{m_1}, \quad \alpha \leq z < \infty, \quad t - r_2 = z,$$

$$t - r_2 = z, \quad r_1 - r_2 = \alpha, \quad r_1 = \frac{-\alpha_3 + \sqrt{D}}{2\delta},$$

$$r_2 = \frac{-\alpha_3 - \sqrt{D}}{2\delta}, \quad D = \alpha_3^2 - 4\delta(\delta + 2),$$

$$m_1 = \frac{1 + \delta}{\delta}\frac{\alpha_3}{\sqrt{D}} - \frac{1 + 2\delta}{\delta},$$

$$m_2 = -\frac{1 + \delta}{\delta}\frac{\alpha_3}{\sqrt{D}} - \frac{1 + 2\delta}{\delta},$$

$$C = \frac{\Gamma(-m_2)}{\alpha^{m_1 + m_2 + 1}\Gamma(m_1 + 1)\Gamma(-m_1 - m_2 - 1)}.$$

The *Type VI* function may be J-shaped or bell-shaped.

TYPE III, $y = Ce^{-\frac{2}{\alpha_3}t}\left(\frac{\alpha_3}{2}t + 1\right)^{\frac{4}{\alpha_3^2} - 1}$,

$$-\frac{2}{\alpha_3} \leq t < \infty, \quad C = \left(\frac{4}{\alpha_3^2}\right)^{\frac{4}{\alpha_3^2} - \frac{1}{2}} \cdot \frac{1}{\Gamma\left(\frac{4}{\alpha_3^2}\right)}.$$

The *Type III* function may be J-shaped or bell-shaped. It is an excellent approximation to the **Bernoulli probability function**

TYPE II, $\alpha_3 = 0$, $-1 < \delta < 0$.

$$y = C(S^2 - t^2)^M, \quad -S \leq t \leq S, \quad S = \sqrt{-(1 + 2/\delta)},$$

$$M = -\left(2 + \frac{1}{\delta}\right), \quad C = \frac{\Gamma(2M + 2)}{(2S)^{2M+1}[\Gamma(M + 1)]^2}.$$

Type II may be bell-shaped or U-shaped.

TYPE VII, $\alpha_3 = 0$, $\delta > 0$.

$$y = C(t^2 + s^2)^{-m}, \quad -\infty > t > \infty, \quad s = \sqrt{1 + \frac{2}{\delta}},$$

$$m = 2 + \frac{1}{\delta}, \quad C = \frac{s^{2m-1}\Gamma(m)}{\sqrt{\pi}\Gamma(m - \frac{1}{2})}.$$

Type VII is always bell-shaped.

TYPE V, $\alpha_3 \neq 0, \quad \delta > 0, \quad \alpha_3{}^2 = 4\delta(\delta + 2)$.

$$y = C(t + r)^{-2m} e^{-\frac{2r(m-1)}{t+r}}, \quad -r \leq t < \infty,$$

$$m = 2 + \frac{1}{\delta}, \quad r = \frac{\alpha_3}{2\delta}, \quad C = \frac{[2r(m-1)]^{2m-1}}{\Gamma(2m-1)}.$$

Type V is always bell-shaped.

The remaining special types *VIII, IX, X, XI,* and *XII* occur infrequently. The normal curve is the special case $\alpha_3 = 0, \delta = 0$. (L.A.A.)

PEARSON'S TYPE III FUNCTION. Pearson System of Probability Functions.

PEAT. Coal.

PEAT MOSS. Bryophytes; Lignite.

PEBA. Mammalia, Edentata. The 9-banded armadillo, *Dasypus novemcinctus,* found from southern Texas and New Mexico to Argentina. (A.W.L.)

PECCARY. Mammalia, Artiodactyla. Animals of two species related to the Old World swine but differing in the 3-toed hind feet and in other anatomical features. The collared peccary or muskhog, *Dicotyles tajaca,* ranges from the southwestern United States to southern South America, and the white-lipped peccary, *D. labiatus,* is found only from British Honduras to Paraguay. The latter species is gregarious, living in bands of large size. Its vicious nature makes it dangerous to encounter, although a single animal is too small to trouble a human being. The collared peccary lives singly or in small groups and is inoffensive. (A.W.L.)

PECORA. An obsolete term for the group of hoofed animals known as **ruminants.** (A.W.L.)

PECTINE. A comb-like organ of a pair found on the under surface of the second abdominal segment of **scorpions.** They are supposed to be accessory organs of reproduction. (A.W.L.)

PEDIATRICS. That branch of medicine which is concerned with the prevention, diagnosis, and treatment of the diseases and disorders of children. (R.S.M.)

PEDICEL. Flower.

PEDICELLARIA. Minute pinchers on the surface of starfishes (**Asteroidea**). Pedicellariae are modified spines formed with two or more apposed jaws. They may be stalked or **sessile** and may have the jaws crossed like the blades of scissors or merely in contact. They vary greatly in form in different species. (A.W.L.)

PEDICULOSIS. Louse infestation. There are several varieties: *pediculosis capitis* refers to head lice, *pediculosis corporis,* to body lice, *pediculosis pubis* to pubic lice. Lice live on the skin and deposit their eggs on hair or, in the case of body lice, on the clothing. Their presence produces an itchy **dermatitis.** Treatment varies with the site of infestation, but the hair should be cut short or removed, and insecticides such as crude coal tar in olive oil, or thymol in olive oil applied. Infested clothing should be destroyed. DDT dusted in clothing when exposure is likely, as with soldiers in campaigns, is a highly effective preventive measure. (D.M.H.)

PEDIPALPI. The **whip scorpions,** an order of Arachnida. (A.W.L.)

PEDOLOGY. The study of the origin and classification of soils. The investigation of the **regolith** as a fundamental natural resource or the basis of terrestrial plant life. (R.M.F.)

PEDUNCLE. A stalk, either of an organ or of an entire animal. The stalks by which **brachiopods,**

crinoids, and similar **sessile** animals are attached to the supporting surface are peduncles. For the use of this term in botany, see **Flower.** (A.W.L.)

PEEWIT. Aves, Charadriiformes. 1. The black-headed gull, *Larus ridibundus.* 2. The common European lapwing, *Vanellus vanellus,* a bird related to the plovers. (A.W.L.)

PEGMATITE. The term pegmatite, derived from the Greek word meaning joined together, was first applied by Haüy in 1822 to a peculiar interpenetrating growth of quartz and **feldspar** sometimes called graphic granite from its resemblance to written characters, particularly those of the Hebrew language. Pegmatite is also used to designate those coarse-grained dikes and sheets, chiefly of **granite** or syenite, that are apophyses of **stocks** or **batholiths,** or of the residual magma, during their congelation. The individual minerals may often reach great size. Granite pegmatites are chiefly composed of alkali **feldspar** and quartz with some **muscovite** or **biotite,** but may carry such minerals as **tourmaline, topaz, beryl, fluorite, apatite, garnet, lepidolite,** etc.

The general characters of pegmatite **dikes** suggest that they are the solidified products of the residue of the **magma** which is rich in volatile matter and water vapor, and because of the abundance of **mineralizers,** remain liquid at relatively low temperatures. These physical-chemical conditions permit the maximum opportunity for the growth of large crystals. (R.M.F.)

PEKAN. Marten.

PEKING MAN. Paleontology of Man.

PELAGIC. Living in the water independent of the bottom and shores. The word applies to both the **plankton** and the **nekton.** It is usually applied only to marine animals but is used to some extent for freshwater forms. (A.W.L.)

PELECANIFORMES. The **pelicans, cormorants, gannets,** and related birds. An order made up of swimming and wading species, many with long necks and legs. (A.W.L.)

PELECYPODA. Lamellibranchiata. The name Pelecypoda for the bivalve mollusks is now widely used but according to the rules of nomenclature the older name, Lamellibranchiata, should be retained. (A.W.L.)

PELE'S HAIR. A fibrous, basic, natural glass (**tachyllite**). The congealed liquid lava blown out of volcanoes. Type locality, the Hawaiian Islands. (R.M.F.)

PELICAN. Aves, Pelecaniformes. Widely distributed birds (**Aves**) of few species. They are large with long

Pelican. (*E. R. Sanborn and N. Y. Zoological Society.*)

necks, short legs, and webbed toes. The beak is very long and the lower mandible bears a flexible pouch which can be distended to accommodate a large amount of food. They live principally on fish but also eat other aquatic animals.

The common pelican of North America, *Pelecanus erythrorhynchos,* is chiefly white with orange beak and feet during the breeding season. It frequents inland waters. (A.W.L.)

PELITE. The general term proposed by Naumann for **argillaceous** sediments containing minute fragments of quartz. Principally used as a textural term, especially when applied to fine-grained volcanic ashes, or **tuff.** Compare and contrast with **bentonite.** (R.M.F.)

PELLAGRA. A deficiency disease resulting from lack of **vitamin** B_2 or G. This disease was first described by Gaspar Casal in 1735. It is a common disease in Europe, Egypt, and in Central America, and in the southern portion of the United States. The largest outbreak in this country was between 1907–1915 and resulted in a high mortality. The disease occurs mostly among the poorer classes whose diet is inadequate.

Pellagra is due to a restricted diet and lack of vitamin B_2 (G), which is not a single substance, but is made up of a number of active principles, including nicotinic acid, riboflavin, and pyridoxine. Those afflicted are accustomed to a diet low in **protein** and made up largely of **carbohydrates.** Predisposing causes are food idiosyncrasies, chronic alcoholism, and diseases which interfere with the assimilation of a proper diet.

The symptoms of the disease appear gradually and affect primarily three of the bodily systems, the skin, digestive tract, and nervous system.

The exposed skin on the neck, hands, and feet becomes dark, raw and scaly, and eventually becomes red with a lifeless scarred appearance. The mouth is sore and ulcerated and the tongue is sensitive, red, and smooth. Diarrhea, severe and persistent, is prominent in acute cases. **Hydrochloric acid** is frequently absent from the stomach.

The symptoms referable to the nervous system are tremor, muscular cramps, and, in marked cases, weakness and paralysis. Mental symptoms are common in the severe form of the disease, and consist of confusion, hallucinations, delusions, and even dementia.

The treatment of pellagra is diet and supplementary vitamins, either as brewer's yeast, or nicotinic acid and riboflavin. In acute cases, supportive treatment with fluids and very large doses of vitamin B_2 are necessary. Liver extract intramuscularly is beneficial. (R.S.M., D.M.H.)

PELORUS. The pelorus, or "dumb compass," is an instrument used on board ship for taking **bearings** of external objects. It consists fundamentally of a circular plate, heavily ballasted and mounted in gymbals. This plate has two pairs of indicators, one pair parallel to the keel of the ship and the other perpendicular to the keel. Concentric with the circular plate, and capable of being rotated independently of each other about a vertical axis, are a graduated dial plate and an alidade or arm for reading angles. The dial plate is graduated in a manner similar to the **compass card** and may be clamped to the circular plate in any desired position. The alidade carries sighting vanes and the line through these vanes passes through the axis of rotation of the instrument carrying indicators at each end which may be read on the dial plate.

The instrument may be used to eliminate **compass errors** from bearings by setting the dial circle relative to the keel indicators on the circular plate so that they give the true **heading.** The external object is then lined up through the sighting vanes on the alidade and the indicator on the alidade will give the true bearing as read on the dial plate. (W.K.G.)

PELUDO. Mammalia, Edentata. The hairy **armadillo** of Argentina, *Tatu pilosa.* (A.W.L.)

PELVIS. Literally, a basin. Commonly applied to the bony pelvis of man, which is a compact pelvic girdle comparable to that of vertebrates generally (**skeletal system**). The human pelvis is properly the basin-like abdominal cavity containing the **viscera,** supported by the bony pelvis, which is composed of the hip bones on either side, and in front, and the **sacrum** and **coccyx.** The pelvis rests upon the lower extremities, and supports the spinal column. Within the pelvis are found the rectum, bladder, and generative organs. There are certain sexual differences in the pelvis. The female pelvis is lighter, more slender, with a cavity that is larger, less funnel-shaped, and shorter. When the female pelvis resembles the male type or is deformed by bony disease such as rickets, childbirth is interfered with and either made difficult or impossible by the vaginal route.

The pelvis of the **kidney** is the principal cavity, which receives the urine from all subordinate divisions and discharges it to the **ureter.** (A.W.L., R.S.M.)

PEN. An internal horny plate found in the **squids.** It is a vestige of a portion of the shell, which was well developed in related **cephalopod** mollusks that are now extinct. (A.W.L.)

PENCIL STONE. Pyrophyllite.

PENDULUMS. We shall here consider only gravity pendulums, leaving **torsion pendulums** and **magnetic pendulums** to be discussed separately. Let a rigid body of mass M swing on an axis which is located at distance r above its center of mass, and with respect to which axis the **moment of inertia** of the body is I. This motion is mathematically not simple, but if the amplitude of swing is small, certain terms in the differential equation of motion may be disregarded and the remaining equation readily solved. The solution gives as the period of the complete oscillation

$$T = 2\pi\sqrt{\frac{I}{Mgr}},$$

in which g is the acceleration of a freely falling body. Huygens found experimentally, what may be proved theoretically, that for any given period of oscillation greater than a certain minimum there are two different distances r for which I/r, and hence T, has the same value. If these two distances are laid off on opposite sides of the center of mass, the two points resulting are "conjugate points" (center of suspension and center of oscillation). Denoting the whole distance between these points by l, the period of swing when the pendulum is suspended at either of them is

$$T = 2\pi\sqrt{\frac{l}{g}}.$$

This is the same as the period for an "ideal simple pendulum," i.e., a single particle of mass m suspended by a weightless thread of length l, for which $r = l$, and $I = ml^2$. Kater utilized this principle in his well-known reversible pendulum (see **Kater's Pendulum**). It was Huygens who first adapted the pendulum to regulate a mechanism for keeping time and thereby gave us the common clock. (L.D.W.)

PENEPLAIN. Meaning nearly a plain. A physiographic term implying a broad flat erosional surface which has been finally developed regardless of the structure, relative hardness, and solubility of the rocks of the region. In the case of widespread **unconformities** the plain of **erosion** which truncates the subjacent de-

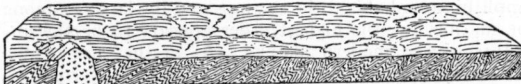

Block diagram showing a peneplain surmounted by a monad-nock of more resistant rock. (*After W. M. Davis.*)

formed rocks and underlies the superjacent formations is an ancient peneplain. Local topographic features which rise above the peneplain are called monadnocks, after the type, Mt. Monadnock, in New England. (R.M.F.)

PENGUIN. Aves, Sphenisciformes. Flightless marine birds (**Aves**) of the south temperate zone and the Antarctic. They have short legs, webbed toes, and paddle-

Galapagos penguin. (*N. Y. Zool. Soc.*)

like wings which are used in swimming. The beak is strong. On land the penguins walk in an erect position with a curiously human air. They swim and dive with exceptional skill and eat fish almost exclusively. The King Penguin is *Aptenodytes longirostris.* (A.W.L.)

PENICILLIN. (Pronounced pĕn-ĭ-sĭl′ĭn.) An antibacterial substance produced by micro-organisms of the *penicillium chrysogenum* group, principally *penicillium notatum NRRL 832* for deep or submerged fermentation, and *NRRL 1249.B21* for surface culture. Penicillin is antibacterial towards a large number of gram-positive and some gram-negative bacteria, and is used in the treatment of a variety of infections. (See below.)

Penicillin was discovered by Alexander Fleming, professor of bacteriology, St. Mary's Hospital, London, in 1929, when he observed that the mold colony—later identified as *penicillium notatum*—inhibited the growth of the pathogenic organism *staphylococcus aureus.* The development of penicillin dates from 1939 when Professor H. W. Florey and associates, working at Oxford University, prepared sufficient material to make chemical tests. In 1941, the Rockefeller Foundation invited Florey and Heatley to visit the United States, with the result that the U. S. Northern Regional Research Laboratory, Peoria, Illinois, undertook the problem of producing sufficient penicillin for clinical testing. A. J. Moyer, microbiologist in the Fermentation Division of the laboratory, found that the addition of corn steeping liquor to the medium on which the mold grows increased the yield of penicillin ten times. This was later improved so that instead of the original 2 Oxford units per milligram a commercial yield of 80 to 100 was obtainable in 4 to 5 days.

The "Oxford unit" is that amount of penicillin which, when dissolved in 50 milliliters of meat-extract broth, just completely inhibits the growth of the test strain of *staphylococcus aureus.* Penicillin is recovered as a mixture of the sodium or calcium salt of most of the organic acids in the original broth, in the form of a pale yellow to dark brown powder, stable at temperatures below 10° C., and generally containing 100 to 500 Oxford units per milligram (8 to 30% sodium penicillin). The finished material is scrupulously ·tested for potency, sterility, toxicity (by injecting into the tail of white mice, four of five mice must survive), and pyrogens (by injecting into the ear of a white rabbit, less than 1° C. rise in temperature must ensue) according to the requirements of the U. S. Food and Drug Administration.

The production of penicillin presents many difficulties which are both new and baffling. How these have been overcome can be comprehended by examining in outline the process of manufacture. During the fermentation there must be complete sterility, a continuous supply of air, controlled agitation, and the maintenance of a temperature of 24° C. The fermentation broth is filtered, and then the liquid is treated with activated carbon to adsorb the active principle, following which the material is again filtered. This time the solid moist residue is recovered and extracted by a solvent, such as acetone 80%, water 20%, and again filtered with the recovery of the liquid. Repeated alternate treatments with activated carbon and solvent may be required to obtain the penicillin in sufficiently concentrated form. In an acid medium the destruction of penicillin is rapid, and it is in an acid medium that the solvent extraction functions best, so the materials must be handled rapidly. This is immediately followed by extraction with an alkaline (sodium bicarbonate) water treatment whereupon the penicillin forms the sodium penicillin salt in water solution. This is the form in which the remaining process of concentrating is performed. The concentration is conducted in an air-conditioned (10% relative humidity), sterile (using ultra-violet lamps) atmosphere by operators wearing surgical caps, gowns, gloves, masks, and foot coverings, all frequently changed. Sterile bottles are charged with say 100,000 units of sodium penicillum solution, and immediately frozen. The final drying is conducted in a high vacuum (pressure of the order of 200 microns of mercury) at a low temperature (−70° C. on the condenser) to obtain sublimation of water below the freezing point of the solution. The final moisture content is less than 1%. The manufacturing process is concluded by stoppering, capping, and labeling the bottles, all in a sterile atmosphere.

The mode of action of penicillin is not completely understood. Apparently it is both bactericidal and bacteriostatic—i.e., it acts by killing organisms and by inhibiting their growth. The bacteriocidal effect does not occur on organisms in the resting phase, but only on those undergoing active multiplication and growth. It is thought that the drug interferes with bacterial metabolism in the early stages of growth.

Penicillin's toxicity is remarkably low even when enormous doses are given. In this respect it has a great advantage over the sulfonamides. The toxic reactions reported have been rare and of mild nature. They include fever, allergic reactions such as urticaria (hives), **dermatitis,** and **conjunctivitis** when the drug is used ·locally on the eye.

The purified drug is administered in a variety of ways. Since it is excreted very rapidly by the kidneys, it is necessary to give repeated doses every few hours in order to maintain an effective blood level. For severe infections the drug is given intravenously or intramuscularly. Oral administration requires 5 times the dosage necessary parenterally and is therefore used only in treating infections with organisms highly susceptible to penicillin action, or where the treatment need not be prolonged. Inhalation therapy with penicillin aerosol is of benefit in certain types of upper respiratory and lung infections, although it is often necessary to combine such treatment with intramuscular injection of the drug. In superficial acute infections penicillin is given locally, sometimes in ointment or powder form, for infections of the skin; it is also used locally for infection of the eye (conjunctivitis). In suppurative conditions, such as empyema and infected joint cavities, it is introduced directly into the affected area.

Not all organisms are susceptible to the action of penicillin. The more common ones against which the drug is effective are staphylococci, streptococci, pneumococci, gonococci, meningococci, Cl. tetani and the other gas-forming bacilli, and the spirochete of syphilis. The penicillin resistant organisms include B. coli, B. typhosus, the dysentery bacilli, Mycobacterium tuberculosis, viruses, yeasts, molds, and the organisms causing undulant fever, plague, cholera, tularemia, and malaria. The degree of susceptibility of each organism is different, and some strains of the same organism are more resistant than others. Occasionally susceptible strains become resistant during therapy, but this has not been common.

The diseases in which penicillin has been used successfully are numerous. The principal ones are severe staphylococcal and streptococcal infections, especially those with septicemia; osteomyelitis, puerperal sepsis, scarlet fever, pneumonia, gonorrhea, syphilis, pneumococcal meningitis, and Vincent's angina (Trench mouth). One of the outstanding successes has been with subacute bacterial endocarditis, which until the advent of penicillin was almost uniformly fatal. Penicillin has also been used extensively in the treatment of war wounds, infected compound fractures, and gas bacillus infections. In many instances penicillin is preferable to the sulfonamides, although in meningococcus meningitis sulfadiazine is still the drug of choice. In some severe infections both sulfonamides and penicillin are used. (R.K.S., D.M.H.)

PENICILLIUM. Ascomycetes.

PENIS.
A projecting, protrusible, or erectile organ of the male animal, used in the act of copulation to introduce the seminal fluid into the genital passages of the female.

In the insects the organ is a fleshy duct enclosed in a chitinous sheath called the aedeagus. In the vertebrates it develops as a fold in the wall of the cloaca and may be no more than a grooved protuberance of this simple origin. The penis of mammals, however, is a complex organ formed by the union of two external folds of tissue. The space between is enclosed as a continuation of the urethra and serves as a common duct for the excretory and reproductive systems. The organ is composed of soft tissue, including some erectile tissue, in three cylindrical bodies. This tissue becomes engorged with blood under certain stimuli and makes the entire organ rigid. When relaxed the terminal portion of the penis is retracted into an enveloping fold of skin called the prepuce, or foreskin.

Special intromittent organs of all animals are known by this name but they are not at all uniform in origin in the different phyla. (A.W.L.)

PENNSYLVANIAN PERIOD.
A period of the Paleozoic era. A systemic term first proposed by H. S. Williams in 1891. Type locality, Pennsylvania. The period began about 250,000,000 years ago. The term Pennsylvanian is roughly equivalent to the more general term Upper Carboniferous. In Britain this system is referred to as the Coal Measures. During this period there were many oscillations of sea level with relatively rapid alternations of marine and terrestrial sediments and fresh-water swamp deposits in which were formed important coal beds. The principal occurrence of the formations of this system are in the Allegheny Plateau region of the eastern United States, westward to the Ohio River. There is an abundance of plant fossils, including the non-flowering plants *Calamites, Equisetum*

("horse tails"), and the progenitors of the modern ferns. These earliest known "forests" harbored the earliest known insects and spiders. The oldest known amphibians inhabited the swamps and lowlands. Among the marine life, the highest order appears to be the ancestors of the modern sharks. As in the case of the Mississippian formations, the essentially terrestrial, clastic sediments of eastern North America grade westward into marine limestone deposits, which in turn grade upward into sediments of Permian Age. The principal marine invertebrate fossils are brachiopods (especially the genus *Productus*), pelecypods, and cephalopods (including nautiloids, goniatites, ceratites, and ammonites). Foraminifera, especially the genus *Fusilina*, were also common. Mountain building began in Europe in early Pennsylvanian time, accompanied with great volcanic activity. In eastern North America these deformative movements culminated, in the late Pennsylvanian and early Permian time, in the folding and uplift of the Paleozoic formations of the Appalachian Geosyncline. (R.M.F.)

PENNYROYAL. Mint Family.

PENSTOCK.
Some considerable surface distance usually separates the intake works and turbines in a medium- or high-head hydroelectric project. Even where an open canal is employed to carry the water from the forebay of a diversion dam to intake works located near the plant, there is still a considerable span to be bridged by a closed water conduit of the pressure type. This water conduit is called the penstock, and is always circular in form because that shape is best adapted to withstand internal pressure. There are many engineering problems, both hydraulic and mechanical, to be met and solved in connection with these penstocks. The hydraulic problems of water hammer and surging are chiefly the result of the inherent momentum of large, rapidly moving masses of water; the mechanical problems are the result of large sizes and weights of pipes used for penstocks, the rugged profiles over which they are laid, and the necessity of making many joints which will be water-tight under high pressure. Penstocks are constructed from wood stave pipe, reinforced concrete, welded and riveted steel pipe, and banded steel pipe, the latter used only for the lower sections of very high-head developments. The penstock may be either buried, partially exposed, or completely exposed. If completely exposed, it rests on concrete or timber saddles. The trend of present practice is toward the use of exposed penstocks because of their greater accessibility and longer life. Exposed pipe is more liable to have ice form in it in the winter, and it also is subjected to larger temperature variation, making it absolutely necessary to provide expansion joints.

It is common practice in this country to proportion the penstock so that the sum of the value of energy lost annually in penstock friction and the annual fixed and operating costs of the pipe line is a minimum. When all the required data are obtainable this does not prove to be a difficult problem.

Penstocks are commonly designed to meet the stress put upon them by designing for the static head, allowing for dynamic conditions by using a factor of safety of two on the elastic limit of the steel used. In other cases it has been the static head plus water hammer introduced by rapid closing of the turbine gates.

Valves to stop the flow of water are always installed in a penstock at its upper end. High-head plants have a valve at the lower end also so that water can be shut off from the turbine in less time than it would take to close the upper valve and drain the penstock. Vacuum relief valves must be provided to avoid subjecting the penstock to full vacuum inside, full air pressure outside, and hence possibility of collapse. The penstock valve will stop the flow of water if a failure is experienced by either the penstock or the turbine, and it also serves to unwater the turbine for inspection and repair. (F.T.M.)

PENTADACTYL APPENDAGE.
The form of vertebrate appendage which is regarded as the fundamental terrestrial limb from which all of the specialized appendages of animals above the fishes have been derived.

The two pairs of vertebrate limbs, pectoral and pelvic, are similar in structure and both are attached to the girdles of corresponding name (**Skeletal System**). Each girdle consists, in the primitive state, of three pairs of bones meeting at the articulation of the limbs. On each side of the body one bone extends toward the back and two toward the middle of the body below. The skeletal structure of the limbs includes a single bone in the segment next to the body, followed by two bones. Then follows a group of small bones, and last five divergent series of moderately long bones which extend into the digits. Of the more constant bones in the pectoral girdle of vertebrates with limbs, the dorsal bone is the scapula and the two ventral are an anterior clavicle (of dermal origin) and a posterior coracoid. This girdle may contain a procoracoid between the last two bones. The bones of the pectoral appendage are, in the order described above, the humerus, the ulna and radius, the carpals, the five metacarpals of the hand and the five series of phalanges in the digits. In the pelvic girdle of vertebrates with limbs, the dorsal bone is the ilium and the two ventral are an anterior pubis and a posterior ischium. The bones of the appendage are the femur, the tibia and fibula, the tarsals, the metatarsals and the phalanges.

Specialized appendages such as the wings of birds, flippers of marine mammals, and the legs of the hoofed species, show either a simplification or a slight increase in complexity of the skeletal system and in some cases a consolidation by webbing of the digits or a still more compact fleshy union between them. In all of these cases, however, the basic pentadactyl structure is still present. (A.W.L.)

PENTAGRID CONVERTER.
This is a **vacuum tube** containing 5 **grids** in addition to the usual **anode** and **cathode** and is used primarily as a combined **oscillator** and mixer or first **detector** in a superheterodyne **receiver**. In operation the cathode and the first two grids form the cathode, grid and anode of a conventional oscillator circuit. The electrons which would all contribute to the plate current of a conventional oscillator tube largely pass through the openings in the grid-like anode and are then acted upon by the incoming signal which is applied to one of the other grids. Upon finally reaching the real anode of the tube the electrons have both the effect of the oscillator and the incoming signal, and thus give a combined effect in the output. A typical connection is shown in the figure. (L.R.Q.)

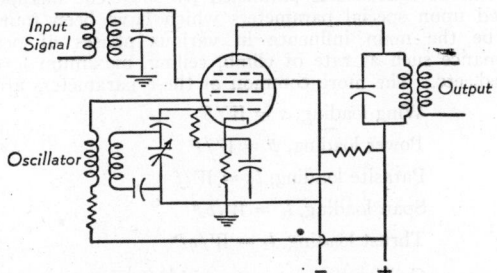

PENTASTOMIDA.
Worm-like **arthropods** of a few parasitic species, sometimes regarded as an order of the class **Arachnida** but more often as a distinct class. The known species live in the **respiratory system** or body cavity of reptiles and mammals. Also called Linguatulida.

The adult has an unsegmented anterior region bearing two pairs of claws and a segmented body whose subdivisions are not metameric. The **larvae** resemble certain mites. They are shorter than the adults and their claws are borne at the ends of leglike prominences. (A.W.L.)

PENTLANDITE.
The mineral **sulfide** of **iron** and **nickel** corresponding to the formula $(Fe,Ni)S$. It is isometric, appears in granular masses; hardness, 3.5–4; specific gravity, 5.0; color, bronze-yellow; opaque. Occurs with **pyrrhotite, millerite, niccolite**, etc. The best known deposit of **pentlandite** is at Sudbury, Ontario, Canada, where it is associated with a nickel-bearing **pyrrhotite**. (E.S.C.S.)

PENTODE.
Tube, Electronic.

PENTOSANS.
Carbohydrates.

PENTOSES.
Carbohydrates.

PENTRAMITES.
Invertebrate Paleontology.

PEONY.
Buttercup Family.

PEPO.
Fruit.

PEPPER.
Piper nigrum. Piperaceae. The pepper plant, *Piper nigrum*, is a woody climbing shrub, which is indigenous in India. It is aided in climbing by the **adventitious roots** which are formed at the nodes. The ovate leaves are evergreen. The flowers are minute, without petals, and borne in slender **spikes**. The **fruit** is a bright red berry less than $1/4''$ in diameter. Each berry contains a single seed. On drying the **pericarp** becomes black and wrinkled.

The berries are gathered before they are ripe and dried, usually by the sun. The dried berries are separated from the stem and ground, producing black pepper. If the pericarp is removed from the berry, leaving the seed and **endocarp**, the ground product is known as white pepper. The pericarp may be removed by using mature berries and soaking them to soften the pericarp. Or machines may rub off the dried pericarp, a method used in western countries. Pepper is one of the most extensively used of all spices. White pepper is less pungent than black, and hence not so conspicuous when used in cooking. Pepper is grown mostly in British India and in the Malaysian regions. Propagation is usually by cuttings, which begin to fruit within four or five years, after which they bear continuously but somewhat irregularly for many years.

Related to *Piper nigrum* is *Piper Betle,* a perennial creeping vine native to Java, but widely grown in tropical Asia. From its leaves is prepared a chew with the nut of the **Areca** palm. *Piper Cubeba* is another species, also native of Java and the Molucca Islands, which yields a volatile oil which is used medicinally.

See also Capsicum in the **Potato Family**. (R.M.W.)

PEPPERMINT, OIL OF.
Volatile Oils.

PEPSIN.
Enzymes.

PEPTIC ULCER.
A common chronic disease of the gastro-intestinal tract characterized by recurrent abdominal pain and indigestion. The symptoms are due to the presence of an **ulcer**, either in the wall of the **stomach**, or in the wall of the **duodenum** just beyond the stomach. Peptic ulcer occurs in all races, in all countries, and at all ages. It is far more common in men than in women, and the peak incidence is between 30 and 50 years. Although it occurs in all constitutional

types, it is seen most often in people of asthenic body build—tall, lean, "nervous," hard-driving individuals. It is estimated that 10% of the population suffer from the disease at some time.

The cause of peptic ulcer is unknown, although many predisposing factors are recognized. Among these are constitutional elements and an inherent failure of the cells of the mucosa to resist acid digestion by the gastric secretions. Neurogenic and psychogenic factors play a large role; it is well known that emotional tensions precipitate and accentuate the disease.

The pain of peptic ulcer is characteristic. It is a gnawing discomfort, felt high in the mid **abdomen**, in the epigastrium. It is associated with hypermotility and a high content of hydrochloric acid in the stomach, and therefore comes on when the stomach is empty, i.e., just before meals. Food dilutes and neutralizes the hydrochloric acid and relieves the pain. Sodium bicarbonate neutralizes the acid and similarly relieves the pain. Other symptoms such as nausea, vomiting, weight loss are not common in uncomplicated ulcer. The outstanding feature, pain, is chronic, but is also periodic. It tends to disappear after a few months, only to reappear again in months or years.

The treatment of simple peptic ulcer consists of rest, diet, and drugs. The latter are used to neutralize the excessive amounts of hydrochloric acid and to reduce the hypermotility of the gastro-intestinal tract. Psychotherapy, especially in individuals suffering from definite psychoneurotic disturbances, is an important phase of treatment (see **psychoneurosis**).

Complications of serious nature are not uncommon. These include perforation of the ulcer into the abdominal cavity with the development of **peritonitis**; **hematemesis** due to massive **hemorrhage** from eroded vessels in the ulcer bed; and chronic obstruction resulting from fibrosis and scarring. Perforation is a surgical emergency which requires prompt treatment. Hemorrhage is treated medically with **transfusions** and iron therapy. Chronic obstruction requires surgical measures usually. (D.M.H.)

PEPTIDES. Aldehydes, Ketones, and Related Compounds.

PEPTONES. Aminoacids, Polypeptides, and Proteins.

PERCENTILE. The percentiles $P_1, P_2, \cdots P_{99}$ are the 99 points dividing a distribution into 100 equal parts. They are determined by this definition in the case of a **serial distribution** and by interpolation in the **cumulative frequency distribution** if a **frequency distribution** has been formed. Sometimes the term first percentile refers to all variates below P_1. (L.A.A.)

PERCH. Pisces, Teleostei. 1. The yellow perch, ringed perch, or common perch, *Perca flavescens,* a fresh-water food fish (Pisces) which lives in lakes and streams from Iowa to South Carolina and northward into Canada. Introduced on the Pacific Coast. 2. The pike-perch, *Stizostedion vitreum,* a related species also known as the wall-eye, glass-eye, wall-eyed pike, and jack salmon. It is a valuable food fish of the Great Lakes basin, the Mississippi valley, and the northeastern states to Virginia. 3. Several marine fishes, some of large size, found in the seas of the Oriental and Australian regions and the Mediterranean. Some ascend rivers. They are related to the sea basses. (A.W.L.)

PERCHED. Used by hydrologists to describe the **ground water** table when it is separated from an underlying layer of ground water by an unsaturated layer of rock. Used by **glaciologists** to describe a glacial boulder, **erratic**, resting on a prominent topographic posi-

tion. Used by **physiographers** and **hydrologists** to describe a stream whose bed is separated from the top of the ground water table by a dry, unsaturated zone. (R.M.F.)

PERCHLORIC ACID AND PERCHLORATES. Perchloric acid ($HClO_4$) is a colorless, fuming, oily liquid, miscible with water, volatile under diminished pressure with safety (at 18 mm., volatilization temperature 16° C.) or by distillation of the dilute solution (70% $HClO_4$ or less) at atmospheric pressure. There is a maximum constant boiling point 203° C. (760 mm.) at 73% $HClO_4$ (distillate) for mixtures of perchloric acid and water. Cold dilute perchloric acid reacts with such metals as **zinc** and **iron**, yielding **hydrogen** gas and the corresponding perchlorate in solution; is stable from the point of view of oxidation and reduction (except that **iodine** is oxidized to **periodic acid,** with liberation of **chlorine, ferrous** salt solutions to **ferric, titanous** salt solutions to **titanic**). Concentrated hot perchloric acid, on the other hand, is a powerful oxidizing agent, exploding violently in contact with charcoal, paper, alcohol; causes serious wounds in contact with the skin.

Prepared by distilling **ammonium** perchlorate with **nitric** and **hydrochloric** acids.

Metallic perchlorates are soluble in water, except that **potassium** perchlorate is slightly soluble. Potassium perchlorate is, however, insoluble in alcohol containing perchloric acid, a property made use of in the qualitative recognition and quantitative estimation of potassium in salt solutions. Perchlorates, when heated, evolve oxygen and leave the chloride as a residue—potassium perchlorate decomposes at 400° C. (R.K.S.)

PERCUSSION. Kinetics.

PERENNIAL. As distinguished from **annuals** and **biennials,** perennials are plants in which at least a part of the vegetative body remains alive for a number of years. Perennial herbs are usually perennial only by underground organs such as roots or rhizomes, the aerial parts dying at the end of each growing season. Some of the species falling in this category have "winter rosettes" of leaves which survive above the ground through the winter months. Examples of such species are the evening primrose and some kinds of asters. Woody perennials include all species of trees and shrubs. In deciduous trees and shrubs the stems are the only aerial parts to remain alive from year to year. In many kinds of evergreen woody plants the leaves are also perennial in the sense that a given crop of leaves may remain on the plant continuously for several years before abscising. (B.S.M.)

PERFORMANCE PARAMETERS, AIRPLANE. The classic method of determining various items of airplane performance from horsepower vs. speed curves at various altitudes is paralleled by short-cut methods based upon special parameters which have been found to be the main influence in various phases of performance such as rate of climb, ceiling, maximum level speed, etc. The more common of these parameters are:

Wing loading, $s = W/S$

Power loading, $p = W/P$

Parasite loading, $l_p = W/f$

Span loading, $l_s = W/b_e^2$

Thrust loading, $l_t = W/\eta P$

Ceiling parameter, $\Lambda = l_s l_t^{4/3}/l_p^{1/3}$.

In the above, W = total weight of airplane and contents, P = rated horsepower, η = propeller efficiency, f = **equivalent parasite area,** b_e = effective monoplane span. (F.T.M.)

PERFUMES. Odorous constituents of plants usually occur as an essential or **volatile oil**, which upon separation is a highly aromatic liquid, and almost always a very complex mixture of substances. Some of these substances are easily destroyed by heating. **Alcohols** and **phenols, aldehyde** and **ketones**, phenolethers and esters—especially esters—are represented among the known constituents. Information of scientific character in the field of odorous chemicals seems to be more specific than general. Very slight differences in constitution often change notably the odorous character of substances, e.g., the contrasting odors of acetic and butyric acids (CH_3COOH, C_3H_7COOH, respectively), and of acetaldehyde and propionaldehyde ($CH_3 \cdot CHO$, $C_2H_5 \cdot CHO$, respectively). Whereas the odors of the paraffin aldehydes from C_3 to C_6 are disagreeable, the paraffin aldehydes from C_7 to C_{14}, inclusive, and the corresponding alcohols are used in traces as desirable odorous additions to a number of perfumes.

The naturally occurring odorous constituents are found distributed in various parts of various plants, as illustrated below:

PART OF THE PLANT	PLANT SUPPLYING ODOROUS MATERIAL
Flowers	Cassia, carnation, clove, hawthorn, hyacinth, heliotrope, jasmin, jonquil, lilac, orange, rose
Flowers and leaves	Lavender, rosemary, violet
Leaves	Bay, cinnamon, eucalyptus, geranium, peppermint, wintergreen
Barks	Cassia, cinnamon
Woods	Camphor, cedar, sandalwood, pine
Roots	Angelica, sassafras
Rhizomes	Ginger, orris, citronella, lemongrass
Fruits	Bergamot, grapefruit, lemon, lime, orange
Seeds	Almond, anise, clove, juniper, nutmeg
Gums, oleoresins	Myrrh, Peru balsam, storax, tolu.

Musk is obtained from the dried secretion of the preputial follicles of the male musk deer, now found only at high elevations in the Himalaya Mountains, and ambergris from the diseased intestines of the whale.

Since most of the odorous substances are volatile and easily vaporized at 100° C., the simplest procedure for their separation from the plant material is distillation with steam in cases where the odorous material is not thereby injuriously affected. Rose, orange blossom, lavender, peppermint and other odorous principles are obtained in this way—but not those of the citrus fruits.

The citrus oils are obtained by expression or squeezing the rinds, thus rupturing the oil cells of the fruits, and collecting the oil. These oils are spoiled by heating.

To obtain the odorous substances of roses and many other flowers, there is a method of absorption by beef and pork fats, called enfleurage when conducted with thin layers of fat at ordinary temperatures, and maceration when fat is melted at about 65° C. The product, called pomade, is treated with alcohol to extract the odoriferous material.

Finally, extraction by the use of a volatile solvent, especially for flowers and leaves, is practiced. After evaporation of the solvent, the residue is known as a concrete. When this concrete is further treated with alcohol, in which solvent the odorless and colored materials are insoluble, and the resulting solution separated from the solvent alcohol, the residue is known as an absolute.

Various substances are used as fixatives, which have little or no odor value themselves, but serve to make odors with which they are mixed less volatile, accordingly available over a longer period of time, and in some cases sweeten the odor.

Some of the individual odorous chemicals that have been identified, and material in which each is contained are as follows:

ODOROUS CHEMICAL	MATERIAL IN WHICH CONTAINED
Benzaldehyde	Bitter almond
Phenylethyl alcohol	Rose
Vanillin	Vanilla
Eugenol	Clove
Citral (geraniol aldehyde)	70% in lemongrass oil 8% in lemon oil
Methyl salicylate	Wintergreen
Ethyl salicylate	Wintergreen, superior
Amyl salicylate	Clover blossom
Benzyl alcohol, formate, propionate	Synthesized
Benzyl acetate	Jasmin
Geraniol esters	Rose
Citronellol esters	Rose
Methylphenyl acetate	Gardenia
Ethyl-, amyl-, benzylphenyl acetates	Synthesized
Linalyl alcohol	Lavender
Linalyl acetate	Bergamot
Phenylacetic aldehyde	Hyacinth
Terpineol	Lilac
Methyl anthranilate	Grapes
Isobutyl-, isoamylsalicylate	Synthesized
Cinnamic aldehyde	Synthesized
Cinnamic alcohol	Storax
Methyl, ethyl, amyl, cinnamyl cinnamate	Synthesized
Anisic aldehyde	Mayblossom
Ionone	Violet
Phenol-, diphenyl oxide	Geranium
Beta-naphthol ethyl ether	Orange blossom
Acetophenone	Synthesized
Para-methylacetophenone	Synthesized
Benzylideneacetone	Sweet pea
Coumarin	Tonka, new mown hay
Isoeugenol	Carnation
Heliotropin	Heliotrope
Hydroxycitronellal	Lily, lilac
Ethyl protocatechnic aldehyde	Vanilla
Gamma-undecalactone	Peach
Methyl, ethyl, isobutyl, amyl benzoates	Synthesized

Odorous materials are widely utilized in perfumes and cosmetics, in flavoring extracts and essences, and in soaps. The blending of natural and artificial odorous substances to produce high grade perfumes is an art. The introduction of traces of a given substance often causes desirable or undesirable results out of all proportion to the amount of material added, and requires for success one who is expert in the art. (See **Volatile Oils**.) (R.K.S.)

PERIANTH Flower.

PERICARDIAL CAVITY. A portion of the body cavity (**coelom**) containing the heart. In the fishes it is the entire **thoracic** part of the cavity but in air-breathing vertebrates the thoracic region also contains the lungs. The cavity surrounding the heart of insects is called the pericardium. It is a division of the **haemo-coele.** (A.W.L.)

PERICARDIAL SINUS. The pericardial cavity or pericardium of invertebrates. (A.W.L.)

PERICARDITIS. Inflammation or infection of the **pericardium** which occurs as the result of bacterial spread through the blood stream, by direct extension from neighboring organs, the lungs and pleura, or as the result of injury such as a stab wound. It occurs most frequently as part of rheumatic fever, as extension from pneumonia, or as tuberculous infection spreading from the lungs. When infection is present the membranous walls of the pericardium become roughened and inflamed and the amount of fluid in the sac is greatly increased. In time the collection of fluid may be so great that it interferes with the function of the heart and results in cardiac tamponade. Excessive fluid may be withdrawn through a needle inserted in the pericardial cavity. If the fluid becomes purulent surgical drainage must be instituted.

Pericarditis without infection occurs as a complication of **coronary occlusion** and infarction of the heart muscle. The muscle which is deprived of its blood supply undergoes **necrosis** and inflammatory changes. The latter frequently involve the adjacent pericardium. On listening over the heart with a stethescope a characteristic rough sound is heard—a "pericardial rub" which represents the movement of the thickened inflamed pericardial sac over the heart with each heart beat.

The symptoms of pericarditis often may be masked by the disease to which it is secondary. Fever, pain in the chest, rapid feeble pulse, and shortness of breath are usually present. X-ray of the chest reveals a characteristic type of enlargement of the heart shadow if fluid is present. (D.M.H.)

PERICARDIUM. 1. The cavity surrounding the **heart** of **insects** and other invertebrates, also called the pericardial sinus. 2. The cellular lining of the **pericardial cavity** in vertebrates and other animals, in which this cavity is a part of the **coelom**. The human pericardium is composed of two layers. The inner layer is adherent to the heart. Between the layers there is about $\frac{1}{2}$ ounce of a thin liquid, the pericardial fluid, which serves as a cushion and prevents friction of the beating heart. (A.W.L., R.S.M.)

PERICARP. Fruit.

PERICLASE. A mineral oxide of magnesium, MgO. Crystallizes in the isometric system. Hardness, 5.5; specific gravity, 3.56±; color, variable with tones of gray, yellow, brown, green and black; transparent. From the Greek, meaning to break around, in reference to the cleavage. (R.M.F.)

PERICYCLE. A layer or cylinder of thin-walled **parenchyma** cells which occurs external to the **xylem** and **phloem.** It is bounded externally by the cells of the cortex, or by the endodermis. In stems it often contains fibers, called pericycle fibers. In roots it gives rise to the meristems which produce lateral roots, and cork cambium which produces the cork and secondary cortex of root bark in most trees. (R.M.W., B.S.M.)

PERIDERM. Lenticels.

PERIDOTITE. The term peridotite is derived from Peridot, the French word for **olivine.**

It is a coarse-grained igneous rock related to *gabbro,* which consists of **olivine** and **proxene** in varying proportions. Certain peridotites contain **spinel, chromite,** or **mica** as accessories. A variety of peridotite made up of **hornblende** and olivine is called cortlandtite because originally described from the township of Cortlandt (not Cortland), N. Y., on the Hudson River south of Peekskill. This rock was formerly called Hudsonite but abandoned as the term hudsonite had been previously applied to a variety of the mineral **pyroxene** found in St. Lawrence County, N. Y.

Rocks consisting essentially of olivine alone are known as dunites, the name coming from the occurrence of this rock in the Dun mountains of New Zealand. In the United States it is found in North Carolina, South Carolina, and Georgia, where corundum is associated with the dunite in commercial quantities. The olivine of peridotites alters readily to the mineral **serpentine,** often to such an extent that the rock itself is called a serpentine. As mentioned above the peridotites may contain **chromite** or other valuable minerals, often to such an extent that they may be commercially exploited, for nickel, platinum, and precious garnet.

Kimberlite from which diamonds are secured is commonly called a mica peridotite, but is more closely related to the **lamprophyres.** (E.S.C.S.)

PERIGYNY. Fruit.

PERIHELION. Perihelion is the point in the **orbit** of any member of the **solar system** when the object is closest to the sun. Since the orbits are all **conic sections** with the sun at one focus, perihelion must lie on the **line of apsides** of the conic.

In orbits of **satellites** and other objects which are referred to primaries other than the sun, terms similar to perihelion are used to indicate points in the orbit closest to the primary. For example, the point in the **moon's** orbit closest to the earth is known as perigee, the corresponding point in an orbit of a satellite of **Jupiter** is known as perijove, etc. (W.K.G.)

PERILLA OIL. Esters.

PERIMORPH. One species of mineral inclosed in another. (R.M.F.)

PERINEUM. The diamond-shaped space at the base of the torso, between the **coccyx** behind and the arch of the pubic bones in front. It is bounded by the thighs on each side. This area comprises the floor of the pelvic cavity. It contains many important muscles which are pierced by the **rectum, urethra** and in the female by the **vagina.** (R.S.M.)

PERIOD. In geology, a major sub-division of an **Era.** The formations which belong in any one period are spoken of as a **System.** (R.M.F.)

PERIOD LUMINOSITY CURVE. Cepheid.

PERIOD LUMINOSITY LAW. Cepheids.

PERIODIC ACID AND PERIODATES. Periodic acid (H_5IO_6) has been isolated as a colorless solid, melting point about 130° C., and at 138° C. begins to decompose, metaperiodic acid (HIO_4) being formed and at higher temperatures iodine pentoxide plus oxygen plus water. $H_4I_2O_9$ and H_3IO_5 have been reported as fairly well established in identity; in solution the evidence points to the presence of HIO_4.

Prepared by reaction of **iodine** and **perchloric acid.** Sodium periodate ($Na_2H_3IO_6$) is formed by reaction of **sodium iodate** plus sodium hydroxide plus **chlorine** (sodium chloride also formed), and the periodate separates as crystals from the medium. In solution, it is stated, periodate gradually forms ozone (see **Oxygen**) and iodate at the ordinary temperatures.

Metallic periodates are solids, slightly soluble in water. Periodates, when heated, evolve oxygen with simultaneous formation of iodate, which is decomposed at higher temperatures. Periodate in acid solution oxidizes **hydrosulfuric acid or sulfurous acid to sulfuric acid,** oxalic acid to carbon dioxide, **manganous** to manganate, and with **hydrogen peroxide** yields oxygen and iodate. (R.K.S.)

PERIODIC FUNCTIONS. If $f(x + p) \equiv f(x)$ for every value of x, where p is a constant, then $f(x)$ is said to be periodic, with period p.

The trigonometric functions are periodic functions; the sine and cosine have a least period of 2π, and the tangent and cotangent have a least period of π.

The **exponential function** has a pure imaginary period $2\pi i$.

The **elliptic functions** are doubly periodic functions. (L.L.S.)

PERIODIC LAW. This "law" states that the chemical and physical properties of elements are a periodic function of the **atomic number** of the elements. (See **Chemical Composition**.) (R.K.S.)

PERIODOGRAM ANALYSIS. Periodogram analysis consists of a statistical technique used to find the periods in time series in which it is assumed that the values of the function repeat themselves after a certain specified time called the period. Naturally such an analysis is needed in the study of wave motion of any kind or of business cycles.

First the trend must be removed in the data. The trend consists of a general movement either upward or downward or a combination of the two on which the periodic movement is superimposed. The removal of trend is accomplished either by use of a polynomial of high degree or by some method of smoothing. After the trend is removed, the search for cycles of various periods is begun. This is done either by **serial correlation**, the curvilinear coefficient of correlation, or by use of simple sums.

Assume $N = jp$ **variates**, N fixed but p and j varying, $y_{11}, y_{12}, \cdots y_{1j}, y_{21}, y_{22}, \cdots y_{2j}, y_{31}, y_{32}, \cdots y_{3j}, y_{p1}, y_{p2}, \cdots y_{pj}$, in that order. Assume a period p, and calculate the j values $M_{j,p} = \sum_{j=1}^{p} y_{pj}$. From these j values for an assumed p calculate $R_p = |$ greatest value of $M_{j,p} -$ least value $M_{j,p}|$, the **absolute value** of the **range** R_p. The values of R_p are plotted against p. Naturally p assumes every value from 1 to N. This forms the periodogram. Wherever the values of R_p are very high comparatively, it is assumed a period of length p has been found. Tests of significance may be applied to R_p under the assumption that the y's are independent. Unfortunately successive y's are generally strongly dependent. Sometimes the technique indicates periods which do not actually occur in the data. Hence care should always be taken in the interpretation of the results of periodogram analysis. In place of R_p we may compute

$$\sigma^2 M_{j,p} = \frac{\Sigma(M_{j,p} - \overline{M}_{j,p})^2}{j}, \quad \overline{M}_{j,p} = \frac{\Sigma M_{j,p}}{j}$$

and construct the periodogram of $\sigma^2 M_{j,p}$ against p. (See **Serial Correlation**.) (L.A.A.)

PERIOPOD. The thoracic appendages of **crustaceans**, behind those which are associated with the mouth. (A.W.L.)

PERIOSTEUM. The tough fibrous membrane which covers the surface of the **bones** of the body except on their joint surfaces. It is tightly adherent to the bone surface and it is from this membrane that new bone regenerates. (R.S.M.)

PERIOSTITIS. Inflammation, acute or chronic, of the **periosteum**. It may occur following injury to the bone or it may be a complication of infection with **septicemia, typhoid fever,** or **syphilis**. In the acute form, abscesses may develop. (R.S.M.)

PERIOSTRACUM. The outer layer of the shell of **brachiopods** and **mollusks**. In the molluscan shell it is formed of a horny organic material called conchiolin, associated with the prismatic middle layer of the shell. It is also organic material in the brachiopods but here it is separated from the prismatic layer of the shell by a thin layer of **calcium** carbonate. (A.W.L.)

PERIPATUS. Onychophora.

PERIPROCT. The surface of the body immediately surrounding the **anus**. It is applied especially to the sea urchins (**Echinoidea**) which have two leathery areas in the otherwise rigid body wall, one the peristome surrounding the mouth and the other the periproct on the upper surface. (A.W.L.)

PERISARC. A chitinous (**chitin**) sheath surrounding the stalks and branches of the colony in the **hydrozoan coelenterates** and in some cases forming cupped expansions around the **polyps**. (A.W.L.)

PERISSODACTYLA. The odd-toed **ungulates,** an order of hoofed **mammals** in which the axis of the foot passes through the middle digit. This digit is larger than the others and is symmetrical. The order contains the **horses** and related species, the **rhinoceroses,** and the **tapirs**. (A.W.L.)

PERISTALSIS. A wave-like series of muscular contractions progressing along the walls of the gastro-intestinal tract, which serve to propel the contents along from the esophagus to the rectum. (R.S.M.)

PERISTOME. The region surrounding the mouth. The term is used chiefly of the ciliated (**cilia**) zone surrounding the mouth in some of the 1-celled animals and the leathery portion of the body wall of sea urchins (**Echinoidea**) in which the mouth opens. (A.W.L.)

PERISTOMIUM. A modified segment or segments behind the mouth in some of the segmented worms (**Polychaeta**). It bears tentacles and other sensory organs and in some forms is modified as a collar. In the species which live in tubes this collar may form a funnel leading to the mouth or a fold which secretes additions to the tube. (A.W.L.)

PERITONEUM. The epithelial lining of the body cavity (**coelom**). In the higher vertebrates the lining of the abdominal cavity in particular. The corresponding tissue in the pleural and pericardial cavities of the thorax is called the pleura and the pericardium, respectively. The peritoneum is a thin layer of tissue made up of flat cells. (See figure at top of page 1062.) (A.W.L.)

PERITONITIS. Inflammation or infection of the **peritoneum**. This is a serious complication of many pelvic and abdominal diseases and is an ever present hazard in abdominal surgery. One of the commonest causes of peritonitis is perforation of some part of the gastro-intestinal tract, such as occurs with a gastric or duodenal ulcer which may break through, or rupture of an inflamed and distended appendix. When this happens, contaminated material gets free in the peritoneal cavity, and severe infection may result. The colon bacillus is frequently the causative agent.

Pelvic peritonitis results from infection of the Fallopian tubes, most often with the gonococcus. Tuberculous peritonitis occurs in the course of tuberculous infection of the bowel or the genito-urinary tract. Pneumococcus peritonitis is seen as a complication of nephrosis (see **nephritis**) in children.

The patient with peritonitis is acutely ill, with fever, rapid pulse and respirations and a distended, tender, rigid abdomen. Treatment depends on the cause. Surgery may be indicated, and **sulfonamides** or **penicillin** are useful if the bacterial agent is sensitive to either of these drugs. (D.M.H.)

PERITREME. A plate surrounding the external opening (**spiracle, stigma**) of an air tube (**trachea**)

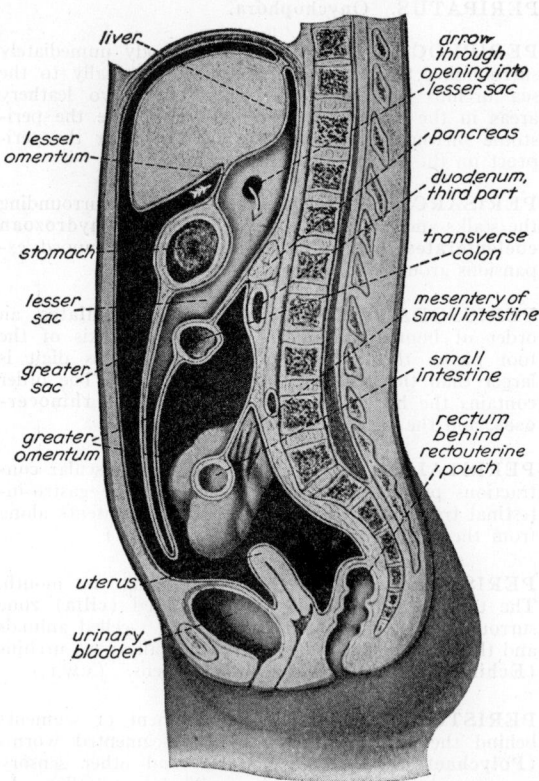

Diagrammatic median section of female body to show the abdominal cavity and the peritoneum on vertical tracing. (*Cunningham, Textbook of Anatomy, Oxford Press.*)

in some of the **arthropods.** The term is applied to certain mites and insects. (A.W.L.)

PERITRICHIDA. Ciliophora.

PERIWINKLE. Mollusca, Gasteropoda. Marine **snails** with a thick conical spiral shell. They live in shallow water, in the tidal zone, and along the shore. Some of the many species are very widely distributed and the genus *Littorina* is represented in all parts of the world. An edible species has been introduced into the United States and is now common on the Atlantic coast north of Delaware Bay. (A.W.L.)

PERKIN MEDAL AWARDS. History and Evolution of Chemistry.

PERLITE. Pearlstone. An unusual form of siliceous lava composed of small spherules of about the size of bird shot or peas. It is grayish in color with a soft pearly luster. The spherules often show a concentric structure and are believed to be formed as a result of a peculiar spherical cracking developed while cooling. They may be confused with oölites which are classified as concretions. (E.S.C.S.)

PERMALLOY. Nickel Alloys; Magnetic Materials.

PERMANENT MOLDS. Casting.

PERMEABILITY. Magnetism.

PERMIAN PERIOD. The name of a geologic period. Type locality, Province of Perm, Russia. The formations of this system were first studied and described

by R. I. Murchison, in 1841. The Permian period began about 230 million years ago, and lasted for about 30 million years. Only the upper Permian appears to be

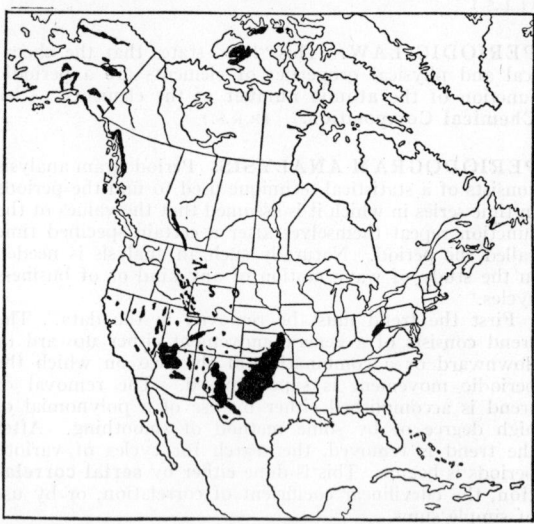

Map showing known areas of outcrops (surface distribution of Permian strata in North America).

represented in North America. Eastern North America was undergoing uplift and erosion during this period, while the seas invaded the West from the Arctic and Gulf. Increasing uplift of the continents and mountain building culminated in the **Applachian Revolution** (U. S.) and the **Hercynian Revolution** (Europe). Increasing aridity and widespread equatorial continental glaciation are disclosed by **tillites** in Australia, Tasmania, New Zealand, India, South America, and Massachusetts. Some of the fossil animals and plants appear to be the progenitors of Mesozoic forms, but a number of marine invertebrates became extinct, including the **trilobites, eurypterids, primitive insects, goniatites,** etc. The fossil fishes are similar to those of the Pennsylvania. The **Amphibia (Stegocephalians)** are approaching extinction. The first fossil reptiles occur in the Permian, including such highly specialized forms as the **Cotylosaurs** (sail-backed lizards). Important **index fossils** are the fern-like genera, *Sphenopteris* and *Glossopteris.* The distinctly latitudinal distribution of animals and plants between Africa and South America has suggested that the North Atlantic was separated from the South Atlantic by a great east-west continent called **Gondwana Land.** Also during this period the Mediterranean was greatly expanded to the west with easterly and south-easterly outlets into the Pacific. (R.M.F.)

PERMINVAR. Nickel Alloys; Magnetic Materials.

PERMUTATIONS. If we have given a set of things, it is often important to know how many different groups or selections can be made by taking some or all of the things of the set, when we take account of the order of arrangement in each group, and when we do not consider the order of arrangement.

Each different arrangement in a definite order which can be made of all or part of a given set of elements is called a permutation.

We are generally concerned with the number of permutations of n things r at a time.

The number of permutations of n things taken r at a time is denoted by many different symbols by different writers; the most commonly used are $_nP_r$, $P_{n,r}$ and $P(n, r)$.

The number of permutations of n different elements taken r at a time is given by $_nP_r = n(n - 1)(n - 2)$ $\cdots (n - r + 1)$, (r factors on the right).

The number of permutations of n different elements taken n at a time is

$$_nP_n = n(n-1)(n-2) \cdots 3 \cdot 2 \cdot 1 = n!$$

If, of n things, p are alike, of one kind, q of another kind, r of another, and so on, the number of permutations of these things taken all together is $\dfrac{n!}{p!q!r!}$. (L.L.S.)

PERNICIOUS ANEMIA. Anemia.

PERONIUM. An ectodermal thickening at the base of a tentacle in some of the jellyfishes. (**Scyphozoa,** order Narcomedusae.) (A.W.L.)

PEROVSKITE. The mineral perovskite is **calcium titanate** $CaTiO_3$, probably **isometric**. It has a cubic **cleavage**; is brittle; hardness, 5.5; specific gravity, 4; luster, adamantine; color, various shades of yellow to reddish-brown or nearly black; transparent to opaque. It is found associated with **chlorite** or **serpentine** rocks occurring in the Urals, Baden, Switzerland, Italy, etc. It was named for a certain Von Perovski. (E.S.C.S.)

PERPETUAL MOTION. The idea of a mechanism which, once started, would operate indefinitely is very old and has been the subject of much discussion and much ridicule. It has never been realized in human invention; yet its impossibility is practical rather than fundamental. It lies in our inability to free moving bodies completely from friction and hence from the dissipation of energy. If one could get away from or neutralize the earth's gravity and other fields of force, so that objects experimented with would not need material support, and if the air and other gases could be completely eliminated from the experimental enclosure, a sphere, for example, set spinning, would continue to spin. Nature approximates these conditions in such mechanisms as the solar system and the atom, but man has hitherto failed to duplicate her arrangements. The only other alternative is to discover an inexhaustible source of energy to supply the loss due to dissipation. The energy from radioactive materials has been used to operate a scientific toy called a "radium clock," but this supply falls off exponentially with time. Solar and cosmic radiation seem inexhaustible, but have not been directly utilized on a large scale. Meanwhile we continue to draw upon apparently limited stores of energy to carry on the activities of what, after all, promises to be only a temporary physical world. (L.D.W.)

PERSEIDS. The Perseids furnish the most reliable of all **meteor showers.** While the **Leonids** have provided some very brilliant displays in the past, about three times each century, their appearance in the intervening years is not at all striking. The Perseids make their appearance during August of each year. Because of the fact that the Perseid showers never are as striking as the Leonids, when at their maximum, we should not expect to find them referred to so frequently in the ancient writings. Extensive search has been made through the old records, however, and we find mention of the Perseids as far back as 830 A.D. The first determination of the **radiant point** in the **constellation** of Perseus was apparently made in 1834.

In appearance the members of the Perseid shower are as striking as those from any other radiant point. Coming as they do during the month of August, when the nights are warm, they are seen by large numbers of people who are always impressed by the relatively slow motion, distinctly reddish appearance, and trails, frequently of several seconds' duration, which characterize the members of this swarm. (W.K.G.)

PERSEUS. (Map, page 380.) This is a rich and brilliant **constellation** of the northern sky. Since it lies right in the milky way it presents many beautiful fields for the opera glass or the small telescope. This is particularly true of the bright star Alpha Persei, which lies in the midst of a very rich and beautiful field.

The star **Algol** (Beta Persei) is the famous **eclipsing variable** whose striking changes in light intensity caused the Arabs to name it the demon star.

In Perseus is to be found the famous double-star **cluster** which is one of the finest objects in the whole sky for an observer with a small telescope. On a clear moonless night, using a relatively low power on the instrument, any observer will be well repaid for his search for this wonderful object. (W.K.G.)

PERSIMMON. Diospyros. Ebenaceae.

PERSON PROCESS. The Person process is a method for controlling the tone separations during photographic printing by means of weak positive masks on film. The masks are made by exposing film under a negative for short intervals so only thin silver deposits or spots are obtained in areas corresponding to the shadow portions of the negative. Usually several masks are made, each differing from the other by a slight change in exposure.

Following exposure, the masks are processed in a manner similar to the negative. One or more of the dried masks are placed in register with the negative, and the edges or corners are fastened to each other with scotch or cellulose tape. Prints are made by exposing through both the negative and the mask, or masks, if more than one is used.

The purpose of the mask is to shade the thin portion of negative so the print will not become blocked, but retain the fine shadow detail of the negative. (S.M.T.)

PERSONAL EQUATION. In making measurements of any character every observer, no matter how skilled he may be, is bound to make certain errors. These errors are of two kinds: accidental errors which will be small in the case of a good observer and which will be distributed in accordance with the **laws of probability;** and systematic errors or errors which are always in the same direction and of approximately the same magnitude. As an example of systematic errors we may cite the case of the observation of the transit of a star across the **reticle** of a **meridian circle.** In this case a good observer will always press the **chronograph** key either slightly too early or slightly too late, depending upon the observer.

The value of the systematic error is known as the personal equation of the observer. Personal equation must be determined empirically for each observer under a variety of different observing conditions. For a good observer the personal equation remains remarkably constant over long periods of time and may be applied directly to any observation. (W.K.G.)

PERSPECTIVE. A method of representing solid objects on a plane surface, as they would appear to an observer's eye when viewed from a given point. The projectors or visual rays converge from the object to the eye, forming a cone or pyramid of rays. The intersection of these rays with the picture plane results in a perspective drawing. Two types of perspective representation are in common use—parallel and angular. In the former, all horizontal lines parallel to the picture plane, and all vertical lines, appear horizontal and vertical; all other parallel lines will intersect, if extended, at one or more common points, called vanishing points. In angular perspective, vertical lines appear vertical; all other parallel lines will intersect, if extended, at one or more vanishing points. (H.C.H.)

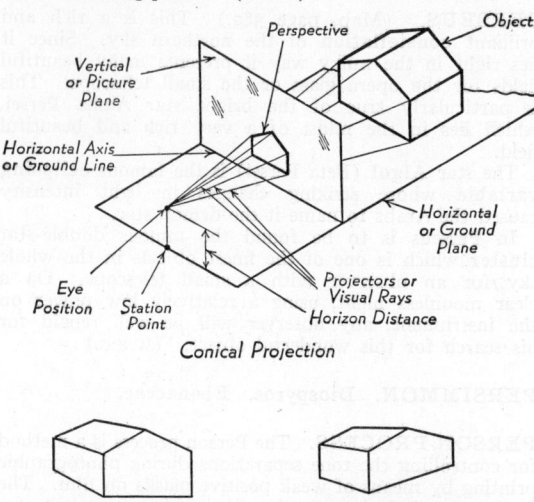

Conical Projection

Parallel Perspective *Angular Perspective*

PERSPIRATION. Sweat. A secretion formed by the **sweat glands** of the skin of some of the mammals. It consists chiefly of water with small quantities of other materials in solution. The dissolved substances include fatty acids, **urea**, **sodium** and **potassium** chloride, phosphates, lactic acid, and cholesterin. By evaporation at the surface of the skin the sweat plays an important part in the regulation of body temperature when the surrounding air is too warm for adequate radiation. It is also important as an excretory medium. (A.W.L.)

PERSULFURIC ACID AND PERSULFATES. Per (mono) sulfuric acid, "Caro's acid" (H_2SO_5 or $HOO \cdot SO_2OH$) and perdisulfuric acid, "persulfuric acid" ($H_2S_2O_8$ or $HOSO_2O \cdot OSO_2OH$) are solids of melting points 45° C. and 65° C., respectively (the latter melting with decomposition). Both acids combine vigorously with a hissing sound with water, and they cause blackening of paper, sugar and paraffin by separation of carbon. Both acids are formed (1) by reaction of chlorosulfonic acid and water, depending upon the ratio, (2) by **electrolysis** of **sulfuric acid** under proper conditions. Better yields of persulfate as **ammonium** persulfate ((NH_4)$_2S_2O_8$) are obtained by electrolysis of ammonium sulfate under proper conditions. Permonosulfuric acid may also be made by reaction of **hydrogen peroxide** and **sulfur** trioxide or anhydrous sulfuric acid.

Permonosulfuric acid oxidizes (1) **hydriodic acid** instantly to **iodine**, (2) **hydrochloric acid** to **chlorine**, (3) **aniline** to nitrobenzene, (4) **ferrous** salt solution to **ferric**, (5) **sulfurous acid** to **sulfuric acid**, but (6) does not reduce **permanganate**, (7) does not give yellow coloration with **titanic** salt solutions, (8) does not give blue coloration with **dichromate** solutions of these reactions. Perdisulfuric acid shows only oxidation of ferrous salt solution to ferric.

Persulfates oxidize (1) potassium iodide slowly to iodine, (2) manganous salt solution to manganese dioxide, (3) manganous salt, in the presence of silver nitrate, to permanganate, but (4) do not give yellow coloration with titanic salt solution, (5) do not give blue coloration with dichromate solutions.

The following properties of persulfates aid in their identification: Persulfates decompose in water to form ozone, which liberates iodine from starch-iodide paper (thus persulfates differ from percarbonates and per-

borates). Persulfates do not react with permanganate or with chromic acid (thus they differ from peroxides). (R.K.S.)

PERTURBATIONS. If there were only two bodies in the universe and if each of these was a homogeneous sphere, the **orbit** of one relative to the other, or the orbits of both relative to the common center of gravity of the system, could be completely determined by the solution of the **two body problem.** In reality, however, there are many more than two bodies in the universe and few of them are homogeneous spheres, with the result that the determination of accurate orbits becomes a problem of tremendous complexity.

Within the **solar system** it fortunately happens that the attraction of one body is dominant, i.e., the attraction of the sun upon any object is far greater than the attractions of all of the other planets combined. In the cases of **satellites,** the attraction of the primary is preponderant. In such cases a close approximation to the true orbit may be obtained by neglecting the attractions of other objects and obtaining a preliminary orbit by the methods of solution of the two body problem.

Using the **Keplerian** ellipse thus obtained it is possible to find at any instant, to a high degree of approximation, the distance of the object from other members of the solar system and hence the attractions of these other objects. The effects of these attractions in changing the motion of the object under consideration may be computed, a second approximation to the true position may be obtained, and this used to obtain more accurate values to the attracting forces. Usually the second approximation is sufficiently accurate for all practical purposes.

The influences which the attractions of the other members of the solar system have on the motions of the object under consideration are known as perturbations. In the case of nearly circular orbits, as in the case of the motions of the planets about the sun, it is possible to obtain, in the form of infinite series, an analytic expression for the perturbations. Such a solution is known as general perturbations. If the orbit is highly eccentric, as in the case of many comet orbits, it is not possible to obtain any general analytic expression and the perturbations are known as special perturbations.

Due to perturbations the orbits of objects have both slow steady changes in the elements, known as secular perturbations, and also relatively short oscillations about an average value, which are known as periodic perturbations. In actual practice it is customary to list both the secular and periodic perturbations in tables from which the accurate position of the planet at any desired instant may be computed. In the case of the perturbations of the Moon, one of the most complicated of all perturbation problems, E. W. Brown's tables of the Moon fill three quarto volumes totalling more than 360 pages.

Perturbations have played an important part in astronomical discovery. After the planet **Uranus** had been discovered and accurately observed, the preliminary orbit, plus perturbations from all known objects, did not accurately represent the observed positions. These deviations could only be explained on the basis of another planet outside of the orbit of Uranus, accurate computations were made and the planet **Neptune** discovered. Deviations of observed positions of Neptune from those computed from the orbit stimulated the search which led to the discovery of the planet **Pluto.** In the case of the planet **Mercury** a perturbation in the longitude of perihelion in the orbit could not be explained on the basis of gravitational theory. This led to a fruitless search for a planet between Mercury and the sun. The perturbation was later explained on the basis of the **theory of relativity** and was one of the early triumphs of this theory. (W.K.G.)

PERTUSSIS. Whooping Cough.

PESSARY. A device made of metal or rubber which is placed in the **vagina** to support the **uterus**, or to correct muscular relaxation or malposition of the uterus. (D.M.H.)

PESTS. Insect Pests.

PETALITE. The mineral petalite, **lithium aluminum silicate**, $LiAl(Si_2O_5)_2$ is **monoclinic**, although crystals are rare, this mineral usually occurring in cleavable, **foliated** masses, whence the name petalite from the Greek meaning a *leaf*. Its hardness is 6–6.5; specific gravity 2.39–2.46; luster, vitreous, colorless to white or gray but may be greenish or reddish; is transparent to translucent. Petalite has been found in Sweden, on the Island of Elba, and in the United States at Bolton, Massachusetts, and Peru, Maine. It is interesting to note that lithium was first discovered in this mineral. (E.S.C.S.)

PETALOID AREA. A flower-like arrangement of pores on the upper (aboral) surface of some of the sea urchins (**Echinoidae**). (A.W.L.)

PETALS. Flower.

PETIOLE. Leaf.

PETIT MAL. Epilepsy.

PETREL. Aves, Procellariiformes. Marine birds (**Aves**) related to the albatrosses. They are powerful fliers and have the feet webbed for swimming. The group includes the fulmars and shearwaters as well as several forms which bear the name petrel. There are many species and many of them bear one or more names in sailors' vernacular. The Cape pigeon, *Daption capensis*, is a well-known petrel of the Southern Hemisphere and the little stormy petrel of the North Atlantic, *Hydrobates pelagicus*, under the name Mother Carey's chicken is probably the most familiar of all to ocean travelers. (A.W.L.)

PETRIFACTIONS. Paleontology.

PETROGENESIS. That branch of **petrology** which deals with the origins of rocks. Practically a synonym for petrology unless, as is usually the practice, confined to the **igneous** (and possibly the **metamorphic**) rocks. (R.M.F.)

PETROGRAPHIC MICROSCOPE. A type of **microscope** especially adapted for use by the mineralogist and petrologist. As compared with the ordinary high-powered microscope, the petrographic microscope has the following additional apparatus:

1. A polarizer, or apparatus for **polarizing** light.
2. An analyzer, or apparatus for analyzing the rays of light after they have passed through the polarizer, and the thin section of rock which is being examined.
3. A rotating stage, which rotates about an axis which is the line-of-sight of the microscope.

The polarizer, sometimes called the **Nicol prism**, is described as "a cleavage rhombohedron of **calcite** (variety, Iceland spar) having four large and two small rhombohedral faces opposite each other, which is modified in the following manner:

"In place of the latter planes, two new surfaces are cut, making angles of 68° (instead of 71°) with the

obtuse vertical edges. These then form the terminal faces of the prism. In addition to this, the prism is cut through in the plane HH', the parts then polished and cemented together again with Canada balsam. A ray of light, a b, entering the prism, is divided into two rays polarized at right angles to each other. One of these, b c, on meeting the layer of balsam (whose refractive index is less than that of the ray b c) suffers total reflection, and is deflected against the blackened sides of the prism and extinguished. The other, b d, passes through and emerges at c, a completely polarized ray of light, that is, a ray with vibrations in one direction only, and that direction of the shorter diagonal prism." Dana. (R.M.F.)

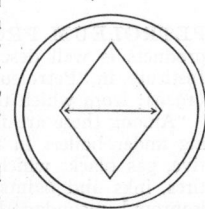

PETROGRAPHY. The branch of **petrology** that has to do with the study and systematic description, both **macroscopic** and microscopic, of the mineral composition, texture and structure of rocks. (R.M.F.)

PETROLEUM. Petroleum and its associated natural gases are mixtures of **hydrocarbons**, or mixtures of many chemical compounds consisting chiefly of the elements, carbon and hydrogen, which have been derived from buried organic matter. The old term, **mineral oil**, used to distinguish this product from whale oil, is a misnomer because petroleum, like **coal**, is of organic origin. In common with coal, however, petroleum is procured from the ground, and is a natural concentration of energy

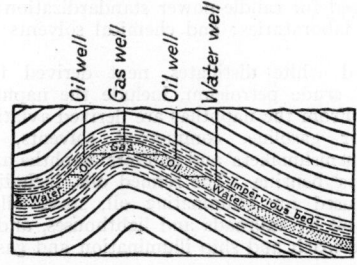

Structure section to show a typical occurrence of gas, oil, and water in an anticlinal structure. (*After U. S. Geological Survey.*)

easily expended but not reproducible except in terms of geologic time. Whatever may be the original sources of materials from which petroleum is formed, it is generally believed that petroleum in commercial amounts from wells, has migrated into the porous formations (**oil sands**) where it is obtained from oil pools in favorable underground structures. Like coal, the products of **oil wells** are extremely variable, there being no such thing as a common **crude**. The physical chemistry of petroleum is exceedingly complex and intimately connected with the technology of the **cracking**. Practically speaking, the product of oil wells may range all the way from **natural gas**, through natural **gasoline** and petroleum, to **asphalt** and **paraffin**. When an oil reservoir is tapped by means of a drilled well, the pressure is relieved and the gas comes out of solution, forcing the oil to the surface. The pressure and gas content are, therefore, exceedingly important in the production of petroleum, the deeper wells usually producing the largest and longest yields. The United States produces approximately $\frac{2}{3}$ of the world production of crude petroleum. First-class powers which have no petroleum, or a great deficiency thereof, within their sovereign territory are: Great Britain, Italy, Japan, France and Germany. (R.M.F.)

Map showing the principal oil fields of the United States. (*Modified after Geological Survey of Kansas.*)

PETROLEUM PRODUCTS. The use of petroleum products is well described by the American Petroleum Institute, in "Petroleum Facts and Figures," pp. 158–160 (1929), from which the following is quoted:

"Among these are fuel gas, which is utilized for burning under boilers in the refineries, and another derivative, gas black, which is used in the making of rubber tires, inks and paints. A series of **alcohols**, including isopropyl, secondary butyl, secondary amyl and secondary hexyl are also recovered from these gases, which are utilized as solvents for the making of **lacquers**, **soaps** and **essential oils**. Liquefied gases, also yielded from the hydrocarbon gases, are utilized in metal cutting and for illumination. Another product, petroleum ether, is used for priming motors and for laboratory work. Natural gasoline, also thus derived, yields light naphthas which, in turn, yield gas-machine gasoline; pentane, used for candle-power standardization; hexane, utilized in laboratories; and chemical solvents for drug extraction.

"So-called white distillates, next derived from the refining of crude petroleum, include the naphthas and kerosene. From the naphthas are derived aviation gasoline, motor gasoline, commercial solvents, blending naphtha, varnishmakers' and painters' naphtha and dyers' and cleaners' benzine. The refined oils, including kerosene, are used for illuminating oil, stove oil, tractor oil, signal oil by railroads and lighthouses, and mineral seal oil for coach and ship illumination and gas absorption.

"The next important product is furnace oil, used as a fuel in oil-burning furnaces in office buildings and homes.

"Intermediate distillates, next derived, yield gas oil and absorber oil. Gas oil is used by gas manufacturing plants in the carburetion of water gas; is an important metallurgical fuel; yields gasoline by the 'cracking' process and also Diesel fuel oil. Absorber oil enters into gasoline and benzol recovery.

"From the heavy distillates next derived come technical heavy oils, waxes and lubricating oil. Technical heavy oils, upon treating, are made into a variety of products. From one of these—white oils—is derived so-called technical oil, used for lubricating special machinery, such as bakers' and candymakers' machinery, and for packing fruit and eggs; also medicinal oil for internal and external use and for the manufacture of salves, creams and ointments. Ink oils are derived from technical heavy oil; also saturating oil, which enters into wool and twine manufacture; emulsifying oil; electrical oils, used for transformers and switches; and flotation oils, used in metal recovery processes.

"From heavy distillates are derived waxes which are used for making candles, chewing gum and candy. Wax

also enters the laundry as a detergent and as an iron wax; has a wide use for sealing purposes, such as for the preserving of fruits and vegetables, and is used by etchers. Saturating wax is applied to cardboard, matches, and paper.

"Lubricating oils derived from petroleum are used wherever there is machinery, special oils or compounds being made for different types.

"From the residues of distillation are derived greases, such as gear grease, switch grease, and cup grease. By a refining process petrolatum is derived from greases. This enters the medicinal field, compounded with other products, in the form of salves, creams and ointments and as petroleum jelly. It is also used for metal coating and lubrication. Residual fuel oil is used for burning under boilers in industrial plants, on ships and in railroad locomotives. Fuel oil is also used for making gasoline by "cracking" processes. Road oil is a residual product of petroleum, as are asphalts and pitches used for roofing, paving, felt saturating, briquetting, rubber making and plastic composition. Another product is coke used for making carbon brushes, carbon electrodes, and also consumed as fuel.

"From refinery sludges are made acid coke used as a fuel; sulfonic acid, utilized as a saponification and de-emulsifying agent; oils and pitches; and **sulfuric acid**, used in **fertilizer** manufacture." (R.K.S.)

PETROLOGY. Petrology is that branch of geology which deals with the rocks forming the **lithosphere** or "crust" of the earth. The term is derived from the Greek, meaning rock and reason, hence a more comprehensive term than **petrography**, which deals only with systematic descriptions of rocks, including their mineral composition, texture, structure, and occurrence. Petrology includes petrography and also methods of classification as founded on systematic description and genetic theories, both experimental and theoretical. Important branches of petrology are **geochemistry** and **geophysics**. (R.M.F.)

PETROMYZONTIA. Cyclostomata.

PETROSILEX. A term frequently encountered in the older literature which refers to those very fine-grained crystalline aggregates which microscopic studies have revealed as devitrified glasses, and may in general be classified as **felsites**. (E.S.C.S.)

PETZITE. A mineral telluride of silver and gold, Ag_3AuTe_2. Crystallizes in the isometric system only at high temperatures. Hardness, 2.5–3; specific gravity, 8.7–9.02; color, gray to black with metallic luster; opaque. Named after W. Petz. (R.M.F.)

pH. Reactions Involving Recombination of Ions.

PHACOLITH. A lens-shaped mass of **igneous** rock found either in the trough or crest of a folded structure. It is believed that the shape and position of a phacolith is determined by the **orogenic** conditions, and that the folding is not a result of their intrusion. The term phacolith comes from the Greek words meaning lens and stone. (E.S.C.S.)

PHAGOCYTE. A cell of the multicellular animal's body which is capable of ingesting foreign particles. In the **sponges** and **coelenterates** and to a limited extent in other more complex animals digestion is carried on wholly or in part by this process. In many animals **bacteria** and particles of dead **tissues** or **cells** are engulfed by such cells. Phagocytes are amoeboid (**Pseudopodium**) in action and some of them move about in the tissue to which they belong by this means. Others are in fixed tissues, carrying on their amoeboid processes at the free end only. Phagocytes may be a part of the endodermal lining of the enteric cavity or wandering cells derived from endoderm, or they may be mesodermal. In the connective tissues and **blood** mesodermal phagocytes occur, and in the lining of the circulatory system cells may act as phagocytes.

Large phagocytes are called macrophages, a term which includes wandering cells of the connective tissues. The opposite, microphage, is rarely met. In the blood of vertebrates all white cells (leukocytes) are phagocytic to some extent, the lymphocytes least of all. (A.W.L.)

PHAGOCYTOSIS. The ingestion of bacteria and other foreign particles by **phagocytes** in the bodies of complex animals. The process is an important part of the defense of animals against microorganisms which cause disease and is also a means of disposing of cells which have served their purpose, such as worn-out red **blood** corpuscles. (A.W.L.)

PHALANGER. Mammalia, Marsupialia. Pouched mammals of an extensive family found only in the Australian and Oriental regions. They are small or medium animals with thick woolly fur. Some resemble mice or squirrels superficially and all are arboreal. With the one exception of the koala they have prehensile tails.

In addition to the true phalangers the group includes the cuscuses, the flying phalangers, long-snouted phalangers, and several other forms. The **koala** is an aberrant species. (A.W.L.)

PHALANGIDA. The **harvestmen** or daddy longlegs, an order of **Arachnida.** (A.W.L.)

Phalangida.

PHALANX, PHALANGES. The small bones of the toes and fingers of **vertebrates.** (A.W.L.)

PHALAROPE. Aves, Charadriiformes. Wading birds (**Aves**) of several species, with long legs, lobed toes, and a moderately long beak. The red (*Phalaropus fulicaris*) and northern (*Lobipes lobatus*) phalaropes breed throughout the northern part of the northern hemisphere, and both the Old and New Worlds have their own species as well. Wilson's phalarope, *Steganopus tricolor,* of North America is one of our most beautiful waders. These birds are unusual in the reversal of ordinary sex behavior and appearance. Females are larger, more brightly colored, and take an aggressive part in courtship, while the males build the nests and incubate the eggs. Phalaropes migrate into the southern hemisphere in winter. (A.W.L.)

PHANATRON. This is a gas-filled hot-cathode type rectifier **tube.** (L.R.Q.)

PHANEROCRYSTALLINE. An adjective applied to an **igneous rock,** the essential constituents of which are distinguishable with the unaided eye. Compare with **macroscopic.** The term is derived from the Greek word meaning visible, plus crystalline. (R.M.F.)

PHANEROZONIA. Starfishes with one or two rows of large plates along the margins of the rays. An order of the class **Asteroidea.** (A.W.L.)

PHANTOM CIRCUIT. Two metallic communication circuits can be made to do the work of three by the addition of certain equipment. Since the third circuit has no wires definitely set aside to its use, it is called a phantom circuit. The phantom circuit may be obtained by using repeating or **impedance coils.** A setup

of a phantom circuit sent over two side circuits by phantom repeating coils is shown in the accompanying figure. The instantaneous direction of current in the

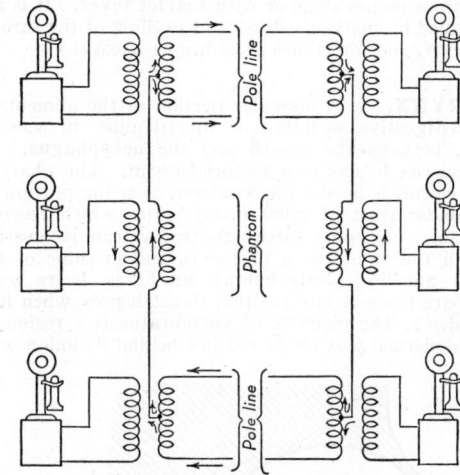

Illustrating a circuit "Phantomed" on two standard circuits (arrows give current flow when phantom is in use.)

phantom circuit is shown by arrows. The repeating coils are **transformers** which are tapped at their electrical center, so that the current in the phantom circuit divides equally in those coils. If the current flow in the side lines (called physical circuits) is equal, and moving oppositely so far as the side line circuit is concerned, their voltages will neutralize, and no induced currents will be present in the side circuits. Thus the side circuit is not affected by the phantom. Good results depend on the use of side circuits which are identical electrically. (F.T.M.)

PHANTOM MEMBER. Graphical Statics.

PHARMACOPEIA. A book containing a list of the **drugs** used in medicine with their formulae, descriptions, chemical tests for identification, purity, and dosage. The United States Pharmacopeia is the standard which is revised and re-issued every 10 years under the supervision of a national committee. (R.S.M.)

PHARYNGEAL CLEFT. Gill Slit.

PHARYNGEAL POUCH. A lateral pouch of the vertebrate **pharynx** which joins a depression in the outer surface of the body wall to form a **gill slit** in all forms which have these openings. In mammals the pouches form in the embryo without breaking through to the exterior.

Pharyngeal pouches are also important as the source of adult anatomical structures. The first pair persist as the cavities of the middle ears and their connections with the pharynx become the **Eustachian tubes.** The palatine **tonsils** arise in the walls of the second pair. The third and fourth pairs give rise to dorsal and ventral outgrowths which become the **parathyroid** and **thymus** glands, respectively. A pair of small postbranchial bodies are interpreted by some anatomists as outgrowths of the fourth pouches and by some as the vestiges of a fifth pair of pouches. (A.W.L.)

PHARYNGITIS ("Sore Throat"). Infection or inflammation of the **pharynx.** Commonly it occurs as part of an upper respiratory infection, a "cold," **tonsillitis,** and hemolytic streptococcal infection.

The principal symptoms are tickling and irritation causing cough. Pain is present on swallowing. The constitutional symptoms of pharyngitis are rarely severe; there is usually only slight fever and malaise. The treatment is similar to that of **coryza** and tonsillitis. Soothing gargles relieve the pain. Severe streptococcal sore throat occurs alone or with **scarlet fever.** It is accompanied by marked redness and swelling of the throat, high fever, and sometimes prostration. (D.M.H.)

PHARYNX. 1. A muscular portion of the alimentary tract (**digestive system**) of invertebrates of various **phyla,** between the mouth and the **oesophagus.** In some species it acts as a suctorial organ. The pharynx of flatworms is lined with ectoderm, as is the portion of the tubular tract of other invertebrates which receives this name. In some species the region can be everted through the mouth as a proboscis, and in some of the marine annelids (**Polychaeta**) its lining bears teeth which are brought into position to catch prey when it is everted. 2. The pharynx of **vertebrates** is a region of the endodermal part of the gut just behind its union with

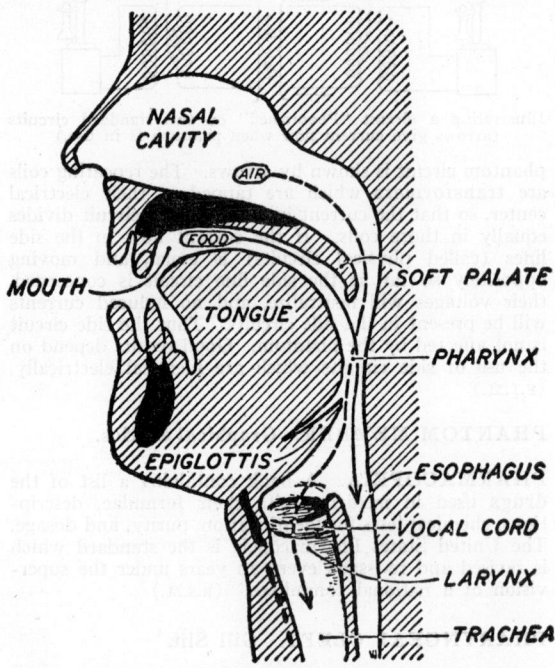

(Carlson and Johnson's The Machinery of the Body, University of Chicago Press.)

the ectoderm of the oral cavity. It is always associated with the **respiratory system.** In the lower **chordates, cyclostomes,** fishes (**Pisces**), and **larval amphibians,** its lateral walls are perforated by gill-slits and in the air-breathing vertebrates the lungs arise from its ventral wall. The nasal passages are dorsal to the mouth, hence the respiratory passages from nostrils to lungs must cross the alimentary tract, and the pharynx remains common to both systems. The **Eustachian tubes** enter in the back portion of the pharynx on either side. The upper portion is called the nasopharynx and is located directly behind the nasal passages. The lower portion is divided into two parts, the oropharynx which is behind and continuous with the mouth, and the laryngo-pharynx which is continuous with the larynx and esophagus.

In man and other vertebrates, the pharynx is important as the source of ductless glands and other structures which arise from the pharyngeal pouches. (R.S.M., A.W.L.)

PHASCOLOGALE. Dasyure.

PHASE. Alternating-Current Circuits.

PHASE CONVERTER. If single-phase voltage is applied to one **phase** of a rotating three-phase induction **motor** it will be found that approximately balanced three-phase voltages may be obtained from the three terminals of the machine. This is because the rotating field induces voltages in the various phase windings which are not much different from the applied voltage. Since the single-phase field is not quite uniform and since the impedance drop in the windings affects the terminal voltage, the output voltages are not quite equal but nearly enough so that they may be used for much three-phase work. This method is used quite widely in a-c railway electrification since the necessity of three conductors makes transmission of three-phase power to the locomotive difficult. (L.R.Q.)

PHASE INVERTER. The push-pull **amplifier** has numerous advantages over single-ended ones but does require a push-pull connection for driving the **grids.** Since the usual amplifier has low-level stages operating single-ended this requires some method of getting two equal voltages 180° out of phase to drive the push-pull grids. One method is to use a **transformer** to couple the single **plate** to the push-pull grids by center-tapping the secondary. However, for resistance coupling a phase inverter may be used. Referring to the figure, part of the output voltage of the upper tube is coupled back

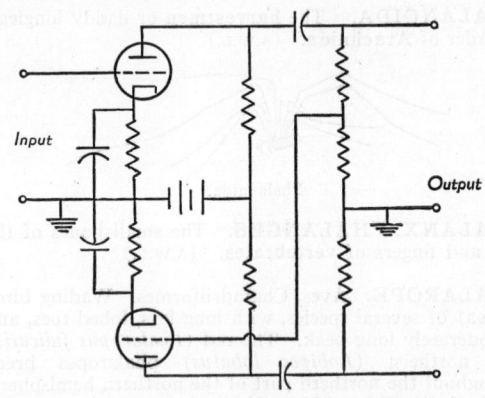

into the grid of the lower tube. Since the output of a tube is 180° out of phase with the input, the input to the second tube is 180° from that of the first tube, hence their outputs are likewise displaced. If the tap on the output resistor of the first tube is adjusted correctly just enough voltage is applied to the grid of the second to cause its output to be equal that of the first. These equal and opposite voltages may then be applied to the push-pull grids of the following stage. (L.R.Q.)

PHASE MODULATION. Modulation.

PHASE RULE. Used by **geochemists** and **petrologists** as a thermodynamic formula in describing the origin of an **igneous** rock in which the **magma** is considered in terms of any system, $P + f = n + z$, when P = number of coexisting chemical phases in the system, f = degrees of freedom, and n = number of independent components in the system. (See **Equilibrium, Chemical; States of Matter.**) (R.M.F.)

PHASE SHIFT CONTROL. This control is commonly applied to **thyratrons** or **ignitrons** operated on a.c. and consists of controlling the breakdown point by shifting the phase of the grid voltage with respect to the anode voltage. Fig. 1a shows the anode voltage

(A), the grid characteristic (B), the d-c grid bias (C) and the a-c grid voltage (D). These are all plotted against time along the horizontal axis. If at any instant

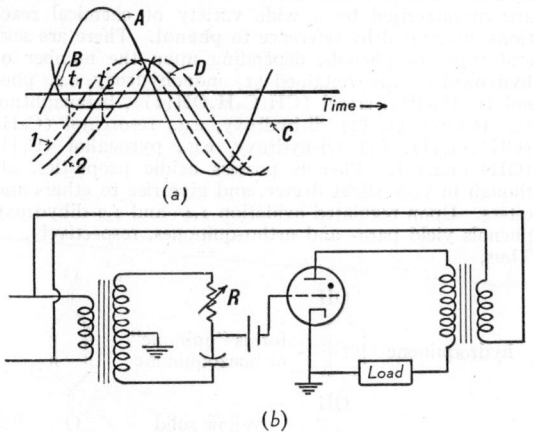

(a)

(b)

the net grid voltage, i.e., the d-c plus the a-c value, goes above the characteristic (B) the tube breaks down and conducts for the rest of that half-cycle. Thus if the grid wave is shifted to position (1) the tube breaks down at time t_1 while if it is shifted to (2) the tube breaks down at t_2. By making the phase continuously variable the average output of the tube may be varied by varying the part of the cycle over which the tube conducts. A simple circuit for doing this is shown at (b). The voltage applied to the grid may be varied over nearly 180° by varying the value of the resistor R. This method of controlling tubes is widely used for motor speed control and for resistance welding control. (L.R.Q.)

PHEASANT. Aves, Galliformes. Game birds (**Aves**) native to the Old World, especially of southeastern Asia and the high altitudes of China and Tibet. The males of many species are gorgeously colored and have much longer tails than the females.

Pheasants are raised in large numbers for game both in Europe and in North America. In some parts of the United States the ring-necked pheasant, *Phasianus colchicis*, has apparently become established after some years of careful protection. The introduction of other species has met with little success although the golden and silver pheasants are listed as introduced species in the fauna of Oregon and Protection Island, Washington. All of these species are Chinese.

The group includes many species with the name pheasant, in addition to the true pheasants of the genus *Phasianus*. Other forms are the tragopans or horned pheasants, the blood pheasants, the monals, the fire-backed pheasants, eared pheasants, golden pheasant, jungle fowls, and argus pheasants. Some of the Indian species bear the names kallege and pukra.

The ruffed **grouse** of North America is known in the southern states as a pheasant but with no scientific reason. (A.W.L.)

PHELLOGEN or **CORK CAMBIUM.** In most dicotyledons this is a tissue of living cells developed either from the cells of the epidermis or from cells lying just beneath the epidermis, or in the case of roots from cells of the pericycle. It is a secondary meristematic tissue, from which cork tissue or phellem and secondary cortex or phelloderm are formed. (R.M.W., B.S.M.)

PHENACITE. The mineral phenacite is an **orthosilicate** of **beryllium** corresponding to the formula Be_2SiO_4.

It is **hexagonal** but the crystals are usually **rhombohedral** in habit. It has a conchoidal **fracture;** is brittle, hardness, 7.5–8; specific gravity, 3; luster, vitreous; colorless to yellowish or reddish, sometimes brown; transparent to translucent. Phenacite is found in **pegmatites** with **topaz, quartz** and **microcline,** and occurs also in **emerald-**bearing **mica schists** of the Ural Mountains. It is found also in France, Norway, Switzerland, Africa, Brazil, and Mexico, and in the United States in Oxford County, Maine; Carroll County, New Hampshire; and in Chaffee and El Paso Counties in Colorado. It derives its name from the Greek meaning deceiver, as it resembles quartz and topaz with which it is associated. It is sometimes spelled phenakite. It has been somewhat used as a gem. (E.S.C.S.)

PHENOCLAST. A textural term proposed by R. M. Field in 1916 for coarsely graded **clastic sedimentary rocks** in which the largest or "show" particles or fragments are referred to as phenoclasts, regardless of their shape or composition. The term implies that the larger constituents of the glomerate have been derived from prelithified rock. Rounded fragments are called pebbles or spheroclasts, which when lithified by means of matrix (sand and clay) and cement form a conglomerate. Angular fragments are called anguclasts (Field), which when lithified by means of matrix (sand and clay) and cement form a **breccia.** (R.M.F.)

PHENOCOPY. A term used in genetics to designate an organic character resembling some inherited Mendelian unit character but developed in the individual in response to some environmental condition and so not transmitted by inheritance. A Lamarckian character superficially like a Mendelian character. (A.W.L.)

PHENOCRYST. A textural term proposed by Iddings in 1892 for macroscopic crystals which are relatively much larger than the crystalline matrix of the **igneous** rock in which they occur. Rocks which have phenocrysts are called **porphyritic.** The term phenocryst is derived from the Greek, meaning show, and crystal. (R.M.F.)

PHENOL. (For phenols, see article below on **Phenols and Quinones.**) Phenol, carbolic acid, hydroxybenzene (C_6H_5OH), is a colorless solid of characteristic odor, poisonous, and corrosive to the skin, melting point 41° C., boiling point 182° C., slightly soluble in water at 20° C. (8.3 grams phenol per 100 grams solution), but at 68.3° C., critical temperature, each layer contains 33.4 grams phenol per 100 grams solution and the layers are miscible; upon cooling the change is reversible; very soluble in ether and miscible in all proportions with alcohol, glycerol, chloroform, absorbs water on exposure to the atmosphere and acquires a reddish color on exposure to air. Phenol reacts (1) with **sodium** hydroxide, yielding sodium phenate (C_6H_5ONa), which with **carbon dioxide** forms phenol plus sodium carbonate (phenol insoluble in **sodium** carbonate or sodium hydrogen carbonate solution), (2) with **phosphorus** pentachloride, yielding chlorobenzene plus phosphorus oxychloride, (3) with **bromine** (or chlorine) in water, yielding 2,4,6-tribromo- (or chloro-) phenol, white precipitate, melting point 96° C. (chloro- 68° C.), (4) with **zinc** heated, yielding **benzene** (C_6H_6) distillate plus zinc oxide residue, (5) with **ferric** chloride solution, yielding violet solution, (6) with **sodium** nitrate plus concentrated **sulfuric acid,** yielding para-nitrosophenol ($C_6H_4(OH)(NO)(1,4)$), brown when cold, blue upon warming, red upon pouring into water, blue upon addition to this of excess sodium hydroxide (Liebermann's reaction), (7) with nitric acid (and concentrated sulfuric acid) yields nitrophenols (para- $C_6H_4(OH)(NO_2)(4)$, ortho- $C_6H_4(OH)$ $(NO_2)(2)$, 2,4-di- $C_6H_3(OH)(NO_2)_2(2,4)$, 2,4,6-tri-$C_6H_2(OH)(NO_2)_3(2,4,6))$, this last substance known as "picric acid" made in this way is an important high explosive. Phenol is obtained (1) from coal tar, in the fraction distilling between 170° C. and 230° C. The reaction with sodium hydroxide is applied in the recovery, followed by treatment with carbon dioxide, and distillation of the separated phenol, (2) from benzene sulfonic acid ($C_6H_5SO_3H$) by heating with sodium hydroxide or calcium hydroxide, (3) from chlorobenzene by heating under pressure with sodium hydroxide in the presence of diphenyl oxide, (4) of scientific interest is the reaction of diazobenzene upon boiling, whereupon phenol is formed. Phenol gives a violet coloration with solutions of ferric salts. Phenol is used (1) as an antiseptic, germicide, and disinfectant, (2) in the manufacture of certain dyes, (3) in the manufacture of picric acid. (For Phenols, see **Phenols and Quinones**.) (R.K.S.)

PHENOLOGY. A biological science which deals with the relations of living things to physical conditions in the environment depending on latitude, longitude, and altitude. Zones in which similar conditions prevail are populated by similar animals and within a given zone the same periodical fluctuations are to be expected. Such zones are indicated on phenological maps by **isophanes**. (A.W.L.)

PHENOLPHTHALEIN. Phthalic Acid.

PHENOLS AND QUINONES. Phenols (containing hydroxyl group, —OH, attached to a benzenoid carbon) are characterized by a wide variety of chemical reactions, illustrated by reference to **phenol**. There are several types of phenols, depending upon the number of **hydroxyl** groups contained (1) monohydroxy, e.g., phenol (C_6H_5OH), cresol ($CH_3C_6H_4OH(2)$), betanaphthol ($C_{10}H_7OH(2)$), (2) di-hydroxy, e.g., resorcinol (C_6H_4 $(OH)_2(1,3)$), (3) tri-hydroxy, e.g., pyrogallol (C_6H_5 $(OH)_3(1,2,3)$). Phenols possess acidic properties, although in very slight degree, and give rise to ethers and esters. Upon regulated oxidation 1,4- and 1,2-dihydroxy phenols yield para- and ortho-quinones, respectively. Thus,

hydroquinone — forms "quinone" or benzoquinone

yellow solid

catechol — forms ortho-benzoquinone

Two forms: green solid, and red solid.

TABLE OF SELECTED REPRESENTATIVE PHENOLS

PHENOL	FORMULA	MELTING POINT, °C.	BOILING POINT, °C.
1. Phenol * (carbolic acid)	$C_6H_5\cdot OH$	41	182
2. Ortho-cresol (2-hydroxyl toluene)	$CH_3C_6H_4OH(2)$	30	191
3. Meta-cresol (3-hydroxyl toluene)	$CH_3C_6H_4(OH)(3)$	12	203
4. Para-cresol (4-hydroxy toluene)	$CH_3C_6H_4(OH)(4)$	34	203
5. Ortho-ethyl phenol (1-ethyl-2 hydroxybenzene)	$C_2H_5C_6H_4(OH)(2)$	−18	207
6. Meta-ethyl phenol (1-ethyl-3-hydroxybenzene)	$C_2H_5C_6H_4(OH)(3)$	−4	214
7. Para-ethylphenol (1-ethyl-4-hydroxybenzene)	$C_2H_5C_6H_4(OH)(4)$	46	219
8. 1,2-Dihydroxybenzene (catechol)	$C_6H_4(OH)_2(1,2)$	105	245
9. 1,3-Dihydroxybenzene (resorcinol)	$C_6H_4(OH)_2(1,3)$	110	276
10. 1,4-Dihydroxybenzene (hydroquinol)	$C_6H_4(OH)_2(1,4)$	170	286
11. 1,2,3-Trihydroxybenzene (pyrogallol)	$C_6H_3(OH)_3(1,2,3)$	133	293 dec.
12. 1,3,5-Trihydroxybenzene (phloroglucinol)	$C_6H_3(OH)_3(1,3,5)$	219	subl. dec.
13. 1,2,4-Trihydroxybenzene	$C_6H_3(OH)_3(1,2,4)$	141	
14. Para-para-biphenol	$(4) OHC_6H_4\cdot C_6H_4OH(4)$	272	subl.
15. Thymol (5-methyl-2-isopropyl phenol)	$C_6H_3(OH)(CH_3)(5)(CH(CH_3)_2)(2)$	51	
16. Carvacrol (2-methyl-5-isopropyl phenol)	$C_6H_3(OH)(CH_3)(2)(CH(CH_3)_2)(5)$		237
17. Guaicol (2-methoxyphenol)	$C_6H_4(OH)(OCH_3)(2)$		
18. Eugenol (2-methoxy-4-allyl phenol)	$C_6H_3(OH)(OCH_3)(2)$ $(CH_2CH:CH_2)(4)$		254
19. Alpha-naphthol	$C_{10}H_7OH(1)$	96	280
20. Beta-naphthol	$C_{10}H_7OH(2)$	122	286
21. 1,2-dihydroxy naphthalene	$C_{10}H_6(OH)_2(1,2)$	60	
22. Anthranol (9-hydroxy anthracene)	$C_6H_4\genfrac{}{}{0pt}{}{COH}{CH}C_6H_4$	170 dec.	
23. Anthrol (1) (1-hydroxyanthracene)	$C_6H_4\genfrac{}{}{0pt}{}{CH}{CH}C_6H_3OH(1)$	151 dec.	
24. Anthrol (2) (2-hydroxyanthracene)	$C_6H_4\genfrac{}{}{0pt}{}{CH}{CH}C_6H_3(OH)(2)$	200 dec.	

* Discussed separately under Phenol.

TABLE OF SELECTED REPRESENTATIVE QUINONES

Quinone	Color	Formula	Melting Point, °C.	Boiling Point, °C.
1. Benzoquinone (quinone).................	Yellow	$C_6H_4(O)_2(1,4)$	116	subl.
2. Ortho-benzoquinone.....................	Red	$C_6H_4(O)_2(1,2)$	60–70 dec.	
3. Naphthaquinone.......................	Yellow	$C_{10}H_6(O)_2(1,4)$	125	100 subl.
4. Ortho-naphthaquinone..................	Red	$C_{10}H_6(O)_2(1,2)$	115 dec.	
5. 2,6-naphthaquinone....................	Red	$C_{10}H_6(O)_2(2,6)$		
6. Anthraquinone, 9, 10*..................	Yellow	$C_6H_4{<}^{CO}_{CO}{>}C_6H_4$	286	380
7. Ortho-anthraquinone,1,2...............	Orange	$C_6H_4{<}^{CH}_{CH}{>}C_6H_2(O)_2(1,2)$	185 dec.	
8. Para-anthraquinone,1,4................	Yellow	$C_6H_4{<}^{CH}_{CH}{>}C_6H_2(O)_2(1,4)$	210 dec.	
9. Alizarin (1,2-dihydroxyanthraquinone).......	Red	$C_6H_4(CO)_2C_6H_2(OH)_2(1,2)$	289 dec.	

* Discussed separately under Anthraquinone

                                                                 (R.K.S.)

The commonly encountered quinones are of the para type, thus,

benzoquinone (1,4), yellow solid ($C_6H_4(O)_2$)

naphthoquinone (1,4), yellow solid ($C_{10}H_6(O)_2$)

anthraquinone (9,10), yellow solid ($C_{14}H_8(O)_2$)

Hydroquinone plus benzoquinone forms the complex known as quinhydrone, dark green solid. Quinhydrone or benzoquinone is readily reduced to hydroquinone (e.g., by sulfurous acid.) Quinhydrone, hydroquinone, or aniline ($C_6H_5NH_2$), which is commonly used to produce benzoquinone, is readily oxidized to benzoquinone (e.g., by sodium dichromate plus sulfuric acid). When naphthalene or alpha-naphthylamine ($C_{10}H_6(NH_2)(1)$) (OH)(4)) is similarly oxidized, naphthoquinone is obtained, and anthracene similarly yields anthraquinone (9,10).

The higher phenols are soluble in water, and in alkaline solution usually absorb oxygen and act as reducing agents, e.g., pyrogallol in alkaline solution is used to estimate the percentage of oxygen in gas mixtures, and in photography as a developer. As reducing agents these phenols generally react with alkaline cupric tartrate (Fehling's solution), and with ammonio-silver nitrate (Tollen's solution). Phenols produce a characteristic color with ferric chloride solution.

Quinones stand alone among the compounds of carbon-hydrogen-oxygen in possessing, as a class, marked color, para are yellow, ortho red. It is believed that the benzenoid structure gives place to the quinoid structure in this case. Quinones, in general, react with hydroxylamine or phenylhydrazine to form mono- and dioximes or mono- and diphenylhydrazones of characteristic properties, e.g., melting point.

PHENOMERIDIAN. Isophane.

PHENOTYPE. Heredity.

PHENYL. Radicals.

PHENYLHYDRAZINE. Hydrazine.

PHENYLHYDRAZONES. Hydrazines, Hydrazones, and Osazones.

PHILLIPSITE. The mineral phillipsite is a zeolite, a hydrous silicate of potassium, calcium, and aluminum, of somewhat uncertain formula. It is monoclinic, forming penetration twins, and sometimes crosses resembling orthorhombic or tetragonal forms. It also may occur in radial groups. Phillipsite is a brittle mineral; hardness, 4–4.5; specific gravity, 2.2; luster vitreous; color, white to light red; translucent to opaque. Like other zeolites phillipsite is found in veins and cavities in basalts, and sometimes in more acidic rocks. It is believed to be a low-temperature mineral. Found in Italy, especially in the lavas of Vesuvius and Monte Somma, and in the basalts of Germany, Ireland, and Australia. Has been reported from Greenland. This mineral was named in honor of the British mineralogist William Phillips. (E.S.C.S.)

PHIMOSIS. An excessively tight foreskin over the head of the penis or clitoris so that it cannot be pushed back over the organ. The condition is corrected by circumcision. (D.M.H.)

PHLEBITIS. Inflammation or infection of the wall of a vein. This is almost invariably associated with some degree of thrombosis. (See Thrombophlebitis.) (D.M.H.)

PHLEBOTOMY. Venesection or withdrawing of blood from a vein in the body in order to reduce the venous pressure. (R.S.M.)

PHLOEM. The phloem is that part of the plant through which foods move from one part of the plant to another. In stems the phloem forms a considerable portion of the bark, and is found outside the cambium.

In angiosperms, the phloem consists of parenchyma cells, fibers, sieve tubes, and companion cells. The parenchyma cells are thin-walled, somewhat irregular in shape, and contain protoplasm. The fibers are long, slender, thick-walled cells having no protoplasm. The sieve tubes are made up of long, thin-walled cells with a peripheral layer of cytoplasm, no nucleus, and a large central vacuole. The walls are perforated in restricted spots called sieve plates. In the more advanced plants these sieve plates occur only at the ends of the cells; in the more primitive plants they are found both at the ends and along the sides of the cells. Sieve tubes thus extend throughout the plant and provide continuous strands of living protoplasm capable of conducting food rapidly from the leaves to its various destinations. A companion cell is a small cell, usually much elongated, which is found adjoining a sieve tube. It has a dense cytoplasm and a distinct nucleus. (R.M.W.)

PHLOGOPITE. The mineral phlogopite is a magnesium-bearing mica, with but little iron, corresponding essentially to the formula $H_2KMg_3Al(SiO_4)_3$. Fluorine is sometimes present. This mica is monoclinic like muscovite, biotite and lepidolite, forming prismatic crystals, occasionally very large, and occurring also in scales and plates. Its cleavage is basal and highly perfect with elastic laminae; hardness, 2.5–3; specific gravity, 2.78–2.85; luster, pearly to submetallic; color, yellowish-brown, green, white and colorless; transparent to translucent; may exhibit asterism, probably due to minute inclusions. Phlogopite is more nearly a characteristic of metamorphic than igneous rocks although occasionally occurring in the latter if they are rich in magnesia and with but little iron. Phlogopite is found especially in Rumania, Switzerland, Italy, Finland, Sweden and Madagascar where it occurs in the crystalline limestones in huge crystals. In the United States it occurs in New York State at Edwards, Hammond, DeKalb, Monroe and, in New Jersey, at Franklin. In Canada it is found at many places in Ontario and Quebec.

The name phlogopite comes from the Greek word meaning like fire, referring to the copper-like reflections often observed in the reddish-brown varieties.

Phlogopite is in demand commercially by the electrical industry for use as an insulator. (E.S.C.S.)

PHOEBE. Aves, Passeriformes. One of the smaller species of North American birds (Aves) belonging with the pewees and kingbirds in the group known as

Phoebe, *Sayornis phoebe*. Dusky brown above, head darker, with a crest. Yellowish white below. Inconspicuous white wing bars.

flycatchers. It builds its nest commonly on beams below bridges and in deserted buildings. Also called the pewit. Both names are descriptive of its call. Two related species of the western half of the continent and the southwestern states respectively, are the Say phoebe, *Sayornis saya,* and the black phoebe, *S. nigricans.* (A.W.L.)

PHOLIDOTA. A small order of mammals containing only the scaly anteaters or pangolins of the Malay archipelago, Africa, and southeastern Asia. (A.W.L.)

PHONODEIK. Musical Sounds.

PHONOGRAPH. Sound Recording.

PHONOLITE. A dense extrusive rock, the equivalent of a nephelite syenite. The name is derived from the Greek word meaning sound, referring to the brilliant metallic ringing sound emitted by certain phonolites when struck. Most if not all of the "clink stones" of the older literature were phonolites. (E.S.C.S.)

PHORONIDEA. A group of marine animals of uncertain relationship. They live in tubes at the bottom of the ocean and are slender animals with a lophophore bearing a series of tentacles about the mouth. In this feature they resemble Bryozoa and Brachiopoda. As in the Bryozoa, the alimentary tract is sharply bent. Phoronids have a closed tubular circulatory system and red blood. They have sometimes been grouped with the primitive chordates but are usually placed as a separate phylum. (A.W.L.)

PHOSGENE, CARBONYL CHLORIDE. Carbon.

PHOSGENITE. The mineral is a rare chloro-carbonate of lead found with cerussite. (E.S.C.S.)

PHOSPHATE. Phosphoric Acid.

PHOSPHINES AND RELATED COMPOUNDS. Phosphines are compounds of phosphorus and hydrogen. By replacement of one or more hydrogen atoms a wide variety of derivatives can be formulated, many of which are known. A table showing the relationship of these compounds is given on the following page, and some of their reactions are described.

Representative reactions of phosphines and their derivatives are as follows:

Mono-substituted phosphines oxidized (by nitric acid) to mono-substituted phosphinic acids.

Di-substituted phosphines oxidized (by nitric acid) to di-substituted phosphinic acids.

Tri-substituted phosphines:

1. Oxidized (by nitric acid) to tri-substituted phosphine oxides (not easily reduced).

2. With chlorine to tri-substituted phosphine dichlorides.

3. With sulfur to tri-substituted phosphine sulfides.

4. With carbon disulfide to tri-substituted phosphine carbon disulfide addition compounds.

5. With methyl (etc.) iodide to tetra-substituted phosphonium iodides.

6. With ethyl azide to tri-substituted phosphine ethyl amine $(CH_3)_3P:NC_2H_5$ plus nitrogen gas (N_2).

Tetra-substituted phosphonium iodides with silver oxide form tetra-substituted phosphonium hydroxides $([(CH_3)_4P]OH)$.

Tetra-substituted phosphonium hydroxides when heated yield tri-substituted phosphine oxides $((CH_3)_3PO)$, plus hydrocarbon (e.g., CH_4).

Benzene plus phosphorus trichloride in the presence of aluminum chloride forms phenyl dichlorophosphine

TABLE SHOWING PROPERTIES OF PHOSPHINES AND RELATED COMPOUNDS.

PHOSPHINES AND THEIR ALKYL-SUBSTITUTION COMPOUNDS	FORMULA	M.P. °C.	B.P. °C.	RELATED OXYGEN COMPOUNDS	FORMULA	M.P. °C.	B.P. °C.
Phosphine.,............	PH_3	-134	-87				
	$PH_2 \cdot PH_2$	-10	57 (735 mm.)				
	$(P_4H_2)_3$	Yellow solid, ignition temperature 160° C.					
Methyl phosphine........	CH_3PH_2		-14	Methyl phosphinic acid...	$CH_3PO(OH)_2$	105	
Ethyl phosphine..........	$C_2H_5PH_2$		25	Ethyl phosphinic acid.....	$C_2H_5PO(OH)_2$		
Phenyl phosphine........	$C_6H_5PH_2$		160	Phenyl phosphinic acid....	$C_6H_5PO(OH)_2$		
Dimethyl phosphine......	$(CH_3)_2PH$		25	Dimethyl phosphinic acid..	$(CH_3)_2PO(OH)$.	76	
Diethyl phosphine........	$(C_2H_5)_2PH$		85	Diethyl phosphinic acid...	$(C_2H_5)_2PO(OH)$		
Trimethyl phosphine......	$(CH_3)_3P$		42	Trimethyl phosphine oxide.	$(CH_3)_3PO$		
Triethyl phosphine........	$(C_2H_5)_3P$		128	Triethyl phosphine oxide...	$(C_2H_5)_3PO$	53	243
Tetramethyl phosphonium iodide................	$[(CH_3)_4P]I$						
Tetraethyl phosphonium iodide................	$[(C_2H_5)_4P]I$						

(phosphenyl chloride ($C_6H_5PCl_2$), (boiling point 225° C.), plus hydrogen chloride.

Phenyldichlorophosphine plus water forms phenyl phosphinous acid ($C_6H_5P(OH)_2$).

Phenyl phosphinous acid plus hydrogen peroxide forms phenyl phosphinic acid ($C_6H_5PO(OH)_2$).

Phenyldichlorophosphine plus hydrogen forms phenyl phosphine $C_6H_5PH_2$.

Phenyldichlorophosphine plus phenyl phosphine forms phosphobenzene ($C_6H_5P:PC_6H_5$), melting point 149° C., yellow solid (similar to azo-compound. Chromophore group —P:P—).

Phosphinobenzene ($C_6H_5PO_2$) similar to nitrobenzene ($C_6H_5NO_2$). (R.K.S.)

PHOSPHORESCENCE. A term commonly applied to the production of light by living things and to the resulting luminosity of ocean water when small luminous animals are abundant in it. Luminosity of animals is only superficially like that of phosphorus, however. It results from the oxidation of organic compounds in some animals, including the common fireflies, and is probably never a true phosphorescence as this term is defined by the physicist. (See **Luminescence.**)

Phosphorescence of the ocean is often due to a multitude of luminous 1-celled animals although a limited show of light may result from larger animals of many forms. Animals capable of producing light are known in every **phylum.** (A.W.L.)

PHOSPHORIC ACID AND PHOSPHATES. Phosphoric acid (H_3PO_4) is a colorless solution, commercially of strength 50% H_3PO_4 (specific gravity at 17.5° C., 1.340); 75% H_3PO_4 (specific gravity at 17.5° C., 1.588); 93% H_3PO_4 (specific gravity at 17.5° C., 1.800). A solution of phosphoric acid of 64% strength has minimum freezing point of —85° C. Phosphoric acid 100% may be made by the proper admixture of **phosphorus** pentoxide and water. Conversely, by concentration of phosphoric acid solution in a vacuum or by heating to 140° C., orthophosphoric acid (H_3PO_4) is obtained as a solid, melting point 42.3° C. When heated at 250° C., pyrophosphoric acid ($H_4P_2O_7$) is obtained (salts tetra- and dipyrophosphates), and at a red heat metaphosphoric acid, "glacial phosphoric acid" (HPO_3)

(salts monometaphosphates) melting point 40° C., which upon cooling yields a glassy solid.

Phosphoric acid reacts with such metals as **magnesium, zinc,** or **iron,** yielding **hydrogen** gas and the corresponding phosphate in solution. Phosphoric acid is (1) a remarkably stable acid from the point of view of oxidation and reduction, and is frequently desirable to use on that account, (2) an esterification agent (see **esters**), e.g., glycerol esterified to glycerophosphoric acid ($C_3H_5(OH)_2 \cdot OPO(OH)_2$), ethyl alcohol to triethyl phosphate (($(C_2H_5O)_3PO$), boiling point 216° C., (3) a non-volatile acid upon heating, e.g., with **sodium** chloride or nitrate, **hydrogen chloride** or nitric acid, respectively, is volatilized and sodium dihydrogen phosphate remains as a residue when slight excess phosphoric acid is used; diethyl phosphoric acid (($(C_2H_5O)_2PO$ (OH)), boiling point 59° C.

In order to obtain phosphoric acid from **calcium** phosphate the latter may be subjected to either of two treatments, (1) by addition of the proper amount of **sulfuric acid,** calcium phosphate is transposed to calcium sulfate, insoluble, plus phosphoric acid solution. This reaction is followed by separation of the solution by sedimentation or filtration, and subsequent evaporation; (2) by addition of **silicon** oxide and **carbon** and subjection to **electric furnace** temperature, phosphorus is volatilized and the vapor may be burned in air to form phosphorus pentoxide. By solution of the oxide, various phosphoric acids are obtained.

The uses of phosphoric acid have been suggested by the chemical reactions previously cited, and the largest quantities are used as an acid to prepare phosphates.

Dilute phosphoric acid reacts with **hydroxides** to form three series of phosphates (the acid is tribasic), e.g., monosodium dihydrogen phosphate (NaH_2PO_4), disodium (mono) hydrogen phosphate (Na_2HPO_4), trisodium phosphate (Na_3PO_4), depending upon the ratio of acid to base reacting. The tribasic and dibasic phosphates, other than those of **sodium, potassium, ammonium,** are insoluble, monobasic phosphates are soluble. Tricalcium phosphate, when treated with the proper ratios of sulfuric acid, yields phosphoric acid (H_3PO_4), monocalcium phosphate ($Ca(H_2PO_4)_2$), dicalcium phosphate ($CaHPO_4$), with accompanying calcium sulfate in each of the three cases. Phosphates are dissolved or transposed by **nitric acid, hydrochloric acid, sulfuric** acid, phosphoric acid (except those of lead, tin, mercury, bismuth), and acetic acid (except those of lead, aluminum, ferric). Upon heating, tribasic phosphates are stable, dibasic phosphates lose water to form pyrophosphates, monobasic phosphates lose water to form metaphosphates, and in such reactions ammonium volatilizes at high temperatures resembling hydrogen, e.g., sodium ammonium hydrogen phosphate (dibasic) upon heating yields sodium metaphosphate residue. Common tests for metaphosphate are as follows:

1. Addition of magnesium salts to boiling solution of metaphosphate produces no precipitate (difference from orthophosphate).

2. Silver nitrate forms a white precipitate (difference from orthophosphate).

3. Albumin is coagulated by aqueous solution of the free acid (difference from pyrophosphate and orthophosphate).

Common tests for orthophosphate are:

1. Silver nitrate gives a yellow precipitate.

2. Ammonium molybdate in nitric acid solution gives a yellow precipitate of ammonium phosphomolybdate, soluble in ammonium hydroxide.

3. Magnesium sulfate in ammonium hydroxide solution gives a white precipitate of magnesium ammonium phosphate. (R.K.S.)

PHOSPHORITE. Phosphorite or phosphate rock is the term applied to accumulations of calcium phosphate in nodular or compact masses often associated with limestones. The phosphates are derived in part from marine invertebrates which secrete shells of phosphate of calcium, and largely from the bones and excrement of vertebrates. While some phosphorite deposits are doubtless original, others are secondary, resulting from the leaching out of the original phosphates and their subsequent deposition elsewhere. Phosphorite is found in the United States in Florida, Idaho, Montana, South Carolina, Tennessee, Wyoming and Utah; also in Russia, France, Algeria, the West Indies and elsewhere. Phosphate rock is economically valuable in the manufacture of fertilizers. (E.S.C.S.)

PHOSPHOROUS ACID AND PHOSPHITES. Phosphorous acid (H_3PO_3, or H_2PO_3H) is a white soluble solid, melting point 74° C.

Phosphorous acid is used in solution, and is usually a reducing agent, e.g., in air changes to phosphoric acid, with hot concentrated sulfuric acid yields phosphoric acid plus sulfur dioxide, with copper sulfate yields finely divided copper metal, with silver nitrate yields finely divided silver metal, with permanganate after some time yields manganous, but occasionally is an oxidizing agent, e.g., zinc plus dilute sulfuric acid yields phosphine.

Phosphorous acid is formed by reaction (1) of phosphorous trioxide and water, (2) of phosphorous trichloride and water (hydrogen chloride evolved, and probably pyrophosphorous acid ($H_4P_2O_5$) first formed). The solution is evaporated to 180° C., and then cooled, whereupon phosphorous acid crystallizes.

Sodium phosphite (disodium phosphite, Na_2PO_3H) and sodium hydrogen phosphite ($NaHPO_3H$) are formed by reaction of phosphorous acid and sodium hydroxide solution in the proper proportions, and then evaporating. Sodium phosphite dry, upon heating, yields sodium phosphate and phosphine.

As an esterification agent (see Ester), phosphorous acid forms, with ethyl alcohol, triethyl phosphite ($(C_2H_5O)_3P$), boiling point 156° C. An unsymmetrical, related compound $(C_2H_5O)_2(C_2H_5)PO$ is known. (R.K.S.)

PHOSPHORUS. Symbol: P. Atomic number: 15. Atomic weight: 31.02. No isotope, but of single atomic form: 31.

The chemical element phosphorus is known in four different forms, namely, (1) yellow phosphorus, density 1.82, melting point 44.1° C., ignition temperature in air 34° C. (preserved under water), boiling point 280° C., reactive with warm sodium hydroxide solution, (2) red phosphorus, density 2.20, melting point 590° C. at 43 atmospheres pressure, ignition temperature 260° C., not reactive with warm sodium hydroxide solution, (3) violet phosphorus, density 2.36, melting point about 600° C., ignition temperature 260° C., insoluble in solvents, (4) black phosphorus, density 2.70, incombustible. Of these varieties, yellow phosphorus is soluble in carbon disulfide, and when the solution is allowed to evaporate spontaneously in air, the residual phosphorus burns spontaneously. Volatilized phosphorus condenses as the yellow variety. Yellow phosphorus reacts with chlorine, sulfur, nitric acid (yielding phosphoric acid), sodium hydroxide solution (yielding sodium hypophosphite plus phosphine gas).

Phosphorus was discovered by Brandt in 1669.

Phosphorus is an important element in plant nutrition, e.g., in the form of phosphate, and in animal nutrition, especially in bone formation. To maintain health, a phosphorus balance must be maintained in the body; that is, intake must equal excretion. Phosphorus plays a part in complex chemical processes in the body about which little is known. It is important in bone formation and here it is linked up with calcium, vitamin D and sunlight. It is present in the blood stream and its concentration remains fixed unless changed by disease such as in rickets, etc. Under nor-

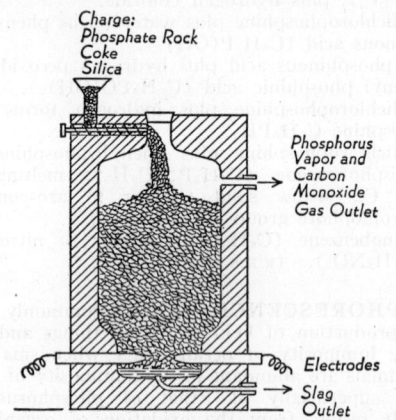

Charge:
Phosphate Rock
Coke
Silica

Phosphorus Vapor and Carbon Monoxide Gas Outlet

Electrodes

Slag Outlet

Manufacture of elementary phosphorus in electric furnace.

mal conditions a varied ordinary diet supplies this element in sufficient quantity. Phosphorus is extensively used as phosphate in fertilizers; as sulfide in ignition tips for friction matches, and as red phosphorus in safety matches. Yellow phosphorus is poisonous, and in contact with the skin causes dangerous wounds. A striking phenomena is the phosphorescence exhibited when yellow phosphorus is heated in darkness in boiling water and the vapors condensed. A phosphorescent region is visible at the place of condensation.

Phosphorus occurs as calcium phosphate (phosphorite, phosphate rock, ($Ca_3(PO_4)_2$) in Florida and Northern Africa, (apatite, $Ca_3(PO_4)_2 \cdot CaCl_2$ or CaF_2) in Quebec, and in the ash of bones. Calcium phosphate is treated in an electric furnace with silicon oxide and carbon (diagram, above), whereupon there is formed calcium silicate fused slag (drawn off as liquid), carbon monoxide gas, and phosphorus vapor, which last is condensed to yellow phosphorus and collected under water. The phosphorus content of iron ore in part determines the value of the ore, an acid-lined furnace such as acid Bessemer does not remove phosphorus, and the basic open hearth furnace operation which does is costly. The phosphorus content of most steel must

SCHEME SHOWING THE INTERRELATIONSHIPS OF PHOSPHORUS-CONTAINING SUBSTANCES

Phosphines Organic phos- phines Phosphonium compounds	PHOSPHORUS		Phosphorus trioxide	Phosphorus tetroxide	Phosphorus pentoxide
		Hypophospho- rous acid Metallic hypo- phosphites	Phosphorous acid Metallic phos- phites	Hypophosphoric acid Metallic hypophos- phates	Phosphoric acid Ortho, pyro, meta Metallic phosphates Insoluble: phosphor- ite, apatite in na- ture Soluble: organic phosphates
			Organic phosphites	Organic hypophos- phates	Organic phosphoric acids Phosphoproteins in proteins Phosphinic acids and phosphine oxides Phosphinobenzene
	Phosphobenzene				

be low, only for certain grades used for free machining is the content appreciable.

Red phosphorus is made by heating yellow phosphorus out of contact with oxygen, under pressure, and in the presence of trace of **iodine** at a temperature of about 200° C.; violet phosphorus, by allowing red phosphorus dissolved in molten **lead** at 580° C. under pressure to crystallize; black phosphorus, by heating yellow phosphorus at 200° C. under very high pressure (12,000 kilograms per sq. cm. or 165,000 lbs. per sq. in.).

Acids: Phosphorus is a constituent of several acids. In increasing order of oxidation the following are those which do not contain carbon: hypophosphorous acid (H_3PO_2); **phosphorous acid** (H_3PO_3); hypophosphoric acid ($H_4P_2O_6$); **phosphoric acid** (H_3PO_4). Glycerophosphoric acid is $C_3H_5(OH)_2PO(OH)_2$.

Bases: Phosphine (PH_3) resembles **ammonia** chemically; phosphonium hydroxide ($PH_4 \cdot OH$) resembles **ammonium** hydroxide chemically.

Bromides: Phosphorus tribromide, phosphorus bromide (PBr_3), colorless liquid, boiling point 173° C., fumes in air, made by reaction of **bromine** and phosphorus, under control of kind of phosphorus, temperature and solvent (e.g., carbon disulfide) to regulate the rate of the reaction; phosphorus pentabromide, phosphoric bromide (PBr_5), yellow solid, fuming in moist air; phosphorus oxybromide ($POBr_3$), white solid, melting point 56° C., boiling point 193° C.

Chlorides: Phosphorus trichloride, phosphorous chloride (PCl_3), colorless liquid, boiling point 73.5° C., fumes in air, made by reaction of **chlorine** and phosphorus, under control of kind of phosphorus, temperature and solvent (e.g., carbon disulfide) to regulate the rate of the reaction; phosphorus pentachloride, phosphoric chloride (PCl_5), yellow solid, fuming in moist air; phosphorus oxychloride ($POCl_3$), colorless liquid, melting point 1° C., boiling point 107° C. These compounds are important reagents in organic chemistry to replace by halogen an hydroxyl (OH) that is attached to **carbon**, e.g., benzoic acid into benzoyl chloride.

Hydrides: Phosphine (PH_3), colorless gas, of characteristic odor, poisonous, boiling point —86° C., burns in air by heating when pure to 100° C. (ordinarily contains diphosphorus tetrahydride inflammable at room temperature); made by reaction of yellow phosphorus and **sodium** hydroxide solution heated, accompanied by diphosphorus tetrahydride (P_2H_4), colorless liquid, boiling point 57° C., spontaneously inflammable in air, accompanies phosphine as prepared above, removable by condensation upon cooling.

Iodide: Phosphorus diiodide (P_2I_4), orange solid, melt-

ing point 110° C.; phosphorus triiodide (PI_3), red solid, melting point 61° C., decomposes upon heating; with water it forms **phosphorous acid** and hydriodic acid.

Nitride: Phosphorus nitride (P_3N_5), white solid, reactive with warm water to yield phosphate plus **ammonia**.

Oxides: Phosphorus trioxide, phosphorous oxide (P_2O_3), white solid, melting point 22.5° C., boiling point 173° C., formed by reaction of phosphorus and a deficiency of air, dissolves in cold water to form **phosphorous acid**; phosphorus tetroxide (P_2O_4), white solid formed by heating phosphorus trioxide in a sealed tube at about 440° C., reactive with water to form a mixture of phosphorus and **phosphoric acids;** phosphorus pentoxide (P_2O_5), white solid, sublimes at 347° C., excellent dehydrating agent, absorbs water avidly, made by burning phosphorus in excess of air or oxygen.

Phosphides: Many metallic phosphides have been described, some seven phosphides of **tin,** and six of **copper.** Formed by reaction of the metal plus phosphorus. Reactive with water or acid to form phosphine or phosphonium salt.

Sulfide: Phosphorus and **sulfur** do not react at low temperatures, nor in **carbon disulfide** solution, but, upon heating with caution, three sulfides of phosphorus have been formed, namely, P_4S_3, P_4S_7, P_4S_{10} (or P_2S_5) of melting points, 172° C., 303° C., 280° C., respectively. Tetraphosphorus trisulfide (P_4S_3), non-poisonous, not attacked by atmospheric moisture, is used to replace yellow phosphorus on match tips.

Other compounds of phosphorus are discussed as follows:

Esters. (See **Phosphoric Acid and Phosphates;** Hypophosphoric Acid and Hypophosphites; Phosphorous Acid and Phosphites.)

Glycerophosphoric acid. (See **Phosphoric acid and Phosphates.**)

Hypophosphates. (See **Hypophosphoric Acid and Hypophosphates.**)

Hypophosphites. (See **Hypophosphorous Acid and Hypophosphites.**)

Hypophosphoric Acid.

Hypophosphorous Acid.

Phosphates. (See **Phosphoric Acid and Phosphates.**)

Phosphines.

Phosphine oxides. (See **Phosphines and Related Compounds.**)

Phosphinic acids. (See **Phosphines and Related Compounds.**)

Phosphinobenzene.   (See **Phosphines and Related Compounds.**)

Phosphites. (See **Phosphorous Acid and Phosphites.**)

Phosphobenzene. (See **Phosphines and Related Compounds.**)

Phosphonium-compounds. (See **Phosphines and Related Compounds.**)

Phosphoproteins. (See **Aminoacids, Polypeptides and Proteins.**) (R.K.S.)

PHOSPHORUS POISONING. This type of poisoning, whether by accident, or intent, is less common since the advent of safety matches which do not contain **phosphorus.** Phosphorus is still used in many of the rat and vermin pastes. The mortality in untreated poisoning is very high since the phosphorus has a special affinity for **liver** cells, which it destroys. The symptoms are early gastrointestinal irritation, abdominal pain and collapse. If the patient lives 2 or 3 days, evidence of liver damage appear, such as marked **jaundice.** Later, drowsiness, delirium and coma develop, and death occurs in a day or two. In the early stages treatment consists of washing out the stomach with dilute solutions of **hydrogen peroxide.** A diet rich in **carbohydrates** and protein should be given to protect the liver. In the acute stage, glucose by vein is indicated. (R.S.M., D.M.H.)

PHOTOCELL. This term is the old designation for all photoelectric devices but is more properly applied only to those which would not be called electronic **tubes.** Thus the two principal types of photocells are the photoconductive cell and the photovoltaic cell. The first of these usually employs selenium as the active element. This is spread across two conductors separated by a short gap (the whole is commonly applied to a base of glass or other insulating material) so the selenium will bridge the gap. After proper heat treatment the conductivity of the selenium will be sensitive to light, being very high when dark and increasing when illuminated by visible light. Such cells are not overly sensitive but are suitable for many applications. The photovoltaic cell is probably most familiar as the exposure meter used in photography. Certain composite surfaces composed of an overlay of one substance on a backing of another will produce a voltage difference across the barrier surface between the two when subjected to light. Cuprous oxide on copper and selenium on iron are two widely used examples of this type cell. Contact is made to the front face by means of an extremely thin layer of metal sputtered over the surface and contacting more substantial connections at the edge. This film is thin enough to allow light to penetrate it and act on the photosensitive materials. When connection is made to an outside circuit current will flow although there is no source of e.m.f. other than that generated by the action of light. While these currents are very small, being of the order of microamperes, they are nevertheless sufficient to operate sensitive current meters and **relays.** In the exposure meter, a microammeter is connected across the terminals of the cell and the deflection is a function of the illumination. By calibrating the instrument in terms of **foot-candles** it may be used as a light meter, or in terms of film exposure as a photographer's aid. (L.R.Q.)

PHOTOCHEMISTRY. When certain substances are subjected to light a chemical change is produced. The production of an image in the photographic plate is possibly the most familiar instance of this, but the reaction of **photosynthesis** in the green leaf of the plant operates on the largest scale. Conversely, the production of light from the heat developed in chemical reactions is a common occurrence. The burning of **magnesium** metal in air produces a high temperature and light of high actinic value which is utilized in photography.

In photochemical reactions **light** supplies the **energy** necessary for the activation of the reacting **molecules**

(Grotthus, 1818, and Draper, 1839). Sometimes the light waves which are absorbed by a body produce only an increase in temperature, sometimes **fluorescence** as in the cases of eosin and fluorescein, and sometimes chemical change. The reaction of **hydrogen** and **chlorine** in light was studied by Bunsen and Roscoe (1862), and they discovered that the amount of chemical change is proportional to the intensity of the light and to the length of time of exposure to the light. The first law of photochemistry (Draper-Grotthus) states that light that is absorbed causes chemical change. The energy of light is measured in quanta

$$E = Nhc/\lambda$$

N is Avagadro's number.
h is Planck's constant.
c is velocity of light.
λ is wave length of light.

Photochemical processes are of two kinds—primary and secondary. The primary process in a photochemical reaction is limited by the Einstein law to the absorption of one quantum by a molecule or atom. A knowledge of the spectrum of the reactants is necessary to determine what happens in this process. The **molecule** may be disrupted into fragments or an **electron** may be excited from a lower orbit to a higher one. Which of these events takes place can often be determined by spectroscopic studies. The secondary process deals with the fate of the molecular fragments or of the excited molecules. The excited molecule may emit its extra energy as light—causing fluorescence; it may lose it by transferring it to other molecules as thermal energy; or it may cause a chemical reaction. On the other hand the molecular fragments may either recombine to give the original reactant or cause further chemical reactions. The study of the quantum yield (which is the number of molecules reacting divided by the number of quanta absorbed), is used as a means of formulating the secondary processes. If the quantum yield is less than one, fluorescence, deactivation or recombination of fragments must take place. If the quantum yield is unity every photon absorbed decomposes one molecule. When the quantum yield is greater than unity (and in some reactions it may be as high as a million) chain reactions are involved. The classical example of such a reaction is the combination of hydrogen and chlorine. The primary reaction is Cl_2 and light $\rightarrow 2Cl$. The chain propagation reactions are

$$Cl + H_2 \rightarrow HCl + H$$

$$H + Cl_2 \rightarrow HCl + Cl$$

creating a cycle which is only stopped by

$$Cl + Cl \rightarrow Cl_2$$

$$H + H \rightarrow H_2$$

Since the last two processes are slow compared to the two before them, one quantum of light can bring about a combination of a million molecules of hydrogen and chlorine. (R.K.S.)

PHOTOCONDUCTIVITY. Many substances exhibit a marked increase in electric conductivity when illuminated. Thus gases may be ionized by light as well as by ultra-violet radiation or x-rays. But the term photoconductivity is commonly applied to crystals which, ordinarily very poor conductors, become distinctly conducting under the action of light.

The most noted example of this phenomenon is found in **selenium,** whose photoconductivity has been known since its discovery in 1873 by May. Unfortunately selenium is far from typical in its manifestations of the property, and the hundreds of researches on it have given many conflicting data. It has finally been recognized, from the work of Gudden and Pohl, that photo-conduction is of two general types: primary or true photoconduction, which is the direct result of radiation penetrating the substance; and secondary effects set up by the photoconduction itself. The case is somewhat analogous to the primary and secondary ionization of a gas, so much utilized in **photoelectric cells.** The fact that in selenium the several secondary effects quite obscure the primary photoconduction is what has occasioned so much confusion. The primary photoconduction current in a crystal is in general proportional to the intensity of the illumination; the secondary is not, and, in the case of "light-negative" selenium, it may actually neutralize the primary and render the crystal less conductive than when in the dark. Some crystals, said to be "idiochromatic," are photoconductive in the pure state, while others, called "allochromatic," acquire the property only by reason of impurities or of previous exposure to suitable radiation.

Films of photoconductive selenium provided with electrodes, and called "selenium cells," have found practical application in the past. They have now, however, been superseded largely by **photoelectric** and **photovoltaic cells.** (L.D.W.)

PHOTOELASTICITY.
This badly chosen term refers to certain changes in the optical properties of isotropic, transparent dielectrics when subjected to stresses. For example, a block of glass, free from optical flaws, exhibits "forced" **double refraction** when put under compression or tension parallel to one of its dimensions. If the block is placed between crossed **Nicol prisms,** the field remains dark so long as the glass is in its normal condition, but as stress is applied, colored fringes appear which are characteristic of the internal deformations of the glass. In 1893 Marsten adapted this principle to the study of elastic stresses in glass or celluloid models of structural parts; and in 1913 Coker developed an apparatus, using circularly- instead of plane-polarized light, for the same purpose. Maris analyzed the altered polarization by means of a **compensator.**

Ewell discovered that torsion imparts to a dielectric cylinder a certain amount of optical activity, i.e., the property of rotating the plane of polarized light. The rotation is in the opposite direction to the torsion, and varies approximately as the fourth power of the torsion. (L.D.W.)

PHOTOELECTRIC CELL. Phototube.

PHOTOELECTRIC PHENOMENA.
Our knowledge of what is known as the photoelectric effect dates from the observation of Hallwachs (1888), who found that a negatively charged body may be discharged when **ultra-violet** radiation falls upon it in a vacuum. We are now mostly interested in the photoelectric effect of visible light, but the general theory applies to radiation of all higher frequencies. Every solid electric conductor is supposed to contain numerous "free" electrons which wander about from atom to atom within the substance. Ordinarily they do not leave the conductor because of the attraction of the positive atoms about them. If, however, such a wandering electron encounters a quantum of radiation, it may receive therefrom sufficient kinetic energy to tear itself loose from the attraction and pass out of the conductor; it then becomes a "photoelectron." (Some physicists have assumed that these electrons are released by the photon energy from the orbital system of the atom instead of being originally free, but this is now considered doubtful.) If there is a suitable electric field between the conductor and a neighboring conductor, the electrons thus released pass across from one to the other, thus setting up a photoelectric current.

The emission of photoelectrons is subject to certain well-known laws. For every conductor there is a photoelectric **work function** p, which represents the kinetic energy which the electron must have in order to escape. Obviously the radiation quantum responsible for the emission must have at least this energy; and if it has more, the electron, taking it all, escapes with a surplus. These facts are embodied in the Einstein photoelectric equation $\frac{1}{2}mv^2 = h\nu - p$, in which ν is the frequency of the radiation, h is the Planck constant, and m and v are the mass and speed of the emitted electron. This expresses the excess kinetic energy after an electron, receiving the quantum $h\nu$, passes the surface with energy loss p. (See **Quantum Theory** and **Planck's Law.**) It follows that radiation of lower than a certain frequency, called the photoelectric threshold frequency, and equal to p/h, cannot cause photoelectric emission from the given conductor. For most metals p is so great that none but the highest frequencies give any result. There are only seven (**sodium, potassium, caesium, rubidium, lithium, barium, strontium**) which react to visible light.

The "photoelectric yield" of a metal, i.e., the emission per unit radiant flux of a given frequency, varies largely with the metal, the condition of its surface, its potential, and the presence of surface films. Careful studies of these factors are necessary in the design or the selection of a **photoelectric cell** for any given purpose. Closely allied phenomena are **photoconductivity** and the **photovoltaic effects.** (L.D.W.)

PHOTOENGRAVING. Photomechanical Reproduction Processes.

PHOTOGENIC ORGAN.
An organ that produces light. In all known cases of light production a complex substance, probably a **protein,** known as luciferin, is oxidized in the production of light. The action may take place inside the cell or outside. The marine protozoan, *Noctiluca,* for example, contains scattered granules which become luminous under the proper stimulus, and the marine worm, *Chaetopterus,* has gland cells at the surface of the body which produce a luminous secretion.

Among the more complex animals, including the **fireflies** and various fishes, photogenic organs are more complex and are extremely varied in form.

The production of light has been interpreted in various ways. It may be incidental to some normal reaction in the body, it may be a warning to other animals, it may serve to attract prey, and it may be for recognition, particularly in connection with mating. (A.W.L.)

PHOTOGRAMMETRIC SURVEYING. Surveying.

PHOTOGRAMMETRY.
Photogrammetry is the application of photographic principles to the science of mapping. In general it consists of taking photographs of small parts of the earth's surface with specially constructed cameras and using these photographs directly to form a picture or indirectly to make a scale map. Terrestrial photogrammetry refers to the use of photographs taken with the optical axis of the camera lens horizontal or oblique from high ground points of known position and elevation overlooking the section to be photographed. Aerial photogrammetry utilizes photographs taken from an airplane with the optical axis of the lens vertical or oblique. Terrestrial photogrammetry is used principally where aerial photographs would not indicate abrupt changes in ground slope such as cliffs. This method is also employed in the preparation of contour maps for sections where it would be extremely difficult to make a topographic survey. (See **Topographic Surveying.**) Since aerial photogrammetry has the greater practical application, the remainder of this article will be devoted to this phase of the subject.

Aerial photographs are pictures taken with an aerial camera while the airplane is in flight. The altitude of the plane should be kept as nearly constant as practicable if more than one exposure is to be made. A series of photographs taken while the plane is proceeding in a fixed direction is called a strip. When the area to be photographed is large, flights must be made in opposite directions. Adjacent exposures must overlap if they are to be satisfactory for practical use.

If it were possible to keep the optical axis truly vertical and all objects were at the same elevation, the aerial photograph would be a true scale map so far as the relative position of the objects is concerned. Since the optical axis usually is tilted a small amount as a result of flight conditions, there is a distortion due to tilt. However, it is normally eliminated by the procedure which is followed in the use of aerial photographs. If this distortion is perceptible, the exposure should be discarded or rectified. Rectification consists of taking a picture of the photograph with a camera having the optical axis of the lens tilted with respect to the plane of the photograph. The main reason why an aerial photograph cannot be used as a scale map is the fact that all points not at the same elevation and not in line with the optical axis (assumed to be vertical) will have a relative displacement on the photograph. This is known as relief displacement.

The scale of an aerial photograph is the ratio of the scaled distance between the images of two ground points on the photograph to the actual distance between these points measured in the same units as those used in scaling the images. The photographic scaled distance should be corrected for relief displacement if large enough to affect the scale.

Aerial photographs may be used to form a mosaic. This is a group of parts of matched photographs which have been reduced or enlarged to a common scale. The part of each common scale photograph used for a mosaic depends upon the required precision. If high precision is desired, the central portion only should be used since there is comparatively little relief displacement and tilt distortion in this area. Before matching, each print is trimmed along natural lines of the photographic detail such as fence lines and boundaries of woods or fields where the match lines will be least likely to show. A mosaic may be constructed without reference to ground points, or its construction may depend on a system of plotted points known as ground control points whose relative position has been established by a ground survey. The latter is a controlled mosaic.

If aerial photographs are to be used for a controlled mosaic or in the construction of a planimetric map or a topographic map, it is necessary to have a series of ground control points which are easily recognizable on the photographs. It is also customary to have a secondary control system because the primary system (ground control) itself is not sufficient for matching the adjacent photographs properly or locating details on planimetric or topographic maps. The secondary control, also known as the photographic control, consists of points which are easily recognizable on two or more photographs and are not a part of the ground control. The first step in making a controlled mosaic or either of the two maps previously mentioned is to plot the ground control points. The second step is the location of the secondary control relative to the ground control. Two methods are used. One is the principal-point traverse method which is applicable only to flat terrain; the other is the radial-line method which can be used for any terrain.

After the secondary control has been located, the plot may be used to match photographs for a mosaic or to make a planimetric or topographic map. Details from aerial photographs are transferred to the control plot by the usual drafting methods or by automatic plotting machines. These machines are also used to establish contours for topographic maps. If drafting procedure is followed, each photograph in turn is placed under the control plot, which must be on tracing paper or acetate film, and properly oriented with respect to the control points. The details are then traced directly on the plot. Due to relief displacement or tilt it may be impossible to orient the photograph so that all control points will match. Under these circumstances parts of the photograph are oriented and traced as separate units.

Automatic plotting machines make use of the principle of stereoscopic observation combined with a drawing attachment. A simple stereoscope is a device which gives a three-dimensional view when applied to a pair of photographs of the same object taken from two different points. It consists of two eye lenses fixed in position above the plane of the photographs. In order to obtain the proper perspective of aerial photographs, a stereoscopic pair is arranged in a certain fixed position with respect to each other. The stereoscope may be used to study photographic details but an additional instrument is required if differences of elevation are to be determined. This is known as the floating-mark system. It may be a part of the stereoscope or a separate unit. The floating-mark system consists of two pieces of glass with a dot at the center of each piece and a micrometer adjustment which makes it possible to move one piece with respect to the other. When viewed through the stereoscope, the two dots appear as one dot which is called the floating mark. When actual elevations are required, the elevation of at least one point on the photographs must be known. The use of additional control points will make it possible to eliminate tilt effects and will permit a more accurate determination of elevations or contours.

In automatic plotting instruments the floating-mark system is connected to a parallel motion having a pencil attachment. Planimetric details are traced by keeping the floating mark in contact with details on the photographs. A contour line corresponding to a given elevation is traced by locking the micrometer at a predetermined reading and then keeping the floating mark in apparent contact with the ground during the tracing.

A theoretical development of the basic principles of these machines and detailed descriptions of the various types are found in texts and pamphlets on photogrammetry. (c.w.c.)

PHOTOGRAPHIC DEVELOPERS.

Photographic developers are chemical solutions containing a number of different compounds so proportioned as to produce the controlled reduction of exposed silver halide grains. During reduction the exposed silver halides are reduced to metallic silver, the invisible latent image formed by exposure being converted into a silver deposit or visible image. Developing solutions normally contain components which can be classified according to their functions into the following heads:

Developing agents.
Preservatives.
Activators.
Restrainers.

Occasionally, developers are encountered which contain less than four components. In these cases, one of the components exhibits more than one function. In other cases additional components are added according to the result desired. These are discussed under developer additions.

Developer Components. 1. Developing agents. Developing agents are selective chemical reducers, usually organic compounds, which possess only sufficient energy to reduce exposed silver halides. Reducers which are too energetic are undesirable as they reduce more than the exposed silver halide grains and thereby fog or block the image. To obtain results desired, many formulae contain two or more reducers. Organic reducers that are good developing agents include:

Amidol.
Glycin.
Hydroquinone.
Metol.
Para-aminophenol.
Paraphenylene diamine.
Pyrogallol. (See **Developing Agents.**)

2. Preservative. Since developing agents in solution are readily oxidized by air, it is necessary to protect the agent from premature oxidation due to contact with the air or dissolved oxygen. Oxidized developing agents lose their reducing power and are thus rendered incapable of development. The preservative most commonly employed in photographic developers is sodium sulfite. Either the desiccated, Na_2SO_3, or the crystal, $Na_2SO_3 \cdot 7H_2O$, form may be used. Sodium bisulfite, $NaHSO_3$, and potassium metabisulfite, $K_2S_2O_5$, are added as extra preservatives when additional sulfite without an increase in the pH of the solution is desired. Many developers containing bisulfite have excellent keeping qualities and are particularly adapted for tank use.

While it is usually assumed that the sulfite is oxidized directly to the sulfate, there is considerable evidence to show that the reaction is more complicated. The exact mechanism is not fully known.

Preservative action may be complicated by a number of effects. While sulfite is normally added to preserve the developing agent, the developing agent may exert a preservative effect on the sulfite as in the case of hydroquinone. Frequently, in solutions containing two developing agents, one agent acts to preserve the other against aerial oxidation. Examples of this is metol preserving amidol and hydroquinone preserving metol.

Sodium sulfite may play more than one role, as protecting the developing agents against aerial oxidation, preventing the formation of certain oxidation products causing stains or dye images, forming complexes which act as silver halide solvents, and acting as a weak alkali to increase the rate of development or the maximum density.

Factors that affect the rate of oxidation and determine the amount of preservative required are:

  a. Susceptibility of developing agent to oxidation.
  b. Alkalinity of the solution.
  c. Concentration of the solution.
  d. Method of development employed.
  e. Temperature at which the developer is kept and used.
  f Keeping quality desired.

Other preservatives, many of which are used with sodium sulfite, are:

  Acetone bisulfite.
  Formaldehyde sodium sulphoxylate.
  Stannous chloride.
  Mannitol.
  Sorbitol.
  Bengoic acid.
  Glycolic acid.
  Salicylic acid.

3. Activators. Activators, also known as accelerators, are compounds that increase the activity of developing agents. Normally, these compounds are alkalis, or compounds which combine with sulfite to form complexes that hydrolyze to liberate a hydroxyl ion and produce the necessary alkali. Experimental evidence shows that the activity of developing agents increases with an increase in alkalinity and that the rate of development is largely a function of the pH of the solution.

The choice of alkalis for a particular formula may be made on considerations other than alkalinity. While it is generally true that increasing the concentration of alkali increases the alkalinity, the concentration of the solution may reach a limit where further additions of alkali lowers the pH, reducing the alkalinity. Alkalis

like the alkali hydroxide are highly ionized whereas salts, as borax and metasilicate, are ionized to a much lower degree. Each developing agent has pH limits within which it functions satisfactorily. Metol develops at a pH of 6, while hydroquinone requires a pH of 9. Low alkalinities are necessary for certain effects, as fine-grain development. The small amount of strong alkalis that can be used in these cases would be such as to produce solutions with short life and poor keeping qualities because of the lack of an alkali reserve. It is better to use larger concentrations of weak alkalis and obtain more stable solutions. Ammonia in ammonium hydroxide is a volatile gas and also a silver halide solvent; properties that may be objectionable. During warm weather it is advisable to use alkalis, like trisodium phosphate and borax, that do not cause blistering due to the generation of a gas, when the emulsion is placed in acid fixing baths. Solutions with strong alkalis produce marked swelling of gelatin and cause softening of the emulsion. Pairs of strong and weak alkalis can be employed to produce buffer combinations, as sodium hydroxide and borax, which produce developers that have a long working life at a constant pH. Some organic acids, as salicyclic, soften gelatin and increase the penetration and diffusion of developer throughout the emulsion.

The alkali in widest use, in most formulae, is sodium carbonate, which is available in either the desiccated, Na_2CO_3, or the monohydrate, $Na_2CO_3 \cdot H_2O$, form. Next in popularity are the borates—sodium metaborate and borax. Others, as trisodium phosphate, sodium sulfite and borax, are used chiefly for fine-grain development. Sodium and potassium hydroxide find application in process or line developers, while paraformaldehyde is restricted to photomechanical processes, as lithography and offset. Organic alkalis, like triethanolamine, are finding some use in color and fine-grain development.

The approximate pH for certain standard developers containing different alkalis are given below:

	pH
Sodium metasilicate	8.5
Borax	9.2
Triethanalamine	10.1
Ammonium hydroxide	10.7
Sodium carbonate	10.4–11.6
Trisodium phosphate	12
Sodium hydroxide	13

Less popular alkalis include: acetone-bisulfite, acetone-formaldehyde, paraformaldehyde, hexmethylenetetramine, sodium aminoacetate, and sodium hexametaphosphate.

4. Restrainers. Unless developing solutions contain restrainers, the developed images are fogged to a greater or lesser degree, depending on the characteristics of the emulsion, the developing agents, and the alkalinity of the developer solution. The selective action of developers for exposed silver halide grains is greatly affected by the above factors. The addition of small amounts of restrainers produces a marked differential in the developer for exposed silver halide grains.

The presence of alkaline restrainers, as potassium bromide, potassium iodide, or sodium chloride, in developers lowers the ionization of the corresponding silver halides, reducing the concentration of the silver ions and consequently restrains development.

Of the restrainers the alkaline bromides are the most common. These are added as fog preventatives because the restraining effect is greater on fog than on the latent image. The addition of alkaline bromide permits longer development, which in turn produces a higher contrast. Where maximum contrast is desired high concentrations of bromide are required. Restraining action can also be accomplished by other ways, e.g., dissolving out the free bromide incorporated in emulsions during development, or by adding quantities of used developer to condition new developers before use.

During recent years, a number of organic nitrogen compounds which are powerful restrainers have been discovered. These antifoggants restrain fog satisfactorily when employed in concentrations as weak as 1:10,000 to 1:100,000, but delay the initial appearance of the image and make longer development necessary.

So far as is known, antifoggants form complexes, through the nitrogen atom, with silver halides. It is believed that the antifoggants are adsorbed at interface between the silver sulfide or the sensitivity speck and the silver halide, and that they form addition products with silver halides, which are stable in alkaline developers. Many of the restrainers are thioanalines. Among the more satisfactory are:

> 6-Nitrobenzimidazole.
> Benzotriazole.
> 5-Chlorbenzimidazole.
> Thioacetanilide.
> Tetrazole.

5. Special agents. Since developers are used for many purposes and under a wide variety of conditions, it often is necessary to add special agents to overcome the difficulties encountered. These include:

Solvents. Methyl, ethyl, isopropyl alcohols, or dioxan, are frequently added in order to dissolve more of the developing agents and obtain solutions with high concentrations, or to keep developing agents in solution, particularly during cold weather.

Water Softeners. Salts producing hardness in water react with the alkalis and sulfite of the developer and reduce its energy, causing the developed images to have lower than normal contrasts. Films processed with developers made up with hard waters have a greater tendency to show water spots, due to salt crust formation around edge of droplet during drying. Calgon (sodium hexametaphosphate) is an interesting water softener because it reacts with the hardness ions, calcium, magnesium, and ion, forming complexes that are soluble.

Wetting Agents. Wetting agents are compounds that reduce the surface tension between a solution and a surface. In photography wetting agents are used to insure instant wetting and uniform processing over entire emulsion surface. Generally speaking, wetting agents are organic compounds having long fatty-acid chains with sulfonic or carboxyl groups attached.

Antiswelling Agents. These compounds prevent gelatin emulsions from absorbing water to the extent that there is danger of excessive softening and frilling. Difficulties of this type are encountered in warm weather and in tropical climates. Sodium sulfate, Na_2SO_4 or $Na_2SO_4 \cdot 10H_2O$, because it is inert and is very soluble, has become an almost universal antiswelling agent for photographic purposes. Additions in amounts sufficient to produce 6–10% concentrations in processing solutions are generally satisfactory.

Hardeners. Alum, chrom alum, aluminum sulfate, formaldehyde and tannic acid have been used as hardeners in developers. Normally special hardening is accomplished in rinse or predeveloper baths.

Penetration Control Agents. Inert materials forming viscous solutions like sugar, corn syrup, dextrin and glycerol, are used to control penetration in both fine-grain and color processing.

Tone Control Agents. The desire for procuring blue-black tones with paper emulsions, particularly the chlorides, bromides, and brom-chlorides, has led to the discovery of numerous agents for the production of blue-black tones. Included in this group are: benzotriazol, nitrobenzimidazol, nitroimidazol, nitrosoguanidine, quinoline, quinone chloride, and thiosemicarbazide. Potassium thiocyanate can also be used but is not as effective.

Silver Halide Solvents. Silver halide solvents have three functions in developers. In fine-grain development, potassium thiocyanate, ammonium chloride, and high concentrations of sodium sulfite exert a solvent action of the silver halide grain. In reversal processing, ammonia, ammonium salts, and thiocyanates prevent blocking of highlights, fogging and insure neutral tones. With developers for positive materials—prints, lantern slides and transparencies—the concentration of the halide solvents control the warmth of the image tones. (S.M.T.)

PHOTOGRAPHIC EMULSIONS. Photographic emulsions are not true emulsions but suspensions of minute silver halide crystals dispered in a protective colloid medium as gelatin, collodion, albumen, casein or agar. Attempts at using cellulose esters have been made recently; however, only gelatin and collodion are of commercial importance.

Gelatin is preferred as a photographic colloid, because the sensitizing bodies present in the gelatin make possible emulsions with greater sensitivity or speed. Gelatin is an excellent emulsifying agent and is readily transformed, from gel to a liquid or the reverse, by changes in temperature. The latter property makes coating of supports, and emulsion processing and working, feasible. The strong protective action of gelatin lowers the rate of reduction of unexposed silver halide crystals in developers so that image formation is readily obtained.

Silver halides employed in emulsions are, the chloride, the bromide and the iodide. Negative emulsions are composed of silver bromide with a small amount of silver iodide. Positive emulsions for films and paper contain silver chloride, or mixtures of silver chloride and silver bromide in varying amounts, according to the tone, speed, and contrast desired.

In **photomicrographs** of negative emulsions, the crystals of silver bromide appear as flat triangular or hexagonal plates with rounded corners. Some globular and needle-shaped crystals are also observed. The thickness of the flat plates is approximately $\frac{1}{10}$ of their diameter. The size of silver bromide crystals range from less than 1 micron to 4 microns. Crystals of silver chloride, or mixtures of silver chloride and silver bromide, as used in positive emulsions, are quite uniform and seldom exceed 0.5 micron in diameter. Multi-layered emulsions contain approximately 1 billion, 10^9, crystals per sq. cm. The areas of individual crystals range from 0.1×10^{-8} cms. for low-speed emulsions to 1.0×10^{-8} cms. for high-speed negative emulsions.

The characteristics of an individual emulsion are primarily dependent on two factors, the size-frequency distribution of the crystals and the composition of the silver halide crystals. The chief problems of the emulsion-maker are the production of uniform suspensions of silver halide crystals with proper size-frequency distribution and correct composition in gelatin, and the ability to reproduce results.

Photographic emulsions are classified as follows:

1. Printing-out emulsions. These emulsions produce images on exposure without development. They are used largely for making portrait proofs which are distinguished by their red or purplish color. Emulsions of this type differ from others in that they usually contain silver nitrate, some free silver, silver salt of an organic acid and a weak free acid. These are known as P.O.P. Proof Papers.

2. Developing-out emulsions. Emulsions for development have an excess of alkaline halides. By varying the composition of the silver halide and treatment, developing-out emulsions may be prepared which are suitable for either negative or positive purposes.

a. Negative emulsions. Negative emulsions are prepared by adding a small amount of a soluble iodide to the bromide used in making the silver halide. The mixed crystals of silver-bromiodide formed are more sensitive to light and produce emulsions with greater speed than silver bromide alone. Negative emulsions are referred to as neutral emulsions if precipitation of the silver halide is carried out in a gelatin solution with an excess of soluble bromide, and ammonia emulsions if the precipitation takes place in a gelatin solution with an excess of soluble bromide in the presence of ammonia or ammoniacal silver. The latter method produces emulsions witht coarser grains which have the highest sensitivity.

b. Positive emulsions. Positive emulsions are prepared by precipitating silver halides containing chloride or mixtures of chloride and bromide in gelatin. The size of the crystals formed are smaller than those of negative emulsions and have a lower sensitivity. Positive emulsions are divided into four classes, according to the composition of the silver halides and their properties.

(1) Chloride emulsions. Because of their slow speed chloride emulsions are used largely for contact printing.

(2) Bromide emulsions. Bromide emulsions are very sensitive and fast. They are used for projection printing exclusively.

(3) Chlor-bromide emulsions. In chlor-bromide emulsions the amount of silver chloride is greater than that of silver bromide. These emulsions are somewhat faster than chloride emulsions and used for contact or slow projection printing. Chlor-bromide emulsions produce warm-toned silver images with a brown or brown-black color.

(4) Brom-chloride emulsions. Brom-chloride emulsions contain more silver bromide than silver chloride. They are faster than chlor-bromide emulsions and used for projection printing where black images and speed printing are desired. Image tones of brom-chloride emulsions are not as warm as chlor-bromide images nor as cold as bromide images.

The preparation of commercial emulsions are trade secrets but the basic procedures are known:

A portion of the gelatin in the formula is swelled by soaking in water and later dissolved with heat. Mixtures of soluble bromides and iodides, or chlorides, are placed in water solution and added to the gelatin solution. Precipitation of silver halides is accomplished by slowly adding a solution of silver nitrate, while stirring, to the mixture. The relative concentration of the solutions, the rate of addition and temperature during mixing, are factors which control the formation, size and dispersion of the crystals in gelatin. The emulsion is then heated or "ripened" at 40–80° C. to recrystallize the silver halides and readjust the size-frequency distribution. Following ripening, more gelatin is added and the emulsion is chilled so it will set quickly. The emulsion is then placed in a press and forced through a screen to break it into shreds or noodles, which are washed, in cold running water to remove the potassium nitrate formed, the excess soluble halides, and certain soluble by-products of the reaction. Chloride emulsions are often prepared without washing or with only a limited washing. After washing, the emulsion is drained, remelted, and additional gelatin and certain agents, such as fog preventatives, are added. The emulsion is then heated, or "after-ripened," to form sensitizing nuclei on the silver halide crystals. This operation increases the sensitivity and contrast of the emulsion and is necessary for the preparation of high-speed negative emulsions. Certain preservatives, or stabilizers, are added so the emulsion can be stored in refrigerated rooms until needed. Before coating the emulsion is melted and sensitizing dyes, hardening agents, wetting agents, etc., are added. After thorough mixing, filtering and heating to coating temperature, it is placed in a coating machine. Supports, as film, paper, or glass, with substratum coatings are fed through machines at proper rates so they become coated with emulsions in uniform layers of desired thickness. The coated supports pass over chill boxes to set the emulsion and then through a series of drying compartments where the rate of drying is carefully controlled so as not to change the sensitivity on the surface. Following drying, the coat-

ings are inspected under proper safelights and the film or paper is cut to desired size and packaged. (S.M.T.)

PHOTOGRAPHIC PRINTING PROCESSES.

Photographic printing processes may be classified, in accordance with the substance comprising the image, as follows:

1. Silver.
    a. Printing-out processes: gelatino-chloride, collodio-chloride.
    b. Developing-out processes: silver chloride, silver bromide, silver chloro-bromide.
2. Iron.
    a. Ferroprussiate (blue print).
    b. Cyanotype (positive blue print).
3. Iron and silver.
    a. Kallitype.
    b. Vandyke.
4. Platinum (platinotype).
5. Palladium.
6. Pigment processes.
    a. Carbon and carbro.
    b. Gum-bichromate.
    c. Fresson, Sury's process.
    d. Glue-print, Resinopigmentype.
    e. Oil-pigment processes (oil, bromoil).
7. Dye processes.
    a. Diazo- dye processes (ozalid).
    b. Dye-mordanting processes.
    c. Dye-bleaching processes.
    d. Dye-development processes.

All of these processes may be used for positive prints on paper, or other supports, to be viewed by reflected light or on glass, film, paper and other transparent supports for viewing by transmitted light.

Silver **printing-out** papers are almost obsolete, having been replaced by developing-out papers.

The only important process employing light-sensitive iron salts is ferro-prussiate (blue print) still widely used in the reproduction of mechanical plans, architect's drawings, etc., but being slowly replaced even in this field by processes employing diazo- dyes (ozalid). The **Vandyke** (silver-iron process) is used for the same purpose and for copies of tracings from which blue prints are to be made. Kallitype, a silver-iron process, is now obsolete.

The platinum and palladium processes employ iron salts as the primary light-sensitive material and from the standpoint of photochemistry should be considered as platinum-iron and palladium-iron processes. Both are now practically obsolete.

Of the various pigment processes listed only carbro is important, being in general use for the making of color prints from color-separation negatives. Bromoil is still used by a few of the older pictorialists for exhibition prints. The other processes mentioned are now obsolete.

Printing processes employing diazo- dyes have come into general use within the last decade and are now serious competitors to the older blue print for copies of tracings, plans, etc.

Dye-mordanting processes have been used principally for the color images required in making color prints from color-separation negatives. In recent years investigation along these lines has been dropped in favor of dye transfer (**imbibition**) processes, e.g., wash-off relief.

The other dye processes listed are confined to color photography, and there only to a limited degree. (C.B.N.)

PHOTOGRAPHIC REDUCTION.

In photography, the term reduction is commonly applied to the processes by which the density of the image is lessened or reduced. Since photographic reduction is not chemically a reduction process, the terms (weakening) used in French and German literature are less open to confusion.

Photographic reducers may be classified as follows:

1. Subtractive. Reducing processes of this type remove the same amount of silver from every density. Thus, while the average density of the image is decreased, the contrast is increased.

2. Proportional. With proportional reducers, the amount of Ag removed from each density is approximately proportional to the original density. In this case, the over-all density is lessened without greatly changing the contrast of the image.

3. Superproportional. With superproportional reducers, the amount of Ag removed from the higher densities is relatively greater than that removed from the lower densities. Thus, the density of the highlights is lessened to a greater degree than the shadow portions of the image.

Photographic reducers are substances capable of converting metallic silver into a compound which is soluble in water, an acid, or other suitable solvent. A typical example is the ferricyanide-hypo reducer, commonly termed Farmer's reducer after its originator Howard Farmer (1883).

(a) $4Ag + 4K_3Fe(CN)_6 = Ag_4(CN)_6 + 3K_4Fe(CN)_6$,

(b) $3Ag_4Fe(CN)_6 + 16Na_2SO_3$

$$= 4Na_5Ag_3(S_2O_3)_4 + 3Na_4Fe(CN)_6.$$

In other words, the silver of the image is oxidized by the ferricyanide, forming silver ferrocyanide which is dissolved by the hypo (sodium thiosulfate). (See also **Superproportional Reducers, Proportional Reducers, Photographic Subtractive Reducers**.) (C.B.N.)

PHOTOGRAPHIC RELIEF ETCHING.
A method of obtaining a dye-absorbing gelatin (or other colloid)-relief image for printing by color photography. The emulsion containing the silver image is treated with a solution which releases nascent oxygen. This oxygen attacks the colloid surrounding the finely divided silver grains and eventually "dissolves" the gelatin. If the silver image is a negative, a positive gelatin-relief image remains. Such a relief image may be dyed for transfer by imbibition. (H.C.C.)

PHOTOGRAPHIC SUBTRACTIVE REDUCERS.
The majority of photographic reducers are of the subtractive type. Some of the more important subtractive reducers are:

Ferricyanide—hypo.
Ferricyanide—thiocyanate.
Ferricyanide—cyanide.
Iodine—hypo.
Iodine—cyanide.
Mercuric chloride—cyanide.
Cupric sulfate—sodium chloride.
Potassium permanganate.
Ferric chloride—hypo.

Some of these are highly poisonous and still others either soften or stain the emulsion. Practically either the ferricyanide-hypo or ferricyanide-thiocyanate reducers meet every requirement. The former is more widely used but the latter deserves to be better known as it is more stable and less likely to stain. (See also **Photographic Reduction** and **Proportional Reducers**.) (C.B.N.)

PHOTOGRAPHY.
The word photography is derived from the Greek roots *photos* (light) and *graphos* (to draw). It is usually attributed to Herschel, although there is some doubt that he was the first to use it publicly. No earlier use of the word photography, however, can be found than in a letter from Herschel to Fox-Talbot dated January 17, 1839. His first published paper using the word photography appeared on March 14 the same year.

Modern photography includes all useful processes for the production of images through the action of radiant energy. Thus all processes employing materials sensitive to light, or to other forms of radiant energy such as the ultra-violet and infra-red, x-rays, the radiation of radium and other **radioactive substances**, are included in the scope of photography. These materials may or may not be used in a camera; the making of a blue print, for example, is a form of photography but does not involve the use of a camera. (C.B.N.)

PHOTOLITHOGRAPHY.
Photomechanical Reproduction Processes.

PHOTOMECHANICAL REPRODUCTION PROCESSES.
These processes include photoengraving, photolithography, photogravure and all photographic processes employed in the reproduction of photographs, drawings, paintings or type mechanically by the printing press.

Printing is usually divided into three classes: (1) relief printing, also called letterpress printing, (2) intaglio printing, and (3) planographic printing. In relief printing, the ink is transferred from the *raised* portions of the printing surface as in printing from ordinary type. In intaglio printing, the ink contained in the *depressed* portions of the printing surface is transferred to the paper to form an image, while in planographic printing the ink adheres only to certain parts of the printing surface and is repelled by the rest. Thus, in printing the ink which adheres to the plate is transferred to the paper to form an image.

Photoengraving. The process employed in the reproduction of photographs, drawings and the like by letter press printing is termed photoengraving. A brief description of the photoengraving process follows. The first step is to make a negative of the photograph, drawing or painting to be reproduced. If the subject is a photograph, or a subject with many different tones from light to dark, a *half-tone screen* is placed directly in front of the film, or plate, on which the photographic copy negative is made. This screen consists of two series of parallel lines at an angle of about 90° to each other, the clear spaces between the lines being approximately the same width as the opaque lines. The number of lines varies in different screens from about 60 to 150 per in., the coarser being used for newspapers, the finer for books and other printed matter employing smooth- or glossy-surfaced papers. The half-tone screen results in a dot image—the highlights being represented by large dots, the shadows by small dots. The accompanying illustration, which represents a portion of a half-tone negative enlarged, shows how the size of the dots vary with the tones of the subject. If this reproduction is examined from a distance, the dots are no longer seen individually but merge to form a continuous image.

Portion of half-tone negative enlarged.

The wet **collodion process** is still widely used in making the half-tone negative but is being superseded by sensitized glass plates and film made especially for the purpose.

To make a printing block for letterpress printing from the half-tone negative, a copper or zinc plate is first cleaned and then coated with fish glue which has been sensitized with ammonium bichromate. When this is dry, the sensitized metal plate is exposed beneath the half-tone negative to a powerful arc light. This renders the exposed portions of the glue insoluble in water; hence when the metal plate is washed in warm water, the unexposed (soluble) glue is washed away leaving parts of the surface bare. The next step is to etch the surface of the metal so as to form an image in relief. When the plate is of copper, ferric chloride is used for etching; while nitric acid is generally used for etching zinc plates. The exposed parts of the metal are attacked by the etching solution and a slight relief is built up. If the process is allowed to continue, however, the etching solution will begin to attack the copper lying beneath the resist. To prevent this, the plate is removed from the etching solution, washed in water and dried, after which a resin, known as Dragon's Blood, is applied to the edges of the relief image and finally this is "burned in" by holding the plate over a gas flame. The plate is now immersed a second time in the etching bath and the process continued until it is necessary to apply the resin a second time. This operation of dusting with the resin and etching may be repeated several times before the relief is sufficiently high to produce a good impression on the press. If part of the image should require deeper etching than the rest, asphaltum paint is applied with a brush to protect the portions which do not require further etching and the plate placed again in the etching bath. This operation of local etching is known as staging. When the work is completed, the asphaltum is removed with suitable solvents.

Zinc, rather than copper, is often employed for line and coarse-screen plates. The procedure is much the same as with copper except that the zinc plate is coated with shellac which has been sensitized with ammonium bichromate. After exposure beneath the negative the unexposed portions of the shellac coating are removed with alcohol leaving the exposed portions to form a resist for the etching process. Dilute nitric acid is used in etching.

Photolithography. Photolithography and offset (offset lithography) are alike except that in photolithography the ink is transferred directly from the plate to the paper while in offset the ink is first transferred from the plate to a rubber roller (blanket) and from this to the paper.

For photolithography and offset a half-tone negative is first made in essentially the same way as for photoengraving. This negative is printed on a zinc or aluminum plate sensitized with a solution of egg albumen and ammonium bichromate. After exposure beneath the negative, the plate is covered with a greasy ink and then washed in warm water to remove the soluble albumen, leaving on the plate an image of hardened albumen and the adhering ink.

To print directly from such a plate, it is first moistened with dampening rollers. The water is taken up by the clear areas but is repelled by the greasy ink which is retained by the hardened albumen. Thus when the water-charged plate is brought into contact with the ink rollers, the hardened albumen accepts the ink but the metal repels it, so that the distribution of ink on the plate reproduces the tone gradations of the image as recorded in the negative. Then when the plate is brought into contact with the paper in the printing press, the ink is transferred to the paper.

In offset (offset lithography) the inked image is first transferred to a rubber roller or "blanket" and then to the printing paper. The offset process enables prints to be made on coarse, rough papers because a more complete transfer of ink can be obtained on such papers from a pliable rubber surface than the harder surface of a metal plate. A further advantage of offset printing is that a larger number of copies can be made since contact with the rubber blanket is less damaging to the image on the metal plate than direct transfer to paper. Offset lithography has spread rapidly in recent years and its use is still on the increase.

For large runs—25,000 copies or more—the so-called deep-etch process is employed. Despite the term "deep etch" the degree of etching is so slight that the process is still properly classified as planographic printing. In the deep-etch process the plate is exposed beneath a positive rather than a negative. After exposure, the plate is washed in warm water to remove the soluble albumen and the exposed parts of the metal plate are lightly etched—grained would perhaps be a more descriptive term—with a weak acid solution. It is then rinsed in water, dried and inked. The hardened albumen is removed by scrubbing the plate with a stiff brush after which the plate is coated with a gum solution. This is retained by the bare parts of the metal (where the hardened albumen had been) but is repelled by the etched areas which are filled with the greasy ink. To print from such a plate, it is first moistened by dampening rollers. The water is absorbed by the areas coated with the gum solution, and repelled by the etched areas which are filled with an ink. Thus the etched portions accept the ink from the inking rollers on the printing press while the water in the gum-coated portions repel it to form an image in which the variations in the quantity of ink correspond with the densities of the positive from which the plate was made.

Gravure. The gravure, or intaglio, processes include photogravure and rotogravure, the latter familiar to most as the process used for the illustrated supplement of Sunday newspapers. Photogravure and rotogravure differ only in minor details. Grained plates are employed in photogravure and the plates printed on a flatbed press while a gravure screen is used in rotogravure and the image is engraved on large copper cylinders from which prints are made in a cylinder (rotary) press. Both photogravure and rotogravure, however, are continuous-tone processes; the screen used does not break up the image into a dot formation as does the halftone screen. Both processes differ from photoengraving and photolithography in requiring a positive rather than a negative. The positive is made directly from the negative (without a screen of any kind) on a slow gravure film which usually has a matt surface to facilitate local retouching. In use, it is exposed and developed much like any other slow, blue-sensitive film emulsion. This positive is next printed on a sheet of gravure tissue which is a gelatin-coated paper sensitized with ammonium bichromate. The effect of exposure beneath the positive is to render the gelatin insoluble to varying depths depending upon the densities of the positive. In other words, the depth of hardening is directly proportional to the amount of exposure under the different densities of the negative, being greatest in the highlight and least in the shadow portions of the subject.

After exposure beneath the positive, the gravure tissue is exposed again but this time beneath the gravure screen. The gravure screen consists of thin transparent lines at right angles to each other with large opaque rectangles between. The object in using the screen is to produce a pattern of fine lines in the image which will enable the wiping blades on the printing press to remove the ink from all but the depressed portions of the printing surface and thus maintain clear, ink-free highlights. The screen has little or no effect on tone rendering and is hardly noticeable.

After this second exposure, the carbon tissue is applied to the copper plate or cylinder so as to obtain intimate contact between the gelatin and the metal plate. Hot water is then poured over the back of the tissue to soften the soluble gelatin. When the gelatin has softened sufficiently, the paper backing is stripped, leaving the gelatin adhering to the metal. Further washing in warm water removes the soluble gelatin, leaving an image in hardened gelatin firmly attached to the surface of the copper.

The plate, or cylinder, is now ready for etching. Ferric chloride is used for this purpose. The action of the etching solution varies with the gelatin image, being directly proportional to its thickness so that the variation of tone from dark to light in the subject is represented by proportional differences in the depth of the etched portions.

In printing from such a plate, the ink taken up from the inking rollers varies with the depth of etching, so that the shadow portions of the image contain the most ink, the half-tones less and the highlights still less. The ink is removed from all but the depressed (etched) areas by wiping or "doctor" blades. Then, on contact with the paper, the ink remaining on the plate is transferred to the paper to form the image. (C.B.N.)

PHOTOMETRY. The measurement of luminous intensity, of luminous flux density, or of **illumination** is known as photometry. The intensity of a light source may be expressed in candles or other arbitrarily defined source-units (see **Candle Power**), while luminous flux density and illumination are expressed in lumens per unit area of cross-section or of surface. A "lumen" is the amount of light or luminous flux received upon a unit surface, all points of which are at a unit distance from a concentrated source of one spherical candle intensity.

Photometers are of many types. Those used for flux-density and candle-power measurement are ordinarily designed to compare the unknown with a known source by balancing in some way the flux densities from the two sources. The most common representatives of this type are the various forms of **bench photometer,** of **wedge photometer,** of **polarization photometer,** and of **integrating photometer.** An important aspect of light-source photometry is the study of the distribution of luminous intensity in different directions,—a variable which the integration photometer is designed to average. Direct indications of luminous flux density or of illumination are afforded by photometers utilizing the **photoelectric cell,** the **selenium cell,** or the photronic cell (see **Photovoltaic Effects**). **Spectral energy distribution** is analyzed by means of various types of **spectrophotometer.**

Since the energy of **radiation** is not at all equally stimulating to the optic nerve, we must recognize two different measures of its intensity: (1) the luminous flux density, in lumens per sq. cm. of cross-section, corresponding to the visual sensation evoked, and (2) the actual flow of power, in watts, per sq. cm., called the radiant flux density. The ratio of the one to the other for any wavelength is the "visibility factor" for that wavelength, while for the whole of any emission (all wavelengths) the corresponding ratio is called the "luminous efficiency" of the emission. The efficiency of a light source is expressed in lumens of visible output per watt of input power. For example, 10 lumens per watt would be typical for modern incandescent lamps such as are in domestic use.

The problem of determination of the brightness of the stars and other objects external to the earth will be found discussed under the heading of **Stellar Photometry.** (L.D.W.)

PHOTOMICROGRAPHY. The photography of objects with the microscope. It should be distinguished

from microphotography which is the term applied to production of small images. (See **Microphotography; Metallography.**)

The apparatus for photomicrography consists of a light source, the microscope and a camera (Fig. 1) all

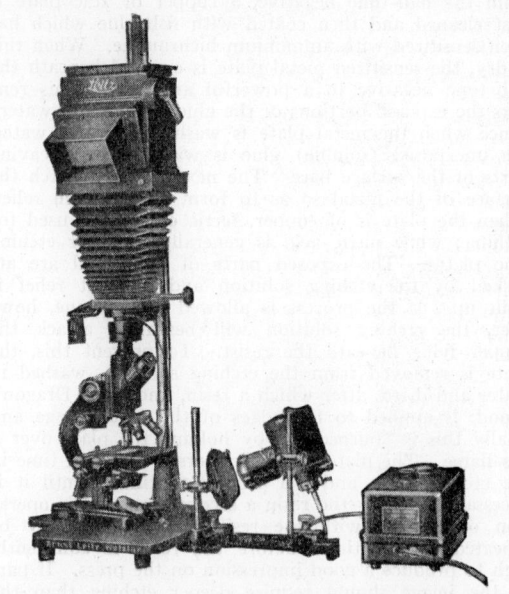

Fig. 1. Leitz vertical photomicrographic camera, with microscope and lamp in position. (*Allen's Photo-Micrography.*)

mounted on a rigid base. The procedures usually employed in preparing specimens for microscopic examination are followed in the preparation of materials for photomicrography. Biological materials are cut into thin sections with a microtome, stained with dyes, if necessary, to disclose their structure, mounted on glass slides and photographed by *transmitted* light. Opaque objects, such as metals, alloys, stones, etc., are ground to a smooth, flat surface, polished, and etched chemically to reveal the structure and photographed by *reflected* light, using a special type of microscope frequently termed a *metalloscope.*

Usually, the advantage of photomicrography over visual examination with the microscope is that a permanent record is obtained. This permanent record is available for study and comparison with others at any time. This is particularly important in the study of changes occurring in an organism or the effect of stress, heat, or chemical treatment on a metal or an alloy. Used with ultra-violet radiation, photomicrography almost doubles the resolving power of the microscope. The increased resolving power is of great value in the study of colloidal particles, the mechanism of filtration and adsorption, pigments and filters, etc. Ultra-violet photomicrography has the further advantage of enabling many organisms and organic structures to be photographed without the staining which is necessary for visual examination. Thus living tissue, which cannot be stained, may be photographed. With certain subjects, the possibility of using infra-red radiation is of special value.

Ciné-photomicrography, motion pictures with the microscope, is of particular value in studying the growth of organisms, movements of colloidal particles, chemical processes, etc., and is rapidly assuming considerable importance.

Photomicrographs in color can be made by most processes although the integral tripack processes (see **Color Photography**) are the most convenient. (C.B.N.)

PHOTON. Light.

PHOTOPERIODISM. This term is applied to the reaction of plants to the daily length of the period of illumination. It is one of the most noteworthy of the reactions of plants to an environmental factor. In most parts of the world marked seasonal variations occur in the length of the daylight period. In temperate zones the length of the daylight period varies from about 8 or 10 hours at the winter solstice to about 14–16 hours at the summer solstice. At higher latitudes the annual variation in day length is greater; at lower latitudes less. In arctic and sub-arctic regions the length of day varies from 24 hours on the longest summer days to zero hours on the "shortest" days; in the equatorial zone day lengths approximate 12 hours the year round.

Although the length of the photoperiod (number of hours of illumination per day) also has effects upon the vegetative development of plants its most significant influences are upon flowering and other phases of the reproductive development of plants. Plants fall into three fairly well defined categories: 1. "Long-day" species, which flower more or less readily in a range of photoperiods longer than a certain critical period, developing only vegetatively under shorter photoperiods. 2. "Short-day" species, which flower more or less readily in photoperiods shorter than a certain critical period, developing only vegetatively at all longer photoperiods. 3. "Indeterminate" species, which exhibit no critical photoperiod, developing both vegetatively and reproductively over a wide range of photoperiods. The length of the critical photoperiod differs according to species but for many plants of both the long-day and short-day types lies in the range of 12 to 14 hours. Examples of short-day species are dahlias, chrysanthemums, asters, cocklebur, and salvia; of long-day species, radish, beets, dill, spinach, lettuce, and grains; of indeterminate species, tomato, cotton, sunflower and buckwheat. Both long-day and short-day varieties may exist even within the same species. Some varieties of soybeans, for example, are short-day plants, while others are long-day plants.

In temperate regions the season of blooming of a plant is largely determined by its photoperiodic reaction. In general short-day plants bloom in the early spring or early fall; long-day plants in the late spring or summer. The geographical distribution of some kinds of plants is at least partly controlled by their photoperiodic reaction. A species cannot maintain itself in a climate in which it is impossible for the cycle of reproductive processes to be completed. Pronounced long-day species, for example, would not as a rule be found in tropical regions.

The leaves are the loci of the reactions leading to the initiation of flower development in plants. If the leaves of short-day Biloxi soybean plants are exposed to long days while the meristems at which flowers are initiated are exposed to short days, no development of flowers takes place. If, however, the leaves are exposed to short photoperiods while the meristems are exposed to long photoperiods, differentiation of flowers soon starts. In like manner, if only the leaves of a long-day dill plant are exposed to photoperiods of suitable length, development of flowers begins very shortly. The mechanism of the photoperiodic reaction is clearly such that the influences of certain processes which occur in the leaves must be transmitted to the meristems, inducing the initiation of flower development. It seems highly probable, therefore, that a hormone-like substance, which is synthesized in the leaves only under suitable photoperiodic conditions, is translocated to the meristems, inducing differentiation of floral parts. The name "florigen" has been suggested for this hypothetical flower-inducing hormone.

If a vegetative short-day plant is transferred from a long day to a short day, exposed to a suitable number of short-day cycles and then returned to a long day, flowers will develop even under long-day conditions.

This phenomenon is called "photoperiodic induction" and occurs in long-day as well as in short-day species. The number of suitable cycles required for photoperiodic induction varies according to species.

A fundamental physiological difference between short-day plants and long-day or indeterminate plants is that the former require a cyclic alternation of light and dark periods if flowers are to develop, while the latter do not. In the short-day Biloxi soybean, for example, flowering is induced only if the plants are exposed to a suitable number of dark periods, each more than 10 hours in length, alternating with light periods of not less than 2 nor more than 20 hours, and optimally about 11 hours in length. Furthermore the dark period must proceed without interruption for at least a certain minimum length of time. For example, in the cocklebur a dark period of sufficient length to induce flower initiation becomes completely ineffective if interrupted at its midpoint by as little as one minute of light of even a relatively low intensity. Long-day and indeterminate plants do not require an alternating cycle of light and dark periods. Most if not all such species will bloom under continuous illumination. Long-day species do, however, require a light period of minimum duration out of each 24-hour day in order for flowers to develop. For dill, a good example of a long-day species, the minimum length of photoperiod which will induce flower differentiation is between 11 and 14 hours. Indeterminate species seem to differ from long-day species chiefly in the fact that they can bloom under shorter photoperiods than the latter.

Practical applications of the principles of photoperiodism have been made in the growing of floricultural greenhouse crops. Short-day species such as chrysanthemums, can be brought into bloom earlier in the fall by decreasing the length of their daily exposure to light. Likewise the time required for long-day floricultural species to attain the flowering stage during the winter months can be shortened by increasing the day length with artificial illumination. (B.S.M.)

PHOTOPHORESIS. Very fine solid or liquid particles suspended in a gas or falling through a vacuum are sometimes given a unidirectional motion by a strong beam of light, and the motion is called photophoresis. If the particles move with the light, the effect is said to be positive; if against the light, it is called negative. Positive photophoresis is usually ascribed to radiation pressure, or regarded as the resultant of radiation pressure and the Crookes radiometer effect of the surrounding gas. Negative photophoresis, observed in fine suspended particles of sulfur or selenium by Ehrenhaft (1917), is thought by some to be due primarily to the radiometer effect. If there is present also a strong electric or magnetic field, the motion may have a component in the direction of the field. (L.D.W.)

PHOTORECEPTOR. A sensory organ which responds to the stimulus of light waves. Eyes are the most familiar organs of this kind but many 1-celled animals are sensitive to light and some of the more complex forms have the surface of the body sensitive to it. True eyes serve for the formation of a visual image whereas the more simple photoreceptors merely indicate the luminosity of the animal's surroundings. In some cases adjustment to light is necessary and the simple light-sensitive organ is adequate for the initial step in the animal's orientation. Eyes serve the very different purpose of enabling the animal to perceive objects about it and are not primarily associated with its adjustment to light. Man depends to an extreme degree on his vision but his skin is sensitive to light in an entirely different way which becomes evident only in the degree of pigmentation. (A.W.L.)

PHOTOSPHERE. The intensely bright portion of the sun which is visible to the unaided eye is known as the photosphere. In reality it is a layer not more than a few hundred miles in thickness which marks the boundary between the dense interior gases of the sun and the cooler, more attenuated gases which go to make up the solar atmosphere. The photosphere radiates with a continuous **spectrum** and the application of the **laws of radiation** indicates that its temperature is about 5750° K.

In appearance the photosphere is brilliantly white, somewhat brighter at the center than at the limb, and is distinctly granular in character. These granules are in reality very large, as much as several hundred miles in diameter, and are relatively short-lived, photographs taken at intervals of less than a minute showing distinct changes. Larger irregular bright areas known as faculae may be frequently seen. Spectroscopic analysis of the light from the faculae indicates that they are masses of heated gases which are rising out through the photosphere to the atmosphere.

Spectroheliographic studies of the photosphere show the presence of areas of hydrogen and calcium vapor, known as flocculi which are somewhat smaller than the faculae and larger than the "granules."

The **sun spots** are relatively dark areas which appear on the photosphere. (w.k.g.)

PHOTOSTAT. Drawing Reproduction.

PHOTOSYNTHESIS. It has long been known that the green parts of plants, when exposed to light under suitable conditions of temperature and water supply, use carbon dioxide from the atmosphere and release oxygen to it. These gaseous exchanges are the opposite of those which occur in **respiration** and are the external manifestation of the process of photosynthesis. In this process hexose **carbohydrates** are synthesized from carbon dioxide and water by the chloroplasts of living plant cells in the presence of light, oxygen being a by-product of the reaction. For each molecule of carbon dioxide used one molecule of oxygen is released. The summary chemical equation for photosynthesis is:

$$6CO_2 + 6H_2O + 673 \text{ kg.-cal.} \rightarrow C_6H_{12}O_6 + 6O_2.$$

As a result of this process radiant energy of sunlight is stored up as chemical energy in the molecules of carbohydrates and other compounds which are derived from them.

In the vascular plants photosynthesis occurs chiefly in the leaves. Carbon dioxide diffuses into the intercellular spaces of the leaf from the atmosphere chiefly through the **stomates**, and then dissolves in the moist walls of the mesophyll cells. The carbon dioxide then diffuses, in solution, to the surface of the **chloroplasts**, which are the actual seat of the process. In most plants the water used in photosynthesis is absorbed by the roots from the soil whence it is translocated to the leaves (see **Ascent of Sap**). In most plants only a very small fraction of the water which reaches the leaves is used in photosynthesis, the bulk of it being lost from the plant in **transpiration**. Except for a usually small portion used in respiration the oxygen set free in the process diffuses out of the leaf into the atmosphere, mostly through the stomates.

Other carbohydrates besides hexoses are synthesized in the leaves, apparently as a result of secondary reactions which follow photosynthesis. Sucrose invariably accumulates in actively photosynthesizing leaf cells. This more complex sugar is built up from the molecules of the simpler hexoses. In most plants insoluble starch also accumulates in leaf cells during photosynthesis. This carbohydrate is synthesized by the condensation of numerous glucose molecules. The sucrose and starch content of leaves decreases during the night as a result of continued **translocation** out of the leaves into other parts of the plant. The sucrose is probably translocated as such, but the starch, being insoluble, must first be converted into simpler, soluble sugars before it can move out of the leaves. The synthesis of starch is not restricted to the green parts of plants but also occurs in many non-green parts as well. A familiar example is the accumulation of starch in potato tubers. The starch in the non-green cells is made out of glucose which comes from the leaves or other photosynthetic organs. Starch occurs in cells in the form of small grains, the type of grain being formed in each kind of plant being more or less characteristic of that species.

Photosynthesis takes place in chlorophyll-containing cells only when carbon dioxide, water and light are available, and when a suitable temperature prevails. Although carbon dioxide constitutes, on the average, only 0.03% of the atmosphere, land plants are entirely dependent upon this source for the carbon dioxide used in photosynthesis. It can be shown experimentally that increasing the carbon dioxide concentration of the atmosphere results, under otherwise suitable conditions, in an increased rate of photosynthesis. If the concentration of this gas in the atmosphere were higher than it is, many plants, much of the time, would photosynthesize at a more rapid rate than they do. In nature sunlight is the source of radiant energy used in photosynthesis although plants will also photosynthesize under artificial light sources of suitable quality and intensity. The intensity, quality, and daily duration of illumination all have an influence on the amount of photosynthesis accomplished per day. The minimum light intensity at which a measurable rate of photosynthesis occurs varies according to species, but is seldom less than 1% of full mid-day summer sunlight. Under natural conditions maximum rates of photosynthesis are attained in single leaves of many species at 25–35% of full sunlight intensity and in some shade species at even lower intensities. The rate of photosynthesis for an entire tree or for a field of grain usually increases with increase in sunlight intensity up to the maximum possible value, which at noon on a clear summer's day in temperate regions is about 1.3 gram-calories per cm.^{2} per min. This relation holds because in a tree or plot of vegetation many of the leaves are shaded by other leaves and hence the higher the incident light intensity the higher the average intensity actually reaching the surface of the leaves. For equal intensities more photosynthesis appears to occur in the orange-short red and blue parts of the spectrum than in the green and yellow. In general the longer the daily period of illumination the more photosynthesis will be accomplished by a plant in the course of a day. A deficiency of water results in a reduced rate of photosynthesis. One of the detrimental effects of prolonged drought upon plants is the marked resulting reduction in the amount of sugar manufactured in photosynthesis. The range of temperatures most suitable for relatively rapid rates of photosynthesis is not the same for all kinds of plants. In general it is higher in tropical than in temperate species, and higher in temperate species than in those of sub-arctic regions. Most photosynthesis in temperate zone plants occurs within the range of 10–35° C. Increase in temperature results in an increase in the rate of photosynthesis up to an optimum which is not the same for all kinds of plants, but which for most temperate zone species lies within the range of 20–30° C. With increase in temperature above the optimum the rate of photosynthesis decreases progressively.

Photosynthesis is the most important of all biological processes. The existence of the entire biological world of plants and animals, with negligible exceptions, hinges upon this process. From a few simple inorganic compounds and from the sugar made in photosynthesis are built up all of the complex kinds of molecules essential in the construction of the bodies of plants and animals or to the maintenance of their existence. Some of these subsequent synthetic processes occur in the plant body; others in the bodies of animals after they have ingested plant materials as foods. Likewise the energy used by plants and animals all represents sunlight energy which was entrapped in sugar molecules during photosynthesis. The entire organic world runs by the gradual expenditure of the energy capital accumulated in photosynthesis.

The magnitude of the photosynthetic process is a question of great biological significance. Corn plants yielding 100 bushels to the acre synthesize about 8700 kg. (nearly 10 tons) of sugar in a growing season. About 33 million kilogram-calories of radiant energy must be transformed into chemical energy in order to accomplish the synthesis of this quantity of sugar. The efficiency of the process is relatively low, however. Only about 1.6% of the radiant energy falling on the acre during the growing season is actually utilized in photosynthesis. The annual world production of sugar by terrestrial plants is estimated at 4×10^{13} kilograms; this represents the transformation of 1.6×10^{17} kg.-cal. of radiant energy into chemical energy. The annual production of sugar by marine plants, mostly **diatoms**, is estimated to be at least twice as great as the production of terrestrial plants and may be considerably greater. (B.S.M.)

PHOTOTROPISM. Movement in Plants; Tropism.

PHOTOTUBE. This is a photosensitive or light-actuated electron **tube**, often popularly referred to as an electric eye. Basically the tube consists of a **cathode** which is photosensitive, i.e., it will emit electrons when illuminated, and an **anode** for collecting the **electrons** emitted by the cathode. There are several materials which are suitable for use as the cathode, some being better suited for a given range of the spectrum than others so the choice is often on the basis of the color light to be used. A widely used cathode for general purposes is a composite of silver which has been coated with caesium and then subjected to an oxygen treatment so the surface is a mixture of silver, caesium and caesium and silver oxides. This surface is one of the most sensitive and at the same time is very sensitive to visible and infra-red light. Other tubes are constructed with **cathodes** which are sensitive to ultra-violet radiation. The exact structure of the tubes varies, some being made by coating the inside of the glass tube with sensitive material and having a center wire for an anode, others by having a semi-cylindrical, a W- or a V-shaped cathode base coated with the sensitive material and a vertical wire located to allow easy collection of the electrons with minimum interference with the light. Since the electrons emitted by the cathode must be drawn to the anode an external source of voltage is required for the operation of the tubes. This voltage normally ranges from a few volts to approximately 100 volts and while usually d.c. it may be a.c., the tube functioning only when the anode is positive. The currents obtained are small, only a few microamperes, but the tube is admirably adapted to vacuum tube amplifier circuits. A typical circuit is shown. The electrons

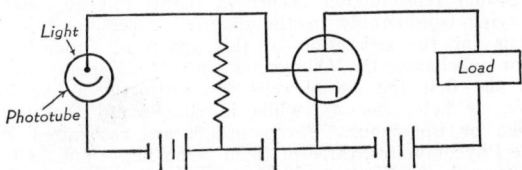

emitted by the action of light cause a current to flow around the phototube loop and thus produce a voltage drop in the resistor in the grid circuit of the vacuum tube. This grid voltage in turn causes a variation in the plate current and for a linear circuit this current will be an amplified image of the phototube current. Gas at a low pressure is often put in the tube to amplify the currents by **ionization** of the gas. When this is done care is necessary to keep the gas from breaking

into a self-maintaining discharge. The intense ionization of a self-maintaining discharge would produce excessive bombardment of the cathode and soon destroy the sensitized surface. The gas-tube currents may also be amplified by vacuum tubes. Phototubes are used in innumerable industrial control applications where interruption of a beam of light is all that is necessary to bring about the desired operation. They are also the basis of the reproducing system of the sound-on-film talking picture systems. (See also **Facsimile.**) (L.R.Q.)

PHOTOVOLTAIC EFFECTS. Several possibly related but very imperfectly understood phenomena are classified under this term. They are characterized by the creation of an **electromotive force** through the incidence of **light.** Two cases only will be described.

The first, discovered by Becquerel and known as the Becquerel effect, employed an electrolytic cell, for example, one with silver electrodes coated with silver iodide and immersed in dilute sulfuric acid. The electrodes being exactly alike, nothing happened so long as the cell was in the dark or even when it was uniformly illuminated. But if one electrode was illuminated more than the other, the cell set up an electromotive force.

A recent type of photovoltaic cell, developed by Lange and much studied, has for one electrode a copper plate, covered by a thin layer of its own oxide (a semiconductor). This "blocking layer" of oxide is, in turn, sputtered with a film of conducting metal, such as aluminum, so thin that it is nearly transparent, but still capable of acting as an electrode. Such an assemblage, in the dark, acts as a **rectifier.** But when light falls on the copper oxide layer, the cell also produces an electromotive force of its own, and if placed in a circuit, causes a current nearly proportional to the illumination. The uses of this device promise to be many. For example, a form having a **selenium** film and known commercially as the "photronic cell," coupled with a sensitive milliammeter, is now used as a direct reading **illumination photometer.** This instrument is familiar to photographers in the rôle of an exposure meter. (L.D.W.)

PHOTRONIC CELL. Photovoltaic Effects.

PHREATIC. The term proposed by Daubree in 1887 for the waters of the **ground water** reservoir, as distinct from the underground waters above the water table, called **vadose.** (R.M.F.)

PHRENIC NERVES. The nerves which control the movement of the **diaphragm.** (D.M.H.)

PHTHALEINS. Phthalic Acid.

PHTHALIC ACID, PHTHALATES, AND PHTHALEINS. Ortho-phthalic acid, benzene-ortho-dicarboxylic acid $(C_6H_4(COOH)_2(1,2))$ is a white solid, melting point 191° C., decomposes, insoluble in cold but soluble in hot water, when heated loses water to form

phthalic anhydride $\left(C_6H_4 \begin{array}{c} CO \\ \diagdown \\ CO \end{array} O \right)$, a white solid, melting point 128° C., boiling point 285° C. Reacts with **phosphorus** pentachloride to form phthalyl chloride $\left(C_6H_4 \begin{array}{c} COCl \\ \diagdown \\ COCl \end{array} \right)$, which further reacts with aluminum chloride to form unsymmetrical phthalyl chloride $\left(C_6H_4 \begin{array}{c} CCl_2 \\ \diagdown \\ CO \end{array} O \right)$. Both chlorides react (1) with **zinc** plus **acetic acid** to form unsymmetrical phthalide $\left(C_6H_4 \begin{array}{c} CH_2 \\ \diagdown \\ CO \end{array} O \right)$, and (2) with **benzene** plus **aluminum** chloride to form unsymmetrical-diphenylphthalide phthalo-

phenone $\left(C_6H_4 \underset{CO}{\overset{C(C_6H_5)_2}{\diagup}} O \right)$.

Phthalic anhydride reacts (1) with **ammonia** to form

phthalimide $\left(C_6H_4 \underset{CO}{\overset{CO}{\diagup}} NH \right)$, white solid, melting

point 238° C., which latter compound reacts (a) with **potassium** hydroxide in alcohol to form potassium phthalimide $\left(C_6H_4 \underset{CO}{\overset{CO}{\diagup}} NK \right)$. When potassium phthalimide is treated with an alkyl halide, e.g., ethyl iodide, ethyl phthalimide $\left(C_6H_4 \underset{CO}{\overset{CO}{\diagup}} N \cdot C_2H_5 \right)$ is formed, from which the primary **amine**, ethyl amine ($C_2H_5NH_2$) may be obtained (plus phthalic acid) by heating with fuming **hydrochloric acid** (Gabriel's synthesis for primary amines), (b) with sodium hypochlorite, to form sodium anthranilate $\left(C_6H_4 \underset{COONa}{\overset{NH_2}{\diagdown}} \right)$, which yields, upon treatment with an acid, anthranilic acid $\left(C_6H_4 \underset{COOH}{\overset{NH_2}{\diagdown}} \right)$; (2) with phenol to form phthaleins, e.g., (a) **phenol** plus phthalic anhydride at 120°, in the presence of sulfuric acid concentrated, forms

phenolphthalein $\left(C_6H_4 \underset{CO}{\overset{C \diagdown^{C_6H_4 \cdot OH(4)}_{C_6H_4 \cdot OH(4)}}{\diagup}} O \right)$, white solid,

melting point 261° C., soluble in alkali to form a red solution, and used as an indicator to determine the neutral point in **titrating** acidic or basic solutions, (b) resorcinol plus phthalic anhydride, similarly, forms resorcinolphtha-

lein, fluorescein $\left(C_6H_4 \underset{CO}{\overset{C \diagdown^{C_6H_3 \underset{O(2)}{\overset{OH(4)}{}}}_{C_6H_3 \underset{O}{\overset{OH(4)}{}}}}{\diagup}} \right)$ dark yellow

solid, decomposes above 290° C., dissolves in alkali to yellow-red solution, which exhibits green fluorescence when dilute. When fluorescein and **bromine** react tetra-bromo-

fluorescein $\left(C_6H_4 \underset{CO}{\overset{C \diagdown^{C_6HBr_2 \underset{O(2)}{\overset{OH(4)}{}}}_{C_6HBr_2 \underset{O}{\overset{OH(4)}{}}}}{\diagup}} \right)$ is formed, the

potassium salt of which is known as eosin, and used as a red dye for wool and silk, (c) N-diethyl-meta-aminophenol plus phthalic anhydride, similarly, forms N-diethyl-meta-amino-phenol phthalein, rhodamine

$\left(C_6H_4 \underset{CO}{\overset{C \diagdown^{C_6H_3 \underset{O(2)}{\overset{N(C_2H_5)_2(4)}{}}}_{C_6H_3 \underset{O}{\overset{N(C_2H_5)_2(4)}{}}}}{\diagup}} \right)$

a red dye.

Ortho-phthalic acid is made by oxidation of **naphthalene** (1) with **sulfuric acid** fuming heated, in the presence of mercuric sulfate, (**sulfur** dioxide also formed, and recovered), (2) with air in the presence of vanadium pentoxide at 450° to 520° C. Ortho-phthalic acid is also formed when **benzene** compounds, containing carbon ortho-substituted groups, are oxidized. Ortho-phthalic acid is used in the manufacture of **indigo** and other dyes. (R.K.S.)

PHTHISIS. Pulmonary **tuberculosis.** (D.M.H.)

PHYCOMYCETES. The Phycomycetes form a group of **fungi** so diverse in habit as to suggest **polyphyletic** origin, quite probably from several different groups

of green **algae,** to which many of them show remarkable similarity. Simpler members of the Phycomycetes consist of but a single cell, while other species have a well-developed branching **mycelium**, always composed of **hyphae** possessing no cross-walls. Many species grow in water and are known as water molds. Others grow out of water. Among the latter are some of the common destructive **parasites.**

In the water-inhabiting species reproduction by asexual **zoöspores** is common. These zoöspores are produced in **zoösporangia**, which are cut off by cross-walls from the ends of the otherwise non-septate

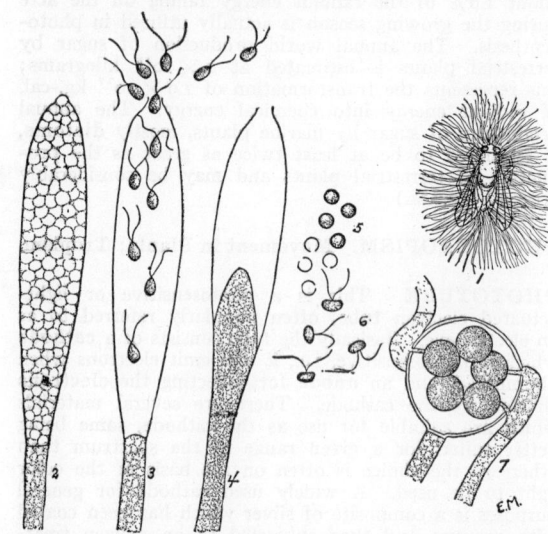

Saprolegnia. Stages in development. 1, fungus growing on a dead fly in water; 2, sporangium with newly formed spores; 3, sporangium discharging zoospores; 4, formation of new sporangium; 5, resting stage following first motile stage 3; 6, second motile stage following resting stage 5; 7, fertilization of eggs in oogonium.

mycelium. Each zoöspore possesses two apical cilia, by means of which it swims about for a time. After this it becomes quiet and secretes around itself a cell wall. From this it may escape as a second type of zoöspore possessing two laterally inserted cilia. After a period of quiescence and wall-formation this second zoöspore puts out a germ tube which develops into an extensive mycelium. The significance of the two kinds of zoöspores, which do not appear in all species of water molds, is not entirely understood. In the non-water inhabiting Phycomycetes asexual reproduction is by means of conidia, minute cells formed in chains or in large masses borne in various ways on the tips of special branches called conidiophores. The conidia are discharged in the air and are borne about by air-currents.

Sexual reproduction occurs in many Phycomycetes, varying considerably in the different orders, and is a basis for the separation of this group of fungi into two subclasses, the Oömycetes and the Zygomycetes. In the first the sexual cells are distinctly unlike in size, or heterogamous, while in the second they are alike or isogamous. Seven orders are recognized in the Phycomycetes, three of them being important. One of these is the Water-molds or Saprolegniales, the members of which are aquatic, and closely resemble certain algae, but are without **chlorophyll.** These fungi include **saprophytes** living on both animal and plant remains, and **parasites,** many of the latter being serious pests in aquaria and fish hatcheries, attacking both fish and eggs. Sexual reproduction is by means of **oögonia** and **antheridia,** although in many species antheridia seem to be functionless and in other species completely lacking. In such cases the oögonia develop without the aid of antheridia, such development being called parthenogenesis. The oögonia are rounded bodies cut off from the tips of hyphae. The antheridia, when present, appear as branches from the hypha which bears the oögonium. These branches grow up around the oögonium into which they develop lateral fertilization tubes penetrating the oögonial wall. When mature, the contents of the oögonium break up to form several spherical eggs, each of which is reached by a fertilization tube. Through this tube a nucleus passes from the antheridium to the egg, where it unites with the egg nucleus. Subsequently a thick wall is formed around each egg, which then passes through a prolonged resting period. Germination of this resting cell is by a germ tube.

The similarity between green algae such as *Vaucheria* species and the Water molds has been frequently noted, and suggests a relationship between them, and lends support to the idea that the Phycomycetes have developed from such green algae through the loss of their chlorophyll and the assumption of a saprophytic habit. According to another theory, the simpler members of the Phycomycetes fall into a line of evolution which starts from non-pigmented unicellular organisms called Flagellates and gradually gives rise to the whole series of Fungi today existing.

A second order of Phycomycetes is made up of parasites on higher plants, many of them being of great importance because of the injury they do to cultivated crops. This order is the Peronosporales, and contains fungi known as Blights and Downy Mildews. In this order the mycelium is composed of hyphae which penetrate between the cells of the host plant tissue, sending into those cells short absorbing tubes called haustoria. Asexual reproduction in this order is by means of conidia, which in some genera are borne in chains in extensive areas beneath the epidermis of the host plant, and liberated by the breaking of this epidermis, and in other genera by conidia borne on the tips of the branches of hyphae which extend outward above the epidermis of the host. These conidia are scattered by the wind and, if they reach a suitable host, give rise to several small laterally biciliate zoöspores; the latter form germ tubes which penetrate the tissues of the host plant, thus causing new infections. Sexual reproduction occurs in the deeper tissues of the infected plant and is very similar to that occurring in the Water-molds.

However, each oögonium develops only one egg. This egg is fertilized within the host tissues and remains there until the latter disintegrate. After this the egg germinates, giving rise to biciliate zoöspores or in some species producing a mycelium directly.

Many members of this order are of great importance, because of their parasitic nature. One of these is the Downy Mildew of Grapes, *Plasmopara viticola,* which was introduced from America into Europe where it became a grave menace to the grape vines. This unfortunate introduction, however, led to the discovery of a control, **Bordeaux mixture.** Originally this mixture was used as a spray to grapevines along the highway as a means of preventing pilfering. It was noticed that the plants so treated were not infected by the fungus, while others were. So the treatment was extended and led to a control of the fungus.

A second serious disease caused by one of the Peronosporales is Potato Blight, due to *Phytophthora infestans.* This fungus attacks the entire plant and in serious cases may completely kill all parts above ground. It appears as a white mold on the under surface of leaves. This whiteness is caused by the abundant coniodiophores. It is this disease which caused the great potato famine in Ireland in 1845. Frequent spraying with Bordeaux mixture is an effective control.

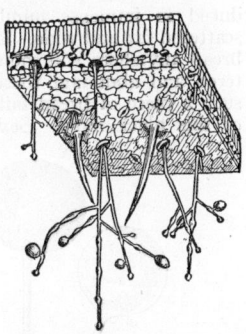

Damping-off is another disease caused by fungi of this order. The causal organisms are usually species of the genus *Pythium,* which attack the stems of seedling plants near the level of the ground and so weaken them that they fall

Perspective view of potato leaf with conidiophores of *Phytophthora infestans* protruding from stomata and bearing conidia.

over and the plant dies. Since moisture favors the growth of this fungus, partial control is obtained by aeration and the avoidance of high humidity.

A third order of Phycomycetes is the Mucorales, and contains the familiar black mold of bread, *Rhizopus nigricans,* a very common species. In this fungus there is an extensive much-branched colorless mycelium which

Left, a potato plant affected with late blight; right, potato leaves showing infected areas.
(*New York Agricultural Experiment Station, Geneva, Bulletin 241.*)

spreads over and within the substratum. Short branches of the mycelium penetrate the substratum and extract from it nutrient materials; erect branches from the mycelium terminate in sporangia, or spore-bearing bodies. The spores of *Rhizopus* are minute spherical bodies pro-

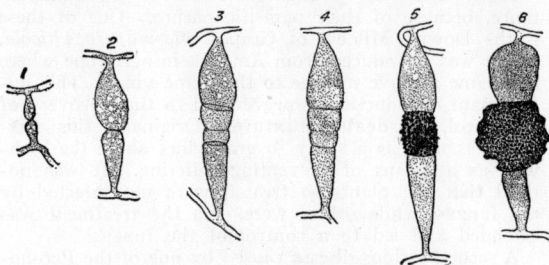

The formation of the zygote of *Rhizopus nigricans*.

duced in immense numbers. When mature they are scattered far and wide. Exposure of a piece of moistened bread anywhere will soon demonstrate how widely scattered and abundant these spores must be: seldom does such a piece of bread fail to nourish a luxuriant growth of this black mold. Sexual reproduction in this order

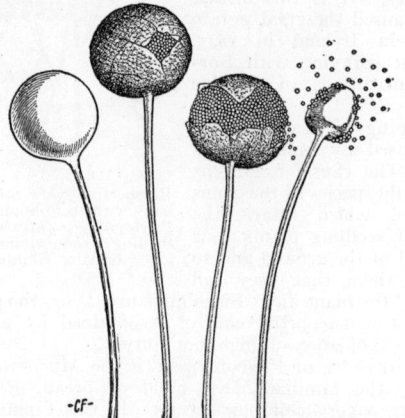

Sporangia of *Rhizopus* of different ages.

is entirely unlike that in the two described earlier. In this group, tips of branches from two different plants come together. Each tip enlarges considerably and becomes densely filled with protoplasm. A wall forms, cutting off the tip of each branch containing several nuclei, after which the walls between the two tips break

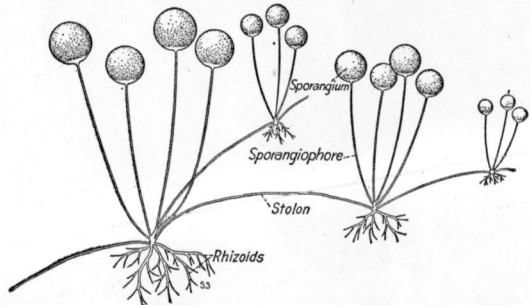

Rhizopus. Plant spreading by stolons and bearing sporangia on sporangiophores.

down and their contents fuse. The resulting cell, called a zygospore, enlarges conspicuously and becomes invested in a thick dark-colored wall. This zygospore usually undergoes a prolonged resting period before germinating. Nuclear fusion in pairs occurs during the

formation of the zygospore. Zygospore formation occurs in a similar fashion in other members of the Mucorales.

In many species, zygospores are formed but rarely. One reason for this is that many species are heterothallic, that is, require the union of hyphae from two plants which look exactly alike but are of very different nature. These hyphae must come from spores which are likewise different, and are usually described as + and − hyphae; often the terms female and male are applied to them, the +, or female, strain showing a more vigorous habit. Species in which the spores are all alike are called homothallic; in these, zygospores are readily formed between hyphae from a single spore, or branches of the same mycelium. Many species in the Mucorales show much-branched conidiophores and are objects of rare beauty when observed with a microscope.

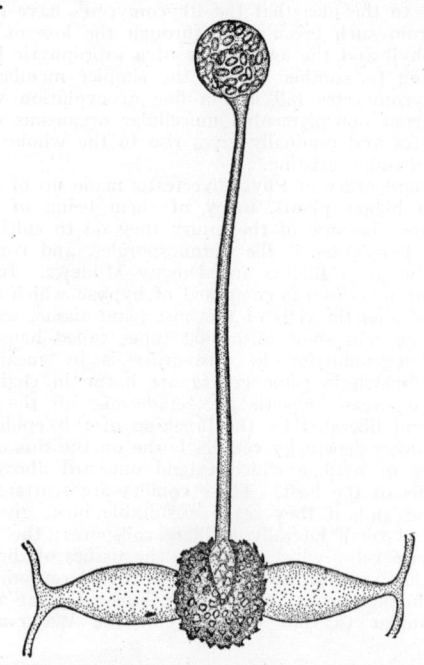

Mucor, a fungus closely related to *Rhizopus.*

One other order of Phycomycetes is worthy of mention. This is the Entomophthorales, many members of which grow parasitically on insects. One species in the genus *Empusa* attacks house flies, resulting in the death of large numbers of flies. Dead flies are often attached to window panes by the numerous hyphae present, and surrounded by a white halo of conidia.

The remaining orders of Phycomycetes are unimportant water-inhabiting forms of little importance to man. (R.M.W.)

PHYLACTOLAEMATA. Ectoprocta.

PHYLLITE. A metamorphic rock intermediate between **slate** and **schist** and characterized by a crinkly surface with the development of much **mica** or micaceous minerals such as **chlorite.** Phyllites may be white due to the presence of **sericite,** green if **chlorite** is present, or red or black due to various organic and mineral substances. Phyllite as a term is derived from the Greek word meaning a leaf, referring to the **cleavage,** though this physical property is much better developed in slates and some schists. Some phyllites are highly **metamorphosed** clay sediments, others represent metamorphosed tuffs, and even felsites, also phyllites have been formed by **mylonitization** of **gray-wackes, granites,** and other rocks. (E.S.C.S.)

PHYLLOPODA. Crustacea.

PHYLLOPODIUM. A form of the biramous appendage of crustaceans in which the general structure is in broad thin plates, with thin walls. They are often associated with the mouth and function in feeding and in respiration. In a few forms they are swimming appendages. (A.W.L.)

PHYLLOTAXY. Leaf.

PHYLLOXERAN, PHYLLOXERID. Insecta, Homoptera. A sucking **insect** related to the plant lice and scale insects. The many species make up a subfamily which, with the adelgids, constitutes the family Phylloxeridae. They differ from the **aphids** in that all females lay eggs and form the scales in their more complex structure, including the four wings of the winged stages.

The most important phylloxerid is a species which attacks grapevines, working on the leaves and roots. It once threatened to ruin the vineyards of France and has destroyed millions of acres of vines. The use of roots of certain American grapes which are not seriously harmed by the pest has greatly lessened the danger from its attack. Tender varieties are grafted onto the resistant roots. (A.W.L.)

PHYLUM. A principal subdivision of the animal kingdom. **Classification.** (A.W.L.)

PHYSICAL CHEMISTRY. Physical chemistry is that branch of chemistry which deals with the theoretical aspects of chemistry, using as its basis the ultimate particles of matter and their fundamental laws as formulated by physics. Whereas the latter science treats of properties which are common to all matter and energy, physical chemistry is primarily concerned with an interpretation of differences found among the various species of matter. The subject matter of physical chemistry is usually divided into the following topics: Properties of **Gases;** Properties of **Liquids;** Properties and Structure of **Solids;** Theory of **Equilibrium;** Theory of Dilute **Solutions;** Theory of Electrical **Conduction** in Liquids; **Ionization; Ionic** Equilibria; **Thermochemistry** and **Thermodynamics; Velocity of Reactions; Photochemistry; Catalysis.** (R.K.S.)

PHYSICAL CIRCUIT. Phantom Circuit.

PHYSICAL DEVELOPMENT. In photography the term physical development is applied to a method of developing the latent image in which the developer contains free silver ions and a reducing agent, such as metol, paraminophenol, etc., in an acid state and deposits silver from the solution on the exposed grains of silver halide. Thus the silver forming the developed image is obtained from the solution and not, as in chemical development, from the reduction of the exposed grains of silver halide.

Physical development has been recommended frequently for miniature work because of the fine grain of the image but has never become popular because in practice it is less reliable than other methods, being more likely to cause general fog and because of the increased exposure required.

It is possible to develop an image after fixation using a physical developer; the exposure, however, must be greatly increased and the process is of theoretical interest only. (C.B.N.)

PHYSICAL MAGNITUDES AND PHYSICAL EQUATIONS. Physics is a quantitative science, dealing primarily with things measurable and expressible in

units. There are hundreds of these physical magnitudes, some simple, some requiring elaborate definition. Many obvious relations exist between them; for example, pressure (or any stress) is the ratio of a force to an area. Careful study reveals that most physical magnitudes have their measures so defined that they may be expressed in terms of not more than three elementary or fundamental magnitudes, combined in various ways. As to which magnitudes should be regarded as fundamental, custom has fixed the choice upon length, mass, and time; to which many physicists add magnetic permeability and dielectric constant, making five in all. In the **c.g.s. system,** for example, the first three of these magnitudes are respectively represented by the centimeter, the gram, and the second, and all other physical units of the system except the magnetic and electric units are expressible in terms of these. Thus the unit of speed is 1 centimeter per second; of area, 1 square centimeter; of force (the dyne), 1 gram-centimeter per second per second; etc. This analysis may be generalized so as not to depend upon any specified system of units. Thus if length be denoted by L, mass by M, and time by T, the "dimension formula" for speed becomes L/T, for area L^2, for force ML/T^2, etc. The derivation of such relationships, called "dimensional analysis," is a highly important item in theoretical physics.

In order that two physical quantities may be equal, or that one may be added to or subtracted from the other, it is obvious that they must have the same makeup and be expressible by the same combination of fundamental units. It follows that in an equation expressing relationship between physical magnitudes, both members and all terms of each member must have the same dimension formula. For example, the total area of a right circular cone of altitude h and having a base of radius r is $a = \pi r^2 + \pi r \sqrt{r^2 + h^2}$, each term of which has the dimension formula L^2 (since π is abstract). Again, the phase angle ϕ of an **alternating current** of frequency n (per second) in a circuit of resistance R (ohms), inductance L (henrys), and capacitance C (farads) is given by

$$\tan \phi = \frac{4\pi^2 n^2 LC - 1}{2\pi nRC}.$$

Since $\tan \phi$ is an abstract quantity, the fraction on the right must also be abstract. Since the 1 in the numerator is abstract, the other term $4\pi^2 n^2 LC$ and the whole numerator must be also; hence the denominator is abstract. That is n, R, and C should have such dimensions that the component fundamental magnitudes cancel when the product nRC is formed. This is true; for the dimension formula of n is $1/T$, and (in electromagnetic measure) that of R is $\mu L/T$, and of C, $T^2/\mu L$; in which μ represents magnetic permeability. Some physical magnitudes are themselves abstract, and therefore independent of the system of units in use; **specific gravity** and **refractive index** are in this class. (L.D.W.)

PHYSICAL MEASUREMENTS. Physical quantities have practical significance only as they are capable of measurement and of expression as bearing definite numerical ratios to appropriate units. In some cases this comparison can be made directly, as by applying a yardstick to the length of a room. More often the quantity to be measured is incapable of such direct attack, and must be determined by means of known relationships to other quantities which are observable. Thus an electric current can be measured only by appealing to certain of its effects; for example, it can be made to form an electrochemical deposit for an observed time and the mass of the deposit then weighed. Likewise, the observable change in volume of mercury is a convenient "measure" of change in temperature.

Most physical units are expressible in terms of certain

primary standards, which are units of the fundamental magnitudes, length, mass, and time; sometimes in connection with the properties of specified substances. In most physical measurements these primary standards are the meter, the kilogram, and the mean solar day. Other "derived" units are defined in terms of these or of multiples or aliquot subdivisions of them (centimeter, gram; see C.G.S. System). Thus the density of a substance may be expressed in g./cm.³, the watt of power is 10^7g.cm.²/sec.³. The centigrade degrees is 1/100 of the temperature interval between the freezing and boiling points of water, which is subdivided on the basis of some specified temperature measure such as the pressure or the volume of a gas, the electrical resistance of a wire, etc., and these in turn must be determined in units appropriate to the respective magnitudes. When the measurement of a physical quantity gives its value in terms of the quantities used in defining its fundamental units, the measurement is said to be "absolute." This is the case, for example, with the measurement of electric current by observing the force with which a magnetic field of known intensity acts upon the conductor carrying the current. (See Abampere.)

In all physical measurements, the instruments used must be "calibrated"; that is, the relation of each subdivision on the instrument scale to the unit of the quantity measured must be ascertained. If each subdivision corresponds to exactly one unit, the instrument is said to be "direct reading"; this is usually true, for example, of ammeters and voltmeters, but seldom of galvanometers.

Various instrumental principles have become standard in physics. We have, for example, many instruments utilizing the vernier, the micrometer, or the optical lever principle. There are also certain well-known general observational methods, some of which are designed to minimize errors. In the method of "substitution," a quantity is determined by substituting for it a known quantity which produces the same effect. In the very common "differential method," the quantity required is the difference between two actually measured quantities. The "null" or "balance" method consists in adjusting the apparatus so that the indicator of the measuring instrument reads zero, as the galvanometer used with a Wheatstone bridge. In the "cumulative method" a large multiple of the quantity sought is measured, e.g., the thickness of a sheet of paper may be found from that of a thousand sheets. The "coincidence method" is useful in measuring periodic phenomena, as in comparing the periods of two pendulums by observing how often the swings coincide. "Compensation" applies to any method in which an error is made to neutralize itself, as in double weighing (see Weighing Methods). Many other schemes, often highly ingenious, are in common use in physical laboratories. (See Errors of Measurement.) (L.D.W.)

PHYSICAL METALLURGY. Metals and Alloys.

PHYSIOGRAPHIC PROVINCES. Physiography.

PHYSIOGRAPHY (GEOMORPHOLOGY). The description and interpretation of the surface features or topographic pattern of the earth. The scientific interpretation of scenery. The science of physiography is one of the major subdivisions of the earth sciences. The term is sometimes loosely used as synonymous with geography, hence the recent tendency to use geomorphology in its place. Since the scenery of any region is fundamentally the present stage of its geologic history, it naturally follows that a discussion of the origin of the topographic or scenic features must include not only an account of the processes of erosion and deposition which are now active, or have been active in the region, but also the manner in which the agents of erosion have been affected or controlled by the stratig-

raphy and structure. The relatively modern science of Geomorphology is peculiarly American in its origin and

Map of North America, showing its main political and physical divisions.

development, and the United States has been divided into a number of physiographic provinces whose natural boundaries have little or no relation to the political or state boundaries. (R.M.F.)

PHYSIOLOGY. A division of biological science which deals with the normal functions of the living body. General physiology is a science which treats of the underlying physical and chemical foundations of vital processes. Physiology in the usual sense is concerned with the more evident vital processes themselve, analyzed to some extent in terms of physics and chemistry but without any attempt to reduce them generally to purely physico-chemical fundamentals.

Physiology, especially the physiology of man, deals with many details of the processes which maintain a normal state of vital activity within the body. Among them the maintenance of a normal internal environment is intimately associated with blood chemistry and the circulation. The maintenance and utilization of reserves of material, the action of enzymes, nervous coordination, and chemical coordination by hormones are other important factors. The response of organs to coordinating influences is also involved in this subject. The all-or-none law bears upon this topic.

Much of the material of physiology is associated with the organ systems. It is briefly considered under contractility, digestion, respiration, circulation, excretion, and reproduction. A limited amount of physiological material is also mentioned under the various organ systems associated with these processes. (A.W.L.)

PHYSIOTHERAPY. An aid to the treatment of the sick or injured by means of physical agents. This includes massage, various forms of light, heat, and electricity, ultra-violet light, air, water, and exercise. Physiotherapy is used extensively in restoring function of wasted, stiff, or contracted muscles after injuries, especially fractures of bones, or after paralysis, particularly infantile paralysis. (R.S.M.)

PHYSOPODA. Thysanoptera.

PHYTOMONADIDA. Mastigophora.

PI SECTION. This is a type network in which the elements are arranged in π shape, i.e., a shunt element across the circuit at each end of a series element. (L.R.Q.)

PICA, PIKA. Mammalia, Rodentia. Small **rodents** related to the rabbits. They live chiefly at high altitudes, ranging from 11,000 to 19,000', and are found only in the northern hemisphere. Two species of the genus *Ochotona* occur in the mountains of western North America and about two dozen in the Old World. All are compactly built, with small ears and a rudimentary tail. In the Old World they are also called tailless hares or mouse-hares. (A.W.L.)

PICHI. Mammalia, Edentata. The pigmy **armadillo,** *Chlamyphorus,* of Argentina. (A.W.L.)

PICIFORMES. An order of birds including the **woodpeckers, toucans,** and related species. They are found in all parts of the world except the Australian region. Most species nest in holes in trees. (A.W.L.)

PICKEREL. Pisces, Teleostei. Fresh-water fishes **(Pisces)** of three species closely related to the pikes and muskellunge. All are North American, living in streams and lakes. They differ from the pike and muskellunge in having both the cheeks and the opercula fully scaled. The little pickerel of the Mississippi Valley and Great Lakes Basin is sometimes called the grass pike, and the true pike is sometimes known as the northern pickerel. All of these species belong to a single genus, *Esox.* (A.W.L.)

PICKLING. Scaling.

PICOLINE. Pyridine.

PICRIC ACID. Phenols.

PICRITE. A rock which, like **peridotite, is** made up chiefly of **femic** minerals, but contains a little **feldspar,** usually **labradorite.** Sometimes **analcite** is present, in which case the rock is called an analcite-picrite. (E.S.C.S.)

PICTORIAL REPRESENTATION. A method of representation based upon oblique representation, in which the effect of perspective representation is obtained.

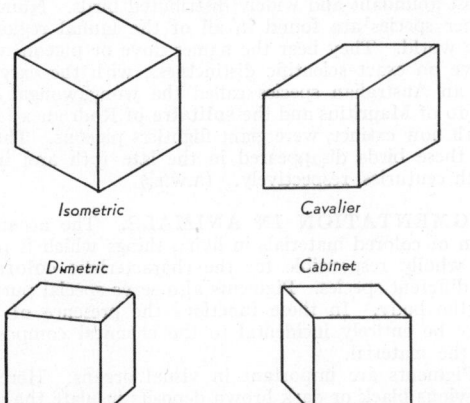

Isometric   Cavalier

Dimetric   Cabinet

In the practical application of pictorial representation, three principal axes are selected, and the actual lengths of the edges of the object are laid off along these axes, resulting in a drawing that is not correct from the standpoint of either orthographic or perspective representation. Since the relative proportions of the object are retained, however, the differences do not detract from the value of the representation. In some forms of representation, however, one or more edges may be drawn to a reduced scale, termed scale reduction, to more closely simulate perspective representation.

The figure shows the appearance of a cube as represented pictorially. Isometric and cavalier drawings correspond to angular and parallel perspective; no scale reduction is used along any axis. Cabinet drawing is similar to cavalier, with scale reduction, usually one-half, along the inclined axis. Dimetric drawing has one vertical and two inclined axes, with one-half scale reduction along the sharply inclined axis. Clinographic representation is used for representing crystals, and employs three reference axes so selected that no face of the crystal will appear as a line. (H.C.H.)

PICTURE PLANE. Perspective.

PIDDOCK. Mollusca, Lamellibranchiata. A **bivalve mollusk** that bores in soft rock and floating wood. Especially a European species of the genus *Pholas,* commonly used as bait and in some localities regarded as a delicacy. The family to which these animals belong is near that containing the **shipworms.** Also spelled piddick. (A.W.L.)

PIE CHART. A method of representing relative magnitudes totaling 100% by means of a circular area divided into sectors whose included angles are proportional to the quantities involved. The pie or 100% circle chart is not as effective as other forms of representation, but has gained favor because of its advertising appeal, particularly when colors are used to differentiate between the area sectors. (H.C.H.)

PIER. A pier is a **masonry** structure acting in compression, the purpose of which is to support, in a suitable manner, some superstructure such as a **bridge, a trestle,** or a wall. Wharves which project into the water to accommodate vessels which berth alongside are also called piers. Piers of this type allow a great many more ships to be docked in a given length of water frontage. The pier may vary from a simple platform-like structure supported on **piles,** to a costly plant which is roofed and walled and provided with elevators, freight storage rooms, passenger waiting rooms, customs inspection facilities, etc. In bridge practice the pier is usually the intermediate support of a multiple-span bridge, its function being to support the intermediate ends of the spans and to do so with minimum obstruction to the stream. Such piers are often made of reinforced concrete, though, in earlier days, they were made of ashlar stone masonry. The pier is built with its long dimension parallel to the stream so that it will receive a minimum of impact of floating debris, ice, and other waterborne loads and will give the least obstruction to the flow of the stream. Thus piers often have to be set on a skew to the bridge center line giving rise to so-called skew bridges. It must support the dead weight of the bridge, and loads due to wind acting on itself and on the bridge and moving load, traction forces, and live and impact loads. In building construction the wall pier is an enlarged section of the wall similar to a **column** which provides lateral rigidity for the wall and may be used to support beams or trusses. (F.T.M.)

PIEZO- AND PYRO-ELECTRIC PHENOMENA. A number of **dielectric** crystalline substances manifest the curious property that, when subjected to strain in suitable directions, they develop electric polarization. Such substances are said to be piezo-electric, and were first observed by Haüy in 1782 and independently by Curie in 1880. Notable examples are quartz (SiO_2) **and**

Rochelle salt (sodium potassium tartarate, $NaKC_4H_4O_6 \cdot 4H_2O$).

The directions in which tension or compression develop polarization parallel to the strain are called the piezo-electric axes of the crystal. Thus the axis of a hexagonal quartz crystal indicated by the arrows in Fig. 1 is known as an "X-axis," and a plate cut, as shown, with its faces perpendicular to this direction is an "X-cut"; while one cut with its faces parallel to the lateral faces of the crystal is a "Y-cut."

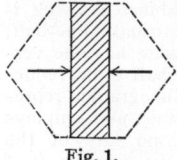
Fig. 1.

The magnitude of the piezo-electric polarization is proportional to the strain and to the corresponding stress, and its direction is reversed when the strain changes from compression to tension. The principal piezo-electric constants of a crystal are the polarizations per unit stress along the piezo-electric axes. While these constants are much greater for Rochelle salt than for quartz, the latter is better adapted to some purposes because of its greater mechanical strength.

Always associated with the piezo-electric property is the inverse property known as electrostriction, namely, the development of mechanical strain as a result of electric polarization. Thus if a piezo-electric plate is placed between metal electrodes and the latter given a difference of potential, the plate becomes either thicker or thinner, depending upon which electrode is positive and which negative.

If a quartz plate is subjected to a rapidly alternating electric field, the inverse piezo-electric property causes it to expand and contract alternately. As an elastic body, the plate has a certain natural frequency of expansion and contraction in the direction of the field, and if the field is made to alternate with the same frequency, the plate responds with a vigorous resonant vibration. This reacts, through the direct piezo-electric property, to augment the electric oscillations. A circuit arranged for this purpose, as in Fig. 2, is known as a piezo-electric or crystal oscillator, the crystal itself, P, being the piezo-electric resonator; T is the oscillation transformer, and C a variable condenser. This device has been much used as a frequency control in radio transmitters. Both X-cut and Y-cut quartz plates are subject to changes of frequency with temperature, due to change of elastic modulus; but certain planes in the crystal have been found, oblique to both X and Y, such that plates cut parallel to them are nearly free from the temperature effect.

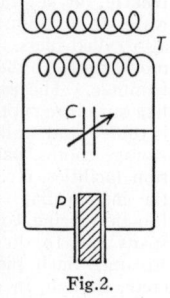
Fig. 2.

The very high piezo-electric response of Rochelle salt makes it adaptable to microphones of the crystal type, now extensively used. In these the crystal is subjected to a variable bending or torsional strain, due to the vibrations of the diaphragm.

The heating or cooling of certain crystals develops polarization similar to piezo-electric polarization. In some cases, at least, this may be a secondary result of thermal expansion or contraction, that is, an indirect piezo-electric effect. This phenomenon is known as "pyro-electricity." (See Crystal.) (L.D.W.)

PIEZOMETER TUBE. A piezometer tube is a tube which is inserted into a region of fluid flow so as to obtain in the connecting tubing a pressure equal to the static pressure existing in the stream. If the piezometer tube be not carefully located, it will receive some impact or suction by Pitot action, and the pressure indicated by it will be in error. For this reason, the piezometer tube location has been the subject of some con-

siderable study. It should open into the stream perpendicular to the thread of the stream, and be made flush with the conduit wall. (F.T.M.)

PIG. Mammalia, Artiodactyla. A group of hoofed animals with two functional and two reduced toes on each foot, all with separate metacarpal or metatarsal bones. They are not ruminants (i.e., they do not chew the cud). The true pigs make up the family Suidae and a closely related family Tayassuidae contains the peccaries.

Pigs are native to the warmer parts of Europe and Asia, the Oriental region, and Africa. None occur in North or South America, although feral races have developed in some areas. In these continents the peccaries are the only native pig-like animals.

Two of the most peculiar species of pigs are the babirusa of Celebes and Boru and the wart-hogs of Africa. The former have four tusks in the male sex, all directed upward and recurved. The enormous heads of the wart-hogs, with their wart-like protuberances and stout tusks, make these animals conspicuously ugly. (A.W.L.)

PIGEON. Aves, Columbiformes. The birds (Aves) of this order are almost exclusively characterized by the name pigeon or dove, although a few related species of the Old World are known as sand grouse. The extinct dodo also belonged here.

Pigeons and doves have the beak swollen at the tip and covered with soft skin at the base, about the nostrils. In North America the group is represented by the band-

Mourning dove, *Zenaidura macroura carolinensis.* Soft olive brown above, buff-gray below, white tips to outer tail feathers.

tailed, *Columba fasciata,* and red-billed pigeons, *Columba flavirostris,* of the western part of the continent, and by several species of doves of similar distribution. The turtle dove, *Streptopelia,* is the only widely distributed species, although the extinct passenger pigeon, *Eclopistes migratorius,* remains a memory of one of the most abundant and widely distributed birds. Numerous other species are found in all of the faunal regions of the world. They bear the names dove or pigeon, which have no exact scientific distinctness, with the exception of an Australian species called the wongawonga. The dodo of Mauritius and the solitaire of Rodriguez Island, both now extinct, were giant flightless pigeons. The last of these birds disappeared in the late 17th and in the 18th centuries, respectively. (A.W.L.)

PIGMENTATION IN ANIMALS. The accumulation of colored materials in living things which is partly or wholly responsible for the characteristic coloration of different species. Pigments also serve special purposes in the body. In these functions the presence of color may be entirely incidental to the chemical composition of the material.

Pigments are important in visual organs. Here impervious black or dark brown deposits insulate the sensitive nerve endings against all light except that which is transmitted by the lens or cornea. In the eyes of some arthropods the pigment is redistributed to admit more light when the surrounding illumination is dim than when it is bright. A similar result is gained in the vertebrate eye by the muscular adjustment of the pigmented iris to change the size of the pupil through which light is admitted. Still another pigment, the visual purple (rhodopsin) is found in the rods of the retina of the vertebrate eye. It is bleached by light and resumes its color in darkness; it is associated with the sensitivity of the eye.

Pigments in the superficial layers of the body are also useful in some animals, independently of the relations discussed under coloration. A familiar example is the protective pigment deposited in the human skin as a protection against **ultra-violet light**. The deposition normally follows excessive exposure to sunlight or to other sources of ultra-violet, and the deposits are lessened when exposure is reduced. These deposits are in the form of granules of melanin in the cells of the innermost layer of the **epidermis** and in branching cells called melanoblasts in the underlying dermis. The pigment loses its granular form and becomes diffuse as the epidermal cells move toward the surface. Melanoblasts are possibly active in the formation of the pigment granules. The pigmentation of hair is not thoroughly understood but both granular and diffuse pigments have been reported.

A definite relation also exists between the normal illumination of the body and pigmentation in other animals, but the nature of the relation is not always known and a definite value to the animal need not exist. A familiar example is the dark upper surface and light lower surface of fishes, whether the upper surface is dorsal, as in most species, or lateral, as in the **flatfishes**. Lack of pigment in fishes of subterranean waters is closely associated.

Pigmentation of **insects** has been shown in several cases to respond to light. Lessened illumination may result in deeper colors, and some observers have secured the same result by moderate increase of light. Extreme changes, however, have resulted in decreased pigmentation in some experiments. Humidity and temperature also affect the depth of pigmentation in some insects, and may modify the pattern.

Incidental colors like the pink flush of human skin result from the presence in the body of the respiratory pigments, haemocyanin and haemoglobin, and waste products. Protein wastes deposited in the superficial tissues of insects are one source of color and the bile pigments of vertebrates, formed by the liver in the modification of haemoglobin, are a source of color in some organs.

The entire subject is closely associated with the chemistry of the living organism on the one hand and with coloration and mimicry on the other. (A.W.L.)

PIGMENTS IN PLANTS. The distinctive green color of leaves and many other plant organs results from the presence in such organs of two pigments called chlorophyll a and chlorophyll b (see **Pyrrole and Related Compounds**). In the higher plants these pigments occur only in the chloroplasts (see **Plastids**). The molecular formula for chlorophyll a is $C_{55}H_{72}O_5$ N_4Mg; for chlorophyll b, $C_{55}H_{70}O_6N_4Mg$. The structural formulae for the molecules of the chlorophylls have also been worked out. The chlorophylls are not water-soluble but can be readily dissolved out of leaf tissues with alcohol, acetone, ether, or other organic solvents. The resulting solutions exhibit the phenomenon of fluorescence; they are deep green when held between an observer and the light, but deep red when viewed in reflected light. By suitable treatments it is possible to obtain pure crystals of chlorophyll from such solutions. Most leaves contain considerably more chlorophyll a than chlorophyll b, often two to three times as much. In the organs of the higher seed plants, with rare exceptions, chlorophyll is synthesized only upon exposure to light. Leaves of grass which develop under a board, for example, contain no chlorophyll. In the leaves of mosses, ferns, and gymnosperms, however, chlorophyll develops

in the dark as well as in the light. The chlorophylls play an indispensable biological role because their presence in plant cells is essential for the occurrence of **photosynthesis**.

Invariably associated with the chlorophylls in the chloroplasts are the yellow pigments carotene and xanthophylls. These pigments are not, however, restricted in their occurrence to the chloroplasts, but may also be present in non-green parts of the plant where they commonly occur in *chromoplasts*. Collectively these pigments, together with certain others which are closely related chemically, are called the *carotenoids*. Carotene is an orange-yellow pigment with the chemical formula $C_{40}H_{56}$. It is especially abundant in the roots of carrots. This compound is of considerable importance because it is the precursor of vitamin A, one molcule of β-carotene being split into two molecules of vitamin A by a simple hydrolytic reaction. Lycopene, a red carotinoid which is isomeric with carotene, is responsible for the red color of the fruits of tomato, pepper, rose and some other species. The commonest xanthophylls found in leaves are lutein and zeaxanthin although others also occur. Both of the above-mentioned xanthophylls have the formula $C_{40}H_{56}O_2$. Another xanthophyll is *fucoxanthin* which imparts to brown **algae** their distinctive color. None of the carotenoids are water-soluble, but all of them can be extracted from plant tissues with suitable organic solvents.

Most of the red, blue, and purple pigments of plants belong to the group of *anthocyanins*. These pigments are chemically related to the **glucosides** and usually occur dissolved in the cell sap, being water-soluble. In general the anthocyanins are red in an acid solution and change in color through purple to blue as the solution becomes more alkaline. Red pigmentation resulting from the presence of the anthocyanins is found in flowers, fruits, bud scales, young leaves and stems, and sometimes even in mature leaves as in those of the red cabbage. Blue and purple pigmentation due to the presence of anthocyanins occurs principally in flowers and fruits.

Another group of cell sap water-soluble pigments is the *anthoxanthins*. These pigments are also chemically related to the **glucosides**. Anthoxanthins often occur in the plant in a colorless form but under suitable conditions their typical yellow or orange color becomes apparent. Some yellow flowers, such as yellow snapdragons, owe their color to the presence of anthoxanthins, but the color of the majority of kinds of yellow or orange flowers is due to carotenoid pigments.

The autumnal coloration of leaves in temperate regions is one of the most spectacular accompaniments of the march of the seasons. Both carotinoid and anthocyanin pigments play an important role in autumnal leaf coloration which is not, contrary to popular opinion, a result of the action of frost. Brilliant development of the anthocyanin pigments in the fall is, however, favored by dry weather during which cool, but not frosty, nights alternate with clear days. During the late summer and early fall the chlorophyll in the leaves gradually decomposes. In many species this simply results in unmasking the yellow carotinoid pigments already present, accounting for the yellow autumnal pigmentation of such species as birch, sycamore, aspen, and tulip trees. In other species synthesis of anthocyanins occurs more or less concomitantly with the disintegration of the chlorophyll; this accounts for the reds or purplish reds characteristic in the autumnal coloration of such species as many oaks, maples, sumacs, and dogwood.

Except in flowers, white is an uncommon color in the externally visible parts of plants, and results from the complete absence of pigments. In some species white streaks or other markings are of common or regular occurrence in leaves and in the leaves of some species such as roses completely white leaves or even entire branches bearing only white leaves sometimes occur. Such branches cannot be propagated because no **photo-**

synthesis can take place in the absence of chlorophyll. As long as such branches remain attached to a plant bearing green leaves they can obtain necessary food from the branches bearing normally pigmented leaves. (B.S.M.)

PIGMENTS, PAINTS, VARNISHES. Solid pigments, usually in admixture with a medium of suspension, are widely used (1) in preservative paint coatings, such as those for wood and steel structures and machinery, (2) in decorative paint coatings, either purely artistic as for paintings, enamels, glass, ceramics, or also utilitarian as for interior decoration of walls, floors, furniture, and for automobiles, cars, trucks, (3) in the manufacture of linoleum oilcloth, window shades, book covers, (4) in the manufacture of paper, wall paper, printing inks, writing inks, (5) in plastics such as rubber, celluloid, bakelite, (6) in cosmetics. This variety of uses suggests a wide variety of materials adapted for each purpose, as well as the many different colors involved. The sources are mainly inorganic, but a number of organic preparations are also important. The properties to be considered for each use are (1) opacity, hiding power, covering power, (2) durability, preservative power, (3) beauty and permanence of color.

A convenient classification for the present purpose is that of color.

White Pigments. Any white powdered substance of sufficient permanence in the vehicle to be used and under the conditions to which the finished material is to be subjected, is available. 1. **Calcium** carbonate is one of the cheapest white pigments, and is used as such, mixed with linseed oil, in putty (gray), and with water for wall coatings (white). 2. **Zinc** oxide (zinc white) is used in compounding rubber, and in paints yielding a rather hard film, not colored by hydrogen sulfide in the air. 3. **Barium** sulfate (**barite**) used as a filler in paints, is not excellent when used alone. 4. **Zinc** sulfide (30%) plus barium sulfate (70%), known as lithopone, is prepared by reaction of zinc sulfate plus barium sulfide in water. 5. **Lead** basic carbonate (white lead) is widely used as a white paint pigment yielding a rather soft film readily colored brown (lead sulfide) by hydrogen sulfide in the air. All lead pigments are poisonous. White lead and zinc white are frequently used in admixture. 6. Lead basic sulfate (sublimed white lead), made directly from lead ores by roasting in air, may contain 20% lead oxide and 5% zinc oxide, is cheaper than genuine white lead and is frequently mixed with other pigments. 7. **Titanium** white (titanium dioxide) is remarkable for its whiteness and high hiding power. It finds important use, either pure or in admixture, in enamels for interior work. For exterior use, titanium oxide alone "chalks," and is accordingly best used in admixture with other pigments, for example, zinc oxide. 8. **Silica**, and china clay, on account of their cheapness, are frequently used in admixture with more expensive pigments.

Yellow Pigments. 1. Cheap, permanent yellow pigments are those of hydrated ferric oxide, such as the mineral **limonite**, frequently with clay, known as ochres, yellow, and siennas, yellowish-brown. These pigments are not affected by an oil medium nor do they affect other pigments, and they furnish a considerable range of tints in the yellow, brown, red range. 2. **Lead** monoxide (litharge), brown-yellow, is used in painting steel, as it is slightly basic, thus increasing resistance to corrosion, causes paint oil to dry quickly. 3. **Lead** chromate (chrome yellow) is excellent in hiding power and brilliance and permanence of tint. Different shades are prepared, depending upon conditions of manufacture, e.g., light shades by use of acidified solutions in the precipitation. Like all lead pigments, the chromate is darkened by hydrogen sulfide. 4. **Zinc** chromate is permanent and unattacked by hydrogen sulfide, but not as excellent in tint and hiding power as lead chromate.

5. **Arsenic** trisulfide (orpiment) either as the natural mineral or artificially prepared. Not permanent to light, affects lead pigments, causing darkening (due to sulfide), and is poisonous. 6. **Cadmium** sulfide, very permanent and brilliant, not to be mixed with lead pigments on account of darkening by lead sulfide. 7. **Gamboge,** a **resin,** is used as a water color pigment, in varnishes, and in lacquers for brass. 8. **Cadmium** lithopones, made by precipitation of cadmium sulfate and **barium** sulfide or **selenide** (or sulfide-selenide) solutions.

Brown Pigments. 1. Cheap, permanent brown pigments are those of hydrated **ferric** oxide and **manganese** dioxide containing clay minerals. Various shades of yellow to brown to red are obtained from a variety of forms found in nature, and darker shades are made by heating these. The yellowish-brown forms are known as siennas, and the darker reddish-brown as umbers. Not affected by the oil medium, do not affect other pigments of good hiding power, cheap, wide variety of shades. 2. Van Dyck brown or Vandyke brown, by heating cork or woody matter to moderate temperatures out of contact with air, and is usually used in admixture with other pigments.

Green Pigments. 1. **Chromic** oxide (chrome green) is permanent under atmospheric conditions and at high temperatures, of excellent tint, of permanent color, can be mixed with other pigments, is unaffected by hydrogen sulfide, alkali, or acids. Other chrome greens (Brunswick green) are made by mixing chrome yellow and Prussian blue. 2. **Copper** greens are copper basic carbonate (**malachite** mineral), copper basic acetate (verdigris), copper basic **arsenite** (Scheele's green), copper acetoarsenite (Paris green, emerald green). These are limited in their satisfactory use due to lack of permanence of color, low hiding power, reaction with sulfide pigments or with moisture, poisonous nature of arsenic compounds. 3. Green lakes, by precipitating certain dyestuffs along with inorganic salts, e.g., of **tin, aluminum.** Used in inks for printing, and for wall finishes.

Red Pigments. 1. **Ferric** oxide (Indian red, Venetian red), either as the mineral **hematite** or prepared by heating ferrous sulfate (alone or with calcium oxide) to the desired temperature. Not affected by oil medium, does not affect other pigments, is of good hiding power, cheap, permanent, used in cosmetics, and in glazes. 2. Trilead tetroxide (red **lead**), orange-red, causes paint oil to dry rapidly, used largely on structural steel alone or as priming coat. 3. **Antimony** trisulfide (antimony vermilion), used in paints (not mixed with lead pigments), and in compounding red rubber. 4. Red lakes, by precipitating certain dyestuffs, e.g., madder, cochineal, aniline dyes, along with inorganic salts, e.g., of tin, aluminum. Used in inks for printing, and for wall finishes.

Blue Pigments. 1. **Iron ferroferricyanide** (Prussian blue, Turnbull's blue), used as a paint pigment, in inks, in laundering, permanent with acids, color destroyed by alkalis. 2. Ultramarine blue (sodium aluminosilicosulfide) used as a paint pigment, discolored by acids, not used where its sulfide content causes color reaction with other pigments, e.g., lead pigments. 3. **Cobalt** blue (cobalt aluminate), by heating to a red heat the precipitate obtained by mixing cobalt salt solution, alum solution, and sodium carbonate solution. Used in oil or water medium and in glazes, not reactive with other pigments, permanent under atmospheric conditions and at high temperatures. 4. **Copper** blue (azurite, copper basic carbonate), color not permanent with sulfides, nor when heated.

Black Pigments. 1. Lamp black (amorphous **carbon**), made by the incomplete combustion of natural gas or petroleum. 2. Bone black or drop black, made by the destructive distillation of bones. Used as a paint pigment, and in stove polishes. 3. Ivory black, made from ivory cuttings as bone black is from bones. Reputed the blackest of pigments, expensive. Used in high-grade varnishes and enamels. 4. Graphite (crystalline carbon), natural or artificial, inert, permanent. Used in protective paints for metals. 5. Asphalt. Used as a protective varnish on metals, concrete, wood—especially roofing materials.

In the manufacture of paints the (1) pigment is suspended in a (2) medium—drying oil, or water—with or without the addition of a (3) thinner or volatile solvent to the drying oil—turpentine, gasoline, benzene are used as thinners—or a (4) drier. Driers are added to hasten the hardening of the drying oil. **Manganese** borate, red **lead**, litharge, finely divided manganese, **cobalt**, **nickel**, **lead**, and metallic soaps, such as linoleates and resinates of manganese, lead, cobalt, are used as driers. Casein increases the adhesiveness of water paints.

Varnishes are similar to paints that contain thinners, in that part (or all) of the medium evaporates. The medium may consist of (1) a drying oil, a thinner, and a drier, with or without pigment. In the use of paints and varnishes a thin film of the material is formed and left exposed on a surface, and the hardening process consists in oxidation of the unsaturated drying oil. (2) **Alcohols, turpentine, benzene,** naphtha, **acetone,** tetrahydronaphthalene, **ethyl acetate,** ethylene **glycol** monoethyl **ether,** in which a **resin,** such as shellac, dammar, copal, sandarac, rosin, synthetic resins of the phenol-formaldehyde type or an **ester,** such as cellulose acetate or cellulose nitrate is incorporated. On exposure, the solvent evaporates, leaving the resin or ester as an adherent film. Some resin films are brittle and not adapted to resist wear, while some ester films are notably permanent and used for such purposes as the coating of automobiles and cloth exposed to the weather.

Paints, varnishes, and lacquers are made for many special purposes. Of special interest may be mentioned (1) shingle stains using creosote oil, or fish oil, (2) luminous paints using radioactive materials such as radium or mesothorium compounds with **barium, strontium,** or **calcium** sulfides or tungstates, (3) linoleum made by mixing specially prepared oxidized linseed oil, rosin, cork dust, and pigments, spreading the mixture on canvas, and allowing to dry. (R.K.S.)

"PIGNOLEA" NUTS. These are the seeds of certain European pines growing in countries bordering the Mediterranean Sea. They are an important food article in European countries; in America they are used mostly as salted nuts and in various candies. (R.M.W.)

PIKE. Pisces, Teleostei. A badly misused name applied in various forms to a number of fresh-water food and game fishes (**Pisces**). They are most abundant in the rivers and lakes of the northern states and Canada. The true pike, *Esox lucius,* is a fish related to the muskellunge. It attains a weight of 40 lbs. and a length of 4′, and is one of the principal game fishes of the north. This species is also called the northern pickerel, but it differs from the closely related pickerels in having the cheeks scaly but the lower half of the opercula bare. Two other species, the wall-eyed pike, *Stizostedion vitreum,* and the sand pike, *Cynoperca canadensis,* are more closely related to the perch than to the true pike. The former is also called the wall-eye, glass-eye, pike-perch, and jack salmon, and in the south is commonly called the salmon. The sand pike is also known as the sauger or gray pike. The wall-eyed pike is an excellent food and game fish, attaining a weight of 25 lbs. The sauger is smaller and less desirable. (A.W.L.)

PIKE-PERCH. Pike.

PILCHARD. Sardine.

PILE. In **foundation** work the pile, often called a bearing pile due to the fact that it carries a direct compressive load, is a post-like member of timber, steel or concrete which is driven into soft ground to support a structure. In case soil conditions are such that an extremely large **footing** or **grillage** is required, piling is used to give sufficient bearing power. Piles are usually necessary for unstable soils and may be used in other types where extremely heavy loads are carried. Formerly, all piling was cut from trees, but, due to the increasing scarcity of timber, especially of the type and size required, reinforced concrete or steel piles are now extensively used.

Many large buildings are constructed on very closely spaced piles driven on 3- or 4-ft. centers. These are surmounted by a monolithic concrete capping which adds to the bearing power by making available whatever bearing power the soil between the piles may possess. Ordinarily, piles are driven deep enough to reach fairly firm soil. Sheet piling differs from bearing piling in that it is used where a continuous wall is required such as in the **cofferdam.** It may consist of vertical wooden planks driven in close contact or specially rolled steel shapes, having interlocking edges, which are driven so that each individual section is firmly united with adjacent sections. Sheet piling is subjected to **lateral** pressures due to earth or water and if driven in water must withstand the impact of floating debris.

Piles are driven by means of a water jet or by a pile driver. The former method, used in sand or gravel, consists of washing the material away from the end of the pile by means of a jet of water. As the material is displaced the pile sinks by its own weight or is driven by light blows from a pile driver. The drop hammer pile driver is made up of a **casting,** which fits into guides on vertical posts, and the machinery and tackle necessary to raise the hammer. The hammer is raised over the end of the pile and then allowed to fall in the guides by gravity. The force exerted by the hammer dropping upon the end of the pile causes it to sink a certain distance. This force is proportional to the weight of the hammer and also to the drop. The single-acting steam-hammer pile driver raises the casting by steam pressure and then allows it to drop by gravity, while in the double-acting type steam pressure raises the casting and increases the force of the blow over that due to gravity. It is preferable to the drop-hammer type since it causes less damage to the pile and can deliver more blows in a given period of time. (F.T.M.)

PILE DRIVER. Pile.

PILES. Hemorrhoids.

PILIDIUM. A ciliated (cilia) larval form of the **nemertine** worms. It forms as a 2-layered saccular structure with a band of cilia around the base and a sensory organ at the apex, resembling in some details the **trochophore** larvae of other animals. The worm develops inside the pilidium and when it breaks out the pilidium perishes. (A.W.L.)

PILL BUG. Crustacea, Isopoda. A small oval terrestrial **crustacean,** commonly found in moist situations at the surface of the ground. They hide in crevices among rocks, under wood, and even invade basements. These forms are also commonly known as sow bugs and wood lice. The pill bugs are properly the species with the power of rolling up into a ball so nearly spherical that it will roll on a slight incline. (A.W.L.)

PILLOW BLOCK. Bearings.

PILLOW LAVA. Effusive volcanic rocks, generally of **basic** composition, which are characterized by pillow-like or bun-like structures formed during the concomitant movement and congelation of the lava. Most pillow lavas are basalts. Frequently the "pillows" have a skin of rock glass called tachylyte. The evidence of the rapidity of the chilling suggests that pillow lavas owe

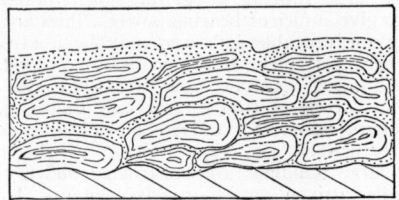

Cross-section of pillow lava. (*Field, Outline, Barnes & Noble.*)

their peculiar structure to having flowed into a body of water or as having originated as aquatic lava flows. (R.M.F.)

PILOCARPINE. Alkaloids.

PILOT BALLOON. Meterological Instruments.

PILOT FISH. Pisces, Teleostei. 1. A small marine fish, *Naucrates ductor,* which accompanies sharks and is sometimes supposed to guide and protect them. It is bluish with several dark vertical bars. 2. One of the **whitefishes** found in fresh water from the northern states to Alaska. (A.W.L.)

PILOTAGE (PILOTING). The term pilotage or piloting is used to describe that type of **navigation in which the positions and motions of a ship are determined by reference to fixed objects on the earth. The landmarks may be natural, such as hills, points of land, small islands, lakes, rivers, etc., or they may be artificial, such as lighthouses, light vessels, **beacons, buoys,** prominent buildings, water towers, railroads, highways, power transmission lines, etc. Two general types of pilotage are recognized by sea navigators: inshore or harbor piloting, and off-shore or coast piloting. Before a ship proceeds up a channel, or into a bay or harbor, the pilot must have a clear mental picture of the locations of all available landmarks, range points, etc. A stranger to the region must obtain this image from a thorough study of **charts**, pilot directions, and similar publications for the region. With the mental picture thoroughly developed, the pilot then guides his ship in much the same manner as an individual finds his way about a city. Aviators use this type of pilotage when operating in the vicinity of a base with which they are thoroughly familiar, or when proceeding by **contact flying**. When a ship is off the coast, with but few recognizable landmarks available, or when a pilot is flying over unfamiliar terrain where prominent landmarks are few and far between, the pilot obtains **lines of position** from the available objects, and then applies geometric constructions to obtain **fixes**.

The **bearing** of an object, taken with the **pelorus** or the **compass**, is the most frequently used line of position in pilotage. If two or more landmarks are available, and are spaced so that their bearings differ by more than 30°, the lines of position will intersect in a fix that is known as a cross-bearing position. A single object will provide only one bearing, but a fix can be obtained if the distance of the object can be measured. In this case the two lines of position will be the bearing line and a circle centered on the object and with radius equal to the distance. Among the various methods for finding the distance we have: the angular height of the object, measured with the **sextant**, if the linear height is known; the use of stadia lines in binoculars; the difference in time of reception of audible and radio signals (see **Radio Beacon**); the time for the echo of a signal from the ship, such as a whistle blast, to return from the shore; etc. None of these methods is as accurate a a bearing

line, and they should not be used when other methods for obtaining a fix are available.

One object may be used for obtaining a running fix by taking two bearings of the object, at different times, provided that the distance and direction run between observations is accurately known. At 1015 the navigating officer of a ship, that is proceeding at 18 **knots** on **heading** 060°, sights a lighthouse about 12 miles away that bears 032° off the port bow. He immediately instructs the helmsman to be very careful with his steering, and proceeds to make a careful study of the **current** and tide tables of the region. From the predicted currents he finds that the ship is making good a **course of** 058° with ground speed 17.8 knots. He then starts a graphical construction (Fig. 1) by drawing the 1015

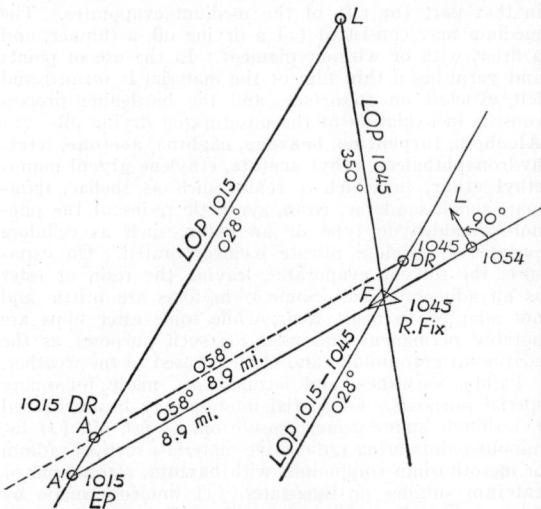

Fig. 1. Scale diagram.

line of position through the lighthouse. This line bears 028° since 060° − 032° = 028°. He then selects a point, A, on this line, usually the point closest to his dead-reckoning position, and draws the course line in direction 058°. At 1045 the lighthouse is 070° off the port bow and a 1045 line of position is drawn through the lighthouse on bearing 350°. The 1015 line must now be advanced to 1045 by measuring off the distance made good in 30 minutes (8.9 miles) along the course line and drawing a line parallel to the 1015 line through the estimated position at 1045. The point of intersection of the 1045 line and the advanced 1015 line is the running fix at 1045. A line drawn through this fix in direction 058° will represent the probable course of the ship. This will intersect the 1015 line in the estimated position at that time. When the ship arrives at point C the ship will have the lighthouse on the port beam. Measurements indicate that at 1045 the ship bears 170° distant 7.2 miles from the light, and that the light will be on the port beam at 1054 and will be 6.7 miles distant at that time.

Proper selection of the second bearing will give a fix without plotting the lines. If, in the above case, the pilot had noted the time when the relative bearing of the light was 064° off the port bow (twice the first value), the triangle A'FL would be isosceles, with the side A'F equal to FL. The side A'F is the distance run in the interval required to "double the angle on the bow" and the bearing and distance of the light at the time of the second observation is obtained without any plotting. An experienced pilot is familiar with a number of similar short-cuts for locating his positions by taking frequent bearings of available objects. Continual use of these, together with careful steering and a thorough knowledge of the speed of the ship and the set and drift of the current, will provide the pilot with a series of successive positions of his ship. In many cases, the scattering of the positions will indicate that the predicted currents are in error. By proper allowance for these abnormal conditions the ship may be saved from disaster. Similar methods are available to the air navigator and they may be used to check wind directions and speeds, as well as providing the pilot with accurate positions of his plane. Either visual or **radio bearings** may be used in flight and a series of running fixes obtained (see **Radio Navigation**).

The contour of the sea bottom may be used for determination of position of a ship. If a ship is held on steady heading and frequent soundings are taken, the depths may be plotted on a sheet of tracing paper, using the distance scale of a chart. The position of the ship is found by placing this tracing over the chart, with the line of soundings parallel to the course line, and moving it about until the observed values correspond with the depths shown. At present, a set of accurate contour maps of the ocean floor is in the process of construction. As these charts become available, and are used with a self-recording fathometer, the method of pilotage by depth will become more common and effective. A similar type of air pilotage will be available when the absolute **altimeter** is completely developed and ready for general use. (W.K.G.)

PILTDOWN MAN (Eoanthropus). Paleontology of Man.

PIN. The structural pin is a solid steel cylinder, usually heat-treated, which is used to connect steel members when freedom of angular movement at the joint is desired. It is threaded at both ends to receive large pressed or cast-steel pin **nuts,** which are used to secure the pin in its final position in the structure. The pin-connected joint is usually assumed to be frictionless but there is always a certain amount of friction present. The pin may be in direct contact with the member, as in the case of the **eye bar,** or it may pass through plates called pin plates which are attached to the members. It is also used to transfer the loads from the bridge **truss** to the end **bearing.** Pins are subjected to bearing, shearing, and bending **stresses.**

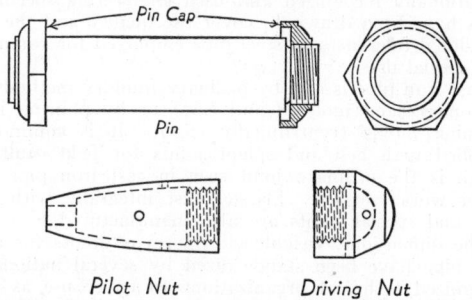

Pilot Nut   Driving Nut

After the members of a pin-connected joint have been assembled in position so that the pin holes line up, the pin is driven through the holes. This driving requires the use of a pilot nut and a driving nut, which are screwed on the ends of the pin. The pilot nut guides the pin through the pin hole and the driving force is applied to the driving nut. These nuts also protect the threads during the driving operation. When the pin is in position the nuts are removed and the pin nuts screwed on. (C.W.C.)

PINACOCYTE. A flattened **cell** of the outer layer of the body wall of **sponges.** These cells are able to change their shape. (A.W.L.)

PINACONE. Acetone.

PINEAL EYE. One of two organs similar to an eye in structure which develop from the roof of the brain. The pineal organ lies behind the parietal organ when both are present, but the evidence of embryology and of their connections with the brain suggests that the two were primitively paired, the pineal organ belonging to the right side, the parietal to the left. In amphibia it is the pineal organ, in reptiles the parietal, which is modified for light reception. The pineal organ undergoes a glandular modification in many vertebrates, including the mammals, leading to the unproved suggestion of an endocrine function. (A.W.L.)

PINEAPPLE. *Ananas sativus.* Bromeliaceae. The pineapple is a native of northern South America. Even in early times it had spread in cultivation to Peru and Mexico; now it is grown throughout tropical lands all over the world. On the short unbranched stem are many thick pale green leaves with sharp-toothed margins. At the top of the stem is borne the **inflorescence,** a cone-like bunch of bluish flowers, each in the **axil** of a **bract.** During ripening of the fruit all parts of the inflorescence, including the axis, bracts, **sepals, petals,** and **ovaries,** coalesce and become very fleshy, forming the familiar multiple fruit. The axis continues beyond this fruit as an elongate rosette of stiff-toothed leaves, which are generally marketed with the fruit. Only rarely do seeds form in this fruit, so propagation is accomplished by cuttings or by means of suckers which are formed quite abundantly.

The fruits are cut and sent to northern markets in a green state, coming into the United States mainly from Porto Rico and Cuba. In Hawaii, where the pineapple is an important plant, immense quantities are canned and put on the market.

From the leaves of the pineapple is obtained a very strong but fine white flexible fiber which is unaffected by water and extremely durable. From this fiber a delicate cloth may be woven. (R.M.W.)

PINHOLE "IMAGE." If a small opening is made in one side of a darkened room or box, an inverted picture of objects outside appears upon the wall opposite the opening. Such a picture differs from a true image in that it is not formed by light from a given point of

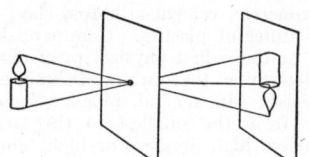

Diagram showing formation of pinhole image.

the source diverging and being reconverged at the corresponding image-point, as by a lens, but is an effect of the rectilinear propagation of light. The only spot on the screen reached by light from a given point of the source is that in direct line with the opening. For this reason, pinhole images are of low intensity. On the other hand, they are free from the distortions to which lens images are subject, and with sufficient exposure, very good photographs can be made by means of them. The pinhole image also affords an excellent means of viewing eclipses of the sun. (L.D.W.)

PINION. Spur Gearing.

PINK BOLLWORM. Insecta, Lepidoptera. A widely distributed enemy of cotton which probably originated in India or Africa. The adult is a small gray-brown **moth,** *Pectinophora gossypiella,* and the **caterpillar** is one-half inch long and is pinkish above. The **larvae** work in the flowers and bolls, causing imperfect development and destroying seeds and lint. It also attacks other plants, including the hollyhock and okra.

This species is found in the western part of the cotton-growing areas of the United States. Vigorous measures have been taken to eliminate it, for no adequate methods of control have been discovered. (A.W.L.)

PINNA. Ferns.

PINNATE. Leaf.

PINNIPEDIA. A division of the mammals containing the **seals, walruses**, and related forms. Sometimes regarded as an order and sometimes as a suborder of the order **Carnivora.** (A.W.L.)

PINNULE. A small lateral branch of the arm of a feather-star (sea lily, **crinoid**). The arms of these animals are sometimes divided into two or more radiating branches but whether branched or simple they bear small lateral branches which give them a feathery form; these are the pinnules. (A.W.L.)

PINO. The column of smoke and ashes emitted by an explosive **volcano**, usually in the beginning or in the early stage of an eruption. The term is of Italian origin signifying the cauliflower-like shape of the cloud as observed during the eruptions of Vesuvius. (R.M.F.)

PIÑON NUTS. Conifers.

PINWORM. Nemathelminthes, Nematoda. A small roundworm (**Nematoda**) which lives in the alimentary tract of man, chiefly in the large intestine. The female is about $2/5''$ long and the male somewhat smaller. Eggs are taken into the mouth in water or from the hands, or on raw vegetables. The entire life cycle takes place in the one host.

Pinworms are usually not harmful but they may cause nervous symptoms and are said sometimes to lead to appendicitis. (A.W.L.)

PIPE. Pipe is, categorically, the physical enclosure of a fluid flow. It is usually circular in cross-section. Pipe is made of steel, wrought iron, cast iron, copper, brass, lead, concrete, cement-asbestos, clay, wood or of a number of different plastics. Commercial sizes available, as well as individual physical properties and characteristics, determine the uses and limitations of the various materials. In general, pipe of carbon or alloy steel in sizes from the smallest to the largest is used exclusively where high pressure or high temperature, or both, are encountered. The other materials are selected for the lower pressure applications (approximately 300 lbs. per sq. in. and less) where their particular properties may be advantageous. Recognized codes having legal standing and covering recognized good practice frequently govern the engineering and technical details of pipe material selection where public interest or safety is involved.

Pipe size is commonly designated in terms of either inside (ID) or outside (OD) diameter.

The largest footage and tonnage are produced in the form of steel pipe which is made from flat-rolled material classified as skelp, plates, and strip. There are three basic methods of manufacture: the welding processes, the seamless process, and shop fabrication.

Welding processes include furnace welding (either butt-welded in sizes $1/8''$ to $4''$ inclusive or lap-welded in sizes $1\frac{1}{4}''$ to $24''$), electric resistance welding (sizes $1/8''$ to $30''$) and fusion welding (sizes $4''$ to $30'$ or larger).

Furnace welded pipe is not welded in a furnace as the term implies. The full length skelp is heated to welding temperature in a furnace and then, as each piece is withdrawn from the furnace, it is welded continuously. The terms "butt-weld" and "lap-weld" refer to the skelp edge forms and method of joining the edges preparatory to welding. In the first case the edges are cut square and upset and butted. The forming and welding of the heated skelp are accomplished by drawing through a bell-shaped die. In the second case they are scarfed and overlapped and passed through welding rolls. The welding and finishing processes leave little or no visible evidence of welds.

Electric resistance welding employs a series of operations, in the first of which the flat-rolled material is cold shaped into tubular form. Union of the seam is then effected by the application of heat and pressure. The welding heat, produced at or near the seam of the material, is generated by resistance to the flow of an electric current introduced through electrodes or by induction.

In the manufacture of tubular products by fusion welding, flat-rolled material which has had its edges suitably prepared is shaped either hot or cold to tubular form and united by a process of fusing with or without simultaneous deposition of filler metal in the molten or vapor state. No mechanical pressure is required with such welding which may be accomplished by either the electric arc or the gas method, or a combination of both. The term "electric welding" ordinarily embraces both the arc-fusion and electric-resistance methods. Fusion welded pipe is made with a welded seam parallel to the axis (straight seam) or winding around the pipe body in the form of a helix (spiral seam).

The seamless process includes the hot piercing method and the cupping method. In the former process, solid steel ingots, blooms, billets or round bars are hot pierced; then by either hot rolling or hot drawing, or a combination of both, the tubes are brought to the desired size. In the latter process, steel plates are hot cupped and hot drawn to the desired size.

Seamless tubes in the smaller sizes, or those with closer dimensional accuracy, are frequently further processed by cold drawing.

Shop fabrication is accomplished by curving flat plates into cylinders to form "cans" which are side-seam welded, then joined end to end to produce the finished pipe. This process is commonly used for the larger diameter pipe.

Nationally recognized and used engineering specifications have been drawn to cover the quality and the dimensional tolerances of steel pipe employed for common and special uses.

Cast iron pipe is cast by ordinary foundry methods in stationary sand molds (sand cast) or by pouring into spinning molds (centrifugally cast). It is commonly supplied with bell and spigot joints for field caulking which is the ordinary joint seen in cast iron pipe for water works service. Flanges cast integrally with the pipe and special joints are also manufactured.

The dimensions, weights, and form of joints for cast iron pipe have been standardized by several nationally recognized technical organizations to cover use as soil pipe, water pipe, and gas pipe.

Copper and brass pipe of seamless manufacture are available in various metal compositions in sizes $12''$ and smaller.

Lead pipe in the smaller diameters is available for plumbing and industrial uses.

Concrete pipe, plain or reinforced, and provided with several types of field joints in sizes from about $4''$ to $84''$, is used for drains and for water pipe under moderate pressures.

Cement-asbestos pipe is made of a material compounded from cement and asbestos fiber as the name indicates. It is mainly used as water pipe in sizes from $4''$ to about $30''$ as well as for various industrial uses for conducting fluids under comparatively low pressures.

Clay pipe in sizes from about $4''$ to $36''$ is manufactured from surface clay, fire clay, or shale. It is ordinarily used for drains and sewers where pressures and strength requirements are negligible.

Wood pipe is made by assembling wood staves and holding them in place by means of a continuous wrapping of steel wire or by individual threaded steel hoops in the same fashion as a wood tank. Continuous wood stave pipe is assembled in the field in such manner that the staves overlap each other with ends of staves staggered along the length of the pipe line. Thus it is distinguished from the ordinary pipe line in which each pipe is an individual unit connected to its neighbor with a field joint.

Plastic pipe has not come into general use but is available in smaller sizes for special purposes. (R. E. BARNARD)

PIPE COATINGS.

Inert materials are applied as films either to the inside or outside of pipe and tubing to prevent or slow down the deterioration or destruction of the metal from chemical action or corrosion. The materials commonly used are paints, enamels, cement, rubber, plastics, or layers of metallic substances such as zinc, tin, lead, copper, nickel, chromium, etc.

In general, thin paint coatings are most adaptable to the least corrosive conditions. Bituminous enamels, usually applied hot and in thicknesses of at least $\frac{1}{16}''$, are far more resistant than thinner coatings. Coal-tar enamel is generally considered best for underground steel water pipe. Cement is used for lining and outside protection. Rubber compounds and synthetic rubber are used most often as linings and the best are vulcanized in place. Plastics are being developed rapidly both as air-drying and as thermosetting types.

Zinc-coated (galvanized) steel pipe is the most widely used pipe with metallic coating. Other metal coatings are used as their resistance and practical economy dictate. (R. E. BARNARD)

PIPE FITTINGS.

A piping system, in fulfilling its function of providing a flow path for liquids or vapors, is rarely a straight run of pipe between two points. Flows are joined, parted, started, stopped, and regulated in the piping system. Only occasionally is it possible to take a "crow flight" path between end connections; the common run of pipe must follow configurations of equipment, walls, floors, beams, etc. Fittings and valves, properly incorporated in the pipe system, enable it to meet these varied service conditions.

In general, fittings consist of the pieces required to make turns, junctions, and reductions. The straight size fittings are the 45° and 90° elbows, the tees, crosses, Y's, laterals, and reducers. These fittings may also be had in special reducing sizes. (F.T.M.)

PIPE FRICTION.

Piping carrying fluid will have a frictional surface with the fluid. The extent to which friction is developed at this surface is a matter governing either the size of pipe which should be used, or the reduction by friction of pressure or velocity head in the pipe. The size of a pipe is not determined alone by the weight or volume of the fluid being transported. For instance, there is no one pipe size that must be selected to carry 500 cu. ft. of steam per min. At 5000 ft. per min. velocity a pipe of 0.1 sq. ft. cross-sectional area is required; at 10,000 ft. per min. it is 0.05 sq. ft. As soon as the velocity as well as the volume is known, the pipe size is determined according to the familiar relation:

Volume rate of flow = Velocity × Cross-sectional area.

The most difficult quantity to determine in this equation is the velocity. Of course, for short pipes it is easy enough to assume a velocity in accordance with actual practice. This assumption is rarely one giving the most economical pipe size, but as the pipe is short the failure to strike an economic balance is of minor importance.

The higher the velocity the smaller the pipe, but, at the same time, the greater the friction loss in the pipe. The most economical pipe size is that for which the annual fixed cost plus the annual cost of friction head is a minimum.

The following symbols are used in the formulae below:

h = Friction head in ft. of the fluid.
f = Coefficient of friction.
L = Length of pipe in ft.
v = Velocity of flow in ft. per sec.
g = 32.2.
d = Inside diameter of pipe in ft.
D = Inside diameter of pipe in inches.
z = Viscosity in centipoises.
S = Specific gravity referred to water as 1.
y = Density in lbs. per cu. ft.
P = Pressure in lbs. per sq. ft.
p = Friction head in lbs. per sq. in.

By assuming that the head lost in friction varies directly as the length of the pipe and as the square of the velocity of flow, and inversely as the hydraulic radius, the formula for feet head lost in friction is obtained:

$$h = \frac{fLv^2}{2gR}.$$

Practically all pipes are circular and $R = d/4$, d being the actual inside diameter of the pipe.

$$h = \frac{4fLv^2}{2gd}.$$

is the common formula for water pipes.

The coefficient for old pipe is to be regarded as highly approximate due to the uncertain nature of the surface of pipe designated merely as "old pipe." The character of the wetted perimeter is liable to change much during the life of the pipe through tuberculation, corrosion, organic growth, scale, silt, etc.

The coefficient f does not remain constant. In order to fit it to the experimental hydraulic data, investigators in this field have found it necessary either to express the equation with fractional exponents on v and d or to express f as a function of v and d. The equations given by Wilson, McAdams, and Setzler are:

For fluid flow in copper, brass, lead pipe

$$f = .00181 + .00662 \left(\frac{z}{DvS} \right)^{0.355}.$$

For fluid flow in clean iron and steel pipe

$$f = .0035 + .00594 \left(\frac{z}{DvS} \right)^{0.424}.$$

These formulae and constants are based on turbulent flow, as opposed to viscous flow before the critical velocity is reached. All ordinary water velocities are higher than the critical velocity and therefore represent turbulent flow—the more desirable of the two from the standpoint of friction drag.

When a compressible fluid such as air or steam flows in a pipe with average velocity v it can be shown that the pressure loss due to friction is

$$P_1 - P_2 = \frac{4fLyv^2}{2gd},$$

from which can be derived

$$p = \frac{fLyv^2}{6gD}.$$

A formula for f in terms of the dimensionless group $\left(\dfrac{Dy^v}{z}\right)$ is:

$$f = .0054 + .375\left(\frac{z}{Dyv}\right),$$

z for steam is

$$z = .0083 + 2 \times 10^{-5}t,$$

t being the steam temperature.

The friction loss through steam fittings and valves can only be approximated. (F.T.M.)

PIPE JOINTS. Individual pipe sections must be joined to form a continuous line. There are many types of joints with numerous variations but all may be included in five general classes:

1. Mechanical joints, restrained or unrestrained against movement along the pipe, which rely for tightness on stuffing box principles or on self-sealing V-shaped flexible gaskets held by steel housings.

2. Screw-joints having standard threads on pipe ends and in couplings.

3. Flanged joints bolted together with sealing gasket between the flanges. Flanges may be made integral with pipe or may be welded or otherwise attached. Flange and bolt dimensions have been standardized for various working pressures.

4. Welded joints made by joining plain- or bevelled-end pipe or fittings by means of electric-arc or gas welding.

5. Caulked joints of the bell and spigot type made tight by filling an annular space with lead or other caulking compound and then manually expanding this material within the bell, using hand tools. (R. E. BARNARD)

PIPERIDINE. Pyridine and Related Compounds.

PIPING CROW. Aves, Passeriformes. Australian birds (**Aves**) of several species related to the crows and jays but not true crows. They are black and white, whence comes their other name, Australian magpie. Unlike the true crows these birds are quite musical and can be taught to whistle tunes and to speak. They are frequently kept as cage birds. (A.W.L.)

PIPISTRELLE. Bat.

PIPIT. Aves, Passeriformes. Small quietly colored birds (**Aves**) of numerous species, related to the wagtails and warblers. They are widely distributed but most species occur in the Old World. In North America the common pipit, *Anthus spinoletta,* nests in the north and at high altitudes in the western mountains, and the one other species, Sprague's pipit, *A. spraguei,* is a bird of the plains. (A.W.L.)

PIRACY (STREAM CAPTURE). A term used by physiographers to designate the manner in which one stream or drainage system grows at the expense of the other. For example, two streams, tributaries to the master stream flowing at the foot of an escarpment, finally tap the head waters of the streams tributary to the master stream on the plateau, diverting the drainage. (R.M.F.)

PIRANHA, PIRAYA. Pisces, Teleostei. *Serrasalmo.* A fish (**Pisces**) of tropical American rivers which is noted for its ferocious attacks on living animals. It has very sharp teeth and is said to be attracted over considerable distances by blood. Although individuals are not large, so many swarm to a feast that a wounded animal in the water is said to be stripped of flesh very quickly. The species is dreaded by man. Similar fishes of African waters are known as dogs of the water. (A.W.L.)

PISCES. As used in astronomy, this is the name of a **constellation** (the Fishes). (Map, page 306.) Pisces is a large constellation which is of importance principally because it is the twelfth sign of the **zodiac.** There are relatively few interesting objects in the constellation although the brightest star (Alpha) is a close double which may be resolved in instruments larger than a four-inch. In spite of the fact that Pisces is the twelfth sign of the zodiac, nevertheless, the **vernal equinox** is located in this constellation at present. This is because **precession** has caused the vernal equinox itself to move back an entire "sign" along the ecliptic since the time when the names were first assigned.

In zoology, Pisces is a class of the phylum **Chordata** made up entirely of aquatic animals which breathe by **gills.** The few species adapted to leave the water either do so for short periods or remain on the wet shores. A few pass longer periods out of water but only in an inert state.

Fishes are characterized by the following structures: 1. The skin is kept moist by glandular secretions and in many species is covered with scales. 2. The appendages are in the form of fins, used in swimming. 3. The **gill slits** persist in the walls of the **pharynx** and serve as the seat of the respiratory gills. 4. The heart has only two principal chambers: an atrium or auricle and a ventricle.

Fishes are widely distributed in the oceans and in fresh waters. Many species are important as food. Billions of pounds are taken each year for this purpose, millions coming from a single species such as the more important salmons and marine species like the herring and cod.

Classification:

Subclass Elasmobranchii. Fishes with **cartilaginous** skeletons.
  Order Plagiostomi. With sharp-pointed (placoid) scales and separate external openings of the gill slits. The **sharks, dogfishes, rays, sawfishes, skates,** etc.
  Order Holocephali. With crushing plates in place of sharp teeth. External openings of the gill slits partly concealed by an **operculum.** The chimaeras.

Subclass Teleostomi. Skeleton composed at least in part of bone. Gill slits covered by an operculum.
  Order Crossopterygii. The air bladder is a paired ventral sac connected with the pharynx and used for breathing air. Many remains of extinct species.
  Order Chondrostei. The **paddle-fishes, sturgeons,** and *Polypterus.* The skeleton is composed largely of cartilage.
  Order Holostei. The **gar-pikes** and **bowfin.** The scales are covered with a hard material called ganoin and the skeleton is bony.
  Order Teleostei. An immense order containing most of the existing fishes. All of the familiar forms such as the **catfishes,** the **trouts** and **bass, salmon, herring, flounders,** and many others belong here. The group is so varied in structure that there is a tendency to elevate the many families that it contains to the rank of orders. When this is done the list is augmented by more than a score of orders in addition to those mentioned here.

Subclass Dipnoi. The **lung fishes.** These animals have slender pointed fins, crushing plates in place of teeth, and an air sac which serves as a lung. The existing species are found in Africa, tropical South America, and Australia. Some live in intermittent streams and pass the dry season coiled up in a cell formed in the dried mud of the stream bed, breathing air. (W.K.G., A.W.L.)

PISCICULTURE. The breeding of fishes (Pisces), together with other methods of maintaining the supply of these animals in heavily fished waters.

Because of the great importance of fishes as food and the consequent enormous removal each season by commercial fisheries, many species would have been depleted long ago but for artificial propagation and protection. In some cases, notably salmon and trout, it is possible to secure the eggs when they are ripe by gentle pressure on the body of the female. This process is called stripping. The germinal secretion of the male, called milt, is then mixed with the eggs to insure fertilization. The eggs are kept under favorable conditions which insure a large percentage of hatching and the young fish are protected until they are able to shift for themselves before being released in open waters. Hatcheries have been erected in many parts of the country for this work.

Some of the important food and game fishes cannot be handled in this way, but by permitting them to breed normally in protected waters it is possible to avoid much of the tremendous destruction which takes place under natural conditions. Waters are stocked with young fishes of a size which guarantees a much higher percentage of survival than can be expected for newly hatched young.

It is possible also to increase the number of fishes in natural waters at low expense by providing breeding stock from the hatcheries, although no protection can be given to the young in such cases other than the usual legal ban on fishing during the breeding season.

Most states now have rigid laws against taking fish during the breeding season or by destructive methods such as unlimited use of nets and the use of explosives. Unfortunately too many persons are ready to violate fishing laws, but these measures have undoubtedly been of great value.

The United States Bureau of Fisheries and the fish and game commissions and bureaus of conservation of the various states are responsible for most of the work in pisciculture being carried on in the United States. (A.W.L.)

PISOLITE. Oölite.

PISTACHIO. *Pistacia vera.* Anacardiaceae. This small tree, with **deciduous pinnate** leaves, is native in southwestern Asia, from which region it has spread in cultivation to the Mediterranean countries. The apetalous flowers are unisexual and borne in panicles and the plants are dioecious. The fruit is a drupe, containing an elongate seed with a greenish kernel, having a very characteristic flavor. The kernels are used in confections and ice cream and also eaten salted.

Related to this true pistachio is *Pistacia Lentiscus,* a shrub or small tree of the Mediterranean region with evergreen, pinnately compound leaves. From it is obtained a **resin,** mastic, which is often chewed by the natives of Turkey. It is used in varnishes and, in medicine, as a mild stimulant. Another species is *Pistacia Terebinthus,* a native of eastern Mediterranean countries, which yields China turpentine. (R.M.W.)

PISTIL. Flower.

PISTON. A piston forms the movable end of a **cylinder,** the volume of which is to be variable. Variable volume needed for expansion or compression of vapors and gases, or for displacement type pumps, is accomplished by the use of a cylinder in which there is a sliding piston. The space enclosed by the cylinder walls, cylinder head, and piston, forms the volume which can be varied by the motion of the piston. Pistons are extremely valuable, for variable volume mechanisms, of which they form an essential part, are the heart of such important machines as **engines, compressors,** and **pumps.**

Pistons may be classified as single-acting, double-acting, disk, plunger, or trunk type; also, of course, by their material. Most steam engines are double-acting, having the piston joined to the **connecting rod** by a piston rod and crosshead. The piston is disk-like in shape, being relatively thin, and having the piston rod rigidly attached to it. The steam engine piston is often made of cast iron, with grooves for two or three rings on its periphery.

The **steam engine** piston is different from the **internal combustion engine** piston because the latter works at

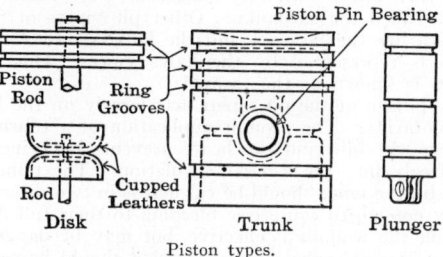

Piston Pin Bearing / Piston Rod / Ring Grooves / Rod / Cupped Leathers / Disk / Trunk / Plunger

Piston types.

much higher temperatures, and often at higher pressures than the steam engine. The internal combustion engine is usually single-acting, so that its piston is of the trunk type. The bore of an internal combustion engine cylinder is smaller, on the average, than the steam engine cylinder. A trunk-type piston is one which has the shape of a short cylinder closed at one end and open at the other. The closed end forms the piston face, in contact with the working medium. The remainder of it is the skirt of the piston, and is provided to align the piston properly in the cylinder, to support the wrist pin, and to provide bearing area on the cylinder wall against the side thrust arising from the angularity of the connecting rod. Since the gasoline engine operates at a much higher rotative speed than the steam engine, more attention is given to reducing the weight of reciprocating parts, such as the piston, in order to secure a better running balance. Many such pistons are made of aluminum alloy. The typical gasoline engine piston has a length about equal to its diameter. It carries internal bosses for the wrist or piston pin, and is ribbed near the top to give strength to the thin sections used. Leakage past the piston is sealed off with cast iron or alloy piston rings which are in the form of slices from a thin-walled cylinder. The rings are split and springy so that in use they press firmly against the cylinder wall, closing the gap between piston and wall. Piston ring grooves are machined circumferentially near the piston face. There are usually two or three of these grooves, the purpose of which is to contain the compression rings. Below them is another groove, which may be of the same size, but sometimes is larger, in which is installed an oil scraper ring, whose function is to prevent excessive amounts of oil being pumped by the piston up to the compression space. The wrist pin may oscillate in either the piston bosses, or the connecting rod, or may float in both.

Some pistons have specially shaped heads which are designed to carry out some idea related to the shape of the combustion chamber. Some are dished, some are crowned, some are very irregular in shape, but in each case the shape chosen was one which seemed especially suitable in conjunction with the shape adopted for the end of the cylinder.

Double-acting designs are occasionally found in internal combustion engine practice, and such pistons are hollow disks, through which cooling water is pumped. The water enters and leaves through a hollow piston rod, to which is attached a swinging or telescoping water inlet tube. (F.T.M.)

PISTON RING. Piston.

PIT VIPER. Reptilia, Sauria. A division of poisonous snakes characterized by the presence of a sharply defined pit between each eye and the adjacent nostril. The numerous species occur only in Asia and the Americas. The **copperhead,** water moccasin or **cottonmouth,** and **rattlesnakes** of North America are pit vipers. Rattlesnakes also occur southward into South America, where the more dreaded species of pit vipers are the large **fer-de-lance** and **bushmaster.** The former attains a length of 7′ and the latter 12′. Their fangs are correspondingly long and their poison glands large, hence their bite is extremely dangerous. The bushmaster is also called the surukuku. Other pit vipers of South America are the jararaca and the labaria. In Asia the group is represented by the halys vipers. One small species is known as the carawila.

The poison of the pit vipers acts largely on the blood (haemotarin). The prompt application of a tourniquet between the bite and the heart prevents its being carried freely into the general circulation. The punctures made by the fangs should be cut across in two directions, deeply enough to cause free bleeding to their full depth. Sucking the wounds is effective, but may be dangerous. In any case of snake bite a physician should be secured promptly if possible and an antivenin serum should be administered. (A.W.L.)

PITCH. Musical Sounds; Resins; Bitumens; Anticline; Chain; Gearing; Screw Thread; Riveted Joints.

PITCH, AIRPLANE. Pitching Moment.

PITCH POINT. Spur Gearing.

PITCH, PROPELLER. Propeller; Air, Pitch of.

PITCHBLENDE. Uraninite.

PITCHER PLANTS. Insectivorous Plants.

PITCHING MOMENT. Rotation of an airplane about a lateral axis passing through the center of gravity is known as *pitch*. Nosing-up and diving motions are the result of moments acting around this axis. These moments are produced by propeller thrust, wing lift and drag, parasitic drags, and tail surface forces. For airplane **trim** these moments must be in equilibrium, whereas for longitudinal stability an increase of angle of attack must produce a diving moment and vice versa. The relation between the moment and the moment coefficient is expressed in the following equation:

$$M = C_m qSc$$

q = Dynamic pressure equivalent to the air speed.

S = Wing area.

c = Wing chord.

(F.T.M.)

PITCHSTONE. Pitchstone is a volcanic glass of a dull pitchy luster. It contains about 5% of water, which distinguishes it from **obsidian,** of which it seems to be a variety.

Pitchstone is found in a variety of colors, including grays, reds, greens, and browns, or it may be black. (E.S.C.S.)

PITH. Stem.

PITHECANTHROPUS. Paleontology of Man.

PITOT TUBE. This is a tube having an opening turned upstream in a fluid flow. It receives the impact of the current against it, which, could it be completely converted into pressure head, would produce in the Pitot tube a pressure head of $\frac{v^2}{2g}$ superimposed on the existing static pressure of the fluid. When the Pitot tube, which is the impact tube, is used in connection with a **piezometer tube,** the static pressure may be subtracted from the total pressure given by the Pitot tube, leaving a difference which is velocity head. This arrangement is frequently used for measuring the velocity of flow of water or air. For example, a Pitot head for measuring the velocity of air is shown in the illustration. The two

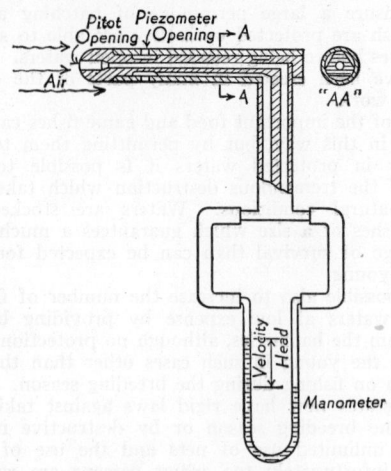

Pitot-static tube. (*Prandtl.*)

leads from this head are carried to either leg of a U-tube manometer. When a current of air flows past the head the liquid is displaced in the manometer by an amount which is proportional to the velocity squared. When this manometer reading is converted into head of the fluid, i.e., air, the velocity is $v = C \sqrt{2gh}$. C is the Pitot coefficient, obtained from a calibration test of the instrument, but is usually assumed to be 1. (F.T.M.)

PITS. Because increase in the thickness of the walls of the **cells** of plants does not take place over the entire surface of the cell, many thin places, called pits, are left in the wall. In cells in any given plant, pits have a very constant appearance, though they do vary greatly from plant to plant. Pits are most abundant in the walls of **tracheids** and **vessels,** and are either entirely absent or of rare occurrence in the walls of **fibers.** The thin places in the walls of two adjoining cells are found always to lie opposite to one another, though they may not have exactly the same size or shape. In many cases the thickened wall of the cell forms an overhanging rim around the pit, which is then called a bordered pit, the surface-view showing a distinct border around the opening. Pits which lack such a margin are called simple pits. In bordered pits the thin membrane has a thickened central portion known as the torus, which is often pressed closely to the overhanging margin, and apparently acts as a valve.

Bordered pits vary greatly in size and shape. Typically both the cavity and the opening of the pit are circular. In much-thickened walls the opening of the pit may become slit-like instead of circular. Often the slits are very numerous and extend far beyond the margin of the actual cavity. Pits may be found irregularly scattered over the entire surface of a cell, or they may occur in regular rows. Frequently they are crowded so closely together as to appear **polygonal** in shape. The function of pits seems to be to facilitate the diffusion of liquids from cell to cell. (R.M.W.)

PITTA. Aves, Passeriformes. *Pitta.* Small, brightly colored birds (**Aves**) of the Old World, also called the ant thrushes, water thrushes, and ground thrushes. They are only superficially like the true thrushes. (A.W.L.)

PITUITARY GLAND (THE HYPOPHYSIS). The pituitary gland is one of the endocrine glands. It is situated at the base of the brain in a depression of the sphenoid bone called the *sella turcica.* Because of its many *hormones* which are vitally concerned in fundamental physiological functions, it is sometimes designated as the "master gland." It is about 1 cm. in size, and composed of three parts, the anterior and the posterior lobes and the pars intermedia, each with different functions.

The posterior lobe of the pituitary gland is of neural origin (derived from brain tissue). Its physiology is not completely understood. Posterior pituitary extract or "pituitrin" possesses three distinct actions. One is the stimulation of the uterus (oxytocic effect) so that the uterine muscle contracts. Another is the constriction of blood vessels (pressor effect); and a third is the action on the kidney tubules to increase water reabsorption (antidiuretic effect). These functions form the basis of the therapeutic uses of posterior pituitary extract. The drug is used to induce labor at the end of pregnancy under certain circumstances. During the third stage of labor it is given to promote contraction of the uterus and thus prevent undue hemorrhage; in the days following delivery it is administered routinely to aid in the involution of the uterus. In *diabetes insipidus,* a disease associated with changes in the hypothalamic region of the base of the brain, it is given because of its antidiuretic effect since patients with this condition excrete enormous quantities of urine. After abdominal operations, pituitrin is given to prevent distension due to relaxation of the bowel musculature. Its pressor substance is probably responsible for this effect.

The anterior lobe originates embryonically from the epithelial lining of the pharynx. This lobe may be divided into two main parts, the anterior and intermediate portions. The function of the latter is not known. The anterior portion has many functions, and several of its hormones have been isolated.

Removal of the anterior lobe in young mammals causes cessation of growth, bone changes, failure of development of the permanent teeth, and failure of sexual development. There are also atrophic changes in many of the endocrine glands, showing that the pituitary is the master control of the endocrine system.

The number of hormones produced by the anterior pituitary is not known. At least nine separate types of activity have been attributed to its secretions, but it is probable that in some instances a single chemical substance is responsible for several effects. The following is a list of hormones as they are commonly designated:

1. Growth. Excessive production results in gigantism and **acromegaly**; decreased production is responsible for certain types of dwarfism.
2. Gonadotrophic. Follicle-stimulating and luteinizing hormones are intimately connected with ovarian function and **menstruation.** In the male, the follicle-stimulating hormone induces the development of spermatozoa in the **testicle** while the luteinizing hormone stimulates the production of the male sex hormone, testosterone.
3. Lactogenic. This hormone is necessary for the initiation of the flow of milk and normal lactation following pregnancy.
4. Thyrotropic. Without thyrotropic hormone, the thyroid gland undergoes atrophy; in the presence of excess, the thyroid increases in glandular tissue and function.
5. Adrenotropic. The adrenal cortex hypertrophies if an excessive amount of this hormone is present, and atrophies in its absence.

About the rest of the principles little is known. They are: 6. Diabetogenic. 7. Parathyrotropic. 8. Ketogenic. 9. Pancreatropic. (R.S.M., D.M.H.)

PIVOT BEARING. Bearings.

PLACENTA. 1. A structure which attaches the unborn **embryo** or fetus to the uterine wall of the mother in the true **mammals.** It is formed from a portion of the **chorion,** and the allantoic or umbilical blood vessels provide communication between it and the circulatory system of the embryo. In the simpler placentas the outer surface of the chorion is merely in close contact with the lining of the **uterus** and the structure is called a contact placenta or a semi-placenta. Many mammals, including man, have a burrowing or true placenta. In this form the chorion bears many root-like processes which invade the lining of the uterus and come into direct contact with the blood of the mother. In no type of placenta do the fetal and maternal bloods mingle, although interchange takes place between them through the thin intervening tissues. Many names have been applied to different forms of placentas, both normal and abnormal.

The human placenta is a flat circular mass of tissue, which serves to establish communication between the mother and the child by means of the **umbilical cord.** It is about 6 or 7″ in diameter, about 1″ in thickness, and weighs about 1 lb. During the third stage of labor, after the birth of the child, the placenta and membranes are passed (in the after-birth). If before the child is due, a separation of a portion of the placenta occurs, uterine hemorrhage results and miscarriage usually takes place (premature separation of the placenta). The placenta is normally attached to the uterus away from the **cervix** or the opening of the uterus into the vagina. When the placenta overlaps or covers this opening a condition known as *placenta previa* exists which may result in fatal hemorrhage with death of the child and mother.

2. Among the viviparous **arthropods** the young are attached to the walls of the genital passages for nourishment in some species. The attaching structure is called a placenta. It occurs in **onychophora, scorpions,** and a few **insects (flies).** The salpians also have an attachment of this kind. (A.W.L., R.S.M.)

PLAGIOCLASE. Feldspar.

PLAGIONITE. A mineral antimony sulfide of lead, $Pb_5Sb_8S_{17}$. Crystallizes in the monoclinic system. Hardness, 2.5; specific gravity, 5.56±; color, gray to black with metallic luster; opaque. From the Greek, meaning oblique in reference to the obliquity of the crystals. (R.M.F.)

PLAGUE. An infectious disease caused by a rod-shaped bacterium, *Pasturella pestis.* Plague probably originated in Central Asia, but it has been known in Europe for many centuries. Its rise and spread have always been associated with wars. The Crusades were responsible for the Black Death of the 14th century which killed approximately ¼ the population of Europe, or 25,000,000 people. Since the middle of the 19th century, the disease has been rare in the Mediterranean basin, even though a pandemic arising in South China has spread along trade routes into the ports of Asia, North and South America, Australia, and Africa and caused many millions of deaths.

Plague is a rat infection transmitted to man by the bite of fleas which have fed on infected rodents. The **epidemic** and pandemic varieties occur in seaports, and spread is accomplished by the rat population of ships. In the western part of the United States, extending from the Pacific to North Dakota, there is a permanent **endemic** plague area. Here, particularly in sparsely populated areas, sylvatic plague is always present in wild rodents, ground squirrels, chipmunks, marmots, prairie dogs, wood rats, hares, and rabbits. At intervals, the disease flares up in these animals and appears as explosive outbreaks which kill off great numbers of them. On rare occasions, man becomes infected—as with the epidemic form—through bites from fleas which have fed on these infected rodents.

The incubation time of plague is from 2 to 5 or more days. The bubonic form is characterized by painful enlargement of the lymphatic glands of the groin, axilla, or neck. Suppurating **buboes** containing the causative organism may occur. The onset is sudden and is characterized by chill, headache, backache, high fever and rapid pulse. Extreme prostration and lethargy develop rapidly and vomiting and delirium may occur. The mortality is around 75% and death usually occurs on the third or fourth day. Even in cases which seem to be recovering, sudden heart failure may develop with immediate death.

The pneumonic form is characterized by rapidly spreading **pneumonia.** Here the onset is sudden, prostration may be extreme and the lungs fill up with frothy blood. Death invariably results in 1 to 4 days. This is the only form of plague which is transmitted from man to man without the intermediatory transmission of fleas. The infection is carried by droplets coughed by the patient. The strictest isolation and quarantine must be followed.

Septicemic plague is a fulminating form which develops so rapidly that death occurs before buboes develop. The signs are those of a fulminating pneumonia accompanied by blood-stained watery sputum. It is invariably fatal within a few days.

The treatment of plague consists of the administration of immune **serum** and **sulfadiazine.** The latter has been effective in reducing the mortality considerably. It is of importance that an early diagnosis be made since treatment is more effective in the early stages. Furthermore, the patient is a dangerous source of infection and the strictest isolation precautions must be used when the diagnosis is suspected.

The prevention of plague is a matter of rodent control. Rat-proofing of ships and ridding cargoes of fleas and rats in port have cut down the incidence tremendously. The control of infected rural rodent populations in the West is also of great importance. Individuals likely to be exposed should be immunized with plague **vaccine** which offers partial protection. (D.M.H.)

PLAICE. Pisces, Teleostei. A European **flatfish,** *Pleuronectes platessa,* related to the flounder but of exclusively marine distribution. (A.W.L.)

PLANCK CONSTANT. Planck's Law.

PLANCK'S EQUATION. An equation developed by Max Planck in 1900 to represent the **spectral energy distribution** of the radiation from a **black body** at a given temperature, in accordance with his then newly conceived **quantum theory** of radiation. It expresses the emissive power of a black body within the infinitely small wavelength range, λ to $\lambda + d\lambda$, as follows:

$$dE_\lambda = 2\pi c^2 h \lambda^{-5} [e^{ch/k\lambda T} - 1]^{-1} d\lambda.$$

T is the absolute temperature of the black body, c is the **electromagnetic constant,** h is Planck's constant (see **Planck's Law**), and k is Boltzmann's constant (see **Ideal Gas Law**). The value of the coefficient $2\pi c^2 h$ is about 3.740×10^{-5} erg cm.^{2}/sec., while the factor ch/k in the exponent is 1.438 cm. deg. An alternative form expresses the radiant energy density, within the same range, of the radiation in thermal equilibrium with a black body as

equal to $8\pi ch\lambda^{-5}[e^{ch/k\lambda T} - 1]^{-1} d\lambda$. The law is found to agree accurately with experiment. (See **Wien's laws.**) (L.D.W.)

PLANCK'S LAW. The fundamental law of the **quantum theory,** expressing the essential concept that energy transfers associated with radiation such as **light** or **x-rays** are made up of definite quanta or increments of energy proportional to the frequency of the corresponding radiation. This proportionality is usually expressed by the quantum formula $q = h\nu$, in which q is the value of the quantum in units of energy and ν is the frequency of the radiation.

h, the constant of proportionality, is known as the elementary quantum of action or more commonly, "Planck's constant." Since q is energy and ν is frequency, h has the dimensions of energy $\times$ time, or **action.** Its value has been found to be about 6.624×10^{-27} erg-second. Thus, if the wavelength of a certain monochromatic green light is 5000 angstroms, its frequency is 6×10^{14} per sec., and its energy is made up of quanta each equal to 6.624×10^{-27} erg-sec. $\times 6 \times 10^{14}$/sec., or 3.97×10^{-12} erg.

Why such a relation should exist between these natural units or quanta of radiant energy and the frequency of the radiation, is still an unsolved problem. (L.D.W.)

PLANE. A plane is a fundamental geometric concept, being one of the basic geometric elements used in the construction of geometric figures in space.

The equation of any plane in **rectangular coordinates** x, y, z is of the first degree (i.e., is a linear equation).

Every linear equation in rectangular coordinates x, y, z is represented by a plane in space.

The normal form of the equation of a plane in rectangular coordinates is

$$x \cos\alpha + y \cos\beta + z \cos\gamma - p = 0,$$

where p is the perpendicular distance from the origin to the plane, and α, β, γ are the direction angles of that perpendicular.

The intercept form of the equation of a plane is

$$\frac{x}{a} + \frac{y}{b} + \frac{z}{c} = 1,$$

where a, b, c are the intercepts of the plane on the axes.

The equation of a plane through 3 given points may be written

$$\begin{vmatrix} x & y & z & 1 \\ x_1 & y_1 & z_1 & 1 \\ x_2 & y_2 & z_2 & 1 \\ x_3 & y_3 & z_3 & 1 \end{vmatrix} = 0.$$

To reduce the general equation $Ax + By + Cz + D = 0$ of a plane to the normal form, divide each coefficient by $\pm\sqrt{A^2 + B^2 + C^2}$.

The coefficients of x, y, z in the equation of a plane are proportional to the **direction cosines** of a line perpendicular to the plane.

The angle θ between two planes $Ax + By + Cz + D = 0$ and $A'x + B'y + C'z + D' = 0$ is given by

$$\cos\theta = \frac{AA' + BB' + CC'}{\sqrt{A^2 + B^2 + C^2} \cdot \sqrt{A'^2 + B'^2 + C'^2}}.$$

These two planes are parallel when and only when $A/A' = B/B' = C/C'$ (i.e., when the corresponding coefficients of x, y, z are proportional).

These two planes are perpendicular when and only when $AA' + BB' + CC' = 0$.

The intercepts of a plane are the distances cut off on each of the coordinate axes; they may be found by putting successively each pair of coordinates equal to 0 and solving the equation of the plane for the other coordinate.

The traces of a plane are the straight lines in which the plane cuts the coordinate planes; they may be found by putting successively each coordinate equal to o in the equation of the plane.

The perpendicular distance of a point $P_1(x_1, y_1, z_1)$ from a plane whose equation is $x \cos \alpha + y \cos \beta + z \cos \gamma - p = o$ is given by

$$d = x_1 \cos \alpha + y_1 \cos \beta + z_1 \cos \gamma - p;$$

the perpendicular distance from the plane $Ax + By + Cz + D = o$ is given by

$$d = \frac{Ax_1 + By_1 + Cz_1 + D}{\sqrt{A^2 + B^2 + C^2}}.$$

If $P_1 = o$ and $P_2 = o$ are the equations of two planes, the equation $P_1 + kP_2 = o$ is a system of planes through the intersection line of the two given planes. (See **Woodworking.**)   (L.L.S.)

PLANE SAILING. This term is applied to the solution of various problems in the **sailings** in which the earth is considered as a plane surface. The particular subject of plane sailing will be found discussed under the topic **Dead Reckoning.**   (W.K.G.)

PLANE SURVEYING. Surveying.

PLANE TABLE. A plane table is a **surveying** instrument which is used for locating and mapping topographical features. A drawing board, accurately made, and arranged so that it may be mounted on a tripod by an adjustable head which allows leveling of the board, is an essential feature of the plane table. Spirit levels are attached to the table in mutually perpendicular directions. The compass, the ruler, and a means for getting a line of sight, such as a telescope or open sights, complete the outfit. A ruler combined with a telescope or with slit sights is called an **alidade.** When the plane table is used for a survey, it is not necessary to take notes of angles or lengths of lines, since they are plotted, at the time of the survey, on the sheet of paper which covers the plane table. Obviously the plane table is not suitable for use in bad weather. When a survey is to be made with this instrument, the table is set up so that some convenient point on the paper is over a selected spot on the ground. The table is leveled and rotated horizontally until it is in **azimuth.** This is accomplished by means of the compass or by sighting back on a known point. It is then clamped in this position and the ruler is brought to the point selected on the paper and swung about it so that the line of sight which parallels the ruler bears on a distant point whose location is desired. A line is drawn in that direction, and after the distance to that point is meas-

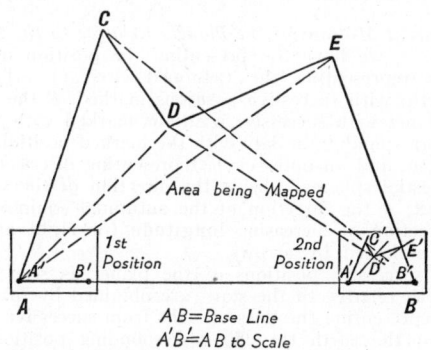

$A B = Base Line$
$A'B' = A B$ to Scale
Plane-table method of survey.

ured, the length of line is plotted to some scale suitable to include the area being mapped on the surface of the plane table.

It is entirely possible to locate the points of a survey, with the exception of the two points specified in the

preceding paragraph without measurement. This method is known as the method of intersections, and requires only one distance to be actually measured. In the method of intersections, the plane table is set over one point, and a line of sight taken on some other selected point, possibly a corner of the area. The distance between those two points is measured, and the other point put in by scale on the plane-table drawing. The alidade is then swung about the initial point, and lines are drawn toward the several other corners which are to be located. The plane table is then set up over the other terminus of the measured base line, and oriented by a compass or by aligning the initially measured line with the initial point by alidade. The alidade is then swung about the second point and lines are drawn again toward the same other corners that are to be located. The intersections will give the location of the points.

Topography may be plotted on a plane table map by either of two methods. If the alidade is telescopic with stadia attachment the general method for topography by stadia may be adapted to plane-table mapping. By this method a small contour interval may be satisfactorily mapped. If the alidade does not have the stadia attachment points, elevation may be secured by the aneroid barometer. When these points are plotted on the map the contours may be secured by interpolation. In either case, sketching contours in the field is preferable to office execution. The use of the aneroid barometer is limited to large contour intervals such as 20 to 100′.   (F.T.M., E.W.S.)

PLANER. For wood planers and shapers, see **Woodworking.**

Planers, shapers and slotters are machine tools that employ single-point tools to generate flat surfaces. In each of these the relative motion of the cutting tool and the work is rectilinear, and either the tool or the work feeds in a direction perpendicular to the cutting stroke. In the planer, the work is held on a horizontal table and moves past a stationary tool; in the shaper, the work is stationary and the tool moves over it. The slotter may be called a vertical shaper, since the tool moves in a vertical direction past the stationary work. All three machines finish surfaces in a similar manner, and their selection depends primarily upon the nature of the work.

The planer is generally used for large work. For comparatively small work the shaper is used, unless a large number of like parts are to be finished. In this case the parts are frequently placed on a planer table in rows, and a number of parts are placed at one setting. This operation is referred to as string planing. The slotter is used for machining flat surfaces which are difficult or inconvenient to machine because they are at right angles to the main dimensions of the part. The slotter is also employed for cutting internal keyways, square holes, and die openings.

Planers and shapers are used for machining surfaces to a high degree of accuracy, and in general require less power per cubic inch of metal removed than machine tools employing multi-toothed cutters. Planer and shaper tools are considerably less expensive than milling cutters; the planer may therefore be used in preference to a milling machine if the castings are poor and subject to hard spots.

The conventional form of crank-driven shaper derives its cutting motion from a pivoted lever, which is driven by an adjustable crank. (See **Quick-Return Mechanism.**) The table which carries the work feeds in a direction perpendicular to the tool motion, and may be adjusted for various heights. Universal shapers are equipped with tilting or adjustable tables so that angular work can be handled. Shaper tools are similar to solid or inserted-bit lathe tools; extension tools are used for cutting keyways and square and splined holes.

Shaper work may be held in a vise or clamped to the table.

There are two standard types of planers that are in extensive use in jobbing and production shops. One of these types is the double-housing planer; the other is the open-side planer. The work is bolted or otherwise securely fastened to a table or platen, the under side of which is provided with two accurately machined guides which slide in guide ways on the planer bed. The table moves against one or more cutting tools, which are held by the rail and side heads, at a speed adapted to the material to be cut. The return stroke, during which no cutting takes place, is usually constant, but is from two to four times as fast as the cutting stroke so as to economize on time.

Planer size is determined by the maximum stroke of the table and the width and height of work that will pass through the housings and underneath the cross-rail. A double housing 30″ x 30″ x 8′ planer, for instance, will machine a part 30″ high, 30″ wide and 8′ long. Open-side planers are classified by the cross-rail height and the length of stroke, and will generally handle work that is somewhat wider than its height.

Planer work may be held in a vise bolted to the planer table or the work may be clamped directly to the table. The table has three or more tee slots running lengthwise and numerous holes for inserting stops and clamping blocks. Castings can generally be clamped in place by using straps or clamps on projecting portions of the work. Planer tools are similar to shaper tools, but are usually larger. In many instances, gang planer tools, carrying three or more tool bits closely adjacent, are used. As each chip is comparatively small, a planer equipped with a gang tool will carry a far greater total feed and depth of cut than are possible with a single-point tool.

A slotter is a shaper with a vertical ram which reciprocates in a ram slide which may be swung to any angle from a vertical position up to 5° from the vertical. The ram receives its motion from a rotating crank which drives a slotted rocker arm through a driving block. A movable saddle is mounted on the bed and dovetail guides permit motion in one direction. The table base is mounted on the saddle and guides integral with the saddle permit perpendicular transverse motion. The rotary table is mounted on a swivel bearing in the table base and may be rotated about its center. The rotary table may be rotated for feeding in either direction by power or by the use of the handwheel; and both the table base and the saddle may be hand-adjusted or power fed in either direction. The rotary table has a graduated edge, and the table handwheel has a micrometer dial subdivided so that the smallest divisions represent two minutes of arc. There are indexing notches provided so that the table can be indexed quickly through the major angles and positioned exactly by a small plunger. (H.C.H.)

PLANETARY MOTIONS. The apparent motions of the planets on the celestial sphere have been observed, recorded and speculated about ever since mankind has existed on the earth. A large part of the pseudo-science of astrology is concerned with these motions, or, more particularly, with the various "aspects," or configurations, of the planets. The motions as seen from the earth are complicated by the fact that the earth is moving about the sun in the same direction as the planets, but with a different rate.

Apparent Motions Relative to the Sun. In Fig. 1, we have S representing the sun, E representing the earth (assumed fixed for the purpose of convenience), P'_1, P', P'_2, P'_3, P'_4 indicating various positions of a planet whose orbit lies between the earth and the sun (inferior planet), and P_1, P, P_2, P_3, P_4, positions of a planet whose orbit is outside of that of the earth (superior planet). The angle between the sun and the

planet (e.g., SEP' or SEP) is defined as the elongation of the planet. For an inferior planet the elongation may have any value from 0 to SEP'_2 or SEP'_4 while

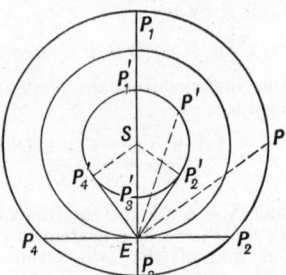

Fig. 1. Motions of planets as seen from earth relative to the sun.

for a superior planet the elongation varies either east or west from 0 to 180°. With elongation 0 we have the planet in the aspect of conjunction.

Inferior planets: It will be noted that the elongation is 0 both at P'_1 and also at P'_3 and hence there are two conjunctions for an inferior planet. To distinguish between them P'_1 is known as superior conjunction and P'_3, inferior conjunction. An inferior planet moves more rapidly in its orbit than does the earth and accordingly from superior conjunction the planet moves out with increasing eastern elongation (evening object) to the point P'_2 (greatest eastern elongation). It then moves in with decreasing elongation, passes the sun and becomes a morning object at inferior conjunction (P'_3) and moves out with increasing western elongation to P'_4 (greatest western elongation) and thence back to superior conjunction again. Hence these planets apparently oscillate back and forth across the direction of the sun. They do not ordinarily pass either between the earth and the sun or directly behind the sun because of the fact that their orbits are not in the plane of the ecliptic (see **Transit of Venus**).

Superior Planets: It must be remembered that these planets are moving more slowly in their orbits than is the earth. These planets apparently move from conjunction at P_1 slowly out to the west of the sun (morning objects) to P_2 where the western elongation is 90° and the aspect is western quadrature. From this point the increase in western elongation increases rapidly to 180° at P_3 (aspect opposition) from which point the elongation becomes east (evening object) and decreases rapidly to 90° eastern elongation at P_4 (eastern quadrature). The decrease in eastern elongation then slows down as the planet moves slowly back to conjunction again.

Apparent Motions of the Planets Relative to the Stars. In Fig. 2, we have S representing the position of the sun, E representing the (assumed circular) orbit of the earth, with successive positions marked, P the orbit of a planet with successive positions marked with numbers corresponding in date with the marked positions for the earth, and an outer circle representing directions on the celestial sphere, V being the direction of the vernal equinox, A the direction of the autumnal equinox and the direction of increasing **longitude** (or **right ascension**) indicated by arrows.

The successive positions of the planet as seen from the earth relative to the stars, are obtained by drawing lines, representing the lines of sight from successive positions of the earth through corresponding positions of the planet. By examining the successive directions, as indicated by numbers on the outer circle, it will be evident that the general trend of the planetary motion is in the direction of increasing right ascension. Such motion in increasing right ascension is known as direct motion. It will be noticed, however, that in the vicinity of opposition of the planet, the motion reverses for a period (numbers 6, 7, and 8) and the planet moves in the direction of decreasing right ascension. Such mo-

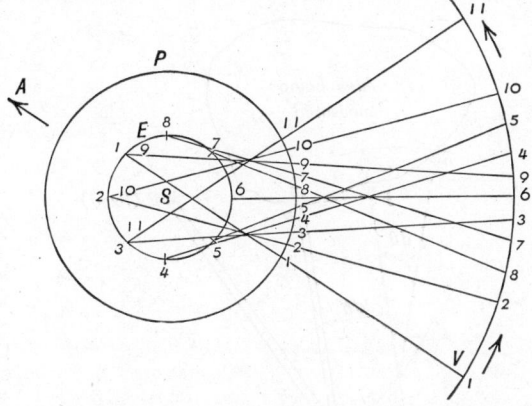

Fig. 2. Motion of planet as seen from the earth relative to the stars.

tion, in the direction of decreasing right ascension, is known as retrograde motion.

Since the diagram is plotted in the plane of the ecliptic with the planet also assumed in this plane, the reversal in direction appears only in longitude. However, there will also be changes in direction of motion both in celestial **latitude** and **declination**, giving the appearance on

the celestial sphere of loops in the motion of the planet. These loops were observed by the early astronomers and it was to account for those on the assumption of the geoconcentric universe that the epicycles of the **Ptolemaic system** became necessary.

The general subject of planetary motions referred to the sun alone will be found discussed under the general topic of orbits. (w.k.g.)

PLANETS. The word planet, which comes from a Greek root meaning "wanderer," was used prior to the 15th century to designate those celestial objects (other than **meteors** and **comets**) which were observed to be in motion relative to the stars. Prior to the 15th century, seven objects were listed as planets: **Sun, Moon, Mercury, Venus, Mars, Jupiter,** and **Saturn.** With the advent of the **Copernican** heliocentric hypothesis for the structure of the **universe**, the sun and moon were removed from the list and the earth added. Since the application of the telescope to astronomy three large planets have been added (**Uranus, Neptune,** and **Pluto**) and over 1000 small planets or **asteroids.** The term as it is used at present applies to any opaque object which is shining by reflected sunlight and travels about the sun in an **orbit.**

In spite of the fact that many of the planets are larger than the earth, their distance is so great that they appear to the naked eye as bright stars. The only certain method for distinguishing a planet from a star without the use of a telescope is to watch it carefully for a considerable period, frequently several days are re-

PLANETARY TABLE I. ORBITAL CHARACTERISTICS

PLANET	MEAN DISTANCE FROM THE SUN		ECCENTRICITY	INCLINATION TO ECLIPTIC, DEGREES	ANGULAR MOMENTUM	PERIODS	
	Millions of Miles	Astr. Units				Sidereal in Years	Synodic in Days
Mercury......	36.0	0.387	0.2056	7.0036	0.02	0.241	115.88
Venus........	67.2	0.723	0.0068	3.3940	0.07	0.615	583.92
Earth........	92.9	1.000	0.0167	0.0000	1.00	1.000	......
Mars........	141.5	1.524	0.0933	1.8501	0.13	1.881	779.94
Jupiter.......	483.3	5.203	0.0484	1.3067	722	11.862	398.88
Saturn........	886.1	9.539	0.0558	2.4909	293	29.458	378.09
Uranus.......	1783	19.182	0.0471	0.7729	64	84.015	369.66
Neptune......	2793	30.057	0.0086	1.7758	94	164.788	367.49
Pluto........	3671	39.518	0.2486	17.1468	1.2(?)	248.430	366.73

PLANETARY TABLE II. PHYSICAL CHARACTERISTICS

PLANET	MEAN DIAMETER		TEMP. IN °K.	ON SCALE OF EARTH = 1			ROTATION PERIOD
	In Miles	Apparent in Seconds of Arc		Mass	Mean Density	Surface Gravity	
Mercury......	3,100	5.45	690	0.037	0.68	0.26	88 (dy.)(?)
Venus........	7,700	30.40	290	0.826	0.94	0.90	(?)
Earth........	7,920	....	287	1.000	1.00	1.00	24.00 (hr.)
Mars........	4,215	8.94	285	0.108	0.71	0.38	24.60 (hr.)
Jupiter.......	85,700	22.65	135	318.4	0.24	2.65	9.9 (hr.)
Saturn........	71,500	9.25	120	95.2	0.12	1.14	10.2 (hr.)
Uranus.......	32,000	1.88	Below 90	14.6	0.25	0.96	10.7 (hr.)
Neptune......	33,000	1.26	Below 90	17.3	0.24	1.00	15.8 (hr.)
Pluto........	2,000(?)	(?)	Below 90	1.0	0.8(?)	(?)	(?)

quired, and if the object is a true planet, it will move relative to the stars. For a quick method of identification, it may be said that usually a planet does not appear to twinkle as do the stars, but this rule is not infallible. With a telescope a planet may be immediately distinguished from a star (with the exception of the planet Pluto or the asteroids) because of the fact that a planet will show an appreciable disk while the stars appear as points of light no matter how much magnification is used.

The planets are classified in two general ways. Mercury and Venus are frequently referred to as the inferior planets, while the others are called the superior planets. Another system of classification considers Mercury, Venus, Earth, and Mars as the minor or terrestrial planets, while Jupiter, Saturn, Uranus, Neptune, and Pluto are called the major planets.

In this article we shall limit ourselves to the material contained in the accompanying tables. Elsewhere in this work will be found discussions of planetary **motions**, descriptions of the individual objects, and an outline of the various theories of planetary evolution under the discussion of the **solar system**. The **satellites** will also be considered elsewhere.

In Planetary Table I will be found the characteristics of the individual planets which have to do with distances and motions of the planets themselves. The various terms used for column headings will be found defined elsewhere in this work either under their individual terms or under the general subject of **orbit**.

Planetary Table II contains data relative to the physical conditions of the planets themselves. The mean diameter in seconds of .arc, given in the third column, is the average diameter as seen from the earth. Since the distances of the planets from the earth, particularly in the case of Venus and Mars, vary through wide limits, the angular diameter at any particular instant may be very different from the value given. The fourth column gives the average temperature of the objects on the Kelvin scale. The values given are those determined by means of **radiometric** observations. In the case of Mercury the value for the sunlit side is given, in the case of Venus the average of the dark and bright sides is used, and for Mars the value obtained for the warmest portion is taken. Further information regarding planetary temperatures will be found in the articles dealing with the individual objects. The mean densities are calculated from the masses and measured diameters. It should be noted that the measured diameter is that of the planet plus atmosphere and probably in the cases of the outer planets the layer of atmosphere is very thick. Possibly this accounts for the low value of the density in comparison with that of the earth. By surface gravity is meant the **gravitational** attraction on unit mass at the surface of the planet. For example, on Jupiter a mass of 10 grams would have a **weight** of 26.5 grams. All values tabulated for the planet Pluto are uncertain. (w.k.g.)

PLANIMETER. The polar planimeter, which is credited to Amsler, is an instrument for the measurement of irregular plane areas. While the area of a regular plane figure may be found by the application of rules for averaging, methods of subdivision, etc., such methods are tedious compared to the ease and accuracy with which the planimeter measures irregular areas. The general form of the polar planimeter is shown in the accompanying figure. There are two arms pivoted together at the point K. The arm PK is free to rotate about the needle point P, and is held in place at that point by a weight. The other arm KF carries a tracing point at F. This is guided around the border of the area to be measured. The arm KF also carries a wheel whose axis must be parallel to KF. The rim of this wheel is in contact with the paper so that any motion of the arm except that in the direction KF will cause the wheel to rotate. A

graduated scale with **vernier** attachment measures the travel of the circumference of the measuring wheel. To use the planimeter, the tracing point is set at some initial

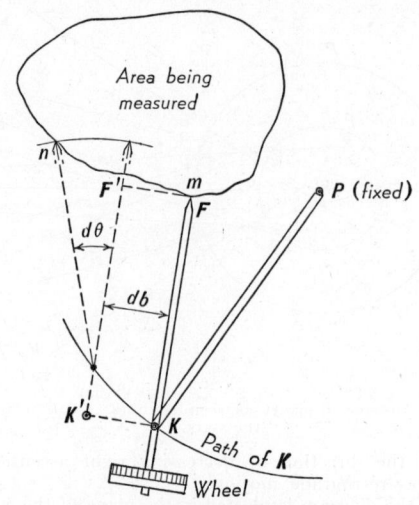

Theory of the Planimeter

point on the circumference of the area, and the reading of the record wheel is next noted. The tracing point is carefully guided around the border of the area clockwise until it has returned to the starting point. A final reading is taken and subtraction of the two readings gives a number which is either the area directly in square inches, or bears a relationship to it as follows:

$$\text{area} = \text{reading} \times \text{a constant}.$$

In some planimeters there is only one constant, but in others the length of the tracing arm may be varied by a micrometer adjustment. This feature is useful when the planimeter is to be used on areas drawn to scale. By employing the proper setting (i.e., the setting constant = scale of area), the area may be obtained directly from the wheel reading.

The explanation of the planimeter rests on a mathematical basis of the integral calculus, since it is primarily an instrument performing a graphical **integration**. It will be shown that the dimensions which are involved in the measurement of area are length of the tracing arm and the movement of the wheel. The length PK has no effect as long as K moves in the arc of a circle. In common use of the planimeter this is true, but large areas sometimes necessitate placing the pole (needle point) inside so that the arm PK describes a complete circle. In this case the area is not read directly from the record wheel. It is the area of the zero circle, plus or minus the wheel reading. The zero circle is that traced by the point F when the plane of the measuring wheel, extended, passes through P. The area of the zero circle, if the graduated scale gives the area directly, is the difference between the readings obtained by traversing the outline of any given area, first with the pole inside and then with the pole outside the area. If the scale does not give the area directly the difference between the readings must be multiplied by a constant to obtain the true value of the zero circle.

The figure above explains the theory of the planimeter. Suppose the tracing point F is moved from m to n along the curve bordering the area. The area swept over by KF can be considered to be made up of a translation dB and of a rotation $d\theta$. The area of translation is the length (L) of the tracing arm KF, multiplied by db. Note also, db is the distance rolled by the wheel. The center K must then be returned to the arc by a motion along the line $K'F'$. This generates no area, and causes no wheel rotation. The area generated in rotation is $\frac{1}{2}L^2d\theta$. The area generated, then, is:

$$A = L \int db + \frac{1}{2}L^2 \int d\theta.$$

When the point F has been carried completely around the closed line, the net area swept through by KF is equal to the area enclosed by that line. By returning point F to its original position, the integral of $d\theta$ becomes zero, and area equals $L \int db$. The $\int db$ is the net distance rolled by the wheel, and is read directly therefrom. Thus it is seen that the area measured by planimeter equals the length $FK \times$ the net distance rolled by the wheel. (F.T.M.)

PLANIMETRIC MAP. Map.

PLANKTON. The portion of the aquatic flora and fauna composed of organisms which merely float and drift and those whose active locomotion is not directional. Most of the organisms of the plankton are small or minute. The terms limnoplankton and haliplankton are sometimes applied to fresh-water and marine organisms, respectively, and other subdivisions have received special names. Floating organisms in the air have been called the aerial plankton, a category which includes no permanently planktonic forms, although some organisms may be carried by air currents in some stages of their development, such as the buoyant seeds of many plants and the pollen of anemophilous species.

The plankton includes many of the smaller invertebrates, such as **protozoans, coelenterates, worms,** and **lower chordates,** many **algae,** and some higher plants. (A.W.L.)

PLANOPHYRITIC. Used by **petrologists** to describe the texture of a **porphoritic igneous** rock in which the **phenocrysts** occur in layers. (R.M.F.)

PLANT ASSOCIATION. Association, Plant.

PLANT CUTTER. Aves, Passeriformes. Birds (**Aves**) of several species found only in Temperate South America. They have a short thick beak with finely serrate edges. Related to the chatterers. (A.W.L.)

PLANT LOUSE. Aphid.

PLANT SUCCESSION. Succession, Plant.

PLANTAIN EATER. Aves, Cuculiformes. An African bird (**Aves**) of an extensive group also called the touracos and louris. They are brightly colored birds and with the exception of the giant plantain eater, which is almost 3′ long, are of moderate size. (A.W.L.)

PLANTAINS. *Plantago* sp. Plantaginaceae. There are some 200 widely distributed species of plantains, many of which are ubiquitous weeds. Some species are stemless plants, the **petioled** leaves forming a rosette which covers a considerable area of ground, from which it excludes other desirable plants. These are the species which cause unsightly patches in lawns.

Plantago major is one of these, with broad ovate leaves on long petioles. *Plantago lanceolata* has **lanceolate** erect leaves. The flowers are borne in elongated **spikes,** the pistil maturing before the **stamens.** Plantains are wind-pollinated plants, although insects occasionally visit them for their pollen. The fruit is a **capsule,** the upper half of which comes off when mature, through the development of a circumferential line of **dehiscence.** The seeds of plantains are often fed to caged birds. In some species the seeds imbibe water and become mucilaginous. One species, *Plantago Psyllium,* native in southern Europe, is sometimes used in medicine under the name Psyllium seed. (R.M.W.)

PLANULA. A **larval** form of the **coelentrates.** It is covered with a ciliated (**cilia**) ectodermal layer and is filled at first with a solid mass of endoderm, which later splits to form the enteric cavity. A free-swimming stage. (A.W.L.)

PLASMA. Blood; Ionized Gases; Chalcedony.

"PLASMOCHIN." (Registered in U. S. Patent Office.) It is also known as pamaquine, and the chemical name is 6-methoxy-8-(1-methyl-4-diethylaminobutylamino)-quinoline. Announced in 1924, this was the first promising synthetic chemical for the treatment of malaria, active against the parasite causing the disease but not directly for the symptoms of the disease itself. (See "Atabrine.") (R.K.S.)

PLASMODESMA. Plasmodesmata are minute threads of **protoplasm** passing through minute pits in the walls of adjoining cells. These threads are so minute that special methods of **histological** preparation are usually needed to make them visible. They are larger and more easily seen in the walls of cells of the **endosperm,** and in cells of **gymnosperms** and ferns. It is possible that they may be found connecting the protoplasts of all living cells in plants, but are too minute to be seen. Their function is not known. (R.M.W.)

PLASMODROMA. A subdivision of the 1-celled animals characterized by the presence of a single kind of **nucleus.** The phylum Protozoa contains two such subdivisions, the subphyla Cilophora and Plasmodroma. This subphylum is divided into three classes and many orders. The classification is briefly summarized below:

Class **Mastigophora.** Animals with one or more **flagella.** Many species have chlorophyll and are intermediate between plants and animals.
Subclass Phytomastigina. Mostly colored species, with chlorophyll.
Order **Chrysomonadida.** Minute species of yellow to brown color, rarely bluish or green.
Order **Cryptomonadida.** With a cytopharynx. Body flattened.
Order **Dinoflagellida.** Body enclosed in a cellulose covering of two parts.
Order **Phytomonadida.** Color green. No cytopharynx.
Order **Euglenoidida.** Cytopharynx present. Color green. Stored food in the form of **carbohydrates.**
Order **Chloromonadida.** Cytopharynx present. Color green. Stored food in the form of oils.
Subclass Zoomastigina. Without chlorophyll. Usually with a single vesicular nucleus.
Order **Pantostomatida.** Animals with both **pseudopodia** and flagella.
Order **Protomonadida.** Small animals with one to three flagella.
Order **Polymastigida.** Minute forms with two to eight flagella and one to many nuclei. Mostly parasitic in the digestive tract of animals.
Order **Hypermastigida.** With more than eight flagella. Parasitic in the intestine of insects.
Class **Sarcodina.** Without a **pellicle** and therefore without constant body form. Pseudopodia are formed for locomotion and to secure food.
Subclass **Rhizopoda.** Species which form temporary pseudopodia of various forms.
Order **Proteomyxa.** Without hard parts. Pseudopodia slender, sometimes branching and anastomosing.
Order **Mycetozoa.** The **slime molds,** formerly regarded as plants. Active stage a plasmodium consisting of a large mass of protoplasm with many nuclei.

Order **Amoebaea.** Without hard parts. Pseudopodia thick protuberances.

Order **Testacea.** With an enclosing shell or test containing a single chamber. Chitinous.

Order **Foraminifera.** With an enclosing test of many chambers. Calcareous.

Subclass Actinopoda. With straight slender radiating pseudopodia which are semipermanent.

Order **Heliozoa.** Chiefly fresh-water species. Body usually spherical, without a central capsule enclosing the inner portion.

Order **Radiolaria.** Marine forms. Usually spherical, with the central portion enclosed in a membranous central capsule.

Class **Sporozoa.** Parasitic species without flagella or cilia and with pseudopodia only in the immature stages.

Subclass Telosporidia. Spores without a polar filament. Spore containing more than one sporozoite or without a resistant capsule. The orders are characterized by peculiarities of the reproductive stages.

Order **Coccidia.** Parasitic chiefly in the lining of the digestive tract of vertebrates and higher invertebrates, and in the associated glands.

Order **Haemosporidia.** Part of the life cycle occurs in the blood of vertebrates and part in the alimentary tract of a blood-sucking invertebrate. The **malarial** parasites and others.

Order **Gregarinida.** Parasitic chiefly in arthropods and annelids.

Subclass Cnidosporidia. Spore with a coiled polar filament.

Order **Myxosporidia.** Mostly parasites of fishes.

Order **Actinomyxidia.** Parasitic in aquatic **annelids.**

Order **Microsporidia.** Typically parasitic in cells of **arthropods** and fishes.

Order **Helicosporidia.** A single peculiar species parasitic in insects.

Subclass Acnidosporidia. Spores without polar filament; with a single sporozoite.

Order **Sarcosporidia.** Parasitic in the muscles of mammals and to some extent of birds and reptiles.

Order **Haplosporidia.** Parasitic in invertebrates and lower vertebrates. (A.W.L.)

PLASMOTOMY. A reproductive process of certain one-celled animals. By this process a multinucleate form of species parasitic in fishes (**Plasmodroma,** Myxosporidia) subdivides into two or more parts, each with several nuclei. (A.W.L.)

PLASTER OF PARIS. Calcium Sulfate.

PLASTIC MOLDING. Plastic molding is analogous to die casting in many of its forms. Plastic molding is a pressure process which is performed either in hydraulic presses with ram capacities of 1000 to 3000 lbs. per sq. in., or in special injection molding machines.

In injection molding the material, in granulated form, is placed in a feed hopper and forced past a spreader through a nozzle into a water-cooled die by a reciprocating plunger. The material is heated to a plastic state as it passes through the injection head. Fast production is the primary advantage of injection molding; one machine used operates at four cycles per minute, and will attain a production rate of 480 parts per hour while using a two-cavity mold. Parts with metal cores can be easily produced by this process, so that the final product will have the surface characteristics of the plastic with the rigidity of metal. The insert need not be carefully finished for plastic covering, as is the case when a metal die casting is prepared for plating.

Compression molding is employed for parts that are too large for successful injection molding, or parts of such thickness that excessive shrinkage must be guarded against. The commonest form of compression mold is the flash mold which permits excess material to escape during final closure of the mold. The mold charge is a previously compressed tablet, or preform, made from powdered or granulated material, and the slight excess flows out between the flash surfaces of the die during final closure. The resulting fin or thin edge must be removed after the part leaves the die, and usually constitutes the only finishing operation required on a plastic molded part.

When bulk material must be employed, a positive mold is used. The material must be accurately weighed or measured before loading since no excess is permissible in the mold. The dies are more subject to wear than flash mold dies. Positive molds may be equipped with strippers for removing deep cup-shaped parts. The floating-chase mold is used for parts whose length is great compared to their diameter. This mold consists of two independent plungers in a tubular die. Both plungers act upon the material so that a uniform density in the part may be attained.

Blowing molds use material in sheet or tube form which is expanded to the finished shape by air, water, or steam pressure. Tube molds consist of two die halves, which are loaded with the stock and closed and the stock is clamped in place by two nozzles. Steam under pressure is admitted through the nozzles and expands the stock to fill the mold. The part is then cooled by circulating water through the nozzles. In sheet molds, the sheets are clamped between two die halves, and a hollow needle is inserted through a groove in the dies. The heated dies render the material plastic, and compressed air is admitted through the needle until the material conforms to the mold shape. The needle is then removed and the finished halves of the part are welded by a pressure sufficient to bring the die edges together, which also cuts off the excess stock. (H.C.H.)

PLASTICITY. However defined by physicists and geophysicists, the effects of plasticity are of interest to the geologist (**stratigrapher**) principally in relation to the development of intraformational structures in **sediments** of the silt or clay grade, and are therefore indicative of the peculiar condition under which the sediments were deposited, deformed or reworked previous to lithification.

Most metals are characterized by high plasticity or ability to undergo permanent deformation in the solid state without fracture. Above the **recrystallization** temperature metals have almost unlimited plasticity because the deformed crystals are transformed into new unstrained crystals spontaneously. (R.M.F., R.H.H.)

PLASTICS. Plastic materials are of two types, namely, (1) bodies which are rigid at ordinary temperatures and pressures, and, when subjected to increased temperature or pressure, become softened so as to permit bending and forming, e.g., glass, (2) bodies which in the process of working soften sufficiently—generally but not always when they are subjected to increased temperature or pressure—so that they can be made to fill molds of forms, e.g., moist clay, moist Portland cement, rubber. The term plastics is loosely used to denote the classes of organic materials represented by "celluloid," "bakelite," "vinylite," "galalith" and rubber. These are used in a great variety of ways, and several are not confined in their uses to the field of plastics but have extensive use—in the form of resins in a solvent—as lacquers, adhesives, textile fibers, fabric treatment, automobile brake linings, impregnated wood, laminated materials such as glass, wood and paper, and finally in various sheets, tubes, filaments and films. (See **Pigments.**)

Of the synthetic plastics, except rubber, the following groups are representative: 1. Nitrocellulose plastics, example Celluloid and its relatives. 2. Phenol-formalde-hyde resins, example Bakelite and its relatives. 3. Urea—formaldehyde resins. 4. Cellulose acetate plastics. 5. Ethyl cellulose. 6. Vinyl polymers, example Vinylite. 7. Acrylic polymers, example methyl methacrylate. 8. Polystyrene. 9. Alkyd resins, example glycerol—phthalic acid resins. 10. **Nylon.**

Nitrocellulose plastics are made by taking 70 to 80 parts by weight of nitrocellulose (11% nitrogen), mixing with non-volatile solvents and plasticizers, e.g., castor oil, camphor, butyl phthalate, diethylphthalate, tricresyl phosphate, at 75–90° C., 20 to 30 parts by weight of camphor, and with 0 to 14 parts by weight of dyes, pigments, fillers. A volatile solvent, e.g., ethyl alcohol, is added, the material is shaped, and the volatile solvent evaporated. Residual volatile solvent remains to the extent of 1 to 5 parts by weight. The working range for molding is 85–120° C. The product, commonly "celluloid," is inflammable and therefore hazardous upon heating. The flash point test shows 160–200° C. The process was perfected and manufacture begun in the United States by J. W. and I. S. Hyatt in 1869.

Phenol-formaldehyde plastics and resins are made by mixing **phenol** and **formaldehyde** solution (37–40% HCHO) with **ammonia** as a **catalyzer.** The mixture is heated to 80–90° C. When the mass has formed a resin, the water is boiled off. The product is immediately drawn off and allowed to cool in thin layers, or a solvent, e.g., ethyl alcohol, acetone, is added to stop further reaction by dilution. Another method is to mix phenol and formaldehyde without adding ammonia, and subject the mixture to a temperature of 140–160° C. in a digestor. Heating slowly to the higher temperature in a period of about 3 hours is necessary to avoid the rapid reaction. Water is allowed to pass off by opening a valve on the digestor. The thin resinous product is drawn off and allowed to cool. When the properly prepared resin is mixed with modifying agents, such as nitrocellulose, resins, casein, and the desired fillers, at 110 to 140° C. under pressure for one hour, a hard molded product is obtained. Products insoluble in most solvents, and infusible, are for use as plastics, and other products soluble in certain solvents for use in lacquers.

Vinyl plastics and resins are made by polymerization of various individuals and mixtures of vinyl chloride, acetate, butyrate, and alcohol.

The natural resins, such as shellac, rosin, and asphalt, are not uniform in composition, and their uses are limited. In contrast, the synthetic resins are very uniform in individual composition, and their uses accordingly vastly outnumber those of the natural resins. On account of the properties being uniform, these properties, when once ascertained, can be depended upon to function regularly. Some of the properties that are important in connection with the uses but shown in different degrees by various synthetic resins are (1) electrical resistance or dielectric strength, (2) tensile strength, (3) flexing strength, (4) dimensional stability, (5) clarity, (6) ability to withstand high and low temperatures, and (7) ability to resist water, acids, oils, and other chemicals. The accompanying diagram PLASTICS COMPARATOR shows many of these results better than could

MOLDING PLASTICS COMPARATOR
(Thermosetting Materials)

Nos. 1–6 represent highest to lowest in order of merit. Relationships are qualitative and variation in formulation may alter position on chart.

	Shock Resistance (Impact—Izod)	Tensile Strength	Flexural Strength	Cold Flow	Hardness (Rockwell)	Heat Resistance (Utility under Continuous Heat)	Dimensional Change on Aging	Thermal Insulation	Specific Gravity	Flammability	Color Possibilities	Color Stability	Utility Around Inserts	Ease of Molding	Water Resistance (Absorption)	Alkali Resistance	Acid Resistance	Organic Solvent Resistance	Loss Factor 60 cycles/sec.	Loss Factor Megacycle/sec.	Resistivity	Dielectric Strength
Phenolics—General Purpose	3	1	3	1	2	2	4	1	2	4	2	2	3	1	4	3	3	2	3	2	3	3
Phenolics—Low Loss	3	2	5	1	2	3	2	2	4	2	4	4	3	3	1	2	2	2	1	1	1	1
Phenolics—Heat Resistant	5	6	6	1	2	1	1	2	4	1	4	3	2	3	2	2	2	2	5	4	5	5
Phenolics—Heat Resistant with Improved Impact	2	3	1	1	2	1	1	2	4	1	4	3	1	4	2	2	2	2	5	4	5	5
Phenolics—Acid Resistant; Alkali Resistant	5	5	4	1	1	3	3	1	1	3	3	3	4	2	3	1	1	1	3	2	2	3
Phenolics—Medium Shock Resistant	2	4	3	1	2	4	5	1	2	5	4	3	1	3	5	3	3	2	4	3	4	4
Phenolics—High Shock Resistant	1	4	4	1	2	4	5	1	2	5	4	3	1	4	5	3	3	2	4	3	4	4
Urea	4	1	2	1	1	5	6	1	3	3	1	1	5	2	4	4	4	2	2	2	2	2

be done with lengthy descriptions. Some specific data are presented in the following table:

MATERIAL	TENSILE STRENGTH LBS. PER SQ. IN.	SOFTENING TEMPERATURE ° C.	ELECTRICAL BEHAVIOR DIELECTRIC STRENGTH VOLTS PER MIL (0.001″)
Hard vulcanized rubber	1,200–2,000	90	300–1,100
Cellulose nitrate	4,000–10,000	75	250–780
Shellac plastics	900–2,000	80	100–400
Phenol-formaldehyde plastics	5,000–10,000	150–250	200–700
Casein-formaldehyde	....	85	125
Vinyl plastics	7,000–9,000	75	350–400
Cellulose acetate	5,000	50–75	720
Urea-formaldehyde	5,000–6,000	....	300–400
Glass	....	....	300–1,500
Porcelain	650–2,300	600–1,000	1,000
Portland cement	2,500	550	40–150

The Comparator gives relative property values, and is only a general summary for quick and easy reference. The ratings may not in all instances confirm published data. The materials are rated according to *usage* experience, as well as laboratory data. For example, urea, on the basis of A.S.T.M. water absorption tests, would warrant a better rating. However, on the basis of continuous exposure to high humidity or water, the rating in the Comparator is 4. In detail, the following is a list upon which the choice of values is based:

Shock Resistance. By this term is meant Izod impact strength in ft.-lb. per in. of milled notch.

Cold Flow. These values were based generally upon actual figures which refer to the distortion of the ½-in. cube of the molded material under a specified load at 122° F.

Hardness. These values were based upon figures obtained by the Rockwell hardness test.

Heat Resistance. These values were obtained from data indicating the highest temperatures at which the molded pieces could be subjected continuously and still maintain their utility.

Dimensional Change on Aging. Plasticizer loss; and water absorption. In choosing these values, the temperature was considered as room temperature, and weathering eliminated as a factor. Since no actual data were available, opinion was based on interpretation of the effect of the foregoing variables on dimensional change with normal indoor use.

Thermal Insulation. These values were based upon thermal conductivity measurements.

Specific Gravity. These values were obtained from several sources and represent definite data.

Flammability. These values were obtained from qualitative data evaluating the relative burning rate of the materials.

Appearance and Color Possibilities. The ratings represent opinion, and were based upon the following, in this order of importance: Transparency, brilliance and depth; translucency and color possibilities; surface smoothness and luster.

Color Change. This represents the intrinsic color change of the natural polymer or filler on aging, obtained from qualitative data and opinions.

Water Resistance. These values were based upon the amount of water absorbed by the material when immersed in water at room temperature for a period of 24 hours.

Alkali Resistance. These values were based upon the resistance to weak alkali, such as household soap solution. Water absorption was also considered when determining these values. Opinion as well as fact was used as basis.

Acid Resistance. These values were based upon the resistance to weak acids generally. Opinion as well as fact was used as basis.

Organic Solvent Resistance. These values were based upon the resistance to the common organic solvents; based on generally qualitative information.

Electrical Properties. These values were based upon experimental data.

Synthetic resin plastics are (1) thermosetting, and (2) thermoplastic. Thermosetting types undergo deep-seated chemical change when subjected to heat and pressure, and set to form a hard infusible mass that does not soften appreciably upon reheating. Thermoplastic types to which all natural resins belong do not undergo such chemical change upon similar treatment but do soften and can then be shaped in molds as desired. Upon cooling, these types harden and retain the acquired shape, but obviously are not temperature-resistant. The heating, molding, cooling, solidifying cycle can be repeated if desired.

Various methods of molding can be utilized depending principally upon the shape of the desired product, e.g., compression, extrusion, casting, injection, jet, and transfer. In compression molding, two dies are heated, the plastic placed between the faces of the dies, and then the dies are closed down over the plastic at the required temperature and pressure; in extrusion the plastic is softened by heat and then forced through a die to produce long continuous articles which are cut into the requisite lengths; and in injection the heated softened plastic is forced into a relatively cool mold for shaping.

Reactions involved in the formation of thermosetting shapes (and synthetic rubber) are believed possible only with certain organic functions, such as those having a double or triple bond in the molecule:

$$>C:C< \qquad >C:O \qquad -C:C- \qquad -C:N$$
$$\text{olefin} \qquad \text{aldehyde; ketone} \qquad \text{acetylene} \qquad \text{nitrile}$$

more markedly if a single atom is completely unsaturated:

$$>C:C:C< \qquad >C:C:O \qquad >C:C:N-$$

and still more so if there is conjugated unsaturation adjacent to a single carbon-to-carbon bond:

$$>C:C-C:C< \qquad O:C-C:O \qquad >C:C-C:N-$$

That there is a definite periodicity throughout the molecule ("high polymer") is shown by x-ray studies. When the molecule is a thread the primary bonds are chain-like in one direction—that of length; when a thin film the chain-like bonds are cross-bound laterally showing two directions of linkage, namely, length and width; and when solid three directions, namely, length, width and thickness, are interlocked. (R.K.S.)

PLASTIDS. Pigments in plants are often located in special bodies called plastids. There are many kinds of plastids: leucoplasts, those which contain no pigment and which are therefore colorless; chloroplasts, those which contain chlorophyll (by far the commonest kind); and chromoplasts, colored plastids which do not contain chlorophyll.

Leucoplasts occur in parts of stems and roots where light fails to penetrate. They absorb glucose and change it to starch.

Chloroplasts occur in cells exposed to light. They manufacture glucose, a simple sugar, out of water and carbon dioxide; may change the glucose to starch or sucrose for temporary storage and these back again to glucose for respiration or translocation. In the algae the shapes of these bodies are many; in a large number of cases the plastid is a thick cup-shaped body occupying the greater part of the volume of the cell; in other algae the plastids have a central mass from which radiating plates or arms extend outward to the cell wall; spiral, net-shaped and ring-shaped plastids are not uncommon in this group of plants. In some algae and in nearly all higher plants, the chloroplasts are small sub-spherical or lens-shaped bodies, varying in number from one to many in a single cell. Always the chloroplasts are found embedded in the **cytoplasm** of the cell. In many plants the continuous movement of the cytoplasm in the cell carries the plastids along with it; in others these bodies have a fixed position. In certain algae and in many cells in higher plants, as for example in the palisade layer of leaves, the chloroplasts may change their position so that they will receive the most favorable amount of light. If the light intensity is low they will present their flat surface to it; while if the light intensity is high, the plastid rotates so that it is placed edge-wise to the light. Chloroplasts contain chlorophyll and other **pigments.**

Chromoplasts have no chlorophyll and no known function. (R.M.W.)

PLASTOGAMY. The union of individuals by which the **plasmodium** of the slime molds is formed. The **cytoplasm** becomes confluent but the **nuclei** remain distinct. In some cases the mass formed is large enough to be conspicuous to the naked eye. (A.W.L.)

PLATE DISSIPATION. This is the power loss occurring at the **plate** of an electronic **tube** and is of considerable importance in high-voltage vacuum tubes. The **electrons** in traversing the space between electrodes under the influence of the applied voltage acquire kinetic energy which must be dissipated when the electron is brought to rest upon striking the plate. This power is equal the product of the plate current and plate voltage (it should be noted that this is the voltage at the plate and not necessarily the voltage applied to the circuit since there is usually a drop in the load). (L.R.Q.)

PLATE EFFICIENCY. Conversion Efficiency.

PLATE GIRDER. Girder.

PLATE RESISTANCE. This is one of the vacuum-tube coefficients which is used in analyzing the behavior of the tube in a circuit. Mathematically it is expressed as

$$r_p = \frac{de_p}{di_p}\Big]\qquad (E_g \text{ constant})$$

where de_p and di_p represent infinitesimal changes of plate voltage and current and E_g is the grid voltage.

The value of the plate resistance, or, more properly, the dynamic plate resistance, is essentially constant over the normal operating range of the tube but is dependent upon the electrode voltages. From the equation it may be seen that it is the reciprocal of the slope of the plate current-plate voltage characteristic curve of the tube. In developing the equivalent circuit of the vacuum tube the plate resistance is the value used for the internal resistance of the equivalent **generator.** (L.R.Q.)

PLATE, STRUCTURAL. Structural plate is a large, flat body of **steel** which is produced by the working of ingots, billets, or slabs in a rolling mill. There are two types of these plates, one known as sheared plate, the other as universal mill plate. The sheared plate is used for square or rectangular shapes, the universal mill product for long strips, like those needed for plate girders. The sheared plate is rolled in a mill having horizontal rollers through which the plate is passed back and forth while the rollers are successively brought closer to each other. During this process the slab or ingot is rolled to a flat plate. Mills having horizontal rollers only are built to roll extremely large plates, some of which are over 100″ wide. The selvage left during rolling is rough and irregular, and the plates must be sheared to standard marketable sizes or customers' specifications

by power shears. Between rolling and shearing operations the plates are straightened, but shearing takes place when the plates are hot, so shrinkage must be allowed for when laying out lines to which the plates are sheared. For this reason, and also for the reason that the shears cannot be manipulated with a great deal of precision, sheared plates must have a large tolerance allowance. However, sheared plates are the only plates available in wide widths.

Universal plate is made on universal mills which have vertical as well as horizontal rolls, the function of the vertical rolls being to edge the plate. Universal plates can be obtained in widths up to 48″. Plates of great length are produced with rolled edges. The allowable tolerances are much smaller, and the purchaser does not need to machine the rolled edge. Structural plate is rolled to several standard thicknesses, beginning at ¼″, and advancing by small increments to 2″.

Another usage of this word describes the structural members with which the tops of walls of buildings are finished off. For example, in the ordinary dwelling built with masonry walls, anchor bolts are built into the masonry at the top of the wall. A substantial wooden timber is bored to receive these bolts and placed on the wall with the bolt projecting through it. The bolts are then capped with nuts which are tightened against washers bearing on the plate. By spacing these anchor bolts at close intervals, the plate is securely anchored to the wall, and forms a base to which rafters or other wooden timbers may be nailed. In wooden wall construction the plate is the horizontal member which is attached to the tops of the studs. (F.T.M.)

PLATEAU. An elevated, relatively flat area or surface of wide extent and underlain by relatively horizontal sedimentary formations. Less correctly used to describe any relatively flat high-level surface of erosion regardless of the structure of the region. (See also **Mesa.**) (R.M.F.)

PLATINOTYPE PROCESS. A process of photographic printing patented by William Willis in 1873. A beautiful process and one of the few producing images of undoubted permanence—it is now obsolete, manufacture of the paper having ceased about 1930. Paper is coated with a ferrous salt (ferrous oxalate), and potassium chloroplatinite. Upon exposure to light the ferrous salt is reduced to the ferric. The faint image is then developed by placing the print in an alkaline oxalate (potassium oxalate), which dissolves the ferric salt and reduces the chloroplatinite to metallic platinum.

$$Fe_2(C_2O_4)_3 + light \rightarrow 2Fe(C_2O_4) + 2CO_2$$

$$6FeC_2O_4 + 3K_2PtCl_4$$

$$\rightarrow 2Fe_2(C_2O_4)_3 + Fe_2Cl_6 + 6KCl + 3Pt.$$

(C.B.N.)

PLATINUM. Symbol: Pt. Atomic number: 78. Atomic weight: 195.23. Density: 21.37. Hardness: 4.3. Melting point: 1773.5° C. (Isotopes: page 290.)

Compact platinum is a grayish-white metal, softer than silver, and hardened by the presence of other metals, **iridium** being frequently used to accomplish this result. Platinum in resistant to attack by air or water at any temperature, by **hydrofluoric acid, hydrochloric acid, nitric acid,** dilute **sulfuric acid;** but is attacked by **aqua regia,** forming chloroplatinic (platinichloric) acid (H_2PtCl_6), by ignition with **sodium** hydroxide, nitrate, peroxide, or cyanide. Probably discovered by Scaliger in the 16th century, and named from its similarity to silver in appearance (*platina* (Spanish), small silver).

Platinum is used in alloys for jewelry (white gold contains small percentages of platinum alloyed with gold); pen points (alloyed with **iridium**); parts of scientific apparatus, especially electrical contacts; surgical tools; standard weights and measures; and containers, such as crucibles and dishes. The change of electrical reistance of a pure platinum wire coil upon the passage of a constant current is used to measure high temperatures. A platinum and platinum-**rhodium** thermocouple is also used for the same purpose.

Finely divided platinum is an important catalyzer for the reaction **sulfur** dioxide gas plus oxygen of air at 450° C., forming **sulfur** trioxide, in the manufacture of **sulfuric acid** by the so-called contact process. The platinum catalyzer must be specially prepared for this purpose. A smooth platinum surface as fine wire gauze, is used as a catalyzer for the reaction **ammonia** gas plus oxygen of air at 600° C., forming nitric oxide which is ultimately recovered as **nitric acid.**

Platinum occurs native but containing other elements, especially **osmium, iridium** and similar metals, **gold,** and **iron.** Found in the Ural Mountains of Russia, and less abundantly in South Africa, Canada, and the United States. By treating with aqua regia, platinum and similar metals are dissolved. After filtration, the filtrate is treated with calcium hydroxide and again filtered. From the resulting filtrate crude platinum is recovered by evaporation to dryness, and ignition.

Chlorides: Platinous chloride ($PtCl_2$); platinic chloride ($PtCl_4$), reddish-brown, hygroscopic, crystalline material. By evaporation of platinic chloride solution with hydrochloric acid, chloroplatinic acid ($H_2PtCl_6 \cdot 6H_2O$), reddish-brown crystals, is obtained, which yields upon ignition at 100° C., chloroplatinic acid ($H_2PtCl_6 \cdot 2H_2O$) as commonly marketed, and, upon ignition in **hydrogen chloride** gas, platinic chloride.

Chloroplatinates: Potassium chloroplatinate (K_2PtCl_6), and ammonium chloroplatinate (($NH_4)_2PtCl_6$) are insoluble in water, and in a mixture of alcohol and ether; sodium chloroplatinate (Na_2PtCl_6) is soluble in the same solvents. This is applied in the separation of potassium in chemical analysis.

Oxides: Platinous oxide (PtO), gray to dark violet; platinic oxide (PtO_2), black.

Sulfides: Platinous sulfide (PtS), black; platinic sulfide (PtS_2), black.

Platinum salts combine with **ammonia,** forming interesting compounds known as platinous and platinic **ammines.** (R.K.S.)

PLATYHELMINTHES. The flatworms, a major division of the animal kingdom containing the most primitive of the triploblastic **Metazoa.** The phylum includes both free-living and parasitic species. Among the latter are the **flukes** and tapeworms, some of which are serious parasites of man.

The phylum is characterized by the following details of structure: 1. The body is bilaterally symmetrical and flattened. 2. The ectoderm is ciliated in free-living forms but forms a **cuticle** in the parasitic species. 3. The mesoderm forms a compact tissue between the various organs, known as a parenchyma. 4. The alimentary tract, when present, has a single opening. 5. The nervous system is a network in which a brain and longitudinal nerve cords are developed. 6. The excretory system consists of large hollow cells with a group of **cilia** extending into the cavity, known as flame cells, connected with tubes.

There are three classes of flatworms:

Class **Turbellaria.** Free-living species. **Planaria.**
Class **Trematoda.** The **flukes.** Parasitic in various parts of the animal body.
Class **Cestoda.** The tapeworms. Parasitic as adults in the alimentary tract of animals. (A.W.L.)

PLATYOSAURUS. Fossil Reptiles.

PLATYPUS. Duckbill.

PLAYA. The flat interior part of an undrained basin on which accumulate fine clastic sediments and chemical precipitates. Playas are formed within desert basins due to intermittent interior drainage, which, during cloudbursts, forms intermittent lakes in which the sediments are deposited. A region remarkable for playas is the Great Basin of the western United States, which covers all of Nevada and Utah. Commercially important mineral salts derived from playa deposits are: **gypsum, sodium** carbonate, the soluble **chlorides,** and **borates.** (R.M.F.)

PLECOPTERA. The **stone flies.** An order of insects with aquatic early stages. The mouth is formed for biting but is usually poorly developed in the adult. They have four wings which fold flat over the back when at rest. Sometimes abundant in the vicinity of water. (A.W.L.)

PLEIADES. The Pleiades is a very famous group of bright stars in the **constellation** of **Taurus.** Probably no one group of stars in the entire sky has received so much notice in classical literature and mythology as has the Pleiades. The Great Pyramid, which was undoubtedly designed for astronomical purposes, is so oriented that in 2170 B.C., when the Pleiades were on the meridian at midnight on the first day of spring, they could be seen through the south passageway. References to this group of stars are to be found in both the Old and New Testaments. One of the seven stars at present is distinctly fainter than the other six and there are many myths regarding this so-called "lost pleiad." In fact, the myths occur in so many different ancient literatures that there is a well-established theory that at one time all seven of the stars were of approximately the same brightness.

The Pleiades is an open **star cluster** with the various members all moving through space together. Long exposure photographs of this group indicate that the space surrounding the stars is filled with a luminous **nebulosity.** (W.K.G.)

PLEISTOCENE. The "great ice age," and first period of the **Quaternary** in the geologic time-scale. The term was proposed by Lyell in 1836. This period, as postulated by the first advance of the continental ice sheets in western Europe and northern North America, is supposed to have begun approximately 2 million years ago. Marine deposits occur in southern California. Both in North America and Europe there was considerable volcanic activity as evidenced by ash beds and basaltic lava flows. The early Pleistocene terrestrial sediments suggest increased humidity in the formerly arid regions. The terrestrial deposits as a whole are classified according to whether they were deposited within or without the glaciated areas. Great areas of **loess** were deposited at the margins of the melting continental glaciers, and the cycles of refrigeration produced not only great ice fields in northern Europe and northern North America, but also caused sympathetic glaciation in the mountainous regions further to the south. The **fossil** plants suggest distinct climatic zones. Both in Europe and North America it has been shown that there were four glacial cycles, or advances and retreats of the continental ice sheets, each with their typical sediments and accompanying floras and faunas. During the Pleistocene there was a great reduction in the number of the larger mammals, except in Asia and Africa. The characteristic types in North America were *Elephas* (**elephant**), as well as the mastodon, *Cervalces* (giant "moose"), a number of *Equidae* (true **horses**), *Smilodon* (last of the great saber-tooth tigers), and *Megalonyx* (giant ground sloth). In South America occur the peculiar Pampean fauna, descended from the earlier indigenous Pliocene forms, including large *Glyptodons, Megatheria* (giant ground sloths), and *Hypidium* (Pampas horse). The interglacial stages were remarkable for the replacement of northern by southern types of plants and animals,

Map of North America showing the maximum extent of glaciers during the Pleistocene Ice Age. The locations and general directions of movement of the great ice sheets are indicated, and regions of local mountain glaciers are shown in black. (*Modified after U. S. Geological Survey.*)

including the progenitors of modern man. The pleistocene is said by some geologists to end with the beginning of the Recent period. Others suggest that the Recent is merely an interglacial epoch preceding the next epoch of continental glaciation. (R.M.F.)

PLENUM. Pressures slightly above atmospheric are known as plenums. Such pressures usually occur in air or gas systems as the result of the action of **fans** or **blowers.** The plenum is measured in small units of pressure, such as ounces per sq. in., or in inches head of a liquid on a differential manometer. (F.T.M.)

PLEOCHROIC HALOES. The term applied to the colored zones concentric to a **radioactive** mineral, such as **zircon,** included in another mineral, such as **biotite.** (R.M.F.)

PLEOPOD. A form of the **biramous appendage** of **crustaceans.** The pleopods occur on the more anterior abdominal segments and are fringed appendages known usually as swimmerets. (A.W.L.)

PLEROCERCOID. A form of the bladder worm **larva** of tapeworms (**Cestoda**). In contrast with the simpler bladder worms which resemble the scolex of the adult, the plerocercoid contains other parts of the body in addition to a scolex. (A.W.L.)

PLEROCERCUS. A form of the bladder worm **larva** of tapeworms (**Cestoda**) which resembles the scolex of the adult. (A.W.L.)

PLESIOSAUR. Fossil Reptiles.

PLEURA. The pleurae are two serous membranes which cover each lung and are reflected against the chest wall. The part covering the lung is called the visceral pleura; it is very thin and firmly adherent to the pulmonary tissue beneath. The layer adherent to the wall of the thorax is the parietal pleura. Between the two layers is the pleural cavity, normally a potential space containing a tiny amount of serous fluid, but in certain pathological conditions constituting a true cavity. The function of the pleurae is to effect excursion of the lungs over the walls of the thorax with a minimum of friction. (See diagram under **Respiratory System.**) (D.M.H.)

PLEURISY. Inflammation or infection of the **pleura,** often producing a characteristic type of chest pain. This is a sharp, piercing sensation which appears during the inspiratory phase of breathing as the inflamed pleural surfaces rub together. The friction produced by this approximation of the parietal and visceral pleura can be heard through the stethoscope as a sound simulating the creaking of leather. It is termed a friction rub.

Pleurisy is often accompanied by effusion, the outpouring of a thin fluid which distends the pleural space. The appearance of fluid causes the disappearance of both the rub and the pain. If the expansion of the lung is sufficiently cut down by a massive accumulation of fluid, **dyspnoea** will develop.

Pleurisy alone, or pleurisy with effusion, is most commonly caused by **tuberculosis.** Other causes are **pneumonia, rheumatic fever,** and **tumors** involving the pleura. (D.M.H.)

PLEUROBRANCHIAE. Gills arising from the sides of the **thorax** in certain **crustaceans.** (A.W.L.)

PLEURODONT. Dentition.

PLEURON. Skeletal System.

PLEUROPODITE. A basal joint of the **biramous appendages** of certain **crustaceans.** It precedes the usually basal coxopodite and is also called the precoxa. (A.W.L.)

PLEXUS. A network. 1. In many animals the processes of **nerve cells** join to form a plexus or nerve net. This is the characteristic form of nervous system in the **coelenterates** and persists with modifications in the flatworms. The nerves of the radially symmetrical **echinoderms** also take on this form. A plexus underlies the ectoderm of these animals and deeper in the body other nerve fibers form plexuses of limited extent. In the vertebrates nerves branch and rejoin in some parts of the body. The brachial plexus, made up of the spinal nerves which enter the arm, and the solar plexus above the stomach are examples. Almost a hundred such plexuses have been named in the human body. 2. A network of blood vessels. The choroid plexuses of the **brain** are the most commonly mentioned examples of this group. They are the very thin and highly vascular roof plates of the most anterior and the most posterior cavities of the brain, which expand into the interior of the cavities. Other vascular plexuses are found elsewhere in the body. (A.W.L.)

PLIOCENE. The last major subdivision of the **Tertiary** in the geologic time-scale. Term proposed by Charles Lyell in 1832 after the type locality in the Paris Basin. The Pliocene Period began approximately 8,000,000 years ago and lasted for about 6,000,000 years. In the United States the principal marine deposits occur on the Pacific Coast, but the outline of the continental margins was approximately what it is at the present day. There was continued mountain-building during the period, and the interior continental terrestrial deposits were relatively thin and unimportant. There was considerable volcanic activity in the Rocky Mountain region, with great extrusions of **rhyolitic** lavas in the Yellowstone Park. The fossil plants are of the modern type, and more abundant in Europe than in North America. Also the marine invertebrates are practically identical with the modern forms. Among the mammals were a number of forms quite similar to those of the **Pleistocene** and recent, including all types of carnivores, horses, browsing camels, antelopes, and mastodons. For a description of the mineral resources of this period see the Tertiary. (R.M.F.)

PLIOPITHECUS. Paleontology of Man.

PLOUGH CUT. Dado.

PLOVER. Aves, Charadriiformes. Wading birds (**Aves**) with moderately long legs and a moderate beak. Although usually found near water, many species fre-

Killdeer. *Oxyechus vociferus.* Grayish above, white below, two black bands across the breast.

quent dry ground. The **killdeer** is the most common of the several North American species. The golden plover, *Pluvialis dominica,* is remarkable for its long migrations between arctic breeding grounds and its winter home in Patagonia. The Bartramian sandpiper is sometimes called the upland plover. True plovers constitute the family Charadriidae. (A.W.L.)

PLUG GAUGE. Measurement.

PLUM. Rose Family.

PLUM CURCULIO. Insecta, Coleoptera. A weevil, *Conotrachelus nenuphar,* which damages plums and other stone fruits, apples, pears, and quinces in the eastern half of the United States. The insects hibernate in fencerows and rubbish in orchards and pupate in the ground, hence the destruction of their hiding places and thorough cultivation of the soil in late July and early August are useful measures of control. Spraying just after the petals drop and again after 10 days with **lead** arsenate (2 lbs. to 100 gal. of water) is effective. For control of the pest on peaches special methods are necessary which differ in various peach-growing regions. (A.W.L.)

PLUMBAGO. Graphite.

PLUMBUM. Lead.

PLUME MOTH. Insecta, Lepidoptera. A small **moth** whose wings are deeply split to form 2–6 slender fringed lobes. In the family Pterophoridae most species have the front wings split for about ⅓ of their length to form two short lobes and the hind wings deeply divided into three lobes. One genus has the wings entire. Members of the Orneodidae have six slender plumes to each wing. (A.W.L.)

PLUMULE. Seed.

PLUNGE (Geological term). For the use of this term in geology, see **Anticline.**

PLUTEUS. A **larval** form of brittle stars and sea urchins (**Echinoidea**). It is bilateral, with a tubular alimentary tract. The body bears several projecting lobes and locomotion is accomplished by **cilia** arranged in a tortuous band. (A.W.L.)

PLUTO. (See tables of planetary data, page 1109.) The planet Pluto, the outermost known **member of the solar system,** was discovered early in 1930 at the Lowell Observatory. The discovery of the planet was announced on March 30, 1930, the anniversary both of Percival Lowell's birth and also of the discovery of **Uranus.** The discovery marked the culmination of a search for a planet outside of the **orbit** of **Neptune** which had been carried on for many years at the observatory of Dr. Percival Lowell at Flagstaff, Arizona. The circumstances which led to the belief that such a planet existed are similar to those which led to the discovery of Neptune. After the **orbit** of Neptune was computed and the motion carried back through the years, it was found that the planet had been observed several times previous to its announcement as a planet, the early observers having recorded it as a star. These early observations were of great value in making an accurate determination of the orbit, and, when all **perturbations** due to known objects had been computed and applied, certain unexplainable differences between observed and computed positions appeared. On the basis of these perturbations Lowell made the necessary laborious computations to determine the positions of a possible planet that might be causing the attractions and predicted Pluto. There is considerable doubt in the minds of many astronomers as to whether Pluto is actually the planet producing the perturbations in the orbit of Neptune or whether the perturbations may not actually be due to accidental errors in the observations of Neptune itself. Whether Pluto is actually the predicted planet itself or not, nevertheless, it is certain that the computations of Dr. Lowell stimulated the search, and that the planet was found as a result of this search.

The name Pluto was selected for the new planet and the first two letters of the name, combined in monogram form ♇, are used as the symbol for the planet. Since these two letters are not only the first two letters of the name of the planet, but also are the initials of Percival Lowell, they are particularly fortunate.

The orbit of Pluto is the most eccentric of all of the orbits of the major planets, and the inclination to the plane of the ecliptic is also the largest. The mean distance of the planet from the sun is slightly less than 40 **astronomical units.** Due to the large value of the eccentricity (0.25) the planet is more than 50 astronomical units from the sun at aphelion and within 30 at perihelion. The latter figure is less than the distance of Neptune from the sun, so at times the planets Pluto and Neptune pass each other. However, the large inclination of the orbit of Pluto makes a collision virtually impossible, the closest approach of the two planets being about 240,000,000 miles.

Comparatively little is known regarding the physical characteristics of this remote member of the sun's family. It appears as a very faint star of about the fifteenth **magnitude,** with a distinctly yellowish color which is in contrast to the greenish appearance of its nearer neighbors in the solar system. In size and mass the planet is apparently much more like the planets Mars and the Earth, than like its closer companions, Uranus and Neptune. (W.K.G.)

PLUTONIC. Plutonic is a general term applied to **igneous rocks** of deep-seated origin as distinguished from volcanic lavas which cool and congeal under the air, or under water. The word is derived from the Roman god Pluto, who ruled the underworld. (R.M.F.)

PLUTONIUM. Chemical Composition, I. Elements.

PNEUMATOLYSIS. Pneumatolysis is the process of alteration of existing rocks and mineral deposits, or the formation of new ones, by means of gases or vapors emanating from the magma. (E.S.C.S.)

PNEUMONIA. An acute infection of the lungs most commonly caused by *Diplococcus pneumoniae*. The disease may be caused by almost any organism, including the **Streptococcus, Staphylococcus,** Friedlander's bacillus, plague bacillus, as well as psitticosis and other viruses, and certain fungi.

The pneumococcus, a small encapsulated coccus occurring in pairs, is one of the most widely distributed of the disease-producing **bacteria.** In a harmless form it is a normal inhabitant of the mouth and throat. Virulent forms of the organism also enter through the upper respiratory passageways, being spread by direct contact with infected secretions. Pneumococci have been divided into more than 33 specific types on the basis of differences in immunological characteristics related to the capsular substance. Different types are of varying virulence. The most prevalent types producing pneumonia are 1, 2, 3, 4, 5, 6, 7, 8, and 14.

Pneumonia is classified as lobar, bronchopneumonia and primary atypical pneumonia. The lobar type is produced, characteristically, by pneumococci and occasionally by Friedlander's bacillus, while the other etiological agents usually produce bronchopneumonia. Primary atypical or "virus" pneumonia is a special form of bronchopneumonia which has been observed to occur in epidemic form since the early 1930's.

Lobar pneumonia is an acute infection localized to one or more lobes of the lung. It occurs most often in the winter months, in young healthy individuals. Pathologically it is characterized by massive **inflammation** and exudation into the smallest divisions of the lung, the alveoli. The normally air-filled lobe is thus converted into a heavy, solid, relatively airless structure. The symptoms are dramatic. The onset is with a chill, and often sharp chest pain on breathing (**pleurisy**). Fever, rapid pulse, "rusty" sputum follow. As the disease progresses, difficulty in breathing (**dyspnoea**), **cyanosis,** and prostration may be marked. The course, duration and mortality of the disease have been altered remarkably since the introduction of the **sulfonamides** and penicillin. Either of these, given early, will control the infection so rapidly that the full-blown picture with a solid lung never develops. The complications of pneumonia have also been altered and reduced to rare occurrences by these drugs. They include **empyema, pericarditis, arthritis,** and **meningitis.**

Primary bronchopneumonia due to staphylococci, streptococci and occasionally pneumococci of the higher types is a much less frequent, less sharply defined clinical entity. It occurs chiefly in children and in the aged. Pathologically it involves the walls of the smaller bronchioles, and to a lesser extent the alveoli. It tends to be patchy in distribution and involves the bases of the lungs. The onset is usually gradual, without a chill. The temperature is elevated, but not very high. Cough and thick sputum are characteristic. The disease responds less dramatically to sulfonamides than does lobar pneumonia.

Secondary Bronchopneumonia occurs as a complication of other diseases, particularly **influenza** and **measles.** It also appears as a terminal event in the course of chronic heart and kidney disease.

Primary atypical pneumonia occurs in early fall and winter epidemics in young individuals. It is chiefly an interstitial pneumonia, involving the tissue between the alveoli; **bronchitis** and bronchiolitis are prominent. Its cause has not been proven definitely, but all evidence points to a **virus** or perhaps several viruses.

Clinically the disease is characterized by a gradual onset of rather mild symptoms, relatively slow pulse, fever, cough, and little sputum. Chills and chest pain are uncommon. There are usually few signs over the chest although x-ray may reveal a widespread involvement of the lungs. The disease does not respond to sulfonamide therapy. The mortality is very low, but in the more severe form, the course may be long with temporary remissions and recurrences. (D.M.H.)

PNEUMOSTOME. 1. The opening leading into the chamber containing the respiratory organs of **scorpions** and **spiders.** 2. A small opening on the right side of a snail between the head and the edge of the mantle. It leads into the mantle cavity of the air-breathing forms. Aquatic forms have a much more extensive opening. (A.W.L.)

PNEUMOTHORAX. The introduction of air into the **pleural cavity,** causing a part or the whole of the **lung** on that side to collapse so that the vesicles of the lung become airless and the lung remains quiet. This occurs following a wound through the chest wall or a rupture of the lung surface through a disease process near the surface, such as tuberculosis.

Artificial pneumothorax is the injection of air into the pleural cavity as a therapeutic measure. It is most frequently used in pulmonary **tuberculosis** to put a diseased lung at rest so that healing may occur. (R.S.M.)

POCHARD. Aves, Anseriformes. A name applied in the Old World to **ducks** of several species related to the scaup ducks, the redhead, and the canvas-back of North America. (A.W.L.)

PODICAL PLATE. Two triangular plates flanking the **anus** in certain **insects.** They lie just behind the tenth abdominal segment and are regarded as the halves of the ventral sclerite of the eleventh segment. They have also been called paraprocts. (A.W.L.)

PODIUM. Tube Foot.

PODOBRANCHIA. A form of gill found in some of the **crustaceans.** It is named from its attachment to the basal segments of the thoracic legs. (A.W.L.)

PODSOL. The name of a leached white or gray **soil.** Supposed to be chiefly characteristic of cold moist climates. (R.M.F.)

POEBROTHERIUM. Oligocene.

POIKILITIC. Ophitic.

POIKILOTHERMY. A condition in which the temperature of the animal body fluctuates according to that of its surroundings. It may be higher or lower, but is always directly conditioned by external temperatures. **Homoiothermy,** in contrast, is the maintenance of a more or less constant body temperature, regardless of the surroundings, and is found only in birds and mammals. Thus a vast majority of animals are poikilothermal.

Poikilothermy has the disadvantage of limiting the activity of the animal to temperatures sufficient for the normal rate of metabolic processes. When it is too cold the animal becomes incapable of movement and its other processes are slowed. As a result poikilotherms must hibernate during the winter as a rule and must avoid the colder regions where their period of active life would be too greatly curtailed.

Animals of this kind are commonly called cold-blooded. (A.W.L.)

POINT OF CONTRAFLEXURE. Bending Moment.

POINT OF INFLECTION ON A PLANE CURVE. Concavity and Convexity of a Plane Curve.

POINTES EN FEUILLE DE LAURIER. Paleontology of Man.

POISEUILLE'S LAW. Viscosity.

POISON CLAW. A jointed appendage of the first body segment of **centipedes**. These appendages are not used in locomotion but are sharp-pointed and work together like a pair of jaws. They bear poison glands and are used in capturing and killing prey. (A.W.L.)

POISONS. Compounds of mercury (-ic), lead, copper, zinc, arsenic, antimony, selenium, barium, thallium, fluorine, are notably poisonous. Soluble metallic cyanides, isocyanic acid esters, **cyanogen**, **hydrogen cyanide**, some glucosides, alkaloids, many **nitrogen** and **sulfur** containing organic substances, yellow **phosphorus**, and **chromates** are extremely poisonous. The poisonous nature of many chemicals, some of which are common and necessary, encountered in laboratory and factory, constitutes a serious hazard to life and health. Careful provision for this hazard should be made. Some poisons, notably **phenol, hydrofluoric acid, sodium** hydroxide, **bromine**, yellow **phosphorus, nitric acid** concentrated, **sulfuric acid** concentrated, are also very corrosive to the flesh.

Among the gases which are extremely hazardous are especially **carbon monoxide**—odorless, and present in all gases of *incomplete* combustion of fuels, whether of coal, oil, or motor fuels—**hydrogen sulfide, nitrogen tetroxide, chlorine, bromine, hydrogen cyanide, cyanogen, cyanogen chloride.** Some poisonous gases are used as fumigants of rooms, foods, and in other ways. Gases thus used include hydrogen cyanide, cyanogen chloride, **sulfur dioxide, carbon disulfide** vapor—inflammable, ethylene oxide, **formaldehyde.** Their removal when used for fumigation and disinfection, before use of the space or materials, is a matter to be carefully confirmed.

Of liquid poisons, carbon disulfide is one commonly used, also phenol and cresols.

Many solids, either as fine powder or solution, are used as insecticides, germicides, fungicides, weed-killers, rodent-killers. Examples are **copper** sulfate—with calcium hydroxide as Bordeaux mixture—copper **arsenite,** copper-acetoarsenite (Paris green), sodium arsenite, calcium arsenate, lead arsenate, sulfur, calcium sulfide (limesulfur), **calcium** hypochlorite, **sodium** hypochlorite, thallous sulfate, mercuric chloride, barium carbonate, sodium fluoride, nicotine, para-dichlorobenzene, pyrethrum, DDT, sodium chlorate—easily ignited in the presence of organic material such as clothing.

In many situations the only substances sufficiently toxic for use against lower organisms are also toxic to man, and the hazard arising from their use cannot be too strongly emphasized.

CODE OF THE STATE OF MASSACHUSETTS FOR MAXIMUM SAFE CONCENTRATIONS OF CERTAIN COMMON TOXIC SUBSTANCES USED IN INDUSTRY

SUBSTANCE	MAXIMUM PARTS PER MILLION OF AIR
1. Ammonia	100
2. Amyl acetate	400
3. Aniline	5
4. Arsine	1
5. Benzene	75
6. Cadmium	0.1
7. Butyl acetate	400
8. Carbon disulfide	15
9. Carbon monoxide	100
10. Carbon tetrachloride	100
11. Chlorine	1
12. Chlorbenzene	75
13. Chlorodiphenyls	1
14. Chloronaphthalene	5
15. Chromic acid	0.1
16. Dichlorbenzene	75
17. Dichlorethyl ether	15
18. Ether	400
19. Ethylene dichloride	100
20. Formaldehyde	20
21. Gasoline	1000
22. Hydrochloric acid	10
23. Hydrogen cyanide	20
24. Hydrogen fluoride	3
25. Hydrogen sulfide	20
26. Lead	0.15
27. Mercury	0.1
28. Methanol	200
29. Nitrobenzene	5
30. Nitrogen oxides	10
31. Ozone	1
32. Phosgene (carbonyl chloride)	1
33. Phosphine	2
34. Sulfur dioxide	10
35. Tetrachlorethane	10
36. Tetrachlorethylene	200
37. Toluene	200
38. Trichlorethylene	200
39. Turpentine	200
40. Xylene, coal-tar naphtha	200
41. Zinc oxide fume	15

(R.K.S.)

POISSON DISTRIBUTION. The Poisson distribution or probability function is defined by $P_x = e^{-m} \frac{m^x}{x!}$, $x = 0, 1, 2, \cdots, \infty$, where P_x is the **probability** of x successes. It may be derived as the limiting case of the **Bernoulli probability function.** $P_x = {}_sC_x p^x q^{s-x}$ when sp remains constant as $s \to \infty$, and $p \to 0$. However, it may occur without the condition that p be small. The **moments** of the Poisson are **mean** $= m$, **variance** $= m$, $\mu_{3:x} = m$, $\mu_{4:x} = 3m^2 + m$, and $\alpha_{3:x} = \frac{1}{\sqrt{m}}$, $\alpha_{4:x} = 3 + \frac{1}{m}$. As $m \to \infty$, the Poisson probability function approaches the **normal probability function** with mean m and variance m. The Poisson probability function occurs in **industrial statistics,** in certain traffic problems, in the number of mutations when certain fruit flies are bombarded by neutrons, or x-rays, and in many other instances where the **universe** is essentially infinite and the probability of success is restricted to a very small portion of the universe. The probability $\sum_{x=a}^{x=b} P_x$ may be found exactly by a χ^2 distribution, or by **Pearson's Type III function,** or from tables of the Poisson probability function. (L.A.A.)

POISSON'S EQUATION. Laplace's Equation.

POISSON'S RATIO. If a rod of elastic material is stretched with sufficient force it can be elongated. The unit elongation (elongation per unit of length) is the strain, and may be denoted by s. At the same time the lateral dimensions will contract, the unit lateral contraction being c. The ratio c/s, which is constant for a given material within the elastic limit, is known as Poisson's ratio. For materials in which there is no directionality to **elasticity,** the value of Poisson's ratio was demonstrated by that celebrated mathematician to be 0.25. The value of 0.30 is generally used for steels, although recent careful determinations indicate 0.28 is a better average value. For aluminum alloys 0.33 is generally used. For values of Poisson's ratio up to 0.50, stretching results in a net increase in volume. At 0.50 the volume remains constant as in the case of plastic deformation of metals. (R.H.H.)

POLAR AIR. Air Masses.

POLAR COMPOUNDS. Valence.

POLAR COORDINATES IN A PLANE.
A point in a plane may be determined in position by two numbers called coordinates. One simple method for fixing the position is by means of polar coordinates.

Let O be a fixed reference point, called the pole (or origin), and let OX be a fixed reference line through O, called the polar axis (or initial line).

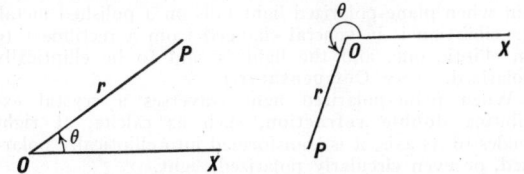

The position of any point P in the plane is determined by the distance OP and by any one of the directed angles having its initial side along OX and its terminal side along OP.

We agree to consider the directed distance OP as positive when measured along the terminal side of angle XOP, otherwise negative; and to regard the angle XOP as positive when it is generated by counter-clockwise rotation of OP from the initial position OX and negative when the rotation is clockwise.

The polar coordinates of the point P are the directed distance OP and the directed angle XOP. The directed distance OP is called the radius vector of P and is usually denoted by r or ρ; the directed angle XOP is called the vectorial angle of P and is usually denoted by θ. A point whose polar coordinates are r and θ is denoted by the symbol (r,θ).

Every pair of polar coordinates is represented by a single point; but conversely, every point has an infinite number of polar coordinates, since any one of a set of co-terminal angles may be used as the vectorial angle and also various combinations of positive and negative coordinates may be used.

Polar coordinate paper is paper ruled into a network of lines by concentric circles and radial lines through the center of the circles, and is useful for plotting points given by polar coordinates.

The distance between the points whose polar coordinates are (r_1, θ_1) and (r_2, θ_2) is given by

$$d^2 = r_1^2 + r_2^2 - 2r_1r_2 \cos(\theta_1 - \theta_2).$$

The relations between rectangular coordinates and polar coordinates in a plane are given by the equations

$$x = r \cos\theta, \quad y = r \sin\theta,$$

and

$$r = \pm\sqrt{x^2 + y^2}, \quad \theta = \arctan(y/x),$$

if the pole coincides with the origin and the polar axis coincides with the positive X-axis. (L.S.S.)

POLAR COORDINATES IN SPACE.
Let P be any point in space, referred in position to a system of rectangular coordinate axes.

The angles α, β, γ which the line OP makes with the X-, Y-, and Z-axes respectively are called the direction angles of the line OP; their cosines are called the direction cosines of OP.

The distance $\rho = OP$ and the three direction angles α, β, γ of OP are sometimes called the polar coordinates of point P.

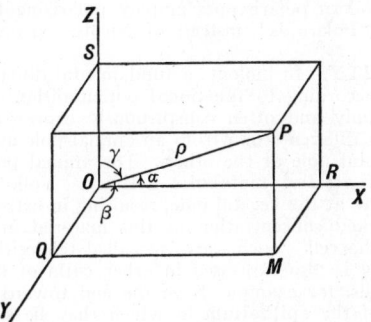

The relation between the rectangular and polar coordinates of a point in space is given by the formulas

$$x = \rho \cos\alpha, \quad y = \rho \cos\beta, \quad z = \rho \cos\gamma, \quad \rho^2 = x^2 + y^2 + z^2.$$

(L.L.S.)

POLAR FRONT. Fronts.

POLAR MOMENT OF INERTIA. Moment of Inertia.

POLARIS.
Polaris (α Ursae Minoris) is the principal star of the **constellation** and, as a matter of fact, about the only claim to recognition which the "little dipper" has. This star has been the closest star to the pole of rotation of the celestial sphere for the past five millenniums and has been used by navigators throughout written history. The antiquity of the use of this star is attested to by the fact that it is found represented on the earliest known Assyrian tablets. At present, Polaris is slightly over 1° from the pole of rotation and hence revolves about the pole in a small circle about 2° in diameter. Twice, and only twice, during every 24 hours Polaris accurately defines the true north **azimuth**.

At present, the star is universally used by navigators and surveyors for the purpose of determining true azimuth and also astronomic **latitude**. Tables have been computed and are to be found in the **Ephemerides** of the various governments for the purpose of reducing the actual position of the star at any particular instant to the actual position of the pole of rotation. (W.K.G.)

POLARISCOPE.
An instrument for ascertaining the properties of **polarized light** or for studying the effects of various agencies upon light of known polarization. It ordinarily consists of a "polarizer" for rendering common light plane-polarized in any desired azimuth, and an "analyzer" for identifying the character of polarized light; between these is usually a mounting for objects whose effect upon the light from the polarizer it may be desired to test. For example, the polarizer may be a **Nicol prism** from which the light emerges vibrating, say, in a horizontal plane; the analyzer may be another, similar Nicol; and between them may be mounted a tube with glass ends in which are placed various liquids to be tested for their rotatory effect. A polariscope of this type for sugar solutions is a **saccharimeter**. Or the object to be tested may be a doubly refracting crystal plate or a metallic reflector, rendering the plane-polarized light elliptically polarized; and the analyzer a Babinet **compensator** or similar device for identifying such light. Some polariscopes use a "half-shade analyzer," for example, a half-wave plate (equal to two quarter-wave plates) covering half the field, the two halves of which look equally bright through a Nicol only when the analyzer is turned into a certain position with reference to the azimuth of the

plane-polarized light entering it. In the simplest polariscopes the polarizer and the analyzer are reflecting plates of opaque glass set at the proper polarizing angle. Many modern polariscopes employ polarizing films, for example "Polaroids," instead of Nicols. (L.D.W.)

POLARITY. In biology, a fundamental differentiation of cells according to functional potentialities. Polarity is commonly and often conspicuously expressed in egg cells by a differentiation from an animal pole at one end to a vegetal pole at the other. The animal pole is the region of greatest **metabolic** activity. Yolk tends to accumulate at the vegetal pole, resulting in extreme cases in a marked concentration of this material in the one part of the cell. Such eggs are called telolecithal.

Polarity is also expressed in other **cells** of the body. Gland cells, for example, have the end toward the free surface of the **epithelium** in which they lie specialized for the accumulation and discharge of the special secretion, while the base of the cell is related to the underlying basement membrane. The phenomenon is also closely associated with the **axial gradients** of the animal body. (See also **Magnetism.**) (A.W.L.)

POLARIZATION. Cell; Electrochemistry; Polarized Light.

POLARIZATION PHOTOMETER. In the main forms of **bench photometer,** the illuminations or luminous flux densities from the two sources to be compared are made equal by regulating the relative distances. In photometers of the polarization type, the same object is accomplished by introducing a pair of Nicol prisms into the beam from the brighter source and turning one of these polarizers until the beam is cut down to equality with that from the other source. The ratio in which the flux density has been reduced, and hence the luminous intensity ratio of the two sources, is readily calculated from **Malus' law,** and the polarizer circle may thus be calibrated to give the ratio directly.

The polarization principle may also be applied to **stellar photometry** for the determination of **stellar magnitudes.** In one type of polarization photometer an "artificial star" is produced in the same manner as is described under **wedge photometer** and a pair of Nicol prisms is introduced into the path of light from this star. The method of use of this instrument is similar to that described for the wedge photometer, the difference being that the angles at which the Nicol prisms are set instead of wedge settings are converted into stellar magnitude difference. In the Pickering type of polarization photometer two telescopic objectives are combined in one tube, so that the images formed by each fall side by side. In front of each objective a plane mirror is placed; one being adjusted so that an image of Polaris is formed in the field of view, and the other adjustable so that the image of any star, close to the meridian, may be brought into the field of view close to the image of Polaris. In the light path from Polaris the pair of Nicol prisms is placed, and the angles at which they are set relative to each other may be calibrated to give the magnitude difference between Polaris and the star under examination. There are many modifications of this same general principle which have been devised by observers for their particular problems. (L.D.W., W.K.G.)

POLARIZED LIGHT. Whenever ordinary light is reflected from a glass plate, a varnished table-top, or other polished **dielectric** surface, we find upon suitable examination that a much larger part of the reflected beam is vibrating at right angles to the plane of reflection than in that plane; whereas originally it gave **no** evidence of any preferential direction of vibration. A little experimenting shows that at a certain angle of incidence (the "polarizing angle," different for different

dielectrics), the component vibrating in the plane of reflection is practically extinguished, all vibration being confined to the plane at right angles to this. The light is then plane-polarized. The effect is more conveniently produced by a **Nicol prism** or by one of the recently invented polarizing films, such as "Polaroid," which polarize by transmission with less loss of light.

When the light passes through two such polarizers in succession, as in a **polariscope,** the fraction of it finally emerging depends upon the angle between the transmission planes of the polarizers, and varies all the way from nearly 100% to zero (see **Malus' law**). The same effect may be produced by two reflections at glass plates turned to reflect in different planes but at the same (polarizing) angle. It seems probable that when the polarizing films above mentioned have been further perfected and cheapened, this intensity-reducing effect will be turned to account in reducing automobile headlight glare.

Metallic reflectors do not produce plane-polarization, but when plane-polarized light falls on a polished metal, its vibration is in general changed from a rectilinear to an elliptic one, and the light is said to be elliptically polarized. (See **Compensator.**)

When plane-polarized light traverses a crystal exhibiting **double refraction,** such as calcite, at right angles to its axis, it is transformed into elliptically polarized, or even circularly polarized, light.

If plane-polarized light is passed through quartz along its axis, or in any direction through one of the many optically active liquids such as turpentine or sugar solution, it undergoes optical rotation, i.e., its vibration plane is twisted around through an angle which steadily increases with the distance traversed in the substance. Different substances have very different rotatory power, and some rotate one way and some the other. The **saccharimeter** is especially adapted to the study of this effect in liquids. Somewhat similar is the **magneto-optical rotation** produced under suitable conditions by a magnetic field. (L.D.W.)

POLARIZING ANGLE. Polarized Light; Brewster's Law.

POLECAT. Mammalia, Carnivora. Long-bodied animals with short legs, related to the weasels and minks. They are known for their disagreeable odor, a characteristic which is responsible for the occasional erroneous application of the name to skunks.

The European polecat, also known as the foumart or foul marten and as the fitchet, fitcher, or fitcheu, is the source of the fur known on the market as fitch. Three other species occur in Europe and Asia, and one, the black-footed polecat, commonly called the black-footed **ferret,** is found in the plains region of North America. The **Cape polecat** of South Africa is more closely related to the skunks than to the true polecats. (A.W.L.)

POLES AND POLARS OF CONICS. For the ellipse $b^2x^2 + a^2y^2 = a^2b^2$, the equation of the **tangent** at the point (x_1, y_1) is $b^2x_1x + a^2y_1y = a^2b^2$. But if (x_1, y_1) is not a point on the ellipse, the preceding equation represents a straight line called the polar of the point $P_1(x_1, y_1)$ with respect to the ellipse, and P_1 is called the pole of the polar line. If the pole P_1 is outside of the ellipse (Fig. 1), the polar of P_1 may be obtained by drawing the two tangents from P_1 to the ellipse, and drawing the chord of contact. If the point P_1 is inside the ellipse (Fig. 2), the polar of P_1 may be obtained by drawing two chords through P_1, drawing the pair of tangents to the ellipse at the ends of each chord, and joining their intersections.

The poles and polars of the other **conic sections** are defined in a similar way.

An important property of poles and polars of conics is: If a point P_2 lies on the polar line of P_1, then P_1 lies on the polar line of P_2. (L.L.S.)

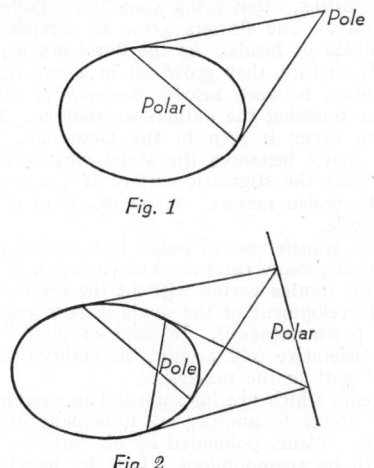

Fig. 1

Fig. 2

POLIAN VESICLE. Water Vascular System.

POLIOMYELITIS (Infantile Paralysis). A common acute virus disease which tends to occur in epidemic form during the late summer months. Poliomyelitis often involves the brain and spinal cord and may result in a flaccid paralysis of various groups of muscles. The disease is caused by a filtrable virus which can be readily isolated from the stools of infected individuals, and identified by animal inoculation. In monkeys and chimpanzees a disease similar to human poliomyelitis can be produced; a few strains will also cause the disease in certain rodents.

The epidemiology of poliomyelitis is not completely understood. It is thought that the virus can enter the body probably through several portals including the mucous membrane of the oral cavity and the pharynx, the lower gastrointestinal tract, and possibly the skin. Once in the body, the virus localizes in (1) the intestinal tract and (2) the brain and spinal cord. There is proof that the gastrointestinal tract serves as one of the major portals of exit for the virus. Patients may continue to excrete active virus in their stools for 6 weeks or longer after the onset of the acute disease, even though they are clinically well. Healthy carriers who have never had any symptoms of the disease are also known to exist. These facts lend support to the theory that the disease is spread by direct human contact or by contamination of food either directly, or through an intermediary agent such as the fly, which is known to carry the virus in epidemic areas.

The pathology of poliomyelitis is primarily an inflammation and destruction of the cells in the gray matter of the spinal cord—the ganglion or motor cells of the anterior horn. The clinical picture is extremely variable. The average incubation period is 10 days. In most instances the illness is mild, of short duration, and not accompanied by muscular weakness. The onset is sudden, with fever, malaise, headache, sometimes sore throat, and often vomiting. Pain in the back and stiffness of the neck and back due to muscular spasm are common. In the abortive type of poliomyelitis, these symptoms disappear after 3–4 days, and the child is well. In the paralytic type, which is much less common, the picture goes on to one of pain and weakness of the muscles and sometimes complete paralysis of the arms and legs and respiratory or other muscles, depending on which parts of the central nervous system are affected. If the respiratory muscles are involved,

death from asphyxiation may occur. In the bulbar type of the disease which involves the cranial nerves and has a high mortality, the voice has a nasal twang, the eye muscles may be paralyzed, and there is difficulty in swallowing; mucus and saliva accumulate, and coughing and choking are common.

The treatment of mild abortive poliomyelitis is symptomatic. In the paralytic form of the disease, the application of hot moist packs to the involved muscles relieves pain and relaxes muscles which are in spasm. Physostigmine and prostigmine are drugs which also seem to be effective in relieving muscular spasm. Early massage, passive and then active exercise of the affected muscles, are now recognized to be beneficial in achieving restoration of function of partially paralyzed limbs. If there is complete destruction of the nerve cells in the spinal cord which supply a group of muscles, nothing will restore the function of these muscles. If the respiratory muscles are involved, the respirator, or iron lung, may be used to give continuous artificial respiration. It may be necessary for a patient to remain in the respirator for weeks or months, until restoration of the function of his respiratory muscles occurs. (D.M.H.)

POLISHING. A surface-finishing process whereby scratches and tool marks are removed by using a polishing wheel made of leather or paper, to the face of which abrasive grains are cemented. (H.C.H.)

POLLACK. Pisces, Teleostei. A marine fish (Pisces) related to the haddock and cod. One species, Gadus pollachius, found in European waters, is also called the whiting-pollack, and another which is taken on both sides of the Atlantic is sometimes called the coal fish. (A.W.L.)

POLLAN. Pisces, Teleostei. A species of whitefish found in Irish lakes, Coregonus pollan. (A.W.L.)

POLLEN TUBE. Flower.

POLLINATION. Pollination is the act of transference of pollen grains to the stigmatic surface of a flower, where the pollen grain will germinate, forming a slender pollen tube, the development of which leads to the process of fertilization.

That the ancients should know anything of the act of pollination is scarcely to be expected, yet they had at least an inkling as to the necessity of the process, since they knew that male flowers of dates must be brought to the female flowers if a crop of fruit was to follow. Picture engravings on Egyptian monuments show this quite clearly. Later the need for caprification in figs was recognized.

But scientific understanding of the significance of the process and the ways in which it was accomplished was delayed until the studies of Joseph Kölreuter in the second half of the eighteenth century and of Christian Sprengel, who followed soon after. From the keen observations of these two men came the first real knowledge of pollination. Later Charles Darwin (1809–1882) made careful observations on pollination, observing the existence of two or more flower types in a single species and pointing out the way in which they might effectively cause cross-pollination, which he believed to be of great importance in maintaining the vigor of a race. He also attentively studied the process of pollination in orchids, observing many of the elaborate developments found in these flowers to insure successful cross-pollination. In the United States, Asa Gray (1810–1888) extended observations to American plants, including many native orchids. In Brazil Fritz Müller (1822–1897) made many interesting additions to the knowledge of the mechanisms and agencies involved in pollination. While these names are those of men whose observations were extensive, many others have also studied the various ways in which flowers are modified seemingly by foresight as to the needs of the process.

Flowers may be self-pollinated or cross-pollinated. In self-pollinated flowers, pollen from the **anthers** is transferred directly to the **stigma** of the same flower, while in cross-pollination pollen is carried to the stigmas of other flowers not on the same plant. Pollen transfer to other flowers on the same plant accomplishes self-fertilization and is called close pollination. That pollen may easily be deposited on stigmas of the same flower does not mean necessarily that self-fertilization shall occur, for in many cases pollen under such a condition fails to develop or develops so slowly that foreign pollen from other flowers soon grows beyond it and brings about fertilization. When pollen from the same flower or plant fails to bring about fertilization, the plant is said to be self-sterile; many varieties of cultivated fruit are of this sort. This explains why it is necessary to plant other varieties of apple among trees of the Mac-Intosh variety; many apple varieties are self-sterile.

Self-pollination must be a fact in certain flowers of the type known as cleistogamous—flowers which never open to allow pollen to be shed into the air or transferred by any means from flower to flower. In such flowers, which are found in many plants, as in several species of violet, the pollen grains may germinate while still in the anther, the pollen tubes growing out to the stigma, and then on to bring about fertilization. In other cleistogamous flowers the pollen is shed from the anthers and falls directly onto the stigma, and there develops. Self-pollination also occurs, though not of necessity, in many perfect flowers. In many cases the stigma is directly beneath the ripened anthers, so that pollen shaken from the anther is likely to fall onto the stigma. In some plants the stigma is always beneath the mature anther, while in other plants movement occurs so that as the part grows older the stigma bends over to a position beneath the anther. In such a case, if cross-pollination has not already occurred, self-pollination may be effected. In other flowers it is the stamen which exhibits movements, curving or bending in such manner that pollination shall be accomplished. In still other plants the filaments gradually elongate so that the anthers are carried upward to the stigma.

Cross-pollination is insured by several means. In many plants, the stamens mature and shed their pollen long before the stigmas are receptive, while in other plants, the stigmas are mature before the pollen of the same flower has ripened. Either case necessarily insures cross-pollination. Equally certain is the occurrence of cross-pollination in those plants which bear unisexual flowers on different plants, or if on the same plant male flowers mature some time before the female, as is the case of the Alder, where the inconspicuous pistillate (see **Pistil**) flowers are mature two to four days before the pendulous **catkins** of staminate (see **Stamen**) flowers begin shedding their pollen.

Another method which tends to effect cross-pollination is the occurrence of flowers of two, or three, or even four different kinds. The purple swamp loosestrife, *Lythrum Salicaria,* shows one such case. Some of the flowers of its spike have anthers borne on long filaments and contain pollen grains of relatively large size; other flowers have stamens with short filaments and contain small pollen grains; while a third flower type is intermediate in habit, the filaments being between the others in length and the pollen grains of intermediate size. In flowers like these, if insect-pollinated, every likelihood is that pollen of one flower type will be carried to stigmas which show the same gradations in length, so that pollen from stamens with long filaments will be deposited on such parts of the insect's body as will be touched by long-styled pistils; and since these long-styled pistils are not found in flowers having long filaments, cross-pollination will probably

occur. Flowers having such differences in style length are called heterostylous flowers.

Many plants, notably those in the **Carrot and Composite Families,** often bring about cross-pollination in another way. The flowers grow in compact groups, either **umbels** or heads. As the floral organs, stamens and pistils, mature, they grow out in a way that brings about contact between nearby flowers, the stigmas of one flower touching the anthers of another. That this does often occur is seen in the **Composite Family,** where in many instances the styles curl back at maturity so that the stigmatic surface is brought directly against the pollen masses. Many plants in this family are self-sterile.

While the transference of pollen from anther to stigma usually occurs, many cases are known in which development of the **ovules** occurs without the presence of any pollen. Development of the ovule in such cases is said to be by parthenogenesis. In some of these it may be that the vegetative cells around the embryo sac play a significant part in the process.

The agents which are instrumental in carrying pollen from one flower to another are principally air-currents and insects. Plants pollinated by air-currents, or wind, are said to be anemophilous; those by insects are entomophilous. In addition to these two main agencies there are several minor ones; in a few plants, water is the agent carrying pollen from flower to flower. A few flowers are considered to be pollinated by snails, others by birds and some even by bats. Man himself may effect pollination by selecting plants of special value to him and transferring pollen from one to the other, taking special precautions to prevent the presence of foreign unwanted pollen; this is artificial pollination to produce special strains of plants. Commonly it is accomplished by gathering mature pollen from unopened anthers of any desired flowers. With a small camel's hair brush, or special pollinating scalpel, this pollen is then placed on the stigmas of another selected flower; the second flower is emasculated, that is, the stamens are removed before maturing, to prevent any possible self-pollination. Flowers to be pollinated are usually kept enclosed in bags to prevent the presence of any unwanted pollen until fertilization has been effected. If accurate results are required all instruments and the hands of the operator are sterilized by immersion in alcohol between successive pollinations.

As a rule wind-pollinated flowers are inconspicuous, and of small size: color, odor, and nectar, all associated with insect-pollinated flowers, are largely lacking. Pollen is produced in immense quantities, since much will be lost. The pollen grains are dry and dust-like, and so float buoyantly in the air. In many plants of this group the stigmas are much-branched, feathery objects, offering a considerable area of sticky surface to catch any pollen that falls on it. To further the ready discharge of pollen, which ordinarily occurs only when the atmosphere is dry, the flowers are often borne in pendulous catkins, or on slender flexible **pedicels,** or have stamens the anthers of which are versatile, that is, attached at the middle and easily moved by any slight disturbance. Examples of wind-pollinated plants are found in the **Gymnosperms,** all of which are so pollinated, in most grasses, in many hardwood trees, such as birches, alders, **oaks,** and beeches, and in the common **cat-tail** of the marshes.

Water-pollinated plants are not numerous. That a plant grows in water is not an indication that its flowers shall be pollinated by water. Indeed, only a very few water-plants are so pollinated. In some cases, as in the marine eel-grass, *Zostera marina,* pollination occurs under water, the heavy pollen grains being carried by the water and by chance reaching the stigmatic surface. The fresh-water plant, *Naias,* is similarly pollinated under water, its pollen grains having a specific gravity of 1. In other water plants, pollination occurs at the surface of the water. Examples of this type are found in the fresh-water eel-grass, *Vallisneria spiralis,* and in the marine *Ruppia maritima.* When the staminate flowers of eel-grass are mature, they break from their rather short stem and, being buoyant because of air-chambers, float to the surface of the water. There the spathes surrounding the flowers open, exposing the stamens within. These float about on the surface of the water, driven by air currents, as well as borne by water currents. The pistillate flower is solitary at the end of a long spiral stalk; when this flower is mature the stalk straightens so that the flower floats on the surface of the water, its three stigmas spreading wide and over-reaching the three-parted perianth. Chance drifting of the staminate flowers brings them near the stigmas which may then be pollinated by contact with the anthers. After pollination and subsequent fertilization, the stalk of the pistillate flower coils into a tight spiral, drawing the developing ovary deep into the water, where it matures.

Plants with flowers pollinated by animals are extremely numerous, and with many variations seemingly calculated to assure cross-pollination. A few tropical plants are said to be pollinated by bats. The flowers have very fleshy petals, open during the evening hours, and are apparently sought by bats, which eat the petals, or perhaps seek any insects which may occur in the flowers. Quite possibly, in going from flower to flower, bats do bring about a transfer of pollen. Pollination by such animals is undoubtedly restricted to a very small number of plants.

Similarly birds, especially hummings-birds and honeysuckers, are held to be the agents pollinating several tropical plants. In these the question arises as to whether the birds are seeking the numerous insects which are to be found in the flowers or after the nectar which is found in the flowers. The flowers visited by birds are rather large, brilliantly colored, often scarlet. Unquestionably humming-birds do seek nectar in flowers, both in the tropics and in temperate regions. Pollination may well result from their visits.

A few flowers are said to be pollinated by snails or slugs. Plants so pollinated have dense masses of flowers borne on a fleshy stock. This mass of sterile tissue attracts these animals, which crawl over it in the search for food, and so are said to carry pollen from flower to flower. Many observers doubt that such animals ever bring about pollination.

All, however, recognize that insects are very important in the pollination of many flowers. In many cases the flowers show remarkable adaptations fitting them to be pollinated by certain insects. The insects which effect pollination are for the most part bees, flies, beetles, and moths and butterflies.

As certain features characterize wind-pollinated flowers, so also insect-pollinated flowers exhibit several common features. In them the pollen grains are usually somewhat adhesive, often sticking together into considerable masses. The surface of each grain is variously sculptured, with knobs, spines, or other protuberances definitely increasing the ability of the grain to stick to the insect body. As in wind-pollinated flowers, so here the pollen would be seriously impaired by water, from which it must be protected during periods when pollination would not occur. Many are the ways in which protection is obtained. In some plants the entire flower bends down at night, while in many others the petals close together over the stamens; often a passing cloud is sufficient stimulus to cause closing, which takes place with surprising speed. In many plants the flowers are located beneath the leaves, as in the common Jewel-weed or Touch-me-not, *Impatiens biflora;* in the common Iris, each stamen is located beneath the broad-petaloid stigma. Often the anther itself is so constructed as to afford considerable protection, the pollen frequently being shed through narrow slits which close tightly during periods of excessive moisture; or small

pores may allow the pollen to escape when advantageous but protect it from water otherwise.

A most obvious characteristic of insect-pollinated flowers is color, which may be found in a single flower with large conspicuous petals, or may result from the massing together of many small flowers, as in the Composite Family. While it is generally assumed that bright color is an aid to pollination because it attracts insects, it should be recognized that insect vision is not necessarily like that of human beings. Often the color of a flower changes with age, many becoming gradually deeper toned, as in the common Lady's Slipper, while others as gradually fade out. Flowers opening at night are almost all white or very light-colored.

The odors of flowers are also assumed to attract insects. Every fragrant flower has an odor which is quite distinctive. The odors of flowers are of many types, from the foul rankness of the Skunk Cabbage and Carrion Flower to the delightful perfume of Verbena, Gardenia, the Roses, and many Lilies. In many cases the odor of the flower is very delicate, being scarcely detectible to many people; in others it is of such penetrating strength as to become objectionable. Often the odor is evident only during certain periods, some flowers being scentless by day but fragrant during the night, while others emit their odors only in broad daylight. All these differences in odors seem designed to attract special insects which will accomplish pollination. The foul odor of carrion calls carrion-flies, and the sweet fragrance of night-blooming flowers attracts moths. After pollination the odor of the flower generally ceases, attraction of insects being no longer of any value.

Many insects undoubtedly visit flowers for the purpose of obtaining nectar, a sweet watery secretion formed in special glands called nectaries. These nectaries are variously located in the flower, usually deep down at the base of the corolla, so that any insect obtaining it must either have a mouth part of sufficient length to reach the nectar or be strong enough to push its way into the flower, passing any obstructions which may serve to protect the nectar from less fortunate insects. Obstructions are found of several sorts. A very common means of excluding such crawling insects as ants, which in all probability would not be efficient pollinators, is by the presence of a barricade of hairs, especially in the throats of flowers having a tubular corolla. Long-tongued bees are able to push through these hairs enough to reach the nectar, while at the same time they are thoroughly powdered with pollen which may be removed later in another flower. The existence of a sticky secretion over the outside of the flower or on the stem of the plant is an effective barrier to many crawling insects, as is also a waxy coating. Especially noteworthy are those flowers in which the nectar is located at the base of a long narrow tube or spur; often the nectar is present in quantities large enough to form a considerable volume. In such cases there is usually an insect with mouth parts just long enough to reach through the tube into the nectar supply, which is sucked up greedily. This becomes the more remarkable when one considers that in some flowers the nectar is at the bottom of a tube which may be 3 or 4″ long, and in the case of one tropical Orchid a nectar-secreting sac over a foot long exists; in such cases insects, usually moths or butterflies, with correspondingly long sucking tubes are found. Often short-tongued insects succeed in obtaining the nectar illegitimately by biting a hole in the wall of the nectar-containing part of the flower, and obtaining the nectar thereby. It is interesting to note, in connection with this problem of nectar-secreting flowers and insects, that the introduction of clover into Australia was not a success until honey-bees were also introduced. The native Australian bees were too short-tongued to reach the nectar and so pollination was not effected. As a consequence the clover crop soon died out, no seed being formed to perpetuate it. With the introduction of suitable bees the plant seeded abundantly.

It is interesting to note that in addition to the existence of colors, odors, and nectar, the structure of the flower and its position on the plant seem designed to facilitate the work of the insect in transferring the pollen. Many flowers are broad and flat, affording convenient support to the insect as it crawls about over the flower. Others, especially those which are visited by long-tongued moths, are borne in such a position as seems most suited to permit the insect to insert its tongue and obtain the nectar. The pollen is collected in quantities by many species of bees, who use it as a food for the developing young.

Several plants have developed a most striking relationship with insects. While the majority of flowers are not visited indiscriminately by all insects but only by certain species or genera, in these special cases the restriction is extreme, both the flower and the insect seeming to be modified especially to serve one another. One such case is seen in the edible fig, pollinated only by a small insect, *Blastophaga grossorum,* which lays its eggs in the ovaries of certain flowers of the fig. Another example of this insect-flower association occurs in a species of Yucca. The creamy-white flowers of this plant are borne in large panicles. They open during the evening and are visited by a small moth, *Pronuba Fuccasella,* which seeks the abundant pollen of the flower. Of this pollen the moth makes a tiny ball which it carries away to another Yucca flower. In the ovary of this flower the female moth lays her eggs, while on the stigmatic surface it deposits its ball of pollen. The developing ovules serve as food for the growing larvae of the insect. However, many ovules grow to maturity to form viable seeds, which perpetuate the plant and so continue the food supply of the moth, which seems to be the sole agent capable of transferring the sticky pollen of the Yucca from plant to plant. Many other cases are known where pollination of the flower depends on the visit of certain insects, which do not, however, themselves depend on the flower for existence. On the other hand, many elaborate devices, such as the keel mechanism in papilionaceous flowers, which ought to assure insect pollination, fail to do so. Garden peas are normally self-pollinated. (R.M.W.)

POLLUCITE. The mineral pollucite is rather rare. It contains caesium, aluminum, silicon, and oxygen, its chemical composition being approximately $H_2O \cdot 2Cs_2O \cdot 2Al_2O_3 \cdot 9SiO_2$. It is isometric, usually in cubic crystals or crystalline masses; conchoidal fracture; brittle; hardness, 6.5, specific gravity, 2.9; luster, vitreous on fresh surfaces; colorless and transparent. Found on the Island of Elba and in the pegmatites of Maine. Pollucite and petalite were found in the granites of Elba and at first named pollux and castorite for the two famous brothers of Roman mythology, Castor and Pollux. Pollucite is derived from the Latin genitive *Pollucis.* (E.S.C.S.)

POLLUX. Pollux (β Geminorum) is the brighter of the "heavenly twins." The two stars are always considered together in ancient writings and in astrology, and, as a matter of fact, are not mentioned individually in literature, but always as the constellation of Gemini. This constellation is of very ancient lineage and is referred to throughout all classical literature. They were always considered of good omen by all peoples and always referred to as twins. Astrologically, the constellation was most favorably regarded, portending genius, goodness, and liberality. (W.K.G.)

POLONIUM. Symbol: RaF (or Po). A radioactive element of the uranium-radium series. (See Radioactive Changes.) (R.K.S.)

POLYBASITE. A mineral antimony sulfide of silver in which copper substitutes for silver to approximately

30 atomic %. Crystallizes in the monoclinic system. Hardness, 2–3; specific gravity, 6.1±; color, black with metallic luster; nearly opaque. From the Greek, meaning many, suggesting the many metallic basis. (R.M.F.)

POLYCHAETA. Chaetopoda.

POLYCHORIC CORRELATION. This is a method, rather complicated, of determining the coefficient of correlation in an $r \times s$ contingency table. (L.A.A.)

POLYCLADIDA. Turbellaria.

POLYCONIC PROJECTION. This form of map projection is essentially a series of simple conical projections of zones on the earth, so stretched as to form a continuous map. While the projection is in reality mathematical rather than geometrical, it may be described in words by considering a series of cones to be constructed which are tangent to the earth at certain selected parallels of latitude. Some meridian is selected as the central longitude for the map. Zones of latitude on the earth, each centered on one of the standard parallels and extending half-way on either side toward the adjacent standard parallels, are projected on the particular cone for the zone. All zones, on the cones, are then cut along the meridian differing by 180° from the central longitude, and developed on a plane.

The resulting map will not be continuous. All parallels will be represented as circles, but the centers of the circles will be different for successive zones. Hence the boundaries of adjacent zones will be tangent along the central meridian, but separate with increase of longitude in both directions from the center. To make the map continuous the zones are stretched, along a north-south direction, until the bounding parallels of adjacent zones are congruent with a curve bisecting the area of discontinuity between the zones. When this is done the meridians in adjacent zones will be discontinuous, and an east-west stretching must be made to bring the meridians into continuous curves. The resulting graticule does not appear to be very different from that of the simple conic, but examination shows that the parallels are not circles and the meridians not converging straight lines.

The advantage of the polyconic projection for maps is that there is virtually no distortion of relative size and shape along the central meridian. The same scale of distance can be used for a considerable distance either side of this line. The polyconic projection is frequently used for maps of river valleys and coast lines when these extend in approximately a north-south direction. For navigational purposes the Lambert projection has completely superseded the polyconic. (W.K.G.)

POLYCYTHEMIA. An abnormal increase in the red cells of the blood so that their total number is in excess of the normal 5 million per cu. mm. *Polycythemia vera* occurs as a primary, slowly progressive disease which is eventually fatal. Patients show reddish cyanosis, enlarged spleen and liver, and they suffer from malaise, digestive symptoms, and often painful thromboses of the superficial veins. Their red counts range from 7 to 11 million.

Secondary polycythemia, as a result of circulatory stasis, occurs with chronic heart disease (particularly of the congenital type), and chronic pulmonary disease, such as emphysema. It is also present in mild degree in individuals who live at high altitudes, the increased numbers of cells compensating for the low oxygen tension. (D.M.H.)

POLYEMBRYONY. The development of more than one individual from a single fertilized egg cell. In this process the egg breaks up during its early development into several to many component parts, each of which becomes a complete animal. It takes place in the phylum **Bryozoa** in connection with the formation of colonies and has been reported in some of the parasitic insects (**Hymenoptera**). Since the host animal defends itself against the efforts of the female parasite to deposit her eggs in its body, the development of many young from each egg successfully placed is an obvious advantage. The process is akin to an asexual reproduction. (A.W.L.)

POLYMASTIGIDA. Mastigophora.

POLYMERIZATION. Association and Polymerization.

POLYMORPHISM. The occurrence of individuals of distinctly different structure or appearance within a species. In many cases two such forms occur and the species is said to be dimorphic rather than polymorphic.

Polymorphism depends upon many different conditions in various groups of animals. The various forms may be adapted for different places in a life cycle, for special parts in a colonial or social organization, or for special stages in a **metamorphosis**. They may also result from the incidence of different environmental conditions due to seasons or to unusual climatic conditions.

The alternation of **polyp** and **medusa** in a reproductive cycle in some species of **coelenterates** is a conspicuous example, as also is the appearance of such varied polyps as the gasterozooids and dactylozooids in colonial species. A differentiation of individuals according to special duties is evident in the castes of social insects. Many insects also vary according to the season or according to geographical range, as in the case of the wet and dry season forms of various tropical butterflies. Seasonal form, geographical race, variety, and subspecies are terms applied to forms of this kind. The influence of unusual conditions sometimes affects an occasional individual, producing an aberration. In chemistry, the term polymorphism is applied to compounds which exhibit **allotrophy**. Thus a substance that can occur in two crystalline forms like calcium carbonate is said to be dimorphous, one that can exist in three crystalline forms like thallous nitrate is said to be trimorphous, etc. The term polymorphous, as stated above, is general for them all. (A.W.L.)

POLYNOMIAL EQUATIONS. A polynomial equation of degree n in one variable or unknown is an equation of the form

$$a_0x^n + a_1x^{n-1} + a_2x^{n-2} + \cdots + a_{n-1}x + a_n = 0 \ (a_0 \neq 0),$$

where n is a positive integer and the coefficients a_0, a_1, $\cdots$, a_n are independent of x. It is also frequently called a rational integral equation.

Polynomial equations are classified into various types according to degree, as: **linear equations, quadratic equations, cubic equations, quartic** (or **biquadratic**) **equations,** quintic equations, etc.

Every **linear equation** and every **quadratic equation** in one unknown can be solved in terms of algebraic expressions involving the coefficients. **Cubic equations** (of third degree) and **quartic equations** (of fourth degree) can also be solved by explicit formulae for the roots in terms of algebraic expressions involving the coefficients. It has been proved, by Abel, however, that a polynomial equation in one unknown of degree higher than 4 cannot be solved in the general case algebraically, that is, for the roots in terms of algebraic expressions involving the coefficients.

In special cases, polynomial equations may be solved by the method of **factoring**.

Approximate values of the real **roots** of polynomial equations may be found by a graphical solution, by constructing the **graph** of the corresponding function and finding the **abscissas** of the intersections of the graph with the X-axis.

If a **polynomial function** $P(x)$ contains $x - r$ as a factor exactly k times and no more, so that $(x - r)^k$ is an exact divisor of $P(x)$ but $(x - r)^{k+1}$ is not, then r is called a multiple root of the equation $P(x) = 0$, or we say that r is a root of multiplicity k or a root of order k. A root of order one is called a simple root.

The so-called "fundamental theorem of algebra" is: Every polynomial equation in one unknown has at least one root (which may be real or complex).

It follows from this that every polynomial equation of degree n in one unknown has n and only n roots, provided a multiple root of order k is counted as k roots.

If r is a simple root of a polynomial equation $P(x) = 0$, (i.e., if $x - r$ is a factor of $P(x)$ once and only once) it is not a root of the equation $P'(x) = 0$, where $P'(x)$ denotes the **derivative** of $P(x)$. If r is a multiple root of order $m > 1$ of $P(x) = 0$, it is a root of order $m - 1$ of $P'(x) = 0$.

If a root of a polynomial equation is known, the corresponding factor may be divided out of the polynomial, leaving a new equation of lower degree, called the depressed equation.

The location of the real roots of a polynomial equation may be investigated by means of **Descartes' rule.** This rule gives more or less definite information about the nature of the roots of a polynomial equation.

The location of the real roots of a polynomial equation may be determined by the following theorem: If $P(x)$ is a polynomial with real coefficients, and if $P(a)$ and $P(b)$ have opposite signs, the equation $P(x) = 0$ has at least one and in fact an odd number of real roots between a and b; if $P(a)$ and $P(b)$ have like signs, the equation $P(x) = 0$ either has no roots between a and b, or it has an even number of roots between a and b.

In locating the real roots of the equation, the following result is also of use: If in substituting in a polynomial $P(x)$ a positive value r by **synthetic division**, all the partial sums are positive, then r is greater than any real root of the equation $P(x) = 0$.

In the process of finding the rational and irrational roots of a numerical polynomial equation, certain **transformations of the equations** are useful.

The determination of the rational roots of a polynominal equation is generally based on the following theorem:

If a polynomial equation, with the coefficient of the highest power of the variable unity, and all the other coefficients integers, has any rational roots, those roots must be integers, and must be divisors of the constant term, if the constant term is not 0.

To find the rational roots of a polynomial equation with first coefficient unity and all other coefficients integers, it is only necessary to test whether the integral factors of the constant term satisfy the equation, usually using synthetic division.

If the coefficient of the highest power of the variable is not 1 or if the coefficients are not integers, we may divide all terms by the coefficient of the highest degree term, and then remove any fractions in the other coefficients by transforming the equation by multiplying the roots by the appropriate factor, and thus reduce the equation to the form of the preceding method.

If a quadratic surd $a + \sqrt{b}$, where a and b are rational but $\sqrt{b}$ is irrational, is a root of a polynomial equation $P(x) = 0$ with rational coefficients, the conjugate surd $a - \sqrt{b}$ is also a root of the equation.

For the calculation of the approximate values of the irrational roots of a polynomial equation, there are available **Horner's method, Newton's method,** interpolation methods, and combinations of these. These methods will give the approximate values of the roots to any desired degree of accuracy.

If a complex number $a + bi$ is a root of a polynomial equation $P(x) = 0$ of degree $n \geqq 2$ with real coefficients, the conjugate complex number $a - bi$ is also a root of the equation $P(x) = 0$.

In other words, complex roots of polynomial equations occur in conjugate pairs.

Approximate values of the roots of a polynomial equation may also be calculated by use of Graeffe's root-squaring method; this method is particularly useful in finding the complex roots of high-degree equations.

The relation between the roots and coefficients of any polynomial equation are given as follows: If $r_1, r_2, \cdots, r_n$ are the roots of the equation

$$x^n + a_1x^{n-1} + a_2x^{n-2} + \cdots + a_{n-1}x + a_n = 0$$

then

$$-a_1 = r_1 + r_2 + \cdots + r_n,$$

$$a_2 = r_1r_2 + r_1r_3 + \cdots + r_{n-1}r_n,$$

$$-a_3 = r_1r_2r_3 + r_1r_2r_4 + \cdots + r_{n-2}r_{n-1}r_n,$$

$$\cdot \cdot \cdot \cdot \cdot \cdot \cdot \cdot \cdot \cdot \cdot \cdot \cdot$$

$$(-1)^na_n = r_1r_2 \cdots r_n.$$

(L.L.S.)

POLYNOMIAL FUNCTION. A rational integral function (or polynomial function) is a **rational function** in which the variable never appears in the denominator of a fraction. It can be expressed in the form

$$P(x) = a_0x^n + a_1x^{n-1} + \cdots + a_{n-1}x + a_n,$$

where x is the variable, n is a positive integer, and the coefficients $a_0, a_1, \cdots, a_n$ are constants (independent of x); in this case, the function is said to be of the nth degree.

Polynomial functions are classified into types according to the degree, into **linear functions, quadratic functions, cubic functions, quartic** (or biquadratic) **functions,** etc. (L.L.S.)

POLYNOMIALS. A polynomial in general is an algebraic **expression** consisting of two or more terms connected by plus or minus signs.

A polynomial in one variable is an expression of the form

$$a_0x^n + a_1x^{n-1} + a_2x^{n-2} + \cdots + a_{n-1}x + a_n,$$

where the coefficients $a_0, a_1, \cdots, a_n$ are **constants** (independent of x), x is the **variable** and n is a positive integer, which is called the degree of the polynomial. The degree of the polynomial is the highest power of the variable that occurs in the expression. A polynomial in a variable x is frequently denoted by a symbol as $P(x)$.

A polynomial in several variables $x, y, z, \cdots$ is an algebraic expression of the form of an algebraic sum of terms of the type $cx^my^nz^p \cdots$, where c is a constant coefficient and each of the **exponents** $m, n, p, \cdots$ is either a positive integer or zero. The degree of such a polynomial term in the variables $x, y, z, \cdots$ is the sum $m + n + p + \cdots$ of the exponents of these variables. The degree of the polynomial is the degree of its term of highest degree. (L.L.S.)

POLYOPISTHOCOTYLINEA. Trematoda.

POLYP. In zoology, a polyp is one of the two types of individuals found in many species of **coelenterates.** The two are the polyp or hydroid and the **medusa.** Polyps are approximately cylindrical, elongated on the axis of the body. One end is usually attached and the other bears the mouth, surrounded by a circlet of tentacles. The wall is relatively thin, due to the thinness of the mesogloea. In the class **Hydrozoa,** polyps are often very simple, like the common little fresh-water species of the genus *Hydra.* Actinozoan polyps, including the **corals** and **sea anemones,** are much more complex, due to the development of a tubular stomodaeum

leading inward from the mouth and a series of radial partitions called mesenteries. Many of the mesenteries project into the enteric cavity but some extend from the body wall to the central stomodaeum.

In medicine, a polyp is a smooth-coated tumor projecting from a mucous surface. It is attached to the surface by a narrow elongated **pedicle.** Polyps are commonly found in the nose, bladder, rectum and large intestine. They may also occur elsewhere in the body. (A.W.L., R.S.M.)

POLYPEPTIDES. Aminoacids, Polypeptides, and Proteins.

POLYPHYLETIC. A group of organisms derived from more than one ancestral stock is said to be polyphyletic. (R.M.W.)

POLYPHYLY. An evolutionary hypothesis which assumes that animals have originated by several lines of descent from as many ancestral forms, rather than from a single primordial form by gradual divergence. Special creation is the extreme of polyphyly but the hypothesis is not in discord with organic **evolution.** Although a great majority of organic forms seem traceable to a common ancestry there are a few peculiarities among the simpler living things which may indicate origin from independent lines. Among them are the cell structure of the **bacteria,** which lack a centralized **nucleus,** and the peculiar **metabolism** of some of the autotrophic species, such as iron and sulfur bacteria. Since modern evolution assumes the origin of life in the beginning from inorganic matter, there is no inherent obstacle to the assumption that living matter in that simple state may have been established independently in more than one place on the surface of the earth. (A.W.L.)

POLYPLACOPHORA. Amphinura.

POLYSACCHARIDES. Carbohydrates.

POLYTROPIC PROCESSES. The expansion or compression of a constant weight of gas may assume a variety of forms, depending on the extent to which heat is added to or rejected from the gas during the process, and also on the work done. There are, theoretically, an infinite number of ways possible in which a gas may expand from an initial pressure p_1, and volume v_1 to a final volume v_2. All these expansions may be grouped generically as polytropic expansions, and all could be represented graphically on the PV plane by the family of curves $pv^n = C$. They are all, in theory, perfectly reversible. n may have any positive value, 0 to ∞, and having been selected numerically it defines the type of expansion. From the infinite number of possible polytropic expansions, it is worth while to isolate four which deserve special attention. When one of the four physical characteristics, to wit, **pressure, temperature, entropy,** or volume, remains constant, expansions of more than ordinary interest are denoted, since they are frequently employed in a practical way, in situations which can be subjected to thermodynamic analysis. The value of the exponent n of the polytropic family for each of these is:

Isobaric	$n = 0$,
Isothermal	$n = 1$,
Isentropic	$n = \gamma$ (γ = ratio of specific heat at constant pressure to that at constant volume)
Isometric	$n = \infty$.

These thermodynamic processes, as they occur in useful machines, are not often of the exact polytropic form desired. For example, an isentropic process which is exemplified, at least theoretically, by expansion of the burned gases after the explosive combustion in the gasoline engine, is modified slightly by the interchange of heat between gases and cylinder wall, whereas a true isentropic has no heat either added or rejected in this way. The particular polytropic curve which would suit these conditions of expansion would depart somewhat from the adiabatic form.

During a polytropic process conditions of the working medium are constantly varying, and analysis may be aimed at determining one of the following: the work done, the heat added, the variation of temperature, and the change of entropy. Some information may be obtained merely by comparing the value of the exponent n with certain other data. For example, if n lies between o and 1, the temperature rises during an expansion and falls during a compression; when n is greater than 1, the temperature falls during expansion and rises during compression. Also, when n is less than γ, heat must be added to obtain an expansion, whereas when it is greater than γ, heat must be expelled. From the above it will be noted that there is a certain range of polytropic expansion in which, although heat is added, the temperature falls. This may seem to some to be paradoxical, but it is readily explained. During these expansions work is being done by the gas at a rate greater than that at which heat is being added, with the result that the deficiency must be made up from within the gas. The only way that this may be accomplished is for the gas to cool and give up some of its internal energy.

The equations for work done and for heat added in the case of the general polytropic expansions are:

$$W = \frac{p_1 v_1 - p_2 v_2}{n - 1}.$$

$$Q = (p_1 v_1 - p_2 v_2)\left(\frac{1}{n-1} - \frac{1}{\gamma-1}\right)$$

Both of these are expressed in foot-pounds. Sometimes a substitution of a definite value of n in one or the other of these equations leads to an indeterminate; for example, with the isothermal,

$$W = \frac{p_1 v_1 - p_2 v_2}{1 - 1}.$$

But since the equation of the isothermal for an ideal gas is

$$pv = C,$$

$$p_1 v_1 = p_2 v_2,$$

and the work equation becomes indeterminate:

$$W = \frac{0}{0}.$$

By approaching the isothermal from a different angle, however, the equation

$$W = pv \log_e \frac{v_2}{v_1}$$

may be deduced for work done. (F.T.M.)

POLYZOA. Bryozoa.

POMACE FLY. Drosophila.

POME. Fruit.

POMEGRANATE. *Punica Granatum.* Punicaceae. The pomegranate is a shrub or small tree which is native to southeastern Europe and southwestern Asia. It has been cultivated since early times, and is now grown extensively in tropical and subtropical regions in both hemispheres. The plant has opposite entire leaves which are elliptical to oblong in shape. The flowers are borne either singly or in small clusters in the axils of the leaves. They are perfect and have a bright red **corolla** of 5–8 **petals,** and many **stamens.** The **fruit** is a many-seeded berry. The outer coat of each seed is the edible portion. It is soft and fleshy and red in color. Pomegranates are used in making jellies and jams. (R.M.W.)

PONS. Brain.

PONTOON. Cottontail.

PONTOON BRIDGE. Bridge.

PONY. Mammalia, Perissodactyla. A **horse** of various small, hardy and agile breeds indigenous to certain islands of the Old World. The Shetland pony is most familiar, from its use for children in the United States. Other breeds occur in the Orkneys and Iceland, northern Europe and Great Britain. (A.W.L.)

PONY TRUSS. Bridge.

POOR-WILL. Aves, Caprimulgiformes. A small North American **goatsucker,** *Phalaenoptilus nuttali.* The species lives in the western half of the United States and Canada. (A.W.L.)

POPPET VALVE. Otto Engine, Valve Gear.

POPPY. *Papaver somniferum.* Papaveraceae. The poppy from which opium is obtained is an annual herb having a smooth branching stem 2–3′ tall, large, dull, green, smooth leaves and solitary single flowers, varying from white to purple in color and rather showy. The flower consists of two **sepals,** which soon fall off when the flower opens, four **petals,** many **stamens** and a single **pistil** with a 1-celled **ovary.** The fruit is a **capsule,** 1–2″ in diameter, containing many small seeds, which escape through a ring of pores which form around the top of the capsule, beneath the persistent stigma.

To obtain opium, the unripe capsules are incised with a knife. From these cuts the milky juice oozes and dries to form a plastic gummy substance, which is scraped off and molded into a ball. This crude opium contains fragments of the plant tissues and considerable dirt. About 10% of the opium is the alkaloid morphine. When first prepared opium is brownish and easily molded. It gradually dies to a hard brittle substance, easily ground to a powder. Besides morphine it contains many other alkaloids.

The opium poppy was known and used by the ancients. It is a native plant in southwestern Asia, but has spread in cultivation over much of Asia as well as parts of Europe and elsewhere. In the Orient opium is much used in smoking, a habit-forming pastime which quickly ruins its follower. To produce opium profitably much cheap labor is necessary, therefore the principal producing countries are in Asia.

Opium and its derivative, morphine, are extremely valuable medicinally, being used to relieve pain and to induce sleep.

Poppy seeds, which contain no harmful substances, are frequently used in bread and cakes. From them is expressed an expensive oil used in cooking and in making artist's paints. (R.M.W.)

POPULATION. A population or universe consists of all the items from which the **sample** is drawn. The population may either be infinite or finite in size; it may consist of qualitative or quantitative variates, and in general does not refer to people. The population may be the yield per acre of a variety of wheat, the weight of a particular animal, or the shape of the head in a human being at a specified age. The population may be mathematically specified by a **probability function**. (L.A.A.)

POPULATION PARAMETER. A population parameter is a value in the **population**, such as the **mean** of the population and the **variance** of the population. In the Bernoulli probability function the parameter is p; in the **normal probability function** there are 2 parameters, the mean m, and the **standard deviation** σ. (L.A.A.)

PORBEAGLE. Pisces, Plagiostomi. A shark, *Lamna cornubica,* found in the warmer parts of the Atlantic, occasionally farther north and in the Pacific. Its length is 8–10′. Also called the mackerel shark, although this name belongs to another larger species. (A.W.L.)

PORCELAIN. Ceramics.

PORCELAIN ENAMEL. A vitreous coating applied to **cast iron** and **steel** sanitary ware, cooking utensils, stove and refrigerator panels, etc. A frit of finely powdered borosilicate glass is applied to the surface by either spraying or dipping. The part is then fired at temperatures near 1500–1600° F. to form an adherent coating. A ground coat and one or more cover coats are generally applied. Coatings on cast iron are generally heavier than on steel and are fired at lower temperatures. Porcelain enamel coatings have excellent corrosion resistance and would have even wider application except for their brittleness. Thinner coatings having improved adherence to the base metal are coming into use. (R.H.H.)

PORCELLANITE. Metamorphosed **marls** and **shales** which look like porcelain. (R.M.F.)

PORCUPINE. Mammalia, Rodentia. An animal with many modified hairs resembling quills in form. These hairs are sharp-tipped rigid spines with finely barbed tips which penetrate flesh very readily and serve as an almost impregnable defense.

North America has two species of porcupines, one ranging over the eastern half of the continent as far south as Virginia and the other, *Erethizon epixanthum,* in the far west, from Alaska to Mexico. The common eastern species, *E. dorsatum,* is also called the hedgehog. Both species are partial to the leaves, twigs and bark of evergreen trees as food.

The tree porcupines differ in having long prehensile tails. They are found in Mexico and South America. Still other species live in Eurasia, Africa, and the Oriental Region. (A.W.L.)

PORGY. Pisces, Teleostei. A fish (**Pisces**) of the family Sparidae, in particular a valuable food fish found along the Atlantic coast from Cape Cod to South Carolina. This species is also called the scup or scuppaug. Among the other species of the family are the red snapper and sheepshead, both among our best marine food fishes. The name is also spelled porgie and porgee. (A.W.L.)

PORIFERA. The sponges. A **phylum** of animals of low organization, related to some of the 1-celled **protozoans** and much more primitive than any other multicellular group. Because of their loosely integrated structure the sponges are regarded as one of three major types of animal organization, designated by the term Parazoa. This group lies between the Protozoa, also a single phylum, and the Metazoa, containing all of the other multicellular phyla.

Sponges develop only two germ layers, the ectoderm and endoderm, but many different cells lie in the mesogloea between the two. The body wall is perforated by many canals leading to a central cavity, the **paragaster**. Some part of these passages is lined with collared **flagellate** cells which produce currents of water flowing inward through the pores and out of a larger opening of the paragaster called the osculum. The body wall contains several kinds of specialized cells. The scleroblasts form hard supporting structures of various forms and materials, called spicules. **Phagocytes** ingest, digest, and transport food. **Porocytes** become perforated to form canals. The outer surface is covered with flattened cells called pinacocytes.

Three kinds of sponges are recognized, according to the plan of the canal system: 1. Ascon sponges have canals leading entirely through the body wall and collared cells (choanocytes) in the lining of the paragaster. 2. Sycon sponges have radial canals lined with choanocytes and opening into the paragaster. Between them inhalant canals lead inward from the outside but do not reach the paragaster. The two types of canals are connected by minute pores called prosopyles through which water must pass to reach the interior. 3. In the rhagon or leucon sponges the canal systems are more intricately branched and the choanocytes are located in small chambers.

Commercial sponges are the skeletal remains of species whose bodies are supported by fibers of a peculiar material, spongin. The organic matter is removed by maceration and washing.

The phylum is divided into three classes:

Class Calcarea. Sponges whose bodies contain calcareous spicules only. The **choanocytes** are large and all three types of canal systems are represented.

Class Hexactinellida. **Spicules** 6-rayed and siliceous. Choanocytes small. Canal system of a simple rhagon type. The glass sponges.

Class Demospongiae. Spicules siliceous but not 6-rayed or an association of siliceous (**silicon**) material and spongin. Rhagon type of canal system. The sponges of commerce are included here. The subfamily Spongillinae includes the only species of sponges found in fresh water. (See also **Invertebrate Paloeontology**.) (A.W.L.)

POROCYTE. A kind of cell found in sponges. They become perforated to form minute pores through which water passes to enter the body of the sponge. (A.W.L.)

POROUS PLUG EXPERIMENT. Joule-Thomson Effect.

PORPHYRINS. Pyrrole and Related Compounds.

PORPHYRITE. A term applied to **hypabyssal igneous** rocks, such as **dikes, sills** and **laccoliths** of andesitic composition and pronounced porphyritic texture. (R.M.F.)

PORPHYROBLAST. A term proposed by Becke, in 1900, for pseudo-**porphyritic** crystals formed by thermodynamic **metamorphism**. (R.M.F.)

PORPHYRY. Porphyry is a textural term applied to **igneous rocks** in which one or more of the mineral constituents present exists as well crystallized individuals in a ground mass that is relatively of much finer grain. The derivation of the word presents an interesting study. The **gasteropods** of the genus Murex were much used for obtaining a purple dye; the Greek name for both the animal and the dye is the same. A certain Egyptian rock which was once much used for building and ornamental purposes displays very prominent crystals in a purplish groundmass and so the same Greek word was applied to it, then later came to mean all rocks of this general appearance. Modern use now restricts the term porphyry to the description of texture alone as in the case of the Egyptian rock. (E.S.C.S.)

PORPOISE. Mammalia, Odontoceti. Small animals related to the toothed **whales.** They reach a length of

Porpoise. (*American Museum of Natural History.*)

about 5′. The muzzle is bluntly rounded and most species have a dorsal fin. Porpoises are found chiefly in the northern oceans although one species ranges from Japan to southern Africa and enters the rivers of China and India. (A.W.L.)

PORPOISING. An undulatory movement of a seaplane consisting of a combination of a vertical oscillation and an oscillation about its transverse axis, which occurs at certain stages of planing. (F.T.M.)

PORT. The word port refers to the opening or means of entry into a certain region. In engineering, the port is that connecting passage or opening by means of which a fluid flow is admitted to such a region as that furnished by a cylinder. The gasoline engine (**Otto engine**), the **Diesel engine,** and the **steam engine,** as well as the **air compressor,** have openings arranged in the cylinder so that the contents may enter and leave properly. In the steam engine, the single port may sometimes serve for both admission and exhaust of the steam. This port leads from the valve chest into the end of the cylinder. The valve will alternately connect the cylinder with either the high-pressure steam or the exhaust region, and the steam flows first in through the port, then out. Some steam engines are double-ported, having separate ports for admission and release of steam. This is an aid to reducing a loss known as **initial condensation.** The **internal combustion engine** usually has two ports, one the inlet, the other the exhaust, since the internal combustion type valve is not suitable for single valve practice. In the gasoline engine, the typical port is a circular opening cored into the cylinder block, terminating at one end in the valve seat, and at the other in the manifold. Usually the valve stem passes through this port for a short distance. If an exhaust port, it must be well water-jacketed or flanged, so that heat may be conducted away rapidly enough to prevent damage to the metal. Smoothness of stream flow through the ports and minimum friction loss are desirable in obtaining the maximum possible inflow to the cylinder, or exhaust from it. The size of a port is fixed by the cylinder bore and piston speed, and the ports should bear a relationship to piston area which has been found to be satisfactory in actual practice. (F.T.M.)

PORTAL. Bridge.

PORTLAND CEMENT. Cement, Portland.

PORTUGUESE MAN-OF-WAR. Coelenterata, Hydrozoa. A beautiful but dangerous colonial **hydrozoan,** *Physalia.* It consists of a large hollow float, the pneumatophore, from which many delicate filaments hang into the water. These filaments are tentacles of a form of individual called a dactylozooid. The stinging cells in them kill the prey and their muscular contractions draw it up within reach of the gastrozooids which digest it. The poison of these animals is virulent and the colony is large enough to be dangerous even to human beings. (A.W.L.)

POSITION ANGLE. The term position angle is used in astronomy to denote the angle between the great circle joining any two celestial objects, and the **hour circle** through one of the objects. In measuring **double stars**

the position angle is the angle between the great circle joining the two stars and the **hour circle** through the brighter of the pair, the angle being measured from the north to the east through 360°. (W.K.G.)

POSITIVE RAYS. Canal Rays.

POSITIVE-RAY ANALYSIS. Mass Spectrograph.

POSITRON. An elementary charged particle, sometimes called the "positive electron" because its mass is apparently of the electronic order and its charge is positive. Positrons were first observed by C. D. Anderson by means of their ionization tracks in the air of a **cloud chamber.** They were thought to arise from the impacts of **cosmic rays** of very high energy upon atoms. Anderson later showed that particles of the same type are emitted when gamma rays of very high frequency are intercepted by a plate of metal. For this purpose the gamma-ray quanta must have energies exceeding a million electron-volts. Blackett and Occhialini have set forth the remarkable theory that in such an encounter the quantum itself, originally neutral, is transformed into a positron and an electron, part of its energy being converted into the masses of these particles (see **Energy** and **Relativity**), and the rest appearing as their kinetic energy. (See **Nuclear Particles.**) (L.D.W.)

POT HOLE. Under favorable conditions where streams flow over the bedrock, swirling eddies will wash sand, gravel or pebbles around and around in the same place with the result that cylindrical holes called pot holes, are worn, often to a considerable depth. These pot holes may be from a few inches to several feet in diameter and rarely as much as 40–50′ deep. Similar features found on the sea shore, the result of wave action, are called sea-mills. Pot holes also have been formed by water from crevasses and ice cliffs and glaciers. (R.M.F.)

POTAMOGALE. Mammalia, Insectivora. *Potamogale.* An amphibious animal of western Africa. It has a slender body and short legs like the mink and otter, and is further characterized by the compressed tail and valves to close the nostrils. (A.W.L.)

POTASH. Term applied to **potassium** compounds. Potash, potassium carbonate; caustic potash, potassium hydroxide. Percentage of potash expressed in analyses of chemicals is for potassium oxide (K_2O). (R.K.S.)

POTASSIUM. Symbol: K (kalium). Atomic number: 19. Atomic weight: 39.096. Density: 0.87. Hardness: 0.5. Melting point: 62.3° C. Boiling point: 760° C. Isotopes: 39 (93.4%), 40 (0.01%), 41 (6.6%).

Potassium is a silver-white metal, can be readily molded, and cut by a knife, oxidizes instantly on exposure to air, and reacts violently with water, yielding potassium hydroxide and **hydrogen** gas, which burns spontaneously in air with a violet flame due to volatilized potassium element, is preserved under kerosene, burns in air at a red heat with a violet flame. Discovered by Davy in 1807.

Potassium occurs as potassium chloride or sulfate in certain salt deposits (**carnallite,** potassium magnesium chloride ($KCl \cdot MgCl_2 \cdot 6H_2O$); **kainite,** potassium magnesium chloride sulfate ($K_2SO_4 \cdot MgSO_4 \cdot MgCl_2 \cdot 6H_2O$)), mainly in Stassfurt, Germany, Alsace-Lorraine, Spain, Poland, California, New Mexico; in common rocks (average of the solid shell of the earth, 2.6%) and the minerals, **feldspar, greensand, alunite, leucite;** present in vegetation (average 1.7%) remaining in the ash when burned.

Potassium metal is obtained by **electrolysis** of fused potassium hydroxide or chloride fluoride mixture in a specially designed cell.

Acetate: Potassium acetate ($KC_2H_3O_2$), white solid, soluble, formed by reaction of potassium carbonate and **acetic acid**, and then evaporating.

Alum: Potash alums are those alums, such as **aluminum** potassium sulfate ($Al_2(SO_4)_3 \cdot K_2SO_4 \cdot 24H_2O$), chromium potassium alum ($Cr_2(SO_4)_3 \cdot K_2SO_4 \cdot 24H_2O$) where potassium sulfate is crystallized with the heavy metal sulfate.

Bromate: Potassium bromate ($KBrO_3$), white solid, soluble, melting point $434°$ C., upon heating oxygen is evolved and the residue is potassium bromide; formed by electrolysis of potassium bromide solution under proper conditions. Used as a source of bromate and bromic acid.

Bromide: Potassium bromide (KBr), white solid, soluble, formed by reaction of potassium hydroxide and **bromine**, and then evaporating and heating to decompose bromate. Use (1) in photography, in engraving and lithographing, (2) in medicine as a sedative.

Carbonate: Potassium carbonate, potash, pearl ash (K_2CO_3), white solid, soluble, formed (1) in the ash when plant materials are burned, (2) by reaction of potassium hydroxide solution and the requisite amount of **carbon dioxide**. Used (1) in making special glasses, (2) in the making of soft soap, (3) in the preparation of other potassium salts (a) in solution, (b) upon fusion; potassium hydrogen carbonate, potassium bicarbonate, potassium acid carbonate ($KHCO_3$), white solid, soluble.

Chlorate: Potassium chlorate, chlorate of potash ($KClO_3$), white solid, soluble, melting point about $350°$ C., powerful oxidizing agent, and consequently a fire hazard with dry organic materials, such as clothes, and with **sulfur**; upon heating oxygen is liberated and the residue is potassium chloride; formed by electrolysis of potassium chloride solution under proper conditions. Used (1) in matches, (2) in pyrotechnics, (3) as disinfectant, (4) as a source of oxygen upon heating. (Hazardous! Use of potassium perchlorate is recommended instead.)

Chloride: Potassium chloride, chloride of potash, muriate of potash (KCl), white solid, soluble, melting point $790°$ C., boiling point $1500°$ C. Common constituent of potassium salt minerals. Volatilized when potassium-bearing silicates, such as **feldspar, leucite**, are heated with **calcium** chloride to a high temperature, as in a cement kiln. Used (1) as an important potassium fertilizer, (2) as a source of other potassium compounds.

Chloroplatinate: Potassium chloroplatinate (K_2PtCl_6), yellow solid, insoluble, formed by reaction of soluble potassium salt solution and chloroplatinic acid. Used in the quantitative determination of potassium.

Chromate: Potassium chromate (K_2CrO_4), yellow solid, soluble, formed by reaction of potassium carbonate and **chromite** at a high temperature in a current of air, and then extracting with water and evaporating the solution. Used (1) as a source of chromate, (2) in leather tanning, (3) in textile dyeing, (4) in inks.

Cobaltinitrite: Dipotassium sodium cobaltinitrite ($K_2NaCo(NO_2)_6 \cdot H_2O$), golden yellow precipitate, formed by reaction of **sodium** cobaltinitrite solution in **acetic acid** with soluble potassium salt solution. Used in the detection of potassium.

Cyanate: Potassium cyanate ($KCNO$), white solid, soluble, formed along with lead metal by reaction of potassium cyanide and **lead** monoxide solids upon heating. Source of cyanate.

Cyanide: Potassium cyanide, cyanide of potash (KCN), white solid, soluble, very poisonous, formed by reaction of calcium **cyanamide** and potassium chloride at high temperature. Used as a source of cyanide and for hydrocyanic acid, but usually replaced by the cheaper sodium cyanide.

Dichromate: Potassium dichromate, chromate of potash ($K_2Cr_2O_7$), red solid, soluble, powerful oxidizing agent, formed by acidifying potassium chromate solution and then evaporating. Used (1) in matches and pyro-

technics, (2) in leather tanning and in the textile industry, (3) as a source of chromate.

Ferricyanide: Potassium ferricyanide, red prussiate of potash ($K_3Fe(CN)_6$), red solid, soluble, formed by reaction of potassium **ferrocyanide** solution and **chlorine**, and then evaporating.

Ferrocyanide: Potassium ferrocyanide, yellow prussiate of potash ($K_4Fe(CN)_6$), yellow solid, soluble, formed by treating "spent oxide" of coal gas works with **calcium** hydroxide to extract ferrous cyanide as soluble calcium ferrocyanide and then treating with potassium carbonate, filtering and evaporating the filtrate. Used (1) as a source of ferrocyanide, but usually replaced by the cheaper sodium ferrocyanide, (2) in blue print paper, (3) in tanning, (4) in tempering steel.

Fluoride: Potassium fluoride (KF), white solid, soluble, formed by reaction of potassium carbonate and **hydrofluoric acid,** and then evaporating. Used in the etching of glass; potassium hydrogen fluoride, potassium bifluoride, potassium acid fluoride (KHF_2), white solid, soluble, formed by reaction of potassium carbonate and excess hydrofluoric acid, and then evaporating.

Glycerophosphate: Potassium glycerophosphate ($(KO)_2PO(C_3H_7O_3)$), pale yellow liquid, formed by warming **glycerol** and **metaphosphoric acid,** and then neutralizing with potassium carbonate. Used in medicine.

Hydroxide: Potassium hydroxide, caustic potash, potassium hydrate (KOH), white solid, soluble, melting point $380°$ C., formed (1) by reaction of potassium carbonate and **calcium** hydroxide in water, and then separation of the solution and evaporation, (2) by electrolysis of potassium chloride under the proper conditions, and evaporation. Used in the preparation of potassium salts (1) in solution, (2) upon fusion.

Hypophosphite: Potassium hypophosphite (KH_2PO_2), white solid, soluble, formed (1) by reaction of **hypophosphorous** acid and potassium carbonate solution, and then evaporating, (2) by reaction of potassium hydroxide solution and **phosphorus** on heating (poisonous phosphine gas evolved).

Iodate: Potassium iodate (KIO_3), white solid, soluble, melting point $560°$ C., formed (1) by electrolysis of potassium iodide under proper conditions, (2) by reaction of **iodine** and potassium hydroxide solution, and the fractional crystallization of iodate from iodide. Used as a source of iodate and iodic acid.

Manganate: Potassium manganate (K_2MnO_4), green solid, soluble, permanent in alkali, formed by heating to high temperature **manganese** dioxide and potassium carbonate, and then extracting with water, and evaporating the solution. The first step in the preparation of potassium manganate and permanganate from pyrolusite.

Nitrate: Potassium nitrate, saltpeter, niter (KNO_3), white solid, soluble, melting point $333°$ C., formed by fractional crystallization of **sodium** nitrate and potassium chloride solutions. Used (1) in matches, explosives, pyrotechnics, (2) in the pickling of meat.

Nitrite: Potassium nitrite (KNO_2), yellowish-white solid, soluble, formed (1) by reaction of nitric oxide plus **nitrogen** tetroxide and potassium carbonate or hydroxide, and then evaporating, (2) by heating potassium nitrate and **lead** to a high temperature and then extracting the soluble portion (lead monoxide insoluble) with water, and evaporating. Used as a reagent (**diazotizing**) in organic chemistry.

Oxalate: Potassium oxalate ($K_2C_2O_4$), white solid, soluble, formed by reaction of potassium carbonate or hydroxide and **oxalic acid,** and then evaporating. Used as a source of oxalate; potassium hydrogen oxalate, potassium binoxalate, potassium acid oxalate (KHC_2O_4), white solid, soluble; potassium tetroxalate ($KHC_2O_4 \cdot H_2C_2O_4 \cdot 2H_2O$), white solid, moderately soluble.

Oxides: Potassium oxide, potassium monoxide (K_2O), white solid, reactive with water to form potassium hydroxide, formed by reaction of potassium and **oxygen**

1133

at reduced pressure, and removal of excess metal by distillation in vacuum; potassium peroxide, potassium tetroxide (K_2O_4), yellow solid, soluble in water with evolution of oxygen, formed by heating potassium at 200° C. in a current of oxygen.

Perchlorate: Potassium perchlorate ($KClO_4$), white solid, very slightly soluble, melting point 610° C., but above 400° C. decomposes with evolution of oxygen gas and formation of potassium chloride residue; formed (1) by electrolysis of potassium chlorate under proper conditions, (2) by heating potassium chlorate at 480° C. and then fractional crystallization. Used (1) as a convenient and safe (preferred to use of potassium chlorate) method of preparing oxygen by heating, (2) in the determination of potassium in soluble salt solution.

Periodate: Potassium periodate (KIO_4), white solid, very slightly soluble, melting point 582° C., formed by electrolysis of potassium iodate under proper conditions.

Permanganate: Potassium permanganate, permanganate of potash ($KMnO_4$), purple solid, soluble, formed by oxidation of acidified potassium manganate solution with **chlorine**, and then evaporating. Used (1) as disinfectant and bactericide, (2) in medicine, (3) as an important oxidizing agent in many chemical reactions.

Persulfate: Potassium persulfate ($K_2S_2O_8$), white solid, slightly soluble, formed by **electrolysis** of potassium sulfate under proper conditions. Used (1) as a bleaching and oxidizing agent, (2) as an antiseptic.

Phosphates: Tripotassium phosphate (K_3PO_4); dipotassium hydrogen phosphate (K_2HPO_4); potassium dihydrogen phosphate (KH_2PO_4); potassium pyrophosphate ($K_4P_2O_7 \cdot 3H_2O$); potassium metaphosphate (KPO_3); white solids, similar in properties and formation to the corresponding and more common **sodium** phosphates.

Phosphites: Dipotassium hydrogen phosphite (K_2HPO_3); potassium dihydrogen phosphite (KH_2PO_3); white solids, similar in properties and formation to the corresponding **sodium** phosphites.

Silicate: Potassium silicate (K_2SiO_3), colorless (when pure) glass, soluble, melting point 976° C., formed by reaction of **silicon** oxide and potassium carbonate at high temperature, similar in properties and uses to the more common sodium silicate.

Sulfates: Potassium sulfate, sulfate of potash (K_2SO_4), white solid, soluble. Common constituent of potassium salt minerals. Used (1) as an important potassium fertilizer, (2) in the preparation of potassium or potash **alums**; potassium hydrogen sulfate ($KHSO_4$), white solid, soluble; potassium pyrosulfate ($K_2S_2O_7$), white solid, soluble, formed by heating potassium hydrogen sulfate to complete loss of water.

Sulfides: Potassium sulfide (K_2S), yellowish to reddish solid, soluble, formed by heating potassium sulfate and **carbon** to a high temperature; potassium hydrogen sulfide, potassium bisulfide, potassium acid sulfide (KHS), formed in solution by reaction of **sodium** hydroxide or carbonate solution and excess **hydrogen sulfide.**

Sulfite: Potassium sulfite ($K_2SO_3 \cdot 2H_2O$); potassium hydrogen sulfite ($KHSO_3$); white solids, similar in properties and formation to the corresponding **sodium** sulfites.

Tartrates: Potassium tartrate ($K_2C_4H_4O_6 \frac{1}{2}H_2O$), white solid, soluble, formed by reaction of potassium carbonate solution and **tartaric acid**, and then evaporating; potassium hydrogen tartrate, potassium bitartrate, potassium acid tartrate, "cream of tartar" ($KHC_4H_4O_6$), white solid, slightly soluble. Obtained from "argols," a by-product of wine fermentation (argols consist of 50–85% potassium hydrogen carbonate and 6–12% calcium tartrate). Used (1) in **baking powder** as a source of acid, (2) as a source of tartrate, potassium sodium tartrate, Rochelle salt ($NaKC_4H_4O_6 \cdot 4H_2O$), white solid, soluble, and **Fehling's solution.** Used (1) in medicine, (2) as a source of tartrate.

Thiocarbonate: Potassium thiocarbonate (K_2CS_3),

yellow solid, soluble, formed by reaction of potassium sulfide and **carbon disulfide.**

Thiocyanate: Potassium thiocyanate, potassium sulfocyanide, potassium rhodanate ($KCNS$), white solid, soluble, melting point about 170° C., formed by fusing potassium cyanide and **sulfur,** and then crystallizing. Used as a source of thiocyanate.

All potassium-containing substances impart a characteristic violet color to the bunsen flame, but small amounts are easily masked by the presence of sodium and its yellow flame. (R.K.S.)

POTATO FAMILY. Solanaceae. This is a relatively small family of some 1700 species, in about 80 genera. The members include herbs, shrubs, and a few trees, most of the last being found in the tropics. The family has many important food and medicinal plants. All of its members contain poisonous **alkaloids,** as solanin, capsaicin, **belladonna,** nicotine, and atropine.

Members of this family have leaves which are usually alternate and variously shaped, often lobed or dissected or **pinnately** compound. The variability frequently occurs in the leaves of a single plant. The flowers occur either singly or in **cymes,** and are regular with an inferior 5-lobed **calyx,** a 5-lobed **corolla** of various shapes; often large and conspicuously colored, five **stamens** and a **pistil** composed of two united **carpels,** a long **style** and a single terminal **stigma.** The **ovary** contains many **ovules.** The fruit is either a **berry** (potato and tomato) or a **capsule** (tobacco).

The most important member of this family is the Potato, *Solanum tuberosum,* commonly called the Irish potato, common potato, White Potato, or English potato, to distinguish it from the sweet potato. The plant is a branched herb, the branches tending to spread out more or less, and the growing 2–4' in height. The green stems are annual, but the tubers, which are modified stems, give to the plant a perennial nature. The leaves are pinnately compound and rather irregular. The flowers are 1–1½″ in diameter, white, often with blue or purple tones or stripes, and with a tubular corolla. The fruit is a globular berry containing many small seeds embedded in a green pulp. Common names given these potato berries are potato balls, potato apples or seed balls. When mature, they are either green or brown; often they contain few seeds, but since they are rarely used in propagating the potato, this is of little consequence. They are used experimentally in the production of new varieties. Once a desirable variety is found, it is perpetuated vegetatively by means of the tubers. These tubers are formed at the tips of modified underground branches, or **rhizomes,** which radiate outward from the basal portion of the stem and to the casual observer resemble roots. The length of these stolons varies from a few inches to a foot or more, depending somewhat on the variety. Rhizome formation and tuber development start soon after the tops appear above ground, and are advanced by darkness and low temperatures. The tubers continue to grow throughout the growing life of the plant. A mature tuber has the structure of a stem, but very much modified. Externally, one may recognize nodes and internodes, the nodes being determined by the eyes, depressions in the surface of the potato, each depression or eye containing a tiny bud. Internally the tuber is largely **parenchymatous** tissue with the cells filled with starch-grains. Near the surface of the potato, cross sections show a faint dark line which is the **vascular** tissue, very much reduced. Potato tubers vary greatly in shape, as well as in size; the better varieties are oblong or oval, with smooth skin and rather shallow eyes. In color, tubers range from brown through yellow to red, according to variety. The length of time required to mature a marketable tuber also varies greatly, some early varieties reaching a desirable size in about two months, while late varieties require five or six months.

The tubers are the principal means of propagation. For this purpose a tuber is cut into irregular pieces, each of which contains two or three eyes. To prevent disease the cut pieces are treated with sulfur and allowed to dry slightly, after which they are planted. Sprouting starts at once, indeed frequently occurs while the potato is still in the storage bin. If the bin is dark the sprouts formed will be long, slender and white. Sprouts formed in light are short, stout and dark green.

In addition to being an important article of food to inhabitants of the temperate regions of the world, the potato has many other uses. **Starch** is made from them, cull or small potatoes being much used for this purpose. To make starch the raw potatoes are cleaned, soaked in water for several hours, then washed thoroughly and reduced to a pulp. This pulp is strained to remove any fibrous material. The strained liquor is allowed to stand, the starch settling to the bottom. This starch layer is drawn off and purified by running slowly down an inclined table where the starch grains once more settle to the bottom. They are then collected and dried thoroughly, then broken up into marketable size.

Another use for potatoes is as a source from which to make **alcohol.** Finally, potatoes are much used as food for domestic animals, especially swine, being fed either raw or cooked or, at times, dried.

The potato is a native plant of cool upland regions of South America, where it has been long cultivated by the natives, who use the tubers for food. From America is was carried to the southern part of Europe, where at first it was grown largely as a curiosity. From Europe the plant was brought back to America and introduced into what is now the United States, thus explaining its name of English or Irish Potato. It is now extensively grown in regions having a cool climate, Maine being particularly noted for its potato crop.

In all green parts of the potato a poisonous alkaloid, solanin, occurs. This substance may also occur in the tubers, particularly if the latter are exposed to light long enough to become green in color.

Another important food plant of this family is the tomato, *Lycopersicum esculentum,* and related species. This is a native of South America, in which continent the tomato is still found wild. The tomato is a coarse branching perennial herb which is usually grown as an annual. It is an important food which in recent years has become immensely popular. Large quantities are consumed raw or canned. Tomato juice is an important article of diet. From the seeds an oil may be obtained, which is used in making soaps.

The tomato was carried to Europe by the early Spanish explorers and became an important food plant in southern European countries. In England and also in the United States it was widely grown before 1830 as a garden ornamental known as "Love Apple," which however was not eaten, but was held by many to be rankly poisonous. The Italians seem to be the first to have braved the danger of tasting the delicious fruit of this "poisonous" plant.

A third member of the family providing food to man is the egg plant, *Solanum Melongena,* a coarse, somewhat woody, branching herb native to India. The plant has a rough stem, 2 or 3′ tall, large sinuate-lobed ovate leaves and purplish oflwers. The fruit is a berry, very large in some varieties. It is eaten either baked or sliced and fried.

Serving rather as a condiment than a food, but still used as a food stuff in some of its varieties, is the **pepper,** *Capsicum annuum,* another native plant of tropical America. Peppers are either annual or biennial plants, with branching stems 1-3′ tall, smooth shining leaves and white flowers. There are many varieties, which bear fruits of a variety of sizes and shapes, as well as degrees of pungency. Some varieties are known as

sweet peppers, and are used in salads or stuffed and baked; others are hot peppers. The pungent taste is due to an acrid compound, capsaicin, which in hot peppers occurs throughout the fruit, but in sweet peppers is largely restricted to the immediate region of the seeds; since only the fleshy **pericarp** is eaten, the seed being removed, this pungent substance is lost.

Peppers are used in many ways other than as food. Small hot peppers are a frequent component of mixed pickles, and also are used in salads. The whole fruit of some varieties is ground up to a powder, and becomes Cayenne pepper, an extremely pungent condiment. Small smooth fruits of var. *conoides* are preserved whole in brine or vinegar, and known as Tabasco sauce or Tabasco peppers. Red peppers are used medicinally. Many varieties of pepper are grown as ornamental plants, the brilliant fruit offering a startling contrast to the dark green leaves.

More widely grown for its ornamental properties is the Petunia, introduced from South America.

Many members of the Potato Family are important drug plants. Here belong *Atropa Belladonna,* yielding **atropine,** and *Datura Stramonium,* the Thorn Apple, called also Jimson Weed or Jamestown Weed. The latter is a tall, rather coarse, branching annual having broad leaves with sinuate margins and large trumpet-shaped white flowers, borne singly in the axils of the leaves (a related species, *D. Tatula,* has purple flowers). The fruit is a prickly capsule. In all parts of the plant, but especially in the seed, are found several drugs. The most abundant is **hyoscyamine,** with small amounts of atropine and scopolamin present. The powdered leaves and seeds are used medicinally, chiefly in treating asthma.

Less important medicinally are several plants such as *Hyoscyamus niger, Solanum Dulcamara,* the Bittersweet, and *Solanum niger,* the garden Nightshade, all held to be poisonous plants if eaten in sufficient quantity by man or domestic animals.

Another member of this family, ranking along with the potato in importance, is the tobacco plant, *Nicotiana Tabacum,* also a native plant of tropical America. It is an annual plant growing 3-6′ or more in height, and stout. The leaves are alternate, simple and rather large and, like the stem, covered with sticky hairs. The rather large flowers are borne in terminal racemes and have a funnel-shaped corolla which is yellow, white, pink or purple in different species and varieties. The fruit is a capsule containing many small seeds. The entire plant contains a poisonous alkaloid, nicotine.

While tobacco plants are sometimes grown as ornamentals, the principal use is made of the large leaves, which are smoked or chewed. Different demands have led to the development of several types of tobacco. One of these is the type used for cigar wrappers in which the leaves are large, thin and have a fine texture with a relatively small amount of vascular tissue. This type of leaf is often obtained by growing the plants in the shade of sheets of cloth. Another type of tobacco has the leaves thicker and with many veins giving a rather coarse structure. The leaves of these coarser sorts are used in making pipe tobacco, chewing tobacco, cigar filler, cigarettes, etc. In all tobaccos considerable differences are produced by cultivation and the nature of the soil on which the plants are grown.

In preparing tobacco for use, the leaves are removed from the plant and dried, either by natural air currents in open ventilated sheds, a process requiring 6–8 weeks, or by artificial heat, which is naturally much quicker. The leaves are then packed tightly in wooden containers and left for several months. During this time fermentation occurs, and important changes take place which lead to the formation of the pleasing aromas of prepared tobaccos. During fermentation the leaves may be repacked several times. The cured leaves are either rolled into cigars or blended, ground up, and used in smoking tobaccos and cigarettes. Mixed with various substances and pressed into cakes it becomes chewing tobacco. Ground fine it becomes snuff.

Tobacco is now extensively grown, not only in tropical regions but also in temperate climates. Its users are very numerous, including a large proportion of the human race. The United States alone grows considerably over a billion lbs. each year. (R.M.W., B.S.M.)

POTENTIAL. This term is used to denote a number of different quantities used in physics, all characterized by their connection with potential energy (hence the name). The most familiar is **electric potential.** One of the simplest is gravitational potential or mass potential. The mass potential at any point is measured by the energy necessary to carry a unit mass from that point to a region of space infinitely removed from all matter. Its value is the integral of $G \dfrac{dm}{r}$, in which G is the **gravitation constant,** dm is any mass element, and r is its distance from the point in question; the integration being extended throughout all existing matter. So far as we know, it always has the same sign; that is, gravitation is always an attraction, and the energy must therefore always be added to the unit mass in the process described; whereas electric potential may be either positive or negative. The magnetic potential at any point is measured by the energy necessary to carry a unit north pole from infinity to that point; or mathematically, it is the line integral of the magnetic intensity between that point and infinity. The "Newtonian potential function" is a mathematical expression occurring as a factor in all these potentials. (See **Laplace's Equation.**)

A characteristic of the mass potential, the electric potential, and the magnetic potential, is that in each case the first derivative of the potential with respect to any direction, as x, at any point in space, is equal in magnitude to the component of the field intensity in that direction at the point in question.

The foregoing potentials all involve inverse-square forces; if the field obeys the inverse-first-power law, the potential involves the logarithm of the distance, and is called the "logarithmic potential."

By analogy, we have also a so-called "thermodynamic potential," which, in reference to a substance in any state, represents the energy which has been required to bring unit mass of the substance to the state in question from some arbitrarily defined, initial state. (L.D.W.)

POTENTIAL DROP, LAW OF. Electric Circuits; Electric Currents.

POTENTIAL ENERGY. Whenever **energy** takes a form not directly associated with motion, it is called potential energy. Energy is apparently able to go into a state of "hibernation" and to remain dormant, sometimes for long periods. We are taught, for example, that there is somehow locked up in coal or petroleum deposits vast stores of energy traceable to the solar radiation of past geological ages. The evidence of such energy is the work which can be derived from burning the fuel in a suitable engine; or more directly, the **heat** produced by its combustion, which we recognize as a form of **kinetic energy.** But a closer analysis shows that, since combustion is a union of carbon and other elements of the fuel with oxygen, and since the process effected by the sun's rays in the growth of carboniferous plants was largely the separation of oxygen from those elements in carbon dioxide and other compounds, there is really no more reason for localizing the energy in the fuel deposit than in the oxygen of the air. A weight lifted from the earth is able to do work in descending, and we say that it has potential energy. But lifting a weight is merely the work of separating two mutually attracting masses, one of which is the earth; so that

the energy may as well be said to belong to the earth as to the lifted weight. These illustrations suffice to show that the "location" of potential energy is quite ambiguous.

A notable characteristic of potential energy is its tendency to transform itself into kinetic energy at every opportunity, and to continue the process until a state of stable equilibrium is reached. This is the well-known **least energy principle.** In many physical processes, such as the swinging of a pendulum or the oscillation of electricity in a conductor, the transformation takes place back and forth from potential to kinetic and vice versa; but in each cycle the total potential energy becomes a little less and the total kinetic energy a little greater (including the heat generated in friction). Ultimately all the energy except a certain minimum corresponding to stable equilibrium takes the form of radiation and is lost. It has thus suffered "degradation" and final "dissipation." (L.D.W.)

POTENTIAL FLOW. The steady flow of an incompressible fluid through a certain region may be created by pressures upon the boundary surfaces of a previously stationary fluid. Study of streamline flows in the x, y plane (two-dimensional flow) is greatly facilitated by the "potential function" describing the flow. Calling the potential function φ,

$$\varphi = f(x, y).$$

Then the component of fluid velocity in the x direction is $u = \dfrac{\partial \varphi}{\partial x}$ and in the y direction, $v = \dfrac{\partial \varphi}{\partial y}$. For the function to be analyzed (i.e., to meet the requirement of incompressibility and continuity),

$$\frac{\partial^2 \varphi}{\partial x^2} + \frac{\partial^2 \varphi}{\partial^2 y^2} = 0.$$

The streamlines have paths described by the stream function ψ, which is also $f(x, y)$. The component velocities in terms of the stream function are:

$$u = \frac{\partial \psi}{\partial y} ; \quad v = -\frac{\partial \psi}{\partial x}.$$

The relation between φ and ψ is:

$$d\psi = \frac{\partial \varphi}{\partial y} dx + \frac{\partial \varphi}{\partial x} dy.$$

The advantage of analysis by potentials is that an analysis of complex flows may be made by adding the potential functions of simple flows which, when superimposed, form the complex flow. Thus the flow around a static cylinder in a rectilinear air stream may be reproduced by superimposing a doublet on a rectilinear stream. while the flow around a rotating cylinder is found by the further addition of a circulation. The potentials are additive, thus:

For the rectilinear stream $\varphi = Vx = Vr \cos \theta$

For the doublet $\varphi = \dfrac{R^2 Vx}{x^2 + y^2} = \dfrac{R^2 V \cos \theta}{r}$

For the circulation $\varphi = \dfrac{\Gamma \tan^{-1} \dfrac{y}{x}}{2\pi} = \dfrac{\Gamma \theta}{2\pi}$

For the complex flow $\varphi = V \cos \theta \left(r - \dfrac{R^2}{r} \right) + \dfrac{\Gamma \theta}{2\pi}$

R = radius of cylinder, V = rectilinear velocity, Γ = **circulation** strength, θ and r are polar coordinates. At any point, r, θ in this flow the normal velocity is $\dfrac{\partial \varphi}{\partial r}$ and the tangential flow is $\dfrac{\partial \varphi}{r \partial \theta}$. (F.T.M.)

POTENTIAL HEAD. Fluid Flow.

POTENTIAL TEMPERATURE.
An air parcel at any given altitude and pressure has a certain temperature which can be determined with proper instruments. If this parcel is lowered or raised dry-adiabatically from its existing pressure to a pressure of 1000 millibars, which is a standard reference pressure, it will assume upon arrival at that level temperature which is designated as its potential temperature. Potential temperature is used to compare the temperature of air parcels at different levels by referring them all to one level. Temperature and pressure of a parcel at its existing level and its potential temperature are related as indicated by the equation,

$$\theta = T\left(\frac{1000}{p}\right)^{.288}.$$

where θ = Potential temperature.
T = Actual temperature.
p = Actual pressure. (P.E.K.)

POTENTIAL TRANSFORMER. Transformer.

POTENTIOMETER.
An instrument used for the measurement or comparison of small potential differences or electromotive forces, based upon the "law of potential drop" (see **Electric Circuits**). One of the simplest potentiometer circuits is shown in the accompanying diagram. Current from a battery B is sent

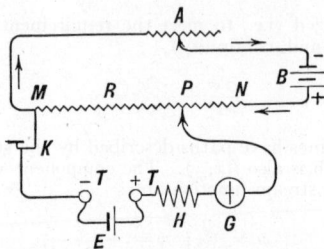

Diagram of potentiometer circuit.

through a resistance MN and is adjustable by means of a rheostat A. From one extremity M of this resistance is taken off a branch circuit containing the potentiometer terminals $+T$, $-T$, between which E, one of the electromotive forces to be compared, is connected. This circuit rejoins the main circuit at a point P which is adjustable so that the partial resistance MP or R can be varied, while MN as a whole remains constant. The $+$ and $-$ leads from E must be connected as shown, and the electromotive force of B must exceed E. The position of P is now adjusted until the galvanometer G shows no current, indicating that the potential drop from P to M just balances the electromotive force E. If two different electromotive forces E_1, E_2, are thus connected and balanced in succession, and if the corresponding values of the resistance MP are R_1, R_2, then since the current through MN is unaltered, the law of potential drop gives

$$\frac{E_1}{E_2} = \frac{R_1}{R_2}.$$

In particular, one of the electromotive forces may be a **standard cell** of accurately known voltage; the other is thereby determined. In such case the standard cell should be safeguarded by a high resistance H, which is gradually reduced as the zero-current adjustment is approached; and the key K should be closed only for an instant. In some potentiometers the whole equipment, including galvanometer and standard cell, is contained in one compact case. (L.D.W.)

POTHEAD.
In electrical distribution networks it is often convenient to run the conductors in a subterranean multi-conductor cable. Where such a cable is brought to the surface and either connected to some equipment or brought up a pole for connection to an overhead

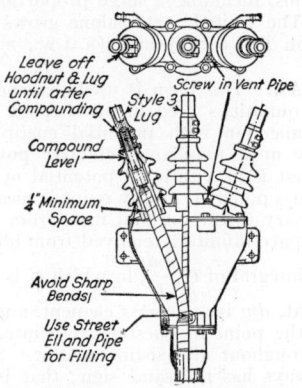

Three-conductor cable pothead.

line, some means for terminating the cable insulation and connecting the conductors to the overhead wires must be provided. This is relatively simple in low-voltage cables, but most cables operate at a voltage which requires a special terminal. This equipment, known as the pothead, is usually unnecessary below 2300 volts. At the cable terminal the separate conductors are fanned out in the pothead, which not only affords the necessary insulation between conductors where they leave the cable, but also seals the end of the cable against the entry of moisture. (F.T.M.)

POTOROO.
Mammalia, Marsupialia. *Potorous.* Any of several species of small animals related to the kangaroos and known also as rat kangaroos. They are found in Australia and Tasmania. (A.W.L.)

POTTER WASP.
Insecta, Hymenoptera. A small **wasp** whose nest is built of mud in the form of a globular pot with a narrow neck. The several species belong to the genus *Eumenes*. All are solitary. (A.W.L.)

POTTERY. Ceramics.

POTTO.
Mammalia, Primates. A name for the African slow **lemurs**, animals which resemble the lorises of Asia in many ways. They are remarkable for the rudimentary index finger, which is a stub without a nail or joints, and for the very short tail. The two species are Bosman's potto and the awantibo, *Perodicticus calabarensis*. (A.W.L.)

POUCHED MOLE.
Mammalia, Marsupialia. A rare pouched mammal of the Australian deserts, *Notoryctes typhlops*, highly specialized for burrowing. Like the true moles it has enormous claws and rudimentary eyes. (A.W.L.)

POUCHED MOUSE. Dasyure.

POULTRY.
A term applied to the domestic breeds of birds, exclusive of cage birds and such ornamental species as the peacocks. The forms embraced by the term are the **chickens, turkeys, ducks, geese,** and **Guinea fowls.** Although pheasants are bred extensively in captivity they are classed as game birds rather than poultry, as are other species which are often raised for the stocking of preserves.

The economic value of poultry varies with the species and with the breed. Most eggs used as food are those of chickens but the flesh of all species is valued. Geese produce the best feathers and down, with ducks a close second. (A.W.L.)

POUT. Pisces, Teleostei. A European fish (**Pisces**), *Gadus luscus,* related to the cod. Also called the bib. (A.W.L.)

POWDER METHOD OF CRYSTAL ANALYSIS. Crystal Structure.

POWDER METALLURGY. Powder metallurgy embraces the production of finely divided metal powders and their union through the use of pressure and heat into useful articles. The temperatures required are below the fusion point of the principal constituent, and bonding depends on interdiffusion of the metal particles in the solid state. It is necessary to provide intimate contact between particles, hence reducing atmospheres are provided in the sintering process to prevent formation of oxide films. Readily oxidized powders such as aluminum require special technique.

Powder metallurgy is taking its place alongside of melting and **casting** and the various processes involving **plastic deformation** of metals as a means of fabrication. While it is inherently costly to produce metal powders, compact them under high pressure in a specially constructed die, sinter in a controlled atmosphere furnace, and in some cases restrike in a finishing die, there are many compensating factors which make such processes economical and for many purposes indispensable.

Among these factors is the great saving in machining in the production of gears, for example, by powder metallurgy as compared with machining them from bar stock or forgings, or even from a rough casting. **Die casting** compares favorably with powder metallurgy in this respect, but in both cases the production must be great enough to justify the high cost of die making. An example of a powder metallurgy product whose success is due to ease of fabrication to high dimensional accuracy is the oil pump gear used in most automobiles. Two involute gears operating in a closed housing provide positive oil pressure. These gears are made from iron powder and require removal of less than 1% of the metal for finishing.

Probably the most important applications of powder metallurgy are those in which a product is made which cannot be duplicated by other methods. There are many examples of this kind. The melting point of tungsten, 6100° F., is much too high for ordinary melting and casting methods and the only way in which filaments for electric lights can be made is to draw them from rods of compacted and sintered tungsten powder. The **cemented carbide** cutting tools are another important product of refractory nature which is readily made by powder metallurgy.

Self-lubricating bronze bearings having controlled porosity are products which can be made only by powder metallurgy. The pores are impregnated with oil, and flow to the bearing surface is maintained by capillary action. Graphite is incorporated with the metal powder in one type of oilless bearing. A material made from powdered copper and graphite is used for electric-current collector brushes, and tungsten-copper or tungsten-silver combinations are used for electric contact points. In contrast to these high-conductivity materials, a high-resistance element is produced from a mixture of copper and porcelain powders, combining a metal with a non-metallic substance. (R.H.H.)

POWER. The time rate of the performance of **work** or of the transfer of **energy.** It may be expressed in units of work per unit time (e.g., foot-pounds per minute or **ergs** per second), or more arbitrarily, as in **horse power** or in **watts.** For most modern purposes, especially where electrically distributed energy is concerned,

power is commonly expressed in watts or kilowatts. Since the product of the power by the time during which it is operative is the work done, the product of a power unit by a time unit, such as the watt-hour or the horse-power-year, is a unit of work or energy. Electrically delivered energy is thus commonly measured in kilowatt-hours, each of which is equal to 3.6×10^{13} ergs or 3.6×10^{6} joules. The average power of large hydro-electric installations subject to wide seasonal variations is often rated in kilowatt-hours per year. For electric circuits having appreciable reactance (see **Alternating Currents**), the power in watts is equal to the product of the voltage, the current in amperes, and a **power factor** dependent upon the reactance.

The measurement of mechanically delivered power is usually effected by some form of power **dynamometer** or "ergometer." This commonly takes the form of a friction brake and a speed indicator, as in the **Prony brake.** Since the power may be considered as the product of the force (of friction) by the linear speed, or the product of the torque by the angular speed, it is necessary only to measure these quantities and to deduce the power therefrom. (L.D.W.)

POWER, AIRPLANE ENGINE. Airplane Engine Performance Characteristics.

POWER FACTOR. Alternating-Current Circuits.

POWER FUNCTION. A power function is an algebraic **function** of the form ax^n, where a and n are constants (independent of x) and x is the **variable.** It has a variable base x and a constant **exponent** n. The exponent may be a positive integer, or it may be a fraction or negative number, or even irrational.

The graphs of the power function are shown in the accompanying figures. (L.L.S.)

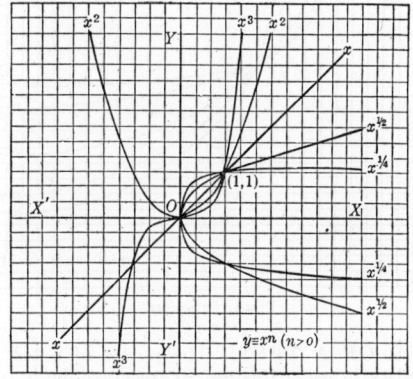

Power function.

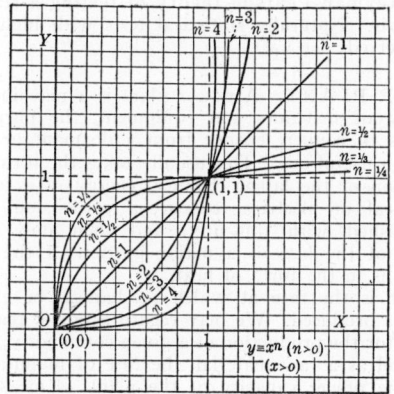

Power function.

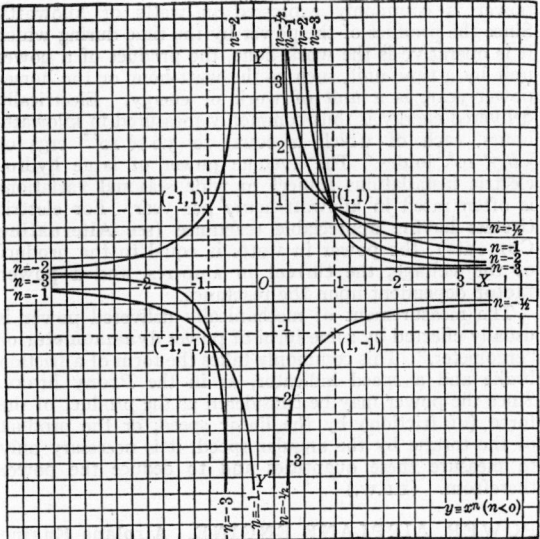

Power function.

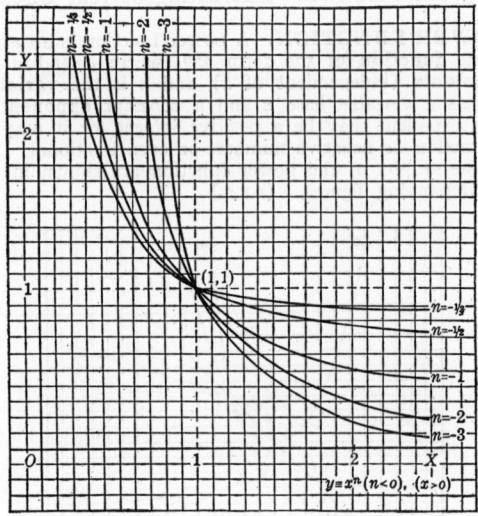

Power function.

POWER OF A TEST. The power of a test is the probability in using this test of rejecting the hypothesis being tested when some alternative hypothesis is true in the **Neyman-Pearson theory of testing** a statistical hypothesis. Naturally statistical tests with high power are preferred. (L.A.A.)

POWER PLANT. A power plant is a machine or an assembly of machines and equipment, the purpose of which is to convert energy from a dormant or useless state to a useful one. This usually takes the form of the conversion of the latent chemical energy of a fuel into mechanical or electrical energy, a conversion that is accomplished with difficulty, and then only at the expense of the loss of a major portion of the energy while it is in the form of heat.

Power is the rate at which mechanical or electrical energy is produced. We may regard a power plant as a factory for the production of a given commodity, energy, from fuel as a raw material, the rate of manufacturing being the power capacity of the plant as measured in hp. or kilowatts.

The possible sources of energy are as follows:

1. Fuels.
2. Streams of water.
3. Ocean tides and waves.
4. Winds.
5. Solar rays.
6. Terrestrial heat.

The **fuels** are most abundant and easily utilized. **Coal** continues the principal fuel for power generation. Coal, oil, gas, and wood all contain **hydrogen** and **carbon** which in rapid chemical union with oxygen from the air liberate heat energy. All fuels have been fired under boilers and gas and oil, in addition, are sources of power in internal combustion engines. The greater portion of the power generated in the United States is produced from fuels.

The hydraulic plant has as its source of energy the kinetic energy of a stream or the potential energy of impounded water. Hydraulic plants are increasing in size and numbers, but it is obvious that a point will be reached beyond which the energy of a stream will not be exploitable in competition with the steam plant.

Power from the winds has served man for many centuries, but the total amount of energy generated in this manner is small. While the expense of installation and the uncertainty of operation have prevented serious consideration of the windmill for other than such intermittent services as pumping, there has been considerable experimentation relative to the generation of electrical energy by moderate-sized wheels for charging storage batteries.

With the exception of terrestrial heat, all of the sources of energy may be traced indirectly to the sun. Evaporation of surface water to form rain clouds which continually replenish the flow of water in streams is accomplished by the heat of the sun; gravitational effects account for tides; warming and cooling of different portions of the earth's atmosphere cause winds and thereby waves; and solar rays, nourishing tropical vegetation through the prehistoric ages, may be held responsible for the deposits of coal, oil, and gas which were formed from that vegetation.

In a few instances the direct rays of the sun have been used to generate energy. The obvious fault of this source is that it is effective during the daylight hours only, and, for continuous service, some reservoir of energy such as a storage battery or a heat-accumulator tank is necessary to carry through the night operation. Also, cloudy weather seriously handicaps the operation of such a plant. In the solar plants that have been constructed, the heat in the rays is absorbed by water in a boiler. The rays are converged upon the boiler by parabolic mirrors set to follow the sun throughout the day; or the same thing may be accomplished by direct absorption on a large flat-surface boiler.

The suggestion of tremendous unleashed power in the daily silent glide of the tides and the evidences of gigantic terrestrial reservoirs of heat to be seen in geysers and volcanoes have, for years, prompted many dreams of harnessing these sources.

Natural steam escapes from surface vents in many portions of the earth. It remained for an Italian, Prince Conti, to develop the use of natural steam. This he did in a successful way at Larderello in Tuscany, Italy, bringing the steam under control by drilling wells.

The heat power plants may be classified as internal or external combustion types. The internal combustion power plant is a self-contained unit using the products of combustion as the working medium. Since it is compact, having all parts included in one machine or engine, and, since its efficiency is, on the average, higher than that of the external combustion cycle, it has been used for automotive service, and where compactness and mobility were desired.

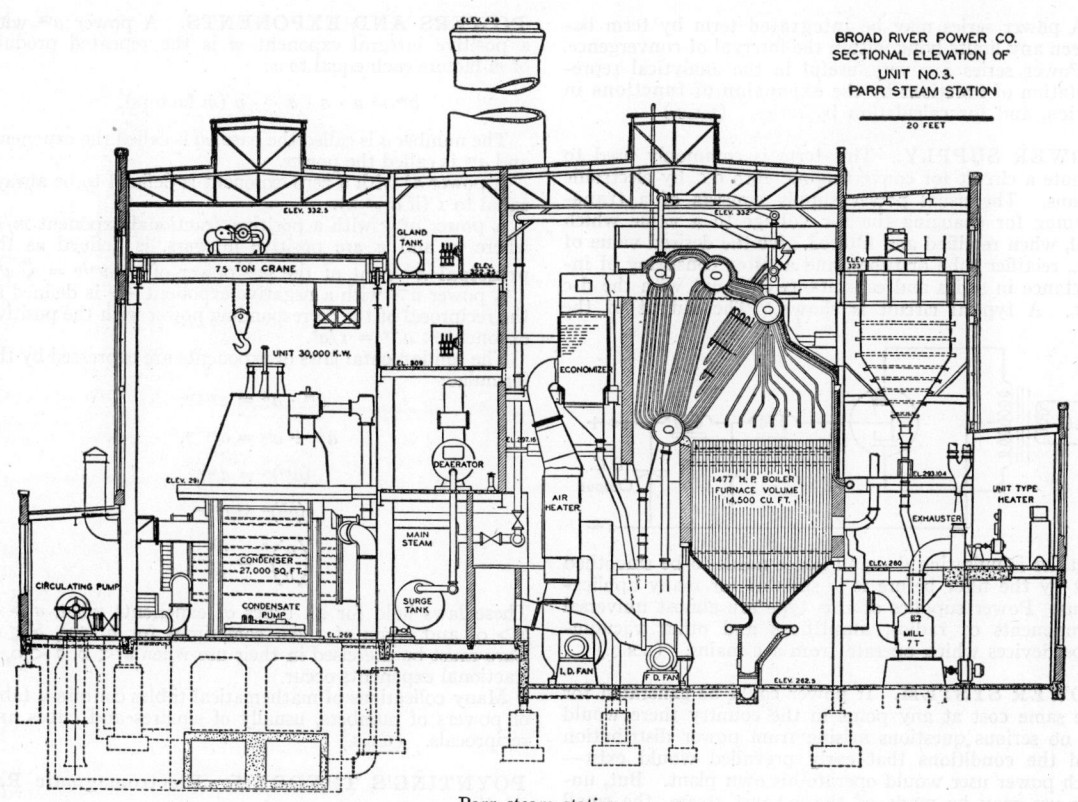

Parr steam station.

Types of fuels which have been required by **internal combustion engines** have been more expensive than those which may be successfully employed in the external combustion cycle, and so there is large power-generating capacity installed in external combustion power plants such as the modern steam plant. A cross-section of a typical steam plant shows the complexity of the modern high-capacity, external combustion cycle. The plant shown in the accompanying figure illustrates the point. The building housing this plant is set off roughly into four regions. Beginning at the left, there is the turbine room; next, a narrow multi-storied section containing electrical galleries and certain of the steam and water auxiliaries; next, the main boiler room with its boilers having fans and other auxiliaries; and lastly, the fuel preparation room, in which fuel is in storage, and from which the fuel is taken to be prepared for burning. (F.T.M.)

POWER PUMP. A power pump is a reciprocating, **piston** and **cylinder** type water **pump** not driven by steam, but by motor, line shaft, belt, or some other mechanical means. The direct-acting steam pump and the steam-driven, flywheel-type pump are thus set off from the power pump. The ordinary form of power pump is known as a triplex pump. It is three-cylindered, single-acting, having three **pistons** driven from a single **crankshaft**. The triplex feature provides fairly uniform discharge, as well as a much steadier load on the motor than would be possible from a single-cylindered pump. The triplex pump has three cylinders in line, and cranks at 120 degrees. The best crankshaft speed is much lower than the usual speeds of motors and gasoline engines, so that it is best to gear drive the pump from the source of power. The efficiency of such pumps is rather high, and there is no limit to the pressures which may be carried. (F.T.M.)

POWER SERIES. An **infinite series** of the form

$$\sum_{n=0}^{\infty} a_n x^n = a_0 + a_1 x + a_2 x^2 + \cdots + a_n x^n + \cdots,$$

where $a_0, a_1, \cdots, a_n, \cdots$ are **constants** and x is a **variable,** is called a power series.

For every power series $\Sigma a_n x^n$ there exists a constant l such that the series is **absolutely convergent** for all values of x such that $|x| < l$, and is **divergent** for all values of x such that $|x| > l$. This number l is called the limit of convergence of the power series, and the interval $(-l, l)$ is called the interval of convergence. The series may converge or diverge at the ends of this interval, for $x = l$ and $x = -l$.

If $\lim\limits_{n \to \infty} \left| \dfrac{a_n}{a_{n+1}} \right|$ exists, it is the limit of convergence.

If $\lim\limits_{n \to \infty} \sqrt[n]{|a_n|}$ exists, it is the **reciprocal** of the limit of convergence.

A power series is **uniformly convergent** within the interval $(-l', l')$, where $l' < l$ and l is the limit of convergence of the given power series.

The function defined as the sum of a power series $\Sigma a_n x^n$ is a **continuous function** of x at all points within the interval of convergence.

For the operations with power series, we have the following theorems:

Two power series may be added together for all values of x for which both series are convergent, i.e., for the smaller of the two intervals of convergence.

Two power series may be multiplied together for all values of x for which both series are absolutely convergent, i.e., for the smaller of the two intervals of convergence.

A power series may be **differentiated** term by term for all values of x within its interval of convergence.

A power series may be **integrated** term by term between any limits lying within the interval of convergence.

Power series are very useful in the analytical representation of functions by the **expansion of functions in series**, and for calculation by series. (L.L.S.)

POWER SUPPLY. This term is commonly used to denote a circuit for converting a.c. into d.c. by electronic means. The usual power supply consists of a **transformer** for changing the a-c voltage to a value which will, when rectified and filtered, give the desired value of d.c., rectifier tube or tubes, and a **filter** consisting of inductance in series and capacitance in shunt with the d-c line. A typical circuit is shown. The output of the

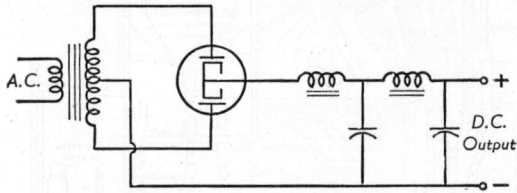

rectifier is pulsating d.c. and hence needs to be smoothed out by the filter before it is suitable for many applications. Power supplies of this type are almost universal components of radios, **amplifiers** and other vacuum-tube devices which operate from a-c mains. (L.R.Q.)

POWER SYSTEM. If power could be generated for the same cost at any point in the country there would be no serious questions arising from power distribution and the conditions that early prevailed would exist—each power user would operate his own plant. But, unless use may be made of the exhaust steam, the small privately owned plant is hardly able to compete with the central system on an economic basis because of the inherently higher efficiency of large generating units and the lower overhead cost of quantity production. So, while large numbers of small isolated plants are in operation at present, the major portion of installed power capacity is to be found in central stations. The bulk of this capacity is in **steam engines** and turbines, mostly **steam turbines**. Power generated by industries may or may not be converted into the electrical form before use, but that generated by central stations is invariably electrical to permit transmission to distant points. The central station is but one link of a chain joining the source of energy and its ultimate user.

The distribution system may be separated into two parts, the primary and secondary systems. The primary distribution system generally consists of a transmission line carrying three-phase current from the switchyard of the plant to a substation located near the load served. The purpose of the substation is to transform the high voltage necessary for economical long-distance transmission to voltages suitable for lines in residential districts and for the primaries of the light pole-top transformers. The secondary distribution system extends from the substation to the customer's meter through the medium of weatherproof pole-top transformers, strategically located with respect to a small group of customers which each one supplies.

Power supply systems are owned and controlled both by municipalities and by public utility companies. Power and light plants, at one time mostly municipally owned, have come more and more into the hands of public utility companies. A utility company may own one or more generating stations and a network of transmission lines serving its territory. The company must be so organized as to weld generating, distributing, and public relations departments into a smooth-working unit. The small company operating one station and servicing one community easily accomplishes this, but the large company requires a complex organization. (F.T.M.)

POWERS AND EXPONENTS. A power a^m with a positive integral exponent m is the repeated product of m factors each equal to a:

$$a^m = a \cdot a \cdot a \cdots a \ (m \text{ factors}).$$

The number a is called the base, m is called the exponent, and a^m is called the power.

A power a^0 with a zero exponent is defined to be always equal to 1 (if $a \neq 0$).

A power $a^{m/n}$ with a positive fractional exponent m/n, where m and n are positive integers, is defined as the **principal n^{th} root** of the mth power of a: $a^{m/n} = \sqrt[n]{a^m}$.

A power a^{-r} with a negative exponent $-r$ is defined as the reciprocal of the corresponding power with the positive exponent r: $a^{-r} = 1/a^r$.

The fundamental laws of exponents are expressed by the formulae:

$$a^m \cdot a^n = a^{m+n},$$
$$a^m \div a^n = a^{m-n},$$
$$(a^m)^n = a^{mn},$$
$$(ab)^n = a^n b^n,$$
$$\left(\frac{a}{b}\right)^n = \frac{a^n}{b^n}.$$

These laws hold for all types of exponents when $a > 0$, $b > 0$, and some of them sometimes for $a < 0$, $b < 0$. Care must be exercised in their use when $a < 0$, $b < 0$, if fractional exponents occur.

Many collections of mathematical tables contain a table of powers of numbers, usually of squares and cubes and reciprocals. (L.L.S.)

POYNTING'S THEOREM. Electromagnetic Radiation.

POZZUOLANE. A **leucite** tuff occurring near Naples and famous for its use in making **cement**. (R.M.F.)

PRAIRIE CHICKEN. Aves, Galliformes. *Tympanuchus*. A **grouse** of the prairie regions of North America. This species was once very abundant over a large area but it is now much more restricted in its range and in numbers. The destruction has been due partly to hunting, partly to the spread of agriculture, and partly to wholesale trapping for the market. In the Dakotas countless birds were taken for sale during the past century. The eastern heath hen, which has recently become extinct, was a variety of the same species, (A.W.L.)

PRAIRIE DOG. Mammalia, Rodentia. *Cynomys*. Burrowing animals of several species, found chiefly west of the Mississippi but introduced into a few eastern localities. They are small stout-bodied animals with shallow cheek pouches. All are plant feeders and in settled regions they sometimes damage crops severely. Also called the prairie marmot. (A.W.L.)

PRANDTL GROUP. Heat Transfer.

PRASE. Chalcedony.

PRASEODYMIUM. Symbol: Pr. Atomic number: 59. Atomic weight: 140.92. Density: 6.48. Melting point: 940° C. No isotope, but of single atomic form: 141. Type of compound: Pr_2O_3, greenish-yellow. Color of salts: Green. Discovered by Welsbach in 1885. A member of the **cerium** sub-group of the rare earth metals. (R.K.S.)

PRATINCOLE. Aves, Charadriiformes. *Glareola*. Birds (**Aves**) of several species found on all continents of the Old World. Their long wings and forked tail give them the appearance of swallows when in flight, but their moderately long legs are a point of resemblance with the plovers. They live chiefly near water but are insectivorous in habits. (A.W.L.)

PRAWN. Crustacea, Decapoda. Small marine **crustaceans** closely related to the shrimps. They differ from the lobsters and crabs in the compressed body and in the use of the abdominal appendages for swimming. Prawns are edible. (A.W.L.)

PRAYING MANTIS. Mantis.

PREAMPLIFIER. This is a class A voltage **amplifier** which receives the signal from a **microphone, pick-up** or other low-level device and amplifies it so it can supply the control and additional amplifier circuits. Thus a preamplifier is commonly used in a radio studio to amplify the microphone output before feeding it into a mixer, line to the transmitter, or other amplifying equipment at the studio. (L.R.Q.)

PREANTENNA. A sensory organ of the pair borne by the first segment of the body in the **Onychophora.** Most arthropods have the first body segment developed in the embryo but lacking in the adult, hence their antennae are appendages of a more caudal segment. The similar organs of Onychophora are named preantennae to distinguish them from the true antennae of the other classes. (A.W.L.)

PRECESSION. An effect manifested by a rotating body when a torque is applied to it in such a way as to tend to change the direction of its axis of rotation. If the speed of rotation and the magnitude of the applied torque are constant, the axis, in general, slowly describes a cone, its motion at any instant being at right angles to the direction of the torque.

A familiar example of precession is an ordinary top. If the axis of spin is not exactly vertical, the force of gravity exerts a torque tending to overturn the top; but instead of tipping over, it "wobbles" with a precessional motion about the vertical through the pivot-point. The gyroscope exhibits similar behavior. A hoop or a coin can roll on edge across the floor because, whenever it tends to tip either way, precession swerves its plane and changes its path, so that it automatically steers itself as a bicycle is steered by the rider.

Precession is due to the fact that the resultant of the angular velocity of rotation and the angular velocity produced by the torque is an angular velocity about a line which makes an angle with the permanent rotation axis; and this angle lies in a plane at right angles to the plane of the couple producing the torque. The permanent axis must turn toward this line, since the body cannot continue to rotate about any line which is not a principal axis of maximum moment of inertia; that is, the permanent axis turns in a direction at right angles to that in which the torque might be expected to turn it. If the rotating body is symmetrical and its motion unconstrained, and if the torque on the spin axis is at right angles to that axis, the axis of precession will be perpendicular to both spin axis and torque axis. Under these circumstances the period of precession is given by $T_p = \dfrac{4\pi^2 I_s}{Q T_s}$; in which I_s is the **moment of inertia** and T_s the period of spin about the spin axis, and Q is the torque. In general the problem is more complicated.

Precession, as it is manifested in astronomy, is a slow, rotary motion of the axis of a rotating body in space. The term is, perhaps, most frequently used in connection with the earth, but all rotating bodies may exhibit the effect. The earth is an oblate **spheroid** with the minor axis the axis of rotation. Hence if we subtract from the earth a sphere with radius equal to the minor axis

we shall have left a shell of continually increasing thickness as we pass from the pole to the equator. Such a rotating shell, with the greatest amount of mass in the plane perpendicular to the axis of rotation, is a characteristic **gyroscope.**

The axis of rotation of the earth is inclined at an angle of 66.5° to the plane of the **ecliptic.** The forces of **gravitational** attraction of the sun and the moon tend to pull the equatorial shell into the plane of the ecliptic and hence a torque is applied tending to change the direction of the axis of rotation. This causes the axis to describe a cone in space (i.e., produces precession).

As the axis of rotation describes a cone, the poles of rotation describe circles in space. The radii of these circles, expressed in angular measure on the **celestial sphere,** are 23.5°. The effect of this motion of the poles is to cause the **vernal equinox** (one point of intersection of the ecliptic with the **equator**) to move along the ecliptic. The motion is slow, about 26,000 years being required for the vernal equinox to make a complete circuit of the ecliptic. The motion is not perfectly regular because of the fact that both the sun and the moon are in different planes and are moving relative to each other, causing a variation in the torque applied to the earth. The variations in the torque produce a slight irregularity in the motion of the poles known as **nutation.**

The motion of the equator among the stars causes slow, and slightly irregular, changes in the **equatorial coordinates** of the **stars.** Since celestial **longitude** is measured from the vernal equinox, the motion of this point produces a change in this coordinate of the stars. The motion of the vernal equinox also changes its location among the **constellations** along the zodiac, the "sign of Aries" (i.e., the vernal equinox) now being located in the constellation of **Pisces** instead of in **Aries.** At present the north pole of rotation of the celestial sphere is close to the star **Polaris** (α **Ursae Minoris**), but 12,000 years hence the star **Vega** (α **Lyrae**) will be close to the pole of rotation and hence be known as the pole star.

Precession enters also in a very important way into the dynamics of **atoms** and **molecules.** (L.D.W., W.K.G.)

PRE-CHELLEAN. Paleontology of Man.

PRECIPITATION. This process involves the separation of a solid from **solution** (1) by evaporation and cooling of certain solutions to **crystallization,** (2) by addition of or to a reagent which forms new insoluble product or products, e.g., **sulfuric acid** to **barium** chloride solution, forming barium sulfate precipitate, (3) by addition or removal of a substance to decrease the solubility of another substance in the medium, e.g., salting out of soap by salt, or removal of **carbon dioxide** from **calcium** hydrogen carbonate solution resulting in the precipitation of calcium carbonate, (4) by heating, e.g., coagulation of certain **proteins.**

The precipitate which is formed in a given case frequently separates in a characteristic form, e.g., **barium** sulfate as fine crystals, difficultly separable from the medium; **lead** chloride as medium crystals, easily separable; **ferric** hydroxide as gelatinous mass or slime, difficultly separable; **silicic** acid as gelatinous mass or as a gel; **silver** chloride as curdy mass. The concentrations used and the temperature frequently affect the form of the precipitate. Very fine crystals are increased in size by digestion, that is, by allowing to stand for some time to take advantage of the fact that small crystals are more soluble than large ones. (See **Hydrometeors.**) (R.K.S.)

PRECIPITATION HARDENING. A very considerable number of alloys are hardenable by a heat-treating procedure based on controlled precipitation throughout the grain structure of minute particles of a metal, **solid solution,** or **intermetallic compound** different from the base metal. It is necessary that the secondary or hardening constituent be soluble in the base metal at an elevated temperature and less soluble or even insoluble at atmospheric temperature. Thus the alloy **duralumin**, for example, is heated to 930° F., which is well under its melting temperature range, until the hardening constituents are in a state of solid solution in the aluminum matrix, whereupon it is quenched in water to prevent reprecipitation and growth of the hardening constituents as their solubility decreases at the lower temperatures. This results in a supersaturated solid solution which does not remain stable at room temperature but tends, over a period of time, to revert to its normal duplex structure. However, because of the low temperature at which the hardening constituents separate from the matrix they remain submicroscopic in size, and being present in very great numbers along the crystallographic **slip** planes of the matrix they effectively interfere with movement along these planes, thus reducing the **plasticity** and increasing the **hardness** and strength of the alloy. The heating and quenching operations are known as the "solution heat treatment," while the spontaneous hardening at room temperature is called age-hardening or precipitation hardening. The latter term is reserved by some for similar treatments carried out on other alloys at slightly elevated temperatures.

In addition to the important aluminum alloys, many magnesium-, copper-, and nickel-base alloys are hardenable by the solution and precipitation heat treatments. Certain copper-bearing steels are also hardenable by this method. (R.H.H.)

PRECISION GAUGE BLOCK. Measurement.

PREFORM. Plastic Molding.

PREGNANCY. (Cyesis, gestation.) The condition of being with child. The duration of pregnancy in humans is usually about 280 days, 9 calendar or 10 lunar months, dating from the time of the last menstrual period. Presumptive signs of pregnancy are absence of the menstrual periods, nausea or vomiting in the morning (**morning sickness**), enlargement of the breasts with pigmentation of the nipples, and enlargement of the abdomen during the last half of pregnancy. Absolute signs of pregnancy are palpitation of the fetal body, movement of the child, and sound of the fetal heart. As early as the 8th day of pregnancy the **Aschheim-Zondek** test is positive.

An extra-uterine or ectopic pregnancy is one in which the fertilized ovum lodges and develops outside the uterine cavity. This usually takes place somewhere along the **fallopian tubes** and, more rarely, free in the abdominal cavity. Such ectopic gestations almost always terminate spontaneously early in the course of the pregnancy, and produce the clinical picture of ruptured ectopic pregnancy. Abdominal pain and shock due to massive hemorrhage into the abdominal cavity make this condition a serious emergency which demands immediate surgical treatment.

A phantom pregnancy, or pseudocyesis, is an hysterical manifestation in which the abdomen enlarges and resembles the enlargement associated with a true pregnancy. It is treated by treating the underlying **psychoneurosis** with psychotherapy. (R.S.M., D.M.H.)

PREHENSION. The flexion of an appendage to grasp an object by folding around it. There are two types of grasping organs among animals: forcipate and prehensile. The human hand illustrates both. Objects may be taken between the thumb and fingers as between the jaws of a forceps, or they may be grasped between the fingers and the palm by the prehensile folding of the digits. Less versatile organs are capable of one or the other type of action, as in the case of the forcipate chela of the lobster and the prehensile tails of some monkeys. (A.W.L.)

PREHISTORIC ART. Paleontology of Man.

PREHNITE. The mineral prehnite is an acid **orthosilicate** of calcium and **aluminum**, $H_2Ca_2Al_2(SiO_4)_3$ It crystallizes in the **orthorhombic** system, usually in masses. It has an uneven fracture; is brittle; hardness 6–6.5; specific gravity, 2.80–2.90; luster, vitreous to pearly; color, various shades of light green to gray or white; translucent. Though not a **zeolite** it is found associated with them and with **datolite, calcite,** etc., in veins and cavities of basic rocks, sometimes in **granites, syenites,** or **gnisses.** It is found in Austria, Italy, the Harz Mountains, France, Scotland, and South Africa where it was originally discovered. In the United States well-known localities are Somerville, Massachusetts; Farmington, Connecticut; Paterson, New Jersey; and Keweenaw County, Michigan. Named for Colonel Prehn its discoverer, who was an early Dutch Governor of the Cape of Good Hope colony in South Africa. (E.S.C.S.)

PREISCHIOPODITE. A segment of the **biramous appendage of crustaceans.** In some species the outer segment (basipodite) of the two that make up the base of the appendage is divided into two parts. The proximal part is then called the probasipodite and the distal is the metabasipodite or preischiopodite. (A.W.L.)

PRELIMINARY SURVEY. Surveying.

PRELOADING. Bearings.

PREMAXILLARY. A small bone lying in front of the **maxillary** bone of the upper jaw in the skull of vertebrates. The two form the median portion of the jaw and bear the incisor teeth, if present. Also called intermaxillary. (A.W.L.)

PREMENTUM. The palpiger of the insect **labium.** (A.W.L.)

PREMOLAR. A tooth lying between the canines and molars of mammals. Premolars are broader teeth than the incisors and canines but are usually less broad than the molars. Among the grazing animals, however, a number of forms have the premolars formed like the molars. The human premolars are also called bicuspids (A.W.L.)

PRENATAL INFLUENCE. Telegony.

PREPUCE. Foreskin.

PRESENTATION. That portion of the fetal body which appears or presents at the cervical opening of the **uterus** during labor. The most common presenting part is the head, usually the occipital portion. This is the normal presentation and is best for an easy delivery Various other portions of the head may also present face, brow, etc.

Breech presentation is presentation of fetal buttocks making labor more difficult. (R.S.M., D.M.H.)

PRESS WORKING. Cold-working metal processes may be divided into four general groups: cutting, bending, drawing, and squeezing. Cutting processes as applied to sheet metal are further subdivided into shearing and blanking processes; the term shearing as used here implies straight cuts on plates and sheets; blanking is the process of cutting parts of almost any shape from sheet metal with a punch and die in a power-actuated press Blanking dies are analogous to the punches and dies used for rivet holes and may be made to cut out parts of almost any shape. They may be made either singly or in multiple so that one, two, three, or more parts may be blanked or cut at one stroke of the press. Piercing dies are small blanking dies that are usually employed for punching circular holes in sheet metal or in previously blanked parts. Progressive dies are those placed in sequence so that successive operations may be performed; combination dies often incorporate blanking and piercing, or blanking and drawing operations in a single die.

To insure accurate work and to eliminate the possibility of chipping the edges of the punches and dies, the two members must be very carefully aligned when they are placed in position on the press ram and bolster. This operation must be performed every time a different die is used, and requires the services of an expert die-setter. To eliminate the care that is necessary in die setting and the undesirable effects of too much freedom in the ram slide, many punches and dies are mounted in a sub-press or die set.

Bending dies are used for bending or folding sheet metal or wire parts into irregular shapes, and function by pressing the stock into cavities in the die by means of a punch of corresponding shape, or by the use of auxiliary attachments such as slides or levers. Drawing dies differ in their action from bending dies in that the deformation produced in the work is usually three-dimensional; a bending die, for example, might be used for one or more parallel folds in a sheet, whereas a drawing die might be used to produce a cup-shaped article from a flat sheet. Either bending or drawing dies may be combined with blanking dies, to serve as combination blanking and bending, or combination blanking and drawing dies. Redrawing dies are second or third operation drawing dies, used where the depth of "draw" is so great that fracture of the material may occur if the shape change is attempted in one operation. In drawing open-end cups from flat circular blanks, for example, one or more redrawing dies are used whenever the depth of draw exceeds one-half the diameter of the cup.

Reducing dies are used for necking the open ends of drawn cups in such applications as cartridge case manufacture. Curling dies, are used for curling or rolling a flange or bead around the open edge of a cup. The cup should be slightly flanged in its final draw to permit the curling punch to start easily. Wiring dies are used for curling and simultaneously enclosing a wire ring inside the flange. The wire is carried by a spring plate on the die which recedes to permit the metal to be rolled entirely around the wire. Bulging dies are special drawing dies that use either a soft rubber pad or a fluid such as oil. They are used for articles that are not straight-sided or tapered inwards. In the hydraulic die, the oil within the blank is compressed by a ram so that it forces the blank to conform to the interior of the die. The die is made in two parts, a stationary half and a movable half, so that it may be opened to permit the finished part to be removed.

Metal spinning is a specialized drawing operation employed for articles which have surfaces of revolution. Metal spinning is usually employed for small-quantity production where the number of parts required does not warrant the use of a drawing die; the process is also used for shapes which are difficult or impossible to produce in dies. Metal spinning lathes are similar to engine lathes. The spinning tool may be either manually or mechanically guided. A wooden (or metal) form is fastened to the faceplate of the spinning lathe, and the sheet disk is placed against it and held in position by a rotating tail center. The material is shaped by forcing the spinning tool against the work as it rotates. The shank of the tool pivots about a stop in the tool rest.

Squeezing operations are the most severe of all cold press processes. Since more or less metal flow in a cold state is involved, squeezing operations are generally performed on hydraulic presses, or on some of the slower motion mechanical presses such as the knuckle-joint press. Sizing is a squeezing operation that is used for flattening or for surfacing parts where a very small amount of metal flow is required. Very close tolerances can be maintained by the sizing process, together with a rate of production much greater than that obtained by milling or grinding. The process is restricted, however, to relatively ductile materials and cannot be used for cast iron and hardened steels. Coining is a squeezing process employed for cold-pressing coins, metals, and emblems and is a restricted-flow process, in contrast to sizing in which the metal can flow in a direction perpendicular to the pressure application. Very high pressures are used in coining, and great care is exercised in the design and operation of the dies since it is probably the most severe of all the squeezing processes. Extrusion press operations are analogous to coining operations in that considerable flow takes place, but a major portion of the die is generally designed so that the flow is unrestricted in a direction perpendicular to the pressure application.

Multiple-slide and multi-slide strip machines are used for complex bending and forming operations on wire and narrow strip stock, for a wide variety of parts such as rings, links, hooks, clips, etc. These machines are allied to the press-working group, and usually consist of four or more cam-actuated slides to which special forming and bending dies may be attached. These machines are also used for combination manufacturing and assembling purposes; a hinge machine, for example, blanks the hinges from strip stock, punches the screw holes, slits and forms the pin bearings, and inserts the hinge pin, producing a completed hinge automatically. (H.C.H.)

PRESSURE. A type of stress, characterized by its uniformity in all directions (as distinguished from compressive stress in one direction). Its measure, as with all other stresses, is the force exerted per unit area; for example, the normal **atmospheric pressure** is about 14.7 lbs. per sq. in. Pressure is usually associated with a decrease in volume; though the opposite stress, accompanying an increase in volume, is sometimes referred to as "negative pressure." This latter must be distinguished from the same term as sometimes used to denote pressures below atmospheric, the pressure of the atmosphere being in such cases taken as an arbitrary zero. It is likewise important, especially in pneumatics, to indicate whether pressure is reckoned from vacuum or from atmospheric pressure as zero. Thus when a tire is inflated to "thirty-five pounds," the actual pressure in the tire is about 50 lbs. per sq. in.

Many different units are used in expressing pressure. The absolute c.g.s. unit is the **barye** or **bar** (1 dyne per sq. cm.). The usual engineering unit is the pound per square inch. Very commonly, pressures are expressed in millimeters or inches of mercury or other liquid, meaning the hydrostatic pressure at the corresponding depth in that liquid. Another common unit in physics is the "atmosphere," standardized as 760 mm. of mercury at 0° C. and equal to about 1,013,246 bars. (See **Pressure Gauges.**) (L.D.W.)

PRESSURE ANGLE. Spur Gearing.

PRESSURE COEFFICIENT, AIRFOIL. This is a parameter often used to describe the variation of pressure over an airfoil or other aerodynamic object immersed in a fluid stream with relative velocity V. Let the free-stream static pressure be designated by p_0, the local airfoil surface pressure by p, and the dynamic pressure corresponding to V by q, then the pressure coefficient is $(p_0 - p)/q$.

Some airfoil pressure coefficients are shown in connection with **compressibility.** The area between the curves of upper and lower surface pressure coefficients when multiplied by q and chord length is the lift per ft. of span. (F.T.M.)

PRESSURE GAUGES. Pressure enters as a variable factor in a vast number of physical phenomena, and there is frequent need for its measurement over a very wide range. For ordinary pressures there are two chief types of gauge: viz., the familiar liquid manometer, of which the mercury barometer is an example, and those instruments dependent upon the deformation of a closed, thin-walled cell of elastic metal, such as the ordinary (Bourdon) steam gauge and the aneroid barometer. Such gauges are secondary instruments in that they must be calibrated in comparison with a primary gauge, such as a manometer. While the mechanical element of a gauge may be a Bourdon tube, a diaphragm, or a bellows, space permits description only of the simple widely used Bourdon type of gauge. As shown in the figure, an oval tube is bent in the form of a portion of a circle

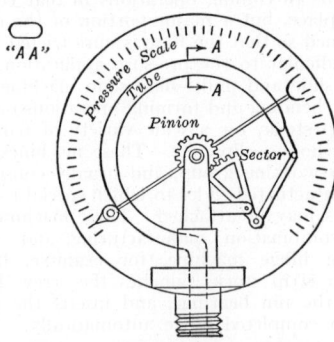

Pressure Gauge (Bourdon Tube).

and sealed at one end. The other end is connected to a region whose pressure is to be measured. If that pressure is above atmospheric, the tube tends to change from oval to circular cross-section, accompanied by an uncurling action. This tendency is resisted by the elastic properties of the metal of the tube, but the tip executes a movement which is proportional to the change of internal pressure. If the internal pressure were below atmospheric, the effect would be reversed and the tube would tend to curve in, moving the sealed tip inward. The extent of this motion is quite small, and it must be amplified by suitable mechanism, so that the limited movement of the tip of the tube may be converted into a full swing of the indicating needle. Modification of the material of the tube, as well as the proportions and wall thickness, will adapt this instrument to a wide range of pressures.

Physical and industrial research often involves very low gas pressures. Gauges for measuring low pressures may be classified under five principal heads: 1. Pressure multipliers, represented by the well-known McLeod gauge. A measured volume of gas at the unknown low pressure is compressed isothermally into a known volume many times less, thus raising the pressure to within the range of an ordinary mercury manometer; the determination is then completed by applying **Boyle's law** or van der Waals' law. 2. Viscosity or "molecular" gauges, dependent upon the fact that the viscosity of a highly rarefied gas is a function of the pressure. In the Langmuir and Dushman design, a horizontal circular disk is rotated rapidly under and near a similar disk suspended in the gas by a quartz fiber; the molecular drag between the disks exerts a measurable torque upon the suspension, which torque is a function of the pressure. 3. Radiometric gauges, like the **Crookes radiometer,** in which the force acting on its vanes depends upon the number of gas molecules present, and hence on the pressure. Knudsen has developed gauges of this class. 4. The hot-wire type, illustrated by the Pirani gauge. A hot filament is cooled by convection, i.e., by the gas molecules coming into contact with it and carrying off its thermal energy, the rate depending, of course, upon

the number of molecules present. The cooling effect is determined by observing the resistance of the filament. 5. Ionization gauges, in which the rarefied gas is subjected to ionization by thermionic emission from a heated filament, and a third electrode, of lower potential than the filament, carries off the positive ions formed. The tube can be so designed that the ionization current is proportional to the gas pressure. Such a gauge was first designed by Buckley, and the Bell Laboratories have perfected a design said to be capable of measuring one-trillionth of an atmosphere.

The highest pressures measurable with the mercury manometer are not over a few hundred kilograms per sq. cm. For very high pressures—thousands of kilograms per sq. cm.—some form of "free piston" gauge is used; that is, one in which the unknown pressure is applied to a small piston, the force exerted on which can be counterbalanced with weights or otherwise. Amagat used a double piston, the larger end balanced against the pressure of a mercury manometer; a sort of inverted hydraulic press. (L.D.W., F.T.M.)

PRESSURE HEAD. Head, as used in the science of **hydraulics,** is the height, actual or imaginary, of a column of fluid which creates a pressure at its base. Viewed in this way, head is a measurement of pressure. It is convertible to the pressure (force per unit area) by multiplying by the density of the fluid. Not only is the static pressure designated in terms of head measured as a linear dimension, but velocity as well, this being distinguished from static head by being called velocity head. This pressure, measured as a head necessary to give a fluid, originally at rest, a velocity V, is $\dfrac{V^2}{2g}$, wherein g is the **acceleration** of gravity. (See **Fluid Flow.**) (F.T.M.)

PRESSURE SHIFT AND PRESSURE BROADENING. **Spectrum** lines are subject to various physical influences brought to bear upon the substance emitting the radiation, and among these is pressure. When the pressure and density of a gas are greatly increased, the lines of its spectrum become less sharp, or broader (without change of total intensity). This broadening is not symmetrical but is greater in the direction of longer wavelength, so that the peak or maximum is shifted toward the red. Thus, when the pressure of mercury vapor is increased from 10 to 50 atmospheres, the 2536-angstrom line has its half-width increased from 0.2 angstrom to 0.8 angstrom, while its peak is shifted about 0.3 angstrom toward the red. These effects are attributed to the influence of atomic **collisions,** which are of course more frequent at high pressure. (L.D.W.)

PRESSURE TENDENCY. Isallobars.

PRESSURE VESSELS. Containers for fluids subjected to pressure may be classified as boilers and fired vessels, or unfired pressure vessels. Vessels, such as stills, autoclaves, and the like, may be fabricated in one piece or made of separate plates formed to shape and riveted or welded at the seams. Pressure vessels of spherical form permit the greatest volume for a given enveloping surface and are uniformly stressed in all directions, thereby affording the most economical utilization of material. Although they may be employed as containers for gases and volatile liquids, they are not in general use. Vessels of cylindrical form, with spherical, semi-spherical, or flat heads at each end, are preferred for liquids and gases because the plates may be more easily preformed, and because such vessels are easier to support than spherical vessels.

Unfired pressure vessels are designed and constructed in accordance with the provisions of the A.S.M.E. Code for Unfired Pressure Vessels; vessels made of carbon steel, used for the storage of petroleum products only, are usually constructed in accordance with a joint code sponsored by the American Petroleum Institute and the A.S.M.E. Boilers are designed in accordance with the A.S.M.E. Power Boiler Code. (See **Boilers**.)

The A.S.M.E. Unfired Pressure Vessel Code recognizes three classifications of vessels, U-68, U-69, and U-70, so denoted from the numbers of the sections in the Code that cover the design and construction. Class U-68 vessels may be used for any purpose, temperature, or pressure; Class U-69 vessels are limited to non-lethal gases and liquids, and have pressure limitations of 400 lbs. per sq. in. and temperature limitations of 700° F.; Class U-70 vessels are limited to pressures of 200 lbs. per sq. in. and operating temperatures of 250° F. In general, class U-68 vessels are more expensive to construct, and require at least one manway to permit access to the interior for fabricating purposes. Class U-70 construction is usually the least expensive, and may be constructed so as to require no manway or access opening.

Unfired pressure vessels may be constructed either with riveted or welded joints; the latter are replacing the former to a considerable extent, since a properly fabricated welded joint requires no calking or pressure sealing. For longitudinal joints, the double-welded vee butt joint is mandatory for most classes of vessels, and is fabricated by welding the seam from both sides. Single-welded vee butt joints and single and double lap joints are used for circumferential and head seams.

The figure shows a cylindrical vessel with various fittings and a dished head; the details above the horizontal

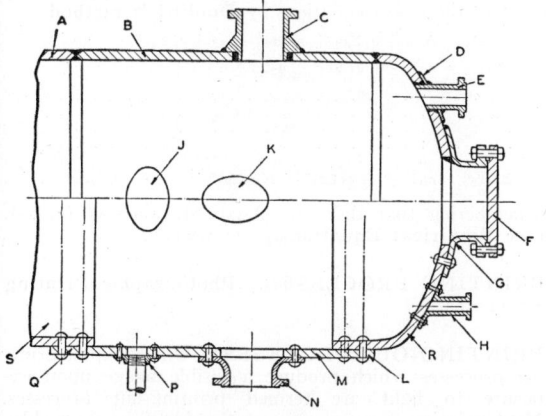

axis are representative of welded joint practice, those below of riveted joint practice. *A* and *B* are successive courses, with the head *R* attached to *B* by a double-welded butt joint. *Q* and *S* are successive riveted courses, with double-riveted chain lap joints at the girth joint, or juncture of *Q* and *R*. *J* and *K* represent elliptical manway openings; the regular-type *J* has the minor diameter of the ellipse in alignment with the vessel axis and is referred to as "long way on sweep"; the irregular-type *K* is referred to as "short way on sweep."

Pressure vessels may be fitted with dished heads, elliptical heads, or flat heads. Dished heads have a shape corresponding to a segment of a sphere, as shown in the figure; the inner radius of the dished portion of the head is termed the crown radius; the radius joining the crown radius to the cylindrical or flanged portion of the head is termed the knuckle radius. Dished heads may be employed as shown, and are then said to be "concave" to the internal pressure; heads placed in the reverse position are said to be "convex" to pressure. The latter type is considered to be only 60% as strong as the former, but will usually permit riveting or welding from the exterior of the vessel. Ellipsoidal heads are actually half oblate ellipsoids, and are somewhat stronger than dished heads. Flat heads may be flanged, with a

knuckle radius joining the cylindrical flange and the head proper, or cut from flat plate, inserted in the vessel and welded around the periphery. Flat heads must usually be considerably heavier than either dished or ellipsoidal heads, for the same internal pressure.

The figure also shows various types of nozzles for attaching pipe and fittings. *C* is a welding-type nozzle, for attachment to a flanged pipe fitting. Welding-type nozzles are available in a range of sizes, and with pressure capacities from 150 to 1500 lbs. per sq. in. The fitting at *E* is a forged steel welding neck, provided with a reinforcing plate *D*, to compensate and provide replacement metal for the opening cut in the head. Straight-neck nozzles, shown at *G* and *H*, are adapted either for riveting or welding, and are obtainable with spherical seats for attachment to the head, or with offset seats as shown at *H*. The nozzle at *G* is used as a manway, with a blind flange *F* serving as a manway cover.

Flared or sweep-type nozzles, shown at *N*, are available in riveting or welding types. These can be used to advantage on all pressure vessels where a smooth orifice is required, since the design minimizes flow turbulence, erosion, and corrosion that may result from sharp-cornered outlet construction.

Forged boiler flanges *P*, are available with plain holes or standard taper pipe threads in three pressure capacities —standard, extra-heavy, and heavy marine types—corresponding to Schedule 40, 80, and 160 pipe. These flanges are usually riveted in place, and can be curved to suit offset or dished head application.

Tanks made of wood are frequently used for water storage and a wide variety of chemical solution and reaction purposes. They usually cost only about one half as much as steel tanks, particularly in the smaller sizes, and can be constructed easily in out-of-the-way places when other material is not available. They are more readily protected from freezing, and do not deteriorate as rapidly, if neglected, as do steel tanks. The life of a wooden tank, if periodically inspected and repaired, is usually fifteen years; cypress tanks often have a useful life of from 20 to 25 years. Cylindrical wooden tanks should have staves dressed on all four sides to a thickness of not less than 2¼" for tanks up to 16' in diameter and depth, or not less than 2¾" for greater diameters and depths. Hoops for cylindrical tanks should be of circular section, bent to fit the outer radius of the tank, connected by malleable iron lugs. The hoops should be located on the tank that the lugs are arranged in a uniform helical line. Wooden cylindrical tanks are usually built with flat bottoms, supported on a wooden or steel grill.

Cylindrical tanks with a conical bottom are frequently specified for applications where suspended solids are present and where rapid drainage may be required. A steel shell for supporting the bottom is necessary, into which the tank is grouted with acid-resisting cement. The steel shell is usually provided with brackets or legs to support the vessel. Dished bottom tanks of similar construction are also available, although dished bottom tanks less than 18" in diameter are usually supplied with steel supporting shells. Support for these sizes usually consists of a wooden false-work supplied by the user.

Plastic molded tanks for internal pressure or vacuum service are available in a range of sizes from 18" to 5' inner diameter and in maximum lengths up to 14'. Vacuum tanks are provided with internal integrally molded stiffening ribs spaced about 1' apart. Drainage slots are provided at the bottom to permit complete removal of the contents of the vessel.

Rectangular plastic molded tanks in a range of sizes varying from 6" to 6' deep, and in lengths up to 15', may be obtained commercially. Such vessels are widely used in the chemical industry as storage and reaction tanks, filters, and for pickling, cleaning, and electroplating applications.

Almost any type of outlet or other fitting can be

molded into tank bottoms or sides. The design and construction of removable heads or covers is essentially similar to that employed for metallic construction. Through bolts are, however, used in preference to cap screws or tap bolts, unless the molded plastic is furnished with molded metal inserts into which the screws may be threaded. (H.C.H.)

PRESSURE-ALTITUDE RELATIONS. Atmospheric Structure.

PREVAILING WESTERLIES. Atmospheric Circulation.

PRIAPULIDA. Gephyrea.

PRIMARY. Transformer.

PRIMARY COLORS. Three colors which, when suitably mixed, will produce all the other colors, including white and black. The colors generally used are an orange-red, green and blue-violet. These colors are sometimes called the additive primaries to distinguish them from the three subtractive or minus-colors, cyan, magenta, and yellow. (H.C.C.)

PRIMARY EMBRYO. An individual formed in the brood pouch or ovicell of some of the bryozoans. It is nourished by the parent through a connection with the wall of the ovicell and produces other individuals by budding. The small masses of cells which arise by this process are called secondary embryos. Each develops into an adult. (A.W.L.)

PRIMARY STRESS. Stress.

PRIMATES. Man, the apes and monkeys, the marmosets, lemurs, and related forms, constituting an order of the class Mammalia.

Most primates are arboreal animals and all species either grasp by opposing the thumb to the fingers or show similarity to grasping appendages of this type in the anatomy of the hand. With the exception of the marmosets, all primaries have nails on at least part of the digits. The lemurs retain a claw on the second toe and the marmosets have a nail only on the great toe. The brain is more highly developed in the primates than in any other animals. (A.W.L.)

PRIMING. Feedwater Treatment, Pumps.

PRIMIPARA. A woman who is giving birth to or has borne but one child. Compare with **Multipara.** (R.S.M.)

PRINCIPAL AXES. Dynamics of Rotation; Moment of Inertia.

PRINCIPAL PLANE OF BENDING. A plane which coincides with a principal axis (see **Principal Axes**) of the cross-section of a flexural member (see **Flexure**) is a principal plane. There are two such planes. The flexure formula is applicable only when the plane of the loads coincides with one of these principal planes. (C.W.C.)

PRINCIPAL PLANES. Principal Stress.

PRINCIPAL ROOT OF A NUMBER. Roots of Numbers.

PRINCIPAL STRAINS. Elasticity.

PRINCIPAL STRESS. A principal stress is a normal stress acting on a plane on which the shearing stress is zero. At any point in a stressed body there

are three mutually perpendicular planes where the shearing stress is zero. Therefore there are three principal stresses. The magnitude and kind of stress (**tension** or **compression**) depend upon the type of applied **load.** The largest of the three, numerically, without regard to the kind of stress, is called the maximum principal stress. The smallest, with due regard to the kind of stress, is called the minimum principal stress. For example, if the three principal stresses have the following values in lbs. per sq. in.: 1000 (tension), 0 and 900 (compression), the maximum is 1000 and the minimum 900, not 0.

Maximum shearing stresses of equal intensity occur on two mutually perpendicular planes which make an angle of 45° with the planes of the maximum and minimum principal stresses.

Principal stresses and maximum shearing stresses at any point in a body are the result of **axial, bending** or **torsional loads** or a combination of these loads.

The planes on which the principal stresses act are called principal planes. (See **Elasticity, Flexure, Stress.**) (C.W.C.)

PRINCIPLE OF LEAST SQUARES. In determining the constants in a function $x_1' = f(x_2, x_3, \cdots x_n)$ the principle of least squares states that the sum of the squared **residuals** $\Sigma(x_1 - x_1')^2$ must be a minimum. By use of calculus and algebra we ordinarily find the constants quite readily. For a polynomial relationship involving only two **variables** x_1 and x_2

$$x_1' = a_0 + a_1 x_2 + a_2 x_2^2 + a_3 x_2^3 + \cdots + a_n x_2^n,$$

we solve the $n + 1$ equations by **Doolittle's method**

$$\Sigma x_1 = N a_0 + a_1 \Sigma x_2 + a_2 \Sigma x_2^2 + \cdots + a_n \Sigma x_2^n$$
$$\Sigma x_1 x_2 = a_0 \Sigma x_2 + a_1 \Sigma x_2^2 + a_2 \Sigma x_2^3 + \cdots + a_n \Sigma x_2^{n+1}$$

$$\Sigma x_1 x_2^n = a_0 \Sigma x_2^n + a_1 \Sigma x_2^{n+1} + a_2 \Sigma x_2^{n+2} + \cdots + a_n \Sigma x_2^{2n+1},$$

remembering that there are N-paired values of (x_1, x_2). (See **Empirical Equations.**) (L.A.A.)

PRINTING PROCESSES. Photographic Printing Processes.

PRINTING-OUT PROCESSES. Photographic printing processes which produce a visible image upon exposure to light are termed printing-out processes. Most printing processes except the familiar silver chloride, bromide or chlorobromide developing-out papers fall into this class.

Of the three silver halides employed in photography, silver chloride, silver bromide and silver iodide, only the first-mentioned produces a strong print-out image. Silver processes designed for printing-out differ from those intended for development in containing free silver. In a printing-out process free silver results in greater sensitiveness to light; its presence in a developing-out material leads to general fog.

All silver printing-out processes are now obsolete, having been superseded by developing papers. The oldest, albumen paper, was supplanted first by collodio-chloride paper and this in turn by gelatinochloride paper. Both of these were similar except in the colloid employed and in both cases the image was toned with gold or platinum, to obtain an image with a more pleasing color, before being fixed in hypo.

The processes employing light-sensitive iron compounds either alone or in combination with silver, platinum or palladium, namely, ferro-prussiate (blue print), cyanotype (positive blue print), Vandyke, kallitype, palladiotype and platinotype (see **Photographic Printing Processes**) are all printing-out processes although the image is faint before processing.

The pigment processes (carbon, gum-bichromate, etc.) are not strictly printing-out processes as the image is not visible after exposure until the water-soluble gelatin is washed away. They are usually considered in this category, however, as the image formed does not require chemical development.

Printing-out processes do not have sufficient sensitivity to adapt them to projection printing which tends to limit their usefulness today. (C.B.N.)

PRIONODESMACEA.
An order of bivalve mollusks (Lamellibranchiata) in one of the classifications now in use. (A.W.L.)

PRISM.
Glass and other transparent materials are cut into many different forms of prism for various optical purposes. Incident light may pass directly through a prism, or may emerge after one or more internal reflections; in some cases it is polarized.

The common triangular prism, familiar in older forms of spectroscope, receives light upon one face and passes

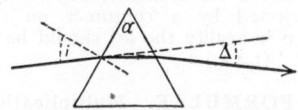

Ray of light passing through triangular prism.

it through another after two refractions, resulting in a total deviation Δ dependent upon the angle of the prism and its refractive index for the light used. If the light is incident at angle i on the first prism face, and if the prism angle is a and the refractive index is n, the total deviation after passage through the prism in a plane at right angles to the prism edge is given by

$$\Delta = i - \alpha + \arcsin\left[n\sin\left(\alpha - \arcsin\frac{\sin i}{n}\right)\right].$$

A little experimenting shows that this deviation has a minimum value when the light traverses the prism symmetrically, entering and emerging at the same angle with the corresponding faces. This angle of minimum deviation is easily shown to be

$$\Delta_{\min} = 2\arcsin\left(n\sin\tfrac{1}{2}\alpha\right) - \alpha,$$

from which the refractive index may be obtained, by experiment, as

$$n = \frac{\sin\tfrac{1}{2}(\Delta_{\min} + \alpha)}{\sin\tfrac{1}{2}\alpha}.$$

The effect of a prism on heterogeneous light may be deduced from the formulae for refractive dispersion. (See Amici Prism, Nicol Prism, Binocular, Monochromatic Illuminator, Total Reflection.) (L.D.W.)

PRISMOIDAL FORMULA.
Earthwork.

PRIVATE EXCHANGE.
PAX and PBX.

PROBABILITY.
The probability that, among several equally likely events, a given event will happen is the ratio of the number of favorable cases to the total number of cases; the probability that this event will fail is the ratio of the number of unfavorable cases to the total number of cases.

It is important to emphasize the assumption in this definition that the events are equally likely.

If the number of favorable cases is a and the number of unfavorable cases is b, the probability that the event will happen is $p = \dfrac{a}{a+b}$, and the probability of failure is $\dfrac{b}{a+b}$.

The probability of certainty is 1 and the probability of impossibility is 0.

In many important cases it is not possible to determine the exact probability of the happening of an event by enumerating all the possible equally likely ways in which an event can happen or fail. This is the case, for instance, in the probabilities involved in life insurance and fire insurance. But it is often possible to make a practical determination of the probability by statistical data based on experience and observation over a large number of cases. If it is found by observation that an event has happened m times out of a total of n cases, we may take m/n as an approximation to the probability of the event, if n is large.

The events of a set are said to be mutually independent or dependent according as the occurrence of one of them does not or does affect the probability of occurrence of others of the set.

The probability that two independent events will both happen is equal to the product of their separate probabilities: $p = p_1 \cdot p_2$. This also holds for three or more independent events.

For dependent events, if the probability of a first event is p_1, and if, after this has happened, the probability of a second event is p_2, then the probability that both events will happen in the order stated is: $p = p_1 \cdot p_2$.

Two events are said to be mutually exclusive if the happening of one excludes the happening of the other.

The probability that one or other of a set of mutually exclusive events will happen is the sum of the probabilities of happening for the separate events: $p = p_1 + p_2 + \cdots$.

The probabilities of happening of the separate events of a set of mutually exclusive events are called the partial probabilities, and the probability that one or other of the events will occur is called the total probability. The total probability is therefore the sum of the partial probabilities.

If p is the probability that an event will happen in any single trial, and $q = 1 - p$ is the probability that the event will fail in any single trial, then the probability that this event will happen exactly r times in n trials is $_nC_r p^r q^{n-r}$; and the probability that it will happen at least r times in n trials is

$$p^n + {}_nC_1 p^{n-1}q + {}_nC_2 p^{n-2}q^2 + \cdots + {}_nC_r p^r q^{n-r}.$$

(L.L.S.)

PROBABILITY CURVE.
In the mathematical theory of probability, it is shown that the probability of committing an error of magnitude x is given by the ordinate of the curve whose equation is

$$y = \frac{h}{\sqrt{\pi}}e^{-h^2 x^2},$$

where h measures the accuracy of the observer. This curve is called the probability curve and is shown in the accompanying figure. (L.L.S.)

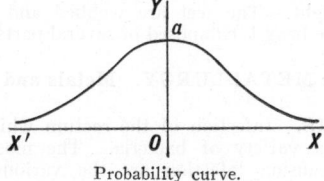

Probability curve.

PROBABLE ERROR.
Least Squares.

PROBABILITY FUNCTION.
A probability function is always positive or zero. In the case of a discrete variable the probability function directly gives the probability that x will occur. An example is the Bernoulli probability function, $P_x = {}_sC_x p^x p^{s-x}$. For a continuous variable probability is given by an element

of area $P_x dx$, and this is generally called the probability function. The ordinates of the probability function are given by P_x. An example is the **normal probability function**

$$P_x dx = \frac{1}{\sigma_x \sqrt{2\pi}} e^{-\frac{(x-m)^2}{2\sigma_x^2}} dx$$

whose ordinates are given by

$$P_x = \frac{1}{\sigma_x \sqrt{2\pi}} e^{-\frac{(x-m)^2}{2\sigma_x^2}}.$$

(L.A.A.)

PROBASIPODITE. A segment of the **biramous appendage** of **crustaceans**. It is the proximal part of the basipodite where this segment is subdivided. In such appendages it usually bears the exopodite. (A.W.L.)

PROBOSCIDEA. The **elephants**. An order of **mammals** including only two species now living. (A.W.L.)

PROBOSCIS. A protruding or protrusible organ associated with the mouth and therefore at the front of the head in most animals. It is used in feeding and in some cases for other purposes.

Among the more primitive animals, such as the flatworms (**Platyhelminthes**) and **annelids**, a portion of the alimentary tract can be everted through the mouth, and is called a proboscis. It may ingest food merely by muscular action but in the annelids it may also be armed with sharp teeth for grasping prey. The proboscis of the flatworm appears near the middle of the lower surface of the body. The **nemertine** worms also have a proboscis but in the phylum it arises independently, although it is associated with the mouth in the adult. Many **arthropods** have an elongate structure which may be termed a proboscis, although there is little uniformity in the different groups. A familiar example is the long sucking tube of the butterflies and moths, formed of the maxillae.

In the **vertebrates** the proboscis consists of the elongated nose and upper lip, and is developed only in the elephants to a conspicuous degree. It serves the animal in respiration, since it is traversed by the nasal passages, and is a delicate and powerful grasping structure. It grasps larger objects by **prehension** and smaller ones by means of small lobes at the tip. (A.W.L.)

PROBOSCIS MONKEY. Mammalia, Primates. A moderately large **monkey** of Borneo, *Nasalis larvatus*, characterized by the long, fleshy, and somewhat drooping nose. This part is largest in adult males and is relatively small and upturned in the young. The nostrils open near each other on the lower surface. The species is closely related to the **langurs**. (A.W.L.)

PROCELLARIIFORMES. The **albatrosses** and **petrels**. An order of marine birds known for their powerful flight. The feet are webbed and the horny sheath of the beak is composed of several parts. (A.W.L.)

PROCESS METALLURGY. Metals and Alloys.

PROCTITIS. Infection of the rectum which may be caused by a variety of bacteria. The most frequent organisms causing infection are the various kinds of **Streptococci, Staphylococci, Gonococci**, etc. Proctitis is characterized by pain and spasm of the rectum. Constitutional symptoms are not usually marked. (R.S.M.)

PROCTODAEUM. A hollow structure formed by the invagination of the outer layer of the body (ectoderm). It associates with the gut of the **embryo** to form the caudal end of the alimentary tract. In the **insects** the proctodaeum is relatively long, forming the entire hind intestine, which consists of the small and large intestines and the rectum. The proctodaeum of **vertebrates** is less extensive, forming only the **anus**. (A.W.L.)

PROCTOLOGY. That division of surgery which deals with the diagnosis and treatment of diseases of the rectum. (R.S.M.)

PROCYON. Procyon (α **Canis Minoris**) is a brilliant star which receives its name from the fact that it precedes the star **Sirius** in its nightly journey across the sky. These two "dog stars" are referred to in the most ancient literatures and were objects of veneration and worship both by the Babylonians and the Egyptians. Astrologically, the star portended wealth, fame, and good fortune.

Procyon, like the other "dog star" Sirius, has a faint companion which is suspected to be a **white dwarf**. (W.K.G.)

PRODUCER'S RISK. The producer's risk is the **probability** that a lot manufactured by the producer should be rejected by a consumer on the basis of sampling when in reality the lot should be accepted by the consumer. (L.A.A.)

PRODUCT FORMULAE. Multiplication.

PROEPIPODITE. A process of the **biramous appendage** of **crustaceans** borne by the pleuropodite. An epipodite. (A.W.L.)

PROFILE. Differential.

PROFILE LEVELING. Differential.

PROFILER. Milling.

PROGESTERONE. Sex Hormones.

PROGLOTTID. A reproductive segment of a tapeworm (**Cestoda**). These segments are budded from the head or **scolex** in a long series which makes up the ribbonlike body of the worm. Each proglottid contains reproductive organs of both sexes and is traversed by the longitudinal nerve cords and the excretory canals. As they mature, the proglottids break away from the caudal end of the worm and pass out of the body of the host with the feces. (A.W.L.)

PROGNATHISM. Protrusion of the lower jaw. Some animals are normally prognathous but the term is usually applied only to abnormal prominence of the jaw in man. Some breeds of bulldogs are conspicuously prognathous; here the term undershot is used with the same meaning. (A.W.L.)

PROGRESSIONS. This term covers the topics of **arithmetic progression, geometric progression**, and **harmonic progression**. (L.L.S.)

PROJECTED AREA. In a bearing or a rivet, the projected area is equal to the product of the diameter of the element and the length along which it is in contact with some other element. (H.C.H.)

PROJECTION, OPTICAL. By far the most common form of projecting apparatus is the stereopticon or magic lantern. This instrument consists essentially of a concentrated source of light L, as an arc, divergent rays from which are caught by a condensing-lens system C (see figure) and made to illuminate the transparent slide S, which thus becomes a secondary source, like the slit of a spectroscope. The objective O, a system of achromatic lenses, forms an enlarged, inverted image of S on a distant white wall at W. For the image to appear right side up, the slide itself must, of course, be inverted. The ratio of enlargement, or magnifying power, M, depends in a simple manner upon the dis-

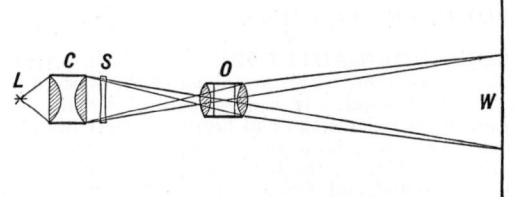

Diagram of stereopticon lens system.

tance D from the slide to the screen and the focal length f of the objective: $M = (D - f)/f$.

In order to exhibit an opaque object, like a postal card or the page of a book, it must be placed at S and illuminated by very strong light focused upon it obliquely from one side by means of lenses or mirrors. Any such picture will appear reversed, right to left, on the screen unless a more complicated optical system is used.

A motion-picture projector differs in no essential way from the stereopticon. The "slides" are, of course, films, and much smaller than standard slides; and the machine is provided with a mechanism for abruptly shifting from one picture to the next on the film while the shutter is closed, and for holding the film stationary while it is open. (L.D.W.)

PROJECTION PRINTING. Projection printing is normally employed to make photographic prints (1) larger than the negative, (2) smaller than the negative, or (3) for correcting the distortion in negatives.

In projection printing, the exposing light passes through the negative and lens and is brought to a focus on a sheet of photographic projection paper placed on the easel. The size of the image formed depends on the focal length of the lens and the relation between the distances from negative to lens and lens to paper. If the latter is greater than the former, the projected image will be larger than that of the negative and the print is an enlargement. If the reverse is true, the projected image is smaller and the print is a reduction.

The projection printer, or enlarger, consists of a light source mounted in a light tight housing, a reflector, set of condensing lenses or diffusion plate, negative carrier, bellows or sliding tube, projection lens, diaphragm, focusing control, support, easel for holding paper, and a light switch.

The illumination may be of various types: tungsten lamp, opal glass enlarger lamp, photoflood lamp, fluorescent tube, or an electric arc. The latter is used in enlargers for large-scale projection, as photomurals or template processes.

Lenses used for projection should be corrected for spherical and chromatic aberration and have an anastigmatic field. All corrections should be made for a plane close to the lens. The better types of enlargers are equipped with projection anastigmats specially designed for projection printing. Some of these are coated to prevent internal reflection. The focal length of the lens is usually equal to, or greater than, the diagonal of the largest negative that the enlarger will accommodate.

Focusing is accomplished either manually—with a screw, which is often equipped with a coarse and fine adjustment, or automatically by means of a system of levers, or rollers, that move along a guide as the projector is raised or lowered.

Projection printing has become very popular, due to the ease with which print sizes can be changed, negatives can be "blown up" so uninteresting areas of negative can be eliminated from print, and because of the

greater convenience for printing-in or shading during exposure. (S.M.T.)

PROJECTIONS. If M_1 is the foot of the perpendicular from P_1 to a line L, then M_1 is called the projection of P_1 on L.

If M_1 and M_2 are the projections of two points P_1 and P_2 on a line L respectively, then the line-segment M_1M_2 is called the projection of the line-segment P_1P_2 on L.

Similarly, the projections of a point and of a line segment upon a plane are defined by drawing perpendiculars to the plane.

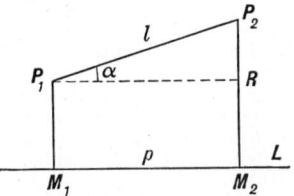

The length of the projection of a line-segment l upon a line L is equal to the length l of the given segment multiplied by the cosine of the angle between the lines: $p = l \cos \alpha$.

The projection of any area upon a plane is equal to the original area multiplied by the cosine of the angle between the given area and the plane.

A fundamental projection theorem is: the projection of a broken line is equal to the sum of the projections of the parts. (L.L.S.)

PROJECTIVE GEOMETRY. Projective geometry may be described as dealing primarily with the properties of figures which are unaltered by central projection. By central projection we mean that, from a fixed point or center of projection, straight lines are drawn through the various points of a given figure, and then a section is taken of these projecting rays by a fixed line or plane, giving a new figure. (L.L.S.)

PROJECTORS, MOTION PICTURE. The projector is similar to the motion picture camera except that it is provided with a light source and condensing lenses for illuminating the film, the illuminated picture being formed on a screen by a projection lens. The intermittent film movement may be obtained either by a pull-down claw of the type used on the camera or the so-called Maltese-cross movement. This is shown in Fig. 1. The cross, which is shown shaded, is attached directly to the sprocket which engages the perforations

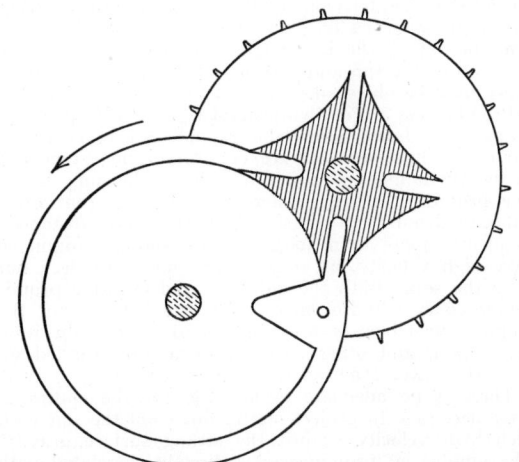

Fig. 1. Maltese-cross movement for intermittent film motion.

along the edges of the film and serves to move the film downwards. The disk with the pin shown entering the cross moves continuously. The pin entering one of the four slots in the cross moves the cross through a 90°-angle before leaving it. The pin does not enter the cross again until it has made a complete revolution. Thus the continuous movement of the disk is converted into intermittent movement of the sprocket and the film is pulled down one picture at a time just as is the case with the claw pull-down used on the camera. (See also **Camera, Motion Picture.**) (C.B.N.)

PROLAMINES. Aminoacids, Polypeptides, and Proteins.

PROLAPSE. The falling or protrusion of an organ or structure due to lack of support, usually secondary to weakness of the ligaments or surrounding muscles. Prolapse of the rectum is a protrusion of a part of the rectal wall externally. Prolapse of the **uterus** is a protrusion of the uterus through the **vagina**, and inverting the vagina so that in an extreme case the uterus hangs between the legs. (R.S.M.)

PROLEG. A fleshy appendage developed in the larvae of certain insects to support the hinder part of the wormlike body. The prolegs bear numerous small hooklets called crochets which grip the supporting surface. Whether these appendages are derived from jointed appendages of the abdomen or not is uncertain; there are evidences to show that such is the case.

The larvae of many sawflies (Hymenoptera) have 6–8 pairs of prolegs and those of most butterflies and moths have five pairs. (A.W.L.)

PROLINE. Aminoacids, Polypeptides, and Proteins.

PROMINENCES. Prominences are regions of the solar **chromosphere** which, for some unexplained reason, extend out to a considerable height from the normal upper surface of this region of the solar atmosphere. In common with the chromosphere the prominences usually appear with a brilliant scarlet color. Up to the middle of the 19th century prominences could be observed only during a total **eclipse** of the sun. In 1868 Lockyer and Janssen discovered a method for observing these interesting phenomena at any time. A high dispersion **spectroscope** will spread out, and hence greatly dilute, the continuous **spectrum** of the sun. The prominences, however, shine by means of hot hydrogen and calcium, elements which have strong isolated bright lines which will appear strongly against the dispersed sunlight. By setting the view telescope of the spectroscope to observe any of these strong lines, usually the red line of hydrogen, the slit of the instrument may be set tangent to the limb of the sun and widened until the entire prominence may be observed.

Prominences are of two general types. The quiescent prominences bear considerable resemblance to clouds in our own atmosphere but are very large and extend out frequently as much as 50,000 miles from the sun. Eruptive prominences are smaller than the quiescent ones, but extend out from the sun to much greater distances, frequently more than 500,000 miles, and are found to have high velocity, as great as 200 miles per sec., out from the sun. Both the quiescent and eruptive prominences contain hydrogen, calcium, and helium, but the eruptive prominences are also found to contain iron, magnesium, and other elements apparently carried up from the lower atmosphere of the sun.

There is no adequate explanation for the enormous force necessary to project matter out from the sun with such high velocity against the strong surface gravity. The number of prominences is directly correlated with the **number** of **sun spots**. The quiescent prominences

may appear at any portion of the sun's disk, but the eruptive type are limited to the latitude zones in which the sun spots are found, and frequently rise from the vicinity of active spots. (W.K.G.)

PROMOTER. Catalysis.

PRONG-HORN ANTELOPE. Mammalia, Artiodactyla. *Antilocapra.* A plains animal of the western half of North America. It is distinguished by the erect horns, hooked at the tip, and bearing a short branch in front. The prong-horn differs from all true **antelopes** in the branching of the horn sheaths and in the periodical shedding and renewal of these sheaths. It is, however, much like the antelopes in appearance. Also called the prongbuck. (A.W.L.)

PRONUBA MOTH, AND YUCCA. Pollination.

PRONY BRAKE. Mechanical horsepower is often conveniently determined by measuring the reaction on a **brake** which bears against a **pulley** or brake drum in such a way as to create a **friction** drag which converts the mechanical **energy** of the rotating shaft into **heat.** Engines and motors are often tested with the use of the Prony brake. The brake which is shown diagrammatically in the figure consists of the brake drum or wheel mounted on the shaft and surrounded by a band of

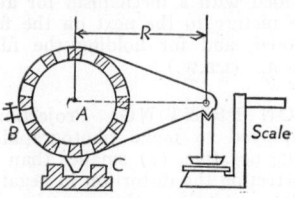

Prony brake.

steel, belting, or some other flexible material, to which are fastened, internally, blocks of wood. A band-tightening mechanism allows the operator to adjust the pressure between the brake blocks and the drum. In this way the frictional drag and the load upon the engine can be varied. To the brake band is rigidly attached an arm having a radius R by means of which the band is kept from rotating. The torque effect of the friction drag is found by measuring the force necessary to keep the brake arm from moving. This scale reading in pounds, multiplied by R, is equal to the pound-feet of **torque** set up as a frictional drag at the surface of the drum. The torque, when multiplied by the angular speed, radians per minute, is the power delivered in foot-pounds per minute. The brake horsepower, then, equals

$$\frac{2\pi RWN}{33,000}.$$

N is the rotative speed of the drum, revolutions per minute, W is the net weight on the scales, net weight meaning the gross recorded weight less the weight necessary to support the dead weight of the unbalanced brake arm. This deduction is known as the tare. The tare could be found by balancing the brake on a knife edge to locate its center of gravity, and then applying the principle of **moments** to determine the reaction at the end of the brake arm. If it is possible to rotate the brake drum slowly backwards, the brake band should be loosened, and the drum rotated slowly forward, then backward. If during this time the weight on the scales is w_1 for forward rotation, and w_2 for backward rotation, the tare is $\frac{1}{2}(w_1 + w_2)$.

The mechanical energy must be absorbed by the brake itself, as this is an absorption type **dynamometer.** For every horsepower 42.4 B.T.U. are generated by friction each minute. The large flywheels of engines have so much windage or fan action that this heat can usually be dissipated to the atmosphere. Small brake drums suitable for motors must be water-cooled to prevent their overheating. To accomplish cooling of this nature, the drum should be provided with internal flanges, and a layer of water will be held against it by centrifugal force. Water can be admitted from a stationary supply tube and removed by a scoop which clears the circumference of the rim. (F.T.M.)

PROOF STRESS. Tension Test.

PROPAGATION; VEGETATIVE REPRODUCTION. It is often very convenient and even essential that plants be propagated by other means than by seeds. One of the principal reasons for this is to perpetuate desirable individual plants, another to increase the stock rapidly. One way of accomplishing this is by **grafting.** Another is by vegetative propagation, by means of cuttings or slips. Various parts of the plant may be used, advantage being taken of the ability of the part used to form **adventitious roots or buds.** Individuals capable of propagation in this manner are patentable in the United States.

A few plants can be propagated by root cuttings. These sections or pieces of the root form buds and also put out roots. The sweet potato is mainly propagated by this method. Roses are also sometimes increased by root cuttings. In many plants the formation of adventitious buds on the roots is advanced by exposure to light. If the roots of a poplar tree are exposed, in a short time adventitious buds grow and form erect stems. Cutting the root which connects this plant with the parent causes the latter to become independent. Such adventitious buds are freely formed on the roots of many thistles. Since the latter become troublesome weeds the habit is one of annoyance, but not of importance to man in this case.

Vegetative propagation is also obtained by means of stem cuttings, often called slips. The stem of the plant to be propagated is cut into sections, each section having several nodes. Usually all or many of the leaves are removed or clipped to reduce transpiration. The cuttings are then placed erect and half buried in moist sand. Pretreatment with **auxin** insures root production in difficult species. In a short time the **parenchyma** cells just back of the cut surface begin to divide rapidly, forming a loose spongy mass of cells, called a callus, which completely covers the wounded end. From this callus tissue, adventitious roots are formed and in some cases adventitious buds also. Not every plant can be rooted with equal facility by this means. Some, like the common *Pelargonium,* usually called geranium, and *Coleus,* root very quickly and easily. The cuttings may be rooted as easily by sticking in water as by burying in sand. Others can be propagated by stem cuttings only when the proper condition obtains. In some plants only young actively growing stems can be used; in others mature hardened wood is necessary. Often such cuttings remain for months before any roots are formed. It has been said that if proper conditions are obtained any plant may be increased by stem cuttings.

Modified stems, such as **rhizomes** and tubers, usually propagate much more readily. The white potato is propagated by cuttings of the ripened tubers, or potatoes. **Bananas** are increased by planting sections of the rhizome. The ubiquitous pest, witch grass, *Agropyron repens,* propagates itself readily by this method. Cultivation by hoe or cultivator breaks the creeping rhizome into sections, each of which becomes a new plant, which seems to grow with increased vigor.

Some plants are propagated by their leaves. Many species of *Sedum,* common rock garden plants, may be thus propagated, the fleshy leaves being removed from the plant and pushed part way down into moist soil. Adventitious roots and an adventitious bud form from the base of the leaf. Some begonias may be readily reproduced by their leaves. The latter are placed flat on the surface of moist sand, partially covered with sand

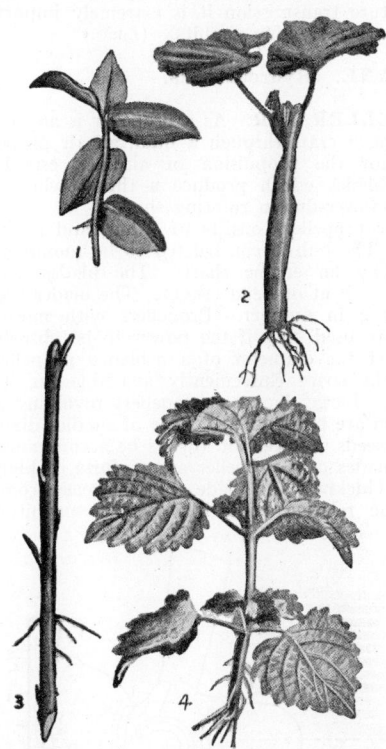

Adventitious roots. 1, growing from node of Wandering Jew; 2, growing from cut end of geranium; 3, growing from lenticels of willow; 4, growing from internode of *Coleus.*

and left undisturbed. Tiny plants soon form on the upper surface of the leaf, along the veins. Cutting across the main veins often helps to cause formation of new plants. *Sanseveria zeylanica,* the common bow-string hemp plant, can be readily propagated by leaf cuttings. The long linear tough leaves are cut into 3- or 4-in. sections and treated like stem cuttings. Species of *Bryophyllum* and the related *Kalanchoë* naturally reproduce by means of their leaves, tiny plants forming in great abundance in the notches or at the apex of the leaves. (R.M.W., B.S.M.)

PROPAGATION CONSTANT. This is a transmission characteristic of a **line** and indicates the effect of the line on the wave being transmitted along the line. It is a complex quantity having a real term, the attenuation constant, and an imaginary term, the wavelength or phase constant. (See **Vector** and associated topics.) The attenuation constant is a measure of the loss in signal strength as the wave travels along a matched line while the wavelength constant is a measure of the phase shift which it undergoes. These relations may be expressed by the following equations:

$$r = a + jB \quad E_r = E_s e^{-(a+jB)}$$

$$|E_r| = |E_s|\,e^{-al}$$

$$|I_r| = |I_s|\,e^{-al}$$

$$\text{Phase displacement} = Bl$$

where r is the propagation constant, a the attenuation constant, B the wavelength constant, l the line length, E_r the received voltage, E_s the sending end voltage and I the corresponding currents. Because of the attenuation on lines used for communication purposes it is necessary to insert **amplifiers** or repeaters at intervals to build the signal back to suitable levels. For sound work the phase shift is usually not important but for **television** and picture transmission it is extremely important and necessitates correcting circuits. (L.R.Q.)

PROPANE. Hydrocarbons.

PROPELLER, AIR. As a propeller is any device for propelling a craft through a fluid, an air propeller is a device for the propulsion of aircraft, especially one having blades which produce a thrust when mounted upon a power-driven rotating shaft.

An air propeller consists of a hub and one or more blades. The hub is constructed to be mounted on and rotated by an engine shaft. The blades are affixed securely to it at one end (root). The blades are usually 2, 3, or 4 in number. Propellers with more than 2 blades are used only if the power to be absorbed is so large that the diameter of a 2-bladed propeller would exceed the space conveniently available for it on the airplane. Large diameter propellers revolving at 1800–2500 rpm are preferable to those of smaller diameters at higher speeds, but cannot always be accommodated.

The blades of a propeller are **airfoils** of high **aspect ratio.** Thickness and blade angle decrease from root to tip. The relative wind increases in magnitude from

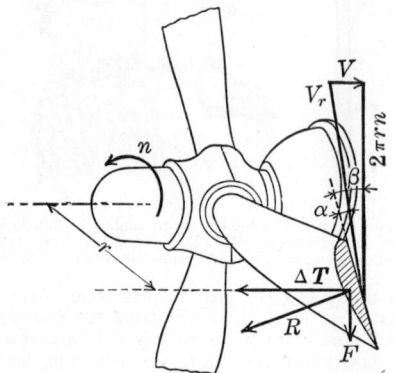

Fig. 1. Aerodynamic action on an element of a propeller blade at radius R.

root to tip because of the increase of tangential component. The twist built into the blade is designed to keep the **angle of attack** at or near the optimum (maximum efficiency) value over the whole blade. The chord of the airfoil usually decreases toward the tip from the mid-blade station. The necessary transition to a thick structural section at the root modifies the inner third of the propeller blade away from an effective airfoil shape.

Propellers may be classified as: 1. Fixed-pitch. 2. Adjustable-pitch. 3. Controllable-pitch. 4. Automatic-pitch. The latter two are described elsewhere. The 2-bladed fixed-pitch propeller has blades and central *boss* in one piece which is bolted between the flanges of a simple steel hub. Since no adjustment of blade angle is possible, this type of propeller must be correctly selected for the average flight condition to be experienced and its performance will be inferior under all other conditions. An adjustable-pitch propeller has separate blades and can therefore be adjusted to new average conditions, but as it is adjustable only at rest, its advantages over the fixed-pitch propeller are limited. It could be adjusted to low pitch to accelerate the take-off, or to high pitch to limit engine speed, but not for both on the same flight. For this feature the more complicated controllable or automatic-speed propellers are necessary.

The action by which thrust and engine counter-torque are produced stems from the aerodynamic pres-

sures on the blades. The resultant of these is resolvable into thrust and torque components. The velocities involved on a small element of the blade at radius

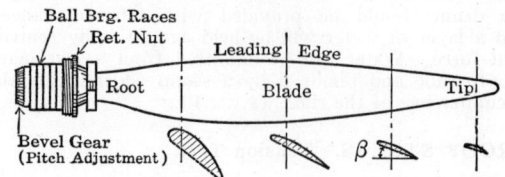

Fig. 2. Typical aluminum alloy propeller blade (with adjustable pitch shank).

r from the center are (1) tangential component $2\pi rn$, (2) airplane forward speed (including any inflow) V. These create a relative wind V_r at α angle of attack. The **propeller blade angle** β will be chosen to make efficiency maximum, so it is near the point for maximum $\dfrac{\text{Lift}}{\text{Drag}}$ of the profile, and much smaller than the angle for maximum lift.

The resultant, R, of the aerodynamic force on the blade element has a thrust component ΔT and a torque component F. The thrust, T, of the whole propeller is the summation of ΔT for all blade elements of which the engine counter torque, Q, is ΣFr. Necessary engine power $= 2\pi nQ$ and thrust power $= VT$. Propeller efficiency is the ratio of thrust power to engine power. Notice that during the first part of a take-off run where V is very small compared to $2\pi rn$, α virtually equals β. Since β at ¾ radius is usually above the stalling angle of attack of a conventional airfoil profile, the thrust is weak until V has increased sufficiently to incline V_r to an α of maximum lift. Then the thrust increases rapidly and a noticeable improvement in take-off acceleration is had. This suggests one of the advantages of being able to control the angle β while propeller is in motion.

A fixed-pitch propeller will be subject to variation of efficiency with air speed. Also, propellers of a geometrically similar family, which would absorb the same engine power at rated rpm, will have different efficiencies, depending on the blade angle. The **propeller coefficient,** known as the speed-power coefficient, is a convenient variable against which to plot efficiency of geometrically similar propellers. The characteristic plot of a group of 3 propellers is shown in Fig. 3 to illustrate why a fixed-pitch propeller of maximum efficiency

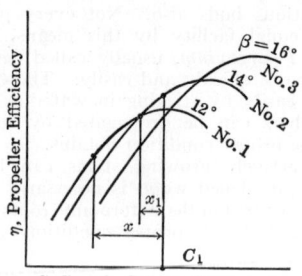

Fig. 3. Speed-power coefficient.

is not the best one (except possibly for a racer). Assume that rated air speed, power, and rpm correspond to a coefficient of C, and that x is the expected operating range of the engine. Propeller No. 2 has the highest efficiency at C, but excels No. 1 only over the partial range x_1. Propeller No. 1 of smaller blade angle is superior in performance over most of the operating range. This aspect of propeller efficiency is confined to the fixed-pitch field as there is no comparable situation if the propeller has controllable or automatic blade angle adjustment.

The efficiency of a propeller at the same *effective pitch* (i.e., V/nd, d being the diameter) is unaffected by changes in altitude since the ratio of $\dfrac{\text{Thrust power}}{\text{Counter-torque power}}$ involves air density in both numerator and denominator. An unsupercharged engine with fixed pitch propeller will not operate at constant V/nd when altitude varies. Lesser air reactions at higher altitudes have a tendency to let n increase, but losses of **volumetric efficiency** of the engine more than overcome this tendency and the net result is a decrease

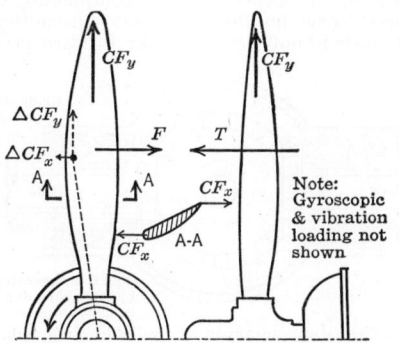

Fig. 4.   Structure loads: T, thrust; F, torque force; CF, centrifugal force.

of n. Supercharged engines with manual or automatic adjustment of blade angle can be operated at rated speed at any altitude and do not lose power until the critical altitude (rated manifold pressure with open throttle) is reached.

Propeller blades are made of laminated wood, aluminum alloy, and steel. The fixed-pitch wooden propeller used to propel almost all light airplanes, as well as many of heavier category, is usually of one-piece construction, being carved from a glued, laminated blank of birch and/or walnut. Except for a regrettable tendency to unbalanced warping, from long exposure to excessive moisture, this material is excellent for low-power propellers. The propellers are inexpensive, and not subject to failure from fatigue cracks. They have a small moment of angular inertia, hence the engine can be "revved" rapidly in emergencies such as an over-shoot of the landing area. The blades are reasonably thin, giving good efficiency, but this virtue diminishes rapidly as horsepower and blade diameter increase.

Aluminum alloy propellers are machined from forgings, then heat treated. Only the blades are of this material, the hubs being steel. The blades are shaped with cylindrical flanged shanks that are clamped into the hub on adjustable propellers and held by ball-bearing races on controllable and automatic types.

Steel propeller blades have been made both solid and hollow. The hollow types must be used for stiffness of the thick section they permit if the propeller is a large one.

Loading forces cause tension, cantilever bending, and torsion. These come from transverse air reaction and radial centrifugal force. As the mass of the blade is not concentrated on the blade center line (although the center of gravity usually is), centrifugal action on the leading and trailing edge mass has transverse components creating torsion. Although not large, these forces, together with aerodynamic moments, if any, need counterbalancing in variable blade angle propellers, otherwise they might impress a continuous load on the blade-setting mechanism. Centrifugal force tends to relieve bending stress on the forward side of the blade and increase it on the rear side (face). A slight rake or tilt ($\tfrac{1}{2}°$) forward of the blade center line causes centrifugal force to produce a moment which alleviates somewhat the transverse air reaction moment.

This effect is amplified by the forward deflection of the blade under the transverse bending moment of the air reaction (thrust).

Summarizing desirable propeller attributes, some of which are conflicting in nature, they are:

1. High aerodynamic efficiency, resulting in converting the maximum practical amount of engine output into propulsive energy.
2. Uniformly high performance over the operating speed range.
3. Small diameter for ground clearance.
4. Low tip speeds for quietness of operation.
5. Propeller weight and angular moment of inertia as low as possible.
6. Resistance to sudden failure, like extended fatigue cracks.
7. Resistance to abrasion.
8. Resistance to moisture absorption and resultant change of shape.
9. Low cost, consistent with reliability. (F.T.M.)

PROPELLER, AIR, CHARACTERISTICS. Except in the field of propeller design, it is not expedient to derive air propeller operating characteristics from basic aerodynamics. A number of special characteristics, whose nature is that of coefficients or parameters, have been found to be of use in judging the over-all performance of propellers. Frequently, propeller data appear to have a random character until plotted against one of these coefficients, whereupon it assumes an orderly character, displayable as a family of curves. Commonly used characteristics include:

Effective pitch ratio, V/nd.
Power coefficient, $P/\rho n^3 d^5$.
Thrust coefficient, $T/\rho n^2 d^4$.
Speed-power coefficient, $V\left(\dfrac{\rho}{Pn^2}\right)^{\!\frac{1}{5}}$.

V = Air speed.
n = Revolutions per sec.
d = Propeller diameter.
P = Engine brake horsepower.
ρ = Mass density of air.
T = Propeller thrust.   (F.T.M.)

PROPELLER, AIR, PITCH OF. *Effective Pitch.* The distance an aircraft advances along its flight path for one revolution of the propeller.

Geometrical Pitch. The distance an element of a propeller would advance in one revolution if it were moving along a helix having an angle equal to its blade angle.

Pitch Ratio. The ratio of geometrical pitch to diameter.

Effective Pitch Ratio. The ratio of effective pitch to diameter, i.e., V/nd.   (F.T.M.)

PROPELLER, AUTOMATIC. Some of the characteristics of an air propeller may be controlled under variable operating conditions by a suitable adjustment of the blade pitch. If manually accomplished by remote control, the propeller is said to be of "controllable pitch." If the control originates in an automatic governor, the propeller is an "automatic propeller."

Generally the characteristic to be controlled is the rotative speed—the "rpm." Engine speed should be held constant at rated rpm for maximum power delivery; also, there are other advantages of automatic control for constant speed. The automatic propeller is built as a *constant-speed propeller* with a centrifugal governor for automatically maintaining constant engine speed through appropriate changes in blade angle. The automatic propeller is therefore a *controllable-pitch propeller* with a speed-responsive governor. Various constant speeds in the range from maximum take-off rpm to minimum cruising rpm may be impressed on the governor by a governor spring-loading control, available to the pilot

Various electrical, hydraulic, and mechanical operating mechanisms have been devised to fit compactly into the hub of the propeller, but all employ the centrifugal governor for primary control. Fig. 1 shows a diagram of the mechanism of an hydraulically operated blade

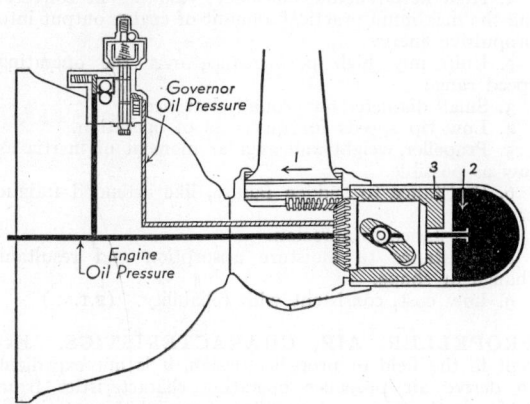

Fig. 1. Diagram of the "Hydromatic" constant speed propeller. When speed decreases the governor acts to drain the cylinder inboard, causing piston (3) to move rearward under engine oil pressure (2), thus causing the cam mechanism to rotate the blade in the direction (1) for pitch decrease. (*Hamilton Standard Propellers.*)

adjustment. The governor, driven by the engine, boosts engine oil pressure and automatically meters oil to and from the propeller to maintain the proper blade angle. The forward section of the propeller hub is an hydraulic cylinder. The straight-line motion of the piston in this cylinder is, by means of a plate-and-roller cam, converted to rotation of a ring gear. This gear's motion will rotate all blades in their sockets, thus effecting a pitch change. For on-speed operation of the propeller the air forces twisting the blade and the net oil pressure acting on the piston leave the mechanism in equilibrium.

When the engine speed falls below the selected rpm, spring force overcomes centrifugal force of the governor weights. The new equilibrium position forces the oil relay valve to open the rear side of the piston to drain, causing a rearward movement of the piston and a decrease in blade pitch. Following this change, an easing of propeller torque allows the engine to pick up speed until desired speed is reached. In case of engine overspeed the flyweight centrifugal force exceeds spring force, with the result that the oil relay moves to admit boosted oil pressure to the rear side of the piston, moving it forward. (F.T.M.)

PROPELLER BLADE ANGLE. The acute angle between the chord of a section of a propeller and a plane perpendicular to the axis of rotation. As the blade angle of an air propeller varies from maximum at hub to minimum at tip it is necessary to localize a section in order to validate a numerical reference to this angle. In the United States, by custom the blade angle of a propeller is the angle at 75% of the radius to the blade tip. (F.T.M.)

PROPELLER, CLUB. A propeller (usually 4-bladed) so called from its short, inefficient, club-like blades. Also it is sometimes called a "test club" or a "test propeller." This is a non-flight propeller of purposely inefficient design. When mounted in place of the normal propeller, the test club provides sufficient torque load to enable power testing of aeronautical engines on fixed test stands. The abnormally small thrust created is of no consequence, but the club must throw enough air back over the engine to accomplish the cooling satisfactorily. (F.T.M.)

PROPELLER COEFFICIENTS. Propeller; Air.

PROPELLER, CONTROLLABLE-PITCH. This is an air propeller constructed to allow adjustment, by the pilot, of the blade angle during flight. This action is called changing the *pitch*. No matter what the airplane speed may be, there will always be a blade angle which will allow rated speed to be attained by the engine. Thus pitch changing allows maximum use of engine power, regardless of airplane speed. In this, it resembles the change speed transmission of an automobile. Usually propeller pitch is continuously variable between extreme positions, whereas automotive transmissions include only three or four forward gear ratios.

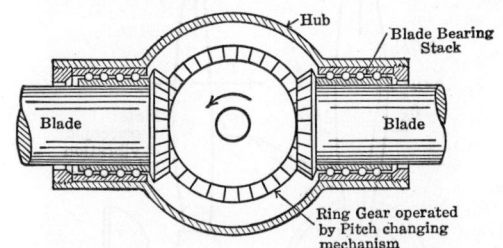

Schematic diagram of controllable pitch hub.

Controllable-pitch propellers have been constructed to operate on hydraulic, electric, and purely mechanical principles. A steel hub is fitted to the engine shaft, this hub having sockets machined to receive the ball- or roller-bearing races which support the shanks of the blades, holding them in proper position but leaving them free to rotate about the blade center line. A gear, fastened to each blade shank, is meshed with another operated by the change pitch mechanism. In an hydraulic mechanism, engine oil pressure is admitted to or released from an operating cylinder which is part of the propeller hub. Motion of the piston is used to effect the change of pitch. The electric propeller has a small reversible electric motor to accomplish the same object. Electric energy is taken from the aircraft electric system, and, after passing through control equipment, enters the hub through insulated slip rings.

Since the controllable propeller is operated mainly with the thought of maintaining some standard engine speed, the automatic propeller, which is a controllable propeller with speed-responsive governor instead of human control, is rapidly replacing it. The blades and the hub mechanism of **automatic propellers** differ but little from the controllable propellers. (F.T.M.)

PROPELLER, COUNTER-ROTATING. An aviation development of considerable interest to the high-power field is the counter-rotating propeller. This is a propulsive arrangement of two propellers for one engine, the two having concentric centers of rotation but being revolved in opposite directions. The planes of rotation are parallel and quite close together, there being about 15″ clear space between the blades of the two propellers. The rear propeller has a bearing on the hub of the other and is driven at the same speed, but oppositely in direction, by gearing in a nose section of the engine crankcase. The hubs are of the automatic-pitch type, the pitch being changed simultaneously and equally on the two propellers.

Counter-rotating propellers were created in response to the increasing power capacity of aviation engines. As these capacities grew first 3-, then 4-bladed propellers were introduced. After the multi-bladed propeller had been developed to the maximum practical diameter the counter-rotating propeller followed. In addition to being capable of absorbing the power of the largest engines developed, the counter-rotating principle possesses some definite aerodynamic advantages. The torque reaction is neutralized. There is no swing to be combatted at take-off. Air flow is smoother over the control surfaces, and lifting surfaces may develop their normal lift, even in the slip-stream of the propeller. (F.T.M.)

PROPELLER THRUST. The thrust of a propeller is the component of fluid reaction on a propeller which is parallel to the direction of advance. The following discussion will relate to the *air propeller*.

The thrust of an air propeller is the result of aerodynamic lift and drag forces on the airfoil-shaped propeller blades, the resultant of which is a thrust normal to the plane of rotation and a *couple* acting counter to the direction of rotation in the same plane.

The magnitude of thrust is related to engine power by "propeller efficiency," η, since

$$\text{Thrust} = \frac{\text{Engine Horsepower} \times 550 \times \eta}{\text{Air Speed, ft. per sec.}}.$$

During flight of an airplane the thrust can be exactly parallel to the relative wind at one air speed only. Since the variation from parallelism over the usual range of flight speed is only about 7–10°, the propeller reactions are generally analyzed as if the plane of rotation were always normal to the relative wind.

Thrust is normally positive, i.e., forwardly directed, but some recent propeller designs have incorporated blade adjustments permitting setting to negative angles of attack. The thrust would then become negative. Uses claimed for negative thrust are (1) to reduce the diving speed of dive bombers, (2) to permit more maneuverability of large flying boats when taxiing, (3) to reduce the length of landing run by decelerating more rapidly than brakes allow, or when wheel friction is abnormally small (icy runways). (F.T.M.)

PROPER ENERGY. Energy.

PROPER MOTION. The individual motion of a star relative to the other stars is known as the proper motion of the star. Up to the early part of the 18th century the belief was current that the stars were all fixed on a sphere commonly known as the **celestial sphere.** Since this sphere was apparently rotating about the earth and possessed other motions such as **precession** and **nutation,** all of the stars had certain motions in common. In 1718, Edmund Halley, while reducing his observations of the positions of the stars, noted that the positions which he obtained for **Sirius** and **Arcturus** differed in position relative to the other stars from the positions given by Ptolemy. Since the differences were greater than could be ascribed to errors in observation, Halley concluded that these two stars were actually not fixed on the celestial sphere but were in motion.

Since Halley's time, many stars have been observed to have proper motion and many long programs are at present under way to study these motions. The standard method of procedure is to compare positions of the stars at two epochs as widely separated as possible. Visual observations made with extreme precision with a **meridian** circle may be used for this purpose, but the photographic methods are far more fruitful since a large number of star positions may be obtained on a single plate. The longer the interval of time between the observations, the more accurate is the determination of the proper motion, and also the smaller the proper motion which can be detected.

Proper motion can only be determined in angular units, and the results are usually expressed in terms of seconds of arc per year. In case the **stellar parallax** of the star is known the velocity in linear units may be computed in accordance with methods discussed under **space velocity** of a star. The largest known proper motion is 10″.25 per year and was found by Barnard for a tenth **magnitude** star. Such a star would require about 200 years to change its position by an amount

equal to the apparent diameter of the moon. There are only about 50 stars known to have proper motions greater than 2″ per year and not more than 1000 with values greater than ½″ per year. Hence, we should not expect the **constellation** figures to have altered appreciably in the 5000 years since they were first described. (W.K.G.)

PROPLIOPITHECUS. Paleontology of Man.

PROPODITE. A segment of the **biramous append-age.** (A.W.L.)

PROPOLIS. A material gathered by **honey-bees** and used for closing crevices in the hive and for filling in sharp angles and attaching loose parts. It is also applied as a varnish to the combs and to the smooth surfaces in the hive. The substance is composed chiefly of resins gathered from plants and has an aromatic fragrance much like that of the leaf buds which furnish some of these resins. (A.W.L.)

PROPORTION. A proportion is a statement of equality between two **ratios.** It may be written in the form $a:b = c:d$ (formerly $a:b::c:d$), or perhaps better $a/b = c/d$.

In a proportion $a:b = c:d$, we call the first and fourth numbers a and d the extremes, and the second and third numbers b and c the means of the proportion. Properties of a proportion, as $a:b = c:d$, are expressed by the following:

1. In any proportion, the product of the means is equal to the product of the extremes: $ad = bc$.

2. In any proportion, the terms are in proportion by inversion, that is, $b:a = d:c$.

3. In any proportion, the terms are in proportion by alternation, that is, $a:c = b:d$.

4. In any proportion, the terms are in proportion by composition, that is, $a + b:b = c + d:d$.

5. In any proportion, the terms are in proportion by division, that is, $a - b:b = c - d:d$.

6. In any proportion, the terms are in proportion by composition and division, that is, $a + b:a - b = c + d: c - d$.

If $a:b = c:x$, then x is called a fourth proportional to a, b, and c.

If $a:b = b:x$, then x is called a third proportional to a and b.

If $a:x = x:b$, then x is called a mean proportional between a and b; this is the same as the **geometric mean** of a and b. (L.L.S.)

PROPORTIONAL LIMIT. The maximum unit stress which can be obtained in a structural material without causing a change in the ratio of the unit stress to the unit **deformation** is called the proportional limit. (See **Ultimate Strength, Modulus of Elasticity, Tension Test.**) (C.W.C.)

PROPORTIONAL REDUCERS. The principal value of proportional reducers is in the reduction of overdeveloped negatives.

Proportional reduction can be obtained with acid solutions of ferric ammonium sulfate (Krauss) and with a mixture of potassium permanganate and ammonium persulfate (Deck). (See also **Photographic Reduction.**) (C.B.N.)

PROPYLITIZATION. Processes, both magmatic and post-magmatic, by which **andesitic** rocks are altered through the action of water, **carbon dioxide,** and **sulfur.** (R.M.F.)

PROSAURIA. Reptilia.

PROSOBRANCHIATA. Synonym of Streptoneura. Gasteropoda. (A.W.L.)

PROSOMA. Cephalothorax.

PROSOPYLE. A minute pore connecting the inhalant and radial canals of sponges. **Porifera.** (A.W.L.)

PROSOSTOMATA. Trematoda.

PROSTATE GLAND. A gland associated with the ducts of the male reproductive system. Its secretion forms part of the seminal fluid.

The name is applied to a gland of some flatworms. The spermiducal glands of earthworms are also called the prostates, although they do not open into the sperm ducts. Some of the **cephalopod** mollusks also have a prostate associated with the sperm duct. In the mammals the gland is well developed, lying at the junction of the urinary bladder and the urethra.

In man, this gland is chestnut-shaped and measures $1\frac{1}{4}$ by $1\frac{1}{2}$". It is composed of glandular and fibrous tissue. The openings of the glandular ducts empty into the urethra. It is not known whether there is any **hormone** secretion by this gland. The function of the prostate is to secrete a fluid which forms a part of the seminal fluid. This gland is subject to infection (prostatitis), **hypertrophy**, and **cancer**.

Prostatitis may occur in an acute or chronic form. It is an exceedingly common disease which affects the adult male. It is usually seen as a complication of **gonorrhea**, but may at times be caused by other organisms such as the *Streptococcus, Staphylococcus,* and *tubercle bacillus.* In an acute infection, an **abscess** may develop. This requires surgical drainage.

The symptoms of prostatitis are many and may be divided into three groups. 1. Sexual. 2. Urinary irritation and obstruction of urinary flow. 3. Referred symptoms. These referred symptoms are usually those of pain, which is often severe, involving the hips, buttocks, legs, thighs, back, or groin. Acute gonorrheal **arthritis** may stem from a focus of infection in this gland.

Hypertrophy (enlargement) of the prostate is exceedingly common in men over 50 years of age. More than 20% of the men past middle life are said to suffer from some degree of prostatic enlargement. The chief symptoms resulting from hypertrophy are due to the obstruction to the outflow of urine. The usual chain of events is as follows: first there is frequency of urination during the night and in the daytime, progressive difficulty in urination, partial retention of urine, and when obstruction is complete, complete retention occurs. Pain referred to the urethra during urination is present in from 75–80% of cases. The most frequent complication of prostatic obstruction is infection which may be limited to the bladder, or may involve the kidneys and ureters as well.

At the present time, excellent results are obtained with surgery of the prostate gland in the treatment of benign hypertrophy, with an exceedingly low mortality. This is remarkable, considering the fact that many of these patients are aged, arteriosclerotic men, who would ordinarily be poor operative risks. The excellent results are due more to the advances made in pre-operative treatment than to the refinements of operative technique. Before a patient is ready for surgery, an attempt is made to relieve the obstruction, and to clear the urinary tract of infection with local installations or with sulfonamides orally. Attention to fluid balance and a good nutritional state are important.

There are two surgical approaches used for removal of the prostate. Before either of them is carried out, the urinary obstruction must be gradually relieved. This is accomplished by one of two methods, depending on the individual case. One method is to drain the bladder through a lower abdominal incision, so that the urine can be removed by suction for a period of time before doing the second stage operation. The other method relieves the obstruction by means of the indwelling catheter. Operation for the removal of the prostate is done at a later date, either (1) through the inside of the bladder by an abdominal incision, or (2) through the perineal route. Both methods give excellent results.

The prostate gland is a common site for the development of cancer. There may be no symptoms from the primary tumor, or the symptoms may be similar to those of benign hypertrophy. If diagnosed sufficiently early, radical surgical removal and the use of radium give the best results. Inoperable cases can only be treated by palliative measures.

In recent years, female **sex hormones** have been used in the treatment of both benign prostatic hypertrophy and cancer of the prostate. The results with benign hypertrophy have not been striking, but in cancer which is inoperable because it has progressed beyond the stage of a solitary tumor mass, diethyl stilbesterol (or similar compounds) causes marked improvement and prolongs useful life in many patients. The use of ovarian hormones is based on the knowledge that the male sex hormone exerts a stimulating effect on the growth of prostatic cancer. By inactivating this hormone with an **estrogen**, the growth of the **tumor** is slowed down, and the tendency to spread or metastasize is diminished. At the same time the patient improves in that his pain disappears and his appetite and weight increase. (A.W.L., R.S.M., D.M.H.)

PROSTATITIS. Prostate Gland.

PROSTOMIUM. A protuberance of the first segment of **annelid** worms, lying in front of the mouth. In the earthworms it is a simple rounded structure. In the **leeches** it enters into the formation of the anterior sucker. Its most remarkable form is the long proboscis, forked at the tip, which appears in the annelids of the genus *Bonellia* (Gephyrea). (A.W.L.)

PROTACTINIUM. Symbol: Pa. Atomic number: 91. Atomic weight: 231. Recognized by the International Committee on Atomic Weights in 1937. A radioactive element of the actinium series. (See **Radioactive Changes.**) (R.K.S.)

PROTAMINES. Aminoacids, Polypeptides, and Proteins.

PROTECTIVE COATINGS. Materials which are usually given protective coatings fall into three general classifications, wood, concrete and metal.

Wood may disintegrate from two causes, weathering and rotting. When unpainted or untreated wood is subjected to the weather it will eventually split and crack due to alternate swelling and shrinking occurring during periods of wet and dry weather. Rotting is caused by living organisms of which there are many types, molds, fungi, bacteria and such parasites as termites. Protection from weathering can usually be obtained by giving the wood a protective coating of paint. This also suffices to keep the wood from rotting, except when in contact with soil. Wood thus employed is usually impregnated with creosote, zinc chloride, or mercuric chloride to prevent rotting.

Most important of the artificial coatings is paint. (See **Pigments, Paints, Varnishes.**)

Although concrete is usually considered quite stable to weathering it will gradually disintegrate, therefore a protective coating can be used to advantage. Alternate wet and dry periods gradually swell concrete to the point where it may crack, and the formation of ice in the surface of concrete chips out pieces of it. One of the best ways to protect concrete against serious disintegration is to give it a coating of tar or asphalt. Concrete roofs, for example, are usually given several coatings of tar and asphalt impregnated felt to make them water-proof. Concrete walls can be made water-repellent by incorporating soap into the concrete mixture during pouring. The soap makes the concrete have a somewhat greasy surface. The greasiness keeps the water from penetrating into the pores of the concrete, thereby keeping the inside of the concrete dry even during a rain. Sodium silicate is also used for the same purpose.

Metals which cannot be given a natural protective oxide coating (see **Corrosion**) must be covered by an artificial coating. (See **Pipe Coatings.**)   (D.E.M.)

PROTEINS. Aminoacids, Polypeptides, and Proteins.

PROTEOMYXA. Sarcodina.

PROTEROZOIC (Algonkian). Next to the oldest of the five **eras** of the earth's history. Separated from the **Archeozoic** (the oldest Era) by a profound **unconformity**. The formations of the Proterozoic contain a preponderance of red sandstones and shales suggesting

Map showing the surface of pre-Cambrian (Archeozoic and Proterozoic) rocks in North America. Largest area shown by dotted pattern; smaller areas by solid black. (*Modified after Willis, U. S. Geological Survey.*)

increasing aridity. **Tillites** also prove that continental glaciers existed in Eastern Canada, Australia, Tasmania, Norway, South Africa and India. The only undoubted forms of life appear to have been low forms of marine plants called calcareous algae. The sedimentary formations of the Proterozoic, especially in North America, contain important ores of **iron** and **copper**. Length of time since the beginning of the Proterozoic, 1000 million years. (R.M.F.)

PROTHALLIUM. Ferns.

PROTOBRANCHIATA. Lamellibranchiata.

PROTOCEREBRUM. The first of the three principal parts of the **arthropod** brain. It is derived from the pair of ganglia in the first segment of the head and innervates the eyes, both simple and compound. In some **arthropods** a median archicerebrum lies in front of this pair of ganglia and the two components together make

up a procerebrum. The remaining two regions of the **brain** are the **deutocerebrum** and the **tritocerebrum**. (A.W.L.)

PROTOCILIATA. Ciliophora.

PROTOCLASTIC STRUCTURE. The type of structure which results from differential flow during fractional and granular differentiation of the crystals from the partly congealed **magma**. (R.M.F.)

PROTOGINE. An old and obsolete term proposed by Turine in 1806 for the central crystalline core of the Alps. Of historic significance because intimately connected with the origin of **granites, gneisses** and mountain building. (R.M.F.)

PROTOHIPPUS. Fossil Mammals.

PROTOMONADIDA. Mastigophora.

PROTON. The proton is most familiar as the nucleus of the **atom** of ordinary **hydrogen** (H^1); that is, as the ionized hydrogen atom. But there is ample evidence that protons exist also in the nuclei of other atoms. Until recently it was supposed that all atomic nuclei are built up of protons and **electrons**; now it appears that there may be other constituents (see **Neutron** and **Positron**), and the existence of electrons as such in the nucleus is in doubt. The mass of the proton is equal to the mass of the hydrogen atom minus the mass of the electron, that is, to about 1.672×10^{-24} gram. (See **Nuclear Particles.**) (L.D.W.)

PROTONEMA. Bryophytes.

PROTOPLASM. The material of which all living things are formed, usually a jelly-like or fluid substance of grayish translucent appearance.

Protoplasm is a complex mixture in which the elements, **carbon, oxygen, hydrogen** and **nitrogen** are always present. **Sulfur** is another constituent of most protoplasm and many other elements, such as **iron, phosphorus, calcium** and **iodine** frequently occur in it. These elements appear in three complex types of chemical compounds and in many others which are intermediate or waste products in the chemical processes of the living body. The three are proteins (see **Aminoacids**), **carbohydrates** and fats (see **Esters**). Proteins are made up of carbon, hydrogen, oxygen and nitrogen and usually sulfur. Nucleo-proteins contain the additional element phosphorus. Both carbohydrates and fats contain carbon, hydrogen and oxygen but in carbohydrates the hydrogen and oxygen are in the same proportions as in water while in fats there is relatively much less oxygen. The proteins are chiefly materials of construction and the carbohydrates and fats are sources of energy.

These materials are associated in a finely divided state known as a **colloid**, in which the extensive surface contacts of particle with particle or with surrounding fluids facilitate all chemical interactions of which they are capable.

Living matter displays several physiological properties which are of paramount importance. First among them is **metabolism**, the power of chemical interchange with the environment by which the organism secures materials and energy. As a result of metabolism **growth** by intussusception is accomplished and the animal adjusts itself to its surroundings by the property of **adaptability**, carried on through the four subordinate properties, **irritability, conductivity, contractility,** and **secretion**. The substance is capable of **reproduction**. By this process any unit characteristic of the **organization** of a given kind of protoplasm is repeatedly produced. These units are not simpler than a single cell; any simpler structure is an end product of the cell and is not capable of complete exercise of these vital properties. (A.W.L.)

PROTOPLAST. The living unit of **protoplasm** contained within the wall of a **cell** forms the protoplast of the cell. This protoplast is an organized unit of protoplasm. (R.M.W.)

PROTOPODITE. A segment of the **biramous appendage.** (A.W.L.)

PROTORE. Enrichment.

PROTOTHERIA. Mammalia.

PROTOZOA. The 1-celled animals, constituting a major division (**Phylum**) of the animal kingdom. They occur in soil or water and many species live as parasites or symbionts in the bodies of other animals. Some protozoans are colonial and in some colonies a division of labor occurs, accompanied by structural specialization of the individuals. Some species are very widely distributed.

Since the body of a protozoan is a single cell it has the subordinate structures of the cell in addition to other specialized parts. It contains one or more nuclei surrounded by cytoplasm. The body may be naked and without permanent form or held in a definite shape by a delicate surface membrane called a pellicle. Some species secrete a shell or test and some form internal hard parts.

The structures which perform special functions are called organelles, since they resemble the multicellular organs of other animals in function but are simpler than the cell itself in structure. Among the more evident and important are the external organelles for locomotion and for securing food. In the various forms of protozoans these are **pseudopodia, cilia, flagella, cirri, membranelles, undulating membranes,** or **tentacles.** Some are associated with a depression in the surface of the body called the cytosome through which food is ingested. Within the body the cytoplasm is differentiated into a clear layer of ectoplasm at the surface and an inner granular endoplasm containing the nucleus. Here also masses of food with a little water from the food vacuoles in which digestion takes place. One or more pulsating or contractile vacuoles are interpreted as excretory organelles. They fill periodically with clear liquid and then discharge. Slender rod-like defensive structures, the trichocysts, lie in the ectoplasm of some species and are discharged when the animal is irritated.

In the bodies of the species which carry on a type of nutrition like that of green plants colored bodies (chromatophores) containing chlorophyll are found in the cytoplasm. These organisms have been interpreted both as plants and as animals, and have in reality some properties of each kingdom. They and some of the parasitic species absorb dissolved foods and do not ingest solids.

The protozoans reproduce commonly by the asexual process known as **fission** although sexual reproduction also takes place at intervals through the process of **conjugation.**

Protozoans are economically important chiefly as the causes of several serious diseases, among them amoebic dysentery, African sleeping sickness, and malaria.

The phylum is divided into two subphyla and five classes as follows:

Subphylum **Plasmodroma. Cilia** never present.
  Class **Mastigophora.** Animals with **flagella.**
  Class **Sarcodina.** Animals which form pseudopodia.
  Class **Sporozoa.** Usually incapable of locomotion as adults but sometimes able to form pseudopodia in immature stages.

Subphylum **Ciliophora.** Ciliated either throughout life or when young.
  Class **Ciliata.** Always ciliated or with related organelles such as membranelles and cirri.
  Class **Suctoria.** Cilia present only in immature individuals. Adults with tentacles. (A.W.L.)

PROTRACTOR. A muscle that draws some part of the body forward. (A.W.L.)

PROTURA. Minute insects of very primitive form. The body is elongate. Neither antennae nor eyes are present. The mouth is formed for sucking. There are three pairs of thoracic legs as in the insects, and a few rudimentary abdominal appendages. Also named Myrientomata and variously considered as a class of **arthropods** and an order of the class **Insecta.** (A.W.L.)

PROUSTITE. A **silver**-bearing **arsenical** mineral corresponding to the formula Ag_3AsS_3 sometimes with a little **antimony.** It rarely is found in **hexagonal** crystals, being usually in compact or disseminated masses or crusts. It is a brittle mineral; hardness, 2–2.5; specific gravity, 5.57–5.64; color, scarlet, vermilion; luster, adamantine, sometimes very brilliant; streak, same as color; transparent to translucent. It is found associated with **pyrargyrite.** Well-known deposits occur in the Harz Mts., Bohemia, Chile and Mexico. In the United States proustite has been found in Colorado, Idaho and Nevada. It was named for the famous French chemist, Louis Joseph Proust. (E.S.C.S.)

PROXIMATE ANALYSIS. The proximate analysis of coal is the determination of moisture, volatile material, fixed carbon, and ash. Much information regarding the firing characteristics and fuel value of coal will be conveyed by the proximate analysis. The test for proximate analysis may be obtained with a small amount of laboratory equipment. For that reason it is useful for checking, from time to time, the quality of coal bought. The proximate analysis is of little use in combustion calculations which are essentially chemical equations. The chemical, or **ultimate, analysis** is required for that work. (F.T.M.)

PRUNE. Rose Family.

PRUNING. The removal of parts of a plant for the purpose of improving it in some desired way is called pruning. Pruning may have as its aim the formation of a plant of more desirable shape. It may be done to reduce the transpiring surface of the plant and so aid it in its struggle to survive when the available water supply is inadequate, a practice of particular value when plants are transplanted and much of the root system is destroyed. Pruning may be done to improve the product of the plant, whether flower or fruit, by throwing greater amounts of food into fewer flower clusters. Finally diseased or damaged portions may be removed to prevent spread of disease and further damage to the plant.

Plants growing in competition for light prune each other. In a dense stand of pines, or other trees, increasing size decreases light penetration to the lower branches. Gradually the latter cease to grow, die, and finally fall off, leaving the trunk straight and unbranched.

Artificial pruning may be done at any time, but is usually best done during the winter or early spring before growth starts. The wound made heals over by means of **callus** tissue which results from increased activity of the **cambium** cells. When a large branch is removed, some time, often years, may be required before the wound is covered. To prevent entrance of disease- and decay-producing organisms, the wound is generally covered with some antiseptic substance, such as white lead paint. When branches are to be pruned, the cut should be made as close as possible to the main trunk and parallel to the surface of the latter, so that the cut surface may be covered as soon as possible by the callus tissue which forms around its edges. Long stubs are rarely covered and provide paths for the entry of fungi which destroy the central portion of the tree trunk. (R.M.W., B.S.M.)

PRURITUS. Intense itching, a symptom of many skin and some constitutional diseases. Pruritus is found at times in the aged due to the degeneration of the skin. It is also present in jaundice.

Pruritus of the anus is very common. In children, the common cause is pinworms; in adults, it is hemorrhoids. Infection and pruritus of the vulva and anus occur in diabetes. (R.S.M., D.M.H.)

PRUSSIAN BLUE. Ferrocyanic Acid and Ferrocyanides.

PSAMMITIC. Arenaceous.

PSEUDOACONITINE. Alkaloids.

PSEUDO-ADIABATIC. Adiabatic Changes in the Atmosphere.

PSEUDOBRANCHIA. A small respiratory structure on the inner surface of the operculum in certain fishes. The Pseudobranchiae serve as supplementary gills. (A.W.L.)

PSEUDOCRYSTALLINE. Crystallography.

PSEUDOEPHEDRINE. Alkaloids.

PSEUDOMORPHINE. Alkaloids.

PSEUDOMORPHS. In mineralogy, defined as minerals having the crystal form of one species and the chemical composition of another. Typical pseudomorphs are malachite in the form of cuprite, barite in the form of quartz, limonite in the form of pyrite. In such cases as these the evidence seems to be that there has been a complete chemical and molecular change but without any change of the original outward form. (R.M.F.)

PSEUDOPHYLLIDEA. Cestoda.

PSEUDOPODIOSPORE. A form of reproductive body or spore produced by some of the 1-celled animals. It is a naked cell like *Amoeba* and is able to form pseudopodia. Also called an amoebula. (A.W.L.)

PSEUDOPODIUM. A cytoplasmic protuberance formed at the surface of the body of 1-celled animals, especially those of the class Sarcodina. Pseudopodia are temporary or semi-permanent structures whose formation and retraction can be observed in many species during brief observation. They serve for securing food and as organs of locomotion.

Four types of pseudopodia are recognized. Lobopodia are temporary and relatively thick structures containing both ectoplasm and endoplasm. The common species of Amoeba form them. Filopodia are also temporary. They are more slender and usually contain only ectoplasm. Rhizopodia are temporary projections which branch and anastomose. Axopodia are very slender and usually straight radiating pseudopodia. They are semi-permanent. This form is characteristic of the heliozoans and radiolarians. (A.W.L.)

PSEUDOSCOLEX. The anterior end of a tapeworm (Cestoda) which lacks a true scolex. The true scolex is provided with one or more suckers or a group of hooks for attachment to the intestine of the host. The pseudoscolex has neither. (A.W.L.)

PSEUDOSCORPION. A small animal resembling the scorpions slightly in appearance. The abdomen is rounded and has no slender posterior part and no sting,

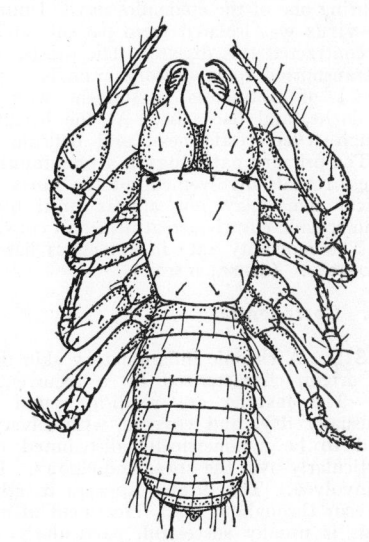

Pseudoscorpion.

but the pair of chelate pinchers at the anterior end of the body resemble those of the scorpions. Pseudoscorpions are commonly found in leaves at the surface of the ground and under the bark of decaying logs. They are only a few millimeters long. The relatively few species make up the order Chelonethida of the class Arachnida. (A.W.L.)

PSEUDOSCORPIONES. Synonym of Chelonethida. **ARACHNIDA.** (A.W.L.)

PSEUDOSTIGMATIC ORGAN. A bristle, often clubbed at the tip and conspicuously large, which occurs in some of the mites. A pair of these structures arise from pits near the anterior end of the body. (A.W.L.)

PSEUDOTROPINE. Alkaloids.

PSEUDOVELUM. A narrow rim of tissue near the margin of the body of jellyfishes. These animals lack the true velum as found in hydrozoan medusae; the pseudovelum differs in its lack of muscle fibers and a nerve ring. (A.W.L.)

PSILOMELANE. Psilomelane is a massive black mineral of indefinite composition, being chiefly manganese oxide with various amounts of barium, potassium and sodium with water.. It may be considered as colloidal manganese dioxide plus impurities. It is commonly associated with pyrolusite and is of secondary origin. Psilomelane is found in Saxony; France; Cornwall, England; India; Brazil and in the United States in Michigan. It is an ore of manganese. The word psilomelane is derived from the Greek words meaning smooth, and black, in reference to the smooth black surfaces so often exhibited. (E.S.C.S.)

PSITTACIFORMES. The parrots and related species, including cockatoos, macaws, paraquets, etc. The strong hooked beak, adapted for opening nuts and seeds, is the most conspicuous character of the group. Two toes are directed forward and two back. An order of the class Aves. (A.W.L.)

PSITTACOSIS (Parrot fever). A very contagious epidemic disease occurring chiefly in parrots but easily transmissible to human beings who handle infected birds. Epidemics of this disease have been reported in Europe, South and Central America. Recently there have been outbreaks in various North American cities. In 1929 during one of the epidemics in the United States a filtrable **virus** was isolated from patients and parrots who had contracted this disease. The disease was successfully transmitted to laboratory animals.

The onset of psittacosis is -sudden, with malaise, chills, headache, and backache. At the height of the disease, high fever, restlessness, and delirum may be present. The primary **pathology** is a **pneumonia**; x-ray of the chest reveals involvement of the lungs although the physical signs may not appear until quite late. There is no specific curative treatment and convalescence is slow. The mortality rate in epidemics has been as high as 30–40%. (R.S.M., D.M.H.)

PSOCID. Corrodentia.

PSORIASIS. A chronic inflammatory skin disease of unknown origin, characterized by a recurrent scaling eruption. The **lesions** are reddish, round or oval plaques, usually dry, and covered with silvery scales. They tend to be symmetrically distributed over the limbs, particularly over the knees and elbows. The scalp is often involved. The disease appears in adults and tends to recur throughout life. Treatment of an attack of psoriasis is usually successful, particularly if begun early, but recurrences cannot be prevented. One of the most effective drugs is chrysarobin, obtained from deposits in the wood of a tree indigenous to India and Brazil. Ammoniated **mercury, salicylic acid** and betanaphthol, applied directly to the lesions are also used. If the disease has been present for several months or longer, it is more resistant to treatment. (D.M.H.)

PSYCHOANALYSIS. A method of eliciting from patients their past emotional experiences, dreams, and subconscious feelings to interpret and explain their present mental state or actions and use them for psychotherapeutic procedures. This form of treatment should only be used by an expert **psychiatrist.** (R.S.M.)

PSYCHONEUROSIS. A disturbance of the emotional life which is based on some conflict between the desires of an individual and what that individual believes is "right," or what society considers to be acceptable behavior. Most psychoneuroses are not associated with abnormal outward behavior, but the inner battle is manifested by a variety of distressing emotional reactions which are expressed in the form of symptoms of physical disease. Quite unconsciously, psychoneurotic patients develop symptoms of "nervous indigestion," "nervous heart," "nervous headache," which they use to protect themselves from conflicts and difficulties which make life too much to bear without some support. Physical disease is an excellent protection because in our society a sick man is the object of sympathy, care, and consideration. And the psychoneurotic is sick; his stomach pain hurts him just as much as such pain arising from an organic disorder.

Some degree of psychoneurosis is an extremely common thing; many individuals who are respected, useful members of society are psychoneurotics. There are no organic causes for the condition, no change in brain structure, or chemical or physical function to account for it. Certain types of personality, however, seem more susceptible. The one who is sensitive, "high-strung," and of asthenic body build, is probably more often affected. These traits are apt to be hereditary, but are influenced greatly by the home environment, particularly in childhood. That environment plays an enormous role is well illustrated by the fact that in wartime, the incidence of psychoneuroses increases tremendously. As many as 60% of medical discharges from the United States Army are on the basis of this type of disease, precipitated by the abnormal stresses and strains which army life imposes on men who, under normal conditions, would have only mild psychoneurotic trends, or none at all.

For convenience, the psychoneuroses are classified as (1) neurasthenia, (2) anxiety states, (3) hysteria, and (4) obsessive-compulsive states.

1. Neurasthenia. In this type of psychoneurosis, the patient complains of mental and physical weakness, irritability, inability to concentrate, insomnia, loss of appetite, and various bodily aches and pains. These symptoms may appear as the cumulative result of early years of maladjustment in an unfortunate home situation, or they may be precipitated by a situation of recent development. Thus the discovery that he dislikes his wife, or his business partner with whom he is forced to work closely, may cause the onset of neurasthenia in a man who has previously been well. Psychotherapy is directed toward uncovering the emotional conflict, explaining its nature to the patient, helping him to adjust his attitude toward the situation, and helping him to face the problem directly rather than take cover behind physical symptoms. In an intelligent, cooperative patient, the prognosis is excellent.

2. Anxiety states. These are the commonest of the psychoneuroses, and the most responsive to treatment. They are manifested by such symptoms as palpitation, breathlessness, lack of appetite, a sensation of fullness in the abdomen, frequency of urination, diarrhea or constipation, tremors, pains, blurring of vision, ringing in the ears, and easy fatigability. Mental symptoms are irritability, depression and fears of all kinds—fears of mental or physical ill-health, fears of heart disease, venereal disease, or any other. Psychiatrists call this abnormal fear of disease with subsequent development of symptoms referable to many diseases, **hypochondriasis.**

The anxiety states are most apt to develop in emotionally unstable individuals who are, or have been, exposed to adverse environment in the home. The influence of a dominant, overanxious parent is a common starting point for an anxiety state. The immediate precipitating factors may be dissatisfaction or disappointment and fears about sexual, financial, domestic or any other matters. The actual form that the symptoms take depends on the cause of the anxiety and the general type of personality, environment, education, etc., that the patient possesses.

Treatment with psychotherapy, as in the neurasthenic, discovers the cause of the anxiety and helps the patient to overcome his fears and make a normal adjustment.

3. Hysteria. An hysterical manifestation, as it is popularly understood, can be part of neurasthenia or an anxiety state. "Crying spells" are hysterical attacks. In medicine, the term usually refers to psychoneurotic symptoms which develop an anatomical expression. Thus fear of developing certain diseases may cause hysterical blindness or hysterical paralyses of limbs in the presence of normal eyes and normal muscles. The patient is unaware of the connection between his fear and the hysterical manifestation. If a visceral organ is involved, hysterical vomiting, retention of urine, or dysmenorrhea may occur. All of these signs are apt to have symbolic significance. For instance, hysterical blindness may be closely related to some distressing visual experience.

In general, hysteria is more apt to occur in young women of unstable emotional makeup. Emotional immaturity, moodiness, self-consciousness, vanity, egocentricity, oversensitiveness, etc., are common personality traits in these individuals.

The treatment of hysteria is psychotherapy. **Hypnosis** and suggestion are often employed successfully.

4. Obsessive-compulsive states. In this form of pychoneurosis, the individual is repeatedly forced against his will, and with no apparent reason or purpose, to think certain thoughts, perform certain acts, or utter certain words. The thoughts may be of a speculative or questioning nature on philosophical subjects, or they may be such an absurd preoccupation as wondering why a chair has four legs instead of one. The action may be a socially prohibitive one, and the words are sometimes obscene. The patient may be tortured by his constantly recurring compulsion, but he is helpless against the insistence of the process.

Compulsion neuroses occur in individuals who are shy, oversensitive, precise, meticulous, and are full of phobias, doubts, fears, and indecisions. Fears of dirt, of bacteria, of insanity, and of cancer are common in them. They translate their own inner conflicts and anxieties into fears of these objects or diseases without being aware of the connection between the two. The compulsive thought or act thus becomes an unconscious symbol, perhaps of some painful experience which has elicited remorse, conflict and repression.

The treatment of the compulsive-obsessive neuroses, as with the other psychoneuroses, is psychotherapy aimed at clarifying the cause of the abnormal behavior and adjusting the patient's attitudes and reactions to avoid recurrence. (D.M.H.)

PSYCHONEUROTIC. Psychoneurosis.

PSYCHOSIS. A mental illness characterized by profound changes in the personality, abnormal behavior and, in certain types, mental deterioration. There are many types of psychoses. The chief ones are (1) the major psychoses, Schizophrenia, Manic-depressive psychoses, and involutional melancholia. These are functional; they are not associated with any demonstrable structural or chemical change in the brain. (2) Those associated with organic disease of the central nervous system such as syphilis, encephalitis, brain tumor, arteriosclerosis, endocrine disease, pellagra, etc. The commonest example of this group is the senile psychosis associated with cerebral arteriosclerosis. (3) Those due to drug intoxication, the alcoholic psychoses, lead, mercury, and carbon monoxide poisoning. (D.M.H.)

PSYCHOSOMATIC MEDICINE. An aspect of medical science which stresses the psychobiological unity of the human being. The close interaction of mind and body in health and disease, and the role that the mind plays in the genesis of organic disease, form the basis of psychosomatic medicine. The common diseases in which psychic as well as organic causes seem to operate are asthma, hypertension, ulcerative colitis, and peptic ulcer. (D.M.H.)

PSYCHROMETER. Hygrometers.

PSYLLIUM SEED. Plantains.

PTARMIGAN. Aves, Galliformes. *Lagopus*. Birds (Aves) related to the true grouse but with the feet and legs fully clothed with feathers. They are found in the far northern parts of Europe, Asia and North America and at high altitudes, above timber line, as far south as Colorado. They are largely mottled gray and brown in the summer but assume white plumage in winter. The change is not, however, always complete but is somewhat conditioned by the climate. The red grouse and the willow grouse or ripa of northern Europe are closely related to the ptarmigans. (A.W.L.)

PTERANODON. Cretaceous.

PTERASPIS. Fossil Fishes.

PTERICHTHYS. Fossil Fishes.

PTERIDOPHYTES. Ferns; Paleobotany.

PTERIDOSPERMS. Paleobotany.

PTEROBRANCHIA. Hemichordata.

PTEROPOD. A marine mollusk whose foot is formed as winglike processes used in swimming. These mollusks belong to the class Gasteropoda. (A.W.L.)

PTEROPOD OOZE. Oceanic Deposits:

PTEROPSIDA. Members of the Pteropsida have large leaves, and definite gaps in the vascular cylinder, where a vascular strand, or leaf trace, passes from the stele to the leaf. It is also characteristic of the Pteropsida that the sporangia are located on the lower surface of the leaf. Most of the vascular plants of today, including all ferns, gymnosperms and angiosperms, belong to this group. (R.M.W.)

PTEROSAUR. Fossil Reptiles.

PTERYGOTA. One of two subclasses into which the class Insecta is divided. The members of this subclass are either winged or closely related to winged forms. Some show evidence of derivation from groups which are typically winged. The great majority of insect orders belong here, only three falling into the suborder Apterygota. (A.W.L.)

PTOMAINE. A class of amines formed by the action of bacteria on proteins or by the metabolism of aminoacids which are broken down into toxic products. (R.S.M.)

PTOMAINE POISONING. A misnomer generally used in describing gastric and intestinal disorders which are usually caused by food poisoning due to the ingestion of food contaminated by bacteria. (R.S.M.)

PTOSIS. 1. A weakness of the muscles of the eyelid causing it to droop over the eye. 2. Falling of any organ or part due to weakness of its supporting tissues—muscle, fat, or ligaments. (D.M.H.)

PTYGMATIC. A term proposed in 1907 by the Swedish geologist, Sederholm, for the primary folding in migmatities. (R.M.F.)

PUBERTY. The period of life when the human reproductive organs begin to function. This usually occurs between the 12th and 17th years, being somewhat later in the male than in the female. Puberty is marked by the development of secondary sexual characteristics, by the appearance of hair in the pubic and axillary regions, enlargement of the breasts and menstruation in the female. The male develops facial and increased body hair, deepened voice, and seminal discharges. In addition to the physical changes, profound psychological alterations and adjustments occur in both sexes. (D.M.H.)

PUBIS. The anterior ventral bone in each half of the pelvic girdle of the vertebrates. Skeletal System. (A.W.L.)

PUBLIC ADDRESS. This is a sound-amplifying system used in connection with large gatherings or where it is desired to cover a large area with sound. The system contains the microphone or phonograph (frequently both), a multistage amplifier to build up the power to a suitable level and finally a loud-speaker or speakers. These systems have come into wide use in recent years as the vacuum tube amplifier has been developed and are now standard equipment in all large auditoriums, stadia, and other places where it is necessary to reach a great number of people with sound announcements or entertainment. (L.R.Q.)

PUDDING WIFE. Pisces, Teleostei. A marine fish (**Pisces**), *Iridio radiatus,* of the wrasse family (Labridae), found in warmer seas. Also called the pudding fish and blue-fish. (A.W.L.)

PUDDINGSTONE. Conglomerate.

PUERPERAL SEPSIS (Child-bed fever). An acute infection of the genital tract which follows childbirth. Long **labor** and manipulations predispose to infection. Its infectious nature was first realized in the middle of the 19th century by the Viennese physician, Semmelweiss, and by Oliver Wendell Holmes, both of whom crusaded for sterile technique in handling obstetrical patients. The etiologic agent is often the hemolytic **streptococcus,** and blood stream invasion (**septicemia**) is common. Treatment is with **sulfonamides** or **penicillin;** the prognosis, with adequate treatment, is good. Before the use of these chemotherapeutic agents, the mortality in severe infections was very high, the disease constituting the major cause of maternal deaths. (D.M.H.)

PUFF-BALLS. Basidiomycetes.

PUFFIN. Aves, Charadriiformes. Marine birds (**Aves**) of several species whose thick compressed beaks are almost as large as the head. They are found on both sides of the Atlantic, nesting in northern latitudes and wintering south to New York and the Mediterranean. Related species of the Pacific are commonly called shearwaters. All belong to the genus *Puffinus.* (A.W.L.)

PUKU. Mammalia, Artiodactyla. A moderately large African **antelope** with fairly long horns which curve forward and are ringed throughout their length. (A.W.L.)

PULASKITE. A variety of **alkali-syenite** containing soda-rich **orthoclase,** usually with **aegirite-augite** and biotite. The type locality is Pulaski County, Arkansas, whence the name. (E.S.C.S.)

PULLEY. Belting; Machines.

PULMONATA. Gasteropoda.

PULSE. The pulse is the expansion and elongation of the arterial walls, produced passively by changes in intra-arterial pressure during contraction (**systole**) and relaxation (**diastole**) of the heart. The pulse is usually felt in the radial artery at the wrist, but it may be felt in any artery which lies near the surface. The heart, with systole, forces blood into the arterial circulation. This blood is accommodated partly by moving the entire arterial column on at greater velocity, and partly by distending the arterial walls. The increase in pressure and distention of the vessel walls is transmitted from one segment of the artery to the next as the pulse wave. The ability of the vessel wall to distend is dependent upon its elasticity. In old age, when the vessel becomes sclerotic and inelastic, it offers increased resistance and this results in elevation of the **blood pressure.**

The examination of the pulse in disease is one of the oldest customs in medicine. Variations in the rate, rhythm and force are significant, particularly in heart disease. The normal rate is 60–80 in adults, and 80–140 in children. With fever the rate increases as a general rule. A disproportionately slow pulse with a high fever is of diagnostic significance in certain infections, notably **typhoid, typhus** and so-called virus **pneumonia.** An increase in the pulse rate is a normal reaction to emotional stimuli, being dependent in this case upon a release of epinephrine (**adrenalin**) by the adrenal glands.

Abnormalities in rhythm may occur as a toxic reaction to **digitalis** or other drugs, or as a result of organic heart disease (**arteriosclerosis, rheumatic fever, diphtheria**) which interferes with the conduction of impulses setting off the contraction of the heart muscle.

Abnormalities in the force of the pulse occur in certain types of heart disease and in severe illness. In shock, the pulse is characteristically weak and thready, in addition to being rapid.

The term pulse is also used in physics. (See **Vibrations and Waves.**) (D.M.H.)

PULSE GENERATOR. Many electronic devices or circuits utilize sharp pulses of current or voltage as a basic part of their operation. Such pulses must frequently be of very short time duration (microseconds) and accurately spaced in time. Others need not be re-

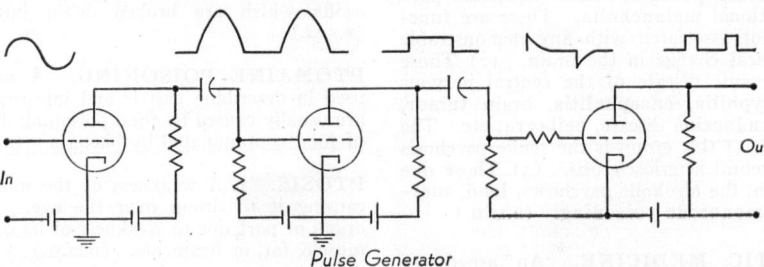

Pulse Generator

peated at regular time intervals but are initiated by some signal and are single narrow pulses each time a signal is received. The various types of circuits for generating these are termed pulse generators. As a simple illustration we might examine a basic pulse generating circuit as shown in the figure. The wave patterns above the circuit diagram represent the wave form at the respective points in the circuit. A sine wave is applied to the **grid** of the first **tube,** which is biased so it clips the wave to the form shown applied to the second tube. This in turn clips the rounded top and gives the square wave shown at its output circuit. By passing this through a condenser-resistor circuit a series of triangular pulses is obtained. These pulses are applied through a grid resistor to the grid of a tube as shown. This tube levels them to produce a series of very narrow pulses spaced at the same intervals as the cycles of the original sine wave. The particular shapes shown in the diagrams are obtained by the proper choice of grid bias and condenser-resistor values in the coupling circuits. When the condenser has a low reactance compared to the resistance in series, the wave is passed without appreciable change of shape but, when the relative values are reversed, the wave form is materially changed as shown in the grid circuit of the final tube. While the circuit shown is relatively simple, it does illustrate some of the possibilities of shaping circuits. (L.R.Q.)

PULVERIZED COAL. The burning of **fuel** in pulverized form has both advantages and disadvantages, but present trends indicate a continually increasing use of pulverized **coal** as fuel costs rise, as supplies of the better grades of coal are exhausted, and as pulverizing practice becomes widespread. The capacity for heat liberation in a pulverized coal furnace is not limited by the capacity of the burners, but by the inability of the **furnace** walls to resist the effects of high temperature, and by the limitations inherent in the physical characteristics of the ash. Pulverized coal therefore can be selected for high-capacity firing. The combustion of pulverized coal is readily adaptable to automatic control, responding rapidly to control manipulation. There is but little loss analogous to the banking loss of stoker fires. High-capacity **boilers** with their high exit-gas temperatures require some heat-saver such as the air preheater if they are to operate efficiently. Pulverized coal firing utilizes preheated air to good advantage and without limitation of the preheated air temperature. Boiler room cleanliness is another asset of the pulverized coal installation.

These advantages are not secured without some undesirable features. Pulverized coal brings with it additional costs and complications of equipment. Forty per cent of the finely divided flocculent ash would pass up the stack and spread itself upon the surroundings were not strenuous efforts made to precipitate most of it. More power is required than for stokers. Low excess air and high capacity raise the temperature of the furnace to where it must be water-cooled to avoid excessive refractory maintenance. Low fusion point ash in particular must be carefully handled or else the boiler will be out of service for slag removal most of the time.

The extremely small size of the pulverized coal particles must be realized. A typical sample will have better than 99% through a 40-mesh screen, 90% through 100-mesh, and 65% through 200-mesh. Thus, it can be seen that the particles are extremely small, and an enormously larger combustion surface is presented than with lump coal. Hence, combustion occurs rapidly—from ¼ to 4 seconds is required with different fuels and burners.

There are two pulverized coal systems. They are called the central system (bin system) and the unit system. A central pulverizing system employs a limited number of larger capacity pulverizers at a central preparation point to prepare coal for all the burners. The driers as well are conveniently installed at this preparation point. From the pulverizers the coal is transported to a central storage bin where it is deposited and its transporting air vented through a "cyclone." The central bin will contain from 12 to 24 hours' supply. The coal is then transported from the central bin to secondary bins, each supplying a burner or group of burners through the medium of "feeders" of varied design. Primary air is added at the feeders.

The unit system is so called from the fact that each burner, or burner group, and pulverizer constitute a unit. Crushed, and sometimes dried, coal is fed to the pulverizing mill at a variable rate governed by the combustion requirements of the boiler and furnace. Preheated primary air is admitted to the mill and is the transport air which carries the coal through the short delivery pipe to the burner.

The pulverizing mills of present design may be classified as (1) impact mills in which a series of swinging hammers or falling balls pulverize the coal, (2) roller mills in which grinding is done by crushing between rollers or balls and a race, (3) chopping or attrition mills.

Pulverized coal can be transported mechanically or pneumatically. Screw conveyors are generally short, straight, and horizontal. The pneumatic system offers more flexibility. The air pressure system is one of inter-mittent delivery, suitable for transporting from bin to bin. In the air mixture system primary air, representing a weight several times that of the coal, is mixed with the coal and carries the coal along in suspension. The air pumping system uses much less air, only enough to render the coal sufficiently fluid to be pumped by pumps.

Type of burners depends on fineness of coal, moisture and volatile content, method of mixing air and fuel, percent total air as primary, and characteristics of the furnace. There are two general types of burners, the long and the short flame. The long flame is produced by moderate-tip velocities coupled with an admission of secondary air through openings in the setting located along the traverse of the flame. These long-flame burners are generally in multiple, directed vertically downwards into the furnace. Short-flame burners are ordinarily set in horizontal position. A short-flame burner produces complete mixture just beyond the burner tip by violently whipping the secondary air through the primary air and fuel. (F.T.M.)

PULVERIZED COAL COMBUSTION. Burner.

PULVINUS. Movement in Plants.

PUMA. Mammalia, Carnivora. A large North American **cat** of uniform tawny to brownish color. The puma proper, *Felis concolor,* is—or was—a species of the eastern half of the continent, ranging from Virginia into Canada. It is now extinct in the settled parts of the country but may still exist in wilder areas. A darker species, *F. coryi,* is found in Florida, a third species, *F. arundivaga,* in Louisiana, and a fourth variable species called the western puma or mountain lion, *F. oregonensis,* ranges from Mexico into Canada. The eastern species has been variously called the cougar, mountain lion, panther, catamount, and painter. It is a menace to stock and sheep but is cowardly in its relations with man. (A.W.L.)

PUMICE. Rhyolitic lavas with a high gas content, when suddenly discharged by volcanic action, congeal in the form of highly **vesicular** natural **glass** called pumice. When ground, mixed with an appropriate binder and pressed into cakes it is the "pumice stone" of commerce which is used as a light abrasive. (E.S.C.S., R.M.F.)

PUMP EFFICIENCY. The static **head** is the height (usually in ft.) of the surface of the water above the gauge point. The pressure head is the static head plus gauge pressure on the water surface plus friction head, all being reduced to the same unit of pressure. Velocity head is the head required to produce a flow of water. The dynamic head is the pressure head plus the velocity head. Except for water velocities considerably above average, or for large volumes handled at low heads, the velocity head can be neglected. For water at atmospheric temperature, 8.33 lbs. = 1 U. S. gal. = 231 cu. in.; also 2.31 ft. head of water = 1 lb. per sq. in. pressure. Hot water weighs less. If the suction head is less than

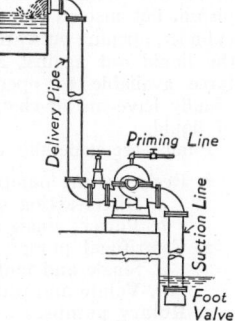

Pump with vacuum suction.

atmospheric, it is given a minus sign. When the suction arrangement is as shown in the accompanying figure, the suction head = − (static head + velocity head + suction pipe and fittings friction loss + loss at entry to suction pipe (which is usually taken at ½$V^2/2g$)).

Hydraulic efficiency

$$= E_h = \frac{\text{operating head}}{\text{operating head} + \text{pump head losses}},$$

Over-all pump efficiency $= E_p = \dfrac{\text{water hp.}}{\text{drive hp.}},$

Mechanical efficiency

$$= E_m = \frac{\text{theoretical pump head} \times \text{capacity}}{\text{drive hp.}},$$

$$E_p = E_h \times E_m.$$

(F.T.M.)

PUMPKIN. *Cucurbita Pepo.* **Gourd Family.**

PUMPKIN-SEED. Pisces, Teleostei. The common **sunfish**, *Eupomotis gibbosus.* A pond fish of the eastern and north central states, found also to a limited extent in streams. A good pan fish. (A.W.L.)

PUMPS, WATER. The function of a pump is to add to the pressure existing in a liquid an increment sufficient for the required service. This service may be the production of a velocity, or the overcoming of friction or external pressure. One of the earliest types of pump to be used was the suction or lift pump (Fig. 1). The

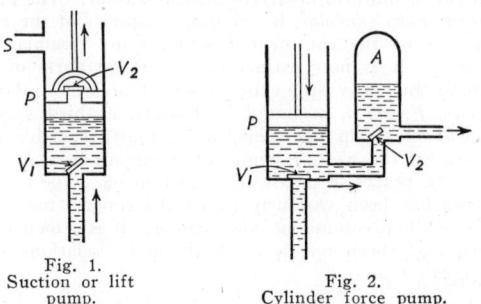

Fig. 1.
Suction or lift
pump.

Fig. 2.
Cylinder force pump.

upward movement of the piston P lowers the pressure in the cylinder and pipe below it, and the liquid is forced up into this space by the atmospheric pressure. On the down-stroke, the valve V_1 closes, imprisoning the liquid in the cylinder; while the piston valve V_2 opens, allowing the piston to plunge into the liquid and draw it up on the subsequent up-stroke, so that it can escape through the spout S. An ordinary cistern pump is of this type.

Another elementary type of pump is the cylinder force pump (Fig. 2). In this the first operation is the same as in the lift pump. There is, however, no valve in the piston, but instead one in the outlet at the side of the cylinder, opening outward, so that the down-stroke forces the liquid out against a pressure limited only by the force available to operate the piston. Such pumps usually have an air-chamber A to equalize the outflow of liquid.

Pumps are generally classed as follows:

1. Reciprocating pumps.
   a. Direct-acting steam; simplex and duplex.
   b. Power; single-acting simplex and triplex.
2. Centrifugal pumps.
   a. Single and multi-stage.
   b. Volute and turbine types.
3. **Rotary pumps.**
   a. Gear and screw pumps (mostly used for pumping oil).
   b. Propeller pumps.
   c. Lobe pumps.
4. Jet pumps.
   a. Steam jet injectors and ejectors.
   b. Water jet ejectors.

Essential data relating to any pump include the head in feet, capacity in gallons per minute, and properties of the liquid such as viscosity, temperature, corrosiveness, grittiness. Secondary data concerning the pump equipment are speed of rotation, power required, and first cost.

The **direct-acting** steam pump is a simple, inexpensive, and reliable piece of equipment—but inefficient as a pumping unit. Its principal use is as a steam boiler auxiliary and as the heat of the exhaust steam can often be recovered in the boiler feed water the low thermal efficiency is not of much importance.

Pumps driven by electric motors, or by steam used expansively, are called **power pumps.** Their over-all efficiencies are high because of the efficient drives.

The triplex pump is often used when the conditions indicate the reciprocating pump in preference to the

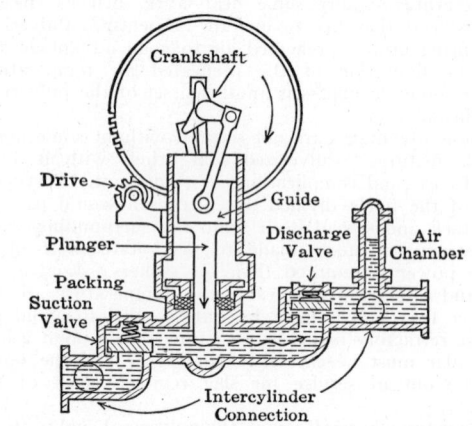

Fig. 3. Diagram of three-cylinder single-acting displacement pump.

centrifugal, and where an efficient pumping unit is required. Calling n the number of pumping strokes per minute, $D \times S$ the diameter $\times$ stroke, and E the volumetric efficiency of the pump (about 90% for pumps in good order) the

$$\text{Capacity} = \frac{ESnD^2\pi}{4 \times 231} \text{ gallons per minute.}$$

The **centrifugal pump** is a velocity machine, that is, its pumping action requires first, the production of a water velocity; second, the conversion of velocity **head** to pressure head. The velocity is given by the rotating impeller, the conversion accomplished by diffusing guide vanes in the turbine type, and in a volute casing surrounding the impeller in the volute type. With few exceptions, all single-stage pumps are of the volute type.

The **specific speed** of a centrifugal pump is $\dfrac{N\sqrt{Q}}{H^{3/4}}$. Ordinarily N is expressed in rpm, Q in gal. per min., and head H in feet. The specific speed of an impeller is an index to its type. Impellers for high heads usually have low specific speeds, while those for low heads have high specific speeds. The specific speed is a valuable index in determining the maximum suction head that may be employed without danger of cavitation or vibration, both of which adversely affect capacity and efficiency.

Allied to the centrifugal pump in several ways are the axial-flow pumps. An **axial-flow** pump, sometimes called a propeller pump, develops most of its head by the propelling action of the vanes in the liquid. It has a single-inlet impeller with the flow entering axially into a guide case. Where the head is developed partly by centrifugal action and partly by vane propulsion, the pump is called a **mixed flow** pump.

The injector is a jet-type feed pump, limited to regular service on small boilers and stand-by service on small and medium-sized boilers. Like the centrifugal pump, the injector operates on the principle of a velocity-pressure conversion but differs from the centrifugal in the manner of creating the velocity. The water acquires its velocity by impact with high-velocity steam leaving an expanding nozzle. The injector will produce a pressure about 50 lbs. per sq. in. higher than the steam pressure used, but the water temperature should not exceed 150° F. The injector is simple, compact, inexpensive, and with no moving parts to wear or require adjustment. As a combined feedwater heater and pump its efficiency is high, but, as a pump alone, very low (less than 5%). Its characteristics recommend it to locomotive service but not to regular service in an efficient stationary plant where feed water is heated by other than live steam.

The selection of any pump is to be made with due regard to the drive. Reciprocating pumps are slow-speed machines, centrifugal pumps high-speed. Power pumps and centrifugal pumps require a driving motor, engine, or turbine. Variable capacity is obtained by throttling the discharge or changing the driver speed.

Mistakes are likely to be made in pump installation by overestimating the possible suction lift. For instance, suppose water is at 140° F. Water will begin to produce steam at this temperature as the pressure is reduced to 2.887 lbs. per sq. in. Hence, the theoretical maximum static head is 2.31 (14.7–2.887), or 27 ft. But, under dynamic operating conditions, steam would be formed at this lift. To be safe against steam binding and separation of the water column, the maximum suction head should be limited. Hot water above 160° F. should be supplied under a positive head—at 200° F. there should be from 5 to 10 ft. head on the pump. Pumps, especially centrifugal, must be filled with water when starting under considerable suction head. A foot valve on the suction intake, if tight, will retain water in the pump, but the foot valves frequently leak slowly if the water contains grit or debris. To prime the pump a check valve by-pass line is satisfactory if there will always be water against the check valve; otherwise an independent fill line can be run to the pump. Steam jet ejectors and other vacuum equipment are often used to prime large pumps. (See **Ram, Hydraulic**.) (F.T.M.)

PUNCHING. Forging.

PUNKIE. Insecta, Diptera. Small biting **midges**, also called sandflies. They are found in abundance at certain times along streams in the eastern mountains, and at some parts of the seashore. (A.W.L.)

PUPA. The third stage of **insects** with complete **metamorphosis**. The pupa is a more or less inert stage but in some insects it retains the power of locomotion to a high degree. The pupae of mosquitoes are an example; they swim as freely as the larvae when disturbed. In contrast the pupae of butterflies and moths can merely move the abdominal segments and those of many flies are quite rigid. Among these inactive pupae some, like those of the **beetles** and **wasps**, have the legs and wings free and are called exarate. Others have the appendages closely attached to the body and are said to be obtected. Those of **Diptera** in some cases are enclosed in a hardened larval skin, the puparium, and are called coarctate pupae. The pupae of butterflies, often brightly colored and strangely shaped, are called chrysalids (singular, chrysalis or chrysalid). (A.W.L.)

PUPIL. The opening in the center of the iris of the **eye** for the transmission of light. The normal pupil is circular and regular in outline and is larger in the young than in adults. Both pupils should be of the same size. The iris acts as a shutter so that the pupil becomes smaller in bright light and when looking at nearby objects, and larger in darkness and upon looking at distant

objects. The characteristics of the pupils, their size and speed of reaction, are of considerable value in the diagnosis of various diseases of the nervous system. In certain forms of late syphilis such as locomotor **ataxia** and **paresis**, the pupils do not dilate or contract in response to light and may be irregular and unequal in size. When morphine is given the pupils contract, and with large doses they become pin-point in size. **Belladonna**, or **atropine**, and **cocaine** dilate the pupil. (D.M.H.)

PURE BENDING. Bending Moment.

PURE CHEMISTRY. Chemistry.

PURE LINE. The descendants of a single self-fertilized individual. The term was established by the botanist, Johanssen, in his work with beans. He found that rigid selection according to size resulted in the establishment of strains in which the seeds varied constantly between certain limits, regardless of further selection. This constancy was most readily obtained by self-fertilization of the plants.

Since this discovery it has been recognized that **clones** and the descendants of genetically identical parents are equally uniform in heritage. These discoveries have an important bearing on **selection** as a factor in organic **evolution**. (A.W.L.)

PURE SHEAR. Shear.

PURINE AND URIC ACID COMPOUNDS. Purine compounds are derivatives of the dicyclodiureide of **malonic** and **oxalic acids**. The dicyclodiureide compound is uric acid

$$\begin{array}{l}\text{HN—CO}\\ \quad |\quad\;\; |\\ \text{OC}\quad\text{C—NH}\\ \quad|\quad\;\; ||\qquad\qquad\text{>CO}\\ \text{HN—C—NH}\end{array}$$

Oxalylurea residue Imidazole ring

Malonylurea residue Pyramidine ring

and purine, the parent compound, is

$$\begin{array}{l}\text{N=CH}\\ |\;1\;\cdot\;6|\\ \text{HC}\;2\;5\text{C—NH}\quad 8\\ ||\;3\;4||\quad\;\;7\;9\;\text{>CH}\\ \text{N—C—N}\end{array}$$

Purine

so that uric acid is 2,6,8-trioxypurine or the keto form of 2,6,8, trihydroxypurine. Caffeine, theobromine, and theophylline are other important purine compounds.

Uric acid ($C_5H_4O_3N_4$—formula above) is a white solid, insoluble in cold water, alcohol or ether, sparingly soluble in hot water. Uric acid is a weak dibasic acid thus forming two series of salts, most of which are very slightly soluble in water (**lithium** urate soluble). Uric acid reacts (1) with nitric acid dilute, forming alloxan

$$\begin{array}{l}\text{HN—CO}\\ \quad|\quad\;\;|\\ \text{OC}\quad\text{CO}\quad\text{plus urea}\\ \quad|\quad\;\;|\\ \text{HN—CO}\end{array}\qquad \begin{array}{l}\text{NH}_2\\ \quad\;\;\text{>CO}\\ \text{NH}_2\end{array}$$

more vigorous treatment yielding oxalylurea

$$\begin{array}{l}\text{OC—NH}\\ \quad\quad\;\;\text{>CO}\\ \text{OC—NH}\end{array}$$

plus **ammonia** plus **carbon dioxide**. The murexide test for uric acid is related to this reaction: Uric acid plus **nitric acid** dilute is evaporated to dryness, yielding red residue, which with **ammonium** hydroxide turns purple or with **sodium** hydroxide blue, (2) with **potassium** permanganate in sodium hydroxide solution, forming allantoin

$$OC—NH$$
$$H_2N—OC—HN—CH—NH \rangle CO,$$

(3) with **phosphorus** oxychloride, forming 2,6,8-trichloropurine

$$\begin{array}{c} N=C\cdot Cl \\ Cl\cdot C \quad C—N \\ N—C—N \rangle C\cdot Cl \\ H \end{array}$$

and 2,6-trichloro-8-hydroxypurine. In these compounds chlorine may be replaced by such groups as methoxy (CH_3O—), ethoxy (C_2H_5O—), hydroxyl (HO—), hydrosulfide (HS—), iodide (I—), hydrogen (H—), and thus many derivatives are obtainable.

Uric acid is found in the urine, blood, and muscle juices of carnivorous animals (herbivorous animals secrete hippuric acid), in the excrement of birds, serpents and insects, and is an oxidation product of the complex nitrogenous compounds of the animal organism.

The following are other important purine compounds, which are also **alkaloids:**

1. Caffeine (theine) 1,3,7-trimethyl-2,6-dihydroxypurine
2. Theobromine 3,7-dimethyl-2,6-dihydroxypurine
3. Theophylline 1,3-dimethyl-2,6-dihydroxypurine
4. Xanthine 2,6-dihydroxypurine
5. Hypoxanthine 6-hydroxypurine
6. Guanine 2-amino-6-hydroxypurine
7. Adenine 6-aminopurine

Caffeine is present in tea leaves (2–5%), in coffee beans (0.75–1.75%), in cola beans (Soudan coffee), (1–2.5%), in cocoa beans (0.1–0.8%); theobromine in cocoa and cola beans, in small amount in tea leaves, and absent from coffee beans; xanthine in beet root juice, in tea leaves, and in sprouting seeds; hypoxanthine and guanine in beet root juice and in tea leaves. (R.K.S.)

PURKINJE EFFECT. This effect, discovered by Abney and studied in some detail by Precht (1899), is perhaps better known as the gamma-wavelength effect. It refers to the variation in **gamma** with the wavelength region used in exposing a photographic material. The variation of gamma with the wavelength of the light makes it necessary to develop 3-color separation negatives for different times to secure three negatives of equal contrasts. It is a source of difficulty in the manufacture of integral tripack materials for color photography. Some present-day panchromatic emulsions show a marked decrease in contrast when exposed to red light, others an increase in contrast. There is as yet no satisfactory explanation of this effect. (C.B.N.)

PURLIN. Bent.

PURPURA. Bleeding into the skin or mucous membranes, producing areas of discoloration which vary in size from pin-point (petechiae) to several inches or more (ecchymoses). Purpura occurs as a symptom of many diseases and its mode of production varies. It is a characteristic feature in the disease **purpura hemorrhagica.** Symptomatic purpura occurs in **scurvy,** as a result of lack of **vitamin** C; in liver disease and disease of the biliary tract due to interference with the production of vitamin K; in some acute infections, particularly meningococcus **meningitis,** due to a toxic factor which increases capillary permeability; in **leukemia** and allied disorders, in toxic reactions to drugs (arsenicals, gold salts, phenobarbital), and acute or chronic **purpura hemorrhagica,** in all of which di-

minished blood platelets are responsible for failure of blood coagulation and abnormal bleeding. Purpura also occurs in various obscure blood diseases of unknown origin. (D.M.H.)

PURPURA HEMORRHAGICA (Thrombocytopenic Purpura). A disease of unknown etiology, characterized by diminution of the blood platelets (see **blood**) and spontaneous hemorrhages into the skin and **mucous membranes.** Females are affected more often than males, and the disease is commonest in children. Acute and chronic forms occur. Treatment of the acute case consists of repeated blood **transfusions.** Remissions and not infrequently permanent recovery occur with this form of purpura. Surgical removal of the spleen is effective in the control of chronic cases, although the mechanism by which such cure is achieved is not understood. (D.M.H.)

PUS. The liquid product of infection in the body. It is made up of cellular debris, micro-organisms, and white **blood** cells (leukocytes). Pus varies greatly in color, consistency and odor. When pus is present as a localized collection in tissue, an abscess is said to be present. Treatment is incision and drainage of the abscess. (R.S.M.)

PUSHBUTTON TUNING. Pushbutton tuning is a semi-automatic method of tuning a radio **receiver** to any one of several preselected stations. The usual types may be divided into two classes, mechanical and electrical. The first is simply a mechanical linkage between the pushbutton on the panel and the tuning condenser or coil in the set. To set this type the linkage is adjusted to vary the tuning just the right amount to set the receiver to the desired station. In the more refined electrical system each button switches in a fixed tuning unit which has been pre-set to the desired frequency. Since these frequencies can be very accurately set and require no mechanical linkages, the results are more satisfactory than the mechanical type. In either type, however, the various buttons and associated circuits or mechanisms are pre-adjusted for various commonly desired stations and then by pushing the proper button the station is tuned in. (L.R.Q.)

PUSULE. A large vacuole opening to the exterior by a canal. It is found in the body of some of the 1-celled animals of the order Dinoflagellida (**Mastigophora**). (A.W.L.)

PYCNIOSPORE. Rust Fungi.

PYCNOGONIDA. The sea spiders, a small number of species constituting a class of **Arthropoda.** All are marine, crawling about on plants and sessile animals. The body is very small and the legs very long, hence the animals seem like clusters of legs attached to each other. They have a sucking mouth. (A.W.L.)

PYCNOMETER. A device for measuring densities of liquids. It is a container, usually in the form of a bottle or a pipette-like tube, the capacity of which is accurately known and which may be completely filled with the liquid. The difference in weight when filled and when empty, together with the known volume of the liquid, gives the density. The pipette form has a mark to show how far to fill it, and is bent into a V-shape to facilitate immersion in a temperature bath. A familiar

Sea spider.

design is the "specific gravity bottle," a small flask with a ground and perforated stopper, and sometimes provided with a thermometer. In one of the most precise forms the stopper has a conical top with the capillary leading to the apex, and both neck and stopper are covered by a tight-fitting ground-glass cap to prevent evaporation.

A preliminary step necessary to precise work with the pycnometer is the determination of its two volume constants; that is, the constants of the linear equation expressing the capacity as a function of the temperature. This is done by filling with distilled water and weighing accurately several times at each of two temperatures near the ends of the range for which the pycnometer is to be used. The bottle form is also adapted to the precise measurement of densities of solids. (See **Density and Specific Gravity.**) (L.D.W.)

PYELITIS, PYELONEPHRITIS. Pyelitis is an infection of the pelvis of one or both **kidneys.** *B. coli* is the commonest causative bacterium, although **streptococci** and **staphylococci** may cause the disease. Infection may occur from the blood stream, or by way of retrograde extension from infection in the **bladder** and **ureters.** Local conditions in the pelvis of the kidney such as congenital anomalies of structure, stones, or stasis due to obstruction to the flow of urine at any point in the urinary tract, predispose to pyelitis. Dilatation of the ureter and kidney pelvis are not uncommonly associated with pyelitis in **pregnancy.**

The symptoms of acute pyelitis are usually sudden in onset; chills, high fever, pain in the flanks and back, and sometimes nausea and vomiting occur. Frequent, painful urination is common. The urine is cloudy, and loaded with **bacteria** and pus cells (white **blood** cells). The symptoms subside rapidly with early treatment with **sulfonamides** in the majority of cases.

Pyelonephritis is an almost invariable complication of pyelitis. It is a more serious disease because the infection spreads to invade the functional part of the kidney, the parenchyma of the organ. The symptoms and signs are similar to those of pyelitis, but more severe. Treatment, as with pyelitis, is with sulfonamides and large amounts of fluid. If pyelonephritis becomes chronic, late cardio-vascular complications with **hypertension** and kidney failure may occur.

Pyelonephrosis, or dilatation of the pelvis of the kidney with pus, occurs as the end-picture of severe pyelonephritis. It is associated with marked destruction of kidney tissue. Surgical removal of the kidney is usually necessary. (D.M.H.)

PYELOGRAM. A diagnostic **x-ray** picture of the **kidney** and **ureter.** This is made by one of two methods. In one, opaque fluid is injected upward through fine catheters inserted into the ureters, after a **cystoscope** has been passed into the bladder. With the other method a radio opaque substance is injected into the blood stream. This material is excreted by the kidneys and shows up the urinary passages when the x-ray is taken. (D.M.H.)

PYGIDIUM. 1. The united caudal segments of the extinct **trilobites,** a group of **arthropods** ancestral to the insects. 2. A caudal portion of the body of some scale insects made up of four fused segments. 3. The terminal segment of the abdomen of a **beetle** when exposed beyond the elytra. (See **Invertebrate Paleontology.**) (A.W.L.)

PYLORUS. The narrow muscular passage from the stomach into the **duodenum.** The pylorus is under a complex **nerve control** and food can only pass from the stomach when this muscular **sphincter** relaxes. It is in the immediate vicinity of the pylorus that **ulcers**

most frequently occur. **Peptic ulcer** or **cancer** in this location may obstruct the passage of food through the pylorus and make surgical interference imperative. Pyloric stenosis is a congenital constriction of the pylorus which results in vomiting and obstruction in early infancy. The situation is remedied surgically by cutting the constricting bands. (R.S.M., D.M.H.)

PYORRHEA. A disease of the gums and structures supporting the teeth, characterized by degeneration and infection of these tissues, resorption of the alveolar bone, and loosening of the teeth. Pyorrhea occurs commonly in old age. Although the primary cause is not known, predisposing causes are manifold. They include local irritation, trauma, malocclusion, ill-fitting fillings and crowns, poor hygiene, faulty tooth-brushing, etc. The role of nutrition and general health is not well understood. Treatment is directed toward clearing the infection and removing badly involved teeth. (D.M.H.)

PYRARGYRITE. An **antimony**-bearing **silver** mineral corresponding to the formula Ag_3SbS_3. It crystallizes in the **hexagonal** system, commonly in **rhombic** prismatic forms. It displays a rhombohedral **cleavage;** fracture, conchoidal to uneven; brittle; hardness, 2.5; specific gravity, 5.77–5.85; luster, metallic; color, grayish-black to black. In thin fragments deep red by transmitted light, otherwise practically opaque; streak, purplish red. Pyrargyrite occurs with **proustite,** other silver minerals, and **galena, sphalerite,** etc. It is found in the Harz Mountains, in Czechoslovakia, Bolivia, Chile, Mexico, and in the United States in Colorado, Idaho, and Nevada. In Canada it is found in the Cobalt region of the Province of Ontario. It derives its name from the Greek words meaning fire and silver. (E.S.C.S.)

PYRHELIOMETER. Meteorological Instruments.

PYRIBOLE. A convenient term for the **hornblende** and **pyroxene** groups of dark minerals which are easily distinguishable with the naked eye from the black **mica, biotite,** but not from each other, when they occur in a fine-grained rock. (R.M.F.)

PYRIDINE AND RELATED COMPOUNDS. Pyridine, monazine (C_5H_5N), contains a ring of 1 nitrogen and 5 carbons with 1 hydrogen attached to each carbon:

Pyridine is a colorless liquid, boiling point 115° C., of unpleasant odor, miscible in all proportions with water, alcohol, or ether, forms salts with acids, the **perchlorate** and **ferrocyanide** sparingly soluble in water and used in the separation and identification of pyridine, reacts only under drastic conditions with **chlorine, bromine, sulfuric acid** (forms beta-pyridine sulfonic acid at 300° C. with concentrated sulfuric acid); non-reactive with **nitric acid, chromic** acid, **potassium** permanganate. Pyridine may be reduced by **sodium** and **alcohol** to piperidine (below), and by heating with **hydriodic acid** to normal-pentane. Pyridine is obtained in the destructive distillation of fat-containing bones—in Dippel's oil—and in small proportions in the similar distillation of coal, peat, carboniferous shales, wood, and in the treatment of various **alkaloids** with **alkalis,** or when they are heated alone or with zinc dust. Pyridine may be detected by the appearance of a red color on addition, to the aqueous solution, of a trace of cyanogen bromide and a few drops of aniline. Pyridine is used as a solvent, especially in the purification of **anthracene**, and as a denaturant of **ethyl alcohol**.

Picolines and Pyridine Carboxylic Acids. Alpha-picoline, alpha-methylpyridine, 2-methylpyridine

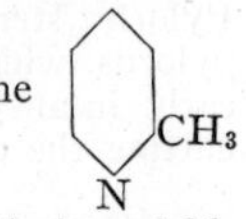

colorless liquid, boiling point 128° C., on oxidation yields picolinic acid, pyridine-alpha-carboxylic acid, sublimes 135° C.

Beta-picoline, beta-methylpyridine, 3-methylpyridine

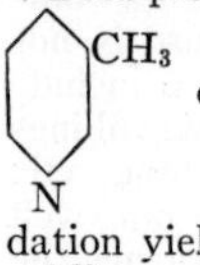

colorless liquid, boiling point 144° C., on oxidation yields nicotinic acid, pyridine-beta-carboxylic acid, sublimes 228° C.

Gamma-picoline, gamma-methylpyridine

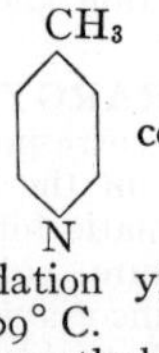

colorless liquid, boiling point 143° C., on oxidation yields isonicotinic acid, white solid, melting point 309° C.

Lutidines. Six dimethylpyridines and three ethylpyridines are known.

Collidines. Collidine, symmetrical-trimethylpyridine (2,4,6), is 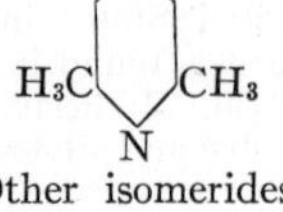 colorless liquid, boiling point 172° C. Other isomerides numbering 21 are theoretically possible.

Hexahydropyridine, piperidine

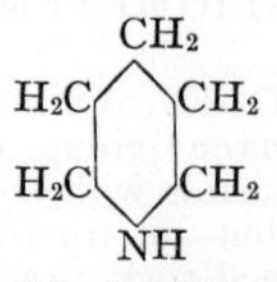

colorless liquid, boiling point 106° C., miscible in all proportions with water, alcohol, ether, or benzene, is formed by the action of sodium and alcohol on pyridine. Piperidine may be obtained by treatment of the **alkaloid** piperine of the pepper plant, by heating with **alkali**.

Azines. When **carbons** of pyridine (monazine) are replaced in succession by **nitrogen**, polyazines are formed, thus:

Orthodiazine (pyridazine)

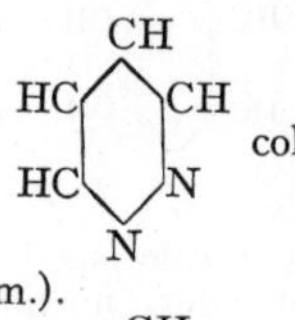

colorless liquid, boiling point 205° C. (755 mm.).

Metadiazine (pyrimidine)

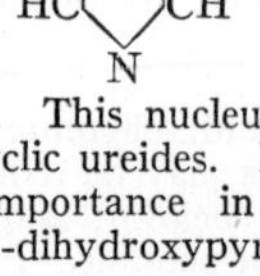

melting point 21° C., boiling point 124° C. This nucleus is contained in purine compounds and cyclic ureides. Derivatives of pyrimidine are of great importance in physiological processes, e.g., 5-methyl-2, 6-dihydroxypyrimidine (thymine) of the **cell** nucleus.

Paradiazine (pyrazine)

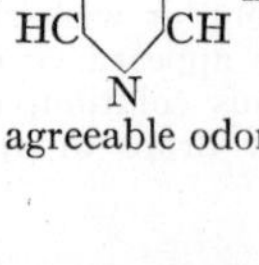

melting point 55° C., boiling point 115° C., of agreeable odor.

Reduction of pyrazine yields hexahydropyrazine (piperazine) 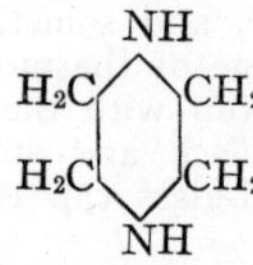 melting point 104° C., boiling point 145° C., soluble in water, and strongly basic.

Quinoline and isoquinoline are benzopyridines (C_9H_7N).

Quinoline

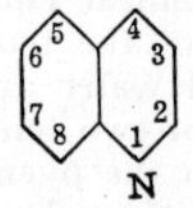

is a colorless liquid, boiling point 238° C., of characteristic odor, slightly soluble in water, miscible with alcohol, ether, and many organic liquids in all proportions, forms **salts** with **acids,** the dichromate sparingly soluble in water, reacts with concentrated sulfuric acid by sulfonation of the **benzene** nucleus, with nitric acid with difficulty (5-nitroquinoline and 8-nitroquinoline), with **chromic** acids in sulfuric acid not reactive, with **permanganate** forms quinolinic acid

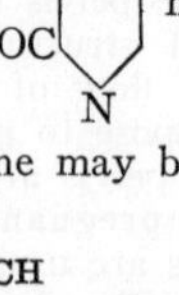

melting point 192° C., plus oxalic acid. Quinoline may be reduced to tetrahydroquinoline

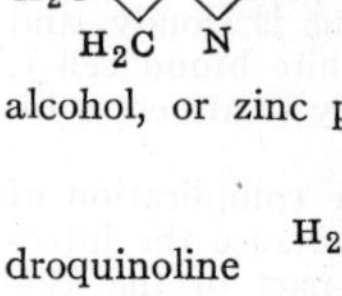

boiling point 245° C., by sodium plus alcohol, or zinc plus hydrochloric acid, and to decahydroquinoline

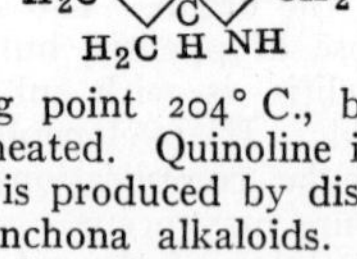

solid, melting point 48° C., boiling point 204° C., by **hydriodic acid** and **phosphorus** heated. Quinoline is found in bone oil and coal tar, and is produced by distillation of many **alkaloids**, e.g., cinchona alkaloids. The classical synthesis of quinoline by Skraup is the reaction of aniline

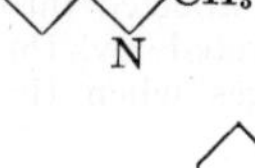

plus **glycerol** ($CH_2OH \cdot CHOH \cdot CH_2OH$), using **nitrobenzene** in **sulfuric acid** heated as oxidizing agent.

Methylquinolines are quinaldine, 2-methylquinoline

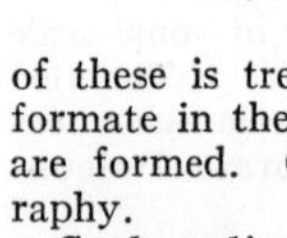

boiling point 247° C., lepidine, 4-methylquinoline

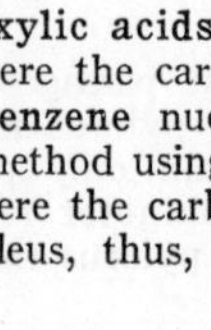

boiling point 257° C. When either of these is treated with ethyl iodide plus ethyl orthoformate in the presence of pyridine, cyanine compounds are formed. Cyanines are dyes used in color photography.

Carboxylic acids of quinoline are of two types, (1) those where the carboxyl group (—COOH) is attached to the **benzene** nucleus. These can be made by the Skraup method using the desired aminobenzoic acid, (2) those where the carboxyl group is attached to the pyridine nucleus, thus, quinoline-2-carboxylic acid, quinaldinic acid melting point 156° C., quino-

line-4-carboxylic acid, cinchoninic acid

melting point 254° C., 6-methoxyquinoline-4-carboxylic

acid, quininic acid melting point 280° C.

(decom.).

Isoquinoline white solid, melting point 23° C., boiling point 243° C., forms salts with acids, the sulfate sparingly soluble in water. Isoquinoline is important as a constituent of certain vegetable **alkaloids**.

Acridine is dibenzopyridine

$$\left(C_{13}H_9N \text{ or } C_6H_4 \left\langle \begin{matrix} CH \\ N \end{matrix} \right\rangle C_6H_4 \text{ or } \right) \text{ white}$$

solid, melting point 108° C., boiling point 346° C. Phenazine is dibenzopyrazine (dibenzoparadiazine)

$$\left(C_{12}H_8N_2 \text{ or } C_6H_4 \left\langle \begin{matrix} N \\ N \end{matrix} \right\rangle C_6H_4 \text{ or } \right) \text{ yellow}$$

solid, melting point 171° C., from which important dyes, e.g., safranines, are derived.

Phenoxazine is white solid, melting point 148° C.

Where **nitrogen** of pyridine is occupied by **oxygen**, pyrone is the compound, and where occupied by **sulfur**, penthiophene; similarly, where nitrogen of quinoline by oxygen chromone is the compound, and where by sulfur, thionaphthalene, and where nitrogen of acridine by oxygen, xanthene, and by sulfur, thioanthracene.

Compounds containing 1 (or more) nitrogens and 4 (or less) carbons in the ring (5-membered) are discussed under **Pyrrole and Related Compounds**. (R.K.S.)

PYRIFORM GLAND. Glands of the **spiders** which secrete silk for the formation of the disks by which threads are anchored. (A.W.L.)

PYRIFORM ORGAN. An organ of unknown function found in the Cyphonautes **larvae** of **bryozoans**. (A.W.L.)

PYRITE. The mineral pyrite or iron pyrites is **iron disulfide**, FeS_2, its **isometric** crystals usually appearing as cubes or **pyritohedrons**. It has a slightly conchoidal to uneven fracture; brittle; hardness, 6–6.5; specific gravity, 4.95–5.10; metallic luster; color, pale to normal brass-yellow; streak, greenish-black; opaque. **Arsenic, nickel, cobalt, copper,** and **gold** may be found in small quantities in pyrite, auriferous pyrite being sometimes a very valuable ore. Pyrite is the commonest of the sulfide minerals, and is of world-wide occurrence. It is

found associated with other sulfides, or with oxides, in **quartz** veins, in **sedimentary** and **metamorphic** rocks, in coal beds, and as the replacement material in fossils. There are many well-known pyrite localities, among which are the Rio Tinto mines in Spain, where copper-bearing pyrite is obtained from huge deposits. In the United States pyrite is found in California, New York, and Virginia in workable deposits. Pyrite is used in the production of sulfur dioxide for the paper industry and the making of sulfuric acid. Its use seems to be on the wane because of the cheapness of Louisiana sulfur. By-product pyrite, obtained from coal, however, is a commercial source of sulfur dioxide in the Illinois region. The name pyrite is derived from the Greek word meaning fire, because of the sparks which result when pyrite is struck with steel. (E.S.C.S.)

PYROCLASTS. A general term for fragmental, volcanic ejectamenta such as **agglomerates**, ashes and **tuffs**. (R.M.F.)

PYROGENETIC MINERALS. A term for the primary **magmatic** minerals of **igneous rocks** as distinguished from those minerals which are the result of special and later processes such as come under the head of pneumatolytic, hydrothermal, etc. (R.M.F.)

PYROLUSITE. The mineral pyrolusite, **manganese** dioxide, MnO_2, appears to be **orthorhombic**, but may be only **pseudomorphous** after **manganite**. It is found massive or in indistinct crystalline aggregates, often acicular. It is soft; hardness, 2–2.5; specific gravity, 4.73–4.86; luster, metallic; color, steel gray to black; streak, black; opaque. Pyrolusite is found as replacement deposits and as residual and sedimentary masses. **Psilomelane** is its usual associate. European localities for pyrolusite are in Bohemia, Saxony, the Harz Mountains, England, and elsewhere. Other deposits occur in India and Brazil. In the United States it is found in Arkansas and Michigan. It is an ore of manganese, is used as a pigment and as an oxidizing agent in glass manufacture. It is from this latter use that it derives the name pyrolusite from the Greek words meaning *fire* and *to wash*. (E.S.C.S.)

PYROMETER. By common usage, the pyrometer is the device for measuring high temperatures. Actually, there are pyrometers which are much used for measuring temperatures in the same range as thermometers, but several of the pyrometers are suitable only for high temperatures. Three types of pyrometers are described as being typical of the entire group. These are (1) the thermoelectric (see **Thermel**) pyrometer, (2) the **optical pyrometer**, (3) the **radiation pyrometer**. (A fourth type, the **resistance pyrometer**, is not commonly used for high-temperature industrial work.)

The thermocouple pyrometer is a convenient meter for measuring a number of high-range temperatures. By means of the rotary switch shown in the diagram a number of thermocouples may be connected in turn to the indicator. The thermocouple is a pair of electrical conductors of different material permanently joined at one end, the other ends being free to connect to an instrument for measuring electromotive force. The indicating instrument is essentially a potential **galvanometer** measuring e.m.f.'s of the magnitude of 50–70 millivolts. The thermocouple itself is delicate and is enclosed in a protecting tube.

An optical pyrometer, as the diagram shows, is one in which the eye compares the radiation emanating from an incandescent object whose temperature is to be measured, with that of a filament electrically heated in the tube of the pyrometer. By interposing between the eye and the hot body a red glass screen, monochromatic light is received by the eye. The current to the bulb is adjusted by rheostat until the intensity of emission from the filament exactly equals that from the light source. It is readily possible to make this adjustment, as with monochromatic light received both from the hot body and the filament, the filament appears black against the image of the hot body when the filament temperature is lower than that of the hot body, white when it is higher than that of the hot body, and disappears when the temperature is the same. In use, the pyrometer is pointed at the hot body, which is viewed through the eyepiece. The pyrometer is held in one hand, and

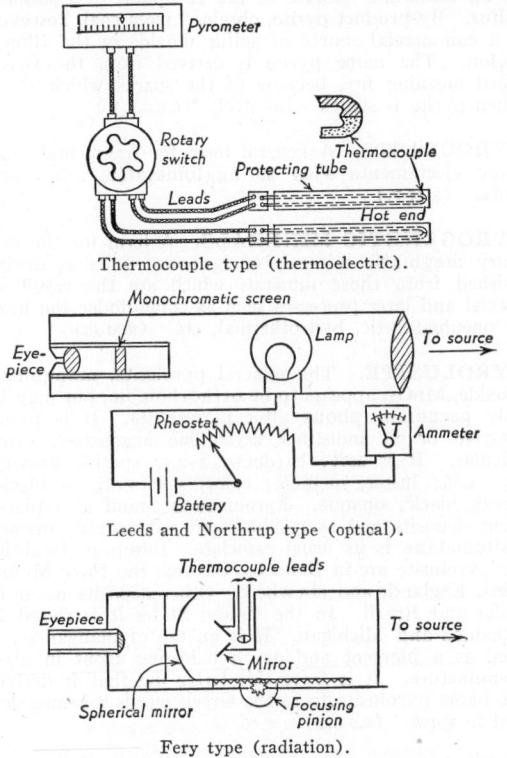

Thermocouple type (thermoelectric).

Leeds and Northrup type (optical).

Fery type (radiation).
Pyrometers.

the rheostat manipulated by the other until the filament disappears against the background. A sensitive ammeter, which measures the current flowing in the filament circuit, is calibrated directly in degrees of temperature.

Total radiation pyrometers are equipped with a mirror which concentrates energy received by radiation from the source upon a thermocouple. This thermocouple is connected to a measuring instrument in a manner similar to that of the thermoelectric pyrometer. Since the total radiation received by a body at T_2 from one at T_1 is $K(T_1{}^4 - T_2{}^4)$, no great error is involved in high-temperature pyrometers by neglecting the term $T_2{}^4$. The heat energy received, then, is proportional to the fourth power of temperature, so the galvanometer in the thermocouple circuit may be calibrated in degrees of temperature of the source which is radiating energy to the mirror, which converges it on the thermocouple.

Radiation and optical pyrometers have this advantage over the thermoelectric pyrometer; no part of them need come into direct contact with the hot body, and no part need be raised to a high temperature. (F.T.M.)

PYROMORPHITE. The mineral pyromorphite is lead chloro-phosphate with a formula corresponding to $(PbCl)Pb_4(PO_4)_3$. The phosphorus is sometimes replaced by arsenic and the lead by calcium. It occurs in prismatic, sometimes hollow, **hexagonal** crystals or may appear in massive forms. It is brittle; hardness, 3.5–4; specific gravity, 6.5–7.1; luster, resinous; color,

green, yellow-green, yellow, brown, and less often gray or white; translucent to opaque. Pyromorphite is a secondary mineral associated with other lead minerals, but is seldom found in large quantities. It has probably resulted from the action of waters bearing phosphoric acid upon the pre-existing lead minerals. Localities for pyromorphite are in the Ural Mountains, Saxony, France, Spain, Cornwall and Cumberland, England; in Scotland, the French Congo, and Australia. In the United States pyromorphite has been found in Chester and Montgomery Counties, Pennsylvania; in Davidson County, North Carolina, and in the Coeur d'Alene mining district of Idaho. The name is derived from the Greek words meaning fire and form. (E.S.C.S.)

PYRONE. Furane and Related Compounds.

PYROPE. Garnet.

PYROPHOSPHATE. Phosphoric Acid.

PYROPHYLLITE. The mineral pyrophyllite is a hydrous **silicate** of **aluminum** corresponding to the formula $H_2Al_2(SiO_3)_4$. **Orthorhombic** with a basal cleavage, it is usually, however, in foliated, lamellar, or fibrous masses, sometimes compact. It is a soft mineral with a greasy feel; hardness, 1–2; specific gravity, 2.8–2.9; luster, pearly to dull; color, white, greenish, grayish, yellowish, and brownish; translucent to opaque. It is found making up **schists** or in **foliated** masses in the Ural Mountains, in Switzerland, Sweden, Brazil, and in the United States in Pennsylvania, North Carolina, Georgia, and California. It is used to some extent as is the mineral **talc,** and also for making slate pencils, hence the name pencil stone sometimes applied to pyrophyllite. (E.S.C.S.)

PYROXENE. This is the name given to a closely related group of minerals, all of which show a distinct **cleavage** angle of 87° or 93° parallel to the fundamental prism. Chemically the pyroxenes are **metasilicates** corresponding to the formula $RSiO_3$, where R may be calcium, magnesium, iron, or less commonly **manganese, zinc, sodium,** or **potassium.** Rarely **titanium, zirconium,** or **fluorine** may be present.

The pyroxenes crystallize in the **orthorhombic, monoclinic,** and **triclinic** systems, like the **amphiboles,** the chief difference between the two groups being the cleavage angles, which for amphibole are 56° and 124°. Pyroxene crystals tend to be short, stout, complex prisms as opposed to the long, slender, and simpler amphiboles.

The pyroxenes are common in the more **basic igneous** rocks, both intrusive and extrusive, and may be developed by the **metamorphic** processes in **gneisses, schists,** and marbles.

The following members of the pyroxene group are described under their own headings: **acmite, aegirite, augite, babingtonite, bronzite, diallage, diopside, enstatite, hypersthene, jadeite, rhodonite,** and **spodumene.**

Pyroxene was so named by Haüy from the Greek meaning fire and stranger, hence a stranger in the domain of fire. Nothing could be more unlike the truth, for the various pyroxenes are typically minerals of the igneous rocks. (E.S.C.S.)

PYROXENITE. A coarse-grained rock related to **gabbro** which consists almost wholly of **pyroxene.** It may also carry small amounts of **quartz** or **olivine.**

Pyroxenite is a rare rock type and of relatively little quantitative importance. (E.S.C.S.)

PYROZALE. Pyrrole and Related Compounds.

PYRRHOTITE.

PYRRHOTITE. The mineral pyrrhotite, sometimes called magnetic **pyrites,** is a sulfide of **iron** with varying amounts of **sulfur.** Analyses indicate formulae from Fe_5S_6 to $Fe_{16}S_{17}$. Pyrrhotite exists in two modifications: it is **hexagonal** below, and **orthorhombic** above 138° C. It is a brittle mineral; hardness, 3.5–4.5; specific gravity, 4.58–4.64; luster, metallic; color, reddish bronze-yellow when fresh, otherwise tarnished; **streak,** grayish-black; **magnetic.** It may carry nickel, generally as **pentlandite,** when it becomes a valuable nickel ore as at Sudbury, Ontario. Pyrrhotite is commonly associated with the basic **igneous rocks** like **gabbro, norite,** etc., and occurs with **chalcopyrite, magnetite, pyrite,** etc. Besides being apparently of **magmatic** origin, it has been found as contact **metamorphic** and as vein deposits. Austria, Italy, Saxony, Bavaria, Switzerland, Norway, Sweden, and Brazil have deposits of more or less importance, and in the United States it has been found associated with **andalusite** crystals at Standish, Maine; also at Brewster, New York; Lancaster County, Pennsylvania, and elsewhere. At Ducktown, Tennessee, it is found together with **copper** and **zinc** minerals. It is mined for its nickel content in Sudbury, Ontario.

Pyrrhotite derives its name from the Greek word meaning reddish in reference to the color of the fresh ore. (E.S.C.S.)

PYRROLE AND RELATED COMPOUNDS.

PYRROLE AND RELATED COMPOUNDS. Pyrrole (monoazole, C_4H_5N or C_4H_4NH), contains a ring of 1 nitrogen and 4 carbons, with 1 hydrogen attached to nitrogen and to each carbon:

Pyrrole is a colorless liquid, boiling point 131° C., insoluble in water, soluble in alcohol or ether. Pyrrole dissolves slowly in dilute **acids,** being itself a very weak **base;** resinification takes place readily, especially with more concentrated solutions of acids; and on warming with acid a red precipitate is formed. Pyrrole vapor produces a pale red coloration on pine wood moistened with **hydrochloric acid,** which color rapidly changes to intense carmine red. Pyrrole may be made (1) by reaction of succinimide

with zinc and **acetic acid,** or with **hydrogen** in the presence of finely divided **platinum** heated, (2) by reaction of **ammonium** saccharate or mucate $(COONH_4 \cdot (CHOH)_4 \cdot COONH_4)$ with glycerol at 200° C. by loss of **carbon dioxide, ammonia,** and water.

When pyrrole is treated with **potassium** (but not with **sodium**) or boiled with solid potassium hydroxide, potassium pyrrole (C_4H_4NK) is formed, which is the starting point for N- derivatives of pyrrole, since reaction of the potassium with halogen of organic compound and with carbon dioxide, readily occurs. When pyrrole is treated with **magnesium** metal and ethyl bromide in ether, pyrrole magnesium bromide plus ethane is formed, which may be used as the starting point for C- derivatives of pyrrole, since reaction with sodium alcoholates readily occurs (with separation of magnesium oxybromide).

The pyrrole nucleus has been shown to be present in the complex substances, **chlorophyll** (the green coloring matter of plants), **haematin** (the red coloring matter of **blood**), and in the coloring matter of **bile.**

Hydropyrrole compounds are 2,5-dihydropyrrole, pyrroline

boiling point 91° C., and tetrahydropyrrole, pyrrolidine

boiling point 81° C.

Keto-pyrrolines are called pyrrolones, e.g., 1-phenyl-4,5,-dimethyl-2-pyrrolone ("antipyrine") and keto-pyrrolidines are called pyrrolidones, e.g., 2-keto-pyrrolidones. Two aminoacids are derivatives of pyrrolidine, namely, proline (pyrrolidine-alpha-carboxylic acid)

melting point 221° C., decomp., and oxyproline (4-hydroxypyrrolidine-2-carboxylic acid)

. Certain **alkaloids** contain the pyrrolidine nucleus.

Azoles. When the **carbons** of pyrrole (monoazole) are replaced in succession by **nitrogen,** di-, tri-, and tetrazoles are formed, thus:

1, 2,–diazole (pyrazole), melting point 70° C., boiling point 188° C.

1, 3,–diazole, iminazole, glyoxaline, melting point 88° C., boiling point 255° C.

1, 2, 3,–triazole

1, 2, 4,–triazole (pyrrodiazole) melting point 121° C., boiling point 260° C.

tetrazole (pyrrotriazole) melting point 155° C., sublimes

Hydrazole compounds are known, 4,5-dihydropyrazole, pyrazoline

boiling point 144° C. and tetrahydropyrazole, pyrazolidine

.

Indole is benzopyrrole (C_8H_7N or $C_6H_4CH{:}CHNH$ or

white, odorous solid, melting point 52° C., boiling point 254° C., soluble in hot water, soluble in ether or alcohol. Indole behaves similarly to pyrrole with acids, and with pine wood moistened with **hydrochloric acid** (cherry-red coloration).

Three methyl indoles are (1) C-methyl indole (3),

skatole white, solid, melting point 95° C., boiling point 268° C., of powerful, disagreeable odor, present in putrefied **albuminous** matter and in human feces, (2) C-methyl indole (2) [structure: CH, CCH₃, NH] white,

odorous solid, melting point 60° C., boiling point 268° C.,

(3) N-methyl indole [structure: CH, CH, NCH₃] colorless liquid, boiling

point 240° C., without unpleasant odor.

Carboxylic acids of indole are (1) indole-3-acetic acid [structure: CCH₂COOH, CH, NH] white solid, melting point 164° C.,

heated above this temperature decomposes into skatole plus **carbon dioxide**, (2) indole-3-alpha-aminopropionic

acid, tryptophane [structure: CCH₂CHNH₂COOH, CH, NH] melting

point 289° C. (decomp.), which gives rise to the derivatives of indole formed during the putrefaction of **proteins**, and yields indole-3-ethyl alcohol, trypophol

[structure: CCH₂CH₂OH, CH, NH] white solid, melting point 59° C.,

by **fermentation** with yeast in the presence of sugar.

Indoxyl is 3-hydroxyindole [structure: C.OH, CH, NH] yellow solid,

melting point 85° C., disagreeable odor, reacts also as

keto-form, pseudo-indoxyl [structure: CO, CH₂, NH] hypothetical,

on exposure to air oxidizes to indigotin.

Indoxylic acid is 3-hydroxyindole-2-carboxylic acid

[structure: C.OH, C.COOH, NH] or [structure: CO, C, H, NH, COOH]

white solid, melting point 178° C. (decom.), and forms an intermediate product in the technical synthesis of **indigo** from phenylglycine-ortho-carboxylic acid

[structure: COOH, NH.CH₂.COOH]

Isatin [structure: CO, CO, NH] or [structure: CO, C.OH, N]

red solid, melting point 199° C., sublimes, is related to indole and indigo.

Indigo blue, indigotin [structure: CO, NH, C : C, OC, HN]

may be made synthetically by reaction of anthranilic

acid (ortho-aminobenzoic acid), [structure: COOH, NH₂] and mono-

chloroacetic acid to form phenylglycocoll ortho-carboxylic acid, which with fused sodium hydroxide yields indoxyl carboxylic acid and then indoxyl. Indoxyl, upon exposure to air, oxidizes spontaneously to indigotin. Since indigotin is insoluble and has no dyeing properties as such, it is customary in dyeing to reduce to soluble indigo white [structure: COH, C—C, COH, HN, HN] by **calcium**

hyposulfite (CaS₂O₄) in alkaline solution, to immerse the cotton fiber in the solution (vat-dyeing), and then withdraw the fiber from the bath and expose to the air, whereupon indigotin is formed in the fiber. In dyeing wool, soluble indigo disulfonic acid, "indigo-carmine," is used.

Monobromo and dibromoindigotin are valuable **dyes**, sometimes used instead of indigo.

Carbazole is dibenzopyrrole, diphenyleneimine (C₁₂H₉N

or C₆H₄·NH·C₆H₄ or [structure: NH]) white solid, melting

point 245° C., boiling point 355° C.

Indazole [structure: CH, NH, N] and isindazole [structure: CH, N, NH] hypothetical

are benzopyrazoles, derivatives of which have been prepared.

Benziminazole is [structure: N, CH, NH] melting point 170° C.

Where **nitrogen** (group —NH) of pyrrole is occupied by **oxygen**, furane is the compound, and where occupied by **sulfur**, thiophene; similarly, where nitrogen of indole by oxygen, coumarone is the compound, and where by sulfur, benzothiophene; and where nitrogen of carbazole by sulfur, diphenylene sulfide.

Compounds containing 1 (or 2) nitrogen and 5 (or 4) carbons in the ring (6-membered rings) are discussed under **Pyridine and Related Compounds**.

Oxygen-nitrogen ring compounds:

Oxazole [structure: HC, N, HC, CH, O] Isoxazole [structure: HC, CH, HC, N, O]

Benzoxazole [structure: N, CH, O] melting point 30° C., boiling point 182° C.

The following tetrapyrrole pigments are known:

1. Chain-compounds. Bile pigments.
   a. Bilirubin ...... (C₃₃H₃₆N₄O₆). The orange pigment of **bile**.
   b. Biliverdin ..... (C₃₃H₃₆N₈O₈). The green pigment of bile.
   c. Mesobilirubin-ogen .... By reduction of bilirubin.
   d. Mesobilirubin . . The brown pigment of **urine**.
   e. Uteroverdin ... The green pigment of eggshells, of the **placenta** of the dog, and of the gallstones of the ox.
2. Ring-compounds. Porphyrins: chlorophyll and blood pigments.
   f. Ooporphyrin . . . (C₃₄H₃₅N₄O₄). The brown pigment of egg-shells.

g. Uroporphyrin . . ($C_{38}H_{38}N_4O_{16}$). Found in the blood and urine in cases of congenital porphyrinuria.

h. Haematopor-
    phyrin ...... ($C_{34}H_{38}N_4O_6$).

i. Aetioporphyrin ($C_{32}H_{38}N_4$).

3. Metallic compounds. Porphyrans.

j. Aetiophyllin ...The magnesium porphyran of chlorophyll.

k. Haematin .....The iron porphyran of haemoglobin

l. Cytochrome ...The respiratory iron porphyran of tissues.

m. Turacin ....The copper porphyran of the feathers of birds.

Aetioporphyrin has been derived from natural sources, and also made synthetically. Chlorophyll, the green pigment of leaves, and haemoglobin, the red pigment of blood, are related to aetioporphyrin. The formula of aetioporphyrin is

(R.K.S.)

PYRROLIDINE. Pyrrole and Related Compounds.

PYTHON. Reptilia, Sauria. Large **snakes** related to the boas. The several species are distributed through Asia, Africa, and Australia. They are largely arboreal, living in forests, usually near the water, which they enter freely. Some of these snakes are very large, the Regal python, *Python reticulatus*, attaining a length of 30′, and one of the African species, *Python sebae,* more than 20′. They are not poisonous, but because of their size and strength they may be dangerous to man under some conditions. (A.W.L.)

Q

Q. This is the figure of merit of a coil and is the ratio of its reactance to its resistance. (L.R.Q.)

QUADRANTAL ANGLES. Angles.

QUADRANT ELECTROMETER. Electroscopes and Electrometers.

QUADRATE BONE. A small bone of roughly quadrangular shape in the vertebrate skull. In the reptiles and lower forms it is included in the articulation of the lower jaw with the skull, and a similar condition persists in the birds. In the mammal the incus or anvil, one of the three small bones of the middle ear, is regarded as the homologue of this bone. (A.W.L.)

QUADRATIC EQUATIONS IN ONE UNKNOWN. A quadratic equation in one unknown is an equation of the form

$$ax^2 + bx + c = 0 \quad (a \neq 0),$$

where a, b, c are constants (independent of x), and x is the unknown.

A quadratic equation may be solved (1) by factoring, (2) by completing the square, (3) by the quadratic formula, or (4) graphically.

If a quadratic equation can be factored by inspection into linear factors, as $a(x - \alpha)(x - \beta) = 0$, then the roots are α and β.

The method of completing the square consists in transposing the constant term, adding a term to both sides to make the left-hand side a perfect square of a binomial, extracting the square root of both sides and solving the resulting linear equations.

The roots of the quadratic equation $ax^2 + bx + c = 0$ may also be found immediately by use of the quadratic formula:

$$x = \frac{-b \pm \sqrt{b^2 - 4ac}}{2a}.$$

One graphical method for solving a quadratic equation $ax^2 + bx + c = 0$ is to plot in rectangular coordinates the graph (a parabola) of the function $y = ax^2 + bx + c$, and then find the abscissas of the points where this parabola cuts the X-axis. This gives only the real roots of the equation.

Another graphical method for the solution of the equation $ax^2 + bx + c = 0$ is obtained by drawing the graph in rectangular coordinates of the equation $y = x^2$ and on the same diagram the graph of the straight line $ay + bx + c = 0$, and finding the abscissas of the points of intersection of these two graphs.

The character of the roots of a quadratic equation is determined by an inspection of the discriminant of the equation, as follows:

If the coefficients a, b, c of the quadratic equation $ax^2 + bx + c = 0$ ($a \neq 0$) are real numbers, then:
  if $b^2 - 4ac > 0$, the roots are real and unequal,
  if $b^2 - 4ac = 0$, the roots are real and equal,
  if $b^2 - 4ac < 0$, the roots are complex numbers.
If the coefficients a, b, c are rational numbers and if $b^2 - 4ac > 0$, then:
  if $b^2 - 4ac$ is a perfect square, the roots are rational,
  if $b^2 - 4ac$ is not a perfect square, the roots are irrational.

For this reason, the expression $b^2 - 4ac$ is called the discriminant of the equation.

The relation between the roots x_1 and x_2 and the coefficients of the equation $ax^2 + bx + c = 0$ are given by:
  the sum of the roots $x_1 + x_2 = -b/a$,
  the product of the roots $x_1 x_2 = c/a$.

An equation is said to be in the quadratic form if it can be transformed into a quadratic equation by means of the substitution of a new variable representing an expression involving the original unknown. Such equations can be solved by use of the methods for quadratic equations. (L.L.S.)

QUADRATIC EQUATIONS, SYSTEMS OF. The usual types of systems of equations involving quadratics are: the linear-quadratic systems, in which one equation is linear and one quadratic, and the quadratic-quadratic systems, in which both equations are quadratic.

Both types of systems of equations may be solved graphically, by constructing the graphs of the equations on the same diagram and finding their intersection points. In the case of the linear-quadratic system, the graphs will be a straight line and a conic, which will intersect in, at most, two points. For the case of the quadratic-quadratic system, the graphs will be two conics, which will intersect in, at most, four points. The coordinates of the intersection points in either case will give only the real solutions.

Linear-quadratic systems of equations may be solved algebraically by solving the linear equation for one unknown in terms of the other, and this result substituted in the quadratic equation gives a quadratic equation in one unknown.

There is no single uniform method for solving quadratic-quadratic systems, but several simple methods will handle the majority of cases.

If both equations of the system are of the form $ax^2 + by^2 = c$, they can be solved as a linear system in x^2 and y^2, and then x and y immediately found.

If both equations are of the form $ax^2 + bxy + cy^2 = d$, we may eliminate the constant terms between the two equations and obtain a single equation of the form $Ax^2 + Bxy + Cy^2 = 0$; this quadratic equation may be solved to obtain one unknown as a linear function of the other. Substitution of the linear expressions in one of the given equations and solving the resulting equation for the remaining unknown enables us to obtain the corresponding values of the other unknown by substitution.

Another method for this case is to substitute $y = vx$ in both equations, solve each resulting equation for x^2 and equate them, and solve the resulting equation for v, then substitute these values of v in one of the equations giving x^2, and thus find x; y can then be found from $y = vx$.

Other special devices can also be used. Systems of equations symmetrical in x and y can often be solved by special devices preserving symmetry, or often by the substitution $x = u + v$, $y = u - v$. (L.L.S.)

QUADRATIC FORM, EQUATIONS IN. Quadratic Equations.

QUADRATIC FORMULA. Quadratic Equations.

QUADRATIC FUNCTION. A quadratic function is a polynomial function of the second degree, and is therefore a function of the form $ax^2 + bx + c$, where a, b, c are constants and x is the variable.

1174

The **graphic representation** in **rectangular coordinates** of a quadratic function $y = ax^2 + bx + c$ is a **parabola** with principal axis vertical (i.e., parallel to the Y-axis).

The value of x for which the quadratic function $ax^2 + bx + c$ takes its least value if $a > 0$, and its greatest value if $a < 0$, is $x = -b/2a$. This extreme value of the function is $\dfrac{4ac - b^2}{4a}$. (L.L.S.)

QUADRATURE. Planetary Motions.

QUADRIC SURFACES. A surface represented by an equation (in **rectangular coordinates**) of the second degree in x, y, z of the form $Ax^2 + By^2 + Cz^2 + Dyz + Ezx + Fxy + Gx + Hy + Iz + J = 0$ is called a quadric surface or a conicoid.

Every plane section of a quadric surface is a **conic section.**

By translation and rotation of axes, the general equation of the second degree in x, y, z may be reduced to one of several type forms, representing the types of surfaces: **ellipsoid** (including the **sphere**), **paraboloid, hyperboloid, cone, cylinder.**

A quadric surface which has a center of symmetry is called a **central quadric.** The ellipsoid and the hyperboloids are surfaces of this character. (L.L.S.)

QUADRUPLE PRODUCTS OF VECTORS. Important formulae for products of four **vectors** are:

$$(\mathbf{a} \times \mathbf{b}) \cdot (\mathbf{c} \times \mathbf{d}) = (\mathbf{a} \cdot \mathbf{c})(\mathbf{b} \cdot \mathbf{d}) - (\mathbf{a} \cdot \mathbf{d})(\mathbf{b} \cdot \mathbf{c}),$$

$$(\mathbf{a} \times \mathbf{b}) \times (\mathbf{c} \times \mathbf{d}) = [(\mathbf{c} \times \mathbf{d}) \cdot \mathbf{a}]\mathbf{b} - [(\mathbf{c} \times \mathbf{d}) \cdot \mathbf{b}]\mathbf{a}.$$

(L.L.S.)

QUAGGA. Mammalia, Perissodactyla. A South African animal, *Equus quagga,* related to the zebras and asses. It is reddish-brown above, blending to white on the legs, and is marked with dark brown stripes on the head, neck, and fore part of the body. A dark stripe also runs along the middle of the back and down the tail, but otherwise the hind quarters are unmarked. Also spelled couagga. (A.W.L.)

QUAIL. Aves, Galliformes. Small compactly built game birds (**Aves**) related to the partridges. Every continent has species of this name, but those of North and South America and those of the Old World belong

Quail (partridge, bob-white) *Colinus Virginianus.* Above, mottled, reddish brown and gray; under parts white barred with black. White spots on head in male; yellow in female.

to different divisions of the family. The widely distributed bob-white, *Colinus virginianus,* is the best known North American species. Some of the western species are known both as partridges and as quails, notably the valley quail of California, *Lophortyx californica,* and in the south the bob-white also receives the name partridge. Quails are swift-footed but their wings are short and they fly only short distances. Although their colors are mostly very quiet, they are beautiful birds.

Both their appearance and their interesting habits make them desirable residents in any locality. (A.W.L.)

QUALITATIVE VARIABLE. A qualitative variable is one which is not measurable but possesses certain characteristics such as curliness of hair or type of body structure. In **industrial statistics** a product may be classified as defective or non-defective, a qualitative variable. (L.A.A.)

QUALITY. Quality is a term used in connection with **musical sounds.** It is also used technically to describe the condition of a **saturated vapor.** A vapor in a condition intermediate between liquid and a dry vapor is said to have a certain quality which may be defined as the ratio of the vaporized portion to the total weight of liquid and vapor. The vaporization of a liquid requires the expenditure upon it of the latent **heat of vaporization.** When heat to this amount is added, the liquid is converted to a dry vapor. If $x\%$ of this heat is added, only $x\%$ of the liquid is vaporized. x is the quality. (F.T.M.)

QUANTIC. A quantic is a **homogeneous algebraic function** of two or more variables, in general containing only positive integral powers of the variables, and so is usually a **polynomial** in several variables. Quantics are classified into quadratic, cubic, quartic, quintic, etc., according to degree, and into binary, ternary, quaternary, etc., according to the number of variables involved. (L.L.S.)

QUANTITATIVE VARIABLE. A quantitative variable is one which possesses measurable characteristics such as length, age, time, etc. (L.A.A.)

QUANTUM MECHANICS. A very general physical theory, or method of physical reasoning, recently recognized as necessary in dealing with entities so small that our laboratory instruments cannot furnish direct observations upon them. The underlying, revolutionary idea, developed primarily by Heisenberg, is that we are not justified in assuming any precisely foreordained and determinate position or motion for such a thing as an atom or an electron. If one throws a baseball, there are instruments which can measure its position and its velocity within limits small as compared with the size of the ball and the speed of its motion; and its subsequent behavior can be calculated with such precision as is afforded by those measurements. But with an atom or an electron, no corresponding proportionately accurate data can be obtained by any method of measurement at our command. Hence any prediction as to the particle's future performance is subject to error or ambiguity, large as compared with the quantities concerned. Instead of saying that an electron will be at such and such a place and have such and such a momentum at a given time, it is possible only to say that the position or momentum in question is more probable than any other, and to calculate how much more probable one assumed value is than another, neighboring value. Thus, the idea of a definite point occupied by an electron is replaced by a whole distribution of possible points, like the spatter of shots on a target. One curious feature, incidentally, is that the more accurately the position of one of these minute particles can be specified, the less determinate is its velocity, and vice versa. To deal with problems involving this "uncertainty principle," Heisenberg has developed a mathematical procedure involving the use of **matrices,** and this phase of the subject is known as "matrix mechanics." **Wave mechanics** is also an important feature of the more general theory. (L.D.W.)

QUANTUM NUMBERS. Quantum Theory.

QUANTUM THEORY.

A general physical theory wherein it is recognized that in many, if not all, processes involving transfer of energy, the energy must be regarded as having an atomistic character, that is, as passing from one system to another only in discrete, finite increments or "quanta." Processes which, on the mass scale directly observable to our senses, appear to be quite continuous, may, when carried to atomic or electronic orders of magnitude, prove to be "quantized," that is, to proceed by small but finite jumps or readjustments. When water is poured from a tub, we observe no discontinuity in its flow, but when allowed to trickle from a pipette, it flows drop by drop. This is only a very crude analogy. A more valid illustration is this: if a spinning grindstone slows down, it appears to do so gradually; but there is reason to believe that when a molecule loses angular speed of rotation, it passes abruptly from one speed to another that is sometimes a little and sometimes much slower, and that the lost energy is emitted as a small or a large quantum of radiation (see **Molecular Spectra**).

The quantum theory had its inception when, in 1900, Max Planck first recognized, in the **spectral energy distribution of thermal radiation,** a statistical distribution of individual entities. This may be illustrated by supposing a whole season's crop of apples graded as to size, and the number of bushels of each size tabulated in order. Likewise the radiation output of a hot body can be graded into different-sized quanta (corresponding to wavelength), and the total amount of energy made up of each tabulated as the spectral energy distribution. It was only by adopting this view that Planck was able to reconcile theory with experiment in the study of **black-body** radiation.

Today we are accustomed to thinking of nearly all phenomena in terms of quantized processes. Radiation is supposed to emanate from atoms or molecules in quanta of definite frequency and wavelength as the result of abrupt changes from one quantum state or energy level to another, the frequency being determined by the amount of energy released in accordance with **Planck's law.** But the mechanism whereby these definite frequencies are produced is as yet unknown. In general, there is associated with each quantum state of an atom or a molecule an integral number, whose physical significance is not fully understood but which changes from one value to another whenever there is an accession or a release of energy. These integers, called quantum numbers, appear in the empirical formulae for the series terms in **atomic spectra,** and each exchange of one quantum number for another corresponds to a radiation quantum or a photon of light of definite frequency. Series of spectral lines are thus built up from different combinations of quantum numbers, corresponding to transitions between different pairs of quantum states or levels. Since an atom or a molecule may be quantized as to different **degrees of freedom,** there are different sets of quantum numbers, characterized, respectively, as azimuthal, radial, rotational, vibrational, etc.

Among the outstanding features of the quantum theory may be mentioned the interpretation of the laws of radiation, of the details of atomic, **molecular,** and **x-ray spectra,** of the Compton effect and the **Raman effect,** of the **Stern-Gerlach experiment,** of certain **magneto-optical** and **electro-optical** phenomena, and of the laws of **specific heat.** (See also **Quantum Mechanics.**) (L.D.W.)

QUAQUAVERSAL.

A term used by structural geologists to describe a dome in which the formations dip outward in all directions. (R.M.F.)

QUARTER-WAVE PLATE. Double Refraction.

QUARTIC EQUATIONS.

A quartic (or biquadratic) equation in one unknown is a **polynomial equation** of the fourth degree and has the general form

$$a_0x^4 + a_1x^3 + a_2x^2 + a_3x + a_4 = 0.$$

The quartic equation $y^4 + py^3 + qy^2 + ry + s = 0$ may be reduced by the substitution $y = x - \dfrac{p}{4}$ to the form $x^4 + ax^2 + bx + c = 0$. Let l, m and n denote the roots of the resolvent **cubic equation**

$$t^3 + \left(\frac{a}{2}\right)t^2 + \left(\frac{a^2 - 4c}{16}\right)t - \frac{b^2}{64} = 0.$$

The required roots of the reduced quartic are then:

$$x_1 = \pm(-\sqrt{l} - \sqrt{m} - \sqrt{n}),$$
$$x_2 = \pm(-\sqrt{l} + \sqrt{m} + \sqrt{n}),$$
$$x_3 = \pm(\sqrt{l} - \sqrt{m} + \sqrt{n}),$$
$$x_4 = \pm(\sqrt{l} + \sqrt{m} - \sqrt{n}),$$

where the upper signs are to be used if $b > 0$, the lower if $b < 0$. (L.L.S.)

QUARTIC FUNCTION.

This is a **polynomial function** of the fourth degree and is therefore of the form $ax^4 + bx^3 + cx^2 + dx + e$, where a, b, c, d, e are **constants** (independent of x) and x is the **variable;** it is also sometimes called a biquadratic function. (L.L.S.)

QUARTILE.

The quartiles, Q_1, Q_2, Q_3, are the three points dividing a distribution into 4 equal parts. They are determined by this definition in the case of a **serial distribution** and by interpolation in the **cumulative frequency distribution** if a **frequency distribution** has been formed. Sometimes the first quartile refers to all **variates** below Q_1. It should be noted that Q_2 is the same as the **median.** (L.A.A.)

QUARTILE DEVIATION.

The quartile deviation, Q.D., is given by the formula

$$\text{Q.D.} = \frac{Q_3 - Q_1}{2},$$

where Q_3 and Q_1 are the third and first **quartiles,** respectively. (L.A.A.)

QUARTZ.

The mineral quartz, oxide of the nonmetallic element **silicon,** is the commonest of minerals, and appears in a greater number of forms than any other. Its formula is SiO_2. Quartz commonly occurs in prismatic **hexagonal** crystals terminated by a pyramid. This pyramid is due to the equal development of two **rhombohedrons,** as may be observed in cases where one rhombohedron predominates. Cleavage is not observed; the fracture is typically conchoidal; hardness is 7; specific gravity, 2.65; luster, vitreous to greasy or dull; colorless to white, pink, purple, yellow, blue, green, smoky brown to nearly black; transparent to opaque. There seem to be two distinct modifications of quartz, depending upon the temperature at which they were formed. The low-temperature variety is formed below 573° C. and is the more common sort, being found in veins, geodes, etc. It is called low-quartz. The high-temperature modification is formed between 573° C. and 870° C., and is found chiefly in granites and **granite** or **rhyolite porphyries.** This is called high-quartz. Above 870° C. **tridymite** is the stable form of SiO_2. The differences between high- and low-quartz are entirely crystallographic, low-quartz having a vertical axis of three-fold symmetry and three horizontal axes of two-fold symmetry, while high-quartz has a vertical axis of six-fold symmetry and six horizontal axes of two-fold symmetry. It is usual to separate the many kinds of quartz into (1) crystalline or vitreous varieties, actual crystals or vitreous crystalline masses, and (2) **crypto-crystalline** varieties, mostly compact non-vitreous sorts, but which may show a crystalline structure under the microscope.

(1) Crystalline or Vitreous: Rock crystal, colorless crystals or masses. Amethyst, clear violet or purple, either crystals or masses. Rose quartz, usually massive but rarely in crystals, delicate shades of pink or rose, sometimes red. Citrine or yellow quartz, sometimes called false or Spanish topaz, light to deep yellow. Smoky quartz, smoky brown to almost black, often called cairngorm stone from Cairngorm, Scotland. Milky quartz, often showing delicate opalescence, transparent to nearly opaque, often with a greasy luster. Aventurine quartz incloses glistening scales of mica or hematite. Rutilated quartz incloses needle-like prisms of rutile called "fleches d'amour." Other acicular minerals such as **actinolite, tourmaline, epidote,** etc., may also be thus inclosed; Cat's Eye shows a peculiar opalescence, probably due to inclosed masses of some fibrous mineral. Tiger's Eye is a siliceous pseudomorph after crocidolite of a golden yellow brown color. (2) Cryptocrystalline: the following cryptocrystalline varieties of quartz are treated under their own headings: **agate, basanite, bloodstone, carnelian, chalcedony, chert, chrysoprase, flint, heliotrope, jasper,** moss agate, **onyx, plasma, prase, sard,** and **sardonyx.** Quartz readily forms **pseudomorphs** after various minerals or structures. Silicified wood is a quartz pseudomorph after the organic material of which it originally consisted. Quartz is often pseudomorphic after **calcite, barite, fluorite,** etc. Quartz is an essential constituent of many **igneous rocks,** for example, **granites, granite porphyries,** and **felsites,** as well as quartz **diorites** and their surface equivalents, the **dacites.** In the **metamorphic** rocks quartz figures very largely in the **gneisses** and **schists,** and, of course, in **quartzite.** In the sedimentary rocks most sandstones are composed chiefly of grains of quartz, and quartz forms veins and nodules in limestones. Of the many foreign localities that have yielded fine specimens of quartz, a few only can be mentioned as: the Swiss Alps, the Piedmont of Italy, the Island of Elba, Dauphine in France, Cumberland in England, Banffshire, Scotland, and Madagascar. Fine amethysts come from the Urals, Ceylon, Madagascar, Uruguay, Mexico, and Brazil. In the United States the following localities are well known: Paris, Maine, especially for rose quartz; Herkimer County, New York, for small but very brilliant crystals found in the Cambrian dolomites or in the soil. Amethyst County, Virginia, furnishes amethysts, as do Lincoln and Alexander Counties, North Carolina. Other localities for amethyst and smoky quartz are South Dakota in the Black Hills, the Pike's Peak district, Colorado; Yellowstone Park, Wyoming; Jefferson County, Montana, and in Canada in the Province of Ontario in the Thunder Bay region. The word quartz is believed to have been originally of German origin. Besides the use of the different varieties of quartz for jewelry and other ornamental purposes, this mineral has extensive industrial uses in the ceramic arts, optical and other sorts of scientific instruments (see **Polarized Light**), abrasive, scouring, polishing materials, and for refractories. (E.S.C.S.)

QUARTZ PORPHYRY. One of the **hypabyssal** or effusive rocks chemically related to the **granite** or alkali family but rich in silica, which occurs as **quartz phenocrysts** in a crypto- or microcrystalline ground mass. (R.M.F.)

QUARTZITE. A hard, tough, and compact metamorphic rock composed almost wholly of quartz sand grains which have been recrystallized to form a particularly massive siliceous rock. The term is also used for non-metamorphosed quartzose **sandstones** and grits whose **clastic** grains have been firmly cemented by silica which has grown in optical continuity around each grain. (R.M.F.)

QUASI-STATIONARY FRONTS. Fronts.

QUATERNARY. Pleistocene.

QUEBRACHO. Tannins.

QUENCH CIRCUIT. Superregenerative Receiver.

QUENCHING. Immersion of hot metals in liquid baths in order to effect rapid cooling. In steel heat-treating practice quenching oils give slower and brine solutions give faster cooling rates than water. Dilute caustic solutions are sometimes used for rapid cooling rates comparable to brine. These baths are maintained at or near room temperature. For special quenching procedures requiring baths held at moderately elevated temperatures, molten metals, such as lead, and fused salts may be used. (See **Steel, Heat Treating.**)

Quenching of ordinary steel is for the purpose of hardening it. Certain non-hardenable steels, for example austenitic stainless steel, and many non-ferrous metals may be cooled rapidly from elevated temperatures for other reasons. (R.H.H.)

QUEZAL. Trogon.

QUICKLIME. Calcium Oxide.

QUICK-RETURN MECHANISM. Machine tools which cut with a straight stroke in one direction, alternated with non-cutting return strokes, are more efficient of operators' time if the return stroke is performed more rapidly than the forward stroke. The **shaper** offers an example of a machine tool in which the cutting is done by moving the work past a fixed cutting tool which cuts in one direction. The work reciprocates, and in order to save as much time as possible on the non-cutting portion of the **cycle,** quick-return mechanisms have been developed, one of which is illustrated in the accompanying figure. This is a variation of a slider

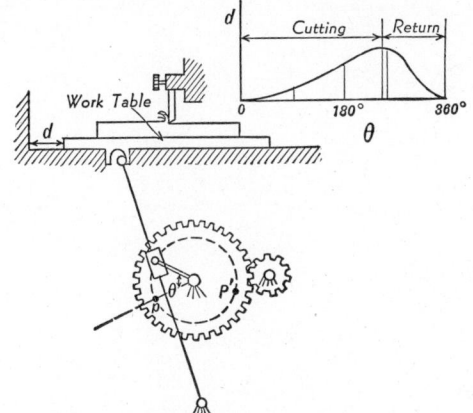

Diagram of the crank shaper.

crank chain, one of the most common of linkages. A pinion engages a large gear to which is affixed a crank arm. The outer end of the crank is pinned to a slider block, which is free to slide on a long swinging arm. As the crank revolves, this arm oscillates back and forth, and by means of a yoke reciprocates a table on suitable ways. The work to be shaped or planed is clamped to this table. The point at which the oscillating arm is **tangent** on either side to the crank pin circle separates the cutting from the return stroke. Since the crank arm turns uniformly, time is proportional to crank angle. The return stroke is accomplished in a much smaller crank angle than the cutting stroke, and consequently consumes less time. A typical displacement diagram for return mechanisms is shown. Many quick-return mechanisms which are patented have features superior to those of the crank mechanism, such as more uniform speed on the cutting stroke, larger quick return ratio, etc. These features are obtained at the expense of more complicated mechanism. (F.T.M.)

QUILL. The portion of a feather which bears none of the slender lateral branches. The hollow shaft which is attached to the skin of the bird. Also the thickened and barbed spines of the porcupines, which are modified hairs. (A.W.L.)

QUILLBACK. Pisces, Teleostei. One of the buffalo fishes, *Carpioides velifer,* of the Mississippi River system. Also called the carp sucker, a name which belongs more properly to a related species, and skimback or river carp. (A.W.L.)

QUILLWORTS. These curious plants, species of the genus *Isoetes,* grow, as a rule, under water. They have short thick corm-like stems and slender dichotomously branching roots. The leaves are slender and somewhat grass-like and are crowded on the upper surface of the short stem. Within the basal portion of the leaf are borne sporangia of two kinds. One, containing large spores or megaspores, is called a megasporangium. The other is a microsporangium and contains numerous very small spores. The megaspores give rise to small multicellular gametophytes on which the archegonia are formed. The microspores become minute multicellular bodies in which are formed the small sperms. A sperm swims to the egg and unites with it to form a zygote. From this a new quillwort is formed. The quillworts are living relics of a once important group. They are of no commercial importance. (R.M.W.)

QUINCE. Rose family.

QUINHYDRONE. Phenols.

QUININE. The chief alkaloid of **cinchona,** the bark of the cinchona tree which is native to certain regions of South America. It was introduced into Europe in 1639 from Peru, for the treatment of ague. The most important use of the drug has been in the therapy of **malaria,** where it exerts a specific action on the parasite. It is used also for the relief of headache, neuralgia, in the treatment of cardiac arrhythmias, as a sclerosing agent in the treatment of varicose veins, and as a stomachic. Occasionally it is employed to induce labor at term; its popular use as an abortifacient in early pregnancy is unwarranted since it has no effect on the **uterus** except just before delivery. (D.M.H.)

QUINNAT. Pisces, Teleostei. A **salmon** of the west coast of North America, ranging from Alaska to the Ventura River. It is among the most important of the food fishes and is the leading salmon of the Columbia River fisheries. Also known as the chinook or king salmon. This fish attains a maximum weight of 100 lbs. and a length of more than 4', but most of those taken weigh only ¼ of this amount. (A.W.L.)

QUINOLINE. Pyridine and Related Compounds.

QUINONES. Phenols and Quinones.

QUINSY. Abscess formation in the tissues around the tonsils, occurring as a complication of acute **tonsillitis.** Symptoms are similar to those of acute tonsillitis. The **lymph glands** in the neck become swollen and tender. Swallowing is extremely painful. Both local and constitutional symptoms are more severe than those in acute tonsillitis. Unless the abscess ruptures spontaneously, incision is required at the proper time. (D.M.H.)

QUINTILE. The 4 quintiles, K_1, K_2, K_3, K_4, are the four points dividing a distribution into five equal parts. They are determined according to the same principles as other **measures of position.** (L.A.A.)

R

RABBET. Dado.

RABBIT. Mammalia, Rodentia. **Rodents** with long ears, large hind legs, and small front legs. Some species burrow and others occupy similar retreats which they do not make for themselves. Many members of the group are called **hares.** There is no sharp distinction between the terms except in their established application to certain species.

Rabbits are found on all continents, although they were introduced into the Australian region. In New South Wales the introduced stock threatened to crowd out even the settlers by its destruction of vegetation. Millions of the animals have been killed per year and the exportation of their hides has somewhat offset their destructiveness.

In North America the common or cottontail rabbit, *Sylvilagus floridanus,* and a few closely related species are widely distributed. One of these species is the brush rabbit, *S. bachmani,* of the Pacific northwest, and two others are the southern marsh rabbit or pontoon, *S. palustris,* and swamp rabbit or cane-cutter, *S. aquaticus.* The large western species are sometimes called hares but more commonly rabbits. The snowshoe rabbit, *Lepus americanus,* also known as the white rabbit or varying hare, lives in the north and in the mountains as far south as Virginia and Colorado. The white-tailed jack rabbit or prairie hare, *L. townsendi,* ranges from the Mississippi River to eastern California. This species becomes white in winter in the northern part of its range. Other species of jack rabbits are found farther west.

The flesh of rabbits is excellent. In the more heavily settled parts of the country they are an important game animal. The fur is thick and soft but the hides are weak, hence they are used chiefly for linings, for cheaper fur garments, and for making felt. After shearing and dyeing rabbit fur reaches the market as northern seal.

Rabbits are bred extensively in captivity as pets, as laboratory animals for use in medicine and bacteriology, to some extent for food, and for the study of heredity. Since they are very prolific they have been among the most useful mammals to the geneticist. (A.W.L.)

RABIES (Hydrophobia). An acute infectious disease of certain animals transmitted to man usually by the bite of an infected dog. Rabies is caused by a **virus** which is present in the saliva of a rabid animal. The disease has been known since 300 B.C. but how it is transmitted was not known until 1804. Pasteur, in 1884, demonstrated that rabies virus is present in the brain of infected animals. Using infected cord and brain tissue he prepared a **vaccine** to be given in repeated injections to individuals bitten by rabid animals. This is the Pasteur treatment which is now widely used in the prevention of the disease.

The period of incubation after a bite from a mad dog is extremely variable, being from 10 days to 2 years; the average is 30–60 days. The nearer the bite to the brain, the shorter is the incubation period. In man, the disease begins with a period of restlessness and depression, followed by gradually developing anxiety, fear, terror, alternating rage and calm. Cerebral irritation may be extreme, and **convulsions** are precipitated by minor noises, lights, or currents of air. The name hydrophobia is derived from the spasmodic contractions of the throat and **larynx** causing agonizing pain when the patient attempts to swallow, or even sees water or hears it mentioned. Extreme thirst and fear of strangling to death drive him to maniacal lengths. Death from cardiac or respiratory failure occurs within a few days.

There is no treatment for rabies. Once the disease develops the patient is doomed to a horrible death. For this reason, preventive treatment with rabies vaccine is imperative whenever an individual is bitten by a rabid animal. If it is not known whether the biting dog is rabid or not, it should be observed for a period of 15 days for signs of the disease, before being released as normal. At the time of a bite from a suspicious animal the wound should be cauterized with pure carbolic acid followed by 95% alcohol. This procedure cannot be relied upon to kill all the virus, but its use does prolong the incubation period and therefore allows more time for adequate vaccine prophylaxis. (D.M.H.)

RACCOON. Mammalia, Carnivora. Stoutly built American animals of moderate size. The head is broad and short, with a sharply pointed muzzle, and the tail is bushy and marked with alternating light and dark rings.

The common raccoon, *Procyon lotor,* ranges over North America east of the Rockies. Two related species live on the Pacific coast and southward into Central America, and one or more additional species have been recorded from South America.

Raccoons are chiefly nocturnal animals. They climb readily and usually nest in hollow trees. In many sections of the country coon hunting at night with specially bred and trained dogs is regarded as among the best sports of the kind.

Raccoon fur is among the better grades, although it is neither very fine nor very beautiful. It is used extensively in making coats. (A.W.L.)

RACE RUNNER. Reptilia, Sauria. Slender lizards, reaching a length of about 10", including the long tapering tail. They are found throughout the United States with the exception of the most northern part. One species is known as the swift, *Cnemidophorus sexlineatus.* (A.W.L.)

RACEME. Flower.

RACEMIC ACID. Tartaric Acid.

RACEMIC COMPOUND. Isomerism.

RACHIS. A shaft. The term may be applied to the vertebral column, but its most common use designates the central shaft of a feather. (A.W.L.)

RACK. Spur Gearing.

RACON. The racon, or **radar** beacon, is a device that has been used for some time for the purpose of identifying ships and planes picked up on radar view scopes during military and naval operations. It is now being developed as a standard aid for both air and sea navigation that may be used by all navigators supplied with radar.

The amount of energy reflected from an object as small as a navigation **buoy,** or even a lighthouse, is frequently too small to produce a signal on the radar scope. The ordinary radar plan of the terrain below an aircraft does not give sufficient detail to permit positive recognition unless some particularly distinctive feature, such as a shoreline, is available.

The racon is a type of **beacon** that contains the equipment for receiving and transmitting **electro-magnetic radiation** of the type used by radar. Whenever a radar pulse from a ship or airborne transmitter arrives at the racon, it "triggers off" a coded pulse of energy. This pulse will arrive at the searching navigator's equipment with sufficient power to set up a signal on the viewing scope. The code characteristic identifies the beacon, and the position of the signal on the PPI gives the bearing and distance of the beacon from the plane. (W.K.G.)

RADAR. Radar (coined from *ra*dio *d*irecting *a*nd *r*anging) is a system for locating reflecting objects by means of radio signals. Anything which will reflect enough energy of a radio signal of high frequency to actuate the **receiver** may be detected and accurately located in space over distances approximating line-of-sight distances from the radar stations. This will be over 100 miles for high-flying planes. Since the radio signals penetrate fog, darkness, rain, and haze, this method of locating planes, ships, and ground objects does not have many of the limitations of the older optical systems. In addition, the ranging by radar means may be much more accurate than that of the optical methods. Although radar was originated a number of years ago, it was not until just before and during World War II that it was brought to the high degree of perfection which permitted its use in detecting enemy equipment and then controlling the gunfire which destroyed that equipment, often without the enemy being visible. Because of the varied reflecting characteristics of many ground installations, bodies of water, types of earth, terrain formations, etc., certain types of radar were used for blind bombing when visual sighting was impossible. These war-developed applications point the way to many obvious peacetime applications such as blind navigation of ships in heavy fogs, blind flying, blind landings, and storm tracking.

Although the details of the various radar systems differ, depending upon the particular type and the use for which it is intended, the basic principles are the same. A series of accurately timed and very short pulses of radio frequency (frequencies range from approximately 100 to several thousand megacycles) are transmitted by a **directional antenna.** Since these pulses represent only a small part of the total time, the power in the pulse

may be quite high without the average power (and hence size of equipment) becoming excessive. If these radio signals strike a conducting object some of the energy is reflected. The reflected pulses may then be picked up by the radar receiving system. Basically the direction of the antenna when a reflection is detected gives the direction to the object, and the time between the transmission of the pulse and its return gives the distance (radio waves travel at 186,000 miles per sec.). The details of making the system sufficiently accurate for practical use are very complex, but the principles may be illustrated by considering a unit used for search and fire control against aircraft. Fig. 1 shows a simplified block diagram of the unit. The timing unit or **pulse generator** controls the sequence of operations by supplying a keying pulse to the transmitting system and at the same time a reference pulse to the receiving system. The pulse is shaped and amplified in the modulator unit and is then applied to the **transmitter.** Here it serves to key the pulse of radio-frequency energy, the high-frequency pulse being fed to the antenna. The field to be observed is scanned by the antenna and its radio signal just as it might be scanned by a searchlight and its beam for visual searching. Various types of antenna structures are used, but all are directional, and the direction may be changed at will. Fig. 2 shows one type: a small dipole of a few

Fig. 2. Radar antenna. (*Electronics.*)

centimeters' length mounted in a parabolic reflector. When a reflected signal is received by the antenna it is fed through the T-R unit to the receiver. Some radars use separate transmitting and receiving antennas, although most use a common antenna for both functions. The T-R unit is merely an electronic switch which cuts out the receiver circuits when a pulse is being transmitted and then opens them between transmitted pulses so they will be ready for the received pulses. The range of the radar is limited by the distance the signal can travel out and back between transmitted pulses, since one pulse must make its complete trip before the next one starts. The received signal is compared in the receiver with the signal originally generated in the pulse generator, the time difference giving the time of travel of the radio wave and hence the distance. There are various methods of doing this, but **cathode ray oscilloscopes** are used as the comparison indicators. As an elementary example, suppose the signal received from the pulse generator initiates the sweep circuit of the

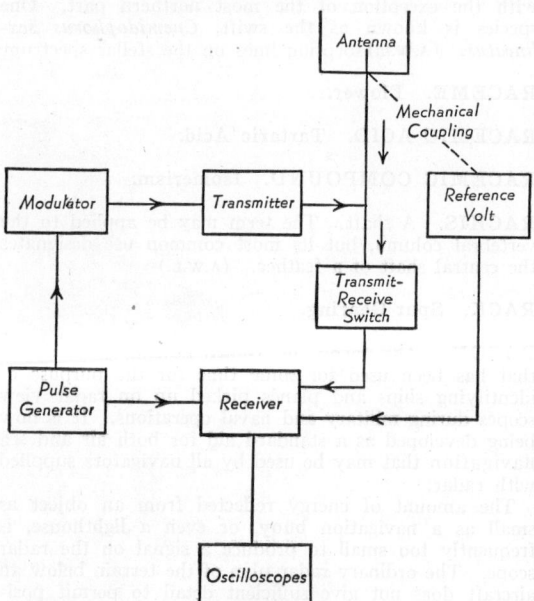

Fig. 1. Block diagram of radar search and control unit.

oscilloscope and that the sweep takes a definite and known time to traverse the oscilloscope screen. This may be accurately adjusted by means of the circuit constants of the sweep circuit. If the returned signal from the receiver is fed to the other set of plates of the oscilloscope, it will appear as a hump or "pip" on the trace. If, for example, the sweep time is 100 microseconds and the pip occurs at a point one fourth of the distance across the scope, the time of travel of the radio wave would be $\frac{1}{4} \times 100$, or 25 microseconds, and the distance to the detected object would be $\frac{1}{2} \times 25 \times 10^{-6} \times 186,000$, or 2.32 miles. This simple illustration neglects, of course, time delays in the circuits, but these are allowed for easily. In many cases they are adjustable to improve the accuracy of ranging. Connected

RADIAL VELOCITY. That component of the space motion of a star that is directed toward the sun is known as the radial velocity of the star, or the velocity of the star in the line of sight. It is measured by spectroscopic methods, employed the Doppler-Fizeau principle, and is determined directly in linear units (i.e., kilometers per sec., or miles per sec.).

Since a comparison spectrum must be available for measurement of the Doppler displacement of the stellar spectral lines, a slit spectrograph must be used for an accurate determination of radial velocity. This instrument is wasteful of light and only one star can be observed at a time. For these reasons the number of stars for which accurate radial velocities are known is small relative to the total number of stars. The ob-

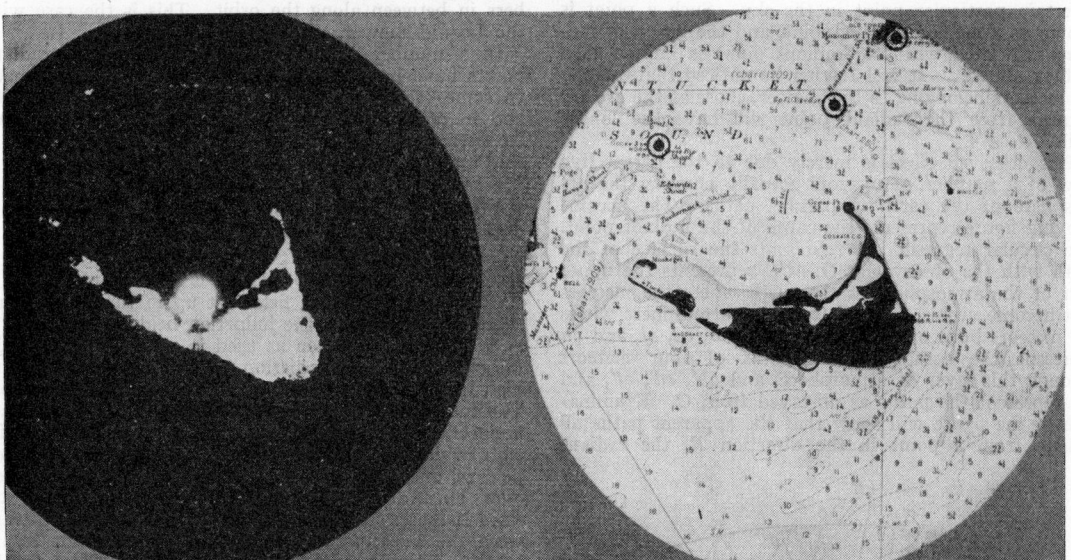

Fig. 3. PPI view compared to map of same area. (*Electronics.*)

with the antenna scanning mechanism is a reference voltage generator whose output is also fed into the receiving system to indicate direction to the reflecting plane when a pulse is received. By electronic means this reference voltage may be compared with the received pulse in such a way as to indicate on a scope to the radar operator when he is pointing directly at the target.

In a PPI (plan position indicator) system the movement of the antenna is tracked by the trace of an oscilloscope tube (the PPI scope). Thus, the position of the trace on the scope corresponds to the direction of the beam from the antenna. A reflection then appears as a bright spot on the oscilloscope, the azimuth of the spot being the azimuth of the reflecting object and the distance of the spot from the center indicating the range of the object. The intensity of the spot is a measure of the reflecting efficiency of the target. Thus, when such a radar is used from a plane, the antenna being used to scan the area below, the oscilloscope will show a shaded pattern of the ground since the reflection is affected by changes of terrain, different building materials, bodies of water, etc. As the radar beam sweeps over points with high-reflection coefficients the spot is bright, whereas points with lesser coefficients give less bright regions on the scope. The result on the screen is a picture of the area being viewed, appearing very much like an aerial photograph. Fig. 3 compares a PPI view with a map of the same area. (L.R.Q.)

RADIAL DRILL. Drilling.

jective prism may be used to determine approximate radial velocities for a large number of stars. In one of the applications of this instrument a comparison spectrum is obtained by interposing a Neodymium screen between the prism and the photographic plate. This produces a few absorption lines on the stellar spectrum relative to which the stellar lines themselves may be measured. Another application of the objective prism to this problem utilizes the fact that the Doppler displacement for a line in the red is greater than that for a line in the violet. Hence the length of the spectrum between these extremes will be changed by an amount proportional to the radial velocity. While the results obtained by the use of the objective prism are only approximate, nevertheless they may be used for statistical study of stellar motions.

From a study of the variations in radial velocity of certain stars, known as spectroscopic binaries, the relative orbits of these objects may be determined. (W.K.G.)

RADIAN MEASURE OF ANGLES. For theoretical purposes, **angles** are generally measured in circular or radian measure. If a circle is drawn with its center at the vertex of an angle, the radian measure of the angle is the ratio of the length of intercepted arc to the radius. The unit angle in this measure is the radian, defined as the angle whose intercepted arc is equal to the radius. The fundamental relation between radian measure and degree measure is given by the equation: π radians = 180°. The symbol (^{r}) is often used after the numerical value of the circular measure to indicate radians, but it is often omitted when no misunderstanding can arise. It follows from the fundamental relation above that:

1 radian = $180°/\pi$ = $57.29578°$ = $57°\ 17'\ 44.6''$ approximately, or $57.3°$ roughly, and

$1°$ = $\pi/180$ radians = $0.017453^{(r)}$ approximately, or $0.017^{(r)}$ roughly.

In terms of radian measure, any circular arc intercepted by an angle with vertex at its center is given in length by the formula: $s = r \cdot \theta$, where r is the radius of the circle and θ is the radian measure of the angle. (L.L.S.)

RADIANT POINT. If the paths of all of the meteors observed from a single station on a given night are plotted on a chart of the sky, it will usually be found that a number of them seem to be coming from a certain particular point in the sky. Such a point is known as a meteor radiant point, and the group of meteors associated with the radiant point is known as a **meteor shower.** It will further be noticed that, among the meteors belonging to the shower, those at the greater distance from the radiant point will have the longer trails.

This observed effect is merely due to the perspective view of a number of meteors actually entering the atmosphere of the earth in parallel paths. The accompanying figure represents the cause of the radiant point. The circular segment AA represents the surface of the earth with the observer at O. CC represents the upper part of the atmosphere of the earth where the meteors first become visible, and BB the lower atmosphere where the meteors burn out and disappear. ab, cd, ef, and gh represent the actual parallel paths of four meteors through this layer of atmosphere, and ab', cd', ef', and gh' represent the paths as observed from O. Examination of the figure will show that the apparent paths all radiate from a point in the direction R, the radiant

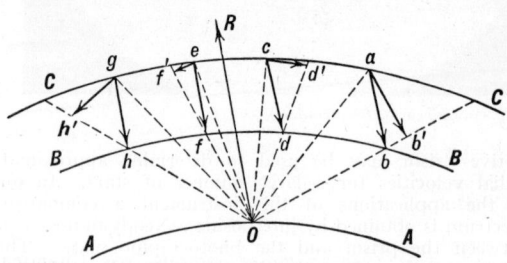

Explanation of radiant point.

point, which is a direction parallel to that in which the meteors are actually entering and traveling through the atmosphere. It will further be noted that the meteors more distant from the radiant point, e.g., ab' and gh', have apparently longer trails than the nearer ones, cd' and ef'.

The location of the radiant point remains approximately fixed with reference to the **constellations** throughout the duration of the shower and is usually named for the constellation in which it appears, e.g., the **Perseid** shower has its radiant in the constellation of **Perseus**, the **Leonids** in **Leo**, etc. Occasionally a shower has a name indicating other characteristics, e.g., the Leonid shower is sometimes referred to as the November meteors because the shower occurs during that month each year, and the **Andromedes** are frequently referred to as the **Bielids** because of their established relation with Biela's **Comet.**

The various showers differ from each other, both in the number of members and in the characteristics of the individual members. Probably one of the most famous showers on record is the Leonid shower of November 12, 1833, during which the number of meteors observed from some stations was estimated as 200,000 per hour for several hours.

Many of the showers occur year after year with definite regularity of date. Such periodic showers may be explained by huge numbers of meteors traveling about the **sun** in an **orbit** which intersects the orbit of the earth. Such a phenomenon has been referred to as a "flying gravel bank," but such a descriptive term is misleading because of the fact that few, if any, of the meteors are large enough to be considered as gravel pebbles. In some cases, the meteors are distributed with fair uniformity all along the orbit, in which case the showers will recur on successive years with approximately the same frequency and appearance. Such is the case with the Perseid shower, which may be observed during the latter part of July and the early part of August each year. In other cases the meteors are concentrated in one or more large swarms with a few scattered members in between along the orbit. This is the case with the Leonid shower.

In a number of cases the orbits of meteor radiant points have been found to agree with orbits of comets. In some cases the comets are still observed as comets, and in other cases the comet itself no longer appears. At present, a large amount of work is being done in this interesting and important field. (W.K.G.)

RADIATION. Electromagnetic Radiation; Thermal Radiation; Solar and Earth Radiation.

RADIATION FIELD. When a conductor carries a.c. there are two types of fields set up in the surrounding space. One of these, the induction field, is predominant at low frequencies such as used in power circuits while the other, called the radiation field, predominates at very high frequencies such as used for **radio communication.** The induction field is responsible for the familiar magnetic effects of **coils** and the interference between circuits which are coupled inductively, i.e., the induction field of one links the other. For radio communication over long distances the radiation field is important since it represents the energy which is radiated outward from the **antenna** system and which does not return to the system but spreads out in space. Close to the transmitting station both types of field may be utilized; in fact, certain wireless record players use the induction field, but at appreciable distances (a few wavelengths) the induction field is negligible. The radiation field consists of an electromagnetic wave traveling at the velocity of light (3×10^{10} cm./sec.). It is this wave which cuts across the receiving antenna and induces the signal which is amplified and demodulated in the **receiver.** This radiation goes out from the antenna in various directions, the exact directions and strengths being dependent upon the antenna characteristics. Thus some of the energy may travel along the earth's surface to the **receiver,** some may go upward to the **ionosphere** and be refracted so it is returned to the earth giving long-distance communication. This radiation field is composed of an electric and a magnetic component, mutually perpendicular and perpendicular to the direction of propagation. Upon leaving the antenna the electric field is parallel to the antenna and the magnetic perpendicular to it. Either or both may contribute to the signal induced in the receiving antenna. At extremely high frequencies the polarization or direction of the electric field or vector has an effect on the character of the received signal. (L.R.Q.)

RADIATION PRESSURE. That electromagnetic **radiation** exerts a pressure upon any surface exposed to it was deduced theoretically by Maxwell in 1871, and proved experimentally by Lebedew in 1900 and by Nichols and Hull in 1901. The pressure is very feeble, but can be detected by allowing the radiation to fall upon a delicately poised vane of polished metal. (See **Nichols' Radiometer.**) It may be shown by the electromagnetic theory, by the **quantum theory,** or by thermodynamic reasoning, making no assumption as to the nature of radiation, that the pressure against a surface exposed in a space traversed by radiation uniformly in all directions is equal to $\frac{1}{3}$ the total radiant energy per unit volume within that space. For **black-body** radiation, in equilibrium with the exposed surface, the energy density is, in accordance with the **Stefan-Boltz-mann law**, equal to $\frac{4\sigma}{c}T^4$; in which σ is the Stefan-Boltzmann constant, c is the **electromagnetic constant**, and T is the absolute temperature of the space. One-third of this energy density is equal to $2.523 \times 10^{-15} T^4$ (ergs/cm.³), which is therefore the pressure in bars. For example, at the boiling point of water ($T = 373.2°$), the pressure amounts to only 0.00005 dyne/cm.² or about 3 lbs. per sq. mile. Such feeble pressures are, nevertheless, able to produce marked effects upon minute particles like gas ions and electrons, and are of importance in the theory of electron emission from the sun, of **cometary** matter, etc. (L.D.W.)

RADIATION PYROMETER. This instrument, of which there are several commercial designs, is based upon the fact that the intensity of the thermal radiation from a heated body depends in a systematic way upon the temperature of its surface. The essential principle involves the focusing of the radiation upon a thermocouple or some other sensitive thermometric detector by means of a suitable mirror. In one form the mirror is concave, producing a real image of a portion of the heated surface; in another it is a hollow cone, in which the radiation converges in an opening at the apex, where the thermocouple is placed. All instruments of this type must be experimentally calibrated, as temperatures computed theoretically from the **Stefan-Boltzmann law** are, for various reasons, found to be in error. A fundamental difficulty arises from the fact that no actual radiator is an ideal **black body**, and the emissive power of one surface differs from that of another at the same temperature. This gives rise to the term "radiation temperature" as different from the actual temperature of the body under examination.

In addition to the foregoing type, sometimes called the "total radiation pyrometer," the **spectrophotometer** or the spectroradiometer (an infrared spectrophotometer) is used in the case of very high temperatures, and the temperature deduced from the "peak" wavelength in accordance with the Wien displacement law. (See **Wien's Laws.**) In this way the surface temperature of the sun, giving out radiation with a maximum at a wavelength of about 5000 angstroms, has been found to be about 5750° C. (L.D.W.)

RADIATION RESISTANCE. This is a fictitious **resistance** which would absorb the same power which a radio **antenna** radiates. Thus, if we say an antenna has a radiation resistance of 73 ohms we mean that it radiates the same amount of power as a 73-ohm resistor would dissipate if it had the same current as the antenna. (L.R.Q.)

RADIATOR. A radiator is a surface especially heated for the emission of heat energy by radiation. Most persons are familiar with the cast-iron sectional-type radiators which are associated with steam and hot-air building-**heating systems.** The hot water or steam on the inside of these sections produces, by **conduction**, a heated outer surface from which the heat is discharged to the air by radiation, aided by **convection** currents which rise over the heated surfaces. It is possible that the greater part of the heat emission from such radiators is by convection rather than radiation. Less artistic in appearance, but equally suitable for heating, are banks of steam pipes, in which the separate pipes are joined at the ends by U-bends. These are usually hung on the wall or suspended from the ceiling in industrial type

buildings. More truly deserving of the name radiator is a domestic electrical appliance consisting of a heater element mounted at the focus of a parabolic reflector. Practically all the heat delivered by this type of radiator is radiant in character. Another type of radiator is the radiant-type gas heater having a number of gas jets producing a flame which heats especially formed fire-clay elements held in it. As a result of the excellent combustion obtained, the fire-clay is heated to incandescence, and emits, radiantly, a great deal of heat **energy**. Of course the products of combustion rising from the heater may deliver a considerable amount of heat by convection, but such heaters are essentially radiators.

On the other hand, there are many devices called radiators in which radiation plays an insignificant part. Outstanding among these, perhaps, is the so-called automobile radiator, by means of which the heat absorbed by the cooling water is dissipated to the atmosphere. If at any time there has been an appreciable amount of heat transfer accomplished by radiation, that is eliminated by recent trends towards enclosed, or nearly enclosed, radiators, in which the radiator core itself is concealed behind polished louvres. The heat transfer in this case takes place mainly by convection, and the design incorporating cellular structure and forced draft is directed towards aiding this type of **heat transfer.** (F.T.M.)

RADICAL AXIS OF TWO CIRCLES. Circle.

RADICAL CENTER OF THREE CIRCLES. Circle.

RADICAL EQUATIONS. A radical equation (or irrational equation) is an **equation** in which the unknown appears under a **radical** sign or with a **fractional exponent.**

Such an equation may usually be solved by squaring both sides of the equation one or more times, or raising both sides to the same other power one or more times, to remove the radicals (or fractional exponents), thus reducing the equation to a **polynomial equation.** Usually one radical should be isolated on one side of the equation before raising to the power.

In this process of solution, **extraneous roots** may be introduced, hence the solution of each radical equation should be checked by substitution in the original equation, and any values not satisfying the equation should be rejected. (L.L.S.)

RADICALS. For chemical radicals, see **Chemical Composition.**

In mathematics, a radical is an indicated **root of a number,** usually a principal root; thus, the radical symbol $\sqrt[n]{a}$ means the principal nth root of a.

Operations with radicals are expressed by the formulae:

$$\sqrt[n]{a \cdot b} = \sqrt[n]{a} \cdot \sqrt[n]{b},$$

$$\sqrt[n]{\frac{a}{b}} = \frac{\sqrt[n]{a}}{\sqrt[n]{b}},$$

if a and b are positive. (L.L.S.)

RADICLE. Seed.

RADIO. Radio Communication.

RADIO AIDS TO NAVIGATION. Radio is of vital assistance to **navigation** in many ways at the present time. The tremendous amount of research in the general field of **electronics** and electromagnetic radiation that has been stimulated by World War II, will undoubtedly bring many new instruments to the aid of the navigator, both sea and air.

The use of radio communication by navigators embraces such important fields as determination of chronometer correction for use in celestial navigation, and for obtaining information regarding present weather at distant landing fields as well as storm warnings and general weather forecasts.

The **loop antenna** has many applications for assisting the air and sea navigators. These will be found discussed under such topics as **Radio Navigation, Radio Compass, Radio Bearing, Radio Range,** and **Radar.** (W.K.G.)

RADIO BEACON. A radio transmitting station, maintained and operated for the purpose of determination of **radio bearings,** is known as a radio beacon. The transmitting frequency, **latitude** and **longitude,** characteristic signal, and times of operation of each radio beacon are listed for sea navigators in the Hydrographic Office Publication "Radio Aids for Navigators," and for aviators in a similar publication of the Civil Aeronautics Administration. In one sense of the term, any radio station might be used as a radio beacon since a radio bearing may be obtained from it. However, only stations listed as beacons can be relied upon, since their positions are carefully determined and their signal characteristics are maintained by responsible agents.

Radio beacons are used for determination of radio bearings and also for "homing" purposes. (See **Radio Compass.**) Sea navigators must exert care not to carry the homing operation too far. The notorious case of the sinking of the Nantucket Shoals Light Vessel by an incoming steamer, that was homing on the radio beacon on the light ship, is an example of perfect and too prolonged homing.

Sky-wave reception at night tends to destroy or decrease the directional characteristics of the **loop,** so for greatest accuracy the bearing must be taken in the daytime or the frequency must be kept low enough so the **ground wave** will be large compared with the sky wave. Accuracies of ½ to 2° may be obtained for a few hundred miles at frequencies around 500 kc. and over much greater distances for lower frequencies. More elaborate **antennae** such as the combination of a loop and vertical, or an Adcock antenna, can give somewhat more accurate results. An Adcock antenna allows the use of higher frequencies and increases the range without interference from the sky wave but does not have as large signal pick-up as the loop.

Many coastal radio beacons transmit audible signals, either through the air or underneath the water, which are synchronized with those transmitted by radio. The difference in time of arrival of the radio and audible signal at the receiving ship will give the distance of the ship from the beacon. (W.K.G., L.R.Q.)

RADIO BEAM. Radio Range.

RADIO BEARING. The **bearing** of a radio transmitter, as determined by a **directional antenna receiver,** is known as the radio bearing of the transmitter.

The first application of radio bearing to navigation was the establishment of radio-compass stations. At first a radio-compass station was a receiving station, whose position was accurately known, that was equipped with a **loop antenna.** A ship within radio range could call the station and request for bearing. On receiving a reply from the station the ship would send a set signal, usually a series of K's, and the land station would determine the radio bearing of the ship and transmit it to the ship. Later the radio-compass station developed into an installation of five or more such individual receivers, each equipped with loop antennae and connected with a central station. In case of a call from a ship, each station would obtain the bearing of the ship and telephone it to the central. At the central, each station was indicated on a large chart and the radio bearings of the ship from each station were plotted. The point of intersection of all bearings gave

the position of the ship which was then radioed out to the ship.

While this procedure is still available, radio bearings of shore transmitters are usually taken from the ship by means of a loop antenna, or the more modern radio direction-finders or radio compasses. Radio bearings as taken are always relative to the **heading** of the ship, and give the **direction** of the station referred to the bow as zero.

Three corrections must be applied to a radio bearing before it can be used as a **line of position** on a **mercator chart** or **small-area plotting sheet.** Loop antennae and radio direction-finders have **deviation** corrections, due to the **magnetic field** of the ship. These must be determined in advance for different headings of the ship and applied to radio bearings as obtained. Then the radio bearing, which is relative, must be changed to true bearing by adding the true heading of the ship. Finally, since radio follows **great circles,** the bearing must be converted from great-circle to mercator, or **rhumb-line,** bearing. Fig. 1 shows two points,

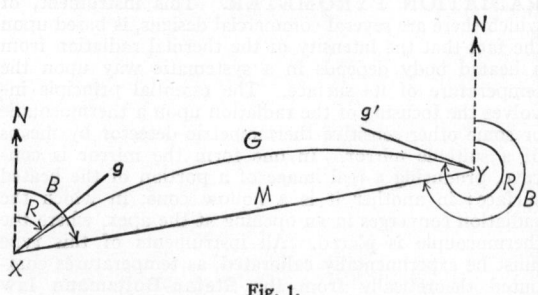

Fig. 1.

X and Y, plotted on a mercator chart, with the rhumb line, XMY, and the great circle, XGY, connecting the two points. The great circle will always be convex toward the nearest pole and we have drawn the figure for the northern hemisphere with true north indicated both at X and Y. The lines gX and $g'Y$ are tangents to the great circle at X and Y respectively, and are the directions in which the signal from Y will arrive at X, and that from X will arrive at Y. Let us consider X to be the receiving station. Then the angle $R(NXg)$ represents the great-circle bearing of Y and the angle $B(NXM)$ the rhumb-line bearing. In this case, it is noted that a correction must be added to the great-circle bearing to obtain the rhumb line. Reversing stations and considering Y the receiver, we see that at this point the correction must be subtracted to obtain rhumb line from great circle.

In case a navigator is working on a **Lambert chart,** as is frequently the case in air navigation, the great-circle bearing is close enough to a Lambert line to be plotted without correction other than for deviation and heading.

The question regarding the value of a fix obtained from radio bearings has long been a matter of much debate. For a number of years the Bureau of Navigation rated a radio **fix** as of less value than an estimated position obtained by dead reckoning. This may well be a correct evaluation at sea, particularly if the dead reckoning is not extended over more than 48 hours. In the air, when we consider the speed of modern planes, the difficulties in holding heading, and the inaccuracies in predicted winds, there is no question but that a radio fix is of tremendous value. With modern equipment, and under best radio conditions, a radio bearing is accurate to within 3°. To allow for possible peculiarity in the reception from some station, it is always advisable to use three radio bearings to plot for fix. In this case the area of the triangle of intersection will indicate the reliability of the fix. The mercator correction should always be applied if it is greater than o°.5. The numerical values of the correction can be quickly obtained from published tables and, if applied with the proper sign, certainly should improve the fix. (See **Radio Navigation**.) (w.k.g.)

RADIO COMMUNICATION.

Communication between distant points not directly connected by an electrical conductor may be accomplished by electromagnetic waves radiated through space. Thus radio communication utilizes radiated energy instead of the conducted energy of the wired methods. (See **Radiation field**.) Obvious benefits of the radio method are the elimination of the expense of installation and upkeep of a wire communication system, and the communication between points difficult of access by wired systems. Furthermore, since energy may be radiated in all directions in space, broadcasting to large numbers of persons is simpler. Radiated electromagnetic waves can be used for communication purposes in three ways. First, they may be used to create a monotone signal of dots and dashes comprising a code. This is **radio telegraphy**. It has certain important commercial and governmental uses, but is not suitable for broadcasting, as it must be interpreted by trained operators. Secondly, electromagnetic waves may be modulated so that they carry the electrical equivalent of sound waves. At the receiver they are caused to reproduce the original sound, whether it was voice or music. This is the field of radio telephony and broadcasting, one which more than any other has entered intimately into the life of the average citizen. Thirdly, it is possible to transmit, by means of these waves, electrical pulses which will recreate at the point of reception, a scene which originated at the transmitter. This **television** process is now the subject of considerable experimentation, and its present approach to commercially satisfactory perfection indicates that it may, in the future, be expected to become an important phase of radio.

There are three essentials to radio communication. These are, first, **transmitter**, or source of electromagnetic radiation; second, the radio waves themselves, travelling between the transmitter and receiver; and thirdly, the **receiver**, which can receive and interpret the electrical pulses either as code, sound, or scene. A requisite to radiation of electromagnetic waves is a source of a.c. of high frequency. The radiation of energy from power lines which carry a.c. at 60 cycles per second is negligible because of the comparatively low frequency. When, however, a current oscillates in a wire suitably arranged for radiation (antenna) at a frequency of, say, 1000 kilocycles (one "kilocycle," by common usage, is 1000 cycles per second), it is possible to radiate considerable power into space in the form of electromagnetic waves, because radiated power increases as the square of its frequency. Furthermore, since at the higher frequencies a smaller antenna is suitable for radiating the energy, it is natural to expect high frequencies to be employed for all radio communication. In any transmitter there must be this source of high-frequency current. This is the **oscillator** and associated **amplifiers**. The frequency of the oscillator is kept under close control by adjusting the characteristics of the circuit incorporating it, so that there may be little or no interference between the waves sent out by different transmitting systems. That is, at the receiving station the electrical circuit employed may be "tuned" to the frequency of one particular transmitter, thus excluding all other wave trains. Between the oscillator and the radiating **antenna,** the transmitter must have a signalling system. For code this might be as simple as a means for interrupting, at will, the oscillating current. (See **Keying**.) In telephony, there will be provided some way of securing **modulation** of the high-frequency carrier wave with the sound wave. The strength of the electric currents created in a **microphone** by the direct action of the sound waves may be amplified and strengthened before being used to modulate the carrier wave, and the modulated wave itself may be strengthened by amplifying. Ultimately the intensified high-frequency currents are sent into the antenna system where they produce the radio waves. For broadcasting purposes these waves flash off into space in all directions, but for commercial telephony or telegraphy, some degree of directional control is possible by the use of **directional antenna**.

Radio-wave propagation from a simple antenna is an action which is analogous to the waves created by dropping a pebble into a quiet pool. Concentric waves travel outwards from the antenna, decreasing in intensity, very rapidly at first, then more slowly, as they travel further from the antenna. Their speed of travel is the same as the speed of light, and they possess the property of being able to pass through and around most obstacles. Atmospheric conditions, however, can affect their travel, as can also sunlight and water. But while the waves may become successively ever more attenuated, they retain their original frequency.

The third essential to radio communication mentioned above, is the receiver. A receiver consists of an antenna suitably arranged to receive the impressed voltages of the electromagnetic waves which reach it from the transmitting antenna. Then, since the receiver must be selective, so that it will reproduce the output of only one of several possible transmitters, means must be provided to "tune" it to the frequency of the selected transmitter. The received signal is very weak, and it must be strengthened by being passed through one or more stages of radio and audio-frequency amplification. Between the radio and audio amplification occurs the action of **detection,** or demodulation, separating the carried wave from the electrical wave which mirrors the original voice or sound wave. Finally, the amplified voice current is fed into a **speaker** which recreates the original sound.

The above discussion, and the related cross-references, can convey only a faint idea of the electrical complexity of modern radio circuits, and give no hint at all of the problems which have arisen from the widespread commercialization of this form of communication. Although in recent years radio has become an important feature of marine and naval operations (see **Radio Beacon** and **Radio Range**) and it is being relied on to an ever increasing extent to secure safety in aerial transportation (see **Blind Flight** and **Landing**), and although there has been developed for public use commercial telephonic service between ship and shore, and between continent and continent, and although radio has been employed for many other interesting and useful services, it is proposed here to dwell only on some of the problems associated with the commercialization of radio broadcasting.

The rise of radio broadcasting in the early 1920's was phenomenal. An important new industry mushroomed into existence in this country almost overnight following the initiation of radio broadcasting service by station KDKA in 1920. Hundreds of firms began the manufacture of receiving sets, thousands of retailers opened up stores for sale and distribution of the product, broadcasting stations were set up everywhere, and the principal chain networks were organized. People everywhere purchased receiving sets, and, with adoption by station WEAF of the toll broadcasting policy, the introductory phase of broadcasting by manufacturers of radio equipment in order to stimulate sales was over, and the sponsored program appeared. The commercial possibilities of the radio broadcasting station as an advertising medium attracted a large amount of capital into the field, and soon there was a chaotic condition of interference between stations.

Radio broadcasting developed using long waves, that is waves having frequencies between 550 and 1500 kc. To avoid interference between stations, a clear **channel** of a band of frequencies of 10 kc. should be provided each station. However, in the broadcast range there are only 96 such channels, and by 1927 there were several hundred broadcasting stations attempting to use them. At that time the situation obviously required more rigid governmental control, and since all radio communication may well be considered interstate, the Government created the Federal Radio Commission to remedy the serious interference which existed among the 600 or more stations. In order to accommodate the large number of stations to the available channels in such a way as to reduce interference, so that the owner of a receiver could obtain satisfactory use from it the Commission had to assign definitely the frequency and power to be used by the different stations, and rigidly control the number of new stations which could be put in service from time to time. Furthermore, it was found necessary where the facilities offered by a station duplicated that of another to cause them to divide their broadcasting time. Also, low-power stations for local broadcasting were frequently required to terminate their activities at sundown.

While the standard broadcast stations still operate on 10 kc. channels in the band from 550 to 1600 kc., much higher frequencies have been developed and are being widely used, both for commercial and entertainment purposes. For commercial services the transmission characteristics of the higher frequencies offer very decided advantages in many cases. In the entertainment field the higher frequencies allow satisfactory international broadcasting and also permit the use of much wider channels than the standard broadcast band. **Television** with its channels of approximately 5 megacycles and frequency modulation with channels of 200 kc. are examples of recent wide-band systems. Obviously these channels would be impractical in the standard broadcast bands because of space limitations. The more or less line-of-sight transmission of very high frequency signals, while a disadvantage for general use, is an advantage for short-range service where the signal is not needed at a distance and is attenuated because of its very high frequency. (F.T.M., L.R.Q.)

RADIO COMPASS. This is probably the most loosely used term in all navigation. When the **loop antenna** was first applied to the determination of **radio bearings**, the term radio-compass station was applied to shore installations that would forward, on request, the bearing of a ship from the station. Next the term was applied to a group of shore installations, each equipped with a loop antenna, from which the navigating officer of a ship within range could obtain the **latitude** and **longitude** of his ship. After the loop antenna and receiving sets had been developed to a state where they could be carried by the ships themselves, the term radio compass was applied to the loop. As new and improved radio equipment became available, the term radio compass was successively applied to any radio device that could be used to determine bearing. A glance through any textbook on navigation, particularly those dealing with air navigation, will yield at least two, and sometimes as many as five, different instruments for radio compass.

At the present time the U. S. Naval Aviation service uses the term radio compass for a direction-finding instrument which has a dial on the instrument panel of the plane that looks in many respects like an ordinary **compass,** and which is used for **heading** the ship in much the same manner as is that instrument. Two radio **antennae** are used with this instrument, a loop and a nondirectional antenna. The volume controls for signal intensity are so adjusted that the signal strength from the loop and the nondirectional antenna are the same when the station is in the plane of the loop. The two antennae are then fed into a single receiver and the signal intensity is illustrated in Fig. 1. The resultant signal intensity from any station is shown by a **galvanometer** on the instrument panel. The loop may now be

rotated and a pointer swings over a compass card on the instrument panel. When the galvanometer needle is in the zero position the reading on the compass dial gives

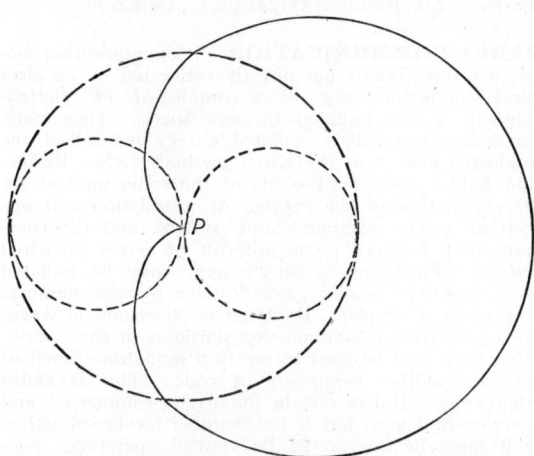

Fig. 1. The length of a line from P to any one of the three curves is proportional to the strength of the signal from a station in the direction toward which the line is drawn. The dotted (figure eight) curve represents the relative intensities in various directions for the loop alone, the dashed (circle) curve the intensity for the nondirectional antenna alone, and the full (catenary) curve that for the combined loop and nondirectional antenna or, in other words, for the radio compass.

the bearing of the transmitting station relative to the ship.

This instrument is frequently used as a "homing" instrument, i.e., an instrument to head a ship toward some selected transmitting station. To be used in this manner the loop is rotated until the compass dial reads zero. The plane is then turned until it is on such a heading that the galvanometer dial indicates zero. The transmitting station is then dead ahead. If the plane remains on this heading it will always be pointing toward the transmitter. However, because of the effect of wind, the plane is not proceeding toward the transmitter, but is spiralling around it. If the wind is known, the loop is turned through the **wind correction angle** and the heading of the plane altered until the galvanometer again reads zero. The plane will then be on **course** toward the transmitter just as long as the galvanometer remains at zero reading and the wind remains constant. If the ship changes heading, the galvanometer dial will show the direction and amount that the plane is off-course, and the pilot can immediately return to proper heading. The pilot must not trust this radio compass too far, for an "off-course" indication may be due to a wind shift. Furthermore, the "homing device" has the pernicious habit of "homing" toward a thunder storm, high-tension line, or some other electromagnetic disturbance, without warning.

New devices, operating on the general principle of **radar,** have been devised which will undoubtedly become known as radio compass when they are available for commercial and private use. (W.K.G.)

RADIO DIRECTION FINDER. Radio Compass.

RADIO FIX. Radio Navigation.

RADIO FREQUENCY. This is the frequency of the carrier signals used for **radio communication,** and while normally thought of as being quite high (many kilocycles), frequencies as low as a few kilocycles are used. The upper limit is fixed by the state of the art, being at present as high as 30,000 megacycles for certain laboratory equipment. The radio frequencies are often subdivided as follows:

FREQUENCY (KC.)	NAME	SYMBOL
10–30	Very low	VLF
30–300	Low	LF
300–3000	Medium	MF
3,000–30,000	High	HF
30,000–300,000	Very high	VHF
300,000–3,000,000	Ultra high	UHF
3,000,000–30,000,000 ...	Super high	SHF
		(L.R.Q.)

RADIO NAVIGATION. The use of radio aids to navigation for checking the **dead-reckoning** position of a ship is known as radio navigation. Following a **radio range**, or using a **radio compass**, is included in the general subject of radio navigation. However, neither of these two methods is completely reliable, and positions of the ship should be checked frequently by using radio bearings for **lines of position**, obtaining what is known as a **radio fix**.

The use of radio bearings as lines of position can best be described by solving two numerical problems. 1. A ship is proceeding on **heading** 330° at 12 **knots**. Three **radio beacons**, A, B, and C, are located in the following positions:

Station........	A	B	C
Latitude......	29° 30′ N	30° 00′ N	28° 40′ N
Longitude.....	83° 20′ W	81° 40′ W	81° 52′ W.

At 0812 the dead-reckoning position of the ship is L = 28° 32′ N & Lo = 82° 42′ W and at that time radio bearings, corrected for deviation of the radio compass, are A = 000°, B = 063°, and C = 117°. These must be changed to true bearings by adding to each the heading of the ship, obtaining: A = 330°, B = 033°, and C = 087°. Since these are **great-circle** bearings, they must be reduced to **rhumb-line** bearings by applying the correction factors for A − 0.05, B + 0°.2, and C + 0°.1. Then, working either on a **mercator** chart, **mercator** plotting sheet, or **small-area plotting** sheet, the corrected rhumb-line bearings are plotted, and the fix determined as the center of the triangle of intersection of the three lines of position. The position of the fix is L = 28° 37′ N & Lo = 82° 44′ W, and, since the sides of the triangle are less than 3 miles, we can assume that the fix is probably correct to within 1 mile. The complete solution is illus-

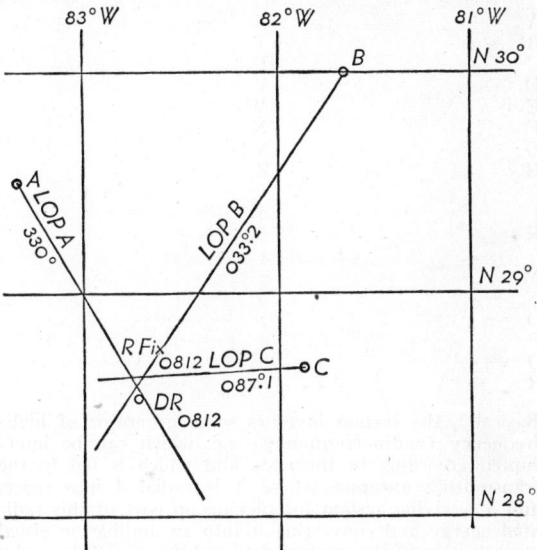

Fig. 1(a). Radio navigation scale diagram.

trated in Fig. 1(a) which is drawn on a small-area plotting sheet and labelled in accordance with standard U. S. Navy procedure.

A plane on routine flight is heading 240° with true air speed of 180 knots. A tail wind of 30 knots is blowing, giving a predicted **course** of 240° and ground speed of 210 knots. The pilot expects that his dead-reckoning position at 1200 will be Latitude 45° 10′ N and Longitude 84° 30′ W. Since he is flying between two layers of overcast, he cannot check his position by either pilotage or celestial navigation. However, there are three radio beacons, A, B, and C, which are available for radio bearings. The positions of the stations are as follows:

Station A, Latitude 44° 17′ N, Longitude 85° 30′ W.
Station B, Latitude 45° 24′ N, Longitude 85° 56′ W.
Station C, Latitude 44° 15′ N, Longitude 84° 38′ W.

At 1155 the radio bearing of A is found to be 336°. At 1200 the radio bearing of B is 043°, and at 1205 the radio bearing of C is 281°. Correcting the bearings for deviation, heading, and rhumb line we have A = 215°.5, B = 282°.6, C = 161°.0. Because of the rapid motion of the plane the 1155 line must be moved forward parallel to itself through the distance that the plane has advanced in 5 minutes, and the 1205 line must be moved back through the distance that the plane moved in 5 minutes. The fix of the 1200 line with the 1155 and 1205 lines, moved to 1200 by dead reckoning, yields L = 45° 15′ N & Lo = 84° 44′ W. The size of the fix triangle indicates that the position is reliable to within 4 miles. The problem, plotted on a small-area plotting sheet and labelled in accordance with U. S. Navy aviation procedure, is shown in Fig. 1(b). (W.K.G.)

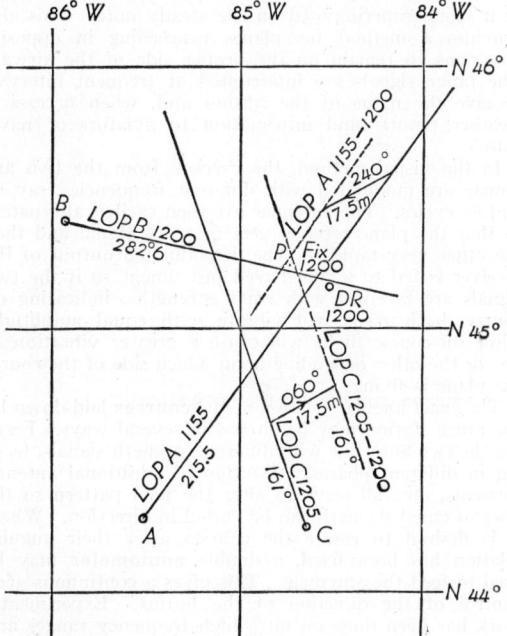

Fig. 1(b). Radio navigation scale diagram.

RADIO RANGE. A radio range is a system of radio signals designed for the purpose of guiding a ship or plane along a designated **track** toward or away from a specified location.

The standard radio range in this country uses relatively low frequency (in the 200–400 kc. band) and, while there are various modifications, the track is indicated by the intersection of two field patterns from the range **antenna** system. The usual antenna arrangement is two pairs of crossed Adcock antennae set 90° in space from one another. This gives two figure-8 field patterns as shown in Fig. 1(a). The patterns overlap in narrow wedge-shaped regions which have their apices at the transmitting station. These overlapping sectors are known as the range or "the beam." Two methods, aural and visible, are used for keeping a ship "on the beam."

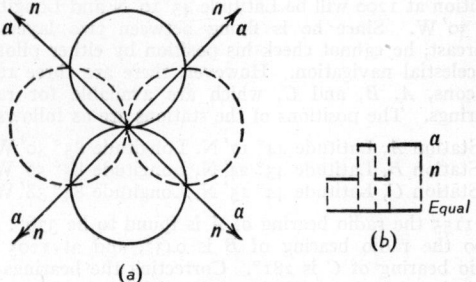

Fig. 1. Radio ranging.

In the aural system the carriers from the two antennae are so modulated with some audio note, say 1000 cycles. The code signal (letter a, dot dash) is transmitted from one antenna, and the code signal (letter n, dash dot) from the other, so timed that on the center of the overlapping region the two signals blend together in a continuous note. This is indicated in Fig. 1(a), while the time sequence of the code characters is shown in Fig. 1(b). To remain directly "on the beam," all that is necessary is for the navigator to head his plane so that the continuous note is heard in his receiver. If he drifts to the right or left of the center, he hears either the a or n signal superimposed on the steady note. This also provides a method for planes proceeding in opposite directions to remain on the proper side of the airway. The range signals are interrupted at frequent intervals to give the name of the station and, when necessary, weather reports and information to aviators or navigators.

In the visual method, the carriers from the two antennae are modulated with different frequencies, say 65 and 85 cycles. The antennae are then excited alternately so that the plane receiver gets first one signal and then the other very rapidly. The demodulated output of the receiver is fed to a tuned reed instrument so if the two signals are received with equal strengths, indicating on course, both reeds will vibrate with equal amplitude, while off-course flight will cause a greater vibration of one or the other depending upon which side of the course the plane is flying.

The exact angular relation of the courses laid down by the range station may be altered in several ways. Feeding the two antennae with different strength signals, feeding in different phases, utilization of additional antenna elements, etc., all serve to alter the field pattern so the lines of equal strength can be varied in direction. Where it is desired to rotate the courses after their angular relation has been fixed, a double goniometer may be used to feed the antennae. This gives a continuous 360° control of the direction of the beams. Experimental work has been done on ultra high-frequency ranges and in the future there will undoubtedly be other systems put into common use.

There is no way with an ordinary receiving antenna for a pilot to tell in which of two opposite directions the range transmitter is located, unless his dead-reckoning position is sufficiently accurate. Several complicated "orientation procedures" are used by aviators to determine the direction of the station. With a modern radio compass installation this difficulty is removed.

To further assist aviation pilots who are navigating by radio ranges a system of markers has been established. With the Adcock antennae now in use no signal is transmitted in a vertical direction. Accordingly, above the transmission station there is a cone-shaped region with apex down in which no signal will be heard, and this is known as the "cone of silence." As a pilot, flying a range, passes above the transmitter the signal strength will drop to zero and then increase to normal strength. The time elapsed between this fading and reappearance will be proportional to the width of the cone, and hence, to the altitude at which the plane is passing above the station. Because of the fact that ordinary fading might produce the same effect, many range stations are now equipped with a Z-type marker. This is a transmitter operating on ultra high frequency (75 megacycles) so designed that its transmission cone fills the cone of silence just mentioned. A special receiver is required to pick up this signal, and, as with the cone of silence, the time during which this signal is heard will indicate the altitude of the plane above the transmitter. To warn the pilot that he is approaching a range station, a fan-type marker, operating on the same frequency as the Z-type, is frequently established at a distance from 3 to 5 miles from the range transmitter. The characteristic signal of such a marker, which is indicated on the chart, will give the pilot a positive location. These are particularly valuable in overcast weather when blind-landing procedures may have to be adopted.

At some points of intersection of radio ranges, M-type markers are established. These are non-directional, low-powered, high-frequency radio stations, transmitting a characteristic signal which will inform the pilot that he is approaching the intersection.

The phrase "on the beam" has become a colloquialism for being safe and proceeding correctly. However, in common with all radio equipment, the ranges cannot be relied upon exclusively, and no pilot should ever proceed blindly along the beam without continually checking his dead reckoning and obtaining fixes whenever possible by pilotage, radio navigation, or celestial navigation. For reasons, far from completely understood at present, radio ranges will split up into several components, or the range may bend from its normal direction, may develop a "kink," or do a number of other things which may lead the incautious navigator into serious trouble, rather than safely into his destination. (W.K.G., L.R.Q.)

RADIO TELEGRAPHY. Radio telegraphy is that form of radio communication which utilizes the dots and dashes of the International code to transmit the intelligence.

A	. —		S	. . .
B	— . . .		T	—
C	— . — .		U	. . —
D	— . .		V	. . . —
E	.		W	. — —
F	. . — .		X	— . . —
G	— — .		Y	— . — —
H	. . . .		Z	— — . .
I	. .		1	. — — — —
J	. — — —		2	. . — — —
K	— . —		3	. . . — —
L	. — . .		4	. . . . —
M	— —		5	. . . . .
N	— .		6	— . . . .
O	— — —		7	— — . . .
P	. — — .		8	— — — . .
Q	— — . —		9	— — — — .
R	. — .		0	— — — — —

Basically, the system involves some generator of high-frequency (radio-frequency) a.c. which can be interrupted according to the code and which is fed to the transmitting antenna where it is radiated into space, and a receiving system for picking up part of this radiated energy and converting it into an audible or visual reproduction of the original dots and dashes. Telegraphy was the first type of wireless communication and was in wide use several years before voice transmission became a practical success. While it is difficult to determine just who was responsible for the invention of radio, or wireless as it was first called, Maxwell is credited with first showing the possibilities mathematically and Hertz with first verifying this theory. Hughes also did valuable work about the same time, demonstrating as early as 1879 the wireless transmission of signals over distances of several yards. He also later extended this range to several hundred yards and discovered the coherer—a device for receiving radio signals. However, these were laboratory experiments and did not constitute a practical system for communication between distant points. It remained for Marconi to develop and patent the first complete practical system. While there were few, if any, basically new points in his system, he deserves full credit for combining various components into a working system. The art developed rapidly from 1897 when Marconi filed for his first patents to 1901 when he successfully spanned the Atlantic. Modulation was invented by Fessenden in 1902 but the first commercial radio telephone service was not inaugurated for several years. The obvious safety aid involved in ship radio led to its immediate adoption by many vessels and shortly to its enforced adoption by all major passenger-carrying vessels. While these early radio systems were universally **damped wave** installations, practically all present day radio telegraph circuits employ **continuous waves**.

At the lower radio frequencies the alternating currents can be generated by special rotating machines, such as the Alexanderson alternator, and in fact were so generated at first. Now nearly all radio signals are generated by vacuum-tube circuits so we shall confine this discussion to them. Regardless of the size or rating of the transmitting station, the high-frequency a.c. originates in the **oscillator**. In the simplest transmitters the output of the oscillator may be connected directly to the antenna but usually this oscillator output feeds an **amplifier** stage or stages where the power is built up by the amplifying action of vacuum **tubes** and their associated circuits. Not only is the power increased beyond that which can satisfactorily be obtained from oscillators directly, but the character of the transmitted signal is improved. The frequency of an oscillator varies with the voltage, with the load, etc., so by using a **buffer** amplifier the frequency **stability** of the station is improved and this is reflected in more reliable reception and decreased interference between stations. Where it is necessary to operate over a considerable range of frequencies the oscillator is self-controlled, i.e., its frequency may be varied by varying the tuning control just as is done in the familiar broadcast **receiver**, while for fixed-frequency operation the oscillator is preferably **crystal**-controlled. Often where only a few fixed operating frequencies are needed the oscillator is still crystal-controlled with several crystals which may be cut in by a selector switch. A simple radio telegraph transmitter is shown in Fig. 1. Not only must the transmitter

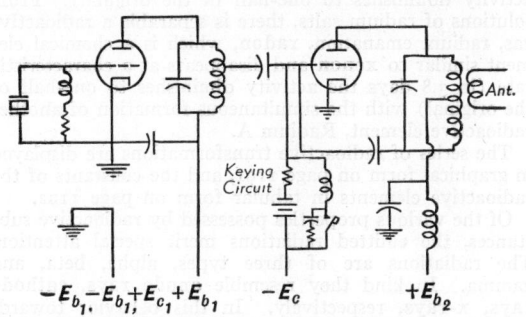

$$-E_{b_1}, -E_{b_1}, +E_{c_1}, +E_{b_1} \qquad -E_c \qquad +E_{b_2}$$

Fig. 1. Simple radio telegraph transmitter.

generate the necessary a-c power for the antenna but suitable means for interrupting this to produce the dots and dashes of the code must be incorporated. While it is necessary in some types of service to stop the oscilla-

tion completely for spaces and hence becomes necessary to key in the oscillator, this produces some frequency drift and often an undesirable chirpy signal unless extreme care is taken in the design and operation of the transmitter. As a consequence of this effect on the output signal, the keying is often done in one of the amplifier stages, the oscillator operating continuously but the keyed amplifier preventing the signals from reaching the antenna. The radio-frequency energy is radiated into space by the antenna, the exact spacial distribution depending upon the antenna system which is often designed for the particular type of service. Many telegraph stations employ **directional antennae** as most of their work is with fixed stations and does not require general coverage.

The receiving antenna intercepts the radio wave and hence has a very small voltage induced in it. This voltage is fed into the receiver which involves some sort of selective circuit for tuning in the desired station, usually one or more stages of amplification, a **detector** and an audio **amplifier**. The reception of continuous wave signals by ear requires some means of converting the high radio frequencies into an audible frequency. This is accomplished by heterodyning the incoming signal with a slightly different frequency of locally generated a.c. to produce an audible beat. This beat, upon **detection**, becomes an **audio-frequency** current which can operate the diaphragm of the headphones or **speaker** to give a sound. This, of course, responds to the keying of the original radio wave. (See **Heterodyne** and **Detection**.) Fig. 2 shows the wave forms produced in this detection process.

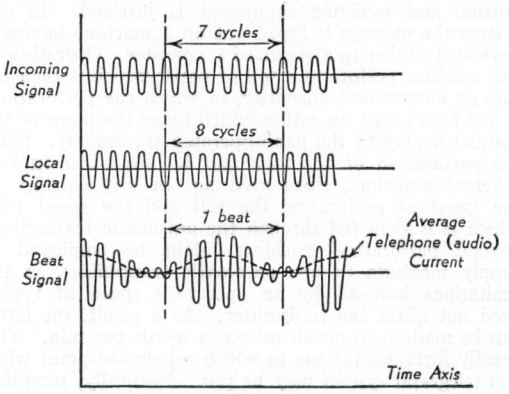

Fig. 2. The heterodyne action.

The simplest form of vacuum-tube **receiver** for this type action is the regenerative detector which serves as an oscillator and detector at the same time. The output may be fed directly into phones. However, this has minor commercial application, much more elaborate circuits being used in modern commercial receivers. Even where the regenerative detector is used it is usually preceded by a stage or two of tuned radio-frequency amplification and followed by one or more audio amplifier stages. Such a receiver is shown in Fig. 3. The various tuned circuits are adjusted so their response is a maximum at the frequency of the station desired. This gives the needed selectivity. After detection the audio signal is built up by the amplifiers to drive a head-set or speaker. Much of the present day reception is done with **superheterodyne** receivers, a beat-frequency oscillator being added to the basic circuit. This oscillator is fix-tuned to the intermediate frequency plus, let us say, 1000 cycles and its output mixed with the intermediate frequency output at the detector. The result is the 1000-cycle beat note. Other refinements include crystal filter circuits to cut out interference from stations on almost the same frequency as the desired one.

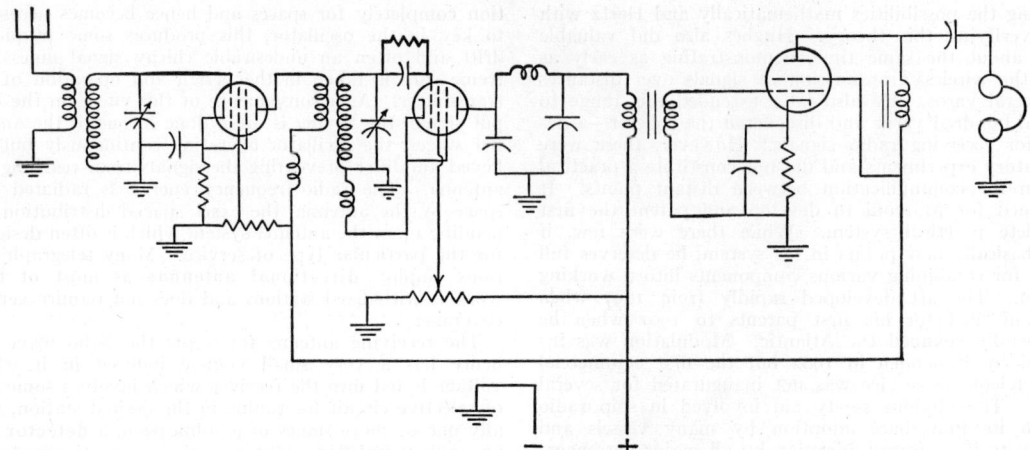

Fig. 3. Commercial radio telegraph receiver.

Manual operation of the radio telegraph employing a hand-operated key at the transmitting end and earphones at the receiver, is satisfactory for a great many services. However, there are instances where limitations in speed of transmission of messages, caused by the inability of even trained and competent operators to receive more than approximately 40 words per min., do not fully develop the possibilities of either the investment or of the channel allotment. Particularly, in point-to-point commercial wireless telegraphy automatic transmitting and receiving equipment is justified. In one system the message is typed out on a machine having a keyboard similar to a standard typewriter. Operation of this machine perforates a roll of paper which is then fed into an automatic transmitter, in which the perforations in the tape create an action which takes the place of the manual keying of the hand-operated transmitter. Since the perforation of the tape and its use occur in two different machines, there need not be any equality of the speed of perforating the roll and the speed with which it is then fed through the automatic transmitter. Several perforating machines might be employed to supply messages to one automatic transmitter, so the limitations imposed by an operator's speed of typing need not affect the transmitter. As a result, the latter can be made to transmit over 200 words per min. This greatly increases the use to which a point-to-point wireless telegraph system may be put. Naturally, reception of such high-speed signals must also be the function of a machine. A commonly used system has an ink recorder in which movement of the pen is controlled by the incoming signal, and a record of dots and dashes is made upon a continuous roll of paper. Operators may then transcribe this ink record into the original message. (L.R.Q.)

RADIOACTINIUM. Symbol: RdAc. A radioactive element of the **actinium** series. (See **Radioactive Changes.**) (R.K.S.)

RADIOACTIVE CHANGES. The phenomenon of radioactivity was discovered by Becquerel in 1896 by the effect produced on a photographic plate by **pitchblende** (**uranium**-containing mineral) while wrapped in black paper in the dark. Soon after this, it was found that uranium minerals and uranium chemicals are radioactive, and strangely the minerals showed more radioactivity than could be accounted for by the uranium content. About the same time there was discovered radioactivity of **thorium** minerals and thorium chemicals. Uranium and thorium are the previously known chemical elements to which is attributable the phenomenon of radioactivity. Later, it was found that **potassium** and **rubidium** pos-

sess a mild degree of radioactivity. There is sufficient thorium element in the oxide of a gas mantle so that, if it is treated as Becquerel treated pitchblende, an image of the mantle web is obtained on the photographic plate.

The excess radioactivity of mineral over chemical uranium led P. and Mme Curie to experiment with the mineral. For detecting the presence of radioactive substance a method was found in the discharge of a charged gold-leaf electroscope, which method also serves for the quantitative estimation of radioactivity by observation of the rate of drop of the gold-leaf. By separating into fractions and examining each fraction by the electroscope, they found in the **bismuth** element fraction the first new radioactive element to be discovered. It was named **polonium** (1898). They found that polonium disappeared rapidly, half of its radioactivity vanished in about six months. The fraction containing **barium** element was also found by them to be radioactive. Repeated fractional crystallizations of the chloride and bromide solutions made possible the recovery by them of practically pure salt of the second new radioactive element. It was named **radium** (1898).

Radium is chemically similar to barium; displays a characteristic **spectrum;** its salts exhibit **phosphorescence** in the dark, a continual evolution of heat taking place sufficient in amount to raise the temperature of 100 times its own weight of water 1° C. every hour; and many remarkable physical and physiological changes have been produced. Radium shows radioactivity a million times greater than an equal weight of uranium, and unlike polonium, suffers no measurable loss of radioactivity over a short period of time (in 1600 years the activity diminishes to one-half of the original). From solutions of radium salts, there is separable a radioactive gas, radium emanation, **radon**, which is a chemical element similar to **xenon** and disappears at a characteristic rate (in 3.8 days the activity diminishes to one-half of the original) with the simultaneous formation of another radioactive element, Radium A.

The series of radioactive transformations are displayed in graphical form on page 1191, and the constants of the radioactive elements in tabular form on page 1192.

Of the various properties possessed by radioactive substances, the emitted radiations merit special attention. The radiations are of three types, alpha, beta, and gamma. In kind they resemble **anode rays, cathode rays, x-rays,** respectively. In this behavior towards **electrical** and **magnetic fields,** the resemblance is qualitatively complete: 1. Alpha rays are positively charged particles of mass 4 and slightly deflected by electrical and magnetic fields. 2. Beta rays are negatively charged **electrons** of mass $1/1800$ of that of the **hydrogen atom,** and largely deflected by electrical and magnetic

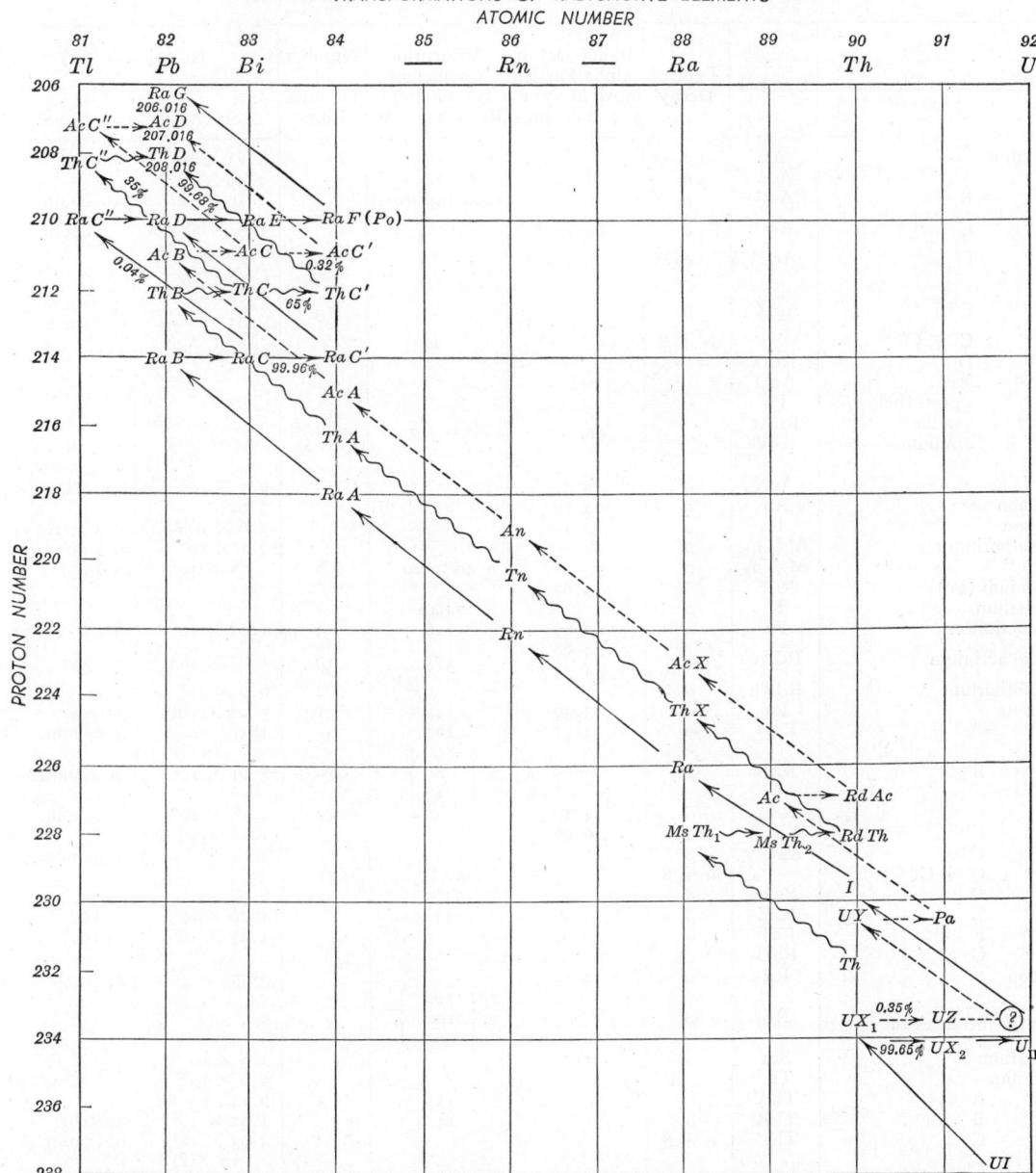

TRANSFORMATIONS OF RADIOACTIVE ELEMENTS
ATOMIC NUMBER

fields. 3. Gamma rays are undeflected by electrical and magnetic fields, and of **wavelength** of the order of 10^{-8} to 10^{-9} cm.

Alpha rays consist of particles shot off from the interior of the atoms of certain **elements** with a definite velocity and a definite range for each element. The velocity is from 5–7% that of **light**. The range is the distance traversed in a homogeneous medium before **absorption,** and is proportional to the cube of the velocity (Geiger and Nuttall, 1911). The penetrating power is the smallest of the three kinds of rays, the beta being of the order of 100 times, and the gamma rays 10,000 times more penetrating. The alpha rays are particles of **helium** carrying two unit positive charges (He^{++}). Ramsay and Royds (1909) experimentally demonstrated that accumulated alpha particles, quite independently of the matter from which they have been expelled, consist of helium. They sealed radon in a

glass tube with wall so thin that the alpha particles passed through the wall into a surrounding vessel and after six days the whole spectrum of helium was observed. Helium itself does not diffuse through such a wall. Therefore alpha particles on losing their charge become ordinary helium. This is the first instance of the *production of a known element* during radioactive transformation. The loss of a single alpha particle by an atom leaves the residual atom four units less in **atomic weight** and two positive charges less in **valence** (Fajans, 1913). The shooting of alpha particles may be visibly registered by Crooke's spinthariscope, in which the tip of a wire, coated by a tiny amount of radium salt, is placed near a screen coated with **zinc** blende. Viewed in the dark with a magnifying eyepiece, each alpha particle striking the zinc blende target emits a visible scintillation. It is *as if* one were seeing atoms. The detection and counting of single alpha particles was

SOME CONSTANTS OF RADIOACTIVE ELEMENTS

Radioactive Element	Symbol	Type of Decay	Range (R) of Alpha Particle in Air at 15° C. 760 mm pres.	Absorption Coefficient (μ) of Beta Rays. cm^{-1}Al	Number of Gamma Lines	Half-period (T)	
						Seconds	Equals
Actinium	Ac	β				4.23×10^8	13.5 yr.
" A	AcA	α	6.58			2.10×10^{-3}	
" B	AcB	β		1000 (approx.)		2.16×10^3	36.0 min.
" C	AcC	α	$\begin{Bmatrix}5.51\\5.09\end{Bmatrix}$			130	2.16 min.
" C'	AcC'	α(?)	(6.5?)			5×10^{-3} (approx.)	
" C''	AcC''	β			3	$\begin{Bmatrix}286\\283\end{Bmatrix}$	$\begin{Bmatrix}4.76\text{ min.}\\4.71\text{ min.}\end{Bmatrix}$
" C' + C''		$\alpha + \beta$		29			
" D	AcD						
" lead	AcD						
" , prot- (below)	Pa						
" , radio-	RdAc						
" uranium	AcV						10^8 to 10^9 yr (approx.)
" X	AcX	α	4.37		5	9.7×10^5	11.2 d.
Actinon	An	α	5.79			3.92	
Ionium	Io	α	3.19			2.6×10^{12}	8.3×10^4 yr.
Mesothorium 1	MsTh$_1$	β				2.1×10^8	6.7 yr.
" 2	MsTh$_2$	β		40 to 20	8	2.21×10^4	6.13 hr.
Polonium (Po)	RaF	α	3.87				
Potassium	K	β		74.49			
Protactinium	Pa	α	3.67	126	3	1.01×10^{12}	3.2×10^4 yr.
Radio actinium	RdAc	α	$\begin{Bmatrix}4.68\\4.34\end{Bmatrix}$	175	10	1.63×10^6	18.9 yr.
Radiothorium	RdTh	α			2	6.0×10^7	1.90 yr.
Radium	Ra	α	3.39	312	1	5.02×10^{10}	1590 yr.
" A	RaA	α	4.72	420		183	3.05 min.
" B	RaB	β		$\begin{Bmatrix}890\\80\\13\end{Bmatrix}$	10	1.61×10^3	26.8 min.
" C	RaC	α	4.1			1.18×10^3	19.7 min.
" C'	RaC'	α	6.96			10^{-6} (approx.)	
" C''	RaC''	β				79.2	1.32 min.
" C' + C''		$\alpha + \beta$		50.13	11		
" D	RaD	β		5500	1	6.94×10^8	22 yr.
" E	RaE	β		45.5		4.26×10^5	4.9 d.
" F	RaF	α				1.21×10^7	140 d.
" G	RaG						
Radon	Rn	α	4.12			3.305×10^5	3.825 d.
Rubidium	Rb	β		$\begin{Bmatrix}700\\190\\900\end{Bmatrix}$			
Samarium	Sm	α	1.2			3.8×10^{19}	1.2×10^{12} yr.
Thorium	Th					5.6×10^{17}	1.8×10^{10} yr.
" A	ThA			153	3	0.14	
" B	ThB	β		14.4		3.82×10^4	10.6 hr.
" C	ThC	$\alpha + \beta$				3.63×10^3	60.5 min.
" C'	ThC'					$\begin{Bmatrix}10^{-9}\ (?)\\<10^{-6}\end{Bmatrix}$	
" C''	ThC''	β		21.6	11	186	3.1 min.
" D	ThD						
" , meso-1 (above)	MsTh$_1$						
" , meso-2 (above)	MsTh$_2$						
" lead	ThD						
" , radio-	RdTh						
" X	ThX	α			2	3.14×10^5	3.64 d.
Thoron	Tn					54.5	
Uranium I	UI	α	$\begin{Bmatrix}2.67\\2.73\end{Bmatrix}$			1.4×10^{17}	4.4×10^9 yr
" II	UII	α	$\begin{Bmatrix}3.12\\3.28\end{Bmatrix}$			9.4×10^{12}	3×10^5 yr.
" X$_1$	UX$_1$	β		460	1	2.12×10^6	24.5 d.
" X$_2$	UX$_2$	β		18		68.4	1.14 min.
" Y	UY	β		300 (approx.)		8.88×10^4	24.6 hr.
" Z	UZ	β		270 to 36		2.4×10^4	6.7 hr.
" lead	RaG						

accomplished by Rutherford and Geiger (1908) by the deflection of an electrometer needle upon the arrival of each alpha particle in a gas at low pressure in an electric field somewhat below the sparking point.

Beta rays are electrons shot off from the interior of the atom of certain elements with velocities varying almost up to that of light. The loss of a single beta particle by an atom leaves the residual atom the same in atomic weight and one negative unit less in valence. (Soddy, Russell, Fleck, Fajans, 1913.)

Gamma rays are like x-rays but more penetrating. The presence of gamma rays from 30 milligrams of radium can be observed in an **electroscope** after passing through 30 cm. of iron (Rutherford). For the protection of the operator, radium is kept in **lead** outer containers or screened by lead sheets. Gamma and beta rays usually accompany each other, and seem to be related in much the same ways as x-rays and cathode rays. The wavelength of gamma rays has been determined by diffraction through crystals, and ranges from about 10^{-8} cm. for the *soft* gamma rays of radium B to 0.7×10^{-9} cm. for the *hard* penetrating rays of radium C.

The end-product of the uranium series indicates lead of atomic weight 206. Actually, uranium lead (RaG) has an atomic weight of 206.016 (from uranium I, 238.14, by loss of 8 alpha particles). The end point of the thorium series indicates lead of atomic weight 208. Actually thorium lead (ThD) has an atomic weight of 208.016 (from thorium 232.12, by loss of 6 alpha particles). The end point of the actinium series indicates lead of atomic weight 207 (stated as 207.016). Atomic weight determinations by Richards (1916) showed lead of North Carolina uranite of atomic weight 206.40, from Colorado carnotite 206.59, and from Joachimsthal pitchblende 206.57.

Radioactive Isotopes. Mesothorium 1 and radium are chemically identical. It is possible to distinguish by chemical methods only 10 different radioactive elements, numbers 81, 82, 83, 84, 86, 88, 89, 90, 91, 92, while from radioactive evidence it appears that there are about 39. All of the elements in each of the 92 possible places of the Mendeleeff-Moseley classification of elements exhibit identical chemical properties. Such elements—isotopic elements—are abundant among the radioactive elements.

Artificial Disintegration of the Elements. In 1919, Rutherford disintegrated the atom of nitrogen by bombardment with alpha particles fired off from radium-C. By means of advancements since then the artificial disintegration of practically any element is possible. The results of these studies have lead (1) to fundamental knowledge of the structure of the atom, and (2) to the application of this knowledge (a) to produce new atoms and atomic particles and (b) to create or absorb energy. All of these results are radically novel and apparently at variance with previous scientific beliefs. Most radical to the layman is probably the annihilation of some of the mass of reactants with the simultaneous creation of a relatively enormous amount of energy. **The disappearance of $1/100$ atomic mass unit** (where the "ordinary" oxygen atom is 16.0000 atomic mass units) **results in the appearance of 9.3 million electron volts (MEV),** of energy, and conversely the appearance of $1/100$ atomic mass unit in the disappearance of 9.3 million electron volts. In the disintegration studied by Rutherford, wherein nitrogen was bombarded with **alpha particles** (helium atoms, atomic mass 4, atomic number 2, electrical charge 2 units $+$), the result is as follows:

Atomic mass:
Atomic number: $\quad {}^{14}_{7}\text{N} \quad + \quad {}^{4}_{2}\text{He} \rightarrow \quad {}^{17}_{8}\text{O} \quad + \quad {}^{1}_{1}\text{H}$

Exact mass: $\quad 14.0075 \quad\quad 4.0039 \quad\quad 17.0045 \quad\quad 1.0081$

Mass sum: $\quad\quad\quad 18.0114 \quad\quad\quad\quad\quad 18.0126$

Appearance of mass: $\quad 18.0126 - 18.0114 = 0.0012$ unit

Disappearance of energy: $\quad 9.3 \times \dfrac{0.0012}{0.0100} = 1.1$ MEV

Curie and Joliot found further, in 1933, that boron, magnesium, or aluminum, when bombarded with alpha particles from polonium, emit positrons (β^{+}) and the emission continues after the polonium is removed. The half-life of this positron emission from aluminum (after the polonium is removed) is 3.25 minutes, and the average energy of ejected positrons is approximately 2.2 MEV by absorption measurements. Radioactive isotopes of many of the chemical elements have been obtained, and some 60 or more eject positrons, some having quite a long half-life. This phenomenon of produced radioactivity is called artificial radioactivity not because there is anything artificial in the radioactivity itself but simply because it is produced artificially. In alpha-particle bombardment the alpha particle is first captured by the atom bombarded and then either a proton (${}^{1}_{1}\text{H}$) or a neutron (${}^{1}_{0}n$) is ejected. The same kind of atom may eject *both* protons *and* neutrons simultaneously from different individual atoms; thus for sodium atoms it is found:

Atomic mass:
Atomic number: $\quad {}^{23}_{11}\text{Na} + {}^{4}_{2}\text{He} \rightarrow \quad {}^{26}_{12}\text{Mg} \quad + {}^{1}_{1}\text{H}$

$\quad\quad\quad\quad\quad\quad {}^{23}_{11}\text{Na} + {}^{4}_{2}\text{He} \rightarrow \quad {}^{26}_{13}\text{Al} \quad + {}^{1}_{0}n$
$\quad\quad\quad\quad\quad\quad\quad\quad\quad\quad\quad$ radioactive

Besides (1) alpha particles (${}^{4}_{2}\text{He}$), it is found that (2) protons (${}^{1}_{1}\text{H}$), (3) neutrons (${}^{1}_{0}n$), and (4) deuterons (${}^{2}_{1}\text{H}$) serve as projectiles for (a) the transmutation of elements, (b) the production of artificial radioactivity, and (c) nuclear fission. Also, (5) photon ($h\nu$) bombardment is effective in a few cases.

In 1932, Cockcroft and Walton bombarded lithium with *protons* having energies of 100 to 700 thousand electron volts (KEV) and found that alpha particles were ejected, as evidenced (a) by scintillations on a fluorescent screen, (b) by photographing the tracks of the particles in a Wilson cloud chamber, and (c) by electrical methods of detection.

Atomic mass:
Atomic number: $\quad {}^{7}_{3}\text{Li} \quad + \quad {}^{1}_{1}\text{H} \quad \rightarrow \quad {}^{4}_{2}\text{He} \quad + \quad {}^{4}_{2}\text{He}$

Exact mass: $\quad 7.0182 \quad\quad 1.0081 \quad\quad 4.0039 \quad\quad 4.0039$

Mass sum: $\quad\quad\quad 8.0263 \quad\quad\quad\quad\quad 8.0078$

Disappearance of mass: $\quad 8.0263 - 8.0078 = 0.0185$ unit

Appearance of energy: $\quad 9.3 \times \dfrac{0.0185}{0.0100} = 17.2$ MEV

Other disintegrations produced by proton bombardment are:

$$ {}^{7}_{3}\text{Li} + {}^{1}_{1}\text{H} \rightarrow {}^{8}_{4}\text{Be} + h\nu $$

$$ {}^{9}_{4}\text{Be} + {}^{1}_{1}\text{H} \rightarrow {}^{6}_{3}\text{Li} + {}^{4}_{2}\text{He} $$

$$ {}^{9}_{4}\text{Be} + {}^{1}_{1}\text{H} \rightarrow {}^{8}_{4}\text{Be} + {}^{2}_{1}\text{H} $$

$$ {}^{11}_{5}\text{B} + {}^{1}_{1}\text{H} \rightarrow {}^{4}_{2}\text{He} + {}^{4}_{2}\text{He} + {}^{4}_{2}\text{He} $$

Since protons with energies as low as 20 KEV are effective in producing some disintegrations, it is believed that there are certain responsive or "resonance" values of energy at which the proton can enter the atomic nucleus readily when its own (proton) energy is relatively small.

When beryllium is bombarded with alpha particles it became clear, after Chadwick (1932) had recognized the neutron as a particle of zero electrical charge and of mass equal to the hydrogen atom or proton, that the reaction involved is:

Atomic mass:
Atomic number: $\quad {}^{9}_{4}\text{Be} + {}^{4}_{2}\text{He} \rightarrow {}^{12}_{6}\text{C} + {}^{1}_{0}n$

wherein a neutron (${}^{1}_{0}n$) was ejected. Due to its bearing no electrical charge, the neutron can readily penetrate the nuclei of atoms, even those that are highly charged and heavy, and accordingly more disintegrations have been produced with neutrons than with any other particle. The disintegrations by neutron bombardment are varied and may result in the ejection of one or more neutrons, of an alpha particle, of a proton, or of a photon, and practically all of the elements have been disintegrated by this method. The following examples are illustrative:

$$\ce{^1_1H + ^1_0n -> ^2_1H} + h\nu$$

ordinary hydrogen into heavy hydrogen

$$\ce{^6_3Li + ^1_0n -> ^3_1H + ^4_2He}$$

production of rare $\ce{^3_1H}$ possessing radioactivity

$$\ce{^{10}_5B + ^1_0n ->\quad ^7_3Li\quad + ^4_2He}$$

$$\ce{^{23}_{11}Na + ^1_0n ->\quad ^{20}_9F\quad + ^4_2He}$$
radioactive

$$\ce{^{24}_{12}Mg + ^1_0n ->\quad ^{24}_{11}Na\quad + ^1_1H}$$
radioactive

$$\ce{^{27}_{13}Al + ^1_0n ->\quad ^{27}_{12}Mg\quad + ^1_1H}$$
radioactive

$$\ce{^{27}_{13}Al + ^1_0n ->\quad ^{24}_{11}Na\quad + ^4_2He}$$
radioactive

$$\ce{^{27}_{13}Al + ^1_0n ->\quad ^{28}_{13}Al\quad} + h\nu$$

$$\ce{^{39}_{19}K + ^1_0n ->\quad ^{38}_{19}K\quad + ^1_0n + ^1_0n}$$
radioactive

Some disintegrations require the use of neutrons of high energy, called fast neutrons, and some require those of low energy, called slow neutrons. Slow neutrons are usually obtained by allowing fast neutrons to pass through some substance containing hydrogen. In the examples cited above, the disintegration by neutron bombardment results in products of decidedly unequal masses, but the neutron bombardment of uranium (1939) splits this atom into two products of almost equal masses. This phenomenon of "equal splitting" is called fission.

By bombardment with *deuterons* ($\ce{^2_1H}$) the lithium atom absorbs deuteron and ejects two alpha-particles, thus:

Atomic mass: Atomic number:	$\ce{^6_3Li}$	$+$	$\ce{^2_1H}$	$\to$	$\ce{^4_2He}$	$+$	$\ce{^4_2He}$
Exact mass:	1.0169		2.0147		4.0039		4.0039
Mass sum:		8.0316			8.0078		

Disappearance of mass: $8.0316 - 8.0078 = 0.0238$ unit

Appearance of energy: $9.3 \times \dfrac{0.0238}{0.0100} = 22.1$ MEV

Also: $\ce{^6_3Li + ^2_1H ->\quad ^7_3Li\quad + ^1_1H}$

$$\ce{^7_3Li + ^2_1H -> ^4_2He + ^4_2He + ^1_0n}$$

$$\ce{^7_3Li + ^2_1H ->\quad ^8_3Li\quad + ^1_1H}$$
radioactive

Two disintegrations by the bombardment of deuterium (heavy hydrogen, $\ce{^2_1H}$) by deuterons are specially interesting:

$$\ce{^2_1H + ^2_1H ->\quad ^3_2He\quad + ^1_0n}$$
rare, stable

$$\ce{^2_1H + ^2_1H ->\quad ^3_1H\quad\quad + ^1_1H}$$
rare, radioactive

The deuterium target is usually "heavy water" frozen on a surface that is cooled by liquid air, or it may be some other solid deuterium compound, say hydrocarbon. Other disintegrations by deuteron bombardment are:

$$\ce{^{10}_5B + ^2_1H ->\quad ^4_2He\quad + ^4_2He + ^4_2He}$$

$$\ce{^{12}_6C + ^2_1H ->\quad ^{13}_6C\quad + ^1_1H}$$

$$\ce{^{12}_6C + ^2_1H ->\quad ^{13}_7N\quad + ^1_0n}$$

$$\ce{^{23}_{11}Na + ^2_1H ->\quad ^{24}_{12}Mg\quad + ^1_0n}$$

$$\ce{^{23}_{11}Na + ^2_1H ->\quad ^{24}_{11}Na\quad + ^1_1H}$$
radioactive

$$\ce{^{27}_{13}Al + ^2_1H ->\quad ^{25}_{12}Mg\quad + ^4_2He}$$

$$\ce{^{196}_{78}Pt + ^2_1H ->\quad ^{197}_{78}Pt\quad + ^1_1H}$$
radioactive

In the last reaction $\ce{^{197}_{78}Pt}$ is radioactive, with a half-life of 18 hours, and is transformed into gold $\ce{^{197}_{79}Au}$ plus an electron. (R.K.S.)

RADIOACTIVE MINERALS. Strutt, one of the earliest workers in radioactive minerals, discovered that a specimen of **thorianite** which he tested contained 280 million times as much helium as the same mineral could generate in a year; thus he assumed that the **igneous rock** in which thorianite formed must be 280 million years old. **Uranium,** another radioactive substance, has been found to break up into **helium** and lead, one atom of uranium yielding eight atoms of helium and one atom of lead. Since these stable atoms of helium and lead can be measured and compared with the amount of uranium which has not disintegrated, it is possible to determine how long it has been since the mineral was formed. Analysis of radioactive minerals, by this method, collected from the oldest known granites of the **Archeozoic** intimate that the age of these rocks is approximately 1 billion, 800 million years old. One of the chief arguments in favor of the "radioactive method" for determining the age of the earth is that it checks when applied to the radioactive minerals collected from successively younger formations. (R.M.F.)

RADIO-FREQUENCY HEATING. Electric Heating.

RADIOGRAPHY. Photography by x-ray radiation or by the gamma rays of radium or **radioactive substances.**

In making radiographs, the x-rays emitted from the anode of the tube (see **X-Rays**) are directed towards the subject to be examined. Upon reaching the subject some pass through while others, meeting parts of the subject which offer greater resistance, are absorbed wholly or partially. Thus a shadow of those parts of the subject which are more opaque to the passage of the rays is cast on the photographic film and, upon development, a shadow image of the variations in the transparency of the subject to x-ray radiation is obtained. This photographic image is termed a radiograph.

While all photographic materials are sensitive to x-ray radiation, special x-ray emulsions are available which are more sensitive, thus permitting shorter exposures, and have the high contrast necessary to differentiate between slight variations in the absorption of x-rays by different parts of the subject. Such films differ from those used in the camera in having much thicker emulsion coatings; sensitivity to x-ray radiation depending largely upon the amount of silver halide present for the formation of an image and not, as with light, on the sensitivity of the different grains of silver halide in the emulsion.

When the exposure must be kept as short as possible as, for example, in photographing organs of the body which cannot be immobilized, or thick sections, intensifying screens are used. These are screens coated with a fluorescent substance, such as calcium tungstate, which convert some of the x-ray energy into ordinary light and so reduces the exposure. Placed in contact with the photographic emulsion an intensifying screen may reduce the exposure to $\frac{1}{25}$ the exposure required by direct exposure to x-ray radiation. With thick metallic castings the reduction in exposure may in some cases reach $\frac{1}{100}$.

Radiographs may also be made by photographing the image on the intensifying screen with an ordinary camera. Since this procedure requires a longer exposure, it is used only when a small image is acceptable. These small radiographs, which may vary in size from $4 \times 5''$ to $1 \times 1\frac{1}{2}''$ do not show as much detail as the larger direct radiograph, which is much better for diagnostic purposes, but meet the requirements of a "screening" examination the object of which is to locate outstanding cases.

This is also the simplest method of making x-ray motion pictures although it is often difficult to obtain sufficient exposure even when using (1) intensifying screens of the highest efficiency, (2) lenses with apertures of $f/1.5$ to $f/1.9$, and (3) emulsions of the highest available speed.

Micro-radiography is the term applied to the radiography of small objects where the detail is too fine to be seen by the unaided eye and the radiograph must either be examined with a low-power microscope or an enlargement must be made by projection. It is a relatively new field and one which is rapidly assuming considerable importance in the study of sections of tissue, leaf structure, insect anatomy, seeds, spray materials, textiles and artificial fibers and even the distribution of the constitutents in an alloy. For this latter the material must be in the form of a thin section. Soft, i.e., low-voltage, x-rays are used and single-coated, fine-grain, high-resolution films or plates. The resulting radiograph may be enlarged from 10 to 100 times, depending on the subject, the film or plate, and the voltage used in exposing the radiograph.

Stereoradiographs are necessary when the distance of an object or malformation below the surface must be determined. They are made by making two radiographs in succession, shifting the position of the tube between exposures a distance equal to the normal interpupillary distance. After processing, the radiographs are placed in a modified Wheatstone type of stereoscope in which the two images are combined to produce a three-dimensional (stereoscopic) image. Stereoradiography is not often used in industrial radiography but at times is of considerable value. (C.B.N.)

RADIOLARIA. Plasmodroma.

RADIOLARIAN OOZE. Oceanic Deposits.

RADIO-METEOROGRAPH. Meteorological Instruments.

RADIOMETER. An instrument for detecting, and usually also for measuring, thermal radiation, especially infra-red. Radiometers are of several different types.

1. *Crookes Radiometer.* The apparatus devised by Sir William Crookes consists of pairs of light vanes mounted at the ends of opposite spokes of a very light, horizontal wheel poised on a pivot like a compass needle. It therefore resembles a miniature paddle-wheel. One side of each vane is a polished metal surface, these surfaces being all directed the same way around the wheel; the other side is blackened. The whole apparatus is mounted inside a glass or quartz bulb from which the air is partially exhausted. When radiation from the sun or other hot source falls on the vanes, the wheel revolves in the direction toward which the polished surfaces are directed. This motion is explained by the higher absorptivity of the blackened surfaces, which causes them to become warmer than the polished surfaces. This, in turn, causes a creeping of gas molecules, known as "thermal transpiration," around the edge of the vane from the polished to the black face, thus building up an excess of pressure against the latter. If a similar vane system is suspended by a fine quartz fiber and provided with a small mirror, the torque caused by the inequality of pressure is made observable and may be used to measure the intensity of the incident radiation.

2. *Nichols Radiometer.* The apparatus used by Nichols and Hull for the measurement of **radiation pressure** (1901) consisted of a pair of small, silvered glass mirrors suspended, in the manner of a **torsion balance,** by a fine quartz fiber within an enclosure in which the air pressure could be regulated. The torsion head to which the fiber was attached could be turned from outside the enclosure by means of a magnet. A beam of light was directed first on one mirror and then on the other, and the opposite deflections observed with mirror and scale. By turning the mirror system around so as to receive the light on the unsilvered side, the influence of the air in the enclosure could be ascertained. This influence was found to be a minimum, and to have an almost negligible value, at an air pressure of about 16 mm. of mercury. The radiant energy of the incident beam was deduced from its heating effect upon a small, blackened silver disk, which was found more reliable than the **bolometer** at first used. With this apparatus the experimenters were able to obtain an agreement between observed and computed radiation pressures within about 0.6 of 1%.

3. Instruments operating on either the Crookes or the Nichols principle, according to the pressure of the residual gas in the enclosure, are in use under the name "vane radiometer." They are more sensitive if the radiation beam is allowed to fall on only one vane, the rest of the vane system being carefully shielded.

4. In addition to these, the **pyrheliometer,** the **bolometer,** the **radiomicrometer,** and the thermopile (see **Thermel)** may be properly classed as radiometers. (L.D.W.)

RADIOMICROMETER. A very sensitive **thermel,** adapted to the detection of feeble thermal radiation; devised by C. V. Boys (1889). It consists of a small bismuth-antimony thermocouple short-circuited by a narrow, oblong loop of copper wire and suspended by a quartz fiber between the poles of a strong permanent magnet. The couple is surrounded by a thick housing of soft iron to shield it from magnetic disturbances and stray radiation. The radiation to be measured is admitted through an opening in this shield and falls on a tiny copper disk soldered to the couple. A very small amount of radiation communicates heat enough to generate a current in the loop and cause it to turn in the magnetic field. The deflection is observed by means of a mirror attached to the loop, which reads on a scale like a **galvanometer.** By using a large lens to concentrate the rays, Boys was able to detect the radiation from a candle more than 2 miles distant. (L.D.W.)

RADIOTHORIUM. Symbol: RdTh. A radioactive element of the thorium series. (See **Radioactive Changes.**) (R.K.S.)

RADIUM. Symbol: Ra. Atomic number: 88. Atomic weight: 225.97. Melting point: 960° C.

Radium metal is white, rapidly oxidized in air, decomposes water, and in the metallic state is a curiosity. Radium is usually handled as the chloride or bromide, either as solid or in solution. Its most remarkable property is its radioactivity, which decreases about 1%

in 25 years. Radium element evolves heat continuously, 0.132 calories per hour per milligram of radium when the decomposition products are retained, and the temperature of radium salts remains about 1.5° C. above the surroundings. Discovered by P. and Mme Curie in 1898. (See **Radioactive Changes**.)

Radium occurs in **pitchblende**, and in **carnotite** along with **uranium**. Radium was first obtained from the uranium residues of pitchblende of Joachimsthal, Czechoslovakia, later from carnotite of southwestern Colorado and eastern Utah. Richer ores have been found in Belgian Congo and in the Great Bear region of northwestern Canada.

Radium is formed by the radioactive transformation of uranium—about 3,000,000 parts of uranium being accompanied in nature by 1 part of radium—and spontaneously generates **radon** gas—about 100 cu. mm. of radon per day per gram of radium.

Chemically related to **barium**, radium is recovered from its ores by addition of barium salt, followed by treatment as for recovery of barium, usually as the sulfate. The sulfates of barium and of radium are insoluble in most chemicals, so they are transformed into **carbonate** or **sulfide**, both of which are readily soluble in **hydrochloric acid**. Separation from barium is accomplished by fractional crystallization of the chlorides (or bromides, or hydroxides). Dry, concentrated radium salts are preserved in sealed glass tubes, which are periodically opened by experienced workers to relieve the pressure. The glass tubes are kept in lead shields.

Radium chloride (RaCl₂) and radium bromide (RaBr₂) are less soluble and radium hydroxide (Ra(OH)₂) more soluble than the corresponding barium compounds.

Radium causes serious flesh burns due to its radioactivity, but under proper control is used in the treatment of certain forms of **cancer**. (R.K.S.)

RADIUS. In zoology, the term radius has two common meanings. 1. It is a principal radiating division of the body of an **echinoderm**. In the starfishes and other forms with projecting arms, the arm or ray is the most conspicuous part of a radius. In more compact forms like the sea urchins each radius is a segment of the body. 2. It also denotes a bone of the forearm or lower foreleg of **vertebrates**.

For the use of the term radius in mathematics, see **Circle** and **Polar Coordinates**. (A.W.L.)

RADIUS OF CURVATURE. In mathematical terms the radius of curvature of a curve is the reciprocal of the curvature k, that is $\frac{1}{k}$, in which k is equal to the change in the direction of a curve per unit length of arc. The radius of curvature of a straight line is infinite. The radius of a **circle** is the radius of curvature of the circular curve. The radius of curvature of an **ellipse, parabola, hyperbola**, etc. (all mathematical curves), is a variable which depends upon the sharpness (curvature) at a particular point on a curve. If these curves are assumed to be made up of a number of infinitely short, connected, circular arcs, the radius of curvature might be defined, in simple terms, as the radius of an infinitely short, circular arc located at the point where the radius of curvature is desired. (See also **Curvature of a Plane Curve**.) (C.W.C.)

RADIUS OF GYRATION. The radius of gyration of an area in respect to a particular axis is the square root of the quotient of the **moment of inertia** divided by the area. It is the distance at which the entire area must be assumed to be concentrated in order that the product of the area and the square of this distance will equal the moment of inertia of the actual area about the given axis. The numerical value of the radius of

gyration, k, is given by the following formula in which I is the moment of inertia and A, the area.

$$k = \sqrt{\frac{I}{A}}$$

The radius of gyration of a **mass** is similar except that the moment of inertia of the mass is involved. (C.W.C.)

RADIUS VECTOR OF A POINT. Polar Coordinates in a Plane.

RADON. Symbol: Rn. Atomic number: 86. Atomic weight: 222. Density: 9.72 grams per liter, 0° C., 760 mm., or 7.5 when air equals 1.00. Melting point: —110° C.

Radon is a colorless gas, radioactive, and formed by **radioactive** transformation of radium. Radon, **thoron** and actinon are isotopes. Discovered by Dorn in 1900 in the radioactive transformation of **radium**, and thoron by Rutherford in 1900 in the radioactive transformation of thorium. Radon decays to ½ its radioactivity in about 4 days. Radon is obtained by bubbling air through a radium salt solution, and collecting the gas plus air. (See **Radioactive Changes**.) (R.K.S.)

RADULA. A horny strip in the floor of the mouth of a **snail**. It bears many rows of minute teeth and is used for scraping up food. The radula is developed in a ventral pouch of the oral cavity known as the radula sac. (A.W.L.)

RAFFIA. *Raphia Ruffia*. Palmaceae. This African **palm** tree has leaves of great length, sometimes as much as 25′. The leaf epidermis may be removed in strips. These strips, known as raffia, have been used in basketwork and for tying up plants. (R.M.W.)

RAFINESQUINA. Invertebrate Paleontology.

RAGLANITE. A term proposed by Adams and Barlow, in 1910, for a **facies** of **nepheline-syenite** also containing abundant **oligoclase feldspar** and the accessory minerals, **mica, calcite, apatite and magnetite**. (R.M.F.)

RAGWEED. *Ambrosia artemisiifolia*. Composite Family.

RAIL. Aves, Gruiformes. Long-legged marsh birds (**Aves**) of wide distribution. They have moderate to long and slender beaks, rather small wings, and a short tail. North America has several species, including the clapper, *Rallus longirostris*, Virginia, *R. limicola*, and sora rail, *Porzana porzana*. The corncrake, *Crex crex*, is a Eurasian species which reaches North America occasionally. In New Zealand the group is represented by the large weka rails which do not fly, although they have wings. (A.W.L.)

RAILS. Steel.

RAILWAY. The railway is a transportation agency which, in spite of the growing importance of other transportation systems (bus, truck, aircraft), is the principal method of transporting goods over long distances. The activities connected with the operation of railways constitute one of the leading industries of the United States by virtue of the services rendered, the capital invested, and the personnel employed. The operation of the railway, furthermore, may and does affect practically all persons living in its sphere of influence, because of the dependence upon transportation for the distribution of goods under the system of manufacture used in a progressive and mechanized civilization such as ours. The services of the railway consist of the movement, for a price as expressed by the current rate, of freight or passenger traffic. From the standpoint of total receipts, freight is the more important. The speed with which a railway can transport goods with a small expenditure of energy means that manufacture can be specialized geographically. Large-scale production, and a considerable division of labor, are possible. These things have contributed in a large degree to making many commodities available at costs which would not be possible under any other system of production. The carrying of passengers is of greater importance to the nation than the net receipts of passenger traffic to the railway would indicate. The availability of definite scheduled transportation to the people as a whole, which transportation is rarely affected by season or weather, has an important economic and political connotation. Suburban passenger railway facilities have grown to be essential to the life of a metropolis, the housing facilities of which are entirely inadequate to its workers, who therefore, by necessity or choice, reside in suburbs which are swiftly and surely connected with the metropolis by the railway. Mobility of labor, rapid transportation of the mail, and carrying of commodities at a premium rate by the express business, are further important branches of railway activity. The railways of the United States are privately owned under the supervision of the Federal body for the control of interstate commerce. Although there are nearly 2000 separate companies divided into three classes by the range of their annual revenue, practically all of the transportation based on ton or passenger mileage is controlled by the class 1 companies, of which there are about 175 organizations. A class 1 railway is that having an annual revenue in excess of $1,000,000.

The railway proper consists of a continuous strip of ground called the right of way, varying from 60 to 150' in width, which extends, with a minimum of curves and departures, in a direct course between centers of population, business, or industry which it serves. On this right of way there is a smooth graded base called the subgrade, upon which is laid **ballast** consisting of cinders or crushed stone. The cross ties which support the rails are usually laid on the subgrade, and afterwards lifted into position when the ballast has been placed. The gradient employed in rolling or mountainous country is, of necessity, a compromise between cost of construction and cost of operation. The steeper the ruling gradient of a line, the more powerful the tractive equipment must be, or the smaller the train size. On the other hand, the cost of reducing the gradient may, in rugged country, become extremely high. One cut-off constructed by a railway company which was able to reduce its ruling gradient from 1.2 to 0.7% is said to have cost approximately $300,000 per mi. This expenditure was justified since it reduced the grade and distance between termini on a trunk line which carried heavy traffic.

The wooden cross tie has become standard in this country, although substitutes in the form of concrete or steel have been tried. New wooden ties are usually creosoted to lengthen their life, and are approximately 7″ x 9″ x 8′ 6″ long. They are spaced a little under 2′ on centers, and on them are laid two parallel steel rails which are firmly spiked to each tie. The hard smooth way provided by the steel rail and the minimum rolling friction of steel wheels on these rails accounts for the extremely low haulage costs per ton mile on railways. The T-rail is the standard for open track work. These rails are rolled in steel rolling mills to standard sections the size of which is denoted by the weight per yd. of run. A 60-lb. rail is a relatively light one, and most of the trackage of this country will be found from this weight upwards to 150 lbs. The present standards of the American Railway Engineering Association are 100-, 110-, and 131-pound rails. The rails are manufactured in lengths of between 30 and

40′, and are coupled together by splice plates of several different designs. These couplings permit **expansion** and contraction of a sufficient amount at each joint.

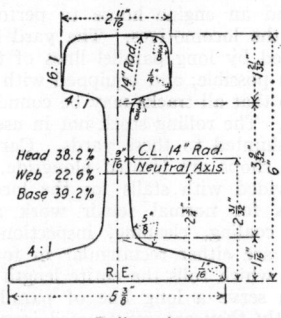

Rail section.

Where the rails bear on the cross ties, there will be wear. When treated ties having a long life expectancy are used, steel tie plates are installed to relieve the cross tie of rail wear. The plate is held securely to the tie and the relative motion, caused by traction or expansion and contraction of the rails, is taken up between the plate and the rail.

Principal routes are nearly all double tracked. The rapid movement of trains, operation of automatic **block signalling**, and freedom from collision hazards, are so superior with the double-tracked arrangement that the additional expense is justified where a large amount of traffic is carried over the railway. Feeder and branch lines are largely single-track, although in many cases where it is thought that a large traffic will eventually develop, the subgrade of the right of way is prepared in sufficient width for double-tracking. Standard practice on double-tracked roads is for the train to advance on the right-hand track. In a few instances trunk lines have triple or quadruple tracks.

Necessary features of the right of way are those structures which are required to carry the railway over obstacles or depressions such as highways, rivers, streams, valleys. The use of **bridges** and trestles, tunnels, **culverts**, and other auxiliary structures, is inevitable in a railway having the trackage implied by membership in the class 1 railway group. Such structures are independently treated in this work.

The railway just described is served by terminal facilities and intermediate stations. The facilities of the intermediate stations will depend upon the importance of the stop. Usually this may be gauged roughly by the population of the locale. The facilities provided consist ordinarily of a station house and side track to which cars may be shunted for unloading or loading of bulk freight, such as sand, lumber, coal, etc. Packaged and crated commodities are handled in a portion of the station set aside for that purpose, or in a separate freight warehouse. The intermediate station contains a waiting room for passengers, and an office from which passenger tickets are dispensed, and which often serves as the base for freight, express, and telegraph services, all of which are concentrated in the hands of one station agent in the small, unimportant stops, but which may be divided among a larger personnel in the case of a city. Express and mail handling facilities and conveniences for passenger loading are also part of the station equipment. Under the head of terminal facilities come also those mentioned for the intermediate stops, although greatly increased and amplified in service to the general public. Larger passenger stations offering diversified services, greater division of labor, and more attention to architectural embellishment, characterize the terminal passenger station. Space is provided also for the executive branch. Long railways are subdivided into divisions with termini intermediate between the divisions as well as at the ends of the line. These termini are important servicing points for the mobile equipment. From the standpoint of the rolling stock, the railway terminal is a "yard" in which the passenger and freight cars may be stored, and an engine house to perform a similar function for the locomotives. The yard is an area of ground covered by long parallel lines of track as close together as is possible, and equipped with switches and cross-overs so that all tracks may be connected with the main railway. The rolling stock not in use, or awaiting use, can be shunted to these yards. Car maintenance and overhaul shops are located alongside. The engine house is provided with stalls for the locomotives and shop facilities for normal repair work and servicing facilities for coaling, cleaning, inspection, and oiling. These houses are either rectangular in form, having a transfer table which rolls the entire length of the house and thus can serve a long line of parallel stalls. Or more frequently they are constructed in circular form; hence the name roundhouse. In a roundhouse the stalls are ranged radially from a central point at which is located a turntable. The incoming **locomotive** is rolled onto this turntable, which is then turned until the turntable track aligns with the short section of track leading to the stall.

The economic operation of a railway system is a very complex affair, reaching into practically all walks of life. The technical maintenance and operation are subjects more appropriate to the nature of this work. This activity consists of the scheduling of trains and the maintenance of them on schedule; servicing of the equipment, both stationary and mobile, so that it will be available for use when needed; and the operation of the physical system, safely and reliably, yet with train speeds as high as possible, for one of the essentials of the service rendered by the railway is fast transportation, whether it be that of freight or of passengers. The activities of technical maintenance must be carried out in a manner to maintain the system against the continuous tendency to deteriorate through such factors as floods, washouts, rust, dirt, rot, snow, friction, wear. It also will include the installation and operation of signalling systems, which have been found necessary, even on double-tracked roads, if heavy traffic schedules are to be maintained. The subject is discussed in more detail under **railway signalling**. Rolling stock must be periodically serviced, at which time work like greasing of bearings, cleaning of upholstery, repair and cleaning of box cars, painting of steel cars, replacement of wheels and brakes, etc., is done. A steam **locomotive**, consisting, as it does, of an entire steam power plant on wheels, is a complicated piece of equipment having many points of wear and many requirements of adjustment. The vibratory nature of its motion when in use, coupled with the presence of high temperatures, high pressures, ash, and smoke, make for expensive but necessary periodic overhauls, and for close inspections at the end of every run. The constant pounding of heavy loads over the track may loosen spikes, crack rail joints, loosen bonds, and these defects can, if unattended, lead directly and rapidly to a serious accident. Hence it is necessary to have a large force going over the right of way constantly, for inspection and repairs. Certain auxiliary services of a special nature are exemplified by the special ballast cleaner trains and by cars fitted with apparatus to detect "rail cancer." Special snow-handling equipment is necessary on railways which are not located in a warm climate. (F.T.M.)

RAILWAY SIGNALLING. The **momentum** arising from the **velocity** and **mass** of a train requires a signalling system for the safety and convenience of operation of a railway train. Railway signalling, which displays signals designed to prevent more than one train occupying a certain section of track at a time, permits the operation of rolling stock at much higher speeds than

would be possible, with safety, if the presence of other trains had to be detected visually by the engineman. Furthermore, passenger trains may be operated more smoothly if the engineman has the information given by the signalling system. The railway automatic **block signal** system has become standard equipment on the railways of this country.

Railway signalling embodies many functions besides the automatic block, but except for the following brief mention, these other classes of signalling are not treated here. A railway system employs signals for advising the engineman when orders from train dispatchers are ready for him. Special signals are employed for controlling railway traffic at junctions, grade crossings, drawbridges, etc., and other fixed or manually operated signals, such as signs along the right of way, indicate yard limits or speed limits. The trainmen make use of torpedoes, flags, or lanterns as signalling devices. Block signals are not always automatic, since many street-car systems have used manually operated block signals for suburban lines. With the disappearance of these extended street railways before the competition of buses, the examples of manually operated block signals have dwindled.

The block signal itself is the physical mechanism employed to signal the condition of the track ahead to the engineman. This it does by a mechanism which conveys the information visually. A pole or tower having a semaphore arm and lights is mounted alongside the track on the right of way, in clear view of the engineman. The trackage of the railway is divided electrically into blocks of from 1 to 5 miles in length. On descending grades the blocks are made longer than on rising grades, since a greater distance is required in which to stop a train. The shorter the block the more trains the system can handle, but the signalling system becomes more expensive, and the average speed of operation may be curtailed. The system used for dividing the line into blocks may be (a) simple blocks (not shown), one beginning where the other ends, each block covered by one block signal, (b) an overlap system, in which the blocks overlap by a distance equal to the space required for a train to stop after application of brakes, (c) the double-semaphore system, in which the condition is being shown not only of the block being entered, but that of the one next ahead. The semaphores in this system are known as the home and distant signals. They are distinguished during the daytime by the position and painting of the semaphore arm, and at night by the relative position of the lights on the pole. The double semaphore is replaceable by a single semaphore which operates at three positions, clear, caution, and stop, shown by vertical, 45°, and horizontal positions of the semaphore. These systems are shown in Figs. 1 and 2.

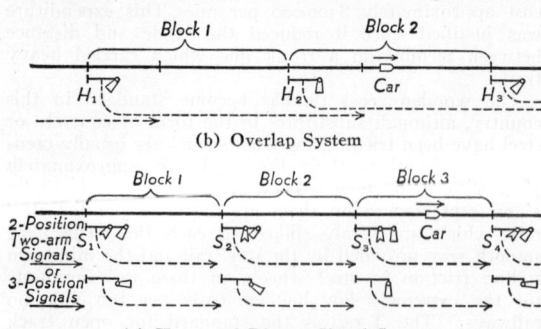

(b) Overlap System

(c) Home and Distant Signal System
Fig. 1. Block signal systems.

The basis of the automatic block-signal system is the track circuit, by means of which the train automatically operates the signals by electricity. As shown in Fig. 3, the block is a section of track in which the rails are insulated from those in the adjoining block by insulating track joints. The rail joints within the block are bonded electrically by bridging them with galvanized wire. For steam roads the track circuit is low-voltage d.c., but in the case of electric railways, in which the track forms a return circuit for the power current, it is not possible to insulate the blocks, and alternating track circuits must be used for automatic block signalling. In the case of a-c electric railways, the blocks must be laid

case of wooden ties by hook-headed spikes which are driven into the latter through holes in the plates. The ties are embedded in a material called **ballast,** usually broken stone. The ballast spreads the load of the train over the earth, holds the track in position and acts as a drainage system. The gauge of the track is the distance measured from inside to inside of rails as shown in Fig. 1. The standard gauge in this country is 4' 8½".

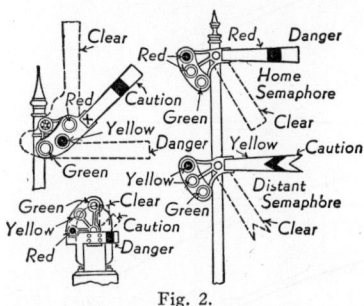

Fig. 2.

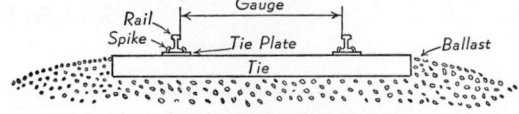

Fig. 1. Railway track.

out with the use of a-c track circuits of much higher frequency than the power current. Fig. 3 illustrates automatic block signal connections for a steam road. A track circuit battery B maintains a current through the track relay R. The armature of this relay closes the semaphore circuit so that the battery B holds the semaphore arm clear. When a train enters the block the

Railway tracks are divided into main and secondary tracks. The secondary system is subdivided into branch, side and spur tracks. A branch is a track for general traffic leading from the main line to communities which are at some distance from the main line. A siding is a track for storing cars at a station or a yard. Side tracks may be located between stations, in which case they are used by slow trains to allow the passing of fast trains. Spur tracks are used for access to quarries, construction projects, mines or industrial plants.

A turnout is a curved track connecting two other tracks. A cross-over is a turnout connecting two parallel tracks which are either straight or curved. The essential features of a turnout are a switch, frogs and guard rails. The switch is a device which causes the train to turn from one track to the other. Where one rail of the turnout crosses a rail of the main track, a device known as a frog is used to allow the flange of the car wheels to pass the intersection. Guard rails are short pieces of rail placed opposite the frog to prevent the flange of the wheels from bearing against the point of the frog and also to prevent them from going in the wrong direction at the point.

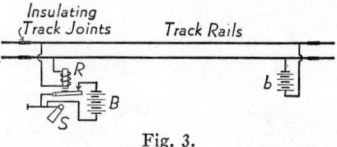

Fig. 3.

wheels and axle short-circuit the relay coil and the battery B is disconnected from the semaphore holding solenoid, causing it to drop to horizontal position. Electrically the track circuits are more complicated when the home and distant signalling is used, and polarized relays are used to distinguish between the impulses received from the two blocks. (F.T.M.)

RAILWAY TRACK. A railway track consists of steel rails which are laid on cross ties. (See **Tie, Railway.**) The rails which are usually separated from the ties by steel tie plates are held in position in the

Fig. 2 shows a diagrammatic illustration of a turnout with a split switch, which is the most common type, in the open and closed position. One rail of the turnout and one rail of the main track are continuous. The transition from one track to the other is made by means of a pair of switch rails which are connected by **tie rods** and pivot about one end. The tie rod at the free end of the switch rails is connected to a switch stand or interlocking tower where the switch is activated.

A stub switch is shown diagrammatically in Fig. 3. The switch rails are free to move as a unit at one end

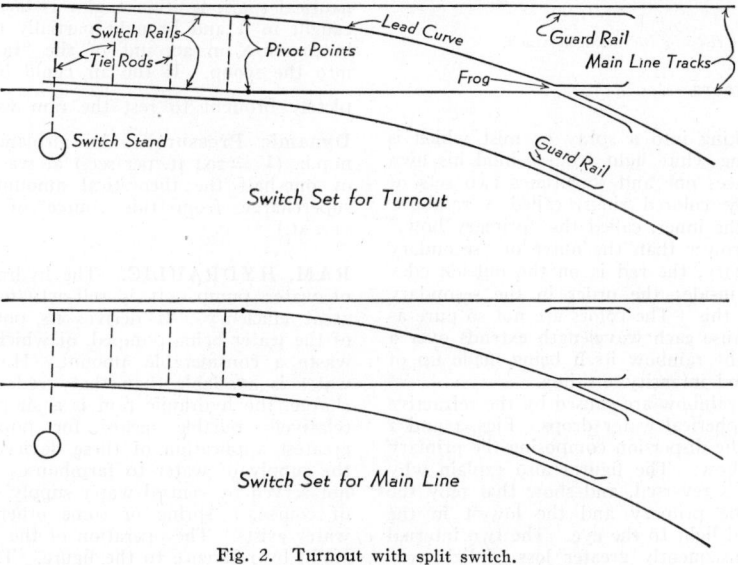

Fig. 2. Turnout with split switch.

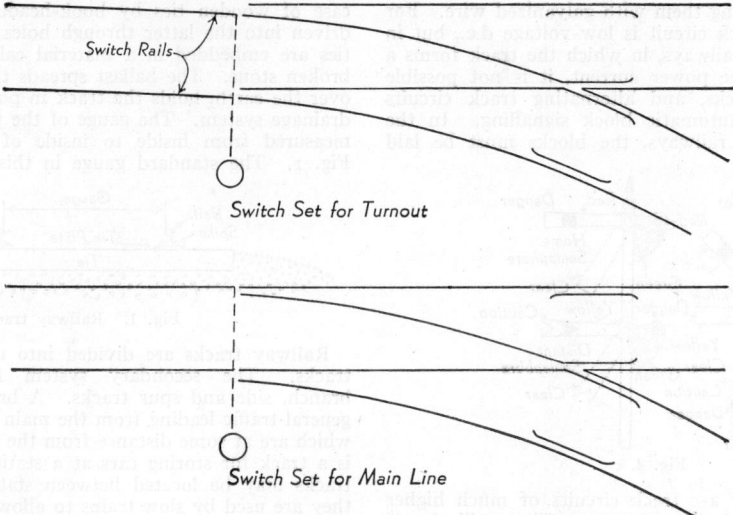

Fig. 3.   Turnout with stub switch.

so that they can be shifted from one track to the other. The free end of the switch rails is connected to the switch stand by a tie rod.

The arrangement of rails where two tracks cross each other is called a crossing. The tracks may be straight or curved but, in any case, four frogs are required. Fig. 4 shows a typical crossing for straight tracks. (c.w.c.)

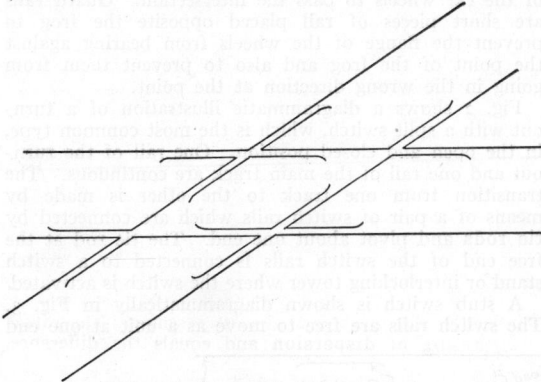

Fig. 4.   Crossing for straight track.

RAIN.   Hydrometeors.

RAINBOW. Looking into a spray or mist which is illuminated by strong white light from behind his own back, an observer sees one and sometimes two sets of concentric, spectrally colored rings, called a rainbow. If two are visible, the inner, called the "primary bow," is brighter and narrower than the outer or "secondary bow." In the primary, the red is on the outside edge and violet on the inside; the order in the secondary being the reverse of this. The colors are not as pure as in a **spectrum,** because each wavelength extends over a wide radial range, the rainbow itself being made up of the fairly pronounced intensity maxima.

The colors of the rainbow are caused by the refractive **dispersion** of the spherical water drops. Figs. 1 and 2 show, respectively, the dispersion composing the primary and the secondary bow. The figures also explain why the order of colors is reversed, and show that only the highest drops in the primary and the lowest in the secondary refract red light to the eye. The two internal reflections, with consequently greater loss of light, ex-

plain why the secondary bow is fainter. The center of the ring system is exactly opposite the source of light; so that natural rainbows are seen only when the

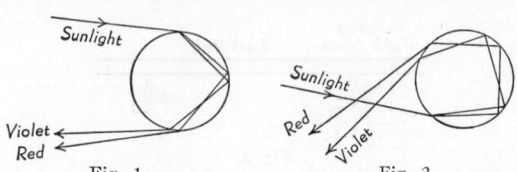

Fig. 1.   Fig. 2.
Formation of primary bow (left) and secondary bow (right). Circles represent raindrops.

sun is near the horizon, unless the observer is elevated high above the surrounding country and can look obliquely downward into the rain. (L.D.W.)

RAKE.   Cutters.

RAM. The male sheep. (See **Press Working; Planer.**)

RAM, AIR. If the carburetor air scoop of an aeronautical spark ignition engine is faced upstream the air caught in it and brought partially to rest will increase in pressure on account of the "ramming" of the air into the scoop. If the air could be ideally and completely brought to rest the ram would be $\frac{\rho}{2} V^2$ (see **Dynamic Pressure**). An airplane travelling at 200 m.p.h. ($V = 294$ ft. per sec.) at sea level, having a ram of one-half the theoretical amount, would receive a supercharge from this source of $1\frac{1}{2}''$ of mercury. (F.T.M.)

RAM, HYDRAULIC. The hydraulic ram is a form of water pump. It is self-actuated, but of no very great efficiency. It derives its power from the flow of the water being pumped, of which it must necessarily waste a considerable amount. However, where more water is available than that needed at the pump discharge, the hydraulic ram is a simple, inexpensive, and relatively reliable means for pumping water. The greatest application of these devices is to be found in the supply of water to farmhouses and other dwellings not served by central water supply systems, and where, of course, a spring or some other source of potable water exists. The operation of the ram may be understood by reference to the figure. The body of the ram is connected with the source of water by a supply pipe, and should be set well below the source so that water will flow rapidly from the supply to the ram. This requirement, of course, rules the ram out as a device for pumping water from wells. The figure shows the ram with its atmospheric valve open. This

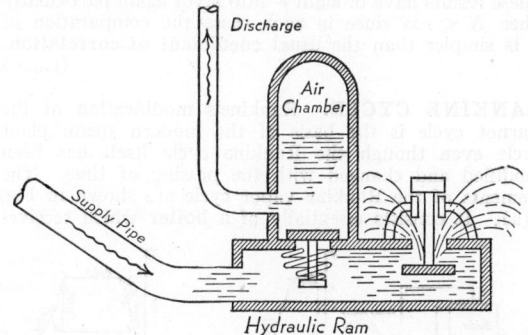

Hydraulic Ram

lets the water flow through the ram at increasing speed until the water flow suddenly drags the valve upwards into the closed position. The momentum of the rapidly moving water in the supply pipe then builds up pressure in the ram sufficient to open the discharge valve and surge some water into the air chamber. After subsidence of the pressure wave, the pressure acting in the discharge closes the discharge valve, and the atmospheric valve drops into open position under the action of its dead weight during the minimum pressure phase of the surge. Flow is resumed through the atmospheric valve, and another cycle of operations begins. Meanwhile the surplus pressure existing in the air chamber due to the incoming surge of water acts to discharge that water into the discharge pipe leading to the water supply system.

In its simplest form the ram is not self-starting; hence it is desirable for the source of water to yield sufficient flow to operate the ram continuously. Where this is not possible, the flow is stored in a reservoir and the ram operated from that periodically. (F.T.M.)

RAMAN EFFECT.
A phenomenon involved in the scattering of light from the molecules of transparent gases, liquids, and solids; discovered by the Indian physicist, C. V. Raman, in 1928. It consists in the appearance of extra spectrum lines in the vicinity of each prominent line of the spectrum of the incident light. For example, if light from a mercury arc shines into carbon tetrachloride (CCl_4, a liquid), some of the violet lines in the scattered light, especially the one at 4358 angstroms, have associated with them a fairly distinct "Raman spectrum," consisting of several lines, some of which are of longer, some of shorter, wavelength than the much brighter, unaltered line. The shorter, usually faint, are called "anti-Stokes" lines, because they violate the law of Stokes (see Luminescence).

The Raman spectrum for a given line is characteristic of the scattering substance, and is made up of lines somewhat more diffuse, or "broader," than the corresponding incident line. Water gives broad bands. The phenomenon extends into the x-ray region, and is somewhat analogous to the Compton effect, in which, however, the scattering particles are electrons instead of molecules. The Raman effect is explained as due to the absorption or contribution of energy from or to the quantum of incident radiation by the scattering molecule, the result being a decrease or an increase in the frequency of the quantum, with corresponding change of wavelength. The phenomenon was predicted by Kramer 4 years prior to its discovery. (L.D.W.)

RAMIE. Boehmeria nivea.

RANDOM FLUCTUATION.
A random fluctuation is a variation due to chance or to random sampling. (L.A.A.)

RANDOMIZED BLOCK.
A randomized block is a design in which there are two sources of variation. Usually it is desired to eliminate one of the sources of variation, either that of rows or columns. We consider s blocks consisting of r treatments in each block forming an rxs table of r columns and s rows. The analysis of variance then would be
Mean square among column means

$$\frac{r\Sigma(\overline{X}_c - \overline{X})^2}{c-1} = V_c$$

Mean square among row means

$$\frac{c\Sigma(\overline{X}_r - \overline{X})^2}{r-1} = V_r$$

Mean square due to error

$$\frac{(rc-1)V_T - (r-1)V_r - (c-1)V_c}{(r-1)(c-1)} = V_e$$

Total mean square

$$\frac{\Sigma\Sigma(X_{ij} - \overline{X})^2}{rc-1} = V_T$$

To test for equality of column means we compute Snedecor's F, $F = \dfrac{V_c}{V_e}$ with $n_1 = c - 1$ degrees of freedom and $n_2 = (r-1)(c-1)$ degrees of freedom. To test for equality of row means, $F = \dfrac{V_r}{V_e}$ with $n_1 = r - 1$, $n_2 = (r-1)(c-1)$. Randomized blocks are used in agriculture and in industry. It is assumed that the variates are distributed normally and independently with the same variance σ^2. Here $\overline{X}$ = mean of all rc items, $\overline{X}_c$ = mean of a column, $\overline{X}_r$ = mean of a row and X_{ij} is the item in the ith row and the jth column. (See Latin Square.) (L.A.A.)

RANG.
The term for a subdivision of the igneous rocks according to their classification by Cross, Iddings, Pirrson and Washington. (R.M.F.)

RANGE.
This term is used in two senses. The range is a measure of dispersion and equals the difference between the largest and the smallest variate in a distribution. The range is used in another closely related sense when we say a variable has a range from $x = a$ to $x = b$. (L.A.A.)

RANGE AND ENDURANCE, AIRPLANE.
The range of an airplane is the distance it can fly in calm air before the fuel supply is exhausted. Its endurance is the time it can remain in flight before its fuel is exhausted. This should be distinguished from refueling endurance, where the airplane is refueled from another while in flight and endurance depends on pilot fatigue or mechanical endurance of the airplane.

The range at a specified speed is somewhat simpler to compute than the maximum range and would be merely the $\dfrac{\text{fuel-carrying capacity}}{\text{specific fuel consumption}}$ multiplied by the air speed, if it were not for the fact that the specific fuel consumption may decrease slightly as fuel is consumed during the flight. The maximum range requires adjustment to the most economical (i.e., maximum $\dfrac{\text{airspeed}}{\text{specific fuel consumption}}$) speed during the entire flight. This can be done either for increments of the flight, assuming fuel consumption constant during each increment, and flight at maximum $\frac{\text{Lift}}{\text{Drag}}$, or one of several formulae developed for this purpose may be used.

Diehl's formula for maximum range is

$$R \text{ (miles)} = 625 \frac{\eta_0}{C_0}\left(\frac{L}{D}\right)_{max}\left[1 - \left(\frac{W_1}{W_0}\right)^{0.6}\right].$$

η_0 = Initial propulsive efficiency (approximately propeller efficiency).
C_0 = Initial engine fuel consumption, lb./hp. hr.
W_0 = Initial gross weight, lbs.
W_1 = W_0 − weight of fuel consumed.

The practical maximum range may be more or less than the theoretical, depending on wind conditions. Conservative viewpoint of the pilot will further reduce the practical maximum range as he allows for (1) unexpected headwinds, (2) time required to search for terminus after long navigational flight over unknown territory, (3) unknown but possible adverse weather conditions, (4) need for circling airport terminus while waiting for landing clearance, (5) small inaccuracies in fuel gauges, (6) inability to pilot continuously at maximum L/D attitude, at constant altitude, and in a direct line between termini, and (7) length of time taken to climb to flight altitude.

Maximum endurance could be of use in estimating time an airplane might stay aloft over a fog-bound airport when meteorological data had indicated a definite time for re-establishment of landing visibility. (F.T.M.)

RANGE OF STRESS. Cycle of Stress.

RANGE MARKS. Two prominent objects, either natural or artificial, which are located along a line which has some particular value for navigators, are known as range marks. Range marks are used for so many different purposes in **navigation** that it would be futile to attempt to list them all. However, one example may be of interest. A channel is entered on **track** 200°, followed for 1100 yards, then the channel turns and track must be altered to 296°. The turning point is marked by a black and white striped **buoy**. A lighthouse on shore is so placed that its **bearing** from the buoy is 200°. Accordingly, when a ship has the lighthouse and the buoy in line, the bearing of the buoy from the ship is 200°, and to follow the first leg of the channel the ship simply keeps the lighthouse "ranging" on the buoy. As the ship approaches the buoy the pilot watches the shore and when a red and white striped target ranges on a white church spire, he alters heading and holds the target on the spire to follow the 296° leg of the channel. Currents may force the pilot to head quite differently from the directions of the channel as given on his chart, but the range marks give a **line of position** and, so long as the ship is on the proper line, the pilot knows he is proceeding in safe water. (See **Radio Range.**) (W.K.G.)

RANGE, MAXIMUM. Range and Endurance; Airplane.

RANK CORRELATION. Consider N pairs of ranked variates (x_1, y_1), $(x_2, y_2) \cdots (x_N, y_N)$ $1 \leq x_{iN} \leq N$, $1 \leq y_{iN} \leq N$, the rank coefficient of correlation ρ^1 is defined

$$\rho^1 = 1 - \frac{6\sum_{i=1}^{N}(x_i - y_i)^2}{N(N^2 - 1)}.$$

The range of variation for ρ^1 is $-1 \leqq \rho^1 \leqq 1$. Tables have been constructed to test the hypothesis that the population value ρ^1 is zero without any assumption as to the distribution of the **variates** (x_i, y_i). It has been shown by H. Hotelling and M. Pabst that as $N \to \infty$, the distribution of ρ^1 approaches normality with **mean** zero and **variance** $\frac{1}{N-1}$. If $N > 10$, the normal approximation seems quite adequate, using $\sigma_{\rho'}{}^2 = \frac{1}{N-1}$. These results have brought ρ' into favor again particularly when $N < 100$ since in such cases the computation of ρ' is simpler than the usual **coefficient of correlation**. (L.A.A.)

RANKINE CYCLE. Rankine's modification of the Carnot cycle is the basis of the modern steam plant cycle even though the Rankine cycle itself has been modified and changed with the passing of time. The elements of the Rankine vapor cycle are shown in Fig. 1(a). It consists essentially of a **boiler** which receives

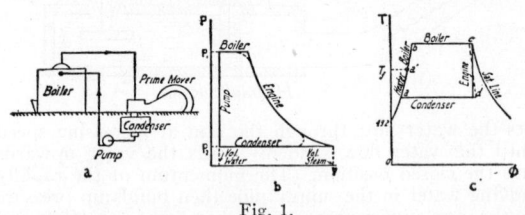

Fig. 1.

feed water from a **pump,** a prime mover to expand the steam **adiabatically,** a **condenser** to receive the exhaust steam from the **engine** and reduce it to water, and a pump to overcome the pressure difference between boiler and condenser. Figs. 1(b) and 1(c) show this cycle on the pressure-volume and temperature-entropy planes. In Fig. 2 is shown a common form of application of the Rankine vapor cycle. This plant is char-

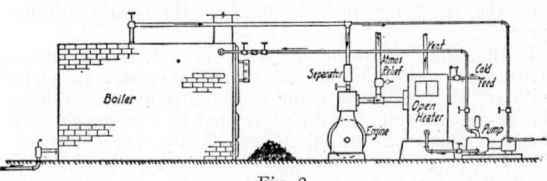

Fig. 2.

acterized by atmospheric exhaust, use of a portion of the exhaust for feedwater heating in an open heater, and a reciprocating steam-driven boiler feed pump, also exhausting to atmosphere. The efficiency of such a plant is necessarily very poor, yet for small amounts of power it represents a type of plant that has the advantage of a minimum investment and can be operated by the nontechnical man in a semi-successful manner.

The Rankine vapor cycle efficiency is (see Fig. 1(c)).

$$E_{rv} = \frac{H_c - H_d}{H'_c - h_{a'}} \quad \text{or} \quad \frac{2545}{S(H'_c - h_{a'})}.$$

where H_c = heat per pound of steam entering prime mover.
H'_c = heat per pound of steam leaving boiler.
H_d = heat per pound of steam leaving prime mover.
$h_{a'}$ = heat per pound of boiler feed water.
S = the steam rate in pounds per horsepower hour

This equation is exact for complete expansion cycles such as may be had by the use of the steam turbine, but is a slightly optimistic approximation for steam engine type of expansion since release pressure is somewhat higher than exhaust. (See **Steam Engine.**) (F.T.M.)

RAOULT'S LAW. The **vapor pressure** of a substance in **solution** is proportional to its **mole fraction.** (See **Solutions and Solubility, Distillation.**) (R.K.S.)

RAPE. Brassica.

RAPESEED OIL. Esters.

RAPFEN. Pisces, Teleostei. A fish (Pisces) of northern and eastern Europe, related to the carps. It lives in quiet waters in lakes and slow streams. (A.W.L.)

RARE EARTH METALS. Cerium; Yttrium.

RASCHIG RING. Gas Absorption.

RASH. A skin eruption of temporary nature which may be caused by sensitivity to **drugs**, to **protein** substances (**allergy**), to chemicals, or may be part of a disease picture. The so-called "exanthems"—**measles**, **german measles, chicken pox, smallpox**, and **scarlet fever**—have distinct and characteristic rashes which are important in differential diagnosis. (R.S.M.)

RASP. File.

RASPBERRY. Rose Family.

RASPBERRY FRUIT-WORM. Insecta, Coleoptera. The **larva** of a small **beetle**, *Byturus unicolor*, which lives on the inside of the fruit on raspberries. It eats the receptacle but is often found in the fruit itself. Since the adults eat the foliage and buds the pest can be held in check by spraying with **lead** arsenate. (A.W.L.)

RASPBERRY-CANE BORER. Insecta, Coleoptera. The **larva** of a beetle which bores in the canes of raspberry plants, ultimately killing them. The adults girdle the tender growth and cause it to wilt. If wilted canes are removed to a few inches below the girdling the development of the larvae is prevented. (A.W.L.)

RASSE. Mammalia, Carnivora. A **civet** of the Oriental region. (A.W.L.)

RAT. Mammalia, Rodentia. In the strict sense an animal of the genus *Rattus,* this term is much more widely applied to many small gnawing animals. The true rats are dull-colored animals with long scaly tails, short legs, small ears, and a pointed muzzle. The common brown rat, *R. norvegicus,* is a typical example. This species was introduced from Europe in Colonial times and has since become a troublesome and sometimes dangerous pest over the entire country. It is much more aggressive than the related black rat, *R. rattus,* and has almost crowded the latter out. The roof rat, *R. alexandrinus,* is a third species found in the southern United States. Still other species of the genus are found in other continents.

Some of the closely related genera include the bandicoot rats, bush rats, bamboo rats, kangaroo rats and cane rats. The name American pouched rats is sometimes applied to a group containing the pocket **gopher.**

Less closely related members of the order are the **Philippine rats, fish-eating rats, wood rats, African crested rat,** and **muskrat.** (A.W.L.)

RATCHET AND PAWL. The ratchet and pawl together comprise an intermittent type of mechanism in which motion derived from a reciprocation or oscillation is converted into intermittent circular motion, constant in direction. This mechanism finds considerable use in lifting **jacks, windlasses,** and similar mechanisms. The ratchet wheel has a circumference covered with ratchet teeth. As shown in the accompanying figure, these are designed to receive a forceful pressure in the counterclockwise direction only. When a ratchet must be reversible, the shape of the teeth necessarily must be modified from that shown in the diagram. The pawl on the oscillating arm slips freely over the teeth in the clockwise direction, but engages and rotates the ratchet when

it moves in the anti-clockwise direction. If the ratchet wheel is under a torque it will rotate backwards when

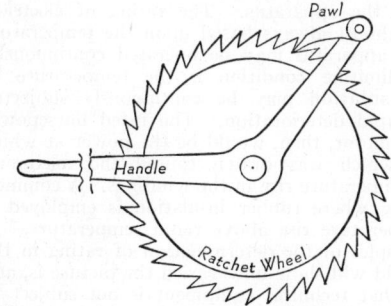

Ratchet and pawl.

the pressure of the driving pawl is released unless a holding pawl is provided. The pawls are held against the ratchet wheel either by dead weight or by spring pressure. (F.T.M.)

RATE OF CHANGE OF A FUNCTION. Derivative of a Function of one Variable.

RATE OF CLIMB, AIRPLANE. When an airplane is in horizontal unaccelerated flight at less than maximum level speed for its altitude, the engine is necessarily throttled somewhat below rated power. If desired, some or all of this unused power could be applied to the airplane without accelerating it if the airplane were directed into a climb. Rate of climb is defined as the vertical component of the velocity of an airplane in a steady unaccelerated climb. If full available engine power is used the rate of climb will become maximum intermediate between minimum and maximum climbing air speeds. If the power required to overcome drag in level flight at various speeds and the thrust power available at full throttle are plotted against air speed, the power intercept between the two represents the power available for producing the rate of climb. The maximum rate of climb may thus be discovered since it equals

$$\frac{550 \ (\text{Maximum excess horsepower available})}{\text{Airplane weight, lbs.}} \ \text{ft./sec.}$$

A high rate of climb is sought in the design of interceptor and fighter military aircraft and in special commercial photographic mapping airplanes. In the military field superiority in rate of climb gives important tactical advantages; in the photographic field it allows the mapping party to reach their working altitude quickly and make maximum use of the occasional superior atmospheric conditions which are necessary for some of this work.

A tangent drawn from the origin to a rate of climb curve (rate of climb vs. air speed) shows the air speed for maximum angle of the flight path. This is the optimum speed for climbing flight to clear fixed obstacles. (F.T.M.)

RATE OF FALL OF RAIN DROPS. Rain drops are commonly accepted as those whose diameter is greater than 0.5 mm. The rate of fall of these drops ranges from about 3 meters per sec. for the smallest to about 8 meters per sec. for the largest. (P.E.K.)

RATEL. Mammalia, Carnivora. A short-legged animal, *Mellivora ratel,* with a broad depressed body and head, resembling the badgers in form. The ratels live in India and Africa. They are nocturnal burrowing animals which eat other animals and insects. From their love of honey they have received the name honey badger. (A.W.L.)

RATING. Mechanical and electrical apparatus is rated on different bases, the particular method being associated with the conditions of usage, or with the historical development of the apparatus. The rating of electrical apparatus is almost always based upon the temperature at which the apparatus may be operated continuously. Usually the limiting condition is the temperature to which the **insulation** may be continuously subjected without its rapid deterioration. The rated horsepower of an electric motor, then, would be the power at which the current which was drawn caused the maximum permissible temperature rise in the windings. A common basis of rating where rubber insulation is employed is a 40° C. temperature rise above room temperature.

A few examples of the determination of rating in the mechanical field will show how varied the picture is, and demonstrate that technical equipment is not subject to uniform systems of rating. Steam **boilers** are rated by a unit based on their heating surface, but progress in the art of generating steam in boilers, and in the field of combustion engineering, has so increased the steam production possible from a given area of heating surface that the actual performance of steam boilers can nowadays be related to their ratings only by the use of a term "per cent rating." A boiler operating at 300 per cent rating would not necessarily be considered overloaded.

The power rating of a **steam engine** depends upon the extent to which steam is used expansively in the engine. The less the expansion, the greater the power rating of a given cylinder volume, and the less economical the engine is of steam. Common practice rates the power as that derived when the **ratio of expansion is 4,** but an engine may, by advancing the cut-off, derive more than this rating, at the expense of steam consumption. A steam turbine has its nozzles, blades, and passages designed for a certain full load at which steam consumption will be minimum. Loads above this can be carried by admitting some high-pressure steam into low-pressure sections of the casing by overload valves. Naturally the **generator** to which the **turbine** is connected is provided with sufficient capacity to absorb this power. Common practice is to rate a turbine at the maximum capacity that can be produced without using overload values, then provide for carrying about 20% overload by secondary or tertiary overload valves. The generator will be rated to absorb the turbine rated power at 80% power factor. To load the turbine over its normal rating the power factor of the generator must be raised. (F.T.M.)

RATIO. The ratio of two numbers is their indicated quotient, frequently expressed as a **fraction.** The ratio of a to b is indicated by the notation $a : b$, or frequently a/b or $a \div b$. (L.L.S.)

RATIO OF EXPANSION. This ratio of certain volumes in the **piston** and **cylinder** engine is of considerable importance, since it is indicative of power output, and other important characteristics of engines. This is also the same as "compression ratio," a term frequently used in describing the modern **Otto cycle engine.** However, ratio of expansion will first be defined for the **steam engine.** In a steam engine, when the piston is at the inner end of its stroke and ready to do work on an outward stroke, there is a small clearance volume created by the necessity for avoiding mechanical contact between piston and cylinder head. As the piston moves under the influence of inflowing steam, the pressure remains nearly constant until the steam is cut off by the valve action. Beyond this point the steam expands with decreasing pressure. The ratio of expansion is the number of times steam is expanded in volume after the point of cut-off. It is equal to the ratio,

$$\frac{\text{volume of the cylinder at the conclusion of a stroke}}{\text{the volume at point of cut-off}}.$$

The greater the ratio of expansion, the smaller the amount of power developed by a cylinder of given size, but the power produced is obtained more efficiently than with small ratios. A common ratio of expansion for steam engines is four.

The cycle of the gasoline engine includes a compression of the charge from maximum cylinder volume into the clearance volume, with a corresponding rise of pressure. After an explosion, the charge re-expands to the maximum volume, so that whether we consider the ratio taken along the compression or expansion portion of the cycle, numerically it will be the same. In the Otto cycle, the ratio of expansion is

$$\frac{\text{The piston displacement} + \text{the clearance volume}}{\text{the clearance volume}}.$$

Since the efficiency of the Otto engine is dependent to a great extent on the ratio of expansion, it is seen that the clearance volume is an important dimension in these engines. Modern engines have "compression ratios" of six or slightly over, as compared to ratios of four obtained in earlier engines. The use of high compression ratios has incurred the problems and results discussed under the head of **detonation.** (F.T.M.)

RATIONAL FUNCTION. A rational function is a **function** which involves the **variable** in only the rational operations of addition, subtraction, multiplication, division and raising to powers with constant integral exponents.

Rational functions are divided into two sub-classes: rational integral functions or **polynomial functions,** and rational **fractional functions.** (L.L.S.)

RATIONAL INTEGRAL EQUATIONS. Polynomial Equations.

RATIONAL INTEGRAL FUNCTION. Polynomial Function.

RATIONAL NUMBERS. Number.

RATIONAL ROOTS OF POLYNOMIAL EQUATIONS. Polynomial Equations.

RATTANS. Species of *Calamus.* Palmaceae. The various **palms** known as rattans are all climbing plants of the Old World tropics. The genus *Calamus,* with its 280 species, contains the greater number of these palms, especially those of commercial importance. As a group the rattans are found in low altitudes, growing in the forests. Many are leaf-climbers, in which the main axis or outer ends of the **pinnae** of the pinnately compound leaves bear stout downward projecting spines. These spines prevent the plants from sliding from any support on which they may rest. The stems are slender (1/8–2½″ in diameter) and very long. Many are from 250–600′ in length, and of uniform diameter throughout this length. The outer portion of the stem is very hard and tough; the inner, softer and rather porous.

In gathering, these stems are cut off close to the base and pulled from their supports. The leaves are then removed and the stem cut into sections 10–20′ long. These may be bent double and thus bundled for shipment. Sometimes they are split before shipping. The outer layer, removed in long narrow strips, is used in weaving mats, making baskets, and for the cane-seats of chairs. The central portion remaining after the cane is stripped off is used in making reed furniture. Coarse brushes are also made from this central portion. Many other minor uses are made of these plants in the eastern countries and islands where they are native. (R.M.W.)

RATTLESNAKE. Reptilia, Sauria. A poisonous **snake** with a series of loosely attached horny rings at the tip of the tail which constitute the rattle. When aroused the snake vibrates its tail, producing a sound between a rattle and a buzz. These snakes have a thick body, tapering at the neck toward the broad triangular head. They are **pit vipers.**

Rattlesnakes occur only in the New World, chiefly in North America. They are most abundant in the warmer parts of the United States. The little massasauga, *Sistrurus catenatus,* which reaches a length of only thirty inches, is the most widely distributed species, ranging from New York to South Dakota and southward into Mexico. It still persists in wild spots, especially swamps, even in heavily settled areas. The ground rattler, *S. miliarius,* of the southeastern coastal area and southern Mississippi Valley is still smaller. These small species are less dangerous than the large rattlesnakes, but even they are not to be disregarded. The common rattlesnake, *Crotalus horridus,* is distributed from Maine to Florida, east of the plains. Another eastern species, the diamond-back rattlesnake, *C. adamanteus,* is found on wet ground from the Carolinas to Louisiana. It is the largest species, with a length of 6', and is correspondingly dangerous. In the western United States several species are found in the plains area and the deserts. Of these the sidewinder, *C. cerastes,* and prairie rattler, *C. confluentus,* are the best known. The latter is found from Canada to Texas over the entire western half of the United States.

These snakes have been ruthlessly destroyed and are now rare except in wild country. Their virulent poison has been the chief cause of economic importance, but the skins and rattles have been attractive as curios in some parts of the country, and in Florida one enterprising snake fancier makes canes of the spinal column and cans the flesh for the market. It is more than probable that filets of rattlesnake will also remain in the class of curios. (A.W.L.)

RAVEN. Aves, Passeriformes. Large **black birds** with more or less iridescent luster. They are closely related to the common crow. The common raven, *Corvus corax,* is found in Europe, Asia, and North America. One variety, the American raven, extends from Canada to Guatemala and from the Rockies to the Pacific. Another, the northern raven, is found from Alaska to Greenland, southward into the northern tier of states, and in the mountains to Carolina. Two species of white-necked ravens are known, one in the southwestern deserts of the United States and the other in Africa. All of these birds eat carrion, eggs, insects, small animals, and to a limited extent vegetable matter. (A.W.L.)

RAY. Pisces, Plagiostomi. Fishes (Pisces) related to the sharks. The body is flattened and the pectoral fins are highly developed to form broad lateral lobes. The tail is slender. Rays are typically marine, but some species ascend rivers and some fresh-water species occur in tropical America. Among the different forms the electric rays (*Torpedo*) and eagle rays (*Myliabatis*) or devil-fishes are noteworthy. The former have electric organs formed of modified muscle tissue and are said to kill their prey by electrocution. The latter attain great size, up to a weight of 800 lbs., and are said to be dangerous to divers. Some of the rays are also known as skates. (A.W.L.)

RAYNAUD'S DISEASE. A vascular disorder characterized by paroxysmal **cyanosis** of the fingers in response to cold or to emotional stimuli. Toes, ears, chin, may also be affected. The disease appears most commonly in women in the third decade of life, and familial cases have been described. The cyanosis is due to interference with the blood supply through constriction of the vessels—vasospasm, resulting from abnormalities of the nervous control of the vessel walls. The cause of these abnormalities is not known. Attacks begin gradually, with mild symptoms, but later there is severe pain, numbness, and discoloration of the affected parts. The attack may last for minutes or hours. In the advanced disease, the fingers may become dark blue or black and gangrenous. Sometimes, a mild chronic form exists, never progressing to **gangrene.**

Treatment is directed toward avoiding the precipitating factors—cold, emotional upsets, etc. Nicotine has a deleterious effect on the vessels, and smoking should be eliminated. Vasodilation may be accomplished with drugs, or surgically by interrupting the sympathetic nerves (see **Autonomic Nervous System**) which carry impulses resulting in vascular constriction. (D.M.H.)

RAYON. Carbohydrates; Cotton.

RAZORBACK. 1. Pisces, Teleostei. One of the less common species, *Xyrauchen texanus,* of buffalo fishes of the Mississippi basin. 2. Mammalia, Artiodactyla. Feral hogs of the southern states. (A.W.L.)

RAZORBILL. Aves, Charadriiformes. A moderately large ocean bird (**Aves**), *Alca torda,* found on both sides of the northern Atlantic. It has a large compressed beak with deep furrows. The species is related to the extinct great auk and to the murres, puffins, and guillemots, and is also known as the razor-billed auk. A number of other vernacular names are indiscriminately applied to various members of this group. (A.W.L.)

REACTANCE. Alternating Currents; Impedance.

REACTANCE TUBE. This is a vacuum tube operated in such a way that it presents the characteristics of a reactance to the rest of the circuit. As reactance takes

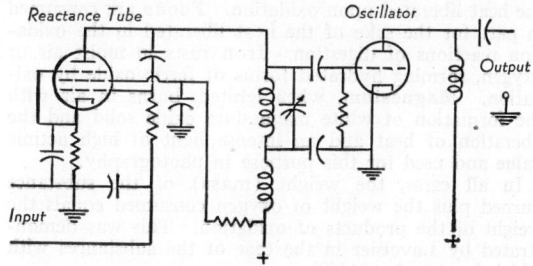

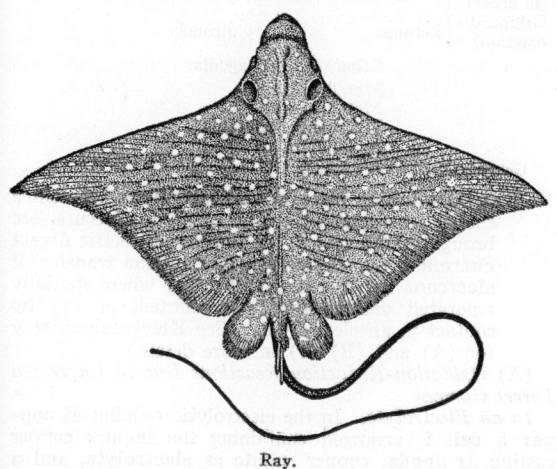

Ray.

a current which is essentially 90° out of phase with the voltage it is necessary that the tube do the same. Reference to the circuit will indicate one method of connecting it to accomplish this. The grid is excited by a voltage obtained from the plate voltage by the resistance-condenser circuit (the condenser connected to the cathode is merely a d-c blocking condenser and is a value which produces no other appreciable effect). The resistance-condenser circuit applies a voltage to the grid which is nearly 90° out of phase with the plate voltage. Since, in a pentode, the plate current is in phase with the grid voltage, this means that the plate current (a-c component) is 90° out of phase with the plate voltage, thus the plate circuit has the desired reactance characteristics. The advantage of the tube over a conventional reactance is that the magnitude of its reactance effect may be easily varied by adjusting the d-c bias applied to the grid. The circuit shows the tube connected as part of the tuning capacity of an **oscillator**. By varying the grid bias (hence the gain) of the tube the tuning of the oscillator is varied. This is used in many frequency-**modulation** transmitters to give the frequency modulation. In this application a suitable audio signal is fed to the grid of the reactance tube so its reactance effect will vary with the audio and hence frequency modulate the oscillator. It is also used in frequency-stabilizing circuits where the deviation from the desired frequency is made to vary the grid bias of the reactance tube and hence correct the oscillator frequency. (L.R.Q.)

REACTION. Apart from the chemical usage of this term (see various **Reactions;** and **Chemistry**), its technical meaning is associated with mechanical action. The reaction force is that which is produced as the result of an **acceleration** of some **mass**, be it a solid or fluid. The recoil of a gun is a reaction force set up by the acceleration of the projectile. In a certain class of **turbine**—both hydraulic and steam—the machines obtain the principal part of their actuating forces by the reaction created when, in the one case, water, and in the other, steam, is accelerated under conditions controlled by the mechanism of the turbine. (See **Jet Propulsion**.)

In structural engineering the term reaction refers to the forces exerted by the supports of a **structure** to counteract the effects of applied loads. (F.T.M.)

REACTIONS INVOLVING OXIDATION-REDUCTION. Originally oxidation referred to the reaction of a substance with **oxygen** gas. The elucidation by Lavoisier of the burning of substances with oxygen is considered by some to mark the beginning of chemistry as a science. When **carbon** burns in excess of oxygen or air, **carbon dioxide** is formed with the accompanying liberation of a definite amount of heat. When **hydrogen** and oxygen of air are burned or subjected to an electric spark, water is formed with the accompanying liberation of a definite amount of heat. When substances containing carbon and hydrogen are similarly burned, carbon dioxide and water are formed with accompanying liberation of a characteristic amount of heat in each case. **Fuels** are burned for the sake of the heat liberated upon oxidation. **Foods** are consumed in part for the sake of the heat liberated in the oxidation reactions of digestion. **Iron** rusts in moist air or oxygen, forming hydrated forms of ferric oxide by oxidation. **Magnesium,** when ignited, burns in air with the formation of white magnesium oxide solid and the liberation of heat and an intense light of high actinic value and used for this purpose in photography.

In all cases, the weight (**mass**) of the substance burned plus the weight of oxygen consumed equals the weight of the products of oxidation. This was demonstrated by Lavoisier in the case of the substances with which he experimented.

Oxidation is a reciprocal process. Oxygen is the oxidizing agent; the substance reacting with oxygen is called the reducing agent. Besides the kinds of simple reactions given in the examples above, there are many others more complex. But the reciprocal principle of oxidation-reduction applies in all cases. Oxidation-reduction, as now understood, includes:

I. Reactions involving oxygen, **chlorine,** and substances supplying such elements on the one hand, with carbon, hydrogen, metals, and other similar substances on the other hand.

II. Reactions taking place in an **electrolyte** wherein a change of electronic charge of an **ion** occurs. Compare **Reactions Involving Recombination of Ions.**

I. Reactions involving oxygen, etc., with carbon, etc., are illustrated as follows:

Reducing Agent	Oxidizing Agent	Resultant	
carbon hydrogen sulfur phosphorus	oxygen, excess	carbon dioxide water sulfur dioxide phosphorus pentoxide	
hydrogen, excess	copper oxides lead oxides tin oxides iron oxides (but not magnesium oxide, aluminum oxide, zinc oxide)	copper lead tin iron	water
magnesium aluminum zinc iron tin lead copper (but not mercury, silver, gold, platinum)	oxygen-furnishing substance	magnesium oxide aluminum oxide zinc oxide ferric oxide or ferroferric oxide stannic oxide lead monoxide cupric oxide	
arsenic antimony aluminum iron copper	chlorine	arsenic trichloride antimony trichloride aluminum chloride ferric chloride cupric chloride or cuprous chloride	
carbon carbon monoxide	copper oxides lead oxides tin oxides iron oxides	copper lead tin iron	carbon dioxide or carbon monoxide or both
aluminum *	iron oxides silicon oxide boron oxide manganese oxides chromic oxide	iron silicon boron manganese chromium	aluminum oxide, glass
silicon	iron oxides	iron	silicon oxide
magnesium zinc iron lead	sulfur	magnesium sulfide zinc sulfide ferrous sulfide lead sulfide	
magnesium (in ether; Grignard's reaction)	aldehydes	secondary alcohols	
	ketones	tertiary alcohols	

Other oxidizing agents:
Metallic nitrates
Metallic chlorates
Metallic perchlorates
Metallic bromates

* Goldschmidt thermit reaction.

II. Reactions taking place in an **electrolyte,** wherein a change of electronic charge of an **ion** occurs, are brought about (A) by an impressed electric **direct current** *in an electrolyte,* (B) by the transfer of **electrons** between ions, either (1) where spatially separated *and* electrically connected, or (2) by contact in an electrolyte. (See **Electrochemistry** for (A) and (B) (2) in more detail.)

(A) *Oxidation-Reduction Reactions Due to Impressed Direct Current*

In an Electrolyte. In the electrolytic refining of **copper** a **cell** is arranged containing the impure copper casting as **anode,** copper sulfate as **electrolyte,** and a thin pure copper sheet as **cathode**. Upon the impression of d.c. of low **electromotive force** in the proper direction, copper of the anode passes into the electrolyte as cupric ion (Cu^{2+}) and pure copper is deposited from the electrolyte onto the cathode. The reaction may be represented thus:

Copper (Cu^0) of anode → Cupric ions (Cu^{2+})

           of electrolyte → copper (Cu^0) of cathode

In a similar manner, the electrolysis of fused **sodium** chloride yields sodium metal at the cathode and **chlorine** gas at the anode (Downs process for sodium, 1920). The reaction may be represented thus:

Sodium cation (Na^+) of electrolyte →

           Sodium metal (Na^0) at cathode

Chlorine ($Cl_2{}^0$) at anode

       ← Chlorine anion ($2Cl^-$) of electrolyte

In all such reactions, there is a *change of charge,* that is, oxidation or reduction, of ions *at* cathode *and* anode. The electrolysis of sodium sulfate solution will clarify this, thus:

Hydrogen ions ($2H^+$) (turning litmus red) plus oxygen gas ($O_2{}^0$) 1 volume, at anode	Sodium cations ($2Na^+$) of electrolyte → ← Sulfate anions ($SO_4{}^{--}$) of electrolyte	Hydroxyl ions ($2OH^-$) (turning litmus blue) plus hydrogen gas ($H_2{}^0$) 2 volumes at cathode

Sodium ions and sulfate ions when discharged react with water at the cathode and anode, respectively, yielding the above products. The total amount of sodium and sulfate ions in the electrolyte remains the same after as before passing the current, the *net* effect (upon stirring the electrolyte) being the decomposition of water to form 1 volume of oxygen at the anode and 2 volumes of hydrogen at the cathode.

Sodium chloride solution electrolyzed with accompanying stirring of the electrolyte yields sodium **hypochlorite** (NaOCl).

Aluminum metal is produced by electrolysis of aluminum oxide dissolved in fused **cryolite** (Na_3AlF_6), and **magnesium** metal by electrolysis of fused magnesium chloride anhydrous ($MgCl_2$).

The charging of a lead storage cell (**Accumulator**) involves reactions in this category, thus:

Lead sulfate ($Pb^{2+}SO_4{}^{--}$) paste of the anode, in contact with the electrolyte, is changed into lead dioxide ($Pb^{4+}O_2{}^0$), brown solid *in situ*. Sulfate ion ($SO_4{}^{--}$) passes into electrolyte	Sulfuric acid-water electrolyte (concentration of sulfuric acid ← increases with the amount of charging →	Lead sulfate ($Pb^{2+}SO_4{}^{--}$) paste of the cathode, in contact with the electrolyte is changed into lead metal (Pb^0) *in situ*. Sulfate ion ($SO_4{}^{--}$) passes into electrolyte

(B) 1. *Oxidation Reduction Reactions by the Transfer of Electrons Between Ions,* where the latter are spatially separated *and* electrically connected. In the Daniell **cell**, one of the oldest known wet batteries, the chemical setup is:

Zinc metal (Zn^0) in zinc sulfate ($Zn^{2+}SO_4{}^{--}$) electrolyte.

Copper sulfate ($Cu^{2+}SO_4{}^{--}$) electrolyte containing copper metal (Cu^0).

The two electrolytes in contact *in the cell*.

The two metals electrically connected *outside the cell* by a metallic conductor.

Results: 1. An electromotive force (1.1 volts) is generated.

       2. Simultaneously, chemical reaction occurs:

Zinc metal is dissolved into the electrolyte as zinc ions at the electrode	Copper ions are precipitated from the electrolyte as copper metal at the electrode

$$Zn^0 \rightarrow Zn^{2+} \quad\xrightarrow{\hspace{2cm}}\quad Cu^{2+} \rightarrow Cu^0$$
Direction of positive
current in the cell

As thus arranged, (1) the chemical reaction generates a definite **electromotive force**, namely, 1.1 volts. The direction of the positive current is from zinc to copper *in* the cell, through the electrolyte, and from copper to zinc *outside* the cell through the metallic conductor. (2) The *amount* of electric current generated is proportional to the *amount* of chemical reaction taking place, namely, 96,500 coulombs per equivalent weight (see **Chemical Composition**) of change. Of zinc, $\frac{65.38}{2}$ = 32.69 grams changed from metallic to ionic; and of copper, $\frac{63.57}{2}$ = 31.79 grams changed from ionic to metallic in generating 96,500 coulombs of electricity.

The discharging of a lead storage cell involves reactions in this category, thus:

Lead dioxide (PbO_2) electrode in **sulfuric acid** electrolyte

Sulfuric acid electrolyte containing lead metal (Pb^0) electrode

The two electrodes electrically connected *outside* the cell by a metallic conductor

Results: 1. An electromotive force (about 2.2 volts) is generated.

       2. Simultaneously, chemical reaction occurs:

Lead dioxide transformed into lead sulfate ($Pb^{2+}SO_4{}^{--}$) at the electrode	Lead sulfate ($Pb^{2+}SO_4{}^{--}$) formed from lead at the electrode

$$Pb^{4+} \leftarrow Pb^{2+} \quad\xleftarrow{\hspace{2cm}}\quad Pb^{2+} \rightarrow Pb^0$$
Direction of positive
current in the cell

As thus arranged, (1) the chemical reaction generates a definite electromotive force, namely, about 2.2 volts when fully charged to about 1.9 volts when almost discharged, the concentration of sulfuric acid *decreases* with the amount of discharging. The direction of the positive current is from lead to lead dioxide *in* the cell through the electrolyte, and from lead dioxide to lead *outside* the cell through the metallic conductor. (2) The amount of electric current generated is proportional to the *amount* of chemical reaction taking place, namely, 96,500 coulombs per equivalent weight (see **Chemical Composition**) of change. Of lead dioxide, $\frac{239}{2}$ = 120 grams changed to lead sulfate, $\frac{303}{2}$ = 152 grams; of lead, $\frac{207}{2}$ = 104 grams changed to lead sulfate, $\frac{303}{2}$ = 152 grams; and of sulfuric acid $\frac{98}{2} \times 2$ = 98 grams removed from electrolyte as lead sulfate in generating 96,500 coulombs of electricity.

SINGLE POTENTIAL DIFFERENCES, 25° C., OF CERTAIN ELEMENTS IN SOLUTIONS OF THEIR IONS WITH REFERENCE TO HYDROGEN GAS AT 1 ATMOSPHERE—PLATINUM—1.00 NORMAL HYDROGEN ION EQUALS 0.00 VOLT

ELEMENT	ION	POTENTIAL DIFFERENCE (In volts, for 1.00 N solution)
Mg	Mg^{++}	+2.40
Al	Al^{+++}	+1.7
Be	Be^{++}	+1.69
Mn	Mn^{++}	+1.1
H_2(Pt)	OH^-	+0.83
Zn	Zn^{++}	+0.76
Cr	Cr^{++}	+0.6
S	S^{--}	+0.51
Fe	Fe^{++}	+0.44
Cd	Cd^{++}	+0.40
Co	Co^{++}	+0.29
Ni	Ni^{++}	+0.22

VARIOUS ELECTRONIC STATES OF OXIDATION-REDUCTION OF SOME OF THE ELEMENTS

Element	Periodic Group	VALENCE −4	−3	−2	−1	0	+1	+2	+3	+4	+5	+6	+7
Chlorine	7				HCl	Cl_2	NaOCl		$KClO_2$		$HClO_3$		$KClO_4$
Bromine	7				HBr	Br_2	NaOBr				$KBrO_3$		
Iodine	7				HI	I_2	NaOI				KIO_3		KIO_4
Fluorine	7				H_2F_2	F_2							
Sulfur	6			H_2S		S				H_2SO_3		H_2SO_4	
Selenium	6			H_2Se		Se				SeO_2		H_2SeO_4	
Tellurium	6			H_2Te		Te		$TeCl_2$		TeO_2		H_2TeO_4	
Nitrogen	5		NH_3			N_2		NO			HNO_3		
Phosphorus	5		PH_3			P	H_3PO_2		H_3PO_3		H_3PO_4		
Arsenic	5		AsH_3			As			$NaAsO_2$		Na_3AsO_4		
Antimony	5		SbH_3			Sb			$SbCl_3$		Sb_2O_5		
Bismuth	5					Bi			$BiCl_3$		$HBiO_3$		
Carbon	4	CH_4		HCH_2OH		C HCHO		CO HCOOH		CO_2 HOCOOH			
Silicon	4					Si				SiO_2 $SiCl_4$			
Tin	4					Sn		SnO $SnCl_2$		SnO_2 $SnCl_4$			
Lead	4					Pb	Pb_2O	PbO $PbCl_2$		PbO_2			
Boron	3					B			H_3BO_3				
Aluminum	3					Al			$AlCl_3$				
Zinc	2B					Zn		$ZnSO_4$					
Cadmium	2B					Cd		$CdSO_4$					
Mercury	2B					Hg	HgCl	$HgCl_2$					
Copper	1B					Cu	CuCl	$CuCl_2$					
Silver	1B					Ag	$AgNO_3$	AgO					
Gold	1B					Au	AuCl		$AuCl_3$				
Nickel	8					Ni		NiO		NiO_2H_2O			
Cobalt	8					Co		CoO	$Co(OH)_3$	CoO_2			
Iron	8					Fe		$FeCl_2$	$FeCl_3$			K_2FeO_4	
Manganese	7B					Mn		$MnCl_2$	$Mn(OH)_3$	MnO_2		K_2MnO_4	$KMnO_4$
Chromium	6B					Cr		$CrCl_2$	$CrCl_3$			K_2CrO_4	
Molybdenum	6B					Mo		$MoCl_2$	$MoCl_3$	MoO_2 $MoCl_4$	$MoCl_5$	MoO_3	
Tungsten	6B					W		WCl_2		WO_2 WCl_4	WCl_5	WO_3 WCl_6	
Uranium	6B					U			UCl_3	UO_2 UCl_4	UCl_5	UO_3	
Vanadium	5B					V		VCl_2	VCl_3	VCl_4	$VOCl_3$		
Titanium	4B					Ti			$TiCl_3$	TiO_2 $TiCl_4$			
Zirconium	4B					Zr				$Zr(SO_4)_2$			
Thorium	4B					Th				$Th(SO_4)_2$			
Cerium	3B					Ce			$Ce_2(SO_4)_3$	$Ce(SO_4)_2$			
Other elements of	3					M			MCl_3				
Elements of	2					M		MCl_2					
Elements of	1					M	MCl						
Elements of	0					E							

ELEMENT	ION	POTENTIAL DIFFERENCE (In volts, for 1.00 N solution)
Sn	Sn^{++}	+0.13
Pb	Pb^{++}	+0.12
H$_2$(Pt)	**H$^+$**	0.00
Bi	Bi^{+++}	−0.2
Cu	Cu^{++}	−0.34
Cu	Cu$^+$	−0.51
Hg	Hg$^+$	−0.80
Ag	Ag$^+$	−0.80
Hg	Hg^{++}	−0.86
O$_2$(Pt)	OH$^-$	−0.40
I$_2$	I$^-$	−0.54
Br$_2$	Br$^-$	−1.07
O$_2$(Pt)	H$^+$	−1.23
Cl$_2$	Cl$^-$	−1.36

Reading down: Each **element** *above loses* electrons more readily than any element below. Accordingly, each element above has a higher *reducing* power than any element below.

Reading up: Each element *below gains* electrons more readily than any element above. Accordingly, each element below has a higher oxidizing power than any element above.

Therefore, any *metal above* will displace, from the salt solution of any metal below, the metal of the latter from its own cation.

(B) 2. *Oxidation-Reduction Reactions by the Transfer of Electrons Between Ions,* where the latter are in contact in an **electrolyte.** Instead of arranging the **electrodes** and electrolytes of the Daniell cell described above wherein the electrode compartments are separated in space and electrically connected through the electrolyte *in* the cell and through the metallic **conductor** *outside* the cell, the materials may be placed in direct contact. To make it still simpler, when zinc metal is placed in copper sulfate solution, a reaction occurs— copper metal is displaced from the solution and an equivalent amount of zinc metals goes into solution as zinc ions. This part of the reaction is identical with that in the Daniell cell. But to generate an electromotive force the conditions laid down for the Daniell cell must be followed.

Reactions of this type may, therefore, be brought about either (1) by the cell method, wherein the electromotive force of each half-cell may be measured by balancing against the electromotive force of a standard half-cell, e.g., a hydrogen electrode, or a calomel (mer-

curous chloride) electrode, or (2) by placing the substances in contact, as in a test tube or a beaker. The resulting transformation of materials, that is, the chemical reaction, is the same in each case. The difference in the electromotive force of the component half-cells is a measure of the tendency to react.

ADDITIONAL SINGLE POTENTIAL DIFFERENCES, 25° C. WITH REFERENCE TO HYDROGEN GAS AT 1 ATMOSPHERE—PLATINUM—1.00 NORMAL HYDROGEN ION EQUALS 0.00 VOLT

FROM	TO	NO. ELECTRONS LIBERATED	VOLTS
Fe(OH)$_2$ + OH$^-$	Fe(OH)$_3$	1	0.65
Ti^{++}	Ti^{+++}	1	0.37
Pb^0 + SO$_4^{--}$	PbSO$_4$	2	0.31
P^0 + 4H$_2$O	5H$^+$ + H$_3$PO$_4$	5	0.3
V^{++}	V^{+++}	1	0.2
Cu$_2$O + 2OH$^-$	2CuO + H$_2$O	2	0.15
Sn^{++}	Sn^{++++}	2	−0.13
H$_2$SO$_3$ + H$_2$O	4H$^+$ + SO$_4^{--}$	2	−0.14
Cu$^+$	Cu^{++}	1	−0.17
H$_2$S	2H$^+$ + S^0	2	−0.17
PbO + 2OH$^-$	PbO$_2$ + H$_2$O	2	−0.3
S^0 + 3H$_2$O	4H$^+$ + H$_2$SO$_3$	4	−0.47
Fe(CN)$_6^{----}$	Fe(CN)$_6^{---}$	1	−0.49
H$_3$AsO$_3$ + H$_2$O	2H$^+$ + H$_3$AsO$_4$	2	−0.49
Ni(OH)$_2$ + 2OH$^-$	NiO$_2$·2H$_2$O	2	−0.49
MnO$_4^{--}$	MnO$_4^-$	1	−0.66
H$_2$O$_2$	2H$^+$ + O$_2$	2	−0.68
C$_6$H$_4$(OH)$_2$ (Hydroquinone)	2H$^+$ + C$_6$H$_4$O$_2$ (Quinone)	2	−0.70
MnO$_2$ + 4OH^{--}	MnO$_4^{--}$ + 2H$_2$O	2	−0.71
Fe^{++}	Fe^{+++}	1	−0.74
Cl$^-$ + 2OH$^-$	OCl$^-$ + H$_2$O	2	−0.94
NO + 2H$_2$O	4H$^+$ + NO$_3^-$	3	−0.94
Cr^{+++} + 4H$_2$O	7H$^+$ + HCrO$_4^-$	3	−1.3
Br$^-$ + H$_2$O	H$^+$ + HOBr	2	−1.33
Mn^{++} + 2H$_2$O	4H$^+$ + MnO$_2$	2	−1.33
Pb^{++} + 2H$_2$O	4H$^+$ + PbO$_2$	2	−1.44
Cl$^-$ + 3H$_2$O	6H$^+$ + ClO$_3^-$	6	−1.45
Cl$^-$ + H$_2$O	H$^+$ + HOCl	2	−1.50
Mn^{++}	Mn^{+++}	1	−1.5
Mn^{++} + 4H$_2$O	8H$^+$ + MnO$_4^-$	5	−1.52
MnO$_2$ + 2H$_2$O	4H$^+$ + MnO$_4^-$	3	−1.63
PbSO$_4$ + 2H$_2$O	4H$^+$ + SO$_4^{--}$ + PbO$_2$	2	−1.7
2H$_2$O	2H$^+$ + H$_2$O$_2$	2	−1.78

Latimer and Hildebrand "Ref. Book of Inorg. Chem.," Macmillan.

Reactions not in solutions of electrolytes but oxidation-reduction in character are of wide scope and of fundamental importance in organic chemistry. They comprise such reactions as transformations (horizontally, not vertically) among the following typical groups of compounds:

Methane	Methyl alcohol	Formaldehyde	Formic Acid	Carbonic Acid
H—C with 3 H	H—C—OH with 2 H	H—C=O with 2 H	H—C(=O)—OH	C with =O, —OH, —OH

Carbon Monoxide
C=O

Carbon Dioxide
O=C=O

Hydrocarbons, primary	Alcohols, primary	Aldehydes	Acids, carboxylic
" secondary	" secondary	Ketones	
" tertiary	" tertiary		
" quarternary			

(See **Hydrocarbons; Alcohols; Aldehydes** (ketones included); **Acids, Carboxylic.**)

Benzene	Phenol
C$_6$H$_6$	C$_6$H$_5$·OH
	Hydroquinone
	C$_6$H$_4$(OH)$_2$(1,4)

Benzoquinone
C$_6$H$_4$(O)$_2$(1,4)

(See **Phenols and Quinones.**)

Ethane	Ethyl amine	Acetamide	Urea
$CH_3 \cdot CH_3$	$CH_3CH_2 \cdot NH_2$	$CH_3 \cdot CONH_2$	$C{=}O$ with NH_2, NH_2
		Methyl cyanide	
		$CH_3 \cdot CN$	

(See **Amines and Amides; Amination.**)

Benzene	Phenylamine (aniline)	Phenylhydrazine	Beta-Phenyl-hydroxylamine	Nitrosobenzene	Nitrobenzene
C_6H_6	$C_6H_5 \cdot NH_2$	$C_6H_5 \cdot NHNH_2$	$C_6H_5 \cdot NHOH$	$C_6H_5 \cdot NO$	$C_6H_5 \cdot NO_2$

(See **Hydrazines; Hydroxylamines; Nitro and Nitroso Compounds.**)

Benzene	Phenylamine	Hydrazobenzene	Azobenzene	Azoxybenzene	Nitrobenzene
C_6H_6	$C_6H_5 \cdot NH_2$	$C_6H_5 \cdot NHNH \cdot C_6H_5$	$C_6H_5 \cdot N{:}N \cdot C_6H_5$	$C_6H_5 \cdot N{:}N \cdot C_6H_5$	$C_6H_5 \cdot NO_2$
				$\overset{..}{O}$	

Benzidine
$(4)H_2N \cdot C_6H_4 \cdot C_6H_4 \cdot NH_2(4)$

(See **Azo- and Related Compounds.**)

Benzene	Thiophenol	Benzene sulfinic acid	Benzene sulfonic acid
C_6H_6	C_6H_5SH	$C_6H_5 \cdot SOOH$	$C_6H_5 \cdot SO_2OH$
	Carbon disulfide	Carbonyl sulfide	
	CS_2	COS	
	Diphenyl sulfide	Phenyl sulfinyl benzene	Phenyl sulfonyl benzene
	$(C_6H_5)_2S$	$(C_6H_5)_2SO$	$(C_6H_5)_2SO_2$

(See **Thioalcohols and Related Compounds.**)

Methane	Methyl chloride	Methylene chloride	Chloroform	Carbon tetrachloride
H_4C	$H_3C \cdot Cl$	$H_2C \cdot Cl_2$	$HCCl_3$	CCl_4

(See **Chlorine, Organic Compounds.**)

Methyl phosphine	Methyl phosphinic acid
$CH_3 \cdot PH_2$	$CH_3PO(OH)_2$
Dimethyl phosphine	Dimethyl phosphinic acid
$(CH_3)_2PH$	$(CH_3)_2POOH$
Trimethyl phosphine	Trimethyl phosphine oxide
$(CH_3)_3P$	$(CH_3)_3PO$
Tetramethyl phosphonium hydroxide	
$(CH_3)_4POH$	

(See **Phosphines and Related Compounds.**)

CLASSIFICATION OF CERTAIN REACTIONS FROM THE STANDPOINT OF OXIDATION-REDUCTION

REDUCTION	OXIDATION
Hydrogenation	Dehydrogenation
Deoxygenation	Oxygenation
Dechlorination	Chlorination
Chloride-formation	Chlorate-formation
Debromination	Bromination
Bromide-formation	Bromate-formation
Nitrite-formation	Nitrate-formation
Amino-formation	Nitro-formation
Cyanide-formation	Cyanate-formation
Sulfide-formation	Sulfate-formation

(R.K.S.)

REACTIONS INVOLVING RECOMBINATION OF IONS. Reactions of acids, bases, salts in water solution are usually reactions of **ions** in one of the following ways:

1. Recombination of anions and cations.
2. Formation of complex anions or cations.
3. Change of the electronic charge of anions or cations or a component part of either.

The third type of reaction is discussed under **Reactions Involving Oxidation-Reduction.** The first two types are considered here.

1. Recombination of anions and cations.

a. Those reactions in which there is *no change of state* involved, that is, no gas and no solid separating from or entering into the solution.

Of these reactions the most common is the neutralization of an acid and a base.

In dilute solution (say 1/20 molar—see **Concentra-**

tion) the electrolytic dissociation data for a few substances is as follows:

Substance	Formula	Percentage Electrolytic Dissociation at M/20
Hydrochloric acid....	H^+Cl^-	100%
Acetic acid.........	$H^+Ac^-(HC_2H_3O_2)$	4%
Sodium hydroxide....	Na^+OH^-	100%
Sodium chloride......	Na^+Cl^-	100%
Sodium acetate.......	Na^+Ac^-	100%
Water..............	H^+OH^-	0.00001%

The following outline represents graphically the state of affairs in two pairs of solutions before mixing and after mixing—in the latter case recombination of ions occurs. $C/$ is the original number of each entity *before* mixing, and $/C$ the final or equilibrium number of each entity *after* mixing and recombination of ions has taken place.

Example 1. Taken: 100 mols M/20 HCl
100 mols M/20 NaOH

$$
\begin{array}{ccccc}
HCl & \rightleftarrows & H^+ & + & Cl^- \\
0/ & & 100/0.00001 & & 100/100 \\
& & + & & + \\
NaOH & \rightleftarrows & OH^- & + & Na^+ \\
0/ & & 100/0.00001 & & 100/100 \\
& & \updownarrow & & \updownarrow \\
& & HOH & & NaCl \\
& & /100 & & /0
\end{array}
$$

Example 2. Taken: 100 mols M/20 HAc
100 mols M/20 NaOH

$$HAc \rightleftharpoons H^+ + Ac^-$$
$$\overset{96/}{} \quad \overset{4/0.00001}{} \quad \overset{4/100}{}$$
$$+ \qquad +$$
$$NaOH \rightleftharpoons OH^- + Na^+$$
$$\overset{0/}{} \quad \overset{100/0.00001}{} \quad \overset{100/100}{}$$
$$\Updownarrow \qquad \Updownarrow$$
$$HOH \qquad HaAc$$
$$\overset{/100}{} \qquad \overset{/0}{}$$

In the first example the real chemical reaction is essentially $H^+ + OH^- \rightleftharpoons HOH$; in the second, $HAc + OH^- \rightleftharpoons HOH + Ac^-$.

In support of this, data from another source is available, namely, the heat of the reaction, which is, per 1 mol of water formed, in example 1, 13,700 calories, and in example 2, 13,400 calories, thus showing that the reactions are not identical. The net *difference* between the two is the **ionization** of **acetic acid**

$$HAc \rightleftharpoons H^+ + Ac^-$$

which evidently absorbs 300 calories.

Acids and bases which are ionized to the degree of **hydrochloric acid** and **sodium hydroxide**, e.g., **hydrobromic acid**, **nitric acid**, **potassium hydroxide**, give the same **heat of reaction**, namely, 13,700 calories. The

HYDROGEN ION CONCENTRATION RANGES (pH) AND COLOR CHANGES OF INDICATORS

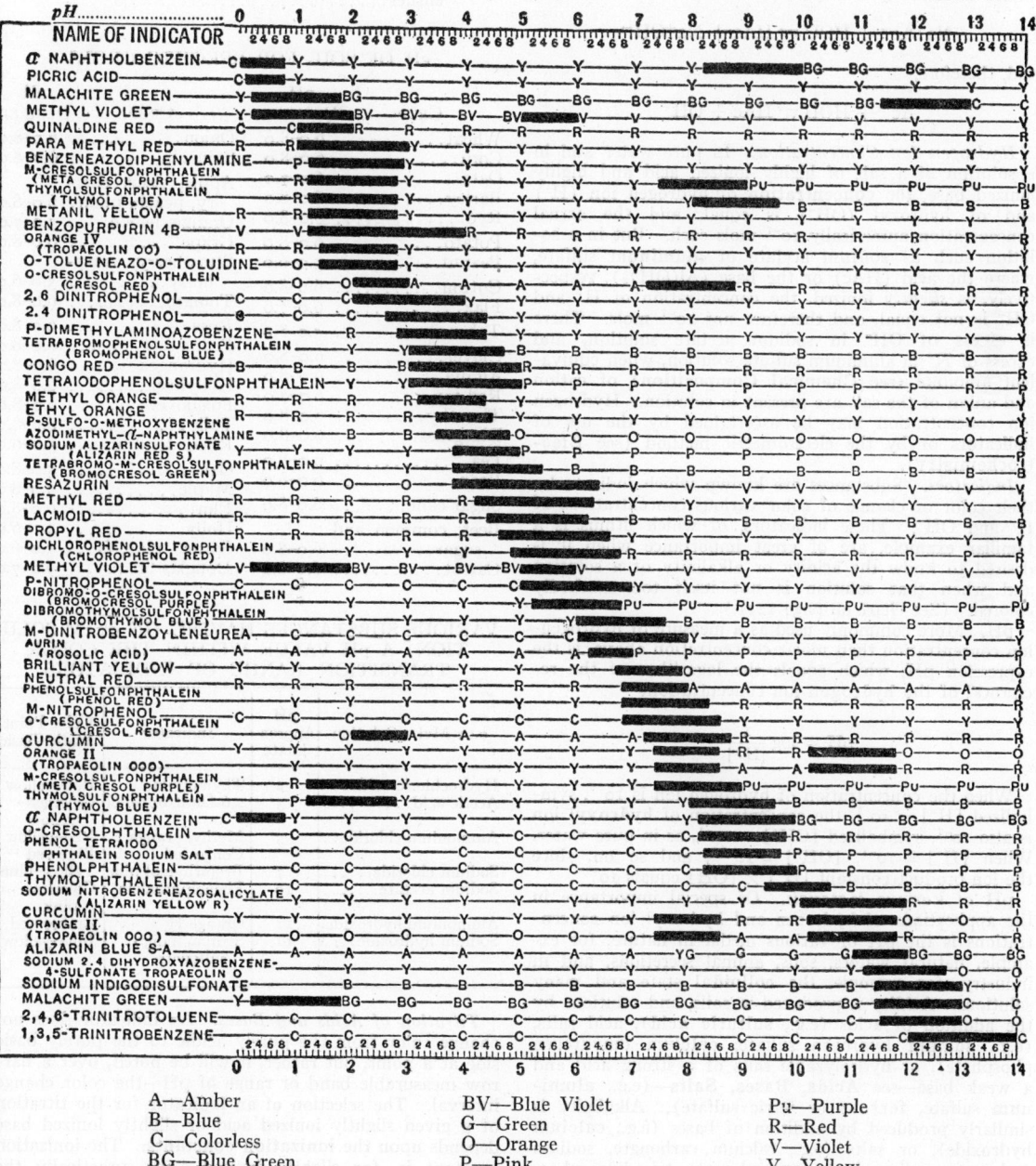

A—Amber
B—Blue
C—Colorless
BG—Blue Green

BV—Blue Violet
G—Green
O—Orange
P—Pink

Pu—Purple
R—Red
V—Violet
Y—Yellow

The pH ranges shown are approximations and are intended to aid in selecting the proper indicator. (*Eastman Kodak Co.*)

heat of reaction of hydrochloric acid and **ammonium** hydroxide (electrolytic dissociation comparable with that of acetic acid) is 12,200 calories—the heat of ionization of ammonium hydroxide is evidently 1500 calories absorbed. The heat of reaction of acetic acid and ammonium hydroxide is 11,900 calories (13.700 − (300 + 1500)).

When solutions of **salts** that are completely ionized are mixed, e.g., **sodium** chloride plus **potassium** nitrate, there is no heat change. No chemical reaction has occurred.

The cases converse to the above are interpreted similarly. Thus, when sodium chloride solution is tested for acidity or alkalinity, it is found to be neutral, while sodium acetate solution, as would be expected from an examination of the above, displays alkalinity, that is, excess of hydroxyl ion (OH−):

$$Na^+Ac^- + HOH \rightleftarrows HAc + Na^+OH^-$$

and, therefore,

$$Ac^- + HOH \rightleftarrows HAc + OH^-$$

Hydrogen Ion Concentration. In pure water and in a solution of a salt of highly ionized acid and highly ionized base, the **concentration** of hydrogen ion (H+) and of hydroxyl (OH−) is equal, and the actual concentration numerically 10^{-7} mols each. But in a solution such as **sodium** acetate or **aluminum** sulfate, where the acid (HAc) or the base (Al(OH)₃), respectively, is *slightly* ionized, the concentration of H+ and OH− is not equal, and therefore not 10^{-7} mols. There is excess of OH− in sodium acetate solution, and excess of H+ in aluminum sulfate solution, when equivalent amounts (see **Chemical Composition**) of cation and anion of the salt are present in solution. Hydrogen ion concentration may be ascertained by the use of indicators or by the electrometric method (see **Electrochemistry**).

Indicators. Substances are known which indicate by their color or change of color various concentrations of H+ and OH−. These indicators, of which litmus is a familiar example, are of great importance when it is desired to know the acidity or alkalinity of a solution and when that solution is not itself too markedly colored. (See chart on page 1211.)

pH. More commonly used as a measure of hydrogen-ion concentration than molar concentration above, is the expression pH, which equals the logarithm of the reciprocal of the **hydrogen** ion concentration:

$$pH = \log \frac{1}{[H^+]}.$$

When the concentration of hydrogen ion is 10^{-7}, symbolized $[H^+] = 10^{-7}$, the concentration of **hydroxyl** ion is also 10^{-7}, symbolized $[OH^+] = 10^{-7}$, as in pure water. When $[H^+] = 10^{-1}$, $[OH^-] = 10^{-13}$, and so on, since the ion product constant $[H^+] \times [OH]$ equals 10^{-14}.

pH *of Various Materials.* Of special importance in the applications of hydrogen and hydroxyl ion concentrations is the pH of various media in nature, for example, natural waters, soils, animal secretions, and in industry, for example, the **colloidal** state and many solutions. Acidity is produced in soils, and solutions by the addition of acids (e.g., **sulfuric acid**), acid salts, (e.g., **sodium** hydrogen sulfate, **calcium** dihydrogen phosphate), or hydrolyzable salts of a strong acid and a weak base—see **Acids, Bases, Salts**—(e.g., **aluminum** sulfate, **ferrous** or ferric sulfate). Alkalinity is similarly produced by addition of bases (e.g., **calcium** hydroxide), or salts (e.g., calcium carbonate, sodium carbonate, sodium hydrogen carbonate, trisodium phosphate, sodium silicate).

pH OF VARIOUS MATERIALS

MATERIAL	pH
Sea water	7.75 to 8.25
Soils	3 to 10
Plant tissues and fluids	About 5.2
Animal tissues and fluids	About 7.0 to 7.5
Blood	7.35–7.5
Urine	5.0–7.0
Milk	6.5–7.0
Gastric juice	1.7
Pancreatic juice	7.8
Intestinal juice	7.7
Internal tissue fluids:	
Minimum, below which acidosis ensues	7.0
Maximum, above which tetany ensues	7.8

pH OF SOIL FOR SPECIFIC CROPS

CROP	pH DESIRED	CROP	pH DESIRED
Wheat	6.0–8.0	Onion	6.0–8.0
Corn	6.0–8.0	Spinach	6.0–8.0
Oats	6.0–7.0	Apple, pear, cherry, peach, plum.	6.0–8.0
Barley	6.0–8.0		
Rye	6.0–8.0	Orange, lemon	5.0–7.0
Potato	6.0–8.0	Grape	6.0–8.0
Peanut	5.0–6.0	Strawberry, raspberry	5.0–6.0
Cotton	6.0–8.0		
Flax	6.0–7.0	Tomato	6.0–8.0
Tobacco	5.0–8.0	Rose	6.0–8.0
Clover	6.0–8.0	Pea, sweet	6.0–8.0
Alfalfa	6.0–8.0	Hydrangea	6.0–8.0
Red top	6.0–7.0	Hollyhock	6.0–8.0
Timothy	6.0–8.0	Foxglove	6.0–8.0
Bluegrass	6.0–8.0	Tulip	6.0–8.0
Pea	6.0–8.0	Rhododendron	5.0–6.0
Bean	6.0–8.0	Oak	5.0–8.0
Sugar cane	6.0–8.0	Maple	5.0–8.0
Beet, common and sugar	6.0–8.0	Elm	6.0–8.0
		Holly	4.0–6.0
Carrot	6.0–8.0	Spruce, pine, fir	5.0–6.0
		Douglas fir	6.0–7.0

VARIOUS SUBSTANCES HAVING (IN 0.01M SOLUTION) A pH VALUE WITHIN THE COLOR TRANSITION RANGE OF INDICATORS

0.01 Molar Solution	pH Approximate	Indicator	Color Point Acidic–Basic
Hydrochloric acid	2	Thymol blue	Red–yellow
Acetic acid	3	2,6-Dinitrophenol	Colorless–yellow
Ammonium chloride	5	Methyl orange	Red–yellow
		Congo red	Blue–red
Sodium chloride	7	Bromthymol blue	Yellow–blue
Sodium acetate	10	Phenolphthalein	Colorless–pink
Ammonium hydroxide	11	Orange II	Amber–red
Sodium hydroxide	12	Sodium indigo sulfonate	Blue–yellow

Titration of Acids and Bases. An indicator does not change color from the purely acidic to the purely basic side at a point, but rather, it will be noted, over a narrow measurable band or range of pH—the color change interval. The selection of an indicator for the titration of a given slightly ionized acid or slightly ionized base depends upon the **ionization constants**. The ionization constant is, for slightly ionized acids, practically the product of the hydrogen ion concentration and of the anion concentration. The following data are in all cases stated for the *first* hydrogen ion dissociation only, thus:

$$\frac{[H^+] \times [Anion^-]}{[Non\text{-}ionized\ acid]} = \text{Ion product constant of acid.}$$

IONIZATION CONSTANTS OF ACIDS (FIRST HYDROGEN IONIZATION) AT 25° C., EXCEPT AS OTHERWISE STATED

ACID	IONIZATION CONSTANT
Acetic	2×10^{-5}
Arsenic	5×10^{-3}
Benzoic	6.5×10^{-5}
Boric	6.5×10^{-10}
Carbonic (18° C.)	3×10^{-7}
Chloroacetic	1.5×10^{-3}
Citric	8×10^{-4}
Dichloroacetic	5×10^{-2}
Formic	2×10^{-4}
Fumaric	1×10^{-3}
Hydrocyanic	7×10^{-10}
Hydroquinone (18° C.)	1×10^{-10}
Hydrosulfuric (18° C.)	9×10^{-8}
Hydrazoic	2×10^{-5}
Hypochlorous (17° C.)	4×10^{-8}
Iodic	2×10^{-1}
Lactic	1.5×10^{-4}
Maleic	1.5×10^{-2}
Malic	4×10^{-4}
Malonic	1.5×10^{-3}
Nitrous (18° C.)	4×10^{-4}
Oxalic	4×10^{-2}
Periodic	2×10^{-2}
Phenol	1×10^{-10}
Phosphoric (18° C.)	1×10^{-2}
Phosphorous	5×10^{-2}
Phenolphthalein	2×10^{-10}
Phthalic	1×10^{-3}
Picric (18° C.)	1.5×10^{-1}
Propionic	1.5×10^{-5}
Pyrophosphoric (18° C.)	1.5×10^{-1}
Salicylic	1×10^{-3}
Succinic	7×10^{-5}
Sulfanilic	6×10^{-4}
Sulfuric	4×10^{-1}
Sulfurous	2×10^{-2}
Tartaric	1×10^{-3}
Thiosulfuric	1×10^{-2}
Trichloracetic (18° C.)	2×10^{-1}
Uric	1.5×10^{-6}

For ionization constants of **nitrogen** bases and nitrogen acids see Amines.

$$\frac{[Cation^+] \times [OH^-]}{[non\text{-}ionized\ base]} = \text{Ionization product of base.}$$

Common Ion Effect. When **sodium** acetate (Na^+Ac^-, ionization large) solution is added to **acetic acid** (H^+Ac^-, ionization small) solution, the concentration of acetate ion is greatly increased with equivalent decrease of hydrogen ion. When **ammonium** chloride (NH_4^+ Cl^-, ionization large) solution is added to ammonium hydroxide ($NH_4^+OH^-$, ionization small) solution, the concentration of ammonium ion is greatly increased with equivalent decrease of hydroxyl ion. This may be demonstrated in such cases by the use of the proper indicator, e.g., methyl orange in the case of acetic acid, phenolphthalein in the case of ammonium hydroxide. The principle is not confined to cases of acids and bases, but is of general application to **electrolytes**. This may be visibly demonstrated by the use of **cupric** bromide solution. The addition of either cupric chloride solution on the one hand, or **potassium** bromide solution on the other hand, causes increase in the cupric and bromide ions, respectively, with resultant color change from the predominant blue of cupric ion to predominant green to brown of non-ionized cupric bromide.

b. Those reactions in which there is *a change of state* involved, that is, a gas or a solid separates from or enters into the solution.

Of these reactions, the most common is the precipitation or the solution of a solid and the evolution or the solution of a gas. In principle, precipitation of solid and evolution of gas are the same, in that in each case the given substance is insoluble (more or less) in the medium (Examples 3 and 4, below). Solution of a solid or of a gas in the medium is the converse of the above in simple cases (Examples 5 and 6, below).

Example 3. Taken: 100 mols M/20 $AgNO_3$
100 mols M/20 HCl

$$
\begin{array}{ccccc}
AgNO_3 & \rightleftarrows & Ag^+ & + & NO_3^- \\
0/ & & 100/0.0013 & & 100/100 \\
& & + & & + \\
HCl & \rightleftarrows & Cl^- & + & H^+ \\
0/ & & 100/0.0013 & & 100/100 \\
& & \Updownarrow & & \Updownarrow \\
& & AgCl & & HNO_3 \\
& & /0 & & /0 \\
& & \Updownarrow & & \\
& & AgCl & & \\
& & (Solid) & & \\
& & /100 & &
\end{array}
$$

Result: AgCl solid separates as precipitate.

Example 4. Taken: 100 mols M/20 Na_2CO_3
200 mols M/20 HCl

$$
\begin{array}{ccc}
 & H_2O & + & CO_2\ (gas) \\
 & /large\ prop. & & /large\ prop. \\
 & & \Updownarrow & \\
2NaCl & & H_2CO_3 & \\
/0 & & /small\ prop. & \\
\Updownarrow & & \Updownarrow & \\
\end{array}
$$

$$
\begin{array}{ccccc}
Na_2CO_3 & \rightleftarrows & 2Na^+ & + & CO_3^{--} \\
0/ & & 200/200 & & 100/small\ prop. \\
& & + & & + \\
2HCl & \rightleftarrows & 2Cl^- & + & 2H^+ \\
0/ & & 200/200 & & 200/small\ prop.
\end{array}
$$

Result: Na_2CO_3 decomposed, CO_2 gas given off.

Example 5. Taken: 100 mols $BaCrO_4$ solid
200 mols M/20 HCl

$$
\begin{array}{ccccccc}
BaCrO_4 & \rightleftarrows & BaCrO_4 & \rightleftarrows & Ba^{++} & + & CrO_4^{--} \\
(solid) & & & & & & \\
100/ & & 0/ & & 0.015/100 & & 0.015/100 \\
& & & & + & & + \\
& & 2HCl & \rightleftarrows & 2Cl^- & + & 2H^- \\
& & 0/ & & 200/200 & & 200/200 \\
& & & & \Updownarrow & & \Updownarrow \\
& & & & BaCl_2 & & H_2CrO_4 \\
& & & & /0 & & /0
\end{array}
$$

Result: $BaCrO_4$ dissolved.

Example 6. Taken: 100 mols CO_2 gas passed into NaOH solution
100 mols M/20 NaOH

$$
\begin{array}{ccccccc}
CO_2 & \rightleftarrows & H_2CO_3 & \rightleftarrows & H^+ & + & HCO_3^- \\
(gas) & & & & & & \\
Large/ & & Small/ & & Small/Small & & Small/Large \\
prop./ & & prop./ & & prop./prop. & & prop./prop. \\
& & & & & & + \\
& & NaOH & \rightleftarrows & OH^- & + & Na^+ \\
& & 0/ & & 100/Small\ prop. & & 100/Large\ prop. \\
& & & & \Updownarrow & & \Updownarrow \\
& & & & H_2O & & NaHCO_3 \\
& & & & /100 & & /Small\ prop.
\end{array}
$$

Result: CO_2 dissolved, and $NaHCO_3$ formed.

2. Formation of Complex Anions or Cations.

The formation or disappearance of complex ions is similar in principle. New combinations are formed without change of electronic charge of the fundamental element.

Example 7. Taken: $Zn(OH)_2$ Solid
Excess NaOH Solution

$$
\begin{array}{c}
Zn(OH)_2 \rightleftharpoons Zn(OH)_2 \begin{array}{l}\nearrow Zn^{++} + 2OH^- \\ \searrow 2H^+ + ZnO_2^{--} \end{array} \\
\text{(solid)} \qquad\qquad + \qquad\qquad + \\
2NaOH \rightleftharpoons 2OH^- \qquad 2Na^+ \\
\qquad\qquad \Updownarrow \qquad\qquad \Updownarrow \\
\qquad\qquad 2H_2O \qquad Na_2ZnO_2 \\
\qquad\qquad\qquad\qquad \text{(solution)}
\end{array}
$$

Result: $Zn(OH)_2$ dissolved. Similar cases: $Na_2Pb^{2+}O_2$, $Na_2Sn^{2+}O^2$, $NaAl^{3+}O_3$, complex zincate anion (ZnO_2^{--}) containing Zn^{2+} formed.

Example 8. Taken: $Zn(OH)_2$ solid.
Excess NH_4OH solution.

$$
\begin{array}{c}
Zn(OH)_2 \rightleftharpoons Zn(OH)_2 \begin{array}{l}\nearrow 2H^+ + ZnO_2^{--} \\ \searrow Zn^{++}\ \}\!- + 2OH^- \end{array} \\
\text{(solid)} \qquad\qquad\qquad + \qquad\qquad\qquad + \\
\qquad 4NH_4OH \begin{array}{l}\nearrow 4NH_3\ \}\!- + 4H_2O \\ \searrow 4OH^- + 4NH_4^+ \end{array} \\
\qquad\qquad\qquad \Updownarrow \qquad\qquad\qquad \Updownarrow \\
\qquad\qquad (NH_3)_4Zn^{++} + 2OH^-
\end{array}
$$

Result: $Zn(OH)_2$ dissolved, complex ammonio-zinc cation $((NH_3)_4Zn^{++})$ containing Zn^{2+} formed. **Similar cases:** $(NH_3)_2Ag^{1+}OH$, $(NH_3)_4Cu^{2+}(OH)_2$, $(NH_3)_2Cu^{1+}OH$, $(NH_3)_4Cd^{2+}(OH)_2$. (R.K.S.)

REACTIONS INVOLVING WATER.

The length of the present discussion is not in proportion to the importance of the subject for the reason that several aspects have conveniently and necessarily been discussed elsewhere. The following articles should be consulted: **Chemical Changes, Reactions Involving Recombination of Ions.**

Reactions involving water are classified as follows:
1. Consumption of water.
2. Production of water.
3. Water as catalyzer.

1. Consumption of Water. This topic is concerned with all reactions in which water is a reactant—hydrolytic and hydration reactions.

a. **Salts** are hydrolyzed in solution to an extent depending upon the strength of the acid of the **anion** and the strength of the base of the **cation**. **Sodium** chloride, a salt of a highly ionized base and a highly ionized acid, is, practically, not hydrolyzed in solution. **Ammonium** acetate, a salt of a slightly ionized base and a slightly ionized acid, is hydrolyzed in solution with the consequent presence of some ammonium hydroxide and some **acetic acid**. One-sided hydrolysis is frequently encountered, e.g., **sodium** carbonate, a salt of a highly ionized base and a slightly ionized acid, is basic (contains excess hydroxyl ions (OH^-)) in solution; **aluminum** chloride, a salt of a slightly ionized base and a highly ionized acid, is acidic (contains excess hydrogen ions (H^+)) in solution.

b. Some salts crystallize from water solution with definite ratios of water called water of **crystallization,** composing the crystal. Water of crystallization is an integral part of such crystals. Spontaneous evolution of water (in cases of efflorescent crystals) or loss of water upon heating, causes change in the crystal form and composition.

c. Some salt crystals, those containing water of crystallization and where the salt is of a slightly ionized non-volatile base and a highly ionized volatile acid (e.g., HCl), are hydrolyzed upon heating, e.g., **magnesium** chloride crystals ($MgCl_2 \cdot 6H_2O$) when heated yield hydrogen chloride (HCl) gas and basic magnesium chloride

$$\left(Mg\begin{array}{l}Cl\\OH\end{array}\right)$$ or, at higher temperature, magnesium oxide (MgO).

d. The hydrolysis of ethyl acetate ester is discussed elsewhere in detail (see **Chemical Changes**). Many classes of organic compounds are hydrolyzable, some preferably in acidic, some in basic, and some in either media. Such compounds include **esters, acetals, acid anhydrides,** polysaccharides (see **Carbohydrates**), **glucosides, tannins, amides** (to form ammonium compounds), **cyanides** (to form **carboxylic acids**), **proteins, oximes, phenylhydrazones,** osazones, semicarbazones, non-benzenoid halogen compounds.

e. **Acetylene** reacts with water in the presence of a catalyzer, e.g., **mercuric salt,** to form **acetaldehyde.** This is a reaction of great importance.

2. Production of Water. This topic is concerned with all reactions in which water is a resultant.

a. In the field of inorganic chemistry, neutralization of acids and bases is a common instance of the production of water. Also, the reduction of oxides, such as **copper, lead, iron** oxides, by heating them in a current of **hydrogen** gas; the burning of hydrogen gas in air or oxygen; and the loss of water of crystallization of crystals.

b. In the field of organic chemistry, the combustion of organic hydrogen compounds (rapid, as in burning, or slow) results in the formation of water from the contained hydrogen.

c. The reaction of **alcohols** plus **acids** to form **esters** (esterification) results in the accompanying formation of water. Esterification is a special case of a general reaction, namely, loss of **hydroxyl** by a compound and loss of **hydrogen** by the same or another compound (chemical dehydration). Examples, ethyl alcohol, 1 molecule, into ethylene ($CH_2:CH_2$) plus water, thus $\frac{H_2C:H}{H_2C:OH}$; ethyl alcohol, 2 molecules, into ethyl ether (($C_2H_5)_2O$), thus $\frac{CH_3CH_2O:H}{CH_3CH_2:OH}$; ethyl alcohol, 1 molecule, plus acetic acid, 1 molecule, into ethyl acetate ($CH_3COOC_2H_5$) thus, $\frac{CH_3CH_2O:H}{CH_3CO:OH}$.

d. Another group of reactions of frequent occurrence in organic chemistry belongs in this category. The reactions are illustrated by the behavior of **nitric** or **sulfuric acid** with benzenoid compounds. These processes, known as **nitration** and **sulfonation,** respectively, result in the formation of water and a **nitro-compound** or a sulfonic acid compound.

3. Water as Catalyzer. In some reactions water plays the role of **catalyzer.**

A mixture of extremely dry hydrogen and oxygen gases is ignited with difficulty, if at all, although water is a resultant when the reaction occurs, and in the presence of water **ignition** is easily accomplished by flame or electric spark. Dixon found that the rate of explosion of **carbon monoxide**-oxygen gas mixtures was definitely affected by the presence of water vapor, thus, when the mixture was saturated with water vapor at 28° C. (3.7% water vapor) the rate of explosion is 1713 meters per sec., when moderately dry, 1305 meters per sec., and when dried with phosphorus pentoxide, 1264 meters per sec.

Baker (1902) showed that extremely dry hydrogen (2 volumes) plus oxygen (1 volume) does not explode even when heated to redness. No combination of the gases occurs when heated below 960° C. (although an electric spark induces an explosion). Previously, it had been demonstrated that extremely dry **ammonia** and **hydrogen chloride** gases do not unite to form ammonium chloride as ordinarily occurs. Burk and Hinshelwood (1927) admitted the dried gases into a glass receiver that had been heated at 200° C. in vacuum to remove all water *except* a film of adsorbed water on the interior surface of the glass receiver. They showed that no ammonium chloride smoke was formed in the gas upon admitting extremely dried ammonia and hydrogen chloride gases, but that solid ammonium chloride was formed on the surface of the receiver *beginning* at the place of entrance of the gases. The reaction probably takes place between ammonium ions (NH_4^+) and chloride ions (Cl^-) in water solution.

An extremely dried mixture of hydrogen and chlorine does not explode in sunlight (as does the moist mixture). Similarly, dried **nitric oxide** and oxygen mixture does not yield **nitrogen** dioxide; **carbon,** when heated to bright redness in dried oxygen, does not burn; **sulfur,** when distilled in dried oxygen, does not burn; **potassium,** when vaporized in dried oxygen, does not burn; **sodium,** burning freely in air, is extinguished when lowered into a jar of dried oxygen. Baker (1922) reported remarkable elevations of the boiling point of liquids that had been subjected to drying for several years by means of phosphorus pentoxide, e.g., bromine, dried 8 years, original boiling point 63° C., new boiling point 118° C., elevation of boiling point 55° C.; benzene, 8.5 years, 80°, 106°, 26° C., respectively; ethyl alcohol, 9 years, 78.5°, 138°, 60° C., respectively. (R.K.S.)

REACTIVE POWER. Alternating-Current Circuits.

REACTOR. Choke Coil.

REALGAR.
The mineral realgar is a **monosulfide of arsenic** corresponding to the formula AsS. It is **monoclinic,** showing short prismatic crystals, or may be in granular or compact masses. It is a soft sectile mineral; hardness, 1.5–2; specific gravity, 3.5; luster, resinous; color, red to orange-yellow; transparent to translucent. Realgar occurs associated with other arsenic minerals and with **gold, silver,** and **lead** ores, although not in great quantities. It has been found as a hotspring deposit and in volcanic sublimations. Realgar has been found in Macedonia, Japan, Switzerland; and in the United States in Yellowstone National Park, as a hot spring deposit, and in Utah and Nevada. The name realgar is derived from the Arabic words *rahj al ghar,* which means the powder of the mine. (E.S.C.S.)

REAMING.
Two-lipped twist and straight fluted drills are satisfactory agents for originating holes, but if a cored hole is to be finished or if a drilled hole is to be enlarged, two lips do not provide sufficient support for the body of the drill and an irregular non-cylindrical hole will result. For this type of work, therefore, drills or reamers with three or more cutting edges must be employed. There are two general classes of reamers, side-cutting and end-cutting types. Fig. 1 shows a rose reamer for enlarging drilled holes or for machining

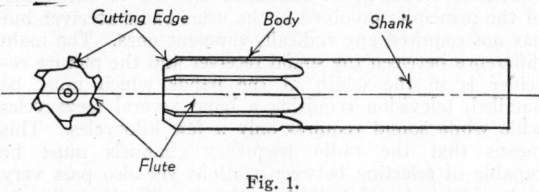

Fig. 1.

punched or cored holes. Such reamers cannot be used for originating holes. The reamer has six cutting edges and is of the end-cutting type. It has no radial clearance on the margin of the flute, but is provided with a back

taper of about .002″ per foot. End-cutting tools can be resharpened frequently since all the cutting is done at the tips. Rose reamers are made somewhat under the nominal size, as they tend to cut oversize. Holes are generally finished to exact size by using either hand or machine side-cutting reamers, with either straight or spiral flutes. Hand reamers are designed to remove only a few thousandths of an inch of metal, and are very slightly tapered on the entering end to facilitate starting.

Adjustable reamers have inserted blades; the main body of the reamer including the straight shank is a solid piece of steel. The blades are made of high-speed steel, and are held by a lock nut at one end and set screws at the other end. The blades can be moved axially along a tapered key which causes them to move radially outward at the same time. The reamer has an adjustment of $\frac{1}{16}$″, and the blades can be replaced by another set when they are worn to the limit of adjustment. The adjustable reamer permits variation in size, but its chief advantage is economy since the blades can be repeatedly adjusted and reground when they become dull.

Hand expansion reamers are used when it is necessary to increase diameters very slightly, especially in repair work. They are intended for light cuts only and are expanded by screwing in the taper expanding plug. The front pilot is generally .010″, and the rear pilot .005″, under the normal size of the reamer.

Shell reamers are hollow cylinders with fluted edges on the outer periphery, and are used with separate arbors to decrease the tool cost.

Taper reamers are used for reaming tapered holes after they have been drilled. Bridge reamers are made in both straight and spiral fluted types, and are used for aligning and reaming punched holes in structural and pressure-vessel work. They are principally used in portable electric or pneumatic tools. Reaming is far superior to drifting for this purpose particularly since it removes the metal around the punched hole, which is likely to be defective on account of the punching process. (H.C.H.)

REARRANGEMENTS.
These are reactions in organic chemistry which involve the transfer of an **atom** or group from one part of the **molecule** to another. **Tautomerism** is a special case of rearrangements in which the two forms are in dynamic **equilibrium.** When such reactions take place the establishment of structural formulae is complicated, since such reactions do not serve as proof for the establishment of structural formulae. (See **Chemical Formulae.**) The following is a list of the more important cases of rearrangements.

1. Allyl Rearrangement:

$$CH_3CH{:}CHCH_2Br \rightarrow CH_2{:}CHCHBrCH_3$$

2. Pinacol Rearrangement: takes place when pinacol is heated with dilute acid, pinacolin formed.

$$(CH_3)_2{-}C{-}C(CH_3)_2 \rightarrow (CH_3)_3C{-}\underset{\underset{OH}{|}}{C}(CH_3)$$
$$\underset{OH\ \ OH}{}$$

3. Benzil Rearrangement: into benzylic acid.

$$\underset{C_6H_5C{-}C{-}C_6H_5}{\overset{O\ \ O}{\|\ \ \|}} \rightarrow (C_6H_5)_2C\!\!\begin{array}{l}OH\\COOH.\end{array}$$

4. Hoffman Rearrangement:

$$RCO\cdot N\!\!\begin{array}{l}H\\H\end{array} \rightarrow RCONHBr \rightarrow RNCO \rightarrow RNH_2$$

5. Beckmann Rearrangement: results when oximes of ketones are treated with certain reagents such as phosphorus pentachloride.

$$R'C—R \quad RC—OH \quad R—C{=}O$$
$$\underset{NOH}{\|} \to \underset{R'N}{\|} \to \underset{R'N—H.}{|}$$

The exchange is between R′ and OH on the opposite sides of the CN bond.

6. Benzidine Rearrangement: Treatment of hydrazobenzene with strong acids.

$$C_6H_5NH—NHC_6H_5 \to H_2N{\cdot}\langle\!\!\bigcirc\!\!-\!\!\bigcirc\!\!\rangle NH_2$$

7. Rearrangement of hydroxylamine derivatives

$$\overset{H}{\underset{|}{C_6H_5NOH}} \to HO\langle\!\!\bigcirc\!\!\rangle NH_2$$

8. Rearrangement in reaction of nitrous acid and alkyl amines

$$CH_3CH_2CH_2{\cdot}NH_2 + HNO_2 \nearrow^{CH_3CH_2—CH_2OH}_{\searrow CH_3CHOH—CH_3} + N_2 + H_2O$$

9. Walden Inversion. A rearrangement takes place in the reaction of optically active substances. The structural formula does not change but the configuration changes due to the interchange of two groups attached to the asymmetrical carbon atom during the course of the reaction. (R.K.S.)

REAUMUR SCALE. Temperature Scales.

RECALESCENCE. A singular phenomenon exhibited by iron and some other ferromagnetic metals. If iron is heated white hot and allowed to cool, it will, at a certain temperature, suddenly evolve enough heat to halt the cooling and even produce a momentary heating. This is easily exhibited by stretching an iron wire against the tension of a spring and arranging a lever index to show slight changes in length. The wire is first heated by an electric current. As it cools and contracts, the index will at a certain point give a perceptible jerk, and then resume its steady motion of contraction. The effect is due to an exothermic change in the crystalline structure. The reverse phenomenon, exhibited on heating, is called "decalescence." For cast iron the re-calescence point is a little below 700° C. Pure iron has two such points, at 780° C. and 880° C.

A somewhat analogous effect is exhibited by some amorphous solids upon **devitrification**, which takes place when the temperature becomes high enough for the substance to crystallize. Non-crystalline sodium silicate, for example, has such a transition point near 500° C., where it suddenly begins to glow. (L.D.W.)

RECAPITULATION. Evolution.

RECEIVER. The radio receiver is the device which picks up the wave from the transmitter and converts it to sound. The simplest form of practical receiver is the crystal set which was once used extensively, then with the advent of satisfactory and cheap vacuum-tube circuits was largely discarded, and which is now coming back as a receiver for ultra high and super high frequencies. Such a set consists of some means of selecting or tuning the desired signal, a rectifying type crystal (galena, silicon, etc.) as a **detector** and a head-set. In its present application for the extremely high frequencies the tuning elements are **lines** and **wave guides.** In the early days of broadcasting the regenerative receiver was almost universally used. However, it has serious limitations, chief among them being its poor audio quality and its radiating ability. As a consequence it is no longer used for regular broadcast reception and its present use is confined to the reception of continuous wave signals. (See **Radio Telegraphy.**) Tuned radio-frequency receivers were also widely used at one time

but are not used much except in the cheaper broadcast receivers and in some long wave commercial stations. This receiver has an antenna-coupling circuit for tuning and coupling the **antenna** to the **grid** of the first amplifier **tube.** This tube may be coupled by a second tuned circuit to another **amplifier** or in smaller sets it may be coupled to the **detector.** Each coupling circuit up to the grid of the detector is tuned and for home receivers the present-day sets have all tuning elements controlled by a single dial. Early sets had each element controlled by a separate dial. While the detector may be any of the conventional types it is usually a triode biased almost to cut-off. This produces rectification and hence demodulation of the radio signal. The audio output of the detector is then amplified by one or more audio-amplifier stages and fed to the **speaker** where it is converted to sound. What might be called the standard receiver today is the superheterodyne which was developed by Armstrong in the early '20's. This circuit differs from others in that it converts all incoming radio-frequency signals to a common carrier frequency. This is accomplished in the first detector, mixer or converter as it is variously called. The signal from the antenna is fed by a tuned coupled circuit to the mixer tube (or in more elaborate sets a tuned radio-frequency stage may be inserted between the antenna and the mixer). In the mixer stage the incoming signal is heterodyned with a locally generated signal so a beat frequency signal, called the intermediate frequency, is produced. This new frequency signal is radio frequency, ranging from around 450 kc. to several megacycles depending upon the purpose for which the receiver is designed. The intermediate frequency has exactly the same modulation as the original signal. In many broadcast receivers the mixer tube combines the functions of mixer and oscillator by using a multiplicity of grids (the **pentagrid converter** is an example of such a tube). However, at higher frequencies it is desirable or even necessary to use a separate tube for **oscillator** and feed its output into the mixer. Regardless of how the oscillator operates, its frequency is always adjusted by the main tuning control of the receiver so the beat frequency output of the mixer is a fixed value. This intermediate frequency signal is then amplified by fixed-tuned radio-frequency amplifiers and then fed to the detector (commonly called the second detector) where it is demodulated. The audio is then further amplified and coupled to the speaker. The simplified-circuit diagram will serve to indicate the various circuits and their relative positions. In some of the cheaper superheterodynes the antenna signal is coupled to the pentagrid first detector, then the intermediate frequency output of this coupled without further amplification to a grid bias or regenerative detector and hence to the final power tube. Various refinements are often added to the higher quality sets. Among these are **automatic volume control, automatic frequency control, noise suppression, tone control,** fidelity and selectivity controls, etc. The superheterodyne gives much greater selectivity by its system of frequency changing and also permits the use of circuits having a more uniform response to the **sidebands.**

The development of television has led to extensions of the principles involved in the usual radio receiver but has not required any radically different ones. The main difference between the sound receiver and the picture receiver is in the width of the bands which must be handled, television requiring a band several megacycles wide while sound requires only a few kilocycles. This means that the radio frequency channels must be capable of selecting between stations yet also pass very wide sidebands. In addition the amplification circuits after the detector (corresponding to the audio amplifiers of the sound set) must satisfactorily amplify over a range of a few million cycles. Frequency **modulation,** on the other hand, has necessitated the use of a some-

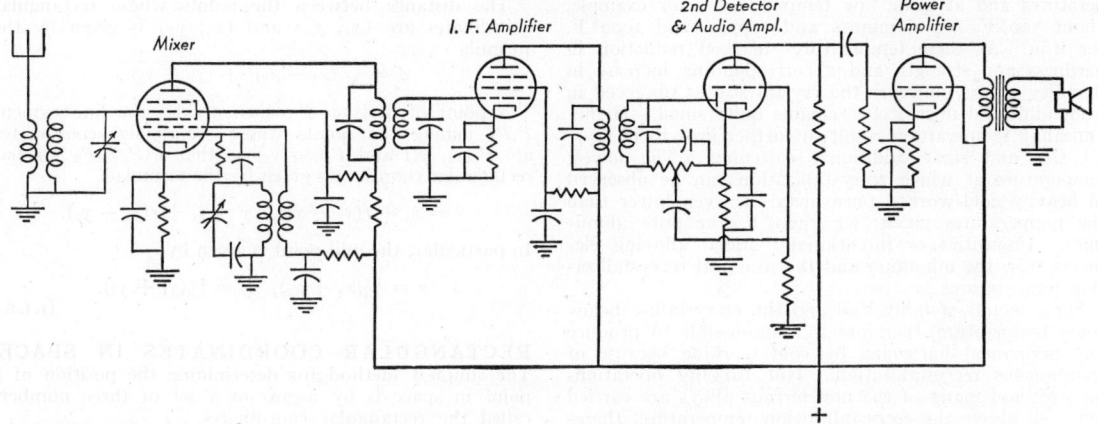

Simplified-circuit diagram of superheterodyne receiver.

what different type of detector. The incoming frequency modulated signal is amplified, converted to the intermediate frequency and this further amplified as in the usual set. However, since the modulation is present as a frequency variation and the loud-speaker responds to an amplitude variation it is necessary to change from one to the other upon detection. This is accomplished in a **discriminator** (a frequency sensitive detector) and then the resulting audio signals, which are amplitude variations, are amplified in a standard audio **amplifier**. However, the radio and intermediate frequency channels of the FM set must be fairly wide band (about 200 kc.) and the audio must be good to about 15,000 cycles to realize the full benefits of this type modulation.

Besides these basic circuit distinctions radios are often classified according to the type **power supply** used. Thus battery receivers for rural and portable use, a-c sets for operating from the 110-volt a-c house-supply circuit, a-c-d-c sets for use with either 110 (nominal) volts a.c. or d.c., etc., are among the various types available.

For special problems involved in the reception of unmodulated signals see **radio telegraphy**.

The common telephone receiver operates by an electromagnetic action. An iron diaphragm is placed in the magnetic field of a permanent steel magnet of U-shape, having soft iron extensions on the pole pieces. When the receiver is not in use, the iron diaphragm is at rest under a constant pull from the permanent **magnet**. On the soft iron extensions are wound coils carrying the receiver current, and which, by electromagnetic action, modulate the constant magnetic **flux** provided by the permanent magnet. The resultant variable magnetic flux vibrates the diaphragm in accordance with the variation of receiver current until it produces sound waves which duplicate those used to vary the current at the **transmitter**. (L.R.Q.)

RECEPTACULUM SEMINIS. A reservoir associated with the genital ducts of the female. The seminal fluid received from the male during coitus is stored in this organ. The term spermatheca is also applied to such reservoirs.

In some of the **insects** a pouch of independent origin receives the seminal fluid. In some species it is connected with the vagina and in others its external opening is independent, but in both forms it is called the bursa copulatrix. It may be associated with a true spermatheca or may itself be the storage reservoir. (A.W.L.)

RECESSIONAL MORAINE. Terminal Moraine.

RECIPROCAL OF NUMBER. Division.

RECIPROCAL SYSTEMS OF VECTORS. From the **quadruple vector product** formula, we obtain:

$$r[abc] = [rbc]a + [rca]b + [rab]c,$$

where $[abc] = (a \times b) \cdot c$, or

$$r = r \cdot \frac{b \times c}{[abc]} + r \cdot \frac{c \times a}{[abc]} + r \cdot \frac{a \times b}{[abc]}.$$

The three **vectors**

$$\frac{b \times c}{[abc]}, \quad \frac{c \times a}{[abc]}, \quad \frac{a \times b}{[abc]},$$

which are perpendicular, respectively, to the planes of b and c, c and a, a and b, are said to form the system reciprocal to the vectors a, b, c.

It is found that if we denote this system of reciprocal vectors by a', b', c', then

$$a = \frac{b' \times c'}{[a'b'c']}, \quad b = \frac{c' \times a'}{[a'b'c']}, \quad c = \frac{a' \times b'}{[a'b'c']}.$$

We also find that

$$[abc] \cdot [a'b'c'] = 1.$$

The system $\hat{i}, \hat{j}, \hat{k}$ of unit vectors is reciprocal to itself. (L.L.S.)

RECONNAISSANCE. Surveying.

RECRYSTALLIZATION. Metals and alloys consist of an aggregate of individual grains or crystals. (See **Metallography**.) When subjected to mechanical deformation as in forging, rolling, or wire-drawing the grains are distorted, usually elongated, and if severely worked as by a 50% reduction in thickness by cold rolling, the grains will be fragmentated. These crystal fragments will, in general, have different **orientations** from the parent crystal but tend to seek certain preferred orientations dependent on the material and the mode of deformation. Plastic deformation of this type tends to harden and strengthen rather than weaken ductile metals. (See **Slip, Plasticity, Cold Working**.) If the final product must meet a maximum hardness or minimum ductility requirement, or if further cold processing is to follow, it may be necessary to recrystallize by means of a process anneal.

Recrystallization is the growth of certain of the grain fragments at the expense of others, resulting in larger, strain-free grains. The recrystallization process is possible because of the increased atomic activity at the elevated annealing temperature.

In the case of severely cold-worked metals recrystallization occurs over a relatively small range of temperatures and at quite low temperatures; for example, about 500° F. for aluminum and copper, and 1000° F. for iron. At these temperatures marked reduction in hardness and strength and a corresponding increase in ductility occurs, however the crystal size as observed in a metallurgical microscope remains quite small. Higher annealing temperatures result in further growth in crystal size and some additional softening. The lowest temperature at which recrystallization can be observed in heavily cold-worked pure metals is even lower than the temperatures given; e.g., 300° F. for pure aluminum. Impurities or intentionally added alloying elements raise the minimum and the practical recrystallization temperatures.

Some metals, notably lead and tin, recrystallize below room temperature, therefore it is impossible to produce any permanent hardening by cold working because of spontaneous recrystallization. Hot forging operations on steel and many of the non-ferrous alloys are carried out well above the recrystallization temperature, therefore spontaneous recrystallization is effective in maintaining the softness and workability of the metal. (R.H.H.)

RECTANGULAR COORDINATES IN A PLANE.
In most mathematical investigations involving geometrical figures, it is necessary to fix the positions of points by means of some system of coordinates, which consists of a set of numbers referred to a framework of reference. For

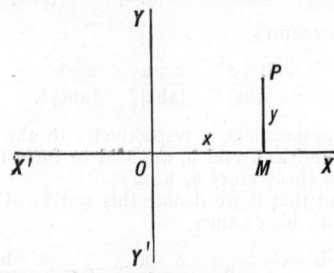

Rectangular coordinates.

a point in a plane, the simplest way to fix its position is by means of a pair of numbers called its rectangular coordinates.

Let $X'X$ and $Y'Y$ be two straight lines perpendicular to each other. These lines are to be used as a framework of reference, and are called rectangular coordinate axes, and their intersection O is called the origin; $X'X$ is called the X-axis and $Y'Y$ the Y-axis. The X-axis is usually drawn horizontally.

Let P be any point in the plane; draw line PM perpendicular to the X-axis. Let the distance OM be regarded as positive when M is to the right of the origin and negative when M is to the left of the origin; let the distance MP be regarded as positive when P is above the X-axis and negative when P is below the X-axis. These distances with the associated algebraic signs are called directed distances.

The rectangular coordinates of the point P are the directed distances OM and MP. The directed distance OM is called the abscissa of P and is usually denoted by x, and the directed distance MP is called the ordinate of P and is usually denoted by y. Together the abscissa x and the ordinate y are the rectangular coordinates of point P. A point whose rectangular coordinates are x and y is commonly denoted by the symbol (x, y).

Every point in a plane has a single pair of rectangular coordinates, and conversely, every pair of rectangular coordinates determines a single point.

Rectangular coordinate paper (or squared paper) is paper ruled into small squares, and is very useful for plotting points given by rectangular coordinates in the plane.

The distance between the points whose rectangular coordinates are (x_1, y_1) and (x_2, y_2) is given by the formula

$$d = \sqrt{(x_1 - x_2)^2 + (y_1 - y_2)^2}.$$

A point of division P which divides the line-segment P_1P_2 joining the points whose rectangular coordinates are $P_1(x_1, y_1)$ and $P_2(x_2, y_2)$ so that $P_1P/P_1P_2 = r$ has rectangular coordinates given by the formulae

$$x = x_1 + r(x_2 - x_1), \quad y = y_1 + r(y_2 - y_1).$$

In particular, the mid-point is given by

$$x = \tfrac{1}{2}(x_1 + x_2), \quad y = \tfrac{1}{2}(y_1 + y_2).$$

(L.L.S.)

RECTANGULAR COORDINATES IN SPACE.
The simplest method for determining the position of a point in space is by means of a set of three numbers called the rectangular coordinates.

Let three mutually perpendicular **planes** be chosen, intersecting in lines $X'X$, $Y'Y$ and $Z'Z$; these lines are also mutually perpendicular. These planes are called rectangular coordinate planes, and are labeled the XY, YZ and ZX planes; the intersection lines $X'X$, $Y'Y$ and $Z'Z$ are called coordinate axes and are labeled the X-axis, Y-axis and Z-axis respectively. The point of intersection of the planes and of the axes is called the origin.

Let us suppose the XY-plane to be horizontal, with the X-axis in the plane of the paper, the ZX-plane to coincide with the plane of the paper and in vertical position, and the YZ-plane therefore perpendicular to the plane of the paper and in vertical position. This is sometimes called a left-handed system. Frequently the X- and Y-axes are interchanged from the preceding position, and we then have a right-handed system.

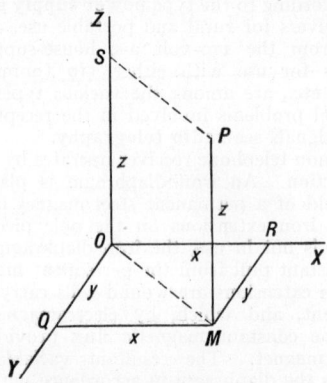

Rectangular coordinates in space.

In the above left-handed system, let P be any point and draw PM perpendicular to the XY-plane, then MR perpendicular to the X-axis and MQ perpendicular to the Y-axis. Let the directed distance MP be regarded as positive when P is above the XY-plane and negative when below; let the directed distance QM (or OR) be considered positive when P is to the right of the YZ-plane and negative when to the left; let the directed distance RM (or OQ) be considered positive when P is in front of the ZX-plane and negative when behind.

The rectangular coordinates of the point P are the directed distances QM (or OR), RM (or OQ) and MP (or OS), and are usually denoted by x, y, z respectively. These rectangular coordinates of a point are therefore the directed perpendicular distances of the point from the coordinate planes.

The distance between the points whose rectangular coordinates are (x_1, y_1, z_1) and (x_2, y_2, z_2) is given by

$$d = \sqrt{(x_1 - x_2)^2 + (y_1 - y_2)^2 + (z_1 - z_2)^2}.$$

A point of division P of a segment joining the points in space whose rectangular coordinates are $P_1(x_1, y_1, z_1)$ and $P_2(x_2, y_2, z_2)$ so that $P_1P/P_1P_2 = r$ has coordinates given by

$$x=x_1+r(x_2-x_1), \quad y=y_1+r(y_2-y_1), \quad z=z_1+r(z_2-z_1).$$

In particular, the mid-point is given by

$$x = \tfrac{1}{2}(x_1 + x_2), \quad y = \tfrac{1}{2}(y_1 + y_2), \quad z = \tfrac{1}{2}(z_1 + z_2).$$

(L.L.S.)

RECTANGULAR HYPERBOLA. Hyperbola.

RECTIFIER. The numerous advantages of a.c. for power transmission and utilization (see **Electric Power Transmission**) have led to its almost universal adoption. There are, however, many applications where d.c. must be used or where it is preferable. Of course, this needed d.c. could be obtained by using a motor-generator set to convert the a-c power into mechanical power and then back into electrical power in the form of d.c. This method is used in many applications, but for many others it is a cumbersome and unnecessary method. The rectifier, in one form or another, is the most used method of converting the a-c supply to d.c. for utilization. Rectifiers may be obtained in several forms but they all depend upon the same effect for their operation, the ability to pass current in one direction and not the other (or much more pronounced in one direction than the other). The thermionic rectifier has a hot **cathode** which emits **electrons** which in turn are attracted to the **anode** when the anode-cathode potential is positive (i.e., the anode is positive) and are repelled by the anode when the potential reverses. Thus the current can flow only when the anode is positive so the applied a.c. is converted to d.c. For low currents and for certain radio applications where gas-filled tubes are objectionable the envelope is evacuated. However, for most industrial applications and for many communications applications the tube is filled with a gas (mercury vapor, neon, argon, xenon, etc.) at low pressure. Such tubes, called **phanotrons,** have much higher efficiencies than vacuum tubes but cannot withstand as high voltages without back-firing, i.e., conducting in the reverse direction. They are not inherently self-limiting in the current they will pass and hence may be seriously damaged or destroyed by excess currents. The **mercury-arc** rectifier is very widely used for large power demands and is rapidly replacing other types of conversion equipment for heavy industrial loads. The copper oxide and selenium disk type rectifiers find many fields of use where the currents required and the voltages applied are not too large. These units depend upon the unidirectional impedance properties of the barrier surface between two dissimilar materials. They are widely used for battery chargers, bias sources for tubes, rectifier type **instruments,** etc., since they have no stand-by losses such as the cathode heating power of the thermionic types, nor any starting requirements like the mercury-arc type. In rating rectifiers various quantities are of importance. Since the applied voltage is alternating, the rectifiers must be rated in terms of the forward peak voltage, i.e., anode positive, and negative peak voltage which they can safely withstand. Current ratings include the peak current (necessary since the current flows in pulses), average current (the d-c value) and sometimes another short-time peak current rating. The various types of rectifiers cover a range of sizes from smaller than a grain of corn to huge tanks as large as a fair-sized room. **Thyratrons** are often called grid-controlled rectifiers. (L.R.Q.)

RECTIGRADATION. Aristogenesis.

RECTILINEAR CHART. A method of representing change by means of a curve drawn on chart paper having rectilinear, or right-angled uniformly spaced coordinates. In general, the independent variable is plotted along a horizontal, and the dependent variable along a vertical axis. (H.C.H.)

RECTUM. A terminal portion of the alimentary tract (**digestive system, intestine**) in which the feces accumulate, pending their periodical discharge. It may be a part of the true gut, lined with endodermal tissue, or a part of the **proctodaeum,** lined with ectoderm. (A.W.L.)

RED BELLY. Pisces, Teleostei. A sunfish of olive color with red belly and lower fins. It is found from Maine to Louisiana. (A.W.L.)

RED CLAY. The most common of deep-sea sediments. A ferruginous clay formed from the alteration products of volcanic ash and other **aeolian** sediments, including **meteoric** material. **Manganese** concretions develop in these muds which are deposited exceedingly slowly and contain little or no organic or **calcareous** matter, due to the solvent power of the sea water under great pressure. (R.M.F.)

RED HARDNESS. Most metals, including tool steels, lose much of their hardness at red heat—about 1000° F. and above. Tool materials such as **high-speed tool steel, Stellite, cemented carbides,** and **diamond,** which retain a considerable part of their hardness at these temperatures are said to have high red hardness or hot hardness. (R.H.H.)

RED LEAD. Lead, Oxides.

RED MUD. A reddish-brown deep-sea mud composed of **aeolian** terrigenous dust or **loess** which is deposited off the seaward end of a large delta or off desert coast lines. (R.M.F.)

RED SHIFT. Spirals; Doppler Effects.

REDFIN. Pisces, Teleostei. The common shiner or **dace,** *Luxilus cornutus,* and another species of the same group which becomes red below during the breeding season. Both are small fishes which are abundant in small streams. (A.W.L.)

REDFISH. Pisces, Teleostei. 1. The Tahoe trout, *Trutta henshawi,* found in streams and lakes on the eastern slope of the Sierras. 2. The channel **bass,** *Sciaenops ocellata,* an important marine food fish taken along the Atlantic and Gulf coasts of the United States. 3. A species of **salmon,** *Oncorhynchus nerka,* second only to the king salmon in the fisheries of the Pacific Coast. Also called the blueback. Its flesh is deep red. (A.W.L.)

REDHORSE. Pisces, Teleostei. Fresh-water fishes (**Pisces**) of the genus *Moxostoma* found in rivers and lakes of the eastern half of the United States. The common redhorse is a good food fish, also called the white sucker or mullet. (A.W.L.)

REDPOLL. Aves, Passeriformes. A small bird (**Aves**) related to the finches. Named for its red crown. One species nests in the northern parts of the northern hemisphere and is known in Europe as the mealy redpoll. Europe has another species, the lesser redpoll, and Asia and North America have the related hoary redpoll, *Acanthis hornemanni,* which only occasionally enters the northern United States. (A.W.L.)

REDSHANK. Aves, Charadriiformes. A Eurasian bird (**Aves**), *Tringa totanus,* related to the snipes and sandpipers, named from the red color of the bare part of the legs. The name is also applied to a larger related species of Europe, *T. erythropus.* (A.W.L.)

REDSTART. Aves, Passeriformes. 1. Birds (**Aves**) of Europe, northern Africa, and palaearctic Asia. The common redstart, *Phoenicurus phoenicurus,* also called the firetail, has the tail and rump chestnut above. This and the allied species are related to the thrushes. 2. The American, *Setophagea ruticella,* and painted, *S. picta,* redstarts of North America are **warblers.** Both are variable in color according to age and sex, adult males showing black and white with red or orange markings. The painted redstart extends into Mexico. (A.W.L.)

REDTAIL. Aves, Falconiformes. The red-tailed hawk, *Buteo borealis.* (A.W.L.)

REDUCED VARIABLES OF STATE. Critical State.

REDUCING MOTION. A reducing motion is one in which a given displacement of rectilinear, rotary, or curvilinear character is converted by the apparatus to a similar motion in which the displacements are at all times proportional to the original motion, but on a smaller scale. The multiplying lever, the large and small pulley, and the inclined plane are a few examples of the many common elements of mechanisms which may be adapted to reducing motions. These are not necessarily exact reducing motions, for frequently very close approximations serve just as well as exactly similar motion. The pantograph, which is described in the article on **parallel mechanisms,** is typical of another group of reducing motions, in which the apparatus is of a more specialized character, and not merely some adaptation of a general element of mechanism. (F.T.M.)

REDUCTION. Reactions Involving Oxidation-Reduction.

REDUCTION OF AREA. Tension Test.

REDUCTION OF PRESSURE TO SEA LEVEL. In order to properly analyze the pressure distribution of the atmosphere it is necessary to refer all atmospheric pressure measurements to some level, and this level is accepted as sea level. All pressure measurements taken above sea level must be reduced to a comparative sea-level value by adding the weight of a fictitious column of air extending from the level of the observatory down to sea level. If the fictitious column of air is not great, no significant error arises in applying the correction, but if the observatory is at several thousands of feet above sea level a considerable error often enters the correction. In general, if the air is very cold, the corrected pressure is too high and if the air is very warm, it is too low. (P.E.K.)

REDUNDANCY. In a structure of the type of a **truss,** redundancy refers to the condition in which there are more members than would be needed to produce stability if the joints act as hinges. Redundant members may be used for the purpose of producing a more rigid structure than that which would be obtained with just enough members to satisfy the conditions for static equilibrium. A flag pole held in a vertical position by 4 guy wires which are spaced equidistant around the pole is not a redundant system, since the wires are incapable of carrying **compression** and only 2 wires act at one time. If the guy wires were replaced by stiff members capable of carrying either **tension** or compression the system would be redundant, as but two are necessary. A truly redundant structure is incapable

of analysis by statical mechanics because the distribution of load between the redundant members and the other members depends upon the elastic properties of the members. The distribution of stress between the stiff members in the above example would depend upon their size and elastic properties. The analysis of such structures can be made by the method of **least work** or other recognized methods for the solution of **indeterminate structures.**

The degree of redundancy or indeterminacy (see **Indeterminate Structure**) of a structure is a number which represents the difference between the number of unknown conditions which must be satisfied and the number of equations of static equilibrium which are applicable. (C.W.C., F.T.M.)

REDUNDANT MEMBER. Indeterminate Structure; Redundancy.

REDWING. Aves, Passeriformes. 1. A European **thrush,** *Turdus musicus,* named from the red markings of the sides and under surface of the wings. 2. The American red-winged **blackbird,** *Agelaius phoeniceus.* The species is black with a conspicuous red patch on the shoulder, margined with whitish or yellowish. (A.W.L.)

REDWOODS. Conifers.

REEDBUCK. Mammalia, Artiodactyla. *Redunca.* A small African **antelope.** The male has relatively small horns which turn forward. Also known by the Dutch equivalent, reitbok. (A.W.L.)

REEVE. Ruff.

REFLECTION. When an emission, such as radiation or sound, traveling in one medium encounters a different medium, part of it in general passes on and undergoes **refraction,** while part is reflected. Even water waves exhibit reflection upon meeting an obstacle, and some of the characteristics of the process are conveniently observed by watching surface ripples. In all cases of "regular" reflection, in which the direction of propagation is sharply defined after reflection, the change takes place in accordance with a very simple law, viz., the reflected and incident wave trains travel in directions making equal angles with the normal to the reflecting surface and lie in the same plane with it. These angles are called, respectively, the angle of reflection and the angle of incidence. For normal incidence, both of these angles are zero. Rough surfaces reflect in a multitude of directions, and such reflection is said to be "diffuse." Only part of the emission or of the energy associated with it is reflected; the ratio of that part to the whole incident emission is called the "reflectivity" of the surface.

Various phenomena may accompany reflection under appropriate circumstances. Sometimes there is a change or even a reversal of phase (see **Vibrations and Waves**); the reflected wave train may be polarized (see **Polarized Light**); or the incident and reflected waves may, through their **interference,** produce stationary waves. If the incident waves are of complex character, the reflection may be selective, due to the difference in reflectivity for the different components. (See **Mirrors and Lenses,** and **Total Reflection.**) (L.D.W.)

REFLECTIVITY. Reflection; Thermal Radiation.

REFLEX. An involuntary secretory or motor response to an external stimulus (see **Nervous System**). Reflex action depends on the passage of impulses from a receptor organ (the skin, muscles, or organs of special sense) to the central nervous system, and then out again to an effector organ (muscle or gland). The route traveled by the impulse is the reflex arc. The contraction of the pupil of the eye in response to light, and the jerking away of the hand when a painful stimulus is applied, are examples of motor reflexes. They are often protective and purposeful. The secretion of saliva and gastric juice when appetizing food is seen or smelled is a secretory reflex.

Conditioned reflexes were first described by Pavlov. He conducted experiments with dogs in which the animals were trained to respond with certain actions to a stimulus not basically related to the action. Thus, if a dog were repeatedly subjected to a certain noise and then fed immediately, he came to associate the noise with food. Eventually he responded to the noise with salivation and gastric secretion, even though the food was not given. This is a conditioned reflex, and illustrates the rôle of higher cortical centers in reflex action. Playing the piano and operating a typewriter are examples of conditioned reflexes. During the learning period, these are voluntary actions, but eventually, the complex motions are carried out instantly, as the eye sees the letter or note, without voluntary thought.

In the physical examination of the nervous system, abnormalities in certain simple reflex movements are valuable aids in testing the integrity of nerve pathways. Normally, tapping the **tendon** just below the flexed knee causes the leg and foot to extend rapidly and come back immediately to the starting position. This reflex movement is called a knee jerk. Overactive, underactive, or unequal knee jerks on the two sides may indicate an abnormal situation somewhere along the reflex arc. Other simple reflexes routinely tested are ankle jerks, biceps and triceps tendon reflexes in the arms. Abnormalities in these and other reflexes occur in such conditions as multiple **neuritis, central nervous system syphilis**, brain and cord **tumors, poliomyelitis**, etc. (D.M.H.)

REFLUX. Distillation.

REFLUX RATIO. Distillation.

REFRACTION. The term refraction properly applies to the change of direction which light or other wave emission experiences on passing obliquely from one medium to another in which its velocity of propagation is different The physical nature of the effect can be visualized by considering a regiment marching in column of platoons across a boundary between smooth turf and freshly plowed ground. If the line of march is perpendicular to the boundary, the platoons are simply slowed up and thus crowded more closely together; but if it is oblique, one end of each platoon is retarded sooner than the other, and the file swings around to a direction nearer the normal (Fig. 1). A train of waves is similarly affected as it passes into a new medium with change of velocity.

Fig. 1. Change of wave length and (in general) of direction of wave train upon entering new medium.

It is easy to show that if i is the angle of incidence and r the angle of refraction at such a boundary (Fig. 2), the refraction is governed by a simple relation known as Snell's law:

$$\sin i = n \sin r;$$

in which n has the same value for various angles of incidence and refraction. This constant n, known as the **refractive index**, depends upon the character of the wave train and of the two media. Physically it represents the ratio of the velocity of the disturbance in the first medium to that in the second. For light passing from one medium to another in which its velocity is greater, so that $n < 1$, we may, for a sufficiently large angle of incidence, encounter the curious phenomenon known as **total reflection**. (L.D.W.)

REFRACTIVE INDEX. When light passes from one medium into another, it in general changes velocity. The ratio of the velocity in the first medium to that in the second is the "relative index of refraction" for the two media. It is the constant of proportionality which appears in Snell's law of **refraction**. If light enters a medium from a vacuum, the ratio is called the "absolute index" for that medium. Let the absolute indices for two media A and B be respectively n_A and n_B. Then it is easy to show that the relative index for light passing from A to B is $n_{AB} = n_B/n_A$. If, for example the absolute indices of water and of glass are respectively $4/3$ and $3/2$, then the relative index from water to glass is $3/2 \div 4/3 = 9/8$; while from glass to water it is $8/9$.

The absolute index for all ordinary transparent substances is greater than 1; but there are some metals (e.g., silver) for which it is less than 1. Since the absolute index for air exceeds unity by less than 0.0003, the relative indices for solids and liquids in air are very nearly equal to their absolute indices. It should be noted that since the refractive index varies with the wavelength (see **Dispersion**), any exact statement of its value must specify the wavelength to which it refers; in tables it is usually given for sodium light (5893 angstroms).

Measurements of refractive index may be made by using a **prism** of the substance at minimum deviation; by the **interferometer**; or by observing the "critical angle" of **total reflection**, as with a **refractometer**. Some typical values are given below (for 5893 angstroms).

Substance	Absolute Index	Substance	Absolute Index
Air............	1.00029	Glycerine.......	1.47
Bromine.......	1.66	Helium.........	1.00007
Carbon dioxide..	1.00097	Ice............	1.3
Diamond.......	2.419	Rock salt......	1.54
Glass..........	1.5 to 1.7	Water..........	1.33

(L.D.W.)

REFRACTOMETERS. Several types of instruments, called refractometers, have been devised for measuring the **refractive index** of any substance. Special forms are used for solids, for liquids, and for gases. Solid and liquid refractometers usually depend upon the principle of **total reflection** and the fact that the sine of the critical angle is equal to the refractive index for light passing from the more to the less refractive medium. The critical angle is what is measured, or deduced from other measured angles.

Suppose that a specimen of the solid or liquid to be tested is brought into optical contact with one face of a glass prism (or "block") of known, higher refractive index and known angle, and that a slightly convergent pencil of light, entering the test substance, is directed at grazing incidence upon the interface between it and the prism. Those rays incident at less than 90° to the normal of the interface enter the prism; the others do not, and the boundary between is sharply defined. The resulting half-pencil traverses the prism and emerges from the other face where the direction of its cut-off edge can be observed (Fig. 1). The angle between the cut-off boundary of the pencil and the first prism face,

Fig. 2. Angles of incidence (i) and refraction (r).

inside the prism, is the critical angle, and can be easily calculated from the observations and the known data. This is the principle of the Pulfrich refractometer.

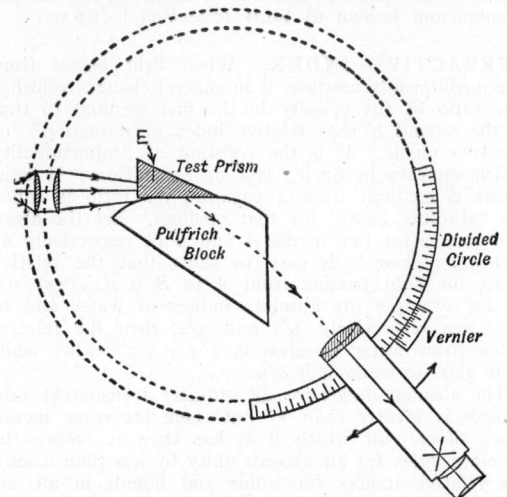

Fig. 1. Pulfrich refractometer (diagrammatic).

Another form, due to Abbe, is used for liquids. A film of the liquid is enclosed between two similar glass prisms (Fig. 2), and the total reflection at the inter-

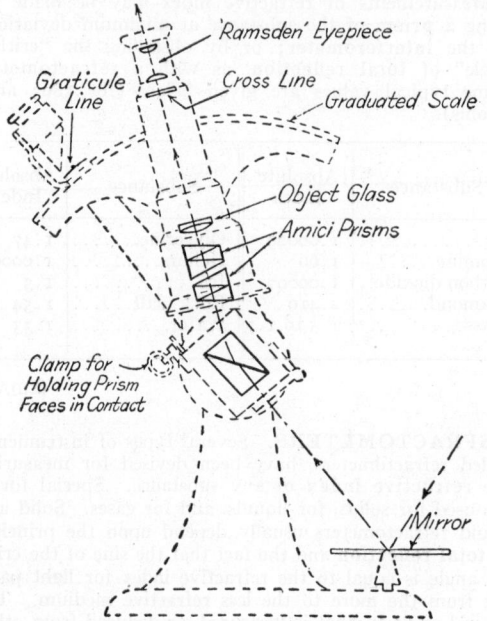

Fig. 2. Optical system of Abbe refractometer.

face observed. Any spectrometer, with a pair of good prisms (preferably right-angled) mounted on the prism table, can be used in this way.

Rayleigh utilized an **interference** method for measuring the indices of gases. Using a **collimator** to render the rays parallel, the stream of light entering one of the slits of the apparatus for **Young's experiment** is passed through a tube of the gas to be tested. The resulting retardation in phase causes a shift of the interference fringes, the amount of which gives the retardation and hence the refractive index of the gas relative to the air outside the tube. (L.D.W.)

REFRACTORIES. Materials which will resist change of shape, weight, or physical properties at high temperatures are known as refractories. The materials which are chiefly used for refractories are fire-clay, silica (see **Silicon**), kaolin, diaspore, alumina (see **Aluminum**), and certain products of the electric furnace, such as silicon carbide. Refractories are used most often in the form of bricks. Refractories are chiefly used to line regions in which **combustion** creates a very high temperature. Also, they are needed to line passages or chambers remote from the combustion region, but which contain high-temperature gases. When refractories are used to line a vessel in which a fluid such as molten steel **or** molten glass is to be contained, the chemical reaction of the refractory is important. Basic and acid refractory brick have their industrial uses. The greatest use of refractory is in **furnaces.** Certain materials (fire-clay, silica, ganister) are resistant to acid slag and are used to contain metallurgical products of acid reaction while others (magnesite, dolomite, alumina) are suitable for basic reaction. Chromite, silicon carbide, and carbon are neutral, but the use of carbon is confined to localities where the atmosphere is reducing. Fire-clay bricks are preferred wherever they give satisfactory service, because of their low cost. Fire-clay brick are classified for temperature duty on the basis of the pyrometric cone scale (see **Seger Cone**). High-temperature duty fire-brick is classed as that which possesses a pyrometric cone rating equivalent to a temperature about 3056° F., whereas low-duty fire-clay must have refractoriness at not less than 2876° F. Between these two limits, the American Society for Testing Materials recognizes two intermediate classifications.

In use, refractories must successfully withstand:

1. *Shrinkage.* A linear shrinkage test of not more than 1.5% when subjected to a temperature of 1400° C. (2550° F.) is recommended. Whenever the shrinkage in use amounts to, say, 3% it is necessary to shut down and repair the installation.

2. *Fusion.* The load test deformation at 1350° C. (2460° F.) should not be more than 1.5%. There is a wide range of refractory materials available through a range of temperatures resistant to fusion, but with their costs approximately in proportion to their temperature resistance. A representative list of these is displayed below. It should be remembered, however, that safe working temperatures are considerably lower than the melting temperatures shown.

3. *Spalling.* Whether due to heating, or to mechanical or structural defects, spalling is a serious matter, since the effective material is thereby lost. A preliminary test is recommended, such as using an air-water mist blast under specified conditions at 1600° C. (2910° F.) when not more than 15% should spall off.

4. *Slag Erosion.* When the face of a refractory material is wetted by a molten slag, and the two are chemically reactive with each other, the refractory is corroded and thereby lost. Acid refractories resist acid slabs best, and basic refractories resist basic slags best.

5. *Abrasion.* Resistance of refractories to the impact of and erosion by solid particles, such as the charge in a pig iron or cupola furnace or the solid ash carried in a furnace flame, is frequently an important matter of concern.

6. *Water Absorption;* 7. *Modulus of Rupture;* 8. *Heat Conductivity;* 9. *Electrical Conductivity* are other points that frequently require consideration in refractory installations.

Refractories play a vital rôle in steam-boiler settings, in pig iron smelters, in open-hearth steel furnaces, in the production of copper, lead, and zinc from their ores, in lime burning, glass, and Portland cement kilns, and in by-product coke ovens.

REFRACTORY MATERIAL	MELTING, TEMPERATURE, °C.	°F.
Kaolin, $Al_2O_3 \cdot SiO_2$.............	1815	3300
Fire-clay (see diagram of Al_2O_3-SiO_2 system below)		
Alumina, fused bauxite..........	1870	3400
Alumina, Al_2O_3, pure sintered....	2050	3720
Silicon carbide, 85–90%.........	1870	3400
Silicon carbide, SiC, pure.......	2250	4080
Silica, quartz, or quartzite.......	1425–1710	2600–3080
Magnesia, 90–95%..............	1970	3980
Magnesia, MgO, pure sintered....	2800	5070
Chromite, $FeO \cdot Cr_2O_3$..........	2180	3960
Chrome, 38% Cr_2O_3.............	1970	3580
Chromium oxide, Cr_2O_3, pure.....	2140	3880
Beryllia, BeO, pure sintered......	2550	4620
Zirconia, ZrO_2, pure sintered.....	2700	4890
Thoria, ThO_2, pure sintered......	3000	5430
Graphite, C, pure (in reducing atmosphere) (softens at 2500° C. (4500° F.))..................	3500	6300

(F.T.M., R.K.S.)

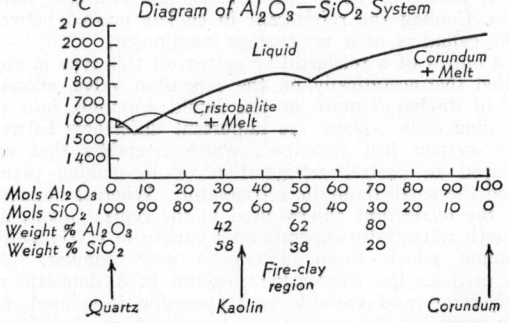

Diagram of Al_2O_3—SiO_2 System

REFRIGERANT. A refrigerant is a substance which is suitable as the working medium of a cycle of operations wherein refrigeration is accomplished. To be satisfactory for this service, the refrigerant should be capable of absorbing heat at a low temperature (i.e., the temperatures associated with ice-making, cold storage, and other forms of refrigeration) and release it at a higher temperature. This may be accomplished only by a suitable expenditure of energy, in accordance with the second law of **thermodynamics**, and through processes involving expansion, evaporation, or chemical change. (The last method is not used.) Refrigerants in actual use can be either gases or vapors.

The use of a gas is but little favored in refrigeration, due to the bulk of the equipment necessary in the cycle; however, refrigeration systems using air as a working medium are entirely successful. Most refrigeration equipment operates with a vaporizable liquid as a working medium. Those refrigerants which are or might be used are shown in table below.

The given data are compiled by W. R. Woolrich, who also states that the properties that the most suitable refrigerant should possess are:

1. It should be readily obtainable.
2. It should condense at normal cooling water temperatures at relatively low pressures.
3. Its boiling point should be sufficiently low not to require vacuum operation.
4. It should have a high latent heat of vaporization.
5. For the service desired the odor should not be objectionable.
6. It should not seriously interfere with lubrication.
(F.T.M.)

REFRIGERATION CYCLE. A refrigeration cycle is any that takes heat at a lower temperature and rejects it at a higher. To do this, the cycle must receive a power input from an external source, and the amount of heat rejected is more than that taken in by the amount of work required to effect the cycle. Theoretically, any power cycle which is reversible could be reversed to create a refrigeration cycle. Actually, practical considerations have caused modification of the reversed power cycle for refrigeration use. Nevertheless, the ordinary vapor compression refrigerating cycle resembles the **Rankine** power cycle to a close degree. In the power cycle, work is done by an **engine**, and in the refrigerating cycle work must be supplied to a reversed engine, which then becomes a pump to pump heat from a lower to a higher temperature. This heat pump is a **compressor**. Most refrigeration cycles make use of a vaporous **refrigerant**, and the following discussion will be confined to that type. In a power cycle the efficiency is the work done divided by the heat supplied. In the refrigeration cycle the efficiency expression is re-

REFRIGERANTS

REFRIGERANT	CHEMICAL FORMULA	BOILING POINT (atmospheric pressure)	MINIMUM PRESSURE REQUIRED (80° F. cooling water)	MELTING POINT	LATENT HEAT OF VAPORIZATION
Ammonia..............	NH_3	− 28° F.	140 lbs. gauge	−107° F.	589.4 B.T.U. per lb.
Carbon dioxide..........	CO_2	Not liquid at atmospheric pressure	960 lbs. gauge	−161° F.	158.6 B.T.U. per lb.
Sulfur dioxide...........	SO_2	+ 14° F.	45 lbs. gauge	−102° F.	172.3 B.T.U. per lb.
Ethyl chloride...........	C_2H_5Cl	+ 54° F.	10 lbs. gauge	−244.5° F.	168.6 B.T.U. per lb.
Methyl chloride..........	CH_3Cl	− 10.6° F.	73 lbs. gauge	−132.5° F.	180.6 B.T.U. per lb.
Butane................	C_4H_{10}	+ 32° F.	25 lbs. gauge	−211° F.	165.2 B.T.U. per lb.
Propane...............	C_3H_8	− 48° F.	130 lbs. gauge	−310° F.	182.6 B.T.U. per lb.
Ethane................	C_2H_6	−127° F.	615 lbs. gauge	−277° F.	234 B.T.U. per lb.
Isobutane..............	C_4H_{10}	+13.6° F.	40 lbs. gauge	−229° F.	158 B.T.U. per lb.
Nitrous oxide...........	N_2O	−128° F.	880 lbs. gauge	−152° F.	163 B.T.U. per lb.
Ether.................	$(C_2H_5)_2O$	+ 94° F.	7 in. vac.	−177° F.	162 B.T.U. per lb.
Carbon bisulfide.........	CS_2	+115° F.	15 in. vac.	−169° F.	153 B.T.U. per lb.
Chloroform.............	$CHCl_3$	+142° F.	21 in. vac.	− 82° F.	109 B.T.U. per lb.
Carbon tetrachloride......	CCl_4	+170° F.	25 in. vac.	− 9° F.	83 B.T.U. per lb.
Dichlorodifluoromethane...	CF_2Cl_2	+ 21° F.	85 lbs. gauge	−247° F.	72 B.T.U. per lb.

placed by "coefficient of performance," which is the reciprocal of efficiency. A coefficient of performance is the heat energy extracted at the low temperature divided by the work which must be supplied to operate the cycle. It measures the cycle performance in that it is an expression of the refrigeration obtained per unit of work supplied to the cycle.

$$\text{Coefficient of performance} = \frac{778Q}{W}$$

Q = heat in B.T.U. absorbed from the ice tank, cold storage room, etc., per pound refrigerant.
W = work in foot-pounds supplied to the compressor per pound refrigerant.

The unit of refrigeration corresponding to the absorption of heat equivalent to the production of 1 ton of ice per day of 24 hours is called a "ton." Normally, about 144 B.T.U. must be abstracted to convert a pound of water to ice. Using this value, together with the proper conversion factors, a ton of refrigeration is found to be equivalent to a heat abstraction of 200 B.T.U. per min. This rate of heat flow is 4.715 hp. As the coefficient of performance measures the ratio of heat abstraction to input, the horsepower which would theoretically be necessary per ton of refrigeration capacity is

$$\frac{4.715}{\text{coefficient of performance}}.$$

An elementary cycle of refrigeration can be operated with a minimum of four pieces of equipment. These are shown in an accompanying figure. Beginning the

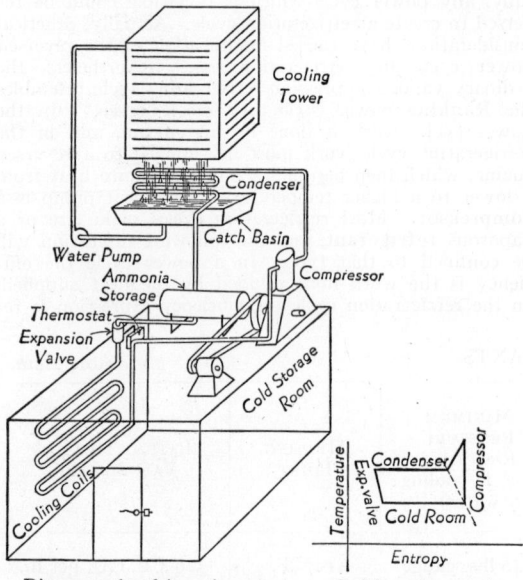

Diagram of refrigerating plant.   Refrigerating cycle.

description with the refrigerating coils, note the bank of pipes in a cold storage room. In these pipes is a liquid whose boiling temperature is lower than the temperature of the cold storage room, so heat is passed through the pipes into this liquid, which absorbs it as heat of vaporization, changing the liquid to a vapor. The vapor is withdrawn from the cooling coils by a motor-driven compressor which increases its pressure until, at the discharge pressure from the compressor, the refrigerant has a saturation temperature high enough so that it may be condensed by some available cooling source such as water. The compressed vapor is discharged into a condenser. This is usually a bank of tubes over which cool water trickles. The heat of vaporization is passed from the vapor to the water,

condensing the former to a liquid. The warmed water is then either wasted, or cooled (by spray pond or cooling tower) and recirculated. The liquid refrigerant then is passed through an expansion valve to the lower pressure of the cooling coils. This is a throttling process. The condition of the refrigerant as it emerges from the expansion valve is liquid, a very small portion of which has been flashed into vapor. The delivery of the refrigerant from the cooling coils to the compressor takes place under the action of natural flow of a vapor in the direction of lowest pressure.

The vapor entering the compressor is dry, or nearly so, and an adiabatic compression raises it to a superheated state. This is shown on the cycle diagram. Known as dry compression, this entails considerably more work from the compressor than is needed merely to raise the vapor to the higher pressure. If the vapor could be brought back along the saturation line instead of the adiabatic line, a large amount of work would be saved, and the coefficient of performance increased. In practice this is sometimes done by:

1. Spraying a small amount of liquid ammonia into the compressor cylinder (wet compression system).
2. Jacketing the compressor cylinder with cold water.
3. Cooling the refrigerant in coolers located between the cylinders of a multi-stage compressor.

Control of a refrigerating system of this type is exercised thermostatically on the expansion valve, allowing it to discharge more or less liquid ammonia into the cooling coils. There are important differences between the system just described, which resembles that employed in central refrigeration or ice-making plants, and the small domestic refrigerator. Whereas ammonia is the refrigerant chiefly used in the central plant, domestic refrigerators operate on a variety of refrigerants, among which Freon (F12) is very popular. Air is used as the condensing medium in a domestic refrigerator, and variable load operation is secured, not by operating the compressor at constant speed and controlling the expansion valve, but by a fixed setting of expansion valve and on-off operation of the compressor. The larger the refrigerating load, the smaller the off intervals of time for the compressor motor. In one leading make, sulfur dioxide vapor is compressed in a small cylinder by a piston whose connecting rod is driven from a crankshaft attached directly to the motor shaft. The compressed vapor is delivered at around 100 lbs. per sq. in. pressure to a condenser which resembles an automobile radiator. Air is drawn by a fan through the openings in the condenser core, and Freon vapor is reduced to a liquid. The liquid flows to a float chamber where a constant level is maintained by a float valve. As more liquid flows into this chamber, the valve passes it into the freezing unit, where it absorbs heat from the refrigerator box and is vaporized. The pressure there is about 20 lbs. per sq. in. gauge. The vapor is drawn by the suction of the compressor into the compressor chamber, where it is ready to begin the cycle again. Despite the small sizes, the coefficient of performance of the domestic refrigerators is not markedly inferior to the large-scale central plant. (F.T.M.)

REGELATION. This curious phenomenon is a direct consequence of the fact that the melting point of ice is measurably lowered by intense pressure. Ice at the normal melting point will, if subjected to great pressure, become liquid, and will "regelate" or refreeze when the pressure is removed. Crushed ice or snow may thus be molded into clear blocks of any desired shape. A small, heavy object placed on a cake of ice not too far below the melting point will gradually bury itself in the ice, the water which results from the pressure escaping from under it and refreezing above it. In the same way a cake of ice resting on a metal grid will hook itself around the bars of the grid. Ice is slippery even when below the freezing point, because the pressure of any hard object, as a skate blade, produces a film of water at the surface of contact. Water extruded under pressure from the terminal wall of ice at the foot of a **glacier** sometimes freezes into snake-like spirals of ice. Regelation is believed to explain the flowing motion of the glacier itself. Only substances which, like water, expand on freezing are capable of exhibiting regelation. (L.D.W.)

REGENERATION. Positive feedback.

In zoology, regeneration is the development of a **tissue** or part of the living body to replace a similar structure that has been damaged or destroyed.

A conspicuous degree of regeneration is possible in some of the simpler animals, including sponges, **coelenterates,** and **worms.** When cut into pieces, the fragments undergo a reorganization of their materials to form complete individuals of smaller size. The process is not unlimited, however, for abnormalities of regeneration take place in some groups when the mutilation is of a certain type. In experiments with flatworms (**Platyhelminthes**), for example, C. M. Child has found that halves of worms or a segment from the middle of the body develop into complete animals, but a head produces only another head and so perishes. T. H. Morgan, in experiments with a species of **earthworm,** found that the amputation of a limited part of the anterior end was followed by complete regeneration but that the removal of more segments resulted in the formation of a minimum number like the original extreme anterior end. Starfishes undergo the regeneration of amputated arms very readily and **mollusks** and **arthropods** are capable of some restoration of lost parts. Insects and crustaceans develop new appendages if the loss occurs before the completion of their growth.

Among the most remarkable cases of regeneration are those of the **bryozoans** and sea cucumbers. The animal (polypide) breaks down within the body wall (zooecium) in the former group to become a disorganized mass called the brown body. From the zooecium a new animal is formed. Sea cucumbers, under extremely irritating stimuli, sometimes discharge the entire intestine, along with the defensive Cuvierian organs. The tract is later replaced.

In complex animals, including man, regeneration is limited to the replacement of parts subject to wear and easy loss, such as hair and nails, and to the renewal of damaged tissues, such as skin. Even the renewal of tissues is limited, some kinds undergoing normal and complete regeneration while others are repaired by the formation of scar tissue of different origin but cannot be replaced.

Researches reported in 1944 indicate that special methods of treatment to prevent the rapid completion of normal healing processes may make it possible for animals of great complexity to regenerate amputated limbs. These studies are in too early a stage to permit an estimate of their ultimate practical importance. (See **Thermal Regeneration.**) (A.W.L.)

REGENERATION, THERMAL. Regeneration is the restoration of a property or quantity to some original state. Thermal regeneration is the restoration of thermal potential by additions of heat.

A furnace so arranged that heat from the flue gases is restored to the furnace by being imparted to incoming air is frequently termed a regenerative furnace. For an example of this, see **Open Hearth.**

Most of the large central power stations of the present time operate on a cycle known as the regenerative cycle, so called because of the method of regenerating the thermal potential of the working medium (feed water) in order to condition it for the boiler. Nearly fifty years ago Cotterill perceived that the extraction of some of the steam from an **engine** for the purpose of bringing

the boiler **feed water** nearer to the saturation temperature of the **boiler,** would result in considerable **thermodynamic** gain over the simple **Rankine cycle.** This idea was first applied to reciprocating steam engine plants, but did not enjoy its present widespread use until the advent of the high capacity steam central station. The reason for the efficiency gain of the regenerative over the Rankine cycle lies in the fact that the steam, as extracted from the turbine, has performed a certain amount of work, but whereas the percentage of the available heat converted into work before extraction is considerable, the percentage decrease in the total heat of the steam is small, and hence, by extracting it before it reaches a low thermal potential, it has remaining in it a high degree of availability as a feed water heating medium. This factor becomes of more and more importance as boiler pressure increases. The elements of the regenerative vapor cycle, shown in the figure, consist

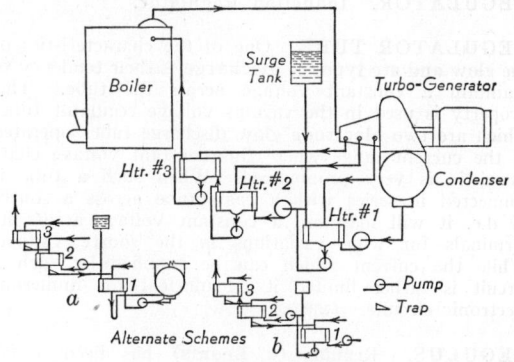

The regenerative vapor cycle.

of the boiler-turbine-condenser combination, and one or more heaters to which steam is extracted, and through which the water flows on its way back to the boiler. Plants using this cycle differ, of course, in the number of heaters employed, and in the manner of handling the heater condensate. The usual arrangement is to employ from two to four extraction heaters, which heat the water to about 80% of the saturation temperature. (F.T.M.)

REGENERATIVE MOTOR. Rockets.

REGIONAL METAMORPHISM. Metamorphism.

REGOLITH. Soil.

REGRESSION. The regression of y on x is given by a functional relation $y = f(x)$. If $f(x) = a_1x + b$, then the regression of y on x is linear. Otherwise it is curvilinear. This result may be extended to m variables $y = f(x_1, x_2, \cdots x_m)$, and if $f(x_1, x_2, \cdots x_m) = a_1x_1 + a_2x_2 + \cdots + a_mx_m$ we have multiple linear regression, otherwise the regression is said to be curvilinear. The constants $a_1, a_2, \cdots a_m$ are usually found by the method of least squares, solving for them by **Doolittle's method.** If the actual points cluster closely about the regression line, curve, plane, or surface, then the dependence of y on $x_1, x_2, \cdots x_m$ is close. The measure of the dependency of y on $x_1, x_2, \cdots x_m$ is given either by the **coefficient of correlation,** the **coefficient of multiple correlation,** the **index of correlation,** or the **index of multiple correlation,** or the index of multiple curvilinear correlation. Tests of significance may be made by the **analysis of variance.** (L.A.A.)

REGULATION, SPEED. Steam engines, steam turbines, internal combustion engines under automatic governing, also certain types of electric motors, vary slightly in speed between no load and full load.

In the case of electric motors, this variation of speed is accounted for by the electrical characteristics. In the case of the prime movers, the variation in speed is due to the characteristics of the governors. The speed regulation is the speed variation between no load and full load expressed as a percentage of the full load. The basic actuation of governors is caused by a change of centrifugal force during change of speed. Usually the centrifugal force is opposed mechanically by springs. The modulus of the springs has a direct bearing upon the speed regulation created by the governor. (F.T.M.)

REGULATION, VOLTAGE. The voltage regulation of a circuit or device is:

$$100 \times \frac{\text{(No load voltage)} - \text{(full load voltage)}}{\text{(full load voltage)}}.$$

(L.R.Q.)

REGULATOR. Induction Regulator.

REGULATOR TUBE. One of the characteristics of the glow and arc types of discharge is their tendency to maintain a constant voltage across the tube. This property is used in the various voltage regulator tubes which are two electrode glow discharge tubes operated in the current range where this constant voltage characteristic is very pronounced. When such a tube is connected in series with a resistance across a source of d.c. it will maintain a constant voltage across its terminals for wide variations in the source voltage. While the current which can be supplied by such a circuit is rather limited it is sufficient for numerous electronic circuits. (L.R.Q.)

REGULUS. Regulus (α Leonis) has been a famous star in all ages. According to the best authority the present name was given to the star by Copernicus. Among the ancients a great variety of different names were used. Regulus has always been regarded as a royal star and among the Persians was considered one of the four rulers of the heavens. As one of the royal stars, Regulus was considered by the astrologers as portending the greatest good fortune. (W.K.G.)

REHEATING. A material, having once been heated, may be found to have undergone important changes; and it may be still further modified when subjected to reheating. In some cases the reheating may have been carried out until the original state was restored, in others to some different temperature. Many industrial processes in which heat plays a part, employ reheating, sometimes to the extent of several "reheats." In annealing, steel is subjected to several reheating processes, which are controlled as to temperature reached, length of time, and cooling, so that treated steels of desired characteristics are produced. When air is compressed to be used in air power tools, it usually cools off between the time it is compressed and the time it is used. Often it is found economical to reheat this air with fuel in order to increase its volume. This is especially true in cases where the air is used expansively in the appliance.

A very interesting application of reheating is to be found in the dozen or so extremely high-pressure (1200 lbs. per sq. in. and up) steam plants operating in this country. An adiabatic heat drop, such as is approximated in steam nozzles, engines, and turbines, implies that the mechanical energy which appears has been produced at the expense of the heat in the steam. The physical properties of steam are such that superheated steam will lose its superheat and become more and more wet as the expansion proceeds to lower pressures. It is found by experience that moisture in expanding steam is undesirable. The inevitable effect of higher pressures is that the saturation line is reached more quickly in an adiabatic expansion and more of the turbine stages operate in the relatively undesirable satu-

rated steam region. In order to relegate the point of saturation to a lower thermal level where high stage efficiency will not be so vital, the steam is resuperheated before it becomes wet. In the resuperheating, or, as

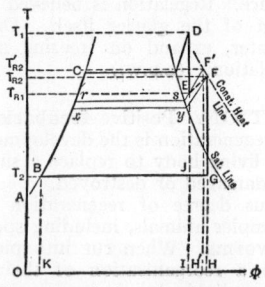

Elements of the reheating cycle.

it is commonly termed, the reheating vapor cycle, the reheating is accomplished by constructing the turbine so that all of the steam may be extracted at a suitable point, resuperheated, then readmitted to the remaining stages for further expansion. This is possible on both single-cylinder and compound turbines but is much simpler on the compound type where the break between high-pressure and low-pressure sections offers a logical point for resuperheating. An additional advantage of reheating is that the reduction in exhaust steam volume per unit of power capacity developed permits some reduction in condenser size.

There are objections to reheating, the principal ones being the added complication of the plant layout and the increased investment cost. Reheat lines are of such size and are so located that concealment is virtually impossible and, furthermore, the whole of the reheat piping is difficult to construct and install. Another objection arises from the energy storage in the steam contained in the reheater and the reheat piping. Close governing will require governor devices on the low- as well as the high-pressure units to overcome the lag in response to normal governor movement. (F.T.M.)

REINDEER. Mammalia, Artiodactyla. A deer of arctic Eurasia. It has antlers in both sexes, set well back on the head and palmately branched at the tips. They are important domestic animals in Lapland and in other parts of the Old World, and have been introduced successfully into Alaska. The two North American species of caribou and several other species of deer are closely related to the true reindeer. (A.W.L.)

REINDEER MOSS. Lichens.

REINFORCED CONCRETE. Briefly described, reinforced concrete is plain concrete in which steel rods or bars are incorporated in such a manner as to reinforce or strengthen the more or less naturally brittle plain concrete. The use of reinforced concrete is of quite recent date, usually being considered as covering about the last 75 years. Its discovery is commonly ascribed to a Parisian gardener named Monier in about the year 1860. Unquestionably the major development of reinforced concrete has taken place since the year 1900 and the United States leads the world in the use of this structural material.

Plain concrete will carry relatively heavy compressive stresses, but any attempt to impose tensile stresses of appreciable magnitude will result in rupture and consequent failure. For this reason plain concrete cannot be used for structural members subjected to a bending action or to a direct tensile action. However, if steel bars are incorporated in such a manner as to carry such tensile action as may develop in the member, then a safe and satisfactory "reinforced concrete" member may be used where otherwise a plain concrete member of the same size, or even larger size, would be wholly inadequate and unsafe.

There are two physical phenomena which are primarily responsible for the successful use of steel and concrete jointly to form structural units or members: 1. The **coefficients of expansion** and contraction for concrete and steel are very nearly the same, thus preventing undesirable, or even disastrous, internal stresses due to differential expansion or contraction. 2. When the concrete hardens it grips the steel bars very tightly and securely and thus permits stress action between the two materials, the result of this joint action being exhibited as the strength of the structural member to resist the imposed load. Usually the steel bars are roughened to give the so-called "corrugated bar" which further assists in strengthening the adhesion between the concrete and steel by affording a sort of mechanical or interlocking bond.

In some structural members, where minimum cross-sectional size is desirable or necessary, steel may be used to carry some of the compressive stress as well as the tensile stress. This condition may occur often in the case of columns and occasionally in the case of beams. In the great majority of cases, however, the steel is used to resist tensile action. Where, as in the case of a continuous girder, the tensile stress alternates between top and bottom of the member, it is customary to bend the steel accordingly, into a zig-zag shape.

The amount of steel required for adequate reinforcement is commonly quite small, varying from 1% more or less for beams and slabs to as much as 6% in some cases for columns. The percentage is usually based on the area in a right cross-section of the member. The bars vary by eighths of an inch from $\frac{1}{4}$–2″, and may be either round or square. The steel may be either mild, medium, or hard, the medium grade being more commonly used with, however, an increasing preference in recent years for the use of the harder grades.

A few of the more important characteristics of reinforced concrete which have made it increasingly popular with architects and engineers as a structural material are as follows: durability in resisting the disintegrating action of the elements, the ease and economy with which it may be cast to any desired form or shape, the fireproof character of the construction, its massiveness and consequent freedom from vibration, and the monolithic character of construction permitted (E.W.S.)

REITBOK. Reedbuck.

REJUVENATION. The restoration of youth, a process which is in the strict sense impossible. Some animals, however, gain increased vigor through special processes. This change is most evident in the one-celled animals as a result of **conjugation,** but the renewal of vigor characterizes the line to which they give rise after conjugation rather than the individuals themselves, and is comparable to the maintenance of diversity by sexual **reproduction** in higher forms.

Rejuvenation in complex animals, particularly man, is a partial restoration of waning functions through various means of stimulating the organs concerned. Since the **endocrine glands** are important in the maintenance of normal functions, the grafting of gland tissue from other animals into the human body has been undertaken to a limited extent. The gonads especially are concerned in such measures. The endocrine functions of these organs have also been found to respond to other operative procedures.

The term rejuvenation is used by geologists (geomorphologists) to describe the increased or renewed erosive power of a river, or river system, by the uplift or tilting of the land. (See **Entrenched Meander.**) (A.W.L., R.M.F.)

RELAPSING FEVER. A group of acute infectious diseases caused by **spirochetes** which are transmitted to man by several species of ticks and lice. Relapsing fever has been known all over the world but the chief centers of spread are Russia, Poland, and the Balkan states. An African form is also known and is called African tick fever. Epidemics of relapsing fever and typhus are often associated, and occur in periods of depression following war when overcrowding, famine, and poor hygienic conditions are prevalent.

The incubation period is usually 7–10 days. The disease is characterized by paroxysms of acute fever lasting several days with intervals between the attacks when the temperature is normal and the patient is apparently well. The diagnosis is easily established by finding the organism in specimens of the patient's blood during a paroxysm of fever.

The disease has been treated with arsenicals such as neoarsphenamine and mapharsen, given during a febrile episode. More recently **penicillin** has been given with strikingly good results. (R.S.M., D.M.H.)

RELATION BETWEEN ROOTS AND COEFFICIENTS OF A POLYNOMIAL EQUATION. Polynomial Equations.

RELATIVE FREQUENCY. If the frequency in a class is f_i, and the total frequency in the distribution is Σf_i, then the relative frequency in the ith class is defined as $\frac{f_i}{\Sigma f_i}$. Similarly the relative frequency of an event is defined as $\frac{x}{s}$ where x is the number of successes and s is the number of trials. (L.A.A.)

RELATIVE GROWTH. Growth of a part in relation to the whole, or in relation to the growth of another part. As living organisms grow, certain parts grow at the same rate as that of the whole organism, some faster, some slower. Growth at the same rate is isogony, at different rates heterogony. The giant chela of the fiddler crab, for instance, is produced by heterogony; the chela grows faster than the crab. In plant seedlings the roots, at first, grow faster than the stem; then the situation is reversed and the stems grow faster than the roots. Some fruits, such as the cucumber, grow faster in length than in width. The final form of organisms and of organs depends upon relative growth.

For the evaluation of data on relative growth a special method has been devised. Growth curves may be obtained by plotting cumulative size against time. This results in a sigmoid curve; growth starts slowly, speeds up, then slows down and ceases. Relative growth is studied by plotting pairs of measurements taken simultaneously against each other on a double logarithmic grid, or the logarithm of one measurement against the logarithm of another on ordinary plotting paper. Time is thus eliminated from the study and the curves turn out to be straight lines for each phase of growth. If the relative growth is isogonic the line forms an angle of 45° with the x-axis, the tangent of which is unity. The equation for the curve is $y = bx^k$, where y is the size of a part at any given moment, x the size of the whole organism (or of another part) at the same moment, b the intercept of the curve on the y-axis on a double logarithmic grid, and k is a constant for the relative growth rate. The constant k is the tangent of the angle which the curve makes with the x-axis, so it is obvious that if $k = 1$ the relative growth rates are isogonic, if greater or less than 1, heterogonic, and that k is a measure of the relative growth rate.

The linear nature of the equation $y = bx^k$ is best revealed by writing it $\log x = \log b + k \log x$. It is used in this form when the logarithms of the data are plotted. The validity of the hypothesis has been well tested. Growth is here considered as a multiplicative process, the growth at any instant being proportional to the size of the organism at that instant. The curve represents a geometric series and, with large amounts of data, grouping is by geometric classes. Time is eliminated because no new data can be plotted until growth has actually occurred. Actual rates of growth are not studied—only relative rate, which permits time to cancel. The chief difficulty with the method lies in the determination of the value of k. It may be roughly approximated by drawing a line on the plotting paper to represent the grouping of the dots and then determining the slope of the line by the usual methods, but if large amounts of data are available and the constant is to be determined with some accuracy it is necessary to resort to rather elaborate equations based on the method of least squares. (P.A.W.)

RELATIVE HUMIDITY. Humidity; Atmosphere; Hygrometers.

RELATIVE VARIABILITY. Relative variability is the ratio of some measure of **dispersion** to some measure of **central tendency**. The only one in use is the **coefficient of variation.** (L.A.A.)

RELATIVE WIND. Velocity of air with reference to a body in it. It is usually determined by measurements made at such a distance from the body that the disturbing effect of the body upon the air is negligible. (F.T.M.)

RELATIVITY. It is impossible, in a brief article, to indicate more than the general trend of the far-reaching domain of scientific thought known as the theory of relativity. This revolutionary philosophy, due largely to the work of Albert Einstein, had its origin in the ambiguities connected with attempts to detect absolute motion, or rather, motion with respect to a supposedly universal and omnipresent medium, the **ether.** The **Michelson-Morley experiment** on the one hand and the **aberration of light** on the other, together with the results of numerous less familiar experiments to similar purpose, had given conflicting evidence. The outstanding conclusion which emerged from them was that, while the motion of bodies relative to one another can be readily observed and has significant results in the physical world, the absolute motion of bodies, or motion with reference to a stationary ether, is not capable of detection and therefore may as well be omitted from physical reasoning.

Proceeding on this basis, Einstein set for himself the problem of formulating the mathematical laws of physics in such a way that they should not depend either upon the positions of the points and lines chosen as a reference system (such as origin and coordinate axes), or upon whether that reference system is supposed to be moving or not. To do this it was necessary, as the reasoning became more general, to give up the cherished Newtonian concepts of absolute length, absolute mass, and absolute time. For the Michelson-Morley experiment, explainable only on some such basis as the **Lorentz-Fitzgerald contraction** hypothesis, had shown that length varies, in a manner not directly observable, with motion in space; so that the absolute or "static" length of a body cannot be determined. Lorentz had concluded, incidentally, that the masses of electric particles, and hence presumably of whole bodies of matter, vary with motion. In the case of high-speed electrons in a vacuum tube, this variation can be detected by their failure to follow the paths prescribed by the Newtonian laws; but the actual, absolute or "rest" mass of a body must be forever unknown because we do not know its absolute motion. Time, also, loses its independence. If we watch things happening on two stars (such as variations in brightness), events which appeal to us to be simultaneous may have actually occurred centuries apart, because of the difference in distance; and the same monochromatic

light from the two stars may reach us with measurably different frequency because of the **Doppler effect.** Since all the magnitudes of physics, such as density, speed, momentum, etc., are analyzable in terms of length, mass, and time, it is clear that all magnitudes are dependent in more or less complex ways upon the forever unknowable absolute motion of the bodies concerned. The outcome of all this was that, in order to avoid incorporating this unknown absolute motion in physical equations as a result of its inseparable connection with the physical magnitudes involved, the equations themselves had to be recast in such a way as to leave out that factor and permit us to forget about absolute motion altogether.

Among the important by-products of this general relativity analysis were the endowment of **energy** with the characteristics of **mass,** the interrelation between **gravitation** and **inertia** (see **Einstein Equivalence Principle**), and the recognition of various fields of force (electric, magnetic, gravitational) as being different aspects of a common phenomenon—the "unified field theory."

In order to facilitate the difficult thinking required for these generalizations, Minkowski devised the plan of describing physical processes in terms of a symbolic space having four coordinates, three being the usual linear coordinates (x, y, z) and the fourth being time. A graph in this 4-dimensional "space-time" or "Minkowski world," called a "world line," represents the continuous history of the particle to which it pertains, any point on the graph corresponding to its position at some particular time. By equations of transformation analogous to those arising upon change of axes in ordinary geometry, Lorentz showed how the description of the performance of a particle can be adapted to the viewpoints of different observers having a uniform motion relative to each other. (L.D.W.)

RELAXATION OSCILLATOR. The relaxation oscillator is an electron tube **oscillator** whose frequency is determined by a condenser-resistance or inductance-resistance circuit rather than the more conventional capacitance-inductance resonant circuits. These oscillators have very distorted wave shapes, giving various outputs such as square waves, trapezoidal waves, triangular waves and pulses of very short duration. These distorted waves have made them ideal for many control and triggering purposes in more elaborate electronic circuits. The relaxation oscillator may be easily synchronized with another oscillator or other source of voltage by injecting some of the other **voltage** into the

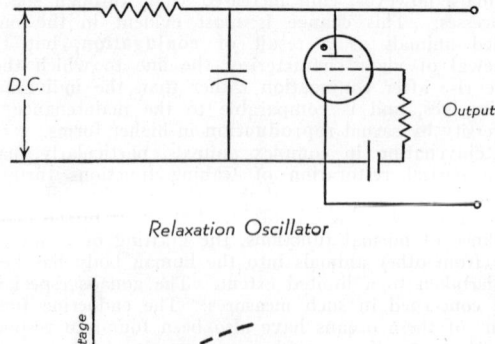

Relaxation Oscillator

Output

relaxation circuit at the proper point. A simple, but widely used, relaxation oscillator is shown in the figure. Here the condenser is charged through the resistance until the voltage across the thryratron reaches the breakdown value when it breaks down and discharges the condenser very rapidly. The process is then repeated and the resultant voltage output is as shown. This oscillator is used for the sweep oscillator of many cathode-ray oscilloscopes (see **Cathode Ray Tube**). The multivibrator is another relaxation oscillator and is widely used for frequency measurements since its output is rich in **harmonics** (see **Frequency Meter**). This oscillator is merely a two-stage resistance coupled **amplifier** which has its output connected back to its input. (L.R.Q.)

RELAY. The electrical relay is a device which utilizes the variation of **current** in an **electric circuit** as a controlling factor in another. For example, a certain change of current in one circuit may cause current to begin to flow in another, by the operation of a relay connected between them. There are numerous types of electrical relays, as they have been widely used in industry, particularly in apparatus of an automatic or semi-automatic nature, or for the protection of electric power equipment, or for communication systems. Protective relays are highly specialized and developed to where they will detect any electrical abnormality, and open the circuit containing that abnormality in any required time interval. Suitable relays will detect over-current, under-current, over-voltage, under-voltage, overload, reverse current, reverse power, abnomal frequency, high temperature, grounds, and phase unbalance.

Usually the relay involves two circuits, the energizing circuit and the relay circuit (the latter variously called the trip circuit, the sounder circuit, etc.). Protective relays may close the trip circuit immediately, or after a definite time interval, or after an inverse time interval. If the trip circuit contacts are normally open, the relay is called circuit closing; if they are normally closed, the relay is called circuit opening.

The automatic protection of electric power circuits is necessary for safety and economy. **Fuses** and automatic **circuit breakers** are the devices most used for opening the circuit. A relay must be used to operate the tripping circuit of the circuit breaker. Fig. 1

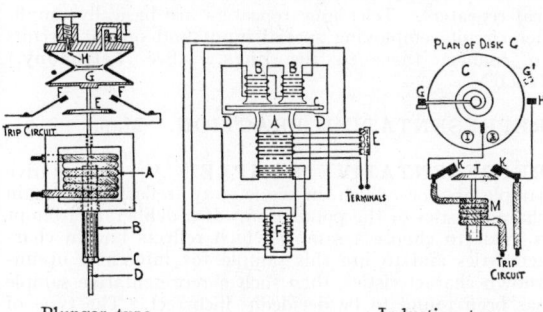

Plunger type.   Induction type.
Fig. 1. Relay.

illustrates the principle of two types of protective relays. They are, respectively, the solenoid and induction relays. Simple inverse-time limit overload relays of the circuit-closing type are shown in each case. Explaining the plunger type, *A* is the operating coil which may be connected in series with the line or through the medium of tripping transformers. *B* is the iron plunger which will be pulled upward against gravity by the "sucking" action of the coil when a predetermined current is reached. The setting for current at which the contacts are closed may be adjusted by moving the plunger up and down on its stem *D* by the adjusting nut *C*. When the plunger starts upwards the pulling force on it increases due to the increased portion within the coil *A*. Thus no floating in an intermediate position is possible. However, the air

compressed by bellows *G* delays the closing until enough can leak out through port *H* to permit the disk *E* to move up to where it will close contacts *FF* of the trip circuit. By adjustment of the leak port needle, the amount of time delay may be changed. *I* is a quick reset valve.

The induction relay receives its current through the connection block *E* by means of which the number of effective turns in the primary winding may be varied. The secondary winding on *A* is connected to the upper pole pieces *BB* through a torque compensator *F*. The magnetic circuit of *F* becomes saturated at high overloads, the strength of the poles *BB* does not increase further and the relay then has a definite-time limit feature. At less severe overloads the relay is inverse-time limit. The interaction of magnetic fields produces a turning torque in the disk *C* which is resisted by a spiral spring. A plan view of the disk is shown in connection with the secondary contacts. The holes *I* are beneath the poles. As they move out under the influence of an overload the torque increases and the contact *G* is swung around to *H*. The holes prevent floating of *G* at some position *G'*, an effect which would destroy the inverse-time feature. The contacts are delicate and often the heavy trip current is passed through secondary contacts, the primary contacts carrying only enough current to energize coil *M*, draw up plunger *J*, and close secondary contacts *KK*.

The telegraph relay has probably served man longer than any other type, and there are few persons indeed who have not listened to the busy clicking in the local telegraph office. The audible signal is made by a sounder which is actuated by a polarized telegraph relay. (See Fig. 2.) This relay consists of soft iron cores on which windings connected to the main line are placed. These cores become magnets under the action of the windings. The relay is polarized by a magnet of the horseshoe type, mounted as shown in the diagram. An armature operates in an air gap between the cores. With no current flowing in the windings, the armature is attracted equally by the electromagnet and the armature tends to remain in a mid position. When the current flows through the windings in a certain direction it will strengthen one electromagnet and weaken the other, so that the armature will be attracted in one direction, carrying with it the contact point, which will close the sounder circuit. Relays have an important place in all wire communication systems.

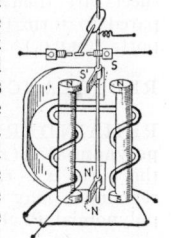

Fig. 2.

The telephone relay is probably the most widely used relay since in its various forms it occurs hundreds of times in the usual telephone system. While there are various specialized types of relays used in telephone circuits, the one commonly referred to as a "telephone" relay has an energizing coil surrounding a soft iron core. This core extends slightly beyond the end of the coil and when the coil is energized attracts a soft iron

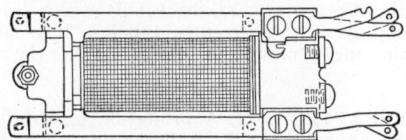

Fig. 3. Telephone relay.

armature, which in turn pushes the contacts open or closed. In many of these relays there are several contacts, some of which may open and others close when the relay is energized. Fig. 3 shows one form of this relay. (F.T.M., L.R.Q.)

RELIABLE SAMPLE. A **sample** is said to be reliable if the **statistics** calculated from it are reliable, that is, such statistics have **confidence limits** which are sufficiently narrow for the purpose at hand. (L.A.A.)

RELIABILITY. The reliability of a **statistic** is determined by tests of significance. If these tests of significance show certain hypotheses are denied, then the statistic is said to be reliable. At the present time more stress is placed on **confidence limits** rather than on computing reliability by use of a standard error of a statistic.

In general the notion of reliability refers to the idea that repeated samples or repeated tests will give essentially the same results. (L.A.A.)

RELIEF. Used by geographers and **physiographers** to define the difference in altitude within a specified or limited region. (R.M.F.)

RELIEF PROCESSES. Methods of color photography classified as relief processes are those in which, instead of an image composed of black silver in gelatin, there is a layer of gelatin of variable thickness. Such a film thus has a variable thickness (gelatin) instead of a variable density (silver). Relief images may be formed from silver images by means of tanning development, by the hardening action of light on silver in a dichromated colloid, or by chemical etching. The relief images may be dyed with appropriate colorants which may then be bound together to form a transparency or they may be transferred to suitable gelatin surfaces by **imbibition.** Films or tissues of gelatin are also obtainable containing pigments of the proper colors. These sheets are then superimposed after the reliefs are prepared to form transparencies or prints on a paper support. (H.C.C.)

RELUCTANCE. Magnetic Circuit.

REMAINDER THEOREM OF ALGEBRA. If a **polynomial** $P(x)$ is divided by a binomial divisor of the form $x - r$, the remainder is $P(r)$; that is, the remainder may be found by substituting $x = r$ in the polynomial $P(x)$. (L.L.S.)

REMORA. Pisces, Teleostei. Marine fishes (**Pisces**) whose anterior dorsal fin is modified to form an oval sucker on the top of the head. This sucker is used to attach the animal to boats, turtles, or other large objects by which the fish may be carried about without effort. From their frequent attachment to sharks they are also called shark suckers, and the name sucking fish is sometimes used. (A.W.L.)

RENAL CORPUSCLE. A structural unit of the vertebrate **kidney.** A renal corpuscle consists of a knot of blood vessels enveloped by a thin-walled expansion of the excretory tubule known as Bowman's capsule. The knot of blood vessels, called a glomerulus, does not lie in the cavity of the tubule but merely bulges into the cavity of the capsule, covered by its thin wall. The portion of the tubule which leads out of the capsule is the secretory tubule. It is also involved in the removal of wastes from the blood. (A.W.L.)

RENAL PAPILLA. 1. A projection extending from the body of **cephalopod** mollusks into the **mantle** cavity. It bears the opening of the excretory duct. 2. The summit of a renal pyramid in the mammalian **kidney.** (A.W.L.)

RENOPERICARDIAL CANAL. A passage connecting the **pericardial cavity** of **mollusks** with the **kidney.** (A.W.L.)

REPEATED LOAD. Load.

REPEATED INTEGRALS. Double Integrals and Triple Integrals.

REPEATER. In communicating over long lines the electrical pulses gradually become attenuated and often distorted so they must be built back to sufficient amplitude and correct shape to satisfactorily operate the receiving equipment. The restoration of these attenuated signals is the function of repeaters. In telegraph circuits various mechanical repeaters are often used. One of the simplest is shown in the diagram. In this simple form

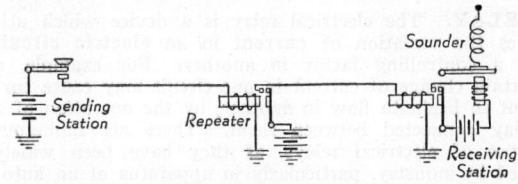

Simple Telegraph Station

communication is possible in one direction only, for the receiving station cannot break the circuit, as it would have to do in order to insert a key. A repeater station, more complicated than that diagrammed, can be installed to provide two-way telegraphy. Two repeaters having holding **coils,** and extra **batteries** are required in this system. These, however, do not restore the original spacing and shape of the pulses and the resulting signals may still give false operation of the receiving devices. In order to overcome this fault regenerative repeaters are used on many telegraph circuits, particularly on **teletype** or printing circuits. These are much more complicated than the simple form shown. They are designed to transmit a restored signal of correct amplitude and shape under the control of only the center portion of each pulse received at the repeater. In this way the outgoing pulse is correct if the center of the incoming one is correct, regardless of what has happened to the incoming pulse outside of this narrow center region. Vacuum-tube repeaters are also used on telegraph circuits and are always used on telephone circuits where the complexity of the signals prevents the use of mechanical repeaters. Telephone repeaters are basically amplifier circuits employing special input and output circuits to connect them to the **lines.** (See **Telephony.**) (L.R.Q.)

REPRESENTATIVE FRACTION. Map.

REPRESENTATIVE SAMPLE. A representative sample is one which in some way reflects the main characteristics of the population. If a deliberate attempt is made to choose a sample which reflects known characteristics and to use this sample for inferences of unknown characteristics, then such a representative sample has been found to be decidedly incorrect. The type of representative sample which is recommended is the **stratified sample.** (L.A.A.)

REPRODUCTION. The production of units of living matter by other usually similar units already in existence. In the ordinary use of the word these units are individual organisms but the power of reproduction is a fundamental property of living matter through which units subordinate to the individual also arise.

Reproduction is at its simplest in unicellular animals (**Protozoa**) such as Amoeba. These animals are single **cells,** composed of a naked mass of **protoplasm** without constant form. In the process of binary fission they divide into two parts, each an independent cell like the parent. Protozoans with constant body form also reproduce by binary fission but in them each half must undergo a reorganization for the development of the complete normal form of its kind. Some protozoans subdivide into more than two parts by multiple fission.

Fission in multicellular animals is limited to simpler forms, such as a few **coelenterates** and flatworms (**Platyhelminthes**). Here it involves a change in the relations of the cells of the body, accompanied by subdivision either longitudinally or transversely and by reorganization within the resulting halves.

In true fission the individuality of the parent merges with that of the resulting offspring.

Multicellular animals also reproduce by budding. In this process the individuality of the parent is retained and its offspring develop gradually from a limited portion of its body. The growth of a coral polyp from the extended margin of the parent is a good example. The process also takes place in sponges, other coelenterates, and some of the worms.

All of these processes are asexual, in contrast with the widespread process of sexual reproduction. In sexual reproduction two reproductive cells (**gametes**, germ cells) normally unite to form a single cell, the zygote, which develops into a new individual. In isogamous reproduction the two cells are similar. More frequently they are distinctly different, a condition known as heterogamy. In the latter case the differences of the cells are the foundation of **sex** in the individuals producing them. One form, the ovum or egg cell, contains a large amount of cytoplasm and often some stored food or yolk. It is produced by the female. The germ cells of the male, known as sperms or spermatozoa, are microscopic in size and are motile, usually through the action of a slender tail resembling a flagellum.

The role of the parents in reproduction may be no more than the production and discharge of the germ cells, involving some special behavior in **mating** and sometimes a close association in **conjugation** or **copulation**. In other cases, notably in the mammals, the young develop in the body of the mother. If the species produces eggs it is said to be oviparous. If the eggs hatch within the body of the parent, it is ovoviviparous. And if the young undergo a long development, nourished by close connection with the body of the mother, it is viviparous.

Specialization of the body of the mother for the nourishment of the young in viviparous species is essential to reproduction, and in the mammals the secretion of milk for their nourishment after birth is equally necessary for their development to an independent stage.

Incubation of the eggs by birds and extensive postnatal care by birds, social insects, and a few other forms are also essential corollaries of reproduction, although not strictly a part of the reproductive process.

Sexual reproduction has the advantage of combining hereditary qualities of different individuals. Through this means it maintains maximum diversity within the species, an end of great value in meeting varied environmental conditions.

A fundamental modification of the sexual process is **parthenogenesis.** (A.W.L.)

REPRODUCTIVE SYSTEM.

REPRODUCTIVE SYSTEM. The assemblage of organs which give rise to the reproductive cells and carry out all accessory functions directly involved in the production of young.

The essential organs of reproduction are the **gonads,** in which the reproductive cells arise, together with the ducts, if any are present, through which these cells are discharged from the body. In the female the gonads are known as ovaries and their ducts as oviducts. The gonads of the male are testes and their ducts are sperm ducts or vasa deferentia.

Regions of the ducts are specialized for various functions in relation to the **germ cells.** In the female they give rise to yolk glands which produce the food

to be stored in the egg, and glands which secrete enclosing membranes or shells, as well as glands which secrete fluids to cover the outside of the egg for various purposes. Other regions become reservoirs (**receptacula seminis**) for the storage of the seminal fluid of the male after **copulation,** and the terminal part of the female genital passages becomes a **vagina** for the reception of the male intromittent organ during that act. The ducts of the male produce glands whose secretions constitute a liquid medium in which the spermatozoa swim.

In many invertebrates the reproductive system is independent of other organic systems, aside from the usual nervous and circulatory connections. In others it is connected with the digestive tract. Vertebrates, on the contrary, have the organs of reproduction usually associated with the **excretory system,** and the common ducts primitively discharge into the **cloaca,** which is also part of the digestive tract. The gonads in this class arise from mesoderm in close association with the mesonephroi, although the germ cells themselves may have another origin. Some of the mesonephric tubules become the slender vasa efferentia of the testes, leading to the principal ducts, the vasa deferentia, which are derived from the mesonephric ducts. Thus in some fishes (**Pisces**) and in **amphibians** the male reproductive and excretory systems have common ducts, and above these classes the loss of the mesonephroi leaves their ducts to the reproductive system alone. In the female a separate pair of oviducts arise, beginning near the ovaries in the funnel-shaped ostia abdominales, which open to the coelom, and extending to the cloaca.

In mammals the union of the posterior parts of the oviducts forms a single terminal passage, the vagina. Anterior to the vagina portions of the ducts are specialized as the uterus, which may remain a forked or paired structure or may, as in man, become a single chamber. In the last type the more slender parts of the ducts leading toward the ovaries persist as the Fallopian tubes. The vagina opens independently to the exterior, but is close to the urethra in an external furrow called the vestibule.

In the male the ventral part of the cloaca separates as a urogenital sinus which forms, in part, a urethra common to the excretory and reproductive systems. Flanking the external opening of this structure in the embryo folds of tissue arise which join to extend the duct and to form a projecting erectile organ, the penis. In more primitive vertebrates such an **intromittent organ** may arise from the wall of the cloaca. As a result of this close association of the two systems, the term urogenital system is often used in vertebrate anatomy.

Various accessory organs of reproduction, not an integral part of the reproductive system, appear in the animal kingdom. Among them the **mammary glands** of mammals are noteworthy. They are developed in two rows along the ventral surface of the trunk or at certain points along these rows. These glands secrete milk for the nourishment of the young and are subject to the intricate regulation of **hormones.** (A.W.L.)

REPTILIA.

REPTILIA. The reptiles, including the **lizards** and **snakes, crocodiles, alligators** and related forms, and **turtles** and **tortoises.** A class of the phylum **Chordata.**

The class is characterized as follows: 1. The **skull** articulates with the spinal column by a single process. 2. The **mandibles** are made up of several bones, joined with the skull by the quadrate bone. 3. The skin is covered with scales. 4. The **heart** is 4-chambered but the separation of the ventricles is incomplete. 5. They are **poikilothermal.** 6. They have extraembryonic membranes during development, a fundamental requirement of terrestrial life in the vertebrates.

Although the group is the lowest class of vertebrates

to attain the capacity for entirely terrestrial life, many reptiles are now amphibious or aquatic. Even the aquatic species, however, come to the land to deposit their eggs.

Reptiles are economically important to a rather limited extent. The flesh of some turtles, lizards, and even snakes is used as food, and the skin of the alligator makes an excellent, if conspicuous, leather. In some regions crocodiles have been known to kill human beings. Usually it seems that the danger from them is very limited. Poisonous snakes also may destroy human life, but here again the danger is usually encountered only under special conditions and snake bites may be regarded as accidental.

The classification of the reptiles is briefly as follows:

Order Prosauria. A single living species, the tuatara of New Zealand, makes up this order. It is a lizardlike animal with a few primitive structural characteristics, including a well-developed **pineal eye.** The order also bears the name Rhynchocephalia.

Order Chelonia (Testudinata). The **turtles, tortoises,** and related species. Characterized by the shell, consisting of an upper **carapace** and a lower plastron, formed of bony plates with a horny sheath.

Order Crocodilia. Large reptiles resembling lizards in form. The jaws are elongated. The thick skin is provided with bony plates. The **crocodiles, alligators, garial,** and related species.

Order Sauria. The lizards and snakes. In some classifications these forms are included in separate orders, and in some the lizards constitute the suborder Sauria and the snakes the suborder Serpentes of the order Squamata. The order including the lizards is also named Lacertilia. All of these animals have elongate bodies. The skin bears horny scales and in some cases bony plates. (A.W.L.)

RESIDUAL. A residual is defined as $x_1 - x_1'$ where x_1 is the actual value of the **variate** and x_1' is the value found from a **regression** equation. The standard error of the residuals is the **standard error of estimate.** In the method of **least squares** the regression equation $x_1' = f(x_2, x_3, x_4)$ is found by making the sum of the squared residuals a minimum. (L.A.A.)

RESIDUAL RADIATION. When light or other radiation falls on the surface of a transparent body, part of it is reflected, part is transmitted, and part is absorbed. The **spectrum** of the transmitted portion, when compared with that of the original radiation, may reveal that certain sharply defined wavelength ranges have failed to get through, but does not indicate whether they have been absorbed or reflected.

Transparent media sometimes reflect very copiously those wavelengths whose absence from the transmitted radiation causes conspicuous absorption bands. Quartz, for example, reflects (or absorbs and re-radiates) **infrared radiation** of 8.5 microns and also that of 20 microns wavelength almost as well as a polished metal. Rock salt does the same at 50 microns. Rubens and Nichols devised an ingenious method of isolating beams of these infrared rays by reflection at polished surfaces of quartz or rock salt, the residue, after several reflections, called by them *Reststrahlen* (residual rays), being almost monochromatic. Analogous properties are exhibited in the visible spectrum by many aniline dyes, the selectively reflected light appearing as a "surface color" complementary to the transmitted portion. (L.D.W.)

RESILIENCE. The resilience of a body measures the extent to which energy may be stored in it by elastic deformation. The implication of the word "stored" in the above definition is that this energy may be released in the form of mechanical work when the force causing the elastic deformation is removed, and that resilience is a property of a material within its proportional limit.

The "modulus of resilience" is the maximum energy storage in a unit volume of the material. In practical units it is the inch pounds of energy stored in a cu. in. of the material stressed to the proportional limit (elastic limit). The modulus of resilience is directly proportional to the square of the stress, and inversely proportional to the modulus of elasticity. (F.T.M.)

RESINS. Natural resins (for synthetic resins, see **Plastics**) are complex compounds composed of **carbon, hydrogen** and relatively small amounts of **oxygen,** which are secreted in various tissues of many plants. In the **pine family,** where resins are very common, they are secreted as oleoresins in resin canal cells, which break down finally, producing resin canals. These canals appear as longitudinal ducts in the sapwood and inner bark, connected laterally by resin canals in the compound wood rays, thus forming an extensive network. A common name given to the oleoresin in this group is pitch, the sticky juice which exudes from the plant wherever it is wounded. On exposure to the air the **volatile oil** in this pitch (oil of turpentine) gradually evaporates, leaving a clear hard glassy substance, the resin, which forms a protective coating over the wound.

Most resins have the same physical properties, being clear, translucent, and of a yellow or brownish color. **Amber,** a fossil resin, is a more or less familiar example. Resins are insoluble in water, but soluble in common organic solvents such as ether and alcohol. All resins burn with a sooty flame. Resins seem to be mainly of value to the plant in that they form protective coverings against the entrance of disease-producing organisms and also prevent excessive loss of water from the thin-walled tissues exposed in the wound.

Resins are separated into several classes. Many of the resins contain almost no volatile oil and are hard, without taste or odor. These are the varnish or hard resins. Other resins, when removed from the plant in which they are formed, and dissolved in volatile oils, form a thick semi-solid mass: these are the oleoresins. In still other cases the resin occurs in combination with a **gum,** forming a gum resin.

Hard Resins. Several of the hard resins, used mainly for making varnishes, are called copals. Most of them come from Africa and are either found in fossil form or obtained from living plants. Other copals come from Australia, New Zealand and East Indian islands. The plants which form them are members of the legume and pine families. The African copals are products of several species of *Trachylobium,* fairly large trees growing in east Africa and Madagascar. The best resin from these trees occurs in a fossil form, often deeply buried in the ground—sometimes in regions where the trees no longer grow. These resins dissolve slowly and are used in making varnishes which are very durable. A South American tree of large size, *Hymenaea Courbaril,* also of the Leguminoseae, yields a very similar resin, which is also found in lumps in the ground around the trees, and used in varnishes.

Another copal is obtained from *Agathis australis,* a very large coniferous tree native in Australia and New Zealand, where it is known as the Kauri pine. Like the other copals, that from the Kauri pine is found in lumps buried in the ground. Most of these lumps are 1 or 2 in. in diameter, but some are much larger, weighing up to 100 lbs. Nearly all of this resin comes from the northern part of North Island of New Zealand. It is frequently called Kauri gum, though it is not a gum, but a true copal resin. Another group of hard resins, known as dammar resins, is obtained from many different trees growing in southern Asia and the East Indian Islands. These resins dissolve readily in alcohol, forming spirit-varnishes.

One of the commonest and most important of the hard resins is rosin, obtained by distilling the pitch, or turpentine, which is a product of several of the native pines of the southeastern United States. This rosin, also known as colophony, is a very important product of that region. Originally the turpentine was obtained by chopping a deep hollow in the base of the trunk of the tree and allowing it to fill up with the turpentine, which was then scooped out. This method was very destructive and wasteful, since much of the oleoresin, turpentine, was lost during the process. The weakened trees were easily blown down.

Now turpentine is obtained by cutting V-shaped gouges in the bark and inserting metal gutters beneath the gouges. These gutters carry the turpentine to a cup placed underneath. As soon as the cut is made, turpentine begins to flow and continues to do so for two or three days, gradually slowing as the drying turpentine allows resin to accumulate and plug the wounds. A new flow is obtained by cutting off a narrow strip of bark from the upper edge of the cut. The process is continued as long as the pitch will flow, which is usually all summer and well along into late fall. Each tree may be turpentined for 6 or 7 successive years or even longer before it ceases to be profitable.

The crude turpentine collected in the cups and the product which has dried on the wound of the tree are removed and carried to the still. Here the turpentine, to which a little water is added, is carefully heated to drive off the oil of turpentine present, together with the water added. The distillate is condensed by passing it through a coil around which cold water is flowing, and collected in a barrel or any suitable container. The two substances, water and oil of turpentine, which make up the distillate, are immiscible and soon separate, the lighter oil of turpentine rising to the top and floating on the water, which is drawn off from the bottom. Oil of turpentine is often called spirits of turpentine, or, in the paint trade, turpentine. In medicine the word turpentine is reserved for the oleoresin which upon distillation yields oil of turpentine and rosin.

The residue remaining in the tank at the end of the distillation is skimmed to remove any impurities such as twigs, bits of bark and dirt, and run into vats to cool. Then it is put into barrels and allowed to harden, forming rosin.

Oil of turpentine is used principally as a solvent for paints and varnishes, because it mixes readily with the various substances used and also because it evaporates quickly, causing the paint or varnish to dry. It is also used in making such things as sealing wax and shoe polish. Very pure grades of turpentine (the oleoresin) are used medicinally.

Large quantities of rosin are used in sizing paper, which makes it take ink without spreading or blotting, gives it a smoother surface and makes it heavier. Rosin is also used in cheaper varnishes, in paints, and in soap making. It is furthermore used as an adulterant of the more expensive resins. Linoleum manufacturers use large amounts of rosin.

In early times large quantities of crude turpentine were used to waterproof the rigging of the sailing vessels and to calk the seams of the hull.

Mastic is a hard resin exuding from the branches of one of the Pistachio trees, *Pistacia Lentiscus*, native of Mediterranean Europe and Southern Asia. Formerly it was extensively used medicinally, for stomach troubles and dysentery, as well as other ailments. Now it is used in making varnishes and in lithographic work. Natives of the region in which it is found chew mastic, which has a pleasant taste.

Since turpentine is a mixture of a volatile substance, spirits of turpentine and a hard resin, it is one of the oleoresins.

Oleoresins. Canada balsam is one of the oleoresins. It is obtained from the bark of *Abies balsamea,* the common balsam fir of northern North America. Canada balsam, because its refractive index is so near that of glass, is much used in optical work and in preparing materials for examination with a microscope.

Little used today is Dragon's blood, an oleoresin obtained from the fruits of *Calamus Draco,* a native palm of southeastern Asia and the Molucca Islands. The resin exudes from the surface of the ripening fruits. It is removed from them by boiling in water. The resin is then moulded into balls or long sticks. It is sometimes used in making varnishes and lacquers.

True lacquer, obtained from the juice of *Rhus vernicifera,* a sumac tree of southeastern Asia, is another oleoresin. To obtain the juice lateral cuts are made in the bark. The exuded sap is collected not only from these cuts but from small branches which are cut off and soaked in water. The juice is cleaned of any foreign substances by straining it through hemp cloth. By slow heating, either artificial or by the sun, the juice is evaporated and stored until used. Lacquer is a poisonous substance, causing intense irritation of the skin in many people. Others seem to be immune. Lacquer is usually applied over some soft wood, commonly soft pine, the pores of which have first been filled by rubbing in a paste of rice and resin, followed by a paste of soft clay and resin. The surface is then covered with cloth and layer after layer of lacquer put over that. Each layer is allowed to dry and rubbed down very smooth before the next layer is added. Any color which is to be added is mixed with the lacquer, with each colored layer covered by a clear layer before another is put on. The final product is a thick covering composed of many thin layers of lacquers. If this is carved the edges of the carving, on careful examination, will show the fine lines separating the different layers. Lacquering is a very old industry, having been carried on in China since the 6th century. Lacquer work is made in many other oriental countries, and the juice of many other trees used as a source for the lacquer used.

Certain resins occur in combination with fragrant volatile oils. One of these is benzoin, obtained from *Styrax Benzoin* by cutting notches in the bark and allowing the resin to collect in them. It is used in making perfumes, in incense, and as a source of benzoic acid, used medicinally.

Another fragrant oleoresin is storax, obtained from *Liquidambar orientalis,* a medium-sized tree growing in southwestern Asia. The resin is obtained by boiling the bark and wood of young branches. It is used medicinally and also in incense.

Gum Resins. Gum resins include myrrh, which exudes from the trunk and branches of *Commiphora Myrrha,* a tree growing in the region around the Red Sea. The lumps of resin are used medicinally, and also in making incense. Another gum resin is frankincense, obtained by cutting notches in the stem of *Boswellia Carterii,* which grows in northeastern Africa and in Arabia. This resin is used in incense. **Asafoetida** is also a gum resin. (R.M.W.)

RESISTANCE. This word describes the opposition offered by a material body to forces which would tend to produce motion, or to the migration of electrons or ions in electric conduction. The mechanical resistance of a body may arise from friction, from stresses set up in rigid anchors or from inertia. Resistance is offered by fluids, such as air or water, to the passage of bodies through them. Aerodynamic resistance is treated elsewhere in this volume. (See **Drag, Aeronautics, Airfoil.** Fluid friction is also treated under **Fluid Friction, Viscosity.** See also **Friction, Mechanics.**)

Electrical resistance is the reciprocal of electrical conductance. Electric conductors are believed to contain free electrons, the movement of which through the substance constitutes electric conduction. In this migration the moving particles evidently meet with some restraint, since their progress is very much slower than in a vacuum tube under the same potential difference. It is natural to visualize this resistance as due to collisions with the atoms, occurring so frequently that the electricity must filter along slowly. If driven by considerable electromotive force, they produce sensible heat; but by what mechanism this comes about is not so evident, and the whole question is complicated by uncertainties as to the nature of the electron and of the atom.

The measure of the resistance of a given conductor is the electromotive force required per unit current, and is usually expressed in ohms. (See Ohm's Law.) The resistance of a wire or other linear conductor of uniform cross-section is proportional to the length l and inversely proportional to the cross-section a: $R = rl/a$. The constant r is the "resistivity" of the substance, usually expressed in ohm-centimeters; and its reciprocal is the "electric conductivity." The dependence of resistivity of metals upon temperature is one of the major problems of electron physics. (See Resistance Thermometers and Superconductivity.)

Pieces of wire may be cut off at such lengths as to have definite resistances, and mounted with convenient connections to form a "resistance box," used in many electrical measurements. A "rheostat" is usually a rugged conductor, often with adjustable resistance, used to introduce a resistance load into a circuit. Resistances are commonly measured by means of some form of bridge, of which the Wheatstone bridge is most familiar.

Some typical resistivities are given in the table below:

RESISTIVITIES OF SOME COMMON MATERIALS
(In ohm-centimeters at 20° C.)

Aluminum...	2.83×10^{-6}	Mercury......	95.78×10^{-6}
Brass.......	7.	Nickel........	7.8
Copper......	1.72	Platinum.....	10.
German silver	33.	Silver........	1.63
Iron (pure)..	10.	Tin..........	11.5
Lead........	22.0	Tungsten.....	5.51

For some purposes it is convenient to express the resistivity as the resistance of one foot of wire of the given metal having a cross-section of one circular mil (a circular mil is the area of a circle 0.001″ in diameter). This value may be obtained by multiplying the resistivity in ohm-centimeters by the factors 6.015×10^6.

The dependence of resistance of many metallic conductors upon temperature is expressed with fair approximation by the linear equation

$$R = R_0 (1 + At),$$

in which R_0 is the resistance at 0° C. and t is the centigrade temperature. The temperature coefficient A is a constant characteristic of the metal.

There remains yet another type of resistance for consideration—heat resistance. This might be said to be the property of offering opposition to the flow of heat. This property is desirable in a heat insulator such as pipe covering, but undesirable in heat transfer equipment. The term resistivity is rarely used in connection with heat flow, as its reciprocal, thermal conductivity, is quite satisfactory, and enjoys the advantage of common usage. (See Heat Transfer, Heat Insulation.) (L.D.W.)

RESISTANCE THERMOMETER. The fact that the electrical resistance of a metal wire increases with rising temperature is the basis of a very useful class of thermometers. One has only to calibrate a given length of wire, as to its resistance in relation to its temperature, enclose it in a suitable protecting tube, and keep it connected with the resistance-measuring bridge, to have a resistance thermometer adapted to a variety of uses over a very wide temperature range. The metal nearly always employed is platinum. The variation of resistivity with temperature of platinum is very

nearly linear, being closely approximated by the formula $r = 0.000000037t + 0.000011$, in ohm-centimeters and centigrade degrees. Callendar found that for any given platinum resistance thermometer there is a slight systematic departure from this formula, characteristic of the particular sample of wire. It is best, therefore, to calibrate each instrument throughout the range for which it is intended. (See Thermometry.)

Care must be taken, in mounting the platinum wire, that it does not come in contact with materials which will contaminate it at high temperatures. Compensation is also necessary for the change of resistance in the wires leading to the platinum spiral. This is commonly effected by balancing against these wires a pair of "dummy" wires similar to and laid alongside them. The instrument is usually provided with a suitably designed resistance bridge, such as the Callendar and Griffiths or the Mueller bridge; which for practical purposes should be portable and self-contained, with battery, galvanometer, balancing rheostat, etc., all in one case, and with a cable leading to the thermometer proper. The bolometer is a highly sensitive special type of resistance thermometer. (L.D.W.)

RESISTIVITY. Resistance.

RESOLVING POWER OF PHOTOGRAPHIC MATERIALS. The resolving power of a photographic material is usually expressed as the number of lines per mm. which are clearly defined as separate lines in the image. Resolving power is usually determined by photographing, with a well-corrected lens, a test object consisting of lines separated by a distance equal to the width of the line.

The resolving power depends principally on the emulsion varying with the range in the sizes of the silver halide grains, the thickness of the coating, its turbidity and color sensitizing. Other factors affecting the resolving power are:

1. The exposure. Resolving power tends to decrease with the exposure because of the increased spread of the image due to the scatter of light in the emulsion.

2. The contrast of the image. The resolving power tends to increase with the contrast of the subject.

3. The color of the exposing light. The resolving power varies with the color of the exposing light but not in the same way for all materials. At present no general statement can be made.

4. The developer. The composition of the developer, in general, does not appear to have a marked effect on the resolving power, but it varies with the developing time increasing to a maximum, then drops slightly and remains approximately constant.

5. Temperature of the developer. The resolving power appears to become less as the temperature of the developing solution increased.

The following values are for a subject with a contrast of 1:30, using a light source of 5400° K and development in a borax-type of developer except as noted.

	LINES PER MM.
High-speed pan film........................	45
Fine-grain film............................	55
Positive motion-picture film..................	70
Contrast-process film (contrast developer).......	130
Photomechanical film (contrast developer)......	200

Special emulsions of much higher resolving power have been made for special purposes. These are very slow, requiring from 10 to 20 times the exposure of bromide paper, but the resolving power is from 600 to 1000 lines per mm. (C.B.N.)

RESONANCE. Every physical system, in general, has one or more natural vibration frequencies characteristic of the system itself and determined by constants pertaining to the system. Thus a flexible string of length l and mass δ per unit length, and subjected to a tension f, will, if struck or plucked and left to itself vibrate with frequencies equal to $\frac{1}{2l}\sqrt{\frac{f}{\delta}}$ and to various integral multiples thereof (overtones). If such a system is given impulses with some arbitrary frequency, it will necessarily vibrate with that frequency even though it is not one of those natural to it. These "forced vibrations" may be very feeble; but if the impressed frequency is varied, the response becomes rapidly more vigorous whenever any one of the natural frequencies is approached, its amplitude often increasing many fold as exact synchronism is reached. This effect is known as resonance. Thus, if one sings a clear note in front of a piano (holding down the "loud" pedal), the string nearest in tune with that note gives an audible resonant response. Applications of the principle are very numerous; among them are the tuning of a radio receiving set to the carrier wave frequency of the broadcasting station, the amplification of musical sounds by resonators, as in a reed organ pipe, and the disproportionate response of **microphone** diaphragms to certain voice frequencies. The resonance is said to be highly selective if impressed impulses in exact synchronism bring much more energetic reponse than those slightly "off key"; sharp tuning of a radio receiver depends upon this condition. (L.D.W.)

RESONANCE DUCT. Jet Engine.

RESONANCE, ELECTRICAL.
In an a-c circuit containing **inductance** and **capacitance** in series the **impedance** is given by:

$$Z = \sqrt{R^2 + \left[(\omega L) - \left(\frac{1}{\omega C} \right) \right]^2}$$

where R is the resistance, ω is 2π times the frequency, L is the inductance and C the capacitance. It can readily be seen that at some frequency the terms in the brackets will cancel each other and the impedance equals the resistance alone. This condition, which gives a minimum impedance (and thus a maximum current for a fixed impressed voltage) and unity **power factor**, is known

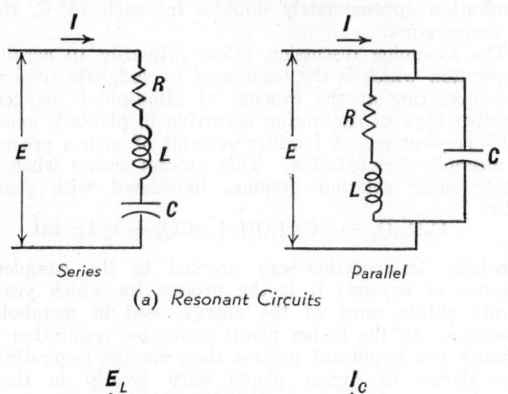

Series   *Parallel*

(a) *Resonant Circuits*

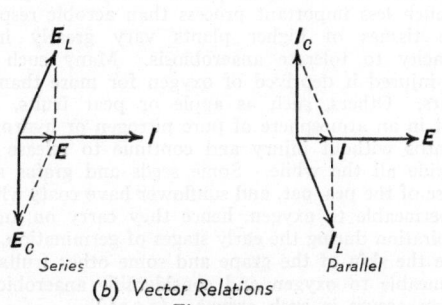

Series   *Parallel*

(b) *Vector Relations*
Fig. 1.

as series resonance. Where the resistance is small the current may become quite large. As the voltage drop across the **condenser** or **coil** is the product of the current and the impedance of that particular unit it may also become very large. The condition of resonance may even give rise to a voltage across one of these units which is many times the voltage across the whole circuit, being, in fact, **Q** times the applied voltage for the condenser and nearly the same for the coil. This is possible since the drops across the coil and condenser are nearly 180° out of phase and thus almost cancel one another leaving a relatively small voltage across the circuit as shown in Fig. 1b. For a circuit composed of an inductance unit in parallel with a capacitance the opposite effects of these two types of reactance will counteract one another at some frequency and produce unity power factor for the circuit. This is parallel resonance or antiresonance as it is sometimes called. In such a circuit the currents in the individual branches may be many times that in the line since they are out of phase and combine vectorially to give the line current. The impedance of a parallel resonant circuit is very high, its behavior being almost identical with that of the current in a series circuit if the Q of the parallel circuit is above 10. Fig. 2 shows a typical frequency-response curve for resonant

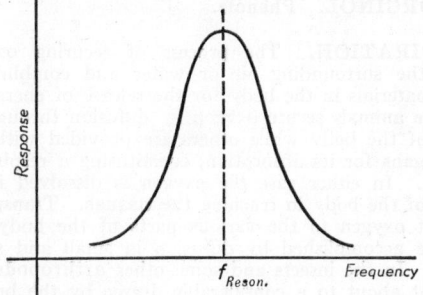

Fig. 2. Typical frequency-response curve for resonant circuit.

circuits, the ordinates being current for a series and impedance for a parallel resonant circuit. In series resonance the resonant condition is given exactly by the following expression which also holds for parallel resonance if the Q is high:

$$\omega L = 1/\omega C.$$

Both types of resonance are widely used in communication circuits to select certain frequencies in preference to others. An example is the tuning circuit of the radio receiver. (L.R.Q.)

RESONANCE RADIATION.
It has long been known that when a gas or a vapor is traversed by light having the same wavelength as that corresponding to some line of the emission **spectrum** of the gaseous substance, a portion of that light is absorbed. It is this fact that gives rise to absorption spectra. (See **Atomic Spectra.**) R. W. Wood observed that if the gas is viewed from one side against a dark ground when traversed by such a beam, the path of the beam becomes visible as a luminous streak. In short, the gas is "excited" and emits this characteristic wavelength; but if the frequency of the incident light is slightly changed, the luminous streak disappears, which would not be so if the phenomenon were ordinary **scattering**. Until recently it was supposed that this re-emission of energy was an example of **resonance,** namely, that the atom was set into resonant oscillation by light waves having one of its own natural frequencies, and that thereupon it became an emitter of the same frequency. Hence the term "resonance radiation."

This view can no longer be supported. For example, Wood used the yellow D radiation of sodium, the wave number (reciprocal wavelength) of which is 16973/cm., and got resonance radiation of the same kind. But Strutt excited the sodium vapor with the second line of the sodium series, wave number 30273 (in the ultra-violet), and found that the resulting secondary radiation consisted of both frequencies. Since the ratio of two wave numbers, 16973 to 30273, is not a simple fraction, the excitation of the former by the latter cannot be explained as resonance. The **quantum theory** avoids the difficulty. From that standpoint a 30273 photon excites the atom to its 30273 or 2Π energy level. In returning to its ground state the atom emits its 30273 spectrum line; but some atoms may first return to the 1Π state, to which the 16973 D radiation would have raised them, and then drop to the ground state from there, emitting the D radiation observed. For some reason the drop from the 30273 to the 16973 level does not result in observable radiation, which would be of wave number 13300, in the extreme red. Instead, that part of the energy seems to be converted directly into heat. Resonance radiation appears to be a species of fluorescence, except that it may take place with no change of frequency. (See **Luminescence**.) (L.D.W.)

RESONANT LINES. Lines.

RESORCINOL. Phenols.

RESPIRATION. The process of securing **oxygen** from the surrounding air or water and combining it with materials in the body for the release of energy.

Some animals secure oxygen by diffusion through the walls of the body while others are provided with special organs for its absorption, constituting a respiratory system. In either case the oxygen is dissolved in the fluids of the body in reaching the **tissues**. Transportation of oxygen to the various parts of the body may also be accomplished by diffusion in small and simple forms. In the insects and some other **arthropods** it is brought about to a considerable degree by the branching air tubes (tracheae) through which air is taken into the body. **Annelid** worms, **phoronids, mollusks,** vertebrates, and possibly a few other invertebrate forms have respiratory pigments in the blood which combine with oxygen for transportation. The more common pigments are a copper compound, haemocyanin, and an iron compound, haemoglobin. They unite with oxygen readily and give up oxygen as readily to the tissues.

In animals with special respiratory systems some muscular movement is necessary to bring fresh supplies of air or water to the surfaces which absorb oxygen. In the terrestrial forms these are the movements of breathing. Insects extend and retract the walls of the abdomen to take in and discharge air, respectively. Mammals elevate the ribs and contract the diaphragm during inspiration, and when the muscles are relaxed expiration takes place. Fishes swallow water and divert it through the **pharyngeal** clefts, where it bathes the gills in passing. Other animals carry on respiratory movements appropriate to their own structure.

The intake and transportation of oxygen constitute external respiration. Once the gas reaches the tissues it is available for the **oxidation** of energy-bearing compounds in the process of internal respiration. This union is accomplished through the action of **enzymes**.

The products of oxidation are partly eliminated from the body by the respiratory system, although this phase of respiration is an excretory function. (A.W.L.)

RESPIRATION IN PLANTS. Except in actively photosynthesizing tissues (see **Photosynthesis**) plant organs are continuously using oxygen from and releasing carbon dioxide to the atmosphere. These gaseous exchanges are the usual external manifestations of the process of respiration which, with respect to plants, is usually defined as the oxidation of foods in plant cells.

Hexose sugars are the foods most commonly oxidized in plant cells, the chemical equation for the process being:

$$C_6H_{12}O_6 + 6O_2 \rightarrow 6CO_2 + 6H_2O + 673 \text{ kg. cal.}$$

This process is the exact opposite of photosynthesis. Green parts of plants also carry on respiration, but when they are exposed to conditions favorable to photosynthesis this process goes on much more rapidly than respiration so that the net gaseous exchanges of the plant are the opposite of those which occur in respiration. In the dark, however, green plant tissues lose carbon dioxide and use oxygen just as non-green tissues do in either light or dark.

Less commonly other kinds of foods, such as fats and proteins, may be oxidized in plant cells. When sugars are oxidized the respiratory ratio (ratio of volume of carbon dioxide released to volume of oxygen absorbed) of the tissue is 1, when fats are oxidized it is about 0.8, and when proteins are oxidized about 0.9. When certain kinds of organic acids are oxidized the respiratory ratio is greater than one. This is also true when part or all of the respiration is of the anaerobic type (see below).

The essential biological significance of respiration is that a part of the energy released is used in the maintenance of metabolic processes and in growth. A considerable part of the energy is usually released as heat, however, and passes out of the plant body without any significant effect upon its metabolism. Release of heat energy in the respiration of plants can easily be demonstrated by placing a mass of germinating seeds in a thermos bottle and determining the temperature rise of the mass with a thermometer. All of the energy released in respiration represents radiant energy of sunlight which was trapped as chemical energy in sugar molecules during the process of photosynthesis.

The rate of respiration varies greatly from one kind of plant tissue to another. In general it is greatest in young, growing tissues such as germinating seeds, root tips, opening buds, or expanding flowers, and least in dormant or senescent tissues such as dry seeds or spores, or ripe fruits. The principal environmental factor affecting the rate of plant respiration is temperature. Within the approximate range of 0°–45° C. the rate of respiration approximately doubles for each 10° C. rise in temperature.

The foregoing discussion refers primarily to aerobic respiration which is the term used to designate respiration occurring at the expense of atmospheric oxygen. Another type of respiration occurring in plants is *anaerobic respiration*. A familiar example of such a process is alcoholic fermentation. This process ensues when a dilute sugar solution becomes inoculated with yeast cells:

$$C_6H_{12}O_6 \rightarrow 2C_2H_5OH + 2CO_2 + 25 \text{ kg. cal.}$$

Alcoholic fermentation can proceed in the complete absence of oxygen; it is the process by which yeast plants obtain most of the energy used in metabolic processes. In the higher plants anaerobic respiration is a much less important process than aerobic respiration. The tissues of higher plants vary greatly in their capacity to tolerate anaerobiosis. Many such tissues are injured if deprived of oxygen for more than a few hours. Others, such as apple or pear fruits, can be kept in an atmosphere of pure nitrogen or hydrogen for months without injury and continue to release carbon dioxide all the while. Some seeds and grains such as those of the pea, oat, and sunflower have coats which are impermeable to oxygen, hence they carry on anaerobic respiration during the early stages of germination. Likewise the skin of the grape and some other fruits is impermeable to oxygen and considerable anaerobic respiration occurs in such organs. (B.S.M.)

RESPIRATORY SYSTEM. The assemblage of organs by which air or water is brought into contact with **tissues** which can absorb part of its contents of **oxygen**.

Many animals absorb oxygen through the surface of the body. This is particularly true of small and simple forms, but the skin of the **earthworms** and that of some **amphibians** absorb all of the oxygen required by the animals, and any moist skin may absorb small quantities. The simplest modification to be introduced as a respiratory system is some extension of the surface to provide for the needs of a more bulky body. Tufted or thin plate-like structures called **gills** project into the water from the surface of many aquatic animals. Such structures are not adapted for air-breathing because their epithelium must be moist for the ready passage of oxygen and their extensive surface favors drying when exposed to air.

In aquatic insects gills contain gas-filled tubes (**tracheae**) and are known as tracheal gills, but in most animals the blood or body fluids circulate through them. Gills of this kind are found in many **annelid** worms and in the **crustaceans.** In the latter group they are sometimes protected by a fold of the body wall. This fold encloses them in a chamber through which water is propelled by special appendages.

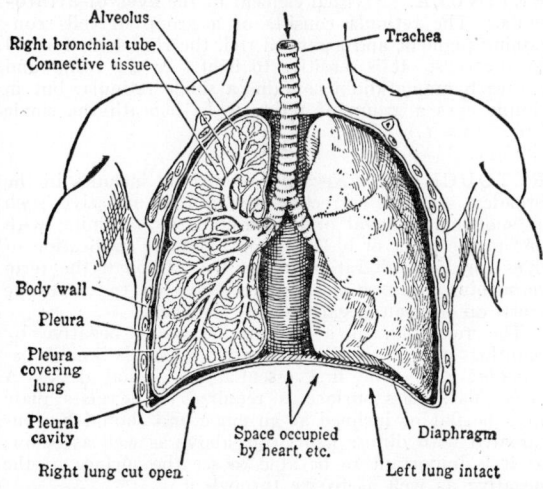

Respiratory system. (*Hegner's College Zoology, Macmillan.*)

In many terrestrial **arthropods** the respiratory system consists of air tubes or tracheae, metamerically arranged. In the primitive state each segment contains a pair of tracheae opening to the surface of the body separately through small pores called spiracles or stigmata. The openings are usually guarded by some closing device or by a grating formed from the **cuticula.** The tracheae have coiled **chitinous** filaments (taenidia) in their walls which keep them distended, and at their inner ends they communicate with finer tracheoles which lack these filaments. These fine tubules lead to the various tissues of the body, although oxygen probably passes from them into the body fluids, rather than directly to the cells. The gas-filled tubes form a closed system in aquatic insects. In these forms the oxygen content is renewed by diffusion from the surrounding water in tracheal gills, as mentioned above.

Spiders have a pair of lung books formed of many thin leaves in depressions in the abdomen. Blood circulates in these leaves and air between them.

The respiratory system of **vertebrates** is associated with the **pharynx.** In primitive **chordates, cyclostomes,** fishes (**Pisces**), and larval **amphibians** the gill slits persist along this passage, so that water taken into the mouth may be expelled from the pharynx without being swallowed. Finely divided blood vessels in the walls of the pharynx or in special outgrowths known as gills along the walls of the pharyngeal clefts receive oxygen from the water as it flows over these surfaces.

In some fishes and in terrestrial vertebrates generally, a saccular outgrowth of the ventral wall of the pharynx forms lungs for the reception of air. In the simplest forms the outgrowth branches to form two sac-like lungs. In more complex lungs the surface is increased by ridges projecting into the cavities from their walls, and in the most highly developed organs of this type there are many minute chambers (alveoli) in a spongy mass of tissue containing muscle and elastic fibers. The original connection with the pharynx persists, leading into a single tube, the trachea, supported by cartilage rings. The principal branches of the trachea are the **bronchi** or bronchial tubes. They lead into finer bronchioles whose branches communicate with the alveoli. (A.W.L.)

RESPIRATORY TREE. A branching diverticulum of the **cloaca** of sea cucumbers (**Holothuroidea**). The structure extends into the body cavity. Contractions of the cloaca pump water through the pair of respiratory trees to the thin-walled ampullae with which they end, thus supplying **oxygen** to the fluid in the body cavity. (A.W.L.)

REST MASS. Relativity.

RESTRAINED BEAM. Fixed Beam.

RESTSTRAHLEN. Residual Radiation.

RESULTANT. In **stress** analysis the resultant is a single vector quantity which will have the same effect on the equilibrium of a body as two or more vector quantities which it replaces. If these latter vectors are parallel the resultant is the scalar sum of these vectors, but if they are not parallel the addition must be made vectorially. (See **Vector and Mechanics.**) (C.W.C.)

RESURGENT. A term proposed by R. A. Daly, in 1908, for **juvenile** gases and vapors which are derived from the assimilated fragments of the intruded rock. These gases may play an important role in the formation of deep-seated **igneous rocks,** or may escape as **magmatic** emanations. (R.M.F.)

RETAINING WALL. A retaining wall is a structure for supporting loose material at an angle greater than its natural **angle of repose.** Usually a retaining wall is of concrete construction, and used to retain a bank of earth. The three principal types of retaining walls are illustrated. They are the simple trapezoidal section with vertical loaded face, the section with earth pressure aiding masonry weight, and the cantilever retaining wall. The successful retaining wall must be built so that it will neither slide on its base under horizontal pressure, nor tip over. The pressure of the earth is, of course, variable, depending on its composition and moisture content. If the earth back of the wall is well drained, so that its moisture content is a minimum, it will possess more cohesion, and exert less overturning pressure against the wall. The retaining wall may, in some cases, have to be designed to withstand pressures of a surcharged embankment having a slope of at least the natural angle of repose. The stability of a solid masonry section, such as that in (a) may be determined in much the same way as that for gravity masonry **dams,** except that earth pressure is not so directly and definitely computable as water pressure. The section shown in (b) will need less masonry because its shape causes it to make use of some of the vertical weight of earth as a stabilizing earth pressure. Also, it is keyed into the ground so that the resistance to sliding is greatly increased. A cantilever retaining wall, shown in (c), is the one with a minimum of masonry, but, because its resistance to overturning is obtained by bending action, internal moments are developed which require that this type of wall be reinforced with steel.

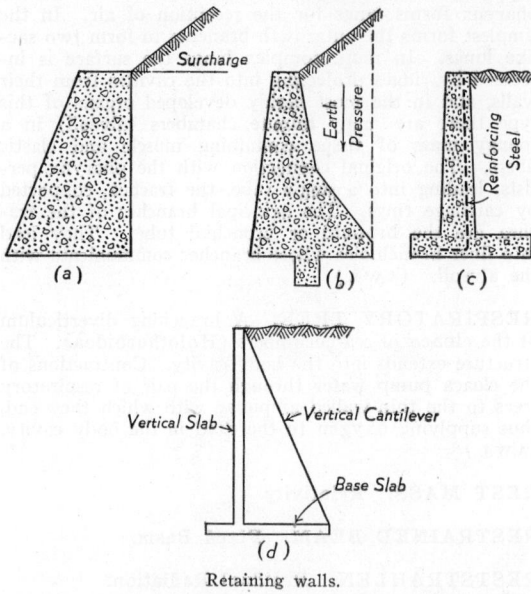

(a) (b) (c)

Vertical Slab Vertical Cantilever

Base Slab

(d)

Retaining walls.

A counterfort retaining wall (d) is made up of a continuous vertical slab which is connected to a continuous horizontal base slab or footing and supported at intervals by vertical cantilevers (counterforts), which rest on the base slab. The earth pressure is carried by the vertical slab, which transfers this load by beam action to the vertical cantilevers. Since concrete is weak in **tension**, the sections of this type of wall must be properly reinforced with steel rods. (F.T.M.)

RETARD COIL. This is the term commonly used in **telephony** to denote a **choke coil.** (L.R.Q.)

RETICLE. A reticle is a set of two or more fine wires placed at the principal focus of a **telescope** lens. The reticle wires are usually sections of spider web or some equally fine fiber. In accordance with the fundamental principle of telescope construction this reticle must be in the principal focus of the **eyepiece,** and when the telescope is directed on an object both the image and the reticle will be in clear view in the eyepiece.

The simplest form of reticle is two perpendicular wires so placed that their point of intersection is on the collimation axis of the telescope. In this case, when an object appears to be on the intersection of the wires the telescope is pointing directly at the object. This is the type of reticle found in telescopic gun sights. Other types of reticles are described under such instruments as the **meridian circle** and the **zenith telescope.**

For convenience in measuring small angles with a telescope the reticle wires are frequently so spaced that two points in a distant object apparently separated by the distance between the reticle wires will subtend some particular angle (e.g., 10 seconds of arc, or 1 minute of arc). Reticle wires so spaced are commonly known as "stadia lines" and are frequently found in the telescopes of sextants, field glasses, or surveyor's instruments. (W.K.G.)

RETICULO-ENDOTHELIAL SYSTEM. A group of mononuclear cells which are primarily **phagocytes.** They are distributed throughout the body in the spleen, liver, lymph nodes and bone marrow. In addition to phagocytosis, the reticulo-endothelial cells are concerned with blood cell formation, bile formation, and iron and pigment metabolism. (D.M.H.)

RETINA. The sensory layer of the camera **eyes** of mollusks and vertebrates. The retina of the human eye includes a layer of nerve fibers on the inner surface which converge to the optic nerve, and various cellular and fibrous layers. Numbering from the inner layer of nerve fibers these are an inner reticular layer, an inner nuclear layer, an outer reticular layer, Henle's fiber layer, the outer nuclear layer, a limiting membrane, the layer of rods and cones, and the pigmented layer. The rods and cones are characteristically shaped sensory cells which are sensitive to light. (See **Vision.**)

The retina is involved in disease by inflammatory processes, especially in **diabetes, kidney** disease, and **syphilis,** and by circulatory changes and hemorrhages in **anemia, arteriosclerosis,** hypertensive cardiovascular renal disease, and **thrombosis.** Detachment of the retina from the underlying layer is very serious, and complete blindness is a usual termination. Tumors may also occur, notably the pigmented tumor, **malignant melanoma** (see **Tumors**). (A.W.L., D.M.H.)

RETINULA. A visual element of the **eyes** of **arthropods.** The retinula consists of a group of cells containing pigment and a central rod, the rhabdom, formed by the cells. It is sensitive to light. In the compound eye each ommatidium contains a single retinula, but in simple eyes a group of retinulae lie beneath the single lens. (A.W.L.)

RETOUCHING. Negative retouching includes in the broadest sense all corrective work on the negative, such as spotting, removal of backgrounds by outlining with an opaque paint or bleaching out, and the application of dyes for local alteration in density, although the term retouching is often in practice restricted to corrective work on portrait negatives with a pencil.

The removing of a background on a negative by painting over it with an opaque paint is known as *blocking-out.* The first essential is a good desk. A large, flat, glass surface is required. The glass plate may be flat or inclined at an angle and should be illuminated with diffuse light from above as well as below, as it is important to be able to see the surface of the negative as well as to see through it.

A few good sable brushes of various sizes, a jar or tube of opaque, a drawing pen, a straight-edge, preferably of metal, and a celluloid draftsman's curve complete the necessary accessories.

Where the subject contains a number of straight lines it is well to rule these with a draftsman's ruling pen, starting with the horizontal lines and ruling all of these first, then the vertical lines. Many of the curves may be ruled in by using the draftsman's curve; those which cannot, must be filled in freehand with a small brush. In using the brush, it is essential, if clean work is to be expected, that the hand should not be cramped. A free, rather flowing motion of the hand from left to right and away from the body, using the fingers underneath the brush as a sliding pivot, leads to smooth, clean work.

To obtain a black background, the subject is covered with a transparent but waterproof varnish which can be obtained from photographic dealers. When the varnish is dry, the negative is immersed in a reducing solution, usually ferricyanide-hypo, or permanganate, and the background removed.

A convenient method of increasing density locally is by the application of a dye solution. A yellow-orange dye, known as neo-coccine, is usually recommended for the purpose as it is readily absorbed by gelatin and may be removed by washing in water if a mistake is made. Naphthalene black has the advantage of corresponding more nearly with the color of the silver deposit, so that it is less difficult to determine when the proper density has been reached. In any case the dye should be applied in a dilute form with a water-color brush and the density built up by successive applications.

Retouching of portrait negatives has as its purpose the removal of freckles and other skin blemishes, subduing over-prominent lines and softening harsh contrasts to improve the modelling. The area of the negative to be retouched is first covered with a resin varnish to retain the lead, after which a moderately hard lead pencil (2H or 3H) sharpened to a fine point is used to fill in the spots. When the spots due to freckles or other skin blemishes have been filled in, any sharp lines are filled in and harsh lines around shadow areas are blended so as to be less abrupt and harsh. (C.B.N.)

RETRACTOR. A muscle which pulls an eversible or extensible part back to its normal resting position in the body. In the starfishes (**Asteroidea**) a pair of muscles in each ray, attached to the pouches of the stomach, serve as retractors for that organ, and in the mussels retractors draw the foot back into the mantle cavity when the animal closes its shell. (A.W.L.)

RETROGRADE MOTION. Planetary Motions.

RETRO-VERSION. Uterus.

REVERBERATION. Acoustics.

REVERBERATORY FURNACE. A metallurgical furnace in which the charge, lying in a shallow hearth, is heated by flame passing over its surface and by radiation from a low roof. (R.H.H.)

REVERSAL OF STRESS. Cycle of Stress.

REVERSE BEARING. Bearing.

REVERSE FAULT. Fault.

REVERSIBLE PROCESSES. Some physical processes are of such a nature that if they are made to take place backward, that is, to go through the same stages in reverse order, the corresponding transfers of energy at each stage are reversed in direction but not in amount. Such processes are briefly characterized as "reversible." Thus a gas may have its density ρ and pressure p increased without any change in temperature (an isothermal compression). This is shown by the curve AB in Fig. 1. It necessitates the removal of heat exactly as

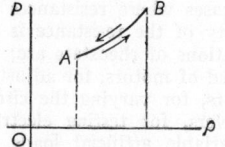

Fig. 1. Reversible pressure-density change in a gas. Fig. 2. Irreversible induction-intensity change in iron.

fast as it is generated by the work of compression. The process may, in theory at least, be duplicated in reverse, so that BA coincides with AB, by allowing the gas to expand and restoring the withdrawn heat energy just as fast as needed to keep the temperature constant. Each change of a **Carnot** cycle is likewise reversible.

On the other hand, consider the magnetization of iron, represented by the B-H (induction-intensity) curve MN in Fig. 2. If, after increasing the magnetization of iron until its condition reaches the stage N, the intensity of the magnetizing field is reduced, no amount of care will avail to make the curve retrace itself. The iron per-

sists in retaining some of the energy that was imparted to it, and returns along NM' to a condition, represented by M', of higher magnetic induction than at first. Magnetizing iron is therefore an irreversible process. (L.D.W.)

REVERSIBLE REACTIONS. Equilibrium.

REVERSING LAYER. The lower part of the sun's atmosphere is frequently referred to as the reversing layer. It is a gaseous layer which is cooler than the **photosphere,** only a few hundred miles in thickness, and gradually merges into the **chromosphere.** The reversing layer receives its name because it is in this region of the sun that the thousands of dark lines in the solar **spectrum,** commonly known as the **Fraunhofer lines,** are produced. As a result of a tremendous amount of research on the identification of the Fraunhofer lines, it is confidently believed that they are all due to elements which are to be found on the earth. These elements are reduced to the gaseous state in the reversing layer because of the tremendous heat from the photosphere. (W.K.G.)

REVERSION. The appearance in an organism of hereditary characters which were not present in its parents but were found in previous generations. The term has been interpreted as the skipping of one generation, a condition more accurately known as atavism, but these two terms are essentially synonymous. The reappearance of characters after one or several generations is made entirely clear by the facts of **heredity.** (A.W.L.)

REYNOLDS CRITERION. Important information as to the behavior of fluid flow was secured experimentally by Reynolds, and the results of his work are important to many fields including heat transfer, hydraulics and aerodynamics. The critical velocity range above which a liquid flow will be turbulent, below which it will be viscous, and in which it may be either viscous or turbulent, depends upon the velocity of flow, the size or shape of the conduit, and the viscosity of the liquid. This latter property varies, of course, with the liquid and with its temperature. Reynolds established the fact that the ratio of

$$\frac{\text{fluid velocity} \times \text{hydraulic radius}}{\text{kinematic viscosity}}$$

was the same for all liquids at the critical velocity. The ratio just given is known as the Reynolds criterion. In the field of aerodynamics, a similar ratio is known as the Reynolds number. In the case of an airfoil, it is

$$\frac{\text{air velocity} \times \text{chord of the airfoil}}{\text{kinematic viscosity of the air}}$$

These two ratios are valuable in their respective fields because tests of models are directly comparable to full scale results of geometrically similar shapes if the Reynolds ratio for the model equals that of the actual or full scale project. This has its practical applications in the field of **hydrodynamics** in the study of water resistance of hulls or floats, and in the study of water velocities, levee problems, etc., of large rivers. It is used also to establish the best proportions of hydraulic turbines through the use of models. Much of the science of **aeronautics** rests upon experimental data obtained in **wind tunnels.** Dangerous inaccuracies might exist in drawing conclusions for actual construction from model tests, unless either the model were tested at a Reynolds number equal to that of the completed project, or due corrections and allowances were made for the Reynolds number. (F.T.M.)

REYNOLDS GROUP. Heat Transfer.

REYNOLDS NUMBER. The **Reynolds Criterion** of hydrodynamics is called Reynolds number in the aerodynamic field.

$$R.N. = \frac{Vl}{\nu}$$

V = Air velocity.

ν = Kinematic viscosity of the air.

l = Some significant dimension of the object, in the direction of air flow such as the chord of an airfoil, diameter of a sphere, length of a plate, etc.

(See **Aerodynamics, Airfoil, Boundary Layer, Burble, Wind Tunnel.**) (F.T.M.)

RF. Radio Frequency.

RH FACTOR. Blood Grouping.

RHABDITE. A crystalline rod-shaped body found in the cells covering the body of **turbellarian** flatworms. They are formed in cells lying between or just below the ectodermal cells, from which they migrate to the outer cells. They have been interpreted as defensive structures but the exact nature of their functions has not yet been established. (A.W.L.)

RHABDOCOELIDA. Turbellaria.

RHABDOM. Eye. Retinula.

RHABDOMERE. A segment of a **rhabdom** when it is made up of clearly distinguishable parts. (A.W.L.)

RHAGON. Porifera.

RHEA. Aves, Rheiformes. *Rhea.* Large flightless birds (**Aves**) of South America. They are called the American ostriches but differ from the true ostriches in the fully feathered head and neck, the longer wings, and the lack of a tail. There are three toes on each foot. (A.W.L.)

RHEBOK. Mammalia, Artiodactyla. *Pelea.* A small **antelope** found in hilly sections of eastern and southern Africa. It has been compared with the chamois in habits. (A.W.L.)

RHEIFORMES. The rheas. An order of birds containing only the few South American species of this name. They resemble the ostriches but differ in details of structure mentioned under **rhea**. (A.W.L.)

RHENIUM. Symbol: Re. Atomic number: 75. Atomic weight: 186.31. Density of metal powder: 10 (approx.). Melting point: 3000° C. Isotopes: 185 (38.2%), 187 (61.8%).

Rhenium is a platinum-white, very hard metal; stable in air below 1000° C.; practically insoluble in **hydrochloric acid or hydrofluoric acid**, but soluble in **nitric acid** with the formation of perrhenic acid; forms sodium rhenate when fused with **sodium** hydroxide and nitrate. Discovered by Noddack and Tacke in 1925 in **tantalite, wolframite,** and **columbite** by the Moseley **x-ray** spectrographic method of analysis, and later found present in **molybdenite,** from which rhenium is obtained. Predicted by Mendeléeff in 1871 as an element to be discovered with properties resembling **manganese,** and named by him dvi-manganese.

Rhenium is obtained from molybdenite by dissolving in nitric acid, precipitating molybdenum as phosphomolybdate, and from the filtrate recovering rhenium as sulfide by **hydrogen sulfide** (1% Re in the product). This product is oxidized and the sublimate of rhenium heptoxide is later reduced to rhenium metal by heating in hydrogen.

Chloride: Rhenium tetrachloride ($ReCl_4$), black liquid, boiling point 500° C.

Oxides: Rhenium dioxide (ReO_2), black solid; trirhenium octaoxide (Re_3O_8), blue solid; rhenium trioxide (ReO_3), red solid; rhenium heptoxide (Re_2O_7), yellow solid; rhenium tetroxide (ReO_4), white solid; all by burning rhenium in a current of **oxygen** gas.

Perrhenate: Potassium perrhenate ($KReO_4$), stable in acid solution, reduced in acid medium by iodide but not by sulfide.

Rhenate: Sodium rhenate (Na_2ReO_4), yellow solid, soluble, stable in alkaline solution.

Sulfides: Rhenium disulfide (ReS_2); rhenium heptasulfide (Re_2S_7), black solid, by reaction of perrhenate solution and hydrogen sulfide.

Rhenium compounds color the bunsen flame pale green. (R.K.S.)

RHEOSTAT. A rheostat is a variable **resistance** used for operation or control of electrical equipment. Rheostats might be classified as metallic, carbon, and electrolytic types. The most common form is the metallic type, in which the resistance is in the form of a metal wire or ribbon, or cast grid, these being made of a metal having poor conductivity, and little deterioration from heating. The variable resistance of metallic rheostats is obtained by bringing out taps from different points of the resistance wire to the points of a multi-pointed switch which can be used to short-circuit different sections of the resistance. Laboratory rheostats are frequently coils of resistance wire wound closely on an insulating cylinder and provided with a sliding contact finger which will bear on the wires themselves, and which can be employed to short-circuit any desired number of turns of the resistance wire.

A carbon rheostat is made of granules or carbon plates held in a frame. The resistance of carbon to flow of **electric current** varies with the pressure on it, and so by providing the rheostat with a screw clamp or other means of changing the pressure, the resistance is made variable. This type of resistor is seldom used with other than small currents. On the other hand, **electrolytic** rheostats are well adapted for large currents. They consist of a tank or barrel containing a solution of salt or acid into which are dipped **electrode** plates. The electrode plates should not be of a metal which is attacked by the solution. The energy is dissipated in heat in the solution, which boils it, so that the heat is ultimately liberated by vaporization of the solution. Since water is capable of absorbing nearly 1000 B.T.U. per lb. if vaporized, it is readily seen why this type of rheostat can be used for large currents. These are sometimes called water rheostats. There are many and varied uses for rheostats, and in many cases where resistances are used some degree of variability of the resistance is desirable. A few of the applications of rheostats are: for starting or controlling the speed of motors, for adjusting the field strength of generators, for varying the circuit characteristics of radio receivers, for testing electrical equipment by interposing variable artificial loads for dimming lights, and for adjusting current flow of any description. (F.T.M.)

RHEUMATIC FEVER (Acute Rheumatic Fever. Acute Articular Rheumatism. Acute Rheumatism). A disease of unknown origin, closely associated with invasion of the body by hemolytic **streptococci**. Rheumatic fever is most prevalent in the north temperate zone. It has been estimated that 170,000 new cases per year occur in the United States, and 840,000 individuals in the United States have **rheumatic heart disease.** The disease attacks children and young adults primarily. It is rare in infancy, appears in children about 5 years of age, and has a peak incidence at 9–11 years, falling off gradually in adult life.

Although the exact cause of rheumatic fever is unknown, certain predisposing factors are well established. The disease tends to be familial and some believe that the predisposition is inherited as a Mendelian recessive (see **Heredity**). Multiple cases in a family are common. The poor are more susceptible than the well-to-do, probably because of such factors as crowding, cold, dampness, and malnutrition which are conducive to the spread of respiratory infections and therefore hemolytic streptococci.

The acute attack of rheumatic fever often begins with sore throat and tonsilitis. Fever, joint pains with redness, swelling and tenderness of the affected joints follow. The arthritis is characteristically migratory, one joint after another being involved acutely but briefly, and healing without residual symptoms or signs. The heart may or may not be involved in the original attack. Rapid pulse rate is common and the appearance of murmurs is indicative of heart damage. Acute **pericarditis** may also occur. Other manifestations of rheumatic fever are **pleurisy, pneumonia** and **subcutaneous nodules**. **Chorea** is generally believed to be a specific rheumatic involvement of the brain.

The pathological pattern of rheumatic fever consists of multiple microscopic nodules showing a characteristic type of **inflammation**. In the heart, these are called Aschoff bodies. They occur throughout the heart muscle and on the heart valves, particularly the mitral and aortic valves. In other organs and tissues, similar tiny granulomas appear.

The course of acute rheumatic fever is dramatically modified by the use of salicylates (aspirin). On such medication, the fever drops, and joint signs disappear rapidly. The patient may then go on to complete recovery, or he may be left with residual **rheumatic heart disease**. In either case he is subject to recurrent attacks of acute rheumatic fever. These can be decreased in number or even eliminated completely if hemolytic streptococcal infections can be prevented. (D.M.H.)

RHEUMATIC HEART DISEASE. It is estimated that approximately 25% of children with **rheumatic fever** develop permanent heart damage. The average life expectancy in this group is roughly 13 to 15 years; if the disease occurred early in childhood, the duration may be much shorter. Rheumatic heart disease is thus an important cause of death in children and young adults.

Rheumatic heart disease which persists after the acute inflammation has subsided is due to fibrosis and scarring of the heart valves. The mitral valve is affected in almost every case; both mitral and aortic valves are involved in about one third of all cases. The patient may have no symptoms of heart disease until years after his acute rheumatic fever. During these years, his heart is under a mechanical disadvantage because of insufficiency of the valves, or constriction (stenosis) of the valve openings, or both. When the strain can no longer be withstood, **heart failure** of the congestive type supervenes. Until the signs and symptoms of failure appear, the patient is said to have inactive rheumatic heart disease. (D.M.H.)

RHEUMATISM. Rheumatic Fever.

RHEUMATOID ARTHRITIS. Arthritis.

RHINENCEPHALON. The portion of the vertebrate **brain** with which the **olfactory** nerves are connected. It is the lower part of the telencephalon. (A.W.L.)

RHINITIS. Inflammation or infection of the mucous membrane of the nose. The common head cold is primarily a rhinitis, as is **hay fever**. (See **Coryza**.) (D.M.H.)

RHINOCEROS. Mammalia, Perissodactyla. *Rhinoceros*. A large animal of the Oriental region or Africa, with rather short legs and a long muzzle bearing one or two conical horns behind the nostrils. There are four species in the Oriental region and two in Africa. The common African species stands almost 6' high at the shoulder and has an anterior horn over 3' long. Because of its thick skin it can be killed only by carefully placed shots. All of the rhinoceroses are described as dull and timid animals but when brought to bay they are aggressive, and because of their great size they are dangerous opponents under some conditions. (A.W.L.)

RHINOPHORE. A posterior tentacle of **olfactory** function in some of the marine mollusks (**Gasteropoda**, Opisthobranchiata). These forms have two pairs of tentacles. Those of the posterior pair have a simple eye at the base and bear organs of the sense of smell. (A.W.L.)

RHIZOIDS. Rhizoids are filamentous outgrowths from the surface, or from epidermal cells, formed of one or many cells, which serve to hold the plants of **mosses** and hepatics or the prothallia of **ferns** to the substratum. Similar structures occur in the **thallophytes**. (R.M.W.)

RHIZOME. A rhizome or rootstock is a horizontal stem growing beneath the surface of the ground or at times, at the surface. It has all the characteristics of a **stem**, such as nodes and internodes, leaves and branches. Often the rhizome is very much enlarged and contains much reserve food material. (R.M.W.)

RHIZOSTOMAE. Scyphozoa.

RHODESIAN MAN. Paleontology of Man.

RHODIUM. Symbol: Rh. Atomic number: 45. Atomic weight: 102.91. Density: 12.44. Melting point: 1985° C. (**Isotopes**: page 290.)

Compact rhodium is a white metal, almost insoluble in acids, including **aqua regia**, and is attacked by **chlorine**, by hot concentrated sulfuric acid, and by fused **potassium** bisulfate. Discovered by Wollaston in 1804. Rhodium metal is used chiefly as an alloy with **platinum**, resulting in a hard, durable alloy. Such an alloy (10% Rh) is used as catalyst in the oxidation of ammonia by oxygen (air), and in conjunction with pure platinum in a **thermocouple** for the measurement of high temperatures. Rhodium itself is used as a plating finish for silverware, and as a mirror surface in searchlights.

Rhodium occurs native in platinum ores, sometimes to the extent of 2%. When **osmium** and **ruthenium** are present, they are removed as volatile oxides; when platinum and **iridium**, by precipitation of the ammonium chloro-compound; insoluble in alcohol, upon addition of ammonium chloride; and when **palladium**, by the precipitation of the ammonium chloro-compound with **ammonium** chloride and **chlorine**. Rhodium remains in the solution in each case, and is precipitated by the addition of a reducing agent, such as **titanium** trichloride, as rhodium metal.

Hydroxide: Rhodium hydroxide ($Rh(OH)_3$), yellow precipitate, by solutions of hydroxides, *e.g.*, **sodium** hydroxide), soluble in excess of the reagent.

Alum: (See Sulfate, below.)

Chloride: Sodium rhodium chloride ($Na_3RhCl_6 \cdot 12H_2O$), dark red crystals, by ignition of oxide with **sodium** chloride and **chlorine** and crystallizing from water.

Oxides: Rhodium sesquioxide (Rh_2O_3), black; rhodium dioxide (RhO_2), brown.

Sulfates: Potassium rhodium alum ($K_2SO_4 \cdot Rh_2(SO_4)_3 \cdot 24H_2O$); potassium rhodium sulfate

$(K_3Rh(SO_4)_3)$, rose-colored crystals, by ignition of oxide with potassium bisulfate and crystallizing from water. (R.K.S.)

RHODOCHROSITE. Rhodochrosite, **manganese carbonate**, $MnCO_3$, is a rose-pink to red **hexagonal** mineral, occurring as small crystals, in cleavable masses, granular or compact. It is a brittle mineral; hardness, 3.5–4.5; specific gravity, 3.3–3.6; luster, vitreous to pearly; color, various shades of pink, red and reddish-brown; transparent to opaque; streak, white. Rhodochrosite has a perfect **rhombohedral cleavage**. Rhodochrosite is formed by replacement and is found as well in sedimentary deposits precipitated like **siderite** by organic matter acting, in the absence of oxygen, upon bicarbonates. It may occur also as a **gangue** mineral. Localities for this mineral are in Rumania, Saxony, Westphalia, and Cornwall, England. In the United States rhodochrosite is found at Franklin, New Jersey; Butte, Montana; and in various localities in Colorado and Nevada. The name rhodochrosite is derived from the Greek meaning rose, and color. (E.S.C.S.)

RHODODENDRON. Heath Family.

RHODONITE. The mineral rhodonite, **manganese metasilicate**, $MnSiO_3$, is a **triclinic** pyroxene forming large, irregular, tabular crystals but usually occurring massive. Prismatic and basal **cleavages** excellent; fracture, conchoidal to uneven; hardness, 5.5–6.5; specific gravity, 3.4–3.7; luster, vitreous to pearly on cleavage faces; color, red to pink and occasionally greenish or yellowish shades; streak, white; transparent to translucent. A variety containing much **calcium** is called bustamite. **Zinc** may replace the manganese in rhodonite; it is then known as fowlerite. Rhodonite is found in the Harz Mountains, Germany; in the Urals of Russia; in Hungary, Italy and Sweden. Bustamite from Mexico, Franklin and Sterling Hill, New Jersey, occurs with **fowlerite**.

Rhodonite has been occasionally used for an ornamental stone. Its name is derived from the Greek meaning a rose, because of the color. (E.S.C.S.)

RHOPALIUM. A tentaculocyst.

RHUBARB, PIE PLANT. *Rheum Rhaponticum*. Polygonaceae. The rhubarb plant is perennial from thick short rhizomes. The large somewhat triangular leaf blades are elevated on long fleshy petioles. The flowers are small, greenish-white and borne in large compound leafy **inflorescences**. The plant is principally grown for its fleshy petioles. These are stewed to yield a tart sauce used as filling for pies and tarts. The plant is indigenous to Asia.

From the rhizomes and roots of another species, *Rheum officinale,* or Medicinal Rhubarb, also native to Asia, is prepared the drug Rheum, used as a strong cathartic, and for its tonic effect on the mucous lining of the nasal cavity. (R.M.W.)

RHUMB LINE. Unless a ship is tacking or executing some other maneuver its course is generally constant for several hours at least. In such a case the ship is said to be following a rhumb line. The rhumb line, or loxodromic curve, may be defined as any curve on the surface of the earth such that the tangent of the curve at any point cuts the meridian through that point at a constant angle. In case this angle has any value other than 0° or 90° it may be proved that the rhumb line is a spiral approaching one of the poles of the earth as a limit.

Obviously the rhumb line course between any two points is the simplest course to follow, for once having set the course it will not have to be changed until the destination is reached. However, except in the particular cases where the two points are either on the same meridian or are both on the equator, the rhumb line will not be the shortest distance between the two points. The **mercator** chart was designed for the purpose of facilitating the laying down of rhumb line courses. On a mercator chart, and only on this chart, the rhumb line appears as a straight line. (W.K.G.)

RHYNCHOBDELLIDA. Hirudinea.

RHYNCHOCEPHALIA. Reptilia.

RHYNCHODAEUM. The part of the **proboscis of nemertine** worms which lies in front of the brain. The proboscis is an eversible structure. When retracted it lies almost entirely behind the brain, but when extended the brain is near its base, thus the greater part of the proboscis constitutes the rhynchodaeum. (A.W.L.)

RHYOLITE. The general term for a group of **acidic igneous rocks**, the **effusive** equivalent of the **granites**. It occurs as lava flows, breccias, and in volcanic necks and dikes. In the porphyritic varieties the **phenocrysts** are frequently quartz or **orthoclase feldspar** imbedded in a highly **felsitic** or glassy ground mass. Rhyolites, including **obsidian**, frequently show **flow, spherulitic, nodular,** and lithophysal structures. (R.M.F.)

RIBBON-FISH. Pisces, Teleostei. Peculiar marine fishes (**Pisces**) with very long compressed bodies. They may reach a length of many feet with a dorsoventral thickness of less than 1′ and a width of only 1″. (A.W.L.)

RIBS. In man, the ribs number twenty-four, twelve on either side. They are attached to the vertebral column behind, and the first seven pairs are connected with the **sternum** in front and are called true ribs. The remaining five are called false ribs. The eighth, ninth and tenth are attached in front to the cartilaginous portion of the next rib above. The lower two, that is the eleventh and twelfth, are not attached in front at all and are called floating ribs. The spaces between the ribs are called intercostal spaces; they contain the intercostal muscles, nerves and arteries. The ribs form the greater part of the bony cage of the **thorax;** they preserve its outline and allow for easy motion in breathing, due to their elasticity. (R.S.M.)

RICE. *Oryza sativa.* Gramineae. Rice is an annual grass which grows wild in tropical Asia and Africa. Cultivated rice is probably derived from an Asiatic species, and is especially adapted to grow in swampy or very wet lowlands. It is a shallow rooted plant, the stems of which tiller abundantly and grow from 2–6′ or more in height. The leaves are long and smooth. The **inflorescence** is a **panicle** the branches of which may occur singly or in pairs. The laterally compressed **spikelets** are 1-flowered, and have a pair of small bristle-like **glumes;** the lemma is tough, parchment-like and sometimes **awned;** the palea resembles the lemma, but is somewhat smaller. A distinctive character of the flower is the presence of six functional **stamens.** Commonly rice is self-pollinated. The grain or karyopsis is enwrapped in the palea, and frequently also in the lemma. In this condition rice grain is known as paddy, or rough rice. The grain itself is smooth and shining, has a pair of longitudinal grooves on its surface, and a glassy endosperm. In structure it is very similar to wheat grain.

The milling of rice involves several processes. First the outer coverings are removed by revolving stones and fans. After this the outer seed-coats and **embryo** are largely removed by rubbing; the remaining grain is scoured and polished by rubbing on leather surfaces. Finally the polished grain is given a coat of glucose and talc, and is ready for market.

The principal use of rice is as a food stuff for human consumption, with China, India and Malaysia consuming the greatest quantity. In the United States rice is used to a slight extent as a breakfast food and in the preparation of other foods. Some rice is used in the manufacture of starch. The hulls and residues from polishing are sometimes fed to stock. Rice straw is used as stock food, for making strawboard and in the Orient for making hats, and many other articles. From the grain the fermented drink, sake, is prepared.

In North America two native grasses are frequently called wild rice. These are *Zizania aquatica* and *Z. miliacea,* tall swamp grasses closely related to *Oryza.* The grain is an important food for wild fowl, and finds a very limited use as a food for man, largely as a novelty. (R.M.W.)

RICE BIRD. Aves, Passeriformes. 1. The American bobolink. 2. The Java sparrow, also known as the paddy bird. (A.W.L.)

RICHARDSON'S EQUATION. Thermionic Phenomena.

RICKETS. A deficiency disease of infancy and early childhood, due to inadequate supplies of vitamin D (see **Vitamins**). The clinical picture is one of skeletal abnormalities due to the failure of lime salts to be deposited in growing cartilage and in newly formed bone. The growing ends of the bones are soft and bend under the stress of weight-bearing. In young infants, the skull is softened along the suture lines (cranio-tabes), and the eminences of the frontal and parietal bones become prominent, resulting in the typical "square head" of the disease. Bowing of the limbs, deformities of the skull, and deformities of the chest persist into adult life. The eruption of the teeth is delayed, and lasting defects in the calcification of the unerupted permanent teeth occur.

The prevention of rickets is accomplished by exposure to sunshine, and by giving vitamin D in the form of cod-liver oil, halibut-liver oil, viosterol, or other antirachitic preparations. The disease is treated by the same means. (D.M.H.)

RICKETTSIAL DISEASES. A group of diseases, world wide in distribution, which are transmitted by arthropods. The causative agents are various species of Rickettsia, microorganisms which stand between viruses and bacteria as to size, cannot be grown outside living cells, and are widely distributed in ticks, mites, spiders, lice and fleas.

The main groups of Rickettsial diseases are the typhus group, the Rocky Mountain spotted fever group, and the Tsutsugamushi disease group. All of these are similar in their pathology: they produce damage of the smaller blood vessels and thrombosis of these vessels; and all produce a clinical picture characterized by fever, lethargy, and a typical rash. Each gives a serological reaction (the Weil-Felix test) which is characteristic. The diseases nevertheless vary considerably in their severity and mortality.

Typhus occurs in epidemic (louse-borne) and endemic (flea-borne) forms. Epidemic typhus is a great scourge in wartime, with a mortality rate of 50 to 70%; its control in World War II by the elimination of the body louse with **DDT** was a major victory. The endemic or murine form occurs chiefly in southern United States. It is a much milder disease than the epidemic variety.

Rocky Mountain spotted fever is transmitted by tick bites. As its name indicates, it is prevalent primarily in the Rocky Mountain region. Its mortality varies greatly with different outbreaks, but is generally high, 5 to 80%—the higher rates occurring in adults.

Tsutsugamushi disease is mite-borne. A variety, if not the identical disease, known as *scrub typhus,* was a major problem in the South Pacific theatre of World War II. The attack rate was very high; the mortality rate was 10 to 25%.

There is no specific treatment for any of the Rickettsial diseases. They are self-limited and run a course of several days to several weeks. (D.M.H.)

RIEBECKITE. The mineral riebeckite, essentially **sodium iron silicate,** $NaFe(SiO_3)_2$, is a monoclinic member of the **amphibole** group, usually in prismatic crystals. It has a prismatic **cleavage;** hardness, 4; specific gravity, 3.4; vitreous luster; color dark bluish to black. It occurs in **granites** and **syenites** chiefly. It is found in Greenland, Portugal, Madagascar and in the United States at Quincy, Massachusetts; near Pikes Peak, Colorado, and the San Francisco Mts., Arizona. (E.S.C.S.)

RIECKE'S PRINCIPLE. The term for **thermodynamic** processes by which recrystallization may take place in **metamorphic rocks.** An explanation of one form of **foliation** and schistosity, whereby tabular minerals such as **hornblende** and **biotite** recrystallize with their longest dimension parallel to the developing planes of foliation. The result of differential pressures and the consequent solution where pressure is greatest, and redeposition and recrystallization where pressure is least. (R.M.F.)

RIFFLEFISH. Pisces, Teleostei. A small fish (**Pisces**) of the genus *Cottus,* related to the miller's thumb of the eastern half of the country. A **sculpin.** It is found from Alaska to the Sacramento River. (A.W.L.)

RIFLE BIRD. Aves, Passeriformes. A **bird of paradise** found in Australia and New Guinea. The several species make up the genus *Ptilorhis.* (A.W.L.)

RIGHI-LEDUC EFFECT. If heat is flowing through a strip of metal and the strip is placed in a magnetic field perpendicular to its plane, a temperature difference develops across the strip. This effect, discovered in 1887 independently by Righi and by Leduc, bears the same relation to the **Nernst effect** that the **Ettingshausen effect** bears to the **Hall effect.** It may indeed be regarded as analogous to the Hall effect, but with a longitudinal flow of heat replacing the electric current and a transverse temperature difference replacing the potential difference. If, to one looking along the strip in the direction of the heat flow, and with the magnetic field downward, the decrease of temperature is toward the right, the effect is said to be positive. It is positive in iron and negative in bismuth. (L.D.W.)

RIGHT ASCENSION. Equatorial Coordinates.

RIGHT TRIANGLES. Triangles.

RIGID FRAME. A rigid frame, also known as a continuous frame, is an **indeterminate structure** in which continuity of action between the intersecting or adjacent members is obtained by means of moment-resisting **joints** (joints capable of resisting **bending moment**).

Rigid frames are made of structural steel or **reinforced concrete.** In the former the joints are either riveted or welded while in the latter continuity is produced by continuing the main reinforcing rods beyond the joint; additional rods may be necessary to develop the required resisting moment.

Rigid frames are used as **bridges,** and as **bents** in mill and multiple-story buildings. The Vierendeel girder bridge is a rigid frame which is similar in outline to the usual bridge **truss.** However, the diagonals are omitted since the **chords** are designed for **flexure.**

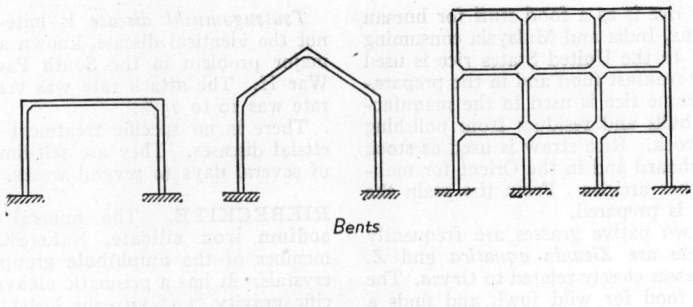

Bents

Vierendeel Girder

Rigid frames.

This truss is very useful for special cases of building framing. Rigid frame action is also utilized in the design of reinforced concrete **culverts** and sewers. (C.W.C.)

RIGIDITY. Elasticity.

RIME ICE. Icing on Aircraft.

RIMMING STEEL. A steel which when cast into ingot molds solidifies with the evolution of considerable quantities of gases. In casting such an ingot, evolution of gas (mainly carbon monoxide) keeps the molten metal turbulent as the thickness of the frozen metal in contact with the ingot mold increases. The solidified ingot has scattered **blowholes** throughout but much less central pipe than **killed steels** and a relatively thick skin free from blowholes. The blowholes weld shut during hot rolling and the thick skin tends to give good surface to the finished product. Rimming steel is characterized by a lack of homogeneity in its chemical composition, especially for the elements carbon, phosphorus, and sulfur. The chemical segregation of the ingot persists in the final product. (R.H.H.)

RING CANAL. 1. In the phylum **Bryozoa**, a part of the body cavity which is incompletely separated from the principal chamber and prolonged into the tentacles. 2. A canal of the water vascular system of starfishes (**Asteroidea**) which forms a complete circle in the disk. It is connected by the stone canal with the madreporite and gives off a radial canal into each arm. The ring canal also bears the Polian vesicles and Tiedemann's bodies. (A.W.L.)

RING GAUGE. Measurement.

RING-DIKE. A more or less circular **dike** controlled by a ring-shaped **fault.** (R.M.F.)

RINGER. This is the bell mechanism of the ordinary telephone. Reference to the figure will show that it differs in principle from the ordinary electric door-bell as no make and break contacts are used. Essentially it consists of two soft iron cores surrounded by coils through which the ringing current (a low-frequency a.c.) is passed, the permanent **magnet** and the soft iron armature which attaches to the bell clapper. The permanent magnet sets up flux equally in the two cores and hence causes no unbalance of attraction for the armature and the bell is in equilibrium. Passage of the alternating ringing current causes the flux due to the coils alternately to add to and subtract from the fixed

flux in the cores and thus first one core and then the other has a maximum flux. This unbalances the attraction of the armature which is pulled to the side having the most flux. Since this alternates, the armature vibrates back and forth, causing the bell to ring. For

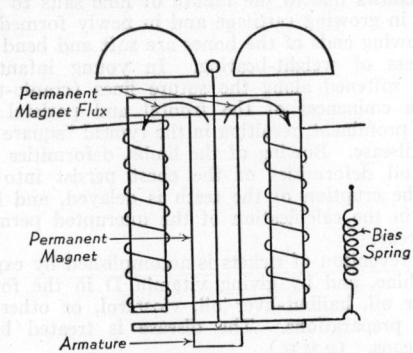

party-line ringing this operation is modified by adding a spring to pull the armature to one side, giving what is known as a biased ringer. Then, if the ringing current is made pulsating d.c. of the right polarity it will cause an increase of flux in the other core and give a pulsating attraction to the armature in that direction. Thus the bell will ring on only one polarity of pulse. The bell at the other party-line phone has the spring attached to the other side so it will ring on the opposite polarity, thereby giving selective ringing on a 2-party line. By proper connection of the biased ringers between line wire and **ground** 4-party selective ringing may be obtained, two ringing over one line and ground and the other two over the other line and ground. (See also **Harmonic Ringing.**) (L.R.Q.)

RING-OILED BEARING. Bearing.

RING-TAILED CAT. Cacomistle.

RINGWORM (TINEA). A contagious disease of the scalp, skin, or nails caused by various **fungi.** Circular reddened patches with crusting and pustules with loss of hair occur. **Epidermophytosis** is a form of ringworm. Treatment varies somewhat with the fungus involved and the site of infection. In general it consists of mechanical removal and chemical destruction of the living fungi, and soothing applications to the irritated skin. Ammoniated mercury, salicylic acid, benozic acid and many other ointments are used. (D.M.H.)

RIPPLE MARK. Corrugations developed in sands and muds by currents in the water which covers them. Ripple marks may be classified as due either to oscillation or translation. The former are the result of oscillation currents set up in the water by the wind; the result of ordinary water waves. The latter are the result of progressive, directional water currents, and the resulting ripple mark is essentially a sub-aqueous **dune**. Ripple mark is helpful in determining the conditions under which aqueous, clastic sediments are deposited. Ripple mark has also been used to help determine the depth of water in which the rippled sediments have been deposited. Ripple mark is also helpful in determining the original position of formations which have been subsequently deformed or overturned. (R.M.F.)

RIPPLE VOLTAGE. When an a.c. is rectified the resultant current or voltage consists of pulsating d.c. In the case of simple half-wave **rectifiers** this output is a series of half-sine waves spaced by equal intervals of no output while for full wave, single-phase rectifiers it consists of half-sine waves with no appreciable space between them. Polyphase rectifiers give outputs which, while they do not vary as markedly as this, consist of a series of adjacent portions of sine waves. In every case the output may be considered as made up of a smooth d-c component and an a-c or ripple component. The first is the value read by a d-c instrument in the circuit and the latter is the component which will produce objectionable hum in communication equipment supplied by the voltage. To reduce this ripple component various **filter** systems are used, the amount of filtering depending upon the equipment being supplied by the rectifier-filter combination. In many industrial applications the output of the rectifier alone is satisfactory, in communications the output must contain an extremely small amount of a.c. This is measured in terms of per cent ripple which is the a-c component of voltage divided by the d-c component of voltage and multiplied by 100. (L.R.Q.)

RITZ PRINCIPLE. Atomic Spectra; Combination Principle.

RIVET. A cylindrical member, with one preformed head, inserted in coincident holes in two or more plates, holding them together by the pressure exerted between the preformed head and a fabricated head at the other end. The fabricated head may be formed, either hot or cold, by ordinary hammering, pneumatic hammering, by a hydraulic pressure die, or by cold-working. Rivets of soft steel or wrought iron are hot-formed for all pressure vessel and structural applications. Rivet holes for pressure vessels are usually drilled to a diameter $\frac{1}{16}''$ greater than the size of the rivet. Punched holes are forbidden by the standard pressure vessel construction codes unless the brittle portion of the plate surrounding the punched hole is removed by reaming after the punching operation. Rivet holes for structural applications are usually punched, $\frac{1}{8}''$ greater than the nominal or shank diameter of the rivet.

Rivet nomenclature is based upon the shape of the heads. Button-head rivets have semi-spherical heads, as shown in the figure; cone head rivets have heads shaped like the frustum of a cone; steeple head rivets have conical heads; countersunk head rivets are formed in countersunk holes, and are flush with the plate in which they are driven. The shank of a rivet is the cylindrical body; rivet grip is the length of the rivet, between the heads, after driving. (For methods of hot riveting, see **Forging**.)

Mass-production rivets are generally headed cold by hammer blows, by the application of pressure, or by spinning over the head of the rivet. Rivets are made of the softer non-ferrous metals such as brass and aluminum alloys, as well as iron and steel. Hammered rivet heads

are formed on helve-type machines; the hammer often has a rotary motion in combination with its vibratory motion. Aluminum alloy rivets may be successfully headed by pressure in a hydraulic press; often a number of rivets are headed in one operation by the use of a suitable fixture or die.

The figure illustrates two methods of rivet head spinning. Spun rivets may be headed by using a coni-

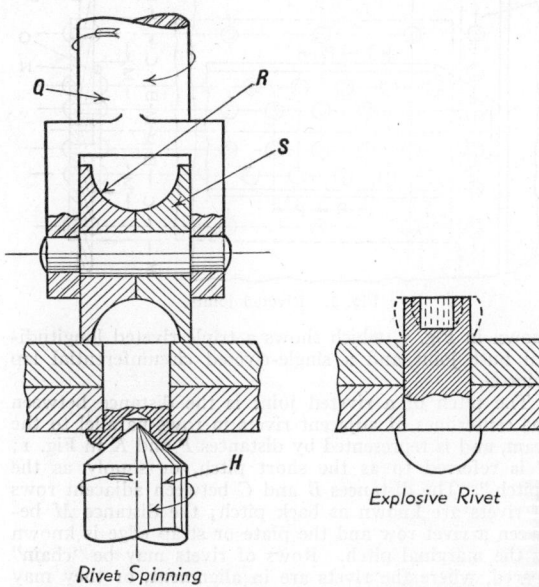

Rivet Spinning

Explosive Rivet

cal-pointed rotating header H in a rivet-spinning machine which resembles a drill press in principle. This process requires a blind hole B in the end of the rivet, but is extensively used on small parts such as typewriter and sewing machine linkages. Spun rivets with button or round heads may be headed by employing a rotating spindle Q with twin rolls R and S, which are free to rotate on an axle A in the spindle fork.

Another type of rivet, shown at the right, is used by aircraft manufacturers for operations where the riveting must be done from one side of the part, thereby dispensing with the customary need of backing up the preformed head during the operation. The projecting shank of the rivet has a small hole filled with an explosive; when the rivet is inserted, the head is touched with a hot iron, which sets off the explosive and effects the deformation shown by the dotted lines. Another type of blind rivet is hollow, and is formed without backing-up by drawing a center pin through the rivet after it has been inserted.

In pressure vessel design and in structural fabrication, rivets are usually placed so that failure will occur by shear of the rivet or by crushing of the rivet body or the plate. (H.C.H.)

RIVETED JOINTS. Seams or joints for pressure vessels may either be riveted or welded, and are referred to as longitudinal or axial joints, and as circumferential or ring joints. (See **Pressure Vessels**.) Riveted joints may be either lap joints or butt joints. In lap joints, the two plate edges to be joined are lapped over each other, and rivets are driven through both plates. Lap joints are subjected to noncollinear forces which form a couple and tend to bend the plate and the rivets, resulting in a tendency to split and tear out the plate. For this reason, lap joints are rarely employed for longitudinal seams if the plate thickness exceeds $\frac{1}{2}''$. They are extensively used, however, for girth and head seams, where the cylindrical form of the telescoping sections provides sufficient rigidity to eliminate this bending action.

In butt joints, the two plates forming the vessel shell are "butted" or placed together, and the connection is effected by butt straps placed on the inside and the outside of the plates. A connection of this character is

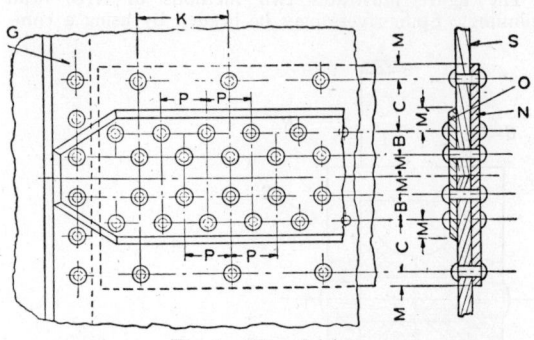

Fig. 1. Riveted joints.

shown in Fig. 1, which shows a triple-riveted longitudinal butt joint and a single-riveted circumferential lap joint G.

The pitch of a riveted joint is the distance between the centerlines of adjacent rivets in rows parallel to the seam, and is represented by distances P and K in Fig. 1; P is referred to as the short pitch, or simply as the "pitch." The distances B and C between adjacent rows of rivets are known as back pitch; the distance M between a rivet row and the plate or strap edge is known as the marginal pitch. Rows of rivets may be "chain" spaced, where the rivets are in alignment, or they may be "staggered," as shown in Fig. 1. (H.C.H.)

RIVETING. Forging.

ROAD-RUNNER. Aves, Cuculiformes. A long-tailed desert bird (**Aves**) of the southwestern United States. The species, *Geococcyx californianus,* is closely related to the cuckoos. It is useful as a destroyer of harmful insects and reptiles. The name comes from its ability to run rapidly, a power which very nearly supplants flight. Also called the snake bird, chaparral cock, and ground cuckoo. (A.W.L.)

ROARING FORTIES. Circulation of the Atmosphere.

ROASTING. Calcination.

ROBBER FLY. Insecta, Diptera. A predacious **fly** of the family Asilidae. Many of these flies are large and all capture living insects as prey, including bees of all kinds. Most of the included species have smooth slender bodies but some are quite hairy and one group is characterized by stout form and dense vestiture. These last mimic bumblebees closely and furnish an apparent case of aggressive mimicry, since the bumblebees are said to be among their victims. (A.W.L.)

ROBIN. Aves, Passeriformes. 1. The European redbreast, *Erithacus rubecula,* a **warbler,** and a related species of the Canary Islands. 2. The American robin, *Turdus migratorius,* a **thrush.** 3. In Australia a species related to the **wheatear** and in New Zealand other birds of the same group. (A.W.L.)

ROBOT PILOT. Automatic Pilot.

ROC, RUC. A semifabulous bird (**Aves**) of enormous size, said to be capable of carrying an elephant in its talons. The tales have been traced to a gigantic bird of Madagascar, somewhat like the ostrich in form, which is now extinct. From the scattered remains and egg shells of these birds, the genus *Aepyornis* has been established, containing several species. (A.W.L.)

ROCHELLE SALT. Tartaric Acid.

ROCHES MOUTONNÉES. The term given by de Saussure, in 1796, to glaciated rock hills resembling the wigs which were fashionable during the late 18th century. Typical roches moutonnées have rounded surfaces sloping gently in the direction of the ice movement with steeper slopes on the lee side, due to the plucking action of the ice. (R.M.F.)

ROCK DRUMLINS. Drumlins.

ROCK SALT. Halite.

ROCKET ASSIST. This term refers to the use of a powder rocket to assist the take-off of an airplane. The thrust of the rocket is taken by the airplane structure and supplements the propeller thrust. Thus heavily loaded fighters and bombers can be assisted to rise off the flight decks of carriers, landplanes may be operated from air ports with more limited runways than would otherwise be needed, and flying boats with their moderate power loadings can be cleared from the water, this being especially helpful when water surface conditions are adverse for take-off.

A rocket containing 25 lbs. of cordite is capable of producing a thrust of over 1000 lbs. for a few seconds. It is claimed that the take-off of a fighter plane equipped with 4 such rockets can be accomplished in about 4 seconds. Rockets are mounted externally and fired electrically. When employed on high-speed military airplanes the supports and rocket tubes are jettisoned after take-off, but this would not necessarily be done where cost was important and speeds moderate. The following precautions must be observed when mounting rockets for assisted take-off.

1. Rocket efflux must miss the empennage and not be reflected from ground or deck back onto any part of the airplane.

2. The position of attachment should be such that the thrust lies in a plane, nearly horizontal (in take-off attitude), which passes through or close to the center of gravity. Thus stability and take-off control are not radically altered.

3. Possibility of shock waves impinging on empennage with adverse results should be investigated.

4. Rockets should be as close as possible to the plane of symmetry of the airplane in order to maintain control if one of the distributed rockets fails to ignite. (F.T.M.)

ROCKET FLIGHT. The flight of a rocket can be conveniently divided into the following phases:

1. Powered ascent during which the fuel is burned and an acceleration given to the rocket.

2. Upward drift as a free projectile, subject to the laws of exterior ballistics. In case the rocket is shot horizontally, as in antitank fire, this phase obviously does not exist.

3. Descent as a freely falling body, modified by air resistance.

4. Final period of deceleration for safety in landing. This also is intentionally non-existent in most rocket flights of the present time.

These phases are all shown in Fig. 1. During the first phase fuel is burned at approximately a constant rate. Where M_1 is the initial mass, M_2 the final mass, the fuel and oxygen combined become $M_1 - M_2$. If the amount of fuel carried is sufficient for T seconds of combustion the average thrust produced is

$$F = \frac{M_1 - M_2}{T} v.$$

In the above equation v is the jet velocity, relative to the rocket, based on complete expansion to atmospheric pressure.

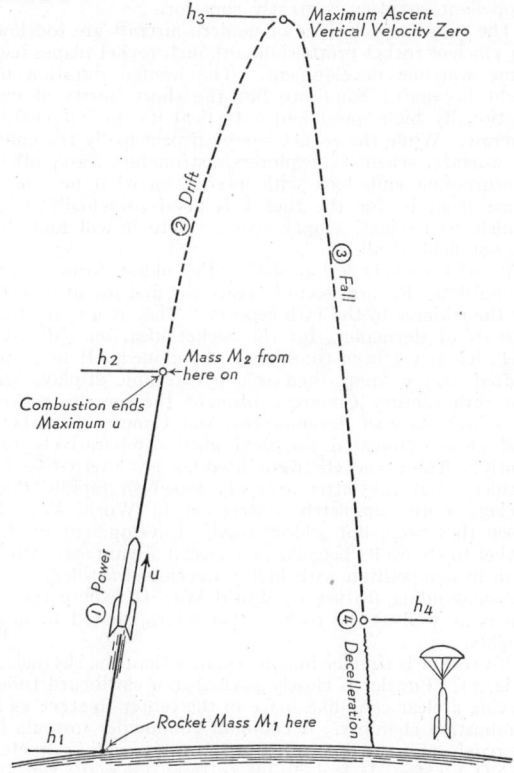

Fig. 1.   Four phases of a rocket flight.

Except for the effect of air drag and gravitation, the maximum rocket velocity attained (relative to earth) is $u_2 = v \log_e \dfrac{M_1}{M_2}$. The actual velocity may sometimes approach half of this. Calculations for speed and altitude reached at the end of phase one are complicated and will not be shown here. In phase two the rocket drifts upward against the deceleration of gravity until brought to rest—or until its vertical velocity component is absorbed. If the end of the first phase has brought the rocket into comparatively thin air, and the trajectory is nearly vertical, then the additional height gained during this phase of the flight is $u_2{}^2/2g$, in which u_2 is the velocity at the beginning of "free" ascent. Future rocket development may see phase one extending to several hundred miles' altitude; then the diminishing acceleration of gravity would have to be taken into account in estimating the free ascent. An equation based on negligible air resistance and a vertical component of rocket velocity of u_2 at the outset of free ascent is:

$$h_3 = \frac{K_1 h_2 + K_2 u_2{}^2 (K_2 + h_2)}{K_1 - u_2{}^2 (K_2 + h_2)}$$

$h_3 =$ Maximum altitude gained by the rocket, ft.
$h_2 =$ Altitude at beginning of "free" ascent, ft.
$u_2 =$ Velocity at beginning of "free" ascent, ft. per sec.
$K_1 = 29 \times 10^{15}$.
$K_3 = 2.12 \times 10^7$.

This same equation would serve to give the velocity at the end of the free fall from height h_3 by replacing h_2 with h_4, the altitude at which phase three of the trajectory goes into the fourth phase. However, since h_4 might be very near sea level (or might be zero for a rocket having no deceleration period), air resistance

should not be neglected. To date small experimental and meteorological rockets have either been allowed to crash or were landed by a parachute. Parachutes are usually opened as soon as descent begins. However, in possible future crew-carrying rockets for extreme altitudes, the deceleration phase will be a topic of major interest—certainly to the crew.

When rockets depart, if ever, from the earth's atmosphere, the navigation problem goes over into astronomical terms. Considerable thought has been given to the problem of decelerating the returning rocket. It might approach earth from a journey into interplanetary space at such terrific velocities that to make a direct approach through the earth's atmosphere, even supposing a successful final deceleration could be accomplished, would be to invite the same disaster suffered by meteors: burning up from friction. It has been theorized that the returning rocket would have to be navigated in a manner to approach the earth's atmosphere tangentially as shown in Fig. 2. By grazing through the upper atmos-

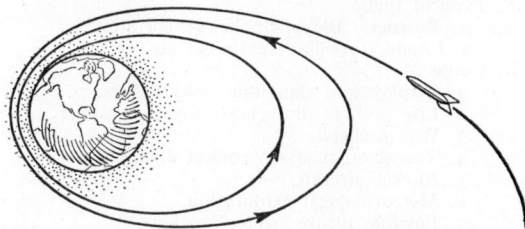

Fig. 2.   Return of space-rocket to earth.

phere on several deceleration loops (based on **Keplerian Laws**) around the earth, each successive one being allowed to pass through air of greater density as the rocket is slowed down, sufficient deceleration might be obtained to let the rocket be safely grounded by parachute, aided by a final burst of reverse rocket power. Some of the problems of interplanetary navigation by rockets have been investigated in great mathematical detail by astrophysicists.

The equation (for h_3) given above will show that, in order to escape the earth's gravitational attraction, i.e., $h_3 = \infty$, the necessary vertical velocity of a projectile is 7 mi. per sec. This, of course, is in the absence of air drag. However, if a rocket were involved instead of a projectile, the rate of combustion might be adjusted to give moderate accelerations until the bulk of the atmosphere was left behind, say at 500 mi. up. The necessary velocity at the beginning of a free ascent to leave earth would still be about 7 mi. per sec. This is called the velocity of escape, or *velocity of liberation*.

The instantaneous ideal efficiency of propulsion of a rocket is zero at the start of a flight, rises to 100% when $u = v$, then begins to decrease. It has been shown by Sänger and others that the average efficiency prevailing from start, when $u = 0$ until speed has reached u, is maximum when $u_1 = 1.6v$. Since the mass ratio of the rocket equals $\log_e \dfrac{u}{v}$, the most efficient use of fuel is therefore when $M_0/M_1 = \log_e{}^{-1} 1.6 = 5$, approximately. This means that only 0.2 lb. of rocket body is permissible per pound of propellant carried.

Using gasoline and liquid oxygen as the propellants, an ideal combustion jet velocity of 14,500 ft. per sec. is indicated, but an actual velocity of $v = 9000$ ft. per sec. would be an achievement. The mass ratio of 5 which yields $u_1 = 1.6v$, neglecting gravitation and air drag, thus is seen to fall short of the 37,000 ft. per sec. necessary to escape from the earth. But if by trimming weight here and there the rocket body should be reduced to, say, .12 lb. per pound of propellants, then, for each pound of propellants, .08 lb. of another "pick-a-back" rocket could be carried along to be fired when the first rocket had reached an optimum altitude. Thus some faint prospect of a space rocket is held out by the step-rocket principle. Alternately, energy sources yielding far higher jet velocities might be sought. When and if earth-science succeeds in dispatching a rocket into outer space it will likely be a one-way unmanned shot to the moon, relying on the crash landing creating some surface disfiguration there to advise us visually that the project was successfully terminated. (F.T.M.)

ROCKETS. Rockets are reaction-propelled bodies which carry all the oxygen needed for combustion along with them on their flight. In contrast, **jet engines** carry only fuel and rely on the surrounding atmosphere for oxygen to support combustion. Jet engines must remain in the atmosphere. Rockets travel best in the absence of an atmosphere; they are the only propulsion means yet developed which may ultimately solve the problem of the "space ship."

Rockets are to be classified according to the fuel type, and by their usage.

A. Type of fuel.
    1. Powder. Black, smokeless, cordite.
    2. Liquid. Alcohol, gasoline, etc.
B. Usage.
    1. Display and signaling. Skyrockets.
    2. Life saving. Breeches buoy line carrier.
    3. War rockets.
    4. Thrust augmenter (**rocket assist**).
    5. Rocket aircraft.
    6. Meteorological exploration.
    7. Possible future "space" rockets.

There have been several ill-considered attempts to employ rockets for purposes foreign to the nature of the rocket. Thus rocket-propelled automobiles are a dismal failure because the operating conditions do not permit an efficiency approaching that of the internal combustion engine. Rockets are high-speed affairs. For them, 50, 100, or 200 m.p.h. is a snail pace. Until speeds are attained exceeding those of our fastest airplanes, rockets will remain comparatively inefficient.

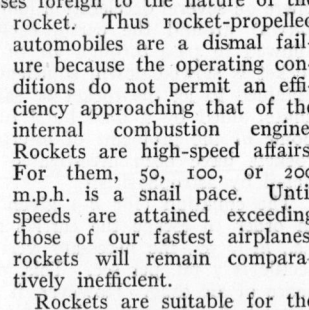

Fig. 1. Skyrocket.

Cap (Air splitter)
"Payload"
Powder charge
Card board tube
Combustion space
Clay nozzle
Paper plug
Fuse
Guide stick

Rockets are suitable for the creation of high velocities without recoil on the launching apparatus. This property gives them peculiar military significance. In rocket form heavy projectiles may readily be fired from small boats, aircraft, and by infantrymen, whereas the same projectile shot, as from artillery, would be accompanied by such a recoil that neither man, boat, nor aircraft could stand up to it. So in World War II small landing craft used in amphibious attacks were packed with fire power equivalent, it was said, to the guns of a cruiser by mounting on them multiple war rocket launchers. Two-man teams delivered the wallop of a heavy piece of antitank ordnance with the "bazooka," a short-range high-velocity rocket of 3½ lb. weight with high explosive war head. Comparatively small warplanes carried several rockets mounted on launching rails usually attached below the wing. This gave them an offensive power nearly equivalent to the mounting of cannon in airplanes —a wartime project that was never completely success-

ful. But portability and simplicity of launching the missile are the war rocket's chief advantages. In long-range aiming accuracy, and efficiency of use of the propellant, artillery is greatly superior.

The subsonic velocities of modern aircraft are too low for efficient rocket propulsion, although rocket planes had some wartime development. The limited duration of flight became a handicap, but the short bursts of exceptionally high speed had a tactical use in interceptor aircraft. While the rocket has been principally the child of warfare, scientists, explorers, astronomers, and other venturesome souls look with interest on what may next come from it, for the rocket is good principally as a vehicle to navigate empty space. Here it will find the biggest field of all.

Powder rockets are probably the oldest form of jet propulsion. Records seem to place the first use of rockets by the Chinese in the 13th century. They reappear after periods of dormancy, for the rocket idea, once discovered, has never been completely abandoned. It first appeared as a weapon, then as a pyrotechnic display. In the 19th century Congreve brought the powder rocket to a high state of development, and Congreve's rockets and ideas dominated the field until comparatively recently. These rockets were used as an alternative to artillery, but the latter arm developed so rapidly that rockets were completely outclassed in World War I, when they were but seldom used. Development of the rocket to secure its peculiar and special advantages rather than in competition with highly developed artillery, was an outstanding feature of World War II, where rocket usage, as well as the rockets themselves, soared to new heights.

No rocket is simpler in construction than the skyrocket (Fig. 1). Powder is closely packed in a cardboard tube, leaving a clear cone-like space in the center to serve as a combustion chamber. A common gunpowder formula is charcoal (C) 15%, sulfur (S) 10%, and saltpeter (KNO$_3$) 75%. It is desirable to slow down the rate of combustion a little in rockets by using less of the oxidizing substance in the mixture. A typical rocket powder is KNO$_3$ 56, C$_{32}$, S$_{12}$. A clay nozzle is formed inside the rocket tube in order to form and guide the jet. When the powder is ignited, by means of a fuse, combustion gases form quickly at a pressure in the central chamber and are violently expelled, carrying along bits of burning carbon. As the powder burns the combustion

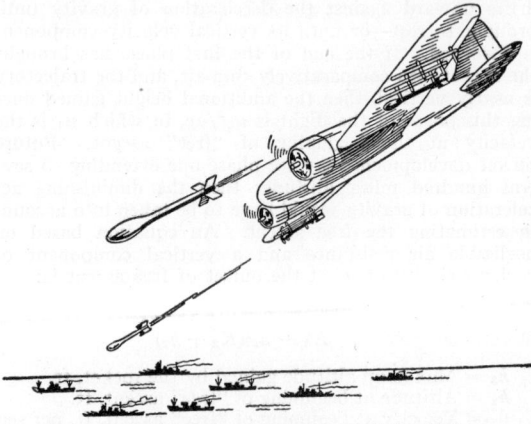

Fig. 2. Aerial rocket bombs launched by airplane.

chamber volume increases and the rate of combustion also. The powder is quickly consumed, but the acceleration produced is quite high. A rocket weighing initially 2 lbs., charged with 0.7 lb. powder, would produce an average thrust of 11 lbs. if it burned uniformly and had a jet velocity of 1500 ft. per sec. This can give the 2-lb. rocket an acceleration of 5.5 g.

The velocity theoretically attained, neglecting air drag and gravitation, is obtained from the differential equation for momentum.

$$mdu + vdm = 0,$$

whence, by appropriate integration,

$$u_2 = v \log_e \frac{M_1}{M_2}.$$

m = Mass of rocket when its velocity is u.
v = Jet velocity relative to the rocket nozzle.
u_2 = Final velocity of the rocket.
M_1 = Initial rocket mass.
M_2 = Final rocket mass.
$M_1 - M_2$ = Weight of powder.
M_1/M_2 = "Mass ratio."

With rocket quantities as mentioned above, the mass ratio is $2/1.3$, with a natural logarithm of 0.432. So

Fig. 3. Rocket plane—1945. Limited range—high speed (550 m.p.h.).

the ideal maximum velocity is 43.2% of jet velocity or $650'$ per sec. Were the rocket aimed for vertical ascent, part of the thrust would overcome gravitation, part would overcome air drag and the remainder would accelerate, with a final speed probably not over $400'$ per sec. Still, this is 270 m.p.h.

The ideal maximum **combustion jet velocities** of gunpowder are low compared with other fuels. Also the ratio of realized to theoretical jet velocity is lower than with liquid fuels because considerable amounts of fuel are thrown out of the rocket in the form of fiery sparks (carbon). This, along with the better control over rate or combustion afforded by liquid fuels, gives the latter a decided advantage and indicates superiority of the liquid fuel rocket for long-distance rocket shots.

Liquid fuel rockets have been built to use gasoline or alcohol as fuels and liquid oxygen as the oxidizing agent. Analine, nitric acid, hydrogen peroxide also have had some use. These liquids are held in separate supply tanks from whence they flow in proper proportions into a combustion chamber. If mixing is adequate, the fuel is almost immediately oxidized, giving up its heat of combustion to the products which thereupon attain a high pressure and temperature. Fig. 4 illustrates the principle of the liquid fuel "motor." Rocket history of the past 2 decades is in large part a record of the trials and tribulations of experimenters seeking to perfect a successful liquid fuel rocket motor. Always fighting weight, these pioneers had to overcome problems of the melting and erosion of combustion chambers, proportioning of fuels, fuel feeds, handling methods suitable for liquid oxygen, the stabilization of flight paths, and a host of others. Liquid fuel rockets are still in the experimental stage. In the United States Dr. Robert Goddard led the way in this research with a succession of liquid fuel rockets of progressively better designs, his work extending over nearly 2 decades, 1920 to 1940.

A diagram of the liquid fuel rocket patterned after some of the early experimental rockets is shown in Fig. 4. A combustion chamber forms the head of the rocket.

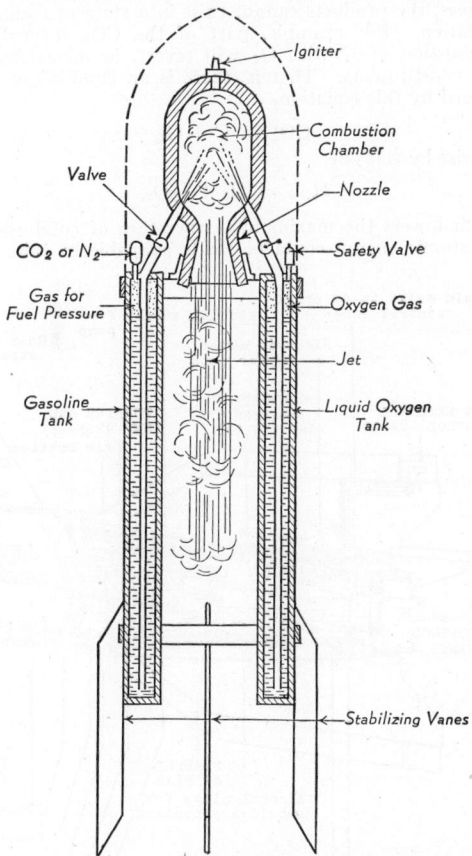

Fig. 4. Early type liquid fuel rocket.

Into this chamber are injected liquid oxygen and gasoline at as steady a rate as possible. Initially ignition is secured from an electrical igniter, after which combustion maintains itself until the fuel or oxygen supply is exhausted. The rate of fuel feed and the tank capacities determine the magnitude of thrust and duration of the powered ascent. Combustion produces high temperature products. The opening through which these may seek release from the combustion chamber will be made sufficiently small so that the products of combustion must build up a considerable fluid pressure before they flow out through the nozzle as rapidly as they are formed. This expansion to atmospheric pressure produces a jet of high velocity which will, however, fail by a large margin of attaining ideal maximum jet velocity. The reasons for this failure are:

1. The oxygen and fuel are not injected in the ideally correct proportions, or else they are not sufficiently mixed before or during combustion.

2. Combustion is not completed before the gas enters the nozzle because of the

a. Time factor, i.e., small chamber and high rate of combustion.

b. Dissociation factor, i.e., incomplete combustion within the chamber because of the thermal equilibrium of a percentage of fuel and oxygen possible at elevated temperatures.

3. Expansion ratio is not great enough to lower the internal energy of the exit gases to the initial state.

With regard to item 2b, long ago it was observed by Dugold Clerk that there was a discrepancy between the actual observed temperature of combustion and the temperature which might be expected to exist by reason of the potential heat, mass, and specific heats involved in the process. When temperatures are several thousand degrees, the products cannot exist in a state of complete oxidation. For example, part of the CO_2 formed by combustion of C with O_2 will revert, or *dissociate*, to free constitutents. That is, there is an equilibrium, expressed by this equation,

$$CO_2 \rightleftarrows C + O_2,$$

likewise by this one,

$$H_2O \rightleftarrows H_2 + \tfrac{1}{2}O_2,$$

which lowers the maximum temperature of combustion. (If atomic energy could be made available to raise the

The rocket in Fig. 4 is poorly shaped for high-velocity flight. It represents a type in use when the possibility of successful liquid fuel rocket flight of any sort was in doubt. The trailing tanks and fins gave it a skyrocket-type of stability. For several reasons, rear end position of the motor is preferable, particularly in high-speed flight. The necessary rocket contents, fuel, controls, etc., can be located in a long, sleek, cylindrical case, sharply pointed at the nose to help penetrate the air at sonic speeds. Stability in flight has been secured by the use of stabilizing fins and gyro-operated control vanes. Examination of the German liquid fuel rocket (V-2) will convey some idea of the progress that already has been made in this field.

As the accompanying illustration shows, the rocket body is cylindrical and elongated. The sharply pointed

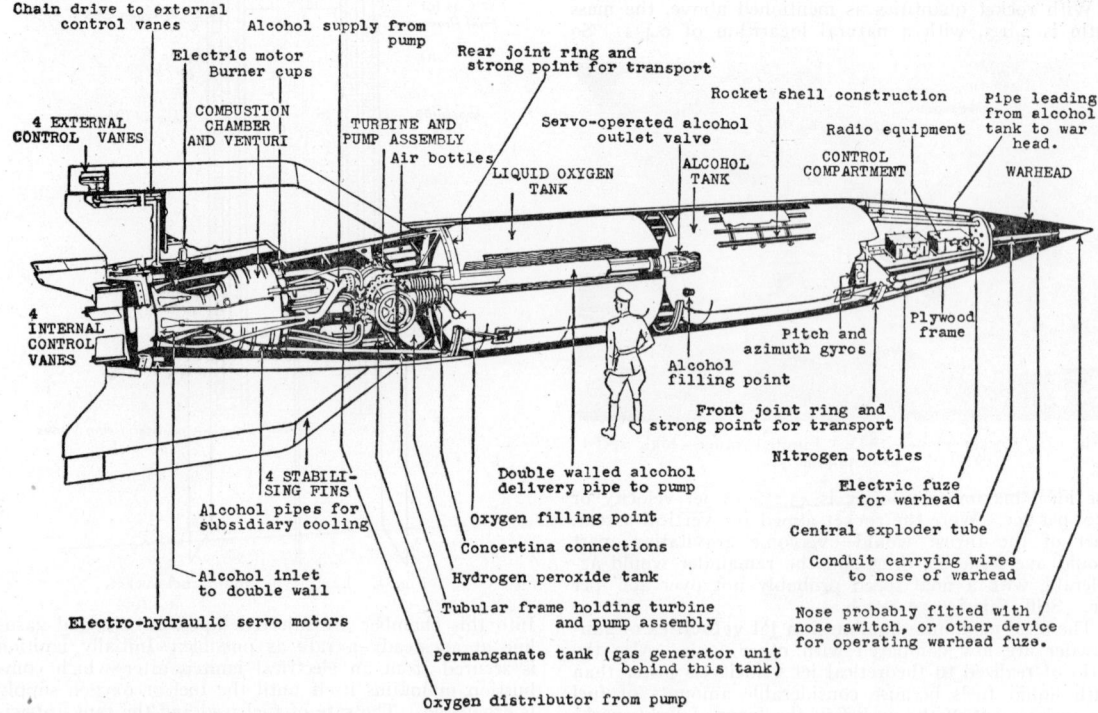

Fig. 5. Liquid fuel rocket—1945. Sectional drawing shows the German V2 rocket bomb. (*British Information Service.*)

temperature and therefore pressure of disintegration products in a rocket chamber, an ultra-high temperature would exist, as the energy is not derived from chemical combination.) However, the dissociation action takes some time for completion—which is not ordinarily allowed to it in the rocket combustion chamber and nozzle. Therefore, high chamber temperatures and nearly adiabatic expansion may be caused to prevail in rocket combustion.

During expansion the temperature falls and the dissociated products may eventually all recombine, liberating the full potential heating value. But since some of the combustion is thus accomplished during the expansion, perfect adiabatic expansion is destroyed, and the nozzle fails to convert the ideal available heat into kinetic energy. For this reason calculations based on equations premised on adiabatic expansion will be in some error for rockets but not as much so as for spark-ignition engines. The expansion through gas turbine and jet engine nozzles is not subject to the same consideration because the use of air rather than oxygen, and in considerable excess, too, usually insures that the temperatures are low enough (1000° F. to 1500° F.) to avoid any substantial amount of dissociation.

nose consists of a 1-ton explosive warhead in back of which are located the radio control equipment and the gyroscopic stability devices used for remote control of the flight. But the interior is mainly filled by tanks containing the liquid propellants. (The data here used are those currently quoted during wartime use of V-2. Although these data are probably a good approximation of the precise figures, it is necessary to point out that these are not the builder's design figures.) The after part of the rocket body is occupied by the combustion equipment. A turbine is used to drive rotary pumps which withdraw the alcohol and liquid oxygen from their tanks and force them into the combustion chamber. The efflux of products of combustion from this chamber takes place through a nozzle about 18 in. in diameter. What the combustion chamber pressure might be has not been publicly stated. It has been estimated that the jet velocity was about 6000 ft. per sec., and this, coupled with the fact that some 8.5 tons of propellants were injected into a hemispherical chamber about 4 ft. in diameter in a little over a minute, implies that the combustion chamber pressure was of a high order. From its pump the fuel is led through jackets surrounding the nozzle and combustion chamber to cool them enough to prevent their destruction. As the heat thus absorbed is retained by the fuel on its way to the burners, this system is called regeneration, and the rocket motor designated a *regenerative motor*. The warm alcohol is mixed with liquid oxygen in some 18 burner cups attached to the forward surface of the combustion chamber. The operation of the rocket is supposed to be about as follows. The fuel and oxygen tanks are filled, their relative capacities corresponding closely to those required for the ideal combustion of methyl alcohol with oxygen. Compressed air from small bottles then forces calcium permanganate into hydrogen peroxide, accompanied by an exothermic chemical reaction which results in the formation of superheated steam. The steam, of course, was produced solely for powering the pump turbine. The pumps transfer the oxygen directly to the burners, and the alcohol to them via the cooling jackets. Ignition is initiated electrically, and thereafter maintains itself as long as the fuel holds out and the turbine continues to turn. As soon as the jet builds up a reaction thrust equal to its weight the rocket takes off. Were the 8.5 tons burned in 70 sec., the exhaust would consist of 7.5 slugs per sec., which, moving at 6000 ft. per sec., would induce a reaction thrust of $\frac{6000 \times 7.5}{2000}$, or about 22.5 tons. The rocket is said to weigh 12 tons ready for launching, consequently a net 10.5 tons is available for acceleration. This would provide an initial acceleration of a little less than 1 g. As the fuel is consumed the rocket mass is diminished and the available thrust can propel the rocket at greatly increased acceleration. The combustion is said to propel this rocket about 25 miles high, after which its momentum carries it upward another 40 miles. U. S. Army postwar experimentation with those rockets increased the range to more than 100 miles altitude.

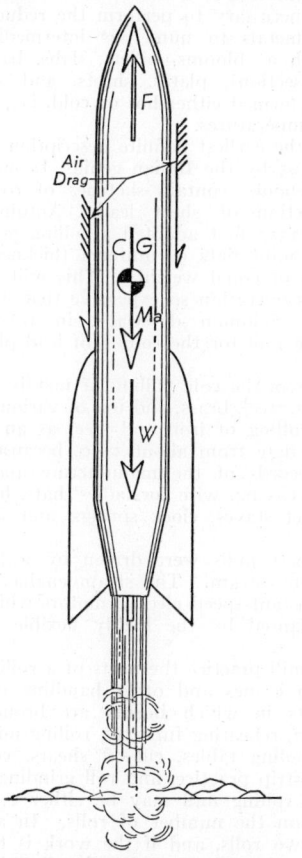

Fig. 6. Forces on the rocket.

The forces acting on a liquid fuel rocket aimed for vertical ascent are shown in Fig. 6. The exhaust jet produces a net reaction of F lbs., which, of course, must equal or exceed the sum of air drag D, and dead weight W. When F exceeds $D + W$, then $F - (D + W)$ is an accelerating force that can increase the velocity, u, of the rocket. An equation balancing the net thrust against inertia yields for vertical rocket acceleration $a = \frac{g}{W} \times [F - (D + W)]$ in which g is the acceleration of gravity. Supposed variations of some of these quantities with extreme altitude are shown by the accompanying graph.

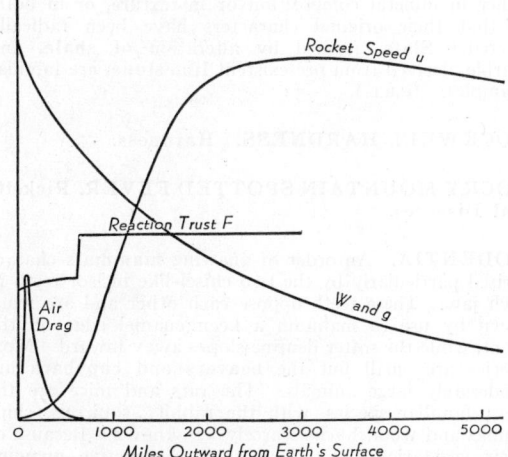

Fig. 7. Supposed variation of factors in rocket repulsion. (Low initial combustion rate to keep speed and air drag limited until the atmosphere is left behind.)

Rocket flights to a few miles of height above the earth's surface are today's wonder. (See **Atmosphere**.) Ascents to such heights that the gravitational field is noticeably weaker will probably come some day. Several books have been written which seriously weigh the pro's and con's of rocket flight into interplanetary space. From a former state of pessimism following analyses of requirements to be met in attempting a rocket-to-the-moon, there now appears a trend of thoughtful reasoning which admits the possibility. (F.T.M.)

ROCK-FLOUR. A peculiar and distinctive white mud the product of the grinding action of **glaciers**. When deposited in lakes rock-flour forms an important constituent of **varves**, or the type of annual cyclic stratification used by glaciologists in determining glacial time. When desiccated and transported by wind this material may form extensive deposits of **loess**. (R.M.F.)

ROCKS. In geologic usage, rock is the material composing the outer part or "crust" of the earth. It is a popular concept that the term rock is restricted to hard or firm substances. To a geologist, however, a body of clay or of volcanic ash is as truly rock as is a mass of hard granite. In general, a rock consists of one or more definite minerals. However, natural glass (**obsidian**), which has no definite composition and structure, makes large rock masses, as in Yellowstone National Park. Rocks composed of definite minerals may be either **simple** or **compound;** for example, the purest **marble** is formed entirely of the one mineral calcite, whereas **granite** consists of feldspar and quartz mixed in varying proportions, usually with minor quantities of mica and other accessory minerals.

All rocks are divided into three major classes, according to mode of origin. **Igneous rocks** are those that have been formed by cooling and solidification of molten masses derived from within the earth. Because of their origin they are sometimes called the primary rocks. In part they are lavas and other products of volcanoes; other masses solidify slowly below the earth's surface to form granite and other crystalline types of igneous rock whose mineral content depends chiefly on the chemical composition of the parent molten mass. **Sedimentary rocks** consist of material that formed a part of pre-existent rocks, and that was moved from its former position, deposited by the action of water, the atmosphere, or glacier ice, and subsequently converted into rock. Examples are conglomerate, sandstone, and limestone. **Metamorphic rocks** are those that were originally igneous or sedimentary, but have been changed either in mineral composition or in texture, or in both, so that their original characters have been radically altered. **Slate**, formed by alteration of **shale**, and marble, derived from pre-existent limestone, are familiar examples. (C.R.L.)

ROCKWELL HARDNESS. Hardness.

ROCKY MOUNTAIN SPOTTED FEVER. Rickettsial Diseases.

RODENTIA.
An order of gnawing **mammals** characterized particularly by the two chisel-like incisor teeth in each jaw. These teeth oppose each other and are worn down by use to maintain a keen enamel edge on the front, while the softer dentine slopes away inward. Most species are small but the **beavers** and **capybara** are moderately large animals. The rats and mice are the most familiar species, with the rabbits, squirrels, chipmunks and woodchucks scarcely less known. Because of their vegetarian habits and their destructive gnawing many rodents are serious crop pests, and as household nuisances they are all too common.

The **rabbits** and their allies are usually included under the Rodentia as a separate sub-order, the Duplicidentata, but are sometimes assigned to an order of their own, the Lagomorpha. (A.W.L.)

RODINAL. Photography.

ROEBUCK.
Mammalia, Artiodactyla. *Caprealus*. The male of the roe **deer** of the Old World. The species is small and the antlers of the male reach a length of little more than a foot. They are rough, and in normal specimens have only three tines. The roe deer is a woodland species. (A.W.L.)

ROENTGEN RAYS. X-Rays.

ROLL, AIRPLANE.
Roll is the angular displacement of an airplane about a longitudinal axis through its center of gravity. As a conventional airfoil is self-damping to roll, rolling instability is rare in aircraft. Roll is intentionally produced by the operation of **ailerons** which increase the effective camber of the wing on one side and decrease that of the other, thus unbalancing the lifting pressures and causing the existence of a rolling moment. The coefficient of rolling moment is defined as the moment directed by the product of span, wing area, and dynamic pressure. Rolls are used by pilots to neutralize "gustiness" and as part of the following common maneuvers—turns, slips, spins. (F.T.M.)

ROLLER.
Aves, Coraciiformes. Birds (**Aves**) of several species found in Africa, the Oriental region, and Europe and Asia. They are brightly colored birds, named from their habit of turning over in flight. (A.W.L.)

ROLLER BEARING. Bearings.

ROLLER CHAIN. Chain.

ROLLER LEVELING.
A method of flattening sheet metal by passing through a succession of bending rolls. A roller leveling machine consists of two parallel trains of small-diameter rolls mounted horizontally in a housing, one train above the other. Top and bottom rows are so off-set or staggered that the sheet passing through the leveler is flexed alternately up and down. The two layers of rollers are movable with respect to each other; so that by means of a screwdown the rollers may be made to mesh loosely or closely as desired, thus controlling the pressure exerted on the piece during the operation. On levelers with the proper amount of meshing of the rolls flatness may be obtained with essentially no increase in the length of the sheet, but the sheet is slightly cold worked by this plastic bending. Roller leveling helps to prevent **stretcher strains** in soft steel sheets. It does not ordinarily produce sheets as flat as **stretcher leveling**, and roller-leveled sheets are used primarily for fabricated articles which are to be stretched considerably, hence the original flatness has no bearing on the shape of the finished article. (R. S. BURNS.)

ROLLE'S THEOREM.
Rolle's theorem is a mathematical result concerning the vanishing of the derivative of a function.

Let $f(x)$ be a **function** which vanishes at $x = a$ and at $x = b$, and which has a finite **derivative** $f'(x)$ at all points in the interval (a, b). Then $f'(x)$ vanishes at some point ξ between a and b. (L.L.S.)

ROLLING; ROLLING MILLS.
Rolling is the process of reducing the cross-sectional area of pieces of metal by passing them between revolving cylinders called rolls. A rolling mill consists essentially of the rolls set in a suitable framework or housing to support them. These rolls are connected to a motor which transmits to them the power necessary to perform the reduction. Rolling mills roll metals to numerous intermediate and final shapes, such as blooms, billets, slabs, bars, rods, rails, structural sections, plates, sheets, and strip. Rolling may be performed either hot or cold, i.e., at elevated or at room temperatures.

Perhaps the earliest definite description of rolling was given in 1495 by the Italian genius, Leonardo da Vinci, whose notebooks contain sketches of rolling mills for the production of sheet lead. Antoine Brulier, of France, in 1552 first adopted a rolling mill to the production of mint flats of uniform thickness in order to make coins of equal weight. This mill was so imperfect and its operation so expensive that its use was discontinued. Solomon de Caus, in 1615, describes a hand-driven mill for the rolling of lead plates for organ pipes.

About 1800, the cold rolling of metals was in general use for iron, steel, brass, and foil of various other metals. The cold rolling of iron and steel as an industry may be said to date from about 1830, because at that time we find records of the manufacture and use of steel reeds for weaving, wire for ladies' hats, hoops for hoop skirts, corset staves, clock springs, and shank steel for shoes.

The earliest mills were driven by water power and later ones by steam. The steam engine, in turn, gave way to constant-speed electric motors which today have been supplanted by the highly flexible individual d-c motor.

In steel-mill practice the parts of a rolling mill are, in addition to cranes and other handling equipment, the soaking pits in which **ingots** are brought to rolling temperature, reheating furnaces, rolling mills, straightening and cooling tables, cut-off shears, coilers and decoilers for strip practice, and roll grinding machines. A stand in a rolling mill may be either 2, 3, or 4 high, depending on the number of rolls. In a 2-high stand there are two rolls, and if the work is to be passed a second time through the rolls, the direction of rotation must be reversed, or the work may be passed back over the top roll for reinsertion in the mill. In a 3-high mill there are three rolls, the lower and upper turning in the same direction. A return pass can be made between the middle and upper rolls without reversing the drive. Several passes may be made in one stand as in the blooming mill illustrated.

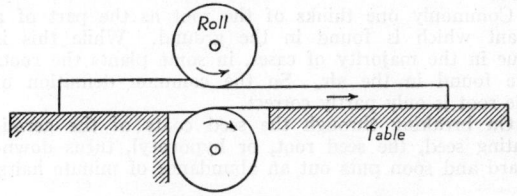

Principle of the rolling mill.

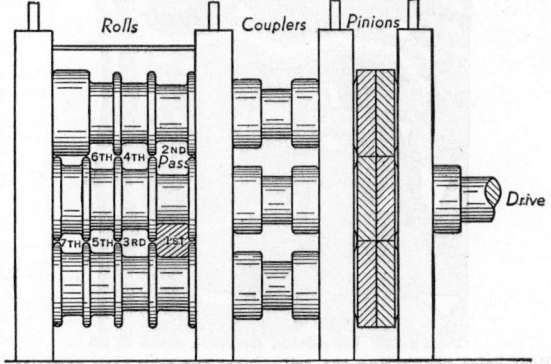

Three-high blooming mill.

Because of the demands created by the automotive, refrigeration, and furniture manufacturers, progress in recent years has probably been most rapid in the production of flat rolled steel, i.e., sheets and strip. This particular phase of rolling, therefore, will be covered in more detail. The production of flat rolled steel products by what is commonly known as "sheet bar practice" prevailed in the early part of the 20th century. This practice consisted in rolling ingots into rectangular bars 8–12″ wide and from ⅜–3″ in thickness and then further reducing by cross-rolling into sheets.

In 1924, strip was obtainable in widths up to 24″ and in lengths as long as demanded, and sheets were obtainable in widths as wide as were demanded (54″) but were limited in length and thickness. As a result of the development of continuous wide strip mills, cold rolled iron and steel sheets up to 90″ wide are available in either cut lengths or coils. The rolling equipment of a modern integrated steel mill now consists of a blooming mill which rolls ingots to slabs or billets, roughing mills, Universal mills and rod and section mills which further reduce these semi-finished products to bars, plates, rods, sections, etc., and continuous hot and cold strip mills which reduce slabs or bars to hot and cold rolled sheet and strip. Hot rolling of ingots to semi-finished products is performed at temperatures ranging from 2300° F. down to 2100° F. Hot rolling of these semi-finished products into other semi-finished and finished products is performed at temperatures between 2200 and 1300° F. In the production of cold-rolled sheet and strip and cold-drawn rods and bars of various dimensions, the oxide is removed from the hot-rolled product by **pickling** and these sections are further reduced on cold rolling mills.

Bars, sheet, and strip are cold rolled or cold finished to final dimensions for several reasons, the most important of which are:

1. To obtain the desired surface finish.
2. Improve dimensional tolerances.

3. To impart desired physical properties.
4. To make gauges lighter than can be produced on hot strip mills.

There are many different kinds of cold reduction mills. The one which enjoys by far the most general use is the conventional 4-high type in which there are two driven work rolls and two larger diameter rolls which back up or support the work rolls. Four-high mills are in use either in tandem or as reversing mills, wherein the strip is coiled on either side of the mill and run back and forth between the rolls until the desired thickness is obtained. Another type of reversing mill which is used to a lesser extent is the Steckel mill in which the work rolls are not driven and all power to perform the reduction is supplied through tension applied to the strip. There are many types of "cluster mills" wherein the supporting or back-up rolls are clustered around small work rolls to give them both vertical and horizontal rigidity. The use of small work rolls is beneficial from the standpoint of making large reductions, particularly on hard materials. (R. S. BURNS.)

ROOF. When a region is covered above to shed rain or snow, the covering is known as a roof. It is an essential feature of all buildings, sheds, etc. A roof consists of a weatherproof covering supported on a smooth surface which, in turn, is supported by certain structural mem-

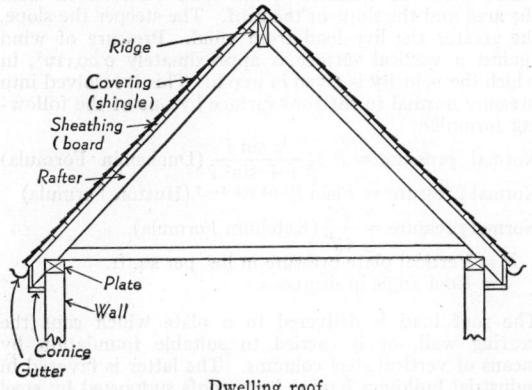

Dwelling roof.

bers. In ordinary building construction, the roof covering is supported by a wood sheathing which is laid on rafters. The latter are structural members of wood which, in the case of ordinary dwellings, form the basic structural members of the roof, but which in the case of large roofs, are themselves laid on purlins (see **Bent**)

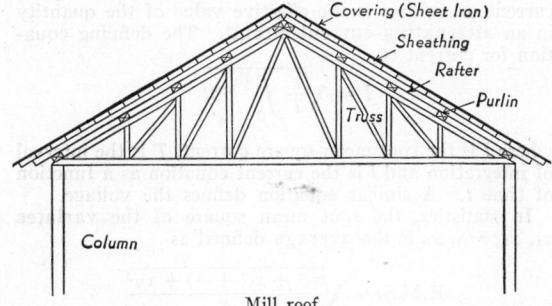

Mill roof.

which are supported by wood or steel trusses. The most common roof covers are slate, composition or wooden shingles, composition roofing paper, galvanized iron sheets (either smooth or corrugated), tinned sheets (smooth), and tile. The type of roof depends upon the expense which the owner is willing to incur in order to secure permanence and also on the kind of building which the roof is to cover. Typical roofs are shown in the accompanying diagram. The loads to which these roofs

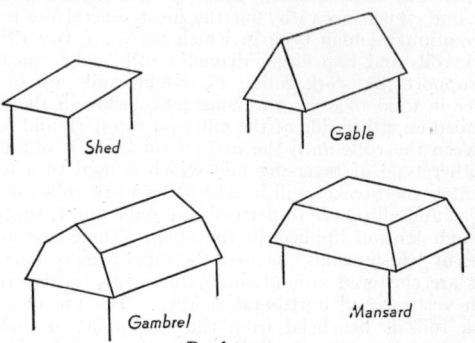

Roof types.

are subjected consist of dead load and live load. The dead load is the weight of the roof itself plus snow, and any structures that may be erected on the roof or suspended from it. The live load is wind or any movable loads which may be suspended from the roof, or which may roll over it. The dead weight varies with the composition of the roof, and the allowance for snow varies with the climate. Wind load varies both with the area and the slope of the roof. The steeper the slope, the greater the live load from wind. Pressure of wind against a vertical surface is approximately $0.0032v^2$, in which the velocity is given in m.p.h. This is resolved into pressure normal to the roof surface by one of the following formulae:

Normal pressure $= P \times \dfrac{2 \sin i}{1 + \sin^2 i}$ (Duchemin Formula)

Normal pressure $= P[\sin i]^{1.84 \cos t - 1}$ (Hutton Formula)

Normal pressure $= \dfrac{Pi}{45°}$ (Ketchum Formula)

P = Vertical plate pressure in lbs. per sq. ft.
i = Roof angle in degrees.

The roof load is delivered to a plate which caps the bearing wall, or is carried to suitable foundation by means of vertical steel columns. The latter is favored in industrial buildings having large roofs supported by steel roof trusses. (F.T.M.)

ROOK. Aves, Passeriformes. A European bird (**Aves**), *Corvus frugilegus*, related to the crows. Its black plumage is glossed with purple and the face of the adult is usually naked and grayish. (A.W.L.)

ROOT MEAN SQUARE. The root mean square current or voltage is the effective value of the quantity in an **alternating-current circuit**. The defining equation for current is:

$$I = \sqrt{\frac{1}{T} \int_0^{T_2} i^2 dt}$$

where I is the root mean square current, T is the interval of integration and i is the current equation as a function of time t. A similar equation defines the voltage.

In statistics, the root mean square of the **variates** $x_1, x_2, \cdots, x_n$ is the **average** defined as

$$\text{R.M.S.} = \sqrt{\frac{x_1{}^2 + x_2{}^2 + \cdots + x_N{}^2}{N}}$$

and in a **frequency distribution** by $\sqrt{\dfrac{\Sigma x^2 f_x}{\Sigma f_x}}$. It is very seldom used. (L.R.Q., L.A.A.)

ROOT OF AN EQUATION. A root of an equation with one unknown is a value of the unknown which satisfies the equation, that is, which makes the two sides of the equation identical in value. (L.L.S.)

ROOT PRESSURE. Ascent of Sap.

ROOTS. In seed plants, the root is generally the first part of the plant to emerge from the germinating seed.

Commonly one thinks of the root as the part of a plant which is found in the ground. While this is true in the majority of cases, in some plants the roots are found in the air. So the common definition of the root is only partly correct.

On breaking through the seed coats of the germinating seed, the seed root, or hypocotyl, turns downward and soon puts out an abundance of minute hairs.

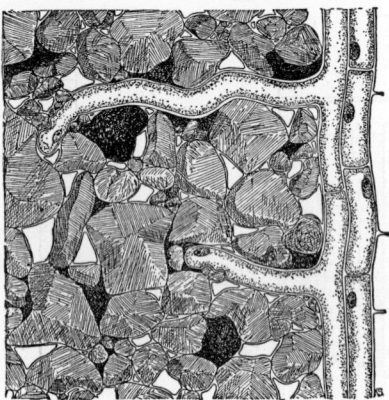

Root hairs penetrating the soil. Note the tiny rock particles, the bits of humus, the films of water, and the air spaces.

These serve the twofold purpose of attaching the hypocotyl firmly in the soil and of absorbing moisture and mineral solutions from the soil. The hypocotyl continues to grow into the soil, elongating and branching, and becomes the root.

Externally there are certain characteristics which distinguish a root from a stem, even when the latter grows underground. The root bears no leaves on its surface and is not separable into nodes and internodes, as is the case in the stem. The apex of the root is covered by a protective structure called the root-cap, a distinctive feature never found in stems. Just back of the tip, the surface of the root is usually provided with root hairs. Branching in roots is quite distinct from that in stems, the branch-roots appearing at irregular intervals; frequently these branch-roots are borne in longitudinal rows, a fact which is correlated with their origin. Branch-roots develop from the pericycle, a tissue deep in the root, and not from the sub-epidermal tissue as do stem branches. In many lower plants the root forks into two equal branches, a method of branching called dichotomous. In higher plants this does not occur.

Several types of roots are recognized. In many plants, the first formed, or primary, root continues to grow downward to form a long root which penetrates deep into the ground. Branch-roots arising from this are commonly much shorter, and of smaller diameter. Such a root is called a tap root, and the entire root system in such cases is known as a tap-root system. Familiar examples are found in the **dandelion**, burdock, and in oak trees. **Monocotyledonous** plants rarely show this form of branching. In them and in many other plants, the primary root soon loses its individuality, the secondary roots becoming larger, and forming an extensive root system, in which no single root is distinguished from the others by its larger size and more obvious downward growth. Such a system is called a fibrous root system, and the individual roots are known as fibrous roots. The common **plantain** has such a root system. In many plants the roots become important places of storage of foods, the roots often being conspicuously enlarged because of this storage. Such roots are called fleshy roots. Their existence is of great value

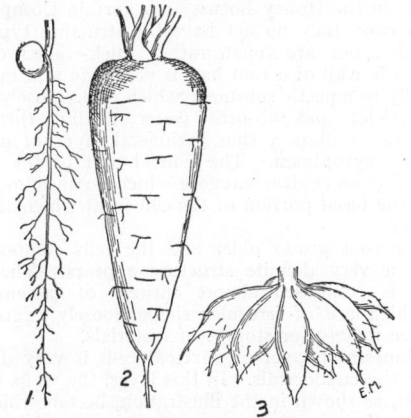

Different kinds of roots. 1, fibrous tap root of pea; 2, fleshy tap root of carrot; 3, fleshy fascicled roots of Dahlia.

to man, since many of them become important crop plants, as for example, carrots, parsnips, turnips and beets. In many plants, several roots are swollen with food so that a group or fascicle of storage roots is formed, as in the *Dahlia*.

In many plants, especially in the tropics, the roots are formed in the air. Such aerial roots are often quite different from ordinary roots, being much coarser, less branched and lacking root hairs. In aerial roots the outer portion is frequently modified to a spongy tissue capable of absorbing moisture rapidly and retaining it tenaciously. This tissue, the velamen, is particularly well developed in epiphytic orchids. Other plants have aerial roots which extend either downward from the branches, or from the lower part of the stem. Such roots are called, respectively, prop roots or brace roots.

A type of modified roots—Aerial roots of poison ivy, *Rhus Toxicodendron.*

The terms are often used interchangeably. These roots soon penetrate the surface of the ground and become like ordinary roots. The **Banyan** tree shows the downward-growing prop roots particularly well, while in the Screw-Pine and in **Corn** plants, brace roots are well-developed. They are especially conspicuous in the **Mangrove** plants, where they form an extensively branching system supporting the plant in the soft mud in which it grows. In many tropical plants, especially in those of large trees, the roots radiate out over the surface of the ground. Often these roots are conspicuously developed vertically, forming thin plates of considerable depth, but only a few inches thick; they are called buttress roots.

In **parasitic plants** the roots may be curiously modified or replaced by absorptive organs which penetrate the tissues of the host plant until they reach the conducting system, from which they obtain their nutrient supply. In some plants, roots are entirely lacking, their functions being taken over entirely by other parts of the plant. In the **saprophytic** Orchid, *Corallorhiza,* for example, the much-branched coarse underground stem, or rhizome, functions as a root; in the **Bladderworts,** *Utricularia,* floating in water, roots are entirely unnecessary and non-existent. In the **Spanish Moss,** *Tillandsia,* growing epiphytically in tropical and subtropical

America, roots are entirely lacking, absorption occurring over the surface of the plant.

Commonly the extent of a root system is very much underestimated. Only when the entire plant is removed from the soil by careful and extensive digging is the great spread of the entire root system seen. Then it is found that the roots may penetrate the soil to a depth much greater than was supposed. In common Red Clover, for example, the roots may go downward to a depth of 8 or 9′, while in **alfalfa** depths of 15–20′ or more are found. The lateral spread of roots is also often very great. In the common Squash the lateral roots may extend outward 10–15′, while other roots of the same plane penetrate the soil to depths of 4–6′, forming a very extensive system. The nature of the soil, and especially the amount of oxygen present in it, are important factors in determining the amount of branching and the extent of the root system. In general, porous well-aerated soil favors extensive branching.

Roots do not seek water, as popularly supposed from the fact that they seem to grow toward water. This tendency is explained by the fact that roots respond very definitely to certain external factors, such as gravity. Most roots react positively to gravity, that is, are positively geotropic, growing directly **downward** into the soil. If a young plant is placed so that its root is horizontal, in a short time it will be found that this root has changed its direction of growth and is growing downward again.

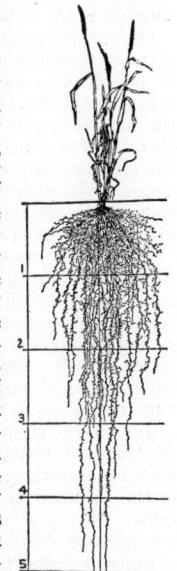

Root system of wheat plant at blossoming time. Figures indicate depths in feet. (*Weaver's Root Development of Field Crops,* McGraw-Hill Book Co., Inc.)

Many roots are also affected by light, from which they turn away, being negatively phototropic. Temperature also affects roots, so that a root encountering a region having a temperature more favorable for its growth will increase more rapidly than other roots. Entirely like this is the response of roots to favorable moisture conditions, which lead to greater growth of the roots extending in that direction. It is such phenomena which lead to the statement that roots seek water. Often they do seem to, especially when they penetrate joints in sewer pipes,

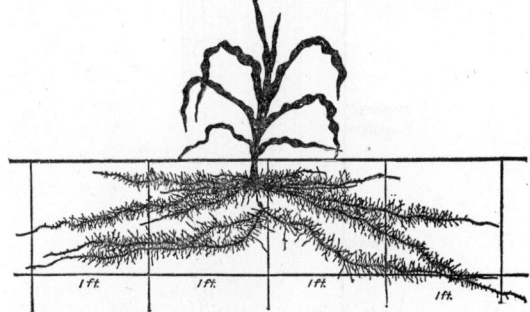

The root system of Iowa Silver Mine corn 36 days old. (*Weaver's Root Development of Field Crops,* McGraw-Hill Book Co., Inc.)

sometimes at distances of 30–60′ or more from the stem of the plant, and form great masses of much branched roots which may completely clog the sewer pipe, and become a source of great inconvenience and expense. Other factors such as the nature of the soil and its contained minerals, also influence the growth of roots.

The internal structure of the root is quite constant and distinct from that of the stem. In the tip of the root an apical meristem, the promeristem, or zone of dividing cells is found. The cells of this region are more or less cubical in shape, thin-walled, and contain a dense protoplasm and a relatively very large nucleus. Because of this last characteristic, the root tip is a particularly suitable place to study nuclei during their divisions.

Over the apical end of the root, and covering the actively dividing cells is the root cap. This is a mass of loosely aggregated cells arising in various ways. In some plants the cells of the root cap are formed from the meristem cells in general. In **dicotyledonous** plants there is frequently a special layer of cells called the calyptrogen, covering the apical surface of the root. These cells divide and form cells which become the root cap. As the root elongates, the protoderm cells give rise to the epidermis of the root.

However formed, the root cap is continuously renewed throughout the life of the root, forming a protective cover over the tip. The outer and consequently older cells of this cap are loosely arranged. Their walls become considerably modified to form a mucilaginous mass, which greatly reduces the friction of the elongating root against the soil particles. This mucilaginous mass is often very conspicuous in the brace roots of corn.

Just back from the apical meristem of the root is the zone of elongation. In this region the cells show very characteristic changes. The most obvious of these is the increase in length which occurs in most of the cells. At the same time there appears in each cell a conspicuous central vacuole which enlarges greatly, the cytoplasm being pushed out to a thin peripheral layer in which is found the nucleus. In roots this elongating region is much shorter than is the corresponding region in the stem, a fact presumably correlated with the denser medium through which it grows. Usually it is less than a centimeter long.

As the root continues elongating the cells gradually change to form a third zone not sharply set off from the second. In this, the zone of maturation, the cells gradually assume their final form. Externally the most

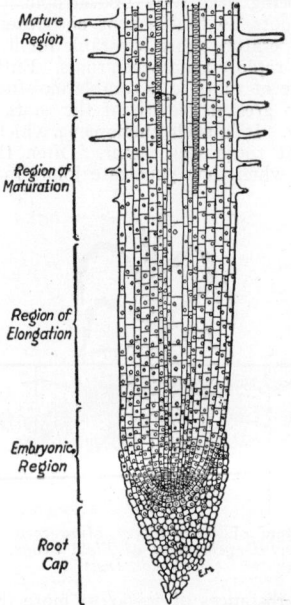

Longitudinal section of onion root showing the regions of development.

conspicuous feature of this zone is the presence of the root hairs. A root hair is a slender outgrowth of an epidermal cell. In size they vary from 0.1 to 10 mm. long, with a diameter averaging about 0.01 mm.; the number

of them formed on the root surface varies from 200 to 400 or more, so that they cause a tremendous increase in the surface of the root. The life of a root hair is not long, being commonly only a few days, after which it disappears. In a few cases root hairs may persist for 2 or 3 years, as in the Honey Locust or in certain **Composites**. In such cases they do not have the structure typical of root hairs, but are tough rather thick-walled objects. Usually the wall of a root hair is very thin and modified externally to a pectic substance which sticks closely to the soil particles and absorbs water readily therefrom. Within the wall is a thin peripheral layer of actively streaming **cytoplasm**. The central part of the hair is occupied by an evident **vacuole** which is continuous with that in the basal portion of the cell from which the hair protrudes.

As the root grows older and the cells composing it mature, a very definite structure appears. The outer portion is composed almost entirely of **parenchyma** cells, which are of irregular shape, loosely aggregated, and serve mainly for storage of materials.

The innermost layer of **cortical** cells is very distinct, forming the endodermis. In this layer, the walls of certain cells, as shown in the illustration, become thickened

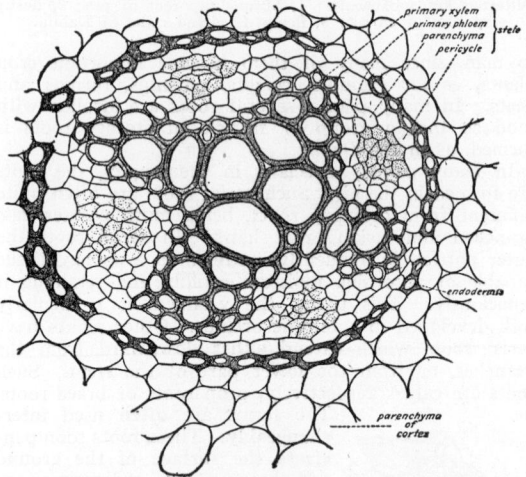

A cross-section of the stele of a young root of the buttercup, *Ranunculus acris.*

and lignified; other cells, called passage cells, retain their thin walls and permit free flow of water to the xylem. In Monocots the walls are all thickened.

Within the endodermis is the stele, or central cylinder. The first cells to differentiate in this region are the water-conducting **xylem** elements, which arise in discrete groups. The number of these groups is usually constant for any given species of plant, cross-sections through the root at this stage showing a number of isolated groups of cells surrounded by unmodified cells. Between these groups of cells and somewhat farther from the center of the root other groups of cells become evident as the root grows older. These are the **phloem** cells. While the walls of xylem cells soon become thickened by the formation of ligno-**cellulose** against the primary wall, those of the phloem cells remain constantly thin. With continued growth of the root, the cells of the central portion gradually become changed into additional xylem cells, so that finally, in most roots, no unmodified cells remain in the center; that is, characteristically the root does not contain pitch cells, though many **monocots** do. Remaining outside the phloem and xylem cells are many only slightly changed cells which form the pericycle. This is the region from which branch roots originate, and also the region which by the division of its cells forms the cork cambium which produces the cork and secondary cortex in the outer bark of tree roots after the first year of growth. The root therefore contains the following tissues derived entirely from the modification of the cells originally resulting from the divisions of the apical meristem; a central cylinder of alternate masses of xylem and phloem cells, surrounded by a sheath of slightly modified cells called the pericycle, these composing the central cylinder. Outside this is the endodermis, with its cells showing characteristic thickenings of the walls. Around the endodermis is the cortex. All these tissues are collectively known as primary tissues.

With further growth of the root, secondary tissues appear. These are formed from a special group of cells known as **cambium** cells. Cambium cells first appear in the region inside the phloem patches, appearing in cross-sections of the root as crescent-like patches which gradually extend radially until they unite to form a continuous sheath around the xylem. By their divisions new cells are formed both inside and outside the cambium band. Those inside gradually develop into secondary xylem cells, while those outside become secondary phloem cells. Continued development of the **cambium** band causes the primary phloem cells and all other cells formed externally to be pushed outward and gradually crushed. Since growth often occurs periodically in the root, especially in regions having alternations of favorable and unfavorable seasons, **annual rings** are found in roots as in stems. As root enlargement occurs, the pericycle cells become meristematic and by their divisions form a mass of cells around the central cylinder as described above.

The functions of the root are primarily anchorage of the plant, absorption of water and mineral salts in solution, and conduction of these to the stem, and also storage of food. Anchorage is obtained by the much-branched far-reaching root system, which penetrates deeply into the ground and resists such forces as wind acting on the top of the plant.

The water of the soil, together with substances in solution, is taken into the plant largely through the portion of the root which is covered by root hairs. The outer wall of the latter soaks up water readily. The wall of the cell is permeable, permitting water to pass through easily. The outer surface of the cytoplasm of the cell is also a membrane, which is semipermeable, that is, readily permits certain substances such as mineral salts in solution to pass through, but does not allow organic substances to pass. A similar condition exists in the membrane of the cytoplasm which separates it from the cell sap in the vacuole. This cell sap has a high concentration of organic solutes dissolved in it, so that it has an osmotic pressure of 4 to 10 atmosphere, which is much greater than that of the soil solution. Since these two solutions are separated by a semipermeable membrane, it is but natural that water molecules should pass more freely into the cell than out, causing water to move from the soil into the plant. Once in the root hair, the water increases the turgor of the latter. This water then passes into the cortical cells and through them to the xylem. There it enters the vessels and moves up through the root into the stem. (See **Ascent of Sap.**)

In many plants, the root, once formed, shortens, apparently by a change in the shape of its cells, and as a result pulls the top down against the ground or even under the surface. Such a phenomenon occurs in the Dandelion.

Food storage, one of the main functions of roots, results from the movement of foods downward from the green leaves into the roots. Often food is stored in such quantities as to cause considerable enlargement of the root. In addition to foods, water is also stored, especially in many desert plants.

Of secondary importance is the habit which many plants have of propagating themselves by means of roots. In this process, **adventitious buds** appear on the root, especially when the latter is exposed to the air. These gradually develop into new plants. This is very important to man, since it is a way of propagating many plants. Sweet potatoes are propagated in this way. (R.M.W., B.S.M.)

ROOTS OF NUMBERS.

An nth root (where n is a positive integer) of a given number a is a number whose n^{th} **power** is equal to a. The number n is called the index of the root.

Thus, x is a square root of a if $x^2 = a$; that is, the square root of a number when multiplied by itself produces the given number. Every positive number a has two real square roots, one of which is positive and the other negative, both having the same **absolute value**. The positive square root of a is called the principal square root of a, and is generally denoted by the radical symbol $\sqrt{a}$.

Every negative number has two **imaginary** square roots; either of them may be chosen as the principal square root.

Every real number has n distinct n^{th} roots. One of these is selected as the principal n^{th} root as follows:

If n is even and a is positive, the one positive n^{th} root of a is called the principal n^{th} root of a.

If n is even and a is negative, any one of the **complex** n^{th} roots of a may be called the principal n^{th} root of a.

If n is odd and a is positive, the one positive n^{th} root of a is called the principal n^{th} root of a.

If n is odd and a is negative, the one negative n^{th} root of a is called the principal n^{th} root of a.

The principal n^{th} root of a is denoted by the **radical** symbol $\sqrt[n]{a}$, or by the **fractional exponent** symbol $a^{1/n}$.

The n n^{th} roots of any number (real or complex) are given by one form of **DeMoivre's theorem.**

Tables of principal roots of numbers, usually square roots and cube roots, are often contained in collections of mathematical tables. (L.L.S.)

ROPE DRIVE.

Ropes passing, in the manner of belts, over **sheaves** are used for the transmission of mechanical power between rotating shafts. Rope drives have been used where conditions were unsuitable for leather belts. Center distances from 50 to 300 or 400 ft. are possible with rope transmission, but not with belts. Ropes are suitable for outdoor location, and for transmission of large amounts of power between shafts which are not parallel. Belting is not suitable for these drives. Extremely long rope drives using wire ropes have been built, but they are rarely constructed today, because the electrification of industry offers more economical transmission of power over long distances. The ropes of rope drives are made of Manila, cotton, or steel; 4- and 6-strand Manila ropes are commonly found in this country, varying in size from $1-1\frac{1}{2}''$ in diameter. These ropes run like belts over grooved sheaves, and any amount of power can be transmitted merely by providing enough rope and sufficient number of grooves on the sheaves. It is readily possible to obtain driving efficiencies of better than 90% in rope drives. (F.T.M.)

RORQUAL. Whale.

ROSE CHAFER.

Insecta, Coleoptera. A medium-sized gray-brown **beetle** which eats the flowers and fruit of roses and various fruit trees. Spraying has not been found an effective method of control although damage to grapes is said to be prevented by the use of **lead** arsenate. The spray recommended is made up of 3 lbs. of lead arsenate and 2 gals. of molasses to 100 gals. of water. It should be applied when the beetles first appear and again after the lapse of a week. (A.W.L.)

ROSE COLD. Hay Fever.

ROSE CURVES. The 3-leafed roses with equations in polar coordinates $r = a \sin 3\theta$ (Fig. 1), and $r = a \cos 3\theta$ (Fig. 2) are shown in the accompanying figures.

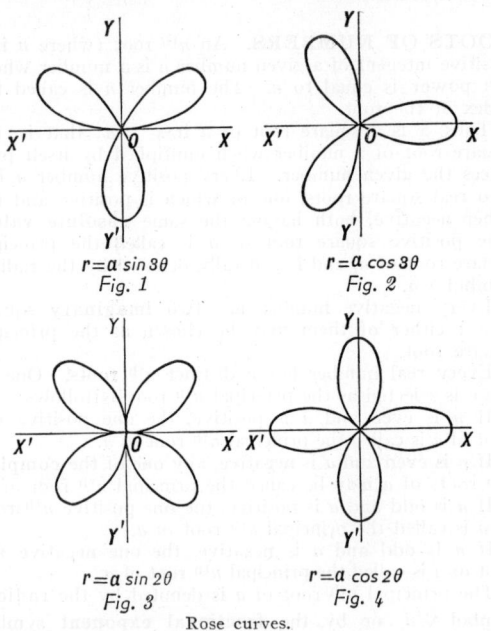

$r = a \sin 3\theta$
Fig. 1

$r = a \cos 3\theta$
Fig. 2

$r = a \sin 2\theta$
Fig. 3

$r = a \cos 2\theta$
Fig. 4

Rose curves.

The 4-leafed rose with equations in polar coordinates $r = a \sin 2\theta$ (Fig. 3), and $r = a \cos 2\theta$ (Fig. 4), are shown in the figures.

The equation $r = a \sin 4\theta$ represents in polar coordinates an 8-leafed rose. (L.L.S.)

ROSE FAMILY. Rosaceae. Comprising this family are some 2000 species of plants, a few of which are of tremendous value to man. Most of them are perennial plants, either trees or shrubs or herbs. Nearly all have alternate leaves, either simple or compound, with stipules present. The flowers are of many forms, and grow in **racemes** or cymes. Frequently the receptacle is more or less hollowed and often forms a part of the mature fruit. The flowers are commonly perfect and have their parts in multiples of five. The fruit may be either dry or fleshy, and in many species is an aggregate of many individual fruits. The flowers are usually conspicuous and insect pollinated. The most important economic members are found in three of the five or more suborders. One of these, the Pomoideae, has the **carpels** often united together and fused with the inner wall of the receptacle, which becomes fleshy; such a fruit is called a pome. In this subfamily are found the apple, pear, and quince.

The apple, *Pyrus Malus,* is a medium-sized tree, usually less than 40′ in height. The wood of the tree is red in color, hard and dense, and frequently used for making tool handles. The flowers are borne on short lateral branches known as spurs, and are pink and pleasantly fragrant. They are self-sterile and insect-pollinated, particularly by the honey-bee. The fruit is a pome, the fleshy part of which is regarded as modified stem tissue, the receptacle. When unripe the fruit contains much malic acid and starch. On ripening the amount of malic acid decreases, while much of the starch changes to sugar. At the same time characteristic aromas and flavors develop. When ripe a change occurs in the middle **lamella** of the cell wall, the substance dissolving away in part so that the cells become more or less separate. On further ripening the flesh in certain varieties becomes mealy and loses most of its tastiness. Each of the five carpels which form the parchment-like core of the apple contains two seeds.

Propagation of the apple tree may be accomplished by seed planting, but with doubtful success since the new plant will usually bear small undesirable fruits. To perpetuate desirable strains, therefore, **grafting** or budding is usually practiced. Many people top-graft old apple trees; frequently a single tree may have a dozen or more varieties grafted thereon.

Apple trees do best only in regions having a fairly cool climate; they are not entirely hardy, temperatures below −20° F. seriously damaging old and well-established trees. Many varieties of apple trees are grown. These fall into three main classes, according to the time when the fruit is mature; summer apples appear first, being ready for use in late July and August; fall apples follow, with Gravenstein one of the most popular varieties of this class; the principal class is the winter apple, the fruits of which keep well so that they are available throughout the winter months. Apples of this class stand storage and shipment through long distances.

Apples are used in several ways. Many are eaten raw; others used as sauce or made into pies and tarts. The high **pectin** content causes them to be much used in the making of jelly. Large quantities of apples, particularly cull fruit, are ground up and the juice pressed out to produce cider. Fermentation causes cider to acquire a considerable percentage of alcohol, it then being known as hard cider. By the action of acetic acid **bacteria,** cider may be changed to vinegar. From the crushed pulp, called pomace, remaining after the juice is expressed, commercial pectin is obtained.

The pear tree, *Pyrus communis,* of Eurasian source, differs in several ways from the apple. Usually it is a taller tree, with a tendency to more upright growth. In contrast to the rough leaves of the apple, those of the pear are smooth and glossy. The wood is very dense and used much as apple wood is used. The flowers are white. The fruit is more juicy and sweeter than is the apple, and the flesh especially when green contains an abundance of stone cells. Pears are used almost entirely as edible fruit, eaten either fresh or canned.

The quince, *Cydonia oblonga,* is a far less important member of this group than are the pear and apple. It is a small, rather shrubby tree. The fruit is large, more or less hairy during growth and hard and yellow at maturity. Each carpel contains several seeds which are invested with a mucilaginous pulp. The main use of quinces is for jams and marmalade.

Another genus included in this group is *Crataegus,* the Hawthorn, many of whose species are planted for ornament. The wood of the larger species is sometimes used in inlay work and for engraving. The genus contains many species, which hybridize freely, offering great difficulty to the taxonomist.

A second subfamily is the Rosoideae, a large one containing many genera and species. Here the fruit is one-seeded and indehiscent, with usually many carpels borne on an enlarged central stalk or carpophore. In some genera the floral axis encloses the carpels in the mature fruit. Shrubs or herbs with simple or compound leaves and various inflorescences comprise the group. Several important plants extensively cultivated.

One of these is the genus *Rubus,* which includes Raspberries, Blackberries, and Dewberries. In these the **fruit** is an aggregate of many drupelets, which cling closely to the axis in the Blackberries and Dewberries, but slip free therefrom in the Raspberries. An erect habit distinguishes Blackberries from the prostrate Dewberries. All have a perennial root system and annual or biennial stems. Propagation is mainly by suckers and root cuttings. The fruits are largely consumed fresh near the place of growing, or canned. Another large genus in this subfamily is *Rosa,* grown widely for its beautiful fragrant flowers. In this genus the fruit or hip is a hollow urn-shaped torus surrounding the many carpels.

From the flowers of certain species, notably *Rosa damascena*, **Rose oil** is obtained.

Another very important genus is *Fragaria*, the Strawberry. Strawberries are low herbs with short thick stems, tough fibrous roots, and trifoliate leaves. The fruit is composed of many small achenes irregularly scattered on the surface of a very much enlarged fleshy receptacle. The hull of the fruit is the persistent calyx and the remains of the stamens. Propagation is mainly by means of the numerous long slender runners, modified branches, which root at their tip and are then severed from the parent plant. Strawberries are used as dessert, for making preserves or jams, and in ice cream. The many varieties are derived from *Fragaria virginiana* of Eastern North America and *Fragaria chiloensis* of the Andean region of South America.

The third subfamily containing genera of importance to man is the Drupoideae, members of which are trees or shrubs with simple leaves and the fruit, a **drupe**, usually one-seeded, in which the **ovary** wall or **pericarp** is differentiated into a fleshy outer portion, the exocarp and mesocarp, and an inner endocarp which is very hard or stony. In these plants the bark contains a gum which frequently exudes in masses. The leaves, bark and seeds are bitter, due to the presence of a glucoside amygdalin which in the presence of an **enzyme** emulsin, produces prussic acid, a deadly poison. Due to this reaction animals eating these parts are frequently killed. The important economic members are *Prunus Persica*, the peach, *Prunus Armeniaca*, the apricot, *Prunus domestica* and related species, the plums and prunes, and *Prunus avium* and *P. Cerasus*, the cherries.

Peaches are derived from a native Chinese tree. The trees are small and usually not very long-lived. They are only semi-hardy, and due to their habit of producing flowers very early in the growing season, before the leaves appear, are frequently badly damaged by late heavy frosts. The flowers are borne singly, are brilliant pink and fragrant. The calyx tube of the flower surrounds the pistil, but is not adnate thereto. The **ovary** is 1-celled, but frequently contains two **ovules** when young, only one of them developing. Peach fruits, the largest of any member of this subfamily, are of two types. In one, the fleshy **mesocarp** slips readily from the stony **endocarp;** these are freestone peaches. In the other the two layers are closely adherent, giving clingstone fruits. Peach fruits are very perishable when mature, and are mostly consumed fresh, canned, or dried. Almond oil may be made from the kernels. During World War I charcoal made from the hard endocarp was used in filters for gas masks.

The apricot likewise is a Chinese plant. The trees are usually rather larger than peach trees, and not so hardy. The flowers are pink, and the fruit smooth-skinned, smaller than peaches, and of the freestone type. Apricots are consumed fresh or dried or canned. Commercial growing is largely restricted to California.

Plums are small to medium trees of several different species, some of which have been cultivated since before the Christian era. Many kinds of plums are consumed fresh. Others are unpalatable when fresh, being used mainly for making preserves and marmalades. Others are very sweet and capable of being dried without souring. When dried these are known as prunes. To prepare prunes, the ripe fruits are picked from the trees and cleaned. The skins are broken to expedite drying. If the climate is suitable the fruits are then spread in the sun to dry. Otherwise drying is done by artificial means. The dried fruit is finally dipped in hot water or scalded in steam, thus destroying any insect eggs which might later develop and cause spoilage of the product. The prunes are now ready to market. California leads the states in prune production.

Cherries are relatively small-fruited trees of medium size, indigeous to Europe. In cultivated kinds, the flowers are borne in small **umbels,** usually opening just as the leaves develop. Sweet cherries are largely consumed fresh. Sour cherries are canned, and used for making pies. Maraschino cherries are artificially colored cherries, the flesh of which has been hardened by soaking in brine and sodium sulfite. Wild cherries, especially black cherries, are frequently used in making jellies and wine.

To this subfamily also belong the **Almonds.** (R.M.W.)

ROSE OF SHARON. Mallow Family.

ROSE REAMER. Reaming.

ROSEMARY. Mint Family.

ROSEMARY, OIL OF. Volatile Oils.

ROSIN. Rosin is a yellow non-crystalline substance of unknown structure produced as a residue on distillation of **turpentine.** (See **Resins.**) (R.K.S.)

ROSS EFFECT. A contraction of the developed photographic image, observed by Ross (1920), occurring in drying and due to the hardening of the gelatin in the immediate vicinity of the image by the by-products of development. As a result, the tanned gelatin in the higher densities dries at a different rate from the unhardened gelatin in the background, setting up strains in the gelatin layer which results in contraction of the image. The effect is of great importance in astronomical photography, particularly in studying double-stars. (C.B.N.)

ROSTELLUM. A projection at the end of the **scolex** of a **tapeworm** which bears the hooks. It serves as an organ of attachment. (A.W.L.)

ROSTRUM. A snout. The term is applied particularly to the elongated snouts of certain fishes and to the **labium** of the **bugs** (Hemiptera). The latter is grooved or folded to form a troughlike support for the slender **mandibles** and **maxillae.** (A.W.L.)

ROTARY DRIER. Drying.

ROTARY ENGINE. At one stage during the course of development of aircraft engines, the rotary engine made its appearance. This engine, although eventually superseded by engines with static cylinders, was sufficiently interesting mechanically to justify some brief description, and had, furthermore, the interest attached of being a type employed on aircraft engaged in military service during World War I. Often today uninformed persons confuse the radial engine with the rotary, which it resembles in appearance. The rotary engine is so called because the cylinders and crankcase rotate. One reason for its early adoption was that this rotation aided in effectively cooling the cylinders. Since that time much has been learned about the scientific dissipation of heat from static finned cylinders. The crankshaft was made stationary on the rotary engine, and in this respect the rotary engine is a kinematic inversion of the modern radial engine. (See **Aeronautical Engine.**) The introduction of a gasoline-air mixture was through a hollow crankshaft into the crankcase. A valve in the piston head allowed this gas to flow from the crankcase into the cylinder during the suction stroke. Castor oil had to be used for lubrication, since ordinary mineral oil would be cut by the gasoline in the crankcase. The crankshaft bearing on the crankcase was more complicated than in static engines, and lubrication was difficult due to the rotation of the crankcase. Also, gyroscopic effects were noticeable when maneuvering an aircraft having the **flywheel** effect provided by these rotating cylinders. (F.T.M.)

ROTARY MILLING ATTACHMENT. Milling.

ROTARY POWER. Polarized Light.

ROTARY PUMP. A rotary pump is one in which the driving element is a rotating shaft, but which, unlike the centrifugal pump, is a positive displacement device. In this class of pumps may be found the gear wheel pumps, in which two spur gears meshing together fit tightly in a casing, one driving the other. The spaces between the casing and the teeth are filled with a liquid being pumped, but on the return the teeth meshing together leave no such space, so that the liquid must be passed out through a discharge opening. The efficiency of this pump depends upon the elimination of clearances between teeth and casing, and upon close fit of the meshing teeth. This is the type of pump most often used to circulate oil in the lubrication systems of automotive engines. Of similar nature is the Roots rotary pump with its 2-lobed impellers. Another type of rotary pump has the rotating element located eccentrically with respect to a circular casing. A number of blades are provided to slide radially in the rotor, and close the gap between rotor and casing. Still other types use screws. These rotary pumps may be direct-connected to electric motors. (F.T.M.)

ROTATION. Kinematics; Dynamics of Rotation.

ROTATION SPECTRUM. Molecular Spectra.

ROTATORIA, ROTIFERA. The wheel animalcules, minute animals with a circlet of cilia at one end of the body, whose movements in some species give the appearance of rotation to the entire disk. They live in water, even in the small quantities found in matted vegetation and temporary pools, and are adapted to withstand long dry periods in such situations. The group is a phylum of minor importance.

Although rotifers are minute their bodies are complex in structure. The body wall consists of an external cuticle and a syncytial ectodermal layer which bounds the internal cavity. There are no muscle layers but bands of muscle are present. The alimentary tract (digestive system) is tubular and includes a pharynx with an elaborate grinding apparatus, known as the mastax, an expanded stomach, with a ciliated (cilia) lining, and a short intestine. Near the anus is an ex-

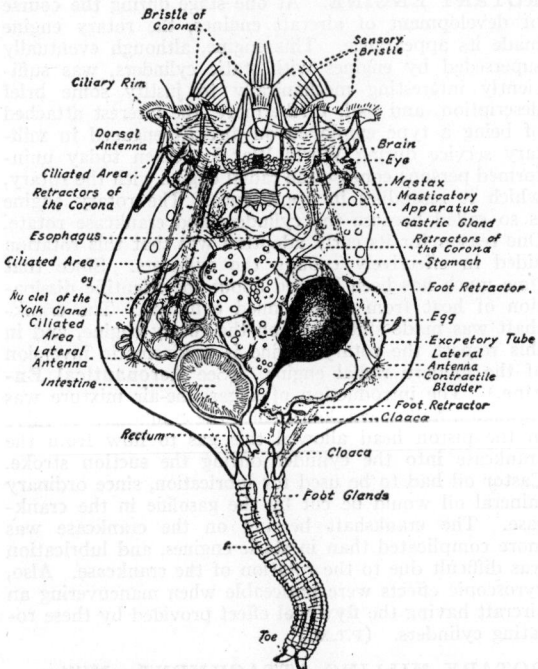

Diagram of a rotifer. *Branchionus rubens* Gosse.
(*After Wesenberg-Lund.*)

panded cloaca which receives the ducts of the excretory and reproductive systems. The essential unit of the excretory system is the flame cell, like that of flatworms. The pair of ducts bearing these cells empty into a contractile vesicle or bladder. Reproduction in this phylum is complex, owing to the frequent occurrence of parthenogenesis and to the adjustment of the life cycle to fluctuating environmental conditions.

Some authorities regard the rotifers as a phylum and others associate with them two other forms of animals, making each of the three groups a class in the phylum Rotifera, also named Trochelminthes. This classification is as follows:

Class Rotatoria. The true rotifers, as described above.
    Order Seisonidea. Ovary paired. The foot, a tapered posterior part of the body, is without subordinate lobes. Both males and females are well developed.
    Order Bdelloidea. Ovary paired. Males unknown. Foot sometimes with subordinate lobes.
    Order Monogononta. Ovary single. Males usually present but very small.
Class Gastrotricha. Minute worm-like creatures.
Class Kinorhyncha. Minute worm-like animals found at the bottom of the sea. (A.W.L.)

ROTATORY POWER. Polarized Light.

ROUND-MOUTHED EEL. Cyclostomata.

ROUNDTAIL. Pisces, Teleostei. A common fish (Pisces), *Gila robusta,* of the Colorado and Gila rivers, related to the chubs and dace. (A.W.L.)

ROUNDWORM. Nemathelminthes.

ROUTE SURVEYING. Surveying.

ROUTER. Woodworking.

ROVE BEETLE. Insecta, Coleoptera. A beetle of the family Staphylinidae, characterized by the long flexible abdomen which is exposed behind the short wing covers (elytra). (A.W.L.)

ROWLAND GRATING. Diffraction Grating.

ROYAL JELLY. The food given by worker honeybees to the young larvae during the first 3 days of their existence and to the larvae of queens until they are fully developed. It is a thick white liquid formed in the stomach of the worker by partial digestion of honey and pollen, and is apparently a highly concentrated food. Queen cells are supplied with the material in excess of the needs of the larva. If conditions within the colony deprive queen larvae of this abundance they fail to become as large, and in some cases they may even develop as intermediate forms between queen and worker. Such individuals may, however, have the instincts of queens and so may mate and lay fertile eggs. The change from royal jelly to a less concentrated food in the case of worker larvae apparently is responsible for the development of worker bees, since both queens and workers may develop from identical eggs. (A.W.L.)

RUBBER. Rubber is a product obtained from latex, a milky secretion which is found in the tissues of a large number of different plants. When the plant is injured, latex flows from the wound. On exposure to air latex coagulates and darkens. Rubber particles occur in the latex in the form of minute spherical globules. While it is true that the latex of many plants will yield rubber, yet few do so in quantities sufficient to give them commercial value.

Rubber was known to early American peoples before the arrival of Columbus. Several tribes used hard balls of rubber in playing a game, other tribes made use of it to render their shoes waterproof, while still others chewed the product. For a long time no use was made of the substance by civilized races. French explorers discovered the source plant and used the native name for the product, changing it, however, to the form caoutchouc, a name by which rubber was long known. The word rubber was a result of its use to rub out pencil marks, discovered by Priestley.

Of the many plants yielding rubber in quantity, the Pará Rubber Tree, *Hevea brasiliensis,* stands in first place. This plant, a member of the **Spurge Family,** is a tree of the Amazonian jungle. It grows as high as 125′, has three-parted leaves of large size, the smooth leaflets being about 2′ long. The flowers are **monoecious** and without petals, and the fruit three-parted, **dehiscent** and containing three large seeds with mottled seed coat very similar to castor beans. These seeds are ejected rather forcibly from the ripened fruit. While attempts have been made to establish plantations of cultivated *Hevea* in South America, little success has attended the attempt, most of the crude rubber being obtained from wild trees. These wild trees grow scattered widely and irregularly in the jungle. They are tapped by ignorant half-breeds who locate the trees. Tapping consists of hacking through the bark, a destructive method which leads the way to infection of the tree by various diseases which finally bring about its destruction. The latex is collected in small cups which are stuck by their edges into the bark below the cut. The flow continues through the morning, but gradually stops when the heat becomes great at noon. The latex is gathered from the cups into pails and rubber-covered bags. If allowed to stand the rubber globules contained in the latex gradually rise. Commercial preparation of rubber is a faster process; a rubber film is fastened around a stick. This is held in the smoke of a fire and additional latex gradually poured over the film, accumulating slowly to form an ever-increasing ball. Kneading and working cause the elimination of any uncoagulated material. Finished balls of crude rubber weigh about 150 lbs. Sometimes the latex is coagulated over the end of a flat wooden paddle, producing a flat pocket-like cake; this is known as knapsack rubber. Besides *Hevea brasiliensis* and other species of *Hevea,* several species of *Castilloa,* of the **Mulberry Family,** are used as a source of rubber. The product obtained from them is known as Gaucho rubber. It is relatively unimportant. In addition to these, many Euphorbiaceous plants are sources from which rubber may be obtained in South and Central America. In Africa, species of *Funtumia* and *Landolphia,* members of the Apocynaceae, are sources of rubber. The first genus contains large tropical trees; the second, vines of immense size growing in the Congo Free State. Methods of obtaining rubber were extremely destructive—mostly the plants were chopped down and cut up, a procedure which soon all but wiped out the African rubber supply. In recent years plantations of *Hevea* have been established in Liberia, mainly by American tire interests.

In Asia, native rubber is obtained from *Ficus elastica,* a member of the Mulberry Family. The plant is familiar to many as the ordinary "Rubber Plant" grown as an ornament in conservatories. In its native home the plant becomes an enormous tree with large spreading branches supported in part by the development of prop roots similar to those of its relative, the **Banyan tree.** It is not an important source of rubber at present time, the product being inferior to that obtained from *Hevea.*

In North America, rubber has been obtained in quantities from various plants. Unlike the tropical rubber sources, those of North America are herbaceous or shrubby plants. Of these the one most used to date is *Parthenium,* a desert shrub occurring in Mexico. The plant has been extensively cultivated in various arid regions of the southwestern United States, and improved strains developed. Chewing of the plant will produce

rubber, but in rather small quantities. In commercial preparation the entire plant is crushed and broken up in water. The particles of rubber thus liberated float to the surface of the water and are removed. They are cleaned, rolled into sheets, washed, and dried, yielding a dark gray product known as guayule rubber.

In the United States several plants have been examined as sources of rubber; among them milkweeds, Indian kelp, and goldenrod. At present none of them produce rubber in any quantity, and the expense of extracting any quantity of latex makes it unlikely that they will compete with the rubber trees.

Most of the immense amount of rubber demanded by present-day civilization is obtained from plantations of *Hevea brasiliensis* grown in the Dutch East Indies, in the Malay Peninsula, and in Ceylon. There the plants are grown from seed, the seedlings being planted out when large enough, or budded with desirable stocks from high-producing trees. Growth is rapid, trees being large enough to tap in 5 or 6 years. In tapping plantation trees, a spiral cut is made in the bark, care being taken not to injure the cambium. A spout is placed at the lower end of the cut and the exuding latex collected in porcelain cups. When exudation from the cut ceases, a thin shaving is removed from the lower edge, a new flow of latex at once resulting. Shaving continues over a period of 3 to 8 weeks, after which the tree is left undisturbed to recover. Complete renewal of bark over the cut area requires about 4 years of growth.

The latex is gathered from the cups and coagulated by different methods. In one of these **acetic acid** is added to the fresh latex, which has been set in shallow pans. The rubber then forms as a sheet on the surface of the liquid. The sheets are removed and squeezed between rollers to remove excess moisture. The resulting product is crepe rubber. Another method of obtaining the rubber is to spray the crude latex into a stream of hot dry air, which quickly evaporates the moisture of the latex, leaving fine particles of rubber, which settle down. Latex has been treated with **ammonia** and shipped in liquid form to the rubber factories.

The Russian dandelion, koksagyz, has been found to grow in the temperate latitude and to yield a satisfactory rubber.

Rubber is a terpene **hydrocarbon,** forms a pseudo-solution with **carbon tetrachloride, chloroform, benzene, turpentine, carbon disulfide,** petroleum naphtha, reacts with ozone to form an ozonide—Harries (1905) deduced the formula of rubber from this reaction as 1,5-dimethylcycloöctadiene-1.5—reacts with sulfur to form the substances commonly known as hard and soft rubber. The hot vulcanization process was discovered by Goodyear in 1839. In this process rubber, alone or dissolved in naphtha, and **sulfur** are intimately mixed, after which the mixture is subjected to a temperature of 113° C. (melting point of sulfur) to 130° C., preferably under pressure, as in an autoclave. The cold vulcanization process was discovered by Parkes in 1846 and is used for thin rubber goods. In this process thin rubber sheets, formed from naphtha solution of rubber and then evaporating the solvent, are treated in a vat with **sulfur** monochloride, dissolved, in order to diminish the violence of the reaction, in carbon disulfide. Rubber may be electrodeposited on metals.

The rate of vulcanization depends upon the percentage of sulfur present, the temperature, and the presence of a **catalyzer.** In the hot cure, **magnesium** oxide was early found to act as a catalyzer or accelerator, and since then organic bases have been found to be excellent accelerators—**aniline** and piperidine were among the first to be used. Para-nitrosodimethyl aniline (0.4 part, by weight) shortens the time of vulcanizing a 90% rubber–10% sulfur mixture at 140° C. from 60 mins. to 20 mins. The activity of many accelerators may be improved by the addition of such substances as **zinc** oxide, **lead** monoxide, **magnesium** oxide, **stearic acid,** oleic acid, pine tar. Something like 80 different rubber accelerators are in use, and in 1940 the quantity of these produced in the United States was over 33,000,000 lbs. The hardness of the product depends upon the percentage of sulfur present and the time of heating. Soft rubber contains 5–10% sulfur, hard rubber 25–40% sulfur, and the latter takes about three times as long as the former for heating. Hard rubber is softened by heating, may then be molded and pressed (an important **plastic**), when cold may be machined and polished, is hard and inelastic, in contrast to the high elasticity of soft rubber. Vulcanized rubber is not reactive with ordinary acid solutions of moderate strength, including **hydrofluoric acid,** but excluding concentrated **nitric** and **sulfuric acids,** nor reactive with alkaline or salt solutions. Certain types of synthetic rubber, for example "Neoprene" (see below), are notably resistant to the softening action of gasoline and other petroleum hydrocarbon fractions.

Vulcanized rubber is subject to deterioration due to light, heat and oxygen of the air, but see the action of antioxidants below. Natural rubber, on the other hand, may be stored for years without noticeable loss of quality.

Rubber may be compounded or have blended with it in the process of manufacture, various materials, such as pigments, fillers, waxes, oils, fibers, to obtain a wide range of properties, and may be combined with various structural materials, e.g., with textile fibers to produce rainproof clothing, machinery **belting,** and automobile tires, with metals, concrete, and wood, to produce acid-proof vessels. Natural hydrocarbon gas black (carbon black) is a reinforcing material imparting high tensile strength to rubber, and is used notably in making automobile tires to increase their resistance to abrasion and cutting.

The permanence of rubber is increased by the use of antioxidants (0.5–2%) in the mix before vulcanizing. Paraminophenol, phenylalphanaphthylamine, phenylbetanaphthylamine, and about 30 other substances are used. Some 20,000,000 lbs. of antioxidants were produced in the United States in 1940. The use of antioxidants has greatly increased the useful life of soft vulcanized rubber, as in automobile tires.

Unvulcanized rubber in pseudo-solution is used as adhesive. Vulcanized rubber is remarkable for its resistance to **abrasion,** its flexibility, **elasticity,** extensibility (10 times the original length for soft rubber bands), resilience, absorption of shock, inertness to water, solutions, gases, high frictional resistance when dry, low frictional resistance when wet, high electrical resistance, low heat conductivity, resistance to moderately high temperature (about 200° C.), plasticity under proper conditions. These properties enable rubber to be utilized in a variety of ways, possibly not exceeded in this respect by any other material.

The uses of rubber hardly need enumeration. They include automobile and airplane tires and tubes, rubber boots, rubber tubing, belting, and rubber hose. Ebonite or hard rubber is used in making battery boxes, combs, and many other products.

The advent of synthetic rubber on a large scale was directly due to the loss of Malaya and the Southwest Pacific plantation rubber producing regions to the Japanese early in 1942. Our economy requires immense amounts of rubber in peace and even larger quantities in war, since automobiles, buses, and trucks run on rubber tires, and to these there must be added during war the rubber for military vehicles and airplanes on a scale hitherto unimagined.

The United States consumed in 1941 approximately 800,000 long tons of plantation rubber, of which 90% came from the Far East region later lost to Japan, and we had on hand, according to the Rubber Survey Committee, on July 1, 1942, only 578,000 long tons of natural and 47,000 long tons of reclaim rubber. Important was the fact that less than one-half of 1% of our total consumption consisted of our own synthetic rubber. Notwithstanding this terrific handicap, in 3 years the synthetic rubber industry was expanded to a production rate greater than our prewar consumption. This is the greatest single development in chemical industry on record.

Chemical and Metallurgical Engineering signalized this accomplishment by awarding its biennial medal in 1943 not to one company as hitherto but to the whole cooperating group of 65 companies directly responsible for the success of the U.S.A. synthetic rubber program. Statistical evidence strikingly supports the Baruch report (Reports of the Rubber Survey Committee) and the "Chem & Met" award, as the following data show:

YEAR	U.S.A. PRODUCTION OF SYNTHETIC RUBBER (long tons)
1935	2,500
1940	5,000
1941	12,000
1942	30,000
1943	325,000
1944	1,000,000

Five principal types of synthetic rubber are recognized, namely, Buna-S, Buna-N, Neoprene, Butyl, and Thiokol Polysulfide. Of these the Buna-S type was selected to be produced on the largest scale, and the name was changed to Government Rubber-S (GR-S). GR-S requires styrene (this accounts for the letter S in the name) and butadiene. Styrene is made by chemically combining benzene and ethylene to form ethyl benzene and then dehydrogenating, thus:

$$C_6H_6 + C_2H_4 \rightarrow C_6H_5 \cdot CH_2 \cdot CH_3 \xrightarrow{-2H} C_6H_5 \cdot CH:CH_2$$

Butadiene ($CH_2:CH \cdot CH:CH_2$) is made from ethyl alcohol or from petroleum fractions. Styrene is a liquid of boiling point 145° C., but butadiene is a gas at ordinary temperature having a boiling point of −4.4° C. so it must be cooled or maintained under pressure if it is to be handled as a liquid as is desirable.

Completed in 20 months under war emergency conditions, the plant at Institute, W. Va., illustrates well the procedure in making GR-S. Before showing this process it would be well to compare its output with the situation if the same quantity of plantation rubber had been contemplated. The land for the plantation product would have to be 400 sq. mi. or 270,000 acres, in the proper climate, and 24,000,000 trees would have to be set out. The trees would have to be tended for 5 years before they would *start* producing rubber. The estimated expenditure would be $80,000,000, and there would be required 90,000 laborers—and a waiting time of 5 years without any rubber. The GR-S plant at West Virginia cost $56,000,000, took 20 months to erect, and requires only about 1500 persons to operate it. Here is the way the Institute plant is arranged:

Unit no.	1	2	3	4	5	6
Long tons (2240 lbs.) per year	20,000	20,000	20,000	20,000	12,500	12,500
Production	80,000 long tons of BUTADIENE of 98.5% purity				25,000 long tons of STYRENE of 99% purity	

Unit no.	7	8	9
Long tons per year	30,000	30,000	30,000
Production	90,000 long tons of GR-S		

As is evident, the process consists in mixing 80 parts by weight of styrene and 25 parts by weight of butadiene. The mixture is put in 175 parts by weight of soapy water, and a colloidal milky liquid results, called the latex. The reaction is purposely stopped before it goes to completion, and the unreacted materials are recovered from the latex by vacuum distillation. The latex contains about 25% rubber. Some antioxidant is added and the latex then *coagulated* by the addition of sodium chloride and sulfuric acid in dilute concentration. The coagulated rubber, or "curds," is filtered on continuous filters, from which it comes as a sheet. The sheet is broken up into particles of uniform size, and these are dried in a vacuum to produce the crude rubber. The plant at Institute produces about one-seventh of our prewar consumption of rubber.

The constitution of various rubbers is shown here:

REACTING UNITS / REACTED UNITS (repeating each other)

$$
\begin{array}{ll}
\text{isoprene} & \text{natural rubber} \\
\text{2-methyl-1;} & \\
\text{3-butadiene} &
\end{array}
$$

1,3-butadiene   styrene   GR-S rubber
phenylethylene

1,3-buta-   acrylo-   GR-N rubber
diene   nitrile

"Chloro-   "Neoprene"
prene"
2-Chloro-1,
3-butadiene

1,3-buta-   isobuty-   Butyl rubber
diene   lene
2-methyl-1-propene

ethylene   sodium   "Thiokol" rubber
dichloride   tetra-
1,2-dichloro-   sulfide
ethane

(R.K.S.)

RUBELLA (GERMAN MEASLES). An acute, very contagious disease probably caused by a filterable virus which has not been definitely isolated. Infection occurs by direct contact with infected individuals. The disease is most prevalent during the first half of the year with a peak incidence in May and June.

The incubation period is longer than that of **measles** or **scarlet fever,** being usually from 16 to 18 days. Children who have had measles are more likely to develop rubella. One attack usually protects against another, but does not protect against measles or scarlet fever.

Symptoms resemble those of a light attack of measles with the following exceptions: the mouth is not involved, the rash lighter, less persistent, and complications are rare. A characteristic and diagnostic finding is the enlargement of the **lymph** nodes at the back of the neck and behind the ear.

The disease can be prevented or modified by the administration of **gamma globulin.** Treatment is symptomatic. (R.S.M., D.M.H.)

RUBELLITE. Tourmaline.

RUBICELLE. Spinel.

RUBIDIUM. Symbol: Rb. Atomic number: 37. Atomic weight: 85.44. Density: 1.53. Melting point: 38.5° C. Boiling point: 700° C. Isotopes: 85 (72.8%) and 87 (27.2%).

Rubidium is a silver-white, very soft metal; tarnishes instantly on exposure to air, soon ignites spontaneously with flame to form oxide; best preserved in an atmosphere of **hydrogen** rather than in naphtha; reacts vigorously with water forming rubidium hydroxide solution and hydrogen gas. Discovered by Bunsen and Kirchhoff in 1860 by means of the spectroscope.

Rubidium occurs in **lepidolite** (lithium aluminosilicate, in amount up to 1% Rb), in certain mineral waters and rare minerals. Rubidium salts may be recovered from the mother liquor upon crystallization of (1) **lithium** salts, (2) **potassium** salts, Stassfurt, Germany. Rubidium metal is obtained by **electrolysis** of the fused chloride out of contact with air.

Chloride: Rubidium chloride (RbCl), white deliquescent solid, melting point 715° C., soluble.

Hydroxide: Rubidium hydroxide (RbOH), white, deliquescent solid, melting point 300° C., soluble.

Oxide: Rubidium oxide (Rb_2O), pale yellow solid, by heating rubidium metal in oxygen or dry air, reactive with water to form soluble rubidium hydroxide.

Other soluble salts: Rubidium sulfate (Rb_2SO_4); rubidium nitrate ($RbNO_3$); rubidium carbonate (Rb_2CO_3).

Slightly soluble salts: Rubidium perchlorate ($RbClO_4$), insoluble in alcohol; rubidium chloroplatinate ($Rb_2Pt\ Cl_6$); rubidium periodate ($RbIO_4$); rubidium permanganate ($RbMnO_4$); rubidium fluosilicate (Rb_2SiF_6).

Volatile rubidium salts, such as the chloride, color the bunsen flame violet. (R.K.S.)

RUBY. Corundum.

RUDACEOUS. A term proposed by A. W. Grabau in 1904 for coarsely graded **clastic** sediments. (R.M.F.)

RUDD. Pisces, Teleostei. A common European freshwater fish (**Pisces**), *Scardinius erythropthalmus,* related to the roach and ide and belonging to the carp family. Also called the red-eye. (A.W.L.)

RUDDER, AIRPLANE. A hinged or movable auxiliary airfoil on an aircraft, the function of which is to impress a yawing moment on the aircraft. (F.T.M.)

RUFF. Aves, Charadriiformes. A European bird (**Aves**), *Machetes pugnax,* related to the sandpipers. The name comes from the seasonal development of a large ruff about the neck of the male. The females are called reeves. (A.W.L.)

RUFFE. Pisces, Teleostei. A small species of **perch,** *Acerina cernua,* found in slow gravelly streams of England. It is also called the pope. (A.W.L.)

RULED SURFACES.

RULED SURFACES. A surface generated by a moving straight line is called a ruled surface. Cylinders and cones are the simplest examples.

The **hyperboloid of one sheet** and the **hyperbolic paraboloid** have two systems of rectilinear generators and are ruled surfaces. (L.L.S.)

RUMINANT. An animal that chews its cud. The term applies to the **oxen, sheep, goats, antelopes** and **deer, camels,** and **chevrotains.** It does not indicate a taxonomic group, although the term Ruminantia has been applied to this part of the order **Artiodactyla** in some classifications. The stomach of the ruminant consists of four chambers. When food is swallowed it passes into the rumen or paunch, where it is stored temporarily while the animal eats. Chewing and digestion are carried out at leisure. The food passes from the rumen to the reticulum or honeycomb, where it is formed into small masses and elevated to the mouth to be chewed. When swallowed the second time it follows a different course through the esophagus to another chamber, the omasum or psalterium, whence it continues to the abomasum. In these two parts gastric digestion is completed and the food is passed on to the intestine. (A.W.L.)

RUNNER. The principal revolving part of a hydraulic reaction turbine is called the runner. It consists of suitably curved blades, the hub, rim, crowns, etc. The runner of a hydraulic turbine is usually quite massive. The corresponding part of an impulse turbine is called the wheel. (See **Hydraulic Turbine.**)

In foundry practice, the channel, made in the sand of the mold, which connects the pouring basin and gate is called a runner. (F.T.M.)

RUNNING FIX. Fix.

RUPTURE. Hernia.

RUSSELL EFFECT. This effect is also known as the Vogel-Colson-Russell effect as it was independently observed in different forms by all three. It is the general term applied to the formation of latent developable images on photographic films and papers by agents other than radiant energy and specifically by resins, metals, printing inks, volatile liquids, gaseous fumes, etc. In many cases the effect appears to be due to the release of hydrogen peroxide which is capable of rendering the silver halides developable without exposure to light. (C.B.N.)

RUST FUNGI. Uredinales. The rust fungi are parasitic **basidiomycetes** which owe their popular name to the reddish color of the **spore** masses in some of the commonest species. Because many of the thousand species found in North America attack important cultivated and wild plants, they are of great economic importance. The group possesses a remarkable variety of spore types. Many species have five kinds of spores. It is notable that any given species has a very limited range of host plants.

One of the best-known species of Rusts is the common Wheat Rust, *Puccinia graminis,* which has long been known. Five spore forms are included in its complex life-history.

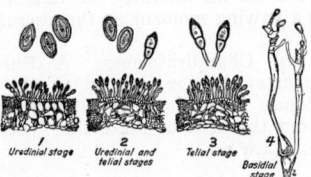

Puccinia graminis in wheat leaves. 1, uredinial stage; 2, uredinial stage changing to telial stage; 3, telial stage; 4, basidiospore germinating and forming basisia and basidiospores. *(After Tulasne.)*

On wheat plants, and on various grasses, there appear during the summer on the stem and leaves reddish spots which on examination are found to contain large numbers of 1-celled spores which are called urediniospores. These are scattered by the wind and reinfect wheat plants continuously during the growing season. Near the end of the growing season, when the host plant is maturing, a new form of spore appears either with the urediniospores or in separate **pustules.** These spores are

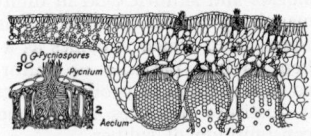

1, Section of barberry leaf infected with *Puccinia graminis* and producing pycnia and aecia; 2, section through pycnium; 3, three pycniospores.

2-celled, of dark color, and have a very thick wall. They are called teliospores, or winter spores, and are able to survive the winter season independent of any host plant, for they can neither attack one nor parasitize it. In the spring each cell of the teliospore puts out a short tube which becomes 4-celled. From each of the four cells a small spore is formed: this is the basidiospore and the 4-celled tube is therefore a basidium. The 1-celled basidiospores are carried by air-currents to suitable host plants, which in this case are not wheat plants, but barberry plants. In contact with the young leaves of the latter plant the basidiospore develops a short tube which penetrates the leaf epidermis and forms a mycelium within the leaf. After a time this mycelium gives rise to a new type of spore which appears on the upper surface of the barberry leaf in small pustules called pycnia or spermagonia. These contain masses of small **hyphae,** from the tips of which are cut off minute 1-celled bodies called pycniospores or spermatia, which seem incapable of reinfecting the host plant. Soon after the formation of the pycnia there appear on the under side of the leaf clusters of orange-colored cups. These are the aecia or cluster cups, in which are formed chains of tightly packed aeciospores. These spores, when released. cannot reinfect barberry plants but must be carried to wheat plants before they can grow. On the wheat plant each aeciospore puts out a short germ tube which penetrates the tissue of the leaf or stem within which it forms an extensive mycelium, from which the urediniospores and later teliospores are formed. So it is apparent that for the completion of its life-history *Puccinia graminis* must have two very different host plants. Many rusts show this character of requiring alternate hosts, and are called heteroecious. Other rusts complete their life cycle on a single host: they are said to be autoecious.

Another rust of great economic importance, especially in the northern United States and Canada, is the white pine blister rust, *Cronartium ribicola,* which, like wheat rust, has two alternate hosts, white pine having the pycnia and aecia, and currants and gooseberries the uredinia and telia stages.

In many rusts one or more of the spore forms may be entirely lacking. For example, in *Gymnosporangium Juniperi-virginianae* there is no uredinial stage, while in the common hollyhock rust, *Puccinia malvacearum,* pycnia, aecia, and uredinia are all lacking, only the teliospores and basidiospores being formed.

A study of the nuclear condition of the various types of rust spores reveals several interesting facts. The basidiospores are uninucleate, as are the minute pycniospores which follow them in the life cycle of wheat rust. For a long time these pycniospores were considered to be without function, they being held to be either degenerate male cells or an asexual type of spore. Recent evidence, however, has been offered that they may actually function as male cells, in some rusts at least.

It is considered that these minute spores, after discharge from the pycnidia, in some way come in contact with the mycelium within the leaf and unite with it so that a binucleate condition results. This binucleate condition then appears in the aeciospores, the urediniospores, and the teliospores. Each nucleus is haploid. Before the basidiospores are formed, nuclear fusion occurs, the basidium having but a single fusion **nucleus.** This undergoes meiosis, producing the haploid nuclei of the basidiospores. This leads to the conclusion that in the complex life cycle of the rusts there is a definite **alternation of generations,** the uninucleate diploid basidium, the haploid basidiospores, and the pycniospores being **gametophytic,** while the rest of the cycle is **sporophytic.**

In the control of rusts, three methods of attack are available. In those rusts which require alternate hosts, the removal of one of the hosts naturally suggests itself. This has been done with considerable success in combating white pine blister rust by destroying all currant and gooseberry plants in regions where white pine is an important timber tree. Similarly, removal of barberry serves to minimize wheat rust infection in regions where the winters are cold enough to prevent continuous infection by urediniospores. When constant reinfection of the same host occurs a second method of control may be found in the breeding of rust-resistant strains. Such strains now exist, for a great many susceptible plants are of great commercial importance. Control of rusts has also been attempted with various sprays with varying results. (R.M.W.)

RUSTING. Corrosion.

RUTABAGA. Brassica.

RUTHENIUM. Symbol: Ru. Atomic number: 44. Atomic weight: 101.7. Density: 12.06. Hardness: 6.5. Melting point: 2450° C. (Isotopes: page 290.)

Compact ruthenium is a grayish-white metal, somewhat harder and more brittle than **platinum,** difficultly soluble in **aqua regia,** and attacked by **chlorine.** Discovered by Claus in 1845.

Ruthenium occurs native in platinum ores, sometimes to the extent of 2%. When **osmium** is present it is removed as volatile osmium tetroxide, and the residue is heated with a mixture of **nitric** and **perchloric acids** to fuming of the latter, whereupon ruthenium tetroxide distils over, and is absorbed in **sodium** hydroxide solution. The resultant orange-red solution is reduced to ruthenium sesquioxide by alcohol. **Ignition** of the sesquioxide in hydrogen forms ruthenium metal. Ruthenium at 600° C., and ruthenium tetroxide at 106° C. yield a sublimate of ruthenium dioxide.

Hydroxides: Ruthenium trihydroxide (Ru(OH)$_3$); ruthenium tetrahydroxide (Ru(OH)$_4$).

Oxides: Ruthenium dioxide (RuO$_2$), green crystals; ruthenium sesquioxide (Ru$_2$O$_3$), black; ruthenium tetroxide (RuO$_4$), golden-yellow crystals. The last compound is volatile upon heating in a mixture of nitric and perchloric acids to fuming of the latter. (R.K.S.)

RUTILE. A mineral composed of **titanium** dioxide which occurs in three distinct forms: as rutile, a **tetragonal** mineral usually of prismatic habit, often twinned; as octahedrite, a tetragonal mineral of **octahedral** habit; and as **brookite,** an **orthorhombic** mineral. Both octahedrite and brookite are relatively rare minerals.

Rutile has a sub-conchoidal fracture; is brittle; luster, metallic-adamantine; color, commonly reddish-brown but sometimes yellowish, bluish or violet; streak, brown; transparent to opaque. Rutile may contain up to 10% of iron.

Experiments in the artificial preparation of titanium dioxide appear to show that rutile is the most stable form and produced at the highest temperature, brookite at a lower temperature, and octahedrite at a still lower temperature.

Rutile is found as an accessory mineral in many kinds of **igneous rocks,** and to some extent in **gneisses** and **schists.** In groups of **acicular** crystals it is frequently seen penetrating quartz as the "fleches d'amour" from Grisons, Switzerland. Rutile is found also in Austria, Italy, Norway, South Australia, and Brazil. In the United States it occurs in Vermont, Massachusetts, Connecticut, New York, Pennsylvania, Virginia, Georgia, North Carolina, and Arkansas.

Rutile derives its name from the Latin *rutilus,* red, in reference to the deep red color observed in some specimens when viewed by transmitted light. (E.S.C.S.)

RYDBERG CONSTANT. A quantity which enters into the frequency or wave number formulae for all **atomic spectra.** It has nearly the same value for all elements, and has only to be multiplied by a factor dependent in a regular way upon the ordinal number of the line to give the wave number of each line in a given spectral series. Rydberg derives the following expression for the constant in the case of any atom, the mass of whose nucleus is M:

$$R = \frac{2\pi^2 m e^4}{c h^3 (1 + m/M)};$$

in which m is the mass and e the charge of the electron, c is the **electromagnetic constant** (speed of light), and h is Planck's constant. Since m/M is in any case very small, it is clear that R cannot vary much from element to element. Its smallest value (for hydrogen) is about 109,678 reciprocal cm., and it can never exceed 109,737 reciprocal cm. If R is multiplied by c, the result is the "Rydberg fundamental frequency" (called by some writers the Rydberg constant); and if this be used in place of R, the spectral series formulae give frequencies instead of wave numbers. (L.D.W.)

RYE. *Secale cereale.* Gramineae. Rye is an annual plant which has a tendency to become perennial. It is a sturdy cereal grass having a much-branched root system which penetrates 4–6' into the ground and tough slender stems which may grow as tall as 6'. The leaves are like those of other **cereal** grasses, but have a definite bluish color, as does the stem. The **inflorescence** is a **spike,** with the individual spikelet 3-flowered and occurring singly at each of the twenty or more joints of the **rachis.** Of the three flowers in a spikelet, only the two lower ones mature, the third aborting. The two **glumes** are narrow, the **lemma** is broad, distinctly keeled, and has a long stiff terminal **awn,** while the **palea** is thin and blunt. Unlike most of the cereal grasses, rye must be cross-pollinated in order to set fruit abundantly. The fruit, a grain, is very similar to that of wheat in structure, and readily separated from the lemma and palea when mature. The grain is long and slender and of much darker color than wheat grains.

Rye apparently originated either in central or southwestern Asia, and is a grain of much more recent **cultivation** by man than are the other cereal grasses. Rye is now extensively cultivated, especially in cold or dry climates, since the plants are capable of growing in much colder regions than wheat, and also on poor soils deficient in moisture content. Russia produces a large part of the world's rye crop, with Germany a second large producer.

Rye is much used in making flour. From this flour is made the dark-colored, somewhat bitter-flavored bread which forms the principal bread of the poorer classes of Europe. Rye grain is also used as a stock food, while the grass is used as forage or dried for hay. The ripened stem, or straw, is much used for bedding, as packing material, and in the manufacture of straw board. Much of the mash used in distilling whiskey and **alcohol** is made from rye.

Rye is subject to attack by a parasitic fungus known as ergot, *Claviceps purpurea,* which causes the grains to hypertrophy. Ergot is a violent poison, and if ground up with rye into flour, may cause serious cases of poisoning. (R.M.W.)

SABATIER EFFECT. This effect, studied by Sabatier, is a reversal phenomenon observed with photographic materials under certain conditions as when the developed image is exposed to diffuse light and redeveloped. Reversal of the image, under these conditions, is usually due to the first negative image acting as a stencil during the second exposure, the positive image being formed by the exposure of the undeveloped silver halide through the negative image.

Reversal of the image, however, may also be produced by the substitution of certain chemicals, such as sodium arsenite, for the exposure to diffuse light. Exposure to x-rays under these conditions does not result in reversal. Students are now agreed that the Sabatier effect is not a simple case of second exposure as described above, but a different phenomenon connected apparently with the development of the images. (C.B.N.)

SABINE'S LAW. Acoustics.

SABLE. Marten.

SACCHARIDES. Carbohydrates.

SACCHARIMETER. This is a special type of polarimeter designed for use in the analysis of sugar solutions. It comprises a polarizing device, commonly a Nicol prism, to effect the polarization of the light, and an equalizer which permits the determination of the amount of rotation of the plane of polarization of this light that has been brought about by passage through a standard thickness of the solution undergoing analysis. Incidentally, such instruments are designed to use or produce monochromatic light since the angle of rotation varies inversely with the wavelength of the light. The specific rotation or specific rotatory power of optically active substances are known and, therefore, if there is only a single optically active substance present in the solution undergoing analysis its concentration may readily be calculated. (R.K.S.)

SACCHARINE. This is an artificial sweetening agent which is not a carbohydrate but a cyclic imide of ortho sulphobenzoic acid.

It is synthesized from **toluene** and is 700 times as sweet as ordinary sugar. The sodium salt is used in **diabetes**. (R.K.S.)

SACCHAROIDAL. Used by petrologists as a textural term meaning granular (sugary). (R.M.F.)

SACCULINA. Crustacea, Cirripedia. A parasitic **crustacean** which lives on crabs. The name is that of the genus to which the animal belongs. When first hatched sacculina is an active **larva** (Nauplius) without a mouth or alimentary tract. It transforms into a Cypris larva which attaches itself to a crab and undergoes a transformation including the development of a perforating organ. By means of this organ the body wall of the crab is penetrated and the parasite, an almost shapeless mass of cells, enters the body cavity. It becomes attached to the alimentary tract of the crab and forms root-like processes which extend ultimately to all parts of the body. As the parasite reaches adult life it bulges from the under surface of the crab's abdomen. A conspicuous effect of the attack is the castration of the host, with accompanying changes in its visible sexual characteristics. (A.W.L.)

SACRUM. The portion of the spinal column of vertebrates, usually formed of several fused vertebrae, with which the pelvis is articulated. **Skeletal system.** (A.W.L.)

SAFELIGHT. In photography the term safelight is applied to a light filter which, when used with a suitable light source, provides illumination by which a photographic material may be examined with reasonable safety. The spectral transmission of the safelight is determined primarily by the spectral sensitivity of the photographic material for which it is designed. Thus a yellow safelight (transmitting green and red) may be used with photographic papers whose sensitivity is low and lies almost wholly within the violet and blue, while a red safelight (absorbing violet, blue and green) may be used with blue-sensitive and orthochromatic negative materials since the sensitivity of these beyond a wavelength of about 550 mμ in the green is so slight as to be of no consequence in practice. Since panchromatic negative materials are sensitive throughout the visible spectrum, the transmission range of the safelight is determined by (1) the wavelength range to which the material is least sensitive, and (2) the wavelength range at which the maximum visibility is obtained. In practice, the second consideration is the more important of the two; hence safelights for panchromatic material are designed to provide the maximum visibility with the smallest amount of light possible. Thus, for panchromatic materials a blue-green safelight provides the maximum visibility consistent with reasonable safety, since at low-intensity levels the maximum sensitivity of the eye is in the blue-green region. No light, however, which is of any value in practice is safe for most high-speed panchromatic materials. These must either be handled in total darkness, examined only briefly when development is almost complete, or desensitized (see **Desensitizing**).

Safelight screens are usually made by coating glass with dyed gelatin although progress has been made in incorporating the dyes in both plastics and in glass to produce screens less subject to fading and to damage by heat and water. (C.B.N.)

SAFETY VALVE. The common form of the safety valve is the pop valve held against its seat by a heavy spring and having a "huddling chamber" to make it open quickly and remain open until a predetermined pressure drop (2-4% of the working steam pressure) has occurred. The A.S.M.E. Boiler Construction Code requires **boilers** having more than 500 sq. ft. of heating surface, or those generating better than 2000 lbs. of steam per hour, to have two or more safety valves. The safety valves should have sufficient relieving capacity to prevent more than 6% pressure rise at maximum rate of **combustion.** Required discharge capacity of a safety valve may be based either on the heat units in the **fuel** consumed or on the amount of steam generated.

In case more than one safety valve is used the smaller one can be set to pop at the desired maximum pressure and the larger at 2 or 3 lbs. higher. The main safety

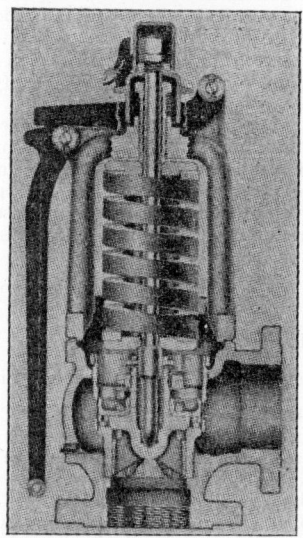

Pop safety valve. (*Crane Co.*)

valves of a large boiler operate to blow down several pounds pressure before closing. A smaller "**vernier**" safety valve giving less pressure drop between pop and close is installed, usually on the **superheater** outlet, though sometimes on the boiler lead, for the purpose of giving partial relief to the high pressure, warning the attendants of high pressure, preventing overheating of superheater tubes, and possibly forestalling popping of the main safety valves and the resultant waste of high-temperature potential heat. Since most safety valves discharge horizontally into a pipe that then turns upward, an impulse force is given to the vent piping which, at least for large valves, needs special anchorage.

The relief valve is a form of safety valve, but usually intended for less severe service and of less importance from the safety viewpoint. Relief valves are applied to air, water, and steam lines, also to tanks, heaters, etc. Among them could be mentioned the back pressure valves and atmospheric relief valves. (F.T.M.)

SAFFLOWER. *Carthamus tinctorus.* **Composite Family.**

SAFFRON. *Crocus sativus.* Iridaceae. *Crocus sativus* is a perennial herb, the native home of which is the eastern Mediterranean region. The stem is an underground flattened **corm**, the surface of which is covered by a few scaly leaves. At the top of the corm is a terminal bud, which develops into linear leaves 5–9″ long, and flowers. The flowers are white or lilac-tinted, with the **perianth** 6-parted and with a very long tube, so that the **ovary** remains below the surface of the ground. The three **stigmas** are bright red. These, when dried, are known as saffron, an orange-yellow dye with a considerable percentage of volatile oil present. Saffron is used in medicines, and in various liqueurs and dishes, to add a pleasant taste and color. (R.M.W.)

SAGE GROUSE, SAGE HEN. Aves, Galliformes. The largest species of **grouse** in North America, *Centrocercus urophasianus.* It ranges from southwestern Canada to Nebraska and central California, inhabiting chiefly the arid plains. It also wanders over grass lands and reaches high altitudes in the mountains. Sometimes called the cock of the plains. (A.W.L.)

SAGEBRUSH. *Artemisia tridentata. Compositae.* A number of other species of this genus are also called sagebrush but *A. tridentata* is more prominent and of wider distribution than the others. This shrub has an extensive root system and may attain heights up to 7′. The gray-green foliage and aromatic odor of this shrub are distinctive. The flowers occur in inconspicuous heads. This plant is distributed from the Black Hills to southern British Columbia to southeastern California to northern Arizona. It reaches its best development, however, in the Great Basin region where it may occur over large areas in nearly pure stands. (B.S.M.)

SAGITTARIUS (The archer) (Map, page 380). This large **constellation** is the ninth sign of the **zodiac**. Lying as it does in a particularly rich portion of the milky way, it contains a large number of star **clusters** and gaseous **nebulae** of great beauty in a moderate-sized telescope. From the large number of faint stars, **cepheid** variables, and globular clusters that seem to congregate in this region, it seems highly probable that the stellar **galactic** system has its greatest extension in this direction. Long-exposure photographs indicate that large numbers of dark or obscuring nebulae lie in this portion of the milky way. (W.K.G.)

SAHUARO. **Cactus Family.**

SAIBLING. Pisces, Teleostei. A fish (**Pisces**) related to the trout. The name has been applied to a species of charr found in mountain lakes of central Europe, *Salvelinus salvelinus,* and in North America to the introduced European trout, *S. alpinus,* a usage also found to some extent in Europe. A related species, *S. aureolis,* found in lakes of the northeastern states, is called the American saibling. The name is also spelled saebling. (A.W.L.)

SAIGA. Mammalia, Artiodactyla. *Saiga.* A small and clumsy **antelope** found on the steppes of western Asia and eastern Europe. Its most conspicuous feature is the peculiarly swollen face. (A.W.L.)

SAILFISH. Pisces, Teleostei. *Istiophorus.* A large marine game fish (**Pisces**) whose dorsal fin is long and

Sailfish, *Istiophorus gladius.* (*American Museum of Natural History.*)

exceptionally high, resembling a sail. The upper jaw is prolonged into a sharp sword like that of the related swordfish. Also called spike-fish. (A.W.L.)

SAILINGS, THE. The position of a vessel at sea, or in the air, is defined by the **latitude** and **longitude**. The position at any particular instant is connected with any other position, either the one just left or the one toward which the vessel is proceeding, by means of the true **course** and distance.

Any given course and distance may be resolved into two components at right angles to each other; the northing or southing and the easting or westing, each expressed in **nautical miles**. The northing or southing may be immediately converted into difference of latitude, expressed in angular units, for the nautical mile is, by definition, approximately equal to a minute of arc along a great circle. However, the conversion from easting or westing, commonly known as **departure**, into difference of longitude, can be accomplished only after taking into account the shape of the **earth** and the approximate latitude of the ship.

The navigator is continually faced by one of two problems. 1. Given the difference of latitude and longitude between two points on the surface of the earth, to find the course and distance between them. 2. Given the course and distance followed by a ship, to find the difference of latitude and longitude between the point of starting and the destination. The different methods of solving these problems are known as the sailings, and include **plane, parallel, middle latitude, mercator, great circle**, and **composite** sailings. (W.K.G.)

SAILPLANE. A sailplane is a highly efficient **glider**. Being designed for the use of expert glider pilots, it is unsuited for primary training. It is characterized by very low sinking speed, nearly flat glide, and perfection of construction. It is capable of rising flight on weak thermal air currents and is the type of aircraft employed for cross-country motorless flights of a sporting nature. The sailplane has high **aspect ratio** (about 20), careful streamlining, clean and smooth external surfaces, and minimum weight consistent with structural safety. The gliding angle in still air is often as flat as 22:1 and the sinking speed as small as 2 ft. per sec. (F.T.M.)

SAINT ELMO'S FIRE. A brush-like discharge from charged objects in the atmosphere is known as St. Elmo's fire. It occurs on ship masts, on aircraft propellers, wings, other projecting parts, and on objects projecting from high terrain. St. Elmo's fire occurs when the atmosphere is charged and a sufficiently strong electrical potential is created between an object and the surrounding air. Aircraft most frequently experience St. Elmo's fire when flying in or near cumulonimbus clouds, thunderstorms, in snow showers, and in dust storms. (P.E.K.)

SAINT HILAIRE. Celestial Navigation.

SAINT VITUS' DANCE. Chorea.

SAKI. Mammalia, Primates. A New World **monkey** of the genus *Pithecia*. Most of the species bear the name of the group, as the white-headed saki and the whiskered saki, but native names have been adopted for some. The hairy saki is called the parauacu and the black saki is also known as the cuxio. No members of the genus have prehensile tails. (A.W.L.)

SAL. Latin term for various salts, much used. Sal ammoniac, **ammonium** chloride; sal soda, **sodium** carbonate; sal volatile, **ammonium** carbonate. (R.K.S.)

SALAMANDER. Amphibia, Urodela. A **vertebrate** with a slender body, short legs, and a long tail. The moist skin of the amphibians limits them to protected habitats, either near water or under some protection on moist ground, usually in the woods. Some species are aquatic throughout life, some take to the water intermittently, and some are entirely terrestrial as adults. The salamanders resemble the lizards superficially but they are easily distinguished by the moist skin, without scales. (A.W.L.)

SALIC. A term proposed by Cross, Iddings, Pirrson, and Washington, in 1906, for the group of relatively common or standard **aluminum-silicate** minerals such as **quartz, feldspars**, and **feldspathoids**. (R.M.F.)

SALICYLIC ACID AND SALICYLATES. Salicylic acid or $C_6H_4(OH)(1)(COOH)(2)$ is a white solid, melting point 159° C., sublimes at 76° C., insoluble in cold water, soluble in hot water, alcohol, or ether. With **ferric** chloride solution, salicylic acid solutions are colored violet (distinction from **benzoic acid**).

Salicylic acid may be obtained (1) from oil of wintergreen, which contains methyl salicylate, (2) by heating dry sodium phenate (C_6H_5ONa) plus **carbon dioxide** under pressure at 130° C., and recovery from the resulting sodium salicylate by addition of dilute **sulfuric acid**. Salicylic acid is a mild disinfectant and antiseptic, and has been used as a food preservative. Salicylic acid and certain salicylates are used in medicine as anti-rheumatics. **Aspirin** is acetylsalicylic acid

$$\left(\begin{array}{c} OOC \cdot CH_3 \\ COOH \end{array}\right)$$, white solid, melting point 135° C.,

used for headaches and colds.

The following are representative esters of salicylic acid:

Methyl salicylate. . . . . $HOC_6H_4COOCH_3$
    Boiling point 222° C.
Ethyl salicylate. . . . . $HOC_6H_4COOC_2H_5$
    Melting point 1° C., boiling point 231° C.
Phenyl salicylate. . . . . $HOC_6H_4COOC_6H_5$
    Melting point 43° C., boiling point 173° C. at 12 mm.
    pressure (R.K.S.)

SALIENTIA. Amphibia.

SALIVARY GLANDS. Glands whose ducts discharge into or near the oral cavity. In the **vertebrates** they are a group of digestive glands lying in various parts of the wall of the oral cavity, chiefly near the bases of the jaws, but among the invertebrates they are an extremely varied assemblage. Thus in the insects some of the salivary glands lie in the **thorax**, and the silk glands of some species are modified salivary glands which may extend through almost the entire length of the body.

The principal salivary glands of man are three pairs: the sublinguals lie in the floor of the oral cavity, the submaxillaries below the angles of the lower jaw, and the large parotids in front of and below the ears. They produce a digestive **enzyme**, ptyalin, which acts on starches, forming maltose, from which glucose is formed by the action of a second salivary enzyme, maltase (see **Carbohydrates**). The presence of digestive enzymes in the salivary secretions is by no means the rule among mammals in general. Among birds it is not common. (A.W.L.)

SALIX. Paleobotany.

SALMON. Pisces, Teleostei. Important game and food fishes (**Pisces**). The true salmons are found only in the northern hemisphere but some closely related forms are found in the Australian region. The family includes the trout also.

Salmon are remarkable for their migration into fresh waters from the sea for the breeding season. Although this is so common as to be almost distinctive, there are fresh-water species which never migrate. These are the so-called landlocked salmon of New England lakes and northward, which are regarded as among our finest game fishes. Two varieties, the ouananiche and the sebago salmon, are classified as subspecies of the Atlantic salmon, *Salmo salar,* an important game fish in the rivers of the Atlantic coast and Europe.

The rivers of the western coast of North America have several species of salmon which are taken in immense numbers to be canned and, to a less extent, to be marketed fresh. The most important species is the chinook, king, or quinnat salmon, *Oncorhynchus tschawytscha*, and next in order come the blueback or redfish, *Oinerka,* and the silver salmon, *O. kisutch.* Other species are of little importance as food.

After the young salmon hatch in the headwaters of the rivers they work their way gradually to the sea, where they remain until mature. Many of the adults perish in their breeding migration but some return to salt water. (See also **Esters.**) (A.W.L.)

SALOL. Drugs.

SALPIAN. Urochordata, Thaliacea. A free-swimming tunicate (**Urochordata**) with a transparent body, marked by opaque muscle bands and internal organs. The salpians live near the surface of the ocean, some as solitary individuals and some in linear series as colonies. They resemble the sessile ascidians in early life, but differ in remaining motile after the transformation of the larva. (A.W.L.)

SALPINGITIS. Infection of the **fallopian tubes.** This common disease is most frequently due to a gonorrheal infection. More rarely, it may be due to tuberculosis, *streptococcus* or *pneumococcus* infection. Gonorrheal salpingitis is not only the most prevalent but is the most disabling in its after-effects. Salpingitis may not occur for months or even years after the original gonorrheal infection.

The symptoms of acute salpingitis are severe lower abdominal pain often colicky in type, with marked tenderness over lower quadrants. There is usually fever and marked leucocytosis. The symptoms and signs may closely resemble those of acute appendicitis and it is often very difficult to distinguish between the two conditions. Treatment of acute gonorrheal salpingitis with **sulfonamides** usually results in prompt control of the infection. If the organism is sulfa-resistant, **penicillin** will effect a cure. Untreated, the disease may run a chronic course lasting for years, lighting up at occasional intervals with a more or less acute attack.

Chronic gonorrheal salpingitis is of great seriousness because of the tenacity of the infection and because it is a common cause of sterility. The most constant symptom is intermittent, dull, dragging pain in the lower abdomen which is usually made worse by exertion. The dull pain often makes the subject irritable, nervous, and neurotic. Menstrual disorders are frequent. Leukorrhea is usually present. Pelvic peritonitis is a common complication, and the formation of peritoneal **adhesions** accounts for many of the direct and associated symptoms.

The recurrent flare-ups of acute salpingitis which occur during the chronic disease are treated with sulfonamides or penicillin. If pyosalpinx or distention of the tube with pus occurs, surgical intervention is necessary. Surgery may also be indicated for late complications, but is never carried out during or immediately after an active infection. (R.S.M.)

SALSIFY, OYSTER PLANT. *Tragopogon porrifolius.* Composite Family.

SALT DOME. Halite.

SALTATION. This term, as proposed by McGee, in 1908, is used by geologists to designate the particular mode of the stream transportation of **clastic** sediments by intermittent leaps or bounds. Probably an important factor in the ultimate transportation of the coarser fragments by streams and rivers. (R.M.F.)

SALTPETER. **Potassium** nitrate. **Sodium** nitrate is Chile saltpeter, and calcium nitrate Norway saltpeter.

SALTS. See individual salts under each metal; also **Acids, Bases, and Salts.**

SALVARSAN. Arsphenamine.

SALVE BUG. Crustacea, Isopoda. A marine **crustacean**, *Aega psora,* parasitic on various fishes. It is elongate oval in form and is a little more than one-half inch long. Found on both sides of the Atlantic. It is said to be used as a salve by fishermen. (A.W.L.)

SALVIA. Mint Family.

SAMARA. Fruit.

SAMARIUM. Symbol: Sm. Atomic number: 62. Atomic weight: 150.43. Density: 7.7. Melting point: >1300° C. (Isotopes: page 290.) Type of compound: Sm_2O_3, white. Color of salts: Pink. Discovered by Boisbaudran in 1879. A member of the **cerium** subgroup of the rare earth metals. (R.K.S.)

SAMPLE. A sample is a portion of the **population** of the universe. Usually some type of random sample should be obtained, although in special circumstances stratified sampling, systematic sampling, or sampling in clusters may be preferable. (L.A.A.)

SAMPLING RELIABILITY. Reliability, Significant Difference.

SAND BOX TREE. Spurge Family.

SAND COLLAR. A thin collar-shaped plate formed of sand glued together with mucus. It is formed by marine **snails** of the family Naticidae in depositing their eggs. (A.W.L.)

SAND CRICKET. Insecta, Orthoptera. *Stenopelmatus.* Thick-bodied clumsy **insects** with large heads and long slender antennae. They live in loose soil, usually under some protective object, in the western United States. They are not true crickets but are more closely related to the long-horned grasshoppers. (A.W.L.)

SAND DOLLAR. Echinodermata, Echinoidea. A sea urchin (**Echinoidea**) with a very thin body, almost

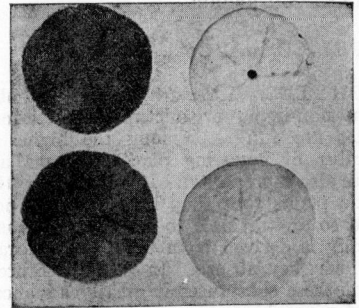

Sand dollars. (*American Museum of Natural History.*)

circular in outline and with a diameter of less than 3″. It is common on both coasts of North America. (A.W.L.)

SAND EEL. Pisces, Teleostei. A long slender marine fish (**Pisces**) found near sandy shores on both sides of the Atlantic. They have a single very long and low dorsal fin and a much shorter median ventral fin. Used chiefly for bait. (A.W.L.)

SAND GROUSE. Aves, Columbiformes. A small group of birds (**Aves**) related to the pigeons but in some ways resembling game birds. They are found chiefly in Africa and Asia but extend to Europe and Madagascar. As the name suggests, they frequent open ground. Their flight is powerful, hence some species migrate over considerable distances. (A.W.L.)

SAND. Sedimentation.

SANDPIPER. Aves, Charadriiformes. Long-legged shore birds (**Aves**) related to the snipes and plovers. They are named from their association with the bare margins of streams and ponds, where their clear piping calls are a familiar sound. North America has a score of species, some bearing other names such as sanderling and knot, and every other continent has representatives of the group during at least part of the year. (A.W.L.)

SANDSTONE. Sand grains cemented by such substances as **silica**, **carbonate** of **lime** or **iron** oxide, so as to form a solid rock is called sandstone. It occurs usually in beds of varying thickness, depending upon the conditions under which the original sediments were laid down. Because it is normally well-jointed and easy to work, sandstone has been much used for building purposes. Unfortunately, however, as most sandstones are quite porous, the weathering action of the atmospheric agencies may have a very deleterious effect upon them. (E.S.C.S.)

SANDSTONE DIKES. Sandstone occurring in fissures which have been filled from above, or from beneath. The latter type are usually the result of earthquake fissures in great flood plain or delta deposits in which the sands have been injected from below. (R.M.F.)

SANIDINE. Feldspar.

SANITARY SEWER. Sewerage System.

SAP. Ascent of Sap.

SAPAJOU. Mammalia, Primates. A **capuchin monkey.** (A.W.L.)

SAPONIFICATION. This is a special case of **hydrolysis** of esters in presence of alkali. (See **Soap.**) (R.K.S.)

SAPPHIRE. Corundum.

SAPROPEL. Cannel Coal.

SAPROPHYTES. These are plants which obtain their food from non-living organic material. Most of the saprophytes are **fungi.** Among the higher plants, a small number of flowering plants and perhaps a few **mosses** are also saprophytes. It is characteristic of these saprophytic plants that they have little or no **chlorophyll,** and so are not able to carry on **photosynthesis.** Their energy is derived from the complex organic substances which they absorb. In many instances the absorption of these substances is greatly advanced by the presence of **mycorhizae.**

Especially is this the case with various species of saprophytic **orchids** which have mycorhizae within the cells of the **roots** or **rhizomes.** The various species of Coral-roots (*Corallorhiza*) are common saprophytic orchids of American woods. These orchids have no roots, absorption occurring in the much-branched fleshy rhizome which gives the plant its name. In this rhizome the mycorhizae are found. These plants have erect stems, leaves reduced to scales, and no chlorophyll. Other saprophytic orchids occur in the continents of the Old World.

Another well-known saprophytic plant is the weird Indian Pipe or Ghost Plant, whose snow-white body is so striking an object, seen in the dark woods it inhabits.

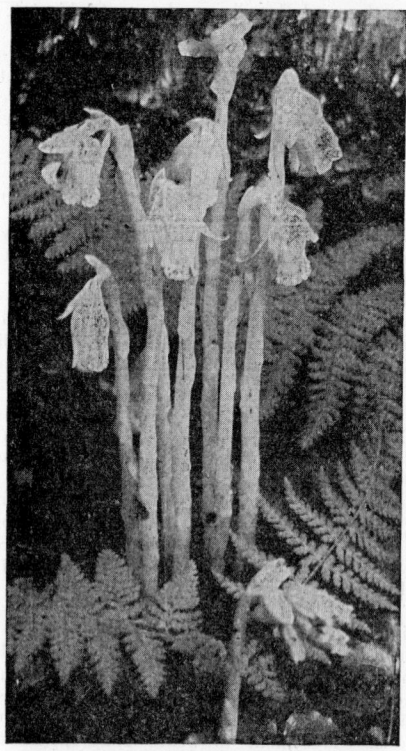

Indian pipe, *Monotropa uniflora;* saprophytic flowering plant.

Beach-drop, *M. Hypopitys;* saprophytic flowering plant.

In this and the related Pinesap, which has a yellowish color, the mycorhizae do not enter deeply into the tissues of the plant roots. The latter form a considerable mass in the decayed vegetable substance on which they grow. On these roots **adventitious buds** appear and develop to the erect flowering shoots.

Other North American saprophytes are found in the Gentian family. These, species of *Bartonia,* have a little chlorophyll, which gives them a greenish color. The leaves are minute scales. Many other saprophytes are found among the **monocotyledonous** plants of the tropical forests both in the New World and the Old (R.M.W.)

SAPSUCKER. Aves, Piciformes. A North American woodpecker which cuts rows of small holes around the trunks of trees and visits them for the sap which exudes and the insects that are attracted by it. The rows of holes are so regular as to be conspicuous. They are said to damage some trees seriously and in some cases to kill them. The yellow-bellied sapsucker, *Sphyrapicus varius,* is the common eastern species. It is represented in the western states by a variety, the red-naped sapsucker, and two other species, the red-breasted and the Williamson sapsuckers, occur in the far west. (A.W.L.)

SAPWOOD. Wood.

SARCODINA. One-celled animals whose organs of locomotion are temporary or semi-permanent protoplasmic processes known as **pseudopodia.** These processes are thrust out and retracted at the surface of the body. A class of the phylum **Protozoa.**

These animals vary from formless masses of **protoplasm** to species which secrete an enveloping test or enclosed hard parts which give them a characteristic shape. Both free and parasitic forms are known. Among the latter the species which causes amoebic dysentery is the most serious to man.

The classification of the class is as follows:

Subclass Rhizopoda. With pseudopodia which are not of the straight semipermanent radial form known as axopodia.
Order Proteomyxa. Without hard parts. Pseudopodia radiating and sometimes branched and united with each other.
Order Mycetozoa. Without hard parts. Pseudopodia rootlike. These animals form a plasmodium consisting of an extensive mass of cytoplasm with many nuclei. Formerly classed as plants and called the slime molds.
Order Foraminifera. With an enclosing test of various colors and forms, calcareous or silicious, and often of several chambers. Most species marine.
Order Amoebaea. Without an enclosing structure. Pseudopodia thick and temporary.
Order Testacea. With a test of one chamber, usually chitinous.
Subclass Actinopoda. With slender radiating pseudopodia.
Order Heliozoa. Usually spherical, with radiating pseudopodia. Body not divided by a central capsule into distinct inner and outer zones. Chiefly in fresh water.
Order Radiolaria. Usually spherical. Body divided into two regions by a central capsule. Marine. (A.W.L.)

SARCOMA. A malignant **tumor** originating in **connective** tissue. These growths are composed of densely packed cells, diffusely embedded in a homogeneous ground substance. Their degree of malignancy varies greatly. They spread by local infiltration and by blood stream invasion. The most frequent sites in which sarcomas develop are bone, lymph nodes, and subcutaneous tissue.

Sarcomas are much less common malignant tumors than are carcinomas or **cancers.** (D.M.H.)

SARCOSINE. Aminoacids, Polypeptides and Proteins.

SARCOSPORIDIA. Sporozoa.

SARD. Chalcedony.

SARDINE. Pisces, Teleostei. *Sardina.* A small fish (Pisces) related to the herrings. It is abundant in the English Channel and the Mediterranean, where it is also known as the pilchard. Immense numbers of these fishes are taken each season, to be marketed in tins. They are delicate, hence they must be prepared promptly in canneries located near the places where the fish are taken. (A.W.L.)

SARDONYX. Agate.

SARGASSUM. Algae.

SARGO. Pisces, Teleostei. Compactly built marine fishes (**Pisces**) of Eurasian waters. They have cutting teeth at the front of the jaws and strong molars. Used as a food fish in Italy under the name dentice. The sheepshead of the Atlantic and Gulf coasts of North America is a member of the same genus. (A.W.L.)

SAROS. The fact that **eclipses** occur in periodic intervals was known to the ancient Chaldeans, and probably even in prehistoric times. This period of 18 years, 11⅓ days (10⅓ days if there happen to be 5 leap years in the interval) is known as the Saros. If an eclipse should occur on January 1st, 1937, at noon, another similar eclipse would occur on January 12th, 1955, at eight o'clock in the evening. The eclipse would not occur at the same point on the earth but would be about 8 hours farther west in **longitude.**

During the course of a Saros there are about 29 lunar and 41 solar eclipses, each repeated during the next Saros, but not at the same portion of the earth. (W.K.G.)

SARRACENIA. Insectivorous Plants.

SARSAPARILLA. *Smilax* sp. Liliaceae. The genus *Smilax* contains some 200 species, most of which are tropical, though a few such as the carrion flower, *Smilax herbacea* and the cat briar, *Smilax rotundifolia,* occur as far north as the New England states. The tropical species are mostly climbing shrubs or vines, usually with prickly stems. The leaves are entire and of oblong to ovate shape. At the base of the leaf is a pair of **tendrils** which are perhaps to be interpreted as modified **stipules,** though such structures are not usually found in **monocotyledons.** The flowers are small, **dioecious** and borne in **umbels.** The fruit is a berry. Some of the South American species are the source of Sarsaparilla, which is obtained from the dried roots. Sarsaparilla is used as a flavoring for beverages and formerly in medicine in the treatment of rheumatism. (R.M.W.)

SASSABI. Mammalia, Artiodactyla. An African **antelope,** *Damaliscus lunatus,* also called the bastard hartebeest. It is almost 4' high, with horns up to 15" long. Their color is deep reddish, blending into black on the back. (A.W.L.)

SASSAFRAS, OIL OF. Volatile Oils.

SATELLITE. The term satellite is usually reserved in astronomy for small planet-like objects that are revolving about the individual **planets** in orbits. The **moon** is the satellite of the earth and has been known from remotest antiquity. The names and dates of discovery of the other satellites of the **solar system** will be found in the accompanying table. In this table there will also be found other data relative to the satellites. The definitions of the various column headings will be found elsewhere in this work. Further information regarding the different satellites will be found in the articles on the individual planets.

Satellites serve a useful purpose to astronomers since the mass of a planet can be determined accurately only if the planet has a satellite. By application of the rigorous expression for the harmonic **Keplerian law of planetary motion** the mass of any planet and satellite may be

THE SATELLITES OF THE SOLAR SYSTEM

NUMBER, NAME AND DATE OF DISCOVERY	MEAN DISTANCE IN MILES FROM PRIMARY	SIDEREAL PERIOD DAYS HOURS	APPARENT STELLAR MAGNITUDE	ON SCALE MOON = 1	
				Diameter	Mass
SATELLITE OF THE EARTH					
Moon....................	238,857	27 7.720	−12.3	2160 mi.	
SATELLITES OF MARS					
1 Phobos..............1877	5,826	0 7.654	11.5	0.0043	
2 Deimos..............1877	14,580	1 6.299	13.0	0.0023	
SATELLITES OF JUPITER					
5 Fifth.................1892	112,600	00 12	13	0.0460	
1 Io....................1610	261,800	01 18	5.5	1.0731	1.09
2 Europa..............1610	461,600	03 13	5.7	0.9062	0.65
3 Ganymede...........1610	664,200	07 04	5.1	1.4816	2.10
4 Callisto..............1610	1,168,700	16 17	6.3	1.4902	0.58
10 Tenth................1938	7,450,000	264	18		
6 Sixth.................1904	7,490,000	266	13.7	0.0374	
7 Seventh.............1905	7,680,000	276	16	0.0115	
11 Eleventh............1938	14,200,000	692	18		
8 Eighth..............1908	14,800,000	739	16	0.0072	
9 Ninth................1914	15,000,000	750	18	0.0072	
SATELLITES OF SATURN					
7 Mimas...............1789	115,300	0 22.618	12.1	0.1870	0.0005
6 Enceladus...........1789	147,800	1 8.885	11.6	0.2445	0.002
5 Tethys...............1684	183,000	1 21.307	10.5	0.3740	0.008
4 Dione................1684	234,400	2 17.686	10.7	0.3452	0.014
2 Rhea.................1672	327,300	4 12.420	10.0	0.5034	0.033
1 Titan................1655	758,800	15 22.691	8.3	1.2083	1.86
8 Hyperion............1848	919,700	21 6.640	13.0	0.1438	0.002
3 Iapetus.............1671	2,210,000	79 7.940	11.0	0.5178	0.08
9 Phoebe..............1898	8,034,000	550.44	14.5	0.0575	
SATELLITES OF URANUS					
1 Ariel................1851	119,100	2 12.489	15.2	0.2589	
2 Umbriel.............1851	165,900	4 3.460	15.8	0.2014	
3 Titania.............1787	272,200	8 16.941	14.0	0.4891	
4 Oberon..............1787	364,000	13 11.118	14.2	0.4315	
SATELLITE OF NEPTUNE					
1 1846	219,800	5 21.044	13.6	1.4384	

found in terms of the mass of the earth-moon system after the distance of the planet from the satellite and its period of revolution are known. The problem of the determination of the masses of the satellites themselves is a more difficult problem. The mass of the moon can be determined in terms of the earth's mass by means of the so-called **barycentric parallax.** Approximate values of the masses of the satellites or Jupiter can be obtained by the mutual **perturbations** which they exert on each other. In the case of Saturn the masses of the satellites may be approximately determined from their mutual perturbations and an approximate check is provided by the positions of the divisions in the rings.

The three outer satellites of Jupiter, the outer satellite of Saturn, the four satellites of Uranus, and the satellite of Neptune all revolve about their primaries in the retrograde sense, i.e., in the direction contrary to that in which all other planets and satellites are revolving and rotating. This retrograde motion can be completely explained on the basis of modern **celestial mechanics.**

The influences which satellites exert on their primaries are very slight. The tidal forces which they exert have some slight effect upon the rotation periods of the primaries but such effects are so small as to be beyond observational measurement. The tidal effects which the planets exert upon the satellites, on the other hand, are in many cases so large that the satellites rotate in approximately the same period as that in which they revolve.

The question as to the origin of the satellite systems is still unanswered. The systems bear so much resemblance to the solar system itself that there is the suggestion that their evolutionary process may be the same as that discussed under the topic of **solar system,** but there are many objections to such a theory. Since the moon is the largest satellite in the solar system in comparison with its primary, and also being larger in proportion to the earth than any other planet is to the sun, it presents some very particular problems. If they ever formed one single mass, that mass must have been rotating with a period of approximately 4 hours, and have been greatly flattened at the poles. Such a mass would tend to break up under the influence of the rapid rotation, but would remain intact unless some external force was present. Such a force is found in the tidal effects of the sun, and the earth-moon system may have been formed by the breaking up of a large parent mass. However, the alternative hypothesis that the earth and moon were formed as two separate bodies at the time that the solar system was formed cannot be disproved. (W.K.G.)

SATURABLE REACTOR. This is a three-legged reactor whose **inductance** is controlled by passing d.c. through the winding on the center leg and thus altering the saturation of the **core.** The windings on the two outer legs are connected in an **alternating-current circuit** and serve as adjustable reactances. This operation is based upon the non-linear characteristic of iron, the permeability of the core and the consequent inductance of the a-c windings being high for low-core flux and decreasing as the saturation is increased. A wide range of reactance values may be obtained by the use of many turns and a small amount of d.c. (L.R.Q.)

SATURATED. In geochemistry this term is used by mineralogists and petrologists to designate minerals, such as **feldspar** which can form in the presence of an excess of free **silica.** The term was proposed by Shand, in 1911, and is frequently used to describe **igneous rocks** which are entirely composed of minerals of this type. (R.M.F.)

SATURATED LIQUID. Liquid which has absorbed the maximum quantity of heat possible, while remaining in the liquid phase, and has, consequently, risen in temperature to the boiling point is said to be "saturated." The quantity of heat contained, as well as the boiling temperature, is a function of the pressure existing on the surface of the liquid. (F.T.M.)

SATURATED VAPOR. A vapor whose temperature corresponds to the boiling temperature at the pressure existing on it, is said to be saturated. Expressing the same thought another way, a vapor is saturated when its temperature is a function of its pressure alone. A saturated vapor may be wet or dry, and the term does not imply, necessarily, a wet vapor. A vapor of 100% **quality,** having no superheat, is said to be dry and saturated. In contrast to a saturated vapor, the temperature of a superheated vapor depends both on the pressure and the degree of superheat. The temperature of a saturated vapor depends on its pressure and increases with increasing pressure. By virtue of the importance of water in both its liquid and vapor phases, its properties have been completely investigated and recorded in tables and charts. The properties of saturated steam will be found among such compilations. The physical attributes of saturated steam are the pressure, temperature, volume, **enthalpy,** and **entropy.** These are always given for steam which is dry and saturated, leaving the reader to apply the **quality** factor when it occurs. The increase of volume on vaporization and the latent heat of evaporation are present in wet steam to the extent of the percent dryness of the steam. One of the important entries in the saturated steam table is that for atmospheric pressure. At 14.7 lbs. per sq. in. absolute pressure, the saturation temperature of steam is 212° F. The heat contained in it as a boiling liquid is 180 B.T.U. (above 32° F.), and its latent heat of evaporation is 970.2 B.T.U. per lb. (F.T.M.)

SATURATION. Color; Magnesium; Vapors; Humidity.

SATURATION CURRENT. Ionization Chamber.

SATURATION OF THE ATMOSPHERE. If a free water surface is introduced into a box from which all gases have been removed, molecules of water will emerge from the liquid surface until the number of molecules escaping from the liquid equals the number returning to the liquid. When a balance is maintained between those evaporating or escaping from the liquid water and those condensing or impinging on and remaining in the liquid, the space within the box is said to be saturated with water vapor. It can hold no more water molecules and, if more are added, these will condense into liquid. If air is admitted into the box there will be no change whatsoever in the rate at which water molecules leave and impinge on the liquid. Component parts of the atmosphere, therefore, have no bearing on the number of molecules of water vapor present. Saturation of space above any liquid water surface depends solely on the number of molecules of water in the air as a vapor when a balance is achieved between the vapor molecules condensing on the liquid and the liquid molecules evaporating into the space above. This molecular balance depends almost entirely on the temperature of the liquid and the water-vapor molecules. The higher the temperature, the more molecules of water vapor can escape from the water surface before they begin to condense and return to the surface. In meteorological practice, as air is the medium in which weather phenomena take place, it is useful to relate saturation of water vapor to the air carrying it. Saturated air as a term is, therefore, used even though the presence of the air does not directly affect the number of molecules of water vapor at saturation. It is customary to speak of saturation specific humidity with particular reference to the weight of air carrying the water-vapor molecules. These saturation specific humidities vary from a few tenths of a gram per kilogram of cold air to 30–40 grams per kilogram of very warm air. (P.E.K.)

SATURATION TEMPERATURE. Saturated Vapor.

SATURN. (See tables of planetary data, page 1109.) Saturn, the "ringed planet," is the sixth major planet in order of distance from the sun and was the outermost planet known to the ancients. In point of size Saturn is the second largest among the planets, having a diameter slightly more than 9 times that of the earth and but slightly less than the diameter of Jupiter.

The physical characteristics of the planet itself are approximately the same as those for Jupiter. The low density, high rotation speed, and variation of rotation period with planetary latitude all point to the probability that the solid core of the planet is relatively small and is surrounded by an atmosphere of very great thickness. As in the case of Jupiter, spectroscopic analysis indicates that the atmosphere is composed to a large extent of methane with some ammonia. All evidence, both observational and theoretical, indicates the surface temperature of the planet to be in the neighborhood of 123° K. (−238° F.). This low temperature coupled with the lack of oxygen in the planet's atmosphere indicates the impossibility of there being any life on Saturn such as we have on the earth.

To the naked eye, Saturn appears comparable to the brighter stars. In a telescope the planet itself has a belted appearance similar to that of Jupiter, but without as many distinctive surface features as are to be seen on the larger planet. From these semipermanent surface features the rotation period of the planet has been determined and found to be but slightly over 10 hours. There is also considerable evidence that the rotation period varies with planetary latitude, being the shorter at the equator.

Undoubtedly, the most remarkable and best known characteristic of Saturn is the ring system which surrounds the planet. These rings have the appearance of circular disks of paper, pierced with a hole in the center to admit the planet. The rings were first noticed by Galileo in 1610 when he first applied the telescope to astronomical observation, but it was not until 40 years later that Huygens was able to accurately describe them. Twenty years later, in 1675, Cassini found that the rings were separated into two parts, and in the middle of the 19th century Bond found a second division of rings, discovering the third or "dusky" ring, close to the planet itself.

Starting from the outside the rings are usually designated by the letters A, B, and C. The outer ring, A, has a diameter of approximately 171,000 miles and is about 10,000 miles wide. Cassini's division, between A and B, is about 3000 miles wide and ring B has a width of 16,000 miles. Between B and C there is a narrow space of about 1000 miles and ring C itself has a width of about 11,500 miles, leaving a space of about 7000 miles between the inner edge of C and the planet itself. In contrast to the great width of the rings their thickness is very slight. Actual measurements of the thickness are very difficult but it certainly is not more than 100 miles. On a model of 10,000 miles to the inch the ring system would have a width of about 17″ while the thickness would be less than that of the thinnest paper.

The rings are parallel to the planet's equator and, since the equator is inclined to the ecliptic at an angle of about 28°, the plane of the rings is inclined to the plane of the ecliptic by the same amount. Saturn revolves about the sun with a period of about 29.5 years and twice during this period the plane of the rings passes through the orbit of the earth. The plane of the rings takes nearly a year to pass the earth's orbit and during this period the earth may pass through the plane of the rings either once or three times. At these times the rings are seen edgewise from the earth and appear as thin needles of light extending out from the planet's equator. Intermediate between the passage of the earth through the plane of the rings they "open out" to their maximum angle at which time they appear as an ellipse produced by tipping a circle by about 27°, or an ellipse with an apparent width about half its maximum length.

The problem as to the constitution of the rings of Saturn was a problem which vexed astronomers from the time when they were first discovered down to the middle of the last century. It was recognized more than 100 years ago that such a wide, thin system could not be in stable equilibrium if it were composed either of a solid or a liquid. After the discovery of the inner ring by Bond and the discovery of a number of very fine divisions in both the A and B rings the mathematician, Clerk Maxwell, proved that the rings are in reality composed of millions of small particles each moving about the primary in orbits. It was further proved that the divisions in the rings are produced by perturbations due to the outer satellites of Saturn. The theoretical proofs were later verified experimentally by Keeler who, using the spectroscope and the Doppler principle, was able to show that the outer edges of the rings were revolving with longer periods than the inner edge. Such a condition could never maintain for a solid and could only be accounted for on the basis of a swarm of satellites moving under the Keplerian laws of planetary motion.

The question of the origin of the ring system of Saturn is still a much debated problem. The two main theories postulate either that a large satellite was "spoiled in the making," or that a large satellite was formed and then exploded under the influence of some unexplained force.

In addition to the millions of satellites that go to make up the ring system Saturn has a system of nine satellites. These outer satellites of Saturn are, on the average, smaller than the satellites of Jupiter, Titan the largest of them having a diameter of about 2600 miles. Titan is visible in a 3-in. telescope and seven of the others may be seen with telescopes of moderately large aperture. Phoebe, the faint outer satellite of Saturn, has a retrograde motion (i.e., revolves about the planet in the direction opposite to the revolution and rotation of the great majority of the other members of the solar system.) (W.K.G.)

SAUERKRAUT. Brassica.

SAUGER. Pisces, Teleostei. A North American fish (Pisces), *Cynoperca canadensis,* related to the walleye. A pike. Found in streams of the St. Lawrence basin and the upper part of the Mississippi River system. (A.W.L.)

SAURIA. Reptilia.

SAUROPSIDA. A term applied collectively to the reptiles and birds. It is not a taxonomic group but indicates the relatively close similarity of birds and reptiles as contrasted with the Ichthyopsida, consisting of fishes and amphibians, and the mammals, which stand alone. (A.W.L.)

SAW. See Milling, Woodworking. Hand sawing is often resorted to in metal work; a hacksaw consists of an adjustable frame for holding separate blades of various lengths. The blades are thin and comparatively narrow, and have teeth along one edge, and holes at each end to fit pins in the frame. Power metal-cutting saws are of three types: band saws, power hacksaws, and circular saws.

A power hacksaw consists essentially of a frame carrying a hacksaw blade and a vise for holding work. The machine has an automatic feed with an arrangement to lift the blade clear of the chips on the return or non-cutting stroke. The vise is adjustable for angular cutting. The machine is usually motor-driven and has a pump to supply coolant for cutting.

A cold saw is a metal-cutting machine which uses a disk saw similar to a milling cutter. One type of machine has an hydraulic feed and a mechanically driven saw which cuts up. Cutting speeds of 35 to 70 ft. per min. can be obtained. Friction saws are used for cutting structural shapes and bars. The saw blades are circular, with vee-shaped peripheral teeth, are generally water-cooled, and are run at speeds up to 20,000 ft. per min. Large sections may be cut in but a few seconds. In some of the larger machines the work is rotated on an axis parallel to the blade axis, at a speed of about *e* ft. per min. to prevent contact with any large area of metal, produce a cleaner cut, and increase the rated capacity of the saw. Thin abrasive wheels are used for cutting extremely hard materials such as hardened high-speed steel, drill rod, glass tubing, or materials which may be abrasive in character.

Metal-cutting *band saws* are essentially similar to wood-cutting band saws. The saw blades are obtainable in widths from $\frac{1}{16}''$ to $\frac{1}{2}''$; as wide a blade as possible is employed; the blade width is generally limited by the minimum radius to be cut. Metal cutting band saws may be used for external or internal cutting. In internal cutting, a starting hole is drilled through the work and the saw band is cut or broken and inserted through the hole. The band is then clamped and welded in a special welding attachment on the frame of the machine. After welding, the joint is annealed and ground on its surface. The joint in the band is a butt-welded joint, can hardly be distinguished from any other part of the saw, and can be made in about one minute. The band saw also finds extensive application in the manufacture of sheet metal blanks where quantities are too large for individual production, but not sufficiently great to warrant the expense of a die set. (H.C.H.)

SAWFISH. Pisces, Plagiostomi. A large marine fish, *Pristis*, one of the rays. It has the upper jaw prolonged into a slender snout set with numerous sharp spines on each side. It is found in tropical America and Guinea. The name is also applied to *Pristiophorus*, a genus of sharks found in Japan and Australia. (A.W.L.)

SAWFLY. Insecta, Hymenoptera. A plant-feeding member of this order, whose more familiar species are the ants, bees, and wasps. The sawflies have four wings,

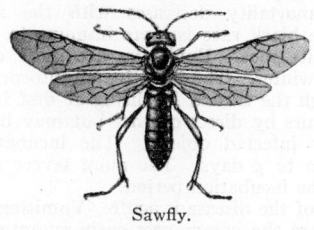

Sawfly.

somewhat like those of the wasps, but the abdomen is broadly connected with the thorax, in contrast with the thin-waisted bodies of the other forms. They are named from the sawlike **ovipositor** with which slits are cut in the tissues of plants to receive the eggs. Some species are of economic importance. Since the **larvae** eat leaves they can be destroyed by the usual arsenical sprays. (A.W.L.)

SAW-TOOTHED WAVE. Cathode-Ray Tube.

SCABIES. "The itch." A skin disease caused by an animal **parasite** (*Sarcoptes Scabiei*) characterized by an intensely itching eruption which involves the hands between the fingers, the **axilla**, nipples, lower abdomen,

buttocks and external genital organs. The itching is always worse at night.

The female parasite burrows beneath the skin where eggs are laid, causing slightly elevated, grayish, tortuous, or dotted lines on the skin surface.

Treatment consists of the external application of various agents which kill the parasites. Sulfur is the most reliable of these, although balsam of Peru, tar, and other compounds are also effective. (D.M.H.)

SCAD. Pisces, Teleostei. A fish (**Pisces**) related to the pompanos, of the family Carangidae. One species, common in the region of the West Indies, is called scad, cigar fish, or round robin. Another, also common in warmer waters although it ranges north to Cape Cod, is known as the big-eyed scad, goggler, or chicharro. The name has also been applied to the related horse mackerel, a European species occasionally taken on the Atlantic coast of North America. (A.W.L.)

SCALAR. Vectors.

SCALAR PRODUCT OF TWO VECTORS. The scalar product (or dot product) of two vectors **a** and **b** is defined as a **scalar** equal in magnitude to the product of the magnitudes of the two vectors by the cosine of the angle between them.

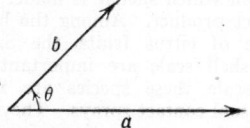

The symbol for the scalar product of **a** by **b** in the Gibbs notation is **a·b**; in another form of vector notation the symbol (**a**, **b**) is used, while in the Hamiltonian notation the symbol S **a b** is used. Then

$$\mathbf{a \cdot b} = ab \cos \theta,$$

where θ is the angle between **a** and **b**.

The scalar product of two vectors may also be interpreted as the length of one vector multiplied by the **projection** of the other on it.

The scalar product of two vectors obeys the **commutative** and **distributive** laws:

$$\mathbf{a \cdot b = b \cdot a}, \quad \mathbf{a \cdot (b + c) = a \cdot b + a \cdot c}.$$

If **a** is perpendicular to **b**, then **a·b** = o; and conversely, if **a·b** = o, then **a** is perpendicular to **b**.

If **a** is parallel to **b**, then **a·b** = ab. Hence, **a·a** = a^2.

For the unit vectors $\hat{\mathbf{i}}, \hat{\mathbf{j}}, \hat{\mathbf{k}}$, we have:

$$\hat{\mathbf{i}} \cdot \hat{\mathbf{i}} = \hat{\mathbf{j}} \cdot \hat{\mathbf{j}} = \hat{\mathbf{k}} \cdot \hat{\mathbf{k}} = 1, \quad \hat{\mathbf{i}} \cdot \hat{\mathbf{j}} = \hat{\mathbf{j}} \cdot \hat{\mathbf{k}} = \hat{\mathbf{k}} \cdot \hat{\mathbf{i}} = 0.$$

If $\mathbf{a} = a_1\hat{\mathbf{i}} + a_2\hat{\mathbf{j}} + a_3\hat{\mathbf{k}}$, $\mathbf{b} = b_1\hat{\mathbf{i}} + b_2\hat{\mathbf{j}} + b_3\hat{\mathbf{k}}$, then

$$\mathbf{a \cdot b} = a_1b_1 + a_2b_2 + a_3b_3. \quad \text{(L.L.S.)}$$

SCALE. A flat structure developed as a covering. 1. The scales of fishes (**Pisces**) are in many cases arranged like shingles to form a complete armor at the surface of the body, and in other forms are small and scattered, merely adding to the resistant qualities of the skin. The elasmobranch fishes have placoid scales, whose form includes a broad base and a projecting enamel-covered point. This form of scale is much like the teeth of these fishes and is supposed to be ancestral to all vertebrate teeth. Ganoid scales, found in relatively few fishes, are regarded as a modification of the placoid type by the loss of the point and the addition of a hard outer layer known as ganoin. Some of the more primitive fishes of other groups have rounded scales with smooth margins, known as cycloid, and others have ctenoid scales, with comblike edges. The two last forms are found in the higher fishes. They bear neither enamel nor ganoin. 2. The scales found over the entire surface of the body in **reptiles**, on the legs of birds (**Aves**), and to a much more limited extent in **mammals**, as on the tails of **rodents**, are quite different from the scales of fishes. Each is a modified area of the skin, thickened and hardened by the development of the horny substance, **keratin**. Scales reach their highest development among the mammals in the **pangolins** and **armadillos**. In the latter they are underlaid by bony plates. 3. The **butterflies** and **moths**, a few **beetles**, and some other insects have the surface of the body and wings more or less covered by flattened scales. These structures are modified **setae**. They often contain pigments and in many species are so formed that they produce iridescent, metallic, or glossy physical colors by breaking up the light rays which they reflect. 4. Scale **insects**, often called scales, are highly specialized sucking insects which live on plants. They belong to the order **Homoptera** and are related to the plant lice and phylloxerans. Most species are minute. The young and the female adults are simplified and remain closely attached to the plant, secreting over themselves a protective covering or scale which gives them their name. This covering is usually characteristic of the species.

Scale insects include many species of economic importance. Among them are the useful cochineal insect, a source of dye, and the lac insect whose scale is the raw material from which shellac is made. China wax is also a scale insect product. Among the harmful species the purple scale of citrus fruits, the San José scale, and the oyster-shell scale are important. Because of the protective scale these species are not easily destroyed by the usual contact sprays. They are controlled by spraying but the concentration of poison must be high and spraying during the dormant stage of the plant is often necessary as a result. Fumigation of citrus trees with cyanide is practiced extensively. This method requires special equipment since the tree must be enclosed in a tent.

In metallurgy scale is the oxide layer that forms on the surface of metals upon heating in air or other oxidizing gases. The heavy scale that forms on steel ingots or billets upon heating for rolling or forging breaks away as the metal is deformed in the mill or under the hammer; however a lighter, often very adherent scale always remains after hot-working operations and after annealing or other heat treatments unless a non-oxidizing protective atmosphere is provided. Scale is removed from steel by pickling, generally in warm dilute sulfuric or hydrochloric acid. The scale which forms on stainless steel is much more resistant and its removal requires strong acids such as mixtures of hot nitric and hydrofluoric. A tight adherent scale is often left on steel for its protective value; for example, steam-blued steel sheets are used for stovepipe. (A.W.L., R.H.H.)

SCALE, BOILER. Water, Utilization.

SCALE EFFECT. Any aerodynamic coefficient which changes with **Reynolds number** is said to be subject to scale effect. (See **Aerodynamics, Airfoil, Drag.**) (F.T.M.)

SCALING. Feedwater Treatment.

SCALLOP. Mollusca, Lamellibranchiata. *Pecten.* Marine **bivalves** of wide distribution in both shallow and deep water. The symmetrical shells are beautifully marked with radiating grooves which give them a scalloped edge. The large muscle which closes the shell of the scallop is eaten but the body is discarded. (A.W.L.)

SCALLOPED BUTT STRAP. Riveted Joints.

SCALP. The soft tissues covering the vertebrate cranium. The term is usually applied only to man, and usually only to the part of the covering of the skull which normally bears hair. (A.W.L.)

SCANDIUM. Symbol: Sc. Atomic number: 21. Atomic weight: 45.10. Melting point: 1200° C. No isotope, but of single atomic form: 45. Type of compounds: Sc₂O₃. Color of salts: Colorless. Discovered by Nilson in 1879, but predicted by Mendeleeff in 1871 as an element to be discovered with properties resembling **boron**. Occurs in a few uncommon minerals, e.g., **wolframite**, sometimes to the extent of 2% oxide. A member of the **cerium** sub-group of the rare earth metals. (R.K.S.)

SCANNING. Television; Facsimile.

SCAPHOPODA. The tooth shells. A small class of **mollusks** with slender tapering shells open at both ends. The foot projects from the larger end of the shell, together with a group of slender tentacles called captacula which arise from the poorly developed head. These captacula have sucker tips and are sensory. In many details of structure the animals are intermediate between the classes **Gasteropoda** and **Lamellibranchiata**. They are marine, burrowing in the bottom by means of the foot. (A.W.L.)

SCAPOLITE. Wernerite.

SCAPULA. The shoulder blade of the vertebrates. **Skeletal system.** (A.W.L.)

SCARAB. Insecta, Coleoptera. A species of **beetle** which was regarded as sacred by the ancient Egyptians. The term is applied especially to the sculptured likenesses of the beetle. From this name the family Scarabeidae has arisen, containing, among the many North American species, the **June bugs** or May beetles, the **tumble bugs,** and the **rose chafer.** (A.W.L.)

SARD. Chalcedony.

SCARLET FEVER (Scarlatina). An acute infectious disease caused by the hemolytic *Streptococcus.* Scarlet fever was first described by Sydenham in 1675. Previous to that time it had been confused with measles and **rubella.**

It is essentially a disease of childhood and the morbidity and mortality decreases with the age of the patient. The black race is more immune to the disease than the white race. Epidemics are more common in the fall and winter. The infective streptococcus usually enters through the throat, localizing at first in this area. Infection occurs by direct contact but may be conveyed indirectly by infected objects. The incubation period varies from 2 to 5 days. The more severe the disease, the shorter the incubation period.

The onset of the disease is acute. Vomiting, fever and sore throat are the commonest early symptoms. Often there is an area of pallor around the mouth and nose. The rash does not appear until 12–24 hours after the onset, and usually spreads over the entire body. The rash is quite characteristic: the skin is suffused with a reddish blush and is uniformly covered by a fine granular eruption. The palate and roof of the mouth are involved in the eruption, and the tongue is red, smooth, and dotted with a few large papillae. The throat symptoms may be severe and swelling of the neck glands may occur. The fever usually remains high for 3–5 days and then decreases gradually. Peeling or desquamation begins within 3–10 days after the disappearance of the rash and is complete in 2 or 3 weeks on the body, longer on the hands and feet.

The rash and the prostration of scarlet fever are due to the toxin produced by the hemolytic streptococci. Against this phase of the disease an effective antitoxin (see **Serum**) which neutralizes the toxin is available. **Penicillin** and **sulfonamides** may be used also, especially in complications such as purulent **sinusitis, otitis media,** etc. (R.S.M., D.M.H.)

SCARLET RUNNER. Bean.

SCARP. The word scarp is an abbreviation of escarpment, and either term may be used to designate a vertical or steep cliff which has resulted from **erosion** or **faulting.** If due to the latter cause, the term fault-scarp is commonly used. An erosional escarpment developed upon beds which tilt at a very gentle angle gives rise to an asymmetric ridge called a **cuesta.** If the beds are of variable hardness several cuestas, one behind the other may develop. Examples of this condition may be seen on the Atlantic and Gulf Coastal plains. (R.M.F.)

SCATTERING. When light enters a body of matter, however transparent, part of it is diffusely reflected or "scattered" in all directions. This is due to the interposition in the light stream of particles of varying size, from microscopic specks down to electrons, and the deflection of light quanta resulting from their encounters with these small obstacles. Similar effects are produced upon infra-red, ultra-violet, x-rays, and other forms of **elestromagnetic radiation,** and upon streams of particles such as **cathode rays** or alpha rays.

The scattering of light was carefully studied by John Tyndall, who used fine suspensions in air and liquids; and also by Ångström and by Lord Rayleigh. The laws of scattering depend upon the nature of the scattering particles. For very fine dust, Rayleigh concluded that the intensity of the light of wavelength λ, scattered in any direction making an angle θ with the incident direction, is directly proportional to $1 + \cos^2 \theta$ and inversely proportional to λ^4. The latter point is noteworthy, in that it shows how much greater is the scattering of the short wavelengths. Thus the sky is blue, and tobacco smoke appears blue, because blue light is scattered more than red. The unscattered light is of course complementary to blue, that is, orange or yellow; which explains the "warm" hues of the sunset. Scattered light is also distinctly plane-polarized (see **Polarized Light**). By far the largest part of scattered monochromatic light is of the same wavelength as the incident; but that scattered by molecules contains faint components of different wavelength; this is known as the **Raman effect.** The scattering of electrons and other particles by **atoms** has given important information as to the structure of atoms. (L.D.W.)

SCAUP. Aves, Anseriformes. A **duck.** The scaup or blue-bill, *Nyroca marila,* is found throughout the northern hemisphere and the lesser scaup, *N. affinis,* is a common North American species. Both are principally black and light gray or white. The head of the scaup duck is glossed with green and that of the lesser scaup with purple. (A.W.L.)

SCHEELITE. The mineral scheelite is **calcium tungstate,** $CaWO_4$. It is a **tetragonal** with an **octahedral** habit although also at times tabular, and may occur massive. It displays an octahedral **cleavage;** is brittle; hardness, 4.5–5; specific gravity, 5.9–6.1; luster, vitreous; color, white to yellowish, reddish, greenish and brownish; white streak; transparent to translucent. Scheelite is found in **pegmatite** and ore veins associated with **granites,** also as a contact **metamorphic** mineral. It is known from Czechoslovakia, Saxony, Italy, Alsace, Finland; Cumberland and Cornwall in England, Mexico; and in the United States in Connecticut, Colorado, South Dakota, Arizona, Nevada, and California. It is an ore of tungsten. The Swedish chemist, Karl Wilhelm Scheele, discovered tungsten in this mineral, which later was named for him. (E.S.C.S.)

SCHICK TEST. A test for immunity to diphtheria. (See **Diphtheria.**) (D.M.H.)

SCHIFF'S REAGENT. Acetaldehyde and Aldehydes.

SCHILLERIZATION. Luster.

SCHIST. The schists form a great group of **metamorphic** rocks chiefly notable for the preponderance of the lamellar minerals such as the **micas, chlorite, talc, hornblende, graphite,** etc. Quartz often occurs in drawn out grains to such an extent that a **quartz** schist is produced. Most schists have in all probability been derived from clays and muds which have passed through a series of metamorphic processes involving the production of **shales, slates** and **phyllites** as intermediate steps. Certain schists have been derived from fine-grained igneous rocks such as lavas and **tuffs.** Most schists are mica schists, but **graphite** and chlorite schists are common. Schists are named for the prominent or perhaps unusual mineral constituent, as **garnet** schist, **tourmaline** schist, **glaucophane** schist, etc. The word schist is derived from the Greek meaning to split, with reference to the easy separation of these rocks in a direction parallel to that in which the platy minerals lie. (E.S.C.S.)

SCHIZOPHRENIA (Dementia Praecox). A mental illness or **psychosis,** characterized by bizarre mental and emotial reactions and oddities of behavior and thought. The disorder is relatively common; approximately 60% of chronically ill patients in state hospitals are schizophrenics. Although its fundamental cause is not known, the disease may be thought of as basically the same type of disturbance which underlies a **psychoneurosis,** except that in schizophrenia the disturbance is deeper and more far-reaching in its effects on the personality. Conflicts between the individual's desires and urges and what his conscience or society consider to be "right," feelings of guilt, insecurity and frustration are all charged with emotional tension which may lead to a psychosis.

Schizophrenia begins most frequently in adolescence or early adult life. It occurs usually in the shy, sensitive, unsociable often imaginative and idealistic individual of asthenic body build. The early symptoms are merely an exaggeration of previous personality traits: the individual tends to be more shy, hypersensitive, quick to take offence. Gradually he retires into his shell and has less and less to do with activities and people. He thus protects himself from painful experiences in what seems to him a confusing, harsh, hostile world. Part of his protection comes from his ability to alter the meaning of things to suit his own needs. Like the child, for whom a broomstick becomes a lively horse, the schizophrenic exercises his imagination to interpret words or actions in a way pleasing to himself. This leads to apparently motiveless, senseless behavior since observers cannot understand the meaning that the actions have to the patient.

The protective shell of the schizophrenic is further strengthened by his ability to get satisfaction by imagining rather than by acting or doing. This is an exaggeration of the pleasure that normal people take in daydreams. For the schizophrenic the world of day-dreams where all of his wishes come true without effort or risk to himself, where he is brilliant, successful, always loved by his beautiful wife, admired by his friends—this world becomes far more attractive and satisfactory than the world of reality. Gradually it supplants the world of reality for him and he comes to live completely in his day-dreams.

By these processes the schizophrenic comes to be remote from the ordinary world and inaccessible to its members. When the disease is far advanced, he may have **hallucinations** and delusions of persecution. These are projections of his own fears and misgivings about himself which he is able to interpret as coming from "other people" or "the world." This is a satisfactory form of defence because it relieves him of self-criticism and any feeling of guilt.

As the disease progresses, psychiatrists speak of deterioration of the individual, but this refers to behavior rather than mental processes. Many schizophrenic patients take no interest in the outside world; they may exhibit **catatonia**: they sit quietly without moving; they stand in rigid positions and stare at a blank wall; they hold their arms or legs in curious attitudes for hours at a time; or they soil themselves and seem completely indifferent. This may be called deterioration of behavior. Yet such patients may come out of their psychotic phase, return to normal behavior and exhibit an intact memory, intelligence, and education.

The prognosis in schizophrenia depends on the duration of the disease. In the early stages, when the patient still has insight into his difficulties, much can be done with **psychotherapy**. Later, when the individual has become accustomed to his private world and finds it pleasing and satisfactory in all respects, it is difficult to persuade him to give it up. The longer the disease has been present, and the more inaccessible the patient becomes to human contact, the less is the likelihood that he can be restored to normal. In cases of certain types, the convulsant drug, metrazol, electric or insulin-shock treatment, producing a profound stimulus to the central nervous system, have given good results. (D.M.H.)

SCHLIER. The term for streaky local segregations which occur in coarse-grained, **intrusive igneous** rocks, usually near the contact of the intrusion with the country rock. The segregations, themselves, appear to be local concentrations of minerals drawn out as lenses or stringers by the movement of the **magma** while still in the viscous state. The term is derived from the German, *Schliere* (singular), *Schlieren* (plural), referring to the streaks produced in artificial cements. (R.M.F.)

SCHMIDT CAMERA. Telescope.

SCHMIDT OBJECTIVE. An objective for reflecting **telescopes,** designed to correct the **aberration** of the spherical mirror without introducing the coma (blurring) to which even a parabolic reflector is subject for wide fields. The results are obtained in somewhat the same way that spectacles correct for defects in vision.

The Schmidt objective as originally designed consists of a concave spherical mirror, functioning in the same way as the objective of any reflecting telescope, but with a plate of glass interposed in front of it perpendicular to its axis at its center of curvature. This glass plate is not plane, but has one surface "figured" in such a way that, as the rays pass through it on their way to the mirror, it so modifies their course as to effect almost perfect correction for the spherical aberration and coma which the mirror would otherwise produce. In a later design, the objective consists of two coaxial cylinders of glass, in contact along a plane perpendicular to the axis. The rear surface of the rear piece is spherically convex and is silvered on the outside, thus presenting a concave spherical mirror to the interior of the glass cylinder. The front surface of the front piece, passing through the center of curvature of the mirror, is the correction surface, serving the same purpose as the glass plate in the older design. The reason for using two pieces is that the final real image *F* is of course produced between the mirror and the correction surface; and it is here that the plane of separation is located, so that the

small photographic film used may be introduced. The general nature of these arrangements will be clear from the figures. (L.D.W.)

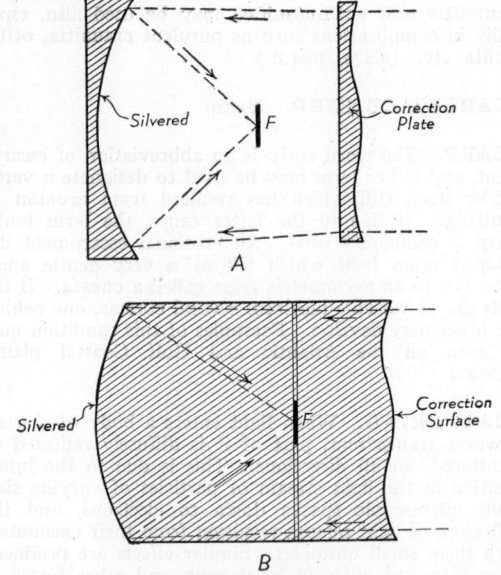

Schmidt objective: *A*, original design; *B*, the "solid Schmidt" (diagrammatic).

SCHORL. Tourmaline.

SCHUMANN REGION. Ultra-Violet.

SCHUPPEN (Literally, Scales). Used by many **structural** geologists to describe an assemblage of thrust slices which overlap like shingles on a roof. The term is used especially for this kind of structure in the Swiss Alps. Essentially synonymous with **imbricate** structure. (R.M.F.)

SCHWARZSCHILD EFFECT. A special phase of reciprocity law failure with photographic materials. (C.B.N.)

SCHWEITZER'S REAGENT. This is an ammoniacal **copper** hydroxide solution used to dissolve cellulose. (See **Carbohydrates.**) (R.K.S.)

SCIATICA. Painful irritation of the sciatic nerves, the large nerve trunks which supply the legs. Sciatica may be due to inflammation of the nerve as in **polyneuritis,** but unilateral sciatica is commonly caused by pressure on the nerve by a protrusion of the intervertebral disk between the lumbar **vertebrae.** This latter condition usually follows injury resulting from lifting. The patient complains of pain down the back of the leg, spasm of the muscles, and a limping gait. Treatment is rest in a hard, flat bed, with boards beneath the mattress, and local heat. Surgery is sometimes indicated. (D.M.H.)

SCIENTIFIC NOTATION FOR NUMBERS IN DECIMAL FORM. Standard Notation for Numbers in Decimal Form.

SCINTILLATION. Radioactive Changes; Spinthariscope.

SCINTILLOSCOPE. Spinthariscope.

SCION. Grafting and Budding.

SCLERENCHYMA. This is a tissue composed of thick-walled **cells** of various forms. These cells have small pits and walls so thick that in many cases the cavity of the cell is nearly obliterated. Mature sclerenchyma cells are dead, containing no **protoplasm.** They serve to strengthen the part of the plant in which they are found or to protect the more delicate structures within. Sclerenchyma cells are of two kinds, stone cells and fibers. Stone cells are small, irregular in shape, and only slightly if at all elongated. They may be found in the **cortex** of the stem or elsewhere in the plant, but are particularly abundant in the endocarp of certain fruits and in seeds. The flesh of pears and blueberries contains many grit particles, which are groups of stone cells. Fibers are very much elongated cells, generally with long pointed ends and with simple pits in the walls. **Hemp** and **flax** are bundles of fibers of great value to man. (R.M.W., B.S.M.)

SCLEROBLAST. Cells of **sponges** which secrete the material of the **spicules.** These cells lie in the middle layer of the sponge (the mesogloea) and are the only source of the hard parts. (A.W.L.)

SCLEROSCOPE HARDNESS. Hardness.

SCOLECITE. This mineral is a **zeolite,** a hydrous **calcium-aluminum silicate,** $CaAl_2Si_3O_{10}3H_2O.$ It occurs in slender **monoclinic** prisms and in fibrous and nodular masses. Hardness is 5–5.5; specific gravity, 2.16–2.4; luster, vitreous to silky; transparent to translucent. When heated, some specimens of scolecite curl up like worms, hence its name, derived from the Greek meaning a worm. This mineral occurs with other **zeolites,** at Baden, Switzerland; Iceland, Greenland; the Deccan region of India; and in the United States, at Golden, Colorado, and Paterson, New Jersey. (E.S.C.S.)

SCOLEX. The portion of an adult tapeworm (**Cestoda**) which is attached to the tissues of the host. It usually consists of a small rounded part at the end of a slender neck leading to the series of segments (proglottids), and usually bears hooks, suckers, or both. The suckers are usually two to four in number. In some species they are replaced by projections or grooves called bothria and in some they are accompanied by accessory suckers. Hooks may be located on a terminal prominence called a rostellum, and in one family they are at the end of slender retractile organs called proboscides. The scolex may be replaced by a modification of the anterior end of the segmented portion of the body (strobila) known as a pseudoscolex. (A.W.L.)

SCOLIOSIS. Abnormal lateral curvature of the spine. Although the curvature is to one side of the body or the other, there is always an element of twisting of the spine. Often it is not curvature of the spine that is complained of by patients, but instead they call attention to a high shoulder or hip, or a projecting shoulder-blade. A curvature that is mild and unnoticed in youth may become troublesome due to its increase after middle-age when there is shrinking of the cartilaginous disks between the vertebrae. Often there are no symptoms connected with this condition unless the scoliosis is of marked degree.

Scoliosis is usually divided into postural or functional, and organic. Many cases of the former condition are due to poor posture. With true or organic scoliosis there exist pathological structural changes in the vertebral bodies. The causes of this are usually congenital abnormalities of the spine, rickets, lung disease, tuberculosis of the spine, infantile paralysis, injuries to the vertebrae, tumor formation, etc.

Treatment of scoliosis depends on the severity, duration and damage to the vertebrae. Exercise and correction of posture play a part in the mild functional forms.

In the more severe forms, various corrective braces are worn. In other cases, surgical fusion of the spine is indicated. (R.S.M., D.M.H.)

SCOLOPHORE. A spindle-shaped organ formed of sensory cells grouped about a nerve ending. It is the fundamental **auditory organ** of **insects,** appearing in a simple form in connection with the body wall and also in the more complex organs of hearing. (A.W.L.)

SCORIA. The term applied to lava which is highly vesicular and slaggy in appearance, due to the escape of the volcanic gases while the lava is still viscous. Scoria may be considered as a very coarse variety of **pumice,** the vesicles occupying approximately the same amount of space as the solid material, and extremely variable in size and shape. (R.M.F.)

SCORPION, SCORPIONIDA. A terrestrial **arthropod** with a conspicuous pair of pinchers and a slender terminal region of the body bearing a clawlike sting, and the order made up of these animals. They are grouped with the **spiders, ticks** and other forms in the class **Arachnida.**

The order is characterized by the conspicuous pinchers and by the sting. The body is divided into **cephalothorax** and abdomen, and the latter consists of a broad anterior portion and a slender postabdomen. The sting, a modified **telson,** bears the opening of the duct of a poison gland. The genital ducts open on the ventral surface of the first abdominal segment, just in front of a pair of comb-like pectines of the second segment which are regarded as accessory reproductive organs.

Scorpions are common in warm dry regions. In the United States they occur as far north as Kentucky. Their poison is not virulent as a rule, although one species found in the region about Durango, Mexico, is said to be dangerous to man. (A.W.L.)

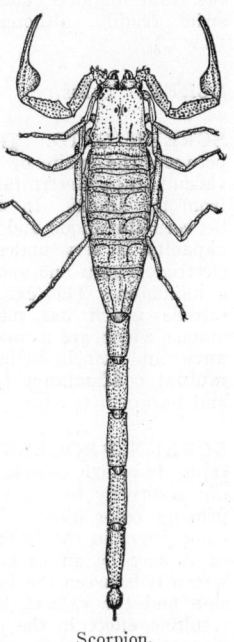

Scorpion.

SCORPION FLY. Insecta, Mecoptera. Moderately large **insects** with four membranous wings. They are named from the peculiar modification of the terminal segments of the abdomen, which fairly resembles that of the scorpions. The apparent sting is, however, made up of the external genital organs. (A.W.L.)

SCORPIUS. (The scorpion.) (Map, page 380.) Scorpius is the eighth sign of the **zodiac.** The **constellation** is rather far south for observation in Europe and North America, but is a beautifully grouped constellation presenting more resemblance to the figure for which it is named than is the case with most of the others.

The brightest star in the group Antares (Alpha Scorpii) is one of the most beautiful stars in the sky. It is distinctly reddish in color and gets its name from the fact that it opposes or rivals **Mars,** the red planet, in color. It is the largest star whose diameter has thus far been measured, having a diameter approximately 450 times that of the sun. The star is so large that if it should replace the sun it would extend out beyond the planet Mars.

Beta Scorpii is a fine **double** for an observer with

a small telescope with the two components distinctly different in color.   (W.K.G.)

SCOTER. Aves, Anseriformes. A duck found chiefly along the seacoasts, and less often on fresh waters. The American scoter, *Oidemia americana*, is a bird of the northern Atlantic, breeding from Alaska to Labrador. The white-winged scoter, *Melanitta deglandi*, breeds from the northern United States northward and is fairly common on the Pacific coast. The surf scoter, *M. perspicillata*, also breeds northward into the Arctic regions and is more generally distributed on both coasts and in inland waters. All of the species are dark colored birds with few lighter areas. (A.W.L.)

SCOTT TRANSFORMER. Transformer.

SCREAMER. Aves, Ciconiiformes. Peculiar South American birds (**Aves**), with moderately long legs and large feet. The beak is like that of the domestic fowl and the wings are provided with two stout spurs on the front margin. The birds are as large as geese and swim readily, although the toes are not webbed. (A.W.L.)

SCREE. Talus.

SCREEN GRID. This is a second **grid** introduced between the main or control **grid** and **anode** of a vacuum **tube** electrostatically to shield the control grid from the anode. In many applications the **feedback** between the anode and grid through the **interelectrode capacity** is very undesirable. By shielding these two electrodes from one another this feedback is reduced to a minimum. This extra grid converts the triode to a tetrode which has markedly different characteristics, among which are a much higher dynamic **plate resistance** and much higher **amplification factor.** The **mutual conductance** is about the same for the triode and tetrode. (L.R.Q.)

SCREEN PROCESSES. Processes of color photography in which color analysis and synthesis are carried out additively by the use of a mosaic screen of minute primary color filters. The screen may be composed of color filters in the form of a regular geometric pattern, or it may be an irregular mosaic. In either case the screen is between the color-sensitive photographic emulsion and the camera lens during analysis so that the resulting effect in the emulsion is that of three color-separation negatives side by side, completely intermingled by means of the screen's pattern.

Color screens may be prepared that are inseparable from the color-sensitive photographic emulsion, or they may be made so that they are on a separate support to be placed in contact with the emulsion. Separable screens are generally geometric and are made by ruling colored lines on the support. In 3-color screens the red, green and blue filter areas must be so adjusted that a satisfactory neutral black or white is obtained with the photographic emulsion and a suitable light source. When a negative is obtained after exposure behind such a color screen it may be reversed or printed onto another emulsion to give a positive image. When this positive is properly registered with the original, or another similar screen, the colors of the subject are reproduced.

Both regular and irregular screens are employed where the screen and emulsion are inseparable. Irregular screens may be made by dusting colored starch grains onto a sticky plate, or by mixing together small globules of colored resins. Such screens often have several thousand dyed particles per sq. in. The photographic emulsion is coated over the screen and the exposure made by light that has passed through the screen elements. The emulsion is reversed to a positive and the original screen is thus used in viewing the final color transparency.

Color screen processes have been widely used in the past but are largely being displaced by subtractive processes in the form of **integral tripacks** (see **Bipacks and Tripacks**). (H.C.C.)

SCREW FASTENINGS. Through bolts and **nuts** are extensively employed as removable fasteners where the bolt has an appreciable amount of clearance, usually $\frac{1}{32}$ or $\frac{1}{16}''$, in the bolt hole. Turned or carefully fitted bolts are frequently used in reamed holes when the bolt is required to resist shearing as well as tensile forces. Bolts applied to unfinished castings or forgings are usually provided with *spot-faced* bolt head and nut seats. Cap and machine screws are used to join parts when one part has an internally threaded hole. These screws are preferable to bolts because they are easier to handle in installations where access to one end of the element is either difficult or impossible. Cap screws are available commercially in sizes from $\frac{1}{4}''$ diameter up. Oval and fillister head cap screws are often preferred to hexagonal head cap screws since the head may be recessed to avoid interference or to facilitate cleaning the part held by the screw. Hexagonal head cap screws may be fastened more tightly than screws with screwdriver slots. The fillister head cap screw with a socket head combines the advantages of the hexagonal head and the slotted fillister screw. This type is fastened by using a special wrench made of hexagonal bar stock. When a screw must be removed frequently, it is often advisable to substitute a stud that may be inserted into the threaded hole and jammed against the bottom so that it is only necessary to remove the nut, thus avoiding wear on the threads in the hole. (In aluminum alloy castings, the aero-thread may be employed instead of using a stud.)

Machine screws are similar in appearance to cap screws but have heads of somewhat smaller proportions. For major diameters see **Gauge Number.** Machine screws have American Standard thread forms in both Coarse-thread and Fine-thread Series.

Stove bolts are employed for assemblies where precision is of no great importance. They are made with either flat or round heads and the screw threads are generally rolled. The square nuts used with them are stamped from common steel. Carriage bolts have a square portion directly under the head to prevent rotation when the nut is tightened and are used for fastening wooden parts together or for fastening metal parts to wood. Expansion and hook bolts are used in semi-permanent fastenings in concrete and masonry. Electric motors and other machinery are usually equipped with one or more eye bolts so that they may be lifted readily and removed with an overhead crane. A turnbuckle is a nut that has both right-hand and left-hand threads, and is used to adjust the length of tie rods and similar devices. The turnbuckle is one of the few devices in which a left-hand thread is employed as a fastener.

Plain washers, Fig. 1, are placed under the heads of hexagonal head screws and under square and hexagonal

PLAIN WASHER   LOCK WASHERS—FOR PREVENTING BOLTS
AND NUTS FROM WORKING LOOSE UNDER VIBRATION
Fig. 1.

nuts to assist in seating the nut or head, or to distribute the pressure exerted. Collar screws, Fig. 2, are square head cap screws with integral washers. Rough washers are punched from common steel; finished washers may be machined from steel bar stock. Lock washers are used to prevent accidental unscrewing of bolts and nuts, either by exerting additional tension on the threads or by biting into the surfaces in contact. It is possible to

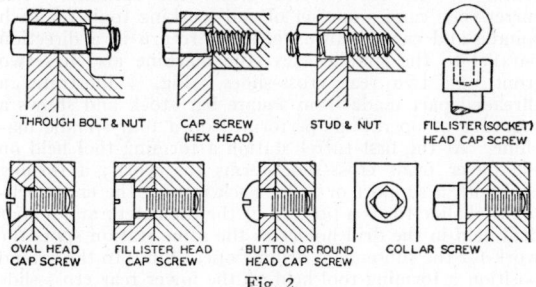

Fig. 2.

obtain button and flat head cap and machine screws with assembled lock washers that cannot drop off, a feature that will be appreciated by anyone who has ever tried to insert a screw with a loose washer in a comparatively inaccessible place.

Castellated and jam nuts, Fig. 3, are representative examples of parts for locking and fixing nuts in place.

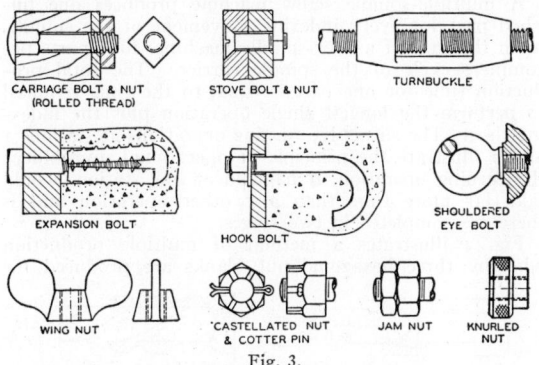

Fig. 3.

The castellated nut is held by a cotter pin and has six locking positions per turn; the jam nut holds the regular nut in position by being screwed against it. Wing and knurled nuts are designed for hand operation. Some forms of fillister head screws are supplied with knurled heads so that they may be screwed into place easily by hand, although the final tightening must be done with a screwdriver or wrench. (H.C.H.)

SCREW GEARING. Worm Gearing.

SCREW MACHINES, AUTOMATIC.

Automatic screw machines are essentially full-automatic bar stock turret lathes. There are two important types: single-spindle and multiple-spindle machines. In one form of single-spindle machine, the work spindle is driven by open and crossed belts so that the spindle rotation may be reversed for threading operations. The turret rotates on a horizontal axis in a plane parallel to or coincident with the spindle axis, and has six tool stations. There are two cross-slides, front and rear, and both cross-slides and the turret slide are actuated by disk or plate cams. The bar stock is held in a collet chuck, and is advanced by a feed finger to a stop, held in the turret, or on a swinging arm. Drills, reamers, threading and knurling tools are carried in the turret; form and cut-off tools are carried on the cross-slides, but some turning tools such as knee tools and hollow mills are held in the turret and are used for heavy cuts, since end cutting imposes less strain on the work and the machine than side cutting. Side cutting and forming is generally performed with circular form tools illustrated in Fig. 1, although straight form tools are also used.

Straight and circular forming cutters are used for form turning and special profiles and can be ground without changing their shape. Fig. 1 illustrates the applica-

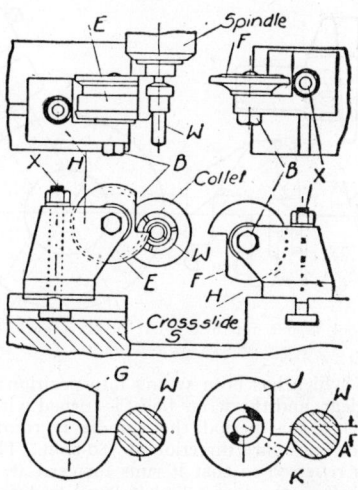

Fig. 1. Circular form tool application.

tion of front and rear circular form tools E and F on the cross-slide. Both cutters are held by clamping screws B in tool holders H; in most instances an auxiliary clamp is used to adjust the cutter to height and to give additional clamping effect. In the illustration, the front cutter is used for forming a shoulder screw blank, and the rear cutter for forming the top of the screw head and for cutting off the finished part. The holders H are keyed to the tee-slots in the cross-slide S and are clamped in place by tee-bolts X.

The cutting face of a circular form cutter is ground along a line parallel to and a distance A from a radial center line, in order to provide clearance for the cutting action and to avoid the insufficient clearance indicated in the radially-faced cutter G, Fig. 1. The center of the cutter J must therefore be set a distance A above the centerline of the spindle. The dotted line at K illustrates correct regrinding procedure.

End-cutting rough turning tools are of two general types: hollow mills and knee tools. Knee tools for automatic screw machines are generally made without an adjusting slide; the tool bit is held by two or three set screws in the solid body of the tool. Knee tools are used chiefly for taking roughing cuts on short work which is finished by a circular forming tool. Hollow mills usually consist of a hollow cylindrical body holding two or more adjustable blades, so arranged that the cutting pressure is axial rather than radial; they are usually, although not always, equipped with back rests.

Box tools are used more generally than any other type for turning straight diameters and are designed primarily for finish turning. The tools have blades which cut tangentially and are equipped with vee or roller back rests. Roller rest tools are preferred for heavy cuts when a diameter is to be turned in one cut; vee rest tools are used for finishing cuts or on free-cutting material such as brass. Box tools with several cutters for turning two or three diameters simultaneously are also employed. Tap holders and die holders are made in two styles: plain drawout type, in which the holder is sufficiently free axially in the body to permit the tap or die to lead itself after being started, and releasing type holders.

Fig. 2 shows a tool layout for machining a hexagon head screw on a single-spindle machine. In this illustration, in which the **front elevation** of the turret and the **plan** of the cross-slide is shown, T represents the turret, H a die holder, B a box tool for turning the body of the screw, *F* a circular form tool for facing and forming the head, *C* a circular cut-off tool, and *S* the swinging stop. In operation, the hexagonal bar stock is brought against

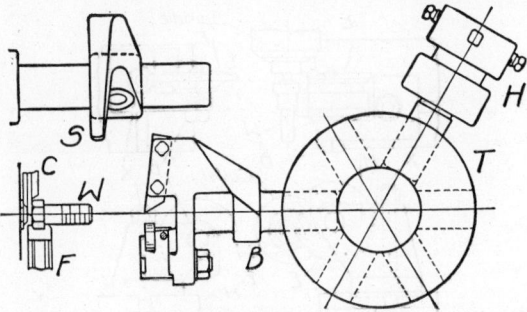

Fig. 2. Tool layout for hexagonal head screw. (*Brown & Sharpe Mfg. Co.*)

the stop *S* which has been swung into position; the stop is swung clear, and the screw body is turned with the box tool. During this period the spindle is rotating in a clockwise or backward direction at 780 rpm. The spindle rotation is reversed so that it runs forward at 165 rpm; during this time the turret had indexed to bring the die in the holder *H* to the work. The screw body is threaded and the spindle is again reversed to 780 rpm backwards, permitting the die to be withdrawn at a rapid rate. The turret slide then moves farther back and indexes to present a blank hole to the work. The circular form tool on the front slide forms the head and moves away from the work, and the rear form tool cuts off the screw, completing the cycle of operations.

Multiple-spindle automatic screw machines are made with two, four, five, six, or eight spindles for holding

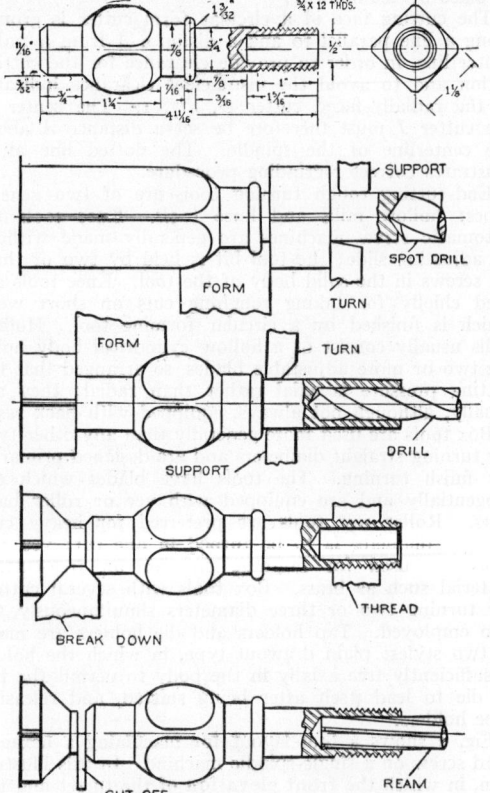

Fig. 3. Operational sequence on a four-spindle automatic. (*Cone Automatic Machine Co., Inc.*)

bar stock, and all stock is machined simultaneously. The spindles rotate in a spindle carrier which indexes around the main drive shaft that passes through the center of the carrier and drives the spindles. The turret slide carries one set of end-working tools for each spindle and can advance, feed and return in a direction parallel to the spindle axes; the machine also has two front and two rear cross-slides. Fig. 3 illustrates a threaded part made from square bar stock and shows a sequence of operations performed on a four-spindle machine. At the first turret station a forming tool held on the lower front cross-slide forms the work; a turning tool with a support or roller back rest on the end-working tool slide turns a portion of the diameter; and a spot drill held in the drill holder at the same station spots the work for the subsequent drilling operation. In the second position a forming tool held on the lower rear cross-slide completes the forming operation; a turning tool completes the turning operation; and the hole is drilled. In the third position a breakdown tool is used for chamfering the work and beveling the forward end for the next piece, and the thread is cut. At the fourth station the drilled hole is reamed to size and the part is cut off. The stock is then advanced against a swinging stop and the cycle is complete.

A multiple-spindle screw machine produces one finished part for every indexing movement of the spindle, or in the case of a four-spindle machine, four parts per complete cycle of the spindle carrier. The total production time for one part is equal to the time required to perform the longest single operation plus the indexing time. The successive turning operations at stations 1 and 2 illustrate the principle of operational divisibility; the turning operation, if handled at one station, would take far more time than any other operation and is therefore completed in two stages.

Fig. 4 illustrates a method of multiple production whereby three hexagonal nut blanks are produced for

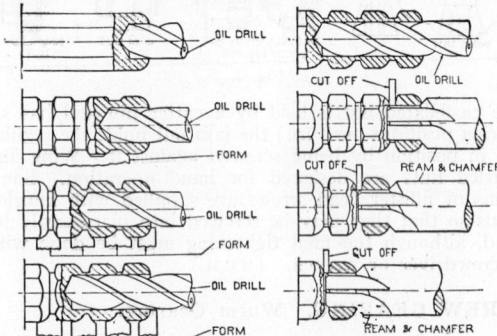

Fig. 4. Producing three nut blanks per index. (*Cone Automatic Machine Co., Inc.*)

every index of the spindle carrier. Successive drilling operations are performed at the first five positions, and the three nuts are rough-formed at stations 2 and 3, and finish-formed at station 4. The first nut is reamed and chamfered and then cut off at station 6; the other two are finished at stations 7 and 8. Multiple-spindle automatics may also be equipped with magazines for second-operation work.

Multiple-spindle automatics are used for extremely high production, complicated work, or where a great deal of metal is removed. Overhead costs on multiple-spindle machines are usually higher than for single-spindle machines. Single-spindle machines are credited with more accurate work. The application of the principle of operational divisibility, however, is confined to multiple-spindle machines, and results in prolonged tool life as well as increased production rates. (H.C.H.)

SCREW PROPELLER. Air Propeller, Water Propeller.

SCREW THREAD. A groove with continuous helicoidal surfaces cut or formed on the external or internal surface of a cylinder. Screw threads are used as removable fasteners, or for the transmission of power. The pitch of a screw thread is the distance between adjacent crests; the lead is the axial distance the screw will advance in one revolution. The lead and pitch of the single-threaded screw shown are identical; in multiple-threaded screws two, three or more threads, 180°, 120°, etc., apart are arranged in parallel. The lead of a double-threaded screw is twice the pitch; of a triple-threaded screw three times the pitch.

Fastening screws have thread profiles with a 60° included angle with either sharp or slightly flattened crests and roots. The former, or sharp V thread, is the older form, but has been largely replaced by the American Standard form shown in the figure. The Whitworth

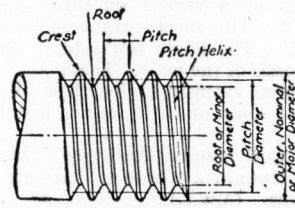

thread, used in Great Britain, has a 55° angle and rounded crests and roots. The Aero-thread system comprises an intermediate insert of spring wire between a threaded hole and the screw, and is useful where high-strength screws are to be fastened in soft non-ferrous parts.

Square, Acme, and Buttress thread form screws are used for power transmission. The surfaces of the Square thread are right helicoids, and the thread depth is equal to one-half the pitch. The Acme thread has an included angle of 29°, with a thread depth equal to half the pitch; it is easier to cut and stronger than the Square thread. The Buttress thread has an included angle of 45°, and one face of the thread is a right helicoid. It is employed for jack screws and gun breech-locks where power is transmitted in one direction only.

Wood screw and self-tapping or self-cutting screw threads permit the screw to cut its own thread as it is inserted in the material. The area of the thread profile on the screw is considerably smaller than the space between adjacent threads, so that greater strength is obtained in the internal threads cut in the weaker mating material.

Pipe threads differ from screw threads in that both internal and external threads are cut on a conical surface, so that the farther the pipe is screwed into the fitting, the tighter the joint becomes. The thread angle is 60°, cut on a conical surface with a taper of ¾″ per ft. (H.C.H.)

SCROFULA. (Formerly called king's evil.) **Tuberculosis** of the lymph nodes of the neck. This was a common disease of childhood in many countries before the 19th century. The organisms are usually transmitted through milk, and tuberculin testing of cows and elimination of tubercle bacilli from milk are responsible for the present very low incidence of the disease in the United States.

The upper deep cervical chain of nodes is most commonly affected. The first sign is the appearance of small lumps in the neck; these gradually swell and become moderately tender; often the nodes form **abscesses** which break through the overlying skin and form chronically draining **fistulas.** Treatment consists of bed rest; at times, sunlight and x-ray; and sometimes surgical excision if abscess formation is extensive. (D.M.H.)

SCROLL CASE. Hydraulic Turbine.

SCROTUM. A sac-like structure containing the two testicles, situated behind and below the penis. It is divided into two compartments by a septum in the middle. (R.S.M.)

SCRUBBER. Gases produced in industrial plants usually originate laden with impurities which render them unfit for the use to which they are to be put. For example, the gas taken from a blast furnace is so dust-laden that, were it used directly in gas engines, the cylinders would soon become badly worn and scored. Scrubbers are means for removing such substances from gas. The scrubbing action is one wherein a liquid, commonly water, is brought in contact with the gas in various ways. Chiefly dust and sulfur dioxide are removed by scrubbers. Other substances in gases are removed by equipment in series with the scrubber; for example, tar extractors. The scrubber has competitors such as electrostatic dust precipitators, centrifugal cleaners, and chemical precipitators. Scrubbing is accomplished in a tower through which the gas is made to pass, generally in counter-current arrangement with respect to the water. The equipment available commercially includes rain-type scrubbers, spray scrubbers, and hurdle scrubbers, the latter being a maze of staggered slats which are wetted, so that the dust impinging upon them will be retained by the water film. (F.T.M.)

SCULPIN. Pisces, Teleostei. A fish (**Pisces**) of the family Cottidae. They are peculiar fishes, usually small, with a broad depressed head and large pectoral fins. They are of no importance as food fishes. Many of the included species bear other names, including the little miller's thumbs of fresh waters and the sea raven which ranges from Cape Cod to the Arctic. One species of the northern Atlantic coast is called the big sculpin, or daddy sculpin. (A.W.L.)

SCURVY. A deficiency disease due to lack of **vitamin** C in the diet. Scurvy was probably first observed as a disease by Hippocrates. In the 13th century it was a curse to Crusaders. In time of war it has killed untold numbers in armies and navies and besieged towns. During the days of sailing ships, often taking months between ports with the resulting lack of fresh food, it affected the crew as a plague. It was even of considerable importance in World War I. In civilized countries its chief importance is in pediatrics and in adult life it is seldom seen. It is rarely seen in breast-fed children but pasteurization of cow's milk destroys the vitamin C and an addition to the diet of this vitamin must be provided for infants under 1 year of age.

The disease can be produced or cured at will by depriving or increasing the vitamin C in the diet either in pure form, as cevitamic acid, or as fruit or vegetables. About 6 months are required to produce scurvy experimentally, as individual susceptibility and quantity of vitamin C previously stored in the body play a part. The earliest sign of scurvy is usually a sallow or muddy complexion, a feeling of tiredness, general weakness, and mental depression. Soon the bones are affected and increasing pain and tenderness develop. The teeth decay easily, become loose and often fall out while the gums bleed easily and are sore. Changes in the blood vessels occur, producing hemorrhages in different parts of the body. In infants, irritability, loss of appetite, fever, and **anemia** also occur. Untreated scurvy is always fatal but this is seldom seen at present due to the prompt curative response obtained from vitamin C. (R.S.M., D.M.H.)

SCUTUM. 1. A shield-like plate on the upper surface of the body of a **tick.** 2. A subdivision of the exoskeleton of an **insect.** The dorsal plate of each segment of the thorax may be divided into as many as four parts. In such a series the scutum is the second from the anterior end. (A.W.L.)

SCYPHISTOMA. A larval stage of the jellyfishes (**Scyphozoa**). The individual first becomes a **planula** larva. This stage develops into the **hydratuba** which is regarded as a greatly reduced polyp stage. The hydratuba is an attached form which may send out a creeping process (stolon) capable of producing other individuals by budding, but sooner or later it undergoes a transverse segmentation to form other individuals. In this stage it is known as a scyphistoma. The segments are saucer-like structures which break away from the scyphistoma as free-swimming ephyra larvae. Each of them develops into an adult jellyfish. (A.W.L.)

SCYPHOMEDUSAE. The jellyfishes. A synonym of **Scyphozoa.** (A.W.L.)

SCYPHOZOA. The jellyfishes. A class of the phylum **Coelenterata** made up entirely of marine animals which are, with very few exceptions, floating forms. The jellyfishes represent the highest development of the **medusa** form of coelenterates, and have lost the **polyp** stage with the exception of the reduced **hydratuba** larva.

Jellyfishes owe their name to the great development of the middle layer of the body (mesogloea), which is a bulky and jelly-like mass. They contain a high percentage of water, sometimes as great as 96%, and are consequently soft-bodied and without rigid support. In the water, however, they are delicate and beautiful. Many are filmy transparent creatures while others are beautifully colored. They are found at various depths and in various seas, and in size they range from species less than 1″ in diameter to the large *Cyanea* with a body 6–7′ in diameter and tentacles 120′ long. They are of no economic importance.

The class is divided into five orders:

    Order Stauromedusae. Body conical, forming a short aboral stalk by which it is temporarily attached. Margin with eight lobes bearing clusters of knobbed tentacles.
    Order Coronatae. Body constricted near the middle. Margin deeply lobed, usually with long tentacles.
    Order Cubomedusae. Somewhat cuboidal, with one tentacle or a group at each angle.
    Order Semaeostomeae. Mouth 4-angled, with a long lip, often frilled, at each angle. Tentacles often long.
    Order Rhizostomae. Mouth with eight long branched lobes. Margin without tentacles. (A.W.L.)

SEA ANEMONE. A complex **polyp** of the class **Anthozoa.** Although closely related to the **alcyonarians** and **corals** the sea anemones are usually **solitary** and in some species are large and beautifully colored. They are without hard supporting structures such as the related forms possess. (A.W.L.)

SEA ARROW. Mollusca, Chephalopoda. Small slender **squids,** *Omma strephes,* which swim very rapidly. Also called flying squids. (A.W.L.)

SEA BAT. Pisces, Plagiostomi. The giant **ray** or devil fish. (A.W.L.)

SEA BEAR. Mammalia, Carnivora. Fur **seal.** (A.W.L.)

SEA BREEZE. When coastal land is heated considerably by the sun a sea breeze springs up, blowing off the cool water onto heated land. A pure sea breeze is usually not more than 1500′ deep, above which there is a weak returning anti-sea breeze. (P.E.K.)

SEA BUTTERFLY. Mollusca, Gasteropoda. **Mollusks** with the foot formed into two wing-like lobes which propel the animal through the sea by slow flapping movements. They make up the order **Pteropoda.** (A.W.L.)

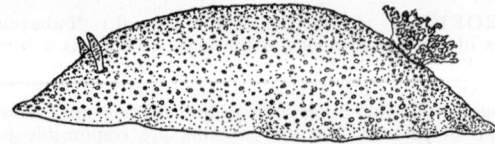

Sea butterfly.

SEA COW. Manatee. Sea Cucumber. Holothuroidea.

SEA ELEPHANT, ELEPHANT SEAL. Mammalia, Carnivora. *Mirunga.* A large **seal** once widely distributed in the southern hemisphere and represented by a variety on the California coast. The species reaches a length of 16′. The name refers to the short proboscis of the male. These seals were hunted extensively for their oil and are no longer found at some of their former haunts. (A.W.L.)

SEA FAN. Coelenterata, Anthozoa. Marine **polyps** of the order **Alcyonaria** whose colonies are in the form of thin lacy fans. (A.W.L.)

SEA FEATHER. Coelenterata, Anthozoa. Marine **polyps** of the order **Alcyonaria.** The colony has a central stalk bearing lateral branches, the whole resembling a feather in appearance. (A.W.L.)

SEA HARE. Mollusca, Gasteropoda. A marine **mollusk** of oval form with two ear-like tentacles near the anterior end which give it this name. The **mantle** almost conceals the shell and the foot forms two lobes by which the animal swims. Sea hares live on seaweed. (A.W.L.)

SEA HEDGEHOG. Pisces, Teleostei. Marine fishes (**Pisces**) with spiny skin. They are able to inflate the body to an almost spherical form by swallowing air, hence they are also called globe fishes. (A.W.L.)

SEA HORSE. Pisces, Teleostei. *Hippocampus.* Marine fishes (**Pisces**) found chiefly among seaweeds in the warmer seas. They swim weakly in a vertical position and attach themselves to some support by the prehensile tail when at rest. The body is marked by ridges in a pattern of rectangles, often prolonged into short spines or long leafy projections resembling the weed in which the animal lives. The elongate snout gives the head a faint resemblance to that of a horse. (A.W.L.)

SEA LEMON. Mollusca, Gasteropoda. A flattened oval marine **mollusk** which feeds on sponges and other

Sea lemon.

sessile animals. The roughened skin and the oval form as seen from above suggest the fruit for which the animal is named. (A.W.L.)

SEA LEOPARD. Mammalia, Carnivora. A large **seal,** *Stenorhynchus leptonyx,* of the southern hemisphere. It is yellowish or tawny with gray spots. Also called the leopard seal. (A.W.L.)

SEA LEVEL. The mean level of the oceans is known as sea level. (P.E.K.)

SEA LILY. Crinoidea.

SEA LION. Mammalia, Carnivora. Animals closely related to the seals. They belong to the same family as the fur seals and with them differ from the other seals in having external ears. The common sea lion, *Zalophus californianus*, is found on the Pacific coast of Mexico and California and Steller's sea lion, *Eumetopias stelleri*, ranges from Bering Strait to southern California. (A.W.L.)

SEA MILL. Pot Hole.

SEA MOUSE. Annelida, Polychaeta. A marine worm, *Aphrodite hastata*, of compact oval form, covered above and on the sides with a felt-like material. It is recorded from Vineyard Sound on the Atlantic coast. (A.W.L.)

SEA PEACH. Chordata, Ascidiacea. An ovoid ascidian, *Halocynthia pyriformis*, found in European waters and along the north Atlantic coast of North America. It is ovoid in form and yellowish to pink or red in color, with a velvety surface. (A.W.L.)

SEA PEN. Coelenterata, Anthozoa. Colonial polyps related to the sea feathers but with much shorter lateral branches or none. (A.W.L.)

SEA SERPENT. Reptilia, Sauria. Although the term has come to mean almost invariably a fantastic myth of some great monster, there are true sea snakes of several species belonging to a group related to the cobras. They are only moderately large. The body is compressed as an adaptation for swimming and the snakes are so thoroughly aquatic that they are either clumsy or helpless when cast ashore.

These snakes are poisonous. The short fangs are near the front of the upper jaw and are grooved. The poison acts on the nervous system, like that of the related cobras.

Sea snakes are confined to the tropical oceans, chiefly the Indian ocean and the western Pacific. A single species, *Pelamydrus platurus*, extends to the eastern Pacific. (A.W.L.)

SEA SLUG. Mollusca, Gasteropoda. Marine mollusks with compact bodies and without shells. Some species have branching processes on the surface of the body by which they breathe. The creeping habits and general form are similar to those of the terrestrial slugs, although there is no closer relationship between the two. The sea slugs make up the order Nudibranchiata. (A.W.L.)

SEA SPIDER. Pycnogonida.

SEA SQUIRT. Ascidiacea.

SEA STAR. Starfish. Asteroidea.

SEA URCHIN. Echinoidea.

SEA WALNUT. Ctenophora.

SEA WOLF. Pisces, Teleostei. One of the wolf fishes, a group of species of fishes (**Pisces**) related to the blennies but much larger. The common wolf fish, *Anarrhichas*, attains a length of 4′ and is an important food fish in Norway and Greenland. Its range extends to Cape Cod. The name wolf fish refers to the large teeth with which the jaws of these fishes are armed. (A.W.L.)

SEAL. Mammalia, Carnivora. Animals whose bodies are highly specialized for life in the ocean although they are still able to move about on land rather clumsily. The fore limbs are paddlelike flippers and the hind limbs are shifted so that they lie close together on opposite sides of the rudimentary tail and serve as a powerful propeller, increasing the effectiveness of the vertical undulating movements by which the animal swims. The body itself is formed to offer little resistance to the water in swimming.

The fur seal represented on the Pacific coast by *Callorhinus alascanus* was once abundant in both hemispheres, but was threatened with extermination by the relentless hunting practiced until the present century. They are killed when they come ashore at their breeding grounds and under present conditions hunting does not interfere with the normal propagation of the species. The Pribilof Islands are an important center for these well-known seals. Fur seals are also called sea bears.

In addition to the fur seals the group includes many other species, among them the harbor seal, *Phoca vitulina*, hair seals, hooded seal, *Crystophora cristata*, and elephant seal, *Mirunga*, in addition to the related sealions. (A.W.L.)

SEAL OIL. Esters.

SEAPLANE. A seaplane is an airplane designed to rise from and alight upon the surface of the water. In common usage a float-type seaplane is meant. Flying boats are not often referred to as "seaplanes." Seaplanes may be classified as:

1. Float seaplanes. Floats attached below fuselage.
2. Flying boats. Hull and fuselage are one and the same.
3. Amphibians. May be either float seaplanes or boats. Auxiliary retractable wheels are provided. These are lowered into position when alighting on land.

The float seaplane is found in single- and twin-float types. Except for some military seaplanes which are single-float, seaplanes generally have 2 floats, these being placed approximately where the wheels would be located on a landplane. Some commercial aircraft have been designed to accommodate either wheels, floats, or skis on the undercarriage structure. Twin floats are always used for the large long-range bimotored seaplanes. The floats are attached rigidly to struts which extend from the fuselage or wing structure. No shock absorption device such as the oleo strut of a landplane, is used. The floats have rudders for use in slow taxiing. Air pilotage of a float seaplane differs but little from that of a wheeled airplane, but taxiing on the surface, taking off, and alighting are quite different and require special technique.

Flying boats replace the float-seaplane type in the larger categories. Although this does not mean that small boats are not built, it is a fact that most small seaplanes are on floats.

The flying boat has a hull which not only provides buoyancy but also houses crew, passengers, and cargo, and fulfills all other requirements of a fuselage. As the lateral stability of a boat is poorer than that of the two-float seaplane, it has wing-tip floats to prevent submergence of a wing tip when in rough water, making a fast turn, etc. Engines and wings need to be elevated well above the water surface to avoid waves and spray. There are different ways of accomplishing this, depending on the size and type of the airplane. Extremely large boats are deep-hulled, and a high-wing position is high enough off the water to allow engine nacelles on the wings. Gull-wing designs are sometimes adopted in order to increase this clearance (see Fig. 1). Smaller hulls will generally have a *cabane* structure to hold the wing above the hull. The engine is then mounted either on the wing or above it. Biplane boats have the lower wing mounted at the top of the hull, or above it, this making the wing cellule high enough so that the engine can be mounted between the wings. When the engine thrust line is well above the center of gravity the thrust produces a nosing down moment which must be corrected by a heavy down-load

Twin-Engined Flying Boat

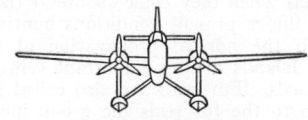

Twin Float Seaplane

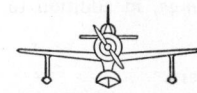

Single Float Seaplane

Fig. 1. Marine aircraft typical flotation systems.

on the tail. For this reason climbing after take-off must be slow and cautious until maneuvering altitude is gained for, if the angle of attack is held near the limit, a sudden partial or complete loss of thrust, due to engine difficulties, would result in a stall before the tail balance load could be relieved.

Since take-off distance of a flying boat is not as limited as for the land airplane, acceleration is of a lower order, permitting use of higher power loading. Comparison of large flying boats with large landplanes finds the boats with wing loadings (lbs. per sq. ft.) approximately the same as landplanes and power loadings (lbs. per hp.) 50% higher.

The hydraulic shapes of floats and hulls are similar. In Fig. 2 a typical flying boat hull is shown. The abrupt

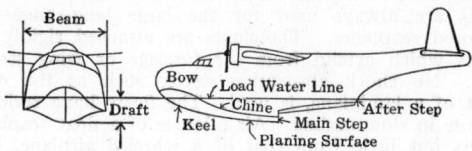

Fig. 2. Elements of the hull.

step on the keel is for the purpose of breaking the suction of the water on the planing surface so that the airplane can take off without the wing lift being greatly in excess of the weight. However, there is always some water suction and take-off speed must exceed stalling speed. The total water displacement of the floats or hull is from 2 to 4 times the airplane weight. This gives a factor of safety against excessive submergence in rough water and when alighting; also, it causes the hull normally to ride high in the water. The under surface must be shaped so that the airplane will "plane" the water at high speed, otherwise the power required to get up take-off speed would be prohibitive.

The take-off run of a seaplane is more often measured by seconds rather than in feet of travel since the water surface available is seldom restricted. Forces acting during the take-off run are pictured in Fig. 3. At the start the weight is supported entirely by buoyancy derived from the displacement of the hull. The propeller thrust moves the hull forward, creating water resistance and air drag. As the thrust is in excess of the combined air and water resistances, the boat continues to accelerate. Although air drag is small and increases

slowly, water resistance increases rapidly until the hull rises and planes the water, thus substituting dynamic reaction of the water for most of the displacement buoyancy. The boat is then riding the water "on the step." Support of the airplane is gradually shifted from water reaction to wing lift as the airplane accelerates to higher speeds. As the hull shifts from buoyancy by displacement to water thrust by planing the resistance of the water to forward motion decreases because of the lower frictional drag of a planing hull and the gradual decrease of water reaction needed as the wings take more of the load. The peak of water resistance is at the

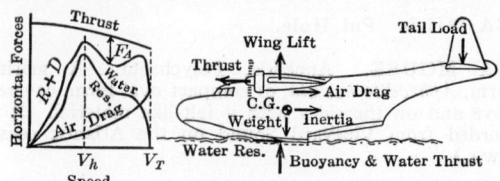

Fig. 3. Flying boat planing "on the step" (forces acting before take-off).

"hump" speed V_h. A hull with a 10-ft. beam will have its hump speed at about 40 m.p.h. Other hulls of similar shape, but different beams, will exhibit a hump speed varying in proportion to the square root of the beam. As Fig. 3 shows, the combined air drag and water resistance decrease for some increase in speed beyond the hump, then may rise if air drag increases faster than water resistance decreases. When the speed has increased to the point V_T, where wing lift is sufficient for air support of the seaplane, take-off occurs and water resistance becomes zero. In order to accelerate continuously the thrust must be in excess of the sum of water resistance and air drag. The difference F_A (Fig. 3) will produce an acceleration of F_A (mass of the seaplane). It is seen that this acceleration is not at all uniform during the take-off run. The take-off time can be estimated as the area enclosed by a curve of the reciprocal of acceleration plotted against speed.

Hulls and floats are constructed of plywood or sheet metal. Customarily they are designed as stressed skin structures, with transverse bulkheads and longitudinal stringers stiffening the skin. The bulkheads divide the interior of floats into water-tight compartments so that punctures caused by accidental contact with floating debris will not result in sinking. (F.T.M.)

SEASICKNESS. This disorder and other similar disorders such as car and airplane sickness are caused by unusual and continuous movements stimulating the equilibrium apparatus of the ear. The unusual stimulation causes reflex symptoms such as headache, nausea and vomiting. The susceptibility of persons varies with the individual, and in certain cases is very marked. (R.S.M.)

SEASON CRACKING. Spontaneous cracking of brass and other metals on standing. Intergranular cracks result from the action of residual internal stresses from cold-working operations aided by surface corrosion. Cold-worked high-zinc brasses sometimes fail during storage in ordinary atmosphere. Ammonia salts and other specific reagents greatly accelerate cracking in brasses subject to this defect. Many other metals, including stainless steels, are subject to cracking under certain conditions of stress and corrosion. (R.H.H.)

SEASONS. The fundamental causes of the seasons are the inclination of the earth's equator to the plane of the ecliptic and the revolution of the earth about the sun. Since the equator is inclined at an angle of approximately 23°.5 to the plane of the ecliptic, and this angle remains approximately fixed, the declination of the sun varies between 23°.5 north on June 21st and

23°.5 south on December 21st. Intermediate between these two dates, on March 21st and September 21st, the declination of the sun is zero. The general characteristics of the earth's motion about the sun are shown in the accompanying figure.

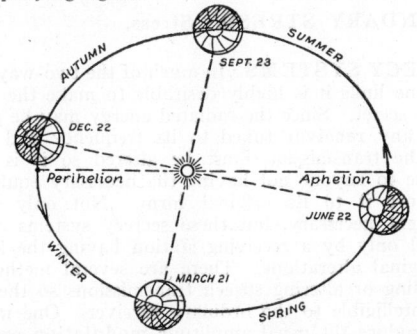

The seasons in the northern hemisphere. This hemisphere is tipped farthest toward the sun at the summer solstice (June 21), and farthest away at the winter solstice (December 21). About the first of January the earth arrives at perihelion.

The amount of heat that a particular spot on the earth receives from the sun in 24 hours is dependent upon three factors: the angle which the line from the earth to the sun makes with the surface of the earth, the time that the sun is above the horizon, and the distance of the earth from the sun. The first two factors depend fundamentally upon the declination of the sun and are the most important effects in the production of seasonal changes. Neglecting for the moment the slight changes in distance of the earth from the sun due to ellipticity of the orbit, and neglecting also the effects of clouds, haze, and dust in the atmosphere, it may be shown that the amount of heat received in one day at any spot on the surface of the earth will be a maximum on that date when the declination of the sun is closest to the latitude of the spot on the earth. Points north of latitude 23°.5 north will receive the maximum amount of heat on June 21 while points south of 23°.5 south will receive the most heat on December 21. The zone on the surface of the earth bounded by the parallels of latitude 23°.5 north and 23°.5 south is known as the torrid zone. The boundaries called the tropics of **Cancer** and **Capricorn**, the names being derived from the signs of the **zodiac** in which the sun is located in June and January. Within the regions north of 66°.5 north and south of 66°.5 south the so-called midnight sun may be observed. The parallels of latitude bounding these zones are known as the arctic and antarctic circles.

Examination of the accompanying figure will show that the earth passes through **perihelion** during the winter season in northern latitudes. In accordance with the **Keplerian law of areas** the motion of the earth is most rapid during this period and the time required for the sun to go from the autumnal to the **vernal equinox** should be shorter than the remainder of the year. Examination of the **calendar** will show that there are 181 days from September 21 to March 21 as compared with 184 days between March 21 and September 21. Furthermore, at perihelion the sun is closer to the earth than at aphelion and hence the heating effect should be greater. While these effects are slight, nevertheless climatic statistics indicate that winter in the northern hemisphere is somewhat milder than the same season in corresponding southern latitude, while the northern summer is somewhat more temperate.

In accordance with the theory of the seasons the hottest portion of the year in the northern hemisphere should be about June 21 while the coldest should be about December 21. Examination of temperature statistics indicates that there is a distinct lag in the seasons, the warmest period coming about August 1 and the

coldest about February 1. The temperature of a certain region on the earth depends upon two factors: the amount of heat received per day and the amount radiated away. The amount received depends upon the factors which we have just discussed while the amount radiated away depends in turn upon the temperature of the radiating substance, the rate of radiation being greater the higher the temperature. On June 21 the amount received is a maximum, but the temperature of the ground has not yet risen to a point sufficiently high to radiate away as much heat as is received. For the next few weeks, the amount received per day decreases, but, since the rate of radiation is not yet as great as the rate of reception of heat, the temperature continues to rise until about the first of August when the balance point is reached. From this date the temperature continues to fall. On December 21, the rate of reception is a minimum and the rate of radiation is in excess of the amount received per day. In spite of the fact that the amount received per day steadily increases from this date, the rate of radiation continues to exceed the rate of reception until about February 1 when the balance point is reached and the earth begins to warm up.

The seasonal conditions in any particular locality are greatly modified by local conditions of altitude, character of the soil, etc. Furthermore, the atmosphere serves both as a blanket and also as a method of transporting heat by convection from one portion of the earth to another. In this connection it is interesting to refer to an old problem of Professor Young in which he asked students to prove that "If weather was used on the spot where it is made, the hottest place on the earth would be the south pole on January 21st." (W.K.G.)

SEBACEOUS CYST. A cystic structure developing in the skin, due to plugging of a duct leading from a sebaceous **gland**. The cyst may increase in size, and some become infected. Sebaceous cysts are commonly called wens and are often seen on the scalp and face as well as other parts of the body. They always contain a white cheesy material.

Treatment is by surgical excision. If the entire cyst wall is not removed recurrence is certain. (D.M.H.)

SEBACEOUS GLAND. A gland of the skin of mammals which secretes an oily substance (sebum). These glands are usually associated with hair follicles and in man are especially abundant in the scalp, although they occur all over the body with the exception of the palms of the hands and soles of the feet. Their secretion keeps the skin pliable and anoints the hair.

In structure these glands are of the compound alveolar type. Their secretory parts are saccular, discharging to a common duct which often opens in the hair follicle. (A.W.L.)

SEBORRHEA. A variety of disorders of **sebaceous glands** are included in this clinical term. The dry, scaling form, "dandruff," is characterized by the presence of fine, branny, slightly greasy scales, which are readily shed. It is common on the scalp but may spread down over the face, neck and ears, and may even be a generalized **dermatitis**. The oily variety is associated with an excess secretion of oil by the sebaceous glands of the scalp and skin. The scalp is covered with a greasy layer which may mat the hair together; this type is associated with permanent baldness. Treatment consists of frequent washing of the hair, the application of olive oil, and of various lotions containing sulfur, resorcin, salicylic acid, or ammoniated mercury. (D.M.H.)

SECOND. Time.

SECONDARY. The secondary winding of a transformer is the winding from which the power is taken. Secondary **electrons** are electrons which have been knocked from an electrode of a **tube** by bombardment by high-speed electrons. (See **Secondary Emission; Transformer; Induction Coil.** (L.R.Q.)

SECONDARY EMBRYO. An individual formed by budding from another that has not yet completed its development. The term is applied to the **bryozoans.** A modified chamber, the **ovicell,** forms a brood pouch in which the fertilized ovum develops into a primary **embryo.** From this individual the secondary embryos are derived. A similar succession of forms occurs in other **phyla,** as in the complex life cycle of the liver flukes, although different terms are used to designate the different stages. (A.W.L.)

SECONDARY EMISSION. This term refers to the result of any of several different processes, in each of which some kind of "primary" emission, when it encounters some form of matter, gives rise to another emission of the same or of different character.

The most familiar example of a secondary emission is the **x-rays,** which have their origin in the impacts of high-speed electrons (cathode rays) upon atoms of matter. The resulting x-rays may themselves act, in turn, as the primary emission and, falling upon solid bodies, cause a secondary x-ray emission. Or they may fall upon a fluorescent substance (see **Luminescence**) and give rise to a secondary radiation of visible light. X-rays, ultra-violet, or light, falling upon a photosensitive metal, may cause a secondary emission of photoelectrons. (See **Photoelectric Phenomena.**) The "recoil" electrons from the Compton scattering of x-rays constitute one form of secondary emission. (See **Compton Effect.**)

A general rule, known as Stokes' law, states that the energy quanta of a secondary emission do not exceed, and are usually less than, those of the primary emission causing it. In case both primary and secondary emission are radiation, this amounts to saying that the secondary radiation is of lower frequency and greater wavelength than the primary radiation. (See **Quantum Theory.**) Thus "hard," short-wavelength, highly penetrating x-rays may pass right through the human body without serious damage to the tissues and then, encountering a metal or other solid object, give rise to much more dangerous secondary rays of longer wavelength, which must be guarded against.

In the investigation of the **cosmic rays,** much uncertainty arises as to what part of the observed effects are due to the cosmic rays themselves and what to secondaries induced by them. (L.D.W.)

SECONDARY ENRICHMENT. The concentration of valuable (ore) minerals in an ore body by the oxidation and leaching of the original ore above the **ground water** level. (R.M.F.)

SECONDARY FLAGELLUM. A small branch of the **antenna** of crustaceans. The whip-like terminal portion of the antenna is called a flagellum. In species with more than one of these structures the smaller flagellum is designated as secondary. The term is not to be confused with the flagella of 1-celled animals. (A.W.L.)

SECONDARY SEXUAL CHARACTERS. The characters of living things whose appearance is definitely associated with the sex of the individual, although the characters have no direct connection with the process of reproduction.

The different colors and patterns of the two sexes in many species of birds and insects are familiar examples of secondary sexual characters. Horns of some males, the manes of many male mammals, and the spurs of cocks are also in this category. In some species the differences resulting from such characters are so great that the two sexes can scarcely be associated by appearance.

Cases are on record among the insects of the classification of males and females of the same species in different genera prior to the discovery of their relationship through other evidence. (A.W.L.)

SECONDARY STRESS. Stress.

SECRECY SYSTEMS. In much of the two-way radio telephone links it is highly desirable to make the transmission secret. Since the radiated energy may be picked up by any **receiver** tuned to its frequency and in its path, the transmission must be altered so it is unintelligible to anyone not having the necessary equipment to restore it to its original form. Not only is the equipment necessary, but these secrecy systems can be decoded only by a receiving station having the key to the original alterations. There are several methods of scrambling or altering speech transmissions so they will be unintelligible to the ordinary receiver. One method is to produce the usual amplitude **modulation** on some frequency other than that of the final **carrier.** All but one **sideband** of this modulation is discarded and this sideband is then modulated upon the desired carrier in such a manner as to invert the frequencies, i.e., the low frequencies now appear as highs and vice versa. A numerical example will perhaps serve to explain the process. Assume a speech range of 200 to 3000 cycles is modulated on a carrier of 100,000 cycles and the upper sideband is utilized. This is a band from 100,200 to 103,000 cycles. If this is now heterodyned with a signal of 103,200 cycles, there will result a band of frequencies from 200 to 3000 cycles but the 200 cycles is due to the original 3000 cycles and the 3000 to the original 200 cycles and in a similar manner all intermediate frequencies have been interchanged. This new band can then be modulated upon the final carrier and transmitted. To obtain the original speech channel at the receiver the reverse process must be followed and it is necessary for the receiving station to know all the intermediate frequencies used at the transmitter. Since as many intermediate steps may be interposed as desired almost any degree of secrecy may be obtained. Another method is to divide the audio band into narrow bands by using **filters** and then interchange them, or invert each band separately. Other modifications involve varying time delays for the various bands. (L.R.Q.)

SECRETARY BIRD. Aves, Falconiformes. A remarkable African bird (**Aves**), *Sagittarius serpentarius*, related to the eagles and vultures. The bird is about four feet tall, with long legs, and is largely terrestrial in habits. It walks and runs very rapidly, and is also a strong flier on the relatively rare occasions when it takes to the air. (A.W.L.)

SECRETION. For the use of this term in geology, see **Concretions.** In biology, secretion is the production and discharge of special products by living **protoplasm.** In the complex body secretion may be carried out by single cells but is more often the function of many associated cells. All of these secreting units are called **glands.**

The details of the entire process are not completely understood. In many cases the special secretion is formed in the **cytoplasm** of the gland cell in granules, or these granules may consist of a foundation substance which is transformed into the secretion characteristic of the gland at the time of discharge. The transformation or the discharge or both may be brought about by a nervous stimulus or by the action of a **hormone.** Thus the salivary glands respond promptly to nerve stimuli resulting from the taste, smell or sight of food, and the pancreas is activated by the hormone secretin produced in the wall of the intestine.

The discharge of secretion may take place through the free end of the gland cell, or through the ruptured end of the cell. In some cases the terminal portion of the cell is converted into the secretion. The existence of these processes is accepted, although the action of many gland cells cannot be classed definitely with any particular one.

The secretions of endocrine glands are not discharged in the usual sense, but are taken up by the blood circulating through the gland. (A.W.L.)

SECTION MODULUS.
An inspection of **flexure** will reveal that the stress in a member subjected to a transverse bending is directly proportional to the external bending moment, and inversely proportional to the ratio of

$$\frac{\text{moment of inertia}}{\text{distance of the farthest stressed element from the neutral axis.}}$$

It is apparent that this ratio is entirely a property of the shape and size of the cross-section of the structural member. This ratio is known as the section modulus, and is an important property of rolled steel sections and other shapes which are used as structural members. When the bending moment to be withstood by a beam or column is divided by this section modulus, the quotient is the maximum bending stress which will exist in that member. (F.T.M.)

SECTOR DISK.
A device much used in physical apparatus to secure an accurately known control of the intensity of a beam of light or other emission. The simplest form is a circular, opaque disk with a sector or sectors of any desired angle cut from it. If the disk is interposed in the path of light rays and rotated rapidly about its center, the resulting intensity, as judged visually, is reduced to a fraction equal to the ratio of the area of the open sectors to that of the whole disk. This arrangement is useful, for example, in a photometer where it is desired to cut down the intensity of one beam to match that of another. By giving the sides of the openings suitable curved shapes instead of cutting them along radii, the intensity may be varied from center to circumference in accordance with any desired law. This affords, for example, a non-selective "wedge" for certain photometric purposes. When a sector disk is used in connection with photographic work, regard must be had for the so-called "intermittency effect," which renders the ratio not strictly accurate. (L.D.W.)

SEDATIVE.
Any therapy that decreases excitement or activity. The sedative **drugs** commonly used are **bromides,** chloral (see **Chloral Hydrate**), barbital and its derivatives.

The chief action of sedatives is obtained by depressing the higher brain centers. The sedative measures other than drugs are hot baths and hot drinks. (D.M.H.)

SEDGE FAMILY.
Cyperaceae. This family of **monocotyledons** is composed of grass-like plants which are found chiefly in marshy places. Most of its members are perennials with creeping **rhizomes** and grass-like leaves. The basal portion of the leaf is a sheath which completely surrounds the stem. The stem is generally solid and triangular in cross-section. The **inflorescence** is a **spike** or a **panicle,** composed of one- to many-flowered spikelets. Each flower, borne in the axil of a bract, has three **stamens** and a single **pistil;** in some sedges there is also a **perianth** of six or many bristles. The fruit is an **achene.** All sedges are wind-pollinated.

Few of the sedges are of any importance. Some of them yield a coarse hay which may be fed to live stock, and is sometimes used for packing material. **Papyrus,** used as **paper** by the ancient peoples, was made from the stems of *Cyperus Papyrus.* The erect stems and leaves of species of *Scirpus* are sometimes dried and woven to form chair seats; chairs finished with this material are called rush-bottomed chairs. The plants are sometimes called bulrushes. Several sedges form tubers which are sometimes used for food. (R.M.W.)

SEDENTARIA.
Chaetopoda.

SEDIMENTARY ROCKS.
Rocks.

SEDIMENTATION.
Filtration.

SEEBECK EFFECT.
Seebeck observed that if a photographic emulsion is exposed to the point at which a visible image appears faintly, and is then exposed to colored light, the emulsion assumes the color of the light to which it is exposed. The colors, however, are weak and mixed with gray. Many attempts have been made to perfect processes of color photography of this character but without success. (See **Thermoelectric Phenomena.**) (C.B.N.)

SEED.
A seed consists of a dormant embryo, together with a quantity of stored food which may be absorbed in the embryo, or may surround it, and one or two seed coats or integuments. The seed develops from an **ovule.** In **Angiosperms** it is completely enclosed by the ovary wall; in **Gymnosperms** it lies exposed on the surface of a scale of the cone.

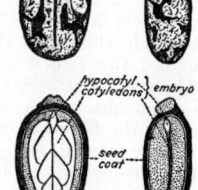

The fertilized egg develops into the embryo which is a young plant contained in a seed. This embryo may be an undifferentiated mass of cells, as it is in the orchid family, but usually is more highly organized. It then consists of a short axis which is called the hypocotyl. At one end of the hypocotyl there is a primitive root called the radicle. At the other end is a terminal bud, called the plumule. This plumule may be nothing more than a small mass of undifferentiated cells, recognizable only as a small bulge at the apex of the hypocotyl, or it may be a well-developed shoot having a short internode and two distinct leaves. Borne laterally at the apex of the hypocotyl there are one or more seed leaves or cotyledons. In many seeds these cotyledons are thin and more

Seed of the castor bean. Above, entire seeds; below, median longitudinal sections.

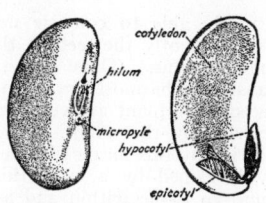

Bean seed. Left, entire seed; right, embryo with seed coat and one cotyledon removed.

or less leaf-like, while in others they are very fleshy, filled with stored food material, and form the greater part of the seed. The number of cotyledons varies. In monocotyledons there is usually only one cotyledon, in dicotyledons there are two, and in gymnosperms there are often many.

In **Angiosperms** the **endosperm** (see **Flower**) is the tissue which results from the triploid endosperm nucleus. It is a tissue which is rich in stored food. The food reserves stored in the seed are **carbohydrates,** especially starches and sugars, **fats** and **proteins.** The latter are present in all seeds, but are particularly abundant in the seeds of the **pea** family. The developing embryo gets its food from the endosperm. In many seeds the embryo uses only a part of the endosperm during its development, so that the mature seed contains much endosperm surrounding the embryo. These are called albuminous seeds. In other plants, the food reserves of the endosperm have been entirely absorbed and restored in the embryo, especially in the cotyledons, which then becomes very fleshy. Such seeds are exalbuminous.

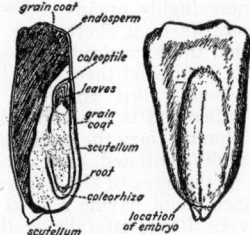

Longitudinal section of a seed of a water lily. (*After Conrad from Curtis, Nature and Development of Plants, Henry Holt & Co.*)

The embryo sac of the ovule is surrounded by a mass of tissue called the nucellus. In most plants this is completely absorbed before the seed reaches maturity. In some seeds it persists and becomes much enlarged. It is then known as the perisperm, and serves as an additional source of stored food.

Surrounding all these are the seed coats, which develop from the integuments of the ovule. The outer coat, called the testa, may be variously modified to aid in the dissemination of the seeds. The inner seed coat is the tegmen. On the seeds of many plants there is a fleshy structure called the aril which grows up around and more or less covers the outer integument. In some seeds, the testa produces an outgrowth called the caruncle, which seems to aid in absorbing water from the soil and passing it on to the seed. Passing through the integuments is a minute hole called the micropyle. It is through this that the pollen tube entered the young ovule; the radicle generally points directly towards it. The seed is attached to the ovary wall by a small stalk or funiculus which, when the seed falls off, leaves a scar called the hilum on the seed coats.

Once started, the plant must remain fixed in its position throughout its life. In higher plants the vegetative

Corn grain. Left, longitudinal section, perpendicular to the broad face of the grain; right, surface view.

parts are very rarely able to colonize new territories. The fruit, or less frequently the seed, is the part which is carried to new regions. There are several agents effecting this transfer. The most important is the wind. Sometimes the seeds of a plant are very small and light, and so easily carried by the wind. The minute seeds of orchids are carried by this means; in these seeds additional buoyancy is gained by a loose thin case which surrounds the embryo tissue within and acts as a float. In other plants the seeds are carried away by hairs which grow from the seed. Milkweed seeds are provided with a tuft of long silky hairs attached at one end of the thin light seed. Cotton fibers serve the same purpose; they completely cover the cotton seed. In other plants the seed is provided with a wing, a flat thin outgrowth from the seed coats. Catalpa seeds are thus equipped, as also are pine seeds. The distances to which the wind carries seeds is considerable. By this means plants are disseminated over many square miles.

Another way in which new land is reached is by ejecting seeds violently from the fruit. When the fruit of the witch hazel is ripe, the dry thick wall of the ovary suddenly snaps and hurls the seeds violently to a distance of many feet. The common Jewel-weed or Touch-me-not scatters its seeds in similar fashion. At maturity the fruit abruptly splits open, and the valves roll back, throwing the seeds many feet away. In similar fashion the pods of many legumes split apart forcefully and scatter the seeds within. Seeds scattered by this means cannot attain the wide dispersal which wind-borne seeds do.

A few plants form seeds which float readily on water for some time without harm. Currents of water may carry the seeds, or the latter, floating on the surface, may be blown to new shores. Other seeds are provided with hooks or barbs or have a sticky surface which causes them to adhere to the bodies of passing animals which scatter them. In most plants, however, it is the fruit which is so carried. Such fruits, as beggar's lice, burdock, goldenrod and many others, are commonly mistaken for seeds.

Having reached an environment where suitable conditions exist, the seed germinates. The seeds of many plants must reach such a place in a very short time or perish, since they remain viable but a very short time. Those of the willow, for example, live only a few days after falling from the parent plant. The seeds of most garden vegetables grow best if planted within a year from the time they ripen, though they may retain their vitality for three or four years with decreasing vigor. On the other hand, some seeds lie dormant for a long time before germinating. Seeds of many weeds, including the common ragweed, the pollen of which causes hay-fever, may live for years before germinating, making it very difficult to eradicate the species by pulling up the plants for a single season. Tests show that the seeds of many plants may remain viable for 20–50 years, though few of them do. It is recorded that the seeds of the Asiatic lotus have germinated after lying dormant for 200 years. But records of viable seeds found in ancient vaults, such as contained the mummies of Egypt, are entirely unfounded.

Certain conditions favor the continued vitality of dormant seeds. Sometimes the seed has a very thick wall which is impervious to water or to oxygen gas, and so excludes the two things necessary to start germination. Until the wall has softened or rotted the seed does not germinate. In other seeds the thick wall resists the pressure of the developing embryo within. Many important crop plants have seeds which germinate slowly because of their thick coats. To hasten germination and to insure a uniform stand of plants the seeds are scarified before planting; that is, the seed coat is rubbed with abrasive substances which break down the impervious wall layers. Often the wall of the seed is sufficiently damaged during mechanical threshing to insure prompt germination when planted. Prolonged soaking sometimes hastens germination.

In other seeds dormancy is inherent in the embryo itself. The embryo may be entirely undeveloped, requiring a long period of slow development before it can break the seed coats. Other seeds germinate only after a period of "after-ripening" which varies from a single winter to many years. The changes occurring during this period are as yet not well understood. The time of "after-ripening" may be considerably shortened by burying the seeds in sand or other suitable material and keeping them cold and moist.

The external conditions necessary to cause germination are adequate water, suitable temperature and oxygen. With some seeds light is an important factor. Most seeds contain very little water, which is one of the reasons why they can survive under adverse conditions such as cold and drought. To germinate they must receive additional water. This added water favors digestion, a process which makes available to the plant the stored food. Both water and oxygen are needed by the germinating embryo, because of the great increase in respiration, the process which frees to the plant the energy stored in the carbohydrates and other compounds.

The temperature at which a seed will germinate varies with different plants. For each there is a considerable range of temperature. The lowest temperature at which germination occurs is called the minimum temperature, and varies from 0–10° C. or even higher. The maximum or highest temperature at which germination takes place is usually between 45 and 59° C. The most favorable temperature, or optimum, is about 30° C. Light favors the germination of many common plants, such as many grasses and troublesome weeds. Other plants, including many common crop plants, are unaffected by light.

Germination is the development of the embryo into a young plant. It becomes completed when the young plant is independent of the food stored in the seed. In most seeds the first visible change is swelling of the seed, which is a result of the increased water content. Often the seed coats are ruptured by the swelling of the contents of the seed. Increased respiration makes available foods which are carried to those regions where active growth occurs, that is, to the hypocotyl and plumule. The radicle pushes out of the seed and attaches itself, by means of root hairs, to the soil particles. These then begin absorbing water from the soil. In some seeds the hypocotyl elongates considerably, often forming an arch which subsequently straightens, lifting the cotyledons and plumule out of the soil and into the air. In other seeds the cotyledons remain permanently underground, the plumule elongating and pushing out into the air. There the first leaves of the plant appear. With their formation the plant becomes independent.

The main foodstuffs of mankind are seeds, especially those of the cereal grains, rice, **wheat, corn, barley,** and **oats.** But seeds are used in many other ways. Many medicinal products are obtained from seeds. **Linseed** oil, **soybean** oil, and **coconut** oil are but a few of the many oils which come from seeds. **Poppy** seeds, caraway seeds, and mustard add flavor to other foods. Clothing is made from the hairy covering of the **cotton** seed. Beads, buttons, and ornaments of various kinds are also often made from seeds. (R.M.W., B.S.M.)

SEED SNIPE. Aves, Charadriiformes. South American birds (**Aves**) resembling quails but classed as intermediate between the true snipes and the gulls. They are closely related to the sheath-bills. (A.W.L.)

SEED-FERNS. Paleobotany.

SEEING. When meterological conditions are such that the atmosphere is very steady (i.e., the atmospheric density decreases with perfect uniformity with increase in height, and there are no strong air currents) the image of a star as seen in a telescope will be a perfectly round disk and very steady. The diameter of the disk depends upon the aperture of the **telescope,** being smaller the larger the aperture of the instrument.

When the atmosphere is unsteady the variations in the **refraction** produce a number of effects on the star images. The images become soft and "fluffy," jump around in the field of view, and are subject to abrupt changes in color. These irregularities in the star images are known as bad seeing. Bad seeing is evident to the unaided eye in the twinkling of the stars. When the stars are twinkling strongly the seeing is bad.

The statement is frequently made that the **planets** do not twinkle. In a sense this statement is true, for to the naked eye the planets do present much more steady images than do the stars. Every point in a planet does twinkle, however, but a planet is made up of a great many such points, since it presents a disk of finite size to the eye. The integrated effect of the twinkling of the individual points in the disk is not apparent to the

unaided eye. In the telescope, when an attempt is made to study individual points on the image of the planet, bad seeing produces such blurring as to make good observations impossible.

Bad seeing is almost as fatal to most types of astronomical observing as actual clouds. Accordingly, in selecting a site for an observatory, expeditions are dispatched to various sites for the purpose of testing the seeing. That site is selected where the seeing averages the best over a long period of time. (W.K.G.)

SEGER CONE. A series of substances having different fusion temperatures might serve roughly to measure the temperature of high-temperature regions such as **furnaces,** since, with a series of substances having progressively increasing fusing temperatures, the temperature naturally lies between the fusion temperature of the last substance fused, and that of the next not yet fused. A series of artificially prepared mixtures, mostly of the oxides such as clays, lime, feldspar, have been designed to form a series of "Seger cones." There are 60 mixtures covering a temperature range from 590 to 2000° C. The variation in fusion temperature between cones of adjacent serial numbers ranges from 20 to 30° C. Such cones have some pyrometric value as control indices in the ceramic industry, and in any case where exact determination of temperature is not necessary, but some simple, inexpensive means for approximating the temperature is desirable. It is apparent that it is necessary to estimate what the temperature to be measured is in order to select the cones for use.

The Seger cones are triangular pyramids about $\frac{1}{2}''$ on each side of the base, and $2''$ high. In use, the temperature is estimated, and four cones of consecutive serial numbers which are thought to include the temperature to be measured are placed on a refractory slab and inserted in the high-temperature region. If the cones have been properly selected, they will exhibit a range of behavior in the furnace varying from complete fusion of some to others remaining unaffected. One will soften until its tip bends over to touch the base, and that one is taken as indicating the temperature. (F.T.M.)

SEISMOGRAPH. An instrument for recording earth tremors; usually housed for the purpose in a suitable seismological observatory. There are two classes of seismograph, one for recording horizontal and the other for recording vertical components of vibration. A well-equipped observatory has three, a north-south horizontal, an east-west horizontal, and a vertical recorder. The instruments are somewhat complicated, but the principle is that of a heavy mass poised in such a way that a vibration of its support, together with the inertia of the mass, causes a relative motion of mass and support; and this motion, suitably amplified, produces the record. In the older forms the recording was done mechanically by a stylus tracing on a revolving drum; in more modern types an electromagnetic current, generated by the motion, operates a **galvanometer** which, by means of a beam of light reflected from its mirror, produces a photographic record of the earth's vibration on a moving film. (See **Earthquakes.**) (L.D.W.)

SEISMOLOGY. Earthquakes.

SEISMOMETER. Earthquakes.

SEISMOSCOPE. Earthquakes.

SEISONIDEA. Rotatoria.

SELACHII. The sharks and dogfishes. A term applied in various classifications to a subclass of the class Pisces and to a suborder of the order Plagiostomi as used in this work. (A.W.L.)

SELAGINELLA. Lycopodiales.

SELECTED AREAS. To determine the form, extent, and general characteristics of the **sidereal universe** as a whole, knowledge of the **magnitudes, spectral types,** and other characteristics of all of the stars would be necessary. To solve completely a problem of such magnitude is obviously impossible, and the process of statistical discussion becomes necessary. In any problem of statistical analysis a sampling of the material under consideration is necessary. For this purpose, Prof. J. C. Kapteyn, a Dutch astronomer, proposed in 1906 a group of 206 "selected areas" distributed all over the sky in accordance with statistical theory, and requested international cooperation in the determination of the characteristics of all stars in these areas. During the past 40 years a large amount of work has been done on these selected areas, and from a statistical discussion of the results most of our information regarding the structure of the sidereal universe has been obtained. (W.K.G.)

SELECTION. A process controlling the reproductive sequence within a **species** either by preserving only a part of the included individuals or by limiting the range of their opportunity for mating. The latter process tends to split the species up into different groups.

The term applies to certain theories of **evolution,** including natural selection and sexual selection, and to methods of improving domestic animals and plants. Most of the varieties of domestic animals have been produced in this way, and the method is supplemented only by hybridization in the production of new breeds.

The simplest application of the process is mass selection, consisting merely of the preservation of the more desirable individuals as breeding stock. Although mass selection is effective and has, indeed, been the original source of many varieties of domestic animals, modern knowledge of heredity shows that it is a relatively crude method. Line selection, based on the study of progenies of single individuals, is a much more exact method of improvement. With the refinements made possible by modern genetics it discloses with reasonable precision the hereditary potentialities of the parents and makes possible the establishment of genetically pure strains embodying characters desired by the breeder. (A.W.L.)

SELECTIVE ASSEMBLY. Interchangeable Manufacture.

SELECTIVE RINGING. Ringer.

SELECTIVITY. This is the ability of a selective circuit to discriminate against undesired signals which are adjacent to the desired ones. The term is commonly applied to radio **receivers** and indicates the receiver's ability to select one station without interference from an adjacent one. It is measured in terms of the amplitude of the input signal at various frequencies necessary to produce a standard output from the set. It is expressed by a curve, usually of V shape, and the steeper the sides the more selective the receiver. (L.R.Q.)

SELENIUM. Symbol: Se. Atomic number: 34. Atomic weight: 78.96. Density: red, 4.50; gray, 4.84. Hardness: 2. Melting point: gray, crystalline, 220° C. Boiling point: 690° C. (**Isotopes:** page 290.)

Selenium exists in several **allotropic** forms, (1) crystalline red selenium alpha and beta, separating in **monoclinic** crystals from solutions of vitreous or amorphous selenium in **carbon disulfide.** The transformation temperature into metallic gray selenium B is 110–120° C. for the alpha and 125–130° C. for the beta, (2) crystalline gray selenium A, formed by heating the vitreous form to 175° C., and changes gradually into metallic gray selenium B, (3) metallic gray selenium B, insoluble in carbon disulfide, and produced when the other forms are heated to 200° C.; metallic luster, malleable, a **conductor** of electricity in proportion to the intensity of the incident light. **Nitric acid** and **aqua regia** dissolve selenium to form selenous acid; attacked by cold concentrated **sulfuric acid** to form a green solution which precipitates free selenium upon dilution with water. Discovered by Berzelius in 1817.

Selenium is used as a decolorizer of glass to counteract the green **ferrous** shade; as selenium or sodium selenite to produce clear red glass, and for red enamels on ceramic ware and steel ware; in vulcanized rubber—the presence of 1–3% selenium notably increases the resistance to abrasion. Selenium is not poisonous, but many of its compounds are exceedingly toxic.

Selenium occurs as selenide in many sulfide ores, especially those of **copper, silver, lead,** and **iron,** and is obtained as a by-product from the **anode** mud of copper refineries and the lead chamber mud of sulfuric acid plants. The mud is (1) fused with **sodium** nitrate and **silica,** or (2) oxidized with nitric acid, and the water extract is then treated with **hydrochloric acid** and **sulfur** dioxide, whereupon free selenium is separated. Chemically related to **tellurium.**

Acids: Selenous acid (H_2SeO_3), soluble, forms metallic selenites; with hydrogen sulfide forms free selenium and **sulfur** which becomes red on heating; is oxidized by **potassium** permanganate to selenic acid; selenic acid (H_2SeO_4), soluble, forms metallic selenates, less stable than selenites.

Chlorides: Selenium monochloride, selenium dichloride (Se_2Cl_2), brown liquid, boiling point 130° C.; selenium tetrachloride ($SeCl_4$), yellowish-white solid, sublimes at 305° C.; selenium oxychloride ($SeOCl_2$), yellowish liquid, boiling point 177° C.

Hydride: Selenium hydride, hydrogen selenide (H_2Se), gas of unpleasant odor, chemically related to **hydrogen sulfide,** hydrogen **telluride,** and **arsenic** hydride (arsine), formed by reaction of selenide with dilute hydrochloric or sulfuric acid.

Oxide: Selenous oxide (SeO_2), white solid, sublimes at 317° C. to yellowish-green vapor, soluble in water to form selenous acid.

Selenides: Metallic selenides are formed by combination of selenium and the metal, or by precipitation of the metallic salt solutions with hydrogen selenide.

Numerous organic compounds of selenium have been prepared. (R.K.S.)

SELENIUM CELL. Photoconductivity.

SELSYN. These devices, also called synchros and autosyns, are used for transferring a mechanical displacement to some distant point electrically. In operation a transmitter or generator selsyn is used with one or more receiver or motor selsyns. They are somewhat similar to wound-rotor induction motors, having a stator winding and a rotor winding. In single-phase machines one of these is a single-phase winding excited from the single-phase power source and the other is a three-phase winding. In three-phase selsyns both are wound for three-phase. In either type the primaries are excited from the common power source while the secondaries are connected in parallel. (See figure.) If the rotors of

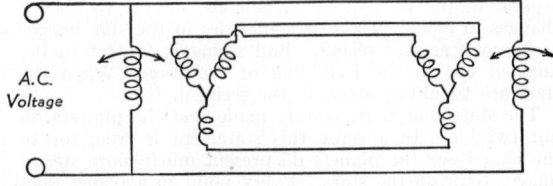

both generator and motor units are in identical positions with respect to the stators the voltages induced in them by transformer action will be identical and there will be no circulating current. However, if one is displaced the induced voltages are no longer the same in magnitude and phase and there will be a consequent circulation of current. This will produce a **torque** in the undisplaced rotor which will cause it to rotate until the voltages are again the same and the rotors will occupy identical positions with respect to their stators. As usually applied the rotor of the generator unit is displaced and the rotor of the motor unit then follows it. A differential synchro is sometimes used for injecting still another control signal. The differential is connected in the circuit between the generator and motor synchros. Remote position indicating is an example of the application of these devices. The generator may be connected to a valve, let us say, while the motor is connected to a dial on the control board. The position of the motor dial will then indicate the position of the valve. Large units may be used to operate devices requiring appreciable torque. They are widely used in sizes ranging from small indicating units to large power units of several horsepower and over distances ranging from a few feet to several miles. (L.R.Q.)

SEMAEOSTOMEAE. Scyphozoa.

SEMEN. The thick grayish-white secretion of the testicles which contains the **spermatozoa**, or fertilizing male cells. It is secreted by the testicles, stored in the seminal vesicles, and ejaculated through the penis in intercourse or during nocturnal emissions. (R.S.M.)

SEMI-AUTOMATIC. Usually a machine, mechanism, or machine tool that requires manual attendance at the beginning and end of each operational cycle, or at definite stages during the cycles. (See **Turret Lathe.**) (H.C.H.)

SEMI-CARBAZIDE. Amines; Amides.

SEMI-CARBAZONES. Amines; Amides.

SEMI-CUBICAL PARABOLA. The semi-cubical parabola is a mathematical curve which is defined as the **locus** of the equation $ay^2 = x^3$ in rectangular coordinates; its form is shown in the accompanying figure. (L.L.S.)

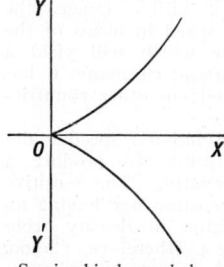
Semi-cubical parabola.

SEMI-DIAMETER. When the **altitude** of a celestial object, that is close enough to the earth to present a finite disk (e.g., **moon, sun, planet**), is measured with reference to the visible sea **horizon** or to an artificial horizon, other than that contained in the bubble **sextant**, it is more convenient to use the upper or lower limb (edge) of the disk, than to estimate the center of the object. When solving the **astronomical triangle** using this measured altitude, as, in determining position at sea, the position of the object given in the **almanac** is that of the center. Hence to obtain the observed altitude of the center a correction must be applied known as the correction for semi-diameter. The correction is to be added or subtracted from that obtained with the sextant, depending upon whether the lower or upper limb of the object is observed. The value of the semi-diameter is given in the almanac for the given date of observation. In using the moon for accurate determination of position, an additional factor, known as **augmentation,** must be considered due to the fact that the distance of the object from the observer varies with the altitude of the object. (W.K.G.)

SEMI-LOGARITHMIC CHART. Logarithmic Chart.

SEMI-LOGARITHMIC PAPER. Logarithmic Paper.

SEMINAL VESICLE. An expanded portion of the genital duct of the male, or a pouch-like derivative of the duct, either for the storage of spermatozoa or the secretion of accessory substances. Structures of this kind are well developed in the **annelid** worms, the **arthropods,** and the **vertebrates.** Often known by the latinized term *vesiculae seminales.* (A.W.L.)

SEMSEYITE. A mineral sulfide of lead and antimony, $Pb_9Sb_8S_{21}$. Crystallizes in the monoclinic system. Hardness, 2.5; specific gravity, 6.08; color, gray to black; opaque. Named after Andor von Semsey, Hungary. (R.M.F.)

SENILITY. The gradual decrease of physical and mental function that accompanies old age. The aging process is a mysterious phenomenon which is not clearly understood. Certain measurable changes occur, but the impetus behind these changes is obscure. Diminution in the circulation to organs and tissues of the body, varying according to the degree of hardening of the arteries, accounts for many of the signs. The age at which senile signs and symptoms begin to appear depends on **heredity,** and to a less extent the amount of abuse and illness that the body has been exposed to during life. (D.M.H.)

SENSE ORGANS. Structures in the animal body which are influenced by certain factors in the environment. Also known as receptors. The action of the environmental factor on the living substance is known as a stimulus. It results in the transmission of a **nerve impulse** to some nerve center and from this point may influence appropriate reactions of the animal or may be stored in memory. Special sense organs are found only in animals with **nervous systems.** Their development involves high specialization in some phase of the general property of living matter called irritability, and to some extent this property persists in all living tissue, whether nervous or sensory or not.

Stimuli arise from contacts with solid objects, from chemical compounds, either dissolved or in the gaseous state, from the incidence of light rays, and from factors which damage the body. We know from various evidences that some animals perceive factors to which our own organs are not sensitive, but as far as we can know, the only stimulating factors are in these several groups.

Contact results in variable pressures to which organs of several kinds are sensitive. In vertebrates **tactile corpuscles** and other similar structures located in surface tissues may be classed as organs of touch. They are sensitive to simple pressure and give rise to images of form through the varying pressures due to uneven surfaces and gross contours. Since sound waves are due to rhythmic compression of the air, the ears and other **auditory organs** such as those of insects are also sensitive to pressures, but only to fluctuations of relatively high frequency (in man 30–30,000 per sec.). Between auditory organs and simple organs of touch are the **lateral line organs** of fishes and the **chordotonal organs** of insects, both related in some anatomical details to the auditory organs of the groups to which they belong. These organs are supposed to be sensitive to fluctuating pressures of lower than auditory frequency.

Tactile organs are also closely allied to sensory organs of insects which apparently enable them to avoid obstacles when flying. Supposedly these organs are sensitive to the changes of air pressure, often extremely delicate, resulting from approach to objects. They are located in the wings.

Dissolved substances stimulate organs of taste and gases or vapors act on organs of smell (**olfactory organs**). In the vertebrates the olfactory organs are associated with the nasal passages or occupy a similar position; nasal structures of fishes are limited to the olfactory function. Vertebrate organs of taste are known as **taste buds** and are located in the oral cavity principally, although aquatic forms may have them also in the skin. Sensory organs of this class in the invertebrates are extremely varied. Some are known as **sensillae.** In addition to organs of taste and smell a general chemical sense is recognized. It is resident in various surface layers and is the least sensitive of the group. These sense organs are known collectively as chemoreceptors.

The varied integumentary sense organs of the human body are known to include some sensitive only to heat, cold, or pain. These organs may be the free nerve endings found in the skin. It has been suggested that pain may also result from overstimulation of other types of sense organs.

Sense organs that are stimulated by light are familiar to us in our own **eyes.** From this stage of complexity they range downward to simple light-sensitive cells. The transition includes organs capable of perceiving fluctuations of light, the direction from which it comes, and movements, as well as organs which form images in varying degrees of precision.

In contrast with sense organs of the kinds mentioned, which are classed as exteroceptors, the body contains others called interoceptors. They are the source of sensations of hunger, thirst, nausea, and pain. Other interoceptors in the muscles, joints, and tendons are associated with the maintenance of equilibrium and are classed as proprioceptors. They are probably subject to varying pressure due to tension of muscles and shifting of the weight of the body. The semicircular canals of the inner ear of vertebrates are also organs of equilibration.

Organs of many invertebrates, such as the **tentaculocysts** of jellyfishes, can be interpreted only by experimental evidence. By testing the animal with different stimuli definite conclusions can often be drawn from its reactions. In most cases there is evidence of functions like those of our own sense organs.

All sense organs consist of nerve endings associated with various specialized cells or tissues. The nerves are not limited to one type of stimulation but their response may be identical under various stimuli. Thus a mechanical shock to the eye produces a sensation of light. The nerve fibers leading from the sense organ toward the central system are sensory or afferent. (A.W.L.)

SENSILLAE. The sense organs of **insects.** The term is usually applied to the **integumentary** sense organs of the group but it is also extended to include the **scolophores** on which organs of hearing and chordotonal organs are based, and the **ommatidia** and **retinulae** of the eyes.

The sensillae of other kinds include some form of cuticular structure, often a projection, associated with a nerve ending and in some cases with gland cells. These organs include some of tactile function and chemoreceptors, both of taste and of smell. In form their external parts are classed as six types: 1. Placoid sensillae end in a thin porous plate or membrane covering a canal. 2. Trichoid sensillae end with a slender **seta.** 3. Basiconic sensillae have a conical protuberance. 4. Styloconic sensillae have a fixed conical base bearing subordinate projections. 5. Coeloconic sensillae end with a depression containing a conical projection. 6. Ampulliform sensillae end with a slender projection in an expanded chamber at the inner end of a long tubule.

Tactile sensillae are distributed over the entire body but are often much more abundant on the legs and sensory appendages such as antennae and palpi. Organs of smell are often abundant on the antennae and in some species appear to be limited to these appendages.

There is some possibility that they may occur on other parts of the body. Organs of taste are undoubtedly associated with the mouth parts, but they are supposed to be present in aquatic insects on the outer surface of the body as well. (A.W.L.)

SENSITIVE PLANT. Movement in Plants.

SENSITIVITY, RECEIVER. This is defined as the signal input necessary for a **receiver** to produce a standard output. The amount of the output, the audio frequency used to modulate the input and the degree of modulation are all specified. The sensitivity of the usual receivers will vary from a few microvolts for highly sensitive ones to several hundred microvolts for smaller ones. (L.R.Q.)

SENSITIVITY DETERMINATION OF PHOTOGRAPHIC MATERIAL. Many different methods of measuring the speed of a plate or film have been proposed, but all may be divided into four groups.

1. The exposure producing the first visible density (**threshold speed**).

2. The exposure required to produce a certain density but greater than the **threshold density** (DIN, Weston speeds and General Electric speeds).

3. The exposure obtained by projecting the straight-line portion of the $D \log E$ curve to the $\log E$ axis (H & D speed).

4. The exposure at the lowest point on the $D \log E$ curve at which the density differences are sufficient to produce a good print (limiting gradient method, Kodak speeds, ASA speeds).

The best known method of speed measurement based on the threshold exposure is that due to Scheiner. The Scheiner speed number was obtained originally by exposing the plate or film in a rotating disk **sensitometer** producing an exposure range of 1:100 in 20 equal steps numbered from 1 to 20, developing under certain prescribed conditions and determining the last visible step. The number of this exposure step—from 1 to 20— became the Scheiner number expressing the speed of the sensitive material. While satisfactory as a means of indicating the speed of a film or plate in terms of the exposure required to produce a visible image, the Scheiner number fails to indicate speed in terms of the exposure required for a negative which will yield a good print. In the land of its origin, Germany, it has been superseded by the DIN speed; in other countries it is practically obsolete.

The DIN (German Industrial Standard) speed is determined from the exposure required to produce a density of 0.1 above the **fog density.** The sensitive material is exposed in a special sensitometer having an intensity scale of 30 steps, ranging in density from 0 to 3.0. The range of exposure is, therefore, 1:1000. The light source is a standardized 40-watt incandescent lamp with a daylight filter and the time of exposure is ½0 second.

After development, the step having a density of 0.1 above fog is determined by means of a special comparison **densitometer.** The density of the exposure step in the sensitometer producing the required density on the film or plate is multiplied by 10 to obtain the DIN speed number.

The DIN method is simple and direct, but has not found wide acceptance outside of Germany.

The speeds employed on the Weston and General Electric exposure meter are determined in the laboratories of the respective companies. After exposure in a continuous time-scale type sensitometer, the sensitive material is developed in the developer recommended by the manufacturer, the densities read and the $D \log E$ curve plotted in the usual way. The Weston speed is determined from the formula speed $= \dfrac{4}{E}$ and the General Electric speed from the formula speed $= \dfrac{5.2}{E} \cdot E$, in both cases, is the exposure required for a density equal to the value of **gamma**.

For convenience in use, the actual laboratory speeds are not published; instead, the speed of the material is given in terms of the nearest speed number on the meter.

The H & D speed is based upon the *inertia* which was defined by Hurter and Driffield as the exposure at the point at which the straight-line portion of the $D \log E$ curve, when extended, intersects the $\log E$ axis. In recent years the practice has been to obtain the H & D speed from the formula speed $= \dfrac{10}{i}$ where i is the inertia as defined. The H & D method of determining speed is no longer used by the manufacturers of sensitized materials in this country, nor in England, and may now be considered obsolete.

In 1943 the American Standards Association gave its official sanction to a method of measuring speed which had been recommended by a committee representing the manufacturers of sensitized materials, the manufacturers of exposure meters, the Optical Society of America and the Photographic Society of America. Speeds determined by this method are designated ASA Speeds. The ASA speed is a value indicating the *minimum* camera exposure which the film must receive in order to produce a negative from which an excellent print may be obtained. This value is determined sensitometrically and is defined as the point on the $D \log E$ curve at which the gradient (slope) of the curve is 0.3 of the average slope over a log exposure range of 1.5. Thus, in the accompanying figure the slope or gradient of the $D \log E$ curve at M is 0.3 of the average

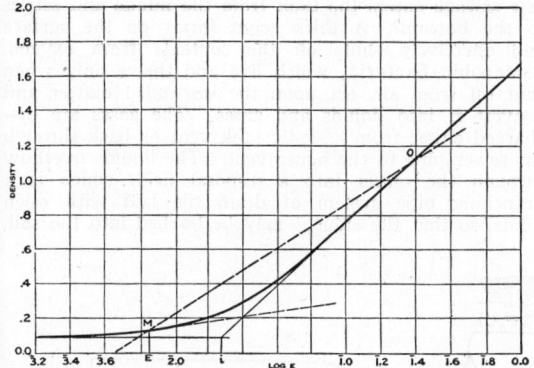

gradient between M and O representing a log exposure range of 1.5. The reciprocal of the exposure at E is the ASA speed. The ASA Speed Number for use on exposure meters and calculators is obtained by dividing the ASA speed by 4. The increased exposure indicated by the ASA speed number is designed to provide a margin of safety in practice sufficient to allow for uncertainties in processing, light conditions, accuracy of exposure meters, camera shutter variations, etc. (C.B.N.)

SENSITOMETRY AND SENSITOMETERS.
Sensitometry, as the term implies, is concerned primarily with the measurement of photographic sensitivity; however, in the broader sense it is concerned with the measurement of the response of photographic materials upon exposure to light or other forms of radiant energy.

Photographic sensitometry involves:

1. The exposure of the sensitive material under measurable and reproducible conditions.
2. The standardization of development.

3. The establishment of standardized and reproducible methods of measuring the product of exposure and development (densitometry).
4. The interpretation of the result.

The instrument used in exposing sensitive materials for sensitometric investigations is known as a sensitometer. A sensitometer consists of (1) a standard light source, and (2) a means of regulating the exposure on the sensitive material in a quantitative manner. Sensitometers may be divided into two types: (1) illumination-scale instruments, and (2) time-scale instruments. In the former, the time of exposure is constant but the illumination on different parts of the sensitive material is varied; in the latter, the illumination is constant and the time of exposure is varied. (See **Density and Densitometers; Gamma; Sensitivity Determination of Photographic Material.**) (C.B.N.)

SEPALS. Flower.

SEPARATING CALORIMETER. Steam Calorimeter.

SEPARATOR, CENTRIFUGAL. Centrifuge, Ultra Centrifuge.

SEPARATION OF SOLIDS. Classification, Filtration.

SEPARATOR, MAGNETIC. The separation of magnetic metals and ores from other non-magnetic material can be accomplished in many ways by varied use of magnets. Sometimes conveyor belts carrying bulk material are discharged over strongly magnetized pulleys which attract such iron or steel particles as bolts, nails and other tramp iron. The iron remains on the belt past the gravity dump point and is finally released by the weakened magnetic field caused by belt travel beyond the pulley. Internal damage to crushers and other machines in which the bulk material is to be processed can thus be prevented. Fine ferrous chips flushed by oil away from cutting operations can be recovered magnetically. In magnetic separation, strong magnetic fields are needed and electric magnets, rather than permanent magnets, must be used. (F.T.M.)

SEPARATOR, STEAM OR OIL. The steam separator is used to remove water from steam. Absolutely dry and saturated steam is a rarity. It is desirable in many cases to have saturated steam, as delivered by the boiler to the prime mover, as dry as possible. Wet steam may cause erosion and **steam engine** knocks. When accentuated by slugs of water from the **boiler**, or from pockets in the steam line, this trouble may be severe enough to knock out cylinder head blow-out plugs. In a **turbine** the damage is blade erosion, even more serious because the damage usually is done before it is suspected. Wet steam not only erodes blading, but also may deposit on it the solids which it has carried from the boiler water. The steam separator is installed to prevent all this. Even when the boiler normally produces a nearly dry steam the separator may be needed, because were high water accidentally carried, or foaming and priming started, the separator would be indispensable. The principles upon which separation is used are (1) reverse current, (2) centrifugal force, (3) wet baffles.

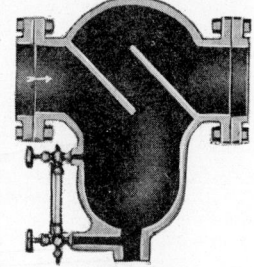

Separator.

The separator should provide for an enlarged path for the steam since it has been found that separation is more effective at lower steam velocities.

Exhaust steam lines from engines and reciprocating steam pumps contain considerable oil as a result of the method of lubricating such equipment. If this steam is to be used in heating systems, to heat feed water, or for industrial processes, the oil content is objectionable. Therefore the oil separator, built much on the same principle as the steam separator, should form part of the exhaust line from such equipment. (F.T.M.)

SEPIOLITE. The mineral sepiolite or meerschaum is soft, white, light in weight, and occurs in clay-like masses. It is a complex, hydrous **magnesium silicate** corresponding to the formula $H_4Mg_2Si_3O_{10}$. It appears to be amorphous; hardness, 2–2.5; specific gravity, 2; color, white, grayish white, sometimes a yellowish- or bluish-green; opaque. It is capable of floating on water, hence the name meerschaum or sea foam. It occurs in Asia Minor associated with **serpentine** and **magnesite,** and may be derived from the latter. Other deposits are in Czechoslovakia, Morocco, and Spain; and in the United States in Pennsylvania and New Mexico. The name meerschaum is from the German. Sepiolite is from the Greek, meaning cuttlefish, referring to the similarity of the bone of that animal to the light, porous sepiolite. (E.S.C.S.)

SEPSIS. Septicemia.

SEPTARIAN STRUCTURE. Mineralized irregular polygonal joints or cracks in certain **concretions.** The structure resembles the pattern of cracks developed by desiccation of mud, and probably resulted from a similar cause—contraction due to desiccation of **colloidal** material. (R.M.F.)

SEPTARIUM. Concretion.

SEPTIBRANCHIATA. Lamellibranchiata.

SEPTICEMIA. Septicemia or bacteremia is said to be present when bacteria are isolated from the blood stream. This is a serious complication of many acute infections, and often carries a grave prognostic significance. Infections caused by hemolytic **streptococci, staphylococci, pneumococci, meningococci,** and colon **bacilli** are most often associated with septicemia, although any organism can find its way from an area of infection into the blood stream. The usual portals of

entry are the lungs, middle ear, mastoid process, the skin, and the genito-urinary tract. Although there are no constant signs of septicemia, the condition is suspected when during the course of an infection, a patient develops unusual fever, hemorrhages into the skin (**purpura**) or joints, signs of **endocarditis, jaundice,** or widespread **abscesses.** In **pneumonia,** meningococcus **meningitis, osteomyelitis,** puerperal or post-abortion infections, the physician is ever on the alert for the development of the signs of septicemia. Whenever the complication is suspected, a blood **culture** is made immediately to verify the diagnosis.

Since the advent of the **sulfonamides** and **penicillin** the mortality from septicemia has been greatly reduced. Although all bacteria are not susceptible to these drugs, fortunately the common ones associated with septicemia are. In addition to chemotherapy, **transfusions** and specific antiserums (see **Serum**) are sometimes useful. If the primary infection is one suitable for surgical drainage, this is carried out, but not until the blood stream is sterilized. (D.M.H.)

SEPTIC TANK. Septic tanks are water-tight compartments generally used for the disposal and purification of residential **sewage.** The tanks are made of concrete or steel. In the simplest form a septic tank is a circular steel tank of 200–600 gals. capacity, fitted with a lid, and with openings for sewage inlet and effluent outlet. The outlet will be baffled so that liquids are drawn from below the surface level to prevent floating solids entering the disposal field. The operation of a septic tank depends on the action of bacteria which convert the solids into liquids, and purify the liquid. The wastes of residences served by septic tanks are conducted to them by soil pipe buried in the ground. The septic tank is also buried; in fact, no part of the sewage system need appear above ground. The raw sewage enters the tank from the intake and settles to the bottom. A thick scum forms on the surface and effectively shuts off the contents from oxygen. **Anaerobic bacteria,** which live and thrive only when shut off from air, act upon the suspended matter and convert it into liquids and gases. The gases are discharged either from a septic tank vent or back through the sewer pipe to the house vent. The liquids overflow through the outlet into a disposal field, which is a branching pipe system of drain tile laid with open joints, so that the effluent may be leached into the soil.

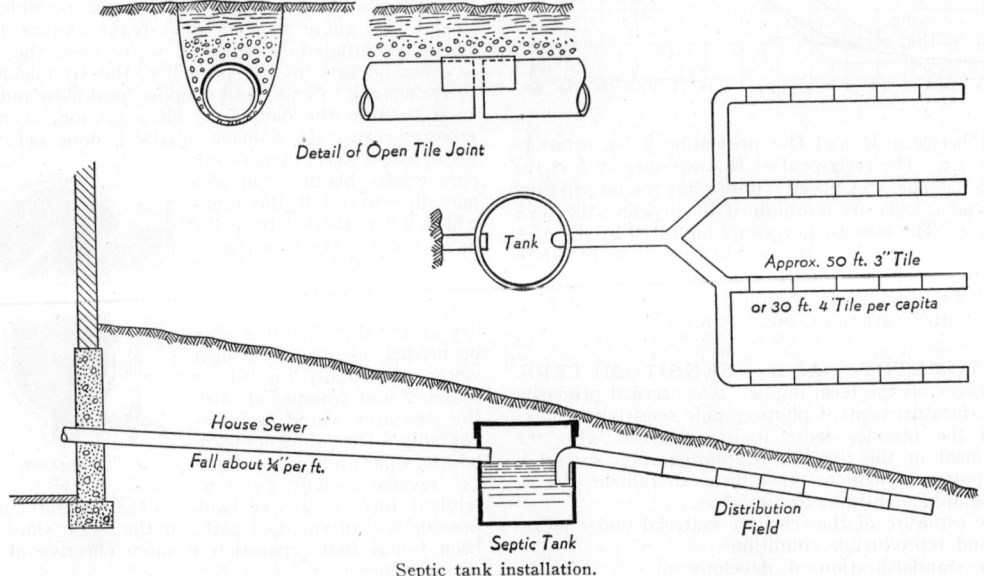

Detail of Open Tile Joint

Tank

Approx. 50 ft. 3" Tile

or 30 ft. 4" Tile per capita

House Sewer

Fall about ¼" per ft.

Distribution Field

Septic Tank

Septic tank installation.

Aerobic bacteria act upon the effluent here to purify by a process of oxidation. The purification of sewage by this method is entirely automatic in character, and the system needs no attention beyond occasional cleaning of digested solids from the bottom of the tank. The septic tank is susceptible of more scientific design than is incorporated in the example just described, but the refinements are not generally considered warranted except for extremely large residences, institutions, or factories. However, where built, the large septic tank is usually of a 2-compartment type, in one of which the settling and anaerobic action take place, and from the other, the effluent is discharged intermittently by special siphons to the drainage field. The intermittent operation permitted by the siphon is much better from the standpoint of disposal of the effluent in a tile field, because when it discharges it floods the field and distributes to all portions, whereas the small outflow from a simple type septic tank is likely to be distributed mainly to one portion of the field, which is thereby liable to become clogged. The intermittent dosing of the disposal field, and the 2-compartment arrangement of the septic tank, permit more sewage to be handled for a given tank volume.

Some of the older municipal sewage disposal plants still make use of the septic tank, which is a horizontal rectangular tank through which the raw sewage is allowed to flow slowly. The solids settle to the bottom, where they are partially reduced to liquids and gases by anaerobic action similar to that which takes place in the residential septic tank. The liquid sewage is discharged at one end while the **sludge** is allowed to accumulate in the bottom of the tank. This sludge is removed periodically by completely emptying the flat-bottom type of septic tanks or by means of sludge discharge pipes in the case of hopper bottom tanks. Occasionally the effluent of septic tanks will contain as much suspended matter as the influent due to the disturbance of the sludge caused by septic action. The **Imhoff tank** is preferred for modern installations, since it eliminates this unsatisfactory condition. A still more recent improvement consists of plain sedimentation tanks, mechanical equipment for continuous sludge collection, and separate sludge digestion tanks. (F.T.M.)

SEPTUM. A thin wall or partition. The term is applied to the radiating plates on the foot of the coral **polyp,** to the transverse partitions which subdivide the body cavity of the **annelid** worms into chambers, and to the partitions between chambers of the shell of **Nautilus,** among the invertebrates. Its most familiar use among the vertebrates is to designate the nasal septum which separates the right and left nasal passages, although it applies also to the partition between the right and left chambers of the heart and to numerous other structures. (A.W.L.)

SEQUENTIAL TEST OF A STATISTICAL HYPOTHESIS. By a sequential test of a statistical hypothesis is meant any statistical test procedure which gives a specific rule, at any stage of the experiment for making one of the following three decisions: (1) to accept the hypothesis being tested (the null hypothesis), (2) to reject the null hypothesis, (3) to continue the experiment by making an additional observation. Such a test procedure is carried out sequentially. This definition and the general theory of sequential tests are due to A. Wald. (L.A.A.)

SEQUOIA. Paleobotany.

SEQUESTRUM. A piece of dead bone which separates from the sound bone. This frequently occurs in chronic infections of bone such as **osteomyelitis.** (R.S.M.)

SERIAL CORRELATION. Serial correlation is the study of **variables** which are ordered in time or space. Such variables occur in the study of time series. Given the **variates** x_i, $i = 1, 2, 3, \cdots, N$, as a measure of the dependence among them we define the serial correlation coefficient for lag L and N observations to be

$$_LR_N = \frac{_LC_N}{V_N} = \frac{x_1x_{L+1} + x_2x_{L+2} + \cdots + x_Nx_L - \dfrac{\left(\sum\limits_{i=1}^{N} x_i\right)^2}{N}}{\sum\limits_{i=1}^{N} x_i^2 - \dfrac{\left(\sum\limits_{i=1}^{N} x_i\right)^2}{N}}$$

in which $x_{N+i} = x_i$. Clearly this is the ordinary **coefficient of correlation** between x_i and x_{L+i}. If we consider the case when x_i's are independently **normally distributed** about the same **mean** with unit **variance** then the distribution of $_LR_N$ has been given by R. L. Anderson. His results correspond to the hypothesis $_LR_N = 0$ in the **population.** Serial correlation has been used to find the lengths of periods in the analysis of cycles and may also be used to test the independence of the variates. Serial correlation has been extended to cases of more than one variable. (L.A.A.)

SERIAL DISTRIBUTION. A serial distribution is one in which the items have not been grouped. It always precedes the formation of a **frequency distribution.** (L.A.A.)

SERIAL TAP. Tapping.

SERIATE FABRIC. A geological term proposed by Cross, Iddings, Pirsson, and Washington in 1906 for the texture of an **igneous rock** whose granular crystals form a complete gradation in size. (R.M.F.)

SERICITE. Muscovite.

SERICITIZATION. Sericite.

SERIEMA. Aves, Gruiformes. *Cariama.* Peculiar South American birds (**Aves**) of several species. Their relationships are doubtful. They have long legs and moderately long necks, with a broad beak, slightly hooked. These birds live in open country and eat small animals and insects. (A.W.L.)

SERIES, BALL BEARING. Bearings.

SERIES, INFINITE. Infinite Series.

SERIES RESONANCE. Resonance.

SERINE. Aminoacids, Polypeptides, and Proteins.

SEROSA. 1. A thin membrane enveloping the developing **embryo** of terrestrial **vertebrates** (reptiles, birds, and mammals) and all of the other **extraembryonic membranes.** It is covered with ectoderm and lined with mesoderm, and in many species is formed simultaneously with the **amnion** as the outer layer of the amniotic folds. It becomes the outer component of the **chorion** and contributes to the formation of the **placenta** of mammals. 2. The lining membrane of any one of the great splanchnic or **lymph** cavities of the vertebrate body. (A.W.L.)

SEROTINE. Bat.

SEROUS GLAND. A gland that produces a watery secretion, in contrast with **mucous glands,** whose secretions are composed of or contain mucus. The term is used in connection with the salivary glands of vertebrates, which are partly serous and partly mixed. The serous gland cells are distinguished in part by their more granular **cytoplasm** and rounded **nucleus,** located near the middle of the cell. (A.W.L.)

SEROW. Mammalia, Artiodactyla. *Capricornis.* The name of animals of several species peculiar to eastern and southeastern Asia. They occur in hilly or mountainous country, sometimes at an altitude of 12,000', and are related to the gorals and goats. They are also called goat-antelopes. (A.W.L.)

SERPENT. Synonym of **snake.**

SERPENTINE. In mathematics, the serpentine is a plane cubic curve whose equation in **rectangular coordinates** is $x^2y + b^2y - a^2x = 0$.

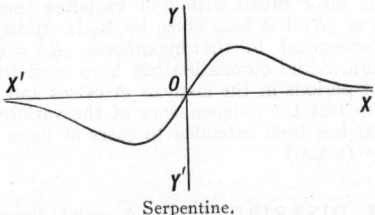

Serpentine.

In mineralogy, serpentine is a hydrous **silicate of magnesium** corresponding to the formula $H_4Mg_3Si_2O_9$. It is a secondary mineral resulting from the alteration of magnesium-bearing silicates such as **enstatite, hypersthene, olivine,** etc., and possibly from the reaction between **feldspar** or other such silicates and waters bearing magnesium. Serpentine forms granular, fibrous, or **foliated** masses occurring as veins, or as rock masses, the latter often mixed with **calcite, dolomite,** or **magnesite** showing delicate clouded tints of green. This variety is known as verd antique or **ophicalcite,** and is used as an ornamental stone. Serpentine is found only as **pseudomorphs,** but with **monoclinic** crystalline structure. Its fracture is usually conchoidal, sometimes splintery; hardness, 2.5–4; specific gravity, 2.50–2.65; luster, greasy, pearly dull; color, usually dark green but may be reddish or yellowish-brown; translucent to opaque. Serpentine is very common and of world-wide distribution. There are many localities in Europe and North America. Decorative verd antique is obtained from Roxbury, Vermont. The name serpentine is derived from the twisting, serpentine markings of the ophicalcite varieties. Besides the use of serpentine as an ornamental stone, much of this mineral is used in the manufacture of heat insulation materials.

Belts of serpentinized **peridotite** are intimately related to the geographical pattern, structure, and origin of island arcs such as the East and West Indies. Belts of serpentinized peridotites also occur in the deeply eroded roots of ancient mountain systems, indicating the pattern and date of the mountain building. (L.L.S., E.S.C.S., R.M.F.)

SERUM. The clear, slightly yellow liquid which is freed when blood is allowed to clot. The blood cells and fibrin are removed and the remaining fluid is serum. The composition of serum in the body shifts, varying with the cell metabolism. The chief constituents are: water; the **proteins, albumin,** and **globulin; hormones; enzymes;** sugar, **cholesterol** and other lipids; non-protein **nitrogen** waste materials; **amino acids;** organic acids; negatively charged **ions** (bicarbonate, **chloride, sulfate** and **phosphate**); and positively charged ions (**sodium, potassium, magnesium** and **calcium**).

The **antibodies,** which are important in protecting against disease, are contained in the globulin fraction of the serum. Various types of sera are therefore collected from individuals who have had a given disease (**convalescent sera**) to use prophylactically or therapeutically in other individuals exposed to, or ill with, the same disease. Effective antisera are also obtained by immunizing animals such as horses and rabbits, bleeding them, and preserving the serum. *Antitoxic* sera are obtained for use against disease caused by toxin-producing bacteria such as **diphtheria, tetanus, botulism, scarlet fever,** etc. *Anti-bacterial* sera are used against **pneumonia, meningitis** anthrax and other bacterial diseases. A *polyvalent* serum is one containing antibodies to several usually related organisms or toxins. (D.M.H.)

SERUM ACCIDENTS. Immediate shock-like reactions which may occur in man and certain animals on the administration of a foreign **protein.** Rarely, such a reaction is fatal.

Serum accidents occur as a result of protein sensitization. Since horse **serum** is the most common offender, individuals who are allergic and have **hay fever** or **asthma** when in contact with horses are the ones most apt to develop a serious reaction.

The symptoms follow the serum injection immediately. Several types occur: **urticaria, asthma, cyanosis** and choking; or chills and extremely high fever; or **shock** with pallor, weakness, rapid pulse, cold moist skin. The prevention of such reactions is accomplished by skin testing and desensitizing sensitive patients by repeated small doses of the serum to be used over 24 hours before giving it in therapeutic amounts. The treatment of the reaction is with **adrenaline** subcutaneously, repeated in a few minutes if necessary. (D.M.H.)

SERUM DISEASE. A delayed **allergic** reaction which occurs in sensitive individuals on the parenteral administration of foreign blood serum. This reaction in some form occurs in about 10% of those treated with small amounts of serum, and up to 90% in individuals receiving 100 cc. or more.

The symptoms begin 6–12 days after the injection of the serum. The onset is marked by intense itching and hives. The skin eruption may be varied and lasts from 1–3 days. The temperature may or may not be elevated; the lymph nodes are usually slightly enlarged. Arthritic symptoms are often present. **Edema** of the face, ankles, and hands may occur.

Serum sickness cannot be prevented and only symptomatic relief can be given when it occurs; fortunately the condition is not serious. (R.S.M., D.M.H.)

SERVAL. Mammalia, Carnivora. An African cat, *Felis serval.* The animal reaches a length of 5' and is light tawny, spotted with black. (A.W.L.)

SERVOMECHANISMS. Probably the earliest recorded use of the word Servo was by a Frenchman, Joseph Farcot, who in 1872 gave the name Servo-Moteur (literally, Slave Motor) to an auxiliary motor on a speed governor for a steam engine. As the term is now used, a servomechanism is a device for automatically controlling some quantity such as position, direction, speed, or temperature, called the output, so that it maintains some desired relationship to another quantity called the input. This desired relationship is maintained by continuously comparing the output with the input, and using the difference between them—which is called the error—to change the output in a direction that tends to decrease the error. This correction of the output is accomplished by control of the servomotor which draws power from a source other than the input. Hence, a servomechanism usually gives over-all power amplification. Since the output may be at some distance from the input, a servomechanism often requires the transmission of information for comparison, or correction, or both.

The input may be constant, as when a ship is steered on a fixed course, or a house is kept at a fixed temperature; or it may be variable, as when a searchlight is pointed at a moving target, or a cutting tool is made to reproduce the contours of a model. Also, the output quantity may be of the same kind as the input, as when the angular position of a repeater compass card is made the same as the angular position of the master compass card; or it may be quite different, as when the speed of a rolling mill is made proportional to a voltage on a controller potentiometer. There is no need to distinguish these various cases since the methods of analysis and dynamic design are the same for all.

In transmitting information to the output from the comparison point where the error is determined, or in transmitting information from the output to the comparison point, there may be delays due to the storage of energy in reactive elements such as mass, compliance, or thermal capacity. This may delay the application of corrective force and cause errors of the type known as lag, or it may delay the removal of corrective force and cause errors known as overshoot. If, in correcting one overshoot, an overshoot in the opposite direction is caused, **hunting** or oscillation may result.

From the comparison point to the output and thence back to the comparison point, the control path is a closed loop, which is an essential characteristic of servomechanisms. In the general case, the input is varying as is, of course, the output, and since any quantity varying with time may, by Fourier analysis, be represented by a series of sinusoidal components of various frequencies, the closed loop of the servo actually carries a group of frequencies. These may be very low— 1 cycle per sec. or even less—but they are still frequencies and are most conveniently treated as such.

One of the major difficulties of servo design is to select the arrangements either so that the various circulating frequencies do not add up in phase at the comparison point, or so that any frequencies that do add up in phase have sufficient attenuation in their path to prevent the steady oscillation that might otherwise occur. This problem of stability in servomechanisms is the same as that in negative-**feedback** amplifiers using vacuum tubes and other simple electrical arrangements. (E. B. FERRELL, Bell Telephone Laboratories.)

SESAME OIL. Fixed Oils.

SET. Deformation.

SET SCREWS. Set screws prevent relative motion by pressure exerted on their points. Several varieties are shown in the figure. Square head set screws have

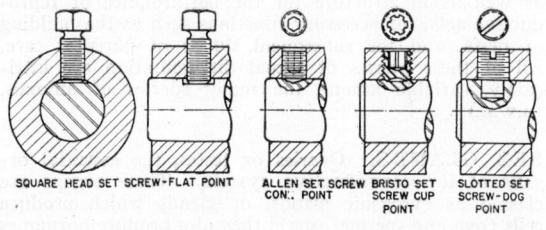

SQUARE HEAD SET SCREW-FLAT POINT   ALLEN SET SCREW CON. POINT   BRISTO SET SCREW CUP POINT   SLOTTED SET SCREW-DOG POINT

heads in which the distance across the flats is equal to the diameter of the screw, with a height equal to three-fourths the screw diameter. Safety or headless set screws may be set so that they do not project above the surfaces of the hub; Allen and Bristo screws require special wrenches, but can be seated more firmly than headless set screws with screwdriver slots. Any of the varieties of screws shown are obtainable commercially with all types of points shown. Cone and cup point

set screws raise burrs on the shaft, and create difficulties in disassembly; shafts are usually made with a "flat," or "spotted" with the point of a drill, to provide a secure seat for the screw and to eliminate burring.

An equation for the size of a set screw is: $d = 0.25 + (D/8)$, where D is the shaft diameter, and d is the set screw diameter, in inches. The transmission capacity of cup point set screws is given by: $F = 2500d^{2.3}$, where F is the resisting force at the surface of the shaft. (H.C.H.)

SETA. Any slender bristle-like structure. The term is usually limited to the hairlike processes of the integument of invertebrates, of which the setae of **annelid** worms are a good illustration. These structures are formed of **chitin** secreted by cells of a pocket in which the base of the seta is lodged. They are also called chaetae. Setae gain high development among the **insects**. Here they are secreted by cells of the **hypodermis** and project from the surface of the cuticula. They vary in form from slender hair-like structures to broad flat scales like those of **butterflies** and **moths.** In some cases they are associated with sense organs or glands. (A.W.L.)

SETTING. This word is commonly used among **boiler** artisans to refer to the brick work which encloses the furnace and partially encloses the boiler itself. Originally this was probably derived from the fact that the **furnace** walls were built, and the boiler set on the brick work. The best modern practice in settings, however, attempts to eliminate boiler supports of that type in favor of independent suspensions because of the unequal rates of expansion of steel and masonry. The boiler setting consists chiefly of walls of a compound type, having **refractory** brick on the hot side, and common brick or steel casing on the exterior, or cool, side. Many types of setting walls are in use today. There are the solid masonry walls which range from a single homogeneous refractory section to one containing special **insulation** between the refractory lining and the exterior casing. This is a common type for steam generating units. Some setting walls are hollow. The air-cooled wall consists of a thin refractory section backed by an air space through which circulate currents of cooling air. Obviously an excellent feature of this construction is that the heat flow which cools the refractory lining may be returned to the furnace by using the cooling air for combustion. Often in settings of this type sections of the setting must be independently supported from a steel or iron skeleton, from which the bricks are hung. Many patented ingenious methods of doing this have appeared, and in the main this is the field of the proprietary setting. In some cases the setting has been partially protected from high temperature by covering a portion of it with water-bearing tubes. This tends to maintain the setting at a lower temperature.

In brick work settings it is necessary to provide for expansion, horizontally and vertically, with joints in the different layers overlapping so as to avoid cracks and infiltration of air. The setting often includes not only the enclosure of the furnace, but virtual enclosure of the boiler itself, which means that different parts of the setting work at widely different temperatures. The brick work is laid with thin joints in a refractory mortar. The exterior surface should be finished for a minimum of heat radiation from it, while the interior should be able to resist high temperature conditions, thermal expansion, spalling, and the action of slag and clinker. (F.T.M.)

SEVENTEEN-YEAR LOCUST. Insecta, Homoptera. A **cicada**, *Tibicina septendecim*, which requires seventeen years to complete its development from egg to adult. The eggs are deposited in slits cut deep into the wood of twigs. After hatching, the young drop to the ground and burrow, feeding on rootlets during their subterranean life. When ready for their transformation into the adults, the **nymphs** leave the ground and climb some plant, where their dry skins can be found after the adults have emerged. The deposition of eggs weakens the twigs and sometimes causes heavy damage in young orchards. Although various broods occur which may overlap, the usual appearance of large numbers is infrequent. By consulting local entomologists the orchardist can learn when to plant trees to avoid trouble. Otherwise young trees can be protected by covering them with cheap cloth.

A variety whose life cycle is completed in 13 years occurs in the south. It is otherwise similar. (A.W.L.)

SEWAGE, SANITARY OR HOUSE. Sewage is a liquid composed of waste from such sources as domestic dwellings, commercial buildings, and factories, together with any ground or surface water which may enter the sewerage system. The average composition of sewage is over 99% water, the solid parts being composed of grease, fats, animal and vegetable matter, both dissolved and undissolved, and some inorganic matter. It is ordinarily swarming with **bacteria** which, in their ability to decompose dead organic matter, become active agents in the process of purification. Among the bacteria there may be some dangerous to human health, making it essential for the sake of general health to practice scientific disposition of the sewage. (E.W.S.)

SEWELLEL. Mammalia, Rodentia. The mountain beaver or **boomer.** A broad stout animal with short legs, found in mountain forests of the western states. It is a burrowing species which feeds on bark, twigs, and leaves. More closely related to the squirrels than to the beavers. (A.W.L.)

SEWEN. Pisces, Teleostei. A **trout** found in the rivers of Wales. (A.W.L.)

SEWERAGE SYSTEM. The term sewerage system refers to any system for the collection and/or disposal of sewage. A sewerage system usually consists of sewers and the necesary auxiliaries for the purification and disposal of the effluent. As sewage is a foul liquid, it must be carried in water-tight underground conduit systems known as sewers. A sanitary sewer is one devoted entirely to carrying sewage, and does not receive storm water which is usually carried by a storm sewer. In a municipality, where a central sewer system serves a large group of homes, industrial buildings, and business houses, the building sewer is connected to a lateral sewer which discharges into a street conduit, called the submain or main sewer. The latter discharges into an outfall sewer, which carries the sewage to the disposal plant. As the average per capita use of water is often assumed to be 100 gals. per day, the sewage load will be about 100 gals. per capita per day. The lateral and main sewers are composed of reinforced concrete or terra cotta pipe with tight joints. The outfall sewer, being larger, is often made of masonry. Sewerage works are plants which receive sewage and convert it into an effluent which may be disposed of without creating a nuisance or a focal point of infection.

A simple method of sewage disposal is described under the head of **septic tank,** but this system is not suitable for central sewerage works by virtue of its imperfect operation. Oxygen is a very important factor in converting sewage into a safe liquid. The oxidizing of sewage takes place either by the **dilution** process or by **aeration** or filtering. Often the discharge of a municipal sewerage works takes place into a nearby stream or river, in which case the dissolved oxygen in the water effects the oxidation required for complete breakdown of the solids. Disposal to a subsurface drainage field, which constitutes a crude form of filter, is illustrated by the tile disposal field, which is an essential part of the septic tank system. Aeration of sewage is illustrated by the activated sludge process. The principal equipment for sewerage works might be briefly enumerated as follows: screens, or racks, for the removal of floating matter, sedimentation tanks for the removal of solids by gravity, **Imhoff tanks,** and activated sludge chambers. In the last-mentioned system, the conversion of the solids in sewage is carried out by forced aeration. Aerobic bacteria are supplied in large quantities by recirculating to the incoming sewage a considerable amount of sludge, which contains large growths of these bacteria. This system is rather extensively used where dilution cannot be practiced, since the effluent is not foul. The operating costs of the activated sludge process may be rather high, because of the necessity for compressing the air for aeration and usage of other mechanical equipment. (E.W.S.)

SEX. The state of an individual as determined by its adaptations for a special part in biparental reproduction and modifications of the process. Also a category of individuals adapted for a special part in reproduction. The usual sexes are male and female. Neuter individuals exist among colonial invertebrates, including both sexless forms and abortive females whose limited reproductive powers are not exercised under normal conditions. Sex is also expressed in hermaphrodite animals which carry on the usual processes of sexual reproduction but have male and female organs in the same individual.

The differentiation of the sexes is associated with the development of two kinds of **gametes** in the process of sexual reproduction. Organs and ducts capable of producing the larger egg cells of the female and providing them with quantities of food material make up a **reproductive system** much different from that which produces the minute male spermatozoa and the seminal fluid in which they are discharged. The development of the external genitalia for internal insemination also results in conspicuous differences since the male has a projecting penis or other **intromittent organ** while the female has the terminal portion of the genital ducts specialized for the reception of this organ. In the mammals the **mammary glands** of the female also constitute a conspicuous sexual distinction. These organs may be classed as essential and accessory organs of reproduction, the former category including the **gonads** and ducts and the latter such parts as the external genitalia and the mammary glands.

Sexes also differ in more or less conspicuous **secondary sexual characters** such as the beard of man, which are definitely associated with sex but have no direct part in reproduction.

The sexes of most animals are specialized in behavior as well as in structure for the performance of reproductive acts, for accessory functions such as the building of nests, and for subsequent duties of parental care. All of these phases of sexual differentiation are intricately variable among the many species of animals. (A.W.L.)

SEX GLANDS. Ovaries or testes, the essential organs of the **reproductive system.** These organs are classed as cytogenic glands, or glands which produce cells (ova and spermatozoa); they also produce hormones and have endocrine functions. (R.S.M., D.M.H.)

SEX HORMONES. The hormones concerned with sexual function, the development of secondary sex characteristics, **menstruation,** and **pregnancy** may be classed as the sex hormones. These are produced by the gonads, the **ovaries** and **testes,** and by the anterior lobe of the **pituitary** gland.

The ovary produces two types of hormones, follicular

and luteal. Follicular hormone is associated with the presence in the ovary of a growing follicle, or egg, with its protective nutrient covering. (See figure page 629.) Since maturation of a follicle occurs each month as part of the menstrual cycle, this hormone undergoes cyclical variations. It is not a single substance but a group of closely related ones called, collectively, estrogens. The most active of the estrogens is estrone or theelin, which is responsible for the development of the sex organs at puberty, and for the development and maintenance of secondary sex characteristics such as the texture and distribution of the hair, the texture of the skin, the distribution of body fat, and the character of the voice. Luteal hormone, or progesterone, is necessary in the preparation and maintenance of the lining of the uterus for the implantation of the fertilized ovum, and the development of the embryo. It acts upon the **mammary gland** during pregnancy to prepare it for lactation, or the secretion of milk.

The estrogens and progesterone are secreted in sequence under the rhythmical stimulation of the anterior **pituitary** acting through its gonadotropic or gonad-stimulating hormones, follicle-stimulating hormone (FSH) and luteinizing hormone (LH). The relationship of these two hormones and the ovarian hormones to the menstrual cycle and pregnancy is discussed under **Menstruation.**

Pharmaceutical preparations of estrogenic substances and progesterone are available for clinical use. A synthetic estrogen called stilbesterol has largely replaced estrone or theelin because it has the advantage of being inexpensive and active orally whereas the natural-occurring substance must be given by injection.

The estrogens are useful in the treatment of menopausal symptoms, senile vaginitis, and in conjunction with luteal hormone, in certain types of amenorrhea (absence of menstruation). In males suffering from inoperable cancer of the **prostate,** the estrogens are effective in controlling the symptoms and slowing the growth of the tumor (see **Prostate Gland**). Progesterone is valuable in the treatment of habitual **abortion,** where it acts by suppressing uterine motility.

The male sex hormones are known as the androgens. The testicular hormone, or testosterone, produced under stimulation by the gonadotropic hormone of the pituitary, is responsible for the development of the sex organs and the secondary sex characteristics at puberty. During adult life, the functions of the seminal vesicles, prostate, and epididymis (see **Reproductive System**) are dependent upon its presence. In the urine of the male a closely related hormone, androsterone, is found in minute quantities. This substance is thought to be testosterone which has been altered in the body. The male and female sex hormones are of strikingly similar chemical structure.

The therapeutic uses of the androgens are in castration syndromes, and in certain cases of cryptorchidism or undescended testicle. (D.M.H.)

SEX LIMITATION.

The conditioning of the expression of a hereditary character in the individual according to its sex. In many known cases exactly the same determining heritage according to Mendelian **heredity** may give different results in the two sexes. A good example is that of color in Ayrshire cattle. Males with determiners for the colors mahogany and red develop the former, while females develop the latter, although both colors may appear in both sexes under proper conditions of inheritance. (A.W.L.)

SEX LINKAGE. Heredity.

SEXAGESIMAL MEASURE OF ANGLES. Degree Measure of Angles.

SEXTANT.

The sextant is a light, portable instrument designed for the purpose of measuring the angular dis-

tance between two objects. It represents the most recent stage in a succession of portable devices for angular measurement advancing from the **astrolabe** through the **cross-staff** down to the modern instrument. The immediate predecessor of the sextant was the quadrant (the "hog-yoke" of the sailing ship era), but the design of this instrument is very similar to that of the modern sextant. Since the sextant is universally used by navigators for the purpose of measuring the apparent **altitude** of celestial objects, the opinion is prevalent that it is purely a navigational instrument. Such is far from the case, and the instrument is of great value to explorers, surveyors, or any person who desires to measure angular distance.

The optical system of the sextant (and quadrant) is diagrammed in the accompanying figure. The mirror A

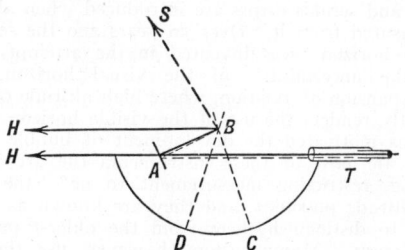

(called the horizon glass) is divided into two sections by a line parallel to the plane of the instrument. The upper section is unsilvered so that an observer looking through the telescope T can see directly through A to an object in the direction H. The mirror B (called the index mirror) may be rotated about an axis perpendicular to the plane of the instrument and coincident with the center of the graduated arc CD. Attached to the index mirror is a vernier arm which sweeps along the arc.

With the mirror B strictly parallel to A the observer will see two superimposed images of the object in the direction H. One of these is the direct image observed through the upper section of the horizon glass, the other is a reflected image with light traveling along the path $HBAT$. Under these conditions the index C should read zero; in case this is not so an "index correction" must be applied to all observations with the instrument.

In case the angular distance between an object in the direction H and another in the direction S (angle SBH) is desired, the observer moves the index arm along the arc until an image of S (light path $SBAT$) is superimposed on the direct image of H. Application of the laws of **reflection** of light will show that the angle CBD through which the mirror is turned is one-half the angle SBH. To obviate the necessity of dividing each reading of the instrument by two, the arc is so graduated that when CBD is actually 60° the index will read 120°.

To obtain the altitude of a celestial object at sea, the instrument is directed toward the visible **horizon** and the index arm moved until the desired object is in the field of view. Now the sextant is rotated back and forth about the optic axis of the telescope (the line HAT in the figure), and the image of the celestial object will apparently swing back and forth along a short arc. The index arm is now carefully adjusted until the celestial object is just tangent to the horizon at the lowest point of its apparent swing. Under these conditions the reading of the sextant will give the sextant altitude measured along the vertical circle through the object.

In case the sextant is to be used for measurement of altitude in a region where the sea horizon is not available, an artificial horizon must be used. This is a level reflecting surface, e.g., a basin of mercury. Simple geometric considerations will show that the angle between a line directly from a celestial object and the line from the image of that object reflected from the level mirror, will be twice the altitude of the object. To measure the altitude with the sextant and artificial horizon, the instrument is pointed toward the level reflector, and the reflected image brought into the field of view directly through the horizon glass of the sextant. Next, the index arm is moved until the direct image of the celestial object (along path *SBAT* in the figure) is brought into the field of view. Now the sextant is rotated back and forth about the optic axis. In this motion the reflected image will remain stationary, but the direct image will apparently oscillate back and forth in a short arc. When the direct image passes through the reflected image at the bottom of its arc, the reading of the sextant will be twice the sextant altitude of the object.

On ships or airplanes the artificial horizon cannot be used. The visible horizon is always more or less ill-defined and serious errors are introduced when altitudes are measured from it. Over 50 years ago the so-called "bubble horizon" was invented in the attempt to remove the uncertainties of the visual horizon. The rapid expansion of aviation, where high altitude of flight frequently renders the use of the visible horizon impossible, has motivated the development of bubble instruments. In most of these instruments the arc is only 45° long, restricting measurement to 90° (the maximum altitude possible), and they are known as bubble octants to distinguish them from the older-type mariner's sextant. Many writers, however, use the term sextant to indicate any type of instrument used for measuring altitude for determination of position by methods of **celestial navigation**. The fundamental principle of all bubble octants is the same. The line of sight of the view telescope is so directed that, when the instrument is held in the proper position for observation of altitude, the image of the bubble of a spherical spirit **level** is in the field of view. A spherical spirit level differs from that used by mechanics in that the upper surface of the chamber containing the liquid is ground to the surface of a sphere so that, when the level is in the plane of the horizon, the bubble is at the "top" of the sphere. With the instrument held so that the bubble is centered in the field of the view telescope, a system of reflecting prisms is adjusted until the image of a selected celestial object is centered on the bubble. The reading of a scale gives the "sextant altitude" of the object. Most octants are so designed that the view telescope can be directed at the visible horizon, when it is available, and the instrument used as a mariner's sextant.

While the use of the bubble removes the uncertainties of the visible horizon, nevertheless other uncertainties are introduced which are of such magnitude that the sea horizon should be used whenever available. **Accelerations** produce a change of position of a level bubble, as may readily be demonstrated by moving a level about over a horizontal surface with anything other than constant velocity. The **Coriolis force** (or acceleration), due to the motion of the ship relative to the surface of the revolving earth, produces a displacement of the bubble of an octant which is proportional to the velocity of the ship and the latitude of the observer. A so-called Coriolis correction may be computed in terms of the speed of the ship, latitude of observation, and relative bearing of the observed object to the track of the ship. This factor is tabulated in the Air **Almanac** and a variety of other publications. However, the random accelerations, due to roughness of the sea or air, are unpredictable and cannot be corrected for. These are usually greater on seaborne ships than in the air, and may introduce errors in a single observation as great as 2° (120 miles in the position of the ship). These accelerations are, as the name implies, random and tend to average out if a number of settings are made in rapid succession. Most modern bubble octants are built with averaging devices so that

a series of settings may be made and the average obtained without reading the scale for each setting.

Before the sextant altitude can be used for obtaining a celestial line of position, two types of corrections, instrumental and observational, must be applied. The instrumental corrections, which are common to both the sextant and the octant, are index correction and eccentricity of the arc. Observational corrections are (1) dip of the horizon, when the visible horizon is used, (2) Coriolis correction, if the bubble is used, (3) atmospheric **refraction**, (4) geocentric **parallax**, (5) **semidiameter**, with **augmentation** when necessary, if the altitude of a limb of an object of finite size (sun, moon, or large planet) is measured.

It is important to consider briefly the accuracy of angular measurements made with these portable instruments. If the instrument is mounted in a sextant holder and used to measure the angle between fixed points an accuracy of 20″ should be attained for a single setting. Such conditions are met in exploratory surveying or in measuring altitudes with an artificial horizon. Such accuracy cannot be expected in observations of altitude taken from a ship or plane. Under the best possible conditions of smooth sea and perfect visibility of the horizon and sky, the fix determined by celestial navigation may be accurate to within ¼ mile. However, under normal conditions an uncertainty of position of from 1 to 2 miles is to be expected. When the bubble is used, the random accelerations, even when a run of ten or more settings is averaged, always introduce an uncertainty of more than 1 mile in the fix, and when the sea or air is rough the error may be as great as twenty odd miles. (W.K.G.)

SHA. Mammalia, Artiodactyla. An Asiatic **sheep** with very wide range. It is found from Persia into India and northward through Tibet. The name more widely used is urial, sha applying to a large variety ranging from northern Tibet to Afghanistan. It is more common at high altitudes, up to 14,000′. (A.W.L.)

SHACKLE. A shackle is a piece used for connecting together two parts. The parts so connected can have some relative motion which is permitted by the shackle, but at the same time the extent of their freedom is limited by the restraining action of the shackle. The connections of the shackle to the part are not required to be as exacting a fit as a **bearing**, but often are of a type needing lubrication. A common example of shackle is the steel shackle employed to connect the ends of the suspension springs of an automotive vehicle to the frame. (F.T.M.)

SHAD. Pisces, Teleostei. A marine fish (**Pisces**) found along the Atlantic coast of the United States. It ascends the coastal rivers to spawn. This species belongs to the same genus as the herring and is ranked as one of our best food fishes. The roe is also regarded as a delicacy. (A.W.L.)

SHAFT. In mining, the term refers to a vertical excavation, sunk to permit access to the interior of the earth. An elevator shaft, used in building construction, refers to a vertical passage for passenger or freight elevators. In mechanical engineering, the term shaft denotes a rotating member used for transmitting power or motion. A spindle is a shaft that supports and drives either a cutting tool or work parts; a lathe spindle is an example of the first classification, and a milling machine spindle an example of the second. Shafts may be classified as transmission shafts and as machine shafts; the former are usually of uniform diameter, and range in size from $^{15}/_{16}$″ diameter to $5^{7}/_{16}$″ diameter or larger. They are usually $^{1}/_{16}$″ under the conventionally accepted fractional sizes, such as $^{15}/_{16}$″ instead of 1″, $1^{7}/_{16}$″ instead of $1^{1}/_{2}$″, $1^{15}/_{16}$″ instead of

2″, etc. Machine shafts are usually, although not always, of non-uniform diameter, such as motor and generator shafts, gear transmission shafting, etc.

Transmission shafting is used for group driving, in which a lineshaft, coupled to and driven from an electric motor, is used to drive countershafts, which in turn drive individual machines. The drive is usually effected by belts and pulleys, although chain drives are sometimes employed. By using tight and loose pulleys between the lineshaft and each of the countershafts, one or more individual machines may be stopped without affecting the others. The drive from the countershaft to the machine may be effected by cone pulleys, to permit variations in the speeds of the individual machines while the lineshaft runs at a constant speed.

The stresses in a transmission shaft are due to the torsion caused by the transmitted power, and the flexure arising from the pulley weight and belt tension. The resultant shearing stress S in lbs. per sq. in., in a rotating shaft subjected to a torque T and a flexural moment M, in inch-lbs., is given by

$$S = \frac{16}{\pi D^3} \sqrt{(K_t T)^2 + (K_m M)^2}$$

where D is the shaft diameter, and K_t and K_m are torsional and flexural service factors. According to the A.S.M.E. Code for the Design of Transmission Shafting, K_t may vary from 1.0 to 3.0, and K_m from 1.5 to 3.0; the lower values are for gradually applied or steady loads, the higher for major shock loads. The permissible shearing stress S may be taken as 8000 lbs. per sq. in. for steels for which no physical properties are specified; where physical specification for the material is introduced, S may be taken as 30% of the tensile elastic limit, or 18% of the ultimate tensile strength, whichever is less. When keyways are cut in the shaft, only 75% of the aforesaid values may be employed. (H.C.H.)

SHALE. Shale is a fine-grained sedimentary rock whose original constituents were clays or muds. It is characterized by thin laminae breaking with an irregular curving fracture, often splintery, and parallel to the often indistinguishable **bedding** planes. (R.M.F.)

SHAMA. Aves, Passeriformes. A jungle bird (**Aves**) of the Oriental region. The several species are found in the Malayan area, in the Philippines, in India, and on various Pacific islands. They are shy birds, but the Indian species is kept as a cage bird for its beautiful song. (A.W.L.)

SHANK. Drilling; Rivet.

SHAPER. Woodworking; Planer.

SHARK. Pisces, Plagiostomi. A carnivorous fish (**Pisces**) of the group with cartilaginous skeletons. The mouth opens on the ventral surface of the head and is

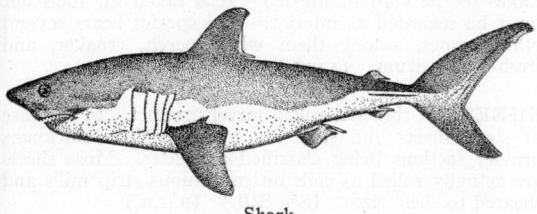

Shark.

armed with many rows of sharp teeth attached to the skin and similar in structure to the placoid **scales** of the body. The tail is of the heterocercal form, having two lobes with the backbone extending into the upper. The openings of the **gill slits** are separate.

The **dogfishes** are the most widely known of the sharks. They are common material for the anatomical

laboratory and are used to some extent as food. Some members of the group, notably the whale shark, *Rhinodon*, attain a length of 30′ or more, and others are large enough to be feared by divers. Although they are reputed to attack man freely, this tendency is also disputed by competent observers. There can be little doubt that under some conditions they do attack human beings.

Among the species whose common names do not indicate their association with this group are the thresher, *Alopias*, the **porbeagle**, the hounds, and the tope. The most peculiar species is the hammerhead, *Sphyrna*, whose head is transversely extended into two processes bearing the eyes at their tips. (A.W.L.)

SHARK SUCKER. Remora.

SHARPNESS OF RESONANCE. Since resonant circuits give a response which varies with frequency, reaching a maximum at the resonant frequency and dropping on either side, it is convenient to have some means of comparing different circuits. The usual purpose of such circuits is to select certain frequencies in preference to others so the common method of defining the sharpness of resonance is to specify the **frequency band** in which the response will exceed an arbitrary value. This arbitrary value is often taken as 70.7% of the maximum as this is the point where the power is half the value at **resonance**. The narrower the frequency band between these two points the sharper the

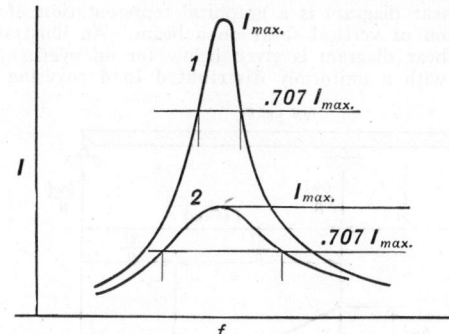

resonance, thus in the figure the circuit having the response curve 1 has much sharper resonance than that of curve 2. (L.R.Q.)

SHEAR. A force which lies in the plane of an area or a parallel plane is called a shearing **force**. It is the force which tends to cause the plane of the area to slide on the adjacent planes.

The vertical shear for any section of a simple beam is the magnitude of the **resultant** of the transverse loads on either side of the section. Transverse loads are those which are at right angles to the length of the beam. If the loads are inclined the vertical components, only, should be used in computing the vertical shear. The resisting shear at any section is the internal force which opposes the shearing action of the external loads. It is numerically equal to the external shear but in the opposite direction. Vertical shear, which is always accompanied by **bending moment** at a section of a beam, is numerically equal to the rate of change of this moment with respect to distance along the beam. This shear is arbitrarily assumed to be positive if the resultant of the vertical loads to the left of a section acts in an upward direction. In a symmetrically loaded simple beam the shear is equal to zero at the center of the beam. (See **Elasticity**.)

In addition to vertical shear in a beam there is always a horizontal shear which is a result of the difference in the flexural stresses (see **Flexure**) between any two vertical planes. The tendency of adjacent horizontal planes to slide upon each other is caused by horizontal shear. The effect may be better understood by visualizing a beam composed of flat planks laid one on top of the other. As the beam bends, due to the applied loads, the bottom of one plank will slide upon the top surface of the one beneath it unless this effect is restrained by friction, nails, bolts, or other **fastenings**.

The unit stresses resulting from vertical shear are called vertical shearing stresses. At any point in a beam these stresses are numerically equal to the horizontal shearing stresses. The variation of the unit shearing stresses over the cross-section of a rectangular beam is parabolic, being equal to zero at the top and bottom surfaces and a maximum at the **neutral axis**. When horizontal and vertical shear, only, act at a point in a body, the body is said to be in a state of pure shear at the point.

Shear is not restricted to beams. It occurs wherever there is bending. The web members of **trusses** and web plates of **plate girders** are designed to carry the shear. Columns, which are subjected to bending caused by eccentric **loads**, or by inclined or **lateral** loads, must be designed to withstand the shearing stresses. **Rivets** and welds (see **Welding**) are also subjected to shear. If the riveted connection is made so that the shear occurs between two plates only, it is called a single shear. When the type of connection is such that the shearing force is opposed by resisting shears acting on two planes, as in the case of three plates riveted together, the condition is called double shear. (See **Rivet**.)

A shear diagram is a graphical representation of the variation of vertical shear on a beam. An illustration of a shear diagram is given below for an overhanging beam with a uniformly **distributed load** covering the

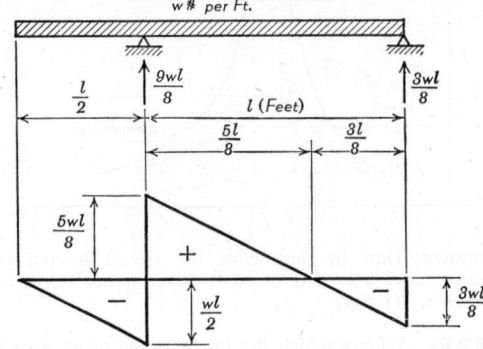

entire length of the beam. The points where the shear changes sign are points of maximum **bending moment**. The area of the shear diagram between any two points is equal to the change of bending moment between these points. (C.W.C.)

SHEAR CENTER. The point through which the external **shear** must act at any cross-section of a **beam** in order to eliminate **torsional stresses** is called the shear center of the particular cross-section in question. The shear center coincides with the **centroid** of the internal shearing forces on the cross-section. If the beam is to bend without twisting, the loads must be applied in such a manner that the external shear at any section will pass through the shear center. The shear center has no meaning for sections where **pure bending** occurs because there is no shear and, therefore, no torsional stress can exist under these conditions. (C.W.C.)

SHEAR RESISTANT. Tension-Field Theory of Beam Analysis.

SHEARING DEFORMATION. Deformation.

SHEARING LOAD. Load.

SHEARING STRESS. Stress.

SHEARWATER. Aves, Charadriiformes. Puffins. A group of marine birds (**Aves**) related to the auks, with large beaks, high at the base and strongly compressed. The name puffin is commonly applied to the Atlantic species and the name shearwater to those of the Pacific. Together they constitute the genus *Puffinus*. (A.W.L.)

SHEATHBILL. Aves, Charadriiformes. *Chionis*. A bird (**Aves**) related to the oyster-catchers. It is named from the horny sheath enclosing the base of the beak. Two species are known, both found in extreme southern latitudes and the Antarctic. (A.W.L.)

SHEAVE. A grooved pulley for use with vee-belts, rope drives, or round leather belts. Sheaves may have single or multiple grooves. Adjustable-diameter sheaves, for varying the diameter of the pulley, and consequently the speed ratio of the drive, are used in small and medium sized vee-belt drives. (See **Belting, Rope Drives**.) (H.C.H.)

SHEEP. Mammalia, Artiodactyla. Animals related to the oxen and goats but more sharply distinguished from the former by their smaller size and the usually spiral horns. They differ from goats in the absence of strong odor in the males and in the lack of a beard.

Aside from the many domestic breeds, *Ovis aries*, sheep are much more highly developed in the Old World. They are mountain animals, often found at high altitudes, and are represented by various species in the Old World, while in North America the mountain sheep or **bighorn**, *O. canadensis*, is the sole representative. Among the other species are the argalis, *O. ammon*, of Tibet, Mongolia, and Kamchatka, the urial or sha, *O. vignei*, the **mouflon**, the **bharal** or blue sheep of Tibet, and several known as the Armenian, Cyprian, and Barbary sheep. (A.W.L.)

SHEEP TICK. Insecta, Diptera. A wingless parasitic **fly**, *Melophagus ovinus*, resembling the true ticks only in its flattened body and leathery texture. Like the other flies it has sucking mouth parts, hence it draws blood from the skin of the host. It is especially harmful to lambs.

Control of the tick requires dipping, preferably twice, after shearing. The commercial preparations for dipping stock are effective against such pests. (A.W.L.)

SHEEPSHEAD. Pisces, Teleostei. 1. An excellent salt-water food fish (**Pisces**), *Archosargus probatocephalus*, related to the snappers. It ranges from Massachusetts to Texas. 2. A fresh-water species, *Aplodinotus grunniens*, one of the drums, found from the Great Lakes to the Gulf of Mexico. It is taken for food but must be regarded as inferior. The species bears several other names, among them white perch, croaker, and fresh-water drum. (A.W.L.)

SHEET. A thin flat-rolled metal product. In the case of steel, sheets and strip are under .25″ in thickness, heavier sections being classified as plates. Most sheets are actually rolled as coils on continuous strip mills and sheared to sheet sizes. (See **Sill**.) (R.H.H.)

SHEET METAL PROCESSES. Many manufactured articles are made from sheet metal. Such articles are usually lighter in weight and are often less expensive than castings or forgings. Such important parts as pipe, elbows, transitions, hoppers, and guards for gears, etc., may be made of sheet iron and steel, galvanized iron, copper, aluminum, or brass.

The shape to which the sheet material is cut while in the flat state is called the stretch-out or blank. The drawing which gives the size and shape of the blank is called the development. In many cases the blank is laid out directly by prick-punching through the corners of the development on the drawing; in other instances, where more than one blank is required, a pattern or template of sheet metal or thin wood is made from the development, and the outline of the blank is scribed on the sheet by following the outline of the pattern.

The blank is cut to the proper shape from the sheet material, and is folded, curved, or stretched until it assumes the desired form. The edges that meet are then joined by seaming, soldering, or riveting, or by a combination of these methods. Hand operations will probably always be necessary in sheet metal fabrication. Snips, which resemble large scissors, are universal and indispensable hand cutting tools. Straight snips may be used for straight or notching cuts; circle snips have curved jaws for cutting circular arcs and other curves. Bench shears are heavy-duty sheet metal cutting tools in which the jaws and handles are longer than in snips. Stakes are the anvils of the sheet metal worker. Such sheet metal operations as tube and taper forming, flanging, seaming, and riveting may be performed with their aid. Most stakes have squared shanks to fit holes in the work bench.

Ball peen, cross peen, and riveting hammers are used by sheet metal workers. Raising hammers have convex faces and are used for producing concave or convex formations in sheet metal by a series of blows. Soft-face hammers, such as plastic-tipped hammers, and hickory mallets are used when hard-faced hammers would deface the work.

A variety of standard machinery is used in sheet metal fabrication to facilitate production, secure a better product, and minimize fatigue of the operator. Foot- or power-actuated squaring shears are used to resquare plates that have been roughly cut on mill shears, and to shear plates into strips, trim edges, and cut square, rectangular, or other straight-sided blanks. Folders and brakes are machines used for bending or folding sheet metal to an angle or lock. Folders are employed for folds of limited width; brakes have unobstructed openings from front to back and are adapted to folds of unlimited width. Roll forming machines are used to curve flat sheets of metal into pipe or other cylindrical formations. The machine usually has three cylindrical rolls, *A*, *B*, and *C*, of which two rolls, *B* and *C*, are hand- or power-actuated. Roll *A* is adjustable vertically to clamp the sheet at the beginning of the bending operation; rear roll *C* is adjustable vertically to produce the required degree of curvature. In hand-operated machines, the sheet is placed between rolls *A* and *B* and roll *A* is adjusted to clamp the sheet. The sheet is then turned up in front in order to give it an initial curvature so that it will pass over the top of roll *C*. Rolls *B* and *C* are then rotated, causing *A* to roll with the sheet, which encircles the upper roll as it is formed. (H.C.H.)

SHEET PILE. Pile.

SHEET-STIFFENER COMBINATION. Stressed Skin Constructon.

SHELDRAKE. Aves, Anseriformes. 1. A duck, *Tadorna tadorna,* of the Old World, marked with green, bay, and white. It is a larger species than most of the ducks but has the characteristic broad beak of the group. 2. The American merganser, *Mergus merganser,* a fish-eating species found throughout North America, although it breeds chiefly north of the United States. (A.W.L.)

SHELL. A hard external covering secreted by folds of the body wall of many animals. The term applies properly to the shells of **Brachiopoda** and **Mollusca**, although it is sometimes used in reference to the hard exoskeleton of **crustaceans**. Also a hard covering of eggs.

Brachiopod shells consist of two valves, one upper and one lower. They have an outer layer of organic matter known as the periostracum, under it a thin layer of calcium carbonate, and a thick inner layer of mixed organic and calcareous matter, deposited in prismatic form. The valves are opened and closed by a complex system of muscles.

Molluscan shells are usually spiral in form, like many common snail shells, or bivalve like those of mussels and the oyster. The valves of such shells are lateral in position. A third rarer form is the shell of **nautilus**, which is spirally coiled in one plane. Internal shells of slugs, chitons, and some cephalopods are in the form of plates of calcareous matter.

The external shells of mollusks have an external horny layer, the periostracum, a smooth lining of nacre or mother-of-pearl, and a thick calcareous middle layer.

Because of their permanence and beauty the shells of many marine mollusks have attracted the attention of collectors, and many have received common names. Among them the spiral staircase or wentletrap, periwinkle, conch, finger shell, cowry, coffee-bean, appleseed, oyster drill, whelk, papal miter, cockle, gem, and others are to be found on the coasts of the United States.

The shells of eggs are calcareous or chitinous coverings secreted by a portion of the female reproductive ducts known as the shell gland. In the birds they are characteristically and often beautifully colored, and in the insects they may be beautifully sculptured. They are sometimes perforated by a minute opening or group of openings called the micropyle for the entry of the sperm. The egg shell of insects is also called the chorion, a term not to be confused with the chorion of vertebrate embryos. (A.W.L.)

SHELL GLAND. A specialized region of the ducts of the female reproductive system which secretes a more or less rigid covering about the egg. (A.W.L.)

SHELL REAMER. Reaming.

SHELL SHOCK. A war neurosis or psychoneurosis which is seen in soldiers far removed from the field of battle as well as in men in combat. Like other psychoneuroses, shell shock has its origin in emotional conflict. The soldier facing battle is forced to repress an agonizing fear for his life. His code of honor and his sense of duty will not permit him to recognize this fear. But, subconsciously, his desire for escape is translated into a physical ill and expressed as a tremor, blindness, deafness or any other disability which renders him unfit for military duty. He thus escapes a dangerous and terrifying situation and at the same time is able to preserve his code of honor and his self-esteem. Shell shock is an hysterical type of psychoneurosis. If early treatment by means of psychotherapy is initiated, the prognosis for complete recovery is good. (D.M.H.)

SHELLAC. A secretion or excretion of the lac insect, *Coccus lacca,* found in the forests of Assam and Siam. Freed from wood it is called "seed lac." It is soluble in alkaline solutions such as ammonia, sodium borate, sodium carbonate and sodium hydroxide, and also in various organic chemicals. When dissolved in acetone or alcohol, shellac yields the familiar shellac varnish of superior gloss and hardness. Orange shellac is bleached with sodium hypochlorite solution to form white shellac. (R.K.S.)

SHEPPARD'S CORRECTION. Sheppard's corrections are applied to the moments of a frequency distribution of more than 100 **variates** which is bell-shaped and whose frequencies taper off to zero gradually at both tails. The variates must be continuous. Then μ_2 corrected $= \mu_2$ uncorrected $- \frac{1}{12}$, where the results are in class interval units. Similarly μ_4 corrected $= \mu_4$ uncorrected $- \frac{1}{2}\mu_2$ uncorrected $+ \frac{7}{240}$, where μ_4 is the fourth central moment and all moments are in class interval units. Similar corrections may be given for a **discrete variable.** Special correction formulae are needed for J-shaped and U-shaped frequency distributions.

Sheppard's corrections are corrections for grouping, and they aid in obtaining somewhat better results if the conditions underlying them are satisfied. (L.A.A.)

SHERARDIZING. Process for applying a thin adherent protective coating of zinc to steel parts by heating at 700° F. in contact with zinc dust in a rotating-drum container. (R.H.H.)

SHICHSHOKIAN DISTURBANCE. Devonian.

SHIELDING. In many circuits, particularly in the fields of communication and electronics where the currents and voltages are often very small it is necessary to protect from external disturbances. This may be largely accomplished by shielding. Disturbing voltages may be induced in a circuit or part of a circuit by electrostatic or electromagnetic (or both) induction. The first is relatively easy to shield against since it is only necessary to surround the circuit with a conducting surface which is grounded. This effectively prevents the electrostatic lines of force from reaching the circuit and thus inducing voltages in it. To shield from electromagnetic effects the magnetic lines of force must be prevented ·from linking the circuit. Since there are no effective magnetic insulators and no perfect conductors this is difficult to do. Fortunately, the number of lines of force which do get to the circuit may be reduced to a small value which will usually not cause trouble. At low frequencies (including the audio range) this is accomplished by surrounding the circuit by a material having high magnetic **permeability,** the higher the permeability the more perfect the shielding. At radio frequencies the effective permeabilities of all magnetic materials approach unity. At these frequencies the eddy currents induced in the shield serve to protect the circuit. As a consequence, the better the conductivity of the shielding the better its protection, aluminum and copper being the materials ordinarily used. (L.R.Q.)

SHINER. Pisces, Teleostei. A small fish (**Pisces**) found in streams east of the Rockies. It occurs in almost all of the states of this region and is sometimes very abundant in the smaller streams. Steel blue above and silvery on the sides. Also called the redfin and dace. The name is applied to other species of the same genus and to less closely related minnows. (A.W.L.)

SHIPWORM. Mollusca, Lamellibranchiata. A peculiar marine **mollusk** which bores into submerged wood and apparently is among the few animals that can digest cellulose and related materials. The shipworms belong to several genera of which *Teredo* is most often cited. All are slender worm-like creatures but they have the characteristic structures of the bivalves. The valves of the shell are small separate parts, located at the anterior end of the worm, and are used for excavating the burrow.

Shipworms do great damage to wooden hulls and marine piling, consequently they have been subject to detailed studies to determine methods of avoiding their destructive attacks. (A.W.L.)

SHOCK. This term is used loosely to describe all types of acute peripheral vascular collapse in which widespread dilatation of the blood vessels institutes a series of circulatory changes. Shock is most commonly associated with **hemorrhage,** severe **trauma, burns,** and **peritonitis.** It is also seen in acute **coronary occlusion.**

The mechanism of shock production is not completely understood. The three theories which explain the situation most satisfactorily are (1) that regional loss of blood and fluid into the involved areas causes reduced blood volume, reduced cardiac output, lowered blood pressure, and tissue **anoxia;** (2) that a toxic substance is released in the injured tissues and this, together with the rapid invasion by bacteria, initiates the shock syndrome; (3) that shock is neurogenic and vasodilation is due to reflex impulses which inhibit the normal constrictor tone of the blood vessels. The first theory is probably the most important and accounts for the majority of instances of traumatic shock.

The clinical features are similar regardless of the mechanism of production. In fully developed traumatic shock, for instance, the patient is apathetic, his skin is clammy and ashen grey in color, and he often feels nauseated and vomits. The pulse rate is rapid, the pulse volume small ("thready pulse"), and the blood pressure low. In burns or crush injuries hemoconcentration occurs, due to loss of plasma from the blood into the damaged tissues. Death due to severe irreversible shock is not uncommon in serious injuries or hemorrhage. Treatment varies somewhat with the type of shock. The general principles governing therapy are restoration of fluid balance, transfusions with whole blood and plasma, elevation of the foot of the bed to help maintain the blood pressure, conservation of body heat, hemostasis by tourniquet application if necessary, and relief of pain and apprehension by the administration of morphine. (D.M.H.)

SHOCK ABSORBER. A shock absorber is a device which absorbs energy of shock either by storage or dissipation. While there are many instances of the use of such equipment on machinery, the two outstanding uses of mechanical shock absorbers are found in the field of automotive and aeronautical engineering.

An automobile shock absorber is provided to absorb the shocks of road irregularity and prevent their transmission to the frame. Strictly speaking, pneumatic tires and springs are in the nature of shock absorbers, but, supplementing these, automobile manufacturers have added shock absorbers which prevent excessive rebound after unusually large spring deflections, or which set up forces aiding the spring in resisting deflection due to the tires passing over a severe bump. Many shock absorbers are simply snubbers, which tighten up when the spring is in its maximum deflected position, and gradually ease it out on the rebound, absorbing energy in mechanical or hydraulic friction, which might otherwise be used in throwing the vehicle.

Shock absorbers are used on **airplanes** to cushion the impact of landing. Although it is possible to land an airplane without impact, this type of landing is rarely accomplished, even by experts, and the usual landing is a **stall** from 3 or 4′ above the ground. Absorbers are installed to absorb the velocity of descent gradually, and prevent large impact stresses in the framework, as well as the unpleasant jarring of the occupants. They used to be of the type which merely stored this energy and then released it. Illustrative of this shock absorber is the type using rubber cords, in which the energy of shock is absorbed by the stretching of a number of loops of rubber shock cord. In their extended position, these cords contain energy which is liberated by a return to the unstretched position. The effect was frequently to throw the aircraft back in the air in a bounce, and such shock absorbers have been largely replaced today by a type in which the energy is dissipated in friction within the absorber.

The oleo shock absorber, used on aircraft, is essentially a smooth-fitting piston and cylinder. The cylinder is attached to the wheel, and the piston to a strut which extends to a fitting on the fuselage or vice versa. The energy is absorbed by compressing this member, thus moving the piston on the cylinder. The cylinder is filled with oil which flows through a small orifice when the strut is compressed. The energy is absorbed in the friction of the liquid flowing through that orifice. The resistance of such a shock absorber automatically adjusts itself to the degree of shock, being comparatively soft for a smooth landing, where the compression in the strut is low. On a hard landing the compression in the strut is greater, and the action of the absorber is harder, although the oil is of course flowing through the orifice faster under the greater pressure developed. For absorbing ground shocks while taxiing some absorbers support the piston on coil springs, others use compressed air.

SHOE. Bearing.

SHOE-BILL. Aves, Ciconiiformes. A large bird (**Aves**) of central Africa, *Balaeniceps rex*, noteworthy for its large beak. The organ is broad and moderately deep, and the upper **mandible** bears a strong hook at the tip. The species stands about 5′ high, has a moderately long neck and long legs, and is gray in color. Also called the whale-headed stork. (A.W.L.)

SHONKINITE. An **igneous rock** term proposed by Pirrson, in 1895, for a basic variety of **syenite** or **monzonite** containing both **orthoclase** and **plagioclase** and small amounts of accessory **nepheline**. (R.M.F.)

SHOOTING STAR. Shooting star is the popular term used to designate **meteors**. These objects bear little if any relation to the **stars** other than that they are seen as bright, rapidly moving objects against the dark sky and hence apparently among the stars. (W.K.G.)

SHOP SPLICE. Splice.

SHORE. Underpinning.

SHORING. Underpinning.

SHORT-CIRCUIT. An electrical circuit is considered to be shorted when the terminals are connected directly together with only the **impedance** of the short connecting leads between them, thus for all practical purposes there is no **resistance** between them, hence no voltage can exist between them. While shorting a circuit which does not contain and is not connected to any source of voltage will produce no harmful effects, shorting a set of terminals across which a voltage normally exists will produce in many instances disastrous current flows. In power circuits protection is often provided by **circuit breakers** or **fuses** which open the circuit under the high values of current which will flow on short-circuits. Even then the transient effects which result from short circuits may cause **generators** to arc over, etc. A short obviously puts the circuit out of use. (L.R.Q.)

SHOT EFFECT. A troublesome phenomenon which gives rise to a sputtering or popping noise in **radio** and **amplifier** apparatus. It was called *Schroteffekt* (small shot effect) by Schottky, who first explained it as due to variations in the number of thermions per sec. emitted from the tube filament. This variation seems to be merely a statistical one, like the variations in the forces acting on particles exhibiting the **Brownian movement**. Its magnitude depends upon several factors, among which is the influence of the space charge (distribution of electrons) within the tube. A somewhat similar effect, produced by random variations in

the velocities of the electrons, and depending upon the temperature of the filament, is manifest in what is called thermal noise, and is superposed on the shot noise. There is also a shot effect in the emission of photoelectrons, observable in the operation of **photoelectric cells**. (L.D.W.)

SHOU. Deer.

SHOVELLER. Aves, Anseriformes. A widely distributed **duck**, *Spatula clypeata*, of the northern hemisphere. The male has a dark green head and is beautifully marked with blue, white, chestnut and orange. The female is brownish but for the blue wing patch. The species is named for the greatly broadened beak. Also called the spoonbill. (A.W.L.)

SHOWER. Precipitation characterized by sudden starting and stopping and by rapid changes in intensity. Shower rain is usually associated with cumulonimbus type clouds and instability in the air. (P.E.K.)

SHREW. Mammalia, Insectivora. Small animals closely resembling the mice in general appearance. They are common in Europe, Asia, Africa and North America. Many species burrow and aquatic habits are not

African jumping shrew. (*American Museum of Natural History.*)

uncommon in the group. Another family, confined to Africa, is characterized by the great development of the hind legs; its members are known as the **jumping shrews**. The **tree shrews** or tupaias of the Oriental region constitute a third family of arboreal habits and somewhat squirrel-like appearance. All of these animals subsist on a diet of worms, insects, and some plant products. (A.W.L.)

SHRIKE. Aves, Passeriformes. *Lanius.* A bird (**Aves**) which is known chiefly for its habit of catching

Shrike.

other birds and small animals and impaling uneaten remnants on thorns. Also known as the butcher bird. The beak is notched and in some species hooked. The numerous species occur on all continents but South America. (A.W.L.)

SHRIMP. Crustacea. 1. The edible shrimps of the order **Decapoda**. Closely related to the prawns and less closely related to the lobsters and crayfishes. The body is compressed, the antennae long and slender, and the chelate appendages small, although the related snapping shrimps have moderately large pinchers. A few species of shrimps and prawns are among the important edible crustaceans. They are found chiefly in warmer waters and are the basis of an important industry on the Gulf Coast of the United States. 2.

The **mantis shrimp**. The name is also applied in other combinations to many of the small crustaceans, including tadpole shrimps, skeleton shrimps, spiny shrimps, and fairy shrimps. (A.W.L.)

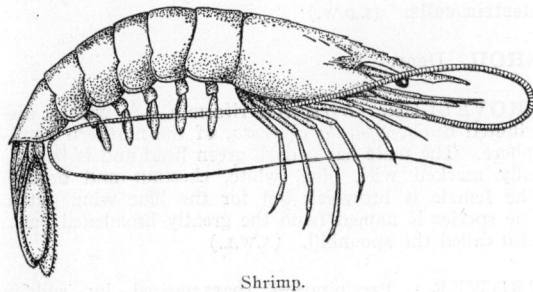

Shrimp.

SHRINKAGE. Foundry Practice.

SHRUBS. Stem.

SHUNT. The shunt is an electrical bypath so arranged that an **electric current** divides and flows partially through the shunt, and partially through the equipment that is shunted. See the accompanying diagram for connections. A shunt may be employed to

extend the range of currents measurable by an **ammeter,** since, when an ammeter is connected around a shunt, the currents flowing in the two branches, i.e., the shunt and the ammeter, are inversely proportional to the respective resistances. Thus, if the resistance of an ammeter is 5 ohms, and of the shunt, 1 ohm, the current through the shunt would be five times that through the ammeter. In such a case, the ammeter reading would have to be multiplied by 6, and 6 would be the multiplying power of this shunt. By the use of shunts of different resistances, a fixed-range ammeter may be adapted to a wide range of currents.

In the above illustration, the shunt has carried the main part of the current. In the shunt-wound **generator,** the field coils are shunted across the armature circuit, but in this case the shunt resistance is very high, and consequently only a very small portion of the current flows through the shunt winding. (F.T.M.)

SHUNT EXCITED. This is a method of exciting tower **antennae** which are not insulated from the **ground** at the base. The feeder is connected to a point about ⅓ of the way up the antenna, the exact location depending on many factors and usually involving some cut-and-try. For proper operation the feeder should slope up to the point of attachment from a point some distance from the base of the antenna, the slope being adjusted experimentally. The term is also applied to the method employed in providing field current for d-c **dynamos.** A shunt-excited machine is one in which the field windings are connected across the **armature** terminals. (L.R.Q.)

SHUNT FEED. In various vacuum-tube **amplifier** circuits the plate voltage may be applied to the **tube** in two ways, in series with the load **impedance or in** parallel (shunt) with it. In audio amplifiers employing transformer coupling the shunt feed circuit keeps **direct** current out of the transformer windings and also permits the resonating of the transformer inductance with the coupling capacitance. Both of these improve the low **frequency response.** In radio-frequency amplifiers

used in **transmitters** shunt feed is often used to avoid the necessity of having **tank** condensers and inductances which will withstand the high d-c voltages in addition to the r-f voltages. Both types of circuits are shown with shunt feed. The **chokes** prevent the by-passing of

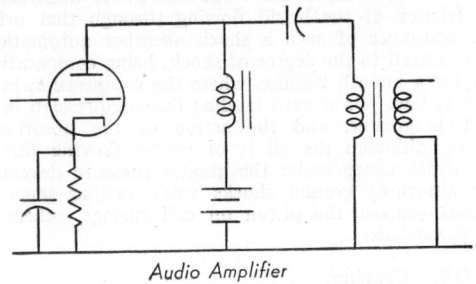

Audio Amplifier

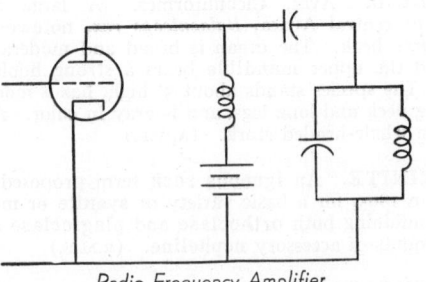

Radio Frequency Amplifier

the useful a.c. through the supply voltage while the condensers block the d.c. from the output circuits and allow the passage of the a.c. (L.R.Q.)

SHUTTERS, PHOTOGRAPHIC. Modern photographic shutters may for all practical purposes be divided into three classes: (1) between-the-lens or diaphragm shutters, (2) focal-plane shutters, and (3) behind-the-lens shutters.

Between-the-lens shutters may be sub-divided into two types: (a) rotary, and (b) leaf shutters. The rotary shutter is found chiefly on inexpensive cameras where only one instantaneous speed is provided in addition to "time" and "bulb." One type of simple rotary shutter is shown in Fig. 1.

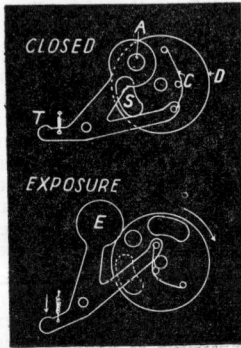

Fig. 1. Simple box camera shutter. *A*—Aperture admitting light to lens. *D*—Shutter disk. *C*—Spring that snaps disk around. *S*—Slot in shutter disk. *T*—Trigger or release lever *E*—Capping disk, preventing double exposure as disk *D* returns after *T* is released. In diagram marked "Exposure," the dotted line shows the position of the shutter disk slot an instant before *T* reaches downward limit.

Leaf shutters are of more complex construction and may have a wide range of instantaneous speeds up to 1/500 second in addition to "time" and "bulb." A typical modern precision shutter of this type is shown in Fig. 2

Presetting the shutter by moving lever (a) to the right sets up tension in spring (b); at $\frac{1}{400}$ sec. an additional spring located under eccentric member (c) is brought into

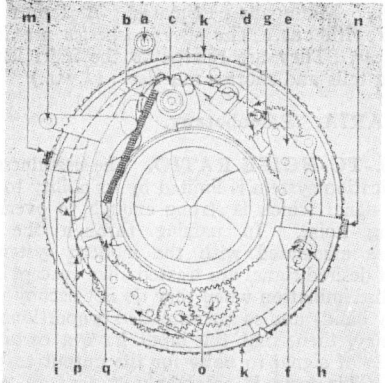

Fig. 2.

action. Shutter speeds are varied by engaged length of gear sector (d) of one member of the gear train retard mechanism (e) and the position of a small oscillating pallet relative to a ratchet wheel (f). Both of these variables are controlled by step-shaped cam shown as a dashed line (g and h). "T" and "B" are determined by position of levers (i), also controlled by a cam. All cams are part of speed-selecting ring (k). Release lever is marked (l) and socket for cable release (m). Moving self-timer lever (n) downward winds up a spring-actuated escapement mechanism (o) which, by means of an oscillating pallet and ratchet wheel (p), delays automatic tripping of shutter through extension of release lever (q) by 10 to 12 secs.

The focal-plane shutter is placed near the focal plane directly in front of the sensitive material. Shutters of this type are found on many miniature cameras, on hand cameras designed particularly for press photography, and on reflex cameras.

Single-curtain shutters contain a long curtain with apertures of various widths. To make an exposure this curtain is wound on a roller at the top, using the wind-key on the side, until the aperture of the desired width is in place. Upon releasing the exposure lever the curtain is freed and the tension on the lower roller causes it to revolve, winding the curtain on it so that the aperture is drawn past the sensitive material, thus making the exposure. A locking device stops the curtain after the aperture has reached the lower roller, so as to prevent a second exposure. The exposure is controlled (1) by the width of the aperture used and (2) by varying the tension which controls the speed at which the aperture is drawn past the film or plate. With this type of shutter it is necessary to cover the lens with a cap or insert the slide of the film or plate holder to prevent exposure of the sensitive material to light when setting the shutter.

This disadvantage is overcome in many cameras by the use of a shutter with two curtains. (See Fig. 3.)

A between-the-lens shutter exposes the entire picture area at one time; with a focal plane shutter, however, the exposure travels from one side of the image to the other as the aperture passes across the face of the sensitive material. This results in a distortion of the image of rapidly moving objects. If the direction of movement of the aperture and the image are the same, the image is elongated; if in opposite directions, the image is contracted; while if the movement of the aperture is at right angles to that of the object, the image is diagonally distorted.

The efficiency of a shutter may be defined as the ratio between the amount of light actually reaching the film for a given time of exposure and the maximum amount which could possibly reach the film for the same time

interval. Thus, a between-the-lens shutter, whose blades opened instantaneously to full opening and remained open for the full exposure time and then closed instan-

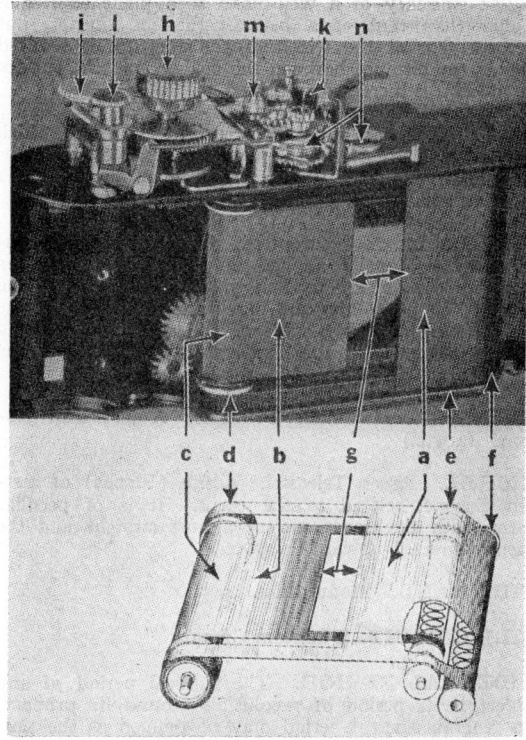

Fig. 3. Focal plane shutter. Operating lever on back of camera first brings edge of curtain (a) over edge of curtain (b), then winds them together across film aperture onto rollers (c and d) setting up spring tension in rollers (e and f). Shutter speeds of $\frac{1}{50}$ to $\frac{1}{1000}$ sec. determined by width of the curtain opening (g), regulated by knob (h) which turns roller (c) in relation to roller (d). Acceleration is compensated for by widening of slit as it travels across film plane due to difference in diameter of rollers (c and d). Speeds 1 to $\frac{1}{25}$ sec. are selected with dial (i) and controlled by an escapement mechanism (k) which varies the delay of curtain (b) after curtain (a) has completed its run. At "B" setting curtain (a) moves across when shutter release button (l) is pressed down, and curtain (b) follows when button is released. Moving self-timer lever (m) in clockwise direction winds up a spring-actuated escapement mechanism (n) which delays automatic tripping of the shutter by 10 to 12 secs. (*Eastman Kodak Co.*)

taneously, would have an efficiency of 100%. However, since this is impossible, it is obvious that while the blades of the shutter are opening and closing, the amount of light reaching the film is reduced. The varia-

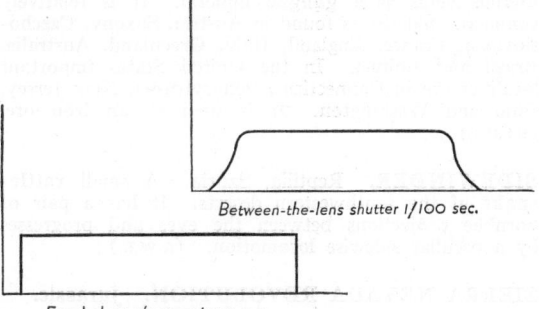

Between-the-lens shutter 1/100 sec.

Focal plane shutter 1/100 sec.

Fig. 4.

tion in illumination during the exposure with a between-the-lens shutter may therefore be represented by a diagram. (See Fig. 4.) The more rapid the opening and closing of the blades of the shutter, the higher its efficiency. Also, the smaller the diaphragm used, the longer the time of exposure at full opening and the greater the efficiency.

The efficiency of a well-made focal-plane shutter is higher than that of a between-the-lens shutter, but becomes less as the width of the aperture is reduced and as the distance between the curtain and the surface of the film is increased. (C.B.N.)

SIAL. That portion of the crust of the earth which forms the continents and, possibly, some of the sub-oceanic **lithosphere.** Rocks predominant in silica (Si) and aluminum (Al). (R.M.F.)

SIALID. Insecta, Neuroptera. A member of the family Sialidae, usually applied only to the **alder flies** although the **dobson fly** and the **fish flies** also belong here. The alder flies are small insects whose larvae live in streams. The habit of the adults of perching on the alders lining the banks gives them their name. (A.W.L.)

SIAMANG. Mammalia, Primates. A species of **gibbon.** (A.W.L.)

SICHEL. Pisces, Teleostei. A fish (**Pisces**) of eastern Europe, related to the carps. It is of peculiar form, with the upper surface almost straight and the mouth directed upward. (A.W.L.)

SIDE BAND. Modulation.

SIDE CIRCUIT. Phantom Circuit.

SIDEREAL PERIOD. The sidereal period of any object is its period of revolution around its primary. In general, sidereal period may be defined as the time required for an object to move from a particular position among the stars back to the same longitude again, as seen from the sun. (W.K.G.)

SIDEREAL UNIVERSE. Milky Way.

SIDERIOLITE. Meteor.

SIDERITE or CHALYBITE. This mineral is a **carbonate** of iron, $FeCO_3$. It is **hexagonal** with **rhombohedral** crystals, and also occurs in various massive forms. It has a rhombohedral **cleavage**; uneven fracture; is brittle; hardness, 3.5–4; specific gravity, 3.83–3.88; luster, vitreous to pearly; color, gray, yellowish-or' greenish-gray, green, reddish-brown and brown. Siderite is found as **concretionary** masses in the **sedimentary rocks;** as a replacement mineral from the action of iron solutions upon limestones; and in metalliferous veins as a **gangue** mineral. It is relatively common. Siderite is found in Austria, Saxony, Czechoslovakia, France, England, Italy, Greenland, Australia, Brazil and Bolivia. In the United States important localities are in Connecticut, Pennsylvania, New Jersey, Ohio and Washington. It is used as an iron ore. (E.S.C.S.)

SIDEWINDER. Reptilia, Sauria. A small **rattlesnake** of the southwestern deserts. It has a pair of hornlike projections between the eyes and progresses by a peculiar sidewise locomotion. (A.W.L.)

SIERRA NEVADA REVOLUTION. Jurassic.

SIEVE TUBES. A sieve tube is an elongate tube occurring only in **phloem.** It is made up of cells, each with a thin wall of **cellulose,** lined with the **protoplast,** and contains a large central **vacuole.** Through its walls there are many perforations, which are localized in those places known as sieve plates. The function of a sieve tube is the conduction of food.

For development, see **cambium.** (R.M.W., B.S.M.)

SIFAKA. Mammalia, Primates. A lemur of the genus **Propithecus.** The several species are found in Madagascar. They are related to the indri lemur but have long tails and shorter muzzles. (A.W.L.)

SIGILLARIA. Paleobotany.

SIGNAL-TO-NOISE RATIO. The usefulness of any communication system is limited by its ability to provide a useful signal which is strong enough to override any interfering noise. In passing through the various **amplifiers** associated with the modern systems both noise and desired signal are treated alike (except for frequency discrimination when they do not occupy the same frequency band) and so are amplified proportionately. It is necessary, then, that the input have a certain minimum ratio of signal to noise for the output to be satisfactory. This term is commonly applied to radio receiving systems. Various types of directional **antennae** will materially improve this ratio in many cases. (L.R.Q.)

SIGNIFICANT DIFFERENCE. Given two statistics θ_1' and θ_2' computed from **samples** of size N_1 and N_2, respectively, both consistent estimates of the **population parameters** θ_1 and θ_2, we wish to test the hypothesis $\theta_1 = \theta_2$ or possibly some other hypothesis. This test is based on the distribution of $\theta_1' - \theta_2'$ or in the general case on some function of θ_1' and θ_2'. If the test gives a point which falls in the critical region, we reject the hypothesis. If this happens, we say there exists a significant difference in θ_1 and θ_2, that is, $\theta_1 \neq \theta_2$. Significant differences may be found in **means, variances, coefficients of correlation,** and other measures. If a difference is shown to be significant statistically, this corresponds then to the statement that the two populations have different parameters. Two parameters may differ significantly in the statistical sense although from a practical point of view the difference may be very small. (L.A.A.)

SILENT CHAIN. Chain.

SILEX. Flint.

SILICATE. Silicon.

SILICATE BOND. A molded grinding wheel made from silicate of soda and abrasives, baked at a high temperature for several days. (H.C.H.)

SILICEOUS SINTER. Geysers.

SILICIC ACID. Silicon.

SILICON. Symbol: Si. Atomic number: 14. Atomic weight: 28.06. Density: 2.4. Hardness: 7. Melting point: 1420° C. Boiling point: 2600° C. Isotopes: 28 (89.6%), 29 (6.2%), 30 (4.2%).

Silicon is (1) a dark gray hard crystalline solid, (2) an amorphous brown powder. Both forms are unaffected by air at ordinary temperatures, but when heated in air to high temperatures a protective layer of oxide is formed; reacts with **nitrogen** at high temperatures to form nitride; with **chlorine** to form chloride; with several metals to form silicides. Crystalline silicon is unattacked by **hydrochloric** or **nitric** or **sulfuric acid,** but attacked by **hydrofluoric acid** to form silicon tetrafluoride gas; soluble in **sodium** hydroxide solution to form sodium silicate and hydrogen gas; reacts with dry chlorine to form silicon tetrachloride. Isolated by Berzelius in 1823.

Silicon occurs abundantly in all ordinary rocks except limestone, is second in abundance of the elements in the earth's crust (27.7% of the solid crust) and exceeded only by oxygen. Present in **igneous rocks** and clays as alumino-silicate; as the oxide (SiO_2) in the minerals **quartz,** sand, **flint,** and the gems **amethyst, jasper, chalcedony, agate, onyx, tridymite, opal, crystobalite;** as silicates, **zircon** (zirconium silicate $ZrSiO_4$), **willemite** (zinc silicate Zn_2SiO_4), **wollastinite** (caicium silicate, $CaSiO_3$), **serpentine** (magnesium silicate $Mg_3Si_2O_7$). Silicon is obtained from the oxide (1) by igniting with **aluminum** powder, (2) by reduction with **carbon** in the **electric furnace,** and is used to reduce many oxides. Silicon is chemically related to **aluminum, boron, titanium** and **carbon.** Silicon is an alloying element in **steel** and in **brass and bronze.**

Acids: Silicic acid (ortho H_4SiO_4 or $SiO_2 \cdot 2H_2O$; meta H_2SiO_3 or $SiO_2 \cdot H_2O$), white gelatinous precipitate, by reaction of solution of silicate and **hydrochloric, nitric** or **sulfuric** acid of proper concentrations, forms silicon oxide upon ignition, forms various silicates by fusion methods; hydrofluosilicic acid (H_2SiF_6), colorless solution, by reaction of silicon fluoride, forms fluosilicates by neutralization methods.

Aluminates: Many complex silico-aluminates or alumino-silicates are formed in nature. Of these, clay in more or less pure form (pure clay, kaolinite; **kaolin,** china clay, $H_4Si_2Al_2O_9$ or $Al_2O_3 \cdot 2SiO_2 \cdot 2H_2O$) is of great importance. Clay is formed by the weathering of igneous rocks, and is used in the manufacture of bricks, pottery, porcelain, Portland cement. (See **Ceramics; Cement, Portland.**) Sodium aluminosilicate is used in water purification to remove dissolved calcium compounds. (See **Calcium** aluminosilicates.)

Carbide: Silicon carbide (SiC), ("carborundum"), bluish-black iridescent crystals, very hard (hardness 9 Mohs scale), used as an **abrasive** and heat refractory material.

Chloride: Silicon chloride, silicon tetrachloride ($SiCl_4$), colorless fuming liquid, boiling point 58° C., by reaction of silicon, or silicon oxide plus **carbon,** heated in **chlorine,** fumes in moist air, and is used for producing white smoke in air.

Fluoride: Silicon fluoride, silicon tetrafluoride (SiF_4), colorless gas, by reaction of silicon oxide, **calcium** fluoride and hot concentrated **sulfuric acid;** reacts with water to form silicic acid plus hydrofluoric acid, and the latter, with excess silicon fluoride, forms hydrofluosilicic acid; combines with ammonia (e.g., $SiF_4 \cdot 2NH_3$).

Fluosilicate: Sodium fluosilicate (Na_2SiF_6), white solid slightly soluble; magnesium fluosilicate ($MgSiF_6$), white solid, soluble.

Hydrides: Silicon tetrahydride, silicon methane, silicane (SiH_4), colorless gas, boiling point 115° C., by reaction of magnesium silicide and **hydrochloric acid,** followed by purification of the gas, reacts with **silver** nitrate yielding silver plus silicon; silicoethane (Si_2H_6), colorless liquid, boiling point 52° C., by reaction of **lithium** silicide (Li_6Si_2) and concentrated hydrochloric acid.

Nitrides: Trisilicon tetranitride (Si_3N_4), by heating silicon oxide plus **carbon** to 1500° C. in a current of **nitrogen** gas.

Oxide: Silicon oxide, silicon dioxide, silica, **quartz, tridymite, crystobalite,** "vitreosil" (SiO_2), white to colorless solid. Ordinary quartz (alpha-quartz), density 2.65, melting point 1425° C., changes into beta-quartz at and above 575° C., these have different optical properties and etching marks with hydrofluoric acid. Beta-quartz is changed into tridymite, density 2.26, melting point 1670° C., at and above 870° C. When silica is maintained some time at the sintering temperature, crystobalite, density 2.32, melting point 1710° C., is formed. Different forms of silicon oxide differ in chemical activity, e.g., quartz is practically unreactive with **sodium** hydroxide solution, whereas amorphous silica,

density 2.20, formed by heating silicic acid, is reactive. Silicon oxide reacts (1) with fused **sodium** carbonate, yielding sodium silicate glass, (2) with sodium and **calcium** carbonate to form ordinary glass, (3) with several metallic oxides to form various colored glasses, (4) with **hydrofluoric acid,** but not with other acids (except hot concentrated **phosphoric acid**). Quartz is commonly found as a mineral in rocks (second in abundance to feldspar among minerals of the earth's crust). Quartz softens gradually like glass when heated somewhat below its melting point, and in the plastic condition may be worked into various shapes like glass (such wares are called by various trade names, e.g., "vitreosil"). Various preparations of silica gel involve the dehydration of silicic acid to silica containing more or less combined water for special purposes of adsorption of gases and liquids.

Silicates: Sodium metasilicate, "water glass" (Na_2SiO_3), colorless (when pure) glass, soluble, melting point 1088° C., formed by reaction of silicon oxide and **sodium** carbonate at high temperature; solution reacts with **carbon dioxide** of the air, or with sodium carbonate solution or **ammonium** chloride solution, yielding silicic acid, gelatinous precipitate. Sodium silicate solution is used (1) in soaps, (2) for preserving eggs, (3) for treating wood against decay, (4) for rendering cloth, paper, wood non-inflammable, (5) in dyeing and printing textiles, (6) as an adhesive (e.g., for paper boxes) and cement. Various fused mixed silicates are used in glasses (sodium calcium silicates in common glass). Various colored silicates may be separated in solution by adding the colored soluble solid (e.g., cobalt nitrate) to sodium silicate solution. (See **Aluminates** above. See **Calcium** silicates.)

Silicones: Compounds containing Si and C and OH, made by reaction of $SiCl_4$ and RMgCl (**Grignard Reagent**) followed by hydrolysis to form $RSi(OH)_3$, $RSi(OH)_2$, $RSi(OH)$. Of these, the first two types are used as baking enamels and resins, and the last two as greases. All show unique stability against heat and chemical action and are of high electrical resistance.

Sulfides: Silicon monosulfide (SiS), yellow solid, somewhat volatile, formed by heating to redness crystalline silicon in sulfur vapor, reactive with water; silicon disulfide (SiS_2), white crystals, formed by heating amorphous silicon and sulfur, and then subliming, reactive with water.

Organic compounds: Silicon tetramethyl ($Si(CH_3)_4$), liquid, boiling point 26° C.; silicon tetraethyl ($Si(C_2H_5)_4$), liquid, boiling point 152° C.; disilicon tetramethyl ($(CH_3)_2Si:Si(CH_3)_2$), boiling point 113° C.; silicon tetraphenyl ($Si(C_6H_5)_4$), solid, melting point 231° C.; methyl silicate, tetramethoxy silicon ($Si(OCH_3)_4$), boiling point 122° C.; ethyl silicate, tetraethoxy silicon ($Si(OC_2H_5)_4$), boiling point 165° C.

In the inorganic world of nature, the element silicon plays a role in the composition of rocks and minerals at least suggestive of the diversity shown by carbon in the organic world of plants and animals. (R.K.S.)

SILICONIZING. Also known as Ihrigizing after the inventor of the process. The surface of iron and steel objects is impregnated with silicon by treatment with silicon carbide and chlorine gas at temperatures between 1700° and 1850° F. A maximum silicon content of about 14% is obtained at the surface, giving the article excellent resistance to acid-corrosion and to scaling at elevated temperatures. (R.H.H.)

SILICOSIS. A chronic disease of the **lungs** seen chiefly in miners. It is caused by inhalation over long periods of time of air containing finely divided particles of **silicon.** Fibrous nodules develop in the alveolar divisions of the lung; after years of exposure, these occupy enough space to cut down seriously the functioning surface of the lung. Decreased vital capacity or possibly total exchange of air with each respiration and **dyspnoea** result. Individuals with silicosis are prone to develop **tuberculosis.** (D.M.H.)

SILICIFICATION. An important geochemical process by which certain sedimentary rocks such as **limestones** and **dolomites,** or calcareous **fossils** are partially or entirely replaced by silica, SiO_2. (See also **Chert** and **Flint.**) (R.M.F.)

SILK. A material produced by many **insects** and **spiders** for the formation of cases in which the **larva** lives, cocoons in which pupation (**pupa**) takes place or eggs are deposited, and webs for the capture of prey. Some **caterpillars** form temporary defenses of silk and both caterpillars and spiders use it to lower themselves from elevated supports **and** sometimes to elevate themselves to such supports.

Silk is formed as a liquid secretion which hardens quickly on exposure to the air. It is spun by being forced through small apertures which mold it into a thread. When so discharged it is a clear, smooth, lustrous material of considerable strength, but spiders form opaque and adhesive threads as well.

The silk of commerce is produced by the silk worm, caterpillar of the moth, *Bombyx mori.* The fine threads of which the cocoon is composed are reeled and spun into threads large enough to be woven. Although the use of pure silk has been encroached upon by manufactured rayon and allied products, it is still produced in large quantities in the Orient and in the Mediterranean countries. Another product of the industry is catgut, produced by removing the silk gland of the worm and artificially drawing it into a thick thread. This product comes chiefly from Spain.

The silk of many other insects has been reeled and spun, and even that of spiders has been made into fabrics. The latter is too fine for practical purposes although the very fine filaments spun by the spider are used as cross-hairs in optical instruments. Pongee silk is the only commercial product of this kind that is derived from another moth than the true silkworm species. (For synthetic fibers such as rayon, see **Carbohydrates.**) (A.W.L.)

SILK GLAND. The gland from which **silk** is produced by the **insects** and **spiders.** Silk glands of most insects are modified salivary glands opening through the mouth parts in the **larvae** alone, but in the order Embiidina some observers report that the silk comes from glands in the front legs, whose ducts open through hairs on the tarsi.

The silk glands of spiders are located in the abdomen, opening through a group of special organs called spinnerets on the under surface near the tip of the body. They are of several forms. Ampulliform glands produce the strong radial lines of the orb webs and aggregate glands secrete the sticky spiral lines. Pyriform glands produce disks by which the anchoring lines of the web are attached to its supports. When prey is caught it is wrapped in fine silk produced from aciniform glands, and still another type, the tubuliform glands, form the silk of which the egg cocoon is made. The last occur only in the female. (A.W.L.)

SILKWORM. Insecta, Lepidoptera. The **caterpillar** of the **moth,** *Bombyx mori,* which produces the silk of commerce. Caterpillars of the family Saturniidae also spin cocoons of silk, and are known as the giant silkworms. This family includes the common luna, cecropia, and polyphemus moths of North America, as well as many other species. (A.W.L.)

SILL. A tabular mass of igneous rock that has been intruded laterally between layers of sedimentary rock, beds of volcanic lava or **ejectmenta,** or even along the direction of the **foliation** in metamorphic rocks. The term sill is synonymous with intrusive sheet. (R.M.F.)

SILLIMANITE. The mineral sillimanite is an **aluminum silicate,** the formula Al_2SiO_5 being like that of **andalusite** and **kyanite.** It is **orthorhombic,** usually in slender prisms, but may be fibrous or massive. Its hardness is 6–7; specific gravity, 3.23–3.24; luster, vitreous; color, various shades of gray, grayish-green, grayish-brown, etc.; transparent to translucent. It occurs in **granites** and **gneisses** as tiny prisms and aggregates, and is often associated with andalusite, **cordierite** and **corundum.** Sillimanite has been found in Bavaria, Czechoslovakia, France, India, Madagascar, Burma and Ceylon, the latter two localities furnishing transparent sapphire-blue gem stones. In the United States sillimanite has been found in Connecticut, New York, Pennsylvania, Delaware, North Carolina and California, where, in Inyo County is the largest deposit in the world. This mineral was named in honor of Benjamin Silliman for many years professor of chemistry and natural science at Yale University. Sillimanite is used in the manufacture of spark plug "porcelains" and laboratory ware. (E.S.C.S.)

SILTSTONE. A term signifying a **clastic** sedimentary rock in which the particles are of silt grade. (R.M.F.)

SILURIAN PERIOD. A major subdivision of the **Paleozoic Era.** Type locality, Wales and Shropshire, England. The formations of this system were first studied and described by R. I. Murchison in 1835. The Silurian Period began 380 million years ago and lasted for 50 million years. In Murchison's time, and for some time afterward, the Paleozoic Era, below the **Devonian,** was divided into the **Cambrian** and Silurian. This practice still holds in parts of Europe. The Silurian formations are well exposed in Eastern North America, especially in New York State, and the length of the **Appalachian geosyncline** where the sediments are principally red **sandstones** and **shales** of delta origin. There was little or no volcanic activity except in southeastern Maine. The Silurian is also well exposed in nearly all of Western Europe, Northern Siberia, Burma, Central Asia, Himalayas, Morocco, Australia, Peru and Bolivia. The system is characterized by all types of sediments, with evidence in certain areas of an arid climate, including thick deposits of salt and **gypsum.** The maximum thickness of sediments is 15,000′, in Britain. The principal types of fossils are **corals, bryozoans** (including primitive reefs), spire-bearing **brachiopods, graptolites, nautiloids, trilobites,** and **ostracods.** The first air-breather, a fossil scorpion, is reported from both Sweden and England. In North America the principal economic products of Silurian Age are rock salt, gypsum, iron ore, petroleum and natural gas. In western Europe the Silurian was brought to a close by a period of mountain building called the Caledonian Disturbance. (R.M.F.)

SILVER. Symbol: Ag (argentum). Atomic number: 47. Atomic weight: 107.880. Density: 10.5. Hardness: 2.5–2.7. Melting point: 960.5° C. Boiling point: 1950° C. Isotopes: 107 (52.5%), 109 (47.5%).

Silver is a white metal, softer than **copper** and harder than **gold;** when molten is luminescent and occludes **oxygen** (20 volumes) but the oxygen is released in solidification ("spitting"); the most malleable and ductile of the metals except gold; as a **conductor** of heat and electricity superior to all other metals; soluble in **nitric acid** containing a trace of nitrate; soluble in hot 80% **sulfuric acid;** reacts with **hydriodic acid** to yield **hydrogen;** insoluble in **hydrochloric** or acetic acid; tarnished by **hydrogen sulfide,** soluble sulfides and many sulfur-containing organic substances (e.g., **proteins**); not affected by air or water at ordinary temperatures, but at 200° C. a slight film of silver oxide is formed; not affected by **alkalis**, either in solution or fused. Discovery prehistoric.

Silver is used (1) as an ornamental metal, either as silverware or silver-plated goods, and deposited on glass for mirrors, (2) in bearings for aircraft engines, (3) as an emergency substitute for copper for bus bars for electric power distribution, (4) in alloys, such as hard solder, and in coins (90% silver with 10% copper as a hardener).

Silver occurs native in Peru, but the chief ores are sulfides (argentite, Ag_2S) and it is frequently associated with gold, lead and copper ores. Silver is obtainable (1) as a by-product in the **electrolytic** refining of copper in the **anode** mud, (2) as a by-product in crude lead by its differential solubility in zinc metal when molten (silver is 3000 times more soluble in zinc than in lead), (3) by leaching the ore with **sodium** cyanide solution and recovery by electrolysis.

Acetate: Silver acetate ($AgC_2H_3O_2$), white solid, only slightly soluble in water.

Bromide: Silver bromide (AgBr), pale yellow precipitate by reaction of silver nitrate solution, and **potassium** bromide solution, insoluble in **nitric acid**, only slightly soluble in ammonium hydroxide, soluble in **sodium** thiosulfate solution, turns dark on exposure to light, melting point of silver bromide 434° C.

Chloride: Silver chloride (AgCl), white precipitate by reaction of silver nitrate solution and **potassium, ammonium,** or **sodium** chloride solution or by **hydrochloric acid**, insoluble in nitric acid, soluble in ammonium hydroxide, and in sodium thiosulfate solution, turns dark on exposure to light, melting point of silver chloride 455° C.

Cyanide: Silver cyanide (AgCN), white precipitate by reaction of silver nitrate solution and **potassium** cyanide solution; silver potassium cyanide ($KAg(CN)_2$), soluble, and recovered as white solid by crystallization.

Chromate: Silver chromate (Ag_2CrO_4), yellow to red to brown precipitate by reaction of silver nitrate solution and **potassium** chromate solution.

Dichromate: Silver dichromate ($Ag_2Cr_2O_7$), red precipitate by reaction of silver nitrate solution and **potassium** dichromate solution, changing to silver chromate upon boiling with water.

Iodide: Silver iodide (AgI), yellow precipitate by reaction of silver nitrate solution and **potassium** iodide solution, soluble in concentrated nitric acid, insoluble in ammonium hydroxide solution, and slightly soluble in sodium thiosulfate solution, turns dark on exposure to light, melting point of silver iodide 552° C. with decomposition.

Nitrate: Silver nitrate, lunar caustic ($AgNO_3$), white colorless crystals, soluble, the source of most insoluble and some soluble silver compounds, formed by the reaction of silver metal and **nitric acid** followed by crystallization, melting point of silver nitrate 212° C., and at 320° C. begins to decompose.

Oxides: Silver monoxide, silver oxide (Ag_2O), brown precipitate, by the reaction of silver nitrate solution and **sodium** or ammonium hydrox'de, soluble in **ammonium** hydroxide (upon standing this solution forms a dangerous explosive), appreciably soluble (1 part in 300) in water, giving a solution that turns litmus blue, decomposes at 300° C. into silver metal and oxygen; silver dioxide, silver peroxide (Ag_2O_2), black precipitate by the reaction of silver nitrate solution and **ammonium** persulfate solution, an active oxidizing agent.

Phosphate: Silver phosphate (Ag_3PO_4), yellow precipitate, by reaction of silver nitrate solution and disodium hydrogen **phosphate** solution, soluble in nitric acid and in ammonium hydroxide, turns dark on exposure to light.

Sulfate: Silver sulfate (Ag_2SO_4), white precipitate, by the reaction of silver nitrate solution and potas-

sium, **sodium** or **ammonium** sulfate solution or sulfuric acid, melting point of silver sulfate 652° C.

Sulfide: Silver sulfide (Ag_2S), black precipitate, by the reaction of silver nitrate solution and **hydrogen sulfide.**

Thiocyanate: Silver thiocyanate (AgCNS), white precipitate by reaction of silver nitrate solution and potassium thiocyanate solution, insoluble in nitric acid, soluble in warm ammonium hydroxide but re-precipitated on cooling.

Silver halides (**iodide, bromide, chloride**) on account of darkening on exposure to light, are applied in photography on the sensitized plate or film, and the solvent action of sodium thiosulfate on silver compounds is applied in the fixing process, the particles of metallic silver remain unattacked.

Colloidal silver and some of its insoluble compounds, are used in medicine for antiseptic and antibacterial action on mucous membranes. Silver nitrate is antibactericidal, astringent, and stimulating.

Silver forms more insoluble salts than any other metal (approached in number of salts by lead and mercury). (R.K.S.)

SILVER THAW. After a period of cold weather and below-freezing temperatures a mass of warm air passing over the region will cause frost or glaze to form on objects that are still at a low temperature. This condition is known as a silver thaw. A silver thaw usually lasts only a few hours as the warm air soon warms all exposed objects above 32° F. (P.E.K.)

SILVERFIN. Pisces, Teleostei. A small fish (Pisces), *Cyprinella whippli,* related to the shiners. Common in clear streams from New York to Alabama and west to Minnesota and Arkansas. It is silvery with a bluish tinge and the paired fins are pale whitish. (A.W.L.)

SILVERFISH. Insecta, Thysanura. *Lepisma.* One of the primitive wingless insects. It is about ½" long, broad in front and tapering behind, with two long slender antennae and three similar processes at the caudal end of the body. It is covered with lustrous grayish scales, whence the common name. The insect eats starchy materials and sometimes defaces book bindings, wall paper, and laundry, but as a rule it is not sufficiently abundant to be a pest. It is often found in damp buildings. Also called the fish moth. (A.W.L.)

SILVERSIDES. Pisces, Teleostei. Small slender fishes (**Pisces**), of numerous species, found chiefly along the coasts of the warmer seas but in a few cases in ponds and streams. They make up the family Atherinidae. (A.W.L.)

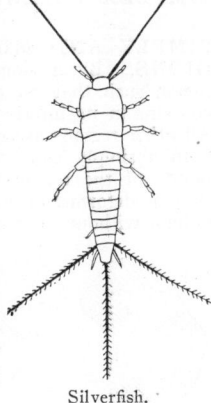

Silverfish.

SIMA. That portion of the crust of the earth, chiefly the sub-Pacific **lithosphere** which is composed chiefly of rocks rich in the ferro-magnesian silicates. (See also **Geosyncline.**) (R.M.F.)

SIMILITUDE. The analysis of an engineering structure may be made by means of mathematical equations or by the use of models. A model is a small-scale reproduction of the prototype in all features which are pertinent to the particular problem under investigation. Model analysis is based on Newton's Law of Similarity. In general this law states that models must be geometrically and dynamically similar to the structures which they represent. No restriction is placed on the size of the model. Consequently the scale and the material of the model may be varied to produce the particular action which is required.

Methods of analysis have been developed which reduce the essential elements for similarity to certain fundamental characteristics. The methods of analysis require a proportionality or scale ratio of the respective forces acting on the model and on the prototype. These forces, depending upon their nature or origin, are expressed in terms of one or several physical characteristics such as **mass**, length and time. The engineer does not always use these characteristics as such but may use **density, velocity, gravity, viscosity,** etc. Often it is impossible to satisfy all of the requirements for similitude simultaneously. In this event skill is required in designing the model in order to reduce the error, which is the result of ignoring some of these requirements, to a negligible value.

Models are used to determine the action of **bridges**, building frames, **airplane** structures and **hydraulic** structures under **load**. The action of some loaded structures is so complex that it is virtually impossible to make a mathematical solution which will have a practical application. Model analysis may then be used advantageously. (C.W.C.)

SIMPLE BEAM. A simple **beam** is one which rests on two end supports in such a manner that the ends of the beam are free to rotate on the supports. Since

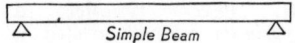
Simple Beam

riveted connections are elastic, the end restraint offered by this type of connection is not very reliable. Therefore, beams and girders which are riveted to supports are generally assumed to be simply supported unless a special type of moment connection is used. (C.W.C.)

SIMPLE CURVE. Circular Curve.

SIMPLEX TELEGRAPH. Telegraphy.

SIMPLY AND MULTIPLY CONNECTED REGIONS. By a simply connected region is meant a region such that no closed curve drawn in the region contains in its interior a boundary point of the region. All other regions are called multiply connected.

In a simply connected region, a curve between two points of the region may always be changed by continuous deformation into any other curve between those points, without passing outside the region.

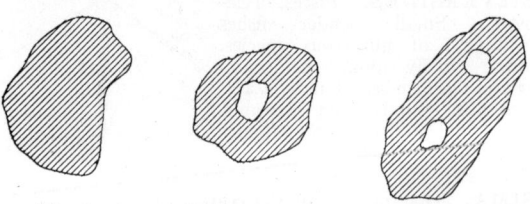

Fig. 1.   Fig. 2.   Fig. 3.
Simply and multiply connected regions.

A region bounded by a simple closed curve is a simply connected region (Fig. 1). A circular ring is an example of a doubly connected region (Fig. 2). A triply connected region is also shown (Fig 3). (L.L.S.)

SIMULTANEOUS EQUATIONS. Systems of Algebraic Equations.

SINANTHROPUS. Paleontology of Man.

SINE CURVE. Trigonometric Curves.

SINEW. A common term applied both to **tendons** and to **ligaments.** (A.W.L.)

SINGING. This is the term used by telephone engineers to designate the result of part of the output of a **repeater** being fed back into the input. This produces oscillations whose frequency is determined by the circuit constants. The frequency may vary over a wide range, but if it is in the audio range and is fed to a **speaker,** the output is a constant note or howl. (L.R.Q.)

SINGLE PHASE. Phase.

SINGLE SIDE-BAND TRANSMISSION. Modulation.

SINGLE SHEAR. Rivet; Shear.

SING-SING. Mammalia, Artiodactyla. An **antelope** of western and central Africa. It is a medium sized species with a white patch on the buttocks. (A.W.L.)

SINGULAR SOLUTION OF A DIFFERENTIAL EQUATION. Ordinary Differential Equations of First Order and Higher Degree than the First.

SINK OR SINK HOLE. Cave.

SINKING SPEED. The vertical component of the velocity of an airplane gliding without acceleration in power-off condition is the sinking speed. It is analagous to "rate of climb" of climbing flight. Methods exist for estimating sinking speed using other data of the airplane, to wit, air speed, altitude, weight, parasite area, wing span, and **airplane efficiency factor.** Gliding at minimum sinking speed should not be confused with the flattest glide. The latter occurs when parasite drag approximately equals induced drag, while for the former the parasite is three times the induced drag. (F.T.M.)

SINTER. The calcareous (see **Calcium**) or siliceous material deposited by mineral springs, usually thermal but not necessarily so, is known as sinter or tufa. Commonly the term sinter is used for the siliceous variety, tufa for the calcareous kind. Siliceous sinter is also known as **geyserite.** Rocks formed from calcareous tufa are known as travertine, and frequently used as decorative stones under the name of "**onyx marble.**" (E.S.C.S.)

SINUS. A pouch or cavity in any organ or tissue, or an abnormal cavity or passage formed by the destruction of tissue. The term is applied to a very large number of such structures in the human body, such as, a dilated portion of a **vein** containing venous blood; a chronically infected tract such as a **fistula;** the air cavities within the cranial bones, especially those located near the nose and connecting with it. They are called accessory sinuses of the nose. They extend from the nasal passages into bones of the skull, and are named according to the bones in which they lie as the frontal, ethmoidal, sphenoidal, and maxillary sinuses. The maxillary sinuses are also called antrums. The maxillary sinuses are found on either side of the nose. They are large in size, and lie between the floor of the eye socket above and the upper teeth below. The frontal sinuses lie in the forehead above the roof of the eye socket, one on either side of the midline of the forehead. The ethmoidal sinuses are three groups with numerous air cells, situated between the eyes in either side of the midline or septum of the nose. The sphenoidal sinuses, two in number, are situated above and behind the nose proper. They are all lined with a delicate mucous membrane continuous with the nasal mucous membrane. When the sinuses become infected, **sinusitis** is said to be present. (A.W.L., R.S.M.)

SINUSITIS. Acute or chronic infection of the paranasal **sinuses.** Infection occurs most commonly during an acute **rhinitis** or common cold or as a complication of allergic rhinitis and hay fever. Infection of the roots of the teeth directly under the floor of the maxillary antra may extend upward and involve these sinuses.

The chief symptoms of sinusitis are pain and headache, usually limited to some portion of the face near the involved sinus. The symptoms are partly due to obstruction of the openings leading from the sinus cavities to the nose, thereby preventing the escape of infected secretion or the entrance of air into the involved sinus. In chronic sinusitis the symptoms are mild. Often an early morning postnasal discharge is the only sign.

The complications of sinusitis are inflammatory diseases of the eye, orbital cellulitis, and **abscess.**

Treatment is local and general, involving the establishment of correct drainage of involved sinuses and correction of any nasal or dental abnormality. **Penicillin** is effective in certain sinus infections. (D.M.H.)

SINUSOID. Trigonometric Curves; Circulatory System.

SIPHON. For the general use of this term in connection with the flow of fluids, see **Hydrokinetics.** In zoology, it refers more specifically to a passage between the **mantle** folds of **bivalve** mollusks through which water enters or leaves the mantle cavity. In some species these passages are developed into long muscular tubes and in others they are no more than poorly marked openings. They are two in number, a dorsal or excurrent siphon and a ventral or incurrent siphon. Water is taken in through the latter, passes through the **gills** to the chambers above them, and flows out through the dorsal siphon.

A part of the mantle border in some of the marine **gasteropods** also forms a tube through which water can be drawn into the mantle cavity. This tube is known as the siphon.

The term oral siphon is applied to the canal leading to the mouth of **ascidians** and the opening from the atrial cavity is sometimes known as the atrial siphon.

Unlike these organs, all of them associated with respiration, the siphon of sea urchins (**Echinoidea**) is a slender tube associated with the alimentary tract. (A.W.L.)

SIPHONAL CANAL. An extension of the lip of the shell surrounding the siphon of many **snails.** (A.W.L.)

SIPHONAPTERA. The fleas. An order of **insects** characterized by highly compressed bodies, sucking mouths, and complete metamorphosis. The eggs are dropped in the quarters occupied by the animals on which the fleas live and the larvae live on organic debris. The great majority of the known species live as adults on mammals but some are found on birds. Species found on the dog and cat are sometimes troublesome to man. (A.W.L.)

SIPHONOGLYPH. A ciliated (Cilia) groove in the angles of the **stomodaeum** of **sea anemones.** It conducts water currents into the enteric cavity for respiratory purposes and as a carrier of food particles, and also carries currents outward. In some species excurrent and incurrent siphonoglyphs occur and in others

a single one serves both purposes by reversal of the effective beat of the cilia. (A.W.L.)

SIPHONOPHORA. Hydrozoa.

SIPHONOZOOID. A specialized **polyp** in the colonial **sea feathers, sea pens,** and related forms. It draws water into the enteric cavity of the colony. (A.W.L.)

SIPHUNCLE. A slender tube extending from the visceral hump of the **nautilus** through perforations in the septa of the shell along the entire spiral. The siphuncle contains blood vessels and its function is supposed to be the maintenance of the gas content of the closed chambers of the shell. By this contained gas the weight of the shell is offset to aid the animal in floating. (A.W.L.)

SIPUNCULIDA. Gephyrea.

SIREN. Musical Sounds.

SIRENIA. The **manatee** and **dugong,** constituting an order of mammals. They are marine animals with fore legs modified as flippers and hind legs lacking. The manatee is found on the east coast of Florida and southward and the dugong is found near Africa, Australia, and the islands of the Oriental region. (A.W.L.)

SIRIUS. Sirius (α **Canis Major**) is the brightest star in the sky and volumes have been written concerning its matchless brilliancy. Historically, it is undoubtedly the most interesting star in the heavens and references to it are found throughout all ancient literatures back to the earliest known writings. Aside from its surpassing brilliancy, the fact that it may be observed from every habitable portion of the earth has served to make it an object of veneration by all peoples. Sirius was worshiped in the valley of the Nile long before Rome was even heard of, and many ancient Egyptian temples were so arranged that the light from this star would penetrate to the inner altars.

Astronomically, Sirius is particularly interesting as being the typical, A, **spectral type.** The companion of Sirius is the first **white dwarf** discovered. (W.K.G.)

SISAL. Agave.

SISKIN. Aves, Passeriformes. A quietly colored **finch.** The numerous species are found throughout the temperate part of the northern hemisphere and in South America. The pine siskin of North America, *Spinus pinus,* also called the pine finch, is finely streaked with brownish on a lighter ground and has a few yellow marks on wings and tail. (A.W.L.)

SITITUNGA. Mammalia, Artiodactyla. An **antelope,** *Limnotragus spekei,* found in swampy ground in central Africa. It is grayish-brown in color and of medium size. Also called the nakong. (A.W.L.)

SIZING. Press Working.

SKARN. Used by **petrologists** to describe that process of contact metamorphism by which certain mineral silicates such as **amphibole, pyroxene** and **garnet** replace limestone and dolomite. (R.M.F.)

SKATE. Pisces, Plagiostomi. Cartilaginous fishes (**Pisces**) characterized by the broad flat body. The pectoral fins are extended at the sides and along the greater part of the head and body and are angled at the outer margin so that the general form is quadrangular. The tail is long and slender and is sometimes armed with spines. The name is applied to the common rays, making up the family Rajidae. Several species are found off the coasts of the United States, including the common skate of the Atlantic coast which also bears the remarkable name, tobacco-box. The family is represented in most of the warm and temperate seas. (A.W.L.)

SKELETAL SYSTEM, SKELETON. An aggregation of rigid or semirigid structures that provide mechanical support for the body and usually a lever system on which the muscles act.

The simplest type of skeletal system is found in the sponges (**Porifera**) and some of the **coelenterates.** Sponges have scattered spicules of calcareous or siliceous matter among their loosely integrated tissues or are held together by a meshwork of fibers composed of the material spongin, so familiar in the sponges of commerce. The commercial sponge is, in fact, the skeleton freed of organic matter. Spicules are variously formed bodies, some straight rods, some with radiating axes from three to six in number, and some expanded at the ends. Approximately similar scattered bodies are found also in alcyonarians, and in some of these animals they are united to form a continuous mass in which the polyps are imbedded or a solid core surrounded by softer material. Rock **corals** lay down a calcareous deposit beneath the base to which each **polyp** of the colony adds.

In many invertebrates the body wall is the sole support of the animal. Even though it is soft, the incorporation of muscular layers in it provides for movement without rigid skeletal structures.

The **echinoderms** differ in having calcareous plates (ossicles) throughout the integument. In the sea urchins they are closely joined to form a shell and in the sea cucumbers they are small and scattered. Other groups show an intermediate condition with closely associated but movable ossicles. These structures in the sea urchin also illustrate lever action in the use of the spines for locomotion and in the peculiar chewing organ known as **Aristotle's lantern.**

The **arthropods** and **vertebrates** differ in the presence of a complex skeletal system which supports the body and forms the foundation of the jointed appendages. In no other phyla are lever-like appendages found.

The arthropods have an exoskeleton made up of hard plates (sclerites) of **chitin** and calcareous matter, formed in circumscribed areas of the **cuticula.** They are separated by flexible zones which provide for freedom of movement, and are moved by muscles attached to their inner surfaces. The **appendages** are formed of rigid segments attached to each other and to the body by the flexible tissue of the joints. The skeleton forms a sheath enclosing the muscles in such appendages.

Among the half-million species of arthropods wide variation is inevitable. In general the exoskeleton of the head forms a compact capsule. The skeleton of the **thorax** also tends to become compact, although its segments are distinctly recognizable in many species. This portion of the exoskeleton also forms a shieldlike **carapace** in many species. The abdomen retains greater mobility, although its segments may be reduced in number and the exoskeleton of each may be consolidated to a ring surrounding the body. A moderately complex segment of the exoskeleton contains a dorsal, a ventral, and two lateral sclerites known respectively as the tergum or tergite, the sternum or sternite, and the pleura or pleurites. In many cases this plan is simplified by fusion or made more complex by subdivision of the sclerites.

An endoskeleton of limited extent in the arthropods is made up of internal projections from the exoskeleton known as apodemes. They serve as places of attachment for muscles.

One important result of exoskeletal support is the provision of wings in the insects as thin-walled sacs of the body wall. By the apposition of the upper and lower walls of these sacs they become thin membranous planes sufficiently strong and rigid to support the body in flight. Thus the jointed appendages are freed from participation in adaptation for flight.

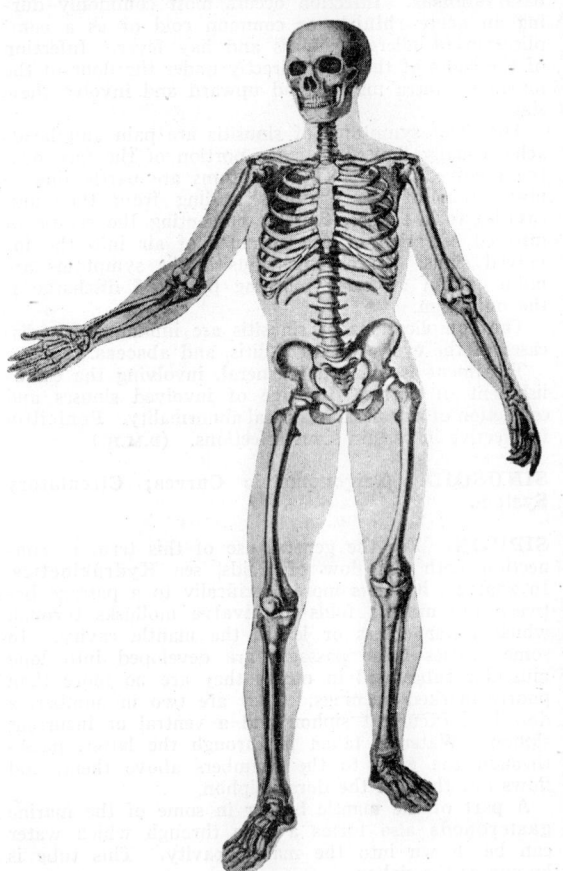

Skeletal system.

Exoskeletal structures also appear in the vertebrates but here they are merely hard parts without the usual supporting functions of the skeleton. They are derived from either or both layers of the skin. The category includes teeth and beaks, horns, claws, hoofs and nails, scales, feathers, and hair. Bony plates associated with scales, as in the alligator and armadillo, are also exoskeletal.

The supporting skeleton of the vertebrates is an endoskeleton composed of **bones** and **cartilage.** It is made up of two chief divisions, an axial skeleton consisting of a vertebral column, ribs, and skull, and an appendicular skeleton including pectoral and **pelvic girdles,** each bearing a pair of appendages. The pharyngeal wall of fishes is supported by the visceral skeleton which persists to a limited extent in the more advanced classes. In the elasmobranch fishes now living, the skeleton is composed entirely of cartilage. In most vertebrates it is made up very largely of bone.

The skull in the elasmobranch fishes consists of a mass of cartilage below, behind, and partially enclosing the brain. This structure is called the chondrocranium. In the bony fishes it is replaced by bones in the same position and is supplemented by superficial bony plates enclosing the remainder of the brain. The jaws of the elasmobranch are also supported by cartilages and in the bony fishes these cartilages are supplemented by bones. Above the fishes a chondro-cranium appears in the embryo and is replaced during development by the ethmoid bone and parts of the sphenoid, temporal, and occipital bones, taken in order from front to back. The remaining bones are not preformed in cartilage. They are more numerous in the lower forms than in man. In the human skull they are the pair of nasal bones in the bridge of the nose, the pair of lachrymals in the orbits, the vomer in the nasal septum, the large frontal bone of the forehead, the parietals in the top and sides of the skull, the

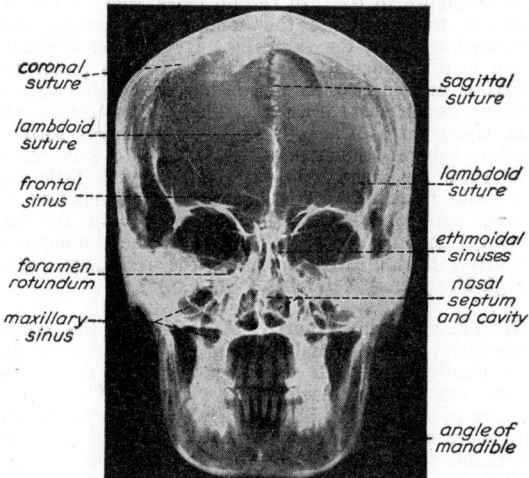

Antero-posterior radiography of a skull. (*Cunningham, Text-book of Anatomy, Oxford Press.*)

upper part of the occipital forming the lower back wall of the skull, and the temporals above the ears. The upper jaw is based on the maxillary bone from which the palatine extends into the hard palate. A slender rod, Meckel's cartilage, supports the lower jaw of the human fetus but the permanent lower jaw is the mandible, a dermal bone formed independently.

The vertebrae that make up the spinal column consist typically of a centrum bearing a neural arch above and a haemal arch below. The neural arch surrounds the spinal cord and is surmounted by a neural spine. In the fishes the haemal arch encloses the dorsal aorta and in the region of the body cavity open haemal arches form ribs known as fish ribs. In other classes and in a few fishes these ribs are replaced by similar slender bones known as true ribs. They develop between the segments of the body lateral to the fish ribs and above them, and articulate with processes of the vertebrae. These ribs join a median ventral structure, the sternum or breast bone, composed of bones and cartilages.

The vertebrae differ in various regions of the body. The first, with which the skull articulates, is the atlas. The second, the axis, is noteworthy for its odontoid process, a solid anterior extension of the centrum. This process is the centrum of the atlas. These two and the remainder of the series in the neck are called cervical vertebrae. The following series with which the ribs articulate are the thoracic vertebrae. The lumbar vertebrae extend to the articulation of the pelvic girdle, where one vertebra or a fused series constitute the sacrum. The following caudal vertebrae lie in the tail or, in the apes and man, are fused to form a mass called the os coccyx.

The pectoral and pelvic girdles are composed of three pairs of bones, two passing downward and toward the median line of the body from the articulations of the appendages and one upward. Of the more constant bones in the pectoral girdle, also called the shoulder girdle, the upper bone is the scapula (shoulder blade),

the anterior of the lower bones is the clavicle (collar bone), and the posterior is the coracoid. These two pairs attach to the sternum. They are supplemented in amphibia and reptiles by other bones. The bones of each half of the pelvic girdle are the dorsal ilium, the anterior pubis, and the posterior ischium. This girdle in many forms is firmly attached to the sacrum.

The girdles are modified in various ways. The principal tendency of the pectoral girdle is toward simplification by the loss of bones, so that in many species only the scapula and clavicle remain, as in man, the coracoid being reduced to a process on the scapula. In the hoofed mammals only the scapula persists. In the moles, however, the girdle is large and strong. The pelvic girdle, in correlation with the stresses that it bears in locomotion, becomes compact. The bones of each side tend to fuse and the pubic symphysis, at the ventral union of the two halves, is very firm. The name pelvis or pelvic bone is often applied to this composite structure.

The primitive appendages are the paired fins of the fishes. These structures are precursors of the **penta-dactyl appendage** of terrestrial vertebrates. From this basic plan the structure of appendages in various groups has come by modification of the proportions and relations of the bones, by loss of digits, and to a limited extent by the fusion of separate bones. Both loss and fusion have occurred in the wings of birds, and maximum loss is found in the legs of the horse and allied species, where a single functional digit remains.

The appendages also contain sesamoid bones, small rounded bones developed in the tendons spanning movable joints. The most constant of these bones is the patella or knee cap.

The visceral skeleton in its primitive condition consists of a series of small bones and cartilages supporting the branchial arches between the gill slits. In the existing vertebrates the embryonic primordia of the hinged jaws are derived from these arches but the visceral skeleton is otherwise greatly reduced except in the fishes. It persists in part as the small bones (ossicles) of the middle **ear**, the hyoid bones supporting the tongue, and the cartilages of the **larynx** and upper part of the **trachea**. (A.W.L.)

SKELETON DIAGRAM. A skeleton or line diagram is a graphical representation of the members of a machine or structure. A simplified diagram of a machine or of some static structure, can be as useful for certain purposes as a complete drawing. In fact, a

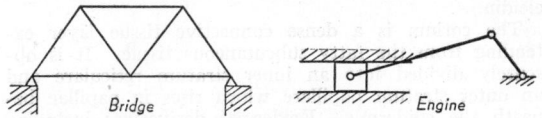

Skeleton diagrams.

skeleton diagram is often more useful than a drawing of the actual equipment it represents, because of the elimination of unnecessary detail which might be confusing. Generally speaking, a skeleton for a static structure is a series of lines which might be thought of as being the center lines of the different structural members. In the case of a machine, a skeleton diagram could be one which would represent the machine kinematically (see **Mechanics**). Such diagrams are useful for study of motions and analysis of stresses. (F.T.M.)

SKEW BRIDGE. Bridge.

SKEW CURVES. Curves in Space.

SKEWNESS. Skewness is that property of a distribution involving its symmetry or asymmetry. As a measure of skewness we define $\alpha_{3:x} = \dfrac{\mu_{3:x}}{\sigma_x^3}$, $\mu_{3:x} = 3rd$

central **moment**, σ_x = **standard deviation.** If the distribution is symmetrical $\alpha_{3:x} = 0$ (Fig. 1); if it possesses a tail to the right $\alpha_{3:x} > 0$ and the distribution is said to

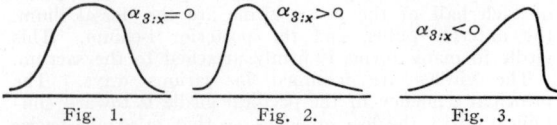

| Fig. 1. | Fig. 2. | Fig. 3. |

be positively skewed (Fig. 2); if it possesses a tail to the left, $\alpha_{3:x} < 0$ and the distribution is said to be negatively skewed (Fig. 3).

The J-shaped distributions are always skewed, and generally the bell-shaped curves are skewed. If in **Pearson's Type III** distribution, $\alpha_{3:x} > 0$, then more **variates** are below the **mean** than above the mean, but the variates above the mean are on the average farther from the mean than the variates below the mean. This is often the case with other skewed distributions. (L.A.A.)

SKIMMER. Aves, Charadriiformes. A bird (**Aves**) related to the terns and gulls but distinguished by its peculiar beak. The entire beak is long and compressed and the lower mandible is much longer than the upper. The bird dips this lower mandible into the water as it flies. Of the few species only one, the black skimmer, *Rhynchops nigra,* is North American. (A.W.L.)

SKIN. The covering of the body of **vertebrates.** It consists of two parts, an outer epidermis and an inner dermis or corium. The former develops in the embryo from the outer **germ layer,** the ectoderm, and the latter from the middle germ layer, the mesoderm.

The epidermis is composed of many layers of cells in two principal strata, the stratum corneum and the stratum germinativum. The flattened cells next the surface are hardened by deposits of pareleidin, a substance related to keratin, and are said to be keratinized. They make up the stratum corneum. Below them are several layers of thicker cells whose active proliferation gives rise to the cells of the stratum corneum. These layers constitute the stratum germinativum. In the outer cells of this stratum granules of keratohyalin appear as forerunners of the pareleidin of the stratum corneum, forming the thin stratum granulosum. A thin clear zone just outside of the granular layer, known as the stratum lucidum, is regarded as the basal layer of the stratum corneum. In its cells the granules of keratohyalin become a diffuse intermediate substance, eleidin.

The corium is a dense connective tissue layer extending from the fatty subcutaneous tissue. It is obscurely divided into an inner stratum reticulare and an outer stratum papillare which rises in papillae beneath the epidermis. Epidermal derivatives including hair follicles, sweat glands, and sebaceous glands extending into the corium and it contains nerve endings, tactile corpuscles, and blood vessels. Smooth muscle fibers attached to the hair follicles lie in it and in some parts of the skin it contains other muscle fibers. The voluntary muscles of the face by which expression is controlled end in it. (A.W.L.)

SKIN EFFECT. The current flowing in an a-c conductor is not uniformly distributed over the cross-section of the conductor but becomes denser the farther the cross-section being considered is from the center of the conductor. This effect, known as skin effect, is caused by the varying inductance of the different parts of the conductor. If we consider a circular conductor, to be specific, and imagine it composed of many elemental filaments in parallel we can readily see why the current flow is more dense as the distance from the center becomes greater. The current in each filament sets up a flux around it which links both it and any other fila-

ments which may be within the lines of flux. It will be realized at once that the filaments in the center can be linked by many more lines than those on the outside, since the outer ones are beyond much of the flux of the inner ones. Thus the inductance of the inner filaments is greater than that of the outer ones and so the **reactance** of the inner ones is greater. This greater **impedance** of the inside causes more of the current to flow in the outer layers and gives rise to the skin effect. As the effective cross-section of the conductor is decreased by this, the **resistance** to a.c. is greater than the resistance to d.c. In certain cases, such as the induction **motor,** where the conductor is largely surrounded by iron, the skin effect is very much greater than in a conductor in free space. In this motor the effect is utilized to obtain high starting torque without serious impairment of the running characteristics. In the layout of a-c buses various arrangements such as hollow squares, channels, etc., are used to save material since the inner part would be of very little use. At radio frequencies the effect is very pronounced (it varies approximately as the square root of the frequency) so only a thin outer layer of the conductor is effective, thus greatly increasing the resistance. (See **High-Frequency Currents.**) (L.R.Q.)

SKINK. Reptilia, Sauria. A **lizard** of the family Scincidae. These animals vary greatly in structure but are usually small with short legs and long tails. The scales are underlaid with bony plates. The several species of the United States are widely distributed but are more common in the south. One of the most remarkable species of the family is the Australian stump-tailed lizard, a thick-bodied species with a very short blunt tail. (A.W.L.)

SKIP DISTANCE. As the frequency of a radio wave is increased the minimum angle of incidence at which the wave will be reflected from the **ionosphere** rather than pass on through becomes greater. This means that the higher the frequency the farther from the **transmitter** the reflected **sky wave** strikes the earth. This distance between the transmitter and the point closest to it at which the sky wave can be received is the skip distance. The **ground wave** is attenuated more rapidly the higher the frequency so that at high frequencies there may be a region in which the ground wave has become too weak for use and in which the sky wave cannot be received because of its skip. Above about 4 megacycles this effect becomes very noticeable, the dead region for higher frequencies running to a distance of a few hundred miles from the transmitter. (L.R.Q.)

SKIPJACK. Pisces, Teleostei. A common fresh-water fish, *Pomolobus chrysochloris,* of moderate size, related to the herring and shad. It is found throughout the Mississippi valley to the Gulf of Mexico and in Lake Erie and Michigan. The back is blue and the sides silvery. Of no value as a food fish. (A.W.L.)

SKIPPER. Insecta, Lepidoptera. An **insect** much like the butterflies but in many ways intermediate between the butterflies and moths. Most species are distinguished by their knobbed and hooked antennae. The few that lack the hook are less easily distinguished from the butterflies except by the veins of the wings or by other details of structure. Most of the skippers of the temperate zones are of moderate size and modest colors but many tropical species are brilliant. They constitute the superfamily Hesperioidea.

The name skipper refers to the vigorous and often erratic flight of these insects. Their small wings and powerful muscles accompany rapid flight. Many species perch readily and make short darting flights from place to place. (A.W.L.)

SKUA. Aves, Charadriiformes. *Catharcta.* Marine birds (**Aves**) related to the gulls. The few species are widely distributed in both hemispheres. (A.W.L.)

SKULL. Skeletal System.

SKUNK. Mammalia, Carnivora. Animals of the Americas, widely known for their foul odor. All of the species are black and white, with a long bushy tail under which are the two glands that produce the defensive secretion.

The skunks of North America have been divided into fifteen species and a number of varieties. These species are grouped as the striped or common skunks and the small spotted skunks, belonging respectively to the two genera, *Mephitis* and *Spilogale*. Both genera are widely distributed.

The fur, especially of the larger striped skunks, is commercially desirable. (A.W.L.)

SKUNK CABBAGE. Aroids.

SKY CONDITION. The state of the sky is also known as the sky condition. In terms of tenths of sky covered, airways' observers in the U.S. recognize four sky conditions.

1. Clear sky is less than $\frac{1}{10}$ cover of clouds.
2. Scattered clouds is $\frac{1}{10}$ to $\frac{5}{10}$ cover.
3. Broken clouds is more than $\frac{5}{10}$ but not more than $\frac{9}{10}$ cover.
4. Overcast is more than $\frac{9}{10}$ cover.

International practice and observations made for synoptic charts in North America recognize 10 states of the sky. They are indicated by code numbers.

0........	no clouds
1........	less than $\frac{1}{10}$
2........	$\frac{1}{10}$
3........	$\frac{2}{10}$ to $\frac{3}{10}$
4........	$\frac{4}{10}$ to $\frac{6}{10}$
5........	$\frac{7}{10}$ to $\frac{8}{10}$
6........	$\frac{9}{10}$
7........	more than $\frac{9}{10}$ but with openings
8........	$\frac{10}{10}$
9........	sky obscured by fog, dust, snow, etc.

Sky condition is also qualified by such remarks as "ugly, threatening sky," "water spout," "tornado to NW," "signs of tropical storm," etc. (P.E.K.)

SKY WAVE. The sky wave is that part of the energy radiated from an **antenna** which does not travel along the surface of the earth but goes upward. This wave does not suffer attenuation by the earth as does the **ground wave.** At the **ionosphere** it passes into the ionized regions and is refracted. If the frequency of the **radiation** is low enough the refraction becomes so great the wave is bent back towards the earth and thus the radio signals are returned and make possible reception great distances from the **transmitter.** As the frequency is increased the wave penetrates successive layers of the ionosphere and if high enough the wave will penetrate all layers and not return to the earth. While the returning sky wave gives reception at distances where the ground wave has died out, it does cause undesirable effects such as **fading,** interference with other signals, etc. (L.R.Q.)

SKYROCKET. Rockets.

SLAG. Slag is a fused product occurring in connection with metallurgical and **combustion** processes. It is composed of the **oxidized** impurities in a metal, and of a fluxing substance, and of ash. In the steel industry, slag is the neutralized product of anhydrous compounds entering into the process. Slag is of great im-portance to the operator of a steel **furnace** or a cupola, in that, through the slag, impurities are separated and removed from the metal. By floating as a molten covering on the pool of metal, slag protects it from oxidation and serves to keep it clean. By controlling the character of slag, and continuous observation, the metallurgist insures that the metal is of the quality desired.

Molten ash is one of the products of **combustion** of coal in certain high-capacity boiler furnaces. It is also called slag. In some plants, the ash is removed from the furnace in this fluid form. Such furnaces are known as slag tap furnaces. Slag has some commercial value as **ballast,** coarse **aggregate** for **concrete,** road metal, etc. (F.T.M.)

SLAKED LIME. Calcium (Hydroxide).

SLATE. A fine-grained homogeneous sedimentary rock composed of **clay** or volcanic ash which has been metamorphosed (foliated) so as to develop a high degree of fissility or **slaty cleavage** which is usually at a high angle to the planes of stratification. This high degree of fissility makes the better grades of slates an extremely useful roofing material which, however, has been somewhat replaced in recent years by synthetic and manufactured substitutes. The finest slates in the world come from Wales, Britain. (R.M.F.)

SLATY CLEAVAGE. That type of **foliation** in fine-grained metamorphosed sedimentary rocks which produces a high degree of fissility. (See **Slate.**) (R.M.F.)

SLEEPER. Dormouse.

SLEEPING SICKNESS. Trypanosomiasis.

SLEET. When raindrops enter a layer of intensely cold air, they become supercooled, that is, cooled below the freezing point but without freezing. In this state they are highly unstable, and upon coming in contact with any object, even a speck of dust, they suddenly freeze into white, spherical pellets of sleet. Larger objects such as twigs or telephone wires, when touched by these supercooled drops, receive a coating of ice, which may result in a landscape of marvelous beauty when the sun appears, but which often proves very destructive if the ice is heavy. Sleet is quite commonly mixed with snow or with rain; but it is always a cold-weather product, and is not, as many suppose, a small form of **hail.** (See **Hydrometeors.**) (L.D.W.)

SLEEVE COUPLING. Coupling.

SLEEVE VALVE. Otto Engine.

SLIDE RULE. The slide rule is an instrument for multiplying, dividing, extracting roots, and obtaining powers of numbers mechanically by logarithmic means. In construction, the simple slide rule consists of two adjacent logarithmic scales which may be so set that a reading on one is added to a reading on another. This represents the addition of logarithms when multiplying numbers. The slide rule for general use is found in several different forms, and there are many special slide rules such as stadia rules, electrical rules, hydraulic computing rules, etc.

The accompanying figure illustrates the principle of multiplication by the slide rule. Two logarithmic scales, A and B, are arranged on a slide rule so that they may be mechanically added. In the illustration given, the process of multiplying three by two is shown. The end of the B scale is aligned over the 3 on the A scale, so that to 3 on the A scale may be added 2 on the B scale. Now if these scales were uniformly divided, the result, of course, would be five units on the A scale, but since they are logarithmically divided, the result on the A scale is the product rather than the sum. Thus, under the 2 on the B scale, one reads the prod-

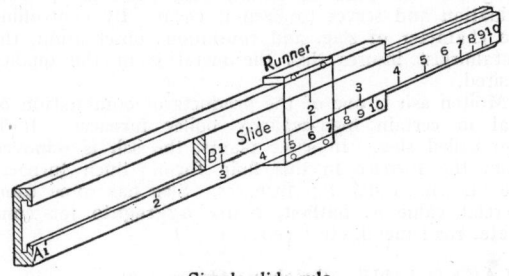

Simple slide **rule.**

uct of $3 \times 2 = 6$ on the A scale. A glass runner is provided so that the alignment of the numbers of the two scales may be facilitated. By halving the divisions, two complete scales could be placed on the rule. Using a full scale, such as A of the figure, and a double scale, one can extract square roots. A triple scale can be used for cube roots. Other scales contained in the ordinary slide rule are a uniform division scale for logarithms, and scales for reading the value of the trigonometric functions of an angle, known as the sine and tangent. (F.T.M.)

SLIDE TOOL. Turret Lathe.

SLIDER. Reptilia. Chelonia. **Turtles** related to the painted turtles and land tortoises. The several species are brownish or greenish, marked with yellow and in some cases also with red. They constitute the genus *Pseudemys.* Most of the species are confined to the southern states but one, known as the red-bellied terrapin, is found near coastal rivers from Florida to Cape Cod and another lives in the Mississippi valley as far north as Iowa. Both of these species are edible. Also called cooters. (A.W.L.)

SLIME MOLDS. Myxomycetes. Slime molds are **saprophytes** found growing on wood, on rocks, or on grass, or wherever there is sufficient moisture to prevent them from drying up. There are some 300 species.

Claimed and rejected by both botanists and zoölogists, this group of organisms occupies a doubtful position in either kingdom. They are simple organisms which have certain characteristics which place them in

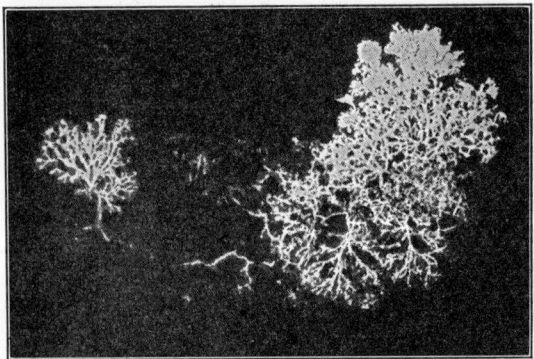

A slime-mold (Myxomycete), *Fuligo septica,* growing on the inner surface of a glass jar. About natural size. (*Gager, General Botany, P. Blakiston's Son & Co.*)

the plant kingdom and others which make them animals. Perhaps they are best treated as organisms which are in a separate classification from both plants and animals.

The vegetative phase of the life history of a typical slime mold starts with the germination of the **spores.** The spores are unicellular non-motile bodies. When sufficient moisture is available the wall of the spore bursts and the **protoplast** escapes. It becomes a uninucleate bit of naked **protoplasm** which moves about in an amoeboid (see **Amoeba**) manner. Soon it forms a single **cilium** at one end and by means of this moves more rapidly in the water. It is now known as a swarm spore. After a time the cilium is lost and it becomes once more like an amoeba, crawling about by forming extensions of its protoplasm, called pseudopodia, and "flowing" into them. In this stage it is called a myxamoeba. Both swarm spores and myxamoebae can increase by simple division.

After a time the myxamoebae move together and coalesce in pairs, forming **zygotes.** Coalescence continues until considerable masses are formed. In these nuclear divisions occur repeatedly, forming a multinucleate mass of naked protoplasm which is not separated into cells. This mass of protoplasm is usually colorless, but sometimes white or yellow or brownish. It has a consistency very similar to that of raw egg white. It is called a plasmodium. The plasmodium is a formless mass which creeps or flows about over the substratum.

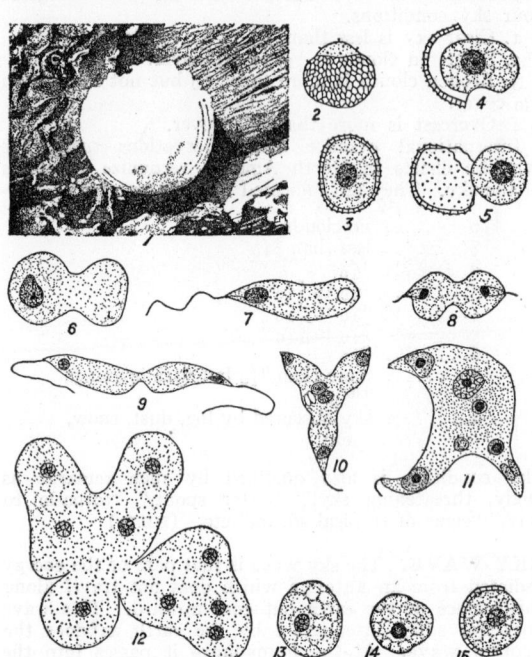

Reticularia Lycoperdon. 1, mature fruiting body on a log; 2, spore, external view; 3, spore, internal view; 4, spore germinating; 5, protoplast which has escaped from spore wall; 6, simple myxamoeba; 7, zoospore; 8, dividing zoospore; 9, fusion of two gametes; 10, 11, three zygotes uniting to form a plasmodium; 12, portion of protoplasm dividing by cleavage to form spores; 13, young spore before wall is formed; 14, spore with young wall; 15, nearly mature spore. (*Malcolm Wilson, Transactions of the Royal Society of Edinburgh, Volume 55.*)

This movement is not a slow regular progress but a series of surges. As it moves about, the plasmodium engulfs food substances such as bacteria, spores of fungi and other organic substances, which are digested in the plasmodium. In some slime molds the plasmodium is replaced by a pseudo-plasmodium consisting of many myxamoebae.

In time the plasmodium, or pseudo-plasmodium, is ready to reproduce. At this time it changes its reactions and turns toward the light. It heaps itself up in certain places. These heaps become the **sporangia.** In some species a long slender stalk is formed by the plasmodium and at the top of the stalk the sporangium is formed. A membrane is formed around the heaped-up mass of protoplasm. The protoplasm and wall form a sporangium. Within this wall the protoplasm becomes separated into numerous small units, each containing a single nucleus. Spores result when walls are formed around each bit of protoplasm. Dispersed among the spores are many slender threads called elaters which may be simple or branched in different species. The aggregate of elaters is called a capillitium. They give strength to the sporangium and also because of their reaction to changes in humidity, which causes them alternately to coil and straighten out, they aid in loosening the spores and pushing them from the sporangium. The spores are capable of resisting prolonged

Fruiting bodies of different kinds of Myxomycetes. (*Kerner's Natural History of Plants, Blackie & Son.*)

desiccation, as well as low and high temperatures. Eventually the spores germinate and the life cycle is repeated.

The sporangia of slime molds show a variety of forms. In some species they are small brown balls appearing singly or in clumps on rotten wood. In other species they are long-stalked structures of great beauty and delicacy. Some resemble small mazes of irregular shape. A few suggest spots of thick brown grease deposited on leaves and stems. The size of the sporangia varies from coarse fusiform bodies $1\frac{1}{2}$–$2''$ long and $1''$ thick, to minute objects scarcely a mm. in diameter.

Slime molds are of no importance to man. (R.M.W., B.S.M.)

SLING PSYCHROMETER. Meteorological Instruments.

SLIP. For the meaning of slip in geology, see **Fault**.

Slip is a characteristic of the induction **motor**. Alternating **current** is used in the field windings of an a-c motor to produce a revolving field. The speed of this revolving field is the synchronous speed, and depends on the frequency of the system. The rotor of the induction type of a-c motor revolves at a speed less than synchronous. The percentage by which the rotor speed falls below synchronous speed is called the slip, and this varies from practically zero at no-load up to maximum at the stalling torque.

The slip of a propeller is the difference between its actual forward motion and that which it would have if it were revolving in a solid medium analogous to a bolt in a nut. The slip of water propellers is less than that of air propellers, because of the relative densities of the two media in which these propellers rotate.

The slip of a **pump** arises in the following way: A

positive displacement type **pump**, such as a piston or plunger pump, should, theoretically, discharge a volume of liquid equal to the piston displacement. Leaky valves, defective piston rings, and other faults, can cause the delivery to be less than the piston displacement. This difference, stated as a percentage, is called slip, and may have a value of about 5% when the pump is in good condition.

In the early stages of the plastic deformation of **metals and alloys** the internal deformation of individual crystals takes place by slip along planes of atoms. The total deformation of a given crystal is the sum of many small lateral displacements in parallel crystallographic planes of a given family. Each slip plane becomes more resistant to further deformation than the remaining potential slip planes; in fact, the crystal as a whole is strengthened by deformation in a manner which has not been adequately explained, thus the theory of hardening metals by **cold working** is not fully developed. (See also **Metallography**.) (F.T.M., R.H.H.)

SLIP FUNCTION. Propeller, Air, Characteristics.

SLIP RINGS. These are conducting rings attached to a rotating part of an electrical machine to make connection through **brushes** with the stationary part of the circuit. They are used where it is not necessary to commutate the current being conducted (see **Commutation**). (L.R.Q.)

SLOPE OF A CURVE. Derivative of a Function of One Variable.

SLOPE OF A LINE. The inclination of a straight line is the least positive angle α from the X-axis to the given line.

The slope of a straight line is the tangent of the inclination angle α of the line: $m = \tan \alpha$.

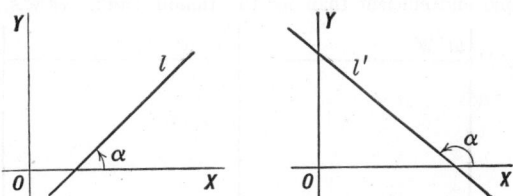

The slope of a horizontal line is o, but a vertical line has no slope.

If a line rises from left to right, its slope is positive, but if the line falls from left to right, its slope is negative.

The slope of the line through the points (x_1, y_1) and (x_2, y_2) is given by the formula

$$m = \frac{y_2 - y_1}{x_2 - x_1}$$

when **rectangular coordinates** are used.

If two lines are parallel, their slopes are equal. If the lines are perpendicular to each other, their slopes are negative **reciprocals** of each other. (L.L.S.)

SLOPE STAKE. Earthwork.

SLOT, AIRFOIL. High Lift Devices.

SLOTH. Mammalia, Edentata. Animals of Central and South America, highly specialized for arboreal life. Their claws are large hooks by which they suspend themselves from branches, and they are so thoroughly adapted to this inverted position that the hair runs from belly to back, opposite to its direction in animals of the usual position. They eat the foliage of cecropia trees by preference. Apparently in coordination with the total lack of competition and risk in their lives, these strange creatures are extremely dull-witted.

The sloths are of two forms: 2-toed and 3-toed, each including several species. The 2-toed sloths, *Choloepus,* are also called unaus, and the native name for the 3-toed species, *Bradypus,* is ai. (A.W.L.)

SLOTTER. Planer.

SLOW-WORM. Blind-Worm.

SLUDGE. When fresh sewage is admitted to settling tanks a certain amount of the solid matter in suspension will settle out, 50% more or less for sedimentation periods of an hour and a half or so. This collection of solids is known as fresh sludge. Such sludge will become actively putrescent in a short time and in modern treatment plants must be passed on from the sedimentation tank before this stage is reached. This may be done in two common ways. The fresh sludge may be passed through the slot in an **Imhoff Tank** to the lower story or digestion chamber. Here decomposition by anaerobic bacteria takes place with considerable liquefaction and reduction in volume. After the decomposition process has run its course (in 6–9 months) the resulting sludge is called "digested" sludge and is relatively inoffensive in character. It may be disposed of by drying on sludge drying beds and spreading on the land. It has little, if any, fertilizing value, being in the nature of humus. The sludge digestion chamber is operated on a periodic schedule of sludge withdrawals.

Alternatively, plain sedimentation basins with mechanical equipment for continuous collection of the fresh sludge may be used. The fresh sludge, so collected, is discharged into separate sludge digestion tanks which operate on the principle of the lower story of the Imhoff Tank except that by means of higher and better temperature control the digestion cycle is much more rapid and efficient than for the Imhoff Tank. (E.W.S.)

SLUG. Mollusca, Gasteropoda. **Mollusks** of the family Limacidae, related to the land snails but without an external shell. The name also appears in the term sea slug, applied to members of the marine group, Nudibranchiata, which are also without external shells. (A.W.L.)

SLUICE. A channel through which water issues is, in some cases, called a sluice. A sluice may be a pressure conduit, or it may be an open flume. For instance, the word may be applied to an artificial channel to lead water from one point to another, especially a temporary wooden channel. Sluices are often incorporated in dam structures. A sluice through a **dam** is a conduit cast in the concrete and equipped with controls called sluice gates. The purpose of the sluice is to empty the reservoir if necessary, to control the water level, and to aid in passing floods. (F.T.M.)

SLURRY. Filtration.

SMALL-AREA PLOTTING SHEET. A small-area plotting sheet is an approximation to a **mercator** plotting sheet, and may be used for solving problems in **dead reckoning** and for plotting **lines of position.**

In constructing the **graticule** for the sheet, the fundamental assumption is made that, within the limit of errors inherent in the problems to be solved, the distance between successive parallels of **latitude,** separated by 1°, is constant all over the sheet. This is equivalent to assuming that the same scale of distance may be used all over the sheet. The ratio between the linear distance, X, between successive **meridians** and the linear distance, Y, between successive **parallels,** is that of **middle-latitude sailing:** $X = Y \cos L_m$, in which L_m is the average latitude of the region covered by the sheet.

Two types of small-area plotting sheet are in common use: (1) the fixed meridian type, and (2) the fixed parallel type. The method for completing the graticules and plotting a point is shown in Figs. 1 and 2 re-

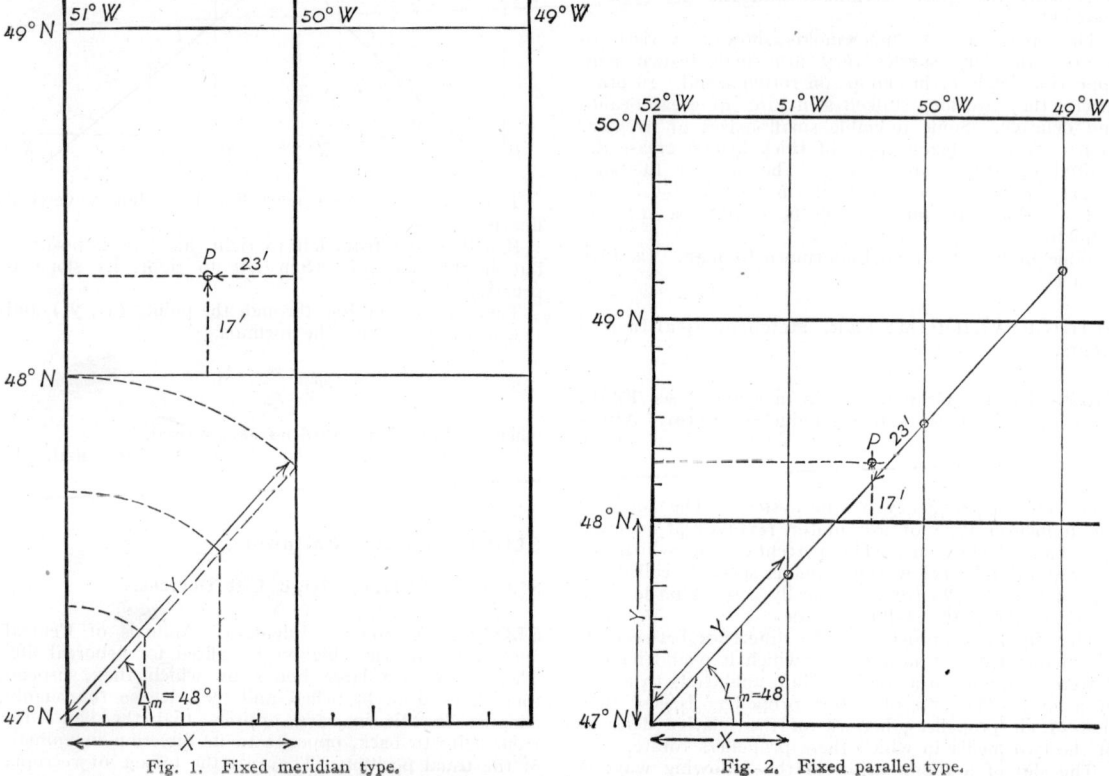

Fig. 1. Fixed meridian type.

Fig. 2. Fixed parallel type.

spectively. In the figures the heavy lines are those printed on the published forms, and the light lines are those drawn to complete the graticule. Dotted lines in the figure are construction lines which need not appear on the completed sheet. The sheets in the figure are completed for $L_m = 48°$ and the point, P, is in latitude 48° 17′N and longitude 50° 23′ W.

In the older or fixed meridian type, Fig. 1, a diagonal line is drawn making an angle L_m with the base line. The distance, X, is that between the fixed meridians, and the distance, Y, is the length of the diagonal, measured between the meridians. Elementary trigonometry shows that $X = Y \cos L_m$. This distance, Y, is then transferred to one of the fixed meridians and becomes 1° of latitude for the small-area plotting sheet. This length, Y, is also equivalent to 60 nautical miles and is used as a scale of distance all over the sheet. The central parallel on the sheet is that of L_m. The central meridian is that of the middle of the region covered by the particular problem to be solved. With this type of sheet the longitude scale is constant, no matter for what region the sheet is drawn. The latitude spacing, and hence the scale of distance, is different for different sheets.

In the more modern, and far more frequently used, fixed parallel type, Fig. 2, the diagonal line making an angle of L_m with the base, is drawn. Distances equal to that between the fixed parallels are laid off along this diagonal and lines are drawn through these points parallel to the latitude scale of the sheet. These become the successive meridians. It will be seen that, as above, $X = Y \cos L_m$. However, on this style of sheet the distance Y is fixed, no matter for what region the sheet is prepared, and a scale for distance ruled on celluloid, or other permanent material, may be used on all sheets. Furthermore, this same scale may be used for measurement of longitude by placing it along the diagonal and then projecting the desired value, parallel to the latitude scale, to the proper latitude.

Plotting sheets of both types and for various scales are published by the U. S. Hydrographic Office, and by several publishing firms. Most of the forms have a compass rose at the center of the sheet so that parallel rulers and dividers may be used, instead of protractor and scale, if the user prefers. However, printed forms are by no means necessary, for the graticule can quickly be drawn on blank paper with the aid of a protractor and scale. Using a blank sheet requires about one minute to obtain the completed graticule, while with the printed forms this time is cut in half.

Either type sheet can be used, without introducing errors greater than those inherent in the problems for which they are designed, for any mid-latitude between the equator and 60° and for areas of about 300 miles sq. For low latitudes the area can be correspondingly increased. (W.K.G.)

SMALL ERRORS. Differentials.

SMALLPOX. (Variola). An acute contagious disease caused by a filtrable **virus.** The disease is characterized by severe constitutional symptoms and a distinctive skin eruption which leaves scarring upon healing. Smallpox has been known since antiquity and lesions have been seen on the skin of an Egyptian mummy of the 20th dynasty. The disease originally spread from India and Central Asia to Europe and became especially prevalent at the time of the Crusades in the 11th century. It was first introduced in America by a Negro slave of Cortez in 1520, producing an epidemic that killed several million people. One century later it appeared in New England. It was first called smallpox in Europe, to differentiate it from grand-pox (**syphilis** of the skin). The discovery of vaccination by Jenner in 1798 is one of the landmarks in the development of medical science.

Since only a small percentage of people have natural immunity, the incidence of smallpox varies with the per cent of the population vaccinated. In the United States, where vaccination is not compulsory, the incidence is relatively high. Permanent immunity usually develops after an attack of smallpox. Epidemics vary greatly in severity. In recent years the disease has been of a mild type.

The incubation period averages between 8–12 days in the majority of cases. The onset of the disease is usually sudden and may begin with a chill. The temperature soon becomes high and headache, generalized pain and vomiting are frequently all present. The patient is prostrated. Delirium, convulsions, and coma occur, primarily in children.

As the skin eruption appears the symptoms and fever abate and the patient feels more comfortable. The eruption first appears as reddish spots which become elevated. After a few days the elevated area is capped by a small blister, which grows bigger and on the 7th or 8th day contains pus. During this period, secondary fever develops, gradually disappearing as healing and desquamation, or peeling, progress. This begins around the 12th or 14th day.

Black or hemorrhagic smallpox is a virulent fulminating type of the disease which invariably causes death in 2–6 days.

Smallpox occurring in those who have been vaccinated is called varioloid, and is usually a mild form.

The deaths in smallpox are often due to complications which are usually infections following secondary invasion by **staphylococci** and **streptococci.** The most common complications are **abscesses, pneumonia, septicemia, nephritis,** and ulcers on the cornea of the eye.

There is no specific treatment for smallpox, and the general treatment is similar to that of any severe acute infection. (R.S.M., D.M.H.)

SMALTITE-CHLOANTHITE. The mineral smaltite-chloanthite is an **isomorphous** mixture of **cobalt diarsenide**, $CoAs_2$ (smaltite), and **nickel** diarsenide, $NiAs_2$, (chloanthite), the proportions varying widely. It occurs in tin-white **isometric** crystals, but is also found massive. It is brittle; hardness, 5.5–6; specific gravity, 6.4–6.6; luster, metallic; color, tin-white to steel gray; streak, nearly black; opaque. It is associated with other nickel and cobalt minerals as well as with **silver** and **copper** ores. It is found in Bohemia, Saxony, Baden, Alsace, Spain, England and New South Wales. In the United States it has been found in Connecticut, New Jersey; and in Canada at Cobalt, Province of Ontario. It is used as an ore of both metals. (E.S.C.S.)

SMEAR METAL. The heat generated by metal cutting is often very great, and may cause a decided change in the nature of the metal adjacent to the cut surface. Under the influence of heat, some of the loose metal particles, and the serrated edges caused by metal cutting, may be welded or combined into an amorphous metal substance known as smear metal. Smear metal has little or no resistance to wear and must therefore be removed from surfaces that are subject to heavy pressures, particularly in bearings. Such processes as **lapping** and **superfinishing** are usually employed for this purpose. (H.C.H.)

SMELL. A chemical sense by which the animal perceives substances in the gaseous state and sometimes in exceedingly small quantities. It is valuable to the individual because the rapid transmission of many volatile substances over long distances permits the recognition of foods or enemies prior to close meeting with them. The sense is resident in **olfactory organs.** (A.W.L.)

SMELT. Pisces, Teleostei. The small fishes (**Pisces**) of the family Osmeridae, including the **candlefish** and the common smelt, *Osmerus eperalnus,* of the Atlantic. Both of these species are marine, ascending the streams to spawn, and both are important food fishes. (A.W.L.)

SMELTING. The process of heating ores to a high temperature in the presence of a **reducing** agent, such as **carbon** (coke), and of a **fluxing** agent to remove the accompanying rock gangue is termed smelting. Iron ore is the most abundantly smelted ore. It contains about 20% gangue (clay and sand). The ore is heated in an air **blast furnace** with coke and **limestone** (fluxing agent) at a temperature above the melting point of iron and **slag** (fusion mixture of impurities and flux). The molten iron (the more dense material) and molten slag (the less dense material) are removed separately from the furnace. (R.K.S.)

SMEW. Aves, Anseriformes. A European name for a small **merganser,** *Mergellus albellus.* Sometimes also applied to widgeons and pochards, and said to be used for the pintail duck, although pintail is by far the more common name. (A.W.L.)

SMILAX. Asparagus; Sarsaparilla.

SMILODON. Pleistocene.

SMITHSONITE. Smithsonite is zinc carbonate, $ZnCO_3$, a **hexagonal** mineral with a **rhombohedral** cleavage. It is a brittle mineral; hardness, 5; specific gravity, 4.3–4.5; luster, vitreous to dull; color, usually white, but may be colored yellowish or brownish or perhaps blue or green due to impurities. It is translucent to opaque. Smithsonite is a secondary mineral after **sphalerite** or may replace **limestone** or **dolomite.** It is sometimes called calamine (but true calamine is a **zinc silicate**) and often associated with it. Smithsonite occurs in Siberia, Greece, Rumania, Austria, Sardinia, Cumberland and Derbyshire, England; New South Wales, South West Africa, and Mexico. In the United States it is found in Pennsylvania, Wisconsin, Missouri, Arkansas, and Utah. This mineral was named in honor of James Smithson, whose generous legacy founded the Smithsonian Institution at Washington, D. C. (E.S.C.S.)

SMOKE. Smoke is the colored product of incomplete combustion, consisting chiefly of particles of unburned carbon. The discharge of smoke to the atmosphere can be considered nothing but an evil—and an unnecessary evil at that. The evil effects of smoke may be considered under three headings, viz., (1) effect on health, (2) financial loss due to incomplete combustion, deleterious effect on plant growth, and begriming of buildings, (3) effect on standard of living.

The smoke nuisance is at its worst, of course, in metropolitan districts. The smoke is produced from both industrial and domestic fires. Smoke abatement workers have found it much easier to render the former class smokeless than the latter.

There has been no definite coordination discovered between diseases of the respiratory tract and smoke density; however, common sense would say that a smoky atmosphere was bound to be less healthful than air alone. Chronic sinus and nose troubles seem to be prevalent in extremely smoky atmospheres. It is a well-known fact that the effect of smoke is deadly to vegetation. One unit of sulfur in the coal gives about three units of **sulfuric acid,** a substance most poisonous to plant life; indeed, the sulfur dioxide content of flue gas has caused more than one large lawsuit to be brought against central station operators. Sulfuric acid plus rain has a detrimental effect upon the limestone of buildings. Smoke corrodes metals, darkens paints, and in many other ways

creates a tremendous economic loss other than that due to loss of heating value of fuel. It is well known by those who have lived in smoky cities that a much lower standard of cleanliness is prevalent. Neither building interiors nor exteriors, clothing, hangings, furniture, etc., can be kept clean.

One of the oldest methods of gauging the smoke emitted by a chimney involves use of the Ringleman smoke comparison chart.

The Ringleman chart is composed of four sets of gratings. When these are placed about 25 ft. from the observer the gratings merge to a solid color ranging from light gray to a dense black. The chart enables a smoke inspector to rate the character of smoke emission from a chimney by comparison. Smoke ordinances are often based on the Ringleman chart, as, for instance, prohibiting the emission of smoke to exceed "No. 2 Ringleman" more than a certain period of time, say, one minute.

Smoke and fog often combine to form a "smog" or fog-smoke mixture. Concentration of smoke particles depends mainly on three factors: 1. The intensity of output at the source. 2. The depth to which smoke is distributed. 3. The strength of winds which carry smoke from the source. Of these three, the second has the greatest bearing on visibility. When air is stable, vertical currents are suppressed; smoke particles, therefore, remain near the ground. Ground inversions preclude any vertical distribution of smoke and these are very common during early morning hours. As the sun rises, the ground inversion or stability is destroyed and air is mixed to a greater vertical depth; smoke is thinned in proportion to the depth of mixing, and visibility improves. On overcast days when stable air lies adjacent to the ground all day, smoke does not thin, and visibility remains considerably reduced. In extreme cases twilight dimness may prevail throughout the day. (F.T.M., P.E.K.)

SMOOTH-HEAD. Pisces, Teleostei. A fish (**Pisces**) related to the salmons. The head is entirely without scales and the median fins, both dorsal and ventral, are located just before the tail. All of the species live in the deep sea. (A.W.L.)

SMUTS. Ustilaginales. Smuts are parasitic **Basidiomycetes** which are so named because of the conspicuous masses of sooty black spores which they form externally on the host plant. Infection by smuts is seldom fatal to the host plant but does seriously reduce its size and may even prevent seed formation completely. The fungus grows as a septate **mycelium** which penetrates between the cells of the host plant; into these cells it sends **haustoria,** which obtain nourishment therefrom. Presently this septate mycelium gives rise to immense numbers of **spores.** Often the presence of the mycelium causes the host tissue to enlarge tremendously, producing irregular tumor-like growths. These are particularly conspicuous in Corn Smut. The spores are thick-walled unicellular objects capable of surviving for some time under unfavorable conditions. On germinating, each spore develops a short germ tube, or promycelium, which becomes from 1–4 cells long. Each of these cells produces a spore. The promycelium becomes a basidium; and the spores, basidiospores. These spores are capable of infecting new host plants, producing therein a mycelium. Conjugation between cells of the mycelium occurs so that each cell comes to have two nuclei. As growth continues, the two nuclei of any cell divide simultaneously so that every cell continues to have two nuclei. When spore formation occurs, the two nuclei fuse, dividing again when the spore germinates. Many variations of this process are found; in many smuts the promycelium buds off from its apex many cells. Often fusion between two of these cells occurs immediately, even before they are separated from the promycelium.

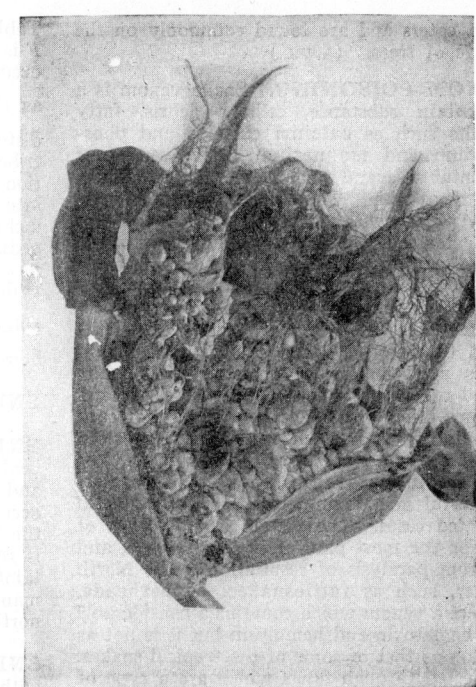

Corn smut. Left, smutted tassel; right, smutted ear.

Because of the rapidity with which smuts may spread and the great reduction of seed production which their presence may cause, smuts are of great economic importance. One species, Corn Smut, *Ustilago zeae,* causes the loss of millions of bushels of corn yearly. Oat smut, *Ustilago avenae,* may cause a 30% reduction in yield, while other smuts are equally important. Control of the parasites may be obtained by rotating crops; but due to the resistant nature of the spores, at least three years should elapse before replanting an infected field to the same crop. Other methods of control consist of soaking seeds in various solutions, such a formaldehyde (see **Aldehyde**) solution in water, or dusting infected seeds with copper compounds. (R.M.W.)

SNAIL. Mollusca, Gasteropoda. A name properly applied to most members of this class but usually given only to the small fresh-water and terrestrial species with

Snail.

coiled shells. The term does not correspond with any part of the scientific classification of mollusks beyond this general application.

Snails are eaten in several countries of Europe, where they are regarded as a delicacy. (A.W.L.)

SNAKE. Reptilia, Sauria. A slender elongate **reptile** with no trace of external appendages and only in a few species with vestiges of the appendicular skeleton. The snakes differ from the lizards also in the large scales of the ventral surface. They are widely distributed and many species are poisonous.

Among the large number of snakes that bear distinctive names are the adders, racers, **vipers** and **pit vipers**, constrictors, including the **python**, **boa**, and **anaconda**, **cobras**, **craits**, **asp**, **bushmaster**, **fer-de-lance** and many others. The poisonous North American species, aside from the relatively unimportant coral snake of the southeast, are the **copperhead**, the **moccasin**, and the **rattlesnakes**. (See also **Fossil Reptiles**.) (A.W.L.)

SNAKE FLY. Insecta, Neuroptera. A peculiar insect found only in the far West. It has four membranous wings and the body is prolonged at the anterior end.

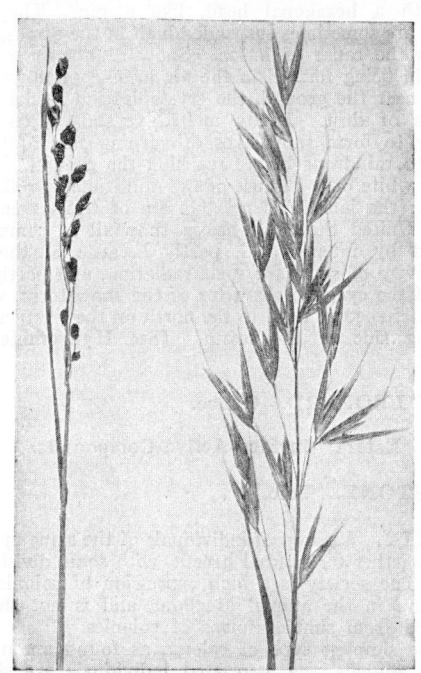

Smutted oat head. Left, smutted plant; right, healthy plant.

They are insect eaters and are found commonly on the bark and foliage of trees. (A.W.L.)

SNAKE VENOM POISONING. Snake venom is a mixture of **protein** substance, cellular debris, fatty matter, and salts such as **calcium** chloride and phosphate, **ammonium** and **magnesium**. The venom of most snakes contains several toxic elements, including a hemotoxin and neurotoxin. The hemotoxin causes damage to the blood vessels with extravasation of blood into the tissues and extensive swelling and discoloration of the soft parts. The neurotoxin attacks nerve centers, causing **paralysis**. Other components are the anti-coagulating factor, which prevents blood **coagulation** and predisposes to hemorrhage; the hemolysin, which causes the red blood cells to dissolve; the agglutinin component which produces agglutination of both red and white blood cells; and other less well established cell poisons.

Various venoms differ in the relative amounts of each of these toxins and in their strength. In the American tropics the **fer-de-lance**, the *mapepire balsayn*, and the *barba amarilla* and the **jararaca** all have a venom that contains a powerful hemotoxin which may cause rapid death in untreated cases. The venom of the cobra of India consists for the most part of the neurotoxin, and death results from paralysis of respiration. The North American snakes, such as **rattlesnakes, copperheads, moccasins**, inject a venom which contains a considerable amount of the hematoxin and hemolysin but it is not as rapid in its action as that of some of the tropical snakes.

Venom is injected through fangs which are a type of highly specialized tooth, and resemble in construction a hypodermic needle.

The mortality from snake bite is very high. In India the annual mortality is around 20,000. Anti-venom serum has been prepared and is easily accessible for use in many countries. In Brazil the Institute at São Paulo, under the auspices of the Government, has made a study of the various venoms, and anti-venom serums are widely distributed. Thus the mortality from snake bites has been reduced from several thousand to less than 50. The annual mortality from this source in the United States is around 150.

In the treatment of snake bite, prompt measures designed to keep the venom from getting into the circulation offer the best method of attack. Proper treatment consists of application of a tourniquet or ligature about the site of the bite. Deep incisions at the site of the fang punctures with the use of some suction device will remove a good deal of venom. Venom is absorbed very rapidly and if several hours elapse before this procedure is carried out the results may be poor. Anti-venom serum should be injected as soon as possible to neutralize the poison. Transfusion and other supportive measures are also important. (R.S.M., D.M.H.)

SNAP GAUGE. Measurement.

SNAPDRAGON. Figwort.

SNEDECOR'S F DISTRIBUTION. Let $F = \frac{s_1^2}{s_2^2}$ where s_1^2 is the estimated **variance** with n_1 **degrees of freedom** and s_2^2 is the estimated variance with n_2 degrees of freedom when the **variates** are **normally distributed** with a common **population** variance of σ_P^2. The distribution of F is given by

$$P_F dF = \frac{n_1^{\frac{n_1}{2}} n_2^{\frac{n_2}{2}}}{B\left(\frac{n_1}{2}, \frac{n_2}{2}\right)} \frac{F^{\frac{n_1}{2}-1}}{(n_1 F + n_2)^{\frac{n_1+n_2}{2}}} dF, \; 0 \leqq F < \infty.$$

Let $\int_{F_\alpha(n_1, n_2)}^{\infty} P_F dF = \alpha$. The value of $F_\alpha(n_1, n_2)$ is given by tables. It should be noted $F_{1-\alpha}(n_1, n_2) = \frac{1}{F_\alpha(n_2, n_1)}$.

Tables for $F_\alpha(n_1, n_2)$ have been constructed. Snedecor's F is related to Fisher's z by $F = e^{2z}$. The F distribution occurs in the **analysis of variance**. We may interpret F as the quotient $\frac{n_2 \chi_1^2}{n_1 \chi_2^2}$, where χ_1^2 is distributed in a χ^2 distribution with n_1 **degrees of freedom** and χ_2^2 is distributed as χ^2 with n_2 degrees of freedom. The F distribution may be transformed into **Fisher's z distribution**, into an **incomplete Beta function**, and, as special cases, includes the **normal probability function, Student's t distribution** and the χ^2 distribution. In the limit as n_1 and n_2 approach infinity in any manner whatever, the F distribution approaches the **normal curve** with mean 1 and variance $2\left(\frac{1}{n_1} + \frac{1}{n_2}\right)$. The approach to normality, however, is very slow. (L.A.A.)

SNELL'S LAW. Refraction.

SNIPE. Aves, Charadriiformes. Long-legged and long-beaked wading birds (**Aves**) related to the sandpipers and woodcock. They are found in marshy ground and occur on all continents. North America has one species, the Wilson snipe, *Capella delicata*, or jack snipe, which is regarded to a limited extent as a game bird. Certainly its erratic flight makes it a severe test of marksmanship. This species breeds from the northern states northward and migrates to South America. (A.W.L.)

SNIPE FLY. Insecta, Diptera. Two-winged **flies** with long legs and a conical abdomen. They belong to the family Rhagionidae, sometimes called Leptidae. Some of the species suck blood and in the western mountains are very annoying to human beings. They receive the name **deer fly** in the west, a term applied to a small horse fly in the east. (A.W.L.)

SNOW. Snow crystals when freshly formed are often almost perfect and exhibit an endless variety of detail. They are commonly flat, 6-sided polygons, stars, or spangles, often of very complicated and beautiful design, but always with the 60° and 120° angles characteristic of the hexagonal system. Sometimes they are needle-like, resembling miniature 6-sided lead pencils, or needle-like with a hexagonal head, like a pin. The finer spicules are sometimes suspended high in the atmosphere, and are the cause of **halos**. Snow differs from **frost** chiefly in being formed in the air instead of upon solid objects near the ground, the crystallization nuclei being particles of dust. Partly melted crystals often cling together to form snowflakes of varying size, and may melt into raindrops before reaching the ground. Snow appears white only because of the multitude of reflecting surfaces; the individual crystals are of transparent ice. In the United States, a heavy snowfall is commonly followed by intense cold, partly because of the low absorptivity of snow for solar radiation, and partly because of the cyclonic character of the snowstorm, which brings a change of wind to the north on the westward or following side of the storm. (See **Hydrometeors**.) (L.D.W.)

SNOW LEOPARD. Ounce.

SOAP. Esters; Surface-Active Compounds.

SOAPSTONE. Talc.

SOCIETY. A group of individuals of the same **species** living together for mutual benefit, with some division of labor. The society is a high expression of colonial organization in the animal kingdom, and is not sharply separated from simpler forms of colonies.

In the simplest type of colony, as found among the 1-celled animals, the associated individuals are similar and each is capable of complete existence in itself. In the same group a slight division of labor appears, accompanied by structural differentiation of the individuals for different tasks. This form of organization persists in the **coelenterates, bryozoans,** and **ascidians,** with varying degrees of structural continuity in the colony and varying degrees of differentiation and division of labor.

Among the more highly organized animals the possibility of association of individuals in a complex society involving division of labor is expressed only among the social insects and man. The insect society, or colony as it is often called, continues the associated principle of structural specialization of the individual for its particular duties, with the resulting castes exemplified in a simple form by the queen, drone, and worker honeybees. Among the **termites** and **ants** the differentiation is much more extreme and complex. To a moderate degree the **honey-bee** colony also shows specialization of behavior among the workers, which may engage in various activities within their powers according to the requirements of the colony at different times.

The human society differs from that of insects in the restriction of inherent fitness for special duties to less evident details of organization. While inherent fitness undoubtedly exists among men, they are structurally of approximately the same form and their specialization is largely a result of training. In other words, specialization of the individual in human society is conspicuously a specialization of behavior. Lack of structural specialization is compensated by the use of tools.

In all cases the society is an extension of the prevailing biological principle that biological units of any degree of complexity can be associated together as component parts of a larger coordinated unit. (A.W.L.)

SOCKET WRENCH. Wrench.

SODA.
Term applied to **sodium** compounds. Soda ash, sodium carbonate; washing soda, sal soda, sodium carbonate decahydrate; baking soda, sodium hydrogen carbonate; caustic soda, sodium hydroxide. Percentage of soda expressed in analyses of chemicals is for sodium oxide (Na_2O). (R.K.S.)

SODA NITRE.
The mineral soda nitre or Chile saltpeter is naturally occurring **sodium nitrate,** $NaNO_3$. Its **hexagonal** crystals are rare, this mineral usually being found in crystalline aggregates, crusts or masses. It is soft; hardness, 1.5–2; specific gravity, 2.24–2.29; vitreous luster; colorless or white to yellow or gray; transparent to opaque. Soda nitre is a most important mineral commercially, being used in the manufacture of nitric acid, other nitrates and fertilizers. The chief soda nitre deposits of the world are those found in the Atacama and Tarapaca deserts of northern Chile, although others exist in the Argentine and Bolivia. Some small deposits have been found in California, New Mexico and Nevada. The origin of these **nitrate** deposits is far from being well understood. They have been regarded as nitrates formed originally by oxidation of organic matter and subsequently leached out. Guano, the excrement of birds, might be the original source of the nitrates. Ground water and ancient marine deposits have been suggested as well as the possibility of derivation from nitric acid produced in the atmosphere during electrical storms. Some invesigators have felt that the **nitrates** may have come from volcanic sources. (E.S.C.S.)

SODALITE.
An isometric mineral, a **sodium aluminum silicate** containing **sodium chloride,** with the chemical composition $3NaAlSiO_4 \cdot NaCl$, **potassium** sometimes replacing a small amount of sodium. It is commonly found as **dodecahedrons** or simply massive. When observed sodalite has a dodecahedral **cleavage;** conchoidal to uneven fracture; brittle; hardness, 5.5–6; specific gravity, 2.14–2.30; luster, vitreous to greasy; color grayish to greenish or yellowish, may be white. It is often a beautiful blue and may sometimes be red. It is transparent to translucent; streak, white. Sodalite is found in **igneous rocks** of **nephelite-syenite** type which have been produced from soda rich **magmas.** It has also been found in the lavas of Vesuvius. Common minerals associated with it are **nephelite** and **cancrinite.** It occurs in the Ilmen Mts. of Russia; at Vesuvius and Monte Somma, Italy; in Norway and Greeland. In Canada, in British Columbia and in Ontario, beautiful blue sodalite is found; and in the United States similar material comes from Kennebec County, Maine. It derives its name from the fact of its soda content. Attempts have been made to use this mineral for ornamental purposes. (E.S.C.S.)

SODIUM.
Symbol: Na (natrium). Atomic number: 11. Atomic weight: 22.997. Density: 0.97. Hardness: 0.4. Melting point: 97.5° C. Boiling point: 880° C. No isotope, but of single atomic form: 23.

Sodium is a silvery-white metal, can be readily molded and cut by knife, oxidizes instantly on exposure to air, and reacts with water violently, yielding sodium hydroxide and hydrogen gas, consequently is preserved under kerosene, burns in air at a red heat with yellow flame. Discovered by Davy in 1807.

Sodium occurs as sodium chloride in the ocean (1.14% Na), in salt deposits (salt, **halite,** NaCl), e.g., in Michigan, New York, Lousiana, in Great Britain, in Germany, in salt lakes, e.g., the Dead Sea (3% Na), Great Salt Lake; in common rocks (average of the solid shell of the earth 2.75% Na) as sodium nitrate (Chile saltpeter, $NaNO_3$) in Chile; as sodium borate (**rasorite, kernite,** $Na_2B_4O_7 \cdot 4H_2O$, in California; **tinkal,** $Na_2B_4O_7 \cdot 10H_2O$, in Tibet); as sodium carbonate (Na_2CO_3) and sulfate (Na_2SO_4) in certain salt lake areas.

Sodium metal is obtained by **electrolysis** of fused sodium chloride or hydroxide out of contact with air. Its uses are limited in extent, but important in particular cases, as in the liberation of a metal from its chloride by reaction of sodium to form sodium chloride, and in certain reactions of organic chemistry.

Acetate: Sodium acetate ($NaC_2H_3O_2 \cdot 3H_2O$), white solid, soluble, formed (1) by reaction of sodium carbonate or hydroxide and **acetic acid,** and then evaporating, (2) by precipitation of **calcium** acetate solution and sodium carbonate solution, followed by filtration, and evaporation of the filtrate. Used (1) as a source of acetate, (2) in the dye industry. Reacts (1) with sulfuric acid, upon heating, yielding sodium hydrogen sulfate non-volatile and acetic acid volatile and condensable, (2) with sodium hydroxide solid, upon heating, yielding sodium carbonate and methane.

Alum: Soda alums are those alums such as **aluminum** sodium sulfate ($Al_2(SO_4)_3 \cdot Na_2SO_4 \cdot 24H_2O$) where sodium sulfate is used instead of the more common potassium or ammonium sulfate.

Aluminate: Sodium aluminate ($NaAlO_2$), white solid, (1) by reaction of **aluminum** hydroxide and sodium hydroxide solution, (2) by fusion of aluminum oxide and sodium carbonate, the solution reacts with carbon dioxide to form aluminum hydroxide. Used as a mordant, and in water purification.

Aluminosilicate: Sodium aluminosilicate is used as a water softener ("Permutite") for the removal of dissolved calcium compounds.

Amide: Sodamide, sodamine ($NaNH_2$), white solid, formed by reaction of sodium metal and dry **ammonia** gas at 350° C. Reacts with **carbon** upon heating, to form sodium cyanide, and with nitrous oxide to form sodium azide (NaN_3).

Arsenate: Sodium arsenate ($Na_3AsO_4 \cdot 12H_2O$), white solid, soluble, formed by oxidation of sodium arsenite. Used (1) as a source of arsenate, (2) as a mordant in dyeing and printing, (3) as an insecticide.

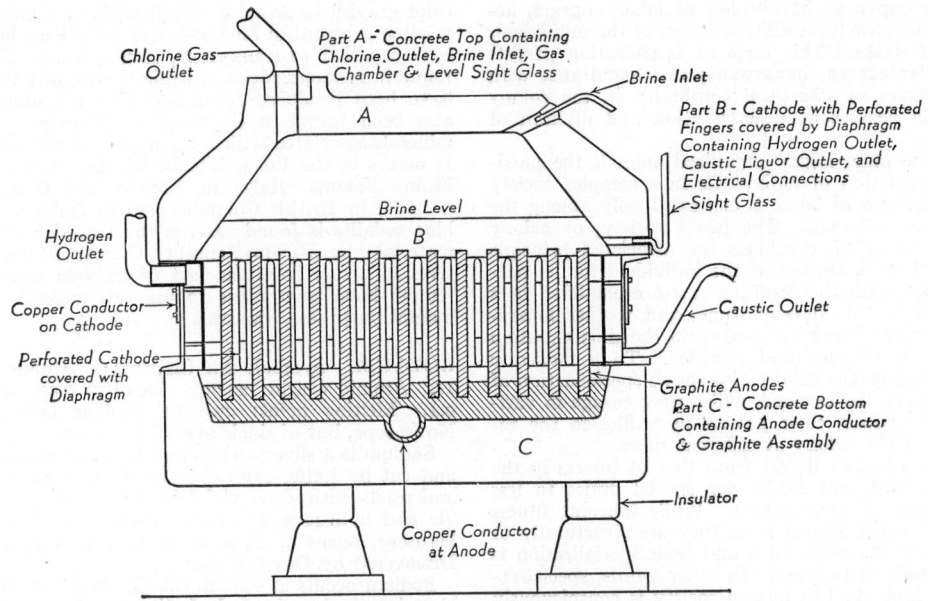

Cross-section of Hooker cell.

Labels in figure:
Chlorine Gas Outlet
Part A - Concrete Top Containing Chlorine Outlet, Brine Inlet, Gas Chamber & Level Sight Glass
Brine Inlet
Part B - Cathode with Perforated Fingers covered by Diaphragm Containing Hydrogen Outlet, Caustic Liquor Outlet, and Electrical Connections
Sight Glass
Brine Level
Hydrogen Outlet
Copper Conductor on Cathode
Caustic Outlet
Perforated Cathode covered with Diaphragm
Graphite Anodes
Part C - Concrete Bottom Containing Anode Conductor & Graphite Assembly
Insulator
Copper Conductor at Anode

Arsenite: Sodium arsenite (NaAsO₂), white solid, soluble, formed by reaction of **arsenic** trioxide and sodium hydroxide or carbonate solution, and boiling. Used (1) as an antiseptic, insecticide, weed-killer, hide preservative, and (2) in dyeing.

Benzoate: Sodium benzoate (NaC₇H₅O₂), white solid, soluble, formed by reaction of **benzoic acid** and sodium carbonate solution, and then evaporating. Used as a food preservative to a limited extent, an antiseptic, in pharmacy, and in dyeing.

Borate: Sodium borate, sodium tetraborate, borax (Na₂B₄O₇·10H₂O), white solid, soluble, formed (1) by reaction of sodium carbonate or hydroxide and **boric acid,** (2) by reaction of sodium carbonate and **calcium** borate (**colemanite**), followed by filtration, and then evaporating the filtrate, or (3) by use or purification of the mineral. Used as a flux in soldering, in ceramics, in textiles and tanning, in laundering and soaps, as a food preservative to a limited extent, and in medicine.

Bromide: Sodium bromide (NaBr), white solid, soluble, melting point 755° C. Used in photography and in medicine.

Carbonates: Sodium carbonate (anhydrous), soda ash (Na₂CO₃), sodium carbonate decahydrate, washing soda, sal soda (Na₂CO₃·10H₂O), white solid, soluble, melting point 851° C., formed by heating sodium hydrogen carbonate, either dry or in solution. Commonly bought and sold in quantity on the basis of oxide (Na₂O) determined by analysis (58.5% Na₂O equivalent to 100.0% Na₂CO₃). Used (1) as a source of carbonate, (2) in laundering, dish-washing, cleansing, and soaps, (3) in water purification for the precipitation of dissolved **calcium** compounds, as for boiler water supplies, (4) in the preparation of sodium salts (a) in solution, (b) upon fusing, at high temperature, for the manufacture of salts, such as sodium silicate from silicon oxide, sodium chromate from chromite, sodium manganate from **manganese** dioxide; sodium hydrogen carbonate, sodium bicarbonate, sodium acid carbonate, baking soda (NaHCO₃), white solid, soluble, when heated, dry or in solution, yields sodium carbonate; formed (1) by precipitation of sodium chloride and **ammonium** hydrogen carbonate cold concentrated solutions, and then **filtering** (ammonia soda process), (2) by reaction of sodium hydroxide or carbonate solution and excess **carbon dioxide.** Used (1) as a leavening agent (with an acid material) for carbon dioxide in baking, (2) in **baking powder,** (3) in ef-

fervescent beverages (with an acid material), (4) in fire extinguishers (with **sulfuric acid**), (5) in medicine as an antiacid. Both sodium carbonate and sodium hydrogen carbonate are relatively cheap and satisfactory mild alkalis. (See **Glass.**)

Chlorate: Sodium chlorate, chlorate of soda (NaClO₃), white solid, soluble, melting point 260° C., powerful oxidizing agent and consequently a fire hazard with dry organic materials, such as clothes, and with sulfur; upon heating oxygen is liberated and the residue is sodium chloride; formed by electrolysis of sodium chloride solution under proper conditions. Used (1) as a weed-killer (above hazard), (2) in matches, and explosives, (3) in the textile and leather industries.

Chloride: Sodium chloride, common salt, rock salt, halite (NaCl), white solid, soluble, melting point 804° C. Source in nature is widely distributed. Formed by reaction of sodium carbonate or hydroxide and **hydrochloric acid.** Used (1) in foods for man and animals, (2) as the original source of most sodium-containing substances, and of chlorine and **hydrogen chloride** (therefore, of most chlorine-containing substances).

Chlorite: (See **Chlorous Acid**).

Chromate: Sodium chromate (Na₂CrO₄·10H₂O), yellow solid, soluble, formed by reaction of sodium carbonate and **chromite** at high temperatures in a current of air, and then extracting with water and evaporating the solution. Used (1) as a source of chromate, (2) in leather tanning, (3) in textile dyeing, (4) in inks.

Citrate: Sodium citrate (Na₃C₆H₅O₇·5½H₂O) white solid, soluble, formed (1) by reaction of sodium carbonate or hydroxide and **citric acid,** (2) by reaction of **calcium** citrate and sodium sulfate or carbonate solution, and then filtering and evaporating the filtrate. Used in soft drinks and in medicine.

Cyanide: Sodium cyanide (NaCN), white solid, soluble, very poisonous, formed (1) by reaction of sodamide and **carbon** at high temperature, (2) by reaction of calcium **cyanamide** and sodium chloride at high temperature, reacts in dilute solution in air with gold or silver to form soluble sodium gold or silver cyanide, and used for this purpose in the cyanide process for recovery of gold. The percentage of available cyanide is greater than in potassium cyanide previously used. Used as a source of cyanide, and for **hydrocyanic acid.**

Dichromate: Sodium dichromate (Na₂Cr₂O₇·2H₂O), red solid, soluble, powerful oxidizing agent, and consequently a fire hazard with dry carbonaceous materials Formed by acidifying sodium chromate solution, and then evaporating. Used (1) in matches and pyrotechnics, (2) in leather tanning and in the textile industry, (3) as a source of chromate, cheaper than potassium dichromate.

Dithionate: Sodium dithionate, "sodium hyposulfate" ($Na_2S_2O_6 \cdot 2H_2O$), white solid, soluble, formed from manganese dithionate solution and sodium carbonate solution, and then filtering and evaporating the filtrate.

Ferricyanide: Sodium ferricyanide, red prussiate of soda ($Na_3Fe(CN)_6 \cdot H_2O$), red solid, soluble, formed by reaction of sodium ferrocyanide solution and chlorine, and then evaporating.

Ferrocyanide: Sodium ferrocyanide, yellow prussiate of soda ($Na_4Fe(CN)_6 \cdot 12H_2O$), yellow solid, soluble, formed by treating "spent oxide" of coal gas works with calcium hydroxide to extract ferrous cyanide as soluble calcium ferrocyanide, and then treating with sodium carbonate, filtering, and evaporating the filtrate. Used (1) as a source of ferrocyanide, (2) in blue print paper, (3) in tanning, (4) in tempering steel.

Fluorides: Sodium fluoride (NaF), white solid, soluble, formed by reaction of sodium carbonate and hydrofluoric acid, and then evaporating. Used (1) as an antiseptic and antifermentative in alcohol distilleries, (2) as a food preservative, (3) as a poison for rats and roaches, (4) as a constituent of ceramic enamels and fluxes; sodium hydrogen fluoride, sodium difluoride, sodium acid fluoride ($NaHF_2$), white solid, soluble, formed by reaction of sodium carbonate and excess hydrofluoric acid, and then evaporating. Used (1) as an antiseptic, (2) for etching glass, (3) as a food preservative, (4) for preserving zoological specimens.

Fluosilicate: Sodium fluosilicate (Na_2SiF_6), white solid, very slightly soluble in cold water, formed by reaction of sodium carbonate and hydrofluosilicic acid. Used (1) in ceramic glazes and opal glass, (2) in laundering, (3) as an antiseptic.

Formate: Sodium formate ($NaCHO_2$), white solid, soluble, formed by reaction of sodium hydroxide and carbon monoxide under pressure at about 200° C. Used (1) as a source of formate and formic acid, (2) as a reducing agent in organic chemistry, (3) as a mordant in dyeing, (4) in medicine.

Hydride: Sodium hydride (NaH), white solid, reactive with water yielding hydrogen gas and sodium hydroxide solution, formed by reaction of sodium and hydrogen at about 360° C. Used as a powerful reducing agent.

Hydroxide: Sodium hydroxide, caustic soda, sodium hydrate, "lye" (NaOH), white solid, soluble, melting point 318° C., an important strong alkali, not as cheap as calcium oxide (a strong alkali) nor sodium carbonate (a mild alkali), but of wide use. Formed (1) by reaction of sodium carbonate and calcium hydroxide in water, and then separation of the solution and evaporation, (2) by electrolysis of sodium chloride solution under the proper conditions, and evaporation. Commonly bought and sold in quantity on the basis of oxide (Na_2O) determined by analysis (77.5% Na_2O equivalent to 100.0% NaOH). Used (1) in the manufacture of soap, rayon, paper ("soda process"), (2) in petroleum and vegetable oil refining, (3) in the rubber industry, in the textile and tanning industries, (4) in the preparation of sodium salts, (a) in solution, (b) upon fusion.

Hypochlorite: Sodium hypochlorite (NaOCl), commonly in solution by (1) electrolysis of sodium chloride solution under proper conditions, (2) reaction of calcium hypochlorite suspension in water and sodium carbonate solution, and then filtering. Used (1) as a bleaching agent for textiles and paper pulp, (2) as a disinfectant, especially for water, (3) as an oxidizing reagent.

Hypophosphite: Sodium hypophosphite ($NaH_2PO_2 \cdot H_2O$), white solid, soluble, formed (1) by reaction of hypophosphorous acid and sodium carbonate solution, and then evaporating, (2) by reaction of sodium hydrox-

ide solution and phosphorus on heating (poisonous phosphine gas evolved).

Hyposulfite: Sodium hyposulfite, sodium hydrosulfite (not sodium thiosulfate) ($Na_2S_2O_4$), white solid, soluble, formed by reaction of sodium hydrogen sulfite and zinc metal powder, and then precipitating sodium hyposulfite by sodium chloride in concentrated solution. Used as an important reducing agent in the textile industry, e.g., bleaching, color discharge.

Iodide: Sodium iodide (NaI), white solid, soluble, melting point 651° C., formed by reaction of sodium carbonate or hydroxide and hydriodic acid, and then evaporating. Used in photography, in medicine and as a source of iodide.

Manganate: Sodium manganate (Na_2MnO_4), green solid, soluble, permanent in alkali, formed by heating to high temperature manganese dioxide and sodium carbonate, and then extracting with water and evaporating the solution. The first step in the preparation of sodium permanganate from pyrolusite.

Nitrate: Sodium nitrate, nitrate of soda, Chile saltpeter, "caliche" ($NaNO_3$), white solid, soluble, melting point 308° C., source in nature is Chile, in the fixation of atmospheric nitrogen nitric acid is frequently transformed by sodium carbonate into sodium nitrate, and the solution evaporated. Used (1) as an important nitrogenous fertilizer, (2) as a source of nitrate and nitric acid, (3) in pyrotechnics, (4) in fluxes.

Nitroprusside: Sodium nitroprusside ($Na_2Fe(CN)_5NO \cdot 2H_2O$), red solid, soluble. Used in testing soluble sulfides.

Nitrite: Sodium nitrite ($NaNO_2$), yellowish-white solid, soluble, formed (1) by reaction of nitric oxide plus nitrogen dioxide and sodium carbonate or hydroxide, and then evaporating, (2) by heating sodium nitrate and lead to a high temperature, and then extracting the soluble portion (lead monoxide insoluble) with water and evaporating. Used as an important reagent (diazotizing) in organic chemistry.

Oleate: Sodium oleate ($NaC_{18}H_{33}O_2$), white solid, soluble, froth or foam upon shaking the water solution (soap), formed by reaction of sodium hydroxide and oleic acid (in alcoholic solution) and evaporating. Used as a source of oleate.

Oxalates: Sodium oxalate ($Na_2C_2O_4$), white solid, moderately soluble, formed (1) by reaction of sodium carbonate or hydroxide and oxalic acid, and then evaporating, (2) by heating sodium formate rapidly, with loss of hydrogen. Used as a source of oxalate; sodium hydrogen oxalate, sodium binoxalate, sodium acid oxalate ($NaHC_2O_4 \cdot H_2O$), white solid, moderately soluble.

Oxides: Sodium oxide, sodium monoxide (Na_2O), white solid, reactive with water to form sodium hydroxide, formed by reaction of sodium hydroxide or peroxide and the requisite amount of sodium metal upon heating; sodium peroxide (Na_2O_2), yellowish-white solid, soluble in water with some evolution of oxygen at 30–40° C., complete at 100° C. Reacts (1) as an oxidizing agent, e.g., nitric oxide converted to sodium nitrate, (2) as a reducing agent, e.g., salt solutions of silver or mercury converted to silver or mercury metal, and oxygen. Formed by heating sodium metal and dry air at 300° C. Used as an important bleaching and oxidizing agent for various materials, e.g., wool, straw, wood pulp, and as a source of oxygen.

Palmitate: Sodium palmitate ($NaC_{16}H_{31}O_2$), white solid, soluble, froth or foam upon shaking the water solution (soap), formed by reaction of sodium hydroxide and palmitic acid (in alcoholic solution) and evaporating. Used as a source of palmitate.

Perborate: Sodium perborate ($NaBO_3 \cdot 4H_2O$) white solid, soluble, stable in air, but in water solution loses oxygen, formed (1) by the electrolysis of sodium borate solution (in the presence of sodium carbonate), (2) by reaction of sodium borate solution, in the presence of sodium hydroxide and excess hydrogen peroxide.

Permanganate: Sodium permanganate, permanganate of soda ($NaMnO_4$), purple solid, soluble, formed by oxidation of acidified sodium manganate solution with chlorine, and then evaporating. Used (1) as disinfectant and bactericide, (2) in medicine.

Phenate: Sodium phenate, sodium phenoxide, sodium phenolate ($NaOC_6H_5$), white solid, soluble, formed by reaction of sodium hydroxide (not carbonate) solution and **phenol**, and then evaporating. Used in the preparation of sodium salicylate.

Phosphates: Trisodium phosphate, tribasic sodium phosphate ($Na_3PO_4 \cdot 12H_2O$), white solid, soluble, formed (1) by reaction of sodium hydroxide and the requisite amount of **phosphoric acid**, and then evaporating, (2) by reaction of disodium hydrogen phosphate plus sodium hydroxide, and then evaporating. Used (1) as a cleansing and laundering agent, (2) as a water softener, (3) in photography, (4) in tanning, (5) in the purification of sugar solutions; disodium hydrogen phosphate, dibasic sodium phosphate ($Na_2HPO_4 \cdot 12H_2O$), white solid, soluble, formed (1) by reaction of dicalcium hydrogen phosphate and sodium carbonate solution, and then evaporating the solution, (2) by reaction of sodium carbonate and the requisite amount of phosphoric acid, and then evaporating. Used (1) in weighting silk, (2) in dyeing and printing textiles, (3) in fireproofing wood, paper, fabrics, (4) in ceramic glazes, (5) in baking powders, (6) to prepare sodium pyrophosphate; sodium dihydrogen phosphate, monobasic sodium phosphate ($NaH_2PO_4 \cdot H_2O$), white solid, soluble, formed (1) by reaction of sodium carbonate and the requisite amount of phosphoric acid, and then evaporating, (2) by reaction of **calcium** monohydrogen phosphate and sodium carbonate solution, and then evaporating the solution. Used (1) in baking powders, (2) in medicine, (3) to prepare sodium metaphosphate; sodium pyrophosphate ($Na_4P_2O_7 \cdot 10H_2O$), white solid, soluble, melting point about 900° C., formed by heating disodium hydrogen phosphate to complete loss of water, followed by crystallization from water solution. Used (1) in electroanalysis; sodium metaphosphate ($NaPO_3$), white solid, soluble, melting point 617° C., formed by heating sodium dihydrogen phosphate or sodium ammonium phosphate to complete loss of water, is an easily fusible phosphate forming colored phosphates with many metallic oxides, e.g., **cobalt** oxide. The hexametaphosphate ($NaPO_3)_6$ is an important water-conditioning agent forming soluble complex compounds with many cations, e.g., Ca^{++}, Mg^{++}.

Phosphites: Disodium hydrogen phosphite ($Na_2HPO_3 \cdot 5H_2O$), white solid, soluble, formed by reaction of **phosphorous acid** and sodium carbonate, and then evaporating at a low temperature, melting point of anhydrous salt is 53° C., at higher temperatures yields sodium phosphate and **phosphine** gas; sodium dihydrogen phosphite ($NaH_2PO_3 \cdot 2\frac{1}{2}H_2O$), white solid, soluble, formed by reaction of phosphorous acid and sodium hydroxide cooled to −23° C. when the crystalline salt separates.

Salicylate: Sodium salicylate ($NaC_7H_5O_3$), white solid, soluble, formed by reaction of sodium phenate and **carbon dioxide** under pressure. Used as a source of salicylate and for **salicylic acid**.

Silicate: Sodium silicate, sodium metasilicate, "water glass" (Na_2SiO_3), colorless (when pure) glass, soluble, melting point 1088° C., formed by reaction of **silicon** oxide and sodium carbonate at high temperature; solution reacts with carbon dioxide of the air, or with sodium carbonate solution or **ammonium** chloride solution, yielding silicic acid, gelatinous precipitate. Sodium silicate solution is used (1) in soaps, (2) for preserving eggs, (3) for treating wood against decay, (4) for rendering cloth, paper, wood non-inflammable, (5) in dyeing and printing textiles, (6) as an adhesive (e.g., for paper boxes) and cement. Sold as granular, crystals, or 40° Baumé solution.

Silicoaluminate: (See aluminosilicate, above.)

Silicofluoride: (See fluosilicate, above.)

Stearate: Sodium stearate ($NaC_{18}H_{35}O_2$), white solid, soluble, froth or foam upon shaking the water solution (soap), formed by reaction of sodium hydroxide and **stearic acid** (in alcoholic solution) and evaporating. Used as a source of stearate.

Sulfates: Sodium sulfate (anhydrous), "salt cake" (Na_2SO_4), sodium sulfate, decahydrate, "Glauber's salt" $Na_2SO_4 \cdot 10H_2O$), white solid, soluble, formed by reaction of sodium chloride and **sulfuric acid** upon heating with evolution of hydrogen chloride gas. Used (1) in dyeing, (2) along with **carbon** in the manufacture of glass, (3) as a source of sulfate, (4) to prepare sodium sulfide; sodium hydrogen sulfate, sodium bisulfate, sodium acid sulfate, "nitre cake" ($NaHSO_4$), white solid, soluble, formed by reaction of sodium nitrate and sulfuric acid, upon heating, with evolution of nitric acid. Used (1) as a cheap substitute for sulfuric acid, (2) in dyeing, (3) as a flux in metallurgy; sodium pyrosulfate ($Na_2S_2O_7$), white solid, soluble, formed by heating sodium hydrogen sulfate to complete loss of water.

Sulfides: Sodium sulfide (Na_2S), yellowish to reddish solid, soluble, formed (1) by heating sodium sulfate and **carbon** to a high temperature. Used (1) as the cooking liquor reagent (along with sodium hydroxide) in the "sulfate" or "kraft" process of converting wood into **paper** pulp, (2) as a depilatory, (3) in sheep dips, (4) in photography, engraving and lithography, (5) in organic reactions, (6) as a source of sulfide, (7) as a reducing agent; sodium hydrogen sulfide, sodium bisulfide, sodium acid sulfide (NaHS), formed in solution by reaction of sodium hydroxide or carbonate solution and excess hydrogen sulfide.

Sulfites: Sodium sulfite (Na_2SO_3), white solid, soluble, dilute solution readily oxidized in air, but retarded by mannitol (**carbohydrates**), formed by reaction of sodium carbonate or hydroxide solution and the requisite amount of **sulfur** dioxide, at high temperature yields sodium sulfate and sodium sulfide. Used (1) as a source of sulfite, (2) as a reducing agent, (3) to prepare sodium thiosulfate, (4) as a food preservative, (5) as a photographic developer, (6) as a bleaching agent and antichlor in the textile industry; sodium hydrogen sulfite, sodium bisulfite, sodium acid sulfite ($NaHSO_3$), white solid, soluble, formed by reaction of sodium carbonate solution and excess sulfurous acid. Uses similar to those of sodium sulfite.

Tartrate: Sodium tartrate ($Na_2C_4H_4O_6 \cdot 2H_2O$), white solid, soluble, formed by reaction of sodium carbonate solution and **tartaric acid**. Used in medicine; sodium potassium tartrate, Rochelle salt ($NaKC_4H_4O_6 \cdot 4H_2O$), white solid, soluble. Used (1) in medicine, (2) as a source of tartrate.

Thiosulfate: Sodium thiosulfate, "Hypo" ($Na_2S_2O_3 \cdot 5H_2O$), white solid, soluble, formed by reaction of sodium sulfite and **sulfur** upon boiling, and then evaporating. Used (1) in photography as fixing agent to dissolve unchanged silver salt, (2) as a reducing agent and antichlor.

Tungstate: Sodium tungstate ($Na_2WO_4 \cdot 2H_2O$), white solid, soluble, by reaction of sodium hydroxide solution and **tungsten** trioxide upon boiling, and then evaporating. Used (1) in fireproofing fabrics, (2) as a source of tungsten for chemical reactions.

Uranate: Sodium uranate, **uranium** yellow (Na_2UO_4), yellow solid, insoluble, formed by reaction of soluble uranyl salt solution and excess sodium carbonate solution. Used (1) in the manufacture of yellowish-green fluorescent glass, (2) in ceramic enamels, (3) as a source of uranium for chemical reactions.

Vanadate: Sodium vanadate, sodium orthovanadate (Na_3VO_4), white solid, soluble, formed by fusion of **vanadium** pentoxide and sodium carbonate. Used (1) in inks, (2) in photography, (3) in dyeing of furs, (4) in inoculation of plant life.

All sodium-containing substances impart a characteristic yellow color to the bunsen flame. (R.K.S.)

SOFTENER, WATER. An apparatus devised to remove from water the dissolved salts producing hardness is known as a softener. Hardness of water is due to dissolved salts that make it difficult to obtain soapsuds in the water. Hardness of water is also indicative of the scale-forming quality of that water as a boiler feed. Consequently, in the power generation field, in the laundry trade, and in many industries making use of water in one way or another, raw water needs to be softened. Water softeners are of two types—precipitation and base exchange. There are several commercial types, but practically all are included in the following classification:

1. Precipitation softeners.
    a. Cold lime and soda.
        (1) Intermittent.
        (2) Continuous.
    b. Continuous hot lime and soda.
    c. Lime and barium (cold).
        (1) Intermittent.
        (2) Continuous.
    d. Lime and sodium aluminate.
2. Base exchange (artificial and natural zeolites).

A precipitation softener embodies the principle of using calculated quantities of soluble reagents to react with the hardness in raw water.

A base exchange softener removes the hardness by a simple filtration of the water through a bed of active material which exchanges its sodium base for the magnesium and calcium in the water. Natural and artificial zeolites are used as the active material. When its softening characteristics are nearly exhausted the zeolite is regenerated by backwashing it with a brine solution. The base exchange softener is simple and effective in that no proportioning of chemicals is required and practically zero hardness is obtained. Base exchange softening may give dangerously high alkalinity when the raw water itself contains large amounts of sodium bicarbonate in addition to calcium and magnesium salts. (F.T.M.)

SOIL. A complex sediment, exceedingly variable as to texture, inorganic and organic composition. Difficult to define except in terms of practical fertility or mode of origin. From the purely geologic point of view, soil is usually the relatively thin upper layer of the unconsolidated mantle of disintegrated and decomposed rock material or regolith which overlies the consolidated bedrock. The upper portion of the regolith is divided into topsoil and subsoil. The topsoil is usually a relatively thin zone of the more highly decomposed mineral constituents of the regolith and contains a varying proportion of organic material called humus. This soil zone is the habitat of the shallow rooted plants, such as most grasses and cereals. The topsoil usually passes gradationally into the subsoil which supplies some of the moisture and food for the deeply rooted plants and trees. The subsoil may or may not pass gradationally into the underlying bedrock. It is important to note that the topsoil is easily destroyed by erosion, when not protected by a mantle of vegetation. Classification of soils founded entirely on the origin of the regolith has been found to be impractical. Since the important topsoils are primarily the result of the interaction of rainfall, temperature and organisms with the regolith, the soil specialist recognizes over 7500 types of soils, irrespective of textural differences, and with particular reference to age and to the climatic and other physical conditions under which each soil has been developed. Soils are therefore usually classified as follows: 1. Young Soils. These usually show their relationship to the parent material and are typical flood plain and hilly land deposits, when the soil surfaces are constantly being replenished or disturbed. 2. Mature Soils. These usually cover relatively flat lands where there are good drainage conditions but relatively little erosion. The development of

these soils has gone so far in some cases, particularly in semi-arid regions, that little relation is shown to the parent material and their nature has therefore been principally determined by climatic and organic factors. 3. Old Soils. These usually cover old flat surfaces which have not been disturbed by erosion or sedimentation for a long time. Such soils, due to the dominance of climatic factors in their formation, have lost many of their original characteristics and have, therefore, developed abnormal features. When soils are intensively cultivated their mineral and organic constituents are rapidly depleted and must be replenished by rotation of crops and the application of natural fertilizers. The method of allowing the land to remain fallow is now known to be inefficient. The complete removal of the vegetable cover, such as may result from over-grazing, deforestation, or dry farming, exposes the soil to rapid erosion and destruction. It has been estimated that already 35 million acres of good soil have already been destroyed in the United States, and that 225 million acres will soon be destroyed if immediate and adequate steps are not taken to conserve them. (R.M.F.)

SOLANINE. Alkaloids.

SOLAR CONSTANT. The rate at which the earth is receiving energy from the sun is known as the solar constant. Technically defined the solar constant is the quantity of **energy** that falls in unit time on a unit area placed perpendicular to the direction of the sun at the mean distance of the earth from the sun. It is usually expressed in units of **heat** energy and many hundreds of observations give a mean value of 1.938 **calories** per sq. cm. per min. Converted into more familiar units this amounts to 1.8 hp. falling upon each sq. meter of the earth, neglecting the effects of atmospheric absorption. If this solar energy had to be paid for at the extremely low rate of one cent per kilowatt hour, it would cost the earth about 478,000,000 dollars each second.

Observations of the solar constant are made by means of the **pyrheliometer** and, of necessity, must be made from the surface of the earth. Corrections must be made to all determinations to allow for variations in the distance of the earth from the sun, due to the **eccentricity** of the earth's **orbit**, and also for the effects of the earth's **atmosphere.** The Smithsonian Institution has established, under the direction of Dr. C. G. Abbot, a number of stations scattered over the northern and southern hemispheres, at which daily observations of the solar constant are made. The determination of the absorptive effects of the earth's atmosphere have been very carefully investigated in a number of painstaking researches.

The results of a long series of observations by Abbot and his associates have shown that the value of the solar contant fluctuates by several per cent. These fluctuations are correlated with the sun-spot number and the solar constant is 2 or 3% higher when the number of sun spots is at a maximum. This variation has been further studied by Pettit at Mount Wilson and he has found that the ratio of the amount of solar radiation in the violet to that in the green varies by as much as 50%, with the intensity of the ultra-violet being greatest at the time of sun-spot maximum. The effects of these variations on the general climate of the earth are under investigation. Some correlations have been definitely established but the whole problem is in too confused a state to warrant definite conclusions. (See **Sun Spots.**) (W.K.G.)

SOLAR MOTION. We know that the so-called fixed stars are actually moving in space and in many cases the **space velocity** has been determined. By 1783, the **proper motions** of thirteen stars had been determined and Sir William Herschel noticed that they seemed to have a preferential character. In the direction of the constellation of **Hercules** he noticed that the stars seemed to be moving apart, while in the opposite direction they appeared to be closing in. He interpreted this phenomenon not as a characteristic of the sidereal system as a whole, but rather a perspective effect caused by the actual motion of the sun in the direction of the constellation of Hercules. During the next 50 years the proper motions of many more stars were determined, and in 1837 Argelander discussed the results statistically and confirmed Herschel's determination. With the rapidly increasing number of proper motion determinations during the past century, numerous statistical discussions of proper motions have been made and all have yielded the same conclusion.

With the application of the **Doppler-Fizeau principle** to the determination of the **radial velocities** of the stars a method for determining the solar motion independently from the proper motions became available. With increasing number of radial velocity determinations, it is found that stars in the general direction of the constellation of Hercules seemed to have a preferential motion toward the sun, while in the opposite part of the sky the preferential motion was one of recession.

Since all methods for the determination of solar motion are purely statistical in character, it must not be expected that results from different methods will agree exactly. The point toward which the sun is apparently moving is known as the solar apex, while the opposite point on the celestial sphere is known as the solar antapex. Results of statistical analysis of proper motions give the position of the solar apex as **right ascension** 18 hr. 03.1m and declination $+27°.0$; while an independent discussion from radial velocities yields 18 hr. 02.4m and $+29°.2$. The value of the velocity with which the sun is moving toward this point in the constellation of Hercules is 19.65 kilometers per sec. (12.3 miles per sec.).

Up to the present time there is no conclusive evidence other than that the sun is moving toward the solar apex in a straight line. Many attempts have been made to employ the solar motion for the determination of the distances of the stars. The complete discussion of the problem is far too complex for discussion here but the results which have been obtained, while not of great accuracy for individual stars, are, nevertheless, of great importance for statistical investigations in problems concerned with the discussions of the structure of the **galactic** system. (W.K.G.)

SOLAR PARALLAX. The mean distance of the earth from the sun is one of the most important constants in astronomical measurement. Known as the **astronomical unit**, it forms the standard of measurement throughout the **solar system**; it is also the base line for the determination of **stellar parallax**. To measure the distance from the earth to the sun directly is obviously impossible and some indirect method must be used. The most common method for the determination of this distance is to determine the angle subtended by an equatorial radius of the earth at the mean distance of the earth from the sun, and this angle is known as the solar parallax.

To measure the solar parallax directly is a very difficult problem for a variety of reasons: the sun is very large and very bright, and, furthermore, when the sun is visible there are no stars visible with reference to which the position of the sun can be measured. Even with these difficulties overcome there remains the smallness of the solar parallax (8".80) which is practically impossible to determine directly with great accuracy.

The most common method for the determination of the solar parallax makes use of the fact that, from the **orbital elements** of planetary orbits, the distance of any of the planets from the sun at any instant can be accurately expressed in terms of the earth's mean distance from the sun as unity. Furthermore, a plane triangle may always be passed through the earth, sun, and any

planet and the angles and sides of this triangle may be computed from the orbit, with distances expressed in terms of the astronomical unit. If any one of the sides of the triangle can be determined in terms of the earth's equatorial radius then the other two may be so expressed and one of these sides will be the distance of the earth from the sun. Hence the problem of determination of solar parallax reduces itself to the determination of the **geocentric parallax** of the planet. For increased accuracy in the determination of the parallax of the planet the object should be stellar in appearance, and should be as close as possible to the earth in order that it may have a large geocentric parallax. The asteroid **Eros** is well suited for this purpose and extensive campaigns for determinations of its geocentric parallax have been carried out at the close oppositions in 1900 and again in 1931. At the time of the close approach in 1931 Eros was only about 16,200,000 miles from the earth and hence the parallax was about 6 times as great as the solar parallax.

Several other indirect methods for the determination of the solar parallax have been devised, but the details of their methods are too complex to be included here. Prior to the discovery of Eros in 1898 the planets **Mercury** and **Venus** were used for the determination of solar parallax. The method of using these objects was to measure the time required for these planets to transit across the disk of the sun and expeditions were dispatched to all parts of the earth to observe this phenomenon.

By international agreement the value 8".80 is used at present for computing the material for various almanacs and ephemerides. A careful investigation of a large number of observations is in progress and it is expected that a third decimal place will eventually be determined. (W.K.G.)

SOLAR RADIATION. The **radiation** from the sun comprises a very wide range of wavelengths from the long **infra-red** rays to the short **ultra-violet** rays, with a maximum intensity in the visible green at about 5000 **angstroms**. However, since the **air** strongly absorbs the wavelengths toward either end of the **spectrum**, the solar radiation received on the surface of the earth is confined, largely, to the visible and near infra-red regions, with a very small proportion of the ultra-violet. This is fortunate, for human beings and many other organisms could not endure the full range of solar radiation. The absorption of the ultra-violet radiation takes place largely in the higher stratosphere, where it probably contributes to the atmospheric ionization (see **Ionosphere**). The longer infra-red is absorbed mainly by dust and water vapor at lower levels, which accounts for the low temperature of the air at high altitudes.

The intensity, or radiant flux density, of the solar radiation is measured by means of various forms of **pyrheliometer** or solarimeter. Its value is known as the **solar constant** and averages about 1.34×10^6 ergs per sq. cm. per sec. The direct **illumination** from the sun approximates 6500 foot-candles. (L.D.W.)

SOLAR SYSTEM. That group of objects which are moving through space with the **sun** is known as the solar system. The following classes of objects are listed as members of the solar system, and the details regarding them as individuals will be found elsewhere: **planets, satellites, asteroids, comets, meteors,** and **meteorites,** and the zodiacal light and gegenschein. The **orbital** and physical data regarding various members of the system may be found tabulated on page 1370. In this article we shall confine ourselves to consideration of the system as a whole.

Examination of the tabular material will indicate several interesting correlations between the various orbital and dynamical factors in the solar system. The mass is overwhelmingly concentrated in the sun, this parent member of the system having nearly 750 times as much mass as all of the rest of the members combined. The distribution of the **moment of momentum**, another important dynamical factor, is interesting in that the four major planets, **Jupiter, Saturn, Uranus,** and **Neptune,** have about 98% of the total for the whole system. With very few, but nevertheless important, exceptions, the members of the solar system rotate on their axes and revolve, either about the sun, or their primary in the case of satellites, in the same directional sense. Furthermore, the orbital planes of the great majority of the members lie within an inclination angle of 20° to the plane of the ecliptic. In so far as we have been able to determine the relative percentages of the various chemical elements which go to make up the various members, the compositions of the different objects bear a remarkable similarity to each other.

For centuries the belief has existed that the solar system is not merely an accidental arrangement of objects in space, but is rather the product of some process of evolution. The mere fact of the common direction of orbital motion of the more than 1300 planets and asteroids is in itself sufficient evidence against any chance arrangement. However, in spite of the labors of the large number of eminent scientists and philosophers, who have worked on the problem during the past three centuries, the origin of the solar system is by no means completely understood.

Since the sun is a typical star, which appears abnormally bright to us merely because of its relatively short distance from the earth, the theories regarding its evolution will be found in the material dealing with the stars. The earliest hypothesis, which is worthy of scientific recognition, is to be found in the writings of Thomas Wright, the theologian Swedenborg, and the philosopher Kant, during the 18th century. None of these gentlemen had much scientific training, with the result that their thories can be regarded as pure hypotheses which violate many of the fundamental principles of dynamics. In the middle of the 19th century the astronomer Laplace attempted to put these hypotheses on a scientific foundation and advanced the so-called Nebular Hypothesis.

In spite of the fact that the Laplacian nebular hypothesis has failed to stand the tests of rigorous analysis and even that Laplace himself gave evidence that he did not regard the theory very seriously, nevertheless, the theory has had such a large popular appeal that a few words regarding it will not be out of place. The theory presumes the existence in space of a large nebulous mass slowly rotating and slowly cooling and condensing. As the mass contracts the angular velocity will increase, since the moment of momentum must be conserved and, with the increase in angular velocity, the centrifugal force at the equator will increase until it becomes greater than the gravitational forces holding the mass together. At this point a ring of matter is split off from the equatorial region of the parent mass. The parent mass continues contracting and increases both the angular velocity and equatorial centrifugal force until another ring is split off. In this way successive rings of matter are produced, each surrounding the equatorial part of the central parent mass. These successive rings of matter split and condense into the major planets with, perhaps, their satellites then formed from the cooling masses in much the same way that the planets themselves were formed. Eventually, the central mass condenses to form the present sun. The common forward motion of all of the planets and their satellites, and the approximately coplanar features of the planetary orbits can all be explained on this theory and it was highly satisfactory to those who did not analyze the mathematics too critically. A careful analysis, however, proves conclusively that the rings thrown off from the primary would not condense into single

planets, but would form swarms of small bodies, such as the asteroids or the rings of **Saturn.**

Other considerations regarding the distribution of angular momentum, etc., completely removed the Laplacian theory from the realm of possibility as an evolutionary process for the solar system. Nevertheless many modifications of the Laplacian theory were proposed during the latter part of the 19th century in the vain attempt to satisfy the dynamics of the observed solar system.

With the dawn of the present century a new idea regarding the birth of the solar system was advanced and the three present theories which are worthy of brief consideration are based upon this new conception. It is known that all of the stars are in motion through space relative to each other and more or less at random. From the observed velocities of the stars, their number, and the volume of space which they occupy, it may be calculated that a close approach, and possibly an actual collision is a probable occurrence during the long life history of the average star. The tidal friction theory of Jeans and Jeffries, and the planetesimal theory of Chamberlin and Moulton both assume the close approach of two stars, while the newer theory of Jeffries postulates a "side-swiping" collision. In either case, one of the stars, or what remains after the side-swiping collision, passes off in a hyperbolic orbit; but either the close approach or the collision will have caused the ejection of material from the star which we shall now refer to as the sun. Three different things may happen to this ejected material: much of it will fall back into the sun due to gravitational attraction, some of it will follow the other star out into space, and some of it will remain revolving about the sun. This latter material is the raw substance of which the planets are constructed.

The fundamental difference between the planetesimal theory and the tidal theory is concerned with what happens to the material ejected from the sun very shortly after it was left behind revolving about the sun. Chamberlin and Moulton in their planetesimal theory postulate that the material condensed and solidified relatively quickly into small objects known as planetesimals, while Jeans believes that the material gathered together in the large masses which now form the major planets. The planetesimal theory then postulates that the planets were formed by the gathering together of the small planetesimals about nuclei and the building up of the planets by a process of accretion. The tidal theory, on the other hand, assumes that the planets were formed by the condensation of large masses of hot diffused material. There are other differences in the theories regarding the distribution of the material about the sun immediately following the catastrophe, but these are too highly technical to be discussed here. The collision theory of Jeffries follows the tidal theory very closely, differing only in the method by which the material was ejected from the sun. None of the theories can be said to be perfect and a great deal of work remains to be done before any positive statement can be made. (W.K.G.)

SOLARIZATION. The term solarization was applied originally to the bronze appearance of the shadows produced on printing-out papers by great overexposure. The term is more generally employed today to describe a condition, resulting from extreme overexposure, which tends to destroy the developability of the image and cause the image to develop as a positive rather than a negative.

The effect varies with different emulsions and with most commercial emulsions requires an increase in exposure of several thousand times. The effect varies also with the time of development, tending to become less pronounced as the time of development is increased.

Solarization does not occur if the film or plate is bathed in a solution of potassium iodide, or a solvent of silver halide, such as hypo, before development. The incorporation of reducing agents, such as p-hydroxyphenylglycin (glycin) in the emulsion diminishes solarization and in some cases prevents it entirely.

Emulsions which solarize readily can be used to produce non-reversed images (for example a positive from a positive). Such emulsions are prepared by pre-exposure to light, by including chemical fogging agents or the addition of a sensitizing dye and a suitable reducing agent. (C.B.N.)

SOLDER. Brazing and Soldering.

SOLDER FITTING. This type of fitting utilizes the phenomenon of capillary action for insuring penetration of molten solder to the entire contact surface. A solder fitting coupling is a sleeve that has a cylindrical bore with an inside shoulder against which the tube ends to be connected fit. Each half of the coupling has an annular groove or solder feed channel, with a solder feed hole entering the groove. The coupling is heated with a blow torch or gas torch, and solder wire is fed through the solder feed hole and melts as it comes into contact with the coupling. The liquefied solder is carried around the entire contact surface by capillary action. This type of fitting is extensively used for non-ferrous pipe and tube connection, and is available in a wide variety of sizes, and in elbow, tee, cross, and coupling form. Valves and other accessories with solder fitting ends can also be obtained. (H.C.H.)

SOLDERING. Joining of metals by adhesion using a metallic bonding alloy. The bonding or soldering alloy has a relatively low melting point and is preferably present as a thin film between the parts to be joined. To obtain good adherence and strength most metals require surface preparation, usually through the application of a flux to remove oxides from the surface. Mechanical surface cleaning is also practiced, particularly in soldering lead, also for aluminum and its alloys whose surface oxide layer is very difficult to remove chemically.

The most common soldering alloys are those consisting of lead and tin, such as 50% lead–50% tin for general purpose work and 60% lead–40% tin for making wiped joints in lead sheet and pipe. A wide melting temperature range is required for this type of soldering. A 40% lead–60% tin composition has a narrower and lower melting temperature range than the higher lead varieties and may therefore be used for soldering tin and other low melting-point alloys. Bismuth is added to lead-tin alloys when further lowering of the melting point is required. (See **Lead and Tin Alloys.**)

Heat for melting the solder and warming the parts to be joined is usually applied by means of the familiar soldering "iron," which is a hand tool with a copper tip of appropriate shape. It is often preheated in a gas or other open flame, or may be continuously heated by an internal electrical heating element. A gas blow pipe or other source of heat may be applied directly to the work.

Solder is relatively soft and soldered joints have relatively low strength compared with **brazing** (sometimes called hard soldering) and **welding**. (R.H.H.)

SOLE. Pisces, Teleostei. A term applied to **flatfishes** of numerous species widely distributed in temperate and tropical seas. Some species ascend rivers. The term is properly applied to members of the group, ranked as a family or as a subfamily, of which the genus *Solea* is typical. Since many other flatfishes are edible, the sole of the inland fish markets is often not true sole. (A.W.L.)

SOLENIA. Tubes of endoderm that pass between the various **polyps** through the middle layer (mesoglea) of **alcyonarians.** (A.W.L.)

SOLENOCYTE. An excretory cell from which a **cilium** extends into the associated excretory tubule. These cells are superficially like the flame cells of **flatworms** but they are smaller and simpler. They occur in connection with the excretory organs of **annelid** worms and in Amphioxus of the lower chordates. (A.W.L.)

SOLENODON. Mammalia, Insectivora. An animal of about the size of a rabbit but with a long slender nose and a long naked tail. The claws are strong, those of the fore feet being much larger than those of the hind feet. The two species occur in the West Indies. (A.W.L.)

SOLENOID. This is an electrically energized **coil** which may consist of one or more layers of windings. It is the basis of all forms of the electromagnet and is thus part of the operating mechanism of many electrically operated devices. One of the simplest forms and at the same time a widely used one is the plunger type solenoid. This is a coil wound on a non-magnetic form in which a magnetic plunger may move. Energizing the coil pulls the plunger up into the coil and thus operates the associated mechanism. The iron clad solenoid is similar except for an iron case surrounding the coil. This increases the magnetic pull on the plunger. Other types use a fixed core and various types of external armatures. Solenoids are widely used for operating **circuit breakers,** track switches, valves, and many other electromechanical devices. (L.R.Q.)

SOLENOIDAL FIELDS IN THE ATMOSPHERE. Circulation Principle.

SOLENOIDAL VECTOR. If the **divergence** of a **vector function** of position vanishes everywhere in a certain region, the function is said to be a solenoidal vector in that region.

If a vector function **v** is the **curl** of a vector function **F**, then **v** is solenoidal. (L.L.S.)

SOLFATARA. Fumarole.

SOLID EXPANSION THERMOMETER. Many devices have employed the expansion of solid bodies as an indicator of temperature change. Wedgwood, a celebrated 18th-century potter, used small blocks of burned clay to estimate, by their expansion, the temperature of his kilns. Because of the low expansion coefficient, any such direct application of solids is necessarily very insensitive; on the other hand the requirement of durability favors their use in certain cases. One rather crude arrangement, sometimes used, is a long wire passing over pulleys, its length being sufficient to insure a measurable expansion.

More commonly, use is made of the warping produced by the differential expansion of two solid strips fastened together. The Breguet spiral is composed of two spiral strips, like watch springs, made of different metals and securely welded together throughout their length. A change in temperature causes the combination to coil or uncoil, a motion which, communicated through gears to a pointer-shaft, serves to give temperature readings on a dial. (L.D.W.)

SOLID OF REVOLUTION. A solid of revolution is a solid bounded by a **surface of revolution,** or by a surface of revolution and certain planes.

Suppose that the curve whose equation in **rectangular coordinates** is $y = f(x)$ is revolved about the X-axis, generating a surface of revolution. The volume of the solid bounded by this surface and the planes $x = a$ and $x = b$ is given by the **definite integral**

$$V = \pi \int_a^b y^2 dx.$$

If the curve whose equation is $x = F(y)$ in rectangular coordinates is revolved about the Y-axis, the volume of the solid bounded by this surface of revolution and the planes $y = c$ and $y = d$ is given by

$$V = \pi \int_c^d x^2 dy.$$

(L.L.S.)

SOLID SOLUTION.

For the general meaning of this term, see **Solution** and **Metals and Alloys**. In geology, as defined by A. Holmes this term is applied to compounds which form minerals and is: a crystalline and homogeneous solid, representing a mixture of two or more substances, sometimes composed of isomorphous compounds. The proportions of the mixture may vary within certain critical limits without destroying the homogeneity of the solid solution. Most of the common silicate minerals which form the **igneous rocks**, such as the **feldspars, amphiboles** and **pyroxenes**, are complex, solid solutions. (Compare with **Isomorphous**.)
(R.M.F.)

SOLIDS.
States of Matter; Statics; Kinetics, Elasticity.

SOLITAIRE.
Aves. 1. Passeriformes. A bird (**Aves**) found in the western mountains from the Black Hills to British Columbia and south into Lower California. It is related to the thrushes and is known for its beautiful song. The full name of this species is Townsend solitaire, *Myadestes townsendi*. 2. Columbiformes. A giant flightless pigeon, *Pezophaps solitarius*, of Rodriguez Island, extinct since the late 18th century. Related to the **dodo**. (A.W.L.)

SOLPUGIDA.
The sun **spiders**, a small order of rare arachnids found chiefly in warm and arid regions. The few North American species have been recorded from Florida and from the region between Kansas, Oregon, and the Rio Grande. Although their appearance is rather formidable they are harmless creatures, nocturnal in habits and very shy. (A.W.L.)

SOLUBILITY.
Solutions and Solubility.

SOLUTIONS AND SOLUBILITY.
Solutions are conveniently classified thus:

1. True Solutions
   a. Ordinary . . . . . Sugar (sucrose) in water, ethyl alcohol in water.
   b. Non-ordinary . **Sodium** chloride in water, sodium hydroxide in water, **hydrochloric acid** in water. Salts, acids, bases.
2. Pseudo-solutions (colloidal) . . Starch emulsion
   Silicic acid gel
   Casein in milk
   Soap in water

Pseudo-solutions are discussed in the article on **Colloidal State**.

It is to be noted that salts, acids, bases are classified as non-ordinary, although commonly encountered, for the reason that such substances in solution are dissociated into two kinds of entities, positively and negatively charged ions. (See **Reactions Involving Recombination of Ions; Reactions Involving Water; Chemical Changes; Reactions Involving Oxidation-Reduction; Electrochemistry**.) A solution consists of one substance (the solute) dispersed in another substance (the solvent) in such a fine degree of comminution that the product is homogeneous, has a greater or less range of concentration of solute, is optically void. Pseudosolutions differ from true solutions in not being opti-

cally void when examined in the ultra-microscope, and suspensions differ from both of these in not being optically void when examined by the naked eye or ordinary microscope. Particles in suspension may be separated from a liquid medium by the mechanical processes of filtration or of sedimentation and decantation.

Solute may be gas, liquid or solid, and solvent liquid or solid. Where there exists a solubility range between solids, that is, there is a series of concentrations of two or more solids, the substances are described as forming a solid solution, as for example in the following systems, namely, (1) **silver-gold**, (2) **cobalt-nickel**, (3) **antimony-bismuth**, (4) **magnesium** sulfate heptahydrate-**zinc** sulfate heptahydrate, (5) **iodine-benzene** crystals. Most solutions exist in the liquid state, and water is the most common solvent encountered. (See **Water; Solvents**).

The solubility of substances is expressed in various units (see **Concentration**). Some illustrative examples of the extent of solubility with various solute-solvent systems and resulting generalizations follow.

Saturated Solutions. The maximum amount of solute dissolvable in a given amount of a stated solvent, under definite conditions of temperature and pressure (pressure is important when dealing with gases, otherwise the effect of pressure is minor) and *in the presence of excess* of the solute, is the solubility of the solute in that solvent. The resulting solution is a saturated solution. (See **Concentration**.)

Unsaturated and Supersaturated Solutions. When less than the saturation concentration of solute is present in a solvent, the solution is unsaturated. When more than the saturation concentration of solute is present in a solvent, the solution is supersaturated. The latter condition is brought about in the case of solids by lowering the temperature of a sufficiently concentrated solution in the absence of solute as such, of dust, of mechanical shock, and of nuclei for **crystallization**. Sodium thiosulfate ($Na_2S_2O_3 \cdot 5H_2O$) is a favored substance for producing supersaturated solution, but many other substances exhibit the phenomenon, e.g., sugar solutions, **calcium** gluconate and many organic compounds. In the case of gases dissolved in liquids, supersaturation may be present upon raising the temperature in the absence of rough solid surfaces, of gases, and of agitation. Water saturated with air at 7° C. is reported as having been kept in this way at 18° C. for 6 days.

Physical Properties of Solutions. Certain physical properties of solutions are of predominant importance. Among these are:

1. The change of **vapor pressure** of a liquid solvent by addition of solute. The vapor pressure is lowered 170 mm. per 1 gram mol of solute in 1000 grams of water for "ordinary solutions," and up to values of about 2 or 3 times this amount for "non-ordinary solutions." For solvents other than water a characteristic value for each is observed.

2. The change of **boiling point** of a solution with the thermometer bulb in the *solution*. The boiling point is raised 0.52° C. per 1 gram mol of solute in 1000 grams of water for "ordinary solutions," and up to values of about 2 or 3 times this amount for "non-ordinary solutions." For solvents other than water a characteristic value for each is observed.

3. The change of **freezing point** of a solution—with the thermometer bulb in the *solution*. The freezing point is lowered 1.85° C. per 1 gram mol of solute in 1000 grams of water for "ordinary solutions," and up to values of about 2 or 3 times this amount for "non-ordinary solutions." For solvents other than water a characteristic value for each is observed.

These three observations were contributed by Raoult (1878).

4. The **osmotic pressure** of a dilute solution (0.5 molar or less). The osmotic pressure at a concentration of 0.1 mol of solute in 1000 grams of water at 0° C. is 2.2 atmospheres, and also proportional to the absolute temperature at constant concentration. This,

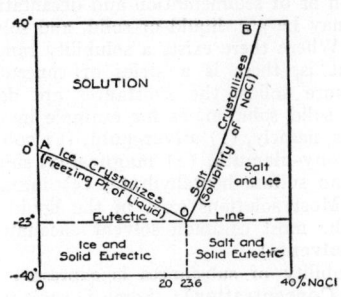

Solubility and freezing point diagram of sodium chloride-water mixtures.

as was pointed out by van't Hoff (1885), is not only analogous to but identical with the behavior of gases:

Equation which expresses the behavior of gases (the more "perfect" the gas, the more exact is the relation):

$\frac{PV}{T} = R$ where P equals pressure, V volume, T absolute temperature, R constant.

Equation which expresses the behavior of solutes in dilute solutions: $\frac{P}{TC} = R$ where P equals osmotic pressure, C concentration of solution, T absolute temperature, R constant.

Solubility of a Substance in the Presence of Two Immiscible Liquids. Distribution or partition of a third substance between two immiscible liquids is illustrated by the following data for several concentrations in each

case. The ratio of concentration of the third substance in each of the two layers is practically constant over a wide range of concentrations.

FORMIC ACID, DISTRIBUTION (13–15° C.) IN

WATER LAYER	BENZENE LAYER
1.016 g. HCOOH per 25 ml.	0.016 g. HCOOH per 150 ml.
1.539	0.023
2.112	0.031
3.826	0.062
7.83	0.138

PHENOL, DISTRIBUTION (20° C.) IN

WATER LAYER	BENZENE LAYER
0.945 g. phenol per liter	2.073 g. phenol per liter
0.711	1.553
0.475	1.036
0.238	0.518

ANILINE, DISTRIBUTION (25° C.) IN

WATER LAYER	BENZENE LAYER
0.0135 g. per 100 ml.	0.1312 g. per 100 ml.
0.0122	0.1282
0.0065	0.0656

OXALIC ACID, DISTRIBUTION (20° C.) IN

WATER LAYER	BENZENE LAYER
0.306 g. (COOH)$_2$ per 100 ml.	0.0653 g. (COOH)$_2$ per 100 ml.
1.064	0.326
3.015	1.148
4.511	1.934

(R.K.S.)

SOLUBILITIES OF REPRESENTATIVE GASES

Solute (Gas)	Solvent (Liquid)	Temperature ° C.	Pressure Atmospheres	Solubility ml. of gas (at 0° C., 760 mm.) in 100 ml. solvent, when the pressure of the gas itself is 760 mm. (with exceptions noted)
Ammonia	Water	0	1	105,000
Hydrogen chloride	Water	0	1	50,000
Sulfur dioxide	Water	0	1	8,000
	Water	20	1	4,000
	Water	40	1	1,900
Hydrogen sulfide	Water	0	1	467
	Water	20	1	258
	Water	40	1	166
Chlorine	Water	20	1	226
Carbon dioxide	Water	0	1	171
	Water	20	1	88
	Water	40	1	53
	Water	0	2	342
	Water	0	0.5	86
	Alcohol (49%)	20	1	98
	Alcohol (99%)	0	1	440
	Alcohol (99%)	20	1	300
	Alcohol (99%)	40	1	220
Nitrous oxide	Water	0	1	130
Acetylene	Water	0	1	173
	Water	20	1	103
	Acetone	0	1	37 grams per liter solution
	Acetone	18	1	21 grams per liter solution
	Acetone	15	1	2,500 ml. per 100 ml. acetone
	Acetone	15	12	30,000 ml. per 100 ml. acetone
	Acetone	−80	1	200,000 ml. per 100 ml. acetone
	Acetone (50%)	0	1	5.7 grams per liter solution
	Acetone (50%)	18	1	1.2 grams per liter solution

SOLUBILITIES OF REPRESENTATIVE GASES

Solute (Gas)	Solvent (Liquid)	Temperature °C.	Pressure Atmospheres	Solubility ml. of gas (at 0° C., 760 mm.) in 100 ml. solvent, when the pressure of the gas itself is 760 mm. (with exceptions noted)
Ethylene..........................	Water	0	1	23
	Water	20	1	12
	Alcohol	0	1	360 vol. per 100 vol. alcohol
	Alcohol	20	1	270 vol. per 100 vol. alcohol
Nitric oxide......................	Water	0	1	7.4
Methane..........................	Water	0	1	5.5
Oxygen...........................	Water	0	1	4.9
	Water	20	1	3.1
	Water	40	1	2.3
Carbon monoxide..................	Water	0	1	3.5
	Water	20	1	2.3
	Water	40	1	1.8
Nitrogen..........................	Water	0	1	2.4
	Water	20	1	1.5
	Water	40	1	1.2
Hydrogen..........................	Water	0	1	2.1
	Water	20	1	1.8
	Water	40	1	1.6

At a constant temperature, the solubility of a given gas in a given solvent is proportional to the pressure, if the gas is of moderate to low solubility (Henry).

In a mixture of gases, the solubility of each gas is determined by its partial pressure (Dalton).

The solubility of gases decreases with increase of temperature. Most gases are expelled gradually but completely from the solvent upon boiling, or by bubbling an inert gas through the solution, or when allowed to remain exposed to the atmosphere. **Hydrogen chloride** is an exception. Boiling hydrochloric acid of any strength gives ultimately a distillate of boiling point 108.58° C. at 760 mm. pressure and of 20.24 percent HCl. At other pressures the boiling point and concentration are different but definite.

SOLUBILITIES OF REPRESENTATIVE LIQUIDS

Solute or Solvent (Liquid)	Solvent or Solute (Liquid)	Temperature °C.	Solubility
Alcohol......................	Water		Miscible in all proportions
Benzene.....................	Water	3	0.030
		23	0.061 } g. H$_2$O per 100 g.
		40	0.114
	Carbon tetrachloride	0	85.3 g. C$_6$H$_6$ per 100 g. solution
		5.5	Miscible
	Chloroform	0	88 g. C$_6$H$_6$ per 100 g. solution
		5	Miscible
	Ethyl alcohol	0	85 g. C$_6$H$_6$ per 100 g. solution
		5.5	Miscible
	Phenol	0	78.3 g. C$_6$H$_6$ per 100 g. solution
		5.1	Miscible
Aniline.....................	Water	30	3.7 g. C$_6$H$_5$NH$_2$ per 100 g. water layer
			5.4 g. H$_2$O per 100 g. aniline layer
		90	6.4 g. C$_6$H$_5$NH$_2$ per 100 g. water layer
			9.9 g. H$_2$O per 100 g. aniline layer
		150	17 g. C$_6$H$_5$NH$_2$ per 100 g. water layer
			24 g. H$_2$O per 100 g. aniline layer
		168	Miscible
	Ethyl alcohol		Miscible
Phenol.....................	Water	20	8.3 g. C$_6$H$_5$OH per 100 g. water layer
			27.9 g. H$_2$O per 100 g. phenol layer
		50	12 g. C$_6$H$_5$OH per 100 g. water layer
			37.3 g. H$_2$O per 100 g. phenol layer
		68.3	Miscible
	Ethyl alcohol		Miscible
Carbon tetrachloride.........	Water	20	0.08 g. CCl$_4$ per 100 g. water
Carbon disulfide.............	Water	20	0.18 g. CS$_2$ per 100 ml. solution
	Ethyl alcohol (97%)	17	100 ml. CS$_2$ per 100 ml. alcohol
Bromine....................	Water	0	4.0
		20	3.5 } g. Br$_2$ per 100 g. solution
		40	3.3

SOLUBILITIES OF REPRESENTATIVE SOLIDS

Solute (Solid)	Solvent (Liquid)	Temperature °C.	Solubility
Sucrose.....................	Water	0	64 ⎫
		20	67 ⎪
		40	70 ⎬ g. $C_{12}H_{22}O_{11}$ per 100 g. solution
		70	76 ⎪
		100	83 ⎭
	Ethyl alcohol (97%)	14	0.36 ⎫
	Ethyl alcohol (50%)	14	47 ⎬ g. $C_{12}H_{22}O_{11}$ per 100 ml. solution
	Water	14	87.5 ⎭
Sodium chloride.............	Water	0	26.3 ⎫
		20	26.4 ⎪
		40	26.7 ⎬ g. NaCl per 100 g. solution
		70	27.3 ⎪
		100	28.1 ⎪
		118	28.5 ⎭
	Hydrochloric acid: 182 g. HCl per liter of solvent	25	7.0 ⎫
	36.5	25	22.3 ⎪
	18.2	25	25.4 ⎬ g. NaCl per 100 g. solution
	9.1	25	25.5 ⎪
	Water	25	26.5 ⎭
Anthraquinone..............	Benzene...........	20	0.26 ⎫ g. $C_{14}H_8O_2$ per 100 g. benzene
		60	0.97 ⎭
	Chloroform	20	0.61 ⎫ g. $C_{14}H_8O_2$ per 100 g. chloroform
		60	1.58 ⎭
Anthracene.................	Methyl alcohol	20	1.8 g. $C_{14}H_{10}$ per 100 g. alcohol
	Benzene	25	1.9 g. $C_{14}H_{10}$ per 100 g. benzene
	Toluene	16	0.9 ⎫ g. $C_{14}H_{10}$ per 100 g. toluene
		100	⎭
	Carbon disulfide	25	2.6 g. $C_{14}H_{10}$ per 100 g. carbon disulfide
	Sulfur dioxide liquid	40	2.1 ⎫
		65	4 ⎬ g. $C_{14}H_{10}$ per 100 g. sulfur dioxide
		99	10 ⎭
Azobenzene.................	Methyl alcohol	10	3.8 ⎫
	Ethyl alcohol	10	5.6 ⎬ g. $(C_6H_5N)_2$ per 100 g. solution
	Propyl alcohol	10	5.7 ⎭
Sulfur.....................	Carbon disulfide	0	18 ⎫
		20	30 ⎬ g. sulfur per 100 g. solution
		40	50 ⎭
	Benzene	15	1.5 ⎫
		26	1.0 ⎬ g. sulfur per 100 g. benzene
		71	4.4 ⎭
	Chloroform	20	1.0 g. sulfur per 100 g. chloroform
	Carbon tetrachloride	25	0.9 g. sulfur per 100 g. carbon tetrachloride
	Sulfur monochloride	0	6 ⎫
		18	10 ⎬ mol % sulfur in mixture
		55	28 ⎭
Phosphorus.................	Carbon disulfide	0	81 ⎫ g. phosphorus per 100 g. solution
		10	90 ⎭
	Benzene	0	1.5 ⎫
		20	3.2 ⎬ g. phosphorus per 100 g. benzene
		40	5.8 ⎭

Increase in temperature generally increases the solubility of a solid in a liquid. Loss of liquid solvent by evaporation at ordinary or elevated temperatures, from a given solution of solid generally results in the recovery of the pure solid as crystals, except when supersaturation occurs.

SOLUTION OF ALGEBRAIC EQUATIONS BY FACTORING. If an equation is in the form of a polynomial in one unknown equated to zero, it may frequently be solved by factoring the polynomial and using the principle: if a product of two or more factors is zero, either one or more of its factors must be zero. (L.L.S.)

SOLUTION OF EQUATIONS. The process of finding the particular values of the unknown for which the two expressions of an equality have identical numerical values is called the solution of the equation.

The process of solution of an equation is therefore the operation of finding the roots of the equation.

The term "solution of an equation" is also sometimes used to mean the same thing as the roots of the equation. (L.L.S.)

SOLUTIONS OF A DIFFERENTIAL EQUATION. Ordinary Differential Equations.

SOLUTRIAN. Paleontology of Man.

SOLVAY PROCESS. Ammonia-soda process for making **sodium** carbonate. (R.K.S.)

SOLVENTS. The most common solvent is water. Water dissolves a great many gases, liquids, and solids, and is much used for this purpose. (See **Solutions; Water.**) Other liquids similarly dissolve many substances without reacting chemically with them. Important considerations in connection with the choice of solvent for a given case are (1) **vapor pressure and boiling point,** (2) solvent power under stated conditions of temperature, (3) ease and completeness of recoverability by **evaporation** and **condensation,** and completeness of separation from dissolved material by evaporation, (4) **heat of vaporization,** (5) miscibility with water or other liquid, if present, (6) inertness to chemical reaction with the materials present, and with the apparatus, (7) inflammability and explosiveness, (8) odor and toxicity, (9) cost of solvent, loss in process, cost of recovering.

Many classes of substances are available as solvents. It is convenient to classify these in the following manner:

1. **Water**
   Various water solutions, e.g., sucrose, glycerol, ethyl alcohol, zinc chloride, zinc bromide, stannous chloride, ammonio-cupric hydroxide, sulfuric acid concentrated.

2. **Hydrocarbons**
   Petroleum fractions
   　　Petroleum ether
   　　Ligroin
   　　Higher boiling fractions
   　　Normal-heptane
   Terpenes
   　　Turpentine
   Benzenoid
   　　Benzene
   　　Toluene
   　　Xylenes
   　　Mesitylene
   　　Cumene
   　　Cymene
   　　Tetrahydronaphthalene
   　　Decahydronaphthalene

3. **Alcohols**
   Methyl alcohol
   Ethyl alcohol
   Normal-propyl alcohol
   Iso-propyl alcohol
   Normal-butyl alcohol
   Amyl alcohol
   Normal-hexanol
   Cyclo-hexanol
   Ethylene glycol
   Propylene glycol
   Diethylene glycol
   Glycerol

4. **Phenols**
   Phenol
   Cresols

5. **Ethers** (See **Alcohols and Ethers**)
   Dimethyl ether
   Diethyl ether
   Dipropyl ether
   Di-iso-propyl ether
   Di-normal-butyl ether
   Ethylene glycol monomethyl ether
   Ethylene glycol monoethyl ether
   Diethylene glycol monomethyl ether
   Diethylene glycol monoethyl ether
   Ethylene oxide
   Propylene oxide
   1,4-Diethylene oxide

6. **Aldehydes**
   Normal-butyl aldehyde
   Crotonaldehyde

7. Ketones   (See **Aldehydes and Ketones)**
   Acetone
   Methyl ethyl ketone
   Methyl iso-butyl ketone
   Furfural

8. **Carboxylic acids** (See **Acids, Carboxylic)**
   Acetic acid

9. **Esters**
   Methyl acetate
   Ethyl acetate
   Iso-propyl acetate
   Normal-butyl acetate

10. **Chloro-organic compounds** (See **Chlorine)**
    Ethyl chloride
    Chloroform
    Carbon tetrachloride
    Ethylene dichloride
    Trichloroethylene
    Propylene dichloride
    1,1,2-Trichloroethane
    Tetrachloroethylene
    Monochlorobenzene
    Acetylene tetrachloride
    Ortho-dichlorobenzene
    Trichlorobenzene
    Epichlorohydrin

11. **Bromo-organic compounds** (See **Bromine)**
    Methyl bromide
    Ethyl bromide
    Bromoform
    Ethylene dibromide
    Ethylene chlorobromide
    Ortho-dibromobenzene

12. **Nitrogen organic compounds** (See **Nitrogen)**
    Pyridine
    Aniline
    Monoethanolamine
    　　(2-hydroxyethylamine)
    Diethanolamine
    　　(di-2-hydroxyethylamine)
    Triethanolamine
    　　(tri-2-hydroxyethylamine)
    Acetamide
    Tetralkyl ammonium compounds

13. **Sulfur organic compounds** (See **Sulfur)**
    Carbon disulfide
    Thiophene

14. **Liquefied gases**
    Ammonia
    Sulfur dioxide
    Hydrogen sulfide
    Hydrogen chloride
    Hydrogen bromide
    Cyanogen
    Nitrogen tetroxide

15. **Inorganic liquids**
    Sulfur monochloride
    Phosphorus trichloride
    Phosphorus tribromide
    Phosphorus oxychloride
    Arsenious chloride
    Hydrogen cyanide

The data available on the subject of solvents is large in amount, but far from complete. The applications are many and growing rapidly. Illustrative selections must necessarily be very limited.

1. **Sulfur, iodine,** yellow **phosphorus** (red phosphorus insoluble) are dissolved by **carbon** disulfide, **and,** upon evaporation of the solvent at room temperature, these solids remain as a residue.

SOLUBILITY OF VARIOUS SUBSTANCES IN CARBON DISULFIDE

SUBSTANCE	TEMPER-ATURE, °C.	GRAMS SUBSTANCE PER 100 GRAMS CARBON DISULFIDE
Sulfur............	0	22.0
	20	41.8
	40	100.0
Iodine............	0	7.9
	20	14.6
	40	25.2
Phosphorus........	0	81.3
	10	89.8

2. Sulfur is dissolved by carbon disulfide, **carbon tetrachloride, chloroform, benzene, toluene, aniline, phenol,** and many other solvents.

SOLUBILITY OF SULFUR IN VARIOUS SOLVENTS

SOLVENT	TEMPER-ATURE, °C.	GRAMS SULFUR PER 100 GRAMS SOLVENT
Aniline............	130	85.3
Benzene...........	15	1.5
Carbon disulfide......	20	41.8
Carbon tetrachloride....	25	0.86
Chloroform.........	19	0.92
Ethylene dichloride.....	25	0.84
Phenol............	174	16.4
Toluene...........	23	1.48
Trichloroethylene......	25	1.63

3. Laszczynski (1894) reported the solubility of about 15 inorganic salts in 7 organic solvents. The following data are illustrative:

SALT	SOLVENT	GRAMS SALT PER 100 GRAMS SOLVENT AT ROOM TEMPERATURE
Mercuric chloride.....	Ether	6.4
(HgCl₂)	Ethyl acetate	30.0
	Acetone	126
Stannous chloride.....	Ether	11.4
(SnCl₂·2H₂O)	Ethyl acetate (77° C.)	73.3
Potassium thiocyanate	Pyridine	6.1
(KCNS)	Acetone	20.7
Silver iodide.........	Pyridine (115° C.)	8.6
(AgI)		

4. In some cases, the solvent, upon evaporation, forms a residue containing solvent of crystallization, e.g., **calcium** chloride tetramethyl alcohol ($CaCl_2 \cdot 4CH_3OH$) below 55° C., calcium chloride triethyl alcohol ($CaCl_2 \cdot 3C_2H_5OH$) above 55° C.

5. The application of solvents is not confined to the solution of solids, but includes also gases and liquids. Nor is it strictly limited to cases where true solutions result, but includes pseudo-solutions, that is, suspensions and emulsions.

6. Common solvents used in pharmacy are water (liquors, aquae, solutions), **ethyl alcohol** plus water (spirits, tinctures, extracts, essences), **sucrose** plus water (syrups), **glycerol** with or without water (glycerites). In the extraction of oils and fats from materials such as seeds, and in dry cleaning of fabrics, such solvents as **carbon** disulfide, carbon tetrachloride, trichloroethylene, **benzene**, ligroin are used. In paints, varnishes and lacquers, **turpentine** is commonly used, also tetrahydronaphthalene, and benzene. Several of the ethers are used in the preparation of **cellulose** nitrate and cellulose acetate lacquers and dopes. In the explosives industry,

acetone is used, and in certain cases ethyl alcohol-ether mixture.

In general, substances of similar chemical constitution dissolve each other, e.g., hydrocarbons dissolve hydrocarbons. By the use of two or more mutually miscible solvents the field of application is greatly extended.

7. Chloro-organic compounds are among the more chemically inert solvents.

8. In the petroleum industry, solvents serve to extract certain portions of the raw material in the process of refining. Liquid sulfur dioxide may be used for differential solution. Liquid propane is also used.

9. Solvent recovery is usually accomplished by evaporation and condensation, but adsorption by such materials as silica-gel is also utilized. (R.K.S.)

SOMITE. One of the longitudinal series of segments into which the bodies of many animals are divided. These segments are clearly shown in a simple form in the **earthworms.** In man they are made evident by the structure of the spinal column and the series of spinal nerves, but they are overshadowed externally by the high development of the appendages. The term is synonymous with metamere.

The segmental masses of mesoderm in the vertebrate embryo are also called somites. They are the primordia of the axial skeleton, voluntary muscles of the body and appendages, and the inner layer of the skin. (A.W.L.)

SONOMETER. Musical Sounds.

SONSTADT SOLUTION. Thoulet Solution.

SORGHUM. *Andropogon Sorghum.* Gramineae. Sorghums are annual grasses of tropical origin. They have an extensive system of wiry roots and solid stems 3–15' tall. The leaves are smaller than those of corn, and capable of rolling up tightly during periods of drought, and quickly unrolling and starting to function when favorable moisture conditions return. Because of this habit sorghums are often grown in regions subject to frequent drought. The **inflorescence** is a **panicle** usually of very compact habit. Ordinarily the **spikelets** are paired, one of the pair being sessile or stemless, the other having a short stem or pedicel. The former is fertile, the latter staminate (see **Stamen**). The grains are enclosed in the glumes and vary considerably in shape in different varieties. There are two main groups of sorghums; the sweet or saccharine sorghums, the juicy pitch of which is a source of syrup; and the grain sorghums, which yield grain, stock food and ensilage. Kaffir is one of the latter group. In Asia grain sorghums are employed in a countless variety of ways, as for fuel, brooms, mats, fences, windbreaks, roof thatch, and in making a fermented drink. (R.M.W.)

SORUS. Ferns.

SOUND. Physically, sound is a longitudinal elastic wave motion propagated by alternate compressions and rarefactions of the medium, usually air. If air is compressed by the sudden movement of some object, as a piece of cardboard, its elasticity causes it to expand and compress the air ahead of it. This generates a wave, in which each molecule of the medium oscillates forward and backward. An excellent analogy is the propagation of a bump or a jerk from the engine at one end of a freight train to the caboose at the other.

There are thermodynamic aspects of the propagation of sound, which in the earlier study of the theory were overlooked. The heating and cooling effects of compression and expansion are not negligible; in deriving the following expression for the speed of propagation of

sound, it is assumed that they are **adiabatic processes.**
The formula commonly used for gases is

$$v = \sqrt{\frac{\gamma p}{\rho}} \, .$$

In this equation p is the pressure of the gas in absolute units and ρ is its density, γ is the ratio of the specific heat at constant pressure to the specific heat at constant volume. Thus, for air at normal temperature and pressure $p = 1,013,250$ dynes per sq. cm. (1 atmosphere), $\rho = 1.293 \times 10^{-3}$ grams per cc., and $\gamma = 1.408$. This gives $v = 33,220$ cm. per sec., which is quite near the measured value. For room temperatures the speed is greater (about 34,400 cm. per sec.), since the density ρ is less.

Modern research in **acoustics** has been greatly aided by the instantaneous photography of sound waves, or rather, of their refraction patterns, cast like shadows upon a photographic plate, by the light from an electric spark. In this way the progress of the wave and its reflection, refraction, and diffraction are permanently preserved for study and measurement. **Sound recording** is also much utilized in the analysis of **musical sounds.**

The mechanism of the ear and of hearing is but imperfectly understood. In some way the sound vibrations reaching the inner ear stimulate the auditory nerves; but the extreme sensitiveness and delicacy of discrimination possessed by the ear are hard to explain. It has been shown that our ability to locate the direction from which a sound comes is due to the slight difference in time at which the waves reach the two ears. (L.D.W.)

SOUND FILM. Attempts to synchronize a phonograph with the motion picture **projector** to produce a talking picture are nearly as old as the motion picture itself. For a long time little progress was made but by 1925 developments in telephony and radio had enabled engineers to overcome the difficulties which earlier inventors had encountered and a few years later the talking picture made its appearance the sound being recorded on a wax disk and reproduced in synchronism with the projected picture. This method of recording sound, however, was soon discarded in favor of a sound record on film.

To record sound on film, it is necessary to convert sound waves into variations in exposure on the film. To do this, the sound waves are first picked up from the air by a microphone. In the microphone the vibrations of the diaphragm result in variations in the current passing through the electrical circuit of which the microphone is a part. This current is amplified by means of vacuum tubes and, through a light-valve or galvanometer, varies or modulates the light reaching the constantly moving film in such a way that either the amount of exposure or the area exposed varies directly with the variations of the current. When the amount of exposure is varied, the sound vibrations are recorded on the film as differences in density (variable density record); in variable width records the sound vibrations are represented by differences in the width of the exposed area on the film.

The light valve referred to in the preceding paragraph consists of a loop of duraluminium tape suspended in a plane at right angles to a magnetic field. When the assembly of magnet and armature is complete, the two sides of the loop constitute a narrow slit, the sides of which lie in a plane at right angles to the lines of force and centered in the air gap. The ends of the loop are connected to the output terminals of the amplifier so that, with the magnet energized, the loop opens and closes in accordance with the variations in the intensity of the current. In the recording machine, the light valve is placed between a constant light source and the moving film so that it functions as a shutter which opens and closes thus varying the exposure in accordance with the alterations in the current from the microphones on the set.

In variable width recording the modulated current produced by the sound vibrations on the microphone is carried to a sensitive oscillograph. This oscillograph galvanometer (see figure) consists of a fine wire loop

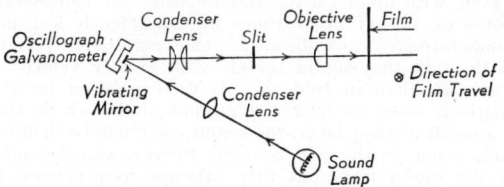

through which the amplified microphone current circulates. A small mirror is cemented to this loop and the loop suspended in a magnetic field. A high-intensity lamp, similar to the ordinary automobile headlight lamp, furnishes the illumination for making the photographic record. Light from this lamp passes through a condenser lens and is focused on the galvanometer mirror, from which it is reflected through another condenser lens to a slit. The resulting slit of light passes through a projector lens which images it on the film. With current flowing through the galvanometer, the wire loop is set into vibration, carrying the mirror with it, and tracing a line of light the width of the slit across the sound track of the film. The result is a sound record of constant density but variable width.

The usual practice is to record the picture and sound on separate films, the camera and recorder being driven by synchronous motors. This enables the cameras to be moved about more freely and permits two or three cameras to shoot the same scene at different angles, all in perfect synchronism with the sound record. There is also the added advantage in being able to use different types of film for each record and to adjust the processing of each record to secure the best results. The films sent to the exhibitor, of course, have the sound record printed in on one side of the pictures.

The complete installation for the projection of talking pictures consists of (1) the projector with its reproducing units for translating the sound record into electric currents, the variations of which correspond exactly to the sound waves recorded, (2) the amplifying equipment by means of which these tiny currents are greatly increased, and (3) the reproducers and horns which convert these currents into the sound one hears from behind the screen.

Sound film is run through a projector which differs from the usual type only in the introduction of the sound-reproducing unit. This is located between the head mechanism of the projector and the lower magazine.

Since the sound-reproducing unit is separated from the picture gate by several inches, it is evident that the sound record cannot be printed directly beside the picture to which it applies. Accordingly the sound record is placed *below* the picture record and a certain amount of slack is allowed between the sprocket which carries the picture record with an *intermittent* motion past the projection lens and the sprockets which carry the sound record *continuously* past the sound-reproducing unit.

In the sound-reproducing unit, light of high intensity is concentrated by an optical system which focuses it in a fine line across the sound track on the film as it passes the aperture. On the other side of the aperture is a photoelectric cell which produces a small electric current the variations of which correspond to the modulated light which reaches it through the film. Thus there is produced an electrical current the variations of which are identical with that which, coming from the microphone on the set, made the sound record on the film. The current produced by the photoelectric cell is strengthened first by a small vacuum-tube amplifier built into the sound-reproducing unit and then by a larger and more powerful amplifying unit. The current, now many times stronger than at first, operates the loud-speakers placed behind the screen. (C.B.N.)

SOUND IN THE ATMOSPHERE.

At times sound travels with great clarity and intensity for considerable distances, but at other times its character is lost in a comparatively short distance. One reason for this lies in the fact that sound travels with greater velocity in warm air than in cold air. If there exists a layer of relatively cold air near the ground above which there is a much warmer layer, the sound wave will be held to a large extent in the cold layer by the downward bending of the sound waves as they attempt to penetrate the warm layer. On the other hand, if the warmer air lies next to the ground and the air progressively becomes colder with altitude, the sound waves will be carried readily aloft. Surface and near-surface inversions favor the propagation of sounds for great distances along the earth's surface by continual refraction from the inversion surface and reflection from the ground or water surface. As sound waves are propagated by the actual motion of air particles, turbulence and strong winds tend to distort and dissipate them. Still cold nights or mornings are, in general, excellent sound propagation periods as opposed to warm windy afternoons. (P.E.K.)

SOUND PRODUCTION.

The incidental production of sound waves in air or similar vibrations in the water by the movements of living things is so common that a sense of hearing serves animals as a valuable means of detecting the presence of enemies, prey, or members of their own species. Together with this sense, the ability to produce sound voluntarily serves many uses, as we know from our own experience. In some cases the act involves only the special use of common organs, as when the beaver slaps the water with his broad tail in a warning signal. Sound production immediately brings to mind, however, organs such as the human larynx with its great range of activity and other organs specially developed for sound production.

Among the vertebrates the production of sound is very commonly associated with the respiratory system, since the air-breathing forms have an opportunity to vibrate special structures by controlled currents of air expelled from the lungs. In the birds the apparatus is the syrinx, located at the inner end of the trachea, and in mammals it is the larynx, near the outer end of the same passage. The frog (Class **Amphibia**) also has a larynx with vocal cords in its very short trachea. Vocal cords are lacking in the reptiles, although some of these animals produce sound by the expulsion of air from the respiratory system, as in the hissing of snakes and the so-called bellow of the alligator. Whispering of human beings is in the same category. Some fishes produce sound by means of the swim bladder. A noteworthy example is the croaker (*Micropogon undulatus* Linn.) of our eastern and southern coastal waters.

Of all vertebrates, the mammals and birds are most generally equipped with elaborate vocal organs, hence the syrinx and larynx of these classes claim special attention. The two organs are quite different in origin and structure.

The syrinx is a chamber located at the junction of the trachea and the two major bronchi. It is supported by modified cartilages and contains a projecting flexible membrane which is thrown into vibration by the expulsion of air from the lungs. Muscular control of the tension of this membrane permits the bird to control the pitch of the sounds produced, which vary from comparatively simple calls to elaborate songs. Some birds are able to mimic sounds very faithfully, hence modulation of the sounds produced by the syrinx is involved, as in the case of human speech. Our mocking bird *Mimus polyglottos*) affords an excellent example of mimicry of the songs of other species, while **parrots** and mynas extend their virtuosity to mimicry of human speech. The mynas, especially, enunciate words with amazing delicacy of inflection.

The larynx is essentially a chamber at the pharyngeal end of the trachea, supported by modified cartilages and in most species of mammals containing vocal folds, or cords, whose tension is regulated by muscles acting on the laryngeal wall. The human larynx is among the most highly developed organs of the kind. A slit-like opening, the glottis, leads into it from the pharynx. This opening is flanked by lobes of the cricoid cartilage and by ligamentous folds, the vocal cords, imbedded in the mucous lining. The larynx is further supported by large right and left quadrangular plates of the thyroid cartilage whose ventral junction forms the "Adam's apple." A flap-like epiglottis closes down over the glottis during the act of swallowing. During ordinary breathing the glottis is V-shaped, becoming more rounded in deep inspiration. When high tones are produced, the tension of the vocal cords closely approximates them and narrows the opening to a mere slit. Although the production of sound by this organ is due simply to the vibration of the cords by the properly controlled ejection of air, sound may be produced, in whispering, without the vibration of the cords and, in either case, an important modulating effect is produced by the mouth and tongue. The oral, nasal and pharyngeal cavities also act as resonators. Individual differences in the pitch of the human voice depend on the length, thickness ,and elasticity of the cords. In any one individual, pitch is varied by changing the tension of the cords. Supplementary resonating structures occur in some mammals, such as the enormously dilated hyoid bone and cavernous diverticula of the larynx in the howler monkeys (*Alouata* sp.). In the **bats** the voice and the highly developed ears serve to detect obstacles in flight by the reflection of sounds.

Among the insects sound production is highly specialized. Here it is of the type called stridulation. The apparatus varies from a comparatively simple roughening of apposed surfaces which produce vibrations when rubbed together, to the elaborate organs of the male **cicada**. A common example of the simple form is found in the milkweed beetles (*Tetraopes*). By moving the prothorax up and down, its ventral sclerite is rubbed against the mesosternum, producing a squeaking sound audible to the human ear at very short distances. A more elaborate stridulation occurs in the order Orthoptera, exemplified by the crickets, katydids and grasshoppers. The crickets have parts called scrapers and files developed on the basal portion of the wings. When rubbed together they throw the entire wing membrane into vibration, producing the shrill singing so familiar to everyone. The short-horned grasshoppers set the wings into vibration by means of a row of fine projections on the inner surface of the hind legs.

The sound-producing organs of the male cicada consist of a pair of deep depressions at the base of the abdomen, covered with thin plates. Within the cavities are membranes which are thrown into vibration by direct muscular action, thus producing sound, and others which serve as resonators. Although the sound produced by cicadas of different species varies greatly in volume and quality, probably no insects exceed in volume or incisiveness the shrill vibrant singing of some of the common cicadas of the United States.

Auditory organs occur in many insects with special stridulating organs, but in others, such as some members of the order Diptera, their presence suggests that the rapid movement of the wings in flight produces the sound that the insect hears, thus providing for recognition within the species. Certainly the wing beat may have such an effect, since it is quite evident to us in the humming of mosquitoes and midges. (A.W.L.)

SOUND RECORDING. The recording and subsequent reproduction of **sound**, somewhat paralleling the development of photography, ranks among the unique achievements of modern times. The fact that its largest application so far has been to entertainment in no way detracts from its noteworthy character or its potential usefulness. The first practical phonograph was constructed by Edison about 1877 and largely improved in the years following. In the Edison apparatus a needle, attached to a diaphragm vibrating with the sound waves, makes indentations in soft tin-foil or wax. When the needle is again drawn over these indentations, it is caused to vibrate as before, and communicates its motion to the diaphragm, so that the original sound is reproduced. In a much-used modification of this device, the needle vibrates sidewise, producing a wavy groove instead of indentations. The nearest modern counterpart of the original Edison phonograph is the "dictaphone" used in lieu of shorthand in some business offices. (See **Sound Film**.)

In another type of recorder now coming into practical use, the sound record is a series of variations in the magnetization along a steel wire or tape, left there by an electromagnet carrying the current output from a microphone. When this magnetic record is drawn between the poles of an inert electromagnet, it generates a varying e.m.f. in the coils thereof which, when amplified, actuates a speaker or earphone. (L.D.W.)

SOUNDER. The sounder is the receiving device used in the aural type of telegraph reception. It consists of an electromagnet whose armature is restrained at each end of its travel by a brass stop. Since the message is transmitted by a code consisting of dots and dashes it is necessary that the receiver be able to tell by sound just how long the armature is pulled down (this being determined by whether the signal is dot or dash). The sounder is constructed to give slightly different sounds when it strikes the two stops so a skilled operator can tell whether it is closing or opening. (L.R.Q.)

SOUNDING. A knowledge of the depth of water under the keel of a vessel is frequently of great importance to a navigator or pilot. The process of obtaining this depth of water is known as taking soundings. The term is also applied directly to the determined depth. All charts of coast lines give the soundings in great detail out to a depth of 100 fathoms. In addition to the depth of water, the character of the bottom is also given. When a vessel is continually taking soundings and using the contour and character of the bottom for the purpose of directing the pilotage, she is said to be "on soundings."

The simplest, and perhaps most effective, method of taking soundings is to use the old-fashioned lead and line. The lead is simply a piece of lead hollowed out at the bottom and "armed" by placing tallow in this hollow for bringing up a sample of the bottom. The line is graduated in measured lengths, due allowance being made for the distance of the person who is "flying the blue pigeon" from the surface of the water. The great difficulty with this method is that the speed of the vessel must be checked in order that the line may be perpendicular to the surface of the water when the lead reaches the bottom.

To avoid the necessity of stopping the ship various types of sounding machines are used. In these machines the lead is attached to a line or wire and allowed to trail out behind the vessel until it reaches the bottom. The depth of the water is determined, not by the length of line payed out, but by a device attached to the lead which measures the maximum pressure of the water. This maximum pressure of the water is proportional to the depth of water and will be attained when the lead reaches the bottom.

Within recent years the so-called sonic depth-recorder has been developed for taking soundings. This instrument operates on the principle that sound travels with a finite velocity in water and the length of time required for a signal to go from the keel to the bottom of the ocean then be reflected back to the ship is proportional to the depth of water under the ship. A device is placed against the skin of the ship near the keel which sends out a sharp signal. A receiver detects the reflected signal from the ocean floor on its return to the ship's bottom. The time between the sending of the signal and the reception of the reflection is converted into the desired sounding. In the so-called fathometer the entire instrument is operated from the pilot house and is so designed that all that is necessary is to turn a switch and the depth of water appears on a dial, all of the operations being fully automatic. (See **Altimeter**.) (W.K.G.)

SOW-BUG. Wood-Louse.

SOYBEAN. *Glycine max.* Bean.

SOYBEAN OIL. Esters.

SPACE CHARGE. The electric charge on a conductor is to be regarded as confined to an infinitely thin layer at the surface and thus, in a sense, as geometrically 2-dimensional. Even when a current is flowing through the conductor, the quantities of positive and negative electricity within any element of volume at any instant are equal, thus neutralizing each other's electrostatic effect. For this reason **Laplace's equation** applies to the interior of a current-bearing conductor as well as to a complete vacuum. This is true whether the conductor be of the metallic or the electrolytic type.

Quite different are the conditions in a so-called vacuum tube, where streams of **electrons** or of positive **ions** (e.g., **canal rays**) may occupy sizable regions to the virtual exclusion of carriers of the opposite sign. (See **Ionized Gases**.) The electricity thus monopolizing such a region is called a "space charge," and may exert marked influence on the performance of the tube.

A typical instance of this occurs in thermionic **rectifiers**. If the cathode is emitting no electrons, the **potential gradient** across from cathode to anode is nearly uniform; that is, the potential increases by the same number of volts for each centimeter from the one electrode to the other. If now a small emission begins from the cathode (because of its being heated), the electron swarm is more dense near the cathode, due to their accelerated motion, much as the stream of water descending from a faucet is broader near the faucet than farther down. The potential gradient is no longer uniform, and the total increase of potential, that is, the "plate voltage," is less than before. Further increase in emission may result in an actual drop of potential near the cathode (followed by a rise farther on), with still further decrease of plate potential. Such a potential minimum is contrary to the condition expressed by Laplace's equation, which therefore does not hold in a region occupied by a space charge. (Another formula, called Poisson's equation, here applies instead. It is $\nabla^2 V = 4\pi\rho$, where V is the potential and ρ is the electric volume density at the point in question.)

In gas-filled discharge tubes the effect of electronic space charge is largely offset by the presence of great numbers of less mobile positive ions. As a result, such tubes not only have much lower **resistance** but have a more uniform potential gradient and give a better approximation to Ohm's law. (L.D.W.)

SPACE FRAME. A space frame or space structure is a three-dimensional **framed structure**. It may be composed of triangles, rectangles or a combination of these forms. Space frames are statically determinate or indeterminate depending on the number of members, support conditions and the rigidity of the **joints**. **Bridges,** framed domes, transmission towers, radio towers and building frames are all space structures.

If the space frame is statically determinate it may be analyzed by means of the equations for three-dimensional statics. These equations state that the summation of all forces parallel to three axes, usually taken as mutually perpendicular, must equal zero and the moments of all forces about these axes must also equal zero.

Indeterminate space frames may be analyzed by methods which are applicable to indeterminate planar structures. Bridges and building frames are highly indeterminate when considered as space frames because the joints have a certain amount of rigidity even though they may not have been designed to resist moment. Thus, all parts do participate to some extent in distributing the **loads** to the foundation. Due to this action some members may be subjected to **torsional stress** as well as **direct** and **bending stresses**. However, it is customary to ignore this action in bridges and buildings and treat the component parts as planar structures. (C.W.C.)

SPACE SHIP. Rocket Flight; Rockets.

SPACE STRUCTURE. Space Frame.

SPACE VELOCITY. The space velocity of a star is its actual velocity in space relative to the sun. The term "fixed star" is one which has been handed down to us from the era when philosophers believed that the stars were fixed on a sphere, rotating about the earth, which they referred to as the celestial sphere. At the present time we believe that practically all of the stars are actually in motion in space with velocities comparable in magnitude to the velocities with which the **planets** are moving about the sun in their orbits. In measuring any velocity it is necessary to have a definite reference point and the space velocity of a star is its velocity referred to the sun.

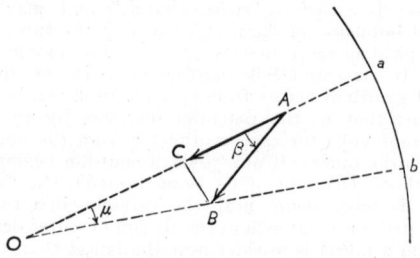

In the figure we have the space velocity of a star represented by the **vector** (directed straight line proportional to the velocity in direction and magnitude) AB. The angle β which this vector makes with the direction of the star from the sun at any particular instant gives the direction of the space velocity at that instant.

As observed from the solar system this space velocity may be resolved into two components: AC the **radial velocity,** and μ the **proper motion.** The radial velocity is determined directly in terms of linear velocity (i.e., in miles or kilometers per sec.) but the proper motion may only be determined in angular units, usually expressed in seconds of arc per year. In order that the space velocity may be known the proper motion must be converted into a linear velocity, commonly known as the transverse velocity of the star. This may be accomplished only if the distance of the star is known. Expressing the distance in terms of the stellar parallax, P'', and calling the transverse velocity T (the line CB in the figure), the following relations may be derived:

$$T = 4.74\,\mu/P'' \text{ kilometers/sec.}$$

or

$$T = 2.94\,\mu/P'' \text{ miles/sec.}$$

With both the transverse velocity, T, and the radial velocity, R, known in the same units, the problem of determining the space velocity S in the same units is merely that of solving the plane right-triangle ACB. This solution yields:

$$S^2 = T^2 + R^2. \qquad \cos\beta = R/S. \qquad \sin\beta = T/S.$$

(W.K.G.)

SPACE-TIME. Relativity.

SPADIX. Flower.

SPAN. Bridge; Truss.

SPAN, EFFECTIVE WING. Wing Span.

SPANDREL BEAM. A **beam** supported by the adjacent exterior **columns** of a multiple-story building which carries the wall **load** of any story is called a spandrel beam. Spandrel beams may also support a part of the floor load. (C.W.C.)

SPANISH FLY. A pharmaceutical preparation consisting of the dried and powdered bodies of a European species of blister beetle. These **beetles,** constituting the family Meloidae, contain a poisonous substance known as cantharidin from the family name Cantharidae formerly applied to the group. Cantharidin is a powerful irritant, dangerous when taken internally. It blisters the surface of the skin and is used as a counter-irritant. (A.W.L.)

SPANISH MOSS. Tillandsia usneoides. Bromeliaceae. Spanish Moss is in fact not a moss at all, but a flowering plant related to the Pineapple. The plant is a perennial herbaceous plant growing as an **epiphyte** on any solid support available. It is found throughout tropical and subtropical America from Brazil to the southern United States.

Spanish Moss is rootless, and has a long, slender, branching stem which reaches a length of 2–6', and narrow tapering leaves. The entire plant surface is covered with grayish scales capable of absorbing water. The flowers are small and yellow; the fruit, a dry **capsule.** The plant propagates itself readily by means of small fragments which are blown about by the wind. Again the plant may be carried by birds, who seek the plant as nest-building material.

The entire plant is frequently used as a packing material or for stuffing furniture. From the stem is obtained a coarse, stiff, black fibrous material, very like horsehair in appearance; this also is used in upholstery. There are several other species in the genus. Other common names applied to this plant are Southern Moss, old man's beard, vegetable horsehair, and long moss. (R.M.W.)

SPANNER. Wrench.

SPAR. In marine parlance a **spar** is a round timber used to extend a sail. Used with this meaning it could be a mast, a boom or a yard.

A structural member used similarly to extend a surface to obtain an air reaction is found in the **wing** of an **airplane.** There, the spar is a principal structural member running the length of the wing. Usually there are two, parallel to one another, but wings have been built having only one spar. In such a design the spar is required to take torsion as well as compression and bending. The spars support the ribs upon which the wing covering is stretched. The reaction of the air upon the latter is transmitted to the spars which in turn are attached to the fuselage. The air reaction may be carried by the spars entirely by bending as in the case of a truly cantilever wing, or it may be carried by spars which are braced outboard of the fuselage by wires or struts. (See **Wing**.)   (F.T.M.)

SPARAGMITE. A general and somewhat local term for late **Proteozoic** poorly graded **clastic,** sedimentary formations including such distinctive types as **arkose** and **graywacke,** intermixed or interbedded with conglomerates and intraformational **breccas.** Some of the pebbles in the conglomerates are **dreikanter,** and the total textural and structural features of the sedimentary suite suggest arid to semi-arid, **torrential,** and **intermontane** conditions of deposition.   (R.M.F.)

SPARK. An electric spark is a sudden breakdown of the insulating strength of the dielectric separating two electrodes, due to the formation of ions by an intense electric field, accompanied by a rush of electricity across the "spark gap," and a flash of light indicating very high temperature. Unlike the arc, glow, and brush discharges, the spark is of very short duration. It may be oscillatory or intermittent, several discharges taking place in quick succession. In gases, the spark takes place only at appreciable pressures, such as normal atmospheric pressure.

If the voltage across a spark gap is progressively raised, a spark passes when it has become sufficiently high. The lowest voltage at which the spark will pass is the "sparking potential"; but there is usually a time interval, called the "spark lag," between the attainment of this voltage and the passage of the spark. Also the voltage may be increased for a moment considerably above this value without producing a spark. These characteristics depend upon the condition of the gas, especially upon the ions and the vapors present in it. After one spark has passed, others follow at lower sparking potential, because of the ions already formed; and this is also true if the "pilot spark" takes place across another neighboring spark gap. Paschen found that for a given pressure the sparking potential is a nearly linear function of the length of the gap, and for a given gap it is a nearly linear function of the pressure. For spherical terminals, the relation is so definite that the "sphere gap" is often used as a rough measure of high voltages. Thus with spherical electrodes 5 cm. in diameter, a 2-cm. spark in air at normal pressure corresponds to a potential difference of 56,300 volts; a 5-cm. gap to 102,250 volts. With 10-cm. spheres, the corresponding voltages are 59,460 and 123,850.

When metallic terminals are used, the light from the spark may exhibit the spark **spectrum** of the vaporized metal. The spark is also the source of a type of ultra-violet radiation known as **Entladungsstrahlen,** which accounts for the effect of the pilot spark, referred to above. (L.D.W.)

SPARK PLUG. The spark plug is a device inserted in the **combustion chamber** of spark ignition engines (**Otto engine**) to provide the insulated electrode and the gap necessary in the high-tension jump spark **ignition system.** As the electrodes are oxidized and pitted by usage, and since insulators may suffer damage, or carbon and oil may accumulate to cause short-circuit, the firing points should be made removable for adjustment, cleaning, or replacement. Universally a threaded hole is provided through the wall of the combustion chamber and the ignition points are built into a small compact "plug" which is provided with a threaded steel shank so that the plug may be screwed into the combustion chamber. Only one of the electrodes leading to the gap needs to be insulated since the electrical circuit is completed through the plug shell and engine body by grounding one end of the high-voltage supply. In operation, parts of the plug remain at high temperatures. The electrodes are made of special alloy steel suitable

for resisting the high temperature and the insulator must be capable of withstanding both high temperature and electric stress of the order of several thousand volts.

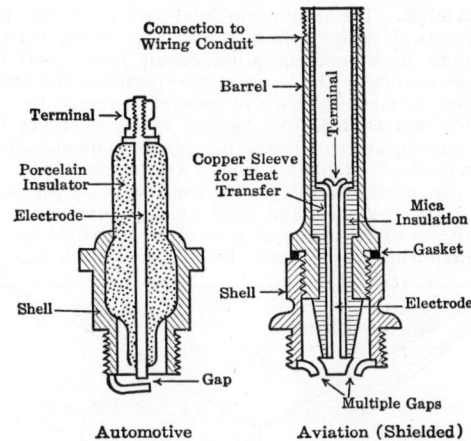

Automotive   Aviation (Shielded)

Molded porcelain and built-up stacks of mica rings are used. Satisfactory operation of the engine is dependent upon proper spark plug temperature.

Plugs can be designed with longer or shorter travel of the heat from the interior tip of the insulator to its point of contact with the shell resulting in higher or lower plug operating temperatures. If the plug is too cool carbonizing and short-circuiting of the gap will be probable; if too hot it may promote pre-ignition and detonation. The length of the gap affects the energy of the spark and the voltage required to produce it. Small gaps reduce electrical stress, make starting easier, but are more easily fouled by bits of carbon and do not ignite the gasoline as efficiently, as has been proved by tests for thermal efficiency under different gap settings. About .03 in. ($\pm$.01) covers the recommendations for plug gaps for all engines except aeronautical, where the gaps are much smaller because of advantages of limiting voltage stress on the plugs and ignition wiring at high altitudes.

A greater variety of spark plugs has been produced for aeronautical service than any other because service conditions are more severe and reliability more desirable than in other internal combustion engine fields. Also the use of radio receiving equipment is an important adjunct to navigation and the ignition system, including spark plugs, will interfere with radio reception unless properly shielded. Operation of aircraft at extremely high altitudes poses further problems in corona loss from the high voltage portions of the ignition system. Electrically shielded and gas-pressurized plugs and ignition wiring harness are employed to meet these special problems of aircraft ignition. (F.T.M.)

SPARROW. Aves, Passeriformes. A common form of small seed-eating bird (**Aves**) of the family Fringillidae. The many species are for the greater part rather quietly colored in browns and grays with streaked and spotted plumage, but some bear conspicuous black or white marks and the browns of some species are very bright. Some of the sparrows have beautiful songs, although none rival such outstanding singers as the brown thrasher and the mockingbird.

The field sparrow, *Spizella pusilla,* chipping sparrow, *S. passerina,* white-throated, *Zonotrichia albicollis,* and white-crowned, *Z. leucophrys,* sparrows, song sparrow, *Melospiza melodia,* and in the west the lark sparrow, *Chondestes grammacus,* are among the well-known North American species. (A.W.L.)

SPASTIC COLON. Constipation.

SPATHE. Flower.

SPATIAL DEGREE. Degree.

SPEAKER. The speaker, or loud-speaker, is the electro-acoustical device which converts electrical current variations in a communication circuit into sound for general coverage. The head-set also performs this function but is used only where general coverage by the sound is not needed. All modern radio **receivers** for home use utilize the speaker but many communications receivers use head phones since reception is usually by only one operator. While various types of speakers have been developed and used at different times, the majority of those in use at present are dynamic speakers. Referring to the figure, the magnet, which may be

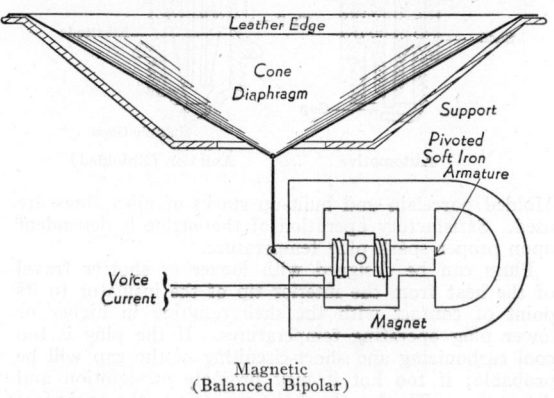

Leather Edge

Cone
Diaphragm

Support

Pivoted
Soft Iron
Armature

Voice
Current

Magnet

Magnetic
(Balanced Bipolar)

electromagnet or permanent magnet, sets up a high flux across the gap containing the voice coil. This coil consists of a few turns of rather light wire wound on a form rigidly attached to the cone. The voice currents, i.e., the output of the audio-frequency **amplifier,** are passed through the voice coil. This current flowing in the presence of the magnetic field produces a force on the coil. As the force varies with the current, the cone is moved back and forth at the frequency of and with amplitudes proportional to the voice currents, thus producing sound waves. An older form of speaker is the magnetic which usually employs a balance armature as shown. This type has been superseded by the dynamic. The basic speaker is usually part of a more complete sound-reproducing apparatus. Thus the ordinary small radio or amplifier unit utilizes the cabinet as a baffle to aid in the reproduction of low-frequency notes. A speaker will not satisfactorily reproduce notes whose wavelength is greater than four times the distance from the cone to the edge of the baffle. More elaborate speaker arrangements involve special acoustical networks, horns, etc., to give a more uniform frequency response over the range desired. For proper operation careful design of these auxiliary components is necessary. Since proper reproduction of low frequencies requires a rather large cone while high frequencies require small cones, may installations use both types in the same unit, sometimes side by side on the same baffle, sometimes as a single coaxial unit. (L.R.Q.)

SPEARMINT, OIL OF. Volatile Oils.

SPECIES. The smallest taxonomic category, excluding the various subdivisions recognized in very detailed study. A species is made up of a group of individuals, but beyond this limitation the definition of the term has varied from the denial of its existence as a natural entity to various attempts at rigid characterization.

As a rule the individuals of a species resemble each

other within established limits of variation, but in some species the sexes are conspicuously different, in some the component individuals differ with the seasons, and in some there are various forms of individuals adapted for different tasks. The principle of fertility as a test of the relationship has been adopted by some biologists, but here again species differ. In some extremely variable species individuals may be found which are incapable of mating, and in some cases different species produce fertile hybrids when crossed. The two criteria, reproductive association and structural resemblance, seem to provide a workable definition when taken together. According to this concept a species may be defined as a natural group of individuals which are either similar in form or associated in a reproductive sequence. Like all other definitions it will be found difficult to apply to

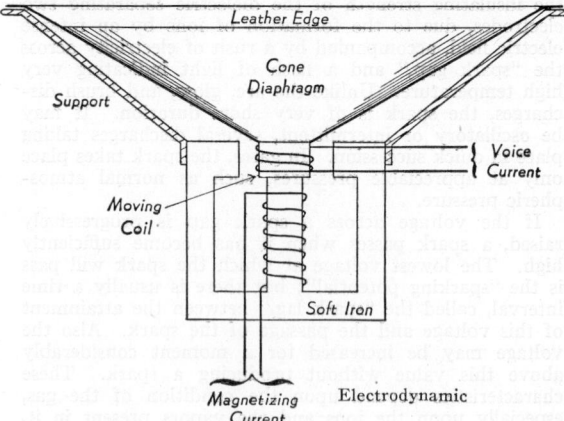

Leather Edge

Cone
Diaphragm

Support

Voice
Current

Moving
Coil

Soft Iron

Magnetizing
Current

Electrodynamic

certain specific cases, but it is more generally applicable than most. (A.W.L.)

SPECIFIC. Any **drug** or treatment which has a special or individual action in the cure of some disorder or disease, such as **quinine** in **malaria** or **salvarsan** in **syphilis.** (R.S.M.)

SPECIFIC GRAVITY. Density and Specific Gravity.

SPECIFIC GRAVITY BOTTLE. Pycnometer.

SPECIFIC HEAT. The specific heat of a substance is the quantity of heat required to impart a unit increase in temperature to a unit mass of that substance; commonly expressed in **calories** per centigrade degree per gram. (Some writers regard the specific heat as the abstract ratio of the quantity just defined to the corresponding value for water). The "thermal capacity" of a body of any mass is the quantity of heat required to raise its temperature one degree (calories per gram), so that the specific heat of a substance may be defined as its thermal capacity per unit mass. Of all known substances, except one of the phases of liquid **helium,** that having the greatest specific heat is hydrogen, for which, at constant volume, it is 2.418 calories per gram per degree; next is water, with specific heat unity (at 15° C.). For gold it is only about 0.031 calories per gram per degree.

These wide variations are qualitatively explained by the fact that, since the atoms of gold are much more massive than those of hydrogen and hence there are much fewer of them in a gram, and since the temperature is determined by the kinetic energy per molecule, the total energy required for a gram of gold is correspondingly less than for a gram of hydrogen. Dulong and Petit (1819) concluded, indeed, that the specific heats of elements are inversely proportional to their atomic weights. (See **Dulong and Petit's Law of Specific Heats.**) The specific heats of crystalline solids have afforded in recent years a fruitful though difficult subject of research. The work of Debye in this field, especially with reference to the variation of specific heats with temperature, is outstanding.

It is necessary to discriminate between the true or temperature specific heat and the apparent or total specific heat which we encounter when the thermal energy imparted is allowed to do more than agitate the molecules or atoms of the substance. Expansion with rise of temperature may require energy because of work done against external pressure or against cohesion. This is notably true with gases, the specific heats of which are, on the average, some 40% greater when they are allowed to expand at constant pressure than when kept at constant volume as the temperature is raised.

SPECIFIC HEATS OF SOME COMMON SUBSTANCES AT ORDINARY TEMPERATURES

Substance	Specific Heat	Substance	Specific Heat
Air (c.v.)	0.171	Iron	0.107
Alcohol (0° C.)	.548	Lead	.031
Aluminum	.214	Mercury	.033
Copper	.090	Nitrogen (c.v.)	.176
Glycerine	.576	Oxygen (c.v.)	.156
Helium (c.v.)	.075	Silver	.056
Hydrogen (c.v.)	2.418	Water	1.000
Ice	.493	Zinc	.090

(L.D.W.)

SPECIFIC SPEED. The specific speed of a hydraulic turbine is the member of rpm at which it would develop 1 hp. under 1-foot head, all dimensions of the turbine being reduced proportionally to enable it to meet these conditions. Mathematically, the specific speed is

$$\frac{rpm\sqrt{hp.}}{(head, in ft.)^{\frac{3}{4}}}$$

The specific speed is the most important single dimension of a turbine. It very fully indicates the characteristics of the **runner.** By making a comparison of specific speeds, the characteristics of turbine runners can be judged without considering their actual speed, head, or power. And yet the specific speed is not completely satisfactory because it is not non-dimensional but is based on the respective dimensional systems, i.e., English or metric.

In general, runners of low specific speeds are suitable for high heads and those of high specific speed are suitable for low heads. The combination of **head** and specific speed is such as to keep the generator speeds within reasonable range for all values of head. This is fortunate since the possible range of head in present-day hydroelectric developments is from 5 to 5000 ft. The range of specific speeds is served by the various types of turbines about as follows:

N_s	Type
10–20	Pelton
20–120	Francis
120–150	Nagler, Kaplan

For an average generator, the cost decreases as the speed increases, at least within certain limits; hence as high a specific speed as can be safely used is chosen.

Specific speed is also a useful reference dimension in the centrifugal pumps field. However, since discharge is more important than power in this field, the specific speed equation has a different form. Centrifugal pump specific speed is

$$\frac{rpm\sqrt{Discharge, gal. per min.}}{(head, in ft.)^{\frac{3}{4}}}$$

(F.T.M.)

SPECTRA. Spectrum.

SPECTRAL CLASS. Examination of stars with the unaided eye will indicate that they are not all of the same apparent color. This fact has been noticed from earliest times, and we find the names of many of the stars carrying an indication of their color, e.g., **Antares** was undoubtedly so named because of its reddish color similar to that of the planet Mars. With the application of the **spectroscope** to astronomical research, several attempts were made to classify the stars according to their spectra. Secchi and Vogel each proposed spectral classification sequences based upon visual observations. The system which is now universally employed was developed by Dr. E. C. Pickering, Director of the Harvard College Observatory from 1877 to 1919. The first results of his classification were published by the observatory with funds made available by Henry Draper, and the classification system is known either as the "Harvard system" or as the "Henry Draper system."

The Harvard system of classification is based upon the relative intensity of certain selected **absorption** lines in the stellar spectra. The various types are designated by letters, and the principal characteristics of the different classes are:

Class B—Helium stars—Absorption lines of helium and hydrogen characteristic features.

Class A—Hydrogen stars—Intensity of helium lines weaker and hydrogen lines the most prominent feature.

Class F—Calcium stars—The *H* and *K* lines of calcium are much stronger than the hydrogen lines, although the latter are still prominent. A few metallic lines are present.

Class G—Solar stars—Calcium lines the most prominent feature, with the hydrogen lines distinctly fainter. Many metallic lines.

Class K—Metallic lines—Calcium lines still strong but spectrum predominated by multitude of lines due to metals. The violet end of the continuous background distinctly weaker.

Class M—Molecular compound lines—Calcium and many metallic lines still prominent, but characteristic feature is the presence of bands due to molecular compounds. The violet end of the continuous background extremely weak.

Accompanying the above changes in the absorption lines there is a progressive change in the position of maximum intensity of the continuous background, the colors of the stars changing from blue for the B-type through yellow for the G-type to red for the M-type. This is an indication of a temperature sequence for the stars, those of the B-type being hottest and those of the M-type coolest. Ninety-nine per cent of all stars thus far examined may be fitted into this sequence of six letters. Stars with spectra intermediate between the various types are designated by a decimal system, e.g., a G5 star is half way between a G- and a K-type star. To provide for the few stars which cannot be fitted into the main sequence there are a number of small classes. For example, certain blue stars, some of which show bright lines, are known as O-type, while red stars, other than M-type, are grouped under N, R, and S.

More than 275,000 stars have been classified at the Harvard Observatory on the foregoing system. The spectra were obtained by **objective prism** photography, and examined and classified by Dr. Annie J. Cannon. These results are published in the Henry Draper Catalogue and its extensions. Other observatories, using objective prisms, or large reflectors and slit spectrographs, have classified large numbers of stars fainter than those included in the Henry Draper Catalogue, and have also made critical examinations of high dispersion spectra to determine slight differences within the individual classes. The correlations between spectral type and the physical characteristics of the stars will be found in the article on **Giant and Dwarf Stars.** (W.K.G.)

SPECTRAL ENERGY DISTRIBUTION. When radiation exhibiting a continuous spectrum, as that from a hot stove or the light from an incandescent lamp, is quantitatively analyzed, it is found that quite different amounts of power are represented by the radiation within equal ranges of wavelength or of frequency having different limits. The proportion in any such range depends upon the character of the source. Thus, in the radiation from a candle, the ratio of the energy output between 6500 and 6600 angstroms (red) to that between 4500 and 4600 angstroms (blue-violet) is greater than the corresponding ratio for the radiation from an arc lamp.

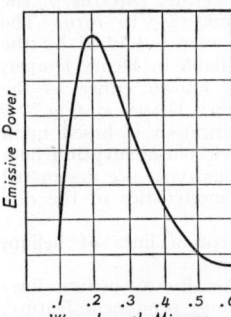

Spectral energy distribution for black body at 1170° C. (1443° Abs.), with peak at 2000 A. (.2 micron).

If we divide the spectrum into small intervals of wavelength, say 10 angstroms, and plot the power output for each range as ordinate with the mean wavelength of the interval as abscissa, the result is a curve showing the distribution of power through the spectrum. When the radiation is due to high temperature, as in the above examples, there is always a wavelength interval having maximum power, that is, the curve has a "peak," from which the ordinates fall off in both directions. Wien pointed out that the higher the temperature of the source, the farther toward the short-wavelength end of the spectrum does this peak lie. (See **Wien's Laws, Planck's Equation,** and **Radiation Pyrometer.**)

An instrument utilizing a prism for dispersing the radiation, together with a thermocouple or similar device for measuring its flux density in different ranges, may be used to analyze infra-red thermal radiation, and is called a "spectroradiometer." The **spectrophotometer** performs a similar service for visible light, except that the results in this case are usually tabulated in terms of the visibility rather than the actual power of the emission. (L.D.W.)

SPECTRAL SERIES. Atomic Spectra; Molecular Spectra; X-Ray Spectra.

SPECTROBOLOMETER. Bolometer.

SPECTROGRAPH. Spectroscope.

SPECTROGRAPHY. The photography of spectra. A spectrograph is a form of spectroscope designed particularly for photographing spectra. (See **Spectrum, Spectroscope.**)

Visual examination in a spectroscope is adequate (1) if the spectrum of the substance is simple and easily recognized, (2) if a permanent record is not required, (3) if the substance can be positively identified by the lines occurring in the visible region. If many of the lines are faint or close together, necessitating exact measurement, or if important identifying lines occur in the ultra-violet or infra-red, the photographic method is the more precise. The procedure is to photograph the spectrum of the sample whose composition is to be determined and then on the same plate, above or below it, the spectrum of a substance whose characteristic lines are well known and can therefore be used in determining the wavelengths of the lines in the spectrum of the unknown substance.

If the substance is known or believed to contain certain elements, the spectra of these may be photographed on the same plate. Then if identical lines are found in both spectra, it is only necessary to determine the wavelengths of the lines which do not correspond to find the other elements present.

Photography extends the range of the spectroscope beyond the visible range into the infra-red and the ultra-violet. No great difficulty is experienced with modern plates in photographing to a wavelength of 1200 millimicrons (12,000 Å) in the infra-red. Ordinary blue-sensitive materials may be used in the ultra-violet up to a wavelength of about 250 millimicrons (2500 Å). Beyond this it is necessary to use (1) plates prepared with a minimum of gelatin (Schumann plates) to overcome the absorption of ultra-violet radiation by gelatin, or (2) to coat the emulsion with materials which fluoresce in the ultra-violet and through fluorescence produce visible light of spectrum lines in the ultra-violet region which can be photographically recorded on regular plates. Various oils have been used by different workers with varying success, the plate being bathed in the oil immediately before use. Lyman in this way was able to reach a wavelength of 580 Å in the extreme ultra-violet. Special plates coated with a fluorescent material for ultra-violet spectrography are now available commercially. The fluorescent coating is removed by washing the plate with acetone before development, or brushing the plate with a soft camel's-hair brush while in the developer.

Spectrography finds many important applications in science and industry. The spectra of the stars show that they are composed of the same elements which we find on the earth. Helium in fact was found in the spectrum of the sun before it was isolated in the laboratory. The spectrum of a star is also an indication of its temperature, age, and the direction and speed in which it is moving.

In the laboratory, spectrography is used in the detection of gases, particularly inert gases such as **krypton, xenon,** etc., studies of full combustion in motors, the presence of impurities in metals and in foods, the composition of alloys, the characteristics of dyes, **pigments,** and coloring matters, etc. (C.B.N.)

SPECTROHELIOGRAPH. As the construction of the term indicates, the spectroheliograph pictures the sun in its **spectrum.** Essentially the instrument consists of a high dispersion spectrograph with a second slit placed directly in front of the photographic plate so that the radiation from only one spectral line is received on the plate. If the instrument is so placed that the first slit is in the principal focus of a telescope directed toward the sun, a narrow strip of the sun's image will be admitted by the first slit and an image of that narrow section of the sun will be formed on the photographic plate in the particular radiation for which the second slit is adjusted. The instrument is so constructed that the first slit may be moved across the image and at the same time the second slit moves across the photographic plate at the same rate. Hence, it is possible to obtain a photograph of the sun in the monochromatic radiation of any particular element, say calcium.

From the **flash spectrum** we can determine the heights to which various elements rise in the solar atmosphere, and thus by means of the spectroheliograph it is possible to obtain photographs of the sun at various levels above the **photosphere.** (W.K.G.)

SPECTROMETER. Spectroscope.

SPECTROPHOTOMETER. An instrument for analyzing the **spectral energy distributions** of sources of light. It is thus a spectroradiometer for visible radiation. Various types are in use. A representative form includes a **monochromatic illuminator,** from the slit of which proceeds an isolated, narrow frequency range of the light under examination. This may be compared, by means of a **wedge photometer** device or otherwise, with the light of the same frequency range from a source whose spectral energy distribution is known. In interpreting the results, it is important to take into account the visibility factor (see **Photometry**), and thus to distinguish between the relative absolute (energy) and apparent (visual) intensities for different wavelengths. There are spectrophotometers which exhibit directly an approximate curve of apparent intensity distribution. A simple type consists of a spectrometer with a neutral absorbing wedge placed with the thick base over the upper end of the slit and the thin edge over the lower. The spectrum then appears as a strip whose upper margin, penetrating visibly to various distances into the region of greater absorption, corresponds to the distribution curve. (L.D.W.)

SPECTRORADIOMETER. Spectral Energy Distribution; Radiation Pyrometer.

SPECTROSCOPE.
Many types of instrument for producing and viewing spectra are included under this term. Variations in form are due, not only to differences in principle, but also to the type of radiation to be examined, which ranges all the way from **infra-red** to **x-rays.**

The earlier spectroscopes, developed by Fraunhofer, Ångström, and others in the early part of the last century, adapted **Newton's** discovery of the **dispersion of** light by a **prism.** The essential features (Fig. 1) are

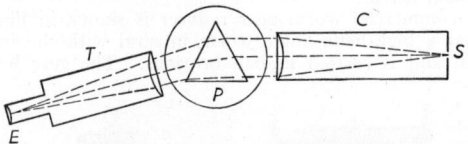

Fig. 1. Diagram of simple prism spectroscope. *C*, collimator with slit *S*; *P*, prism; *T*, telescope for viewing spectrum at eyepiece *E*.

a slit, S, a **collimating** lens, C, for rendering the light from the slit parallel before entering the prism, one or more dispersing prisms, P, and a **telescope,** T (or a **camera**), for forming images of the slit in the various wavelengths and thus providing a method for viewing or photographing the **spectrum.** The light passes through these in the order named, being deviated by the prism through various angles according to the wavelength. When a spectroscope is provided with a graduated circle for measuring deviations, it is called a "spectrometer." The "direct-vision" or non-deviation spectroscope, employing an **Amici prism,** is a compact instrument for qualitative purposes. The photographic spectroscope, known as a "spectrograph," is now almost universally used in spectral research. (See **Spectrography.**)

Many modern spectroscopes employ the **diffraction grating** instead of the prism. In the concave grating spectroscope, developed by Rowland, the collimating lens and telescope or camera objective are unnecessary be-

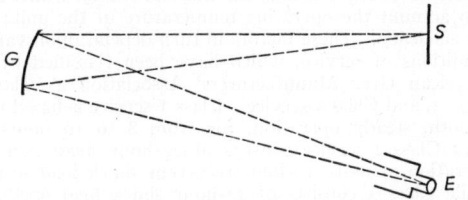

Fig. 2. Concave grating spectroscope. *S*, slit; *G*, grating; *E*, eyepiece (or plate-holder). (Diagrammatic.)

cause of the focusing effect of the grating itself (Fig. 2). Other types of instruments developed for particular purposes will be found described under **X-Ray Spectrometer** and **Objective Prism.** (L.D.W.)

SPECTROSCOPIC BINARIES.
Within the past 50 years we have become acquainted with a class of **binary stars** which are not **double stars** in the ordinary sense of the term, because of the fact that the components are too close together for them to be observed separately even in **telescopes** of the highest **resolving power.** In such binaries the period is usually short and the orbital velocities high. Unless the orbit plane happens to be perpendicular to the line of sight, the orbital velocities will have components in the line of sight and the observed **radial velocity** of the system will vary periodically. Since radial velocity is measured with the spectroscope, employing the **Doppler-Fizeau principle,** the binaries so observed are known as spectroscopic binaries. In some spectroscopic binaries the spectra of both stars are visible and the lines are alternately double and single. Such stars are known as double-line binaries. In others the spectrum of only one component is seen and the lines in this spectrum move periodically from violet toward the red and back again.

The determination of the **orbit** of a spectroscopic binary is made from a long series of observations of the radial velocity of one or more components of the system. The observations are first plotted against time and from the resulting curve the period may be obtained. With this period determined observations are then reduced to a single epoch and the best possible curve drawn through the points, obtaining what is known as the velocity curve of the system. If the orbit is circular the velocity curve will be a sine curve, if elliptical, the shape of the curve will depend upon the eccentricity of the ellipse and the orientation of the major axis with reference to the line of sight. From the shape of the velocity curve the orbit of the system in space may be determined. In the solution of the spectroscopic orbit it is impossible to determine individually the semimajor axis, a, and the inclination of the orbit plane, i. However, the product of the semimajor axis by the sine of the inclination (i.e., $a \sin i$) may be determined directly in linear units (i.e., in either miles or kilometers). If either a or i can be obtained from other types of observations, as in the case of **eclipsing binaries,** a complete solution for the orbit can be made. (W.K.G.)

SPECTROSCOPIC PARALLAX.
The term spectroscopic parallax of a star is applied to a determination of the distance of a star in which the **stellar parallax** is determined from observations of spectral peculiarities of the star together with determinations of the apparent brightness of the object.

The apparent brightness of a star depends upon two fundamental factors: the intrinsic brightness of the star and its distance from the observer. Expressed on the **stellar magnitude** scale, we find the apparent magnitude, mg, the **absolute magnitude,** M, and the stellar parallax, π'', to be connected by the analytical expressions: $M = mg + 5 + 5 \log_{10} \pi''$. The apparent magnitude of a star may be determined by a variety of methods of **stellar photometry,** and if a method is available for the determination of the absolute magnitude the value of the stellar parallax may be determined.

The relative intensities of certain **spectral lines** are different in **giant and dwarf stars** of the same spectral type. The relative intensities of selected pairs of lines may be compared in stars of the same spectral type and known absolute magnitudes and a "calibration curve" obtained. The relative intensities of the same pairs may then be found in stars of unknown absolute magnitudes and the calibration curves used to determine the absolute magnitude. Thus, the parallax may be determined from a study of spectra. The accuracy of the determinations of spectroscopic parallax compares very favorably with parallaxes obtained from the relative trigonometric methods. (W.K.G.)

SPECTRUM. This term usually refers to the array of wavelengths or frequencies resulting from the dispersion of light or other radiation, as by a prism or a diffraction grating, and revealed by the spectroscope. The emission spectrum of a substance is that of the radiation which it emits when "excited," while the result of passing white light through the substance is its absorption spectrum. Incandescent solids, liquids, or gases under high pressure give a continuous spectrum, while gases under a relatively low pressure give either an atomic spectrum made up of lines, or a molecular spectrum of bands, or both. The spectrum of the same substance may differ according to the method of excitation. Thus we may have the flame spectrum, the arc spectrum, or the spark spectrum of, say, iron. That is, iron salts may be vaporized in a flame, or an arc or a spark may be passed between iron electrodes. X-ray spectra, while in many essential respects similar to those of light, belong in a distinct class, because of the differences in technique required by the much shorter wavelengths employed. (L.D.W.)

SPECULAR IRON. Hematite.

SPEED, AIRCRAFT DESIGN GLIDING. Airplane Speeds.

SPEED, AIRCRAFT DESIGN LEVEL. Airplane Speeds.

SPEED, GROUND. Airplane Speeds.

SPEED, HUMP. Seaplane.

SPEED, LANDING. Airplane Speeds.

SPEED, MINIMUM FLYING. Airplane Speeds.

SPEED, MAXIMUM VERTICAL. Airplane Speeds.

SPEED RATIO. Chain; Belting; Gearing.

SPEED REDUCERS. Geared speed reducers afford all the advantages of toothed gearing and require very little attention or maintenance other than periodic inspection of the oil supply, if they are properly selected and installed. Practically all types of gearing are employed in their construction; the most important forms of reducers are those with parallel input and output shafting, for which spur, helical, and herringbone gearing are used; and those with perpendicular shaft axes, which employ spiral bevel, hypoid, or worm gearing. One of the simplest forms of speed reducers is the gearmotor shown in Fig. 1. The unit illustrated is an "offset" re-

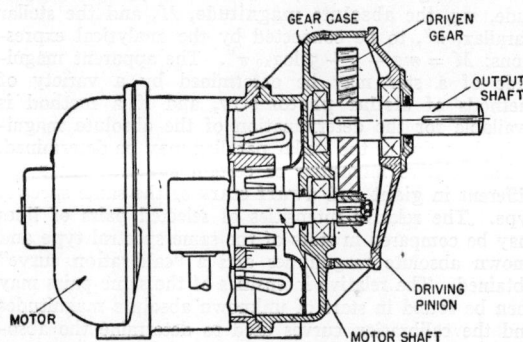

Fig. 1. Gearmotor speed reducer.

ducer, in which a helical pinion on the end of the motor shaft drives a helical gear on the output shaft. The gear case is supported by the motor housing, and can be

adjusted so that the output shaft is in the same horizontal plane as the motor shaft, or in several intermediate positions as well as in the position shown, thereby making the unit adaptable to various conditions of installation. Double and triple reduction units are available for high-speed ratios; in some of these the motor is flange-mounted on the reducer case. Planetary gear motors, with the motor and output shaft axes in alignment, are available in single and double reduction units and for a wide range of ratios and power capacities up to about 75 hp.

Parallel shaft reducers with a single set of helical or herringbone gears are used for reductions up to 10:1, and are known as single-stage reducers; two-stage units, with two sets of gears, are used for reductions from 10:1 to 60:1; and three-stage reductions over 60:1.

Right-angle, two-stage speed reducers with a spiral bevel gear set for the high-speed stage, and a helical gear set for the low-speed stage, are common. Single and triple reduction units are also available. Units of this character can also be procured with vertical output shafts, projecting either above or below the gear case, for agitator and mixer drives.

Worm gear reducers are frequently employed for high velocity ratios and heavy power demands. The driving motor is generally connected to the input or worm shaft by means of some form of flexible coupling; the drive from the worm gear shaft, or output shaft, to the driven machine can be effected either through a flexible coupling for direct-connected drives, or by spur gearing, belt or chain drives.

A commercial worm gear reducer is shown in Fig. 2. It has a high-helix angle worm integral with the input shaft, and is carried in ball bearings. The gear has a

Fig. 2. Worm gear reducer.

bronze rim bolted to a cast iron spider mounted on the slow-speed or output shaft and is carried in roller bearings. Horizontal units are also available in which the worm is above the gear; there are also double reduction units in which two sets of worm gearing are incorporated in a single case.

Worm gear reducers should be selected upon consideration of three important factors: the mechanical rating of the unit, which involves the strength and wear load capacities of the gearing; the thermal rating, which takes into account the operating temperature of the unit; and the efficiency. These factors in turn depend upon various conditions of service, which have been classified by the American Gear Manufacturers' Association as Class 1, Class 2, and Class 3 service. Class 1 service is based upon smooth, steady operation, for from 8 to 10 hours per day; Class 2 service consists of 24-hour steady, smooth operation or 8- to 10-hour recurrent shock load service; while Class 3 consists of 24-hour shock load operation. (H.C.H.)

SPEED REGULATION. Control, Motor; Governor.

SPEED RING. Hydraulic Turbine.

SPEED, STALLING. Airplane Speeds.

SPEED, TAKE-OFF. Airplane.

SPELTER. Galvanizing.

SPENOLITH. A term proposed by Burckhardt in 1906 for wedge-shaped igneous intrusions. (R.M.F.)

SPERM. Gamete.

SPERM DUCT. Any of the ducts of the testes. Reproductive System. (A.W.L.)

SPERM SAC. Seminal Vesicle.

SPERM VESICLE. A dilation of the terminal portion of the vasa deferentia of the arrow worms. Unlike the seminal vesicles of most animals, which are nearer to the testes, these reservoirs open directly to the exterior. (A.W.L.)

SPERMACETI. Esters.

SPERMATHECA. Receptaculum seminis.

SPERMATOPHORE. A packet of spermatozoa. In some of the invertebrates, including **annelid** worms, **arthropods,** and **mollusks,** instead of swimming freely in a seminal fluid the reproductive cells of the male are transferred to the female in these small bundles. The formation of the packets is a function of the male genital ducts, which secrete a retaining envelope about the masses of sperm cells. (A.W.L.)

SPERMATOPHYTA. Seed Plants. This is the largest division of the plant kingdom, comprising over 130,000 species. Members of this division, including nearly all plants of economic value, are of the utmost importance to man, for from them come practically all his food and that of his domestic animals, his clothing, and, to a great degree, his shelter.

Seed plants are much more complex than are those in the other divisions. Their common distinguishing feature, of course, is the **seed.** A seed is composed of an **embryo** plant which has developed from a mature egg, usually after the latter is fertilized, together with a certain amount of reserve food which may surround the embryo or be stored within its tissues, and one or more seed coats which completely surround the embryo plant and the food reserves.

In seed plants there occurs a definite alternation of generations. The **gametophyte** generation, however, is very much reduced, and entirely dependent on the **sporophyte** throughout its brief existence. The familiar seed plants are the sporophyte generation.

Seed plants are divided into two groups, the **gymnosperms** and the **angiosperms,** of which the gymnosperms are the more primitive. In them the seed is formed on the surface of an ovuliferous scale. Pines, spruces, cedars, larches, and cycads are all gymnosperms. The angiosperms are more advanced, and are generally called flowering plants. Two groups of angiosperms are recognized, the **dicotyledons** and the **monocotyledons.** Plants in these two groups are separated by the number of seed leaves or cotyledons present in the embryo. Roses, grapes, lilies, maples, and orchids are all angiosperms. (See also **Paleobotany.**) (R.M.W.)

SPERMATOZOA. The male sex **cells,** which are formed in the testicles and are the main element of the semen or seminal fluid. With coitus they are ejaculated into the vagina, where under favorable conditions a single spermatozoon unites with an **ovum.** In a single ejaculation several million spermatozoa are contained in the semen. The spermatozoon is a small cell about 1/500″ in length; in shape it resembles a tadpole in that it has a head, a long body, and a long tail which is responsible for its motility. (R.S.M., D.M.H.)

SPERMOPHILE. Mammalia, Rodentia. A small slender animal with short legs, small external ears, a short hairy tail, and large cheek pouches. Also called ground squirrel and gopher. The name spermophile is derived from the genus *Spermophilus,* in which these animals were formerly grouped.

Most of the several species of spermophiles are confined to limited areas in the arid lands of the west. The striped species, commonly called the gopher, ranges from central Ohio to the Rockies and from Canada to Texas. This species is brown with alternating clayyellow stripes and rows of spots on the back and sides. It is destructive to crops and damages lawns. (A.W.L.)

SPERRYLITE. A mineral diarsenide of platinum, $PtAs_2$. Crystallizes in the isometric system. Hardness, 6–7; specific gravity, 10.58; color, white; opaque. Named after Francis L. Sperry, Sudbury, Ontario. (R.M.F.)

SPESSARTITE. Garnet.

SPHAGNUM. Bryophytes.

SPHALERITE — ZINC BLENDE — BLENDE. Sphalerite is zinc sulfide, ZnS, crystallizing in the isometric system frequently as **tetrahedrons,** sometimes as cubes or **dodecahedrons,** but usually massive with easy **cleavage,** which is dodecahedral. It is a brittle mineral with a conchoidal fracture; hardness, 3.5–5; specific gravity, 3.9–4.1; luster, adamantine to resinous, commonly the latter. It is usually some shade of yellow brown or brownish-black, less often red, green, whitish, or colorless; streak, yellowish or brownish, sometimes white; transparent to translucent. Certain varieties are phosphorescent or fluorescent. Sphalerite is the commonest of the zinc-bearing minerals, and is found associated with **galena, chalcopyrite, tetrahedrite, barite, fluorite,** etc., as a result of contact **metamorphism,** and as replacements, and **vein** deposits. There are very many foreign localities, including Saxony, Bohemia, Switzerland; Cornwall, in England; Spain, Sweden, Japan, and elsewhere. In the United States sphalerite is found in Arkansas, Iowa, Wisconsin, Illinois, Colorado, New Jersey, Pennsylvania, Ohio, and especially in the so-called Tri-State area which includes parts of Kansas, Missouri, and Oklahoma, the largest zinc-producing district in the world at the present time. The word sphalerite is derived from the Greek, meaning treacherous, and its older name, blende, meaning blind or deceiving, refers to the fact that it was often mistaken for lead ore. (E.S.C.S.)

SPHENE. Titanite.

SPHENISCIFORMES. The penguins. An order of birds (**Aves**) characterized by the paddle-like wings, webbed feet, and upright posture when out of the water. They are remarkable swimmers, using both wings and feet under the water. Penguins are confined chiefly to the more southern parts of the southern hemisphere, including the Antarctic regions, but one species is found on the Galapagos Islands. (A.W.L.)

SPHENOID. A bone located in the anterior part of the base of the skull. It forms part of the floor of the brain cavity and is of complex shape, articulating with most of the other bones of the skull. (A.W.L.)

SPHENOPHYLLALES. Paleobotany.

SPHERE.

A spherical surface is a **surface** all points of which are at a fixed distance, its radius, from a fixed point, its center. It is often referred to as a sphere, although this term frequently means the solid bounded by the spherical surface.

The equation in **rectangular coordinates** of a sphere with center at (h, k, l) and radius r is:

$$(x - h)^2 + (y - k)^2 + (z - l)^2 = r^2.$$

Any equation of the form

$$x^2 + y^2 + z^2 + Ax + By + Cz + D = 0$$

represents a sphere.

For a sphere of radius R, the area of the surface is $4\pi R^2$ and its volume is $\frac{4}{3}\pi R^3$. (L.L.S.)

SPHERE PHOTOMETER. Integrating Photometer.

SPHERICAL ABERRATION.

If the surfaces of a **lens** or the reflecting surface of a mirror are spherical, the rays refracted through or reflected from the outer portions will be brought to a focus in a different plane than those from the center, thus producing a blurring of the resultant image known as spherical aberration. This effect is more pronounced in short-focus lenses or mirrors than in long-focus instruments, for the curvature of the surfaces of the short-focus instruments is greater.

The decrease of spherical aberration with increase in focal length was discovered very early in the history of optical instruments, and during the 17th century we find telescope builders increasing the focal lengths of their instruments to tremendous proportions. Telescopes with focal lengths between 100 and 200' were not uncommon during this period, and the problem of supporting the long thin tubes so that they could be used for astronomical purposes and remain straight was one calling for great ingenuity. Descartes in 1637 published the theory of spherical aberration and showed that theoretically it could be corrected by grinding the surfaces of lenses and mirrors in curves other than spheres. The difficulties of grinding the required lens curves were apparently greater than those of operating the long-focus telescopes. About the middle of the 18th century John Dolland published the fact that spherical aberration could be corrected by the same method employed for the treatment of **chromatic aberration**, i.e., by using two lenses, one convergent and the other divergent. All modern telescopic lenses of good quality now employ the double object glass with the figures of the lenses and the separation between the components depending upon the ideas of the makers. For wide-angle, short-focus lenses, such as are used in modern hand **cameras** and in astrographic cameras, the simple pair of lenses does not provide sufficient correction for all of the aberrations, and three or more lenses are used in combination. Perhaps the most common is the so-called doublet, in which two pairs of lenses are used with a considerable separation between the pairs.

Spherical aberration may be corrected in the case of a mirror by grinding the concave surface in the form of a **paraboloid** of revolution instead of in the form of a sphere. This provides almost complete correction for rays entering the mirror parallel to the axis of revolution of the paraboloid, i.e., along the principal axis of the mirror; but for rays entering at a moderately large angle, spherical and various other aberrations make their appearance in the resultant image. Hence, while the reflecting type of telescope can be more easily corrected for the aberrations along the axis, nevertheless it cannot be used to obtain photographs of large areas with good definition throughout. (See **Telescope**.) (W.K.G.)

SPHERICAL ANGLE.

A spherical angle is formed by two intersecting arcs on the surface of a **sphere**; it is measured by the plane angle formed by the **tangents** to the arcs. (L.L.S.)

SPHERICAL COORDINATES.

The spherical coordinates of a point in space are three numbers which determine the position of the point and which are particularly adapted to problems involving **spheres**. In case the point is assumed to be located on the surface of a sphere of known or unknown radius two angular coordinates are sufficient for locating the position of the point on the surface of the sphere.

The accompanying figure represents reference frames for both spherical and also 3-dimensional rectangular

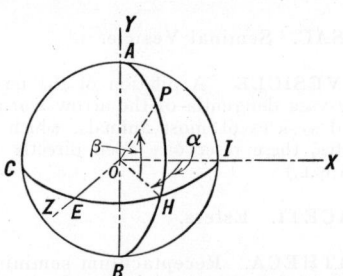

coordinates. In setting up the spherical system an origin, O is first selected, and assumed to be the center of the sphere of reference. A line, AB, is then passed through the origin and must intersect the sphere in diametrically opposite points. This line is known as the fundamental line for the spherical system. A plane is then passed through the origin perpendicular to the fundamental line, and this plane becomes known as the fundamental plane. Since the fundamental plane passes through the center of the sphere it must intersect the sphere in a great circle, $CEHI$. Since the fundamental plane is perpendicular to the fundamental line, AB, the points A and B must be poles of the great circle on the sphere. For reference purposes some fundamental direction, say OI in the figure, must be selected in the fundamental plane.

To determine the position of the point, P, on the sphere a plane, $AOBHP$, is passed through the fundamental line, intersecting the surface of the sphere in the great circle, $APHB$. One angular coordinate, α in the figure, is measured in the fundamental plane from the fundamental direction to the line of intersection of the plane passed through the point with the fundamental plane. The other angular coordinate, β in the figure, is measured in the plane, $AOBHP$, through the point from the fundamental plane to the radius of the sphere through the point P. Both angular coordinates may be measured by the arcs of the great circles cut out on the sphere by the planes involved, e.g., α may be measured by the arc, IH, of the fundamental plane, and β by the arc, HP, of the plane through the point and the fundamental line.

To define the position of P in 3-dimensional space referred to the point O as origin the radius of the sphere, ρ, is also necessary. The transfer from spherical to **rectangular coordinates** may readily be made by passing the XY, XZ, and XY planes through the sphere as shown in the figure. Then we have at once:

$$x = \rho \cos \beta \cos \alpha; \quad y = \rho \sin \beta; \quad z = \rho \cos \beta \sin \alpha,$$

$$\rho^2 = x^2 + y^2 + z^2; \quad \tan \alpha = \frac{z}{x}; \quad \tan^2 \beta = \frac{y^2}{x^2 + z^2}.$$

(W.K.G.)

SPHERICAL HARMONICS.

A spherical harmonic is frequently defined as any homogeneous **polynomial** V_n in the variables x, y, z of the nth degree which satisfies **Laplace's equation**

$$\frac{\partial^2 V}{\partial x^2} + \frac{\partial^2 V}{\partial y^2} + \frac{\partial^2 V}{\partial z^2} = 0.$$

If this equation is transformed to **spherical coordinates** r, θ, ϕ, the solution is called a solid spherical harmonic, U_n. If this function be divided by r^n, we obtain a function of θ and ϕ only, which is called a surface spherical harmonic. When the spherical harmonic is independent of ϕ, it is called a surface zonal harmonic and satisfies Legendre's equation, and is therefore a **Legendre function**. (L.L.S.)

SPHEROCLAST. Phenoclast.

SPHEROIDAL STATE. Ebullition.

SPHEROMETER. An instrument for measuring the curvature of solid spherical surfaces, either convex or concave, such as those of lenses; a measurement in which high precision is not easily attained. The most familiar mechanical device for this purpose is a form of **micrometer.** It resembles a small three-legged stool, the sharp steel points of whose legs form an equilateral

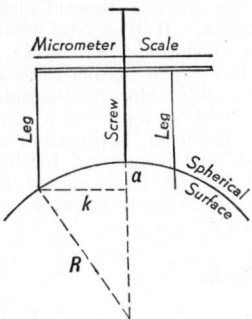

Diagrammatic section of spherometer.

triangle. The micrometer screw, also with a sharp point, is mounted at the center of this supporting trivet, and is adjusted to read zero when all four points are in one plane, as determined by standing the instrument on a flat plate of glass and screwing the micrometer point down to it. The distance from each of the legs to the central axis must be accurately known. If it is called k, and if the elevation (or depression) of the micrometer point to fit a given spherical surface is a, then the radius of curvature of the surface is readily calculated as

$$R = \frac{k^2 + a^2}{2a}.$$

The chief sources of error are in determining just when contact takes place between micrometer point and surface, and in measuring k. (L.D.W.)

SPHERULITES. In lavas there are frequently to be found radial or concentric aggregates of **mineral** crystals from microscopic size to an inch or two in diameter, although rarely their dimensions may be much greater. These radial aggregates are believed to be the result of rapid growth of small **crystals** at favorable points in the quickly cooling lava. (E.S.C.S.)

SPHINCTER. A ring of muscle fibers surrounding a hollow organ or the opening of a duct. By their contraction sphincters constrict or close the passages with which they are associated. Muscles of this kind are found in the iris of the **eye** around the margin of the pupil, in the wall of the **anus**, at the union of the urethra with the urinary **bladder**, at the **pylorus** between the stomach and the duodenum, and in other similar relations. (A.W.L.)

SPHYGMOMANOMETER. An apparatus for measuring the **blood pressure.** It consists of a rubber-bag cuff which is wrapped around the upper arm. This is inflated by a hand bulb. The cuff is connected by rubber tubing to a measuring device which is either a sealed column of mercury or a spring scale. Sufficient pressure is pumped into the rubber cuff to compress the brachial artery in the upper arm. A stethoscope is applied over the artery below the cuff and air is gradually allowed to escape from the cuff until the pulse can be heard. The reading on the scale or column of mercury at this point indicates the systolic pressure or the highest pressure in the arteries during contraction of the **heart.** The deflation of the cuff is continued, and that point on the scale when the last sound of the disappearing pulse is heard is the diastolic pressure, or lowest pressure in the artery during diastole, or relaxation of the heart muscle between beats. The normal systolic reading of an adult varies from 110 to 130 or 140 mm. of mercury. Normal diastolic readings vary from 60 to 90 mm. of mercury. (R.S.M., D.M.H.)

SPICA. Spica (α **Virginis**) is a brilliant first **magnitude** star. Spica is particularly interesting in that it is believed to be the star that provided Hipparchus with the data which enabled him to discover the **precession** of the equinoxes. The temple at Thebes was oriented with reference to Spica in about 3200 B.C. Later, temples which were oriented to this same star indicated the motion of the star due to precession and provided the necessary data. (W.K.G.)

SPICULE. A small hard body formed within the tissues of animals. Spicules are formed in all sponges (**Porifera**) by cells called scleroblasts in the mesogloea. They vary in form from simple rods to complex structures radiating from a common center on three to six axes. Some are straight with expanded ends. In the **coelenterates** spicules appear only in some of the **alcyonarians,** where they are formed in the mesogloea by cells that migrate from the ectoderm. The term has also been applied to the radiating hard structures of the 1-celled **radiolarians.** The formation of bone in the embryos of vertebrates begins with the deposition of spicules which are later consolidated into larger masses of bone. (A.W.L.)

SPIDER. Arachnida, Araneina. **Arthropods** of almost exclusively terrestrial habits, known commonly for their ability to spin silken webs. They differ from other arthropods in one or more of the following characters: The body is divided into **cephalothorax** and abdomen, and the latter is unsegmented. The head bears a group of simple eyes. Four pairs of legs are present. The jaws are perforated by the ducts of poison glands. The ventral surface of the abdomen bears the openings of the **lung books,** respiratory organs of peculiar form, and a group of **spinnerets** through which the ducts of the silk glands open.

Spiders are widely known and unaccountably repulsive. In fact, they are among the most interesting of all animals, and with a single known exception, the black widow spider, are harmless. Even though they secrete poison most of them are too small to bite a human being unless on a very thin fold of tissue, and most seem to have no inclination to bite. Even the large hairy species commonly called banana spiders or tarantulas are mild-mannered creatures.

To what extent the bad reputation of the black widow, *Latrodectus mactans,* is deserved seems difficult to establish. Apparently its bite is severely poisonous and occasionally fatal, and apparently it is vicious in habits. There is contradictory evidence, however, so the case is not wholly settled.

Spiders vary greatly in habits. Some spin funnel-like webs in which they hide to await their prey, others form irregular webs, and still others make the orb webs so beautifully demonstrated in our gardens on dewy mornings. Other forms spin no web but capture their prey by pouncing on it from concealment or by open chase. Among these forms are the crab spiders, named from their short broad form, which lie in wait on plants and are sometimes almost perfectly hidden in flowers by their concealing coloration. The wolf spiders are stout hairy species, often black in color. They hunt like the predators for which they are named.

The spiders that capture prey without the use of webs have other uses for silk, such as the formation of cocoons or egg-sacs in which the eggs are deposited, and the construction of a smooth lining for their hiding places. The most remarkable example of the latter use is the nest of the trapdoor spiders of warm regions. These nests consist of a silk-lined burrow with a beveled margin at the surface of the ground. A lid hinged with tough silk fits perfectly into this beveled depression and can be held shut by the spider, which provides in its lining two depressions to be gripped by the claws.

The mating habits of spiders are also remarkable. The male, in many species much smaller than the female, goes through a courting procedure as complex as that of the birds, and is often killed by his consort. Reproductive adaptations are also peculiar in the males, in that the palpi are modified to convey the seminal fluid to the genital passages of the female. When sexually mature the male spins a web in which the contents of the reproductive organs are discharged, to be taken up into the cavities in the palpi. When the individual is successful in securing a mate he thrusts the palpi one at a time into her genital aperture. (A.W.L.)

SPIKE. Flower.

SPIKE, HOOK-HEAD RAILWAY. Railway Track. Spline. Key.

SPIKELET. Flower.

SPILITE. A term used to define a **basaltic** rock whose **feldspars** are chiefly **albite** or soda-rich **oligoclase.** (R.M.F.)

SPILLWAY. One of the important adjuncts of a **dam** of the overflow type is a spillway, which is simply an opening through or over which excess water may flow when the reservoir is full. The spillway may be a certain overflow section of the dam, or it may be cut in rock along the normal reservoir line at one side of the dam. It is necessary for the crest to be sufficiently long that maximum expected floods may be passed through it without the depth of water on the spillway crest exceeding a predetermined depth. A plain spillway fixes the maximum depth of water in the reservoir. The discharge over a spillway is something like that over a weir, and the rate of flow is proportional to the length of crest and the ⅗ power of the **head** over the crest. It is possible to raise the water level several feet above the spillway, and yet preserve the safety features, by equipping it with crest gates, which normally will be lowered onto the crest of the spillway, obstructing it and raising the water level, but which may be raised out of the way and leave the stream unobstructed during the flood seasons. (See **Gates.**) (F.T.M.)

SPINAL ANAESTHESIA. Anaesthesia.

SPINAL CORD. The large axial structure of the central **nervous system** of **vertebrates,** extending from the brain back through the body. It lies in the dorsal body wall and is surrounded by the neural arches of the spinal column. It consists of nerve tracts or aggregations of fibers for communication of impulses to and from other parts of the nervous system, and of nuclei, which are aggregates of cell bodies of **neurons.**

In man the spinal cord is a cylindrical mass of nervous tissue about 18″ long within the vertebral canal of the spine. Thirty-one pairs of nerves are attached to the **spinal cord.** The spinal cord diminishes in size from above downward, and is suspended within the spinal canal by certain ligaments and membranes, and bathed in **spinal fluid.**

The functions of the spinal cord are exceedingly complex, and cannot be understood without a thorough knowledge of the equally complex anatomy. In general, the cord may be said to be the central receiving and distributing plant for all nervous impulses going to and coming from all parts of the body. This includes sensory and motor impulses, voluntary and involuntary impulses, and **reflex** impulses. Some of the fiber tracts running in the cord have elaborate connections with the higher centers in the **cerebrum,** where conscious control is exerted. (A.W.L., R.S.M., D.M.H.)

SPINAL FLUID. The spinal canal contains spinal fluid which is in free communication with the fluid surrounding the brain and filling its ventricles. This fluid is produced by the vascular choroid plexus of the ventricles of the **brain.** It is crystal clear and contains minute amounts of chlorides and other inorganic salts, plus traces of glucose and plasma proteins. The spinal fluid serves as a fluid cushion for the **spinal cord.** Removal of spinal fluid by means of a spinal tap is commonly used for diagnostic information and therapeutic procedures. The fluid is cultured for bacterial growth, examined chemically, the cells counted, and its pressure within the canal measured, as well as other specific tests. (A.W.L., R.S.M., D.M.H.)

SPINDLE. Shaft.

SPINE. The vertebral column or backbone, in man, composed of thirty-three vertebrae which, grouped according to regions, are as follows: seven cervical, twelve thoracic, five lumbar, five sacral, and four coccygeal vertebrae. (R.S.M.)

SPINEL. The mineral spinel is one of a group of minerals which crystallize in the **isometric** system with an **octahedral** habit, and whose chemical compositions are analogous. These minerals are combinations of bivalent and trivalent oxides of **magnesium, zinc, iron, manganese, aluminum,** and **chromium,** the general formula being represented as $R''O \cdot R''_2'O_3$. The bivalent oxides may be MgO, ZnO, FeO, and MnO, and the trivalent oxides Al_2O_3, Fe_2O_3, Mn_2O_3, and Cr_2O_3. The more important members of the spinel group are spinel, $MgAl_2O_4$; gahnite, zinc spinel, $ZnAl_2O_4$; franklinite $(Fe \cdot Mn \cdot Zn)$, $(Fe \cdot Mn)_2O_4$, and **chromite,** $(Fe \cdot Mg)Cr_2O_4$. True spinel has long been found in the gem-bearing gravels of the Island of Ceylon and in the limestones of Burma and Siam. Spinel usually occurs in isometric crystals, octahedrons, often twinned. It has an imperfect octahedral **cleavage;** conchoidal fracture; is brittle; hardness, 8; specific gravity, 3.5–4.1; luster, vitreous to dull; transparent to opaque; streak white; may be colorless, rarely through various shades of red, blue, green, yellow, brown, or black. These colors are doubtless due to small amounts of impurities. The clear red spinels are called spinel-rubies or balas-rubies and were often confused with genuine rubies in times past. Rubicelle is a yellow spinel. A violet-colored manganese-bearing spinel is called almandine. Spinel is found as a **metamorphic** mineral, also as a primary mineral in **basic rocks,** because in such **magmas** the absence of alkalies prevents the formation of feldspars, and any aluminum oxide present will form corundum or combine with **magnesia** to form spinel. This fact accounts for the finding of both ruby and spinel together. Besides the localities mentioned above, in Burma and Siam, the Ceylon yields beautiful specimens. Spinel is found in Italy and Sweden and on the Island of Madagascar. Also in the United States in Orange County, New York, and in Sussex County, New Jersey, are many well-known spinel localities. Spinel is found also in Macon County, North Carolina, and in Canada in Quebec and Ontario. The name spinel is derived from the Greek, meaning a spark, in reference to the fire-red color of the sort much used for gems. Balas ruby is derived from Balascia, the ancient name for Badakhshan, a country of central Asia situated in the upper valley of the Kokcha River, one of the principal tributaries of the Oxus. (E.S.C.S.)

SPINNERET. A spinning organ of the **spiders**. The spinnerets are located on the ventral surface of the abdomen, near or at its tip, and vary from one to three pairs. They are conical to cylindrical in form. Each has a membranous terminal portion called the spinning field, through which run many minute spinning tubes from the silk glands. The nature of the spinning tubes varies, different tubes producing different kinds of silk.

In the act of spinning the liquid silk is forced through the spinning tubes, to harden on exposure to the air as silk. The spinnerets bear spines, some of them apparently tactile, and are moved by muscles, so that the spider is able to form and place its threads with precision. See page 243 for use of spinneret in formation of rayon filaments. (A.W.L.)

SPINNING. Ductile sheet metals that can be cold-drawn can also be fabricated by spinning into cylinders and other shapes having rotational symmetry. The sheet blank is held in a spinning lathe between a former or chuck, which establishes the inside contour, and a rotating tailstock. The sheet is worked by means of pressure of a tool against the outer surface. Several chucks of progressively deeper contour may be required for deep vessels, and the metal may require process annealing between operations to overcome the work-hardening. Soft aluminum, copper, brass, and low-carbon steel are readily shaped by spinning, while harder metals such as stainless steel can be spun with proper tooling and annealing practice.

Spinning is a relatively inexpensive process especially suited to the production of small numbers of parts for which the cost of drawing dies would be excessive. It is sometimes used in combination with drawing and other fabricating operations. (See **Press Working.**) (R.H.H.)

SPINTHARISCOPE. A simple apparatus, also called a scintilloscope, designed by Sir William Crookes for observing the scintillations produced by alpha rays (see **Radiactive Emissions**). It consists of a short metal tube with a screen of phosphorescent zinc sulfide closing one end, and a strong magnifying lens at the other, focused upon the screen. Projecting out in front of the screen, inside the tube, is a narrow metal point, the tip of which bears a trace of radium. The alpha particles emitted by the radium fall upon the screen and produce scintillations which, under suitable conditions, are plainly visible through the lens, looking like minute sparks. To see them clearly, one must have been in a dark room for some minutes; also care should be taken not to expose the screen to a strong light beforehand, or its phosphorescence may obscure the scintillations. A large-scale modification of the instrument has been used for counting scintillations and hence estimating the emission of alpha particles from radioactive substances. (L.D.W.)

SPINULOSA. Asteroidea.

SPINY ANTEATER. Echidna.

SPIRACLE. 1. A small opening on the surface of the insect body, leading into an air tube of the **respiratory system.** Insect spiracles are simple openings in some species, but usually they are guarded by some structure which prevents the entrance of foreign particles. This guard varies from a fringe of hairs to an elaborate sieve-like plate. Some insects depend on a closing apparatus of the associated **trachea** for the exclusion of particles.

In the primitive condition spiracles apparently were paired and **metameric,** one opening on each side of each segment. This arrangement is only slightly modified in some species, which are said to be peripneustic. In other cases the number of openings is greatly restricted, only one or two pairs providing the sole entrance to the respiratory apparatus. Insects with a single pair at the anterior end of the body are called propneustic, those with a pair at the posterior end metapneustic, and those with a pair at each end amphipneustic. The spiracles may also be associated with tubes, either extending the tracheae beyond the general surface of the body or forming external conduits for air, by which the insect may reach the air from within the water or decaying matter in which it lives. 2. The greatly reduced external opening representing the first of the series of **gill slits** in elasmobranch fishes. (A.W.L.)

SPIRAL BEVEL GEARING. Bevel Gearing.

SPIRAL CURVE. In railway or highway alignments a spiral curve, sometimes called an easement or transition curve, is one which provides a gradual change of curvature when passing from a tangent (straight line) to a circular curve. The cubic parabola and cubic spiral are curves well suited for this purpose. The former is used when spirals are to be laid out by offsets from the

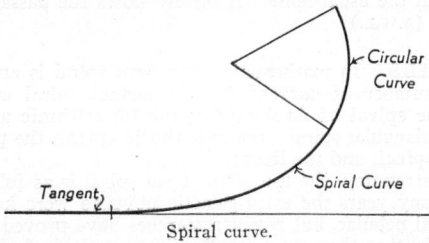

Spiral curve.

tangent (see **Tangent Offset**), since the offset to any point on the curve varies as the cube of the distance along the tangent from the point where the curve begins. When the curve is to be laid out by **deflection** angles the cubic spiral is used. In this curve the offset from the tangent is proportional to the cube of the distance along the curve measured from the point of tangency. Spiral curves are particularly well adapted for the use of **superelevation** since they furnish a means for gradually increasing this quantity from zero to the amount required on the circular portion of the curve. Railroads use a combination of spiral and circular curves to connect the tangents on their main lines. The value of the spiral is being recognized in modern highway design for high-speed traffic. (C.W.C.)

SPIRAL GEARING. Helical gears used for transmitting power between shafts whose axes are neither parallel nor intersecting are commonly, although incorrectly, called spiral gears. They may be adapted to any shaft axes angle, although they are usually employed for shaft axes at 90° to each other. The tooth elements are similar to those of parallel-shaft helical gears; the usual method of tooth measurement is by normal diametral pitch P_n. In 90° shaft-angle gearing, both pinion and gear have helical teeth of like hand, with complementary helix angles H_p and H_g. Spiral gear pitch diameters, like parallel-shaft helical gear diameters, may be adjusted in design to accommodate a special or unusual center distance. Power-transmitting capacity of spiral gearing is limited. Although some operating characteristics are analogous to worm gearing, the contact area between the teeth of spiral gears is theoretically a point, and in actual practice is confined to a very small area. Spiral gears are generally used when motion rather than power is of major importance. (See **Helical Gearing.**) (H.C.H.)

SPIRAL MILLING. Milling.

SPIRAL OF ARCHIMEDES. The spiral of Archimedes is a mathematical curve of spiral form, as shown in the accompanying figure. Its equation in polar coordinates is $r = a\theta$. (L.L.S.)

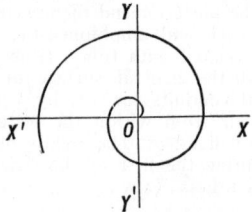

Spiral of Archimedes.

SPIRAL VALVE. A thin ridge projecting from the lining of the intestine of elasmobranch fishes to the center of the cavity and following a spiral course through the length of the tube. The structure provides greatly increased surface for the digestion and absorption of food. It is so arranged that the food must follow a spiral course, but beyond this regulating effect is not a valve in the usual sense. It merely slows the passage of food. (A.W.L.)

SPIRALS. In mathematics, the term spiral is applied to a number of curves. The important spiral curves are: the **spiral of Archimedes**, the **logarithmic spiral** (or equiangular spiral), the **hyperbolic spiral**, the **parabolic spiral**, and the **lituus**.

In astronomy, the use of the term spiral is as follows. For many years the extra-galactic **nebulae** were known as spiral nebulae, but recent researches have proved conclusively that they are a totally different class of objects from the galactic nebulae. Their **spectra**, instead of being characterized by the bright lines commonly found in the galactic nebulae, are of the absorption type such as would be produced by a large number of stars of various **spectral classes.** Long-exposure photographs of the extra-galactic nebulae, taken with instruments of long focal length, have definitely proved the hypothesis that these objects are in reality large groups of very distant stars.

In at least five of the extra-galactic nebulae **Cepheids** have been discovered, and the application of the period-luminosity relation for Cepheids gives distances of between 100,000 and 1,000,000 **light years** from the sun. By various statistical methods, largely developed by Hubble at the Mount Wilson Observatory, the distances of a number of other extra-galactic objects have been obtained. Distances of the order of magnitude of one billion (1,000,000,000) light years have been obtained; which seems to be about the limiting distance at which these objects can be distinguished with the 100-in. telescope. Researches on the distances of the extra-galactic nebulae will be one of the first problems for the new 200-in. telescope.

The extra-galactic nebulae are classified by Hubble as: 1. Spirals (with subdivisions of normal spirals and barred spirals. 2. Elliptical. 3. Irregular. A more detailed subdivision of each main type has been suggested, but is too complex to be considered here. Of the extra-galactic nebulae which are bright enough to

have been classified thus far, about 77% belong to the first type, 20% to the second type, and 3% to the third.

The normal spirals are characterized by two main whorls, each of which may have several branches, which come out from opposite sides of a large central condensation and wind about this condensation in the same plane and in the same direction. The central condensation is distinctly ellipsoidal in form, being much more extensive perpendicular to the plane of the whorls than are the whorls themselves. The spirals are found in all types of orientation, some with the plane of the spiral perpendicular to the line of light, and in others with the whorls seen edge on. In the latter cases a dark bank is observed passing across the nucleus as though there was a dark extension in the plane of the whorls which would otherwise be unobserved. The barred spirals differ from the normal spirals in that the arms begin at the ends of a long bar which extends outward from the central condensation or nucleus.

The elliptic extra-galactic nebulae show no structural forms even when examined with the most powerful instruments. They are ellipsoidal in form, ranging from spheres to thin lens-shaped volumes. The elliptic nebulae which apparently have the largest size, are the companions to the Andromeda spiral.

The irregular extra-galactic nebulae are relatively rare objects. As the name implies, they have no definite form, and appear like clouds in space. The largest one known is found in the constellation of Sagittarius.

A spectrographic study of some of the brighter spirals indicates that these objects are in rotation about their minor axes. From the rotational speeds and sizes of these objects Hubble estimates their total masses to be of the order of magnitude of 600,000,000 to 1,000,000,000 times the mass of the sun.

The absorption lines in the spectra of the extra-galactic nebulae show strong displacement to the red, with the amount of displacement increasing with the distance of the object from the sun. If this "red shift" is interpreted as a **radial velocity** effect it implies that all of the extra-galactic nebulae are receding from the sun, with the more distant ones receding the more rapidly. The implication from these results is that the entire universe of nebulae is expanding, and several attempts have been made to explain this theoretically. However, Hubble and other observational workers are by no means convinced that the red shift of the lines in extra-galactic nebular spectra is due to a radial velocity effect, although they are unable as yet to find any other satisfactory explanation. Until the question as to whether these observations prove the expansion of the universe is answered, it does not seem desirable to enter upon a discussion of so controversial a problem.

Statistical studies of extra-galactic nebulae indicate that there are between 50 and 100 million of these objects within the limiting distance of observation with the 100-in. telescope. Studies of the distribution of these objects throughout the immense volume of space (approximately 1,000,000,000 light years in extent) indicate clustering of the objects in certain regions. The surveys are still in progress and results are uncertain, but it seems probable that there are galaxies of galaxies in space. The suggestion has been made that our own galactic system may be such a group. (L.L.S., W.K.G.)

SPIRE. The portion of a conical **snail** shell above the large whorl into which the aperture leads. (A.W.L.)

SPIRIFIER. Invertebrate Paleontology.

SPIRILLUM. Bacteria.

SPIROGYRA. Algae.

SPITTLE BUG. Frog Hopper.

SPLEEN. An organ of vertebrates derived from mesenchyme and lying in the mesentery. It is closely associated with the circulatory system. The organ consists of masses of tissue of granular appearance, known as lymphoid tissue, located around fine terminal branches of veins and arteries. According to one interpretation, these vessels are connected through the spleen pulp by modified capillaries called splenic sinuses. The pulp is supported by a reticular tissue foundation and contains blood cells of all kinds in addition to the characteristic mesenchymal cells. The functions of the organ are the formation of blood cells, the destruction of old red corpuscles, the removal of other debris from the blood stream, and as a reservoir for blood.

The human spleen is situated in the left upper part of the abdomen, behind the stomach, just below the diaphragm. This organ, in the normal individual, measures about 5″ by 3″ by 2″ in size. In certain diseases it often increases in size, and it may even fill a large portion of the left side of the abdomen. The spleen enlarges in malaria, bacterial endocarditis, leukemia, pernicious anemia, Hodgkins' disease, Banti's disease, tumors, and cysts of the spleen.

The spleen is classified as a ductless gland. Its presence is not necessary for life. It may be removed surgically and often is, following abdominal injuries with rupture and hemorrhage from the spleen, or in the treatment of certain blood diseases (hemorrhagic purpura, familial jaundice, etc.), or for the removal of splenic tumors or cysts. Congenital anomalies such as accessory spleens occur, and rarely the spleen has been found to be completely absent. The ancients are said to have removed the spleen from runners, believing that this increased their speed. (A.W.L., D.M.H.)

SPLICE. A splice is a connection which is used to transfer stress in a direct line between two adjacent members or between the separate, adjacent parts of the same member. In general, a splice consists of splice material and fastenings, each of which must be strong enough to carry the stress. Structural steel is spliced by plates or angles which are fastened by means of rivets, bolts or welding. A gusset plate is not ordinarily considered a splice plate unless it is used in conjunction with other splice material to transfer stress between adjacent chord members at a joint. Structural timber is spliced by steel plates and bolts or by additional splice timbers which are connected by bolts or by timber connectors. The reinforcing rods of reinforced concrete are spliced by welding or by overlapping them far enough to assume that bond (see Bond Stress) will produce the stress transfer.

Splices are made for two reasons. First, there is a limiting length to all manufactured material. When the length of the member exceeds the limiting length of any of its parts, it is necessary to use two or more pieces for each of these parts. Splices are used to form one continuous piece. Secondly, transportation conditions and erection equipment limit the length of structural members or parts of structures which can be completed at a fabricating plant. This necessitates a field connection. A splice which is made at a fabricating plant is a shop splice while one made in the field is a field splice. (C.W.C.)

SPLIT SWITCH. Railway Track.

SPLIT-TAIL. Pisces, Teleostei. A small silvery minnow, *Pogonichthys macrolepidotus,* of central California. (A.W.L.)

SPODUMENE. The mineral spodumene is a lithium aluminum silicate corresponding to the formula $LiAl(SiO_3)_2$ and occurs in monoclinic prismatic crystals, occasionally of very large size. It also occurs massive.

Spodumene has a perfect prismatic cleavage often very noticeable; uneven to splintery fracture; brittle; hardness, 6.5–7; specific gravity, 3.13–3.20; luster, vitreous to pearly; color, grayish- to greenish-white, green, yellow and purple. Its streak is white; it is transparent to translucent. Spodumene is characteristically a mineral of the pegmatites, and it is found in Sweden, Ireland, Madagascar and Brazil. In the United States it is found especially in the pegmatites of Oxford County, Maine; in the towns of Goshen, Huntington and Chesterfield in western Massachusetts; at Branchville, Connecticut; in North Carolina; in South Dakota in huge crystals and in San Diego and Riverside Counties in California. The name spodumene is derived from the Greek meaning ash-colored, particularly appropriate for the slightly weathered varieties. Hiddenite, the beautiful emerald-green or yellow-green spodumene that is used as a gem, was named for W. E. Hidden. Kunzite, named in honor of the late George F. Kunz, is a transparent lilac to rose-colored spodumene from Madagascar and California, also used as a gem stone. Spodumene alters rather readily to a mass of albite and muscovite. The commercial use of spodumene is chiefly as a source of lithium compounds. (E.S.C.S.)

SPONDYLITIS. Infection of the vertebrae of the of the spine. Tuberculous and pyogenic osteomyelitis are the common causes. (D.M.H.)

SPONGE. An animal of the phylum Porifera. In the vernacular the word refers to the sponge of commerce, which is the skeleton of a sponge from which the animal matter has been removed by maceration and washing. The material of which these sponges are composed is spongin. Calcareous (calcium) and siliceous (silicon) sponges are, of course, too harsh for similar use.

Commercial sponges are derived from various species and come in many grades, from the fine soft lambs' wool sponges to the coarse grades used for washing cars. They come from fisheries in the Mediterranean and in the West Indies and nearby waters. The industry has reached a value of about $3,000,000 annually but the manufacture of substitutes from rubber and cellulose is destined to reduce its importance. Cellulose sponges, in particular, have the good qualities of natural sponges with greater uniformity. (A.W.L.)

SPONGIN. A horny material found in the skeletons of sponges. In some species it unites spicules of other materials and in some it forms fibers in which the spicules are imbedded. The sponges of commerce have a skeleton formed wholly of a fibrous mass of spongin. (A.W.L.)

SPONGIOBLAST. A form of cell in the mesogloea of sponges by which the fibrous structures of the body are formed. (A.W.L.)

SPOONBILL. 1. Pisces, Chondrostei. The paddlefish of the Mississippi River system. 2. Aves, Ciconiiformes. Large long-legged wading birds related to the herons and flamingoes. The beak is very long and broad, with the terminal portion still further broadened to give it a spoonlike form. 3. Aves, Anseriformes. The shoveller ducks. (A.W.L.)

SPORANGIUM. Flower.

SPORE. A spore is a special type of reproductive cell which develops directly into a new plant. Spores are of many kinds. In the Thallophytes they may be asexual cells or zygotes. Thick-walled spores formed after the union of isogametes are called zygospores. Oöspores are fertilized eggs. In the higher plants spores are always produced by diploid plants and develop into haploid gametophytes. In Selaginella and the seed plants small spores, microspores, produce male gametophytes; large spores, megaspores, produce female gametophytes. Pollen grains in the seed plants are microspores. (See also **Paleobotany.**) (R.M.W., B.S.M.)

SPOROBLAST. A stage in the life cycle of some of the 1-celled animals of the class **Sporozoa.** After fertilization, the nucleus of the zygote subdivides. Each of the daughter nuclei, together with a surrounding mass of **cytoplasm,** becomes a sporoblast which secretes an enveloping membrane and becomes a spore. (A.W.L.)

SPOROCYST. A stage in the life cycle of some 1-celled animals, characterized by the presence of a heavy protecting wall enclosing a number of separate cells, the spores. (A.W.L.)

SPORONT. The stage in the **reproductive** cycle of 1-celled parasites of the class **Sporozoa** which subdivides to produce the active **sporozoites.** The sporozoites begin a new life cycle. (A.W.L.)

SPOROPHORE. Fungi.

SPOROPHYLL. A sporophyll is a **spore**-bearing leaf. It actually bears the sporangia which contain the spores. It may be greatly modified in structure and appearance. (R.M.W., B.S.M.)

SPOROPHYTE. In plants in which there is a distinct **alternation of generations,** the generation which bears the spores is called the sporophyte generation. A spore-bearing plant is a sporophyte. Its cells all contain the diploid or double number of **chromosomes.** (R.M.W.)

SPOROSAC. A sessile medusoid (**medusa**) that remains attached to the parent colony. Since the medusoids produce reproductive cells, these individuals, without need for locomotion, retain no other conspicuous function. (A.W.L.)

SPOROTRICHOSIS. A fungus infection caused by the genus *Sporothrichium.* It is characterized by tumor-like nodules, abscess formation and ulcers of the skin. Most of the cases in the United States have been reported in the Middle West where the parasite has been found on vegetable matter.

Treatment by **iodine** compounds orally is usually successful in arresting the disease. (R.S.M., D.M.H.)

SPOROZOA. One-celled animals of parasitic habits which are usually incapable of locomotion. The immature stages of some species form pseudopodia but neither **cilia** nor **flagella** are ever present. A class of the phylum **Protozoa.**

The sporozoans absorb nutriment from the body of the host, a type of nutrition rarely possible save in parasites. In connection with their parasitic habits also they undergo intricate cycles of **reproduction,** often associated with life in alternating hosts of different species. They attack animals of almost every phylum, sometimes with fatal results or with more or less serious disturbances of normal functions. In man these relations are illustrated by the **malarial** parasites whose complex life cycle is completed in Anopheles **mosquitoes** and the human blood stream. Malaria, the result of their presence in the human body, varies from the periodic chills and fever of the milder forms to an extremely virulent disease.

The classification of the sporozoans is work for an expert, depending upon highly technical distinctions. **In** brief summary the class is divided as follows:

> Subclass Telosporidia.
> Order Coccidia. Intracellular parasites in digestive **epithelium.** Some are transmitted by blood-sucking animals. Parasitic in invertebrates and vertebrates.
> Order Haemosporidia. Parasitic in the blood of vertebrates and in the alimentary tract of blood-sucking invertebrates. The malarial parasites belong here.
> Order Gregarinida. Parasitic in various stages in the cells and in the cavity of the invertebrate intestine.
> Subclass Cnidosporidia.
> Order Myxosporidia. Mostly parasitic in fishes; a few in amphibians and reptiles.
> Order Actinomyxidia. In the body cavity or the lining of the alimentary tract of aquatic **annelids.**
> Order Microsporidia. Mostly intracellular parasites in **arthropods** and fishes.
> Order Helicosporidia. A single known species, parasitic in insects.
> Subclass Acnidosporidia.
> Order Sarcosporidia. Parasitic chiefly in the muscles of mammals, some in birds and reptiles.
> Order Haplosporidia. Found in invertebrates and lower vertebrates. (A.W.L.)

SPOROZOITE. A stage in the reproductive cycle of the 1-celled parasites of the class **Sporozoa.** The sporozoites are the active individuals formed by the subdivision of the zygote or sporont. (A.W.L.)

SPORULATION. The process of **spore** production, a type of reproduction more common in plants than in animals, but highly developed in some of the 1-celled forms. In this process the parent cell subdivides to form a number of smaller active cells which begin a new life cycle. (A.W.L.)

SPOT FACE. Counterbore.

SPOT WELDING. Welding.

SPOT-TAIL. Pisces, Teleostei. A small **minnow,** *Hybopsis hudsonius,* widely distributed and abundant in the fresh waters of the eastern half of the United States. Also called the **shiner.** It usually has a black spot at the base of the tail. A species with a similar mark in a closely related genus, *Cyprinella stigmatura,* is called the spotted tail. The latter is found only in the southeast. (A.W.L.)

SPRAIN. A variety of injuries in or about a joint which occur when the movement of the joint is carried beyond its normal range, or forcibly in a direction where its range is limited. Sprains occur in the ankle, knee, wrist, elbow, and spine, in that order of frequency. An injury in which a sudden wrenching or twisting produces tearing of the **ligaments** or **tendons** is a common form of sprain. Displacement of **cartilages** between joints, tearing of muscles around the joint, and tearing of the synovial membrane are others. The symptoms of a sprain are those of **inflammation** with pain, swelling, and limitation of motion of the part.

Treatment consists of immobilization of the joint with an adhesive strapping or an elastic bandage, elevation of the extremity, and sometimes aspiration of the joint cavity if an effusion or outpouring of inflammatory fluid occurs. (D.M.H.)

SPRAT. Pisces, Teleostei. A small marine food fish (**Pisces**), *Clupea sprattus,* related to the herring. It occurs on the Atlantic coast of Europe. (A.W.L.)

SPRAY PATTERN. Foundry Practice.

SPRING INDEX. The ratio between the mean diameter of the coil of a helical compression spring and the diameter or radial thickness of the spring wire or strip. The spring index is used as a modifier in computing the safe load a spring may carry. (H.C.H.)

SPRING PEEPER. Amphibia, Anura. A tree frog, *Hyla crucifer,* about an inch in length, with an elongate **X** on its back. The color varies greatly, including shades of brown, yellow, green and gray. The species is found through most of the eastern half of North America. It is named from the early appearance of the animals in the spring, when their not unmusical tones are one of the first indications that hiberating animals are active again. (A.W.L.)

SPRING TAIL. Collembola.

SPRINGBOK. Mammalia, Artiodactyla. A small South African **antelope,** *Antidorcas marsupialis,* of striking appearance. One of the **gazelles.** The animal is cinnamon and white with a dark brown streak on the sides. (A.W.L.)

SPRINGS. Used to absorb energy or shock as in automobile chassis springs; to serve as a source of power as in clocks or watches; and to provide a force to maintain pressure between contacting surfaces as in friction clutches. Springs with ground ends are generally more

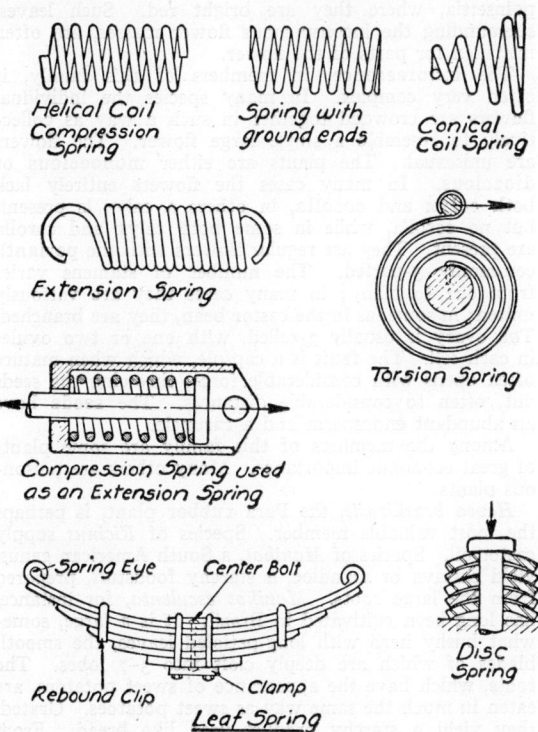

Helical Coil Compression Spring

Spring with ground ends

Conical Coil Spring

Extension Spring

Torsion Spring

Compression Spring used as an Extension Spring

Spring Eye *Center Bolt*

Rebound Clip *Clamp*

Disc Spring

Leaf Spring

satisfactory than those with plain ends; compression springs are more desirable for heavy loads than extension springs, because of the possibility of stress concentration in the loop of the extension spring. Compression springs can, however, be employed for tensile loading. Conical coil springs, if properly designed, may be compressed flat under load. Disk springs represent a recent development that is being extensively employed for heavy loads. Laminated or leaf springs are used **in** vehicles of various types, although coil springs are

now being used in automotive applications. Coil springs may be made of square, rectangular, or round wire. (See also **Ground Water.**) (H.C.H.)

SPRUCE. Conifers.

SPRUE. A disease of tropical countries, marked by anemia, wasting, gastrointestinal disturbances, sore tongue, and diarrhea with the passage of voluminous, fatty, frothy, foul-smelling stools. An exactly similar clinical picture known as **celiac disease** occurs in children in tropical and non-tropical countries. It is believed that sprue and celiac disease are manifestations of the same pathological entity.

The exact cause of sprue is unknown, but there is evidence to suggest that it is a deficiency disease. The deficiency may be in the nature of dietary lack, or digestive tract disturbances which interfere with the proper absorption of food. Treatment is especially effective in early cases. A high protein and low fat diet, injections of liver, brewers' yeast or other forms of **vitamin B** complex by mouth are followed by improvement and remission. Celiac disease is treated similarly; it tends to undergo permanent remission as maturity is reached.

The anemia of sprue is of the macrocytic type, i.e., the red cells are pale and large. Recently it has been discovered that synthetic folic acid, one of the vitamin B complex, causes an improvement in the anemia of sprue and a marked improvement also in the symptoms of the disease. (D.M.H.)

SPUR GEARING. This form of toothed gearing is used for transmitting power between shafts whose axes are parallel; the figure illustrates the important elements of a set of spur gears. The velocity ratio of a spur

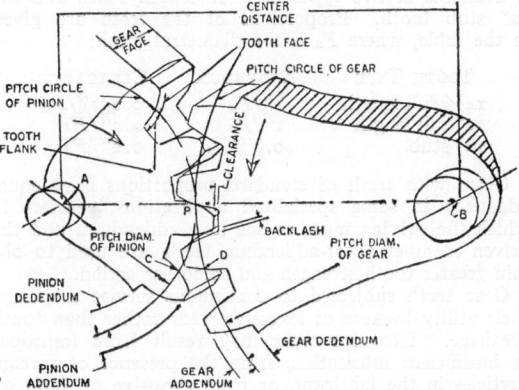

gear set is the ratio of the number of revolutions of one gear to the number of revolutions of the other. The **pitch circles** of a pair of spur gears are those imaginary circles that are equivalent to the peripheries of a pair of friction wheels that would operate without slipping at the same velocity ratio and center distance as the gears themselves. The point of tangency *P* of the pitch circles on the line of centers is the pitch point of the gearing. The smaller of the two gears is usually referred to as the pinion.

Circular pitch is the distance from a point on the profile of one tooth to a corresponding point on the profile of an adjacent tooth measured on the pitch circle. In the figure, the arc *CP* or *DP* is the circular pitch of the pinion or the gear. The relation of circular pitch and pitch diameter is as follows:

$$\text{Pitch diameter} = \frac{\text{Number of teeth} \times \text{Circular pitch}}{\pi}.$$

Diametral pitch is the ratio of the number of teeth to the pitch diameter and is expressed as follows:

$$\text{Pitch diameter} = \frac{\text{Number of teeth}}{\text{Diametral pitch}}.$$

Circular pitches may be given as ½″, ¾″, 1¼″, etc.; diametral pitches as 4, 6, 8, 16, etc. Diametral pitch is commonly employed for commercial gearing since pitch diameters and center distances can be thereby expressed in whole numbers or commonly-used fractions.

It may be seen from the above that the velocity ratio of a gear set is dependent upon, and inversely proportional to, the numbers of teeth in the pinion and gear, or:

$$\frac{\text{rpm pinion}}{\text{rpm gear}} = \frac{\text{Number of teeth in gear}}{\text{Number of teeth in pinion}}.$$

In order that spur gears may operate at a constant angular velocity of the driven member for each increment of rotation of the driving member, the tooth curves must be such that the common perpendicular to the tooth profiles at any point of contact will pass through the pitch point. Theoretically, almost any curve may be used for the profile of the teeth of one gear; if the tooth profile for the mating gear is constructed so that the common normal to the point of contact passes through the pitch point, the gearset will operate at a constant angular velocity, and the gears are said to be conjugate. In practice, only one curve, the involute of a circle, is used. The directrix of a gear tooth involute is the base circle tangent to an angular line passing through the pitch point. The angle between this line and a perpendicular to the line of centers of the gear axes is known as the pressure angle, or angle of obliquity, of the gearing. Two angles, 14½° and 20°, are in common use; the 20° gear is available in two types, the full length tooth and the 20° stub tooth. Proportions of the teeth are given in the table, where P_d is the diametral pitch:

Tooth Type	Addendum	Clearance
14½ full length	$1''/P_d$	$0.157''/P_d$
20° full length	$1''/P_d$	$0.2''/P_d$
20° stub	$0.8''/P_d$	$0.2''/P_d$

Gears with teeth of standard proportions have equal addenda; in some specialized applications, gearsets in which the driving member has long-addendum, and the driven member short-addendum teeth, are used to obtain greater tooth strength and smoother action.

Gear teeth subjected to continuous service may lose their utility because of excessive wear rather than tooth breakage. Excessive wear may result from improper or insufficient lubrication, from the presence of foreign particles in the lubricant, or from excessive pressure on the surfaces in contact. Suitable lubrication and encasing or housing the gearset may eliminate some of this trouble, but the teeth must be of such proportions that compressive fatigue failure of the materials cannot occur.

An internal gear or annular is a ring in which gear teeth are cut on the inner periphery; it is literally a spur gear turned inside out. Compared to spur gearing, an internal gear operating with a pinion has several advantages: shafts can rotate in the same direction without the use of an idler gear; annular teeth are stronger than spur gears of the same size and number of teeth; and the center distance between shaft axes is equal to the difference of the pitch radii of the gear and pinion, instead of to the sum. Tooth interference is more likely in internal gearing than in spur gearing, however, and somewhat more limited methods of cutting annular gears are available.

A rack is a straight-line spur gear, and may also be defined as a spur gear whose pitch radius approaches infinity as a limit. A rack and pinion may be used in combination to convert rotary to reciprocating, or reciprocating to oscillatory motion. (H.C.H.)

SPUR-FOWL. Aves, Galliformes. Long-tailed Indian and Ceylonese partridges which resemble pheasants. (A.W.L.)

SPURGE FAMILY. Euphorbiaceae. Members of this large family are found in many tropical and temperate regions. The greater number are trees or shrubs, with a few herbaceous species, especially in cooler regions. Many of the tropical species are interesting xerophytes, plants capable of enduring the driest climates. Often these have a habit very similar to that of species of *Cactus,* with which they may easily be confused, especially if not in flower. However, nearly all members of the spurge family contain a milky juice which exudes from them when the surface is cut or broken. This milky juice, or latex, will readily distinguish them from cacti, which lack latex. Leaves, when present, are usually alternate and have **stipules.** In many species, such as the frequently cultivated *Euphorbia splendens,* or "Crown of Thorns," the leaves soon drop off, leaving a spine-covered stem. In one genus, *Phyllanthus,* leaves are frequently reduced to minute scales, and the stem flattened and green; in these the small pinkish flowers are borne around the edge of the flattened stem. In some species of *Euphorbia* the leaves near the top of the stem become brilliantly colored, as in *Euphorbia pulcherrima,* the poinsettia, where they are bright red. Such leaves, surrounding the inconspicuous flower masses, are often mistaken for parts of the flower.

The **inflorescence,** in members of this family, is often very complex. In many species the individual flowers are crowded together in such a way as collectively to resemble a single large flower. The flowers are unisexual. The plants are either **monoecious** or **dioecious.** In many cases the flowers entirely lack both **calyx** and **corolla,** in others a calyx is present, but no corolla, while in some both calyx and corolla are present. They are regular flowers with the **perianth** commonly 5-parted. The number of stamens varies from one to many; in many cases they are variously united; in some, as in the castor bean, they are branched. The ovary is usually 3-celled, with one or two ovules in each cell. The **fruit** is a capsule, which when mature often opens with considerable force, throwing the seeds out, often to considerable distances. The **seeds** have an abundant endosperm and a caruncle.

Among the members of this family are some plants of great economic importance. Many others are poisonous plants.

Hevea brasiliensis, the Para **rubber** plant, is perhaps the most valuable member. Species of *Ricinus* supply castor oil. Species of *Manihot,* a South American genus, yield cassava or mandioc, a starchy foodstuff, prepared from the large roots. *Manihot esculenta,* for instance, has long been cultivated in Brazil. It is a large, somewhat bushy herb with long-petioled leaves, the smooth blades of which are deeply cleft into 3-7 lobes. The roots, which have the appearance of sweet potatoes, are eaten in much the same way as sweet potatoes. Grated, they yield a starchy product used like bread. From the roots tapioca may be prepared. The poisonous principle which is present in many species of *Manihot* is removed by squeezing or destroyed by heating. From other species of this genus may be obtained, by tapping, a milky juice which is a source of rubber.

Hura crepitans, the sand box tree, is another member of the family of some slight commercial value. The plant is a fairly large tree the stem of which is covered with short, sharp spines, and bears long-petioled, toothed leaves. The fruit is composed of numerous hard carpels which, when mature, explode violently, throwing the seeds out forcibly. These fruits, about 3″ in diameter, were formerly gathered and wired to prevent bursting. When dry they were used as containers for the fine sand which was then used to blot ink—hence the common name of the tree. The wood of the tree is used locally, but rarely exported. The milky juice of the tree is very poisonous. This juice, mixed with meal or similar substances, and thrown into the waters of a stream or lake, stupefies the fish present therein, so that they may be readily captured. The poison does not render the fish unfit for human consumption.

Several species of *Croton* yield important purgative drugs. *Croton eluteria* gives Cascarilla bark, used as a tonic.

Jatropha Curcas, a small shrub or tree, bears eggshaped green fruits which contain a high percentage of an odorless oil called "curcas" oil, used in making paints, as a lubricant, and in soap-making. From the leaves of this tree, natives of the Philippine Islands prepare a fish poison used in much the same way as that obtained from *Hura*.

Aleurites species, natives of the East Indian region, are important sources of oil. (B.S.M.)

SPURIOUS CORRELATION. Spurious correlation is correlation which is misleading. Take two variables *x* and *y* which are independent. Consider a third variable *z* which is correlated with *x* and with *y*. Form $v = f(x, z)$ and $w = f(z, y)$, then frequently *v* and *w* will show considerable dependence which may be traced to the correlation of *x* with *z* and *y* with *z*. In general, then, spurious correlation may be defined as correlation which is introduced by other variables rather than the ones under study.

The real question at issue in correlation is actually this. Are the variables in which we are interested *x* and *y* or *v* and *w*? If the variables are *x* and *y* then the correlation of *v* with *w* is spurious. If the causal variables, however, are *v* and *w* and not *x* and *y*, then the correlation of *v* with *w* is valid and not spurious. If two variables have no logical connection of any kind, naturally any correlation found between them will be spurious. (L.A.A.)

SPUTTERING. A result of the disintegration of the metal cathode in a vacuum tube due to bombardment by positive ions. Atoms of the metal are ejected in various directions, leaving the cathode surface in an abraded and roughened condition. The ejected atoms alight upon and cling firmly to the tube walls and other adjacent surfaces, forming a blackish or lustrous metallic film. This effect is often utilized to form very fine-grained coatings of metal upon surfaces of glass, quartz, etc., purposely exposed to the sputtering. Films of different metals can be obtained by using cathodes made of these metals. Glass plates may be thus silvered, or suspension fibers of spun quartz rendered conducting for use in electrometers, etc. (L.D.W.)

SQUALL. A sudden, strong wind which may or may not be accompanied by a wind shift is a squall. Rain and snow squalls are showers accompanied by strong gusts. (P.E.K.)

SQUALL LINE. Thunderstorms, rain showers or squalls, and snow showers or squalls often appear in a long line sometimes reaching hundreds of miles but only one squall in depth. Such lines of squalls or squall lines normally move perpendicular to their leading edge. Squall lines are of two types, cold-front and prefrontal. Cold-front squall lines are composed of numerous thunderstorms, squalls, or showers occurring at the cold front, as it moves across the terrain, because the cold air behind the front is underrunning and lifting the warm air ahead of the front. It is necessary that the warm

and lifted air be unstable or potentially unstable and possess sufficient moisture to permit the development of cumulonimbus clouds. Along a normal cold-front squall line there are breaks in the squall clouds. Thus it is possible for a cold front to pass a given locality with nothing more than broken clouds as evidence (also a wind shift), whereas a locality 5 miles away may experience a severe thunderstorm and considerable damage.

Prefrontal squall lines originate from a combination of several factors, probably three:

1. A tongue of moist air in advance of and parallel to a cold front.

2. Strong southerly winds at low levels ahead of the cold front carrying warm air northward at a rapid rate, and westerly winds at high levels carrying comparatively cold air over the northward-moving warm air, thus creating instability.

3. Lifting of the air well in advance of the cold front because of its fluid properties, and because of convergence in the warm sector of a cyclone. These three conditions are most often satisfied in the warm sector of a temperate-zone cyclone just ahead of the cold front.

Prefrontal squall lines may begin with a few scattered showers or thunderstorms located in the moist tongue, but some develop rapidly into a solid line of showers or thunderstorms located from 50–200 miles ahead of the cold front. A prefrontal squall line, once developed, travels like the cold-front squall line, perpendicular to its leading edge and with the cold front. Prefrontal squall lines are usually violent and destructive, often accompanied by tornadoes, hail storms, gusty surface winds, and severe turbulence. Velocity of a prefrontal squall line is usually greater. (P.E.K.)

SQUAMATES. Fossil Reptiles.

SQUARE. Screw Thread; Measurement.

SQUARE WAVES. Any periodic **wave**, regardless of its shape, may be analyzed into a series of sine and cosine components whose frequencies are harmonically related. The number of these components will be determined by the shape of the wave, but in general the sharper the corners of the original wave the more component terms. Thus a square wave will require a wide range of frequencies to express it. These components are not mere mathematical fictions but are true electrical components in the case of an electric wave. They may be separated and examined by means of proper filter circuits. Since a square wave will contain a long series of frequencies it may be used for rapidly determining the frequency response of a piece of equipment by applying the wave to the input and noting the distortion of the output wave. The distortion is due to certain frequencies of the original wave being attenuated or amplified out of proportion in passing through the circuit. Thus the necessity of making a laborious series of tests at various frequencies using sine waves is avoided. When an operator is properly trained in interpreting the results of such testing it offers a rapid means of checking amplifiers, networks, etc. These square waves may be generated by a variety of electronic circuits. (L.R.Q.)

SQUASH. *Cucurbita maxima* (*pepo*) and (*C. moschata*). **Gourd Family.**

SQUASH BORER. Insecta, Lepidoptera. The **larva** of a **moth,** *Melittia satyriniformis,* which bores in the root and stem of squash vines, often killing the plant. The moth is a beautiful species with olive fore wings, transparent hind wings, and legs tufted with orange and black.

The larva thrusts waste material out of holes in the stem. When detected by this means it can be cut out of the plant and killed. In large fields where hand control is impossible, deep plowing when the vines are dead kills many of the insects. Other methods depend on an accurate knowledge of the time of deposition of the eggs. Spraying is effective during that period. (A.W.L.)

SQUASH BUG. Insecta, Hemiptera. A true bug, *Anasa tristis,* about ⅝″ long, gray-brown above and mottled with yellow below, showing red in flight from areas beneath the wings. It sucks the juices of pumpkins, squash, and related vines and spreads a bacterial wilt.

The adults hibernate beneath debris on the ground and may be trapped under pieces of board and destroyed. Colonies of young are said to be destroyed by dusting with calcium cyanide (see **Calcium**) (50% or 25% cyanide), and the masses of eggs are large enough to be seen and destroyed by hand. They are often deposited on the underside of leaves, however, where they are difficult to see. (A.W.L.)

SQUID. Mollusca, Cephalopoda. Marine animals with usually elongate bodies bearing a pair of lateral fins. The head is provided with large eyes and ten tentacles, two much longer than the rest and expanded at their tips.

The squids vary in size from the little butterfly squid of the Maine coast, scarcely more than an inch long, to the giant squids of the deep seas with bodies, exclusive of the tentacles, 20′ long. The common squids of the genus *Loligo* are also called calamaries.

Squids are eaten by the Chinese, who regard them as a delicacy. In North America they are important as bait in the marine fisheries of the northern Atlantic coast. (A.W.L.)

SQUIRREL. Mammalia, Rodentia. An arboreal or terrestrial **rodent** with a long bushy tail. These animals, together with the chipmunks, ground squirrels, woodchucks, and flying squirrels, make up the family Sciuridae. Representatives of the true squirrels are found on all continents but Australia.

Flying squirrels belong to a subfamily distinguished from the true squirrels and related forms by the presence of folds of skin stretching from the front to the hind legs along the sides of the body. These folds, held extended by the legs, support the animal in the air during long gliding leaps.

Still another small division of the family contains the pigmy squirrels of Africa, Borneo and the Philippines. These few species are about as large as mice.

North America has ten species of true squirrels, some widely distributed and extremely variable, and two species of flying squirrels, *Glaucomys.* The little red squirrel, *Sciurus hudsonicus,* is the most widely known, with the gray, *S. carolinensis,* and fox, *S. niger,* squirrels very close. The two last are valued as small game; their flesh is excellent. (A.W.L.)

SQUIRREL-CAGE MOTOR. Motor, Electric.

SQUIRTING CUCUMBER. *Ecballium Elaterium.* Gourd Family.

STABILITY. Equilibrium of Forces; Least Energy Principle; Buoyancy; Stability, Mechanical.

STABILITY, AIRPLANE. The aerodynamic stability of an airplane in flight is a form of **mechanical stability.** As the airplane has three degrees of freedom, it possesses three components of stability, viz., longitudinal, lateral, and directional.

Longitudinal stability is stability around the lateral axis. For normal flight this is an important type of stability. An airplane which is stable longitudinally will, if the controls are released, recover from a dive or a

climb and resume normal flight automatically. This will not happen instantly. In most cases, if the ship is diving and the controls are released, it will come out of the dive and begin to climb until the speed drops appreciably. It then will dive again but much less steeply, and once more begins to climb, but at a smaller angle than in the first case. Any alternating motions are known technically as *oscillations,* those from dive to climb being called *phugoid oscillations.* Most airworthy airplanes will resume normal flight after 3 to 5 oscillations.

Longitudinal stability is the most important of the three stabilities because small angular rotations about the lateral axis create large variations of aerodynamic reaction on the wing. Each angular degree of pitch changes the angle of attack a degree. (The range of angle of attack from no lift to maximum lift is often no more than 16°). Tendencies to longitudinal instability would be extremely hazardous during landing maneuvers. Lateral and directional instability are undesirable, but while they might cause piloting to be tedious, they do not pose the same risks as longitudinal instability. Longitudinal stability is more difficult to achieve than the others, because an airplane is unsymmetrical about its pitching axis but symmetrical about the rolling axis.

An airfoil is inherently unstable but the combination of wing and stabilizer can have positive stability provided the center of gravity is located at about the 30% chord point of the wing. As shown in Fig. 1, a stable

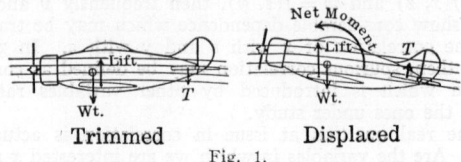

Trimmed	Displaced

Fig. 1.

monoplane is flying in "trim" at comparatively low angle of attack. Center of pressure is well back on the wing and the stabilizer is operating at negative angle of attack. Neglecting the small, and nearly balanced, thrust and drag moments, the moment of the stabilizer lift equals the moment of the wing lift. Now assume that the pilot introduces a disturbance by operating the controls momentarily to "nose-up" the airplane. The higher angle of attack moves the lift center forward so that the wing, of itself, would move to an even larger angle. However, the stabilizer lift will have diminished, or even reversed in direction in an amount sufficient to overcome adverse wing moment and produce a net stabilizing moment which returns the airplane to its original attitude. While the angular momentum may carry the airplane through the neutral position, the direction of the stabilizing moment correspondingly reverses and after a few reversals the phugoid oscillation is completely damped. If the center of gravity is displaced forward or backward far enough, this positive stability will disappear, therefore it is necessary to control disposable load (baggage, fuel, etc.) and keep center of gravity within predetermined stability limits.

Lateral or Rolling Stability. Lateral motions involve one translation, side slipping, and two rotations: rolling, about the longitudinal axis; and yawing, about an axis that is normally vertical. Each of these rotations affects the position of the axis of the other one. So lateral motions are more complicated than longitudinal. For one motion, let us assume that the pilot keeps the airplane from yawing, by the use of the rudder. It may roll and slip then, but it cannot yaw. Stability in this motion will be called *rolling stability.*

Rolling stability may be obtained in two ways. The most common procedure is to build or rig the wings with a **dihedral** varying from 1–3°. When a ship is banked without turning, it tends to sideslip or slide downward toward the low side. If the wings have dihedral, the air then strikes the low wing at a much greater angle of attack than the high. This, of course, increases the lift on the low wing, and restores the ship to its normal position. Rolling stability may also be secured by the use of *sweep-back*, though the effect of sweep-back is not as pronounced as dihedral. An airplane with an ordinary amount of dihedral or sweep-back will stay more or less right side up if kept to a fixed heading with the rudder, without any use of ailerons.

Directional or Yawing Stability. An airplane has directional stability if forces, called into action when it is yawed, tend to restore it to its original direction of flight. Positive stability of this nature is secured by the use of a large vertical *fin* at the tail. The directional stabilizing action is similar to that of a weathervane. Wing sweep-back also secures some measure of directional stability through the unbalance of drag when such a wing is yawed.

General. Stability is always desirable, but not necessarily in the maximum possible degree. Excessive stability reduces maneuverability and renders the airplane "stiff" to the controls. The degree of stability secured by dihedral must be carefully balanced against that secured by fin, else there may result one of two types of lateral-directional instability.

1. "Dutch roll" from too small a fin for the dihedral used.

2. Spiral instability from too large a fin.

Most airplanes that are statically stable are dynamically stable. Even when dynamically unstable, this will usually occur at high speed and show up during altitude tests rather than during takeoff or landing. Dynamic analysis is tedious and costly and is not ordinarily performed, except for unconventional or large and costly airplanes. Most wind-tunnel tests are devised to obtain static stability characteristics and the model needs to be a small replica only of the external shape of the airplane. For dynamic stability the model must also be dynamically similar, that is, have a mass and moment of inertia which mirrors the prototype. (F.T.M.)

STABILITY, FREQUENCY.

This term is commonly used to denote the accuracy with which the carrier frequency of a radio **transmitter** maintains a constant value. Various methods are used to improve this, among them being **crystal** control of the **oscillator,** electron coupling of the oscillator, high **Q** tuned circuits, etc. (See also **Frequency Deviation.**) (L.R.Q.)

STABILITY, INSTABILITY, AND VERTICAL CURRENTS.

Everywhere that air is in motion some vertical perturbations are present. Isolated parcels and currents of air are thus started upward or downward in a layer of surrounding air. The action of the environment on the displaced parcels is a measure of the sta-

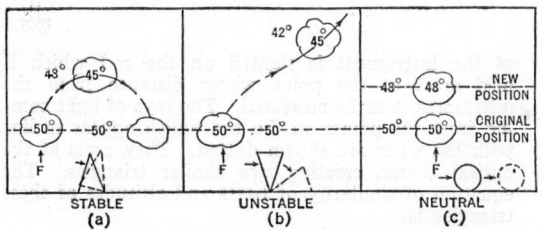

Fig. 1. Equilibrium (mechanical and atmospheric).

bility or instability of the air. If the parcel is forced back to its original position, the air is stable and does not favor vertical motions; if the parcel is accelerated in its vertical movement then the air is unstable and favors vertical motions; if the parcel comes to rest at a new position, neither rising nor falling, then the air is in an equilibrium state. See Fig. 1. Thus, vertical currents occur only if favored by unstable air. Otherwise, displaced parcels will oscillate with a definite period and

displacement until the oscillations are damped out. Stability criteria are directly related to the amount and distribution of water vapor in the air.

1. Dry air moves upward and downward with a temperature change of approximately 5.5° F. per 1000'.

a. When the temperature in the surrounding air decreases less rapidly than 5.5° F. per 1000', a rising parcel will grow colder than its environment, will be more dense, and will sink to its original position. The air is then said to be stable for such displacements.

b. When the temperature in the surrounding air decreases more rapidly than 5.5° F. per 1000', a rising parcel will grow warmer than its environment, will be less dense, and will accelerate in its vertical path. Air is then said to be unstable for such displacements.

c. If the temperature in the surrounding air decreases exactly 5.5° F. per 1000', the parcel will remain the same temperature as its environment, will remain of the same density, and will slowly come to rest due to friction. Air is then said to be in a neutral state or in equilibrium with regard to rising and sinking parcels.

2. Saturated air moves upward with a temperature drop equal to the moist adiabatic rate of cooling for the particular temperature and pressure of the parcel. These rates of cooling vary from 1.8° F. per 1000' for very warm air to nearly 5.5° F. per 1000' for very cold air. Cooling rates for rising saturated air also change continuously as the parcel rises, becoming greater as the parcel rises.

a. If saturated air parcels move upward and remain always colder than their environment, they will be more dense and will tend to sink to the somewhat lower position. The air is then stable for saturated rising parcels. Because sinking parcels do not long remain saturated, they are governed by the laws of unsaturated but water-containing air.

b. If saturated air parcels move upward and remain always warmer than their environment, they will be less dense and will rise with accelerated velocity. The air is then unstable for saturated rising parcels.

3. Unsaturated but water-containing air is governed by the stability criteria for dry air while it remains unsaturated, and for saturated air after it becomes saturated due to cooling in ascent. In most cases sinking air is governed by the rules of dry air because such parcels soon become unsaturated.

a. If the lapse rate is greater than the dry-adiabatic rate of 5.5° F. per 1000', the air is absolutely unstable for both the unsaturated and the saturated states because any rising parcel will be warmer than its environment.

b. If the lapse rate is less than the moist-adiabatic rate for any saturated parcel, the air is absolutely stable because any rising parcel will cool with a greater rate than the lapse rate in its environment. It will be more dense than its environment at all points during the ascent with a constant tendency to sink.

c. If the lapse rate lies between the dry-adiabatic rate and the moist-adiabatic rate then the air is conditionally unstable. It is stable for all parcels so long as they remain unsaturated, but can become unstable if the parcels become saturated and then are carried upward into a region where their temperature is greater than their environment.

In all cases of stability and instability for parcels, the ultimate criteria are solely these: 1. If the parcel is and will remain warmer than its environment, it will rise with accelerated velocity and the environment is unstable for that parcel. 2. If the parcel is and will remain colder than its environment, it will tend to sink and the environment is stable for that parcel.

Oftentimes a whole layer of air is lifted bodily along a front or sloping terrain. Lifting of a layer of air is nearly always a destabilizing process and if the layer grows unstable after lifting, it is said to have been convectively unstable, i.e., convection of the layer created an unstable state.

1. All stable air which remains unsaturated during the total lift remains stable after lift, but its stability is slightly reduced by the process.

2. Unsaturated stable layers which become saturated uniformly through the layers during lift remain stable after lift, but their stability is slightly reduced by the process.

3. Unsaturated conditionally unstable layers which become saturated uniformly through the layers during lift become unstable upon saturation.

4. All layers of air which are more nearly saturated at the top than at the base will remain stable during lift because of the difference in cooling rates of unsaturated air and saturated air.

5. All layers of air which are more nearly saturated at the base than the top will soon become unstable if lifted because the top continues to cool at the rate for unsaturated air after the base becomes saturated and cools at the rate for saturated air.

6. All saturated stable layers of air which are lifted remain stable but their margin of stability is slightly reduced by the process.

Stable air is unfavorable to vertical currents and is therefore conducive to stratification of clouds and smoke, to comparatively poor visibility, and to smooth flying. Unstable air favors vertical currents and therefore is conducive to billowed cumuloform clouds, to thunderstorms and showers, to good visibility, and to rough flying air. (P.E.K.)

STABILITY, MECHANICAL. Stability is that property of a body which will cause it to develop forces in opposition to any position or motion disturbing influence. The subject may be divided into *static stability* and *dynamic stability*. The former is concerned with the production of the restoring forces, the latter with the oscillations that are set up in the system as a result of the restoring forces.

Another classification is into (1) *positive stability* when the displaced object returns to an initial state of equilibrium after a temporary disturbance, (2) *neutral stability* when the object tends to remain in a definite position but when disturbed may come to rest in a new position, (3) *negative stability* (i.e., instability), when the object assumes an entirely new position when disturbed from its initial state. A simple damped pendulum illustrates the first; a sphere the second; while a slender cylinder standing vertically on end is a case of negative stability.

Let it be assumed that an object at rest or in a state of uniform motion receives a disturbing force. Depending on the kind of stability possessed, it might react with one of the motions shown in the accompanying figure.

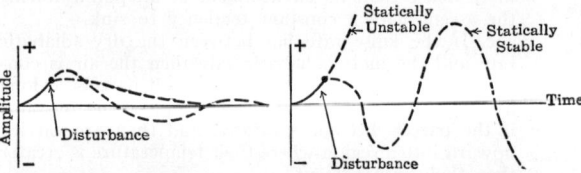

Left: Positive stability (both statically and dynamically stable).
Right: Negative stability (both dynamically unstable).

If it is dynamically stable as well as statically stable, its motion-time history may be one of diminishing oscillation or of simple subsidence, depending on the magnitude of damping, and inertial effects. Dynamic instability may occur with either static stability or static instability. These lead to divergent oscillation, or to complete divergence. (F.T.M.)

STABILIZER, AIRPLANE. Any airfoil whose primary function is to increase the stability of an aircraft. It usually refers to the fixed horizontal tail surface of an airplane, as distinguished from the fixed vertical surface. (F.T.M.)

STABLE FLY. Insecta, Diptera. A true fly, *Stomoxys calcitrans,* similar in appearance to the house fly but with a mouth fitted for piercing and sucking. It is common in barns, breeding in manure and other organic refuse, but it sometimes enters houses. The impression that house flies bite is due to the occasional attacks of this similar species. (A.W.L.)

STACK. In geology a rock pillar or monument which occurs relatively close to marine cliffs which have usually been developed in hard horizontally bedded but jointed sedimentary rocks. In some cases a stack is the remaining pillar of an arch. Both sea arches and stacks are well developed on the coast lines of the Gaspé Peninsula and the North East Highlands of Scotland. (R.M.F.)

STADIA. This term refers to a method of obtaining distances and elevations of points on the earth's surface with respect to some known point by means of a **transit** equipped for stadia readings and an ordinary **level rod.** Stadia measurements are most useful when applied to measurement of inaccessible lines or some limited area such as a farm. It is not applied to precise control work such as the triangulation surveys of the Coast and Geodetic survey. Since contour data for all surrounding points may be taken from one instrument station, the stadia method is particularly adaptable to topographical surveying.

The stadia transit is equipped with two parallel horizontal wires which appear in the field. These are in addition to the ordinary centering hairline. The principle of distance measurement by stadia is explained in connection with the diagram. The telescope

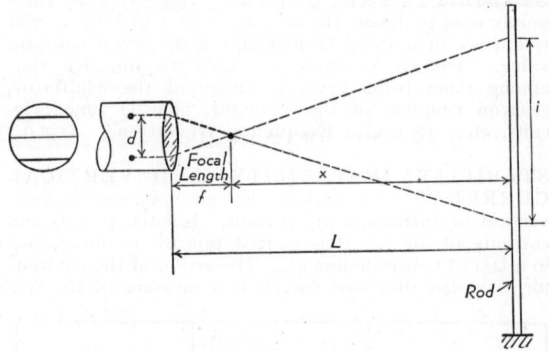

of the instrument is sighted on the rod which is held erect at the point whose distance from the instrument is to be measured. The rays of light coming from the points on the rod which appear in line with the wires are shown dotted. They cross at the optical center, creating two similar triangles. The equation of similarity of bases and altitudes of these triangles is:

$$\frac{x}{f} = \frac{i}{d},$$

which reduces to:

$$x = i\frac{f}{d}.$$

The above shows that to measure distance L one needs to read the intercept on the rod of the stadia wires, multiply it by a factor which is a characteristic of the instrument, and add the focal length. This gives the distance from the point to the objective lens of the telescope. To this must be added the distance from the objective lens to the center of the instrument in order to get the total distance from the point to the transit setup. This last quantity is variable, depending upon the focusing of the telescope. However, an average value may be used without appreciable error. If c is this average value and H is the total distance from center of instrument to rod, then,

$$H = i(f/d) + (f + c).$$

If the ground elevation at the rod is different from that at the transit certain corrections must be applied to the above method. The corrections involve trigonometric functions derived from the angle of elevation read on the vertical circle of the transit. This angle is also employed to find the elevation of the distant point above the instrument station. (F.T.M., E.W.S.)

STAGE. Steam Turbine.

STAGE EFFICIENCY. By stage efficiency is meant the fraction of **isentropic** heat drop in a **steam turbine** stage that is transferred to the rotor as mechanical energy. The remainder goes to reheating the steam at the lower pressure. The failure of any stage to convert all the heat drop into mechanical energy may be accounted for in:
1. Friction of steam on nozzles, blades, and casing.
2. Throttling flow of steam through clearance spaces.
3. Shock loss.
A well-designed turbine may have a stage efficiency as high as 85%. (F.T.M.)

STAGNATION POINT, FLUID FLOW. A point in the **boundary layer** of fluid flow where viscous friction has brought part of the boundary layer to rest. Stagnation sometimes allows reverse flows along the wall, resulting in an unstable fluid surface which breaks up into vortices and random fluid motions. As long as this point is well to the rear on an airfoil the profile drag is low and lift is sustained, but let the stagnation point advance towards the leading edge and large drag increases accompany a dwindling lift. (See **Fluid Flow.**) (F.T.M.)

STAINLESS AND HEAT-RESISTING STEELS. The development of stainless steels began in 1913 when Harry Brearley in England observed the rust-resistance of alloys containing about 12% chromium, and Benno Strauss in Germany discovered that the addition of nickel to high-chromium steels produced remarkable changes in the mechanical properties.

The original chromium-type steel is hardenable by heat treatment and its first important commercial applications were for cutlery (No. 3 in the table). Modifications with higher carbon and chromium contents, such as Nos. 4 and 5, are now used for the best quality cutlery. By lowering the carbon content (No. 1) a heat treatable steel is produced having strength and hardness suitable for machine parts and stressed structures. A modification having a high sulfur (or selenium) content has improved machinability (No. 2).

By further reduction in carbon content and increase in chromium content, steels are produced which harden only slightly by heat treatment (Nos. 6 and 7). The 17% chromium alloy is widely used as sheet or strip which can be fabricated by cold drawing and forming. The 27% chromium alloy is used primarily for heat-resisting applications.

All of the high chromium and chromium-nickel alloys have good strength and resistance to scaling at elevated temperatures compared with ordinary steels, hence they have many applications in processing equipment in the food, chemical, oil, synthetic rubber, and munitions industries where both corrosion and heat resistance are required. In general, the scaling resistance increases with the chromium content and the chromium-nickel compositions have the highest tensile and **creep** strengths.

Austenitic grain structure of 18-8 stainless steel at 100 times magnification. Etched by electrolysis in 10% perchloric acid. (See **Metallography.**)

By the addition of 6½% or more nickel, alloys are obtained having physical and mechanical characteristics not found in the carbon or low-alloy steels. The basic alloy (No. 9) is known as 18-8 for its chromium and nickel content. In these alloys the austenitic crystal structure, which is the normal structure for carbon steels at high temperatures, is retained at room temperature. (See **Steel.**) This structure has a high capacity for plastic deformation and develops considerable hardness and strength by **cold working.** Because of the absence of a structural transformation range at elevated temperatures, these alloys are not capable of hardening by **heat treatment,** thus they are dependent on the cold-working processes such as rolling and drawing of sheets, and drawing of rods, wire, and shapes to develop high strength. Even in the soft condition, which is obtained by rapid cooling from a high temperature such as 2000° F., these austenitic alloys have a relatively high tensile strength of the order of 90,000 lbs. per sq. in. One grade (No. 8) develops up to 200,000 lbs. per sq. in. tensile strength in the full hard condition. (See **Temper.**) Its **hardness** is about 42 Rockwell "C" or 400 Brinell, which is not high enough to retain a good cutting edge. These steels are used principally for aircraft, railway cars, and other structural applications requiring a high strength-weight ratio.

Because of their high combination of ductility and strength, the austenitic stainless steels can withstand severe cold drawing operations. Another aid in fabricating these materials is their good weldability, especially in spot welding. When welded structures are to be subjected to highly corrosive media, precautions must be taken to prevent intergranular **corrosion.** This type of failure can occur if the material has been "sensitized" by migration of chromium carbides to the grain boundaries. This occurs during exposure to temperatures in the range 900–1600° F. or during slow cooling through this temperature range. Such exposure can occur in material adjacent to welds, in heating operations during fabrication, or in service. Corrosion failure of this nature can be prevented by the use of "stabilized" grades (Nos. 10 and 11) in which sufficient titanium or columbium is present to combine with the carbon and prevent the formation of chromium carbides at the grain boundaries.

The corrosion resistance of stainless steels is believed to be the result of a protective oxide film which is stable, tightly adherent, and submicroscopically thin. When broken, the film regenerates itself provided oxygen is available. This accounts for the exceptional resistance to nitric and other oxidizing acids and the lesser resistance to hydrochloric acid and other reducing agents. Materials which depend on protective films for corrosion resistance, including aluminum and nickel alloys in addition to stainless steels, are subject to corrosion by pitting

plane that has "stalled" has had the organized stream-line flow over the upper wing surface partly destroyed. The lift is partially lost so weight and lift are out of equilibrium. Since the forward end of a conventional airplane is the heaviest it drops faster than the tail, hence a stall is followed by a nosing down motion accompanied by rapid loss of altitude. (See **Stagnation Point, Airfoil, Auto-Rotation.**) (F.T.M.)

STAMEN. Flower.

STAMPING. Press Work.

REPRESENTATIVE STAINLESS STEELS

No.	AISI * TYPE	CARBON %	CHROMIUM %	NICKEL %	OTHER ELEMENTS †	BRINELL HARDNESS
1	410	.15 max.	11.50–13.50	....	....	400‡
2	416	.15 max.	12.00–14.00	....	P, S, or Se .07% min. Mo or Zr .60% max.	400‡
3	420	Over .15	12.00–14.00	....	....	525‡
4	440A	.60–.75	16.00–18.00	....	....	577‡
5	440C	.95–1.20	16.00–18.00	....	....	688‡
6	430	.12 max.	14.00–18.00	....	....	150
7	446	.35 max.	23.00–27.00	....	....	175
8	301	.08–.20	16.00–18.00	6.00– 8.00	....	160
9	302	.08–.20	17.00–19.00	8.00–10.00	....	150
10	321	.10 max.	17.00–19.00	8.00–11.00	Ti 4 × C min.	160
11	347	.10 max.	17.00–19.00	9.00–12.00	Cb 8 × C min.	160
12	316	.10 max.	16.00–18.00	10.00–14.00	1.75–2.50% Mo	170
13	309	.20 max.	22.00–24.00	12.00–15.00	....	170

* American Iron and Steel Institute.

† The maximum sulfur and phosphorus contents are 0.04% except as noted. The maximum silicon content is 1%. The maximum manganese content is 1% for the chromium grades and 2% for the chromium-nickel grades.

‡ Quenched to full hardness and given a stress-relieving tempering treatment at a temperature under 700° F. All others are in the annealed state.

when the film is permanently broken at a given point on the surface. This is a particularly dangerous type of corrosion since a single hole through a sheet or plate usually makes the part unserviceable. Small particles of foreign substances which attach themselves to the surface are believed to be a cause of pit or contact corrosion under certain conditions. Proper cleaning and maintenance of processing equipment is a safeguard against this form of corrosion. The molybdenum-bearing composition (No. 12) is especially resistant to pitting.

For high strength and resistance to scaling up to temperatures as high as 2000° F., high chromium-nickel content steels such as No. 13 are used. (See also **Nickel Alloys.**)

The stainless-steel types that have been selected for discussion represent less than ½ the total number recognized by the industry. Other modifications are available in which special properties such as machinability, **spinning** quality, or scaling resistance are developed to a high degree. (R.H.H.)

STALACTITE. A stalactite is a deposit of **calcium carbonate** which hangs icicle-like from the roof or wall of a **limestone** cavern, and is formed by the dripping of mineralized **solutions.** Corresponding columnar structures built upward from the floors of caves beneath the stalactites in a similar manner, are called stalagmites. Stalactite is derived from the Greek, meaning to fall in drops; stalagmite, from the Greek, meaning that which drops. (R.M.F.)

STALAGMITE. Stalactite.

STALL. A term in common use to describe the condition of an aerodynamic **burble** upon a wing. An air-

STANDARD ATMOSPHERE. A standard atmosphere is defined as one in which the sea-level temperature is 15° C.; the sea-level pressure, 1013.2 millibars; and the lapse rate 6.5° C. per km. up to 11 km. (the stratosphere). In the ft.-lb.-sec. system this amounts to very nearly a sea-level temperature of 59° F., a sea-level pressure of 29.92″ of mercury, and lapse rate of very nearly 3.5° F. per 1000′. The real atmosphere only occasionally duplicates the standard atmosphere. (P.E.K.)

STANDARD CELL. Precise measurements of electromotive force, as with a **potentiometer**, require a constant, known source of voltage. The only practicable means of supplying this is by some special type of **electrolytic cell.** Of these, two have been highly developed and much used.

The older is the Clark standard cell, which, as improved by Carhart, has a mercury **cathode** and an amalgamated zinc **anode.** The cathode is submerged in a **mercurous** sulfate paste, and the **electrolyte** is a solution of zinc sulfate saturated at 0° C. The whole is sealed in a suitable glass tube. This cell has an electromotive force of 1.440 volts at 15° C., with a decrease of 0.00056 volt per degree rise in temperature.

In 1891 Weston substituted **cadmium** amalgam for zinc, and a continuously saturated solution of cadmium sulfate for the zinc sulfate electrolyte, the whole in an H-shaped tube. The Weston normal cell has a somewhat lower voltage than the Clark (1.0183 volts at 20° C.), but its temperature coefficient is practically negligible. The official adoption of the voltage above given for the Weston normal cell by the standards laboratories of Great Britain, the United States, and Germany (1911) gave in effect an alternative definition of the volt, i.e., 0.98203 of the electromotive force of the

Weston normal cadmium cell at 20° C. An unsaturated type of Weston cell is much used commercially in the United States. (See also **Reactions Involving Oxidation-Reduction.**) (L.D.W.)

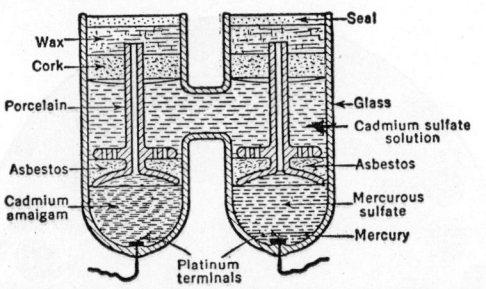

Cross-section of standard Weston normal cadmium cell.

STANDARD DEVIATION. The standard deviation is a measure of **dispersion** and **variability** and is undoubtedly the most important of any measure of variability. If the standard deviation is large, the **variates** are dissimilar; if the standard deviation is small the variates are more alike. Consider the variates $x_1, x_2, \cdots, x_N$, then the standard deviation of x, σ_x is defined by

$$\sigma_x = \sqrt{\frac{(x_1 - \overline{X})^2 + (x_2 - \overline{X})^2 + \cdots + (x_N - \overline{X})^2}{N}} = \sqrt{\frac{\Sigma(x_i - \overline{X})^2}{N}}$$

where $\overline{X}$ is the **mean** $= \dfrac{\Sigma x}{N}$. In the case of a **frequency distribution** with class marks x_i and class frequencies f_i the formula for σ_x becomes

$$\sigma_x = \sqrt{\frac{\Sigma f_i (x_i - \overline{X})^2}{\Sigma f_i}}$$

where $\overline{X}$ is the mean as before. Usually, however, an arbitrary origin, a, is used and class interval c, $d' = \dfrac{x - a}{c}$ and for computational purposes

$$\sigma_x = c \sqrt{\frac{\Sigma f_i d'^2}{\Sigma f_i} - \left(\frac{\Sigma f_i d'}{\Sigma f_i}\right)^2}.$$

Using the definition of $\mu_{2:x}$, $\sigma_x = \sqrt{\mu_{2:x}}$. The square of the standard deviation is the **variance**. The quantity $\sqrt{\dfrac{\Sigma(x - a)^2}{N}}$ where a is any value is a minimum when $a = \overline{X}$, an important property of the standard deviation.

Given r sets with $N_1, N_2, \cdots N_r$ variates, respectively, and respective means and variances $\overline{X}_1, \overline{X}_2, \cdots X_r$, $\sigma_1^2, \sigma_2^2, \cdots \sigma_r^2$, then the variance of the combined group σ^2 may be found from either of the formulas

$$N\sigma^2 = N_1\sigma_1^2 + N_2\sigma_2^2 + \cdots + N_r\sigma_r^2 + N_1(\overline{X}_1 - \overline{X})^2 + \cdots + N_r(\overline{X}_r - \overline{X})^2,$$

or

$$\sigma^2 = \frac{N_1(\sigma_1^2 + \overline{X}_1^2) + \cdots + N_r(\sigma_r^2 + \overline{X}_r^2)}{N} - \overline{X}^2,$$

where
$$N = N_1 + N_2 + \cdots + N_r,$$
and
$$\overline{X} = \frac{N_1\overline{X}_1 + N_2\overline{X}_2 + \cdots + N_r\overline{X}_r}{N},$$

the mean of the combined sets. The variance of the combined set is always greater than or equal to the smallest variance in all the sets. The **expected value** of the **sample** variance in a sample of N variates is $\dfrac{N - 1}{N}$

times the **population** variance, and is not the population variance.

The standard deviation is used in the **coefficient of correlation**, in **regression** lines, in the **normal curve**, in finding **standard units**, in the measure of **skewness** a_3, and the measure of **kurtosis** a_4, in the **coefficient of variation** and in many other cases. When squared it is used in the **analysis of variance**. Two sample variances from the same normal population may be tested by use of **Snedecor's F distribution**. (L.A.A.)

STANDARD ERROR OF ESTIMATE. Error of Estimate.

STANDARD ERROR OF SAMPLING. Consider a statistic θ' calculated from **samples** of N variates. Its standard error, $\sigma_{\theta'}$, is defined as the **standard deviation** of θ'. (L.A.A.)

STANDARD METER. The fundamental unit of the metric system is called the meter. The standard platinum meter bar, known as the **mètre des archives**, deposited at Sèvres, was supposedly so constructed that at 0° C. its length is one ten-millionth of the earth's meridian quadrant at sea level. It is one of three similar bars constructed at the same time (1793) and kept in different places. This standard superseded a provisional meter based on the length of a seconds pendulum. Its ratio to the standard yard of the **English system** is about 1.0936 (1 U. S. yard equals 0.9144 meter). The international prototype meter is defined as 1,553,164.13 times the wavelength of the red line of **cadmium** in air at 760 mm. pressure and 15° C. (This method of fixing the standard was suggested by Professor A. A. Michelson.) (L.D.W.)

STANDARD INTEGRALS. Integration, Technique of.

STANDARD NOTATION FOR NUMBERS IN DECIMAL FORM. The use of positive and negative **exponents** finds useful practical application in the so-called standard (or scientific) notation for numbers in decimal form. In much scientific work and elsewhere, very large and very small numbers need to be expressed in convenient form. For this purpose, a number between 1 and 10 is multiplied by the appropriate power (positive or negative) of 10. Thus, 93,000,000 would be written 9.3×10^7, and 0.000,000,065 would be written 6.5×10^{-8}. (L.L.S.)

STANDARD TIME. With the adoption of mean solar time and the improvement in the manufacture of time pieces, each locality set its clocks to indicate its own local mean time. As railway systems developed, an intolerable confusion resulted from so many different local times being used along the lines. The first step toward standard time was the introduction of so-called "railway time" which was the local civil time of some important station on a particular railroad. As interstate and international communication expanded, particularly after the invention of the telephone and telegraph, it became apparent that a universal standardization of time must be adopted.

The modern plan developed from a suggestion made by Sandford Fleming in 1878 and is now used by the armed services and international sea and air transport systems. This **Zone Time** is described in a separate article but has never been adopted for civilian life, although it is approximated to in the Standard Time system which is in general use. A few of the smaller nations still use the local civil times of their individual capitals as standard everywhere within their borders. The larger nations employ a system on which the earth is divided into standard time zones. Each zone is 15° of **longitude**, or one hour of time, in width, with the center of each zone an integral number of hours east or west of Greenwich, England.

In Canada and the United States five standard time zones are employed. These are known as Atlantic (Maritime), Eastern, Central, Mountain, and Pacific, and the centers of each are 4, 5, 6, 7, and 8 hours west of Greenwich, respectively. The boundaries of each zone are not set along meridians of longitude, but are determined by national, state, municipal, or commercial convenience. In regions close to the zone boundaries there is still some confusion remaining, but the conditions are infinitely better than those of 50 years ago.

In 1916 the so-called "daylight saving" or "summer time" practice was introduced. Originally the use of daylight saving time was left to local option and much confusion resulted, particularly since the railroads were required by interstate commerce commission ruling to operate on the regular standard time for each region. During World War II, the use of daylight saving time was enforced for the entire country under the name of "war time." On this system each zone was shifted 1 hour toward Greenwich with Eastern War Time being actually Atlantic Standard Time, Central War Time being Eastern Standard, and so on across the continent. (W.K.G.)

STANDARD UNIT.

A variable may be changed to standard units by the transformation $t = \dfrac{x - X}{\sigma_x}$, where t is in standard units, x is the **variate,** $\overline{X}$ is the **mean,** and σ_x is the **standard deviation** of the distribution. In such cases t is said to have a mean zero and a standard deviation of 1.

A standard unit in metrology is a unit whose definition is fixed by law or by some recognized authority; e.g., the U. S. standard yard, the international ohm, etc. (L.A.A., L.D.W.)

STANN(IC), (OUS). Tin.

STANNITE.

The mineral stannite is a **sulfo-stannate** of **copper** and **iron,** sometimes with some **zinc,** corresponding to the formula $Cu_2S \cdot FeS \cdot SnS_2$. It is **tetragonal;** brittle with uneven fracture; hardness, 3.5; specific gravity, 4.3–4.5; metallic luster; color, gray to black, sometimes tarnished by **chalcopyrite; streak,** black; opaque. It occurs associated with **cassiterite,** chalcopyrite, **tetrahedrite** and **pyrite,** probably the result of deposition by hot alkaline solution. This mineral occurs in Bohemia: Cornwall, England; Tasmania, Bolivia and in the United States in South Dakota. It derives its name from the Latin word for tin, *stannum.* (E.S.C.S.)

STANNUM. Tin.

STAPHYLOCOCCI.

A group of organisms which are the commonest agents causing suppurative or pus-forming **lesions.** They are spherical cocci which usually grow in grape-like clusters. On solid media, they produce different pigments, a property which is the basis of their classification as *staphylococcus aureus, albus,* and more rarely, *citreus,* the yellow, white and green pigment-producers respectively.

Staphylococci are constantly in the air, water, and on the surface of the body. Different strains vary widely in their ability to produce disease in man. The most virulent infections are produced by *hemolytic staphylococcus aureus. Staphylococcus albus* is relatively non-pathogenic. Boils, furuncles, and **abscesses** in any organ or tissue are most frequently caused by staphylococci. **Osteomyelitis,** or infection of bone, is due to staphylococcus infection in 75% of cases. If the organisms gain entry into the blood stream (**septicemia**) widespread abscesses in such sites as the brain, meninges, kidney, liver, lung, etc., may result. Sometimes **pneumonia** is a primary staphylococcal infection.

The treatment of staphylococcal infections depends on their nature and site. The organisms are highly sus-

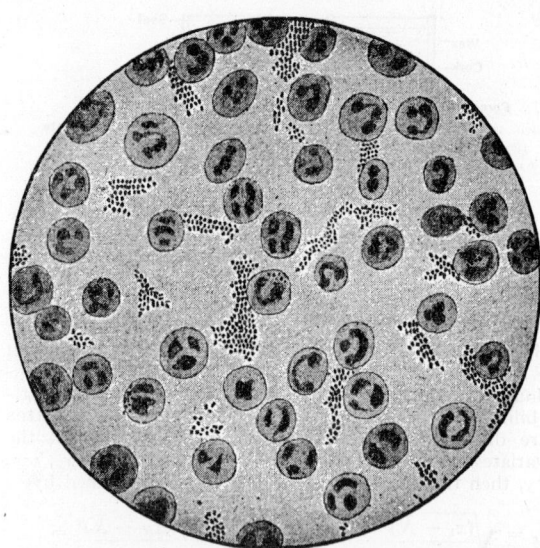

Staphylococci from an abscess of the parotid gland (Jakob). The staphylococci are seen as clumps of dots between pus cells. (*Todd and Sanford, Clinical Diagnosis by Laboratory Methods, W. B. Saunders Co.*)

ceptible to the action of **sulfonamides** and **penicillin.** Penicillin particularly has resulted in the control and cure of many fulminating infections that would otherwise have been fatal. (D.M.H.)

STAR ALTITUDE CURVES. Celestial Navigation.

STAR CATALOGUES.

Any listing of stars, usually arranged in order of increasing **right ascension,** is known as a star catalogue. Originally, star catalogues were intended merely for the purpose of providing accurate positions of the stars for use by navigators, but many modern catalogues are designed to provide particular characteristics of the stars.

The oldest existing star catalogue is contained in the **Almagest** of Ptolemy issued about 137 A.D. This is undoubtedly a reissue of the catalogue of Hipparchus of which no copy is known to exist. The Almagest catalogue was the only one of any real value until the 15th century when an Arabian catalogue made its appearance. Tycho Brahe's catalogue of 1580 marks the dawn of the modern era of star catalogues and since that time many others have appeared. Probably the most comprehensive catalogue issued is the **Bonner Durchmusterung,** which first appeared about 1850, together with the various extensions which have since been published. During the last half of the 19th century the Astronomische Gesellschaft sponsored a catalogue of accurate positions of the majority of the stars contained in the Bonner Durchmusterung.

With application of photography to astronomy several projects have been launched for obtaining comprehensive star catalogues. By far the most ambitious of all of the photographic catalogues is the so-called Astrographic Catalogue which was started as a cooperative effort of 18 observatories in 1887. When completed the catalogue will contain positions of between 3 and 4 million stars measured on about 44,000 plates. As well as the positions of the stars, there will be also issued prints from the plates forming the atlas frequently referred to as the Carte du Ciel. At present the astrographic catalogue is somewhat more than $\frac{2}{3}$ finished.

Many particular types of catalogues are issued for particular purposes, such as catalogues of **stellar magnitudes, spectral class, proper motions, double stars, variable stars,** etc. (W.K.G.)

STAR CONNECTION. This is the connection of the various phases of an a-c machine or circuit in which one end of all phases is connected to a common point, the other end of each phase going to a line. The **Y** is the three-phase star and is the most common example of this type connection. (L.R.Q.)

STAR COUNTS. Milky Way.

STAR FINDER. An instrument designed for the purpose of quickly determining the **horizontal coordinates** of a star, at a certain place and time, with an accuracy sufficient to locate it for observational purposes, is known as a star finder. A star finder may also be used for the reverse process of determining the name of an unknown star whose horizontal coordinates have been approximately measured.

Various mechanical types of star finders have been designed and placed on the market. The simplest of these types is a tube mounted in the same manner as an **equatorial telescope,** but having the altitude of the polar axis easily adjustable for **latitude.** The **declination** circle has the values for the 22 **navigator's stars** indicated, and has the hour circle arranged with various movable dials so that the **hour angle** of any navigator's star may be quickly set for local **time** in a given longitude. All types of mechanical star finders must be properly oriented and kept level while in use, hence they are far more practical for use on land than on a ship at sea or in the air.

For use on a moving ship several types of star finders have been devised which have a map of the celestial sphere, usually drawn on the **stereographic projection,** and with a set of transparent disks to be placed over the map. The disks have curves marked on them showing horizontal coordinates and the set covers a wide latitude range. To use such a star finder the disk most nearly corresponding to the latitude of the observer is placed over the star map and oriented in accordance with printed directions. The observer's local time and longitude govern the orientation in most cases. With the disk in proper position the horizontal coordinates of objects on the star map may be immediately determined. (W.K.G.)

STAR STREAMING. Moving Cluster.

STARCH. Carbohydrates.

STARFISH. Asteroidea.

STARK EFFECT The effect of a strong, transverse electric field upon the **spectrum** lines of a gas subjected to its influence. In many respects it resembles the more complicated types of **Zeeman effect,** but is subject to different laws, and may change radically in character and in multiplicity of component lines with increasing field intensity. The phenomenon, first observed by Stark in 1913, is conveniently studied by means of a **canal-ray** tube having behind the cathode a third electrode which may be given a high positive potential, in order to impose the desired field upon the radiating canal-ray particles. (L.D.W.)

STARLING. Aves, Passeriformes. A bird (**Aves**) of any of several species native to Europe, Asia, and northern Africa. The African glossy starlings belong to a family containing also the Asiatic grackles or hill mynas, while the true starlings are placed in a closely related family.

The one species which concerns us in North America is the common European starling, *Sturnus vulgaris,* which was introduced into New York in 1890 and has

Starling. *Sternus vulgaris.*

since spread a third of the way across the continent. The males are black with green and blue iridescence and light tips on many of the smaller feathers, and the females are brownish-gray. The beak is rather large. In Europe the starlings are valued as destroyers of insects and for their song. They are able mimics. The same good qualities are worthy of consideration in America but the birds are also destructive of native species and in the fall they become a nuisance by gathering in great flocks to roost in countless thousands in the trees of residential districts. The net verdict is against them, and they are commonly regarded as another of our imported pests. (A.W.L.)

STARS. There is probably no one class of physical objects that has attracted more popular and scientific attention throughout the ages than have the stars. A small portion of the mass of mythological material that is associated with these bodies will be found in the various articles dealing with the individual **constellations** and a few of the brighter stars.

The methods for determining the physical characteristics of the stars will be found under such titles as: **Stellar Parallax, Stellar Magnitude, Spectral Classification, Binary Stars, Variable Stars, Cepheids, Giant and Dwarf Stars,** etc. The characteristics of a typical main sequence G-type star will be found in the articles dealing with the **Sun** and various solar characteristics. See table, p. 1370. (W.K.G.)

STARTER, ENGINE. The cycle of the **internal combustion engine** requires that the engine be initially revolved from an external source of power. The mechanism supplying this initial effort is known as the starter. Certain small semi-portable or stationary engines, also tractor engines, truck engines, etc., are frequently started by manual cranking, but a mechanical starter is standard equipment on automobile engines. **Diesel engines** are rarely hand cranked, although they are started by more varied types of starters than are employed with the gasoline engine. **Aeronautical engines** can be started by swinging the propeller by hand. However, many are now equipped with starters, the commercial plane because of the size of the engine and the inconvenient location of the propeller viewed from the standpoint of manual starting, the small plane for convenience of the private owner, or in emulation of the standard practice on automobiles.

Starters for automobile engines are electric motors of a series wound type, deriving their supply of electricity from a storage battery. The relation of motor output to battery capacity is such that the use of the motor must be restricted to short intervals of time, between which the battery is recharged. Normally, the starting motor is disengaged from the **engine.** During the starting cycle it is connected by a reduction gear which multiplies the **torque** produced by the motor. A small pinion on the motor engages with a large annular gear, usually placed on the circumference of the flywheel of the engine. The speed reduction is between 10:1 and 15:1. The method of engaging the pinion with the gear for starting used most frequently is the Bendix drive. This is shown in

PHYSICAL CHARACTERISTICS OF TYPICAL STARS *

STAR	SPECTRAL CLASS	TEMPER-ATURE IN °K.	DENSITY IN TERMS OF WATER	REFERRED TO SUN AS UNITY		
				Luminosity	Mass	Diameter
GIANTS						
Antares....................	M_0	3,100	0.0000003	3500	30	480
Aldeberan..................	K_5	3,300	0.00002	90	4	60
Arcturus...................	K_0	4,100	0.0003	100	8	30
Capella....................	G_0	5,500	0.002	150	4.2	12
MAIN SEQUENCE						
β Centauri.................	B_1	21,000	0.02	3100	25	11
Vega......................	A_0	11,200	0.1	50	3	2.4
Sirius A...................	A_0	11,200	0.4	26	2.4	1.8
Altair.....................	A_5	8,600	0.6	9.2	2	1.4
Procyon...................	F_5	6,500	1.2	5.4	1.1	1.9
α Centauri A...............	G_0	6,000	1.1	1.12	1.1	1.0
The Sun...................	G_0	6,000	1.4	1	1	1.0
70 Ophiuchi A..............	K_0	5,100	0.9	0.42	0.9	1.0
61 Cygni A.................	K_7	3,800	1.3	0.21	0.5	0.7
Krueger 60A................	M_3	3,300	9	0.002	0.3	0.3
WHITE DWARFS						
Sirius B...................	F	7,500	27,000	0.10	0.96	0.034
O₂ Eridani B...............	A_0	11,000	64,000	0.003	0.44	0.019

* See article on **Stars**, p. 1369.

---

the accompanying figure. The motor shaft is extended to form support for an externally threaded sleeve and two collars. One collar is fixed to the shaft and to one end of a coil spring. The other end of the spring is at-

Starter rotor.

tached to the remaining collar, as is also the threaded sleeve. When the shaft revolves, the sleeve and collar revolve as well, being driven by the spring. The pinion has an internal thread corresponding to that on the sleeve, and is unbalanced by having a weight cast on one side of the circumference. When the starter is connected to the battery, the armature immediately starts to revolve, but not so the pinion, due to its inertia and the effect of the unbalanced weight. Consequently, it moves endwise along the thread. The starting motor is so placed that this movement engages the pinion with its gear. When the end of the thread is reached, the pinion jams against a stop, and must then, perforce, turn with the armature. The shock of connecting a moving with a stationary body is eased by the spring connection between the pinion and the armature shaft.

Small Diesel engines are started by electric starters, but this is not usual in the sizes in which these engines are usually built. Because of the high compression pressure, a heavy torque must be exerted to revolve the Diesel engine during starting. Some engines have small one- or two-cylinder gasoline engines as an integral auxiliary for starting. The gasoline engine is started by hand cranking, then mechanically clutched to the Diesel for

the purpose of starting the latter. After the Diesel starts firing the starter engine is de-clutched and stopped. The largest Diesels generally use compressed air from storage tanks which are recharged after a start. Some manufacturers supply the compressed air to one cylinder, only, of the engine, and have the exhaust valves propped open on the other cylinders so as to reduce the torque required. Others suuply compressed air to all cylinders at the time of the normal power impulse by means of a rotary distributor valve.

Electric starters have been used on aeronautical engines, and it is probable that as starters continue to be applied in that field, the electrical starter will be the more important type. Its use was at one time restricted because of the aversion of manufacturers to including a wet storage battery in an airplane, and to the lack of suitable aircraft types of storage batteries. With the development of the latter, and the increasing desire for storage battery current to operate lights, radio, etc., there is a noticeable trend to the use of electric starters. (F.T.M.)

STARTER, MOTOR. The current in all types of electric motors is limited largely by the back voltage of the armature. Since this voltage is zero at starting, the current drawn by the machine under this condition may be many times that drawn when running at rated speed. For small motors this current may still not be too large for the system so such motors are frequently started by connecting them across the line without any auxiliary starting equipment. This method of line-starting is ordinarily used only for fractional horsepower d-c motors but a-c motors up to several hp. are started this way. Under these conditions it is necessary to provide only overload protection for the motor. The current drawn by large machines would impose too much of a strain on the supply system so such motors are provided with special starters. For d-c motors these starters consist of some arrangement for starting with extra resistance in the armature circuit and then for cutting this out automatically or manually as the motor comes up to speed. The starter usually provides overload protection and under voltage release protection so if the voltage fails the motor is not connected directly to the line when power is restored. A four-point manual starting box is illustrated. Squirrel cage induction motors and synchronous motors (which usually start as induction motors) are commonly started on reduced voltage. Since these are a-c machines the **transformer** offers an efficient means of reducing the voltage. Auto-transformers are commonly used for this purpose, the motor being started on a tap position, then the voltage increased to full value in one or more steps as the machine comes up to speed. At the running position the transformer is disconnected and the motor is connected directly to the line. This operation may be manual or automatic. An example of this type starter is illustrated (the starter is frequently called a compensator). Wound rotor induc-

STATES OF MATTER.

STATES OF MATTER. The term refers to the solid, liquid, and gaseous forms in which matter presents itself. If we started with any pure element at the lowest attainable temperature and, keeping it at some fixed pressure, as that of the atmosphere, raised the temperature gradually to the highest point known to the laboratory, the substance would be observed to pass through certain stages and to undergo changes from one stage to another at more or less definite transition points. If the pressure were different, similar changes would take place, but in general at somewhat altered transition temperatures. The same statements may be made concerning a pure compound, except that there are some compounds which are so unstable as to decompose before attaining all the states. (For example, if one attempts to melt nitrogen iodide, the result is an explosion, the products of which are nitrogen gas and iodine vapor.)

The stages referred to are the solid, the liquid, and the vapor states; which, for any pure substance, are determined by the two variables, temperature and pressure,

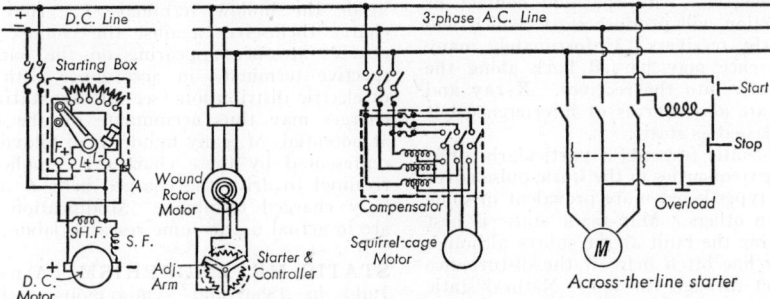

Some methods of starting electric motors.

tion motors offer an additional method of reducing the starting current. This may be done by inserting resistance in the rotor circuit, the effect being analogous to the armature resistance of the d-c starter, and then cutting this out as the machine speeds up. This also offers a means of speed control of the motor. In large plants where many machines are used a special starting bus supplying reduced voltage may be used. The motors are connected to this for starting and switched to the full voltage bus as they come up to speed. (L.R.Q.)

STARTING BOX. Starter, Motor.

STARVATION. A state of existence without food or with inadequate food. When animals are completely deprived of food their only source of energy for the essential processes of **metabolism** is the material already present in the body. The **carbohydrate** stored as glycogen is quickly used up, leaving only the stored fat, the circulating protein (see **Amino Acids**), and the protein of the tissues. Studies of the progress of starvation in mammals have shown that the fat is almost completely used up and that the greatest loss of tissue proteins is from the muscles, due to their great bulk. In percentages, however, the **liver, spleen,** and **gonads** lose more of their bulk than other parts of the body, and the **heart** and central **nervous system** are maintained at the expense of the other parts with very little loss of substance. Extensive observations of the details of metabolism and bodily changes during starvation allowed to continue to the death of the animal have been recorded.

A remarkable result of starvation in some of the lower invertebrates is a progressive shrinkage of the body as a whole, in contrast with the emaciation which results in vertebrates. Observations on the flatworm, *Planaria*, have shown that ultimately it even retraces its development to assume an embryonic form. (A.W.L.)

with the density as a third, related variable. If the temperature and pressure are under complete control, it is possible to pass from any one to any other of the three states, either directly by a single transition, or by passing through the third state with two transitions. This will be clear from the accompanying phase diagram,

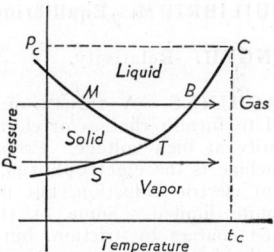

Temperature-pressure curves for substance like water. *C* represents the critical state, *T* the three-phase equilibrium point.

which represents the temperature-pressure equilibrium curves for a substance converging at the **three-phase equilibrium** or "triple point" T. (It is understood that each phase of the substance is to be strictly pure; for example, no air is to be mixed with the vapor.) Thus, if a mass of pure ice were kept completely enclosed by itself at a fixed pressure below 4.6 mm. (its "triple point"), and its temperature raised, it would turn directly into vapor at the **sublimation** point S; but if the pressure exceeded 4.6 mm., the ice would first become liquid at the **melting point** M, and the water would then vaporize at the **boiling point** B, corresponding to the existing pressure. The volume of the enclosure would, of course, have to be varied to keep the pressure constant during these changes.

When the temperature has exceeded the critical temperature t_c (see **Critical State**), neither the solid nor the liquid phase is longer possible, even with greatly increased pressure. The liquid-vapor curve must be considered as terminating at C, since it pertains specifically to two phases in equilibrium, while beyond the point C the substance does not exhibit two phases. The one apparently homogeneous phase now remaining is said to be a true "gas." (L.D.W.)

STATIC. This is the name commonly applied to all the various random electrical disturbances which are picked up by a radio **receiver**. These can be divided into two general classes, natural and man-made static. The first is caused by various types of natural electrical discharges, the most pronounced being those of lightning. However, a static-producing discharge is not necessarily, or even usually, a visible lightning discharge. Various static charges are often continually building up and discharging in the atmosphere and hence inducing disturbances in the receiver. Cosmic radiations are also responsible for static. These types of natural static are often called atmospherics. The types of man-made static are almost as numerous as the electrical machines which man has developed. Any sparking contact or poor electrical connection will produce static which will be picked up by nearby receivers. Unfortunately many types of this interference may be fed back along the power lines and directly into the receiver. **X-ray** and diathermy machines are also sources of interference but cannot be properly classed as static.

The elimination of static presents a particularly difficult problem since the frequencies in the static pulse cover a wide band, certain types being more prevalent in some frequency ranges than others. Man-made static is best eliminated by correcting the fault at the source although a **filter** in the power line often helps if the disturbance is coming into the set through the line. Natural static can be minimized, but not eliminated. For amplitude **modulation** systems limiting the **frequency band** to which the receiver responds will reduce the noise. Various types of limiters will also reduce the effect since the static signal is frequently greater than the desired one. Frequency modulation is inherently less susceptible to interfering noise and offers almost noise-free reception. (L.R.Q.)

STATIC EQUILIBRIUM. Equilibrium of Forces.

STATIC LENGTH. Relativity.

STATIC MACHINES. A variety of devices have been employed to furnish charges of electricity of considerable quantity at high voltage. Probably the simplest static machine is the **electrophorus**, operating on the principle of electric induction, but its output and voltage are quite limited. Some of the older machines generated charges by friction, but modern machines are of the induction type. Among the latter, the well known Toepler-Holtz and Wimshurst machines have rotating glass or mica plates bearing metal "carriers" on which the charges are induced as on the metal plate of the electrophorus. Recently some very powerful induction machines, embodying all the principles of the older types, have been developed by Van de Graaff and others and have come into use in nuclear research. One of the simplest of these is illustrated in the diagram.

Two endless belts, B, of silk or paper revolve on insulated drums D, the upper of which are mounted inside two hollow, rounded metal tanks T forming the terminals of the machine. R represents the positive terminal of an ordinary static machine or of a high-voltage rectifier-transformer, which keeps the system from the inductor plate to the "comb" of charging points P_1, positively charged at about 10,000 volts. The negative inductor plate I_1 and the charging comb P_2, opposite P_1 and I_2, respectively, are grounded at one end, their other ends being maintained negative

by induction. As the belts revolve, the sharp teeth of the combs P, under the influence of the inductors I, discharge upon the belts, which carry the charge

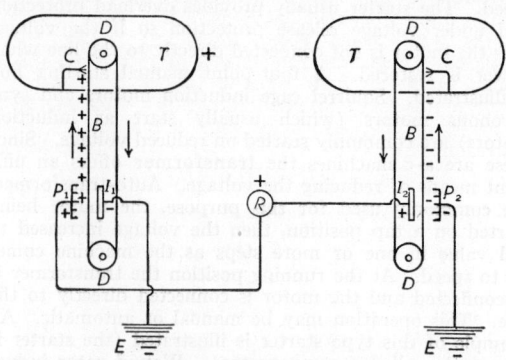

Diagrammatic sketch of a modern electrostatic generator.

inside the hollow terminals T. Here the belts discharge themselves against the contact brushes C, the charges at once appearing on the outside of the respective terminals, in accordance with Faraday's law of electric distribution (see **Electrostatics**). Very large charges may thus accumulate on the terminals T, T, at potentials of many hundreds of kilovolts. The energy represented by these charges is supplied by the work required to drive the charged belts B toward the similarly charged terminals. Modifications of this design are in actual use in some research laboratories. (L.D.W.)

STATIC METAMORPHISM. A term proposed by Judd in 1889 and synonymous with the German, *Belastungmetemorphismus.* That form of regional metamorphism which may be considered as due to "load," and which is primarily induced by vertical or gravity pressure, as opposed to regional metamorphism which is caused by **orogenic** pressures. (R.M.F.)

STATICAL MOMENT. Moment.

STATICS. Statics is that branch of mechanics which deals with particles or bodies in **equilibrium** under the action of **forces** or of **torques**. It treats of the composition and resolution of forces, the equilibrium of bodies under balanced forces, and such properties of bodies as center of gravity and moment of inertia.

A set of forces may be exerted along lines which all lie in the same plane, in which case they are said to be coplanar. Again, the lines of action may all intersect at one point, so that the forces are "concurrent." A torque, or moment, is that which tends to produce rotation about some axis. The measure of the torque of a given force about an axis not parallel to its line of action is the product of the force, the perpendicular distance from its line of action to the axis, and the sine of the angle between the axis and the direction of the force. Two forces of equal magnitude, acting along parallel lines in opposite directions, constitute a couple, the torque of which, about any axis perpendicular to its plane, is the product of either force by the perpendicular distance between their lines of action. Such a pair of forces has no resultant and no equilibrant, since it can neither be replaced nor balanced by a single force. It is possible to have a system of more than two forces, not necessarily parallel or even coplanar, but which is equivalent to a couple in that it has a torque without having a resultant. Only another couple, or its equivalent, can balance such a system. One of the basic propositions of statics is that any system of forces is, in general, equivalent to a single force acting along a definite line and a couple whose torque axis is in a definite direction; and that for equilibrium, this force and this couple must both become zero. These ideas are most conveniently expressed by means of the notation of vector analysis.

A force F, whose line of action lies at θ degrees to one of two mutually perpendicular axes, may be resolved into components of $F \cos \theta$ and $F \sin \theta$, parallel to the axes. When several forces are concurrent, and coplanar as well, the components of their resultant are found as the algebraic sums of the components of the separate forces. The resultant of two coplanar forces P and Q having an included angle θ, is expressed by the equation

$$R = \sqrt{P^2 + Q^2 + 2PQ \cos \theta}.$$

The angle between the resultant and the force P is

$$\text{arc tan } \frac{Q \sin \theta}{P + Q \cos \theta}.$$

If there are more than two forces whose resultant is to be found, any two may be combined to find the resultant which may then be combined with a third, and so on until the last resultant found is that for the complete array of forces. The same problem may be solved graphically by the vectorial addition of the forces. In vectorial addition, lines whose lengths are equal to the magnitudes of the forces are drawn in the directions of the forces so that the force representing one line follows another, forming part of a polygon, the arrows of which, as shown in Fig. 1, are in succession.

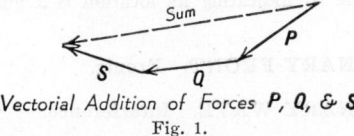

Vectorial Addition of Forces P, Q, & S
Fig. 1.

cession. The resultant is the line drawn from the beginning of the first force to the end of the last. This resultant force, with direction of arrow reversed, is the equilibrant, or neutralizing force. When forces are concurrent, the line of application of the resultant is known to pass through the point of concurrence of the forces; but with non-concurrent forces, neither the force polygon nor the analysis by components is sufficient to establish the line of application of the resultant force. To locate this line, another condition of statics must be introduced. It is that the moment of the resultant about any moment axis is equal to the algebraic sum of the moments of the forces about the same axis. The funicular polygon may also be used to establish the line of action of non-concurrent forces. Rays are drawn to the corners of the force polygon from some assumed pole, and the sides of the funicular polygon drawn parallel to these rays. The corners of the funicular

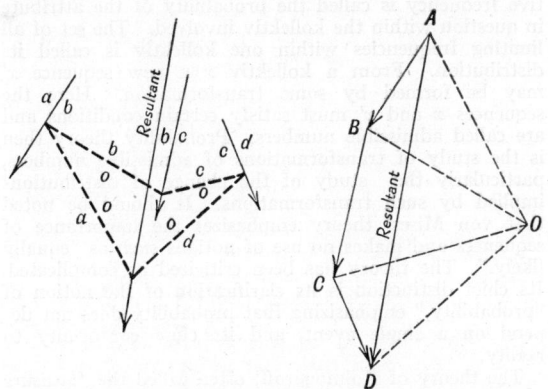

Forces in space.   Force polygon.
Fig. 2,   Funicular polygon,

polygon lie on the forces themselves, and the closing lines, oa and od, intersect on the line of action of the resultant. (See **Bow's Notation**.)

When forces are not coplanar, the solution, in general, involves simultaneous solution of six equations, as follows:

Let F_x, F_y, and F_z be components parallel to x, y, and z axes, respectively.

R_x, R_y, R_z are the corresponding components of the resultant.

$$\Sigma F_x = R_x,$$
$$\Sigma F_y = R_y,$$
$$\Sigma F_z = R_z,$$
$$\Sigma M_{Fx} = M_{Rx},$$
$$\Sigma M_{Fy} = M_{Ry},$$
$$\Sigma M_{Fz} = M_{Rz} \text{ in which the } M\text{'s}$$

are the moments of the respective components about either axis perpendicular to them.

Static equilibrium exists if the algebraic sum of the components of the forces in any direction is zero, and if the algebraic sum of the moments of the forces about any axis is zero. When a body is in equilibrium under forces, some of which are unknown, an analysis may be made by applying these conditions of equilibrium, provided that, of the quantities required to specify the forces, the number which are unknown does not exceed the number of equations afforded by the given conditions. In the graphical analysis of static equilibrium, the conditions of equilibrium are that the force polygon and the funicular polygon close.

A common application of the laws of equilibrium of coplanar forces is the calculation of the reactions of a simply supported beam loaded with parallel forces which are perpendicular to the length of the beam. The reactions are determined by considering the whole system to be one in equilibrium, so that the moment equation may be applied by selecting the moment center on

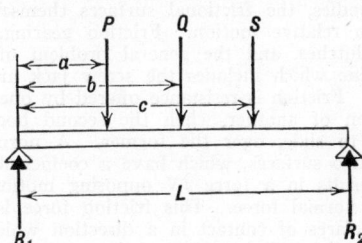

Beam loaded with parallel forces
Fig. 3.

one of the unknown reactions. The moment of this reaction thus becomes zero, and the moment equation has only one unknown reaction. If a beam of span L (Fig. 3) is loaded with forces P, Q, and S, at distances, a, b, and c from R_1, the sum of the moments equals

$$R_2 L - Pa - Qb - Sc.$$

For equilibrium, this quantity must equal zero; hence

$$R_2 = \frac{Pa + Qb + Sc}{L}.$$

Triangularly framed structures may be analyzed by algebraic resolution and composition of forces, or by the graphical method of funicular and force polygons. Since the latter is simpler and entirely as useful, it is the one most frequently employed. Consider a simple derrick, framed as shown in Fig. 4, loaded with weight W. This condition of static equilibrium may be analyzed for the forces acting in the structural members by drawing a force polygon for each joint, beginning with that which has only two forces unknown in magnitude, then proceeding to some other joint, where the same condition exists. In the example given, it is plain that the joint

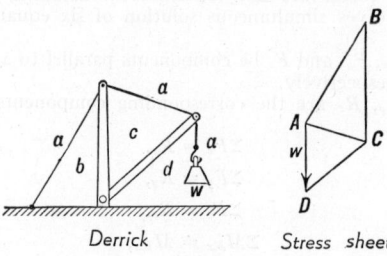

Derrick   Stress sheet

Fig. 4.

at the load must be analyzed first. The equilibrium polygon for this joint is *CDA*, and is drawn by laying off *DA* in the direction of *W*, and of the magnitude *W*, and locating point *C* as the intersection of lines *DC* and *AC*, drawn in the directions which those forces actually have in the derrick. Using the magnitude of *CA* thus determined, the joint at the top of the mast may next be analyzed, locating the point *B*. The magnitudes and character of the forces acting on triangularly framed structures are thus determined by the method of analysis of joints graphically by force polygons which close. These force polygons begin at a point where all corners of the polygon are known except one, and proceed from there in succession to other joints, which will offer the same conditions. Each joint has a separate force polygon, but it is usual to place these polygons in juxtaposition along the lines where they have lines in common, for convenience and simplification. The resulting diagram, which is a composite force polygon, is known as a stress diagram or stress sheet. (See **Graphical Statics**.)

Certain cases of **friction** lie properly in the realm of statics, since, though the friction may be that of moving bodies, the frictional surfaces themselves may not be in relative motion. Friction gearing, wedges, friction clutches, and the general problem of the inclined plane which includes the screw jack are typical examples. Friction is resistance offered by one body to the motion of another when the second body slides, or tends to slide, over the former. A normal force between two surfaces, which have a coefficient of friction *f*, results in a force *fN* opposing motion, where *N* is the normal force. This friction force is tangent to the surfaces of contact in a direction which would oppose motion. The resulting force, acting between a body and its supporting surface, takes the direction shown in Fig. 5. The horizontal component of the

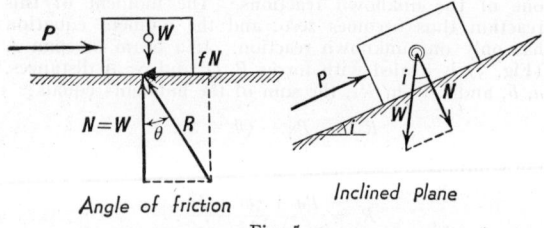

Angle of friction   Inclined plane

Fig. 5.

reaction *R* is equal to *P*. As *P* increases up to the point of sliding, the angle θ increases until its tangent equals the coefficient of friction of the surfaces in contact. When *P* exceeds *fN* there will be relative motion between the body and the supporting surface. Should a body having weight *W* rest on a plane inclined at angle *i*, the coefficient of friction between the surfaces being *f*, the force necessary to start the body up the plane must be greater than $W \sin i + fW \cos i$ where

the force *P* is parallel to the plane. If the force *P* is inclined to the plane, only its component parallel to the plane will be available for starting the load, and its perpendicular component may either tend to increase or decrease the friction resistance.

The centers of gravity of areas and masses come under the head of statics, since they may be considered as made up of a number of elementary areas or masses, which could be treated as proportional to a set of parallel forces acting through their individual centers. (See **Centroid, Moment, Moment of Inertia, Friction, Equilibrium of Forces**.)   (L.D.W., F.T.M.)

STATION. A wood stake or any other type of marker which is used to locate accurately a point on a **traverse** line or the intersection of traverse lines is called a station. The stake is driven flush with the ground. The exact point or the point of intersection is indicated by a tack.

Any station which is used for a transit setup is a transit station. Lines which connect transit stations are transit lines. The transit lines, taken collectively, form what is known as the horizontal control of the survey.

Open traverses are marked by stations every 100' along the line. These are known as full stations. Intermediate stations, called plus stations, are placed wherever it is necessary to locate adjacent artificial or natural features with respect to the transit line.

A stake which is driven at an angle over a station for the purpose of indicating its location is a guard stake. (C.W.C.)

STATIONARY FRONT. Fronts.

STATIONARY WAVE. Interference.

STATISTIC. A statistic is a function of **sample** values such as the **mean** of a sample, the **variance** of a sample, the **coefficient of correlation** in a sample, etc. (L.A.A.)

STATISTICAL MOMENT. Moment.

STATISTICAL PROBABILITY. Two modern theories of probability have been used in mathematical statistics, that of von Mises and that of Kolmogoroff. The theory of von Mises, usually called the "frequency" theory, depends upon the notion of a "kollektiv." A kollektiv *x* in one dimension consists of an endless sequence $x_1, x_2, x_3, \cdots$, of observations or experiments. Briefly a kollektiv corresponds to a **population** or universe. If x_i represents a success we replace x_i by 1, or a failure we replace it by 0. We may form the **relative frequency** of success in a finite sequence and in the case of a kollektiv the limiting value of the relative frequency is called the probability of the attribute in question within the kollektiv involved. The set of all limiting frequencies within one kollektiv is called its distribution. From a kollektiv *x* a new sequence *x'* may be formed by some transformation. Here the sequences *x* and *x'* must satisfy certain conditions and are called admissible numbers. Probability theory then is the study of transformations of admissible numbers, particularly the study of the change of distributions implied by such transformations. It should be noted that von Mises' theory emphasizes the importance of sequences and makes no use of notions such as "equally likely." The theory has been criticized as complicated. Its chief distinction is its clarification of the notion of "probability," emphasizing that probability does not depend on a single event, and its close conformity to reality.

The theory of Kolmogoroff, often called the "measure theory of probability," has been extensively used. Roughly speaking, the probability of an event is a definite number found by the theory of measure and the axioms merely express the rules for operating with such numbers. Up to this point the general theories of probability have been discussed. It is important to define statistical probability. In case the probability of an event may not be determined by mathematical reasoning, then it is necessary to have a series of experiments or to record a series of observations. From this series we may by application of **Bernoulli's theorem** be practically certain that the value of the probability so found will differ little from the true probability. Great care must be exercised to see that the series of observations constitutes a random **sample** or, if properly planned, is a **stratified sample.** This method must be used in insurance and in the social, physical, and biological sciences. (L.A.A.)

STATISTICAL QUALITY CONTROL.

A tool of industrial management comparable with production control and cost control. In the past it was usually a routine of inspections, investigations, corrections, and more inspections designed to insure the passage of only good products from one department to another or to the shipping room. At present the words "quality control" almost automatically imply "by statistical methods."

It is necessary to unlearn certain common misconceptions and to accept new views in order to understand how statistical methods can apply to manufacturing processes, assembly processes, typing of letters, writing of orders, and research in agriculture.

1. Inspection of manufactured products is performed not just to show whether these parts are acceptable; its more important function is to show whether the process, machine, or operator who made the parts is *capable* of working within tolerances and whether he is *actually* doing it.

2. No machine can make two articles exactly alike; all "identical" products or processes differ among themselves and tolerances must be allowed.

3. It is not necessary that a process be turning out thousands of parts a day in order to use statistical methods; these have been used where only one large assembly was completed in a day.

4. It does not require advanced mathematics to use these methods; tables in handbooks provide constants and formulae, and the arithmetic computations are simple.

All items or products fall into three classifications. There are (1) those items which have measurable characteristics or variables, (2) those items which are acceptable or rejectable because of either good or bad attributes, and (3) those which *can* have an infinite number of defects, but which *usually* have only a few observable ones.

As an example of the inspection for variables, consider an inspector who is weighing boxes of a powder. The specification states that the net weight must be 340 grams (about 12 oz.) plus 10 grams minus nothing. Each half-hour the inspector takes a sample of five boxes from the packaging machine, weighs the contents, and records the values. Then he computes the average of the five and notes the range of the sample (the difference between the lightest and the heaviest contents). Either from previous data or from six such samples, he computes the average range and multiplies by a constant to find the control limits within which the average weight of each sample must lie. He multiplies the average range by another constant and learns what extreme individual weights might be expected if he were to weigh several hundred boxes. If these lightest and heaviest anticipated weights lie within the specified figures, the machine is *capable* of doing its work and, if the average of each sample lies within the control limits, the machine is *actually* doing its work. In this example the grand average of all samples should remain very close to 345 grams.

When a process is not capable of working within the tolerances, either the process or the tolerances must be changed. If a machine or process does not work within its control limits, there is some assignable cause of variation which must be found and eliminated.

As an example of the inspection for attributes, consider an inspector who is checking small electrical devices for "grounds" or electrical paths between the windings and the metal frame. This defect could result from either faulty material or careless workmanship in assembly. During each 2-hour period, the inspector takes 100 units from the assembly line and makes a quick check on each to determine the presence or absence of a ground fault. He records the total number of items tested and the number of defectives found. This enables him to compute the average per cent defective for the process over a period when the rate of production is steady. By the application of statistics, he then computes control limits within which the per cent defective for each group of 100 items should lie. Any group showing more or less defective items must be investigated because it indicates an assignable cause of deviation. This cause should be discovered whether it makes the process better or worse.

The inspector may be checking only $\frac{1}{10}$ of the total production but he learns more about the *process* by the use of control limits than he would learn by checking all items and avoiding the use of statistics. He learns precisely its capabilities.

As an example of the indeterminate sample size where the possible number of defects is large, consider the "squawk sheet" for a large airplane. The final inspectors of the finished plane fill out this sheet which lists all the faults that are observed. The statistical department analyzes these sheets to determine whether the over-all assembly process is getting better or worse. It also notes the frequency with which various faults occur. The total number of faults must follow a random fluctuation within certain limits which are computed by statistical methods. Lack of this random fluctuation may reveal improper performance of the inspectors. These limits may shift slowly with time if the quality of the assembly work changes.

In each of these examples which portray the elementary quality control operations, it can be seen that the use of an appropriate statistical method makes it possible for an inspector to learn more about the capabilities and the performance of the processes than he could learn by any other means. Usually this can be accomplished by using *less* inspection in a *more* efficient manner. (J. A. FIZZELL.)

STATISTICALLY SIGNIFICANT.

A result is said to be statistically significant if the **null hypothesis** has been rejected. For example, if we say the difference in sample **means** is significant we imply we have rejected the hypothesis of the equality of the means in the two **populations** from which the two **samples** are drawn. While we have used sample means as an example, the theory is perfectly general and may be applied to **statistics** of any kind. Sample results should never be used unless statistical significance has been established, i.e., results have been determined for the populations from which the samples were drawn. Of course, sample statistics may differ statistically and yet from an engineering point of view the difference may be too small to be of any importance. (L.A.A.)

STATISTICS.

Statistics is essentially a modern subject combining the mathematical theory of **probability** and the techniques of testing hypotheses. It is primarily concerned with the development of theories which have immediate applications and depends heavily on mathematics while deriving much of its inspiration from applied problems. Modern statistical methods arose from the classical theory of probability and the theory of least squares. Formally statistics may be defined in various ways. Statistics is the subject concerned with the collection, tabulation, and analysis of data. This definition emphasizes descriptive statistics. Statistics is the study of variation. This definition implies the analysis of variance. At the higher level, statistics is that subject in which we make inferences concerning the whole population by use of a sample, usually a random sample. Statistics is of importance in the biological, physical, and social sciences as well as the fields of finance, business, industry, and insurance. Topics treated in statistics are descriptive statistics, analysis of variance, Bernoulli's theorem, law of large numbers, the central limit theorem, the chi-square test, correlation, De Moivre-Laplace theorem, distribution functions, factor analysis, design of experiments including Latin squares, and randomized blocks, multiple correlation, normal probability function, Pearson system of probability functions, Gram-Charlier series, Bernoulli probability function, Poisson probability function, likelihood, Lexis distribution, partial correlation, Neyman-Pearson theory of testing hypotheses, periodogram analysis, serial correlation, sequential test of a statistical hypothesis, Fisher's z distribution, Snedecor's F distribution, Student's t distribution, hypergeometric probability function, consistent, efficient, and sufficient statistics, stratified samples, index numbers, confidence limits, quality control, industrial statistics, the test of independence in a contingency table, and tetrachoric correlation. (L.A.A.)

STATOBLAST. Asexual reproductive bodies of freshwater Bryozoa. They are disk-shaped buds formed in the parent colonies with a protective test to enclose them. When the colony dies in the winter the statoblasts remain alive and inactive, to produce new colonies in the spring. (A.W.L.)

STATOCYST. A form of sense organ found at the margin of hydrozoan medusae. It is a vesicular structure filled with a liquid containing small calcareous granules. The ectodermal lining is sensory and is apparently stimulated by the impacts of the granules as the animal moves. Varying stimuli in the eight statocysts around the margin of the animal bring about the proper movements to maintain its normal position in the water. Lithocyst. (A.W.L.)

STAUROLITE. The mineral staurolite is a complex silicate of iron and aluminum corresponding to the formula $HFeAl_5Si_2O_{13}$, but somewhat varying and may carry magnesium or manganese. It is orthorhombic, prismatic, twins common, often producing cruciform crystals. It is a brittle mineral; fracture, sub-conchoidal; hardness, 7–7.5; specific gravity, 3.6–3.7; luster, sub-vitreous to resinous; color, dark brown, sometimes reddish to nearly black; grayish streak; translucent to opaque. Staurolite is a metamorphic mineral usually the result of regional rather than contact metamorphism, and is common in schists, phyllites and gneisses together with garnet, kyanite, tourmaline, etc. Well-known foreign localities are in Switzerland and Brittany; and in the United States this mineral is common in the schists of New England, and those of the southern Alleghenies. Frequently the crystals are found loose in the soil after the disintegration of the country rock. The name staurolite is derived from the Greek meaning a cross, in reference to the twin crystals, the more nearly perfect crosses being somewhat in demand as curios. Quite frequently used as a baptismal stone, for according to early legend when the fairies heard of the crucifixion of Christ their falling tears when they reached the earth became crosses. (R.M.F.)

STAUROMEDUSAE. Scyphozoa.

STAY PLATE. Compression Member.

STAYBOLT. The surfaces of pressure vessels, such as boiler drums and tanks, which are not of a natural bulged shape, such as the cylinder or the sphere, must be stayed against bulging by special tension rods called staybolts. An example of a stayed surface will be found in the horizontal-return tubular boiler, which is simply a cylinder with flat ends having tubes extending from one end to the other, parallel to the axis of the cylinder. The tubes act to stay the flat surfaces, but the portion of the surface above the water line is free of tubes, and must therefore be supported by staybolts. Staybolts may be simple tie rods threaded on the ends to receive nuts which are screwed on the outside of the surface, or they may be hollow or flexible, the latter type having ball or socket joint at one or both ends of the staybolt. (F.T.M.)

STEADITE. A hard microconstituent found in cast iron consisting of a eutectic mixture rich in phosphorus. In high-phosphorus irons it may cause brittleness and impair machinability. (R.H.H.)

STEADYREST. Lathe.

STEAM. Steam is the vapor of water at or above its boiling temperature. The universal occurrence of water, its cheapness, its solvent properties, its suitability as a medium for a vapor cycle, and general familiarity with it and its properties, all have contributed to make of steam the most important vapor used by man. If a closed vessel is partially filled with water, and heat applied at a high temperature, the water will absorb the heat and rise in temperature until its molecular activity (which depends on its temperature) is so increased that the internal molecular attractions are no longer able to maintain it in the form of a liquid, and molecules leave the surface of the liquid and occupy the space above it in the rarefied condition of a vapor. Water has a definite vapor pressure at every temperature; when the vapor pressure becomes equal to the pressure above the liquid, the boiling point has been reached. If the steam is produced in a closed vessel, it is invisible. If it is released to the atmosphere, it partly condenses to form a cloud of mist (small drops of liquid water), which is visible. The temperature at which the boiling takes place is dependent upon the pressure which is maintained in the boiler. Steam which is in contact with the boiling water is known as saturated steam. If it is led away from the boiler, and subjected to further heating, it can become superheated, that is, its temperature may be raised above the saturation temperature. This elevation of temperature is known as superheat. (See Quality; Saturated Vapor; Superheat.)

The saturation temperature of steam increases under pressure, at first rapidly, then more slowly, with uniform increments of pressure, until a temperature of 706.1° F. is reached, at a pressure of 3226 lbs. per sq. in. The heat required to boil off a pound of water at the saturation temperature into dry saturated steam is greater at the lower pressures, and decreases as pressures increase, until it disappears at 3226 lbs. per sq. in. This high pressure is the critical pressure at which steam and water are identical in physical properties. Steam at a temperature above 706.1° F. is certain to be superheated, no matter what the pressure. The heat contained in saturated water increases with increase of pressure until it becomes equal to the full enthalpy at the critical pressure. The rate of increase of heat of a liquid does not correspond with the rate of decrease of the heat of evaporation, and their sum, i.e., the enthalpy of the dry steam, increases until it becomes maximum between 400 and 450 lbs. per sq. in., after which it decreases. At the critical pressure a pound of saturated steam contains less heat than at any other pressure. (See **Critical State**.) (F.T.M.)

STEAM CALORIMETERS. Dry saturated **steam** is rarely produced by a **boiler**. If unequipped with a **superheater**, the boiler will generate steam which contains droplets of moisture suspended in the steam. The per cent of steam in a unit quantity of steam and water is known as the **quality**. Steam exhausted from an **engine** or **turbine** has a considerable amount of moisture, the quality often being as low as 85%. It is not possible to judge the quality by inspection of steam, or of pipes in which steam is flowing. A thermometer is of no avail, and yet, there often arises the need for knowing the quality of steam flowing in a pipe line. To meet this need, there have been devised so-called steam calorimeters—devices for experimentally determining the quality of steam. Such calorimeters are made in two forms, each of which has its own particular sphere of usefulness. These instruments, which differ in principle, range of quality measurable, and technique of operation, are known as the *separating* and *throttling* calorimeters. (See **Calorimetry**.)

The separating calorimeter operates on a mechanical principle. It separates the water from the steam by centrifugal force, or difference of density of water and steam, and the principle is simply one of effecting the separation, and then separately measuring the water and dry steam. If A lbs. of water are separated from B lbs. of dry steam, the quality of the mixture is $\dfrac{B}{A + B}$.

The calculations required in this calorimeter are thus seen to be simple, but the steam must be comparatively wet, because an extremely small amount of moisture cannot be effectively separated from the stream of dry steam. The calorimeter is provided with a catch chamber in which the separated water collects. The amount collected is measured by a graduated gauge glass connected to the calorimeter. The dry steam is measured by being passed through an **orifice**. A peculiar property of the orifice is that the weight of steam passed is proportional to the steam pressure. Using a fixed orifice, a scale of steam flow can be incorporated on the dial of an ordinary steam pressure gauge. However, the flow through the orifice is proportional to the calorimeter steam pressure only if the final pressure after passing the orifice is less than 58% of the initial pressure. This is equivalent to a requirement that the steam to be tested be at a pressure of over 10 lbs. gauge. For testing atmospheric exhaust steam, the calorimeter would have to discharge into a region of 12″ mercury vacuum or better.

The throttling calorimeter is not suitable for other than the measurement of steam of very high quality. Its action depends upon the fact that the throttling of high-pressure steam through an orifice to a low pressure leaves the steam with the same heat content at the low pressure that it had at the high. Unless the high-pressure steam exceeds 1800 lbs. per sq. in. (a rare case), the heat it contains exceeds that of dry saturated steam at atmospheric pressure, so that it may be slightly wet, and yet have enough heat to produce superheated steam after being throttled. Now while the thermometer is of no avail in ascertaining the state of saturated steam, it does, in connection with pressure, definitely determine the state of superheated steam. That is, a temperature and a pressure together fix the state of steam and numerical data such as total heat, **entropy**, volume, etc., are readily obtained from compilations of steam data. Unless the steam is initially too wet to superheat the throttled product, the pressure and temperature readings of the low-pressure steam in the calorimeter are sufficient to determine its heat. Taking this, also, as

the heat of the high-pressure steam, the quality is computed by the relationship: heat of wet steam = heat of liquid + quality × heat of evaporation. The quality

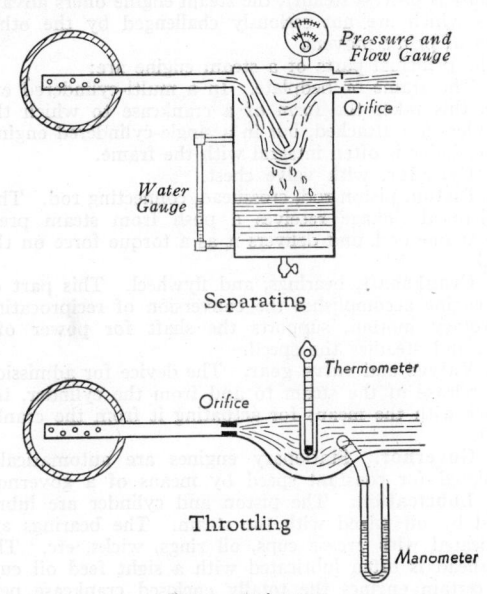

Steam calorimeters.

and heat of evaporation are taken from steam tables at the pressure of the pipe line. The accompanying diagrams are intended to be illustrative of the principle, rather than of the actual arrangement of the calorimeter equipment. A very different apparatus, also known as a steam calorimeter, and used for measuring *specific heats* of substances, is described under calorimetry. (F.T.M.)

STEAM ENGINE. A steam engine is a positive displacement piston and cylinder machine which, when supplied with steam at a pressure above its exhaust pressure, uses that steam expansively for the production of power which it makes available as a rotating **torque** at a crankshaft or flywheel. The steam engine is, with few exceptions, double acting, and at present is most

Ames automatic engine.

frequently built and used in sizes of less than 500 **hp.** Larger power units are generally installed as turbines. The steam engine is characterized by moderate or low speeds (100–500 rpm), the use of atmospheric exhaust (or a moderate vacuum), high starting torque, and ease of conversion to reversible operation if desired. The excellence of the **steam turbine** when a large amount of power is to be generated, especially where the exhaust is at a high vacuum, coupled with the high efficiency of the **Diesel engine** as a prime mover, has greatly restricted the field in which the steam engine is economically a superior prime mover. However, where a boiler must be supplied anyway, as for the generation of heating steam, the steam engine is usually superior to the Diesel engine as a source of power, and where exhaust pressures are high (often the case in industry, where the exhaust is process steam), the steam engine offers advantages which are not seriously challenged by the other types of prime movers.

The principal parts of a steam engine are:

1. The frame or bedplate. In a multi-cylindered engine, this takes the form of a crankcase to which the cylinders are attached, but in a single-cylindered engine, the cylinder is often integral with the frame.

2. Cylinder, with valve chest.

3. Piston, piston rod, crosshead, connecting rod. This mechanical linkage receives a push from steam pressure at one end, and delivers it as a torque force on the crank.

4. Crankshaft, bearings, and flywheel. This part of the engine accomplishes the conversion of reciprocating to rotary motion, supports the shaft for power offtake, and steadies the speed.

5. Valves and valve gear. The device for admission and release of the steam to and from the cylinder, together with the means for actuating it from the crankshaft.

6. Governor. Stationary engines are automatically regulated for constant speed by means of a governor.

7. Lubrication. The piston and cylinder are lubricated by oil mixed with the steam. The bearings are lubricated with grease cups, oil rings, wicks, etc. The crosshead is often lubricated with a sight feed oil cup. On certain engines the totally enclosed crankcase permits the use of a splash system of oiling.

Unlike the gasoline engine, in which the valves have become standardized on the poppet type, the steam engine is built with many different valve types. The

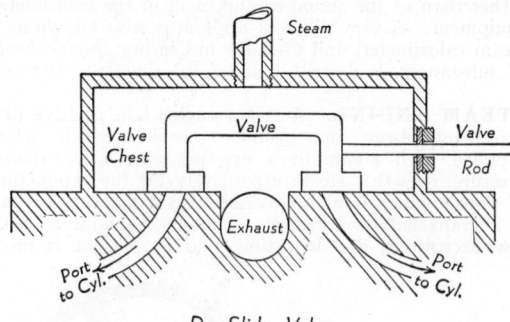

D Slide Valve

original valve was known as the D slide valve, because in cross-section it resembled the capital letter D. Only one valve was needed to admit and release the steam at both ends of the cylinder. This valve had definite disadvantages, to wit: the unbalanced steam pressure on it was difficult to handle except in low-pressure installations; its motion could be no faster than that of the piston, and consequently "wire drawing" of the steam existed during closure of the ports. Governing, by changing the point of cut-off, was possible only at the expense also of changing the other events of the cycle. The use of the D slide valve necessitated dual-flow engines, that is, steam entering and leaving by the same port. The alternate heating and cooling of the ports caused a large thermal loss known as initial condensation. To overcome some or all of these disadvantages, a century of engine development has evolved the following improvements:

1. The piston valve—no unbalanced pressure.

2. The Corliss—no wire drawing. Partially balanced pressure, partially reduced condensation. Cut-off independently controlled.

3. The poppet valve—balanced pressure, partially reduced condensation, and cut-off independently controlled.

4. Multi-port balanced valves—reduction of wire drawing. Balanced pressure.

5. Uniflow engine—elimination of initial condensation. Adaptable to any of the improved valves.

A steam engine converts from 5 to 15% of the heat supplied to it into work, depending on the state of the steam supplied, and on the exhaust pressure. The heat unconverted is composed: first, of heat remaining in the exhaust steam; second, initial condensation; third, incomplete expansion; fourth, wire drawing; fifth, friction; and sixth, radiation (negligible). The first of these is the largest, and is reducible only within certain limits. Incomplete expansion results from the release of the steam at the end of the stroke at a pressure higher than the exhaust. By using a longer stroke, this could be eliminated, but there is a point beyond which the increased cost of the engine more than offsets the gain derived by eliminating this loss.

The cycle upon which the engine operates is briefly described as follows: Slightly before the piston reaches the dead-center position corresponding to minimum cylinder volume, the valve connects the cylinder with the

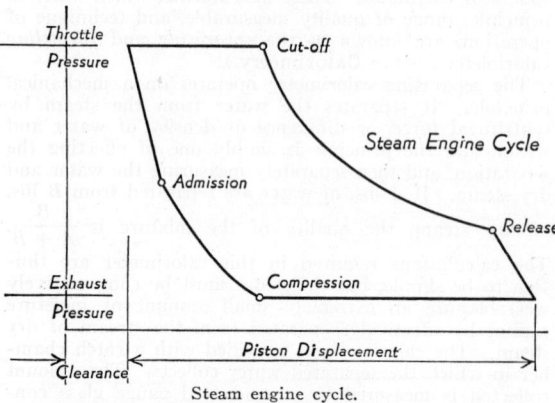

Steam engine cycle.

steam line so that as the piston starts on its outward travel, the full steam pressure is acting on it. The beginning of this action is known as the event of *admission*. When some 20 to 30% of the stroke has been completed, the valve closes the port on the event known as *cut-off*, and during the remainder of the stroke, the steam is expanded adiabatically to the accompaniment of decreasing pressure. Near the end of its stroke, the valve again opens the port, this time connecting the cylinder with the exhaust line. This event is known as *release*. The cylinder remains connected with the exhaust during the return stroke of the piston, and the steam is expelled until approximately 2/3 of the return stroke has been completed. The valve then closes the port, and the remaining steam is trapped in the cylinder and compressed. The beginning of this process is known as the event of *compression*. The four events just described govern the form of the steam engine cycle. The engine using it will be able to develop a horsepower hour using from 10 to 25 lbs. of steam, depending upon the expansion permitted by the terminal conditions of the steam.

The steam engine may be mechanically controlled to give variable output so that when it is connected to a load which varies, it maintains nearly constant speed. There are two methods of accomplishing this result. In one, called cut-off governing, the per cent of the stroke during which the valve connects the cylinder with the boiler, is varied, and in this way different amounts of steam are admitted to the cylinder at one pressure. The mechanism to effect this type of control is incorporated in the valve drive. The other method, called throttling governing, consists of interposing an artificial resistance to create a pressure drop between the boiler and the engine, so that although the same volume is admitted on each stroke (the cut-off being constant), the weight of steam admitted will vary because of the variation in density created by throttling. The governor, in this case, operates on a throttle valve located at the steam inlet. (F.T.M.)

STEAM ENGINE INDICATOR. Thermodynamics; Indicator.

STEAM PURIFIER.

Steam purifiers are inserted in boiler drums to cleanse the steam of suspended moisture entrained in it during the process of generation. Boilers that steam slowly and have ample steam space in the drum do not need purifiers. At high rates of steaming, especially at high pressures, there are possibilities of wet steam generation and carry-over of entrained solids into the steam outlet. Clean steam is always desirable, especially when the steam is passed through turbines. Special cleaning equipment mounted inside the steam drum may be a simple *dry pipe* (a perforated pipe extending through the steam space to the outlet nozzle) or a more complex *washer*.

The steam washer is built into the steam drum so that incoming feed water will wash the entrained boiler water from the steam by bubble or spray action. If there is any moisture carry-over with the steam it is of the relatively pure feed water and not boiler water which is usually saturated with impurities. (F.T.M.)

STEAM RATE.

The rate at which a **steam engine** or **turbine** consumes **steam** per unit power output is its steam rate. Originally, this quantity was called water rate, but since an enigne operates on steam, the term steam rate seems more appropriate. It is usually given in units of pounds of steam per horsepower hour, or per kilowatt hour in the case of a direct-connected or turbo-generator unit. The steam rate varies with the load, being minimum at a load known as the most economical load. This is the rated load in the turbine, and somewhat less than rated in the engine. Steam rate is affected by initial pressure and degree of superheat; also, by exhaust pressure. The higher the initial pressure, or temperature, and the lower the final pressure, the smaller the steam rate will be. Thermal efficiency is inversely proportional to steam rate. Steam rate, multiplied by load, is steam consumption, and when this product is plotted against load, the result is often nearly a straight line, up to the most economical load. This is the well-known **Willans Line.** (F.T.M.)

STEAM RECEIVER.

The steam receiver is a tank or vessel providing a volume for the storage of **steam**, so that sources which use the steam in a fluctuating manner may be supplied from the storage in the receiver without causing a large fluctuation of pressure in whatever supplies the steam to the receiver. For example, the use of steam by an engine is pulsating, and where the boiler drum which supplied that steam had insufficient volume of steam storage, a cyclical variation of pressure was created, which, in some cases, caused excessive vibration of steam pipes and boilers. This was remedied by setting up between the **boiler** and the engine a receiver of suitable volume to augment the vapor storage of the boiler. Cross-compounded **steam engines** require a steam receiver intermediate between the exhaust of the high-pressure cylinder and the inlet of the low. The reason for this is that in this compounding arrangement, the low-pressure cylinder does not take in steam simultaneously with the exhaust of the high-pressure cylinder. A receiver of volume about equal to that of the high-pressure cylinder must be provided to prevent fluctuations of pressure on the inlet to the low-pressure cylinder. For much the same reason, receivers are used between the stages of multi-stage air compressors of the piston-cylinder type, though a cooling action is also incorporated in these receivers. (F.T.M.)

STEAM TRAP.

A device which automatically operates to allow the discharge of water from a certain region, and prevent the escape of **steam** along with it, is known as a trap. The trap is used when condensate is to be drained from a vessel occupied by condensing steam, without the loss of steam. It traps the steam in the vessel and passes the water of condensation. Traps are used with steam heaters and cookers of all types, wherein a surface is interposed between the steam and that which is heated, to drain the condensation from steam lines, and for many similar services. A great deal of heat which is otherwise wasted by a partially opened drain valve will be saved by a trap if its discharge is connected to some point where the heat in the hot condensate can be used. The principal methods of operating a steam trap are expansion, float, tilting under the influence of accumulated condensate, and sinking bucket.

In an expansion trap, advantage is taken of the fact that condensate is usually a little cooler than the steam, and a metal expansion element is placed where it may be covered either with steam or condensate, depending on the amount of condensate in the trap. When it is covered with condensate it is cooler, and shrinks, opening a valve to pass the condensate from the trap. As the condensate is discharged, it uncovers the expansion element, and the hot steam causes it to expand and close the discharge valve. Such traps are frequently used on steam radiators. In a float-type trap, the float is attached to the end of a pivoted lever, from the other end of which a link bar extends to open or close a discharge valve, depending on the position of the float. These two types tend to produce a continuous discharge through a valve which is cracked open just sufficiently to maintain a condition of equilibrium in the water level of the trap. The tilting- and bucket-type traps are intermittent in operation, which is much better from the standpoint of wear on the discharge valve. In the tilting trap, water collects in a pivoted chamber until the weight of it overcomes the counterbalance and the whole chamber tilts until a valve is opened and the discharge of water through it relieves the trap so that the counter-balance can return it to the closed position. A pivoted bucket trap is shown in the accompanying diagram. As the condensate is collected in the trap, it fills the bucket, which otherwise would tend to float on the surface of the water in the bucket chamber. When the bucket is filled it

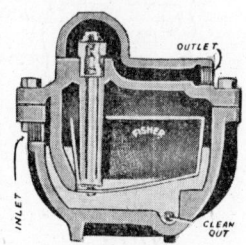

Steam trap.

sinks, thereby opening a discharge valve. The pressure of the steam forces the water out through the discharge valve, and as soon as the bucket is empty, it again floats on the water in the chamber, closing the discharge valve. (F.T.M.)

STEAM TURBINE.

Prime movers utilizing **steam** are the reciprocating engine and the turbine. One is essentially a pressure machine, the other a flow machine. The reciprocating parts of engines limit their speed to a comparatively low value, but turbine energy is obtained from a number of small forces working at high velocity. Smaller dimensions and freedom from vibration give the turbine an advantage in first cost, space, and foundation requirements. Both engine and turbine are reliable prime movers.

Probably the most significant factor bearing on the importance of the steam turbine is that it is the only prime mover available in the largest sizes desired. While many of the larger turbines are specially made to the purchaser's specifications, manufacturers have developed the practice of arranging the blading within a number of standard sizes of casings so that turbines of almost any capacity can be built at reasonable cost and fairly good efficiency. Advantages of a turbo-generator are its (1) low first cost per kilowatt capacity, (2) low maintenance cost, (3) economy of foundation and building cubical content, (4) high efficiency when operated far into the low-pressure range, (5) uniform angular velocity with freedom from vibration, (6) oil-free character of the condensate.

The action in a steam turbine is the transfer of energy from the heat form first to kinetic energy of a high velocity steam jet, then to the energy in the rotating shaft. Its principal parts are:

1. Nozzles to change heat energy to work energy, and to direct the course of steam onto blades.

2. Blades, which change the kinetic energy of the jet of steam into shaft horsepower.

3. Rotating shaft, to which the blades are affixed.

4. A casing, which encloses the steam path and supports fixed parts.

5. Governor, bearings, lubrication, and other auxiliary devices.

When steam is expanded **adiabatically** through a stationary nozzle, it does not retain all the heat it originally had. The heat released during expansion does not disappear. According to the first law of thermodynamics, it must reappear as work energy in equivalent amounts (778 foot-pounds for every B.T.U.). In the case of steam expanding through a stationary nozzle, the moving steam must gain this mechanical energy, with the result that its speed is considerably increased. An ordinary expansion involving a heat drop of approximately 100 B.T.U. of steam has the capacity to increase its speed to the amazingly large magnitude of 30 miles per min.

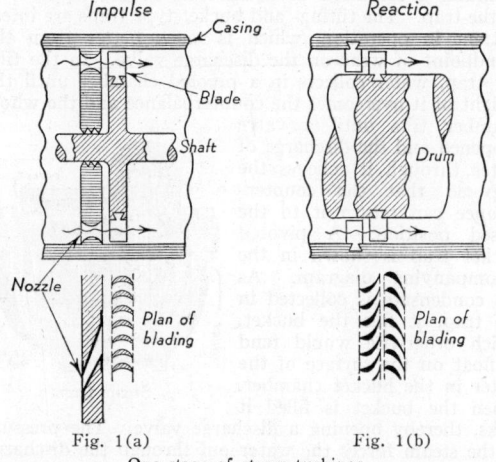

Fig. 1(a)   Fig. 1(b)

One stage of steam turbines.

This may give some inkling as to the reason why so light a fluid medium as steam is capable of producing so much power in a machine of moderate dimensions. There are two principles of action employed in turbines, known as the impulse and reaction principles. The impulse principle involves stationary nozzles and moving blades which absorb the mechanical energy from the steam as it flows over the blades. In the reaction turbine the nozzles are themselves attached to the shaft. In one case the motivating force is one of impulse of a stream against a blade; the other, one of a reaction force created by the acceleration of the steam in the moving nozzles. A hydraulic analogy of the reaction turbine is the common revolving lawn sprinkler. The machines using these two principles exhibit some dissimilarity. For example, in the impulse turbine, the blades

are rather heavy, of steel, attached to wheels which are mounted on the shaft. The nozzles have definite nozzle form, and are fixed in the casing. Reaction turbine blades are lighter, of bronze, and are directly affixed to a drum, which takes the place of the shaft. The moving nozzles themselves are made of blades so shaped as to give nozzle action. Further comparison of the impulse and reaction principles may be had by reference to the accompanying figure, which illustrates diagrammatically the nozzle and blade arrangement of both types. In each case one stage is represented. The nozzle of the impulse turbine speeds up the steam and guides it onto the blades which move past. The pressure drop is entirely consummated in the nozzle. The velocity increases in the nozzle and decreases on the blade. In the reaction stage, the row of fixed blades gives a nozzle effect which expands the steam and directs it onto a row of moving blades, also composed of nozzles. These, in turn, expand the steam, receiving thereby a reaction force from the acceleration of the expanding steam. The pressure drops in both fixed and moving nozzles. The fixed nozzles give the steam sufficient velocity to glide into the moving blades at their speed, or slightly faster. The expansion in the moving blades is sufficient to increase the relative velocity of the steam so that when it leaves them there will be no component of velocity in their direction.

Very high rotative speeds are necessary if the entire adiabatic heat drop for the turbine is released in one set of nozzles. For example, an impulse turbine with a conservative expansion would create a steam speed which, if absorbed on one row of blades revolving in an 18-in. circle, would require a rotative speed at the shaft of better than 10,000 rpm. This gives a clue to the desirability of staging a turbine by subdividing the heat drops, the energy being absorbed after each incremental heat liberation. Thus, in a five-stage impulse turbine, each stage may be called on to absorb only one-fifth of the total heat drop. The subdivision is even greater in the case of reaction turbines, where the B.T.U.'s liberated per stage rarely exceed 10. In consequence, the number of stages found in the reaction type turbine is greater than in the impulse. Except for areas of nozzles, size of blades, and blade angles, the stages resemble one another.

Large steam turbines fall into three classes:

1. Straight reaction.

2. Straight impulse.

3. Impulse reaction.

The straight reaction may have fifty or more stages, the straight impulse rarely more than twenty (half that number is more common). One or two impulse stages preceding a straight reaction section enable reduction of the number of reaction stages to about twenty.

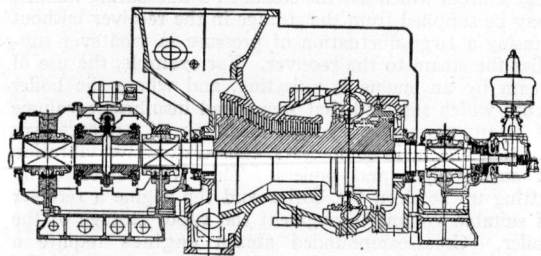

Fig. 2. High-pressure section of 110,000-kw. Westinghouse turbine.

An impulse turbine is pressure staged, i.e., the pressure drop is subdivided among a number of stages, any one stage of which may be, in addition, velocity staged. By this is meant that, following one set of nozzles there may be two rows of moving blades which, together, effect the reduction of velocity and the absorption of energy created in those nozzles. The two rows of moving blades are separated by a row of fixed reversing blades, which receives the steam from the first row and directs it at the proper angle for the second row. A velocity stage within a pressure stage is known as a Curtis stage. A simple pressure stage is a Rateau stage, and the reaction stage often is called Parson's staging. As steam expands through a multi-stage turbine it increases in volume after each stage, making it necessary either that the diameter of the circle in which the blades travel be increased, or that the height of the blades be increased to provide sufficient area for the increased volume of flow. Usually both of these expedients are adopted, so that the turbine exhibits, roughly, a somewhat conical shape, being smallest at the high-pressure end, and largest at the exhaust.

Essential to the operation of a turbine are a number of auxiliary devices. The impulse turbine rotor receives a moderate end thrust, due to fluid friction on the blades. The reaction turbine has a large end thrust, arising from the drop of steam pressure, across the moving blades, acting on the blading annulus. While the end thrust of an impulse turbine can be accommodated by special thrust bearings, the large forces set up in a reaction turbine necessitate the use of a special device known as a dummy piston. The dummy piston is a circular plate mounted concentric with the axis of the turbine, and having, on one side, high, and on the other, low steam pressure. The pressures are so chosen that the direction of the resultant force on the plate is counter to the end thrust on the blades. A separate piston is used to balance each different diameter of the drum, so that although steam pressure and end thrusts may vary with different loads, the equalizing pressures also similarly vary. A thrust bearing is provided to absorb the small amount of unbalance which may still exist.

The moving parts of the turbine are few. Principally, there are two bearings, one at each end of the turbine, which support the rotor. These are rather heavily loaded so that plain babbitted bearings are usual. Oil is pumped to these bearings and wasted from them to a sump, from which it is withdrawn, filtered, and cooled before being again supplied to the bearings. An auxiliary pump maintains oil pressure during the starting and stopping cycle of the main unit. Where the shaft of the turbine projects through the casing, means are provided for packing against leakage of the steam outward at the high-pressure end, and infiltration of air at the low-pressure end. Packed **stuffing boxes** are used only on small turbines. This service is performed on large turbines by sealing **glands** which are built into the turbine, and which seal the shaft with steam or water. The outward leakage from the glands is minimized by labyrinths.

Governing of steam turbines is accomplished by three methods, viz.: (1) throttling at inlet, (2) varying number of inlet nozzles in action, (3) varying duration of full pressure puffs (blasts), of which there are several per second. In addition, some turbines are provided with hand-operated by-pass valves which, by admitting high-pressure steam to low-pressure stages, enable the turbine to carry more overload though, of course, at reduced economy. Of these methods, the first is widely used on small turbines. In the large turbine field, the second is applied to the Curtis type, and the third to the Parsons type.

The losses in a steam turbine, as is the case with the engine, are topped in magnitude by the heat in the exhaust. Unfortunately, this is difficult to reduce because of the troubles attending the use of steam of **quality** lower than about 85% in the turbine. Wet steam erodes turbine blades, due to the high velocity with which the particles of moisture strike them. The other losses which occur in steam turbines are:

1. Thermodynamic losses.
    a. Leakage. Past shaft gland packings, dummy pistons, diaphragms and blade tips.
    b. Blade and disk friction.
    c. Throttling at the control valves.
    d. Non-stream line flow at other than design conditions.
    e. Leaving loss, and pressure losses in exhaust nozzle.
2. Mechanical losses.
    a. Bearing and stuffing box friction.
    b. Windage action of idle blades.
    c. Gland water-seal power.
    d. Oil pump and governor power.

All of the thermodynamic losses of one stage are returned to the steam as it enters the next lower stage, so that one definite advantage of multi-staging is to make available to the next stage the thermal losses of the preceding one. This is one factor in the superior performance of multi-stage turbines. To illustrates the point, consider the case of friction on the nozzles. If the turbine casing is well insulated, the heat that is generated by friction of the steam against the nozzles simply elevates the nozzle temperature until it returns, by conduction to the steam, as much heat as is generated by friction. The greater part of the stage loss is returned to the steam at the low pressure, and in turbine design it is considered that all of the reheat occurs at the low pressure of the stage. Fig. 3 shows how an adiabatic heat drop, as in nozzles, followed by a reheat, in which

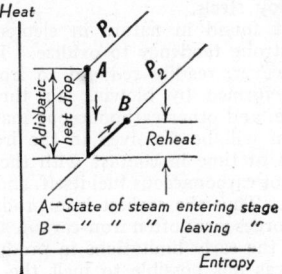

Fig. 3.   Steam conditions in one stage of a turbine.

the stage losses are added at the low pressure, results in a steadily increasing **entropy** in a multi-stage turbine. This explains the typical expansion line of the steam turbine, which, as pressure is reduced, increases in entropy, whereas an ideal turbine would expand the steam at constant entropy. The increase of entropy corresponds to a decrease of energy availability, and that turbine which has the minimum entropy increase shows the best performance. (See **Stage Efficiency**.)

It is obvious from the foregoing analysis that all of the heat taken out of the steam during its passage from the throttle to the exhaust must have been put onto the rotating shaft as mechanical energy. This is true because of the negligible radiation loss from the casing. Friction and power required to drive the governor and oil pump take their toll of this energy, but they rarely amount to as much as 5% of the heat drop, so that practically all of the heat drop appears at the turbine shaft coupling as useful work. (F.T.M.)

STEARIC ACID AND STEARATES. Stearic acid ($H \cdot C_{18}H_{35}O_2$ or $C_{17}H_{35} \cdot COOH$ or $CH_3(CH_2)_{16} \cdot COOH$) is a white solid, melting point 69° C., boiling point 383° C., insoluble in water, slightly soluble in alcohol, soluble in ether. Stearic acid may be obtained from glyceryl tristearate, present in many solid fats, such as tallow, and in smaller percentage in semi-solid fats (lard) and liquid vegetable oils (cottonseed oil, corn oil), by **hydrolysis**. The crude stearic acid, after separation of the water solution of glycerol, is cooled to fractionally crystallize the stearic and **palmitic acids,** which are then separated by **filtration** (oleic **acid** in the liquid), and fractional **distillation** under diminished pressure. With **sodium** hydroxide, stearic acid forms sodium stearate, a soap. Most soaps are mixtures of sodium stearate, palmitate and oleate.

The following are representative esters of stearic acid: Methyl stearate ($C_{17}H_{35}COOCH_3$), melting point 38° C., boiling point 215° C. at 15 mm. pressure; ethyl stearate ($C_{17}H_{35}COOC_2H_5$), melting point 35° C., boiling point 200° C. at 10 mm. pressure; glyceryl tristearate [tristearin ($C_3H_5(COOC_{17}H_{35})_3$)], melting point 70° C. approx.

Stearic acid is used (1) in the preparation of metallic stearates, such as aluminum stearate for thickening lubricating oils, for waterproofing materials, and for varnish driers, (2) in the manufacture of "stearin" candles, and is added in small amounts to paraffin wax candles. As the glyceryl **ester,** stearic acid is one of the constituents of many vegetable and animal oils and fats. (R.K.S.)

STEATITE. Talc.

STEEL. The essential elements in steel are iron and carbon, although manganese, silicon, copper, phosphorus, and sulfur may be present in greater amounts than carbon even in the so-called plain carbon or unalloyed steels. In alloy steels certain of these elements and many others including nickel, chromium, molybdenum, tungsten, and vanadium may be present in substantial amounts but carbon rarely exceeds 1.5% in either plain carbon or alloy steels.

Iron is not found in nature in elemental form because of its strong tendency to oxidize. The iron oxide ores are, however, readily reduced to iron by hot reducing gases formed by blowing air through burning charcoal, coke, and other carbonaceous materials. More or less carbon will be dissolved in the iron, depending on the length of time in contact with the hot reducing gases or the hot carbonaceous fuel itself, and the temperature attained. Thus the earliest iron products produced in primitive forges were often iron-carbon alloys or steels.

Because of the early limitations in producing a strong air blast it was not possible to melt the metal, but in the red-hot pasty condition in which it was withdrawn from the fire it could be readily welded together and shaped by hammering.

The art of hardening steel was undoubtedly discovered independently by many a smith who, by chance, cooled his red-hot "iron" by immersing it in water or other liquids. If the metal was too low in carbon, it did not respond to this treatment and some artisans apparently learned to increase the carbon content by methods similar to modern **carburizing.**

The irons and steels produced by direct reduction of iron ores inevitably contained as impurities some of the silica, clay, and other minerals present in the ore. Although considerable purification can be achieved by merely melting the steel and casting, it is doubtful if the required temperatures were attainable. However, in the course of development of improved bellows and of stack-type furnaces of increasing height in which the ores and fuels reached higher temperatures, a molten product of very high carbon content was finally obtained directly from the ore. The higher carbon content is the result of increased rate of absorption of carbon by solid iron at the higher temperatures. Since the melting point of iron-carbon alloys decreases progressively as the carbon content increases, it is possible to obtain molten metal without ever reaching the high temperatures necessary to melt pure iron or low-carbon steels. Castings from such a melt must have been disappointing in that they could not be welded or forged when heated to a red heat and were quite brittle at normal temperatures. While the new product found uses in cast form, only within the past century has it been used as the basis for making steel and other wrought or workable products. The indirect process of first making a high-carbon **pig iron** in a **blast furnace** and then refining it in a **Bessemer, open hearth,** or **electric furnace,** or by duplex processes involving more than one of these furnaces, is the basis of the modern steel industry. The most recent attempts to revert to the direct process of making steel from ore have not been economically successful.

The early history of the production of high-carbon **cast iron** from ore, the first stage in the indirect process for making steel, is not definitely known.

A cast-iron stove made at least 15 centuries ago has been discovered in China. Several other castings made in China before 1000 A.D. have been found, one weighing about 1500 lbs. The modern cast-iron period began in the 14th century when molten iron flowed from blast furnaces in the Rhine Valley. The process was used in England in about 1500, and in 1619 coke was first used instead of charcoal as the fuel and carburizing agent. Cast iron found many applications in the industrial revolution in England and later in America. Modern cast irons are made by remelting pig iron, the product of the blast furnace, in order to control its composition and adapt it to the product to be cast, often using alloying elements as in steel-making practice.

The principal developments in the second phase of the indirect process, the production of low-carbon iron and steel from pig iron, are of more recent occurrence. In 1766 a British patent was granted on a process for refining pig iron in a **reverberatory furnace** under oxidizing conditions which removed carbon and other readily oxidized elements. Improvements by Henry Cort in 1784 and Joseph Hall in 1830 virtually completed the development of the hand "puddling" process for making **wrought iron,** a low-carbon iron which could be welded and forged in the manner of the direct-process irons. Although this was a rather difficult product to make, requiring much arduous labor, wrought iron was produced in sufficient amounts to be used for many structural purposes in bridges and buildings and for machine parts, often replacing the relatively weak and brittle iron castings formerly used.

In 1856 the Bessemer process was introduced, making it possible for the first time to produce steel in relatively large quantities, thus introducing the age of steel. In 1864 the even more important open hearth steel-making furnace was invented by Siemens and Martin. The electric furnace followed in about 1890.

It will be recalled that small amounts of iron and steel have been made directly from the ore for many centuries. By means of Huntsman's "crucible process" (1740) it became possible to melt the crude "blister steel" made by cementation or carburization of direct-process irons, thereby liberating the slag and producing a casting of improved purity and uniformity which could then be forged into tools or weapons and heat treated to suitable hardness. The **crucible** process of melting steel is slow and very expensive compared with modern steel-making processes; however, until electric furnaces became well established the crucible process was used for melting high-quality tool and alloy steels from selected melting stock of high purity. In modern practice the electric induction furnace can be compared to the crucible process since both are ordinarily used to melt the charge without working or refining the metal.

The terminology of the iron-carbon alloys is, unfortunately, often confusing in that the term iron is applied to both very low and very high carbon materials. Thus wrought iron produced by direct reduction from ore under noncarburizing conditions or by the modern indirect methods has a very low carbon content. Of more industrial importance is open hearth iron, also widely known as Armco iron, produced in the open hearth furnace under highly oxidizing conditions in order to remove practically all the carbon, manganese, silicon, and phosphorus present in the charge of pig iron and steel scrap. This material generally has 0.025% carbon or less and under 0.10% total of carbon, manganese, silicon, phosphorus, and sulfur. On the other hand, pig iron, the product of the blast furnace, and **cast iron,** generally made by remelting pig iron in a cupola, have high carbon contents, usually over 3.0%. Furthermore, **malleable cast iron,** the product of annealing white cast iron, has a carbon content usually between 2.0 and 3.0%. White-heat malleable iron, a product of British origin not generally made in this country, is a white cast iron whose carbon content has been practically all removed by an oxidizing heat treatment.

In many respects steels are intermediate between the very low carbon, and often very pure irons, such as wrought iron and open hearth iron, and the high-carbon cast irons which generally have high contents of silicon, phosphorus, sulfur, etc. The limits for carbon in steels are about 0.04% for low-carbon rimming steels such as are used for deep drawing sheets, and 2.25% in certain alloy die steels. Within this range there are certain fairly well-defined classifications as follows:

% CARBON	REPRESENTATIVE APPLICATIONS
.04– .12	Low-carbon rimming and killed steels for drawing and stamping sheets, wire, nails, rivets.
.08– .25	Low-carbon steel for structural purposes, boilers and fireboxes, pipe and tubing, machine parts to be surface carburized and heat-treated.
.25– .50	Medium-carbon steels for high-strength structural purposes, steel castings, forgings, machine parts to be hardened by heat treatment.
.45– .75	Intermediate-carbon steels for railroad car axles, hard-drawn spring wire, heat-treated alloy steel springs, rails, railroad car wheels, shock-resisting tools.
.75–1.25	High-carbon steels for cold-drawn music wire, drill rod, heat-treated carbon steel springs, cutting tools and dies, drills, reamers, ball bearings.
1.25–2.25	High-carbon tool steels for drawing dies, finishing tools, files, etc., in which high abrasion-resistance is more important than toughness.

Open hearth iron and the low-carbon drawing-grade steels consist, structurally, of crystals or grains of ferrite containing elements such as manganese, phosphorus, and copper in solid solution. (See **Metallography.**) A very small amount of carbon is also soluble in the ferrite and any excess will be present as particles of **cementite,** Fe_3C, embedded in the ferrite, or as patches of pearlite, an intimate mechanical mixture of ferrite and Fe_3C having a characteristic microstructure. Also present in iron and steels of all catagories are certain non-metallic particles such as iron and manganese sulfides, and various silicates and oxides. (See **Inclusions.**)

In the annealed or soft state the structures of all steels consist of various mixtures of ferrite, pearlite, and cementite. Ferrite is soft and ductile, with a tensile strength of about 40,000 lbs. per sq. in. Cementite is very hard and brittle. Pearlite has intermediate characteristics with a tensile strength of about 100,000–140,000 lbs. per sq. in., depending on the fineness of its structure. As the carbon content increases the proportion of ferrite decreases and the pearlite increases until at about 0.80% carbon the structure is completely pearlitic. This is also known as the **eutectoid** structure. Steels with higher carbon contents may consist of pearlite and cementite, the latter generally being present as massive particles embedded in the pearlite or as a continuous film around the pearlite grains. However, dead soft annealed high-

carbon tool steels do not contain pearlite, all of the carbon being present as rounded cementite particles embedded in a matrix of ferrite.

When carbon steels are used in the annealed or normalized condition, their structures are essentially as described above and their properties change gradually

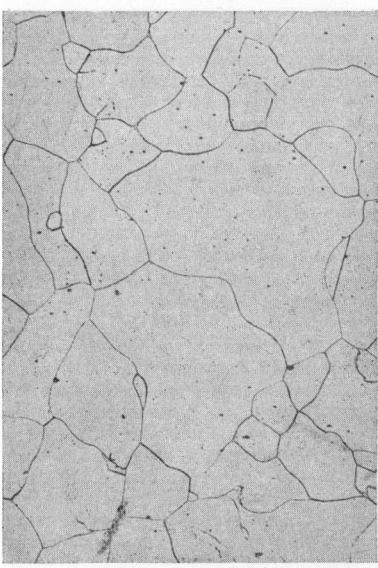

Ferrite grains in ingot iron at 100 times magnification. Etched in a solution of 3% nitric acid in alcohol (nital). (See **Metallography.**)

from those of ferrite to those of pearlite, with further increase in hardness and wear-resistance when massive cementite is present. However, for many applications steels are **heat-treated** by quenching in oil, water, or brine, which results in profound changes in their struc-

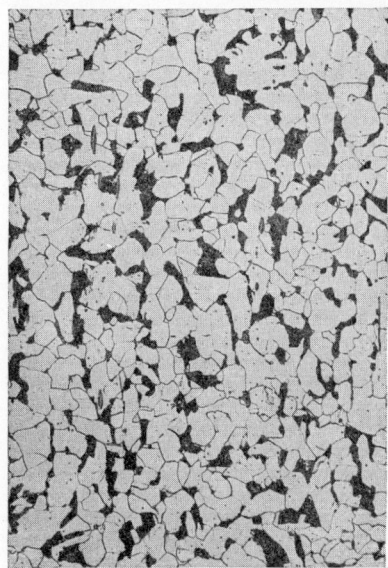

Ferrite (light) and pearlite (dark) in a 0.15% carbon steel at 100 times magnification. Etched in nital.

tures and properties. These changes are related to a property of iron-carbon alloys known as allotropy, or change in crystal structure with temperature. (See page 292.) When an annealed steel is heated above a certain temperature, e.g., 1400° F. for a 0.50%-carbon steel, its structure is changed from ferrite and pearlite to austenite. Whereas the original ferrite had a **crystal structure** known as body-centered-cubic, austenite has a face-centered-cubic structure of somewhat higher density. Furthermore, all of the carbon present is soluble in the solid austenite at this and higher temperatures so that the material is homogeneous rather than a

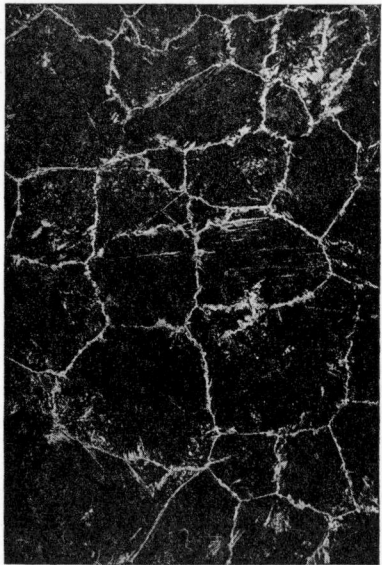

Pearlite (dark) with cementite in the grain boundaries of a 1.2% carbon steel at 100 times magnification. Etched in nital.

mechanical mixture of ferrite and pearlite. Although the transformation is reversible under equilibrium conditions, upon rapid cooling the austenite does not revert to ferrite and pearlite. Instead, a new structure known as martensite is produced and some of the austenite may

Pearlite in a 0.75% carbon steel at 500 times magnification showing intimate mixture of ferrite and cementite. Etched in solution of picric acid in alcohol.

remain untransformed, especially in the case of certain alloy steels. Martensite has a body-centered-tetragonal crystal structure which is very similar to that of ferrite; however, it contains the carbon atoms in a state of supersaturated solid solution. Martensite is very hard and quite brittle.

For nearly all purposes hardened steel must be reheated or **tempered** to relieve internal stresses and reduce brittleness. (See **Heat Treating.**) This is accomplished at relatively low temperatures such as 400° F. If a greater measure of toughness and ductility is required at some sacrifice of hardness and strength, higher tempering or drawing temperatures up to about 1200° F. are used. The structural changes occurring during tempering include transformation of any retained austenite to martensite and the decomposition of martensite at the higher temperatures to a mixture of ferrite and cementite known as sorbite. The cementite particles are first precipitated in a very finely divided form throughout the matrix of ferrite, and as the tempering temperature in-

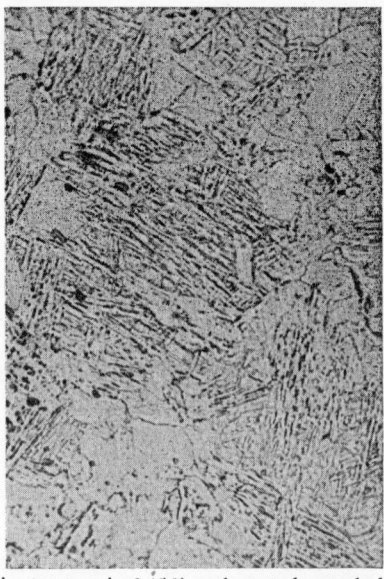

Martensitic structure in 0.15% carbon steel quenched in water from 1660° F. 500 times magnification. Etched in solution of picric acid in alcohol.

creases the particles grow by agglomeration until they become easily distinguishable in the microstructure as spheroids. Prolonged treatment at about 1300° F., known as spheroidizing, produces the largest cementite particles. Concentration of the hard cementite particles in this manner leaves a continuous matrix of soft ferrite. In this condition the indentation hardness and strength are at a minimum, and in the case of the higher carbon steels the machinability is very good, but not necessarily at a maximum.

In the foregoing discussion the direct quench and temper type of hardening treatment has been described. There are several alternative treatments, some of quite recent invention, for the heat-treatment of steel. These include time quenching, austempering, and martempering. All of these treatments are based on the fact that in the quenching of steel the cooling rate down to a black heat (about 900° F.) is very critical, but below this temperature a lower rate can be used. The actual transformation to the hard martensitic structure takes place in the lower temperature range, a few hundred degrees F. above room temperature. This transformation is accompanied by an increase in volume. In a large section or a piece having both large and small sections, temperature gradients will occur in any rapid-cooling process, thus the transformation and volume change will occur first in the small sections and the outer portions of large sections. Since the steel is at a black heat it has lost its high-temperature plasticity and the differential volume changes cause high internal stresses which can result in cracking. Special treatments have, therefore, been devised to avoid rapid cooling in the low-temperature range.

Time-quenching consists of withdrawing the piece from the quenching medium after a predetermined time while it is still hot and cooling it to room temperature in air or in a less drastic quenching medium. This method has been used to prevent cracking in the hardening of tools and dies.

Martempering, developed during the war, has the same purpose as time-quenching—full hardening without cracking or distortion. In the martempering process rapid cooling is interrupted at a temperature just above the martensite transformation temperature and the piece is held in a constant-temperature bath until the temperature is essentially equalized throughout the piece before final cooling to room temperature. This is followed by tempering in the usual manner.

Austempering consists of quenching into a bath of molten salt or metal held at a fixed temperature such as 750° F., and holding at this temperature until the hardening transformation is complete before cooling in air to room temperature. Martensite is not formed and no tempering treatment is required. The structure produced in this process by transformation at a constant elevated temperature is known as Bainite. Austempered steels develop a combination of strength, ductility, and toughness which is superior to that of conventionally quenched and tempered steels. Because of the necessity for rapid cooling down to the transformation temperature and the relatively poor cooling capacity of the elevated temperature bath, the process is applicable only to small sections and to steels having good **hardenability.**

The following list of steel compositions has been selected from hundreds now included in the specifications of the various technical societies, the government, and large users of steel. While the alloy and heat-treating grades far outnumber the non-heat treating grades, the latter make up the greatest tonnage. (See **Tool Steels, Steel Castings, Stainless Steel, Magnetic Materials.**)

STEEL CASTINGS. A relatively small proportion of the total domestic steel production is made into steel castings. (See **Casting, Steel.**) These range in size from small machine parts to hammer base foundation castings weighing 250,000 lbs. Most steel castings are a medium carbon (0.25–0.35% carbon) steel; however, the entire range of carbon and alloy contents ordinarily produced as wrought steels can be used for steel castings with slight modification to insure soundness. Steel castings are often heat-treated by normalizing or quenching, followed by a tempering or drawing treatment. The mechanical properties of heat-treated castings approach those of corresponding wrought steels and the ductility and toughness are much greater than that of cast iron.

Steel castings have many applications for railroad car and locomotive construction, including frames, trucks, couplings, and driving wheels. There are many other applications in rolling mills, machine tools, farm implements, automobiles, road-building machinery, cranes, valves and fittings, etc.

Although the rapidly increasing use of **welding** has made it possible to fabricate certain parts from wrought-steel shapes and forgings instead of using castings, the inherent ease of production of castings insures their continued use for parts having complicated shapes. (R.H.H.)

STEFAN-BOLTZMANN LAW. An important law of **thermal radiation,** discovered empirically by J. Stefan in 1879 and deduced theoretically 5 years later by L. Boltzmann. It states that the total emissive power of a **black body** is proportional to the fourth power of the absolute temperature of the black body:

$$E = \sigma T^4.$$

The constant of proportionality, σ, called the Stefan-Boltzmann constant, has been determined experimentally as 5.672×10^{-5} erg/sec. cm.^{2} deg.^{4}; but the measurements

TYPICAL COMPOSITIONS OF REPRESENTATIVE WROUGHT STEELS AND IRONS
(Excluding Tool Steels and Stainless Steels)

C	Mn	Si	P	S	Ni	Cr	Mo	OTHERS	HEAT TREATMENT *	DESIGNATION
.015	.025	.003	.008	.025	....	....	....		None	Open hearth iron (Armco Iron)
.08	.03	.18	.12	.015	....	(2.85% slag)	....		None	Wrought iron (Aston process).
.06	.35	.003	.008	.030	....	....	....		None	Low-carbon sheets and strip.
.12	.75	.....	.11	.20	....	....	....		None	Bessemer free machining screw stock.
.15	.45	.20	.030	.035	....	....	....		C.Q.T.	Carburizing steel for machine parts.
.20	.40	.10	.030	.035	....	....	....		None	Mild steel for pipe, plate, shapes.
.35	.75	.20	.030	.035	....	....	....		Q.T.	Medium-carbon machine steel.
.45	.65	.04	.030	.035	....	....	....		None	Railway car axle steel.
.70	.75	.18	.030	.035	....	....	....		None	Rails (100 lb./yd.)
.90	.40	.20	.015	.015	....	....	....		None	Music wire.
1.00	.40	.20	.030	.035	....	....	....		Q.T.	Carbon steel springs.
.40	1.75	.20	.030	.035	....	....	....		Q.T.	Manganese steel for machine parts.
.60	.85	2.00	.030	.030	....	....	....		Q.T.	Silicon-manganese spring steel.
.40	.75	.20	.030	.035	3.5	....	....		Q.T.	Nickel structural and machine steel.
.40	.75	.20	.030	.035	1.25	0.65	....		Q.T.	Nickel-chromium machine steel.
.35	.45	.20	.030	.035	3.50	1.50	....		Q.T.	Nickel-chromium machine steel.
.30	.60	.20	.025	.030	1.50	0.75	0.30		Q.T.	Nickel-chromium-molybdenum machine steel.
.15	.55	.20	.025	.030	1.80	....	0.25		C.Q.T.	Nickel-molybdenum carburizing steel.
.40	.75	.20	.030	.035	....	1.00	....		Q.T.	Chromium machine steel.
1.00	.35	.25	.020	.025	....	1.35	....		Q.T.	Chromium ball bearing steel.
.30	.50	.20	.030	.035	....	1.00	0.20		Q.T.	Chromium-molybdenum aircraft tubing.
.50	.75	.20	.030	.035	....	1.00	....	0.18 Va	Q.T.	Chromium-vanadium spring steel.
.35	.45	.25	.020	.025	....	1.20	0.20	1.2 Al	N	Nitriding steel.
.35	1.05	.30	.030	.035	.45	.40	0.12		Q.T.	National Emergency machine steel.
.08	.60	.03	.06	.025	.70	....	....	0.60 Cu	None	Low alloy high-strength structural steel.

* None—Used as supplied by the steel mill. Special cooling or **annealing** treatments may be required in manufacture.
C.Q.T.—Carburized, quenched, and tempered. (See **Carburizing.**)
Q.T.—Quenched and tempered. The quenching medium will depend on the nature of the part, especially the section size. The higher carbon and alloy grades are generally quenched in oil, the others in water or brine. (See **Heat Treatment.**)
N—Nitrided. (See **Nitriding.**)

(R.H.H.)

are difficult and the precision is doubtful. Boltzmann's deduction was based upon the thermodynamic theory of **radiation pressure.**

In accordance with this law and the principle of exchange of radiant energy, and since the emissivity and the absorptivity are equal, the net rate at which a black body of area a (sq. cm.) loses energy by radiation when placed in an enclosure of temperature T_m is

$$\frac{dw}{dt} = a\sigma(T^4 - T_m{}^4) \text{ (ergs per sec.).}$$

Newton's law of cooling approximates the Stefan-Boltzmann law for small differences of temperature only. (See also **Heat Transfer.**)   (L.D.W.)

STEGANOPODES, PELECANIFORMES. Birds with long legs, webbed feet, and long beaks and necks. The **pelicans** are typical of the group, which also contains the **cormorants** and some related marine forms. (A.W.L.)

STEGOCEPHALIA. Fossil Amphibia.

STEGODON. Fossil Mammals.

STEGOSAURUS. Fossil Reptiles.

STEGOTHERIUM. Miocene.

STEINBOK. Mammalia, Artiodactyla. A small African **antelope,** any member of several species. The group to which they belong includes several species named as steinboks, *Raphiceros,* and in addition the royal antelope, *Neotragus pygmaeus,* the grysbok, *Nototragus melanotus,* and the **oribi.** The name is also applied to the alpine **ibex.** (A.W.L.)

STELE. The **vascular** tissue of the axis of a plant is called the stele or central cylinder. The principal tissues composing it are the **xylem** and the **phloem.** Others are pith, pith rays, cambium, and pericycle. The most primitive type is the protostele, which consists of a solid central mass of xylem surrounded by a cylinder of phloem. There is no pith in a protostele. Protosteles are found in the roots of all plants and in the stems of some of the ferns. Many would set off the stele of the root as a separate type known as a radial stele, since the central mass of xylem is not a cylinder but has several arms projecting outward from its surface, with the phloem concentrated between these arms. In stems, only those of certain Pteridophytes like *Lycopodium,* have radial steles.

A siphonostele is composed of concentric cylinders of xylem and phloem enclosing a central pith. If the xylem cylinder is next the pith and is surrounded by the phloem, the stele is an ectophloic siphonostele; if there are two cylinders of phloem, one inside and one outside the xylem, it is an amphiphloic siphonostele. The most common type in **Gymnosperms** and **dicotyledon** stems is the ectophloic siphonostele. The amphiphloic siphonostele is found in members of the gourd family and in some ferns. Commonly there are many gaps in the siphonosteles, where vascular strands pass outward into a leaf. These gaps are called leaf gaps; the strand, a leaf trace.

A dictyostele is a siphonostele which is so broken up by numerous leaf gaps that it appears to be made up of a number of separate strands. It is really a very much dissected siphonostele. Dictyosteles are found in some **ferns** and in many **dicotyledons.** (R.M.W., B.S.M.)

STELLAR MAGNITUDE. In the first star catalogues issued by Hipparchus and Ptolemy the relative apparent brightness of the stars was designated by a system of six numbers referred to as the magnitudes of

the stars. Twenty of the brightest stars were referred to as first magnitude, while those at the limit of visibility were called sixth magnitude. The stars with brightness intermediate between the two extremes were assigned to a magnitude number with the numbers increasing with faintness of the stars. With the application of the **telescope** to astronomy many faint stars were discovered and the need for additional magnitude numbers became evident. Unfortunately for modern astronomers, the attempt was made to amplify the ancient magnitude system not only to include the fainter stars, but also to indicate finer gradations of brightness by a decimal system. The result is that astronomers are now using a system which was started about 2000 years ago and has all of the clumsiness and inconvenience for modern observers which is characteristic of so many of the ancient scientific instruments.

There is no definite evidence that Hipparchus or Ptolemy had any idea in mind at the time that they first used the magnitude system other than to provide a rough descriptive term for the stars. In the early part of the 19th century Sir John Herschel found that the apparent brightness of a first magnitude star is about 100 times that of a sixth magnitude. In 1850 Pogson proposed a fixed scale of stellar magnitudes based upon the original scale of Hipparchus and Ptolemy, but so adjusted that it would agree at the sixth magnitude with the system employed by Argelander in his famous **Bonner Durchmusterung.** Adopting the announcement of Herschel that the ratio of brightness of a first and sixth magnitude star is approximately 100, Pogson proposed that the ratio between successive magnitudes should be $\sqrt[5]{100}$ or approximately 2.512. This leads to an analytical expression for the magnitude scale as follows:

Call B_1 the apparent brightness of a star of magnitude H and B_2 the apparent brightness of a star of magnitude J, then $B_1/B_2 = 2.512^{(J-H)}$ or, expressed in logarithmic form, $\log_{10} B_1 - \log_{10} B_2 = 0.4\,(J - H)$.

Since the magnitude scale is a scale of relative brightness, it is necessary to establish a system of standards. For this purpose a group of stars in the immediate vicinity of the north celestial pole has been selected. The magnitudes of the stars in this "north polar sequence" have been very carefully determined and agreed upon by the International Astronomical Union. All magnitudes determined at the present time should be referred, either directly or indirectly, to this standard sequence.

The magnitude scale as originally established referred to the relative apparent visual brightnesses of the stars. With the application of photography to astronomy difficulty with the magnitude scale immediately became evident. If we have two stars of the same visual magnitude, one of them blue and the other red, the photographic image of the blue star will be much stronger than the photographic image of the red star. The colors of the stars in the sky vary with the different spectral types, and the visual magnitude differences between a number of stars of different spectral types will differ considerably from the magnitude differences obtained by photographic means. Furthermore, the photographic magnitudes, so-called, will be different, depending upon the type of plate used and the characteristics of different telescopes, and it becomes necessary to be very explicit in defining the particular range of wavelengths of **spectral** energy that are to be used in any magnitude scale. The difference between the photographic and visual magnitude of a star is known as the **color index** of the star, the term arising from the fact that the color is the determining factor in the magnitude scale difference.

With the application of various other types of radiation measuring instruments, such as **bolometers** and **radiometers,** to the measurement of the apparent brightnesses, of the stars the necessity has arisen for various different magnitude scales such as bolometric magnitude, radiometric magnitude, etc. The problem of the intercorrelation of the different systems is at present in a very confused state and much research is being carried on in this important field. It is devoutly to be hoped for that in the future some system of expressing the apparent brightnesses of the stars may be devised that will replace the present complicated inverse logarithmic scale of magnitudes.

For the purpose of expressing the intrinsic brightness of a star, independent of the distance of the star from the earth, a system of **absolute magnitudes** has been devised which will be discussed in more detail elsewhere. (W.K.G.)

STELLAR PARALLAX.

STELLAR PARALLAX. The term stellar parallax is used by astronomers as a means for expressing the distance of a given star. Technically defined, stellar parallax is the angle that would be subtended by the mean distance of the earth from the sun (one **astronomical unit**) at the distance of the star from the sun.

From the earliest days of the Pythagoreans any theory of the structure of the **universe** which postulated that the earth might move about the sun was objected to on the ground that such motion should produce an apparent motion of the stars. **Copernicus**, in proposing his heliocentric theory, met this objection by postulating that the distances of the stars were so incomparably greater than the distance of the earth from the sun that no instrumental methods would be capable of detecting the motion even if it did exist. The attempts to test the Copernican doctrine by searching for this so-called stellar **parallax** gave a tremendous impetus to the design of accurate instruments, but even with the improvement in instrumental equipment the effect was not observed and the Copernican theory lost ground. It was not until 1838 that **Bessel** was able to definitely prove that the effect is present.

The type of effect to be looked for is illustrated in the figure. The type of curve which the stars should

Parallaxes of the stars. Owing to the earth's revolution the nearer stars describe parallax orbits annually with respect to the remote stars.

apparently follow due to the earth's motion about the sun varies from an ellipse with eccentricity equal to that of the earth's orbit for stars at the pole of the ecliptic, to oscillations back and forth along a straight line for stars in the plane of the ecliptic.

The problem of determination of stellar parallax is theoretically very simple. All that is necessary is to make a series of observations of the positions of a star on any system of **spherical coordinates** (e.g., **right ascension** and **declination**) and from the observed changes in position throughout the year determine the stellar parallax. This so-called absolute method was attempted many times but failed to reveal any definite value because of the fact that the instrumental corrections were larger than the effect sought for. This is not surprising when we consider that the largest stellar parallax which has ever been found (for the star Proxima Centauri) has a value of 0″.783, or equivalent to the angle subtended by a ten cent piece at a distance of approximately 3 miles.

With the failure of the absolute method to yield values for the parallaxes of the stars, Bessel and Struve decided upon an indirect or relative method for determining the desired quantity. This method is based upon the assumption that certain stars are at such a great distance that their parallaxes are too small for detection, but that there are other stars closer to the sun which should show motion relative to the distant background. Bessel selected the star 61 **Cygni**, which was assumed to be relatively close to the earth from a large **proper mo-**

tion, while Struve selected the star **Vega** which has an appreciable proper motion and is also so bright as to imply closeness to the earth. Proceeding by different methods, in 1838 both Bessel and Struve were able to show that the stars which they had selected showed parallactic motion relative to the background of stars.

Until the application of photography to astronomy the problem of determination of stellar parallaxes was very tedious and laborious and up to 1880 distances of less than 25 stars had been determined. With the application of photography the progress of parallax determination became very much more rapid, and at present many long programs of observations both in the northern and southern hemisphere are nearing completion.

The photographic method consists in first selecting stars which are suspected, either from proper motion, **spectral type**, or other characteristics, to be relatively close to the sun. Plates are taken of these stars, great care being exercised in the guiding, and the brightness of the "parallax star" is reduced until its photographic image compares favorably with the images of the fainter "background stars." The plates are all taken at the same **hour angle**, either east or west, to eliminate so far as possible atmospheric effects, and the dates on which the plates are taken are separated as much as possible to make the effect of the earth's motion as large as possible. Twenty or thirty plates, extending over several years, are taken and the position of the parallax star carefully measured with reference to half a dozen background stars. **A least squares** solution will yield the motion of the star relative to the background. This motion will consist both of the proper motion and the parallactic shift. The former may be separated from the latter because proper motion is linear in character while the parallactic shift is periodic. For stars within 5 million times the sun's distance from the earth (parallax 0″.04) the mean of two or three determinations will be correct within 20%. For twice this distance the results are only accurate enough for statistical purposes, while beyond this distance the trigonometric method is practically valueless. Occasionally, due to an unfortunate choice of parallax star or of comparison stars, the value of the stellar parallax comes out to be a negative quantity. Such a "negative parallax" simply means that the star under observation is more distant than those selected for comparison purposes. For the more distant stars beyond the range of the trigonometric method certain other methods are available such as: parallaxes of members of **moving clusters**, mean parallaxes, dynamical parallaxes of double stars, and **spectroscopic parallaxes**.

The "General Catalogue of Stellar Parallaxes," published by Dr. Frank Schlesinger at the Yale University Observatory in 1935, lists 7534 parallaxes determined either by the relative trigonometric or spectroscopic methods, and 2482 parallaxes determined by the dynamical method. (W.K.G.)

STELLAR PHOTOMETRY.

STELLAR PHOTOMETRY. The problem of determining the **stellar magnitude**, or brightness, of a star is known as stellar **photometry**. Since the magnitude scale is a purely arbitrary one, all methods of stellar photometry consist fundamentally in the comparison of the brightness of one star with the brightness of a star of standard magnitude. The simplest method is to make direct visual comparison between the two stars and estimate directly the difference in magnitude. In the so-called "Argelander method" two stars of standard brightness are selected, one brighter and the other fainter than the star under consideration. The difference in magnitude between these "comparison stars" should not be greater than one magnitude. The observer then mentally divides the magnitude difference between the comparison stars into, say, ten, light steps and estimates the brightness of the unknown star in terms of these steps.

It is a well-known fact that the eye can more accurately determine when two objects are of equal brightness than it can determine the difference of brightness of two unequal objects. A number of different devices, such, for example, as the **wedge photometer** and the **polarizing photometer,** are employed for equalizing the brightness either of an artificial star or a star of known magnitude with that of the star under consideration.

The above methods all employ the human eye as the instrument of measurement and the characteristics of different eyes vary not only between different observers, but also, to some extent, from day to day for an individual observer. In the attempt to remove the human element from the determination of stellar magnitudes various stellar photometers have been devised which measure directly the intensity of the radiation from star images formed by telescopes. Such photometers employ devices such as **photoelectric cells, selenium cells, thermopiles,** etc., by means of which the radiant energy is converted into a **galvanometer** deflection, with the deflection a complicated function of the amount of radiant energy. The technique of using such instruments is difficult to master and the process of obtaining results is slow and tedious, but in the hands of skilled observers the results with these physical photometers are more accurate than those obtained by any other method. These types of instruments are selective for particular wavelength bands of radiation, and great care must be taken in comparing results obtained with one instrument and telescope with the results obtained by others.

The determination of stellar magnitudes either by direct visual comparison, by the use of visual photometers, or by the use of physical photometers is slow and fatiguing for the observers, with ten magnitudes per hour being a fair average rate of measurement. With the photographic plate and modern astrographic **cameras,** images of several hundred stars may be obtained in one hour of observing with the telescope. The size and blackness of a star image are complicated functions of the magnitude of the star. A plate taken of a region of the sky for the determination of the magnitudes of the stars must be standardized in some manner. There are many different methods of plate standardization in use, for example, photographing a standard sequence of stars, such as the polar sequence, on the same plate, or exposing the plate to a number of small areas with light of different intensity in each area. The plates are measured in a **densitometer,** usually of the physical type of microdensitometer, and the photographic densities of the various star images and standarization images are obtained. From the densities of the standarization images a calibration curve for the plate may be obtained and the magnitudes of the stars obtained. In the photographic method, as in all methods of stellar photometry, color effects are very troublesome and the intercomparison of results obtained by different observers, using different photographic plates and cameras, is a difficult task.

In all methods of stellar photometry a great many precautions must be taken to insure results of any value. After all known corrections for observer, method, and telescope have been applied, there always remains the troublesome and uncertain error introduced by the effects of atmospheric absorption. (W.K.G.)

STELLITE. A non-ferrous alloy containing 25–40% chromium, 40–80% cobalt, 0–25% tungsten, 0.75–4% carbon, and smaller amounts of iron, manganese, and silicon. It is hard and unmachinable as case and must be finished by grinding. In addition to its high hardness and **red hardness** which make it useful as a cutting tool material, it has excellent corrosion-resistance which is a requirement for applications such as surgical instruments and polished mirrors in optical instruments. (See **Alloys.**) (R.H.H.)

STEM. The stem of a plant is that part which bears the **leaves** and **flowers** and later **fruits.** Commonly it grows erect, lifting these various organs up above the ground. It is readily distinguished from the root by being separated into joints or nodes and internodes, and by bearing leaves or by having on its surface scars left by the falling of leaves.

Stems may be classified in several ways. If one considers the duration of the stem, it becomes an annual lasting but one year, a biennial living 2 years, or a perennial stem which grows for several years. If one considers the internal structure of the stem, he finds it to be herbaceous or woody. An herbaceous stem is one which is largely made up of **parenchymatous** cells, without a great mass of woody tissue. In temperate regions such stems last but a single year, at the end of which they die. In annuals the entire plant dies, while in herbaceous perennials the top dies, but the basal portion including the root and the lower stem lives on. Woody plants are those in which the stem is predominantly composed of **vascular** tissues. Such plants are either trees or shrubs or vines. Trees are commonly distinguished by the existence of a single stem or trunk which does not branch at its base, whereas in shrubs no single trunk exists, but several of equal size result from basal branchings. Branching is either excurrent or deliquescent. When excurrent, the trunk is distinctly recognizable throughout its length, the branches coming from it being much smaller, as in many **conifers** such as spruce or fir. Usually such trees have a conical shape, due to the progressively smaller branches from bottom to top of the tree. In deliquescent branching, the main trunk branches into several large branches which in turn divide, as in the elm. Vines are distinguished by their long relatively slender stems which usually require external support. Another classification separates erect stems from those which are procumbent or trailing on the ground, from scandent stems which clamber over supports, and from twining stems, which wind tightly around any supporting object.

In many plants the stem is very much reduced in size, appearing as a small often flattened ball, as in the common Cyclamen or in certain Cactus plants. Other plants are said to be stemless, the stem existing only as a small object at the top of the root, the leaves arising from it seeming to come directly from the top of the root. A familiar example is the dandelion. The first year of growth in many biennials results in a similarly stemless plant; carrots, beets, and parsnips are common examples.

As previously noted, one of the functions of the stem is to elevate the leaves into a position where they may function most efficiently and the flowers to a position where they may become more conspicuous and where the resulting fruits may be better scattered. Not only does the stem perform this function, but it also permits a great increase in the number of leaves and flowers which may be borne. The stem is the organ through which sap ascends from the root to the leaves and through which organic materials elaborated in the leaves pass to the place of storage. The stem itself may be the place in which materials are stored.

The stem develops from a **bud.** In the **seeds,** the embryo has a terminal rudimentary bud, the plumule, from which the first stem develops. The tip of this stem bears a terminal bud, from which further increase in the stem is developed. In the axils of the leaves lateral buds are formed which develop into branches. Elongation of the stem takes place only in the tip, extending downward therefrom a few inches, and is caused by the cellular changes like those that occur also in the elongating **root.** In some plants, as in Grasses, intercalary growth also occurs. This is growth in the region of the older nodes of the stem. Increase in diameter of the stem results from the divisions of special cells called **cambium** cells.

When the extreme variations of stem are considered, with their range from tiny plants less than half an inch in height to forest giants towering 300′ and over, and from vines with slender wiry stem less than $\frac{1}{16}''$ in diameter through succulent herbs to sturdy trunks 40′ or more through, it is not surprising that stem structure should be very variable and often complex. Yet they are all composed of the same types of cells and are all arranged on two fundamental patterns, one found in **dicotyledonous** plants, the other characterizing the **monocotyledons**.

In dicotyledons, the growing tip of the young stem has cells which are all alike, having a dense cytoplasm, and large nuclei, which, if the stem is growing, will be dividing frequently. These cells comprise the promeristem, the region where active cell division occurs, but little change in cell form.

As these cells increase in number, some are carried ahead, while others remain unchanged in position. The latter gradually show very evident changes in size and shape, and in the nature of their walls. The outermost layer, called the dermatogen or protoderm, is made up of somewhat flattened cells which will become the epidermis, a protective covering against entrance of disease-producing organisms and against excessive loss of water. Within the body of the stem tip certain strands of cells become distinct by their elongate shape and dense protoplasmic content. These procambium strands are the beginnings of the vascular tissue presently to appear. In cross-sections of the stem the procambium appears as a ring of separate masses of cells. The remaining cells of the growing tip are parenchymatous cells, called the ground meristem, changed but little from the promeristem condition.

As the procambial cells grow older, they gradually change in form. Those cells which are nearest the center of the stem become xylem cells, those towards the circumference of the stem become phloem, with **cambium** cells separating the two types. In some stems the differentiation of cells continues until the procambial strands have united to form a continuous cylinder which gives place to concentric rings of xylem and phloem cells separated by a band of cambium cells. In other stems, the strands remain distinct, forming separate vascular bundles. The ground meristem or parenchyma in the center of the stem becomes the pith, that surrounding the strands becomes the cortex and pericycle, while the radiating masses of parenchyma cells between the separate strands make up the pith rays. All these tissues, derived indirectly from the differentiation of the promeristem cells, form the primary body, composed of primary tissues.

The epidermal cells are often somewhat elongated in the direction of the length of the stem. The outer wall of an epidermal cell is frequently much thickened and cutinized, so that it becomes impervious to water. Many **stomata** are found in the epidermis. As the diameter of the stem increases, the epidermal cells gradually become stretched until they finally break apart and are lost.

The cortex inside the epidermis is comprised of several kinds of cells. Those nearest the surface are the collenchyma cells. These are modified parenchyma cells, the walls of which are thickened in their angles. These cells usually contain chloroplasts. Because of their thickened walls, collenchyma cells give support to the stem, while at the same time they manufacture food. The parenchyma cells of the cortex are thin-walled, and

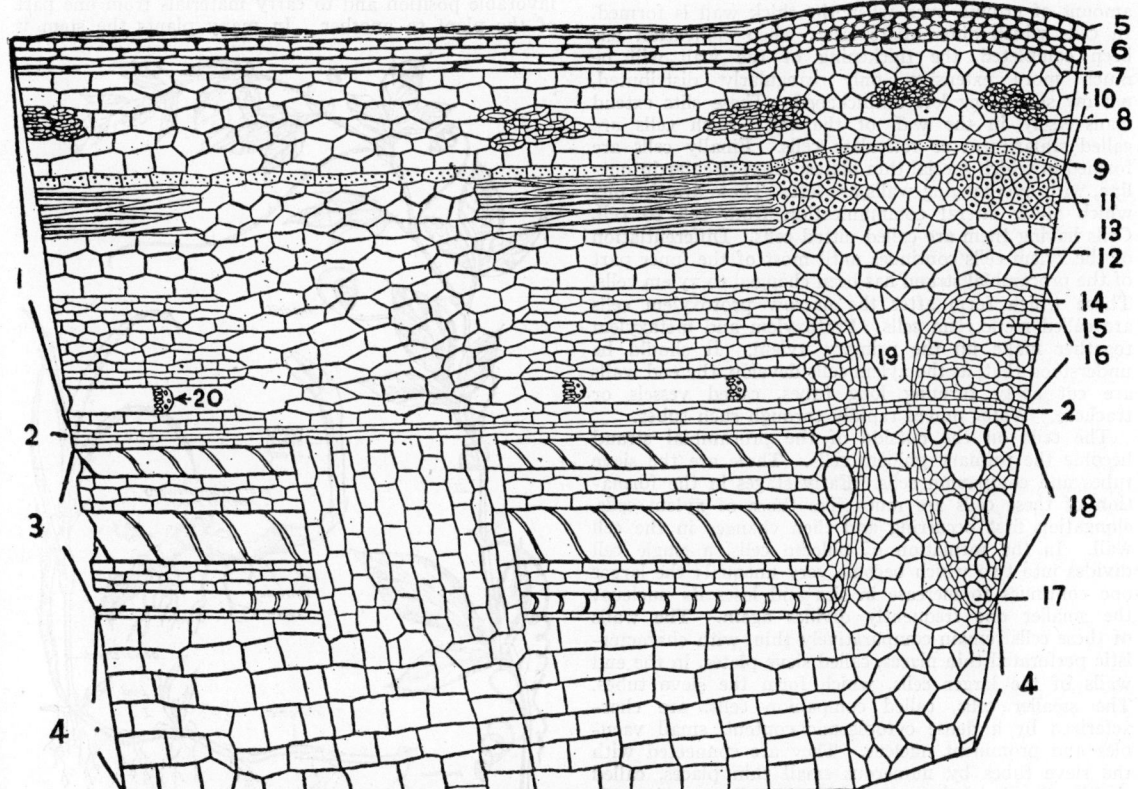

1, bark; 2, cambium; 3, wood; 4, pith; 5, epidermis; 6, collenchyma; 7, parenchyma of cortex; 8, stone cells; 9, starch sheath; 10, primary cortex; 11, pericyclic fibers; 12, pericyclic parenchyma; 13, pericyclic; 14, phloem parenchyma; 15, sieve tube; 16, phloem; 17, vessel; 18, xylem; 19, pith ray; 20, sieve plate.

Diagram showing the tissues derived from the primary meristens. (*Redrawn by permission, from Stevens Plant Anatomy, The Blakiston Co.*)

either rounded in shape or, through mutual pressure, more or less angular. Those near the surface of the stem often contain chloroplasts, and carry on photosynthesis. They give rigidity to the stem because of their turgor pressure, and also serve as storage tissue. In some stems there are also found in the cortex thick-walled sclerenchyma cells. These may be either long slender fibers or short stone cells. The innermost cells of the cortex, the endodermis, are rarely conspicuously filled with starch grains, which give to these cells collectively the name of starch sheath. In a few plants the innermost cells of the cortex form a definite endodermis, the walls of each cell being much thickened.

The tissues inside the cortex include the pericycle, the vascular bundles, pith rays, and the pith. This is the stele.

The cells of the pericycle are very similar to those of the cortex, so much so that it is often very difficult to distinguish one from the other. In the pericycle of stems, which usually is much thicker than that in roots, sclerenchyma cells, both fibers and stone cells, may be found along with the parenchyma, just as in the cortex.

The cells comprising the vascular system—xylem and phloem—are very much modified. The procambial cells towards the center of the stem first elongate greatly, without appreciably increasing in diameter. Soon changes appear in the wall, secondary deposits of cellulose being laid down against the primary wall. The manner in which the wall is thickened varies in different cells. In some the thick deposits are in rings: these are formed while the cell is still elongating and so are gradually separated. Such cells are called annular cells. In other cells of the first-formed xylem, which is called protoxylem, the wall thickenings are in the form of spirals, producing spiral cells, which allow a certain amount of growth even when the thick wall is formed. In cells which differentiate later, when elongation has been completed, the thickening of the wall will be much more extensive, only irregularly distributed, narrow slits being left unthickened. These slits extend transversely in the wall of the cell. Such cells are called scalariform or reticulate cells. Finally cells are formed in which special thin plates, often circular in outline, with overhanging walls are left. These are the pits which allow lateral communication from cell to cell. Cells having them are called pitted cells. Differentiation of the xylem cells continues until most of the inner part of the procambial strand has been changed to xylem cells. Those which form after the narrow protoxylem cells are called metaxylem cells. Protoxylem and metaxylem together make up the primary xylem. It should be understood that, as the xylem cells develop, the end walls are cut away forming long tubes, called vessels or tracheae. Water moves rapidly through such tubes.

The cells on the outside of the procambial strand become the primary phloem cells. These are the sieve tubes and companion cells. Early stages in the formation of these cells are much like those of xylem cells, elongation first occurring and then changes in the cell wall. In the formation of phloem cells, a single cell divides into two which become very unequal; the larger one continues to increase in size and loses its nucleus; the smaller one frequently divides again. The walls of these cells remain comparatively thin, with characteristic perforated thin places, called sieve plates, in the end walls of the larger cells, which form the sieve tubes. The smaller cells, called companion cells, are characterized by a dense cytoplasmic content, small vacuoles and prominent nucleus. They are connected with the sieve tubes by numerous small thin places, called simple pits, in their walls. Phloem cells are channels in which food passes through the stem. In addition to these various cells, the phloem contains parenchyma cells which are used for storage of materials and in some plants long thick-walled fibers, called phloem or bast fibers.

Parenchyma and fibers also occur in the xylem elements. Between the xylem and phloem elements there is a band of cells which remain unmodified and become an important tissue in many stems. This is the cambium, which by its divisions gives rise to the secondary tissues which compose the bulk of the stems of woody plants. These secondary tissues are the secondary xylem and phloem, and differ only slightly from the primary xylem and phloem, being unlike mainly in their origin. Secondary rays, which provide for lateral translocation of food, are also produced by the cambium.

In the center of the stem is the pith, composed of large, thin-walled cells arranged in irregular fashion. They function principally as places of storage of food.

In nearly all monocotyledons, no cambium is formed, therefore the monocot stem is composed entirely of primary tissues. The arrangement of these tissues is vastly different from that of the stems of dicotyledons. The vascular tissues occur in the form of separate small bundles which are scattered throughout the stem. It is impossible to distinguish any limit separating cortex from pericycle and pith. In many monocots the central portion of the stem is entirely free from bundles and recognized as a pith. Often the pith breaks up, forming a hollow stem.

All Gymnosperms have woody stems. The development and structure of these are quite similar to that of dicotyledons, but, except in a few uncommon species, the xylem is composed entirely of tracheids, distinguished from the tracheae of the angiosperms by the fact that they are elongated cells instead of long tubes formed from many cells, and no companion cells are formed in the phloem.

In most plants the function of the stem is to display the leaves and reproductive organs in the most favorable position and to carry materials from one part of the plant to another. In many plants the stem is

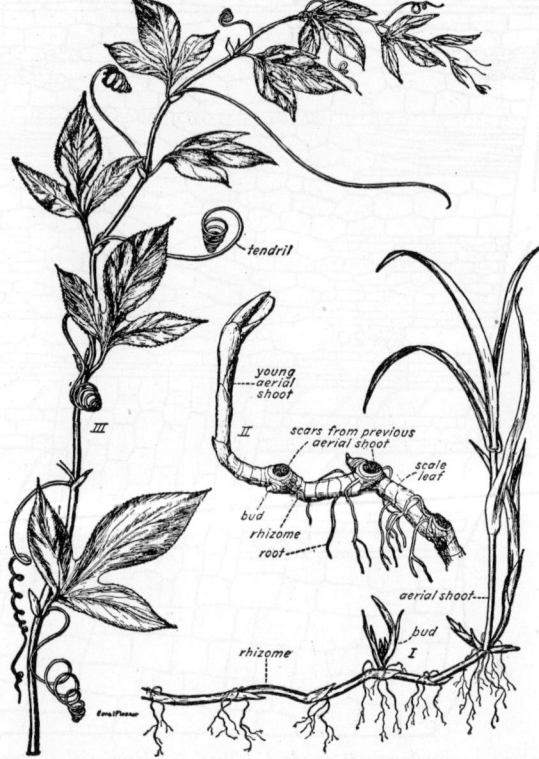

Types of modified stems. I, rhizome of quack grass (*Agropyron repens*); II, rhizome of Solomon's seal (*Polygonatum commutatum*); III, stem of passion flower (*Passiflora incarnata*) with tendrils which are modified stems.

a highly specialized structure with different functions. Often these specialized stems take over the function of one of the other organs of the plant.

The outer tissues of the stems of nearly all plants are green. Therefore some **photosynthesis** takes place in these tissues. There are many plants in which the stem is the principal, if not the only, place where photosynthesis occurs. In many cases, as in some of the **spurges** and **cacti**, the appearance of the stem differs very little from that of any other plant; but leaves are very much reduced or entirely lacking, all photosynthesis occurring in the stem. In other species of cactus, such as the Prickly Pear and the widely cultivated crab or Christmas cactus, the stem is very much flattened, but still distinctly recognizable as a stem. In some plants, however, the modification has become extreme. The ultimate branches of the stem have become very much flattened and have a shape which gives them every appearance of a leaf. Only their position in the axil of a tiny scale, the real leaf, betrays their true nature. The dainty Smilax, *Asparagus asparagoides*, of the florist, has branches of this kind. So also does the Butcher's Broom, a marsh plant of Europe, which is widely cultivated and appears during the Christmas season, stained a brilliant scarlet. The inconspicuous greenish-white flowers of this plant occur in the center of that part which is commonly assumed to be a leaf. The tiny needle-like "leaves" of the garden asparagus are really branches.

In a few plants the stem becomes a very important reproductive part. Runners, long slender branches from the base of the stem, grow out horizontally over the surface of the ground, and root at their tip. There a new plant is formed. With death and disintegration of the connecting stem, the young plant becomes separate from its parent. This is a form of vegetative reproduction. A stolon differs very little from a runner; it is a prostrate branch which regularly roots at its nodes, and sends up new plants not only at its tip but also from the nodes. Rootstocks or **rhizomes** are spreading underground branches which produce **adventitious roots** at their nodes and often spread widely.

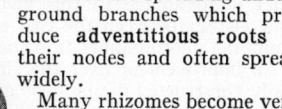

Many rhizomes become very fleshy because of an accumulation in them of food materials. Their principal function then is storage. Storage occurs in the stems of many different plants. If only a portion of the rhizome becomes enlarged, it is called a tuber. The common white potato is a very familiar tuber, which is formed at the tip of a slender rhizome. The "eyes" of the potato are really the nodes of the stem; from them buds will give rise to branches when the potato grows. More modified than the tuber is the corm.

Potato tuber with scale-like leaf at "eye."

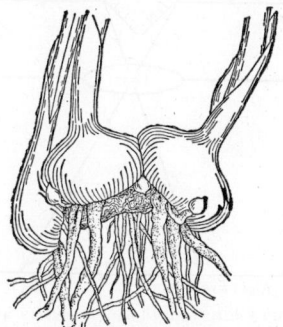

Corms of *Gladiolus* consisting chiefly of fleshy stems.

This is a short thick erect rootstock. Often it is much broader than it is long. Buds are formed on the upper surface of the corm. Each of these buds grows into a new plant, exhausting the substance of the old corm. New corms form at the base of the old one. Corms known to all are those of *gladiolus* and *crocus*. These are usually incorrectly called bulbs. A bulb is a fleshy bud, composed of a short thick stem and many fleshy or scaly leaves or leaf bases. Onions form true bulbs, as do tulips, hyacinths and many lilies.

The materials stored in the stems so far considered are mainly reserve food. Other stems become swollen with stored water. Many cacti and Euphorbias have stems of this sort.

Thorns are usually stems or branches which have become stiff and pointed and serve to protect the plant. Tendrils are organs which support a plant as it grows up through other plants. Not all tendrils are stems. Some, like those of the grape, are definitely so; others are leaves or parts of leaves. Usually plants which have tendrils have slender stems which lack sufficient strength to support themselves. Plants of this type are vines. There are several types of vines, or climbing plants. One of them supports itself by twining tightly around any available support. The direction of twining is very constant for any species, some invariably turning in a clockwise direction and others counterclockwise.

Other climbers support themselves by **adventitious roots** which form in abundance and cling tightly to any support. Other climbing plants are supported solely by the presence of many spines or prickles, often hooked, or pointed backwards so that the stem does not easily slide off any object on which it rests. Climbing roses illustrate this type of climber. But many tropical vines are much better illustrations. Often these grow to great lengths, hanging in long festoons from the tops of tall trees, or growing in tangled masses over low shrubby plants. In these tropical climbers, which are commonly called lianas, the stems often assume curious flattened or fluted or irregular shapes. In diameter they vary from a fraction of an inch to many inches; they may attain a length of 400 or 500 feet. They are one of the most characteristic and annoying features of the tropical rainforest. (See also **Cambium, Wood**.) (R.M.W., B.S.M.)

STENOPODIUM. The biramous appendage of **crustaceans** in its slender form. (A.W.L.)

STEPHANITE. The mineral stephanite, silver **antimony sulfide**, $5Ag_2S,Sb_2S_3$, is found in short prismatic or tabular **orthorhombic** crystals. It is a brittle mineral; hardness, 2–2.5; specific gravity, 6.2–6.3; metallic luster; color, black; **streak**, black; opaque.

Stephanite occurs associated with other silver minerals and is believed to be primary in character. Foreign localities are in Czechoslovakia, Saxony, the Harz Mountains, Sardinia; Cornwall, England; Chile and Mexico. In the United States it is found in Nevada, where it is an important silver ore. It was named for the Archduke Stephan of Austria, mining director of that country at the time this mineral was first described. (E.S.C.S.)

STEPHENSON GEAR. Valve Gear.

STEPTOE. A hill or mountain whose top projects above a lava flow which has surrounded its lower flanks. (R.M.F.)

STEREOGRAPHIC PROJECTION. This type of **map projection** is used to some extent by navigators, but more commonly for maps of the **celestial sphere**, particularly in constructing the basic map for **star finders**. This projection has a valuable property in that circles on the sphere appear as circles on the projected map. Because of this property, several attempts have been made to use the stereographic projection for drawing lines of position obtained from celestial objects. In the proposed methods, the substral point would be plotted on the chart and the circle of position drawn about this point with a radius equal to the observed zenith distance of the object. This method is proposed for use by aviators, where speed in drawing lines of position and obtaining a fix is particularly desirable, and where extreme accuracy is never possible because of errors inherent in measuring altitude with a bubble sextant in a moving plane. Thus far this use of the stereographic projection has not been developed for practical use, but it shows promise of success. (W.K.G.)

STEREOISOMERISM. Isomerism and Stereoisomerism.

STEREOPTICON. Projection; Optical.

STEREOSCOPE. Binocular Vision.

STEREOSCOPIC PHOTOGRAPHY. A stereoscopic photograph is one which reproduces in the mind of the observer the same sensation of three dimensions produced by observation of the object itself.

Students of optics have long been aware that the two eyes perceive dissimilar views of an object.

The two views received by the eyes differ in two significant respects. Since the angle of vision is different

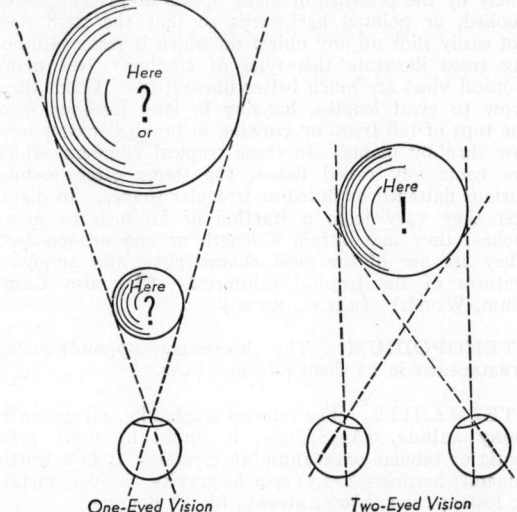

Fig. 1. In one-eyed vision, the brain may interpret the retinal image as a large object far away or a small object near by. In two-eyed vision, the object is fixed in direction, size, shape, and distance.

for each eye, the objects appear in a slightly different guise. If there are several objects in the field of vision, they appear in slightly altered relations to each other. Both the angles and distances of the objects in the scene are sufficiently changed to present a different picture to each eye.

A second important difference in the two views is a change in light distribution. As the angles of view change, highlights appear at different positions and with varying degrees of intensity.

The brain does not record a genuine sensation of depth unless the two eyes convey to it two complete and somewhat different pictures of the object seen. It is only by a rapid comparison and combination of both views that the brain is able to perceive depth directly.

All versions of the stereoscope are devices for duplicating the visual process of depth perception. They are all instruments that present two somewhat dissimilar

2-dimensional pictures, one to each eye of the observer. The brain reacts to these views as it would to the object itself and therefore "sees" a 3-dimensional image. The two pictures must be positioned so that the two eyes perceive mutually exclusive views.

In 1833 the British scientist, Sir Charles Wheatstone, stated the general theory of binocular vision, and in the course of his paper suggested a method of building a stereoscope. It is he who is generally credited with the invention of the device. A year later Professor Elliott of the Department of Logic at the University of Edinburgh decided to illustrate his theories on perception of distance with a practical demonstration. Since photography was unknown at that time Professor Elliott drew a pair of simple pictures containing a cross, a tree and a crescent moon. He placed these on transparent paper and designed a viewer.

Another investigator, Sir David Brewster, built a box-type stereoscope which he equipped with geometric drawings. In 1849 the Abbé Moigno equipped the Brewster stereoscope with daguerreotypes. A large variety of scenes could now be reproduced in pairs with ease and accuracy. By 1860 the stereoscope was a popular parlor amusement and a London stereoscope company was doing a thriving business.

In this country, Oliver Wendell Holmes became interested in the device and he designed a hand model for parlor use. He saw great possibilities in the device and predicted that the new invention would find many applications.

Since the original Wheatstone invention, stereoscopic photography has gone through a large number of improvements, most of which were developed for technical and scientific applications of the instrument.

The simplest method of providing a pair of stereoptic photographs is to take two successive pictures of the same object by shifting the camera a small distance laterally, generally equal to the distance between the two eyes. The camera may be mounted on a slide bar and the change of angle between the scenes controlled. The chief disadvantage to this method is that a period of time elapses between the two exposures. This difficulty limits the photographer to scenes of still life.

An improvement over the single-lens system is a specially constructed double-lens camera. Here the shutter

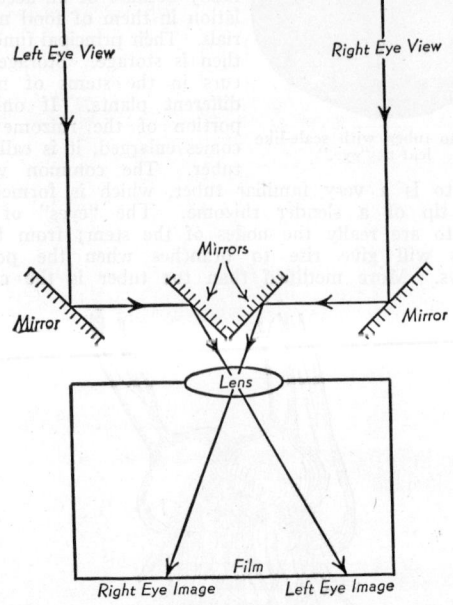

Fig. 2. A beam-splitter attachment converts any small single-lens camera into a one-shot camera for taking 3-dimensional views.

mechanism is synchronized for simultaneous exposures, and the pairs of scenes are recorded, one on each half of the film. After development the prints may be separated for stereoscopic use. A variation of this system is the use of mirrors which direct the two images through the single lens. In this manner two separate images may be obtained simultaneously with only one lens.

The problem of viewing stereoscopic pairs is more difficult. Ideally the two images should appear in the same location, yet each image must be seen only by the eye for which it is intended. Three general methods of image separation have been devised: geometric optics, color optics and polarization optics.

The earliest viewers of Wheatstone and Elliott used complicated systems of mirrors to achieve the proper picture separation. Brewster's model and its later development by Holmes use lenses to permit the eyes to look straight ahead in a normal manner with the focus relaxed as though viewing a distant scene. The modern stereoscopic viewer, permitting only one observer at a time, is a compact and inexpensive device. It can be used for viewing both transparencies and reflection-type prints.

A different approach to the problem of viewing was made in 1853 by Rollman and by D'Almeida a few years later. By their process, the two prints, called color anaglyphs, one reproduced in red, the other in blue-green or green, are superimposed. The system makes use of matching viewers. Since the red filter prevents the red reproduction from being seen by one eye, while the blue print is clearly visible, and the blue filter acts similarly, the effect is a stereoscopic presentation. The system has the great advantage of being capable of display to a large number of people at one time. However, the color filters greatly reduce contrast and brightness. They preclude any possibility of representation in natural colors. In addition, the strain caused by the conflicting colors restricts the viewing of color anaglyphs to short periods.

In 1901, F. E. Ives developed the parallax system of viewing which has the advantage of requiring no viewing devices. But the viewing positions are limited, there is a loss of detail, and the photographic technique required for the Ives system is complicated.

In 1891, the British scientist Anderton first experimented with polarized light as a means of image separation using tourmaline and other crystalline polarizing materials. But lack of an efficient and commercially practical light-polarizer prevented the development of

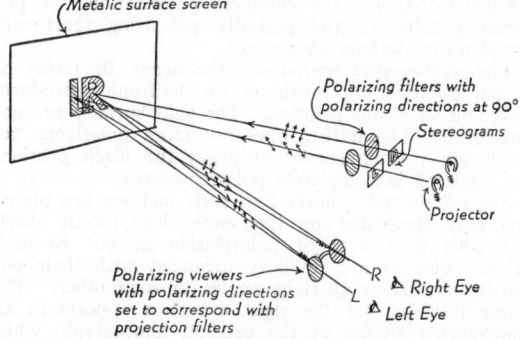

Fig. 3.   The principle of 3-dimensional pictures in full color.

these techniques. The invention of a practical sheet-polarizer by E. H. Land reopened the entire field of picture separation by means of polarizing filters.

In 1934, J. Mahler invented the photo plasticon, a complicated device which permits only one observer at a time. It is the only instrument yet devised which presents a full-color stereoscopic image without loss of definition and yet requires no viewers of any kind.

The application of Polaroid sheet-polarizers to the problems of stereoscopic photography gave great impetus to the art of 3-dimensional representation. None of the foregoing procedures was suitable for presentation to a large audience without sacrificing much of the detail of the picture. None of the above methods could produce a satisfactory print that could be handled and observed as any other photographic print. In 1940, however, Land and Mahler developed the vectograph which met both these needs.

If two polarizing sheets are placed across each other, with their polarizing directions at right angles, no light will pass through them. If their polarizing directions are parallel, light will penetrate. If, however, one of the polarizers has only a partially polarizing effect, light

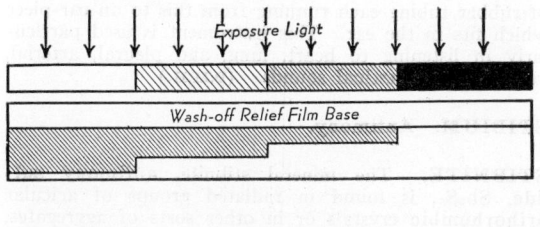

Exposure: Varying amounts of light transmitted through the light and dark areas of a stereo negative create latent image of varying depth in dyed relief film emulsion.

Development: Where latent image is developed the surrounding gelatin of emulsion is hardened by tanning developer.

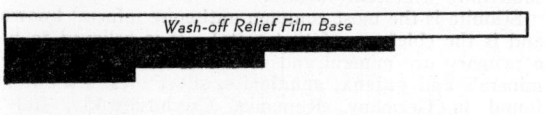

Wash-off: Unhardened gelatin washes off, leaving relief image which soaks up varying quantities of printing solution.

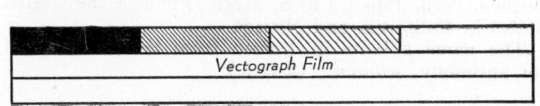

Vectograph Printing: Varying densities of vectograph image set up by solution transferred from relief image. Similar image set up on other side of vectograph film.

Fig. 4.   Vectograph process in diagram.

passing through it will appear gray to a polarizer placed across it. By controlling the degree of polarization, the observer can, through a crossed polarizer, see many shades or tones from very light, at zero polarization, to completely opaque for complete polarization. A polarizer parallel to the partially polarizing sheet will reveal no gradations whatsoever.

The vectograph reproduces the scene in terms of partial polarization, much as the halftone reproduces light and shade for printing. The light surfaces are not polarizing. The darker ones are more polarizing, reproducing varying shades of gray; the black portions are rendered in completely polarizing areas.

For stereoscopic effects two such pictures are placed one over the other on the same clear plastic sheet, but with their axes of polarization at 90° to each other. Viewers are similarly arranged with their polarizing directions at right angles to each other. The image intended for the right eye then appears in all the varying shades of the original photograph, while the picture intended for the left eye remains invisible. The left eye receives its exclusive image, and the brain combines the two to form a 3-dimensional scene.

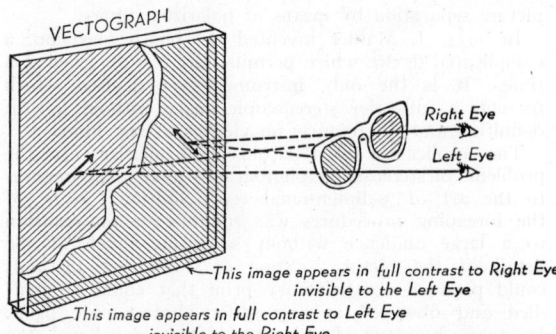

VECTOGRAPH

Right Eye

Left Eye

This image appears in full contrast to Right Eye invisible to the Left Eye

This image appears in full contrast to Left Eye invisible to the Right Eye

Fig. 5. Two-eyed vision. Three-dimensional vectographs have one image on each side of the film, each image being rendered in terms of degree of polarization, with their polarization directions at right angles. The front image appears in full contrast to the right eye because its vibration direction is crossed with the "optical slots" of the right eyepiece.

The applications of 3-dimensional photography vary from military and scientific uses to such recreational fields as motion pictures and the family snapshot album. In photomicrography, surface structures and minute objects can be observed in all the detail of their actual dimensions. Aerial photographs faithfully reproduce contour changes for relief maps. Instruction in solid geometry and celestial navigation no longer require an explanation of 3-dimensional phenomena from flat projections. The possibilities are far-reaching in all commercial fields where display is used. Museums and scientific exhibits can achieve far more effective representation of their materials by the use of depth perception. (Prepared by Polaroid Corporation, Cambridge, Mass.)

STERILITY. Inability to reproduce; barrenness. Sterility may be the fault of husband or wife or both. Sterility in either sex may be due to (1) anatomical defects, congenital or acquired, (2) disease or the results of disease in the reproductive organs, especially that due to **gonorrhea**, (3) glandular disturbances especially those concerning the **thyroid, pituitary, adrenal, ovarian** and **testicular** glands, (4) poor health, due either to chronic disease or deficiency in diet. (R.S.M., D.M.H.)

STERILIZATION. Any procedure by which an individual is made incapable of reproduction. Sterilization of the feeble minded, insane, or criminal insane, or for therapeutic measures, is a simple procedure. No mutilating operations are done and there is no loss of sexual urge, no change in appearance, behavior, or sensation. In the male a small incision is made in the upper part of the scrotum and a small piece of the vas deferens is removed, thereby preventing the passage of sperm cells to the outside. In the female, through an incision into the peritoneal cavity, small pieces of both fallopian tubes are removed. (R.S.M.)

STERLET. Pisces, Chondrostei. A **sturgeon,** *Acipenser ruthenus,* of moderate size found in the Black and Caspian seas and in rivers of Siberia and Europe. As a food fish it is one of the most desirable members of the group, both for its flesh and for its caviar. (A.W.L.)

STERN-GERLACH EXPERIMENT. An experimental test by O. Stern and W. Gerlach (Germany, 1924) of the magnetic moment of **atoms.** A stream of metallic atoms, issuing from a vaporizing furnace through a narrow slit, entered a strong magnetic field. The magnetic intensity was perpendicular to the atom stream, and had a strong gradient in its own direction. If magnetic moments of atoms are due to revolving electrons, the atoms should, according to classical theory, begin to precess at all angles about the field direction, and the atomic beam should simply broaden into a band. According to the quantum theory, they should precess at certain angles only, and the original stream should be divided into several distinct streams. Actually it proved to be divided into two, oppositely deflected streams. From this the experimenters concluded that the atoms tested have but one "magneton" each. (See **Magnetism** and **Precession.**) (L.D.W.)

STERNITE. Skeletal System.

STERNUM. The breast bone. Skeletal System. (A.W.L.)

STEROLS. Alcohols and Ethers.

STETHOSCOPE. An instrument used to transmit and to amplify various sounds produced by certain organs of the body. The stethoscope in its common form is composed of a cup-shaped device and two pieces of rubber tubing each running from this to an ear-piece which fits in the ear. This instrument is used particularly in listening to heart, lung, and pleural, arterial, venous, and intestinal sounds. (D.M.H.)

STIBIUM. Antimony.

STIBNITE. The mineral stibnite, **antimony sulfide,** Sb_2S_3, is found in radiated groups of acicular **orthorhombic** crystals or in other sorts of aggregates, as well as blades, also as columnar or granular masses. It shows a highly perfect pinacoidal cleavage; conchoidal fracture; hardness, 2; specific gravity, 4.5–4.6; luster, metallic and very brilliant on cleavage faces or freshly fractured surfaces. Its color is a steely gray; the streak very similar in color, may be covered with a black, sometimes iridescent tarnish.

Stibnite is the most common antimony mineral known and is the chief ore of that metal. It is believed to be a primary ore mineral and occurs with other antimony minerals and **galena, sphalerite,** silver ores, etc. It is found in Germany, Rumania, Czechoslovakia, Italy, Bornéo, Peru, Japan, China, Mexico; and in the United States in California and Nevada.

The name stibnite is derived from the Latin word for antimony, *stibium.* (E.S.C.S.)

STICK INSECT. Walking-Stick.

STICKLEBACK. Pisces, Teleostei. Small marine and fresh-water fishes (**Pisces**) of the family Gasterosteidae, characterized by a series of strong spines along the back in front of the dorsal fin. (A.W.L.)

STIFF DIAGONAL. Diagonal.

STIFFENER. Bridge, Monocoque.

STIGMA. 1. A secondary sexual mark of **insects**. In many species of **butterflies** it consists of a patch on the wing of the male bearing modified scales on a more or less modified area of the wing membrane. 2. Used by some entomologists in place of **spiracle**. 3. In botany, the receptive part of the gynoecium or pistil of the **flower**, to which the pollen grain is carried in the act of **pollination**. In many flowers the surface of the stigma is sticky and so receives and holds the pollen grains more surely. (See **Flower**.) (A.W.L., R.M.W.)

STILBITE (DESMINE). The mineral stilbite, $(Na_2Ca)O \cdot Al_2O_3 \cdot 6SiO_2 \cdot 6H_2O$, is a **zeolite**, the compound **monoclinic** crystals of which are usually grouped in approximately parallel positions, forming sheaf-like aggregates, which have a soft pearly luster, whence the name stilbite from the Greek, meaning luster. The less commonly used term desmine is likewise from the Greek, meaning a bundle. Stilbite has one perfect cleavage; uneven fracture; is brittle; hardness, 3.5–4; specific gravity, 2–2.2; luster, vitreous to pearly; color, usually white but may be brownish, yellowish, red or pink. Its streak is white, and it is transparent to translucent. Like the other zeolites stilbite occurs in cavities in **basalts** and **traps**, rarely in **granites** and **gneisses**. Of the many foreign localities may be mentioned Trentino, Italy; the Harz Mountains; Valais, Switzerland; Arendal, Norway; the Ghats Mountains of India; and Mexico. The Triassic traps of New Jersey and Pennsylvania furnish specimens as do also rocks of the same age in Nova Scotia. (E.S.C.S.)

STILL-BIRTH. The birth of a dead fetus. (D.M.H.)

STILLSON WRENCH. Wrenches.

STILT. Aves, Charadriiformes. *Himantopus*. A bird (**Aves**) with unusually long legs, a long neck, and a long slender beak, slightly upcurved. Related to the avocets. In spite of their attenuated lines the stilts are beautiful and graceful birds. Their plumage is chiefly black or gray and white, but the beak and legs of some species are pink or red. There are only a few species but they are represented in all parts of the world. (A.W.L.)

STILT BUG. Insecta, Hemiptera. A **bug** with a slender body and very long slender antennae and legs. The few species make up the family Neididae. They are sluggish insects of no economic importance. (A.W.L.)

STIMULANT. Any agent or **drug** which increases the activity of an organ. Such drugs are **caffeine, strychnine, adrenaline, camphor, ammonia, digitalis** and its allies, etc. These drugs act principally on the nervous, respiratory or cardiac systems. Various remedial measures which are stimulating are cold air and water, contrast baths with hot, then cold water, and regulated exercise. (R.S.M., D.M.H.)

STINK BUG. Insecta, Hemiptera. A flattened **bug** of generally ovate form, in many species with an angular outline. Most species are moderately large, reaching a length of about ½″. The many species, constituting the family Pentatomidae, are also characterized by the fetid odor of the secretion discharged from glands opening on the lower surface of the body.

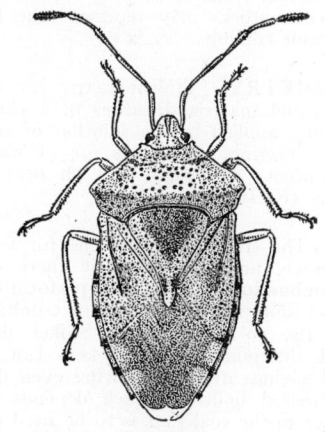

Stink bug.

One species, the **harlequin cabbage bug** or **calico-back**, is a troublesome pest. It is best controlled by clean cultivation of fields, hand picking of bugs and their eggs, and the use of trap crops, planted early to attract the insects.

Some members of the group eat other insects and are probably beneficial in destroying pests, but unfortunately even the harmless species may contaminate berries with their unpleasant odors. (A.W.L.)

STINKSTEIN. Limestone.

STIPES. Maxilla.

STIPULES. Leaf.

STOAT. Mammalia, Carnivora. A slender short-legged animal, *Mustela erminea*, related to the weasels and sometimes called the greater weasel. It has been regarded by some zoologists as common to both hemispheres, living in northern latitudes, but authorities now consider the common North American weasel to be a distinct, although closely related, species. Except in the extreme southern parts of their range, both of these animals turn white in the winter, retaining a black tip on the tail. In this stage they are often called ermine. (A.W.L.)

STOCHASTIC VARIABLE. A stochastic variable is a chance variable or one whose value is determined by a **probability function**. Examples of chance **variables** are the height of a person, measurements of some kind in a laboratory, the number of defectives in a lot, and the intelligence of a person. Generally all variables in **statistics** are chance variables. If x and y are stochastic variables then $x + y$, $x - y$, xy, and x^2y^2 are chance variables also. If x and y are chance variables, so is x/y, provided $y \neq 0$.

Some well known theorems on stochastic variables follow. The **expected value** of the sum of two chance variables x and y equals the expected value of x plus the expected value of y; i.e., $E(x + y) = E(x) + E(y)$. The expected value of the product of two chance variables x and y is the expected value of x times the expected value of y if x and y are independent, i.e., $E(xy) = E(x) \cdot E(y)$. The expected value of a constant times a chance variable equals the constant times the expected value of the variable, i.e., $Ec(x) = cE(x)$. (L.A.A.)

STOCK. In geology, a mass of **igneous rock** not of great areal extent, generally with a circular or elliptical cross-section, which has been intruded vertically into previously existing formations.

In some cases stocks may represent the lower portions of volcanic conduits. (R.M.F.)

STOICHIOMETRY. Stoichiometry is a study of heat, energy, and material balances of a chemical system, commonly applied to the solution of problems in Chemical Engineering. It makes use of many of the laws and concepts of general and physical chemistry, and physics. (D.E.M.)

STOKER. The stoker is a machine for feeding **coal** into a **furnace,** and supporting it there during the period of **combustion.** It may also perform other functions, such as supply of air, control of **combustion,** distillation of the volatile matter. Stoker development has reached the point where stokers can offer substantial fuel savings over hand firing even under small and medium-sized boilers. Much depends upon suiting the stoker to the coal that is to be fired on it. The character of the volatile matter is studied and arrangement made for burning all of it through the introduction of over-fire air when necessary. The ash characteristics also influence stoker selection. The stoker ought to be responsive to variable load. Control of combustion necessitates a variable drive of the stoker. The power to drive stokers is usually furnished by electric motors or steam engines.

Modern stokers may be classified as overfeed, underfeed, and conveyor.

The sprinkler or spreader stoker is a type of overfeed, but with horizontal grates. A rotor is driven at high speed and so located that any coal dropped on it will be thrown onto the grates. A feeder delivers the coal at the proper rate from the hopper into the range of action of the rotor blades. Some sprinkler stokers have a pneumatic or steam jet delivery of the coal to the grates.

Stokers most in use nowadays are the conveyor (**Fig. 1**) and underfeed types. A chain grate stoker

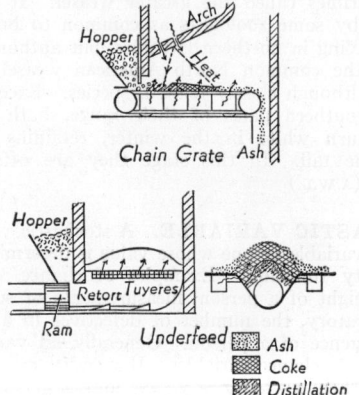

Fig. 1. Diagram of combustion in coal stokers.

(one type of conveyor) is a broad endless belt composed of short connected links. This belt passes over two sprockets so that it makes a flat upper surface upon which coal may rest. One of the sprockets is power driven, so that there is a very slow motion of the grate in a direction which will drag coal from the bottom of the hopper, carry it into the furnace, and finally dump the ash into a hopper. The rate of combustion is varied by simultaneous control of air pressure, velocity of the grates, and thickness of the fuel bed. Heat from the incandescent zone is radiated to the

ignition **arch** and reflected back to the green fuel bed at its entry to the furnace. The volatile matter that is driven off is mixed with air admitted through and over the green fuel bed, and, being confined to the arch, is passed through the hottest zone of the furnace. The carbon left behind then reaches the ignition point and burns, giving up heat, part of which is radiated back to the incoming green coal. There is no means of breaking up a crust formed by a coal which will soften and fuse together, but the chain grate stoker will burn coals that would be too fine or easily packed for use in an underfeed stoker, as well as coals that clinker badly.

The natural draft chain grate stoker is adaptable to small or medium-sized boilers, operated at fairly steady loads. This type of stoker is being used less and less, more favor being given to the single retort underfeed stoker. At their best point they will use from 70% to 80% excess air. Coal of fine sizes sifts through between the grate bars and causes losses unless a system is installed for its recovery. However, they have burned lignite in a very successful manner. The chain grate type is applicable to burning fine coals of high ash content and clinkering characteristics. A large floor area is required. This stoker is hard to bank and is very sensitive to furnace design.

The efficient combustion of certain classes of coal requires that the coal be gasified under conditions which will prevent the coal from fusing into an air-tight crust. A coal that has a tendency to fuse together into a solid cake of coke must be burned in an underfeed stoker, in which the fresh coal is fed to the fire zone by being pushed up from underneath. This type of feed creates a heaving action at the fire line, which tends to prevent the formation of an air-blanketing crust. An underfeed stoker consists of a retort, usually trough-shaped in industrial stokers, but resembling a pot in the domestic type. The coal is forced into this retort by an endless screw or ram in such a way as to accomplish the feed action outlined above. Air is supplied at the sides of the retort under pressure sufficient to enable it to penetrate the zone of incandescence. Heat from an incandescent zone of burning carbon is radiated and conducted downwards into the green fuel which is being crowded up to the ignition line from below. The volatile matter which is given off is mixed with air supplied from tuyeres, also below the ignition line. The inflammable mixture then passes directly through the hottest zone of the furnace—the incandescent fuel bed—insuring that all parts of it reach the ignition point. The ever-upward motion of the fuel keeps the bed well broken up, allowing free passage of gases and oxygen to all parts of it.

The single retort underfeed stoker is well adapted to a wide range of coals and to all sizes up to 1000 boiler horsepower. As it sets low and requires no ash basement, it is easily applied to almost any existing boiler. For these reasons, and because of its inherent smokelessness, it is especially adaptable to industrial service. Good operation is secured in small as well as large sizes with everyday boiler efficiencies ranging from 65 to 70%. The most efficient types have auxiliary moving grates to break up the fuel bed that overflows the retort. The size of a retort is limited by the ability of the air to penetrate from the tuyeres to the zone of burning. Large underfeed stokers cannot be built with single large retorts, for this reason; rather, they must be built with a number of parallel retorts. However, the multiple retort underfeed stoker does not resemble a number of single retort stokers placed side by side. Instead, the retorts are sloped downwards slightly, and the coal emerges from the retort on to an overfeed section, where the residual carbon is burned.

The multiple retort underfeed and forced draft chain grate types are essentially large, high-capacity boiler types having large initial cost and installation expense. A basement ashpit is required for both types. Skilled operators are needed, especially for the chain grate type, but they can readily obtain from 70 to 80% boiler efficiency, using about 50% excess air. From the fore-

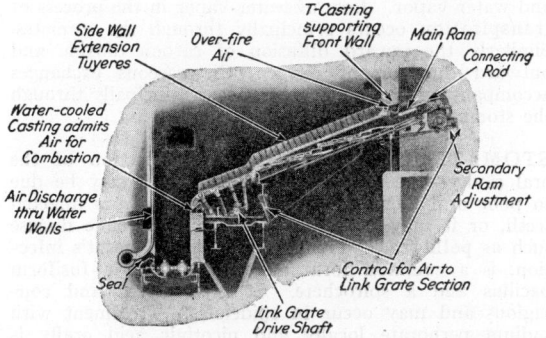

Fig. 2. Side view multiple retort underfeed stoker.

going remarks, it is evident that these stokers are central station types. The forced draft chain grate stoker is conceded to be the only type for anthracite or coke breeze. The ignition arches required increase the cost of the furnace. The complication of air zoning the plenum chamber must be accepted in order to control air in an efficient manner. Its flexibility and banking characteristics are better than the natural draft type but not so good as the underfeed.

Considerable similarity is apparent in multiple retort underfeed stokers as far as the underfeed portion is concerned, but examination of the overfeed portions discloses major differences. Ash is handled by dump plates or by clinker grinders. (See **Combustion; Combustion Control; Furnace.**) (F.T.M.)

STOKES' LAW. Luminescence; Viscosity.

STOKES' THEOREM. Stokes' theorem is an important mathematical result used in mathematical physics particularly. It transforms a **line integral** in space into a **surface integral** or vice versa. It was given by the great English mathematician and physicist Sir George Gabriel Stokes (1819–1903).

Let P, Q, R be **functions** of x, y, z, which, together with their first **partial derivatives**, are **continuous** in a region V of space, and let S be a **surface** lying in V and bounded by the curve C; let a, β, γ be the **direction angles** of the normal to S. Then Stokes' theorem is:

$$\int_C (P\,dx + Q\,dy + R\,dz) = \iint_S \left\{ \left(\frac{\partial R}{\partial y} - \frac{\partial Q}{\partial z} \right) \cos \alpha \right.$$
$$\left. + \left(\frac{\partial P}{\partial z} - \frac{\partial R}{\partial x} \right) \cos \beta + \left(\frac{\partial Q}{\partial x} - \frac{\partial P}{\partial y} \right) \cos \gamma \right\} \, dS.$$

In **vector** language and vector notation, this theorem may be stated: The **surface integral** of the **curl** of a **vector function** F of position taken over any surface region S is equal to the **line integral** of $\mathbf{F}$ taken around the boundary C of the region:

$$\iint_S (\Delta \times F) \cdot \hat{n} \; dS = \int_C \mathbf{F} \cdot d\mathbf{r}.$$

(L.L.S.)

STOLON. In botany, a stolon is a branch which grows out horizontally from the base of the **stem**, takes root and gives rise to a new plant at the **nodes** or at the tip.

In zoology, a shoot growing from the base of an animal or a **colony**, from which other individuals arise by budding. (R.M.W., A.W.L.)

STOMACH. The pouch-like structure, part of the **digestive system**, which serves as the temporary collecting place for the food as it leaves the **esophagus.** It lies high up on the left side of the abdomen just beneath the **diaphragm.** It lies obliquely extending from the extreme left side beneath the ribs across the midline of the abdomen about 3″ above the navel. The size of the stomach varies depending on the contraction of its muscular walls, the amount of food that it contains, and the pressure on its outside walls due to the adjacent small and large intestine. The entrance to the stomach from the esophagus is known as the cardiac orifice and the exit of the stomach is known as the **pylorus** or pyloric orifice. Both of these openings are closed by circular muscle fibers known as **sphincters,** except when food is passing through them. The walls of the stomach are made up of oblique longitudinal and circular smooth muscle fibers not under voluntary control. This arrangement in its walls allows for progressive wave-like movement of the food toward the pyloric opening, and also for a thorough mixing of the food. The glands of the stomach secrete mucus, **hydrochloric acid** and a substance which is converted into **pepsin** when it mixes with the acid.

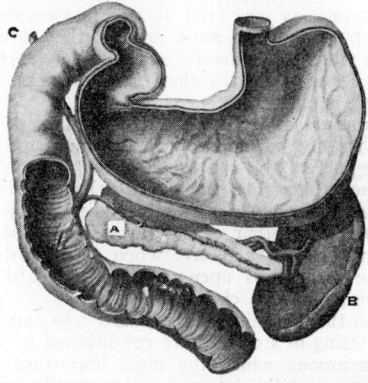

View of stomach and duodenum with part of anterior portion removed. *A*, pancreas; *B*, spleen; *C*, common bile duct.

Functions of the stomach are to mix the food with the gastric juices, and to pass it into the **duodenum** in small portions at intervals. The movements of the stomach also serve to reduce the size of the swallowed food particles. Very little actual **digestion** takes place in the stomach. **Protein** digestion begins in the stomach but is incomplete. In general, the food is prepared for the digestive action in the small intestine. The acid of the stomach has further a definite germicidal action on the swallowed food.

Since the secretion and movement of the stomach are controlled by the involuntary or sympathetic nervous system, anger, fear or worry interfere with or can alter the activity of the stomach. Exercise taken too soon after meals has a similar effect.

The common disorders affecting the stomach are (1) peptic ulcer, (2) cancer, (3) gastritis. (R.S.M., D.M.H.)

STOMACH ULCER. Peptic Ulcer.

STOMATE OR STOMA. The name applied to the minute pores which occur abundantly in the epidermis of leaves and, less abundantly, in the epidermis of young stems, flower parts and fruits. Each stomate is located between two distinctive epidermal cells called *guard cells.* The size and shape of the guard cells varies considerably from one species to another. Unlike other epidermal cells the guard cells contain chloroplasts. In many species of plants stomates occur in both the upper and lower epidermis of the leaf. In many other species, especially of woody plants, stomates occur only in the

lower epidermis. Even in those species in which the stomates are present in both epidermises there are commonly, although not invariably, more per unit area in the lower epidermis. In floating leaves such as those of the water lily stomates are present only in the upper epidermis. In many kinds of plants the stomates are restricted to grooves or furrows in the leaf.

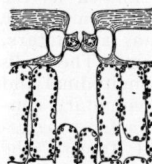

A portion of the lower epidermis of a geranium leaf.

Sunken stoma of carnation.

The number of stomates per unit area varies with the kind of plant and also, within limits, with the conditions under which the plant has developed. The range is from a few thousand to about a hundred thousand per sq. cm. of leaf surface. A single corn plant has been estimated to bear from 140 to 240 million stomates and the number on a large tree could be represented only by a figure of astronomical dimensions. The size of the individual stomates also varies greatly from species to species, their dimensions being expressed in microns. In some species the fully open stomates may be as large as 8 to 10 x 30 to 40 microns, as measured along the two axes of the elliptical pore, but in most species they are smaller. Species in which the stomates are relatively small usually have more per unit area than those which have relatively large stomates.

The structure of the cell wall of a guard cell is quite complex, varying considerably in thickness and elasticity from one part of the cell to another. This wall structure is such that when the guard cells increase in turgidity their inner walls—those bounding the pore—bow away from each other causing a widening of the stomate. In general, therefore, when the guard cells are turgid the stomate is open; when the guard cells are flaccid the stomate is closed.

Shifts in the turgidity of the guard cells causing opening and closing of stomates are conditioned by a number of factors among which the most important are light, temperature, and the internal water supply of the leaf. In general stomates are open in the light and closed in the dark although there are many exceptions to this statement. Exposure of the guard cells to light appears to activate the conversion of starch (a secondary product of photosynthesis) in those cells to sugars. This results in an increase in the osmotic pressure of the cell sap of the guard cells which in turn causes osmotic movement of water into the guard cells from adjoining epidermal cells in which there has been no increase in osmotic pressure. The influx of water into the guard cells causes the increase in their turgidity which in turn results in stomatal opening. Concomitantly there is a reduction in the turgidity of the contiguous epidermal cells. With a failure of illumination the opposite train of events apparently takes place in the guard cells. In general, low temperatures are unfavorable to stomatal opening and when the temperature falls below the optimum for stomatal opening for a given species the stomates will remain closed or open only incompletely even if light conditions and water supply are favorable. Similarly drought conditions, by resulting in a reduction in the water content of the leaves, are usually unfavorable to stomatal opening even if light and temperature conditions are favorable. During prolonged droughts the stomates of many species remain nearly or completely closed most of the time. Night opening of the stomates is of regular occurrence in some species such as certain cacti, in which they are not usually open in the daytime, and may occur in some other species under certain conditions.

When the stomates are open they serve as the principal pathways through which gases diffuse into or out of the leaf; when the stomates are closed all gaseous exchanges between a leaf and its environment are greatly retarded. The gases of greatest physiological importance which enter or depart from a leaf principally through the stomates are oxygen, carbon dioxide, and water vapor. Loss of water vapor in the process of **transpiration** occurs principally through the stomates. Similarly the inward diffusion of carbon dioxide and outward diffusion of oxygen, the gaseous exchanges accompanying **photosynthesis,** occur principally through the stomates. (B.S.M.)

STOMATITIS. Inflammation and infection of the oral cavity—the gums, lips, cheeks. This may be due to poor oral hygiene, infection spreading from infected teeth, or it may be associated with a systemic disease such as **pellagra.** "Trench mouth," or Vincent's infection, is a common form due probably to a fusiform bacillus and a spirochete. It is recurrent and contagious and may occur in epidemics. Treatment with sodium perborate locally and nicotinic acid orally is usually effective, but the disease may be quite resistant. Penicillin has proved to be the most effective drug employed so far. (D.M.H.)

STOMATOPODA. Synonym of Hoplocarida. **Crustacea.** (A.W.L.)

STOMODAEUM. A tube lined with tissue derived from the surface of the body, associated with the alimentary tract (**Digestive System**). The tube forms as a depression in the outer surface and deepens until it meets the lining of the enteric cavity and breaks through into it. The gullet of sea anemones is one form of stomodaeum. In animals with a tubular alimentary tract it forms the anterior part of the tube, consisting of the oral cavity in **vertebrates** and of the entire fore gut of **insects.** This region of the insect alimentary tract includes the oral cavity, **pharynx, oesophagus, crop,** and **gizzard.** (A.W.L.)

STONE CANAL. Water Vascular System.

STONE FLY. Plecoptera.

STONE ROLLER. Pisces, Teleostei. A bottom-feeding fish, *Campostoma anomalum,* of moderate size found in small streams from Wyoming to New York and south to the Gulf. It lives on plant matter. (A.W.L.)

STONECHAT. Aves, Passeriformes. A bird (**Aves**), *Saxicola torquata,* of central and northern Europe. The male has a black head and back, a white collar, and reddish under parts; the female is more brownish. (A.W.L.)

STOP BATHS. The term stop bath is applied to solutions used to stop development. Stop baths may be divided into two classes: (1) acid stop and (2) hardening stop baths.

The acid stop bath in widest use is composed of

| Water | .... 64 oz. | 1 liter |
| Acetic acid, 28% | ..... 3 oz. | 48.0 cc. |

This is used most frequently in the processing of papers where it not only stops development but tends to prevent stains arising from the oxidation of the developer or the formation of reduced colloid silver from the continued action of the developer.

Other stop baths in less common use include:

Sodium bisulfite
Citric acid

Hardening stop baths are particularly useful when processing at high temperatures. A 2–3% solution of chrome alum is the most widely used hardening stop bath. In using a solution of chrome-alum as a stop bath, the following precautions must be observed:

1. The solution must be fresh. Used solutions lose their hardening properties rapidly.

2. The film should be rinsed briefly, but thoroughly, before being placed in the stop bath.

3. The film must be kept in motion for 30 seconds to 1 minute after immersion in the stop bath.

If these precautions are not observed, a deposit of aluminum sulfite will be formed on the surface of the film and the hardening of the gelatin will be adversely affected. Stains of aluminum sulfite are readily removed when wet but very difficult to remove when dry.

The addition of a small quantity of acid makes the formation of aluminum sulfite stains less likely and increases the useful life of the solution but reduces its hardening properties.

Baths of chrome alum should be discarded when they turn green or turbid in appearance. At temperatures above 85° F., the efficiency of a chrome-alum stop bath may be increased by the addition of sodium sulfate (*not* sulfite). The sodium sulfate tends to prevent the swelling of the gelatin and thus increases the rate of hardening. The following formula is suggested.

(Kodak SB-4)

Water	32 oz.	1 liter
Potassium chrome alum	1 oz.	30 grams
Sodium sulfate (dess)	2 oz.	60 grams

This solution when fresh is a violet-blue by tungsten light, turning a yellow-green with use. When the blue-violet color disappears, the solution should be discarded as it will no longer harden the emulsion. (C.B.N.)

STORAGE BATTERY. The storage battery is one or more electric cells having a reaction which is reversible to a high degree. Not only is it necessary that the chemical reaction be reversible but the plates and products of the reaction must be insoluble in the electrolyte. There are two widely used commercial types of storage battery, the lead and the nickel-iron-alkaline (Edison) cells.

The lead cell consists of positive and negative plates, separated by wood, hard rubber, or fiberglass separators, and the electrolyte, the whole being in a hard rubber or glass container. The active material of the positive plates is lead peroxide either in a thin film on the supporting lead plate or as a filler in the holes of a cast lead-antimony alloy grid. The negative plate usually consists of a lead-antimony grid containing sponge lead as the active material. The electrolyte is sulfuric acid and water, producing chemical reactions as follows:

$$PbO_2 + Pb + 2H_2SO_4 \rightleftarrows 2PbSO_4 + 2H_2O.$$

The equation is read from left to right for discharge and the opposite for charge. It will be noted that the discharge converts the electrolyte to water so the state of charge may be determined by the specific gravity of the electrolyte. This is preferable to voltage readings which are very unreliable since the cell has a nearly constant voltage for a wide range of charge conditions. The specific gravity of a fully charged cell is about 1.300 and the voltage is approximately 2 volts.

Lead cells gradually lose their usefulness from a variety of causes. Loss of active material from the plates due to normal dislodging in service, excessive sulphation, local action in the plates, buckling of plates, deterioration of separators, etc., all contribute to the decay of the cell.

The nickel-iron-alkaline cell which was commercially perfected in this country by Edison and often called the Edison cell is composed of a nickel oxide positive plate, an iron negative plate, and potassium and lithium hydroxide solution as the electrolyte. The chemical reactions are considerably more involved than those in the lead cell and the structure of the plates is more complicated, the Edison plates usually consisting of perforated tubes of nickel containing the active ingredients, the tubes being supported in a grid structure. The state of charge of the Edison cell is usually determined by voltage readings, these varying from about 1.4 for fully charged to 1.0 for a practically discharged cell. The ability of this battery to withstand abuse makes it preferable to the lead type for many applications, especially in railroad signal service. The life is also considerably greater than for the lead cell.

In spite of the widespread availability of a-c power there are many applications of the storage battery. Among these may be listed automobile service, railway signalling, railway coach lighting, farm lighting, telephone service, industrial trucks and submarine service. (L.R.Q.)

STORK. Aves, Ciconiiformes. Birds of the Old World and South America. They are large, attaining a length of about 4′, and have long legs and a long straight beak. The common white stork of Europe, *Ciconia alba,* which nests on the tops of chimneys is the best-known species. Others occur in Europe, Asia, Africa, Australia and South America. Among the last are some of the giant storks, also called **jabirus**. The **adjutant** or marabou storks of Africa and India are peculiarly untidy-looking birds, due to the almost bare skin of the head and neck with its few straggling feathers.

The name stork is confused to a slight extent with that of the related herons, as in the case of the whale-headed or shoe-billed species, of the White Nile. This species, called both heron and stork, has an enormous and powerful beak, both broad and deep and provided with a strong hook at the tip. (A.W.L.)

STORM SEWER. Sewerage System.

STRABISMUS (SQUINT OR CROSS-EYE). The inability to fix the sight of both eyes on a given point at the same time; one eye or the other alone sees the object at any given moment, but never both at the same time. Strabismus can be corrected by non-operative measures such as glasses and training. The operative treatment is very successful and is done by the shortening and lengthening of certain eye muscles. (R.S.M., D.M.H.)

STRAIGHT LINE, IN A PLANE. The equation of any straight line is of the first degree, i.e., is a **linear equation**, in the **rectangular coordinates** x and y.

The **locus** of any linear equation (first degree equation) in rectangular coordinates x and y is a straight line.

A straight line parallel to the Y-axis has an equation of the form $x = a$, where a is the X-intercept; a line parallel to the X-axis has an equation of the form $y = b$, where b is the Y-intercept.

The fundamental type forms of the equation of a straight line in rectangular coordinates are the following:

The equation of a line with **slope** m and Y-intercept b is

$$y = mx + b \quad \text{(Slope-intercept form)}.$$

The equation of a line with slope m and passing through the point (x_1, y_1) is

$$y - y_1 = m(x - x_1) \quad \text{(Point-slope form)}.$$

The equation of a line passing through the two points (x_1, y_1) and (x_2, y_2) is

$$\frac{y - y_1}{y_2 - y_1} = \frac{x - x_1}{x_2 - x_1} \quad \text{(Two-point form)}.$$

The two-point form may also be written in the determinant form:

$$\begin{vmatrix} x & y & 1 \\ x_1 & y_1 & 1 \\ x_2 & y_2 & 1 \end{vmatrix} = 0.$$

The equation of the straight line with intercepts a and b is

$$\frac{x}{a} + \frac{y}{b} = 1 \quad \text{(Intercept form).}$$

The equation of the line whose normal (perpendicular) from the origin makes an angle ω with the X-axis and is of length p is

$$x \cos \omega + y \sin \omega - p = 0 \quad \text{(Normal form).}$$

To reduce an equation $ax + by + c = 0$ of a line to the normal form, divide all the terms of the given equation by $\sqrt{a^2 + b^2}$.

In **polar coordinates**, the equation of a line may be written

$$r \cos (\theta - \omega) = p,$$

where p is the length of the perpendicular to the line from the origin and ω is the angle which this perpendicular makes with the X-axis.

Special cases are: If $\omega = 0$, the line is perpendicular to the polar axis and the equation becomes $r \cos \theta = p$. If $\omega = \dfrac{\pi}{2}$, the line is parallel to the polar axis and the equation is $r \sin \theta = p$.

A straight line through the origin (pole) in polar coordinates has the equation $\theta = \alpha$, where α is its inclination angle.

The condition that the three lines whose equations in rectangular coordinates are $a_1x + b_1y + c_1 = 0$, $a_2x + b_2y + c_2 = 0$, $a_3x + b_3y + c_3 = 0$ shall intersect in a common point is

$$\begin{vmatrix} a_1 & b_1 & c_1 \\ a_2 & b_2 & c_2 \\ a_3 & b_3 & c_3 \end{vmatrix} = 0.$$

The angle between two lines whose slopes are m_1 and m_2 is given by the formula

$$\tan \theta = \frac{m_1 - m_2}{1 + m_1 m_2}.$$

The perpendicular distance from a line whose equation is in the normal form $x \cos \omega + y \sin \omega - p = 0$ to the point (x_1, y_1) is given by the formula

$$d = x_1 \cos \omega + y_1 \sin \omega - p.$$

The perpendicular distance from the line whose equation is $ax + by + c = 0$ to the point (x_1, y_1) is given by

$$d = \frac{ax_1 + by_1 + c}{\pm \sqrt{a^2 + b^2}}.$$

If $L_1 = 0$ and $L_2 = 0$ are the equations of two given straight lines, then $L_1 + kL_2 = 0$ is the equation of the system of all straight lines passing through the intersection of the given lines, when k takes all positive and negative values (including o).

If the equations of two given lines are $x \cos \omega_1 + y \sin \omega_1 - p_1 = 0$ and $x \cos \omega_2 + y \sin \omega_2 - p_2 = 0$, the equations of the bisectors of the angles between them are:

$$x \cos \omega_1 + y \sin \omega_1 - p_1 = \pm (x \cos \omega_2 + y \sin \omega_2 - p_2).$$

(L.L.S.)

STRAIGHT LINE, IN SPACE. Consider any line l in space. Through the origin O of a system of rectangular coordinate axes in space draw a line OP parallel to the given line l, and let the direction angles of the line OP be α, β, γ. Then the **direction cosines** of the line l are the cosines of these angles α, β, γ.

If α, β, γ are the direction angles of any line, then

$$\cos^2 \alpha + \cos^2 \beta + \cos^2 \gamma = 1.$$

If a, b, c are numbers proportional to the direction cosines of a line, then the direction cosines are given by

$$\cos \alpha = \frac{a}{\pm\sqrt{a^2 + b^2 + c^2}}, \quad \cos \beta = \frac{b}{\pm\sqrt{a^2 + b^2 + c^2}},$$

$$\cos \gamma = \frac{c}{\pm\sqrt{a^2 + b^2 + c^2}}.$$

If two lines have direction angles α, β, γ, and α', β', γ', then the angle θ between them is given by

$$\cos \theta = \cos \alpha \cos \alpha' + \cos \beta \cos \beta' + \cos \gamma \cos \gamma'.$$

Two lines are parallel when their corresponding direction angles (and direction cosines) are equal.

Two lines are perpendicular when the sum of the products of their corresponding direction cosines is o.

The locus of two simultaneous **linear equations**

$$A_1x + B_1y + C_1z + D_1 = 0, \quad A_2x + B_2y + C_2z + D_2 = 0$$

is a straight line, unless the coefficients of x, y, z are proportional.

A parametric form of the equations of a straight line are:

$$x = x_1 + t \cos \alpha, \quad y = y_1 + t \cos \beta, \quad z = z_1 + t \cos \gamma$$

where P_1 is a point on the line, α, β, γ are direction angles of the line, and t is a variable parameter.

The symmetric form of the equations of a straight line is:

$$\frac{x - x_1}{\cos \alpha} = \frac{y - y_1}{\cos \beta} = \frac{z - z_1}{\cos \gamma},$$

where (x_1, y_1, z_1) is a point on the line, and $\cos \alpha$, $\cos \beta$, $\cos \gamma$ are the direction cosines of the line. Another form of the symmetric equations is:

$$\frac{x - x_1}{a} = \frac{y - y_1}{b} = \frac{z - z_1}{c},$$

where a, b, c are numbers proportional to the direction cosines of the line.

The two-point form of the equations of a straight line is:

$$\frac{x - x_1}{x_2 - x_1} = \frac{y - y_1}{y_2 - y_1} = \frac{z - z_1}{z_2 - z_1}.$$

The plane $Ax + By + Cz + D = 0$ and the line

$$\frac{x - x_1}{a} = \frac{y - y_1}{b} = \frac{z - z_1}{c}$$

are perpendicular if

$$\frac{a}{A} = \frac{b}{B} = \frac{c}{C},$$

and are parallel if $aA + bB + cC = 0$.

A plane passing through a given line and perpendicular to one of the coordinate planes is called a projecting plane of the given line. (L.L.S.)

STRAIGHT LINE MOTION. If a **mechanism** has links designed so that a point on one of them moves in a straight line, that mechanism is one of a particular class known as the straight line mechanism if it is devoid of sliding links. The accompanying figure shows the Watt motion, an approximate straight-line mechanism, and the Peaucellier, an exact straight-line motion. The point P, which is located on the intermediate length BC in the Watt straight-line mechanism, moves, within limits set by the mechanism, in what is nearly a straight line. For best results, this point P should be located by the ratio

$$\frac{AB}{CD} = \frac{PC}{PB}.$$

In the Peaucellier mechanism, line *OD* is fixed, links *OA* and *OB* are equal, link *CD* equals *OD*, and links *PB*,

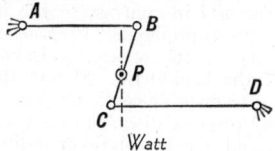

Watt

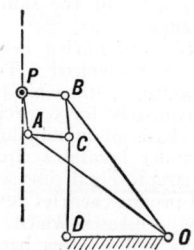

Peaucellier
Straight line mechanism.

BC, CA, and *AP* are equal. Within the limits of its motion, this mechanism traces at point *P* a straight line perpendicular to the line *OD*. (F.T.M.)

STRAIN. Deformation.

STRAIN AXES. Elasticity.

STRAIN-HARDENING. Cold Working.

STRAMONIUM. Potato Family.

STRAND. Wire Rope.

STRANGLING FIG. Fig. Moraceae.

STRAIGHT-EDGE. One of three, any two of which when placed together, coincide throughout their length. (See also **Lapping, Origination.**) (H.C.H.)

STRAIN. Deformation and **Elasticity.**

STRATH. Geographically defined as a broad alluviated valley. Geomorphologically defined as a valley or confluent valleys which represent a local base level of erosion or local and incipient **Peneplain.** (R.M.F.)

STRATIFICATION. Bedding.

STRATIFIED DRIFT. Glacial Deposits.

STRATIGRAPHIC THROW. Fault.

STRATIFIED SAMPLE. Consider a **population** which may be divided into mutually exclusive portions, each part called a stratum, and from each stratum a random **sample** of **variates** is chosen. The resulting sample is called a stratified sample. The stratification may be determined by some logical classification such as geographical location, income group, political affiliation, racial origin, or any other characteristic. Stratified sampling is usually more precise with fewer items needed in the sample, and if it is possible to achieve the stratification with sufficient ease so as not to make the process too expensive, it is quite advantageous. The theory of stratified sampling has been developed chiefly by J. Neyman. Let $\overline{X}$ be an estimate of the **mean,** *m* of the population, then

$$\overline{X} = \frac{n_1\overline{X}_1 + n_2\overline{X}_2 + \cdots + n_r\overline{X}_r}{n},$$

where $\overline{X}_i$ is the sample mean of the *i*th stratum, $n = \sum_{i=1}^{r} n_i$, n_i = number of variates in the sample from the *i*th stratum, when we know T_i = number of variates in the *i*th stratum, and $T = \Sigma T_i$, $\sigma_i^2 = $ **variance** in the *i*th stratum and

$$\sigma_{\overline{X}^2} = \sum_{i=1}^{r} \frac{T_i - n_i}{T_i - 1} \cdot \frac{\sigma_i^2}{n_i}.$$

It is readily proved that $\sigma_{\overline{X}^2}$ is always less than the variance of the mean when using random sampling provided that not all the means of the strata are equal. In all these results it is assumed that T_i and ΣT_i are known or may be very accurately estimated. If this cannot be done it is advisable not to use stratified sampling. To estimate the σ_i^2 in the formula for $\sigma_{\overline{X}^2}$ we may use

$$\sigma_{s.i}^2 \frac{n_i}{n_i - 1} \cdot \frac{T_i - 1}{T_i},$$

where $\sigma_{s.i}^2$ is the sample variance.

An important question is the size of n_i. In general to obtain the optimum precision for $\overline{X}$, it is advisable to make n_i proportional to the standard deviation σ_i and to the number of variates T_i. The advantage of stratified sampling is that $\overline{X}$ is unbiased and possesses a minimum variance. On the whole stratified sampling is readily adapted to many purposes. (L.A.A.)

STRATIGRAPHY. The study of the origin and chronological successions of the observable rocks of the **lithosphere,** in which each lithologic unit is considered to be a formation. The term formation is usually confined to bedded or stratified rocks, including lava flows and volcanic ashes. The major principles involved in the correlation (dating) of formations are: 1. The law of superposition, or that the chronological sequence of any stratigraphic section depends upon the original order in which the formations were laid down; thus the fundamental basis of Stratigraphy is **Structural Geology.** 2. Index Fossils, or those species of fossils whose stratigraphic age are already known. 3. **Lithology.** Igneous rocks may be dated by the age of the sedimentary rocks which they intrude, or overlie; or by **radioactive minerals.** (R.M.F.)

STRATOCUMULUS. Clouds.

STRATOSPHERE. That portion of the earth's atmosphere above the tropopause. This air is free of all weather phenomena being practically without moisture and having in general an isothermal structure. (See **Atmosphere.**) (P.E.K.)

STRATOSTAT. Balloon.

STRATUS. Clouds.

STRAWBERRY. Rose Family.

STREAK. This term, as used by the mineralogist, denotes the color of a powdered **mineral.** Usually determined by rubbing the mineral on a piece of unglazed porcelain, called a streak plate. All metals, and most minerals showing a metallic **luster,** show the same color whether in the solid or powdered form. The **silicates** and most of the minerals having a non-metallic luster, show different colors according to whether each is viewed in mass or as a powder. (R.M.F.)

STREAMLINE. If a fluid has steady flow without large-scale turbulence the paths of elements of the fluid are streamlines. If a body is "streamlined" its shape allows the streamline pattern to follow the surface closely, in contrast to a "bluff body" where the streamlines break away from the rear portions of the body. The more nearly a body conforms to the pattern of natural streamline flow, the less will be its drag. As low drag is usually an asset, streamlining is a mark of excellence in shape design for aeronautics.

Through popular interest in aviation, 'streamlined" was appropriated to mean, broadly, excellence in design, smoothness in organization, elimination of uneconomical appendages, and similar ideas. (See **Drag; Fluid Flow.**) (F.T.M.)

STREAMLINES IN THE ATMOSPHERE. Lines drawn everywhere tangent to wind vectors. They show the instantaneous flow-pattern of the air at a given time only and do not indicate trajectories of air parcels. Streamlines are used for 2-dimensional flows (horizontal plane), although actually they are a 3-dimensional function. The vertical velocity is usually considered insignificant in comparison with the horizontal flow. Streamline charts are useful and quickly prepared forecast tools for obtaining wind forecasts at a given flight altitude. Trajectory charts, however, are more useful for projecting atmospheric phenomena. (P.E.K.)

STREPSIPTERA. An order of peculiar parasitic insects called stylopids by entomologists. They live in the bodies of other insects, including **wasps** and **Homoptera**, and are not likely to be seen without special efforts to find them. The female remains in the host as a grub-like **individual** but the male becomes a winged adult whose hind wings alone are developed for flight. The **metamorphosis** begins with an active **larval** form which seeks and enters a host and undergoes a gradual transformation through several molts leading to the ultimate grub-like form. (A.W.L.)

STREPTOCOCCI. A group of small, round, bacterial organisms of many strains, differing widely in pathogenicity. They are frequently the cause of many acute,

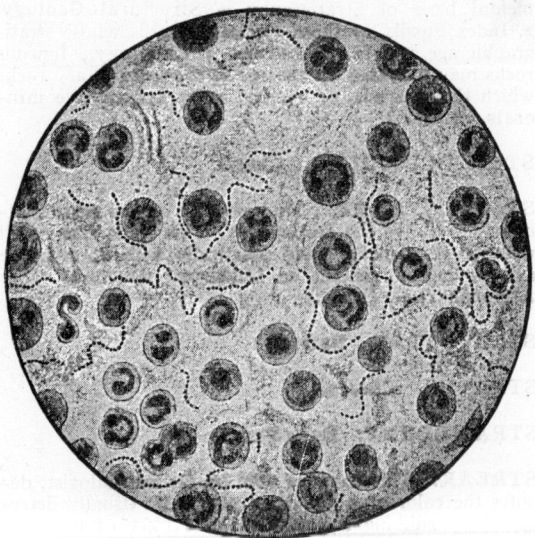

Streptococci from a case of empyema (Jakob). The *streptococci* appear in chain formation lying among pus cells. (*Todd and Sanford, Clinical Diagnosis by Laboratory Methods, W. B. Saunders Co.*)

chronic and often fatal diseases. This group of cocci is characterized by multiplying by division in one plane of space only. Therefore, in the stained preparation on a slide the cocci appear in chain-formation.

Pasteur recognized the *Streptococcus* in 1878 in studies on sepsis following childbirth. Koch saw the organisms in pus from wound infections about the same time.

The classification of streptococci is based on their ability to break up or hemolyze red blood cells. Strains producing a hemolytic **toxin** are thus spoken of as *hemolytic streptococci* in contrast to the non-hemolytic type, and the green-producing *streptococcus viridans*. The pathogenicity of the organisms is correlated with the presence of the hemotoxin. Hemolytic streptococci are the chief disease-producers; *streptococcus viridans* occasionally may produce disease (e.g., subacute bacterial **endocarditis**), and non-hemolytic or indifferent streptococci are harmless. The green and indifferent varieties are normal inhabitants of the skin, throat and other mucous membranes.

Among the disease-producing hemolytic streptococci, several groups are recognized. To *group A* belong most of the organisms pathogenic to man, and within this group approximately forty specific types have been identified on the basis of immunological characteristics. Any type can, under favorable circumstances, produce any one of the streptococcal diseases. Group-A hemolytic streptococci produce **scarlet fever, erysipelas,** and acute **tonsillitis.** Diseases which may be caused by these organisms as well as other bacteria include **osteomyelitis, puerperal sepsis, meningitis, pneumonia, otitis media, impetigo** and any skin or tissue infection. **Rheumatic fever** and **nephritis** are closely related to group-A hemolytic streptococcal infections but the mechanism of this relationship is obscure.

Hemolytic streptococci are very sensitive to **sulfonamides** and **penicillin.** These chemotherapeutic agents have been highly effective in controlling streptococcal disease and curing infections which formerly would have resulted fatally. In the treatment of scarlet fever antitoxic **serum** is available to combat the effects of the potent rash-producing toxin. (R.S.M., D.M.H.)

STREPTOMYCIN. An **antibiotic** produced by the ray **fungus,** *Actinomyces griseus.* As this organism grows in culture media, it liberates a substance which inhibits the growth of certain bacteria. This substance, isolated from the culture filtrate, purified, and concentrated, is streptomycin. It has a low degree of toxicity and, while less effective against the coccal organisms than **Penicillin,** is effective against some organisms against which penicillin is inactive. These include H. influenza, Friedländer's bacillus, B. tularensis, and some gram negative (see **Bacteria**) bacilli. Clinical trial has been limited, but results so far indicate that streptomycin exerts a therapeutic effect in influenzal **meningitis, tularemia,** Friedländer's bacillus **pneumonia,** and **peritonitis,** and urinary tract infections due to gram negative rods. **Tuberculosis** in the experimental animal (guinea pig) improves strikingly with streptomycin therapy. Use of the drug in human tuberculosis suggests that certain forms of the disease may respond, but investigations must be carried much further before its value in tuberculosis can be determined. Further investigation is necessary also to determine its usefulness in typhoid fever, undulant fever, and intestinal infections.

STREPTONEURA. Gasteropoda.

STRESS. A stress is the quantitative expression of a condition within an elastic material due to deformation, or strain, brought about by external forces, inequalities of temperature, or otherwise. Its measure is always the ratio of a force to an area. By some, stress is interpreted as a force distributed over an area, and the above ratio is called the "unit stress." Central, torsional and bending **loads** cause stress. The total resisting force acting at any section of the body divided by the area of the section is the average stress, commonly expressed in points per sq. in. The unit stress is the resisting force on a unit of area. The component of a stress which acts at right angles to a surface is known as the normal stress. If this stress is produced by a load whose **resultant** passes through the **center of gravity** of the area, it is called an axial or direct stress and is always uniformly distributed over the area. A normal resisting force which causes the fibers to increase in length is a tensile stress, while one which shortens the fibers is a compressive stress. The latter is often called a bearing stress. The component of any stress which lies in the plane of the area is a shearing stress. (See **Elasticity**.)

Direct tensile or compressive stresses are known as primary stresses. The bending stress, resulting from deflection, is called a secondary stress. The stresses developed in a column due to the lateral deflections are of a secondary nature. The rigidity of the riveted or welded joints of a truss which has deflected due to the axial **deformation** of its members causes bending stresses in the members which are classified as secondary stresses. The resistance offered by a body to a combination of direct and bending loads is frequently called a combined stress. A normal stress which occurs at a point in a plane on which the shearing stress is zero is known as a principal stress. If this normal stress is tensile it is often called a diagonal tension stress; if compressive it is known as a diagonal compression stress.

The internal resisting force which arises in a restrained body due to temperature changes is a thermal stress. The adhesive resistance which is developed in the concrete surrounding the steel reinforcing rods when a reinforced concrete member is subjected to load is known as **bond** stress. Safe unit resisting forces which are used in design are called working stresses. These are usually taken as a percentage of the **ultimate stress** or the elastic limit of the material.

The stress developed in **bridge** members as a result of **traction** between the wheels of the **live load** and the supporting surface is called a traction stress. The effect of these stresses is usually neglected in highway bridge design but must be considered in railway bridges.

Any stress which is developed in a body before the application of an initial external load is an initial stress. These stresses are usually the result of a manufacturing process or a manufacturing and erection procedure. Shrinkage of **concrete** or other cast materials produce initial stresses. Initial **tension** is a term frequently used to denote the following initial stress conditions:

1. The stress **in the shank of a hot-driven rivet** after it has cooled. Contraction causes compression in the connected materials which, in turn, produces tension in the rivet.

2. The stress **in the shank of a bolt,** used to connect two or more parts, after the nut has been tightened.

3. The stress in riveted or bolted diagonals of bracing systems of bridges and buildings as the result of pulling them into position with **drift pins.** These diagonals are usually fabricated slightly shorter than the theoretical length in order to produce a rigid bracing system after being pulled into their proper place.

4. The stress **in tie rods** of bracing systems caused by tightening the nuts at the end connections. If the tie rod is equipped with a turnbuckle, initial tension is produced by tightening the turnbuckle. (c.w.c.)

STRESS CONCENTRATION. Fatigue.

STRESS CONCENTRATION FACTOR. Laboratory tests and field experience as well as theoretical analysis indicate large concentrations of stress at those points where the cross-section of a member changes abruptly. This condition is not particularly serious except for **repeated loads** because the material will yield, if stressed beyond the **elastic limit.** The yielding causes a redistribution of stress in the vicinity of the discontinuity which will not affect the member as a whole.

In the case of repeated loads a factor is introduced to evaluate the effect of the abrupt change or other discontinuity. This factor, which is called the stress concentration factor, is the ratio between the **endurance limit** of the member without the abrupt change (the same as the endurance limit of the material) and the endurance limit of the member with the abrupt change. Both of these may be found by experimental means. The magnitude of the localized stress may be predicted by multiplying the stress obtained from the conventional formulae by the stress concentration **factor.** (c.w.c.)

STRESS DIAGRAM. Graphical Statics.

STRESSED-SKIN CONSTRUCTION. This term refers to a type of load-carrying structure whose covering, casing, or "skin," carries all or part of internal stresses resisting some applied external load. It is an alternative to a **truss** as a load-carrying structure. Stressed-skin construction may be advantageously employed where a skin or covering must be supplied for reasons other than structural, for by giving appropriate consideration to load-carrying qualities, the skin will serve a dual purpose. An an example, several components of airplanes must have a surface covering for aerodynamic reasons. If this covering can, by proper design, take some of the tension, compression, or shear stress that would otherwise be borne by a structure supporting the skin, then some of that structure may be eliminated with a corresponding saving in weight.

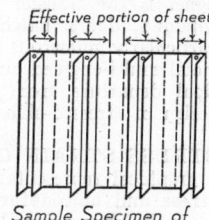

Sample Specimen of Flat Stiffened Sheet

In no other field of construction has stressing of the covering been so widely employed as in the construction of the *airframe* of airplanes. Here costly construction that saves some weight is more readily justified than in most other fields.

In practice, the gauge of metal sheets used as the skin is liable to be so thin that a light network or grid of stiffening members will be employed to give the skin support, shape, and stability. In long tube-like structures (i.e., an airplane fuselage) this grid is composed of longitudinal stringers and transverse stiffening rings or bulkheads. The relative position of these is illustrated in the figure, which shows a simple beam created

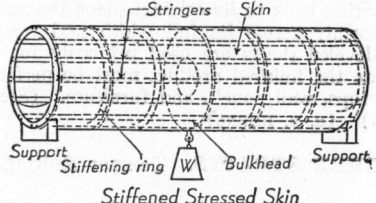

Stiffened Stressed Skin

by a tube whose covering is reinforced by a network of stiffeners.

Stressed skin will resist tension very well, up to the ultimate strength of the material, but shear and compression limits are met at considerably reduced values because of the early failure, under such stresses, due to local instability, wrinkling between stiffeners, etc. The presence of wrinkles under compression does not necessarily mean structural failure in the case of thin ductile material such as aluminum alloy. It has been found that a stiffened sheet will carry some additional load even after buckling starts. However, even where some wrinkling would be safe structurally, it is not considered feasible in aeronautic design to permit it because of the loss of optimum aerodynamic shape of the covering, the misalignment of parts which may be attached to the covering, and the psychological effect upon passengers. The design of stiffened stressed-skin structures has developed in both theoretical and research phases of structural investigation. The irregular behavior of structural elements subject to local instability has prevented correlation of theory and test, as has also the dependence of test results upon dimensions, support, etc., of the specimens. However, general methods of attack have been devised which are effective if supplemented by strength tests of specimens whose construction is similar to that which it is proposed to use in the completed design.

A common method of estimating the compressive strength of a stiffened sheet is to consider only part of the sheet as working with the stiffener, but that part to carry a stress equal to that of the stiffener. There have been several research investigations of this problem and reasonably reliable equations have been developed for computing the effective width of the sheet. The strength of the stiffener depends greatly on its gauge, size, and shape; consequently it is well to construct specimens and test to determine the critical buckling stress.

One of the most difficult problems faced by the designer is the transfer of load around cutouts such as windows, doors, hatches, attachments, etc. These are particularly numerous in the fuselage and wing of all-metal multi-engined airplanes and account for the principal departure from rational and definite structural design of stressed-skin airplanes. (F.T.M.)

STRESS-STRAIN CURVE. A stress-strain curve is a graphical representation of the relation between unit **stress** and unit **deformation** in a stressed body as a gradually increasing **load** is applied. (See **Tension Test.**) (C.W.C.)

STRETCHER. Masonry.

STRETCHER LEVELING. Flattening of sheet metals by stretching. A stretcher leveler consists primarily of a set of stationary jaws to hold one end of a sheet and a set of movable jaws attached to a hydraulic piston to hold the other end of the sheet. By means of hydraulic pressure the sheet is stretched gradually, taking up all of the slack that may exist in the form of waves or buckles and continuing until the sheet is stretched very taut and flat. The stretching proceeds, theoretically, past the elastic limit of the steel, in the neighborhood of 1–2% elongation, so that the sheet will not spring back to its original shape but will spring back uniformly across its full width and remain flat. Stretcher-leveled sheets are used primarily for applications where the finished product must be perfectly flat, i.e., table tops, wall panels, and furniture. (See **Roller Leveling.**) (R. S. BURNS)

STRETCHER STRAINS. Temper Rolling.

STRIAE. The scratches on bedrock or on pebbles and boulders which are the result of glaciation. These are called glacial striae to distinguish them from the striae which occur on the surfaces of fault planes. (R.M.F.)

STRIDULATION. Sound Production.

STRIGIFORMES. The owls. An order of birds with the hooked beak and talons of birds of prey but differing from the hawks and related forms in having the eyes directed forward and in their nocturnal habits. (A.W.L.)

STRIKE. For the use of this term in geology, see **Anticline.**

STRIKE-FAULT. Fault.

STRING PLANING. Planer.

STRINGER. Bridge.

STRIPPING. Gas Absorption.

STROBILATION. An asexual process of reproduction by terminal budding. Some of the **annelid** worms reproduce in this manner and the proglottids of tapeworms (**Cestoda**) are formed by a similar process, leading to the application of the term strobila to the segmented portion of the body. The production of jellyfishes by their scyphistoma larvae is another example. (A.W.L.)

STROBILUS. A strobilus is a group of **sporophylls** or **spore**-bearing leaves aggregated together to form a compact cone-like structure. (R.M.W.)

STROBOSCOPE. The stroboscope is an instrument for viewing moving objects so they appear stationary. In its simplest form it may consist of a revolving disk with holes spaced around the edge, the moving object being observed through these holes. Since the object can be seen only when a hole is opposite the eye, a cyclic motion can be made to appear stationary if the speed of the disk is adjusted so a hole comes opposite the eye only when the moving object reaches the same point in each cycle; at other times the disk blocks out the view. As the object is thus always seen in the same position the persistence of vision makes it appear to be still. The principal drawback to this type stroboscope is the blurring of rapidly moving objects since they can be seen for a short time while the disk hole is moving before the eye. Much more accurate stroboscopes utilize a flashing light to illuminate the moving machine. Again, if the light flashes once for each cycle of the motion the eye sees it only in one position, i.e., that at which the flash occurs. Even though the machine may have general illumination so the observer sees the usual blur of a rapidly moving machine, the flash of stroboscopic light causes it to appear stationary. Since the flash can be made of extremely short duration by using gaseous discharge lamps the machine being observed moves a negligible distance while the light is on and thus appears sharply defined. If the light is not quite synchronized the machine appears to move slowly in one direction or the other depending upon whether the light is fast or slow. The greatest value of the stroboscope is in the study of machine parts while they are under the stresses due to motion. As an example, a part may be badly distorted by its motion. If the stroboscope is synchronized with this part's motion, the part will appear stationary, but any distortion will be clearly visible. In the study of wear, distortion, vibration, etc., of rapidly moving machine parts the stroboscope has been an invaluable aid and has enabled investigators to solve experimentally many difficult problems. (L.R.Q.)

STROKE. Apoplexy.

STROMATOLITH. A term proposed by Foye, in 1916, for banded **gneisses** composed of alternate layers of igneous and metamorphic (**schistose**) rocks. Compare with **lit-par-lit.** (R.M.F.)

STROMEYERITE. A mineral sulfide of silver and copper, $AgCuS$. Crystallizes in the orthorhombic system. Hardness, 2.5–3; specific gravity, 6.2–6.3; color, gray to blue with metallic luster; opaque. Named after Fr. Stromeyer (1776–1835), Göttingen. (R.M.F.)

STRONTIANITE. The mineral strontianite is **strontium carbonate**, $SrCO_2$, usually occurring in whitish-yellow or whitish-green masses of radiated acicular crystals, or in fibrous or granular form. When distinctly crystallized it is obviously **orthorhombic**, but such crystals are rare. It has a nearly perfect prismatic **cleavage**;

uneven fracture; brittle; hardness, 3–3.5; specific gravity, 3.68–3.72; luster, vitreous; color, as above, also green, gray and colorless; streak, white; transparent to translucent. Strontianite occurs in veins chiefly in limestones, occasionally in the crystalline rocks, and usually associated with calcite and celestite. It is found in the metalliferous veins in the Harz Mountains and Saxony. It is commercially important in Westphalia where it is mined for use in the beet sugar industry. In the United States, crystalline masses and geodes of strontianite are found in Schoharie County, New York, long a famous locality for this mineral. (E.S.C.S.)

STRONTIUM. Symbol: Sr. Atomic number: 38. Atomic weight: 87.63. Density: 2.6. Melting point: 800° C. Boiling point: 1150° C. (Isotopes: page 290.)

Strontium is a silver-white metal, soft as lead, malleable, ductile, oxidizes rapidly on exposure to air, burns when heated in air emitting a brilliant light and forming oxide and nitride, reacts with water yielding strontium hydroxide and hydrogen gas. Discovered by Hope and by Klaproth in 1793, and isolated by Davy in 1808.

Strontium occurs chiefly as sulfate (celestite, SrSO$_4$) and carbonate (strontianite, SrCO$_3$), although widely distributed in small concentration. The commercially exploited deposits are mainly in England. The sulfate or carbonate is transformed into chloride, and the electrolysis of the fused chloride yields strontium metal.

Acetate: Strontium acetate (Sr(C$_2$H$_3$O$_2$)$_2$), white crystals, soluble, by reaction of strontium carbonate or hydroxide and acetic acid, and then crystallizing.

Carbide: Strontium carbide (SrC$_2$), black solid, by reaction of strontium oxide and carbon at electric furnace temperature, decomposes water yielding acetylene gas and strontium hydroxide.

Carbonates: Strontium carbonate (SrCO$_3$), white solid, insoluble, (1) by reaction of strontium salt solution and sodium carbonate or bicarbonate solution, (2) by reaction of strontium hydroxide solution and carbon dioxide, decomposes at 1200° C. to form strontium oxide and carbon dioxide, is dissolved by excess carbon dioxide, forming strontium bicarbonate; strontium bicarbonate (Sr(HCO$_3$)$_2$), solution, by excess carbon dioxide and strontium hydroxide solution.

Chloride: Strontium chloride (Sr(Cl$_2$·6H$_2$O), white crystals, soluble, by reaction of strontium carbonate or hydroxide and hydrochloric acid, and then crystallizing. Anhydrous strontium chloride (SrCl$_2$) absorbs dry ammonia gas.

Chromate: Strontium chromate (SrCrO$_4$), yellow precipitate, by reaction of strontium salt solution and potassium chromate solution.

Cyanamide: Strontium cyanamide (SrCN$_2$), mixed with cyanide (Sr(CN)$_2$), by heating strontium carbide at 1200° C. with nitrogen gas.

Hydride: Strontium hydride (SrH$_2$), white solid, by heating strontium metal or amalgam in hydrogen gas at 250° C., reactive with water, yielding strontium hydroxide and hydrogen gas.

Hydroxide: Strontium hydroxide (Sr(OH)$_2$), white solid, (1) by reaction of strontium oxide and water, (2) by precipitation of strontium salt solution with sodium hydroxide solution, yields hydrate (Sr(OH)$_2$·8H$_2$O) on crystallizing, decomposes upon heating at about 850° C. to form oxide (SrO) and water.

Nitrate: Strontium nitrate (Sr(NO$_3$)$_2$), white crystals, soluble, by reaction of strontium carbonate or hydroxide and nitric acid, and then crystallizing, used in pyrotechnics for the production of red light.

Oxides: Strontium oxide (SrO), white solid, melting point about 2400° C., reactive with water to form strontium hydroxide; strontium peroxide (SrO$_2$·8H$_2$O), white precipitate, by reaction of strontium salt solution and hydrogen or sodium peroxide yields anhydrous strontium peroxide (SrO$_2$) upon heating at 130° C. in a current of dry air.

Oxalate: Strontium oxalate (SrC$_2$O$_4$), white precipitate, by reaction of strontium salt solution and ammonium oxalate solution.

Sulfate: Strontium sulfate (SrSO$_4$), white precipitate by reaction of strontium salt solution and sulfuric acid or sodium sulfate solution, insoluble in acids, by heating with carbon yields strontium sulfide, by boiling with sodium carbonate solution yields strontium carbonate.

Sulfides: Strontium sulfide (SrS), grayish-white solid, by heating strontium sulfate and carbon, reactive with water to form strontium hydrosulfide solution; strontium hydrosulfide (Sr(SH)$_2$) solution, (1) by reaction of strontium sulfide and water, (2) by saturation of strontium hydroxide solution with hydrogen sulfide; strontium polysulfides are formed by boiling strontium hydrosulfide with sulfur.

Volatile compounds of strontium, such as the chloride, color the bunsen flame a brilliant red. (R.K.S.)

STROPHOID. The strophoid is a plane curve which may be defined geometrically as follows: Through a

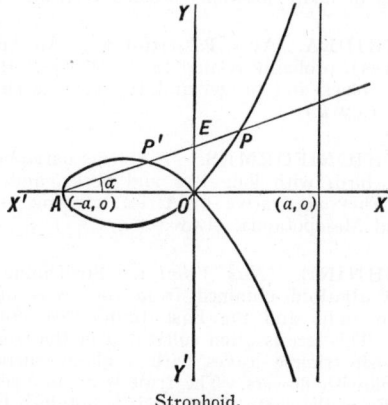

Strophoid.

fixed point $A(-a,o)$ on a set of rectangular axes draw any line AE and on this line locate points P and P' such that $OE = EP = EP'$. As AE rotates about A, points P and P' describe the strophoid.

The parametric equations of the curve are:

$$x = a \sin \alpha, \quad y = a \tan \alpha(1 + \sin \alpha).$$

The equation in rectangular coordinates is

$$x^3 + xy^2 + ax^2 - ay^2 = 0.$$

(L.L.S.)

STRUCTURAL COLOR. Interference.

STRUCTURAL CONTOURS. Sub-surface contours which represent the structure-in-depth of orogenic belts of the lithosphere. Compare with isopach and isobath. (R.M.F.)

STRUCTURAL GEOLOGY. The study of the arrangement, or rearrangement, of the materials which form the lithosphere or outer rocky portion of the earth. The general subject of structural geology is related to a number of other phases of geology, including metamorphism and the structure of the earth as a whole. Ordinarily the term is confined to the study of the deformational or structural features of the observable portion of the lithosphere, including the making of geologic maps (geologic surveying). Structural geology is, therefore, intimately related to stratigraphic geology (Stratigraphy). While the primary object in a problem in structural geology is to determine the character as well as the space, time relationship of the deformations, the primary object in stratigraphy is to determine the order of geological events. From a practical point of view there is a close relation between structural and stratigraphic geology. In **petrology**, the structure of a rock refers to the arrangement of the mineral constituents. A number of different types of folds and faults are described under the subject of structural geology. During recent years the science of **geophysics** has supplied a number of methods and techniques which are highly essential to modern research in structural geology. (R.M.F.)

STRUCTURAL MOUNTAINS. Mountains.

STRUT. A strut is a structural member subjected to **compression.** Its conditions of loading and analysis are the same as for a **column.** If there is any difference between strut and column, it rests on the following points. A column is usually thought of as being a fairly large compression member, vertical in position. Small columns are frequently called struts; also, struts are compression members which are incorporated into structures in many positions besides vertical. (F.T.M.)

STRUTHIDEA. Aves, Passeriformes. An Australian bird (**Aves**), probably related to the shrikes. Its colors are gray, black and brown and the beak is short and curved. (A.W.L.)

STRUTHIONIFORMES. The true **ostriches.** An order of birds with long legs and necks, and vestigal wings. They are native to Africa, ranging as far as Syria and Mesopotamia. (A.W.L.)

STRYCHNINE. *Nux Vomica.* Strychnine is the principal **alkaloid** obtained from the seeds of a tree native to India and the East Indies, *Strychnos Nux Vomica.* This tree is often cultivated in the tropics. It has opposite simple leaves with a glossy surface, and small yellowish flowers. The fruit is an orange-colored berry. From the coats of the seeds is obtained the alkaloid strychnine, one of the most bitter substances known. Strychnine is used as a poison for rodents and vermin; it has long had wide use in medicine but without sound pharmacological reason. Its chief action is stimulation of the nervous system. All parts are affected, but the **spinal cord** is the site of the most striking alterations. Strychnine produces "spinal convulsions," i.e., tonic contractions of all the muscles in response to the slightest stimuli. In therapeutic doses it has no effect on the heart, or on the gastrointestinal tract, and it is not a "tonic," contrary to popular belief. Its only rational use is as a respiratory stimulant when the central nervous system is so depressed that effective doses can be used with safety. This occurs in poisoning with **barbital** and its derivatives.

Strychnine poisoning, either through accident or with suicidal intent, is not uncommon; death from this poisoning is particularly violent and agonizing. Following a poisonous dose of strychnine the muscles twitch, muscular spasms appear and are soon followed by convulsion. During the convulsions, the mind remains acutely clear and there is great apprehension on the part of the patient. The convulsions increase in rapidity and degree and are initiated by any slight stimulus such as turning on a light, or any sudden sound. Death takes place from asphyxia due to paralysis of respiratory centers in the brain and **anoxemia** as a result of continuous spasm of the respiratory muscles.

Immediate treatment for this poisoning is the intravenous administration of a depressant drug, particularly one of the rapid-acting barbital derivatives such as amytal. As soon as the patient is quieted, gastric lavage with tannic acid or strong tea, or **potassium** permanganate can be carried out without danger of setting off another convulsion.

A related South American tree, *Strychnos toxifera,* is a source of **curare,** used by the native Indians as an arrow poison. To obtain the poison the bark is scraped off and macerated in water. (D.M.H.)

STUB SWITCH. Railway Track.

STUB TOOTH. Spur Gearing.

STUBS. Gauge Number.

STUBS, LINE. Line stubs are relatively short lengths of **line,** either open or closed at the end, which are connected across the main line. In telephone practice stubs are usually short lengths of **cable** which are connected at cable junctions to build out the line capacitance to the desired value so a better match (see **Impedance, Match**) can be obtained at the junction. In radio frequency work, particularly at the very high and ultra high frequencies, the stub line serves as a valuable matching aid. In these applications the stub is an appreciable fraction of a wavelength long (order of 1/4 to 1/2 wavelength). Such stubs are frequently used to match a transmission line to an **antenna.** By proper location of an open or shorted stub near the junction of the line and load a match can be obtained, the location and length of the stub depending upon the characteristics of the line and load but being subject to calculation in any given case. For many such problems curves or tables are available to aid in the solution. In microwave work where lines are coaxial or **wave guides** the matching stub is widely used to match the various junctions since the dimensions at these frequencies make them entirely feasible and other more conventional matching means are not satisfactory. (L.R.Q.)

STUD. Screw Fastenings.

STUDENT'S *t* DISTRIBUTION. Student's *t* distribution is given by

$$P_t dt = \frac{\Gamma\left(\dfrac{n}{2}\right)}{\sqrt{n\pi}\,\Gamma\left(\dfrac{n}{2} - 1\right)} \left(1 + \frac{t^2}{n}\right)^{-\frac{n+1}{2}} dt.$$

$-\infty < t < \infty$, $n =$ the **degrees of freedom.** Student's *t* distribution is needed in testing the significance of a **sample mean**, the difference of two sample means, the significance of a **coefficient of correlation**, the significance of regression coefficients and **confidence limits** in these cases. Most of these results are due to R. A. Fisher who gave a rigorous proof of the distribution in 1925, although Student gave the result after an empirical demonstration in 1908. In all **cases** it is assumed that the variates are **normally distributed** with the same **population variance.**

To test the hypothesis that the population mean is *m* we use

$$t = \frac{\overline{X} - m}{s} \sqrt{N}$$

with $N - 1$ degrees of freedom, where $\overline{X}$ is the mean of the sample and s^2 is

$$\frac{\displaystyle\sum_{i=1}^{N} (x_i - \overline{X})^2}{N - 1}.$$

$N =$ number of **variates** in the sample

To test the hypothesis $m_1 = m_2$ we use

$$t = \frac{\overline{X}_1 - \overline{X}_2 - (m_1 - m_2)}{s \sqrt{\dfrac{1}{N_1} + \dfrac{1}{N_2}}}$$

with $N_1 + N_2 - 2$ degrees of freedom, $\overline{X}_1 =$ mean of the first sample, $\overline{X}_2 =$ mean of the second sample, $N_1 =$ number of items in the first sample, $N_2 =$ number of items in the second sample,

$$s^2 = \frac{N_1\sigma_1^2 + N_2\sigma_2^2}{N_1 + N_2 - 2}, \quad \sigma_1^2 = \frac{\sum_{i=1}^{N_1}(x_i - \overline{X}_1)^2}{N_1},$$

$$\sigma_2^2 = \frac{\sum_{i=1}^{N_2}(x_i - \overline{X}_2)^2}{N_2}.$$

We assume the variates are not paired. If the variates are paired, naturally we use $d_i = x_i - y_i$, and test the hypothesis $m_1 = m_2$ by

$$t = \frac{\overline{d}_i}{s}\sqrt{N}$$

with $N - 1$ degrees of freedom where

$$s^2 = \frac{\Sigma(d_i - \overline{d}_i)^2}{N - 1} \quad \text{and} \quad \overline{d}_i = \frac{\Sigma d_i}{N}.$$

To find confidence limits at the α level of significance in all cases above we assume m unknown, $m_1 - m_2$ unknown or finally m_d unknown in

$$t = \frac{\overline{d}_i - m_d}{s}\sqrt{N},$$

and since these other quantities are known we can solve for m, $m_1 - m_2$ or m_d. Thus we get three cases:

1. $$m = \overline{X} \pm \frac{st_\alpha}{\sqrt{N}},$$

where $\overline{X}$ and s is found from the sample and t_α is given by tables of Student's t distribution.

2. $$m_1 - m_2 = \overline{X}_1 - \overline{X}_2 \pm st_\alpha\sqrt{\frac{1}{N_1} + \frac{1}{N_2}}$$

3. $$m_d = \overline{d}_i \pm st_\alpha\sqrt{N}.$$

To test the hypothesis that the population value of a **coefficient of correlation** ρ is zero we set

$$t = \frac{\sqrt{N - 2}\,r}{\sqrt{1 - r^2}}$$

and use the tables of Student's distribution with degrees of freedom $N - 2$, where N is the number of pairs in the correlation, and r is the sample coefficient of correlation.

To test whether a **coefficient of partial correlation** $r_{12 \cdot 34 \ldots k}$ from N sets of values is significant or not we use

$$t = \frac{\sqrt{N - k}\,r_{12 \cdot 34 \ldots k}}{\sqrt{1 - r^2_{12 \cdot 34 \ldots k}}}$$

with degrees of freedom $N - k$ and the tables of Student's distribution.

It may be asked what should be done if the variates are not distributed normally but form a distribution which is moderately skewed. In such a case, if the degrees of freedom are 8 or more it seems safe to use Student's t distribution. If the variates are distributed in a J-shaped distribution it would be wise to use the 1% level of significance before rejecting a hypothesis unless the degrees of freedom are 50 or more in which case the usual t tables may be used. In case $\sigma_1^2 \neq \sigma_2^2$, Student's t test does not apply. There is no general agreement in this case, but if N_1 and N_2 are both large (more than 100) we may use

$$t = \frac{\overline{X}_1 - \overline{X}_2 - (m_1 - m_2)}{\sqrt{\dfrac{s_1^2}{N_1} + \dfrac{s_2^2}{N_2}}}$$

where

$$s_1^2 = \frac{\sum_{i=1}^{N_1}(x_i - \overline{X}_1)^2}{N_1 - 1}, \quad \text{and} \quad s_2^2 = \frac{\sum_{i=1}^{N_2}(x_i - \overline{X}_2)^2}{N_2 - 1},$$

and refer t to the tables of the normal curve.

Student's t distribution approaches a normal curve as the degrees of freedom increase. (L.A.A.)

STUDENT'S t TEST. Student's t Distribution.

STUFFING BOX. A device for preventing leakage or transfer of fluid between moving parts, usually consisting of a relatively soft packing compressed or confined by an adjustable member called a gland. Stuffing box packings differ from gaskets in that they are used in confined spaces, and do not of themselves withstand stresses due to fluid pressure. In the usual form, the stuffing box consists of a hollow cylinder surrounding the moving (reciprocating or rotating rod or shaft) member; the space between the hole and the rod or shaft is filled with packing compressed by the gland. Packings may consist of relatively plastic material, bonding such substances as cotton fabric, rope, or asbestos, but packings of rubber, leather, pressed graphite, or molded plastics are also used.

In high-speed machinery, such as turbines and rotary compressors, where no adequate cooling is available, and in high-pressure equipment where small clearances are required, and for which packings are inadequate, devices known as labyrinths may be used. A labyrinth consists of a series of projections on the rotating element, running in close contact with grooves on the stationary element. To enable a labyrinth to function as a seal, there must be some fluid flow; the fluid first passes a restriction and then expands into a chamber, which consumes a certain amount of energy. After a series of such expansions, a considerable pressure drop will exist between the initial and terminal points of the labyrinth. In order to render the device operable, there must always be some leakage at the terminal point. The effective operation of a labyrinth is less marked for liquids than for vapors. (See **Fluid Seals**.) (H.C.H.)

STUPOR. Partial unconsciousness. In stupor the patient can be aroused as a rule by various stimulating measures. Stupor is seen in certain central nervous system diseases, psychoses, and severe infections, but is more often associated with overdoses of drugs or alcohol. (D.M.H.)

STURGEON. Pisces, Chondrostei. Fishes (**Pisces**) of moderate to very large size, found in the oceans and fresh waters of the entire northern hemisphere. Their chief external characteristic is the series of bony plates arranged in rows along the back and sides, separated by wide spaces containing only small hard elements.

The sturgeons are excellent food fishes and are the source of caviar.

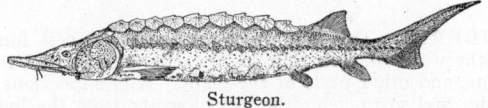
Sturgeon.

Among the true sturgeons the small **sterlet**, *Acipenser ruthenus*, and the giant hausen, *A. huso*, of Asiatic and European waters are noteworthy. The group also includes the shovel-beaked sturgeons, *Scaphirhynchus*, of the Mississippi River system and the rivers of central Asia, and the paddle fishes, *Polyodon*, of the same order are sometimes called toothed sturgeons. (A.W.L.)

STURM'S THEOREM. A method for locating the **real roots** of a **polynomial equation** is given by Sturm's theorem, which may be stated as follows:

Let $P(x) = 0$ be a polynomial equation with real coefficients and without multiple roots. Modify the usual process (Euclid's algorithm) for finding the **highest common** factor of $P(x)$ and its first **derivative** $P_1(x)$ by exhibiting each remainder as the negative of a polynomial P_k, thus:

$$P = q_1 P_1 - P_2, \quad P_1 = q_2 P_2 - P_3, \quad P_2 = q_3 P_3 - P_4, \cdots,$$

$$P_{n-2} = q_{n-1} P_{n-1} - P_n,$$

where P_n is a constant $\neq 0$. If a and b are real numbers, with $a < b$, neither of which is a root of $P(x) = 0$, then the number of real roots of $P(x) = 0$ between a and b is equal to the excess of the number of variations of sign of

$$P(x), P_1(x), P_2(x), \cdots, P_{n-1}(x), P_n,$$

for $x = a$ over the number of variations of sign for $x = b$. Terms which vanish are to be dropped out before counting the variations of sign. (L.L.S.)

STYE. A staphylococcal infection of a **sebaceous gland** on the margin of the eyelids. Styes are frequently seen with errors of refraction of the eye. (D.M.H.)

STYLE. Flower.

STYLET. Small sharp structures used for piercing. The term applies to the calcareous **spines** of the **proboscis** of some **nemertine** worms and to the slender piercing organs of some insect mouths. The latter are modified **mandibles** and **maxillae**. (A.W.L.)

STYLOLITE. A columnar-like structure, at right angles, or highly inclined to the bedding planes of certain **limestones**, believed to be produced by differential vertical movements induced under great pressure. The term is derived from the Greek words meaning column and a stone. (E.S.C.S.)

STYLOMMATOPHORA. Gasteropoda.

STYLOPID. Strepsiptera.

SUB-LAND DRILL. Counterbore.

SUBAERIAL FAN. Alluvial Fan.

SUBCUTANEOUS. Situated under the skin, as a subcutaneous **tumor** or **abscess**, etc. (R.S.M.)

SUBCUTICULA. The **tissue** beneath the **cuticula** of **flukes** and tapeworms (**Cestoda**). In these parasitic forms of flatworms (**Platyhelminthes**) the body is covered with a resistant noncellular cuticula and the ectodermal cells have sunk into the **parenchyma,** so that the layer immediately within the cuticula is composed of a mixture of ectodermal and mesodermal cells and muscle fibers. (A.W.L.)

SUBERIN. This is a waxy substance which is found in the walls of cork cells in the outer tissues of stems, roots, and other parts of the plant. It is impervious to water and so prevents the loss of water from the living cells of the plant. Suberinization also prevents the entry of food, thus causing the early death of cork cells. The walls are comparatively thin. It is very similar to cutin, which forms on the walls of epidermal cells and makes them waterproof. (R.M.W., B.S.M.)

SUBLIMATION. This term applies to the transition, under suitable conditions, directly between the vapor and the solid state of a substance. If solid iodine is placed in a tube and slightly warmed, it vaporizes and the vapor reforms into crystals on the cooler parts of

the tube. Many crystalline substances, both metallic and non-metallic, may be similarly sublimated in a vacuum; fairly large crystals of selenium have been thus prepared. The most familiar sublimates are **frost** and **snow.** As in the case of other changes of state, sublimation is accompanied by the absorption or evolution of heat, the quantity of which per unit mass is called the "heat of sublimation" of the substance. At pressures near the triple point (see **Three-Phase Equilibrium**) the heat of sublimation is approximately equal to the sum of the heats of fusion and vaporization. (See **Vapor Pressure;** and **Distillation, Evaporation, and Drying.**) (L.D.W.)

SUBMAXILLARY GLAND. Salivary Gland.

SUBMENTUM. Labium.

SUBNEURAL GLAND. A gland found below the brain in the **ascidians.** Its duct leads to the ciliated funnel, an organ supposed to be sensory. The function of the gland is unknown. (A.W.L.)

SUBNORMAL DISPERSION. If the **Lexis ratio,** D, is less than 1, the distribution is said to possess subnormal dispersion. (See **Coefficient of Dispersion.**) (L.A.A.)

SUBNORMAL OF A CURVE. Tangents and Normals to a Plane Curve.

SUBSEQUENT STREAMS. Consequent Streams.

SUBSIDENCE. Subsiding air is sinking air and is associated with lateral divergence. Subsidence in the atmosphere is a stabilizing influence; it also decreases relative humidity within the sinking air as it warms the air. Atmospheric pressure usually rises under the influence of subsidence which is normally associated with anticyclones. Clear or partially clouded skies are the usual weather in a region of subsidence. (P.E.K.)

SUBSILICIC. A term proposed by Clarke, in 1911, to replace "basic." (R.M.F.)

SUBSOIL. Soil.

SUBSTATION. An electrical substation is a point in the transmission system, outside of the main generating station, where the power may be transformed and switched to render more flexible and reliable service. Depending upon their purpose in the transmission scheme, substations may be classified as transmission, distribution and industrial. The transmission substation consists merely of **transformers** and **circuit breakers** for connecting the secondaries to various outgoing lines. For more reliable service, by offering better switching facilities, the station is usually somewhat more complex than this, having various combinations of breakers and feeders by using auxiliary bus systems. The equipment and its arrangement in the distribution substation varies widely between different stations, being determined by the reliability of service deemed necessary, the type of distribution system served, etc. The primary function of such a substation is to receive the power from the high-voltage transmission line and transform it to the distribution voltage (usually of the order of a few thousand volts) and distribute this power to the various distribution feeders. The substation may contain voltage-regulating equipment and always contains certain fault-protecting devices, the complexity being governed by the type load being served from the station. A typical outdoor substation is shown. Industrial substations are designed to meet the needs of a particular plant and contain the same general components as the others. The various pieces of equipment composing the substation may be located in a substation building or may be of weatherproof construction and located out-of-doors. The latter arrangement offers certain advantages, among them the avoiding of arranging high-voltage lines inside the building, avoiding of the high-voltage entrance connec-

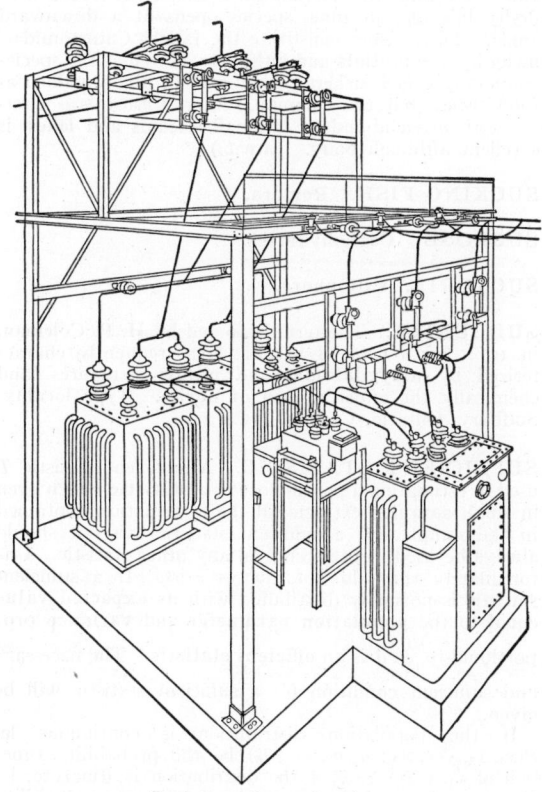

Typical outdoor substation.

tions, the lesser cost since no building is required, etc. Outdoor substations have the breakers and transformers in weatherproof encloures with the **relays**, operating mechanisms, etc., usually in cabinet type structures along the side. The connecting buses are arranged in a neat system of parallel and perpendicular buses supported from steel framework. Distribution substations are frequently located in residential sections of cities so considerable care is taken to make them pleasing in appearance, the electrical apparatus being neatly arranged, the grounds landscaped, etc. (L.R.Q.)

SUBSTITUTE MEMBER. Graphical Statics.

SUBSTITUTION METHOD. Physical Measurements.

SUBSURFACE WATER. Ground Water.

SUBTANGENT TO A CURVE. Tangents and Normals to a Plane Curve.

SUBTRACTION. Subtraction is the **inverse operation to addition**. The result of subtracting one number from another is called the difference of the two numbers. The difference $a - b$ between two numbers a and b is that number c such that $b + c = a$. It follows that $(a - b) + b = a$ and $(a + b) - b = a$. That is, the operations of addition and subtraction are mutually destructive: each undoes the effect of the other.

To subtract one number from another, change the

sign of the number to be subtracted and proceed as in addition. (L.L.S.)

SUBTRACTIVE COLOR PROCESS. A method of photographic color synthesis using two or more superimposed colorants which selectively absorb their complementary colors from white light.

Most modern processes of color photography make use of a subtractive synthesis to yield prints or transparencies. In a 3-color process the colorants cyan, magenta,

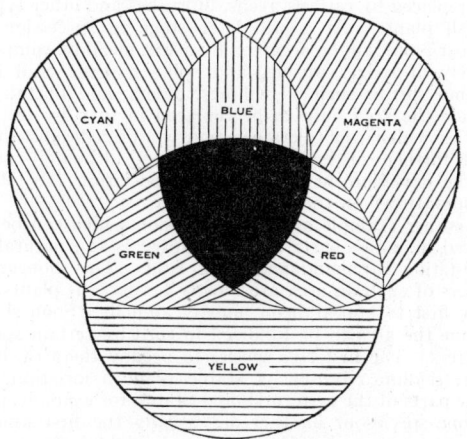

(*Neblette's Photography.*)

and yellow are used to control the amounts of red, green and blue in a beam of white light. This beam of white light may be either that of a projector with its color transparency, or the light reflected from a white support, such as paper, on which the color reproduction is printed. In the first case, the light passes through the colorants once, while in the print viewed by reflection the light must traverse the colorants twice.

The colorants are positive or negative images which have been prepared by one of the methods discussed under **color image formation.** A cyan positive image (a cyan colorant controls red light), for example, may be prepared from the negative that recorded the red present in the subject. The magenta and yellow images are likewise made from green and blue record negatives. These colorant images are superimposed in register to yield the final reproduction. The three colorants may be in separate removable layers or they may be physically inseparable as in the modern integral tripacks (see **Bipacks and Tripacks**). The contrast of the colorant images must be approximately double for a picture to be viewed by transmitted light as compared to one to be viewed by reflection. The accuracy of color reproduction by a subtractive synthesis as compared to an additive is chiefly dependent on how satisfactorily the three colorants cyan, magenta, and yellow fulfil their role, as red, green and blue absorbers respectively. Color correction is often adopted to improve the accuracy of reproduction when using the colorants generally available. (H.C.C.)

SUBTRACTIVE REDUCERS. Photographic Reduction.

SUBUMBRELLA. Medusa.

SUCCESSION, PLANT. This is the term used to designate the gradual replacement of one plant association by another. Succession is caused by slow changes in the environmental factors which influence the establishment, development and survival of plants. Many of these changes, such as shading, increase in the humus content or porosity of the soil, decrease in soil temperature, etc., are brought about by plants themselves.

A good example of a plant succession occurs in the filling of a shallow pond. At first the deeper parts of such a pond are occupied only by submersed aquatic species of plants. As the pond becomes shallower, as a result of the accumulation of plant remains and the deposition of the silt caught in such remains, floating-leaf species such as water lilies invade the area and largely shade out the underwater species. With further decrease in the depth of the water the floating leaf aquatics are replaced by cattails, reeds, bulrushes and other typical marsh plants. At a still later stage in succession the former pond will be occupied by a wet meadow composed largely of sedges, rushes, and spike rushes. Still later swamp shrubs such as willows, alders, and buttonbush invade the area only to be replaced in turn by trees of species which can tolerate a poorly drained substratum. Still further successional stages may ultimately result in the development of a climax association (see **Plant Association**) on the area.

Another example of plant succession which can be observed in many parts of the country is the natural reforestation of abandoned farm land. The pioneer invaders of such an area are mostly herbaceous plants, the very first to appear being mostly annuals. Soon shrubs invade the area to be followed in turn by certain species of trees. The first tree species to occupy the area, however, seldom represent a stable plant association. In large parts of the eastern United States, for example, pines of one species or another are usually the first kind of tree to occupy abandoned fields, but they are in time replaced by other species, most commonly by oaks or hickories or both. In many regions still further successional stages will be passed through before a stable climax association is attained.

Other examples of kinds of areas on which plant succession is initiated are sand dunes, burned over forest lands, cliffs or raw talus slopes, and the bottoms of lakes which have been exposed by drainage. Important consideration should be given to the principles of plant succession in forest and range management and in other land utilization problems. (B.S.M.)

SUCCESSIVE DERIVATIVES. Higher Derivatives.

SUCCESSIVE INTEGRATIONS. Double Integrals and Triple Integrals.

SUCCULENTS. Succulent plants are those which are fleshy, that is, have their stems or leaves greatly enlarged to serve as water-storage organs. Succulents occur wild in regions in which there is a very limited supply of available water, as in the desert lands of western America, central Asia, and many parts of Africa. Among stem succulents the outstanding representatives are the members of the **Cactus** family, with the milkweed and **spurge** families also ranking high. Leaf succulents include many plants such as *Sedums, Crassulas, Bryophyllums,* and *Gasterias.* Many of the leaf succulents are plants of extremely curious habit. In some, the leaves are so swollen with stored water that they form a nearly spherical mass growing almost buried in the ground; in others the masses of leaves form compact rosettes at the tips of the branches; while in others the swollen leaves appear like small green beads along a slender stem.

Because of their odd and often ornamental habit, succulent plants are frequently grown in cultivation, especially as house plants. Many are especially desirable for this purpose, since they require little care and are remarkably tolerant to all conditions except too much water, which will cause quick rotting and death to many of them. Many of this group of plants produce large gaudy flowers which add to their popularity as house plants. (R.M.W.)

SUCKER. Pisces, Teleostei. A fish (**Pisces**) which lives near the bottom of streams, feeding on vegetation and small animals. The mouth is usually provided with fleshy lips and in some species opens at a downward angle. These fishes constitute the family Catostomidae, including the **mullets** and **buffaloes** as well as the species commonly called suckers. They are not highly valued as food fishes, but the common sucker, *Catostomus commersoni,* often abundant in small streams and lakes, is excellent although bony. (A.W.L.)

SUCKING FISH. Remora.

SUCROSE. Carbohydrates.

SUCTORIA. Ciliophora.

SUDBURITE. The term proposed by H. P. Coleman, in 1912, for a type of basaltic lava frequently characterized by **amygdaloidal** and **pillow** structures, and chemically the effusive form of **norite.** Type locality, Sudbury, Ontario, Canada. (R.M.F.)

SUFFICIENT STATISTIC. A sufficient statistic T used in estimating a parameter θ is a **statistic** which even in small **samples** extracts all the information contained in the sample. If a sufficient statistic exists it should always be used in preference to any other statistic. Unfortunately they do not always exist. If a sufficient statistic is normally distributed with its **expected value** equal to the **population parameter** and **variance** proportional to $\frac{1}{N}$, it is an **efficient statistic.** The necessary and sufficient condition for a sufficient statistic will be given.

If the population distribution is continuous let $P(x_1, x_2, \cdots, x_N; \theta_1, \theta_2, \cdots, \theta_h)$ be the probability function of $x_1, x_2, \cdots, x_N$; if the distribution is **discrete**, let $P(x_1, x_2, \cdots, x_N; \theta_1, \theta_2, \cdots, \theta_h)$ be the discrete probability of obtaining $x_1, x_2, \cdots, x_N$. In either case a necessary and sufficient condition that T be a sufficient statistic for estimating θ_1 is that P factor as follows: $P(x_1, x_2, \cdots, x_N; \theta_1, \theta_2, \cdots, \theta_h) = g_1(T; \theta_1, \theta_2, \cdots, \theta_h) \cdot g_2(x_1, x_2, \cdots, x_N; \theta_2, \theta_3, \cdots, \theta_h)$. (L.A.A.)

SUGAR. Sugar is the name given to a group of chemical compounds included in the **carbohydrates.** They are composed of carbon, hydrogen, and oxygen, usually in the proportions of $C_6H_{12}O_6$ or $C_{12}H_{22}O_{11}$. Sugars are colorless compounds, either solid or liquid, and many form aqueous solutions having a sweet taste. Commercial sugar is sucrose or saccharose, which has the chemical formula $C_{12}H_{22}O_{11}$, and occurs in many plants, mostly in amounts too small to be of importance commercially. A few plants, however, contain this sugar in their sap in considerable quantity. The most important sugar-containing plants are sugar cane and the **sugar beet.** Sweet **Sorghums,** several **Palms,** and Sugar Maples also yield sugar.

SUGAR CANE. *Saccharum officinarum.* Gramineae. This plant is a perennial grass, growing best in regions having a continuous hot climate and abundant moisture. The plant possesses a shallow fibrous root system and thick underground rhizomes from which rise the erect stems 1-2" in diameter and 6-15' in height. These stems are many-jointed, solid, and bear at the nodes broad leaves with clasping bases. Each leaf is 3 or more feet long. The inflorescence is a loose panicle, which is rarely produced on plants grown outside the tropics. The **spikelets** of the **panicle** are two-flowered, the lower one sterile, the upper fertile. At the base of the spikelet is a tuft of long silky hairs. The grain is very small, rather silky, and rarely viable, especially in plants grown in less favorable regions.

Propagation of sugar cane is therefore done by means of stem cuttings rather than seed. In this propagation, young stems are cut up into sections, each containing three buds. These sections are planted in holes in the ground and covered with about 2″ of soil. Sprouts appear above ground in about two weeks and grow rapidly. The crop is ready for the first harvest in 12-15 months; successive crops may be gathered yearly thereafter for a number of years, varying greatly with the nature of the soil.

In harvesting the stout stems are cut down, stripped of their leaves and top, and hauled to the factory as soon as possible. There the canes are passed between heavy variously toothed rollers, which crush and shred them. The crushed material is passed through a series of heavy rollers which squeeze out much of the juice. Hot water is sprayed over the stem mat as it passes through the rollers, ensuring extraction of greater amounts of the juice. The residue remaining after extraction is called bagasse, and is used as a fuel to run the mill. The extracted juice is strained to remove solid materials, then heated and allowed to stand in tanks so that any dirt present in the liquid may settle out. The clear liquid is then treated with lime, or other chemicals, which remove additional impurities, which unite with the lime and rise to the surface of the liquid as a scum, or settle to the bottom as a sediment. The clear liquid is then evaporated under partial vacuum to a sirup. Boiling is continued in vacuum pans until the sirup becomes very thick and concentrated, at which time small crystals begin to appear. These crystals increase in size as more liquid is added and boiled down. When the desired crystal size is reached, the concentrated sirup is removed to centrifugal machines consisting of perforated cylindrical baskets within heavy jackets. Rapid revolving of these baskets causes the liquid portion of the sirup to be thrown out through the holes to the sides of the container. The crystalline portion remains in the basket. This is washed to remove any sirup which remains as a film over the crystal surface, then removed and dried. The use of sugar is too well known to need mention. It is an important source of energy and also much used to impart a sweet flavor to substances which might be unpleasant otherwise. As a by-product of this process a thick viscous brown liquid is obtained. This is molasses, which is used for cooking, for stock feeding, and for the manufacture of rum and alcohol.

Sugar Maple. *Acer saccharum*. Another source of sugar, of limited importance commercially, is the sugar maple, *Acer saccharum*, a deciduous hardwood tree of eastern North America. The sugar is obtained from the sweet sap of the tree. To obtain this sap, small holes are bored into the sapwood near the base of the trunk, a channeled spout driven into each hole, and a bucket hung on the spout. Clear, snappy, cold nights and warm quiet days favor the flow of sap, which may run 3 or 4 gallons to a tree daily for a period of 3 or 4 weeks. The colorless sap is poured into slightly tilted flat pans so corrugated that the sap gradually flows downward. Wood fires under the pan cause the thin layer of sap to evaporate rapidly, producing a thin brownish sirup which is mostly sucrose, but characteristically flavored by substances contained in the sap. This sirup may be further concentrated until it crystallizes as a brown sugar, which is usually prepared in small cakes. From a good tree 2 lbs. of sugar or a quart of thick sirup may be obtained each year. Maple products are much used to flavor tobacco, and in candy manufacturing. Maple sirup is frequently mixed with other sirups before marketing. The principal producing regions are Vermont and New York, with other New England states and the states along the Great Lakes producing the rest.

Honey. Another source of sugar is the nectar found in many flowers. The amount found in each flower is too minute to be directly obtained by man. Certain insects gather this nectar in quantities. One insect in particular, the honey-bee, is an especially industrious collector of nectar. This they store in waxen cells, first having partially digested the nectar. Bees can be made to store much more honey than will be needed for carrying them through the winter. This surplus honey is then available to man. Certain flowers, notably clover, buckwheat, and lindens, give to honey obtained from them a particularly attractive fragrance. Honey is used most commonly as a table luxury. (R.M.W.)

SUGGESTION. Producing a condition in which activity, conduct, or thought is determined by impressions or ideas imparted by others. (See Hypnosis.) (D.M.H.)

SUINA. A division of the hoofed animals including the pigs and peccaries. They differ from the other members of the order Artiodactyla, to which they belong, in the absence of horns, in the thick skin and bristly hair, and in the fact that they are not ruminants. (A.W.L.)

SULFA DRUGS. Thioalcohols and Related Compounds.

SULFADIAZINE. Sulfonamide Drugs.

SULFAGUANADINE. Sulfonamide Drugs.

SULFAMERAZINE. Sulfonamide Drugs.

SULFANILAMIDE. Sulfonamide Drugs.

SULFAPYRIDINE. Sulfonamide Drugs.

SULFATE. Sulfuric Acid.

SULFATHIAZOLE. Sulfonamide Drugs.

SULFIDE. Hydrosulfuric Acid.

SULFINIC ACIDS. Thioalcohols and Related Compounds.

SULFINYL COMPOUNDS. Thioalcohols and Related Compounds.

SULFITE. Sulfurous Acid and Sulfites.

SULFONAMIDE. Thioalcohols and Related Compounds.

SULFONAMIDE DRUGS. A group of chemotherapeutic agents capable of curing serious generalized or systemic bacterial infections in man. In 1935, Domagk, a German investigator, was the first to observe the clinical value of *prontosil*, a red compound derived from azo dyes. Para-amino-benzene-sulfonamide was shown to be the effective portion of the prontosil molecule, and this substance was given the name sulfanilamide.

Sulfanilamide was the first of the group to receive wide clinical trial. It was found to be effective in the treatment of hemolytic streptococcal and staphylococcal infections. Within the space of a few years the following related drugs were the most important ones to be synthesized and given clinical trial: *sulfapyridine, sulfathiazole, sulfaguanadine, sulfadiazine* and *sulfamerazine*. They act by inhibiting the growth of bacteria rather than by killing organisms. All of them are now in use for the treatment of various infections. Sulfadiazine at present is the most widely used because it is highly effective and less toxic than most of the others. It is used in the treatment of hemolytic streptococcal and staphylococcal infections, in **puerperal sepsis, pneumonia, gonorrhea,** meningococcus, **meningitis,** the **dysenteries, cystitis, pyelitis** and many other acute infections.

Until the advent of **penicillin,** the sulfonamides were supreme in chemotherapy. Penicillin and the sulfonamides are, in general, effective against the same organisms, but penicillin has the advantage of being nontoxic. The current widespread use of sulfonamides without the direction of a physician is decried because of the serious and sometimes fatal toxic reactions which may ensue. Hemolytic **anemia, agranulocytosis,** hyperleukocytosis, **purpura, hemorrhagica,** various skin eruptions, **hepatitis, psychoses,** kidney stones, **hematuria,** complete suppression of kidney function and anuria may occur during the course of therapy. (D.M.H.)

SULFONIC ACIDS. Thioalcohols and Related Componds.

SULFONYL COMPOUNDS. Thioalcohols and Related Compounds.

SULFUR. Symbol: S. Atomic number: 16. Atomic weight: 32.06. Isotopes: 32(96%), 33(1%), 34(3%). The chemical element sulfur is known in two different forms, namely, (1) alpha-**rhombic** sulfur, yellow, density 2.07, melting point 112.8° C. to form gamma-sulfur; (2) beta-**monoclinic** sulfur, yellow, density 1.96, melting point 119° C. to form gamma-sulfur. Below 96° C. beta-monoclinic changes to alpha-rhombic. The boiling point of sulfur is 444.6° C., as a brownish-red vapor turning deep red at 500° C. and straw yellow at 650° C. The discovery of sulfur is prehistoric. Sulfur is largely consumed (1) in insecticides, either as "flowers of sulfur" or as calcium sulfide, "lime-sulfur," (2) in making sulfur dioxide gas and sulfites, such as calcium hydrogen sulfite plus sulfurous acid solution used as the cooking liquor for making sulfite paper pulp from wood (see **Paper**), (3) in making sulfur trioxide, **sulfuric acid,** and sulfates.

Free sulfur occurs in the alpha-rhombic form; brittle; fracture conchoidal; hardness 1.5–2.5; specific gravity 2; luster, resinous; color, usually yellow, but may be brownish, reddish, or greenish, transparent to translucent.

Alpha-rhombic and beta-monoclinic sulfur are soluble in **carbon** disulfide, and when the solution is allowed to evaporate spontaneously in air, the residual sulfur crystallizes as alpha-rhombic. Sulfur burns with a beautiful violet transparent flame of characteristic odor when ignited in air at 250° C. to form sulfur dioxide, and combines with many metals when heated, e.g., **iron** powder to form ferrous sulfide. Color and **viscosity** of sulfur change with change of temperature approximately as follows: maximum of viscosity (50,000 times that of water at 17° C.) is at about 200° C.; at this temperature liquid sulfur is dark red, having arrived at this by passing through stages from pale yellow. Viscosity at 300° C. is about 2000. At about 200° C. the liquid darkens almost to black. If, at a temperature near the boiling point, liquid sulfur is poured into cold water, it forms a soft, elastic, ductile mass like rubber, which in a few days at ordinary temperatures changes to alpha-rhombic sulfur. When liquid sulfur is kept at a temperate only slightly below the melting point for a sufficient time for part of the liquid to crystallize, and the remaining liquid poured off, beta-monoclinic crystals are obtained, which within a few days at ordinary temperatures change to alpha-rhombic sulfur.

Sulfur occurs as free sulfur in many volcanic districts, and may have been formed in part by **sublimation,** by decomposition of hydrogen sulfide, or metallic sulfides, or by organic agencies. It is often associated with **limestones** and **gypsum.** Sulfur is found in Spain, Iceland, Japan, Mexico, and Italy (especially Sicily), which was the producer for the world until about the beginning of the twentieth century, when Herman Frasch, by inventing the superheated water method of mining sulfur, made available the great Louisiana and Texas deposits. This method of mining is at the same time a method of purifying sulfur, as in the process of heating, accompanying materials remain unmelted at the temperature at which sulfur melts and is drawn off; In the Louisiana and Texas deposits the sulfur is associated with gypsum, occurring in the caprock overlying the salt plugs that have pierced the strata underlying the Gulf coastal plain. In the United States sulfur is also found in California, Colorado, Nevada, and Wyoming. Sulfur

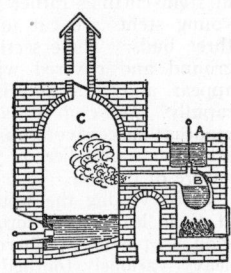

Diagram of furnace for the sublimation of sulfur. *A*, sulfur as charged to furnace; *B*, sulfur melted by heating and vaporized; *C*, chamber maintained below the melting point of sulfur for the direct condensation of sulfur as finely divided solid (sublimed sulfur).

also occurs as (1) sulfides, e.g., iron disulfide, **pyrite** (FeS₂), lead sulfide, **galenite** (PbS), copper iron sulfide, copper pyrite (CuFeS₂), zinc sulfide, **zinc blende** (ZnS), mercury sulfide, **cinnabar** (HgS); and (2) as sulfates, e.g., calcium sulfate, **gypsum** (CaSO₄·2H₂O), barium sulfate, **barite** (BaSO₄).

Acids: Sulfur is a constituent of several acids. In increasing order of oxidation the following are those which contain sulfur and oxygen but not carbon: hydrosulfuric acid (H₂S); thiosulfuric acid (H₂S₂O₃); tetrathionic acid (H₂S₄O₆); hyposulfurous acid (H₂S₂O₄); sulfurous acid (H₂SO₃); dithionic acid (H₂S₂O₆); persulfuric acid (H₂S₂O₈); chlorosulfonic acid (Cl·SO₂OH), liquid, boiling point 155° C., reactive with explosive violence with water forming sulfuric acid plus **hydrochloric acid,** and formed by distillation of concentrated sulfuric acid and **phosphorus** pentachloride or oxychloride; fluosulfonic acid (F·SO₂OH); those which contain carbon (thioic acids): thiocarbonic acid (H₂CS₃); ethane thiolic acid (CH₃COSH); ethane thionic acid (CH₃·CSOH); ethane thionthiolic acid, methane carbodithioic acid (CH₃·CSSH); methyl sulfonic acid (CH₃·SO₂OH); methyl sulfinic acid (CH₃·SOOH); benzene sulfonic acid (C₆H₅·SO₂OH).

Bases: Sulfur is a constituent of some bases, e.g., trimethyl sulfonium hydroxide ((CH₃)₃SOH).

Bromide: Sulfur bromide (S₂Br₂), dark red liquid, boiling point 56° C. at 0.2 mm. pressure.

Chlorides: Sulfur chloride, sulfur monochloride (S₂Cl₂), yellowish-brown liquid, of irritating odor, boiling point 139° C., formed by reaction (1) of **chlorine** and sulfur heated, (2) of chlorine and **carbon** disulfide, and separated from **carbon** tetrachloride by fractional distillation. It dissolves sulfur readily, and is used in the process of vulcanization of rubber; sulfur dichloride (SCl₂), dark red fuming liquid, boiling point 59° C., formed by reaction of sulfur monochloride and chlorine in the presence of a trace of iodine; sulfur tetrachloride (SCl₄), unstable above −20° C., formed by saturating sulfur monochloride with chlorine at −22° C.; thionyl chloride (SOCl₂), colorless liquid, boiling point 78° C., fumes in moist air to yield **hydrochloric acid** and sulfurous acid, formed by reaction of sulfur trioxide and sulfur monochloride at about 80° C.; sulfuryl chloride (SO₂Cl₂), colorless, fuming liquid, boiling point 69° C., with water yields hydrochloric acid plus sulfuric acid, formed by reaction of chlorosulfuric acid heated with a **catalyzer,** e.g., **mercuric** sulfate; thiocarbonyl chloride ("thiophosgene," CSCl₂), red fuming liquid, boiling point

73° C., formed by reaction of carbon disulfide and chlorine followed by treatment with **stannous** chloride.

Fluorides: Sulfur hexafluoride (SF_6), colorless, odorless, tasteless gas, melting point $-55°$ C., remarkably stable, unaffected by fused alkalis, formed by reaction of sulfur and **fluorine**; thionyl fluoride (SOF_2) and sulfuryl fluoride (SO_2F_2) are gases which react with water to give hydrofluoric acid in both cases plus sulfurous and sulfuric acids, respectively.

Hydride: Hydrogen sulfide (H_2S), colorless gas, odor of rotten eggs, poisonous, density 1.539 grams per liter, 0° C., 760 mm., or 1.190 when air equals 1.000, melting point $-83°$ C., boiling point $-60°$ C., not decomposed at 350° C., burns in air to form sulfur dioxide and water, preferred reagent in solution for the preparation of many metallic sulfides, e.g., copper sulfide. Made by action of dilute **hydrochloric** or sulfuric acid on a sulfide, such as **ferrous** sulfide or **calcium** sulfide, and used either as gas or solution. As a solution in water, see Hydrosulfuric acid, below.

Oxides: Sulfur dioxide (SO_2), colorless gas, sharp characteristic odor, density 2.9269 grams per liter, 0° C., 760 mm., or 2.264 when air equals 1.000, melting point $-73°$ C., boiling point $-10.0°$ C., critical temperature 157° C., critical pressure 78 atmospheres, very soluble in water forming sulfurous acid, formed by the burning in air of (1) sulfur, (2) sulfur-containing organic material, e.g., coal, protein, (3) sulfides, and (4) by the reaction of sulfites and an acid. Used (1) as a refrigerant,

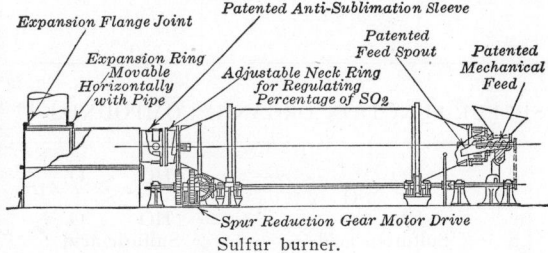

Expansion Flange Joint

Patented Anti-Sublimation Sleeve

Expansion Ring Movable Horizontally with Pipe

Patented Feed Spout

Adjustable Neck Ring for Regulating Percentage of SO_2

Patented Mechanical Feed

Spur Reduction Gear Motor Drive

Sulfur burner.

(2) in the preparation of sulfur trioxide, sulfuric acid, sulfites, (3) as a disinfectant, (4) in bleaching; sulfur sesquioxide (S_2O_3), by dissolving sulfur in sulfur dioxide, or by reaction of sulfur trioxide and hydrazine; sulfur trioxide (SO_3), (1) alpha-sulfur trioxide, colorless liquid, boiling point 45° C., crystallizes to needle crystals of melting point 17° C., fumes in air, reacts with water vigorously, (2) beta-sulfur trioxide (probably S_2O_6), white solid, looks like asbestos, fumes in air, reacts with water somewhat less vigorously than the alpha form yielding sulfuric acid. Sulfur dioxide and oxygen of air react when heated in the presence of a catalyzer, e.g., platinized asbestos or vanadium pentoxide, to yield sulfur trioxide; sulfur heptoxide (S_2O_7), sublimes at 10° C., formed by the reaction of ozone and sulfur dioxide or trioxide, unstable; sulfur tetroxide (SO_4), solid, formed by the action of the silent electric discharge on sulfur dioxide plus oxygen, unstable.

Sulfides: Sulfides of many elements are known. Metallic sulfides are (1) soluble, e.g., **sodium, potassium, ammonium** sulfides (Na_2S, K_2S, (NH_4)$_2S$), (2) reactive with water, e.g., **calcium** sulfide (CaS) to form calcium hydrogen sulfide and calcium hydroxide, **aluminum** sulfide to form hydrogen sulfide and aluminum hydroxide, (3) insoluble, e.g., **silver** sulfide (Ag_2S) brown, **lead** sulfide (PbS) brown, **copper** sulfide (CuS) black, **ferrous** sulfide (FeS) black, **zinc** sulfide (ZnS) white, **arsenious** sulfide (As_2S_3) yellow, and (4) organic sulfides are noteworthy, e.g., methyl sulfide (dimethyl sulfide (CH_3)$_2S$), phenyl sulfide (diphenyl sulfide (C_6H_5)$_2S$), carbon disulfide (CS_2), carbonyl sulfide (COS).

Hydrosulfuric acid and soluble sulfides are important chemical reagents for the precipitation of insoluble sulfides, which are produced and separated from each other by means of their characteristic solubilities in various acids, and in excess sodium or ammonium sulfide.

Carbon disulfide is a colorless, odorous liquid, boiling point 46° C., inflammable, explosive when mixed with

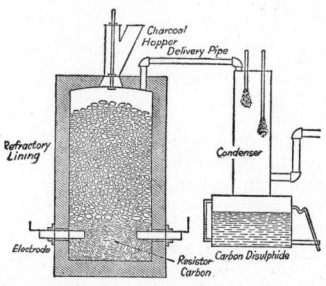

Charcoal Hopper Delivery Pipe

Refractory Lining

Condenser

Electrode

Resistor Carbon

Carbon Disulphide

Electric furnace for the manufacture of carbon disulfide.

air and ignited, and poisonous. Carbon disulfide dissolves sulfur, yellow phosphorus, iodine, bromine, chlorine, hydrocarbons, oils, fats. Made by reaction of carbon (coke) and sulfur in an electric furnace, and recovered by condensation of the vapor. Used (1) as a solvent, (2) as an insecticide, (3) in the manufacture of carbon tetrachloride.

Other compounds of sulfur are discussed as follows:

Benzothiophene. (See **Thiophene and Related Compounds.**)

Dithionates. (See **Dithionic Acid and Dithionates.**)

Dithionic acid.

Hydrosulfides. (See **Hydrosulfuric Acid and Sulfides; and Thioalcohols and Related Compounds.**)

Hydrosulfuric acid.

Hyposulfites. (See **Hyposulfurous Acid and Hyposulfites.**)

Hyposulfurous acid.

Isothiocyanates. (See **Thiocyanic Acid and Thiocyanates.**)

Esters. (See **Sulfuric Acid and Sulfates; Sulfurous Acid and Sulfites; Thiocyanic Acid and Thiocyanates.**)

Mercaptans. (See **Thioalcohols and Related Compounds.**)

Mustard gas. (See **Thioalcohols and Related Compounds.**)

Organic Compounds. See **Thioalcohols and Related Compounds.**

Persulfates. (See **Persulfuric Acid and Persulfates.**)

Persulfuric acid.

Sulfates. (See **Sulfuric Acid and Sulfates.**)

Sulfides. (See **Sulfur, Sulfides, Hydrosulfuric Acid and Sulfides; Thioalcohols and Related Compounds.**)

Sulfinic acids. (See **Thioalcohols and Related Compounds.**)

Sulfinyl-compounds. (See **Thioalcohols and Related Compounds.**)

Sulfites. (See **Sulfurous Acid and Sulfites.**)

Sulfones. (See **Thioalcohols and Related Compounds.**)

Sulfonic acids. (See **Thioalcohols and Related Compounds.**)

Sulfonium-compounds. (See **Thioalcohols and Related Compounds.**)

Sulfonyl-compounds. (See **Thioalcohols and Related Compounds.**)

Sulfoxides. (See **Thioalcohols and Related Compounds.**)

Sulfuric acid.

Sulfurous acid.

Sulfuryl chloride. (See **Sulfur, Chlorides.**)

Tetrathionates. (See **Tetrathionic Acid and Tetrathionates.**)

Tetrathionic acid.

SCHEME SHOWING THE INTERRELATIONSHIPS OF SULFUR-CONTAINING SUBSTANCES

	SULFUR In nature		
Hydrogen sulfide		Sulfur dioxide	Sulfur trioxide
Hydrosulfuric acid		Sulfurous acid	Sulfuric acid Most abundantly used acid
Metallic sulfides Pyrite, galenite, sphalerite, chalcopyrite in nature		Metallic sulfites	Metallic sulfates Gypsum, barite, strontianite, magnesium sulfate in nature
Insoluble sulfides Of silver, lead, mercury; arsenic, antimony, tin, bismuth, copper, cadmium, iron, cobalt, nickel, manganese, zinc		Insoluble sulfites Except sodium, potassium, ammonium	Insoluble sulfates Of lead, mercurous, barium, strontium, calcium
Soluble sulfides Of sodium, potassium, ammonium		Soluble sulfites Of sodium, potassium, ammonium	Soluble sulfates Of sodium, potassium, ammonium
Reactive with water Of magnesium, calcium, strontium, barium			
Organic sulfides Sulfonium-compounds		Organic sulfites	Organic sulfates
		Sulfinic acids Sulfinyl-compounds	Sulfonic acids Sulfonyl-compounds

Thiophene Penthiophene Thioazole Cystine In proteins	Thioic acids Thiocyanic acid Thiourea Thiocarbonic acid	Thiosulfuric acid Tetrathionic acid Hyposulfurous acid	Dithionic acid	Persulfuric acid

SCHEME SHOWING INTERRELATIONSHIPS OF SULFUR-FUNCTION ORGANIC COMPOUNDS

Related to	$\begin{smallmatrix}H\\ \\H\end{smallmatrix}\!\!>\!S$ Hydrogen sulfide	$\begin{smallmatrix}HO\\ \\HO\end{smallmatrix}\!\!>\!SO$ Sulfurous acid	$HO\!\!>\!S\!\!<^{O}_{O}$ (HO) Sulfuric acid
$>$CH of benzene	Thiophene Penthiophene		
—CH_2OH $>$CHOH $\geqslant$COH of alcohols and phenols	Thioalcohols (—SH) Thioethers ($>$S) Tertiary sulfonium compounds Thiophenols		
—CHO $>$CO of aldehydes and ketones	Thioaldehydes Thioketones		
H—COOH Carboxylic acids	Thioic acids: Thiolic $\left(-C\!\!<^{O}_{SH}\right)$ Thionic $\left(-C\!\!<^{S}_{OH}\right)$ Thionthiolic $\left(-C\!\!<^{S}_{SH}\right)$	Sulfinic acids $\left(-S\!\!<^{O}_{OH}\right)$ Sulfinyl (Sulfoxides) ($>$SO)	Sulfonic acids $\left(-S\!\!<^{O}_{OH}\right)$ Aminosulfonic Phenolsulfonic Sulfonyl (Sulfones) $\left(>S\!\!<^{O}_{O}\right)$
HO—COOH Carbonic acid	Thiocyanates Isothiocyanates Thioureas Thiocarbonic acid (H_2CS_3)		

Thioacids. (See **Sulfur, Acids.**)
Thioalcohols.
Thioaldehydes.
Thiocarbonates.   (See **Thiocarbonic Acid and Thio-carbonates.**)
Thiocarbonic acid.
Thiocyanates.   (See **Thiocyanic Acid and Thiocyanates.**)
Thiocyanic acid.
Thioethers.   (See **Thioalcohols and Related Compounds.**)
Thioketones.   (See **Thioaldehydes and Thioketones.**)
Thiols. (See **Thioalcohols and Related Compounds.**)
Thiolic acid. (See **Sulfur, Acids.**)
Thionic acids. (See **Sulfur, Acids.**)
Thionyl chloride. (See **Sulfur, Chlorides.**)
Thiophene.
Thiophenols.   (See **Thioalcohols and Related Compounds.**)
Thiosulfates.   (See **Thiosulfuric Acid and Related Compounds.**)
Thiosulfuric acid.
Thiourea.

(R.K.S.)

SULFUR DIOXIDE. Sulfur.

SULFURIC ACID AND SULFATES. Sulfuric acid, "oil of vitriol" (H_2SO_4), a colorless oily liquid, is encountered as a colorless solution, commercially of strength 60° Baumé (specific gravity 60° F., water at 60° F., 1.7059, 77.67% H_2SO_4, 22.33% water), 66° Baumé (specific gravity at 60° F., water at 60° F., 1.8354, 93.19% H_2SO_4, 6.81% water); specific gravity 1.835 (approximately 95% H_2SO_4); oleum 20% (20% free SO_3, 80% H_2SO_4, 85.3% total SO_3, specific gravity at 20° C., 1.927); oleum 40% (40% free SO_3, 60% H_2SO_4, 89.0% total SO_3, specific gravity at 20° C., 1.966). Sometimes colored brown to black by carbon. With the exception of the varieties of oleum (or "fuming sulfuric acid"), the highest strength of sulfuric acid ordinarily used is 66° Baumé (93.19% H_2SO_4), but in the manufacture of sulfuric acid and oleum by the contact process a strength of 98% to 99% is produced. Sulfuric acid 100% may be made by the proper admixture of sulfur trioxide and water, or by crystallization from sulfuric acid 98% upon cooling. There is a maximum constant boiling point 317° C. (768 mm.) at 98% H_2SO_4 (distillate) for mixtures of H_2SO_4 and water, and of H_2SO_4 and SO_3.

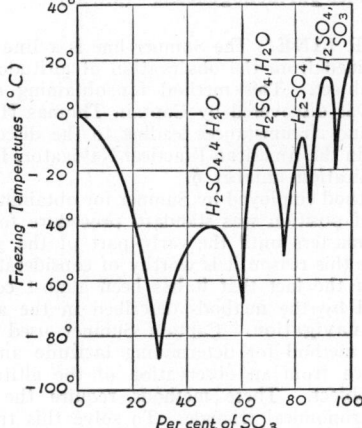

Freezing-point curve of system: sulfuric acid—water—sulfur trioxide. (*Mellor, Modern Inorganic Chemistry, Longmans, Green & Co.*)

The freezing point relations of water-sulfur trioxide mixtures is shown in the diagram. Sulfuric acid when mixed with water generates much heat. A commonly

used strength for dilute sulfuric acid is 24.5 grams H_2SO_4 per 100 milliliters of solution (5 normal).

Dilute sulfuric acid reacts (1) with many hydroxides, e.g., **sodium** hydroxide, to yield two series of sulfates (the acid is dibasic), e.g., sodium sulfate or sodium hydrogen sulfate, depending upon the ratio of acid to base reacting, (2) with many ordinary oxides, e.g., **magnesium** oxide, to yield the corresponding sulfate, e.g., magnesium sulfate solution, (3) with some carbonates, e.g., **zinc** carbonate, to yield the corresponding sulfate, e.g., zinc sulfate solution plus **carbon dioxide** gas (calcium carbonate is soon coated by a layer of calcium sulfate, which prevents further reaction), (4) with some sulfides, e.g., **ferrous** sulfide, to yield the corresponding sulfate, e.g., ferrous sulfate plus hydrogen sulfide gas, (5) with many metals, e.g., zinc, if not too pure (but not copper), to yield the corresponding sulfate, e.g., zinc sulfate solution plus **hydrogen** gas, (6) with solutions of some salts to yield the corresponding sulfate, e.g., **barium** chloride, changed to barium sulfate precipitate, calcium citrate, malate, tartrate to calcium sulfate precipitate and the free organic acid in solution.

Higher strengths of sulfuric acid react similarly in kind to the cases of (1), (2), (3), (6) above, but not, in general, in the remaining cases. (4) Sulfides react to yield the corresponding sulfates, but accompanied by **sulfur**; (5) reactions of metals depend upon the metal and the strength of sulfuric acid. **Copper** and concentrated sulfuric acid yield copper sulfate and **sulfur** dioxide gas. Iron reacts similarly, yielding ferric sulfate in the place of copper sulfate. Concentrated sulfuric acid is (7) thus an oxidizing agent, and a further example is the oxidation of **sulfur** to sulfur dioxide (the reacting sulfuric acid is reduced to sulfur dioxide), (8) a sulfonating agent, e.g., **naphthalene** sulfonated to naphthalene sulfonic acids (mono- alpha or beta, di- several), (9) an esterification agent, e.g., methyl **alcohol** esterified to dimethyl sulfate ((CH_3)$_2SO_2$), melting point −32° C., boiling point 189° C., or methyl hydrogen sulfate ($CH_3 \cdot SO_2OH$), ethyl alcohol esterified o diethyl sulfate ((C_2H_5O)$_2SO_2$), melting point −26° C., boiling point 208° C., or ethyl hydrogen sulfate ($C_2H_5O \cdot SO_2OH$), (10) a dehydration agent, e.g., **formic acid** into carbon monoxide, sugar blackened with separation of carbon, (11) an addition agent, e.g., ethylene into ethyl hydrogen sulfate, (12) a non-volatile acid upon heating, e.g., with **sodium** chloride or nitrate, **hydrogen chloride** or nitric acid, respectively, is volatilized and sodium sulfate or sodium hydrogen sulfate remains as a residue.

In order to obtain sulfuric acid, two sources of sulfur are available for conversion into sulfur dioxide, which is the first stage. These are (1) sulfur, (2) metallic sulfides, namely, **pyrite** (iron disulfide (FeS_2)), and **copper, lead, zinc** sulfides of smelter operations. In all these cases sulfur element is converted into sulfur dioxide gas by roasting in a current of air. In the second stage of the process sulfur dioxide is converted into sulfur trioxide by either of two methods, (1) "contact process," (2) "chamber process," both of which are catalytic. In the contact process sulfur dioxide, carefully purified, is mixed with air, and passed over a specially prepared **catalyzer.** The catalyzer commonly used is finely divided platinum (derived from platinum salts) on the surface of **asbestos, magnesium** sulfate, or **silica** gel, which act as porous, large surface supports. Vanadium pentoxide distributed on such a support as silica gel is also used. The third stage of the process involves the absorption of sulfur trioxide in water to form sulfuric acid. Experiment has demonstrated that the most satisfactory absorbent is 98% to 99% sulfuric acid. Acid of this strength is maintained in the absorber. When "oleum" is desired, water is not added.

In the chamber process, the above second and third stages are combined by introducing into large lead-lined chambers the required ratios of **sulfur** dioxide, water mist, and **nitrogen** oxides. Carefully designed towers are arranged (a) Glover tower, *before* the chambers, to introduce the nitrogen oxides previously dissolved in sul-

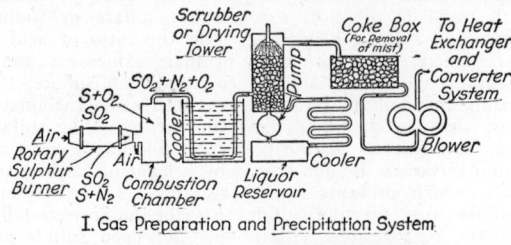

I. Gas Preparation and Precipitation System

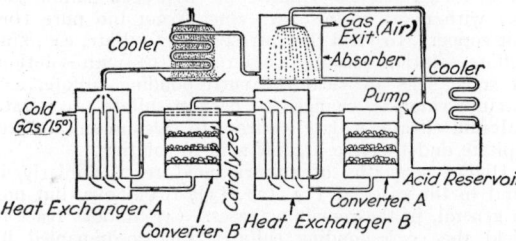

II. Heat Exchanger, Converter, and Absorption System

Sketch—Contact process for sulfuric acid (platinum catalyst).

furic acid ("Gay-Lussac acid"), and to cool the sulfur dioxide-air mixture from the burners, (b) Gay-Lussac tower, *after* the chambers, to recover the reduced and unused nitrogen oxides by dissolving them in sulfuric acid ("Glover acid"). An incidental stage in the chamber process is the concentration of "chamber acid" (62–70% H_2SO_4) by evaporation in acid-resistant pans, such as fused quartz, or by spray evaporation to the desired concentration.

The uses of sulfuric acid have been suggested by the chemical reactions previously cited, and the largest quantities are used (1) in the production of **explosives**, where it is used with nitric acid as "mixed acid," (2) in refining **petroleum** distillates, usually "oleum" used, (3) in the production of superphosphate **fertilizer** and phosphoric acid, ordinary "chamber acid" used, (4) of hydrochloric and nitric acids, (5) of sulfates, and (6) in the cleaning of metals, e.g., iron.

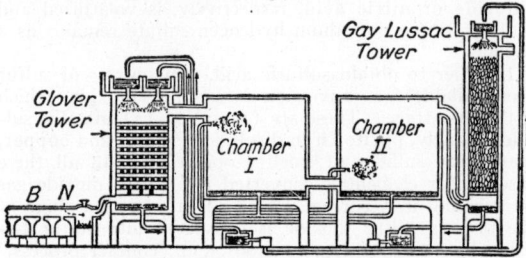

Chamber process for sulfuric acid.

All metallic sulfates, except **barium** sulfate, **strontium** sulfate, **lead** sulfate, **mercurous** sulfate, are soluble in water, but **calcium** sulfate and mercuric sulfate are only slightly soluble.

Metallic sulfates, upon heating, behave in an individually characteristic manner, e.g., sodium sulfate stable, sodium hydrogen sulfate into sodium pyrosulfate, barium sulfate stable, ferric sulfate into ferric oxide plus sulfur trioxide (an early method for obtaining sulfur trioxide).

Solutions of sulfates give a white precipitate, insoluble in hydrochloric acid, with barium chloride solutions. (R.K.S.)

SULFUROUS ACID AND SULFITES. Sulfurous acid (H_2SO_3) is a colorless solution formed when **sulfur**

dioxide gas is dissolved in water, and the presence of sulfur dioxide imparts its own odor. Sulfur dioxide may be completely expelled from sulfurous acid solution by boiling.

As a bleaching agent, sulfurous acid is used for whitening wool, silk, feathers, sponge, straw, wood; as a bleaching and preservative agent for dried fruits, and decolorizes a solution of rosaniline (fuchsine, magenta).

Sulfurous acid is a strong reducing agent, being oxidized to **sulfuric acid** (1) on standing in contact with air, (2) by **chlorine, bromine, iodine** yielding **hydrochloric, hydrobromic, hydriodic acids**, respectively, (3) by **nitric** or **nitrous acid** yielding nitric oxide, (4) by **permanganate**. Sulfurous acid is itself reduced by **zinc** and dilute sulfuric acid, to **hydrogen sulfide**. Sulfurous acid is formed (1) by dissolving sulfur dioxide gas in water, (2) in effect, by reaction of sulfite or bisulfite solution and an acid.

Sodium sulfite (Na_2SO_3) and sodium hydrogen sulfite ($NaHSO_3$) are formed by reaction of sulfurous acid and sodium hydroxide or carbonate in the proper proportions and concentrations. Sodium sulfite dry, upon heating, yields sodium sulfate and sodium sulfide. Sodium pyrosulfite, "sodium metabisulfite" ($Na_2S_2O_5$), is a common sulfite. Crystalline sulfites are obtained by warming the corresponding bisulfite solutions. **Calcium** hydrogen sulfite ($Ca(HSO_3)_2$) is an important substance used in conjunction with excess sulfurous acid in converting wood to paper pulp. Sodium sulfite and **silver** nitrate solutions react to yield silver sulfite, white precipitate, which upon boiling decomposes forming silver sulfide, brown precipitate. **Barium** sulfite is soluble (a white precipitate formed by the addition of barium chloride to sulfite solution containing hydrochloric acid is barium sulfate in consequence of the previous oxidation of sulfite).

As an esterification agent, sulfurous acid forms dimethyl sulfite ($(CH_3O)_2SO$), boiling point 126° C., and diethyl sulfite ($(C_2H_5O)_2SO$), boiling point 161° C. Sulfites give a white precipitate with barium chloride, soluble in hydrochloric acid with evolution of sulfur dioxide. Sulfites decolorize iodine in acid solution. (R.K.S.)

SUMMER EGG. A form of egg produced by some species of **crustaceans**, usually during the summer. These eggs have thin shells and are parthenogenetic (i.e., they develop without being fertilized), in contrast with the eggs of biparental generations. The significance of such specialization in the reproductive processes is considered under **parthenogenesis**. (A.W.L.)

SUMNER LINE. The Sumner line is a **line of position** obtained from the observation of altitude of some celestial object. This method for obtaining a line of position was discovered by Captain Thomas H. Sumner in 1837, and circumstances leading to the discovery are described in the American Practical Navigator, Bowditch, H.O. publication number 9.

The method employed by Sumner for obtaining a celestial line of position was standard procedure for American ship masters until the early part of the 20th century. For this reason it is worthy of consideration here, in spite of the fact that it has been almost completely superseded by the methods described in the article on **celestial navigation**. Captain Sumner used the old-fashioned method for determining **latitude** and **longitude** at sea from an observation of the altitude of a celestial object. These methods require the solution of the astronomical triangle. To solve this triangle at least three parts must be known. At sea, the altitude of the object, as obtained from **sextant** observations, will give one part, and the declination of the object will give another. To obtain the third part, either the latitude or the longitude of the observer must be known. Since latitude and longitude are the coordinates for which the navigator is seeking, it would seem at first glance that a solution of the problem is impossible. However, if the object observed is approximately due east or west, a change in latitude of several miles will produce but slight effect on the computed longitude. On the other hand, if the object is nearly due north or south, a change in longitude will produce but slight effect in the computed latitude. An approximate value of latitude and longitude can be obtained by methods of **dead reckoning** (DR). If the observed object is within 45° of east or west, the DR latitude is used and the longitude is computed. When the object is within 45° of the meridian, the DR longitude is used in the computation of the latitude. If the DR position is known to within 20 miles, the computed latitude or longitude will be accurate to within ¼ of a mile, unless the observed object is close to the **zenith.**

In Captain Sumner's case, his ship had experienced gales and fog for a number of days, and the DR position was very uncertain. When the clouds broke away, he obtained an altitude of the sun in the forenoon. Since his DR position was uncertain, he assumed several values for latitude separated by about 20 miles, and computed the corresponding longitudes. On plotting these positions he found that they lay along a straight line on a **mercator chart.** He made the assumption, which has since been established as sound, that his ship must be on the line.

After the publication of this discovery, the use of Sumner lines became the standard procedure for most navigators. If two objects, differing in bearing by at least 45°, are available for observation, two Sumner lines can be obtained and a **fix** determined. If, as is frequently the case during daylight hours, only one object is available, this one object is observed twice with an interval of time between the two observations sufficient to produce a change in **bearing** of at least 45°. The fix of these two lines is obtained by moving one line to the time of the other by the method of running fix.

A Sumner line is, in reality, a small circle on the earth, with a point on the earth directly under the observed object as center. Unless the object is within 10° of the zenith, the curvature is so slight as to be negligible in drawing the line on a mercator chart. (W.K.G.)

SUN. The sun is without question the most important of all of the celestial objects, not only to the earth, but also to all other members of the **solar system.** Whatever life exists on the earth, or elsewhere in the solar system, is absolutely dependent upon the **radiation** from the sun for its existence. All forms of energy which we employ on the earth come from the sun, either by its present radiations, as in the case of water power or wind power, or from its radiations in the past, as in the case of coal or oil. Even the **tidal energy,** which comes principally from the moon, is influenced to a considerable extent by the position of the sun relative to the moon. From the purely scientific point of view, the sun is of tremendous importance to the astronomer, since it is a typical **dwarf** star of the G₀ **spectral class,** and is close enough to the earth to permit of careful analysis.

Somewhat detailed descriptions of the different portions of the sun and its surrounding atmosphere will be found in articles on the **photosphere,** the **reversing layer, chromosphere, corona, prominences,** and **sun spots.** The mean distance of the sun, as determined from many measurements of **solar parallax,** is 149,680,000 kilometers, or 93,005,000 miles. The mean diameter is 1,393,700 kilometers, or 866,000 miles; about 109.1 times the diameter of the earth. Since volume is proportional to the cube of the diameter, we find that the volume of the sun is 1,300,000 times that of the earth. In order to get some concept of these sizes and distances consider the sun as a globe 2′ in diameter; the earth on this same scale would be a sphere only 0.22″ in diameter, and would be distant from the sun 215′. On this same scale the nearest **star** would be 11,000 miles away! The mass of the sun may be determined from the gravitational attraction which it exerts upon the earth and is found to be 331,950 times that of the earth, or 1.982×10^{33} grams (or 2×10^{27} tons). From the mass and the volume the mean density of the sun is found to be about ¼ that of the earth, or 1.4 times that of water. The gravitational force on the surface of the photosphere is 27.6 times that on the earth, which is equivalent to saying that a person weighing 100 lbs. on the earth would weigh nearly 1½ tons on the sun.

From a large number of observations of sun spots and also by spectroscopic observations the rotation period of the sun has been found to be different at different distances from the sun's equator. At the equator the sidereal period of rotation is about 24.65 days; in latitude 30°, 25.85 days; in latitude 60°, 30.93 days; and at the poles about 34 days. Such a varying rotation period indicates certainly that the sun is not a solid, but there is no adequate explanation for the differing rotation periods, even for a gaseous object. The sun's equator is inclined to the plane of the ecliptic by 7° 10′.5.

Observations of the **continuous spectrum** from the sun and the application of the **laws of radiation** indicate that the effective temperature of the surface of the sun is about 5750° K. From the value of the **solar constant** we calculate the rate of radiation from the surface of the sun as 89,500 **calories** per sq. cm. per min.; equivalent to about 84,000 hp. per sq. meter. Residents of northern climates will appreciate this amount of energy better when they realize that it would melt a sheet of ice 40′ thick in about 1 minute! The problem as to the source of this tremendous amount of energy is a vexing one, as is the whole problem of radiation from the stars. (W.K.G.)

SUN BITTERN. Aves, Gruiformes. A South American bird (**Aves**), *Europyga helias,* of moderate size, related to the cranes. (A.W.L.)

SUN COMPASS. The sun compass is a device utilizing the direction of the sun for direction or orientation purposes. The instrument operates on much the same principle as that of the **sun dial.** In the sun dial the gnomon for casting the shadow is set accurately parallel to the earth's axis of rotation and the direction of the shadow indicates local apparent **time.** In the sun compass the dial is set for local apparent time and the direction of the shadow is used in connection with a **compass card.** The instrument is quite complicated, for it must be set for terrestrial **latitude, longitude,** and local apparent time. It has been of great service in connection with flights in the polar regions of the earth, where the weakness and uncertainty of the horizontal component of the earth's magnetic field render the use of the magnetic **compass** very dangerous. (W.K.G.)

SUN DIAL. It is logical to suppose that from the earliest times mankind has used the apparently moving sun as a means for reckoning **time.** As the sun apparently moves across the heavens during the day the position and length of the shadow cast by an opaque rod will continually change. The positions or lengths of this shadow may be used for the purpose of subdividing the period between sunrise and sunset. Any device which utilizes the shadow cast by the sun for the purpose of subdividing the day into equal parts is known as a sun dial.

It is difficult to say just when the first sun dial was constructed. The earliest written record that we have is found in Isaiah XXXVIII:8, which was written approximately 700 years before the Christian era. The earliest instrument which has come down to us is a device which was built in Egypt but for which the exact date of construction is unknown. The first dial was constructed for Rome in about 146 B.C. Sun dials came into general use during the 13th century, and the development of the different types advanced rapidly following this period. By the time that mechanical clocks and watches made their appearance in the 15th century a multitude of different types of sun dials had been constructed and many volumes written regarding the theory of the various devices.

There are two fundamental types of sun dials. The most common type of fixed dial is that which marks the divisions of the day by the direction which the shadow of the sun has at any particular instant. The dial itself may be set at any desired angle, but the most common is the type in which the plate is horizontal and the style, which casts the shadow, is so placed as to be parallel to the axis of rotation of the earth. The second fundamental type of dial makes use of the fact that the length of the shadow of the sun varies throughout the day, being the shortest at noon and the longest at sunrise and sunset. Practically all

if the brilliant photosphere were not present, the sun spots themselves would appear intensely brilliant. A typical sun spot has a dark irregularly shaped central portion known as the umbra, surrounded by a lighter region known as the penumbra. Because of the distance of the sun, the smallest sun spots which can be studied must be at least 150 miles in diameter. Spots with diameters of 40,000 to 50,000 miles are quite common, and instances have been recorded where a number of spots were so close together that the penumbra blended into one area nearly 150,000 miles across.

Sun spots are usually relatively short-lived. About a quarter of them last but a single day, and as many again from 2–4 days. In a few cases large spots have persisted for over a month, the longest case on record being a large group of spots which lasted for nearly 18 months.

Sun spots confine themselves almost exclusively to the solar latitude zones between 5° and 40° north and south of the solar equator. The total number of spots on the surface of the sun varies, with a somewhat regular periodicity of approximately 11 years. The accompanying figure shows the total number of spots observed during each single year from 1885 to 1936. It

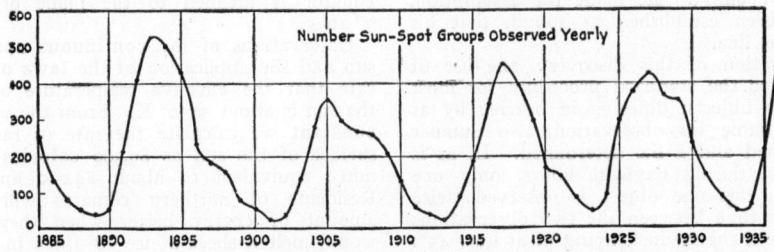

The sun-spot number cycle. The point for each year represents the number of sun-spot groups observed during that year. The curve shows the roughly periodic variation in the numbers. (*From data by S. B. Nicholson.*)

portable sun dials are of this type. The great difficulty with this type of dial is that, because of the change in **declination** of the sun with **season**, it is necessary to have different scales of time for different periods of the year.

It is impossible in a work of this character to discuss the multitude of ingenious and beautiful types of sun dials that have been used in the past for the purpose of keeping time and are in use at present as ornaments or items of curiosity. In adjusting the horizontal type of sun dial, such as may be purchased from a number of dealers in garden supplies or curios, it is important to remember that the style should be parallel to the axis of the earth. That is, it should lie exactly in the true north-south plane, and the north end should be so elevated that the angle which the style makes with the horizontal plate is equal to the latitude of the observer. When properly adjusted the sun dial will read local apparent time. This time will differ from that ordinarily kept by watches both by the **longitude** difference between the position of the dial and the **standard time meridian** and also by the **equation of time.** (W.K.G.)

SUN SPIDER. Solpugida.

SUN SPOTS. As the term implies, sun spots are spots on the surface of the sun which make their appearance on the **photosphere.** There is no record of when these phenomena were first observed. Frequently, they are so large as to be seen with the unaided eye when the brilliancy of the sun is cut down either by thin clouds or by darkened glass. There are Chinese records of sun spots long prior to the early part of the seventeenth century, when Galileo first observed them through his telescope.

In the first place, it must be clearly understood that sun spots are not really dark. They are merely darker than the surrounding regions of the photosphere, and,

will be noted that the shape of the curve and the interval between maxima and minima is not the same for each cycle. Along with the variation in number of sun spots there goes a shift in the average location of the spots. At the beginning of a cycle the spots are located at the outer regions of the latitude zones, i.e., between 30° and 40° solar latitude. As the number increases the maximum number of spots is located in about 16° latitude, and by the end of the cycle the spots are about 5° from the solar equator. The beginning of the new cycle is heralded by the appearance of a few spots at considerable distance out from the equator.

Many attempts have been made during the last two centuries to correlate terrestrial phenomena with the number of spots on the surface of the sun. The correlation between sun-spot number and electromagnetic phenomena on the earth is positive, and we find that at times of sun-spot maximum magnetic storms, with the accompanying shifts in compass variation, and interference with radio and telegraphic communication, are most prevalent. Maximum and minimum auroral displays follow slightly after times of maximum and minimum sun spots. The correlation between solar disturbances and weather conditions on the earth is both weak and puzzling. It has been shown that the solar constant varies with sun-spot number, being a maximum when the number of spots is greatest. However, there is definite evidence that the temperature of the air at the earth's surface is lower at the times of sun-spot maximum. Studies of the width of tree rings, made by Douglass in Arizona, indicate direct correlation between the periods of wide rings and the sun-spot periodicity.

Sun spots are observed to have a distinctly whirling character, not unlike cyclonic storms in the atmosphere of the earth. They are distinctly magnetic in character, spots in which the whirling motion is in one direction, all having one magnetic polarity, while spots whirling in the opposite direction will have the opposite magnetic polarity. When two spots are observed relatively close together, as is frequently the case, the two are whirling in opposite directions, and hence have opposite magnetic polarity.

It is believed that sun spots are caused by a whirling mass of gas just below the surface of the photosphere. Hot gases are brought up until they break through the photosphere, where the sudden reduction in pressure causes them to cool and spread out over the surface, causing a relative darkening. The influence of spots is felt clear out through the atmosphere of the sun, for at the time of sun-spot maximum there is also a maximum number of **prominences**. The shape of the solar **corona** is also influenced by the number of spots on the surface of the sun. In spite of the tremendous amount of attention that sun spots have received, there is no adequate explanation as to their origin. (w.k.g.)

SUN STROKE. HEAT STROKE. This condition is due to exposure to high temperatures which overwhelm the heat-regulating mechanism of the brain. The onset may be sudden, with immediate loss of consciousness, or this event may be preceded by headache, nausea, dizziness, and visual disturbances. The face is flushed, the skin hot and dry. The temperature in severe cases can rise to 109° F. Rapid feeble pulse and shallow respirations precede death. The latter may occur within a few minutes after onset of symptoms, but if the patient survives 48 hours, recovery is probable. The milder degrees of reaction to intense heat are classified as heat cramps, due to loss of **sodium chloride** in perspiration, and heat exhaustion. In heat exhaustion there are pallor, weakness, dizziness, profuse perspiration, slight elevation of temperature, but rarely loss of consciousness.

Heat cramps are prevented by the ingestion of extra sodium chloride during hot weather. The treatment of heat exhaustion requires only rest in a cool place and perhaps mild stimulants. Heat stroke is an emergency which demands prompt treatment. Cold baths, cold enemas, ice packs, and stimulants are used. (d.m.h.)

SUNDEWS. *Drosera* species. **Insectivorous Plants.**

SUNFISH. Pisces, Teleostei. A name of varied uses, applied to fresh-water and marine fishes (**Pisces**) of three distinct forms and many species, all superficially similar in the short, high, and compressed body. One species, *Lampris luna*, is found in the Mediterranean and the northern Atlantic. It attains a length of 4′ and is bluish with silver spots and red fins. Another group of several species related to the globe fishes is characterized by the very short truncated tail and by the enormous size that they attain. One species is known to reach a weight of 500 lbs. These giant sunfishes are widely distributed in temperate and tropical seas.

In marked contrast the familiar fresh-water sunfishes of North America are among the small pan fish related to the basses. They reach a maximum length of 5–10″. The common sunfish or **pumpkin-seed** reaches about 8″. The various species are found in streams and ponds, especially the latter, and rank as superior food fish and fair game fish. They take dry or wet flies and on light tackle are a valuable supplement to the larger bass in the waters of heavily settled areas. (a.w.l.)

SUNFLOWER. *Helianthus annuus.* **Composite Family.**

SUNSTONE. Feldspar.

SUNU. Mammalia, Artiodactyla. *Nesotragus.* A large African **antelope** related to the sing-sing. The species

is black with white rings around the eyes and white under parts and ears. The horns are long and thin. (a.w.l.)

SUPERCHARGER. The performance of an internal combustion engine is indicated, among other things, by the brake hp. output. A review of the factors affecting power indicates that atmospheric conditions have a significant effect. (See **Engine Performance**.) A naturally aspirated (unsupercharged) engine is able to draw into the cylinders on suction strokes only from 70-85% of the fuel charge which it is theoretically capable of inducing (see **Volumetric Efficiency**). Consequently, the mean effective pressures are not as large as they might be, and power output per cu. in. of piston displacement does not reach its maximum possible value. Compression of the incoming air, or air-fuel mixture, somewhat above ambient pressure is a natural way of increasing output at sea level or of regaining it at altitudes. A compressor used for this purpose is designated a *supercharger*.

The reader should be cognizant of the fact that raising the sea-level power of an existing engine by adding a supercharger is not ordinarily feasible, as the original design may be insufficient to withstand the increased structural stresses created by higher cylinder pressures. However, supercharging for the purpose of regaining sea-level rating at higher altitudes may be readily added provided the supercharger is not used at sea level and is employed only to offset the effect of altitude. Thus it is seen that supercharging falls into two categories: (1) supercharging that contemplates increasing capacity per cu. in. of piston displacement under approximately constant atmospheric conditions; and (2) supercharging that will keep power of an engine up near its rated output at high altitudes. The first category is sometimes termed "ground boosting," and the second "altitude supercharging." The latter predominates in the aeronautical field. The former will be found occasionally in stationary, marine, and surface-vehicle usage of engines and in the two-cycle aircraft engine.

Either centrifugal compressors (see **Air Compressor**) or positive displacement blowers can be used to supercharge. For ground boosting the latter are excellent, since boost varies linearly with speed, and so cylinder pressures and shaft torque hold up well under slow-speed operation, whereas the pressure boost given by a centrifugal compressor varies as the square of the speed and drops off rapidly as speed is cut. On the other hand, the displacement type suffers from excess bulk and weight compared to the centrifugal—a fact not likely to recommend it to the aeronautical field.

Sea-level gasoline engines are not often supercharged because they are most frequently found in small capacities where, by relatively high rotative speeds, engine size may be kept small. Furthermore, unless special high-octane fuels are used, supercharging may create detonation troubles. Diesels, however, are frequently supercharged, thus allowing the burning of larger amounts of fuel without creating excessive combustion-chamber temperatures. The higher mean effective pressures, coupled with the use of two-cycle principle of operation, has done much to remove from the Diesel the stigma of excessive weight and bulk. Displacement blowers used in this field have been of the lobe and eccentric-vane types. These are often driven from the crankshaft by means of V-belts. The Buchi supercharge system uses a centrifugal compressor driven by an exhaust gas turbine.

High-altitude flights in airplanes would have been impossible without supercharging of the engine. Beginning at the time of World War I, research and experiment in airplane-engine supercharging has made great strides and accomplished remarkable results. The tactical advantage attained by the use of military aircraft with superior flight ceilings, and airplane critical altitudes, has provided the greatest stimulus for development in this field, since comparatively simple supercharging (or none at all) will fulfill most civil aviation needs.

Airplane-engine superchargers are, universally, high-speed centrifugal types because of the emphasis on minimum size and weight. The principal types are (1) the internal gear-driven type, receiving its drive from the crankshaft, and (2) the external exhaust gas turbine-driven type. Since supercharging may be obtained by staging the over-all compression as well as by single stage, and as different gear ratios to drive the same impeller are sometimes employed, many different combinations of equipment are possible. The variations mainly represent individual ideas of how to obtain maximum power at altitudes without sacrificing too much power driving surplus compressor capacity at sea level. Three

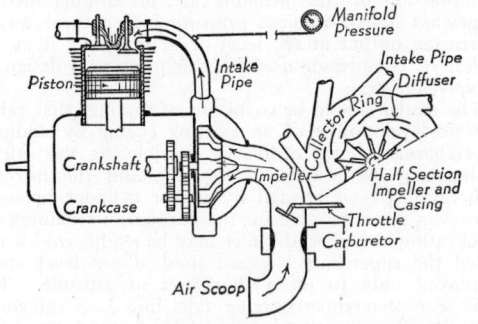

Fig. 1. Single stage supercharger. Diagram emphasizes induction system.

of the many systems developed are shown in Fig. 2, while Fig. 1 shows the standard simple single-stage, single-speed, internal gear-driven supercharger.

The standard gear-driven supercharger is an integral part of large-capacity engines. As Fig. 1 shows, the carburetor is located on the suction side of the impeller, so a combustible mixture is being compressed. The duralumin impeller is geared up to make from 10 to 15 thou-

sand rpm, and is capable of boosting atmospheric pressure approximately 1 atmosphere. It would not be permitted to accomplish this at sea level as the cylinder explosion pressures would be too high; also, **detonation** would probably occur. By throttling the intake at the **carburetor**, the cylinder pressures can be kept within safe limits. This is gauged, by the pilot, by the *manifold pressure;* that is, the mixture pressure just before it enters the cylinder. As an airplane climbs (engine rpm constant), it is necessary to open the throttle gradually in order to keep manifold pressure constant. When the throttle is full open a *critical altitude* is reached, and further gain in altitude will be attended by decrease of output since there is no more supercharger reserve left. All this is diagrammed for engine A in the graph, Fig. 2.

In the effort to reduce the required low-altitude supercharger drive power and at the same time increase the power available at high altitudes, additions are made to the standard system, three of which appear in Fig. 2. They are: (B) single-stage, two-speed, (C) two-stage, two-speed, and (D) two-stage with exhaust gas turbine driving the auxiliary stage.

In system B, a gear-shifting mechanism is incorporated which makes possible a low and high gear ratio. Although it adds but little extra weight or complexity to the installation, it produces an increase in power at medium altitudes (not much gained at extreme altitudes). Also, the low blower ratio is not as wasteful of power near sea level. Its application is limited because of low compression efficiency and high manifold temperatures (detonation), which are produced where extra high gear ratio is attempted. The attached graph shows that take-off is with low blower ratio engaged. After reaching critical altitude, some more altitude is gained before changing to high blower.

The two-stage, two-speed supercharger has a main stage like that of engine A, although usually with smaller ratio, thus giving lower main stage critical altitude. An auxiliary stage, also engine-driven, has an impeller whose speed is made selective by a gear-shifting mechanism, incorporating clutches. The auxiliary stage, therefore, adds a low-blower and high-blower ratio to the main

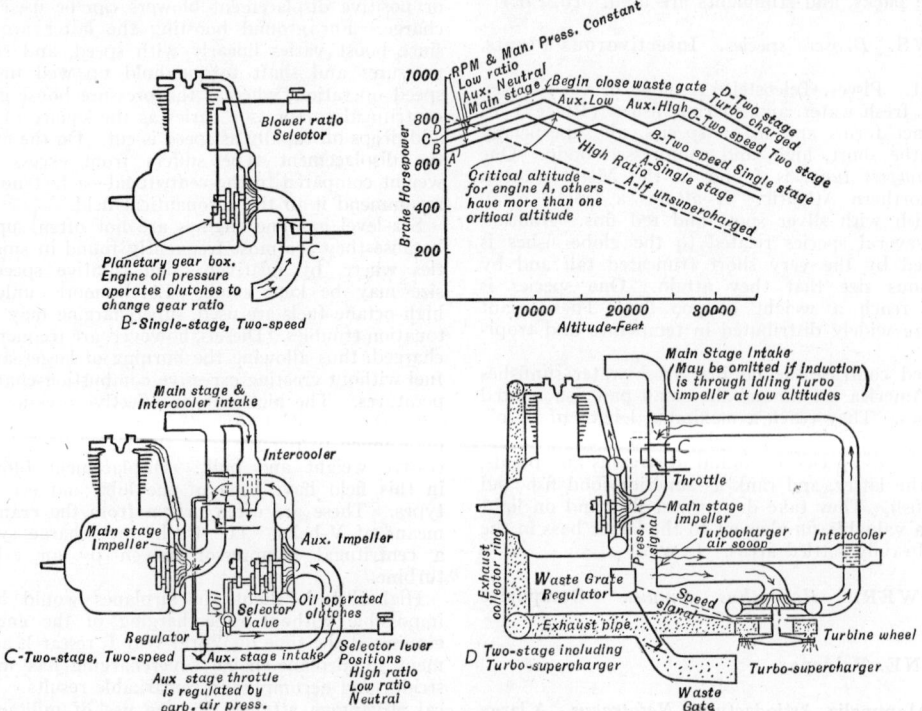

Fig. 2.

stage supercharge. This system is superior to a single-stage high-ratio supercharger for high altitudes, for the auxiliary stage may remain idle (neutral position) until the supercharge of the main stage is completely developed. Furthermore, an intercooler is included between the two stages which not only improves compression performance, but opposes detonation, since otherwise the adiabatic temperature rise of both stages in series would be very great at high altitudes.

Advantages of this extra complication are shown in the C curve of power output. Disadvantages are cost, weight, and complication of the extra stage, the ducts, and the intercooler, as well as the added drag of the larger air scoop area.

The exhaust gas turbo-supercharger consists of a single-stage turbine wheel directly connected to the compressor impeller. The engine exhaust is collected and led to a nozzle box where it arrives at a pressure of several lbs. per sq. in. gauge. Pressure ratio across the nozzles of the supercharger varies with altitude, but is typically 1.5 to 2.5. However, this can be controlled by releasing part of the exhaust to the atmosphere through a waste gate. The turbine speed and air-pressure boost can be controlled in this manner. Advantages of this method of supercharging are: (1) partial use of the energy in the incompletely expanded exhaust gases; (2) more latitude in arrangement of the auxiliary stage in an airframe (gear-driven stages must be *on* or *in* the engine); (3) turbine tends to run faster and give compensating supercharge as altitude is increased because atmospheric pressure decreases more rapidly than exhaust manifold pressure; (4) the power output is maintained up to much higher altitudes than for other types. Disadvantages are: (1) weight and complication of extra ducts, intercooler, controls, etc.; (2) some loss of power at low altitude when waste gate is open, probably caused by induction pressure drop through turbo-charger impeller and intercooler (avoidable by use of extra air scoop for main stage); (3) structural conditions imposed on turbine wheel are severe, viz., high centrifugal stresses contemporary with high temperature; (4) higher back pressure is created on cylinder exhaust than would otherwise exist.

Fig. 2-D diagrams a turbo-supercharger installation. Here it becomes the auxiliary stage; in some engines, especially in-line types, it may be the main stage. Exhaust gas from the collector ring flows to the nozzle box, some of it being diverted through the partly open waste gate. Gas expanding through the nozzles to atmospheric pressure flows across the blades, turning the turbine wheel at high speed. The direct-connected impeller draws in air from the intake duct, where it may have some ram, and adiabatically boosts it in pressure and temperature. Passing through the intercooler the temperature is lowered, after which fuel is added at the carburetor and the mixture given its second stage of compression by the gear-driven impeller. Waste-gate position is controlled by two signals, one derived from carburetor air pressure, one from turbine shaft speed. The latter includes both speed anti-surge and maximum safe-speed components. The power output curve shows that as the main stage reaches critical altitude, the turbo-supercharger is brought into play (by waste-gate closure) and maintains the power nearly constant until altitudes are reached where a combination of factors, chiefly attainment of maximum allowable turbine speed and loss of compression efficiency, create a normal type of altitude power loss.

The power necessary to drive a supercharger is considerable but, of course, it is more than repaid by increased engine output. The power required to operate the supercharger of a single-stage type is a function of the pressure ratio, the air temperature, the thermal efficiency, and the compressor efficiency.

Let

P = Net engine brake hp.
W = Air-fuel ratio.
f = Specific fuel consumption, lbs. per brake hp. hour.
e = Adiabatic compression efficiency. 65-75% for the centrifugal supercharger.
T = Air temperature at supercharger inlet.
R = Pressure ratio ($R > 1$).

Supercharger horsepower = $\dfrac{PfWT}{10,000e} [R^{.186} - 1]$.

Consider the power required to drive the supercharger of an engine having a critical altitude of 8000'. R is theoretically 1.35, T is 490° R. For $f = .6$, $W = 16$, $e = .65$, the equation above shows that the supercharger would consume about 6% of the output. ((F.T.M.)

SUPERCONDUCTIVITY. An abnormally high electrical **conductivity** appearing quite abruptly in certain metals when cooled through a very low, characteristic transition temperature. In 1911 Onnes, at Leyden, found that a column of frozen mercury with which he was experimenting, and which had a resistance of 0.084 ohm at 4.3° K., acquired a vanishingly small resistance (less than 0.000003 ohm) when cooled to 3° K. The resistance was measured by the potential drop while carrying a known current. The more familiar metals exhibiting this property are magnesium, zinc, cadmium, mercury, aluminum, tin (tetragonal only), and lead. Alloys of these metals also show the effect, as do some alloys of metals not mentioned; e.g., the alloy composed of two parts of gold and one of bismuth; also some compounds, as lead sulfide and tungsten carbide (though neither tungsten nor carbon is superconductive). The transition points are always within a very few degrees of absolute zero. For columbium it is 9.2° K., for columbium carbide, 10.1° K.; for magnesium it is only 0.7° K. The transition is sharper for metallic monocrystals than for microcrystalline masses.

One of the most curious aspects of the phenomenon, discovered by Onnes and Tuyn, is the apparent "perpetual motion" of a current in a superconducting circuit, such as a lead ring immersed in liquid helium. The current may be started inductively by cooling the metal in a magnetic field and then withdrawing the field, whereupon the current continues to flow indefinitely. If two points on such a ring are connected to a galvanometer and the ring parted between these points while it is thus conducting, the current stops, but not until it causes a throw in the galvanometer. The superconductive state may be removed, not only by heating but by applying a magnetic field of above a certain threshold intensity, or by using too strong a current. Another curious phase may be described as the almost perfect diamagnetism of metals when superconductive, the magnetic **permeability** being reduced practically to zero. These phenomena are not explained on any simple theory. There is no evidence of any structural change, and the relations observed between superconductivity and thermal conductivity, specific heat, etc., at low temperatures do not throw much light on the question. (L.D.W.)

SUPERCONTROL TUBE. Variable Mu Tube.

SUPERCOOLED LIQUID. Supersaturated Vapor.

SUPERCOOLED LIQUID IN THE ATMOSPHERE. Drops of rain and cloud droplets often are cooled to temperatures well below 32° F. and remain in liquid form. They are then said to be supercooled. If such drops and droplets are disturbed, they freeze almost instantly, partially or entirely, depending on their temperature and size. Passage of aircraft through supercooled drops and droplets in the atmosphere results in rapid and severe icing. (P.E.K.)

SUPERELEVATION.

SUPERELEVATION. When the plane of a roadway is tilted on a curve (commonly known as banked), it is said to be superelevated. The purpose of superelevation is to permit a vehicle to round a curve on a roadway at high speed without danger of overturning or skidding. The superelevation can be made so that the resultant of dead weight and centrifugal force passes through the vertical plane of symmetry of the vehicle. In this condition, no side sway would be felt by the occupants. However, the superelevation necessary to accomplish this is different for each vehicle speed, so that it is apparent that the superelevation of a highway presupposes an average vehicle speed. The same is true of railways, although the variation of speeds with which the trains round curves is less than in the case of highway traffic.

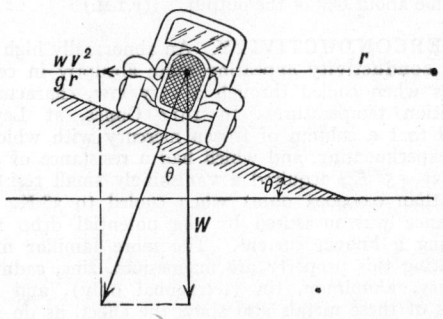

To illustrate how the superelevation depends upon vehicle speed, let it be assumed that an automobile approaches a curve on a highway at a speed of V (ft. per sec.). If the radius of the turn is r, the centrifugal acceleration is $\dfrac{V^2}{r}$. Furthermore, assume that the weight is W pounds. While negotiating the curve, the car is subject to two forces, one, the weight vertically downward, the other, centrifugal force acting horizontally away from the center of curvature, and having a magnitude $\dfrac{WV^2}{gr}$.
The surface of the road must be perpendicular to the resultant for no "side sway." If superelevation is given as the angle of bank (see figure), the angle of superelevation θ has a tangent equal to centrifugal force divided by weight. This tangent is $\dfrac{V^2}{gr}$, demonstrating that the superelevation must be made with respect to the radius of curvature and the velocity of the vehicle. It is independent of the dimensions and weight of the vehicle. (F.T.M.)

SUPERFICIAL VELOCITY. Gas Absorption.

SUPERFINISHING.

SUPERFINISHING. An abrasive process for removing smear metal, scratches and ridges produced by machining and grinding operations, and other surface irregularities, from parts that are to have a highly finished surface. The process resembles lapping in that a lubricated abrasive stone is applied to the surface at comparatively low speeds and light pressures. A superfinishing head whose base is attached to the cross-slide of an engine lathe may be used. The base supports two vertical cylindrical guides on which the head proper may be manually adjusted to the work by the hand lever shown. The abrasive stone is carried in a vertical slide which is subjected to the action of a spring for applying pressure to the stone. The stone pressure may be regulated to suit the requirements of the work by turning the screw at the top of the slideway.

In operation, the work rotates between centers at a surface speed of about 30 ft. per min. for roughing to from 70 to 100 ft. per min. for finishing operations. The stone is moved axially along the surface of the work by using the coarsest feed of the lathe carriage. As the carriage feeds, the stone is subjected to short, frequent, axial oscillations by the motor at the rear of the superfinishing head. For average work, a spring pressure of from 12 to 20 lbs. per sq. in. is employed, using kerosene as a lubricant. The cutting action cycle is as follows: when the stone is first applied to the work, it comes in contact with the peaks of the minute serrations with a high unit pressure because of the small area of contact, causing the abrasive to tear out comparatively large metal particles and effecting a comparatively large loss of abrasive particles. These particles are immediately washed away by the lubricant. This tearing tendency is rapidly reduced as the serrations wear flat, and the metal particles finally become so small that they immediately oxidize and begin to fill the pores of the stone. This metallic oxide is in itself a polishing agent and contributes to the finishing action. When the combination of improvement in surface condition of the work and the dulling and glazing of the stone face have reached a certain point, the decrease in unit pressure allows the lubricant to prevent further contact between the stone and the work. The surfaces in contact have such a comparatively large area, and consequently such a low unit pressure, that the lubricant is drawn between the surfaces and forms an oil film as in perfectly lubricated bearings. For repetitive work the process therefore ceases at the same point in each cycle, and surfaces of like degrees of finish are obtained on duplicate parts. After a superfinished part is removed and an unfinished part is substituted, the stone is automatically dressed sharp by its application to the peaks of the serrations on the new part and the cycle is repeated. (H.C.H.)

SUPERHEAT.

SUPERHEAT. Superheat is the addition of heat to produce **vapor** at a higher temperature than saturation. Superheat is possible when the vapor is led away from the liquid from which it was boiled. For this reason, superheaters are installed in **boilers** so that the final product may be elevated in temperature from 50 to 200 or 300° F. above the saturation temperature. The temperature added is called the degree of superheat, and the equipment to superheat is known as a superheater.

The effectiveness with which water vapor may be employed as a working medium in a power cycle is enhanced by superheating it. The less erosive character of dry steam and the lower heat losses from pipes carrying dry steam, have made superheating very desirable, so that many boilers are equipped with **superheaters** at present. (F.T.M.)

SUPERHEATER, STEAM.

SUPERHEATER, STEAM. The superheater is heat-transfer surface, tubular in character, arranged to receive saturated steam at the inlet of the tubes and deliver superheated steam steadily at the outlet. Heat is received from combustion of fuels by direct radiation to the surface of the superheater tubes, by convection of hot furnace gases made to flow over the tube surfaces, or both. Superheater tubes are made small in diameter to promote maximum contact of the steam with the hot tube wall. Usually the superheater is placed within the boiler setting and thus uses heat from the boiler furnace, but separately set and fired superheaters have occasionally been employed.

Superheaters may be classified as *convection* or *radiant* types; also as *interdeck*, which is primarily convective, but receives considerable radiant energy as it is screened from the furnace by only a few rows of tubes. Both types are usually constructed of small smooth tubes in single or multiple hairpin loop, with sufficient loops in parallel between headers to pass the required quantity of steam.

On account of the alloy steels employed in their construction, and because of the low coefficient of heat transfer to dry vapor, superheating surface tends to become expensive. By using higher temperature differences, surface can be saved, but at the risk of encountering slagging difficulties, whereas conservatively low temperature differences not only yield expensively larger superheaters, but also incur the difficulty of disposing of them and supporting them. The convection-type superheater has a rising superheat characteristic while the radiant type has just the reverse. At higher ratings the furnace gas is at higher temperatures, but that does not affect the energy level of the combustion quanta, which are the source of heat for the radiant superheater. Therefore an increased flow through the radiant superheater will have less superheat per pound. But the heat supply to the convection superheater is actually increased, and increased more rapidly than the steam flow, hence there is more superheat per pound.

Since the maximum steam temperatures now employed are close to the working limit of metals employed, close control over the superheat is essential in the steam generator of advanced design. This is true not only of the superheater itself, but also for the turbine. Superheat may be controlled in a number of ways, of which the following are the more important:

1. Follow the superheater with a water-spray desuperheater operated by a temperature regulator.

2. Gas by-passing. This method is employed for regulating convection superheaters. It is necessary to employ an oversize superheater, one which will give the required superheat at the lowest specified load. At higher loads the by-pass is opened, allowing some of the gas to flow around rather than over the superheater. Although the gas passing the superheater is hotter, its volume is less, hence the compensation.

3. Combine convection and radiant types in series or place an interdeck type so as to receive the proper proportions of radiant and convection energy. Neither of these methods is adequate to maintain uniform temperature over a load range, but the variation will be substantially less than for either type alone. May be combined with gas by-passing.

4. Other methods employing a combination of radiant and convection superheaters with twin furnaces which may be differentially fired are said to have given excellent results over a wide range of superheat control.

5. By-pass a varying portion of the steam, at an intermediate point in the superheater, through a tubular desuperheater cooled by boiler water or steam. (F.T.M.)

SUPERHETERODYNE. Receiver.

SUPERIMPOSED RIVER VALLEY. A river valley which is independent of present structural control may be described as either superimposed or antecedent.

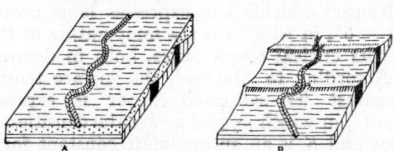

Diagrams illustrating the development of a superimposed river valley.

In the former case it is implied that the river has been able to maintain its course across resistant structures such as ridges, because it started as a consequent stream and has been "let down" on the underlying or non-conformable structure. (R.M.F.)

SUPERLINGUA. Paragnatha.

SUPERNORMAL DISPERSION. If the Lexis ratio, D, is greater than 1, the distribution is said to possess supernormal dispersion. Ordinarily when D is significantly greater than 1, we presume the variable consists of sets which have constant probabilities within the set but probabilities which differ from set to set. (See **Coefficient of Dispersion.**) (L.A.A.)

SUPERPOSITION, LAW OF. The fundamental law in **stratigraphy** and historical geology stating that underlying strata are older than overlying strata unless the formations have been inverted by folding or by low-angle thrusts. (R.M.F.)

SUPERPROPORTIONAL REDUCERS. Superproportional reducers are useful in the reduction of negatives of extremely high contrast due to the contrast of the subject as well as overdevelopment. Negatives of subjects of normal contrast which have been overdeveloped should be reduced in a proportional reducer.

The alkaline persulfates are the only oxidizing substances which form practical superproportional reducers. (See also **Photographic Reduction, Photographic Subtractive Reducers, Proportional Reducers.**) (C.B.N.)

SUPER-REGENERATIVE RECEIVER. In the ordinary regenerative **receiver** the sensitivity goes up as the **feedback** is increased but if the feedback is increased to produce the maximum amplification (just before oscillations start) the circuit is unstable and breaks into oscillation. The super-regenerative circuit utilizes this high gain point without the instability by introducing a voltage of low radio frequency in the plate supply lead. Since this voltage subtracts from the plate supply voltage every half-cycle it will lower the net plate voltage to the point where any started oscillations die out. The circuit is adjusted so oscillations actually start to build up, giving very high gain, but are killed off at the low radio frequency (quench frequency) rate and so do not reach an objectionable amplitude. The quenching frequency may be generated by a separate oscillator **tube** or may be generated by the regular detector tube in a so-called self-quenching circuit. The gain of these detectors is enormous but they are subject to several limitations such as: poor quality, radiation and subsequent interference with other receivers, strong interchannel hiss, poor selectivity, etc. They are, nevertheless, quite widely used for reception in the **very high frequency** region. (L.R.Q.)

SUPERSATURATED AIR. Air devoid of salt, dust, and ions can be supersaturated with respect to a free water surface. Inasmuch as the atmosphere contains numerous salt particles, dust, and ions, as well as microscopic plants and animals, supersaturation is seldom if ever a reality because these materials, particularly salt particles and ions, serve as nuclei for condensation. Actually many clouds form at less than saturation with respect to open water surfaces. (P.E.K.)

SUPERSATURATED SOLUTION. Solutions.

SUPERSATURATED VAPOR. A supersaturated vapor is one which remains dry, although its heat content is less than that of dry and saturated steam at the pressure. Supersaturation is an unstable condition, and is found in the **steam** emerging from the nozzles of a steam turbine. The abnormality of the phenomenon is similar to that of supercooling. An experiment often performed in physics laboratories is the careful, slow cooling of water in a vessel absolutely free from vibration or motion. In a perfectly quiescent state, the water may be slowly cooled to below the normal freezing temperature without appearance of ice. In the supercooled state a slight jar on the containing vessel will immediately create a return to normal condition, with a rise of temperature to the freezing point, and the appearance of crystals of ice in the water. Supersaturation probably results from the very rapid expansion of steam in the nozzle, permitting the traverse of a short distance before the condensation of moisture is completed. At a certain definite point, however, known as the Williams limit, the supersaturation vanishes, and the steam regains the wet state which would be normal in view of the pressure and the heat content. Supersaturation of vapor is impossible in the presence of numerous changed ions or dust particles. (F.T.M.)

SUPERSONICS. Ultrasonics.

SUPPRESSED CARRIER. Modulation.

SUPPRESSOR GRID. This is a third grid introduced in the thermionic vacuum tube between the screen grid and the plate to suppress secondary emission from the plate. By doing this the undesirable dip in the plate characteristic of the tetrode is eliminated. (L.R.Q.)

SUPPURATION. The formation of pus. Any infection in which pus develops is suppurative. The common pyogenic or pus-forming organisms are staphylococci and streptococci. (D.M.H.)

SUPRARENAL GLAND. Also called adrenal. Endocrine Gland. (A.W.L.)

SURCHARGE. Retaining Wall.

SURFACE GAUGE. Measurement.

SURFACE INTEGRAL OF VECTOR FUNCTION.
Let $\mathbf{F}(\mathbf{r})$ be a vector function of the position vector $\mathbf{r}$, let S be a region on a curved surface, and let $\hat{\mathbf{n}}$ be a unit normal vector, normal to S at any point of the region, and let θ be the angle between $\mathbf{F}$ and $\hat{\mathbf{n}}$. Then the surface integral $\iint_S F \cos\theta\, dS$ is called the surface integral of the vector function $\mathbf{F}$, and is denoted by

$$\iint_S \mathbf{F} \cdot \hat{\mathbf{n}}\, dS.$$

(L.L.S.)

SURFACE INTEGRALS. Let $f(x, y, z)$ be a function which is continuous within and on the boundary of a region S on a given curved surface. Let this region be divided up into n sub-regions ΔS_k, and let (x_k, y_k, z_k) denote any point in ΔS_k. Form the sum

$$\sum_{k=1}^{n} f(x_k, y_k, z_k)\Delta S_k = f(x_1, y_1, z_1)\Delta S_1 + \cdots$$
$$+ f(x_n, y_n, z_n)\Delta S_n.$$

The limit of this sum as each $\Delta S_k \to 0$ (and $n \to \infty$) is called the surface integral of $f(x, y, z)$ over the region S, and is denoted by

$$\iint_S f(x, y, z)dS.$$

Let $R(x, y, z)$ be a continuous function in a region S on a curved surface; divide S into sub-regions ΔS_k as before, and let $\Delta\sigma_k$ be the projection of ΔS_k on the XY-plane. Then the limit of the sum

$$\lim_{\Delta\sigma_k \to 0} \sum_{k=1}^{n} R(x_k, y_k, z_k)\Delta\sigma_k$$

is defined as a surface integral of $R(x, y, z)$ and may be denoted by $\iint_\sigma R(x, y, z)d\sigma$. In rectangular coordinates, the element $d\sigma$ becomes $dxdy$, and the integral is written $\iint_S R(x, y, z)dxdy$. Similarly we may define surface integrals $\iint_S Q(x, y, z)dzdx$ and $\iint_S P(x, y, z)dydz$.

These surface integrals usually occur in combination as a sum, usually written in the form

$$\iint_S (Pdydz + Qdzdx + Rdxdy).$$

(L.L.S.)

SURFACE OF REVOLUTION. The surface generated by revolving a plane curve about a line lying in its plane is called a surface of revolution.

Examples are: the sphere, right circular cylinder and cone, ellipsoid of revolution, hyperboloid of one and of two sheets of revolution, and paraboloid of revolution.

To find the equation of a surface generated by revolving a curve in one of the coordinate planes about one of the axes in that plane: Substitute in the equation of the curve the square root of the sum of the squares of the two variables not measured along the axis of revolution for that one of these two variables which occurs in the equation of the curve. (L.L.S.)

SURFACE TENSION. Fluid surfaces exhibit certain features resembling the properties of a stretched elastic membrane; hence the term surface tension. Thus, one may lay a needle or a safety-razor blade upon the surface of water, and it will lie at rest in a shallow depression caused by its weight, much as if it were on a rubber air-cushion. A soap bubble, likewise, tends to contract, and actually creates a pressure inside, somewhat after the manner of a rubber balloon. The analogy is imperfect, however, since the tension in the rubber increases with the radius of the balloon, and the pressure inside, which would otherwise decrease, remains approximately constant; while the liquid "film tension" remains constant and the pressure in the bubble falls off as the bubble is blown.

Whenever two dissimilar substances make contact at an interface, the inequalities of molecular attraction (cohesion), together with other forces in operation, tend to change the shape of the interface until, in accordance with the least energy principle, the potential energy of the whole molecular system attains a minimum value. If both substances are fluid, the surface does actually adjust itself to this condition. For example, a drop of oil suspended at rest in another liquid of the same density assumes a spherical form because the minimum-energy curvature is the same for all points of the surface. But if the drop is rotating, centrifugal forces alter the equilibrium and the drop becomes spheroidal; or in the case of a drop hanging from the end of a pipette, gravity enters as a component and the drop becomes pear-shaped. (See also Capillarity.)

In any case, the value of the surface tension at any interface is determined by the nature and the physical condition of the two substances in contact. The surface tension of a liquid (against air) decreases with rising temperature. An empirical formula known as the Eötvös-Ramsey-Shields law expresses it as proportional to $t_c - t - 6°$, in which t is the temperature of the liquid and t_c is its critical temperature, both in degrees centigrade. According to Macleod, the surface tension of a liquid against its saturated vapor is expressed by $K(\rho_L - \rho_V)^4$, in which ρ_L and ρ_V are the densities of liquid and vapor and K is an approximate constant for a given substance. At the critical point, when the densities become equal, the surface tension should be zero and thus no longer impede diffusion; though this conclusion is in slight disagreement with the Eötvös-Ramsey-Shields law above. (L.D.W.)

SURFACE-ACTIVE COMPOUNDS. Those substances that lower the surface tension of a liquid when exposed to a gas, e.g., water in air, or reduce the interfacial tension between two immiscible liquids, e.g., kerosene and water. Effective substances in this respect are the sodium soaps of the fatty acids, namely, lauric with C_{12}-chain, myristic C_{14}-, palmitic C_{16}-, stearic C_{18}-, oleic C_{18}- and 1 olefin bond, and linoleic C_{18}- and 2 olefin bonds. These reduce the surface tension of pure water in air at 20° C. from 73 dynes per cm. to the order of 25 upon dissolving $\frac{1}{10}\%$ soap in the water. The laurate and myristate are more soluble than the higher -C soaps, and are thus better for use with sea water. They are made from such oils as coconut, babassu, and palm-kernel. Olefin bonds in the C-chain bring about greater lowering of the surface tension, and hydroxyl groups, as in recinoleate (hydroxyoleate from castor oil) cause less lowering than the plain chain. Soaps of higher than C_{18}- are practically insoluble in water, and consequently ineffective in lowering the surface tension of water.

Some simple interfacial tension values are the following:

SYSTEM	INTERFACIAL TENSION dynes per cm. at 20° C.
n-Hexane—water	51
Carbon disulfide—water	48
Carbon tetrachloride—water	45
Chlorobenzene—water	37
Benzene—water	35
Nitrobenzene—water	26
Diethyl ether—water	11
Aniline—water	6

Although detergency is a complicated phenomenon, certainly among the important factors are surface and interfacial tensions, since the lowering of these assists in the penetration and wetting of the material to be cleansed, and in the emulsification and suspension of the "dirt" by the action of the cleansing solution. Another factor is the orientation of the molecules of the detergent. These are made up of two oppositely charged groups, one of which is hydrophilic (water-loving) and the other hydrophobic (water-fearing). In ordinary soap the —COONa group is the former and the hydrocarbon residue (—$C_{17}H_{35}$ of stearate) is the latter, and the result is that the —COONa group faces into the water *at its surface* and the hydrocarbon residue faces away from the water. Oily dirt is displaced from fabric by the wetting action of the soap solution to form droplets that can be easily detached from the fabric by stirring, and are thereupon emulsified in the soap solution. Rinsing completes the separation.

The chief disadvantage of soap as a detergent is that it forms insoluble soaps with the calcium, magnesium, and iron salts of hard water, and not until after this reaction has taken place does the residual soap function as a cleansing agent. This causes a loss of soap equivalent to the hardness of the water—a loss which may be ½ of the soap used—and it also causes clots of the insoluble soap which must be removed as so much more dirt. This handicap in the use of soap in hard water can be overcome (1) by the addition of sodium hexametaphosphate or of tetrasodium pyrophosphate, which form water-soluble complex ions with the hardness-producing metals, or (2) by the use of sulfates of the C_{12}- and C_{14}-chain alcohols instead of soap. These compounds, in the form of their sodium salts, possess a double advantage (which may, however, be offset economically by their higher cost) in that (a) the calcium, magnesium, and iron of the hard water do not produce insoluble compounds, and (b) they may be used in acid solution in contrast to soap which cannot be so used since the free fatty insoluble acid is thereby formed. These sulfates, such as Dreft, Drene, Gardinol, are made by hydrolyzing the oil, reducing the free fatty acid so formed to the corresponding alcohol, reacting this alcohol with sulfuric acid to produce the organic hydrogen sulfate, and then neutralizing with sodium carbonate to form the sodium salt desired, thus:

R—COOH → R—CH₂OH → R—CH₂OSO₂OH →

Free fatty   Alcohol   Organic hydrogen
acid   sulfate

R—CH₂OSO₂ONa.

Sodium salt
of organic
hydrogen sulfate

Other detergents useful with hard water and for wetting purposes are the sodium salt of a monogylceride monosulfate, as in the case of sodium glyceryl monolaurate sulfate (Syntex M), and a sodium secondary alcohol sulfonate, as in the case of sodium dioctylsulfosuccinate (Aerosol OT). Sodium dibutylsulfosuccinate is stated to be soluble even in 75% solution of zinc chloride, and a 0.1% solution of this agent in 20% sodium sulfate has a surface tension of only 37.5 dynes per cm.

Surface-active compounds are numerous and widely used as wetting agents and detergents in kier-boiling, bleaching, dyeing and scouring baths in the textile industry. A tabulated list of 350 such materials may be found in *Silk and Rayon* in various issues in the years 1937-39. These compounds are also important in the leather industry for wetting dry hides and skins; in the insecticide spray industry; in glass- and metal-cleaning compounds; in the wet processing of clay and cement; in flame-proofing textile fibers and wood; and in materials for removing wall-paper. (R.K.S.)

SURFACES. Metal products, particularly sheet, plate, and bars, are supplied in various surface finishes. Hot-rolled or forged surfaces generally have a layer of oxidized metal, or scale, which may be loosely to very tightly adherent. This layer may be removed by chemical action resulting in a pickled finish.

Hot-rolled and pickled products are often finished to size by cold-rolling flat products or cold-drawing bars and certain shapes. This gives a smooth, dense surface. For some purposes it is desirable to have a roughened surface on cold-rolled products. This can be accomplished by grit-blasting the finishing rolls. Where high-dimensional accuracy and smooth surface are required, bar stock may be finished by grinding or machining; however, cold-drawn bars are the most widely used.

In addition to these surface finishes, stainless steel, bronze, and certain other non-ferrous alloys may be ground and polished in various degrees up to a high luster or mirror finish. Many alloys may also be brightened by **electropolishing.**

Bearing surfaces of machine parts are sometimes prepared by **superfinishing,** a mechanical lapping operation which gives a mirror finish and permits very small dimensional tolerances.

Instruments for measuring the roughness of commercial finishes are usually of the tracer type in which a finely pointed diamond stylus is drawn across the surface. The motion of the stylus is highly magnified and recorded by electrical means. Indications of surface roughness are measured in microns or in micro-inches. Several types of optical instruments are also used to evaluate surface roughness.

In addition to finishes in which the base-metal surface is retained, protective films and coatings are widely used on metal products. These range from oils and other liquid rust-preventatives to relatively heavy metallic coatings applied by electrodeposition or other means. (See **Anodizing, Bonderizing, Calorizing, Chromizing, Galvanizing, Parkerizing, Sherardizing, Siliconizing, Terne Plate, Tinplate, Vitreous Enamel.**)

Very hard surfaces on steel are produced by processes known as **carburizing, nitriding,** and **hard surfacing.**

A surface may be represented analytically by (1) an **explicit** equation form $z = f(x, y)$, or (2) the **implicit** equation form $F(x, y, z) = o$, where x, y, z denote **rectangular coordinates,** or **spherical coordinates,** or **cylindrical coordinates,** or (3) by a set of **parametric equations** $x = f(u, v)$, $y = g(u, v)$, $z = h(u, v)$.

To study the surface defined by a given equation, find the traces of the surface on the coordinate planes, obtained by putting each coordinate x, y, z equal to o in turn, and then find sections of the surface by planes parallel to the coordinate planes, obtained by putting each coordinate equal to constant values in turn. By putting together these plane sections, one obtains an idea of the nature of the surface. (R.H.H., L.L.S.)

SURGE IMPEDANCE. Characteristic Impedance.

SURGE TANK. The surge tank is a water tank employed to absorb irregularities in flow. It may be used where the total amount of water flowing around the closed cycle is constant, but where the volume passing one point in the cycle varies from that at another. For example, in a condensing power plant, the rate at which feedwater is pumped back to the boiler may be different from that at which steam is supplied to the turbine, although the integrated flows over a definite time interval would be the same. A surge tank interposed between the point of discharge of condensate from the condenser, and the intake to the boiler feed pump, would have a water level which would rise and fall to take care of these irregularities, and in this surge of water in the tank there would be compensation for the different rates of flow.

A hydroelectric plant is frequently provided with a surge tank which is attached to the **penstock** near the plant, by means of a vertical stand pipe. The use of the surge tank is to cushion the penstock **water hammer** which would otherwise arise when turbine gates are suddenly closed. The surge tank is doubly valuable to a penstock because it will not only absorb energy during the deceleration, but will also provide a ready reservoir from which the turbines can draw temporarily, as when they are started during normal operations, or when the sudden heavy demand causes rapid opening of the gates. (F.T.M.)

SURICATE. Mammalia, Carnivora. An animal of southern Africa, *Suricata tetradactyla,* related to the monogooses and with them belonging to the civet group. It is of moderate size, gray with transverse dark bands on the back and a whitish crown, and has rather short legs. It makes an interesting pet. This species shares the name meerkat with one of the mongooses. (A.W.L.)

SURVEYING. This term covers the art of determining the shape, contour, position, or dimensions of any part of the earth's surface, and further, of representing this information on paper. Maps and profiles are the usual method of representing the results of a survey. The data from which these drawings are constructed is obtained by field work, which consists in measuring distances and angles both horizontal and vertical. When the area surveyed is less than 100 sq. mi. in extent, it is considered a plane surface, and the surveying of it is called plane surveying. Surveys of larger tracts of the earth's surface must recognize the fact that the earth's surface is curved. This class of surveying is known as geodetic surveying, and of necessity is more difficult and exacting. However, in no case is a survey to be considered as an exact representation of the area which was surveyed. All surveying is, of course, accurate precise work, necessitating the use of specially designed instruments. However, the degree of precision obtainable varies with the class of survey and the need for precise results. The precision attained in a survey is represented by percent of difference allowable in two measurements of the same distance, or allowed error of closure of a closed survey.

Land surveying consists of the measurement of distances and angles for the purpose of establishing new boundary lines or reestablishing old boundaries when the original corners have been obliterated.

Route surveying is a general term used to cover all surveys made subsequent to and in connection with the final location of a highway, railroad or similar cross-country project. The first step in route surveying is a general study and investigation of all territory along the proposed route. This is known as a reconnaissance.

The second step is the preliminary survey of one or more tentative locations which are selected as a result of the reconnaissance. The preliminary survey is made in order to secure information from which a **topographic map** and **profile** of each of these locations may be made. The establishing of a trial center line of the project on the route finally selected after studying the topographic maps and profiles is called the paper location. The location survey which is the final step is the field work involved in establishing the final center line on the ground and in running a set of profile levels (see **Profile Leveling**) for this center line.

Any survey made with a **transit** is a transit survey.

A construction survey is any survey made in connection with the preparations for the actual construction of a highway, railroad, bridge, building, sewer or any other engineering structure. In general it consists of locating stakes for lines and grades.

Surface surveys of mining claims and underground surveys of mines are classed under the heading of mine surveying.

City surveying applies to surveys made within the boundaries of a city for mapping and municipal construction purposes. It is usually based on a higher degree of accuracy than ordinary land surveying.

Hydrographic surveying refers to surveys of large bodies of water made for the purpose of determining the configuration of the bottom and other pertinent information. The depth at various points, which are located by **triangulation** from the shore, is found by soundings.

Photogrammetric surveying is a term applied to the use of photographs of the earth's surface in connection with the preparation of maps. (See **Photogrammetry.**)

The many divisions of this subject are treated separately in this volume. (See **Maps, Azimuth, Bearings, Metes and Bounds, Land Subdivision, Level, Transit, Declination, Plane Table, Level Rod, Stadia, Differential, Curve, Tangent Offset, Topography, Contour, Compass, Triangulation, Traverse.**) (F.T.M., C.W.C.)

SURVIVAL OF THE FITTEST. Evolution.

SUSLIK. Gopher.

SUSCEPTANCE. Admittance.

SUSPENSION BRIDGE. Bridge.

SUSSEXITE. The term proposed by J. Kemp in 1892 for an igneous rock composed chiefly of **nepheline** and **aegirine**, and essentially free from **feldspar**. Type locality, Sussex County, New Jersey. (R.M.F.)

SUSU. Dolphin.

SUTURE. 1. The line of union of the adjacent flat bones making up the skull. 2. The surgical sewing-up of a wound or incision. Suture material is generally classified either as absorbable or non-absorbable. The absorbable type is made of either plain or **chromic** catgut. Non-absorbable material is silk, linen, fine wire, or strands of synthetic composition.

A fascial suture is one fashioned of a strip of **fascia** which is usually removed from the thigh where it covers the external muscles. Such sutures are often described as living sutures as they act similarly to a graft. They are used principally in the repair of large **herniae** where the tissues are weak. (R.S.M., D.M.H.)

SWAGE. Forging.

SWALLOW. Aves, Passeriformes. An insect-eating bird (**Aves**) with a short wide beak, long and relatively narrow wings, weak feet and legs, and usually a forked tail. The distribution of the swallows is worldwide. They constitute the well-marked family Hirundinidae.

The swallows' nests are built in burrows, holes in trees, and about human dwellings, hence some of the species are familiar friends. In North America the purple martin, *Progne subis,* is widely known from the large colonies that nest in bird houses year after year. Among the true swallows of this continent the barn swallow, *Hirundo erythrogaster,* is among the most beautiful and is undoubtedly the most widely known. The bank, *Riparia riparia,* cliff, *Petrochelidon albifrons,* and rough-winged, *Stelgidopteryx ruficollis,* swallows are also widely distributed and locally common, though less beautiful than the tree swallow and the western violet-green swallow, *Tachycineta thalassina.* The glossy blue and green shades of the upper parts of the last two species are very striking, but both species are found in wild areas, hence they are less familiar than those mentioned above. (A.W.L.)

SWALLOWTAIL. Insecta, Lepidoptera. A large **butterfly,** usually with slender tails extending from the hinder angles of the hind wings. The many species of swallowtails belong to the family Papilionidae. They are widely distributed in the tropical and temperate zones, especially in the former where some are very beautiful and brilliantly colored. Twenty-one species occur in North America. Most of them are yellow with black markings or vice versa but the common pawpaw swallowtail of the eastern and southern states is greenish-white with black bands and some red marks. Some of our species have metallic blue or green scales on the hind wings. Although some species lack the tails, they are also called swallowtails by association with the typical forms. (A.W.L.)

SWAMP. Where the flatness of the land, the presence of impervious soils or bed rock, or abnormal amounts of plant material obstruct or entirely prevent the normal drainage of an area, an excess of moisture will accumulate to the point of saturation and a swamp will come

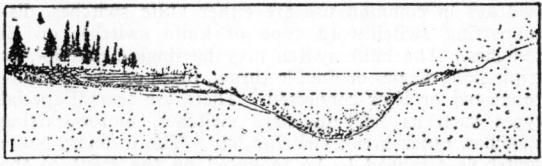

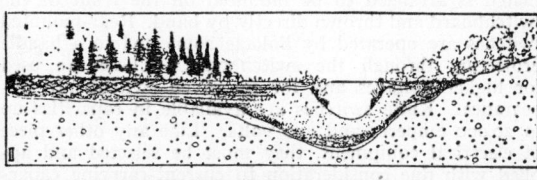

Origin and evolution of a peat bog. (*After Dachnowski.*) I, II and III, illustrating the successive stages in the filling of a pond by the growth of the peat bog. (*Field, Outline of Geology, Barnes & Noble.*)

into existence. While most swamps are level this is not a necessary condition, for hillside swamps are by no means uncommon, due to a constant supply of percolating ground water which maintains the swampy condition. Lake basins are occasionally filled with vegetation and sediment, thus becoming swamps; these are frequently referred to as muskegs, a word of American Indian origin.

Swamps may be formed on the flood plains of rivers as well as upon their deltas; they are characteristic of the flat ill-drained areas of the Atlantic Coastal Plain, examples of which are the Great Dismal Swamp which covers about 2000 sq. miles in the states of Virginia and North Carolina, and the Everglades of Florida, covering about 4000 sq. miles.

Coastal salt-water swamps may develop in the zone between high and low tides or extend up river estuaries; examples of these are common along the Atlantic and Gulf coasts of the United States. In certain northern latitudes swamps develop into peat bogs. Peat bogs are an important source of fuel in Northern Europe, and also serve as an interesting illustration of the origin of **coal,** as exemplified in the **peat, lignite,** bituminous coal series. The accompanying figure illustrates the formation of a peat bog. (R.M.F.)

SWAN. Aves, Anseriformes. A large bird (**Aves**) related to the geese and of similar form and habits. They differ in the length of the neck, which is at least as long as the body in the swans. Although the more familiar species are white, some swans are marked with black and in Australia a species with almost entirely black plumage occurs. The two North American species, the whistling, *Cygnus columbianus,* and trumpeter, *C. buccinator,* swans, breed far to the north and are not often seen. The whistling swan also bears the name whooper in Europe. (A.W.L.)

SWAY BRACING. Bridge.

SWEAT GLAND. A coiled tubular **gland** derived from the outer layer of the skin of **mammals** but extending into the inner layer. These glands are distributed over almost the entire surface of the body in man and many other species but are lacking from the skin of some marine and fur-bearing species.

The secretion of the sweat glands varies greatly. Human sweat is composed chiefly of water, with various salts and organic compounds in solution. It contains minute amounts of fatty materials, **urea,** and other wastes. In certain parts of the body the sweat glands are modified and produce wholly different secretions, including the wax of the outer ear. The sweat of other animals is normally different in composition from that of man.

The sweat glands perform an important function in the maintenance of body temperature by drawing heat from the surface of the body when the temperature of the air is too high to permit adequate radiation. Animals without sweat glands, such as the dog, accomplish the same result by panting, and so evaporating water from the moist lining of the oral cavity and **pharynx.** (A.W.L.)

SWEEP OSCILLATOR. This is the timing axis oscillator for a cathode ray oscilloscope. (See **Cathode-Ray Tube.**) (L.R.Q.)

SWEET FLAG. Aroids.

SWEET POTATO. *Ipomoea Batatas.* Convolvulaceae. The plant is a trailing perennial, the stems of which twine in counter-clockwise direction around supporting objects. These stems arise from much-thickened roots which are rich in **starch.** In cultivation many varieties have been developed, with many different leaf shapes. Dark green, heart-shaped leaves with shining surface occur in several varieties, while in others the leaves are variously lobed and dissected. The flowers, seldom produced in plants grown in northern latitudes, are about 2″ across, purple, and borne either singly or in small axillary **cymes.** The fruit is a **capsule.**

Various methods of propagation are employed. Small roots may be planted whole, **adventitious buds** soon forming and giving rise to shoots which appear above the ground in about four weeks. Root cuttings from growing plants may also be used, especially in regions where the growing season is long. On occasion stem cuttings may be used, but necessarily demand a long growing season and reach maturity very late in the season. Seed may be grown, but germination is slow and uneven, and the product not uniform.

The sweet potato is an American plant, a native of the West Indies and Central America. In cultivation it has gradually spread out of tropical lands, new varieties being developed to suit new localities. At present the crop is grown as far north as Cape Cod.

The principal use of the sweet potato is for human consumption, although some are fed to swine. The vines are frequently used as stock food. From the roots starch, flour, **glucose** and **alcohol** are extracted, to a limited extent.

Sweet potatoes are frequently called yams. This application of the name yam to the sweet potato is confusing, since the true yam is an entirely different plant, *Dioscorea Batatas* (Dioscoreaceae), widely grown in tropical lands for its edible tubers, which are rich in sugar, watery, and soft when cooked. The flowers are white. Propagation is mainly by cuttings of tubers, each containing one or more eyes, or small buds, such as are found in the white potato tuber. Yams are widely used as food.

Another species of Ipomoea, *I. purpurea,* is the Morning Glory, frequently cultivated for its showy flowers. (R.M.W.)

SWEETBREAD. Originally the **thymus** gland of the calf as an article of food, and secondarily applied to the **pancreas.** The term now indicates the pancreas more often in popular usage. Both of these glands are to be regarded as delicacies which are not common enough to be of great dietary interest. The pancreas contains a fairly large amount of **vitamin B.** Otherwise neither gland is a valuable food save for its energy content. (A.W.L.)

SWIFT. 1. Reptilia, Sauria. **A lizard.** The name is applied without scientific accuracy to some of the **iguanas,** including small species of

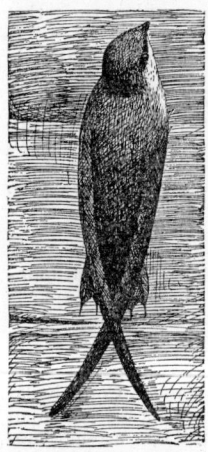

Chimney swift, *Chaetura pelagica.* Sooty brown. Tail feathers sharply pointed. Long narrow wings.

two different genera. Most of these lizards occur in the southwestern United States and Mexico but one species, the pine or fence lizard, *Sceloporus undulatus,* is found as far north as Michigan, New Jersey, and Oregon. 2. Aves, Micropodiformes. Small birds (**Aves**) with short wide beaks and long slender wings. They are superficially like the swallows but are more closely related to the hummingbirds. The numerous species of swifts are widely distributed in both hemispheres, five occurring in North America. The common chimney swift, *Chaetura pelagica,* which has abandoned its original habit of nesting in hollow trees to occupy our chimneys, is both widely distributed and abundant, while the remaining species are found only in the far West and southward. The swifts make their nests of various materials cemented together and fastened to their support with saliva, and one species of the Oriental region uses the secretion alone, without foreign materials. The nests of this species are attached to the walls of caves and are the famous edible bird nests of Chinese epicures. (A.W.L.)

SWIMMERET. The **biramous appendages** of the abdomen of a **crustacean.** Also called pleopods. (A.W.L.)

SWINE. Pig.

SWING BRIDGE. Bridge.

SWINGING CHOKE. This is a variable inductance **choke** often used as the input choke for a smoothing **filter** of a **power supply.** The requirements for the input choke vary with the load on the filter so one value of inductance is needed for no load (other than the bleeder across the filter output) and a much lower value is needed when load is applied. By proper adjustment of the air gap in the **core** of an iron-cored choke this variation in inductance can be made automatic as the current through it varies with the load demand. (L.R.Q.)

SWITCH, ELECTRIC. The switch is a circuit making and breaking device used in electrical circuits for manual operation under normal circuit conditions. It is distinguished from the **circuit breaker** in function since the breaker performs these operations under abnormal conditions such as overloads. The breaker is usually more complex in mechanical and electrical construction since it must be capable of handling the abnormal circuit conditions usually present when it functions. The same apparatus, however, often is used to serve both as an ordinary switching means, i.e., opening and closing the circuits under normal conditions at the will of the operator, and as a circuit breaker, i.e., operating automatically when the circuit ceases to function normally. This is true both of the many small breakers which are replacing fuses in many applications and of the large power oil breakers used for protecting and controlling high-voltage transmission lines.

There are many different kinds of switches. Not considering the numerous kinds of small snap switches, push button switches, instrument and control switches, the switches in common use are either knife switches, disconnecting switches (a type of knife switch), or oil switches. The knife switch may be single-, double-, or triple-pole, single- or double-throw, fused or plain, front-connected or back-connected. There are several special types such as field-discharge switches, motor-starting switches, quick-break switches, etc. The ordinary knife switch is arranged to be mounted on the front of the switchboard and thrown directly by hand. Rear-mounted switches are operated by linkage connected to a handle projecting through the switchboard. These are used for higher voltages and for improved switchboard appearance. Knife switches are available in capacities up to 20,000 amperes and 750 volts. Like any other piece of electrical equipment, they must be selected and applied with due consideration to current-carrying capacity and voltage rating.

The disconnecting switch (often called a "disconnect") is a form of knife switch used primarily to isolate apparatus for inspection or repair, or as a transfer switch so that connections may be changed. It is a switch for opening the circuit only after the current has been interrupted by other means.

Live parts of the oil switch are surrounded by oil retained in a tank. Oil switches are manually operated and not intended to be opened under other than normal load. (F.T.M., L.R.Q.)

SWITCH RAILS. Railway Track.

SWITCHBOARD. Originally the power switchboard consisted of a panel of some insulating material with various knife **switches** mounted on it, and it was from this that the modern switchboard derived its name.

However, the present-day board has little in common with the earlier ones other than being the center of the control system for the switching of electrical energy. The primary function of the board is to permit the proper distribution and control of the energy to various loads. In performing this function it should meet certain fundamental requirements, among which are:

1. Safety to operating personnel.
2. Economic operation of apparatus.
3. Flexibility consistent with economics involved.
4. Rugged mechanical construction and arrangement of components. The exact construction of the power switchboard will be governed by its particular application but such boards may be grouped into the following classes:

   a. Direct-control panel type.
   b. Remote mechanical-control type.
   c. Direct-control truck type.
   d. Electrically operated.

Direct-control boards are generally used for low- and medium-capacity installations where cost is to be kept to a minimum. Such boards have the circuit breakers, disconnect switches, instruments, etc., mounted directly on the board. Sometimes low-voltage boards are of the live-front type, i.e., the switches, etc., are mounted on the front of the board, but the trend is to the dead-front type where breakers and switches are mounted on the rear. Remote mechanical-control boards have the breakers mounted separately from the board and operated through mechanical linkages by controls on the board, thus giving greater safety for personnel. The truck type boards are used in many industrial applications for 15,000 volts and less. These boards have the equipment mounted in steel compartments, completely assembled by the manufacturer, and capable of being withdrawn for inspection and servicing. All high-voltage parts are enclosed and interlocks are provided for safety when the trucks are withdrawn. The electrically operated board has the breakers and other high-voltage equipment located separately from the board and controlled by solenoids or motor operating devices. Such boards may be of the panel type or truck type.

Several materials are more or less standard for use as the panel, among them marble, slate, ebony asbestos and steel. The first three are insulating materials but are being rapidly replaced in modern boards by steel because of its greater strength and improved appearance.

The switchboard equipment varies with the application and type of board, but will include one or more of the following components: breakers, disconnect switches, voltmeters, ammeters, watt-hour meters and other operating instruments, relays, and the necessary bus construction for connecting the various circuits. The present trend is for the board to be assembled in the manufacturer's plant and be shipped as a unit or as several completely wired units requiring a minimum of connecting in the field. Such construction usually results in a neater and more practical board than would be obtained if built up entirely in the field.

The telephone switchboard bears even less resemblance to its namesake than does the power board. The modern telephone board is a jack and plug system for connecting any line with any other line in the office or for connecting to other offices or exchanges (see Telephony). If switches were used (as was the case in very early practice) the complications would be beyond all reason. The board has jacks for all lines being served, various signalling lights, patch cords for connecting from one line to another, and certain associated switches and circuits for the operator's use in handling the connections. The exact components and arrangement are determined by the type service being rendered by the board, i.e., common battery, local battery, A boards, B boards, long distance, etc.   (L.R.Q.)

SWITCHING. Switching refers to the practice or policy used in connection with the subdivision and distribution of power from a central point to a number of radiating power lines. Switching is practiced in most power stations, and in certain substations or switching centers. There are three possible situations to be met, viz.:

1. Power may be distributed at the same voltage at which it is generated.
2. Power may be distributed at a higher voltage than that generated.
3. Power may be distributed at several voltages, one of which is the same as the generator.

Naturally, switching employed to get the first-named condition is simpler because it does not involve the use of transformers for changing voltage. However, this switching arrangement is confined to small stations because ordinary transmission line voltages are very much higher than the usual generator voltage.

The simplest switching arrangement possible is known as the single-bus system. A single bus is provided, to which are connected the generators and the feeder lines. Both generators and feeders are connected to the bus through automatic circuit breakers and disconnecting switches. For normal switching conditions, this simple arrangement will meet every requirement. However, there is no flexibility, and the failure of any generator circuit requires the withdrawal of the corresponding machine and breaker from service. A double-bus single-breaker system provides more flexibility at very little additional cost. This switching arrangement is shown in Fig. 1. Such an arrangement will eliminate the possi-

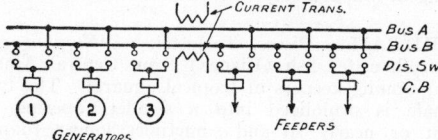

Fig. 1. Double bus, single circuit breaker system.

bility of a long shut-down resulting from a bus failure. It also permits maintaining service while working on either bus, such as cleaning the insulators, etc. It does not, however, eliminate the necessity of withdrawing apparatus from service in case of trouble on the circuit breaker, and in some cases this system is extended to include a circuit breaker in the line to each bus.

In larger stations, or in more important switching centers, different types of switching systems have been developed, some of which are extremely complicated, but are justified by virtue of the service rendered. A double-bus double-breaker system, in which will be found both low and high tension buses, is shown in Fig. 2. Here

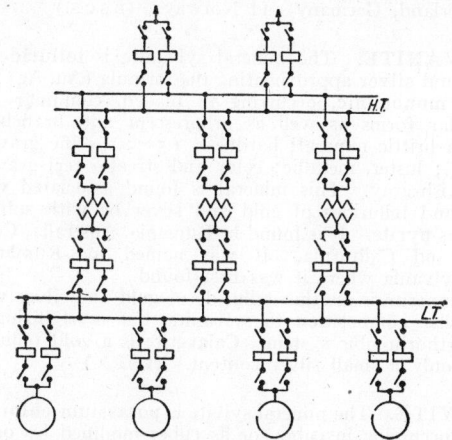

Fig. 2. Bus system using double buses and double breakers.

the outgoing lines (two in number) operate at a voltage higher than that of the generators (four). The increase of voltage is accomplished by transformers located between the high- and low-tension buses. There are three transformers in parallel. In this system the failure of any one bus, or any one circuit breaker, or any one disconnecting switch, will cause only momentary interruption of activity on that line, as in each case there is a spare unit. The installation of duplicate equipment, or duplicate circuits, is very expensive, but the indirect cost to a power and light company of loss of customer good-will through temporary interruptions of service, is considered great enough to justify such a switching system in the more important centers. (F.T.M.) •

SWORD-BEARER. Insecta, Orthoptera. A long-horned **grasshopper** of the group known as cone-headed grasshoppers from the conical prolongation of the head. The sword-bearer is named from the long **ovipositor** of the female. This organ is a slender slightly curved blade longer than the entire body. The species is found in the northern states east of the Rockies. (A.W.L.)

SWORDFISH. Pisces, Teleostei. A large marine fish (**Pisces**), *Xiphias gladius,* whose upper jaw is prolonged into a flat pointed process, the sword, about ⅓ the length of the body. The dorsal fin is very large and scales are lacking. These fishes reach a maximum length of about 16′ and a weight of several hundred lbs. They are related to the sailfish.

The sword is said to be used to spear larger prey and cases are on record of its being driven deep into the planking of wooden ships. (A.W.L.)

SWORDTAIL. Pisces, Teleostei. *Xiphophorus.* A small fresh-water fish (**Pisces**) from Central America, popular among keepers of tropical aquaria. The tail of the male is prolonged into a slender tapering lobe, straight or nearly so and sometimes longer than the body. They belong to the **killifish** family. (A.W.L.)

SYCON. Porifera.

SYENITE. Syenite is a coarse-grained, granular, therefore intrusive, **igneous rock** of the general composition of **granite** except that **quartz** is either absent or present in relatively small amount. The **feldspars** are **alkaline** in character and the dark mineral is usually **hornblende.** Soda-lime feldspars may be present in small quantities. The term syenite was originally applied to hornblende granite like that of Syene in Egypt from whence the name is derived. Syenite is not a common rock, some of the more important occurrences being in New England, Arkansas, Montana, New York State (syenite gneisses), Switzerland, Germany and Norway. (E.S.C.S.)

SYLVANITE. The mineral sylvanite is **telluride** of gold and silver approximating the formula (Au, Ag) Te₂. It is **monoclinic,** occurring in bladed, columnar and granular forms as well as arborescent and branching. It is a brittle mineral; hardness, 1.5–2; specific gravity, 7.9–8.3; luster, metallic; color and streak, steel gray to yellowish-gray. This mineral is found associated with gold and tellurides of gold and silver or with sulfides such as **pyrite.** It is found in Rumania, Australia, Colorado and California. It was named for Rumanian Transylvania where it was first found.

Krennerite is another telluride of gold and silver with a similar composition to sylvanite, but crystallizing in the orthorhombic system. Calaverite is a gold telluride with only a small silver content. (E.S.C.S.)

SYLVITE. The mineral sylvite is **potassium chloride,** KCl, occurring in cubes, or as cubes modified by **octahedra.** It is therefore **isometric.** It has a perfect cubic cleavage; uneven fracture; is brittle; hardness, 2; specific gravity, 1.9; luster, vitreous; colorless when pure but may be white, bluish, yellowish or reddish due to impurities. It is soluble. It is much rarer than **halite** and has been found as sublimates at Mt. Vesuvius and as bedded deposits at Stassfurt, Germany. It is used as a source of potash salts. Potassium chloride was called by the early chemists *Sal digestivus Sylvii,* whence the name of the mineral. (E.S.C.S.)

SYMBIOSIS. This name is applied to an association of two organisms in which each derives some benefit from the association. The two organisms may both be plants. Or one may be a plant; the other, an animal. It is often difficult to determine exactly what each component gains when living together in this way. When careful study shows that one organism gains much more than the other, the association is really one of parasitism. A frequently cited example of true symbiosis is the **lichen,** a composite plant formed of a **fungus** and an **alga** growing together to produce an organism entirely unlike either component. Here the fungus gains nutrient from the alga; presumably the alga gains protection and an increased supply of water. However, it seems as though the fungus gained the greater benefit from its association with the alga. Another example of symbiosis in which the two organisms are both plants is found in **mycorhizae.** Here fungus **hyphae** grow closely around or within the root tissues of some higher plant, which seemingly receives an increased supply of water and inorganic food materials from the presence of the fungus. The latter obtains food. Nodule-forming **bacteria** in roots may be considered an example of symbiosis.

The dependence of plants on insects to effect **pollination** is not usually to be regarded as a case of symbiosis. Symbiosis does, however, seem to exist in such plants as the **Yucca** and the **Fig,** which are entirely dependent on insects for pollination, while the **larva** of the insects live in the **ovary** and feed on the **ovules** of the plants pollinated. Here obviously the two organisms are mutually benefited and dependent each on the other. (See **Animal Association.**) (R.M.W.)

SYMBOLS, CHEMICAL. Chemical Composition.

SYMMETRICAL or UNSYMMETRICAL FOLDS. Anticline.

SYMMETRY. The significance of this term in mathematics is treated in the article on **Locus of an Equation.** In zoology, symmetry is the arrangements of the parts of animal bodies in relation to centralized axes. The bodies of some 1-celled animals are asymmetrical and of others, notably the Heliozoa, spherically symmetrical with the hard parts of the skeleton radiating in various directions from a common center. By far the most common forms of symmetry, however, are those known as radial and bilateral.

Radial symmetry is especially common among the sessile animals such as **sea anemones** and the related **jellyfishes** whose movements are weak. These animals have a principal axis passing through the mouth from which similar structures extend on several radii. The same form of symmetry appears in the echinoderms although these animals begin life as bilaterally symmetrical larvae. The radial symmetry of the adult accompanies sluggish movement and in some forms food-securing habits like those of sessile animals.

Bilaterally symmetrical animals have similar halves flanking a median plane in the principal axis of the body. Sense organs are concentrated near the end that goes first in locomotion, forming a **head** in which the mouth opens as a rule. This end of the body is the cephalic end, in contrast with the opposite caudal end where the tail is attached in the vertebrates. The originally upper and lower or dorsal and ventral surfaces are also differentiated, since the animal rests on the latter while the former is exposed to surrounding influences, and the sides of the body are known as right and left. This type of symmetry prevails in all actively moving animals.

The value of each type of symmetry is clearly correlated with the mode of life in which it is found. Sessile animals receive food and are subjected to dangers only when the responsible factors approach under their own powers of locomotion or on currents in the water. It is an advantage to the animal to be able to perceive such factors as easily in one direction as another. Bilaterally symmetrical animals move about in search of food, hence the end of the body that normally goes first has the chief need of powers of perception, while the upper and lower surfaces are exposed to different environmental conditions and the sides are similar in their contacts. (A.W.L.)

SYMPATHETIC NERVOUS SYSTEM. Autonomic Nervous System.

SYMPHYLA. Small and rare animals living in moist debris at the surface of the ground. They are related to primitive **insects** and **centipedes** and are usually regarded as a class of the phylum **Arthropoda.** They have a pair of antennae but no eyes. The segments of the body are well marked, bearing eleven or twelve pairs of legs. The animals breathe by **tracheae.** (A.W.L.)

SYNAPSE. The association between nerve cells of animals above the **coelenterates.** In coelenterates the nerve net is made up of cells whose processes are structurally connected but in the higher **nervous systems** the fine terminal branches of nerve processes merely come into close contact with those of adjacent cells. While some structural continuity persists in these animals, the synaptic association is regarded as an important foundation for the highly specialized type of nervous coordination that reaches its most complex state in man. The synapse has the important property of allowing impulses to pass in only one direction. (A.W.L.)

SYNAPSIS. Meiosis.

SYNCARIDA. Crustacea.

SYNCHRO. Selsyn.

SYNCHRONIZING, POWER CIRCUITS. Parallel Operation.

SYNCHRONIZING, RADIO CIRCUITS. There are various types of synchronizing requirements in the field of radio and television engineering. In certain applications two or more **oscillators** need to be synchronized, i.e., tied together so their frequencies remain the same. This may be accomplished by injecting into each circuit a synchronizing signal from some common source (which may be one of the oscillators being synchronized). It is characteristic of many oscillator circuits that they will lock in frequency with an injected signal if its frequency is near the natural frequency of the oscillator or is a **harmonic** of the natural frequency. In **television** it is essential that the reproducing circuits of the receiver be synchronized with the transmitter camera device so the reconstructed scene will have its components in the proper places. This is accomplished by transmitting synchronizing pulses from the **transmitter.** In many of the recently developed electronic devices the synchronization of various components of the system plays an extremely important part but the basic method is that used for two oscillators. (L.R.Q.)

SYNCHRONOUS CONDENSER. This is a synchronous **motor** operated over-excited and without load.

When built only for this application the shaft does not have any means provided for taking off motor power and is consequently much lighter than the corresponding motor. A synchronous motor will draw a leading current when over-excited, the magnitude of the current and its angle of lead being determined by the load and the amount of over-excitation on the field. When there is no load (as is the case of the synchronous condenser) the current is almost 90° out of phase with the voltage and thus gives the same effect as a condenser. Such operation not only tends to counteract the usual lagging current taken by most industrial loads but also serves to stabilize the voltage at the end of a long transmission **line.** The advantages are so pronounced that many large power systems have synchronous condensers across them. The trend at present is towards totally enclosed machines having hydrogen as the cooling medium as it provides better cooling and gives less windage loss. (L.R.Q.)

SYNCHRONOUS CONVERTER. Converter.

SYNCHRONOUS IMPEDANCE. This is a fictitious **impedance** used to replace the combined actual impedance of the **armature** of an **alternator** and the effect of armature reaction. By its use the voltage **regulation** of the alternator may be calculated from no-load tests. (L.R.Q.)

SYNCHRONOUS MOTOR. Motor, Electric.

SYNCLINE. The syncline is a structure in which the strata are bent downward in an inverted arch, the sides of which are designated the limbs. The syncline may be a broad open fold or tightly compressed with steep dips, and **pitch** either upward or downward. (R.M.F.)

SYNCLINORIUM. Anticlinorium.

SYNCOPE. A fainting spell, in which the unconsciousness is due to a temporary cerebral anemia; i.e., insufficient circulation of blood in the **brain.** (R.S.M.)

SYNCYTIUM. A mass of **protoplasm** containing many **nuclei,** not separated by cell boundaries. In some cases the syncytium is a network in which partially distinct cell bodies are connected by protoplasmic strands while in others the mass is broadly continuous. The nerve net of the jellyfishes is an example of the former type and the plasmodium of Mycetozoa (**Sarcodina**) is a conspicuous illustration of the latter. The plasmodium is formed by the joining of separate cells. The formation of syncytia by repeated nuclear subdivision without accompanying division of the **cytoplasm** has also been observed. (A.W.L.)

SYNDROME. A group of symptoms characterizing or occurring in any abnormal state or disease. (R.S.M.)

SYNECOLOGY. Autecology.

SYNGAMY. A synonym of **conjugation.**

SYNGENETIC. Epigenetic.

SYNODIC PERIOD. The synodic period of any member of the **solar system** is the time required for the object to go from some particular position relative to the sun as seen from the earth back to the same position again. In the case of the **moon** the synodic period is the time required for the moon to go from **conjunction,** or new moon, back to conjunction again. This period of approximately 29.5 days is the original **month** as used by ancient astronomers in the construction of the **calendar.**

Since a **planet** is best observed at **opposition,** the synodic period of the planet gives the interval of time between successive positions of favorable observation. The synodic period is related to the sidereal period, i.e., the actual period of revolution of an object about the sun, by a simple relationship:

Let P be the sidereal period of the object.

S the synodic period of the same object.

E the sidereal period of the earth (approximately 365.25 days).

Then

$1/S = 1/P - 1/E$ for planets with orbits inside that of the earth.

$1/S = 1/E - 1/P$ for planets with orbits outside that of the earth.

(W.K.G.)

SYNOPTIC REPORTS AND CHARTS. In meteorology, observed weather conditions are placed in brief coded form as a synopsis of the conditions. Such briefs of conditions are known as synoptic reports. Likewise, a map upon which numerous weather reports are placed and upon which an analysis of the data appears is known as a synoptic chart. Synoptic charts may show conditions at sea level, any other level, or in cross-section.

the isobars shown in millibars on one end and inches of mercury on the other. Winds blow clockwise about regions of high pressure and counterclockwise about regions of low pressure. The high cell over the southeastern part of the U. S. is therefore feeding warm moist air over the eastern part of the country. A double low center is located just east of the Rockies in the Saskatchewan-Dakota region. The fronts are the heavy lines on which are attached teeth and lobes, according to the symbols listed in the lower left corner of the map. All of the storminess of the day is associated with the low centers or the frontal zones. During the summertime, fronts and cyclonic systems are comparatively weak. In winter, they are well defined and vigorous. The daily weather map is useful to the layman in interpreting forecasts and in amateur forecasting. (P.E.K.)

SYNOVIA. The transparent sticky fluid contained in a joint cavity or tendon sheath. It is secreted by the synovial membrane, the shiny smooth lining membrane of a joint or tendon sheath. The fluid serves as a lubricant to the joint or tendon so that no friction occurs between opposing surfaces. (R.S.M.)

SYNOVITIS. Inflammation of a **synovial** membrane characterized by an increase in synovial fluid, swelling, pain about the involved joint. If the fluid in a joi'

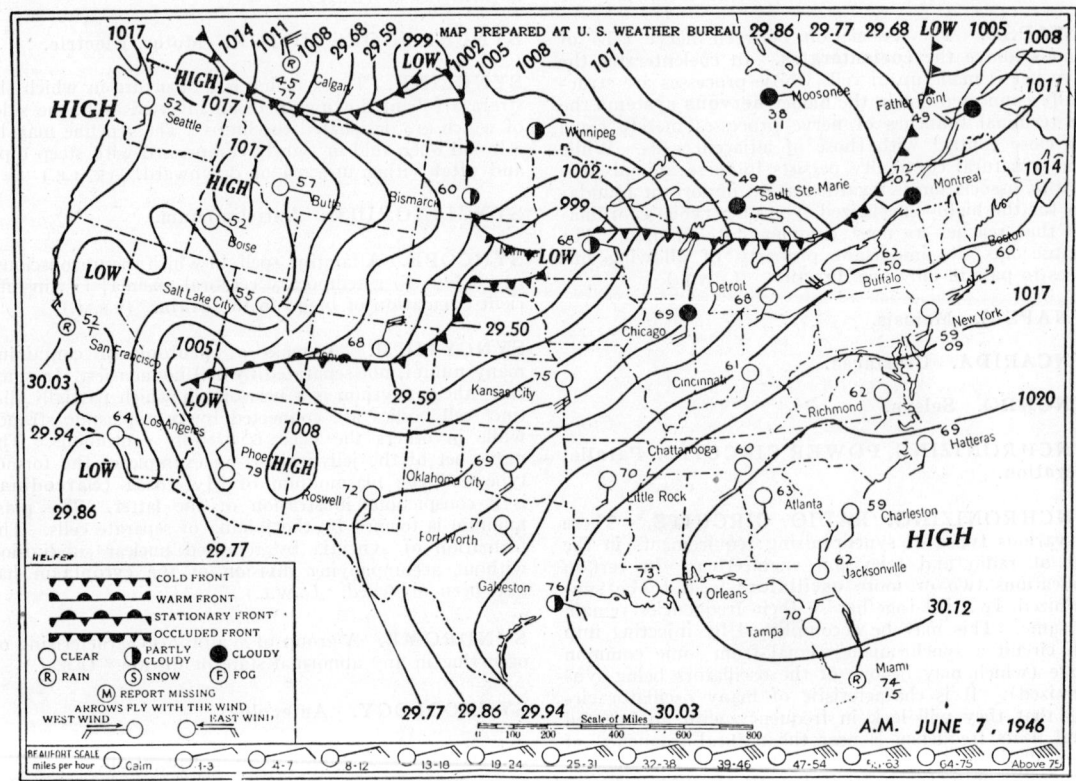

Weather maps of the United States and adjacent areas are issued daily by the United States Weather Bureau. A sample map is illustrated. Narrow black lines are isobars. Sea level pressure is the same everywhere along an isobar. In addition, the isobars show the direction and velocity of the wind. Winds blow nearly parallel to the isobars. On the average, they blow at an angle of 20° across the isobars, directed away from high pressure and toward low pressure. The distance in miles between isobars divided into 2500 will yield the approximate surface wind velocity ($V = 2500/D$). The numbers at the ends of the isobars are pressure values of

becomes infected, suppuration may occur; surgical treatment to drain the pus from the joint may be necessary. (D.M.H.)

SYNTEXIS. The term proposed by Loewinson-Lessing, in 1899, for the generation of **magmas** either by remelting or assimilation of portions of the **lithosphere** regardless of the type or variation of its lithology. (Contrast with **Anatexis.**) (R.M.F.)

SYNTHETIC DIVISION. Synthetic division is an abbreviated process using detached coefficients for finding the **quotient** of a **polynomial** in one variable x by a divisor of the form $x - r$, where r is a constant. It may be indicated thus:

To divide the polynomial $a_0x^4 + a_1x^3 + a_2x^2 + a_3x + a_4$ by $x - r$, the following scheme is computed:

$$
\begin{array}{ccccc|l}
a_0 & a_1 & a_2 & a_3 & a_4 & \underline{r} \\
 & a_0r & A_1r & A_2r & A_3r & \\
\hline
\end{array}
$$

$a_0, \ A_1 = a_0r + a_1, \ A_2 = A_1r + a_2, \ A_3 = A_2r + a_3, \ R = A_3r + a_4$

Then the quotient is $a_0x^3 + A_1x^2 + A_2x + A_3$ and the remainder is R. A similar process applies to a polynomial of any other degree.

The synthetic division process may be described in words by the following rule:

To divide a polynomial in one variable x by a binomial divisor of the form $x - r$, arrange the polynomial in descending powers of x, as $a_0x^n + a_1x^{n-1} + \cdots + a_{n-1}x + a_n$.

Arrange the detached coefficients $a_0, a_1, \cdots, a_n$ in order in the first line, supplying any missing power of x with a zero coefficient, and write r at the right.

Bring down a_0 in the first place in the third line. Multiply a_0 by r, write the product in the second line under a_1, and write their sum in the third line directly underneath; multiply this sum by r, add the product to a_2, and write the sum underneath in the third line, etc., continuing in the same way until finally a product is added to the last coefficient a_n.

Then the last sum in the third line is the remainder, and the preceding sums are the coefficients of the powers of x in the quotient, beginning with x^{n-1} and arranged in descending order.

If the remainder in this synthetic process is zero, the divisor is shown to be a factor of the given polynomial. (L.L.S.)

SYNTHETIC SUBSTITUTION. If the **synthetic division** process is carried out with a **polynomial** $P(x)$ and a divisor $x - r$, by the **remainder theorem** the last result R is the value $P(r)$ of the polynomial $P(x)$ when we substitute r for x; in this case, if we are concerned only with the remainder R (and not with the quotient), we may call the process synthetic substitution. (L.L.S.)

SYNTHETIC RESINS. Plastics.

SYPHILIS. An infectious disease, congenital or acquired, caused by the spirochete, *Treponema pallidum*. Syphilis is characterized by primary and secondary stages during which it is highly contagious, and a noncontagious, late or tertiary stage, marked by involvement of many organs and tissues. In the congenital form there is no primary lesion and late manifestations predominate.

The early history of syphilis is not entirely clear. It is thought by some historians that the disease was first introduced into Europe by Columbus' returning sailors, and subsequently spread through Italy where it became a great scourge, by the soldiers of Charles VIII. Another view is that the disease has existed in civilized man since antiquity. Early Egyptian and Assyrian inscriptions as well as bony changes found in mummies are interpreted as supporting this theory. In the Middle Ages, syphilis occurred in severe, widespread epidemics with enormous mortality rates. Later it became a milder disease, and its venereal nature was recognized. It was not until 1905, when Schaudinn discovered the *Treponema pallidum* to be the causative organism, that syphilis and **gonorrhea** were recognized as two distinct diseases. The next landmarks were the development of the diagnostic **Wassermann Reaction** in 1906 and the discovery in 1908 by Ehrlich of salvarsan (**Arsphenamine**, "606") which is a highly effective remedy against the disease.

It is estimated that 10% of the population of the United States is infected with syphilis. The incidence varies widely in different groups, from less than 1% to 25% or higher. It is especially high in Negroes.

1. *Acquired Syphilis.* Syphilis is usually transmitted during sexual intercourse. Occasionally an extragenital sore occurs—on the lip or finger—from kissing or from contact with contaminated material, drinking glasses, etc. The causative agent is a motile "corkscrew" spiral organism which abounds in primary and secondary lesions. The spirochetes gain entry through an abraded surface, usually on the genital organs. In the male, the primary sore is commonly on the penis, where it is easily seen. In the female, it is found either on the mucous surface of the **vulva** or within the **vagina** where it may remain unnoticed.

The primary stage is marked by the appearance of the **chancre**, which develops at the site of invasion, usually 2–4 weeks after exposure. Before the chancre develops, the organisms often have invaded the body tissues by way of the lymph channels and blood stream. A fully developed chancre appears as a clean, slightly raised, hard, circumscribed **ulcer**, which exudes a thin, highly infectious secretion. By dark-field microscopic examination, this secretion can be seen to teem with spirochetes. The lesion is painless, but tender lymph nodes may develop in the groin. The blood Wassermann test is usually negative throughout the primary stage and the diagnosis can be made only by finding the spirochetes in the chancre.

The secondary stage begins 6 weeks to 6 months after the primary. By the time of onset of this period, the organisms have invaded the tissues, and the body's defense reactions are active. **Antibodies** are being produced, and the Wassermann reaction is positive. The second stage may be so mild as to pass unnoticed, but it is usually ushered in with a rash over the skin and **mucous membranes**. The eruption is extremely variable and may imitate any skin disease. The commonest type is the macular, in which the lesions consist of flat, or slightly raised, rose-colored spots, most prominent over the abdomen and chest. Characteristic features of the secondary eruption are its symmetrical distribution, painless, non-itching nature, and its tendency to appear on the palms and soles. Its duration is variable from weeks to months, and it may fade and leave a faint pigmentation for a time. With treatment the rash disappears promptly. Spirochetes are present in skin and mucous-membrane lesions, and are particularly easy to demonstrate in the soft moist sores on the genetalia, the condylomata.

Constitutional symptoms in the secondary stage are usually mild. They consist of sore throat, slight fever, headache, and some enlargement of the superficial lymph nodes.

The latent period is the interval between the secondary and tertiary phases of the disease. During this time, the patient may enjoy excellent health, and be unaware that he has syphilis. The duration is variable—from a few weeks to many years—even 25 or 30. During this time, the various organs and tissues are harboring spirochetes, but these are inactive, hibernating as it were.

The tertiary stage is the late picture, which is usually seen only in untreated syphilis. It is marked by the development of **gumma**, tumor-like masses in the skin and visceral organs, and inflammatory changes in the cardiovascular and central nervous systems. Since any organ or tissue may be attacked in tertiary syphilis, the signs and symptoms are extremely varied.

The most common types of tertiary syphilis are those involving the cardiovascular and central nervous systems. Cardiovascular involvement usually appears earlier—10 to 15 years after infection. It occurs as **aortitis**, aortic **aneurysm** and sometimes **coronary occlusion**. In syphilitic heart disease, aortitis is commonly followed by insufficiency of the aortic valve. This mechanical defect puts a tremendous strain on the heart, which compensates by **hypertrophy** of the muscle. Eventually the strain is too great for the reserve, and **heart failure** and finally death are the result.

Central nervous system syphilis may not develop for 30 years after the primary infection. It occurs as a **meningitis,** as an arteritis or **inflammation** of the small cerebral vessels with secondary changes in the brain tissue, and as involvement of the cells of the brain and cord in **tabes dorsalis** and **paresis.** The latter two are discussed elsewhere under their respective headings.

2. *Congenital Syphilis.* The congenital type of the disease is transmitted only through the mother who may have acquired her disease before or after conception. The spirochetes reach the fetus through the **placenta,** and disseminate widely through the fetal organs. In the most severe forms of infection, the child may show at birth an extensive skin eruption, fissures about the angles of the mouth and nose, the characteristic nasal discharge, "snuffles," bone lesions, and enlargement of the liver and spleen. Underdevelopment and poor nutrition are conspicuous; such a child rarely lives longer than a few days. In other instances, the infant may appear normal at birth, yet after a few months develop the typical signs and symptoms of the disease.

Children who survive the active congenital disease, or those in whom the infection remains latent, often show certain permanent stigmata which make the diagnosis apparent at a glance. "Saddle nose" and deformed teeth (Hutchinson's teeth)—peg-shaped notched incisors, which are widely spaced—are highly characteristic. Bony lesions, inflammation of the cornea of the eye (interstitial keratitis), deafness, and central nervous system syphilis such as occurs in the acquired form, are also seen in children with congenital syphilis.

3. *Prognosis and Treatment.* Syphilis can be cured, and congenital syphilis can be prevented. The success of treatment depends largely on the duration of the disease. If begun during the first few weeks or months after infection, proper treatment results in permanent cure in almost all instances. In the late, or tertiary stage when irreparable tissue damage has occurred, cure is not possible in the sense of a return to normal structure and function, although arrest of the disease and control of the symptoms can often be accomplished.

For many years, the arsenical drugs, **arsphenamine,** neoarsphenamine, Mapharsen, etc., used alternately with **bismuth,** have been the drugs of choice. Neoarsphenamine intravenously at weekly intervals for 6 weeks, alternating with 6 weekly injections of bismuth intramuscularly continued over a period of a year and a half, is the usual routine. Continuation of treatment for at least a year after the Wassermann test becomes negative is also included in most schemes of treatment. Congenital syphilis is treated similarly but more cautiously. Pregnant women found to be syphilitic are given more intensive arsenical therapy, especially just prior to delivery.

The accepted scheme for treatment is now undergoing considerable change because of the introduction of massive arsenotherapy and the advent of **penicillin.** Only for primary and secondary stage cases of syphilis is massive therapy used. The entire treatment consists of a total of 1.2 grams of Mapharsen given slowly by continuous intravenous drip 7–10 hours per day for several days. So far, good results and few toxic effects have been found. Penicillin is also being used in the treatment of acute primary and secondary syphilis. The drug is given in large doses over a period of several days, and the early results have been excellent. Further evaluation of the rapid treatments must await the passage of time.

In tertiary syphilis, the problems of treatment are somewhat different. Less intensive therapy is recommended. In central nervous system diseases, tabes and paresis, tryparsamide and fever therapy, either with malaria or a "hot box" are used. Iodides, neoarsphenamine, and bismuth are used cautiously, especially in cardiovascular syphilis. (D.M.H.)

SYRINX. Sound Production.

SYSTEM. For the use of this term in geology, see **Period.**

SYSTEMIC HEART. The principal **heart** of the **cephalopod** mollusks, in contrast with the supplementary **branchial hearts.** The systemic heart includes two or four auricles which receive blood from the gills. The blood passes from the thin-walled auricles into a single muscular ventricle whose contractions propel the liquid through the arteries to the body. (A.W.L.)

SYSTEMS OF ALGEBRAIC EQUATIONS. In the consideration of **equations** with more than one unknown, we may deal with single equations in several unknowns or with systems of such equations.

An **algebraic equation** with more than one unknown will, in general, be satisfied by an unlimited number of sets of values of the unknowns; such an equation is called an indeterminate equation.

A set of algebraic equations with more than one unknown is said to form a system when the equations are considered together with the object of determining whether they are satisfied by one or more sets of values of the unknowns. The term "simultaneous equations" is often used to mean a system of equations.

A solution of a system of equations with more than one unknown is any set of corresponding values of the unknowns which satisfy the equations of the system. (L.L.S.)

SYSTEMS OF CIRCLES. Circle.

SYSTEMS OF EQUATIONS INVOLVING QUADRATICS. Quadratic Equations, Systems of.

SYSTEMS OF LINEAR ALGEBRAIC EQUATIONS. Linear Equations, Systems of.

SYSTEMS OF LINEAR DIFFERENTIAL EQUATIONS. Linear Differential Equations.

SYSTEMS OF LINES. Straight Line in a Plane.

SYSTEMS OF LOGARITHMS. Logarithms.

SYSTOLE. That period during which the **heart** muscle is contracting and the blood is expelled from the ventricles of the heart into the aorta and arterial system (Compare **Diastole**). (R.S.M.)

T SECTION. This is a network connection which has two elements in series in one line and a third element in shunt from the junction of the two series elements to the opposite line, thereby giving a circuit diagram which looks like a T. (L.R.Q.)

TAB, AIRCRAFT CONTROL. An auxiliary airfoil attached to a control surface for the purpose of reducing the control force or trimming the aircraft. (F.T.M.)

TABES DORSALIS (Locomotor ataxia). A chronic syphilitic infection of the **spinal cord**, occurring as a manifestation of **tertiary** syphilis. It is the end-picture in about 5% of untreated cases. Pathologically, there is a degeneration of the cells of the posterior columns of the cord, and interference, therefore, with the sensations of touch and pain. The nerves of the organs of special sense, particularly the optic nerve, are often involved.

The signs and symptoms of tabes are numerous, and do not make their appearance in any set order. Sometimes well-marked neurological signs may be present in the absence of symptoms, or symptoms may be striking long before neurological signs can be detected. So-called lightning pains—severe shooting, lightning-quick pains in the legs are often an early symptom. Others are difficulty in urination, incontinence of urine, loss of sexual power, and diminished vision. Early signs are loss of the tendon reflexes, particularly the knee and ankle jerks, weakness of the eye muscles, changes in the pupils—small, unequal, irregular pupils which react sluggishly to light if at all. Abnormalities of gait are common; they are due to muscular incoordination and loss of the position sense. As the disease progresses, the walk is often characteristic: the foot is raised high, thrown forward forcibly, and slapped to the ground.

Visceral crises are characteristic of tabes. Of these, gastric crises are the most common. In their most severe form they consist of recurrent attacks of excruciatingly severe abdominal pain, accompanied by violent and continuous vomiting. Even **morphine** gives no relief. The crises may last several hours or several days. They may be the only manifestation of central nervous system syphilis.

The trophic disturbances, chronic ulcers, Charcot joints, which result from loss of sensation and therefore absence of reactions which normally protect the part, usually occur late in the disease. Charcot joints (named for the neurologist who described them) are tremendously deformed joints with marked destruction and abnormal mobility. The knees, hips and ankles are the most commonly affected.

The diagnosis of tabes is established on the basis of the history, physical findings, and a positive spinal-fluid Wassermann test. Tabes often occurs in combination with **paresis**; the picture is then spoken of as taboparesis.

Treatment is less effective in tabes than in other forms of neurosyphilis. Early cases sometimes do well on neo-arsphenamine, but more fully developed ones require tryparsamide and fever therapy. The latter is given artificially in a specially constructed cabinet or "hot box," or by the inoculation of malaria parasites and the production of tertian or vivax **malaria.** The prognosis for cure is never very hopeful in advanced tabes, but even then improvement and remission may occur. (D.M.H.)

TABLING. Classification.

TACHYLYTE or TACHYLITE. Pure tachylite is a natural, **basic** black glass which may form along the chilled contacts of **dikes** or **sills.** It also occurs as a rind on basic **pillow lavas** which have been suddenly chilled by plunging into water. Occasionally it forms entire flows from certain Hawaiian volcanoes. (R.M.F.)

TACONIC REVOLUTION. Ordovician.

TACONITE. Term for the ferruginous **cherts** associated with the iron ores of Lake Superior district. (R.M.F.)

TACTILE ORGANS. Organs of touch. Sensory organs located at the surface of the body which are stimulated by pressure. The chief tactile organs of the human body are known as tactile corpuscles. They consist of an elliptical bulb about $\frac{1}{300}''$ or less in length, enclosed in a connective tissue sheath and divided incompletely by plates of the same tissue. From one to several coarse nerve fibers enter the corpuscle and follow a winding course inside, ending between or on the connective tissue cells. These corpuscles are especially abundant in the skin of the finger tips and in other parts where the sense of touch is highly developed, hence there can be no reasonable doubt that they serve this sense. They are related to other corpuscles of more limited occurrence which may also be tactile, and are supplemented in the skin by free nerve endings, some of them expanded into disklike nets, which are regarded as tactile.

In many invertebrates hairlike projections at the surface of the body serve for ready reception of tactile stimuli and in the vertebrates true hairs, often bristle-like, transmit such stimuli to the nerve endings at their bases. From the sensitiveness of the human scalp to light contacts transmitted through erect hairs it is easy to judge the value of these tactile hairs. They are well illustrated by the whiskers or vibrissae of the cat and other mammals. (A.W.L.)

TAFFRAIL LOG. Patent Log.

TAGMA. Body regions of **Arthropoda.** The bodies of these animals consist typically of the head, **thorax**, and abdomen but many forms show a confluence of parts or further subdivision. The **centipedes** and millipedes (**Diplopoda**), with a few related forms, have only a head and a long segmented body also called the trunk. **Arachnida** and **Crustacea** have a **cephalothorax** or prosoma in which the segments belonging to the head and thorax are not sharply separated, and a segmented abdomen or opisthosoma. Subdivisions of the abdomen such as appear in the centipedes are the anterior **mesosoma** and the posterior **metasoma**, sometimes called the postabdomen. The plural of tagma is tagmata. (A.W.L.)

TAHR. Mammalia, Artiodactyla. A wild **goat,** *Hemitragus jemlaicus,* of the Himalayas. This species, together with two other Asiatic species, differs from the typical goats in the lack of a beard. It is found in forested regions at high altitudes but does not enter the open country of Tibet. (A.W.L.)

TAIL. A solid prolongation of the axis of the vertebrate body at the end opposite to the head. The tail contains a portion of the spinal column together with blood vessels, nerves and muscles, but it is entirely posterior to the body cavity which is confined to the trunk. By association with the tail, this end of the body is known as the caudal end, a term applied to all bilaterally symmetrical animals. Many caudal appendages are called tails by a similar association, although they are not true tails. Thus the tail of the fish in the ordinary use of the term is properly the tail fin.

The tails of many **vertebrates** are without apparent value but in some of the **primates** they are prehensile appendages, used in climbing, and the hoofed animals use them in driving away insects. The tail of the **alligator** is an effective weapon. (A.W.L.)

TAIL, AIRPLANE. The rear part of an airplane, usually consisting of a group of stabilizing planes, or fins,

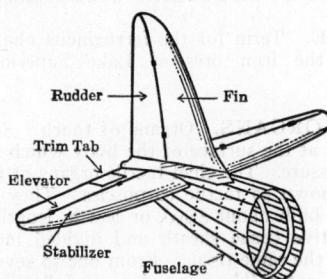

Rudder — Fin
Trim Tab
Elevator
Stabilizer
Fuselage

Typical cantilevered empennage.

to which are attached certain controlling surfaces such as elevators and rudders; also called "empennage." (F.T.M.)

TAILSTOCK. Lathe.

TAKE-OFF. Airplane.

TAKIN. Mammalia, Artiodactyla. A hoofed animal, *Budorcas taxicolor,* found in eastern Tibet and adjacent regions. It is a moderately large animal of heavy build, with strong curved horns. Related to the goats. (A.W.L.)

TALC. The mineral talc is a **magnesium silicate** corresponding to the formula $H_2Mg_3(SiO_3)_4$ which occurs as **foliated** to fibrous masses, its **monoclinic** crystals being so rare as to be almost unknown. It has a perfect basal cleavage the folia non-elastic although slightly flexible; it is sectile and very soft; hardness, 1; specific gravity, 2.5–2.8; luster, waxlike or pearly; color, white to gray or green; translucent to opaque. It has a distinctly greasy feel. Talc is a **metamorphic** mineral resulting from the alteration of silicates of magnesium like **pyroxenes, amphiboles, olivine** and similar minerals. It is found chiefly in the metamorphic rocks, often those of a more basic type due to the alteration of the minerals above mentioned. Of the many foreign localities may be mentioned the Austrian Tyrol, the St. Gotthard district of Switzerland, Bavaria and Cornwall, England. In Canada talc is found in Brome County, Quebec and Hastings County, Ontario. In the United States well-known localities are to be found in Vermont, New Hampshire, Massachusetts, Rhode Island, New York, Pennsylvania, Maryland and North Carolina.

A coarse grayish-green talc rock has been called soapstone or steatite and was formerly much used for stoves, sinks, electrical switchboards, etc. Talc finds much use as a cosmetic, for lubricants and as a filler in paper manufacturing. Most tailor's "chalk" consists of talc. The origin of the word talc is not definitely known. (E.S.C.S.)

TALLOW. Esters.

TALUS. The mass of coarse rock fragments which accumulate at the foot of a cliff as a result of the processes of weathering and gravity. In Great Britain the term scree is used for such material. (R.M.F.)

TAMANDUA. Mammalia, Edentata. The lesser **anteater** of Central and South America, a species related to the great anteater and of similar form. It differs in the short-haired prehensile tail, which is used in climbing. The name is that of the genus, derived from the Portuguese. (A.W.L.)

TAMARAO. Mammalia, Artiodactyla. A small **buffalo** native to the island Mindora in the Philippines. It is related to the Indian buffalo and to the anoa. (A.W.L.)

TANAGER. Aves, Passeriformes. An American bird (**Aves**) of brilliant coloration, related to the finches. The tanagers are chiefly birds of the Central and South American tropics, where about 400 species are found. Of the few species that enter the United States only the scarlet tanager, *Piranga erythromelas,* is widely known. The brilliant red of its body, contrasting with the black wings and tail, makes it one of our most striking species. The dull olive plumage of the female exemplifies a common contrast between the sexes in the entire group. (A.W.L.)

TANAIDACEA. Crustacea.

TANG. Drilling.

TANGENT OFFSET. A method occasionally used by surveyors to reduce field notes or area or route surveys to map form is known as plotting by "tangent offsets." A small-scale sketch of the **traverse** made with a protractor and scale, will greatly facilitate the plotting since it serves as a guide for locating the initial line so that the traverse will fall within the limits of the paper.

The use of tangent offsets is illustrated in Fig. 1. The initial line ab is located by means of the sketch men-

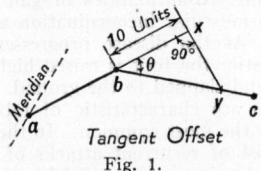

Tangent Offset
Fig. 1.

tioned above and prolonged a distance of 10 units to a point x where a perpendicular is erected. The line xy, equal to the natural tangent of angle θ multiplied by 10 units, is then plotted on the perpendicular. The angle θ may be obtained by direct measurement of the **deflection** angles in the field or by taking the difference of the **bearings** or **azimuths** of the two lines. A line connecting b and y will give the direction of bc and serves as a base upon which to lay off the distance from b to c as measured in the field. Each successive course may be plotted in relation to the preceding course by the method just described. To avoid errors in plotting it is well to occasionally check the bearing of a particular line in relation to a fixed meridian by means of scaled distances.

A method known as "offsets from the tangent" is frequently used for laying out **circular curves** in the field. The geometry of the method is illustrated in **Fig. 2.**

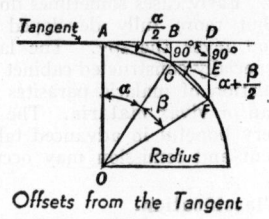

Offsets from the Tangent
Fig. 2.

In order to lay out the curve it is necessary to locate points such as C and F in respect to the tangent through A which is the point where the curve begins. A transit is set up at A, sighted in the direction of the tangent AD, and points B and D located by taped distances. The transit is then set up over B, sighted back on A and a 90° angle turned. The tangent offset BC is taped which locates point C. Point F may be located in a similar manner. The coordinates used in the field operations are calculated from assumed **chord** lengths such as AC and CF which fix the value of the angles α and β respectively when the radius of the curve is given. (c.w.c.)

TANGENT PLANE TO A SURFACE.

A straight line is said to be tangent to a surface at a given point P if it is the limiting position of a secant line through P and a neighboring point P′ on the surface when P′ is made to approach P along the surface.

All tangent lines to a surface at a given point lie in a plane, which is called the tangent plane to the surface at the given point.

If the equation of a surface in rectangular coordinates is $F(x, y, z) = 0$, the tangent plane at the point (x_1, y_1, z_1) has the equation

$$F_x(x_1, y_1, z_1)(x - x_1) + F_y(x_1, y_1, z_1)(y - y_1)$$
$$+ F_z(x_1, y_1, z_1)(z - z_1) = 0,$$

where $F_x(x_1, y_1, z_1)$ is the value of the **partial derivative** $\frac{\partial F}{\partial x}$ at the point (x_1, y_1, z_1), and similarly for the other coefficients. The equation of the tangent plane is often written

$$\frac{\partial F}{\partial x}(x - x_1) + \frac{\partial F}{\partial y}(y - y_1) + \frac{\partial F}{\partial z}(z - z_1) = 0,$$

but it must be understood that the values of $\frac{\partial F}{\partial x}$, etc., must be taken at the point of tangency.

If the equation of the surface in rectangular coordinates is of the form $z = f(x, y)$, the equation of the tangent plane at (x_1, y_1, z_1) is

$$f_x(x_1, y_1)(x - x_1) + f_y(x_1, y_1)(y - y_1) = z - z_1.$$

A normal line to a surface at a point P_1 is the line through P_1 which is perpendicular to the tangent plane to the surface at P_1.

If the equation of the surface in rectangular coordinates is of the form $F(x, y, z) = 0$ or of the form $z = f(x, y)$, the equation of the normal (line) at a point (x_1, y_1, z_1) is, respectively,

$$\frac{x - x_1}{F_x(x_1, y_1, z_1)} = \frac{y - y_1}{F_y(x_1, y_1, z_1)} = \frac{z - z_1}{F_z(x_1, y_1, z_1)},$$

or

$$\frac{x - x_1}{f_x(x_1, y_1)} = \frac{y - y_1}{f_y(x_1, y_1)} = \frac{z - z_1}{-1}.$$

(L.L.S.)

TANGENTS AND NORMALS TO PLANE CURVES.

Consider a plane curve (Fig. 1); take a point P_1 on the curve, also take a second point P_2 on the curve and draw the secant line P_1P_2. If as P_2

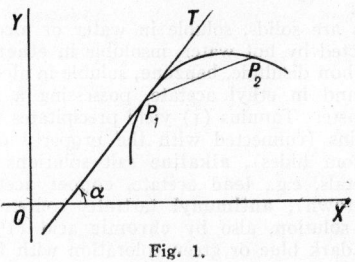

Fig. 1.

approaches P_1 along the curve, the secant P_1P_2 approaches a limiting position P_1T, this line P_1T is called the tangent to the curve at P_1.

Another form of definition of tangent to a curve, which can be shown to be equivalent to the preceding, is the following (Fig. 2): Let a line l be drawn cut-

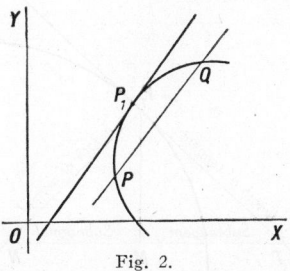

Fig. 2.

ting the curve in two points P and Q, and let this line move parallel to itself so that P and Q move closer together: then if when P and Q approach coincidence, the secant line PQ approaches a limiting position P_1T, this line P_1T is called the tangent to the curve at P_1.

The **slope** of the tangent to a curve at a point P_1 is often called the slope of the curve at P_1.

If the equation of a curve in rectangular coordinates is $y = f(x)$, then the slope of the tangent to the curve at a point (x_1, y_1) is given by the value of the **derivative** $f'(x_1)$ at the value $x = x_1$. The equation of the tangent at (x_1, y_1) is then

$$y - y_1 = f'(x_1)(x - x_1).$$

If the equation of a curve is given in **polar coordinates** as $r = f(\theta)$, the direction of the tangent at a point (r_1, θ_1) is determined by $\tan \psi_1 = r_1/f'(\theta_1)$, where ψ_1 is the angle between the **radius vector** to the given point and the tangent at this point; the angle α, which the tangent makes with the X-axis (or polar axis) is then $\alpha_1 = \theta_1 + \psi_1$. (Fig. 3).

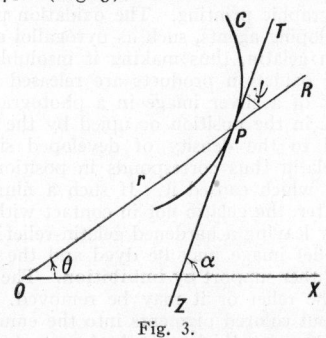

Fig. 3.

A normal to a curve is a straight line perpendicular to a tangent to the curve; its slope is the negative reciprocal of the slope of the tangent. If the equation of the curve in rectangular coordinates is $y = f(x)$, then the equation of the normal at the point (x_1, y_1) is

$$y - y_1 = -\frac{1}{f'(x_1)}(x - x_1).$$

If the tangent and normal at P_1 intersect the X-axis at T and N respectively, then we define: P_1T = length of tangent at P_1, and P_1N = length of normal, at P_1 (Fig. 4).

The projections on the X-axis of the length of tangent P_1T and the length of the normal P_1N are called the subtangent and subnormal at P_1. Then subtangent =

TM_1, subnormal $= M_1N$, where M_1 is the projection of P_1 on $X'X$.

If m is the slope of the tangent at $P_1(x_1, y_1)$, then subtangent $= -y_1/m$, subnormal $= my_1$. (L.L.S.)

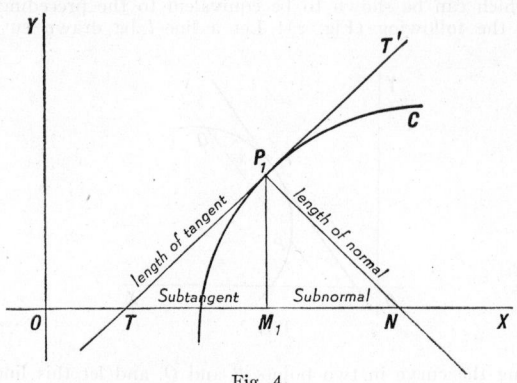

Fig. 4.

TANGERINE. Citrus Fruits.

TANIPAN. Carapato.

TANK. Pressure Vessels.

TANK CIRCUIT. This is the name often given to the inductance-capacitance parallel circuit in the grid or plate circuit of a vacuum tube **oscillator** or class C **amplifier.** It is the resonant frequency of this circuit which determines the output frequency of the vacuum-tube stage. In this type operation the tube is operating in pulses rather than continuously as in Class A service but the tank circuit provides the electrical fly-wheel action needed to keep the circuit operating between pulses. (L.R.Q.)

TANNING DEVELOPMENT. The use of a tanning developer to obtain a hardened gelatin relief for use in color photographic printing. The oxidation products of certain developing agents, such as pyrogallol and hydroquinone, tan gelatin, thus making it insoluble in warm water. The oxidation products are released during the development of a silver image in a photographic emulsion and are in the position occupied by the image and proportional to the density of developed silver. The hardened gelatin thus corresponds in position with the silver image which caused it. If such a film is treated with hot water, the gelatin not in contact with the silver washes away leaving a hardened gelatin-relief image.

Such a relief image may be dyed and the dye transferred to another support by **imbibition.** The silver may be left in the relief or it may be removed. It is also possible to put colored pigments into the emulsions during manufacture in which case the image obtained after washing in hot water will be properly colored. Such a pigmented relief image may be stripped off onto another support in register with the other colorant images making up the color reproduction or it may be used as part of a color transparency. (H.C.C.)

TANNINS. Tannins are substances generally related to one of the **phenols,** pyrogallol or catechol, and they are found in many plants. By their action on animal skins they cause changes which make the skins resistant to decomposition and at the same time leave them flexible and very strong, greatly improved in wearing qualities. Skins so treated are said to be tanned, and are called leather. Tanning is a very old art, having been practiced in China since long before the Christian era. It was also known to the American Indian before the arrival of the white man.

Tannins are found in various parts of the plant, appearing frequently in leaves, and in the cortical tissues of stems. Tannins may be found in the walls of cells or in the **vacuoles;** often their presence causes the cell to appear dark-colored. Many fruits, such as the persimmon, contain large amounts of tannin, especially before they are ripe. Wound tissues, and especially the **hypertrophied** tissues known as **galls,** which result from the bites of certain insects, are particularly rich in tannins. Tannins appear to be by-products of the **metabolism** of the plant. When present in the epidermal cells they are seen as a deterrent to snails, which might injure the leaf by feeding on it, to parasitic fungi which might otherwise enter the leaf tissue, and as a protection against desiccation, since they form substances impervious to water.

An important source of tannin is the bark of various trees, especially that of the hemlock, and several species of **oaks.** The bark is removed from the tree in sheets approximately 4' long. Stripping from the tree is usually done in the spring, when the **cambial** cells are most active and the bark separates easily. To remove the bark two rings are cut completely through the bark and round the tree. A longitudinal slit is made through the bark from one ring to the other. Using a blunt long-handled implement the bark is then pried loose from the tree and allowed to dry. By felling the tree, the entire trunk may be stripped of its bark in this way. The dried bark is shipped to mills which extract the tannin. Tannins from these barks are used to tan leather for shoe-soles and other heavy leathers. The wood of the chestnut tree yields a tannin similarly used. The available supply of tannin from these sources does not meet the demand; so many foreign plants are now being used.

Of these the most important are small trees of the genus *Schinopsis,* natives of the southern part of South America, including southern Brazil, Bolivia and other southern countries. These trees are known by the name "quebracho," which means "ax-breaker," because of their very hard, dense, heavy, dark-red wood, which is cut with difficulty. The heartwood of the tree contains 20-27% tannin, which is obtained by cutting the wood into small chips and extracting with water. This tannin is often used in combination with tannins from other plants.

The bark of many other trees yields large amounts of tannins. Among these are the **mangrove,** and several species of *Acacia,* known as wattles, natives of Australia. Fruits also may be a source of tannin. The fruits of *Terminalia chebula,* called myrobalans, are an important tannin source. The tree is a native of tropical Asia. Another fruit rich in tannin is divi-divi, the pods of a legume, *Caesalpinia coriaria,* which is native in tropical America and the West Indies. Sumac leaves, especially those of *Rhus coriaria,* a shrub or small tree native in Mediterranean Europe, are rich in tannins. To obtain the tannin, the plants are cut down and spread out to dry. The leaves are then removed from the stems and packed into bags which are shipped to the mills. There the leaves are first cleaned and then ground up. The tannins from this source are used in manufacturing fine leathers, like glove leathers. Leaves of other species of sumac, including the various American sumacs, also contain tannins which, however, are not so valuable and are little used.

Tannins are solids, soluble in water or alcohol, usually extracted by hot water, insoluble in **ether, chloroform, carbon** disulfide, **benzene,** soluble in alcohol-ether mixture, and in ethyl acetate, possessing a bitter astringent taste. Tannins (1) yield precipitates with gelatin, proteins (connected with the property of making leather from hides), **alkaline** salt solutions of many heavy metals, e.g., **lead** acetate, **copper** acetate (precipitate brown), **antimonyl** tartrate, concentrated dichromate solution, also by **chromic** acid (1% CrO_3), (2) yield **dark** blue or green coloration with **ferric** salt solutions, (3) in alkaline solution, absorb **oxygen** and yield dark colored solution, (4) with **iodine** in potassium iodide plus small proportion of ammonium hydroxide, yield red color, (5) with dilute solution potassium **ferricyanide** in ammonium hydroxide, yield a red to brown coloration (care not to use excess reagent).

fluorides, tantalum concentrating in the crystals and columbium in the mother liquid. Tantalum metal may be obtained by **electrolysis** of fused potassium tantalum fluoride, or by reduction of the pentoxide with **carbon** in the electric furnace. On heating, the metal absorbs large volumes of **nitrogen** and **hydrogen,** which are

REACTIONS OF TANNINS

REAGENT	PYROGALLOL TANNINS	CATECHOL TANNINS	PHLOROGLUCINOL TANNINS
	1. Oak wood 2. Chestnut wood 3. Galls 4. Sumac 5. Myrobalans 6. Divi-divi	1. Pine barks 2. Oak barks (but not oak wood, fruit or galls) 3. Acacia 4. Quebracho wood 5. Cassia bark 6. Mangrove bark 7. Cutch 8. Gambier	1. Gambier
With ferric salt solution	Dark blue color	Greenish-black color	
Bromine water	No precipitate	Yellowish to brownish precipitate (excess reagent)	
Sulfuric acid concentrated		Water extract yields dark red or crimson layer at the junction, on diluting turns pink	
On leather	Forms a bloom	No bloom formed	
Acid, boiled		Red insoluble phlobaphenes	
Moisten pine wood with water extract, add concentrated hydrochloric acid.			Red to purple color

Tannins are used (1) in making hides and skins into leather, after preparation of the hide, e.g., dehairing, (2) in the manufacture of inks, (3) in dyeing, as a mordant, (4) in the clarification of solutions, e.g., wine, and (5) as a source of gallic, pyrogallic, and tannic acids. (R.K.S., R.M.W.)

TANTALITE. A mineral oxide of iron, manganese, columbian, and tantalum. Crystallizes in the orthorhombic system. Hardness, 6–6.5; specific gravity, 5.20±; color, brown to black; transparent in thin sections. (Compare with **Columbite.**) (R.M.F.)

TANTALUM. Symbol: Ta. Atomic number: 73. Atomic weight: 181.4. Density: 16.6. Melting point: 2850° C. No isotope, but of single atomic form: 181.
Tantalum is a slightly bluish metal; ductile, malleable, and when polished resembles **platinum;** burns upon being heated in air; insoluble in **hydrochloric** or **nitric acid,** but soluble in hydrofluoric acid or a mixture of hydrofluoric and nitric acids. Tantalum metal was used as a filament for the incandescent electric lamp, but has been superseded by **tungsten.** Ductile tantalum and tantalum-clad steel are used for chemical apparatus as a substitute for platinum; for surgical instruments; as a resistant **electrode;** in electric current rectifiers; in vacuum technique because of its high capacity for absorbing carbon monoxide, nitrogen, hydrogen, and oxygen. Tantalum carbide is an important *abrasive.* Discovered by Ekeberg in 1802.
Tantalum occurs, usually with **columbium,** in **tantalite** (Fe(TaO$_3$)$_2$, 85% Ta$_2$O$_5$), **samarskite** (20% Ta$_2$O$_5$) chiefly found in Western Australia, and South Dakota. Recovered along with columbium by fusion with **potassium** hydrogen sulfate in the residue after subsequent extraction with water. Tantalum and columbium are separated by fractional crystallization of the **potassium**

only removed by heating to fusion in a vacuum. Chemically related to **vanadium** and columbium.
Chloride: Tantalum pentachloride (TaCl$_5$), white crystals, melting point 221° C., boiling point 242° C.
Fluoride: Tantalum potassium fluoride (K$_2$TaF$_7$).
Oxide: Tantalum dioxide (TaO$_2$), brown solid; tantalum pentoxide (Ta$_2$O$_5$), white solid.
Tantalates: (tantalum of valence plus 5).
Tantalic acid: (HTaO$_3$), gelatinous precipitate by addition of water to tantalum pentachloride. (R.K.S.)

TAPACULO. Aves, Passeriformes. *Pteroptochus.* A South American bird (**Aves**) superficially like the wrens. The several species are distributed over the entire continent. (A.W.L.)

TAPE WORM. Tape worms (**cestodes**) are parasitic flat worms, segmented in form, and having a head portion (**scolex**) which can continually give rise to new segments. As long as this head remains in the human intestine the worm has not been destroyed, for from this an entire new body can grow. These worms do not possess a digestive canal and are usually **hermaphrodites.** Nourishment is absorbed from the host through the surface of the worm's body. Fertilization is accomplished between segments of different worms or between different segments of the same worm.
Certain tape worms apparently do not greatly harm the host while others may produce marked symptoms such as severe anemia. Larval forms of the worm may invade the various tissue of man in certain forms of cestode infestation and cause serious trouble.
The kinds of tape worms commonly found in man are as follows: 1. The fish tape worm (*Diphyllobothrium latum*) is found throughout the world and infests humans when infected fish, which has not been thoroughly cooked, is eaten. Serious symptoms may not develop

for years, but often a condition resembling pernicious anemia develops. Recovery follows the expulsion of the worm. 2. The Dwarf tape worm (*Hymenolopsis nana*) is also found throughout the world and is the most common of tape worms occurring in the United States. It is of small size, being only a few centimeters long, but when present in large numbers the worms cause diarrhea, loss of weight and appetite, and nervous

Dwarf Tapeworm, head, middle segments, and terminal segments. From stained and mounted specimens (photographs ×30). (*Todd and Sanford, Clinical Diagnosis by Laboratory Methods, W. B. Saunders Co.*)

manifestations. The symptoms disappear when the parasites are expelled. 3. The pork tape worm (*Taenia solium*) is large, measuring 9 to 12′ in length. Most of the symptoms and dangers of this worm are due to the invasion of the tissue by its larval forms. This form is rare in the United States. 4. The beef tape worm (*Taenia saginata*) is common in the world wherever beef is eaten. Cold storage will kill the parasite. The adult forms are extremely large worms, some having been found that measure 30 to 50′ in length. Colic and pain may be present in patients infected with this worm. Spontaneous discharge of a long piece of worm may inform the patient of his guest.

The diagnosis of worm infestation is made by finding either segments of worms or eggs in the feces. The white blood count often reveals an eosinophilia.

Treatment with various antihelminths is usually successful if correctly employed. The drugs most frequently used are oleoresin aspidium, carbon tetrachloride, and hexylresorcinol. (D.M.H.)

TAPER. In conical parts, taper is the difference in diameter per unit of length. A part having diameters of 3″ and 2⅝″ at either end of a 6″ length, is said to have a taper of ¾″ per ft. Twist drills and some shank-type milling cutters are equipped with self-holding taper shanks where the degree of taper is so small that the frictional force between the shank and the socket is sufficient to drive the tool. The Morse Taper is used for lathe centers and twist drill shanks, and is approximately ⅝″ per ft.; the Brown and Sharpe Taper is used for milling cutters, and is approximately ½″ per ft. The Jarno Taper is used on profiling and die-sinking machines, and has a taper of .600″ per ft. Standard taper pins or dowels have a taper of ¼″ per ft. Taper sizes are usually designated by numbers, as a No. 7 B&S, or a No. 3 Morse. Since there is little or no uniformity, it is necessary to refer to tables for machine tool taper dimensions. (H.C.H.)

TAPER TURNING. Lathe.

TAPIOCA. Spurge Family.

TAPIR. Mammalia, Perissodactyla. *Tapirus.* Moderately large animals found in the neighborhood of water in the forests and Central and South America and the Malayan region. They reach a length of 8′ and a height of a little over 3′ and are stoutly built. Their most conspicuous characteristic is the short trunk into which the snout is prolonged. Only one of the five species occurs in the Old World.

The natives of the American tropics hunt tapirs for their flesh and hides, although the value of the latter for leather is limited. (A.W.L.)

TAPPING. Tapping is an internal threading process and may be accomplished by hand or by machine. The figure shows the necessary tools and the sequence of operations in threading a blind hole (one which does not

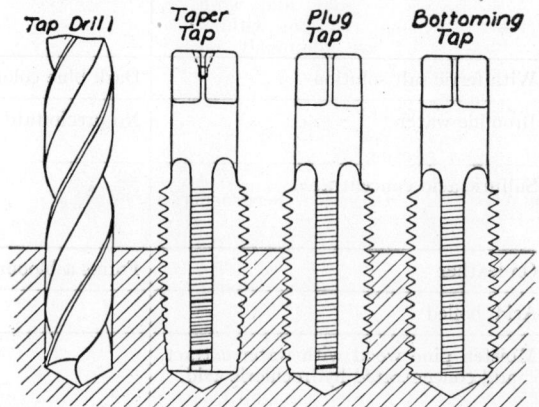

go through the work). The bulk of the metal in a threaded hole is removed by a tap drill which has a diameter equal to or slightly greater than the root diameter of the thread. The threads may be cut by using several taps in succession. The taper tap has from seven to ten chamfered threads. The long chamfer serves as a guide in aligning the axes of the tap and the hole. The plug tap has from three to five chamfered threads and is the most frequently used tap, particularly in machine tapping. In hand tapping it is not as easy to start as the taper tap. The bottoming tap has a chamfer on the first thread only and is never used as a starting tap. It is only employed for finishing blind holes which require threading to the bottom.

Taps are made of carbon or high-speed steel. High-speed steel taps permit more rapid production and hold their cutting edge longer. Carbon steel taps give satisfactory service for occasional jobs and are much less expensive. Taps are made with two, three, or more flutes. The flute serves to form the cutting edge, and provides room for chips. Taps are made with squared shanks for use with a tap wrench, or they may have straight cylindrical shanks for use in chucks in machine tapping.

Through holes may be tapped with a taper tap if the length of the hole does not exceed twice its diameter; in such an instance the taper tap has too much cutting edge in contact at one time, which requires excessive tapping torque and may result in tap breakage. Plug taps or so-called serial taps may be used for long holes. Serial taps are made in sets of three, the first two of which are undersize in outer diameter. The first tap roughs out the thread; the second cuts the thread a little fuller; and the third is used for finishing and bringing the hole to size, thus distributing the work of tapping among three taps. (H.C.H.)

TARANTULA. 1. Arachnida, Araneina. A name applied indiscriminately to many of the large hairy **spiders** of the warmer parts of the Americas, perhaps most commonly to the forms so often imported into temperate latitudes in bunches of bananas. The name comes from the generic name *Tarantula,* which has been applied to an entirely different group of spiders. 2. Arachnida, Pedipalpi. A genus of **whip scorpions.** The application of this name is dependent upon rules of nomenclature which have apparently received no authoritative interpretation. It will undoubtedly remain in popular usage as a name for many large spiders, whatever its scientific disposition. (A.W.L.)

TARDIGRADA. The bear animalcules, a group of minute animals with four pairs of short legs bearing claws. The mouth has piercing stylets and is formed for sucking. Neither respiratory nor circulatory organs are present. Bear animalcules occur in water and in moist places on land, and while they are evidently arthropods, their relations within this **phylum** are uncertain. They are commonly regarded as a separate class. (A.W.L.)

TARN. Cirque.

TARO. Aroids.

TARPAN. Mammalia, Perissodactyla. A wild **horse** of the steppes of central Asia. This species has been regarded as feral rather than a natural species, but this interpretation has been disputed by other authorities. It is in any case closely related to the domestic horse and may be the ancestral form. (A.W.L.)

TARPON. Pisces, Teleostei. *Tarpon.* A large marine fish (**Pisces**) of silvery color, found in the waters of the West Indies, along the Gulf Coast, and northward to a limited extent along the Atlantic coast. It reaches a length of 6′ and is known for its very large scales, which may be 3″ across. Also known as the silver fish, silver king, savanilla, and sabalo.

The tarpon is among the great game fishes but is not commonly regarded as desirable for food. It is eaten in Central America. (A.W.L.)

TARRAGON. Artemisia.

TARSIER. Mammalia, Primates. *Tarsius.* A peculiar animal found on some islands of the Oriental region. It is about as large as a rat and has a very short muzzle and enormous eyes. The hind legs, particularly the ankles, are long, and the tips of the digits are expanded into fleshy disks. The tarsier moves about on its hind legs by springy leaps. It is one of the extreme forms related to the lemurs. (A.W.L.)

TARSUS. 1. The terminal division of the leg of an insect. It consists of five segments in the typical form, the terminal segment bearing a pair of claws. Tarsi are modified in various species by the fusion or loss of segments, and in one genus, *Bittacus,* the terminal joint folds back on the next to form a grasping organ. 2. The shank of the leg of a bird (**Aves**). 3. The proximal portion of the foot of **vertebrates,** containing several tarsal bones. 4. The framework of connective tissue which gives shape to the eyelid. (A.W.L.)

TARTAR EMETIC. Tartaric Acid.

TARTARIC ACID AND TARTRATES. Tartaric acid ($H_2 \cdot C_4H_4O_6$ or $COOH \cdot CHOH \cdot CHOH \cdot COOH$) is a white solid, melting point of dextro or laevo 170° C. (of racemic, dextrolaevo, about 205° C., of mesotartaric, inactive, 140° C.) (see **Isomerism**), soluble in water, slightly soluble in alcohol, insoluble in ether. Tartrates (like **citrates**) in solution change silver of ammonio-silver nitrate into metallic silver. **Potassium** hydrogen tartrate, and **calcium** tartrate, on account of their solubility characteristics, are of importance in the separation and recovery of tartaric acid. The former salt is readily converted into the latter, and the resulting calcium tartrate plus dilute sulfuric acid yields tartaric acid plus calcium sulfate, and the latter may be separated by filtration. Tartaric acid may be obtained by evaporation of the filtrate. Ester: Diethyl tartrate ($COOC_2H_5$ ($CHOH$)$_2COOC_2H_5$), melting point 17° C., boiling point 280° C. Tartaric acid may be obtained (1) from some natural products, e.g., in the juice of grapes and acid fruits, often in conjunction with citric or **malic acid**; potassium hydrogen tartrate, "argol," in the residue of wine vats, (2) by synthesis. Tartaric acid is a dibasic acid, that is, two series of salts and **esters** are known. Tartaric acids is used (1) in **baking powders** as potassium hydrogen tartrate ("cream of tartar") with **sodium** hydrogen carbonate ("baking soda"), (2) in medicine, e.g., potassium antimonyl tartrate ("tartar emetic"), (3) in effervescent medicinal salts, (4) in **blue printing** as ferric tartrate, (5) in silvering mirrors—ammonio-silver nitrate yielding smooth deposit of silver. Sodium potassium tartrate ("Rochelle salt," $NaKC_4H_4O_6 \cdot 4H_2O$) is used in medicine, and in the preparation of **Fehling's solution,** which is an alkaline cupric solution made by mixing **copper** sulfate solution, sodium potassium tartrate solution and **sodium** hydroxide solution, and is used as an oxidizing reagent in the case of many organic compounds, such as **glucose** and reducing sugars, and **aldehydes,** with which cuprous oxide, red to yellow precipitate, is formed. (R.K.S.)

TASMANIAN DEVIL. Mammalia, Marsupialia. A Tasmanian animal, *Sarcophilus ursinus,* resembling the badger in its stout build and large head. It is nocturnal in habits, hiding during the day in a burrow or in natural crevices. It eats all kinds of living animals, even killing forms much larger than itself. The species is grouped with the **dasyures** of Australia. (A.W.L.)

TASMANIAN WOLF. Mammalia, Marsupialia. *Thylacinus.* A pouched animal closely resembling the wolves and of similar habits. It is related to the **dasyures.** Also called the thylacine.

Like the wolves of the northern hemisphere, this species has been killed in large numbers for its attacks on domestic animals and is now restricted to the wilder mountainous parts of Tasmania. (A.W.L.)

TASTE BUD. A sensory organ of the **vertebrates,** sensitive to contact with substances in solution and those in a liquid state. A taste bud consists of a spindle-shaped group of cells in the **epithelium** of the vertebrate tongue and in some species in the lining of the mouth and **pharynx.** Taste buds have been reported in the skin of some aquatic animals. The bud includes supporting cells of thick spindle shape and slender taste cells ending with a short taste hair which projects into a minute pit at the free end of the bud. The interpretation of these two types of cells is a subject of disagreement; it is possible that they represent stages in the development of a single form of cell.

The action of taste buds results in four fundamental taste sensations: sweet, sour or acid, salty, and bitter. Perception of these chemical properties is localized in the tissues containing organs of taste, but differentiation of the organs accompanying this localization has not been demonstrated. (A.W.L.)

TATLER. Aves, Charadriiformes. A north American **sandpiper.** *Heteractitis incanus,* the wandering tatler, is found along the Pacific coast of North America, where it breeds in northern latitudes, and on some of the Pacific islands. (A.W.L.)

TATOUAY. Mammalia, Edentata. The broad-banded **armadillo** of South America, *Cabassous unicinctus*. It is similar to the 6-banded armadillo in many ways. The largest of the group with the exception of the giant armadillo of the same region. (A.W.L.)

TAURUS (The bull) (Map, page 380). Taurus, the second sign of the **zodiac,** is a constellation of very great antiquity, two of its open **clusters,** the **Pleiades** and the **Hyades,** being frequently referred to in the Bible, and **Aldebaran,** its brightest star, is mentioned by both Homer and Hesiod.

Aldebaran (Alpha Tauri) is the standard first **magnitude** star of the northern hemisphere. The star is distinctly yellowish in appearance and has a measured diameter about 60 times that of the sun. It is a **double** star, but difficult to resolve except with moderately large instruments.

The two asterisms, the Pleiades and the Hyades, are both open clusters, i.e., groups of stars moving through space together. The Pleiades group is also noteworthy in that it is filled with diffuse **nebulous** material.

There are a number of double stars avaible for observers with small telescopes. With a wide-field instrument, such as an opera glass, the two doubles Sigma and Theta can both be seen at once and, with the surrounding stars, make a very interesting spectacle. (W.K.G.)

TAUTOMERISM. This is a phenomenon observed in organic chemistry when a substance has one molecular formula but must be assigned two structural formulae. (**Chemical Formulae.**) Two sets of reactions characteristic of two mutually incompatible groups in the **molecule** and two sets of derivatives can be produced by reactions depending on but slight variations of the experimental conditions. Consequently a unique structural formula cannot be assigned to such a compound. This difficulty is solved in theoretical organic chemistry by assuming the existence of two compounds with different structural formulae and also an existence of a dynamic **equilibrium** between these two forms. The latter differ in their structural formulae in general by the position of a hydrogen **atom** and that of a double bond. The following are the more important cases of tautomerism.

1. Ethyl ester of acetoacetic acid.

The equilibrium mixture at room temperatures consists of 7% keto and 93% enol form. The enol form can be obtained in the free state by treating the **sodium** derivative with dry **hydrogen chloride** and distilling. The enol form can be preserved in the pure state at low temperatures but at room temperatures it is converted into the equilibrium mixture in two weeks. The enol form gives characteristic reactions with **ferric chloride.** The keto form can be obtained by **crystallization** at low temperatures of the **ester** from an **alcohol,** ether, or hexane solution. It does not give the ferric chloride reaction and at room temperatures is gradually converted into the equilibrium mixture.

2. Acetone and similar ketones containing a hydrogen on the carbon atom next to the carbonyl group.

3. Acetaldehyde.

4. Phloroglucinol.

5. Nitroparaffin compounds.

$$R_2C-NO_2 \rightleftharpoons R_2C=N\underset{OH}{\overset{O}{\lessgtr}}$$

6. Nitrous acid.

$$HONO \rightleftharpoons HNO_2$$

7. Hydrogen cyanide.

$$HCN \rightleftharpoons HNC$$

Two sets of derivatives can be formed CH_3CN methyl cyanide and CH_3NC methyl isonitrile.

8. Diazo compounds.

9. Isatin.

10. Sugars.

The existence of this tautomerism is invoked to explain mutarotation. [See **Carbohydrates** (glucose)].

(R.K.S.)

TAXIS. **Tropism,** and also **Movement in Plants.**

TAXONOMY. The science of **classification.** In dealing with the many details involved in classifying the half-million known species of animals, it has been found necessary to adopt rules of procedure in order to secure approximate uniformity and stability of results. The work of the earlier naturalists was without such restrictions, hence much of the confusion of modern taxonomy has arisen from the interpretation of these early contributions.

Of the several codes of taxonomic procedure the International Rules of Zoological Nomenclature, drawn up by the International Commission on Zoological Nomenclature, is commonly regarded as authoritative in its field. These rules establish many principles relating to scientific names of all ranks. Among them is the provision that all zoological nomenclature shall start with the tenth edition of Linnaeus' "Systema Naturae," published in 1758. All prior works are discarded. A second important provision is the law of priority, which states that the valid name first applied to a group or species shall stand, all names subsequently applied to the same category becoming synonyms. All names, to be accorded consideration, must be published by printing or by a similar permanent form of reproduction and distributed to other scientists either in a recognized periodical or by private enterprise.

All scientific names are latinized, although the range of choice of names is not limited to the Latin; it is, in fact, practically unlimited. The Rules make several specific provisions. Restrictions are placed on the formation of names of superfamilies, families, and subfamilies, which are formed of the stem of the type genus plus the ending -oidea, -idae, or -inae, respectively. The type genus is the included genus regarded by the author of the greater group as typical.

The rules relating to genera have been the source of most confusion. Once applied, a generic name is not to be used again in the animal kingdom, and if inadvertently used a second time, the later usage becomes a homonym of the name as first applied and is replaced by a new name. In naming a genus the author is supposed to indicate a type species which shall be the embodiment of the characteristics of the genus. Types were designated long before the formulation of these rules but many genera were erected without stated types before their value was appreciated. To meet these cases rules for the citation of types of genera described without them are available. The type species must be among those included in the original description of the genus, and the first designation of such a type takes precedence over all others. As a result historical research has been necessary to fix the older generic names and some are still in a state of confusion. Such cases may be submitted to the Commission for special ruling.

The International Rules of Botanical Nomenclature agree in the main with zoological procedure. The essential differences are summarized under **nomenclature.** See also **Paleobotany.**

In naming species the specimens available, or a selection of them, should be labeled as types and preserved, preferably in the permanent collections of a museum. The species bears the name of the genus to which it belongs, followed by its own name, and for completeness the name of the author of the species should be appended. The generic name is capitalized and both it and the specific name are underlined or in printing are italicized. The use of two names for each species is the principle of binomial nomenclature.

Many other details of taxonomic procedure, based on written and unwritten laws, are of interest only to specialists. A minimum of knowledge in this field is essential to any well-rounded biological training. (A.W.L.)

TAYLOR'S THEOREM. Expansion of Functions in Series.

TAYRA. Mammalia, Carnivora. *Galera.* A South American animal of the weasel family. It is among the larger species, comparing with the otter in size, and has the characteristic long body and short legs of the weasels, with a long and rather heavily furred tail. These animals range from white to black in color. (A.W.L.)

TCHEBYCHEFF'S INEQUALITY. Let x be a variable, either discrete or continuous, whose probability function possesses a **mean** $\overline{X}$ and **standard deviation** σ_{xi} then the probability of obtaining a **deviation** as much as or greater in absolute value than $t\sigma_x$ is less than or equal to $1/t^2$. In symbols

$$P[\ |\ x - \overline{X}\ | \geqq t\sigma_x] \leqq \frac{1}{t^2}.$$

This inequality is often useful in testing for **significant differences.** A better inequality is due to B. H. Camp. (L.A.A.)

TEA. *Thea sinensis.* Theaceae. The tea plant is an evergreen shrub probably indigenous to China, where it has been cultivated since early times. The plant possesses alternate elliptical leaves which when mature are tough, and vary in length from 2 to 5 in. The flowers are axillary (see **Axil**) and appear singly or in small groups. They are white, slightly fragrant, and about an inch in diameter. Each flower has numerous **stamens** and a single **pistil** composed of three **carpels.** The fruit is a woody **capsule** containing three large seeds.

For successful growth tea must be planted in regions having abundant rainfall. In China most of the tea plants are grown on small farms. The plants are grown from seed or from nursery stock and begin to yield crops when 3 or 4 years old, continuing to do so thereafter for 50 years. The young shoots appear in flushes, growing rapidly for a time. From these flushes are picked the young leaves used for tea. Several flushes occur each year. Picking is done entirely by hand, necessitating an abundant supply of cheap labor.

After picking the leaves are spread out and allowed to wilt for some time. The limp leaves are then rolled and crushed in machines, so that the cells are bruised and certain enzymes freed. The crushed leaves are again spread out and allowed to ferment, after which they are again rolled and dried until crisp. They are then sorted and graded. While still warm the leaves are packed in lead-lined chests partly to conserve the aroma of the leaf and partly to prevent absorption of any odor which might spoil the product. This method of preparation produces black tea, which when brewed yields a rich orange-colored drink. Black tea is principally obtained from China and Ceylon. The best grade is known as Orange-Pekoe.

The preparation of green tea differs in several particulars from that of black tea. After picking, the leaves are at once steamed so that no oxidative **enzymes** remain. Steaming tends to bring out the aroma. After steaming, the leaves are rolled and dried, usually by machine. They are then sifted, graded and packed. Green teas are principally produced by Japan and largely exported to the United States.

Oolong is a slightly fermented tea produced in Formosa, and is intermediate between green and black tea.

In China tea is often delicately scented by exposing the leaves to the odors of flowers. Large quantities of Jasmine flowers are used for this purpose, the product being known as Jasmine Tea.

Tea is principally used as a beverage. Great Britain and its dominions consume by far the greatest amount used outside the producing countries. (R.M.W.)

TEAK. *Tectona grandis.* Verbenaceae. The teak tree is tall, with very rough-surfaced oblong leaves from 10–20″ long and 8–15″ broad, and small white or blue-tinted flowers borne in large panicles. The tree is native in the tropics of Asia and is frequently grown in plantations in India, Java, and other Asian countries for its hard wood. This wood is very durable and much used in shipbuilding and in the making of fine furniture. The wood is very heavy and dries slowly. Drying is hastened somewhat by girdling the tree at the base and leaving it standing for a year or more. During this time it dies and dries out, after which it is felled and floated to the shipping port. (R.M.W.)

TEAL. Aves, Anseriformes. A small **duck** whose beak has almost parallel sides. The several species are beautifully marked. Teals are found throughout Europe, Asia, and North America. The common European species migrates into Africa and is sometimes found in northern and eastern North America, while other species have a much more limited range. In North America the blue-winged teal, *Querquedula discors,* is common east of the Rockies, the green-winged teal, *Nettion carolinense,* throughout the continent, and the cinnamon teal, *Q. cyanoptera,* west of the Mississippi. (A.W.L.)

TEASER TRANSFORMER. Transformer.

TECTIBRANCHIATA. Gasteropoda.

TECTOGENE. A hypothetical complete down-fold of the **lithosphere** or **sial** as intimated by the sinuous linear strip of strong negative gravity anomalies which follows the East Indian and West Indian island arcs. (See also **Geosyncline.**) (R.M.F.)

TECTONITE. A term proposed by Backland, for a variety of **mylonite** produced from **para-schists.** (R.M.F.)

TEE BOLT. A bolt with a carefully finished square head designed to fit tee-shaped slots in milling machine or planer tables and in lathe faceplates; used for holding machine vises or clamps for work parts. (H.C.H.)

TEGMEN. For the use of this term in botany, see **Seed.** In zoology, it has two meanings. 1. A thickened forewing of the kind found in the **grasshoppers** and related forms. These wings are less thickened than the **elytra** of many beetles but they also serve as wing covers under which the hind wings are folded when at rest. Most tegmina vary from scarcely thickened membranes to leathery appendages. 2. The leathery top of the **calyx** of crinoids. (A.W.L.)

TEJU. Reptilia, Sauria. *Tupinambis.* A large **lizard** of the West Indies and South America. Lives in the forests near water but is not aquatic. Reaches a length of a yard, with long slender tail and heavy forequarters. It is chiefly olive and black. (A.W.L.)

TELEGONY. A supposed influence of a scrub sire on the offspring of a blooded dam by later mating with a blooded sire. Many animal breeders suppose that the accidental mating of a choice female with a mongrel makes it impossible to be sure of securing pure-bred young at subsequent births. There is nothing in the findings of geneticists to support this view. Influence of the sire is limited to the hereditary potentialities of his **germ cells,** and once his own progeny have been born, the female carries no residual influence of the mismating. (A.W.L.)

TELEGRAPHY. Communication by telegraph, whether the older manual type or the more recent automatic or printing type, is done by a code of electrical pulses. In the manual type the operator sends a certain combination of pulses for each letter of his message and the receiving operator then transcribes them into the characteristic letters. In the more complicated automatic types the sending operator uses a keyboard similar to that of a typewriter, and the equipment transforms the striking of a key into the proper signals (not the same code as for manual operation) and a machine at the receiving end selects the proper letter and the message is typed. The message may be transcribed to a punched tape and then transmitted automatically.

In its simplest form the telegraph consists of **battery,** **key,** and **sounder,** with connecting wire or wires. The telegraph key is a type of switch designed for easy operation by hand. The sounder is similar to a **relay,** the essential difference being that the sounder has no electrical contacts to close, but its armature strikes a stop at each end of its travel, giving a sound characteristic of the stop. Thus a skilled operator can tell the position of the armature by the sound and can interpret the code message.

A system commonly used in this country is the closed-circuit, so named because the electrical circuit is normally closed, current flowing when no message is being sent. Several stations are usually connected in series, with two common batteries serving all. A circuit is shown in Fig. 1. Since the stations are all in series the

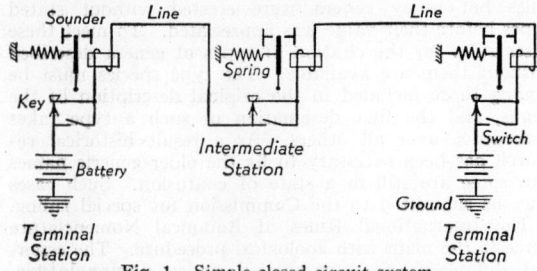

Fig. 1. Simple closed circuit system.

keys must be short-circuited when not in use, hence a switch is attached to each key. To operate, the switch is opened and the key is tapped in accordance with the code. This sends pulses of current through the wire and ground circuit, causing all sounders to operate. These hit the stops on either side of the armature and give the message to the receiving operator.

While the single-current system just described is older and somewhat simpler, a better arrangement is the double-current system of Fig. 2. In this system cur-

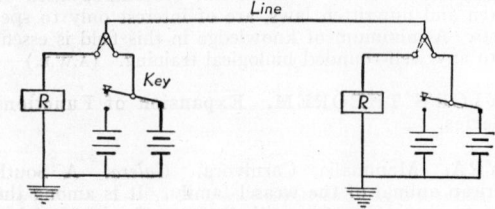

Fig. 2. Double current system.

rent flows in one direction for a signal and in the opposite direction for a space. The operation is more positive than with single-current circuits. These currents may be sent as illustrated with two batteries oppositely connected or a means may be provided for reversing the connections to a single battery. The system requires a special relay, known as a polarized relay, between the line and the sounder. A polarized relay is one so designed that current flowing in its windings in one direction will close the contacts in one direction, and current flowing in the opposite direction will close the contacts in the other direction. The combination of relay and sounder is shown as R in Fig. 2. The connections are set for transmission of signals from left to right, the operation of the key at the left station sending

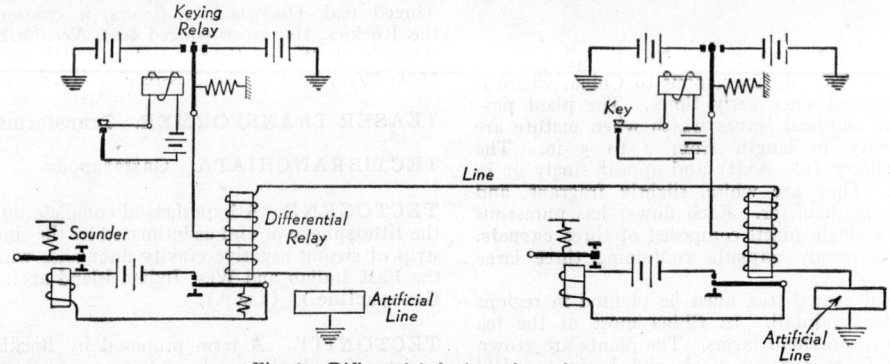

Fig. 3. Differential duplex telegraph system.

pulses of current to the receiver R at the right station. There may be intermediate stations.

Both circuits shown allow transmission in only one direction at a time, but there are several arrangements providing simultaneous transmission in both directions. This is known as duplex service as opposed to the previously described simplex service. The differential duplex system is shown in Fig. 3. The duplex operation depends upon a split-winding relay and a balancing of the actual line with an artificial one. An artificial line is a circuit designed so that its electrical characteristics are the same as those of the actual line. Therefore, when it is connected to the relay at one end and the line at the other, the two halves of the relay winding and associated connections are electrically identical. Operation of the key at the left station causes the keying relay to send current in first one direction and then the other into the differential relay. Since the two sides of the relay are equal electrically, the current divides, half flowing upward and half downward. The **magnetic flux** of the windings oppose and cancel. This prevents the relay picking up and operating the sounder. However, current coming into the relay from the line all flows through the winding in the same direction, the flux is not neutralized, and the relay picks up, operating the sounder. Thus both transmission and reception may occur at the station at the same time, and only the incoming signals operate the sounder. The half of the transmitted current which flowed upward in the relay goes out over the line to the right station and operates its relay as just described. (See also **Telephony** and **Teletype**.)   (F.T.M., L.R.Q.)

TELEODESMACEA. Bivalve mollusks, including the soft-shelled **clam**, the **shipworms**, and many other marine species. The group is included as an order in one of the classifications now widely used. Equivalent to the order Eulamellibranchiata. **Lamellibranchiata.** (A.W.L.)

TELEOSTEI. Bony fishes of the vast majority of existing species. The group is regarded in most classifications as a very large order of the class **Pisces**, divided into many suborders and families, but a tendency also exists to elevate it to a higher rank and to make its principal subdivisions orders. The principal subdivisions are well established, whatever the rank accorded to them. In North America they number about a score, containing above 3000 species of fresh-water fishes. The principal forms among these many species and the more important food and game species are treated under their vernacular names in this work, as **bass, eel, killifish, salmon, shad, trout,** etc.   (A.W.L.)

TELEOSTOMI. The true fishes, a subclass of the class **Pisces**. (A.W.L.)

TELEPHONE EXCHANGE. Exchange, Telephone.

TELEPHONY. Telephony is the science of communicating speech by electrical means over wire circuits. The complete system has at least three fundamental components, the transmitter which converts the sound variations into electrical variations, the transmission circuits and the receiver which converts the electrical variations back into sound.

A successful system must meet certain requirements in order to reproduce the speech with a satisfactory degree of fidelity. Ordinary sound may cover a range of frequencies from 20 cycles per sec. to almost 20,000 cycles per sec., the exact range varying from person to person. To transmit this complete band would be a complicated and expensive process so considerable research has been applied to the problem of just how much is really necessary for different services. For simple speech transmission it has been found that a band from about 250 to 2800 cycles is ample. This gives good intelligibility with enough of the characteristic overtones of the voice to make most voices easily recognizable over the phone and yet it is not wide enough to offer serious technical problems. The absence of the frequencies below 250 cycles greatly reduces the amount of power necessary in the system since an abnormal proportion of the total sound power is included in the lower end of the audio band. When telephone circuits are used for special purposes such as program lines for radio networks this range must be considerably extended, introducing many special problems in designing the transmission circuits. Wired circuits for **television** networks are still more difficult to obtain, but can be worked out satisfactorily if cost is not a factor.

The transmitting equipment for the ordinary telephone consists of a single-button carbon **microphone**, called a transmitter. This is connected to a **battery** whose current passes through the transmitter and is modified by the sound waves. It is this modified current which is transmitted over the wire circuits, often through several intermediate pieces of equipment, to the receiver where the sound is reproduced.

Since no special line problems are involved in the local telephone circuits we may start a discussion of the various telephone systems with this relatively simple one. The local service is handled through central offices where the various lines are arranged so they may be interconnected at will to make the desired connections between the subscribers. The exact nature of the equipment in the office and at the subscriber's location depends upon whether the source of d.c. is a battery located at the customer's instrument or whether it is located at the office. Many rural telephones where fairly long **lines** are involved use the first method in order to avoid excessive losses which would result if the d.c. were transmitted from a central location. All modern city service is of the common battery type, all telephones served by a given central office being supplied from the same main battery located at the office. Special connections are used to avoid interference between the many circuits using this same battery. A brief outline of the major operations occurring in making a call through a manual office (this is an office in which the connections are made by the telephone operator rather than automatically as in a machine switching office) will illustrate the principles. At the subscriber's premises the telephone equipment (commonly called a subscriber's subset) consists basically of the transmitter, the receiver, a signal bell and an induction coil or **transformer**. The latter is used to connect the receiver to the line so better matching can be secured and so d.c. can be kept out of the receiver windings. (See **Anti-Side Tone** for typical connections.) Removing the receiver from the hook, or, for hand sets, the removing of the hand set from the cradle, closes switch contacts connecting the instrument across the line, the circuit between line wires being completed through the induction coil primary and the transmitter. This connection causes a signal lamp associated with this particular phone to light on the switchboard. The operator responds by plugging into the corresponding answering jack on the board, asking for the number, and then completing the connection by means of another plug cord. Ringing is then usually done automatically until the called party answers or the calling party hangs up. When the call is over and the parties hang up other supervisory lamps on the board light and indicate to the operator that the connection should be removed.

Since complete service requires the connection of any phone with any other phone a limit in the number of phones which can be handled is obviously reached. For rapid service this limit is the point at which the operator cannot reach the jack for every phone served by the office. With present equipment the board can hold 10,-

ooo jacks within the reach of the operator. When a community has more than this number of phones additional offices must be provided to handle them. The various offices serving a community constitute an exchange or exchange area. Calls from one office to another can be

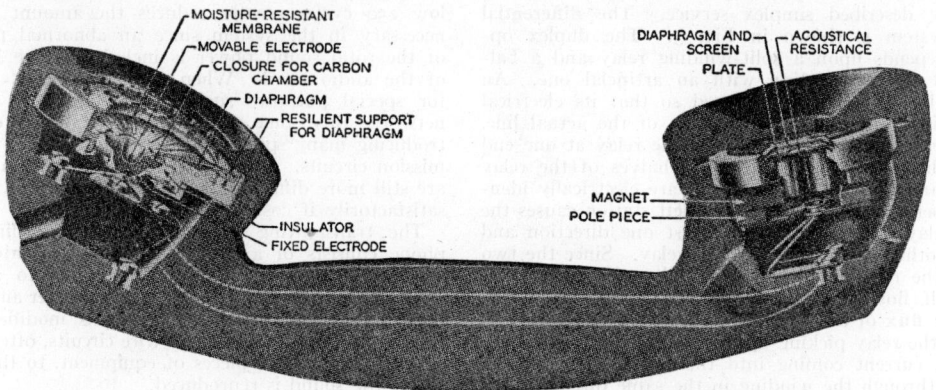

Fig. 1. Sectional view of a modern telephone hand set.

MOISTURE-RESISTANT MEMBRANE
MOVABLE ELECTRODE
CLOSURE FOR CARBON CHAMBER
DIAPHRAGM
RESILIENT SUPPORT FOR DIAPHRAGM
INSULATOR
FIXED ELECTRODE

DIAPHRAGM AND SCREEN
PLATE
ACOUSTICAL RESISTANCE
MAGNET
POLE PIECE

handled by an extension of the simple process given above, the calling party's operator passing his call on to an operator in the called office who completes the connection. Separate switchboards are used for taking calls, i.e., answering a calling party and requesting the number, and for connecting to the called line. The first is called the A board and the second the B board.

Dial or, more accurately, machine switching performs the connecting functions automatically under the direction of the dial of the calling telephone. This type service is replacing the manual equipment in many localities as it gives quicker and more reliable service at a lower operating cost. It is, however, higher in initial cost. The transmitting equipment is modified to include a dial which is used by the subscriber to indicate the desired number. The dialing operation consists of placing the finger in the dial at the desired numbers and rotating the dial to a stop, then allowing it to return freely, repeating this for each letter or digit. This operation, through the action of the freely returning dial, sends pulses to the central office, these pulses being used to control the necessary switching. The simplest dial system is the rotary selector or step-by-step system which will be outlined to illustrate the principles involved. When the calling party removes his receiver from the hook, the line circuit is closed through the dial contacts. Instead of a lamp lighting on a board as in the manual system, a line finder starts searching for the calling line. This searching is initiated by the closing of the hook switch contacts at the calling subscriber's phone and is terminated when the finder locates the line (indicated by the connection at the phone). The subscriber then hears the dial tone indicating that the office circuits are ready to receive his call. Reference should be made to Fig. 2. The subscriber then dials his number, one digit at a time. For each digit a number of pulses are sent over the line, the number corresponding to the number being dialed. These pass through the line finder connections to a switching device called a selector. This is a notching relay mechanism, which will be notched up one step for each pulse, there being ten steps to handle the range of possible values for each digit. When the even sequence of pulses stops (this happens when the subscriber rotates the dial for his second digit), the selector automatically rotates by steps until it connects to a second selector which is not busy. This second selector responds to the second digit pulses in a similar manner. This process is repeated for all but the last two digits. For the next to the last digit the selector (which in this special application is com-

monly called a connector) notches up, but does not rotate automatically. However, when the last digit is dialed the connector rotates step by step to the dialed position. Connection is now all but complete to the called line, and, if it is not busy the connection is com-

pleted automatically and the called phone is rung. Upon completion of the conversation the parties hang up and all switches return to normal. For large exchange areas a system more complicated in the electrical sense but more rapid in service is used. This is the panel system. A more recent development, the cross-bar system, is also

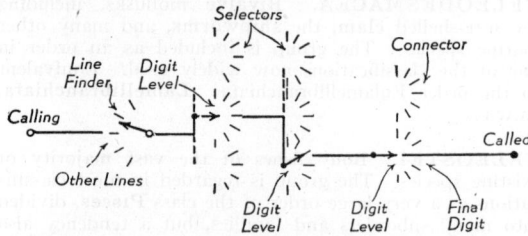

Fig. 2. Simplified dial connections.

Selectors
Line Finder
Digit Level
Calling
Other Lines
Digit Level
Connector
Called
Digit Level
Final Digit

designed for large service areas. Neither of these latter systems does its switching under the direct control of the dial pulses but rather through an intermediate mechanism.

Long distance, or inter-city, service introduces many new problems. The electrical signals traversing the lines are subject to attenuation, frequency distortion, interference and various other undesirable effects. Additional equipment must be used to avoid these. The lines themselves are expensive and much has been done to utilize them to the fullest, phantom and carrier circuits being means of doing this. The lines may consist of the familiar open wire lines, of lead sheathed cables containing many lines or of lead sheathed coaxial cables. The cables may be overhead or underground, the coaxial being universally underground. Cables are often operated filled with dry nitrogen under a slight pressure, thus decreasing the losses. While open-wire lines are no longer loaded, cable lines for voice frequencies (lines not having carrier channels) are usually loaded. This process, developed by Pupin, consists of inserting at fairly frequent intervals (ranging from a fraction of a mile to a little over a mile) lumped inductances to increase the over-all inductance of the lines. This loading improves transmission by decreasing distortion and attenuation but it also puts a definite upper limit to the frequency which the line will pass. It is this latter limit which prevents its use with the usual carrier circuits. Even for radio program circuits it becomes necessary to insert the loading coils at very close intervals, since the closer the spacing the higher the cut-off frequency.

To extend further the range of communication over lines it becomes necessary to insert **amplifiers**, or repeaters, at intervals along the line to build the signal strength back up. The spacing of these amplifiers varies with the line and its use, ranging from around 200–300 miles for open-wire lines, 50–100 miles for many cables, down to 5 miles for coaxial cables used for wired **television** programs. Basically these repeaters are vacuum-tube amplifiers but they must be inserted in the circuits in such a way that no energy can feed back from the output to the input. In some cases separate lines are used for the conversation in the two directions so no special problem is encountered except at the terminals where both directions must be brought together on a single line for transmission to the subscribers. In many other cases, however, the same line serves for both directions just as it does in local service. Here special care must be exercised to prevent feedback and consequent distortion. This is accomplished by the bridging transformer or hybrid coil in a manner shown in Fig. 3. The circuits are

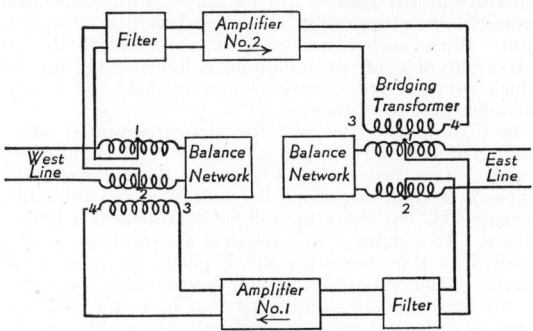

Fig. 3.   Repeater station.

so balanced that no signal from the output of one amplifier is introduced in the input of the other. The **filters** limit the range of signals which may go into the amplifiers and hence make the balancing problem less severe since balance need not hold over such a wide range of frequencies. The amplifiers are conventional vacuum-tube circuits and the balancing networks are combinations of capacity, resistance and inductance which will present the same **impedance** as the line over the usable frequency range.

Transpositions are used extensively to decrease interference from various sources. (See **Inductive Interference**.) (L.R.Q.)

TELESCOPE. The name telescope implies that it is a device for "seeing at a distance." There are two modern methods for accomplishing this purpose: (1) the refracting telescope which uses a **lens** for gathering light and forming a real image of a distant object, and (2) the reflecting telescope which employs a concave mirror for the same purpose. In both of these types an **eyepiece** is employed to "magnify" and study the image. In this article the development of the modern instrument will be briefly treated from the historical point of view.

It is probable that experiments on the accomplishment of "seeing at a distance" were carried out from the time of introduction of lenses into Europe during the thirteenth century for the purpose of correcting defective vision. However, it is not until 1608 that we find definite evidence of the discovery of telescopic vision, for in October of that year Jan Lippershey applied to the States-General of Holland for a patent to an instrument for seeing at a distance. The States-General appreciated the importance of the instrument for military and naval purposes and refused to grant the patent, but did commission Lippershey to continue his experi-

ments and purchased the rights to his instrument. The news of Lippershey's discovery spread like wildfire over Europe, and by May, 1609, Galileo had heard of the discovery and set about to construct an instrument for himself based upon Lippershey's design. Galileo immediately applied his instrument to astronomical observing, and before the middle of 1610 had made and announced a number of important discoveries.

The type of instrument first built by Galileo is shown diagrammatically in Fig. 1 and is now known as the

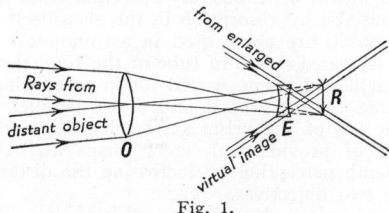

Fig. 1.

Galilean telescope. A converging (convex) lens, O, known as the objective, forms at its principal focus a real inverted image, R, of a distant object. Before reaching the focal plane of the objective the converging rays are intercepted by a divering (concave) lens, E, known as the eyepiece. If the eye is placed behind E an enlarged virtual image of the distant object is observed. This image is in the same orientation as the distant object (i.e., it is an erect image), and the **magnifying power** of the instrument depends upon the relative focal lengths of O and E.

The Galilean type of telescope now survives only in the form of opera and field glasses, usually of from 2- to 5-power. The principal advantages of this type are (1) the erect image, (2) the brilliant illumination of the field of view, (3) the diminished distance between O and E which gives a more compact instrument, and (4) the partial correction of spherical and chromatic **aberrations** by the combination of convex and concave lenses. Set against these advantages there are the important difficulties that (1) no **reticle** (cross-hairs) can be used, and (2) the field of view with such an instrument is very limited. In Galileo's instrument the field of view was only $7' 15''$, less than ¼ the diameter of the moon.

To overcome the disadvantages of the Galilean form of telescope, Johann Kepler in 1611 described the type of instrument shown diagrammatically in Fig. 2, which

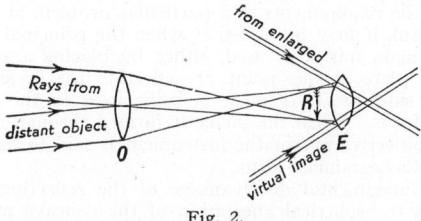

Fig. 2.

is now known as the Keplerian telescope, or, more commonly, as the astronomical telescope. The real image R of a distant object is formed by the objective O as is the case with the Galilean telescope. The difference between the two types of instruments is found in the eyepiece, E, which, in the original Keplerian instrument, was a relatively short-focus convergent lens set in such a position that R is close to its principal focus. The eye is placed behind E and an enlarged, inverted, virtual image of the distant object is observed. This instrument overcomes the fundamental disadvantages of the Galilean instrument, for a reticle or a **filar micrometer** may be used, with the wires placed at R, and the field of view is large. The chromatic and spherical aberrations are very severe when simple convex lenses are used, and the correction of these defects requires the construction of complicated objectives and eyepieces.

In addition to its use in astronomical observing, the Keplerian form of telescope has a multitude of other uses such as fixing a line of sight (as in surveying instruments or in gunsights) and for reading scales on leveling rods, **galvanometers**, etc. The inverted image is frequently a disadvantage for terrestrial observing, and erecting eyepieces of various designs have been invented. These require the addition of extra lenses with consequent loss of light, both by reflection from the lens surfaces and also by absorption in the glass itself. Such erecting systems are never used in astronomical observing. The increased length of tube of the Keplerian form over the Galilean is compensated for in prism binoculars by reflecting the rays back and forth parallel to the tube. The use of the prism system has the additional advantages of providing an erect image and also increasing depth perception by increasing the distance between the two objectives.

For astronomical observing an object glass of large diameter is of fundamental importance both for the purpose of gathering more light, and hence increasing the brightness of faint objects, and also to diminish the size of the **diffraction** rings and thus increase the resolving power. Telescope builders of the 17th and 18th centuries were unable to obtain glass disks of sufficient clearness and homogeneity for the construction of large object glasses, and turned to the reflecting type of telescope. In such instruments a concave mirror is used to form the real image to be examined with the eyepiece. This image will be formed in the open end of the tube, and to examine it the observer would have to put his head directly in the path of the incoming rays and seriously reduce the light-gathering power of the instrument. To avoid this difficulty a number of different schemes have been evolved. Newton placed a small mirror in the path of the rays, slightly inside the focus and inclined at an angle of 45° so that the image would be formed at one side of the tube where the eyepiece was located. Cassegrain, and others, adopted the scheme of reflecting the light from the objective mirror back through a hole cut out of the center of the mirror. The relative advantages, in the Cassegrainian type, of using convex hyperbolic mirrors inside the focus of the objective, concave elliptical mirrors outside the focus, etc., are far too complex to be considered in a work of this character. However, by the use of different types of secondary mirrors, the focal length and efficiency of a telescope using one objective mirror may be altered to suit the requirements of a particular problem at hand. In general, it may be said that when the principal focus of the main mirror is used, either by placing a photographic plate at this point or reflecting the image out to one side, the instrument is being used in the Newtonian focus. When the image is formed through a hole in the objective mirror the instrument is said to be used in the Cassegrainian focus.

One fundamental disadvantage of the reflecting telescope is the spherical aberration of the concave mirror. Even though this may be corrected along the axis by the use of a concave paraboloidal mirror instead of a spheroidal mirror, nevertheless, the field of good definition is limited. Recognizing this difficulty, glass makers and lens designers continued their endeavors to produce large aperture refracting telescopes, culminating in 1897 with the dedication of the Yerkes Observatory telescope, with an objective diameter of 40″. It seems highly improbable that any refractors will ever be made with lens diameters greater than this. Not only are telescope builders faced with the difficulty of obtaining the large, clear glass disks and grinding the four surfaces of the achromatic objective, but also there is the difficulty of maintaining the lens figure as the telescope is moved to different positions, since, of necessity, the glass must be supported at the edges. For photographing large areas of the sky the refractor still has its place, and compound objectives are continually being designed and built for the purpose of improving definition over large angular fields.

Another difficulty with reflecting telescopes has been the tarnishing of the reflecting surface. In the original reflectors, in which the mirrors were built of metal, this tarnishing of the surface required the complete regrinding of the figure of the mirror at frequent intervals. These metal mirrors were also subject to serious changes in figure due to expansion and contraction of the metal with changes of temperature. Both of these difficulties have been overcome by making the mirror of glass and depositing a metallic coating on the figured concave surface. The glass does not have to be clear or transparent, can be supported at the back, and has such a low temperature coefficient of expansion that the figure does not change seriously with temperature variation. Originally, the glass mirrors were coated with silver, which was chemically deposited on the surface. This could readily be dissolved off and replaced after it had tarnished. The 200-in. reflector, and probably all the reflectors of the future, will be coated with **aluminum** deposited by evaporation and recondensation in a vacuum. When such a surface is first exposed to the air a transparent oxide of aluminum is immediately formed which protects the surface from tarnishing for a considerable period of time.

In 1930, a new type of telescope was invented which has revolutionized certain phases of astronomical research. This instrument, known as the Schmidt telescope, is a combination of the reflecting and refracting instruments. Its reflecting mirror is ground as a hollow sphere. To eliminate the spherical **aberrations** of such a mirror, a thin correcting lens is placed in front of the mirror. The curvature of the mirror and the figure of the lens may be so adjusted that a working field as large as 20° may be used without appreciable aberrations at the edges, and the effective focal length may be reduced to such a point that the f-ratio (see **Camera**) may be as low as 0.5. The image is formed between the mirror and the lens, and the photographic plate, or film, must be curved by pressing it against a convex spherical surface. Since the telescope cannot be used for visual work, it is sometimes referred to as the Schmidt camera. During World War II, the Schmidt camera was successfully used for high-altitude reconnaissance photography. (W.K.G.)

TELESCOPING GAUGE. Measurement.

TELETYPE. Teletype is the transmission and reception printing of messages by electrical means, being in principle a receiving typewriter whose keys are operated from the transmitting station, either by means of a keyboard or previously prepared tape. While there are two systems in use in this country, the simplex and the multiplex, they differ more in details than in principle.

Both systems utilize the Baudot code which employs 5 equal length pulses in various combinations to give the letters, numerals, punctuation, etc., necessary for operation of such a system. The nature of this code will become clear as we discuss the operation of the simplex or start-stop teletypewriter. Referring to Fig. 1, we have a sending device, the connecting line and a receiving device. In addition to the schematic parts shown, there are, of course, many other components such as transmitting key board, **relays**, receiving printer, etc. When a key is pressed at the transmitter the sending brush arm is released and starts revolving (clockwise in the diagram), sweeping over successive segments of the sending commutator. Pressing of the key (this is not the conventional telegraph key, but rather a key of a typewriter-like key board) also causes certain of the transmitting contacts to close, the ones selected being determined by the letter being transmitted, i.e., "a" closes contacts 1 and 2, "b" closes 1, 4 and 5, etc. As soon as the sending arm starts revolving it operates the receiving end relays so the receiving brush arm starts revolving at practically the same speed as the transmitting one. Thus the two arms cross corresponding commutator segments together. If the letter "a" is being transmitted, battery is connected to the line as the sending arm crosses segments 1 and 2. This in turn causes the receiving magnets of 1 and 2 to operate. Other segments are dead for this letter.

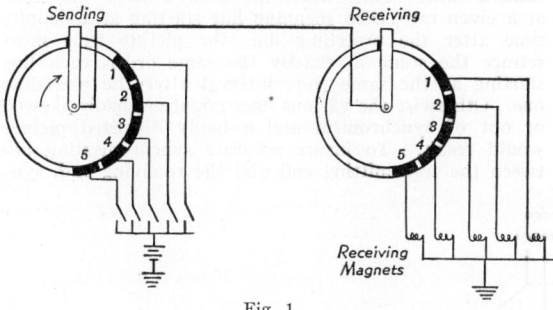

Fig. 1.

After crossing the 5 code segments, the transmitting arm is stopped and the receiving arm operates a printing relay and then is stopped, ready for starting on another letter. The receiving magnets, operating a mechanical selector mechanism set up the selection of the "a" type bar and when the printing circuit is actuated by the receiving arm, the letter is printed. Other letters cause other magnet combinations to pick up and thus select the proper type bars. The start-stop method allows the two arms to be synchronized for each revolution and thus the corresponding segments of the commutators are crossed at the same time. The multiplex system is much faster and does not stop and start the arms but uses other means of synchronizing.

A great part of the telegraph traffic is handled by various kinds of printers and there is also available to private subscribers a teletypewriter service on a basis somewhat similar to the familiar telephone service. In this service any teletype subscriber can be connected through teletype switchboards to any other teletype subscriber just as a telephone subscriber can connect with any other phone. Business houses operating on a national basis make extensive use of the system for transmitting orders, etc., from branch to home office or home to branch. (L.R.Q.)

TELEVISION. Television is the transmission of scenes, either still or motion, by electrical means, commonly by radio, for instantaneous viewing without permanent recording. For a practical system certain fundamental components or functions are necessary:

1. Camera device to pick up the scene.
2. Transducer to convert the light impulses of the scene to corresponding electrical pulses.
3. Transmitter to convert the electrical pulses into proper form to be transmitted to the receiver.
4. Receiver to pick up the transmitted signals and convert them to the proper form to apply to a transducer.
5. Transducer to convert the electrical pulses back into light in a reproduction of the original scene.

While in both still and motion pictures the entire scene is utilized at the same time to produce the film image, in all the present practical systems of television it is necessary to break up the scene into minute elements and utilize these elements in an orderly sequence. This process is called scanning and is very similar to the process of reading line by line the printed page. In reading the eye progresses along the line word by word,

relatively slowly, and then returns rapidly to the beginning of the next line, and repeats this process on down the page. In scanning the picture must be broken down into lines of successive elements just as the eye breaks down the printed page into words. While there have been various methods developed for performing this function all those in use at the present time utilize some electrical means. However, since it gives a rather clear idea of the basic process, we might consider one of the earlier mechanical methods using a revolving disk. This disk had a series of small holes along a spiral near its outer edge as shown in Fig. 1. Suppose, now, the

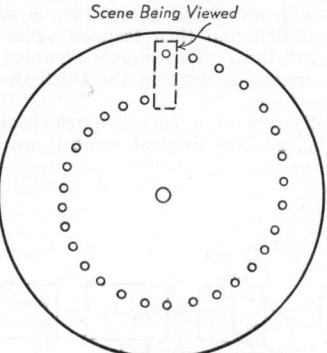

Fig. 1.   Mechanical scanning disk.

picture is viewed through these holes as the disk is rapidly revolved. The outermost hole passes across the top of the picture so if the observer is looking through it he sees successively the parts of the picture opposite the hole as it passes across the scene. As the first hole moves beyond the edge of the picture, the second comes to the other edge, and since it is one hole diameter nearer the center of the disk than the first, it covers a second line across the picture. Then the third hole crosses a line just below the second and so on until the disk has made a complete revolution when the entire picture has been covered. This is repeated at the rate of revolution of the disk. If these successive small sections of the picture as seen through the disk can be made to produce corresponding electrical effects the entire picture can be reproduced in an orderly sequence of electrical pulses. When the disk was the standard means of scanning this was accomplished by replacing the eye of the above discussion by a phototube. Modern television, as has been said, utilizes an electrical scanning method.

The received scene must be reconstructed from these electrical pulses, or their received equivalents. If we assume that the original scanner broke the picture down into ten lines and each line had ten elements side by side (this is determined by the width of the hole, being the width of the picture divided by the hole width) then we have 10 times 10 or 100 elements in the scene and our reproduced picture must be built of 100 blocks. It can readily be seen that this would give a very coarse mosaic effect to the picture since each element is fixed in intensity of light. More lines would give more elements and finer detail in the receiver scene. A close comparison may be drawn between this and the printed pictures of newspapers and magazines. The relatively coarse newspaper pictures are lacking in detail while the usual magazine half-tones give very good detail. The difference between these pictures is the number of dots or elements (clearly visible if the printed picture is viewed through a magnifying glass) of which they are composed. Various technical considerations dictate the exact number of lines into which the television scene may be broken, but at present it is of the order of 500, while the future holds promise of much higher numbers and consequently much better detail.

Besides the number of lines per scene, the engineer must decide upon the repetition rate. In moving pictures at present the action effect is produced by projecting upon the screen successive scenes taken and projected at the rate of 24 per sec. The persistence of vision of the eye then gives the effect of smooth motion. The television pictures must also be repeated at a rate high enough to give the illusion of smooth motion. While the motion picture rate of 24 frames per sec. would be satisfactory for this the standard power supply frequency of 60 cycles dictates the use of 30 frames per sec. for television. Further to improve the quality of the reproduced picture, the scanning is not done for adjacent lines in order, but the picture is scanned over alternate lines first and then scanned again over those missed the first time. This double scanning, known as interlaced scanning, is done in the thirtieth-of-a-second period of one frame.

A block diagram of a complete television system is shown in Fig. 2. The original scene is focused on the picture carrier. Both are then picked up by the receiving antenna, amplified and fed through the first detector and intermediate frequency amplifiers of a **superheterodyne receiver.** The two types of signal are then separated and each is fed to its proper **detector** or demodulator. The sound signals, now at audio frequency, are further amplified and drive the **loudspeaker.** The picture signal circuits are much more complex. In order to reproduce the scene at the receiver, it is necessary for the receiver transducer, whatever its nature, to follow exactly the operation of the camera tube. Thus when the camera scans the scene at a given rate, each scanning line starting at a definite time after the preceding one, the picture tube must retrace the scene in exactly the same order, each line starting at the same time interval after the preceding one. Otherwise the various lines might get badly skewed or out of synchronism and a badly distorted picture would result. To insure accurate synchronization between the transmitting end and the receiving end, syn-

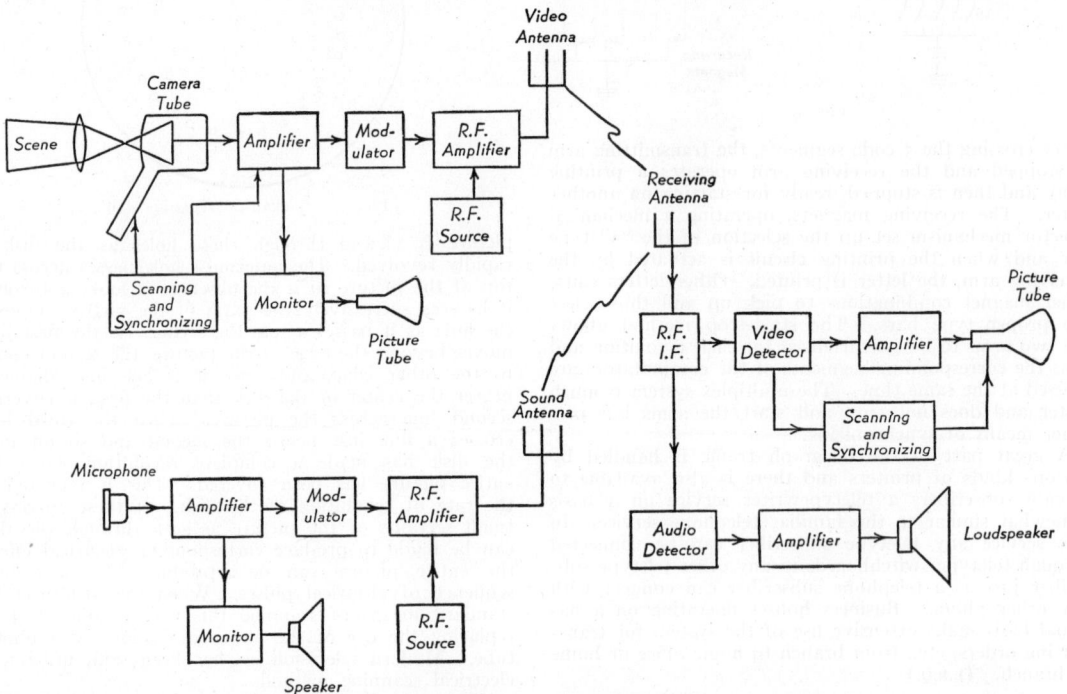

Fig. 2. Block diagram of television system.

camera tube by a light lens system. The camera tube converts this light picture into the sequence of electrical elements necessary for transmission. (For the operation of two of the various types of camera tube see **Iconoscope** and **Image Dissector.**) The very minute electrical signals coming from the camera tube are amplified by wide band **amplifiers,** the wide band being necessitated by the great range of frequencies produced by the modern multi-line systems. This wide band amplifier feeds a monitor circuit which reproduces the televised scene on a picture tube so the operator can check the camera circuit operation continuously. It also drives the modulator which modulates the picture signals on a radio-frequency **carrier in a manner very** similar to that of the audio **modulation** of conventional broadcasting. The modulated radio frequency is then fed to the **antenna** and radiated into space. At the same time the **microphone** is picking up the sound associated with the scene. This signal is amplified and impressed on the radio frequency sound **carrier** just as for any amplitude modulated **transmitter.** The sound-modulated carrier is radiated simultaneously with the

chronizing pulses are transmitted at the end of each scanning line. In addition synchronizing pulses to govern the return of the scanning to the top of the scene are also transmitted. These various pulses are impressed on the signal in the transmitter during the short time interval while the scanning is returning from the end of one line to the beginning of the next. In the receiver these synchronizing pulses must be separated from the detected signal and routed into the proper channels. They are then used to synchronize the sweep oscillators which give the reversed scanning at the receiver or picture tube. The video signals without the synchronizing pulses are amplified in the proper channels of the circuit and also fed to the picture tube. The electronic picture tube is the **cathode ray tube.** The electron beam issuing from the gun is modulated in intensity by the picture or video signal so the intensity of the fluorescent spot produced by it on the tube screen is a reproduction of the intensity of the corresponding part of the original scene. The synchronizing and scanning circuit produces a sweep signal which is applied to the picture tube by plates or coils just as in the oscilloscope discussed in the section on cathode ray tubes. This sweep action carries the electron beam relatively slowly across the screen, then blanks it and returns it rapidly, moves it down and repeats, the time for each operation being the same as for the corresponding operation in the camera tube, the two operations being linked together by the synchronizing pulses. After completing the scanning of alternate lines of the picture, the beam is deflected back to the top of the picture and repeats this operation, now filling in the alternate lines which were skipped on the first scanning, again in exact synchronism with the same process in the camera tube. It can be seen, then, that since the intensity of the spot corresponds at each instant to the intensity of the original scene and the position of the spot corresponds with the position of the original scanning position at the original scene, the reproduced effect on the screen of the picture tube is the scene which was picked up by the camera.

While the various systems of television in current use are all electronic in nature and differ in details rather than principle, the systems are highly complicated and are subject to certain limitations. The frequency band of the video signal is proportional to the number of lines, the width of the elements in any line and the number of frames per second, giving a band width of several megacycles for satisfactory reproduction. Since this requires a correspondingly wide radio-frequency **channel** it is entirely impracticable to use carrier frequencies in the ordinary broadcast band, both because the entire band could accommodate less than one station, but also because of complications in designing suitable selective circuits at these frequencies. These factors have forced television to the very high frequency region. Further to aid in the problem of getting the most out of the available spectrum space, both **sidebands** are not transmitted, but one is eliminated except for the components nearest the carrier. This is known as vestigial transmission and is a compromise between conventional double sideband and single sideband transmission. The audio signal band is transmitted on adjacent frequencies, the spectral arrangement being shown in Fig. 3. Because of the nature of propagation at

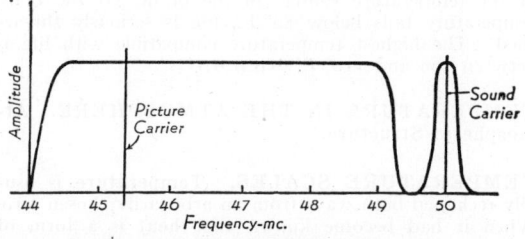

Fig. 3.   Spectral distribution of signal.

these high frequencies the range of television stations is very limited, reliable reception being possible over distances very little greater than **line-of-sight** distances. Under abnormal atmospheric conditions reception can be secured over greater distances, sometimes even running into hundreds of miles, but this is extremely erratic and occurs only rarely. The reflection and diffraction problems which accompany these high frequencies also enter the problem and limit the satisfactory reception of television. It seems doubtful if reliable reception will be available for years for any but the residents of urban and suburban areas. The conventional chain service of standard broadcasting is also unlikely since the wide band of video frequencies introduces grave problems in wire transmission. The **coaxial cable** is the only possibility for wired program connections and it is very expensive. Automatic relay transmitters and receivers operating on ultra high frequencies offer the possibility of radio-linked networks. However, the entire trans-

mitting procedure is so expensive at present it is doubtful if smaller communities will justify the expense so television service for these and rural areas must wait for technical advances to bring down the costs. In spite of these limitations, the service which can be rendered in large population centers is at present satisfactory and within the very near future bids fair to be greatly improved, both by development of systems using more lines and even giving color television.   (L.R.Q.)

TELIOSPORE.   Rust Fungi.

TELLURIUM.   Symbol: Te.   Atomic number: 52. Atomic weight: 127.61.   Density: 6.25.   Hardness: 2.3. Melting point: 452° C.   Boiling point: 1390° C.   (Isotopes: page 290.)

Tellurium is a silver-white brittle semi-metal; stable in air, and in boiling water; when heated in air burns with a greenish flame to form the dioxide; insoluble in **hydrochloric acid**, but dissolved by nitric acid or **aqua regia** to form telluric acid; dissolved by **sodium** hydroxide solution; combines with chlorine upon heating to form tellurium tetrachloride. Discovered by Reichenstein in 1782.

Tellurium occurs chiefly as telluride in **gold, silver, copper, lead,** and **nickel** ores in Colorado, California, Ontario, Mexico, and Germany, and infrequently as free tellurium and **tellurite** (tellurium dioxide, TeO_2). The **anode** mud from copper and lead refineries, or the flue dust from roasting telluride gold ores is treated by fusion with sodium nitrate and carbonate and the melt extracted with water. The resulting solution is acidified carefully with **sulfuric acid**, whereupon tellurium dioxide is precipitated, and the dioxide reduced to free tellurium by heating with carbon. Chemically related to **selenium.**

Acids: Tellurous acid (H_2TeO_3), white crystals, only slightly soluble in water; telluric acid (H_2TeO_4), white crystals, soluble in water, forms telluric oxide upon heating to a red heat.

Chlorides: Tellurium dichloride ($TeCl_2$), black solid; tellurium tetrachloride ($TeCl_4$), white solid, melting point 224° C., formed by reaction of tellurium and excess chlorine.

Hydride: Tellurium hydride, hydrogen telluride (H_2Te), by reaction of aluminum telluride with water or hydrochloric acid, gaseous, unstable, of unpleasant odor, chemically related to hydrogen selenide.

Oxide: Tellurium dioxide (TeO_2), white solid, sublimes at 450° C., only slightly soluble in water, soluble in acids, and in sodium hydroxide; tellurium trioxide (TeO_3), orange crystals, insoluble in water, in nitric acid, and in cold **hydrochloric acid**, soluble in hot concentrated sodium hydroxide solution.

Tellurides: Metallic tellurides are formed by combination of tellurium and the metal, or by precipitation of the metallic salt solutions with hydrogen telluride.

Numerous organic compounds of tellurium have been prepared.   (R.K.S.)

TELOGONIA.   Nematoda.

TELPHER.   An electric hoist suspended from a wheel carriage which rolls on an overhead track is called a telpher. The operator rides on a platform, also supported by the carriage, and controls the hoisting as well as the horizontal movement of the telpher. The telpher can be employed to lift and transplant all material which can be suspended from a hook, and is suitable for a variety of service in industrial plants, being able to transport lumber, pipe, iron (usually with electromagnet), partially or completely manufactured parts, packages, and bulk material (with grab bucket). The carriage may roll on an I-beam suspended from an overhead structure. In building interiors it may be suspended from the roof trusses. Outside, of course, a special supporting structure must be built. (See **Crane**.) (F.T.M.)

TELSON. The hind part of the body of an **arthropod**, beyond the segmented abdomen. The **anus** opens in this structure, which is otherwise extremely varied. In the **lobster** it is a flap associated with a pair of broadened appendages to form a powerful fanlike swimming organ and in the **scorpions** it is the sting. The telson of the **horseshoe crab** is the spine-like terminal appendage. (A.W.L.)

TEMPER. Temper has two distinct meanings in metallurgy. Hardened steels are tempered after quenching to increase their toughness. (See **Heat Treatment**.) In the heat-treatment of small tools the "temper" may be drawn by observing the "temper color" of a brightened area as it becomes oxidized during heating. The first indication is a light straw color which appears at about 425° F. The piece may be quenched to arrest the tempering process. At higher temperatures brown, purple, blue, and dark blue to black oxides form. At the higher temperatures the hardness decreases appreciably. In the production heat-treatment of both tool and machine steels, tempering is carried out in batch or continuous-type furnaces in which the heating cycle is carefully controlled by pyrometers. Under these conditions much more uniform results are obtained than by the old temper color method.

Temper also refers to the degree of hardness induced by a **cold-working** process such as sheet rolling or wire drawing. Tempers are expressed as soft, ¼ hard, ½ hard, etc., or in **gauge numbers** reduction in thickness or diameter. The following table applies to copper alloy sheet and strip.

Commercial Temper Designation	B. and S. Nos.	Reduction of Hard Thickness, %
Soft (annealed)	0	0.0
Quarter hard	1	10.9
Half hard	2	20.7
Three-quarter hard	3	29.4
Hard	4	37.1
Extra hard	6	50.0
Spring	8	60.5
Extra spring	10	68.7

Reduction in thickness of an annealed brass sheet from 0.0808″ (12 gauge) to 0.0508″ (16 gauge) by cold rolling would result in a hard temper. (R.H.H.)

TEMPER CARBON. Cast Iron.

TEMPER ROLLING. Soft annealed low-carbon steel and ingot iron sheets tend to form stretcher-strains, a rough surface condition, when drawn into autobody components or similar parts. Stretcher-strains, also called Lüder lines or orange peel, are formed when the material reaches the **yield point**. In areas where the total stretch is greater than the yield point elongation, the surface again becomes quite smooth. In order to prevent stretcher-straining, annealed sheets are temper-rolled to eliminate the yield point elongation. A reduction by cold rolling of only 0.5–2.0% is sufficient. In the case of iron or steel made by the rimming process the yield point elongation gradually returns, making it necessary to draw or stamp the sheet within a limited time after temper rolling. **Roller leveling** and **stretcher leveling** also aid in the prevention of stretcher-straining. (R.H.H.)

TEMPERATURE. Fundamentally, temperature is a manifestation of the average translational kinetic energy of the molecule of a substance due to heat agitation, and is measurable by any one of many physical effects

due to changes or differences in this energy. Thus, substances expand, their electrical resistivity changes, gases and vapors exert varying pressure, the viscosity of fluids alters, etc., as the temperature varies; and the state of aggregation of any substance (whether solid, liquid, or gaseous), under a fixed pressure, depends primarily upon the temperature. Very imperfectly, also, our special temperature sense is able to judge whether one body with which we come into contact is warmer or colder than another. Heat energy always transfers itself spontaneously from the warmer to the cooler parts of any body or system of bodies, never in the reverse direction, and the transfer ceases when the temperatures become equalized. The temperature of a vacuum may be defined as the temperature of a small body placed in it and in thermal equilibrium with it. All measurements of temperature, upon whatever principle they are based, are comprised under **thermometry**. (See also **Temperature Scales**.) (L.D.W.)

TEMPERATURE GRADIENT IN THE ATMOSPHERE. The maximum rate of decrease of temperature with distance in the atmosphere is known as the thermal gradient. Temperature gradients directed vertically are known as lapse rates. (P.E.K.)

TEMPERATURE, HUMAN. The normal temperature of the human body is usually said to be 98.6° F. Actually this normally varies in the course of 24 hours, so that variation between maximal and minimal temperatures during this period is about 1.8° F. The above figures are those obtained when the temperature is taken by mouth. The rectal temperature is 1° F. higher.

The body temperature is determined by the relation of two factors: the amount of heat produced and the amount of heat eliminated. The amount of heat produced depends on the basal **metabolism**. Additional heat is produced by muscular activity. Extra heat is eliminated by increase in radiation, condensation, or evaporation on the skin surface, and by more rapid and deeper breathing. When insufficient heat is eliminated the temperature rises and **fever** is said to be present. Normally the delicate adjustment in the body between heat production and heat elimination is under control of the temperature centers in the brain. If the body temperature falls below 86° F., life is seriously threatened. The highest temperature compatible with life is between 109 and 110° F. (D.M.H.)

TEMPERATURE IN THE ATMOSPHERE. Atmospheric Structure.

TEMPERATURE SCALES. Temperature is usually reckoned both ways from an arbitrarily chosen zero. When it had become known that heat is a form of **energy** and that temperature is determined by the mean transitional energy of molecular agitation, the concept of an **absolute zero** of temperature became possible; but since such a condition is apparently unattainable and is inconveniently far removed from temperatures of everyday experience, a more practicable reference point, such as the melting point of ice, is preferable for ordinary purposes.

Starting at such a zero, it would seem logical to base the temperature scale upon equal quantities of some selected effect caused by temperature change. For example, a change of one Fahrenheit degree in the temperature of ice-cold mercury increases its volume by almost exactly one ten-thousandth, so that each ten-thousandth might be taken as indicating one degree of temperature. However, a different plan has been followed. Instead of one fixed point, two are chosen, and the temperature interval between them divided into aliquot parts.

Thus, for the centigrade scale, the fixed points are the freezing and the boiling points of water, and the interval is divided into 100 parts, so that the freezing point is 0° C. and the boiling point is 100° C. On the Fahrenheit scale, these points are marked 32° and 212° respectively, the zero point having no obvious physical significance. The ratio of the Fahrenheit to the centigrade degree is therefore 5:9; and if t_F and t_C are respectively the Fahrenheit and the centigrade value of the same temperature, their relation is easily seen to be

$$t_F = \tfrac{9}{5}t_C + 32°, \quad \text{or} \quad t_C = \tfrac{5}{9}(t_F - 32°).$$

The original of the centigrade scale was the Celsius scale, on which the boiling point was zero and the freezing point 100°. The Réaumur scale, with freezing as zero and boiling at 80°, is still used in parts of Central Europe. Special limited scales have been adopted for certain purposes; for example, the Leyden low-temperature scale (or low-temperature range of the centigrade scale), taking the boiling point of hydrogen at −252.74° and that of oxygen at −182.95° as its fixed points and using the centigrade degree.

Since the centigrade value of absolute zero has been determined as −273.16°, it follows that if we wish to express an "absolute temperature" in centigrade degrees, we have but to add 273.16° to the centigrade temperature:

$$T = t_C + 273.16°.$$

This is the basis of the Kelvin scale, and therefore 0° C. is 273.16° K. (degrees Kelvin). (See **Thermochemistry**.)

It will be noted that nothing is here said as to the basis upon which equal subdivisions or "degrees" of a fixed temperature interval are established. For this the reader is referred to **Thermometry**. (L.D.W.)

TEMPLATE. Press Working; Sheet Metal Processes.

TEMPORAL. A bone of the vertebrate skull. In the human skull it lies at the side, centering about the ear. It forms the posterior part of the cheek bone with the articulation of the lower jaw, and bears the mastoid process behind the ear. It includes the squamosal bone and the bones of the ear. (A.W.L.)

TENCH. Pisces, Teleostei. A European fresh-water fish (**Pisces**), *Tinca tinca*. It is found in ponds, lakes, and other quiet waters, particularly those with a muddy bottom. The species is known to reach a weight of 5 lbs. in some cases. (A.W.L.)

TENDON. Connective tissue structures connecting muscles with their skeletal supports. They are composed of parallel white fibers, closely bound into bundles between which the cells of the tendon are compressed. The entire tendon is surrounded by a fibrous sheath known as the vagina fibrosa which is split in some cases to form a cavity containing a mucoid liquid. Such a sheath is called a vagina mucosa. It develops where a wide range of movement is necessary.

Aponeuroses, ligaments, and **fasciae** are fibrous structures resembling tendons in structure. The first are broadly expanded and usually bind down muscles. The second connect bones. The third are fibrous coverings of muscles and other organs. When tendons are severed they must be sutured together again. At least 6 weeks are required for union between the cut ends. (A.W.L., R.S.M.)

TENDRIL. A tendril is a slender elongated structure which either twines around any supporting object or is attached thereto by means of small disks. Tendrils may be modified **stems**, as in the grape, **leaves**, or leaflets as in various peas, or **stipules**. (R.M.W.)

TENON. Dado.

TENORITE. A mineral oxide of copper, CuO. Crystallizes in the monoclinic system. Hardness, 3.5; specific gravity, 5.8–6.4; color, gray to black with metallic luster. Named after M. Tenore (1780–1861), Naples. (R.M.F.)

TENOSYNOVITIS. An infection of the membranous sheaths which envelop tendons in various parts of the body. The sheaths most commonly involved are those which surround tendons controlling movement of the fingers, wrist, and hand, and provide frictionless movements of these parts. Normally there is a small amount of fluid within the sheaths to provide lubrication. When infection takes place, more fluid accumulates with the production of pus. The only treatment that will save the function of the tendon is very early surgical drainage.

The symptoms of tenosynovitis are acute pain, swelling, and redness over the affected part. (D.M.H.)

TENREC. Mammalia, Insectivora. *Centetes*. An animal of Madagascar. This species resembles the other members of the order in its compact body, short legs, and long sharp muzzle. It is the largest of the group, attaining a length of 16″, and is clothed with a mixture of spines, bristles, and hair. Tenrecs are nocturnal burrowing animals. (A.W.L.)

TENSILE LOAD. Load.

TENSILE STRESS. Stress.

TENSION. In structural engineering tension is used to denote the longitudinal force which causes the fibers of a member to elongate, thus giving rise to tensile stress. (C.W.C.)

TENSION MEMBER. Any member of a structure which is subjected to a primary tensile **stress** is called a **tension** member. The analysis or design depends on that part of the cross-sectional area which is effective in resisting the tensile stress. The effective area at any section is called the net section. It is the gross area minus the area of any holes including those filled by rivets or **bolts**. The net section may be based on a right section or a zig-zag section. The net section of a threaded member such as a bolt or **tie rod** is the area at the root of the thread. The strength of a tension member is a function of the minimum net section.

If the member is composed of two or more independent parts other than **eye bars**, the parts may be connected by **tie plates** or by tie plates and **lacing bars** to equalize the stress and prevent excessive distortion during fabrication, shipping or erection. (C.W.C.)

TENSION TEST. Next to **hardness** tests, tension tests are the most frequently used to determine the **mechanical properties of metals.** Tension test specimens necessarily vary in form with the product to be tested. A machined cylindrical specimen with threaded or shouldered ends for gripping is used when the material is sufficiently thick. Standard flat specimens are used for flat-rolled products. While both types of specimens have a reduced central section to insure breaking within a measured gauge length, wires and certain special shapes such as steel reinforcing bars for concrete are tested in full section without preparation. Special cast test specimens are often attached to castings, or cast separately, and these are generally tested without machining.

The significant loads determined in the test are reported as unit stresses based on the area of the original section. (Stress = load ÷ area.) The elongation is expressed as % increase in length of the gauge-marked section. The initial gauge length is generally 2 in., although an 8-in. gauge length is used for certain flat specimens and other gauge lengths are used for special specimens. The gauge length should be specified when elongation is reported since % elongation values are higher for short than for long gauge lengths.

The elongation measured over a fixed gauge length, and the reduction of area of the section at the fracture are measures of ductility. In cylindrical specimens the area is readily determined from the final diameter at the fracture. The % reduction of area is then determined as (original area — final area) ÷ original area.

Autographic load-deformation curves are often drawn during the test. From such a curve the modulus of elasticity, proportional limit, and yield strength can be determined.

A typical curve has an essentially linear portion (*OA*) in which the deformation is proportional to the applied load. It follows that the unit stress (load ÷ original area) is proportional to the unit strain (deformation ÷ original gauge length) in accordance with Hooke's Law. The numerical value of this ratio (e.g., in lbs. per sq. in.) is known as Young's Modulus or Modulus of Elasticity. (See **Elasticity**.)

The maximum stress that is developed without deviation from proportionality of stress to strain is the pro-

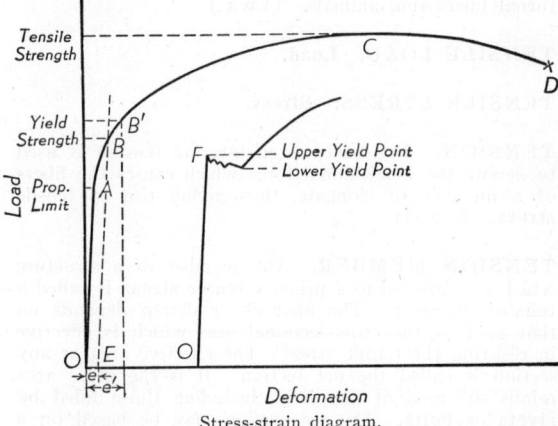

Stress-strain diagram.

portional limit (the stress corresponding to load *A*). The maximum stress that can be applied without causing permanent deformation upon release of the load is the elastic limit. Usually there is little difference between the proportional limit and the elastic limit. Both are dependent on the sensitivity of the measuring devices used and certain details of testing technique. For this reason the yield strength is generally used as a practical measure of the elastic properties of metals.

The yield strength is the stress at which the stress-strain curve deviates from the initial straight line by a specified increment of strain. The yield strength corresponding to the load at *B* is based on the specified strain deviation or offset *e*. The value of *e* may be as low as 0.0001 in. per in. of gauge length but the most commonly used value is 0.002 in. or 0.2% strain.

If the load should be released after reaching *B*, the load-deformation relationship will follow the line *BE*, or a curved line terminating between *O* and *E*. Thus the permanent strain will be *e* or a somewhat smaller value. When the final or permanent strain is specified, the stress is known as the proof stress.

An alternate type of yield strength is based on a specified total extension under load, such as 0.5%. If the specified extension is *e'* the load *B'* determines the "extension-under-load" yield strength. Load *B'* may be greater or less than *B*.

The tensile strength, or ultimate tensile strength, is the maximum stress developed in the tension test. (Load *C* ÷ original area.)

The breaking stress, corresponding to load *D*, is seldom determined or reported.

In loading tension specimens of many soft irons and steels a point is reached where stretching continues without increase in load. The unit stress obtained by dividing this load, *F*, by the original area of the section is called the yield point. The elongation of the specimen at the yield point may reach 8% in some instances, after which the load will again increase to a maximum in the normal manner. Upper and lower yield points are indicated at *F*; both are used but the upper yield point is influenced by variations in testing technique such as alignment of the specimen in the testing machine and speed of test. Yield points occur only rarely in the non-ferrous metals.

Conventional stress-strain curves are necessarily similar to the load-deformation curves from which they are derived. True stress-strain curves can also be derived in which the stress is based on the actual or instantaneous area of the cross-section. Such curves do not have a maximum corresponding to *C* but increase continuously to the breaking load. (R.H.H.)

TENSION-FIELD THEORY OF BEAM ANALYSIS. The analysis of **plate girders** and other built-up beams of structural engineering is based on the assumption that the web does not buckle along diagonal lines under the action of the diagonal compressive stresses. (See **Stress, Principal Stress.**) Although these stresses vary through the depth of web (see **Web Plate**) between **flanges** at any section, they are usually assumed to be constant and equal to the value at the **neutral axis**. The diagonal compressive stress at a point on the neutral axis is equal numerically to the **shearing stress** at that point. Design procedure is such that these webs will not buckle if the **intermediate stiffeners** are properly spaced.

In **airplane** structures the webs of built-up beams are very thin. Consequently the possibility and the effect of diagonal buckling must be investigated even though the webs are supported by closely spaced stiffeners. These stiffeners divide the web into a series of panels in which buckling may occur. After a web has buckled there is a complete redistribution of stress at the section. The web within the panel acts as a series of inclined strips subjected to **tensile stress**. The flanges must resist both vertical and horizontal components of this diagonal tension at the junction of the flange and web. The horizontal component causes **direct stress** in the flange and the vertical component produces flexural stress (see **Flexure**) in the flange which acts as a **beam** continuous over the stiffeners. Buckled webs which carry the shear by means of inclined tension only are called tension-field webs. Beams whose webs are designed to buckle are known as Wagner Beams.

If the shear at the neutral axis is less than a certain critical value at which buckling takes place, the web is said to be shear resistant and the usual formulae for beam analysis are applicable. The critical stress at which a braced panel will buckle may be obtained by analytical or experimental means. Formulae which are applicable to beams with shear resistant webs cannot be used for beams with tension-field webs. (C.W.C.)

TENSION-FIELD WEB. Tension-Field Theory of Beams.

TENSORS. The tensor concept is a generalization of that of a **vector**, and requires for its specification more than three components.

A tensor may be defined as a set of n^r components which are functions of the coordinates of any point in space of n dimensions, which is transformed linearly and homogeneously, according to certain rules, when a transformation of coordinates is made. Tensors are called covariant, contravariant or mixed, according to the law of transformation. The number r is called the rank or order of the tensor. (L.L.S.)

TENTACLE. A slender fleshy protuberance of the body wall, in the region of the mouth and usually arranged as a group surrounding the mouth. Tentacles are well developed among the **coelenterates,** the marine **annelids,** and the **cephalopod** mollusks, and are present in all **Bryozoa, Brachiopoda,** and **Phoronidea.** In the coelenterates they contain stinging cells and are capable of wrapping around the prey to bring it to the mouth. They also act to secure food by ciliary movement; in the Bryozoa the cilia covering them apparently carry a food-bearing current down the funnel that they form about the mouth. Cephalopod tentacles also enfold the prey but in addition they are provided with cuplike suckers which enable them to grip even very smooth surfaces. In this group and to a limited extent in the coelenterates they are used for locomotion.

In the **echinoderms** a sensory appendage at the end of each radial water vessel is called a terminal tentacle and, in the sea cucumbers, tube feet in the region of the mouth are developed into tentacles of various forms, sometimes finely branched. (A.W.L.)

TENTACULOCYST. A sense organ of the jellyfishes and a few allied forms. It consists of a reduced tentacle at the margin of the body, hooded by a small projection of the margin. The tentacle bears a pigment spot which may be a light-sensitive organ and is flanked by two pits regarded as olfactory organs. Tentaculocysts lie at the ends of the interradial and perradial canals. (A.W.L.)

TEOSINTE. Corn.

TEPHRITE. A term proposed by Cordier, in 1816, for a variety of **basalt** containing both **plagioclase feldspar** and **nepheline** or other soda-feldspathoids. Tephrite differs from **basanite** because of the absence of **olivine.** (R.M.F.)

TERBIUM. Symbol: Tb. Atomic number: 65. Atomic weight: 159.2. No isotope, but of single atomic form: 159. Type of compound: Tb$_2$O$_3$. Color of salts: Colorless. Discovered by Mosander in 1842. A member of the **yttrium** sub-group of the rare earth metals. (R.K.S.)

TERGITE. Skeletal System.

TERMINAL EFFICIENCY. The structural efficiency with which stresses may be transmitted through the terminal connection of a structural member is called "terminal efficiency." This is of more importance to tension than compression members. Near the point of connection of a member such as a tie-rod, a cable, or a rolled or drawn section, the connecting device will usually alter the distribution of stress over the cross-section of the member. Threads or rivet holes may be cut to develop the type of connection desired. Eccentric loadings and short radius bends may have to be introduced. The terminal efficiency may be defined as the safe working stress of the member, including terminals, divided by the safe working stress of the section used for the member. It is possible to have terminal efficiencies greater than 100%. For example, by swaging the ends of a rod its diameter at the terminal fitting can be increased sufficiently to provide greater strength than that possessed by the normal section of the rod. (F.T.M.)

TERMINAL MORAINE (END MORAINE). When balance is maintained between the melting of a **glacier** and its forward advance, the debris carried on (super-glacial); within (englacial); and dragged along the bottom (subglacial); is dumped at that point and builds up a heterogeneous mass of the transported material called the terminal moraine. If a glacier is slowly retreating and makes successive halts farther and farther up the valley, a series of terminal moraines are formed which are spoken of as recessional moraines. (R.M.F.)

TERMINAL TEMPERATURE DIFFERENCE. Two fluids not having the same temperature and being on opposite sides of a heat-conducting surface cause a temperature difference to exist across the surface, a difference that may vary from point to point on the surface. If one of the fluids is in motion the temperature difference where it joins or leaves the conducting surface is the terminal temperature difference. Water flowing through a tube which is surrounded by steam whose temperature exceeds that of the water will receive heat by conduction through the tube and become heated. The temperature difference between steam and water dwindles as the water traverses the tube, becoming minimum at the discharge end of the tube. The minimum temperature difference is often called the terminal temperature difference. (F.T.M.)

TERMITE. Isoptera.

TERMITOPHILE. An insect of another form living in the nest of **termites.** Both ant and termite colonies are inhabited by other insects, some apparently living as scavengers and profiting by the supply of food and by the protection afforded by the colony while others produce secretions used by their hosts and so live in a commensal relationship. Even in the latter cases the termitophiles may eat the young of the termites, but since they render some return they are not to be regarded as parasitic on the colony. Some, however, are nourished by food supplied by the termites in return for the desired secretion. Insects of such habits are often very strangely formed, differing conspicuously from other members of the orders to which they belong.

Among the known termitophiles are **flies, larvae of moths,** one species of **Homoptera,** and many **beetles.** (A.W.L.)

TERN. Aves, Charadriiformes. A bird (**Aves**) resembling the gulls but with a more slender and tapering beak. The tail is forked. Terns are very like gulls in habits and are usually seen in flight over water, both on the coasts and near ponds and lakes. About 10 species are normally found in the United States. (A.W.L.)

TERNE PLATE. An iron or steel sheet coated by the hot-dip process with a lead-tin alloy containing up to 25% tin.

There are two classifications known as "short ternes" and "long ternes." Short ternes are produced in certain small sheet sizes in light gauges only and usually carry heavy weights of coating. They are used principally for roofing purposes.

Long terne sheets are produced in sizes 24" x 60" and larger in a wide range of thicknesses and usually carry much lighter weight coatings than short ternes. Long terne sheets are often supplied in deep drawing grades, in which case the coating acts as a carrier for lubricating oils or compounds in the drawing dies and also provides protection from rust during processing and storage. The alloy coating also greatly facilitates soldering and is a good base for paint. Among the principal applications of long ternes are metal caskets and automotive parts such as gasoline tanks, radiator side walls, oil filters, and mufflers. (R.H.H.)

TERRA COTTA. Terra cotta is the name given to a product made of burnt clay which has been fired at a high temperature. It is employed in building construction as a decorative feature, and is to be found in copings, cornices, facades, and other trim. It is used both on exteriors and interiors. Terra cotta is one of the most permanent of building materials, and a great deal of distinctive individual ornamentation of buildings of present and past times has been achieved by liberal use of terra cotta. A standard building material used to construct interior partitions, to back up exterior face brick in a masonry wall, or to form a complete exterior wall for an industrial building, is known as hollow building tile. The aim in manufacturing this product, of course, is to produce a cheap fireproof building material, but not an especially decorative one.

A great deal of terra cotta work is special, i.e., in its shape and surface decoration, as well as color and method of attachment. Models of the separate pieces are made and plaster casts taken. These casts are then used to mold the clay body of the terra cotta. When the molded clay has dried it is removed from the mold and fired at high temperature until it has become hard and unaffected by moisture. It is known now as a bisque, and is ready for the application of liquid glazes, known as slips. The minerals which produce the color are incorporated in this slip. The bisque is either sprayed or dipped in the slip, and returned to the kiln for firing of the glaze. Terra cotta is set in mortar, with carefully pointed joints, and in some cases may be reinforced by metal ties or by shelves or caps. (F.T.M.)

TERRA ROSSA. A red **ferruginous**, residual earth derived from the surface alteration of limestones. A characteristic soil of the *Karst* lands surrounding the Adriatic Sea. (R.M.F.)

TERRAPIN. Reptilia, Chelonia. Certain of the pond and land **turtles**, especially those of the genera *Malaclemmys* and *Pseudemys*. The related genus *Terrapene* includes the box turtles, and the common painted turtles are also closely connected forms. The species whose flesh

Diamond back terrapin. (*N. Y. Zool. Soc.*)

is so highly esteemed is the diamond-backed terrapin or salt-marsh turtle found in salt marshes along the Atlantic coast from Massachusetts to Florida. This and a related species of the Gulf coast make up the genus *Malaclemmys*. The species of *Pseudemys* are commonly called sliders or cooters, one species bearing the name red-bellied terrapin. They are edible but are less highly valued than the diamond-back. Terrapins are widely distributed in the northern hemisphere. Elsewhere they are limited to a few species in Central and South America. (A.W.L.)

TERRESTRIAL COORDINATES. The position of a point on the surface of the earth may be defined to a high degree of precision by considering the earth as a sphere and establishing on that sphere a reference frame

for spherical coordinates. The most common system is to consider the line joining the poles of rotation of the earth as a fundamental line. The plane perpendicular to this line, and passing through the center of the earth, is the plane of the **equator** and cuts out on the surface of the earth a great circle known as the terrestrial equator. Great circles on the earth perpendicular to the plane of the equator and passing through the poles of rotation are known as terrestrial **meridians**. The fundamental direction selected in the plane of the equator is the direction of the point of intersection of the meridian through Greenwich, England, with the equator.

Terrestrial **longitude** of a point on the earth is the spherical coordinate measured in the plane of the equator either east or west from the point of intersection of the meridian through Greenwich to the point of intersection of the meridian through the point in question. **Latitude** is the spherical coordinate measured from the equator north or south along the meridian to the point in question. (See **Latitude, Longitude, Celestial Navigation**, etc.). (W.K.G.)

TERRESTRIAL MAGNETISM. The outstanding facts of the earth's magnetism were investigated by William Gilbert about 1600. He recognized that the earth acts as a huge bipolar magnet, and anticipated by many years the actual discovery of its magnetic poles. These are located, respectively, on the Boothia peninsula about 500 miles northwest of Hudson's Bay and on the Antarctic Continent some 750 miles northwest of Byrd's "Little America." (The pole north of Hudson's Bay is, of course, one of the negative or south-seeking kind, since the positive poles of compasses point toward it.) Our direct knowledge of the terrestrial magnetic field is confined near the earth's surface, and is usually represented by maps showing lines of equal magnetic **variation** (departure from the true north), lines of equal **inclination** or dip (see **Dip Needle**), and lines of equal magnetic intensity (see **Magnetometer**). The inclination at middle latitudes approximates 70°, while the total intensity in this region is usually something like 0.6 **oersted**, with a horizontal component of about 0.2 oersted. There is a line of zero variation (agonic line) at all points of which the compass indicates the true north; this line passes down through the United States and traverses the globe in each hemisphere from northwest to southeast.

The terrestrial field has certain variations, chief among which are: 1. The diurnal variation, in which the compass swings back and forth through an angle of many minutes once a day. 2. The secular variation, a slow periodic movement of many degrees with a period of several centuries. 3. Random variations, sometimes very great, when they are called "magnetic storms." These last are unquestionably associated in some way with those solar activities which result in **sun spots**, since their occurrence is almost simultaneous with the appearance of sun spots, and they have the same 11-year cycle. The **aurora borealis** seems to be similarly involved. The mechanism behind terrestrial magnetic phenomena has not been fully explained; though the rotating earth's electrical condition (see **Lightning**), together with thermoelectric earth-currents, doubtless contribute to some of the effects. (L.D.W.)

TERRESTRIAL PHOTOGRAMMETRY. Photogrammetry.

TERTIARY. A major subdivision of the **Cenozoic**, or last geologic era. The periods of the Tertiary from the oldest to the youngest are: **Paleocene, Eocene, Oligocene, Miocene, Pliocene.** (R.M.F.)

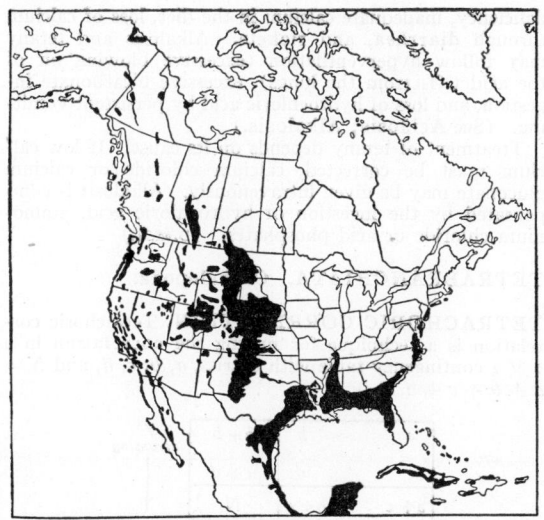

Map showing the surface distribution (areas of outcrops) of Tertiary strata in North America.

TESCHENITE. Theralite.

TESLA COIL. A type of **induction coil** in which the primary has a high-frequency spark gap instead of the usual interrupter, and whose secondary yields an intense high-frequency discharge. A typical arrangement is shown in the figure. A high-voltage transformer T, or sometimes an ordinary induction coil, sends sparks across the primary gap G_1, which is in circuit with a condenser C_1 and the primary air-core coil P, composed of a few turns of heavy copper wire or tubing. Because of the oscillatory nature of the condenser discharges, this circuit is the seat of powerful high-frequency oscillations. The secondary, S, consists of many turns of fine wire. The secondary circuit may be "tuned" by means of a variable condenser C_2, and when it is in **resonance** with the primary, the oscillations in it are very intense. A torrent of high-frequency sparks plays across the secondary gap G_2; or one terminal may be grounded and sparks drawn from the other. (L.D.W.)

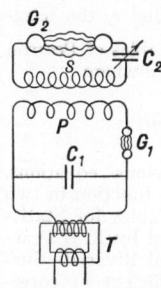

Diagram of Tesla-coil circuits.

TEST. A non-cellular structure enclosing the body of an animal and formed by the outer layers of cells. A test is found in some 1-celled animals, giving the name Testacea to one division of the phylum. In these forms it is composed of various materials, including **chitin** and calcareous (see **Calcium**) deposits. It may be either rigid or flexible. Foreign particles such as grains of sand are incorporated into some of these tests.

A tough enveloping test is also formed about the body of the **tunicates**, giving them their name through the synonymous term tunic. In this group the structure is formed of a material closely related to cellulose, which is a common plant product. (A.W.L.)

TEST OF INDEPENDENCE IN A CONTINGENCY TABLE. It is frequently desired to test whether items in contingency tables of r rows and c columns are independent. If we represent the actual number of **variates** in the cell formed by the intersection of the rth row and the cth column by n_{rc} and the theoretical number by n'_{rc} we know that on the hypothesis of independence

$$n'_{rc} = \frac{n_c n_r}{N}$$

where N is the number of items in the whole table and n_c is the number of items in the cth column and n_r is the number of items in the rth row. Then we compute

$$\chi^2 = \sum \frac{(n_{rc} - n'_{rc})^2}{n'_{rc}}$$

summed over all cells, $n'_{rc} \geqq 5$. For a test of significance we use the χ^2 distribution with $(n-1)(r-1)$ **degrees of freedom.** If the hypothesis is rejected we may if desired compute the **coefficient of mean square contingency.** If $n'_{rc} < 5$, special methods must be used or certain cells must be combined. If $n'_{rc} > 5$, χ^2 is found somewhat more easily by

$$\chi^2 = N\left[\left(\sum \frac{n^2_{rc}}{n_r n_c}\right) - 1\right],$$

the summation being extended over all cells. (L.A.A.)

TEST OF THE DIFFERENCE OF TWO MEANS. Student's t Distribution.

TESTA. Seed.

TESTACEA. Sarcodina.

TESTICARDINES. Brachiopoda.

TESTICLE. Testis. The testicles are the two male sex glands which are suspended from the groin by the spermatic cords, and are supported and enclosed by the **scrotum.** Each gland measures about $1\frac{1}{2}''$ in length, $1''$ in width and about $\frac{3}{4}''$ in thickness from side to side. The testicle is joined to the spermatic cord by means of the epididymis which is a coiled duct lying at the upper portion of the testicle. When uncoiled this duct measures $20'$ in length. The function of the testicle is to produce **spermatozoa** and the male **sex hormone.** If both testicles are removed before puberty, secondary sex characteristics fail to develop, due to the absence of testosterone. The skin remains smooth, the voice is high-pitched, fat develops around the breasts and buttocks, and the pubic hair is scanty. Erections are feeble, and no ejaculation occurs. The individual is timid, lacks ambition, normal combativeness and aggressiveness. Such an individual is known as a eunuch. Eunuchoidism usually results from failure of development of the testes, which is usually secondary to a **pituitary** disorder.

In early fetal life both the ovaries and the testicles lie in front of and below the kidneys. During fetal growth they descend. The ovaries finally lodge on the side wall of the pelvic cavity. The testicles normally continue downward and descend through and out of the abdomen in the region of the groin to the scrotum. This descent may be arrested along any portion of this pathway, and one or both testicles may remain in the abdominal cavity, in its wall or in the groin. The condition is known as **cryptorchidism.** Men with undescended testicles are sterile, but are sexually normal otherwise. The situation is corrected by an operation in which the undescended testicle is brought down to its normal scrotal position. This operation is best done at or shortly after puberty. In some cases glandular injections will cause the testicle to descend without operative interference.

Diseases affecting the testicles are tumors, which are rare, and infections. These include **tuberculosis, syphilis,** and acute staphylococcal, gonococcal, or other pyogenic infections. The testes are involved as a complication of **mumps,** and atrophy and sterility may be the end result. (R.S.M., D.M.H.)

TESTIS. Testicle.

TESTUDINATA. The turtles and tortoises, constituting an order of the class **Reptilia**, also known as the order Chelonia. These animals have a short broad body enclosed in a shell composed, in most species, of closely joined bony plates covered with thin plates of horn or tortoise shell. The bony plates are flattened ribs and vertebrae, with supplementary dermal plates. The upper part of the shell is the **carapace** and the plate below the body is the plastron; the two are firmly united along the sides in most species. The skin is covered with scales and the jaws are without teeth, forming a horn-sheathed cutting beak.

Most species of the order are partially aquatic but some are strictly terrestrial. The marine turtles are the most thoroughly aquatic species, but even they come to shore to lay their eggs. These species have all four limbs developed as paddlelike flippers while the other members of the group merely have webbed toes. (A.W.L.)

TETANUS (Lock-Jaw). An infectious disease caused by *Clostridium tetani* which liberate a toxin having a peculiar affinity for nervous tissue.

Tetanus has been known since ancient times. No specific treatment was available until the development of an antitoxin by Behring and Kitasato in 1890. In 1922 in the United States, 1480 persons died from tetanus, but in World War I, when tetanus antitoxin was given to all wounded American soldiers, among 224,089 wounded, only 36 developed the disease.

Tetanus bacilli are **anaerobic bacteria, spore**-forming rods which are found in the excreta of man and many domestic animals. They thrive well in soil and manure; the spores are extremely resistant to heat and chemicals. The disease usually follows an injury of the puncture or penetrating type. Blank cartridge wounds are particularly dangerous in this respect. Infection occurs by the introduction of spores into the wound. The organisms will not grow unless there is devitalized tissue present. If they gain a foothold, they multiply deep in the tissues and produce an extremely potent neurotoxin which is responsible for the signs and symptoms of the disease.

Following an incubation period of 5–10 days, the earliest symptoms appear. They are restlessness, irritability, stiffness and tightness of the various muscles. The jaw muscles are usually affected first, and later become fixed in a spasm which gives the disease its name—lockjaw. In severe cases, extreme rigidity of the muscles and painful **convulsions** set off by the slightest noise or other stimuli occur. Death may result from spasm of the respiratory muscles and asphyxia. Often, however, the immediate cause of death is not apparent.

All wounds of the dirty, deep lacerated type should be given immediate surgical care and tetanus antitoxin should be given at once. After the disease once develops, the important factor in treatment is to control the convulsions and to give sufficient antitoxin to neutralize the effect of the poison produced by the tetanus organisms. Before the use of antitoxin the mortality from tetanus was from 80–100%. At present the mortality is only about 25%.

Recently a tetanus toxoid has been produced. This is a preparation of treated toxin which produces active immunity when given in small doses subcutaneously. It is given to those who work in the soil, and may also be given to children routinely as a prophylactic measure. (D.M.H.)

TETANY. Increased neuromuscular irritability manifested by spasm of the muscles, particularly of the extremities, and sometimes convulsions. It is commoner in children than in adults. Tetany is due to an upset in the inorganic salt metabolism which may be caused by a number of factors. There are two common types: one resulting from low serum **calcium**, and the other by an upset in acid-base balance with resultant alkalosis. Conditions resulting in low calcium include **parathyroid** deficiency, inadequate calcium in the diet, loss of calcium through **diarrhea, and rickets.** Alkalosis and tetany may follow hyperventilation (excessive blowing off of the acid CO_2 from the lungs), excessive bicarbonate ingestion, and loss of hydrochloric acid by persistent vomiting. (See **Acidosis; Alkalosis.**)

Treatment of tetany depends on its cause. If low calcium must be corrected, calcium chloride or calcium gluconate may be given intravenously. Alkalosis is compensated by the ingestion of hydrochloric acid, ammonium chloride or acid phosphates. (D.M.H.)

TETRABRANCHIATA. Cephalopoda.

TETRACHORIC CORRELATION. Tetrachoric correlation is a technique for finding the **correlation** in a 2×2 contingency table with entries, a, b, c, d, and $N = a + b + c + d$.

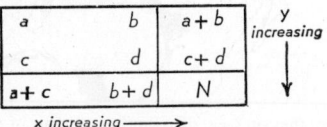

In the table above we have the standard form. We determine h and k by the following equations

$$\int_h^\infty \frac{e^{-\frac{t^2}{2}}}{\sqrt{2\pi}} dt = \frac{b+d}{N}, \quad \int_k^\infty \frac{e^{-\frac{t^2}{2}}}{\sqrt{2\pi}} dt = \frac{c+d}{N}$$

such that $h \geq 0$ and $k \geq 0$. This can always be accomplished if the contingency table is put into the standard form above. We then calculate $\frac{d}{N}$ and find r, the tetrachoric coefficient of correlation from the tables of Pearson for volumes of the normal bivariate surface where

$$\frac{d}{N} = \int_h^\infty \int_k^\infty P(t_1, t_2, r) dt_1, dt_2$$

having found h and k by the previous equations, $P(t_1, t_2, r)$ being the normal probability function in two variables. These tables are given in "Tables for Statisticians and Biometricians, Part II" edited by Karl Pearson in the section Volumes of the Normal Bivariate Surface, pp. 78–137. The tetrachoric coefficient of correlation is based on the assumption of a normal distribution in both variables. Hence, it cannot be compared to the usual **coefficient of correlation.** (L.A.A.)

TETRADYMITE. The mineral tetradymite is a **bismuth tellurium sulfide** corresponding to the formula $Bi_2(TeS)_3$. It is **rhombohedral.** Occurs usually in gold **quartz** veins. Found in Norway, Sweden, England, Bolivia, British Columbia, and in the United States in Virginia, North Carolina, Georgia, Montana, Colorado and elsewhere. It derives its name from the Greek word meaning fourfold in reference to the double twin crystals occasionally developed. (E.S.C.S.)

TETRAGONAL SYSTEM. Crystallography.

TETRAHEDRITE. The mineral tetrahedrite is a **copper-antimony sulfide,** the formula being $Cu_8Sb_2S_7$. The copper may be replaced in part by iron, lead, zinc, mercury, etc., and the antimony by arsenic. This mineral is **isometric** and the crystals are **tetrahedrons** often highly modified. It is also found massive. It has no **cleavage;** uneven fracture; brittle; hardness, 3–4; metallic luster; color, gray to black; streak, gray to black but may be brown or red; practically opaque. There are many European localities for tetrahedrite among which might be mentioned Rumania, Czechoslovakia, France, and Cornwall, England. It is found in

Algeria and Bolivia; and in the United States in Colorado, Arizona and Utah. While chiefly an ore of copper it may be also mined for **silver** or other replacing metals. (E.S.C.S.)

TETRAPHYLLIDEA. Cestoda.

TETRAPLASY. A theory of the origin and **evolution** of living things formulated by Henry Fairfield Osborn. According to this theory, also known as the tetrakinetic or tetraplastic theory, four complexes are the foundation of all that exists and occurs in the organic world. These were designated as the inorganic environment, the organism, the Heredity-germ, and the life environment. In more recent publications they have been discussed under the terms physical environment, internal environment, heritage, and organic environment. The theory is distinctly mechanistic, since it postulates that vital phenomena are adequately explained as the results of interaction among these complexes. (A.W.L.)

TETRARHYNCHIDEA. Cestoda.

TETRATHIONIC ACID AND TETRATHIONATES. Tetrathionic acid ($H_2S_4O_6$) is a colorless solution formed by reaction of **barium** tetrathionate and dilute **sulfuric acid,** and filtering off barium sulfate. Dilute solution of tetrathionic acid may be boiled without decomposition but the concentrated solution decomposes yielding sulfuric acid, **sulfur** dioxide and **sulfur.**

Sodium tetrathionate ($Na_2S_4O_6$) is made (1) by reaction of sodium thiosulfate and **iodine** with accompanying formation of iodide, (2) by reaction of sodium thiosulfate and **ferric** salt solution, lead dioxide, barium peroxide, or sodium peroxide, (3) by **electrolysis** of sodium thiosulfate. No visible reaction takes place when sodium tetrathionate is mixed with (1) ammoniacal **silver** nitrate, (2) **ammonium** sulfide, (3) sodium hydroxide (although thiosulfate, trithionate, sulfite and sulfide may be formed). (R.K.S.)

TETRAXONIDA. A division of the sponges (**Porifera**) included in some classifications as an order characterized by the presence of spicules with four axes. These sponges belong to the **Demospongiae** of this work. (A.W.L.)

TETRODE. Tube, Electronic.

TEXTURE. This term, as used by **petrographers,** denotes primarily the absolute and relative size and shapes of the visible constituents of a rock. In an igneous rock the texture of the mineral aggregate depends upon its crystallinity, **granularity,** and **fabric.** Texture, when applied to clastic sedimentary rocks should include the surface characteristics of the constituent particles, or clasts. (R.M.F.)

THALAMUS. A great nuclear center in the **brain,** formed just behind the portion that gives rise to the cerebral hemispheres. Communication between the outlying parts of the body and the cerebral cortex is established by the routing of nerve impulses through the thalamus, where the redistribution of impulses is accomplished. Thus the center is important in the types of nervous control that are highly developed in man. (A.W.L.)

THALIACEA. The salpians or salps. A group of transparent marine animals related to the ascidians but drifting freely in the water in contrast with the sessile habits of that group. They constitute a class of the subphylum Urochordata (**Chordata**).

This class is closely related in fundamental characters to the ascidians, showing a high specialization of structure with an enclosing test of material similar to cellulose. It is divided into two orders:

Order Multistigmatea. Barrel-shaped animals with muscle bands forming complete rings about the body.
Order Astigmatea. Body usually flattened. Muscle bands incomplete. (A.W.L.)

THALLIUM. Symbol: Tl. Atomic number: 81. Atomic weight: 204.39. Density: 11.86. Melting point: 303.5° C. Isotopes 203 (29.4%), 205 (70.6%).

Thallium metal is bluish-gray upon fresh exposure, changing to dark gray on standing, this oxidation increased with temperature above 25° C.; soft, and may be easily cut with a knife; malleable but of low tenacity so that it must be extruded to form wire; **nitric acid** is the best solvent; forms alloys with many metals, e.g., **mercury, cadmium, zinc, silver, copper, magnesium.** Discovered by Crookes in 1861.

Thallium occurs in small amounts in **pyrite, zinc blende,** and **hematite** of certain localities, and in a few rare minerals in Sweden and Macedonia. For the recovery of thallium from flue dust of pyrite burners, the dust is boiled with water, allowed to stand some time, filtered, and **hydrochloric acid** added to the filtrate, whereupon crude thallous chloride is precipitated. This is purified by further treatment, and thallium metal obtained (1) by **electrolysis** of the sulfate solution or (2) by fusion of the chloride with **sodium** cyanide and carbonate. Thallium compounds are poisonous, and as such used as rat poison, as insecticide, as a depilatory.

Hydroxides: Thallous hydroxide (TlOH), chemically similar to sodium hydroxide, reacts characteristically with traces of ozone to give brown coloration, when on filter paper; thallic hydroxide (Th(OH)₃), brown precipitate (possibly TlO·OH) by reaction of **sodium** hydroxide and thallic salt solution, insoluble in excess sodium hydroxide.

Oxides: Thallous oxide (Tl₂O), black solid, melting point 300° C., when molten attacks glass and porcelain; thallic oxide (Tl₂O₃), brown to black solid.

Salts: Thallous. Thallous sulfate (Tl₂SO₄), soluble; thallous carbonate (Tl₂CO₃), soluble; thallous **alum** (Tl₂SO₄·Al₂(SO₄)₃·24H₂O). These compounds are chemically similar to those of sodium. Thallous chloride (TlCl), white resembling silver chloride in appearance and formation and used in an electrical **cell.** Thallous chromate (Tl₂CrO₄) yellow precipitate; thallous sulfide (Tl₂S) black precipitate in acetic acid, neutral or alkaline solution. These compounds are chemically similar to those of lead. Thallous chloroplatinate (Tl₂PtCl₆), pale orange precipitate. Thallous cobaltinitrite (Tl₃Co(NO₂)₆), pale red precipitate. These compounds are chemically similar to those of potassium.

Thallic. Thallic chloride (TlCl₃), nitrate (Tl(NO₃)₃), sulfate (Tl₂(SO₄)₃) all white to colorless soluble solids.

Volatile thallium salts, such as the chlorides, color the bunsen flame green. (R.K.S.)

THALLOPHYTES. This division, the second largest of the plant kingdom (the seed plants being the largest) is considered to be the oldest and to contain the most primitive types of plants.

Plants of this division vary from minute forms composed of a single cell to seaweeds 200′ or more in length. These plants do not have any root, stem or leaves, though some have structures which resemble these parts in form. However, with a few exceptions in the brown **algae,** all these plants are composed of aggregations of **cells** which show very little differentiation. A plant body thus not differentiated into true roots, stems and leaves is called a thallus.

Several methods of reproduction are found in the thallophytes. The simplest is **cell division,** which occurs in the unicellular forms. A second method is by spores. These are single cells set apart for reproductive purposes. Many spores are provided with protective cell walls which keep them alive in unfavorable environments. Reproduction by spores is either sexual or asexual. Several kinds of spores are found in this division. In many thallophytes, asexual reproduction is by means of zoospores; these are reproductive cells provided with one or more cilia, thread-like objects which by their beating propel the spore through the water.

Another form of reproduction is sexual. This occurs in many thallophytes. The cells which function in sexual reproduction are called **gametes**. As a rule gametes are incapable of developing until they unite in pairs, forming cells called zygotes. From the latter new plants develop. Such zygotes usually acquire thick walls and become resting spores. If they are produced by fusion of gametes of equal size (isogametes) they become zygospores; if by fertilization of an egg by a sperm (heterogametes) they become oöspores.

There are two main groups of thallophytes, the **algae** and the **fungi**. Algae have **chlorophyll** in their cells and so are separated from fungi, which always lack chlorophyll. In addition to these two groups, there is a third, composed of plants which are formed by an intimate union of an alga with a fungus. These are the **lichens**, which in spite of their dual nature are usually described as separate plants, and classified as such. (See also **Paleobotany**.) (R.M.W.)

THALLUS. A thallus is a plant body having no differentiation into roots, stems or leaves. (R.M.W.)

THEBAINE. Alkaloids.

THECA. The outer ridge surrounding the depression in the hard deposit of a stony **coral** in which the **polyp** lives. The deposit is formed by the basal disk of the polyp, first as a flat plate and later in a series of ridges as the ectoderm of the disk is thrown into folds. The theca is a circular ridge in most species. (A.W.L.)

THEELIN. Sex Hormones.

THEOBROMA CACAO. Sterculiaceae. A medium-sized tree growing wild in the lowlands from Mexico to northern South America. It has shining evergreen leaves about a foot long and small flowers, which grow from buds on the trunk or large branches of the tree. The fruits are 6–12″ long and about 4″ in diameter, and have a ribbed rough surface. Each contains from 20–50 flattened seeds, or beans, embedded in a gelatinous pulp. The tree is extensively cultivated in low humid climates in regions where there is a rich soil. It has been introduced into various Old World countries, and is grown extensively in tropical Africa. The cultivated tree is somewhat smaller than the wild one and begins to bear in 4 or 5 years when grown from seed.

The mature pods are cut from the tree and split open. The seeds are then scooped out and fermented for a week or so. During fermentation the color of the seeds darkens to a reddish tone and a rich aromatic odor develops. The pulp surrounding the seeds liquefies and runs off. After fermentation the seeds are dried and shipped to the manufacturers, located chiefly in the United States and Europe.

In the factories, the seeds are cleaned and then roasted for a short time (1–2 hours). After roasting, the seeds are cracked and the shell separated from the **cotyledons**. The shells may be ground up and used in the manufacture of cheap grades of cocoa, or they may be burned as fuel. From the cotyledons is ground out by heated mills an oily liquid which hardens into the familiar chocolate. If part of the oil is squeezed out and the residue ground to a powder, it is cocoa. When chocolate is mixed with sugar and flavored with **vanilla**, it becomes sweet chocolate.

The vegetable fat removed from the pressed beans is known as cocoa butter, and is used in the manufacture of various pharmaceutical preparations, and also in the preparation of confections. (R.M.W.)

THEOBROMINE. Purines.

THEODOLITE. Altazimuth; Meteorological Instruments.

THEOPHYLLINE. Alkaloids.

THEORETICAL PLATE. Distillation.

THEORY OF EQUATIONS. The name "Theory of Equations" usually is applied to the study of the properties of **polynomial equations** and of methods for the numerical solution of such equations. (L.L.S.)

THEORY OF ESTIMATION. Given the **statistics** in a **sample**, it is necessary from these to make some inference concerning the **population parameters** or values. Two methods of estimation will be discussed, that of Markoff and that of R. A. Fisher. Markoff stated that estimates of population parameters which are unbiased and have minimum **variance** should be used. Ordinarily this principle cannot be utilized for all values of the parameter.

R. A. Fisher stated that an estimate should at least be a **consistent statistic** and if possible should be efficient and sufficient. Usually such statistics which are at least consistent and efficient may be found by the method of maximum **likelihood**. It appears then that a particular estimate depends on the **probability function** of the variates. Examples of some estimates are the **mean** of a sample for the mean of a population and the **variance** of a sample for the variance of a population. These are point estimates. Estimation by intervals is accomplished in the theory of **confidence limits**. (L.A.A.)

THEORY OF SAMPLING. In the theory of sampling we are interested in the study of **samples** drawn from a population. Among topics covered by such a theory are the **theory of estimation**, methods of finding the **moments** of **statistics**, methods of obtaining the distribution of statistics, **stratified samples**, random samples, Snedecor's F distribution, Fisher's z distribution, Stuent's t distribution, confidence limits, the χ^2 test, and the **Neyman-Pearson theory of testing a statistical hypothesis**. The problems involved in sampling theory are sometimes very difficult. (L.A.A.)

THEORY OF SIMILARITY. Similitude.

THERALITE. Theralite is a granular intrusive **igneous** rock composed chiefly of **labradorite**, **nephelite** and **augite**. Duppau, Bohemia, is the type locality. If **analcite** is present instead of nephelite the rock is called teschenite from Teschen, in Moravia. The name theralite is derived from the Greek meaning eagerly sought for, because such a rock type was believed to exist and sought for before its actual discovery. (E.S.C.S.)

THERIODONTIA. Fossil Reptiles.

THERMAL CAPACITY. Specific Heat.

THERMAL CONDUCTION. Every substance is in some measure a conductor of heat, though liquids are generally poor conductors and gases almost non-conductors. The best conductors are metals. The flux of heat through a layer of any substance by conduction is proportional to the temperature gradient (fall of temperature per unit thickness), and to a factor called the "thermal conductivity" of the substance, defined as the quantity of heat transmitted per unit time per unit cross section per unit temperature gradient. The thermal conductivities of a few solids are given below, in calories per cm. per sec. per °C.:

Aluminum........0.480	Iron (cast)........0.161
Copper............918	Lead.............083
Cork.............0001	Paraffin..........0006
Glass............002	Quartz....0.033 or 0.017
Ice..............005	Silver...........1.006

(It will be noted that quartz, a highly birefringent crystal, has different conductivities along and perpendicular to its optic axis, as do such crystals in general.)

The mechanism of thermal conduction is probably at least three-fold. Thermally agitated atoms and molecules doubtless actually jostle each other and thus mechanically pass along the heat energy. Thermal radiation between neighboring atoms or molecules should have a similar result. But neither of these explains the enormous difference in conductivity between, say, copper and glass, both of which are dense, fine-grained solids; or between silver and lead, both soft, crystalline metals of similar chemical properties. If, however, we examine the electric conductivity of these substances (see **Electric Conduction and Resistance**), we discover that good thermal conductors are also good electrical conductors, and we are led to suspect that thermal as well as electrical conduction may depend upon the activity of electrons. The relationship is brought out quantitatively by the **Wiedemann-Franz law.**

The speed with which a temperature wave progresses by thermal conduction depends upon the "thermal diffusivity" of the conductor, which is its thermal conductivity divided by its thermal capacity per unit volume. (L.D.W.)

THERMAL CONVECTION. This is a familiar phenomenon, consisting in the transfer of heat by the automatic circulation of a fluid (liquid or gas) due to differences in temperature and density. While water and liquids generally are poor conductors, a kettle of water is quickly heated throughout by applying heat at the bottom. The warmer water, being less dense, is compelled to rise by the colder, which, sinking to the bottom, is warmed in its turn. The process is more clear-cut when a definite circuit is provided, as in the heating coil attached to a hot-water tank or radiator system. Gasoline engines are cooled by a similar circulation, either entirely automatic or augmented by a small rotary pump. Gases likewise exhibit convection. A chimney "draws" when the air inside it is warmer than that outside, so that the greater pressure difference outside forces the air inward at the bottom. (The term "draw" is manifestly misleading.) The motion of the air in hot-air furnaces and in the **winds** of the atmosphere are good examples. There is a type of **pressure gauge** which depends upon the cooling effect of convection currents in the gas upon a hot filament immersed in it, the rate of cooling being a function of the gas density. (L.D.W.)

THERMAL DEGREE. Temperature Scales.

THERMAL DIFFUSIVITY. Thermal Conduction.

THERMAL EFFICIENCY. Thermal efficiency is output, in heat units, divided by the heat supplied or chargeable. Thermal efficiency may partially define the operating condition of both a machine or a static piece of equipment. In the case of static equipment, if it is well insulated it may have very high thermal efficiencies. A machine which converts heat supplied into work output (the **steam engine, Diesel engine**, etc.) is always characterized by low thermal efficiencies because the conversion of the low-grade type of heat energy into high-grade energy of mechanical work is accomplished with considerable difficulty. The thermal efficiencies of the best

prime movers today rarely exceed 35%, and are not found higher than 40% even when optimum conditions of loading, maintenance, and fuel employed are present. Thermal efficiency of a prime mover may be based on the output at the shaft per unit of heat supplied, or upon the electrical output in the case of a generator drive. It may also be based on cylinder hp. in piston and cylinder prime movers. Heat supplied and chargeable to internal combustion engines is the heat in the fuel. Heat supplied to steam prime movers is the heat in the **steam** at the throttle; the heat chargeable is the heat supplied after deducting the heat of condensate in the exhaust. (F.T.M.)

THERMAL EQUATOR. The belt of maximum temperature surrounding the earth which moves north and south with (but lagging) the sun's motion. It is also spoken of as the center of the area bounded by the yearly mean isotherms of 80° F. (P.E.K.)

THERMAL METAMORPHISM. Metamorphism.

THERMAL RADIATION. All bodies that are not at absolute zero emit radiation excited by the thermal agitation of their molecules or atoms, whether there are other causes of excitation or not. This thermal radiation ranges in wavelength from the longest **infra-red** to the shortest **ultra-violet** rays, its **spectral energy distribution,** however, depending upon the nature of the body and upon its temperature. The total emissive power of a surface at any temperature is the rate at which it emits energy of all wavelengths and in all directions, per unit area of radiating surface. The flux density (per unit solid angle) in various directions obeys the **cosine emission law** approximately; but strictly only in the case of a **black body.** Thermal radiation is observed and measured by means of different types of **radiometer,** by the **bolometer,** and by the **radiomicrometer;** also, in the shorter wavelengths, by its photographic and photoelectric effects.

In ordinary surroundings bodies not only emit but receive radiation. The net transfer of energy depends upon the relative rate of emission and absorption, and this, in turn, upon the temperature of the body and that of its surroundings. The dependence of this rate upon the two temperatures was investigated by Newton, by Dulong and Petit, by Stefan, and by Boltzmann. (See **Newton's Law of Cooling** and **Stefan-Boltzmann Law.**) **Wien's laws** have to do with the dependence of the spectral energy distribution of black-body radiation upon temperature, the last word upon which at present, however, seems to be **Planck's equation** developed through the **quantum theory.** Kirchhoff pointed out that the absorptivity of a surface (ratio of absorbed to incident radiation) and its emissivity (ratio of emissive power to that of a black body at the same temperature) are equal. It follows that good radiators are good absorbers. The reflectivity (ratio of reflected to incident radiation) is equal to one minus the absorptivity; hence good radiators and absorbers are poor reflectors.

Most surfaces exhibit "selective" emission, absorption, and reflection; that is, their emissivities, absorptivities, and reflectivities for different wavelengths are not proportional to those of a black body, and the proportion differs for different surfaces. Thus copper has abnormally high reflectivity and low absorptivity in the red; silicon, in the ultra-violet.

A substance transparent to thermal radiation is said to be "diathermanous." A material could be completely described in this respect by giving its absorption coefficient in the various parts of the spectrum. Thus, rock salt is very diathermanous throughout the infra-red and visible ranges; water and glass only in the visible, filtering out much of the infra-red. Well-defined absorption bands often occur in different regions of the spectrum. (See **Absorption Spectrum** and **Residual Radiation.**) (L.D.W.)

THERMAL RATING. Speed Reducers.

THERMAL WAVES. Long Waves in the Prevailing Westerlies. (P.E.K.)

THERMEL. The well-known Seebeck effect (see Thermoelectric Phenomena) is the principle underlying a large class of thermoelectric thermometers. The essential feature is a circuit composed of two different metals, the two junctions of which are at different temperatures, and in which a net electromotive force develops as a result of this temperature difference. Any device which uses this electromotive force, or the current due to it, as a measure of temperature is called a thermel (a term introduced by W. P. White).

The simplest form is a "thermocouple," composed of two pieces of metal, or wires, soldered or welded together at their ends, the other ends being connected to a **galvanometer** or a **potentiometer**. Various pairs of metals are used, for example, antimony and bismuth, copper and iron, or copper and constantan (an alloy of copper and nickel). High-temperature thermocouples are commonly of platinum, with some other refractory metal such as iridium or an alloy of platinum and iridium, rhodium or chromium. One of the junctions may be enclosed in a protecting tube, the other being kept at zero by means of melting ice. In cases where a temperature difference only is desired, the two copper or platinum lead-wires may be attached to the opposite ends of a single wire of the other metal, and the two junctions placed at the two points to be compared.

A "thermopile" is composed of a number of thermocouples in series, the alternate junctions being assembled in two bunches which are used like the junctions of a single thermocouple. This arrangement multiplies the thermoelectromotive force and gives greater sensitivity. An exceedingly sensitive form of thermel is the **radio-micrometer**. The "thermoelement" utilizes a thermocouple to detect and measure very feeble currents by their heating effect on a fine wire. (L.D.W.)

THERMIONIC PHENOMENA. In view of the commotion among the atoms and electrons of a heated substance, it is not surprising that electric particles, both positive ions and electrons, should be projected from a highly heated body. If the body is electrically charged, particles of the same sign as the charge, when once through the surface barrier (see **Work Function**), are repelled into the surrounding space, where they can be detected. Electric particles thus emerging, either positive or negative, are called thermions. The heated body may be a filament of pure metal, electrically heated, or a layer of some chemical substance spread over and heated by such a filament. The usual experimental arrangement is to enclose the thermionic emitter as an electrode in a tube or bulb, along with another electrode of opposite sign, so that the field between the two will set up a stream of the released thermions, called a thermionic current. With sufficient voltage, this current reaches a maximum or "saturation" value, the ions being then swept away as fast as they are released.

In the early experiments of Becquerel, Guthrie, Edison, Elster and Geitel, and others, it was found that at lower temperatures (up to a red heat) the thermionic emission from metals is predominantly positive, but that at much higher temperatures (white heat) the negative or electronic emission rapidly surpasses the positive and becomes all-important. The positive emission from a pure metal like platinum or tungsten appears to be due to impurities such as potassium, and falls off with prolonged heating; for metallic salts it consists of ions of the metal composing the salt. Some metals emit electrons much more copiously than others; a notable example is thorium, an adsorbed film of which on tungsten gives very copious electron emission at high temperatures.

The saturation current for any metal is an exponential function of the temperature, expressed by Richardson's equation

$$I = AT^n e^{-B/T},$$

in which T is the absolute temperature and A and B are constants dependent on the metal. The exponent n appears in some cases to be $\frac{1}{2}$, while in others the value 2 fits better. B, and perhaps A, involves the thermionic work function. The emission is not quite steady, being subject to the so-called **shot effect**. In the various forms of **triode**, the thermionic current is controlled by means of a grid, an arrangement of great practical importance in radio and elsewhere. (L.D.W.)

THERMITE. Aluminum.

THERMO SIPHON. The term thermo siphon refers to a method of circulation of a liquid arising from the slight difference of density of the hot and cool liquid. At one time automobile engines were provided with thermo siphon action in the cooling water system. Although the principal portion of circulation of modern automobile engines is achieved by a pump, there is some thermo siphon action aiding the pump. Thermo siphon circulation is used to considerable extent on stationary engines which have an external radiator for cooling the jacket water. When the water leaves the radiator and enters the engine jacket, it is heated and expands slightly. Its decrease in density causes the column of hot water in the engine to weigh less than the equivalent column of cool water in the radiator, so there is a continuous displacement of heated water from the jacket by cool water flowing in from the radiator. This system requires large hose, unobstructed hose connections from engine to radiator, and large radiators having considerable height. (F.T.M.)

THERMOCHEMISTRY. Every chemical reaction is accompanied by a definite change in the heat content of the system. Most reactions evolve heat and are described as exothermic reactions, but some, such as the reaction of **nitrogen** plus **oxygen** to form **nitric oxide** absorb heat and are described as endothermic.

The amount or quantity of heat change in a given reaction is measured in large or small **calories** (number of kilograms or grams, respectively, of water * multiplied by the temperature change in °C.) or in B.T.U. (number of lbs. of water multiplied by the temperature change in °F.). The amount of heat is calculated in various ways, for example, per 1 gram or per 1 lb. of one reacting substance, as in the case of solid and liquid **fuels** and **foods**, per 1 cu. meter or per 1 cu. ft. of gaseous fuels, per 1 equivalent in the neutralization of **acids** and **bases**, or per 1 mol in expressing heats of formation of compounds.

The intensity of heat is measured by the degrees of temperature on the centigrade scale (melting point 0° C., boiling point 100° C. of water), the Fahrenheit scale (melting point 32° F., boiling point 212° F. of water) or the absolute scale + 273° C. (which is also the Kelvin scale) and + 491° F. As stated, 1° C. equals 1.8° F. It is estimated that the temperature of the **hydrogen**-oxygen flame may attain 2000° C., the **acetylene**-oxygen flame 2500° C., and the thermite reaction (**aluminum** plus **iron** oxide yielding aluminum oxides plus iron) even higher temperatures. Coker and Scoble (1913) estimated the temperature attained in coal gas-air mixture in the ratio of 1 to 5.66, as 2250° C. and the pressure observed 433 lbs. per sq. in.

Changes of energy always accompany chemical reaction, and most common of these is heat energy; these

---

* Temperature of water from 3.5–4.5° C., ordinary calorie; from 14.5–15.5°, normal calorie; from 0–100° C. divided by 100, mean calorie.

have been stated to be "of hardly less importance to the modern chemist than the material changes themselves" (Huddeston).

Thermochemical considerations may be classified as follows:

1. When no change in composition is involved.
   a. No change of state takes place.
      Heat capacity or atomic heat of solid elements (Dulong and Petit, 1819).
      Heat capacity or molecular heat of solid compounds (Kopp, 1864). (See **Chemical Composition**.)
      Heat of dilution of solutions.
      Heat of wetting and adsorption.
   b. A change of state takes place.
      Heat of fusion or solidification (melting point, freezing point).
      Heat of vaporization or condensation (boiling point, condensing point).
      Heat of sublimation (sublimation point).
      Heat of transition (transition point of allotropic substances).
      Heat of solution or crystallization of solids, liquids, gases.

2. When a change in composition is involved.
   a. Temperature effects.
      (1) Temperature of reaction.
          Reactions at ordinary temperatures.
          Reactions at high temperatures.
          The temperature attained depends upon the rate of heat evolution in the system less the rate of heat loss from the system.
      High temperatures attained by combustion of fuels:
          Local, by use of hydrogen—or acetylene—oxygen flame.
          General, by use of fuel gas and pre-heated air, in metallurgical furnaces, ceramic, cement and glass kilns and furnaces, by-product coke ovens. (See **Fuels**.)
      Very high temperatures attained in the electric furnace, producing such products as calcium carbide, silicon carbide, phosphorus, graphite, carbon disulfide, silicon, aluminum oxide fused, silicon oxide fused.

HEAT OF FORMATION OF OXIDES, CHLORIDES, SULFIDES

ELEMENT	OXIDE	CALORIES PER 16 GRAMS OXYGEN	CHLORIDE	CALORIES PER 35.5 GRAMS CHLORINE	SULFIDE	CALORIES PER 32 GRAMS SULFUR
Hydrogen	H_2O, gas	58	HCl, gas	22	H_2S, gas	5
Sodium	Na_2O	99	NaCl	98	Na_2S	90
Potassium	K_2O	86	KCl	104	K_2S	88
Magnesium	MgO	146	$MgCl_2$	77	MgS	82
Calcium	CaO	152	$CaCl_2$	95	CaS	114
Strontium	SrO	141	$SrCl_2$	99	SrS	113
Barium	BaO	133	$BaCl_2$	103	BaS	111
Boron	B_2O_3	93	BCl_3, liquid	31	B_2S_3	25
Aluminum	Al_2O_3	130	$AlCl_3$	56	Al_2S_3	115
Carbon	CO_2	47	CCl_4, liquid	8	CS_2, liquid	−11
	CO	27				
Silicon	SiO_2, fused	99	$SiCl_4$, liquid	37	SiS_2	16
Titanium	TiO_2	109	$TiCl_4$, liquid	46		
Nitrogen	NO, gas	−22				
Phosphorus	P_2O_5	73	PCl_3, liquid	26	P_4S_3	26
Oxygen					SO_2, gas	69
Sulfur	SO_2, gas	35	S_2Cl_2, liquid	7		
Vanadium	V_2O_5	87	VCl_4, liquid	40		
Chromium	Cr_2O_3	89	$CrCl_3$	47		
Manganese	MnO	97	$MnCl_2$	57	MnS, ppt.	47
Iron	Fe_2O_3	66	$FeCl_3$	32		
	Fe_3O_4	67				
	FeO	64	$FeCl_2$	41	FeS	23
Cobalt	CoO	57	$CoCl_2$	39	CoS, ppt.	20
Nickel	NiO	58	$NiCl_2$	37	NiS, ppt.	21
Copper	CuO	35	$CuCl_2$	26	CuS	12
	Cu_2O	40	CuCl	33	Cu_2S	19
Silver	Ag_2O	7	AgCl	31	Ag_2S	5
Gold	Au_2O_3	−4	$AuCl_3$	9		
Zinc	ZnO, fused	84	$ZnCl_2$	50	ZnS	46
Cadmium	CdO	65	$CdCl_2$	46	CdS	34
Mercury	HgO	22	$HgCl_2$	27	HgS	11
	Hg_2O	22	HgCl, ppt.	32		
Tin	SnO_2	69	$SnCl_4$, liquid	32		
	SnO	68	$SnCl_2$	41	SnS	23
Lead	Pb_3O_4	44				
	PbO	52	$PbCl_2$	43	PbS, ppt.	22
Arsenic	As_2O_3	49	$AsCl_3$, liquid	24	As_2S_2	10
Antimony	Sb_2O_3	55	$SbCl_3$, liquid	29	Sb_2S_3	12
Bismuth	Bi_2O_3	45	$BiCl_3$	30		

(2) Temperature coefficient of reaction.
About 2 for a rise in temperature of
10° C. (See **Chemical Changes.**)
b. Heat effects.
Heat of reaction, general.
Specific cases—heat of neutralization of
acids and bases; heat of dissociation; heat
of formation.

The initial and final conditions of a reaction or of a
series of reactions determine the total heat effect without
regard to the intermediate steps. This is the formulation of Hess' Law of Constant Heat Summation. Its
usefulness is evident by a simple example. The heat of
reaction of carbon to carbon dioxide $(C + O_2 \rightarrow CO_2)$ is
94,400 calories obtained by burning a weighed amount
of carbon with excess oxygen in a bomb calorimeter.
The heat of reaction of carbon monoxide to carbon
dioxide $(CO + 0.5\ O_2 \rightarrow CO_2)$ is 68,000 calories obtained
by burning a measured volume of carbon monoxide in
a gas calorimeter. It is not easy to ascertain the heat of
reaction carbon to carbon monoxide $(C + 0.5\ O_2 \rightarrow CO)$
directly, but since the heat evolved by the two routes

$$C \longrightarrow CO_2$$
$$\searrow\ \nearrow\qquad \text{is the same, then the difference between}$$
$$CO$$

the first and second reactions ascertained above is the
desired result, namely, 26,400 calories.

The heat of formation of three grand groups of chemical substances is shown on preceding page. (R.K.S.)

THERMOCOUPLE. Thermel.

THERMODYNAMIC POTENTIAL. Potential.

THERMODYNAMICS.
This branch of physics had
its origin in the classical discoveries of Rumford, Davy,
Joule, and others early in the nineteenth century, which
identified **heat** as a form of **energy.** Gradually the
mechanism of heat and the statistics of molecular motion
were revealed by the researches of such men as Maxwell,
Kelvin, Clausius, and Boltzmann, until now, with the
added assistance of the **quantum theory,** we discuss the
dynamics of **molecules** almost as confidently as that of
visible bodies.

The first law of thermodynamics is embodied in the
fact of the **mechanical equivalent of heat,** and need be
touched upon here only by expressing it in its customary
algebraic form $W = JQ$; meaning that Q heat units are
equivalent to JQ work units. Our best present value of
J is 41,855,000 ergs per calorie. With increasing emphasis upon the fact that heat is kinetic energy, has grown
the tendency to express heat directly in ergs or joules and
to do away with the **calorie** and the constant J; but the
old notation still persists.

The second law of thermodynamics is really a con-
fession of our helplessness in making molecules do what
we wish, because of their inconceivable numbers and their
submicroscopic size. It states, in effect, that the only
way in which we can utilize any of the supply of heat
energy in a body, and make it do mechanical work, is to
find another body whose molecules have less average heat
energy (i.e., a body at lower temperature), set up between them a mechanism (engine) through which the
molecules of the warmer body can contribute some of
their excess energy to those of its cooler neighbor, and
capture part of this donated energy on the way. (No
such engine will run on heat energy given by a cooler to
a warmer body; it would have to be run from outside as
a refrigerating machine.) It is shown that if the warmer
and cooler bodies are at the respective absolute temperatures T_1 and T_2, the maximum fraction of the transferred
energy that even an ideal engine could capture is $(T_1 -
T_2)/T_1$, which is thus the maximum ideal heat-engine
efficiency.

The variables commonly chosen in thermodynamic rea-

soning are temperature, entropy, pressure, and volume,
and in terms of these we write the **characteristic equations** of the substances, such as air, steam, etc., used in
engines. The purely dynamic aspects are especially concerned with volume and pressure, and thermodynamic
diagrams are often drawn with these as coordinates. For
example, if a gas expands and its pressure diminishes,
this change may be represented by the curve AB and the
corresponding work by the area $AB\beta a$ (Fig. 1); while
if, as in an engine, the change is a cyclic one, the net

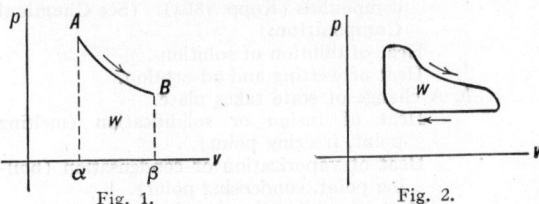

Fig. 1.   Fig. 2.
Representation of work during a unidirectional or a cyclic
change in volume and pressure.

work derived from each cycle on one side of the piston
is represented by the area W enclosed by the curve (Fig.
2). A steam engine indicator, for example, automatically
draws such a curve at each stroke of the engine, the efficiency of whose performance can thus be deduced. (See
**Carnot Cycle, Rankine Cycle, Entropy, Joule-Thom-
son Effect, Reversible Processes,** etc.) (L.D.W.)

THERMOELECTRIC PHENOMENA.
If two
strips of different metals are closely joined at one end,
and the junction kept at a different temperature from
the rest of the strips, an electromotive force develops in
the two-part conductor. This thermoelectromotive force,
sometimes called the Seebeck effect from its discoverer,
depends upon the metals and the temperature distribution
in them. The pair of metals is called a thermocouple,
and the change in the electromotive force per degree
change in temperature at the junction is the thermoelectric power of the thermocouple.

The thermoelectric power is a linear function of the
junction temperature; while the thermoelectromotive
force itself is a quadratic function, having a maximum
value at some point (for iron and copper about 275° C.).

Peltier discovered that when a feeble battery current
is sent through a thermocouple, the junction is thereby
either warmed or cooled, according as the direction of
the current is from the − to the + or from the + to
the − metal. (This is altogether apart from the ordinary resistance heating effect.) Kelvin found that a
potential difference develops even in a single metal (except in the case of lead) if one end is warmer than the
other. The thermoelectromotive force of a thermocouple
appears to be a combination of a "Peltier electromotive
force" at the junction and the Kelvin or "Thomson
electromotive forces" in the two strips.

If several metals are joined to form a circuit of non-
uniform temperature, the resultant thermoelectromotive
force is the algebraic sum of the several Peltier and
Thomson electromotive forces, and gives rise to a thermoelectric current, such as that utilized in any **thermel.**

Kelvin discovered that if a weak current is sent
through a wire which is heated at one point, the current
causes a flow of heat, sometimes one way and sometimes
the other, depending upon the metal.

None of these phenomena is fully understood; but they
are believed to be due to the activity of **electrons** which
are either free or related in some peculiar way to the
crystal structure of the metals. (L.D.W.)

THERMOGRAPH.
A recording thermometer. A
common and very simple form employs a Breguet differential expansion spiral (see **Solid Expansion Thermometer**). One end of the spiral is securely fixed to

a metal post, while to the other is attached a long rod bearing at its end a pen or pencil. As the spiral coils and uncoils with changes of temperature, this marker moves up and down on a drum or disk covered with paper and slowly revolved by clockwork; thus leaving a graph of the temperature changes during a period of several hours or days. Another arrangement is to use a **thermel** connected with a **galvanometer**, the mirror of which reflects a spot of light upon a slowly moving photographic film. Thermographs are in extensive use at meteorological observing stations, for the exploration of the upper air, in deep sea soundings, etc. (See **Meteorological Instruments.**) (L.D.W.)

THERMOLUMINESCENCE. Luminescence.

THERMOMETRY. The measurement of **temperature** and of changes in temperature has been based upon many different heat effects. Among those which have been extensively developed are: (1) the expansion of solids, liquids, and gases, illustrated by **solid expansion thermometers, liquid expansion, thermometers,** and the constant pressure **gas thermometer;** (2) the change of pressure in a gas kept at constant volume (constant volume gas thermometer), or of the saturated vapor pressure of a liquid (see **Vapors**); (3) the change in resistivity of metals (metallic resistance thermometer); (4) the Seebeck thermoelectric effect (**thermel**); (5) the brightness of very hot bodies (**optical pyrometer**); and (6) the character of the **thermal radiation** from the heated body (**radiation pyrometer**).

With so many temperature indices, it is necessary to have a standard of temperature measure. For various reasons, it was once found desirable to fix upon gas pressure at constant volume as the practical standard measure of temperature. That is, equal changes of temperature were defined as those corresponding to equal changes of pressure in a selected gas (hydrogen) kept at constant volume. The constant volume hydrogen thermometer thus became the reference instrument for the calibration of other types. In 1927, however, the United States, Great Britain, and Germany proposed, and thirty-one nations represented at the Seventh General Conference of Weights and Measures unanimously adopted, what is now called the international temperature scale. From $-190°$ to $+660°$ C., the measure of temperature is based upon the indications of a standard platinum resistance thermometer, specified and used in accordance with certain formulas. From $+660°$ C. to the melting point of **gold** a platinum-platinrhodium thermel is the reference instrument; and above the gold point, the optical pyrometer is used as standard. The basic fixed points of this scale are the boiling point of **oxygen** ($-182.97°$ C.), the freezing and the boiling points of water, the boiling point of **sulfur** ($+444.60°$ C.), the melting point of **silver** ($+960.5°$ C.), and the melting point of gold ($+1063°$ C.).

Kelvin was long ago impressed with the dependence of such thermometric methods upon characteristic properties of certain arbitrarily chosen substances. Even constant-volume gas thermometers, using different gases, do not quite agree. Kelvin therefore proposed an ideal absolute temperature scale on which changes of temperature, whatever the substance concerned, are strictly proportional to the quantities of heat converted into mechanical work during **Carnot cycles** bounded by the respective temperatures and by the same two adiabatic **entropy** limits. A constant-volume gas pressure thermometer using a gas obeying the **ideal gas law** would perform in exact agreement with this standard. The subsequent researches of Joule and Kelvin on the **Joule-Thomson effect** enabled them to calculate the corrections necessary to convert the hydrogen constant volume standard into this "thermodynamic scale" (more properly, this thermodynamic standard of temperature measure) proposed by Kelvin, which is now used as the basis of the international scale above mentioned. (L.D.W.)

THERMOPILE. Thermel.

THERMOSTAT. The thermostat is a device, the purpose of which is to regulate, automatically, the temperature of a body or of an enclosure, or to maintain the temperature at some predetermined value, or within a certain range. Thermostats are employed widely in heating systems in order to adjust or regulate the temperature of the heating medium. They also have a great many uses in industry, where fluids, chambers, or material must be maintained at predetermined temperatures in order to accomplish the industrial process satisfactorily. There are other scattered uses of thermostats, many of which are very important, and among these might be cited the control of engine cooling water temperature in the modern automotive vehicle, the control of heat-treating furnaces, etc.

Thermostats are actuated mainly either by the expansion of a fluid, or by expansion of a metallic element. A very common example of the former is the wafer-type thermostat, in which thin disk-shaped shells, called sylphons, are filled with gas, or partially filled with a liquid of suitable boiling temperature. Actuated by the developed internal pressure, the expansion and contraction of these sylphons, the cases of which are flexible, constitute a motion which can be mechanically transmitted to a regulating **valve or relay.** While the motion of any one disk is relatively small, a number of them may be made into a group, the motions of which are additive.

Metallic element thermostats employ the well-known principle of linear expansion with temperature change. The element is not always built as a straight rod; satisfactory thermostats are constructed of a bimetallic strip wound in spiral form, which tends to coil or uncoil with changes of temperature.

Thermostats are sometimes actuated by thermoelectric currents or by the varying resistance of conductors, the effects of which, suitably amplified, may be utilized to control heating elements. (F.T.M.)

THICKENING. Classification.

THICK-FILM LUBRICATION. Lubrication.

THICK-KNEES, THICKNEES. Aves, Charadriiformes. A long-legged European bird (**Aves**) whose habits are similar to those of our killdeer. The same genus is represented by African, Indian, South American, and Australian species. The European bird is also called the stone curlew, but there is some question about its inclusion in this order. (A.W.L.)

THIGMOTROPISM. Movement in Plants.

THIN SECTIONS. The term used by **petrographers** and **mineralogists** for slices of rocks or minerals which are cut and ground to the approximate thickness of a thousandth of an inch, and mounted on glass slides for use with the **petrographic microscopic.** The average practical thickness for normal petrographic or mechanical analysis is 0.03 mm. (R.M.F.)

THIN-FILM LUBRICATION. Lubrication.

THINNER. Pigments, Paints, Varnishes.

THIOALCOHOLS AND RELATED COMPOUNDS. Thioalcohols and thioethers are organic derivatives of **hydrogen sulfide** containing the groups —SH and ≡S respectively. The thioalcohols (and thiophenols) may also be named as mercaptans, hydrosulfides, sulfhydrates, and thiols. The last of these names (thiols) is given priority in the accompanying table. Sulfinic and sulfonic acids are organic compounds containing the groups —SO_2H and —SO_3H respectively.

TABLE SHOWING INTERRELATIONSHIP BETWEEN VARIOUS ORGANIC SULFUR COMPOUNDS

THIOALCOHOLS	SULFINIC ACIDS	SULFONIC ACIDS
Boiling Point		
(Hydrogen sulfide HSH −62° C.)		
Methanethiol CH₃SH 6° C.	Methyl sulfinic acid CH₃SOOH	Methyl sulfonic acid CH₃·SO₂OH
(Methyl hydrosulfide, methyl mercaptan)		167° C. dec.
Ethanethiol C₂H₅SH 37° C.	Ethyl sulfinic acid C₂H₅SOOH	Ethyl sulfonic acid C₂H₅SO₂OH
(Ethyl hydrosulfide, ethyl mercaptan)		
Propanethiol C₃H₇SH 67° C.		
(Normal-propyl hydrosulfide, normal-propyl mercaptan)		
Phenylmethanethiol C₆H₅CH₂SH		
(Benzyl hydrosulfide)		
Dithioglycol C₂H₄(SH)₂		
(Ethylene mercaptan)		
Benzenethiol C₆H₅SH	Benzene sulfinic acid C₆H₅SOOH	Benzene sulfonic acid C₆H₅SO₂OH
(Phenyl hydrosulfide, thiophenol)		Benzene disulfonic acid
		$C_6H_4(SO_2OH)_2(1,3)$
		Benzene trisulfonic acid
		$C_6H_3(SO_2OH)_3(1,3,5)$

NAPHTHYLAMINE SULFONIC ACIDS	NAPHTHOL SULFONIC ACIDS	SULFONIC ACIDS
Naphthionic acid	Naphthol sulfonic acid (Alpha acid)	Alpha-naphthalene sulfonic acid
$C_{10}H_6(NH_2)(1)(SO_2OH)(4)$	$C_{10}H_6(OH)(1)(SO_2OH)(2)$	$C_{10}H_7SO_2OH(1)$
Laurent's acid	Naphthol sulfonic acid (Beta acid)	Beta-naphthalene sulfonic acid
$C_{10}H_6(NH_2)(1)(SO_2OH)(5)$	$C_{10}H_6(OH)(2)(SO_2OH)(6)$	$C_{10}H_7SO_2OH(2)$
Tobias' acid	Naphthol disulfonic acid (R-acid)	Naphthalene disulfonic acids
$C_{10}H_6(NH_2)(2)(SO_2OH)(1)$	$C_{10}H_5(OH)(2)(SO_2OH)_2(3,6)$	$C_{10}H_6(SO_2OH)_2(2,6);(2,7);(1,5);(1,6)$
Bronner's acid	Naphthalmine disulfonic acid (S-acid)	Methylene disulfonic acid
$C_{10}H_6(NH_2)(2)(SO_2OH)(6)$	$C_{10}H_5(NH_2)(1)(SO_2OH)_2(4,8)$	$CH_2(SO_2OH)_2$
		Ethylene disulfonic acid
		$C_2H_4(SO_2OH)_2(1,2)$

SULFONYL CHLORIDES		
Methyl sulfonyl chloride	Ethyl sulfonyl chloride	Benzene sulfonyl chloride
CH_3SO_2Cl	$C_2H_5SO_2Cl$	$C_6H_5SO_2Cl$

THIOETHERS SULFIDES	SULFINYL-COMPOUNDS SULFOXIDES	SULFONYL-COMPOUNDS SULFONES
Methylene sulfide CH₂S	Methyl sulfinylmethane (dimethyl	Methylsulfonylmethane (dimethyl
(thiomethylene)	sulfoxide) (CH₃)₂SO	sulfone)(CH₃)₂SO·
Dimethyl sulfide (CH₃)₂S		
(methylthiomethane) M.P. 38° C.		
Diethyl sulfide (C₂H₅)₂S	Ethylsulfinylethane (diethyl	Ethylsulfonylethane (diethyl
(ethylthioethane) M.P. 92° C.	sulfoxide)(C₂H₅)₂SO	sulfone)(C₂H₅)₂SO₂
Divinyl sulfide		
(CH₂:CH)₂S		
Diallyl sulfide		
(CH₂:CHCH₂)₂S		
Dichloroethyl sulfide (beta, beta prime)		
("mustard gas")(ClCH₂·CH₂)₂S		
Dibenzyl sulfide	Benzylsulfinylphenylmethane	Benzylsulfonylphenylmethane
(C₆H₅CH₂)₂S	(dibenzyl sulfoxide) (C₆H₅CH₂)₂SO	(dibenzyl sulfone)(C₆H₅CH₂)₂SO₂
Diphenyl sulfide (C₆H₅)₂S	Phenylsulfinylbenzene (diphenyl	Phenylsulfonylbenzene (diphenyl
(phenylthiobenzene)	sulfoxide)(C₆H₅)₂SO	sulfone)(phenyl sulfene)(C₆H₅)₂SO₂
		Diethyl sulfone dimethylmethane
		(acetone diethyl sulfone, sulfonal)
		$(CH_3)_2C(SO_2C_2H_5)_2$
		Trional
		$\left.\begin{array}{l}CH_3\\C_2H_5\end{array}\right\rangle C(SO_2C_2H_5)_2$
		Tetronal
		$(C_2H_5)_2C(SO_2C_2H_5)_2$
Cyanogen sulfide	Carbon disulfide	Diethylene disulfide
(CN)₂S M.P. 60° C.	CS₂ B.P. 46° C.	$C_2H_4S·SC_2H_4$
Carbonyl sulfide COS B.P. −48° C.	Methyldithiomethane (dimethyl	Phenyldithiobenzene (diphenyl
	disulfide) (CH₃)₂S₂	disulfide)(C₆H₅)₂S₂
Acetyl disulfide	Ethyldithioethane (diethyl	Diphenylene disulfide (thianthrene)
(CH₃CO)₂S₂	disulfide)(C₂H₅)₂S₂	$C_6H_4(S)_2C_6H_4$
Benzoyl disulfide	Benzyldithiophenylmethane	
(C₆H₅CO)₂S₂	(dibenzyldisulfide)(C₆H₅CH₂)₂S₂	
Carbonyl disulfethyl		
CO(SC₂H₅)₂		
Allyl trisulfide		
(C₃H₅)₂S₃		

SULFONIUM COMPOUNDS		
Trimethyl sulfonium iodide ((CH₃)₃S)I	Trimethyl sulfonium hydroxide ((CH₃)₃S)OH	(R.K.S.)

THIOALDEHYDES AND THIOKETONES. Thio-aldehydes and thioketones are produced by the action of **hydrogen sulfide** on **aldehydes** and **ketones.** They are ill-smelling liquids, but change on standing to odor-less compounds, trithioaldehydes or trithioketones, e.g., thio-acetaldehyde, ethanethial ($CH_3 \cdot CHS$) changes to trithio-acetaldehyde ($CH_3 \cdot CHS)_3$. Trithioacetone ((($CH_3)_2CS)_3$), alpha, melts at 101° C.; beta melts at 125° C.; gamma melts at 81° C. **Potassium** per-manganate causes oxidation yielding sulfonyl compounds. (R.K.S.)

THIOCARBONIC ACID AND THIOCARBO-NATES. Thiocarbonic acid (H_2CS_3) is a yellow oily liquid, melting point 20° to 30°, with decomposition into **carbon disulfide** plus **hydrogen sulfide,** made by addition of acid, e.g., **hydrochloric acid,** to sodium thiocarbonate.

Sodium thiocarbonate (Na_2CS_3) is made by reaction of sodium hydrogen sulfide and carbon disulfide (similar to the reaction of sodium hydroxide and **carbon dioxide** to form **sodium** carbonate). (R.K.S.)

THIOCYANIC ACID AND THIOCYANATES. Hydrogen thiocyanate (HCNS) is a gas, unstable, and upon cooling it solidifies to an odorous solid, melting point 5° C. to a liquid that soon changes to a yellow solid. Hydrogen thiocyanate is soluble in water, and the solution is thiocyanic acid, moderately stable when dilute and cold, but not otherwise.

Thiocyanic acid is formed by reaction of **barium** thiocyanate solution and dilute **sulfuric acid,** and fil-tering off barium sulfate.

Sodium, potassium, barium, or **calcium** thiocyanate may be made by reaction of **sulfur** and the correspond-ing **cyanide** upon heating to fusion. Ammonium thio-cyanate (plus ammonium sulfide) may be made by reac-tion of **ammonia** and **carbon disulfide,** a reaction which probably accounts for the presence of ammonium thiocyanate in the products of the destructive distilla-tion of coal.

Silver, lead, cuprous, and **thallous** sulfocyanates are insoluble, and **mercuric** and **stannous** sulfocyanates slightly soluble. All of these are soluble in excess of soluble (e.g., ammonium) thiocyanate forming complexes. Ferric thiocyanate is a red solution, used in detecting either ferric or thiocyanate in solution, and is extracted from water by amyl alcohol.

When thiocyanic acid is treated with oxidizing agents, e.g., **nitric acid, sulfuric acid, hydrocyanic acid** is formed; when treated with reducing agents, e.g., **alumi-num** and dilute hydrochloric acid, hydrogen sulfide plus carbon plus ammonium chloride are formed.

Esters: Ethyl thiocyanate ($C_2H_5 \cdot SCN$), colorless liquid, boiling point 142° C. Formed by reaction (1) of potassium thiocyanate and potassium ethyl sulfate, (2) of cyanogen chloride and ethanethiol. Oxidizable with fuming nitric acid to ethyl sulfonic acid ($C_2H_5 \cdot SO_2OH$), and reducible with zinc and dilute sulfuric acid to ethane thiol (C_2H_5SH). Ethyl isothiocyanate ($C_2H_5 \cdot NCS$), colorless, odorous liquid, boiling point 132° C. Formed by reaction of ethyl **amine** and carbon disulfide. Re-ducible to ethyl amine ($C_2H_5NH_2$) plus methylene sul-fide (CH_2S). Allyl isothiocyanate ("mustard oil," $C_3H_5 \cdot NCS$) liquid, boiling point 151° C., odor of mus-tard, and causes blisters in contact with the skin.

Soluble thiocyanates give a characteristic red colora-tion with ferric salts. (R.K.S.)

THIOIC ACIDS. Sulfur.

THIOLS. Thioalcohols and Related Compounds.

THIOPHENE AND RELATED COMPOUNDS. Thiophene (($CH)_4S$) is a liquid, boiling point 84° C., resembling **benzene** in odor and properties. Thiophene

is present in coal tar and is recovered in the benzene distillation fraction (up to about 0.5% of the benzene present). Its removal from benzene is accomplished by mixing with concentrated **sulfuric acid,** soluble thio-phene sulfonic acid being formed. Thiophene gives a characteristic blue coloration with isatin in concentrated sulfuric acid.

Thiophene may be formed (1) by passing ethyl sul-fide (diethyl sulfide) through a red-hot tube, (2) by reaction of **sodium** succinate and **phosphorus** trisul-fide. **Chlorine** and **bromine** yield chloro- and bromo-substitution products, respectively, cold fuming **nitric acid** yields thiophene sulfonic acid. Thiophene aldehyde ($C_4H_3S \cdot CHO$), liquid, boiling point 198° C., resembles **benzaldehyde** chemically rather than furfural. The cor-responding primary **alcohol,** and **carboxylic acid** are known. Where sulfur of thiophene is occupied by oxy-gen, **furane** is the compound, and where by nitrogen (group —NH), **pyrrole.**

Benzothiophene ($C_6H_4 \cdot (CH)_2S$) is a solid, melting point 31° C., boiling point 221° C., and resembles naph-thalene. Where sulfur of benzothiophene is occupied by oxygen, coumarone is the compound, and where by nitro-gen (group —NH), indole.

Thiophene   HC $\begin{smallmatrix}4&3\\5&2\end{smallmatrix}$ CH   (C_4H_4S)
             HC   CH
                  S

Benzothiophene
(Thionaphthene)   ($C_6H_4(CH)_2S$)
                         S

Penthiophene   HC $\overset{CH_2}{\diagup}$ CH   (C_5H_6S)
                HC   CH
                     S

Diphenylene sulfide

                S
         melting point 97° C.
         boiling point 332° C.

Nitrogen-sulfur ring compounds:
             Thiazoles:
           HC——N
Thiazole
           HC   CH
              S

        boiling point 117° C.

Benzothiazole

                  N

                  CH
               S

            Thiazines:
Phenthiazine, thiodiphenylamine
             NH

              S
       melting point 150° C.
       boiling point 370° C.

Methylene blue, 2, 8, tetramethyldiamino-thiazonium chloride

(R.K.S.)

THIOPHENOLS. Thioalcohols and Related Compounds.

THIOSULFURIC ACID AND THIOSULFATES.
Thiosulfuric acid ($H_2S_2O_3$) is possibly formed upon addition of an acid to **sodium** thiosulfate solution, but immediately decomposes into **sulfur** dioxide gas and sulfur, yellow precipitate, the latter appearing gradually (more rapidly with higher concentrations).

Sodium thiosulfate, "hypo" (misnamed hyposulfite) ($Na_2S_2O_3 \cdot 5H_2O$), is used (1) to dissolve **silver** chloride, bromide, iodide in the photographic "fixing" bath, soluble sodium silver thiosulfate being formed plus sodium chloride, bromide, iodide, (2) in reaction with **iodine** in solution, sodium tetrathionate and sodium iodide being simultaneously formed, or with **ferric** salt solution, sodium tetrathionate and ferrous being simultaneously formed, (3) in reaction with **chlorine** as an "antichlor" forming sulfate and chloride. Sodium thiosulfate reacts with silver nitrate solutions yielding silver sulfide, brown precipitate, and with permanganate yielding **manganous**. **Sodium** amalgam changes sodium thiosulfate to sodium sulfide plus sodium sulfite.

Sodium thiosulfate is formed (1) by reaction of sodium sulfite solution and sulfur upon warming, (2) by reaction of sodium sulfite solid and sulfur upon heating, (3) by complex reaction of sulfur and sodium hydroxide solution upon warming—sulfur yields sodium sulfide plus sodium sulfite and the latter reacts with excess sulfur forming sodium thiosulfate, and the sodium sulfide present may be converted into sodium thiosulfate by passing in sulfur dioxide until the solution changes from yellow to colorless.

Thiosulfates are commonly identified as follows:

1. Dilute acids precipitate sulfur from thiosulfates (difference from sulfides and sulfites).

2. Zinc sulfate and sodium nitroprusside give no color (difference from sulfites). (R.K.S.)

THIOUREA.
Thiourea, "thiocarbamide" ((NH_2)$_2CS$), is a white solid, melting point 172° C.; easily hydrolyzed to **ammonia** plus **carbon dioxide** plus **hydrogen sulfide**; chemically analogous to urea; oxidized to **urea** by cold potassium permanganate solution.

Thiourea is formed by heating ammonium **thiocyanate** at 170° C. After about an hour 25% is converted. With **hydrochloric acid** thiourea forms thiourea hydrochloride; with **mercuric** oxide thiourea forms a salt, and with **silver** chloride it forms a complex salt.

Symmetrical diphenyl thiourea, "thiocarbanilide" ((C_6H_5NH)$_2CS$), is a solid, melting point 154° C., and when heated with concentrated **hydrochloric acid** yields aniline plus phenylisocyanate. Formed by reaction of aniline and **carbon disulfide**. Symmetrical-diethylthiourea ((C_2H_5NH)$_2CS$) is a solid, melting point 77° C. (R.K.S.)

THOMSON EFFECT. Thermoelectric Phenomena.

THOMSON E.M.F. Thermoelectric Phenomena.

THONGALLEN. Types of concretions which occur
in **loess**. When these concretions simulate the forms of human figures they are called *Loessmännchen*. (R.M.F.)

THORACENTISIS.
Removal of fluid or pus from the chest or pleural cavity by inserting a needle or trocar through the chest wall between the ribs. (D.M.H.)

THORACOPLASTY. Tuberculosis.

THORAX.
1. The division of the **arthropod** body between the head and abdomen. It consists of three or four metameric segments of the body, often closely united so that their boundaries are difficult to determine. It bears a pair of jointed appendages on each segment, developed for walking or swimming, and in some species for grasping, and in the insects the two posterior segments may bear a pair of wings each. The three segments of the insect thorax are known as the pro-, meso-, and metathorax. 2. A division of the trunk of **vertebrates**, just behind the head and neck. This region is supported by the ribs and contains the thoracic cavity, a division of the **coelom**. In the **mammals** it is separated from the abdominal cavity by the diaphragm and is further subdivided into the pleural cavities containing the **lungs** and the pericardial cavity containing the **heart**. It is a bony cage made of the sternum in front, the vertebral column behind, and the ribs connecting the two. (A.W.L., R.S.M.)

THORIANITE.
The mineral thorianite is chiefly composed of **thorium uranium oxide**, occurring in black nearly opaque cubic crystals in Ceylon and Madagascar. It is radioactive, and as such is particularly valuable in helping to date the absolute, as well as the relative, ages of the rocks in which it occurs. (See **Radioactive Minerals; Chronology**.) (R.M.F.)

THORITE.
The mineral thorite is a **silicate** of the rare element **thorium** and corresponds to the formula $ThSiO_4$. It is **tetragonal** and exhibits a prismatic **cleavage**. The original thorite was black in color with a specific gravity of 4.4–4.8. A variety orangite, so called from its orange-yellow color, has a specific gravity of 5.19–5.40. It has been found partly altered to thorite. Uranothorite contains **uranium** oxide. Thorite occurs in Norway in **augite syenites**. Thorite and orangite occur in Sweden, and orangite and uranothorite are found in Madagascar. Uranothorite is found in Ontario. (E.S.C.S.)

THORIUM.
Symbol: Th. Atomic number: 90. Atomic weight: 232.12. Density: 11.5 (11.3–11.7). Melting point: 1845° C. No isotope, but of single atomic form: 232.

Thorium metal is dark gray, dissolves in **hydrochloric acid**, is made passive in **nitric acid**, not affected by fusion with alkalis; at 450° C. combines with **chlorine** and with **sulfur**, at 650° C. combines with **hydrogen** and with **nitrogen**. An outstanding property is the **radioactivity** of all thorium-containing substances. Discovered by Berzelius in 1828.

Thorium occurs in **monazite** sand in Brazil, India, North and South Carolina, which sand contains 3–9% thorium oxide, and is the chief source; also found in thorite containing about 60% oxide and in thorianite, about 80% oxide. When heated with concentrated **sulfuric acid** the minerals form thorium sulfate, from which, by a series of reactions, thorium nitrate, the chief commercial compound, is obtained. Upon ignition the nitrate yields oxide which finds extensive use in the manufacture of gas mantles, where a mixture of 99% thorium oxide and 1% **cerium** oxide gives maximum luminosity at a comparatively low temperature. The oxides are formed in place by ignition of the mantle fiber, which has been previously impregnated with the corresponding nitrates and dried.

Hydroxide: Thorium hydroxide ($Th(OH)_4$), white, gelatinous precipitate, by treating thorium salt solutions with sodium hydroxide solution.

Nitrate: Thorium nitrate ($Th(NO_3)_4$), crystallizes with 5, 6, and 12 molecules of water, and is the chief commercial salt.

Oxide: Thorium oxide (ThO_2), white solid, prepared by ignition of the nitrate, oxalate, or hydroxide. (R.K.S.)

THORN APPLE. Potato Family.

THORNY-NOSE. Pisces, Teleostei. A peculiar marine fish (**Pisces**) found near New Zealand. It has very long serrate dorsal and ventral fins and the nose bears two strong spines. (A.W.L.)

THORON. Symbol: Tn. A radioactive element of the thorium series. (See **Radioactive Changes.**) (R.K.S.)

THOULET SOLUTION. An aqueous solution of **potassium mercuric** oxide the specific gravity of which is 3.19. Used by petrologists to separate mineral particles according to their specific gravities. Synonym, Sonstadt Solution. (R.M.F.)

THRASHER. Aves, Passeriformes. Moderately large North American birds (**Aves**) related to the wrens, the mockingbird, and the catbird. They have a long curved beak. Several of the species are among our finest singers.

The brown thrasher, *Toxostoma rufum,* is a familiar bird east of the Rockies. Our 6 other species are western and southwestern. (A.W.L.)

THREAD GAUGE. Measurement.

THREE-BODY PROBLEM. If we assume that three or more objects exist in the universe, each of them attracting every other in accordance with the law of **gravitation,** the problem of predicting subsequent positions and motions is commonly referred to as the three-body problem, or the *n*-body problem.

The **two-body problem** has been completely solved and the full solution may be expressed in comparatively few words and symbols. However, if we add one or more other bodies the solution becomes one of exceeding complexity and has never been accomplished in any form which is at all suitable for computational purposes. In fact, only one complete solution has ever been made, in spite of the labors of practically all of the great mathematicians of the past 3 centuries.

Even though no general solution of the problem is available, nevertheless, there are several practical computational methods for determining the positions of **planets** and other members of the **solar system,** taking into account the gravitational attraction of all effective members. Such solutions are all made by successive approximations and various methods of computing **perturbations,** rather than by the application of any general solution.

A number of particular solutions of the three-body problem have been made by mathematicians, notable among them being the solution by Lagrange. He showed that it is possible for an **asteroid** to be stable in a position such that it is equidistant from both the sun and **Jupiter.** In this case the three objects would be on the vertices of an equilateral triangle and the asteroid **orbit** would have the same period as that of Jupiter. This case is illustrated in nature by the members of the so-called **Trojan group.** (W.K.G.)

THREE-HINGED ARCH. Arch.

THREE-PHASE EQUILIBRIUM. For every pure, chemically stable substance there is a certain temperature and pressure at which it can exist in all three states or phases, solid, liquid, and vapor, each phase being in equilibrium with each of the others. At higher temperatures and pressures than those at this so-called "triple point," the liquid and vapor states may attain equilibrium; solid-vapor equilibrium (see **Sublimation**) is possible at lower temperatures and pressures; while solid-liquid equilibrium can be obtained at higher pressures and at lower or higher temperatures according as the substance contracts or expands upon melting (see **Fusion**). These three equilibria may be represented by three temperature-pressure graphs which

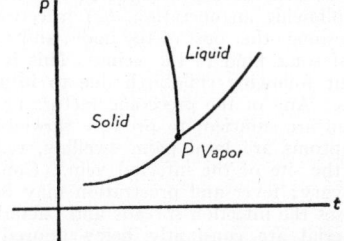

Triple point (*P*) on temperature-pressure diagram.

converge at the triple point. The figure illustrates the case of water, which contracts on melting, and for which the triple point is at +0.072° C. and 4.6 mm. of mercury. (L.D.W.)

THREE-WIRE SYSTEM. Much of the domestic and small commercial electric power loads are supplied by a dual-voltage, three-wire system, the voltage between two of the wires being approximately 220 while that between the other two wire combinations is approximately 110 volts. As usually shown schematically the neutral wire (the one common to the two 110-volt combinations) is placed in the center, giving 110 volts between the neutral and either outside wire while the outside wires have 220 between them. In applying the load to this type system an attempt is made to keep it balanced on either side of neutral. For a perfectly balanced load no current flows in the center wire, while for an unbalanced load only the unbalance current (the difference between the currents in the two 110-volt loads) is carried by the neutral wire. Thus there is a reduction in power loss and voltage drop in the line, giving a better, more efficient service. (See **Balance Coil.**) (L.R.Q.)

THRESHOLD SPEED, THRESHOLD DENSITY, THRESHOLD EXPOSURE. In photography the threshold density is the smallest density visible—above the **fog density**—when the image is placed in contact with a sheet of white paper. Threshold exposure is the exposure necessary to produce the threshold density. Threshold speed may be defined in several ways but, basically, it is the reciprocal of the threshold exposure. The use of the term threshold in photographic technology arose from the use of "Schwellenwert" in a similar sense in German literature.

The Scheiner method of measuring and expressing emulsion speeds is the best known of the methods based on the threshold speed. (C.B.N.)

THRIPS. Thysanoptera.

THROAT. Internally, the pharynx: the cavity behind the mouth into which the nasal passages open and from which the esophagus and trachea lead. Externally, the ventral part of the neck. (A.W.L.)

THROMBOPHLEBITIS. Thrombosis associated with inflammation or **phlebitis** of a vein. There are two general types: (1) Superficial thrombophlebitis in which veins near the surface of the body are involved; and (2) deep thrombophlebitis in which the deep internal veins are involved. Any vein in the body may be affected, but those of the legs and pelvis are most commonly involved. Varicose veins are more subject to phlebitis than normal ones. With infection of the vein a soft friable clot forms in the involved vessel, blocking the blood flow in this portion of the vein. The condition is then termed thrombophlebitis.

The causes of thrombophlebitis are (1) stasis, or slowing of the blood stream, as occurs when a patient is confined to bed for a prolonged period, as in severe illness or following an operation, (2) infection, chronic or acute, in some other part of the body, and (3) trauma or injury of some kind to the veins. This is especially apt to occur following childbirth due to injury to the pelvic veins. Any of the preceding factors or combination of them are sufficient to produce thrombophlebitis.

The symptoms are local pain, swelling, redness, and heat, over the site of the infected vein. Constitutional symptoms vary; fever and prostration may be marked. In some cases the infection spreads and bacteria and infected material are constantly being poured into the blood stream, resulting in septicemia. At times a piece of clot (embolus) breaks loose from the infected vein and is carried into the general circulation to the lung. If it is large enough to shut off the main artery to the lung, death results very quickly—at times instantly. If smaller, the clot may travel in a smaller arterial branch to the lung. If the clot is infected, pneumonia results.

The treatment of thrombophlebitis is mainly rest in bed with such general treatment as is given any acute infection. If an extremity is involved, heat is applied, and the part is elevated. Injection of the parasympathetic ganglia along the spine with novocaine is one of the newer and more effective treatments. It acts by dilating the blood vessels of the involved area, improving the circulation and thus clearing up the infection more rapidly. (D.M.H.)

THROMBOSIS. The forming of a clot in a vessel, thereby shutting off the circulation beyond the point of thrombosis. This usually results from infection, injury to the vessel, or almost complete stasis in the vessel. (See **Coronary Occlusion.**) (R.S.M.)

THROMBUS. A clot or plug forming in a blood vessel and attached to the vessel wall. When the thrombus becomes dislodged and is caried by the blood current to some other part of the circulation, it is called an **embolus.** (R.S.M.)

THROTTLE. To throttle means to choke. Throttling a fluid flow is to decrease the cross-sectional area of flow at some point, as by partly closing a valve, damper, or gate in the conduit. (See **Throttling Process.**)

This word is in common usage to denote the adjustable control on the quantity of the working medium entering prime movers and hence has become synonymous, in common usage, with the control of prime mover power. (F.T.M.)

THROTTLING PROCESS. Throttling is the reduction of pressure of a fluid passing a restriction or **orifice.** During a throttling expansion of a fluid there must be no work done, i.e., the enthalpy contained in the fluid after throttling must be unchanged from the original. Throttling as a thermodynamic process may be used to advantage as in the control of a steam engine, where the weight of steam drawn into the engine per cycle is varied by varying the pressure with a throttle valve. The automobile engine is governed in speed by variable throttling of the intake, so that the cylinders receive more or less charge, depending on the extent to which the pressure of the incoming gas is reduced by throttling. (See **Steam Calorimeter.**) (F.T.M.)

THROUGH BRIDGE. Bridge.

THROW. For the use of this term in geology, see **Fault.**

THRUSH. (In zoology) Aves, Passeriformes. In the broad sense a bird (**Aves**) of the family Turdidae, but the term is most widely applied to the members of this family that retain a spotted breast as adults, while other species which lose the spots as their adult plumage develops receive other names. Among the latter are the **robin** and **bluebirds** of North America. The North American thrushes are of moderate or small size, brown, gray, or olive above and white below, with spots similar to the back. The wood thrush, *Hylocichla mustelina,* is widely distributed in the eastern half of the United

Wood thrush, *Hylocichla mustelina.* Bright brown above, white below, with large round spots on the breast and sides.

States, and the hermit thrush, *H. guttata,* is even more widely distributed, though less commonly known.

The family includes the European fieldfare, **redwing,** and **blackbird,** the ring-ouzel, and the Old World chats. In North America it is also represented by the Townsend solitaire and the **wheatear,** in addition to the robin and the several species of bluebirds.

The song-thrush of Europe, *Turdus philomelus,* is also called the mavis, and in America the term thrush is misapplied to the Louisiana water thrush, which is a warbler, because of its similar color and pattern.

In medicine, thrush is a contagious parasitic infection of the mouth due to a **fungus** called *Oidium albicans.* The organism produces white patches (apthae) on the mucous membrane of the mouth and tongue. It occurs primarily in nursing infants who are debilitated from some other infection, but may occur in adults when the oral hygiene has been neglected, or where the gums have been injured by too violent cleansing.

The disease is treated by adopting the proper methods of oral hygiene, and swabbing the mouth with mild antiseptic solutions such as boracic acid, or 1% aqueous gentian violet. (A.W.L., R.S.M.)

THRUST BEARING. Bearings.

THRUST COEFFICIENT, PROPELLER. Propeller Coefficients.

THULIUM. Symbol: Tm. Atomic number: 69. Atomic weight: 169.4. No isotope, but of single atomic form: 169. Type of compound: Tm_2O_3. Color of salts: Green. Discovered by Cleve in 1879. A member of the **yttrium** sub-group of the rare earth metals. (R.K.S.)

THUMBLESS MONKEY. Mammalia, Primates. A group of African **monkeys** constituting the genus *Colobus.* They are named from the reduction of the thumb, which is either entirely absent or reduced to a small projection, with or without a vestigial nail. Some of the included species are called colobs, one is a guereza, and one is the king monkey. (A.W.L.)

THUNDER. An electrical discharge in the atmosphere which sets up a sound wave all along the column of air in which the discharge occurred. If the discharge occurs close at hand, there is one sharp burst or "explosion" but if the discharge occurs some distance away then portions of the sound wave reach an observer over a period of time. Echoes of thunder also occur when the sound wave or parts of it are reflected from objects projecting from the earth such as hills and buildings. (P.E.K.)

THUNDERSTORMS. Cumulonimbus clouds accompanied by lightning and thunder. Normally thunderstorms are accompanied by torrential rain for brief moments during the passage of the storm, but occasionally no precipitation reaches the ground. Often they cause hail and gusty surface winds of considerable velocity. Sometimes they are attended by tornadoes which cause great damage. Vertical velocities inside thunderstorms are extremely erratic and as high as 120 m.p.h. Thunderstorms originate as a result of the occurrence of three factors and the absence of any of the three will not permit their formation.

1. Sufficient number of vertical perturbations in the air to cause parcels of air to start upward.
2. Sufficiently unstable air to permit the vertical perturbations to rise with accelerated velocity up to great enough heights to insure penetration of the freezing level.
3. Sufficient moisture to insure the development of cloud droplets and eventually snow and rain in the rising parcels of air.

The first requirement may be fulfilled by mechanical lifting of air over terrain or up a frontal surface, or by heating of the surface air over warm ground or water. The second requirement may be fulfilled by one of four processes in the atmosphere.

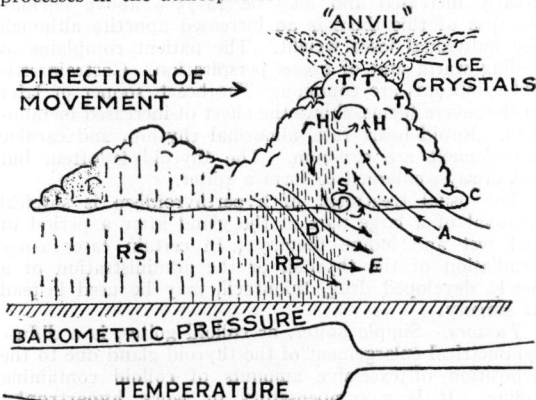

A cross-section of a typical cumulo-nimbus (thunderstorm) cloud showing the air currents involved as well as the formation of hail and areas of rain. *A* is the inflowing-ascending air; *D*, the outflowing-descending air; *C*, the storm collar; *S*, roll of "scud," where *A* and *D* meet; *E*, wind gust; *H*, hail; *T*, thunderheads; *RP*, primary rain; *RS*, secondary rain.

a. Heating from below, i.e., along the ground or water.
b. Cooling from above, i.e., advection of cold air at high altitude.
c. Simultaneously heating from below and cooling from above.
d. Differential cooling of a layer of air in such a manner that the top grows colder than the bottom.

Necessary moisture is not always present in the atmosphere to assure thunderstorm development even though the other two requirements are met. Moisture content of the air must be such that saturation can be caused by the two other requirements and its distribution must be such that lower levels reach saturation first. Because moisture is distributed horizontally by large-scale air currents in tongues and islands in the atmosphere, the distribution of thunderstorms often assumes these patterns.

Heating from below normally occurs over land during daylight hours and over water whenever cold air flows from cold continents over warm water, or from cold water over warmer water. Thermal or air-mass thunderstorms of this type are at a maximum over land in the late afternoon and are restricted almost entirely to the spring and summer of the year in countries as far north as the United States and Canada. Thermal or air-mass thunderstorms occur over oceans at any time of day or night and generally are at a maximum during winter and spring. Air-mass thunderstorms are isolated and distributed at random.

Cooling from above by nocturnal radiation from the moist air takes place over ocean areas where the surface water temperature is maintained virtually constant. This creates instability and results in early morning thunderstorms at sea.

Cooling aloft may also occur by advection of cold air over a layer of warm air. Thunderstorms of this type are usually high level with bases 1 or 2 miles above the earth. They occur over both land and water. In North America they are common to the region just east of the Rockies during summer.

On some occasions both heating from below and cooling from above occur simultaneously. Thunderstorms which result may lie as low as a few thousand feet or as high as several miles. Heating from below may arise from any of the processes that raise the temperature of the lower levels of the atmosphere, but high-level cooling is due to advection of cold air.

Differential cooling of an air layer occurs when an entire layer of air is lifted. If the air remains unsaturated the top will cool slightly more rapidly than the base. If the layer remains unsaturated at its top, but becomes saturated at its base, lifting will cool the top rapidly but the base only slowly, causing a rapid trend toward instability. Lifting occurs along cold and warm fronts, along mountains, and at a relatively slow rate in air that is undergoing convergence. (See Cumulonimbus.) (P.E.K.)

THYLACINE. Tasmanian Wolf.

THYME. Mint Family.

THYMUS. A ductless gland situated in the upper anterior portion of the chest cavity. It reaches its maximum activity during puberty and thereafter it diminishes in size and activity in most individuals. Its function in the body is not known. It is known, however, that certain children who have an enlarged thymus are subject to sudden attacks of unconsciousness, convulsions, sometimes sudden death. At autopsy a large thymus is found. The condition is described as *status thymicolymphaticus.*

Tumors of the thymus are found in a certain per cent of patients with myasthenia gravis, a disease of the muscles which makes them fatigue easily. (D.M.H.)

THYRATRON. Tube, Electronic.

THYROID GLAND. This important gland of internal secretion is made up of two flattened lobes lying beneath the superficial muscles of the lower anterior part of the neck on either side of the trachea. The two portions of the gland are connected by a small bridge of thyroid tissue lying across part of the trachea. Other masses of accessory thyroid tissue may sometimes be present along the length of the trachea. During pregnancy and menstruation the thyroid may temporarily increase in size. Complete removal and abnormal secretion of the gland causes grave systemic disturbances. The gland is also subject to **tumor** formation, which may be **benign,** causing enlargement only (adenoma), or toxic (toxic adenoma), or **malignant (cancer).**

The function of the thyroid is to serve as a storehouse for **iodine** and to secrete into the blood stream thyroid hormone, which has a stimulating effect on growth and metabolism. The thyroid affects other ductless glands and the sympathetic nervous system. In a reverse manner, other endocrine glands in turn influence the thyroid; this is particularly true of the pituitary gland, which has a multiplicity of influences on the endocrine system. The interglandular relationship of the thyroid as well as other endocrine glands is not well understood at present.

Diseases of the thyroid may be classified as those due to:

1. Hypofunction: cretinism and myxedema.
2. Hyperfunction: exophthalmic goiter and toxic nodular goiter (toxic adenoma).
3. **Tumors:** simple or colloid goiter; **adenoma; nodular** goiter; malignant disease, **cancer.**
4. Inflammatory disease: acute and chronic thyroiditis.

Diseases due to hypofunction: Cretinism is a condition originating in fetal life, or soon after birth, due to a more or less complete lack of thyroid secretion, and sometimes congenital absence of the gland. It is not incompatible with long life. The condition is rare in

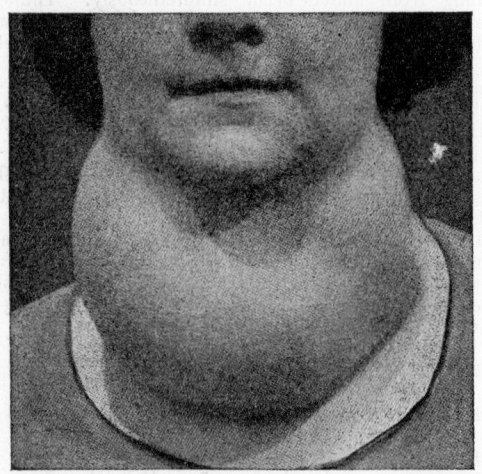

Large simple goiter in girl seventeen years old. (*Nelson Loose-Leaf Surgery, Thomas Nelson & Sons.*)

the United States. It is most common in iodine-deficient districts, where there is a lack of iodine in the maternal organism.

Often the disease is not recognized until the child has reached the age of 1 year, when a failure of physical and mental development becomes apparent. The skin is coarse, scaly, the tongue is thickened, and the child is dull and slow, with an intelligence far below normal.

The most important treatment is prophylactic and consists of proper treatment of pregnant women in goiter districts with iodine compounds. Cretins must be treated early in life with thyroid gland substance if relief of symptoms is to be obtained.

Hypothyroidism or **myxedema** is a disease of adults which may occur spontaneously, or as a result of the removal of too much of the gland in surgical treatment of Graves' disease. The disease occurs four to five times more frequently in women than in men. A goiter of simple type may be present although this is not usually the case.

The onset is gradual and occurs most frequently after 25 years of age. There is slowing of mental and bodily activities, an increase in weight, decreased appetite, and loss of a feeling of well-being. The individual complains of feeling cold. Blood pressure is lowered and the pulse is slowed. Mental sluggishness, forgetfulness, and lack of ambition may be marked. There is a tendency to increased sleep and drowsiness. The skin is thinned, coarsened, dry, and scaly, and the nails and hair may be brittle. Pains in the joints are common. **Anemia** is a frequent finding. The **basal metabolism** varies from minus 5 to minus 40.

Brilliant and dramatic results are obtained by administration of thyroid gland substance or thyroxin. The symptoms entirely disappear as long as substitution of thyroid substance is maintained to counteract the deficiency. The basal metabolism rises to normal limits and this test provides a means of estimating the proper maintenance dose of thyroid substance.

Diseases Due to Hyperfunction. Hyperfunction of the thyroid results in exophthalmic goiter (Graves' disease, hyperthyroidism, Basedow's disease, Thyrotoxicosis). The cause of this condition is not known, but constitutional predisposition seems to be a factor. It is a disease of civilized countries and is most common where conditions tend toward mental activity and strain. More women than men are affected; frequently a nervous shock or strain seems to start the process. The clinical manifestations are the result of an increased metabolic rate, increased **metabolism** of cells, and disturbances in the **autonomic nervous system.** Exophthalmus, a protrusion of the eyeballs giving a peculiar staring expression to the eyes, is usually present in marked cases; the cause of this disturbance is not known.

Hyperthyroidism begins insidiously, with early symptoms of fatigue, loss of weight, marked nervousness. The patient's disposition may change, with the development of irritability and inability to get along with people. A fine tremor, seen best when the hands are extended, is characteristic. The basal metabolism is greatly increased and may be 40–75% above normal. Because of this there is an increased appetite although the individual loses weight. The patient complains of feeling warm and of excess perspiration. Gastrointestinal symptoms are common. The heart, sooner or later in the severe forms, shows the effect of increased metabolism. Rapid heart rate, abnormal rhythm, and cardiac enlargement are common. The thyroid is often, but not always, enlarged to form a goiter.

The usual treatment of hyperthyroidism is surgical removal of a large part of the gland after a period of bed rest and iodine therapy. In certain cases x-ray irradiation of the thyroid, or the administration of a newly developed drug, thiouracil, may be used instead of surgery.

Tumors. Simple goiter, or colloid goiter, is a diffuse symmetrical enlargement of the thyroid gland due to the deposition of excessive amounts of colloid containing iodine. It is a compensatory or work **hypertrophy** which is a response to inadequate supplies of **iodine** in food and water. The disease usually appears soon after puberty; it is commoner in females than in males. There are no symptoms other than enlargement of the neck, and, if the tumor reaches very great size, secondary pressure symptoms. The metabolic rate is not disturbed. There are certain districts in the world where colloid goiter is common. These in general are in mountainous regions and the sites of glacial ice fields—that is, districts deficient in iodine. Near the sea coast, where sea-water iodine is present, the disorder is relatively uncommon. In the United States, goiter districts are found around the Great Lakes, Mississippi Valley, and the Northwest. In these areas supplementary iodine added to the water to make up the deficiency is important in order to correct this disorder and prevent cretinism in offspring.

Many colloid goiters gradually decrease after puberty. Others may respond to iodine treatment, and a few require surgery if they reach great size.

Adenoma of the thyroid seldom causes constitutional symptoms unless the **adenoma** becomes toxic and then hyperthyroidism develops. These adenomatous growths develop from fetal rests of thyroid tissue. Surgical removal is the treatment for both the simple and toxic varieties; it gives excellent results.

Malignant tumors, **carcinoma** and **sarcoma** are rare and difficult to distinguish from benign adenomas. Surgical and x-ray treatment are used palliatively.

Thyroiditis is infection of the thyroid gland. This condition is very uncommon. (D.M.H.)

THYROIDITIS. Thyroid Gland.

THYSANOPTERA. The thrips, an order of minute **insects,** with or without wings. Their mouths are formed for sucking, the tarsi have an expanded tip, and the metamorphosis is gradual. When wings are present there are two pairs, both formed of slender membranes with very long fringes.

In spite of their minute size a number of species of thrips are serious pests, attacking cultivated plants.

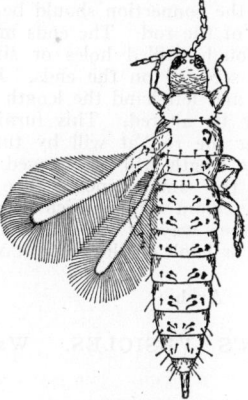

Thrips.

One of these is the onion thrips, introduced from Europe over sixty years ago. It attacks onions, cabbage, melons, and other plants and is sometimes troublesome in greenhouses, although it is not the common greenhouse thrips. Spraying with nicotine sulfate is an effective check, although onions are difficult to spray because the insects cannot easily be reached between the leaves. In greenhouses fumigation is desirable.

Some thrips live on other small insects, eggs, and mites, and so are of some value in checking other pests. (A.W.L.)

THYSANURA. Small to moderately large insects of the primitive wingless group, commonly called bristletails. The more common forms are the **silverfish** and the **fire-brat,** both grayish with silvery luster. The body is about ⅓″ long in both species, bearing long slender antennae and at the opposite end a pair of similar cerci and a slender median filament. The order also includes some species with forceps-like appendages at the tip of the body in place of the slender cerci. (A.W.L.)

TIC. A recurrent, spasmodic, involuntary muscular reaction taking place in a group of muscles. Simple tic is of psychogenic origin and is apt to develop in nervous people; it may accompany psychoneurotic disorders. It is a severe form of the so-called "habit-spasm," and twitching of a part of the face is its most common form. Tic douloureux is a form of neuralgia

of the facial nerve which is characterized by sudden severe pain over one side of the face, accompanied by reflex contraction of the muscles in this area. (D.M.H.)

TICK. Arachnida, Acarina. A blood-sucking parasite related to the spiders and mites. Its body is leathery and saclike, with piercing mouth parts forming a small protuberance called the capitulum and a shield or scutum marking the dorsal surface in most species. The ticks are larger than their relatives, the mites, ranging from three to six millimeters in length. When filled with blood the female sometimes reaches a length of ½″.

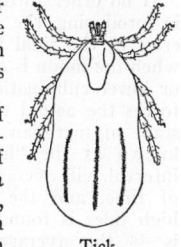

Tick.

The common dog or wood tick of the eastern and central states, *Dermacentor variabilis,* often attaches itself to human beings in the woods and is a common pest of domestic animals, especially dogs, during the warm months. A closely related species in the western states, *D. andersoni,* transmits the dangerous disease, **Rocky Mountain spotted fever.** The Texas cattle tick, *Margaropus annulatus,* transmits Texas fever of cattle, also a destructive disease. (A.W.L.)

TIDES. The periodic rise and fall of the oceans, or other large bodies of water, relative to the surrounding land is commonly known as the tides. Since the earth itself is not a perfectly rigid body, there are tides in the earth itself, and observations of the land tides have provided useful information regarding the rigidity of the interior of the earth.

Tides are produced by the combination of a number of external forces, with the principal force being the **gravitational** attraction of the **moon.** Owing to the differences of distance of the moon from various portions of the earth the amount of the attractive force will be different in different places and tend to produce a deformation. The accompanying diagrams indicate (on a very much exaggerated scale) the deformation forces, and

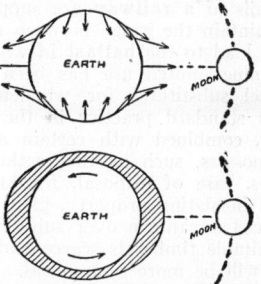

Tides in a very deep ocean. If the whole earth were covered by very deep water, and if the earth's rotation were slower, high tides would occur under and opposite the moon. The upper figure, after Darwin, shows the tide-raising force of the moon at different places.

the resultant tides in a very deep ocean surrounding a rigid earth.

The tidal force due to the gravitational attraction of the sun ranks second in importance to the tide-raising forces from the moon. Twice during each month, at times of full and new moon, the forces from the sun and moon are acting in parallel directions and produce the maximum tide range known as "spring tides." At times of quarter-moon the forces are acting at right angles to each other and the resultant minimum range of tides is known as "neap tides." Other tidal forces due to the rotation of the earth, the revolution of the earth-moon system, and the revolution of the sun-earth system, make the problem of tide prediction one of great complexity. Detailed **harmonic analysis** of the observed tides have indicated the presence of tidal forces due to the attractions of several of the major **planets**. In the open ocean, from observations taken on isolated islands, the average range of ocean tides relative to the land is about 2½'. Configuration of shore line and contour of the ocean floor greatly increase this range for most stations along the coasts of the continents. The maximum range of tides in the land is about 9".

If no other forces than lunar attraction were effective in producing the tides, the time of high water (the crest of the tidal bulge) would occur at a given point when the moon is on the local **meridian** (either at upper or lower culmination). Due to the other effective tidal forces the actual time of high tide differs from the instant of meridian passage of the moon by an amount known as the "lunar interval." The effect of lunar interval will average zero in the course of a long period of time, and the average interval between successive high tides is found to be 12 hours 25.5 minutes, which is ½ the average interval between successive upper culminations of the moon.

The difference between the actually observed time of high tide and the time calculated from the instant of transit of the moon across the meridian, combined with the lunar interval, is known as the "establishment of the port." The values of the "establishment" are obtained observationally for various ports and are tabulated in tide tables. The values obtained for two ports, which may be separated by only a few miles, may be very different owing to the configuration of the coast line between the ports.

The tides represent expenditure of energy, and a large part of this must come from the **kinetic energy** of rotation of the earth. This should tend to check the earth's rotation and thus increase the length of the day, but the effect is too small to have been observed. Considerable effort has been directed towards utilizing tidal energy for useful purposes to man, but so far the success has been very limited. Tidal forces also play an important part in the evolutionary processes of the various astronomical bodies. (W.K.G.)

TIE. The rails of a **railway** are supported by cross ties, which maintain the gauge between rails and transmit the wheel load to the **ballast** in which the ties are embedded. Some limited use has been made of concrete and steel substitutes for wooden ties, but the wooden tie is standard practice in the United States. The low cost, combined with certain advantages that wooden ties possess, such as the method of fastening rails by spikes, ease of disposal, freedom from corrosion, electrical insulating property, give the wood tie a very pronounced advantage over substitutes. In countries where suitable timber is scarce and expensive, the substitute tie will be more economical. Most cross ties in this country are made of oak or Southern pine. Locust, hickory, and beech are also accepted. The size of a class A tie is 7" × 9" × 8' 6". The ties must be made from sound live timbers, free from soft or decayed spots, shakes, worm holes, or other imperfections which would impair the usefulness. Ties may be made from a log by sawing or hewing slabs from the surface, or by sawing or splitting larger logs. The ties are barked before use. Ties are spaced about 21" on centers, and are embedded nearly to their upper surface, with rock or gravel thoroughly tamped around them.

At the present time, practically all wooden ties are creosoted before use, as it has been amply proved that creosote extends the life of the tie sufficiently to more than pay for the cost of creosoting. This fact is not likely to be altered with time, because of the increasing cost of ties, and the improvements of creosoting processes. In a treating plant, the tie is first seasoned and steamed. A vacuum is maintained for a sufficient time to draw out the sappy moisture content of the timber,

and then after this is drained off the cylindrical retort is filled with hot creosote oil. A pressure is then applied, and this drives the oil into the pores of the wood. Then the pressure is released, and the excess oil is drained from the retort, after which a quick vacuum is applied to complete the extraction of the excess creosote. (F.T.M.)

TIE PLATE. Compression Member; Railway Track.

TIE ROD. A tie rod is a slender structural rod capable of carrying tensile loads only. Since the ratio of its length to the **radius of gyration** of its cross-section is usually very large, it would buckle (bend) under the action of compressive forces. Tie rods are used for airplane structures and in steel structures such as bridges, industrial buildings, tanks, towers, and cranes. Tie rods known as sag rods are sometimes used in connection with purlins (see **Bent**) to take the component of the loads which is parallel to the **roof**. The strength of a tie rod is the product of the allowable working stress and the minimum cross-sectional area. In a rod which is threaded at the end and not **upset** to allow for the reduction of area caused by the thread, the minimum area will be that which occurs at the root of the thread. Tie rods are connected at the ends in various ways, but the strength of the connection should be, at least, equal to the strength of the rod. The ends may be threaded and passed through drilled holes or shackles and retained by nuts screwed on the ends. If the ends are threaded right- and left-hand the length between points of loading may be altered. This furnishes a method for pre-stressing the rod at will by turning it in the nuts so that the length will be changed. A turnbuckle will accomplish the same purpose. The ends may also be swaged to receive a fitting which is connected to the supports. Another way of making end connections is to forge an eye or hook on the rod. (C.W.C.)

TIED ARCH. Arch.

TIEDEMANN'S VESICLES. Water Vascular System.

TIGER. Mammalia, Carnivora. One of the largest of the **cats**, *Felis tigris*, ranking with the lion in size and strength. The fur normally varies from reddish to brownish yellow, with transverse black stripes and a black-ringed tail. The total length of adult males, including the tail, is 9–10'. The tiger is an Asiatic animal, found chiefly in the warm southern countries, but also northward into Turkestan and southern Siberia. It is by no means a tropical species.

Tigers have been hunted extensively for sport, but they have also had to be destroyed in India, Java, and Sumatra because of their destruction of domestic animals. Occasionally also tigers have become man-eaters. Tales of the killing of these great beasts are numerous in the records of big game hunting. (A.W.L.)

TIGER BEETLE. Insecta, Coleoptera. A small long-legged **beetle** of predacious habits. They are usually found on exposed earth or sand in the glare of the sun, where they both run and fly very rapidly. The numerous species are blue or green, reddish, or white, with or without a characteristic pattern of spots and dashes on the elytra. The **larvae** live in burrows in the ground, lying with the head at the entrance to the burrow ready to seize any victim that comes near. (A.W.L.)

TIGER-CAT. Ocelot.

TILE FISH. Pisces, Teleostei. A marine fish (**Pisces**), *Lopholatilus chamaeleonticeps*, found in the warmer seas.

These fishes are covered with small scales and are brightly colored. They are chiefly remarkable for the discovery of a new species off the Massachusetts coast in the seventies which promised to be a valuable food fish. The fishes were taken for a time in the same way as cod, but they disappeared abruptly. (A.W.L.)

TILL. A general term for coarsely graded and extremely heterogeneous sediments of glacial origin. Till is generally classified as unstratified drift which may vary from clays to mixtures of clay, sand, and boulders. A particularly sticky form of clay till is called gumbo. (R.M.F.)

TILLERING. Grass Family.

TILLITE. Conglomerate.

TIMARAU. Tamarao.

TIMBER. Wood.

TIME. In the purely physical sense of the term, time is defined as measured duration. In accordance with this definition we say that two intervals of time are equal if a body, moving in **equilibrium,** moves over equal distances in the two intervals. The moving object is frequently referred to as the clock.

From the earliest recorded history the apparently moving sun has been adopted as the clock for regulating human affairs, and the apparent solar day is the interval of time between successive passages of the true sun across a local meridian. For many centuries upper culmination, or apparent noon, marked the beginning of an apparent day and local apparent time was the **hour angle** of the true sun. In 1925, to bring the apparent day into synchronism with that used in civil life, the beginning of the apparent day was transferred to lower culmination, or midnight, and the local apparent time defined as the hour angle of the true sun plus 12 hours.

The various hour angles, or local apparent times, were formerly recorded by the position of the shadow of a rod on a graduated dial known as a **sun dial.** Such an instrument is useful only when the sun is shining, and the need for mechanical clocks arose. As the development of these timekeepers progressed, with ever increasing accuracy, from burning candles, water clocks, sand glasses, and other such contrivances, down to the modern mechanical and electric clocks, it became evident that the apparent solar day is not constant in length throughout the year. These variations in length may be traced to the fact that the earth is not only rotating on an axis, but is also revolving about the sun in an **orbit.** The plane of the orbit is not perpendicular to the axis of rotation of the earth, and the angular motion of the earth in its orbit is not constant but in accordance with the elliptical motion described by **Kepler's Law of Areas.**

To assist in the development of mechanical clocks, the mean solar day was introduced. This is a day which is the average, in length, of all apparent solar days in a given year. A purely fictitious object, known as the **mean sun,** was defined and a mean solar day is the interval between successive passages of this fictitious object across any local meridian. The mean solar day begins at lower culmination of the mean sun and local mean time is the hour angle of the mean sun plus 12 hours. The difference between local apparent time and local mean time, taken in the algebraic sense of apparent minus mean, is known as the **equation of time.**

The term local civil time is sometimes used as synonymous with local mean time. Prior to 1925 the mean solar day began at upper culmination (mean noon) and the civil day began at midnight.

All three of the different kinds of time thus far discussed are measured from the local meridian. Accord-

ingly, only those people living in the same terrestrial **longitude** would have synchronous clock readings. To avoid this confusion the surface of the earth has been divided into a series of **standard time** zones. The different sorts of standard time will be discussed elsewhere, but standard time is defined as the civil time of some standard meridian.

While the sun provides the most convenient reference point for measuring time for everyday life, nevertheless it is not convenient for stellar astronomy because of the fact that the sun is continually moving eastward through the stars. Sidereal time is defined as the hour angle of the **vernal equinox.** Since the sun is apparently moving to the eastward through the stars, due to the revolution of the earth about the sun, there is one more sidereal day than solar day in the course of a **year.** A clock keeping sidereal time gains approximately 4 minutes each day on a mean solar clock, the sidereal clock agreeing with the civil time clock on approximately September 21. The **right ascension** of a star is measured from the vernal equinox in a direction contrary to the direction of apparent rotation of the **celestial sphere** and the sidereal time is the hour angle of the vernal equinox, and hence measured in the direction of apparent rotation of the celestial sphere. Therefore, sidereal time minus right ascension is equal to the hour angle of a star.

The standard unit of time for the physical sciences is the mean solar day as defined above. The practical unit is $1/86,400$ part of the mean solar day and is known as the mean solar second. This is one of the three basic units of the so-called **c.g.s. system.** It should be carefully noted that the mean solar day is based upon a purely fictitious object known as the mean sun. Hence the unit of time is just as arbitrary in character as are the other two units of the c.g.s. system. (See **Relativity.**) (W.K.G.)

TIME CONSTANT. This is a measure of the time of response of a circuit or device. In electrical circuits containing inductance and resistance it is L/R, L being the inductance in henries and R the resistance in ohms. For a capacitance and resistance circuit is it RC, C being the capacitance in farads and R as before. For both of these it is the time required for the circuit variables (current, charge, voltage) to reach to within $1/e$ of the final value, where e is the base of the natural logarithms (2.72). (L.R.Q.)

TIME QUENCHING. Steel.

TIME-GAMMA CURVE. This curve represents the relationship between gamma and the time of development for a particular sensitive material, developing formula and manner of development (temperature, agitation, etc.). The time-gamma curve is obtained by plotting the value of gamma (as determined from several sensitometric exposures which have been developed for different times) on the vertical axis against the time of development along the base of the chart. (See **Exposure-Density Relationship.**) From the time-gamma curve can be obtained (1) the time of development for a given gamma, (2) the gamma which will be produced by a given time of development under the same conditions, and (3) an indication of the range of gamma to which the particular film or plate should be developed for best results, since it is generally advisable to develop to a gamma which is at least 75% of the maximum gamma obtainable on that particular film or plate. (C.B.N.)

TIMING AXIS OSCILLATOR. Sweep Oscillator.

TIN. Symbol: Sn (stannum). Atomic number: 50. Atomic weight: 118.70. Density: white, 7.31 at 20° C.;

gray, 5.75 at 20° C. Hardness: 1.5–1.8 (white). Melting point: 231.85° C. Boiling point: 2260° C. (Isotopes: page 290.)

Tin is a silver-white metal with a bluish tinge, softer than zinc and harder than lead; malleable, ductile at 100° C.; can be powdered at 200° C., and upon exposure to temperatures below 18° C. crumbles to a grayish powder due to the "tin pest," which is caused by the transformation of white to gray tin (the reverse transformation may be brought about by heating gray tin to about 100° C.); when a bar of tin is bent a marked creaking sound is emitted due to the friction of the crystals; not oxidized on exposure to air at ordinary temperatures; burns to stannic oxide when heated to high temperatures in air or oxygen; soluble in hydrochloric acid to form stannous chloride; converted by nitric acid into insoluble beta-stannic acid; soluble in aqua regia to form stannic chloride; soluble in sodium hydroxide solution slowly to form sodium stannite and hydrogen gas; reacts with chlorine to form volatile stannic chloride. Discovery prehistoric.

Tin is used (1) as a protective coating on iron and steel and on copper. Tin coating on steel sheet is largely used for "tin can" containers, the coating being applied by dipping the sheet in molten tin or by electrolysis in acid or alkaline tin salt solution (very thin and continuous layers of tin are thus produced), (2) in alloys, such as solder, bronze, pewter, and bearing metals.

Tin occurs as oxide (cassiterite, tin stone, stannic oxide, SnO_2), obtained commercially in Federated Malay States, Dutch East Indies, and Bolivia. The ore is concentrated and then roasted to oxide (83–88% stannic oxide). The product is treated in a blast furnace and crude tin recovered. Refining is conducted by electrolysis, or by fractional fusion.

Chlorides: Stannous chloride ($SnCl_2 \cdot 2H_2O$), white solid, soluble, formed by reaction of tin metal and hydrochloric acid and then crystallizing, used as a mordant and reducing agent in dyeing and printing textiles; stannic chloride ($SnCl_4 \cdot 5H_2O$), (1) white crystals, soluble by reaction of stannous chloride with chlorine and then crystallizing, (2) colorless liquid, soluble, boiling point 114° C., formed by heating tin metal in chlorine and condensing the distillate.

Hydroxide: Stannous hydroxide ($Sn(OH)_2$), white gelatinous precipitate, by reaction of stannous chloride solution and alkalis, soluble in acids and in sodium hydroxide, insoluble in ammonium hydroxide.

Nitrate: Stannous nitrate ($Sn(NO_3)_2$), white solid, by reaction of tin metal and dilute nitric acid and crystallization, soluble in water with slight excess of nitric acid.

Oxalate: Stannous oxalate (SnC_2O_4), white precipitate, by reaction of stannous chloride solution and oxalic acid or ammonium oxalate solution, upon heating yields stannous oxide.

Oxides: Stannous oxide (SnO), black solid, insoluble, (1) by heating stannous chloride solution and sodium carbonate solution several hours, (2) by ignition of stannous oxalate out of contact with air; stannic oxide (SnO_2), white solid, insoluble, (1) by heating tin metal to a high temperature in air or oxygen, (2) by reaction of tin metal and concentrated nitric acid and ignition of the precipitate, (3) by reaction of stannic salt solution with alkalis, and ignition of the precipitate.

Stannates: Sodium stannate (Na_2SnO_3) and potassium stannate (K_2SnO_3), colorless solutions, by reaction of stannic salt solutions with excess of sodium or potassium hydroxide, respectively.

Stannic acids: Stannic acid (H_2SnO_3), white solid, insoluble (1) alpha, by reaction of stannic salt solution and alkalis, or by reaction of stannate solutions with acids, (2) beta, by reaction of tin metal and nitric acid.

Stannites: Sodium stannite (Na_2SnO_2) and potassium stannite (K_2SnO_2), colorless solutions, by reaction of

stannous salt solutions with excess of sodium or potassium hydroxide, respectively. Powerful reducing agents.

Sulfates: Stannous sulfate ($SnSO_4$), white solid, soluble; stannic sulfate ($Sn(SO_4)_2$), white solid, soluble.

Sulfides: Stannous sulfide (SnS), dark brown precipitate, by reaction of stannous salt solution and hydrogen sulfide, insoluble in sodium sulfide solution but soluble in sodium polysulfide solution, forming sodium thiostannate; stannic sulfide (SnS_2), yellow precipitate, by reaction of stannic salt solution and hydrogen sulfide, soluble in sodium sulfide solution, forming sodium thiostannate.

Stannous chloride solution, when treated with mercuric chloride solution, yields a precipitate white to gray, depending upon the relative amounts of mercurous chloride (white) and mercury metal (black) formed. (R.K.S.)

TIN STONE. Cassiterite.

TINAMOU, TINAMU. Aves, Crypturiformes. A bird (**Aves**) resembling the partridges superficially. Characterized by the vestigial tail and by details of anatomy resembling the ostriches. The several species live only in South America. They are like game birds in many respects.

Tinamous are remarkable for their courting habits. The females court the males and the males fill the usual role of mother in the care of eggs and young. The nest is a scantily lined depression in the ground and the eggs are unusually smooth and glossy, resembling porcelain of a bluish green or wine red color. Some species have been reported as singers of exceptional ability. (A.W.L.)

TINCTURE. An alcoholic (**ethyl alcohol**), or water and alcohol, extract of a **drug.** The proportion of drug in a tincture varies, but is usually 10%. (R.S.M., D.M.H.)

TINGUAITE. A term proposed by Rosenbusch, in 1887, for a usually **porphyritic dike** rock chemically related to **aegirine-phonolite.** Similar to sölvsbergite but containing **nepheline.** (R.M.F.)

TINPLATE. Thin sheet steel with a protective coating of pure tin applied either by a hot dipping process or by electrodeposition. The principal use is for tin cans and other containers. (R.H.H.)

TISSUE. An aggregation of **cells** of characteristic form, specialized for the performance of some limited function or functions. All cells of the multicellular animal body take part in the formation of tissues, and tissues in turn are to a great extent incorporated in organs. The cells in a tissue may be all alike, or several kinds may be present, and in some tissues the cells are supplemented by a conspicuous bulk of intercellular materials.

Tissues are divided into five classes: epithelial, muscular, nervous, connective, and vascular. (A.W.L.)

TISSUE CULTURE. A method for the study of the behavior of single **cells** and isolated bits of tissues of the multicellular body independent of their usual surroundings. Minute pieces of tissue are separated from a growing mass, as in an **embryo** or a **tumor,** and placed under aseptic conditions in a nutrient medium in small glass cells, where they can be observed microscopically. Tissues taken from birds or mammals must be kept at the body temperature of the species from which they are derived. Under proper conditions the cells of these isolated fragments continue to grow and divide, behaving to some degree as in the normal organism. Cells in cultures, however, do not grow old but persist at a uniform level of metabolism, apparently indefinitely.

This method of study has been especially important in cancer research, in studies of the physiology of senility, and in the observation of cellular phenomena in the living units.

Some progress has also been made in the attempt to cultivate, under aseptic conditions, isolated tissues and organs of plants. Very young embryos have been grown in culture media based on the liquid endosperm of the coconut (coconut milk). Excised root-tips and portions of embryos have been grown in culture media based on cane sugar solution to which are added vitamins and salts containing the chemical elements essential for plant metabolism. (A.W.L., P.A.W.)

TIT. Aves, Passeriformes. A small insect-eating bird (**Aves**) of agile habits and friendly nature, exemplified by the **chickadees**, *Penthestes*, and titmice, *Baeolophus*, of North America. They are chiefly birds of the northern hemisphere, but the group is represented in the Australian region. The colors of most species are quiet, ranging from white to black through bluish grays, relieved by a limited amount of buff or chestnut. The blue tit and the azure tit of the Old World, however, are much more brightly colored, as their names imply.

In North America some species of chickadee is to be found in almost every locality, while titmice occur in the far west and in the states east of the Mississippi.

These species are unusually fearless, and although they have no reputation as singers their cheery calls are always welcome. In the winter they visit feed boxes readily and soon become accustomed to close observation if the observer avoids sudden movements. (A.W.L.)

TITANIUM. Symbol: Ti. Atomic number: 22. Atomic weight: 47.90. Density: 4.5. Melting point: 1800° C. (Isotopes: page 290.)

Compact titanium is a while metal, when cold it is brittle and may be powdered, but at a red heat may be forged and drawn into wire. At 610° C. titanium reacts with **oxygen** to form titanium dioxide; at 800° C. with **nitrogen** to form titanium nitride; and upon heating with **chlorine** to form titanium tetrachloride. Cold, dilute **sulfuric acid** readily dissolves titanium metal to form titanous sulfate, and the hot, concentrated acid yields titanic sulfate. Discovered by Gregor in 1791.

Ferrotitanium is used as a "scavenger" for the removal of oxygen and nitrogen from molten iron and steel, and copper-titanium and manganese-titanium similarly for brass and bronze.

Titanium is used in some **stainless steels** in order to stabilize the carbon as carbide. Titanium carbide is an important **abrasive.**

Titanium occurs in practically all rocks, estimated by Clarke as ninth in abundance of the elements of the earth's crust (0.58% Ti), and in two important ores, rutile (titanium dioxide, TiO_2), and ilmenite (ferrous titanate, $FeTiO_3$). Ilmenite is a usual component of monazite sand of India, Brazil, and southeastern United States. The **magnetites** (ferroferric oxide, Fe_3O_4) of New York State contain titanium oxide.

Titanium metal is obtained by reduction of the oxide (1) with **carbon** in the **electric furnace**, or (2) with **aluminum** powder upon ignition.

Chlorides: Titanium dichloride ($TiCl_2$), black, deliquescent crystals; titanium trichloride ($TiCl_3$), violet crystals; titanium tetrachloride ($TiCl_4$), liquid, boiling point 136° C., which fumes in moist air, and used for producing white smoke screens in air.

Hydroxides: Titanous hydroxide ($Ti(OH)_3$), dark color; titanic hydroxide ($Ti(OH)_4$), white precipitate, soluble in acids, soluble in bases to form titanates.

Oxalate: Titanium potassium oxalate ($TiOC_2O_4 \cdot K_2 C_2O_4 \cdot 2H_2O$), greenish-white crystals, used as a mordant in textile and leather dyeing.

Oxides: Titanium monoxide (TiO) of slight importance; titanium sesquioxide (Ti_2O_3), black, lustrous crystals by heating the dioxide in hydrogen; titanium dioxide (TiO_2), white powder, used as a paint pigment of high covering power and stability, and in glass and ceramic ware; titanium peroxide (TiO_3), formed in solution as a yellow to red color upon the addition of **hydrogen peroxide** to titanic salt solution and an important test for titanic.

Sulfates: Titanous sulfate ($Ti_2(SO_4)_3$), violet solution by reduction of titanic solutions with **zinc** metal; titanic sulfate ($Ti(SO_4)_2$), colorless solution used as a textile mordant. (R.K.S.)

TITANIUM DIOXIDE, TiO_2. Octahedrite.

TITEL. Mammalia, Artiodactyla. An African **antelope** related to the hartebeest. Also called the bubaline antelope. It is smaller than the hartebeest and has relatively short thick horns, ringed and black in color. (A.W.L.)

TITI. Mammalia, Primates. An American **monkey** of the genus *Callithrix*. They are small animals closely related to the squirrel monkeys but differing in their more rounded heads, smaller eyes, and in the longer hair of the tail. The several species are found chiefly in Brazil but to some extent in other parts of the Amazon valley. (A.W.L.)

TITMOUSE. Tit.

TITRATION. Reactions Involving Recombination of Ions; Analysis.

TOAD. Amphibia, Anura. An animal related to the frogs and of similar form. The term is not a scientific one and its application lacks precision. As a rule the toads are better adapted for life away from the water and are found in merely moist situations, such as woods and gardens. The arboreal species are called both tree toads and tree frogs. The skin of the toads is glandular and more or less warty, a circumstance that is no doubt responsible for the false idea that handling toads will produce warts. Like the frogs, most toads have an aquatic larval stage.

The surinam toad, *Pipa pipa,* is probably the most famous member of the group because of the peculiar habits of reproduction. During the breeding season the skin of the female's back becomes thick and soft. The male imbeds each egg, as it is laid, in this soft skin, and development proceeds to completion in a resulting pouch. The young issue as small toads.

Anderson's tree toad. (*American Museum of Natural History, Photo by Mary C. Dickerson.*)

The common toads belong to the family Bufonidae. These species contrast strikingly with frogs in their warty skin and shorter and weaker hind legs. They are largely insectivorous and are of some slight value in the garden. Unlike the frogs, they deposit their eggs in long strings of jelly. (A.W.L.)

TOAD BUG. Insecta, Hemiptera. A small broad **bug** found on the muddy shores of ponds and streams. It is peculiarly like a toad in appearance and habits, capturing insects as prey and burrowing at times. Only a few species are known. (A.W.L.)

TOADSTONE. An old English term for **amygdaloidal basalts** interbedded with **carboniferous limestones** of Derbyshire, England. The rock probably takes its name from the resemblance of the amygdales to the warts or spots on the skin of a toad. (R.M.F.)

TOADSTOOLS. Agarics and Basidiomycetes.

TOBACCO. Potato Family.

TOBACCO WORM. Insecta, Lepidoptera. The **caterpillar** of a large **sphinx moth,** *Protoparce sexta,* which eats the leaves of tobacco, tomato, and other plants. The moth is gray with a row of six orange spots on each side of the abdomen and the larva is green, about 3″ long, with a stout horn near the caudal end of the body. Its sides are marked with oblique whitish lines.

Crop rotation as prescribed for the region involved, and dusting with lead arsenate (see **Lead**), are effective methods of control on large fields. The **larvae** are so easily seen when they become large enough to cause severe damage that hand picking is effective on truck crops of smaller extent.

Tobacco is attacked by other caterpillars, hence the term tobacco worm is not always restricted to this form. (A.W.L.)

TODY. Aves, Coraciiformes. A small green and red insect-eating bird (**Aves**), *Todus viridis,* of the West Indies. The legs are relatively small and the beak is long and flattened. They nest in tunnels along the banks of streams like their relatives, the kingfishers. (A.W.L.)

TOEPLER-HOLTZ MACHINE. Static Machines.

TOGGLE. The toggle is the particular class of **mechanism** commonly used to apply heavy pressure. The toggle effect which will be described is incorporated in mechanisms known variously as toggle mechanisms, toggle joints, etc. The principle of the toggle is the straightening out of a flat angle between two joined

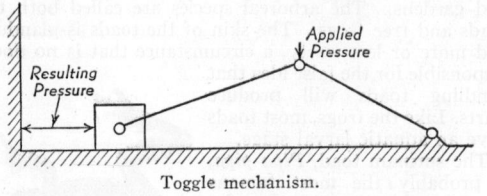

Toggle mechanism.

members, one of which is affixed at its outer end, the other connected at its outer end to an anvil, jaw, hammer, die, or whatever is used at the pressure face. The diagram shows how two links, one pivoted and the other connected to a sliding block, have a common joint at which pressure is applied. Due to the flat angle between the two links, a comparatively small force effectively applied in a direction tending to straighten them out still more is capable of overcoming a large resistance at the moving end. The force is applied at the joint by means of a threaded screw, a cam, a crank and connecting rod, or any other suitable means. Examples of the use of the toggle are found in stamping machines, presses, crushers, etc.

A toggle bolt is one having a short bar pivoted to the nut so that the bolt, with nut not attached, can be pushed through a hole in a wall, then reversed. The bar, which is laid parallel to the bolt when entering the hole, swings to a perpendicular position, thus retaining the bolt in the hole. (F.T.M.)

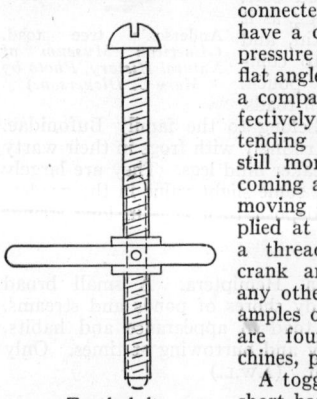

Toggle bolt.

TOGGLE JOINT. Machines; Toggle.

TOLERANCE. Interchangeable Manufacturing; Fit.

TOLLEN'S SOLUTION. Formaldehyde.

TOLUENE. Toluene, "toluol," methyl **benzene,** phenyl methane $\left(C_6H_5 \cdot CH_3 \text{ or } \right)$ is a colorless, odorous **hydrocarbon,** boiling point 110.5° C., insoluble in water, miscible in all proportions with alcohols, ether, chloroform, and many organic liquids, dissolves **iodine, sulfur,** oils, fats, resins, **phosgene,** burns when ignited with a smoky flame. Toluene reacts (1) with **chlorine,** to form substitution products (one-half of the chlorine forms **hydrogen chloride**), such as (a) ortho- and para-chlorotoluenes ($(1)CH_3 \cdot C_6H_4 \cdot Cl(2)$ and $(1)CH_3 \cdot C_6H_4 \cdot Cl(4)$) at ordinary temperatures, moist, and in the presence of a catalyzer (e.g., **iodine, iron**). Both products are colorless, pleasant-smelling liquids, heavier than and insoluble in water, unchanged by sodium hydroxide solution, and (b) benzyl chloride ($C_6H_5CH_2Cl$), benzal chloride ($C_6H_5CHCl_2$), benzotrichloride ($C_6H_5CCl_3$), at boiling temperature, dry, and in the absence of a catalyzer. All three products are colorless, heavy liquids separable by fractional distillation, of pungent and irritating odors, insoluble in water, reactive with sodium hydroxide solution to form benzyl alcohol ($C_6H_5CH_2OH$), benzaldehyde (C_6H_5CHO), sodium benzoate (C_6H_5COONa), respectively, (2) with oxidizing agents, e.g., **sodium** dichromate plus **sulfuric acid,** to form benzoic acid (C_6H_5COOH). Toluene, benzyl chloride, benzotrichloride, benzyl alcohol, benzaldehyde form benzoic acid upon oxidation, (3) with concentrated **nitric acid,** to form ortho- and para-nitrotoluene ($(1)CH_3 \cdot C_6H_4 \cdot NO_2(2)$ and $(1)CH_3 \cdot C_6H_4 \cdot NO_2(4)$), dinitrotoluene ($(1)CH_3 \cdot C_6H_3(NO_2)_2(2,4)$, trinitrotoluene ("T.N.T.") ($(1)CH_3 \cdot C_6H_2(NO_2)_3(2,4,6)$), (4) with concentrated **sulfuric acid,** to form ortho- and para-toluene sulfonic acids ($(1)CH_3 \cdot C_6H_4 \cdot SO_3H(2)$ and $(1)CH_3 \cdot C_6H_4 \cdot SO_3H(4)$). From the former, **saccharin** $\left(C_6H_4 {<}^{CO}_{SO_2}{>}NH \right)$, which is some 500 times sweeter than sucrose and not a sugar, is made, and from the latter, chloramine T ($(1)CH_3 \cdot C_6H_4 \cdot SO_2 \cdot NClNa(4)$), an important antiseptic. Toluene is obtained from coal tar and coal gas, as described under **benzene.**

Toluene is used (1) as a solvent for various substances, (2) in the manufacture of nitrotoluenes for toluidines and explosives such as T.N.T., of sulfonic acids for saccharin and chloramine T, of benzoic acid; and of other organic chemicals, especially dyes and perfumes. (R.K.S.)

TOMATO. Potato Family.

TOMATO WORM. Insecta, Lepidoptera. A large **caterpillar,** *Protoparce quinquemaculata,* similar to the tobacco worm in appearance and habits and belonging to a closely related species. The moths of the two species are similar but are easily distinguished by comparison. (A.W.L.)

TOMBOLO. A type of sand bar which connects one island with another, or an island to the mainland. (R.M.F.)

TONALITE. Diorite.

TONE CONTROL. This is the control for regulating the **frequency response** of an audio **amplifier.** In its most common form as found in most radio **receivers** it consists of a condenser in series with a variable resistance shunted across the circuit at some point. Since such a combination passes high frequencies more easily than low values, the highs will be attenuated, the degree being determined by the setting of the variable resistance. Thus by varying the resistance more or less attenuation may be given the high frequencies and the effect is as if the bass response were being varied. In more elaborate tone controls separate bass and treble controls are provided, thus allowing a balanced control of the response. (L.R.Q.)

TONGUE. For the use of this term in geology, see **Apophysis.** In anatomy an organ associated with the floor of the oral cavity, usually projecting or protrusible. The true tongue is a vertebrate structure, occurring in all classes above the fishes. It is made up largely of voluntary muscle fibers, so distributed that it can be protruded and withdrawn and swung in every possible direction. It arises embryonically from the floor of the anterior part of the pharynx, extending forward into the oral cavity.

The tongue is used in man and other species for the manipulation of foods in chewing, and in some of the amphibians, reptiles and birds it aids in capturing food, usually by adhesion. In man it is an important organ of speech, aiding in the modulation of sounds produced by the larynx.

The word is sometimes applied to the **radula** of the snails and to parts of the insect mouth, but only from superficial similarity with the true tongue of the vertebrates. (A.W.L.)

TONGUE SHELL. A common name for animals of the phylum **Brachiopoda,** also called lamp shells. (A.W.L.)

TONKA. Bean.

TONSILLITIS. An acute infection of the tonsils and surrounding throat tissues, usually caused by the hemolytic streptococcus. Tonsillitis may be a mild disease of short duration or a severe one with high fever and constitutional symptoms. In follicular tonsillitis both tonsils may be red, tremendously swollen and covered with patches of pus forming a thick exudate. The throat is sore and exceedingly painful, especially on swallowing. Either mild or severe tonsillitis may be followed immediately by scarlet fever, or after a latent period by **rheumatic fever** or **nephritis.** Faucial **diphtheria** or diphtheria involving chiefly the tonsillar areas must be distinguished from streptococcus tonsillitis.

Treatment of the severe type of infection consists of irrigation with warm saline or glucose, and in some instances, **sulfonamides** or penicillin. (D.M.H.)

TOOL. An appliance, device, or machine used by a worker in pursuing his art. In the metal-working trades, tools include hand tools such as **hammers, saws, chisels,** etc., tools for gauging and **measurement,** and power and machine tools. (H.C.H.)

TOOL STEELS. Steels used in the manufacture of all types of tools are many, diversified in composition and properties and made under strict quality controls to meet rigid specifications and withstand rigorous service conditions. These steels, comprising less than 1% of the tonnage of all steels produced, are now made by the basic electric melting process to the almost complete exclusion of the crucible process formerly used. They are, however, still made in comparatively small lots and orders for one size or one composition are often of only a few hundred lbs. in weight.

Tool steels require, in almost all instances, a potential of high hardness and strength and, therefore, generally possess a carbon content in excess of .50%. The plain carbon tool steels, still in extremely wide use, contain little or no alloying elements as the name implies. All other tool steels contain one or more alloying elements (manganese, silicon, chromium, tungsten, molybdenum, vanadium, nickel) in quantities from less than 1% total alloy content in certain types to over 45% total alloy content in other types. The composition of typical tool steel types is shown in the table on p. 1480.

Carbon tool steels may be used for innumerable purposes covering the entire range of machine and hand tool operations. They are adapted to meet the requirements of certain applications involving repeated shock yet requiring higher than average wear resistance because of their peculiar quality of hardening to a limited depth, readily controllable. Internal stresses remaining in the steel are compressive at the surface in this case and result in greatly increased fatigue strength and service life. Carbon tool steels are manufactured with .50–1.40% carbon, the carbon content selected being high where wear resistance is of prime importance and being low where toughness is a major consideration.

Chromium and chromium-vanadium tool steels containing .75–2.50% chromium have the ability to harden to a greater depth than carbon tool steels and are, therefore, applicable for large size tools and dies (see **Hardenability of Steel**). A large range of applications from shock-resisting tools to cutting tools and dies, small- or intermediate-size rolls, bearings and bearing races is encountered with these steels based on their exceptional strength and toughness, their variable hardness and hardenability. As with the straight carbon tool steels the service requirements dictate the carbon content.

The low alloy chromium-nickel-molybdenum die steels are essentially high-carbon varieties of the Engineering S.A.E. alloy steels. The higher carbon increases the depth of hardening and the maximum hardness obtainable in large sizes. These oil-hardening steels when hardened to 58 to 62 Rockwell "C" combine an excellent degree of toughness, strength and resistance to wear and find use in dies and machine parts which require a good combination of hardness and toughness.

The *manganese oil-hardening steels* are among the most important of the tool steel types because they can be used for many types of tool and die work. This group of steels exhibits low dimensional changes in hardening, a high as-quenched hardness combined with a great depth of hardening, and a freedom from cracking when intricate sections are quenched. These steels do not possess any **red hardness** properties which would enable cutting at high speeds or which would make them applicable for hot-working operations. The low distortion properties enable this group of steels to find considerable application for tools which cannot be ground or machined after heat-treating.

Silicon tool steels originally used for springs are valuable primarily because of their high resistance to shock, very high fatigue-resistance and good wearing ability. They can be treated to a fairly high degree of hardness, making them applicable to many punch and die applications, as well as to shock-resisting tools. They are well adapted for chisels, punches, rivet sets, coal cutters and for shear blade where good wearing qualities and a comparatively low-cost steel are required. They are especially useful as shanks for tipped carbide or high-speed steel tools for which purpose they have largely supplanted many other types in the last several years.

Tungsten-finishing steels are generally used in cases where extreme wear resistance or abrasion resistance is necessary and where the ability to retain a keen cutting edge is desired. The wear resistance of tungsten-finishing steels is 4 to 10 times that of plain carbon steels at maximum hardness. In addition to applications for cutting tools the tungsten-finishing steels are extensively used for small tube and wire drawing dies as well as for sizing plugs and gauges and forming and piercing punches.

A recent addition to the steels used for tools is the group of materials known as *graphitic steels* which contain a relatively high carbon content of which a portion is graphitized during heat treatment. The remaining ungraphitized carbon is generally in an amount equal to the eutectoid content of the steel. Such a structure results in a hardenable free-machining steel to which can be added various alloy additions to change the hardening characteristics in accordance with the need of different applications.

Tungsten chisel steels are used for much the same purpose as silicon tool steels, namely, shock-resisting tools, but find additional use in tools for hot-working purposes. For the latter purposes they are used when the dies for hot-working are subject to considerable shock, such as in heading dies for square-headed bolts, tools with deep recesses or grooves, and hot shear blades having notched edges. In certain instances the tungsten chisel steels will be carburized to produce a tool having a very high toughness and at the same time a surface near file hardness.

The *air-hardening die steels* containing 1% carbon and 5% chromium and 1% molybdenum compete with both the manganese oil-hardening steels and the high-carbon high-chromium steels as cold-work die materials. Their wear-resistance is midway between these two types of competing steels but their toughness is greater than either. They possess a very high hardenability and can be hardened in air with a distortion amounting to only about ¼ that found in oil-quenched manganese oil-hardening steels. They are, therefore, used where fair abrasion-resistance must be combined with exceptional toughness, such as in dies for blanking, forming, drawing and thread rolling and in rolls and punches. They also have a limited application for certain types of shear blades.

High-carbon high-chromium steels first developed commercially as a substitute for high-speed steel cutting tools during World War I have received very little use for this purpose because of their exceptional brittleness but their high wear-resistance combined with remarkable non-deforming qualities make them extremely useful for cold-working dies. They may be obtained in either the air- or the oil-hardening varieties, the latter having a slightly greater wear-resistance than the former. The wear-resistance of the high-carbon high-chromium steels is approximately 8 times that of carbon tool steels, and

by virtue of their high chromium content resist oxidation at high temperatures and have an appreciable resistance to staining when hardened and polished. Properly designed tools made from these steels can punch steel plate up to ¼″ in thickness and they also are used for shear blades, slitting cutters, cold-forming rolls, gauges and drawing dies.

Die steels used for hot-working operations are readily divided into two classes: (1) those containing high chromium contents, and (2) those containing high tungsten contents. These steels must possess a high resistance to deformation at the temperature of working and should resist fire or heat checking. The 1% carbon–4% chromium steels have excellent wear-resistance and sufficient rigidity for applications where hot-working is performed mainly by compressive action, finding considerable use for gripper dies in bolt and rivet plants. The most widely used of all die steels for general hot-working operations are the *low-carbon 5%-chromium steels* with molybdenum or tungsten additions which were originally developed for the die casting of aluminum alloys. These steels are air-hardening and in addition to their die casting die applications are used for forging dies, punches, piercers and mandrels for hot work, shear blades for hot work and all types of hot-working dies which involve shock. The *tungsten hot-work steels,* sometimes called semi-high-speed steels, containing low percentages of carbon and 9–15% tungsten plus additions of other alloying elements, will withstand the highest working temperatures without failure and also possess better wearing qualities than any of the other hot-work die steels generally in use. Their main disadvantage lies in the fact that they will not withstand shock and cannot generally be rapidly cooled in operation with water without danger of breakage.

High-speed steels, regardless of type, have a striking similarity in their physical make-up. Practically all of the high-speed steels possess a very high alloy content combined with sufficient carbon to permit hardening to file hardness; they are all hardened from temperatures just under their fusion point; they harden so deeply that almost any size encountered commercially will have a uniform hardness throughout the cross-section; and all

Type	C	Si	Mn	W	Cr	V	Mo	Ni	Co
Carbon tool steels	.60–1.40	.25	.25			Opt.			
Chromium tool steels	.50–1.10	.25	.25		.75–2.50	Opt.			
Low alloy chromium-nickel-molybdenum die steels	.70	.25	.50		1.00–2.00			.20–.50	1.00–2.00	
Manganese oil-hardening steels	.95	.25	1.20	0.50	.50	.20			
Silicon tool steels	.50	2.00	.80		Opt.	Opt.	Opt.		
Tungsten-finishing steels	1.30	.25	.25	4.00	Opt.	Opt.	Opt.		
Graphitic steels	1.50	1.00	.30	Opt.	Opt.		Opt.	Opt.	
Tungsten chisel steels	.50	.25	.25	1.00–2.00	1.00	Opt.			
Air-hardening die steels	1.00	.30	.50	5.00		Opt.	1.00		
High-carbon high-chromium steels, oil-hardening type	2.25	.25	.30	Opt.	12.00	Opt.			
High-carbon high-chromium steels, air-hardening type	1.00–1.50	.25	.30		12.00	Opt.	.80		Opt.
Chromium hot-work steels (1)	1.00	.25	.25		4.00	Opt.	Opt.		
(2)	.35	.90	.25	1.05	5.00	Opt.	1.00		
Tungsten hot-work steels (1)	.35	.25	.25	11.00	3.00	Opt.			
(2)	.30–.45			15.00	3.00–4.00	Opt.	Opt.	Opt.	
High-speed steels (1)	.50–.80	.25	.25	18.00	4.00	1.00			
(2)	.80	.25	.25	18.00	4.00	2.00	1.00		
(3)	.75	.25	.25	18.00	4.00	1.00–2.00			5.00–8.00
(4)	.75	.25	.25	20.00	4.00	2.00			12.00
(5)	.80	.25	.25	Opt.	4.00	1.00–2.00	8.50		
(6)	.80	.25	.25	6.00	4.00	2.00	5.00		
(7)	1.25	.25	.25	6.00	4.00	4.00	5.00		
(8)	.80	.25	.25	6.00	4.00	2.00	5.00		8.00

high-speed steels can be hardened to near maximum hardness by cooling in still air. But perhaps the most important property common to all high-speed steels is the ability to retain a considerable hardness at elevated temperatures.

The basic type of high-speed steel which has been developed over a period of many years and the one which has received the most study contains 18% tungsten, 4% chromium, 1% vanadium and .50–.80% carbon. It is used for both cutting and die applications. The war emergency period caused the commercial development of molybdenum high-speed steels, the most popular of which are the 6% tungsten-5% molybdenum type and the 8% molybdenum types with vanadium or tungsten in small amounts. To each of these types cobalt in amounts up to 12% may be added to increase the hot hardness and cutting efficiency in certain instances.

Actually no one composition of high-speed steel can always meet all of the requirements. The standard tungsten and the standard molybdenum types can be more generally applied than any others. These steels have good toughness and cutting ability and the tungsten type can be very readily fabricated and heat treated. The addition of vanadium in amounts over 1.50% to the tungsten and molybdenum types offers an advantage of greater wear-resistance and hot-hardness and these steels are, therefore, more suited for fine and finishing cuts on hard and soft materials. The cobalt types are slightly more brittle than the non-cobalt types but will give better performance on cutting hard materials at high speeds.

High-speed steels of any type have a hardness at 1100° F. of between 450 and 520 Brinell and have a hardness at 1200° F. of between 300 and 400 Brinell. At 1100° F. they are superior in hot-hardness to the cast cutting materials but at or above 1200° F. their hardness is lower than that of the cast cutting materials. (GEORGE A. ROBERTS, VANADIUM ALLOY STEEL CO.)

TOOTH. 1. A hard structure projecting from the wall of the mouth or the anterior part of the alimentary tract (**digestive system**) and used for grasping and breaking up food. Teeth vary from the chitinous projections on the radula of **mollusks** to the complex structures of the vertebrates. Chitinous teeth of **annelid** worms, located on the walls of the **pharynx,** are also known as jaws.

Vertebrate teeth are formed of two layers of hard materials over a living papilla, the pulp, which contains blood vessels and nerves. They form in the **embryo** from ingrowths of the outermost layer of the body, the ectoderm, associated with masses differentiated in the middle layer, the mesoderm. The mesoderm forms the dental papilla and around its outer surface, except at the end that is to remain connected with the body, lays down a layer of dentine. This material is similar to bone in composition but not in minute structure. It is harder, contains less organic matter than bone, and is the ivory of commerce. The ectodermal cap over the dental papilla develops into an enamel organ and deposits a layer of enamel on the dentine. Enamel is made up of minute prisms and is the hardest substance produced by the animal body, containing only 2–4% organic matter. When the tooth takes its place in the jaw in the process of eruption the enamel organ is destroyed, hence no more of this substance can be produced. Dentine may be deposited later in life, however, encroaching on the pulp cavity.

The teeth of sharks are the most primitive vertebrate teeth. They are formed like the placoid scales of the body, with a principal point and sometimes smaller supplementary points, and are superficially attached in rows on the jaws. Both scales and teeth contain dentine covered with enamel.

In other fishes (**Pisces**) and **amphibians** teeth are distributed in various parts of the oral cavity and are associated with the bones of the jaws and skull. In reptiles and **mammals** they are limited to the jaws and are seated in sockets (alveoli) in the bones by means of a third hard substance, cementum, deposited between the dentine of the roots and the bone of the jaw.

Primitive teeth are simple conical structures. In mammals this type persists only in the canines. The other teeth are sharp-edged incisors, broad molars with projecting cusps on the apposed surfaces, and premolars of intermediate form. The premolars of man are also called bicuspids. An assemblage of teeth of these kinds, variously specialized for cutting, tearing and holding, and grinding and crushing, is known as a heterodont dentition, and is the fundamental plan of **dentition** in the mammals. The various forms of teeth are the basis of specialization in different groups of mammals for the use of limited types of food. Thus species that depend on plant tissues have greater need of crushing teeth while carnivorous forms need cutting teeth for their tougher food. The molars of the former are broad and are so folded that the worn apposed (occlusal) surfaces show alternating bands of dentine, enamel, and cementum in patterns characteristic of the kind of animal. In the elephants these materials form transverse ridges. The incisors of grazing animals are also flattened, forming clipping, rather than cutting, teeth. The canines are reduced or lacking. In carnivores the incisors are greatly reduced, the canines are highly developed, and the molars are sharp-edged. Molars of the two jaws work together like the blades of shears. Some mammals, notably certain **anteaters,** have lost all trace of teeth.

The tusks of various animals are greatly elongated teeth projecting from the jaws for use in fighting and digging. They are usually devoid of enamel. Those of the elephants are upper incisors and in early growth are provided with enamel tips. Those of **walruses** and **swine** are canines. Single tusks of **elephants** weighing almost 200 lbs. have been recorded but the average is less than 100. These tusks provide the finest ivory.

Some teeth grow constantly as they are worn away, while others attain a fixed form within a short period and still others undergo a partial compensation for wear. The incisors of **rodents** are of the first type. These chisel-like teeth wear away in gnawing but are constantly renewed by growth at their bases to preserve a uniform length. The molars and premolars of horses are elevated in the jaws by lengthening of the roots, but the amount available for use during the life of the animal is regulated by the height of the crown at the initial formation of the tooth. Human teeth assume a fixed form. The only compensation for wear is the partial filling in of the pulp cavity by the deposition of dentine.

A compensation for breakage and wear is also found in the replacements that occur in lower vertebrates. Mammals normally have no more than two sets, milk and permanent, and when once the permanent teeth of adult life have assumed their functional positions in the jaws, loss through accident is permanent. Rarely a third set of teeth develops.

The word tooth also denotes:

2. The interlocking projections at the hinged edges of the valves of the bivalve molluscan shell.

3. Tooth-like projections on a hard structure or on the surface of the body of an animal. (A.W.L.)

TOP MINNOW. Pisces, Teleostei. Any small fish (**Pisces**) of several species, belonging to several genera, which live in quiet water of streams and in pools, swamps, and ditches, feeding on mosquito larvae and other small insects. Some are also called mosquito fish and one species is known as the rain-water fish. These fishes vary from less than 1″ to 3″ in length. They are most abundant in the southern states and southward into Central America, living in fresh and brackish water.

For introduction into lily pools to keep down mosquitoes these fishes are the most desirable. (A.W.L.)

TOPAZ. The mineral topaz is a **silicate** of **aluminum** and **fluorine** corresponding to the formula $(AlF)_2SiO_4$. It is **orthorhombic** and its crystals are mostly prismatic terminated by pyramidal and other faces, the **basal pinacoid** being often present. Massive varieties are known. It has an easy and perfect basal **cleavage** hence for this reason gems or fine specimens should be handled with care to avoid developing cleavage flaws. The fracture is conchoidal to uneven; hardness, 8; specific gravity, 3.4–3.6; luster, vitreous; color, of typical topaz, wine or straw-yellow but may be colorless, white, gray, green, blue or reddish-yellow; transparent to translucent. When heated, yellow topaz often becomes a reddish pink. Topaz is found associated with the more **acid rocks** of the **granite** and **rhyolite** type and may occur with **fluorite** and **cassiterite**. Topaz comes from many foreign localities, a few of which are: Russia in the Urals and the Ilmen Mountains; Czechoslovakia, Saxony, Norway, Sweden, Japan, Brazil and Mexico. In the United States topaz has been found in Oxford County, Maine; Carroll County, New Hampshire; Fairfield County, Connecticut; El Paso and Chaffee Counties, Colorado; and in Texas, Utah and California. The name topaz is derived from the Greek meaning to seek, which was the name of an island in the Red Sea that was difficult to find from which a yellow stone, now believed to be a yellowish-olivine, was obtained in ancient times. In the Middle Ages any yellow stone was called topaz, but now the name is properly applied only to the species here described. (E.S.C.S.)

TOPAZOLITE. Garnet.

TOPE. Pisces, Elasmobranchii. *Galeus.* A small **shark** found in all temperate and tropical seas. More commonly called the dog shark or dogfish in North America, and sometimes the hound shark. It attains a length of 6′ but is usually much smaller. It is used as food to some extent. (A.W.L.)

TOPIOLITE, $FeTa_2O_6$. Mossite.

TOPOGRAPHIC MAP. Map; Topographical Mapping.

TOPOGRAPHIC SURVEYING. Topographical Mapping.

TOPOGRAPHICAL MAPPING. Topographical mapping consists of representing on a map the physical features of a given section of land, by showing thereon **contours** or hachures representing the elevation, and noting various other physical features by conventionalized signs. Thus trees, streams, marshes, and roads may be part of a topographic survey. The method of topographic surveying and mapping of large areas consists of establishing points for horizontal and vertical control, and surveying the adjacent area from these points by the use of direct leveling or **stadia.** (F.T.M.)

TOPOGRAPHY. Topography is a term used in a broad sense to include all details, both natural and artificial, which are required for a topographic map. The details cover such points as the disposition of the parts of the earth's surface; namely, hills, valleys, plains, plateaus, etc., and the location of waterways, bridges, highways, railroads, cultivated fields, forests, buildings, etc. The delineation of these features is also known as topography. (C.W.C.)

TOPOLOGY. Topology may be described as a study of those properties of a geometrical figure unaffected by any deformation not involving tearing or joining. (L.L.S.)

TOPSOIL. Soil.

TORNADOES. Some thunderstorms, particularly the line-squall type, occasionally develop a violent whirl of air which extends down from the base of the cloud and touches the earth. It often draws up into the cloud again and may strike some distance away or never reappear. Very low pressure prevails inside a tornado because of its great vorticity. When a tornado passes over a building, therefore, the building virtually blows up because of the low pressure surrounding it. Tornadoes in the U. S. are most common to the South and Middle West. (P.E.K.)

TORQUE. A torque, often called a torsional or twisting moment, is a **moment** which tends to twist a body about an axis of rotation. In the accompanying illustration a **shaft** of diameter d is connected to a rigid

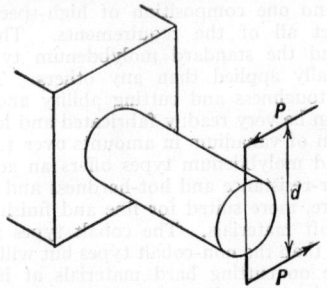

support. The forces P tend to rotate the shaft in a counter-clockwise direction causing a torque $(T) = Pd$.

For a body in free rotation the torque $= I\alpha$, in which I is the **moment of inertia** of the mass of the body and α is the angular acceleration. (See **Statics.**) (C.W.C.)

TORSION. Elasticity.

TORSION BALANCE. The torsion balance is an instrument for measuring very feeble forces of attraction or repulsion. It has played an important rôle in the demonstration of **Coulomb's laws** and in measuring the **gravitation constant** and **radiation pressure**. It consists essentially of a light horizontal rod suspended by a slender elastic wire or fiber and carrying at each end a small ball, vane, or other object upon which the unknown force is to act tangentially. The resulting torsional deflection is measured by an **optical lever**, the small mirror of which is attached at the base of the suspension. To calculate the corresponding torque (and hence the force), the "torsion coefficient" of the suspension must have been previously determined. This is the torque required to twist the suspension through an angle of 1 radian, and is conveniently obtained by using the same suspension as the support for a **torsion pendulum** of known **moment of inertia**, and measuring the period. In the Cavendish method for determining the gravitation constant, small balls of gold or platinum are used and the torque applied by the attractions of much more massive balls or cylinders of lead in the same horizontal plane. The Eötvös balance (see **Gravitation and Gravity**) is a highly specialized form of the torsion balance, used in gravity measurements. (L.D.W.)

TORSION PENDULUM. The torsion pendulum consists of a body suspended by a fine wire or elastic fiber in such a manner that it will execute rotational oscillations as the suspension twists and untwists. If I is the **moment of inertia** of the body with respect to the axis of oscillation, and if K is the **torsion coefficient** of the fiber (torque required to twist it through an angle of one radian), the period of oscillation is given by the simple formula

$$T = 2\pi\sqrt{\frac{I}{K}} \qquad (1)$$

Both I and K may have to be determined by experiment in actual laboratory practice. This is easily done by measuring the period T and then adding to the suspended body another of known moment of inertia I', giving a new period of oscillation T', which is also measured. From (1),

$$T' = 2\pi \sqrt{\frac{I + I'}{K}} \; ; \qquad (2)$$

and the solution of the two equations now gives $K = 4\pi^2 I'/(T'^2 - T^2)$, $I = T^2 I'/(T'^2 - T^2)$.

The oscillating balance wheel of a watch is, in effect, a torsion pendulum, with the suspending fiber replaced by hairspring and pivots. The watch is regulated, first roughly by adjusting I (for which purpose screws are set radially into the rim of the wheel), then accurately by varying the free length of the hairspring, and hence its torsion coefficient K. (L.D.W.)

TORSIONAL LOAD. Load.

TORSIONAL STRESS.
The **shearing stress** which occurs at any point in a body as the result of an applied **torque** or **torsional load** is called a torsional stress. If the body is a circular shaft, the stress is found from the formula:

$$s_s = \frac{Tc}{J}$$

in which s_s = Required **unit stress**.
T = Torque.
c = Radial distance from the center of the shaft to the point.
J = **Polar moment of inertia** of the cross-sectional area.

This formula is based on the assumption that a plane section before twisting remains a plane after the torque is applied; also that the radii remain straight. The theory of **elasticity** shows that these assumptions are true for all sections except those adjacent to the applied torque and the supports.

The cross-section of a non-circular body becomes decidedly warped after twisting, so the stress must be obtained by formulae developed from the theory of elasticity, by experimental means or by approximate formulae which are given in textbooks on advanced strength of materials. (C.W.C.)

TORSK.
Pisces, Teleostei. A marine fish (**Pisces**), *Brosme brosme*, found in northern waters. Reported as abundant near the Shetland and Orkney islands. It is related to the cod. (A.W.L.)

TORTOISE.
Reptilia, Chelonia. A **turtle**. There is no accurate scientific distinction between the two terms, although some species are called turtles and some tortoises. This is true even of members of the same family.

In North America the term turtle is commonly used, although a few species, including the gopher turtle of the south, belong to the genus of land tortoises. This group is known especially for the giant land tortoises once so abundant on the Galapagos Islands. These animals attain a weight of 500 lbs. and live a century or more.

One species of northeastern South America is noteworthy for the angular prominences on the shell and head and for peculiar projections fringing the long neck. It is called the matamata. (A.W.L.)

TORTOISE SHELL.
The mottled horny plates of the shell of a marine **turtle**, *Eretmochelys imbricata*. This species is found in warm seas, extending as far north as Massachusetts occasionally. It is known as the hawksbill or tortoise-shell turtle.

Tortoise shell was once highly valued for ornamental

use, as in toilet articles and the handles of pocket knives, but it has been largely replaced by synthetic materials. (A.W.L.)

TORTOISES. Fossil Reptiles.

TORUS.
In mathematics, a torus (or anchor ring) is the figure generated by revolving a circle about an axis in its plane but not intersecting it.

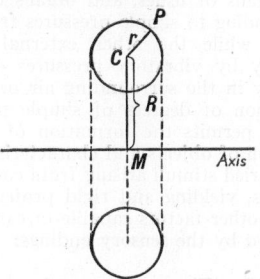

Generation of torus.

The volume of a torus is $2\pi^2 R r^2$ and the area of its surface is $4\pi^2 R r$, where r is the radius of the generating circle and R is the distance of the center of the generating circle from the axis of revolution.

In zoology, a torus is a pad at the tip of a **digit** of the **vertebrates**. These structures form the tough resilient bearing surfaces of the appendages of most vertebrates, and are nicely illustrated by the pads on which the dog and cat walk. (L.L.S., A.W.L.)

TOTAL DERIVATIVE. Partial Derivatives.

TOTAL DIFFERENTIAL. Differentials.

TOTAL REFLECTION.
Total reflection occurs when light is reflected in the more refractive of two media from the interface between them, at any angle of incidence exceeding the so-called "critical angle." This is the angle whose sine is equal to the relative **refractive index** for light attempting to pass from the more to the less refractive medium. (See **Refraction**.) Thus, for water against air, for which the index is 0.75, the critical angle is 48° 35'; so that if light is incident in the water at an angle of 50°, it will be totally reflected. This phenomenon may be easily observed by holding a glass of water with the surface slightly above the eye and inserting the finger or a matchstick into the water. The under side of the surface is then seen to reflect this object like a perfect mirror, and gives the impression of a mercury surface inverted.

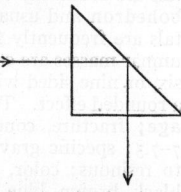

Total reflection in a right-angled prism.

The principle is often utilized in reflecting prisms such as those used in the prism **binocular**. Here the refractive index is about 0.6 and the critical angle, therefore, only 36° 52', while the angle of incidence is commonly 45° (see figure). **Refractometers** utilize this principle. For some metals whose refractive index for light is less than unity, total reflection occurs on the outside of the surface. The same is true of glass for x-rays. (See **Mirage**.) (L.D.W.)

TOUCAN.
Aves, Coraciiformes. A bird (**Aves**) of South and Central America, characterized by the enormous beak. Most species are brilliantly colored and of large size. The beak, several times as large as the head, is of very light structure, with serrated edges used in cutting up fruit. It is covered with a thin horny shell and contains a fine bony reticulum. Toucans are also remarkable for their peculiar habit of tossing bits of

food into the air and catching them in the mouth to be swallowed. (A.W.L.)

TOUCH. A special sense that responds to contacts. The tactile organs through which the stimuli of contact are received are sensitive to varying pressures, sometimes of exceedingly slight degree. They are related to other sense organs which respond to pressure, including internal receptors, chordotonal organs of insects, lateral line organs of fishes, and organs of hearing, but differ in responding to simple pressures from the outside of the body, while the other external receptors are stimulated only by vibratory pressures of varying frequency, usually in the surrounding air or water.

The perception of degrees of simple pressure by organs of touch permits the formation of mental images of conformation of objects and characteristics of surfaces through the varied stimuli arising from contact with high and low points, yielding and rigid projections, adhesive materials, and other factors capable of carrying the pressure transmitted by the sensory endings. (A.W.L.)

TOUGHNESS. The property of a material which enables it to absorb energy when stressed beyond the elastic limit is called toughness.

The modulus of toughness is the total energy per unit of volume which must be absorbed in order to produce failure by fracturing (breaking). It is equal to the area between the stress-strain curve and the axis of unit deformation.

A ductile (see Ductility) material is tougher than a brittle material; therefore the former should be used for members which are subjected to impact or energy loads since it is impossible to predict an exact value for these loads. (c.w.c.)

TOURACO. Plantain-Eater.

TOURMALINE. The mineral tourmaline is a complex silicate of aluminum and boron, but because of isomorphous replacements this mineral varies widely in chemical composition, iron, magnesium, and lithium entering into combination to a greater or less extent with the aluminum and boron. Tourmaline belongs to the hexagonal system, its crystals are usually prismatic, tending to be long and slender, often acicular. The crystals are ordinarily terminated with three faces of a rhombohedron and usually hemimorphic. The smaller crystals are frequently found in radial arrangement, and columnar masses are common. The prisms are usually three, six, or nine sided with heavy vertical striations producing a rounded effect. Tourmaline is essentially without cleavage; fracture, conchoidal to uneven; brittle; hardness, 7–7.5; specific gravity, 2.9–3.2; luster, vitreous inclining to resinous; color, in common tourmaline black, bluish-black, brown, blue, green, red or pink, and in the transparent varieties colorless (rare), various shades of rose and pink, greens, blues and browns. The color arrangement in tourmaline is of considerable interest; bi-colored crystals are common and may be green at one end and pink at the other, or green on the outside, and pink within, which, in the case of transparent or translucent crystals, is very attractive. The opaque black tourmaline is called schorl, which term was applied to all tourmaline until 1703 when tourmaline was introduced, it being a corruption of the Ceylonese word, *turamali*. The origin of the word schorl is not known, but is perhaps Scandinavian. The rose or pink tourmalines are called rubellite (from ruby); the dark blue, indicolite (from Indigo); the lighter, Brazilian sapphire; the green, Brazilian emerald; the brown, dravite (from the Drave district of Carinthia); and the colorless, achroit from the Greek meaning without color. Small tourmalines are found in granites and some gneisses. Due, no doubt, to the mineralizing action of magmatic vapors, tourmaline is found particularly well developed in pegmatites, and

as a contact **metamorphic** mineral. A few of the important foreign localities are: The Ural Mountains, Bohemia, Saxony, the Island of Elba, Norway; Devonshire and Cornwall, England; Greenland, Madagascar, and Brazil. In the United States in Oxford and Androscoggin Counties, Maine; Grafton and Sullivan Counties, New Hampshire; Hampshire County, Massachusetts; Haddam and Fairfield Counties, Connecticut; St. Lawrence County, New York; Sussex County, New Jersey; Delaware County, Pennsylvania; and San Diego County, California. (E.S.C.S.)

TOWHEE. Aves, Passeriformes. A North American bird (Aves) related to the finches and sparrows. The common eastern species, *Pipilo erythrophthalmus,* is a black and white bird with red-brown sides. It is seen chiefly on the ground and nests chiefly beneath tangled thickets. From its call this bird is also known as the chewink. Four other species are found in the west, three congeneric with the eastern towhee and a fourth, the green-tailed towhee, *Oberholseria chlorura,* is placed in a related genus. (A.W.L.)

TOXEMIA. The varying degrees of prostration which result from the presence of specific toxins liberated by the bacteria associated with such diseases as scarlet fever, diphtheria, tetanus, etc. The term is also used loosely and incorrectly to describe various vague symptoms in any severe illnesses. (D.M.H.)

TOXEMIAS OF PREGNANCY. A variety of complications occurring during the course of pregnancy. The most important are pernicious vomiting, acute yellow atrophy of the liver, pre-eclampsia, eclampsia, and several forms of nephritis.

Pernicious vomiting occurs during the first trimester, and is an exaggeration of the normal nausea and vomiting, or "morning sickness" of pregnancy. Its cause is unknown. Although it occurs more often in neurotic individuals, it has an organic basis which is little understood. In severe cases, unless the pregnancy is terminated, the patient may die. At autopsy, the liver shows toxic necrotic changes.

Acute yellow atrophy (see Hepatitis) is a rare and calamitous disease, which occurs in the last half of pregnancy. It is characterized by sudden and extensive necrosis of the liver so that the organ shrinks in a few days to half its normal size. There is deep jaundice and prostration. Recovery is unusual and death usually supervenes promptly.

Pre-eclampsia and eclampsia occur in the latter part of pregnancy, or at term. Pre-eclampsia is characterized by elevation of the blood pressure, albuminuria, and edema. Headache, visual disturbances, diminished urinary output and abdominal pain are the common symptoms. Pre-eclampsia may go on to eclampsia just before or just after delivery. In this case the patient develops convulsive seizures and coma. Unless immediate treatment is given, death is the outcome. Morphine, magnesium sulfate and other sedatives are used effectively. The cause of eclampsia is unknown. At autopsy characteristic lesions in the kidney and liver are seen but their pathogenesis is still mysterious.

Glomerulonephritis, pyelonephritis, and nephrosclerosis may first become manifest during pregnancy; it may be difficult to distinguish them from pre-eclampsia if they appear late in the pregnancy. Their persistence after delivery is a characteristic feature.

Pre-natal care is directed toward the prevention of toxemias, or their early recognition and treatment. This is the reason that regular examinations are made by the physician throughout pregnancy. (D.M.H.)

TOXIN. A soluble poison produced and liberated by certain bacteria, insects, snakes, and plants. Toxins are usually protein substances which may be destroyed by heat. They vary in potency, and in their tissue affinities. The toxins produced by **tetanus** and *botulinus bacilli* are extremely potent and affect only nervous tissue; **diphtheria** toxin attacks the nervous system and the heart. Other organisms liberating toxins are hemolytic streptococci (notably in scarlet fever), **dysentery** bacilli, and staphylococci causing food poisoning.   (D.M.H.)

TRACHEA.   1. An air tube of the **arthropod** respiratory system. These tubes open at the surface of the body through **spiracles** and lead inward, branching extensively and in some species expanding to form air sacs. At their inner extremities they connect with very fine tubules called tracheoles, of independent origin. Tracheae are ingrowths of the outer layer of the body and are lined with a continuation of the cuticula, which forms spiral rods in their walls. Near the spiracle they are often provided with a muscular closing device by which air can be excluded if it contains harmful materials. The tracheoles form within cells of the lining of tracheae, later breaking out and assuming their connection with the larger tubes. They have no spiral supporting rods (taenidia).

2. The principal air tube of the vertebrate **respiratory system**, also called the wind pipe. It leads from the **pharynx** to the major branches (bronchi) connecting it with the **lungs, and** its wall is supported by rings of cartilage. At the pharyngeal end in **amphibians** and **mammals** it forms an expanded larynx containing vocal cords, and at the point where it forks to form the bronchi in birds the remarkable vocal organ called the syrinx is developed. (See **Vessel.**)   (A.W.L.)

TRACHEIDS.   A tracheid is an elongate xylem cell (angular in transverse section) with tapering ends; when mature it contains no **protoplasm.** In the thickened, lignified walls of a tracheid there are many bordered **pits.** Functionally a tracheid serves both for conduction of water and for support.

For their development, see **cambium.**   (R.M.W.)

TRACHEOPHYTES.   All plants with true **vascular** system composed of **xylem** and **phloem** are known as tracheophytes. **Pteridophytes** and **spermatophytes** are the tracheophytes. The alternative group is the atracheata, the **thallophytes** and **bryophytes,** which lack these special modified conducting cells.   (R.M.W.)

TRACHOMEDUSAE.   Hydrozoa.

TRACHYTE.   The name of an **extrusive, igneous,** fine-grained or **porphyritic rock,** the surface equivalent of **syenite.** Trachyte is predominant in alkali **feldspar** and usually contains **biotite** and **augite.** Trachyte is an old name, proposed by Brongiart in 1813, and has never been altered or supplanted. It is derived from the Greek, meaning rough.   (R.M.F.)

TRACING.   Drawing Reproduction.

TRACK.   The term track is used in **navigation** to describe the direction and distance a ship has moved in a given length of time. If a ship is known to be in a definite position at a given time and known to be in another position after the elapse of a definite period of time, then the direction of the **rhumb line** between the two points is the track and the length of the line gives the track distance. Other lines such as **Lambert, great circle,** etc., might be computed between the two points, yielding the Lambert track, etc., but the term track is usually restricted to the rhumb line. It should be carefully noted that the positions at the extremes of the track must be accurately known. These may be determined by any reliable **fix** or, in the case of airplanes, by landmarks beneath the ship. (See **Course** and **Heading.**)   (W.K.G.)

TRACKING.   In radio circuits this term is used to designate the following of two or more circuits tuned from the same shaft. Thus in the superheterodyne **receiver** the **oscillator** tuning must track with the **tuned radio frequency** circuits so the difference in frequency will always be the constant intermediate frequency.

In gunnery the term is used to designate the following of the target by the director or gun.   (L.R.Q.)

TRACTION.   In a narrow sense, traction refers to the **friction** developed between a powered surface and one in contact with it. The most common example of traction is the resistance to slipping developed at the point of contact of a driven wheel with the surface on which it rolls. The locomotive driver on its rail, the pneumatic tire on the highway, or the traction engine wheel on earth, illustrate this case. But traction is not confined to wheels, as may be proved by citing another example. The traction sheave of an elevator is a grooved pulley, power-driven, over which passes a rope which is driven by the sheave. The friction between the rope and pulley constitutes the traction.

In a larger sense, traction includes the act of pulling a load over a surface by overcoming the resistance to motion, and the word may also be used descriptively of any vehicle which by its excess of power over its own tractive needs is able to pull other vehicles. In the case of an automobile rolling on a highway, traction is applied through the rear wheels by means of a live axle which delivers a torque to the wheel. The load on the wheel presses it against the road surface enough to develop sufficient friction so that the resistance to motion will be overcome before the wheel slips on the road. If the roadbed be covered with a surface of low frictional power, for example, wet clay, there will be little traction developed because of the low coefficient of friction. The traction of an automobile must overcome wind resistance, wheel bearing friction, rolling resistance, and grade.

The railway train is drawn by the locomotive by means of traction developed between the drivers and the rail. The weight on the drivers, multiplied by the coefficient of friction of steel on steel is equal to the traction. In winter sand is sprinkled on icy rails in order to increase the coefficient of friction and so get more tractive power. The train resistance which traction must overcome is composed of the rolling resistance of wheels on rails, journal friction (static), air resistance, grade, and friction of wheel flanges on rails. The tractive efforts needed for starting a train are much higher than those required to keep it moving. While standing, oil is squeezed out of the bearing surfaces at the wheels, and metal to metal contact exists for the first few revolutions when the train has started. After the journal has revolved, oil is drawn into the surface between the journal and bearing, and a lower coefficient of friction results. Rolling resistance would be zero if the rail and wheels were perfectly inelastic, but rails deform slightly ahead of the wheels, making the reaction of the rail on the wheel inclined somewhat from the vertical in the direction of increased drag.   (F.T.M.)

TRACTION STRESS.   Stress.

TRADE WINDS.   Air Circulation of the Atmosphere; Winds.

TRAGOPAN.   Aves, Galliformes. A large game bird **(Aves)** found in wooded country at high altitudes in China and northern India. These birds are related to the pheasants and are known as horned pheasants and, improperly, as Argus pheasants. The last name belongs to a different group. The head bears a pair of fleshy projections, the horns, and the plumage of all of the several species is beautifully colored.   (A.W.L.)

TRAJECTORY OF AIR PARCELS.

A parcel of air located in a given pressure field will move with the gradient wind of the field (assuming steady flow). At the end of a few hours, the parcel will locate in some new region where it has been carried by the wind. If, however, the pressure field and therefore the wind is changing, the parcel will not move into a position indicated by the existing gradient flow. It will follow a

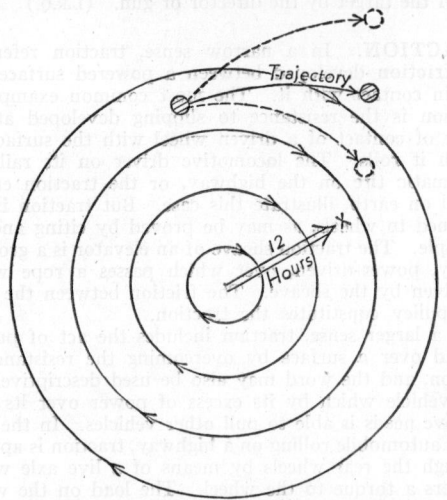

Trajectory of air parcels about anticyclone.

trajectory or path dictated by successive gradient directions and velocities as indicated by **synoptic charts.** An approximation to its trajectory can be had by extrapolating the parcel's indicated movement for as small a time interval as practicable (usually 3 or 6 hours between synoptic charts) using successive synoptic charts. The average of the velocity vectors at the beginning and the end of a given time interval would be taken as the true velocity and direction of the parcel over that time interval. Obviously, the smaller the time interval, the more accurate the trajectory. One of the charts may be a prognostic chart for computing future trajectories. Air trajectories are valuable in estimating the influence the earth will have on air as it flows over varied earth surfaces. (P.E.K.)

TRANSCENDENTAL EQUATIONS.

A transcendental equation is an equation which is not an algebraic equation; it contains one or more **transcendental functions.**

Transcendental equations include as special types: **trigonometric equations, exponential equations, logarithmic equations,** etc. (L.L.S.)

TRANSCENDENTAL EQUATIONS, SOLUTION OF.

Transcendental equations may be solved approximately by graphical methods, by successive interpolation, or by **Newton's method.**

One graphical method is to write the equation in the form of a single **function** equated to zero, plot the **graph** of this function and find the **abscissas** of the point or points of intersection of this graph with the X-axis; this gives only the real roots of the equation.

Another graphical method is to write the equation in the form of an equality of two appropriately chosen functions, construct the graphs of the two functions on the same diagram, and find the abscissas of their intersection points. (L.L.S.)

TRANSCENDENTAL FUNCTION.

A transcendental function is a **function** which is not an **algebraic function.**

Among the transcendental functions are the following types: **trigonometric functions, inverse trigonometric functions, exponential functions, hyperbolic functions, logarithmic functions, gamma functions, beta functions, elliptic functions, Bessel functions,** etc. (L.L.S.)

TRANSCENDENTAL NUMBERS.

A transcendental number is a **number** which is not an **algebraic number,** i.e., is not a root of a **polynomial equation** in one unknown with integral coefficients.

Two well-known transcendental numbers are $e \approx 2.71828$ and $\pi \approx 3.14159$. (L.L.S.)

TRANSFER OF SKILLS. Measurement.

TRANSFORMATION OF COORDINATES.

In **rectangular coordinates in the plane,** the transformations of coordinates are: translation of axes and rotation of axes.

If a set of rectangular axes are moved from a first position OX, OY to a new position $O'X'$ and $O'Y'$ such that $O'X'$ and $O'Y'$ are respectively parallel to OX and OY, the axes are said to be translated from the first to the second position.

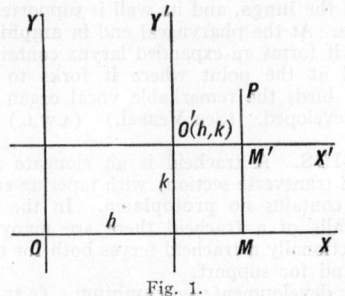

Fig. 1.

If the axes are translated to a new origin $O'(h,k)$, then (Fig. 1) the old coordinates are given in terms of the new by the formulae:

$$x = x' + h, \quad y = y' + k.$$

If a set of rectangular axes are rotated about the origin through an angle θ, the old coordinates in terms of the new are given by (Fig. 2)

$$\begin{cases} x = x' \cos \theta - y' \sin \theta, \\ y = x' \sin \theta + y' \cos \theta. \end{cases}$$

In **rectangular coordinates in space,** we have also translation of axes and rotation of axes.

If the origin is translated to a new origin $O'(h, k, l)$, the equations for the old coordinates in terms of the new are:

$$x = x' + h, \quad y = y' + k, \quad z = z' + l.$$

If $(\alpha_1, \beta_1, \gamma_1)$, $(\alpha_2, \beta_2, \gamma_2)$, $(\alpha_3, \beta_3, \gamma_3)$ are respectively the **direction angles** of three mutually perpendicular axes OX', OY', OZ' with respect to a given set of per-

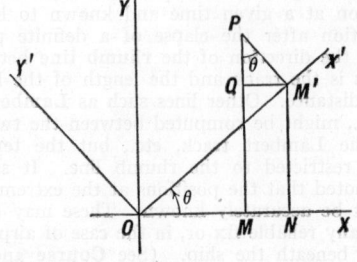

Fig. 2.

pendicular axes OX, OY, OZ, then the equations for rotating the old axes into the new position $O - X'Y'Z'$ are:

$$x = x' \cos \alpha_1 + y' \cos \alpha_2 + z' \cos \alpha_3,$$
$$y = x' \cos \beta_1 + y' \cos \beta_2 + z' \cos \beta_3,$$
$$z = x' \cos \gamma_1 + y' \cos \gamma_2 + z' \cos \gamma_3.$$

These nine **direction cosines** satisfy the following six equations:

$$\cos^2 \alpha_1 + \cos^2 \beta_1 + \cos^2 \gamma_1 = 1,$$
$$\cos \alpha_1 \cos \alpha_2 + \cos \beta_1 \cos \beta_2 + \cos \gamma_1 \cos \gamma_2 = 0,$$
$$\cos^2 \alpha_2 + \cos^2 \beta_2 + \cos^2 \gamma_2 = 1,$$
$$\cos \alpha_2 \cos \alpha_3 + \cos \beta_2 \cos \beta_3 + \cos \gamma_2 \cos \gamma_3 = 0,$$
$$\cos^2 \alpha_3 + \cos^2 \beta_3 + \cos^2 \gamma_3 = 1,$$
$$\cos \alpha_3 \cos \alpha_1 + \cos \beta_3 \cos \beta_1 + \cos \gamma_3 \cos \gamma_1 = 0.$$

(L.L.S.)

TRANSFORMATIONS OF POLYNOMIAL EQUATIONS.

To transform a **polynomial equation** of the nth degree into an equation each of whose **roots** is m times the corresponding root of the original equation, we multiply the successive coefficients of the equation, beginning with that of the next to the highest degree term, by m, m^2, $\cdots$, m^n. If the given equation is

$$a_0 x^n + a_1 x^{n-1} + \cdots + a_{n-1} x + a_n = 0,$$

the new equation will be

$$a_0 x^n + m a_1 x^{n-1} + m^2 a_2 x^{n-2} + \cdots$$
$$+ m^{n-1} a_{n-1} x + m^n a_n = 0.$$

If any power of the variable below the highest is missing, it must be written with a zero coefficient.

To transform a polynomial equation into one whose roots are changed in sign from those of the original, we change the signs of the coefficients of the odd powers of the variable in the equation.

To transform a polynomial equation of the nth degree

$$P(x) \equiv a_0 x^n + a_1 x^{n-1} + \cdots + a_{n-1} x + a_n = 0$$

into an equation each of whose roots is less by h than a corresponding root of the given equation, we proceed as follows: Divide the polynomial $P(x)$ by $x - h$ synthetically and denote the remainder by R_n; divide the quotient by $x - h$ and denote the remainder by R_{n-1}; etc.; continuing this process n times, obtaining a_0 as the last quotient and R_1 as the last remainder. Then the coefficient a_0 and the remainders R_1, R_2, $\cdots$, R_n, in order, are the coefficients of the transformed equation, which is therefore

$$a_0 x^n + R_1 x^{n-1} + R_2 x^{n-2} + \cdots + R_{n-1} x + R_n = 0.$$

The divisions should be performed synthetically. (L.L.S.)

TRANSFORMER.

Without doubt, adoption of a.c. in favor of d.c. by the growing electric industry was due as much to the ease and efficiency with which power could be transferred from low to high potential, and vice versa, as to the advantages of **alternators** over d-c **generators**. The power transformer is used to increase alternator voltage to economical transmission voltage or to decrease alternator or line voltage for station power service. It can be defined as a device for transferring electric energy from a circuit to another by magnetic induction, usually with a change of voltage. There are no moving parts, nor is there any electrical connection of the two circuits (except in the case of the auto-transformer); the energy is transferred through magnetic linkage. Regardless of the voltage, the energy supply circuit is termed the primary, and the energy receiving circuit the secondary.

Transformers may be classed as follows:

1. According to purpose.
   a. Constant voltage ratio.
   b. Constant current.
   c. Feeder voltage regulation.
2. According to use.
   a. Heavy duty. Power type as used in substations.
   b. Distribution. At customer end of the distribution system.
   c. Instrument. Potential and current transformers of small capacity and light weight, but having accurate ratios.
3. According to arrangement of magnetic circuit.
   a. Shell type. Large, low-voltage units.
   b. Core type. Small, high-voltage units.
4. According to arrangement of electric circuit.
   a. Single- or three-phase. Small transformers are often three-phase, but large capacity is met by the three-phase connection of three single-phase transformers.
   b. Connection of three-phase windings. Δ to Δ, Δ to Y, Y to Y, open Δ.
5. According to cooling.
   a. Air-cooled.
   b. Oil-filled. Oil-cooled by direct radiation or by heat transfer to separate water-cooling system.
   c. Pyranol-filled.

The ordinary single-phase transformer consists of the magnetic circuit, the windings, the leads, the insulating bushings (solid dielectric, condenser, or oil-filled types),

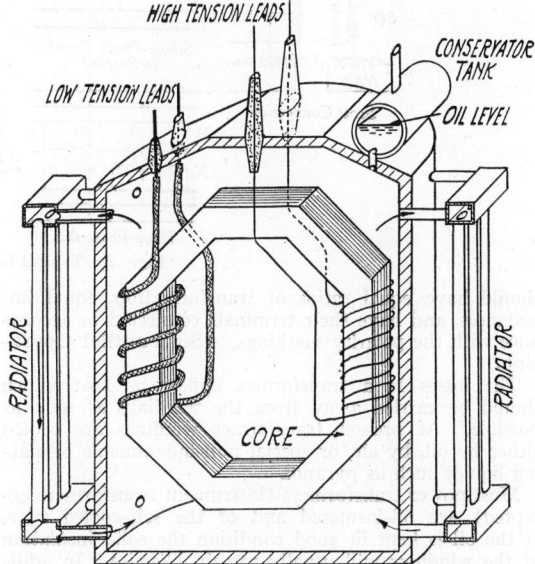

Fig. 1. Schematic sectional view of a single-phase, self-cooled transformer.

the insulating oil and its cooling system, and the tank to contain the oil and provide support for the other components. The auto-transformer is a modification of this, having only one winding which is tapped to give a low voltage. Reference to Fig. 2 will indicate that power can flow from primary to secondary by conduction. However, the two parts are also linked by the magnetic flux, thus allowing transformer action also. The use of part of the winding for both primary and secondary, and the fact that all of the power is not transformed makes the auto-transformer more efficient and cheaper than the conventional type. It is, however, not suitable for many applications because of the direct connection between the high- and low-voltage sides. This, together with the fact that its advantage decreases with increased voltage ratio, gives it a very limited use.

In an ideal transformer the voltage ratio is the same as the turn ratio. Except for transformer losses, $N_1I_1 = N_2I_2$, in which N is number of turns and I is effective current. Due to the excellent efficiency of transformers, especially in the larger sizes (where the efficiency is 99% or better), the ratio E_1/E_2 of voltage is very nearly N_1/N_2. The losses which occur in a transformer are iron losses and copper losses. The iron losses are nearly constant and are present as long as the transformer is connected to the supply circuit. Transformer regulation is much superior to alternator, rarely exceeding 5%. A transformer is rated at its maximum continuous kva., on the secondary side, as limited by heating.

When single-phase transformers are operated alone there is no obstacle to correct connections. When single-phase or polyphase operate in parallel, or when three single-phase transformers are connected into a three-phase bank, the terminals must be joined correctly on the basis of polarity. The standard marking for polarity is an H_1 on the primary lead and an X_1 on the secondary. When the current is flowing towards the transformer in the H_1 lead it will be flowing away from it in X_1. A few typical connections for transformers are shown in Fig. 2. Transformers which are operated in parallel

latter with measuring instruments. Small transformers are of importance because their applications in the central station are exceedingly numerous. They are used for two very good reasons: to protect station operators from contact with high voltage and to permit the use of trip coils, instruments, etc., of moderate current and voltage capacity. Through the use of small transformers low-voltage circuits can be obtained which reflect the characteristics of high-voltage ones to a practical degree of accuracy. (F.T.M., L.R.Q.)

TRANSFUSION. Transfer of blood or plasma from one person to another, usually by intravenous drip through a sterile rubber tube connected with a needle placed in a vein. The person from whom the blood is taken is known as the donor. The patient receiving the blood is the recipient. Both the donor and recipient must be of the same blood group and, furthermore, must be tested to make sure that the two bloods will be compatible. In emergencies when there is no time for grouping, a donor of the universal type (Group I, Jansky. Group IV, Moss. Group O, International) may be used, although this is not desirable. If plasma is used, no typing or cross matching is necessary since the antigens have been removed with the red cells.

At present, transfusions are used in many conditions while formerly they were used only as a measure of last resort. After any severe hemorrhage, transfusion may be life-saving. In anemia, in the course of severe

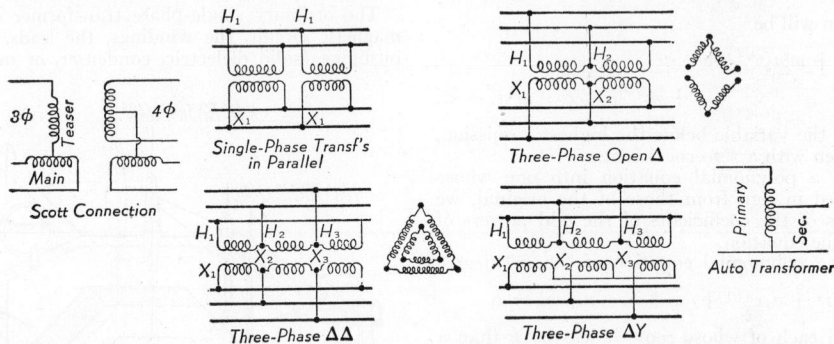

Fig. 2. Typical transformer connections.

should have equal ratios of transformation, equal impedances, and have their terminals connected in accordance with the polarity markings. (See **Parallel Operation**.)

The losses in a transformer appear as heat which should be carried away from the windings as soon as possible. At present transformer windings are cooled either by oil, by air, or special non-inflammable insulating liquids such as pyranol.

Most power transformers (instrument transformers excepted) are oil-insulated and of the self-cooled type. If the oil is kept in good condition the solid insulation of the windings will usually remain likewise. In addition to its insulating qualities, the oil should have low viscosity and high coefficient of expansion for good circulation. The expansion of the oil, while essential to good circulation, has a disadvantage in that air is exhaled or inhaled due to the total increase or decrease of oil volume in the tank. This phenomenon, called "breathing," is the principal source of water and sludge in the oil. Water comes from the condensation of atmospheric moisture "breathed" into the tank; sludge, from the oxidation of the surface of the oil. A conservator tank is used on some transformers to expose the minimum of oil surface to the air.

Small current and potential transformers may be divided into two classes, i.e., tripping transformers and instrument transformers. The former are used primarily with relays and trip coils of switching equipment; the

infections, preoperatively to prepare a patient for operation, post-operatively if blood loss has been great, and in many other conditions transfusions are a very important therapeutic measure.

There are many methods of transfusion. Since blood banks have become a part of most hospitals, the so-called indirect method has come to be used almost exclusively. With this technique, the total amount of blood to be given is withdrawn from the donor at one time into a suitable container containing sodium citrate which prevents clotting. The blood is then typed, cultured, Wassermann tested, and stored at 0° C. It may be kept so for as long as one week before being strained and given to a recipient of the same type. Great care is necessary in the handling and typing of blood, for serious and even fatal transfusion reactions can occur due to contaminated or incompatible blood. (See **Blood Groups**.) (D.M.H., R.S.M.)

TRANSIENT CURRENTS. Transients.

TRANSIENTS. Everyone has noticed that when an electric heater is turned on or off, the light from lamps on the same or closely connected circuits gives a slight flicker, and, if a radio receiver is operating at the time, it produces an audible click. A lightning flash or the sudden readjustment of connections in a transmission line may produce a similar "surge" of current. This is due to the fact that when the current in any part of a network changes, because of a change in the resistance of that part or the electromotive force operative in it, there is a readjustment of the potential differences and currents in the other parts. (See **Kirchhoff's Laws of Networks.**) Such readjustments take place very quickly, and the corresponding momentary fluctuations of current are called transients.

Mathematically, the general expression for the current in any circuit as a function of the time contains terms which rapidly become negligible; these are the transient terms. A very common and comparatively simple example is found in the case of an unbranched circuit having resistance R, inductance L, and series capacitance C, to which is suddenly applied an alternating electromotive force of low frequency n and maximum voltage E_0. If the switch is closed at a voltage maximum, the voltage at any time t (sec.) thereafter is $E_0 \cos 2\pi nt$, and the current is given by the equation

$$I = \frac{E_0}{\sqrt{R^2 + \left(2\pi nL - \frac{1}{2\pi nC}\right)^2}} \cos(2\pi nt + \Delta) + k_1 e^{m_1 t} + k_2 e^{m_2 t}$$

where Δ is the phase difference between electromotive force and current. (See **Alternating Currents.**) In the last two terms, k_1, k_2, m_1, and m_2 are constants depending upon the circuit characteristics R, L, and C. The quantities m_1 and m_2 are both negative; as a result, these exponential terms "decay" or fade away after a few cycles. Only the first term is therefore usually written. (L.D.W.)

TRANSIT. The transit is an instrument used by engineers and surveyors for measuring horizontal angles. Modern transits are usually equipped with a level attached parallel to the telescope, and a graduated vertical circle as well, so that the transit may be employed as a

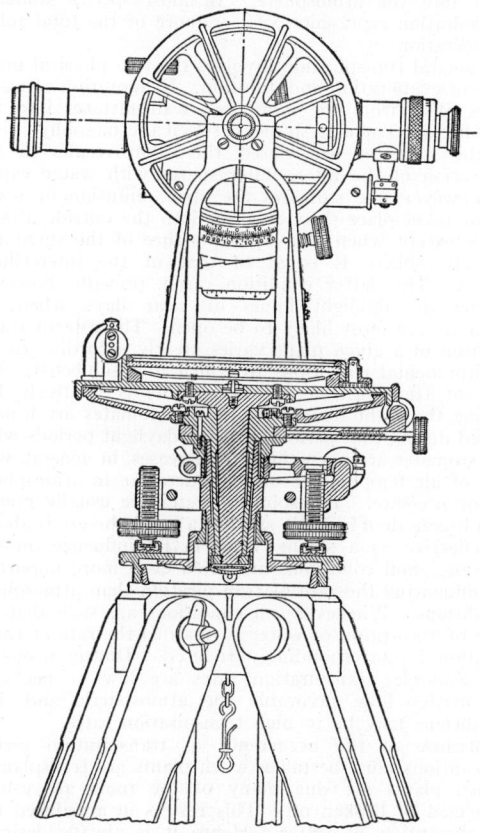

Transit.

level, and may also be used to read vertical angles. It may also be equipped with stadia wires. An exceptionally accurate transit of the type used for precise surveying is known as a theodolite. A transit is constructed largely of metal, and is mounted upon a wooden tripod. A tripod plate (foot plate) is screwed to a tripod head provided on the tripod. This tripod plate contains the socket of a ball and socket joint, and has a smooth surface upon which leveling screws may rest. The base of the instrument is carried on a vertical conical shaft which turns in a socket, on the lower end of which is a ball which is a part of the ball and socket joint just mentioned. The conical socket carries also projections into which the leveling screws are threaded. By operating these leveling screws, the shaft may be accurately aligned in vertical position. The shaft carries at its top a horizontal plate on which are mounted at right angles two levels which are used by the instrument man to guide and operate the leveling screws, so that the plate may be horizontal, and the shaft vertical. A compass is also mounted on this plate. The telescope is mounted on horizontal trunnions which are supported on standards attached to the above mentioned plate (limb). The telescopes are usually erecting, although many favor the inverting telescope. The motion of the telescope around the vertical axis can be accomplished by loosening clamping screws and swinging it at will until it is brought to sight on the desired object. It can then be clamped, and small additional motions obtained by the use of a **vernier** adjustment. The horizontal plate is usually graduated to ½° and the vernier permits readings to single minutes of arc. (F.T.M.)

TRANSIT LINE. Station.

TRANSIT OF VENUS OR MERCURY. Both **Venus** and **Mercury** revolve about the **sun** in **orbits** which lie inside of the orbit of the **earth** about the sun. Accordingly, if the planes of the planetary orbits coincided with the plane of the ecliptic, once during the **synodic** period of each planet the object should pass between the earth and the sun. Because of the small angular diameter of the planets relative to that of the sun, at the time they pass between the earth and sun they appear as small black spots moving across the brilliant disk of the sun. Such a phenomenon is known as a transit of Venus or a transit of Mercury.

The orbits of both Mercury and Venus are inclined to the plane of the ecliptic with the result that transits do not occur during each synodic period, but only when the planet happens to come to inferior conjunction close to the passage through a **node**. Transits of Mercury can occur only in May and November, when the earth crosses Mercury's line of nodes; those of Venus occur only in June and December. Since Mercury is closer to the sun than Venus, transits of Mercury are more common than those of Venus. The last transit of Venus came in 1882 and the next will take place in 2004.

The first recorded observation of a transit of Venus was made in England in 1639 by Horrocks. Four transits have been observed since that time. Since observations of the duration of transit of the planet across the disk of the sun made from widely separated positions on the earth provide a method for determination of the important astronomical constant known as **solar parallax,** expeditions were always dispatched to observe the phenomena. Within recent years more accurate methods for determination of solar parallax have been devised, and henceforth, transits of planets across the sun will be of value only for the purpose of obtaining accurate positions of the objects and thus improving the orbital **elements.** (W.K.G.)

TRANSIT STATION. Station.

TRANSIT SURVEY. Surveying.

TRANSLOCATION. This is the term most commonly used to designate the movement of solutes from one part of a plant to another. The principal conductive tissues in plants are the xylem and the phloem (see **Stems**) which constitute continuous tissue systems from just back of the tip of every root into all parts of every leaf and other lateral organs. One of the most important types of translocation in plants is the downward movement of the carbohydrates made in the leaves into the other parts of the plant. This movement occurs in the phloem and, in woody plants, only in the youngest layers of phloem. In tall trees and some vines the carbohydrates which reach the root tips may have moved hundreds of feet through the continuous phloem from the leaves in which they were made. Most of the translocation of solutes through the phloem occurs in the sieve tubes. The exact mechanism of downward movement of solutes is not known. The rate of movement is in some cases so great that it appears likely that the metabolic activity of the cells operates in some manner to speed up movements of the solute molecules.

Under some conditions considerable upward movement of carbohydrates and other soluble foods occurs in plants. This type of translocation also occurs in the phloem. The new shoots formed on woody plants when the buds resume growth in the spring are largely made at first from foods which move in an upward direction through the stems from the older existing stems. A similar upward translocation of foods occurs in the phloem towards the aerial portions of young seedlings from the seeds, towards fruits or flowers which are located at a terminal position on stems, and during the early stages of the development of leaves or shoots from bulbs, corms, tubers, or other underground organs. It is not known with certainty whether solutes can move simultaneously in both directions through the phloem, or whether they can only move in an upward direction under certain conditions, and in a downward direction under other conditions.

The mineral salts absorbed by the roots are moved mostly, and perhaps entirely, in an upward direction through the xylem. A large proportion of the mineral salts are translocated to the leaves and hence may be moved in some plants for distances of several hundreds of feet. The dissolved mineral salts are present in relatively low concentration in the water in the xylem conduits and appear to be passively carried through the plant in an upward direction in the ascending columns of water (see **Ascent of Sap**). Once translocated into a leaf the mineral salts do not necessarily remain there permanently. Some of them may subsequently be translocated out of the leaf in essentially the same form in which they enter it. Others may be used in the synthesis of complex compounds and subsequently be re-translocated out of the leaf in the form of such compounds. Shortly before the abscission of leaves or floral parts, for example, a considerable proportion of the nitrogen, phosphorus, potassium, sulfur, and magnesium are translocated back into the stems in one form or another. On the other hand there seems to be little or no such reverse translocation from the leaves of such elements as calcium, iron, boron, or manganese. All movement of minerals out of the leaves into other parts of the plant appears to occur in the phloem. It is possible that there may be a periodic daily circulation of at least some of the mineral elements in plants. There are indications, for example, that mobile phosphorus compounds can move from the root to the leaf through the xylem and return from the leaf to the root in the phloem within a 24-hour period. (B.S.M.)

TRANSMISSION LINE. Electric Power Transmission.

TRANSMITTER. In telephony the transmitter is the **microphone,** universally a single-button carbon type. The radio transmitter is the complete group of equipment utilized in converting the incoming audiofrequency signal from the studio into the modulated radio frequency signal fed to the **antenna.** The usage varies somewhat and may mean just the modulators and radio frequency stages or it may mean all the audio as well as radio-frequency stages. For **radio telegraphy** the transmitter consists of the **oscillator,** various radio frequency **amplifiers** and the **power supply** with necessary operating auxiliaries such as **keying** circuits. For radio telephony the transmitter consists of the same essential components as for telegraphy plus the necessary audio equipment and modulators. (See **Radio Communication** and **Radio Telegraphy.**) (L.R.Q.)

TRANSPIRATION. All plants require water for their continued existence and most kinds of plants require it in considerable quantities. In spite of the absolute indispensability of water for their growth and metabolism plants in general are very inefficient in their use of water. An overwhelmingly large proportion of the water absorbed by most kinds of terrestrial plants escapes from them as water vapor. This process of the loss of water vapor from plants is called *transpiration*. Although some water vapor loss may occur from any organ of a plant which is exposed to the atmosphere, in the vast majority of plants most transpirational water loss occurs from the leaves. There are two kinds of foliar transpiration: 1. Stomatal transpiration in which water vapor loss occurs through the stomates. 2. Cuticular transpiration, in which evaporation of water takes place directly through the surface of the epidermal cells through the cuticular layer into the atmosphere. In most species stomatal transpiration represents 90% or more of the total foliar transpiration.

Stomatal transpiration involves the two physical processes of evaporation and diffusion. Evaporation of water takes place from the moist cell wall surfaces into the air-filled intercellular spaces between the mesophyll cells. If the stomates are closed this soon results in the saturation of the intercellular spaces with water vapor. If, however, the stomates are open, diffusion of water vapor takes place through them into the outside atmosphere except when the vapor pressure of the surrounding atmosphere is equal to that of the intercellular spaces. This latter condition rarely prevails, however, during the daylight hours of clear days when the stomates are most likely to be open. The rate of transpiration of a given plant varies greatly according to the environmental conditions to which it is subjected. The rate of transpiration of most plants is relatively low during the night hours because the stomates are usually closed during that period. During daylight periods when the stomates are open the rate increases, in general, with rise of air temperature or with decrease in atmospheric vapor pressure. Transpiration rates are usually greater in a breeze than in quiet air, but a gentle breeze is almost as effective as a strong wind in its influence on this process. Soil conditions are often even more important in influencing the rate of transpiration than atmospheric conditions. Whenever soil conditions are such that the rate of absorption of water is retarded the rate of transpiration is correspondingly reduced. During droughts, for example, transpiration rates are low or negligible no matter how favorable the atmospheric and light conditions may be to high transpiration rates.

Because of the occurrence of transpiration certain precautions must be taken when plants are transplanted. When plants are dug many of the roots are usually damaged or broken off. This results in a reduced rate of absorption of water. Hence it is always desirable and often imperative to reduce the transpiration rate of the plant by removing leaves, pruning, or covering the plant in some suitable manner. Newly transplanted plants should also be copiously watered. Gradual development of new roots on the transplanted plants sooner or later permits sufficient absorption of water to compensate for normal transpirational water loss.

Transpiration is not an essential plant process but a necessary consequence of the structure of plants. The structure of green terrestrial plants must be such that carbon dioxide gas will have access to moist cell wall surfaces in which it can dissolve (see **Photosynthesis**). The principal carbon dioxide absorbing surfaces of vascular land plants are the mesophyll cell walls bounding the intercellular spaces of the leaves. Most carbon dioxide diffuses into the leaves through the open stomates. When the stomates are open, outward diffusion of water vapor is unavoidable and most of the water lost from plants escapes in this way. Some of the incidental effects of transpiration upon the plant are beneficial but none is indispensable. Likewise some of the incidental effects of the process are detrimental to plants but plants have existed through countless centuries in spite of transpiration.

Rates of transpiration rarely, if ever, exceed 5 grams per sq. decimeter of leaf surface per hour and are usually considerably less. Under drought conditions the rate of transpiration may fall to an almost zero value. Only approximate computations can be made of the loss of water from vegetation covered areas but considerable interest from the standpoint of water conservation and meteorology attaches to such figures. It has been calculated, for example, that an acre of corn in central Illinois transpires water equivalent to 15 acre-inches in the course of a growing season. The corresponding figure for an acre of young apple orchard in central New York is 9 acre-inches. (See also **Guttation**.) (B.S.M.)

TRANSPOSITION. Inductive Interference.

TRANSVERSE LOAD. Load.

TRAP. A general term for fine-grained, **basic** and therefore dark-colored **igneous** rocks, having a relatively high specific gravity. The term is derived from an old Scandinavian word meaning a stair, because these rocks, as particularly well displayed in the 3000-ft. cliffs of the Faroe Islands, develop architectural forms simulating colossal stairways. The term is also used as a synonym for **basalt**, especially by contractors who use it for road metal. (Compare with **Whinstone**.) (R.M.F.)

TRAUMA. Any injury to the body caused by an outside physical force such as a fall, blow, or weapon, etc. Traumatic surgery is surgery that deals with injuries, and is usually of an emergency nature. (R.S.M.)

TRAVERSE. A traverse is a series of connected straight lines, forming the outline of a plot of land or the center line of a roadway or railway. A traverse which forms a closed figure is a closed traverse. A traverse which is not closed is an open traverse. The traverse is obtained by means of a survey. Ordinarily, the transit and tape are used, the transit being successively set up over each point constituting the junction of two adjacent lines. The distances between transit stations (the length of the straight lines) are measured by taping with a graduated tape or **chain**. The bearings of the lines or the angles between the lines, as well as the length of the lines, are noted by the instrument man at the time of the survey in a field book. These field notes are later employed to map the traverse by one of the several methods available, such as **tangent offsets, latitudes and departures**, coordinates, etc. A traverse which follows the shore line of a body of water is a meander line. The general outline of the

shore is found by taking offsets from the transit lines at regular intervals and also at points where there is an abrupt change of shape.

In navigation, the term traverse or traverse sailing is sometimes used to determine, by **dead-reckoning** methods, the course and distance made good by a ship, when the ship has been on several different headings in the time interval involved. Cases of this sort commonly arise in determining the day's run from noon to noon. An example of traverse sailing is discussed in detail in the article on dead reckoning. (F.T.M., C.W.C., W.K.G.)

TRAVERSE TABLES. In many problems in surveying and navigation (e.g., the important navigational problem of **dead reckoning**) the solution of a plane right-triangle becomes necessary. In traverse tables the plane right-triangle is solved without the necessity of using tables of **trigonometric functions**. The tables are constructed by tabulating for successive values of one apex angle the two sides of the triangle as functions of the hypotenuse. The traverse tables as published for navigational purposes solve the dead-reckoning problem by tabulating difference of **latitude** and **departure** as functions of the distance for each course either in quarter points of the compass or every degree. Since the tables must contain pages for each value of the apex angle they are more bulky than ordinary trigonometric tables. However, the rapidity with which the plane triangle may be completely solved more than compensates for the time lost in turning pages, particularly when extreme accuracy is not required in the solution. (W.K.G.)

TRAVERTINE. Carbonated waters dissolve large amounts of **calcium** carbonate, especially under high temperature. Such waters reaching the earth's surface as hot springs often deposit the calcium carbonate, in great quantities. This material is called travertine from the ancient name for Tivoli, Italy, where a very thick deposit occurs. Travertine may be compact, crystalline, fibrous or, if rapidly deposited, spongy and porous. The less compact varieties are known as tufa. Travertine is being formed at the Mammoth Hot Springs, Yellowstone National Park and at many other localities. A banded travertine used as an ornamental stone is called **onyx marble** or Mexican onyx. (R.M.F.)

TREE HOPPER. Insecta, Homoptera. An **insect** of moderate size with a peculiarly formed **thorax**, prolonged over the hinder part of the body and in many species with other projections of bizarre form. These insects make up the family Membracidae, related to the **leaf hoppers** and spittle bugs.

The buffalo tree hopper, *Ceresa bubalus*, sometimes damages twigs of fruit trees by laying eggs in the bark and is occasionally troublesome in the garden through sucking the juices of young seedlings. Clean cultivation is an effective control. (A.W.L.)

TREE OF HEAVEN. Ailanthus.

TREE OF LIFE. Phylogenetic tree. A diagrammatic scheme for showing the probable evolutionary relationship of the various kinds of organisms. Arising from primordial living matter and now found at their simplest in the 1-celled forms, all forms of plants and animals may be traced through graded relations to their present condition. (A.W.L.)

TREE SHREW. Mammalia, Insectivora. *Tupaia*. Small arboreal animals of the Oriental region. They resemble squirrels closely in general form and in the long tail but the muzzle is long and sharp as in other **shrews**. One species is remarkable for the feather-like form of the tip of the tail, which is elsewhere clothed with short hair. The group is interesting chiefly because it has been regarded by one school of thought as probably ancestral to the primate stock. (A.W.L.)

TREE TOAD, TREE FROG. Amphibia, Anura. Small animals like the other **frogs** and **toads** in form but largely of arboreal habits. The toes are provided with adhesive disks at the tips. The spring peeper and the cricket frog, each about an inch in length, are well known North American species. Our tree frogs deposit their eggs in the water but in some species the larval stage is passed in the egg, which is attached to a leaf. The extensive family Hylidae includes tree frogs of every continent. Species of another family occur in Madagascar and South America. (A.W.L.)

TREE-CREEPER. Aves, Passeriformes. A small dull-colored bird (**Aves**) with a long curved beak. It seeks its prey, consisting of insects and other small creatures, in the crevices of bark, moving about the trunks and branches of trees in any position. The group is represented in North America by the brown creeper, *Certhia familiaris,* and its several subspecies. (A.W.L.)

TREE-FERNS. Paleobotany.

TREE-PIE. Aves, Passeriformes. A bird (**Aves**) of the Oriental region related to the magpies. The colors of these birds are shades of brown, black, and gray and their beaks are relatively short, but in habits they re semble the magpies. (A.W.L.)

TRELLIS DRAINAGE. The type of drainage pattern which is similar to a trellis or fireman's ladder. Trellis drainage is characteristic of the canoe-shaped valleys of the Appalachians. (R.M.F.)

TREMATASPIS. Fossil Fishes.

TREMATODA. The flukes, a class of flatworms (Phylum **Platyhelminthes**). They are very variable in size, are flattened or rounded in cross-section, and are exclusively parasitic. Flukes resemble the free-living flatworms (Turbellaria) more closely than tapeworms, but in addition to their parasitic habits they differ in having the mouth usually near the anterior end of the body and the intestine usually forked and sometimes provided with anal openings. The body is covered with a **cuticle** and bears suckers and in some species hooks or spines by which the worm attaches itself to the host.

Many flukes hatch from the egg as free-swimming **larvae** and pass through a life cycle in two or more hosts before attaining maturity. The passage from host to host is made with the changes in the metamorphic cycle.

Some flukes attach themselves to the gills or skin of fishes while others live in the internal organs of various animals. Man is affected by four types, the liver, lung, blood, and intestinal flukes, found chiefly in tropical and Oriental countries. Since some of these parasites pass some stages of their **metamorphosis** in fishes, the eating of uncooked or imperfectly cooked fish is dangerous in regions where they occur.

The class is divided into two subclasses and five orders:

Subclass Monogena. External parasites with a single host.
Order Monopisthodiscinea. Small species on the skin and gills of fishes. Body without suckers but with a posterior disk bearing hooks.
Order Monopisthocotylinea. With or without anterior suckers. No oral sucker. Posterior end with a sucking disk, usually bearing hooks. Parasitic on marine fishes.
Order Polyopisthocotylinea. Posterior lobe with at least two suckers. On gills and in mouths of fishes, and on skin or in bladder of amphibians and reptiles.

Subclass Digena. Internal parasites with two or more hosts. With one or two suckers.
Order Gasterostomata. Mouth at the middle of the body. Parasitic in fishes and marine mollusks.
Order Prosostomata. Mouth at the anterior end. A large order. Many species, attacking many invertebrate and vertebrate hosts. (A.W.L.)

TREMOLITE. The mineral tremolite is a **calcium-magnesium silicate** corresponding to the formula $H_2Ca_2Mg_5(SiO_3)_8$ belonging to the **amphibole** group. The replacement of magnesium by ferrous **iron** causes tremolite to approach **actinolite** in composition. Tremolite is **monoclinic**, developing bladed prismatic crystals, but it is frequently found in compact columnar, granular, or fibrous masses. The perfect prismatic cleavage at angles of 56° and 124° typical of this group is to be noted; hardness, 5–6; specific gravity, 2.9–3.1; luster, vitreous to silky; color, varies from white or whitish-gray through shades of green or greenish-yellow; transparent to opaque. Tremolite is formed as a result of contact **metamorphism** and occurs in **marbles**, **dolomites**, and **schists**. It may alter to **talc**. Tremolite is found in Switzerland, in the St. Gotthard region, being named for the Tremola Valley, and is common elsewhere in Europe. In the United States it occurs in Maine, Pennsylvania, and New York. In Canada tremolite has been found in Quebec and Ontario.

Hexagonite is a pinkish-purple variety of tremolite which contains a small amount of manganese. So called because it was at first believed to be **hexagonal**. It has been since shown to be monoclinic, and is found in St. Lawrence County, New York. Some nephrite and asbestos is tremolite. (E.S.C.S.)

TRENCH MOUTH. Stomatitis.

TREPAN. Trephine.

TREPANG. Holothuroidea.

TREPANNING. Originating a hole by removing an annular ring of material, instead of reducing to chip form the entire volume of the material originally within the hole. Contrast **boring, drilling.** (H.C.H.)

TREPHINE. 1. A hollow, circular, saw-tooth, drill-like instrument that is used for removing a disk of bone from the skull. This opening may then be enlarged or several of these drill holes may be made or connected so as to raise a piece of bone to expose a portion of the brain. 2. To open the skull with a trephine. The skull is usually operated upon for removal of an intracranial **tumor,** to relieve intracranial pressure, or to drain a brain abscess. (R.S.M., D.M.H.)

TRF. TUNED RADIO FREQUENCY. Receivers.

TRIANGLES. It is shown in **geometry** that the three sides and three angles of a plane triangle are so related that when any three parts, excluding the case of three angles, are given, the other three parts are determined, so that the triangle is fixed in shape and size. Geometry also gives methods for constructing the triangle by geometrical constructions with straightedge and compasses when three appropriate parts are given.

The process of finding the unknown parts of a triangle from a set of given parts is called the solution of the triangle.

If the known parts of a triangle are given by their numerical values, we may draw the triangle to scale, carrying out the geometric constructions and measurements with a graduated ruler, compasses, and protractor, and then on the completed figure we may measure off the required parts with ruler and protractor, and obtain the numerical measures of these parts. This is called the graphic solution of triangles. This method is limited in accuracy; however, many important practical problems in which only a moderate degree of accuracy is required can be readily solved by the graphic method; it is also of use as a check on the trigonometric solution of triangles.

The trigonometric solution of triangles is based on calculations with the use of formulae. This method is capable of any desired degree of accuracy.

To solve a right triangle, ABC:

Choose one of the formulae expressing the definition of one of the trigonometric functions of one of the acute angles of the triangle which involves an unknown part (side or angle) and two known parts, and solve for the unknown part.

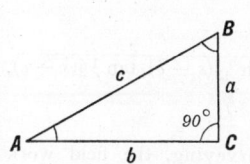

We also use the geometric relations $A + B = 90°$, and the Pythagorean theorem $a^2 + b^2 = c^2$.

Every oblique triangle can be solved by means of the methods for solving right triangles, by division into two right triangles by drawing a perpendicular from one vertex to the opposite side or opposite side extended. Special formulae are also available for the solution of oblique triangles.

The law of sines is: Any two sides of a plane triangle are proportional to the sines of the opposite angles:

$$\frac{a}{b} = \frac{\sin A}{\sin B}, \quad \frac{b}{c} = \frac{\sin B}{\sin C}, \quad \frac{c}{a} = \frac{\sin C}{\sin A};$$

also

$$\frac{a}{\sin A} = \frac{b}{\sin B} = \frac{c}{\sin C} = D,$$

where D is the diameter of the circumscribed circle of the given triangle.

The law of cosines is: The square of any side of a plane triangle is equal to the sum of the squares of the other two sides minus twice the product of these two sides and the cosine of their included angle:

$$a^2 = b^2 + c^2 - 2bc \cos A,$$
$$b^2 = c^2 + a^2 - 2ca \cos B,$$
$$c^2 = a^2 + b^2 - 2ab \cos C.$$

The law of tangents is: The tangent of half the difference of any two angles of a plane triangle is equal to the difference of the two opposite sides divided by their sum, multiplied by the cotangent of half their included angle:

$$\tan \frac{1}{2}(A - B) = \frac{a - b}{a + b} \cot \frac{1}{2} C,$$
$$\tan \frac{1}{2}(B - C) = \frac{b - c}{b + c} \cot \frac{1}{2} A,$$
$$\tan \frac{1}{2}(C - A) = \frac{c - a}{c + a} \cot \frac{1}{2} B.$$

The half-angle formulae for a plane triangle are:

$$\tan \frac{1}{2} A = \frac{r}{s - a}, \quad \tan \frac{1}{2} B = \frac{r}{s - b}, \quad \tan \frac{1}{2} C = \frac{r}{s - c},$$

where $s = \frac{1}{2}(a + b + c)$, and

$$r = \sqrt{\frac{(s - a)(s - b)(s - c)}{s}}.$$

The best formulae for use in checking the solution of an oblique triangle are the Mollweide's formulae:

$$\frac{a + b}{c} = \frac{\cos \frac{1}{2}(A - B)}{\sin \frac{1}{2}C},$$
$$\frac{a - b}{c} = \frac{\sin \frac{1}{2}(A - B)}{\cos \frac{1}{2}C},$$

and four others obtained by cyclic interchange of letters.

The fundamental formula for the area of a triangle is:

$$\text{area} = \frac{1}{2} \text{ base} \times \text{altitude}.$$

If the angles of the triangle are denoted by A, B, C and the opposite sides by a, b, c, then:

$$\text{area} = \sqrt{s(s - a)(s - b)(s - c)}, \text{ where } s = \frac{1}{2}(a + b + c);$$

$$\text{area} = \frac{1}{2}ab \sin C = \frac{1}{2}bc \sin A = \frac{1}{2}ca \sin B;$$

$$\text{area} = \frac{1}{2} a^2 \frac{\sin B \sin C}{\sin (B + C)} = \frac{1}{2} b^2 \frac{\sin A \sin C}{\sin (A + C)}$$
$$= \frac{1}{2} c^2 \frac{\sin A \sin B}{\sin (A + B)}.$$

In the solution of oblique triangles, four cases need to be distinguished:

I. Given one side and two angles, a, A, B.
II. Given two sides and an angle opposite one of them, as a, b, A.
III. Given two sides and the included angle, as a, b, C.
IV. Given the three sides, a, b, c.

All oblique triangles can be solved, by use of natural functions (i.e., the actual values of the trigonometric functions), by use of the law of sines, the law of cosines, and the angle formula $A + B + C = 180°$, as follows:

Case I, given a, A, B: Angle C may be found by the angle formula, then b and c may be found by use of the law of sines used twice.

Case II, given a, b, A: Angle B may be found by use of the law of sines, angle C from the angle formula, and c by the law of sines again.

Case III, given a, b, C: Side may be found by the law of cosines, and angles A and B from the law of sines used twice, or angle A by the law of sines and angle B from the angle formula.

Case IV, given a, b, c: The angles may all be found by the law of cosines, or angle A from the law of cosines and angles B and C from the law of sines, or angle A from the law of cosines, angle B from the law of sines, and angle C from the angle formula.

In all cases, the solution may be checked by Mollweide's formulae.

Case II is called the ambiguous case, as there may be one solution, or two solutions, or no solution. Given a, b, A:

1. If $A < 90°$ and $a < b \sin A$, no solution.
2. If $A < 90°$ and $a = b \sin A$, one solution, a right triangle.
3. If $A < 90°$ and $b > a > b \sin A$, two solutions, oblique triangles.
4. If $A < 90°$ and $a \geq b$, one solution, an oblique triangle.
5. If $A < 90°$ and $a \leq b$, no solution.
6. If $A > 90°$ and $a > b$, one solution, an oblique triangle.

All oblique triangles can be solved, by the use of logarithms, with the law of sines, the law of tangents, the half-angle formulae, and the angle formula, as follows:

Case I, given a, A, B: Angle C is found from the angle formula, then the law of sines is used to find b, c.

Case II, given a, b, A: Use the law of sines to find angle B, the angle formula to find angle C, then the law of sines to find c.

Case III, given a, b, C: Use the law of tangents to find $\frac{1}{2}(A - B)$, from this and the angle formula find A and B, then use the law of sines to find c.

Case IV, given a, b, c: Use the half-angle formulae to find $\frac{1}{2}A$, $\frac{1}{2}B$, and $\frac{1}{2}C$.

Mollweide's check formulae are adapted to logarithmic calculation.

The solution of right spherical triangles is based on the following formulae (where C is the right angle):

$$\sin a = \sin A \cdot \sin c,$$
$$\sin a = \tan b \cdot \cot B,$$
$$\cos A = \cos a \cdot \sin B,$$
$$\cos A = \tan b \cdot \cot c,$$
$$\cos c = \cot A \cot B,$$
$$\sin b = \sin B \sin c,$$
$$\sin b = \tan a \cot A,$$
$$\cos B = \cos b \cdot \sin A,$$
$$\cos B = \tan a \cdot \cot c,$$
$$\cos c = \cos a \cdot \cos b.$$

Let the five values a, b, co-A, co-B, co-c (where co-means complement of) be arranged in order as indicated in the accompanying figure. Denote any one of these as a middle part, then two of the other parts are adjacent to it and the other two parts are opposite to it. The above ten formulae are summarized in the following rules, known as Napier's rules of circular parts:

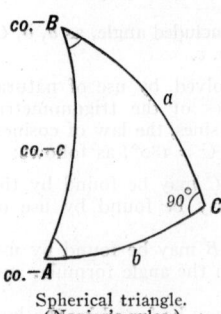

Spherical triangle.
(Napier's rules.)

1. The sine of a middle part is equal to the product of the tangents of the adjacent parts.
2. The sine of a middle part is equal to the product of the cosines of the opposite parts.

The solution of oblique spherical triangles is based on the following formulae:

Law of sines: $\dfrac{\sin A}{\sin a} = \dfrac{\sin B}{\sin b} = \dfrac{\sin C}{\sin c}.$

Law of cosines: $\cos a = \cos b \cos c + \sin b \sin c \cos A,$
$$\cos A = -\cos B \cos C + \sin B \sin C \cos a.$$

Half-angle and half-side formulae:

$$\tan \frac{1}{2} A = \sqrt{\frac{\sin (s - b)\,\sin (s - c)}{\sin s \sin (s - a)}}$$

where $s = \frac{1}{2}(a + b + c),$

$$\tan \frac{1}{2} a = \sqrt{\frac{-\cos \sigma \cdot \cos (\sigma - A)}{\cos (\sigma - B) \cos (\sigma - C)}}$$

where $\sigma = \frac{1}{2}(A + B + C).$

Napier's Analogies (Formulae):

$$\frac{\tan \frac{1}{2}(a - b)}{\tan \frac{1}{2}c} = \frac{\sin \frac{1}{2}(A - B)}{\sin \frac{1}{2}(A + B)},$$

$$\frac{\tan \frac{1}{2}(a + b)}{\tan \frac{1}{2}c} = \frac{\cos \frac{1}{2}(A - B)}{\cos \frac{1}{2}(A + B)},$$

$$\frac{\tan \frac{1}{2}(A - B)}{\cot \frac{1}{2}C} = \frac{\sin \frac{1}{2}(a - b)}{\sin \frac{1}{2}(a + b)},$$

$$\frac{\tan \frac{1}{2}(A + B)}{\cot \frac{1}{2}C} = \frac{\cos \frac{1}{2}(a - b)}{\cos \frac{1}{2}(a + b)}.$$

Delambre's Analogies:

$$\frac{\sin \frac{1}{2}(A - B)}{\cos \frac{1}{2}C} = \frac{\sin \frac{1}{2}(a - b)}{\sin \frac{1}{2}c},$$

$$\frac{\sin \frac{1}{2}(A + B)}{\cos \frac{1}{2}C} = \frac{\cos \frac{1}{2}(a - b)}{\cos \frac{1}{2}c},$$

$$\frac{\cos \frac{1}{2}(A - B)}{\sin \frac{1}{2}C} = \frac{\sin \frac{1}{2}(a + b)}{\sin \frac{1}{2}c},$$

$$\frac{\cos \frac{1}{2}(A + B)}{\sin \frac{1}{2}C} = \frac{\cos \frac{1}{2}(a + b)}{\cos \frac{1}{2}c}.$$

The area of a spherical triangle is given by the formula

$$\Delta = \frac{\pi R^2 E}{180},$$

where $E = A + B + C - 180°$ is the spherical excess of the triangle and R is the radius of the sphere, and E is also given by L'Huillier's formula:

$$\tan \frac{1}{4} E =$$
$$\sqrt{\tan \frac{1}{2}s \cdot \tan \frac{1}{2}(s - a) \cdot \tan \frac{1}{2}(s - b) \cdot \tan \frac{1}{2}(s - c)},$$

where $s = \frac{1}{2}(a + b + c)$. (L.L.S.)

TRIANGULATION. In surveying, the field work necessary to obtain the angular measurements between the sides of a series of connected triangles, and the length of one or more of the sides, is known as triangulation. The system of connected triangles is called a triangulation system. The topographic surveying of land requires the establishment of control points of known position and elevation. This control usually takes the form of a series of surveyed triangles, with the apices located on prominent points in the area, such as peaks, mounds, cliffs. In laying down a network of triangles, a fairly level region is selected on which to measure a straight **base line**. The transit is then set over one end of this base line, and the angles between pairs of selected control points taken. The transit is then taken to the other end of the base line, and the angles are again taken to the same points. By occupying, progressively, advancing triangulation points the network may be extended over as large an area as necessary. These angles are read to a high degree of precision by very accurate instruments, using the method of repetition. Even then the angular data as taken in the field will not be consistent, one triangle with another or with the whole. Such small errors or inconsistencies as exist must be adjusted, or balanced, by distributing the error. This is a fairly specialized and intricate process but must be executed to obtain a consistent system of triangles of highest probable accuracy. Upon the completion of this process the position of any triangulation point with respect to any other or with respect to the base line may be very accurately computed. The United States Coast and Geodetic Survey has covered the country with a triangulation system, the sides of the triangles often being many miles in length. The accuracy of triangulation depends upon the accuracy with which the instrument is constructed and read, and the precision with which the base line is measured. The precision of triangulation is classified as first order, second order, third order, etc., first order triangulation being that in which the base line is required to be measured to an accuracy of 1 part in 25,000, and the average triangle closure being only one second. The geodetic surveying calls for first and second order triangulation, but ordinary intermediate work is satisfactory with third or fourth order triangulation. (F.T.M., E.W.S.)

TRIASSIC PERIOD. The earliest, major subdivision of the **Mesozoic Era** of the geologic time-scale. One of the oldest **systemic** terms and denoting a three-fold division of the German formations into the lower or Bunter sandstones, the middle or Muschelkalk limestones, and the upper or Keuper copper-bearing shales. The term Triassic was proposed by Alberti in 1834. The period began about 200 million years ago and lasted for about 50 million years. The maximum thickness of formations, 25,000', occurs in the Alps. In eastern North America, especially in Massachusetts, Connecticut, and New Jersey, occur a thick series of red sandstones, arkoses, shales, and argillites of fresh water origin, containing the footprints of the earliest known

Map showing the surface distribution (areas of outcrops) of Triassic and Jurassic strata in North America. Some areas of doubtful age and extent not shown in British Columbia. All Atlantic Coast Areas are Triassic. In much of the western United States the Triassic and Jurassic have not yet been satisfactorily separated.

dinosaurs. Interbedded with the sedimentary formations are numerous basalt lava flows and sills. This eastern facies of the Triassic is called the Newark Series after the type locality in New Jersey. In Arizona and New Mexico occur fresh-water clastic sediments containing the prostrate fossil trunks of ancestral Sequoias in which the original body structures have been perfectly replaced by silica. Some of the tree trunks are over 150′ long and 3–6′ in diameter. In the Cordilleran region the Triassic formations are marine. Triassic rocks also occur in South America, British Isles, western Europe, Asia, Africa, and Australia. During this period there were great changes in the plants and animals, as disclosed by the fossils. Ferns, cycads, and conifers predominated among the plants. The modern corals, hexacoralla, predominate over the earlier tetracoralla of the Paleozoic. Cystoids and Blastoids have become extinct. The brachiopods are less abundant, their place being taken by the pelecypods. Modern insects begin their development in this period, and the modern (bony) fishes are ascendant. Among the terrestrial animals, the Paleozoic amphibia (Stegocephalia) are replaced by the dinosaurs. The highest types of marine invertebrates are ammonites. Marine reptiles, called Ichthyosaurs, and flying reptiles called Pterosaurs, first appeared in Europe. Small reptilian-like mammals also made their first appearance. The economic products of this period are chiefly salt, gypsum, and copper. In eastern North America the period closed with relatively slight uplift and block-faulting of the Newark Series, called the Palisades Disturbance. In the Pacific Coast region there was a withdrawal of the marine waters to mark the close of the period. (R.M.F.)

TRIBO-ELECTRIFICATION. Frictional Electricity.

TRIBOLUMINESCENCE. Luminescence.

TRICERATOPS. Fossil Reptiles.

TRICHINA. Nemathelminthes, Nematoda. A small roundworm parasitic in the intestine and muscles of mammals. The species is now classified as *Trichinella spiralis* but it commonly retains the name trichina from that of an older genus.

Adult worms develop in the intestine and bear living young which migrate through the intervening tissues and encyst in voluntary muscle fibers, where they perish unless the flesh is eaten by another animal and the cysts are dissolved in the process of digestion. (A.W.L.)

TRICHINIASIS. An infection caused by a worm of the class called **nematodes.** The larvae of the worms are found in pigs and man acquires the disease by eating infected pork. After ingestion, the larvae develop to the adult stage in the small intestine. Countless **embryos** are then given off which penetrate the walls of the intestine, enter the circulation, and finally lodge in muscle tissue, where they become encysted. (See **Trichina.**)

The symptoms depend on the number of larvae ingested and the number of worms that mature and discharge embryos into the intestine. The first symptoms accompany the period of invasion. These consist of abdominal pain, cramps, vomiting, and diarrhea. On the ninth or tenth day after eating the infected pork, the period of dissemination, corresponding to the migration of embryos to the muscles of the body begins. This is marked by high fever, tenderness, swelling, and pains in the body muscles. A characteristic feature is edema about the face, particularly around the eyes. Mild cases last for 2 weeks, more severe may last for 6 weeks, and are followed by many months of weakness. Convalescence lasts as long as worms are invading muscles and becoming encysted there.

The mortality varies in different outbreaks from 1–25%. The average mortality is about 6%.

The treatment is mainly directed to relief of symptoms and general supportive measures. Once present, the encysted forms remain in the muscle for life. Prophylaxis is important. Thorough cooking of pork destroys all the infecting parasites. In many countries pork is examined in the slaughter house for their presence. (D.M.H.)

TRICHINOSIS. Trichiniasis.

TRICHOCYST. A slender structure found in the outer layer of the body in many species of 1-celled animals (ciliates). They are discharged from the body under stimulation, forming a fine thread with a cap at the base. Although they have been interpreted as defensive structures they have also been regarded as organelles of attachment. They are varied and are not yet wholly understood. (A.W.L.)

TRICHOGYNE. A trichogyne is a beak-like or thread-like projection from an oögonium which functions to receive the **sperm** cells and to convey the sperm nucleus to the egg. (R.M.W.)

TRICHOPTERA. Caddis fly. An order of **insects** closely related to the more primitive moths. Many species are moth-like but all differ from the great majority of moths in their rudimentary biting mouth parts. (A.W.L.)

TRICKLE CHARGER. This is a rectifier-type battery charger which is kept connected across the storage battery at all times, charging it at a slow rate and thus keeping it in a well charged condition. (L.R.Q.)

TRICLADIDA. Turbellaria.

TRICLINIC SYSTEM. Crystallography.

TRIDYMITE. The mineral tridymite is, like quartz, **silicon** dioxide, SiO_2, but is a high-temperature variety, probably stable above 870° C. It has a conchoidal fracture; is brittle; hardness, 7; specific gravity, 2.28–2.33; vitreous luster; usually colorless and transparent. It is found chiefly in volcanic rocks of the more acidic types like **rhyolite, trachyte,** and **andesite.** It is not a particularly uncommon mineral, occurring in Germany, France, Italy, Japan, the Island of Martinique, Mexico, and in the United States in Wyoming and Washington. Tridymite is **hexagonal** but when heated to about 1470° C. passes into an **isometric** form, cristobalite, which was first noted in the andesitic lavas of the Cerro San Cristobal, Pachuca, Mexico, together with tridymite. Cristobalite has been found also in California and in Germany. (E.S.C.S.)

TRIGGER CIRCUIT. Many of the recent developments in the application of electron **tubes** have utilized various types of triggering action. By this is meant a circuit which suddenly changes its electrical condition, just as if a trigger had tripped it. While there are many varieties of such circuits, they represent some form of unstable electrical equilibrium so the triggering disturbance, whatever it may be, causes it to break over into a new condition. These circuits range from very simple **thyratron** devices to elaborate vacuum-tube circuits which give some special type of output. Many of the modern developments would be impossible without these devices. (L.R.Q.)

TRIGONOMETRIC CURVES. The graphs of the trigonometric functions are shown in Figs. 1–7.

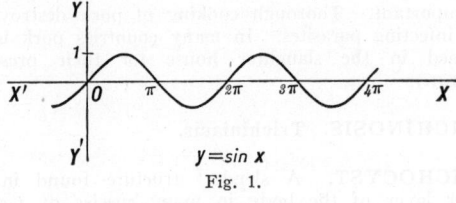

$y = \sin x$
Fig. 1.

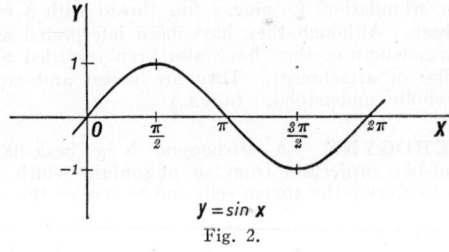

$y = \sin x$
Fig. 2.

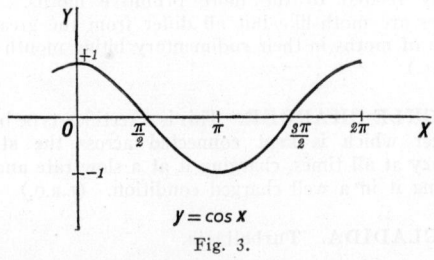

$y = \cos x$
Fig. 3.

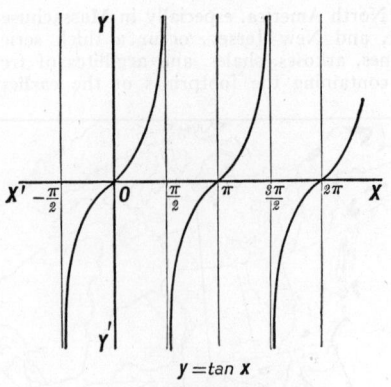

$y = \tan x$
Fig. 4.

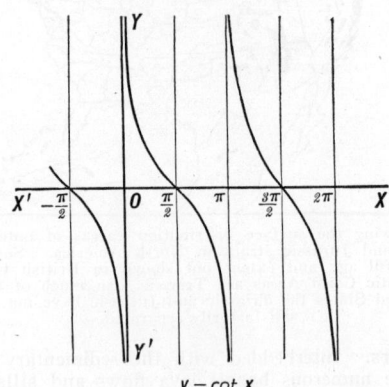

$y = \cot x$
Fig. 5.

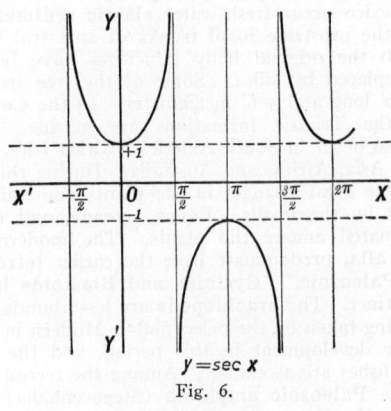

$y = \sec x$
Fig. 6.

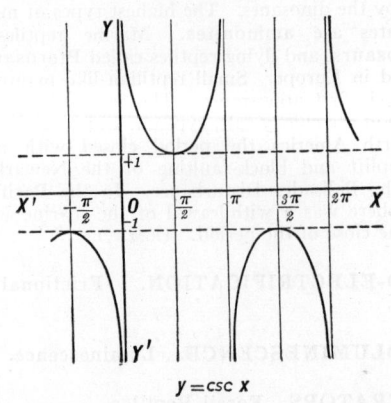

$y = \csc x$
Fig. 7.

The graph of $y = \sin x$ is a wave-curve. Its amplitude is its greatest height 1, its wavelength is 2π. The graph of $y = a \sin x$ has amplitude a and wavelength 2π. The graph of $y = \sin bx$ has amplitude 1 but wavelength $2\pi/b$. The graph of $y = a \sin bx$ has amplitude a and wavelength $2\pi/b$. The graph of $y = a \sin (bx + c)$ has phase difference of c from that of $y = \sin x$. The cosine graph differs from the sine graph in phase by $90°$ or $\frac{\pi}{2}$.

By adding corresponding ordinates of several simple sine wave-curves, we obtain new types of wave-curves, as for example that in Fig. 8.

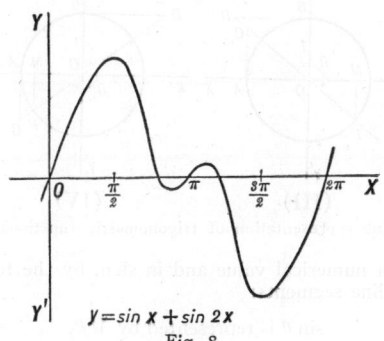

$y = \sin x + \sin 2x$
Fig. 8.

The graph of the equation $y = ae^{-kt} \cos(at + \beta)$ is shown in Fig. 9. It represents a damped vibration curve. (L.L.S.)

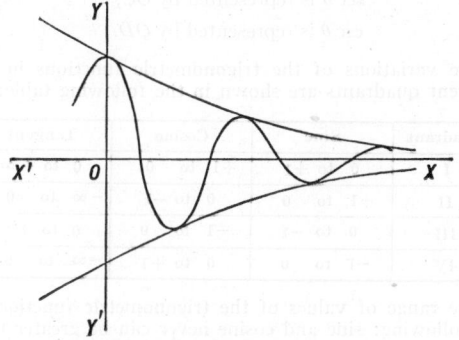

Fig. 9. Damped vibration wave.

TRIGONOMETRIC DIFFERENTIALS. Integration, Technique of.

TRIGONOMETRIC EQUATIONS. A trigonometric equation is a conditional equation which involves one or more trigonometric functions of an unknown angle.

Trigonometric equations have an endless number of solutions, due to the fact that trigonometric functions are periodic functions.

In solving trigonometric equations there are many methods which can be used, but one useful method in simple cases is to express all functions occurring in the equation in terms of one function, and then solve this equation algebraically for that function; from this the value of the unknown angle can then be found at once. The fundamental trigonometric identities will often be needed in transforming equations to simpler forms. (L.L.S.)

TRIGONOMETRIC FUNCTIONS. Let θ be any angle placed in standard position, with initial side along OX, and let P be any point on its terminal side; drop a perpendicular PM from P to the X-axis. Let x, y, r be the **abscissa, ordinate,** and **radius vector** of P. The **ratios** of these three directed distances are called the trigonometric ratios for angle θ; they are:

$$y/r, \ x/r, \ y/x; \ r/y, \ r/x, \ x/y.$$

Each of these six trigonometric ratios for an angle depends only on the angle and not on the position of the point chosen on the terminal side of the angle and each

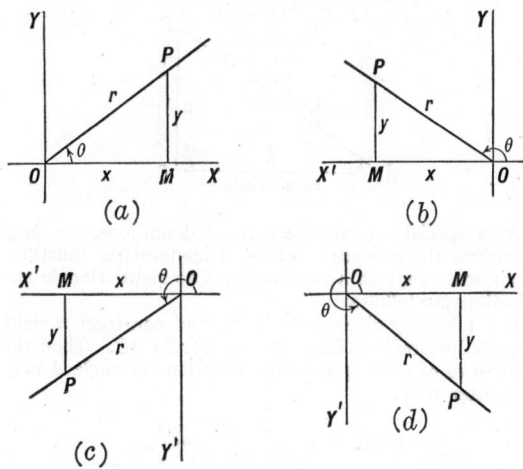

ratio changes when the angle changes. In other words, each trigonometric ratio is a **function** of the angle. The trigonometric ratios are therefore called the trigonometric functions.

The trigonometric ratios or trigonometric functions are named as follows:

$y/r =$ sine of θ, written $\sin \theta$,
$x/r =$ cosine of θ, written $\cos \theta$,
$y/x =$ tangent of θ, written $\tan \theta$,

and the reciprocals of these:

$r/y =$ cosecant of θ, written $\csc \theta$ (or cosec θ),
$r/x =$ secant of θ, written $\sec \theta$,
$x/y =$ cotangent of θ, written $\cot \theta$ (or ctn θ).

The functions are undefined when θ is such that a **zero** denominator occurs in these definitions.
$(\sin \theta)^2$ is written $\sin^2 \theta$, $(\cos \theta)^3$ is written $\cos^3 \theta$, $(\tan \theta)^n$ is written $\tan^n \theta$, etc.
Other functions sometimes used are:

versed sine of θ: vers $\theta = 1 - \cos \theta$,
coversed sine of θ: covers $\theta = 1 - \sin \theta$,
haversine of θ: hav $\theta = \frac{1}{2}(1 - \cos \theta) = \frac{1}{2}$vers θ,
exsecant of θ: exsec $\theta = \sec \theta - 1$.

The algebraic signs of the trigonometric functions in the various quadrants are:

Quadrant	sin	cos	tan	csc	sec	cot
I	+	+	+	+	+	+
II	+	−	−	+	−	−
III	−	−	+	−	−	+
IV	−	+	−	−	+	−

When any one of the trigonometric functions is given, the others are determined and may be found by use of the fundamental relations between the functions, or by a geometric method.

The trigonometric functions of quadrantal angles are given by the table:

	0°	90°	180°	270°	360°
sin	0	1	0	−1	0
cos	1	0	−1	0	1
tan	0	∞	0	∞	0

As $\theta \to 90°$ through smaller values, $\tan \theta \to +\infty$, and as $\theta \to 90°$ through larger values, $\tan \theta \to -\infty$, etc.

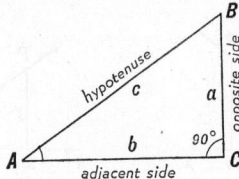

As a special case of the general definitions, we may formulate the definitions of the trigonometric functions of an acute angle in a form adapted to right triangle applications as follows:

Let A be any given acute angle and construct a right triangle by drawing BC perpendicular to AC. Then the definitions of the trigonometric functions of angle A may be written thus:

$$\sin A = \frac{\text{opposite side}}{\text{hypotenuse}} = \frac{a}{c},$$

$$\cos A = \frac{\text{adjacent side}}{\text{hypotenuse}} = \frac{b}{c},$$

$$\tan A = \frac{\text{opposite side}}{\text{adjacent side}} = \frac{a}{b},$$

$$\csc A = \frac{\text{hypotenuse}}{\text{opposite side}} = \frac{c}{a},$$

$$\sec A = \frac{\text{hypotenuse}}{\text{adjacent side}} = \frac{c}{b}$$

$$\cot A = \frac{\text{adjacent side}}{\text{opposite side}} = \frac{b}{a}$$

Each trigonometric function of the complement of an acute angle is equal to the co-function of the angle:

$$\sin (90° - A) = \cos A, \qquad \cos (90° - A) = \sin A,$$

$$\tan (90° - A) = \cot A, \qquad \cot (90° - A) = \tan A,$$

$$\sec (90° - A) = \csc A, \qquad \csc (90° - A) = \sec A.$$

The trigonometric functions of 30°, 45°, 60° are given in the following table:

Angle	sin	cos	tan
30°	$\frac{1}{2}$	$\frac{1}{2}\sqrt{3}$	$\frac{1}{3}\sqrt{3}$
45°	$\frac{1}{2}\sqrt{2}$	$\frac{1}{2}\sqrt{2}$	1
60°	$\frac{1}{2}\sqrt{3}$	$\frac{1}{2}$	$\sqrt{3}$

The trigonometric functions may be represented graphically, both in magnitude and sign, by certain geometric line segments. Draw a unit circle (of radius 1) with the vertex of the given angle at the center of the circle. Let P be the point where the terminal side of angle θ cuts the circle, and draw PM perpendicular to $X'X$, AC perpendicular to $X'X$, BD perpendicular to $Y'Y$. Then for each quadrant, the trigonometric functions are repre-

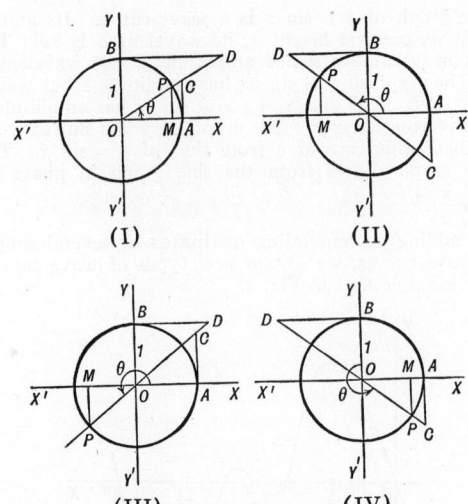

Line representation of trigonometric functions.

sented, in numerical value and in sign, by the following directed line segments:

$\sin \theta$ is represented by MP,

$\cos \theta$ is represented by OM,

$\tan \theta$ is represented by AC,

$\cot \theta$ is represented by BD,

$\sec \theta$ is represented by OC,

$\csc \theta$ is represented by OD.

The variations of the trigonometric functions in the different quadrants are shown in the following table:

Quadrant	Sine	Cosine	Tangent
I	0 to +1	+1 to 0	0 to +∞
II	+1 to 0	0 to −1	−∞ to 0
III	0 to −1	−1 to 0	0 to +∞
IV	−1 to 0	0 to +1	−∞ to 0

The range of values of the trigonometric functions is the following: sine and cosine never can be greater than +1 nor less than −1, the secant and cosecant never can be between +1 and −1, while the tangent and cotangent may take all positive and negative values.

Reduction to the first quadrant may be made as follows:

To find any trigonometric function of any angle θ, take the same function of the acute angle A between the terminal side of θ and the X-axis, and prefix the algebraic sign given by the rule of signs for the quadrants.

For the trigonometric functions of supplementary angles: The sine of an obtuse angle is equal to the sine of the supplementary acute angle; the cosine of an obtuse angle is equal to minus the cosine of the supplementary acute angle; and the tangent of an obtuse angle is equal to minus the tangent of the supplementary acute angle.

The trigonometric functions of a negative angle are given by the formulae:

$$\sin (-\theta) = -\sin \theta,$$

$$\cos (-\theta) = \cos \theta,$$

$$\tan (-\theta) = -\tan \theta.$$

An odd function $f(x)$ is one for which $f(-x) = -f(x)$, and an even function is one for which $f(-x) = f(x)$, for all values of x. The sine, tangent, cosecant, and cotangent are odd functions, and the cosine and secant are even functions.

Reduction formulae for the trigonometric functions:

$$\begin{cases} \sin\ (\ 90° - \theta) = & \cos\theta \\ \cos\ (\ 90° - \theta) = & \sin\theta \\ \tan\ (\ 90° - \theta) = & \cot\theta \end{cases} \quad \begin{cases} \sin\ (180° - \theta) = & \sin\theta \\ \cos\ (180° - \theta) = & -\cos\theta \\ \tan\ (180° - \theta) = & -\tan\theta \end{cases}$$

$$\begin{cases} \sin\ (\ 90° + \theta) = & \cos\theta \\ \cos\ (\ 90° + \theta) = & -\sin\theta \\ \tan\ (\ 90° + \theta) = & -\cot\theta \end{cases} \quad \begin{cases} \sin\ (180° + \theta) = & -\sin\theta \\ \cos\ (180° + \theta) = & -\cos\theta \\ \tan\ (180° + \theta) = & \tan\theta \end{cases}$$

$$\begin{cases} \sin\ (270° - \theta) = & -\cos\theta \\ \cos\ (270° - \theta) = & -\sin\theta \\ \tan\ (270° - \theta) = & +\cot\theta \end{cases} \quad \begin{cases} \sin\ (270° + \theta) = & -\cos\theta \\ \cos\ (270° + \theta) = & \sin\theta \\ \tan\ (270° + \theta) = & -\cot\theta \end{cases}$$

$$\begin{cases} \sin\ (360° - \theta) = & -\sin\theta \\ \cos\ (360° - \theta) = & \cos\theta \\ \tan\ (360° - \theta) = & -\tan\theta \end{cases} \quad \begin{cases} \sin\ (360° + \theta) = & \sin\theta \\ \cos\ (360° + \theta) = & \cos\theta \\ \tan\ (360° + \theta) = & \tan\theta \end{cases}$$

When n is an even integer, any trigonometric function of $n \cdot 90° \pm \theta$ is numerically equal to the same function of θ; when n is an odd integer, any trigonometric function of $n \cdot 90° \pm \theta$ is numerically equal to the co-function of θ; the algebraic sign of the result is that of the given function of $n \cdot 90° \pm$ an acute angle.

The six trigonometric functions are related to each other by various relations. The fundamental relations are:

1. The reciprocal relations:

$$\csc\theta = 1/\sin\theta, \quad \sin\theta = 1/\csc\theta,$$
$$\sec\theta = 1/\cos\theta, \quad \cos\theta = 1/\sec\theta,$$
$$\cot\theta = 1/\tan\theta, \quad \tan\theta = 1/\cot\theta.$$

2. The square relations:

$$\sin^2\theta + \cos^2\theta = 1,$$
$$1 + \tan^2\theta = \sec^2\theta,$$
$$\cot^2\theta + 1 = \csc^2\theta.$$

3. The quotient relations:

$$\tan\theta = \frac{\sin\theta}{\cos\theta}, \quad \cot\theta = \frac{\cos\theta}{\sin\theta}.$$

Formulae for the trigonometric functions of the sum and difference of two angles are:

$$\sin\ (x \pm y) = \sin x \cos y \pm \cos x \sin y,$$
$$\cos\ (x \pm y) = \cos x \cos y \mp \sin x \sin y,$$
$$\tan\ (x \pm y) = \frac{\tan x \pm \tan y}{1 \mp \tan x \tan y}.$$

These are often called the addition theorems for the trigonometric functions.

From these are obtained the following formulae for transforming sums and differences of trigonometric functions into products, and products into sums and differences:

$$\sin x + \sin y = 2 \sin \tfrac{1}{2}(x + y) \cos \tfrac{1}{2}(x - y),$$
$$\sin x - \sin y = 2 \cos \tfrac{1}{2}(x + y) \sin \tfrac{1}{2}(x - y),$$
$$\cos x + \cos y = 2 \cos \tfrac{1}{2}(x + y) \cos \tfrac{1}{2}(x - y),$$
$$\cos x - \cos y = -2 \sin \tfrac{1}{2}(x + y) \sin \tfrac{1}{2}(x - y);$$

and

$$2 \sin x \cos y = \sin\ (x + y) + \sin\ (x - y),$$
$$2 \cos x \sin y = \sin\ (x + y) - \sin\ (x - y),$$
$$2 \cos x \cos y = \cos\ (x + y) + \cos\ (x - y),$$
$$-2 \sin x \sin y = \cos\ (x + y) - \cos\ (x - y).$$

Formulae for the trigonometric functions of a double angle are:

$$\sin 2x = 2 \sin x \cos x,$$
$$\cos 2x = \cos^2 x - \sin^2 x,$$
$$= 1 - 2 \sin^2 x,$$
$$= 2 \cos^2 x - 1,$$
$$\tan 2x = \frac{2 \tan x}{1 - \tan^2 x}.$$

Half-angle formulae for the trigonometric functions are:

$$\sin \tfrac{1}{2}x = \pm\sqrt{\frac{1 - \cos x}{2}},$$
$$\cos \tfrac{1}{2}x = \pm\sqrt{\frac{1 + \cos x}{2}},$$
$$\tan \tfrac{1}{2}x = \pm\sqrt{\frac{1 - \cos x}{1 + \cos x}},$$
$$\tan \tfrac{1}{2}x = \frac{1 - \cos x}{\sin x} = \frac{\sin x}{1 + \cos x}.$$

Transformation of $a \cos x + b \sin x$ is given by:

$$a \cos x + b \sin x = c \cos\ (x - \alpha),$$

where $c = \sqrt{a^2 + b^2}$, $\tan \alpha = b/a$;

$$a \cos x + b \sin x = c \sin\ (x + \beta),$$

where $c = \sqrt{a^2 + b^2}$, $\tan \beta = a/b$.

All co-terminal angles have the same values for their trigonometric functions:

$$f(\alpha \pm n \cdot 360°) = f(\alpha),$$

or

$$f(\alpha \pm 2n\pi) = f(\alpha),$$

where f denotes one of the trigonometric functions.

Hence, all six trigonometric functions are **periodic functions** with the period 2π (radians), or $360°$. This is the least period for sine, cosine, secant and cosecant, but tangent and cotangent have also a smaller period π (or $180°$) since

$$\tan\ (x + \pi) = \tan x, \quad \cot\ (x + \pi) = \cot x$$

and

$$\tan\ (x + k\pi) = \tan x, \quad k \text{ any integer, etc.}$$

The trigonometric functions all repeat their values in cycles of $360°$.

When the angle is small and expressed in **radian measure**, the sine, tangent and the angle have important relations, expressed by:

$$\lim_{\theta \to 0} \frac{\sin\theta}{\theta} = 1 \quad \text{or} \quad \sin\theta \approx \theta,$$

$$\lim_{\theta \to 0} \frac{\tan\theta}{\theta} = 1 \quad \text{or} \quad \tan\theta \approx \theta,$$

when θ is small and is expressed in radians.

A function expressible in the form $y = c \sin\ (kx + \phi)$, where c, k, ϕ are constants, is a simple harmonic function; c is the amplitude, ϕ is the phase angle and k determines the period: $k/2\pi$ is the frequency and $2\pi/k$ is the period. It may also be written in the form $y = a \cos kx + b \sin kx$. (L.L.S.)

TRIGONOMETRIC IDENTITIES. Trigonometric Functions.

TRIGONOMETRIC INTEGRALS. Integration, Technique of.

TRIGONOMETRIC RATIOS. Trigonometric Functions.

TRIGONOMETRIC TRANSFORMATIONS. Trigonometric Functions.

TRIGONOMETRY. The name "trigonometry" is derived from two Greek words meaning measurement or solution of **triangles**.

While the solution of triangles forms an important

part of modern trigonometry, it is by no means the only part or even the most important part. In the development of methods for the solution of triangles by computation, certain **functions** of **angles** occur, and the study of the properties of these functions and their applications to various mathematical problems, including the solution of triangles, constitutes the subject matter of trigonometry.

In plane trigonometry, the solution of plane triangles is considered; spherical trigonometry treats of the solution of spherical triangles.

An idea of the subject matter usually assigned to trigonometry may be gained by consulting the following topics: **Angles, Degree Measures of Angles, Radian Measure of Angles, Trigonometric Functions, Trigonometric Curves, Trigonometric Equations, Inverse Trigonometric Functions, DeMoivre's Theorem, Logarithms, Triangles.** (L.L.S.)

TRILOBITES. Invertebrate Paleontology.

TRIM, AIRPLANE. The trim of an airplane denotes the balance of aerodynamic forces around a lateral axis through the center of gravity. An airplane is symmetrical about the other two axes of motion, and consequently would normally always be in trim with respect to those axes. An airplane would be in trim if it had no tendency to nose up or down in steady unaccelerated flight with the elevator control neutralized. Although an airplane which is not thus trimmed can be held in a steady attitude by the use of elevator controls, it is convenient for the pilot on cross-country journeys to be relieved of the steady balancing pressure, which he might otherwise have to maintain, by being able to make trimming adjustments. The two methods commonly employed to make the trim adjustment are: 1. Adjustment of the leading edge of the stabilizer up or down so as to change the tail angle. 2. An adjustment of a **tab** located in the trailing edge of the elevator. In stabilizer adjustment the angle of attack of the tail is altered until the air reaction required for balance is obtained. Tab adjustment produces a change in the effective camber of part of the elevator surface, which causes the elevator to ride in a displaced position, thus changing the effective camber of the whole tail surface, and achieving a desired balancing load.

When perfectly trimmed, the airplane should maintain steady flight at a constant angle of attack with hands off the controls. In this condition the moment of the air reaction on the horizontal tail surfaces is equal in magnitude and opposite in nature to the sum of the moments of the thrust, wing lift and drag, and parasitic drag, all moments being taken with reference to the center of gravity of the complete airplane. Due to the change of center of pressure on a wing at different angles of attack, the magnitude of tail moment necessary for balance varies with the speed of the airplane. Consequently, when the forward speed is changed a new position of stabilizer or tab is required for trim.

The trimming control on an airplane is usually a small crank or hand wheel located accessible to the pilot, but placed in a subordinate position, as it is one of the less frequently operated and less essential controls. (F.T.M.)

TRIMMING MOMENT. Trim, Airplane.

TRIODE. A 3-electrode, thermionic **vacuum tube.** The thermionic current in a hot-cathode tube is unidirectional; that is, reversal of the voltage stops the current, so that the tube acts as a "valve" or rectifier. If between the hot cathode and the anode or "plate" of such a tube there is interposed a wire mesh or grid, and if this third electrode is given a slight negative potential or bias, the effect is to reduce the flow of electrons somewhat, by an amount dependent upon this potential. For a suitable adjustment of the voltages and the filament temperature, the thermionic current may prove very sensitive to changes in the grid potential. Very feeble electric oscillations communicated to the grid may then appear as large fluctuations in the main thermionic current, and the triode becomes an **amplifier.** This device, originated by DeForest, is now a most important factor in **radio** reception and in many other types of apparatus requiring high amplification of electric impulses.

By so connecting the circuits that the current fluctuations are made to "feed back," that is, to affect the grid voltage, the amplifier becomes regenerative, tending to build up the fluctuations to a maximum. Large tubes so operating are used to produce the electric oscillations necessary for the emission of the carrier wave in radio broadcasting, sometimes radiating energy at the rate of many kilowatts. The details of the circuits used for these purposes often involve highly complicated radio technique.

To secure the greatest sensitivity, the potential of the grid is adjusted to a value dependent upon the par-

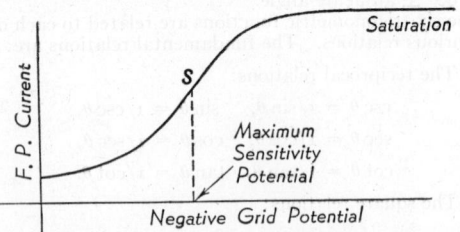

Characteristic curve for typical triode, which is most sensitive at S.

ticular triode. This value is indicated on the "characteristic curve" of the triode as the point S of maximum slope (see figure), where the filament-to-plate current varies most rapidly with the grid potential. (See **Tube, Electron.**) (L.D.W.)

TRIP HAMMER. Forging.

TRIPACK. Bipacks and Tripacks.

TRIPLE INTEGRAL. Let $f(x, y, z)$ be a **continuous function** in a region V of space. Let this region be divided up into n sub-regions ΔV_k, and let (x_k, y_k, z_k) be any point in ΔV_k. Then form the sum $\sum_{k=1}^{n} f(x_k, y_k, z_k)\Delta V_k$. Let $n \to \infty$ and let the greatest diameter of each $\Delta V_k \to 0$. The **limit** of this sum is then called the triple integral of $f(x, y, z)$ extended over the region V, and we write

$$\lim_{\Delta V_k \to 0} \sum_{k=1}^{n} f(x_k, y_k, z_k)\Delta V_k = \iiint_V f(x, y, z)dV.$$

An iterated triple integral

$$\int_{x_1}^{x_2}\left[\int_{y_1}^{y_2}\left(\int_{z_1}^{z_2} f(x, y, z)dz\right) dy\right] dx$$

means a series of successive integrations; z_1 and z_2 will in general be functions of x and y, and y_1 and y_2 will in general be functions of x, and x_1 and x_2 will be constants. Such an iterated (or repeated) triple integral is sometimes written

$$\int_{x_1}^{x_2} \int_{y_1}^{y_2} \int_{z_1}^{z_2} f(x, y, z)dz\, dy\, dx,$$

or

$$\int_{x_1}^{x_2}dx \int_{y_1}^{y_2}dy \int_{z_1}^{z_2} f(x, y, z)dz.$$

A triple integral may be interpreted geometrically as a **volume.**

The fundamental theorem for triple integrals may be stated: A triple integral $\iiint_V f(x, y, z)dV$ extended over a region V is equal to an iterated triple integral of the form

$$\int_{x_1}^{x_2}\left[\int_{y_1}^{y_2}\left(\int_{z_1}^{z_2} f(x, y, z)dz\right)dy\right]dx,$$

where the limits z_1, z_2, y_1, y_2, x_1, x_2 depend on the boundary of the region V.

In **spherical coordinates**, the element of volume $dx\,dy\,dz$ in a triple integral becomes $r^2 \sin\phi\,d\phi\,d\theta\,dr$.

In **cylindrical coordinates**, the element of volume $dx\,dy\,dz$ becomes $r\,d\theta\,dr\,dz$. (L.L.S.)

TRIPLE POINT. Three-Phase Equilibrium.

TRIPLE PRODUCTS OF VECTORS. The triple scalar product of three vectors $(a \times b)\cdot c$ is a scalar, which may be represented by the volume of the parallelepiped on **a, b, c** as adjacent edges.

An important formula for the triple scalar product is:

$$(a \times b)\cdot c = \begin{vmatrix} a_1 & a_2 & a_3 \\ b_1 & b_2 & b_3 \\ c_1 & c_2 & c_3 \end{vmatrix}.$$

This triple scalar product $(a \times b)\cdot c$ is sometimes denoted by $[a\,b\,c]$.

The condition that three vectors **a, b, c** lie in one plane is $(a \times b)\cdot c = 0$, or $[a\,b\,c] = 0$.

The triple vector product $(a \times b) \times c$ is represented by the important formula

$$(a \times b) \times c = (c\cdot a)b - (c\cdot b)a.$$
(L.L.S.)

TRIPLOBLASTIC ORGANIZATION. A form of organization based on three embryonic **germ layers**. With the exception of the phyla **Protozoa, Porifera, Coelenterata**, and possibly **Ctenophora**, all animals begin their development with the formation of these three layers, ectoderm, mesoderm, and endoderm. The ectoderm and endoderm develop first as outer and inner layers of the body wall, and the development of **diploblastic** animals proceeds by the differentiation of adult structures from these layers. In triploblastic groups a third layer forms between the two; from this mesodermal layer other parts of the **adult** are formed. (A.W.L.)

TRIPOLI. Abrasives.

TRISECTRIX OF MACLAURIN. This curve is the **locus** of the equation in rectangular coordinates

$$x^2 + xy^2 + ay^2 - 3ax^2 = 0.$$

It can be used for the trisection of an angle. (L.L.S.)

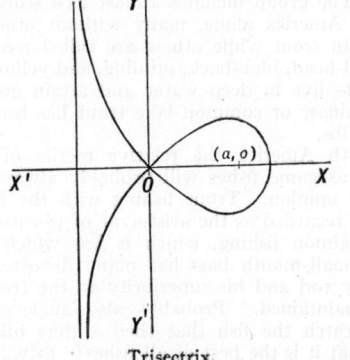
Trisectrix.

TRITOCEREBRUM. The third (posterior) division of the brain of **arthropods**. It is derived from the pair of ganglia of one segment of the head. (A.W.L.)

TROCHELMINTHES. Rotatoria.

TROCHOID. A trochoid is a mathematical curve which may be regarded as a generalization of the cycloid.

If a circle of radius a rolls along a straight line, the locus of a point P on a radius of the circle at a distance

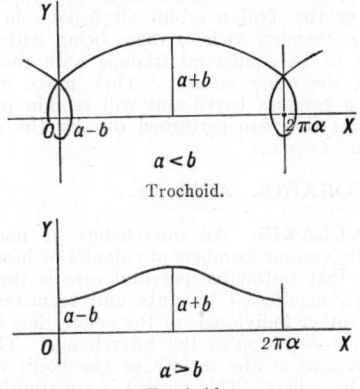
Trochoid.

Trochoid.

b from the center is called a trochoid. If $b > a$ (point outside the circle), the curve is frequently called a prolate cycloid, and if $b < a$ (point inside the circle), the curve is called a curtate cycloid.

Parametric equations of the trochoid are:

$$x = a\theta - b\sin\theta, \quad y = a - b\cos\theta.$$
(L.L.S.)

TROCHOPHORE. An invertebrate **larva** with a bent or curved tubular alimentary tract (**Digestive System**) lined with ciliated (**Cilia**) tissue. The surface of the body bears an encircling band of cilia and a tuft at the apex of the body. These larvae develop from the fertilized egg and after drifting and swimming for a time undergo a transformation to become adults.

Trochophore larvae are found in the phyla **Bryozoa, Brachiopoda, Phoronidea, Annelida**, and **Mollusca**. They are also called trochospheres. (A.W.L.)

TROCHUS. The ciliated (**cilia**) disk at the end of the body of a **rotifer**. In many species the end of the body is flattened and is surrounded by two rows of cilia, with the mouth opening between. The outer part is called the cingulum and the disk surrounded by the inner row of cilia the trochus. (A.W.L.)

TROCTOLITE. Gabbro.

TROGON. Aves, Cuculiformes. A brilliantly colored and beautifully feathered bird (**Aves**) of the Oriental region, tropical Africa, and Central and South America. A single species, the Coppery-tailed trogon, *Trogon ambiguus*, enters southern Texas and Arizona. These birds have a moderately large beak, strong and curved. Crests and plumes are highly developed in some of the males, and in this sex the tail is sometimes very long.

The **quezal** of Central America is one of the most widely known of the trogons, probably because it has been pictured frequently on postage stamps of Guatemala and partly because its brilliant colors persist in museum specimens. The male is brilliant green with blood-red under parts below the breast. It has a very long tail, drooping plumes over the wings, and the head bears a rounded crest. (A.W.L.)

TROJAN GROUP. This is the name applied to a group of 11 asteroids which carry the names of the various heroes of the Trojan wars. This group of asteroids all have their periods and mean distances nearly identical with the planet **Jupiter.** The group is of great theoretical interest because of the fact that they represent examples of the solution of the **three-body problem** proposed by Lagrange. Lagrange proved theoretically that an object so located that it is equidistant from both Jupiter and the sun would be in a stable position, i.e., would remain there and continue to go about the sun with the same period as Jupiter. The members of the Trojan group all behave in approximately this manner, each of them being within 20° of the vertex of an equilateral triangle with the sun and Jupiter at the other vertices. They move about this vertex in a complex curve and will remain in this vicinity unless they are perturbed out by the attraction of **Saturn.** (W.K.G.)

TROPACOCAINE. Alkaloids.

TROPHALLAXIS. An interchange of nourishment between the various members of colonies of insects. The older idea that instinctive parental care is the basis of the concern manifested by **ants** and **termites** for the welfare of other individuals of the colony has been substituted by recognition of this interchange. The insects exude secretions at the surface of the body which are licked off by others. This reward is apparently responsible for much of the solicitude. Even the insects of other species living within such colonies are welcome guests in some cases because of a similar contribution to their hosts. The relationship has complex ramification which explain other aspects of colonial organization, such as cannibalism, that are difficult to reconcile with the older interpretations of mutual aid within the colony. (A.W.L.)

TROPHI. Mastax.

TROPHOSOME. One of several terms used to designate individuals that take in and digest food in colonies of primitive animals, or the stage of development of an individual during which it eats and grows. Trophosome is applied to the **polyp** stage of **hydrozoa,** also called gasterozooids. The term trophozooid is applied to an early stage in the development of the **coral** polyps of the family Fungidae, known as mushroom corals, and the term autozooid to individuals that take food in **alcyonarian** colonies. The root appears again in trophozoite, applied to the growing stage of some parasitic Protozoa (**Sporozoa**). (A.W.L.)

TROPHOZOITE. Trophosome.

TROPHOZOOID. Trophosome.

TROPICAL AIR MASS. Air Masses.

TROPICAL CYCLONE. Hurricanes.

TROPINE. Alkaloids.

TROPINONE. Alkaloids.

TROPISM. For the use of this term in botany, see **Movement in Plants.** In zoology, a tropism is an unavoidable response of an animal to some environmental stimulus involving the orientation of the body in relation to the causative factor. In very simple animals this type of reaction is common. **Protozoans,** for example, may always move toward or away from light, and even animals with organized nervous systems may have some nerve paths associated in such a way that a given condition always evokes the same reaction. This type of response is also called a taxis, a term applied to the responses made without a nervous system and at the other extreme to simple reflexes, which may be nervously intricate although they are automatic.

Both tropism and taxis are combined with various prefixes for special cases. Thus a topotropism or topotaxis is a reaction toward the inciting stimulus and a phobotropism is a withdrawal from the stimulus. These reactions are also designated as positive and negative. Reactions to chemical stimuli are sometimes called chemotropisms. Light evokes phototaxis. By such terms any special orientation of this nature may be concisely expressed. (A.W.L.)

TROPOPAUSE. The discontinuity surface separating the **stratosphere** from the **trophosphere.** It varies in height from about 55,000′ at the equator to 25,000′ over the poles. (P.E.K.)

TROPOSPHERE. The air below the tropopause in which the temperature, on the average, decreases with altitude. All storms occur in the troposphere. (P.E.K.)

TROUGH LINE. A line drawn in a pressure field along which the isobars are symmetrical and curved cyclonically. A V-shaped trough normally contains a front; a U-shaped trough generally contains no front or a very weak one. Usually there is considerable weather associated with a trough line of the V variety. Trough line movements can be computed and a forecast made of future positions. (P.E.K.)

TROUPIAL. Aves, Passeriformes. A name derived from the French and applied variously to members of the family Icteridae, including the **orioles, blackbirds,** and New World **grackles.** The name has been used by various writers for the grackles, for the orioles, and for all members of the family. Also spelled troopial. (A.W.L.)

TROUT. Pisces, Teleostei. A leading game fish (**Pisces**) found in cold waters, including lakes and streams and, some species, in salt water. The various species commonly called trout have been classified with the **salmons** in the genus *Salmo,* and separately in the genera *Trutta, Salvelinus,* and others. They vary greatly in size and to a slight extent in game qualities. The scales are minute to small and the flesh is superior.

Trout are common in streams of the northern hemisphere and have been widely transported for stocking other than their native waters. The brook or speckled trout, *Salvelinus fontinalis,* is the most widely distributed species, ranging from Maine to the Dakotas and northward throughout the continent. Rainbow trout, *Trutta iridea,* from the west coast, are now widely stocked in other waters, and, in places where trout fishing is important, imported species, including the European brown trout, *T. trutta,* and the Lochleven trout, are to be expected. The group includes almost two score of species in North America alone, many without other common names than trout while others are called redfish, greenback, steel-head, blue-back, saibling, and yellow-fin. The lake trouts live in deep water and attain great weight. The mackinaw or common lake trout has been recorded up to 90 lbs.

In North America the relative merits of the trout and bass as game fishes will probably always remain a matter of opinion. Trout fishing with the fly rod has long been regarded as the aristocrat of piscatorial sports, next to salmon fishing, which is less widely available, but the small-mouth bass has many devotees as a fish for the fly rod and his superiority to the trout is often loyally maintained. Probably also anglers will continue to catch the fish that their waters offer, and to be sure that it is the best of all fishes! (A.W.L.)

TROUTON-NOBLE EXPERIMENT. Ether.

TROUTON'S LAW. Heat of Vaporization.

TROUT-PERCH. Pisces, Teleostei. A common fish (Pisces) of the Great Lakes and the rivers of the eastern half of the United States, rare in the south. It attains a maximum length of about 10″, with scales like the perches and otherwise similar to the trout. It is one of two species of the family Percopsidae. (A.W.L.)

TRUE BEARING. Bearing.

TRUEING. Grinding.

TRUFFLE. Ascomycetes.

TRUMPETER. Aves, Gruiformes. *Psophia*. A long-legged and long-necked bird (**Aves**) of South America. The few species are characteristically terrestrial in habits, living in the forests and flying poorly. They live in flocks and are said to be tamed in Brazil for the protection of domestic fowls, with which they live contentedly. The word also appears in the names of the trumpeter-hornbills of Africa and the trumpeter swan of North America, both species of other orders. (A.W.L.)

TRUNCUS ARTERIOSUS. The great arterial vessel leading from the **heart** of **vertebrates** in the primitive unpaired condition and in the **embryo**. All blood leaving the heart passes through this vessel, which is differentiated to form the conus arteriosus and the ventral aorta in the fishes and persists as a short trunk in the **amphibians**. In the **reptiles** its subdivision to form the aorta and the pulmonary artery is begun and in the birds and mammals this splitting is complete. The truncus is divided by the growth of a pair of ridges in a spiral course along opposite walls. These ridges unite to form a partition. (A.W.L.)

TRUNK. 1. The body of a **vertebrate**, bearing the neck and head at one end and the tail at the other, with the two pairs of appendages attached laterally or ventrolaterally near the two ends. The trunk contains the body cavity in which the viscera lie, and is supported by the vertebral column and the ribs. It is divisible into an anterior thorax and a posterior abdomen. 2. The proboscis of the elephant. This is a muscular organ formed of the elongated nose and upper lip. It contains the greatly elongated nasal passages, which are used for raising water and small particles of food, such as grain, to the mouth by inhaling them into the terminal parts of the tubes. The appendage is also used for grasping by prehension and by the opposition of two finger-like processes at its tip. (A.W.L.)

TRUNK, TELEPHONE. This is the name of a **line** connecting two or more telephone switching points. Thus trunks connect the various central offices within an exchange area. They also connect private exchanges with the telephone company's office. It takes relatively few trunk lines since many calls passing through a switchboard will be for other phones connected to the same board and hence will require no trunk. (See **Telephony**.) (L.R.Q.)

TRUSS. A truss is a framed structure composed of a series of adjoining triangles which are formed by straight members, all lying in one plane. The members are connected at their points of intersection by **pins** or **gusset** plates or by welding. The point of intersection is called a panel point or joint. Since rigidity of the truss is secured by triangles which cannot deform without changing the length of the sides, it is generally assumed that loads applied at the panel points will produce direct stress only. This will be true if the gravity axes of the

members meet at the panel points and the effect of secondary bending stresses, caused by the **deflection** of the truss, is disregarded. If a load is applied to a member between the panel points it must distribute the load by beam action to the adjacent panel points. The top line of members of a truss is called the top **chord** and the lower line is known as the lower or bottom chord. The members connecting the top and bottom chords are

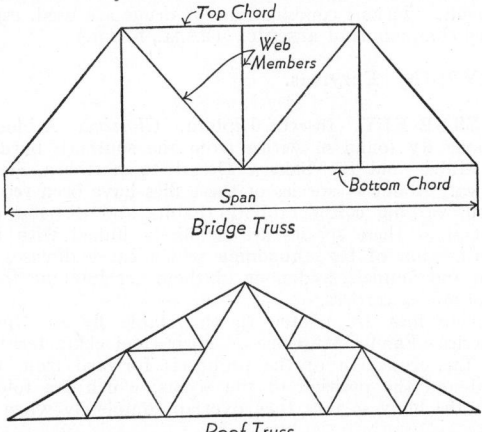

Bridge Truss

Roof Truss

called web members. The span of the truss is the horizontal distance center to center of end bearings.

Framed bridges consist of two or more parallel vertical trusses which support the floor system. The lateral bracing in the plane of the top and bottom chords of bridges form the web members of horizontal trusses. A roof truss is one which supports the roof of a building. Trusses are also used over auditoriums to carry the loads from the floors above. The members of a truss are usually made of steel or wood, although concrete trusses have been used for bridges. Steel has the advantage over wood in that it has far greater strength and is less subject to deterioration, but the latter is frequently used for temporary structures and is becoming increasingly popular for reasonably permanent structures. (See **Fuselage; Wing Truss**.) (C.W.C.)

TRYPANOSOME. Protozoa, Mastigophora. A one-celled parasitic animal found in the **blood** plasma of **vertebrates**. The body is characteristically long and slender, with a single **nucleus** and one **flagellum** arising from a small basal body and running along the edge of an undulating membrane, sometimes through the greater part of the length of the body, forming a short free portion at the end. These parasites constitute the genus *Trypanosoma*.

Some species of trypanosomes cause serious diseases of man and domestic animals. Among them *Trypansoma gambiense*, the cause of African sleeping sickness, has been widely publicized. This species is transmitted by the **tsetse fly**, a blood-sucking insect related to the stable fly of our own continent. In South America Chagas' disease is caused by a trypanosome transmitted by a bug. Still another species causes the very destructive disease, nagana, which affects both wild and domestic animals in Africa. (A.W.L.)

TRYPANOSOMIASIS (SLEEPING SICKNESS). A chronic parasitic disease of African and South American regions characterized by wasting, fever, lassitude, and profound lethargy. The parasites responsible for the disease are known as *Trypanosoma gambiense* and *Trypanosoma rhodesiense* (African forms) and *Trypanosoma cruzi* (South American form). The parasites are transmitted by the **tsetse fly**, in which they undergo various changes.

The incubation period in man varies from 10 to 20 days. The acute stage is characterized by irregular fever, enlargement of the **lymph nodes**, and skin rashes. Later, after a latent period of months or years, brain symptoms occur (sleeping sickness). This chronic stage is characterized by headache, mental dullness, paralysis, coma, and convulsions, usually with eventual slow death. A form of this disease occurs which is acute throughout its course, resulting in death. Treatment. of all types is difficult. Various combinations of drugs are used, especially compounds of **arsenic**. (R.S.M., D.M.H.)

TRYPSIN. Enzymes.

TSETSE FLY. Insecta, Diptera. *Glossina.* A blood-sucking fly found in Africa from the southern borders of Arabia and the Sahara Desert to northern South Africa. About 15 species of these flies have been recognized, varying considerably in habits and distribution. Several of these species are definitely linked with the transmission of **trypanosomes** which cause diseases of man and animals, and none of them are immune from suspicion as carriers.

These flies are related to the stable fly of North America (Family Muscidae). Their chief characteristics are the projection of the proboscis forward from the head and the position of the wings, which are folded over the body so that they overlap completely when at rest.

Tsetse flies suck plant juices as well as blood, and are known to attack animals of many species, both warm- and cold-blooded. They show many interesting reactions. Moving objects attract them more readily than others, and they see dark colors more readily than light. White persons are said to be bothered very little by them when in the company of Negroes. Some species frequent watercourses while others are found in dry areas.

The problem of controlling tsetses in Africa has been important because of the great virulence of the diseases that they transmit. (A.W.L.)

TSUNAMI (SUNAMI). Earthquakes.

TUATARA, TUATERA. Reptilia, Prosauria. A primitive **reptile**, *Sphenodon punctatus,* of New Zealand, superficially like the lizards but different in several anatomical details. It is chiefly noteworthy for the high development of the pineal or median eye. (A.W.L.)

TUBE. The tube is a hollow cylindrical body having a length much greater than its diameter. Generally the tube is used to conduct a fluid, though it may also be used as a container or structural member. Tubes are made of many and varied materials; in fact, almost anything that may be worked into tubular shape has a use somewhere as a tube: steel, iron, copper, brass, glass, cardboard. Metal tubes may be made by rolling a piece of skelp to tubular shape and welding or soldering the joint. This type of tubing has been superseded by seamless tubing, by virtue of the development of processes for producing seamless tubes from a billet. A billet is heated to a plastic state, then pierced by forcing it over the point of a mandrel with revolving rolls which work the metal from the center towards the outside, so that a short thick tube is produced. This intermediate product is then passed between several other mandrels and rolls, each one of which effects a thinning of the tube wall and an elongation. After passing through finishing rolls, the tubes are discharged to cooling and straightening tables. Seamless tubes may also be produced by cupping a circular plate. The plates are first pressed into bowl-shape and then drawn through several dies, each succeeding one being of smaller diameter. The dies eventually bring the tube down to the final shape. A steel tube is differentiated from a steel **pipe** by the seamless character, and by the nominal size. The tube is sized by its external diameter, the pipe by its nominal internal diameter. (F.T.M.)

TUBE, ELECTRON. The electron tubes form the heart of many of the modern developments in communication and industry. It is safe to say that without them we would have no **radio, television,** satisfactory long-distance telephone service, and many of our common every-day necessities could not be produced as cheaply and as well without the tubes. The electron tube is a device in which **electrons** are freed from the restraints of a solid conductor, pass across a free space (vacuum or gas at low pressure) and are again collected by a solid conductor, but during this passage in free space are controlled in manners which would be impossible if they had not been temporarily freed. Thus we may control the flow of current magnitude in a wire by controlling the voltage across its terminals, but when the electrons are passing between the electrodes of a tube we can superimpose many kinds of complicated controls if desired.

Tubes may be classified in various ways and no one way gives a complete picture. As a first general classification we might use vacuum and gas-filled, the first being a tube which has been as highly evacuated as possible and the latter being one which has been evacuated and then partially refilled with a gas at low pressure. The characteristics of the two groups are usually quite different. Then they may be classified according to the number of elements, two-electrode tubes being diodes (the simplest electron tube must have at least two electrodes), three-electrode tubes being **triodes,** four being tetrodes, five being pentodes and so on. To give anything approaching a satisfactory picture of the tube it is necessary to use both types of classification and this is usually done.

Let us examine the vacuum-tube family first, starting with the simple diode. This has a thermionic cathode which may be directly (filament type) or indirectly heated to produce electron emission. The other electrode is the anode which collects the electrons. Since electrons are negative charges they are attracted by positive-charged electrodes and repelled by negatively charged ones. Thus if the anode is made positive with respect to the cathode the emitted electrons will be attracted to it, the current usually varying as the 3/2 power of the applied voltage. If the anode is negative it repels the electrons and no current flows from one electrode to the other since the electrons are the carriers of the charge. It should be noted that the direction of the electron flow is opposite the conventional current direction, the result of an unfortunate error in understanding the nature of electricity in the early days of the science. We can see, then, that if an alternating voltage is impressed across a diode current will flow only during the part of the cycle in which the anode is positive, thus giving rectifier action. This is the major use of this tube.

If a third electrode is inserted in the form of a grid between the **cathode** and **anode** its potential will also affect the electron flow across the tube. As this new electrode, the grid, is much closer to the cathode than the anode is, a given voltage on it will have much more effect on the electrons than the same voltage wou'd if impressed on the plate. This is the basis of the amplifying action of all the grid-controlled tubes and the amplification factor of the tube is a measure of this relative effectiveness, varying from two or three to approximately a hundred in triodes and much higher for tetrodes and pentodes. The electrons which are controlled by the voltage on the grid pass on through the open grid structure and go to the plate (when it is positive as it normally is), thus giving the grid control over the plate current. Another advantage of this control is that the grid takes little or no power from the controlling circuit, yet the controlled power in the plate circuit may be quite large (there is no creation of energy involved since the additional power comes from the direct current operating source).

The triode has undesirably high capacity between the anode and grid so a second grid may be put between these two electrodes to shield them electrostatically from one another. This gives the screen grid or tetrode tube. The screen grid is operated at a positive potential and ordinarily serves no purpose except shielding although in certain applications it is used for other effects. This tube is characterized by a high **amplification factor**, often many hundred, and a high dynamic **plate resistance**, running well over a megohm in some cases. Its big disadvantage is a **secondary emission** effect when the plate potential is lower than that of the screen. This is overcome in the pentode by inserting a third grid, the suppressor, between the screen grid and the plate, the suppressor being connected to the cathode. This new grid effectively suppresses secondary emission and at the same time the tube retains the desirable characteristics of the tetrode. Another method of accomplishing the same result is the beam tube which is often used for power **amplifiers**. In this tube the electron stream across the tube is focused in such a way that it serves as its own suppressor.

Certain special-purpose vacuum tubes have still other electrodes, the pentagrid converter tube, for example, has five grids which are utilized to make the tube serve both as the **oscillator** and mixer tube of a superheterodyne **receiver**. Other multigrid tubes are used for mixing two or more signals to give a combined effect in the output. Other tubes have effectively two distinct tubes within the same envelope, e.g., a diode or duodiode and triode or pentode are often in the same envelope to serve as **detector** and first audio **amplifier** of a receiver.

The vacuum-tube family ranges from extremely small tubes used for **ultra high frequency** work to the giants of many kilowatt capacity used in radio **transmitters** and large industrial applications. Many of these larger sizes are water-cooled to dissipate the large amount of heat generated by the losses in the tube. The diodes are often called kenotrons and the grid-controlled tubes pliotrons, particularly in industrial applications.

The second group of tubes, the gas-filled, have in general quite different characteristics due to the ionization of the gas during operation. The diode is used primarily as a rectifier and is similar in general construction to the vacuum diode. It is possible to have a cold cathode in a gas-filled tube and in some low power applications such a cathode is often used. Besides the passage of electrons as in the kenotrons, the ionized gas gives positive ions as additional current carriers. The passage of the original electrons across the tube will produce ionization by colliding with the gas molecules if the voltage applied across the tube is sufficient to give the electrons the necessary velocity. This ionization gives rise to more electrons and the positive ions, the latter being attracted to the negative cathode where they neutralize the space charge sheath around the cathode. The result of this is a greatly lowered tube voltage drop with the tube current going up tremendously and no longer following the $\frac{3}{2}$ power law. In fact, if the current is not limited by external **resistance** it reaches destructive values in a small fraction of a second. Hence gas-filled tubes must always be used in circuits with enough resistance to limit the current to safe values. Gas-filled diodes are widely used for rectifiers where large currents are desired since the drop across the tube is low and consequently the efficiency is high. The usual hot cathode tube has a drop between 8 and 15 volts. The **mercury pool tube** is a special type of gas-filled diode. The hot cathode tubes are often called phanotrons.

If the grid is inserted between the cathode and anode of a gas-filled tube the effect is decidedly different from the vacuum tube. At low grid voltages the tube behaves as a vacuum tube, but as the grid voltage is raised, the gas suddenly breaks into a discharge, the voltage drop falling to a low value and the current increasing to a high value. Thus a voltage on the grid can trigger a large current in the plate circuit. This grid-controlled, hot cathode, gas-filled tube is the thyratron. The behavior of the thyratron can be forecast from its grid control characteristic curve, which shows the relation between the grid voltage and the anode voltage for breakdown of the gas. Thus for any given anode voltage the critical grid voltage to cause the tube to trigger may be obtained. Once that tube has broken down the grid no longer has control. For the grid to regain control it is necessary for the plate voltage to fall below the tube drop value and remain below this long enough for the gas ions to recombine. This deionization time puts a definite limit to the frequency with which the tube can be triggered and controlled. On a.c. the anode voltage falls every cycle and if the cycle lasts long enough the tube will not start on the next positive half-cycle unless the grid voltage exceeds the critical value. Thus while the control is not continuous as it is in vacuum tubes, it is possible to control within one cycle which is sufficient for many applications. While it has limited use in communication work it is very valuable for industrial control where a minute actuating power may control, through the thyratron, kilowatts of power in the plate circuit. For many industrial applications the triggering effect is exactly what is desired, the operation being equivalent to a switch without any arcing contacts, mechanical time lag, etc. By using special circuits, such as the **phase shift control**, various special effects may be obtained with the thyratron. Many of these tubes have other grids for producing the desired characteristics. The **ignitron** is another type of controlled gas-filled tube. The grid-glow tube is a cold cathode tube similar to the thyratron.

Besides the tubes discussed here the **phototube** is also an electron tube. (L.R.Q.)

TUBE FOOT. A characteristic appendage of **echinoderms**, associated with the **water vascular** system. In the starfishes (**Asteroidea**) the tube feet are rounded protuberances in the ambulacral grooves along the under surface of the rays. They are cupped at the tips and grip surfaces by adhesion and suction. The tube feet of brittle stars are without suckers. Those of sea urchins (**Echinoidea**) form respiratory organs, and a group of ten around the mouth are sensory. Tube feet of **sea cucumbers** are suctorial with the exception of a group around the mouth that form the **tentacles**. (A.W.L.)

TUBER. Stem.

TUBERCULIN TEST. A skin test which indicates whether or not an individual is sensitive to the **protein** of the *tubercle bacillus*. The test is performed by injecting a small amount of the protein as tuberculin into the skin and observing the injection site for redness and swelling. In a positive test, the reaction is characteristic and persists for several days.

A positive tuberculin test means that the individual has at some time been infected with *tubercle bacilli* and has become sensitized or allergic to the organism. It does not necessarily mean active infection at the time of the test (see **tuberculosis**). Approximately 40% of young adults have positive tuberculin tests. (D.M.H.)

TUBERCULOSIS. A chronic or acute infectious disease caused by an invasion of the body by the *Bacillus tuberculosis*. It may exist without causing symptoms (inactive tuberculosis) or with symptoms (active tuberculosis). The symptoms of tuberculosis depend on the organ involved, the virulence of the strain of tubercle bacilli and the resistance of the individual infected. Almost any organ or tissue of the body may be attacked by the tuberculous process, although the commonest site is the lungs.

Tuberculosis was known to the Greeks and excellent descriptions of the disease date from the time of Hippocrates. Galen regarded it as infectious. Robert Koch in 1882 demonstrated the causative organism.

The disease is exceedingly widespread and is found in animals as well as in man. It is common in bovine animals, and there is also an avian form. Man is subject to disease from both human and bovine types of organisms. It is estimated that tuberculosis causes about ⅛ of all deaths. In civilized portions of the world there has been a progressive reduction in the morbidity and mortality from this disease. This is due to (1) better economic and social conditions, (2) the present campaign against tuberculosis with its educational aspect toward hygiene and opportunity for diagnosis, (3) earlier diagnosis of the disease both by clinical and x-ray examination, (4) protection of the healthy from active cases, (5) better forms of treatment, especially artificial **pneumothorax** and surgical collapse therapy.

All ages are susceptible, but tuberculosis is most likely to occur in young adult life. Certain races, particularly those who have never before been exposed, are more subject to the disease than others. This was true of the American Indians, who died in great numbers from the new disease which the white race brought to them. The Negro and Irish people also seem to suffer more from the disease than others. Certain occupations have a predisposing effect in that unsanitary surroundings, and particularly exposure to dust, increase its incidence. Also malnutrition, debility from overwork, mental and physical strain, the stresses of puberty and particularly pregnancy seem to increase the susceptibility to tuberculosis.

Tubercle bacilli (Mycobacterium tuberculosis) are slender, non-motile rods which are extremely resistant to heat and chemicals. Human, bovine and avian types are distinguishable by cultural characteristics. All can withstand drying for long periods, and the human type may remain alive in dried sputum for as long as 10 months. The human and bovine organisms enter the body through several portals. Ingestion of contaminated milk is the route for the entrance of bovine tuberculosis, while pulmonary infection with the human type of organism usually occurs through inhalation. Occasionally, infection enters through the skin.

Once in the body, the reaction set up around the invading organisms is similar, no matter what tissue is attacked. It consists of an accumulation of inflammatory cells including "giant cells" around the tubercle bacilli, forming a tiny, at first microscopic nodule or tubercle. As the tubercle grows, its center breaks down, becomes necrotic, and caseates (i.e., becomes "cheesy"). Several small tubercles may coalesce to form gradually larger ones. The lesion may reach several inches in diameter, and show up as a shadow on an x-ray film. When the center of such a large tubercle caseates, a cavity is formed. The necrotic contents of the cavity may be coughed up in copious amounts. Such sputum is highly infectious and menaces all the individuals exposed to it. As time goes on, the healing stage of tuberculosis begins, with gradual fibrosis and sometimes calcification of the healed tubercles.

Bovine tuberculosis occurs primarily in children. It is a disease of the **lymphatic system** caused by drinking milk from infected cows. Since the advent of pasteurization of milk and inspection of cattle herds, this form of the disease has decreased strikingly. The lymph nodes serve as filters for all foreign material and in this capacity they collect tubercle bacilli as these enter the body through the mouth and gastrointestinal tract. Lymph-node tuberculosis has long been known as **scrofula**. The nodes most commonly involved are the cervical in the neck, and the mesenteric nodes, which drain the intestinal tract. These nodes may also be involved in tuberculosis due to the human type of bacillus. Cervical-node tuberculosis is characterized by swelling, and often suppuration or the discharge of pus from the node. Mesenteric-node tuberculosis is usually without symptoms, although in some children general debility, weight loss, fever, and diarrhea occur. Occasionally acute, abdominal pain causes the disease to be mistaken for appendicitis. Rupture of a node may be followed by tuberculous **peritonitis**.

The treatment of tuberculosis of the lymph nodes is conservative. Surgical removal of infected neck nodes is avoided if possible, because of the tendency of such an infection to ulcerate and form a chronic draining **sinus** to the outside. X-ray treatment and ultra-violet light, in addition to attention to diet and general supportive measures, are employed successfully.

Pulmonary tuberculosis, due to the human strain of tubercle bacilli, is by far the commonest type of the disease. It occurs as a childhood primary complex, consisting of a localized, usually single tubercle in the lung with simultaneous involvement of the regional lymph nodes at the root of the lung, and as a secondary or adult type of infection. The ordinary opportunities for infection are such that approximately 50% of young adults give evidence of having had at some time a primary tuberculous infection. This evidence includes a positive skin test (**tuberculin test**) and the minute shadow of a healed calcified tubercle on an x-ray film of the lungs. The primary infection is usually symptom-free and passes unnoticed. In early childhood, however, it may be a severe disease, spreading rapidly from the original tubercle throughout the lungs, and resulting fatally. Usually the primary complex heals and the individual never has a recurrence of the disease.

Secondary, or the adult type of, pulmonary tuberculosis occurs only in the presence of an old primary lesion. The two theories of its pathogenesis are: (1) it represents a breaking down of the primary tubercle with dissemination in the lung; (2) it is a re-infection following exposure to a large dose of organisms. In either case, it is thought that the primary infection is responsible for an altered reaction in the host's tissues. He has been sensitized and is allergic to the protein of the tubercle bacillus. At the same time, he has developed a certain amount of **immunity**. As the result of the interaction of these two factors, the secondary, or adult, disease tends to be a chronic, low-grade process, with gradual development of thick-walled cavities, and eventually fibrosis and healing.

The symptoms of pulmonary tuberculosis depend on the stage of the disease. If the infection is near the **pleura**, and the latter is involved, the onset may be with acute **pleurisy** and effusion of inflammatory fluid into the pleural cavity. Most commonly in an early active case, the patient notices malaise, weight loss, loss of appetite, easy fatigability, afternoon fever, and slight cough. The sputum usually contains tubercle bacilli by the time such symptoms have developed, and x-ray will reveal an area of infiltration in the lung, most often in the apex of the right lung. As the disease goes on, fever, night sweats, cough, and sputum may increase. Hemorrhage from the lungs and **hemoptysis** occur in about 50% of cases. Sometimes an acute pneumonic episode or tuberculous **pneumonia** develops. Gradually, with proper rest and treatment, the symptoms subside, and the disease settles down to a chronic and slow-healing stage. At this time, there may be extensive involvement of both lungs with cavitation and large amounts of sputum, yet the patient may have little or no fever and feel relatively well. The clinical picture is extremely variable, depending on the natural resistance of the patient to the organism and his degree of acquired immunity. Some individuals are unable to handle the infection, and in spite of good treatment early in the disease they go on to develop severe symptoms and widespread, often fatal, tuberculosis. Others are able to keep a small lesion localized, and in the course of a year of treatment complete healing may be accomplished.

The complications of pulmonary tuberculosis are associated with spread of the disease to near and distant organs. In some instances the pulmonary disease may be quite minor, and the first manifestation may occur when urinary tract, or abdominal, tuberculosis begins to cause symptoms.

Miliary tuberculosis occurs when the organisms gain access to the blood stream and are suddenly spread throughout the body to all the organs and tissues. The lungs, brain, meninges, liver and spleen usually show great numbers of small miliary tubercles. This form of tuberculosis is universally fatal, usually within a few weeks.

Tuberculosis of the larynx is common in advanced pulmonary disease. It results from direct contact with the organisms as the infectious sputum is coughed up and spat out. Tuberculous **bronchitis** and **bronchiectasis** also occur by direct extension of the infection within the lung. Tuberculous meningitis, which is usually fatal, may be a part of the picture of miliary disease, or may occur from the breakdown of a single tubercle which formed when a few organisms reached the brain through the blood. Intestinal tuberculosis is most often a complication of tuberculosis of the lungs. Spread of the tubercle bacilli to bones, genito-urinary tract, or any other organ or tissue through the blood may occur. Blood-stream invasion is the common way for the disease to reach these organs, although primary disease does occasionally occur in them. Serous-cavity tuberculosis involving the **pleura** and the **pericardium** usually is the result of extension from the lungs, or in the case of the pericardium, lungs or tracheobronchial lymph nodes. Tuberculous peritonitis is a complication of intestinal or mesenteric node tuberculosis by extension, or of pulmonary disease. In the latter case, the peritoneal infection is blood borne.

The treatment of pulmonary tuberculosis always involves a period of several years at least. For best results certain fundamentals must be adhered to: a strict regime of bed rest, good food, and fresh air, in cheerful surroundings, are all essential. High, dry, sunny climates have long been thought to be desirable. Although these may be the most pleasant, they have no special curative value. Early treatment is one of the most important features in achieving a cure.

In recent years, collapse therapy in tuberculosis has been developed and carried out with considerable success. It is based on the fact that healing of a diseased lung is more rapid if the organ can be put at rest. Several types of collapse therapy are used. **Pneumothorax** is employed in unilateral lung disease which has not responded to conservative measures after adequate trial. It consists of introduction of air into the pleural cavity between the chest wall and the lung until the pressure in the cavity is sufficient to force the expanded lung to give up its air and collapse down to a fraction of its normal size. Since the air is absorbed rapidly, it must be reintroduced at intervals—once or twice weekly. The treatment is continued for a minimum of 2 years as a rule, before the lung is allowed to re-expand.

When pneumothorax fails because of **adhesions** between the chest wall and the lung, in such a position that they cannot be cut, or when there is a stiff-walled cavity which remains open in spite of pneumothorax, thoracoplasty, or permanent surgical collapse of the chest by removal of the ribs in two or three operative stages, saves many patients who would otherwise die of their disease.

Crushing one of the phrenic nerves, which temporarily paralyzes half of the **diaphragm**, is another means of decreasing the activity of the diseased lung. It may be used alone or in conjunction with pneumothorax or thoracoplasty.

The various forms of abdominal tuberculosis are treated with x-ray and ultra-violet light, as well as the usual general measures. Tuberculosis of the genito-urinary tract is largely a surgical problem. Bone disease, or tuberculous **osteomyelitis,** is treated by immobilization and sometimes by surgical fusion of diseased joints.

With recent developments in chemotherapy (**sulfonamides, penicillin,** etc.) a new spurt has been given to the search for a drug which will cure tuberculosis. **Streptomycin** has proved promising especially in the treatment of early exudative lesions of the lungs. Early effects suggest that it is effective also in laryngeal or tracheobronchial tuberculosis, and even in tuberculosis meningitis. So far none has been found, although a few compounds still under investigation appear promising.

The prognosis in tuberculosis depends on many factors. The type, duration and extent of disease when treatment is begun, the resistance of the patient to the tubercle bacillus are of prime importance. Early treatment increases the per cent of cures enormously. The importance of continuation of treatment, usually for a minimum of 2 years, cannot be overestimated. Since relapses are relatively common even after apparent cure, restriction of activities and regular check-up examinations for a period of years are essential. (D.M.H.)

TUBULARIAE. Hydrozoa.

TUBULIDENTATA. The aard-varks. An order of mammals found only in Africa. Based chiefly on the form of the teeth, which are composed of numerous subordinate parts penetrated by radiating tubules. The animals themselves are thick-bodied clumsy creatures with long snouts. The name means earth-pig and refers to their somewhat pig-like appearance. (A.W.L.)

TUBULIFORM GLANDS. Silk glands of **spiders** whose secretion is used in forming the cocoon to contain the eggs. They occur only in the female. (A.W.L.)

TUCOTUCO. Mammalia, Rodentia. Rat-like burrowing animals of South America. They have gray fur and red incisor teeth. The tail is only moderately long and is clothed with short fur. A closely related form with vestigial ears is known in Chili by the name cururo. (A.W.L.)

TUCUXI. Mammalia, Odontoceti. A fresh-water **dolphin** found in the Amazon river system. It belongs to a family differing from that which contains the bouto. (A.W.L.)

TUFA. Sinter.

TUFF. Tuff or volcanic tuff is a sedimentary rock, resulting from the partial or complete consolidation of the products of explosive volcanic eruptions. Tuffs may be well sorted and stratified, due to the action of wind or water, or may have an unsorted, heterogeneous character. As particles making up a tuff become coarser the rock grades into an **agglomerate.** (E.S.C.S.)

TUKON HARDNESS. Hardness.

TULAREMIA. An infectious disease caused by *Bacterium tularense*. It occurs primarily as a fatal disease of rodents, mainly rabbits and hares. It may be transmitted to man by the bite of certain flies or ticks, or by handling the raw meat of infected rodents, usually wild rabbits. Most of the cases have been described in the United States, 43 states having reported them. The only other countries reporting cases are Norway,

Russia and Japan. The infection is not transmitted from man to man.

The incubation period of tularemia averages about 3 days. The onset is acute, with headache, vomiting, body pains and fever. The lymph nodes draining the area of the portal of entry on the skin often become infected and ulcerated. In certain cases the eye has been primarily infected. Severe systemic reactions, pneumonia, and cardiac involvement may occur. Convalescence is slow and it may require 6 months to a year before return to normal health occurs. One attack confers immunity. The mortality rate is about 4%. The disease is often erroneously diagnosed as influenza, typhoid, tuberculosis or undulant fever. Accurate diagnosis can be made by culturing the organism from the lymph nodes or other lesions, and by the detection of antibodies in the blood serum of the patient.

Prevention in cooks, market men, and hunters may be accomplished by the wearing of rubber gloves when dressing wild rabbits. Treatment with **streptomycin** is effective in the acute infection. (D.M.H.)

TULIP. Lily Family.

TULLIBEE. Pisces, Teleostei. A common fish (Pisces), *Leucichthys tullibee,* of the small lakes from Minnesota to New York and northward into Canada. One of the ciscoes or lake herrings. Also recorded as the mongrel whitefish and said to occur in small numbers in the Great Lakes. (A.W.L.)

TUMBLE-BUG. Insecta, Coleoptera. A beetle of the family Scarabaeidae which forms and buries balls of dung. Numerous species of scavengers in this family have the same habit. They are said to use the dung as a supply of food during periods when they remain underground, and also to bury a ball with an egg attached, to provide food for the developing larva. The balls of dung are often much larger than the beetles themselves, and their clumsy maneuvers in rolling their booty to the place where it is to be buried are responsible for the name tumble-bug. (A.W.L.)

TUMOR. 1. Any swelling or abnormal enlargement of a part of the body. 2. A growth or mass of tissue which develops in the body, serves no use, and grows independently of surrounding tissues. Such growths are divided into two classes, **benign** and **malignant** tumors. Benign tumors do not spread over the body; they cause symptoms only when large enough to interfere mechanically with the functions of surrounding structures and they rarely recur after complete excision. Examples of benign tumors are fibroids of the uterus, fatty and fibrous tumors, and simple cysts. Malignant tumors, **cancer** and **sarcoma**, do invade surrounding tissues, disseminate throughout the body (metastasize) by means of the blood and lymph vessels, may recur after removal, cause constitutional symptoms, and usually terminate fatally.

There are many theories regarding the cause of tumors, but as yet no adequate explanation has been produced. Every tissue in the body is capable of developing tumors; their number and variety are tremendous. They are classified on the basis of cellular structure, the tumor being named according to the tissue from which it originated. The classification is difficult and too complex to be given in full detail here. A simple classification for the more common types according to tissue composition is given below.

I. Connective-Tissue Tumors.
  A. Benign.
    1. Fibroma or fibroid tumor—fibrous tissue.
    2. Chondroma—cartilaginous tissue.
    3. Osteoma—osseous or bone tissue.
    4. **Lipoma**—fatty tissue.

  B. Malignant.
    1. **Sarcoma.** One of the large groups of tumors composed of any one of the varieties of connective tissue mentioned under benign tumors.
II. Muscle-Tissue Tumors.
  A. Benign.
    1. Leiomyoma—smooth muscle.
III. Epithelial Tumors.
  A. Benign.
    1. Papilloma—made up of surface epithelium—skin or mucous membrane.
    2. **Adenoma**—made up of glandular epithelium.
  B. Malignant.
    1. Carcinoma or **cancer.**
      a. Squamous-cell carcinoma or epithelioma, composed of flat epithelial cells.
      b. Adenocarcinoma—composed of glandular epithelial cells.
IV. Endothelial Tumors. Composed of cells that line blood vessels, lymph vessels, joints, and the body cavities.
  A. Benign.
    1. Hemangioma—blood vessels.
    2. Lymphangioma—lymph vessels.
  B. Malignant.
    1. Endothelioma—an ill-defined group of tumors, some of which are malignant.
V. Pigmented Tumors.
  A. Benign.
    1. Pigmented moles or naevi.
  B. Malignant.
    1. Melanoma. A tumor arising in pigmented moles of the skin or in the pigmented coat of the eye, the retina.
VI. Nervous Tissue Tumors.
  A. Benign.
    1. Meningioma—from cells covering the arachnoid villi.
    2. Acoustic neuroma—from the sheath of the acoustic or 8th cranial nerve.
  B. Malignant.
    1. Glioma—from the several types of specialized connective tissue in the brain and spinal cord.
    2. Neuroblastoma—from primitive nerve cells in the adrenal gland.
    3. Ganglioneuroma—from adult nerve cells and fibers.

(D.M.H.)

TUNA, TUNNY. Pisces, Teleostei. A large marine fish, *Thunnus thynnus,* of the mackerel family. It attains a maximum weight of more than 1500 lbs. and is highly valued both as a game fish and for food. About 15,000,000 lbs. are canned annually in California.

As a game fish the tuna has no superior. It is an unusually vigorous fighter, readily taken with rod and reel on flying fish bait. Specimens over 100 lbs. are frequently caught and a record fish, more than 10′ long, was taken off the coast of Nova Scotia only a few years ago.

The tuna is also called the albacore and horse-mackerel. (A.W.L.)

TUNDRA. The tundras are the Arctic plains which, while supporting mosses and lichens in profusion, and, locally, various flowering shrubs, are treeless. The top soil is usually a black muck; the subsoil is perpetually frozen. (R.M.F.)

TUNED CIRCUIT. Resonance.

TUNED RADIO FREQUENCY. This is a type of amplifier circuit which uses resonant circuits in the grid and/or plate circuits. As commonly used it is used to denote a type receiver which has one or more such stages preceding the detector, the condenser or inductance of the resonant circuit being varied to select the particular station desired. (See **Receiver.**) (L.R.Q.)

TUNG OIL. Fixed Oils.

TUNGSTEN. Symbol: W. Atomic number: 74. Atomic weight: 183.92. Density: 19 (18.6–19.1). Melting point: 3370°. (Isotopes: page 290.)

Tungsten is a silver-white to steel-gray, brittle, hard metal; not oxidized by air at ordinary temperature but burns at high temperature, best dissolved by a mixture of **hydrofluoric and nitric acids.** Tungsten is used in the production of special alloy tool steels for cutting purposes (16–20% W); in electric lamp filaments and radio tubes; and in special hard alloys, e.g., "carboloy" (tungsten carbide and cobalt) and "stellite" (cobalt, 55%; chromium, 35–40%; tungsten, 3–10%); chemically related to **chromium, molybdenum,** and **uranium** elements. The chemical, paint pigment, and tanning industries use tungsten compounds.

Discovered by d'Elhujar brothers in 1783. Occurs as **scheelite** (calcium tungstate, $CaWO_4$) and **wolframite** (iron and manganese tungstate) chiefly obtained in China. The United States production in Idaho, Nevada, and California is significant. Mexico, Bolivia, Peru, Ecuador, and Argentina are potential producing countries.

Fusing the ore with sodium carbonate and nitrate yields sodium tungstate (Na_2WO_4), which is extracted with water and later acidified, whereupon tungstic oxide (tungsten trioxide, WO_3) is obtained. The oxide is reduced by heating with **carbon** or with **hydrogen** to form tungsten metal, from which by "swaging" (rapid mechanical hammering at 1500° C. in an electric furnace in an atmosphere of hydrogen) ductile wire is obtained.

Chlorides: Chlorides of the following composition are reported: WCl_4, WCl_5, WCl_6, $WOCl_4$, WO_2Cl_2.

Oxides: Tungsten dioxide (WO_2), brown solid, by reduction of the trioxide by hydrogen below 700° C., which upon continuation of the same treatment above 780° C. forms tungsten metal; ditungsten pentoxide (W_2O_5), blue solid, by reduction of the trioxide by hydrogen at 250–300° C.; tungstic oxide, tungsten trioxide (WO_3), lemon-yellow solid, converted by alkalis to soluble tungstates, by heating with **chlorine** to tungsten oxychloride (WO_2Cl_2), by heating with **hydrogen** sulfide or sulfur to tungstic sulfide (WS_3); by alkalis to tungstates. Ammonium phosphotungstate is insoluble, likewise the potassium compound, but the sodium compound is soluble in water. (R.K.S.)

TUNIC. The tough enveloping layer or test of the **ascidians.** It is formed of a material similar to cellulose, and is the source of the name tunicate applied to some of these animals. (A.W.L.)

TUNICATA. Urochordata.

TUNING FORK. The tuning fork is a convenient device for preserving a comparatively pure harmonic vibration frequency at nearly constant value. It is a U-shaped bar of elastic material, usually steel (but in some modern forks, of fused quartz), the prongs of which vibrate alternately toward and away from each other, with two nodes near the bend of the U. The fork may be set vibrating by striking one prong with a mallet, and will, after a moment to allow some high overtones to die out, emit a nearly pure musical tone. Large forks are often made to be driven electrically, like an electric bell or buzzer, and will then vibrate continuously for an indefinite time. Tuning forks are used in many experiments on **musical sounds,** as standards of pitch, and also for the control of electric oscillations and electric timing devices. A fork may be tuned by grinding off the ends of the prongs or by means of sliding weights attached to the prongs. Once tuned, the frequency varies only with changes in the elastic modulus of the material. This is slightly dependent upon temperature; hence, for precise work, a fork should be kept in a thermostatically controlled enclosure. (L.D.W.)

TUNING INDICATOR. In radio circuits with **automatic volume control** the output is held constant for an appreciable amount of mistuning of the circuits. Thus the loudness of the output is not a satisfactory indication of when the set is correctly tuned. To remedy this various types of tuning indicators are used. These respond to the carrier amplitude and hence are not affected by the avc. Formerly various instruments were used for this purpose, but now instruments are used only in communications receivers while ordinary home receivers use a special vacuum tube. This tube, often called a magic eye, has a willemite-coated cone-shaped plate upon which the electrons impinge, causing the coating to fluoresce. By connecting the grid so the carrier magnitude can control the number or pattern of the electrons hitting this coating, it can be used to indicate relative carrier strength. In any tuning, the correct setting for a given station is that which makes the carrier amplitude maximum. (L.R.Q.)

TUNNEL DRIER. Distillation, Evaporation, and Drying.

TUR. Mammalia, Artiodactyla. A wild **goat** of the Caucasus mountains. It is a large species with strong, slightly spiral horns. Although closely related to the sheep it has the characteristic beard of the goats. (A.W.L.)

TURBELLARIA. The free-living flatworms, a class of the phylum **Platyhelminthes.** Unlike the parasitic members of this phylum, these worms have a cellular ectodermal covering bearing **cilia,** with the exception of a few parasitic members which resemble the flukes and tapeworms in the absence of cilia. The free-living members of the class also have sense organs located at the anterior end of the body, including a pair of eyes and tentacles. The mouth opens on the ventral surface, either near the head or near the end of the body, and the alimentary tract (**Digestive System**) is branched in many forms to extend widely through the body. The terminal portion of the tract forms a protrusible proboscis. The turbellarians are **hermaphrodite** with very few exceptions.

Flatworms of this group are often common in small streams and ponds, and some are marine. They glide over surfaces by the action of the cilia of the lower surface, aided by the secretion of a trail of mucus, or move by undulations produced by muscular action. Their food consists of small animals, living or dead.

The class is divided into four orders:

Order Acoela. Without a hollow gut. The endodermal cells are fused together. These worms live in **symbiosis** with algae contained in the cells of the body.
Order Rhabdocoelida. Mouth anterior, leading into an unbranched gut. Chiefly fresh-water species. Some small marine forms.
Order Tricladida. Mouth approximately central. Gut with three main branches. This order includes the common fresh-water forms, *Dendrocoelum* and *Planaria,* and some species that live in moist situations on land. Some are large and brightly colored.
Order Polycladida. Mouth behind the middle of the body. Gut with many branches from a small main portion. Marine species. Some large and leaf-like. (A.W.L.)

TURBINE. Gas Turbine; Hydraulic Turbine; Steam Turbine.

TURBOT. Pisces, Teleostei. A large flatfish, *Psetta maxima,* of the European side of the Atlantic. It reaches a length of 3′ and is regarded as the best food fish of the group. On the Atlantic coast of North America a small and useless related species is commonly called the window pane. (A.W.L.)

TURBULENCE. This is a term used to describe a characteristic quality of fluids in motion. There are several concepts of turbulence which, though broadly related, differ greatly in detail. Thus turbulence is employed in the literature of such dissimilar fields as the hydraulics of liquid flow in conduits, the flow of steam in turbines, the injection of fuel into Diesel engine **combustion chambers,** the design and arrangement of burners for steam boiler **furnaces,** the **aerodynamics** of **airfoils,** and the accuracy of **wind-tunnel** research. It affects many phases of these fields, i.e., **heat transfer, thermal efficiency, lift** and **drag** forces, skin friction, comparison of experimental data, etc. Definitions of turbulence are by necessity arbitrary and depend on the purpose for which they are to be applied.

Turbulence consists of a superimposing of irregular eddy currents on a uniform flow. This would include the classical vortices of Helmhotz, with velocity gradients in the turbulent area, as well as isotropic turbulence.

The latter type is the common "turbulence" of applied aerodynamics, and will be amplified in the following discussion.

A turbulent flow might be considered to be a main flow of the centers of gravity of elementary masses of the fluid with turbulence being the irregular motions of these elementary masses about their individual centers of gravity. Turbulence is sometimes quantitatively described as the root-mean-square speed fluctuation as a percentage of the mean speed. Turbulence in air streams is characterized by high-frequency fluctuations about the mean air speed, the magnitude of fluctuation being usually less than 3% of the mean speed. (Low-frequency pulsations of greater magnitude are usually called "gusts.") These fluctuations are irregular in magnitude and direction. The magnitude has been measured directly by the *hot wire anemometer,* developed at the Bureau of Standards, and indirectly by comparison of sphere drags. Turbulence arises from the flow of air over obstacles and the breaking up of the vortices formed. Given sufficient time the viscous properties of air will damp turbulence components. Thus turbulence of the free atmosphere ought to be zero. This has been experimentally confirmed.

Until the qualifying effects of wind-stream turbulence were understood it was impossible to correlate the experimental results at different **wind tunnels** because the initial turbulence was different in each case. After measuring the characteristic tunnel turbulence and allowing for it (especially on C_{Lmax} and C_D) very good correlation was obtained. Low turbulence air streams can be produced by placing fine screens across the tunnel immediately after the turbulence is produced (by models, fixed vanes, and propellers) so as to break up the vortices and produce turbulence of small "grain," and then give the viscosity of the air a chance to damp out turbulence by a slow-speed return to the test section. Increasing the effective viscosity by increasing the air density through compression is of further assistance. (See **Boundary Layer, Heat Transfer, Friction Fluid.**) (F.T.M.)

TURBULENT FLOW. Turbulence.

TURKEY. Aves, Galliformes. *Meleagris.* A large game bird (**Aves**) of North and Central America. Wild turkeys are now abundant in some of the protected areas of the eastern states, and in the southwest they may be found in wild areas. In comparison with their former abundance, however, they are now rare. The eastern species is easily distinguished from the western by the absence of white tips on the feathers of the tail and rump. A third species occurs in Central America. (A.W.L.)

TURMERIC. *Curcuma longa.* Zingiberaceae. The name turmeric has been given to both the plant, *Curcuma longa,* and to its derivatives, a dye and a drug, which are obtained from the swollen **rhizomes** of the plant.

Curcuma, a native of southern Asia, is widely grown in India and other tropical Asian and East Indian lands, where it is used as a drug as well as a condiment and dyestuff. The plant has long smooth pointed leaves, and dull yellow flowers. To prepare it for use, the yellow-brown aromatic rhizome is cleaned and cut into pieces 1–2″ long. The cut surfaces show the bright yellow color of the substance of the rhizome. These cut pieces are dried in an oven and then ground. In western countries turmeric is sometimes used as a dye; as an ingredient to curry powder; and in chemistry, where it is a test for alkali. Other species of *Curcuma* yield products used in making Oriental tonics and medicines. (R.M.W.)

TURNBULL'S BLUE. Ferricyanic Acid and Ferricyanides.

TURNING. Lathe.

TURNING POINT. Differential.

TURNIP. Brassica.

TURNSTONE. Aves, Charadriiformes. *Arenaria.* Shore birds (**Aves**) related to the plovers and oystercatchers. They are found chiefly in the far north, migrating southward, chiefly along the coasts, in both the Old and New Worlds. They winter as far south as Patagonia. (A.W.L.)

TURPENTINE. Resins; Pistachio; Terpenes.

TURQUOIS. The mineral turquois is a hydrated **phosphate** of **aluminum** and **copper.** Its exact composition is doubtful, the formula may be expressed $H_5Al(OH)_2 \cdot 6Cu(OH)(PO_4)_4$; iron is often present. This mineral is found in minute **triclinic** crystals, but chiefly massive as seams and crusts. The fracture is conchoidal; hardness, 5–6; specific gravity, 2.6–2.8; luster, soft waxy; color, may be various shades of blue, bluish-green and green; essentially opaque. It takes a good polish and the sky-blue varieties have long been used as a gem material. Unfortunately many beautiful blue stones in time change their color to some greenish hue, usually not attractive, rendering them practically valueless. For hundreds of years turquois has been mined in Persia where it is found with **limonite** filling crevices in a brecciated **trachyte-porphyry;** and because it found its way into Europe through Turkey it became known as turquois, from the French word *turque,* Turkish. Other mines were worked by the Egyptians in ancient times on the Sinai Peninsula. Turquois is also found in Siberia, Turkestan, Saxony and France; and in the United States in Arizona, California and New Mexico. A blue stone that has passed for turquois is in reality odontolite, from the Greek meaning tooth, usually fossil teeth or bones colored with iron phosphate. Odontolite is softer than true turquois, has a somewhat higher specific gravity, 3.0–3.5, and may be distinguished by chemical tests, or by a microscopical examination which will reveal its organic structure. (E.S.C.S.)

TURRET LATHE. Mass-production turning and facing equipment may be classified as hand-operated, semi-automatic, and completely automatic machinery. Hand-operated equipment generally requires the attention of an operator while the work is in process; semi-automatic equipment usually requires an operator to load and unload the parts to be machined, after which the machine performs a cycle of operations and stops at its conclusion without further attention; automatic machinery requires no service from the operator except gauging the parts as they are produced, resharpening and resetting the tools when required, and filling a hopper with blanks, or in the case of bar machines, inserting fresh bar stock when the original bar has been processed. To illustrate, one operator is required for every turret lathe in use; one man can handle from two to four semi-automatic chucking machines; but six or even more automatic screw machines can, in some instances, be taken care of by a single operator.

Turret lathes are used in the mass-production system of manufacture to care for the wide, diversified field of production between the engine lathe on one hand and the automatic or semi-automatic bar and chucking machine on the other. Parts which are machined in an engine lathe by being held in a chuck can be drilled and reamed by using tools held in the tailstock sleeve; and the part is subsequently bored, turned, and faced by a series of tools held in the tool post. Turret lathes operate on precisely the same principle. The single hole tailstock, however, is replaced by a six-station turret which holds six tools of such character that from six to fifteen cutting and finishing operations may be successively performed on the part without the necessity of changing the tooling equipment; and the single operation tool post is replaced by a four- or six-station turret which permits four or more cross-slide turning and facing operations.

Modern turret lathes have 5- or 6-station turrets for end cutting operations, and a four-station cross-slide turret for side cutting. The machine has an all-geared headstock, similar to that of a modern engine lathe, and a carriage which carries the turret and is free to move on the bed of the machine parallel to the spindle axis. The bed also supports and guides the turret saddle which carries the turret. Both the turret saddle and the cross-slide carriage may be power fed longitudinally or parallel to the spindle axis; the cross-slide can also be power fed at right angles to the spindle.

The figure shows a turret lathe tooled for turning rocker bolts from hexagonal bar stock. The rocker bolt is shown at the top of the figure, and the spindle and work are represented at each station as though the spindle were indexed around the turret. In operation 1, the turret is moved forward to a stop, and the bar stock B in the collet chuck is moved forward until it meets the work or stock stop shown in the turret. The turret is indexed, and the $1\frac{1}{4}''$ diameter shank of the bolt turned by using a knee tool with a roller back rest. The turret is indexed, and operation 3, forming the head of the bolt by means of the front cross-slide form tool, and operation 4, turning the $1\frac{1}{2}''$ diameter body of the bolt by a knee tool with a roller back rest, are performed simultaneously at this station. The turret is indexed, and the end of the bolt is formed by an end-facing tool with a roller back rest held in the turret. The turret is indexed, and the bolt is threaded, operation 6, by using an automatic opening die head; the thread-cutting dies can be retracted so that it is unnecessary to reverse the direction of rotation of the turret spindle to run off the die. The turret is indexed, and the bolt is cut off by means of the rear cross-slide tool T. The turret is again indexed, bringing it to the initial position shown in operation 1. (H.C.H.)

TURTLE. Reptilia, Chelonia. A reptile with a broad flattened body enclosed in a shell formed of a dorsal carapace and a ventral plastron, united at the sides.

Atlantic green turtle. (*N. Y. Zool. Soc.*)

Most species are able to withdraw the head, legs, and tail into the shell for protection. The exposed parts of the skin are scaly.

Many turtles are partially aquatic and some marine

ROCKER BOLT

$1\frac{1}{4}''$-7-USS

OPER. 5 FORM END OF SCREW

OPER. 3—FORM HEAD OF SCREW

OPER. 6 THREAD END OF SCREW

OPER. 4 TURN $1\frac{1}{2}''$ DIA.

AUTOMATIC OPENING DIE HEAD

R—ROLLER BACK REST

OPER. 7 CUT OFF AND CHAMFER STOCK END

T—TOOL BIT

TURRET

OPER. 2 TURN $1\frac{1}{4}''$ DIA.

BAR STOCK-B

OPER. 1 FEED HEXAGONAL BAR STOCK TO STOP, AND CLAMP IN C

P—TOOL POST-P

S—CROSSLIDE-S

SPINDLE BAR CHUCK

Soft-shelled turtle. (*N. Y. Zool. Soc.*)

species leave the water only to deposit their eggs. In the latter the legs are developed as broad flippers. Turtles are to be found along almost any stream and are often abundant. The group includes both herbivorous and carnivorous species. The flesh of some species is regarded as an unusual delicacy.

Some species of turtles are known by distinctive vernacular names, as slider, cooter, and terrapin, and a number of them are called tortoises. (See also Fossil Reptiles.) (A.W.L.)

TURTLE STONE. Concretion.

TWILIGHT. Twilight is produced primarily by reflection of the light of the sun from the upper atmosphere of the earth, but effects of scattering of light, and refraction also enter into the period of duration of twilight. Morning twilight begins and evening twilight ends when the sun is about 18° below the horizon, although this value varies somewhat with the purity of the atmosphere.

The angle which the path of the sun at setting makes with the horizon depends upon the latitude, the sun setting perpendicular to the horizon at the equator. The apparent velocity of the sun on the equator is practically constant for all latitudes and all seasons, but the time required for the sun to get 18° below the horizon will be much shorter at the equator than in high latitudes. Accordingly, the duration of twilight in the tropics is much shorter than it is in high latitudes such as Scotland.

The so-called twilight arch may be observed above the eastern horizon on a clear evening as the sun sets in the west. This is a blue segment bounded by a faintly reddish arc, and is in reality the shadow of the earth cast on the upper atmosphere. (W.K.G.)

TWILIGHT SLEEP. A light anaesthesia produced by the hypodermic use of morphine and scopolamine. It has been used in labor and also pre-operatively and post-operatively. In this state the patient is not totally unconscious, but does not remember the occurrence of any pain. It is not often used in labor at present as more satisfactory drugs are available. (R.S.M.)

TWIN BANDS. Twin bands in crystals are zones in which has occurred a change in orientation of the atomic crystal structure from that of the parent crystal. Although twinning is very noticeable in the microstructure of certain metals, it has relatively little effect on the physical, mechanical, or chemical properties.

Profuse twinning occurs in many metals and alloys having face-centered cubic crystal structures when these metals are annealed after mechanical deformation (for example copper and most of its alloys, nickel and its alloys, and austenitic stainless and heat resisting steels). These are known as annealing twins.

Mechanical twins are sometimes produced in iron and low-carbon steels when subjected to shock. These are also known as Neumann bands. (R.H.H.)

TWIN CRYSTALS. Those crystals in which one or more parts regularly arranged are in reverse position with reference to the other part or parts. They often appear externally to consist of two or more crystals symmetrically united, and sometimes have the form of a cross or star (Dana). (R.M.F.)

TWINS. Two individuals born at the same birth. The incidence of twins is about 1 to 2% of births. Heredity is a factor in the etiology of multiple pregnancies. Identical twins are twins developing from the cleavage of one fertilized ovum. They are always of the same sex and similar in appearance. Fraternal twins are twins developing from two fertilized ova. These

may be of either sex, and may or may not be of similar appearance. (D.M.H.)

TWIST DRILL. Drilling.

TWISTED CURVES. Curves in Space.

TWITCHELL'S REAGENT. This is a catalyst for the hydrolysis of fats. (R.K.S.)

TWO-BODY PROBLEM. The so-called two-body problem is the foundation of celestial mechanics. The solution of the problem requires two fundamental assumptions: (1) that two and only two objects exist in the universe, and (2) that some law of force between the two objects is given. With these assumptions admitted the two-body problem may briefly be stated as follows: given the relative positions of two objects at any instant, together with their motions and masses at that instant, to predict the positions and motions of the objects at any subsequent instant.

The two-body problem was first solved by Newton by considering the motions of the individual planets about the sun. To satisfy the first assumption of the problem Newton assumed that the force between the sun and an individual planet was so much greater than the force between any two planets, that the sun and the planet could be considered, as a first approximation, to be isolated in space. The inclusion of the forces between the different planets as well as the force between the sun and the individual planet leads to the three-body problem. When Newton first attacked the two-body problem no law of force between objects in space was known, but the Keplerian laws of planetary motion had already been empirically derived. Newton was familiar with the characteristics of centrifugal force and realized that no object could revolve about another (as the moon does about the earth) unless there is some force of attraction between them to counteract the centrifugal force. Considering as a first approximation that the orbit of the planet about the sun is circular he found that if the force of attraction between the sun and the individual planets varies inversely as the square of the distance then, and only then, does the so-called harmonic law of Kepler result.

Newton realized that in his establishment of a theoretical foundation for the third of Kepler's laws of planetary motion he had violated the first, for Kepler specified that the motion must always be elliptical and Newton had used the circle (a very particular ellipse). The mathematics of Newton's time was not sufficiently developed to permit of a solution of the problem of elliptical motion and Newton was forced to develop the theory of fluxions, the parent of the modern calculus, to accomplish his solution. Using the Cartesian system of geometry and his own theory of fluxions, Newton was able to show that all three of the Keplerian laws of planetary motion were consequences of the planets moving about the sun under the influence of a force emanating from the sun and varying inversely as the square of the distances of the objects from the sun. In a similar manner Newton was able to explain the motions of Jupiter's satellites, and the motion of the moon about the earth. Newton firmly believed that the force involved was the so-called force of gravitation which had been described some time previous in connection with the laws of falling bodies on the earth. He stated the law in its familiar form that: every particle of matter in the universe attracts every other with a force which varies directly with the product of the masses of the two objects and is inversely proportional to the square of the distance between them.

With the law of force between the two objects thus stated Newton was able to completely solve the two-body problem as stated in the opening paragraph of this article. Publication of his solution was delayed for many years, because Newton was unable to verify his hypothesis numerically, since the distance between the earth and the moon was imperfectly known. Subsequent applications of the two-body problem have led to modern methods for orbit computation and the multitude of other problems in the field of celestial mechanics. Within recent years the two-body problem has had a number of applications in the treatment of problems in atomic structure. (W.K.G.)

TWO-CYCLE. Two-cycle refers to a sequence of operations by means of which the cycle of the **internal combustion engine** is performed. It is a shortened form of "two-stroke cycle," which may be taken to imply that the cycle is completed in two strokes of the piston. The two-stroke actions to be described here should be compared with the four-stroke described under **four-cycle.**

A two-cycle engine must be so arranged that it can be supplied with a fresh charge when the **piston** is in the extreme outward position. In general, any internal combustion engine must first introduce the fresh charge, compress it, then ignite and expand it after combustion, obtaining power, then exhaust the products of combustion. To obtain all these functions in two strokes of the piston, it is necessary to shorten the period of time that can be allotted to induction and exhaust. For this reason, the **volumetric efficiency** of the two-cycle engine is inferior to that of the four, and, although it receives a power impulse every revolution, instead of every two revolutions, the power falls short of being double that of the four-cycle engine of corresponding size and speed.

In a simple type of two-cycle engine, valves are replaced by ports in the cylinder, which are uncovered by the piston as it nears crank end dead center. One cylinder port leads to the atmosphere, and through it the burned gases are expelled by the pressure remaining in the cylinder. The other port opens from a by-pass to the crankcase, and through it a slightly compressed charge is delivered to the cylinder when the piston uncovers the port. In some types, the gas is compressed in the crankcase by the piston, but in others an external compressor of a piston or rotary type is employed The extremely short time which the ports are open renders the two-cycle engine less suitable for high speeds than the four-cycle engine. However, with certain modifications, the speed may be considerably increased without great sacrifice of efficiency. One method is to place auxiliary poppet exhaust valves in the cylinder head, mechanically operated, to aid in clearing the cylinder of burned gas and this will also aid indirectly in obtaining a better induction of the fresh charge. However, the introduction of an exhaust valve nullifies one important advantage of the two-cycle principle, namely, the absence of such valves. Crankcase compression, while simple enough in the single-cylinder gasoline engine, is complicated in multi-cylinder engines because the crankcase must be separated into sections by air-tight diaphragms so that compression can occur. This offers an obstacle to the smooth supply of fuel to all cylinders from a single carburetor. This problem has been solved in two-cylinder engines by using a special rotary **valve** in the wall between the crankcase sections. As Diesels do not mix the fuel with the air until after it is compressed in the cylinder, crankcase compression in them poses only the problem of subdividing the crankcase. (F.T.M.)

TWO-HINGED ARCH. Arch.

TYLOSAURUS. Cretaceous.

TYLOSES. Tyloses are balloon-like outgrowths which develop from the living **parenchyma** cells in the woody tissues of older portions of stems. The protrusions push through the **pits** in the cell wall and gradually fill the lumen, or cavity, of the cell until the latter is completely clogged. (R.M.W.)

TYMPANUM. A thin membrane associated with organs of hearing. The tympanum vibrates in response to sound waves and its vibrations are communicated to the nerve endings of the auditory organ. In **insects** the tympanum is a modified area of the integument. In **vertebrates** the tympanum or **ear** drum develops as a specialized area of the skin in the region of the ear, and persists in this condition in the **amphibians**. In the higher classes of vertebrates this area is depressed until it lies at the inner end of the canal of the outer ear. (A.W.L.)

TYPE METAL. Lead Alloys.

TYPHOID FEVER. An acute generalized infection due to a small bacillus, *E. Typhosa*. Typhoid fever is world wide in distribution. Before the days of water and milk inspection, great epidemics of the disease occurred in the summertime and took many lives. Typhoid is spread by feces-contaminated water, milk, or food. Chronic **carriers** who work as food handlers are common sources of dissemination. The organism gains entrance through the mouth and localizes in areas of the intestinal tract where lymphoid tissue occurs. Peyers' patches in the terminal part of the small intestine, or ileum, are the primary sites of localization. From here the organisms spread to the mesenteric lymph nodes, the thoracic duct, and into the blood stream.

Typhoid is usually a severe, long drawn-out disease, and complications are common. The onset is often with chills; **fever** is high and sustained for several weeks, but the **pulse** is slow. A rose-colored eruption is characteristic. The spleen is large and soft. Complications are **pneumonia, cholecystitis, thrombophlebitis,** hemorrhage from the bowel, and perforation of the bowel; the latter two carry a high mortality rate.

Treatment of the uncomplicated case is supportive, with special attention to a high caloric diet. **Sulfonamides, penicillin** and **streptomycin** seem to be inactive against the typhoid bacillus. The disease can be prevented by the injection of a **vaccine.** Immunization programs have been effective in preventing outbreaks among troops in wartime. (D.M.H.)

TYPHOON. Hurricanes.

TYRANNOSAURUS. Cretaceous; Fossil Reptiles.

U

UDAD. Mammalia, Artiodactyla. The Barbary sheep, *Ammotragus lervia*, a large African species with a growth of long hair extending from the throat down over the fore legs. The horns are long, heavy, and strongly curved. This species has been called the udad, aoudad, or audad, but according to some writers the name arui is applied to it by the Arabs. (A.W.L.)

UHF. Radio Frequency.

UINTAITE. Gilsonite.

UINTATHERIUM. Eocene; Fossil Mammals.

ULCER. A shallow, open sore which penetrates the skin or mucous membrane. On the skin, a variety of bacteria may be found in the ulcer, but staphylococci commonly predominate. Ulcers are usually sluggish in character, showing little tendency toward healing. They occur on the surface of the body, especially the legs, as a result of impaired circulation, and are most commonly associated with varicose veins, diabetes, and arteriosclerosis. They may also occur in specific forms in syphilis, tuberculosis, and other diseases. Cancerous growths may become ulcerated, especially in the later stages. Ulcers are prone to form following a trivial injury in areas of the body where circulation is poor.

Peptic ulcer, or ulcer of the stomach or duodenum, is a common disease characterized pathologically by a circumscribed area of destruction involving the lining mucous membrane and the muscle coats beneath.

Ulcers can occur lower in the intestinal tract. They accompany typhoid fever, and in certain forms of colitis are a constant finding in the large intestine. Tuberculosis of the bowel is accompanied by ulceration of the wall, characteristically in the ileum. (D.M.H., R.S.M.)

ULTIMATE ANALYSIS. One of the methods of reporting the analysis of a fuel is on the basis of the chemical elements present and their proportions by weight. Since such an analysis reports the composition of a substance in terms of its ultimate elements, it has been called the ultimate analysis. Essentially, it is a chemical analysis, as contrasted to the physical basis of the proximate analysis. The analysis of a fuel, solid, liquid, or gaseous, as the case may be, for the purpose of resolving it into an ultimate analysis, is a process requiring the trained knowledge of the chemist, the apparatus of a well-equipped chemical laboratory, and no inconsiderable perfection of technique. For this reason, the taking of an ultimate analysis has become a specialized subject. Through several years of fuel research and experimentation, many data on the ultimate analyses of coals from different seams have been accumulated, and it is safe to say that very few commercial seams of the present time are without a published analysis of their typical product.

Combustion calculations are essentially calculations of chemical reactions. Any quantitative work starting with chemical reactions must rest on a knowledge of weights of the elements entering into the reaction. It is to be expected, therefore, that the ultimate analysis would be required for combustion calculations. It is in the field of combustion calculations that the engineer is to be found making extensive use of the ultimate analysis of his fuel. (F.T.M.)

ULTIMATE STRENGTH. Tension Test.

ULTRA HIGH FREQUENCY. Radio Frequency.

ULTRABASIC. A term proposed by Judd in 1881 for exceedingly mafic igneous rocks composed largely, if not entirely, of the ferro-magnesium minerals such as olivine and pyroxene. The limiting figure of total silica is approximately 45%, or barely sufficient to supply the needs of the basic silicates. (R.M.F.)

ULTRAMICROMETER. Condenser.

ULTRAMICROSCOPE. The ultramicroscope is not an instrument of extraordinary magnifying power, as its name might suggest. The term has reference rather to a special system of illumination for very minute objects. Such objects as colloidal particles, fog drops, or smoke particles are held in liquid or gaseous suspension in an enclosure with an intensely black background (usually of the black-body type). They are illuminated by a convergent pencil of very bright light entering from one side and coming to focus in the field of view—the so-called "Tyndall cone" familiar in experiments on scattering. With this arrangement, objects too small to form visible images in the microscope produce small diffraction ring systems, which appear as minute bright specks on a dark field. The device is used in studying the Brownian movement, in the Millikan droplet method of measuring the electronic charge (see Electron), in observing ionization tracks in the cloud chamber, etc. (L.D.W.)

ULTRASHORT WAVES. Electric Oscillations and Electric Waves.

ULTRASONICS. Elastic waves of frequencies far beyond the range of audibility, called ultrasonic or supersonic waves, present some interesting aspects. Such waves are conveniently produced by quartz crystal oscillators (see Piezo-Electricity) designed for frequencies ranging up to 200 or 300 kilocycles per sec. Various acoustic phenomena may be demonstrated in these high ranges, and the waves are useful in illustrating principles of optics. For example, it is possible to construct a coarse, concave diffraction grating to form the spectrum of such sounds, or an ultrasonic interferometer to measure their wavelengths.

Some curious effects are observed when the oscillator is immersed in a vessel of oil. The surface of the liquid bulges up into an agitated heap and emits a spray. If the oscillator is placed at the bottom of the vessel and a horizontal metal plate is lowered into the oil, the plate experiences a distinct upward thrust which has pronounced maxima and minima as the plate is pushed downward, corresponding to the interference nodes and antinodes of "stationary" waves. A glass rod held between the fingers and dipped into the oil is so violently (though silently) agitated that its friction may burn the fingers. Small animals in water thus agitated quickly die, and blood corpuscles are destroyed, which suggests caution in exposing the body to such high-frequency vibrations. The waves have been used in sounding the depth of water. (L.D.W.)

ULTRA-VIOLET RADIATION. A range of radiation of frequencies next higher than those of the visible violet. If light from an open arc is passed through a quartz prism and allowed to fall on a white wall, the familiar continuous spectrum appears, ranging from the extreme red to the extreme violet. But if we substitute for the white wall a suitable fluorescent screen, the spectrum is seen to extend considerably beyond the violet, that is, into the region of shorter wavelengths known as ultra-violet. This spectral region has been observed over more than three "octaves" of the radiation frequency scale, roughly from 4000 A (angströms) at the extremity of the violet to below 400 A on the border of the x-ray region. The ultra-violet range has pronounced photographic and ionizing effects, and so is easily detected. The chief hindrance to its study is its rapid absorption in most forms of matter; even air is a serious obstacle to the shorter ultra-violet waves. The sun is an intensely hot source of radiation, but its observable spectrum ceases quite abruptly just below 3000 A because, it is believed, of absorption by atmospheric gases, chiefly oxygen and ozone. (This is fortunate, as the shorter radiations may be very injurious to living tissues; sunburn is attributed in large measure to them. Like x-rays, they should be applied to the body only under the direction of a doctor.) It is therefore necessary to turn to artificial sources, chief among which are solid-electrode arcs and, especially, the mercury arc. Since quartz and fluorite are much more transparent to ultra-violet than is glass, it is necessary that plates, lenses, and prisms for this region be made of these materials. Silver is a much poorer reflector of ultra-violet rays than certain alloys, so that mirrors and reflection gratings are made of the latter. Schumann developed the technique of spectroscopy in the far ultra-violet (the "Schumann region") and prepared plates especially adapted to its photography; so that now, with the vacuum spectrograph and Schumann plates, the ultra-violet spectra of substances are studied almost as thoroughly as the visible.

In medicine, ultra-violet radiation is used therapeutically as a stimulating measure in certain conditions, to promote healing of indolent and sluggish wounds and for disinfectant purposes. In certain diseases it is curative, for example, in rickets. It is also used in lymph node tuberculosis and in certain skin diseases. In 1924 it was shown by Hess and Steenbock that various foods became activated with vitamin D, showing marked antirachitic powers after being exposed to ultra-violet rays. Milk is commonly treated in this way, as well as other foods. Viosterol is made by the ultra-violet irradiation of ergosterol. (L.D.W., R.S.M.)

UMBEL. Flower.

UMBILICAL CORD.
A tube-like structure extending from the navel of the fetus to the placenta of the mother. It contains two arteries and one vein and serves as an excretory, digestive, and respiratory mechanism for the fetus. (R.S.M.)

UMBILICUS.
The scar on the ventral surface of the abdomen where the umbilical cord is severed at birth in the placental mammals. The navel. (A.W.L.)

UMBO.
A rounded prominence near the hinged edge of a bivalve shell. In the shells of mollusks an umbo appears on each valve, representing the point where the initial deposits of shell were formed. Many shells show lines of growth where the margin has extended little by little from the umbo. The lower valve of the shell of brachiopods extends behind the upper in a beak which is also called the umbo. This structure is perforated by the stalk of the animal. (A.W.L.)

UNADJUSTED MOMENT.
An unadjusted moment is a sample moment in a frequency distribution to which no correction for grouping has been made. (L.A.A.)

UNAFLOW. Uniflow Engine.

UNAU. Sloth.

UNCERTAINTY PRINCIPLE.
Quantum Mechanics.

UNCONFORMITY.
If deposition in a given area is interrupted for a time by erosional processes, then renewed, a dissected surface will separate the two groups of beds, which are then said to be unconformable, and the erosion surface marking their contact is called an unconformity. If this unconformable relationship of two rock groups is of limited extent, the unconformity is then said to be local, if of wide extent it is called a regional unconformity. If two groups of sedimentary rocks are separated by an unconformity on tilted beds,

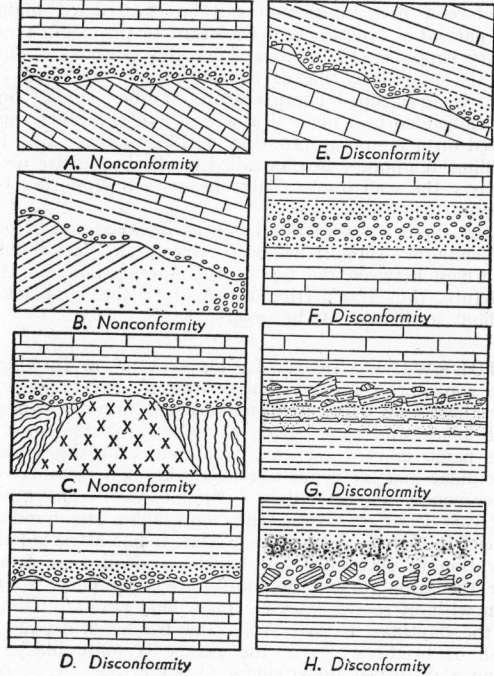

Types of unconformities. (Field, Outline, Barnes & Noble.)

it is then called a nonconformity or angular unconformity. When the plane of erosion occurs between relatively horizontal formations it is called a disconformity. Principal types of unconformities are illustrated in this column. A represents a nonconformity having an irregular erosion surface, developed on the upturned edges of different formations, and covered with a basal conglomerate which grades through sandstone and shale into limestone. In B the nonconformable series has also been deformed together with the second wave of deformation which affected the already deformed beds beneath the unconformity. C represents a major nonconformity in which the erosion surface truncates highly metamorphosed formations and batholithic intrusives. The basal conglomerate rests upon a plane of erosion which, in itself, must represent a great physical hiatus or lack of stratigraphic record for this region. D and E represent disconformities which have been developed upon a prelithified surface. In E the disconformable formations have been subsequently deformed (tilted). F represents a gradational contact, or one in which erosion and depositions have been relatively continuous. G represents a gradational contact developed in plastic sediments, with the development of mud cracks and intraformational conglomerates. H represents a disconformity in which the basal breccia is composed of the lithified fragments of the older formations, plus foreign clastic material. As the fossils above and below the disconformity are the same, the disconformity, though pronounced, signifies slight, if any, hiatus (**diastem**). Disconformities of types *F*, *G*, and *H* may represent either slight or great hiatus. (Note) The true amount of hiatus can only be measured by paleontological means. With the exception of a major nonconformity, the physical evidence of hiatus (unconformity) is seldom a safe criterion alone. Thus the terms nonconformity and disconformity have structural but not necessarily stratigraphic, or time, significance. (R.M.F.)

UNDERPINNING. Underpinning is a term used to denote all work done in connection with the construction of a new **foundation** under an existing structure or under the old foundation itself.

Before a new foundation can be constructed under any part of a structure it is necessary to provide a temporary support. In building work this temporary support is secured by means of shoring or needling. A shore is an inclined **compression member** having one end embedded in a niche in the wall and the other supported on a temporary **footing** outside of the wall. In needling, temporary **beams** called needle beams are used to carry the wall **load**. These beams are supported by temporary footings.

After the new foundation has been installed and before the temporary support has been removed, wedges should be driven between the top of this foundation and the structure, thus transferring the load to the foundation without settlement of the superstructure. (C.W.C.)

UNDERWING. Insecta, Lepidoptera. A moth whose fore wings are colored in dull shades of gray or brown, but whose hind wings are in most species brightly banded with black and some shade of yellow, orange or red. The name is used for the large moths of the family Noctuidae and the genus *Catocala*, although members of other genera are colored in a similar way. The genus *Catocala* is so extensive, with more than one hundred species in North America alone, that it has attracted the attention of specialists and has been studied in detail. For the same reason the moths are very likely to be seen without special search. They hide about buildings in the daytime, as well as on the trunks of trees. A flash of bright wings in the garage is quite likely to mean a disturbed underwing. (A.W.L.)

UNDULANT FEVER (Brucellosis, Malta Fever). A specific fever found in man, cattle, goats and other animals, caused by organisms called *Brucella melitensis* and *Brucella abortus*. Human infection usually occurs from the drinking of unpasteurized milk from infected animals or from handling infected animals. The infection in animals causes abortion.

Infection in man gives a varied picture. The onset may be acute or gradual. Fever may be slight and of intermittent nature, persisting for weeks or months. Weakness, malaise, gastrointestinal symptoms, are the rule. The mortality is low but recurrences and relapses are quite common.

Pasteurization or boiling of milk is a sure preventive. In some districts 20% of cattle are infected.

There is no specific treatment although **serum and vaccine** therapy may be tried. Otherwise the disease is treated with the same measures as are used in any general systemic infection. (R.S.M.)

UNDULATING MEMBRANE. A form of *organelle* found in some of the 1-celled animals. Undulating membranes of two forms appear, one in the ciliates and the other in certain flagellates. The former is composed of numerous **cilia** associated in a row near the oral structures of the animal. The flagellate membrane is developed in the **trypanosomes** as a delicate membrane along one side of the body, bordered by the **flagellum**. It vibrates in an undulating movement as an organ of locomotion. (A.W.L.)

UNGUICULATA. A division of the class **Mammalia** used in the older classifications for the forms with claws. If used at all in modern classifications it is regarded as a division of the subclass Eutheria containing the orders **Insectivora, Dermoptera, Chiroptera, Carnivora, Rodentia, Edentata, Pholidota,** and **Tubulidentata.** (A.W.L.)

UNGULATA. An older division of the class **Mammalia** whose standing is like that of the Unguiculata. It includes the hoofed animals and related forms whose feet bear heavy nails or nail-like hoofs, embracing the orders **Artiodactyla, Perissodactyla, Proboscidea, Sirenia,** and Hyracoidea. Although the term is almost obsolete it persists commonly in reference to the members of the first two orders as the even-toed and odd-toed ungulates, respectively. (A.W.L.)

UNIAXIAL CRYSTALS. Double Refraction.

UNIFIED FIELD THEORY. Fields of Force; Relativity.

UNIFLOW ENGINE. The term uniflow designates one-way flow of a fluid through a cylinder. While it is sometimes employed to differentiate between the flow of gases in two- and four-stroke cycle engines, uniflow has come principally to designate that improved type of **steam engine** in which the exhaust is arranged so that the steam flows from the end of the cylinder to exhaust ports located near the center, and does not reverse its direction of flow during exhaust, as is the case in the dual flow engine. This elimination of exhaust steam flow over inlet ports is the major advance in steam engine design in recent years, because it eliminates **initial condensation** in ports and cylinder head. The overcoming of this rather large loss places the uniflow engine in a favorable position to compete with other types of prime movers. It is more expensive to construct, but so marked are its advantages that it has come to be the only type of steam engine considered where a large efficient engine is wanted.

Except for the cylinder construction, the uniflow engine is very similar to any other steam engine. The cross-

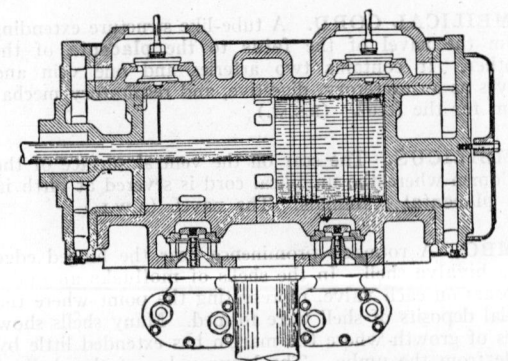

Section of cylinder and valves of universal uniflow engine.

section through the cylinder of a uniflow (sometimes written unaflow) engine shows how a ring of exhaust ports situated at the center of the cylinder will be uncovered by the **piston** as it nears the end of an expansion stroke. It also shows why the piston must be much longer in this type engine than in the dual flow engine.

On the return stroke the piston covers the ports fairly early in the stroke, and a considerable amount of work would be drawn from the **flywheel** in compressing the steam during the long back stroke were it not for the fact that this uniflow is equipped with auxiliary exhaust valves which remain open a portion of the exhaust stroke, thus delaying the point of compression and preventing the reabsorption of considerable work from the flywheel. This delayed compression does not produce higher thermal efficiencies than a full compression, but increases the power which may be developed per cubic inch of piston displacement. Some manufacturers build full compression uniflow engines. (F.T.M.)

UNIFORM CONVERGENCE OF SERIES. An infinite series $\sum_{1}^{\infty} u_n(x)$, each of whose terms is a function of x defined in an interval I, is said to be uniformly convergent in that interval if it **converges** for every value of x in I and if, for any arbitrary $\epsilon > o$, an index N independent of x exists such that

$$\left| \sum_{i=n+1}^{\infty} u_i(x) \right| < \epsilon$$

for every $n > N$ and for every value of x in I.

The most important test for uniform convergence is the so-called Weierstrass' M-test:

If the terms of $\Sigma u_n(x)$ are **continuous functions** of x in an interval I, and if $|u_n(x)| \leq M_n$ for every n and for all values of x in I and the M_n are positive constants, then if ΣM_n is convergent, the series $\Sigma u_n(x)$ is uniformly convergent in I and is also **absolutely convergent** in I.

Some of the important uses of uniform convergence are indicated by the following theorems:

If $\Sigma u_n(x)$ is uniformly convergent in an interval I and if its terms are continuous functions of x in that interval, then its sum is itself a continuous function of x in I.

A series of continuous functions which converges uniformly in an interval I may be integrated term by term, provided the limits of integration are finite and lie in the interval I.

If a uniformly convergent series be integrated term by term, the resulting series will be uniformly convergent.

Any convergent series of functions may be differentiated term by term if the resulting series is uniformly convergent. (L.L.S.)

UNIFORMITARIANISM. Cataclysm.

UNILATERAL TOLERANCE. Interchangeable Manufacture.

UNIRAMOUS LIMB. A crustacean appendage consisting of a single unbranching series of segments. It is derived from the **biramous appendage** by the loss of the **exopodite** and is very well illustrated by the walking legs of the **thorax** of lobsters and related forms. The **chela** of these animals is a modified uniramous appendage. (A.W.L.)

UNIT DEFORMATION. Deformation.

UNIT HEATER. When a means for producing heat is combined with one for circulating it within a building, the combination is known as a unit heater. There are several manufacturers who supply unit heaters commercially; usually the product takes the form of a gas- or a steam-heated surface such as a cellular core, finned tube, or pipe coil. Air is blown over this heating surface under the influence of a fan or blower, generally electrically driven. Formerly unit heaters were employed principally in industrial buildings or mercantile buildings, but in recent years the desire of owners of steam or hot-water heated homes to eliminate radiators from the rooms, and to obtain a measure of air conditioning, has led to the development of similar equipment for the home. The domestic equipment is considerably different in design from the heavy, rougher, industrial types, but in it, also, is generally employed a heating surface over which a silent propeller type fan blows the air which, when heated, is passed through ducts or grills into the room. (F.T.M.)

UNIT OPERATIONS. Unit operations is a study of the physical changes which take place in materials undergoing chemical processing, in order to design equipment in which to conduct these changes. Its major classifications are flow of fluids, heat transfer, evaporation, drying, air conditioning, gas absorption, distillation, extraction, mixing, separation of solids, crystallization, filtration, and crushing and grinding. (D.E.M.)

UNIT PROCESSES. Unit processes is the study of the chemical changes which take place in materials undergoing processing in order to find the best conditions for the reactions, and to design suitable equipment. The major classifications are sulfonation, nitration, reduction, ammonation, oxidation, halogenation, hydrolysis, esterification, alkylation, polymerization, and various special reactions such as the Friedel and Crafts reaction. (D.E.M.)

UNIT STRESS. Stress.

UNITED STATES STANDARD. Gauge Number.

UNIT-PRODUCTION. Measurement.

UNITS. Physical Measurements; Physical Magnitudes and Physical Equations; Metric System; C. G. S. System; M. K. S. System; Electric and Magnetic Units.

UNIVERSAL CHUCK. Chuck.

UNIVERSAL GRINDER. Grinding.

UNIVERSAL JOINT. Coupling.

UNIVERSE. The term universe, in its complete physical sense, should apply to all matter in existence. We know that this matter is not uniformly distributed, but rather that it is gathered into aggregates ranging in size from the smallest atom to the largest collection of matter known to astronomers.

We have considerable knowledge regarding the small aggregates which make up objects on the surface of the earth. We know something regarding the larger aggregates known as the **planets, stars, nebulae,** and other astronomical objects contained within the boundaries of the **milky way,** or galactic system. We have observational data from which theories may be formed regarding the size and structure of our own local system and the milky way as a whole, but we are now approaching the region where we have nothing more than hypotheses which await more complete confirmation.

What is beyond our milky way? Is this galactic system the largest aggregate of matter which exists in the universe? Such are questions upon which much research is being carried on at present and for which answers can by no means be considered as found. It is in search of answers for such questions that the 200-in. telescope is constructed.

This much can be said at present: we do know that there are many aggregates of matter which lie outside of the confines of our milky way. Descriptions of these extra-galactic objects will be found discussed under the title of **spirals.**

Statistical discussions indicate that the number of these objects may be as great as 10^{10}—a number comparable with the number of stars in our own galactic system. Very recent studies have shown quite conclusively that the extra-galactic objects are not uniformly distributed through space. The term "galaxies of galaxies" has been applied to these, the largest aggregates of matter thus far discovered. Far too little is known of them to permit of any discussion of their sizes and forms. (See **Population**.)   (W.K.G.)

UNSTRATIFIED DRIFT. Till.

UNSYMMETRICAL BENDING. The condition which exists at any cross-section of a flexural member (see **Flexure**) when the plane of the **loads** contains the **shear center** but does not coincide with either of the two **principal planes of bending** is known as unsymmetrical bending. Under these conditions the **flexure** formula is not applicable because the **neutral axis** is not perpendicular to the plane of the loads although it does pass through the **center of gravity** of the cross-section.

The flexural stress due to unsymmetrical bending is found from the formula.

$$s = \frac{M \cos\theta y}{I_x} + \frac{M \sin\theta x}{I_y}$$

in which s = Flexural unit stress at a point.
    M = **Bending moment.**
    x, y = Coordinates of the point taken with respect to the **principal axes.**
  I_x, I_y = **Moments of inertia** referred to the principal axes.
    θ = Angle which the plane of the loads and hence the plane of the bending moment makes with the y-axis.

It should be noted that this formula, which is applicable if the stress does not exceed the **proportional limit,** may be obtained by resolving the bending moment into components parallel to the two axes. The flexure formula is then applied to each of the principal axes and the results combined by the theory of superposition. The simplest way to determine the sign of the resultant stress (+ for **tension** and − for **compression**) is to note the sign for bending about each axis and use this sign when combining the results in the formula.   (C.W.C.)

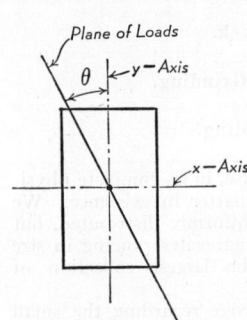

UPLIFT. If water should find its way from a **reservoir** to the surface between the base of a **dam** and its foundation, it would exert a pressure upwards against the base of the dam. It would, in the extreme case, equal the full hydrostatic head corresponding to the height of the water surface above the base of the dam. This water pressure is known as uplift, and could possibly overturn a gravity type dam if it were not prevented, or allowed for in the design. Sometimes $\frac{2}{3}$ of the static head is assumed to be acting as uplift. Uplift might be considered to be maximum at the upstream edge of the base, decreasing from that to zero at the downstream edge. Where measures are taken to prevent uplift, the foundation must be very thoroughly grouted, cut-off walls must be let into the foundation near the upstream edge of the base, and drains provided to relieve any pressure which might be built up by a slow seepage.   (F.T.M.)

UPSET. Forging.

URALITE. This is a metamorphic mineral. It is a well-established fact that **pyroxene** rocks may be **metamorphosed** into **hornblende** rocks. If the hornblende thus produced is fibrous and retains the original form of the pyroxene, it is called uralite, and the process by which the change is brought about is called **uralitization.** It seems quite clear that uralitization is a chemical process which in many cases is accompanied by the generation of new minerals such as **calcite, epidote,** and **magnetite.** Uralite was first observed in rocks from the Ural Mountains, hence its name.   (E.S.C.S.)

URALITIZATION. Uralite.

URANINITE. A mineral approximating the composition UO_2, but containing besides the higher oxide of **uranium,** UO_3, and oxides of **lead, thorium,** and rare earths. The uraninite may occur as black **octahedral** crystals of high specific gravity (9.0–10.63); when in masses of pitchy luster is called **pitchblende.** All uraninites and pitchblende contain a minute amount of radium. It was in pitchblende obtained from the Joachimsthal in Czecho-Slovakia that Mme Curie discovered radium. Other localities for uraninite are in Saxony, Rumania, Norway, Cornwall, East Africa, and in the United States in the pegmatites of Connecticut, North Carolina, and South Dakota, and in Gilpin County, Colorado. An important occurrence of pitchblende is at Great Bear Lake, Northwest Territories, Canada, where it has been found in large quantities associated with silver, and is now a commercial source of radium.   (E.S.C.S.)

URANIUM. Symbol: U. Atomic number: 92. Atomic weight: 238.14. Density: 18.7. Melting point: < 1850° C. (Isotopes: 234 (trace), 235 (0.7%), 238 (99.3%).)

Uranium is a white metal, ductile, malleable, and capable of taking a high polish, but tarnishes readily on exposure to the atmosphere. Finely divided uranium takes fire on exposure to air, and the compact metal burns when heated in air at 170° C. Uranium metal slowly decomposes water at ordinary temperatures and rapidly at 100° C.; is soluble in **hydrochloric acid** and in **nitric acid;** and is unattacked by **alkalis.** Chemically related to **chromium, molybdenum,** and **tungsten;** and, like **thorium,** is radioactive. In the radioactive decomposition radium is formed. (See **Radioactivity.**) Discovered by Klaproth in 1789.

Uranium occurs in **pitchblende** (75–90% U_3O_8) and **carnotite** (62–65%). The oldest and most celebrated deposit of pitchblende is that of Joachimsthal in Czecho-Slovakia, known since early in the sixteenth century, in the ore from which mine the discovery of radioactivity was made. Carnotite is a **potassium** uranium **vanadate,** and occurs in southwestern Colorado and eastern Utah as an extensive deposit. Richer ores have been found in recent years in Belgian Congo, and Great Bear Lake region of northern Canada. Uranium, isotope of mass 235, has been the subject of much investigation because of its disintegration into barium and krypton when bombarded by slow neutrons. (See **Radioactivity.**) The so-called atomic bomb is really a nuclear bomb making use of the above disintegration to release its enormous energy.

Oxides: Oxides of the following composition are reported: UO_2, U_2O_5, UO_3, U_3O_8, UO_4. UO_2 and UO_3 have been well established; U_3O_8 is believed to be a mixture, $UO_2 \cdot 2UO_3$. Uranium dioxide, brown to black solid, is obtained when uranium metal is burned in air at 170° C., and when U_3O_8 is heated in **hydrogen** at 650° C.; U_3O_8, green solid, by heating any other oxide of uranium in air to 700° C.; uranium trioxide, brick-red solid, by heating uranic acid, ammonium diuranate or ammonium uranyl carbonate to a maximum temperature of 300° C.

Salts: 1. Uranous, e.g., chloride UCl₄, green color, strong reducing agents, and not fluorescent in violet light. 2. Uranyl, e.g., chloride UO_2Cl_2, yellow color, markedly fluorescent in violet light, readily reduced in solution to uranous by zinc metal, tin metal, copper metal, iron metal, and ferrous salts. Small amounts of sodium uranate are used in ceramics to produce yellow glazes, and in the dyeing industry as a mordant. (R.K.S.)

URANOTHORITE. Thorite.

URANUS. (See tables of planetary data, page 1109.)
Uranus, the first planet to be "discovered," was found accidentally by Herschel in 1781 while sweeping the sky with a 7-in. reflecting telescope of his own manufacture. His discovery stirred up a tremendous amount of popular interest in astronomy, and history relates that during the weeks following the discovery, the streets in front of Herschel's house were crowded with people eager to get a view of the telescope and a glimpse of the discoverer. Herschel named the planet Georgium Sidus (star of the Georges) in honor of the then reigning king of England, George III, but the name was never adopted on the continent. Many Europeans called the planet Herschel, in honor of the discoverer, but the name Uranus, proposed by Bode, is the one which has survived.

To the naked eye Uranus is barely visible as a sixth magnitude star, but in a telescope of moderate aperture the object appears as a disk. The disk has a bluish-green appearance, and no surface markings have ever been observed. Early observations showed that the planet is very much flattened at the poles, which gives evidence of high rotational speed, but it was not until 1912 that Lowell and Slipher were able to prove, by the Doppler principle, that the period of rotation is about 10.75 hours. The relatively high albedo and low densities both give evidence of a thick layer of atmosphere about the planet. Reasoning from purely theoretical grounds, on the basis of the mean distance of the planet from the sun, we find that the temperature of the surface of Uranus should be about 63° K. (−346° F.). At such a low temperature all gases except possibly hydrogen, helium, and argon should be condensed out of the atmosphere. However, within recent years, Dunham has shown the existence of a large amount of methane in the atmosphere of Uranus, which indicates that the temperature must be considerably higher than that given by purely theoretical reasoning.

One strikingly interesting characteristic of Uranus is found in the fact that the axis of rotation lies almost in the plane of the ecliptic, being inclined to it by an angle of less than 10°. Furthermore, the direction of rotation is opposite to that of all of the other members of the solar system.

Uranus has four small satellites less than 1000 miles in diameter. The orbit planes of these satellites lie close to the plane of the planet's equator, and hence are nearly perpendicular to the plane of the ecliptic. The satellites revolve about the primary in the same directional sense as that in which the planet rotates, i.e., in the retrograde direction. (W.K.G.)

URCHIN. 1. Mammalia, Insectivora. The European
hedgehog. 2. Echinodermata, Echinoidea. The sea-urchins and the related flat forms called cake urchins and sand dollars. Although usually accompanied by a prefix, the name is commonly shortened, as in the case of the common green and purple urchins of the Atlantic. (A.W.L.)

UREAS. Amines and Amides.

UREASE. Enzyme.

UREDINALES. Rust Fungi.

UREDINIOSPORE. Rust Fungi.

UREIDES. Purine and Uric Acid Compounds.

UREMIA. An acute or chronic disorder resulting from
retention of nitrogenous waste products in the body due to impairment of kidney function. The majority of cases complicate the latter stages of severe nephritis, but uremia may occur in kidney failure from any cause. The usual ones are various kidney diseases, mercury poisoning, reflex suppression of kidney function, etc.

The symptoms of uremia are headache, vomiting, dizziness, and diarrhea. Psychic disorders are frequently present, delirium, stupor, and coma sometimes occur. Convulsions or, more frequently, muscular twitching are frequent. Other symptoms are those of the underlying disease and the accompanying acidosis.

The treatment of uremia is directed toward the elimination of accumulated nitrogenous substances by aiding renal and cardiac function. The prognosis is usually poor. (D.M.H., R.S.M.)

URETER. The two tubes, each of which connects a
kidney with the bladder. The urine drains from the kidney pelvis into the ureter, down to the bladder. Each ureter is about twelve to fourteen inches long and is extremely small in diameter. (See Excretory System.) (R.S.M.)

URETHANES. Amines and Amides.

URETHRA. The narrow passageway through which
the urine flows from the bladder to the outside. In the female, the urethra is about 1½″ long and the external opening (meatus) lies above the vagina, between it and the clitoris. In the male, it measures about 8″, runs the entire length of the penis; it also serves to carry the seminal fluid to the outside. The urethra is often the site of acute and chronic infection. (See Excretory and Urogenital Systems.) (D.M.H., R.S.M.)

URIAL. Sha.

URIC ACID. Purine and Uric Acid Compounds.

URINALYSIS. The chemical and microscopical ex-
amination of the urine. (D.M.H.)

URINARY BLADDER. Excretory and Urogenital
Systems.

URINE. A transparent amber fluid, slightly acid in
reaction, with a characteristic odor and a specific gravity varying from 1.005 to 1.030. Urine is excreted by the kidneys, passes through the ureters into the bladder, where it is stored and is discharged through the urethra. The average amount excreted by the kidneys in 24 hours in health is about three pints. The amount varies somewhat according to the quantity of fluid drunk and the amount lost through the skin, lungs, and intestinal tract.

The urine may become cloudy on a vegetable diet due to precipitation of phosphates. In disease cloudiness may develop from the presence of pus.

Urine is the medium by which the body gets rid of a large portion of the end-products resulting from food metabolism. The waste material is composed principally of urea, an end-product of protein metabolism, ammonia, hippuric acid, purine bodies, and salts, mostly sodium chloride.

The important abnormal constituents of the urine which, when constantly present, signify some disease process are **albumin, glucose, acetone, pus,** red blood cells, **bacteria,** and casts of the kidney tubules.   (D.M.H.)

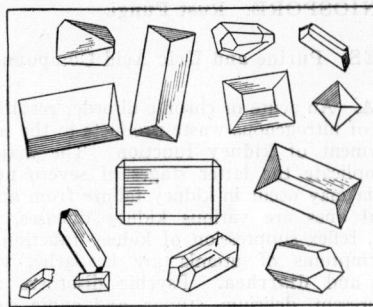

Prismatic forms of triple phosphate crystals from urine (×250). *(Todd and Sanford, Clinical Diagnosis by Laboratory Methods, W. B. Saunders Co.)*

UROCHORDATA. Chordata. A subphylum containing the marine animals known as **ascidians, tunicates, salpians,** and appendicularians. They represent an extreme adaptation of primitive chordate structure for **sessile** or drifting life.

Urochordates are characterized by the presence of a cuticular covering called the **test** or **tunic,** a word associated with the name Tunicata, also applied to the group. They have a very large **pharynx,** occupying most of the body and perforated by many openings. In some species these openings lead to the exterior and in others to a peribranchial chamber opening to the exterior by a constricted **cloacal** aperture. Ordinarily, water bearing particles of food is taken into the pharynx by way of the mouth through ciliary (**cilia**) action, passes into the peribranchial space, and is discharged by the cloacal aperture. When irritated, however, the animal may contract spasmodically and discharge jets of water from both mouth and cloacal aperture. In the pharynx food is caught by an endostyle as in the **Cephalochordata.**

The Urochordates are classified as follows:

Class Larvacea. Appendicularians. Forms resembling larval tunicates, with a trunk and tail. Minute, free-swimming, and transparent.

Class Ascidiacea. The ascidians or sea-squirts. Tailed and free-swimming as larvae but sessile as adults.

Class Thaliacea. The salpians. Floating transparent forms, sometimes colonial. Moderate in size. (A.W.L.)

URODELA. The **salamanders, newts,** and related forms of tailed **Amphibia.** An order, also named Caudata.   (A.W.L.)

UROGENITAL SYSTEM. A term sometimes applied to the closely associated **excretory** and **reproductive** systems of the vertebrates. The gonads and the kidneys arise from the same mass of embryonic tissue, the intermediate **mesoderm,** and as the ducts of the excretory system develop they are appropriated in part by the reproductive organs. In the male the mesonephric or Wolffian ducts persist as the vasa deferentia, connected with the testes by modified mesonephric tubules, the vasa efferentia. These parts do not persist in the female, where separate Müllerian ducts arise in association with the ovaries. Both male and female genital ducts join the **cloaca** during embryonic development, and here further intimate association is evident.

The cloaca persists in all classes, but in only a few of the primitive mammals. In most mammals a more or less complete separation takes place.

This region of the gut bears the **allantois,** an extra-embryonic membrane, as a ventral appendage, and receives dorsolaterally the mesonephric ducts, from which the metanephric ducts or ureters branch to the kidneys. The cloaca expands and the angle between the allantoic stalk and the intestine progresses in mammals until the chamber is subdivided into the dorsal rectum and the ventral urogenital sinus. The latter receives the mesonephric ducts. By its continued growth the common portion of these ducts is absorbed into its wall, so that the mesonephric ducts and ureters join it separately, the latter in front of the former. The anterior part, bearing the ureters, together with a portion of the allantoic stalk, becomes the urinary bladder. The slender posterior portion becomes the female urethra, the duct of the bladder, and in the male forms the proximal portion of the urethra. External folds flanking the opening of the urethra form the labia minora of the female, between which the vagina, derived from the united parts of the Müllerian ducts, and the urethra open separately. In the male these folds unite to form part of the penis, with an enclosed tube which becomes the distal or phallic portion of the urethra. Thus in the male the urethra remains a common duct of the excretory and reproductive systems. (A.W.L.)

UROPOD. A swimmeret of certain **crustaceans,** modified as a broadly expanded swimming appendage. The pair of uropods associated with the flattened **telson** form the fan-like tail of the **lobsters** and **crayfishes.** By powerful contractions of the abdomen this structure is used to scull the animal rapidly backward through the water. (A.W.L.)

UROSTYLE. A long rod of bone articulated with the last vertebra of the **frog** and related forms, in the skeletal axis. Although this bone is not segmented, it shows evidences of being formed from coalesced vertebrae, like the os coccyx of **primates.** (A.W.L.)

URSA MAJOR. (The greater bear.) (Map, page 380.) This **constellation** is probably best known for the asterism known in this country as the big dipper and in England as the plough or the wagon. The constellation is circumpolar for both Europe and North America, and two of the stars in the dipper, known as the pointers, are very useful in locating the star **Polaris,** since the line joining them, if extended, will pass close to the celestial pole.

The star Mizar (Zeta Ursae Majoris) at the bend of the handle of the dipper is an easy visual **double star.** Tradition says that this object was used by the American Indians as a test of vision, a person able to see this star as double being credited with good vision. The star is certainly one of the earliest doubles known, having been so named by the ancient Arabs. (W.K.G.)

URSA MINOR. (The smaller bear.) (Map, page 380.) This **constellation** is best known because of the fact that the bright star at the end of the handle of the asterism, frequently referred to as the little dipper, is at present the closest bright star to the north celestial pole of rotation. This star **Polaris** (Alpha Ursae Minoris) will be described elsewhere. Other than this star the constellation contains very few objects of interest or importance. (W.K.G.)

URTICARIA. Hives. An allergic response characterized by the appearance on the skin of firm, elevated, circumscribed lesions varying from the size of a pea to several inches in diameter. The eruption is characterized by intense itching. The patches may appear and disappear or change their position during the course of the day. Usually the hives disappear as quickly as they develop, but sometimes they may persist from 12 to 24 hours. Hives may occur from a variety of causes, some of which are not well understood. In general, urticaria occurs when a hypersensitive individual comes in contact with a specific **protein** substance. Certain protein foods, serum or **blood** injections, or certain drugs may cause this type of eruption.

Treatment consists, when possible, of the avoidance of the offending protein. At times, desensitization with graded doses of this protein subcutaneously is successful. During an attack **adrenaline** by hypodermic injection gives prompt relief usually. (D.M.H.)

USTILAGINALES. Smuts.

UTERINE TUBES. Fallopian Tubes.

UTERUS. A portion of the female reproductive passages in which the eggs or young are retained during all or part of **embryonic** development. A uterus is found in some invertebrates, as the roundworms and **arthropods.** In the mammals, accompanying their highly specialized reproductive processes, it is important. It is formed in these animals either as a pair of chambers in the two Müllerian ducts or as a portion of the fused region of the two ducts. Uteri of the latter kind are of three fundamental forms. A duplex uterus is forked through most of its length, and is in reality a pair of chambers united at one end. In the uterus bicornis there is an unpaired chamber of considerable extent, branching in front to form two diverticula bearing the oviducts. A uterus simplex is an unpaired chamber.

The wall of the uterus contains heavy layers of involuntary muscle and its lining is glandular and partially ciliated (cilia). The lining is specialized for the reception of the fertilized ovum in pregnancy, and takes part in the formation of circulatory connections with the developing **embryo** through the **placenta.**

The human uterus is a hollow, muscular, pear-shaped organ situated in the lower pelvic cavity between the **bladder** and the **rectum.** The broader portion of the uterus is the upper portion, at each end of which the **fallopian tubes** enter. The lower portion is called the cervix and projects downward into the **vagina.** The virginal uterus is small, measuring about 3″ long, 2″ broad, and 1″ in thickness. During pregnancy the uterus enlarges enormously and may extend upward to the ribs. After pregnancy the uterus is always a little larger than its original state. After the **menopause** the organ becomes smaller.

Since the uterus is suspended and held in position by ligaments, its position may become disturbed. Anteversion indicates that the uterus is inclined too far forward, while retroversion indicates that the organ is displaced backward toward the rectum.

The function of the uterus is to receive the ova from the ovaries through the fallopian tubes. If the ovum is fertilized it is retained and development takes place within the uterine cavity. When development is complete the fetus is expelled from the uterus by contractions of the uterine muscular wall, aided by contractions of the abdominal muscles.

The uterus may become involved in infectious processes, especially after childbirth and **abortions,** particularly of the criminal variety. **Tumors** are common in the uterus, both of the benign and malignant variety. The most common benign tumor that occurs in the uterine wall is a fibroid. Removal of the uterus is called hysterectomy. The operation to correct malposition is called suspension of the uterus. (A.W.L., R.S.M.)

UTILIZATION FACTOR. When a **transformer** is used to supply a **rectifier** system the current wave forms are normally irregularly shaped and hence the losses are different than would be expected at first glance, the transformer running hotter for a given current than if it were supplying a **resistance** load. The ratio of the d-c power output to the normal a-c rating for the same heating losses is the utilization factor of the transformer. (L.R.Q.)

U-VALLEY. A river valley whose transverse profile has been changed from the V- to the U-shape by the work of a glacier. While the ice tongue is occupying the valley, any tributary glaciers or streams are unable

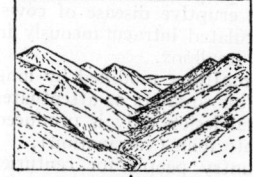

Diagrams showing a stream-cut valley (A), and as it appears after glaciation (B). *(After U. S. Geological Survey.)*

to cut below the surface of the ice lying in the main valley. When the glacier finally melts the tributary valleys do not enter the main valley at grade but appear as hanging valleys. The stream which flows in a U-valley is called a misfit stream because it could not have transported the large glacial boulders among which it flows. (R.M.F.)

UVAROVITE. Garnet.

UVULA. A soft mass of tissue which hangs downward from the posterior part of the soft palate above the base of the tongue. When it is too long and causes irritation or tickling in the throat it may be shortened by astringent solutions, or in extreme cases by surgery. (R.S.M.)

V

VACCINATION. The process of introducing a vaccine into the body for the purpose of immunization against disease. The immunity is the result of the stimulation of **antibody** production against the organism contained in the vaccine. Vaccination was first used against smallpox, but is now carried out as a method of immunizing against many diseases including **whooping cough, typhoid fever, cholera, yellow fever** and **rabies.** Vaccination against the common cold, dysentery, meningococcus meningitis, streptococcal infections, etc., have met with little or no success. (D.M.H.)

VACCINE. A preparation which on injection will induce an active **immunity** in the body. Vaccines are made up of dead or attenuated infectious agents—**bacteria** or **viruses** and each one is specific: the intracutaneous inoculation of tetanus vaccine protects the individual against tetanus, and rabies vaccine protects against rabies, etc. (D.M.H.)

VACCINIA (Cowpox). An eruptive disease of cows, the virus of which, when inoculated intracutaneously in man, gives protection against smallpox.

Vaccinia is believed to be **smallpox** modified by long passage through an animal. This so alters or attentuates the smallpox **virus** that a local lesion only is produced when the virus is inoculated into man (vaccination).

It has been known by country people for centuries that cowpox will protect against smallpox, but it remained for Jenner in 1798 to put vaccination on a practical basis.

Revaccination should be done (1) at intervals of 7 to 10 years, (2) repeatedly when previous vaccinations fail to "take," (3) when exposed to a smallpox patient.

If smallpox occurs in a patient who has been vaccinated many years before so that its effect has worn off, the disease assumes a less serious form than without previous vaccination. (R.S.M.)

VACUOLE. A small globule of clear fluid in the cytoplasm of a **cell.** In preparations for microscopic study the contents of the vacuole are usually dissolved away, so that an open space alone remains, but even vacuoles whose contents are undisturbed usually appear vacant because of their transparency. In the multicellular body fat cells afford a good illustration of vacuoles as globules of fat accumulate in them. One-celled animals also offer a good example in the contractile or pulsating vacuole. This is a globule of clear liquid that forms and discharges periodically, sometimes at a fixed point in the cell. It is interpreted as an organ for the removal of surplus water with dissolved wastes from the **protoplasm.** (See Cell.) (A.W.L.)

VACUUM. A perfect vacuum would be a region entirely devoid of matter. Because of diffusion and the volatility even of solids, this condition is merely an ideal. Even if we imagine it to be attainable, the old concept of a vacuum as mere emptiness is modified by evidence leading to the conclusion that so-called "space" is capable of participation in processes of a physical nature. Faraday's notion of **fields of force,** involving stresses in space, and the undulatory theory of **light** (not to mention the more recent relativistic aspects of the subject), have profoundly influenced scientific thinking in regard to space by endowing it with physical properties.

Experimentally a vacuum is simply a region of very low pressure, usually attained by some form of **air pump.** A pressure of a mm. or so of mercury would be called a "rough vacuum," while one of 0.0001 mm. is a "high vacuum." The best vacuum obtainable by artificial means at present is probably of the order of 10^{-8} mm.; though such pressures cannot at present be measured with certainty. (L.D.W.)

VACUUM THERMOCOUPLE. A device for measuring very feeble electric currents, either alternating or direct, by means of their heating effect. The current to be measured is passed through a short, very fine platinum wire enclosed in a small, evacuated glass bulb. In good thermal (but not electric) contact with the center of this wire is placed one junction of a small thermocouple composed of platinum platinum-rhodium wires (or other metals), the other junction being kept at the constant temperature for which the instrument has been calibrated. When the unknown current is turned on, it heats the fine platinum wire slightly, and sets up a change of electromotive force in the couple which bears a definite relation to the heating current, as determined by calibration with known currents. This voltage change may, if necessary, be amplified by a d-c **amplifier** in order to make it measurable. The glass bulb containing the thermocouple must be housed in a heat-insulating, opaque box or case to prevent heat from reaching the couple by either radiation or conduction, and thoroughly evacuated to avoid convection. (L.D.W.)

VACUUM TUBE. Tube, Electron; Crookes Tube; Ionized Charge; Triode.

VACUUM-TUBE VOLTMETER. Instruments, Electrical.

VADOSE. A geologic term referring to a type of circulating underground water. (R.M.F.)

VAGINA. A sheath. The term is most familiar in reference to the terminal portion of the female genital passages, which receives the intromittent organ of male during **copulation.** This canal extends from the **vulva** upward to the **cervix** of the **uterus,** and is 3–3½″ long. The term vagina is also used in anatomy to designate other sheath-like structures, such as the fibrous sheath of tendons. This structure is called a vagina fibrosa when solid and a vagina mucosa when it contains a fluid-filled cavity surrounding the tendon. (A.W.L., R.S.M.)

VALENCE. Valence is the capacity of an **atom** to combine with other atoms to form a **molecule.** It is specified as the number of **hydrogen** atoms or twice the number of **oxygen** atoms with which one atom of the element under question will combine. Thus nitrogen has the valence 3,2,4,5 in the compounds NH_3, NO, NO_2, N_2O_5. A further distinction is made by considering positive and negative valences. If the hydrogen is assigned the valence of plus one, and oxygen that of minus two, and if the valences in a compound are made to total up to zero, we have a formal scheme of positive and negative valences. In ammonia, NH_3, the three hydrogen atoms each with a valence of plus one exactly balance the one nitrogen atom with the valence of negative three. Many atoms possess more than one valence,

1522

but the principal valence is correlated with the periodic table and the **atomic structure** of the atom. (See **Chemical Composition.**) The inert gases have the valence zero. The principal positive valence is the number of the group in which the element falls in the periodic table. Thus hydrogen is one, **lithium** also one, **boron** three, etc. The negative valence is eight minus the number of the group in the periodic table. Negative valences greater than four do not occur. For example, oxygen has the valence of eight minus six, that is two negative in H_2O (water); and nitrogen has the valence of eight minus five, that is three negative in NH_3 (ammonia).

On the basis of modern electronic theory of atomic structure we can classify the different types of valence. The guiding principle is that the atoms tend to assume an inert gas electronic structure of eight **electrons** in the outer shell (in the case of hydrogen it is two). To do this the atom either loses to, gains from, or shares with other atoms, electrons. This process leads to molecule formation. The following are the principal types of valences and their electronic interpretation.

Electrovalence or polar valence is associated with a transfer of an electron from one element to the other in order to complete by such a transfer the octet of each element. Thus in sodium chloride the sodium atom has one valence electron outside a closed octet of eight. By loss of this electron the **sodium** atom becomes positively charged sodium **ion** because the nuclear positive charge exceeds that of the electrons by one. On the other hand, the **chlorine** atom has a grouping of seven electrons in the outer shell. It picks up another electron to complete its outer shell to an octet, but in so doing obtains a total charge of one minus, becoming a chloride ion. The result is that in sodium chloride we are not dealing with sodium atoms and chlorine atoms but with sodium and chloride ions. This is experimentally substantiated. The forces holding the ions together are the **electrostatic forces,** which are equal to the product of the electronic charges on the ions divided by the product of the separation squared times the dielectric constant of the medium. Thus when the sodium chloride **crystal** is placed in solvent of high **dielectric constant** such as water, the forces between the ions are weakened and the ions float away from each other. In other words, electrolytic **dissociation** takes place. It must be noted that polar valences have no specific directional effects in space. The electrostatic attraction is best satisfied by a close packing of the ions. Inasmuch as there are large stray electric fields present in polar compounds, they possess a high melting point and considerable hardness.

Homopolar or covalent bonds are formed by a different mechanism. Here again we have as the basis the tendency of each atom to complete its outer shell of electrons to eight, or in the case of hydrogen to a doublet. In contrast to polar valence, in covalence we have no direct transfer of electrons, but merely a sharing. In the case of molecular hydrogen each hydrogen atom with its one electron shares this electron with the other hydrogen. The result is that each atom in the molecule has at least part of the time a complete shell of two electrons. The electrons can be visualized as traveling in orbits encompassing the two hydrogen nuclei. It is a property of the covalent bond that it is not weakened by electrolytic solvents and that it has a definite direction in space. These directional effects of covalent bonds are expressed in stereochemistry. Thus, for example, the four valence bonds of the carbon atoms are arranged to extend from the center of a tetrahedron to the four corners. Furthermore, since there is a one-to-one saturation of the electron forces, the stray electric fields are negligible, the melting points are low, and the crystals are soft.

Another type of bond which occurs in solids is the metallic bond. It can be considered as an extreme case of sharing of electrons in that an electron gas (present in the crystal lattice) is shared not by two ions but by all the ions in the lattice. This electron gas is responsible for the metallic properties of certain solids, especially for thermal and electrical conductivity. (R.K.S.)

VALLEY BREEZE. On hot days, uneven terrain gives rise to uphill breezes, i.e., from the valley up mountain or hill slopes. This breeze is known as a valley breeze; it is an anabatic wind. With sunset, the breeze dies. Valley breezes are seldom strong. (P.E.K.)

VALVE. For the meaning of this term as used in engineering, see **Valves.** For the meaning of valve as used in radio, see **Rectifiers.**

In zoology, a valve is a structure that regulates the flow of materials through a tubular organ. It may consist either of a muscular band in the wall of the organ or of a system of flaps that close together to impede flow in one direction.

The valves of the alimentary tract (**Digestive System**) are of the former type, consisting of a **sphincter** muscle encircling the tract within a projecting ridge of its lining. The chief valves are the **pyloric** at the union of the stomach with the **duodenum** and the ileo-colic at the junction of the small and large intestines.

In the **circulatory system** valves are highly developed in the vertebrate **heart** and appear in some tubular vessels. The latter, in **vertebrates,** are limited to the **lymphatic** vessels and veins of the extremities. They consist of paired folds of the lining with their convex surfaces toward the passage. Pressure against this surface separates the flaps of the valves, while pressure against their concave faces forces them together and blocks the passage.

In the heart the passages from veins to heart, from auricles to ventricles, and from heart to arteries are guarded by valves composed of two or three flaps, and in the fishes (**Pisces**) the proximal part of the ventral aorta is in some species developed as a valvular conus arteriosus. In the **mammalian** heart the openings of the pulmonary artery and aorta are guarded by valves of three parts, known as the semilunar valves. The valve between the right auricle and ventricle is also 3-parted, and is called the tricuspid, while the mitral valve between the left auricle and ventricle has only two flaps. These last valves are reinforced by slender chordae tendineae extending from their edges to the opposite walls of the ventricles. The heart also has a valve of Thebesius or coronary valve at the opening of the coronary sinus and a Eustachian valve at the opening of the inferior vena cava. (A.W.L.)

VALVE GEAR. The mechanical linkage by which the valves of an engine derive their motion and timing from the crankshaft rotation is called, collectively, the valve gear. The term is not restricted to any type of engine but is in more frequent use, perhaps, in the steam engine field. Here, both the internal combustion and steam engine gear will be described.

With very few exceptions, the internal combustion engine uses a poppet valve, which is actuated by a short reciprocation, the amplitude of which is well within the ability of a cam to produce. The cam is also an economical method of converting rotation to reciprocation for multi-cylinder engines because of the comparative ease with which a number of cams may be machined on a single shaft. The poppet valve is retained against its seat by a strong spring. When it is to be opened, a rod or lever is pressed against it, and moved far enough to create the necessary valve lift. This lifting motion is derived from the cam through a cam follower, and tappet rods or push rods, as illustrated in Figs. 1 and 2. The camshaft itself is driven from the crankshaft. Camshaft drive must be positive, so gears or chains are always used.

Two variations of valve gear suitable for in-line engines are pictured in Fig. 1. As both types are drawn

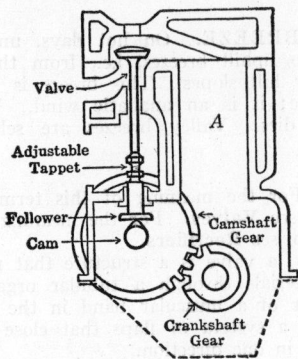

Fig. 1A. L head valve arrangement.

for the same bore and stroke, the figures demonstrate the effect of valve placement and valve gear space needs upon the engine shape. These sketches are intended to be illustrative of valve gear only and many essential engine details (i.e., valve springs, piston, etc.) are purposely omitted. Fig. 1A represents a common arrange-

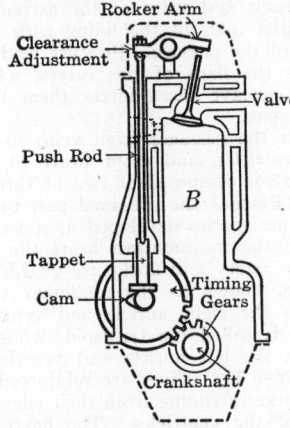

Fig. 1B. Overhead valve arrangement.

ment of the gear in an L head **four-cycle** gasoline engine. The camshaft is gear-driven from the crankshaft at half speed. The cam bears against a flat follower which raises a tappet, whose pressure against the valve stem then accomplishes the valve opening. The beginning of valve motion is a function of the mesh of the gears and can be altered by placing different teeth in mesh. The height and duration of valve lift are determined by the cam profile. Because of the likelihood of unequal thermal expansion of the valve gear and the cylinder barrel a slight clearance is introduced between tappet and valve stem to prevent the valve from continuously riding open when hot. This clearance should be precisely maintained within a thousandth of an inch of the builder's recommendation. Adjustments, either at the tappet or rocker arm (case B) are provided for this purpose. To prevent valve gear noise arising from the cyclical closure of this clearance, special devices such as hydraulic tappets have sometimes been used. These keep the above-mentioned clearance closed when the engine is running but yield and compensate for expansion before the valve will ride from its seat in closed position. The overhead valve (*B*) requires a few more parts in its valve gear but is often used by engine builders on ac-

count of thermodynamic benefits associated with this location.

Some radial engines have been built with a short two-cam camshaft for each cylinder. However, a more common gear for these engines is the multiple lobe cam mounted on a drum (Fig. 2). A radial engine is built

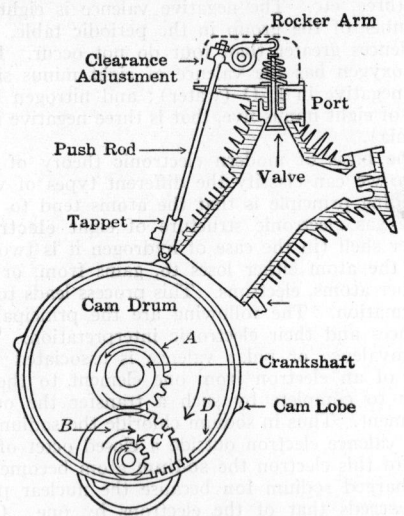

Fig. 2. Radial engine valve gear.

with an odd number of cylinders for each row so that cylinders may be fired alternately (i.e., firing order 1-3-5-7-9-2-4-6-8 for nine cylinders) in the interest of good dynamic balance. A plate cam with four lobes, revolving oppositely to the crankshaft rotation at one-eighth speed will accomplish the operation of the inlet valves necessary for this type of firing order. Another group of four lobes offset from the inlet group will attend to the requisite exhaust valve motion. The gear *A* is mounted tightly on the crankshaft and turns gears *B* and *C* which are integral and mounted on a fixed pivot. *C* meshes internally with gear *D* which is cut on the inside of the cam drum. The cam drum has a bearing on the crankshaft. The combined reduction of *A*, *B*, *C* and *D* is 8:1 and it will be observed that *D* has rotation opposite to *A*.

The **steam engine** valve gear (Fig. 3) is found in varied forms of many mechanical details. Furthermore,

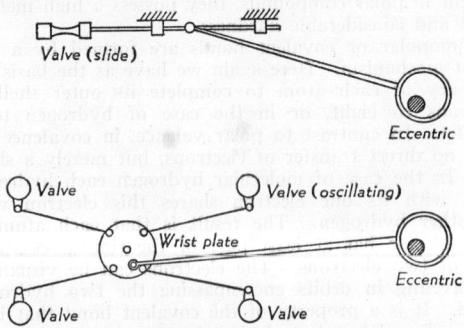

Fig. 3. Steam engine valve gear.

the valve gear of a steam engine may be designed to give reversible operation, and to effect the governing necessary for constant speed at variable load. Steam engine valves can be plain sliding, oscillating, or poppet. The throw of the valve, or the amplitude of its reciprocation, is too large to be produced by a **cam** mounted on the **crankshaft**, so the drive usually originates from an **eccentric**. In a simple slide valve engine this eccentric is encircled by an eccentric strap, to which is fastened an eccentric rod. The eccentric rod is pinned to the valve stem so that the rotation of the crankshaft will reciprocate the valve. Steam engines which are governed by the cut-off method have the governor action interposed between the crankshaft and the eccentric, that is, the eccentric is not keyed to the shaft, but is free thereon, being maintained in a position relative to the engine crank, determined by the load and controlled by the governor.

Engines with oscillating valves derive their motivation from an eccentric on the crankshaft, but there is interposed between the eccentric rod and the valve a wrist plate and valve reach rods. The Corliss engine is like this. In a simple slide valve engine the eccentric angle leads the crank angle in the direction of rotation by a definite amount. If two eccentrics are provided, one on either side of the crank by this same angle, then the engine could be operated forward or reverse if the valve could be driven from either of the eccentrics. This idea is incorporated in the Stephenson valve gear. In locomotive practice, more flexibility of control than is possible in the simple valve gear just described is desired, and the gear used in most locomotives is a type

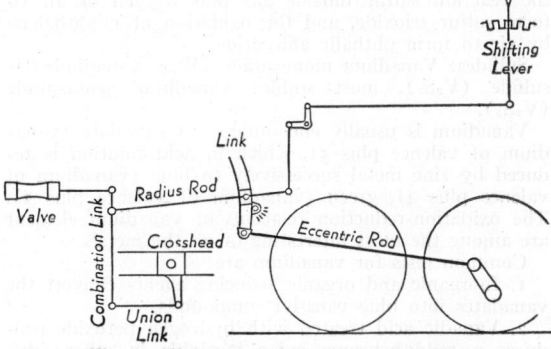

Fig. 4. Walschaert's locomotive valve gear.

which derives part of its motion from the crosshead, and part from the crankshaft. A detailed description of this gear is not in order; however, a skeleton diagram of the linkage is given in Fig. 4. (F.T.M.)

VALVES, ENGINE. Engine valves are the closures of the ports of entry or exit of the working medium of engines—steam, Otto, or Diesel. Correct execution of the thermodynamic cycle of operation of a heat engine is dependent upon precision timing of the valve motion, admitting and releasing the working medium:

Valve types may be classified as:

1. *Sliding.* D slide valves, piston valves, and other steam engine types.
2. *Oscillating.* Corliss valve.
3. *Poppet.* Reciprocating, but not sliding. Mushroom and tulip valves of internal combustion engines. Balanced poppet valves for steam engines.

Valves are actuated, and timed relative to the cycle of operation, by the kinematics of the valve gear. (See **Otto Engine, Steam Engine, Valve Gear.**) (F.T.M.)

VALVES, PIPE. A pipe valve is a pipe line accessory used to start, stop, and regulate fluid flows. It consists of a body to house it and to give a means of connecting it to the pipe, a valve seat, a valve, a valve stem, a top, a means of operating the stem—usually screw and hand wheel, and a packing to prevent leakage around the stem as it emerges from the top. Valves are furnished for either screwed or flanged connection. They are made of brass, bronze, malleable iron, cast iron, cast steel, and forged steel. Cast and forged steel are employed in high pressure-temperature service. Valve sealing materials (sometimes constituting the valve seat,

sometimes attached to the valve face) include rubber for cold water, brass and bronze for ordinary temperatures, stainless steel, monel metal, and various other alloys for high temperatures.

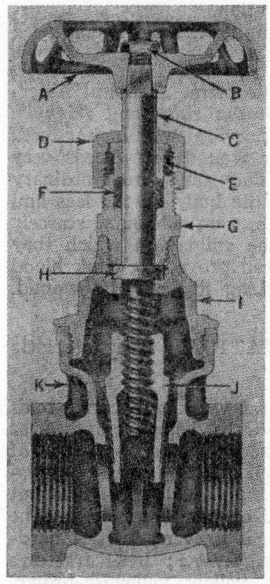

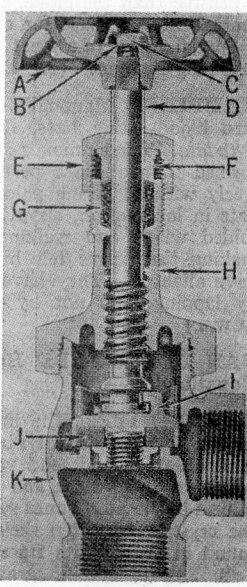

(*Jenkins Brothers.*)

Fig. 1.   Fig. 2.

Fig. 1. Section through a gate valve. A. Hand wheel; B. Wheel nut; C. Spindle; D. Packing nut; E. Gland; F. Packing box; G. Packing box nipple; H. Spindle collar; I. Bonnet; J. Solid wedge; K. Body.

Fig. 2. Section through angle type of globe valve. A. Hand wheel; B. Index plate; C. Hand wheel nut; D. Spindle; E. Packing nut; F. Gland; G. Packing box; H. Bonnet; I. Disk holder; J. Composition disk or washer; K. Body.

Valves most used are the ordinary hand-operated globe and gate valves, and the check valves. These are classified thus:

1. Globe valves (straight and angle).
   a. Inside screw; outside screw.
   b. Screw bonnet top; bolted yoke top.
2. Gate valves (straight and angle).
   a. Rising stem; non-rising stem.
   b. Wedge valve (split and solid); parallel seat valve.
3. Check valves (lift and swing types).
   a. For vertical pipe.
   b. For horizontal pipe.

The globe valves do not allow a line to drain completely; also, they offer more frictional resistance than gate valves. They are frequently used in very small

Fig. 3. Horizontal swing check valve. (*Jenkins Brothers.*) A. Cap; B. Hanger pin; C. Hanger; D. Hanger nut; E. Disk; F. Body.

lines (both water and steam) and where the valve is to be used for throttling, as they can be closely regulated and the seats, which are liable to be cut away in throttling service, can be more easily replaced than in gate valves. Gate valves are used in large pipe lines, in high-pressure steam lines, and in all service where small friction loss is wanted. The gate valves are more expensive than the globe, and their longer stems require more clearance space for operation.

Special valves of a great many types are also employed in piping systems. (F.T.M.)

VAMPIRE. Mammalia, Chiroptera. Tropical American blood-eating bats, *Desmodus rufus* and *Diphylla ecaudata*. These animals are provided with incisor teeth formed to produce a peculiar wound that bleeds very freely without giving pain. They differ from ordinary bats in locomotion, elevating the body on the legs and folded wings and walking thus on all-fours, scarcely making themselves felt by the animal on which they have alighted. They have been proved to drink blood by lapping, and not by sucking it from the wound. (A.W.L.)

VAN DER WAALS' EQUATION. Characteristic Equations.

VANADINITE. The mineral vanadinite corresponds to the formula $(PbCl)Pb_4(VO_4)_3$, being composed of lead chloride and lead vanadate in the proportion of 90.2% of the former and 9.8% of the latter. It crystallizes in the hexagonal system, is usually prismatic, but the crystals are often skeletal or cavernous; it may be found in crusts. Its fracture is uneven; brittle; hardness, 2.75–3; specific gravity, 6.7–7.2; fresh fractures show a resinous luster; color, yellow, yellowish-brown, reddish-brown, and red; streak, white to yellowish; translucent to opaque. Vanadinite, by no means a common mineral, occurs as an alteration product in lead deposits. It is found in the Urals, Austria, Spain, Scotland, Morocco, the Transvaal, Argentina, and Mexico. In the United States it occurs in Arizona, New Mexico, and South Dakota. It is used as an ore of vanadium and to some extent of lead as well. It is interesting to note that this mineral was first described as a chromate upon its discovery in Mexico in 1801. It was not until the discovery of the metal vanadium in 1830 that the true nature of this compound was known. (E.S.C.S.)

VANADIUM. Symbol: V. Atomic number: 23. Atomic weight: 50.95. Density: 5.69. Melting point: 1710° C. No isotope, but of single atomic form: 51.

Vanadium is a silver-white, very hard (Mohs' scale of hardness 7) metal; oxidizes upon being exposed to air, and upon ignition forms the pentoxide; insoluble in hydrochloric acid, slowly dissolves in hydrofluoric, nitric, or sulfuric acids (hot, concentrated), or aqua regia; insoluble in sodium hydroxide solution. Discovered by Del Rio in 1801. Subsequent investigations made by Sefstrom, Wohler, and Berzelius, all about 1830, and by Roscoe in 1868, are of great interest.

Most of the vanadium of commerce is used in the form of ferro-vanadium (30–40% V) in the manufacture of special alloy steels to which vanadium (not usually exceeding 1%, and as low as 0.2%) imparts hardness and toughness, and increases the elastic limit and resistance to shock or impact. Vanadium steel is used for tools and in the construction of vehicles.

Vanadium occurs as patronite, containing vanadium pentasulfide, in Peru, as carnotite, potassium uranyl vanadate, in Colorado and Utah, as vanadinite, lead vanadate, in Arizona, New Mexico, South East Africa, and northern Rhodesia. Ships burning Venezuelan or Mexican petroleum fuel oil recover vanadium oxide from the boiler and stack dust. In Italy the refining of bauxite ore (for aluminum) yields vanadium, and in Germany some iron ores contain vanadium. Vanadium and radium ore is found in southwestern Colorado and southeastern Utah; in Arizona a complex ore of gold, silver, and lead contains vanadium and molybdenum;

and extensive deposits of phosphate rock in Idaho may prove to be our principal source of vanadium. The sulfide ore is roasted to remove sulfur, and the residue fused with sodium carbonate, forming sodium vanadate. This last is extracted with water and excess of sulfuric acid is added, causing precipitation of vanadium pentoxide, which is later reduced by carbon or aluminum at high temperatures.

Chlorides: Vanadium dichloride (VCl_2), green crystalline solid, a strong reducing agent; vanadium trichloride (VCl_3), pink crystalline solid; vanadium tetrachloride (VCl_4), reddish-brown liquid, boiling point 148° C.

Hydroxides: Vanadium dihydroxide $(V(OH)_2)$, brown precipitate by reaction of sodium hydroxide solution with hypovanadous acid (one of the most powerful of reducing agents) lavender solution; vanadous hydroxide $(V(OH)_3)$, green precipitate by reaction of sodium hydroxide solution with vanadous salt green solution.

Oxides: Vanadium monoxide (VO), gray solid; vanadium trioxide (V_2O_3), black solid; vanadium dioxide (VO_2), dark blue solid; vanadium pentoxide (V_2O_5), orange to red solid. The last is the most important oxide; formed by the ignition in air of vanadium sulfide, or other oxide, or vanadium; used as a catalyzer, e.g., the reaction sulfur dioxide gas plus oxygen of air to form sulfur trioxide, and the oxidation of naphthalene by air to form phthalic anhydride.

Sulfides: Vanadium monosulfide (VS); vanadium trisulfide (V_2S_3), most stable; vanadium pentasulfide (V_2S_5).

Vanadium is usually encountered as vanadate (vanadium of valence plus 5), which in acid solution is reduced by zinc metal successively to blue (vanadium of valence plus 4), green (vanadium of valence plus 2). The oxidation-reduction relations of vanadium element are among the most interesting of all the metals.

Common tests for vanadium are:

1. Inorganic and organic reducing agents convert the vanadates into blue vanadyl compounds.

2. Vanadic acid treated with hydrogen peroxide produces a reddish-brown color insoluble in ether (difference from chromates). (R.K.S.)

VANDYKE. Drawing Reproduction.

VANILLA. Orchidaceae. Vanilla is a climbing orchid native to Central America and Mexico. Vanilla extract is obtained from the fruit. (See Bean.) (R.M.W.)

VANILLIN. Aldehydes, Ketones, and Related Compounds.

VANISHING POINT. Perspective.

VAN'T HOFF LAW. Equilibrium.

VAPOR CYCLE. A vapor cycle is so named from the fact that it uses the same vapor over and over again, passing it around a closed loop and subjecting it to various thermodynamic changes by means of which useful energy is produced from raw heat.

In the process of conversion of the heat of a fuel into a useful form of energy there are three essentials: First, a heat absorber where the heat liberated by combustion of the fuel is absorbed by the working medium; second, a heat utilizer in which as much of the heat as is available is taken from the working medium and converted into a useful form of energy; third, the working medium itself passing back and forth between the first two—a carrier of heat energy. The thermodynamic aspects of its passage from the heat absorber to the heat utilizer and back again constitute the vapor cycle. (See Rankine Cycle, Regenerative Cycle, Reheating, Mercury Vapor Cycle.) (F.T.M.)

VAPOR LOCK. Volatility of gasoline makes for easier starting of an engine using it, but creates one undesirable feature, namely, vapor lock. This phenomenon is associated with the fuel supply to internal combustion engines. It occurs mainly in the fuel lines conveying a volatile fuel such as gasoline. If the fuel line from supply tank to engine has many bends and fittings, if it is exposed to engine heat, or if parts of it are under vacuum, bubbles of fuel vapor may form in the lines, causing irregular engine operation, or even stoppage. While this condition might sometimes occur initially in a fixed engine supply system, once remedied it is not liable to recur. On the other hand, vehicular engines may be operated under more varied conditions of temperature and altitude; hence are more frequently troubled with vapor lock. Manufacturers of automobiles and other surface vehicles had to design gasoline supply systems with possible vapor lock in mind. Except for unusual extremes of operating conditions, such vehicles seldom encounter vapor lock nowadays. When an automobile engine is used at high altitude or in desert heat without special modification of the fuel system, some trouble may develop from vapor lock. Temporary relief may be secured by chilling the fuel lines and insulating them from engine heat.

Airplane builders also encounter vapor lock difficulties, due principally to altitude. Except for small, low-ceiling airplanes, fuel pumps have superseded gravity supply. When fuel pumps were first introduced vapor lock troubles disappeared—only to reappear as airplanes began to be operated at altitudes above approximately 15,000'. Stoppages persisted in occurring after experimental elimination of the possibility of fuel vaporization. Then it was found that considerable air was dissolved in the fuel, and began to separate from it and airbind pumps and other parts of the supply systems at high altitudes. Although this might have been called "air lock," it remained customary to refer to all such phenomena as vapor lock. Upon installation of separators which allowed removal of gas and vapor from strategic points in the system, engines could be operated to very high altitudes without vapor lock trouble. (F.T.M.)

VAPOR PRESSURE. The vapor pressure of a substance (solid or liquid) is the pressure exerted by its vapor when in **equilibrium** with the substance. For pure substances it depends only on the temperature. The simplest way to measure the vapor pressure of a substance is to introduce a small amount of it into the closed end of a barometer tube and note the decrease in the height of the barometer. (See table below.)

The vapor pressure of a **solvent** is lowered on dissolving the solute in it. This lowering for dilute solutions is proportional to the mole fraction of the solute (**Raoult's Law**). The lowering of the vapor pressure of the solution can be related to the lowering of the freezing point and the elevation of the boiling point. These phenomena serve as a basis for **molecular weight** determinations. If both components of the solution are volatile, each lowers the vapor pressure of the other and the ratios of the two substances in the liquid and vapor phase are not necessarily the same. Use is made of this fact to separate the two substances by **distillation.** (See **Deliquescence, Efflorescence,** and **Water.**) (R.K.S.)

VAPOR PRESSURE THERMOMETER. Thermometry.

VAPORS. A substance in the gaseous state, but below its critical temperature, is called a vapor. If a pure liquid partly filling a closed container is allowed to stand, the space above it becomes filled with the vapor of the liquid, which develops a pressure. This **vapor pressure** increases up to a certain limit, depending upon the temperature, where it becomes constant, and the space is then said to be saturated.

Such a body of vapor is not subject to all of the laws of gases. If the space occupied by it is diminished without change of temperature, there is no increase in pressure, but instead part of the vapor condenses. And if the temperature is raised, the pressure goes up not at a uniform but at an increasing rate, because of both the expansion of the liquid and the further **evaporation** from it. The relation of vapor to liquid takes on a curious aspect as the **critical state** is approached, in which the vapor and the liquid have equal density.

The **characteristic equations** applying to vapors are of various forms, differing somewhat from those for gases. A typical empirical equation of the sort, proposed by Callendar to represent the behavior of steam, is

$$v = \frac{RT}{p} + f(T) + c,$$

VAPOR PRESSURE OF SOME SUBSTANCES AT VARIOUS TEMPERATURES IN MILLIMETERS OF MERCURY EXCEPT AS STATED

	−20° C.	0° C.	+20° C.	100° C.
Water......................	0.8 (Ice)	4.6	17.5	760
Ethyl alcohol...............	2.5	12.2	43.9	1690
Diethyl ether...............	37.6 (−30° C.)	185.3	442.2	4860
Acetone...................	11.2 (−30° C.)	89.1 (+5° C.)	184.8	2790
Acetic acid.................			11.7	417
Carbon disulfide............	46.5	127.3	298	3360
Carbon tetrachloride........	9.8	32.9	91	1460
Benzene...................			77	760 (79.6° C.)
Toluene...................			37 (30° C.)	557
Aniline....................			2.4 (50° C.)	45.7
Nitrobenzene...............			7.5 (80° C.)	20.9
Ethylene glycol.............				39 (120° C.)
Ammonia..................	1.88 atm.	4.24 atm.	8.46 atm.	
Sulfur dioxide..............	0.63 atm.	1.53 atm.	3.23 atm.	
Carbon dioxide.............	19.4 atm.	34.4 atm.	56.5 atm.	
Oxygen...................	49.7 atm. (−118° C.)			
Nitrogen..................	33.5 atm. (−147° C.)			
Sulfur....................				0.010
Iodine....................		0.030	0.20	45.5
Mercury...................			0.0012	0.273
Naphthalene...............				19
Camphor..................				380 (180° C.)

in which R is the ideal gas constant, c is a constant to be obtained empirically, and $f(T)$ is a slowly varying function of the absolute temperature T. (L.D.W.)

VARIABILITY. Dispersion.

VARIABLE. A symbol which represents any one of a set of numbers (or elements of any kind) is said to be a variable. The elements of the set are called values of the variable, and the set of elements itself is called the range of the variable. (L.L.S.)

VARIABLE MU TUBE. This is a vacuum tube with a grid so constructed as to give a variable amplification factor. To accomplish this the spacing of the grid wires is not constant but varies regularly along the grid. The result is that the mutual conductance characteristic of the tube approaches the grid-voltage axis very gradually rather than abruptly as in tubes of other types. The figure shows the characteristics of both conventional and variable mu tubes. Since the slope

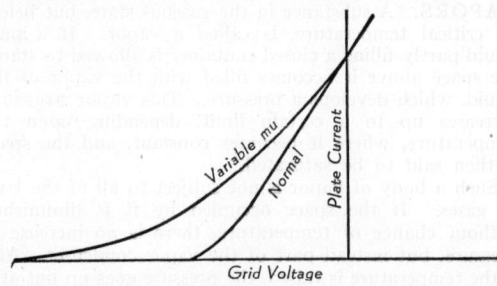

of this curve is the mutual conductance of the tube, it will be apparent at once that the mutual conductance of the variable mu tube may be varied over a wide range by varying the operating bias of the tube. This property is utilized in all standard **automatic volume control** circuits, the gain of the tube being controlled by the bias voltage which is obtained from the rectifier carrier. (L.R.Q.)

VARIABLE SPEED TRANSMISSION. Speed Changers.

VARIABLE STARS. Any star whose light is known to fluctuate is called a variable star. There is evidence that the ancients noticed certain temporary stars, now known as **novae**, and that the star **Algol** was recognized by them as a variable. The first definite record that we have of observation of a variable star is in connection with observations of **Mira** during the late 16th and early 17th centuries. In 1844, Argelander published the first catalogue of variable stars which contained eighteen entries. The publication of this catalogue stirred up a great deal of interest in variable stars and they became objects of much search and study.

At the present time most variable stars are discovered from examination of photographic plates, taken at different times, of the same region of the sky. At present over 10,000 variables are known, and it is estimated that at least 5% of all stars are at least slightly variable. When a variable star is first observed it is assigned a serial number in order of discovery within the year, together with the year of discovery (e.g., 256,1923 designates the 256th variable discovered in 1923). After a sufficient number of observations have been taken to confirm the variability of the star and the fact that it is really a discovery, the star is assigned letters, in accordance with a system established by Bayer, followed by the genitive case of the constellation within which the object is located (e.g., RZ Herculis). Novae are not included in this system of classification, being referred to by "nova" followed by the genitive

case of the constellation and the year of discovery (e.g., Nova Herculis 1934).

The observation of variable stars consists essentially in systematic **photometric** observations of the objects over long periods of time. The methods employed are essentially those of stellar **photometry** and these will be found discussed elsewhere in this volume. The observations are carried on not only by professional astronomers at many observatories all over the world, but also by an enthusiastic group of amateurs who have banded themselves together in societies for the purpose of carrying on this valuable work. In the United States, the American Association of Variable Star Observers, commonly known as AAVSO, has its headquarters at the Harvard College Observatory, holds two general meetings each year, and its membership contributes thousands of observations annually. Its membership is open to anyone who is interested in astronomical observing, whether or not he has any equipment other than eyesight and enthusiasm.

When a sufficient number of observations of the magnitude of a variable have been obtained, a **light curve** is plotted and the study of the characteristics of the object is undertaken. Variable stars divide themselves into two natural groups: the periodic and non-periodic variables. In the accompanying figure the number of periodic variable stars within certain limits of period is

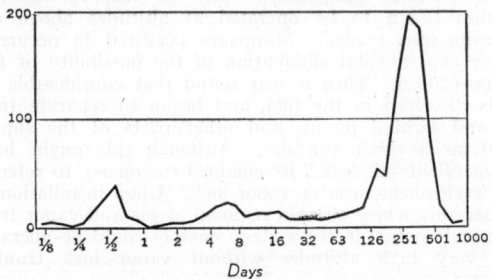

Distribution of periods of variable stars. (*Russell, Dugan and Stewart, Astronomy, Ginn & Co.*)

plotted as ordinate against the logarithm of the period in days. Examination of the figure indicates that there are three principal groups of periodic variables: one group with periods of less than 1 day, a second with periods between 3 and 50 days, and a large group with periods between 120 and 750 days. Periodic variables with periods under 50 days are known as **Cepheids**, while those with periods greater than 100 days are known as **long-period variables**. The non-periodic variables are divided into two groups: the novae and the **irregular variables**. We sometimes find **eclipsing binaries** referred to as eclipsing variables, but they are not variables in the exact sense of the term, since the apparent variation in light is not due to any intrinsic variability within the stars themselves, but is due to changes in configuration of the objects as seen from the earth. (W.K.G.)

VARIABLES OF STATE. Characteristic Equations.

VARIANCE. The variance is the square of the standard deviation; also the number of **degrees of freedom** of a system. (L.A.A., L.D.W.)

VARIATE. A variate is a particular value of the variable. The variable may be the temperature, but $70°$, $72°$, $73°$, $\cdots$, etc., are variates. (L.A.A.)

VARIATE DIFFERENCE CORRELATION. Variate difference correlation consists of the correlation of the first differences of the two **variables** instead of correlating the variables themselves. It is sometimes used in time series. (L.A.A.)

VARIATION.

This term has a number of uses. It is applied in connection with the compass; and this topic is discussed in the article on **compass correction**. The term variation is also used in mathematics and biology, and these usages will be discussed in that order in this article.

Variation is a simple and important type of mathematical relationship between variables, which is of very frequent occurrence in physical problems.

If two **variables** x and y are related by the **power function** relationship $y = kx^n$, where k and n are constants, we say that y varies as x^n, or y varies as the nth power of x, or that y is proportional to x^n. The factor k is called the constant of variation, or the constant of proportionality, or the proportionality factor. Special cases are:

1. ($n = 1$). If two variables are so related that their **ratio** is always constant, so that $y/x = k$ or $y = kx$ (k a constant), we say that y varies directly as x (or that y is proportional to x). The notation $y \propto x$ is sometimes used to express the fact that y varies directly as x.

2. ($n = -1$). If two variables are so related that their **product** is always constant, so that $xy = k$ or $y = k/x$ (k a constant), we say that y varies inversely as x (or that y is inversely proportional to x).

Other types of variation are defined as follows:

If a variable z varies directly as the product of two variables x and y, so that $z = kxy$ (k constant), we say that z varies jointly as x and y.

If a variable z varies jointly as x and the **reciprocal** of y, so that $z = kx/y$, we say that z varies directly as x and inversely as y.

A common type of variation is one where a variable varies directly as the product of two variables and inversely as the square of another variable (as in the law of gravitation).

Instead of expressing variation statements by equations, we may express them as **proportions**. Thus, if y varies directly as x, we may write $y_1 : y_2 = x_1 : x_2$; and if y varies inversely as x, we may write $y_1 : y_2 = x_2 : x_1$.

In biology, variation is the deviation of organisms from the mean development of their kind. Very few individuals are even approximately identical in form, although some species show a much wider range of variation than others.

The probable cause of variation is the interaction of diverse heritages with varying environmental conditions in the formation and maintenance of individuals. The study of **heredity** has disclosed an intricate reproductive mechanism for the maintenance of diversity in the heritage as one source of variation, and environmental conditions have been found correlated with variation in many cases.

As actual difference in living things variation is particularly interesting in the field of **evolution**. Here the heritability of variations is important, since evolutionary change sufficient to account for the origin of species can only be attained through a long succession of generations. Characteristics due to the variable heritage are obviously heritable. They are subject to change only through the process of **mutation**, as far as we know, hence this process becomes an important source of variations. Efforts to determine the cause of mutations have shown that they may result from the incidence of unusual environmental conditions, such as x-rays, terrestrial radiation, and radium emanations, and from more usual factors such as heat and moisture. Thus the most advanced information on the subject leads to the initial premise that variations are due to the intricacies of reaction between variable heritages and environments.

The actual variation of existing species has been studied extensively and is the subject of an extensive literature. Differences between individuals may be sexual, seasonal, or due to specialization within colonies. These types of variation are often extreme, resulting in individuals of very different structure and appearance. On a lesser scale the similar individuals of any species show incidental differences of no fundamental importance. (L.L.S., A.W.L.)

VARIATION OF LATITUDE.

In the latter part of the 18th century the mathematician Euler predicted, from purely theoretical considerations, that the **latitude** of every point on the surface of the earth should be varying. The amount of this variation depends upon the shape, the elasticity, and other physical characteristics of the earth, together with the attractions of external objects, such as the **moon**, for our rotating **planet**. The first actual observations of variation of latitude were made in 1888, nearly a century after Euler's prediction, by the careful observations of the astronomer Kustner.

The effect may be described by considering the motion of a wheel, the bearings of which have become so worn that the axle fits loosely in the hub. When considering variation of latitude alone we consider the axis of the earth as fixed in space in spite of the fact that we know that it is actually slowly moving due to **precession** and nutation. As the earth slips slightly about on this axis the position of the pole of rotation wanders about over a small area in the vicinity of the average position of the pole. By definition the plane of the **equator** is perpendicular to the axis of rotation of the earth, and if this line moves, so also must the equator move. Since latitude is measured from the equator, the motion of the equator must produce a variation of latitude.

The actual amount of variation of latitude is a small quantity, the total area over which the pole moves being about the size of a baseball diamond. Nevertheless, small as the effect may be, it is appreciable and the date should be very carefully specified whenever defining boundary lines by means of latitude.

During the past 50 years a great mass of observational material has been gathered, but as yet the problem has not been completely solved. At present an international campaign is under way in connection with which a number of stations all over the earth are constantly at work observing the latitudes of their localities by use of the **zenith telescope.** (W.K.G.)

VARICELLA. Chickenpox.

VARICOCELE.

Enlargement and dilatation of the veins surrounding the spermatic cord in the male. When this condition is marked the knot of varicose veins forms a swelling in the upper scrotum. Surgical interference is not wise unless the condition causes severe symptoms.

Ovarian varicocele is a similar condition in the female of the veins in the ligaments near the ovary. (R.S.M.)

VARICOSE VEINS.

An enlarged tortuous condition of the superficial veins of the body. This disorder is most often seen in the legs and thighs.

Varicose veins are directly caused by any disorder that destroys or weakens the **valves in the veins**, whether it be infection or increase in the venous pressure such as from a **tumor**, pregnant uterus, or constriction of the leg from tight circular garters, or from constant standing, etc. A tendency toward varicose veins seems to be hereditary, and is probably due to the inheritance of weak valve structure in the veins, the basic cause of varicosities.

The symptoms caused by varicose veins are pain, cramps, or tired feeling in the legs, swelling of the ankles and legs when the condition is marked. Complicating the later stages of varicose veins are **thrombophlebitis**, varicose **ulcers**, and **eczema**. Varicose ulcers over the legs are chronic and show little tendency to heal because of the poor circulation. The eczema results from poor skin nutrition, also on a circulatory basis.

The treatment of varicose veins includes elevation of the extremities and rubber bandages to improve the venous return and prevent edema. Surgical ligation or excision of the veins obliterates the superficial venous channels and forces the blood to return by the deep veins. This objective is also accomplished by the injection of chemical sclerosing solutions into the veins, causing occlusion of the vessel lumen, fibrosis, and final complete obliteration of the diseased veins. (D.M.H., R.S.M.)

VARIEGATION. Albinism.

VARIOLA. Smallpox.

VARIOLES. Variolite.

VARIOLITE. A fine-grained basic rock that contains spherulites made up of fibers of feldspar and augite in radial development.

The spherulites themselves are known as varioles, and the texture of such rocks is said to be variolitic. The term is derived from the Latin *variola,* smallpox, given to the rock because of resemblance of the pea-like or pustular forms of the varioles on weathered surfaces to smallpox pustules. (E.S.C.S.)

VARLEY LOOP. This is a bridge-type measurement used by telephone engineers to locate line faults. The circuit is shown in the figure. The method differs

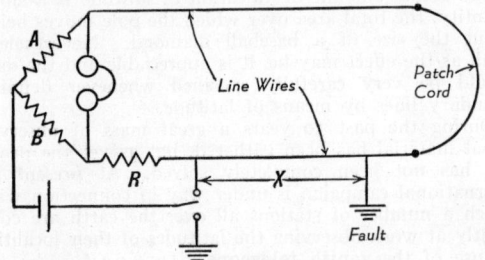

Circuit for Varley loop.

from the **Murray loop** in that the **resistance** of the line loop may be measured by throwing the switch up and balancing the bridge without changing any connections. When the switch is thrown down and the bridge again balanced, the distance X to the fault may be determined from the equation

$$X = \frac{BR_L - AR}{r(A + B)}.$$

Where A, B, and R are shown on the figure, R_L is the loop length times the resistance per ft., r. (L.R.Q.)

VARNISHES. Pigments, Paints, Varnishes.

VAS DEFERENS. Urogenital System.

VASCULAR ELEMENTS. The vascular elements of a plant are the tissues which serve to conduct materials in the plant. The principal tissues of the vascular system are the xylem and phloem. (R.M.W.)

VASCULAR SYSTEM. This is a complex system of cells and tissues called xylem and phloem, and serving to conduct water, mineral salts, and food through the plant. It also gives strength and support to the plant. It composes the bulk of the tissues of roots and stems of woody plants. (See **Translocation;** see also **Circula-**

tory **System** for a discussion of the vascular system of animals.) (R.M.W.)

VASCULAR TISSUE. Blood, lymph, and related fluids of the body. These liquids are regarded as tissues because they consist of characteristic cells lying in an intercellular substance. The liquid condition of the intercellular substance is responsible for their fluidity. (A.W.L.)

VECTOGRAPH. Stereoscopic Photography.

VECTOR ADDITION. Let **a** and **b** be any two vectors (Fig. 1). To add vector **b** to vector **a**, we place the origin of **b** at the terminus of **a**, then the vector **c** extending from the origin of **a** to the new terminus of **b**

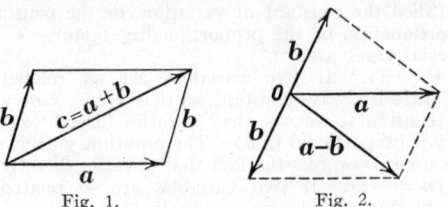

Fig. 1.   Fig. 2.

is defined as the sum (or resultant) of **a** and **b**, and is denoted by **a** + **b**. This operation is often called composition of vectors.

Vector addition obeys the **commutative** and **associative** laws of algebra: **a** + **b** = **b** + **a**, (**a** + **b**) + **c** = **a** + (**b** + **c**).

To subtract a vector **b** from a vector **a**, we take the negative of **b** and add —**b** to **a**. (Fig. 2.)

The components of a vector **a** are any vectors whose sum is **a**. The components most frequently used are parallel to a set of rectangular coordinate axes; these are called rectangular components. The operation of finding components of a given vector is often called the decomposition of the vector.

In order to be able to express a vector in terms of its rectangular components in convenient form, we take unit vectors parallel to the X, Y, Z axes of a **rectangular coordinate** system and denote them by $\hat{i}$, $\hat{j}$, $\hat{k}$, respectively.

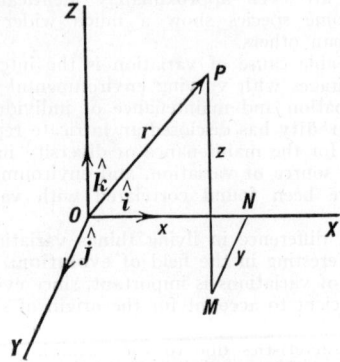

Fig. 3. Vector components.

In terms of the unit vectors $\hat{i}$, $\hat{j}$, $\hat{k}$, we may express any given vector **r** with origin at the origin of the coordinate system by the form (Fig. 3)

$$\mathbf{r} = x\hat{i} + y\hat{j} + z\hat{k},$$

where x, y, z are the rectangular coordinates of the terminus of **r**. Then $x\hat{i}$, $y\hat{j}$, $z\hat{k}$ are the rectangular components of **r**.

If $\mathbf{r} = x\hat{i} + y\hat{j} + z\hat{k}$, then the magnitude of **r** is $r = \sqrt{x^2 + y^2 + z^2}$, and the direction of **r** is given by the direction cosine relation $\cos \alpha : \cos \beta : \cos \gamma = x : y : z$.

If two vectors **a** and **b** have rectangular components a_1, a_2, a_3 and b_1, b_2, b_3 in magnitude, then to add **a** and **b**, we have

$$\mathbf{a} + \mathbf{b} = (a_1 + b_1)\hat{\mathbf{i}} + (a_2 + b_2)\hat{\mathbf{j}} + (a_3 + b_3)\hat{\mathbf{k}}.$$

Let a vector **a** have rectangular components a_1, a_2, a_3 (in magnitude) with respect to a first set of rectangular coordinate axes X, Y, Z, and components a'_1, a'_2, a'_3 (in magnitude) with respect to any other set of rectangular axes, with the same origin. Let the relations of the axes be given by the following scheme of direction cosines:

	X	Y	Z
X'	l_1	l_2	l_3
Y'	m_1	m_2	m_3
Z'	n_1	n_2	n_3

where l_1, l_2, l_3 are the cosines of the angles which the X, Y, Z axes make with the X' axis, and similarly for the m's and n's. Then the formulae for **transformation of coordinates** are:

$$x' = l_1x + l_2y + l_3z, \qquad x = l_1x' + m_1y' + n_1z',$$
$$y' = m_1x + m_2y + m_3z, \qquad y = l_2x' + m_2y' + n_2z',$$
$$z' = n_1x + n_2y + n_3z, \qquad z = l_3x' + m_3y' + n_3z'.$$

Then the vector components must transform in a similar way:

$$a'_1 = l_1a_1 + l_2a_2 + l_3a_3, \qquad a_1 = l_1a'_1 + m_1a'_2 + n_1a'_3,$$
$$a'_2 = m_1a_1 + m_2a_2 + m_3a_3, \qquad a_2 = l_2a'_1 + m_2a'_2 + n_2a'_3,$$
$$a'_3 = n_1a_1 + n_2a_2 + n_3a_3, \qquad a_3 = l_3a'_1 + m_3a'_2 + n_3a'_3.$$

(L.L.S.)

VECTOR ANALYSIS.

The name Vector Analysis is usually applied to the study of the properties and applications of **vectors**. It may be described also as the Algebra, the Geometry, and the Calculus of Vectors.

The principal topics relating to Vector Analysis are: **Vectors, Vector Addition, Vector Multiplication, Scalar Product of Two Vectors, Vector Product of Two Vectors, Triple Products of Vectors, Vector Geometry, Vector Derivatives, Vector Integrals, Line Integral of Vector Function, Surface Integral of Vector Function, Gradient of a Scalar Function, Divergence of a Vector Function, Curl of a Vector Function, Linear Vector Function, Dyadics.** (L.L.S.)

VECTOR DERIVATIVES.

Let **r** be a **vector function** of a scalar variable t. If the origin of **r** is kept fixed and t is varied, the terminus of **r** will describe a

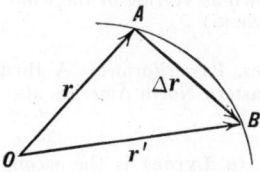

curve. Let A and B be two neighboring points on this curve, and let **r** and **r**′ be their position vectors, then **r**′ − **r** = Δ**r** is a vector increment of **r**; it is a vector having the direction of the secant AB, which approaches the **tangent to the curve** at A as B approaches A. Let Δt be the change (increment) in t corresponding to the increment Δ**r** of **r**. Form the ratio Δ**r**/Δt. If $\lim\limits_{\Delta t \to 0} \dfrac{\Delta \mathbf{r}}{\Delta t}$ exists, it is called the **derivative** of **r** with respect to t, and is denoted by $\dfrac{d\mathbf{r}}{dt}$. It is a vector tangent to the curve described by the terminus of **r**.

If $\mathbf{r} = x\hat{\mathbf{i}} + y\hat{\mathbf{j}} + z\hat{\mathbf{k}}$, and if x, y, z are functions of a scalar variable t, then

$$\frac{d\mathbf{r}}{dt} = \hat{\mathbf{i}}\frac{dx}{dt} + \hat{\mathbf{j}}\frac{dy}{dt} + \hat{\mathbf{k}}\frac{dz}{dt},$$

$$\frac{d^2\mathbf{r}}{dt^2} = \hat{\mathbf{i}}\frac{d^2x}{dt^2} + \hat{\mathbf{j}}\frac{d^2y}{dt^2} + \hat{\mathbf{k}}\frac{d^2z}{dt^2}.$$

If **r** is the position vector of a particle and if t is the time, then $\dfrac{d\mathbf{r}}{dt}$ will represent the vector velocity and $\dfrac{d^2\mathbf{r}}{dt^2}$ the vector acceleration of the particle.

If **r** is constant in magnitude but variable in direction, then $\dfrac{d\mathbf{r}}{dt}$ is perpendicular to **r**.

Let **u**, **v**, **w** be vector functions of a scalar variable t. Then the vector derivatives of various types of **products of vectors** are given by the following formulae:

$$\frac{d}{dt}(\mathbf{u} \cdot \mathbf{v}) = \frac{d\mathbf{u}}{dt} \cdot \mathbf{v} + \mathbf{u} \cdot \frac{d\mathbf{v}}{dt},$$

$$\frac{d}{dt}(\mathbf{u} \times \mathbf{v}) = \frac{d\mathbf{u}}{dt} \times \mathbf{v} + \mathbf{u} \times \frac{d\mathbf{v}}{dt},$$

$$\frac{d}{dt}[\mathbf{u} \cdot (\mathbf{v} \times \mathbf{w})] = \frac{d\mathbf{u}}{dt} \cdot (\mathbf{v} \times \mathbf{w}) + \mathbf{u} \cdot \left(\frac{d\mathbf{v}}{dt} \times \mathbf{w}\right) + \mathbf{u} \cdot \left(\mathbf{v} \times \frac{d\mathbf{w}}{dt}\right),$$

$$\frac{d}{dt}[\mathbf{u} \times (\mathbf{v} \times \mathbf{w})] = \frac{d\mathbf{u}}{dt} \times (\mathbf{v} \times \mathbf{w}) + \mathbf{u} \times \left(\frac{d\mathbf{v}}{dt} \times \mathbf{w}\right) + \mathbf{u} \times \left(\mathbf{v} \times \frac{d\mathbf{w}}{dt}\right).$$

If we have given a vector function of several independent scalar variables, its **partial derivatives** may be defined in a manner similar to that for scalar functions of several variables.

The symbolic vector expression

$$\hat{\mathbf{i}}\frac{\partial}{\partial x} + \hat{\mathbf{j}}\frac{\partial}{\partial y} + \hat{\mathbf{k}}\frac{\partial}{\partial z}$$

is denoted by Δ and is usually called "del" (occasionally "nabla"). When applied directly to a scalar function of position, we obtain the **gradient** of the function; when applied by the **scalar product** process to a vector function of position, we obtain the **divergence** of the function; and when applied by the **vector product** process to a vector function of position, we obtain the **curl** of the function. (L.L.S.)

VECTOR GEOMETRY.

Let **r** be a **variable vector**, with origin at O and let s be a **scalar** variable. Then $\mathbf{r} = s\mathbf{a}$ is the equation of a straight line through O, parallel to **a**; and $\mathbf{r} = \mathbf{b} + s\mathbf{a}$ is the equation of a line through the terminus of **b** and parallel to **a**. The equation

$$\mathbf{r} = \mathbf{a} + s(\mathbf{b} - \mathbf{a}) = s\mathbf{b} + (1 - s)\mathbf{a}$$

is the equation of a line through the ends of two vectors **a** and **b**. If we have given three vectors, **a**, **b**, and **c**, with the same origin, they terminate in the same straight line if

$$x\mathbf{a} + y\mathbf{b} + z\mathbf{c} = 0 \text{ and } x + y + z = 0.$$

The equation of a straight line through the end of **b** and parallel to **a** may also be written $\mathbf{a} \times (\mathbf{r} - \mathbf{b}) = 0$.

Let **r** be a variable vector, and let s and t be scalar variables. Then the equation of a plane through the end of **c** and parallel to **a** and **b** is $\mathbf{r} = \mathbf{c} + s\mathbf{a} + t\mathbf{b}$. The equation of a plane through the ends of three noncoplanar vectors **a**, **b**, **c** is $\mathbf{r} = s\mathbf{a} + t\mathbf{b} + (1 - s - t)\mathbf{c}$. The condition that four vectors **a**, **b**, **c**, **d** with the same origin terminate in the same plane is

$$x\mathbf{a} + y\mathbf{b} + z\mathbf{c} + w\mathbf{d} = 0 \text{ and } x + y + z + w = 0.$$

The equation of a plane perpendicular to a and through the end of **b** may be written $\mathbf{a} \cdot (\mathbf{r} - \mathbf{b}) = 0$. The equation of a plane through the end of **b** and parallel to **c** and **d** may be written $(\mathbf{c} \times \mathbf{d}) \cdot (\mathbf{r} - \mathbf{b}) = 0$. The equation of a plane through the ends of three vectors **a**, **b**, **c** with the same origin is $(\mathbf{r} - \mathbf{a}) \cdot (\mathbf{a} - \mathbf{b}) \times (\mathbf{b} - \mathbf{c}) = 0$, or $(\mathbf{a} \times \mathbf{b} + \mathbf{b} \times \mathbf{c} + \mathbf{c} \times \mathbf{a}) \cdot (\mathbf{r} - \mathbf{a}) = 0$.

The vector to a point of division of the segment joining the ends of **a** and **b** is $\mathbf{r} = \mathbf{a} + s(\mathbf{b} - \mathbf{a})$, where s is the ratio of the partial segment AP to the total segment AB, where A and B are the ends of **a** and **b**.

If r is a variable vector, the equation of a circle (or of a sphere), with center at the origin of $\mathbf{r}$, is $\mathbf{r} = \mathbf{a}$ or $\mathbf{r} \cdot \mathbf{r} = \mathbf{a} \cdot \mathbf{a}$ (written $\mathbf{r}^2 = \mathbf{a}^2$). If the center of the circle is at the end of **c**, the equation of the circle is $(\mathbf{r} - \mathbf{c})^2 = \mathbf{a}^2$ or $\mathbf{r}^2 - 2\mathbf{r} \cdot \mathbf{c} = \mathbf{a}^2 - \mathbf{c}^2 = \text{constant}$. If the origin is on the circumference of the circle, $\mathbf{c} = \mathbf{a}$, and the equation of the circle is $\mathbf{r}^2 - 2\mathbf{r} \cdot \mathbf{a} = 0$. (L.L.S.)

VECTOR INTEGRALS. Vector integration, as an operation, may be defined as in the case of a scalar function, as the **inverse operation** to vector differentiation.

Corresponding to ordinary **definite integrals** of functions of one variable, and to **double integrals** of functions of two variables, we have vector integrals of the following types: **line integrals of vector functions** and **surface integrals of vector functions**. (L.L.S.)

VECTOR MULTIPLICATION. There are two distinct kinds of products of two **vectors**: the **scalar product** and the **vector product**.

Combining these two types of products, we get two kinds of **triple products of vectors**, and also several products of four vectors, giving rise to **reciprocal systems of vectors**. (L.L.S.)

VECTOR PRODUCT OF TWO VECTORS. The vector product (or cross product) of two **vectors** is defined as a vector perpendicular to their plane in the sense of advance of a right-handed screw rotated from the first to the second of these vectors through the smaller angle between them, and having a magnitude equal to the product of the magnitudes of the two vectors by the sine of the angle between them.

The vector product is represented in magnitude by the area of the parallelogram of which the two given vectors are adjacent sides.

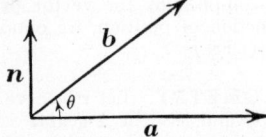

The vector product of **a** by **b** is denoted by $\mathbf{a} \times \mathbf{b}$ in the Gibbs notation, by $[\mathbf{a}, \mathbf{b}]$ in another common form of notation, and by $V\ \mathbf{ab}$ in the Hamiltonian notation.

We have:

$$\mathbf{a} \times \mathbf{b} = (ab \sin \theta)\hat{\mathbf{n}},$$

where θ is the angle between **a** and **b**, and $\hat{\mathbf{n}}$ is the unit normal vector to **a** and **b**.

The vector product $\mathbf{a} \times \mathbf{b}$ does not obey the **commutative law**, but we have

$$\mathbf{a} \times \mathbf{b} = -\mathbf{b} \times \mathbf{a}.$$

It is, however, **distributive**:

$$(\mathbf{a} + \mathbf{b}) \times \mathbf{c} = \mathbf{a} \times \mathbf{c} + \mathbf{b} \times \mathbf{c}.$$

If **a** is parallel to **b**, then $\mathbf{a} \times \mathbf{b} = 0$, and conversely. It follows that $\mathbf{a} \times \mathbf{a} = 0$ for any vector **a**.

For the unit vectors $\hat{\mathbf{i}}, \hat{\mathbf{j}}, \hat{\mathbf{k}}$, we have:

$$\hat{\mathbf{i}} \times \hat{\mathbf{i}} = \hat{\mathbf{j}} \times \hat{\mathbf{j}} = \hat{\mathbf{k}} \times \hat{\mathbf{k}} = 0,$$

$$\hat{\mathbf{i}} \times \hat{\mathbf{j}} = \hat{\mathbf{k}}, \quad \hat{\mathbf{j}} \times \hat{\mathbf{k}} = \hat{\mathbf{i}}, \quad \hat{\mathbf{k}} \times \hat{\mathbf{i}} = \hat{\mathbf{j}}.$$

If $\mathbf{a} = a_1\hat{\mathbf{i}} + a_2\hat{\mathbf{j}} + a_3\hat{\mathbf{k}}$, $\mathbf{b} = b_1\hat{\mathbf{i}} + b_2\hat{\mathbf{j}} + b_3\hat{\mathbf{k}}$, then $\mathbf{a} \times \mathbf{b}$
$= (a_2b_3 - a_3b_2)\hat{\mathbf{i}} + (a_3b_1 - a_1b_3)\hat{\mathbf{j}} + (a_1b_2 - a_2b_1)\hat{\mathbf{k}}$

$$= \begin{vmatrix} \hat{\mathbf{i}} & \hat{\mathbf{j}} & \hat{\mathbf{k}} \\ a_1 & a_2 & a_3 \\ b_1 & b_2 & b_3 \end{vmatrix}.$$

(L.L.S.)

VECTOR PRODUCTS. Vector Multiplication.

VECTORIAL ANGLE. Polar Coordinates in a Plane.

VECTORS. Line-segments, as AB and BA, in which opposite senses are distinguished, are called directed line-segments.

A directed line-segment is also called a vector. A vector then has magnitude (length), direction and sense.

A physical quantity which has the three attributes of magnitude, direction and sense may be called a vector quantity. Examples of vector quantities are displacement of a particle, velocity and acceleration of a particle, force acting on a body, etc.

If a vector is a directed line-segment AB, in which the sense is from A to B, we call A the origin and B the terminus of the vector and sometimes denote it by the symbol $\overrightarrow{AB}$.

A vector is frequently denoted by a single letter, printed in bold-face type, as **a**.

A scalar is a number.

By a scalar quantity we shall mean a physical quantity whose measure is completely expressed by a scalar or number; it has only magnitude. Examples of scalar quantities are temperature, weight, potential, etc.

Scalars are denoted by light-face type.

The magnitude (or length) of a vector **a** is often denoted by a or by $|\mathbf{a}|$. The unit vector having the same direction and sense as **a** but of unit length is sometimes denoted by â.

To multiply a vector **a** by a positive scalar m, we multiply the magnitude of **a** by m, leaving the direction and sense unchanged.

The negative of a vector **a** is the vector of the same magnitude and direction but of opposite sense.

There is great diversity of notation in vector analysis, particularly with regard to products of vectors. (L.L.S.)

VEE-BELT. Belting.

VEERING WIND. Any clockwise change in wind direction is known as veering of the wind. It is opposite to backing. (P.E.K.)

VEERY. Aves, Passeriformes. A thrush, *Hylocichla fuscescens*, of eastern North America, also called Wilson's thrush. (A.W.L.)

VEGA. Vega (α Lyrae) is the second brightest star visible in northern latitudes. It is visible during some portion of every clear night throughout the year, and dominates the summer skies. Because of its distinctly bluish tinge it is one of the most beautiful stars of the northern skies and references to it are found in all of the ancient literatures. At one time Vega was the pole star and, because of **precession**, the pole will be close to Vega about 11,500 years hence. Lockyer claims that the temples at Denderah in Egypt were oriented to this star as early as 7000 B.C.

Vega is of interest astronomically for it is the brightest star in the general vicinity of the **solar apex** or the point on the celestial sphere toward which the **sun** is moving, carrying all other members of the **solar system** along with it. (W.K.G.)

VEGETABLE IVORY. Vegetable ivory is obtained from the fruits of various **palms** of the genus *Phytelephas*, known as ivory palms. These plants, natives of the wet forests of tropical America, have short thick stems and erect, **pinnate** leaves of very large size. The plants are **dioecious.** The staminate (see **Stamen**) flowers are borne on an elongate fleshy **spadix;** while the pistillate (see **Pistil**) flowers occur in groups of six or seven. Each flower gives rise to a berry, the six or seven berries of a group being united. This fruit has a hard outer layer, covered with woody protuberances, and contains several seeds. During development these seeds contain a milky-white fluid, which hardens as the seed matures, and which is the endosperm of the seed. It becomes almost as hard as ivory, and is used in making billiard balls, buttons, and a variety of small articles. (R.M.W.)

VEIN. For the use of this term in anatomy, see **Circulatory System.** For its use in botany, see **Leaf.** In geology the term vein is understood to refer to small or large fissures which have been filled with mineral matter by deposition from aqueous solutions, including "liquors" and gaseous emanations from magmas. Lode means essentially the same thing as vein, being an old Cornish mining term referring to the formations which would "lead" or direct the miner to the desired minerals. (E.S.C.S.)

VELARIUM. A flange of thin tissue projecting from the margin of a jellyfish over the concave surface of the body. Unlike a true **velum** it contains neither muscles nor a nerve ring. This structure is well developed in the order Cubomedusae (**Scyphozoa**) and increases the resemblance of these animals to the **hydrozoan** medusae. Also called the pseudovelum. (A.W.L.)

VELIGER. A form of **larva** found in the phylum **Mollusca.** Some mollusks hatch as a **trochophore** and later develop into a second free-swimming larval stage, the veliger. In this stage a ciliated (**cilia**) ring called the **velum** is developed and the foot becomes larger. The shell becomes coiled in the veliger stage of some **gasteropods.** (A.W.L.)

VELOCITY. The time rate of change of position. (Unless angular velocity is specified, this term is understood to refer to linear motion, which may be emphasized by the expression "linear velocity.") Strictly, the velocity of a moving point must specify both the speed and the direction of the motion, and is therefore a **vector;** though the term is sometimes more loosely used as merely synonymous with speed. The velocity of a point is the time rate of the distance s from a fixed origin O, expressed as the **vector derivative** of s with respect to the time, ds/dt (Fig. 1); while the speed is the magnitude of the velocity and is not a vector. If the direction of motion is constant, so that the motion is in a straight line (but not necessarily with constant speed), and if the line of motion is clearly understood, it is convenient to treat the distance s and the velocity ds/dt as scalars with respect to some zero point on that line and with appropriate algebraic signs (Fig. 2); otherwise they must be regarded as vectors. If the velocity is variable, account must be taken of the **acceleration.** Examples of both curved and rectilinear motion are treated under **kinematics.** (L.D.W.)

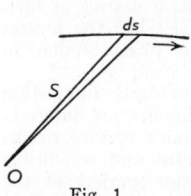

Fig. 1.

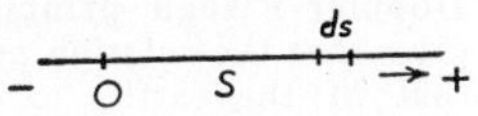

Fig. 2.

VELOCITY CURVE. A plot of **radial velocity** of a star as ordinates against time as abscissae is known as the velocity curve for the star. The method of formation of a mean velocity curve is similar to that described for the determination of a mean **light curve.** The use of the velocity curve for determining the **orbit** of a **spectroscopic binary** is discussed elsewhere. (W.K.G.)

VELOCITY HEAD. Fluid Flow.

VELOCITY OF LIBERATION. Rocket Flight.

VELOCITY MODULATION. This is a form of electron **modulation** in which the electrons passing through a resonant cavity in a tube such as the **klystron** are acted upon by a modulating field in such a manner that their velocities cause them to pass through the collector cavity in groups. (L.R.Q.)

VELOCITY RATIO. Belting, Chain, Gearing.

VELUM. 1. In **hydrozoan** medusae, a thin flange of tissue projecting from the margin of the body over its concave surface, leaving a restricted circular opening. The velum contains a nerve ring and muscles. It aids in locomotion by the contraction of the body to force jets of water from the concave subumbrellar surface. 2. In the **veliger** larva of **mollusks,** a ciliated (**cilia**) organ adjacent to the mouth. It is formed of the postoral ciliated ring of the trochophore larva together with the preoral ring, and serves as an organ of locomotion and to bring food to the mouth. 3. The double ciliated ring surrounding the oral end of the body of a **rotifer.** (A.W.L.)

VELVET ANT. Insecta, Hymenoptera. A **wasp** of the family Mutillidae. The females of these insects are wingless and consequently are antlike in form, but they differ in the absence of the dorsal prominence on the slender waist that characterizes the ants. They are densely hairy insects, usually brightly banded with some shade of red or yellow, black, and sometimes white. The males differ in other details than the presence of wings, hence it is difficult to associate the sexes unless they are taken together. Velvet ants are parasitic in the nests of other insects and some have been reported as parasites on the tsetse fly. (A.W.L.)

VENA CAVA. A main venous trunk returning blood to the **heart** in vertebrates. Above the fishes the paired condition of the venous system is modified, and in the **reptiles,** birds (**Aves**), and **mammals** the posterior parts of the body are drained by the tributaries of a single trunk vein, the posterior vena cava. In the mammals the paired condition of the anterior venous system also disappears, giving rise to an anterior vena cava. Owing to the erect posture of man these vessels are also known as the superior and inferior venae cavae. The name is also applied to a venous trunk in the **cephalopod** mollusks. It drains the head and subdivides to form the **branchial veins** to the gills. (A.W.L.)

VENDACE. Pisces, Teleostei. A fresh-water fish related to the salmon, found in lakes of Ireland. (A.W.L.)

VELAMEN. Root.

VENEER. Wood.

VENTILATION. This is the process of changing the air in any space by natural or mechanical means. In its simplest form ventilation results from the effect upon buildings of winds, or from the natural levity of heated air. Air blown horizontally against a building creates a small **plenum** to windward and a small **vacuum** to leeward. Through openings intentionally created, such as opened windows, doors, ventilators, etc., this pressure difference will cause a ventilating current to flow. Also through cracks in walls, clearances around window sash, and the like, unintentional and unwanted ventilation may be brought into existence. Natural ventilation can be secured in the absence of atmospheric motion by providing elevated outlets (roof ventilators, stacks, air shafts) so that cool air may displace warm air upwards and out of the vents.

Mechanical systems of ventilation are more adaptable to continuous positive control of air flow. The vitalizing air pressure of these systems is usually created by a **centrifugal fan,** although axial flow and propeller fans are sometimes used. Ducts of sheet metal, wood, and other materials of construction carry the air to and/or from the space to be ventilated. Quietness in operation and even distribution of air (i.e., no drafts) are especially sought in these systems when installed in public buildings, dwellings, offices, and stores. Vibration absorbent mountings for fan and motor, low fan feeds, moderate air velocities, and acoustical duct linings are some of the means of reducing noise level. This level should not be above that usually existing in the spaces being ventilated, for example, from 30 to 50 decibels in dwellings and offices.

The minimum quantity of air in circulation consistent with human comfort has been the subject of extensive observations by the American Society of Heating and Ventilating Engineers and by others. In the usual space allotment of dwellings (approx. 400 cu. ft. per person) about 1½ changes of air per hour will satisfactorily ventilate from both the chemical and thermal standpoint. This amounts to about 10 cu. ft. air per person per min. In workrooms, auditorium, schoolrooms and other places of congregation the free space allotment per person is less and the required number of air changes greater.

In conclusion, be it noted that ventilation is one aspect of **air conditioning,** for the latter term is concerned not only with quantity of air circulated, but also with its quality. (F.T.M.)

VENTRICLE. 1. A strongly muscular chamber of the **heart** from which **blood** is pumped to various parts of the body. In the **mollusks** it is a single chamber and in the vertebrates it is single in the more primitive forms, including fishes, amphibians, and some reptiles. In the reptiles its subdivision is begun, and in the crocodilians right and left ventricles appear. This division is complete in the birds and mammals. 2. A cavity of the vertebrate **brain.** The first and second ventricles are in

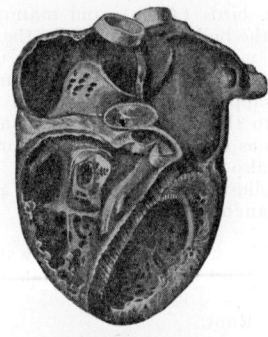

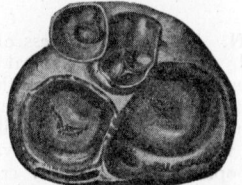

Cavities of the heart, showing valves between the heart chambers.

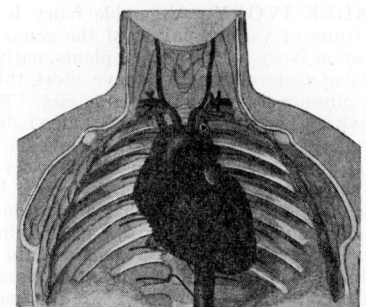

Heart and main vessels. (Lungs removed.)

the cerebral hemispheres. The third is formed of the persisting median cavity of the first primitive brain vesicle. The fourth ventricle is the cavity of the third primary vesicle, and all other remnants of the original cavities become narrowed passages. The ventricular system contains cerebrospinal fluid. (A.W.L.)

VENTURI METER. Flow Meter.

VENUS. (See tables of planetary data, page 1109.) Venus may be called the earth's twin **planet,** for in size, density, and general constitution, if not in all physical characteristics, Venus is much like the earth. This planet is one of the most conspicuous objects in the sky and may easily be viewed in full daylight with the naked eye, provided attention is directed to the proper point in the heavens. For this reason Venus is frequently used by navigators for obtaining a **Summer line** in the daytime to use in conjunction with an observation of the sun.

Venus, like **Mercury,** has its orbit between the earth and the sun and never appears very far away from the sun. Since at maximum elongation Venus appears either in the eastern sky at sunrise, or in the western sky at sunset, this is the planet that is most commonly referred to as the morning or evening "star." Like Mercury the ancient astronomers had two names for it; calling it Phosphorus as the morning star, and Hesperus as the evening. As Venus passes about in its apparent path from inferior conjunction back to inferior conjunction again it goes through a series of **phases** exactly as does the moon. In fact, this change in phase of Venus was one of the tests proposed by the anti-**Copernicians** prior to the invention of the telescope. At inferior conjunction Venus is closer to the earth than any other astronomical object except the moon (and an occasional asteroid or comet), having a distance of only 26,000,000 miles. Since at superior conjunction the planet has a distance of nearly 160,000,000 miles the changes in apparent diameter are very great and, as a matter of fact, Venus appears the brightest to us (i.e., has the largest apparent area) when in the crescent phase similar in shape to the moon when about 5 days old.

Telescopically Venus is not a particularly interesting object except insofar as the phase changes are interesting. There are very few and very faint surface markings; in fact, the markings are so faint and so illusive that the determination of the rotation period of the planet by direct observation is practically impossible. The application of the spectroscope, by applying the **Doppler-Fizeau principle** to the spectral lines, indicates that the rotation period is considerably longer than that of the earth, 30 days being a fair compromise between various discordant results.

The high reflecting power of the planet, together with a value for **surface gravity** comparable with the value for the earth, lead to the conclusion that Venus has a dense **atmosphere.** A tremendous amount of research has been applied to the problem of the constitution of the atmosphere of Venus. Delicate and exacting tests have failed to indicate the presence of either water vapor or oxygen in the atmosphere of Venus, although there is considerable evidence of the existence of carbon dioxide. The absence of water vapor precludes the possibility of any large bodies of water on the planet and makes the existence of any forms of life such as we know them on the earth highly improbable.

Comparatively little is known concerning the surface conditions of Venus because the dense atmosphere of the planet prevents observations being made of the surface. Measurements of the surface temperature of the planet indicate a range of from 333° K. (140° F.) on the sunlit portions, down to 253° K. (−4° F.) for the dark regions.

At rare intervals Venus passes through inferior conjunction at a time when the sun is close to the node of its apparent path. At such times the planet will transit the disk of the sun and, since the planet will then be quite close to the earth, the transits may be used for the determination of **solar parallax.** The last transits occurred in 1872 and 1882 and the next ones will come in 2004 and 2012. (W.K.G.)

VENUS' FLY-TRAP. *Dionaea muscipula.* **Insectivorous Plants.**

VENUS' GIRDLE. Ctenophora, Cestida. A transparent marine animal of ribbon-like form. The longitudinal axis of the body lies across the width of the ribbon, hence the body is short and is greatly elongated on one transverse axis and very short on another. (A.W.L.)

VERATRINE. Alkaloids.

VERDET CONSTANT. Faraday Effect.

VERMIFUGE. A drug used to rid the intestinal tract of worms or intestinal **parasites.** (D.M.H., R.S.M.)

VERNAL EQUINOX. The point of intersection of the **ecliptic** and the equator where the sun apparently passes from south to north of the earth's equator is known as the vernal equinox. The direction of the

vernal equinox is the fundamental direction in the equatorial and ecliptic systems of **spherical coordinates** from which right ascension and celestial longitude are measured. The sun is in the direction of the vernal equinox on approximately March 21 and this date indicates the beginning of the spring season in the northern hemisphere. On the ancient **calendars** the passage of the sun through the vernal equinox indicated the beginning of the new year. Some 2000 years ago the vernal equinox was in the **constellation of Aries** and was frequently referred to as the first of Aries. Astrologers frequently use this same terminology for the position of the sun on March 21 in spite of the fact that, due to **precession,** the vernal equinox has shifted into the constellation of **Pisces.** (W.K.G.)

VERNALIZATION. This term is applied to the low-temperature treatment of seeds before sowing which shortens the time to flowering of the plants which develop from them. This technique is employed principally with cereals and especially with winter wheats. It is used as a widespread agricultural practice chiefly in Russia. Vernalization of the grains of a winter wheat so shortens its life cycle that it can be grown as a spring wheat. Grains of the Turkey Red variety of winter wheat, for example, can be vernalized by soaking them to a moisture content of about 60% and exposing them to a temperature of 1–3° C. for 9–10 weeks while holding them at approximately that water content. The grains are then sowed, the plants developing from them reaching the heading stage in 110–120 days from the beginning of the cold treatment. Unvernalized seeds of the same variety require about 150 days from sowing to reach the heading stage under the same conditions. (B.S.M.)

VERNIER. Measurement.

VERNIER CALIPER. Measurement.

VERONAL. Barbital.

(See Vertebrate Paleontology.)

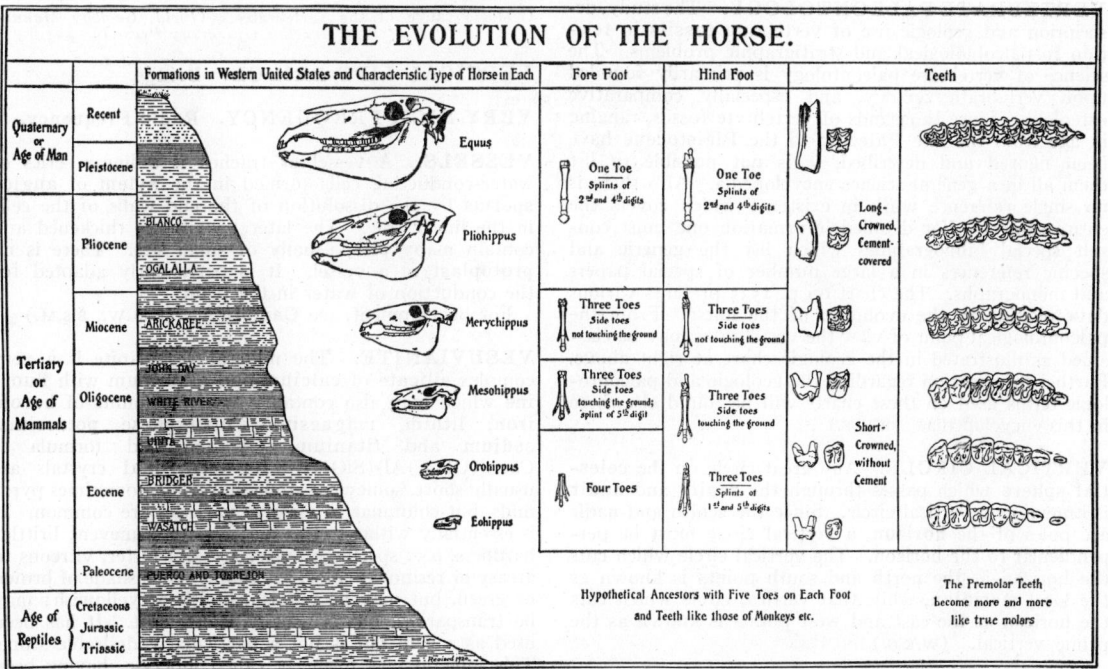

THE EVOLUTION OF THE HORSE.

		Formations in Western United States and Characteristic Type of Horse in Each	Fore Foot	Hind Foot		Teeth
Quaternary or Age of Man	Recent		One Toe Splints of 2ʳᵈ and 4ᵗʰ digits	One Toe Splints of 2ʳᵈ and 4ᵗʰ digits		
	Pleistocene	Equus				
	Pliocene	BLANCO / OGALALLA — Pliohippus			Long-Crowned, Cement-covered	
Tertiary or Age of Mammals	Miocene	ARICKAREE — Merychippus	Three Toes Side toes not touching the ground	Three Toes Side toes not touching the ground		
	Oligocene	JOHN DAY / WHITE RIVER — Mesohippus	Three Toes Side toes touching the ground; splint of 5ᵗʰ digit	Three Toes Side toes touching the ground	Short-Crowned, without Cement	
	Eocene	UINTA / BRIDGER — Orohippus / WASATCH — Eohippus	Four Toes	Three Toes Splints of 1ˢᵗ and 5ᵗʰ digits		
	Paleocene	PUERCO AND TORREJON				
Age of Reptiles	Cretaceous Jurassic Triassic		Hypothetical Ancestors with Five Toes on Each Foot and Teeth like those of Monkeys etc.			The Premolar Teeth become more and more like true molars

Evolution of the horse. (*American Museum of Natural History.*)

VERTEBRA. One of the thirty-three bones making up the **spinal column**. In man, there are seven cervical, twelve dorsal, five lumbar, five sacral and four coccygeal vertebrae. (R.S.M.)

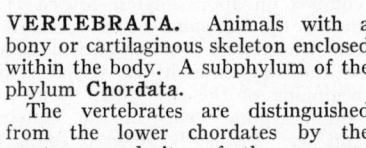

Vertebral column. (*Cunningham, Textbook of Anatomy, Oxford Press.*)

VERTEBRATA. Animals with a bony or cartilaginous skeleton enclosed within the body. A subphylum of the phylum **Chordata**.

The vertebrates are distinguished from the lower chordates by the greater complexity of the **nervous system**, including a well-developed **brain**, as well as by the replacement of the **notochord** by a spinal column in most forms.

The group is divided into six classes:

Class **Cyclostomata.** The **hag-fishes** and **lampreys**, or round-mouthed eels. A group with a persistent notochord and without hinged jaws.

Class **Pisces.** The fishes. Animals of aquatic habit, relatively few species leaving the water at all. They breathe by gills.

Class **Amphibia.** Frogs, salamanders and related forms. Moist-skinned animals, usually with an aquatic larval stage in which gills appear. Some are permanently aquatic and some exclusively terrestrial.

Class **Reptilia.** Lizards, snakes, turtles, crocodiles, etc. Never with gills, even though aquatic. Skin scaly.

Class **Aves.** The birds. Skin clothed with feathers and scales. Mostly flying species.

Class **Mammalia.** Skin clothed with hair in most species. Young nourished with milk secreted by the mother. (A.W.L.)

VERTEBRATE PALEONTOLOGY. The study, description and geologic use of vertebrate **fossils** in relation to paleobiological and stratigraphic problems. The science of vertebrate paleontology is primarily founded upon vertebrate zoölogy, and especially comparative osteology. Since thousands of vertebrate fossils, ranging in age from the late **Palezoic** to the **Pleistocene** have been figured and described, it is not possible to list them all in a general science encyclopedia. Also there is no single reference work in existence which covers the entire subject. For detailed information one must consult special bibliographies which list the generic and specific references in a large number of special papers and monographs. The chart on p. 1535 presents various data relating to the evolution of the horse. From the paleontological point of view the vertebrates may be classified as illustrated in the geologic chart at right above. Further information regarding the geologic and paleontologic terms used on these charts will be found elsewhere in this encyclopedia. (R.M.F.)

VERTICAL CIRCLE. Any great circle on the **celestial sphere** which passes through the **zenith** and **nadir** is known as a vertical circle. Since the zenith and nadir are poles of the **horizon**, a vertical circle must be perpendicular to the horizon. The vertical circle which cuts the horizon at the north and south points is known as the local **meridian**, while that vertical circle which cuts the horizon at the east and west points is known as the prime vertical. (W.K.G.)

VERTICAL CURVE. Grade.

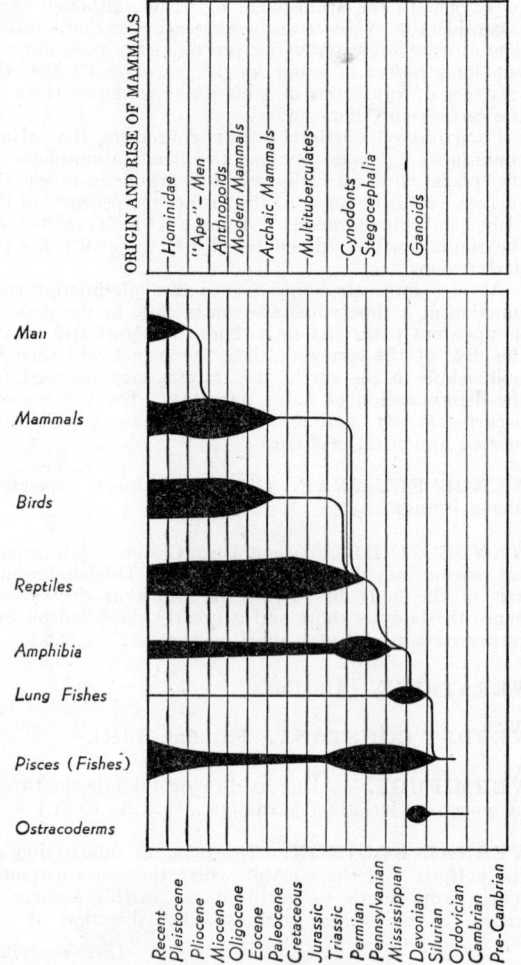

Geologic range of the vertebrates. (*Field, Geology Manual, Part II, Princeton University Press.*)

VERTICAL LIFT BRIDGE. Bridge.

VERY HIGH FREQUENCY. Radio Frequency.

VESSELS. A vessel or trachea is a linear series of water-conducting cells formed in the **xylem** of **angiosperms** by the dissolution of the end walls of the cells in the linear row. The lateral walls are thickened and contain many **pits**, usually of small size. There is no **protoplast** in a vessel. It is particularly adapted for the conduction of water in the plant.

For development, see **Cambium**. (R.M.W., B.S.M.)

VESUVIANITE. The mineral vesuvianite is a very complex **silicate** of **calcium** and **aluminum** with **fluorine** which may also contain varying amounts of **boron, iron, lithium, magnesium, manganese potassium, sodium** and **titanium**. A suggested formula is $Ca_6Al(OH,F)Al_2(SiO_4)_5$. Its **tetragonal** crystals are usually short, somewhat stoutish, prisms, sometimes pyramids, but columnar to massive varieties are common. It is essentially without cleavage; fracture, uneven; brittle; hardness, 6.5; specific gravity, 3.3–3.5; luster, vitreous to greasy or resinous; color, commonly some shade of brown or green, but may be reddish, bluish or yellowish; may be transparent, but is usually translucent. It has been used as a gem but is not a particularly desirable stone. This mineral was formerly called idocrase, having been named by Haüy from the Greek words meaning *form* and *mixture* because it resembled crystals of other species, scarcely a valid distinction. Werner gave it the name vesuvianite from Mt. Vesuvius where it was first found in blocks of limestone appearing as inclusions in the **lava.** Vesuvianite is not a constituent of the **igneous rocks,** but rather a contact **metamorphic** mineral resulting from the alteration of impure **limestones** and **dolomites.** It is usually associated with **diopside, wollastonite, epidote, grossularite, garnet, etc.** There are many localities worthy of mention among which are: The Urals, Czechoslovakia, Rumania, Trentino and Monzoni in Italy, as well as at Mt. Vesuvius and Mt. Somma, Switzerland, Mexico and Japan. In the United States vesuvianite is found in Androscoggin and York Counties in Maine; Orange County, New York; Sussex County, New Jersey; Garland County, Arkansas; and Riverside and Tulare Counties in California. (E.S.C.S.)

V-FLAT. Belting.

VHF. Very High Frequency.

VIBRACULUM. A modified form of individual in the **bryozoan** colony. These colonies contain the complex polypides and two reduced forms whose functions appear to be the protection of the entire group against the lodging of **sessile** animals. These forms are the **avicularium** and the vibraculum. The latter is no more than a bristle or filament that moves back and forth, sweeping across the surface of the colony. Numerous vibracula sometimes act synchronously in this movement. (A.W.L.)

VIBRATION SPECTRUM. Molecular Spectra.

VIBRATIONS AND WAVES. These terms are used in very broad senses and apply to a large variety of phenomena and processes. "Vibration" commonly refers to a to-and-fro motion, but we shall have to extend the meaning to include any periodic physical process, such, for example, as a cyclic variation in electric or magnetic field intensity. When an elastic body is deformed and released, it is in general set into oscillation such that the displacement of any particle from its equilibrium position is a more or less complicated harmonic function of the time (see **Harmonic Motion** and **Harmonic Analysis**). The vibration may or may not be symmetrical with respect to the neutral position; in any case the maximum displacement is called the amplitude of the vibration. By analogy, the same terms and the same analysis are applied to vibrations of any type.

If a vibratory disturbance occurs at any point in a medium having sufficient continuity to transmit displacements from one part to another, a train of waves is propagated outward from the seat of the disturbance. The speed of propagation depends upon the closeness of coupling between adjacent particles of medium and the consequent magnitude of the restoring forces; and upon whatever reaction of the medium corresponds to mechanical inertia. In some cases also the speed varies with the frequency, as with light in a material medium. In any case the wavelength, viz., the distance traversed during a complete vibration period, is related to the frequency of vibration and the speed of propagation by the simple equation $v = \nu\lambda$ in which v is the speed, ν the frequency, and λ the wavelength. Thus, if sound waves of frequency 250 vibrations per sec. are traveling with a speed of 1000 ft. per sec., the wavelength is 4'. In case the vibrations are of complex character and the different components travel with different speeds, the resulting "wave group," traveling with its characteristic **group velocity,** may be very sharply defined and may thus constitute a "wave packet," resembling a single pulse or unrepeated wave. (See **Interference.**)

The theorem of Fourier states that any vibration or wave train, however complex, can be resolved into simple harmonic components of various amplitudes and frequencies and in various phases. Of these components, the one of lowest frequency (and in the case of elastic vibrations, usually of greatest amplitude) is the "fundamental"; the others are "overtones."

The character of a wave process may be described mathematically by means of a "wave equation" which specifies the condition at any point of the wave field in terms of the position and of the time; or graphically by one or more "wave form" curves, of which the ordinates represent the periodically variable displacements at any point, and the abscissas the time. (See **Wave Propagation.**)

In applied dynamics, vibrations might be classified as natural and artificial. An outstanding example of the former is the earthquake. Almost any rotating machinery furnishes examples of the latter. Civilized man comes into contact with artificial vibrations constantly, but only occasionally with those of a natural type. Sometimes vibration can be an asset; witness the use of vibrators to jar a **pattern** loose from molding sand, vibrators in therapeutical work, and a few others. In the main vibration is a nuisance, possibly destructive, indicative of wear and inefficiency, likely to produce weakening of a structure, and fatiguing physically and mentally. Man-made vibrations are much more likely to be encountered in damaging amounts in cities than in rural areas. There is considerably more machinery in and about the cities. Most of the streets are hard-paved. They carry a high average density of traffic, and further, there are subways, elevated railroads, and a concentration of commercial vehicles which produce all sorts of disturbances on the crust of the earth. In a region of large modern buildings these vibrations might enter the structural steel skeleton of the building below ground level, and be felt throughout the whole structure. Concrete in monolithic construction transmits vibrations exceptionally well. This condition has reached such proportions that the prevention of vibration, or a very effective damping of it, is of definite financial value, and hence the economic incentive to attack vibration from an engineering standpoint is provided. By the use of suitable vibration-absorbing materials, and with proper isolation of machines having vibration, or, better still, the more scientific balancing of rotating apparatus, much has been done to reduce vibration to a point where it is not objectionable. This work has appeared in many different fields, as is brought out by illustrations of vibration control taken at random. In the automotive field, the number of cylinders of engines has been increased so that there will be a smoother flow of power due to overlapping power impulses, vibration dampers have been designed and applied to crankshafts, and engines have been mounted in resilient supports like rubber. The widespread use of pneumatic tires has helped more than anything in reducing vibration from passing vehicles. The spring suspension of many pieces of rotating machinery, which were once rigidly bolted to their foundations, is a further example. Most domestic refrigerators will be found to have the rotating refrigerating unit flexibly supported so that a minimum of vibration will be transmitted. Buildings may be insulated partially from vibration if the problem is considered at the time of laying the foundations, but not much can be done about it once the building is finished. (See **Mechanics, Balancing, Noise and Vibration, Critical Speed.**) (L.D.W., F.T.M.)

VIBRATOR. The vibrator is a rapidly acting automatic **switch** for alternating the polarity on a **transformer** primary fed from a d-c source, thus in effect converting the d-c supply to a.c. so the transformer can function. It is widely used in automobile and other portable radio or electronic apparatus. Formerly a double-type vibrator was widely used, the second part being connected to the transformer secondary and serving as a **rectifier** for the output voltage. The tendency now is towards the use of a single or non-synchronous vibrator and some other type rectifier. A typical application is shown in the diagram. The vibrator is pulled

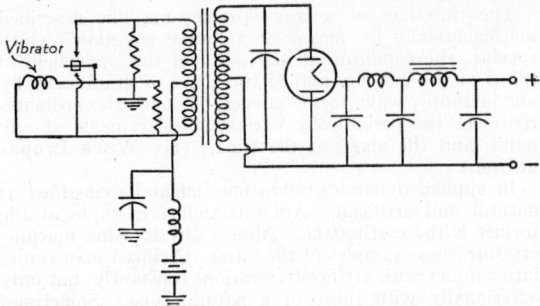

Vibrator

against the lower contacts by the vibrator winding. This shorts the vibrator winding so it releases and the vibrator reed throws over to the other contacts due to the spring action of the reed. This removes the coil short and thus the coil is again energized and pulls the reed back and so the cycle is repeated. It can be seen that this operation connects the ground terminal of the battery to first one end and then the other of the transformer winding. This produces current flowing first one way and then the other in the primary (actually in the two halves of the winding) so there is induced in the secondary a voltage determined by the turns ratio of the transformer. This secondary voltage is then rectified and filtered to give the high-voltage plate supply needed in most electronic applications. (L.R.Q.)

VICKERS HARDNESS. Hardness.

VICUNA, VICUGNA, VICUNIA. Mammalia, Artiodactyla. A wild South American animal, *Lama vicunia*, related to the guanaco and the domestic llama and alpaca. The vicunia is the smaller of the two wild species. It lives in the mountains of Ecuador, Peru, and Bolivia. (A.W.L.)

VIDEO FREQUENCY. This is the frequency band of the **television** picture signal, i.e., it corresponds to the audio frequency of conventional radio. These frequencies range from a few cycles per second to several million, imposing severe requirements on any **amplifier** circuits passing them. (L.R.Q.)

VIERENDEEL GIRDER BRIDGE. Rigid Frame.

VILLARD EFFECT. This is a special phase of the Clayden effect and was discovered by Villard in 1900, who found that a latent image on a photographic plate, or film, which had been produced by x-ray is partially destroyed upon exposure to diffused white light. (C.B.N.)

VINCENT'S ANGINA. Stomatitis.

VINCENT'S INFECTION. Stomatitis.

VINE. Stem.

VINEGAR EEL. Nemathelminthes, Nematoda. A minute roundworm, *Anguillula aceti*, found in the "mother" of vinegar. It reaches a length of 2 mm. The worms have been found in other situations, including the human bladder. (A.W.L.)

VINEGARONE. Whip Scorpion.

VIOSTEROL. Vitamin D.

VIPER. Reptilia, Sauria. A poisonous **snake** with a pair of long tubular fangs near the front of the upper jaw. Most species of vipers are also characterized by the relatively short and thick body and the broad triangular head. The group includes the typical vipers of the Old World and the **pit vipers** of North and South America.

The Old World vipers include a number of Asiatic, European, and African species bearing the name viper and in addition the two species known as the asp and eja. The Egyptian horned viper is sometimes known as the cerastes from the name of the genus to which it belongs. The African puff-adder is a viper, named from its habit of inflating the body when disturbed. This name has unfortunately been borrowed for an entirely harmless snake of the eastern states, related to the hog-nosed snake. It is variously known as the puffing or spreading adder or blowing viper. When disturbed it flattens its body and makes an impressive bluff, but it is quite harmless.

Many of the Old World vipers are dangerous. Their poison is similar in nature to that of the pit vipers. (A.W.L.)

VIREO. Aves, Passeriformes. *Vireo*. A small quietly colored bird (**Aves**) of a family related to the warblers. The vireos are mostly gray to olive gray above and white below. Some species show traces of contrasting black or yellow. They build beautiful cupped nests, suspended in the crotch of a twig. (A.W.L.)

VIRGINIUM. Chemical Composition, I. Elements.

VIRGIN'S BOWER. Buttercup Family.

VIRGO. (See map, page 380.) Virgo, the sixth sign of the **zodiac,** is one of the earliest-named among the constellations. In every known literature we find references to this constellation and always connected in some manner with a maiden and the harvest. Among the Egyptians, Virgo was associated with Isis and was said to have formed the milky way by dropping innumerable wheat heads in the sky.

Astronomically, the constellation is famous for the large number of **nebulae** found in it. Sir William Herschel found no less than 323 of these objects in this part of the sky and more recent observations have raised the number to over 500. A large number of variable stars are also to be found in the constellation. The brightest star in the constellation is the well-known star **Spica**.

Since Virgo is a feminine sign, it is generally considered **astrologically** as unfortunate, although those born under this sign are supposed to be thrifty and ingenious. (W.K.G.)

VIRTUAL TEMPERATURE. Air free from water vapor has a certain density for fixed temperature and pressure. If, however, some water enters the air as vapor and the pressure held constant, the density will decrease slightly because the molecular weight of water is 18, compared to 28.97 for air. Also, if the air remains water-free and the temperature increases slightly, the density will also decrease. Virtual temperature of air is that temperature required to maintain its density constant if its water-vapor content is removed. Virtual temperature in air containing water is always greater than existing temperature.

$$T_v = \frac{T}{1 - 0.379e/P}$$

where T_v = Virtual temperature.
T = Existing temperature.
e = Vapor pressure of water vapor.
P = Total pressure of the air. (P.E.K.)

VIRTUAL WORK PRINCIPLE. Equilibrium of Forces; Least Energy Principle.

VIRULENCE. This term is used in describing pathogenic organisms. It is indicative of the disease-inciting power of these organisms. Variations in virulence occur not only among different species of microorganisms but also among strains of the same species. **Bacteria** gain or lose their virulence according to environmental conditions outside and within the body. This accounts in part for the presence of **epidemics** at certain times and the rise and fall and severity of epidemics. (R.S.M.)

VIRUS. An ultra-microscopic infectious agent. Diseases caused by viruses include **measles, mumps, chicken pox, smallpox, rabies, poliomyelitis, encephalitis,** the common cold, **herpes, dengue, influenza** and many others.

Viruses are filterable; that is, they are so small that they will pass through the pores of a porcelain filter. Information about their size—which varies between 10 and 150 millimicra in different species, has been collected by indirect means, using filtration and sedimentation data; recently with the development of the **electron microscope,** certain of the larger viruses have been photographed and their actual shapes seen.

Viruses are thought to be living organisms, capable of multiplication. They cannot be grown outside living tissue, and therefore must be identified indirectly. This is done by studying the **pathology** produced in experimental animals on inoculation of virus-infected tissue; by determining the host range, that is the animals which are susceptible to a given virus; and by testing immunological responses, i.e., specific **antibodies** in infected animals. These characteristics are more or less specific for each virus. Different tissue specificities are also characteristic. Thus the dermatropic agents attack primarily skin: smallpox, chicken pox, and measles viruses are included in this group. Rabies, encephalitis and poliomyelitis viruses are primarily neurotropic, and attack nervous tissue. (D.M.H.)

VIRUS DISEASES. Many diseases, both plant and animal, are called virus diseases. The organisms causing these diseases are too small to be seen except with the aid of the **electron microscope** by means of which the larger viruses have been photographed. Recent investigations indicate that they are individual giant protein molecules.

Virus diseases attack herbaceous plants more frequently than they do woody plants, and often cause serious damage, especially to such cultivated plants as **tobacco, potato** and **sugar-cane.** The number of virus diseases is great, but the host range for any one virus is usually rather strictly limited. Virus diseases are classified by the effects they produce rather than by their properties. These effects vary greatly with the different diseases. Leaves of the tobacco plant when infected with tobacco mosaic disease become spotted with light areas, and are rather stunted. Mosaic disease of the cucumber causes the leaves to become irregularly mottled and somewhat wrinkled. Virus diseases may not be fatal but do greatly reduce the vigor of the plant and also diminish the value of the crop produced by such plants.

These diseases are transmitted from plant to plant through wounds. Very commonly sucking insects serve as carriers, transferring the virus, which seems able to exist within the insect body, to another plant which becomes infected through the insect bite. Once within the plant the virus spreads rapidly. In some forms the virus seems to be carried into the seed and to infect the new plant when the latter develops from the seed. A great deal of experimental work has been done on the causative agents of virus diseases, which has shown among other things that they are remarkably tolerant

to substances which would destroy other organisms, to high temperatures, and prolonged desiccation. (B.S.M., R.M.W.)

VISCACHA. Mammalia, Rodentia. *Viscaccia.* A large stoutly built burrowing animal of the South American pampas, related to the chinchillas. The contrast between these animals has been likened to that between our related squirrels and woodchucks, the one gracefully built and the other a clumsy burrowing form. (A.W.L.)

VISCERA. Organs lying more or less freely in the cavities of the body. Usually applied to the **heart** and **lungs** as thoracic viscera and to the **stomach, intestines, spleen, liver** and **pancreas,** and some of the **reproductive** organs as abdominal viscera. The singular form, viscus, is rarely met. (A.W.L.)

VISCERAL ARCH. The column of tissues persisting between adjacent **gill slits** in the wall of the vertebrate **pharynx.** The arch is lined with endodermal tissue and covered outside with ectodermal. It contains a bony or cartilaginous support belonging to the **visceral skeleton** and an **aortic arch.** In the fishes the gills are supported by these arches. (A.W.L.)

VISCERAL CLEFT. Grill Slit.

VISCERAL MASS. A compact mass of tissue containing some of the internal organs of the **mollusks.** It forms the main part of the body. (A.W.L.)

VISCERAL SKELETON. A portion of the vertebrate skeleton supporting the walls of the **pharynx** and forming the jaws and hyoid apparatus. In the primitive condition it consists of a series of bony or cartilaginous arches extending from the ventral wall of the pharynx upward in the **visceral arches.** Seven pairs of these structures persist in the fishes. The most anterior of the visceral arches forms part of the upper and all of the lower jaw, supported by skeletal structures derived from this source. The skeleton of the second arches and variable derivatives of those following constitute a hyoid apparatus supporting the base of the tongue. Other components of the visceral skeleton become the cartilages of the larynx and upper part of the trachea.

One of the most remarkable transformations of this part of the skeleton is the migration of small bones of the first and second arches into the middle **ear** as the hammer, anvil, and stirrup which bridge the cavity in the mammals. (A.W.L.)

VISCOMETER. Viscosity.

VISCOSITY. A property of fluids, either liquid or gaseous, which may be briefly described as a resistance to flow. When a solid is subjected to shear, a **stress** develops in it, increasing with the shear until a condition of static equilibrium is reached by virtue of the **elasticity** of the solid. On the other hand, when a fluid starts to flow, while there is likewise a **shearing** stress, the opposing reaction is of the nature of an internal friction, and the equilibrium attained when the flow becomes steady is brought about by the viscosity of the fluid opposing the shearing motion. As one layer of the fluid moves past an adjacent layer, the interaction of molecules at the boundary and the passage of molecules across the boundary both ways results in the transmission of energy from the faster to the slower, and hence tends to check the relative motion. This reaction is proportional to the rate of shear, thus:

$$\text{shearing stress} = \eta \times \text{rate of shear.}$$

The constant η is called the viscosity coefficient of the fluid. For liquids, it decreases, in general, with rise of temperature; with gases, it increases. Thus, molasses becomes more fluid but air becomes less so, at a higher temperature.

When a fluid flows through a capillary tube because of a pressure difference Δp at the opposite ends, the volume per unit time is given by Poiseuille's law:

$$\frac{V}{t} = \frac{\pi r^4 \Delta p}{8\eta l},$$

in which r is the radius and l the length of the capillary. The viscosity coefficient may thus be measured, since it is easy to measure r, l, and Δp and to observe the rate of flow V/t. Several types of viscometer depend upon this principle.

Stokes derived an expression for the force necessary to keep a small sphere of radius r moving at a uniform speed v through a given fluid: $f = 6\pi\eta vr$; an expression very useful in studying drops or other spherical particles falling freely in the air. (L.D.W.)

VISCOSITY MANOMETER. Pressure Gauges.

VISE. Clamping and holding tools used in machine shops. Essentially, a vise consists of one stationary and one movable jaw, adjusted by means of a screw and nut. Bench vises are used for hand operations, such as chipping and filing, and are usually fastened to a shop bench. Hand vises are small tools fitted with a handle, and are designed to be held in one hand while the other holds a small file, engraving tool, etc. Machine vises are used in milling, drilling, and other machine tool processes, and are usually constructed so that they may be clamped to the machine table by clamps or **tee bolts.** Machine vises are usually made with replaceable jaw plates, so that special or false jaws may be substituted. (See **Milling.**)

Machine vises used for mass-production work are sometimes made with cam-actuated jaws, to permit quick-acting operation; vises actuated by compressed air are also extensively used. (H.C.H.)

VISIBILITY. That greatest distance toward the horizon at which an unaided normal eye can clearly distinguish prominent objects without aid of optical devices. In measuring visibility, officially, it is necessary that the value given exist over more than half the horizon. (P.E.K.)

VISIBILITY FACTOR. Photometry.

VISION. The formation of mental images of the shape and color of objects through the reception of light rays reflected from their surfaces to sensory organs known as **eyes.** While some organs included in this category are supposed to be merely sensitive to light and capable of perceiving its intensity and the direction of its source, others are either known or supposed to form sharply defined images like our own.

The compound eyes of **arthropods** are supposed to form fairly definite images by a process known as **mosaic vision.** The interpretation of such organs is necessarily theoretical since they are conspicuously different from our own. Their action is certainly very different from that of the camera eye of man.

For, from the optical standpoint, the human eye is merely a **camera,** with the retina substituted for a photographic plate. The optical system, however, extends continuously from the cornea back to the retina, instead of being localized in an objective. It consists of four transparent media: cornea, aqueous humor, "crystalline" lens, and vitreous humor, bounded by curved interfaces. The crystalline lens is of non-uniform refractive index, and is flexible like rubber, so that its shape and focal length can be controlled by a set of muscles; it is in this way that the focusing or "accommodation" is accomplished. The lens casts a sharp inverted image on the sensitive **retina.** Each rod or cone

affected by the light sends an impulse to the visual centers in the brain. Here the mental image resulting from the total stimulation of the retina is formed.

The normal human eye forms clear images of objects 18' away or more without effort. Closer objects require accommodation by the contraction of the ciliary muscle, which permits the lens to become more convex. For this reason close vision is more tiring than distant. Cephalopods and fishes on the contrary must accommodate for distant vision. The change is accomplished by the contraction of muscles which draw the retina and lens closer together. In amphibians and reptiles the lens is moved farther from the retina for close vision.

In the human eye a blind spot occurs at the point of entrance of the optic nerve where only nerve fibers exist. A depression in the retina at the axis of the eyeball, called the fovea, is the point of most acute vision. Here very few nerve fibers intervene between the nerve endings (rods and cones) and the source of light. Certain areas of the retina are also different in their sensitiveness to light of various lengths. Although color vision is conditioned by a number of variations in stimuli, the retina has certain normally characteristic visual fields. Red and green are perceived by a limited central area and blue and yellow by this area and a surrounding extension. The marginal field perceives only black and white and the grays.

The perception of size, shape, and distance depends partly on experience and partly on comparison with other known objects in view. The relative distances of objects are determined by stereoscopic vision through the mental association of slightly different images formed by the two eyes. Animals whose eyes are directed outward from the sides of the head are incapable of this type of vision. The same principle is used in stereoscopic photography. Two pictures made simultaneously from slightly separated points of view are examined through the stereoscope so that each eye sees one member of the pair. The difference in images may also be noted by closing or covering one eye at a time while looking at some object. (See **Binocular Vision.**)

Defects of vision in man include various forms of color-blindness, due to hereditary abnormality of the cones in the retina, and malformations of the eyeball that interfere with the formation of sharp images. When a normal eye is relaxed the image focuses directly on the retina; such an eye is called emmetropic. If the ball is slightly elongated the rays focus slightly in front of the retina during relaxation and the individual is said to be short-sighted or myopic. In the opposite condition light rays come to a focus behind the retina of the short eye, which is said to be long-sighted or hyperopic.

The commonest form of color-blindness is the red-green type. The condition varies but in general red and green register as different degrees of yellow. A more extreme form called monochromatic color-blindness results in the perception of all colors as shades of gray. (See **Color.**)

Other adjustments of vision related to the intensity of illumination have been recorded. They involve various structural adaptations of the eyes to secure maximum stimulation for dim light and the familiar contractile iris to cut down the amount of light entering the eye. (A.W.L., R.S.M.)

VISUAL BINARIES. A visual **binary star** is one for which the angular separation between the two components is great enough to permit the system to be observed as a **double star** in a telescope. The **resolving power** of the **telescope** employed is an important factor in the detection of a visual binary and as telescopes of larger and larger aperture are built there will be an ever increasing number of visual binaries discovered. Also, the brightness of the objects is an important factor in the detection of the double character of a star, it being easier to see as separate objects two faint stars separated by a small angular distance, than two bright stars separated by the same angular distance. Adopting certain arbitrary definitions as to what shall be considered a visual binary, Aitken estimates that about 1 star out of every 18 is a visual binary.

Visual binary stars are studied from observations taken either with a **filar micrometer** or a **stellar interferometer.** The brighter star of the pair is known as the primary and the fainter as the secondary. The **position angle** of the secondary with respect to the primary is measured, together with the angular distance between the two components. The time of the observation is also recorded. After a sufficient number of observations have been obtained they are plotted in polar **coordinates,** using the primary star as origin. Through these plotted points the most probable ellipse is drawn, the only restriction on the ellipse being that the **Keplerian Law of Areas** must be satisfied. The ellipse thus drawn is known as the apparent ellipse.

This apparent ellipse is the projection of the actual elliptical **orbit,** of the secondary with reference to the primary, on the plane perpendicular to the line of sight of the observer. From this projected ellipse the complete **elements** of the orbit may be computed, the semimajor axis, a, being expressed in **angular units** unless the **stellar parallax** of the system is known. (w.k.g.)

VITAMINS. A group of substances which are present in varying amounts in certain animal and plant tissues. Vitamins are necessary for normal nutrition, growth, and function of the body. The quantities needed are often extremely minute, but since the body cannot (with a few relative exceptions) manufacture its own supply, vitamins must be supplied in the diet. Their absence or partial deficiency produces various characteristic diseases and disturbances of body growth and function.

Biological, rather than chemical, methods are responsible for the discovery of vitamins, and for much of our early knowledge of them. These methods proved the necessity for their supply to the organism, as well as the properties of vitamins, a summary of which is given in the accompanying table.

In this country, extreme deficiency states of a particular vitamin are not as common as they are in other parts of the world. But milder forms of deficiency diseases, especially rickets and pellagra, are not infrequent. Deficiency in diet is not the only factor involved, for disturbances in absorption from the gastrointestinal tract may result in deficiency states even though adequate quantities of vitamins be supplied in the diet. The requirements for certain vitamins are enormously increased during certain periods of life—during the ages of active growth, and during pregnancy and lactation.

Vitamin A is a fat-soluble vitamin necessary for growth. Another function is concerned with vision in dim light. Such vision is mainly a function of specialized cells, the rods of the retina, and is dependent on an adequate supply of the pigment, visual purple, a compound of vitamin A and a protein. (See **Vision.**) One of the earliest manifestations of vitamin A deficiency is, therefore, night blindness, or diminished vision in dim light. The vitamin is also necessary for the maintenance of the epithelial tissues (**epithelium**) in a normal state. In its absence, specialized epithelial cells undergo **atrophy,** and change to a more primitive type; this may seriously interfere with the function of the involved part. Thus the surface of the **conjunctiva** and **cornea** of the eye

VITAMINS

Designation	Date of Discovery	Properties	Main Effects of Deprivation in Man	Dietary Sources
A	1913	Fat-soluble, relatively Heat stable, relatively Well stored by the body	Night blindness, xerophthalmia, dermatosis	Fish oils, liver, egg yolk, butter, carrots and leafy green vegetables
The B complex B₁ thiamin	1926	Water-soluble Heat Stable Poorly stored by the body	Beri-beri, polyneuritis, cardiac enlargement and failure, loss of appetite	Lean pork and beef. Peanuts, dried peas and beans, oatmeal, wheat, green peas, and beans
Nicotinic acid	1867 1939	Water-soluble Heat stable	Pellagra, dermatitis, stomatitis, gastro-enteritis, psychosis	As thiamin, plus milk, egg yolk
B₂ (G) riboflavin	1933	Water-soluble Heat stable	Cheilosis and angular stomatitis, glossitis, dermatitis	Liver, lean meats, egg yolk, peanuts, pears, peaches, milk, whole grain wheat, carrots, peas, spinach
C Ascorbic or cevitamic acid	1932	Water-soluble Heat stable	Scurvy	Citrus fruits, tomatoes, turnips, sweet potatoes, leafy green vegetables
D	1922	Fat-soluble, relatively Heat stable, relatively Well stored in the body	Rickets	Fish liver oils, egg yolk, butter, cream
E Alpha-tocopherol	1922	Fat-soluble Heat stable	Habitual abortion, neuro-muscular disorders	Wheat germ oil, leafy vegetables, eggs
K	1939	Fat-soluble Heat stable. Not well stored	Hemorrhagic disease of the newborn, bleeding associated with obstructive jaundice, sprue, etc.	Kale, spinach, carrot tops, tomatoes, liver

become hard, dry, and may ulcerate. Xerosis, xeroph-thalmia, and keratomalacia are the diseases which represent various stages in this process. The epithelium of the skin and its specialized structures are affected too: the skin becomes rough, dry, scaly, and the sweat glands dry up. The epithelial cells of the respiratory and genito-urinary tracts may also be involved. Vaginitis of the so-called senile type may be a manifestation of A deficiency.

Vitamin A is absorbed from the intestine and transported by the blood and lymph to the liver, where it is stored. In the liver the precursors of the vitamin, chiefly alpha- and beta-carotene, are converted into vitamin A.

The B group of vitamins is a complex which contains three well-established compounds (thiamin, nicotinic acid, and riboflavin) and several recently discovered constituents whose functions and requirements for man are not yet completely understood—(pyridoxine, folic acid, pantothenic acid, biotin, inositol, anti-gray hair factor, etc.). Only the three whose properties and functions are well known will be discussed here.

Vitamin B_1, thiamin, the deficiency of which results in beri-beri was the first to be differentiated from other members of the B complex. It has been isolated and synthesized in pure crystalline form. In practice it is used in the form of the hydrochloride. It is present in both blood cells and serum in small quantities, and is excreted in the urine in amounts which reflect intake and storage. In animals, it has been demonstrated to be abundant in liver, heart, kidneys and voluntary muscle.

Vitamin B_1 plays an important role in the fundamental processes of oxidation in the body. In the oxidation of carbohydrate it is active as a coenzyme—carboxylase. Although much is known of the chemistry of thiamin in the body, little is known of the mechanism by which the clinical picture of B_1 deficiency is produced. The vitamin seems necessary for the maintenance of the health of nerve tissue, intestinal and cardiovascular function, appetite, and growth. Lack of thiamin produces beri-beri, polyneuritis, dilatation and hypertrophy of the heart, edema, and muscular atrophy. Alcoholics are prone to thiamin deficiency, because of their tendency to substitute alcohol for the normal diet. Peripheral neuritis is seen most commonly in alcoholics. All of these manifestations respond promptly to adequate therapeutic doses of thiamin.

Nicotinic acid, although discovered in 1867, was not known to be important for man until 1939. Deficiency of nicotinic acid produces pellagra in humans, and "black tongue" in dogs. Nicotinic acid in the form of the amide forms a part of one of the essential enzyme systems, used for transferring oxygen in oxidations concerned with fundamental life processes in the cell. How this function is related to the physiological and structural changes (dermatitis, glossitis, proctitis, dementia) resulting from its deficiency is not known. The vitamin is widespread in body tissues and is stored relatively well. It is available in pure form as nicotinic acid and nicotinic acid amide for clinical use. In addition to its use as a specific for pellagra, the drug is now being used because of its vasodilating action in angina pectoris, intermittent claudication (cramping pain in the legs on exercise, due to a diminished blood supply), and thrombophlebitis.

Vitamin B_2 (G) is a complex pigment with a green fluorescence called riboflavin or lactoflavin. The clinical picture resulting from deficiency of this vitamin was first described in 1938; the outstanding findings are cheilosis, stomatitis at the angles of the mouth, seborrheic dermatitis, glossitis, and vascularization of the cornea. The lip and mouth lesions have been familiar to clinicians for many years, but their relation to a vitamin deficiency is recent knowledge. Riboflavin deficiency frequently accompanies pellagra and the typical lesions of both nicotinic acid and riboflavin deficiency are often found in that disease.

Riboflavin, like nicotinic acid, forms an oxidation enzyme and, as such, acts as an oxygen carrier to the cell. The mode of production of the clinical picture of deficiency is again unknown. The vitamin is widely distributed in the cells, and rather large amounts are excreted in the urine. Little is known of its absorption and storage. Riboflavin is available in pure form for clinical use. It is, however, very unstable to light and air, and solutions must be used soon after preparation.

Vitamin C is ascorbic or cevitamic acid. Its deficiency results in scurvy, a disease characterized by weakness, anemia, swelling of the gums and multiple hemorrhages. The vitamin is a powerful oxidation-reduction agent. It has been isolated in pure form. It cannot be synthesized by the human body, although there is considerable difference among species in this respect: rats, chickens, and dogs can synthesize it, while guinea pigs cannot. When supplied in the diet, it is completely absorbed in the intestine and distributed widely throughout the body. Its concentration in the blood reflects the concentration in the tissues and forms the basis of a laboratory test to measure its deficiency.

Vitamin C is necessary for the maintenance of the so-called intercellular cement substance of mesothelial tissue or structures derived from it, such as connective tissue, bone, capillaries, etc. In C deficiency these tissues become abnormally fragile, and hemorrhages are common, especially in the gums and into the joints. All signs and symptoms respond promptly to treatment with ascorbic acid.

Vitamin D is a fat-soluble vitamin which is necessary for the absorption of calcium and phosphorus, and is concerned directly in the metabolism of bone. It is a specific in the prevention and treatment of rickets, childhood tetany, and osteomalacia. It is necessary for the formation of bone and teeth, and large amounts are therefore required in childhood. In some unknown way, it is related to the action of the parathyroid glands. Vitamin D is not a single substance, but only 2 of the 10 or 11 sterols known to have antirachitic properties are of importance medically. These are activated ergosterol (Viosterol) and 7-dehydro-cholesterol. The D substances are activated sterols, complex substances often closely asociated with fats in plants and animals. Cholesterol is an important sterol which is present in the skin; activation of cholesterol by ultraviolet light changes the compound and produces vitamin D. This is the way in which man obtains the greater part of his vitamin D, relatively little being taken in with food. The vitamin can be stored in the liver. It is toxic in very large doses far outside the range for therapeutic use.

Vitamin E is a fat-soluble vitamin essential for reproductive processes in certain animals; its necessity for man has not been established. In the female white rat, deficiency of vitamin E results in premature death and resorption of the fetus in utero. The possible corresponding condition in humans is habitual abortion. The evidence that the two situations are related by a common etiology is now less convincing than formerly, and there is considerable doubt that vitamin E plays any role in humans. The relationship to certain of the neuro-muscular disorders is still under investigation.

Vitamin E is a tocopherol, a solid alcohol present in relatively large amounts in wheat germ oil, from which it was first isolated. It has been isolated in pure form and has been synthesized. There are several tocopherols, alpha being the most active of the naturally occurring ones. The mechanism by which vitamin E produces its effect is not known, nor is it known whether it is important in biologic processes in the cell.

Vitamin K is a recently described factor necessary for the adequate production of prothrombin, a substance which is present in the blood and plays an active role in the clotting mechanism. Vitamin K deficiency results in a low prothrombin level and is therefore associated with several hemorrhagic states: the bleeding tendency accompanying obstructive **jaundice**, hemorrhagic disease of the newborn, and bleeding associated with diseases of the gastrointestinal tract, such as **sprue**, **celiac disease**, etc.

Vitamin K is a substituted derivative of naphthoquinone and occurs in several forms. It has been isolated in pure form and synthesized and is available for clinical use. As a preventive of hemorrhagic disease of the newborn, it is highly effective and is now often given routinely to mothers just before delivery.

STRUCTURAL CONFIGURATIONS OF VITAMINS

Vitamin A:

Vitamin B₁ (Thiamine Chloride (the hydrochloride)):

Nicotinic acid:

Vitamin B₂ (G) (Riboflavin):

Vitamin C (Ascorbic or Cevitamic acid):

Vitamin D (Irradiated Ergosterol, or Calciferol):

Vitamin E (Alpha-tocopherol):

Vitamin K:

(R.K..S, D.M.H.)

VITELLARIUM. A yolk-forming organ. The term is applied in the flatworms (**Platyhelminthes**) to a long series of glandular bodies associated with the **oviducts**. They form both the yolk and the shells of the eggs. The same name designates a portion of the ovary of **rotifers**. This organ is divided into a germarium where the egg cells are produced and the larger vitellarium. (A.W.L.)

VITRAIN. A term proposed by M. Stopes, in 1919, for a glassy variety of **coal** which occurs in bituminous coal as bright narrow and easily friable bands which may be distinguished from **clarain**, especially with the aid of the microscope. (R.M.F.)

VITRELLA. A crystal cell of the group between the cornea and the retinula in the compound **eye** of **arthropods.** (A.W.L.)

VITREOUS ENAMEL. Porcelain Enamel.

VITRIFIED BOND. A molded grinding wheel made from clay or flint, abrasive, and water, burned at a high temperature for several days. (H.C.H.)

VITRIOL. Term applied to **sulfates**. Oil of vitriol, concentrated **sulfuric acid**; blue vitriol, **copper** sulfate crystals; green vitriol, **ferrous** sulfate crystals; white vitriol, **zinc** sulfate crystals. (R.K.S.)

VITROPHYRE. A vitrophyre is a volcanic glass carrying sporadic distinct crystals of **feldspar** and other minerals; in short, a **porphyritic** glass. (E.S.C.S.)

VIVIANITE. The mineral vivianite is a hydrous **iron phosphate**, $Fe_3P_2O_8 \cdot 8H_2O$, its **monoclinic** crystals are usually prismatic or bladelike but may be in massive forms. Vivianite has one perfect **cleavage**; hardness 1.5–2; specific gravity, 2.58–2.68; luster, pearly on cleavage faces, otherwise vitreous; colorless, when freshly exposed, but becoming blue or brownish with the alteration of the ferrous to ferric iron; transparent to translucent. Vivianite is an associate of **pyrrhotite**, **pyrite** and **copper** and **tin** ores. It is found also in clay beds forming the so-called "blue iron earth" which is common and of wide distribution in peat bogs. Vivianite is found in Rumania, Bavaria, Cornwall in England, Australia, Bolivia, Greenland and elsewhere in Europe. In the United States it occurs in New Jersey, Delaware and Colorado. This mineral was named by Werner after the English mineralogist J. G. Vivian, its discoverer. (E.S.C.S.)

VIVIPARITY. A reproductive process involving the internal nourishing of the young by the body of the mother during the early stages of development and their birth when sufficiently advanced to carry on essential processes of life. It contrasts with oviparity, in which reproduction is accomplished by the formation and discharge of eggs, and ovoviviparity, in which the eggs hatch in the body of the mother, but there are no special adaptations for the direct prenatal nourishing of the young.

Both viviparity and ovoviviparity necessitate internal **insemination.** They are not limited to particular groups of animals but occur in many forms of invertebrates and vertebrates, including roundworms, rotifers, insects, fishes, reptiles and mammals. Although most of the forms below the mammals are probably ovoviviparous, there is good evidence for the interpretation of some of the parasitic flies as truly viviparous, since the larvae are produced just prior to their transformation into pupae. True viviparity is at its maximum, however, in the mammals, but even here a transition occurs from the oviparous monotremes to the true mammals, including man. The young are nourished prior to birth by interchange with the blood stream of the mother through the **placenta.** (A.W.L.)

VIVISECTION. The dissection of living animals. The practice of vivisection has aroused so much emotional opposition that it has been widely publicized in the daily press. In the strict sense, scientists do dissect living animal to learn of processes taking place in their bodies more accurately than is possible without this procedure. Such studies are conducted, however, with the greatest possible humanity. Animals are anesthetized for operative procedures and are killed painlessly at the end of experimental study. If the destruction of life is to be regarded as cruelty, then our use of domestic animals as food is on a par with vivisection.

The study of processes and relations within the body of the animal while it is still a living organism has been of incalculable benefit to medicine and has been directly responsible for the saving of many human lives. As an example among recent discoveries, the functions of the adrenal cortex were discovered through experiments with cats from which the glands had been removed. The animals were kept alive by administering extracts of the adrenal cortex of other species, and the final perfection of these extracts so that they could be administered safely to human beings provided the first alleviation for Addison's disease. (A.W.L.)

VOGESITE. A term proposed by Rosenbusch, in 1887, for a **syenitic lamprophyric igneous rock** in which the characteristic minerals are generally **hornblende** or **augite** and **oligoclase** or **andesine** feldspar. (R.M.F.)

VOLATILE. Coal, Proximate Analysis.

VOLATILE OILS. The volatile oils are distinguished from the **fixed oils** by the fact that a drop of one of these oils does not leave a spot on paper. Members of certain plant families, such as the **Mint Family,** contain a larger percentage of such oils than do other families. But volatile oils are in no sense restricted to any small group, nor are they found only in certain tissues. Sometimes, certain parts may be principally used for the oils, as the seeds of the **Carrot Family.**

Various methods are used in extracting the oils from the plant tissue. Many are distilled with water or steam, the oil being carried over with the distillate. In others, as for example oil of bitter almonds, the oil develops in the tissues only after **fermentation.** It is then obtained by **distillation.** Another method, and one especially used for more delicate and valuable oils, is called "enfleurage." In this method the flowers containing the oil are spread as a thin layer over a layer of lard or olive oil. The latter absorbs the delicate

oil in the flowers, after which distillation may separate the volatile oil from the other.

Volatile oils are much used as perfumes, flavorings, drugs and solvents. Attar of Roses, or Rose oil, from **Roses,** is one of the most valuable. Jasmine oil, from *Jasminum grandiflora* (Oleaceae) petals is somewhat less valuable. It is obtained by absorbing the oil from the petals of the plant in olive oil. Geranium oil, from several species of *Pelargonium,* cultivated mainly in Southern France and Spain, is much used to adulterate rose oil. Bay oil, from the leaves of *Pimenta acris* (Myrtaceae), a native of the West Indies, is much used in perfumes, toilet preparations and bay rum. From the leaves and flowers of many members of the Mint Family fragrant perfume oils are obtained: *Lavendula vera,* used by the Romans to scent their baths, gives lavender oil, an expensive perfume oil, while *L. spica* yields Spike oil, a cheaper oil than lavender. Dried lavender plants are frequently used in bedding and clothing to impart to them a delicate fragrance. Another mint, *Rosmarinus officinalis,* yields oil of rosemary. Certain grasses also yield fragrant oils. Among these are *Cymbopogon citratus,* a native of India and Ceylon, from which is obtained lemon grass oil, used as an adulterant for lemon oil, as well as for its own fragrance, and *Cymbopogon Nardus,* from which citronella oil is obtained. The latter is used as a repellent for mosquitoes and other insects. These two grasses are cultivated to a limited extent for their oils. Verbena oil is a valuable product from *Verbena triphylla* and other species (Verbenaceae). Neroli oil is obtained from the flowers of the orange tree (Rutaceae); this oil is used in cologne and in liquors. From another member of this family comes **Bergamot oil.** The Laurel Family (Lauraceae) has many members which yield volatile oils. From *Cinnamomum zeylanicum* come cinnamon oil and cinnamon-leaf oil, used not only in perfumes but also in medicines and as a flavoring. The green leaves of *Cinnamomum Cassia,* a native of the East Indies and India, give cassia oil, an expensive and consequently much adulterated oil used in perfumes and medicines. The bark and roots of *Sassafras officinale,* a tree native in southern United States, yield sassafras oil, used in making cheaper perfumes. From *Cananga Odorata* (Anonaceae) comes oil of cananga, called also Ylang ylang, "Flowers of Flowers," a very expensive oil. The tree is a native of southeastern Asia. Various species of Iris, particularly *Iris germanica,* yield orris root and orris oil, the dry powdered rhizome being used. The powder has an odor suggestive of violets, and is used in compounding medicines, sachets and tooth powders. Southern Italy leads in the production of orris root.

Among volatile oils are many primarily used in flavoring and as drugs. Many of these come from members of the Mint Family. One of them is oil of peppermint from the leaves and stems of *Mentha piperita.* The states of Michigan, Indiana, and New York grow much of the native crop, but are far behind Japan in production. Oil of peppermint is mostly menthol, and is used is flavoring chewing gum, tooth pastes, and as an inhalant and perfume. It is commonly adulterated. From *Mentha spicata,* or Spearmint, comes oil of spearmint, used as a drug, as a flavoring in cooking, and in preparing mint sauce, in chewing gums and in cheaper perfumes. The principal producing regions are New York, Michigan, India and Russia. Clove oil is obtained from the flower buds of the Clove tree, *Eugenia aromatica.* It is used as a drug and for flavoring and in perfumes. *Artemisia absinthium,* one of the **Composite Family,** yields oil of wormwood, used in medicine, as a worm repellent, and in the preparation of absinthe. Oil of anise seed is obtained from the seeds of *Pimpinella Anisum,* a member of the **Carrot Family** indigenous to Egypt and now widely cultivated. The oil is used in perfumes, as a drug, and in liqueurs. Anise seed oil is also used to make a trail which will be followed by foxhounds. Many other members of the **Carrot Family** have aromatic oils in the seeds. Star anise, so named because of the star-shaped fruit, from *Illicium verum* (Magnoliaceae), is a similar oil. Camphor oil is obtained from the **Camphor** Tree, *Cinnamomum Camphora.* Lemon oil comes from the skin of the lemons, *Citrus Limonium.* Bitter almond oil is extracted from *Prunus communis,* of the **Rose Family.** Juniper oil is obtained from *Juniperus communis,* a Conifer. It is used in medicine, also in varnish making, and in gin. Among leguminous plants the genus *Eucalyptus,* especially *E. globulus,* yields from the leaves oils known as eucalyptus oils, used in making perfumes, as antiseptics, in scented soaps and toilet preparations, also in concentrating ores by the flotation process. The trees are natives of Australia, and have been introduced into California.

Oil of turpentine, also called spirits of turpentine, is obtained by the distillation of exudates of *Pinus ponderosa* and *P. Taeda* (Coniferae). It is used as a solvent in making paints and varnishes, and also in medicine. Venetian turpentine, obtained from the European Larch, *Larix europea,* is a similar product. (See **Hydrocarbons.**) (R.M.W.)

VOLCANIC BOMB. Lapilli.

VOLCANITE. A term proposed by Hobbs, in 1893, for a volcanic rock from the Lipari Islands largely composed of the minerals **anorthoclase** and **augite.** (R.M.F.)

VOLCANO. A volcano is a conical mountain built up around a vent in the crust of the earth. It is formed of lava and fragmental material which has flowed out in a highly heated and liquid state, or from matter ejected by explosive eruptions, or both. It is convenient to classify volcanoes as of three types: 1. The explosive type from which solid fragmental material and gases

Pelée on the Island of Martinique, West Indies, which occurred in 1902. The latter eruption was without lava and consisted of a great cloud of incandescent gases and dust, which destroyed the city of St. Pierre and almost its entire population of about 28,000 inhabitants. Mauna Loa and Kilauea in the Hawaiian Islands are examples of the effusive type of volcano and Mt. Vesuvius is an excellent example of the intermediate type. Most of the active Alaskan volcanoes are of the intermediate type as were doubtless the now-extinct volcanoes of the northwestern part of the United States which include: Mt. Shasta, Mt. Rainier, Mt. Hood, Mt. Baker, and others.

The explosive type may be recognized by the steep cone (often seen with the intermediate type as well); the effusive type of volcano with its long lava flows builds a low cone of great areal extent as compared with its height. (E.S.C.S.)

VOLE. Mammalia, Rodentia. The meadow or field mice, constituting the genus *Microtus.* The genus is limited to the northern hemisphere and in Asia does not extend south of the Himalayas. Voles are characterized by rootless molar teeth formed of two rows of alternating triangular prisms. About 20 species occur in North America. (A.W.L.)

VOLT. The volt, which is the practical unit of **electromotive force** and **electric potential,** may be defined in different ways, some of which are equivalent, while others give rise to slightly different values. Perhaps the simplest definition of the "absolute volt" is that electromotive force or potential difference against which one watt of power is necessary to maintain an electric current of 1 (absolute) **ampere;** or against which one joule of energy is necessary to transfer an electric charge of one (absolute) **coulomb.** The "international volt," on the other hand, is defined in terms of the international ampere and international **ohm** in accordance with **Ohm's**

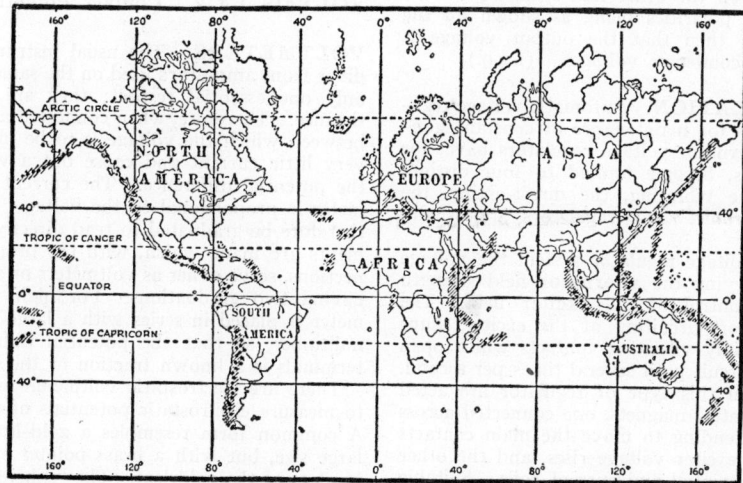

Map showing the distribution of active and recently extinct volcanoes. (*Tarr's New Physical Geography,* The Macmillan Co.)

are erupted. This material may consist of blocky pieces, often ejected in a partly fluid condition, *lapilli,* or dust, the latter often incorrectly called ash. 2. The effusive type, characterized by quiet eruptions of liquid lava with little or no explosive violence. Such lavas are very fluid and frequently flow for many miles. 3. The intermediate type which may at times erupt explosively with accompanying flows of lava. Typical highly explosive eruptions have been: that of Krakatao in the Straits of Sunda, Dutch East Indies, which in 1883 blew up about a cubic mile of rock which rose as dust over 15 miles in the atmosphere; and that of Mt.

law (one volt maintains a current of 1 ampere through a 1-ohm resistance). The international volt exceeds the absolute volt by about 34 parts in 100,000. (See also **Standard Cell.**) One may get a fair idea of the magnitude of 1 volt by noting that the electromotive force of a 3-cell automobile battery is about 6 volts, and that of the ordinary electric light supply is 110 volts. (L.D.W.)

VOLTAGE DIVIDER. The ordinary three-terminal resistance may be used as a voltage divider. If no current is taken from the intermediate tap the voltages will be in proportion to the resistance included between the taps. In many applications a potentiometer is used as a voltage divider so the voltage may be adjusted by

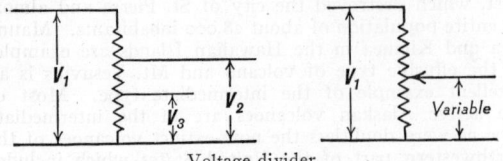

Voltage divider.

varying the position of the tap. If current is drawn from the tap the exact distribution of the voltage is altered but the total voltage applied across the divider is still divided between the various sections. The figure shows some typical arrangements. (L.R.Q.)

VOLTAGE DOUBLER. This is a connection of condensers and rectifiers across an a-c source which gives a d-c output voltage approximately twice that of the normal connection. The exact value of the output voltage depends upon the load and for very high loads may not even approach the theoretical double value. A circuit is shown in the figure. On one half-cycle C_1 charges

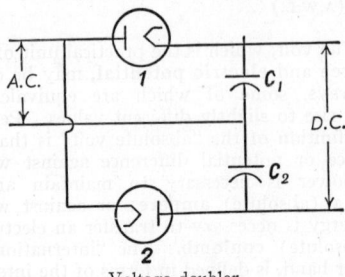

Voltage doubler.

through tube 1 and on the other half-cycle C_2 charges through tube 2, the polarities being as shown on the diagram. It is seen then that the output voltage is the sum of the two condenser voltages. (L.R.Q.)

VOLTAGE REGULATION. Automatic voltage regulators are relied upon for maintenance of constant **generator** voltage. Alternator-voltage regulators have, for all practical purposes, become limited to four distinct mechanical types, the vibrating, the direct-acting, the rheostatic, and electronic. D-c regulators are usually rheostatic.

In the vibrating voltage regulator system the voltage is maintained by varying the **alternator** field strength indirectly through control of the **exciter** field. The basic idea is the short-circuiting of the exciter shunt field rheostat by rapidly vibrating contacts which open and close the short-circuit path several times per second. The main contacts in this type of regulator are acted on by two sets of control magnets, one connected across the exciter bus and tending to move the main contacts farther apart as the exciter voltage rises, and the other acted upon by a-c potential and current coils. Suitable springs and counterweights allow adjustment to be made. When the main contact closes it energizes the relay magnet, thus closing the relay contact, short-circuiting the exciter rheostat, raising the exciter voltage, and consequently, the alternator voltage. The use of the exciter voltage as one of the main control circuits prevents the alternator voltage "overshooting."

Compensating current winding of the a-c **solenoid** is provided with a dial switch to give any amount of compensation required for the feeder circuit in which the current transformer is located. The vibrating-type regulator may be applied to d-c as well as a-c systems. The exciters should be selected with characteristics known to function satisfactorily with the type of regu-

lators used. The regulators may be operated in parallel, also one regulator may control more than one generator through the use of special exciter rheostats. Successful operation of individually regulated exciters in parallel depends upon control of the wattless current which may circulate between the alternators as a result of momentary differences in excitation. Control of this feature is worked out on the basis of alternator power factor.

In direct-acting regulators, an induction motor principle furnishes the actuating impulse. The torque produced is counteracted by a spring (and the exciter field rheostat is an integral part of the regulator). The rheostat arm has pure rolling motion, hence very little effort is required for the voltage regulating motion. A damping mechanism consists of a disk and magnets.

Unlike the first two types, the rheostatic regulator can be used in plants where the excitation is taken normally from a constant-potential bus. The rheostatic regulator does what an operator would do, except that it does it more quickly and provides instantaneous correction to standard voltage. Rheostatic regulators should be used in the case of large slow-speed exciters, the magnitude of whose field current would prove embarrassing to the vibrating-type regulator. It is also applied where exciter field control would give too slow a response to the control impulse.

The electronic regulator employs electron tubes to control the field current of the generator directly or the exciter field. These tubes are controlled by the output voltage so a rise of voltage decreases the current through them and a decrease of voltage increases the current. By proper adjustment of the circuit the control may be made sensitive enough to maintain essentially constant bus voltage. (F.T.M., L.R.Q.)

VOLTAMETER. Coulombmeter.

VOLT-AMPERE. Alternating Currents.

VOLTA'S LAW. Contact Potential Difference.

VOLTMETERS. The usual instruments of this class differ from **ammeters** used on the same type of service in only one essential respect: they are of very high resistance. Therefore, when connected across the terminals between which the voltage is to be measured, they take very little current and cause but a very slight drop in the potential difference. The current through the voltmeter is proportional to the voltage, and the scale may therefore be graduated to read directly in volts. Instruments are made which, with the proper change in connections, serve either as voltmeters or ammeters, the scale having two graduations. For high voltages, the voltmeter is placed in series with a large resistance, called a multiplier, so that the potential difference between its terminals is a known fraction of the voltage under test.

There are electrostatic voltmeters which may be used to measure electrostatic potentials of thousands of volts. A common form resembles a gold-leaf **electrometer** of large size, but with a brass pointer swinging on a scale in place of the gold-leaf. The "sphere gap" (see **Spark**) may also be used for approximate high-voltage measurements. (See **Instruments, Electrical.**) (L.D.W.)

VOLUME EXPANDER. Expander.

VOLUME INTEGRAL. Triple Integral.

VOLUMES BY DOUBLE INTEGRALS. Consider the solid bounded below by a region S in the XY-plane, with boundary $x = a$, $x = b$, the X-axis and a curve $y = \phi(x)$, bounded above by a surface $z = f(x, y)$, and laterally by planes $x = a$, $x = b$, $y = o$ and the cylindrical surface formed by parallels to the Z-axis through the points of the curve $y = \phi(x)$. The volume of this solid is given by the double integral

$$V = \int_a^b \left(\int_0^{\phi(x)} z\,dy \right) dx = \int_a^b \left(\int_0^{\phi(x)} f(x, y)\,dy \right) dx.$$

(L.L.S.)

VOLUMES BY PARALLEL SECTIONS. If the plane perpendicular to the X-axis at a distance x from the origin cuts from a given solid a section whose area is $A(x)$, then the volume of that part of the solid between $x = a$ and $x = b$ is given by

$$V = \int_a^b A(x)dx.$$

(L.L.S.)

VOLUMES BY TRIPLE INTEGRALS. A volume in general may be found by the triple integral

$$V = \int_{x_1}^{x_2} \left[\int_{y_1}^{y_2} \left(\int_{z_1}^{z_2} dz \right) dy \right] dx,$$

where the limits of the successive integrations are determined by the boundary of the given solid. (L.L.S.)

VOLUMETRIC ANALYSIS. Analytical Chemistry.

VOLUMETRIC EFFICIENCY. Volumetric efficiency is a term applicable to a **piston** and **cylinder** mechanism in which an outward stroke of the piston induces a **vacuum** which draws a gas into the cylinder. This efficiency is of special importance in the **internal combustion engine,** and no complete explanation of engine action is possible without invoking it. Volumetric efficiency may be defined as the weight of gas actually drawn in on an induction stroke, divided by the weight which would occupy the piston displacement under standard conditions of atmospheric pressure and 60° F. If an engine revolved very slowly, and the induction passages were large and unobstructed, the cylinder might be filled with a gas at practically atmospheric pressure, but still the volumetric efficiency could be less than 100% by the heating of this fresh charge through contact with warm manifold and cylinder walls. Since internal combustion engines rotate at speeds from 300 to 3000 rpm, a definite pressure decrement must be expected as necessary to overcome inertia and friction in order to get the cylinder filled with gas in so short an interval of time. Of course, the above refers to normal operation, as it is possible to obtain volumetric efficiencies higher than 100% by **supercharging.**

From the above it will be realized that volumetric efficiency has little in common with **thermal efficiency,** but depends on such factors as the rotative speed of the engine, the fraction of the cycle which is given over to induction, the shape of the ports and valves, and the temperature of the gas. The latter is affected by heating in manifolds, carburetor air heaters, or cylinders, though this may be partially offset by some refrigeration obtained in the vaporization action of the carburetor. (F.T.M.)

VOLUTE PUMP. A volute is a spiral or scroll. A volute **centrifugal pump** has a spiral casing surrounding the impeller, so that as water is discharged uniformly around the periphery of the impeller it will be collected in a chamber of increasing cross-sectional area. In this way the discharge from the rim of the rotating impeller is delivered to the discharge outlet without the necessity of the water velocity near the outlet being higher than average. (F.T.M.)

VOMER. A bone of the vertebrate skull. In mammals it is a thin vertical plate in the posterior part of the nasal septum. (A.W.L.)

VOMITING. The expulsion of the stomach contents through the mouth. This is accomplished by reversal of the direction of the normal waves of **peristalsis** in the gastrointestinal tract. Vomiting is caused by irritation of the stomach or bowel by local irritation, or by stimulation of the vomiting center in the brain by **drugs,** pressure (as by a brain tumor or intracranial pressure) ; or by obstruction in the gastrointestinal tract in the esophagus, stomach, or in the bowel. (D.M.H.)

VORTEX. Aerodynamics.

VORTEX, BOUND. The assumption of a vortex whose filament coincides with the spanwise axis of an airfoil is a useful expedient in deriving aerodynamic equations for lift and drag. This vortex is considered to replace the airfoil and create an equivalent streamline pattern in a rectilinear velocity field. It is not a physical reality over the span but is detected as the two tip vortices which trail downstream and assist in creating the **downwash.** (F.T.M.)

VORTEX LAWS. A vortex is a mass of fluid in which the flow is circulatory. See **Circulation, Aerodynamic.**) The filament or thread of the vortex is the locus of the centers of circulation. Hydrodynamic analysis of fluid flow, attributed to Helmholtz, has produced the following laws governing vortex flow:

1. The strength Γ of a vortex is constant along the filament.

2. The identity of the fluid in a vortex does not alter during the life of the vortex.

3. Filaments have no ending. They are either closed paths, or the ends extend to infinity. (F.T.M.)

VORTEX SHEET, TRAILING. The strength of the circulation component of air flow around an airfoil of finite span cannot be accounted for by the bound vortex alone. Rather, it must be assumed that there are many vortices whose filaments coincide along the span but which occupy varying proportions of the span. As each vortex filament must trail downstream to infinity the vortex pattern takes on something of the appearance sketched in the figure. The numerous filaments

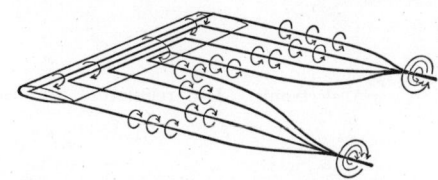

streaming from the trailing edge create a vortex sheet. Derivations based on this assumption of multiple vortices are verified by experiment. The assumption, moreover, disposes of what was at one time a flaw in the circulation theory of lift. At the airfoil tips where the tip vortices turn downstream the downwash velocity induced by the vortex should be infinity since the radial distance from the vortex filament is zero there. But that the downwash is everywhere of small magnitude was well known. Since the strength of multiple vortices is additive the proximity of numerous other trailing vortices to the tip vortex will minimize the downwash at the tips.

The spanwise halves of the vortex sheet represent a surface of discontinuity which is unstable and each half rolls up into a powerful single vortex at somewhere about a span length behind the airfoil. (F.T.M.)

VUG. A rock cavity lined, but incompletely filled, with mineral matter so that a part of the available space remains empty. (E.S.C.S.)

VULTURE. Aves, Falconiformes. A large flesh-eating bird (**Aves**) with a hooked beak but with claws less strongly developed than those of the eagles, hawks, and

Turkey buzzard, *Cathartes aura septentrionales.* Black with brown edging to feathers. Skin of head and neck bare and red.

owls. They feed largely on carrion, but many species are also known to attack living animals.

The Old World vultures belong to a family distinct from the New World species. The latter differ in having the nostrils confluent, so that the beak is perforated transversely. With the exception of the lammergeier, an

Old World species, all vultures have the head and neck almost bare of feathers and sometimes brightly colored.

The turkey vultures or turkey buzzards, *Cathartes aura,* of North America are our most widely distributed representatives of the group, and the condor of South America is probably most widely known for its enormous size. It has been recorded with a length of 4' and a wingspread of 9'. The California vulture or condor, *Gymnogyps californianus,* of the southwestern states and Lower California has also been recorded with a maximum length of 4' or more, and its wingspread is said to reach almost 11'. Although their habits are repulsive, all of these birds are magnificent fliers, soaring for long periods without flapping a wing. (A.W.L.)

VULVA. The area containing the external genitalia of the female corresponding to the **penis** and scrotum in the male. They comprise the two labia majora and the parts lying between them. They are as follows:

The larger labia are two large fatty folds of skin lying between the thighs. Posteriorly they end at the **anus,** anteriorly they end in the *mons Veneris,* a fatty prominent elevation over the pubic bone. They correspond to the **scrotum** in the male. The lesser labia are two smaller similar folds of skin within the fold of the larger labia. Anteriorly they are prolonged over the **clitoris** to form the foreskin or **prepuce.** Just below the clitoris within the labial folds is the opening of the **urethra,** above the **vagina.** Opening into the area about the vagina are the ducts of several glands whose special purpose is lubrication. (R.S.M.)

W

WACKE. An old English term for a dark, greenish-brown clay, a decomposition product of **basalts** and **tuffs.** (R.M.F.)

WAD. The mineral wad, sometimes called bog manganese, occurs in amorphous masses, and consists of mixtures of **manganese** oxides, MnO_2 and MnO and oxides of other metals such as **copper, lead, cobalt, iron,** etc. It is bluish- to brownish-black, usually soft enough to soil the fingers and often porous and light. It is not a distinct mineral species. (E.S.C.S.)

WAGNER BEAM. Tension-Field Theory of Beam Analysis.

WAGTAIL. Aves, Passeriformes. Slender insect-eating birds (**Aves**) of the Old World, related to the larks and pipits. There are numerous species in both hemispheres. (A.W.L.)

WALKING-STICK. Insecta, Orthoptera. A slender wingless insect related to the **grasshoppers.** The walking-sticks are elongate in every part and closely resemble the twigs or stalks of the vegetation on which they live. Together with winged species found in the warmer regions of the world they make up the family Phasmidae. Some of the winged species resemble leaves. (A.W.L.)

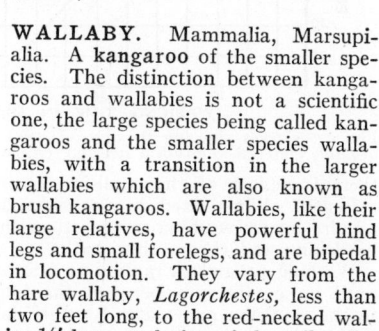

Walking-stick.

WALLABY. Mammalia, Marsupialia. A **kangaroo** of the smaller species. The distinction between kangaroos and wallabies is not a scientific one, the large species being called kangaroos and the smaller species wallabies, with a transition in the larger wallabies which are also known as brush kangaroos. Wallabies, like their large relatives, have powerful hind legs and small forelegs, and are bipedal in locomotion. They vary from the hare wallaby, *Lagorchestes,* less than two feet long, to the red-necked wallaby whose body is $3\frac{1}{2}'$ long, exclusive of the tail. The spur-tailed wallabies, *Onychogale,* are peculiar in having the tail tipped with a horny spur. (A.W.L.)

WALLAROO. Mammalia, Marsupialia. A stoutly built **kangaroo,** *Macropus robustus.* It is one of the large species, thickly furred and gray in color. (A.W.L.)

WALNUT. Wood.

WALNUT OIL. Esters.

WALRUS. Mammalia, Carnivora. A giant marine animal related to the seals but constituting a distinct family. Adults reach a length of more than $12'$ and a weight of a ton to 3000 lbs. The feet of the walrus are adapted for swimming but they are used for clumsy locomotion on land, as in the seals. In early life the body is covered with thick light brown fur but after middle age this vestiture tends to disappear. The muzzle bears a number of very thick bristles.

Walruses have the canine teeth of the upper jaw prolonged as tusks. The ivory of these tusks is used extensively by the Eskimos.

These animals are confined to the Arctic seas and are commonly regarded as constituting an Atlantic, *Odobaenus rosmarus,* and a Pacific, *O. obesus,* species, the latter with longer tusks. (A.W.L.)

WALSCHAERT GEAR. Valve Gear.

WANDEROO, WANDERU. Mammalia, Primates. The purple-faced monkey of Ceylon, one of the **langurs.** The name has also been applied incorrectly to the lion-tailed monkey of western India, belonging to the **macaques.** (A.W.L.)

WAPITI. Mammalia, Artiodactyla. The American **elk,** *Cervus canadensis,* a member of the red deer group. It is a large species, attaining a height of more than $5'$ at the shoulder, with gracefully branched antlers $4-5\frac{1}{2}'$ long. The species once ranged entirely across the continent but is now restricted to the western mountains. (A.W.L.)

WARBLE FLY. Insecta, Diptera. *Hypoderma.* A **bot fly** whose **larva** migrates through the connective tissues of cattle to complete their development in small abscesses called warbles opening through the skin of the animal's back. The adult flies attach their eggs to the hairs of cattle and the newly hatched larva enters the skin by way of the hair follicle. During its development it migrates extensively before reaching its final position under the skin of the back. The perforations leading into the warbles damage the best part of the hide, and the insect is sometimes a source of economic loss as a cause of illness in cattle. The maggots can be pressed out of the warbles when they once become evident or can be destroyed by smearing an ointment over the openings in the skin. A mixture of one ounce of iodoform (see **Iodine**) to five of vaseline has been recommended for this purpose. (A.W.L.)

WARBLER. Aves, Passeriformes. A small bird (**Aves**) related to the thrushes. The warblers of the Old World are an extensive family (Sylviidae) represented in North America only by the kinglets and gnat-catchers. The birds commonly called warblers in America are more accurately distinguished as wood warblers and make up the family Mniotiltidae, more closely related to the vireos.

Both groups include species whose common names do not indicate their association. Among European examples are the whitethroat, the hedge sparrow, the firecrest, and among the American warblers are the **oven bird,** water thrushes, **shats,** and **redstarts.**

Many of these birds are beautiful. Their numerous species are a delight to bird lovers during the spring migration in the United States, and in the fall, due to the great variation of patterns and colors between the sexes and the immature individuals, they are as much a puzzle as a pleasure. (A.W.L.)

Hooded warbler, *Wilsonia citrina.* Olive green above, yellow below. The male has a black hood covering the top of the head and running around the neck to the throat, giving the effect of a yellow mask across the face.

WARM FRONTS. Fronts.

WARM SECTOR. That sector of a wave cycle occupied by the warmer air mass is known as the warm sector. A warm sector is always smaller than the cold sector which is occupied by the colder air mass. (P.E.K.)

WARMOUTH. Pisces, Teleostei. A species of **sunfish**, *Chaenobryttus gulosus,* also called the red-eyed bream. It is an olive-green fish, marked with red and blue especially in northern waters. Maximum length ten inches. The species ranges from the Great Lakes to Florida and Texas and is common in quiet waters in the South. (A.W.L.)

WART. Verrucae, or warts, are circumscribed, raised skin **lesions**, gray, tan, brown, or black in color, having a piled up or horny surface. Verruca vulgaris commonly occurs on the hands of children, as multiple, small warts over the fingers and back of the hand. Other forms are plantar warts on the soles of the feet, and venereal warts on the genitalia. Warts are believed to be due to an infectious agent, as yet unidentified.

Treatment consists of removal of the lesions by the application of **salicylic acid,** glacial **acetic acid,** or trichloracetic acid. Freezing with solid **carbon dioxide,** the application of **radium,** and **x-ray** are also used. In many instances the warts clear spontaneously without treatment. (D.M.H.)

WART-HOG. Mammalia, Artiodactyla. A very ugly **pig** of Africa. It has a large head bearing excrescences which add the prefix to its name. The broad muzzle also bears strong upturned tusks. Two species are recognized, one, *Phachochoerus africanus,* ranging from Abyssinia southward through eastern Africa and the other, *P. aethiopicus,* confined to the southeastern portion of the continent. (A.W.L.)

WASHBURN AND MOEN. Gauge Number.

WASHER. Screw Fastenings.

WASHING. In photography, washing operations are employed to remove the excess reagents and soluble by-products, which affect later processing or impair the permanency of the photographic image.

Final washing following fixation removes the soluble silver thiosulfate complexes and fixing solution left in emulsion. Normally, the complexes are removed by washing with water or by repeated soaking in fresh-water baths. Under certain conditions, washing times are reduced or residual products are removed with **hypo eliminators.**

Failure to wash out the soluble complexes produces reactions which in time stain both highlight and unexposed areas. Incomplete removal of thiosulfate (hypo) and tetrathionates (oxidation products of thiosulfate) produces gradual sulfurization of image and fading.

Water fit for human consumption is usually satisfactory for washing. Sulfur water should be avoided because of its effect on the silver image. Hard water containing large amounts of calcium, iron, magnesium, sodium and potassium salts has a tendency to leave deposits on negative surfaces that are almost impossible to remove when the negative is dry. Hard water also shows water marks when droplets have dried.

Time of Washing. The rate of washing is dependent on the speed of diffusion of thiosulfate from the emulsion and on its removal from the emulsion surface. Since the diffusion of thiosulfate from the emulsion varies exponentially with time, the quantity of thiosulfate removed from the emulsion at any moment is proportional to that present in the emulsion. The rate of diffusion also depends on the difference between the

hypo concentrations in the emulsion and the wash water. The lower the concentration of hypo in water, the faster the rate of diffusion. In practice, the washing time, or removal of hypo from the emulsion surface, depends on the time required for a complete change of water and on the degree of agitation. In ideal conditions of washing, hypo is removed from the surface faster than it diffuses out of the emulsion. The volume of water used should be sufficient to cover the material. The more rapidly the water is renewed, the faster will be the rate of washing. The effect of temperature on hypo elimination is shown in the curve.

At temperatures below 60° F. the rate of washing is retarded, while at temperatures higher than 70° F.

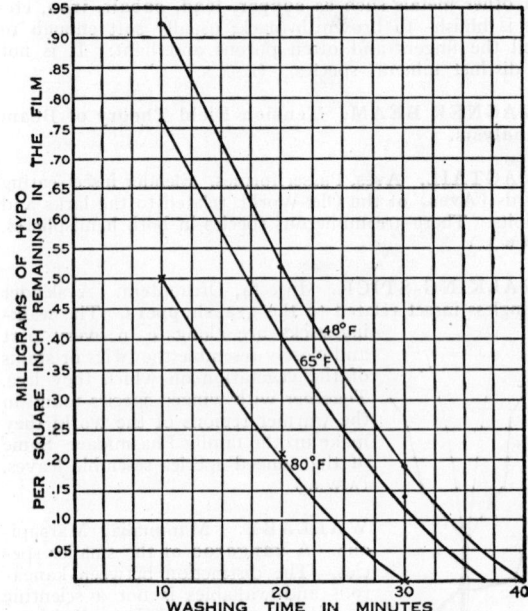

Fig. 1. Curves showing the rate of hypo elimination from a negative material to an ideal stream. (*Neblette's Photography.*)

there is danger of the emulsion becoming swelled and softened. Films fixed in potassium-alum fixing baths require longer washing periods than those fixed in non-hardening and chrom-alum baths.

Washing progresses rapidly at first and then decreases until an equilibrium between the hypo concentrations in the emulsion and water is established. (See Fig. 2.)

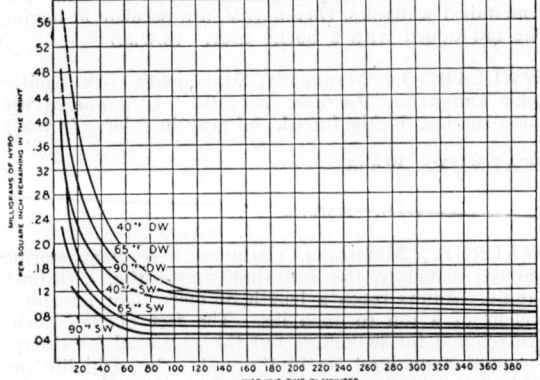

Fig. 2. Curves showing the rate of hypo elimination from papers. (*Neblette's Photography.*)

The average time required for establishment of equilibrium during washing for different photographic materials is:

Film .... 7 minutes
Single-weight papers ...... 15–60 minutes
Double-weight papers ...... 35–120 minutes

It is impossible to completely wash out the last traces of hypo from prints because of the adsorption of thiosulfate to the paper fiber and the baryta coating. Hypo so retained is termed residual hypo and can only be removed with the aid of hypo eliminators.

The minimum washing time for any material is the sum of the time for a complete water change plus the time for the establishment of equilibrium. The time for a complete water change can be determined by adding a dye or ink solution to the washer and noting the time until the color of the wash water becomes as clear as that from the tap.

Washing Tests. Tests to determine the completeness of washing (hypo tests) may be of two types: those used to detect the presence of hypo diffusing from emulsions and those which determine the presence of residual hypo retained by negative or print. For films, the alkaline permanganate test is widely employed to detect diffusible hypo and determine satisfactory washing. A stock solution is prepared by mixing:

Potassium permanganate .... 0.5 gram
Sodium hydroxide .... 1.0 gram
Distilled water to make ...... 1.0 liter

Films thought to be washed are removed from tray or tank and allowed to drain into 10–20 cc. of a test solution prepared by diluting 1 part of the stock solution with 20 parts of water. If traces of hypo are present, the color of the solution changes from violet to orange in 30 seconds. An objection to the test is the fact that oxidizable organic matter present in water produces the same color changes as hypo. The water is tested for interference by adding a volume of tap water equal to that drained from film to another sample of the test solution. Hypo is absent if both samples have the same shade.

For prints the only reliable test for hypo elimination is silver nitrate. While common hypo tests, as alkaline permanganate, sodium azide, and mercuric chloride, show the presence of diffusible hypo in wash water, they do not indicate the amount of hypo retained by the paper base and baryta coating—a quantity sufficient to cause image fading.

Tests for residual hypo are made by taking a strip of undeveloped paper that has been fixed and washed along with a batch of prints, and partially immersing it in a 1% silver nitrate bath for 3 minutes. Any yellow to brown coloration found on the immersed portion after rinsing shows the presence of residual hypo. (S.M.T.)

WASHING SODA. Sodium.

WASP. Insecta, Hymenoptera.

An **insect** related to the ants and bees. Some species are solitary and some social. Most species have four membranous wings, the front wing much larger than the hinder pair and both

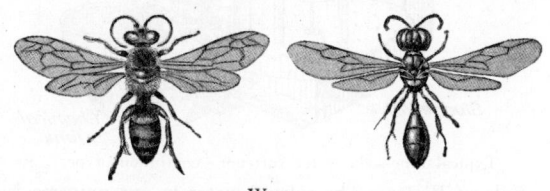

Wasp.

with few veins, joined to form closed cells. Some wasps burrow, some build nests of mud, and some use a coarse paper made by chewing wood from weathered surfaces. The last species are more commonly called hornets. Wasps are well known for their ability to sting; the more commonly known species inflict painful wounds because of their large size.

From the scientific point of view the term wasp is almost without value, since it applies to many different groups of insects. The members of one superfamily, the Vespoidea, include the spider wasps, the cuckoo wasps, the velvet ants, the true ants, the true wasps, and the potter wasps, each making up a different family, while the hornets and yellow-jackets, most familiar of wasps, are only one family of the group. Scientifically another superfamily, Sphecoidea, is made up entirely of insects called wasps according to some writers, but includes the bees according to others. The wasps of this division are represented by the giant cicada-killer, the largest of North American wasps, and the thread-waisted wasps.

Wasps have been extensively studied because of their complex behavior in making and provisioning nests for their young. In the works of the Peckhams and the Raus some remarkable and interesting records are preserved. (A.W.L.)

WASSERMANN REACTION.

A serological test used in the diagnosis of **syphilis**. It was described by Wassermann in 1906 and has been of great importance in detecting latent cases as well as active syphilitic disease ever since. The test is performed on blood serum or spinal fluid, and depends upon the presence of **antibodies**. A positive result indicates that antibodies are present, and the individual has been infected with syphilis at some time, even though at the time of the test he may be asymptomatic. The spinal fluid Wassermann may be positive even though the blood is negative; it always indicates central nervous system syphilis. The test on blood or spinal fluid is also a guide as to the effectiveness of treatment, although the early change from a positive to a negative result during therapy does not alter the program of prolonged treatment in the least.

Usually the blood Wassermann does not become positive until several weeks after onset of the infection. Characteristically, at about the time that the chancre begins to heal, a positive blood test is obtained. If treatment is begun before this serological reaction takes place, the prognosis for early cure is better than if the Wassermann reaction is already positive when therapy is undertaken.

The Wassermann reaction is not completely specific. It may be positive in yaws, and occasionally in infectious mononucleosis and leprosy. (D.M.H.)

WATER.

Water (H_2O) is a colorless (blue in thick layers), odorless, tasteless liquid, melting point $0°$ C. (one of the standard temperature points), boiling point $100°$ C. at 760 mm. pressure (another standard temperature point). At 770.0 mm. pressure, the boiling point is $100.366°$ C.; at 750.0 mm., $99.360°$ C.; at 740.0 mm., $99.255°$ C.; at 730.0 mm., $98.877°$ C.; at 380 mm., $81.7°$ C.; at 76 mm., $46.1°$ C.; at 1520 mm., $120.6°$ C.; at 7600 mm., $180.5°$ C. Density, 1.000000 gram per milliliter (or 1.000073 gram per cu. cm.) at $3.98°$ C. (one of the standard density points). At $0°$ C., the density is 0.99987 gram per milliliter; at $8°$ C., 0.99988; at $15°$ C., 0.99913; at $16°$ C., 0.99897; at $17.5°$ C., 0.99871; at $20°$ C., 0.99823; at $25°$ C., 0.99707; at $40°$ C., 0.99224; at $50°$ C., 0.99807; at $75°$ C., 0.97489; at $100°$ C., 0.95838; at $120°$ C., 0.9434. **Critical temperature** $374°$ C., **critical pressure** 217.7 atmospheres, **critical density** 0.4 gram per cu. cm. **Viscosity**, 0.01792 poise (dyne-second per sq. cm.) at $0°$ C. (specific viscosity 1.000). At $20°$ C., the viscosity is 0.01005 poise (specific viscosity 0.561); at $50°$ C., 0.00549 (specific viscosity 0.307); at $75°$ C., 0.00380 (specific viscosity 0.212), at $100°$ C., 0.00284 (specific viscosity 0.158). **Surface Tension** against air, at $0°$ C., 75.6 dynes per cm.; at $10°$ C., 74.22; at $20°$ C., 72.75; at $30°$ C., 71.18; at $60°$ C., 66.18; at $100°$ C., 58.9. **Specific heat**, 1.00000 at $15°$ C. (standard of specific heat). At $0°$ C., the specific heat is 1.00874; at $25°$ C., 0.99765, at $35°$ C.,

0.99743 (minimum); at 50° C., 0.99829; at 65° C., 1.00001; at 80° C., 1.00239; at 100° C., 1.00645; at 120° C., 1.016; at 180° C., 1.04. Electrical **conductivity** 0.04 × 10⁻⁶ reciprocal ohms at 18° C. (Kohlraush and Heydweiller, 1902), of pure water in **equilibrium** with air 0.8×10^{-6}, of ordinary distilled water about 5×10^{-6}. **Dielectric constant** (specific inductive capacity), 81.07 at 18° C. (compare ethyl alcohol 25.8 at 20° C., carbon disulfide 2.6 at 20° C.).

The chemical composition of water has been the subject of intensive studies from the early years of the science of chemistry. E. W. Morley, of Western Reserve University, Cleveland, Ohio, in 1895, reported the weight ratio of **oxygen** to **hydrogen** in water as 7.9395 to 1.00000, and the volume ratio, 1.00000 to 2.00288. F. P. Burt and E. C. Edgar, of England, in 1916, considered, on the basis of their experiments, 7.9387 to 1.00000 the most exact weight ratio. The present value accepted by the International Union of Chemistry, Committee on Atomic Weights, is 8.0000 to 1.0080. The H_2O molecules in ice and water are associated together by hydrogen bonding. This phenomenon accounts for the abnormally high **heat of vaporization** (and condensation), which is 585 calories (15° C.) per gram of water at 20° C., and 540 at 100° C. (compare butane 88, **ethyl alcohol** 204, **acetone** 125, **acetic acid** 97, **carbon tetrachloride** 46), and high **heat of fusion** (and solidification), which is 80 calories (15° C.) per gram of water (compare **benzene** 30, ethyl alcohol 25, acetone 21, acetic acid 44, carbon tetrachloride 4).

For water as **catalyzer** see **Reactions Involving Water.**

Pure water, especially when free from dissolved gases, may be heated above 100° C., even to 180° C., without boiling, but on further heating boiling with explosive violence may occur. Steam at 100° C. occupies a volume 1700 times greater than water at 100° C. Pure water, when not agitated, may be cooled somewhat below 0° C. without freezing, but on further cooling congeals with increase of volume (density of ice 0.917) exerting great force, when confined, but if in intimate contact with water at atmospheric pressure the temperature is 0° C. Vapor pressure of ice and of water 4.579 mm. at 1 atmosphere pressure, 0° C. Triple point, ice-water-water vapor, +0.007° C. in vacuum. When water is compressed to say 20,000 atmospheres and then cooled, other varieties of ice, all denser than water, are formed (ice II 12% denser than water ice, III 3% denser. Six varieties of ice are known).

Pure water may be obtained (1) by distillation and condensation of water, (2) by partial freezing of water followed by separation of the pure ice and melting, (3) by burning of hydrogen or reduction of heated oxides

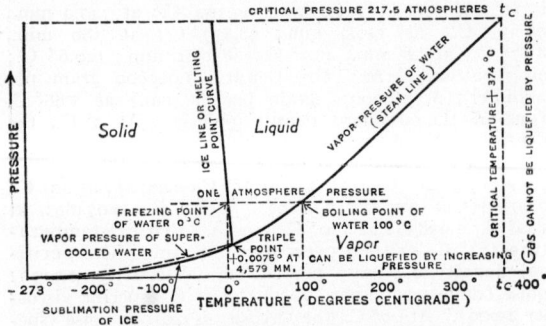

Pressure-temperature diagram for water.

by hydrogen and collecting the product. Hydrogen containing organic substances, e.g., **hydrocarbons,** when burned in air or heated with **copper** oxide, form water.

"Heavy water," deuterium oxide (D_2O, molecular weight 20, that of ordinary water 18), is concentrated in the residual water of **electrolytic** cells that have

been operated a long time, ordinary water being more readily decomposed than heavy water. The physical constants of deuterium oxide are: melting point 3.8° C., boiling point 101.42° C., density at 25° C. 1.1056, temperature of maximum density 11.6° C., solubility of sodium chloride 15% less than in ordinary water.

Water occupies a distinct position among liquids in the matter of dissolving gases, liquids and solids. While there are other liquids that exceed it in solvent power in specific cases, no other shows such a wide range and general intensity of solvent power. Many chemical reactions take place in water as the solvent medium, and many of these occur instantaneously. (See **Solutions and Solubility.**)

Water of the ocean, containing dissolved salts, covers 73% of the surface of the earth; water of lakes (fresh and salt) and rivers forms an important portion of the land surface; water vapor of the atmosphere is a constituent affecting climate and plant growth; snow and ice of mountain tops serve upon melting as regional water supplies, and of the polar regions determine ocean currents and climate; and underground water is an industrial and agricultural source of water supply. The effect of water in changing the earth's surface is mainly due (1) to the disintegration of rocks when ice forms in the interstices, (2) to the mechanical carrying of particles of various sizes from higher to lower levels, (3) to the solution of parts of the rocks, (4) to its beneficial effect on plant growth. Water stands alone in its importance to plant and animal life, and to industry.

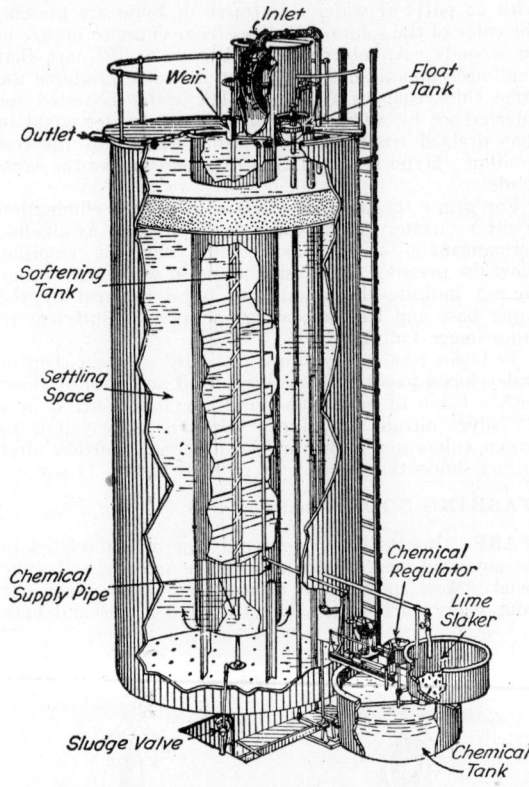

Typical lime-soda water softener—continuous type.

Rôle of Water. The rôle of water in our universe is on a par with, if not surpassing, that of any known substance. The scope is briefly as follows:

In nature
  Geochemical
    The hydrosphere (See below)
    Water as vapor, liquid, ice and snow
    Water at rest and in motion

In nature (*Continued*)
  Biochemical
    Water and plant organisms
    Water and animal organisms
In industry
  As solvent, and medium of reaction (See **Solutions**)
  Humidity
  Extinguisher of fires by lowering the temperature
  Drying and wetting
In living
  Beverages, foods and their preparation (See **Foods**)
  Washing, cleansing and sanitation
In science
  Standard of reference for many data (See above)
  Deliquescence and water absorption
  Efflorescence and loss of water
  Drying and desiccation
  Reactions involving water
    1. Consumption of water
    2. Production of water
    3. Water as catalyzer

Natural waters may be contaminated with (1) insoluble suspended material. This settles out upon standing, or may be filtered, (2) soluble inorganic matter, (3) soluble organic matter. This may impart color or acidity. The latter may be readily neutralized by addition of a base. The former may be removed by precipitation of gelatinous **aluminum** hydroxide (aluminum sulfate or alum plus sodium carbonate), a process which also removes practically all the bacteria which may be present.

In ocean and salt lake waters the principal content is **sodium** chloride, with small amounts of **calcium, magnesium, potassium,** and **sulfate, carbonate;** in fresh lake, river and underground water the content is variable in amount over a wide range, some surface waters in contact with igneous rocks are practically pure water except for dissolved air, others are notable for their high calcium content in **limestone** regions, usually containing calcium and magnesium as hydrogen carbonate and chloride, respectively ($Ca(HCO_3)_2$, $MgCl_2$). When the hydrogen carbonate is boiled calcium carbonate ($CaCO_3$) is precipitated and can be separated. In order to separate dissolved calcium and magnesium, treatment with calcium hydroxide and sodium carbonate has long been practiced. Other treatments include passage of the water over artificial sodium zeolites or over various ionic resins, addition of trisodium phosphate.

The cycle of water in nature is evaporation from the oceans, then condensation upon cooling by contact with colder bodies of land, e.g., mountains, or cold currents of air from the polar regions. This may fall as rain or snow, which are practically pure water, except for dissolved air. Rain runs off the surface into lakes or rivers or seeps into the earth. Snow and ice form a storage

ANALYSES OF SEA WATERS

DISSOLVED SOLIDS		ANALYSES OF DISSOLVED SOLIDS					
		Chloride	Sodium	Potassium	Sulfate	Calcium	Magnesium
Atlantic Ocean......	3.30 to 3.74%	55.3%	30.6%	1.1%	7.7%	1.2%	3.7%
Great Salt Lake, Utah	14.99 to 23.04	56.0	33.2	1.6	6.6	0.2	2.5
Owens Lake, Cal....	7.2	25.7	37.4	2.2	10.0	0.02	0.01
Dead Sea..........	19.22 to 26.00	65.8	11.7	1.9	0.3	4.7	13.3
Baltic Sea..........	0.3 to 0.8						
White Sea..........	2.6 to 3.0						
Black Sea..........	1.8 to 2.2						
Red Sea...........	5.1 to 5.9						

ANALYSES OF RIVER WATERS
PARTS PER MILLION, AVERAGE

Place	Silica	Calcium	Magnesium	Sodium plus Potassium	Sulfate	Chloride	Total Dissolved Solids
Mississippi							
Minneapolis, Minn., 1 yr (1906–07)......	15	40	14	10	18	2	200
Memphis, Tenn., 1 yr (1908)............	24	36	12	19	43	9	202
New Orleans, La., 1 yr (1905–06)........	11	32	8	13	24	10	166
Ohio (calculated from tributaries).........	12	18	5	8	17	7	
Missouri							
Kansas City, Kans., 1 yr (1906–07)......	37	62	18	44	135	13	426 (1909)
Lake Superior (Aver. of 11 anal.)..........	13	22	5	6	4	2	60
St. Lawrence (Aver. of 11 anal.)..........	5	24	5	5	9	6	134
Penobscot							
Bangor, Me., 3 yr. (1909–12)............	2	4	1	....	6	2	62
Hudson							
Hudson, N. Y., 1 yr. (1906–07).........	11	21	4	8	16	4	108
Rio Grande							
Laredo, Tex., 1 yr (1905–06)...........	29	104	23	119	228	164	791
Colorado							
Yuma, Ariz., 1 yr. (1893)...............	19	66	13	194	231	183	706
San Joaquin							
Lathrop, Cal., 1 yr. (1906).............	16	18	8	27	26	30	161

(*Continued on next page*)

ANALYSES OF RIVER WATERS—*Continued*

PARTS PER MILLION, AVERAGE

Place	Silica	Calcium	Magnesium	Sodium plus Potassium	Sulfate	Chloride	Total Dissolved Solids
Sacramento							
Sacramento, Cal., 1 yr. (1906)..........	19	15	7	15	13	9	124
Columbia							
Cascade Locks, Wash. (Bonneville, Ore.)							
1 yr. (1911–12)......................	14	17	4	9	12	3	97
Willamette							
Salem, Ore., 1 yr. (1911–12).............	15	5	1	4	4	2	51
Thames, England, 1906-13...............	.....	.....	.....	.....	.....	17	227
Rhone, France.......................	2	26	4	4	27	1	64
Rhine, Cologne......................	0.2	26	6	3	13	4	52
Elbe...............................	.....	.....	.....	.....	.....	123	414
Danube							
Budapest...........................	1	27	7	1	14	1	51
Nile, Egypt							
Cairo..............................	17	13	7	16	4	3	60
Amazon							
Obidos.............................	29	15	1	9	2	7	63

TURBIDITY OF RIVER WATERS

PARTS PER MILLION, AVERAGE

Place	Turbidity	Place	Turbidity
Mississippi		Columbia	
Minneapolis, Minn. 1 yr. (1906–07)........	10	Cascade Locks, Wash. (Bonneville, Ore.),	
Memphis, Tenn., 1 yr. (1908).............	556	1 yr. (1911–12)........................	27
Hudson		Willamette	
Hudson, N. Y., 1 yr. (1906–07)............	13	Salem, Ore., 1 yr. (1911–12)..............	8

CHEMICAL DENUDATION IN THE UNITED STATES

Drainage Area	Area Drained (Square miles)	Dissolved Solids to Ocean (Short tons per square mile per annum)
North Atlantic.............................	159,400	130
South Atlantic.............................	123,900	94
Eastern Gulf of Mexico......................	142,100	117
Western Gulf of Mexico....:................	315,700	36
Mississippi River...........................	1,265,000	108
Laurentian Basin (U.S.A.)....................	175,000	116
Colorado River of Arizona....................	230,000	51
South Pacific..............................	72,700	177
North Pacific..............................	270,000	100
Sum...................................	2,753,800	Average 98
Great Basin...............................	334,700	
Total...................................	3,088,500	

TOTAL DENUDATION OF THE COLUMBIA RIVER BASIN

Above Cascade Locks, Wash. (Bonneville, Ore.)
Observations, Aug., 1911, to Aug., 1912

Drainage Area: Columbia River above Cascade Locks, Wash.

Area Drained: 175,200 Square miles
Mean Discharge: Minimum 60,600 Second-feet, Jan. 1, 1912
Maximum 624,900 Second-feet, June 10, 1912
Dissolved Matter: 17,000,000 Short tons in 1 year
Suspended Matter: 7,000,000 Short tons in 1 year

CHEMICAL DENUDATION OF THE LAND SURFACE OF THE EARTH BY CONTINENTS

	Land Surface Million square miles	DISSOLVED SOLIDS TO OCEAN	
		Metric tons per square mile	Million metric tons per annum
North America............	6	79	474
South America.............	4	50	200
Europe....................	3	100	300
Asia......................	7	84	588
Africa....................	8	44	352
Total..................	28	Aver. 68.4	Total 1,914

AVERAGE HARDNESS OF WATER FROM PUBLIC SUPPLY SYSTEMS IN THE UNITED STATES IN 1923

State	Average Hardness as Calcium Carbonate in Parts per Million	Population Served Thousand	% of Total Population of State	State	Average Hardness as Calcium Carbonate in Parts per Million	Population Served Thousand	% of Total Population of State
Alabama..........	53	283	12	Missouri.........	148	1,245	37
Arizona...........	221	49	15	Montana..........	91	66	12
Arkansas.........	149	94	5	Nebraska.........	239	247	19
California.........	172	1,800	52	Nevada..........	74	16	21
Colorado.........	144	330	35	New Hampshire...	9.7	107	24
Connecticut.......	25	826	60	New Jersey.......	48	1,983	63
Delaware....../....	51	114	51	New Mexico.......	126	22	6
District of Columbia	80	438	100	New York........	47	7,576	73
Florida............	296	204	21	North Carolina....	22	157	6
Georgia...........	27	421	15	North Dakota.....	141	36	6
Idaho............	91	36	8	Ohio............	153	2,767	48
Illinois...........	156	3,435	53	Oklahoma........	400	194	10
Supplied from Lake				Oregon..........	9.6	276	35
Michigan.......	131	2,824	44	Pennsylvania......	69	3,556	41
Not supplied from				Rhode Island.....	12	475	79
Lake Michigan...	274	611	9	South Carolina....	31	105	6
Indiana...........	264	873	30	South Dakota.....	503	40	6
Iowa..............	298	412	17	Tennessee......,..	57	416	18
Kansas...........	316	223	13	Texas............	136	841	18
Kentucky.........	90	387	16	Utah............	158	151	34
Louisiana.........	54	431	24	Vermont..........	39	38	11
Maine.............	18	127	17	Virginia...........	45	489	21
Maryland.........	53	792	55	Washington.......	41	570	42
Massachusetts.....	14	2,668	69	West Virginia.....	76	174	12
Michigan..........	134	1,722	47	Wisconsin........	145	760	29
Minnesota........	158	714	30	Wyoming.........	119	25	13
Mississippi........	14	46	3				
				United States.....	99	38,757	37

PERCENTAGE OF POPULATION OF THE UNITED STATES SERVED BY PUBLIC WATER SUPPLY

DATA FOR 1930, EXCEPT AS STATED

State	Percentage of Population Served by Public Water Supply	State	Percentage of Population Served by Public Water Supply
Massachusetts................	97.0	Followed by 25 other states, and the following states, making a total of 32 below the average.	
New Jersey..................	94.7		
New York...................	90.0		
Rhode Island (1933).........	89.4	Kentucky..................	34
Connecticut.................	89.0	Alabama..................	31
California...................	84.6	North Carolina.............	31
New Hampshire..............	82.8	Arkansas..................	25
Nevada (1933)...............	81.5	South Carolina.............	24
Followed by 8 other states, making a total of 16 above the average.		North Dakota..............	22
		Mississippi.................	18
Average United States........	63.1		

AVERAGE CONSUMPTION OF WATER IN VARIOUS CITIES

City	Gallons Per Capita Per Day	City	Gallons Per Capita Per Day
London (1924)	43	Tokio (1913)	32
Paris (1913)	38	Sydney (1913)	50
Madrid (1913)	84	Toronto (1913)	118
Rome (1913)	120	New York City	115
Berlin (1913)	35	Chicago	235
Cairo (1913)	25	Baltimore	130
Calcutta (1913)	62	Milwaukee	85

The consumption of water in European cities is usually about 20 to 50 gals. per capita per day, but in the cities of the United States is much larger, from 75 to 150. In the United States, the consumption of water is distributed about 35% domestic, 40% industrial, and the remainder for municipal use plus waste.

SEASONAL EVAPORATION OF WATER AT VARIOUS PLACES
Data in Inches of Water Evaporated

	Berkeley, California, 1905	Minidoka Dam, Idaho, 1909–10	Boston, Massachusetts, 1875–90	Nebraska Interstate Canal, 1909–10	Lee Bridge, England, 1860–73
Jan.	1.0	2.2	1.0	2.0	0.8
Feb.	1.4	2.5	1.0	2.2	0.6
Mar.	2.1	4.0	1.7	3.5	1.1
Apr.	3.1	7.0	3.0	6.0	2.1
May	4.7	11.2	4.5	8.5	2.8
June	5.7	12.3	5.5	11.0	3.1
July	5.5	15.0	6.0	14.7	3.4
Aug.	5.1	13.5	5.5	12.7	2.8
Sept.	4.6	11.0	4.1	10.0	1.6
Oct.	4.3	8.5	3.2	7.6	1.1
Nov.	2.7	5.8	2.2	5.2	0.7
Dec.	1.3	3.5	1.5	3.0	0.6
1 year	41.6	96.5	39.2	86.4	20.7

source of water which is liberated when and as the temperature rises. Underground water gradually evaporates through porous soil, flows to lower levels, or remains as a storage source. Plants and animals enter the cycle to use such water as is necessary and available, and returning to the atmosphere or the surface of the earth such as is not retained. Animal waste and animal remains furnish a food source for **bacteria** which in some cases are hazardous to animal life. Such water is purified (1) by **filtration** through sand filters, usually accompanied by precipitation of **aluminum** hydroxide, (2) by disinfection, most frequently by chlorine or a hypochlorite, and, in many cases (3) by treatment with activated **carbon** to remove odors. Industry enters the cycle (1) to use the water for power, usually hydroelectric, (2) to generate steam, in which case the purity or impurity of the water used is an important consideration in connection with steam boiler practice, (3) to use the water for cleansing purposes, in cases where soap is used the extent to which **calcium** or **magnesium** ("hardness") is present represents loss of soap, (4) to use the water for dissolving or separating solids, liquids, gases. The effluents from such applications represent a wide range of impurities, e.g., from coal-gas works, woodpulp mills, tanneries, packing-house plants. These effluents, along with those from domestic sewage, present a serious problem in sanitation. Rivers complete the cycle of water in nature by returning to the oceans the excess of water plus such natural, industrial, and domestic additions of insoluble and soluble materials as are contained.

Volume and Mass of Hydrosphere

Data of Kossinna (1921), accepted by Clarke (1924)

Volume of hydrosphere, including the oceans, the mediterranean seas, the border seas, and the gulfs:

1,370,323,000 cu. kilometers or 327,672,000 cu. miles.

Mass of hydrosphere, described as above:

$1,411.4 \times 10^{15}$ metric tons

Density of sea water of normal salinity (35 parts per 1000) at 0° C.:

1.028 (1.03, Clarke)

Mass of salt in hydrosphere:

49.4×10^{15} metric tons

The hydrosphere is estimated to be 7% of the total mass of the earth. (R.K.S.)

WATER BEAR. Bear animalcule. **Tardigrada.** (A.W.L.)

"WATER BLOOM." Algae.

WATER BOATMAN. Insecta, Hemiptera. An aquatic bug of the family Corixidae. These insects are flattened, broad at the head and tapering bluntly at the opposite end of the body. The fringed posterior legs project like a pair of oars and are used in swimming. Water boatmen breathe air but they are able to descend to considerable depths, carrying a film of air on the ventral surface of the body. They feed on ooze containing plant matter and minute animals. (A.W.L.)

WATER BRAKE. Dynamometer.

WATER BUCK. Mammalia, Artiodactyla. *Kobus.* A large **antelope** of southern and eastern Africa. It frequents rocky hills in the vicinity of rivers. The horns are more than 2′ long, slightly curved and ringed almost to the tips. (A.W.L.)

WATER DEER. Mammalia, Artiodactyla. *Hydropotes.* A small **deer** found along the margins of the Yangtse Kiang river in China. The male has long curved tusks in the upper jaw and neither sex has antlers. The species is also remarkable in producing 3–6 young at a time. (A.W.L.)

WATER DOG. Amphibia, Urodela. The **mud puppy.** (A.W.L.)

WATER FLEA. Crustacea, Cladocera. Minute aquatic **crustaceans** of compact form, usually transversely compressed and provided with a bivalve **carapace.** They are superficially like fleas in form. (A.W.L.)

WATER GAP. Gap.

WATER HAMMER. Sudden stoppage of water flow in a long pressure conduit caused by the closing of **valves** can, if the rate of closure be rapid enough, cause the conduit to be subjected to a sharp, hammer-like blow from a steep front pressure **wave.** Water moving in a long pipe-line has considerable mass. To decelerate a mass requires a force equal to the mass times the deceleration (negative **acceleration**). If the rate of deceleration is large, the force will be large. Hence if a valve or gate is suddenly closed, the water has high deceleration, and a large force is set up. Due to the elastic nature of conduit material, this force acts expansively, slightly stretching the pipe. When the **inertia** force has disappeared, the pipe regains its original girth and produces secondary pressure waves. A calculation shows that the power required to decelerate water in a 5-ft. pipe 2000′ long is 1400 hp. This calculation was based on an assumption of 10 ft. per sec. water velocity, and 5 seconds was the time taken to close the valve, and it should acquaint one with the magnitude of power behind the water hammer. To cushion all or parts of the water conduit against the destructive effect of water hammer forces, relief valves, bursting plates, and surge tanks have been used. Water hammer may also be caused by the sudden collapse of steam bubbles upon entering cold water, as when the steam is turned into a cold radiator partly filled with water. (F.T.M.)

WATER HEATING. Feedwater Heating; Heating.

WATER HYACINTH. *Eichhornia crassipes.* Pontederiaceae. This plant occurs widespread in tropical and subtropical regions, where it often becomes a troublesome weed. In Florida it sometimes forms floating masses so dense as to become a serious hindrance to river navigation. Very noticeable are the leaves, the **petioles** of which are swollen in bladder-like enlargements containing many air-spaces. These cause the plant to remain floating at the surface of the water. Because of the broad shining green blades the plants are easily blown about on the surface by the wind. The dark-colored roots form a dense mass beneath the water surface. The root cap at the tip of each rootlet is a very conspicuous structure. The flowers are showy and pale lavender in color. They are trimorphic, there being three different lengths of **styles.** The plant is frequently found in cultivation in northern regions, where, however, it is not hardy. (R.M.W.)

WATER LEVEL REGULATOR. Feedwater Regulator.

WATER MEASURER. Insecta, Hemiptera. A long slender **bug** that creeps slowly on the surface of water. Also called the marsh treaders. These are not the common insects that skate rapidly on the water, although they are closely related. The water measurers make up the family Hydrometridae, and the other insects are water striders of the family Gerridae. (A.W.L.)

WATER MOLDS. Phycomycetes.

WATER PENNY. Insecta, Coleoptera. A small flattened oval **insect** found chiefly on the underside of rocks in running water. Water pennies are the **larvae** of **beetles** of the family Psephenidae. They resemble crustaceans and were originally described as such. (A.W.L.)

WATER PHEASANT. Aves, Charadriiformes. A large and beautiful water bird (**Aves**) of India and Ceylon. Its nest floats on the water or is anchored to water plants. Related to the jacana. (A.W.L.)

WATER PRESSURE. Head.

WATER PROPELLER. A water propeller is a very short section of an endless screw used for propulsion of power-driven vessels. Although when first developed, a short section of an actual endless screw was used, the modern propeller uses only segments of the screw which are arranged radially on a hub, there being usually two, three, or four blades, depending on the type of vessel. The water propeller depends largely upon push exerted on the water by its astern face, and in this respect is unlike the air propeller, in which thrust is developed by vacuum on its forward face. For this reason, also, and because of the greater density of water as compared to air, the water propeller has a relatively small diameter and large blade area. This means that the disk area, i.e., the area of the circle swept by the tips of the blade, is largely occupied by blade. Water propellers are usually made of cast iron or manganese bronze, the latter being more suitable than iron for salt water. The propeller is securely attached to a propeller shaft, which passes through the hull at the stern, and from there is connected to the engine by couplings, and possibly also by universal joints. The propeller shaft is given a slight slant downwards to the rear, and leaves the hull through a stuffing box which reduces water leakage to small value. The pitch of the propeller is the distance it would advance in a complete turn, considering it a screw revolving in a solid medium. In use, it does not actually advance this distance, the difference being **slip.** This slip averages about 15% for a properly selected propeller. In any application of the water propeller, it is important to select the propeller with due regard to the speed and power of the engine. The weight and size of the boat are of much less importance. If a propeller is improperly selected, it will either permit the engine to race, churning the water and producing little thrust, or it will hold the engine speed below rated, and prevent development of its full output. (F.T.M.)

WATER PURIFICATION. Water purification methods may be classified into processes for (1) removing material in solution, (2) removing material in suspension. The former processes are employed principally where the water contains objectionable mineral matter in solution, rendering it unsuitable for domestic or industrial uses. Various objectionable mineral forms in solution may occur but the principal process employed under this heading is that of water-softening which removes the calcium salts typical of hard or limestone water.

The principal problem of water purification is the removal of matter in suspension from the larger particles, which will settle out readily, right on down to the most minute inorganic matter and including bacteria as well.

Plain sedimentation in tanks or basins will remove the larger, readily-settling particles satisfactorily but requires too long a period of detention for a high removal of the finely divided material. Plain sedimentation as a principal or sole means of purifying water is now seldom employed although in earlier days it was often so used.

Certain chemicals such as alum, and certain of the iron salts when added in very small quantities to water naturally hard, or made so by the addition of an alkali, react to produce a sticky, gelatinous precipitate or "floc." This precipitate readily settles, forming larger masses as it settles by agglomeration, and as it passes down through the water it picks up the finer material by surface adhesion and its sticky and gelatinous character. This method of treating water with chemicals, followed by settling, is spoken of as coagulation and sedimentation. By this means clarification by sedimentation is enormously speeded up, especially for the more difficult turbidities caused by finely divided matter in suspension.

Modern sanitary standards, however, usually require that coagulation and sedimentation be followed by filtration for more complete removal of all matter in suspension. In this case the coagulated and settled water is passed through a rapid sand filter, the combined treatment processes removing 98–88½% of the matter in suspension, including the bacteria.

As a final process, the effluence from the filters is treated with chlorine to eliminate the remaining bacteria and produce sterile water. The amounts of chlorine so used are small and should not normally affect the taste or odor of the water.

With a high rate of chlorine application, settled water or even unsettled water may be rendered sterile and safe for consumption, but under these conditions so much chlorine has to be used for safety that it commonly produces a distinct taste and odor.

The rapid sand filter, with its auxiliary treatment of coagulation and sedimentation, was especially developed to handle very turbid waters for which, by reason of the very finely divided nature of the matter in suspension, plain sedimentation was ineffective and slow sand filters were unsatisfactory.

Historically, slow sand filters were developed first. They consist of very large enclosed sand beds through which the water passes very slowly. In time, a thin organic coating of bacterial jelly forms on top of the sand bed and acts as a very fine strainer to screen out bacteria or very minute inorganic particles. Slow sand filters will not operate satisfactorily for very turbid waters because the necessarily frequent cleaning of the filter bed, in itself a large job, disturbs and destroys the organic coating on the beds too often for efficient operation. Thus slow sand filters have a field of use limited to the less turbid waters of infrequent occurrence.

In the case of the rapid sand filter, some of the floc passes on into the filter and forms a thin, inorganic, gelatinous coating on top of the sand bed which acts as a fine straining medium to take out the fine material and bacteria. The rapid sand filter is readily washed, as frequently as may be required, by forcing water at a high rate back through the filter and washing off the upper dirty layer on the sand bed. This would be impractical in the case of the slow sand filter because of the large volumes of water required. In the latter case, the dirty layer is skimmed off the top of the sand by means of flat spades, washed, returned to the bed, and re-spread.

Rapid sand filters operate at a daily rate of about 125,000,000 gals. per acre of bed, and slow sand at daily rates of from 3,000,000–10,000,000 gals. per acre. (E.W.S.)

WATER SCORPION. Insecta, Hemiptera. Moderately large water bugs of oval or slender and elongate form, with a long breathing tube at the end of the body. The front legs are adapted for catching small prey. (A.W.L.)

WATER SPOUT. Funnel-shaped tornado cloud at sea or over water of lakes or rivers. It is the equivalent of a tornado but does not usually reach the same violence. Sea craft in its path will be damaged and the sea surface only slightly confused. (See **Tornadoes.**) (P.E.K.)

WATER STRIDER. Insecta, Hemiptera. A moderately large **bug** whose two posterior pairs of legs are modified for locomotion on the surface film of water. The claws are set back from the tip of the leg and the hairs of the dense covering are turned under so that the tip of the leg rests on their curved surfaces. These bugs are common on streams and ponds. The members of one genus, *Halobates,* are the only truly marine insects. They are found in tropical waters, often far from land, and their eggs have been found attached to floating feathers. (A.W.L.)

WATER TABLE. Ground Water.

WATER TANK. Tank.

WATER VAPOR. Steam and Vapors.

WATER VASCULAR SYSTEM. A tubular system of **echinoderms** through which sea water is circulated. It is formed from the **coelom** and is associated chiefly with locomotion, respiration, and the securing of food.

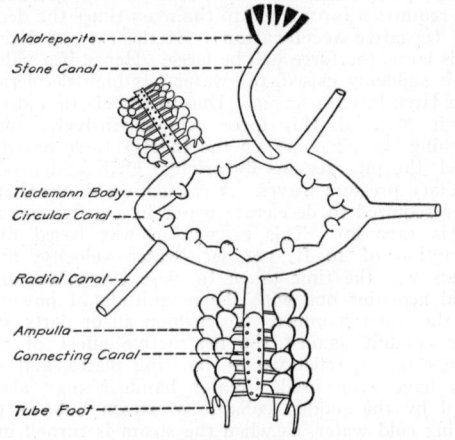

Madreporite ----------
Stone Canal ----------
Tiedemann Body ----
Circular Canal ----
Radial Canal ----
Ampulla ----------
Connecting Canal ----
Tube Foot ----------

Water vascular system of the starfish. (*Drawn by W. J. Moore.*)

In the starfishes (**Asteroidea**) the madreporite on the surface of the disk is a round perforated plate leading into this system of tubes. Directly below it the **stone canal** extends into the disk and joins a ring canal encircling the axis of the body. From the ring canal a radial canal passes out along the axis of each ray, and various appendages arise from all of these canals. The appendages of the ring canal are a series of Polian vesicles and a series of minute Tiedemann's bodies (or vesicles). The former are saccular reservoirs and the latter glandular bodies. The radial canals bear transverse branches with a small vesicular ampulla inwardly and a slender **tube foot,** cupped at the end, extending out through an opening in the hard wall of the animal and projecting in the ambulacral groove on the under surface of the ray. These tube feet grip by suction and adhesion.

The madreporite of sea cucumbers projects into the cavity of the body; several madreporites and stone canals may also occur. The body cavity is in communication with the exterior by way of the branching respiratory trees that arise from the cloaca. A similar condition exists in the crinoids, which have several stone canals opening by single pores into the cloaca, and other pores from the cloaca to the surrounding water. (A.W.L.)

WATER WALL. Water walls are installed on most modern boilers that are designed to operate at high rating. They are also often added to old boilers which, by the further installation of high-capacity combustion equipment, economizers, and air preheaters, may be operated at greatly increased capacity. When pulverized coal was first used refractory walls were standard, but pulverized coal could be burned by 10 to 20% excess air, whereas 50% was considered good in the existing stoker-fired furnaces. The results of the higher furnace temperatures accompanying pulverized coal were destruction of refractories and slagging of walls and tubes. The furnace temperature was above the softening point of the ash and particles of it flying about in a sticky or molten state adhered to the first cool surface they touched. Unless the advantages gained from the reduction of flue gas loss were to be sacrificed, furnace designs had to be changed. This led to the introduction of water-cooled furnaces in which the combustion space was partially or completely surrounded by tubes carrying water which, by absorbing radiant heat directly as soon as it was evolved from the combining molecules of fuel and oxygen, prevented the attainment of destructively high furnace temperature. (F.T.M.)

WATER-LILIES. Nymphaeaceae. The water-lilies form a small family of water or marsh plants. The leaves may be submerged or floating or carried well above the surface of the water on stiff petioles, as in the Lotus. The flowers are usually large and solitary. The principal genera are *Cabomba, Nuphar, Nymphaea,* or *Castalia,* and *Nelumbium.*

Cabomba is a genus of tropical American water-lilies having two types of leaves; some are submerged and much-divided into linear segments, while others are entire and floating with the petiole centrally attached. The small flowers are borne on long peduncles, and have their parts in threes. These plants are frequently used in aquaria, both for ornament and to oxygenate the water.

Nuphar, or *Nymphozanthus,* is a genus of yellow-flowered plants occurring in the northern hemisphere. *Nuphar advena* is the common yellow water-lily or spatterdock of the marshes.

Nymphaea or *Castalia* contains the showy-flowered water-lilies so frequently grown in artificial ponds. Northern hardy forms are white- or sometimes pink-flowered and fragrant. Many tropical species have red, yellow, blue, or pink flowers of great beauty. These flowers float on the surface of the water, as do the large cleft leaves. The fruit is a berry containing many seeds, each enveloped in a spongy aril. The fruit ripens under water, the mature seeds floating upward from the fruit and drifting about, by means of the air bubbles contained in the aril. Eventually each seed sinks to the bottom.

Nelumbium is a genus which contains but two species, *N. lutea,* a native of the southern half of North America, and *N. Nelumbo,* of Asia and the East Indian Islands. The American species is pale yellow-flowered, the flowers and also the large peltate leaves standing well above the surface of the water. The Asiatic species is the Sacred Lotus, which has showy fragrant pink flowers of great beauty. The fruit of the lotus is a curious obconical receptacle in the top of which are embedded the many carpels. At maturity the receptacle is very light and dry, so that when broken from its stalk it floats on the water, carrying the seeds about until it breaks apart. The seeds of the lotus are used as food by many peoples, especially in Asia.

Victoria regia, the giant water-lily of the Amazon, is related to *Nymphaea.* It is a plant of tremendous size, the floating leaf with its upturned rim often having a diameter of six feet or more. The flowers are likewise very large. The seeds are used as food in the Amazon valley, where the plant is native. (R.M.W.)

WATERMELON. *Citrullus Vulgaris;* Gourd Family.

WATERSHED. The area which supplies water to a stream and its tributaries by direct runoff and by ground water runoff is the drainage area or watershed for the stream. The yield of the watershed is the direct runoff plus the ground water runoff for a given period of time. That part of yield which is used for industrial or domestic purposes is called the draft. When the draft approaches the yield, the latter may be adjusted by the construction of an impounding reservoir. (C.W.C.)

WATKINS-FACTOR. Factorial Development.

WATT. A metric unit of power. The absolute watt is equivalent to 10^7 ergs (one joule) of work per sec. The watt is especially convenient in electrodynamics, because the practical electrical units are so chosen that the product of the current (amperes) by the electromotive force (volts) at any instant equals the power in watts. The international watt is so defined in terms of the international ampere and the international volt, and thus differ slightly from the absolute watt. One horsepower is equal to about 746 watts, so that the kilowatt (1000 watts) is approximately 1.34 hp. (L.D.W.)

WATT-HOUR METER. Integrating Meters.

WATTMETERS. A wattmeter has two coils, one fixed, the other capable of turning in the field of the first, both coils being without iron cores. The fixed coil is connected in series with the main circuit, so as to carry the whole current (or, with d-c instruments, a known fraction of it, as determined by a shunt). The movable coil, which is of high resistance, is connected across the terminals of the "load," that is, that portion of the circuit in which the power is to be measured, and the small current in the coil is therefore proportional to the voltage between these terminals. This coil turns against a hairspring, and since the torque is proportional to the product of the currents in the two coils, it is proportional to the product of the main current by the terminal voltage, that is, to the required power. The scale may therefore be graduated directly in watts. The wattmeter may be replaced by an ammeter (in series with the load) together with a voltmeter (across the load terminals). To obtain the power it is merely necessary to multiply their readings together; except that in the case of alternating currents with reactance in the load, this product must also be multiplied by a "power factor" equal to the cosine of the phase angle. (L.D.W.)

WAVE CYCLONES. Cyclones which develop in the temperate zone are waves on frontal surfaces in contrast with tropical cyclones in which fronts are not prominent features. Cyclones develop more frequently along a Polar front than any other place, but they also appear on the inter-tropical front and on other minor fronts. It is on the Polar front, however, that wave cyclones develop into major storms and climax their lives as large-scale vortices. Wave cyclone development occurs in the following sequence of events.

1. A fairly prominent front exists between two air masses of some density contrast. The frontal slope is approximately in equilibrium.

2. A perturbation of wave form with the major motion nearly horizontal, rather than vertical (as in gravity

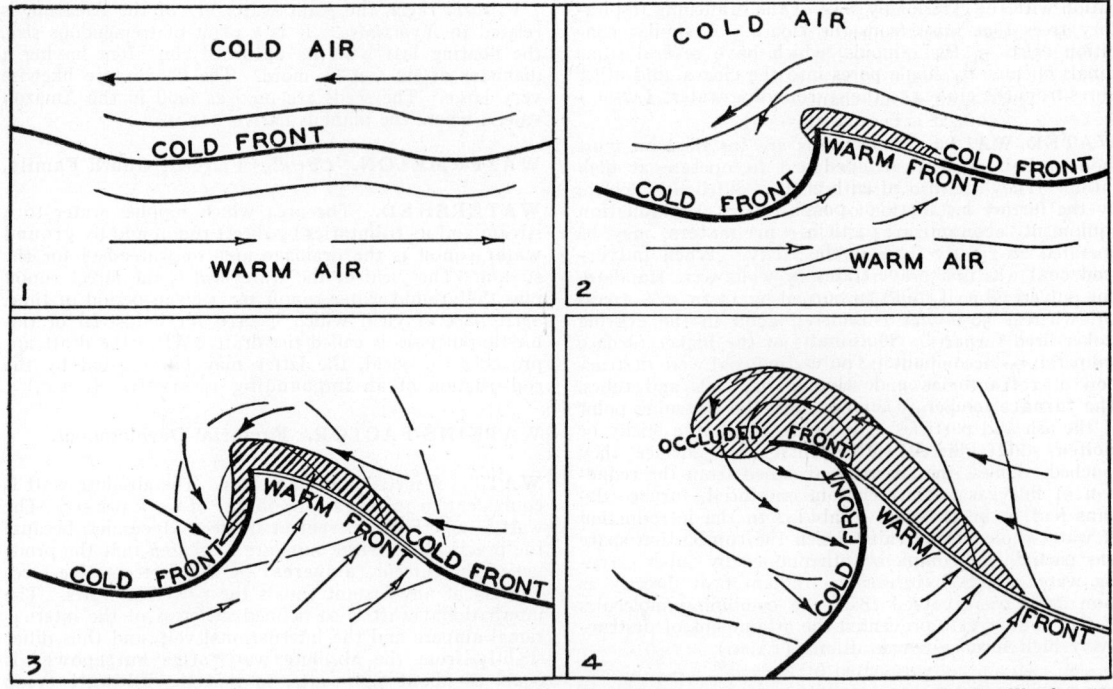

History and development of an extra-tropical cyclone. (Forming occlusion.) (*Gillmer and Nietsch, Clouds, Weather and Flight.*)

waves), develops on the frontal surface. One part of the front is bent toward the cold air as a warm front and an adjacent part is bent from equilibrium toward the warm air as a cold front. A deep indentation into the cold sector (cold air mass) occurs with the crest of the wave at the maximum point of indentation.

3. The wave moves along the frontal zone, but while it moves it also begins to occlude, i.e., the warm sector begins to close because the cold front moves more rapidly than the warm front. Pressure falls about the center but most rapidly in the direction toward which the crest of the wave is moving.

4. As the cold front overtakes the warm front, occlusion continues and the warm-sector air is squeezed aloft between the frontal surfaces.

After the fronts have occluded and the warm sector is gone, only an approximate vortex remains which slowly loses intensity. Typical warm and cold frontal weather are associated with both fronts respectively. In the fall, winter, and spring months of the year, most of the rain and snow falling over the United States and Canada is due to wave cyclones. Movement of the cyclones is generally easterly with a north or south component but it should be remembered that wave cyclones tend to move along the frontal zone in which they developed. Rate of movement varies from less than 100 miles per day to as much as 1000 miles; on the average, about 400–500 miles in summer and 600–700 miles in the winter. Velocity of the wave crest (center of the cyclone) is greatest during the early stages of the storm and normally decreases as it approaches maturity. Wave cyclones often occur in a series with a definite wavelength, thus have a uniform spacing. Under the influence of a well-established major circulation pattern the paths of cyclones can be predicted and the general aspect of weather estimated for several days. With transient general circulation patterns, however, the paths of cyclones become quite variable and accurate prediction from one group of cyclones to the next becomes next to impossible. (P.E.K.)

WAVE FILTER. Electric Oscillations and Electric Waves.

WAVE GUIDE. While it is ordinarily considered necessary to have a conductor surrounded by an insulating medium to conduct electrical energy from one point to another, it can be done in the reverse fashion, i.e., the insulating medium conducts the energy and is surrounded by a conductor which guides the energy flow. This is the wave-guide principle. Since the success of the method depends upon the correct dimensions of the guide and these vary as an inverse function of frequency the method is not feasible except at ultra high and higher frequencies, but at these ranges it is the standard means of conducting the energy. The guide may be considered a coaxial line without a center conductor or may be regarded as a pipe or tube down which the energy flows. Electrically the energy is carried by a radio wave which is reflected from the guide surfaces so the result is guided transmission along the tube. These guides have a variety of uses at the higher frequencies, being used to form resonant cavities, to conduct radio-frequency power from point to point, to serve as radiators or collectors by shaping them as open-ended horns, etc. (L.R.Q.)

WAVE MECHANICS. Wave mechanics is a more or less direct outgrowth of the **quantum theory,** and an integral part of **quantum mechanics.** The fact that radiant **energy** (light, x-rays, etc.) is certainly emitted by atoms or molecules and is as certainly done up in parcels, called quanta, the magnitude of each of which is definitely associated with a vibration or wave frequency of some kind (see **Planck's Law**), leads one to inquire what there is about an atom or the electrons in it that has to do with vibrations or waves. The now classic **Davisson-Germer experiment** gave most conclusive evidence that electrons actually do have wave characteristics even when flying freely through space (or at least when they strike and rebound from something like a crystal), and that, again, the energy of their motion is expressible in terms of a wave or vibration frequency. Even whole atoms are reflected by crystals as if they were waves, as shown by the experiments of Ellett, Olson, and Zahl.

Such facts have given rise to the idea that perhaps all physical processes are, in the last analysis, wave processes, with frequencies or wavelengths appropriate to the quanta into which the energy divides itself. Indeed it seems not impossible that the very atoms of which matter is composed are complex wave patterns, and that when an atom changes from one "quantum state" to another, it is because this wave pattern changes to one of different frequency. (A useful analogy is found in a metal plate clamped at the center and covered with sand; when stroked with a violin bow it shows one or another of a variety of possible complex wave patterns.) Instead of being particles which revolve in orbits like planets, the electrons in the atom, according to this conception, become wave trains reverberating like sound in a closed room, and setting up stationary interference patterns corresponding to the stationary quantum states. It is of such boldly revolutionary concepts that the new wave mechanics is built. The mathematical formulation of the theory has been developed largely by de Broglie and by Schroedinger. (L.D.W.)

WAVE PROPAGATION. In the propagation of a train of waves, each particle of the medium undergoes some sort of periodic variation, represented by the departure of some periodic variable from a neutral or zero value. This variable may be a position (geometrical coordinate), a pressure or other stress, a magnetic intensity, an electric intensity, a temperature, etc. Let the departure of the variable from its zero or equilibrium value at any instant be represented by d. If the variation is harmonic, it may, for any one particle of the medium, be represented by the equation $d = a \cos 2\pi\nu t$; in which a is the amplitude and ν the frequency of the periodic variation, and t is the time reckoned from an instant when d is at its maximum value. (See **Harmonic Motion.**) But if we consider different particles, we must also provide for differences in phase, by adding an adjustable phase term Δ:

$$d = a \cos (2\pi\nu t + \Delta). \tag{1}$$

Now if a train of waves is moving in a homogeneous medium in the direction, let us say, of the X-axis, this phase term Δ is a linear function of x; so that, if we could arrest the process for a moment and examine conditions along the X-axis, the phase of d would be found to differ by equal amounts at equal intervals of distance. This linear function has the form

$$\Delta = \Delta_0 - \frac{2\pi\nu}{V} x;$$

in which Δ_0 is the value of Δ at the origin, and V is the speed of the wave propagation. Also, the amplitude a is in general some function of x, called a "wave function"; let it be represented by $\psi(x)$. Substituting these expressions in (1), we obtain the simple harmonic "wave equation":

$$d = \psi(x) \cdot \cos \left[2\pi\nu \left(t - \frac{x}{V} \right) + \Delta_0 \right]. \tag{2}$$

According to Fourier's theorem of **harmonic analysis,** any periodic variable can be expressed as the sum of a number of simple harmonic variables, so that any wave equation in one dimension (x) may be written by equating d to a series of terms similar to (2) but with different values of $\psi(x)$, ν, Δ_0, and sometimes also of V. (See **Group Velocity** and **Vibrations and Waves.**) (L.D.W.)

WAVELENGTH CONSTANT. Propagation Constant.

WAVELLITE. The mineral wavellite is a hydrous phosphate of **aluminum,** formula $(Al \cdot OH)_3 (PO)_4 \cdot 5H_2O$. It is **orthorhombic** but crystals are of rare occurrence as it is ordinarily found in crusts or radial aggregates, sometimes fibrous. Its hardness is 3.5–4; specific gravity, 2.3–2.4; may be of various colors, gray,

blue, green, yellow, black, or colorless. It has a vitreous luster, and is translucent. This mineral is of secondary origin, probably formed by waters bearing phosphoric acid which have acted on aluminum minerals. Wavellite is found in Saxony, Bavaria, Devonshire, from whence it was originally described; and in the United States in Chester and Cumberland Counties, Pennsylvania; and Montgomery and Garland Counties, Arkansas. It was named after its discoverer, Dr. Wavel. (E.S.C.S.)

WAVEMETER. Frequency Meters and Wavemeters.

WAVES. Vibrations and Waves; Wave Propagation.

WAX MOTH. Insecta, Lepidoptera. A small **moth** whose **larva** lives in the combs of bee hives, spinning silken tunnels as it burrows through them. Although it eats the wax of which the combs are built and will attack the pure wax in stored comb foundation, careful studies have shown that it does not thrive on a pure wax diet. The other materials in old combs are a necessary source of nitrogen, without which the **caterpillar** may live but cannot grow and develop normally.

The moth does not become a serious pest in strong colonies of bees but it is sometimes a cause of serious damage in weak colonies and stored combs. It may be killed by fumigating stored supplies with **carbon disulfide.** This fumigant is highly explosive and must be used with due precautions to prevent ignition. The vapor is heavier than air and may collect in dangerous quantities in low places, but fortunately its disagreeable odor makes it easy to detect. (A.W.L.)

WAXES. Esters.

WAXWING. Aves, Passeriformes. A bird (**Aves**) of the northern hemisphere, smoothly gray and brown, with limited yellow shading and black marks, and red tips of horny material on some of the wing feathers. The name waxwing refers to these tips. The head bears a sharp crest. One species, the cedar waxwing or cedar bird, *Bombycilla cedrorum,* is peculiar to the United States. The Bohemian waxwing, *B. garrula,* nests in northern latitudes in Europe and North America and is occasional through the northern half of the states. A third species, the Japanese waxwing, *B. japonica,* breeds in eastern Asia. (A.W.L.)

Cedar waxwing, *Ampelis cedrorum.* Soft cinnamon-brown. A yellow band across the end of the tail. A crest on the head. The end of each secondary feather in the wing has a red tip like a drop of sealing-wax, whence the name of the bird.

WEAKFISH. Pisces, Teleostei. A food fish, *Cynoscion regalis,* belonging to the **drum** family. It attains a length of $2\frac{1}{2}'$ and is found from Cape Cod to Florida. Also called the squeteague. Two related species of more southern distribution are the white and the spotted weakfish. (A.W.L.)

WEASEL. Mammalia, Carnivora. A small carnivorous animal, slim and short-legged, related to the minks, ferrets, and martens. Several species of weasels occur in Eurasia and a dozen North American species have been described. The common or long-tailed weasel, *Mustela noveboracensis,* ranges from Illinois to Carolina and northward into Canada. It is brown above and yellowish below in the summer, becoming white in winter, with a black-tipped tail, in the northern part of its range. The winter phase has also been called **ermine.**

The name long-tailed weasel applies also to another species, *M. longicauda,* found only on the plains from Kansas northward. The short-tailed weasel, *M. cicognanii,* resembles the common species but is white below in summer; it ranges from the northern states to Alaska. Other species are of limited distribution or are not commonly known. (A.W.L.)

WEATHER. All the atmospheric phenomena occurring for short periods such as hourly or daily. Thus, existing phenomena constitute weather while the average weather over long periods of time constitutes climate. (P.E.K.)

WEATHER ANALYSIS. Synoptic weather reports received from a relatively large area, i.e., part of a continent, or part of the world, are plotted on a weather chart in a systematic manner. The chart is then analyzed in such a manner that fronts, isobars, isallobars, isotherms, precipitation areas, clouded areas, and air masses are indicated. Also, forecast movements of prominent features are usually indicated. This is an analysis of current weather. It is a "moon-view" of the weather of the area of interest on a single chart or map. (P.E.K.)

WEATHER FORECASTING. Anticipation of the next hour's, the next day's, or the next week's weather is the work of a weather forecaster. Forecasts may be divided into several categories.

1. Micro-forecasts or highly detailed forecasts for small regions.

2. Macro-forecasts or general forecasts for large areas.

3. Short-period forecasts covering generally 24 hours or less, but sometimes 48 hours.

4. Long-period forecasts covering more than 48 hours and up to and including a week.

5. Seasonal forecasts.

An example of a micro-forecast is one issued for aircraft operation for one flight in which the period will be only a few hours. Cloud bases and tops are described to within a few hundred feet, depth and degree of roughness of air are outlined, temperature and winds within very small limits of variation are given the pilot for computation of a flight time and drift. Location of thunderstorms and icing conditions also are cited when necessary.

An example of a macro-forecast is found in one issued for farming operations covering a few days, i.e., for cutting and curing a crop of hay in a state or several states. In this forecast the amount of cloudiness, the general drying characteristics of the atmosphere, and the probability of dew or rain are outlined. Forest-fire hazard forecasts also are macro-forecasts covering humidity conditions in general for a large area for several days.

Forecast accuracy depends on two factors, (1) the required detail of the forecast, (2) the length of the forecast period. Micro-forecasts are usually inaccurate for more than 24 hours and often at the end of a 12-hour period some adjustments must be made. Macro-forecasts can be fairly accurate for several days. Forecasts for periods of more than a few days must be very generalized. The reason for failing accuracy as the forecast period grows longer lies in the fact that accelerations in atmospheric activities are ever-present in varying quantity and, as such, are not capable of being analyzed. Varying accelerations alter considerably the velocity of atmospheric phenomena until computed or estimated velocities on which a forecast is based are no longer valid. Forecasts then go off schedule. Another factor is introduced in the unknown quantity of vertical wind currents. Such currents are known to exist but there is no direct method of computing or estimating them. These currents alter considerably the wind structure of the atmosphere, which in turn changes circumstances on which a forecast is based.

A short-period forecaster works with many synoptic charts. He uses a surface map as the basic chart. Also available are maps of 5000, 10,000, 15,000, and 20,000', cross-sections of the atmosphere, temperature and humidity soundings up to the stratosphere, temperature difference between selected level maps, moisture pattern maps, streamline and temperature maps, as well as a variety of innovations he has found useful. Each one of these charts contains some pertinent data taken from synoptic observations. The forecaster transposes the current synoptic charts into charts as they are expected to appear at some future time. These are called prognostic charts. Features of the futuristic charts are then interpreted in terms of atmospheric phenomena which is the forecast for that future time.

A long-period forecaster uses other charts which are ordinarily mean-value fields of atmospheric properties, synoptic patterns of a generalized nature, and the mean wind blowing at some chosen altitudes such as 10,000'. Long-period forecasts must, of necessity, be stated in broad and general terms. Such forecasts are useful for some activities which are seriously affected by abnormal or large-scale changes in the weather. (P.E.K.)

WEATHER MAPS. Synoptic Reports and Charts.

WEATHER OBSERVATIONS. Throughout the world, at pre-selected sites on continents, on ships at sea, on aircraft in flight, certain properties and features of the atmosphere are observed at regular intervals of 3, 6, 12, or 24 hours. Also, along airways throughout the world properties and features of the atmosphere useful in aircraft operation are observed hourly. Usually included in these synoptic observations are: form of low, middle, and high clouds; state of weather, i.e., snow, rain, thunderstorms; visibility; height of lower clouds; total sky cover; direction and velocity of the wind; atmospheric pressure, temperature and dew point; relative humidity; tendency of the barometric pressure; total change in pressure in previous 3 hours; past state of the weather. Several other items, such as maximum or minimum temperature, time of onset or ending of rain, total snow on ground also are included in some reports. (P.E.K.)

WEATHER SYMBOLS. Observations of atmospheric conditions are stated in numerical values or in terms of symbols. Transmission of a weather observation always is partly in code, with symbols or numbers representing specific values, and partly in direct numerical values. Temperature and dew point and velocity of the wind are values that can be observed and transmitted directly. Type, amount, and direction of movement of clouds usually must be coded with a symbol or a number representing a certain value of each. Many meteorological quantities also are plotted in symbol form on a synoptic chart. Weather-symbol systems in use in the United States and Canada are two in number, (1) the airways' system, used for hourly reports, which consists of symbols and numbers, (2) the international system which consists of numbers only and is used for 6- and 12-hourly observations. (P.E.K.)

WEATHERING. The processes by which the atmospheric agencies, commonly associated with the weather, mechanically disintegrate or chemically decompose the rocks at or near the earth's surface. Mechanical weathering includes the effects produced by changes of temperature, the action of frost, etc.; chemical weathering includes the solvent action of water, the union of atmospheric oxygen with rock materials—oxidation, union with atmospheric carbon dioxide—carbonation, and the chemical combination of substances with water—hydration. (E.S.C.S.)

WEAVER. Pisces, Teleostei. Marine fishes (Pisces) with poisonous spines on the dorsal fins and opercula. They are found in European and South American waters. The greater weaver, *Trachinus draco,* or sting-bull of British seas, is an excellent food fish. (A.W.L.)

WEAVER BIRD. Aves, Passeriformes. A bird (Aves) of a large group found in Africa, Australia, and tropical Asia. They build remarkable nests, weaving their materials intricately and sometimes in complex forms. Some species are gregarious, building large nests for the entire colony. Many species of weavers are brilliantly colored. The group includes the ox birds, whydah birds, bishop birds, munias, and weaver finches. (A.W.L.)

WEB MEMBER. Bridge, Truss.

WEB PLATE. Bridge, Girder.

WEBWORM. Insecta, Lepidoptera. A caterpillar that surrounds the site of its work on plants with a mixture of silk and debris. Several insects belonging to different families are known as webworms. The burrowing webworms (Acrolophidae) eat the roots of **grass** and may damage **corn** when planted on sod ground. The sod webworms work at the base of the stem and damage **grasses, cereals,** and other plants. They belong to the subfamily Crambinae of the family Pyralidae, which also contains the cabbage and garden webworms, members of a different subfamily, Pyraustinae. The former affects cabbage and related plants and the latter attacks corn, cotton, and various species of garden plants. The European corn borer is closely related to the last two species.

Cultural methods are the most important in controlling these pests. They vary according to the species, the crop, and the conditions. (A.W.L.)

WEDGE. Machines.

WEDGE LINE. A line drawn symmetrically in a wedge of pressure along which the isobars are curved anticyclonically and symmetrically. Wedge lines or wedges of pressure are elongated anticyclones and are always U-shaped in contrast to trough lines which often are V-shaped. (P.E.K.)

WEDGE PHOTOMETER. In the many forms of bench photometer, the illuminations or luminous flux densities from the two sources to be compared are made equal by regulating the relative distances. In photometers of the wedge type, the same object is accomplished by pushing into the beam from the brighter source a graduated wedge of absorbing material until its intensity is cut down to equality with that from the other source. The scale reading on the wedge indicates the ratio in which the flux density has been reduced, and hence the luminous intensity ratio of the two sources. It is highly important that the absorbing wedge shall be "neutral," that is, not selective as to wavelength, in its absorption; otherwise there will be an alteration of color as well as of total intensity.

This same principle may be applied to **stellar photometry.** By means of a complicated system of diaphragms, lenses, and color screens, an "artificial star" image may be formed and reflected into the eyepiece of a telescope close beside the image of a star whose **stellar magnitude** is desired. The source of light for this artificial star is usually a small incandescent lamp whose brightness is maintained as nearly constant as is possible by using a storage battery of large capacity. In the path of the light from this lamp is placed the neutral optical wedge so that the brightness of the artificial star may be varied, with the amount of variation proportional to the position of the wedge. In determining magnitude with this instrument the observer sets the telescope on a star whose brightness is desired and adjusts the wedge until the artificial star and the star image formed by the telescope have the same ap-

parent brightness. The telescope is then turned to a star of known magnitude, e.g., a star of the north polar sequence, and the wedge readjusted. The difference in wedge positions on the two stars may be converted into the difference of magnitude between the stars by means of a calibration curve previously obtained for the instruments and the observer. (L.D.W., W.K.G.)

WEEDS. Any plant growing where it is not wanted and seeming to have no usefulness may be called a weed. Many plants become weeds when introduced to new regions, where they can grow rapidly. Cultivated land, in which competition with other plants is very much reduced, is such a region. Here small plants such as purslane, which would be choked out in competition with other plants, can spread rapidly. Attempted eradication by hoeing often fails to remove such plants, because of the ease with which they put out branching stems which root at the nodes and become reestablished. Other plants become weeds when introduced into countries in which they do not naturally occur and where they meet with little competition from native plants. The common white daisy, introduced into America in colonial times as a garden flower, grows rapidly in hay fields and by its presence greatly reduces the value of the crop. Another weed causing even greater loss in such situations is the common hawkweed or devil's paintbrush. The prickly-pear cactus, introduced into Australia and into Mediterranean countries, has become a troublesome weed. In the western plains the Russian thistle, *Salsola Kali,* becomes a pest which, however, may be of value at times because of its great drought-resisting ability. When young the plants are fairly good forage. When mature they are utterly worthless, breaking loose from their roots and rolling about as tumbleweeds, scattering seeds far and wide. Other plants such as dandelion and plantain, because of the coarse unsightly habit of their leaves, become weeds when they get into lawns.

Various means are used to control weeds. Whenever possible they may be kept down or eliminated by thorough hoeing, which prevents them from getting established. In other cases the tops may be prevented from forming, thus gradually starving the plants by preventing **photosynthesis.** Sometimes they may be prevented by covering the ground with paper or other materials. This method is successfully used in the pineapple fields of the Hawaiian Islands. Recently some success has been obtained by using certain lethal sprays, such as iron sulfate solution, which sticks to the broad leaves of such plants as kale or mustard, but does not materially harm the grain among which the weeds are growing. The plant hormones themselves have also been used to kill broad-leaf weeds, which is accomplished by accelerating the normal life cycle. The plant hormone widely used for this purpose is the so-called 2,4-D, which is the chemical compound 2,4-dichloro-phenoxyacetic acid. Weed killers have also been used to spray poison ivy, where the latter has become an obnoxious and unwanted weed. The ingredients of these sprays are selected for their capacity to produce ultimate death by interfering with the metabolism of the weed. (R.M.W.)

WEEVIL. Insecta, Coleoptera. A **snout beetle,** member of a large division of the order known as the Rhynchophora. These insects, with some exceptions, have the head prolonged into a snout which bears the small mandibles at its tip. The snout is most conspicuous in the nut weevils, where it exceeds the length of the body and is very slender. Some of the weevils are important pests, notably the cotton boll weevil and the granary weevil. The former lives as a **larva** in the squares and bolls of the cotton plant, preventing the normal development of seeds and lint, and the latter is found chiefly in stored grains. (A.W.L.)

WEIGHING METHODS. The use of a **balance** for measuring masses is something more than the mere placing of equal weights on the two pans. The chief reasons are that the refinement required often goes beyond the smallest weights in any set, and that the arms of a balance are never exactly equal in length.

The former condition is commonly met by the use of a "rider," a small weight (usually 1 milligram) sliding along a scale on the beam. The most approved procedure, however, makes use of the known sensibility of the balance, expressed in divisions of the pointer scale per unit excess weight on one pan. The arm inequality may be allowed for by the substitution method, in which the object to be weighed is first counterpoised, or nearly so, and then replaced by known weights on the same pan, the small difference in pointer reading being noted and the difference in weight computed from it.

Both difficulties can be met at once by the method of "double weighing." The object is first placed on the left pan and weights nearly equal to it on the right, the resulting pointer reading being r_1. The object and weights are now interchanged, with pointer reading r_2. Then if the sensibility of the balance is s, and if the weights used total a value w, the weight of the object is given by

$$W = w + \frac{r_1 - r_2}{2s}.$$

The sensibility is, in general, a function of the load, and before precise weighing is attempted, a table or a graph should be prepared from which s can be obtained from the value of w. When the pointer is used, it should not be allowed to come to rest, but its equilibrium position should be deduced from the extremes of its small oscillation (see **Damping**).

It is of course presumed in any case that the errors of the weights themselves have been accurately determined. Due allowance must also be made in precise weighing, for the buoyancy of the air, or rather for the difference of the buoyant force on the object to be weighed and that on the weights in the opposite pan. If the weights are mainly of brass (density 8.4 g./cm.^{3}), the corrected or "vacuum" weight of an object of volume V (cm.^{3}), weighed in air of density ρ (g./cm.^{3}), is

$$W_0 = \left(1 - \frac{\rho}{8.4}\right)W + V\rho,$$

where W is the uncorrected result of the weighing. Under ordinary laboratory conditions, the value of ρ is approximately 0.00119 g./cm.^{3}, giving 0.99986 as the coefficient of W in the above formula. (L.D.W.)

WEIGHT. Gravitation and Gravity; Weighing Methods; Errors.

WEIGHT, AIRPLANE. Airplane weight is the total weight of an airplane and its contents. As number of passengers, fuel, and baggage are variables, airplane weight actually is a variable, ranging between minimum and maximum values. The minimum weight employed for commercial designing in the United States is the weight empty with standard equipment, plus crew, plus full capacity of lubricating oil, plus 0.25 lb. of fuel per hp. maximum continuous engine rating. The maximum weight of an airplane is the gross weight which can be lifted from the longest available runway or water surface at sea level and raised to a safe flying altitude. Conservative aeronautic practice does not sanction flight with this maximum weight except in emergencies. Safety allowances are provided by specifying airworthiness requirements for commercial design. Then the maximum practical weight becomes that for which the airplane is certificated as complying with all the officially promulgated airworthiness requirements for normal operations.

The (empty) weight of an airplane is a matter of far greater importance to aeronautics than dead structural weights are in almost any other field. Insufficient structure is a hazard that is not countenanced, but unnecessary weight represents a continuous loss of pay load during the operating life of a commercial airplane. Therefore **weight control** is given great emphasis in airplane design and construction. (F.T.M.)

WEIGHT CONTROL, AIRPLANE. Adequacy of an airplane structure for the loads imposed on it and ability of the airplane to carry the maximum weight of fuel, passengers, and cargo at high speeds are of the greatest importance to aviation—to the designer, the builder, and the user of airplanes. Unfortunately any unnecessary margin of structural weight employed in meeting the first requirement will penalize the second, and vice versa.

Airplane weight affects the tactical performance of military airplanes. Commercial airlines are interested primarily in pay load, so weight control in design and construction is imperative for the airliner. The airplane of superior performance (speed, range, etc.), and which maintains adequate structural safety will be more attractive to the private owner. Here again, weight control is important.

After an airplane is built, and while it is in use, weight control continues to demand attention. If accessories such as flares or radio are added, there must be a corresponding reduction of freight or baggage capacity. Also, the effect on center of gravity needs investigation in order to maintain predetermined stability limits. Weight control must be practiced by every scheduled airline. Cargo compartments are provided fore and aft and cargo suitably apportioned to each in order to maintain trim.

All phases of aviation are "weight conscious," but none more so than manufacturing. In this field the specialized service of "weight control engineering" is becoming more and more essential to manufacturing as airplane sizes increase. The ultimate object is to establish the methods and control, both in design and production, whereby the lightest weight possible is achieved in a structure of adequate strength. The weight-control engineer alone cannot hope to control the final airplane weight closely. Designers, fabricators, and many others must be weight-conscious, with the weight-control engineer in continuous touch with all phases of construction and continuously cognizant of the weight being incorporated into the structure. Subassemblies are weighed and tested for center of gravity so that the practical weight control is not delayed until the airplane is finished. However, the actual weight and center of gravity of the completed airplane is the final test of successful weight control.

Enough has been said to point out the importance of technical weight control in aviation. Weight must be closely watched and controlled from the inception of design, through construction, and during the life of the craft. A carefully compiled record of weight and balance must accompany an airplane throughout its commercially airworthy life. (F.T.M.)

WEIL'S DISEASE (Spirochetal jaundice). An acute infectious disease caused by a spirochete (*Spirochaeta icterohaemorrhagiae*). This spirochete is present in rats, and is excreted in their urine. Such urine, contaminating water supplies, is a major source of infection. The agent is also found in dogs, cats, mice, pigs, and horses.

The disease is characterized by a sudden onset after an incubation period of 10 days, of fever, chills, gastrointestinal symptoms, **jaundice,** enlargement of the spleen, and a tendency to hemorrhage into the skin and mucous membranes.

Immune serum and **bismuth** have been used in the

treatment with some success, although the main treatment is still supportive. (D.M.H.)

WEIR. A weir is any **dam** or **bulkhead** over which water flows, or it may be a bulkhead containing a notch through which water flows, the notch at no time becoming completely submerged. A weir is usually employed to measure the volume in a flow of water. This it accomplishes through the fact that a discharge through a weir bears a certain definite relationship to the **head** of water over its crest, and this head is comparatively easy to measure. Uses of a weir, then, are for the precise measurement of a large flow of water, or of one where peculiar conditions eliminate other methods of volume measurement. The weir is frequently used for measuring the flow of small streams, and the discharge from all sorts of hydraulic apparatus. It also is used for measuring hot water. When used with water which is not at atmospheric pressure and temperature, the general weir formulae do not apply, and the weir must be calibrated by a primary meter.

There are many different forms of weirs, such as sharp-crested and flat-crested, rectangular and V-notch, trapezoidal, broad-crested and submerged. Also the distance from the crest to the bottom of the channel and from the edge of the weir to the side of the channel

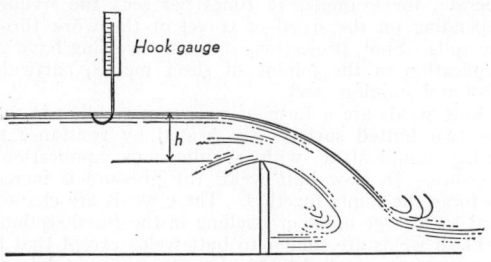

Sharp-crested weir.

distinguishes a weir. If the edge of the channel is also the edge of the weir, it is said to have no end contraction, and this type is recommended, since no corrections for end contraction need be made. The crest, however, should be high enough over the bottom that a standard crest contraction exists. The figure shows a cross-section through a trapezoidal weir, with sharp-edged crest. This type is better than the flat crest, as results may be duplicated with greater accuracy. The height of water over the crest should be measured at least $2\frac{1}{2}$ crest heads upstream, so that the water surface contraction will not be included in the reading of head. The head is measured by a hook gauge, which is a hook submerged in the water and raised until the point just touches the surface. To the stem of the hook is attached a scale, so that the head may be read. Where it is possible to flood the weir, and then shut off the water supply, the hook gauge can be used to give a zero reading, since the water will stand back of the weir at exactly the crest elevation. Where this is not possible, a surveyor's level can be employed to give the elevation of the crest in comparison with some reference point on the hook-gauge support.

If the flow is highly variable, it is better to use a triangular-notch weir than rectangular, because at small discharges the head does not become so minute. Both 90° and 60° V-notch weirs are used. A trapezoidal-shaped notch whose sides slope four on one theoretically gives the discharge without "end contraction" corrections, because the decrease of flow resulting from end contractions is compensated for by the flow through the small triangles on either side of the weir. The bottom width is used as crest length in the regular rectangular weir formula. This is known as the Cipolletti weir.

Weirs having flat or rounded crests can be used to measure discharge of water, and often the overflow section of a dam can be used in this way. The discharge is proportional to the length of crest, and to the $\frac{3}{2}$ power of the head over the crest, but the factor of proportionality must be established by experiments with models which duplicate hydrodynamically the proposed weir.

The most common formula for the discharge of water through a suppressed rectangular weir is the Francis formula:

$$Q = 3.33Lh^{\frac{3}{2}}.$$

Q is the discharge, cu. ft. per sec., L is the crest length in ft., and h the crest head in ft. Specific weir installations may call for corrections to this formula for end contractions or velocity of approach, or both. Furthermore, if the crest is very close to the bottom of the channel of approach, a correction for suppression of crest contraction is used. The discharge through V-notch weirs is given by the following formula:

$$Q = Ch^{\frac{5}{2}}.$$

Q and h have the meanings given above, and C is a constant depending upon the angle of the notch. It is approximately 2.5 for a 90° notch. (F.T.M.)

WELDED JOINTS. Fig. 1 shows various types of welded joints. Double-welded butt joints are those welded from both sides of the plate; single-welded butt

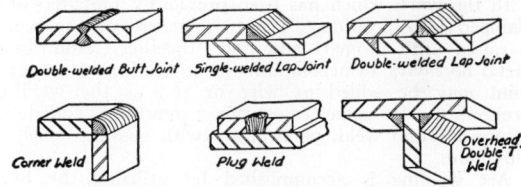

Double-welded Butt Joint   *Single-welded Lap Joint*   *Double-welded Lap Joint*

Corner Weld   *Plug Weld*   *Overhead Double T Weld*

(Hesse's Engineering Tools and Processes.)

joints are welded from one side only. In **pressure vessel** design, double-welded butt joints are considered to have a joint efficiency of 80%; single-welded butt joints 70%; double-welded fillet lap joints and single-welded fillet lap joints with plug welds have a joint efficiency of 65%. In structural practice, butt welds may be stressed to 13,000 lbs. per sq. in. in tension and 11,300 lbs. per sq. in. in shear. The allowable shearing stress in full-fillet lap joints is 1000 lbs. per lineal in. of length per $\frac{1}{8}''$ of fillet leg for ordinary strength welds; that is, a $\frac{1}{2}''$ weld has a strength of 4000 lbs. per in. of length; a $\frac{7}{8}''$ weld a strength of 7000 lbs. per in. of length. For welds made with coated rods, the strength is about 20% greater. (H.C.H.)

WELDING. Welding is a method of joining metals by means of fusion or by interdiffusion in the solid state under the influence of high temperature and pressure as in forge welding. Metals having similar composition may be united in one homogeneous piece by fusing together the edges in contact, or by additional molten metal of the proper characteristics deposited where it will form a fused joint with each piece. In any machine or structure there are numerous cases where permanent junctions between the component parts are required. Before the advent of fusion welding, these were made by riveting or bolting and forge welding.

In a number of instances within comparatively recent years, structures have been built with welded joints at a lower cost than was possible formerly. Typical of such practice might be mentioned the welding of joints in pipe lines, eliminating flanges, couplings, and cumbersome fittings. Bridges and buildings of structural steel have been fabricated by arc welding, replacing the older riveted connections. Modern shipbuilding practice is turning to welding as the principal means of fabrication. Where weight saving is as important as strength, welding is especially suitable because of the elimination of the weight of rivets and bolts, and often of the flanges in which they are seated. Many complicated machine parts, which at one time could be produced only by casting, are now built up from structural steel shapes by welding. To the above illustrations may be added the aircraft structures, where, especially in the fuselage tubing, welding has become standard practice. Machinery frames and bases, tanks, and steel frames are welded. Even high-temperature and pressure vessels, such as boiler drums, are now accepted if welded in conformance with standard procedures.

Welding can be practiced on castings, on rolled and forged metals, and on thin sheet metals, using techniques appropriate for the thickness of section being welded. Steel, iron, copper, nickel, aluminum, duraluminum, and even magnesium may be welded. The principal methods of welding are gas welding, arc welding, resistance welding, pressure welding, and combination methods such as atomic-hydrogen welding.

Gas welding is usually accomplished by the oxyhydrogen or oxyacetylene methods, the latter being the most common, with oxyhydrogen welding being used primarily on aluminum. High pressure oxygen and acetylene are led through reducer valves to reduce the pressure to a torch where they are mixed and burned at the tip. The flame, either oxidizing, reducing, or neutral, is adjusted to the proper size and brought into contact with the work which has been previously prepared and clamped into position. The temperature of the work is raised until it melts and flows together. Additional metal necessary to accomplish the formation of a sound joint may be added as wire or rod as the welding progresses. In some cases fluxes must be added to insure a sound weld, particularly with stainless steel or aluminum.

Arc welding is accomplished by utilizing the heat generated by an electric arc, either a.c. or d.c., to fuse the metals together. A common method is to use a motor-generator set to supply d.c. at relatively low voltage. One electrode is fastened to the work, and the welding rod, usually coated with a flux, is made the other electrode, the arc being struck between the rod and the work. Metal from the rod melts off and is fused into the joint. Automatic machines for arc welding use coils of wire to feed into the arc, flux being supplied from a hopper as the work travels under the arc. This process is also known as submerged arc welding. Another method of arc welding is the carbon arc in which one electrode is a rod of carbon. It is generally used on automatic machines. The wandering of the arc is minimized by placing a magnetic field near the arc to stabilize it. Extra filler material is fed into the arc if needed, and it is often shielded by gas generated from a chemically treated string or cord.

Atomic-hydrogen arc welding is a combination of gas welding and arc welding. In this case heat is supplied to the work by a stream of hydrogen passing through an arc struck between two tungsten electrodes. As the hydrogen passes through the arc, it is dissociated into its atomic form. When it recombines to form molecular hydrogen, a large amount of heat is generated in a small space, making an intensely hot flame. The hydrogen blankets the weld as it is made and largely prevents loss of alloying materials. This method also has been applied to automatic machines, no flux being necessary. A large variety of alloy steels for exacting specifications are welded by this method.

Heliarc welding is a modification of the atomic hydrogen method for the welding of magnesium; here helium gas is blown along a single tungsten electrode, the work being the other terminal as in ordinary electric arc welding. The helium serves to protect the highly reactive magnesium from burning.

Electric-resistance welds may be divided into a number of different categories but all have the common characteristic, that the metal is heated by its own resistance to a semi-fused or fused state by the passage of very heavy currents for very short lengths of time and then welded by the application of pressure.

Spot welding is accomplished by clamping the lapped edges of the work between two water-cooled copper electrodes usually having a small tip area. Pressures up to 50,000 lbs. per sq. in. are applied and currents up to 60,000 amperes. The heavy current which is applied for a fraction of a second melts the metal at the interface of the lapped pieces of the work, resulting in complete welding of the two pieces under the contact area of the electrodes.

Projection welding is a modification of spot welding and is much used in mass-produced articles. Small projections are stamped into the work at the desired positions. The electrodes are applied opposite this projection and thus many welds may be made simultaneously.

Seam welds are another type of resistance weld. Here the electrodes are in the form of rollers which oppose each other on opposite sides of the work. Current is passed usually intermittently between the electrodes as the work moves along. The current may operate, for example, 12 times per sec., the frequency depending on the speed of travel of the work through the rolls. Spot, projection, and seam welding have wide application in the joining of sheet metals, particularly steel and stainless steel.

Butt welds are a form of resistance welding in which the two butted surfaces are heated by resistance to a fusing temperature with simultaneous application of pressure. In slow butt welds the pressure is increased to forge the joints together. These welds are characterized by a large upset or swelling in the fused section.

Flash welds are similar to butt welds except that high pressures are applied suddenly after the metal is quite hot, resulting in a small upset or swelling in the heated area.

Pressure welding is also quite similar to butt welding except that the heat is usually applied by some source other than resistance to electric current. The parts to be welded are cleaned and carefully fitted together to exclude as much air as possible. Pressure is applied and the parts heated to a forging temperature where upsetting and welding occur. This method has been used extensively in the joining of pipe in the field. It is of interest to note that such an operation is a mechanized form of forge welding as practiced by the blacksmith long before fusion welding methods were known. Some of the most modern welding techniques make use of the principle of interdiffusion in the solid state under pressure without melting or fusion of any part of the joint. (F.T.M and C. R. TAYLOR)

WELS. Pisces, Teleostei. A catfish, *Siluris glanis,* found in European rivers east of the Rhine. It is a large fish, attaining commonly a length of 6', and a maximum of twice that figure. (A.W.L.)

WELSBACH MANTLE. This is a gauze made out of thorium and cerium oxides which is used for gas mantles. The heat derived in the combustion of the gas is converted by this mantle into light energy. (R.K.S.)

WENTLETRAP. Mollusca, Gasteropoda. A marine mollusk with a white shell of elongated conical form, with many convex whorls bearing prominent ribs. Also called spiral staircases. The 200 species are widely distributed. (A.W.L.)

WERNERITE. The mineral wernerite is a silicate of calcium and aluminum which contains also some soda and chlorine. It can be considered as an isomorphous mixture of two molecules corresponding to the following compositions:

$CaCO_3 \cdot 3CaAl_2Si_2O_8$, called meionite, and
$NaCl \cdot 3NaAlSi_3O_8$, called marialite.

Its **tetragonal** crystals are coarse and thick, often very large. It occurs also in massive forms. It has a distinct prismatic **cleavage;** subconchoidal fracture; is brittle; hardness, 5–6; specific gravity, 2.66–2.73; luster, vitreous to rather dull; color, white to gray, red, green, or blue; translucent to nearly opaque. Wernerite is found in the **metamorphic rocks,** particularly those rich in calcium, also in contact metamorphic deposits in limestones. It has been found in **basic** igneous rocks, probably as a secondary mineral. Notable localities are Lake Baikal, Siberia, Arendal, Norway, and Madagascar. In the United States it is found in Massachusetts, New York, and New Jersey. Grenville, in the Province of Quebec, Canada, is an important locality. Wernerite is named in honor of A. G. Werner, a famous German mineralogist (1749–1817). (E.S.C.S.)

WESTON CELL. Standard Cell.

WET-AND-DRY-BULB THERMOMETER. Hygrometers.

WET-BULB TEMPERATURE. Humidity.

WETTING AGENTS. Surface-Acting Compounds.

WHALE. Mammalia, Odontoceti and Mystacoceti. A marine animal of completely aquatic habits. The whales vary to a considerable degree in structure, as is indicated by their classification in two orders. Whales 20' long are among the smaller species. Individuals 60 to 80 or 90' long have been recorded among several of the large kinds.

The whalebone whales (**Mystacoceti**) live on small marine animals which are separated from the water by the sievelike whalebone fringes of the jaws. The toothed whales (**Odontoceti**) include some actively predacious species. Species of both orders were once widely sought for their oil and whalebone, but the latter has long been supplanted by manufactured products and the oil has given way for many uses to petroleum oils. Sperm oil is still valued as a fine lubricant and sperm whales are killed also for a waxy material, spermaceti, used in the manufacture of cosmetics. The peculiar substance called ambergris is formed in the intestine of the whale and is used to make the odor of **perfumes** more persistent.

Whales are specialized for life in the ocean by the formation of the pectoral appendages into paddlelike flippers and by the complete loss of hind limbs. The tail is expanded horizontally into a pair of broad lobes, the flukes, which serve as a powerful swimming organ. The **respiratory system** is also highly specialized. The lung capacity is great and the nostrils are located high on the head, so that very little of the animal need be exposed to enable it to breathe. In some species they are combined to form a single opening. The whales are without hair, which would be useless to animals living always in the water, but they have a thick layer of fat, the blubber, as an insulation against the cold of the surrounding water. Much of the oil secured from their bodies is from the blubber, but the sperm whale also has an enormous cavity in the head filled with oil.

The various kinds of whales are briefly mentioned under the two orders. (See **Esters.**) (A.W.L.)

WHEAT. *Triticum sativum* and other species. Gramineae. Wheat is an annual plant producing the most valuable of cereal grains used by the white race. The plants grow either as summer annuals, seed being planted in the spring and the harvest gathered in the fall of the same year, or winter annuals, the seed then being planted in the fall and growing until stopped by cold weather, developing during that time an abundant root system which insures rapid growth in the springtime, the mature crop being ready for harvest in early summer. (See **Vernalization.**)

When wheat seeds germinate a small primary root system is formed by the development of the radicle or seed root. This primary root system lasts but a short time, being soon replaced by a system of **adventitious roots** arising from the lowermost nodes of the stem and extending outward and downward to fill the soil with an extensive fibrous root system. The stem of the wheat plant is 2–4' tall and usually hollow, although some species have solid stems. The dried stems form wheat straw, frequently used in the manufacture of straw board. The leaves are of the ordinary grass type. The **inflorescence** is composed of very short-stemmed **spikelets** attached alternately on opposite sides of a zigzag axis. Each spikelet has 2–8 flowers. In certain varieties known as bearded wheats the **lemma** of the flower bears a long bristle or **awn.** Most species of wheat, particularly those grown in temperate climates, are close-pollinated. Durum wheat and primitive species are cross-pollinated. Evidence indicates, also, that wheats grown in hot dry climates are cross-pollinated. The fruit or grain of cultivated wheat varies somewhat according to the species. In many kinds of wheat the fruit separates readily from the surrounding floral **bracts,** while in a few kinds the lemma and **palea** tightly enwrap the grain. The grain bears at its apex a tuft of short hairs called the brush and has a distinct groove along the side which was against the palea. In section, a wheat grain shows several very distinct parts. Externally there is a layer several cells thick called the **pericarp,** or ovary wall. Within this is a layer two cells thick, the tegmen, which is formed from the inner integument. The outer integument was absorbed during the development of the grain. Next is the **nucellus, a** single cell in thickness. These three layers constitute some 8% of the grain, and make up the substance which is called bran. Within these is the **endosperm,** which forms the bulk of the grain; the outermost layer of cells of the endosperm is the aleurone layer. In the basal portion of the seed next to the endosperm is the so-called germ, or embryo, from which the new plant may develop.

There are several ways of classifying wheats. Botanically they are separated according to the structure of the spikelet, the number of **florets** and the nature of the parts of the flower. Among the kinds of wheat recognized in this classification are einkorn, spelt, emmer, durum, and common wheat. Again, if the palea and lemma adhere to the grain, the wheat is classified as spelt wheat, while if the grain readily separates from these two parts it is naked wheat. If one turns to the grain itself there are hard wheats, in which the grain is horny and has a high **protein** content, and soft wheats with starchy grains. The nature of the soil in which the wheat is grown and the climate have considerable effect on the nature of the grain. Hard wheats are separated into hard spring wheat, hard winter wheat, and durum, the latter being especially rich in protein content.

While a considerable quantity of wheat is used directly as food for domestic animals, the greater part is ground into flour. This milling of wheat is a very carefully controlled process. The first step is the thorough cleaning of the grain, removing therefrom any other substance. During this process the brush of the grain is removed. The cleaned grain is then moistened slightly in order to soften the outer layers so that they may be more easily removed subsequently. The moistened grain is then passed between iron rollers. The first of these are corrugated and break up the grains. Each successive pair of rollers grind the grain into finer and finer particles. Early in this grinding the coarse flakes of bran and the germ are removed. These are disposed of as such or ground up separately. The ground grain is passed through fine bolting silks which insure a very even grade of fineness of the flour particles. Every precaution is taken in flour mills to prevent the accumulation of dust particles in the atmosphere, since these may form very dangerous explosive mixtures.

Flour is classified according to the amount of the grain included in the final product, into Graham flour, which contains the entire grain; whole wheat flour, which contains all the grain except about half of the bran; and straight bread flour, which results when all the bran is removed early in the grinding process. Wheat flour is used extensively in making breads, crackers, and pastries. Because it is so highly glutinous that pastes made from it will support their own weight, durum wheat is used in making spaghetti and macaroni. In making this product a thick viscous paste or dough is prepared. This is forced, under great pressure, through holes in metal dies. Metal pins may project into the hole in the die, causing the dough which is forced through to emerge as a hollow tube. This is macaroni. As they emerge, the tubes or strings are cut into suitable lengths and hung up to dry. Durum wheat alone has the necessary properties for making macaroni and similar substances. If any other wheat were used the product would break apart from its own weight. In addition to flour and macaroni, much wheat is used in making breakfast foods. The familiar puffed wheat results from heating wheat grains under pressure and suddenly releasing the pressure; the grain expands rapidly. Some wheat is used for making whiskey and certain varieties of beer. (R.M.W.)

WHEAT MIDGE. Insecta, Diptera. A minute fly whose **larva** develops in the growing kernel of wheat. Introduced from Europe into Canada early in the nineteenth century, the species was troublesome in New York about the middle of the century but has not been serious since. Cultural methods are an adequate protection. They include crop rotation, fall plowing to bury and destroy the larvae, and the destruction of all debris from infested fields. (A.W.L.)

WHEATEAR. Aves, Passeriformes. A bird (**Aves**), *Oenanthe oenanthe*, related to the thrushes and bluebirds. It nests in the northern part of Europe and in Alaska, and is widely distributed in the Old World and occasionally in the United States during its southern migrations. (A.W.L.)

WHEATSTONE BRIDGE. One of the simplest and best-known **bridge** networks for measuring electrical resistances. Referring to Fig. 1, let R_1 be the unknown resistance and R_2 a known resistance, preferably not very different from R_1, in terms of which R_1 is to be measured. R_3 and R_4, called the ratio arms, are two other resistances which may be varied continuously or by very small stages and the values of which are known, either in ohms or relatively to each other. From C to D

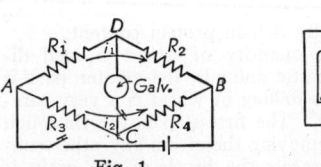

Fig. 1.

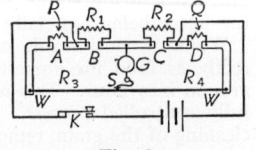

Fig. 2.

Sketches of simple and four-gap Wheatstone bridge circuits.

is the bridge proper, containing a **galvanometer**. To measure R_1, the resistances R_3 and R_4 are adjusted until the galvanometer shows no current, which means that C and D are at the same potential. It then readily follows from the **law of potential drop** that $R_1:R_2 = R_3:R_4$,

and hence $R_1 = R_2R_3/R_4$. Fig. 2 shows a "slide-wire" form having a graduated resistance wire and four gaps. The **Carey-Foster bridge** is a special type of Wheatstone bridge. (L.D.W.)

WHEEL-AND-AXLE. Machines.

WHELK. Mollusca, Gasteropoda. A moderately large marine **mollusk** with a spirally coiled shell. It is used as a bait in the cod fisheries and in Europe is eaten. (A.W.L.)

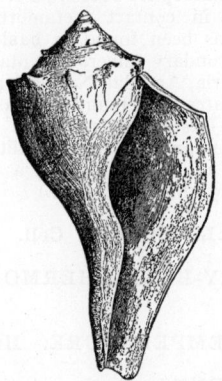

Fulgur, the whelk. (*U.S.B.F. Report, 1897.*)

WHIMBREL. Aves, Charadriiformes. Northern shore birds (**Aves**) with long legs and a long curved beak. Related to the curlews and plovers. The name is European; the species that occur in North America are called here the Eskimo, *Phaeopus borealis,* and Hudsonian, *P. hudsonicus,* curlews. (A.W.L.)

WHINCHAT. Aves, Passeriformes. A small European bird (**Aves**) related to the bluebirds and thrushes. It nests in the far north and winters in Africa. (A.W.L.)

WHINSTONE. A popular British term for **basic**, fine-grained **igneous rocks** belonging to the basaltic group and including **diabase, dolorite, spidiorite,** also greenstone and the **lamprophyres.** Synonym for **trap** or trap rock. (R.M.F.)

WHIP SCORPION. Arachnida, Pedipalpi. A large **arthropod** whose abdomen bears a whiplike posterior portion, although some species of the same order lack this terminal filament. The **pedipalps** are large and strong, either chelate or simple. The first pair of legs are modified as slender many-jointed sensory appendages. They are tropical animals, only a few species entering the southern part of the United States.

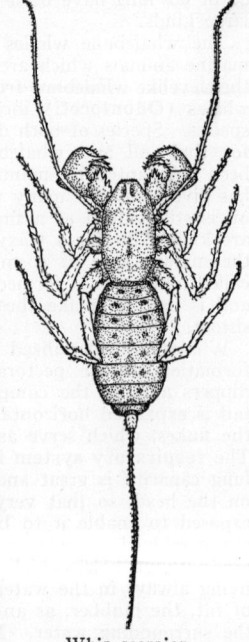

Whip scorpion.

In the Southwest many fears and superstitions are associated with these unpleasant looking animals, which are called vinegarones (also spelled vinegar roan and vinegaroon). They discharge a disagreeable, sour-smelling secretion, but are entirely harmless to man. (A.W.L.)

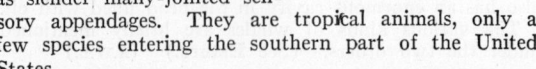

WHIPPOORWILL. Aves, Caprimulgiformes. A nocturnal bird, *Anthrostomus vociferus,* with a short beak and wide mouth, adapted for taking insects in flight.

Whippoorwill, *Antrostomus vociferous.* Mottled grayish, reddish, and white. Long wings, conspicuous bristles around the mouth.

Its call has been likened to the words used in its name. Often the three syllables are repeated over and over scores of times without cessation. One of the **nightjars.** (A.W.L.)

WHIRLWIND. Small cyclonic whirls of air usually not more than a few hundred feet high and several tens of feet in diameter. If the whirls pick up dust they are more commonly known as dust-devils. A few whirlwinds grow larger than the average and do some damage, particularly to grain fields during harvest. Whirlwinds originate in a shallow layer of very unstable air lying immediately above the ground and are therefore primarily afternoon phenomena. (P.E.K.)

WHISTLER. Mammalia, Rodentia. An animal related to the woodchuck. A **marmot.** Its range extends from Alaska to Montana and Washington. (A.W.L.)

WHITE ANT. Isoptera.

WHITE DWARFS. In the article on **giant and dwarf stars** it was shown that more than 99% of the observed stars may be classified either as super giants, giants, or main sequence stars. A few stars do not fit into this general scheme at all and form a unique and little understood class known as the white dwarfs. The first star of this class to be discovered was the companion star to **Sirius.** The orbit of the pair has been computed, and from it the mass of the companion was computed and found to be about the same as that of the sun. Its **absolute magnitude** was known and the **luminosity** found to be about 1/360 that of the sun. On the ordinary giant and dwarf hypothesis such a star should be in the dwarf M-type **spectral class** having low brightness per unit area. However, when its spectral class was finally determined it was found to be of class F and hence hotter than the sun and brighter per unit area. From the brightness per unit area and the intrinsic brightness, the diameter can be computed, and it is found to be about 30,000 miles, or of planetary dimensions. This means that we have an object of stellar mass and planetary dimensions, and an unbelievably great density, of the order of magnitude of 30,000 times the density of water. At such a density a cu. ft. of this stellar material would weigh 935 tons.

This star is one of the general class known as white dwarfs, all with spectral classes between A and F. The total number of such stars is not known, for they are all apparently very faint and cannot be detected as white dwarfs except by spectrographic analysis, which is difficult for faint objects. Furthermore, their intrinsic brightness is so small that only the closer ones would be observed at all (the few known at present are all within 15 **light years** of the sun).

The problem of the internal constitution of objects of such great density is as yet far from completely solved.

The best hypothesis is that the material consists of atomic nuclei stripped of all external **electrons** and tightly packed together by gravitational compression. Such matter would be in a state much like a single huge molecule, or crystal. (W.K.G.)

WHITE FLY. Insecta, Homoptera. A minute sucking **insect** related to the phylloxerans and scale insects. The adult has four wings and is mealy white. In spite of their small size these insects are important pests. One species, *Asterochiton packardi,* attacks strawberry plants, one, *Dialeurodes citri,* is a pest of citrus fruits in Florida, and two attack both flowers and vegetables in greenhouses.

Cyanide fumigation is recommended for the control of the greenhouse species, while spraying with oil emulsions is effective against others. In Florida an ingenious method of spraying with an infusion of the spores of fungi has been devised. Three species of fungi live on the immature insects and destroy them in large numbers when the colonies are once inoculated. (A.W.L.)

WHITE LEAD. Lead (Carbonates).

WHITE LIGHT. Color.

WHITE WHALE. Mammalia, Odontoceti. A small species related to the **narwhal.** The **beluga.** (A.W.L.)

WHITEFISH. Pisces, Teleostei. A lake fish of the **salmon** family, genus *Coregonus.* The several species are excellent food fishes and the common whitefish of the Great Lakes is among the most important in the inland fisheries of the United States. Several species are known by various distinctive names, among them pilotfish, whiting, cisco, bluefin, and tullibee. (A.W.L.)

WHITENOSE. Pisces, Teleostei. A common fish (Pisces), *Moxostoma anisurum,* of the Great Lakes and Ohio River basins, related to the redhorse. It reaches a length of seventeen inches and is a good but not an important food fish. (A.W.L.)

WHITETHROAT. Aves, Passeriformes. A European **warbler.** The common species, *Sylvia cinerea,* is gray-brown above and whitish below, and the lesser whitethroat, *S. curruca,* is gray above with dark brown ear coverts and tinged with pinkish on the breast. (A.W.L.)

WHITING. Pisces, Teleostei. 1. Marine fishes, *Menticirrhus,* of the **drum** family. One occurs on the Atlantic coast from Maryland to Brazil and another, called the silver or surf whiting, is found from Virginia to Texas. 2. A member of the **cod** family found on the Atlantic coast of North America. Also called the silver **hake** and stock-fish. (A.W.L.)

WHITWORTH. Screw Thread.

WHOOPING COUGH (Pertussis). An acute contagious disease of childhood, caused by a small bacillus, *Hemophilus pertussis,* first identified by Bordet and Gengou in 1906. Nearly 50% of the cases occur during the first two years of life and most of the deaths are in this age group. During the first year of life the mortality rate is at least 25%. In children over 5, the disease is rarely fatal.

The incubation period of whooping cough is from 5–15 days. The disease may run a mild brief course, or it may be a severe illness, especially in the very young child, lasting 6 weeks or more.

The symptoms of the first week and a half are mild and are those of **bronchitis.** Although the disease is most contagious at this stage, diagnosis is difficult unless there is a history of exposure. Instead of subsiding, the coughing becomes more frequent, occurring in paroxysms, terminating in the characteristic whoop. Several exhausting paroxysms may occur together until a small amount of mucus is coughed up, giving immediate relief. Vomiting may occur after these paroxysms. **Convulsions** are not uncommon in infancy. Nose bleeds and hemorrhages beneath the **conjunctivae** of the eye appear when the coughing is very severe. Hemorrhages in the lungs and brain may also occur. This active stage of the disease averages between 3 and 6 weeks, the intensity and severity of attacks slowly diminishing.

The most important complication of whooping cough is bronchopneumonia, which causes most of the deaths from the disease. **Atelectasis** and **bronchiectasis** may occur. **Otitis media** due to secondary invading organisms is frequently encountered.

The prevention of whooping cough by immunization with **vaccine** has been very successful. It should be carried out in all children between the ages of 6 and 12 months. If an infant has been exposed to the disease, he should be given immune or convalescent **serum** immediately, to protect him against the disease and modify the severity of its course should it develop.

Treatment of the active case is general and supportive. Bed rest, sedation, and medication to control the cough are useful. (D.M.H.)

WHYDAH BIRD. Aves, Passeriformes. A brilliantly colored and long-tailed **weaver bird** of Africa. The several species constitute the genus *Vidua*. (A.W.L.)

WICHERT TRUSS. Bridge.

WIDE FLANGE BEAM. I-Beam.

WIDGEON, WIGEON. Aves, Anseriformes. A **duck** with a relatively small beak, widest at the base. The European species is occasionally taken in the United States, in addition to a native species called the baldpate and sometimes the American widgeon. A third species occurs in South America. All belong to the genus *Mareca*. (A.W.L.)

WIDMANSTAETTEN STRUCTURE. A microstructure often observed in steels and certain other alloys. It is a duplex structure in which one phase forms sets of parallel plates embedded in the matrix. (See **Metallography**.) Typical examples are found in medium carbon cast steels and in meteoric iron. (R.H.H.)

WIEDEMANN-FRANZ LAW. The most casual observation reveals that at ordinary temperatures the metals which are the best electrical conductors are also the best conductors of heat. (See **Thermal Conduction**.) In fact, if we calculate the ratio of the thermal conductivity (in calories per cm. per sec. per degree) to the electrical conductivity (in reciprocal ohms-centimeters) for a number of metals at 0° C., we get results like the following: copper, 0.00000156; platinum, 0.00000143; lead, 0.00000169; etc.; all in calorie-ohms per degree per sec. The ratio appears to be nearly constant.

The Wiedemann-Franz law states that the ratio of the thermal to the electrical conductivity for all metals is proportional to the absolute temperature T, and in the above units equal to $6.11 \times 10^{-9}\ T$. For 0° C. (273° absolute) this gives 0.00000147 calorie-ohm per sec. per degree, which is very close to the observed value for platinum. The Wiedemann-Franz formula is theoretical and its coefficient involves both the **Boltzmann constant** and the electronic charge. For most metals the ratio as observed is a little higher than that given by the formula, doubtless because of thermal conduction due to other causes than electronic activity. (L.D.W.)

WIEN'S LAWS. From a study of the **spectral energy distribution** of **thermal radiation**, W. Wien, in 1896, arrived at three laws relating to the radiation from a **black body**.

1. The wavelength λ_m of the spectral distribution, for which the radiation has greatest intensity, is inversely proportional to the absolute temperature T of the black body:

$$\lambda_m T = c_1.$$

Thus as the temperature rises, the "peak" of the distribution curve is displaced or shifted toward the short-wavelength end of the spectrum. This is commonly called Wien's "displacement law." The value of the "displacement constant" c_1 is about 0.2897 centimeter-degree.

2. The emissive power of the black body within the maximum-intensity wavelength interval $d\lambda$ is proportional to the fifth power of the absolute temperature:

$$dE_m = c_2 T^5 d\lambda.$$

Subsequent work by Planck and others gives the value of the constant c_2 as about 1.288×10^{-4} erg/cm.^{3} sec. deg.^{5}

3. Wien's third law is an attempt to express the spectral energy distribution of the radiation from the black body at temperature T, as follows:

$$dE_\lambda = A\lambda^{-5}e^{-B/\lambda T}d\lambda,$$

in which $dE\lambda$ is the emissive power within the wavelength interval $d\lambda$ and A and B are constants to be empirically determined.

The first and second laws are in accord with thermodynamic theory and with **Planck's equation**, and also agree very accurately with experiment. The third law is empirical, but is almost identical in form with **Planck's equation** and agrees with it closely except for short wavelengths. (L.D.W.)

WILD RICE. Rice.

WILDCAT. Mammalia, Carnivora. *Lynx.* Any of five species of **cats** found in North America, related to the Canada lynx but of slightly smaller size. One species, also called the bobcat, is found in forested regions throughout the United States and southern Canada, and extends into Mexico. It is now confined to wilder areas. (A.W.L.)

WILDEBEEST. Gnu.

WILLANS LINE. A characteristic curve for the **steam turbine** is that of steam consumption per kilowatt-hour versus load in kilowatts. From this curve the Willans line, conveying information of great value, can be plotted. The Willans line shows total steam consumption at each load. In conjunction with the load curve it can be used to find total **steam** required per day or any other period of time. The steam-rate curve for all turbines will exhibit a node corresponding to the most efficient load. This point will be observed on the Willans line as a break on the nearly straight slopes on each side of it. This is the load point at which the first overload valve opens. The slope of the Willans line is a result of the combined effect of the various losses in the steam turbine.

For the sake of comparison with the ideal turbine, the Willans line of an ideal turbine is shown in line *OC* in the figure. The line *OC′*, parallel to *AB*, would be the Willans line of a turbine having stage efficiencies less than those of the ideal turbine (100%) but, at the same time, having no no-load losses. The steam rate of such a turbine would be constant at all loads up to the most economical. The line *AB* represents the steam consumption of an actual turbine under specified operating conditions. Practically all steam consumption curves are straight lines between the no-load and most economical load points. Above the most economical load performance varies. In general computers assume that the steam rate at full rated load is 5% greater than at the most

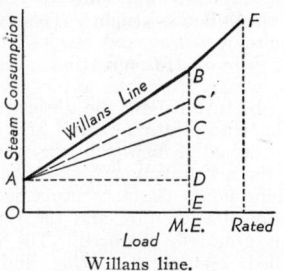

Willans line.

economical load, when the latter is about 80% of the rated load. The end point of the Willans line, F, is usually joined to B by a straight line unless the overload characteristics of the particular turbine are known. (F.T.M.)

WILLEMITE. The mineral willemite is an ortho-silicate of zinc, Zn_2SiO_4, occurring in hexagonal prisms, as masses or scattered grain. It has a good basal cleav-age; conchoidal fracture; is brittle; hardness, 5.5; spe-cific gravity, 3.9–4.2; subvitreous luster; usually some shade of yellow, yellowish-green, green, or reddish-brown, but may be colorless, white, or blue to nearly black; transparent to opaque. Much willemite is strongly fluorescent in yellow or yellowish-green hues. Willemite occurs associated with other zinc materials in Belgium, Algeria, the French Congo, South West Africa, and Greenland. In the United States, except for three oc-currences, one in Colorado, one in New Mexico, and one in Utah, Sussex County, New Jersey, is the only locality in this country for willemite and is the only one in which that mineral is found in quantity. Here it is found associated with zincite and franklinite, forming an important ore of zinc. It was named by the French mineralogist, Michel Lévy, in honor of King William the First of the Netherlands. (E.S.C.S.)

WILLET. Aves, Charadriiformes. A North American wading bird, *Catoptrophorus semipalmatus*, long legged and with a long straight beak. They are mottled grayish above and white below. Closely related to the sand-pipers. (A.W.L.)

WILLIOT DIAGRAM. A Williot diagram is a graphi-cal representation (drawn to any convenient scale) of the relative movements of the **panel points** of a loaded planar **truss**. Longitudinal **deformation** of the truss members produces horizontal and vertical **deflections** of all **joints** except the end panel points which are re-strained against movement in one or both of these directions. These longitudinal deformations are given by the formula:

$$e = \frac{PL}{AE}$$

in which e = Total deformation.
    P = Total **stress.**
    L = Length.
    A = Cross-sectional area.
    E = **Modulus of elasticity.**

If it were possible to draw a diagram of the deflected truss itself to scale, the relative position of the panel points could be found by plotting the truss as a series of connected triangles, using the deformed length of the members. Since the longitudinal deformations are small compared with the theoretical length of the unstressed members, such a diagram is not practical.

However, it is possible to represent the longitudinal deformations to scale in a diagram which will give the

correct relative movement of the panel points. In this case one of the joints is assumed to be fixed in position and one of the members at the joint fixed in direction.

The undeformed shape of a symmetrical truss, sym-metrically loaded, is shown in Fig. 1. The tension and

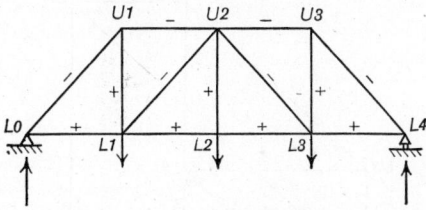

Fig. 1. Undeformed shape of a symmetrical truss, sym-metrically loaded.

compression members are indicated by $+$ and $-$ signs respectively.

Fig. 2 shows a Williot diagram for this truss which is based on the correct assumption that U_2L_2 is vertical before and after loading. As a basis for plotting the deformations it is also assumed that one end of this member, L_2, is fixed with relation to the movement of all other panel points. Tensile deformations should be shown as acting away from the joint and compressive deformations toward the joint. The plotting is begun by establishing the joint L_2 at L_2'. The relative position U_2 is found by laying off the deforma-tion of U_2L_2 to scale away from the joint in the assumed fixed direc-tion of this member. The relative position of the third joint of the triangle $L_1L_2U_2$ is obtained by drawing the deformations of the members L_1U_2 and L_1L_2 from the points U_2' and L_2' respectively in the correct direction and parallel to these members in the unloaded truss. The intersection of perpen-diculars from the end of these de-formations locates the position of L_1 relative to U_2 and L_2 because joint L_1 in the loaded truss lies at the intersection of the deformed

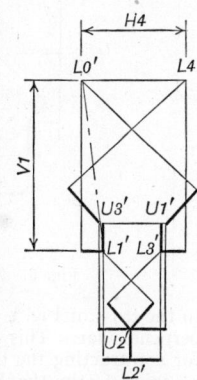

Fig. 2. Williot dia-gram for truss in Fig. 1.

lengths of L_1U_2 and L_1L_2. The relative positions of the other joints are located in a similar manner. Since the truss is symmetrical in all respects, U_2L_2 will be vertical in the loaded truss and the diagram will give the correct relative and actual displacements. The relative dis-placement of any two joints is obtained by connecting the corresponding points with a straight line. The dash-dot line $L_0'L_1'$ indicates that joint L_1 has moved downward and to the right with respect to L_0. The vertical deflection of L_1 is the vertical projection of this line which is marked V_1 on the diagram. The hori-zontal movement of L_4, which is restrained against downward movement by the support, is the distance H_4. Because of symmetry it is not necessary to actually plot points U_3', L_3', and L_4'.

The full lines of Fig. 3 show a Williot diagram based on the incorrect assumption that L_0L_1 is horizontal in the loaded structure. L_0 is the joint which was as-sumed to be fixed in position. Since L_1 does not lie on a horizontal line through L_0, it is necessary to apply a correction diagram called a Mohr correction diagram. The points U_1', L_1', etc., of Fig. 3 represent the posi-tion of the panel points of the loaded structure relative to the unloaded structure assuming L_0L_1 to be hori-zontal. To correct for the fact that L_4 is at the same elevation as L_0, the truss must be rotated. In rotating, each joint moves vertically a distance proportional to its horizontal distance from the fixed joint and horizontally proportional to its vertical distance from this joint. Plotting these distances with their signs changed produces the truss shown by dash lines in Fig. 3. It should be noted that the correction diagram is similar

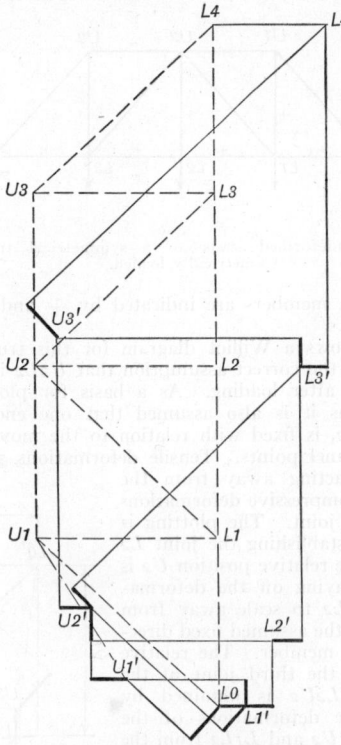

Fig. 3. Mohr correction diagram.

to the truss of Fig. 1 and the corresponding members are perpendicular. This similarity forms the practical basis for constructing the diagram. The vertical displacement of joint $L4$ in the Williot diagram is divided into as many parts as the lower chord of the real truss; then the remainder of the diagram can be constructed from similarity. The actual movement of any joint such as $U1$ is shown in direction and magnitude on the combined Williot-Mohr diagram by a dash-dot line joining $U1$ and $U1'$. In this case the diagram indicates that $U1$ has deflected downward and to the right. A Mohr correction diagram will always be necessary for any unsymmetrical truss with or without symmetrical loading. (C.W.C.)

WILSON CHAMBER. Cloud Track.

WILSON EXPERIMENT. The theory of the electromagnetic field in dielectrics requires that if a dielectric move across a magnetic field, an electric polarization should be produced in it at right angles to the field and to the motion; just as in a conductor, likewise moved, electric induction is set up. The question was tested by H. A. Wilson in an experiment which has become classic. A hollow cylinder of dielectric material, coated on its inner and outer cylindrical surfaces with metal, was rotated about its axis in a magnetic field whose lines of force were parallel to the axis. The metal coatings were connected to a sensitive electrometer, which registered a charge, reversing with the reversal of the field, and having both the sign and the magnitude required by the theory. (L.D.W.)

WILTING. This phenomenon, observed most commonly in leaves, results from a complete or nearly

complete loss of turgor by the cells. Drooping, rolling, or folding are the common external manifestations of the wilting of leaves. During hot summer days plants often wilt during the daylight hours only to regain their turgidity during the following night. This type of wilting is called *temporary wilting* and results from a daytime excess of the rate of transpiration over the rate of absorption of water from the soil. Even when the soil water supply is adequate the absorption rate often lags behind the transpiration rate if the latter is high, resulting in a gradual reduction in the water content of the plant and the accompanying phenomenon of wilting. During the night hours the absorption rate exceeds the usually low transpiration rate and the turgidity of the plant tissues is gradually restored. The other type of wilting is called *permanent wilting* and occurs only when the available soil water supply has been exhausted. Plants can recover from permanent wilting only when the soil water supply is replenished. Permanent wilting is of common occurrence in most kinds of plants whenever prolonged drought conditions prevail. Some kinds of plants can endure this condition for long periods without lethal effects, others will die within a few days or a few weeks if subjected to continuous permanent wilting. (B.S.M.)

WIMSHURST MACHINE. Static Machines.

WINCH. A mechanism arranged to hoist by manual or mechanical power through the medium of a rope, cable, etc., winding on a drum is a winch. The winch is provided with means for stepping up the torque obtainable and a brake for lowering the load. However, the brake is sometimes omitted in the simplest winches. The setup of a winch used for hoisting is shown in the accompanying figure. The hoisting rope passes up and

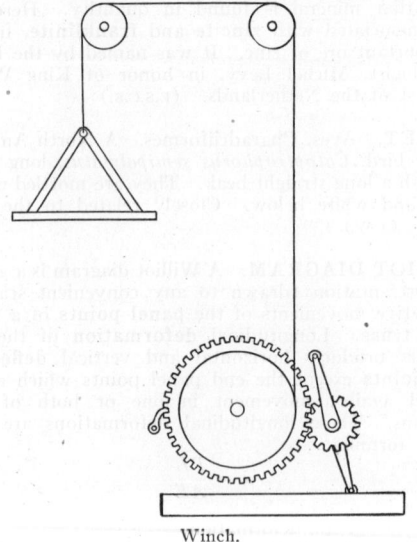

Winch.

over two sheaves, so that a load may be hoisted from a position adjacent to the winch. The hand power is applied at a crank or cranks, which are attached to a shaft bearing a pinion gear. This pinion engages with a large gear attached to the winding drum. On the winding drum there is a brake, generally of the band type, which is brought into play when a load is to be lowered. A high-capacity winch may have a double reduction gear, instead of the single gear illustrated. Winches are necessarily slow-moving, and they are not ordinarily economical for other than intermittent hoisting jobs. (F.T.M.)

WIND CORRECTION ANGLE. The angle between the **heading** of a plane and the **course** made good is known as the wind correction angle. The angle is always measured from the heading to the course and is named right (+) or left (−) to indicate the direction of drift. The signs permit of immediate algebraic addition of the wind correction angle to the heading to obtain the course being made good.

The wind correction angle must always be determined if a plane is to make good a specified course whenever there is a wind blowing. The method of determination is always graphical, either by actual drawing or by use of any one of the numerous types of **dead-reckoning computers**. The **vector** diagram constructed is similar to that used in graphical methods of **dead reckoning** and from the diagram not only the wind correction angle (WCA), but also the predicted ground speed (PGS) of the plane can be determined.

The procedure may best be described by consideration of a specific situation. A plane is to proceed to a point on rhumb-line bearing 105° distant 180 miles. The plane is to cruise at 130 knots and the wind is from 345° speed 25 knots. The pilot wishes to know the wind correction angle, the heading of the plane, and the estimated time of arrival (ETA). The velocity-vector diagram is shown in Fig. 1, which is constructed and

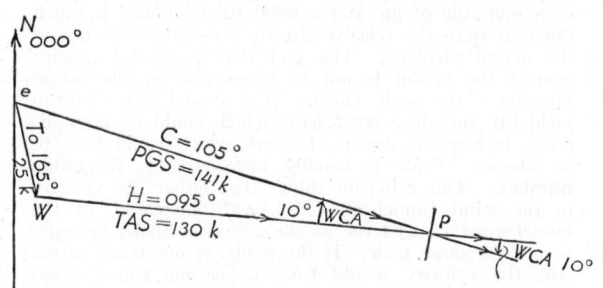

Fig. 1. Velocity-vector diagram for determining wind correction angle and predicted ground speed.

labelled in accordance with the standard procedure of U. S. Naval Aviation. From a point e a vector ew is drawn representing the velocity in which the wind is moving the plane. From e a line is drawn in the direction of the course to be made good. Then from w an arc is struck, using either dividers or a scale, with length equal to the true air speed of the plane (TAS). This arc cuts the course direction line in the point p. This completes the velocity-vector diagram with ep representing course to be made good and PGS, wp the heading and TAS, and the angle at p, measured from wp toward ep, representing WCA. Measurement with scale and protractor on the completed diagram gives PGS = 141 knots and WCA = 10° Right. Since, by definition H + WCA = C we have in this case H + 10° = 105° or H = 095°. Finally with PGS = 141 knots and distance to be made good 180 miles, we have an expected elapsed time of 1hr 17m for the flight. This interval added to the time of departure will give the ETA.

Throughout flight the navigator must keep careful watch with the **driftmeter** and if the drift is not equal to the predetermined wind correction angle, the wind must be determined and a new heading computed to make good the desired course. This procedure is illustrated under **Air Plot**. (w.k.g.)

WIND EQUATIONS. Many factors influence the movement of air in any quasi-horizontal plane at or above the surface of the earth. In the friction layer, frictional forces play an important role. Convergence and divergence also play their part in determining air flow. Isallobaric fields have some influence. But, in general, three major forces play the main roles in establishing the winds. They are:

1. The Coriolis or Deflective Force due to the earth's rotation which is given by $2\omega V\rho \sin \phi$.

2. The pressure gradient of the atmosphere which is given by $\dfrac{dp}{dx}$.

3. Centrifugal force due to curvature of path of air parcel, which is given by $\dfrac{V^2\rho}{r}$.

If air movement is unaccelerated, these three forces are the only factors involved in determining the wind. When other forces are active, accelerations and decelerations alter equilibrium conditions among the three principal forces. However, equilibrium conditions hold, except under the excessive influence of one or more of the minor forces, and winds are generally determined by the balance among the three main forces.

There are several cases possible for establishing a balance of forces among the three. The following are for the northern hemisphere.

Case 1. If the parcel of air is moving in a field of pressure which is higher at the center of the field than at the exterior, the pressure gradient is directed outward from the center. In this case, the flow is clockwise and the deflective or Coriolis Force balances the pressure gradient and the centrifugal force.

$$\underset{\left(\substack{\text{pressure} \\ \text{gradient}}\right)}{\frac{dp}{dx}} + \underset{\left(\substack{\text{centrifugal} \\ \text{force}}\right)}{\frac{V^2\rho}{r}} = \underset{\left(\substack{\text{Coriolis} \\ \text{force}}\right)}{2\omega V \sin \phi}$$

where dp/dx = The pressure gradient.
  V = The velocity of the parcel.
  ρ (rho) = The density of the air.
  ω (omega) = The angular velocity of the earth.
  r = The radius of curvature of the parcel's path.
  ϕ (phi) = The latitude.

Solving for this equation we obtain:

$$V = r\omega \sin \phi \pm \sqrt{(r\omega \sin \phi)^2 - \frac{r}{\rho}\frac{dp}{dx}}.$$

But $v = 0$ when $\dfrac{dp}{dx} = 0$, and we have

$$V = r\omega \sin \phi - \sqrt{(r\omega \sin \phi)^2 - \frac{r}{\rho}\frac{dp}{dx}}.$$

It is not possible to give meaning in this case to a negative value for the radical, therefore, the quantity under the radical must always be positive. This places a limiting value in wind velocity in a clockwise or anticyclonic system of winds which is given by the value of zero for the quantity under the radical.

We then have

$$V_{\max} = r\omega \sin \phi.$$

For a given latitude the quantity $\omega \sin \phi$ becomes fixed and $V_{\max}/r$ is constant. At the equator the quantity ($\sin \phi$) becomes zero and the velocity also becomes zero. Anticyclonic winds at the equator, therefore, are **not** possible.

For a given pressure gradient, the velocity of an anticyclonic wind increases toward the center of the system until the critical value of velocity is reached.

Case 2. If the parcel of air is moving in a field of pressure such that the pressure is less at the center than the exterior, the pressure gradient is directed inward toward the center. In this case, the flow is counterclockwise and the pressure gradient balances the deflective force and the centrifugal force.

$$\frac{dp}{dx} = \frac{V^2\rho}{r} + 2\omega\rho V \sin \phi.$$

The solution to this equation is

$$V = \pm \sqrt{(r\omega \sin \phi)^2 + \frac{r}{\rho}\frac{dp}{dx}} - r\omega \sin \phi.$$

Again, when $\frac{dp}{dx} = 0$, $V = 0$ and the $+$ sign before the radical is necessary.

$$V = \sqrt{(r\omega \sin \phi)^2 + \frac{r}{\rho}\frac{dp}{dx}} - r\omega \sin \phi.$$

The quantity under the radical is always positive so there can be no limiting value on the velocity. At the equator, where $\sin \phi$ is zero, the pressure gradient is still present under the radical to give the velocity a positive value. Winds blowing under these conditions are known as cyclonic winds and include both tropical and wave cyclones as the main types.

Case 3. If the flow of air is in a straight line, the centrifugal-force term drops out and the balance of forces is between the pressure gradient and the deflective force only.

$$\frac{dp}{dx} = 2\,\omega\rho V \sin \phi.$$

$$V = \frac{dp/dx}{2\,\omega\rho \sin \phi}.$$

This wind is known as the geostrophic wind in contrast to Cases 1 and 2, which are gradient wind cases. Geostrophic winds increase southward for a given pressure gradient and theoretically become infinitely great at the equator where the denominator becomes zero. They are directly proportional to the pressure gradient at any given latitude.

Case 4. At the equator where the deflective force drops out entirely, a balance between the pressure gradient and centrifugal force is possible.

$$\frac{\rho V^2}{r} = \frac{dp}{dx}.$$

$$V = \sqrt{\frac{r}{\rho}\frac{dp}{dx}}.$$

Winds of this type are cyclostrophic winds and blow cyclonically or counterclockwise about a center of low pressure. Hurricanes are the best example of cyclostrophic winds and, even though they develop and migrate from the equatorial zone, the deflective force plays no significant role until the center is considerably removed from low latitudes.

In the southern hemisphere, wind relations are the mirror image of the northern hemisphere, i.e., they blow counterclockwise about centers of high pressure, clockwise about centers of low pressure, and with high pressure to the left. (P.E.K.)

WIND GAP. Gap.

WIND INDICATOR. A device that indicates direction and/or velocity of the surface wind. Examples: Wind "sock," swivelling wind tee, smoke pot, weather vane, anemometer. The wind sock, or cone, is a tapered fabric sleeve pivoted on a standard. When exposed to a wind stream it "weathervanes" to point the wind direction. Being large and bright colored, it is easily observed from aircraft, and gives needed information to the pilot planning a landing in the vicinity. The small end is usually unsupported, other than by the wind, hence a rough indication of the magnitude as well as direction of the wind, is afforded by the cone, since it will be limp in calm air, and partially so in light winds. (F.T.M.)

WIND ROSE. Any diagram showing features of wind direction and velocity in the form of a hub, with spokes of a wheel representing direction or velocity values or both. The most common wind rose is one in which the length of the spokes of the wheel extending into the cardinal directions represents the frequency of winds

from that direction. Wind roses are used primarily in climatology to show the wind features of a particular area. (P.E.K.)

WIND TUNNEL. Aerodynamic data upon which are founded the rational design of heavier-than-air craft, are obtained in large measure from tests made by blowing air past a stationary model which is supported so that the air forces acting on it may be measured. The conduit which contains and directs the air across the model, together with the auxiliaries required for its operation, is known as a wind tunnel. The auxiliaries mentioned are a **blower** or **propeller**, a return duct, a honeycomb, and corner vanes. Besides heavier-than-air research, the wind tunnel is used to obtain data on **airship** hulls, automobile shapes, and other cases where air reactions are important. Theoretically, the reaction of air upon any shape or object is the same, whether the object be moving through still air, or whether the object is at rest, and the air moving past it. Actually, however, there are some differences, arising largely from the fact that the air in a wind tunnel has a certain degree of **turbulence** not present in a calm atmosphere. Furthermore, model tests may not be applied directly to full-scale designs without certain corrections. Chief among these is the scale correction. The relative size of a molecule of air and a wind tunnel model is much different than the relative size of a molecule of air and the actual airplane. The fact that molecular density against the airfoil is not in proportion in the model introduces the scale effect. If a model test were to yield lift and drag coefficients which could be used directly in airplane design, the scale effect would have to be absent. Scale is usually measured by **Reynolds number.** This criterion shows that either the velocity in the wind tunnel must be vastly increased, or the model must be as large as the actual airplane, in order to get the same scale. If the model is one-tenth actual size, the velocity would have to be ten times actual flying velocity. As this is impracticable, models have been tested at lower scales than **airplanes** operate, and corrections for scale have had to be applied. Considerable knowledge as to the magnitude of scale effect, and means for correcting it, have been devised from tests in a variable density wind tunnel. This special type is arranged so that compressed air having pressure of several atmospheres, and a corresponding high density, can be blown over the model. The advantage here is that by increasing the density of air sufficiently, a small model may be tested at reasonable tunnel velocities, and yet have the Reynolds number of a magnitude comparable to that of a full-sized airplane. Besides scale and turbulence, corrections must be made for shape of the wind tunnel walls, tunnel wall interference, and aspect ratio.

The wind tunnel is used to measure the drag of all sorts of objects, particularly components of aircraft, and to measure the drag and lift forces of complete model aircraft. It is also employed to test for stability and balance of the same. The principal classifications of tunnels are once-through and return-circuit types. Another classification is open or closed test section. The air is usually blown by a propeller driven by a variable speed electric motor. The models most frequently tested are those having a wing span varying from 2–3'. This means that the tunnel must have a 30–50" width at the test section. However, tunnels are built in a variety of sizes, ranging anywhere from 6" to 60' at the throat.

Ordinarily, a tunnel has a nozzle to give something like constant acceleration to the air, and direct it into the test section, which is a constant area section in which the models are mounted. Passing the test section, the tunnel is slowly enlarged in order to recover the pressure head. In a once-through tunnel, the propeller is mounted at the end of the diverging section. In a return-circuit tunnel it may be mounted there or in the

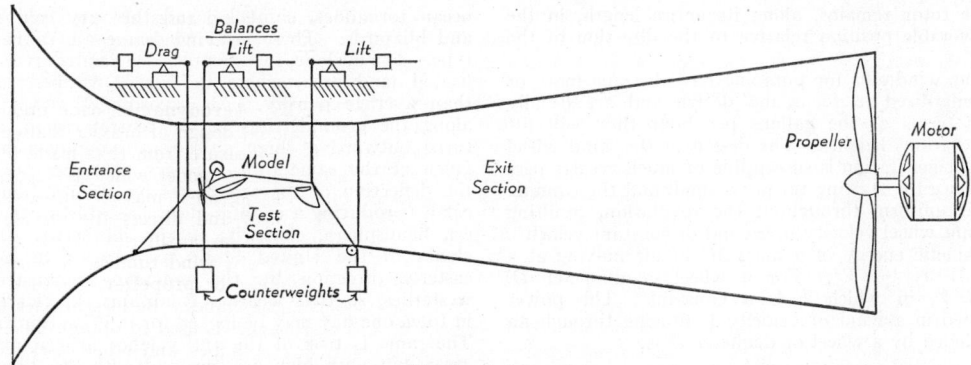

Simple wind tunnel.

return circuit. The materials from which tunnels have been constructed are wood, plaster, concrete, and steel. The sections employed have been circular, octagonal, hexagonal, square, and rectangular. The models are supported from a balance, which can be of a suspension wire type (Goettingen), or a rigid post-like support (National Physical Laboratory). In either case the purpose of the balance is to maintain the model in a predetermined attitude with respect to the wind stream, and quantitatively measure the air forces which arise therefrom.

The "atmospheric" wind tunnel illustrates most of the foregoing remarks. Varied tests of great value are performed in tunnels of this character upon solid models held by the balances in a horizontal airstream. Unfortunately this fails to cover all the needed testing and research and other designs of tunnels are required.

Pressure Tunnels. Closed-throat tunnels when made air-tight and structurally strong enough to contain air above atmospheric pressure are used in research when Reynolds number and viscosity effects are important since the increase of mass density accentuates the viscosity effects.

Vertical Tunnels. Generally these are atmospheric but having upwardly directed air jets. They are most frequently employed to test airplane spinning characteristics. The model is a replica of the prototype, being similar not only in shape but also in mass disposal. It is launched into the vertical jet and spins freely if a light-weight flying model is used, or it is mounted free to spin on a post if the model is heavy. The light-weight models are effective in discovering the spin recovery properties produced by the control surfaces.

Other special purpose tunnels are (a) smoke tunnels for visualizing air flows at low speeds, (b) supersonic tunnels for compressibility effects, (c) refrigerated and altitude tunnels for test work on engines and aircraft auxiliaries.

With the increasing size and speed of the modern airplane, two tendencies have appeared in wind tunnel construction. One has been the tendency to increase of the working section, with a maximum width of 80′, so that full-scale airplanes can be tested in the aerodynamic laboratory. The other has been in the increase of working speeds, with speeds running as high as 600 m.p.h. so that compressibility effects can be adequately investigated. The Mach number, which is the ratio of the tunnel speed to the speed of sound, is now as important as the Reynolds number. (F.T.M.)

WIND VANE. Meteorological Instruments.

WINDHOVER. Falcon.

WINDMILL. Any arrangement of sails or blades which will turn the shaft to which they are attached when exposed to a wind may be called a windmill. The practical use of the windmill is to obtain for local use some of the energy represented by atmospheric motion.

Power from the winds has served man for many centuries, but the total amount of energy generated in this manner is small. While the expense of installation and the uncertainty of operation have prevented serious consideration of the windmill for other than such intermittent services as pumping, there has been considerable experimentation relative to the generation of electrical energy by moderate-sized wheels for charging storage batteries. 35-volt third-brush generators which may be set to produce maximum charge at the best wheel speed have been proposed for farm lighting plants. A 15-ft. American type wheel has, under test, produced an average of 2.5 kw-hr. per day.

There are four well-differentiated types of windmills, viz.:

1. The multi-bladed turbine wheel, or American type.
2. The Dutch type.
3. The propeller-type high-speed wheel.
4. The rotor.

The American type was given its name in Europe to distinguish it from the Dutch type. Its wheel is kept into the wind by a tail or rudder. The regulation in high winds is usually accomplished by automatic devices which swing the wheel out of the wind when it becomes excessive.

Although its large 4-sailed wheel is efficient, the Dutch type, being difficult to regulate and operate, and having a higher first cost, has been used mainly in its place of origin, the Low Countries of Europe. The rpm is lower than the American wheel, but the tip speed is relatively higher because of the length of the sails. Regulation is accomplished by turning the movable top of the tower into the wind.

Developed during World War I, for aircraft generator drive, the propeller type of windmill has possibilities of development into a convenient source of motive power for farm electric systems. It is inherently a high-speed wheel enabling the generator to be coupled directly to its shaft. Regulation can be accomplished by adjusting the pitch of the blades.

The most recently perfected mechanism for obtaining energy from the winds is the rotor. The phenomenon which causes a rapidly rotating baseball or tennis ball to take a curved path was first investigated by Professor Magnus and has been named the Magnus effect. The industrial application of this was successfully accomplished by Flettner in his rotorship.

The characteristics of the rotor windmill are:

1. Ability to start in very low winds.
2. With constant rotor rpm the wheel is automatically regulated, since an increase in wind speed above the designed condition leaves the rotative force on the rotor unchanged unless the rpm of the rotor also increases.
3. After the designed value of wind velocity is exceeded, the delivered power remains a constant.

4. The rotor remains, along its entire length, in the most favorable position relative to the direction of the wind.

Turbine windmills for pumping have become more or less standardized in form and design and are usually rated in terms of the gallons per hour they will lift through various heights. The design of the wind wheel for electric generation is susceptible of much greater perfection since the starting torque is small and the running torque is uniform throughout the revolution, resulting in constant wheel velocity in a wind of constant velocity.

The kinetic energy in a mass M' of air moving at a velocity V is $\frac{1}{2}M'V^2$. For a wheel of diameter D, $M' = kD^2V$, in which k is a constant. The power represented in a wind of velocity V flowing through an area included by a wheel of diameter D is:

$$P = kD^2V^3.$$

This equation shows that the power of a windmill is proportional to the cube of the wind velocity and the square of the wheel diameter. Using 0.0763 lb. as the weight of 1 cu. ft. of air, it can be shown that the theoretical wind hp. is:

$$\text{Hp.} = 5.36 \times 10^{-6} D^2 V^3$$

Hp. = the theoretical wind hp.
D = the wheel diameter in ft.
V = the wind velocity in m.p.h.

The power actually developed will be but a fraction of this wind hp.—about 10% for the American type wheel, and 20% for the Dutch type. The ability of a wheel to start in light winds is an unquestioned advantage.

The starting torque, and therefore the wind speed at which the wheel will start, is largely determined by the blade angle; the greater its pitch, the greater the starting torque, but the lower the running speed. This means a heavier wheel for the same hp. than one designed to start at a higher value of wind velocity. An adjustment must be made between cost of construction and percentage of time in service. The velocity duration for a station 100′ above ground is very little higher in value than that for the 50-ft. point; however, it must be remembered that the wind hp. increases as the cube of the velocity and an increase from 10 to 12 m.p.h. is equivalent to a 73% increase in power. (F.T.M.)

WINDOW PANE. Turbot.

WINDPIPE. The trachea, in the **respiratory system** of air-breathing vertebrates. (A.W.L.)

WINDS. Since only slight horizontal variations in the density of the air can occur ordinarily, and the total volume of the atmosphere is substantially constant, it is evident that a wind is necessarily a circulation; that is, any movement of the air in one direction must be offset by a return current elsewhere. All such movements are the result of **thermal convection** due to the heat of the sun.

The winds may be roughly classified into three types: 1. Winds which are in a sense permanent, including the easterly trade winds of the tropics and the westerly winds prevailing in the temperature zones. 2. Seasonal winds, such as the monsoons of the Indian ocean, which blow northward toward the warm Asiatic continent in summer, bringing heavy rainfall to India, and which pour southward over the Himalayas from the cold plateaus in winter. (To such may be added the diurnal land and sea breezes, which have similar origin but on a much smaller scale.) 3. Local winds and storms, which temporarily interrupt the more general air movements prevailing at the time. Among these are the great **cyclones** and **anticyclones** characteristic of North American climate, ocean cyclones (including hurricanes or typhoons), **thunderstorms, tornadoes,** waterspouts or

ocean tornadoes, chinooks, and the very cold mistrals and blizzards. These local winds are all of the vortex type. The cyclone, for example, is generated over a large, heated land area, where the air is warmed to lower than average density. Air accordingly moves inward along the ground, rises in the central region, and returns outward at high altitudes; the movement being given at the same time a rotary character because of the deflection of the air masses by the rotation of the earth (producing a counter-clockwise whirl in the northern hemisphere, clockwise in the southern). The cyclones of the United States are carried in a northeasterly direction by the prevailing north-temperate westerlies, so that weather conditions experienced, say, in Iowa one day may be looked for in Wisconsin the next. The same is true of the anticyclones formed over cold areas, but with high pressure and with the direction of the local air currents reversed. Hurricanes, tornadoes, waterspouts, etc., are similarly formed but in varying circumstances and on different scales.

The winds are practically all confined to a very few miles vertically above the earth's surface. Above this lies the **stratosphere,** believed to be a region of almost perpetual calm. (See **Atmosphere.**) Air currents at high altitudes are commonly studied by means of small pilot balloons, released for the purpose and watched through telescopes. (L.D.W.)

WING. For the use of this term in aeronautics, see **Wing, Airplane.** In zoology, a wing is a broad thin appendage used to support the animal in the air, either through its resistance to air currents or by muscular movement against the air. The support may be accompanied by the progress called **flight** or may sustain the animal in one spot. The latter action is called hovering.

Wings and flight have been developed in only four groups of animals: the extinct pterosaurs, a group of **reptiles;** the **insects;** the **birds;** and the **bats,** an order of mammals. Other animals progress through the air to a limited extent, but they merely coast, supported by extended surfaces of various kinds; this form of locomotion is known as gliding and is considered with flight.

Only the insects have developed wings independent of their other appendages. In this class there are two pairs at the maximum, attached dorsolaterally to the second and third segments of the thorax. They are developed as saclike outgrowths of the body wall whose upper and lower walls become closely apposed and relatively very thin to form a light membrane. Some rigidity is conferred by the cuticula of the wing membrane and additional support is provided by thick-walled tubes called veins or nervures running through the structure. The principal veins follow the courses of tracheae which are functional during the developmental stages, but others of a supplementary nature may be present. The number and arrangement of the veins have been of the greatest importance in the classification of the insects.

Insect wings are modified in some orders beyond the changes of shape, venation, texture, and vestiture which do not interfere with their use in flight. Examples are the **elytra** of beetles, the **tegmina** of grasshoppers, and the **halteres** of flies.

The wings of bats are thin folds of skin extending between elongated digits of the front legs, along the sides of the body to the hind legs, and in some species thence to the tail. They are admirably adapted to flight but they have the one great weakness of presenting a greatly extended surface from which the radiation of heat may take place, hence the activities of bats are limited to warmer climates and warmer seasons of temperate regions. The **pterodactyls** had similar wings, but the problem of radiation was not the same to them since they were presumably cold-blooded.

The wings of birds, like those of bats and pterodactyls, are supported by the skeleton of the pectoral appendages. In this group, however, they are no more than modified pectoral appendages. The extended surface of the bird's wing is made up of large stiff feathers and the supporting skeleton is reduced in size and complexity. The resulting advantage in the lightness of the wings and in their presenting no radiating surface greatly extends the radius of activity of birds. No climate is too cold for them, and no season too severe. They are among the most widely distributed of all living things.

The wings of bats retain short clawed appendages on the anterior margin, but those of birds are wholly transformed into organs of flight except in the young **hoatzin**, which uses similar clawed digits in climbing. The only other noteworthy modification of birds' wings is found in the **penguins**, where they are small paddles with greatly reduced feathers, useless for flight but effectively developed for swimming. Reduction to a vestigial state has also occurred in various flightless species, such as the ostrich and the kiwis.   (A.W.L.)

WING, AIRPLANE.

Wing is a general term applied to the airfoil, or one of the airfoils, designed to develop a major part of the lift of an airplane. An airfoil is an aerodynamic shape which yields useful lift for the price of relatively small drag. An airplane wing is an airfoil whose lifting power is utilized to make flight possible. The reader is referred to **airfoil** for wing aerodynamics.

The most common wing arrangements are biplane and monoplane. The biplane was once the principal type, but now airplanes usually have monoplane wings.

The wing is composed of a surface, of members supporting that surface in the external shape of the desired airfoil, and of an underlying structural framework which receives the air reactions from the above-mentioned components and transmits the lift to the fuselage, nacelle, etc. Typical construction of a light airplane wing is shown in Fig. 1. The load-carrying structure consists of two

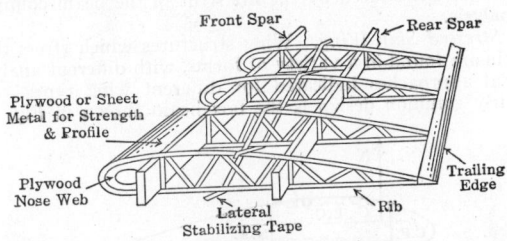

Fig. 1.   Simple wing structure.

spars to which ribs are fastened at intervals of about 1'. The ribs form the profile of the airfoil desired. In stressed metal skin types, the ribs and attached skin and the stiffeners (see Fig. 3) provide for stiffness and rigidity, while the spars, webs, and stiffeners take the bending load.

The forward edge or "nose" section of a fabric-covered wing is usually strengthened with sheet metal or plywood because here the unit air forces are greatest; also here it is most necessary to maintain the true airfoil profile. The fabric is stretched over the ribs and sewed firmly to each. Upon application of the fabric finish (airplane "dope") the fabric shrinks to a taut, smooth wing-covering. Air reactions on this surface are transmitted to the numerous ribs, and from the ribs to the spars. This structure requires internal support against drag forces in the plane of the spars. If the spars are not cantelevered from rigid root fittings, external struts or wires must support the spars. The wing truss therefore must be duplex in character, carrying both normal and chord loads.

Spars are constructed of wood and of metal. Usually wood is preferred for small and medium airplanes, while aluminum alloy in the form of drawn shapes, tubes, corrugations, etc., is used for the built-up spars of the larger airplanes. Some typical spar cross-sections are

shown in Fig. 2. Ribs are (1) built up from spruce strips by gluing and nailing, (2) stamped one-piece from metal sheet, (3) built up of metal shapes by riveting

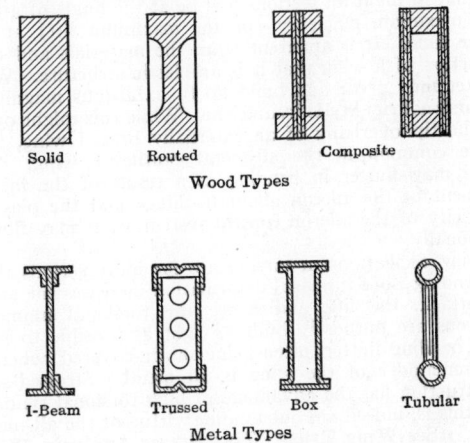

Fig. 2.   Typical wing spar sections.

or welding. Fabric covering is limited to smaller airplanes. Aluminum alloy sheet such as 24ST, or plywood, is preferred for covering large wings. Although fabric does nothing except provide an airfoil profile, a plywood or sheet metal skin may be counted on to carry tensile, compressive and shear loads. The "stressed-skin" construction is lighter than flexible two-spar construction for large wings. A stressed-skin wing may employ the skin to resist torsion and the internal structure consisting of spars, shear webs, corrugated sheet, etc., to resist bending and shear.

The light-plane wing is occasionally a simple structure, similar to Fig. 1, with constant chord and profile, containing, internally, only wing structure and aileron controls. In contrast, the wing of the large multi-engined airplane is not only structurally more complex, but may also be required to provide space and fittings for some or all of the following:

1. Power cell assembly, consisting of engine mount, nacelle, cowling, controls, fuel, and oil supply.
2. Landing gear, retraction mechanism, wheel wells.
3. Flaps and control.
4. De-icer inflation air tubing.
5. Fuel tanks and piping.
6. Fire-extinguishing equipment.
7. Emergency "ditching" equipment (life rafts, etc.) and automatic release for same.
8. Electrical equipment. Position and landing lights, flap and wheel position indicators, etc.
9. Military equipment—guns, bomb and rocket racks.
10. Miscellaneous equipment such as Venturis, engine and propeller control and instrument leads, air scoops and ducts.

These large wings are tapered to secure aerodynamic advantage and the **airfoil classification** is frequently

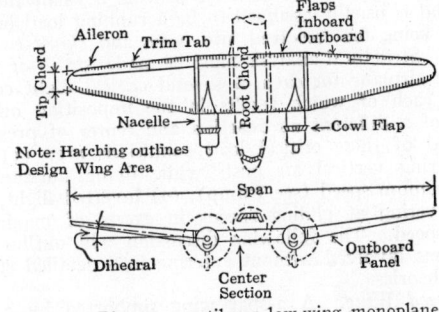

Fig. 3.   Bi-motor cantilever low-wing monoplane.

different at the root and tip because the taper in thickness is not the same as the taper in chord. The wing is often built with allowance for twist due to torsion load, so that in flight all sections will be at the angle of attack for maximum efficiency (or for maximum lift) at the same time. It is apparent from the materials and construction of a wing that it is an elastic structure. Wing flutter may occur and build up to hazardous magnitude if torsional rigidity is insufficient. While this might occur in the tip overhang of an externally braced wing, it is more common in the all-metal cantilever wing. The wing may flutter in bending as a result of the inertia moment of the aileron about its hinge and the play or elasticity of the aileron control system, or it may flutter torsionally.

Wing oscillations in torsion are far more serious than in bending since torsional deformation increases the angle of attack, the lift, and hence the torsional moment. Ailerons are purposely built as light as possible to prevent bending flutter, often being fabric-covered, whereas the remainder of the wing is all metal. Stressed-skin construction has the advantage of high torsional rigidity, and this is, indeed, an outstanding virtue of the all-metal wing. (See **Wing Truss, Wing Stress Analysis, Wing Area, Airfoil, Airplane, Aspect Ratio, Chord, Controls, Airplane Flight, Dihedral, Equivalent Monoplane, Flap, Monoplane, Tab.**) (F.T.M.)

WING AREA. The only area of an airplane wing to have been officially specified is the "design wing area." This is the area enclosed by the projection of the wing outline, including ailerons and flaps, but ignoring fairings and fillets, on a surface containing the wing chords. The outline is assumed to extend through nacelles and through the fuselage to the plane of symmetry. (See **Wing, Airplane.**) (F.T.M.)

WING SPAN. Maximum distance measured parallel to lateral axis from tip to tip of airfoil of an airplane wing inclusive of ailerons, or of a stabilizer inclusive of elevator.

Effective. True span of wing less corrections for tip loss. (F.T.M.)

WING STRESS ANALYSIS. Of all structural designs, none presents more difficult problems of analysis, nor requires as perfect a design, as that of the airplane structure. In that structure the *wing* analysis involves the greatest variety of problems. The nature of aviation causes an airplane structure to receive large increments of dynamic load, sometimes attaining two or three times the static loading. Intentional acrobatics may extend the dynamic load even more. The reader is referred to **load factor** for description of dynamic load determination. When the load on and size of the wing have been fixed by aerodynamics, structural design must follow with a wing design that has adequate strength to meet the contemplated load factor, but which has no surplus strength or unnecessary weight.

Air loads received from the ribs are conveyed at close intervals to the spars. The rib spacing is small and the air load is usually assumed to be a running load on the spars, being first resolved into *beam* and *chord* components. Structural analysis must be made for all or parts of the airplane for each of several critical flight conditions, each of which results in the imposition on the wing of a different air reaction and center of pressure. Typical of these critical load conditions are: (1) encountering vertical air gusts while flying horizontally at maximum speed (see **Bump**), (2) inverted flight, (3) maneuvering at gliding speeds in excess of maximum level speed. The following discussion will outline the problems involved without entering into detailed structural theories.

Trussed Wing. A 2-spar wing supported by a lift truss and made rigid by a drag truss offers the least

complication in design since, mostly, it is solvable by the methods of statical mechanics and simple structural theory. The **wing truss** has axial compression or tension in all members except the spars themselves.

The beam component of air reaction is apportioned between front and rear spar by decomposition from the center of pressure position. The spar is considered to receive a distributed air load, the spanwise variation of which must have been determined by airfoil aerodynamics. As is shown in Fig. 1, this results in reactions at

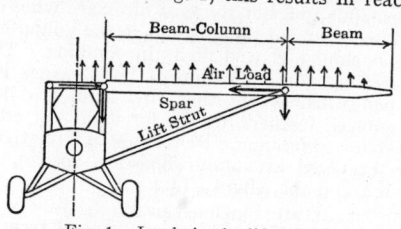

Fig. 1. Loads in the lift truss.

the root fitting and outboard strut fitting. Since the strut is inclined to the spar, it not only supports the beam reactions, but places the spar under compression. Therefore, although the other members of the lift truss have axial loads only, the spar is a **beam-column** and must be analyzed by methods too complicated to admit of inclusion here. Basic formulae were worked out for beam-columns by Müller-Breslau and Berry.

The chord component loads the drag truss which consists of the spars, compression struts, and wire braces arranged in the form of a Warren truss with the spars for chords. If the lift truss strut is not in the plane of the beam component, it will produce a chordwise component to load the drag truss. The average spar compression produced by the drag truss must be included with compression introduced by the lift strut in the beam-column analysis.

Stressed Skin Wing. Wing structures which stress the skin are found in a variety of forms, with different analytical approaches indicated for different wing types. A fairly common design is the box beam shown in Fig. 2.

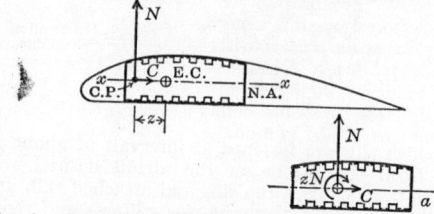

Fig. 2. Box beam with stressed-skin flanges. Bending and torsion loading separated.

Here the wing is shown to be composed of a skin reinforced laterally with "hat section" stiffeners. Ribs stiffen transversely and establish the airfoil profile. Two (several in some wings) shear webs extend from upper to lower surfaces and, together with the reinforced skin between them, form a box beam of considerable torsional rigidity. This beam is considered to resist bending caused by the components of air reaction N (normal or beam component) and C (chord component). Since N acts through an aerodynamic center of pressure, it introduces a twist. To separate bending and torsion, N is moved to the elastic center $E. C.$ of the beam by introducing a torsional moment or torque zN. The elastic center is that point on the neutral axis about which the sum of the first moments of elementary moments of inertia of the beam material is zero, i.e., where $\Sigma x I_{xx} = 0$. The principal axes of bending through $E. C.$ may not coincide with the direction of N and C, but usually are near enough to be so considered. However, in cases of considerable unbalance of distribution of flange material horizontally it may be necessary to find the angle of the principal axes to XX, and resolve the bending moments normal to these new axes before computing the bending stress. The torsion produces a shearing stress which can be analyzed by treating the box beam as a shell in torsion from torque zN. The nose section of the wing may contribute appreciably to torsional rigidity and, if so, the torsional analysis is that of connected shells, treatment of which is beyond the present scope, but may be found in texts devoted to aircraft structural design. The flange material of the box beam is thus seen to be in combined shear (from torsion) and compression or tension (from bending). The resulting stress must be determined by some of the analytical or graphical methods of combining stresses, such as the Culmann Circle of Stress. (F.T.M.)

WING TRUSS. A wing which obtains rigidity by truss action is said to be trussed. The resultant air reaction on an **airfoil** is inclined to the plane of the chord, hence two truss systems are necessary—a lift truss and a drag truss. The lift truss is composed of the spar and external bracing in a plane normal to the wing chord, or approximately so. The drag truss is in the plane of the spars. The high-wing monoplane used for illustration has a lift truss connecting an outboard point on the spar

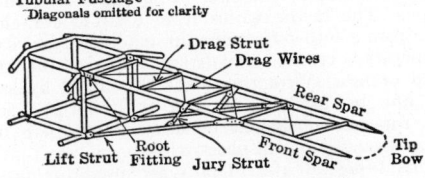

Tubular Fuselage
Diagonals omitted for clarity

Drag Strut
Drag Wires
Rear Spar
Lift Strut   Root   Jury Strut   Front Spar   Tip Bow
Fitting

Diagrammatic view of trussed wing.

with the lower fuselage longeron. During flight this strut is in tension. The stress changes to compression when air lift ceases and dead weight of the wing is supported by the truss. The strut is a long slender column and hard landings may impose critical compression stresses on it. It is strengthened by a jury strut which serves to reduce the unsupported length. The spar is in combined compression and bending.

The drag truss system consists of compression struts (sometimes extra heavy "bulkhead" ribs are used) which divide the plane of the spars into "bays." Each bay is crossed with drag wires and counter wires which absorb the shear of the chord component of air reaction. At high angle of attack the character of this shear may reverse, hence the need for counter-diagonals. In stressed-skin designs, no drag truss is needed. (F.T.M.)

WING WALL. Abutment.

WINTER EGG. A form of egg produced by some invertebrates, usually in the fall. It has a thick protective shell and in species with a complex life cycle, such as the **aphids**, it is produced by the sexual generation. Such eggs usually pass through the winter before hatching, but similar eggs may be produced at the beginning of other unfavorable seasons, as periods of drought. (A.W.L.)

WINTERGREEN. Heath Family.

WIRE. A wire is defined as a slender rod or filament of drawn metal. If covered with electrical **insulation**, it is called insulated wire. There are manifold uses of wire in the present day and time. Among the more important uses are fencing, woven and barbed type; binding of extra strong nature for heavy packages; spring stock; ropes and hawsers; electrical **conductors** or resistors; and stock for forming various small light-weight parts, such as mouse traps, egg beaters, triggers, latches,

etc. The materials from which wires are drawn are wrought iron, steel, annealed copper, aluminum, and phosphor bronze. Wires are sized by the areas in circular mills or American wire gauge. (See **Gauge**.) Despite the arguments in favor of one system using the diameters in mills as the size numbers, the American wire gauge is commonly used in gauge numbers from 40 to 0000. Descending gauge numbers indicate increasing wire sizes. Electrical conductor is usually annealed copper wire, although aluminum is used for transmission lines, and iron for resistance wires. Most low-voltage electrical conductors are insulated. Rubber compound and varnished cambric are the principal materials for insulating wires. The standardizing agency in the field of electric wiring is the National Electric Code. (F.T.M.)

WIRE DRAWING. The term "wire drawing" has two separate and distinct meanings. First, it is descriptive of the action whereby a rod is reduced to a wire by being pulled forcibly through a round die which reduces its diameter and increases its length (see figure). Secondly, wire drawing refers to a similar action occurring in fluid flow through a small aperture. The usage in the second case is derived from the first. Wire drawing as applied to a **fluid** is usually employed in discussions of the flow of **steam** or **gas** through the **valves** of engines. When the valves are partially open, the small area through which a fluid flows causes a reduction of pressure which is known as wire drawing. The description below refers entirely to the production of wires.

To make a wire, the metal is first prepared as a rod by rolling. It is then treated to give it a surface coating

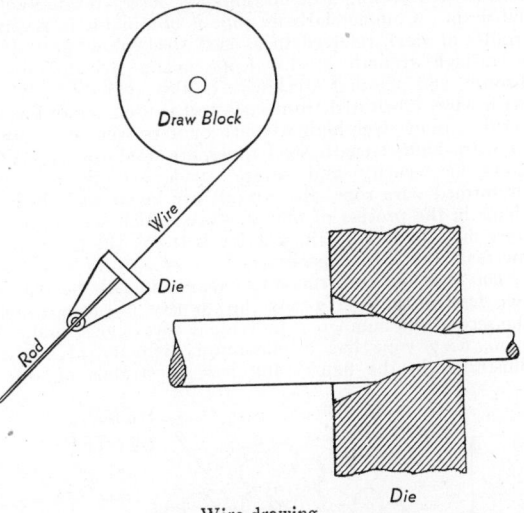

Draw Block
Wire
Die
Rod
Die

Wire drawing.

which will act as a lubricant during its passage through the die. The method of applying the lubricant gives rise to the wet and dry process of wire drawing. In the dry process the wire is dipped in an emulsion of hydrated lime (see **Calcium**), and dried in baking ovens. Then just before it is passed into the die it passes through a greasing operation. The combination of lime film and grease provides the lubricant between the rod and the die during the drawing process. In the wet process the rod is first dipped in a solution of copper sulfate, which results in a thin coating of copper in the wire. It is then dipped in a vat containing a fermented mixture made from meal and yeast. Upon being withdrawn from this vat, it is fed wet to the die. The liquor, together with the thin copper deposited in the salt bath, lubricates the rod in the wet drawing process.

Drawing consists of pointing the rod and threading it through the die. It is then gripped by pinchers and drawn a short distance by hand power applied through a draw bar. Then the end is clamped in a vise on a revolving drum called the drawblock. This drum is revolved slowly by power applied through a geared shaft, and in so revolving continues to pull the wire through the die until the rod has been completely reduced. When the stock is passed through the die it is elongated and reduced in cross-section. The elongation is known as draft, and may amount to from 10 to 40 times the length of the original rod. Sometimes more than one draft is necessary to finish the wire. In such cases it is removed from the block and started through another smaller die.

The dies are the most essential part of wire drawing. They are made of chilled cast iron, steel, cemented carbides, or diamond. The hard cemented carbide type dies are widely used. Diamond dies are used only for fine wire. A typical cross-section of a die is shown in the figure. The hole through the die is tapered, and the part where the reduction in wire size occurs is reamed to a very smooth finish. Two tapers are employed, one in which most of the reduction takes place, the other one of smaller taper, in order to ease the wire down to the exact size at exit from the die. The initial part of the tapered hole is not machined, as it serves only as a region for application of the lubricant to the rod. A plan of a draw bench which is the production unit in wire drawing, is also given. (F.T.M.)

WIRE ROPE. Wire rope is in wide use for hoisting and haulage, and for static loads, such as guy and supporting wires for stacks, masts, and towers. Wire rope consists of cold-drawn steel wires wrapped into strands and twisted around a hemp center or core saturated with lubricant. Commercial wire rope is obtainable in several grades of steel, referred to as cast steel, plow steel, and extra-high-strength steel. Rope made of aluminum, bronze, and stainless steel wire is also available. Rope with wires fabricated from cast steel is used for ordinary service; plow steel, high-strength or improved plow steel, or extra-high-strength steel ropes are used for high degrees of security and severe service conditions. In preformed wire rope, the strands are given their helical shape in the process of manufacture, which serves to reduce metallic fatigue and friction between the parts, and increases the load capacity.

The type and construction of wire rope is indicated by two figures, the first giving the number of strands, and the second the number of individual wires per strand. A 2-in., 6 x 7 rope has a maximum diameter of 2", as illustrated in the figure, and has six strands of seven

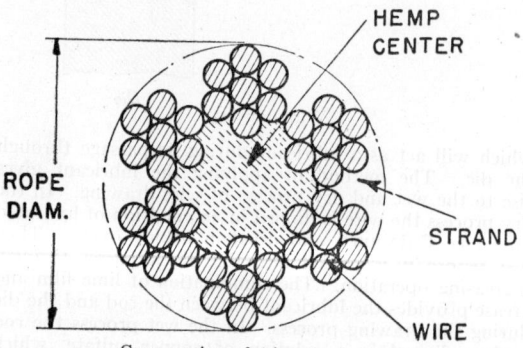

Cross-section of 2-in., 6 x 7 rope.

wires each. For a given diameter, ropes made up of many wires are more flexible than those with few wires, and may therefore be used on sheaves of smaller diameter. The common rope types are 6 x 7 coarse, 6 x 19 flexible, and 6 x 37 and 8 x 19, extra-flexible. The 6 x 7 rope is used for mine and yard haulage and guy wires; 6 x 19 is the standard hoisting rope, and is used for mine and ore hoists, car pullers, cranes, and elevators. Extra-

flexible rope is used for small diameter sheave applications, and for steel mill ladles, cranes, and high-speed elevators.

A wire rope under load is subjected to stresses caused by the load lifted (which includes the weight of the carrier and the rope length between the carrier and the drum), to stresses caused by sudden starting and stopping and taking up slack in the rope, and to stresses caused by bending the rope around the sheave. If the load is lifted suddenly, the force F, induced by acceleration, is given by

$$F = Wa/32.2$$

where W is the total load and a the acceleration in ft. per sec. per sec. Impact loads, caused by taking up rope slack suddenly, are difficult to evaluate but may cause an increase in the actual stress in the rope by as much as 100%. Impact stresses are usually accounted for by a reasonable increase in the factor of safety used for rope selection. When a wire rope is bent around a sheave, the outer wires tend to increase in length. This action induces a tensile stress in the wires, over and above that caused by the direct load, and is dependent upon the wire size and rope construction. The total tensile force induced by bending varies with the cube of the diameter of the rope and inversely as the diameter of the sheave.

The factor of safety of a wire rope is the ratio of the breaking strength to the sum of the stresses induced in the rope. The factor varies from 8 to 12 for elevator service, from $2\frac{1}{2}$ to 5 for mine hoists, derrick service, and hand-operated cranes, and from 4 to 6 for power-actuated cranes. Wire rope which is joined by splicing should not be used for severe or important service. Spliced rope is considered to have about 75% of the breaking strength of unspliced rope.

Various types of attachments are used for fastening wire rope to crane hooks and other devices; details and necessary dimensions may be found in suppliers' and manufacturers' catalogs. The rope socket attachment is the only one that will develop 100% of the full rope strength. It is made of forged steel with a tapered socket into which the separated wires of the rope are anchored by means of high-grade zinc poured into the socket in molten state. Another attachment uses a steel thimble for the loop or eye of the rope; the rope itself is fastened by clips, clamps, or by splicing. Rope connections made with clips or clamps are not recommended as permanent fastenings since they develop only 50 to 75% of the strength of the rope. (H.C.H.)

WIRELESS TELEGRAPHY. Radio Telegraphy.

WIREWORM. Insecta, Coleoptera. Slender hard-bodied larvae of the **click-beetles.** They live under bark and in logs or in the ground. Among the latter species are some important crop pests which damage the roots of grass and grains, and a few other crops including potatoes and cotton. The corn wireworm is one of the most widely known, often destroying entire fields.

The hard bodies of these worms render them immune from repellent substances that can be used in the soil, but crop rotation including clover, soy beans, buckwheat, flax, or some other species not subject to attack, is an effective method of avoiding serious loss. (A.W.L.)

WIRING, ELECTRIC. Wiring consists of running the necessary electrical **conductors** from distribution panels to such lamps, plug-in sockets, motors, ovens, etc., as may be served from the distribution center. The interpretation given here to wiring includes the field of interior wiring, only. Wiring must meet so widely differing service conditions that different types have been developed. The conditions which may vary in different installations are: average quantity of **current** flowing, light or heavy power service, need for finished appearance or concealment of wiring, presence of moisture or fumes, different structural types of buildings in which wiring is installed, and time of installation, i.e., during construction or in old buildings. Other factors which bear on the development and use of different types of wiring are the money which can be expended on wiring, and the possible loss which might result from electrical fires. These factors have brought about the creation of the following methods of wiring:

1. Exposed knob and cleat wiring (mill type).
2. Concealed knob and tube wiring.
3. Raceway wiring.
4. Armored conductor wiring (BX).
5. Rigid iron conduit wiring.
6. Non-metallic protected cable.

The first two of these methods involve suspension of the conductors a certain distance apart (minimum 5″), or their enclosure in flexible loom when, as at switches and fixtures, they must converge. However, knob wiring, wherein the insulators are supported on porcelain knobs or cleats, or in tubes where they pass through holes in beams and walls, very definitely insulates the conductors from one another, and eliminates contact with anything but the porcelain insulation. Of course, when poorly installed, and not taut, the conductors may sag between insulators and may create a very definite fire hazard. Even when installed perfectly, little can be said for the decorative appearance of open wiring, so that it is limited to industrial buildings, garages, sheds, and other places where appearance is not a primary consideration. Knob and tube wiring installed in residences is inexpensive, although it must be done during construction of the building. If well installed, it is as good as any method of wiring, except possibly the rigid iron conduit.

Raceway wiring is to be thought of as a type whose maximum usefulness is in the field of wiring installation subsequent to construction, where the opening of partitions or ceilings for the purpose of introducing concealed wiring is impossible or undesirable. Raceway wiring is surface wiring, with the wires concealed in wooden or metal raceway which is attached to the surface of walls or ceilings. A wooden raceway is provided with a molded decorative cover which may be worked into a semblance of decoration in a room. The same cannot be said for a metal raceway, but it is of small size as compared to the wooden raceway, less conspicuous, and is frequently used for surface wiring of offices and mercantile establishments. In raceway wiring, the base is installed and the wires laid therein, complete from outlet to outlet, then the cover is applied.

Flexible armored conductor is probably used more nowadays for residence wiring than any other type. The armored conductor consists of two rubber-covered wires (three-conductor cable is also commercially produced) covered with a spiral serving of craft paper over which is laid a double spiral steel sheath. The flexibility of the latter permits the armored cable to be bent, although if too small a radius of bending is attempted, the sheath will be damaged. Armored cable is not moisture-proof unless it is of a special type having a lead sheath between armor and the conductor. However, it is approved for wiring in dry locations, such as residences. The cable must be continuous from outlet to outlet, and is generally employed in connection with iron outlet boxes, to which it is attached by special clamping connectors. It does not need to be held by insulators, and may be passed through holes bored in beams or joists without the aid of insulating bushings. The non-metallic protected cable is very similar to this, the major difference being in the protective sheath which is woven.

Rigid iron conduit is the best method of wiring; also the most expensive. It is superior in durability, is moisture-proof, and fireproof, reliable, safe, and mechanically strong. While it is not the intention here to give details of conduit wiring (there are standard handbooks treating that subject), some of the general principles are stated. Although the conduit is an ordinary water pipe smoothed on the inside, there is little in common between the installation of conduit and of water pipe, unless it be that junction of adjacent lengths is made by screwed couplings in both cases. To understand the differences let it first be noted how wires are installed in a conduit. A run of conduit having been installed complete and unbroken between the terminal points, the wire is drawn in as follows. A fish tape or wire, a tempered steel wire of rectangular cross-section, is pushed through the conduit until its end appears at the farther end. A draw line is then attached to it and by withdrawing the fish tape, the line is drawn through the conduit. The wires are in turn attached to the draw line and drawn into position. This method of installation requires (1) that the conduit interior be smooth and uninterrupted, (2) that the bends be of long radius and limited in number. If the conduit were not smooth internally there might be difficulty in pushing the fish tape through it; moreover, the roughness would doubtless damage the insulation on wires being drawn in. When conduit is cut and threaded the ends should be reamed to remove burrs. Ordinary pipe elbows are not used in conduit wiring. The pipe itself is bent to a long radius or long radius elbows are used. Due to the snubbing action of bends on wire being drawn through the conduit, not more than four equivalent 90°. bends are permitted between pulling points, and many prefer to limit the number to three. In size, conduits smaller than ½″ iron pipe size are not to be used, while conduits larger than 4″ are seldom required. Fiber ducts or tunnels are favored over the large iron conduit. Manufacturers have developed, in place of ordinary pipe elbows, tees, etc., lines of special conduit fittings designed to satisfy all requirements for outlets, junctions, etc. (F.T.M., L.R.Q.)

WISENT. Mammalia, Artiodactyla. The European bison. (A.W.L.)

WITCH OF AGNESI. The witch (of Agnesi) is a plane curve which may be defined geometrically as follows: Place a circle of radius a with its center on the Y-axis and passing through the origin; draw any line

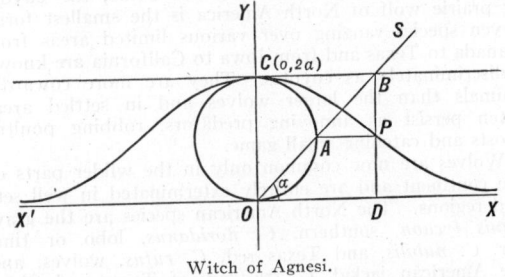

Witch of Agnesi.

OS cutting the circle at A and cutting the tangent at C at B; draw BD perpendicular to the X-axis and AP parallel to the X-axis, and let BD and AP intersect at P. As OA rotates about O, the point P describes the witch.

The **parametric equations** of the witch are

$$x = 2a \cot \alpha, \quad y = 2a \sin^2 \alpha$$

The equation in **rectangular coordinates** is:

$$x^2 y + 4a^2 y - 8a^3 = 0,$$

or

$$y = \frac{8a^3}{4a^2 + x^2}.$$

(L.L.S.)

WITCHES' BROOMS. The malformations known as witches' brooms are formed in many different kinds of woody plants. Many of them are caused by the presence of *Exoascus,* an **ascomycete,** which grows parasitically in the tissues of the plant. A bud infected by *Exoascus* is stimulated to rapid growth. So are the lateral buds of this shoot. The result is a bush-like mass of branches. The leaves of these infected branches are commonly dwarfed and fall earlier than the other leaves of the plant. The stems contain a much greater amount of **parenchymatous** tissue than is found in the normal branches. No reproductive parts are found on the brooms.

Witches' brooms are formed on fir trees; here they are caused by *Aecidium elatinum,* one of the **rusts.** A rust, *Gymnosporangium,* also causes the formation of witches' brooms on white cedars. Hackberry trees are often conspicuously covered with witches' brooms. In this plant they are caused by a mite, *Phytoptus,* which attacks the plant. (R.M.W.)

WITHERITE. The mineral witherite is **barium** carbonate, BaCO₃, crystallizing in the **orthorhombic** system. It is interesting to note that at 811° C. it changes to the **hexagonal** system, and at 982° C. it appears to become **isometric.** It has a rather imperfect prismatic **cleavage;** uneven fracture; hardness 3–3.7; specific gravity, 4.2–4.35; luster, vitreous to resinous; color, white to yellowish or grayish; streak, white; transparent to translucent. Witherite is not a common mineral, being found in veins, and often associated with **galena** as at Alston Moor, Cumberland, England. Associated with barite at Freiberg, Saxony, and at Lexington, Kentucky. Named in honor of Dr. William Withering, an English botanist. (E.S.C.S.)

WOLF. Mammalia, Carnivora. A large dog-like animal belonging to the family Canidae, containing the jackals and domestic dogs, the various wild dogs, and the foxes. The wolves are closely related to the domestic dogs and are supposed to be the ancestral stock of the latter. The relatively few species are found in Europe, Asia and North America, with the one exception of the Antarctic wolf of the Falkland Islands, a species somewhat smaller than the coyote, with a less bushy tail, black at the middle and tipped with white. The kaberu, also called the Abyssinian wolf, is related to the jackals.

Of the true wolves of the genus *Canis,* the **coyote** or prairie wolf of North America is the smallest form. Seven species ranging over various limited areas from Canada to Texas and from Iowa to California are known indiscriminately as coyotes. They are more cowardly animals than the larger wolves and in settled areas often persist as annoying predators, robbing poultry roosts and catching small game.

Wolves are now common only in the wilder parts of the continent and are entirely exterminated in well settled regions. The North American species are the gray, *Canis lycaon,* southern, *C. floridanus,* lobo or timber, *C. nubilis,* and Texas red, *C. rufus,* wolves, and the American jackal, *C. frustron,* of Texas and Oklahoma. (A.W.L.)

WOLFFIAN DUCT. Urogenital System.

WOLFRAM. Tungsten.

WOLFRAMITE. The mineral wolframite, **tungstate** of iron and manganese, is an **isomorphous** mixture of tungstate of iron, FeWO₄, and tungstate of manganese, MnWO₄, the amounts being variable. The pure iron tungstate is called ferberite and the manganese tungstate, hübnerite. It has been proposed that the name ferberite be applied to mixtures of not less than 80% FeWO₄ and not more than 20% MnWO₄, and that the term hübnerite be given to mixtures of not less than 80% MnWO₄ but not more than 20% FeWO₄. Wolframite would thus include the minerals of intermediate composition, and its formula would be written (Fe,Mn)WO₄. Wolframite is **monoclinic,** usually appearing in tabular, columnar or bladed crystals, sometimes quite large; also may be massive. Its hardness is 5–5.5; specific gravity, 7.1–7.5; color, gray, reddish-brown, brown or black; streak, reddish-brown to black; luster, submetallic; opaque, occasionally magnetic. Wolframite is found associated with **apatite, cassiterite, quartz, fluorite,** etc.; **in granites** and **pegmatites.** Often with **scheelite,** CaWO₄, and sometimes as a **pseudomorph** after that mineral. Wolframite is found in Czechoslovakia, Rumania, Saxony, Cornwall in England, New South Wales, Bolivia; and in the United States at Trumbull, Connecticut; Luna and Lincoln Counties, New Mexico. It also occurs in small quantities in Nevada and Utah.

Ferberite, which has monoclinic tabular crystals, sometimes massive, resembles wolframite, and occurs in Spain and in Boulder County, Colorado.

Hübnerite crystals are monoclinic, often long fibrous or bladed, may be massive; resembles wolframite and is found in Peru, the Black Hills, South Dakota; San Juan County, Colorado; White Pine County, Nevada, and Lemhi County, Idaho. (E.S.C.S.)

WOLLASTONITE. The mineral wollastonite is **calcium** metasilicate, CaSiO₃, which is found as tabular or short prismatic **monoclinic** crystals. This mineral has a hardness of 4–5; specific gravity, 2.8–2.9; color, yellowish, reddish or brownish, to gray, white or colorless; luster, vitreous to pearly; transparent to translucent. Wollastonite in the main is formed by the action of contact **metamorphic** processes on limestones at relatively high temperatures (600° C.+). Its common associates, **diopside, vesuvianite, garnet** and **epidote,** suggest this origin. Some of the more important foreign localities are in the copper mines of Rumania, in the lavas of Monte Somma and Vesuvius, in Finland and Mexico. In the United States it is found in Essex and Lewis Counties, New York; Keweenaw County, Michigan; and Riverside County, California. Wollastonite was named in honor of the English chemist, William Hyde Wollaston. (E.S.C.S.)

WOLVERINE. Mammalia, Carnivora. A stoutly built short-legged animal, *Gulo luscus,* of northern North America, ranging from the mountain forests of Colorado and Pennsylvania to the Arctic. It is over 3' long, with large feet, a short bushy tail, and shaggy fur. Also known as the glutton, and by the Canadian French name carcajou.

The wolverine is noted as a vicious killer and as a despoiler of camps. It not only damages foods that it cannot eat but steals objects that it cannot use and so makes itself a general nuisance to campers and trappers in the northern woods. Since it climbs, gnaws, and digs with great facility, protection of supplies against its attacks is difficult. Wolverines eat dead animals as well as prey that they kill, and often rob trap lines.

A second species of wolverine, *G. luteus,* of slightly smaller size and different proportions, has been recognized, extending southward in the Sierra Nevada of California. (A.W.L.)

WOMB. The mammalian **uterus. Reproductive System.** (A.W.L.)

WOMBAT. Mammalia, Marsupialia. *Phascolomys.* Stoutly built pouched animals of Australia and Tasmania, broad of body and with short thick legs. They have a pair of chisel-like incisor teeth in each jaw, like the rodents, and have similar food habits. They live in burrows or in crevices among rocks. Several species are known. (A.W.L.)

WOOD. (TIMBER. LUMBER.) The word wood is frequently applied to fuel, which in this case means forest products cut to a size suitable for burning in stoves. Again, especially in the plural form, the word refers to a stand of growing trees, especially if the stand covers a considerable area. The word may also apply to finished products such as boards, joists, and beams, all called wood. In botanical language, the word wood means that part of the stem, trunk, or branches, which is composed of the water-conducting **xylem** and associated fibers, especially in perennial plants in which these cells form a considerable mass, as in trees, shrubs and vines.

The word timber is also applied to standing trees, especially in the phrase standing timber, meaning trees large enough to be commercially valuable. Another common phrase is timber-land. Timber also means prepared forest products, especially any large beams, or coarse products. **Lumber** has a similar meaning, but is used in a rather broader sense, applying generally to any forest products such as boards, planks, beams, etc.

Practically all plants yielding wood of any value are **gymnosperms** or **dicotyledonous angiosperms.** Among **monocotyledons** only certain **bamboos** and **palms** have any appreciable use, and that largely restricted to the regions in which these plants grow.

It is worthy of note that forests have a very great value to man in addition to their yields of wood. For a forest is a very important factor in conserving water, preventing it from rapid running-off, and so tending greatly to reduce the possibility of disastrous floods. This retention of water is due to several factors; one because the roots themselves tend to form hollows in which water stands and settles, to move slowly through the ground, another because the increased supply of humus accumulating in the ground under the trees acts as a sponge, absorbing and retaining water. Vast amounts of water are absorbed by trees and given off to the air by transpiration.

In North America there are several types of forests, distinguished mainly by the kinds of trees which are most numerous in any kind. In eastern North America three main regions are recognized. There is the Northern Forest, occupying the northernmost tier of states from Minnesota east, and reaching downward in the mountains to Virginia and Tennessee. The dominant trees of this forest are **coniferous**, with white pine the most valuable species. Other coniferous species are hemlock, fir, and spruce. Hardwoods such as oaks, maples and birches are important trees of this northern forest. The southern forest occupies the region bordering the Atlantic and Gulf of Mexico and extending from the southeastern corner of Virginia into the eastern part of Texas. This also is a forest dominated by coniferous trees, mostly species of pine. Hardwood trees of this forest include sweet gum, ash, and tupelo. Between these two forests and extending west into the prairie region is the hardwood forest, in which are found oaks, black walnut, hickory, and basswood. In the western half of the country there are two regions, one, the Pacific forest, occupying the states bordering the Pacific ocean, and composed largely of Douglas fir, redwoods and several species of pine, all trees frequently attaining tremendous size. Occupying higher elevations in the western mountain states is the Rocky Mountain forest, containing such trees as Douglas fir, Engelman spruce, several pines, aspen, red cedar, and many others. Including the southernmost tip of Florida and the southern borders of Louisiana and Texas is a forest of an entirely different type, the tropical forest, characterized by such trees as magnolia, and live oak with an undergrowth of palmetto, which occupies vast regions of land in Central and South America, as well as the West Indies.

Woods are divided into two main classes, softwoods and hardwoods, distinguished not so much by the nature of the wood as by the trees. Softwoods are those obtained from **coniferous** trees such as pines, spruces,

hemlocks and firs; hardwoods come from **deciduous** trees, and include such trees as oaks, ashes, maples, basswood, poplars, gums, as well as many tropical trees. It will be seen that the relative hardness of the wood is not an indication of its classification, since many hard pines, which are softwoods, are much harder than the soft wood of the bass, the poplar and the tulip tree, all of which are classed as hardwoods. Softwoods are much more extensively used by man, being employed widely in building construction. Hardwoods are used for furniture, interior finish, and for products demanding special wood structure.

The structure of wood is a very important factor in determining its ultimate use. The woody tissue of any plant is composed entirely of cells or vessels of various kinds, the majority of them being elongated and thick-walled. In **gymnosperms**, from which come the softwoods, these cells are **tracheids**, elongated cells in the walls of which there are many **pits**. These tracheids vary in size and in the thickness of their walls; those which are formed during the season of most active growth, in the spring, being larger and having thinner walls than those formed later. The hollow center, or lumen, of the spring-formed wood cells is much larger than that of the summer wood. As a result of this variation in cell size and structure, the wood of the tree is made up of distinct concentric layers of cells, often very sharply distinct. These are the annual rings. By counting them one may obtain an accurate knowledge of the age of the tree. In addition to the tracheids there are in the gymnosperm wood other cells which extend in narrow radial bands outward towards the surface of the stem. These are the ray cells which collectively form the rays of the wood. In softwoods the rays are not conspicuous; in many they are very minute. In hardwoods the cellular structure of the wood is more complex. Tracheids are present and resemble those of the softwood in appearance. In addition there are many vessels. These are composed of many cells arranged in vertical rows extending considerable distances along the stem. It is characteristic of vessels that there is no cross wall separating the component cells. Pits are found in the lateral walls of the vessels. In hardwoods, there are also many fibers, slender elongate cells with thick walls, in which there are only a few small pits. Compared with those of softwoods, the rays of hardwoods are very large and often form a conspicuous feature of the wood. All these cells in both hard and soft woods are formed by the **cambium** cells. When first formed, all wood cells contain **protoplasm.** Very early in their development, the protoplasm of the vessels, fibers and tracheids is lost, the cell becoming void of any living contents. Only the ray cells and any **parenchyma** cells which may remain retain their protoplasm. Therefore the greater part of the woody tissue of the tree is dead. For some time this wood functions, it being the region through which water and dissolved mineral matter passes up through the stem. Wood thus functioning is called sapwood. It is usually pale in color. As the stem increases in size, changes occur in the cells, and there is less and less **ascent of sap.** Their walls frequently become much darker in color; often the lumina of the cells becomes filled up with various solid substances. When this condition obtains the wood is known as heartwood. The separation between sapwood and heartwood is usually quite distinct; it does **not** necessarily coincide with an annual ring, but may form a very irregular region. The appearance of the heartwood differs greatly in different woods; in some it is never distinct from the sapwood, in others it forms a small mass in the center of the stem, while in others it includes almost the entire stem. In color heartwood ranges from white through yellows, reds, greens, browns, and even black in different trees; often it is streaked and mottled with different shades of color.

The first step in the preparation of wood is cutting

it into suitable dimensions. Usually this means that the felled trunks or boles, known as logs, are cut into lengths varying from 2 or 3' to 16' or more, depending on the use to which the wood will be put. These logs are then sawed into boards, planks, or larger pieces. The manner in which a log is sawed is often a very important factor in the value of the product, since many woods are valuable for their grain, which differs according to the way the wood is cut. The word grain, applied to wood, has many meanings. Commonly it refers to the appearance of the wood when finished. This is determined by the nature of the cells composing the wood and their arrangement. In many woods, especially the softwoods, the cells are all parallel so that the wood splits easily. Such wood is said to have a straight grain. In other woods, the cells are oriented in various directions. Such woods split with great difficulty, and are called cross-grained woods. Wood composed mostly of small thick-walled cells, with few vessels, is called fine-grained wood, in contrast to that having many vessels, usually of large diameter, known as coarse-grained wood. The rays of the wood often form an important feature of the grain, since when they are large they may appear in the form of distinct flakes or streaks in the surface of the wood. The size and appearance of these flakes is greatly changed by the method of sawing the wood.

The commonest way of sawing is called plain sawing. In this, successive boards or planks are cut from the log, beginning at one side and cutting successive pieces through the center to the opposite side. In those woods in which grain is of little value, especially those with very small rays, this method of cutting is entirely satisfactory, for it is quickly and easily done. In other woods, especially those in which the rays are large and form a conspicuous feature of the finished wood, it is not so satisfactory, since variation results. This variation is largely due to the nature of the rays, which extend directly outward from the center to the surface of the wood, and have little thickness compared to their other two dimensions. Consequently the first cuts will be across the rays, which will then appear as narrow linear streaks in the wood. When boards near the center are cut, the ray will be nearly parallel to the cut and so appear as an irregular broad patch on the surface, adding greatly to the beauty of the wood. Methods of cutting have been devised which increase the number of boards which show these radial cuts. These methods are known as quarter-sawing. There are many of these, all having as their object the cutting of the greatest number of radial cuts most economically. Another important feature of the grain of wood, which can be changed by the method of cutting, is found in the annual growth increment, which forms the annual rings of the cross-section. In sawed woods these annual layers appear as lines or streaks, often elaborately twisting.

Another important step in the preparation of wood is drying or seasoning. This may be done either before or after the log is cut into boards. During this process much of the water which is present in the wood, both in the lumina of the cells and in the cell walls, themselves, is lost. As a consequence there is a certain amount of shrinkage of the wood, which is greater tangentially than in the other dimensions. As a result, the wood on drying tends to check or crack longitudinally, especially in large pieces of wood. For most purposes, undried wood, known as green wood, is entirely unsatisfactory; it will twist, warp, and crack as it dries; it cannot easily be glued, or finished, and it is more subject to attacks of fungi and boring insects. There are two methods of seasoning woods. The older method was to pile the wood in such a manner as to secure the maximum exposure to air, separating each piece from the next by narrow strips of wood, or by other means, and leaving the piled-up wood to dry slowly. This is called air-drying or air-seasoning. To dry wood

in this way requires a month or more for boards an inch thick and several years for very large sticks, especially if the wood be a hardwood containing much water. Naturally the climate greatly affects the time required, wet seasons causing very slow drying. Recently much of the wood used, especially that used in making furniture, interior finish, shingles, etc., has been dried artificially in large rooms, heated artificially. By this method it is possible to dry wood in a few days instead of months or years. The product is also much drier and better fitted for use. This method is known as kiln-drying or kiln-seasoning. Properly prepared wood has the following properties which give it its special value. It possesses great strength and incompressibility, yet is flexible, giving without breaking, and elastic, recovering after bending. Many woods have special properties which particularly fit them for special uses.

The uses made of wood are so numerous that only those of the greatest importance can be named. Immense quantities of wood, both hard and soft, are used in the pulp and paper industry. In construction work other large quantities are used, and in a variety of ways. Poorer grades of lumber are much used in making forms in which concrete work is poured. Scaffoldings and other framework also use quantities of the cheaper kinds of lumber. Wood is still a most important material for the construction of dwellings and other buildings. For the framework and for the rough finish of the walls and floors, the poorer kinds of wood are mostly used. For the outer parts, such as shingles and clapboards, and for the interior finish materials and floors, better grades, often of special kinds of wood, are used. Shingles, for example, are mostly cut from white or red cedar, which is very resistant to the weather. Floors are made from hard pine, from maple or oak, and a few other woods which do not splinter easily and which resist heavy wear favorably. Nearly all the woods used in construction work are softwoods, which are light in weight, durable and easy to work.

For cabinet work and furniture an entirely different group of woods is used. Most of these are hardwoods, selected for their fine grains, or for their color, or because fashion and the whims of fancy happen to give them popularity. Many of the woods used for this purpose are tropical woods difficult to obtain and so of great value. In early days furniture was made from solid pieces; later it became the custom to use thin pieces of wood glued to a background of a different wood. These thin pieces were called veneer. There were several reasons for using veneer. One of course was economy, since an expensive block of wood would yield many more pieces of veneer than it would boards. Another reason was found in the fact that it was often difficult to fashion things from the hard dense woods used. They would split or crack easily, whereas when glued to a more easily handled more suitable wood they became usable. Finally, and very important, was the fact that often veneers could be obtained showing a beautiful grain which could rarely be obtained otherwise. So veneering became important.

There are three ways of cutting veneer wood. They may be sawed like any board: this method is wasteful since much material is wasted in the cut, as sawdust; it cannot produce very thin pieces, and it does not give the best grain in many cases. An advantage is found in the high polish which may be given veneers cut by this method. A second method of cutting veneers is slicing. In this method the short piece of wood, called a bolt, from which the veneers are to be cut, is moved up and down against a heavy stationary blade. This method has an advantage over sawing in that it allows much thinner veneers to be cut. As in sawed veneers, sliced veneers are limited to the size of the bolt from which they are cut, which is frequently a factor against them. The third method and the one used in cutting nearly all the veneers in the United States, is rotary-cutting. In this method the bolt is revolved against the knife, the veneer coming off in a thin sheet of any width desired. In order that the wood may be more easily cut and the veneer handled better, the bolt from which it is to be cut is usually soaked in water or steamed for some time before cutting, both by the slicing and the rotary method. After cutting, veneers are pressed flat and dried, in which condition they may be held indefinitely. They are commonly glued to some hardwood when they are used.

Another way in which veneers are used is in the making of plywood. This is made by gluing together three or more layers of veneer, the grain of each layer being at right angles to the one above it. Plywood is very resistant to blows and not easily cracked or broken. For which reason it is much used in making boxes, table tops, crates, and panels designed for various purposes.

In addition to these major uses of wood, there are many others, each of which uses immense quantities. Cooperage is one of these. There are two kinds, slack and tight, each producing a barrel or container suited for a special purpose. Slack cooperage produces barrels, kegs, tubs, etc., which are used as containers for vegetables, for nails, and for many other things. Formerly flour was mostly shipped in barrels of this sort. Tight cooperage produces containers which are tight, the staves fitting very close together, and the heads being very closely fitted. Woods used in tight cooperage must have large pores filled. White oak, in which the pores are completely plugged, is especially suitable. If other woods are used, they must be coated with a film of paraffin which will plug the pores. Quantities of wood are used in the manufacture of charcoal, which is extensively used in making steel, explosives, and carbon dioxide gas, and also as a filter in the manufacture of sugar. Charcoal is made by allowing a closely packed pile of wood to burn with insufficient air, as a result of which the volatile materials are driven off and the carbon of the wood left. Millions of cords of wood are used for fuel each year in the United States; much of this is waste wood which would be of little value for any other purpose. Large quantities of light softwoods, especially baswood, are used in making excelsior; in this the wood is first scored and then stripped off in thin pieces by knives. Excelsior is largely used to pack around glassware and dishes, and as a stuffing for mattresses and upholstery of the cheaper grades. Other uses of wood are for railroad ties, poles, and posts, piling, all of which call for woods which may be treated in such a way as to resist rotting and the attacks of various animals.

The following woods have properties which particularly fit them for special uses. Lignum-vitae, the heaviest of all woods, is a dark, greenish-brown, fine-grained wood with an oily appearance. It is obtained from *Guaiacum officinale,* a small tree native in tropical America. The wood of another species, *Guaiacum sanctum,* is also used. Both trees have compound leaves and showy flowers borne in clusters in the axils of the leaves. The fine-grained wood with its much-crossed fibers is used in making pulley sheaves, and as bushing around the propeller shafts of steamships, as well as for bowling balls and mallet heads. The wood contains a resin, guaiacum, which was formerly highly esteemed as a valuable medicine used in the treatment of social diseases. Today it is of slight importance as a drug. Rosewood is obtained from *Dalbergia nigra* and other species, natives of South America. It is a hard close-grained wood with a pleasant fragrance. Formerly it was much used as a cabinet wood. Sandalwood is obtained from trees native in the East Indian region. It is a firm-textured wood of dull yellow color, which darkens with age. Like rosewood, sandalwood has a characteristic aromatic odor. The tree, *Santalum album,* is interesting because it is a parasite on the roots of other plants. Satinwood is obtained from *Chloroxylon*

Swietenia, and other species, natives of India and Ceylon. The wood has a golden-yellow color and a brilliant luster which makes it a valuable cabinet wood. Several species of walnut, particularly *Juglans regia* of Europe and Asia and *Juglans nigra* of North America, yield dark brown woods of great value. They are much used in furniture making, and in interior finish in houses, and also because when once dried they are not subject to any changes through swelling or shrinkage, they are particularly favored as material from which to make gun and rifle stocks. Walnut is capable of taking a very high polish. Another dark heavy wood which is much used in furniture making is ebony, the wood of *Diospyros Ebenum,* and other species. The sapwood of the tree is soft and creamy white and of little value; the heartwood is very dark, often black, and hard. There are many other less well known woods, which are frequently used in furniture making.

(See also **Cedar, Mahogany, Oak.**) (R.M.W.)

WOOD DISTILLATION. Destructive Distillation.

WOOD HOOPOE. Aves, Piciformes. An African bird (**Aves**), of a family closely related to the true hoopoes. The wood hoopoes differ in the metallic gloss of the plumage and the long wedge-shaped tail. Also called chatterers. (A.W.L.)

WOOD PULP. Cellulose, under **Carbohydrates.**

WOOD RAT. Mammalia, Rodentia. A moderately large rodent, more common in the western states but represented by a few species in other sections. They are well and unfavorably known from their habit of invading houses and camps and carrying away anything edible. They differ from true rats in the shorter furry tail and the larger eyes and ears. Ten species have been described, all in the genus *Neotoma.*

These animals are also called pack rats and trade rats, the latter from their habit of replacing what they take with some other object. (A.W.L.)

WOODCHUCK. Mammalia, Rodentia. A large heavy bodied animal of the northern hemisphere. Woodchucks have short stout legs and are powerful burrowing animals, penetrating many feet into the ground. They also climb readily, although somewhat clumsily. They eat vegetation of many kinds and sometimes become troublesome in fields and gardens.

North America has three species of woodchucks, the common woodchuck, *Marmota monax,* the yellow-bellied woodchuck, *M. flaviventris,* and the whistler, *M. caligata.* The first occurs from Kansas to Georgia, northward to Alaska and Hudson Bay. The second ranges from the Rocky Mountains to the Pacific and the third is also western, ranging from Montana and Washington to Alaska. The names **marmot** and groundhog are also generally applied to them. The Old World species are widely distributed in Europe and Asia, where they are more commonly called marmots.

The flesh of woodchucks is eaten but it is inferior to that of other common rodents. While its flavor is good it is coarse in texture as compared with that of squirrels and rabbits. (A.W.L.)

WOODCOCK. Aves, Charadriiformes. A woodland bird (**Aves**), *Scolopax minor,* of the northern and eastern part of the United States, related to the snipes. It has a long straight beak and short neck, tail and legs, and its mottled brown plumage is a fine example of protective coloration. Of limited interest as a game bird. (A.W.L.)

WOOD-HEWER. Aves, Passeriformes. A small brown bird (**Aves**), of a family found only from Mexico to southern South America. The family includes more than 200 species, mostly limited to the temperate parts of the continent. Among them are the **oven birds** and several other groups. (A.W.L.)

WOODLOUSE. Crustacea, Isopoda. Small terrestrial **crustaceans** of oval form, somewhat flattened. They live under bark and debris near the surface of the ground, in decaying vegetation, and in other sufficiently moist situations, and are common in gardens. Also called sowbugs. These forms constitute the families Porcellionidae and Oniscidae, and in the closely related family Armadillididae are found the pill bugs. The latter are able to roll themselves into almost perfect spheres when disturbed. The numerous species vary in color from gray to brownish and blackish. (A.W.L.)

WOODPECKER. Aves, Piciformes. A bird **(Aves)** whose beak is adapted for chipping wood and whose feet are formed for gripping the bark of trees. The tail is composed of stiff feathers and is used as a brace against the surface to which the bird clings. Woodpeckers excavate deep holes in trees as nests and deposit white eggs whose shells are like translucent china. They also dig into decaying wood for the insects contained in it. The capture of insects is facilitated by the sharp barbed tongue and sticky saliva. Both the barbed tongue and the stiff tail are lacking in a few genera.

Woodpeckers are well represented in all regions except the Australian. They vary from the diminutive piculets of South America and Asia to the great ivory-billed species, which attain a length of 18". North America has more than a score of species, including the flickers and sapsuckers. Of these the red-headed, *Melanerpes erythrocephalus,* and downy, *Pyrobates pubescens,* woodpeckers and the flickers are widely known and the large pileated woodpecker, *Ceophloeus pileatus,* or cock-of-the-woods of northern and western forests and ivory-billed woodpecker, *Campephilus principalis,* of southern localities are among the rarest. (A.W.L.)

WOODRUFF KEY. Key.

WOOD'S METAL. Alloys.

WOODWORKING. A cutting process performed with the aid of both hand and mechanically actuated tools. Hand saws are used for bringing stock to size, for roughing out grooves, and for numerous other purposes. There are two important types of hand **saw** teeth: cross-cut saw teeth, which are pointed and are employed for cutting across grain; and ripsaw teeth, which have chisel edges and are used for cutting with the grain of the wood. The backsaw is a fine-tooth cross-cut saw with a thin blade and is designed for accurate work. As its name indicates, it is reinforced with a heavy steel rib. A compass saw is a cross-cut saw with a narrow tapered blade, and is used to cut curves and circles after a preliminary hole has been bored in the stock. A coping saw has a C-shaped steel frame and a very narrow blade and is used for cutting curves in thin stock.

Chisels are cutting tools used in joint construction and in fitting and shaping. Chisels are used to cut **mortises**, grooves, bevels and chamfers. Gouges are chisels of circular section. Outside bevel gouges are used for wood-turning roughing cuts and for cutting flutes with spherical ends (blind flutes). Inside bevel gouges are used for hollows and for grooving and edge shaping. Veining or carving tools are gouges of vee or circular shape, and are employed in ornamental carving.

A hand plane is essentially a chisel carried in a frame. The jack plane is so named because it may be used to do the work of several other planes. The plane iron cap is clamped to the plane iron or cutter, and breaks the shavings as they enter the throat of the plane. The lever cap locks the plane iron to the body of the plane. The cutter is advanced or retracted for a coarse or fine cut by the adjusting nut. The jack plane is about 14" long; the smoothing plane, fore plane and jointer planes are almost identical with the jack plane except that they are respectively 8", 18", and 30" long. The block plane is a small plane for planing end grain. It has no cap iron or handle. The plane iron in a block plane is set at a smaller angle to the plane bottom than the jack plane iron and its bevel is turned up instead of down.

Drills and bits are used for boring holes in wood. An auger bit is a cutting tool of helicoidal form with a screw point which serves to pull the tool into the wood. A pair of spurs outline the hole and cut off the fibers, which are picked up by the lips of the bit. Taper head or twist drills are extensively employed in wood boring because the greater part of the fluted portion of the drill can be ground away in resharpening. It requires pressure to make it cut, does not elevate the chips as readily as the double-twist auger bit, and is therefore not as satisfactory for deep hole boring. It is very satisfactory for end grain service and is extensively used for cross-grain where absolutely smooth holes are not required.

Woodworking machinery is extensively used in modern practice. A circular saw consists of a horizontal spindle carrying the saw, which projects above the surface of the saw table. A ripping fence or guide can be set at various distances from the saw, and is employed as a gauge for cutting boards to definite widths. The cut-off fence is employed for end cutting where the long dimension of the board moves at right angles to the saw.

Saws with either cross-cut or ripping teeth are employed for circular saws. Grooves for notched joints may be cut by using a series of saws properly spaced by collars or washers. Dado heads, for dado, plough, and rabbet cutting, may also be used on saw tables.

A bandsaw is a continuous steel ribbon with teeth along one edge, which runs over a pair of wheels whose diameters denote the size of the saw. The work is placed on a horizontal table and fed by hand against the front edge of the saw. The thrust of the work is borne by roller guides which are set as close as possible to the work. Bandsaws are obtainable in widths from ⅛" up, and are generally made with a brazed lap joint. Bandsaws with tilting tables, and saws in which the frame tilts, thereby permitting angular or beveled cuts, are available. The bandsaw is used for cutting lumber to shape, and for curved work where its narrow blade permits small radii to be sawed.

Scroll saws or jig saws are used for cutting internal and external irregular curves. The saw blade has a reciprocating motion, and cuts on the down stroke. The scroll saw can be used for irregular holes in the work where a starting cut from the outer edge to the hole, which is required in bandsawing, is not desirable.

The wood planer is employed for surfacing rough lumber and planing wide boards. The cutter head is composed of a cylindrical body with two, three, four, or six knives which are clamped to the body but may be adjusted to compensate for resharpening. The wood jointer is used for preparing edges to be joined accurately. The table is adjustable for various depths of cut, and a fence is adjustable for finishing bevels as well as squared stock. The wood shaper is employed for moldings and similar work. It has a vertical rotating spindle, on which cutters or knives of different shapes may be placed and clamped.

Lathes and drill presses are extensively used in woodworking. (See **Lathe, Drill Press.**) Many other machines for woodworking are in common use. The molder is used for cutting moldings, hexagons, full- and semi-rounds, and other strips of irregular cross-section. It resembles a combination of planer and jointer with two additional vertical cutter-heads. The matcher is a machine with six cutter-heads for simultaneously finishing all sides of matched flooring stock. The tenoner is a machine with several cutter-heads for machining in one operation tenons of varied shapes. (H.C.H.)

WOOLLY BEAR. Insecta, Lepidoptera. The densely hairy **caterpillar** of a **moth** of the family Arctiidae. Most hairy caterpillars belong to this family and most species of the family have hairy larvae. (A.W.L.)

WORK. Work may be regarded as the transfer of **energy** from one body to another; or, in a slightly different sense, as energy in process of transfer. While there are numerous instances of energy transfer whose mechanism is unknown (by radiation, for example), in all cases open to direct observation the process appears to involve two essential factors: (1) the exertion of a force by one body A upon another body B, and (2) the motion or displacement of B in a direction in which the force has an effective component (not necessarily in the direction of the force itself). Thus, the wind may exert a force upon a sail in the direction north-northeast, while the boat actually moves straight north, that is, at an angle of 22° 30' with the force. But if the force is F, it has a component in the direction of motion, equal to $F \cos 22° 30'$ or $0.9239 F$; and work is therefore done.

It would be logical to define the measure of work as the quantity of energy transferred. But custom has it just reversed; we fix upon a measure of work and define the unit of energy as that transferred when unit work is done. The quantity of work itself is defined as the product of the displacement (say, in feet) by the component of the force (say, in pounds) in the direction of that displacement; thus giving rise to composite work units such as the **foot-pound**, or the **erg** and the **joule.** Thus, it comes about that we also commonly express energy in foot-pounds, ergs, or joules.

It should be made clear that the displacement is not necessarily caused by the force when work is done. Thus, if one throws a tennis ball against the rear of a rapidly receding truck, it cannot be said that this causes the truck to move. Nevertheless the ball does work on the truck which it would not do were the truck standing still; and as a result the ball rebounds with less energy than from a stationary surface. (L.D.W.)

WORK FUNCTION. The general meaning of this term, much used in connection with electronic phenomena, may be made clear from the following application. Imagine an electron trying to escape from the heated, negatively charged filament of a radio tube. It moves about readily enough within the metal, where there is no general electric field; but upon leaving the interior and receding from the positive atomic nuclei of the metal, the electron must necessarily experience a backward jerk which robs it of some of its kinetic energy. The work function, which expresses this lost energy, is characteristic of the metal. Once outside the surface, the negative surface charge helps the fugitive electron along; but it always has less speed and less kinetic energy than it would have had but for the parting jerk. Photoelectric emission involves the same thing. A somewhat analogous case is that of molecules escaping from a liquid in evaporation. They are likewise hindered (by cohesion), the loss of kinetic energy being in this case sustained by the liquid itself (which is actually cooled) and represented by the **heat of vaporization.** Electronic work functions are commonly expressed in **electron volts.** (L.D.W.)

WORKING STRESS. Factor of Safety; Stress.

WORLD LINE. Relativity.

WORM. A word without exact scientific limitations. Applied to creeping animals of the invertebrate phyla

with long slender bodies, but also to some flatworms with broad thin bodies and only inaccurately to worm-like forms such as some of the insects. Scientifically it embraces four phyla: **Platyhelminthes** or flatworms, **Nemertea** or ribbon worms, **Nemathelminthes** or roundworms, and **Annelida** or segmented worms. (A.W.L.)

WORM GEARING. Screw gearing or worm gearing is used to transmit power between shafts with perpendicular, non-intersecting axes. The worm is usually of cylindrical form, and resembles a screw; a section through the worm thread shows that the teeth are straight-sided and analogous to those of an involute rack. Worms are cut on a lathe or a thread milling machine and are often ground and polished after cutting and hardening to obtain surface precision and finish.

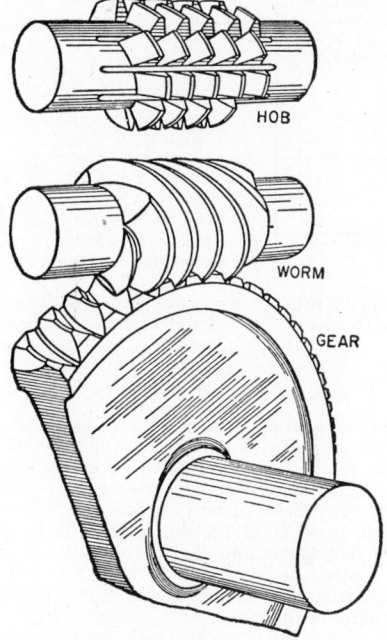

Fig. 1. Worm gearing and hob.

The worm wheel is essentially a helical gear with a face curved to fit a portion of the worm periphery. The tooth form and shape are obtained by cutting the wheel with a special form cutter known as a *hob,* which is essentially a replica of the worm, furnished with longitudinal flutes to provide cutting edges. In cutting the worm wheel teeth, the hob and the wheel blank are rotated at a speed ratio exactly that of the finished set; the hob is properly located with respect to the plane of the wheel and fed in radially until the teeth have been cut to full depth. This cutting action generates worm wheel teeth that are of involute form at the midplane of the wheel, and are conjugate to the hob and consequently to the worm.

Tooth measurement in worm gearing is generally based on circular pitch, although diametral pitch gearing is manufactured and stocked by gear manufacturers. Circular pitch is measured in the diametral plane of the wheel and in a plane passing through the axis of the worm. If D_g represents the pitch diameter of the wheel, P_c the circular pitch, and N_g the number of teeth in the wheel, then

$$D_g = \frac{N_g P_c}{\pi}.$$

The lead L of the worm is the distance that a thread advances in one turn, or the distance that a point on the pitch circle of the worm wheel will advance during one revolution of the worm. If N_w represents the number of threads or "starts" in the worm, then:

$$L = N_w P_c.$$

A triple-threaded worm has a lead equal to three times the pitch; in a single-threaded worm the lead and pitch are alike.

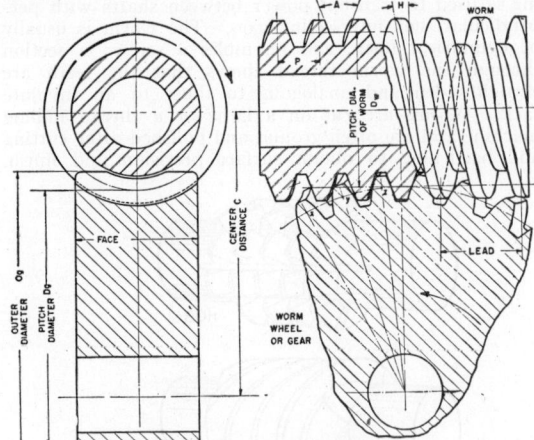

Fig. 2. Worm gearing nomenclature.

The velocity ratio R of a worm gear set depends upon the lead of the worm and the pitch diameter of the wheel, or,

$$R = \frac{\text{rpm Worm}}{\text{rpm Wheel}} = \frac{N_g}{N_w}.$$

Unlike most gearing, the velocity ratio is independent of the pitch diameter of one of the elements—the worm. The worm pitch diameter can therefore be selected to suit a particular center distance, or to make use of a stock hob and thereby dispense with the cost of a special cutter.

The lead angle H of the worm threads is the angle between a line tangent to the thread helix at the pitch line and a plane perpendicular to the axis of the worm. It is found from the following, where D_w represents the pitch diameter of the worm.

$$\text{Tan } H = \frac{P_c N_w}{\pi D_w}.$$

The tooth pressure angle is measured in a plane passing through the axis of the worm, and is equal to one-half the thread profile angle A. Pressure angles of $14\frac{1}{2}°$ are commonly used for single- and double-threaded worms, and $20°$ for triple- and for quadruple-threaded worms. However, in many modern worm gear reducer sets, pressure angles as high as $30°$ are employed.

The nature of the tooth engagement in worm gearing causes greater sliding action between the surfaces in contact than in the case of spur gearing. The amount of this sliding action varies with the helix angle, and affects the efficiency of the gearing although it contributes to the smoothness of the drive. Efficiency depends not only on the material of the worm and worm wheel, the amount and character of the lubricant, the velocity of rubbing, but also upon the size of the helix angle of the worm. The efficiency may be estimated from the following:

$$E, \% = 100 - \frac{R}{2}.$$

The above expression refers to commercial worm gear reducers properly mounted and lubricated.

The power-transmitting capacity of worm gearing is based upon the strength of the teeth, the ability to resist wear and abrasion, and the heat-radiating capacity. Since the teeth on the gear are usually weaker than the threads on the worm, the design for strength and for wear is analogous to the method used for spur gearing. The heat-radiating capacity is a function of the efficiency and the square of the pitch diameter of the gear.

Worm gear sets should be carefully aligned in the axial plane of the worm, with the shaft axes at 90°. If the set is arranged so that the worm is underneath the wheel, the former may be run in an oil bath to insure adequate lubrication. Installations should preferably be enclosed to retain the lubricant and to prevent the admission of dust or foreign matter.

Globoidal worm gearing is a form of worm gearing in which a worm of "hour-glass" shape envelops a gear with straight-sided teeth. The capacity of this form of gearing is considerably greater than that of the conventional cylindrical type, and its efficiency is somewhat higher. (H.C.H.)

WORMWOOD, OIL OF. Volatile Oils.

WOUND. Any break in the continuity of a bodily surface, internal or external, caused by an outside injury or force.

An operative wound is one that is made surgically to expose an organ or portion of the body.

An aseptic wound is a clean wound which is not infected by pathogenic organisms. Such a wound is said to heal by primary intention.

An infected wound is one which is infected with pyogenic organisms. Such a wound heals by secondary intention, by adhesion of granulating surfaces, or by third intention, by filling of the wound with granulation tissue.

A lacerated wound is one in which the tissues are torn.

A contused wound is one in which crushing and bruising have occurred.

A punctured wound is a deep wound, small in diameter, made by a pointed object.

An incised wound is one made cleanly, as by a knife or other cutting instrument. (R.S.M.)

WOU-WOU. Mammalia, Primates. The gray or silver gibbon of Java. (A.W.L.)

WRASSE. Pisces, Teleostei. Marine fishes (**Pisces**) found among rocks and coral reefs in tropical and temperate seas. Many are beautifully colored. Some of the larger wrasses are excellent food fishes. The name applies generally to members of the family Labridae. Two species of the Atlantic coast are known as the cunner (blue perch, chogset, bergall), *Tautogolabrus adspersus,* and the tautog (oyster-fish, black-fish), *Tautoga onitis.* (A.W.L.)

WREN. Aves, Passeriformes. A small bird (**Aves**), related to the larger thrashers. Distinguished by its long curved beak and often sharply erected tail. Several

Carolina wren, *Thryothorus ludovicianus.* Bright reddish-brown above; wings and tail barred with black. A long white line over eye. Underparts, creamy buff.

species habitually build their nests about dwellings or in bird houses. From this habit the widely distributed and vociferous little house wren, *Troglodytes gedon,* has become widely known and the Bewick wren, *Thryomanes bewicki,* a more musical species, has introduced himself to many residents of the eastern states. The Carolina wren, *Thryothorus ludovicianus,* also frequents human habitations. North America is the home of a dozen species and Europe and Asia also have representatives of the group, but it attains its greatest diversity in South America. (A.W.L.)

WRENCHES. Tools for turning and tightening bolts, nuts, and pipe. A variety of wrenches are shown in the figure. Socket and box wrenches are usually preferred to open-end wrenches, since there is less likelihood of slipping over the corners of the nut. Open-end wrenches can be used in many places, however, where socket wrenches will not enter. "S" wrenches are particularly applicable where the swing of the wrench handle is limited. Adjustable open-end wrenches, such as the type shown, or the familiar "monkey" wrench, are used

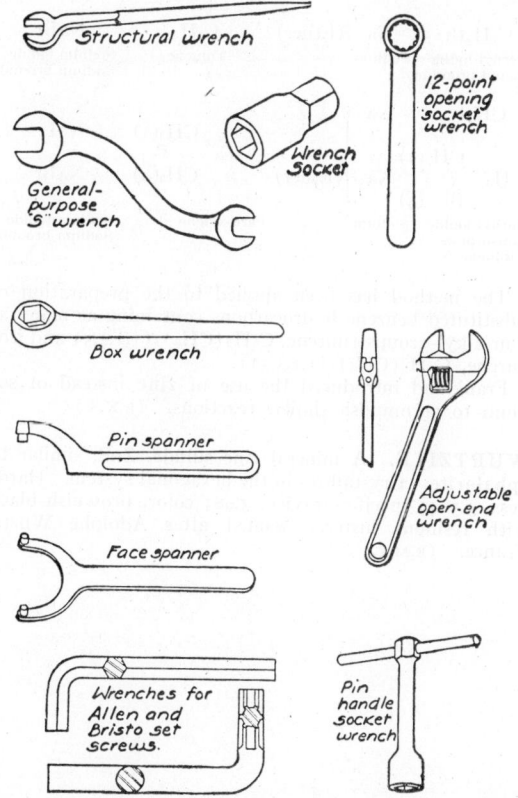

(Hesse's Engineering Tools and Processes.)

for general purpose applications, where it is not feasible to provide a set of solid jaw wrenches. Pin spanners and face spanners are used for round nuts and chucks which have drilled holes either in the periphery or in the face to fit the projecting pins of the wrench. Set screw wrenches are used for setting and tightening socket head set screws, or socket head screws.

Alligator and Stillson wrenches are used for cylindrical work, such as piping and rods. An alligator wrench has a vee-shaped jaw; one side of the jaw is smooth, the other is furnished with ratchet teeth, to enable the wrench to "bite" into the part to be turned, effecting a wedging action that increases the grip of the wrench as the turning force increases. The Stillson wrench is an adjustable wrench fitted with a pivoted jaw with ratchet teeth which tends to wedge on the work. Alligator and Stillson wrenches should not be used for bolts and nuts, except in emergencies, since they deform the heads and make it difficult to use ordinary wrenches. (H.C.H.)

WRIST. The slender part of the forearm at its attachment with the hand, especially the region containing the group of small carpal bones between the radius and ulna of the arm and the metacarpals of the hand. (A.W.L.)

WROUGHT IRON. Wrought iron is a ferrous material aggregated from a solidifying mass of pasty particles of highly refined metallic **iron,** with which, and without subsequent fusion, is included a minutely and uniformly distributed quantity of **slag.** This definition of wrought iron indicates that it is a material made of two components; one, iron of a high degree of purity, the other, slag (chiefly silicate of iron). In the finished product the slag is distributed through the iron in threads and fibers, of which there is an enormous number. The slag imparts to the wrought iron a fibrous structure, quite different from the crystalline structure of cast metals. Wrought iron has made for itself a name as a metal which has resistance to corrosion, and which is exceptionally suitable for structural purposes where the structure is subject to shock. Wrought iron also can be readily worked, forged, machined, welded, galvanized, etc. Among the many applications of wrought iron might be mentioned tubes, pipes, and tanks.

Wrought iron has been made for centuries, but until comparatively recently, its production involved a large amount of hand labor. In former years, wrought iron was produced in the puddling furnace, wherein molten iron was refined with the aids of oxidizing agents, which resulted in the production of an iron silicate, slag. In this process the furnace temperature towards the end of the heat run is maintained high enough to keep the slag in molten condition, but low enough for the iron to become pasty. The workman thoroughly mixes the slag with a spongy iron, producing a sponge of iron and slag. This working is known as rabbling. When this process is complete, the sponge is taken from the furnace with tongs, and put in a squeezing machine, which presses out the surplus slag. The resulting product is then collected and hammered into blooms or rough bars of the wrought iron. Much arduous and tedious work has to be applied to produce a comparatively small sponge with the puddling process. The rabbling process requires a large amount of labor, yet this system continued to be used to produce wrought iron until comparatively recent years.

In the Byer's process, a modern wrought-iron manufacturing method capable of large output, the pig iron is melted in a cupola, then refined in a **Bessemer** converter. The slag is independently produced in a tilting **open hearth** furnace. When the molten slag is ready it is poured into a large ladle and carried to a processing machine where the liquid refined iron from the converter is slowly poured into it. The temperature of the slag is maintained at a value which is sufficiently lower than the melting point of iron that, as the iron is poured into it, the iron rapidly assumes a semi-solidified form. The pouring of iron into the slag is attended by certain mechanical oscillations and relative movement which is designed to produce a uniform distribution of iron in the slag ladle. When all the iron has been poured into the slag, it lies in the bottom of the ladle as a spongy slag-iron mixture over which the excess slag floats. This excess is poured off, and the sponge is dumped into a press which ejects still more slag and squeezes the sponge into a bloom which can be finished in a **rolling mill.** (F.T.M.)

WRYBILL. Aves, Charadriiformes. A New Zealand bird (**Aves**), whose beak is asymmetrical. The terminal half of the organ bends to the right. The bird is said to seek its food, consisting of insects and other small animals, by reaching under the edges of stones. This habit is apparently well served by the peculiar adaptation. (A.W.L.)

WRYMOUTH. Pisces, Teleostei. A marine fish (**Pisces**), related to the blennies. Found off Cape Cod. The name applies to all members of the family Cryptacanthodidae. (A.W.L.)

WRYNECK. Aves, Piciformes. A bird (**Aves**) related to the woodpeckers but with soft tail feathers. The name is from the curious habit of turning and extending the head displayed by the European species. The few known members of the group are found in Europe, Asia, and Africa. (A.W.L.)

WULFENITE. The mineral wulfenite is **lead molybdate** corresponding to the formula $PbMoO_4$, analyses showing that a part of the lead may be replaced by calcium. Wulfenite crystallizes in the tetragonal system usually in thin tabular forms, but is also found massive. It is a brittle mineral; hardness, 2.75–3; specific gravity, 6.5–7; luster, adamantine to resinous; color, yellowish to green or red, may be whitish or grayish; transparent to translucent. Wulfenite is a secondary mineral found in association with other lead minerals such as **galena**, **pyromorphite**, etc. It is believed to have been formed, at least in part, by the action of waters containing molybdenum salts on cerussite, anglesite, pyromorphite, etc. Especially important foreign localities are in Yugoslavia, Czechoslovakia, Morocco, French Congo, New South Wales and Mexico. In the United States it has been found in Phoenixville, Pennsylvania, and in the Organ Mountains, New Mexico; Yuma County, Arizona; Box Elder and Salt Lake Counties, Utah; and in Clark and Eureka Counties, Nevada. Wulfenite was named in honor of F. X. von Wülfen, an Austrian mineralogist of the 18th century. (E.S.C.S.)

WURTZ-FITTIG-FRANKLAND REACTION. Sodium metal was used by Wurtz as reagent for the preparation of paraffin **hydrocarbons** by treating alkyl iodide in ethereal solution, thus:

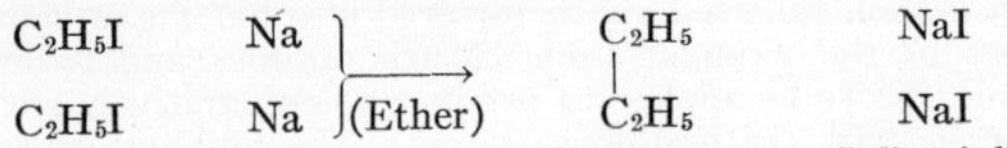

$$
\begin{array}{llll}
C_2H_5I & Na & C_2H_5 & NaI \\
C_2H_5I & Na \ (Ether) & C_2H_5 & NaI \\
\text{Ethyl iodide} & \text{Sodium} & \text{Normal-butane} & \text{Sodium iodide}
\end{array}
$$

The method has been applied to the preparation of paraffin hydrocarbons as high in the series as hexacontane ($C_{60}H_{122}$). The alkyl radicals may be the same or different in the **iodide** or iodides taken.

Sodium metal was also used similarly by Fittig as reagent for the preparation of hydrocarbons by treating aryl **bromide** or iodide in the presence of dry ether, thus:

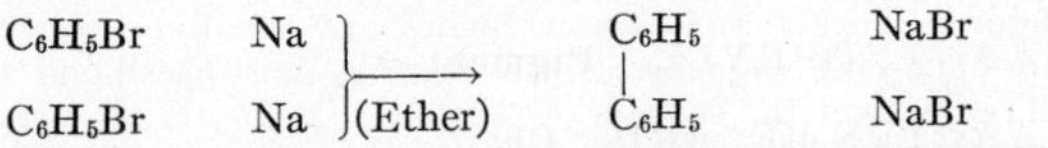

$$
\begin{array}{llll}
C_6H_5Br & Na & C_6H_5 & NaBr \\
C_6H_5Br & Na \ (Ether) & C_6H_5 & NaBr \\
\text{Phenyl iodide} & \text{Sodium} & \text{Biphenyl} & \text{Sodium bromide}
\end{array}
$$

When alkyl iodide and aryl bromide are taken, the hydrocarbon is of the mixed alkyl-aryl type, thus:

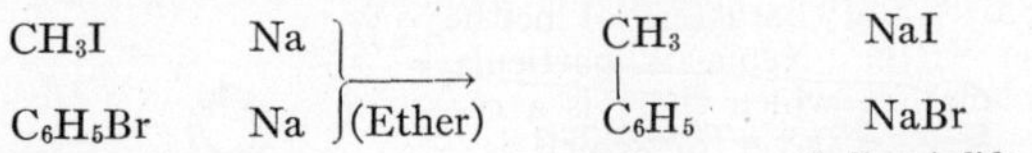

$$
\begin{array}{llll}
CH_3I & Na & CH_3 & NaI \\
C_6H_5Br & Na \ (Ether) & C_6H_5 & NaBr \\
\text{Methyl iodide} & \text{Sodium} & \text{Toluene} & \text{Sodium iodide} \\
\text{Phenyl bromide} & & & \text{Sodium bromide}
\end{array}
$$

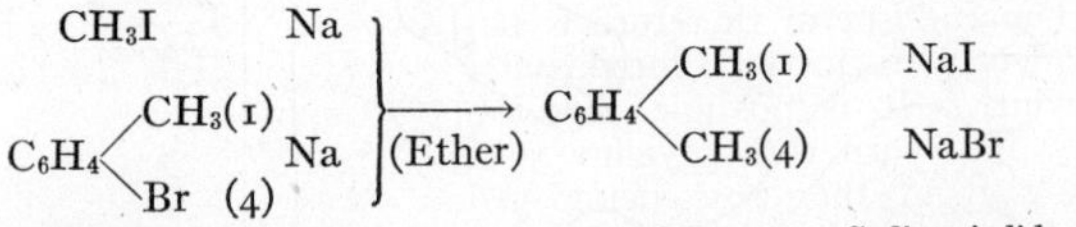

$$
\begin{array}{llll}
CH_3I & Na & & \\
C_6H_4\!\!\begin{array}{l}CH_3(1)\\Br\,(4)\end{array} & Na \ (Ether) & C_6H_4\!\!\begin{array}{l}CH_3(1)\\CH_3(4)\end{array} & \begin{array}{l}NaI\\NaBr\end{array} \\
\text{Methyl iodide} & \text{Sodium} & \text{Para-xylene} & \text{Sodium iodide} \\
\text{Para-bromo-} & & & \text{Sodium bromide} \\
\text{toluene} & & &
\end{array}
$$

The method has been applied to the preparation of substituted benzene hydrocarbons containing as many as four alkyl-groups (durene, $C_6H_2(CH_3)_4(1,2,4,5)$ and isodurene, $C_6H_2(CH_3)_4(1,2,3,5)$).

Frankland introduced the use of **zinc** instead of sodium to accomplish similar reactions. (R.K.S.)

WURTZITE. A mineral zinc sulfide, ZnS, similar to **sphalerite**. Crystallizes in the hexagonal system. Hardness, 3.5–4; specific gravity, 3.98; color, brownish-black with resinous luster. Named after Adolphe Wurtz, France. (R.M.F.)

X

XANTHINE. Alkaloids.

XANTHOPHYLL. Pigments in Plants.

XANTUS BECARD. Chatterer.

XENIA. The name xenia is given to that phenomenon in which, two plants having been crossed, the characters of the male parent appear at once in the seed. Ordinarily the characters of neither parent are recognizable so early. Xenia is particularly well shown in **corn** plants, in which there is a conspicuous **endosperm,** the outermost layer of which is the aleurone layer. This endosperm results from the fusion of a sperm **nucleus** with two polar nuclei in the process of **fertilization.** The endosperm therefore is triploid. In corn it is frequently distinctly colored, with yellow dominating over white. It is possible therefore to pollinate a white-grained corn with a yellow parent, all the grains resulting from the cross being yellow since the gene (or factor) which produces the yellow color is in every cell of the triploid endosperm. Red or purple colors often occur in corn, usually in the aleurone layer, and are dominant over colorless aleurone. So again crosses may be made showing directly the effect of genes from the male parent on the endosperm of the seed.

In some plants the influence of the male element is more extensive, involving tissues which are of female origin. This phenomenon is known as metaxenia. (R.M.W., B.S.M.)

XENOBLAST. A term proposed by Becke in 1903 for metamorphic crystals with undeveloped **crystal faces.** (Compare with **Idioblast.**) (R.M.F.)

XENOCRYST. A term proposed by Sollas in 1894 for crystals, usually corroded, which are foreign to the **magma** from which the **igneous rock** in which they occur has crystallized. (R.M.F.)

XENOLITH. A fragment, large or small, of a foreign rock included in an igneous mass. The term is derived from the Greek, meaning stranger and stone. Xenoliths, both large and small, are best displayed at the contacts or margins of **batholiths.** (R.M.F.)

XENON. Symbol: Xe. Atomic number: 54. Atomic weight: 131.3. Density: 5.851 grams per liter, 0° C., 760 mm., or 4.53 when air equals 1.00. Melting point: −112° C. Boiling point: −107.1° C. (Isotopes: page 290.)

Xenon is a colorless, odorless gas, of negative chemical properties with ordinary materials. Discovered by Ramsay and Travers, in 1898, in ordinary **air** to the extent of 1 part xenon in about 11,000,000 air. (R.K.S.)

XEROPHYTES. These are plants which are adapted to grow in regions in which there is a decided lack of water, as in deserts. Plants growing in such localities have become modified in various ways which enable them to survive. Since the greatest problem is the lack of water, xerophytes must be so formed as to avoid excessive loss of water. There are several factors which reduce water loss. In many of these plants the leaves are greatly reduced in size or are completely lost, as in most Cacti and many Euphorbias. Most of the water lost to the plant passes through the **stomata** in the leaves and stem epidermis. In many xerophytes the stomata are sunk deep in small pits. The number of stomata is often reduced and also their size. As a further protection against loss of water the epidermis may be covered with a very thick cuticle.

The sap of xerophytic plants has properties which make it hold water tenaciously, tending to reduce loss. Many xerophytes have a fleshy habit, so that greater space is available for storage of water. Sometimes the root is tremendously swollen, in others the stem becomes the enlarged part, while in many it is the leaves. Often the leaves form dense tufts which cause great reduction of free surfaces.

Xerophytes are extremely tolerant of prolonged drought. Some may be kept free from water for several years, yet revive and grow when supplied with water. Most of these plants cannot stand too much water, quickly rotting under such conditions. Many species of xerophytes are covered with thorns. This is commonly assumed to protect them against grazing animals. (R.M.W.)

XIPHOSURA. The horseshoe or king crabs, a class of the phylum **Arthropoda** containing only a single genus, *Limulus,* with five species, one found along the Atlantic coast of North America from Florida to Nova Scotia and the other four on the eastern coast of Asia. They are animals of ancient lineage, showing no important change from **fossils** of the **Triassic.**

King crabs are distinguished by the following characters: 1. The **cephalothorax** is covered by a continuous arched plate of horseshoe shape. 2. Six segments of the abdomen are fused to form a continuous piece, hinged to the cephalothorax. 3. The abdomen bears a long caudal spine, the **telson.** 4. The appendages of the cephalothorax include six pairs associated with the mouth, five of them chelate. The five posterior pairs have the basal segments formed for crushing food. 5. The appendages of the abdomen are six pairs of broad plates bearing gills and used in swimming. 6. There are two large compound eyes and two smaller median eyes.

These animals burrow in sand and mud near the shore, feeding on small animals. They come to shore to deposit their eggs in sand. They are of little economic importance but have been used as food for pigs and domestic fowls, and to a limited extent as fertilizer. (A.W.L.)

X-RAY SPECTRA. We know that it is possible, by means of the **diffraction-grating** effect of a crystal, to analyze a beam of non-homogeneous **x-rays** into its various wavelengths, thus forming a **spectrum,** and to measure the relative intensities of its various components (see **X-Ray Spectrometer).** When **cathode rays** of sufficient energy fall upon a specimen of some element, as the metal "target" of an x-ray tube, the resulting x-rays, thus analyzed, are in general found to consist of a continuous spectrum (somewhat analogous to the radiation from a heated solid) with an intensity maximum and an upper frequency limit at frequencies dependent upon the speed of the cathode rays. Upon this may be superposed certain groups of much sharper maxima, which may be regarded as rather diffuse spectrum lines. These are characteristic of the material of the target, not of the incident cathode rays; except that if the cathode rays are produced below a certain voltage, some groups of lines do not appear at all.

As in the case of ordinary diffraction-grating spectra with light, each spectrum line may appear in two or three different orders, of which only the first need be considered. There are then recognized, in the spectrum of each element, several distinct line series; viz., the K, L, M, and N series (perhaps more); which are believed to be due to electron transitions ending in as many distinct quantum or energy states. The lines in any series correspond to "drops" to these respective states from various higher states or levels; the greater the drop, the greater the frequency of the resulting line (see **Quantum Theory**). There is a definite relation between the frequencies of the lines for a given element and the lower frequency limits or edges of the absorption bands of the same element for x-rays (see **Absorption Spectrum**).

If attention is fixed upon the strongest line of any one series (say the K series) as produced by different heavy elements, it is observed that the higher the **atomic number** of the element, the higher the frequency of this line; and if one plots a curve with these variables as coordinates, the result is a **parabola**. In fact, as we pass from one element to the next in the atomic-number series, the square root of the frequency of the selected line always increases by the same amount. This is known as Moseley's law. (L.D.W.)

X-RAY SPECTROMETER. X-ray spectroscopes or spectrometers, of which there are several designs, all utilize the **diffraction-grating** effect of crystals in analyzing x-rays. The Bragg spectrometer, one of the earliest types, is quite simple (see figure). The rays

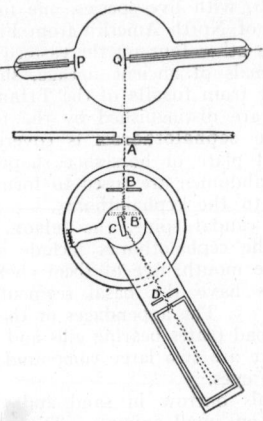

A Bragg x-ray spectrometer.   (Diagrammatic.)

enter through a pair or a succession of slits *A*, *B*, which limit them to a narrow parallel beam, and fall upon a suitable crystal *C* mounted on a prism table at the center of a graduated circle. The table, with the crystal, can be rotated so as to vary the angle of incidence. On another arm is mounted an **ionization chamber**, which the reflected rays enter through a set of slits *D*. As the crystal is rotated slowly, this second arm is rotated just twice as fast, so as to keep the axis *CD* always at the same angle with the reflecting planes of the crystal as the fixed incident beam *AC*. Thus, when any of the x-rays are reflected, they enter the chamber and produce an electrometer deflection proportional to their intensity. By plotting the electrometer readings against wavelengths as computed from Bragg's law, it is possible to exhibit the distribution curve and the contours of the spectrum lines. In more modern forms, the spectrum is progressively recorded on a photographic plate or film as the crystal rotates, and the instrument is then an x-ray spectrograph. (L.D.W.)

X-RAYS. It was in 1895 that Professor Wilhelm Roentgen, of Würzburg, Germany, discovered the x-rays, quite by accident, while experimenting with a **Crookes tube**. He observed the fluorescence of a barium platino-

cyanide screen which happened to lie near the tube, and traced the effect to something which emanated from the spot where the **cathode rays** struck the tube wall. Putting the tube in a pasteboard box made no difference; so it was not **light or ultra-violet radiation** that caused the fluorescence. Investigations followed rapidly, and within only a few weeks the x-rays were being used by surgeons to examine the bones of living people.

It was soon found that x-rays arise wherever cathode rays encounter solids; that "targets" of high **atomic weight** yield more copious x-rays; and that the greater the speed of the cathode particles, the more penetrating, or the "harder," the x-rays are. Special tubes were designed for producing x-rays. The earlier tubes were of the Crookes type, depending on the conduction of ionized gas. Those most used now are thermionic, of the Coolidge type, with a hot-wire cathode operating in a high vacuum; a construction which permits the passage of very high-speed electrons under voltage control, the quantity of them, and the intensity of the resulting x-rays, being regulated by the filament temperature.

Early experiments indicated that x-rays are something essentially different from light, an erroneous conclusion based upon the failure to observe regular reflection, refraction, or diffraction. We now know that this failure was due to the extremely short wavelength of the rays, the range of which extends from the extreme ultra-violet into the **gamma-ray** region, that is, from 10^{-7} to 10^{-9} cm. The early theories attributed x-rays to the sudden stopping of the cathode-ray electrons; but the discoveries of Barkla, Moseley, Bragg, and others have shown that they arise in the atoms of the bombarded substance and that they have wavelengths and spectra quite analogous to visible **atomic spectra**. (See **X-Ray Spectra**.)

There are three principal means of detecting x-rays: the fluorescent effect, the photographic effect, and the ionizing effect. The only method at first available for distinguishing radiations of different wavelength was to measure their penetration or their **absorption coefficient** in various substances. The discovery of the x-ray **diffraction** or grating effect of crystals, by von Laue, Friedrich, and Knipping, in 1912, made it possible to analyze the rays and measure their wavelengths very much as light is studied with the **spectroscope**. When x-rays of given wavelength are incident upon a crystal turned in various directions, the layers of atoms, at certain angles of incidence, reflect wave trains in phase with each other which, if caught on a photographic plate, produce a "Laue pattern." While the matter is not as simple as in the case of light incident on a **diffraction grating**, it is nevertheless possible to interpret such patterns in somewhat the same way as a line spectrum, and to deduce the wavelength from it. A unit convenient for expressing x-ray wavelengths is the "x-unit," which is 10^{-11} cm. or 0.001 angstrom. (See **X-Ray Spectrometer, Bragg's law**, and **Crystal Structure**.)

X-rays are used in medicine to photograph various organs and tissues of the body, or to view them directly (by projecting the x-rays upon a fluorescent screen—fluoroscopy). Because of their destructive action on cells, x-rays are also used in the treatment of certain forms of **cancer** and other malignant **tumors, leukemia**, certain skin diseases, and other pathological conditions. (L.D.W., D.M.H.)

X-UNIT. X-Rays.

XYLEM. The **cells** composing the woody tissues of higher plants are xylem cells. They have undergone extreme modification during their development, becoming greatly elongated and having much-thickened walls. Xylem tissue includes **tracheids, vessels, fibers**, and **parenchyma**. (See **Stem**.) (R.M.W.)

XYLENE. Hydrocarbons.

Y

YAGUARONDI. Jaguarondi.

YAK.
Mammalia, Artiodactyla. A species of **ox**, *Poëphagus grunniens,* found in Tibet and as a domestic animal in other parts of Asia. It is a large animal, attaining a height of well over 5′ and a weight of more than 1000 lbs. The horns are large, and are of the characteristic curved form usually found in oxen, with smooth surfaces. The animals are marked chiefly by the long hair that clothes the flanks, legs, and tail, drooping almost to the ground.

Like other domestic cattle, yaks are important as a source of meat, milk, hides, and hair. (A.W.L.)

YAM. Sweet Potato.

YAPOK.
Mammalia, Marsupialia. *Chironectes.* An **opossum** found from Guatemala to Brazil. It differs from other members of the group in its aquatic habits, resembling the mink and otter in this respect. The hind toes are webbed. (A.W.L.)

YARD. Measurement.

YAW, AIRCRAFT.
This is the angular displacement about an axis parallel to the normal axis of an aircraft. The normal axis of an airplane in horizontal flight is vertical. Intentional yaw is produced principally by the rudder. (F.T.M.)

YAWS (Frombesia Tropica).
An infectious disease common in the tropics, affecting principally the skin. Yaws is caused by a spirochetal organism, *Treponema pertenue,* which is closely related to the spirochete of **syphilis.** Unlike syphilis, however, yaws is non-venereal.

Infection usually occurs in childhood. Widespread, raised, rough, cauliflower lesions containing the infectious organism develop on the skin. Improvement and relapses may occur over a period of years.

The blood **Wassermann reaction** is positive in yaws. Treatment with arsenicals, such as are used in syphilis, is highly effective, as is **penicillin.** (D.M.H.)

Y-CONNECTION.
Three-phase a-c equipment is wound with three wires whose currents differ 120° electrically in **phase.** The windings can be connected either in **Y or delta.** In balanced electrical condition, voltages and currents are the same in all coils. In the Y-connections one end of all three coils is connected in a common joint, and leads from each of the other ends constitute the three-phase line. The Y-connection is preferred for alternators because of the usefulness of the neutral point, and because the line voltage is the $\sqrt{3}$ times the phase voltage. The ability to bring out a neutral point and ground it either through resistance or reactance, is advantageous because it aids in working out protection and selectivity of control of parallel alternators. (F.T.M.)

YEAR.
The year is the longest natural unit for measuring **time.** Several different kinds of year are in use, but all of them depend upon the same phenomenon—the revolution of the earth about the sun. The variations between the different kinds of year arise from the use of different reference points external to the earth relative to which the revolution period is measured.

The sidereal year is the revolution period of the earth about the sun from a given star back to the same star again. Its length, expressed in mean solar **days,** hours, minutes, and seconds, is 365^d 6^h 9^m $9.^s5$ $(365.^d25636)$ and from a purely mechanical point of view this is the true revolution period of the earth.

For general living purposes the sidereal year has no particular significance, and for this purpose the time required for the earth (or apparently the sun) to pass from the vernal equinox back to the vernal equinox again is used. This is known as the tropical year and is 365^d 5^h 48^m $46.^s0$ $(365.^d24220)$ in length, the difference between this and the sidereal year being due to **precession.** This is the year which is the basis of practically all ancient and modern **calendars.**

A third type of year, which is seldom used, is the period for the earth to pass from some point in its **orbit** (e.g., perihelion) back to the same point again. This anomalistic year is 365^d 6^h 13^m $53.^s0$ $(365.^d25964)$ in length, differing from the other two because of the fact that the line of apsides of the earth's orbit is slowly moving at the rate of $11''$ per year. (W.K.G.)

YEASTS.
Saccharomycetales. Yeasts are unicellular **Ascomycetes** in which no **mycelium** develops, and presumably are degenerate forms. They possess two methods of reproduction; one is the asexual process known as **budding;** the other a reduced type of **ascus** formation. In this the contents of a single cell divide to form four or eight cells; this would then be an ascus resulting from the transformation of a single cell. In some species a fusion of two cells occurs prior to ascus formation.

Yeasts are of great economic value, because of their ability to cause fermentations when present in solutions of **carbohydrates.** As a result of their activities the carbohydrates are broken down to ethyl **alcohol** and **carbon dioxide.** The final step in the process corresponds to anaerobic **respiration.** Yeast is therefore used in producing wine from grape juice, cider from apple juice, and beer from sprouted barley, as well as grain (ethyl) alcohol from various sources.

A similar process occurs in bread-making. Yeast plants, mixed with starch and in the form of yeast cakes, are mixed in the dough; immediately the yeast acts on the starch present, producing first glucose, then alcohol and carbon dioxide. After the process has gone on for a sufficient time the dough is kneaded and presently baked. Kneading serves to break up the bubbles of carbon dioxide which have formed. These are the cause of the many small holes present in raised bread. Any alcohol present is driven off in the baking process. (R.M.W., B.S.M.)

YELLOW FEVER.
An acute infectious disease which at present exists only in West Africa and certain parts of South America. Its geographical extent has been gradually limited by proper sanitary measures. Yellow fever is not contagious but is transmitted by the bite of a **mosquito,** most commonly the female *Aëdes aegypti,* which has bitten a yellow fever patient during the first few days of his disease. Twelve days after biting an infected patient, the mosquito is then able to infect a non-immune person. During the 18th and 19th centuries yellow fever was seen in North America and Europe, being transmitted by mosquitoes breeding in the open water tanks of sailing vessels.

The organism causing yellow fever is a filterable **virus.** Once in the body, it produces **hepatitis,** an inflammation and fatty degeneration of the liver cells; to a less extent, similar degeneration occurs in the cells of the spleen, kidneys and heart.

The incubation period is from 3 to 6 days. The symptoms in most cases are mild and consist of fever, slight headache, and perhaps jaundice. A severe case, however, is characterized by a sudden onset, acute prostration, fever, slow pulse rate, generalized aches and pains, and jaundice; the kidneys are particularly involved and large amounts of albumin in the urine is a constant feature. Vomiting may be uncontrollable. Due to bleeding throughout the gastrointestinal tract the "black vomit" characteristic of this disease is produced.

Failure of the kidneys, with complete suppression of their function, or excessive bleeding from the gastrointestinal tract indicate a rapid termination. Death usually occurs between the 6th and 9th days. The mortality rate in the severe form of the disease is about 60%.

Malaria, infectious hepatitis and Weil's disease may be difficult to distinguish from yellow fever by the clinical picture.

There is no specific treatment for yellow fever. In preventing the disease, a vaccine is used successfully. The destruction of Aedes mosquitoes in their breeding places is an essential control measure. (D.M.H.)

YELLOWBIRD. Aves, Passeriformes. The American goldfinch, *Astragalinus tristis,* and the yellow warbler, *Dendroica aestiva.* Only the male of the former species is yellow, and it has the crown and wings black and the tail marked with black. The yellow warbler is more generally yellow in both sexes. The name yellowbird is not commonly used. (A.W.L.)

YELLOW-FIN. Pisces, Teleostei. A species of trout, *Trutta macdonaldi,* found in the upper part of the Arkansas River. (A.W.L.)

YELLOW-JACKET. Insecta, Hymenoptera. A small black and yellow wasp that builds its nest of paper, either in a hole in the ground or under some object near the ground. The use of partly decayed wood in making the paper gives these nests a brownish color. Several species are known. (A.W.L.)

YELLOW-LEGS. Aves, Charadriiformes. An American bird (Aves) related to the snipes and sandpipers. Two species, the greater, *Totanus melanoleucus,* and lesser, *T. flavipes,* yellow-legs, are widely distributed over North America and migrate into South America. The latter is occasionally seen in Europe. (A.W.L.)

YELLOW-THROAT. Aves, Passeriformes. *Geothlypis.* A North American warbler distinguished by its bright yellow front with a black patch through the eyes and along the side of the head. Known as the Maryland yellow-throat in the eastern and central states and in its western varieties as the western, Pacific, tule, and salt-marsh yellow-throat. It frequents low thickets and marshes. (A.W.L.)

YERBA. Maté.

YIELD. Stress Concentration Factor, Watershed.

YIELD POINT. The minimum unit stress at which a structural material will deform without an increase in the load is called the yield point. Some materials do not have a yield point and in others it is not a well-defined value. Consequently, in these cases it has become common practice to use a quantity called the yield strength. The yield strength is the unit stress corresponding to a specific amount of permanent unit deformation. (See Elasticity; Tension Test.) (C.W.C.)

YIELD STRENGTH. Yield Point; Tension Test.

YOKE. The yoke of a motor or generator is the supporting and magnetic structure back of the poles, i.e., it is the part of the machine through which the flux passes in going between poles at the ends away from the armature. (L.R.Q.)

YOLK GLAND. A portion of the reproductive system of the female by which yolk is secreted. In some animals called the vitellarium. (A.W.L.)

YOLK SAC. An accessory embryonic membrane formed in the vertebrates as an enveloping structure around the yolk of the egg. It is connected with the mid gut of the embryo and serves for the absorption of nourishment during embryonic life, and in some species, notably the fishes, after the individual has become active. The wall of the structure is composed of the same germ layers that form the gut, a lining endoderm, and a covering of splanchnic mesoderm. In the latter blood and blood spaces develop at an early period, later forming a network of vitelline vessels from which blood flows into the body of the embryo by way of a pair of large omphalomesenteric veins. Branches of the arterial system of the body extend into this plexus, completing a cycle for the transportation of the absorbed food to the developing body.

The yolk sac persists even in the mammals, where yolk is usually not present. In these forms it serves for the absorption of materials from the surrounding uterus during early development and is involved in the development of the circulatory system. It soon becomes a vestige, however, as its functions are taken over by other membranes. (A.W.L.)

YOUNG'S INTERFERENCE EXPERIMENT. In 1801 Thomas Young made the epochal discovery of the interference of light waves, by means of an experiment which has become classic. Light from a narrow slit L falls on a plate in which are two parallel slits, A, B, very close together, so that from the further side of the latter there emerge two exactly similar wave trains. (See figure.) These overlap in the region beyond AB

Light from single source L gives rise to two wave trains at A and B, which produce interference fringes on screen F.

and produce interference. If a screen F is placed at some distance from AB, alternate bright and dark bands or fringes appear on it, parallel to the two slits. If a translucent screen is used (or in the case of white light, a plate of colored glass acting as a color filter), these bands may be viewed by means of a magnifier beyond it at E, or better, a low-power micrometer eyepiece, with which the width of the band-interval can be measured.

It is easy to show from the elementary theory of interference that if $AB = s$, if the distance from AB to F is x, and if the wavelength of the light is λ, the distance on the screen between any two consecutive dark bands or any two consecutive bright bands is $b = x\lambda/s$. Therefore if b, s, and x are measured, we have at once a means of determining the wavelength: $\lambda = bs/x$.

Other devices have proved more satisfactory than the pair of slits, such as Young's "biprism," Fresnel's mirrors, or Lloyd's mirror; each of which produces a double virtual image of the slit L to serve as the two wave-train sources A and B. (L.D.W.)

YOUNG'S MODULUS. Elasticity.

YPSILOID CARTILAGE. A cartilaginous structure associated with the pelvic girdle in some of the salamanders. It is a Y-shaped projection from the pubis. (A.W.L.)

YTTERBIUM. Symbol: Yb. Atomic number: 70. Atomic weight: 173.04. Melting point: 1800° C. (Isotopes: page 290.) Type of compound: Yb_2O_3. Color of salts: Colorless. Discovered by Urbain in 1907. A member of the **yttrium** sub-group of the rare earth metals. (R.K.S.)

YTTRIUM. Symbol: Y. Atomic number: 39. Atomic weight: 88.92. Density: 3.80. Melting point: 1490° C. No isotope, but of single atomic form: 89. Type of compound: Y_2O_3. Color of salts: Colorless. Discovered by Mosander in 1842.

Yttrium occurs in a few uncommon minerals, e.g., **gadolinite** (35–45% Y_2O_3), **xenotine** (55–65%), **fergusonite** (25–45%), **euxenite** (15–35%).

The yttrium sub-group of the rare earth metals consists of the elements yttrium, terbium, dysprosium, holmium, erbium, thulium, ytterbium, and lutecium. The **potassium** sulfate compounds of all of these elements are relatively soluble in water. Yttrium is the most abundant member of the sub-group but not so abundant as are nine of the ten members of the **cerium** sub-group. (R.K.S.)

YUCCA. Pollination.

YUCCA BORER. Insecta, Lepidoptera. A giant skipper, *Megathymus yuccae*, whose **larva** bores in the stem and root of yucca plants. Several species of these insects are known, expanding from 2–3½″ in the adult stage. While yucca seems to be the prevailing food plant, some species are known to attack agave. They are limited in distribution to the southern states and south into Central America with the exception of one that has been taken in central Colorado and western Nebraska. (A.W.L.)

YUCCA MOTH. Insecta, Lepidoptera. A small **moth,** also commonly called the pronuba moth from the name of its genus. It lives in the flowers of yucca, depositing its eggs in the ovary of the plant and then **pollinating** the flower. The developing larva eats the seeds, but since many more are formed than it is capable of consuming, the exchange is beneficial to the plant. Several species of these moths are known. All belong to the same genus, whose early name, *Pronuba,* has been supplanted by another, *Tegeticula.* Yuccas are also frequented by moths of the genus *Prodoxus,* which are of no service in pollinating the flowers. They are called false yucca moths. (A.W.L.)

ZEBRA. Mammalia, Perissodactyla. An animal of the genus *Equus* related to the asses and quagga, distinguished by the complete or nearly complete transverse striping of the body and legs. The stripes vary in the several species from white to yellow-brown, alternating with dark brown to black. All of the zebras occur in the southern half of Africa. (A.W.L.)

ZEBU. Mammalia, Artiodactyla. The common domesticated cattle, *Bos indicus,* of India, characterized by the highly developed dewlap and by the sharply defined hump on the withers. Similar cattle are found in China and Africa, and they have been introduced into the Americas for hybridization with range cattle. In southern Texas and Brazil these crosses are promising, since the humped cattle are more tolerant of heat and more resistant to insect- and tick-borne diseases. In America the zebu has been known also as Brahman cattle. (A.W.L.)

ZEEMAN EFFECT. An effect of a moderately intense magnetic field upon the structure of the spectrum lines of a gas when subjected to its influence. The phenomenon, sought unsuccessfully by Faraday and finally observed by Zeeman in 1896, consists in the splitting up of each line into two or more components. In the simpler cases, when the source is viewed at right angles to the field, there are three components, of which the middle one has the same frequency as the unmodified line. This component is plane-polarized to vibrate parallel with the field, while the two side components vibrate at right angles to the field. When the source is viewed in the direction of the field, there are only two components, displaced in opposite directions, and circularly polarized in opposite senses. (See **Polarized Light.**) These phenomena constitute the so-called "normal" Zeeman effect.

With most lines, however, the number of components is greater, in some cases reaching twelve or fifteen. They are symmetrically arranged and symmetrically polarized. The displacements, as in the simpler case, are proportional to the magnetic field intensity H, and are always expressible, in wave numbers, as rational multiples of the displacement in the normal effect, which is $4.67 \times 10^{-5}H$ (reciprocal centimeter), a quantity known as the "Lorentz unit." The Zeeman effects observed in sun spots give valuable information as to the magnetic conditions in those areas.

Closely related to the Zeeman effect are two others, the Paschen-Back effect, produced by very strong magnetic fields, and the Back-Goudsmit effect, observed with the spectra of elements having a nuclear magnetic moment, such as bismuth. (See also **Stark Effect.**) (L.D.W.)

ZENITH. The point on the celestial sphere directly overhead is the observer's zenith. The astronomical zenith is defined as the point where the plumb line extended up from the surface of the earth will intersect the celestial sphere. Owing to the fact that the plumb line may be affected by local gravitational effects, such as large mountains in the vicinity, the geographical zenith is the point where a line perpendicular to the surface of a smooth earth would intersect the celestial sphere. The angular distance between the astronomical and geographical zenith is the station error of the point on the surface of the earth. Because of the fact that the earth is an oblate spheroid rather than a perfect sphere, neither the plumb line nor a perpendicular to the surface of the earth will pass through the geometrical center of the earth unless the observer is either at one of the poles or is on the equator. The geocentric zenith is defined as the point where a line extended from the center of the earth through the observer will intersect the celestial sphere. The angular distance between the astronomical and geocentric zenith is the reduction of latitude for the observer. (W.K.G.)

ZENITH TELESCOPE. This instrument, as the name implies, is designed for use at or very close to the zenith. A telescope is mounted in the same manner as the merdian circle, i.e., in the altazimuth form with the azimuth so fixed that the instrument is always in the plane of the meridian. In place of the accurate circles for measuring altitude which are to be found on the meridian circle, this instrument carries a very accurate level so adjusted that the bubble is in the center of the tube when the telescope is pointing at the zenith. The reticle of the instrument carries in addition to the set of wires parallel to the meridian, as in the meridian circle, two wires parallel to the axis of rotation, i.e., perpendicular to the meridian. One of these wires is fixed and, when the instrument is in proper adjustment with the level bubble in the center of the tube, the fixed wire is in the plane of the prime vertical, i.e., passes through the zenith. The other wire may be moved by means of a fine screw which is parallel to the meridian. The head of this screw is accurately calibrated so that the distance of the movable wire from the zenith wire may be determined in seconds of arc.

The zenith telescope is used primarily for the accurate determination of terrestrial latitude by what is commonly known as Talcott's Method. In the figure we have a representation of the celestial sphere drawn in

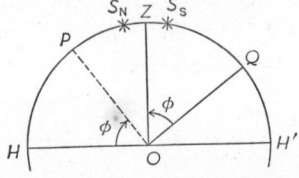

the plane of the observer's meridian $HPZQH'$. HOH' is the plane of the horizon, Z is the astronomic zenith, P the pole of rotation, and Q the direction of the equator. HP is the altitude of the pole and hence is, by definition, the astronomic latitude of O. Inspection of the figure shows that QZ, the declination of the zenith, is also equal to φ.

For the purpose of determining QZ a pair of stars, S_s and S_n, are selected which have approximately the same right ascension and hence will cross the meridian within a short time of each other. The declinations of the stars are so selected that one will pass just south of the zenith and the other just north. Such pairs of stars, known as Talcott pairs, have been selected and their declinations accurately determined by various observatories. Of course for the selection of the proper pairs to be used for a particular station the latitude of that station must be approximately determined in advance by any of the methods discussed under latitude.

At the time that one of the stars is approaching the meridian the observer watches in the eyepiece of the zenith telescope and, when the star appears, he sets the movable wire upon the star and keeps it there until the star has crossed the meridian. The reading of the screw gives the distance that the star is north or south of the zenith, either S_nZ or S_sZ, depending upon which star is observed first. The second star is then observed and the other zenith distance determined in the same manner. From the value of the measured zenith distances and the declinations of the stars the value of the declination of the zenith can be accurately determined and hence the latitude.

The great advantage of this method is that both stars are observed close to the zenith and hence the correction for **astronomical refraction** is very small in either case. Furthermore, since one star passes north and the other south of the zenith, the refraction corrections practically neutralize each other. The measurement of the zenith distance by means of the screw can be very accurately made. The disadvantages of the method are that it requires the use of the accurately adjusted fixed instrument, and the declinations of the stars must be very accurately known.

Observations of this character are constantly being made at various stations all over the world for the purpose of determining the **variation of latitude**. This is also the method used by the Coast and Geodetic Survey for the determination of latitude of their fundamental stations. (w.k.g.)

ZEOLITE. Water.

ZEOLITE GROUP. To the zeolite group of minerals belong a number of hydrous **silicates** of **aluminum** which also ordinarily contain **sodium** or **calcium**, but rarely they may carry **barium, strontium, magnesium, potassium**, etc. These minerals are not related crystallographically as they occur in the **isometric, orthorhombic, hexagonal**, and **monoclinic** systems, but they are all characterized by the presence of water, up to 10 or 20%, which is easily released with the application of heat. They are all rather soft minerals, hardness, 3.5–5.5; of low specific gravity, 2.0–2.5, and they will decompose readily upon treatment with acid, most of them yielding a gelatinous mass. The easy fusion, together with the rapid expulsion of water, is responsible for the name of this interesting group; it is derived from the Greek words to boil and a stone, hence zeolite, "a boiling stone." The zeolites are secondary minerals, usually found filling fissures and cavities in the more **basic igneous rocks** as basalt, gabbro, etc., but occasionally in the more **acidic** types as granite or in gneisses. The following members of the zeolite group are described under their own headings: **Analcite, chabazite, heulandite, natrolite, phillipsite, stilbite, thomsonite, scolecite,** and **harmotone.** (e.s.c.s.)

ZERO. The number zero has the fundamental properties expressed by the formulae

$$a + o = a, \quad a \cdot o = o,$$

and

$$o/a = o, \quad \text{if } a \neq o,$$

where a is any number.

Division by zero is not defined and is not a permissible operation. (l.l.s.)

ZERO BEAT. Beat Frequency.

ZERO EXPONENT. Powers and Exponents.

ZERO OF A FUNCTION. A zero of a **function** of one variable is a value of the **variable** which gives the value zero to the function.

A zero of a function $f(x)$ is a **root** of the corresponding **equation** $f(x) = o$. (l.l.s.)

ZEUGLODON. Eocene.

ZINC. Symbol: Zn. Atomic number: 30. Atomic weight: 65.38. Density: 7.1. Hardness: 2.5. Melting point: 419.4° C. Boiling point: 907° C. (Isotopes: page 290.)

Zinc is a bluish-white metal, malleable and ductile at 150° C., but at 180° C. it changes rapidly so that at 205° C. it may be easily powdered; remains lustrous in dry air but is slightly tarnished in moist air or in water; burns upon heating to vaporization with a bluish flame, forming zinc oxide; soluble in acids—slowly when pure but rapidly on contact with **copper** or **platinum**; soluble in alkalis. Discovery prehistoric.

Zinc is one of the four most largely produced and utilized metals. Used (1) as a protective coating for iron and steel ("galvanized iron"), (2) as a constituent of various **alloys**, especially brass, (3) as plates for primary **batteries**, e.g., dry cell, (4) in zinc base die castings, (5) as a reducing agent of wide range of application in inorganic and organic chemical reactions, (6) as a source of zinc for zinc compounds, e.g., zinc oxide and salts.

Zinc occurs chiefly as sulfide (**sphalerite**, zinc blende, ZnS), carbonate (**smithsonite**, $ZnCO_3$), or oxide (**franklinite**, iron-zinc-manganese oxide). The sulfide ore is roasted to form the oxide. The oxide is mixed with **carbon** (coal) and heated to 1200° C. Zinc vapor is condensed outside the reaction chamber, and cast into blocks called spelter. A low-temperature process employs roasting of the sulfide preferably to the sulfate, which is later extracted with water, and zinc metal is obtained by electrolysis.

Acetate: Zinc acetate ($Zn(C_2H_3O_2)_2 \cdot 2H_2O$), white crystals, soluble. Used as a mordant in dyeing textiles, as a wood preservative, and medicinally.

Carbonate: Zinc carbonate ($ZnCO_3$), white insoluble powder, by grinding smithsonite, or by reaction of zinc salt solution and **sodium** hydrogen carbonate solution. Used as a pigment and medicinally.

Chloride: Zinc chloride ($ZnCl_2$), white, soluble, deliquescent, fusible (melting point 365° C.) solid, formed by reaction of zinc oxide, carbonate, or metal with **hydrochloric acid** solution, and evaporating. Used in a great variety of ways, such as a wood preservative, in soldering fluxes, as a mordant in dyeing textiles, in adhesives and cements, in fluids for embalming and taxidermy; zinc ammonium chloride ($ZnCl_2 \cdot 5NH_3 \cdot H_2O$), white solid.

Chromate: Zinc chromate, "zinc yellow" ($ZnCrO_4$), yellow solid, insoluble, formed by reaction of zinc salt solution and **sodium** chromate solution. Used as a pigment.

Dichromate: Zinc dichromate ($ZnCr_2O_7$), orange-red solid, insoluble, formed by reaction of **chromic acid** and zinc hydroxide. Used as a pigment.

Hydroxide: Zinc hydroxide ($Zn(OH)_2$), white, gelatinous precipitate, formed by reaction of zinc salt solution and alkali hydroxide solution, but soluble in excess of either **sodium** or **ammonium** hydroxide, and soluble in acids.

Nitrate: Zinc nitrate ($Zn(NO_3)_2 \cdot 6H_2O$), white, deliquescent crystals, soluble, formed by reaction of zinc oxide, carbonate, or metal with **nitric acid**.

Oxide: Zinc oxide (ZnO), white, insoluble powder, formed (1) by burning zinc vapor in air, (2) by igniting zinc carbonate, hydroxide, or nitrate. Used (1) as a paint pigment, (2) in compounding rubber, (3) in pharmaceutical and cosmetic preparations, (4) in white printing inks, (5) in ceramic glazes, (6) in dental cements; zinc peroxide (ZnO_2), white precipitate, formed by reaction of zinc chloride solution and **barium** peroxide, followed by recovery of the precipitate. Used as an antiseptic in medicine and in cosmetics.

Phosphate: Zinc phosphate ($Zn_3(PO_4)_2 \cdot 4H_2O$), white precipitate, formed by reaction of zinc salt solution and **sodium** triphosphate solution. Used in dental cements, and in medicine.

Sulfate: Zinc sulfate, white vitrol ($ZnSO_4 \cdot 7H_2O$), white crystals, soluble, formed by reaction of zinc oxide, carbonate, or metal with **sulfuric acid,** or by roasting the sulfide ores at a relatively low temperature, and later extracting with water and crystallizing. Used (1) in preserving wood, glue, and skins, (2) in the manufacture of lithopone pigment, (3) as electrolyte for zinc plating, (4) as a mordant in dyeing textiles, (5) in medicine as an emetic and mild disinfectant.

Sulfide: Zinc sulfide (ZnS), white precipitate, by reaction of zinc salt solution and a soluble **sulfide.** This reaction is applied in the preparation of lithopone by mixing solutions of zinc sulfate and barium sulfide, the product formed being a mixture of zinc sulfide and barium sulfate. Lithopone is used as an important paint pigment.

Stearate: Zinc stearate ($Zn(C_{18}H_{35}O_2)_2$), white insoluble powder, formed by the reaction of zinc salt solution and **sodium** stearate solution. Used in medicine (1) as a non-irritant powder and in skin diseases, (2) in paints and linoleum as a dryer, (3) in cosmetics.

When zinc salt solutions are ignited with **cobalt** nitrate solution, a green cobalt zincate is formed.

REPRESENTATIVE ALLOYS CONTAINING ZINC

Zinc die
casting.....	95% Zn	3.9% Al	1% Cu	0.06% Mg
Battery plate.	64% Zn	21% Sn	12% Pb	3% Cu
White solder..	60% Zn	40% Cu		

Alloy melting
at 327° C... 50% Zn   50% Cd
Alloy melting
at 295° C... 30% Zn   70% Cd
Alloy melting
at 280° C... 10% Zn   90% Cd

(See **Copper** for other alloys of zinc.)

(R.K.S.)

ZINC ALLOYS. "Wrought zinc and zinc alloys may be obtained in the form of rolled strip and sheet and drawn rod and wire. Wrought zinc has desirable corrosion-resistant properties for many types of service, and as corrosion products that may form are colorless, they do not stain other materials. Wrought zinc has chemical characteristics particularly adapted to certain uses such as dry batteries and offers combinations of desirable physical and mechanical properties at relatively low cost. Practically all commercial wrought zinc contains small amounts of natural impurities, mainly lead, iron and cadmium, while the present commercial alloys usually contain additions of copper up to about 1 per cent and sometimes smaller amounts of other elements."

"Of the various metals suitable for pressure die castings, zinc has the advantage of low material cost. Furthermore, because of the low melting point of zinc alloys, good die life may be obtained with dies of plain carbon steel and it is not necessary to give the dies special heat treatment.

"The following established uses give a general idea of the wide range of suitable applications. In the automotive field body hardware, carburetors, fuel pumps, radiator ornaments, speedometer frames and taxi meter housings and parts are commonly die cast from zinc alloys. Gears for many uses including washing machine and electric hoist gears have proven successful. Refrigerator hardware, door checks, and other interior hardware in the building field; small motor frames, radio chassis and parts in the electrical industry and various housings and brackets, in business equipment offer further example." (*The Metals Handbook,* American Society for Metals.)

ZINC BLENDE. Sphalerite.

ZINC POISONING (Brass founders' ague). This industrial disease is caused by inhalation of fumes from zinc oxide. It is characterized by chills and fever accompanied by profuse sweating. It develops in individuals working with zinc, such as smelters, metal refiners, galvanizers, etc. There is no treatment, and the disorder lasts only a day or two and apparently leaves no after-effects. (R.S.M.)

ZINCITE. The mineral zincite is a **zinc** oxide corresponding to the formula ZnO. Its **hexagonal** crystals are rare, as it usually occurs massive, foliated, or in coarse to fine grains. When the crystals are observable it reveals a perfect cleavage parallel to the base of the prism. The fracture is conchoidal. Its hardness is 4–4.5; specific gravity, 5.4–5.7; luster, subadamantine to vitreous; orange-yellow streak; color, red to orange-yellow; translucent to opaque. Zincite occurs in considerable quantities at Franklin Furnace, New Jersey, with **willemite** and **franklinite.** Zincite has been found in Poland, Tuscany, Spain, Saxony, Tasmania, and elsewhere. Except at Franklin Furnace, it is not an important ore of zinc. (E.S.C.S.)

ZIRCON. The mineral zircon, **zirconium** silicate is commonly found in square **tetragonal** prisms, although sometimes assuming pyramidal or irregular forms. It may be found as grains in sands and gravels. It is without good **cleavage;** is brittle, with a conchoidal fracture; hardness, 7.5; specific gravity, variable from 4.2 to 4.8; luster, adamantine, brilliant; color, green, yellow-green, golden-yellow, red, red-brown, brown, and blue. It is said that the colorless stones have been produced by heating the brown ones, to imitation diamonds. The name zircon comes from an Arabic word *zarqun,* meaning vermilion, or perhaps from the Persian *zargun,* meaning golden-colored. These words are corrupted into jargoon, a term applied to the light-colored zircons. The yellow zircon is called hyacinth, from a word of East Indian origin. In the Middle Ages all yellow stones of Indian origin were called hyacinth, but today it is restricted to the yellow zircons.

Zircon occurs in the **igneous rocks** of **acid** type, such as **granites** and **syenites,** e.g., the zircon syenites of southern Norway. It is sometimes found in **gneisses** and **schists,** and is common in river gravels. Foreign localities are the Ural Mountains; Trentino, Monte Somma, and Vesuvius; Arendal, Norway; Ceylon, India; Siam; at the Kimberley mines, South Africa; Madagascar; and in Canada in Renfrew County, Ontario, and Grenville, Quebec. In the United States zircon is found at Litchfield, Maine; Chesterfield, Massachusetts; in Essex, Orange, and St. Lawrence Counties, New York; Henderson County, North Carolina; the Pike's Peak district, Colorado; and Llano County, Texas. (E.S.C.S.)

ZIRCONIUM. Symbol: Zr. Atomic number: 40. Atomic weight: 91.22. Density: 6.44. Melting point: 1900° C. (Isotopes: page 290.)

Crystalline zirconium of high purity is a white, soft, ductile, and malleable metal, but that of 99% purity, when obtained at high temperature, is hard and brittle. Amorphous zirconium is a bluish-black powder. At about 500° C. zirconium burns in air; heated in **hydrogen** forms hydride; heated in **nitrogen** a nitride; and heated in **chlorine** the tetrachloride. A mixture of **hydrofluoric** and **nitric acids** dissolves the metal. On the laboratory scale, zirconium metal may be produced by the reduction of the chloride, oxide, or potassium zirconium fluoride with **sodium** metal. Discovered by Klaproth in 1789.

Zirconium occurs in **zircon** (zirconium silicate, $ZrSiO_4$), and **baddeleyite** (zirconium oxide, ZrO_2). Zirconium is used as a deoxidizer and grain refiner in **certain** steels.